CHILTON

IMPORT AUTO REPAIR MANUAL

1998-2002 Edition

THOMSON

DELMAR LEARNING ™

Chilton Import Car Repair Manual 1998-2002

Vice President, Technology and Trades SBU:
Alar Elken

Channel Manager:
Beth A. Lutz

Production Editor:
Elizabeth Hough

Executive Director, Professional Business Unit:
Greg Clayton

Marketing Coordinator:
Brian McGrath

Custom Publishing Coordinator:
Claudette Corley

Publisher, Professional Business Unit:
David Koontz

Production Director:
Mary Ellen Black

Technology Project Manager:
David Porush

Editorial Assistant:
Kristen Shenfield

Production Manager:
Larry Main

Editors:
Rich Rivele
Christine L. Sheeky
Tim Crain
Thomas A. Mellon

ISBN: 0-8019-93633-6

NOTICE TO THE READER

Table of Contents

Model Index

EDITORIAL POLICY

Manufacturer and Model Coverage

This Manual does not seek to cover every make and model that is currently available on the market. Rather, the Chilton Editorial Staff makes judicious decisions as to which makes and models warrant coverage. Those that are included herein represent Chilton's judgement as to the makes and models that make up 90% of vehicles that will be presented to the average technician for diagnosis and repair. In general, this Manual does not cover:

- Exotics (e.g. Rolls-Royce, Dodge Viper; Alfa Romeo, etc.)
- OEM's with no U.S. presence (e.g. Fiat)
- OEM's that have not sold enough units to be a factor in the repair market.

Model Year Information

Every effort is made to gather current data for use in this Manual. Data is acquired from manufacturers at the time when each OEM chooses to release it. Different manufacturers choose to release their new model information at different times of the year. Indeed, the same manufacturer can be early one season with information, and then late the next season. As a result, not all models are equally current when each edition of this Manual goes to press. You will note that the Editorial Staff has taken care to indicate the currency of coverage for each model.

Safety Notice

Proper service and repair procedures are vital to the safe, reliable operation of all motor vehicles, as well as the personal safety of those performing the repairs. This manual outlines procedures for servicing and repairing vehicles using safe effective methods. The procedures contain many NOTES, WARNINGS and CAUTIONS which should be followed along with standard safety procedures to eliminate the possibility of personal injury or improper service which could damage the vehicle or compromise its safety.

It is important to note that repair procedures and techniques, tools and parts for servicing vehicles, as well as the skill and experience of the individual performing the work vary widely. It is not possible to anticipate all of the conceivable ways or conditions under which vehicles may be serviced, or to provide cautions as to all of the possible hazards that may result. Standard and accepted safety precautions and equipment should be used when handling toxic or flammable fluids, and safety goggles or other protection should be used during cutting, grinding, chiseling, prying, or any other process that can cause material removal or projectiles.

Some procedures require the use of tools specially designed for a specific purpose. Before substituting another tool or procedure, you must be completely satisfied that neither your personal safety, nor the performance of the vehicle will be endangered.

Although information in this manual is based on industry sources and is as complete as possible at the time of publication, the possibility exists that some vehicle manufacturers made later changes which could not be included here. Information on very late models may not be available in some circumstances. While striving for total accuracy, the Publisher cannot assume responsibility for any errors, changes, or omissions that may occur in the compilation of this data.

LOCATING AND USING INFORMATION

Organization

The Table of Contents, located at the front of the book, lists each Unit Repair Section and Model Specific section in this manual.

To find where a particular model specific section is located in the book, you need only look in the Table of Contents. Once you have found the proper section, you may wish to find where specific procedures are located in that section. Turn to the Index at the front of the model specific section. At the upper left-hand side is a listing of the main topics within that section and the page number on which they may be found. Following the main topics is an alphabetical listing of all of the procedures within the section and their page numbers.

The Model Index, located just after the Table of Contents in the beginning of this manual, may also be used to locate the specific section for any vehicle model covered in this manual.

Specifications

Specifications charts for all models covered in this book are located in Chapter 1. They include: Vehicle and Engine Identification, General Engine Specifications, Engine Tune-Up Specifications, Capacities, Valve Specifications, Crankshaft & Connecting Rod Specifications, Piston & Ring Specifications, Engine Fastener Torque Specifications, Brake Specifications, Maintenance Interval Specifications, Ball Joint Specifications, Wheel & Tire Specifications, and Wheel Alignment Specifications.

Unit Repair Sections

The three Unit Repair Sections are written to cover all applicable models for the specific system or component, unless specifically noted otherwise. The procedures covered in the URS are not repeated in the model specific sections, therefore, refer to the URS for the service procedures for the applicable systems or components. Refer to the Table of Contents for URS coverage.

Model Specific Sections

The model specific sections are grouped by manufacturer and arranged in alphabetical order. The text and illustrations that comprise the service procedures in each model specific section are arranged in the following order of systems and components: Engine Repair (Gasoline, then Diesel if applicable), Fuel System (Gasoline, then Diesel if applicable), Drive Train, Steering and Suspension.

All illustrations are located as close as possible to the applicable procedure. Procedures are for all models in the particular section unless specifically noted otherwise.

Part Numbers

Part numbers listed in this book are not recommendations by the Publisher for any product by brand name. They are references that can be used with interchanges manuals and aftermarket supplier catalogs to locate each brand supplier's discrete part number.

Special Tools

Special tools are recommended by the vehicle manufacturer to perform their specific job. Use has been kept to a minimum, but where absolutely necessary, they are referred to in the text by the part number of the tool manufacturer. These tools may be purchased, under the appropriate part number, from your local dealer or regional distributor, or an equivalent tool can be purchased locally from a tool supplier or parts outlet. Before substituting any tool for the one recommended, read the previous Safety Notice.

ACKNOWLEDGMENTS

This publication contains material that is reproduced and distributed under a license from Ford Motor Company. No further reproduction or distribution of the Ford Motor Company material is allowed without the expressed written permission from Ford Motor Company.

Portions of the material contained herein have been reprinted with permission of General Motors Corporation, Service Technology Group.

SPECIFICATIONS

1

ACURA
2.3CL • 2.5TL • 3.0CL • 3.2TL • 3.5RL • INTEGRA • NSX

ENGINE AND VEHICLE IDENTIFICATION

Engine								Model Year	
Code	Liters (cc)	Cu. In.	Cyl.	Fuel Sys.	Engine Type	Eng. Mfg.		Code ①	Year
B18B1	1.8 (1834)	112	4	PGM-FI	DOHC	Honda		W	1998
B18C1	1.8 (1797)	110	4	PGM-FI	DOHC	Honda		X	1999
B18C5	1.8 (1797)	110	4	PGM-FI	DOHC	Honda		Y	2000
C30A1	3.0 (2977)	183	6	PGM-FI	DOHC	Honda		1	2001
C32A6	3.2 (3206)	196	6	PGM-FI	SOHC	Honda		2	2002
C32B1	3.2 (3206)	196	6	PGM-FI	DOHC	Honda			
C35A1	3.5 (3474)	211	6	PGM-FI	SOHC	Honda			
F23A1	2.3 (2254)	138	4	PGM-FI	SOHC	Honda			
G25A4	2.5 (2451)	150	5	PGM-FI	SOHC	Honda			
J30A1	3.0 (2997)	183	6	PGM-FI	SOHC	Honda			
J32A1	3.2 (3210)	196	6	PGM-FI	SOHC	Honda			

PGM-FI: Programmed Fuel Injection

DOHC: Double Overhead Camshaft

SOHC: Single Overhead Camshaft

① 10th digit of the Vehicle Identification Number (VIN)

93471C01

GENERAL ENGINE SPECIFICATIONS

Year	Model	Engine Displacement Liters (cc)	Engine ID	Fuel System Type	Net Horsepower @ rpm	Net Torque @ rpm (ft. lbs.)	Bore x Stroke (in.)	Compression Ratio	Oil Pressure @ rpm
1998	Integra	1.8 (1834)	B18B1/①	PGM-FI	140@6300	127@5200	3.19x3.50	9.2:1	50@3000
	Integra GSR	1.8 (1797)	B18C1/②	PGM-FI	170@7600	128@6200	3.19x3.43	10.0:1	50@3000
	Integra Type R	1.8 (1797)	B18C5/②	PGM-FI	195@8000	130@7500	3.19x3.43	10.6:1	50@3000
	2.3CL	2.3 (2254)	F23A1/YA3	PGM-FI	150@5700	152@4800	3.39x3.82	9.3:1	50@3000
	2.5TL	2.5 (2451)	G25A4/UA2	PGM-FI	176@6300	170@3900	3.35x3.40	9.6:1	50@3000
	NSX	3.0 (2977)	C30A1/NA1	PGM-FI	252@6600	210@5300	3.54x3.07	10.2:1	50@3000
	3.0CL	3.0 (2997)	J30A1/YA2	PGM-FI	200@5000	195@4800	3.39x3.39	9.4:1	71@3000
	3.2TL	3.2 (3206)	C32A6/UA3	PGM-FI	200@5300	210@4500	3.35x3.40	9.6:1	50@3000
	NSX	3.2 (3206)	C32B1/NA2	PGM-FI	290@7100	224@5500	3.66x3.07	10.2:1	50@3000
	3.5RL	3.5 (3474)	C35A1/KA9	PGM-FI	210@5200	224@2800	3.54x3.58	9.6:1	50@3000
1999	Integra	1.8 (1834)	B18B1/①	PGM-FI	140@6300	127@5200	3.19x3.50	9.2:1	50@3000
	Integra GSR	1.8 (1797)	B18C1/②	PGM-FI	170@7600	128@6200	3.19x3.43	10.0:1	50@3000
	2.3CL	2.3 (2254)	F23A1/YA3	PGM-FI	150@5700	152@4800	3.39x3.82	9.3:1	50@3000
	NSX	3.0 (2977)	C30A1/NA1	PGM-FI	252@6600	210@5300	3.54x3.07	10.2:1	50@3000
	3.0CL	3.0 (2997)	J30A1/YA2	PGM-FI	200@5600	195@4800	3.39x3.39	9.4:1	71@3000
	NSX	3.2 (3206)	C32B1/NA2	PGM-FI	290@7100	224@5500	3.66x3.07	10.2:1	50@3000
	3.2TL	3.2 (3206)	J32A1/UA5	PGM-FI	225@5500	216@5000	3.50x3.39	9.8:1	71@3000
	3.5RL	3.5 (3474)	C35A1/KA9	PGM-FI	210@5200	224@2800	3.54x3.58	9.6:1	50@3000
2000	Integra	1.8 (1834)	B18B1/①	PGM-FI	140@6300	127@5200	3.19x3.50	9.2:1	50@3000
	Integra GSR	1.8 (1797)	B18C1/②	PGM-FI	170@7600	128@6200	3.19x3.43	10.0:1	50@3000
	NSX	3.0 (2977)	C30A1/NA1	PGM-FI	252@6600	210@5300	3.54x3.07	10.2:1	50@3000
	NSX	3.2 (3206)	C32B1/NA2	PGM-FI	290@7100	224@5500	3.66x3.07	10.2:1	50@3000
	3.2TL	3.2 (3206)	J32A1/UA5	PGM-FI	225@5500	216@5000	3.50x3.39	9.8:1	71@3000
	3.5RL	3.5 (3474)	C35A1/KA9	PGM-FI	210@5200	224@2800	3.54x3.58	9.6:1	50@3000
2001	Integra	1.8 (1834)	B18B1/①	PGM-FI	140@6300	127@5200	3.19x3.50	9.2:1	50@3000
	Integra GSR	1.8 (1797)	B18C1/②	PGM-FI	170@7600	128@6200	3.19x3.43	10.0:1	50@3000
	NSX	3.0 (2977)	C30A1/NA1	PGM-FI	252@6600	210@5300	3.54x3.07	10.2:1	50@3000
	NSX	3.2 (3206)	C32B1/NA2	PGM-FI	290@7100	224@5500	3.66x3.07	10.2:1	50@3000
	3.2TL	3.2 (3206)	J32A1/UA5	PGM-FI	225@5500	216@5000	3.50x3.39	9.8:1	71@3000
	3.2CL	3.2 (3210)	J32A2/YA4	PGM-FI	260@6100	232@3500	3.50x3.39	10.5:1	71@3000
	3.5RL	3.5 (3474)	C35A1/KA9	PGM-FI	210@5200	224@2800	3.54x3.58	9.6:1	50@3000

PGM-FI: Programmed Fuel Injection

① DB7: 4 door
  DC4: 3 door

② DB8: 4 door (Except Type R)
  DC2: 3 door

93471C02

For complete service labor times, order the Chilton Labor Guide

ENGINE TUNE-UP SPECIFICATIONS

Year	Engine Displacement Liters (cc)	Engine ID/VIN	Spark Plug Gap (in.)	Ignition Timing (deg.) MT	AT	Fuel Pump (psi)	Idle Speed (rpm) MT	AT	Valve Clearance In.	Ex.
1998	1.8 (1834)	B18B1/①	0.039-0.043	16B	16B	40-47②	700-800	700-800	0.003-0.005	0.006-0.008
	1.8 (1797)	B18C1/③	0.051	16B	16B	48-55②	700-800	700-800	0.006-0.007	0.007-0.008
	1.8 (1797)	B18C5/③	0.039-0.043	16B	—	47-54②	750-850	—	0.006-0.007	0.007-0.008
	3.0 (2977)	C30A1/NA1	0.043-0.047	15B	15B	47-53②	750-850	730-830	0.006-0.007	0.007-0.008
	3.2 (3206)	C32A6/UA3	0.039-0.043	—	15B	38-45②	—	590-690	HYD	HYD
	3.2 (3206)	C32B1/NA2	0.043-0.047	—	15B	47-53②	—	750-850	0.006-0.007	0.007-0.008
	3.5 (3474)	C35A1/KA9	0.039-0.043	—	15B	43-50②	—	600-700	HYD	HYD
	2.3 (2254)	F23A1/YA3	0.039-0.043	12B	12B	47-54②	650-750	650-750	0.009-0.011	0.011-0.013
	2.5 (2451)	G25A4/UA2	0.039-0.043	—	15B	43-50②	—	650-750	0.009-0.011	0.011-0.013
	3.0 (2997)	J30A1/YA2	0.039-0.043	—	10B	41-48②	—	700-800	0.008-0.009	0.011-0.013
1999	1.8 (1834)	B18B1/①	0.039-0.043	16B	16B	40-47②	700-800	700-800	0.003-0.005	0.006-0.008
	1.8 (1797)	B18C1/③	0.051	16B	16B	48-55②	700-800	—	0.006-0.007	0.007-0.008
	3.0 (2977)	C30A1/NA1	0.043-0.047	15B	15B	47-53②	750-850	730-830	0.006-0.007	0.007-0.008
	3.2 (3206)	C32B1/NA2	0.043-0.047	—	15B	47-53②	—	750-850	0.006-0.007	0.007-0.008
	3.5 (3474)	C35A1/KA9	0.039-0.043	—	15B	43-50②	—	600-700	HYD	HYD
	2.3 (2254)	F23A1/YA3	0.039-0.043	12B	12B	47-54②	650-750	650-750	0.009-0.011	0.011-0.013
	3.0 (2997)	J30A1/YA2	0.039-0.043	—	10B	41-48②	—	700-800	0.008-0.009	0.011-0.013
	3.2 (3210)	J32A1/UA5	0.039-0.043	—	10B	41-48②	—	630-730	0.008-0.009	0.011-0.013

93471C03

ENGINE TUNE-UP SPECIFICATIONS

Year	Engine Displacement Liters (cc)	Engine ID/VIN	Spark Plug Gap (in.)	Ignition Timing (deg.)		Fuel Pump (psi)	Idle Speed (rpm)		Valve Clearance	
				MT	AT		MT	AT	In.	Ex.
2000	1.8 (1834)	B18B1/①	0.039-0.043	16B	16B	40-47②	700-800	700-800	0.003-0.005	0.006-0.008
	1.8 (1797)	B18C1/③	0.051	16B	16B	48-55②	700-800	—	0.006-0.007	0.007-0.008
	3.0 (2977)	C30A1/NA1	0.043-0.047	15B	15B	47-53②	750-850	730-830	0.006-0.007	0.007-0.008
	3.2 (3206)	C32B1/NA2	0.043-0.047	—	15B	47-53②	—	750-850	0.006-0.007	0.007-0.008
	3.5 (3474)	C35A1/KA9	0.039-0.043	—	15B	43-50②	—	600-700	HYD	HYD
	2.3 (2254)	F23A1/YA3	0.039-0.043	12B	12B	47-54②	650-750	650-750	0.009-0.011	0.011-0.013
	3.0 (2997)	J30A1/YA2	0.039-0.043	—	10B	41-48②	—	700-800	0.008-0.009	0.011-0.013
	3.2 (3210)	J32A1/UA5	0.039-0.043	—	10B	41-48②	—	630-730	0.008-0.009	0.011-0.013
2001	1.8 (1834)	B18B1/①	0.039-0.043	16B	16B	40-47②	700-800	700-800	0.003-0.005	0.006-0.008
	1.8 (1797)	B18C1/③	0.051	16B	16B	48-55②	700-800	—	0.006-0.007	0.007-0.008
	3.0 (2977)	C30A1/NA1	0.043-0.047	15B	15B	47-53②	750-850	730-830	0.006-0.007	0.007-0.008
	3.2 (3206)	C32B1/NA2	0.043-0.047	—	15B	47-53②	—	750-850	0.006-0.007	0.007-0.008
	3.5 (3474)	C35A1/KA9	0.039-0.043	—	15B	43-50②	—	600-700	HYD	HYD
	2.3 (2254)	F23A1/YA3	0.039-0.043	12B	12B	47-54②	650-750	650-750	0.009-0.011	0.011-0.013
	3.0 (2997)	J30A1/YA2	0.039-0.043	—	10B	41-48②	—	700-800	0.008-0.009	0.011-0.013
	3.2 (3206)	J32A1/UA5	0.039-0.043	—	10B	41-48②	—	630-730	0.008-0.009	0.011-0.013
	3.2 (3210)	J32A2/YA4	0.039-0.043	—	10B	41-48②	—	700-800	0.008-0.009	0.011-0.013

NOTE: The Vehicle Emission Control Information label reflects specification changes during production and must be used if they differ from this chart.

B: Before Top Dead Center

HYD: Hydraulic

① DB7: 4 door
   DC4: 3 door

② At idle, pressure regulator vacuum hose disconnected

③ DB8: 4 door (Except Type R)
   DC2: 3 door

93471C04

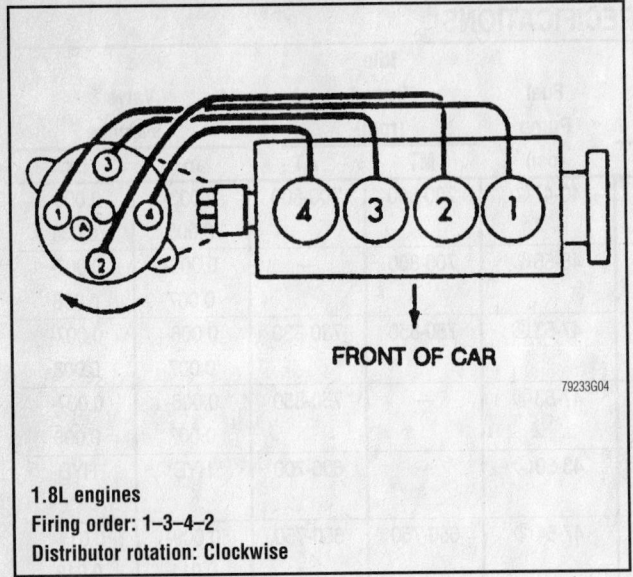

1.8L engines
Firing order: 1–3–4–2
Distributor rotation: Clockwise

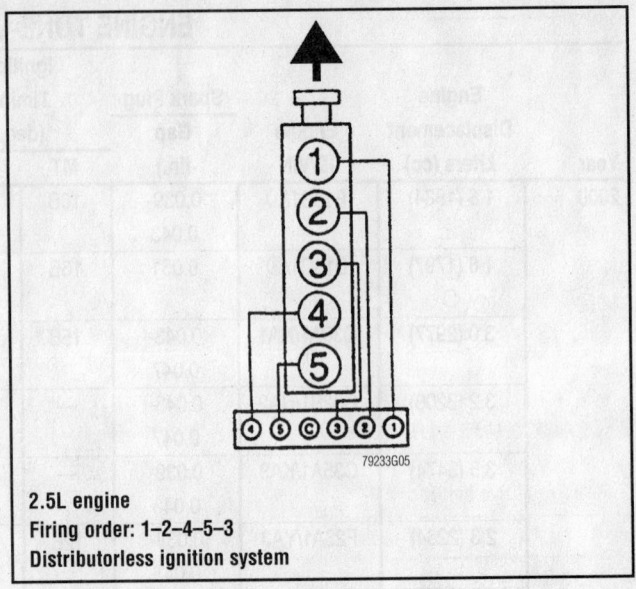

2.5L engine
Firing order: 1–2–4–5–3
Distributorless ignition system

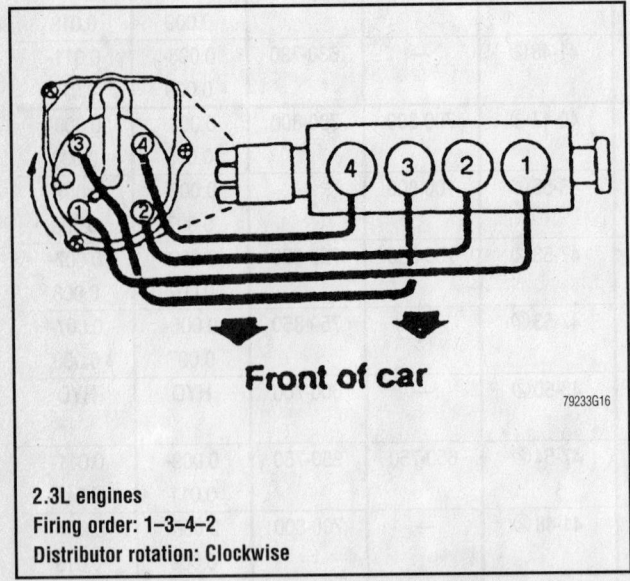

2.3L engines
Firing order: 1–3–4–2
Distributor rotation: Clockwise

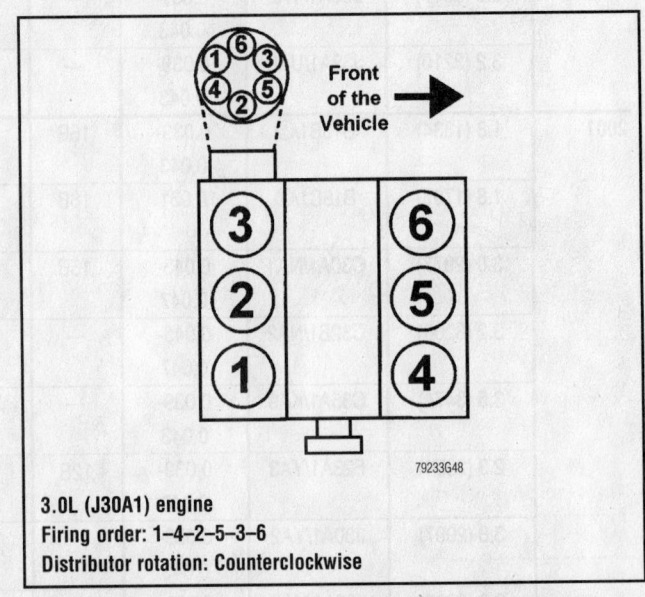

3.0L (J30A1) engine
Firing order: 1–4–2–5–3–6
Distributor rotation: Counterclockwise

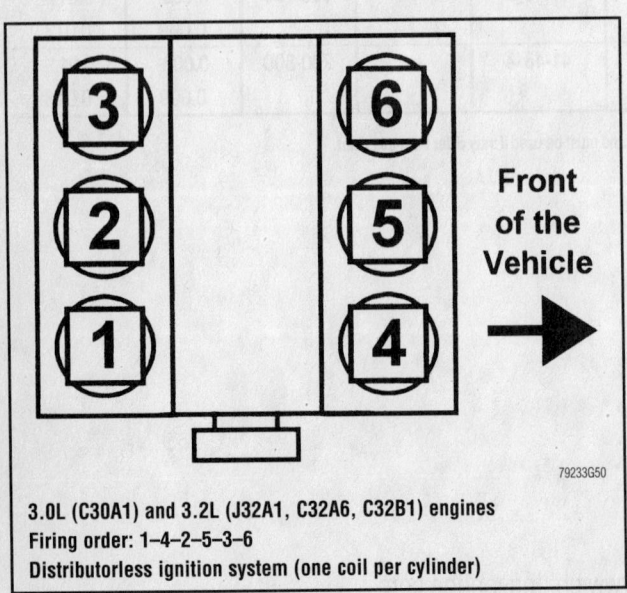

3.0L (C30A1) and 3.2L (J32A1, C32A6, C32B1) engines
Firing order: 1–4–2–5–3–6
Distributorless ignition system (one coil per cylinder)

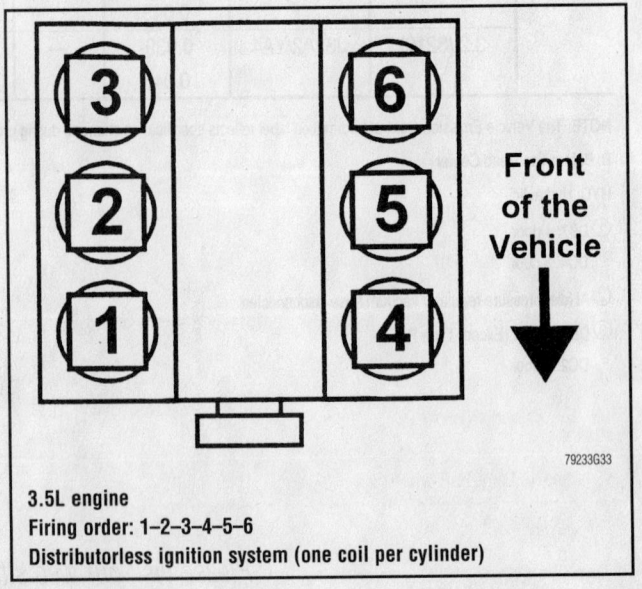

3.5L engine
Firing order: 1–2–3–4–5–6
Distributorless ignition system (one coil per cylinder)

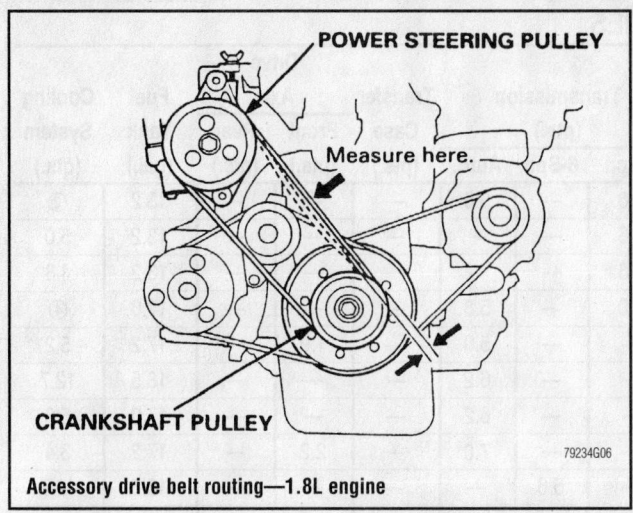

Accessory drive belt routing—1.8L engine

79234G06

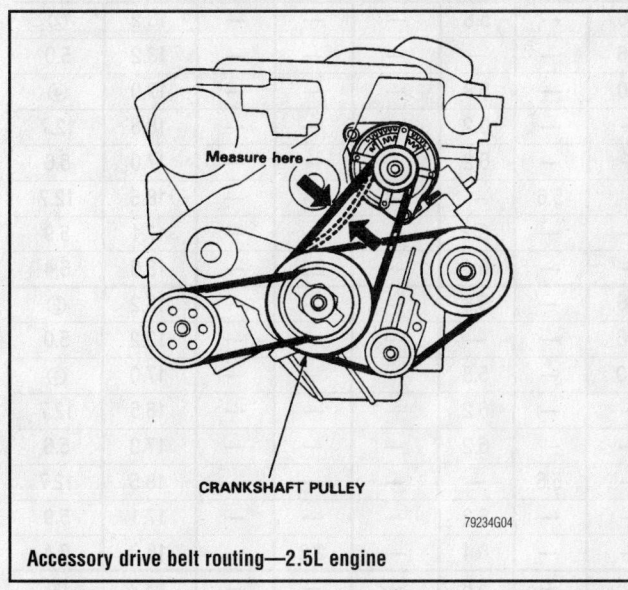

Accessory drive belt routing—2.5L engine

79234G04

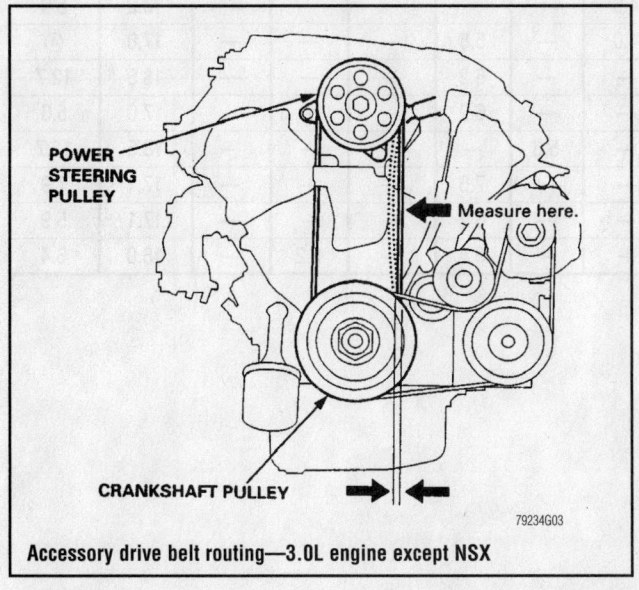

Accessory drive belt routing—3.0L engine except NSX

79234G03

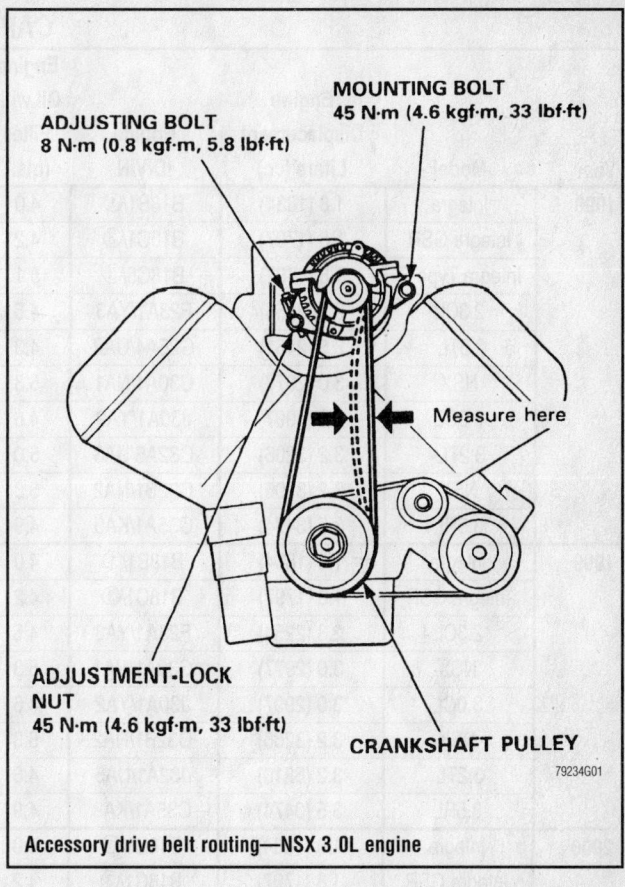

Accessory drive belt routing—NSX 3.0L engine

79234G01

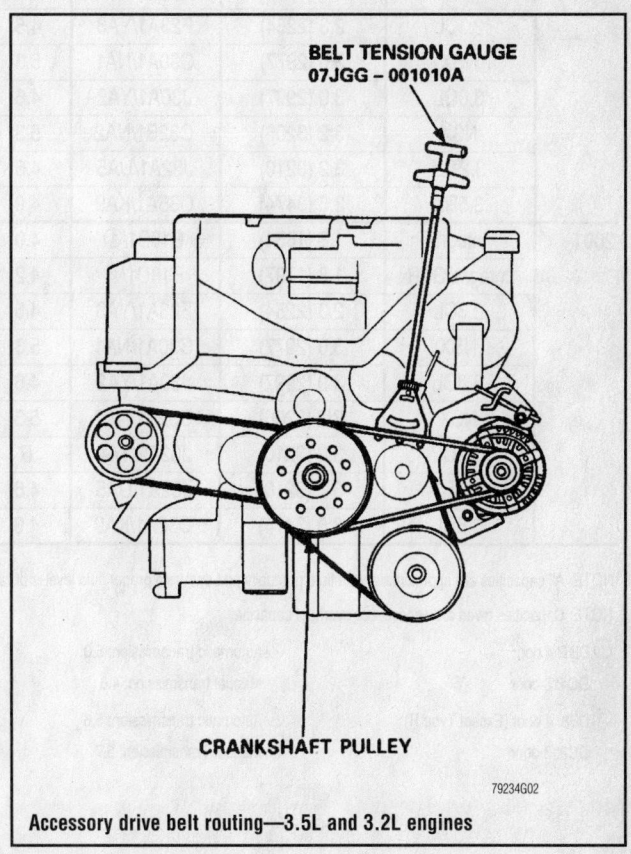

Accessory drive belt routing—3.5L and 3.2L engines

79234G02

Timing belt service is covered in Section 3 of this manual

CAPACITIES

Year	Model	Engine Displacement Liters (cc)	Engine ID/VIN	Engine Oil with Filter (qts.)	Transmission (pts.)			Transfer Case (pts.)	Drive Axle		Fuel Tank (gal.)	Cooling System (qts.)
					5-Spd	6-Spd	Auto.		Front (pts.)	Rear (pts.)		
1998	Integra	1.8 (1834)	B18B1/①	4.0	4.6	—	5.8	—	—	—	13.2	②
	Integra GSR	1.8 (1797)	B18C1/③	4.2	4.6	—	—	—	—	—	13.2	5.0
	Integra Type R	1.8 (1797)	B18C5/③	5.1	4.6	—	—	—	—	—	13.2	4.8
	2.3CL	2.3 (2254)	F23A1/YA3	4.5	4.0	—	5.8	—	—	—	17.0	④
	2.5TL	2.5 (2451)	G25A4/UA2	4.3	—	—	5.0	—	1.9	—	17.2	5.2
	NSX	3.0 (2977)	C30A1/NA1	5.3	—	—	6.2	—	—	—	18.5	12.7
	3.0CL	3.0 (2997)	J30A1/YA2	4.6	—	—	6.2	—	—	—	17.0	5.6
	3.2TL	3.2 (3206)	C32A6/UA3	5.0	—	—	7.0	—	2.2	—	17.2	3.4
	NSX	3.2 (3206)	C32B1/NA2	5.3	—	5.6	—	—	—	—	18.5	12.7
	3.5RL	3.5 (3474)	C35A1/KA9	4.9	—	—	6.4	—	2.2	—	18.0	6.4
1999	Integra	1.8 (1834)	B18B1/①	4.0	4.6	—	5.8	—	—	—	13.2	②
	Integra GSR	1.8 (1797)	B18C1/③	4.2	4.6	—	—	—	—	—	13.2	5.0
	2.3CL	2.3 (2254)	F23A1/YA3	4.5	4.0	—	5.8	—	—	—	17.0	④
	NSX	3.0 (2977)	C30A1/NA1	5.3	—	—	6.2	—	—	—	18.5	12.7
	3.0CL	3.0 (2997)	J30A1/YA2	4.6	—	—	6.2	—	—	—	17.0	5.6
	NSX	3.2 (3206)	C32B1/NA2	5.3	—	5.6	—	—	—	—	18.5	12.7
	3.2TL	3.2 (3210)	J32A1/UA5	4.6	—	—	6.2	—	—	—	17.1	5.9
	3.5RL	3.5 (3474)	C35A1/KA9	4.9	—	—	6.4	—	2.2	—	18.0	6.4
2000	Integra	1.8 (1834)	B18B1/①	4.0	4.6	—	5.8	—	—	—	13.2	②
	Integra GSR	1.8 (1797)	B18C1/③	4.2	4.6	—	—	—	—	—	13.2	5.0
	2.3CL	2.3 (2254)	F23A1/YA3	4.5	4.0	—	5.8	—	—	—	17.0	④
	NSX	3.0 (2977)	C30A1/NA1	5.3	—	—	6.2	—	—	—	18.5	12.7
	3.0CL	3.0 (2997)	J30A1/YA2	4.6	—	—	6.2	—	—	—	17.0	5.6
	NSX	3.2 (3206)	C32B1/NA2	5.3	—	5.6	—	—	—	—	18.5	12.7
	3.2TL	3.2 (3210)	J32A1/UA5	4.6	—	—	6.2	—	—	—	17.1	5.9
	3.5RL	3.5 (3474)	C35A1/KA9	4.9	—	—	6.4	—	2.2	—	18.0	6.4
2001	Integra	1.8 (1834)	B18B1/①	4.0	4.6	—	5.8	—	—	—	13.2	②
	Integra GSR	1.8 (1797)	B18C1/③	4.2	4.6	—	—	—	—	—	13.2	5.0
	2.3CL	2.3 (2254)	F23A1/YA3	4.5	4.0	—	5.8	—	—	—	17.0	④
	NSX	3.0 (2977)	C30A1/NA1	5.3	—	—	6.2	—	—	—	18.5	12.7
	3.0CL	3.0 (2997)	J30A1/YA2	4.6	—	—	6.2	—	—	—	17.0	5.6
	NSX	3.2 (3206)	C32B1/NA2	5.3	—	5.6	—	—	—	—	18.5	12.7
	3.2CL	3.2 (3210)	J32A2/YA4	5	—	—	7.6	—	—	—	17.1	7.9
	3.2TL	3.2 (3210)	J32A1/UA5	4.6	—	—	6.2	—	—	—	17.1	5.9
	3.5RL	3.5 (3474)	C35A1/KA9	4.9	—	—	6.4	—	2.2	—	18.0	6.4

NOTE: All capacities are approximate. Add fluid gradually and ensure a proper fluid level is obtained.

NOTE: Capacities given are service, not overhaul capacities

① DB7: 4 door
　 DC4: 3 door

② Automatic transmission: 5.0
　 Manual transmission: 4.6

③ DB8: 4 door (Except Type R)
　 DC2: 3 door

④ Automatic transmission: 5.6
　 Manual Transmission: 5.7

93471C05

VALVE SPECIFICATIONS

Year	Engine Displacement Liters (cc)	Engine ID/VIN	Seat Angle (deg.)	Face Angle (deg.)	Spring Test Pressure (lbs. @ in.)	Spring Installed Height (in.)	Stem-to-Guide Clearance (in.)		Stem Diameter (in.)	
							Intake	Exhaust	Intake	Exhaust
1998	1.8 (1834)	B18B1/①	45	45	NA	NA	0.0010-0.0020	0.0020-0.0030	0.2591-0.2594	0.2579-0.2583
	1.8 (1797)	B18C1/②	45	45	NA	NA	0.0010-0.0022	0.0020-0.0031	0.2156-0.2159	0.2146-0.2150
	1.8 (1797)	B18C5/②	45	45	NA	NA	0.0010-0.0022	0.0020-0.0031	0.2156-0.2159	0.2146-0.2150
	3.0 (2977)	C30A1/NA1	45	45	NA	NA	0.0010-0.0020	0.0020-0.0030	0.2156-0.2159	0.2146-0.2150
	3.2 (3206)	C32A6/UA3	45	45	NA	NA	0.0010-0.0020	0.0020-0.0030	0.2157-0.2161	0.2146-0.2150
	3.2 (3206)	C32B1/NA2	45	45	NA	NA	0.0010-0.0029	0.0020-0.0030	0.2156-0.2159	0.2146-0.2150
	3.5 (3474)	C35A1/KA9	45	45	NA	NA	0.0010-0.0020	0.0020-0.0030	0.2157-0.2161	0.2146-0.2150
	2.3 (2254)	F23A1/YA3	45	45	NA	NA	0.0008-0.0018	0.0022-0.0031	0.2159-0.2163	0.2146-0.2150
	2.5 (2451)	G25A4/UA2	45	45	NA	NA	0.0008-0.0018	0.0022-0.0031	0.2156-0.2159	0.2146-0.2150
	3.0 (2997)	J30A1/YA2	45	45	NA	NA	0.0008-0.0018	0.0022-0.0031	0.2159-0.2163	0.2146-0.2150
1999	1.8 (1834)	B18B1/①	45	45	NA	NA	0.0010-0.0020	0.0020-0.0030	0.2591-0.2594	0.2579-0.2583
	1.8 (1797)	B18C1/②	45	45	NA	NA	0.0010-0.0022	0.0020-0.0031	0.2156-0.2159	0.2146-0.2150
	3.0 (2977)	C30A1/NA1	45	45	NA	NA	0.0010-0.0020	0.0020-0.0030	0.2156-0.2159	0.2146-0.2150
	3.2 (3206)	C32B1/NA2	45	45	NA	NA	0.0010-0.0029	0.0020-0.0030	0.2156-0.2159	0.2146-0.2150
	3.5 (3474)	C35A1/KA9	45	45	NA	NA	0.0010-0.0020	0.0020-0.0030	0.2157-0.2161	0.2146-0.2150
	2.3 (2254)	F23A1/YA3	45	45	NA	NA	0.0008-0.0018	0.0022-0.0031	0.2159-0.2163	0.2146-0.2150
	3.0 (2997)	J30A1/YA2	45	45	NA	NA	0.0008-0.0018	0.0022-0.0031	0.2159-0.2163	0.2146-0.2150
	3.2 (3210)	J32A1/UA5	45	45	NA	NA	0.0008-0.0018	0.0022-0.0031	0.2159-0.2163	0.2146-0.2150

93471C06

Heater Core replacement is covered in Section 2 of this manual

VALVE SPECIFICATIONS

Year	Engine Displacement Liters (cc)	Engine ID/VIN	Seat Angle (deg.)	Face Angle (deg.)	Spring Test Pressure (lbs. @ in.)	Spring Installed Height (in.)	Stem-to-Guide Clearance (in.)		Stem Diameter (in.)	
							Intake	Exhaust	Intake	Exhaust
2000	1.8 (1834)	B18B1/①	45	45	NA	NA	0.0010-0.0020	0.0020-0.0030	0.2591-0.2594	0.2579-0.2583
	1.8 (1797)	B18C1/②	45	45	NA	NA	0.0010-0.0022	0.0020-0.0031	0.2156-0.2159	0.2146-0.2150
	3.0 (2977)	C30A1/NA1	45	45	NA	NA	0.0010-0.0020	0.0020-0.0030	0.2156-0.2159	0.2146-0.2150
	3.2 (3206)	C32B1/NA2	45	45	NA	NA	0.0010-0.0029	0.0020-0.0030	0.2156-0.2159	0.2146-0.2150
	3.5 (3474)	C35A1/KA9	45	45	NA	NA	0.0010-0.0020	0.0020-0.0030	0.2157-0.2161	0.2146-0.2150
	2.3 (2254)	F23A1/YA3	45	45	NA	NA	0.0008-0.0018	0.0022-0.0031	0.2159-0.2163	0.2146-0.2150
	3.0 (2997)	J30A1/YA2	45	45	NA	NA	0.0008-0.0018	0.0022-0.0031	0.2159-0.2163	0.2146-0.2150
	3.2 (3210)	J32A1/UA5	45	45	NA	NA	0.0008-0.0018	0.0022-0.0031	0.2159-0.2163	0.2146-0.2150
2001	1.8 (1834)	B18B1/①	45	45	NA	NA	0.0010-0.0020	0.0020-0.0030	0.2591-0.2594	0.2579-0.2583
	1.8 (1797)	B18C1/②	45	45	NA	NA	0.0010-0.0022	0.0020-0.0031	0.2156-0.2159	0.2146-0.2150
	3.0 (2977)	C30A1/NA1	45	45	NA	NA	0.0010-0.0020	0.0020-0.0030	0.2156-0.2159	0.2146-0.2150
	3.2 (3206)	C32B1/NA2	45	45	NA	NA	0.0010-0.0029	0.0020-0.0030	0.2156-0.2159	0.2146-0.2150
	3.5 (3474)	C35A1/KA9	45	45	NA	NA	0.0010-0.0020	0.0020-0.0030	0.2157-0.2161	0.2146-0.2150
	2.3 (2254)	F23A1/YA3	45	45	NA	NA	0.0008-0.0018	0.0022-0.0031	0.2159-0.2163	0.2146-0.2150
	3.0 (2997)	J30A1/YA2	45	45	NA	NA	0.0008-0.0018	0.0022-0.0031	0.2159-0.2163	0.2146-0.2150
	3.2 (3210)	J32A2/YA4	45	45	NA	NA	0.0008-0.0018	0.0022-0.0031	0.2159-0.2163	0.2146-0.2150
	3.2 (3210)	J32A1/UA5	45	45	NA	NA	0.0008-0.0018	0.0022-0.0031	0.2159-0.2163	0.2146-0.2150

NA: Not Available

① DB7: 4 door
DC4: 3 door

② DB8: 4 door (Except Type R)
DC2: 3 door

93471C07

CRANKSHAFT AND CONNECTING ROD SPECIFICATIONS

All measurements are given in inches.

Year	Engine Displacement Liters (cc)	Engine ID/VIN	Crankshaft Main Brg. Journal Dia.	Main Brg. Oil Clearance	Shaft End-play	Thrust on No.	Connecting Rod Journal Diameter	Oil Clearance	Side Clearance
1998	1.8 (1834)	B18B1/①	②	③	0.0040-0.0180	4	1.7707-1.7717	0.0008-0.0015	0.0060-0.0160
	1.8 (1797)	B18C1/④	⑤	⑥	0.0040-0.0180	4	1.7707-1.7717	0.0013-0.0020	0.0060-0.0160
	1.8 (1797)	B18C5/④	⑤	⑥	0.0040-0.0180	4	1.7707-1.7717	0.0015-0.0020	0.0060-0.0160
	2.3 (2254)	F23A1/YA3	⑦	⑧	0.0040-0.0180	4	1.7707-1.7717	0.0008-0.0019	0.0060-0.0120
	2.5 (2451)	G25A4/UA2	2.1644-2.1654	0.0007-0.0019	0.0040-0.0140	4	1.7707-1.7717	0.0006-0.0017	0.0060-0.0120
	3.0 (2977)	C30A1/NA1	2.5187-2.5197	0.0009-0.0019	0.0040-0.0110	3	2.0463-2.0472	0.0016-0.0025	0.0060-0.0120
	3.0 (2997)	J30A1/YA2	2.8337-2.8346	0.0008-0.0017	0.0040-0.0140	3	2.0857-2.0866	0.0008-0.0017	0.0060-0.0140
	3.2 (3206)	C32A6/UA3	2.6762-2.6772	0.0008-0.0017	0.0040-0.0110	3	2.1248-2.1257	0.0009-0.0018	0.0060-0.0120
	3.2 (3206)	C32B1/NA2	2.5187-2.5197	0.0009-0.0019	0.0040-0.0110	3	2.0463-2.0472	0.0016-0.0025	0.0060-0.0120
	3.5 (3474)	C35A1/KA9	2.6762-2.6772	0.0008-0.0017	0.0040-0.0110	3	2.1248-2.1257	0.0009-0.0018	0.0060-0.0120
1999	1.8 (1834)	B18B1/①	②	③	0.0040-0.0180	4	1.7707-1.7717	0.0008-0.0015	0.0060-0.0160
	1.8 (1797)	B18C1/④	⑤	⑥	0.0040-0.0180	4	1.7707-1.7717	0.0013-0.0020	0.0060-0.0160
	2.3 (2254)	F23A1/YA3	⑦	⑧	0.0040-0.0180	4	1.7707-1.7717	0.0008-0.0019	0.0060-0.0120
	3.0 (2997)	J30A1/YA2	2.8337-2.8346	0.0008-0.0017	0.0040-0.0140	3	2.0857-2.0866	0.0008-0.0017	0.0060-0.0140
	3.0 (2977)	C30A1/NA1	2.5187-2.5197	0.0009-0.0019	0.0040-0.0110	3	2.0463-2.0472	0.0016-0.0025	0.0060-0.0120
	3.2 (3206)	C32B1/NA2	2.5187-2.5197	0.0009-0.0019	0.0040-0.0110	3	2.0463-2.0472	0.0016-0.0025	0.0060-0.0120
	3.2 (3210)	J32A1/UA5	2.8337-2.8346	0.0008-0.0017	0.0040-0.0140	3	2.1644-2.1654	0.0008-0.0017	0.0060-0.0140
	3.5 (3474)	C35A1/KA9	2.6762-2.6772	0.0008-0.0017	0.0040-0.0110	3	2.1248-2.1257	0.0009-0.0018	0.0060-0.0120
2000	1.8 (1834)	B18B1/①	②	③	0.0040-0.0180	4	1.7707-1.7717	0.0008-0.0015	0.0060-0.0160
	1.8 (1797)	B18C1/④	⑤	⑥	0.0040-0.0180	4	1.7707-1.7717	0.0013-0.0020	0.0060-0.0160
	2.3 (2254)	F23A1/YA3	⑦	⑧	0.0040-0.0180	4	1.7707-1.7717	0.0008-0.0019	0.0060-0.0120
	3.0 (2997)	J30A1/YA2	2.8337-2.8346	0.0008-0.0017	0.0040-0.0140	3	2.0857-2.0866	0.0008-0.0017	0.0060-0.0140
	3.0 (2977)	C30A1/NA1	2.5187-2.5197	0.0009-0.0019	0.0040-0.0110	3	2.0463-2.0472	0.0016-0.0025	0.0060-0.0120
	3.2 (3206)	C32B1/NA2	2.5187-2.5197	0.0009-0.0019	0.0040-0.0110	3	2.0463-2.0472	0.0016-0.0025	0.0060-0.0120
	3.2 (3210)	J32A1/UA5	2.8337-2.8346	0.0008-0.0017	0.0040-0.0140	3	2.1644-2.1654	0.0008-0.0017	0.0060-0.0140
	3.5 (3474)	C35A1/KA9	2.6762-2.6772	0.0008-0.0017	0.0040-0.0110	3	2.1248-2.1257	0.0009-0.0018	0.0060-0.0120
2001	1.8 (1834)	B18B1/①	②	③	0.0040-0.0180	4	1.7707-1.7717	0.0008-0.0015	0.0060-0.0160
	1.8 (1797)	B18C1/④	⑤	⑥	0.0040-0.0180	4	1.7707-1.7717	0.0013-0.0020	0.0060-0.0160
	2.3 (2254)	F23A1/YA3	⑦	⑧	0.0040-0.0180	4	1.7707-1.7717	0.0008-0.0019	0.0060-0.0120
	3.0 (2997)	J30A1/YA2	2.8337-2.8346	0.0008-0.0017	0.0040-0.0140	3	2.0857-2.0866	0.0008-0.0017	0.0060-0.0140
	3.0 (2977)	C30A1/NA1	2.5187-2.5197	0.0009-0.0019	0.0040-0.0110	3	2.0463-2.0472	0.0016-0.0025	0.0060-0.0120
	3.2 (3206)	C32B1/NA2	2.5187-2.5197	0.0009-0.0019	0.0040-0.0110	3	2.0463-2.0472	0.0016-0.0025	0.0060-0.0120
	3.2 (3210)	J32A1/UA5	2.8337-2.8346	0.0008-0.0017	0.0040-0.0140	3	2.1644-2.1654	0.0008-0.0017	0.0060-0.0140
	3.2 (3210)	J32A2/YA4	2.8337-2.8346	0.0008-0.0017	0.0040-0.0140	3	2.1644-2.1654	0.0008-0.0017	0.0060-0.0140
	3.5 (3474)	C35A1/KA9	2.6762-2.6772	0.0008-0.0017	0.0040-0.0110	3	2.1248-2.1257	0.0009-0.0018	0.0060-0.0120

① DB7: 4 door
  DC4: 3 door

② Nos. 1, 2, 4 and 5: 2.1644-2.1654
  No. 3: 2.1642-2.1651

③ Nos. 1,2,4, and 5: 0.0009-0.0017
  No. 3: 0.0012-0.0019

④ DB8: 4 door (Except Type R)
  DC2: 3 door

⑤ Nos. 1, 2, 4 and 5: 2.1644-2.1654
  No. 3: 2.1643-2.1653

⑥ Nos. 1, 2, 4 and 5: 0.0009-0.0017
  No. 3: 0.0012-0.0019

⑦ Nos. 1, 2 and 4: 2.1646-2.1655
  No. 3: 2.1644-2.1654
  No. 5: 2.1650-2.1660

⑧ Nos. 1 and 4: 0.0005-0.0015
  No. 2: 0.0008-0.0018
  No. 3: 0.0010-0.0019
  No. 5: 0.0004-0.0013

93471C08

Brake service is covered in Section 4 of this manual

PISTON AND RING SPECIFICATIONS
All measurements are given in inches

Year	Engine Displacement Liters (cc)	Engine ID/VIN	Piston Clearance	Ring Gap			Ring Side Clearance		
				Top Compression	Bottom Compression	Oil Control	Top Compression	Bottom Compression	Oil Control
1998	1.8 (1834)	B18B1/①	0.0004-0.0016	0.0080-0.0140	0.0160-0.0220	0.0080-0.0200	0.0018-0.0028	0.0018-0.0026	NA
	1.8 (1797)	B18C1/②	0.0004-0.0016	0.0080-0.0140	0.0160-0.0220	0.0080-0.0200	0.0018-0.0028	0.0016-0.0026	NA
	1.8 (1797)	B18C5/②	0.0004-0.0016	0.0080-0.0140	0.0160-0.0220	0.0080-0.0200	0.0018-0.0028	0.0016-0.0026	NA
	2.3 (2254)	F23A1/YA3	0.0008-0.0020	0.0080-0.0140	0.0160-0.0220	0.0080-0.0280	0.0014-0.0024	0.0012-0.0022	NA
	2.5 (2451)	G25A4/UA2	0.0004-0.0016	0.0080-0.0120	0.0140-0.0200	0.0080-0.0280	0.0018-0.0028	0.0012-0.0022	NA
	3.0 (2977)	C30A1/NA1	0.0002-0.0014	0.0100-0.0160	0.0140-0.0200	0.0080-0.0280	0.0012-0.0022	0.0012-0.0022	NA
	3.0 (2997)	J30A1/YA2	0.0006-0.00160	0.0080-0.0140	0.0160-0.0220	0.0080-0.0280	0.0014-0.0024	0.0012-0.0022	NA
	3.2 (3206)	C32A6/UA3	0.0010-0.0020	0.0100-0.0160	0.0160-0.0220	③	0.0022-0.0031	0.0012-0.0022	NA
	3.2 (3206)	C32B1/NA2	0.0002-0.0012	0.0080-0.0120	0.0140-0.0200	0.0080-0.0280	0.0014-0.0026	0.0012-0.0024	NA
	3.5 (3474)	C35A1/KA	0.0010-0.0020	0.0100-0.0160	0.0160-0.0220	③	0.0022-0.0031	0.0012-0.0022	NA
1999	1.8 (1834)	B18B1/①	0.0004-0.0016	0.0080-0.0140	0.0160-0.0220	0.0080-0.0200	0.0018-0.0028	0.0018-0.0026	NA
	1.8 (1797)	B18C1/②	0.0004-0.0016	0.0080-0.0140	0.0160-0.0220	0.0080-0.0200	0.0018-0.0028	0.0016-0.0026	NA
	2.3 (2254)	F23A1/YA3	0.0008-0.0020	0.0080-0.0140	0.0160-0.0220	0.0080-0.0280	0.0014-0.0024	0.0012-0.0022	NA
	3.0 (2977)	C30A1/NA1	0.0002-0.0014	0.0100-0.0160	0.0140-0.0200	0.0080-0.0280	0.0012-0.0022	0.0012-0.0022	NA
	3.0 (2997)	J30A1/YA2	0.0006-0.00160	0.0080-0.0140	0.0160-0.0220	0.0080-0.0280	0.0014-0.0024	0.0012-0.0022	NA
	3.2 (3206)	C32B1/NA2	0.0002-0.0012	0.0080-0.0120	0.0140-0.0200	0.0080-0.0280	0.0014-0.0026	0.0012-0.0024	NA
	3.2 (3210)	J32A1/UA5	0.0006-0.00160	0.0080-0.0140	0.0160-0.0220	0.0080-0.0280	0.0014-0.0024	0.0012-0.0022	NA
	3.5 (3474)	C35A1/KA	0.0010-0.0020	0.0100-0.0160	0.0160-0.0220	③	0.0022-0.0031	0.0012-0.0022	NA
2000	1.8 (1834)	B18B1/①	0.0004-0.0016	0.0080-0.0140	0.0160-0.0220	0.0080-0.0200	0.0018-0.0028	0.0018-0.0026	NA
	1.8 (1797)	B18C1/②	0.0004-0.0016	0.0080-0.0140	0.0160-0.0220	0.0080-0.0200	0.0018-0.0028	0.0016-0.0026	NA
	3.0 (2977)	C30A1/NA1	0.0002-0.0014	0.0100-0.0160	0.0140-0.0200	0.0080-0.0280	0.0012-0.0022	0.0012-0.0022	NA
	3.0 (2997)	J30A1/YA2	0.0006-0.00160	0.0080-0.0140	0.0160-0.0220	0.0080-0.0280	0.0014-0.0024	0.0012-0.0022	NA
	3.2 (3206)	C32B1/NA2	0.0002-0.0012	0.0080-0.0120	0.0140-0.0200	0.0080-0.0280	0.0014-0.0026	0.0012-0.0024	NA
	3.2 (3210)	J32A1/UA5	0.0006-0.00160	0.0080-0.0140	0.0160-0.0220	0.0080-0.0280	0.0014-0.0024	0.0012-0.0022	NA
	3.5 (3474)	C35A1/KA	0.0010-0.0020	0.0100-0.0160	0.0160-0.0220	③	0.0022-0.0031	0.0012-0.0022	NA
2001	1.8 (1834)	B18B1/①	0.0004-0.0016	0.0080-0.0140	0.0160-0.0220	0.0080-0.0200	0.0018-0.0028	0.0018-0.0026	NA
	1.8 (1797)	B18C1/②	0.0004-0.0016	0.0080-0.0140	0.0160-0.0220	0.0080-0.0200	0.0018-0.0028	0.0016-0.0026	NA
	3.0 (2977)	C30A1/NA1	0.0002-0.0014	0.0100-0.0160	0.0140-0.0200	0.0080-0.0280	0.0012-0.0022	0.0012-0.0022	NA
	3.0 (2997)	J30A1/YA2	0.0006-0.00160	0.0080-0.0140	0.0160-0.0220	0.0080-0.0280	0.0014-0.0024	0.0012-0.0022	NA
	3.2 (3206)	C32B1/NA2	0.0002-0.0012	0.0080-0.0120	0.0140-0.0200	0.0080-0.0280	0.0014-0.0026	0.0012-0.0024	NA
	3.2 (3210)	J32A1/UA5	0.0006-0.00160	0.0080-0.0140	0.0160-0.0220	0.0080-0.0280	0.0014-0.0024	0.0012-0.0022	NA
	3.2 (3210)	J32A2/YA4	0.0006-0.00160	0.0080-0.0140	0.0160-0.0220	0.0080-0.0280	0.0014-0.0024	0.0012-0.0022	NA
	3.5 (3474)	C35A1/KA	0.0010-0.0020	0.0100-0.0160	0.0160-0.0220	③	0.0022-0.0031	0.0012-0.0022	NA

NA; Not Applicable

① DB7: 4 door
   DC4: 3 door

② DB8: 4 door
   DC2: 3 door (Except Type R)

③ RIKEN: 0.0080-0.0280 inches
   TEIKOKU: 0.0080-0.0200 inches

93471C09

TORQUE SPECIFICATIONS
All readings in ft. lbs.

Year	Engine Displacement Liters (cc)	Engine ID/VIN	Cylinder Head Bolts	Main Bearing Bolts	Rod Bearing Bolts	Crankshaft Damper Bolts	Flywheel Bolts	Manifold Intake	Manifold Exhaust	Spark Plugs	Lug Nut
1998	1.8 (1834)	B18B1/①	②	56	③	130	④	17	23	13	80
	1.8 (1797)	B18C1/⑤	②	⑥	⑦	130	76	17	23	13	80
	1.8 (1797)	B18C5/⑤	②	⑥	⑦	130	76	17	23	13	80
	3.0 (2977)	C30A1/NA1	56	⑧	⑨	181	④	16	25	13	80
	3.2 (3206)	C32A6/UA3	⑩	⑪	33	174	④	16	22	13	80
	3.2 (3206)	C32B1/NA2	56	⑫	29	181	76	16	22	13	80
	3.5 (3474)	C35A1/KA9	56	⑬	33	181	54	16	22	13	80
	2.3 (2254)	F23A1/YA3	⑩	⑪	⑭	181	④	16	23	13	80
	2.5 (2451)	G25A4/UA2	⑮	⑯	24	181	54	16	23	13	80
	3.0 (2997)	J30A1/YA2	⑰	⑱	⑲	181	54	16	23	13	80
1999	1.8 (1834)	B18B1/①	②	56	③	130	④	17	23	13	80
	1.8 (1797)	B18C1/⑤	③	⑥	⑦	130	76	17	23	13	80
	3.0 (2977)	C30A1/NA1	56	⑦	⑨	181	④	16	25	13	80
	3.2 (3206)	C32B1/NA2	56	⑫	29	181	76	16	22	13	80
	3.5 (3474)	C35A1/KA9	56	⑬	33	181	54	16	22	13	80
	2.3 (2254)	F23A1/YA3	⑳	㉑	⑭	181	④	16	23	13	80
	3.0 (2997)	J30A1/YA2	⑰	⑱	⑲	181	54	16	23	13	80
	3.2 (3210)	J32A1/UA5	⑰	⑱	⑲	181	54	16	23	13	80
2000	1.8 (1834)	B18B1/①	②	56	③	130	③	17	23	13	80
	1.8 (1797)	B18C1/⑤	②	⑥	⑦	130	76	17	23	13	80
	3.0 (2977)	C30A1/NA1	56	⑧	⑨	181	④	16	25	13	80
	3.2 (3206)	C32B1/NA2	56	⑫	29	181	76	16	22	13	80
	3.5 (3474)	C35A1/KA9	56	⑬	33	181	54	16	22	13	80
	2.3 (2254)	F23A1/YA3	⑳	㉑	⑭	181	④	16	23	13	80
	3.0 (2997)	J30A1/YA2	⑰	⑱	⑲	181	54	16	23	13	80
	3.2 (3210)	J32A1/UA5	⑰	⑱	⑲	181	54	16	23	13	80
2001	1.8 (1834)	B18B1/①	②	56	③	130	④	17	23	13	80
	1.8 (1797)	B18C1/⑤	②	⑥	⑦	130	76	17	23	13	80
	3.0 (2977)	C30A1/NA1	56	⑧	⑨	181	④	16	25	13	80
	3.2 (3206)	C32B1/NA2	56	⑫	29	181	76	16	22	13	80
	3.5 (3474)	C35A1/KA9	56	⑬	33	181	54	16	22	13	80
	2.3 (2254)	F23A1/YA3	⑳	㉑	⑭	181	④	16	23	13	80
	3.0 (2997)	J30A1/YA2	⑰	⑱	⑲	181	54	16	23	13	80
	3.2 (3210)	J32A2/YA4	⑰	⑱	⑲	181	54	16	23	13	80
	3.2 (3210)	J32A1/UA5	⑰	⑱	⑲	181	54	16	23	13	80

① DB7: 4 door
　DC4: 3 door

② Step 1: 22 ft. lbs.
　Step 2: 63 ft. lbs.

③ Step 1: 14 ft. lbs.
　Step 2: 23 ft. lbs.

④ Manual transmission: 76 ft. lbs.
　Automatic transmission: 54 ft. lbs.

⑤ DB8: 4 door (Except Type R)
　DC2: 3 door

⑥ Step 1: 22 ft. lbs.
　Step 2: Cap Nos. 1, 5: 56 ft. lbs.
　　Cap Nos. 2, 3 and 4: 49 ft. lbs.

⑦ Step 1: 14 ft. lbs.
　Step 2: 33 ft. lbs.

⑧ Cap bolts: 29 ft. lbs.
　Cap bridge bolts: 48 ft. lbs.

⑨ 14 ft. lbs. plus 116 degrees

⑩ Step 1: 29 ft. lbs.
　Step 2: 56 ft. lbs.

⑪ 11 mm: 57 ft. lbs.
　9 mm: 29 ft. lbs.
　Side: 36 ft. lbs.

⑫ Step 1: Cap bolts 48 ft. lbs.
　Step 2: Side bolts 36 ft. lbs.

⑬ Step 1: Outer (9mm) 29 ft. lbs.
　Step 2: Inner (11mm) 56 ft. lbs.
　Step 3: Side (10mm) 36 ft. lbs.

⑭ Step 1: 14 ft. lbs.
　Step 2: Rotate 90 degrees

⑮ Step 1: 29 ft. lbs.
　Step 2: 51 ft. lbs.
　Step 3: 72 ft. lbs.

⑯ Step 1: 22 ft. lbs.
　Step 2: 54 ft. lbs.

⑰ Step 1: 29 ft. lbs.
　Step 2: 51 ft. lbs.
　Step 3: 72.3 ft. lbs.

⑱ Step 1: Cap bolts 56 ft. lbs.
　Step 2: Side bolts 36 ft. lbs.

⑲ Step 1: 14 ft. lbs.
　Step 2: Rotate 90 degrees

⑳ Step 1: 22 ft. lbs.
　Step 2: Rotate 90 degrees
　Step 3: Rotate 90 degrees
　Step 4: If new bolt, plus 90 degrees

㉑ 11 mm bolts:
　Step 1: 22 ft. lbs.
　Step 2: 51 ft. lbs.
　6mm bolts: 104 inch lbs.

93471C10

For complete Engine Mechanical specifications, see Section 1 of this manual

BRAKE SPECIFICATIONS
All measurements in inches unless noted

Year	Model		Brake Disc Original Thickness	Brake Disc Minimum Thickness	Brake Disc Maximum Runout	Brake Drum Diameter Original Inside Diameter	Brake Drum Diameter Max. Wear Limit	Brake Drum Diameter Maximum Machine Diameter	Minimum Lining Thickness Front	Minimum Lining Thickness Rear	Brake Caliper Bracket Bolts (ft. lbs.)	Brake Caliper Mounting Bolts (ft. lbs.)
1998	2.3CL	F	0.910	0.830	0.004	—	—	—	0.06	—	80	54
		R	0.390	0.310	0.004	—	—	—	—	0.06	28	18
	3.0CL	F	0.980	0.910	0.004	—	—	—	0.08	—	80	54
		R	0.390	0.310	0.004	—	—	—	—	0.08	28	18
	3.5RL	F	0.910	0.830	0.004	—	—	—	0.06	—	80	36
		R	0.350	0.300	0.004	—	—	—	—	0.06	28	17
	2.5TL	F	0.910	0.830	0.004	—	—	—	0.06	—	—	36
		R	0.350	0.300	0.004	—	—	—	—	0.06	28	17
	3.2TL	F	0.910	0.830	0.004	—	—	—	0.06	—	—	36
		R	0.350	0.300	0.004	—	—	—	—	0.06	28	17
	Integra	F	0.830	0.750	0.004	—	—	—	0.06	—	80	23
		R	0.350	0.310	0.004	—	—	—	—	0.06	28	17
	Integra R	F	0.900	0.830	0.004	—	—	—	0.06	—	80	36
		R	0.350	0.310	0.004	—	—	—	—	0.06	28	17
	NSX	F	1.100	1.020	0.004	—	—	—	0.06	—	80	36
		R	0.830	0.750	0.004	—	—	—	—	0.06	80	36
1999	2.3CL	F	0.910	0.830	0.004	—	—	—	0.06	—	80	54
		R	0.390	0.310	0.004	—	—	—	—	0.06	28	18
	3.0CL	F	0.980	0.910	0.004	—	—	—	0.08	—	80	54
		R	0.390	0.310	0.004	—	—	—	—	0.08	28	18
	3.5RL	F	0.910	0.830	0.004	—	—	—	0.06	—	80	36
		R	0.350	0.300	0.004	—	—	—	—	0.06	28	17
	2.5TL	F	0.910	0.830	0.004	—	—	—	0.06	—	—	36
		R	0.350	0.300	0.004	—	—	—	—	0.06	28	17
	3.2TL	F	1.100	1.020	0.004	—	—	—	0.06	—	80	36
		R	0.350	0.310	0.004	①	6.693②	②	—	③	41	17
	Integra	F	0.830	0.750	0.004	—	—	—	0.06	—	—	24
		R	0.350	0.310	0.004	—	—	—	—	0.06	—	24
	NSX	F	1.100	1.020	0.004	—	—	—	0.06	—	80	36
		R	0.830	0.750	0.004	—	—	—	—	0.06	80	36
2000	3.5RL	F	0.910	0.830	0.004	—	—	—	0.06	—	80	36
		R	0.350	0.300	0.004	—	—	—	—	0.06	28	17
	3.2TL	F	1.100	1.020	0.004	—	—	—	0.06	—	80	36
		R	0.350	0.310	0.004	①	6.693②	②	—	③	41	17
	Integra	F	0.830	0.750	0.004	—	—	—	0.06	—	—	24
		R	0.350	0.310	0.004	—	—	—	—	0.06	—	24
	NSX	F	1.100	1.020	0.004	—	—	—	0.06	—	80	36
		R	0.830	0.750	0.004	—	—	—	—	0.06	80	36

93471C11

BRAKE SPECIFICATIONS

All measurements in inches unless noted

Year	Model		Brake Disc			Brake Drum Diameter			Minimum Lining Thickness		Brake Caliper	
			Original Thickness	Minimum Thickness	Maximum Runout	Original Inside Diameter	Max. Wear Limit	Maximum Machine Diameter	Front	Rear	Bracket Bolts (ft. lbs.)	Mounting Bolts (ft. lbs.)
2001	3.5RL	F	0.910	0.830	0.004	—	—	—	0.06	—	80	36
		R	0.350	0.300	0.004	—	—	—	—	0.06	28	17
	3.2TL	F	1.100	1.020	0.004	—	—	—	0.06	—	80	36
		R	0.350	0.310	0.004	①	6.693②	②	—	③	41	17
	3.2CL	F	1.100	1.020	0.004	—	—	—	0.06	—	80	36
		R	0.350	0.310	0.004	①	6.693②	②	—	③	41	17
	Integra	F	0.830	0.750	0.004	—	—	—	0.06	—	—	24
		R	0.350	0.310	0.004	—	—	—	—	0.06	—	24
	NSX	F	1.100	1.020	0.004	—	—	—	0.06	—	80	36
		R	0.830	0.750	0.004	—	—	—	—	0.06	80	36

NA: Not Available

F: Front

R: Rear

① Rear parking brake drum: 6.693 inches

② Rear parking brake drum maximum diameter: 6.732 inches

③ Rear pad: 0.06 inches

Rear parking brake shoes: 0.04 inches

93471C12

For Accessory Drive Belt illustrations, see Section 1 of this manual

WHEEL ALIGNMENT

Year	Model		Caster		Camber		Toe-in (in.)	Steering Axis Inclination (Deg.)
			Range (+/-Deg.)	Preferred Setting (Deg.)	Range (+/-Deg.)	Preferred Setting (Deg.)		
1998	2.3CL	F	—	+3.16	—	-0.10	0	—
		R	—	—	—	-0.57	0.06	—
	2.5tl	F	1.00	+2.12	1.00	0	0 +/- 0.05	—
		R	—	—	1.00	+0.50	0.06 +/- 0.06	—
	3.0cl	F	1.00	+3.28	1.00	0	0 +/- 0.06	—
		R	—	—	0.50	+0.50	0.06 +/- 0.06	—
	3.2tl	F	1.00	+2.80	1.00	0	0 +/- 0.06	—
		R	—	—	0.50	-0.50	0.06 +/- 0.06	—
	3.5rl	F	1.00	+2.81	1.00	0	0 +/- 0.06	—
		R	—	—	1.00	-0.50	0.06 +/- 0.06	—
	Integra ①	F	1.00	+1.16	1.00	+0.16	0 +/- 0.16	—
		R	—	—	1.00	+0.75	0.16 +/- 0.16	—
	Integra ②	F	1.00	+1.16	1.00	+0.16	0 +/- 0.06	—
		R	—	—	1.00	+0.50	0.06 +/- 0.06	—
	NSX ③	F	0.25	+8.00	0.17	-0.33	0.14 +/- 0.04	—
		R	—	—	0.50	-1.50	0.18 +/- 0.06	—
	NSX ④	F	0.25	+8.23	0.17	+0.50	0.14 +/- 0.04	—
		R	—	—	0.50	-2.00	0.18 +/- 0.06	—
1999	2.3CL	F	—	+3.16	—	-0.10	0	—
		R	—	—	—	-0.57	0.06	—
	3.0cl	F	1.00	+3.28	1.00	0	0 +/- 0.06	—
		R	—	—	0.50	+0.50	0.06 +/- 0.06	—
	3.2tl	F	1.00	+2.80	1.00	0	0 +/- 0.06	—
		R	—	—	0.50	-0.50	0.06 +/- 0.06	—
	3.5rl	F	1.00	+2.81	1.00	0	0 +/- 0.06	—
		R	—	—	1.00	-0.50	0.06 +/- 0.06	—
	Integra ①	F	1.00	+1.16	1.00	+0.16	0 +/- 0.16	—
		R	—	—	1.00	+0.75	0.16 +/- 0.16	—
	Integra ②	F	1.00	+1.16	1.00	+0.16	0 +/- 0.06	—
		R	—	—	1.00	+0.50	0.06 +/- 0.06	—
	NSX ③	F	0.25	+8.00	0.17	-0.33	0.14 +/- 0.04	—
		R	—	—	0.50	-1.50	0.18 +/- 0.06	—
	NSX ④	F	0.25	+8.23	0.17	+0.50	0.14 +/- 0.04	—
		R	—	—	0.50	-2.00	0.18 +/- 0.06	—
2000	3.2tl	F	1.00	+2.80	1.00	0	0 +/- 0.06	—
		R	—	—	0.50	-0.50	0.06 +/- 0.06	—
	3.5rl	F	1.00	+2.81	1.00	0	0 +/- 0.06	—
		R	—	—	1.00	-0.50	0.06 +/- 0.06	—
	Integra ①	F	1.00	+1.16	1.00	+0.16	0 +/- 0.06	—
		R	—	—	1.00	+0.75	0.16 +/- 0.16	—
	Integra ②	F	1.00	+1.16	1.00	+0.16	0 +/- 0.06	—
		R	—	—	1.00	+0.50	0.06 +/- 0.06	—
	NSX ③	F	0.25	+8.00	0.17	-0.33	0.14 +/- 0.04	—
		R	—	—	0.50	-1.50	0.18 +/- 0.06	—
	NSX ④	F	0.25	+8.23	0.17	+0.50	0.14 +/- 0.04	—
		R	—	—	0.50	-2.00	0.18 +/- 0.06	—

93471C13

WHEEL ALIGNMENT

Year	Model		Caster Range (+/-Deg.)	Caster Preferred Setting (Deg.)	Camber Range (+/-Deg.)	Camber Preferred Setting (Deg.)	Toe-in (in.)	Steering Axis Inclination (Deg.)
2001	3.2tl	F	1.00	+2.80	1.00	0	0 +/- 0.06	—
		R	—	—	0.50	-0.50	0.06 +/- 0.06	—
	3.2cl	F	1.00	+2.80	1.00	0	0 +/- 0.06	—
		R	—	—	1.00	-0.50	0.06 +/- 0.06	—
	3.5rl	F	1.00	+2.81	1.00	0	0 +/- 0.06	—
		R	—	—	1.00	-0.50	0.06 +/- 0.06	—
	Integra ①	F	1.00	+1.16	1.00	+0.16	0 +/- 0.16	—
		R	—	—	1.00	+0.75	0.16 +/- 0.16	—
	Integra ②	F	1.00	+1.16	1.00	+0.16	0 +/- 0.06	—
		R	—	—	1.00	+0.50	0.06 +/- 0.06	—
	NSX ③	F	0.25	+8.00	0.17	-0.33	0.14 +/- 0.04	—
		R	—	—	0.50	-1.50	0.18 +/- 0.06	—
	NSX ④	F	0.25	+8.23	0.17	+0.50	0.14 +/- 0.04	—
		R	—	—	0.50	-2.00	0.18 +/- 0.06	—

① Except Type R
② Type R
③ Except Type S
④ Type S

93471C14

For Tire, Wheel and Ball Joint specifications, see Section 1 of this manual

TIRE, WHEEL AND BALL JOINT SPECIFICATIONS

Year	Model	OEM Tires Standard	Optional	Tire Pressures (psi) Front	Rear	Wheel Size	Ball Joint Inspection
1998	2.3CL	205/55VR16	None	32	29	6-JJ	NS
	3.0CL	205/55VR16	None	32	29	6-JJ	NS
	2.5TL	205/60HR15	None	30	29	6-JJ	NS
	3.2TL	205/65R15	None	29	29	6.5-JJ	NS
	3.5RL	P215/60VR16	None	29	29	6.5-JJ	NS
	2.2CL	205/55VR16	None	32	29	6-JJ	NS
	NSX	Fr. 215/45ZR16 Rr. 245/40ZR17	None	33	40	6-JJ	NS
	Integra	P195/60/HR14	P195/55VR15	Std: 29 Opt: 35	Std: 29 Opt: 33	Std: 5.5-JJ Opt: 6-JJ	NS
1999	Integra	P195/55VR15	None	32	30	6-JJ	NS
	3.2TL	P205/60VR16	None	29	29	6-JJ	NS
	2.3CL	205/55VR16	None	32	29	6-JJ	NS
	3.0CL	205/55VR16	None	32	29	6-JJ	NS
2000	Integra	P195/55VR15	None	32	30	6-JJ	NS
	3.2TL	P205/60VR16	None	29	29	6-JJ	NS
	2.3CL	205/55VR16	None	32	29	6-JJ	NS
	3.0CL	205/55VR16	None	32	29	6-JJ	NS
2001	Integra	P195/55VR15	None	32	30	6-JJ	NS
	3.2TL	P205/60VR16	None	29	29	6-JJ	NS
	3.2CL	P205/60VR16	None	29	29	6-JJ	NS
	2.3CL	205/55VR16	None	32	29	6-JJ	NS
	3.0CL	205/55VR16	None	32	29	6-JJ	NS

OEM: Original Equipment Manufacturer

PSI: Pounds Per Square Inch

STD: Standard

OPT: Optional

NS: Not Specified by manufacturer

93471C15

SCHEDULED MAINTENANCE INTERVALS
ACURA 3.0CL

TO BE SERVICED	TYPE OF SERVICE	7.5	15	22.5	30	37.5	45	52.5	60	67.5	75	82.5	90	97.5	105	112.5	120
		— VEHICLE MILEAGE INTERVAL (x1000)															
Engine oil	R	✓	✓	✓	✓	✓	✓	✓	✓	✓	✓	✓	✓	✓	✓	✓	✓
Engine oil filter	R		✓		✓		✓		✓		✓		✓		✓		✓
Air cleaner element	R				✓				✓				✓				✓
Valve clearance	S/I	Inspect only if noisy, and every 30,000 miles thereafter.															
Spark plugs	R														✓		
Timing Belt	R														✓		
Water pump	S/I														✓		
Accessory drive belts ①	S/I						✓		✓				✓				✓
Brake fluid	R								✓				✓				
Parking brake	A		✓		✓		✓		✓		✓		✓		✓		✓
Rotate tires	S/I	✓	✓	✓	✓	✓	✓	✓	✓	✓	✓	✓	✓	✓	✓	✓	✓
Tie rod ends, steering gear box and boots	S/I		✓		✓		✓		✓		✓		✓		✓		✓
Suspension components	S/I		✓		✓		✓		✓		✓		✓		✓		✓
CV-joint boots	S/I		✓		✓		✓		✓		✓		✓		✓		✓
Brake hoses and lines	S/I		✓		✓		✓		✓		✓		✓		✓		✓
Fluid levels and condition	S/I		✓		✓		✓		✓		✓		✓		✓		✓
Cooling system hoses and connections	S/I		✓		✓		✓		✓		✓		✓		✓		✓
Exhaust system	S/I		✓		✓		✓		✓		✓		✓		✓		✓
Fuel lines and connections	S/I		✓		✓		✓		✓		✓		✓		✓		✓
Supplemental Restraint System (SRS)	S/I	Ten years after production															

R: Replace   S/I: Inspect and service, if needed   A: Adjust

① Inspect accessory drive belt tension and general condition.

FREQUENT OPERATION MAINTENANCE (SEVERE SERVICE)

If a vehicle is operated under any of the following conditions it is considered severe service:

- Towing a trailer or using a camper or car-top carrier.

- Repeated short trips of less than 5 miles in temperatures below freezing, or trips of less than 10 miles in any temperature.

- Extensive idling or low-speed driving for long distances as in heavy commercial use, such as delivery, taxi or police cars.

- Operating on rough, muddy or salt-covered roads.

- Operating on unpaved or dusty roads.

- Driving in extremely hot (over 90°) conditions.

Engine oil & filter: replace every 3750 miles or 6 months, whichever occurs first.

Air cleaner element: clean every 30,000 miles, starting with the 30,000 mile interval.

Timing belt: replace at the 60,000 mile interval, if the vehicle is driven in very high temperatures (over 110°F/43°C)

or in very low temperatures (under -20°F/-29°C). Otherwise, refer to the scheduled maintenance chart.

Transmission fluid: replace every 30,000 miles.

Front and rear brakes: inspect every 7500 miles or 6 months, whichever occurs first.

CV-joint boots: inspect every 7500 miles or 6 months, whichever occurs first.

Lights and controls: inspect every 15,000 miles.

Vehicle underbody: inspect every 15,000 miles.

93471C16

For Wheel Alignment specifications, see Section 1 of this manual

SCHEDULED MAINTENANCE INTERVALS
ACURA 2.3CL, 2.5TL, 3.2TL, 3.5RL, INTEGRA & NSX

TO BE SERVICED	TYPE OF SERVICE	VEHICLE MILEAGE INTERVAL (x1000)												
		7.5	15	22.5	30	37.5	45	52.5	60	67.5	75	82.5	90	97.5
Engine oil	R	✓	✓	✓	✓	✓	✓	✓	✓	✓	✓	✓	✓	✓
Rear brake discs, calipers & pads	S/I		✓		✓		✓		✓		✓		✓	
Rotate tires	S/I	✓	✓	✓	✓	✓	✓	✓	✓	✓	✓	✓	✓	✓
A/C filter	R		✓		✓		✓		✓		✓		✓	
A/C filter (3.5RL)	R				✓				✓				✓	
Brake hoses & lines (including ABS)	S/I		✓		✓		✓		✓		✓		✓	
Cooling system hoses & connections	S/I		✓		✓		✓		✓		✓		✓	
Driveshaft boots	S/I		✓		✓		✓		✓		✓		✓	
Exhaust system	S/I		✓		✓		✓		✓		✓		✓	
Front brake discs & calipers	S/I		✓		✓		✓		✓		✓		✓	
Fuel pipes, hoses & connections	S/I		✓		✓		✓		✓		✓		✓	
Suspension components	S/I		✓		✓		✓		✓		✓		✓	
Suspension mounting bolts	S/I		✓		✓		✓		✓		✓		✓	
Tie rods, steering gear box & boots	S/I		✓		✓		✓		✓		✓		✓	
Steering operation, tie rod ends, steering gearbox & boots	S/I		✓		✓				✓				✓	
Valve clearance (2.5TL)	S/I		✓		✓		✓		✓		✓		✓	
Valve clearance (2.2CL & NSX)	S/I				✓				✓				✓	
Parking brake	S/I		✓		✓		✓		✓		✓		✓	
Air cleaner element	R				✓				✓				✓	
Automatic transmission fluid	R				✓				✓				✓	
Brake fluid (including ABS) (Integra & 2.5TL)	R				✓				✓				✓	
Brake fluid (including ABS) (3.5RL)	R								✓				✓	
Brake fluid (including ABS) (2.2CL, 3.2TL & NSX)	R						✓				✓			
Front differential oil (2.5TL)	R				✓				✓				✓	
Front differential fluid (3.2TL & 3.5RL)	R								✓				✓	
Manual transmission fluid	R				✓				✓				✓	
ABS operation	S/I				✓				✓				✓	
Drive belt(s)	S/I				✓				✓				✓	
Spark plugs (2.2CL, 2.5TL, Integra except GSR)	R				✓				✓				✓	
Spark plugs (3.2TL, Integra GSR & NSX)	R								✓				✓	
Spark plugs (3.5.RL)	R								✓					
Engine coolant	R						✓				✓			

93471C17

SCHEDULED MAINTENANCE INTERVALS
ACURA 2.3CL, 2.5TL, 3.2TL, 3.5RL, INTEGRA & NSX

TO BE SERVICED	TYPE OF SERVICE	VEHICLE MILEAGE INTERVAL (x1000)												
		7.5	15	22.5	30	37.5	45	52.5	60	67.5	75	82.5	90	97.5
ABS high pressure hose (NSX)	R								✓					
Fuel filter	R								✓					
PCV valve	S/I								✓					
Timing belt (except as noted below)	R												✓	
Timing belt & timing balancer belt (2.2CL)	R												✓	
Timing belt & timing balancer belt (3.5RL) ①	R													
Transmission fluid	R												✓	
Distributor, ignition cap & rotor (2.2CL, 2.5TL & Integra)	S/I								✓					
Idle speed (2.2CL, 2.5TL, 3.2TL, Integra & NSX)	S/I								✓					
Idle speed (3.5RL) ②	S/I		✓		✓		✓		✓		✓		✓	
Ignition wires	S/I		✓		✓								✓	
TWC converter heat shield	S/I		✓		✓		✓		✓		✓		✓	
Water pump	S/I				✓				✓				✓	
Water pump (3.5RL) ②	S/I		✓		✓		✓		✓		✓		✓	

R: Replace   S/I: Service or Inspect

① Replace at 105,000 miles.

② Service or inspect at 105,000 miles.

FREQUENT OPERATION MAINTENANCE (SEVERE SERVICE)

If a vehicle is operated under any of the following conditions it is considered severe service:

- Extremely dusty areas.

-50% or more of the vehicle operation is in 32°C (90°F) or higher temperatures, or constant operation in temperatures below 0°C (32°F).

-Prolonged idling (vehicle operation in stop and go traffic).

-Frequent short running periods (engine does not warm to normal operating temperatures).

-Police, taxi, delivery usage or trailer towing usage.

Oil & oil filter: change every 3750 miles.

Brake hoses & lines (including ABS) (3.5RL): service or inspect every 7500 miles.

Cooling system hoses & connections (3.5RL): service or inspect every 7500 miles.

Driveshaft boots (2.2CL, 2.5TL, 3.2TL & 3.5RL): check every 7500 miles.

Exhaust system (3.5RL): check every 7500 miles.

Brake discs, calipers & pads: service or inspect every 7500 miles.

Fuel pipes, hoses & connections (3.5RL): check every 7500 miles.

Power steering system: service or inspect every 7500 miles.

Suspension components: service or inspect every 7500 miles.

Tie rod ends, steering gear box & boots (2.2CL, 2.5TL, 3.2TL & 3.5RL): service or inspect every 7500 miles.

Air cleaner element (NSX): service or inspect every 7500 miles.

Air cleaner element (except NSX): service or inspect every 15,000 miles.

Front differential fluid (2.4TL, 3.2TL, 3.5RL): replace every 15,000 miles.

Transmission fluid (NSX): replace every 15,000 miles.

Transmission fluid (2.2CL, 2.5TL, 3.2TL & 3.5RL): replace every 30,000 miles

Timing belt (2.5TL & 3.2TL): replace every 60,000 miles.

Water pump (2.5TL & 3.2TL): service or inspect every 60,000 miles.

93471C18

For Maintenance Interval recommendations, see Section 1 of this manual

SCHEDULED MAINTENANCE INTERVALS
ACURA
INTEGRA, NSX
2.3CL, 2.5TL, 3.2TL, 3.5RL

The following should be used as a guide when determining the amount of work required for a particular service. In estimating how long a particular Scheduled Maintenance Service should take, please observe the following:

- Labor Time is time based on field research and data supplied by the vehicle manufacturer.
- Labor time operations are given in hours and tenths of an hour.
- All labor operations are to be used as a guide.

Mechanic Skill Level Codes:
(A) PRECISION: Highly skilled with multiple certification.
(B) GENERAL: Normally skilled with certification.
(C) MAINTENANCE: Semi-skilled working on certification.

	LABOR TIME
7500 Mile Service (C)	
All Models	.9
Inspect front brake pads add	
NSX	.1
15000 Mile Service (B)	
2.3CL	2.1
2.5TL	2.2
3.5RL	1.8
Integra,NSX	1.8
22500 Mile Service (C)	
All Models	.9
Inspect front brake pads	
add NSX	.1
30000 Mile Service (B)	
2.3CL, 3.5RL	3.1
2.5TL	4.3
3.2TL	2.9
Integra	3.9
NSX	3.5
w/AT add	.5

	LABOR TIME
37500 Mile Service (C)	
All Models	.9
Inspect front brake pads add NSX	.1
45000 Mile Service (B)	
2.3CL, 3.2TL	2.8
2.5TL, 3.5RL	2.5
Integra	2.4
NSX	2.9
52500 Mile Service (C)	
All Models	.9
Inspect front brake pads add NSX	.1
60000 Mile Service (B)	
2.3CL, 3.5RL	5.1
2.5TL	5.6
Integra, 3.2TL	5.4
NSX	5.7
w/AT add	.5
67500 Mile Service (C)	
All Models	.9
Inspect front brake pads add NSX	.1

	LABOR TIME
75000 Mile Service (B)	
2.3CL, 3.2TL	2.8
2.5TL, 3.5RL	2.5
Integra	2.4
NSX	2.9
82500 Mile Service (C)	
All Models	.9
IInspect front brake pads add NSX	.1
90000 Mile Service (B)	
2.3CL	7.2
2.5TL	8.5
3.2TL	7.4
3.5RL	4.6
Integra	7.6
NSX	7.1
w/AT add	.5
97500 Mile Service (C)	
All Models	.9
Inspect front brake pads add NSX	.1

93471C19

SCHEDULED MAINTENANCE INTERVALS
ACURA
3.0CL

The following should be used as a guide when determining the amount of work required for a particular service.
In estimating how long a particular Scheduled Maintenance Service should take, please observe the following:

● Labor Time is time based on field research and data supplied by the vehicle manufacturer.
● Labor time operations are given in hours and tenths of an hour.
● All labor operations are to be used as a guide.

Mechanic Skill Level Codes:
(A) PRECISION: Highly skilled with multiple certification.
(B) GENERAL: Normally skilled with certification.
(C) MAINTENANCE: Semi-skilled working on certification.

	LABOR TIME		LABOR TIME		LABOR TIME
7500 Mile Service (C)		**45000 Mile Service (B)**		**90000 Mile Service (B)**	
3.0CL	.9	3.0CL	2.8	3.0CL	4.6
15000 Mile Service (B)		**52500 Mile Service (C)**		w/AT add	.5
3.0CL	2.1	3.0CL	.9	**97500 Mile Service (C)**	
22500 Mile Service (C)		**60000 Mile Service (B)**		3.0CL	.9
3.0CL	.9	3.0CL	5.1	**105000 Mile Service (C)**	
30000 Mile Service (B)		w/AT add	.5	3.0CL	7.2
3.0CL	3.1	**67500 Mile Service (C)**		**112500 Mile Service (C)**	
w/AT add	.5	3.0CL	.9	3.0CL	.9
37500 Mile Service (C)		**75000 Mile Service (B)**		**120000 Mile Service (B)**	
3.0CL	.9	3.0CL	2.8	3.0CL	2.8
		82500 Mile Service (C)			
		All Models	.9		

93471C19A

For Tune-up, Capacities and Firing orders, see Section 1 of this manual

AUDI
A4 • A6 • CABRIOLET

ENGINE AND VEHICLE IDENTIFICATION

Engine							Model Year	
Code	Liters (cc)	Cu. In.	Cyl.	Fuel Sys.	Engine Type	Eng. Mfg.	Code ①	Year
AEB	1.8 (1781)	107	4	MFI-Turbo	DOHC	Audi	W	1998
AFC	2.8 (2771)	169	6	MFI	SOHC	Audi	X	1999
AHA	2.8 (2771)	169	6	MFI	DOHC	Audi	Y	2000
							1	2001
							2	2002

MFI: Multi-point Fuel Injection

SOHC: Single Overhead Camshaft

DOHC: Double Overhead Camshaft

① 10th digit of the Vehicle Identification Number (VIN)

93471C20

GENERAL ENGINE SPECIFICATIONS

Year	Model	Engine Displacement Liters (cc)	Engine ID	Fuel System Type	Net Horsepower @ rpm	Net Torque @ rpm (ft. lbs.)	Bore x Stroke (in.)	Compression Ratio	Oil Pressure @ rpm
1998	A4 Sedan 1.8 T	1.8 (1781)	AEB	MFI	150@5700	155@1750	3.18x3.40	9.5:1	72-101@3000
	A4 Avant 2.8	2.8 (2771)	AHA	MFI	190@6000	207@3200	3.25x3.40	10.3:1	44-73@3000
	A4 Sedan 2.8	2.8 (2771)	AHA	MFI	190@6000	207@3200	3.25x3.40	10.3:1	44-73@3000
	A6 Sedan	2.8 (2771)	AHA	MFI	200@6000	207@3200	3.25x3.40	10.3:1	44-73@3000
	Cabriolet	2.8 (2771)	AFC	MFI	172@5500	184@3000	3.25x3.40	10.3:1	44-73@3000
1999	A4 Avant 1.8 T	1.8 (1781)	AEB	MFI	150@5700	155@1750	3.18x3.40	9.5:1	72-101@3000
	A4 Sedan 1.8 T	1.8 (1781)	AEB	MFI	150@5700	155@1750	3.18x3.40	9.5:1	72-101@3000
	A4 Avant 2.8	2.8 (2771)	AHA	MFI	190@6000	207@3200	3.25x3.40	10.3:1	29@2000
	A4 Sedan 2.8	2.8 (2771)	AHA	MFI	190@6000	207@3200	3.25x3.40	10.3:1	29@2000
	A6 Avant	2.8 (2771)	AHA	MFI	200@6000	207@3200	3.25x3.40	10.3:1	29@2000
	A6 Sedan	2.8 (2771)	AHA	MFI	200@6000	207@3200	3.25x3.40	10.3:1	29@2000
2000	A4 Avant 1.8 T	1.8 (1781)	AEB	MFI	150@5700	155@1750	3.18x3.40	9.5:1	72-101@3000
	A4 Sedan 1.8 T	1.8 (1781)	AEB	MFI	150@5700	155@1750	3.18x3.40	9.5:1	72-101@3000
	A4 Avant 2.8	2.8 (2771)	AHA	MFI	190@6000	207@3200	3.25x3.40	10.3:1	29@2000
	A4 Sedan 2.8	2.8 (2771)	AHA	MFI	190@6000	207@3200	3.25x3.40	10.3:1	29@2000
	A6 Avant	2.8 (2771)	AHA	MFI	200@6000	207@3200	3.25x3.40	10.3:1	29@2000
	A6 Sedan	2.8 (2771)	AHA	MFI	200@6000	207@3200	3.25x3.40	10.3:1	29@2000
2001	A4 Avant 1.8 T	1.8 (1781)	AEB	MFI	150@5700	155@1750	3.18x3.40	9.5:1	72-101@3000
	A4 Sedan 1.8 T	1.8 (1781)	AEB	MFI	150@5700	155@1750	3.18x3.40	9.5:1	72-101@3000
	A4 Avant 2.8	2.8 (2771)	AHA	MFI	190@6000	207@3200	3.25x3.40	10.3:1	29@2000
	A4 Sedan 2.8	2.8 (2771)	AHA	MFI	190@6000	207@3200	3.25x3.40	10.3:1	29@2000
	A6 Avant	2.8 (2771)	AHA	MFI	200@6000	207@3200	3.25x3.40	10.3:1	29@2000
	A6 Sedan	2.8 (2771)	AHA	MFI	200@6000	207@3200	3.25x3.40	10.3:1	29@2000

MFI: Multi-point Fuel Injection

93471C21

For complete service labor times, order the Chilton Labor Guide

ENGINE TUNE-UP SPECIFICATIONS

Year	Engine Displacement Liters (cc)	Engine ID/VIN	Spark Plug Gap (in.)	Ignition Timing (deg.) MT	AT	Fuel Pump (psi)	Idle Speed (rpm) MT	AT	Valve Clearance In.	Ex.
1998	1.8 (1781)	AEB	0.039	①	①	50-58	820-900	820-900	HYD	HYD
	2.8 (2771)	AFC	0.039	—	①	46-55	—	650-750	HYD	HYD
	2.8 (2771)	AHA	0.039	①	①	55-61	700-800	700-800	HYD	HYD
1999	1.8 (1781)	AEB	0.039	①	①	50-58	820-900	820-900	HYD	HYD
	2.8 (2771)	AHA	0.039	①	①	55-61	700-800	700-800	HYD	HYD
2000	1.8 (1781)	AEB	0.039	①	①	50-58	820-900	820-900	HYD	HYD
	2.8 (2771)	AHA	0.039	①	①	55-61	700-800	700-800	HYD	HYD
2001	1.8 (1781)	AEB	0.039	①	①	50-58	820-900	820-900	HYD	HYD
	2.8 (2771)	AHA	0.039	①	①	55-61	700-800	700-800	HYD	HYD

NOTE: The Vehicle Emission Control Information label reflects specification changes made during production and must be used if diferrent from this chart.

NOTE: Fuel pump pressure specifications with the fuel pressure regulator vacuum hose attached.

HYD: Hydraulic

① The basic setting is controlled by the ECU and is not adjustable

93471C22

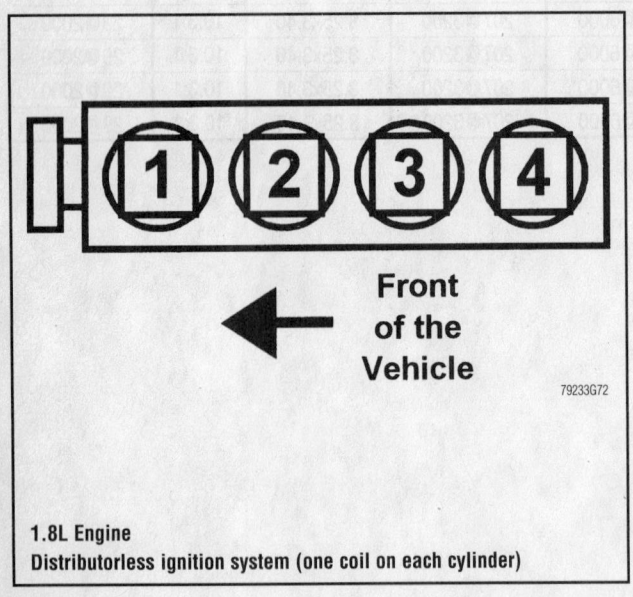

1.8L Engine
Distributorless ignition system (one coil on each cylinder)

79233G72

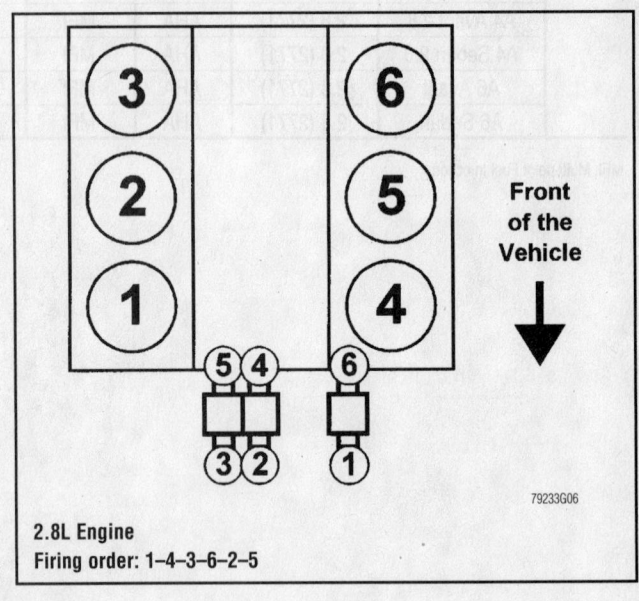

2.8L Engine
Firing order: 1–4–3–6–2–5

79233G06

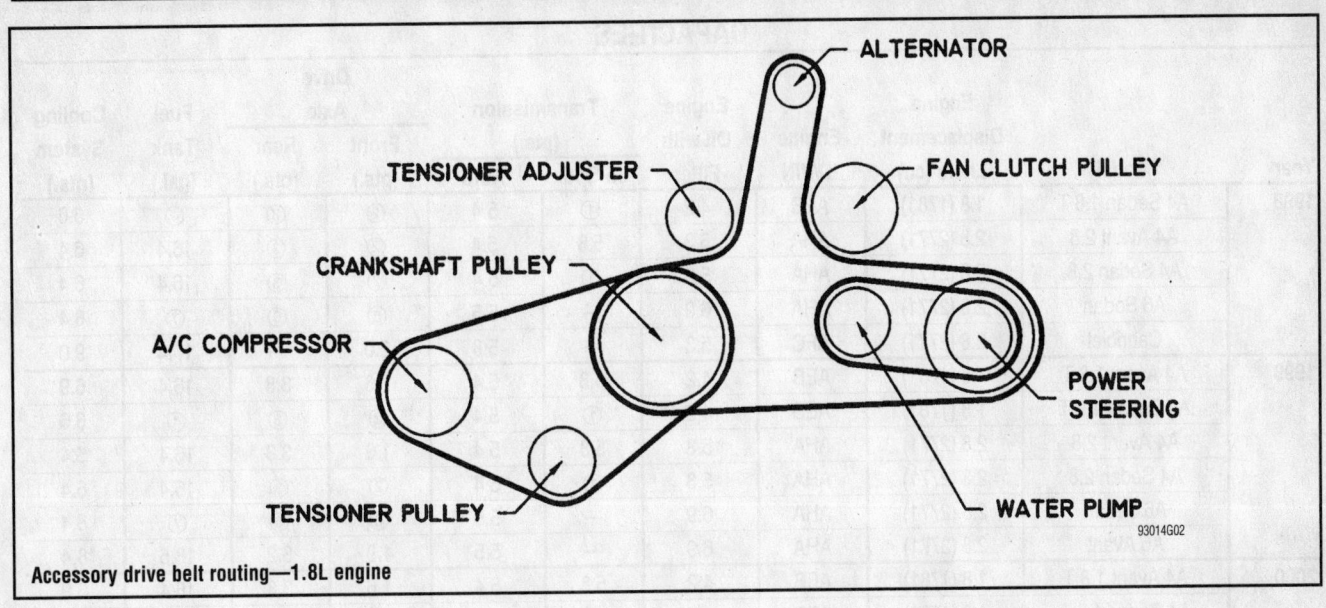

Accessory drive belt routing—1.8L engine

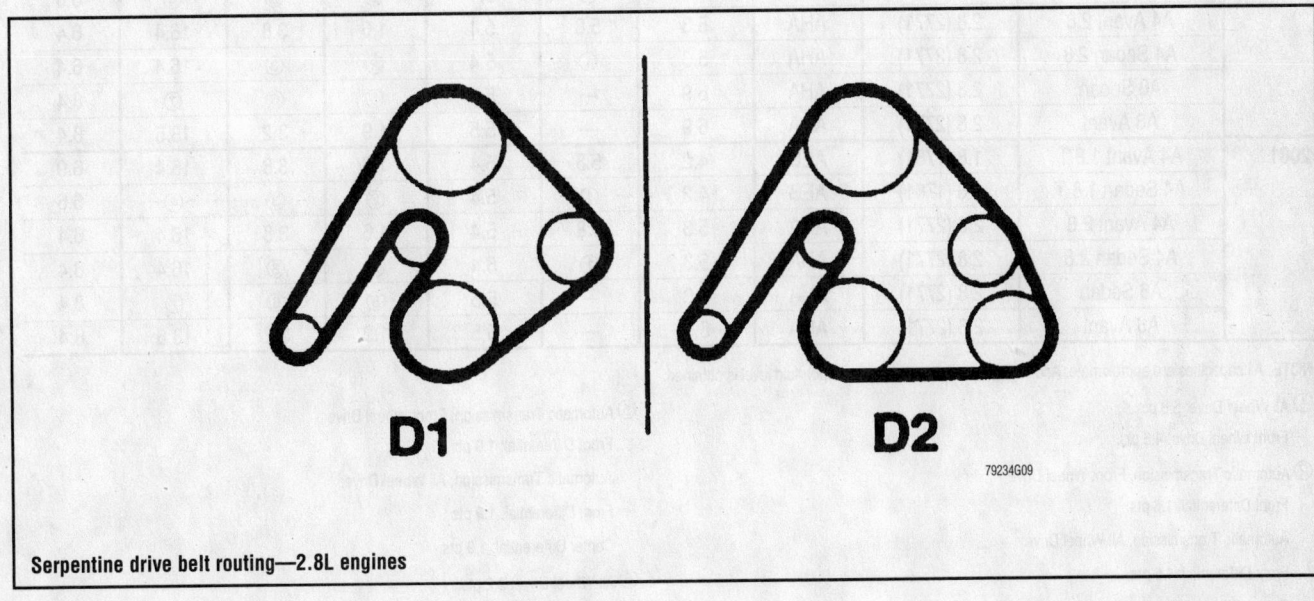

Serpentine drive belt routing—2.8L engines

CAPACITIES

Year	Model	Engine Displacement Liters (cc)	Engine ID/VIN	Engine Oil with Filter	Transmission (pts.)		Drive Axle		Fuel Tank (gal.)	Cooling System (qts.)
					5-Spd	Auto	Front (pts.)	Rear (pts.)		
1998	A4 Sedan 1.8 T	1.8 (1781)	AEB	4.2	①	5.4	②	③	④	6.6
	A4 Avant 2.8	2.8 (2771)	AHA	5.3	5.8	5.4	②	③	16.4	6.4
	A4 Sedan 2.8	2.8 (2771)	AHA	5.3	①	5.4	②	③	16.4	6.4
	A6 Sedan	2.8 (2771)	AHA	6.9	—	5.5	⑤	⑥	⑦	8.4
	Cabriolet	2.8 (2771)	AFC	5.3	—	5.8	2.0	—	17.4	9.0
1999	A4 Avant 1.8 T	1.8 (1781)	AEB	4.2	5.8	5.4	1.6	3.8	16.4	6.9
	A4 Sedan 1.8 T	1.8 (1781)	AEB	4.2	①	5.4	②	③	④	6.6
	A4 Avant 2.8	2.8 (2771)	AHA	5.3	5.8	5.4	1.6	3.8	16.4	6.4
	A4 Sedan 2.8	2.8 (2771)	AHA	5.3	①	5.4	②	③	16.4	6.4
	A6 Sedan	2.8 (2771)	AHA	6.9	—	5.5	⑤	⑥	⑦	8.4
	A6 Avant	2.8 (2771)	AHA	6.9	—	5.5	1.9	3.2	18.5	8.4
2000	A4 Avant 1.8 T	1.8 (1781)	AEB	4.2	5.8	5.4	1.6	3.8	16.4	6.9
	A4 Sedan 1.8 T	1.8 (1781)	AEB	4.2	①	5.4	②	③	④	6.6
	A4 Avant 2.8	2.8 (2771)	AHA	5.3	5.8	5.4	1.6	3.8	16.4	6.4
	A4 Sedan 2.8	2.8 (2771)	AHA	5.3	①	5.4	②	③	16.4	6.4
	A6 Sedan	2.8 (2771)	AHA	6.9	—	5.5	⑤	⑥	⑦	8.4
	A6 Avant	2.8 (2771)	AHA	6.9	—	5.5	1.9	3.2	18.5	8.4
2001	A4 Avant 1.8 T	1.8 (1781)	AEB	4.2	5.8	5.4	1.6	3.8	16.4	6.9
	A4 Sedan 1.8 T	1.8 (1781)	AEB	4.2	①	5.4	②	③	④	6.6
	A4 Avant 2.8	2.8 (2771)	AHA	5.3	5.8	5.4	1.6	3.8	16.4	6.4
	A4 Sedan 2.8	2.8 (2771)	AHA	5.3	①	5.4	②	③	16.4	6.4
	A6 Sedan	2.8 (2771)	AHA	6.9	—	5.5	⑤	⑥	⑦	8.4
	A6 Avant	2.8 (2771)	AHA	6.9	—	5.5	1.9	3.2	18.5	8.4

NOTE: All capacities are approximate. Add fluid gradually and ensure a proper fluid level is obtained.

① All Wheel Drive: 5.8 pts.
   Front Wheel Drive: 4.8 pts.

② Automatic Transmission, Front Wheel Drive:
   Front Differential: 1.6 pts.
   Automatic Transmission, All Wheel Drive:
   Front Differential: 1.6 pts.
   Center Differential: 1.6 pts.

③ All Wheel Drive: 3.8 pts.

④ Front Wheel Drive: 16.6 gal.
   All Wheel Drive: 16.5 gal.

⑤ Automatic Transmission, Front Wheel Drive:
   Front Differential: 1.9 pts.
   Automatic Transmission, All Wheel Drive:
   Front Differential: 1.9 pts.
   Center Differential: 1.9 pts.

⑥ All Wheel Drive: 3.2 pts.
   Front Differential: 1.9 pts.

⑦ All Wheel Drive: 21.1 gallons
   Front Wheel Drive: 18.5 gallons

93471C23

VALVE SPECIFICATIONS

Year	Engine Displacement Liters (cc)	Engine ID/VIN	Seat Angle (deg.)	Face Angle (deg.)	Spring Test Pressure (lbs. @ in.)	Spring Installed Height (in.)	Stem-to-Guide Clearance (in.)		Stem Diameter (in.)	
							Intake	Exhaust	Intake	Exhaust
1998	1.8 (1781)	AEB	45	45	NA	NA	0.039 ①	0.051 ①	0.2339-0.2350	0.2339-0.2343
	2.8 (2771)	AFC	45	45	NA	NA	0.039 ①	0.051 ①	NA	NA
	2.8 (2771)	AHA	45	45	NA	NA	0.031 ①	0.031 ①	0.2339-0.2350	0.2339-0.2343
1999	1.8 (1781)	AEB	45	45	NA	NA	0.031 ①	0.031 ①	0.2339-0.2350	0.2339-0.2343
	2.8 (2771)	AHA	45	45	NA	NA	0.031 ①	0.031 ①	0.2339-0.2350	0.2339-0.2343
2000	1.8 (1781)	AEB	45	45	NA	NA	0.031 ①	0.031 ①	0.2339-0.2350	0.2339-0.2343
	2.8 (2771)	AHA	45	45	NA	NA	0.031 ①	0.031 ①	0.2339-0.2350	0.2339-0.2343
2001	1.8 (1781)	AEB	45	45	NA	NA	0.031 ①	0.031 ①	0.2339-0.2350	0.2339-0.2343
	2.8 (2771)	AHA	45	45	NA	NA	0.031 ①	0.031 ①	0.2339-0.2350	0.2339-0.2343

NA: Not Available

① To measure: Insert a new valve into guide with end of valve flush with end of guide. Use a dial indicator to measure axial valve head movement.

93471C24

CRANKSHAFT AND CONNECTING ROD SPECIFICATIONS

All measurements are given in inches.

Year	Engine Displacement Liters (cc)	Engine ID/VIN	Crankshaft				Connecting Rod		
			Main Brg. Journal Dia.	Main Brg. Oil Clearance	Shaft End-play	Thrust on No.	Journal Diameter	Oil Clearance ①	Side Clearance
1998	1.8 (1781)	AEB	2.1267-2.1275	0.0008-0.0016	0.0030-0.0090	3	1.8811-1.8837	0.0004-0.0020	0.0040-0.0140
	2.8 (2771)	AFC	2.5573-2.5598	0.0007-0.0018	0.0027-0.0091	3	2.1243-2.1268	0.0006-0.0024	0.0060-0.0140
	2.8 (2771)	AHA	2.5573-2.5598	0.0007-0.0018	0.0028-0.0091	3	2.1243-2.1268	0.0006-0.0024	0.0060-0.0140
1999	1.8 (1781)	AEB	2.1267-2.1275	0.0008-0.0016	0.0030-0.0090	3	1.8811-1.8837	0.0004-0.0020	0.0040-0.0140
	2.8 (2771)	AHA	2.5573-2.5598	0.0007-0.0018	0.0028-0.0091	3	2.1243-2.1268	0.0006-0.0024	0.0060-0.0140
2000	1.8 (1781)	AEB	2.1267-2.1275	0.0008-0.0016	0.0030-0.0090	3	1.8811-1.8837	0.0004-0.0020	0.0040-0.0140
	2.8 (2771)	AHA	2.5573-2.5598	0.0007-0.0018	0.0028-0.0091	3	2.1243-2.1268	0.0006-0.0024	0.0060-0.0140
2001	1.8 (1781)	AEB	2.1267-2.1275	0.0008-0.0016	0.0030-0.0090	3	1.8811-1.8837	0.0004-0.0020	0.0040-0.0140
	2.8 (2771)	AHA	2.5573-2.5598	0.0007-0.0018	0.0028-0.0091	3	2.1243-2.1268	0.0006-0.0024	0.0060-0.0140

① To measure oil clearance torque as follows:

AFC and AHA engine connecting rods: 15 ft. lbs.

AEB engine connecting rods: 22 ft. lbs.

93471C25

Timing belt service is covered in Section 3 of this manual

PISTON AND RING SPECIFICATIONS
All measurements are given in inches

Year	Engine Displacemen Liters (cc)	Engine ID/VIN	Piston Clearance	Ring Gap			Ring Side Clearance		
				Top Compression	Bottom Compression	Oil Control	Top Compression	Bottom Compression	Oil Control
1998	1.8 (1781)	AEB	0.0014-0.0022	0.0078-0.0157	0.0078-0.0157	0.0098-0.0197	0.0023-0.0035	0.0019-0.0315	0.0011-0.0023
	2.8 (2771)	AHA	0.0008-0.0012	0.0140-0.0200	0.0200-0.0280	0.0100-0.0200	0.0010-0.0030	0.0010-0.0030	0.0010-0.0030
	2.8 (2771)	AFC	0.0008-0.0012	0.0140-0.0200	0.0200-0.0280	0.0100-0.0200	0.0010-0.0030	0.0010-0.0030	0.0010-0.0030
1999	1.8 (1781)	AEB	0.0014-0.0022	0.0078-0.0157	0.0078-0.0157	0.0098-0.0197	0.0023-0.0035	0.0019-0.0315	0.0011-0.0023
	2.8 (2771)	AHA	0.0008-0.0012	0.0140-0.0200	0.0200-0.0280	0.0100-0.0200	0.0010-0.0030	0.0010-0.0030	0.0010-0.0030
2000	1.8 (1781)	AEB	0.0014-0.0022	0.0078-0.0157	0.0078-0.0157	0.0098-0.0197	0.0023-0.0035	0.0019-0.0315	0.0011-0.0023
	2.8 (2771)	AHA	0.0008-0.0012	0.0140-0.0200	0.0200-0.0280	0.0100-0.0200	0.0010-0.0030	0.0010-0.0030	0.0010-0.0030
2001	1.8 (1781)	AEB	0.0014-0.0022	0.0078-0.0157	0.0078-0.0157	0.0098-0.0197	0.0023-0.0035	0.0019-0.0315	0.0011-0.0023
	2.8 (2771)	AHA	0.0008-0.0012	0.0140-0.0200	0.0200-0.0280	0.0100-0.0200	0.0010-0.0030	0.0010-0.0030	0.0010-0.0030

93471C26

TORQUE SPECIFICATIONS

All readings in ft. lbs.

Year	Engine Displacement Liters (cc)	Engine ID/VIN	Cylinder Head Bolts	Main Bearing Bolts	Rod Bearing Bolts	Crankshaft Damper Bolts	Flywheel Bolts	Manifold		Spark Plugs	Lug Nut
								Intake	Exhaust		
1998	1.8 (1781)	AEB	①	②	③	④	⑤	7	18	22	89
	2.8 (2771)	AFC	①	⑥	③	④	⑤	15	18	22	89
	2.8 (2771)	AHA	①	⑥	③	④	⑤	15	18	22	89
1999	1.8 (1781)	AEB	①	②	③	④	⑤	7	18	22	89
	2.8 (2771)	AHA	①	⑥	③	④	⑤	15	18	22	89
2000	1.8 (1781)	AEB	①	②	③	④	⑤	7	18	22	89
	2.8 (2771)	AHA	①	⑥	③	④	⑤	15	18	22	89
2001	1.8 (1781)	AEB	①	②	③	④	⑤	7	18	22	89
	2.8 (2771)	AHA	①	⑥	③	④	⑤	15	18	22	89

① Step 1: 44 ft. lbs.
   Step 2: 90 degrees
   Step 3: 90 degrees

② Step 1: 44 ft. lbs.
   Step 2: 180 degrees

③ Step 1: 22 ft. lbs.
   Step 2: 90 degrees

④ Center Bolt, installed with oil:
   Step 1: 148 ft. lbs.
   Step 2: 180 degrees
   Damper Bolts: 15 ft. lbs.

⑤ Flywheel MT:
   Step 1: AFC engine: 30 ft. lbs. all others, 44 ft. lbs.
   Step 2: 90 degrees
   Step 3: 90 degrees
   Flexplate AT:
   Step 1: 44 ft. lbs.
   Step 2: 90 degrees

⑥ Step 1: 48 ft. lbs.
   Step 2: 90 degrees

93471C27

Heater Core replacement is covered in Section 2 of this manual

BRAKE SPECIFICATIONS
All measurements in inches unless noted

Year	Model		Brake Disc Original Thickness	Brake Disc Minimum Thickness	Brake Disc Maximum Runout	Minimum Lining Thickness Front	Minimum Lining Thickness Rear	Brake Caliper Bracket Bolts (ft. lbs.)	Brake Caliper Mounting Bolts (ft. lbs.)
1998	A4 Sedan	F	①	②	0.002	③	—	92	④
		R	0.390	0.310	0.002	—	0.080	⑤	26
	A4 Avant	F	①	②	0.002	③	—	92	④
		R	0.390	0.310	0.002	—	0.080	⑤	26
	A6 Sedan	F	0.984	0.906	0.002	0.078	—	89	18
		R	0.394	0.315	0.002	—	0.079	70	26
	Cabriolet	F	0.980	0.910	0.002	0.080	—	92	18
		R	0.390	0.310	0.002	—	0.080	48	26
1999	A4 Sedan	F	①	②	0.002	③	—	92	④
		R	0.390	0.310	0.002	—	0.080	⑤	26
	A4 Avant	F	①	②	0.002	③	—	92	④
		R	0.390	0.310	0.002	—	0.080	⑤	26
	A6 Sedan	F	0.984	0.906	0.002	0.078	—	89	18
		R	0.394	0.315	0.002	—	0.079	70	26
	A6 Avant	F	0.984	0.906	0.002	0.078	—	89	18
		R	0.394	0.315	0.002	—	0.079	70	26
2000	A4 Sedan	F	①	②	0.002	③	—	92	④
		R	0.390	0.310	0.002	—	0.080	⑤	26
	A4 Avant	F	①	②	0.002	③	—	92	④
		R	0.390	0.310	0.002	—	0.080	⑤	26
	A6 Sedan	F	0.984	0.906	0.002	0.078	—	89	18
		R	0.394	0.315	0.002	—	0.079	70	26
	A6 Avant	F	0.984	0.906	0.002	0.078	—	89	18
		R	0.394	0.315	0.002	—	0.079	70	26
2001	A4 Sedan	F	①	②	0.002	③	—	92	④
		R	0.390	0.310	0.002	—	0.080	⑤	26
	A4 Avant	F	①	②	0.002	③	—	92	④
		R	0.390	0.310	0.002	—	0.080	⑤	26
	A6 Sedan	F	0.984	0.906	0.002	0.078	—	89	18
		R	0.394	0.315	0.002	—	0.079	70	26
	A6 Avant	F	0.984	0.906	0.002	0.078	—	89	18
		R	0.394	0.315	0.002	—	0.079	70	26

① Teves/Ate Calipers:
  Venetilated Disc: 0.984 inches
  Non-ventilated Disc: 0.590 inches
  Lucas Calipers:
  Venetilated Disc: 0.870 inches
  Non-ventilated Disc: 0.590 inches
  Double Piston Calipers:
  Venetilated Disc: 1.180 inches

② Teves/Ate Calipers:
  Venetilated Disc: 0.905 inches
  Non-ventilated Disc: 0.510
  Lucas Calipers:
  Venetilated Disc: 0.790 inches
  Non-ventilated Disc: 0.430 inches
  Double Piston Calipers:
  Venetilated Disc: 1.100 inches

③ Teves/Ate and Lucas Calipers: 0.080 inches
  Double Piston Calipers: 0.12 inches

④ Teves/ATE: 18 ft. lbs.
  Lucas: 22 ft. lbs.
  Double Piston Calipers: 148 ft. lbs.

⑤ Front Wheel Drive Models (Ribbed Bolt): 70 ft. lbs.
  All Wheel Drive Models (Socket-head Bolt): 44 ft. lbs.

93471C28

WHEEL ALIGNMENT

Year	Model		Caster Range (+/-Deg.)	Caster Preferred Setting (Deg.)	Camber Range (+/-Deg.)	Camber Preferred Setting (Deg.)	Toe-in (in.)	Steering Axis Inclination (Deg.)
1998	A4 ①	F	—	—	0.42	-0.58	0.08 +/- 0.02	—
	Standard Suspension	R	—	—	0.33	+0.50	0.08 +/- 0.04	—
	A4 ①	F	—	—	0.42	-0.83	0.08 +/- 0.02	—
	Sport Suspension	R	—	—	0.33	+0.50	0.13 +/- 0.04	—
	A4 ①	F	—	—	0.42	-0.33	0.08 +/- 0.02	—
	Heavy Duty Suspension	R	—	—	—	—	—	—
	A4 ②	F	—	—	0.42	-0.42	0.08 +/- 0.02	—
	Standard Suspension	R	—	—	—	—	—	—
	A4 ②	F	—	—	0.42	-0.66	0.08 +/- 0.02	—
	Sport Suspension	R	—	—	—	—	—	—
	A4 ②	F	—	—	0.42	-0.50	0.08 +/- 0.02	—
	Heavy Duty Suspension	R	—	—	—	—	—	—
	A6 ③	F	—	—	0.42	-0.75	0.08 +/- 0.02	—
	FWD	R	—	—	0.30	+0.70	0.16 +/- 0.08	—
	A6 ④	F	—	—	0.42	-1.00	0.08 +/- 0.02	—
	FWD	R	—	—	0.30	+0.70	0.24 +/- 0.08	—
	A6 ⑤	F	—	—	0.42	-0.58	0.08 +/- 0.02	—
	FWD	R	—	—	0.30	+0.70	0.12 +/- 0.08	—
	A6 ③	F	—	—	0.42	-0.75	0.08 +/- 0.02	—
	AWD	R	—	—	0.50	+0.70	0.06 +/- 0.04	—
	A6 ④	F	—	—	0.42	-1.00	0.08 +/- 0.02	—
	AWD	R	—	—	0.50	+0.70	0.06 +/- 0.04	—
	A6 ⑤	F	—	—	0.42	-0.58	0.08 +/- 0.02	—
	AWD	R	—	—	0.50	+0.70	0.06 +/- 0.04	—
1999	A4 ①	F	—	—	0.42	-0.58	0.08 +/- 0.02	—
	Standard Suspension	R	—	—	0.33	+0.50	0.08 +/- 0.04	—
	A4 ①	F	—	—	0.42	-0.83	0.08 +/- 0.02	—
	Sport Suspension	R	—	—	0.33	+0.50	0.13 +/- 0.04	—
	A4 ①	F	—	—	0.42	-0.33	0.08 +/- 0.02	—
	Heavy Duty Suspension	R	—	—	—	—	—	—
	A4 ②	F	—	—	0.42	-0.42	0.08 +/- 0.02	—
	Standard Suspension	R	—	—	—	—	—	—
	A4 ②	F	—	—	0.42	-0.66	0.08 +/- 0.02	—
	Sport Suspension	R	—	—	—	—	—	—
	A4 ②	F	—	—	0.42	-0.50	0.08 +/- 0.02	—
	Heavy Duty Suspension	R	—	—	—	—	—	—
	A6 ③	F	—	—	0.42	-0.75	0.08 +/- 0.02	—
	FWD	R	—	—	0.30	+0.70	0.16 +/- 0.08	—
	A6 ④	F	—	—	0.42	-1.00	0.08 +/- 0.02	—
	FWD	R	—	—	0.30	+0.70	0.24 +/- 0.08	—
	A6 ⑤	F	—	—	0.42	-0.58	0.08 +/- 0.02	—
	FWD	R	—	—	0.30	+0.70	0.12 +/- 0.08	—
	A6 ③	F	—	—	0.42	-0.75	0.08 +/- 0.02	—
	AWD	R	—	—	0.50	+0.70	0.06 +/- 0.04	—
	A6 ④	F	—	—	0.42	-1.00	0.08 +/- 0.02	—
	AWD	R	—	—	0.50	+0.70	0.06 +/- 0.04	—
	A6 ⑤	F	—	—	0.42	-0.58	0.08 +/- 0.02	—
	AWD	R	—	—	0.50	+0.70	0.06 +/- 0.04	—
2000	A4 ①	F	—	—	0.42	-0.58	0.08 +/- 0.02	—
	Standard Suspension	R	—	—	0.33	+0.50	0.08 +/- 0.04	—
	A4 ①	F	—	—	0.42	-0.83	0.08 +/- 0.02	—
	Sport Suspension	R	—	—	0.33	+0.50	0.13 +/- 0.04	—

93471C29

Brake service is covered in Section 4 of this manual

WHEEL ALIGNMENT

Year	Model		Caster Range (+/-Deg.)	Caster Preferred Setting (Deg.)	Camber Range (+/-Deg.)	Camber Preferred Setting (Deg.)	Toe-in (in.)	Steering Axis Inclination (Deg.)
2000 (cont.)	A4 ①	F	—	—	0.42	-0.33	0.08 +/- 0.02	—
	Heavy Duty Suspension	R	—	—	—	—	—	—
	A4 ②	F	—	—	0.42	-0.42	0.08 +/- 0.02	—
	Standard Suspension	R	—	—	—	—	—	—
	A4 ②	F	—	—	0.42	-0.66	0.08 +/- 0.02	—
	Sport Suspension	R	—	—	—	—	—	—
	A4 ②	F	—	—	0.42	-0.50	0.08 +/- 0.02	—
	Heavy Duty Suspension	R	—	—	—	—	—	—
	A6 ③	F	—	—	0.42	-0.75	0.08 +/- 0.02	—
	FWD	R	—	—	0.30	+0.70	0.16 +/- 0.08	—
	A6 ④	F	—	—	0.42	-1.00	0.08 +/- 0.02	—
	FWD	R	—	—	0.30	+0.70	0.23 +/- 0.08	—
	A6 ⑤	F	—	—	0.42	-0.58	0.08 +/- 0.02	—
	FWD	R	—	—	0.30	+0.70	0.12 +/- 0.08	—
	A6 ③	F	—	—	0.42	-0.75	0.08 +/- 0.02	—
	AWD	R	—	—	0.50	+0.70	0.06 +/- 0.04	—
	A6 ④	F	—	—	0.42	-1.00	0.08 +/- 0.02	—
	AWD	R	—	—	0.50	+0.70	0.06 +/- 0.04	—
	A6 ⑤	F	—	—	0.42	-0.58	0.08 +/- 0.02	—
	AWD	R	—	—	0.50	+0.70	0.06 +/- 0.04	—
	S4	F	—	—	0.42	-0.58	0.08 +/- 0.02	—
		R	—	—	0.33	+0.50	0.08 +/- 0.04	—
	TT Coupe	F	0.50	7.97	0.50	-0.75	0.07 +/- 0.05	—
	AWD	R	—	—	0.20	-1.2	0.06 +/- 0.04	—
	TT Coupe	F	0.50	7.97	0.50	-0.75	0.07 +/- 0.05	—
	FWD	R	—	—	0.20	-2.00	0.08 +/- 0.04	—
2001	A6 ①	F	—	—	0.42	-0.75	0.08 +/- 0.02	—
	FWD	R	—	—	0.30	+0.70	0.16 +/- 0.08	—
	A6 ②	F	—	—	0.42	-1.00	0.08 +/- 0.02	—
	FWD	R	—	—	0.30	+0.70	0.23 +/- 0.08	—
	A6 ③	F	—	—	0.42	-0.58	0.08 +/- 0.02	—
	FWD	R	—	—	0.30	+0.70	0.12 +/- 0.08	—
	A6 ①	F	—	—	0.42	-0.75	0.08 +/- 0.02	—
	AWD	R	—	—	0.50	0.70	0.06 +/- 0.04	—
	A6 ②	F	—	—	0.42	-1.00	0.08 +/- 0.02	—
	AWD	R	—	—	0.50	+0.70	0.06 +/- 0.04	—
	A6 ③	F	—	—	0.42	-0.58	0.08 +/- 0.02	—
	AWD	R	—	—	0.50	+0.70	0.06 +/- 0.04	—
	S4	F	—	—	0.42	-0.58	0.08 +/- 0.02	—
		R	—	—	0.33	+0.50	0.08 +/- 0.04	—
	TT Coupe	F	0.50	7.97	0.50	-0.75	0.07 +/- 0.05	—
	AWD	R	—	—	0.20	-1.2	0.06 +/- 0.04	—
	TT Coupe	F	0.50	7.97	0.50	-0.75	0.07 +/- 0.05	—
	FWD	R	—	—	0.20	-2.00	0.08 +/- 0.04	—

① With aluminum mounting brackets
② Without aluminum mounting brackets
③ With standard suspension
④ With sport suspension
⑤ With heavy duty suspension

TIRE, WHEEL AND BALL JOINT SPECIFICATIONS

| Year | Model | OEM Tires | | Tire Pressures (psi) | | Wheel | Ball Joint |
		Standard	Optional	Front	Rear	Size	Inspection
1998	A8	225/60HR16	225/55WR17	34	34	Std: 7-J Opt: 8-J	NS
	Cabriolet	195/65HR15	205/55HR16	34	34	7-J	NS
	A4	195/65VR15	205/55HR16	34	34	7-J	NS
	Avant	195/65VR15	205/55HR16	34	34	7-J	NS
1999	A8	225/60HR16	225/55WR17	34	34	Std: 7-J Opt: 8-J	NS
	Cabriolet	195/65HR15	205/55HR16	34	34	7-J	NS
	A4	205/60HR15	205/55ZR16	34	34	7-J	NS
2000	A8	225/60HR16	225/55WR17	34	34	Std: 7-J Opt: 8-J	NS
	Cabriolet	195/65HR15	205/55HR16	34	34	7-J	NS
	A4	205/60HR15	205/55ZR16	34	34	7-J	NS
2001	A8	225/60HR16	225/55WR17	34	34	Std: 7-J Opt: 8-J	NS
	Cabriolet	195/65HR15	205/55HR16	34	34	7-J	NS
	A4	205/60HR15	205/55ZR16	34	34	7-J	NS

OEM: Original Equipment Manufacturer

PSI: Pounds Per Square Inch

STD: Standard

OPT: Optional

NS: Not Specified by manufacturer

93471C31

For complete Engine Mechanical specifications, see Section 1 of this manual

SCHEDULED MAINTENANCE INTERVALS
Audi

TO BE SERVICED	TYPE OF SERVICE	VEHICLE MILEAGE INTERVAL (x1000)												
		7.5	15	22.5	30	37.5	45	52.5	60	67.5	75	82.5	90	97.5
Engine oil & filter ①	R	✓	✓	✓	✓	✓	✓	✓	✓	✓	✓	✓	✓	✓
Automatic shiftlock operation	S/I	✓	✓	✓	✓	✓	✓	✓	✓	✓	✓	✓	✓	✓
Cooling system	S/I	✓	✓	✓	✓	✓	✓	✓	✓	✓	✓	✓	✓	✓
Passenger compartment air filter	R		✓		✓		✓		✓		✓		✓	
Automatic transmission fluid, filter & final drive	S/I		✓		✓		✓		✓		✓		✓	
Battery electrolyte level	S/I		✓		✓		✓		✓		✓		✓	
Brake system (brake pads & fluid level)	S/I		✓		✓		✓		✓		✓		✓	
Drive axle shaft boots	S/I		✓		✓		✓		✓		✓		✓	
Engine (check for leaks)	S/I		✓		✓		✓		✓		✓		✓	
Exhaust system	S/I		✓		✓		✓		✓		✓		✓	
Idle speed ②	S/I		✓		✓		✓		✓		✓		✓	
Manual transmission fluid	S/I		✓		✓		✓		✓		✓		✓	
ODB System check for codes	S/I		✓		✓		✓		✓		✓		✓	
V-belts ③	S/I			✓		✓				✓		✓		
Air cleaner element	R				✓				✓				✓	
Spark plugs	R				✓				✓				✓	
Power steering fluid level	S/I				✓				✓				✓	
Automatic transmission fluid A1④	R						✓						✓	
Timing belt	R												✓	
Brake fluid ⑤	R													
Front axle dust seals on ball joints & tie rod ends	S/I								✓					
Poly-ribbed belt	R												✓	
Rotate tires	S/I	✓												

R: Replace   S/I: Service or Inspect

① Reset service interval display, if equipped.

② Except California models.

③ Replace at 45,000 & 90,000 miles.

④ Replace at mileage interval or every 2 years, whichever comes first.

⑤ Replace every 2 years regardless of mileage.

FREQUENT OPERATION MAINTENANCE (SEVERE SERVICE)

If a vehicle is operated under any of the following conditions it is considered severe service:

- Extremely dusty areas.

- 50% or more of the vehicle operation is in 32°C (90°F) or higher temperatures, or constant operation in temperatures below 0°C (32°F).

- Prolonged idling (vehicle operation in stop and go traffic).

- Frequent short running periods (engine does not warm to normal operating temperatures).

- Police, taxi, delivery usage or trailer towing usage.

Oil & oil filter: change every 3750 miles.

Automatic transmission fluid: replace every 30,000 miles.

93471C32

SCHEDULED MAINTENANCE INTERVALS
AUDI
A4, A6, A8, CABRIOLET

The following should be used as a guide when determining the amount of work required for a particular service. In estimating how long a particular Scheduled Maintenance Service should take, please observe the following:

● Labor Time is time based on field research and data supplied by the vehicle manufacturer.
● Labor time operations are given in hours and tenths of an hour.
● All labor operations are to be used as a guide.

Mechanic Skill Level Codes:
(A) PRECISION: Highly skilled with multiple certification.
(B) GENERAL: Normally skilled with certification.
(C) MAINTENANCE: Semi-skilled working on certification.

	LABOR TIME
7500 Mile Service (C)	
A4	1.1
A6, A8	1.3
Cabriolet	1.1
15000 Mile Service (B)	
A4	1.9
A6	2.3
A8	1.8
Cabriolet	2.1
22500 Mile Service (C)	
A4, A6, A8	.9
Cabriolet	.6
30000 Mile Service (B)	
A4	2.4
A6	2.7
A8	2.8
Cabriolet	2.5

	LABOR TIME
37500 Mile Service (C)	
A4, A6, A8	.9
Cabriolet	.6
45000 Mile Service (B)	
A4	1.6
A6	1.8
A8	2.1
Cabriolet	1.8
52500 Mile Service (C)	
A4	1.1
A6, A8	1.3
Cabriolet	1.1
60000 Mile Service (B)	
A4	2.4
A6	2.7
A8	2.8
Cabriolet	2.5
67500 Mile Service (C)	
A4, A6, A8	.9
Cabriolet	.6

	LABOR TIME
75000 Mile Service (B)	
A4	1.9
A6, Cabriolet	2.1
A8	1.8
82500 Mile Service (C)	
A4, A6, A8	.9
Cabriolet	.6
90000 Mile Service (B)	
A4	2.9
A6	3.2
A8	3.3
Cabriolet	2.8
Replace timing belt (2.8L) add	1.5
97500 Mile Service (C)	
A4	1.1
A6, A8	1.3
Cabriolet	1.1

93471C33

For Accessory Drive Belt illustrations, see Section 1 of this manual

BMW
318i • 318iS • 318iC • 323is • 323iC • 325i • 325iS • 325iC • 328i • 525i • 525ti • 528i • 530i • 530ti • 540i • 740i • 740iL • 840Ci • 750iL • 850Ci • 850CSi • M3 • Z3

ENGINE AND VEHICLE IDENTIFICATION

Engine								Model Year	
Code	Liters (cc)	Cu. In.	Cyl.	Fuel Sys.	Engine Type	Eng. Mfg.		Code ①	Year
M44B19	1.9 (1895)	116	4	②	DOHC	BMW		W	1998
M52B25	2.5 (2494)	152	6	③	DOHC	BMW		X	1999
M52TUB25	2.5 (2494)	152	6	③	DOHC	BMW		Y	2000
M52B28	2.8 (2793)	170	6	③	DOHC	BMW		1	2001
M52TUB28	2.8 (2793)	170	6	③	DOHC	BMW		2	2002
M62B44	4.4 (4398)	268	8	④	DOHC	BMW			
M73B54	5.4 (5379)	328	12	⑤	SOHC	BMW			
S52B32	3.2 (3152)	192	6	⑥	DOHC	BMW			

DOHC: Double Overhead Camshaft

SOHC: Single Overhead Camshaft

① 10th digit of the Vehicle Identification Number (VIN)

② Bosch ML-Motronic w/knock control (2 sensors)

③ Siemens MS 41.0 with knock control

④ Bosch HFM-Motronic M5.2 with adaptive knock control

⑤ Bosch HFM Motronic M5.2 (dual DME) with adaptive knock control

⑥ Siemens MS 41.1 with knock control (2 sensors)

93471C34

GENERAL ENGINE SPECIFICATIONS

Year	Body Type	Model	Engine Displacement Liters (cc)	Engine ID/VIN	Fuel System Type	Net Horsepower @ rpm	Net Torque @ rpm (ft. lbs.)	Bore x Stroke (in.)	Compression Ratio	Oil Pressure @ rpm
1998	E36	318i	1.9 (1895)	M44B19	①	138@6000	133@4300	3.35x3.29	10.0:1	7.4@900
	E36	318ti	1.9 (1895)	M44B19	①	138@6000	133@4300	3.35x3.29	10.0:1	7.4@900
	E36	323is	2.5 (2494)	M52B25	②	168@6000	181@4300	3.31x2.95	10.5:1	7.4@800
	E36	323iC	2.5 (2494)	M52B25	②	168@6000	181@4300	3.31x2.95	10.5:1	7.4@800
	E36	Z3 1.9	1.9 (1895)	M44B19	①	138@6000	133@4300	3.35x3.29	10.0:1	7.4@900
	E36	Z3 2.8	2.8 (2793)	M52TUB28	②	189@5300	203@3950	3.31x3.31	10.2:1	7.4@700
	E36	328i	2.8 (2793)	M52B28	②	190@5300	206@3950	3.31x3.31	10.2:1	7.4@700
	E36	328is	2.8 (2793)	M52B28	②	190@5300	206@3950	3.31x3.31	10.2:1	7.4@700
	E36	328iC	2.8 (2793)	M52B28	②	190@5300	206@3950	3.31x3.31	10.2:1	7.4@700
	E36	M3	3.2 (3152)	S52B32	③	240@6000	236@3800	3.40x3.53	10.5:1	7.4@700
	E39	528i	2.8 (2793)	M52B28	②	190@5300	206@3950	3.31x3.31	10.2:1	7.4@700
	E39	540i	4.4 (4398)	M62B44	④	282@5700	310@3900	3.62x3.26	10.0:1	7.4@580
	E38	740i	4.4 (4398)	M62B44	④	282@5700	310@3900	3.62x3.26	10.0:1	7.4@580
	E38	740iL	4.4 (4398)	M62B44	④	282@5700	310@3900	3.62x3.26	10.0:1	7.4@580
	E38	750iL	5.4 (5379)	M73B54	⑤	322@5000	361@3900	3.35x3.11	10.0:1	7.4@600
1999	E36	318ti S	1.9 (1895)	M44B19	①	138@6000	133@4300	3.35x3.29	10.0:1	7.4@900
	E36	323i	2.5 (2493)	M52B25	②	168@5500	181@3950	3.31x2.95	10.5:1	7.4@700
	E46	323i Sedan	2.5 (2494)	M52TUB25	②	168@5500	181@3950	3.31x2.95	10.5:1	7.4@700
	E36	323iC	2.5 (2494)	M52B25	②	168@5500	181@3950	3.31x2.95	10.5:1	7.4@700
	E36	Z3 2.3	2.5 (2494)	M52TUB25	②	170@5500	181@3500	3.31x2.95	10.5:1	7.4@700
	E36	Z3 2.8	2.8 (2793)	M52TUB28	②	193@5500	206@3500	3.31x3.31	10.2:1	7.4@700
	E36	Z3 Coupe	2.8 (2793)	M52TUB28	②	193@5500	206@3500	3.31x3.31	10.2:1	7.4@700
	E46	328i Sedan	2.8 (2793)	M52TUB28	②	190@5300	206@3950	3.31x3.31	10.2:1	7.4@700
	E36	328is	2.8 (2793)	M52B28	②	190@5300	206@3950	3.31x3.31	10.2:1	7.4@700
	E36	328iC	2.8 (2793)	M52B28	②	190@5300	206@3950	3.31x3.31	10.2:1	7.4@700
	E36	M3	3.2 (3152)	S52B32	③	240@6000	236@3800	3.40x3.53	10.5:1	7.4@700
	E39	528i	2.8 (2793)	M52B28	②	193@5500	206@3500	3.31x3.31	10.2:1	7.4@700
	E39	528i SW	2.8 (2793)	M52B28	②	193@5500	206@3500	3.31x3.31	10.2:1	7.4@700
	E39	540i SW	4.4 (4398)	M62B44	④	282@5400	324@3600	3.62x3.26	10.0:1	7.4@580
	E39	540i	4.4 (4398)	M62B44	④	282@5400	324@3600	3.62x3.26	10.0:1	7.4@580
	E38	740i	4.4 (4398)	M62B44	④	282@5400	324@3700	3.62x3.26	10.0:1	7.4@580
	E38	740iL	4.4 (4398)	M62B44	④	282@5400	324@3700	3.62x3.26	10.0:1	7.4@580
	E38	750iL	5.4 (5379)	M73B54	⑤	326@5000	361@3900	3.35x3.11	10.0:1	7.4@600

93471C35

For Tire, Wheel and Ball Joint specifications, see Section 1 of this manual

GENERAL ENGINE SPECIFICATIONS

Year	Body Type	Model	Engine Displacement Liters (cc)	Engine ID/VIN	Fuel System Type	Net Horsepower @ rpm	Net Torque @ rpm (ft. lbs.)	Bore x Stroke (in.)	Compression Ratio	Oil Pressure @ rpm
2000	E36	318ti S	1.9 (1895)	M44B19	①	138@6000	133@4300	3.35x3.29	10.0:1	7.4@900
	E36	323i	2.5 (2493)	M52B25	②	168@5500	181@3950	3.31x2.95	10.5:1	7.4@700
	E46	323i Sedan	2.5 (2494)	M52TUB25	②	168@5500	181@3950	3.31x2.95	10.5:1	7.4@700
	E36	323iC	2.5 (2494)	M52B25	②	168@5500	181@3950	3.31x2.95	10.5:1	7.4@700
	E36	Z3 2.3	2.5 (2494)	M52TUB25	②	170@5500	181@3500	3.31x2.95	10.5:1	7.4@700
	E36	Z3 2.8	2.8 (2793)	M52TUB28	②	193@5500	206@3500	3.31x3.31	10.2:1	7.4@700
	E36	Z3 Coupe	2.8 (2793)	M52TUB28	②	193@5500	206@3500	3.31x3.31	10.2:1	7.4@700
	E46	328i Sedan	2.8 (2793)	M52TUB28	②	190@5300	206@3950	3.31x3.31	10.2:1	7.4@700
	E36	328is	2.8 (2793)	M52B28	②	190@5300	206@3950	3.31x3.31	10.2:1	7.4@700
	E36	328iC	2.8 (2793)	M52B28	②	190@5300	206@3950	3.31x3.31	10.2:1	7.4@700
	E36	M3	3.2 (3152)	S52B32	③	240@6000	236@3800	3.40x3.53	10.5:1	7.4@700
	E39	528i	2.8 (2793)	M52B28	②	193@5500	206@3500	3.31x3.31	10.2:1	7.4@700
	E39	528i SW	2.8 (2793)	M52B28	②	193@5500	206@3500	3.31x3.31	10.2:1	7.4@700
	E39	540i SW	4.4 (4398)	M62B44	④	282@5400	324@3600	3.62x3.26	10.0:1	7.4@580
	E39	540i	4.4 (4398)	M62B44	④	282@5400	324@3600	3.62x3.26	10.0:1	7.4@580
	E38	740i	4.4 (4398)	M62B44	④	282@5400	324@3700	3.62x3.26	10.0:1	7.4@580
	E38	740iL	4.4 (4398)	M62B44	④	282@5400	324@3700	3.62x3.26	10.0:1	7.4@580
	E38	750iL	5.4 (5379)	M73B54	⑤	326@5000	361@3900	3.35x3.11	10.0:1	7.4@600

PGM-FI: Programmed Fuel Injection

① Bosch ML-Motronic w/knock control (2 sensors)

② Siemens MS 41.0 with knock control

③ Siemens MS 41.1 with adaptive knock control (2 sensors)

④ Bosch HFM-Motronic M5.2 with adaptive knock control

⑤ Bosch HFM Motronic M5.2 (dual DME) with adaptive knock control

93471C36

ENGINE TUNE-UP SPECIFICATIONS

Year	Engine Displacement Liters (cc)	Engine ID/VIN	Spark Plug Gap (in.)	Ignition Timing (deg.)		Fuel Pump (psi)	Idle Speed (rpm)		Valve Clearance	
				MT	AT		MT	AT	In.	Ex.
1998	1.9 (1895)	M44B19	①	②	②	41-47 ③	850-950	850-950	HYD	HYD
	2.5 (2494)	M52B25	①	②	②	48-54 ③	700-800	650-750	HYD	HYD
	2.5 (2494)	M52B25	①	②	②	48-54 ③	700-800	650-750	HYD	HYD
	2.8 (2793)	M52B28	①	②	②	48-54 ③	700-800	650-750	HYD	HYD
	2.8 (2793)	M52TUB28	①	②	②	48-54 ③	750-850	650-750	HYD	HYD
	3.2 (3152)	S52B32	①	②	②	48-54 ③	650-750	650-750	HYD	HYD
	4.4 (4398)	M62B44	①	②	②	48-54 ③	530-630	530-630	HYD	HYD
	5.4 (5379)	M73B54	①	—	②	48-54 ③	—	550-650	HYD	HYD
1999	1.9 (1895)	M44B19	①	②	②	41-47 ③	850-950	850-950	HYD	HYD
	2.5 (2494)	M52B25	①	②	②	48-54 ③	700-800	650-750	HYD	HYD
	2.5 (2494)	M52TUB25	①	②	②	48-54 ③	700-800	650-750	HYD	HYD
	2.8 (2793)	M52B28	①	②	②	48-54 ③	700-800	650-750	HYD	HYD
	2.8 (2793)	M52TUB28	①	②	②	48-54 ③	700-800	650-750	HYD	HYD
	3.2 (3152)	S52B32	①	②	②	48-54 ③	650-750	650-750	HYD	HYD
	4.4 (4398)	M62B44	①	②	②	48-54 ③	700-800	530-630	HYD	HYD
	5.4 (5379)	M73B54	①	—	②	48-54 ③	—	550-650	HYD	HYD
2000	1.9 (1895)	M44B19	①	②	②	41-47 ③	850-950	850-950	HYD	HYD
	2.5 (2494)	M52B25	①	②	②	48-54 ③	700-800	650-750	HYD	HYD
	2.5 (2494)	M52TUB25	①	②	②	48-54 ③	700-800	650-750	HYD	HYD
	2.8 (2793)	M52B28	①	②	②	48-54 ③	700-800	650-750	HYD	HYD
	2.8 (2793)	M52TUB28	①	②	②	48-54 ③	700-800	650-750	HYD	HYD
	3.2 (3152)	S52B32	①	②	②	48-54 ③	650-750	650-750	HYD	HYD
	4.4 (4398)	M62B44	①	②	②	48-54 ③	700-800	530-630	HYD	HYD
	5.4 (5379)	M73B54	①	—	②	48-54 ③	—	550-650	HYD	HYD

NOTE: The Vehicle Emission Control Information label reflects specification changes during production and must be used if they differ from this chart.

B: Before Top Dead Center

HYD: Hydraulic

① Three mass and four-mass electrodes cannot be adjusted

    Dual mass electrodes: 0.035-0.039 inches

    All models except M models: 0.028-0.031 inches

    M models: 0.024-0.028 inches

② Controlled by the Engine Control Module (ECM) and cannot be adjusted

③ At idle, pressure measured at injectors

93471C37

For Wheel Alignment specifications, see Section 1 of this manual

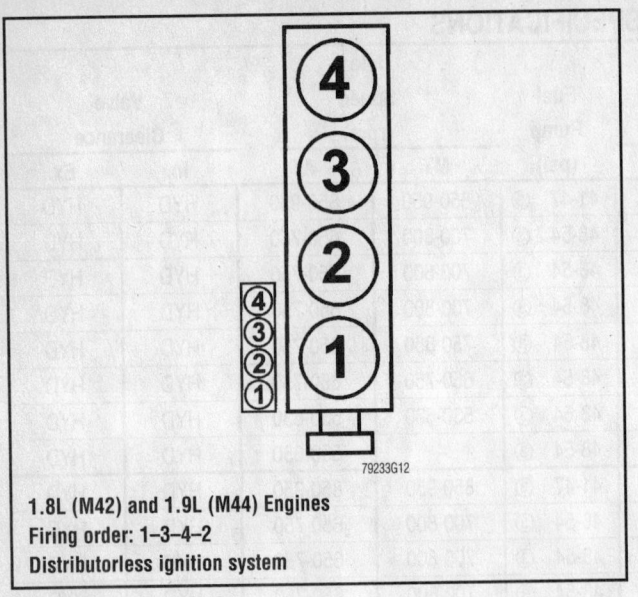

1.8L (M42) and 1.9L (M44) Engines
Firing order: 1–3–4–2
Distributorless ignition system

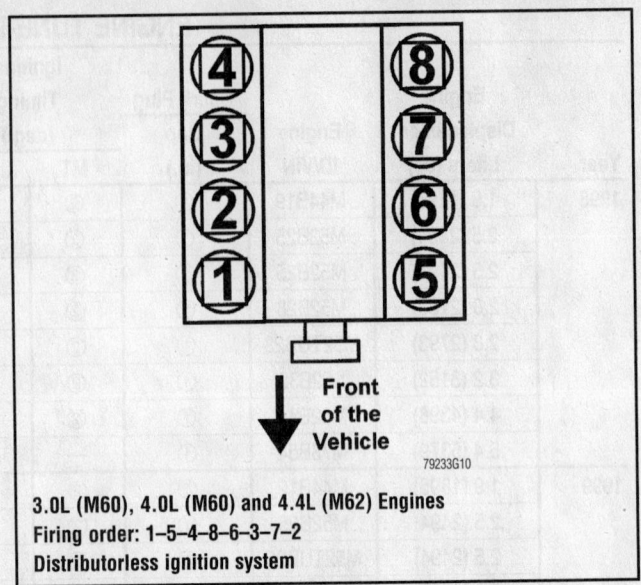

Front of the Vehicle

3.0L (M60), 4.0L (M60) and 4.4L (M62) Engines
Firing order: 1–5–4–8–6–3–7–2
Distributorless ignition system

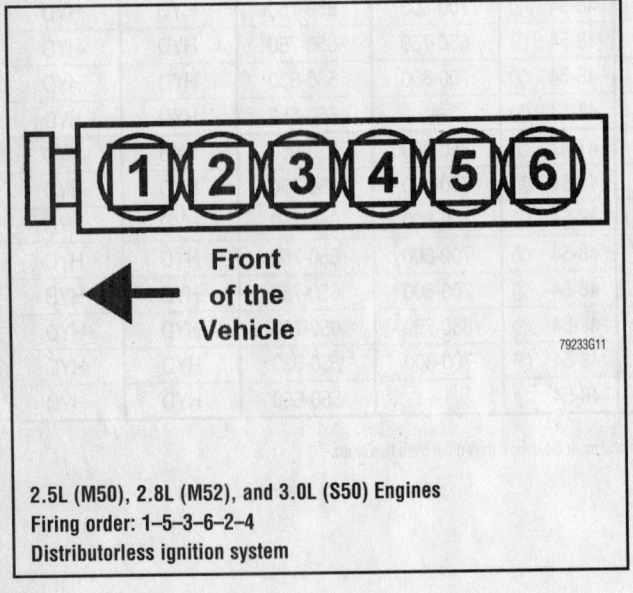

Front of the Vehicle

2.5L (M50), 2.8L (M52), and 3.0L (S50) Engines
Firing order: 1–5–3–6–2–4
Distributorless ignition system

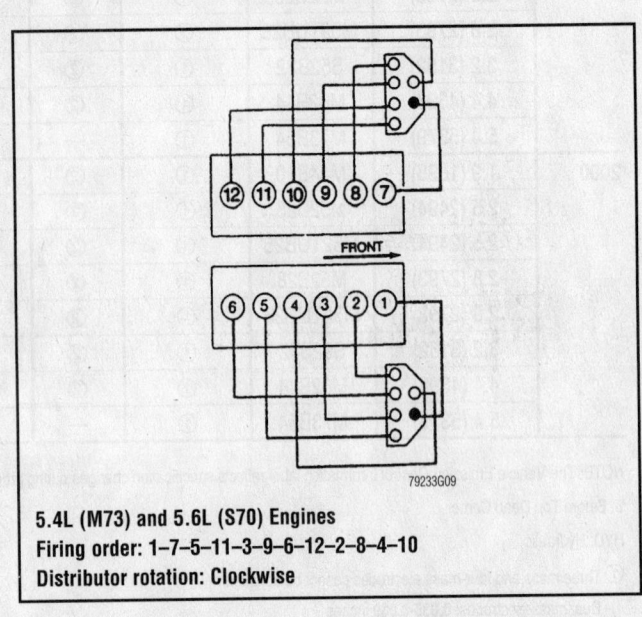

FRONT

5.4L (M73) and 5.6L (S70) Engines
Firing order: 1–7–5–11–3–9–6–12–2–8–4–10
Distributor rotation: Clockwise

CAPACITIES

Year	Body Type	Model	Engine Displacement Liters (cc)	Engine ID/VIN	Engine Oil with Filter (qts.)	Transmission (pts.)			Drive Axle		Fuel Tank (gal.)	Cooling System (qts.)
						5-Spd	6-Spd	Auto.	Front (pts.)	Rear (pts.)		
1998	E36	318i	1.9 (1895)	M44B19	5.3	2.7	—	6.4	—	3.6	13.2	6.9
	E36	318ti	1.9 (1895)	M44B19	5.3	2.7	—	6.4	—	3.6	13.2	6.9
	E36	323is	2.5 (2494)	M52B25	6.9	3.2	—	6.4	—	3.6	13.2	8.9
	E36	323iC	2.5 (2494)	M52B25	6.9	3.2	—	6.4	—	3.6	17.0	8.9
	E36	Z3 1.9	1.9 (1895)	M44B19	5.3	3.2	—	6.4	—	3.6	13.5	6.9
	E36	Z3 2.8	2.8 (2793)	M52TUB28	6.9	3.2	—	6.4	—	3.6	13.5	8.9
	E36	328i	2.8 (2793)	M52B28	6.9	3.2	—	6.4	—	3.6	16.4	8.9
	E36	328is	2.8 (2793)	M52B28	6.9	3.2	—	6.4	—	3.6	16.4	8.9
	E36	328iC	2.8 (2793)	M52B28	6.9	3.2	—	6.4	—	3.6	16.4	8.9
	E36	M3	3.2 (3152)	S52B32	6.9	3.2	—	6.4	—	3.6	16.4	10.6
	E39	528i	2.8 (2793)	M52B28	6.9	3.2	—	7.0	—	3.4	18.5	11.1
	E39	540i	4.4 (4398)	M62B44	7.9	—	4.8	7.0	—	3.4	18.5	13.2
	E38	740i	4.4 (4398)	M62B44	7.9	—	—	11.7	—	3.4	22.5	13.3
	E38	740iL	4.4 (4398)	M62B44	7.9	—	—	11.7	—	3.4	22.5	13.3
	E38	750iL	5.4 (5379)	M73B54	8.5	—	—	11.7	—	3.4	25.1	13.8
1999	E36	318ti S	1.9 (1895)	M44B19	5.3	2.7	—	6.4	—	3.6	14.5	6.9
	E36	323i	2.5 (2493)	M52B25	6.9	3.2	—	6.4	—	3.6	16.4	8.9
	E46	323i Sedan	2.5 (2494)	M52TUB25	6.9	3.2	—	6.4	—	3.6	16.4	8.9
	E36	323iC	2.5 (2494)	M52B25	6.9	3.2	—	6.4	—	3.6	16.4	8.9
	E36	Z3 2.3	2.5 (2494)	M52TUB25	6.9	3.2	—	6.4	—	3.6	13.5	8.9
	E36	Z3 2.8	2.8 (2793)	M52TUB28	6.9	3.2	—	6.4	—	3.6	13.5	8.9
	E36	Z3 Coupe	2.8 (2793)	M52TUB28	6.9	3.2	—	6.4	—	3.6	13.5	8.9
	E46	328i Sedan	2.8 (2793)	M52TUB28	6.9	3.2	—	6.4	—	3.6	16.4	8.9
	E36	328is	2.8 (2793)	M52B28	6.9	3.2	—	6.4	—	3.6	16.4	8.9
	E36	328iC	2.8 (2793)	M52B28	6.9	3.2	—	6.4	—	3.6	16.4	8.9
	E36	M3	3.2 (3152)	S52B32	6.9	3.2	—	6.4	—	3.6	16.4	10.6
	E39	528i	2.8 (2793)	M52TUB28	6.9	3.2	—	6.4	—	3.4	18.5	11.1
	E39	528i SW	2.8 (2793)	M52TUB28	6.9	3.2	—	6.4	—	3.4	18.5	11.1
	E39	540i SW	4.4 (4398)	M62B44	7.9	—	—	7.0	—	3.4	18.5	13.2
	E39	540i	4.4 (4398)	M62B44	7.9	—	4.8	7.0	—	3.4	18.5	13.2
	E38	740i	4.4 (4398)	M62B44	7.9	—	—	11.7	—	3.4	22.5 ①	13.3
	E38	740iL	4.4 (4398)	M62B44	7.9	—	—	11.7	—	3.4	22.5 ①	13.3
	E38	750iL	5.4 (5379)	M73B54	8.5	—	—	11.7	—	3.4	25.1	13.8

93471C38

For Maintenance Interval recommendations, see Section 1 of this manual

CAPACITIES

Year	Body Type	Model	Engine Displacement Liters (cc)	Engine ID/VIN	Engine Oil with Filter (qts.)	Transmission (pts.) 5-Spd	6-Spd	Auto.	Drive Axle Front (pts.)	Rear (pts.)	Fuel Tank (gal.)	Cooling System (qts.)
2000	E36	318ti S	1.9 (1895)	M44B19	5.3	2.7	—	6.4	—	3.6	14.5	6.9
	E36	323i	2.5 (2494)	M52B25	6.9	3.2	—	6.4	—	3.6	16.4	8.9
	E46	323i Sedan	2.5 (2494)	M52TUB25	6.9	3.2	—	6.4	—	3.6	16.4	8.9
	E36	323iC	2.5 (2494)	M52B25	6.9	3.2	—	6.4	—	3.6	16.4	8.9
	E36	Z3 2.3	2.5 (2494)	M52TUB25	6.9	3.2	—	6.4	—	3.6	13.5	8.9
	E36	Z3 2.8	2.8 (2793)	M52TUB28	6.9	3.2	—	6.4	—	3.6	13.5	8.9
	E36	Z3 Coupe	2.8 (2793)	M52TUB28	6.9	3.2	—	6.4	—	3.6	13.5	8.9
	E46	328i Sedan	2.8 (2793)	M52TUB28	6.9	3.2	—	6.4	—	3.6	16.4	8.9
	E36	328is	2.8 (2793)	M52B28	6.9	3.2	—	6.4	—	3.6	16.4	8.9
	E36	328iC	2.8 (2793)	M52B28	6.9	3.2	—	6.4	—	3.6	16.4	8.9
	E36	M3	3.2 (3152)	S52B32	6.9	3.2	—	6.4	—	3.6	16.4	10.6
	E39	528i	2.8 (2793)	M52TUB28	6.9	3.2	—	6.4	—	3.4	18.5	11.1
	E39	528i SW	2.8 (2793)	M52TUB28	6.9	3.2	—	6.4	—	3.4	18.5	11.1
	E39	540i SW	4.4 (4398)	M62B44	7.9	—	—	7.0	—	3.4	18.5	13.2
	E39	540i	4.4 (4398)	M62B44	7.9	—	4.8	7.0	—	3.4	18.5	13.2
	E38	740i	4.4 (4393)	M62B44	7.9	—	—	11.7	—	3.4	22.5 ①	13.3
	E38	740iL	4.4 (4398)	M62B44	7.9	—	—	11.7	—	3.4	22.5 ①	13.3
	E38	750iL	5.4 (5379)	M73B54	8.5	—	—	11.7	—	3.4	25.1	13.8

NOTE: All capacities are approximate. Add fluid gradually and ensure a proper fluid level is obtained.

NOTE: Capacities given are service, not overhaul capacities

① With optional self leveling rear suspension 25.1 gallons

93471C39

VALVE SPECIFICATIONS

Year	Engine Displacement Liters (cc)	Engine ID/VIN	Seat Angle (deg.)	Face Angle (deg.)	Spring Test Pressure (lbs. @ in.)	Spring Installed Height (in.)	Stem-to-Guide Clearance (in.)		Stem Diameter (in.)	
							Intake	Exhaust	Intake	Exhaust
1998	1.9 (1895)	M44B19	①	45	NA	NA	② 0.0197	② 0.0197	0.2372-0.2340	0.2579-0.2583
	2.5 (2494)	M52B25	①	45	NA	NA	② 0.0197	② 0.0197	0.2372-0.2340	0.2378-0.2384
	2.8 (2793)	M52B28	①	45	NA	NA	② 0.0197	② 0.0197	0.2372-0.2340	0.2378-0.2384
	2.8 (2793)	M52TUB28	①	45	NA	NA	② 0.0197	② 0.0197	0.2372-0.2340	0.2378-0.2384
	3.2 (3152)	S52B32	①	45	NA	NA	② 0.0197	② 0.0197	0.2372-0.2340	0.2378-0.2384
	4.4 (4398)	M62B44	①	45	NA	NA	② 0.0197	② 0.0197	0.2157-0.2161	0.2146-0.2150
	5.4 (5379)	M73B54	①	45	NA	NA	② 0.0197	② 0.0197	0.2766-0.2772	0.2740-0.2734
1999	1.9 (1895)	M44B19	①	45	NA	NA	② 0.0197	② 0.0197	0.2372-0.2340	0.2579-0.2583
	2.5 (2494)	M52B25	①	45	NA	NA	② 0.0197	② 0.0197	0.2372-0.2340	0.2378-0.2384
	2.5 (2494)	M52TUB25	①	45	NA	NA	② 0.0197	② 0.0197	0.2372-0.2340	0.2378-0.2384
	2.8 (2793)	M52B28	①	45	NA	NA	② 0.0197	② 0.0197	0.2372-0.2340	0.2378-0.2384
	2.8 (2793)	M52TUB28	①	45	NA	NA	② 0.0197	② 0.0197	0.2372-0.2340	0.2378-0.2384
	3.2 (3152)	S52B32	①	45	NA	NA	② 0.0197	② 0.0197	0.2372-0.2340	0.2378-0.2384
	4.4 (4398)	M62B44	①	45	NA	NA	② 0.0197	② 0.0197	0.2156-0.2159	0.2146-0.2150
	5.4 (5379)	M73B54	①	45	NA	NA	② 0.0197	② 0.0197	0.2766-0.2772	0.2740-0.2734

93471C40

For Tune-up, Capacities and Firing orders, see Section 1 of this manual

VALVE SPECIFICATIONS

Year	Engine Displacement Liters (cc)	Engine ID/VIN	Seat Angle (deg.)	Face Angle (deg.)	Spring Test Pressure (lbs. @ in.)	Spring Installed Height (in.)	Stem-to-Guide Clearance (in.)		Stem Diameter (in.)	
							Intake	Exhaust	Intake	Exhaust
2000	1.9 (1895)	M44B19	①	45	NA	NA	② 0.0197	② 0.0197	0.2372-0.2340	0.2579-0.2583
	2.5 (2494)	M52B25	①	45	NA	NA	② 0.0197	② 0.0197	0.2372-0.2340	0.2378-0.2384
	2.5 (2494)	M52TUB25	①	45	NA	NA	② 0.0197	② 0.0197	0.2372-0.2340	0.2378-0.2384
	2.8 (2793)	M52B28	①	45	NA	NA	② 0.0197	② 0.0197	0.2372-0.2340	0.2378-0.2384
	2.8 (2793)	M52TUB28	①	45	NA	NA	② 0.0197	② 0.0197	0.2372-0.2340	0.2378-0.2384
	3.2 (3152)	S52B32	①	45	NA	NA	② 0.0197	② 0.0197	0.2372-0.2340	0.2378-0.2384
	4.4 (4398)	M62B44	①	45	NA	NA	② 0.0197	② 0.0197	0.2156-0.2159	0.2146-0.2150
	5.4 (5379)	M73B54	①	45	NA	NA	② 0.0197	② 0.0197	0.2766-0.2772	0.2740-0.2734

NA: Not Available

① Valve seat angle: 45 degrees

   Correction angle outside: 15 degrees

   Correction angle inside: 60 degrees

② To measure: Insert a new valve into guide

   with end of valve flush with end of guide.

   Use a dial indicator to measure axial valve head movement.

93471C41

CRANKSHAFT AND CONNECTING ROD SPECIFICATIONS

All measurements are given in inches.

	Engine		Crankshaft				Connecting Rod		
Year	Displacement Liters (cc)	Engine ID/VIN	Main Brg. Journal Dia.	Main Brg. Oil Clearance	Shaft End-play	Thrust on No.	Journal Diameter	Oil Clearance	Side Clearance
1998	1.9 (1895)	M44B19	①	0.0007-0.0018	0.0031-0.0064	4	1.7720-1.7706	0.0004-0.0020	0.0060-0.0160
	2.5 (2494)	M52B25	①	0.0007-0.0029	0.0031-0.0064	5	1.7720-1.7706	0.0007-0.0022	0.0060-0.0196
	2.8 (2793)	M52B28	①	0.0007-0.0029	0.0031-0.0064	5	1.7720-1.7706	0.0007-0.0022	0.0060-0.0196
	2.8 (2793)	M52TUB28	①	0.0007-0.0029	0.0031-0.0064	5	1.7720-1.7706	0.0007-0.0022	0.0060-0.0196
	3.2 (3152)	S52B32	①	0.0007-0.0029	0.0031-0.0064	5	1.7720-1.7706	0.0007-0.0022	0.0060-0.0120
	4.4 (4398)	M62B44	②	0.0007-0.0018	0.0033-0.0101	3	1.8901-1.8887	0.0007-0.0022	0.0060-0.0196
	5.4 (5379)	M73B54	③	0.0010-0.0020	0.0033-0.0101	5	1.7720-1.7706	0.0006-0.0023	0.0060-0.0196
1999	1.9 (1895)	M44B19	①	0.0007-0.0018	0.0031-0.0064	4	1.7720-1.7706	0.0004-0.0020	0.0060-0.0160
	2.5 (2494)	M52B25	①	0.0007-0.0029	0.0031-0.0064	5	1.7720-1.7706	0.0007-0.0022	0.0060-0.0196
	2.5 (2494)	M52TUB25	①	0.0007-0.0029	0.0031-0.0064	5	1.7720-1.7706	0.0007-0.0022	0.0060-0.0196
	2.8 (2793)	M52B28	①	0.0007-0.0029	0.0031-0.0064	5	1.7720-1.7706	0.0007-0.0022	0.0060-0.0196
	2.8 (2793)	M52TUB28	①	0.0007-0.0029	0.0031-0.0064	5	1.7720-1.7706	0.0007-0.0022	0.0060-0.0196
	3.2 (3152)	S52B32	①	0.0007-0.0029	0.0031-0.0064	5	1.7720-1.7706	0.0007-0.0022	0.0060-0.0120
	4.4 (4398)	M62B44	②	0.0007-0.0018	0.0033-0.0101	3	1.8901-1.8887	0.0007-0.0022	0.0060-0.0196
	5.4 (5379)	M73B54	③	0.0010-0.0020	0.0033-0.0101	5	1.7720-1.7706	0.0006-0.0023	0.0060-0.0196
2000	1.9 (1895)	M44B19	①	0.0007-0.0018	0.0031-0.0064	4	1.7720-1.7706	0.0004-0.0020	0.0060-0.0160
	2.5 (2494)	M52B25	①	0.0007-0.0029	0.0031-0.0064	5	1.7720-1.7706	0.0007-0.0022	0.0060-0.0196
	2.5 (2494)	M52TUB25	①	0.0007-0.0029	0.0031-0.0064	5	1.7720-1.7706	0.0007-0.0022	0.0060-0.0196
	2.8 (2793)	M52B28	①	0.0007-0.0029	0.0031-0.0064	5	1.7720-1.7706	0.0007-0.0022	0.0060-0.0196
	2.8 (2793)	M52TUB28	①	0.0007-0.0029	0.0031-0.0064	5	1.7720-1.7706	0.0007-0.0022	0.0060-0.0196
	3.2 (3152)	S52B32	①	0.0007-0.0029	0.0031-0.0064	5	1.7720-1.7706	0.0007-0.0022	0.0060-0.0120
	4.4 (4398)	M62B44	②	0.0007-0.0018	0.0033-0.0101	3	1.8901-1.8887	0.0007-0.0022	0.0060-0.0196
	5.4 (5379)	M73B54	③	0.0010-0.0020	0.0033-0.0101	5	1.7720-1.7706	0.0006-0.0023	0.0060-0.0196

① Standard yellow 2.3615-2.3618 inches
  Standard green: 2.3613-2.3615 inches
  Standard white: 2.3611-2.3613 inches

② Standard yellow 2.7553-2.7555 inches
  Standard green: 2.7550-2.7552 inches
  Standard white: 2.7548-2.76550 inches

③ Standard yellow 2.9521-2.9523 inches
  Standard green: 2.9518-2.952 inches
  Standard white: 2.9516-2.9518 inches

93471C42

PISTON AND RING SPECIFICATIONS
All measurements are given in inches

| Year | Engine Displacemen Liters (cc) | Engine ID/VIN | Piston Clearance | Ring Gap | | | Ring Side Clearance | | |
				Top Compression	Bottom Compression	Oil Control	Top Compression	Bottom Compression	Oil Control
1998	1.9 (1895)	M44B19	0.0004-0.0016	0.0078-0.0394	0.0078-0.0394	0.0157-0.0551	0.0007-0.0078	0.0007-0.0039	NA
	2.5 (2494)	M52B25	0.0004-0.0016	0.0004-0.0118	0.0078-0.0157	0.0098-0.0197	0.0007-0.0024	0.0019-0.0026	0.0007-0.0024
	2.8 (2793)	M52B28	0.0004-0.0016	0.0039-0.0118	0.0078-0.0157	0.0098-0.0197	0.0008-0.0024	0.0012-0.0026	0.0007-0.0024
	2.8 (2793)	M52TUB28	0.0004-0.0016	0.0039-0.0118	0.0078-0.0157	0.0098-0.0197	0.0008-0.0024	0.0012-0.0026	0.0007-0.0024
	3.2 (3152)	S52B32	0.0010-0.0023	0.0098-0.0157	0.0078-0.0157	0.0098-0.0197	0.0012-0.0026	0.0007-0.0022	0.0007-0.0022
	4.4 (4398)	M62B44	0.0002-0.0015	0.0039-0.0118	0.0078-0.0157	0.0078-0.0354	0.0008-0.0024	0.0008-0.0024	NA
	5.4 (5379)	M73B54	0.0010-0.0020	0.0059-0.0138	0.0078-0.0157	0.0157-0.0551	0.0007-0.0022	0.0007-0.0022	NA
1999	1.9 (1895)	M44B19	0.0004-0.0016	0.0078-0.0394	0.0078-0.0394	0.0157-0.0551	0.0007-0.0078	0.0007-0.0039	NA
	2.5(2494)	M52B25	0.0004-0.0016	0.0004-0.0118	0.0078-0.0157	0.0098-0.0197	0.0007-0.0024	0.0019-0.0026	0.0007-0.0024
	2.5 (2494)	M52TUB25	0.0004-0.0016	0.0004-0.0118	0.0078-0.0157	0.0098-0.0197	0.0007-0.0024	0.0019-0.0026	0.0007-0.0024
	2.8 (2793)	M52B28	0.0004-0.0016	0.0039-0.0118	0.0078-0.0157	0.0098-0.0197	0.0008-0.0024	0.0012-0.0026	0.0007-0.0024
	2.8 (2793)	M52TUB28	0.0004-0.0016	0.0039-0.0118	0.0078-0.0157	0.0098-0.0197	0.0008-0.0024	0.0012-0.0026	0.0007-0.0024
	3.2 (3152)	S52B32	0.0010-0.0023	0.0098-0.0157	0.0078-0.0157	0.0098-0.0197	0.0012-0.0026	0.0007-0.0022	0.0007-0.0022
	4.4 (4398)	M62B44	0.0002-0.0015	0.0039-0.0118	0.0078-0.0157	0.0078-0.0354	0.0008-0.0024	0.0008-0.0024	NA
	5.4 (5379)	M73B54	0.0010-0.0020	0.0059-0.0138	0.0078-0.0157	0.0157-0.0551	0.0007-0.0022	0.0007-0.0022	NA
2000	1.9 (1895)	M44B19	0.0004-0.0016	0.0078-0.0394	0.0078-0.0394	0.0157-0.0551	0.0007-0.0078	0.0007-0.0039	NA
	2.5(2494)	M52B25	0.0004-0.0016	0.0004-0.0118	0.0078-0.0157	0.0098-0.0197	0.0007-0.0024	0.0019-0.0026	0.0007-0.0024
	2.5 (2494)	M52TUB25	0.0004-0.0016	0.0004-0.0118	0.0078-0.0157	0.0098-0.0197	0.0007-0.0024	0.0019-0.0026	0.0007-0.0024
	2.8 (2793)	M52B28	0.0004-0.0016	0.0039-0.0118	0.0078-0.0157	0.0098-0.0197	0.0008-0.0024	0.0012-0.0026	0.0007-0.0024
	2.8 (2793)	M52TUB28	0.0004-0.0016	0.0039-0.0118	0.0078-0.0157	0.0098-0.0197	0.0008-0.0024	0.0012-0.0026	0.0007-0.0024
	3.2 (3152)	S52B32	0.0010-0.0023	0.0098-0.0157	0.0078-0.0157	0.0098-0.0197	0.0012-0.0026	0.0007-0.0022	0.0007-0.0022
	4.4 (4398)	M62B44	0.0002-0.0015	0.0039-0.0118	0.0078-0.0157	0.0078-0.0354	0.0008-0.0024	0.0008-0.0024	NA
	5.4 (5379)	M73B54	0.0010-0.0020	0.0059-0.0138	0.0078-0.0157	0.0157-0.0551	0.0007-0.0022	0.0007-0.0022	NA

NA: Not Available

93471C43

TORQUE SPECIFICATIONS
All readings in ft. lbs.

Year	Engine Displacement Liters (cc)	Engine ID/VIN	Cylinder Head Bolts	Main Bearing Bolts	Rod Bearing Bolts	Crankshaft Damper Bolts	Flywheel Bolts	Manifold Intake	Manifold Exhaust	Spark Plugs	Lug Nut
1998	1.9 (1895)	M44B19	①	②	③	243	88	④	⑤	⑥	80
	2.5 (2494)	M52B25	①	②	③	302	⑦	④	⑧	⑥	80
	2.8 (2793)	M52B28	①	②	③	302	⑦	④	⑧	⑥	80
	2.8 (2793)	M52TUB28	⑨	⑩	③	302	⑦	④	⑧	⑥	80
	3.2 (3152)	S52B32	①	②	③	302	⑦	④	⑧	⑥	80
	4.4 (4398)	M62B44	⑨	⑩	⑪	⑫	⑦	④	⑤	⑥	80
	5.4 (5379)	M73B54	⑨	⑩	③	⑬	88	④	⑤	⑥	80
1999	1.9 (1895)	M44B19	①	②	③	243	88	④	⑤	⑥	80
	2.5 (2494)	M52B25	①	②	③	302	⑦	④	⑧	⑥	80
	2.5 (2494)	M52TUB25	⑨	⑩	③	302	⑦	④	⑧	⑥	80
	2.8 (2793)	M52B28	①	②	③	302	⑦	④	⑧	⑥	80
	2.8 (2793)	M52TUB28	⑨	⑩	③	302	⑦	④	⑧	⑥	80
	3.2 (3152)	S52B32	①	②	③	302	⑦	④	⑧	⑥	80
	4.4 (4398)	M62B44	⑨	⑩	⑪	⑫	⑦	④	⑤	⑥	⑭
	5.4 (5379)	M73B54	⑨	⑩	③	⑬	88	④	⑤	⑥	80
2000	1.9 (1895)	M44B19	①	②	③	243	88	④	⑤	⑥	80
	2.5 (2494)	M52B25	①	②	③	302	⑦	④	⑧	⑥	80
	2.5 (2494)	M52TUB25	⑨	⑩	③	302	⑦	④	⑧	⑥	80
	2.8 (2793)	M52B28	①	②	③	302	⑦	④	⑧	⑥	80
	2.8 (2793)	M52TUB28	⑨	⑩	③	302	⑦	④	⑧	⑥	80
	3.2 (3152)	S52B32	①	②	③	302	⑦	④	⑧	⑥	80
	4.4 (4398)	M62B44	⑨	⑩	⑪	⑫	⑦	④	⑤	⑥	⑭
	5.4 (5379)	M73B54	⑨	⑩	③	⑬	88	④	⑤	⑥	80

① Cast iron block. Replace, wash and oil bolts
Step 1: 22 ft. lbs.
Step 2: 90 degrees
Step 3: 90 degrees

② Cast iron block. Replace, wash and oil bolts
Step 1: 14.8 ft. lbs.
Step 2: 50 degrees

③ Replace, wash and oil connecting rod bolts
Step 1: 14.8 ft. lbs.
Step 2: 70 degrees

④ All M6 fasteners: 88 inch lbs.
All M7 fasteners: 11 ft. lbs.
All M8 fasteners: 16 ft. lbs.

⑤ Coat with Molykkote HSC compound
All M6 fasteners: 88 inch lbs.
All M7 fasteners: 11 ft. lbs.

⑥ M12x1.25: 14.8-19.1 ft. lbs.
M14x1.25: 21.4-24.3 ft. lbs.

⑦ New micro-encapsulated screws:
Automatic transmission: 88 ft. lbs.
Manual transmission: 77.4 ft. lbs.

⑧ Coat with Molykkote HSC compound or equivalent
All M6 fasteners: 88 inch lbs.
All M7 fasteners: 14.8 ft. lbs.

⑨ Aluminum block. Replace bolts. Do not remove coating
Step 1: M52: 29.5 ft. lbs. M62, M73: 22 ft. lbs.
Step 2: M52: 90 degrees. M62: 80 degrees. M73: 60 degrees
Step 2: M52: 90 degrees. M62: 80 degrees. M73: 60 degrees

⑩ Aluminum block. Replace bolts. Do not remove coating
Step 1: 14.8 ft. lbs.
Step 2: 70 degrees

⑪ Replace, wash and oil connecting rod bolts
Step 1: 14.8 ft. lbs.
Step 2: 80 degrees

⑫ Step 1: 73.7 ft. lbs.
Step 2: 60 degrees
Step 3: 60 degrees
Step 4: 30 degrees

⑬ Step 1: 73.7 ft. lbs.
Step 2: 60 degrees
Step 3: 60 degrees

⑭ 740i with Sport Package 18 inch wheels: 100 ft. lbs

93471C44

BRAKE SPECIFICATIONS
All measurements in inches unless noted

Year	Body Type	Model		Brake Disc			Brake Drum Diameter			Minimum Lining Thickness		Brake Caliper	
				Original Thickness	Minimum Thickness	Maximum Runout	Original Inside Diameter	Max. Wear Limit	Maximum Machine Diameter	Front	Rear	Bracket Bolts (ft. lbs.)	Mounting Bolts (ft. lbs.)
1998	E36	318i	F	0.803	①	0.007	—	—	—	0.118	—	81	26
			R	②	①	0.007	③	④	④	—	⑤	48	26
	E36	318ti	F	0.803	①	0.007	—	—	—	0.118	—	81	26
			R	②	①	0.007	③	④	④	—	⑤	48	26
	E36	323is	F	0.803	①	0.007	—	—	—	0.118	—	81	26
			R	②	①	0.007	③	④	④	—	⑤	48	26
	E36	323iC	F	0.803	①	0.007	—	—	—	0.118	—	81	26
			R	②	①	0.007	③	④	④	—	⑤	48	26
	E36	Z3 1.9	F	0.803	①	0.007	—	—	—	0.118	—	81	26
			R	②	①	0.007	③	④	④	—	⑤	48	26
	E36	Z3 2.8	F	0.803	①	0.007	—	—	—	0.118	—	81	26
			R	②	①	0.007	③	④	④	—	⑤	48	26
	E36	328i	F	0.803	①	0.007	—	—	—	0.118	—	81	26
			R	②	①	0.007	③	④	④	—	⑤	48	26
	E36	328is	F	0.803	①	0.007	—	—	—	0.118	—	81	26
			R	②	①	0.007	③	④	④	—	⑤	48	26
	E36	328iC	F	0.803	①	0.007	—	—	—	0.118	—	81	26
			R	②	①	0.007	③	④	④	—	⑤	48	26
	E36	M3	F	0.803	①	0.007	—	—	—	0.118	—	81	26
			R	②	①	0.007	③	④	④	—	⑤	48	26
	E39	528i	F	0.803	①	0.007	—	—	—	0.118	—	81	26
			R	⑥	①	0.007	⑦	④	④	—	⑤	48	26
	E39	540i	F	1.118	①	0.007	—	—	—	0.118	—	81	26
			R	⑥	①	0.007	⑦	④	④	—	⑤	48	26
	E38	740i	F	1.118	①	0.007	—	—	—	0.118	—	81	26
			R	0.409	①	0.007	⑧	④	④	—	⑤	48	26
	E38	740iL	F	1.118	①	0.007	—	—	—	0.118	—	81	26
			R	0.409	①	0.007	⑧	④	④	—	⑤	48	26
	E38	750iL	F	1.196	①	0.007	—	—	—	0.118	—	81	26
			R	0.724	①	0.007	⑧	④	④	—	⑤	48	26
1999	E36	318ti S	F	0.803	①	0.007	—	—	—	0.118	—	81	26
			R	②	①	0.007	③	④	④	—	⑤	48	26
	E46	323i Sedan	F	0.803	①	0.007	—	—	—	0.118	—	81	26
			R	②	①	0.007	③	④	④	—	⑤	48	26
	E36	323is	F	0.803	①	0.007	—	—	—	0.118	—	81	26
			R	②	①	0.007	③	④	④	—	⑤	48	26
	E36	323iC	F	0.803	①	0.007	—	—	—	0.118	—	81	26
			R	②	①	0.007	③	④	④	—	⑤	48	26
	E36	Z3 2.3	F	0.803	①	0.007	—	—	—	0.118	—	81	26
			R	②	①	0.007	③	④	④	—	⑤	48	26
	E36	Z3 2.8	F	0.803	①	0.007	—	—	—	0.118	—	81	26
			R	②	①	0.007	③	④	④	—	⑤	48	26
	E36	Z3 Coupe	F	0.803	①	0.007	—	—	—	0.118	—	81	26
			R	②	①	0.007	③	④	④	—	⑤	48	26

93471C45

BRAKE SPECIFICATIONS
All measurements in inches unless noted

Year	Body Type	Model		Brake Disc Original Thickness	Brake Disc Minimum Thickness	Maximum Runout	Brake Drum Diameter Original Inside Diameter	Max. Wear Limit	Maximum Machine Diameter	Minimum Lining Thickness Front	Minimum Lining Thickness Rear	Brake Caliper Bracket Bolts (ft. lbs.)	Brake Caliper Mounting Bolts (ft. lbs.)
1999 (cont.)	E46	328i Sedan	F	0.803	①	0.007	—	—	—	0.118	—	81	26
			R	②	①	0.007	③	④	④	—	⑤	48	26
	E36	328is	F	0.803	①	0.007	—	—	—	0.118	—	81	26
			R	②	①	0.007	③	④	④	—	⑤	48	26
	E36	328iC	F	0.803	①	0.007	—	—	—	0.118	—	81	26
			R	②	①	0.007	③	④	④	—	⑤	48	26
	E36	M3	F	0.803	①	0.007	—	—	—	0.118	—	81	26
			R	②	①	0.007	③	④	④	—	⑤	48	26
	E39	528i	F	0.803	①	0.007	—	—	—	0.118	—	81	26
			R	⑥	①	0.007	⑦	④	④	—	⑤	48	26
	E39	528i SW	F	0.803	①	0.007	—	—	—	0.118	—	81	26
			R	⑥	①	0.007	⑦	④	④	—	⑤	48	26
	E39	540i SW	F	1.118	①	0.007	—	—	—	0.118	—	81	26
			R	⑥	①	0.007	⑦	④	④	—	⑤	48	26
	E39	540i	F	1.118	①	0.007	—	—	—	0.118	—	81	26
			R	⑥	①	0.007	⑦	④	④	—	⑤	48	26
	E38	740i	F	1.118	①	0.007	—	—	—	0.118	—	81	26
			R	0.409	①	0.007	⑧	④	④	—	⑤	48	26
	E38	740iL	F	1.118	①	0.007	—	—	—	0.118	—	81	26
			R	0.409	①	0.007	⑧	④	④	—	⑤	48	26
	E38	750iL	F	1.196	①	0.007	—	—	—	0.118	—	81	26
			R	0.724	①	0.007	⑧	④	④	—	⑤	48	26
2000	E36	318ti S	F	0.803	①	0.007	—	—	—	0.118	—	81	26
			R	②	①	0.007	③	④	④	—	⑤	48	26
	E46	323i Sedan	F	0.803	①	0.007	—	—	—	0.118	—	81	26
			R	②	①	0.007	③	④	④	—	⑤	48	26
	E36	323is	F	0.803	①	0.007	—	—	—	0.118	—	81	26
			R	②	①	0.007	③	④	④	—	⑤	48	26
	E36	323iC	F	0.803	①	0.007	—	—	—	0.118	—	81	26
			R	②	①	0.007	③	④	④	—	⑤	48	26
	E36	Z3 2.3	F	0.803	①	0.007	—	—	—	0.118	—	81	26
			R	②	①	0.007	③	④	④	—	⑤	48	26
	E36	Z3 2.8	F	0.803	①	0.007	—	—	—	0.118	—	81	26
			R	②	①	0.007	③	④	④	—	⑤	48	26
	E36	Z3 Coupe	F	0.803	①	0.007	—	—	—	0.118	—	81	26
			R	②	①	0.007	③	④	④	—	⑤	48	26
	E46	328i Sedan	F	0.803	①	0.007	—	—	—	0.118	—	81	26
			R	②	①	0.007	③	④	④	—	⑤	48	26
	E36	328is	F	0.803	①	0.007	—	—	—	0.118	—	81	26
			R	②	①	0.007	③	④	④	—	⑤	48	26
	E36	328iC	F	0.803	①	0.007	—	—	—	0.118	—	81	26
			R	②	①	0.007	③	④	④	—	⑤	48	26
	E36	M3	F	0.803	①	0.007	—	—	—	0.118	—	81	26
			R	②	①	0.007	③	④	④	—	⑤	48	26

93471C46

Timing belt service is covered in Section 3 of this manual

BRAKE SPECIFICATIONS
All measurements in inches unless noted

Year	Body Type	Model		Brake Disc Original Thickness	Brake Disc Minimum Thickness	Brake Disc Maximum Runout	Brake Drum Diameter Original Inside Diameter	Brake Drum Diameter Max. Wear Limit	Brake Drum Diameter Maximum Machine Diameter	Minimum Lining Thickness Front	Minimum Lining Thickness Rear	Brake Caliper Bracket Bolts (ft. lbs.)	Brake Caliper Mounting Bolts (ft. lbs.)
2000 (cont.)	E39	528i	F	0.803	①	0.007	—	—	—	0.118	—	81	26
			R	⑥	①	0.007	⑦	④	④	—	⑤	48	26
	E39	528i SW	F	0.803	①	0.007	—	—	—	0.118	—	81	26
			R	⑥	①	0.007	⑦	④	④	—	⑤	48	26
	E39	540i SW	F	1.118	①	0.007	—	—	—	0.118	—	81	26
			R	⑥	①	0.007	⑦	④	④	—	⑤	48	26
	E39	540i	F	1.118	①	0.007	—	—	—	0.118	—	81	26
			R	⑥	①	0.007	⑦	④	④	—	⑤	48	26
	E38	740i	F	1.118	①	0.007	—	—	—	0.118	—	81	26
			R	0.409	①	0.007	⑧	④	④	—	⑤	48	26
	E38	740iL	F	1.118	①	0.007	—	—	—	0.118	—	81	26
			R	0.409	①	0.007	⑧	④	④	—	⑤	48	26
	E38	750iL	F	1.196	①	0.007	—	—	—	0.118	—	81	26
			R	0.724	①	0.007	⑧	④	④	—	⑤	48	26

NA: Not Available

F: Front

R: Rear

① Minimum thickness is stamped in the brake disk shell
   Maximum machining limit per side: 0.031inches

② Solid brake rotor: 0.331 inches
   Ventilated brake rotor: 0.685 inches

③ Parking brake drum diameter: 6.299 inches

④ Parking brake drum maximum runout: 0.004 inches
   Wear limit and machining specifications are not available

⑤ Rear brake pad wear limit: 0.118 inches
   Parking brake shoe wear limit: 0.059 inches

⑥ Solid brake rotor: 0.331 inches
   Ventilated brake rotor: 0.724 inches

⑦ Parking brake drum diameter: 7.283 inches

⑧ Parking brake drum diameter: 7.086 inches

93471C47

WHEEL ALIGNMENT

Year	Model		Caster Range (+/-Deg.)	Caster Preferred Setting (Deg.)	Camber Range (+/-Deg.)	Camber Preferred Setting (Deg.)	Toe-in (in.)	Steering Axis Inclination (Deg.)
1998	3 series ①	F	0.50	+0.68	0.50	-0.50	0.15 +/- 0.06	—
		R	—	—	—	—	0.20 +/- 0.05	—
	3 series ②	F	0.50	+0.83	0.50	+0.85	0.15 +/- 0.06	—
		R	—	—	0.25	-2.00	0.20 +/- 0.05	—
	528i	F	0.50	+6.47	0.50	-0.22	0.05 +/- 0.08	—
		R	—	—	0.30	-2.19	0.13 +/- 0.08	—
	5 series touring edition ③	F	0.50	+0.30	0.50	-0.22	0.12 +/- 0.08	—
		R	—	—	0.33	+0.83	0.13 +/- 0.08	—
	5 series touring edition ④	F	0.50	+0.45	0.50	-0.50	0.08 +/- 0.08	—
		R	—	—	0.25	+0.83	0.08 +/- 0.06	—
	5 series touring edition ⑤	F	0.50	+0.57	0.50	-0.62	0.12 +/- 0.08	—
		R	—	—	0.33	+0.83	0.13 +/- 0.08	—
	5 series sedan ⑥	F	0.50	+0.30	0.50	-0.22	0.12 +/- 0.08	—
		R	—	—	0.08	+0.07	0.18 +/- 0.03	—
	5 series sedan ⑤	F	0.50	+0.57	0.50	-0.62	0.12 +/- 0.08	—
		R	—	—	0.08	+0.83	0.13 +/- 0.08	—
	5 series sedan ⑦	F	0.50	+0.57	0.50	+0.22	0.12 +/- 0.08	—
		R	—	—	0.08	+0.07	0.18 +/- 0.03	—
	5 series sedan ⑧	F	0.50	+0.30	0.50	-0.22	0.12 +/- 0.08	—
		R	—	—	0.08	+0.07	0.18 +/- 0.03	—
	5 series sedan ⑦⑧	F	0.50	+0.57	0.50	+0.22	0.12 +/- 0.08	—
		R	—	—	0.42	+0.83	0.13 +/- 0.08	—
	7 series ⑨	F	0.50	+0.97	0.50	-0.22	0.12 +/- 0.08	—
		R	—	—	0.08	+0.45	0.15 +/- 0.08	—
	7 series ⑩	F	0.50	+0.20	0.50	-0.60	0.12 +/- 0.08	—
		R	—	—	0.08	+0.45	0.15 +/- 0.08	—
	7 series ⑪	F	0.50	+0.77	0.50	+0.30	0.12 +/- 0.08	—
		R	—	—	0.08	+0.07	0.15 +/- 0.08	—
	8 series ⑫	F	0.50	+0.98	0.50	-0.18	0.16 +/- 0.04	—
		R	—	—	0.25	+0.38	0.16 +/- 0.04	—
	8 series ⑬	F	0.50	+0.13	0.50	-0.42	0.16 +/- 0.04	—
		R	—	—	0.25	+0.38	0.16 +/- 0.04	—
	M3 ⑭	F	0.50	+0.63	0.50	+0.92	0.08 +/- 0.04	—
		R	—	—	0.17	+0.75	0.25 +/- 0.04	—
	M3 ⑮	F	0.50	+0.58	0.50	+0.77	0.08 +/- 0.04	—
		R	—	—	0.17	+0.75	0.25 +/- 0.04	—
	Z3 ⑯	F	0.50	+0.68	0.50	-0.73	0.16 +/- 0.06	—
		R	—	—	0.50	-0.33	0.30 +/- 0.09	—
	Z3 ⑰	F	0.50	+0.80	0.50	-0.07	0.16 +/- 0.06	—
		R	—	—	0.50	-0.83	0.37 +/- 0.12	—
1999	3 series ①	F	0.50	+0.68	0.50	-0.50	0.15 +/- 0.06	—
		R	—	—	—	—	0.20 +/- 0.05	—
	3 series ②	F	0.50	+0.83	0.50	+0.85	0.15 +/- 0.06	—
		R	—	—	0.25	-2.00	0.20 +/- 0.05	—
	528i	F	0.50	+6.47	0.50	-0.22	0.05 +/- 0.08	—
		R	—	—	0.30	-2.19	0.13 +/- 0.08	—

93471C48

Heater Core replacement is covered in Section 2 of this manual

WHEEL ALIGNMENT

Year	Model		Caster Range (+/-Deg.)	Caster Preferred Setting (Deg.)	Camber Range (+/-Deg.)	Camber Preferred Setting (Deg.)	Toe-in (in.)	Steering Axis Inclination (Deg.)
1999 (cont.)	5 series touring edition ⑥	F	0.50	+0.30	0.50	-0.22	0.12 +/- 0.08	—
		R	—	—	0.33	+0.83	0.13 +/- 0.08	—
	5 series touring edition ④	F	0.50	+0.45	0.50	-0.50	0.08 +/- 0.08	—
		R	—	—	0.25	+0.83	0.08 +/- 0.06	—
	5 series touring edition ⑤	F	0.50	+0.57	0.50	-0.62	0.12 +/- 0.08	—
		R	—	—	0.33	+0.83	0.13 +/- 0.08	—
	5 series sedan ⑥	F	0.50	+0.30	0.50	-0.22	0.12 +/- 0.08	—
		R	—	—	0.08	+0.07	0.18 +/- 0.03	—
	5 series sedan ⑤	F	0.50	+0.57	0.50	-0.62	0.12 +/- 0.08	—
		R	—	—	0.08	+0.83	0.13 +/- 0.08	—
	5 series sedan ⑦	F	0.50	+0.57	0.50	+0.22	0.12 +/- 0.08	—
		R	—	—	0.08	+0.07	0.18 +/- 0.03	—
	5 series sedan ⑧	F	0.50	+0.30	0.50	-0.22	0.12 +/- 0.08	—
		R	—	—	0.08	+0.07	0.18 +/- 0.03	—
	5 series sedan ⑦⑧	F	0.50	+0.57	0.50	+0.22	0.12 +/- 0.08	—
		R	—	—	0.42	+0.83	0.13 +/- 0.08	—
	7 series ⑨	F	0.50	+0.97	0.50	-0.22	0.12 +/- 0.08	—
		R	—	—	0.08	+0.45	0.15 +/- 0.08	—
	7 series ⑩	F	0.50	+0.20	0.50	-0.60	0.12 +/- 0.08	—
		R	—	—	0.08	+0.45	0.15 +/- 0.08	—
	7 series ⑪	F	0.50	+0.77	0.50	+0.30	0.12 +/- 0.08	—
		R	—	—	0.08	+0.07	0.15 +/- 0.08	—
	8 series ⑫	F	0.50	0.98	0.50	-0.18	0.16 +/- 0.04	—
		R	—	—	0.25	+0.38	0.16 +/- 0.04	—
	8 series ⑬	F	0.50	+0.13	0.50	-0.42	0.16 +/- 0.04	—
		R	—	—	0.25	+0.38	0.16 +/- 0.04	—
	M3 ⑭	F	0.50	+0.63	0.50	+0.92	0.08 +/- 0.04	—
		R	—	—	0.17	+0.75	0.25 +/- 0.04	—
	M3 ⑮	F	0.50	+0.58	0.50	+0.77	0.08 +/- 0.04	—
		R	—	—	0.17	+0.75	0.25 +/- 0.04	—
	Z3 ⑯	F	0.50	+0.68	0.50	-0.73	0.16 +/- 0.06	—
		R	—	—	0.50	-0.33	0.30 +/- 0.09	—
	Z3 ⑰	F	0.50	+0.80	0.50	-0.07	0.16 +/- 0.06	—
		R	—	—	0.50	-0.83	0.37 +/- 0.12	—
2000	3 series ①	F	0.50	+0.68	0.50	-0.50	0.15 +/- 0.06	—
		R	—	—	—	—	0.20 +/- 0.05	—
	3 series ②	F	0.50	0.83	0.50	+0.85	0.15 +/- 0.06	—
		R	—	—	0.25	-2.00	0.20 +/- 0.05	—
	528i	F	0.50	+6.47	0.50	-0.22	0.05 +/- 0.08	—
		R	—	—	0.30	-2.19	0.13 +/- 0.08	—
	5 series touring edition ⑥	F	0.50	+0.30	0.50	-0.22	0.12 +/- 0.08	—
		R	—	—	0.33	+0.83	0.13 +/- 0.08	—
	5 series touring edition ④	F	0.50	+0.45	0.50	-0.50	0.08 +/- 0.08	—
		R	—	—	0.25	+0.83	0.08 +/- 0.06	—
	5 series touring edition ⑤	F	0.50	+0.57	0.50	-0.62	0.12 +/- 0.08	—
		R	—	—	0.33	+0.83	0.13 +/- 0.08	—

93471C49

WHEEL ALIGNMENT

Year	Model		Caster Range (+/-Deg.)	Caster Preferred Setting (Deg.)	Camber Range (+/-Deg.)	Camber Preferred Setting (Deg.)	Toe-in (in.)	Steering Axis Inclination (Deg.)
2000 (cont.)	5 series sedan ⑥	F	0.50	+0.30	0.50	-0.22	0.12 +/- 0.08	—
		R	—	—	0.08	+0.07	0.18 +/- 0.03	—
	5 series sedan ⑤	F	0.50	+0.57	0.50	-0.62	0.12 +/- 0.08	—
		R	—	—	0.08	+0.83	0.13 +/- 0.08	—
	5 series sedan ⑦	F	0.50	+0.57	0.50	+0.22	0.12 +/- 0.08	—
		R	—	—	0.08	+0.07	0.18 +/- 0.03	—
	5 series sedan ⑧	F	0.50	+0.30	0.50	-0.22	0.12 +/- 0.08	—
		R	—	—	0.08	+0.07	0.18 +/- 0.03	—
	5 series sedan ⑦⑧	F	0.50	+0.57	0.50	+0.22	0.12 +/- 0.08	—
		R	—	—	0.42	+0.83	0.13 +/- 0.08	—
	7 series ⑨	F	0.50	0.97	0.50	-0.22	0.12 +/- 0.08	—
		R	—	—	0.08	+0.45	0.15 +/- 0.08	—
	7 series ⑩	F	0.50	+0.20	0.50	-0.60	0.12 +/- 0.08	—
		R	—	—	0.08	+0.45	0.15 +/- 0.08	—
	7 series ⑪	F	0.50	+0.77	0.50	+0.30	0.12 +/- 0.08	—
		R	—	—	0.08	+0.07	0.15 +/- 0.08	—
	M3 ⑭	F	0.50	+0.63	0.50	+0.92	0.08 +/- 0.04	—
		R	—	—	0.17	+0.75	0.25 +/- 0.04	—
	M3 ⑮	F	0.50	+0.58	0.50	+0.77	0.08 +/- 0.04	—
		R	—	—	0.17	+0.75	0.25 +/- 0.04	—
	Z3 ⑯	F	0.50	+0.68	0.50	-0.73	0.16 +/- 0.06	—
		R	—	—	0.50	-0.33	0.30 +/- 0.09	—
	Z3 ⑰	F	0.50	+0.80	0.50	-0.07	0.16 +/- 0.06	—
		R	—	—	0.50	-0.83	0.37 +/- 0.12	—
	Z8 ⑱	F	0.50	+0.98	0.50	-0.18	0.15 +/- 0.04	—
		R	—	—	0.25	+0.17	0.15 +/- 0.04	—
	Z8 ⑲	F	0.50	+0.13	0.50	-0.42	0.15 +/- 0.04	—
		R	—	—	0.25	+0.38	0.15 +/- 0.04	—
	Z8 ⑳	F	0.50	+0.13	0.50	-0.40	0.21 +/- 0.04	—
		R	—	—	0.25	+0.38	0.16 +/- 0.04	—

① Standard suspension
② Sport suspension
③ With m-technic suspension
④ M chasis
⑤ With low slung sport suspension
⑥ With standard suspension
⑦ With rough road package
⑧ With air suspension
⑨ With standard suspension
⑩ With low slung sport suspension
⑪ With rough road package
⑫ With standard suspension
⑬ With sport suspension
⑭ With 3.0L engine
⑮ With 3.2L engine
⑯ With standard suspension
⑰ With sport suspension
⑱ With standard suspension
⑲ With low slung sports suspension
⑳ 850 csi

93471C50

Brake service is covered in Section 4 of this manual

TIRE, WHEEL AND BALL JOINT SPECIFICATIONS

Year	Model	OEM Tires		Tire Pressures (psi)		Wheel Size	Ball Joint Inspection
		Standard	Optional	Front	Rear		
1998	318i Sedan	185/65R15	none	29	34	6-J	4-36 in. ①
	318is	205/60HR15	225/55VR15 225/50ZR16	29	32	7-J	4-36 in. ①
	318ti	205/60HR15	225/55VR15 225/50ZR16	29	32	7-J	4-36 in. ①
	328i Coupe	205/60VR15	225/55VR15 225/50ZR16	29	32	7-J	4-36 in. ①
	328i Convertible	205/60HR15	255/50ZR16	29	32	7-J	4-36 in. ①
	328is Coupe	205/60HR15	225/50ZR16	29	34	7-J	4-36 in. ①
	7-Series	235/60WR16	245/55WR16	32	29	Std: 7.5-J Opt: 8-J	4-36 in. ①
	M3	Fr: 225/45ZR17 Rr: 245/40ZR17	none	41	48	Fr: 7.5-J Rr: 8.5-J	4-36 in. ①
	M Roadster	Fr: 225/45ZR17 Rr: 245/40ZR17	none	35	38	Fr: 7.5-J Rr: 8.5-J	4-36 in. ①
	528i	205/65R15	225/60R15 235/45R17 255/40R17	33	41	Std: 6.5-J Opt: 8-J Opt: 8-J	4-36 in. ①
	540i	225/55R16	235/45R17 255/40R17	35	42	Std: 7-J Opt: 8-J Opt: 9-J	4-36 in. ①
	Z3	225/50ZR16	205/60R15 225/45ZR17	29	29	7-J	4-36 in. ①
1999	318is	205/60HR15	225/55VR15 225/50ZR16	29	32	7-J	4-36 in. ①
	323i	205/60HR15	255/50ZR16	29	32	7-J	4-36 in. ①
	328is Coupe	205/60HR15	225/50ZR16	29	34	7-J	4-36 in. ①
	528i	205/65R15	225/60R15 235/45R17 255/40R17	33	41	Std: 6.5-J Opt: 8-J Opt: 8-J	4-36 in. ①
	540i	225/55R16	235/45R17 255/40R17	35	42	Std: 7-J Opt: 8-J Opt: 9-J	4-36 in. ①
	740i	225/55R16	235/45R17 255/40R17	35	42	Std: 7-J Opt: 8-J Opt: 9-J	4-36 in. ①
	Z3	225/50ZR16	205/60R15 225/45ZR17	29	29	7-J	4-36 in. ①
2000	318is	205/60HR15	225/55VR15 225/50ZR16	29	32	7-J	4-36 in. ①
	323i	205/60HR15	255/50ZR16	29	32	7-J	4-36 in. ①
	328is Coupe	205/60HR15	225/50ZR16	29	34	7-J	4-36 in. ①
	528i	205/65R15	225/60R15 235/45R17 255/40R17	33	41	Std: 6.5-J Opt: 8-J Opt: 8-J	4-36 in. ①

93471C51

TIRE, WHEEL AND BALL JOINT SPECIFICATIONS

| Year | Model | OEM Tires | | Tire Pressures (psi) | | Wheel Size | Ball Joint Inspection |
		Standard	Optional	Front	Rear		
2000 (cont.)	540i	225/55R16	235/45R17 255/40R17	35	42	Std: 7-J Opt: 8-J Opt: 9-J	4-36 in. ①
	740i	225/55R16	235/45R17 255/40R17	35	42	Std: 7-J Opt: 8-J Opt: 9-J	4-36 in. ①
	Z3	225/50ZR16	205/60R15 225/45ZR17	29	29	7-J	4-36 in. ①

OEM: Original Equipment Manufacturer

PSI: Pounds Per Square Inch

STD: Standard

OPT: Optional

① Torque required in inch lbs. to rotate ball joint when removed from the knuckle

93471C52

For complete Engine Mechanical specifications, see Section 1 of this manual

SCHEDULED MAINTENANCE INTERVALS
1998-00 BMW

TO BE SERVICED	TYPE OF SERVICE	SERVICE INTERVALS			
		INITIAL 1200 MILES	OIL SERVICE	INSPECTION I	INSPECTION II
Oil level	S/I	✓			
Engine oil	R	①			
Engine oil & filter	R②		✓	✓	✓
Engine air cleaner element	R③				✓
Spark plugs	R				✓
Fuel filter	R④				✓
Fuel, vapor lines & fuel cap	S/I	✓		✓	✓
Cooling system	S/I	✓		✓	✓
Exhaust pipe & muffler	S/I	✓		✓	✓
Catalytic converter & shielding	S/I	✓		✓	✓
Throttle linkage	S/I			✓	✓
Engine (check for leakage)	S/I	✓			
Engine drive belts	S/I				✓
Maintenance Indicators	RE		⑤	✓	✓
Engine coolant	R			⑥	⑥
Oxygen sensor	R⑦				
Intake air dust separators	S/I⑧				
Brake & clutch fluids ⑥	S/I			✓	✓
Brake pads & discs	S/I			✓	✓
Parking brake system	S/I			✓	✓
Power steering system	S/I			✓	✓
Rear axle fluid	S/I			✓	✓
Steering play, suspension track rods, front axle joints, steering linkage & joint disc	S/I			✓	✓
Transmission fluid/oil	S/I			✓	⑨
Wheel centering hubs	S/I			✓	✓
Rear axle fluid ⑩	R		✓		✓
OBD system for codes	S/I	✓		✓	✓

R: Replace   S/I: Service or Inspect   RE: Reset

Note: BMW does not rely solely on vehicle mileage to determine service intervals. An on-oboard diagnostic center, monitors engine operating conditions, along with mileage, to determine the most effective maintenance intervals. The information is then conveyed to the driver through the service indicator lights, located in the center of the instrument p[anel.

① Service is not required for 318 & 325 models.

② On vehicles operated less than 6200 miles per year, more frequent service may be required.

③ Replace more frequently if vehicle is operated in dusty conditions.

④ Recommended service for California models, required for all other models.

⑤ Reset the oil service indicator lights only.

⑥ Replace every 2 years with inspection service.

⑦ Replace every 100,000 miles on all models.

⑧ Required service for 850 models only.

⑨ Change fluid (A/T) or oil (M/T) at inspection.

⑩ At first oil service, then at each inspection.

FREQUENT OPERATION MAINTENANCE (SEVERE SERVICE)

If a vehicle is operated under any of the following conditions it is considered severe service

- Extremely dusty areas.

- 50% or more of the vehicle operation is in 32°C (90°F) or higher temperatures, or constant operation in temperatures below 0°C (32°F).

- Prolonged idling (vehicle operation in stop and go traffic).

- Frequent short running periods (engine does not warm to normal operating temperatures).

- Police, taxi, delivery usage or trailer towing usage.

93471C53

SCHEDULED MAINTENANCE INTERVALS
BMW
3 SERIES, 5 SERIES, 7 SERIES, 8 SERIES
M3, Z3

The following should be used as a guide when determining the amount of work required for a particular service.
In estimating how long a particular Scheduled Maintenance Service should take, please observe the following:

● Labor Time is time based on field research and data supplied by the vehicle manufacturer.
● Labor time operations are given in hours and tenths of an hour.
● All labor operations are to be used as a guide.

Mechanic Skill Level Codes:
(A) PRECISION: Highly skilled with multiple certification.
(B) GENERAL: Normally skilled with certification.
(C) MAINTENANCE: Semi-skilled working on certification.

	LABOR TIME
Initial 12000 Mile Service (B)	
All Models	1.4
Engine Oil Service (B)	
All Models	.8
Inspection I (B)	
All Models	3.2
Inspection II (B)	
All Models	4.9

93471C54

For Accessory Drive Belt illustrations, see Section 1 of this manual

DAEWOO
Lanos • Nubira • Leganza

ENGINE AND VEHICLE IDENTIFICATION

		Engine						Model Year	
Code ①	Liters (cc)	Cu. In.	Cyl.	Fuel Sys.	Engine Type	Eng. Mfg.		Code ②	Year
6	1.6L (1597)	97.5	4	MFI	DOHC	Daewoo		X	1999
Z	2.0L (1997)	122	4	MFI	DOHC	Daewoo		Y	2000
2	2.2L (2156)	134	4	MFI	DOHC	Daewoo		1	2001
								2	2002

MFI: Multi-port Fuel Injection

SOHC: Single Overhead Camshaft

DOHC: Double Overhead Camshafts

① 8th digit of VIN

② 10th digit of VIN

93471C55

GENERAL ENGINE SPECIFICATIONS

Year	Model	Engine Displacement Liters (cc)	Engine ID/VIN	Fuel System Type	Net Horsepower @ rpm	Net Torque @ rpm (ft. lbs.)	Bore x Stroke (in.)	Compression Ratio	Oil Pressure @ rpm
1999	Lanos	1.6L (1597)	6	MFI	105@5800	105.8@3800	3.11 X 3.21	9.5:1	30 @ idle
	Nubira	2.0L (1997)	Z	MFI	128@5400	135@4400	3.38 X 3.38	9.6:1	30 @ idle
	Leganza	2.2L (2156)	2	MFI	131@5200	147@2800	3.4 0X 3.70	9.6:1	4.35 @ idle
2000	Lanos	1.6L (1597)	6	MFI	105@5800	105.8@3800	3.11 X 3.21	9.5:1	30 @ idle
	Nubira	2.0L (1997)	Z	MFI	128@5400	135@4400	3.38 X 3.38	9.6:1	30 @ idle
	Leganza	2.2L (2156)	2	MFI	131@5200	147@2800	3.40 X 3.70	9.6:1	4.35 @ idle
2001	Lanos	1.6L (1597)	6	MFI	105@5800	105.8@3800	3.11 X 3.21	9.5:1	30 @ idle
	Nubira	2.0L (1997)	Z	MFI	128@5400	135@4400	3.38 X 3.38	9.6:1	30 @ idle
	Leganza	2.2L (2156)	2	MFI	131@5200	147@2800	3.40 X 3.70	9.6:1	4.35 @ idle

NA: Not Available

MFI: Multi-port Fuel Injection

93471C56

ENGINE TUNE-UP SPECIFICATIONS

Year	Engine Displacement Liters (cc)	Engine ID/VIN	Spark Plug Gap (in.)	Ignition Timing (deg.) MT	Ignition Timing (deg.) AT	Fuel Pump (psi)	Idle Speed (rpm) MT	Idle Speed (rpm) AT	Valve Clearance Intake	Valve Clearance Exhaust
1999	1.6L (1597)	6	0.043	10 B	10 B	41-47	825	825	HYD	HYD
	2.0L (1997)	Z	0.031	8 B	8 B	41-47	825	825	HYD	HYD
	2.2L (2156)	2	0.031	6 B	6 B	41-47	850	850	HYD	HYD
2000	1.6L (1597)	6	0.043	10 B	10 B	41-47	825	825	HYD	HYD
	2.0L (1997)	Z	0.031	8 B	8 B	41-47	825	825	HYD	HYD
	2.2L (2156)	2	0.031	6 B	6 B	41-47	850	850	HYD	HYD
2001	1.6L (1597)	6	0.043	10 B	10 B	41-47	825	825	HYD	HYD
	2.0L (1997)	Z	0.031	8 B	8 B	41-47	825	825	HYD	HYD
	2.2L (2156)	2	0.031	6 B	6 B	41-47	850	850	HYD	HYD

NOTE: The Vehicle Emission Control Information label often reflects specification changes made during production. The label figures must be used if they differ from those in this chart

B: Before top dead center

HYD: Hydraulic

93471C57

For Tire, Wheel and Ball Joint specifications, see Section 1 of this manual

CAPACITIES

Year	Model	Engine Displacement Liters (cc)	Engine ID/VIN	Engine Oil with Filter (qts.)	Transaxle (pts.)		Fuel Tank (gal.)	Cooling System (qts.)
					Manual	Auto.		
1999	Lanos	1.6L (1597)	6	4.0	4.0	24.4	12.7	7.4
	Nubira	2.0L (1997)	Z	4.1	4.0	24.4	13.7	7.4
	Leganza	2.2L (2156)	2	4.2	3.8	15.4	15.8	7.4
2000	Lanos	1.6L (1597)	6	4.0	4.0	24.4	12.7	7.4
	Nubira	2.0L (1997)	Z	4.1	4.0	24.4	13.7	7.4
	Leganza	2.2L (2156)	2	4.2	3.8	15.4	15.8	7.4
2001	Lanos	1.6L (1597)	6	4.0	4.0	24.4	12.7	7.4
	Nubira	2.0L (1997)	Z	4.1	4.0	24.4	13.7	7.4
	Leganza	2.2L (2156)	2	4.2	3.8	15.4	15.8	7.4

NOTE: All capacities are approximate. Add fluid gradually and ensure a proper level is obtained.

93471C58

VALVE SPECIFICATIONS

Year	Engine Displacement Liters (cc)	Engine ID/VIN	Seat Angle (deg.)	Face Angle (deg.)	Spring Free Length (in.)	Valve Guide Inside Diameter (in.)		Stem Diameter (in.)	
						Intake	Exhaust	Intake	Exhaust
1999	1.6L (1597)	6	45	45	NA	0.236-0.237	0.236-0.237	0.236	0.236
	2.0L (1997)	Z	45	45	NA	0.236-0.237	0.236-0.237	0.2755-0.2760	0.2747-0.2752
	2.2L (2156)	2	40	44	NA	0.235-0.236	0.235-0.236	0.235-0.234	0.234-0.236
2000	1.6L (1597)	6	45	45	NA	0.236-0.237	0.236-0.237	0.236	0.236
	2.0L (1997)	Z	45	45	NA	0.236-0.237	0.236-0.237	0.2755-0.2760	0.2747-0.2752
	2.2L (2156)	2	40	44	NA	0.235-0.236	0.235-0.236	0.235-0.234	0.234-0.236
2001	1.6L (1597)	6	45	45	NA	0.236-0.237	0.236-0.237	0.236	0.236
	2.0L (1997)	Z	45	45	NA	0.236-0.237	0.236-0.237	0.2755-0.2760	0.2747-0.2752
	2.2L (2156)	2	40	44	NA	0.235-0.236	0.235-0.236	0.235-0.234	0.234-0.236

NA: Not Available

93471C59

CRANKSHAFT AND CONNECTING ROD SPECIFICATIONS
All measurements are given in inches.

Year	Engine Displacement Liters (cc)	Engine ID/VIN	Crankshaft				Connecting Rod		
			Main Brg. Journal Dia.	Main Brg. Oil Clearance	Shaft End-play	Thrust on No.	Journal Diameter	Oil Clearance	Side Clearance
1999	1.6L (1597)	6	2.1640-2.1650	0.0004-0.0010	0.003	NA	1.6910-1.6920	0.0007-0.0028	0.0027-0.009
	2.0L (1997)	Z	2.2824-2.2832	0.0005-0.0023	0.0027-0.0118	NA	1.9283-1.9286	0.0050-0.0016	NA
	2.2L (2156)	2	2.2824-2.2832	0.0005-0.0023	0.0027-0.0118	NA	1.9283-1.9286	0.0050-0.0016	NA
2000	1.6L (1597)	6	2.1640-2.1650	0.0004-0.0010	0.003	NA	1.6910-1.6920	0.0007-0.0028	0.0027-0.009
	2.0L (1997)	Z	2.2824-2.2832	0.0005-0.0023	0.0027-0.0118	NA	1.9283-1.9286	0.0050-0.0016	NA
	2.2L (2156)	2	2.2824-2.2832	0.0005-0.0023	0.0027-0.0118	NA	1.9283-1.9286	0.0050-0.0016	NA
2001	1.6L (1597)	6	2.1640-2.1650	0.0004-0.0010	0.003	NA	1.6910-1.6920	0.0007-0.0028	0.0027-0.009
	2.0L (1997)	Z	2.2824-2.2832	0.0005-0.0023	0.0027-0.0118	NA	1.9283-1.9286	0.0050-0.0016	NA
	2.2L (2156)	2	2.2824-2.2832	0.0005-0.0023	0.0027-0.0118	NA	1.9283-1.9286	0.0050-0.0016	NA

NA: Not Available

93471C60

PISTON AND RING SPECIFICATIONS
All measurements are given in inches.

Year	Engine Displacement Liters (cc)	Engine ID/VIN	Piston Clearance	Ring Gap			Ring Side Clearance		
				Top Compression	Bottom Compression	Oil Control	Top Compression	Bottom Compression	Oil Control
1999	1.6L (1597)	6	0.0012	0.019	0.019	NA	0.0008	0.0008	NA
	2.0L (1997)	Z	0.00039-0.0011	0.019	0.019	NA	NA	NA	NA
	2.2L (2156)	2	0.0011-0.0020	0.019	0.019	0.0015-0.055	NA	NA	NA
2000	1.6L (1597)	6	0.0012	0.019	0.019	NA	0.0008	0.0008	NA
	2.0L (1997)	Z	0.00039-0.0011	0.019	0.019	NA	NA	NA	NA
	2.2L (2156)	2	0.0011-0.0020	0.019	0.019	0.0015-0.055	NA	NA	NA
2001	1.6L (1597)	6	0.0012	0.019	0.019	NA	0.0008	0.0008	NA
	2.0L (1997)	Z	0.00039-0.0011	0.019	0.019	NA	NA	NA	NA
	2.2L (2156)	2	0.0011-0.0020	0.019	0.019	0.0015-0.055	NA	NA	NA

NA: Not Available

93471C61

For Wheel Alignment specifications, see Section 1 of this manual

TORQUE SPECIFICATIONS
All readings in ft. lbs.

Year	Engine Displacement Liters (cc)	Engine ID/VIN	Cylinder Head Bolts	Main Bearing Bolts	Rod Bearing Bolts	Crankshaft Damper Bolts	Flywheel Bolts	Manifold		Spark Plugs	Lug Nuts
								Intake	Exhaust		
1999	1.6L (1597)	6	①	②	③	④	⑤	18	18	18	88
	2.0L (1997)	Z	⑥	②	⑦	15	⑧	13	11	15	88
	2.2L (2156)	2	⑥	②	⑦	15	⑧	13	11	15	88
2000	1.6L (1597)	6	①	②	③	④	⑤	18	18	18	88
	2.0L (1997)	Z	⑥	②	⑦	15	⑧	13	11	15	88
	2.2L (2156)	2	⑥	②	⑦	15	⑧	13	11	15	88
2001	1.6L (1597)	6	①	②	③	④	⑤	18	18	18	88
	2.0L (1997)	Z	⑥	②	⑦	15	⑧	13	11	15	88
	2.2L (2156)	2	⑥	②	⑦	15	⑧	13	11	15	88

① Step 1: 18 ft. lbs. (25 Nm)
Step 2: Plus 60 degrees
Step 3: Plus 60 degrees
Step 4: Plusl 60 degrees
Step 5: Plus 10 degrees

② Step 1: 37 ft. lbs. (50 Nm)
Step 2: Plus 45 degrees
Step 3: Plus 15 degrees

③ Step 1: 18 ft. lbs. (25 Nm)
Step 2: Plus 30 degrees
Step 3: Plus 15 degrees

④ Step 1: 70 ft. lbs. (95 Nm)
Step 2: Plus 30 degrees
Step 3: Plus 15 degrees

⑤ Step 1: 25 ft. lbs. (35 Nm)
Step 2: Plus 30 degrees
Step 3: Plus 15 degrees

⑥ Step 1: 18 ft. lbs. (25 Nm)
Step 2: Plus 90 degrees
Step 3: Plus 90 degrees
Step 4: Plus 90 degrees

⑦ Step 1: 26 ft. lbs. (35 Nm)
Step 2: Plus 45 degrees
Step 3: Plus 15 degrees

⑧ Step 1: 48 ft. lbs. (60 Nm)
Step 2: Plus 30 degrees
Step 3: Plus 15 degrees

93471C62

BRAKE SPECIFICATIONS
All measurements in inches unless noted

| Year | Model | | Brake Disc | | | Brake Drum | | | Minimum Lining Thickness | Brake Caliper | |
			Original Thickness	Minimum Thickness	Maximum Run-out	Original Inside Diameter	Max. Wear Limit	Maximum Machine Diameter		Bracket Bolts (ft. lbs.)	Mounting Bolts (ft. lbs.)
1999	Lanos	F	0.950	0.900	0.0040	—	—	—	0.280	20	70
		R	—	—	—	7.87	7.91	7.91	0.020	—	—
	Nubira	F	0.950	0.870	0.0010	—	—	—	0.280	16-24	70
		R	0.410	0.330	0.0010	7.87	7.91	7.91	①	41	23
	Leganza	F	0.950	0.870	0.0010	—	—	—	0.430	20	70
		R	0.410	0.314	0.0010	—	—	—	0.080	48	23
2000	Lanos	F	0.950	0.900	0.0040	—	—	—	0.280	20	70
		R	—	—	—	7.87	7.91	7.91	0.020	—	—
	Nubira	F	0.950	0.870	0.0010	—	—	—	0.280	16-24	70
		R	0.410	0.330	0.0010	7.87	7.91	7.91	①	41	23
	Leganza	F	0.950	0.870	0.0010	—	—	—	0.430	20	70
		R	0.410	0.314	0.0010	—	—	—	0.080	48	23
2001	Lanos	F	0.950	0.900	0.0040	—	—	—	0.280	20	70
		R	—	—	—	7.87	7.91	7.91	0.020	—	—
	Nubira	F	0.950	0.870	0.0010	—	—	—	0.280	16-24	70
		R	0.410	0.330	0.0010	7.87	7.91	7.91	①	41	23
	Leganza	F	0.950	0.870	0.0010	—	—	—	0.430	20	70
		R	0.410	0.314	0.0010	—	—	—	0.080	48	23

① Rear Disc brake pad: 0.08 inch
   Rear Brake drum shoe: 0.08 inch

93471C63

WHEEL ALIGNMENT

Year	Model		Caster Range (+/-Deg.)	Caster Preferred Setting (Deg.)	Camber Range (+/-Deg.)	Camber Preferred Setting (Deg.)	Toe-in (in.)	Steering Axis Inclination (Deg.)
1999	Laganza	F	1.00	+3.00	1.00	-0.20	-0.05 +/- 0.04	—
		R	—	—	1.00	-0.80	0.08 +/- 0.06	—
	Lanos ①	F	0.91	+2.60	0.70	+1.20	0 +/- 0.08	—
		R	—	—	1.00	-1.60	0 +/- 0.08	—
	Lanos ②	F	1.00	+1.50	0.70	+1.20	0 +/- 0.08	—
		R	—	—	1.00	-1.60	0 +/- 0.08	—
	Nubira	F	0.75	+3.00	0.75	-0.40	0 +/- 0.08	—
		R	—	—	0.75	-0.90	0.04 +/- 0.06	—
2000	Laganza	F	1.00	+3.00	1.00	-0.20	-0.05 +/- 0.04	—
		R	—	—	1.00	-0.80	0.08 +/- 0.06	—
	Lanos ①	F	0.91	+2.60	0.70	+1.20	0 +/- 0.08	—
		R	—	—	1.00	-1.60	0 +/- 0.08	—
	Lanos ②	F	1.00	+1.50	0.70	+1.20	0 +/- 0.08	—
		R	—	—	1.00	-1.60	0 +/- 0.08	—
	Nubira	F	0.75	+3.00	0.75	-0.40	0 +/- 0.08	—
		R	—	—	0.75	-0.90	0.04 +/- 0.06	—
2001	Laganza	F	1.00	+3.00	1.00	-0.20	-0.05 +/- 0.04	—
		R	—	—	1.00	-0.80	0.08 +/- 0.06	—
	Lanos ①	F	0.91	+2.60	0.70	+1.20	0 +/- 0.08	—
		R	—	—	1.00	-1.60	0 +/- 0.08	—
	Lanos ②	F	1.00	+1.50	0.70	+1.20	0 +/- 0.08	—
		R	—	—	1.00	-1.60	0 +/- 0.08	—
	Nubira	F	0.75	+3.00	0.75	-0.40	0 +/- 0.08	—
		R	—	—	0.75	-0.90	0.04 +/- 0.06	—

① With power steering
② Manual steering

93471C64

TIRE, WHEEL AND BALL JOINT SPECIFICATIONS

| Year | Model | OEM Tires | | Tire Pressures (psi) | | Wheel Size | Ball Joint Inspection |
		Standard	Optional	Front	Rear		
1999	Lanos	185/60R14	None	32	32	5.5-J	①
	Leganza	205/60R15	None	29	29	6-J	①
	Nubira	185/65R14	None	30	28	5.5-J	①
2000	Lanos	185/60R14	None	32	32	5.5-J	①
	Leganza	205/60R15	None	29	29	6-J	①
	Nubira	185/65R14	None	30	28	5.5-J	①
2001	Lanos	185/60R14	None	32	32	5.5-J	①
	Leganza	205/60R15	None	29	29	6-J	①
	Nubira	185/65R14	None	30	28	5.5-J	①

OEM: Original Equipment Manufacturer

PSI: Pounds Per Square Inch

STD: Standard

OPT: Optional

① Replace if any measurable movement is found.

93471C65

For Tune-up, Capacities and Firing orders, see Section 1 of this manual

SCHEDULED MAINTENANCE INTERVALS
NUBRIA

TO BE SERVICED	TYPE OF SERVICE	VEHICLE MILEAGE INTERVAL (x1000)																
		6	12	18	24	30	36	42	48	54	60	66	72	78	84	90	96	102
Drive belt	S/I			✓			✓			✓			✓			✓		
Engine oil & filter ①	R	✓	✓	✓	✓	✓	✓	✓	✓	✓	✓	✓	✓	✓	✓	✓	✓	✓
Cooling system hose & connections	S/I		✓		✓		✓		✓		✓		✓		✓		✓	
Engine coolant	S/I	✓	✓	✓	✓			✓		✓		✓		✓	✓	✓		✓
	R					✓					✓					✓		
Fuel filter	R					✓					✓					✓		
Fuel line & connections	S/I		✓		✓		✓		✓		✓		✓		✓		✓	
Air cleaner element	S/I	✓	✓	✓	✓		✓	✓	✓	✓		✓	✓	✓	✓		✓	✓
	R					✓					✓					✓		
Ignition Timing	S/I		✓		✓		✓		✓		✓		✓		✓		✓	
Spark plugs	S/I			✓						✓				✓				
	R					✓					✓					✓		
Ignition cables	R	Replace every 60,000 miles (96,000 km)																
Evaporative emission canister & vapor lines	S/I					✓					✓					✓		
PCV system	S/I			✓			✓			✓			✓			✓		
Timing belt (exc. Calif.)	R	Replace every 72,000 miles (115,000 km)																
Timing belt (Calif.)	S/I	Inspect every 60,000 miles (96,000 km) and 90,000 (144,000 km)																
	R	Replace every 102,000 miles (163,200 km)																
Air filter (A/C)	R	✓	✓	✓	✓	✓	✓	✓	✓	✓	✓	✓	✓	✓	✓	✓	✓	✓
Exhaust pipes & mountings	S/I		✓		✓		✓		✓		✓		✓		✓		✓	
Brake/clutch fluid ②	S/I	✓	✓		✓	✓		✓		✓		✓		✓	✓		✓	✓
	R			✓			✓			✓			✓			✓		
Front brake pads & discs	S/I	✓	✓	✓	✓	✓	✓	✓	✓	✓	✓	✓	✓	✓	✓	✓	✓	✓
Rear brake drums & linings	S/I		✓		✓		✓		✓		✓		✓		✓		✓	
Parking brake	S/I		✓		✓		✓		✓		✓		✓		✓		✓	
Brake line & connections (inc. booster)	S/I	✓	✓	✓	✓	✓	✓	✓	✓	✓	✓	✓	✓	✓	✓	✓	✓	✓
Rear hub bearing & clearance	S/I		✓		✓		✓		✓		✓		✓		✓		✓	
Manual transaxle oil	S/I		✓		✓		✓		✓		✓		✓		✓		✓	
Clutch & brake pedal free-play	S/I	✓	✓	✓	✓	✓	✓	✓	✓	✓	✓	✓	✓	✓	✓	✓	✓	✓

93471C66

SCHEDULED MAINTENANCE INTERVALS
NUBRIA

TO BE SERVICED	TYPE OF SERVICE	VEHICLE MILEAGE INTERVAL (x1000)																
		6	12	18	24	30	36	42	48	54	60	66	72	78	84	90	96	102
Automatic transaxle fluid ③	S/I		✓		✓		✓		✓				✓		✓		✓	
Chassis & underbody bolts & nuts, tighten/secure	S/I	✓	✓	✓	✓	✓	✓	✓	✓	✓	✓	✓	✓	✓	✓	✓	✓	✓
Tire condition & inflation pressure	S/I	✓	✓	✓	✓	✓	✓	✓	✓	✓	✓	✓	✓	✓	✓	✓	✓	✓
Wheel alignment	S/I	Inspect when abnormal condition is noted.																
Steering wheel & linkage	S/I	✓	✓	✓	✓	✓	✓	✓	✓	✓	✓	✓	✓	✓	✓	✓	✓	✓
Power steering fluid & lines	S/I		✓		✓		✓		✓		✓		✓		✓		✓	
Driveshaft boots	S/I	✓	✓	✓	✓	✓	✓	✓	✓	✓	✓	✓	✓	✓	✓	✓	✓	✓
Seat belts, buckles & anchors	S/I	✓	✓	✓	✓	✓	✓	✓	✓	✓	✓	✓	✓	✓	✓	✓	✓	✓
Lubricate locks, hinges & hood latch	S/I		✓		✓		✓		✓		✓		✓		✓		✓	
Tire rotation	R	Rotate every 6,000 miles (9,600 km)																

R: Replace   S/I: Inspect and service, if needed

① Change engine oil and filter every 3,000 miles (4,800 km) or 3 months under severe conditions

② Change brake/clutch fluid every 9,000 miles (14,400 km) or 9 months under severe conditions

③ Change automatic transaxle fluid every 30,000 miles (48,000 km) or 30 months under severe conditions

FREQUENT OPERATION MAINTENANCE (SEVERE SERVICE)

If a vehicle is operated under any of the following conditions it is considered severe service:

- Towing a trailer or using a camper or car-top carrier.

- Repeated short trips of less than 5 miles in temperatures below freezing, or trips of less than 10 miles in any temperature.

- Extensive idling or low-speed driving for long distances as in heavy commercial use, such as delivery, taxi or police cars.

- Operating on rough, muddy or salt-covered roads.

- Operating on unpaved or dusty roads.

- Driving in extremely hot (over 90°) conditions.

93471C67

SCHEDULED MAINTENANCE INTERVALS
LANOS

TO BE SERVICED	TYPE OF SERVICE	VEHICLE MILEAGE INTERVAL (x1000)																
		6	12	18	24	30	36	42	48	54	60	66	72	78	84	90	96	102
Drive belt (1.5L)	S/I			✓			✓			✓			✓			✓		
	R						✓						✓					
Drive belt (1.6L)	S/I			✓			✓			✓			✓			✓		
Engine oil & filter ①	R	✓	✓	✓	✓	✓	✓	✓	✓	✓	✓	✓	✓	✓	✓	✓	✓	✓
Cooling system hose & connections	S/I		✓		✓		✓		✓		✓		✓		✓		✓	
Engine coolant	S/I	✓	✓	✓	✓		✓	✓	✓	✓		✓	✓	✓	✓		✓	✓
	R					✓					✓					✓		
Fuel filter	R						✓				✓					✓		
Fuel line & connections	S/I		✓		✓		✓		✓		✓		✓		✓		✓	
Air cleaner element	S/I	✓	✓	✓	✓		✓	✓	✓	✓		✓	✓	✓	✓		✓	✓
	R					✓					✓					✓		
Ignition Timing	S/I		✓		✓		✓		✓		✓		✓		✓		✓	
Spark plugs	S/I			✓						✓				✓				
	R					✓					✓					✓		
Ignition cables	R	Replace every 60,000 miles (96,000 km)																
Evaporative emission canister & vapor lines	S/I					✓					✓					✓		
PCV system	S/I			✓			✓			✓			✓			✓		
Timing belt (exc. Calif.)	R	Replace every 60,000 miles (96,000 km)																
Timing belt (Calif.)	S/I	Inspect every 60,000 miles (96,000 km) and 90,000 (144,000 km)																
	R	Replace every 102,000 miles (163,200 km)																
Exhaust pipes & mountings	S/I		✓		✓		✓		✓		✓		✓		✓		✓	
Brake/clutch fluid ②	S/I	✓	✓		✓	✓		✓	✓		✓	✓		✓	✓		✓	✓
	R			✓			✓			✓			✓		✓			
Front brake pads & discs	S/I	✓	✓	✓	✓	✓	✓	✓	✓	✓	✓	✓	✓	✓	✓	✓	✓	✓
Rear brake drums & linings	S/I		✓		✓		✓		✓		✓		✓		✓		✓	
Parking brake	S/I		✓		✓		✓		✓		✓		✓		✓		✓	
Brake line & connections (inc. booster)	S/I	✓	✓	✓	✓	✓	✓	✓	✓	✓	✓	✓	✓	✓	✓	✓	✓	✓
Rear hub bearing & clearance	S/I		✓		✓		✓		✓		✓		✓		✓		✓	
Manual transaxle oil	S/I		✓		✓		✓		✓		✓		✓		✓		✓	
Clutch & brake pedal free-play	S/I	✓	✓	✓	✓	✓	✓	✓	✓	✓	✓	✓	✓	✓	✓	✓	✓	✓

93471C68

SCHEDULED MAINTENANCE INTERVALS
LANOS

TO BE SERVICED	TYPE OF SERVICE	VEHICLE MILEAGE INTERVAL (x1000)																
		6	12	18	24	30	36	42	48	54	60	66	72	78	84	90	96	102
Automatic transaxle fluid ③	S/I		✓		✓		✓		✓				✓		✓		✓	
Chassis & underbody bolts & nuts, tighten/secure	S/I	✓	✓	✓	✓	✓	✓	✓	✓	✓	✓	✓	✓	✓	✓	✓	✓	✓
Tire condition & inflation pressure	S/I	✓	✓	✓	✓	✓	✓	✓	✓	✓	✓	✓	✓	✓	✓	✓	✓	✓
Wheel alignment	S/I	Inspect when abnormal condition is noted.																
Steering wheel & linkage	S/I	✓	✓	✓	✓	✓	✓	✓	✓	✓	✓	✓	✓	✓	✓	✓	✓	✓
Power steering fluid & lines	S/I		✓		✓		✓		✓		✓		✓		✓		✓	
Driveshaft boots	S/I	✓	✓	✓	✓	✓	✓	✓	✓	✓	✓	✓	✓	✓	✓	✓	✓	✓
Seat belts, buckles & anchors	S/I	✓	✓	✓	✓	✓	✓	✓	✓	✓	✓	✓	✓	✓	✓	✓	✓	✓
Lubricate locks, hinges & hood latch	S/I		✓		✓		✓		✓		✓		✓		✓		✓	
Tire rotation	R	Rotate every 6,000 miles (9,600 km)																

R: Replace   S/I: Inspect and service, if needed

① Change engine oil and filter every 3,000 miles (4,800 km) or 3 months under severe conditions

② Change brake/clutch fluid every 9,000 miles (14,400 km) or 9 months under severe conditions

③ Change automatic transaxle fluid every 50,000 miles (80,000 km) or 50 months under severe conditions

FREQUENT OPERATION MAINTENANCE (SEVERE SERVICE)

If a vehicle is operated under any of the following conditions it is considered severe service:

- Towing a trailer or using a camper or car-top carrier.

- Repeated short trips of less than 5 miles in temperatures below freezing, or trips of less than 10 miles in any temperature.

- Extensive idling or low-speed driving for long distances as in heavy commercial use, such as delivery, taxi or police cars.

- Operating on rough, muddy or salt-covered roads.

- Operating on unpaved or dusty roads.

- Driving in extremely hot (over 90°) conditions.

93471C69

SCHEDULED MAINTENANCE INTERVALS
LEGANZA

TO BE SERVICED	TYPE OF SERVICE	VEHICLE MILEAGE INTERVAL (x1000)																
		6	12	18	24	30	36	42	48	54	60	66	72	78	84	90	96	102
Drive belt	S/I			✓			✓			✓			✓			✓		
Engine oil & filter ①	R	✓	✓	✓	✓	✓	✓	✓	✓	✓	✓	✓	✓	✓	✓	✓	✓	✓
Cooling system hose & connections	S/I		✓		✓		✓		✓		✓		✓		✓		✓	
Engine coolant	S/I	✓	✓	✓	✓							✓	✓	✓	✓		✓	✓
	R					✓					✓					✓		
Fuel filter	R					✓					✓					✓		
Fuel line & connections	S/I		✓		✓		✓		✓		✓		✓		✓		✓	
Air cleaner element	S/I	✓	✓	✓	✓		✓	✓	✓	✓		✓	✓	✓	✓		✓	✓
	R					✓					✓					✓		
Ignition Timing	S/I		✓		✓		✓		✓		✓		✓		✓		✓	
Spark plugs	S/I			✓						✓				✓				
	R					✓					✓					✓		
Ignition cables	R	Replace every 60,000 miles (96,000 km)																
Evaporative emission canister & vapor lines	S/I					✓					✓					✓		
PCV system	S/I			✓			✓			✓			✓			✓		
Timing belt (exc. Calif.)	R	Replace every 72,000 miles (115,000 km)																
Timing belt (Calif.)	S/I	Inspect every 60,000 miles (96,000 km) and 90,000 (144,000 km)																
	R	Replace every 102,000 miles (163,200 km)																
Air filter (A/C)	R	✓	✓	✓	✓	✓	✓	✓	✓	✓	✓	✓	✓	✓	✓	✓	✓	✓
Exhaust pipes & mountings	S/I		✓		✓		✓		✓		✓		✓		✓		✓	
Brake/clutch fluid ②	S/I	✓	✓		✓			✓		✓		✓		✓			✓	✓
	R			✓			✓			✓			✓			✓		
Front brake pads & discs	S/I	✓	✓	✓	✓	✓	✓	✓	✓	✓	✓	✓	✓	✓	✓	✓	✓	✓
Rear brake drums & linings	S/I		✓		✓		✓		✓		✓		✓		✓		✓	
Parking brake	S/I		✓		✓		✓		✓		✓		✓		✓		✓	
Brake line & connections (inc. booster)	S/I	✓	✓	✓	✓	✓	✓	✓	✓	✓	✓	✓	✓	✓	✓	✓	✓	✓
Rear hub bearing & clearance	S/I		✓		✓		✓		✓		✓		✓		✓		✓	
Manual transaxle oil	S/I		✓		✓		✓		✓		✓		✓		✓		✓	
Clutch & brake pedal free-play	S/I	✓	✓	✓	✓	✓	✓	✓	✓	✓	✓	✓	✓	✓	✓	✓	✓	✓
Automatic transaxle fluid ③	S/I		✓		✓		✓		✓				✓		✓		✓	
	R										✓							

93471C70

SCHEDULED MAINTENANCE INTERVALS
LEGANZA

TO BE SERVICED	TYPE OF SERVICE	VEHICLE MILEAGE INTERVAL (x1000)																
		6	12	18	24	30	36	42	48	54	60	66	72	78	84	90	96	102
Chassis & underbody bolts & nuts, tighten/secure	S/I	✓	✓	✓	✓	✓	✓	✓	✓	✓	✓	✓	✓	✓	✓	✓	✓	✓
Tire condition & inflation pressure	S/I	✓	✓	✓	✓	✓	✓	✓	✓	✓	✓	✓	✓	✓	✓	✓	✓	✓
Wheel alignment	S/I	Inspect when abnormal condition is noted.																
Steering wheel & linkage	S/I	✓	✓	✓	✓	✓	✓	✓	✓	✓	✓	✓	✓	✓	✓	✓	✓	✓
Power steering fluid & lines	S/I		✓		✓		✓		✓		✓		✓		✓		✓	
Driveshaft boots	S/I	✓	✓	✓	✓	✓	✓	✓	✓	✓	✓	✓	✓	✓	✓	✓	✓	✓
Seat belts, buckles & anchors	S/I	✓	✓	✓	✓	✓	✓	✓	✓	✓	✓	✓	✓	✓	✓	✓	✓	✓
Lubricate locks, hinges & hood latch	S/I		✓		✓		✓		✓		✓		✓		✓		✓	
Tire rotation	R	Rotate every 6,000 miles (9,600 km)																

R: Replace   S/I: Inspect and service, if needed

① Change engine oil and filter every 3,000 miles (4,800 km) or 3 months under severe conditions

② Change brake/clutch fluid every 9,000 miles (14,400 km) or 9 months under severe conditions

③ Change automatic transaxle fluid every 30,000 miles (48,000 km) or 30 months under severe conditions

FREQUENT OPERATION MAINTENANCE (SEVERE SERVICE)

If a vehicle is operated under any of the following conditions it is considered severe service:

- Towing a trailer or using a camper or car-top carrier.

- Repeated short trips of less than 5 miles in temperatures below freezing, or trips of less than 10 miles in any temperature.

- Extensive idling or low-speed driving for long distances as in heavy commercial use, such as delivery, taxi or police cars.

- Operating on rough, muddy or salt-covered roads.

- Operating on unpaved or dusty roads.

- Driving in extremely hot (over 90°) conditions.

93471C71

Timing belt service is covered in Section 3 of this manual

SCHEDULED MAINTENANCE INTERVALS
DAEWOO
NUBRIA

The following should be used as a guide when determining the amount of work required for a particular service.
In estimating how long a particular Scheduled Maintenance Service should take, please observe the following:

- Labor Time is time based on field research and data supplied by the vehicle manufacturer.
- Labor time operations are given in hours and tenths of an hour.
- All labor operations are to be used as a guide.

Mechanic Skill Level Codes:
(A) PRECISION: Highly skilled with multiple certification.
(B) GENERAL: Normally skilled with certification.
(C) MAINTENANCE: Semi-skilled working on certification.

	LABOR TIME		LABOR TIME		LABOR TIME
6000 Mile Service (C)		**42000 Mile Service (C)**		**78000 Mile Service (C)**	
All Models	1.1	All Models	1.0	All Models	1.1
12000 Mile Service (C)		**48000 Mile Service (B)**		**84000 Mile Service (C)**	
All Models	2.0	All Models	2.5	All Models	2.1
18000 Mile Service (C)		**54000 Mile Service (C)**		**90000 Mile Service (C)**	
All Models	1.8	All Models	1.7	All Models	2.6
2400 Mile Service (B)		**60000 Mile Service (C)**		**96000 Mile Service (C)**	
All Models	2.2	Nubria	2.5	All Models	2.0
30000 Mile Service (C)		**66000 Mile Service (C)**		**102000 Mile Service (C)**	
All Models	1.7	All Models	1.0	All Models	2.0
36000 Mile Service (C)		**72000 Mile Service (B)**			
All Models	2.0	Nubria	4.0		

93471C72

SCHEDULED MAINTENANCE INTERVALS

DAEWOO
LANOS

The following should be used as a guide when determining the amount of work required for a particular service. In estimating how long a particular Scheduled Maintenance Service should take, please observe the following:

- Labor Time is time based on field research and data supplied by the vehicle manufacturer.
- Labor time operations are given in hours and tenths of an hour.
- All labor operations are to be used as a guide.

Mechanic Skill Level Codes:

(A) PRECISION: Highly skilled with multiple certification.
(B) GENERAL: Normally skilled with certification.
(C) MAINTENANCE: Semi-skilled working on certification.

	LABOR TIME		LABOR TIME		LABOR TIME
6000 Mile Service (C) All Models	1.1	42000 Mile Service (C) All Models	1.0	78000 Mile Service (C) All Models	1.1
12000 Mile Service (C) All Models	2.0	48000 Mile Service (B) All Models	2.5	84000 Mile Service (C) All Models	2.1
18000 Mile Service (C) All Models	1.8	54000 Mile Service (C) All Models	1.7	90000 Mile Service (C) All Models	2.6
2400 Mile Service (B) All Models	2.2	60000 Mile Service (C) Lanos	4.0	96000 Mile Service (C) All Models	2.0
30000 Mile Service (C) All Models	1.7	66000 Mile Service (C) All Models	1.0	102000 Mile Service (C) All Models	2.0
36000 Mile Service (C) All Models	2.0	72000 Mile Service (B) Leganza, Nubria	4.0	

Heater Core replacement is covered in Section 2 of this manual

93471C73

SCHEDULED MAINTENANCE INTERVALS
DAEWOO
LEGANZA

The following should be used as a guide when determining the amount of work required for a particular service. In estimating how long a particular Scheduled Maintenance Service should take, please observe the following:

● Labor Time is time based on field research and data supplied by the vehicle manufacturer.
●● Labor time operations are given in hours and tenths of an hour.
●●● All labor operations are to be used as a guide.

Mechanic Skill Level Codes:
(A) PRECISION: Highly skilled with multiple certification.
(B) GENERAL: Normally skilled with certification.
(C) MAINTENANCE: Semi-skilled working on certification.

	LABOR TIME
6000 Mile Service (C)	
All Models	1.1
12000 Mile Service (C)	
All Models	2.0
18000 Mile Service (C)	
All Models	1.8
2400 Mile Service (B)	
All Models	2.2
30000 Mile Service (C)	
All Models	1.7
36000 Mile Service (C)	
All Models	2.0

	LABOR TIME
42000 Mile Service (C)	
All Models	1.0
48000 Mile Service (B)	
All Models	2.5
54000 Mile Service (C)	
All Models	1.7
60000 Mile Service (C)	
Leganza	2.5
66000 Mile Service (C)	
All Models	1.0
72000 Mile Service (B)	
Leganza	4.0

	LABOR TIME
78000 Mile Service (C)	
All Models	1.1
84000 Mile Service (C)	
All Models	2.1
90000 Mile Service (C)	
All Models	2.6
96000 Mile Service (C)	
All Models	2.0
102000 Mile Service (C)	
All Models	2.0

93471C74

HONDA
ACCORD • CIVIC • DEL SOL • PRELUDE

ENGINE AND VEHICLE IDENTIFICATION

Engine						Model Year	
Code	Liters	Cu. In. (cc)	Cyl.	Fuel Sys.	Eng. Mfg.	Code	Year
B16A2	1.6	97 (1595)	4	PGM-FI	Honda	W	1998
D16Y5	1.6	97 (1590)	4	PGM-FI	Honda	X	1999
D16Y7	1.6	97 (1590)	4	PGM-FI	Honda	Y	2000
D16Y8	1.6	97 (1590)	4	PGM-FI	Honda	1	2001
D17A1	1.7	101.7 (1668)	4	PGM-FI	Honda	2	2002
D17A2	1.7	101.7 (1668)	4	PGM-FI	Honda		
F20C1	2.0	121.9 (1997)	4	PGM-FI	Honda		
F23A1	2.3	137 (2254)	4	PGM-FI	Honda		
F23A4	2.3	137 (2254)	4	PGM-FI	Honda		
F23A5	2.3	137 (2254)	4	PGM-FI	Honda		
H22A4	2.2	132 (2157)	4	PGM-FI	Honda		
J30A1	3.0	183 (2997)	6	PGM-FI	Honda		

PGM-FI: Programmed Fuel Injection

93471C75

Brake service is covered in Section 4 of this manual

GENERAL ENGINE SPECIFICATIONS

Year	Model	Engine Displacement Liters (cc)	Engine ID/VIN	Fuel System Type	Net Horsepower @ rpm	Net Torque @ rpm (ft. lbs.)	Bore x Stroke (in.)	Compression Ratio	Oil Pressure @ rpm
1998	Civic	1.6 (1590)	D16Y5	PGM-FI	115@6300	104@5400	2.95x3.54	9.6:1	50@3000
	Civic	1.6 (1590)	D16Y7	PGM-FI	106@6200	103@4600	2.95x3.54	9.4:1	50@3000
	Civic	1.6 (1590)	D16Y8	PGM-FI	127@6600	107@5500	2.95x3.54	9.4:1	50@3000
	Prelude	2.2 (2157)	H22A4	PGM-FI	①	158@5500	3.43x3.57	10.0:1	50@3000
	Prelude SH	2.2 (2157)	H22A4	PGM-FI	①	158@5500	3.43x3.57	10.0:1	50@3000
	Accord Coupe (EX, LX)	2.3 (2254)	F23A1	PGM-FI	150@5700	152@4900	3.39x3.82	9.3:1	50@3000
	Accord Coupe (EX, LX)	2.3 (2254)	F23A4	PGM-FI	150@5700	152@4900	3.39x3.82	9.3:1	50@3000
	Accord Sedan (DX)	2.3 (2254)	F23A5	PGM-FI	150@5700	152@4900	3.39x3.82	9.3:1	50@3000
	Accord Sedan (EX, LX)	2.3 (2254)	F23A1	PGM-FI	150@5700	152@4900	3.39x3.82	9.3:1	50@3000
	Accord Sedan (EX, LX)	2.3 (2254)	F23A4	PGM-FI	150@5700	152@4900	3.39x3.82	9.3:1	50@3000
	Accord Coupe (EX, LX)	3.0 (2997)	J30A1	PGM-FI	200@5500	195@4700	3.39x3.39	9.4:1	50@3000
	Accord Sedan (EX, LX)	3.0 (2997)	J30A1	PGM-FI	200@5500	195@4700	3.39x3.39	9.4:1	50@3000
1999	Civic	1.6 (1590)	D16Y5	PGM-FI	115@6300	104@5400	2.95x3.54	9.6:1	50@3000
	Civic	1.6 (1590)	D16Y7	PGM-FI	106@6200	103@4600	2.95x3.54	9.4:1	50@3000
	Civic	1.6 (1590)	D16Y8	PGM-FI	127@6600	107@5500	2.95x3.54	9.4:1	50@3000
	Civic	1.6 (1595)	B16A2	PGM-FI	160@7600	111@7000	3.19x3.05	10.2:1	50@3000
	Prelude	2.2 (2157)	H22A4	PGM-FI	①	158@5500	3.43x3.57	10.0:1	50@3000
	Prelude SH	2.2 (2157)	H22A4	PGM-FI	①	158@5500	3.43x3.57	10.0:1	50@3000
	Accord Coupe (EX, LX)	2.3 (2254)	F23A1	PGM-FI	150@5700	152@4900	3.39x3.82	9.3:1	50@3000
	Accord Coupe (EX, LX)	2.3 (2254)	F23A4	PGM-FI	150@5700	152@4900	3.39x3.82	9.3:1	50@3000
	Accord Sedan (DX)	2.3 (2254)	F23A5	PGM-FI	150@5700	152@4900	3.39x3.82	9.3:1	50@3000
	Accord Sedan (EX, LX)	2.3 (2254)	F23A1	PGM-FI	150@5700	152@4900	3.39x3.82	9.3:1	50@3000
	Accord Sedan (EX, LX)	2.3 (2254)	F23A4	PGM-FI	150@5700	152@4900	3.39x3.82	9.3:1	50@3000
	Accord Coupe (EX, LX)	3.0 (2997)	J30A1	PGM-FI	200@5500	195@4700	3.39x3.39	9.4:1	50@3000
	Accord Sedan (EX, LX)	3.0 (2997)	J30A1	PGM-FI	200@5500	195@4700	3.39x3.39	9.4:1	50@3000
2000	Civic	1.6 (1590)	D16Y5	PGM-FI	115@6300	104@5400	2.95x3.54	9.6:1	50@3000
	Civic	1.6 (1590)	D16Y7	PGM-FI	106@6200	103@4600	2.95x3.54	9.4:1	50@3000
	Civic	1.6 (1590)	D16Y8	PGM-FI	127@6600	107@5500	2.95x3.54	9.4:1	50@3000
	Civic	1.6 (1595)	B16A2	PGM-FI	160@7600	111@7000	3.19x3.05	10.2:1	50@3000
	Prelude	2.2 (2157)	H22A4	PGM-FI	①	158@5500	3.43x3.57	10.0:1	50@3000
	Prelude SH	2.2 (2157)	H22A4	PGM-FI	①	158@5500	3.43x3.57	10.0:1	50@3000
	Accord Coupe (EX, LX)	2.3 (2254)	F23A1	PGM-FI	150@5700	152@4900	3.39x3.82	9.3:1	50@3000
	Accord Coupe (EX, LX)	2.3 (2254)	F23A4	PGM-FI	150@5700	152@4900	3.39x3.82	9.3:1	50@3000
	Accord Sedan (DX)	2.3 (2254)	F23A5	PGM-FI	150@5700	152@4900	3.39x3.82	9.3:1	50@3000
	Accord Sedan (EX, LX)	2.3 (2254)	F23A1	PGM-FI	150@5700	152@4900	3.39x3.82	9.3:1	50@3000
	Accord Sedan (EX, LX)	2.3 (2254)	F23A4	PGM-FI	150@5700	152@4900	3.39x3.82	9.3:1	50@3000
	Accord Coupe (EX, LX)	3.0 (2997)	J30A1	PGM-FI	200@5500	195@4700	3.39x3.39	9.4:1	50@3000
	Accord Sedan (EX, LX)	3.0 (2997)	J30A1	PGM-FI	200@5500	195@4700	3.39x3.39	9.4:1	50@3000
	S2000	2.0 (1997)	F20C1	PGM-FI	240@8300	153@7500	3.43X3.31	11:01	85@3000

93471C76

GENERAL ENGINE SPECIFICATIONS

Year	Model	Engine Displacement Liters (cc)	Engine ID/VIN	Fuel System Type	Net Horsepower @ rpm	Net Torque @ rpm (ft. lbs.)	Bore x Stroke (in.)	Compression Ratio	Oil Pressure @ rpm
2001	Civic DX	1.7 (1668)	D17A1	PGM-FI	115@6100	110X4500	2.95X3.72	9.5:1	50@3000
	Civic GX	1.7 (1668)	D17A1	PGM-FI	115@6100	110X4500	2.95X3.72	9.5:1	50@3000
	Civic LX	1.7 (1668)	D17A1	PGM-FI	115@6100	110X4500	2.95X3.72	9.5:1	50@3000
	Civic EX	1.7 (1668)	D17A2	PGM-FI	127@6100	117X4500	2.95X3.72	9.9:1	50@3000
	Prelude	2.2 (2157)	H22A4	PGM-FI	①	158@5500	3.43x3.57	10.0:1	50@3000
	Prelude SH	2.2 (2157)	H22A4	PGM-FI	①	158@5500	3.43x3.57	10.0:1	50@3000
	Accord Coupe (EX, LX)	2.3 (2254)	F23A1	PGM-FI	150@5700	152@4900	3.39x3.82	9.3:1	50@3000
	Accord Coupe (EX, LX)	2.3 (2254)	F23A4	PGM-FI	150@5700	152@4900	3.39x3.82	9.3:1	50@3000
	Accord Sedan (DX)	2.3 (2254)	F23A5	PGM-FI	150@5700	152@4900	3.39x3.82	9.3:1	50@3000
	Accord Sedan (EX, LX)	2.3 (2254)	F23A1	PGM-FI	150@5700	152@4900	3.39x3.82	9.3:1	50@3000
	Accord Sedan (EX, LX)	2.3 (2254)	F23A4	PGM-FI	150@5700	152@4900	3.39x3.82	9.3:1	50@3000
	Accord Coupe (EX, LX)	3.0 (2997)	J30A1	PGM-FI	200@5500	195@4700	3.39x3.39	9.4:1	50@3000
	Accord Sedan (EX, LX)	3.0 (2997)	J30A1	PGM-FI	200@5500	195@4700	3.39x3.39	9.4:1	50@3000
	S2000	2.0 (1997)	F20C1	PGM-FI	240@8300	153@7500	3.43X3.31	11:01	85@3000

PGM-FI: Programmed Fuel Injection

① Manual transaxle: 195@7000

Automatic transaxle: 190@6600

93471C77

For complete Engine Mechanical specifications, see Section 1 of this manual

ENGINE TUNE-UP SPECIFICATIONS

Year	Engine Displacement Liters (cc)	Engine ID/VIN	Spark Plugs Gap (in.)	Ignition Timing (deg.) MT	AT	Fuel Pump (psi)	Idle Speed (rpm) MT	AT	Valve Clearance In.	Ex.
1998	1.6 (1590)	D16Y5	0.039-0.043	12B	12B	28-36	620-720	650-750	0.007-0.009	0.009-0.011
	1.6 (1590)	D16Y7	0.039-0.043	12B	12B	28-36	620-720	650-750	0.007-0.009	0.009-0.011
	1.6 (1590)	D16Y8	0.039-0.043	12B	12B	28-36	620-720	650-750	0.007-0.009	0.009-0.011
	2.3 (2254)	F23A1	0.039-0.043	12B	12B	40-47	650-750	650-750	0.009-0.011	0.011-0.013
	2.3 (2254)	F23A4	0.039-0.043	12B	12B	40-47	650-750	650-750	0.009-0.011	0.011-0.013
	2.3 (2254)	F23A5	0.039-0.043	12B	12B	40-47	650-750	650-750	0.009-0.011	0.011-0.013
	2.2 (2157)	H22A4	0.039-0.043	15B	15B	47-54	650-750	650-750	0.006-0.007	0.007-0.008
	3.0 (2997)	J30A1	0.039-0.043	—	10B	41-48	—	630-730	0.008-0.009	0.011-0.013
1999	1.6 (1595)	B16A2	0.047-0.051	16B	—	31-38	650-750	—	0.006-0.007	0.007-0.008
	1.6 (1590)	D16Y5	0.039-0.043	12B	12B	28-36	620-720	650-750	0.007-0.009	0.009-0.011
	1.6 (1590)	D16Y7	0.039-0.043	12B	12B	28-36	620-720	650-750	0.007-0.009	0.009-0.011
	1.6 (1590)	D16Y8	0.039-0.043	12B	12B	28-36	620-720	650-750	0.007-0.009	0.009-0.011
	2.3 (2254)	F23A1	0.039-0.043	12B	12B	40-47	650-750	650-750	0.009-0.011	0.011-0.013
	2.3 (2254)	F23A4	0.039-0.043	12B	12B	40-47	650-750	650-750	0.009-0.011	0.011-0.013
	2.3 (2254)	F23A5	0.039-0.043	12B	12B	40-47	650-750	650-750	0.009-0.011	0.011-0.013
	2.2 (2157)	H22A4	0.039-0.043	15B	15B	47-54	650-750	650-750	0.006-0.007	0.007-0.008
	3.0 (2997)	J30A1	0.039-0.043	—	10B	41-48	—	630-730	0.008-0.009	0.011-0.013
2000	1.6 (1595)	B16A2	0.047-0.051	16B	—	31-38	650-750	—	0.006-0.007	0.007-0.008
	1.6 (1590)	D16Y5	0.039-0.043	12B	12B	28-36	620-720	650-750	0.007-0.009	0.009-0.011
	1.6 (1590)	D16Y7	0.039-0.043	12B	12B	28-36	620-720	650-750	0.007-0.009	0.009-0.011
	1.6 (1590)	D16Y8	0.039-0.043	12B	12B	28-36	620-720	650-750	0.007-0.009	0.009-0.011
	2.0 (1997)	F20C1	0.039-0.043	5B	5B	47-54	750-850	—	0.008-0.010	0.010-0.011
	2.3 (2254)	F23A1	0.039-0.043	12B	12B	40-47	650-750	650-750	0.009-0.011	0.011-0.013

93471C78

ENGINE TUNE-UP SPECIFICATIONS

Year	Engine Displacement Liters (cc)	Engine ID/VIN	Spark Plugs Gap (in.)	Ignition Timing (deg.) MT	AT	Fuel Pump (psi)	Idle Speed (rpm) MT	AT	Valve Clearance In.	Ex.
2000 (cont.)	2.3 (2254)	F23A4	0.039-0.043	12B	12B	40-47	650-750	650-750	0.009-0.011	0.011-0.013
	2.3 (2254)	F23A5	0.039-0.043	12B	12B	40-47	650-750	650-750	0.009-0.011	0.011-0.013
	2.2 (2157)	H22A4	0.039-0.043	15B	15B	47-54	650-750	650-750	0.006-0.007	0.007-0.008
	3.0 (2997)	J30A1	0.039-0.043	—	10B	41-48	—	630-730	0.008-0.009	0.011-0.013
2001	1.7 (1668)	D17A1	0.039-0.043	8B	8B	40-47	650-750	650-750	0.007-0.009	0.009-0.011
	1.7 (1668)	D17A2	0.039-0.043	8B	8B	40-47	650-750	650-750	0.007-0.009	0.009-0.011
	2.0 (1997)	F20C1	0.039-0.043	5B	5B	47-54	750-850	—	0.008-0.010	0.010-0.011
	2.3 (2254)	F23A1	0.039-0.043	12B	12B	40-47	650-750	650-750	0.009-0.011	0.011-0.013
	2.3 (2254)	F23A4	0.039-0.043	12B	12B	40-47	650-750	650-750	0.009-0.011	0.011-0.013
	2.3 (2254)	F23A5	0.039-0.043	12B	12B	40-47	650-750	650-750	0.009-0.011	0.011-0.013
	2.2 (2157)	H22A4	0.039-0.043	15B	15B	47-54	650-750	650-750	0.006-0.007	0.007-0.008
	3.0 (2997)	J30A1	0.039-0.043	—	10B	41-48	—	630-730	0.008-0.009	0.011-0.013

NOTE: The Vehicle Emission Control Information label often reflects specification changes made during production. The label figures must be used if they differ from those in this chart.

B: Before Top Dead Center

93471C79

For Accessory Drive Belt illustrations, see Section 1 of this manual

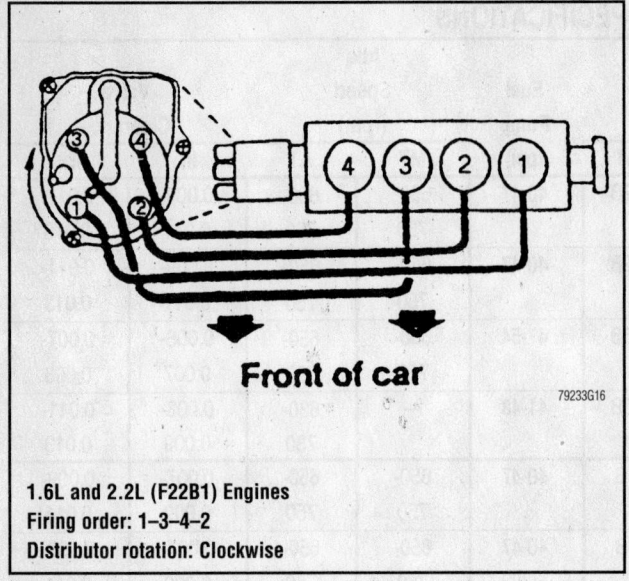

1.6L and 2.2L (F22B1) Engines
Firing order: 1–3–4–2
Distributor rotation: Clockwise

79233G16

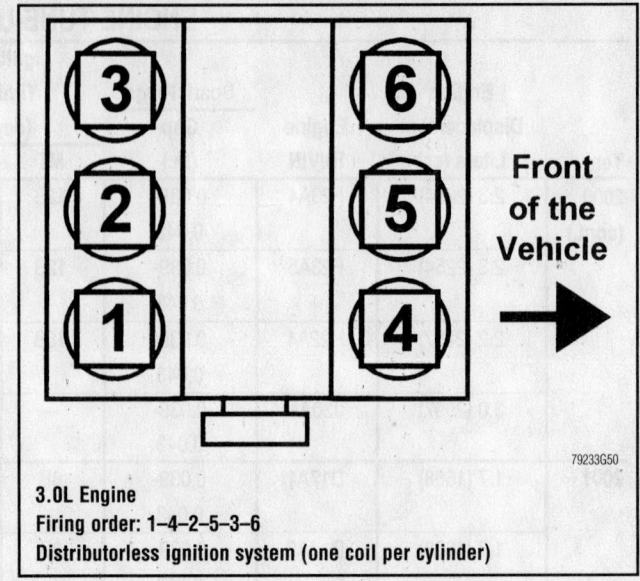

Front of the Vehicle

3.0L Engine
Firing order: 1–4–2–5–3–6
Distributorless ignition system (one coil per cylinder)

79233G50

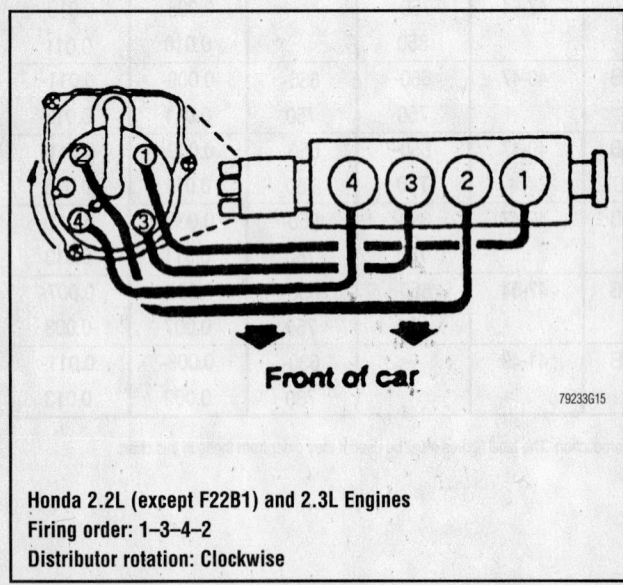

Honda 2.2L (except F22B1) and 2.3L Engines
Firing order: 1–3–4–2
Distributor rotation: Clockwise

79233G15

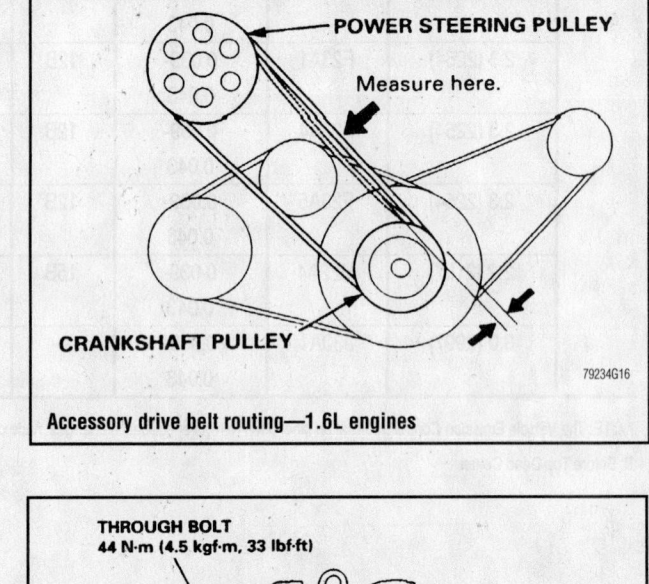

POWER STEERING PULLEY

Measure here.

CRANKSHAFT PULLEY

79234G16

Accessory drive belt routing—1.6L engines

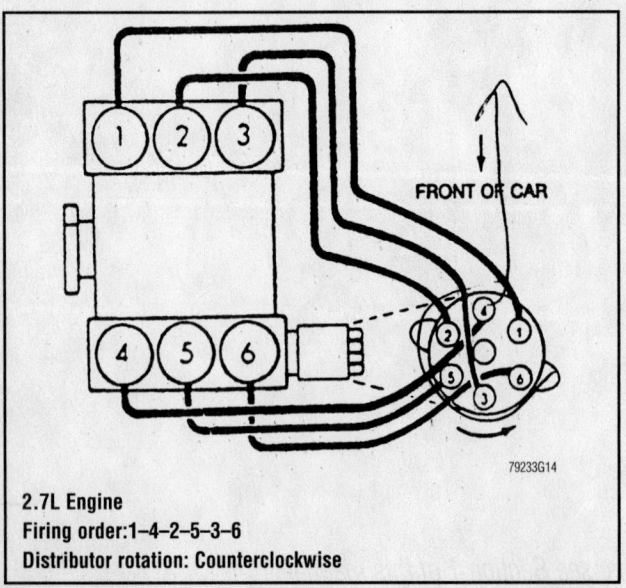

FRONT OF CAR

2.7L Engine
Firing order: 1–4–2–5–3–6
Distributor rotation: Counterclockwise

79233G14

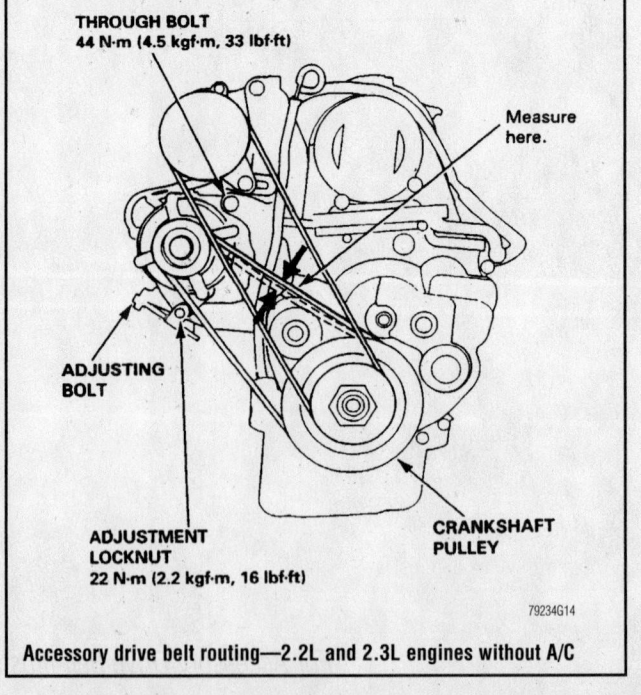

THROUGH BOLT
44 N·m (4.5 kgf·m, 33 lbf·ft)

Measure here.

ADJUSTING BOLT

ADJUSTMENT LOCKNUT
22 N·m (2.2 kgf·m, 16 lbf·ft)

CRANKSHAFT PULLEY

79234G14

Accessory drive belt routing—2.2L and 2.3L engines without A/C

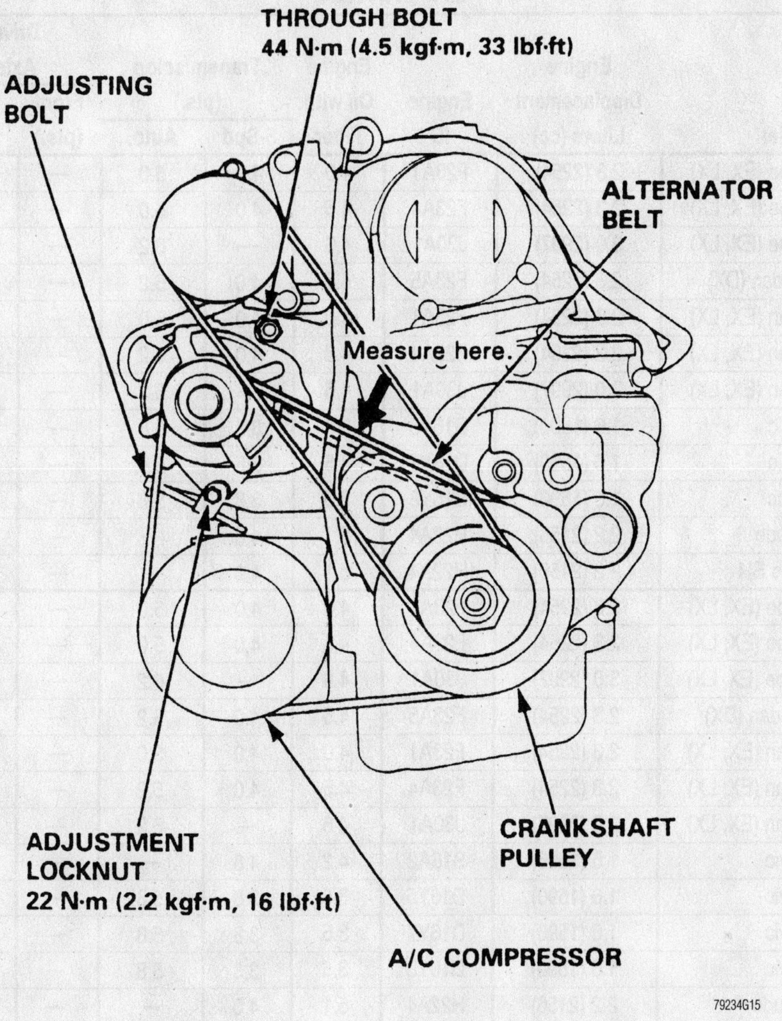

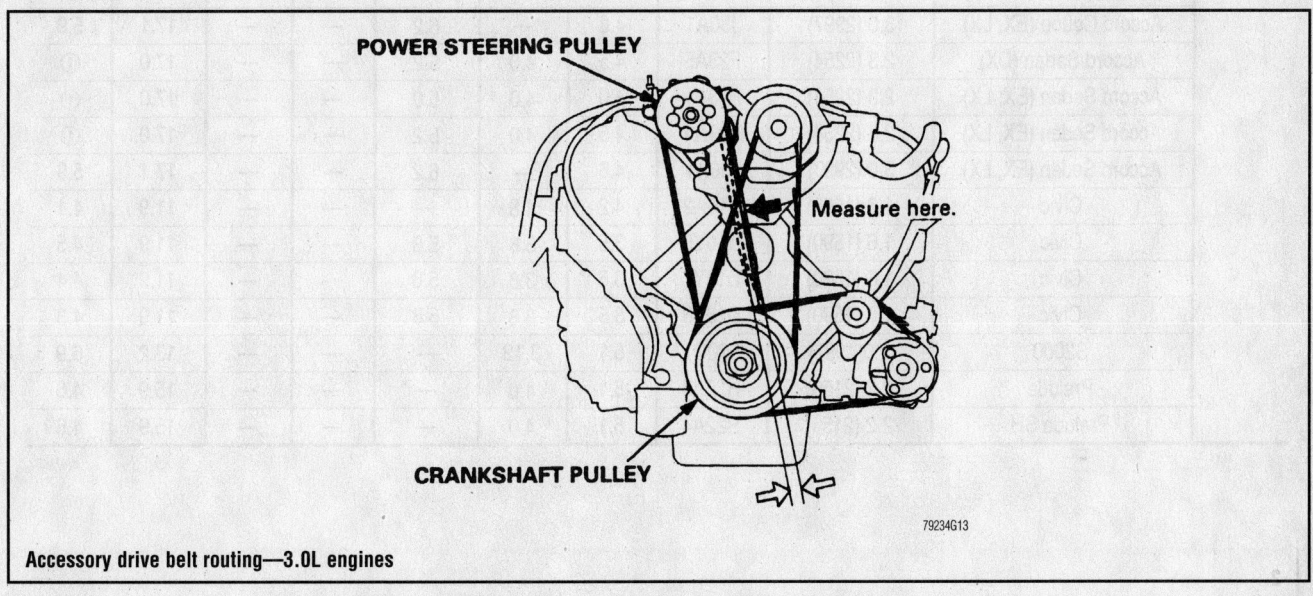

THROUGH BOLT
44 N·m (4.5 kgf·m, 33 lbf·ft)

ADJUSTING BOLT

ALTERNATOR BELT

Measure here.

ADJUSTMENT LOCKNUT
22 N·m (2.2 kgf·m, 16 lbf·ft)

CRANKSHAFT PULLEY

A/C COMPRESSOR

79234G15

Accessory drive belt routing—2.2L and 2.3L engines with A/C

POWER STEERING PULLEY

Measure here.

CRANKSHAFT PULLEY

79234G13

Accessory drive belt routing—3.0L engines

For Tire, Wheel and Ball Joint specifications, see Section 1 of this manual

CAPACITIES

Year	Model	Engine Displacement Liters (cc)	Engine ID	Engine Oil with Filter	Transmission (pts.) 5-Spd	Transmission (pts.) Auto.	Drive Axle Front (pts.)	Drive Axle Rear (pts.)	Fuel Tank (gal.)	Cooling System (qts.)
1998	Accord Coupe (EX, LX)	2.3 (2254)	F23A1	4.0	4.0	5.0	—	—	17.0	①
	Accord Coupe (EX, LX)	2.3 (2254)	F23A4	4.5	4.0	5.0	—	—	17.0	①
	Accord Coupe (EX, LX)	3.0 (2997)	J30A1	4.6	—	6.2	—	—	17.1	5.9
	Accord Sedan (DX)	2.3 (2254)	F23A5	4.5	4.0	5.2	—	—	17.0	①
	Accord Sedan (EX, LX)	2.3 (2254)	F23A1	4.0	4.0	5.0	—	—	17.0	①
	Accord Sedan (EX, LX)	2.3 (2254)	F23A4	4.5	4.0	5.2	—	—	17.0	①
	Accord Sedan (EX, LX)	3.0 (2997)	J30A1	4.6	—	6.2	—	—	17.1	5.9
	Civic	1.6 (1590)	D16Y5	3.5	3.8	5.8	—	—	11.9	4.5
	Civic	1.6 (1590)	D16Y7	3.5	3.8	5.8	—	—	11.9	4.4
	Civic	1.6 (1590)	D16Y8	3.5	3.8	5.8	—	—	11.9	4.3
	Prelude	2.2 (2156)	H22A4	5.1	4.0	—	—	—	15.9	4.6
	Prelude SH	2.2 (2156)	H22A4	5.1	4.0	—	—	—	15.9	4.6
1999	Accord Coupe (EX, LX)	2.3 (2254)	F23A1	4.0	4.0	5.0	—	—	17.0	①
	Accord Coupe (EX, LX)	2.3 (2254)	F23A4	4.5	4.0	5.0	—	—	17.0	①
	Accord Coupe (EX, LX)	3.0 (2997)	J30A1	4.6	—	6.2	—	—	17.1	5.9
	Accord Sedan (DX)	2.3 (2254)	F23A5	4.5	4.0	5.2	—	—	17.0	①
	Accord Sedan (EX, LX)	2.3 (2254)	F23A1	4.0	4.0	5.0	—	—	17.0	①
	Accord Sedan (EX, LX)	2.3 (2254)	F23A4	4.5	4.0	5.2	—	—	17.0	①
	Accord Sedan (EX, LX)	3.0 (2997)	J30A1	4.6	—	6.2	—	—	17.1	5.9
	Civic	1.6 (1595)	B16A2	4.2	4.8	—	—	—	11.9	4.1
	Civic	1.6 (1590)	D16Y5	3.5	3.8	5.8	—	—	11.9	4.5
	Civic	1.6 (1590)	D16Y7	3.5	3.8	5.8	—	—	11.9	4.4
	Civic	1.6 (1590)	D16Y8	3.5	3.8	5.8	—	—	11.9	4.3
	Prelude	2.2 (2156)	H22A4	5.1	4.0	—	—	—	15.9	4.6
	Prelude SH	2.2 (2156)	H22A4	5.1	4.0	—	—	—	15.9	4.6
2000	Accord Coupe (EX, LX)	2.3 (2254)	F23A1	4.0	4.0	5.0	—	—	17.0	①
	Accord Coupe (EX, LX)	2.3 (2254)	F23A4	4.5	4.0	5.0	—	—	17.0	①
	Accord Coupe (EX, LX)	3.0 (2997)	J30A1	4.6	—	6.2	—	—	17.1	5.9
	Accord Sedan (DX)	2.3 (2254)	F23A5	4.5	4.0	5.2	—	—	17.0	①
	Accord Sedan (EX, LX)	2.3 (2254)	F23A1	4.0	4.0	5.0	—	—	17.0	①
	Accord Sedan (EX, LX)	2.3 (2254)	F23A4	4.5	4.0	5.2	—	—	17.0	①
	Accord Sedan (EX, LX)	3.0 (2997)	J30A1	4.6	—	6.2	—	—	17.1	5.9
	Civic	1.6 (1595)	B16A2	4.2	4.8	—	—	—	11.9	4.1
	Civic	1.6 (1590)	D16Y5	3.5	3.8	5.8	—	—	11.9	4.5
	Civic	1.6 (1590)	D16Y7	3.5	3.8	5.8	—	—	11.9	4.4
	Civic	1.6 (1590)	D16Y8	3.5	3.8	5.8	—	—	11.9	4.3
	S2000	2.0 (1997)	F20C1	5.1	3.12	—	—	—	13.2	6.9
	Prelude	2.2 (2156)	H22A4	5.1	4.0	—	—	—	15.9	4.6
	Prelude SH	2.2 (2156)	H22A4	5.1	4.0	—	—	—	15.9	4.6

93471C80

CAPACITIES

Year	Model	Engine Displacement Liters (cc)	Engine ID	Engine Oil with Filter	Transmission (pts.)		Drive Axle		Fuel Tank (gal.)	Cooling System (qts.)
					5-Spd	Auto.	Front (pts.)	Rear (pts.)		
2001	Accord Coupe (EX, LX)	2.3 (2254)	F23A1	4.0	4.0	5.0	—	—	17.0	①
	Accord Coupe (EX, LX)	2.3 (2254)	F23A4	4.5	4.0	5.0	—	—	17.0	①
	Accord Coupe (EX, LX)	3.0 (2997)	J30A1	4.6	—	6.2	—	—	17.1	5.9
	Accord Sedan (DX)	2.3 (2254)	F23A5	4.5	4.0	5.2	—	—	17.0	①
	Accord Sedan (EX, LX)	2.3 (2254)	F23A1	4.0	4.0	5.0	—	—	17.0	①
	Accord Sedan (EX, LX)	2.3 (2254)	F23A4	4.5	4.0	5.2	—	—	17.0	①
	Accord Sedan (EX, LX)	3.0 (2997)	J30A1	4.6	—	6.2	—	—	17.1	5.9
	Civic	1.7 (1668)	D17A1	3.7	3.2	5.8	—	—	11.9	5.4
	Civic	1.7 (1668)	D17A2	3.7	3.2	5.8	—	—	11.9	5.4
	S2000	2.0 (1997)	F20C1	5.1	3.12	—	—	—	13.2	6.9
	Prelude	2.2 (2156)	H22A4	5.1	4.0	—	—	—	15.9	4.6
	Prelude SH	2.2 (2156)	H22A4	5.1	4.0	—	—	—	15.9	4.6

NOTE: All capacities are approximate. Add fluid gradually and ensure a proper fluid level is obtained.

NOTE: Capacities given are service, not overhaul capacities

① Automatic Transaxle: 5.7
Manual Transaxle: 5.8

93471C81

For Wheel Alignment specifications, see Section 1 of this manual

VALVE SPECIFICATIONS

Year	Engine Displacement Liters (cc)	Engine ID/VIN	Seat Angle (deg.)	Face Angle (deg.)	Spring Test Pressure (lbs. @ in.)	Spring Installed Height (in.)	Stem-to-Guide Clearance (in.) Intake	Stem-to-Guide Clearance (in.) Exhaust	Stem Diameter (in.) Intake	Stem Diameter (in.) Exhaust
1998	1.6 (1590)	D16Y5	45	45	NA	NA	0.0010-0.0020	0.0020-0.0030	0.2157-0.2161	0.2146-0.2150
	1.6 (1590)	D16Y7	45	45	NA	NA	0.0010-0.0020	0.0020-0.0030	0.2157-0.2161	0.2146-0.2150
	1.6 (1590)	D16Y8	45	45	NA	NA	0.0010-0.0020	0.0020-0.0030	0.2157-0.2161	0.2146-0.2150
	2.3 (2254)	F23A1	45	45	NA	NA	0.0008-0.0018	0.0022-0.0031	0.2159-0.2163	0.2146-0.2150
	2.3 (2254)	F23A4	45	45	NA	NA	0.0008-0.0018	0.0022-0.0031	0.2159-0.2163	0.2146-0.2150
	2.3 (2254)	F23A5	45	45	NA	NA	0.0008-0.0018	0.0022-0.0031	0.2159-0.2163	0.2146-0.2150
	2.2 (2157)	H22A4	45	45	NA	NA	0.0010-0.0022	0.0020-0.0031	0.2156-0.2159	0.2156-0.2159
	3.0 (2997)	J30A1	45	45	NA	NA	0.0008-0.0018	0.0022-0.0031	0.2159-0.2163	0.2146-0.2150
1999	1.6 (1595)	B16A2	45	45	NA	NA	0.0010-0.0022	0.0020-0.0031	0.2156-0.2159	0.2146-0.2150
	1.6 (1590)	D16Y5	45	45	NA	NA	0.0010-0.0020	0.0020-0.0030	0.2157-0.2161	0.2146-0.2150
	1.6 (1590)	D16Y7	45	45	NA	NA	0.0010-0.0020	0.0020-0.0030	0.2157-0.2161	0.2146-0.2150
	1.6 (1590)	D16Y8	45	45	NA	NA	0.0010-0.0020	0.0020-0.0030	0.2157-0.2161	0.2146-0.2150
	2.3 (2254)	F23A1	45	45	NA	NA	0.0008-0.0018	0.0022-0.0031	0.2159-0.2163	0.2146-0.2150
	2.3 (2254)	F23A4	45	45	NA	NA	0.0008-0.0018	0.0022-0.0031	0.2159-0.2163	0.2146-0.2150
	2.3 (2254)	F23A5	45	45	NA	NA	0.0008-0.0018	0.0022-0.0031	0.2159-0.2163	0.2146-0.2150
	2.2 (2157)	H22A4	45	45	NA	NA	0.0010-0.0022	0.0020-0.0031	0.2156-0.2159	0.2156-0.2159
	3.0 (2997)	J30A1	45	45	NA	NA	0.0008-0.0018	0.0022-0.0031	0.2159-0.2163	0.2146-0.2150
2000	1.6 (1595)	B16A2	45	45	NA	NA	0.0010-0.0022	0.0020-0.0031	0.2156-0.2159	0.2146-0.2150
	1.6 (1590)	D16Y5	45	45	NA	NA	0.0010-0.0020	0.0020-0.0030	0.2157-0.2161	0.2146-0.2150
	1.6 (1590)	D16Y7	45	45	NA	NA	0.0010-0.0020	0.0020-0.0030	0.2157-0.2161	0.2146-0.2150
	1.6 (1590)	D16Y8	45	45	NA	NA	0.0010-0.0020	0.0020-0.0030	0.2157-0.2161	0.2146-0.2150

93471C82

VALVE SPECIFICATIONS

Year	Engine Displacement Liters (cc)	Engine ID/VIN	Seat Angle (deg.)	Face Angle (deg.)	Spring Test Pressure (lbs. @ in.)	Spring Installed Height (in.)	Stem-to-Guide Clearance (in.)		Stem Diameter (in.)	
							Intake	Exhaust	Intake	Exhaust
2000 (cont.)	2.0 (1997)	F20C1	45	45	NA	NA	0.0010-0.0020	0.0020-0.0030	0.2157-0.2162	0.2146-0.2150
	2.3 (2254)	F23A1	45	45	NA	NA	0.0008-0.0018	0.0022-0.0031	0.2159-0.2163	0.2146-0.2150
	2.3 (2254)	F23A4	45	45	NA	NA	0.0008-0.0018	0.0022-0.0031	0.2159-0.2163	0.2146-0.2150
	2.3 (2254)	F23A5	45	45	NA	NA	0.0008-0.0018	0.0022-0.0031	0.2159-0.2163	0.2146-0.2150
	2.2 (2157)	H22A4	45	45	NA	NA	0.0010-0.0022	0.0020-0.0031	0.2156-0.2159	0.2156-0.2159
	3.0 (2997)	J30A1	45	45	NA	NA	0.0008-0.0018	0.0022-0.0031	0.2159-0.2163	0.2146-0.2150
2001	1.7 (1668)	D17A1	45	45	NA	NA	0.0008-0.0020	0.0020-0.0031	0.2157-0.2161	0.2146-0.2150
	1.7 (1668)	D17A2	45	45	NA	NA	0.0008-0.0020	0.0020-0.0031	0.2157-0.2161	0.2146-0.2150
	2.0 (1997)	F20C1	45	45	NA	NA	0.0010-0.0020	0.0020-0.0030	0.2157-0.2162	0.2146-0.2150
	2.3 (2254)	F23A1	45	45	NA	NA	0.0008-0.0018	0.0022-0.0031	0.2159-0.2163	0.2146-0.2150
	2.3 (2254)	F23A4	45	45	NA	NA	0.0008-0.0018	0.0022-0.0031	0.2159-0.2163	0.2146-0.2150
	2.3 (2254)	F23A5	45	45	NA	NA	0.0008-0.0018	0.0022-0.0031	0.2159-0.2163	0.2146-0.2150
	2.2 (2157)	H22A4	45	45	NA	NA	0.0010-0.0022	0.0020-0.0031	0.2156-0.2159	0.2156-0.2159
	3.0 (2997)	J30A1	45	45	NA	NA	0.0008-0.0018	0.0022-0.0031	0.2159-0.2163	0.2146-0.2150

NA: Not Available

93471C83

For Maintenance Interval recommendations, see Section 1 of this manual

CRANKSHAFT AND CONNECTING ROD SPECIFICATIONS

All measurements are given in inches.

Year	Engine Displacement Liters (cc)	Engine ID/VIN	Crankshaft Main Brg. Journal Dia.	Crankshaft Main Brg. Oil Clearance	Crankshaft Shaft End-play	Thrust on No.	Connecting Rod Journal Diameter	Connecting Rod Oil Clearance	Connecting Rod Side Clearance
1998	1.6 (1590)	D16Y5	2.1644-2.1654	0.0007- ① 0.0016	0.004-0.018	3	1.7707-1.7717	0.0008-0.0017	0.0006-0.0016
	1.6 (1590)	D16Y7	2.1644-2.1654	0.0007- ① 0.0016	0.004-0.018	3	1.7707-1.7717	0.0008-0.0017	0.0006-0.0016
	1.6 (1590)	D16Y8	2.1644-2.1654	0.0007- ① 0.0016	0.004-0.018	3	1.7707-1.7717	0.0008-0.0017	0.0006-0.0016
	2.3 (2254)	F23A1	②	③	0.004-0.018	4	1.7707-1.7717	0.0008-0.0024	0.0006-0.0016
	2.3 (2254)	F23A4	②	③	0.004-0.018	4	1.7707-1.7717	0.0008-0.0024	0.006-0.018
	2.3 (2254)	F23A5	②	③	0.004-0.018	4	1.7707-1.7717	0.0008-0.0024	0.0006-0.0016
	2.2 (2157)	H22A4	④	⑤	0.004-0.018	4	1.8888-1.8898	0.0011-0.0020	0.0006-0.0016
	3.0 (2997)	J30A1	2.8337-2.8346	0.0008-0.0020	0.004-0.018	3	2.0857-2.0866	0.0008-0.0020	0.006-0.018
1999	1.6 (1595)	B16A2	2.1644- ⑥ 2.1354	0.0009- ⑦ 0.0020	0.004-0.018	3	1.7707-1.7717	0.0013-0.0022	0.0006-0.0016
	1.6 (1590)	D16Y5	2.1644-2.1654	0.0007- ① 0.0016	0.004-0.018	3	1.7707-1.7717	0.0008-0.0017	0.0006-0.0016
	1.6 (1590)	D16Y7	2.1644-2.1654	0.0007- ① 0.0016	0.004-0.018	3	1.7707-1.7717	0.0008-0.0017	0.0006-0.0016
	1.6 (1590)	D16Y8	2.1644-2.1654	0.0007- ① 0.0016	0.004-0.018	3	1.7707-1.7717	0.0008-0.0017	0.0006-0.0016
	2.3 (2254)	F23A1	②	③	0.004-0.018	4	1.7707-1.7717	0.0008-0.0024	0.0006-0.0016
	2.3 (2254)	F23A4	②	③	0.004-0.018	4	1.7707-1.7717	0.0008-0.0024	0.006-0.018
	2.3 (2254)	F23A5	②	③	0.004-0.018	4	1.7707-1.7717	0.0008-0.0024	0.0006-0.0016
	2.2 (2157)	H22A4	④	⑤	0.004-0.018	4	1.8888-1.8898	0.0011-0.0020	0.0006-0.0016
	3.0 (2997)	J30A1	2.8337-2.8346	0.0008-0.0020	0.004-0.018	3	2.0857-2.0866	0.0008-0.0020	0.006-0.018
2000	1.6 (1595)	B16A2	2.1644- ⑥ 2.1354	0.0009- ⑦ 0.0020	0.004-0.018	3	1.7707-1.7717	0.0013-0.0022	0.0006-0.0016
	1.6 (1590)	D16Y5	2.1644-2.1654	0.0007- ① 0.0016	0.004-0.018	3	1.7707-1.7717	0.0008-0.0017	0.0006-0.0016
	1.6 (1590)	D16Y7	2.1644-2.1654	0.0007- ① 0.0016	0.004-0.018	3	1.7707-1.7717	0.0008-0.0017	0.0006-0.0016

93471C84

CRANKSHAFT AND CONNECTING ROD SPECIFICATIONS

All measurements are given in inches.

Year	Engine Displacement Liters (cc)	Engine ID/VIN	Crankshaft				Connecting Rod		
			Main Brg. Journal Dia.	Main Brg. Oil Clearance	Shaft End-play	Thrust on No.	Journal Diameter	Oil Clearance	Side Clearance
2000 (Cont.)	1.6 (1590)	D16Y8	2.1644-2.1654	0.0007- ① 0.0016	0.004-0.018	3	1.7707-1.7717	0.0008-0.0017	0.0006-0.0016
	2.0 (1997)	F20C1	2.1644-2.1654	0.0007-0.0016	0.004-0.014	4	1.8888-1.8898	0.0012-0.0020	0.0006-0.0016
	2.3 (2254)	F23A1	②	③	0.004-0.018	4	1.7707-1.7717	0.0008-0.0024	0.0006-0.0016
	2.3 (2254)	F23A4	②	③	0.004-0.018	4	1.7707-1.7717	0.0008-0.0024	0.006-0.018
	2.3 (2254)	F23A5	②	③	0.004-0.018	4	1.7707-1.7717	0.0008-0.0024	0.0006-0.0016
	2.2 (2157)	H22A4	④	⑤	0.004-0.018	4	1.8888-1.8898	0.0011-0.0020	0.0006-0.0016
	3.0 (2997)	J30A1	2.8337-2.8346	0.0008-0.0020	0.004-0.018	3	2.0857-2.0866	0.0008-0.0020	0.006-0.018
2001	1.7 (1668)	D17A1	2.1644-2.1654	0.0007- ① 0.0014	0.004-0.014	4	1.7707-1.7717	0.0009-0.0017	0.0006-0.0016
	1.7 (1668)	D17A2	2.1644-2.1654	0.0007- ① 0.0014	0.004-0.014	4	1.7707-1.7717	0.0009-0.0017	0.0006-0.0016
	2.0 (1997)	F20C1	2.1644-2.1654	0.0007-0.0016	0.004-0.014	4	1.8888-1.8898	0.0012-0.0020	0.0006-0.0016
	2.3 (2254)	F23A1	②	③	0.004-0.018	4	1.7707-1.7717	0.0008-0.0024	0.0006-0.0016
	2.3 (2254)	F23A4	②	③	0.004-0.018	4	1.7707-1.7717	0.0008-0.0024	0.006-0.018
	2.3 (2254)	F23A5	②	③	0.004-0.018	4	1.7707-1.7717	0.0008-0.0024	0.0006-0.0016
	2.2 (2157)	H22A4	④	⑤	0.004-0.018	4	1.8888-1.8898	0.0011-0.0020	0.0006-0.0016
	3.0 (2997)	J30A1	2.8337-2.8346	0.0008-0.0020	0.004-0.018	3	2.0857-2.0866	0.0008-0.0020	0.006-0.018

① Journals 1 and 5
 Journals 2, 3 and 4: 0.0009 - 0.0019
② Journals 1, 2 and 4: 2.1646 - 2.1655
 Journal 3: 2.1644 - 2.1654
 Journal 5: 2.1650 - 2.1660
③ Journals 1, 2 and 4: 0.0008 - 0.0020
 Journal 3: 0.0010 - 0.0022
 Journal 5: 0.0004 - 0.0016

④ Journals 1 and 2: 1.9676 - 1.9685
 Journal 3: 1.9674 - 1.9683
 Journal 4: 1.9679 - 1.9688
 Journal 5: 1.9680 - 1.9690
⑤ Journals 1 and 2: 0.0008 - 0.0020
 Journal 3: 0.0010 - 0.0022
 Journal 4: 0.0005 - 0.0020
 Journal 5: 0.0004 - 0.0016

⑥ Journals 1, 2, 4 and 5
 Journal 3: 2.1642 - 2.1651
⑦ Journals 1, 2, 4 and 5
 Journal 3: 0.0012 - 0.0021

93471C85

For Tune-up, Capacities and Firing orders, see Section 1 of this manual

PISTON AND RING SPECIFICATIONS
All measurements are given in inches.

Year	Engine Displacement Liters (cc)	Engine ID/VIN	Piston Clearance	Ring Gap			Ring Side Clearance		
				Top Compression	Bottom Compression	Oil Control	Top Compression	Bottom Compression	Oil Control
1998	1.6 (1590)	D16Y5	0.0004-0.0020	0.006-0.024	0.012-0.028	0.008-0.031	0.0014-0.0050	0.0012-0.0050	N/A
	1.6 (1590)	D16Y7	0.0004-0.0020	0.006-0.024	0.012-0.028	0.008-0.031	0.0014-0.0050	0.0012-0.0050	N/A
	1.6 (1590)	D16Y8	0.0004-0.0020	0.006-0.024	0.012-0.028	0.008-0.031	0.0014-0.0050	0.0012-0.0050	N/A
	2.3 (2254)	F23A1	0.0008-0.0020	0.008-0.024	0.016-0.028	0.008-0.031	0.0014-0.0050	0.0012-0.0050	N/A
	2.3 (2254)	F23A4	0.0008-0.0020	0.008-0.024	0.016-0.028	0.008-0.031	0.0014-0.0050	0.0012-0.0050	N/A
	2.3 (2254)	F23A5	0.0008-0.0020	0.008-0.024	0.016-0.028	0.008-0.031	0.0014-0.0050	0.0012-0.0050	N/A
	2.2 (2157)	H22A4	0.0002-0.0020	0.010-0.024	0.024-0.035	0.008-0.031	0.0022-0.0050	0.0016-0.0050	N/A
	3.0 (2997)	J30A1	0.0006-0.0030	0.008-0.024	0.016-0.028	0.008-0.031	0.0014-0.0050	0.0012-0.0050	N/A
1999	1.6 (1595)	B16A2	0.0004-0.0020	0.008-0.024	0.016-0.028	0.008-0.028	0.0018-0.0050	0.0016-0.0050	N/A
	1.6 (1590)	D16Y5	0.0004-0.0020	0.006-0.024	0.012-0.028	0.008-0.031	0.0014-0.0050	0.0012-0.0050	N/A
	1.6 (1590)	D16Y7	0.0004-0.0020	0.006-0.024	0.012-0.028	0.008-0.031	0.0014-0.0050	0.0012-0.0050	N/A
	1.6 (1590)	D16Y8	0.0004-0.0020	0.006-0.024	0.012-0.028	0.008-0.031	0.0014-0.0050	0.0012-0.0050	N/A
	2.3 (2254)	F23A1	0.0008-0.0020	0.008-0.024	0.016-0.028	0.008-0.031	0.0014-0.0050	0.0012-0.0050	N/A
	2.3 (2254)	F23A4	0.0008-0.0020	0.008-0.024	0.016-0.028	0.008-0.031	0.0014-0.0050	0.0012-0.0050	N/A
	2.3 (2254)	F23A5	0.0008-0.0020	0.008-0.024	0.016-0.028	0.008-0.031	0.0014-0.0050	0.0012-0.0050	N/A
	2.2 (2157)	H22A4	0.0002-0.0020	0.010-0.024	0.024-0.035	0.008-0.031	0.0022-0.0050	0.0016-0.0050	N/A
	3.0 (2997)	J30A1	0.0006-0.0030	0.008-0.024	0.016-0.028	0.008-0.031	0.0014-0.0050	0.0012-0.0050	N/A
2000	1.6 (1595)	B16A2	0.0004-0.0020	0.008-0.024	0.016-0.028	0.008-0.028	0.0018-0.0050	0.0016-0.0050	N/A
	1.6 (1590)	D16Y5	0.0004-0.0020	0.006-0.024	0.012-0.028	0.008-0.031	0.0014-0.0050	0.0012-0.0050	N/A
	1.6 (1590)	D16Y7	0.0004-0.0020	0.006-0.024	0.012-0.028	0.008-0.031	0.0014-0.0050	0.0012-0.0050	N/A
	1.6 (1590)	D16Y8	0.0004-0.0020	0.006-0.024	0.012-0.028	0.008-0.031	0.0014-0.0050	0.0012-0.0050	N/A
	2.0 (1997)	F20C1	0.0002-0.0011	0.010-0.024	0.024-0.035	0.008-0.031	0.0018-0.0035	0.0016-0.0028	N/A
	2.3 (2254)	F23A1	0.0008-0.0020	0.008-0.024	0.016-0.028	0.008-0.031	0.0014-0.0050	0.0012-0.0050	N/A
	2.3 (2254)	F23A4	0.0008-0.0020	0.008-0.024	0.016-0.028	0.008-0.031	0.0014-0.0050	0.0012-0.0050	N/A

93471C86

PISTON AND RING SPECIFICATIONS
All measurements are given in inches.

Year	Engine Displacement Liters (cc)	Engine ID/VIN	Piston Clearance	Ring Gap			Ring Side Clearance		
				Top Compression	Bottom Compression	Oil Control	Top Compression	Bottom Compression	Oil Control
2000 (cont.)	2.3 (2254)	F23A5	0.0008-0.0020	0.008-0.024	0.016-0.028	0.008-0.031	0.0014-0.0050	0.0012-0.0050	N/A
	2.2 (2157)	H22A4	0.0002-0.0020	0.010-0.024	0.024-0.035	0.008-0.031	0.0022-0.0050	0.0016-0.0050	N/A
	3.0 (2997)	J30A1	0.0006-0.0030	0.008-0.024	0.016-0.028	0.008-0.031	0.0014-0.0050	0.0012-0.0050	N/A
2001	1.7 (1668)	D17A1	0.0004-0.0016	0.006-0.024	0.012-0.024	0.008-0.031	0.0014-0.0024	0.0012-0.0022	N/A
	1.7 (1668)	D17A2	0.0004-0.0016	0.006-0.024	0.012-0.024	0.008-0.031	0.0014-0.0024	0.0012-0.0022	N/A
	2.0 (1997)	F20C1	0.0002-0.0011	0.010-0.024	0.024-0.035	0.008-0.031	0.0018-0.0035	0.0016-0.0028	N/A
	2.3 (2254)	F23A1	0.0008-0.0020	0.008-0.024	0.016-0.028	0.008-0.031	0.0014-0.0050	0.0012-0.0050	N/A
	2.3 (2254)	F23A4	0.0008-0.0020	0.008-0.024	0.016-0.028	0.008-0.031	0.0014-0.0050	0.0012-0.0050	N/A
	2.3 (2254)	F23A5	0.0008-0.0020	0.008-0.024	0.016-0.028	0.008-0.031	0.0014-0.0050	0.0012-0.0050	N/A
	2.2 (2157)	H22A4	0.0002-0.0020	0.010-0.024	0.024-0.035	0.008-0.031	0.0022-0.0050	0.0016-0.0050	N/A
	3.0 (2997)	J30A1	0.0006-0.0030	0.008-0.024	0.016-0.028	0.008-0.031	0.0014-0.0050	0.0012-0.0050	N/A

NA: Not Available

93471C87

For complete service labor times, order the Chilton Labor Guide

TORQUE SPECIFICATIONS
All readings in ft. lbs.

Year	Engine Displacement Liters (cc)	Engine ID/VIN	Cylinder Head Bolts	Main Bearing Bolts	Rod Bearing Bolts	Crankshaft Damper Bolts	Flywheel Bolts	Manifold Intake	Manifold Exhaust	Spark Plugs	Lug Nut
1998	1.6 (1590)	D16Y5	①	②	23	134	87	17	23	13	80
	1.6 (1590)	D16Y7	①	②	23	134	87	17	23	13	80
	1.6 (1590)	D16Y8	①	②	23	134	87	17	23	13	80
	2.2 (2157)	H22A4	③	54	34	181	76④	16	23	13	80
	2.3 (2254)	F23A1	⑤	⑥	⑦	181	76④	16	23	13	80
	2.3 (2254)	F23A4	⑤	⑥	⑦	181	76④	16	23	13	80
	2.3 (2254)	F23A5	⑤	⑥	⑦	181	76④	16	23	13	80
	3.0 (2997)	J30A1	③	⑧	⑦	181	76④	16	23	13	80
1999	1.6 (1585)	B16A2	⑨	⑩	30	130	76	17	23	13	80
	1.6 (1590)	D16Y5	①	②	23	134	87	17	23	13	80
	1.6 (1590)	D16Y7	①	②	23	134	87	17	23	13	80
	1.6 (1590)	D16Y8	①	②	23	134	87	17	23	13	80
	2.2 (2157)	H22A4	③	54	34	181	76④	16	23	13	80
	2.3 (2254)	F23A1	⑤	⑥	⑦	181	76④	16	23	13	80
	2.3 (2254)	F23A4	⑤	⑥	⑦	181	76④	16	23	13	80
	2.3 (2254)	F23A5	⑤	⑥	⑦	181	76④	16	23	13	80
	3.0 (2997)	J30A1	③	⑧	⑦	181	76④	16	23	13	80
2000	1.6 (1585)	B16A2	⑨	⑩	30	130	76	17	23	13	80
	1.6 (1590)	D16Y5	①	②	23	134	87	17	23	13	80
	1.6 (1590)	D16Y7	①	②	23	134	87	17	23	13	80
	1.6 (1590)	D16Y8	①	②	23	134	87	17	23	13	80
	2.0 (1997)	F20C1	⑤	⑪	⑫	181	94	16	23	13	80
	2.2 (2157)	H22A4	③	54	34	181	76④	16	23	13	80
	2.3 (2254)	F23A1	⑤	⑥	⑦	181	76④	16	23	13	80
	2.3 (2254)	F23A4	⑤	⑥	⑦	181	76④	16	23	13	80
	2.3 (2254)	F23A5	⑤	⑥	⑦	181	76④	16	23	13	80
	3.0 (2997)	J30A1	③	⑧	⑦	181	76④	16	23	13	80
2001	1.7 (1668)	D17A1	①	②	24	⑦	87	16	23	13	80
	1.7 (1668)	D17A2	①	②	24	⑦	87	16	23	13	80
	2.0 (1997)	F20C1	⑤	⑪	⑫	181	94	16	23	13	80
	2.2 (2157)	H22A4	③	54	34	181	76④	16	23	13	80
	2.3 (2254)	F23A1	⑤	⑥	⑦	181	76④	16	23	13	80
	2.3 (2254)	F23A4	⑤	⑥	⑦	181	76④	16	23	13	80
	2.3 (2254)	F23A5	⑤	⑥	⑦	181	76④	16	23	13	80
	3.0 (2997)	J30A1	③	⑧	⑦	181	76④	16	23	13	80

① Step 1: 14 ft. lbs.
Step 2: 36 ft. lbs.
Step 3: 49 ft. lbs.
Step 4: Bolts 1-2, retorque to 49 ft. lbs.

② Step 1: 18 ft. lbs.
Step 2: 38 ft. lbs.

③ Step 1: 29 ft. lbs.
Step 2: 51 ft. lbs.
Step 3: 72.3 ft. lbs.

④ Automatic transaxle: 54 ft. lbs.

⑤ Step 1: 22 ft. lbs.
Step 2: Rotate 90 degrees
Step 3: Rotate 90 degrees
Step 4: If new bolt rotate additional 90 degrees

⑥ Step 1: 11mm bolts, 29 ft. lbs.
Step 2: 11mm bolts, 58 ft. lbs.
Step 3: 6mm bolts, 8.7 ft. lbs.

⑦ Step 1: 14 ft. lbs.
Step 2: Rotate 90 degrees

⑧ Step 1: Cap bolts, 56 ft. lbs.
Step 2: Side bolts, 36 ft. lbs.

⑨ Step 1: 22 ft. lbs.
Step 2: 61 ft. lbs.

⑩ Step 1: 18 ft. lbs.
Step 2: 54 ft. lbs.

⑪ Step 1: Bearing cap bolts to 22 ft. lbs.
Step 2: Bearing cap bolts plus 60 degrees
Step 3: 8mm bolts to 16 ft. lbs.

⑫ Step 1: 18 ft. lbs.
Step 2: Plus 90 degrees

93471C88

BRAKE SPECIFICATIONS
All measurements in inches unless noted

Year	Model		Brake Disc Original Thickness	Brake Disc Minimum Thickness	Brake Disc Maximum Runout	Brake Drum Diameter Original Inside Diameter	Brake Drum Diameter Max. Wear Limit	Brake Drum Diameter Maximum Machine Diameter	Minimum Lining Thickness Front	Minimum Lining Thickness Rear	Brake Caliper Bracket Bolts (ft. lbs.)	Brake Caliper Mounting Bolts (ft. lbs.)
1998	Accord	F	0.910	0.830	0.004	—	—	—	0.060	—	—	①
		R	0.358	0.310	0.004	8.66	8.70	8.70	—	0.080 ②	—	17
	Civic	F	0.840	0.750	0.004	—	—	—	0.060	—	—	①
		R	—	—	—	7.87	7.91	7.91	—	0.080	—	—
	Prelude	F	0.910	0.830	0.004	—	—	—	0.060	—	83	①
		R	0.390	0.320	0.004	—	—	—	—	0.060	—	17
1999	Accord	F	0.910	0.830	0.004	—	—	—	0.060	—	—	①
		R	0.358	0.310	0.004	8.66	8.70	8.70	—	0.080 ②	—	17
	Civic	F	0.840	0.750	0.004	—	—	—	0.060	—	—	①
		R	—	—	—	7.87	7.91	7.91	—	0.080	—	—
	Prelude	F	0.910	0.830	0.004	—	—	—	0.060	—	83	①
		R	0.390	0.320	0.004	—	—	—	—	0.060	—	17
2000	Accord	F	0.910	0.830	0.004	—	—	—	0.060	—	—	①
		R	0.358	0.310	0.004	8.66	8.70	8.70	—	0.080 ②	—	17
	Civic	F	0.840	0.750	0.004	—	—	—	0.060	—	—	①
		R	—	—	—	7.87	7.91	7.91	—	0.080	—	—
	Prelude	F	0.910	0.830	0.004	—	—	—	0.060	—	83	①
		R	0.390	0.320	0.004	—	—	—	—	0.060	—	17
	S2000	F	0.990	0.910	0.004	—	—	—	0.060	—	80	24
		R	0.476	0.390	0.004	—	—	—	—	0.060	41	17
2001	Accord	F	0.910	0.830	0.004	—	—	—	0.060	—	—	①
		R	0.358	0.310	0.004	8.66	8.70	8.70	—	0.080 ②	—	17
	Civic	F	0.840	0.750	0.004	—	—	—	0.060	—	—	①
		R	—	—	—	7.87	7.91	7.91	—	0.080	—	—
	Prelude	F	0.910	0.830	0.004	—	—	—	0.060	—	83	①
		R	0.390	0.320	0.004	—	—	—	—	0.060	—	17
	S2000	F	0.990	0.910	0.004	—	—	—	0.060	—	80	24
		R	0.476	0.390	0.004	—	—	—	—	0.060	41	17

NA: Not Available

F: Front

R: Rear

① Calipers with long pins beyond bolt threads, 54 ft. lbs.
  Calipers with no pin beyond threads, 20 ft. lbs.

② With rear disc: 0.060

93471C89

WHEEL ALIGNMENT

Year	Model		Caster Range (+/-Deg.)	Caster Preferred Setting (Deg.)	Camber Range (+/-Deg.)	Camber Preferred Setting (Deg.)	Toe-in (in.)	Steering Axis Inclination (Deg.)
1998	Accord	F	1.00	+2.80	1.00	+0.06	0 +/- 0.03	—
		R	—	—	0.50	-0.50	0.03 +/- 0.03	—
	Civic	F	—	+1.66	—	0	0.03	—
		R	—	—	—	-1.00	0.03	—
	Prelude	F	1.00	+2.66	1.00	0	0 +/- 0.03	—
		R	—	—	1.00	-0.45	0.03 +/- 0.03	—
1999	Accord	F	1.00	+2.80	1.00	+0.06	0 +/- 0.03	—
		R	—	—	0.50	-0.50	0.03 +/- 0.03	—
	Civic	F	—	+1.66	—	0	0.03	—
		R	—	—	—	-1.00	0.03	—
	Prelude	F	1.00	+2.66	1.00	0	0 +/- 0.03	—
		R	—	—	1.00	-0.45	0.03 +/- 0.03	—
2000	Accord	F	1.00	+2.80	1.00	+0.06	0 +/- 0.03	—
		R	—	—	0.50	-0.50	0.03 +/- 0.03	—
	Civic	F	—	+1.66	—	0	0.03	—
		R	—	—	—	-1.00	0.03	—
	Prelude	F	1.00	+2.66	1.00	0	0 +/- 0.03	—
		R	—	—	1.00	-0.45	0.03 +/- 0.03	—
	S2000	F	0.75	+6.00	0.50	-0.50	0 +/- 0.03	—
		R	—	—	0.50	-1.50	0.12 +/- 0.03	—
2001	Accord	F	1.00	+2.80	1.00	+0.06	0 +/- 0.03	—
		R	—	—	0.50	-0.50	0.03 +/- 0.03	—
	Civic	F	—	+1.66	—	0	0.03	—
		R	—	—	—	-1.00	0.03	—
	Prelude	F	1.00	+2.66	1.00	0	0 +/- 0.03	—
		R	—	—	1.00	-0.45	0.03 +/- 0.03	—
	S2000	F	0.75	+6.00	0.50	-0.50	0 +/- 0.03	—
		R	—	—	0.50	-1.50	0.12 +/- 0.03	—

93471C90

TIRE, WHEEL AND BALL JOINT SPECIFICATIONS

Year	Model	OEM Tires		Tire Pressures (psi)		Wheel Size	Ball Joint Inspection
		Standard	Optional	Front	Rear		
1998	Accord DX	P195/70SR14	None	32	32	5-J	NS
	Accord EX, LX 4-cyl.	P195/65HR15	None	32	32	5-J	NS
	Accord LX V6	P205/65VR15	None	30	30	5-J	NS
	Prelude	P205/50VR16	None	32	32	6.5-JJ	NS
	Civic	P185/65SR14	None	30	29	5-J	NS
1999	Civic	P185/65R14	P195/55VR15	30	29	5-J	NS
	Accord DX	P195/70SR14	None	32	32	5-J	NS
	Accord EX, LX 4-cyl.	P195/65HR15	None	32	32	6.5	NS
	Accord LX V6	P205/65VR15	None	30	30	5.5-J	NS
	Prelude	P205/50VR16	None	32	32	6.5-JJ	NS
2000	Civic	P185/65R14	P195/55VR15	30	29	5-J	NS
	Accord DX	P195/70SR14	None	32	32	5-J	NS
	Accord EX, LX 4-cyl.	P195/65HR15	None	32	32	6.5	NS
	Accord LX V6	P205/65VR15	None	30	30	5-J	NS
	Prelude	P205/50VR16	None	32	32	6.5-JJ	NS
	S2000	P225/50VR16	None	32	32	5-J	NS
2001	Civic	P185/65R14	P195/55VR15	30	29	5-J	NS
	Accord DX	P195/70SR14	None	32	32	5-J	NS
	Accord EX, LX 4-cyl.	P195/65HR15	None	32	32	6.5	NS
	Accord LX V6	P205/65VR15	None	30	30	5-J	NS
	Prelude	P205/50VR16	None	32	32	6.5-JJ	NS
	S2000	P225/50VR16	None	32	32	5-J	NS

OEM: Original Equipment Manufacturer

PSI: Pounds Per Square Inch

STD: Standard

OPT: Optional

NS: Not specified by manufacturer

93471C91

Timing belt service is covered in Section 3 of this manual

SCHEDULED MAINTENANCE INTERVALS
Honda Civic, Accord, Prelude, S2000

TO BE SERVICED	TYPE OF SERVICE	VEHICLE MILEAGE INTERVAL (x1000)												
		7.5	15	22.5	30	37.5	45	52.5	60	67.5	75	82.5	90	97.5
Engine oil & filter	R	✓	✓	✓	✓	✓	✓	✓	✓	✓	✓	✓	✓	✓
Front brake pads	S/I	✓	✓	✓	✓	✓	✓	✓	✓	✓	✓	✓	✓	✓
Rotate tires	S/I	✓	✓	✓	✓	✓	✓	✓	✓	✓	✓	✓	✓	✓
Cooling system, hoses & connections	S/I		✓		✓		✓		✓		✓		✓	
Driveshaft boots	S/I		✓		✓		✓		✓		✓		✓	
Exhaust system	S/I		✓		✓		✓		✓		✓		✓	
Front brake discs & calipers	S/I		✓		✓		✓		✓		✓		✓	
Front wheel alignment	S/I		✓		✓		✓		✓		✓		✓	
Front & rear wheel alignment (Prelude w/4WS)	S/I		✓		✓		✓		✓		✓		✓	
Fuel pipes, hoses & connections	S/I		✓		✓		✓		✓		✓		✓	
Parking brake adjustment	S/I		✓		✓		✓		✓		✓		✓	
Power steering system	S/I		✓		✓		✓		✓		✓		✓	
Rear brake discs, calipers & pads	S/I		✓		✓		✓		✓		✓		✓	
Suspension components	S/I		✓		✓		✓		✓		✓		✓	
Suspension mounting bolts	S/I		✓		✓		✓		✓		✓		✓	
Tie rods, steering gear box & boots	S/I		✓		✓		✓		✓		✓		✓	
Valve clearance (Prelude VTEC) ①	S/I		✓		✓		✓		✓		✓		✓	
Valve clearance (Accord L4, Civic & Prelude non-VTEC)	S/I				✓				✓				✓	
Parking brake	S/I		✓		✓				✓				✓	
Air cleaner element	R				✓				✓				✓	
Transmission fluid (Civic CVT)	R				✓		✓		✓		✓			
Transmission fluid (A/T or M/T) (except as noted below)	R				✓				✓				✓	
Transmission fluid (Prelude L4)	R												✓	
Brake fluid (including ABS) (Accord V6)	R				✓				✓				✓	
Brake fluid (including ABS) (Accord L4, Civic, & Prelude)	R						✓						✓	
Spark plugs (non-VTEC)	R				✓				✓				✓	
Spark plugs (VTEC) ①	R								✓				✓	
ABS operation	S/I				✓				✓					
Alternator drive belt	S/I				✓				✓					

93471C92

SCHEDULED MAINTENANCE INTERVALS
Honda Civic, Accord, Prelude, S2000

TO BE SERVICED	TYPE OF SERVICE	VEHICLE MILEAGE INTERVAL (x1000)												
		7.5	15	22.5	30	37.5	45	52.5	60	67.5	75	82.5	90	97.5
Power steering pump belt	S/I				✓				✓				✓	
Rear brake drums, wheel cylinders & linings (except Prelude)	S/I				✓				✓				✓	
Engine coolant	R						✓				✓			
ABS high pressure hose	R								✓					
Fuel filter	R								✓					
Timing belt	R												✓	
Timing balancer belt	R												✓	
Distributor, ignition cap & rotor	S/I								✓					
Idle speed	S/I								✓					
Ignition wires	S/I								✓					
PCV valve	S/I								✓					
TWC converter heat shield	S/I								✓					
Water pump	S/I												✓	

R: Replace   S/I: Service or Inspect

① S2000: 105,000 miles

FREQUENT OPERATION MAINTENANCE (SEVERE SERVICE)

If a vehicle is operated under any of the following conditions it is considered severe service:

- Extremely dusty areas.

- 50% or more of the vehicle operation is in 32°C (90°F) or higher temperatures, or constant operation in temperatures below 0°C (32°F).

- Prolonged idling (vehicle operation in stop and go traffic).

- Frequent short running periods (engine does not warm to normal operating temperatures).

- Police, taxi, delivery usage or trailer towing usage.

Oil & oil filter: change every 3750 miles.

Driveshaft boots: service or inspect every 7500 miles.

Front brake discs & calipers, & rear brake discs, calipers & pads: service or inspect every 7500 miles.

Power steering system: service or inspect every 7500 miles.

Suspension components: service or inspect every 7500 miles.

Tie rods, steering gear box & boots: service or inspect every 7500 miles.

Air cleaner element: service or inspect every 15,000 miles.

Transmission fluid (Accord V6 & Civic CVT): replace every 15,000 miles.

Transmission fluid (Accord L4, Civic, & Prelude): replace every 30,000 miles.

Timing balancer belt: replace every 60,000 miles.

Timing belt: replace every 60,000 miles.

Water pump: service or inspect every 60,000 miles.

93471C93

Heater Core replacement is covered in Section 2 of this manual

SCHEDULED MAINTENANCE INTERVALS
HONDA
ACCORD, CIVIC, PRELUDE, S2000

The following should be used as a guide when determining the amount of work required for a particular service.
In estimating how long a particular Scheduled Maintenance Service should take, please observe the following:

- Labor Time is time based on field research and data supplied by the vehicle manufacturer.
- Labor time operations are given in hours and tenths of an hour.
- All labor operations are to be used as a guide.

Mechanic Skill Level Codes:
(A) PRECISION: Highly skilled with multiple certification.
(B) GENERAL: Normally skilled with certification.
(C) MAINTENANCE: Semi-skilled working on certification.

	LABOR TIME		LABOR TIME		LABOR TIME
7500 Mile Service (C)		**45000 Mile Service (B)**		**75000 Mile Service (B)**	
All Models	1.2	All Models	2.1	All Models	2.7
15000 Mile Service (B)		*replace trans fluid add*	.5	*adjust valves add*	1.1
All Models	2.1	*replace brake fluid add*	.6	*replace trans fluid add*	.5
adjust valves add	1.1	*replace engine coolant add*	.7	*replace engine coolant add*	.7
22500 Mile Service (C)		**52500 Mile Service (C)**		**82500 Mile Service (C)**	
All Models	1.2	All Models	1.2	All Models	1.2
30000 Mile Service (B)		**60000 Mile Service (B)**		**90000 Mile Service (B)**	
All Models	3.4	All Models	3.6	All Models	3.6
adjust valves add	1.1	*adjust valves add*	1.1	*adjust valves add*	1.1
replace trans fluid add	.5	*replace trans fluid add*	.5	*replace trans fluid add*	.5
replace spark plugs add	.5	*replace spark plugs add*	.5	*replace spark plugs add*	.5
replace brake fluid add	.6	*replace brake fluid add*	.6	*replace brake fluid add*	.6
37500 Mile Service (C)		*replace fuel filter add*	.3	*replace timing belt add*	2.2
All Models	1.2	**67500 Mile Service (C)**		*replace timing belt V6 add*	3.5
		All Models	1.2	**97500 Mile Service (C)**	
				All Models	1.2

93471C94

HONDA
INSIGHT

ENGINE AND VEHICLE IDENTIFICATION

		Engine					Model Year	
Code ①	Liters (cc)	Cu. In.	Cyl.	Fuel Sys.	Engine Type	Eng. Mfg.	Code ②	Year
ECA1	1.0 (995)	61	3	PGM-FI	SOHC	Honda	Y	2000
SOHC: Single Overhead Camshaft							1	2001
① 8th position of VIN							2	2002
② 10th position of VIN								

93471C95

GENERAL ENGINE SPECIFICATIONS

Year	Model	Engine Displacement Liters (cc)	Engine Series (ID/VIN)	Fuel System	Net Horsepower @ rpm	Net Torque @ rpm (ft. lbs.)	Bore x Stroke (in.)	Compression Ratio	Oil Pressure @ rpm
2000	Insight	1.0 (995)	EAC1	PGM-FI	73@5700	91@2000	2.83x3.21	10.8:1	50@3000
2001	Insight	1.0 (995)	ECA1	PGM-FI	73@5700	91@2000	2.83x3.21	10.8:1	50@3000

PGM-FI: Programmed Fuel Injection

93471C96

ENGINE TUNE-UP SPECIFICATIONS

Year	Engine Displacement Liters (cc)	Engine ID/VIN	Spark Plug Gap (in.)	Ignition Timing (deg.)	Fuel Pump (psi)	Idle Speed (rpm)	Valve Clearance In.	Valve Clearance Ex.
2000	1.0 (995)	ECA1	0.039-0.043	12	30-37	900-950	0.007-0.009	0.008-0.010
2001	1.0 (995)	ECA1	0.039-0.043	12	30-37	900-950	0.007-0.009	0.008-0.010

NOTE: The Vehicle Emission Control Information label often reflects specification changes made during production. The label figures must be used if they differ from those in this chart.

93471C97

CAPACITIES

Year	Model	Engine Displacement Liters (cc)	Engine ID/VIN	Engine Oil with Filter (qts.)	Manual Transaxle (qts.)	Rear Drive Axle (pts.)	Fuel Tank (gal.)	Cooling System (qts.)
2000	Insight	1.0 (995)	ECA1	2.6	1.59	—	10.6	4.6
2001	Insight	1.0 (995)	ECA1	2.6	1.59	—	10.6	4.6

NOTE: All capacities are approximate. Add fluid gradually and check to be sure a proper fluid level is obtained.

93471C98

VALVE SPECIFICATIONS

Year	Engine Displacement Liters (cc)	Engine ID/VIN	Seat Angle (deg.)	Face Angle (deg.)	Spring Test Pressure (lbs. @ in.)	Spring Installed Height (in.)	Stem-to-Guide Clearance (in.)		Stem Diameter (in.)	
							Intake	Exhaust	Intake	Exhaust
2000	1.0 (995)	ECA1	45	44.5-45.0	NA	NA	0.001-0.0020	0.002-0.0030	0.2157	0.2146
2001	1.0 (995)	ECA1	45	44.5-45.0	NA	NA	0.001-0.0020	0.002-0.0030	0.2157	0.2146

NA: Not Available

93471C99

CRANKSHAFT AND CONNECTING ROD SPECIFICATIONS

All measurements are given in inches.

Year	Engine Displacement Liters (cc)	Engine ID/VIN	Crankshaft				Connecting Rod		
			Main Brg. Journal Dia.	Main Brg. Oil Clearance	Shaft End-play	Thrust on No.	Journal Diameter	Oil Clearance	Side Clearance
2000	1.0 (995)	ECA1	1.5741-1.5750	0.0006-0.0013	0.004-0.0140	3	1.4164-1.4173	0.0008-0.0014	0.004-0.0140
2001	1.0 (995)	ECA1	1.5741-1.5750	0.0006-0.0013	0.004-0.0140	3	1.4164-1.4173	0.0008-0.0014	0.004-0.0140

93471C00

For complete Engine Mechanical specifications, see Section 1 of this manual

PISTON AND RING SPECIFICATIONS
All measurements are given in inches.

Year	Engine Displacement Liters (cc)	Engine ID/VIN	Piston Clearance	Ring Gap			Ring Side Clearance		
				Top Compression	Bottom Compression	Oil Control	Top Compression	Bottom Compression	Oil Control
2000	1.0 (995)	ECA1	0.0002-0.0017	0.006-0.012	0.014-0.020	0.008-0.028	0.0022-0.0031	0.0012-0.0022	NA
2001	1.0 (995)	ECA1	0.0002-0.0017	0.006-0.012	0.014-0.020	0.008-0.028	0.0022-0.0031	0.0012-0.0022	NA

93471CA1

TORQUE SPECIFICATIONS
All readings in ft. lbs.

Year	Engine Displacement Liters (cc)	Engine ID/VIN	Cylinder Head Bolts	Main Bearing Bolts	Rod Bearing Bolts	Crankshaft Damper Bolts	Flywheel Bolts	Manifold		Spark Plugs	Lug Nuts
								Intake	Exhaust		
2000	1.0 (995)	ECA1	①	②	③	④	33	16	16	17	80
2001	1.0 (995)	ECA1	①	②	③	④	33	16	16	17	80

① Step 1: 29 ft. lbs.
   Step 2: Additional 90 degrees.
   Step 3: 6 mm bolts 8.7 ft. lbs.

② Step 1: 18 ft. lbs.
   Step 2: Plus an additional 60 degrees.

③ Step 1: 7.2 ft. lbs.
   Step 2: Plus an additional 90 degrees

④ Step 1: 14 ft. lbs.
   Step 2: Plus an additional 90 degrees.

93471CA2

BRAKE SPECIFICATIONS
All measurements in inches unless noted

Year	Model		Brake Disc			Brake Drum Diameter			Min. Lining Thickness	Caliper Guide▲ Pin Bolts (ft. lbs.)
			Original Thickness	Minimum Thickness	Maximum Run-out	Original Inside Diameter	Max. Wear Limit	Maximum Machine Diameter		
2000	Insight	F	0.665-0.673	0.590	0.004	—	—	—	0.313	16①
		R	0.170	—	0.005	7.083-7.087	7.13	NA	0.040	—
2001	Insight	F	0.665-0.673	0.590	0.004	—	—	—	0.313	16①
		R	0.170	—	0.005	7.083-7.087	7.13	NA	0.040	—

NA: Not Available

F: Front

R: Rear

① Lbs. lbs.

93471CA3

WHEEL ALIGNMENT

Year	Model		Caster		Camber		Toe-in (in.)	Steering Axis Inclination (Deg.)
			Range (+/-Deg.)	Preferred Setting (Deg.)	Range (+/-Deg.)	Preferred Setting (Deg.)		
2000	Insight	F	1.00	+2.00	1.00	0	0 +/- 0.03	—
		R	—	—	1.00	-1.00	0.06 +/- 0.06	—
2001	Insight	F	1.00	+2.00	1.00	0	0 +/- 0.03	—
		R	—	—	1.00	-1.00	0.06 +/- 0.06	—

93471CA4

For Accessory Drive Belt illustrations, see Section 1 of this manual

TIRE, WHEEL AND BALL JOINT SPECIFICATIONS

| Year | Model | OEM Tires | | Tire Pressures (psi) | | Wheel | Ball Joint |
		Standard	Optional	Front	Rear	Size	Inspection
2000	Insight	P165/65R14	None	32	32	5-J	NS
2001	Insight	P165/65R14	None	32	32	5-J	NS

OEM: Original Equipment Manufacturer

PSI: Pounds Per Square Inch

STD: Standard

OPT: Optional

NS: Not specified by manufacturer

93471CA5

HYUNDAI
ACCENT • ELANTRA • SONATA • TIBURON • XG 300

ENGINE AND VEHICLE IDENTIFICATION

Engine								Model Year	
Code	Liters (cc)	Cu. In.	Cyl.	Fuel Sys.	Engine Type	Eng. Mfg.		Code	Year
K	1.5 (1495)	91.17	4	MPFI	SOHC	Hyundai		W	1998
K	1.5 (1495)	91.17	4	MPFI	DOHC	Hyundai		X	1999
M	1.8 (1795)	109.54	4	MPFI	DOHC	Hyundai		Y	2000
F①	2.0 (1975)	120.52	4	MPFI	DOHC	Hyundai		1	2001
F②	2.0 (1997)	121.90	4	MPFI	DOHC	Hyundai		2	2002
D	2.4 (2351)	143.46	4	MPFI	DOHC	Hyundai			
E	2.5 (2493)	152.13	6	MPFI	DOHC	Hyundai			
G	3.0 (2972)	181.40	6	MPFI	DOHC	Hyundai			
T	3.0 (2972)	181.40	6	MPFI	SOHC	Hyundai			

MPFI: Multi-Point Fuel Injection

SOHC: Single Overhead Camshaft

DOHC: Double Overhead Camshafts

① Elantra and Tiburon

② Sonata

93471CA6

For Tire, Wheel and Ball Joint specifications, see Section 1 of this manual

GENERAL ENGINE SPECIFICATIONS

Year	Engine Displacement Liters (cc)	Engine ID/VIN	Fuel System Type	Net Horsepower @ rpm	Net Torque @ rpm (ft. lbs.)	Bore x Stroke (in.)	Compression Ratio	Oil Pressure @ rpm
1998	1.5 (1495)	K	MFI	92@5500	97@4000	2.97 x 3.29	10.0:1	21@Idle
	1.8 (1795)	M	MFI	124@6000	116@5000	3.23 x 3.35	10.0:1	24@Idle
	2.0 (1975)	F ①	MFI	140@6000	133@4800	3.23 x 3.68	10.3:1	24@Idle
	2.0 (1997)	F ②	MFI	137@6000	129@4000	3.35 x 3.46	9.0:1	12@Idle
	3.0 (2972)	T	MFI	142@5000	168@2500	3.59 x 2.99	8.9:1	12@Idle
1999	1.5 (1495)	N	MFI	92@5500	97@4000	2.97 x 3.29	10.0:1	21@Idle
	2.0 (1975)	F	MFI	140@6000	133@4800	3.23 x 3.68	10.3:1	24@Idle
	2.4 (2351)	D	MFI	137@6000	129@4000	3.41 x 3.94	10.0:1	12@Idle
	2.5 (2493)	E	MFI	142@5000	168@2500	3.59 x 2.99	10.0:1	12@Idle
2000	1.5 (1495)	N	MFI	92@5500	97@4000	2.97 x 3.29	10.0:1	21@Idle
	2.0 (1975)	F	MFI	140@6000	133@4800	3.23 x 3.68	10.3:1	24@Idle
	2.4 (2351)	D	MFI	137@6000	129@4000	3.41 x 3.94	10.0:1	12@Idle
	2.5 (2493)	E	MFI	142@5000	168@2500	3.59 x 2.99	10.0:1	12@Idle
2001	1.5 (1495)	N	MFI	92@5500	97@4000	2.97 x 3.29	10.0:1	21@Idle
	2.0 (1975)	F	MFI	140@6000	133@4800	3.23 x 3.68	10.3:1	24@Idle
	2.4 (2351)	D	MFI	137@6000	129@4000	3.41 x 3.94	10.0:1	12@Idle
	2.5 (2493)	E	MFI	142@5000	168@2500	3.59 x 2.99	10.0:1	12@Idle
	3.0 (2972)	G	MFI	192@6000	178@4800	3.59 x 2.99	10.0:1	12@Idle

MFI : Multi-Port Fuel Injection

① Tiburon

② Sonata

93471CA7

GASOLINE ENGINE TUNE-UP SPECIFICATIONS

Year	Engine Displacement Liters (cc)	Engine ID/VIN	Spark Plugs Gap (in.)	Ignition Timing (deg.)		Fuel Pump (psi)	Idle Speed (rpm)		Valve Clearance	
				MT	AT		MT	AT	In.	Ex.
1998	1.5 (1495)	K	0.039-0.043	6-16B	6-16B	43	700-900	700-900	HYD	HYD
	1.8 (1795)	M	0.039-0.043	5-15B	5-15B	43	700-900	700-900	HYD	HYD
	2.0 (1975)	F①	0.039-0.043	5-15B	5-15B	43	700-900	700-900	HYD	HYD
	2.0 (1997)	F②	0.039-0.043	3-7B	3-7B	48	650-850	650-850	HYD	HYD
	3.0 (2972)	T	0.039-0.043	3-7B	3-7B	48	600-800	600-800	HYD	HYD
1999	1.5 (1495)	K	0.039-0.043	6-16B	6-16B	43	700-900	700-900	HYD	HYD
	2.0 (1975)	F	0.039-0.043	5-15B	5-15B	43	700-900	700-900	HYD	HYD
	2.4 (2351)	D	0.039-0.043	3-7B	3-7B	48	650-850	650-850	HYD	HYD
	2.5 (2493)	E	0.039-0.043	7-17B	7-17B	48	600-800	600-800	HYD	HYD
2000	1.5 (1495)	K	0.039-0.043	6-16B	6-16B	43	700-900	700-900	HYD	HYD
	2.0 (1975)	F	0.039-0.043	5-15B	5-15B	43	700-900	700-900	HYD	HYD
	2.4 (2351)	D	0.039-0.043	3-7B	3-7B	48	650-850	650-850	HYD	HYD
	2.5 (2493)	E	0.039-0.043	7-17B	7-17B	48	600-800	600-800	HYD	HYD
2001	1.5 (1495)	K	0.039-0.043	6-16B	6-16B	43	700-900	700-900	HYD	HYD
	2.0 (1975)	F	0.039-0.043	5-15B	5-15B	43	700-900	700-900	HYD	HYD
	2.4 (2351)	D	0.039-0.043	3-7B	3-7B	48	650-850	650-850	HYD	HYD
	2.5 (2493)	E	0.039-0.043	7-17B	7-17B	48	600-800	600-800	HYD	HYD
	3.0 (2972)	G	0.039-0.043	3-7B	3-7B	48	600-800	600-800	HYD	HYD

HYD: Hydraulic Valve Lifters

B: Before Top Dead Center

① Tiburon

② Sonata

93471CA8

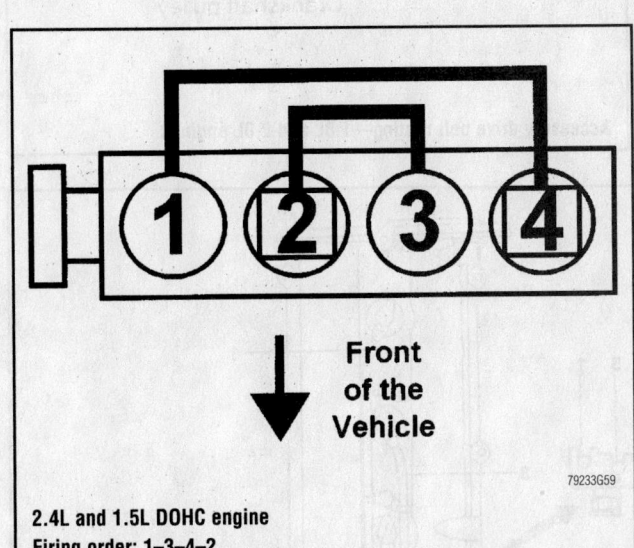

2.4L and 1.5L DOHC engine
Firing order: 1–3–4–2
Distributorless ignition system

79233G59

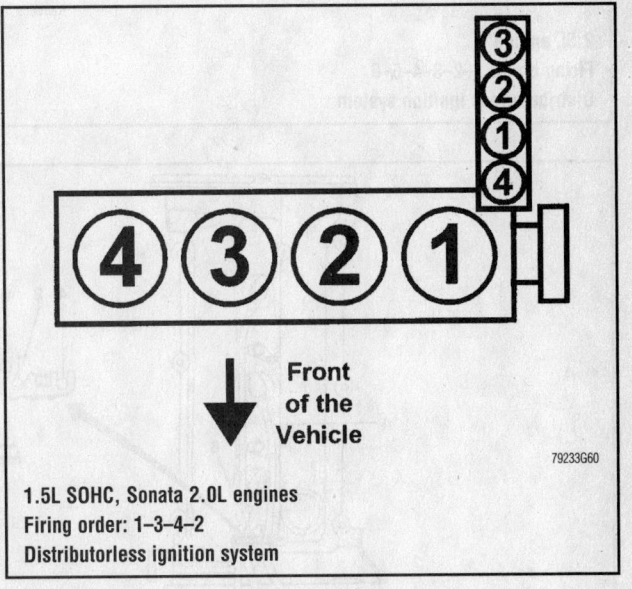

1.5L SOHC, Sonata 2.0L engines
Firing order: 1–3–4–2
Distributorless ignition system

79233G60

For Wheel Alignment specifications, see Section 1 of this manual

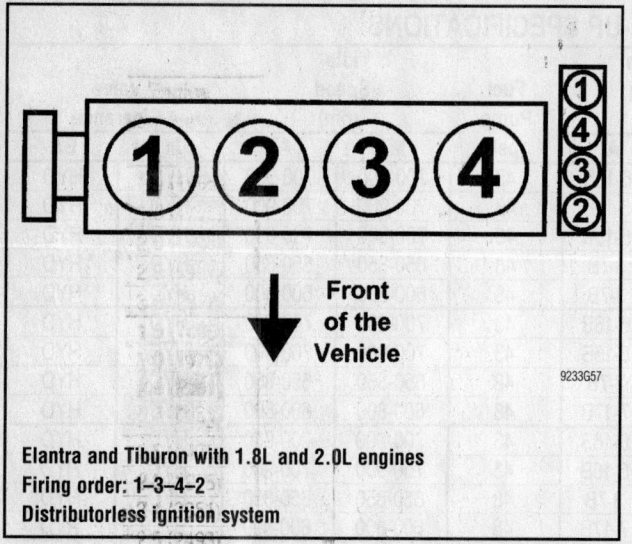

Front of the Vehicle

Elantra and Tiburon with 1.8L and 2.0L engines
Firing order: 1–3–4–2
Distributorless ignition system

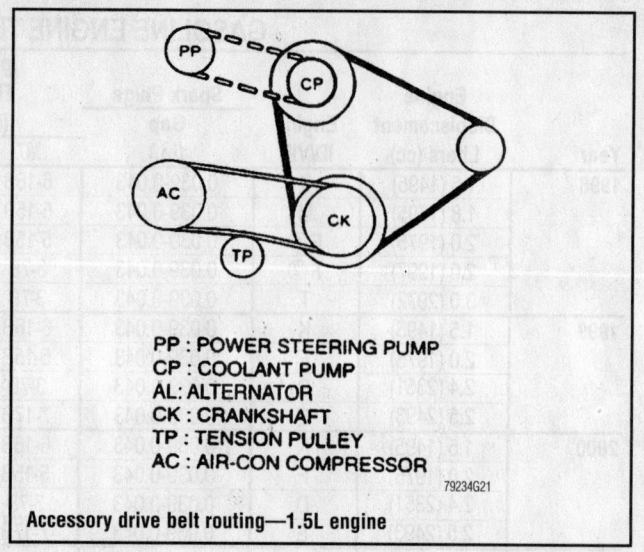

PP : POWER STEERING PUMP
CP : COOLANT PUMP
AL: ALTERNATOR
CK: CRANKSHAFT
TP : TENSION PULLEY
AC : AIR-CON COMPRESSOR

Accessory drive belt routing—1.5L engine

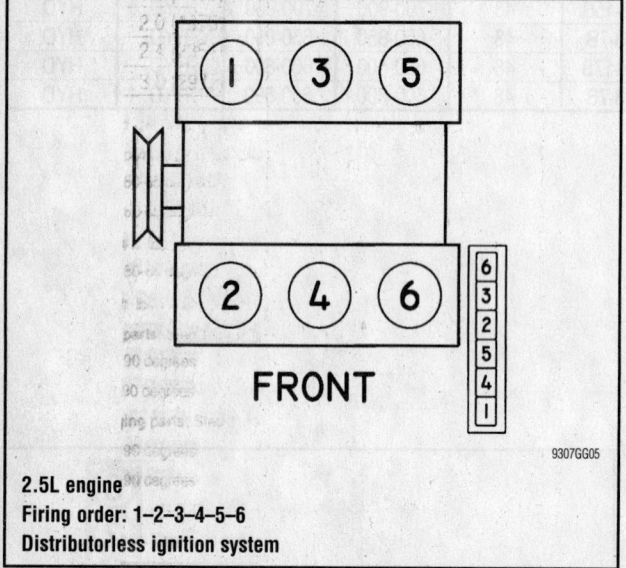

FRONT

2.5L engine
Firing order: 1–2–3–4–5–6
Distributorless ignition system

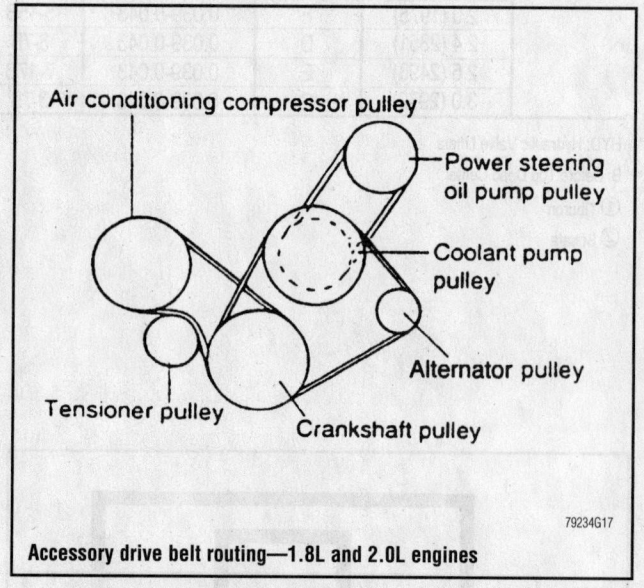

Air conditioning compressor pulley

Power steering oil pump pulley

Coolant pump pulley

Alternator pulley

Tensioner pulley

Crankshaft pulley

Accessory drive belt routing—1.8L and 2.0L engines

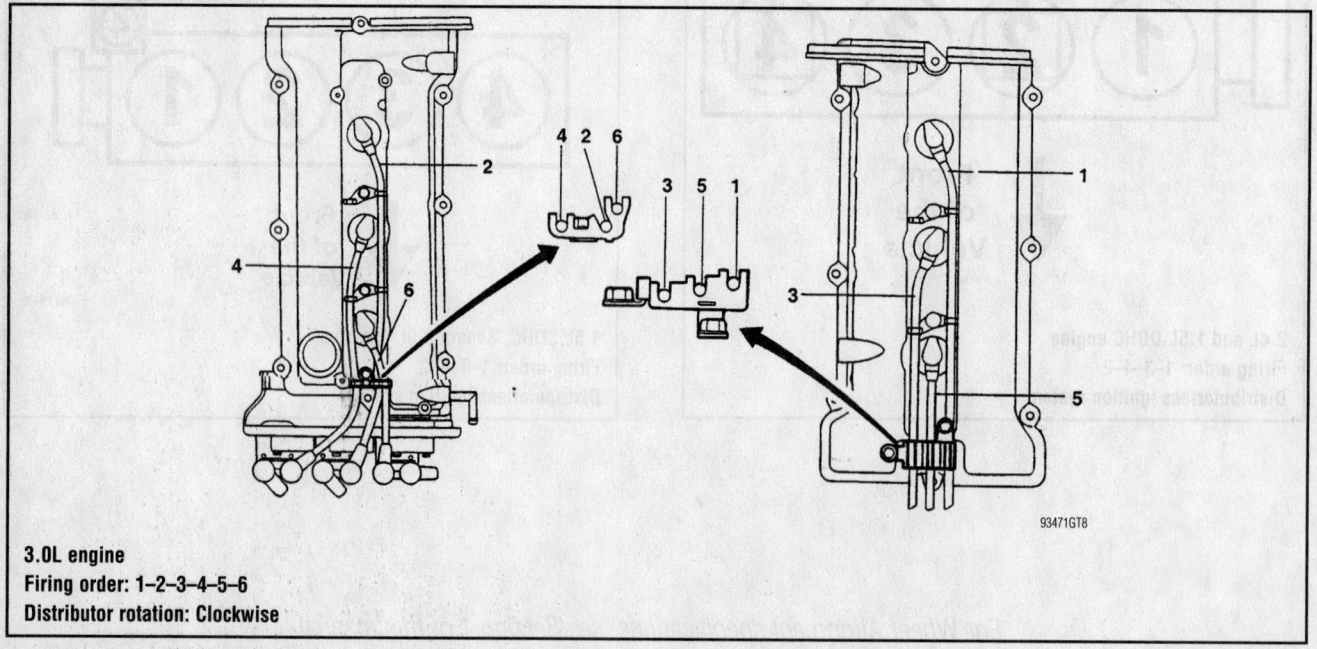

3.0L engine
Firing order: 1–2–3–4–5–6
Distributor rotation: Clockwise

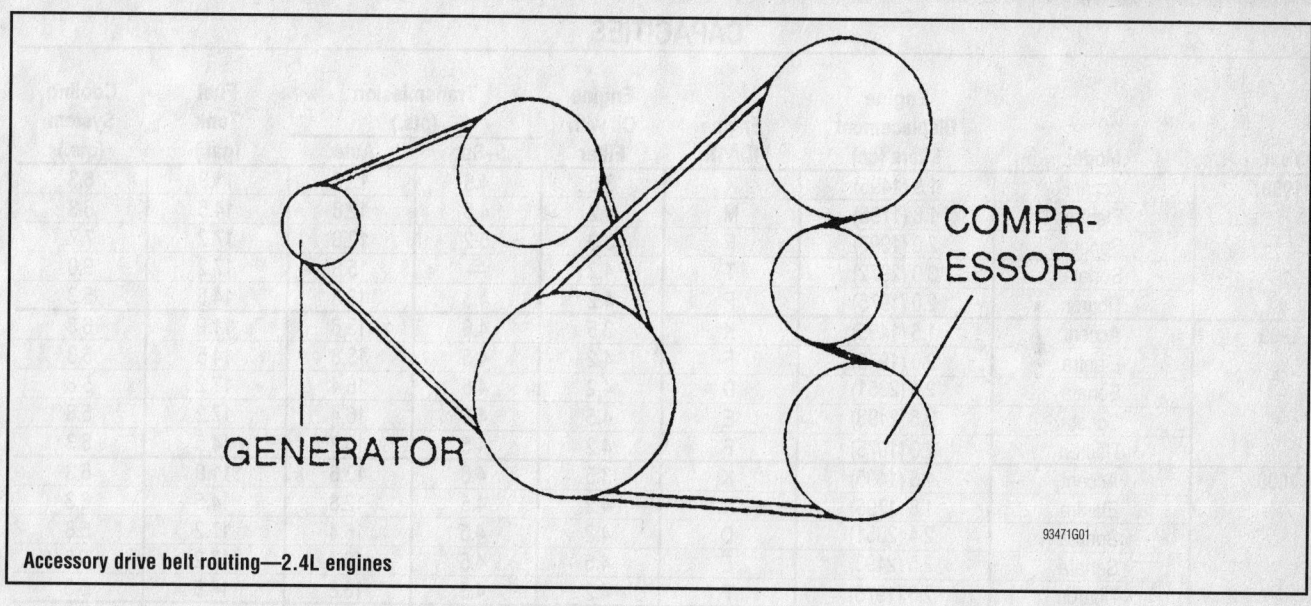

Accessory drive belt routing—2.4L engines

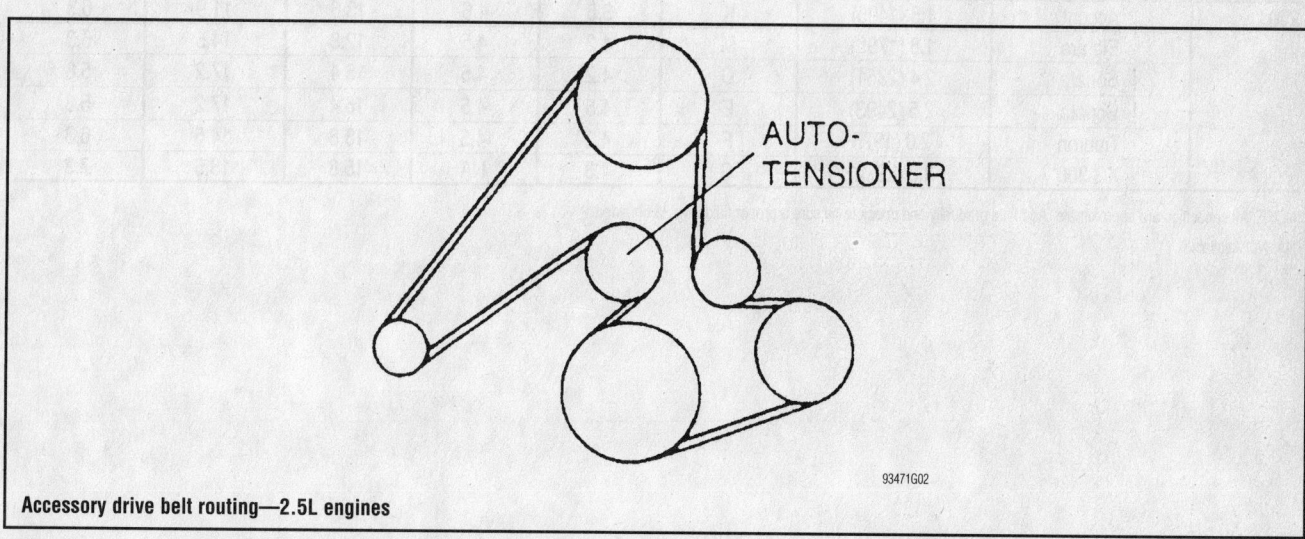

Accessory drive belt routing—2.5L engines

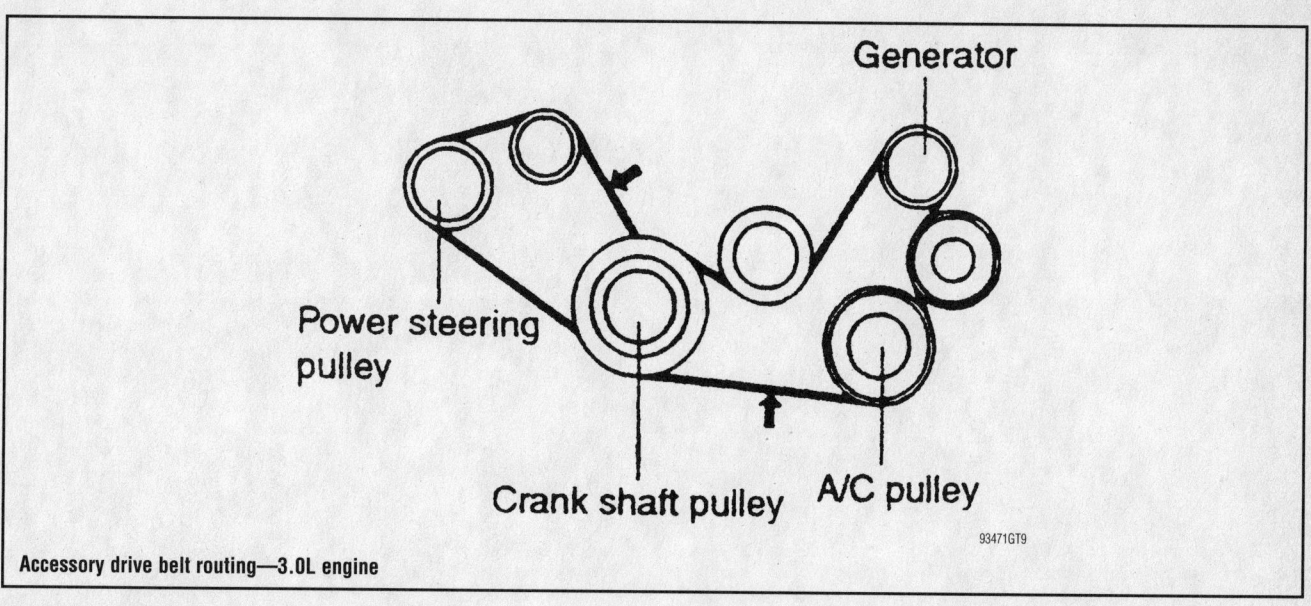

Accessory drive belt routing—3.0L engine

CAPACITIES

Year	Model	Engine Displacement Liters (cc)	Engine ID/VIN	Engine Oil with Filter	Transmission (pts.)		Fuel Tank (gal.)	Cooling System (qts.)
					5–Spd	Auto.		
1998	Accent	1.5 (1495)	K	3.5	4.6	13.6	11.9	6.3
	Elantra	1.8 (1795)	M	4.2	4.5	12.8	14.5	6.3
	Sonata	2.0 (1997)	F	3.9	5.2	12.8	17.2	7.7
	Sonata	3.0 (2972)	T	4.2	—	15.8	17.2	9.0
	Tiburon	2.0 (1975)	F	4.2	4.5	13.8	14.5	6.3
1999	Accent	1.5 (1495)	K	3.5	4.6	13.6	11.9	6.3
	Elantra	2.0 (1975)	F	4.2	4.5	13.8	14.5	6.3
	Sonata	2.4 (2351)	D	4.2	4.5	16.4	17.2	5.8
	Sonata	2.5 (2493)	E	4.5	4.5	16.4	17.2	5.8
	Tiburon	2.0 (1975)	F	4.2	4.5	13.8	14.5	6.3
2000	Accent	1.5 (1495)	K	3.5	4.6	13.6	11.9	6.3
	Elantra	1.8 (1795)	M	4.2	4.5	12.8	14.5	6.3
	Sonata	2.4 (2351)	D	4.2	4.5	16.4	17.2	5.8
	Sonata	2.5 (2493)	E	4.5	4.5	16.4	17.2	5.8
	Tiburon	2.0 (1975)	F	4.2	4.5	13.8	14.5	6.3
2001	Accent	1.5 (1495)	K	3.5	4.6	13.6	11.9	6.3
	Elantra	1.8 (1795)	M	4.2	4.5	12.8	14.5	6.3
	Sonata	2.4 (2351)	D	4.2	4.5	16.4	17.2	5.8
	Sonata	2.5 (2493)	E	4.5	4.5	16.4	17.2	5.8
	Tiburon	2.0 (1975	F	4.2	4.5	13.8	14.5	6.3
	XG 300	3.0 (2972)	G	4.3	NA	15.8	18.5	7.3

NOTE: All capacities are approximate. Add fluid gradually and check to be sure a proper fluid level is obtained.

NA: Not Available

93471CA9

VALVE SPECIFICATIONS

Year	Engine Displacement Liters (cc)	Engine ID/VIN	Seat Angle (deg.)	Face Angle (deg.)	Spring Test Pressure (lbs. @ in.)	Spring Installed Height (in.)	Stem-to-Guide Clearance (in.)		Stem Diameter (in.)	
							Intake	Exhaust	Intake	Exhaust
1998	1.5 (1495)	K	45	45	54@1.358	1.358	0.0012-0.0024	0.0014-0.0026	0.3920	0.3960
	1.8 (1795)	M	45	45	56@1.458	1.358	0.0008-0.0019	0.0019-0.0033	0.2348-0.2354	0.2334-0.2342
	2.0 (1975)	F①	45	45	56@1.457	1.358	0.0008-0.0019	0.0019-0.0033	0.2348-0.2354	0.2334-0.2342
	2.0 (1997)	F②	45-45.5	45-45.5	66@1.575	③	0.0008-0.0019	0.0020-0.0033	0.2585-0.2891	0.2571-0.2579
	3.0 (2972)	T	44-44.5	45	74@1.591	1.590	0.0012-0.0024	0.0020-0.0035	0.3150	0.3134
1999	1.5 (1495)	K	45	45	54@1.358	1.358	0.0012-0.0024	0.0014-0.0026	0.3920	0.3960
	2.0 (1975)	F	45	45	56@1.457	1.358	0.0008-0.0019	0.0019-0.0033	0.2348-0.2354	0.2334-0.2342
	2.4 (2351)	D	45-45.5	44-44.5	56@1.457	1.358	0.0008-0.0019	0.0020-0.0033	0.2585-0.2891	0.2571-0.2579
	2.5 (2493)	E	45	45	49@1.378	1.378	0.0009-0.0020	0.0014-0.0026	0.2348-0.2354	0.2343-0.2348
2000	1.5 (1495)	K	45	45	54@1.358	1.358	0.0012-0.0024	0.0014-0.0026	0.3920	0.3960
	2.0 (1975)	F	45	45	56@1.457	1.358	0.0008-0.0019	0.0019-0.0033	0.2348-0.2354	0.2334-0.2342
	2.4 (2351)	D	45-45.5	44-44.5	56@1.457	1.358	0.0008-0.0019	0.0020-0.0033	0.2585-0.2891	0.2571-0.2579
	2.5 (2493)	E	45	45	49@1.378	1.378	0.0009-0.0020	0.0014-0.0026	0.2348-0.2354	0.2343-0.2348
2001	1.5 (1495)	K	45	45	54@1.358	1.358	0.0012-0.0024	0.0014-0.0026	0.3920	0.3960
	2.0 (1975)	F	45	45	56@1.457	1.358	0.0008-0.0019	0.0019-0.0033	0.2348-0.2354	0.2334-0.2342
	2.4 (2351)	D	45-45.5	44-44.5	56@1.457	1.358	0.0008-0.0019	0.0020-0.0033	0.2585-0.2891	0.2571-0.2579
	2.5 (2493)	E	45	45	49@1.378	1.378	0.0009-0.0020	0.0014-0.0026	0.2348-0.2354	0.2343-0.2348
	3.0 (2972)	G	45-45.5	45-45.5	74@1.591	1.826	0.0009-0.0020	0.0020-0.0033	0.258-0.2594	0.257-0.2580

① Tiburon
② Sonata
③ Free length: 1.902

93471CA0

For Tune-up, Capacities and Firing orders, see Section 1 of this manual

CRANKSHAFT AND CONNECTING ROD SPECIFICATIONS
All measurements are given in inches.

Year	Engine Displacement Liters (cc)	Engine ID/VIN	Crankshaft				Connecting Rod		
			Main Brg. Journal Dia.	Main Brg. Oil Clearance	Shaft End-play	Thrust on No.	Journal Diameter	Oil Clearance	Side Clearance
1998	1.5 (1495)	K	2.2440	0.0011-0.0018	0.0019-0.0068	3	1.7700	0.0009-0.0016	0.0039-0.0098
	1.8 (1836)	M	2.2400	0.0011-0.0018	0.0023-0.0100	3	1.7700	0.0009-0.0016	0.0039-0.0098
	2.0 (1975)	F①	2.2400	0.0011-0.0018	0.0023-0.0100	3	1.7700	0.0009-0.0016	0.0039-0.0098
	2.0 (1997)	F②	2.2433-2.2439	0.0008-0.0020	0.0020-0.0070	3	1.7709-1.7715	0.0008-0.0020	0.0040-0.0100
	3.0 (2972)	T	2.3614-2.3622	0.0008-0.0020	0.0020-0.0070	3	1.9677-1.9685	0.0006-0.0018	0.0039-0.0098
1999	1.5 (1495)	K	2.2440	0.0011-0.0018	0.0019-0.0068	3	1.7700	0.0009-0.0016	0.0039-0.0098
	2.0 (1975)	F	2.2400	0.0011-0.0018	0.0023-0.0100	3	1.7700	0.0009-0.0016	0.0039-0.0098
	2.4 (2351)	D	2.2434-2.2442	0.0007-0.0014③	0.0020-0.0098	3	1.7709-1.7717	0.0008-0.0020	0.0040-0.0098
	2.5 (2493)	E	2.4402-2.4409	0.0002-0.0009	0.0028-0.0098	3	1.8891-1.8898	0.0007-0.0014	0.0039-0.0098
2000	1.5 (1495)	K	2.2440	0.0011-0.0018	0.0019-0.0068	3	1.7700	0.0009-0.0016	0.0039-0.0098
	2.0 (1975)	F	2.2400	0.0011-0.0018	0.0023-0.0100	3	1.7700	0.0009-0.0016	0.0039-0.0098
	2.4 (2351)	D	2.2434-2.2442	0.0007-0.0014③	0.0020-0.0098	3	1.7709-1.7717	0.0008-0.0020	0.0040-0.0098
	2.5 (2493)	E	2.4402-2.4409	0.0002-0.0009	0.0028-0.0098	3	1.8891-1.8898	0.0007-0.0014	0.0039-0.0098
2001	1.5 (1495)	K	2.2440	0.0011-0.0018	0.0019-0.0068	3	1.7700	0.0009-0.0016	0.0039-0.0098
	2.0 (1975)	F	2.2400	0.0011-0.0018	0.0023-0.0100	3	1.7700	0.0009-0.0016	0.0039-0.0098
	2.4 (2351)	D	2.2434-2.2442	0.0007-0.0014③	0.0020-0.0098	3	1.7709-1.7717	0.0008-0.0020	0.0040-0.0098
	2.4 (2351)	D	2.2434-2.2442	0.0007-0.0014③	0.0020-0.0098	3	1.7709-1.7717	0.0008-0.0020	0.0040-0.0098
	3.0 (2972)	G	2.3617-2.3620	0.0007-0.0014	0.002-0.0098	3	1.9677-1.9685	0.0009-0.0020	0.0039-0.0098

① Tiburon

② Sonata

③ No. 3: 0.0009 - 0.0016

93471CB1

PISTON AND RING SPECIFICATIONS
All measurements are given in inches.

Year	Engine Displacement Liters (cc)	Engine ID/VIN	Piston Clearance	Ring Gap			Ring Side Clearance		
				Top Compression	Bottom Compression	Oil Control	Top Compression	Bottom Compression	Oil Control
1998	1.5 (1495)	K	0.0008-0.0016	0.008-0.020	0.008-0.020	0.008-0.039	0.0016-0.0033	0.0016-0.0033	snug
	2.0 (1975)	F①	0.0008-0.0016	0.009-0.015	0.013-0.019	0.008-0.024	0.0015-0.0031	0.0012-0.0027	snug
	2.0 (1997)	F②	0.0004-0.0012	0.010-0.018	0.014-0.020	0.008-0.028	0.0012-0.0028	0.0012-0.0028	snug
	3.0 (2972)	T	0.0008-0.0016	0.012-0.018	0.010-0.016	0.008-0.028	0.0012-0.0035	0.0008-0.0024	snug
	1.8 (1836)	M	0.0008-0.0016	0.010-0.016	0.018-0.024	0.008-0.028	0.0012-0.0028	0.0012-0.0028	snug
1999	1.5 (1495)	K	0.0008-0.0016	0.008-0.020	0.008-0.020	0.008-0.039	0.0016-0.0033	0.0016-0.0033	snug
	2.0 (1975)	F	0.0008-0.0016	0.009-0.015	0.013-0.019	0.008-0.024	0.0015-0.0031	0.0012-0.0027	snug
	2.4 (2351)	D	0.0008-0.0012	0.010-0.014	0.006-0.022	0.004-0.016	0.0008-0.0024	0.0008-0.0024	snug
	2.5 (2493)	E	0.0008-0.0016	0.009-0.015	0.013-0.019	0.008-0.024	0.0015-0.0031	0.0012-0.0027	snug
2000	1.5 (1495)	K	0.0008-0.0016	0.008-0.020	0.008-0.020	0.008-0.039	0.0016-0.0033	0.0016-0.0033	snug
	2.0 (1975)	F	0.0008-0.0016	0.009-0.015	0.013-0.019	0.008-0.024	0.0015-0.0031	0.0012-0.0027	snug
	2.4 (2351)	D	0.0008-0.0012	0.010-0.014	0.006-0.022	0.004-0.016	0.0008-0.0024	0.0008-0.0024	snug
	2.5 (2493)	E	0.0008-0.0016	0.008-0.014	0.015-0.020	0.008-0.028	0.0016-0.0031	0.0012-0.0028	snug
2001	1.5 (1495)	K	0.0008-0.0016	0.008-0.020	0.008-0.020	0.008-0.039	0.0016-0.0033	0.0016-0.0033	snug
	2.0 (1975)	F	0.0008-0.0016	0.009-0.015	0.013-0.019	0.008-0.024	0.0015-0.0031	0.0012-0.0027	snug
	2.4 (2351)	D	0.0008-0.0012	0.010-0.014	0.006-0.022	0.004-0.016	0.0008-0.0024	0.0008-0.0024	snug
	2.5 (2493)	E	0.0008-0.0012	0.008-0.014	0.015-0.020	0.008-0.028	0.0016-0.0031	0.0012-0.0028	snug
	3.0 (2972)	G	0.0008-0.0016	0.012-0.018	0.010-0.016	0.008-0.028	0.0012-0.0035	0.0008-0.0024	snug

① Tiburon
② Sonata

93471CB2

For complete service labor times, order the Chilton Labor Guide

TORQUE SPECIFICATIONS
All readings in ft. lbs.

Year	Engine Displacement Liters (cc)	Engine ID/VIN	Cylinder Head Bolts	Main Bearing Bolts	Rod Bearing Bolts	Crankshaft Damper Bolts	Flywheel Bolts	Manifold Intake	Manifold Exhaust	Spark Plugs	Lug Nut
1998	1.5 (1495)	K	①	40-44	25-28	110-118	94-101	11-14	11-14	18	65-80
	1.8 (1795)	M	②	③	34-39	125-133	88-95	13-18	22-30	18	65-80
	2.0 (1975)	F	②	③	34-39	125-133	88-95	11-14	17-22	18	65-80
	2.0 (1997)	F	76-83	47-51	36-38	80-94	94-101	18-22	18-22	18	65-80
	3.0 (2972)	T	④	55-61	36-38	109-115	65-70	11-14	11-16	18	65-80
1999	1.5 (1495)	K	①	40-44	25-28	110-118	94-101	11-14	11-14	18	65-80
	2.0 (1975)	F	②	③	34-39	125-133	88-95	11-14	17-22	18	65-80
	2.4 (2351)	D	⑤	18⑥	⑦	80-94	94-101	⑧	⑨	15-22	67-81
	2.5 (2493)	E	⑩	⑪	⑫	130-138	53-55	14-15	18-22	15-22	67-81
2000	1.5 (1495)	K	①	40-44	25-28	110-118	94-101	11-14	11-14	18	65-80
	2.0 (1975)	F	②	③	34-39	125-133	88-95	11-14	17-22	18	65-80
	2.4 (2351)	D	⑤	18⑥	⑦	80-94	94-101	⑧	⑨	15-22	67-81
	2.5 (2493)	E	⑩	⑪	⑫	130-138	53-55	14-15	18-22	15-22	67-81
2001	1.5 (1495)	K	①	40-44	25-28	110-118	94-101	11-14	11-14	18	65-80
	2.0 (1975)	F	②	③	34-39	125-133	88-95	11-14	17-22	18	65-80
	2.4 (2351)	D	⑤	18⑥	⑦	80-94	94-101	⑧	⑨	15-22	67-81
	3.0 (2972)	G	75-82	65-72	36-39	NA	NA	9-10	20-40	15-22	67-81

① Cold: 51-54 ft. lbs.; Warm: 58-61 ft. lbs.

② Step 1: M10 bolts to 22 ft. lbs. and M12 bolts to 26 ft. lbs.
   Step 2: Plus 60-65 degrees
   Step 3: Plus 60-65 degrees

③ Step 1: 20-24 ft. lbs.
   Step 2: Plus 60-65 degrees

④ Cold: 65-72 ft. lbs.; Warm: 72-80 ft. lbs.

⑤ If changing parts: Step 1: 14 ft. lbs. then back off completely
   Step 2: Plus 90 degrees
   Step 3: plus 90 degrees
   If not changing parts: Step 1: 14 ft. lbs.
   Step 2: Plus 90 degrees
   Step 3: Plus 90 degrees

⑥ Plus 90-94 degrees

⑦ 13-16 ft. lbs. plus 90-94 degrees

⑧ M8: 11-14 ft. lbs.
   M10: 13-18 ft. lbs.
   Nuts: 22-30 ft. lbs.

⑨ M8: 18-22 ft. lbs.
   M10: 25-40 ft. lbs.

⑩ Step 1: 18 ft. lbs.
   Step 2: plus 60-64 degrees
   Step 3: plus 45-49 degrees

⑪ M10: 20-24 ft. lbs. plus 90-94 degrees
   M8: 10-13 ft. lbs. plus 90-94 degrees

⑫ 10-13 ft. lbs. plus 90-94 degrees

93471CB3

BRAKE SPECIFICATIONS
All measurements in inches unless noted

| Year | Model | | Brake Disc | | | Brake Drum Diameter | | | Minimum Lining Thickness | | Brake Caliper | |
			Original Thickness	Minimum Thickness	Maximum Run-out	Original Inside Diameter	Max. Wear Limit	Maximum Machine Diameter	Front	Rear	Bracket Bolts (ft. lbs.)	Mounting Bolts (ft. lbs.)
1998	Accent	F	0.750	0.670	0.002	—	—	—	0.039	—	50	①
		R	—	—	—	7.100	—	7.165	—	0.039	—	—
	Elantra	F	0.750	0.670	0.002	—	—	—	0.039	—	50	①
		R	—	—	—	8.000	—	8.079	—	0.039	—	—
	Elantra w/rear disc	F	0.750	0.670	0.002	—	—	—	0.039	—	50	①
		R	0.354	NA	NA	—	—	—	—	0.031	—	23
	Sonata	F	0.866	0.787	0.002	—	—	—	0.079	—	55	23
		R	—	—	—	8.858	—	8.936	—	0.031	—	—
	Sonata V6	F	0.866	0.787	0.002	—	—	—	0.079	—	55	23
		R	—	—	—	9.000	—	9.080	—	0.059	—	—
	Sonata w/rear disc	F	0.866	0.787	0.002	—	—	—	0.079	—	55	23
		R	0.472	0.413	0.005	—	—	—	—	0.079	—	23
	Tiburon	F	0.866	0.787	0.002	—	—	—	0.079	—	55	①
		R	—	—	—	8.000	—	8.079	—	0.059	—	—
	Tiburon w/rear disc	F	0.866	0.787	0.002	—	—	—	0.079	—	55	①
		R	0.354	NA	NA	—	—	—	—	0.031	—	23
1999	Accent	F	0.750	0.670	0.002	—	—	—	0.039	—	50	①
		R	—	—	—	7.100	—	7.165	—	0.039	—	—
	Elantra	F	0.750	0.670	0.002	—	—	—	0.039	—	50	①
		R	—	—	—	8.000	—	8.079	—	0.039	—	—
	Elantra w/rear disc	F	0.750	0.670	0.002	—	—	—	0.039	—	50	①
		R	0.354	NA	NA	—	—	—	—	0.031	—	23
	Sonata	F	0.945	0.787	0.002	—	—	—	0.079	—	51-63	16-24
		R	—	—	—	9.000	—	8.936	—	0.059	—	—
	Sonata w/rear disc	F	0.945	0.880	0.003	—	—	—	0.079	—	51-63	16-24
		R	0.390	0.350	0.005	—	—	—	—	0.079	—	23
	Tiburon	F	0.866	0.787	0.002	—	—	—	0.079	—	55	①
		R	—	—	—	8.000	—	8.079	—	0.059	—	—
	Tiburon w/rear disc	F	0.866	0.787	0.002	—	—	—	0.079	—	55	①
		R	0.354	NA	NA	—	—	—	—	0.031	—	23

93471CB4

BRAKE SPECIFICATIONS
All measurements in inches unless noted

Year	Model		Brake Disc Original Thickness	Brake Disc Minimum Thickness	Brake Disc Maximum Run-out	Brake Drum Diameter Original Inside Diameter	Brake Drum Diameter Max. Wear Limit	Brake Drum Diameter Maximum Machine Diameter	Minimum Lining Thickness Front	Minimum Lining Thickness Rear	Brake Caliper Bracket Bolts (ft. lbs.)	Brake Caliper Mounting Bolts (ft. lbs.)
2000	Accent	F	0.750	0.670	0.002	—	—	—	0.039	—	50	①
		R	—	—	—	7.100	—	7.165	—	0.039	—	—
	Elantra	F	0.750	0.670	0.002	—	—	—	0.039	—	50	①
		R	—	—	—	8.000	—	8.079	—	0.039	—	—
	Elantra w/rear disc	F	0.750	0.670	0.002	—	—	—	0.039	—	50	①
		R	0.354	NA	NA	—	—	—	—	0.031	—	23
	Sonata	F	0.945	0.787	0.002	—	—	—	0.079	—	51-63	16-24
		R	—	—	—	9.000	—	8.936	—	0.059	—	—
	Sonata w/rear disc	F	0.945	0.880	0.003	—	—	—	0.079	—	51-63	16-24
		R	0.390	0.350	0.005	—	—	—	—	0.079	—	23
	Tiburon	F	0.866	0.787	0.002	—	—	—	0.079	—	55	①
		R	—	—	—	8.000	—	8.079	—	0.059	—	—
	Tiburon w/rear disc	F	0.866	0.787	0.002	—	—	—	0.079	—	55	①
		R	0.354	NA	NA	—	—	—	—	0.031	—	23
2001	Accent	F	0.750	0.670	0.002	—	—	—	0.039	—	50	①
		R	—	—	—	7.100	—	7.165	—	0.039	—	—
	Elantra	F	0.750	0.670	0.002	—	—	—	0.039	—	50	①
		R	—	—	—	8.000	—	8.079	—	0.039	—	—
	Elantra w/rear disc	F	0.750	0.670	0.002	—	—	—	0.039	—	50	①
		R	0.354	NA	NA	—	—	—	—	0.031	—	23
	Sonata	F	0.945	0.787	0.002	—	—	—	0.079	—	51-63	16-24
		R	—	—	—	9.000	—	8.936	—	0.059	—	—
	Sonata w/rear disc	F	0.945	0.880	0.003	—	—	—	0.079	—	51-63	16-24
		R	0.390	0.350	0.005	—	—	—	—	0.079	—	23
	Tiburon	F	0.866	0.787	0.002	—	—	—	0.079	—	55	①
		R	—	—	—	8.000	—	8.079	—	0.059	—	—
	Tiburon w/rear disc	F	0.866	0.787	0.002	—	—	—	0.079	—	55	①
		R	0.354	NA	NA	—	—	—	—	0.031	—	23
	XG 300	F	0.413	0.096	0.002	—	—	—	0.079	—	51-63	16-24
		R	0.390	0.080	NA	—	—	—	—	—	51-63	16-24

NA: Not Available

F: Front

R: Rear

① Upper: 28 ft. lbs.
 Lower: 19 ft. lbs.

93471CB5

WHEEL ALIGNMENT

Year	Model		Caster Range (+/-Deg.)	Caster Preferred Setting (Deg.)	Camber Range (+/-Deg.)	Camber Preferred Setting (Deg.)	Toe-in (in.)	Steering Axis Inclination (Deg.)
1998	Accent	F	0.50	+1.80	0.50	0	0 +/- 0.12	—
		R	—	—	0.50	-0.68	0.12 +/- 0.08	—
	Elantra	F	0.50	+2.35	2.00	+4.00	0 +/- 0.12	—
		R	—	—	0.50	-0.70	0.14 +/- 0.02	—
	Sonata	F	1.00	+3.25	0.50	0	0 +/- 0.12	—
		R	—	—	0.50	-0.50	0.08 +/- 0.08	—
	Tiburon	F	0.50	+2.45	2.00	+4.00	0 +/- 0.12	—
		R	—	—	0.50	-0.90	0.14 +/- 0.02	—
1999	Accent	F	0.50	+1.80	0.50	0	0 +/- 0.12	—
		R	—	—	0.50	-0.68	0.12 +/- 0.08	—
	Elantra	F	0.50	+2.35	2.00	+4.00	0 +/- 0.12	—
		R	—	—	0.50	-0.70	0.14 +/- 0.02	—
	Sonata	F	1.00	+3.25	0.50	0	0 +/- 0.12	—
		R	—	—	0.50	-0.50	0.08 +/- 0.08	—
	Tiburon	F	0.50	+2.45	2.00	+4.00	0 +/- 0.12	—
		R	—	—	0.50	-0.90	0.14 +/- 0.02	—
2000	Accent	F	0.50	+1.80	0.50	0	0 +/- 0.12	—
		R	—	—	0.50	-0.68	0.12 +/- 0.08	—
	Elantra	F	0.50	+2.35	2.00	+4.00	0 +/- 0.12	—
		R	—	—	0.50	-0.70	0.14 +/- 0.02	—
	Sonata	F	1.00	+3.25	0.50	0	0 +/- 0.12	—
		R	—	—	0.50	-0.50	0.08 +/- 0.08	—
	Tiburon	F	0.50	+2.45	2.00	+4.00	0 +/- 0.12	—
		R	—	—	0.50	-0.90	0.14 +/- 0.02	—
2001	Accent	F	0.50	+1.80	0.50	0	0 +/- 0.12	—
		R	—	—	0.50	-0.68	0.12 +/- 0.08	—
	Elantra	F	0.50	+2.35	2.00	+4.00	0 +/- 0.12	—
		R	—	—	0.50	-0.70	0.14 +/- 0.02	—
	Sonata	F	1.00	+3.25	0.50	0	0 +/- 0.12	—
		R	—	—	0.50	-0.50	0.08 +/- 0.08	—
	Tiburon	F	0.50	+2.45	2.00	+4.00	0 +/- 0.12	—
		R	—	—	0.50	-0.90	0.14 +/- 0.02	—
	XG 300	F	1.00	+3.15	0.50	0	0 +/- 0.12	—
		R	—	—	0.50	-0.50	0.08 +/- 0.08	—

93471CB6

Timing belt service is covered in Section 3 of this manual

TIRE, WHEEL AND BALL JOINT SPECIFICATIONS

Year	Model	OEM Tires		Tire Pressures (psi)		Wheel Size	Ball Joint Inspection
		Standard	Optional	Front	Rear		
1998	Accent	P155/80R13	P175/70R13 P175/65R14	30	30	5-J	①
	Elantra	P175/65R14	P195/60R14	30	30	5.5-JJ	①
	Sonata	P195/70R14	P205/60HR15	30	30	Std: 5.5-JJ Opt: 6-JJ	①
1999	Accent	P155/80R13	P175/70R13 P175/65R14	30	30	5-J	①
	Elantra	P195/60R14	None	30	30	5.5-JJ	①
	Sonata	P195/70R14	P205/60HR15	30	30	Std: 5.5-JJ Opt: 6-JJ	①
	Tiburon	P195/60R14	None	30	30	5.5-JJ	①
2000	Accent	P155/80R13	P175/70R13 P175/65R14	30	30	5-J	①
	Elantra	P195/60R14	None	30	30	5.5-JJ	①
	Sonata	P195/70R14	P205/60HR15	30	30	Std: 5.5-JJ Opt: 6-JJ	①
	Tiburon	P195/60R14	None	30	30	5.5-JJ	①
2001	Accent	P155/80R13	P175/70R13 P175/65R14	30	30	5-J	①
	Elantra	P195/60R14	None	30	30	5.5-JJ	①
	Sonata	P195/70R14	P205/60HR15	30	30	Std: 5.5-JJ Opt: 6-JJ	①
	Tiburon	P195/60R14	None	30	30	5.5JJ	①
	XG 300	P205/65R15	None	30		6.0J	①

OEM: Original Equipment Manufacturer

PSI: Pounds Per Square Inch

STD: Standard

OPT: Optional

① Replace if any measurable movement is found.

93471CB7

SCHEDULED MAINTENANCE INTERVALS
ACCENT, ELANTRA, SONATA, TIBURON AND XG300

TO BE SERVICED	TYPE OF SERVICE	VEHICLE MILEAGE INTERVAL (x1000)												
		7.5	15	22.5	30	37.5	45	52.5	60	67.5	75	82.5	90	97.5
Engine oil & filter	R	✓	✓	✓	✓	✓	✓	✓	✓	✓	✓	✓	✓	✓
Automatic transaxle fluid	S/I		✓		✓		✓		✓		✓		✓	
Brake pads, calipers & rotors	S/I		✓		✓		✓		✓		✓		✓	
Driveshafts & boots	S/I		✓		✓		✓		✓		✓		✓	
Wheel bearing grease	S/I				✓				✓				✓	
Air cleaner filter	R				✓				✓				✓	
Automatic transaxle fluid & filter	R				✓				✓				✓	
Brake fluid	R				✓				✓				✓	
Engine coolant	R				✓				✓				✓	
Fuel hose, vapor hose & fuel filter cap	S/I							✓						
Spark plugs	R				✓				✓				✓	
Spark plugs (Sonata 3.0L V6)	R								✓					
Bolts & nuts on chassis & body (Accent)	S/I				✓				✓				✓	
Drive belts	S/I				✓				✓				✓	
Exhaust pipe connections, muffler & suspension bolts	S/I				✓				✓				✓	
Manual transaxle oil	S/I				✓				✓				✓	
Rear brake drums, linings & parking brake	S/I				✓				✓				✓	
Steering gear rack, linkage & boots	S/I				✓				✓				✓	
Suspension ball joints & dust covers (Accent)	S/I				✓				✓				✓	
Timing belt (Accent & Elantra)	S/I				✓				✓				✓	
Timing belt (except Accent & Elantra)	R								✓					
Fuel filter	R							✓						
Fuel lines & connections	S/I								✓					
Vacuum & crankcase ventilation hoses	S/I								✓					

R: Replace   S/I: Service or Inspect

FREQUENT OPERATION MAINTENANCE (SEVERE SERVICE)
If a vehicle is operated under any of the following conditions it is considered severe service
- Extremely dusty areas.
- 50% or more of the vehicle operation is in 32°C (90°F) or higher temperatures, or constant operation in temperatures below 0°C (32°F).
- Prolonged idling (vehicle operation in stop and go traffic).
- Frequent short running periods (engine does not warm to normal operating temperatures).
- Police, taxi, delivery usage or trailer towing usage.

Oil & oil filter: change every 3000 miles.
Brake pads, calipers & rotors: service or inspect every 7500 miles.
Driveshaft boots: service or inspect every 7500 miles
Steering gear rack, linkage & boots: service or inspect every 7500 miles.
Air cleaner filter: service or inspect every 15,000 miles.
Automatic transaxle fluid & filter: replace every 15,000 miles.
Rear brake drums & linings: service or inspect every 15,000 miles.
Spark plugs: replace every 24,000 miles.

93471CB8

Heater Core replacement is covered in Section 2 of this manual

SCHEDULED MAINTENANCE INTERVALS
HYUNDAI
ACCENT, ELANTRA, SONATA, TIBURON, XG300

The following should be used as a guide when determining the amount of work required for a particular service.
In estimating how long a particular Scheduled Maintenance Service should take, please observe the following:

- Labor Time is time based on field research and data supplied by the vehicle manufacturer.
- Labor time operations are given in hours and tenths of an hour.
- All labor operations are to be used as a guide.

Mechanic Skill Level Codes:
(A) PRECISION: Highly skilled with multiple certification.
(B) GENERAL: Normally skilled with certification.
(C) MAINTENANCE: Semi-skilled working on certification.

	LABOR TIME		LABOR TIME		LABOR TIME
7500 Mile Service (C)		**37500 Mile Service (C)**		**67500 Mile Service (C)**	
All Models	.4	All Models	.4	All Models	.4
15000 Mile Service (C)		**45000 Mile Service (C)**		**75000 Mile Service (C)**	
Accent, Elantra	.7	Accent, Elantra	.8	Accent, Elantra	.8
Sonata	.8	Sonata	.8	Sonata	.8
Tiburon, XG300	.7	Tiburon, XG300	.8	Tiburon, XG300	.8
w/AT add	.1	*w/AT add*	.1	*w/AT add*	.1
22500 Mile Service (C)		**52500 Mile Service (C)**		**82500 Mile Service (C)**	
All Models	.4	All Models	.8	All Models	.4
30000 Mile Service (B)		**60000 Mile Service (B)**		**90000 Mile Service (B)**	
Accent, Elantra	3.5	Accent, Elantra	4.4	Accent, Elantra	3.7
Sonata	3.3	Sonata	5.6	Sonata	3.3
Tiburon, XG300	3.5	*w/3.0L add*	.5	Tiburon, XG300	3.7
w/AT add	.6	Tiburon, XG300	5.2	*w/AT add*	.6
		w/AT add	.6	**97500 Mile Service (C)**	
				All Models	.4

93471CB9

INFINITI
G20 • I30 • Q45

ENGINE AND VEHICLE IDENTIFICATION

		Engine						Model Year	
Code ①	Liters (cc)	Cu. In.	Cyl.	Fuel Sys.	Engine Type	Eng. Mfg.		Code ②	Year
C	2.0 (1998)	122	4	MFI	DOHC	Nissan		W	1998
C	3.0 (2988)	182	6	MFI	DOHC	Nissan		X	1999
B	4.1 (4130)	252	8	MFI	DOHC	Nissan		Y	2000
								1	2001
								2	2002

MFI: Multi-port Fuel Injection

DOHC: Double Overhead Camshaft

① 4th digit of the Vehicle Identification Number (VIN)

② 10th digit of the Vehicle Identification Number (VIN)

93471CB0

GENERAL ENGINE SPECIFICATIONS

Year	Model	Engine Displacement Liters (cc)	Engine ID/VIN	Fuel System Type	Net Horsepower @ rpm	Net Torque @ rpm (ft. lbs.)	Bore x Stroke (in.)	Compression Ratio	Oil Pressure @ rpm
1998	I30	3.0 (2988)	C	MFI	190@5600	205@4000	3.66x2.89	10.1:1	63-80@3000
	Q45	4.1 (4130)	B	MFI	266@5600	278@4000	3.66x2.99	10.2:1	67-81@3000
1999	G20	2.0 (1998)	C	MFI	140@6400	132@4800	3.39x3.39	9.5:1	46-57@3200
	I30	3.0 (2988)	C	MFI	190@5600	205@4000	3.66x2.89	10.1:1	63-80@3000
	Q45	4.1 (4130)	B	MFI	266@5600	278@4000	3.66x2.99	10.2:1	67-81@3000
2000	G20	2.0 (1998)	C	MFI	140@6400	132@4800	3.39x3.39	9.5:1	46-57@3200
	I30	3.0 (2988)	C	MFI	190@5600	205@4000	3.66x2.89	10.1:1	63-80@3000
	Q45	4.1 (4130)	B	MFI	266@5600	278@4000	3.66x2.99	10.2:1	67-81@3000
2001	G20	2.0 (1998)	C	MFI	140@6400	132@4800	3.39x3.39	9.5:1	46-57@3200
	I30	3.0 (2988)	C	MFI	190@5600	205@4000	3.66x2.89	10.1:1	63-80@3000
	Q45	4.1 (4130)	B	MFI	266@5600	278@4000	3.66x2.99	10.2:1	67-81@3000

MFI: Multi-port Fuel Injection

93471CC1

Brake service is covered in Section 4 of this manual

GASOLINE ENGINE TUNE-UP SPECIFICATIONS

Year	Engine Displacement Liters (cc)	Engine ID/VIN	Spark Plug Gap (in.)	Ignition Timing (deg.)		Fuel Pump (psi) ①	Idle Speed (rpm)		Valve Clearance	
				MT	AT		MT	AT	Intake	Exhaust
1998	3.0 (2988)	C	0.039–0.043	15B	15B	34	625	700	HYD	HYD
	4.1 (4130)	B	0.039–0.041	—	15B	34	—	650	HYD	HYD
1999	2.0 (1998)	C	0.031–0.035	15B	15B	34	800	800	HYD	HYD
	3.0 (2988)	C	0.039–0.043	15B	15B	34	625	700	HYD	HYD
	4.1 (4130)	B	0.039–0.041	—	15B	34	—	650	HYD	HYD
2000	2.0 (1998)	C	0.031–0.035	15B	15B	34	800	800	HYD	HYD
	3.0 (2988)	C	0.039–0.043	15B	15B	34	625	700	HYD	HYD
	4.1 (4130)	B	0.039–0.041	—	15B	34	—	650	HYD	HYD
2001	2.0 (1998)	C	0.031–0.035	15B	15B	34	800	800	HYD	HYD
	3.0 (2988)	C	0.039–0.043	15B	15B	34	625	700	HYD	HYD
	4.1 (4130)	B	0.039–0.041	—	15B	34	—	650	HYD	HYD

NOTE: The Vehicle Emission Control Information label often reflects specification changes made during production. The label figures must be used if they differ from those in this chart.

B: Before top dead center

① 43 psi with regulator vacuum hose disconnected

93471CC2

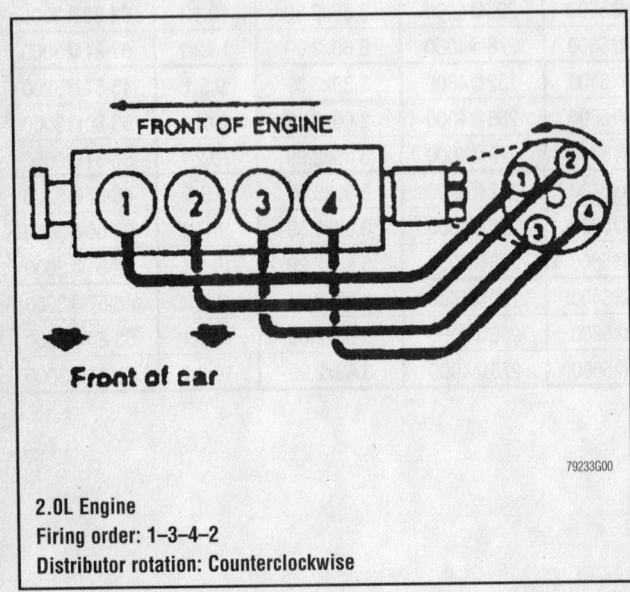

2.0L Engine
Firing order: 1–3–4–2
Distributor rotation: Counterclockwise

79233G00

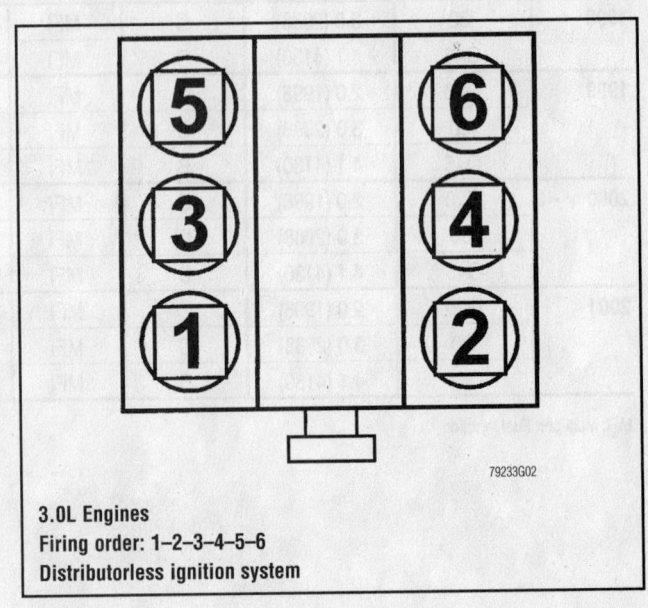

3.0L Engines
Firing order: 1–2–3–4–5–6
Distributorless ignition system

79233G02

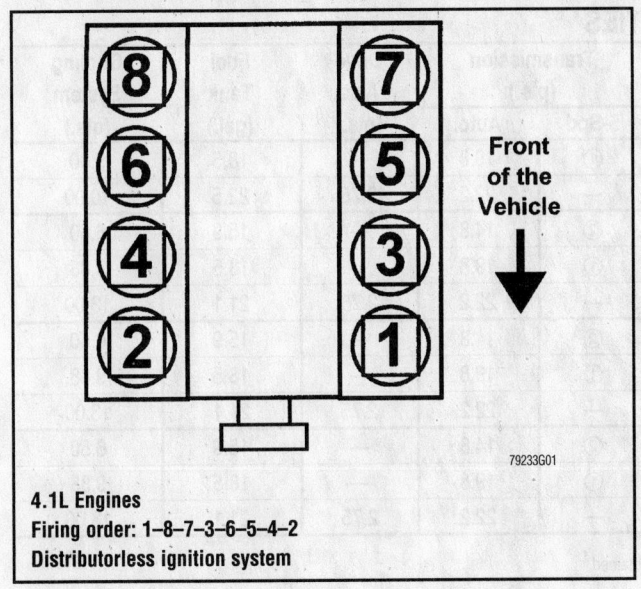

4.1L Engines
Firing order: 1–8–7–3–6–5–4–2
Distributorless ignition system

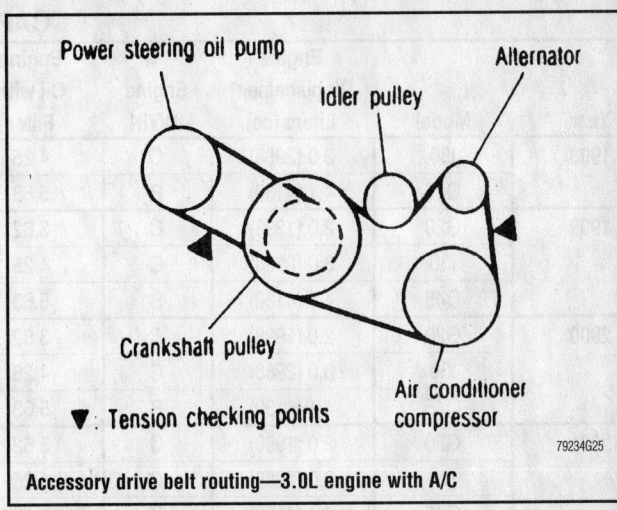

Accessory drive belt routing—3.0L engine with A/C

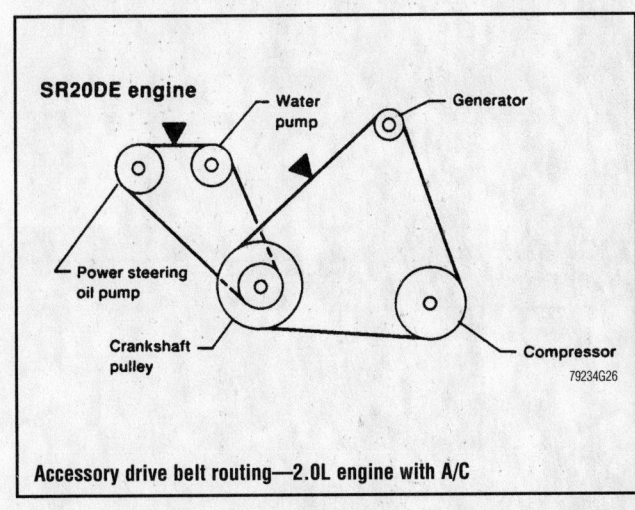

Accessory drive belt routing—2.0L engine with A/C

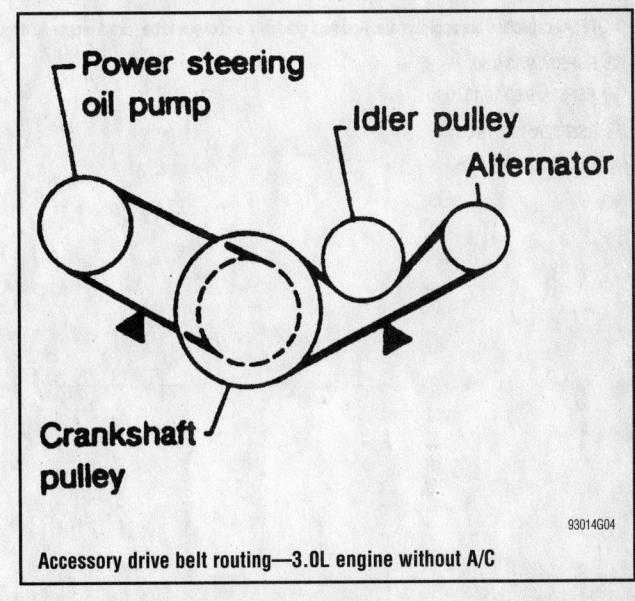

Accessory drive belt routing—3.0L engine without A/C

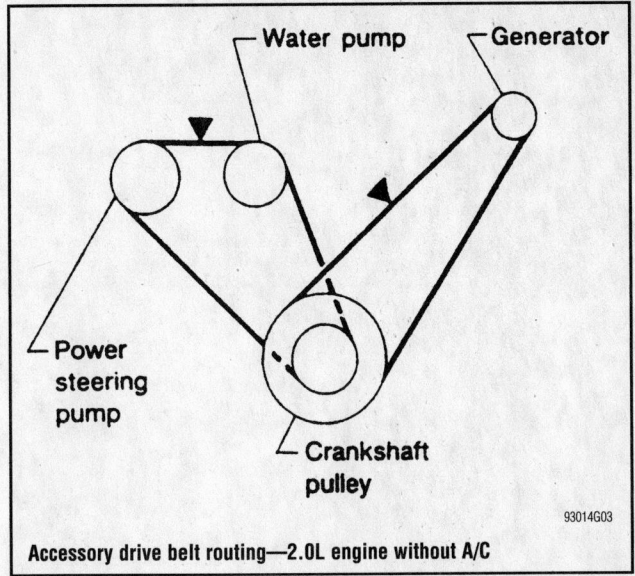

Accessory drive belt routing—2.0L engine without A/C

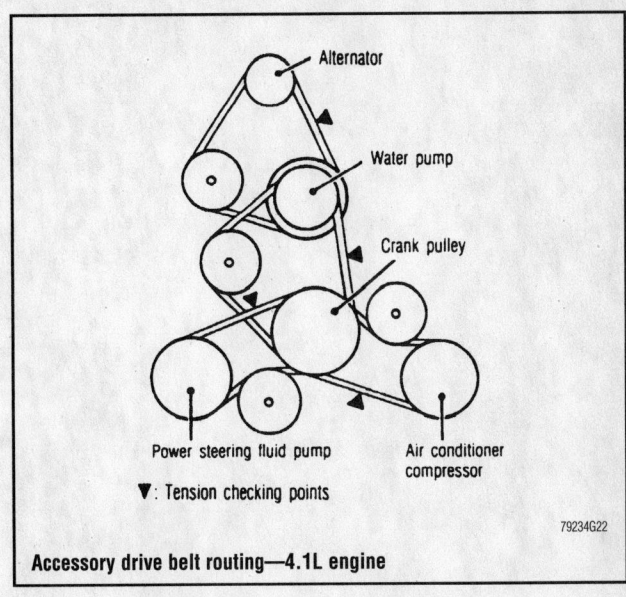

Accessory drive belt routing—4.1L engine

For complete Engine Mechanical specifications, see Section 1 of this manual

CAPACITIES

Year	Model	Engine Displacement Liters (cc)	Engine ID/VIN	Engine Oil with Filter	Transmission (pts.)		Drive Axle (pts.)	Fuel Tank (gal.)	Cooling System (qts.)
					5-Spd	Auto.			
1998	I30	3.0 (2988)	C	4.25	①	19.8	—	18.5	9.00
	Q45	4.1 (4130)	B	5.63	—	22.2	2.75	22.5	10.90
1999	G20	2.0 (1998)	C	3.62	②	14.8	—	15.9	6.50
	I30	3.0 (2988)	C	4.25	①	19.8	—	18.5	9.88
	Q45	4.1 (4130)	B	5.63	—	22.2	2.75	21.1	13.00
2000	G20	2.0 (1998)	C	3.62	②	14.8	—	15.9	6.50
	I30	3.0 (2988)	C	4.25	①	19.8	—	18.5	9.88
	Q45	4.1 (4130)	B	5.63	—	22.2	2.75	21.1	13.00
2001	G20	2.0 (1998)	C	3.62	②	14.8	—	15.9	6.50
	I30	3.0 (2988)	C	4.25	①	19.8	—	18.5	9.88
	Q45	4.1 (4130)	B	5.63	—	22.2	2.75	21.1	13.00

NOTE: All capacities are approximate. Add fluid gradually and check to be sure a proper fluid level is obtained.

① RSF50V: 9.13-9.50
RSF50A: 9.50-10.13

② RS5F32A: 7.62 - 8.00
RS5F32V: 7.86 - 8.25

93471CC3

VALVE SPECIFICATIONS

Year	Engine Displacement Liters (cc)	Engine ID/VIN	Seat Angle (deg.)	Face Angle (deg.)	Spring Test Pressure (lbs. @ in.)	Spring Free Height (in.)	Stem-to-Guide Clearance (in.)		Stem Diameter (in.)	
							Intake	Exhaust	Intake	Exhaust
1998	3.0 (2988)	C	45.25-45.75	NA	120.1@1.085	1.845	0.0008-0.0021	0.0016-0.0029	0.2348-0.2354	0.2341-0.2346
	4.1 (4130)	B	45.25-45.75	44.85-45.10	120.4@1.055	1.946	0.0011-0.0020	0.0014-0.0020	0.2743-0.2744	0.3134-0.3136
1999	2.0 (1998)	C	45.25-45.75	44.85-45.10	127.9-144.3@1.181	1.943	0.0008-0.0021	0.0016-0.0029	0.2348-0.2354	0.2341-0.2346
	3.0 (2988)	C	45.25-45.75	NA	120.1@1.085	1.845	0.0008-0.0021	0.0016-0.0029	0.2348-0.2354	0.2341-0.2346
	4.1 (4130)	B	45.25-45.75	44.85-45.10	120.4@1.055	1.946	0.0011-0.0020	0.0014-0.0020	0.2743-0.2744	0.3134-0.3136
2000	2.0 (1998)	C	45.25-45.75	44.85-45.10	127.9-144.3@1.181	1.943	0.0008-0.0021	0.0016-0.0029	0.2348-0.2354	0.2341-0.2346
	3.0 (2988)	C	45.25-45.75	NA	120.1@1.085	1.845	0.0008-0.0021	0.0016-0.0029	0.2348-0.2354	0.2341-0.2346
	4.1 (4130)	B	45.25-45.75	44.85-45.10	120.4@1.055	1.946	0.0011-0.0020	0.0014-0.0020	0.2743-0.2744	0.3134-0.3136
2001	2.0 (1998)	C	45.25-45.75	44.85-45.10	127.9-144.3@1.181	1.943	0.0008-0.0021	0.0016-0.0029	0.2348-0.2354	0.2341-0.2346
	3.0 (2988)	C	45.25-45.75	NA	120.1@1.085	1.845	0.0008-0.0021	0.0016-0.0029	0.2348-0.2354	0.2341-0.2346
	4.1 (4130)	B	45.25-45.75	44.85-45.10	120.4@1.055	1.946	0.0011-0.0020	0.0014-0.0020	0.2743-0.2744	0.3134-0.3136

NA: Not Available

93471CC4

For Accessory Drive Belt illustrations, see Section 1 of this manual

CRANKSHAFT AND CONNECTING ROD SPECIFICATIONS

All measurements are given in inches.

Year	Engine Displacement Liters (cc)	Engine ID/VIN	Crankshaft				Connecting Rod		
			Main Brg. Journal Dia.	Main Brg. Oil Clearance	Shaft End-play	Thrust on No.	Journal Diameter	Oil Clearance	Side Clearance
1998	3.0 (2988)	C	2.3610-2.3612	0.0014-0.0021	0.0039-0.0098	3	1.7704-1.7706	0.0013-0.0023	0.0079-0.0138
	4.1 (4130)	B	2.5181-2.5183	0.0002-0.0008	0.0039-0.0102	3	2.0460-2.0462	0.0008-0.0018	0.0079-0.0138
1999	2.0 (1998)	C	2.1643-2.1646	0.0002-0.0009	0.0039-0.0102	3	1.8885-1.8887	0.0008-0.0018	0.0079-0.0138
	3.0 (2988)	C	2.3610-2.3612	0.0014-0.0021	0.0039-0.0098	3	1.7704-1.7706	0.0013-0.0023	0.0079-0.0138
	4.1 (4130)	B	2.5181-2.5183	0.0002-0.0008	0.0039-0.0102	3	2.0460-2.0462	0.0008-0.0018	0.0079-0.0138
2000	2.0 (1998)	C	2.1643-2.1646	0.0002-0.0009	0.0039-0.0102	3	1.8885-1.8887	0.0008-0.0018	0.0079-0.0138
	3.0 (2988)	C	2.3610-2.3612	0.0014-0.0021	0.0039-0.0098	3	1.7704-1.7706	0.0013-0.0023	0.0079-0.0138
	4.1 (4130)	B	2.5181-2.5183	0.0002-0.0008	0.0039-0.0102	3	2.0460-2.0462	0.0008-0.0018	0.0079-0.0138
2001	2.0 (1998)	C	2.1643-2.1646	0.0002-0.0009	0.0039-0.0102	3	1.8885-1.8887	0.0008-0.0018	0.0079-0.0138
	3.0 (2988)	C	2.3610-2.3612	0.0014-0.0021	0.0039-0.0098	3	1.7704-1.7706	0.0013-0.0023	0.0079-0.0138
	4.1 (4130)	B	2.5181-2.5183	0.0002-0.0008	0.0039-0.0102	3	2.0460-2.0462	0.0008-0.0018	0.0079-0.0138

93471CC5

PISTON AND RING SPECIFICATIONS
All measurements are given in inches.

Year	Engine Displacement Liters (cc)	Engine ID/VIN	Piston Clearance	Ring Gap			Ring Side Clearance		
				Top Compression	Bottom Compression	Oil Control	Top Compression	Bottom Compression	Oil Control
1998	3.0 (2988)	C	0.0004-0.0012	0.0087-0.0161	0.0197-0.0291	0.0079-0.0272	0.0016-0.0031	0.0012-0.0028	SNUG
	4.1 (4130)	B	0.0004-0.0012	0.0106-0.0181	0.0154-0.0248	0.0079-0.0272	0.0016-0.0031	0.0012-0.0028	SNUG
1999	2.0 (1998)	C	0.0004-0.0012	0.0079-0.0118	0.0138-0.0197	0.0079-0.0236	0.0018-0.0031	0.0012-0.0026	SNUG
	3.0 (2988)	C	0.0004-0.0012	0.0087-0.0161	0.0197-0.0291	0.0079-0.0272	0.0016-0.0031	0.0012-0.0028	SNUG
	4.1 (4130)	B	0.0004-0.0012	0.0106-0.0181	0.0154-0.0248	0.0079-0.0272	0.0016-0.0031	0.0012-0.0028	SNUG
2000	2.0 (1998)	C	0.0004-0.0012	0.0079-0.0118	0.0138-0.0197	0.0079-0.0236	0.0018-0.0031	0.0012-0.0026	SNUG
	3.0 (2988)	C	0.0004-0.0012	0.0087-0.0161	0.0197-0.0291	0.0079-0.0272	0.0016-0.0031	0.0012-0.0028	SNUG
	4.1 (4130)	B	0.0004-0.0012	0.0106-0.0181	0.0154-0.0248	0.0079-0.0272	0.0016-0.0031	0.0012-0.0028	SNUG
2001	2.0 (1998)	C	0.0004-0.0012	0.0079-0.0118	0.0138-0.0197	0.0079-0.0236	0.0018-0.0031	0.0012-0.0026	SNUG
	3.0 (2988)	C	0.0004-0.0012	0.0087-0.0161	0.0197-0.0291	0.0079-0.0272	0.0016-0.0031	0.0012-0.0028	SNUG
	4.1 (4130)	B	0.0004-0.0012	0.0106-0.0181	0.0154-0.0248	0.0079-0.0272	0.0016-0.0031	0.0012-0.0028	SNUG

93471CC6

For Tire, Wheel and Ball Joint specifications, see Section 1 of this manual

TORQUE SPECIFICATIONS
All readings in ft. lbs.

Year	Engine Displacement Liters (cc)	Engine ID/VIN	Cylinder Head Bolts	Main Bearing Bolts	Rod Bearing Bolts	Crankshaft Damper Bolts	Flywheel Bolts	Manifold Intake	Manifold Exhaust	Spark Plugs	Lug Nuts
1998	3.0 (2988)	C	①	②	③	④	61-69	20-23	21-24	14-22	72-87
	4.1 (4130)	B	⑤	⑥	⑦	260-275	61-69	12-15	20-23	14-22	72-87
1999	2.0 (1998)	C	⑧	⑨	⑩	105-112	61-69	13-15	27-35	14-22	72-87
	3.0 (2988)	C	①	②	③	④	61-69	20-23	21-24	14-22	72-87
	4.1 (4130)	B	⑤	⑥	⑦	260-275	61-69	12-15	20-23	14-22	72-87
2000	2.0 (1998)	C	⑧	⑨	⑩	105-112	61-69	13-15	27-35	14-22	72-87
	3.0 (2988)	C	①	②	③	④	61-69	20-23	21-24	14-22	72-87
	4.1 (4130)	B	⑤	⑥	⑦	260-275	61-69	12-15	20-23	14-22	72-87
2001	2.0 (1998)	C	⑧	⑨	⑩	105-112	61-69	13-15	27-35	14-22	72-87
	3.0 (2988)	C	①	②	③	④	61-69	20-23	21-24	14-22	72-87
	4.1 (4130)	B	⑤	⑥	⑦	260-275	61-69	12-15	20-23	14-22	72-87

① Step 1: 72 ft. lbs.
Step 2: Loosen bolts completely
Step 3: 25-33 ft. lbs.
Step 4: Tighten an additional 90-95 degrees
Step 5: Repeat Step 4

② Step 1: Shift crankshaft to align the bearing beam
Step 2: Tighten all bolts to 24-28 ft. lbs.
Step 3: Tighten an additional 90-95 degrees

③ Step 1: Tighten all nuts to 15 ft. lbs.
Step 2: Tighten an additional 90-95 degrees

④ Step 1: 29-36 ft. lbs.
Step 2: Tighten an additional 60-66 degrees

⑤ Step 1: 22 ft. lbs.
Step 2: 69 ft. lbs.
Step 3: Loosen bolts completely
Step 4: 18-25 ft. lbs.
Step 5: Tighten an additional 90-95 degrees or 69-72 ft. lbs.

⑥ Step 1: Shift crankshaft back and forth to seat bearing caps
Step 2: Tighten inner cap bolts to 27-31 ft. lbs.
Step 3: Tighten outer cap bolts to 20-24 ft. lbs.
Step 4: Tighten No. 1-3, 5 inner cap bolts an additional 60 degrees
Step 5: Tighten No. 4 inner cap bolt and additional 35 degrees
Step 6: Tighten outer cap bolts an additional 35 degrees
Step 7: Tighten bearing cap side bolts to 34-38 ft. lbs.

⑦ Step 1: 10-12 ft. lbs.
Step 2: Tighten an additional 60-65 degrees or 43-48 ft. lbs.

⑧ Step 1: 29 ft. lbs.
Step 2: 58 ft. lbs.
Step 3: Loosen bolts completely
Step 4: 25-33 ft. lbs.
Step 5: Tighten an additional 90-100 degrees
Step 6: Repeat Step 5.

⑨ Step 1: 24-28 ft. lbs.
Step 2: Tighten an additional 45-50 degrees or 54-61 ft. lbs.

⑩ Step 1: 10-12 ft. lbs.
Step 2: Tighten an additional 60-65 degrees or 28-33 ft. lbs.

93471CC7

BRAKE SPECIFICATIONS
All measurements in inches unless noted

| Year | Model | Front Brake Disc | | | Rear Brake Disc | | | Minimum Lining Thickness | | Brake Caliper | |
		Original Thickness	Minimum Thickness	Maximum Run-out	Original Thickness	Minimum Thickness	Maximum Run-out	Front	Rear	Bracket Bolts (ft. lbs.)	Mounting Bolts (ft. lbs.)
1998	I30	0.870	0.787	0.003	0.350	0.310	0.003	0.079	0.059	①	16-23
	Q45	1.100	1.024	0.003	0.350	0.551	0.003	0.079	0.079	②	24-31
1999	G20	0.870	0.787	0.003	0.350	0.310	0.003	0.079	0.059	①	16-23
	I30	0.870	0.787	0.003	0.350	0.315	0.003	0.079	0.059	①	16-23
	Q45	1.100	1.024	0.003	0.350	0.551	0.003	0.079	0.079	②	24-31
2000	G20	0.870	0.787	0.003	0.350	0.310	0.003	0.079	0.059	①	16-23
	I30	0.870	0.787	0.003	0.350	0.315	0.003	0.079	0.059	①	16-23
	Q45	1.100	1.024	0.003	0.350	0.551	0.003	0.079	0.079	②	24-31
2001	G20	0.870	0.787	0.003	0.350	0.310	0.003	0.079	0.059	①	16-23
	I30	0.870	0.787	0.003	0.350	0.315	0.003	0.079	0.059	①	16-23
	Q45	1.100	1.024	0.003	0.350	0.551	0.003	0.079	0.079	②	24-31

① Front: 53-72
  Rear: 28-38
② Front: 103-118
  Rear: 28-38

93471CC8

For Wheel Alignment specifications, see Section 1 of this manual

WHEEL ALIGNMENT

Year	Model		Caster		Camber		Toe-in (in.)	Steering Axis Inclination (Deg.)
			Range (+/-Deg.)	Preferred Setting (Deg.)	Range (+/-Deg.)	Preferred Setting (Deg.)		
1998	I30	F	0.75	+2.75	0.75	-0.25	0.04 +/- 0.04	14.25
		R	—	—	0.75	-1.00	0.04 +/- 0.15	—
	Q45	F	0.75	+6.42	0.75	-0.66	0.08 +/- 0.04	—
		R	—	—	0.50	-0.75	0.09 +/- 0.10	—
1999	G20	F	0.75	+1.92	0.75	0	0.04 +/- 0.04	—
		R	—	—	0.75	-1.03	0.16 +/- 0.16	—
	I30	F	0.75	+2.75	0.75	-0.25	0.04 +/- 0.04	14.25
		R	—	—	0.75	-1.00	0.04 +/- 0.15	—
	Q45	F	0.75	+6.42	0.75	-0.66	0.08 +/- 0.04	—
		R	—	—	0.50	-0.75	0.09 +/- 0.10	—
2000	G20	F	0.75	+1.92	0.75	0	0.04 +/- 0.04	—
		R	—	—	0.75	-1.03	0.16 +/- 0.16	—
	I30	F	0.75	+2.75	0.75	-0.25	0.04 +/- 0.04	14.25
		R	—	—	0.75	-1.00	0.04 +/- 0.15	—
	Q45	F	0.75	+6.42	0.75	-0.66	0.08 +/- 0.04	—
		R	—	—	0.50	-0.75	0.09 +/- 0.10	—
2001	G20	F	0.75	+1.92	0.75	0	0.04 +/- 0.04	—
		R	—	—	0.75	-1.03	0.16 +/- 0.16	—
	I30	F	0.75	+2.75	0.75	-0.25	0.04 +/- 0.04	14.25
		R	—	—	0.75	-1.00	0.04 +/- 0.15	—
	Q45	F	0.75	+6.42	0.75	-0.66	0.08 +/- 0.04	—
		R	—	—	0.50	-0.75	0.09 +/- 0.10	—

93471CC9

TIRE, WHEEL AND BALL JOINT SPECIFICATIONS

Year	Model	OEM Tires Standard	OEM Tires Optional	Tire Pressures (psi) Front	Tire Pressures (psi) Rear	Wheel Size	Ball Joint Inspection
1998	I30	P205/65R15	P215/55R16	35	35	6.5-JJ	①
	Q45	P215/60VR16	None	35	35	7-JJ	①
1999	G20	P195/65HR15	None	35	35	6-JJ	①
	Q45	P215/60VR16	P225/50VR17	35	35	Std: 7-JJ Opt: 7.5-JJ	①
	I30	P205/65R15	P215/60HR15	35	35	Std: 6-JJ Opt: 6.5-JJ	①
2000	G20	P195/65HR15	None	35	35	6-JJ	①
	Q45	P215/60VR16	P225/50VR17	35	35	Std: 7-JJ Opt: 7.5-JJ	①
	I30	P205/65R15	P215/60HR15	35	35	Std: 6-JJ Opt: 6.5-JJ	①
2001-02	G20	P195/65HR15	None	35	35	6-JJ	①
	Q45	P215/60VR16	P225/50VR17	35	35	Std: 7-JJ Opt: 7.5-JJ	①
	I30	P205/65R15	P215/60HR15	35	35	Std: 6-JJ Opt: 6.5-JJ	①

OEM: Original Equipment Manufacturer

PSI: Pounds Per Square Inch

STD: Standard

OPT: Optional

① Replace if any measurable movement is found.

93471CC0

SCHEDULED MAINTENANCE INTERVALS
Infinity G20, I30, Q45

TO BE SERVICED	TYPE OF SERVICE	7.5	15	22.5	30	37.5	45	52.5	60	67.5	75	82.5	90	97.5
Engine oil & filter	R	✓	✓	✓	✓	✓	✓	✓	✓	✓	✓	✓	✓	✓
Automatic transaxle fluid	S/I		✓		✓		✓		✓		✓		✓	
Brake lines & cables	S/I		✓		✓		✓		✓		✓		✓	
Brake pads & discs	S/I		✓		✓		✓		✓		✓		✓	
Differential gear oil (Q45)	S/I		✓		✓		✓		✓		✓		✓	
Driveshaft boots (I30 & G20)	S/I		✓		✓		✓		✓		✓		✓	
Full-active suspension fluid (Q45) ①	S/I		✓		✓		✓		✓		✓		✓	
Manual transaxle oil (G20 & I30)	S/I		✓		✓		✓		✓		✓		✓	
Air cleaner filter	R				✓				✓				✓	
Exhaust system	S/I				✓				✓				✓	
Fuel lines	S/I				✓				✓				✓	
Steering gear linkage axle & suspension parts	S/I				✓				✓				✓	
SUPER HICAS linkage (J30 & Q45)	S/I				✓				✓				✓	
Vapor lines	S/I					✓			✓				✓	
Engine coolant	R								✓					
Spark plugs	R								✓					
Timing belt	R								✓					
Drive belts	S/I								✓					

R: Replace S/I: Service or Inspect

① Replace at 60,000 miles (if not previously replaced).

FREQUENT OPERATION MAINTENANCE (SEVERE SERVICE)

If a vehicle is operated under any of the following conditions it is considered severe service

- Extremely dusty areas.

- 50% or more of the vehicle operation is in 32°C (90°F) or higher temperatures, or constant operation in temperatures below 0°C (32°F).

- Prolonged idling (vehicle operation in stop and go traffic).

- Frequent short running periods (engine does not warm to normal operating temperatures).

- Police, taxi, delivery usage or trailer towing usage.

Oil & oil filter: change every 3750 miles.

Brake pads & discs: service or inspect every 7500 miles.

Driveshaft boots (G20 & I30): service or inspect every 7500 miles

Exhaust system: service or inspect every 7500 miles.

Steering gear, linkage, axle & suspension ball joints: service or inspect every 7500 miles.

Steering linkage, ball joints & front suspension ball joints: service or inspect every 7500 miles.

SUPER HICAS linkage (Q45): service or inspect every 7500 miles.

93471CD1

SCHEDULED MAINTENANCE INTERVALS
INFINITI
G20, I30, Q45

The following should be used as a guide when determining the amount of work required for a particular service. In estimating how long a particular Scheduled Maintenance Service should take, please observe the following:

- Labor Time is time based on field research and data supplied by the vehicle manufacturer.
- Labor time operations are given in hours and tenths of an hour.
- All labor operations are to be used as a guide.

Mechanic Skill Level Codes:
(A) PRECISION: Highly skilled with multiple certification.
(B) GENERAL: Normally skilled with certification.
(C) MAINTENANCE: Semi-skilled working on certification.

	LABOR TIME		LABOR TIME		LABOR TIME
7500 Mile Service (C)		**37500 Mile Service (C)**		**75000 Mile Service (C)**	
All Models	.6	All Models	.6	G20, Q45	1.0
15000 Mile Service (C)		**45000 Mile Service (C)**		I30	.9
G20, Q45	1.0	G20, Q45	1.0	**82500 Mile Service (C)**	
I30	.9	I30	.9	All Models	.6
22500 Mile Service (C)		**52500 Mile Service (C)**		**90000 Mile Service (B)**	
All Models	.4	All Models	.6	G20, I30	2.2
30000 Mile Service (B)		**60000 Mile Service (B)**		Q45	2.3
G20, I30	1.7	G20, I30	7.3	**97500 Mile Service (C)**	
Q45	1.6	Q45	7.9	All Models	.6
		67500 Mile Service (C)			
		All Models	.6		

93471CD2

For Tune-up, Capacities and Firing orders, see Section 1 of this manual

KIA
SEPHIA

ENGINE AND VEHICLE IDENTIFICATION

			Engine					Model Year	
Code ①	Liters (cc)	Cu. In.	Cyl.	Fuel Sys.	Engine Type	Eng. Mfg.		Code ②	Year
1	1.8 (1793)	109	4	EGI	DOHC	KIA		W	1998
								X	1999
EGI: Electronic Gasoline Injection								Y	2000
DOHC: Double Overhead Camshafts								1	2001
① 8th digit of VIN								2	2002
② 10th digit of VIN									

93471CD3

GENERAL ENGINE SPECIFICATIONS

Year	Model	Engine Displacement Liters (cc)	Engine VIN	Fuel System Type	Net Horsepower @ rpm	Net Torque @ rpm (ft. lbs.)	Bore x Stroke (in.)	Com- pression Ratio	Oil Pressure @ rpm
1998	Sephia	1.8 (1793)	1	EGI	125@6000	120@4500	3.19x3.43	9.5:1	43-57@3000
1999	Sephia	1.8 (1793)	1	EGI	125@6000	120@4500	3.19x3.43	9.5:1	43-57@3000
2000	Sephia	1.8 (1793)	1	EGI	125@6000	120@4500	3.19x3.43	9.5:1	43-57@3000

EGI: Electronic Gasoline Injection

MFI: Multi-port Fuel Injection

93471CD4

ENGINE TUNE-UP SPECIFICATIONS

Year	Engine Displacement Liters (cc)	Engine VIN	Spark Plug Gap (in.)	Ignition Timing (deg.)		Fuel Pump (psi)	Idle Speed (rpm)		Valve Clearance	
				MT	AT		MT	AT	Intake	Exhaust
1998	1.8 (1793)	1	0.028-0.032	3-13B	3-13B	64	750-850	750-850	HYD	HYD
1999	1.8 (1793)	1	0.028-0.032	3-13B	3-13B	64	750-850	750-850	HYD	HYD
2000	1.8 (1793)	1	0.028-0.032	3-13B	3-13B	64	750-850	750-850	HYD	HYD

NOTE: The Vehicle Emission Control Information label often reflects specification changes made during production. The label figures must be used if they differ from those in this chart

B: Before top dead center

HYD: Hydraulic

93471CD5

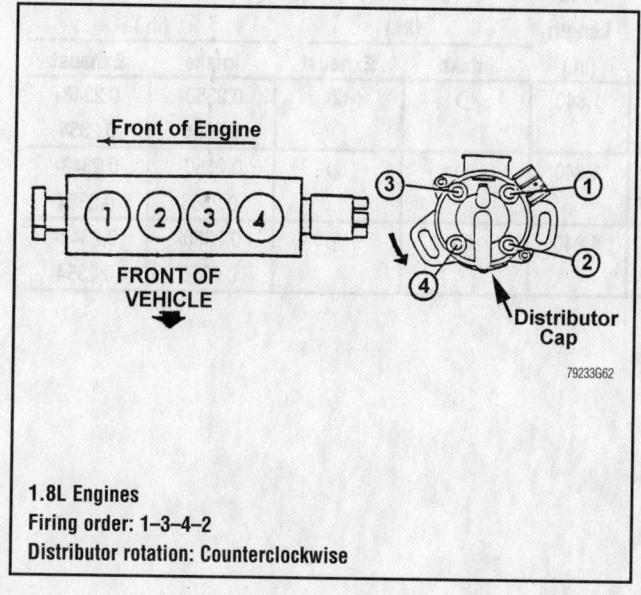

Front of Engine

FRONT OF VEHICLE

Distributor Cap

79233G62

1.8L Engines
Firing order: 1–3–4–2
Distributor rotation: Counterclockwise

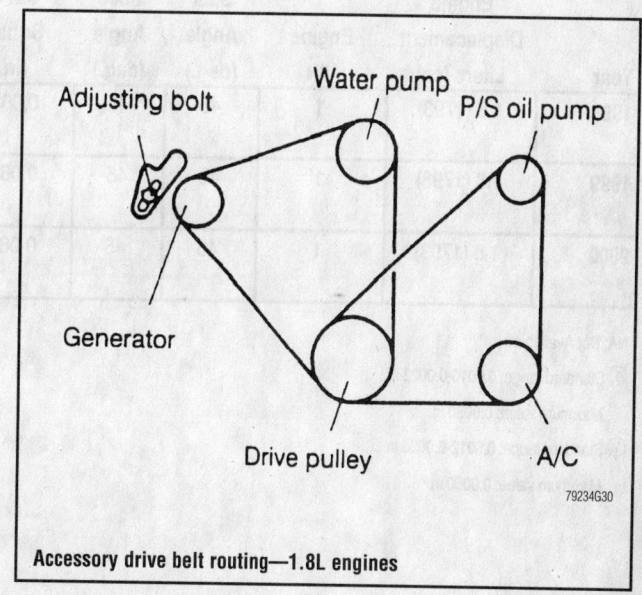

Adjusting bolt

Water pump

P/S oil pump

Generator

Drive pulley

A/C

79234G30

Accessory drive belt routing—1.8L engines

CAPACITIES

Year	Model	Engine Displacement Liters (cc)	Engine VIN	Engine Oil with Filter	Transaxle (pts.)		Fuel Tank (gal.)	Cooling System (qts.)
					Manual	Auto.		
1998	Sephia	1.8 (1793)	1	4.0	5.6	11.4	13.2	6.3
1999	Sephia	1.8 (1793)	1	4.0	5.6	11.4	13.2	6.3
2000	Sephia	1.8 (1793)	1	4.0	5.6	11.4	13.2	6.3

NOTE: All capacities are approximate. Add fluid gradually and ensure a proper level is obtained.

93471CD6

VALVE SPECIFICATIONS

Year	Engine Displacement Liters (cc)	Engine VIN	Seat Angle (deg.)	Face Angle (deg.)	Maximum out of Square (in.)	Spring Free Length (in.)	Stem-to-Guide Clearance (in.)		Stem Diameter (in.)	
							Intake	Exhaust	Intake	Exhaust
1998	1.8 (1793)	1	45	45	0.0638	1.840	①	②	0.2350-0.2356	0.2348-0.2354
1999	1.8 (1793)	1	45	45	0.0638	1.840	①	②	0.2350-0.2356	0.2348-0.2354
2000	1.8 (1793)	1	45	45	0.0638	1.840	①	②	0.2350-0.2356	0.2348-0.2354

NA: Not Available

① Standard range: 0.0010-0.0023 in.

  Maximum value: 0.0080 in.

② Standard range: 0.0012-0.0025 in.

  Maximum value: 0.0080 in.

93471CD7

CRANKSHAFT AND CONNECTING ROD SPECIFICATIONS
All measurements are given in inches.

Year	Engine Displacement Liters (cc)	Engine VIN	Crankshaft				Connecting Rod		
			Main Brg. Journal Dia.	Main Brg. Oil Clearance	Shaft End-play	Thrust on No.	Journal Diameter	Oil Clearance	Side Clearance
1998	1.8 (1793)	1	2.1629-2.1636	0.0010-0.0017	0.0032-0.0111	3	1.7693-1.7700	0.0008-0.0019	0.0044-0.0019
1999	1.8 (1793)	1	2.1629-2.1636	0.0010-0.0017	0.0032-0.0111	3	1.7693-1.7700	0.0008-0.0019	0.0044-0.0019
2000	1.8 (1793)	1	2.1629-2.1636	0.0010-0.0017	0.0032-0.0111	3	1.7693-1.7700	0.0008-0.0019	0.0044-0.0019

93471CD8

PISTON AND RING SPECIFICATIONS
All measurements are given in inches.

Year	Engine Displacement Liters (cc)	Engine VIN	Piston Clearance	Ring Gap			Ring Side Clearance		
				Top Compression	Bottom Compression	Oil Control	Top Compression	Bottom Compression	Oil Control
1998	1.8 (1793)	1	0.0015-0.0021	0.006-0.011	0.012-0.018	0.008-0.027	0.0020-0.0030	0.0010-0.0030	SNUG
1999	1.8 (1793)	1	0.0015-0.0021	0.006-0.011	0.012-0.018	0.008-0.027	0.0020-0.0030	0.0010-0.0030	SNUG
2000	1.8 (1793)	1	0.0015-0.0021	0.006-0.011	0.012-0.018	0.008-0.027	0.0020-0.0030	0.0010-0.0030	SNUG

93471CD9

TORQUE SPECIFICATIONS
All readings in ft. lbs.

Year	Engine Displacement Liters (cc)	Engine VIN	Cylinder Head Bolts	Main Bearing Bolts	Rod Bearing Bolts	Crankshaft Damper Bolts	Flywheel Bolts	Manifold		Spark Plugs	Lug Nuts
								Intake	Exhaust		
1998	1.8 (1793)	1	①	②	35-37	9-13 ③	71-76	14-19	28-34	11-17	65-87
1999	1.8 (1793)	1	①	②	35-37	9-13 ③	71-76	14-19	28-34	11-17	65-87
2000	1.8 (1793)	1	①	②	35-37	9-13 ③	71-76	14-19	28-34	11-17	65-87

① Step 1: 36 ft. lbs.
  Step 2: Loosen fully
  Step 3: 29 ft. lbs.
  Step 4: Tighten 90 degrees
  Step 5: Additional 90 degrees

② Step 1: 29 ft. lbs.
  Step 2: Loosen fully
  Step 3: 14.5 ft. lbs.
  Step 4: Tighten 90 degrees
  Step 5: Tighten 60 degrees

③ Crankshaft pulley

93471CD0

BRAKE SPECIFICATIONS
All measurements in inches unless noted

Year	Model		Brake Disc			Brake Drum			Minimum Lining Thickness	Brake Caliper	
			Original Thickness	Minimum Thickness	Maximum Run-out	Original Inside Diameter	Max. Wear Limit	Maximum Machine Diameter		Bracket Bolts (ft. lbs.)	Mounting Bolts (ft. lbs.)
1998	Sephia	F	0.940	0.710	0.0040	—	—	—	0.080	33-49	19-21
		R	0.400	0.320	0.0039	7.87	7.91	7.91	0.079	33-49	22-29
1999	Sephia	F	0.940	0.710	0.0040	—	—	—	0.080	33-49	19-21
		R	0.400	0.320	0.0039	7.87	7.91	7.91	0.079	33-49	22-29
2000	Sephia	F	0.940	0.710	0.0040	—	—	—	0.080	33-49	19-21
		R	0.400	0.320	0.0039	7.87	7.91	7.91	0.079	33-49	22-29

F: Front

R: Rear

93471CE1

Timing belt service is covered in Section 3 of this manual

WHEEL ALIGNMENT

Year	Model		Caster Range (+/-Deg.)	Caster Preferred Setting (Deg.)	Camber Range (+/-Deg.)	Camber Preferred Setting (Deg.)	Toe-in (in.)	Steering Axis Inclination (Deg.)
1998	Sephia	F	0.75	+2.45	0.50	0	0.11 +/- 0.12	12.58
		R	—		0.50	-0.52	0.07 +/- 0.12	—
1999	Sephia	F	0.75	+2.45	0.50	0	0.11 +/- 0.12	12.58
		R	—	—	0.50	-0.52	0.07 +/- 0.12	—
2000	Sephia	F	0.75	+2.45	0.50	0	0.11 +/- 0.12	12.58
		R	—	—	0.50	-0.52	0.07 +/- 0.12	—

93471CE2

TIRE, WHEEL AND BALL JOINT SPECIFICATIONS

Year	Model	OEM Tires		Tire Pressures (psi)		Wheel Size	Ball Joint Inspection
		Standard	Optional	Front	Rear		
1998	Sephia	P175/70SR13	P185/60HR14	29	29	Std: 5-JJ Opt: 6-JJ	①
1999	Sephia	P175/70SR13	P185/60HR14	29	29	Std: 5-JJ Opt: 6-JJ	①
2000	Sephia	P175/70SR13	P185/60HR14	29	29	Std: 5-JJ Opt: 6-JJ	①

OEM: Original Equipment Manufacturer

PSI: Pounds Per Square Inch

STD: Standard

OPT: Optional

① Replace if any measurable movement is found.

93471CE3

Heater Core replacement is covered in Section 2 of this manual

SCHEDULED MAINTENANCE INTERVALS
1998-00 Kia Sephia

TO BE SERVICED	TYPE OF SERVICE	VEHICLE MILEAGE INTERVAL (x1000)																
		7.5	15	22.5	30	37.5	45	52.5	60	67.5	75	82.5	90	97.5	105	112.5	120	120
Accessory drive belts	S/I				✓				✓				✓				✓	✓
Air cleaner element	R				✓				✓				✓				✓	✓
Air conditioner system	S/I	Inspect the system operation and refrigerant amount annually.																
Brake lines, hoses and connections	S/I				✓				✓				✓				✓	✓
Chassis and body fasteners	T				✓				✓				✓				✓	✓
Clutch pedal height, freeplay and operation	S/I				✓				✓				✓				✓	✓
Cooling system hoses and coolant level	S/I				✓				✓				✓				✓	✓
CV-joint boots	S/I				✓				✓				✓				✓	✓
Engine coolant	R				✓				✓				✓				✓	✓
Engine oil and filter	R	✓	✓	✓	✓	✓	✓	✓	✓	✓	✓	✓	✓	✓	✓	✓	✓	✓
Exhaust system heat shields	S/I				✓				✓				✓				✓	✓
Front and rear brakes	S/I				✓				✓				✓				✓	✓
Front ball joints S/I	S/I				✓				✓				✓				✓	✓
Fuel filter	R								✓									
Fuel lines and hoses	S/I				✓				✓				✓				✓	✓
Idle speed	A				✓				✓				✓				✓	✓
Locks and hinges	L	✓	✓	✓	✓	✓	✓	✓	✓	✓	✓	✓	✓	✓	✓	✓	✓	✓
Spark plugs	R				✓				✓				✓				✓	✓
Steering operation and linkage	S/I				✓				✓				✓				✓	✓
Timing belt (California models)	R														✓			
Timing belt (California models)	S/I								✓				✓					
Timing belt (except California models)	R								✓								✓	✓

R: Replace   S/I: Inspect and service, if needed   L: Lubricate   A: Adjust   T: Tighten

FREQUENT OPERATION MAINTENANCE (SEVERE SERVICE)

If a vehicle is operated under any of the following conditions it is considered severe service

- Towing a trailer or using a camper or car-top carrier.
- Repeated short trips of less than 5 miles in temperatures below freezing, or trips of less than 10 miles in any temperature.
- Extensive idling or low-speed driving for long distances as in heavy commercial use, such as delivery, taxi or police cars.
- Operating on rough, muddy or salt-covered roads.
- Operating on unpaved or dusty roads.
- Driving in extremely hot (over 90°F) conditions.

Engine oil and filter: replace every 5000 miles or 5 months, whichever occurs first.

Air cleaner element: inspect ever 15,000 miles or 15 months and replace every 30,000 miles or 30 months, whichever occurs first.

Fuel system hoses (California models only): replace every 105,000 miles.

Emission system hoses (non-California models): inspect every 55,000 miles or 55 months, whichever occurs first.

Emission system hoses (California models): inspect every 60,000 miles or 60 months, whichever occurs first.

Front and rear disc brakes: inspect every 15,000 miles or 15 months, whichever occurs first.

Chassis and body fasteners: tighten every 15,000 miles or 15 months, whichever occurs first.

Locks and hinges: lubricate every 5000 miles or 5 months, whichever occurs first.

93471CE4

SCHEDULED MAINTENANCE INTERVALS
KIA
SEPHIA

The following should be used as a guide when determining the amount of work required for a particular service. In estimating how long a particular Scheduled Maintenance Service should take, please observe the following:

- Labor Time is time based on field research and data supplied by the vehicle manufacturer.
- Labor time operations are given in hours and tenths of an hour.
- All labor operations are to be used as a guide.

Mechanic Skill Level Codes:
(A) PRECISION: Highly skilled with multiple certification.
(B) GENERAL: Normally skilled with certification.
(C) MAINTENANCE: Semi-skilled working on certification.

	LABOR TIME		LABOR TIME		LABOR TIME
7500 Mile Service (C)		**52500 Mile Service (C)**		**90000 Mile Service (B)**	
All Models	.6	All Models	.6	All Models	2.4
15000 Mile Service (C)		**60000 Mile Service (B)**		**97500 Mile Service (C)**	
All Models	.6	All Models	4.1	All Models	.6
22500 Mile Service (C)		Calif.	2.3	**105000 Mile Service (B)**	
All Models	.6	**67500 Mile Service (C)**		All Models	.6
30000 Mile Service (B)		All Models	.6	*Calif. models replace timing*	
All Models	2.1	**75000 Mile Service (C)**		*belt add*	2.0
37500 Mile Service (C)		All Models	.6	**112500 Mile Service (C)**	
All Models	.6	**82500 Mile Service (C)**		All Models	.6
45000 Mile Service (C)		All Models	.6	**120000 Mile Service (B)**	
All Models	.6			All Models	4.1

93471CE5

Brake service is covered in Section 4 of this manual

LEXUS
ES300 • GS300 • GS400 • IS300 • LS400 • LS430 • SC300 • SC400

ENGINE AND VEHICLE IDENTIFICATION

Engine								Model Year	
Code ①	Liters (cc)	Cu. In.	Cyl.	Fuel Sys.	Engine Type	Eng. Mfg.		Code ②	Year
1MZ-FE	3.0 (2995)	183	6	SFI	DOHC	Toyota		W	1998
1UZ-FE	4.0 (3969)	242	8	SFI	DOHC	Toyota		X	1999
2JZ-GE	3.0 (2997)	183	6	SFI	DOHC	Toyota		Y	2000
3UZ-FE	4.3 (4264)	262	8	SFI	DOHC	Toyota		1	2001
								2	2002

SFI: Sequential Multi-port Fuel Injection

DOHC: Double Overhead Camshaft

① Located on the timing belt cover

② 10th digit of the VIN

93471CE6

GENERAL ENGINE SPECIFICATIONS

All measurements are given in inches.

Year	Model	Engine Displacement Liters (cc)	Engine ID/VIN	Fuel System Type	Net Horsepower @ rpm	Net Torque @ rpm (ft. lbs.)	Bore x Stroke (in.)	Compression Ratio	Oil Pressure @ rpm
1998	ES 300	3.0 (2995)	1MZ-FE	SFI	188@5200	203@4400	3.44x3.27	10.5:1	43-78@3000
	GS 300	3.0 (2997)	2JZ-GE	SFI	220@5800	210@4800	3.39x3.39	10.0:1	47-84@3000
	GS 400	4.0 (3969)	1UZ-FE	SFI	260@5300	270@4500	3.45x3.25	10.4:1	43-85@3000
	LS 400	4.0 (3969)	1UZ-FE	SFI	260@5300	270@4500	3.45x3.25	10.4:1	43-85@3000
	SC 300	3.0 (2997)	2JZ-GE	SFI	220@5800	210@4800	3.39x3.39	10.0:1	47-84@3000
	SC 400	4.0 (3969)	1UZ-FE	SFI	260@5300	270@4500	3.45x3.25	10.4:1	43-85@3000
1999	ES 300	3.0 (2995)	1MZ-FE	SFI	210@5800	220@4400	3.44x3.27	10.5:1	43-78@3000
	GS 300	3.0 (2997)	2JZ-GE	SFI	225@6000	220@4000	3.39x3.39	10.5:1	47-84@3000
	GS 400	4.0 (3969)	1UZ-FE	SFI	300@6000	310@4000	3.44x3.25	10.5:1	43-85@3000
	LS 400	4.0 (3969)	1UZ-FE	SFI	290@6000	300@4000	3.44x3.25	10.5:1	43-85@3000
	SC 300	3.0 (2997)	2JZ-GE	SFI	225@6000	220@4000	3.39x3.39	10.5:1	47-84@3000
	SC 400	4.0 (3969)	1UZ-FE	SFI	260@5300	270@4500	3.45x3.25	10.4:1	43-85@3000
2000	ES 300	3.0 (2995)	1MZ-FE	SFI	188@5200	203@4400	3.44x3.27	10.5:1	43-78@3000
	GS 300	3.0 (2997)	2JZ-GE	SFI	220@5800	210@4800	3.39x3.39	10.0:1	47-84@3000
	GS 400	4.0 (3969)	1UZ-FE	SFI	260@5300	270@4500	3.45x3.25	10.4:1	43-85@3000
	LS 400	4.0 (3969)	1UZ-FE	SFI	260@5300	270@4500	3.45x3.25	10.4:1	43-85@3000
	SC 300	3.0 (2997)	2JZ-GE	SFI	220@5800	210@4800	3.39x3.39	10.0:1	47-84@3000
	SC 400	4.0 (3969)	1UZ-FE	SFI	290@5300	270@4500	3.45x3.25	10.4:1	43-85@3000
2001	ES 300	3.0 (2995)	1MZ-FE	SFI	210@5200	220@4400	3.44x3.27	10.5:1	43-78@3000
	GS 300	3.0 (2997)	2JZ-GE	SFI	220@5800	220@3800	3.39x3.39	10.0:1	47-84@3000
	GS 430	4.3 (4264)	3UZ-FE	SFI	300@5300	325@4500	3.58x3.25	10.4:1	43-85@3000
	IS 300	3.0 (2997)	2JZ-GE	SFI	215@5800	218@3800	3.39x3.39	10.0:1	47-84@3000
	LS 430	4.3 (4264)	3UZ-FE	SFI	290@5300	320@4500	3.58x3.25	10.4:1	43-85@3000

SFI : Sequential fuel injection

93471CE7

For complete Engine Mechanical specifications, see Section 1 of this manual

ENGINE TUNE-UP SPECIFICATIONS

Year	Engine Displacement Liters (cc)	Engine ID/VIN	Spark Plug Gap (in.)	Ignition Timing (deg.)	Fuel Pump (psi)	Idle Speed (rpm)	Valve Clearance	
							Intake	Exhaust
1998	3.0 (2995)	1MZ-FE	0.043	8-12B①	44-50	650-750	0.006-0.010	0.010-0.014
	3.0 (2997)	2JZ-GE	0.043	8-12B②	44-50	650-750	0.006-0.010	0.010-0.014
	4.0 (3969)	1UZ-FE	0.043	8-12B②	44-50	700-800	0.006-0.010	0.010-0.014
1999	3.0 (2995)	1MZ-FE	0.043	8-12B①	44-50	650-750	0.006-0.010	0.010-0.014
	3.0 (2997)	2JZ-GE	0.043	8-12B②	44-50	650-750	0.006-0.010	0.010-0.014
	4.0 (3969)	1UZ-FE	0.043	8-12B②	44-50	700-800	0.006-0.010	0.010-0.014
2000	3.0 (2995)	1MZ-FE	0.043	8-12B①	44-50	650-750	0.006-0.010	0.010-0.014
	3.0 (2997)	2JZ-GE	0.043	8-12B②	44-50	650-750	0.006-0.010	0.010-0.014
	4.0 (3969)	1UZ-FE	0.043	8-12B②	44-50	700-800	0.006-0.010	0.010-0.014
2001	3.0 (2995)	1MZ-FE	0.043	8-12B①	44-50	650-750	0.006-0.010	0.010-0.014
	3.0 (2997)	2JZ-GE	0.043	8-12B②	44-50	650-750	0.006-0.010	0.010-0.014
	4.3 (4264)	3UZ-FE	0.043	8-12B③	44-50	700-800	0.006-0.010	0.010-0.014

NOTE: The Vehicle Emission Control Information label often reflects specification changes made during production. The label figures must be used if they differ from those in this chart.

B: Before top dead center

① Terminals TE1 and E1 of check connector must be connected

② Terminals TC and E1 of check connector must be connected

③ LS 430: Terminals TC and CG of check connector must be connected

    GS 430/300: Terminals TC and E1 of check connector must be connected

93471CE8

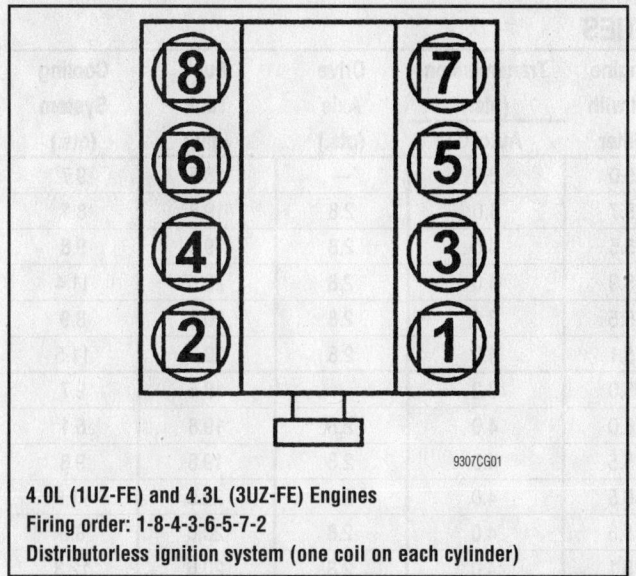

4.0L (1UZ-FE) and 4.3L (3UZ-FE) Engines
Firing order: 1-8-4-3-6-5-7-2
Distributorless ignition system (one coil on each cylinder)

9307CG01

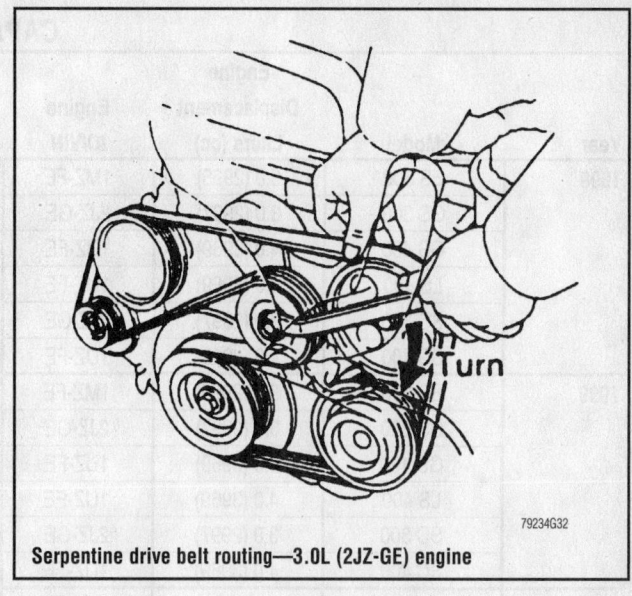

Serpentine drive belt routing—3.0L (2JZ-GE) engine

79234G32

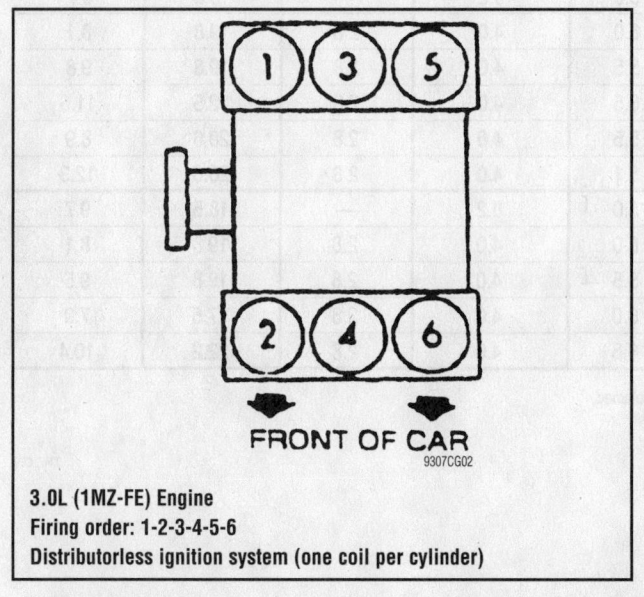

FRONT OF CAR

9307CG02

3.0L (1MZ-FE) Engine
Firing order: 1-2-3-4-5-6
Distributorless ignition system (one coil per cylinder)

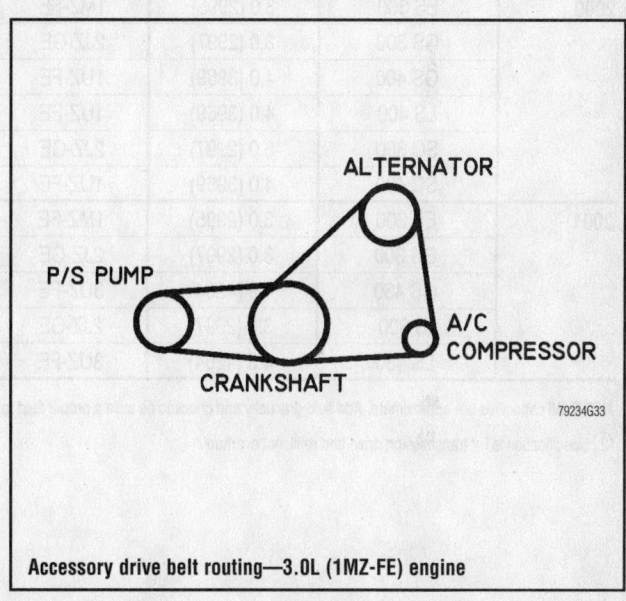

ALTERNATOR

P/S PUMP

A/C COMPRESSOR

CRANKSHAFT

79234G33

Accessory drive belt routing—3.0L (1MZ-FE) engine

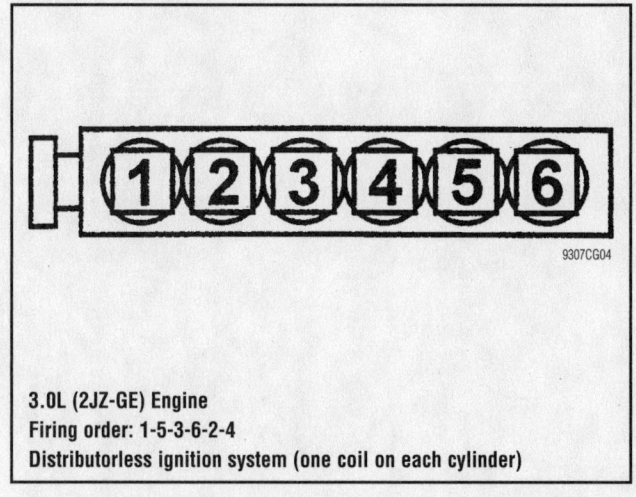

9307CG04

3.0L (2JZ-GE) Engine
Firing order: 1-5-3-6-2-4
Distributorless ignition system (one coil on each cylinder)

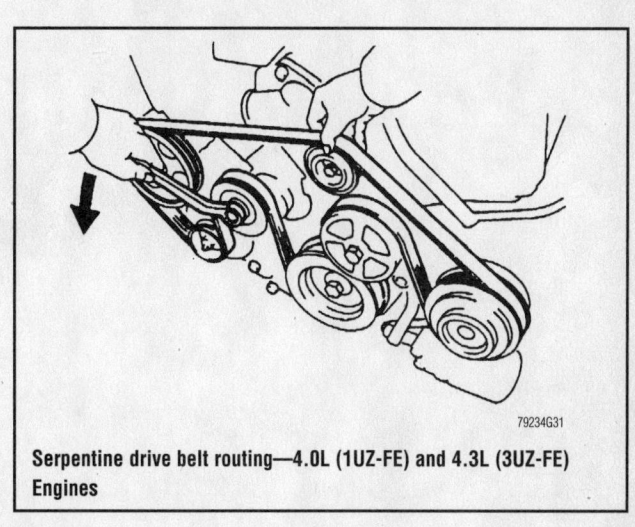

79234G31

Serpentine drive belt routing—4.0L (1UZ-FE) and 4.3L (3UZ-FE)
Engines

For Accessory Drive Belt illustrations, see Section 1 of this manual

CAPACITIES

Year	Model	Engine Displacement Liters (cc)	Engine ID/VIN	Engine Oil with Filter	Transmission (pts.) Auto. ①	Drive Axle (pts.)	Fuel Tank (gal.)	Cooling System (qts.)
1998	ES 300	3.0 (2995)	1MZ-FE	5.0	8.2	—	18.5	9.7
	GS 300	3.0 (2997)	2JZ-GE	5.7	4.0	2.8	19.8	8.1
	GS 400	4.0 (3969)	1UZ-FE	5.5	4.0	2.8	19.8	9.8
	LS 400	4.0 (3969)	1UZ-FE	5.9	4.0	2.8	22.5	11.4
	SC 300	3.0 (2997)	2JZ-GE	5.5	3.4	2.8	20.6	8.9
	SC 400	4.0 (3969)	1UZ-FE	5.1	4.0	2.8	20.6	11.5
1999	ES 300	3.0 (2995)	1MZ-FE	5.0	8.2	—	18.5	9.7
	GS 300	3.0 (2997)	2JZ-GE	6.0	4.0	2.8	19.8	8.1
	GS 400	4.0 (3969)	1UZ-FE	5.5	4.0	2.8	19.8	9.8
	LS 400	4.0 (3969)	1UZ-FE	6.5	4.0	2.8	22.5	11.6
	SC 300	3.0 (2997)	2JZ-GE	5.5	4.0	2.8	20.6	8.9
	SC 400	4.0 (3969)	1UZ-FE	5.1	4.0	2.8	20.6	12.3
2000	ES 300	3.0 (2995)	1MZ-FE	5.0	8.2	—	18.5	9.7
	GS 300	3.0 (2997)	2JZ-GE	6.0	4.0	2.8	19.8	8.1
	GS 400	4.0 (3969)	1UZ-FE	5.5	4.0	2.8	19.8	9.8
	LS 400	4.0 (3969)	1UZ-FE	6.5	4.0	2.8	22.5	11.6
	SC 300	3.0 (2997)	2JZ-GE	5.5	4.0	2.8	20.6	8.9
	SC 400	4.0 (3969)	1UZ-FE	5.1	4.0	2.8	20.6	12.3
2001	ES 300	3.0 (2995)	1MZ-FE	5.0	8.2	—	18.5	9.7
	GS 300	3.0 (2997)	2JZ-GE	6.0	4.0	2.8	19.8	8.1
	GS 430	4.3 (4264)	3UZ-FE	5.5	4.0	2.8	19.8	9.5
	IS 300	3.0 (2997)	2JZ-GE	6.0	4.0	2.8	17.5	7.9
	LS 430	4.3 (4264)	3UZ-FE	6.5	4.0	2.8	22.2	10.4

NOTE: All capacities are approximate. Add fluid gradually and check to be sure a proper fluid level is obtained.

① Specification is for transmission drain and refill, not overhaul.

93471CE9

VALVE SPECIFICATIONS

Year	Engine Displacement Liters (cc)	Engine ID/VIN	Seat Angle (deg.)	Face Angle (deg.)	Spring Test Pressure (lbs. @ in.)	Spring Free-Length (in.)	Stem-to-Guide Clearance (in.)		Stem Diameter (in.)	
							Intake	Exhaust	Intake	Exhaust
1998	3.0 (2995)	1MZ-FE	NA	44.5	41.9-46.3@ 1.331	1.791	0.0010- 0.0024	0.0012- 0.0026	0.2154- 0.2159	0.2152- 0.2157
	3.0 (2997)	2JZ-GE	NA	44.5	41.9-46.3@ 1.358	①	0.0010- 0.0024	0.0012- 0.0026	0.2350- 0.2356	0.2348- 0.2354
	4.0 (3969)	1UZ-FE	NA	44.5	②	2.130	0.0010- 0.0024	0.0012- 0.0026	0.2154- 0.2159	0.2152- 0.2157
1999	3.0 (2995)	1MZ-FE	NA	44.5	41.9-46.3@ 1.331	1.791	0.0010- 0.0024	0.0012- 0.0026	0.2154- 0.2159	0.2152- 0.2157
	3.0 (2997)	2JZ-GE	NA	44.5	41.9-46.3@ 1.358	①	0.0010- 0.0024	0.0012- 0.0026	0.2350- 0.2356	0.2348- 0.2354
	4.0 (3969)	1UZ-FE	NA	44.5	②	2.130	0.0010- 0.0024	0.0012- 0.0026	0.2154- 0.2159	0.2152- 0.2157
2000	3.0 (2995)	1MZ-FE	NA	44.5	41.9-46.3@ 1.331	1.791	0.0010- 0.0024	0.0012- 0.0026	0.2154- 0.2159	0.2152- 0.2157
	3.0 (2997)	2JZ-GE	NA	44.5	41.9-46.3@ 1.358	①	0.0010- 0.0024	0.0012- 0.0026	0.2350- 0.2356	0.2348- 0.2354
	4.0 (3969)	1UZ-FE	NA	44.5	②	2.130	0.0010- 0.0024	0.0012- 0.0026	0.2154- 0.2159	0.2152- 0.2157
2001	3.0 (2995)	1MZ-FE	NA	44.5	41.9-46.3@ 1.331	1.791	0.0010- 0.0024	0.0012- 0.0026	0.2154- 0.2159	0.2152- 0.2157
	3.0 (2997)	2JZ-GE	NA	44.5	41.9-46.3@ 1.358	①	0.0010- 0.0024	0.0012- 0.0026	0.2350- 0.2356	0.2348- 0.2354
	4.3 (4264)	3UZ-FE	45	44.5	45.9-50.7@ 1.3795	2.130	0.0010- 0.0024	0.0012- 0.0026	0.2154- 0.2159	0.2152- 0.2157

NA: Not Available

① Pink: 1.7209
  Yellow: 1.7362

② GS400: 45.9-50.7@1.3795
  SC400 & LS400: 45.9-50.7@1.378

93471CE0

For Tire, Wheel and Ball Joint specifications, see Section 1 of this manual

CRANKSHAFT AND CONNECTING ROD SPECIFICATIONS

All measurements are given in inches.

Year	Engine Displacement Liters (cc)	Engine ID/VIN	Crankshaft				Connecting Rod		
			Main Brg. Journal Dia.	Main Brg. Oil Clearance	Shaft End-play	Thrust on No.	Journal Diameter	Oil Clearance	Side Clearance
1998	3.0 (2995)	1MZ-FE	2.4011	①	0.0016-0.0095	2	2.0863-2.0866	0.0015-0.0025	0.0059-0.0118
	3.0 (2997)	2JZ-GE	2.4403-2.4409	0.0010-0.0016	0.0008-0.0087	4	2.0465-2.0472	0.0009-0.0016	0.0098-0.0158
	4.0 (3969)	1UZ-FE	2.6373-2.6378	②	0.0008-0.0087	3	2.0465-2.0472	0.0011-0.0021	0.0063-0.0138
1999	3.0 (2995)	1MZ-FE	2.4011	①	0.0016-0.0095	2	2.0863-2.0866	0.0015-0.0025	0.0059-0.0118
	3.0 (2997)	2JZ-GE	2.4403-2.4409	0.0010-0.0016	0.0008-0.0087	4	2.0465-2.0472	0.0009-0.0016	0.0098-0.0158
	4.0 (3969)	1UZ-FE	2.6373-2.6378	②	0.0008-0.0087	3	2.0465-2.0472	0.0011-0.0021	0.0063-0.0138
2000	3.0 (2995)	1MZ-FE	2.4011	①	0.0016-0.0095	2	2.0863-2.0866	0.0015-0.0025	0.0059-0.0118
	3.0 (2997)	2JZ-GE	2.4403-2.4409	0.0010-0.0016	0.0008-0.0087	4	2.0465-2.0472	0.0009-0.0016	0.0098-0.0158
	4.0 (3969)	1UZ-FE	2.6373-2.6378	②	0.0008-0.0087	3	2.0465-2.0472	0.0011-0.0021	0.0063-0.0138
2001	3.0 (2995)	1MZ-FE	2.4011	①	0.0016-0.0095	2	2.0863-2.0866	0.0015-0.0025	0.0059-0.0118
	3.0 (2997)	2JZ-GE	2.4403-2.4409	0.0010-0.0016	0.0008-0.0087	4	2.0465-2.0472	0.0009-0.0016	0.0098-0.0158
	4.3 (4264)	3UZ-FE	2.6373-2.6378	②	0.0008-0.0087	3	2.0465-2.0472	0.0008-0.0019	0.0063-0.0138

① Journal No. 1 and 4: 0.0006 - 0.0013 inch
   Journal No. 2 and 3: 0.0010 - 0.0018 inch

② Journal No. 1 and 5: 0.0007 - 0.0013 inch
   Remaning journals: 0.0011 - 0.0018 inch

93471CF1

PISTON AND RING SPECIFICATIONS

All measurements are given in inches.

Year	Engine Displacement Liters (cc)	Engine ID/VIN	Piston Clearance	Ring Gap			Ring Side Clearance		
				Top Compression	Bottom Compression	Oil Control	Top Compression	Bottom Compression	Oil Control
1998	3.0 (2995)	1MZ-FE	0.0033-0.0042	0.0098-0.0138	0.0138-0.0177	0.0059-0.0157	0.0008-0.0028	0.0008-0.0024	SNUG
	3.0 (2997)	2JZ-GE	0.0014-0.0027	0.0118-0.0185	0.0138-0.0205	0.0051-0.0177	0.0004-0.0028	0.0012-0.0028	SNUG
	4.0 (3969)	1UZ-FE	0.0033-0.0041	0.0098-0.0177	0.0197-0.0276	0.0059-0.0197	0.0008-0.0028	0.0004-0.0020	SNUG
1999	3.0 (2995)	1MZ-FE	0.0033-0.0042	0.0098-0.0138	0.0138-0.0177	0.0059-0.0157	0.0008-0.0028	0.0008-0.0024	SNUG
	3.0 (2997)	2JZ-GE	0.0014-0.0027	0.0118-0.0185	0.0138-0.0205	0.0051-0.0177	0.0004-0.0028	0.0012-0.0028	SNUG
	4.0 (3969)	1UZ-FE	0.0033-0.0041	0.0098-0.0177	0.0197-0.0276	0.0059-0.0197	0.0008-0.0028	0.0004-0.0020	SNUG
2000	3.0 (2995)	1MZ-FE	0.0033-0.0042	0.0098-0.0138	0.0138-0.0177	0.0059-0.0157	0.0008-0.0028	0.0008-0.0024	SNUG
	3.0 (2997)	2JZ-GE	0.0014-0.0027	0.0118-0.0185	0.0138-0.0205	0.0051-0.0177	0.0004-0.0028	0.0012-0.0028	SNUG
	4.0 (3969)	1UZ-FE	0.0033-0.0041	0.0098-0.0177	0.0197-0.0276	0.0059-0.0197	0.0008-0.0028	0.0004-0.0020	SNUG
2001	3.0 (2995)	1MZ-FE	0.0033-0.0042	0.0098-0.0138	0.0138-0.0177	0.0059-0.0157	0.0008-0.0028	0.0008-0.0024	SNUG
	3.0 (2997)	2JZ-GE	0.0014-0.0027	0.0118-0.0185	0.0138-0.0205	0.0051-0.0177	0.0004-0.0028	0.0012-0.0028	SNUG
	4.3 (4264)	3UZ-FE	0.0033-0.0041	0.0118-0.0157	0.0157-0.0197	0.0059-0.0157	0.0012-0.0031	0.0008-0.0024	SNUG

93471CF2

For Wheel Alignment specifications, see Section 1 of this manual

TORQUE SPECIFICATIONS
All readings in ft. lbs.

Year	Engine Displacement Liters (cc)	Engine ID/VIN	Cylinder Head Bolts	Main Bearing Bolts	Rod Bearing Bolts	Crankshaft Damper Bolts	Flywheel Bolts	Manifold		Spark Plugs	Lug Nuts
								Intake	Exhaust		
1998	3.0 (2995)	1MZ-FE	①	②	③	159	61	11	36	13	76
	3.0 (2997)	2JZ-GE	④	⑤	⑥	239	61	20	29	13	76
	4.0 (3969)	1UZ-FE	⑤	⑥	③	181	61	13	32	13	76
1999	3.0 (2995)	1MZ-FE	①	②	⑦	159	61	11	36	13	76
	3.0 (2997)	2JZ-GE	④	⑤	⑥	243	61	21	30	13	76
	4.0 (3969)	1UZ-FE	③	⑧	③	181	61	13	32	13	76
2000	3.0 (2995)	1MZ-FE	①	②	⑦	159	61	11	36	13	76
	3.0 (2997)	2JZ-GE	④	⑤	⑥	243	61	21	30	13	76
	4.0 (3969)	1UZ-FE	③	⑧	②	181	61	13	32	13	76
2001	3.0 (2995)	1MZ-FE	①	②	⑦	159	61	11	36	13	76
	3.0 (2997)	2JZ-GE	④	⑤	⑥	243	61	21	30	13	76
	4.3 (4264)	3UZ-FE	⑨	⑧	⑦	181	⑩	13	32	13	76

① Head bolt:
Step 1: 40 ft. lbs.
Step 2: Plus 90 degrees
Recessed head bolt: 13 ft. lbs.

② 6-point bolts: 20 ft. lbs.
12-point bolts:
Step 1: 16 ft. lbs.
Step 2: Plus an additional 90 degrees

③ Step 1: Tighten to 29 ft. lbs.
Step 2: Plus 90 degrees

④ Step 1: 25 ft. lbs.
Step 2: Tighten an additional 90 degrees
Step 3: Tighten an additional 90 degrees

⑤ Step 1: 33 ft. lbs.
Step 2: Plus 90 degrees

⑥ Step 1: 22 ft. lbs.
Step 2: Plus 90 degrees

⑦ Step 1: 18 ft. lbs.
Step 2: Plus 90 degrees

⑧ Nuts:
Step 1: 20 ft. lbs.
Step 2: Plus 90 degrees
Bolts: 36 ft. lbs.

⑨ Step 1: 44 ft. lbs.
Step 2: Plus 90 degrees

⑩ Step 1: 36 ft. lbs.
Step 2: Plus 90 degrees

93471CF3

BRAKE SPECIFICATIONS
All measurements in inches unless noted

| Year | Model | Front Brake Disc | | | Rear Brake Disc | | | Minimum Lining Thickness | Brake Caliper | |
		Original Thickness	Minimum Thickness	Maximum Run-out	Original Thickness	Minimum Thickness	Maximum Run-out		Bracket Bolts (ft. lbs.)	Mounting Bolts (ft. lbs.)
1998	ES 300	1.102	1.024	0.0020	0.394	0.354	0.0059	0.0390	①	②
	GS 300	1.260	1.181	0.0020	0.472	0.413	0.0020	0.0390	③	25
	GS 400	1.260	1.181	0.0020	0.472	0.413	0.0020	0.0390	③	25
	LS 400	1.102	1.024	0.0020	0.630	0.591	0.0020	0.1180	③	25
	SC 300	1.102	1.024	0.0020	0.630	0.591	0.0020	0.0390	③	25
	SC 400	1.260	1.181	0.0020	0.630	0.591	0.0020	0.0390	③	25
1999	ES 300	1.102	1.024	0.0020	0.394	0.354	0.0059	0.0390	①	②
	GS 300	1.260	1.181	0.0020	0.472	0.413	0.0020	0.0390	③	25
	GS 400	1.260	1.181	0.0020	0.472	0.413	0.0020	0.0390	③	25
	LS 400	1.102	1.024	0.0020	0.630	0.591	0.0020	0.1180	③	25
	SC 300	1.102	1.024	0.0020	0.630	0.591	0.0020	0.0390	③	25
	SC 400	1.260	1.181	0.0020	0.630	0.591	0.0020	0.0390	③	25
2000	ES 300	1.102	1.024	0.0020	0.394	0.354	0.0059	0.0390	①	②
	GS 300	1.260	1.181	0.0020	0.472	0.413	0.0020	0.0390	③	25
	GS 400	1.260	1.181	0.0020	0.472	0.413	0.0020	0.0390	③	25
	LS 400	1.102	1.024	0.0020	0.630	0.591	0.0020	0.1180	③	25
	SC 300	1.102	1.024	0.0020	0.630	0.591	0.0020	0.0390	③	25
	SC 400	1.260	1.181	0.0020	0.630	0.591	0.0020	0.0390	③	25
2001	ES 300	1.102	1.024	0.0020	0.394	0.354	0.0059	0.0390	①	②
	GS 300	1.260	1.181	0.0020	0.472	0.413	0.0020	0.0390	③	25
	GS 430	1.260	1.181	0.0020	0.472	0.413	0.0020	0.0390	③	25
	IS 300	1.260	1.181	0.0020	0.472	0.413	0.0020	0.0390	③	25
	LS 430	1.181	1.102	0.0020	0.630	0.571	0.0020	0.0390	—	④

F: Front

R: Rear

① Front: 79 ft. lbs.
  Rear: 34 ft. lbs.

② Front: 25 ft. lbs.
  Rear: 14 ft. lbs.

③ Front: 87 ft. lbs.
  Rear: 77 ft. lbs.

④ Front: 81 ft. lbs.
  Rear: 58 ft. lbs.

93471CF4

For Maintenance Interval recommendations, see Section 1 of this manual

WHEEL ALIGNMENT

Year	Model		Caster Range (+/-Deg.)	Caster Preferred Setting (Deg.)	Camber Range (+/-Deg.)	Camber Preferred Setting (Deg.)	Toe-in (in.)	Steering Axis Inclination (Deg.)
1998	ES 300	F	0.75	+2.30	0.75	-0.62	0 +/- 0.08	13.06
		R	—	—	0.75	-0.80	0.16 +/- 0.08	—
	GS 300	F	0.50	+7.55	0.50	-0.27	0.06 +/- 0.08	8.83
		R	—	—	0.50	-0.78	0.06 +/- 0.08	—
	LS 400 ①	F	0.75	+7.00	0.75	+0.34	0.12 +/- 0.08	8.42
		R	—	—	0.75	-0.83	0.09 +/- 0.09	—
	LS 400 ②	F	0.75	+7.42	0.75	+0.08	0.04 +/- 0.08	8.75
		R	—	—	0.75	-1.42	0.12 +/- 0.08	—
	SC 300	F	0.75	+3.01	0.75	+2.00	0.04 +/- 0.08	9.03
		R	—	—	0.75	+1.08	0.08 +/- 0.24	—
	SC 400	F	0.75	+3.00	0.75	+2.00	0.04 +/- 0.08	9.03
		R	—	—	0.75	+1.08	0.08 +/- 0.24	—
1999	ES 300	F	0.75	+2.30	0.75	-0.62	0 +/- 0.08	13.06
		R	—	—	0.75	-0.80	0.16 +/- 0.08	—
	GS 300	F	0.50	+7.55	0.50	-0.27	0.06 +/- 0.08	8.83
		R	—	—	0.50	-0.78	0.06 +/- 0.08	—
	LS 400 ①	F	0.75	+7.00	0.75	+0.34	0.12 +/- 0.08	8.42
		R	—	—	0.75	-0.83	0.09 +/- 0.09	—
	LS 400 ②	F	0.75	+7.42	0.75	+0.08	0.04 +/- 0.08	8.75
		R	—	—	0.75	-1.42	0.12 +/- 0.08	—
	SC 300	F	0.75	+3.01	0.75	+2.00	0.04 +/- 0.08	9.03
		R	—	—	0.75	+1.08	0.08 +/- 0.24	—
	SC 400	F	0.75	+3.00	0.75	+2.00	0.04 +/- 0.08	9.03
		R	—	—	0.75	+1.08	0.08 +/- 0.24	—
2000	ES 300	F	0.75	+2.30	0.75	-0.62	0 +/- 0.08	13.06
		R	—	—	0.75	-0.80	0.16 +/- 0.08	—
	GS 300	F	0.50	+7.55	0.50	-0.27	0.06 +/- 0.08	8.83
		R	—	—	0.50	-0.78	0.06 +/- 0.08	—
	LS 400 ①	F	0.75	+7.00	0.75	+0.34	0.12 +/- 0.08	8.42
		R	—	—	0.75	-0.83	0.09 +/- 0.09	—
	LS 400 ②	F	0.75	+7.42	0.75	+0.08	0.04 +/- 0.08	8.75
		R	—	—	0.75	-1.42	0.12 +/- 0.08	—
	SC 300	F	0.75	+3.01	0.75	+2.00	0.04 +/- 0.08	9.03
		R	—	—	0.75	+1.08	0.08 +/- 0.24	—
	SC 400	F	0.75	+3.00	0.75	+2.00	0.04 +/- 0.08	9.03
		R	—	—	0.75	+1.08	0.08 +/- 0.24	—
2001	ES 300	F	0.75	+2.30	0.75	-0.62	0 +/- 0.08	13.06
		R	—	—	0.75	-0.80	0.16 +/- 0.08	—
	GS 300	F	0.50	+7.55	0.50	-0.27	0.06 +/- 0.08	8.83
		R	—	—	0.50	-0.78	0.06 +/- 0.08	—
	GS 430	F	0.50	+7.55	0.50	-0.27	0.06 +/- 0.08	8.83
		R	—	—	0.50	-0.78	0.06 +/- 0.08	—
	IS 300	F	0.50	+6.12	0.50	-0.50	0.04 +/- 0.08	9.42
		R	—	—	0.50	-0.07	0.08 +/- 0.08	—
	LS 430 ①	F	0.75	+6.75	0.75	-0.08	0.04 +/- 0.08	9.00
		R	—	—	0.75	-1.00	0.12 +/- 0.08	—
	LS 430 ②	F	0.75	+7.25	0.75	-0.25	0.04 +/- 0.08	9.25
		R	—	—	0.75	-1.55	0.12 +/- 0.08	—

① Except with air suspension

② With air suspension

TIRE, WHEEL AND BALL JOINT SPECIFICATIONS

Year	Model	OEM Tires		Tire Pressures (psi)		Wheel Size	Ball Joint Inspection
		Standard	Optional	Front	Rear		
1998	GS 400	225/55VR16	235/45ZR17	Std: 32 Opt: 33	Std: 32 Opt: 33	Std: 7.5-JJ Opt: 8-JJ	U: 9-30 in. ①
	SC 300	225/55VR16	None	32	32	6.5-JJ	U: 9-30 in. ①
	SC 400	225/55VR16	None	32	32	6.5-JJ	U: 9-30 in. ①
	GS 300	P215/60VR16	225/55VR16	30	30	7.5-JJ	U: 9-30 in. ①
	LS 400	P225/60VR16	None	29	29	7-JJ	U: 9-30 in. ①
	ES 300	P205/65VR15	None	26	26	6-JJ	U: 9-30 in. ①
1999	GS 400	225/55VR16	235/45ZR17	Std: 32 Opt: 33	Std: 32 Opt: 33	Std: 7.5-JJ Opt: 8-JJ	U: 9-30 in. ①
	SC 300	225/55VR16	None	32	32	6.5-JJ	U: 9-30 in. ①
	SC 400	225/55VR16	None	32	32	6.5-JJ	U: 9-30 in. ①
	GS 300	P215/60VR16	225/55VR16	30	30	7.5-JJ	U: 9-30 in. ①
	LS 400	P225/60VR16	None	29	29	7-JJ	U: 9-30 in. ①
	ES 300	P205/65VR15	None	26	26	6-JJ	U: 9-30 in. ①
2000	GS 400	225/55VR16	235/45ZR17	Std: 32 Opt: 33	Std: 32 Opt: 33	Std: 7.5-JJ Opt: 8-JJ	U: 9-30 in. ①
	SC 300	225/55VR16	None	32	32	6.5-JJ	U: 9-30 in. ①
	SC 400	225/55VR16	None	32	32	6.5-JJ	U: 9-30 in. ①
	GS 300	P215/60VR16	225/55VR16	30	30	7.5-JJ	U: 9-30 in. ①
	LS 400	P225/60VR16	None	29	29	7-JJ	U: 9-30 in. ①
	ES 300	P205/65VR15	None	26	26	6-JJ	U: 9-30 in. ①
2001	GS 430	225/55VR16	235/45ZR17	Std: 32 Opt: 33	Std: 32 Opt: 33	Std: 7.5-JJ Opt: 8-JJ	U: 9-30 in. ①
	GS 300	P215/60VR16	225/55VR16	30	30	7.5-JJ	U: 9-30 in. ①
	LS 4300	P225/60VR16		29	29	7-JJ	U: 9-30 in. ①
			P225/55R17 95H	32	35		
			225/55R17 97W	35	36		
	IS 300	P205/55R16 89V	215/45ZR17	33	33	NA	U: 9-30 in. ①
	ES 300	P205/65VR15	None	26	26	6-JJ	U: 9-30 in. ①

NA: Not Available

OEM: Original Equipment Manufacturer

PSI: Pounds Per Square Inch

STD: Standard

OPT: Optional

L: Lower

U: Upper

① Torque required in inch lbs. to rotate ball joint when removed from the knuckle

93471CF6

For Tune-up, Capacities and Firing orders, see Section 1 of this manual

SCHEDULED MAINTENANCE INTERVALS
Lexus ES300, IS300, GS300, GS400, GS430, SC300, SC400, LS400, LS430

TO BE SERVICED	TYPE OF SERVICE	VEHICLE MILEAGE INTERVAL (x1000)												
		7.5	15	22.5	30	37.5	45	52.5	60	67.5	75	82.5	90	97.5
Engine oil & filter	R	✓	✓	✓	✓	✓	✓	✓	✓	✓	✓	✓	✓	✓
Air conditioning filter (LS 400) ①	S/I	✓	✓	✓	✓	✓	✓	✓	✓	✓	✓	✓	✓	✓
Automatic transaxle fluid & filter	S/I		✓		✓		✓		✓		✓		✓	
Ball joints & dust covers	S/I		✓		✓		✓		✓		✓		✓	
Bolts & nuts on chassis & body	S/I		✓		✓		✓		✓		✓		✓	
Brake fluid ②	S/I		✓		✓		✓		✓		✓		✓	
Brake line pipes & hoses	S/I		✓		✓		✓		✓		✓		✓	
Brake linings & drums	S/I		✓		✓		✓		✓		✓		✓	
Brake pads & discs (front & rear)	S/I		✓		✓		✓		✓		✓		✓	
Differential oil	S/I		✓		✓		✓		✓		✓		✓	
Driveshaft boots (ES 300)	S/I		✓		✓		✓		✓		✓		✓	
Steering gear housing oil	S/I		✓		✓		✓		✓		✓		✓	
Steering linkage	S/I		✓		✓		✓		✓		✓		✓	
Air filter	R				✓				✓				✓	
Exhaust pipes & mountings	S/I				✓				✓				✓	
Fuel lines & connections	S/I				✓				✓				✓	
Engine coolant	R						✓					✓		
Fuel tank cap gasket	R								✓					
Spark plugs	R								✓					
Charcoal canister	S/I								✓					
Drive belts	S/I								✓					
Valve clearance	S/I								✓					

R: Replace   S/I: Service or Inspect

① Replace at 15,000 miles.

② Replace at 30,000 miles (unless previously replaced).

FREQUENT OPERATION MAINTENANCE (SEVERE SERVICE)

If a vehicle is operated under any of the following conditions it is considered severe service

- Extremely dusty areas.

- 50% or more of the vehicle operation is in 32°C (90°F) or higher temperatures, or constant operation in temperatures below 0°C (32°F).

- Prolonged idling (vehicle operation in stop and go traffic).

- Frequent short running periods (engine does not warm to normal operating temperatures).

- Police, taxi, delivery usage or trailer towing usage.

Oil & oil filter: change every 3750 miles.

Ball joints & dust covers: service or inspect every 7500 miles.

Bolts & nuts on chassis & body: service or inspect every 7500 miles.

Brake linings & drums: service or inspect every 7500 miles.

Brake pads & discs (front & rear): service or inspect every 7500 miles.

Driveshaft boots (ES 300): service or inspect every 7500 miles.

Brake linings & drums: service or inspect every 7500 miles.

Steering linkage: service or inspect every 7500 miles.

Air filter: service or inspect every 15,000 miles.

Automatic transmission fluid & filter: replace every 15,000 miles.

Differential oil: replace every 15,000 miles.

Exhaust pipes & mountings: service or inspect every 15,000 miles.

Drive belts: service or inspect at 60,000 miles & every 7500 miles thereafter.

Timing belts: replace every 60,000 miles.

93471CF7

SCHEDULED MAINTENANCE INTERVALS
LEXUS
ES300, IS300, GS300, GS400, GS430, SC300, SC400, LS400, LS430

The following should be used as a guide when determining the amount of work required for a particular service. In estimating how long a particular Scheduled Maintenance Service should take, please observe the following:

● Labor Time is time based on field research and data supplied by the vehicle manufacturer.
● Labor time operations are given in hours and tenths of an hour.
● All labor operations are to be used as a guide.

Mechanic Skill Level Codes:
(A) PRECISION: Highly skilled with multiple certification.
(B) GENERAL: Normally skilled with certification.
(C) MAINTENANCE: Semi-skilled working on certification.

	LABOR TIME
7500 Mile Service (C)	
All Models	.4
LS400/430 add	.2
15000 Mile Service (C)	
ES300, IS300, SC300	1.6
GS300/400/430	1.5
LS400/430	1.7
SC400	1.6
22500 Mile Service (C)	
All Models	.4
LS400/430 add	.2
30000 Mile Service (B)	
ES300, IS300, SC300	2.4
GS300/400/430	2.3
LS400/430	2.5
SC400	2.4

	LABOR TIME
37500 Mile Service (C)	
All Models	.4
LS400/430 add	.2
45000 Mile Service (B)	
ES300, IS300, SC300	2.1
GS300/400/430	2.0
LS400/430	2.2
SC400	2.1
52500 Mile Service (C)	
All Models	.4
LS400/430 add	.2
60000 Mile Service (B)	
ES300, IS300, SC300	3.9
GS300/400/430	3.8
LS400/430	4.2
SC400	3.9
67500 Mile Service (C)	
All Models	.4
LS400/430 add	.2

	LABOR TIME
75000 Mile Service (C)	
ES300, IS300, SC300	1.6
GS300/400/430	1.5
LS400/430	1.7
SC400	1.6
82500 Mile Service (C)	
All Models	.4
LS400/430 add	.2
90000 Mile Service (B)	
ES300, IS300, SC300	2.4
GS300/400/430	2.3
LS400/430	2.5
SC400	2.4
97500 Mile Service (C)	
All Models	.4
LS400/430 add	.2

93471CF8

For complete service labor times, order the Chilton Labor Guide

MAZDA
626 • MIATA • MILLENIA • MX6 • PROTEGE

ENGINE AND VEHICLE IDENTIFICATION

Engine							Model Year		
Code ①	Liters (cc)	Cu. In.	Cyl.	Fuel Sys.	Engine Type	Eng. Mfg.	Code ②		Year
BP	1.8 (1839)	112.2	4	MPFI	DOHC	Mazda	W		1998
FP	1.8 (1839)	112.2	4	MPFI	DOHC	Mazda	X		1999
FS	2.0 (1991)	121.5	4	MPFI	DOHC	Mazda	Y		2000
KJ	2.3 (2254)	137.2	6	MPFI	DOHC	Mazda	1		2001
KL	2.5 (2496)	152.3	6	MPFI	DOHC	Mazda	2		2002
Z5	1.5 (1489)	90.8	4	MPFI	DOHC	Mazda			
ZM	1.6 (1597)	97.4	4	MPFI	DOHC	Mazda			

MPFI: Multi-Point Fuel Injection

DOHC: Double Over Head Cam

① Located above the starter

② 10th digit of the Vehicle Identification Number (VIN)

93471CF9

GENERAL ENGINE SPECIFICATIONS

Year	Model	Engine Displacement Liters (cc)	Engine ID/VIN	Fuel System Type	Net Horsepower @ rpm	Net Torque @ rpm (ft. lbs.)	Bore x Stroke (in.)	Compression Ratio	Oil Pressure @ rpm
1998	626 DX	2.0 (1991)	FS	EFI	114@5500	124@4500	3.27x3.62	9.0:1	57-71@3000
	626 ES-V6	2.5 (2497)	KL	EFI	160@5500	156@5000	3.33x2.92	9.2:1	49-71@3000
	626 LX	2.0 (1991)	FS	EFI	114@5500	124@4500	3.27x3.62	9.0:1	57-71@3000
	626 LX-V6	2.5 (2497)	KL	EFI	160@5500	156@5000	3.33x2.92	9.2:1	49-71@3000
	Miata	1.8 (1839)	BP	EFI	133@6500	114@5500	3.27x3.35	9.0:1	43-57@3000
	Millenia	2.5 (2497)	KL	EFI	170@5800	160@4800	3.33x2.92	9.2:1	49-71@3000
	Millenia S	2.3 (2255)	KJ	EFI	210@5300	210@3500	3.16x2.92	10.0:1	44-66@3000
	Protege DX	1.5 (1489)	Z5	EFI	①	96@4000	2.96x3.29	9.4:1	43-57@3000
	Protege ES	1.8 (1839)	BP	EFI	122@6000	117@4000	3.27x3.35	9.0:1	43-57@3000
	Protege LX	1.5 (1489)	Z5	EFI	①	96@4000	2.96x3.29	9.4:1	43-57@3000
1999	626 ES	2.0 (1991)	FS	EFI	125@5500	127@3300	3.27x3.62	9.0:1	57-71@3000
	626 ES-V6	2.5 (2497)	KL	EFI	170@6000	163@5000	3.33x2.92	9.5:1	49-71@3000
	626 LX	2.0 (1991)	FS	EFI	125@5500	127@3300	3.27x3.62	9.0:1	57-71@3000
	626 LX-V6	2.5 (2497)	KL	EFI	170@6000	163@5000	3.33x2.92	9.5:1	49-71@3000
	Miata	1.8 (1839)	BP	EFI	②	③	3.27x3.35	9.5:1	43-57@3000
	Millenia	2.5 (2497)	KL	EFI	170@5800	160@4800	3.33x2.92	9.2:1	49-71@3000
	Millenia S	2.3 (2255)	KJ	EFI	210@5300	210@3500	3.16x2.92	10.0:1	44-66@3000
	Protege DX	1.6 (1597)	ZM	EFI	④	⑤	3.07x3.29	9.0:1	43-57@3000
	Protege ES	1.8 (1839)	FP	EFI	⑥	⑦	3.27x3.35	9.1:1	43-57@3000
	Protege LX	1.6 (1597)	ZM	EFI	④	⑤	3.07x3.29	9.0:1	43-57@3000
2000	626 ES	2.0 (1991)	FS	EFI	125@5500	127@3300	3.27x3.62	9.0:1	57-71@3000
	626 ES-V6	2.5 (2497)	KL	EFI	170@6000	163@5000	3.33x2.92	9.5:1	49-71@3000
	626 LX	2.0 (1991)	FS	EFI	125@5500	127@3300	3.27x3.62	9.0:1	57-71@3000
	626 LX-V6	2.5 (2497)	KL	EFI	170@6000	163@5000	3.33x2.92	9.5:1	49-71@3000
	Miata	1.8 (1839)	BP	EFI	②	③	3.27x3.35	9.5:1	43-57@3000
	Millenia	2.5 (2497)	KL	EFI	170@5800	160@4800	3.33x2.92	9.2:1	49-71@3000
	Millenia S	2.3 (2255)	KJ	EFI	210@5300	210@3500	3.16x2.92	10.0:1	44-66@3000
	Protege DX	1.6 (1597)	ZM	EFI	④	⑤	3.07x3.29	9.0:1	43-57@3000
	Protege ES	1.8 (1839)	FP	EFI	⑥	⑦	3.27x3.35	9.1:1	43-57@3000
	Protege LX	1.6 (1597)	ZM	EFI	④	⑤	3.07x3.29	9.0:1	43-57@3000

EFI: Electronic Fuel Injection

① California, New York and Massachusetts: 90@5500
Except California, New York and Massachusetts: 92@5500

② LEV states: 138@6500
Except LEV states: 140@6500

③ LEV states: 117@5000
Except LEV states: 119@5500

④ LEV states: 103@5500
Except LEV states: 105@5500

⑤ LEV states: 106@4000
Except LEV states: 107@4000

⑥ LEV states: 120@6000
Except LEV states: 122@6000

⑦ LEV states: 119@4000
Except LEV states: 120@4000

93471CF0

ENGINE TUNE-UP SPECIFICATIONS

Year	Engine Displacement Liters (cc)	Engine ID/VIN	Spark Plug Gap (in.)	Ignition Timing (deg.)		Fuel Pump (psi)	Idle Speed (rpm)		Valve Clearance	
				MT	AT		MT	AT	In.	Ex.
1998	1.5 (1489)	Z5	0.041	6-18B	6-18B	29-34	650-750	700-800	0.011	0.011
	1.8 (1839)	BP	0.041	6-18B	6-18B	39-45	700-800	700-800	HYD	HYD
	2.0 (1991)	FS	0.041	11-13B	6-18B	37-46	650-750	650-750	HYD	HYD
	2.3 (2255)	KJ	0.030	6B	6B	39-48	600-700	600-700	0.011	0.011
	2.5 (2497)	KL	0.041	9-11B	9-11B	39-45	600-700	600-700	HYD	HYD
1999	1.6 (1597)	ZM	0.040-0.043	6-18B	6-18B	30-36	650-750	650-750	0.010-0.012	0.010-0.012
	1.8 (1839)	BP	0.041	6-18B	6-18B	39-45	700-800	700-800	HYD	HYD
	1.8 (1839)	FP	0.040-0.043	6-18B	6-18B	30-36	650-750	650-750	0.009-0.012	0.009-0.012
	2.0 (1991)	FS	0.041	11-13B	6-18B	37-46	650-750	650-750	HYD	HYD
	2.3 (2255)	KJ	0.030	6B	6B	39-48	600-700	600-700	0.011	0.011
	2.5 (2497)	KL	0.041	9-11B	9-11B	39-45	600-700	600-700	HYD	HYD
2000	1.6 (1597)	ZM	0.040-0.043	6-18B	6-18B	30-36	650-750	650-750	0.010-0.012	0.010-0.012
	1.8 (1839)	BP	0.040-0.043	6-18B	6-18B	53-61	750-850	750-850	0.008-0.009	0.012-0.013
	1.8 (1839)	FP	0.040-0.043	6-18B	6-18B	30-36	650-750	650-750	0.009-0.012	0.009-0.012
	2.0 (1991)	FS	0.040-0.043	11-13B	6-18B	30-38	650-750	650-750	0.009-0.012	0.009-0.012
	2.3 (2255)	KJ	0.030	6B	6B	39-48	600-700	600-700	0.011	0.011
	2.5 (2497)	KL	①	9-11B	9-11B	30-36	600-700	600-700	②	③

NOTE: The Vehicle Emission Control Information label often reflects specification changes made during production. The label figures must be used if they differ from those in this chart.

B: Before top dead center

HYD: Hydraulic

① 626 models: 0.028-0.031
Millenia models: 0.039-0.043

② 626 models: 0.010-0.012
Millenia models: HYD

③ 626 models: 0.010-0.013
Millenia models: HYD

93471CG1

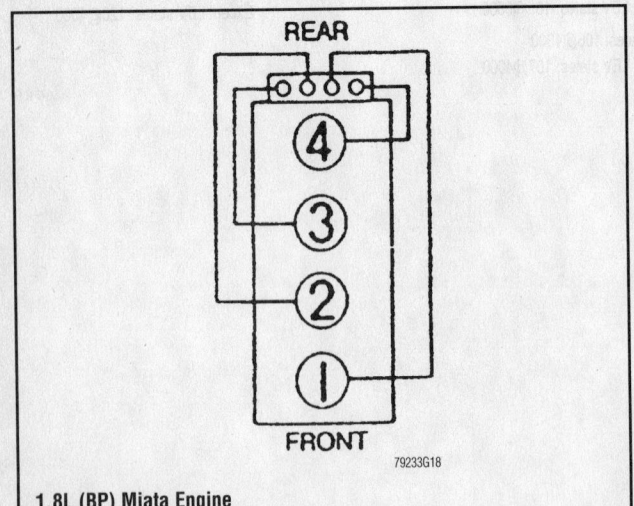

1.8L (BP) Miata Engine
Firing order: 1–3–4–2
Distributorless ignition system

79233G18

1.5L and 1.8L Non-Miata Engines
Firing order: 1–3–4–2
Distributor rotation: Counterclockwise

79233G66

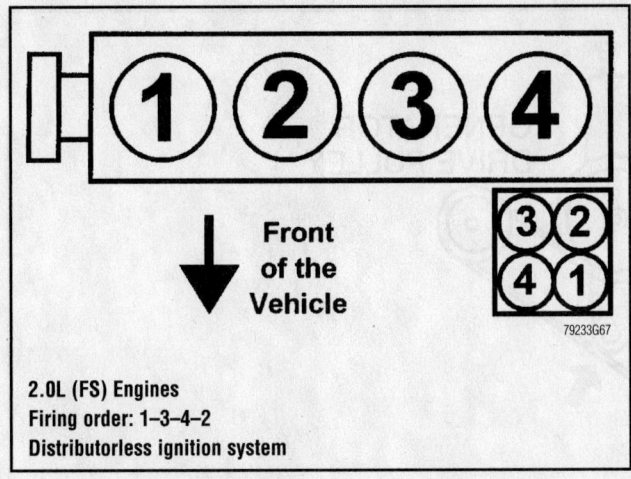

2.0L (FS) Engines
Firing order: 1–3–4–2
Distributorless ignition system

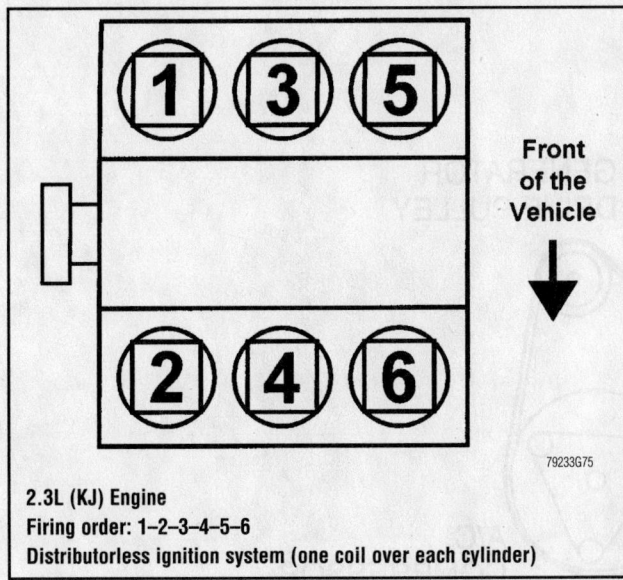

2.3L (KJ) Engine
Firing order: 1–2–3–4–5–6
Distributorless ignition system (one coil over each cylinder)

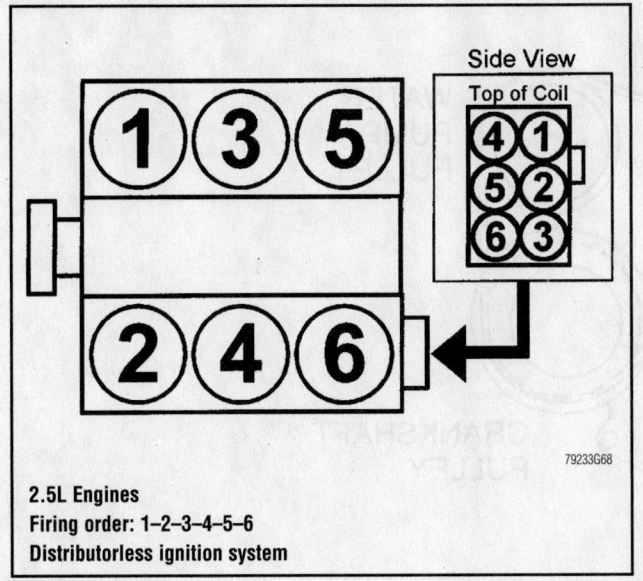

2.5L Engines
Firing order: 1–2–3–4–5–6
Distributorless ignition system

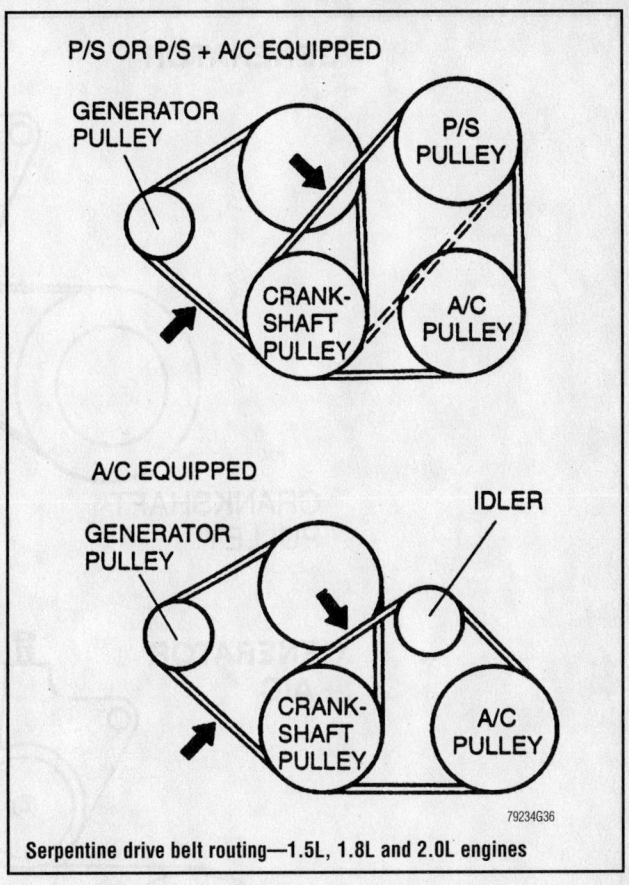

Serpentine drive belt routing—1.5L, 1.8L and 2.0L engines

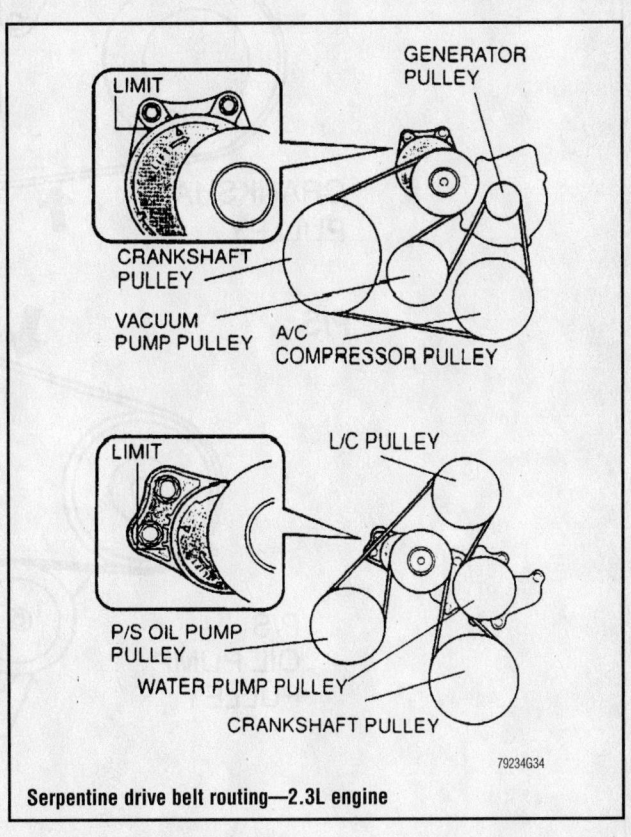

Serpentine drive belt routing—2.3L engine

Timing belt service is covered in Section 3 of this manual

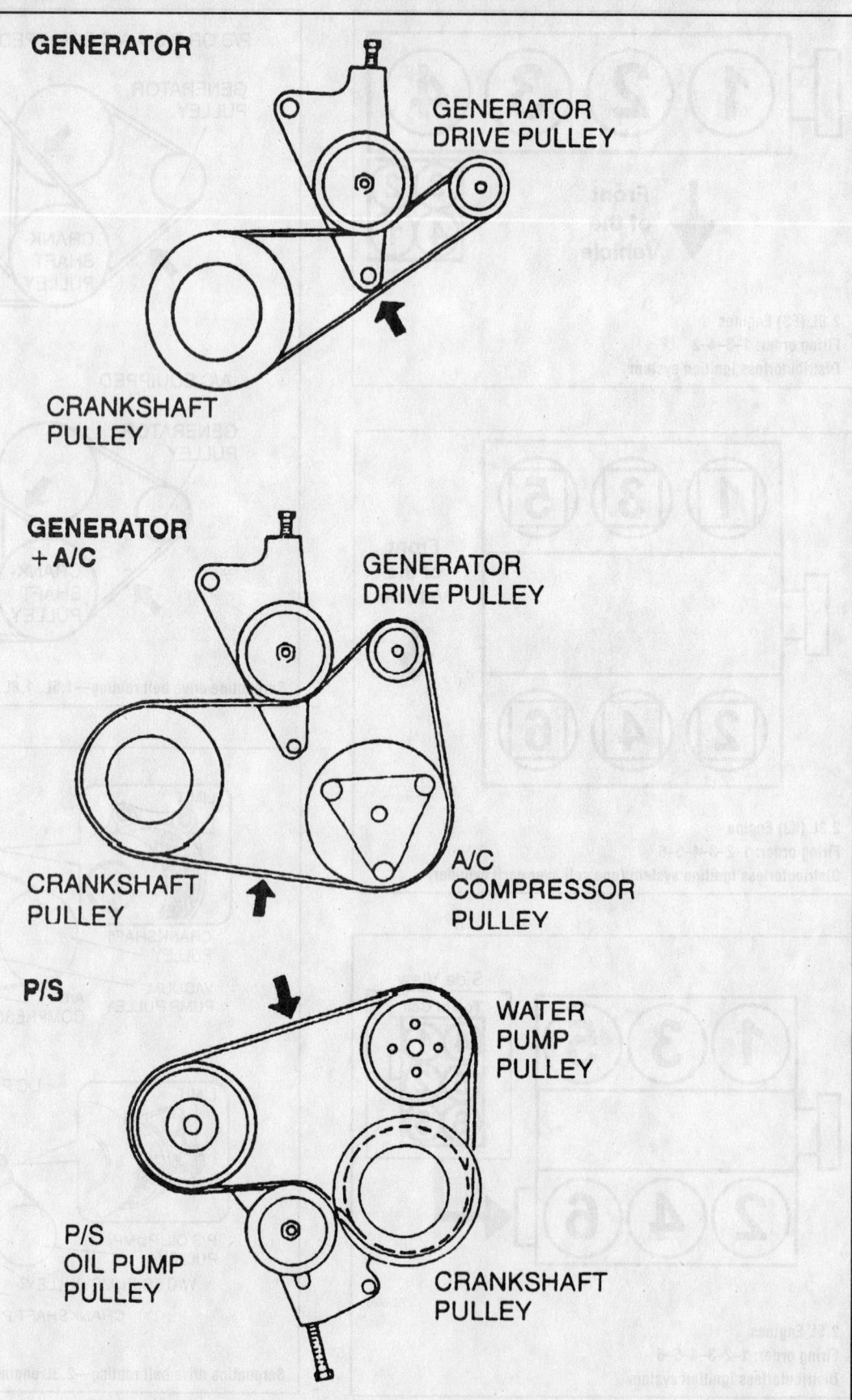

GENERATOR

GENERATOR DRIVE PULLEY

CRANKSHAFT PULLEY

GENERATOR + A/C

GENERATOR DRIVE PULLEY

CRANKSHAFT PULLEY

A/C COMPRESSOR PULLEY

P/S

WATER PUMP PULLEY

P/S OIL PUMP PULLEY

CRANKSHAFT PULLEY

79234G35

Serpentine drive belt routing—2.5L engines

CAPACITIES

Year	Model	Engine Displacement Liters (cc)	Engine ID/VIN	Engine Oil with Filter	Transmission (pts.)		Drive Axle		Fuel Tank (gal.)	Cooling System (qts.)
					5-Spd	Auto.	Front (pts.)	Rear (pts.)		
1998	626 DX	2.0 (1991)	FS	3.7	5.8	18.4	①	—	15.5	7.4
	626 ES-V6	2.5 (2497)	KL	4.2	5.8	18.6	①	—	15.5	7.9
	626 LX	2.0 (1991)	FS	3.7	5.8	18.4	①	—	15.5	7.4
	626 LX-V6	2.5 (2497)	KL	4.2	5.8	18.6	①	—	15.5	7.9
	Miata	1.8 (1839)	BP	4.0	4.2	13.5	—	2.1	12.7	6.3
	Millenia	2.5 (2497)	KL	4.2	—	16.9	①	—	18.0	7.4
	Millenia S	2.3 (2255)	KJ	4.3	—	16.9	①	—	18.0	7.4
	Protege DX	1.5 (1498)	Z5	3.7	5.6	11.3	①	—	13.2	6.3
	Protege ES	1.8 (1839)	BP	4.0	5.6	11.3	①	—	13.2	6.3
	Protege LX	1.5 (1498)	Z5	3.7	5.6	11.3	①	—	13.2	6.3
1999	626 ES	2.0 (1991)	FS	4.0	6.0	17.6	①	—	16.9	7.9
	626 ES-V6	2.5 (2497)	KL	4.2	6.0	17.0	①	—	16.9	7.9
	626 LX	2.0 (1991)	FS	4.0	6.0	17.6	①	—	16.9	7.9
	626 LX-V6	2.5 (2497)	KL	4.2	6.0	17.0	①	—	16.9	7.9
	Miata	1.8 (1839)	BP	4.0	4.2	13.5	—	2.1	12.7	6.3
	Millenia	2.5 (2497)	KL	5.2	—	16.9	①	—	18.0	7.4
	Millenia S	2.3 (2255)	KJ	5.9	—	16.9	①	—	18.0	7.4
	Protege DX	1.6 (1597)	ZM	5.0	5.6	15.2	①	—	13.2	7.9
	Protege ES	1.8 (1839)	FP	5.0	5.6	15.2	①	—	13.2	9.9
	Protege LX	1.6 (1597)	ZM	5.0	5.6	15.2	①	—	13.2	7.9
2000	626 ES	2.0 (1991)	FS	3.7	6.0	17.6	①	—	16.9	7.9
	626 ES-V6	2.5 (2497)	KL	4.2	6.0	17.0	①	—	16.9	7.9
	626 LX	2.0 (1991)	FS	3.7	6.0	17.6	①	—	16.9	7.9
	626 LX-V6	2.5 (2497)	KL	4.2	6.0	17.0	①	—	16.9	7.9
	Miata	1.8 (1839)	BP	4.0	4.2	13.5	—	2.1	12.7	6.3
	Millenia	2.5 (2497)	KL	5.2	—	16.9	①	—	18.0	7.4
	Millenia S	2.3 (2255)	KJ	5.9	—	16.9	①	—	18.0	7.4
	Protege DX	1.6 (1597)	ZM	5.0	5.6	15.2	①	—	13.2	7.9
	Protege ES	1.8 (1839)	FP	5.0	5.6	15.2	①	—	13.2	9.9
	Protege LX	1.6 (1597)	ZM	5.0	5.6	15.2	①	—	13.2	7.9

NOTE: All capacities are approximate. Add fluid gradually and check to be sure a proper fluid level is obtained.

① Included in transaxle

93471CG2

Heater Core replacement is covered in Section 2 of this manual

VALVE SPECIFICATIONS

Year	Engine Displacement Liters (cc)	Engine ID/VIN	Seat Angle (deg.)	Face Angle (deg.)	Maximum out of Square (in.)	Spring Free Length (in.)	Stem-to-Guide Clearance (in.)		Stem Diameter (in.)	
							Intake	Exhaust	Intake	Exhaust
1998	1.5 (1489)	Z5	45	45	0.0520	1.240	0.0010-0.0023	0.0012-0.0025	0.2154-0.2159	0.2152-0.2157
	1.8 (1839)	BP	45	45	0.062	①	0.0010-0.0023	0.0012-0.0025	0.2351-0.2356	0.2349-0.2354
	2.0 (1991)	FS	45	45	0.061	1.732	0.0010-0.0023	0.0012-0.0025	0.2351-0.2356	0.2349-0.2354
	2.3 (2255)	KJ	NA	45	0.062	1.413	0.0010-0.0023	0.0012-0.0025	0.2351-0.2356	0.2349-0.2354
	2.5 (2497)	KL	45	45	0.064	1.847	0.0010-0.0023	0.0012-0.0025	0.2351-0.2356	0.2349-0.2354
1999	1.6 (1597)	ZM	NA	NA	NA	NA	NA	NA	NA	NA
	1.8 (1839)	BP	45	45	0.062	①	0.0010-0.0023	0.0012-0.0025	0.2351-0.2356	0.2349-0.2354
	1.8 (1839)	FP	NA	NA	NA	NA	NA	NA	NA	NA
	2.0 (1991)	FS	45	45	0.061	1.732	0.0010-0.0023	0.0012-0.0025	0.2351-0.2356	0.2349-0.2354
	2.3 (2255)	KJ	NA	45	0.062	1.413	0.0010-0.0023	0.0012-0.0025	0.2351-0.2356	0.2349-0.2354
	2.5 (2497)	KL	45	45	0.064	1.847	0.0010-0.0023	0.0012-0.0025	0.2351-0.2356	0.2349-0.2354
2000	1.6 (1597)	ZM	NA	NA	NA	NA	NA	NA	NA	NA
	1.8 (1839)	BP	45	45	0.062	①	0.0010-0.0023	0.0012-0.0025	0.2351-0.2356	0.2349-0.2354
	1.8 (1839)	FP	NA	NA	NA	NA	NA	NA	NA	NA
	2.0 (1991)	FS	45	45	0.061	1.732	0.0010-0.0023	0.0012-0.0025	0.2351-0.2356	0.2349-0.2354
	2.3 (2255)	KJ	NA	45	0.062	1.413	0.0010-0.0023	0.0012-0.0025	0.2351-0.2356	0.2349-0.2354
	2.5 (2497)	KL	45	45	0.064	1.847	0.0010-0.0023	0.0012-0.0025	0.2351-0.2356	0.2349-0.2354

NA: Not Available

① Intake: 1.80 in.
Exhaust: 1.903 in.

93471CG3

CRANKSHAFT AND CONNECTING ROD SPECIFICATIONS

All measurements are given in inches.

Year	Engine Displacement Liters (cc)	Engine ID/VIN	Crankshaft				Connecting Rod		
			Main Brg. Journal Dia.	Main Brg. Oil Clearance	Shaft End-play	Thrust on No.	Journal Diameter	Oil Clearance	Side Clearance
1998	1.5 (1489)	Z5	1.9661-1.9667	0.0008-0.0014	0.0032-0.0111	4	1.5725-1.5730	0.0012-0.0018	0.0044-0.0103
	1.8 (1839)	BP	1.9661-1.9667	0.0008-0.0014	0.0032-0.0111	4	1.7693-1.7699	0.0008-0.0017	0.0044-0.0103
	2.0 (1991)	FS	2.2022-2.2029	①	0.0031-0.0111	4	1.8874-1.8880	0.0005-0.0015	0.0043-0.0103
	2.3 (2255)	KJ	2.4385-2.4391	0.0015-0.0022	0.0032-0.0111	4	2.0843-2.0848	0.0010-0.0016	0.0071-0.0157
	2.5 (2497)	KL	2.4385-2.4391	0.0015-0.0022	0.0032-0.0111	4	2.0843-2.0848	0.0010-0.0016	0.0071-0.0157
1999	1.6 (1597)	ZM	NA	NA	NA	NA	NA	NA	NA
	1.8 (1839)	BP	1.9661-1.9667	0.0008-0.0014	0.0032-0.0111	4	1.7693-1.7699	0.0008-0.0017	0.0044-0.0103
	1.8 (1839)	FP	NA	NA	NA	NA	NA	NA	NA
	2.0 (1991)	FS	2.2022-2.2029	①	0.0031-0.0111	4	1.8874-1.8880	0.0005-0.0015	0.0043-0.0103
	2.3 (2255)	KJ	2.4385-2.4391	0.0015-0.0022	0.0032-0.0111	4	2.0843-2.0848	0.0010-0.0016	0.0071-0.0157
	2.5 (2497)	KL	2.4385-2.4391	0.0015-0.0022	0.0032-0.0111	4	2.0843-2.0848	0.0010-0.0016	0.0071-0.0157
2000	1.6 (1597)	ZM	NA	NA	NA	NA	NA	NA	NA
	1.8 (1839)	BP	1.9661-1.9667	0.0008-0.0014	0.0032-0.0111	4	1.7693-1.7699	0.0008-0.0017	0.0044-0.0103
	1.8 (1839)	FP	NA	NA	NA	NA	NA	NA	NA
	2.0 (1991)	FS	2.2022-2.2029	①	0.0031-0.0111	4	1.8874-1.8880	0.0005-0.0015	0.0043-0.0103
	2.3 (2255)	KJ	2.4385-2.4391	0.0015-0.0022	0.0032-0.0111	4	2.0843-2.0848	0.0010-0.0016	0.0071-0.0157
	2.5 (2497)	KL	2.4385-2.4391	0.0015-0.0022	0.0032-0.0111	4	2.0843-2.0848	0.0010-0.0016	0.0071-0.0157

NA: Not Avilable

① No. 1, 2, 4 & 5: 0.0009-0.0020 in.
No. 3: 0.0012-0.0022 in.

93471CG4

Brake service is covered in Section 4 of this manual

PISTON AND RING SPECIFICATIONS
All measurements are given in inches.

Year	Engine Displacement Liters (cc)	Engine ID/VIN	Piston Clearance	Ring Gap			Ring Side Clearance		
				Top Compression	Bottom Compression	Oil Control	Top Compression	Bottom Compression	Oil Control
1998	1.5 (1489)	Z5	0.0012-0.0016	0.006-0.011	0.010-0.015	0.003-0.006	0.0014-0.0025	0.0012-0.0025	0.0030-0.0060
	1.8 (1839)	BP	0.0010-0.0014	0.006-0.011	0.006-0.011	0.008-0.027	0.0012-0.0026	0.0012-0.0027	0.0030-0.0060
	2.0 (1991)	FS	0.0015-0.0020	0.006-0.012	0.006-0.012	0.008-0.028	0.0014-0.0026	0.0014-0.0026	SNUG
	2.3 (2255)	KJ	0.0004-0.0014	0.006-0.010	0.010-0.014	0.008-0.030	0.0014-0.0025	0.0012-0.0025	0.0028-0.0062
	2.5 (2497)	KL	0.0012-0.0022	0.006-0.012	0.010-0.016	0.008-0.028	0.0008-0.0025	0.0012-0.0025	0.0008-0.0020
1999	1.6 (1597)	ZM	NA	NA	NA	NA	NA	NA	NA
	1.8 (1839)	BP	0.0010-0.0014	0.006-0.011	0.006-0.011	0.008-0.027	0.0012-0.0026	0.0012-0.0027	0.0030-0.0060
	1.8 (1839)	FP	NA	NA	NA	NA	NA	NA	NA
	2.0 (1991)	FS	0.0015-0.0020	0.006-0.012	0.006-0.012	0.008-0.028	0.0014-0.0026	0.0014-0.0026	SNUG
	2.3 (2255)	KJ	0.0004-0.0014	0.006-0.010	0.010-0.014	0.008-0.030	0.0014-0.0025	0.0012-0.0025	0.0028-0.0062
	2.5 (2497)	KL	0.0012-0.0022	0.006-0.012	0.010-0.016	0.008-0.028	0.0008-0.0025	0.0012-0.0025	0.0008-0.0020
2000	1.6 (1597)	ZM	NA	NA	NA	NA	NA	NA	NA
	1.8 (1839)	BP	0.0010-0.0014	0.006-0.011	0.006-0.011	0.008-0.027	0.0012-0.0026	0.0012-0.0027	0.0030-0.0060
	1.8 (1839)	FP	NA	NA	NA	NA	NA	NA	NA
	2.0 (1991)	FS	0.0015-0.0020	0.006-0.012	0.006-0.012	0.008-0.028	0.0014-0.0026	0.0014-0.0026	SNUG
	2.3 (2255)	KJ	0.0004-0.0014	0.006-0.010	0.010-0.014	0.008-0.030	0.0014-0.0025	0.0012-0.0025	0.0028-0.0062
	2.5 (2497)	KL	0.0012-0.0022	0.006-0.012	0.010-0.016	0.008-0.028	0.0008-0.0025	0.0012-0.0025	0.0008-0.0020

NA: Not Available

93471CG5

TORQUE SPECIFICATIONS
All readings in ft. lbs.

Year	Engine Displacement Liters (cc)	Engine ID/VIN	Cylinder Head Bolts	Main Bearing Bolts	Rod Bearing Bolts	Crankshaft Damper Bolts	Flywheel Bolts	Manifold Intake	Manifold Exhaust	Spark Plugs	Lug Nut
1998	1.5 (1489)	Z5	①	40-43	22-25	116-122	71-76	14-18	12-17	11-16	65-87
	1.8 (1839)	BP	56-60	40-43	35-36	116-122	71-76	14-18	29-34	11-16	65-87
	2.0 (1991)	FS	①	②	③	116-122	71-76	14-18	15-20	11-16	65-87
	2.3 (2255)	KJ	④	⑤	③	116-122	45-49	14-18	14-18	11-16	65-87
	2.5 (2496)	KL	④	⑥	③	116-122	45-49	14-18	14-18	11-16	65-87
1999	1.6 (1597)	ZM	⑦	NA	NA	116-122	71-76	14-18	15-20	11-16	65-87
	1.8 (1839)	BP	56-60	40-43	35-36	116-122	71-76	14-18	29-34	11-16	65-87
	1.8 (1839)	FP	⑦	NA	NA	116-122	71-76	14-18	15-20	11-16	65-87
	2.0 (1991)	FS	①	②	③	116-122	71-76	14-18	15-20	11-16	65-87
	2.3 (2255)	KJ	④	⑤	③	116-122	45-49	14-18	14-18	11-16	65-87
	2.5 (2497)	KL	④	⑥	③	116-122	45-49	14-18	14-18	11-16	65-87
2000	1.6 (1597)	ZM	⑦	NA	NA	116-122	71-76	14-18	15-20	11-16	65-87
	1.8 (1839)	BP	56-60	40-43	35-36	116-122	71-76	14-18	29-34	11-16	65-87
	1.8 (1839)	FP	⑦	NA	NA	116-122	71-76	14-18	15-20	11-16	65-87
	2.0 (1991)	FS	①	②	③	116-122	71-76	14-18	15-20	11-16	65-87
	2.3 (2255)	KJ	④	⑤	③	116-122	45-49	14-18	14-18	11-16	65-87
	2.5 (2497)	KL	④	⑥	③	116-122	45-49	14-18	14-18	11-16	65-87

NA: Not Available

① Step 1: 16 ft. lbs.
  Step 2: Tighten each bolt 90 degrees
  Step 3: Repeat Step 2

② Step 1: 16 ft. lbs.
  Step 2: Tighten each bolt 90 degrees

③ Step 1: 19 ft. lbs.
  Step 2: Tighten each bolt 90 degrees

④ Step 1: 17-19 ft. lbs.
  Step 2: Tighten each bolt 90 degrees
  Step 3: Repeat Step 2

⑤ Step 1: Inner bolts: 17-19 ft. lbs.
  Step 2: Outer bolts: 13.5-15.5 ft. lbs.
  Step 3: Inner bolt Nos. 1-3: Tighten each bolt 70 degrees
  Step 4: Inner bolt No. 4: Tighten each bolt 80 degrees
  Step 5: Outer bolts: Tighten each bolt 60 degrees
  Step 6: Repeat Step 3-5

⑥ Step 1: Inner bolts: 17-18 ft. lbs.; Outer bolts: 13-15 ft. lbs.
  Step 2: Inner bolt Nos. 1-3: Tighten each bolt 70 degrees
  Step 3: Inner bolt No. 4: Tighten each bolt 80 degrees
  Step 4: Outer bolts: Tighten each bolt 60 degrees
  Step 5: Repeat Step 2

⑦ Step 1: 13-16 ft. lbs.
  Step 2: Tighten 85-95 degees
  Step 3: Repeat step 2

93471CG6

For complete Engine Mechanical specifications, see Section 1 of this manual

BRAKE SPECIFICATIONS
All measurements in inches unless noted

| Year | Model | | Brake Disc | | | Brake Drum | | | Minimum Lining Thickness | Brake Caliper | |
			Original Thickness	Minimum Thickness	Maximum Runout	Original Inside Diameter	Max. Wear Limit	Maximum Machine Diameter		Bracket Bolts (ft. lbs.)	Mounting Bolts (ft. lbs.)
1998	626	F	0.940	0.870	0.002	—	—	—	0.080	33-36	45-49
		R	0.390	0.310	0.002	9.00	NA	9.06	0.040	22-28	30-41
	Miata	F	0.790	0.710	0.004	—	—	—	0.040	58-65	78-88
		R	0.350	0.310	0.004	—	—	—	0.040	33-36	45-49
	Millenia	F	1.100	1.020	0.002	—	—	—	0.080	47-62	63-84
		R	0.370	0.290	0.002	—	—	—	0.080	12-17	16-23
	Protege	F	0.870	0.790	0.002	—	—	—	0.040	29-36	40-49
		R	0.354	0.276	0.002	7.87	7.91	NA	0.040	33-44	46-60
1999	626	F	0.940	0.870	0.002	—	—	—	0.080	33-36	45-49
		R	0.390	0.310	0.002	9.00	NA	9.06	0.040	22-28	30-41
	Miata	F	0.790	0.710	0.004	—	—	—	0.040	58-65	78-88
		R	0.350	0.310	0.004	—	—	—	0.040	33-36	45-49
	Millenia	F	1.100	1.020	0.002	—	—	—	0.080	47-62	63-84
		R	0.370	0.290	0.002	—	—	—	0.080	12-17	16-23
	Protege	F	NA	①	0.002	—	—	—	②	33-36	58-75
		R	—	—	—	7.87	7.91	NA	0.040	—	—
2000	626	F	0.940	0.870	0.002	—	—	—	0.080	22-28	58-75.2
		R	0.390	0.310	0.002	9.00	NA	9.06	0.040	26-28	34-49
	Miata	F	0.790	0.710	0.004	—	—	—	0.040	33-39	37-50
		R	0.350	0.310	0.004	—	—	—	0.040	26-28	34-49
	Millenia	F	1.100	1.000	0.002	—	—	—	0.080	47-62	63-84
		R	0.370	0.300	0.002	—	—	—	0.080	12-17	16-23
	Protege	F	NA	①	0.002	—	—	—	②	33-36	58-75
		R	—	—	—	7.87	7.91	NA	0.040	—	—

F: Front

R: Rear

① With 1.6L engine: 0.780 in.
   With 1.8L engine: 0.870 in.

② With 1.6L engine: 0.060 in.
   With 1.8L engine: 0.080 in.

93471CG7

WHEEL ALIGNMENT

Year	Model		Caster Range (+/-Deg.)	Caster Preferred Setting (Deg.)	Camber Range (+/-Deg.)	Camber Preferred Setting (Deg.)	Toe-in (in.)	Steering Axis Inclination (Deg.)
1998	626 ①	F	1.00	+2.13	1.00	-0.70	0.12 +/- 0.16	15.06
		R	—	—	1.00	-0.10	0.12 +/- 0.16	—
	626 ②	F	1.00	+2.15	1.00	-0.70	0.12 +/- 0.16	15.06
		R	—	—	1.00	-0.10	0.12 +/- 0.16	—
	626 ③	F	1.00	+2.05	1.00	-0.70	0.12 +/- 0.16	15.06
		R	—	—	1.00	-0.10	0.12 +/- 0.16	—
	Miata	F	1.00	+5.75	1.00	+0.05	0.12 +/- 0.16	11.63
		R	—	—	1.00	-0.75	0.12 +/- 0.16	—
	Millenia	F	1.00	+2.23	0.75	-0.19	0.12 +/- 0.16	9.09
		R	—	—	1.00	-0.31	0.12 +/- 0.16	—
	Protégé	F	1.00	+1.88	1.00	-0.75	0.08 +/- 0.16	—
		R	—	—	1.00	-0.52	0.08 +/- 0.16	—
1999	626 ①	F	1.00	+2.13	1.00	-0.70	0.12 +/- 0.16	15.06
		R	—	—	1.00	-0.10	0.12 +/- 0.16	—
	626 ②	F	1.00	+2.15	1.00	-0.70	0.12 +/- 0.16	15.06
		R	—	—	1.00	-0.10	0.12 +/- 0.16	—
	626 ③	F	1.00	+2.05	1.00	-0.70	0.12 +/- 0.16	15.06
		R	—	—	1.00	-0.10	0.12 +/- 0.16	—
	Miata	F	1.00	+5.75	1.00	+0.05	0.12 +/- 0.16	11.63
		R	—	—	1.00	-0.75	0.12 +/- 0.16	—
	Millenia	F	1.00	+2.23	0.75	-0.19	0.12 +/- 0.16	9.09
		R	—	—	1.00	-0.31	0.12 +/- 0.16	—
	Protégé	F	1.00	+1.88	1.00	-0.75	0.08 +/- 0.16	—
		R	—	—	1.00	-0.52	0.08 +/- 0.16	—
2000	626 ①	F	1.00	+2.13	1.00	-0.70	0.12 +/- 0.16	15.06
		R	—	—	1.00	-0.10	0.12 +/- 0.16	—
	626 ②	F	1.00	+2.15	1.00	-0.70	0.12 +/- 0.16	15.06
		R	—	—	1.00	-0.10	0.12 +/- 0.16	—
	626 ③	F	1.00	+2.05	1.00	-0.70	0.12 +/- 0.16	15.06
		R	—	—	1.00	-0.10	0.12 +/- 0.16	—
	Miata	F	1.00	+5.75	1.00	+0.05	0.12 +/- 0.16	11.63
		R	—	—	1.00	-0.75	0.12 +/- 0.16	—
	Millenia	F	1.00	+2.23	0.75	-0.19	0.12 +/- 0.16	9.09
		R	—	—	1.00	-0.31	0.12 +/- 0.16	—
	Protégé	F	1.00	+1.88	1.00	-0.75	0.08 +/- 0.16	—
		R	—	—	1.00	-0.52	0.08 +/- 0.16	—

① With 14 in. wheels
② With 15 in. wheels
③ With 16 in. wheels

93471CG8

For Accessory Drive Belt illustrations, see Section 1 of this manual

TIRE, WHEEL AND BALL JOINT SPECIFICATIONS

Year	Model	OEM Tires		Tire Pressures (psi)		Wheel Size	Ball Joint Inspection
		Standard	Optional	Front	Rear		
1998	626 2.0L	P185/70R14	None	32	36	5.5-JJ	8-43 in. ①
	Miata	P185/60HR14	P195/50VR15	26	26	6-JJ	3-16 in. ① ②
	626 2.5L	P205/60HR15	None	32	36	6-JJ	8-43 ①
	Millenia	P205/65HR15	P215/55VR16	32	29	Std: 6-JJ Opt: 6.5-JJ	2-30 in. ①
	MX-6, base	P195/65SR14	None	32	36	5.5-JJ	8-43 in. ①
	MX-6 LS	P205/55VR15	None	32	36	6.5-JJ	8-43 in. ①
	Protégé, exc ES	P175/70SR13	P185/65SR14	32	32	5-J	8-43 in. ①
	Protégé ES	P185/65SR14	None	32	32	5.5-JJ	8-43 in. ①
1999	Millenia	P205/65HR15	P215/55VR16	32	29	Std: 6-JJ Opt: 6.5-JJ	2-30 in. ①
	Miata	P185/60HR14	P195/50VR15	26	26	6-JJ	3-16 in. ① ②
	Protégé 1.6L	P185/65R14	None	32	29	6-JJ	8-43 in. ①
	Protégé 1.8L	P185/65R14	P195/50VR15	32	32	Std: 5.5-JJ Opt: 6-JJ	8-43 in. ①
	626 2.0L	P185/70R14	None	32	36	5.5-JJ	8-43 in. ①
	626 2.5L	P205/60HR15	None	32	36	6-JJ	8-43 in. ①
2000	Millenia	P205/65HR15	P215/55VR16	32	29	Std: 6-JJ Opt: 6.5-JJ	2-30 in. ①
	Miata	P185/60HR14	P195/50VR15	26	26	6-JJ	3-16 in. ① ②
	Protégé 1.6L	P185/65R14	None	32	29	6-JJ	8-43 in. ①
	Protégé 1.8L	P185/65R14	P195/50VR15	32	32	Std: 5.5-JJ Opt: 6-JJ	8-43 in. ①
	626 2.0L	P185/70R14	None	32	36	5.5-JJ	8-43 in. ①
	626 2.5L	P205/60HR15	None	32	36	6-JJ	8-43 in. ①

OEM: Original Equipment Manufacturer

PSI: Pounds Per Square Inch

STD: Standard

OPT: Optional

① Torque required in inch lbs. to rotate ball joint when removed from the knuckle

② Applies to uppper and lower ball joints

93471CG9

SCHEDULED MAINTENANCE INTERVALS
Mazda

TO BE SERVICED	TYPE OF SERVICE	VEHICLE MILEAGE INTERVAL (x1000)												
		7.5	15	22.5	30	37.5	45	52.5	60	67.5	75	82.5	90	97.5
Engine oil & filter	R	✔	✔	✔	✔	✔	✔	✔	✔	✔	✔	✔	✔	✔
Air cleaner element	R				✔				✔				✔	
Engine coolant ①	R				✔				✔				✔	
Spark plugs (Millenia KJ engine)	R				✔				✔				✔	
Automatic transaxle fluid	S/I				✔				✔				✔	
Bolts & nuts on chassis & body	S/I				✔				✔				✔	
Brake lines, hoses & connections	S/I				✔				✔				✔	
Cooling system	S/I				✔				✔				✔	
Disc brakes	S/I				✔				✔				✔	
Drive belts (Millenia ②)	S/I				✔				✔				✔	
Drive shaft dust boots	S/I				✔				✔				✔	
Exhaust system heat shield	S/I				✔				✔				✔	
Front & rear suspension ball joints	S/I				✔				✔				✔	
Fuel lines & hoses	S/I				✔				✔				✔	
Idle speed	S/I				✔				✔				✔	
Steering operation & linkages	S/I				✔				✔				✔	
Engine timing belt ③	R								✔					
Fuel filter	R								✔					
Manual transmission	R								✔					
Hose & tube for emission	S/I								✔					

R: Replace   S/I: Service or Inspect

① (Millenia): replace initially at 45,000 miles & every 30,000 miles thereafter.

② (Millenia KJ engine): replace every 105,000 miles

③ (Calif.): inspect every 30,000 miles & replace at 105,000 miles (if not replaced previously).

FREQUENT OPERATION MAINTENANCE (SEVERE SERVICE)

If a vehicle is operated under any of the following conditions it is considered severe service

- Extremely dusty areas.

- 50% or more of the vehicle operation is in 32°C (90°F) or higher temperatures, or constant operation in temperatures below 0°C (32°F).

- Prolonged idling (vehicle operation in stop and go traffic).

- Frequent short running periods (engine does not warm to normal operating temperatures).

- Police, taxi, delivery usage or trailer towing usage.

Oil & oil filter: change every 5000 miles.

Oil & oil filter (Puerto Rico): change every 3000 miles.

Air cleaner element: service or inspect every 15,000 miles

Automatic transaxle fluid: service or inspect every 15,000 miles.

Bolts & nuts on chassis & body: tighten every 15,000 miles.

Disc brakes: service or inspect every 15,000 miles.

93471CG0

For Tire, Wheel and Ball Joint specifications, see Section 1 of this manual

SCHEDULED MAINTENANCE INTERVALS
MAZDA
626, MIATA, MILLENIA, PROTEGE

The following should be used as a guide when determining the amount of work required for a particular service. In estimating how long a particular Scheduled Maintenance Service should take, please observe the following:

- Labor Time is time based on field research and data supplied by the vehicle manufacturer.
- Labor time operations are given in hours and tenths of an hour.
- All labor operations are to be used as a guide.

Mechanic Skill Level Codes:
(A) PRECISION: Highly skilled with multiple certification.
(B) GENERAL: Normally skilled with certification.
(C) MAINTENANCE: Semi-skilled working on certification.

	LABOR TIME		LABOR TIME		LABOR TIME
7500 Mile Service (C)		**37500 Mile Service (C)**		**75000 Mile Service (C)**	
626	.4	626	.4	626	.4
Miata	.4	Miata	.4	Miata	.4
Millenia	.4	Millenia	.4	Millenia	.4
Protege	.4	Protege	.4	Protege	.4
15000 Mile Service (C)		**45000 Mile Service (C)**		**82500 Mile Service (C)**	
626	.4	626	.4	626	.4
Miata	.4	Miata	.4	Miata	.4
Millenia	.4	Millenia	.4	Millenia	.4
Protege	.4	Protege	.4	Protege	.4
22500 Mile Service (C)		**52500 Mile Service (C)**		**90000 Mile Service (B)**	
626	.4	626	.4	626	3.0
Miata	.4	Miata	.4	Miata	2.8
Millenia	.4	Millenia	.4	Millenia	3.0
Protege	.4	Protege	.4	Protege	2.5
30000 Mile Service (B)		**60000 Mile Service (B)**		**97500 Mile Service (C)**	
626	3.0	626	4.9	626	.4
Miata	2.8	Miata	5.3	Miata	.4
Millenia	3.0	Millenia	4.9	Millenia	.4
Protege	2.5	Protege	5.1	Protege	.4
		67500 Mile Service (C)			
		626	.4		
		Miata	.4		
		Millenia	.4		
		Protege	.4		

93471CH1

MITSUBISHI
3000GT • DIAMANTE • ECLIPSE • GALANT • MIRAGE

ENGINE AND VEHICLE IDENTIFICATION

Engine							Model Year	
Code ①	Liters (cc)	Cu. In.	Cyl.	Fuel Sys.	Type	Eng. Mfg.	Code ②	Year
4G15/A	1.5 (1468)	87	4	MFI	SOHC	Mitsubishi	W	1998
4G93/C	1.8 (1834)	112	4	MFI	SOHC	Mitsubishi	X	1999
420A/Y	2.0 (1996)	122	4	MFI	DOHC	Mitsubishi	Y	2000
4G63/F	2.0 (1997)	122	4	MFI-Turbo	DOHC	Mitsubishi	1	2001
4G64/G	2.4 (2351)	143	4	MFI	SOHC	Mitsubishi	2	2002
6G72/H	3.0 (2972)	181	6	MFI	SOHC	Mitsubishi		
6G72/J	3.0 (2972)	181	6	MFI	DOHC	Mitsubishi		
6G72/K	3.0 (2972)	181	6	MFI-Turbo	DOHC	Mitsubishi		
6G72/L	3.0 (2972)	181	6	MFI	SOHC	Mitsubishi		
6G74/P	3.5 (3497)	213	6	MFI	SOHC	Mitsubishi		

MFI: Multiport fuel injection

SOHC: Single overhead camsha

DOHC: Double overhead camshafts

① Engine ID / 8th digit of the VIN

② 10th digit of the VIN

93471CH2

For Wheel Alignment specifications, see Section 1 of this manual

GENERAL ENGINE SPECIFICATIONS

Year	Model	Engine Displacement Liters (cc)	Engine ID/VIN	Fuel System Type	Net Horsepower @ rpm	Net Torque @ rpm (ft. lbs.)	Bore x Stroke (in.)	Compression Ratio	Oil Pressure @ rpm
1998	Mirage	1.5 (1468)	4G15/A	MFI	92@6000	93@3000	2.97x3.23	9.2:1	54@2000
	Mirage	1.8 (1834)	4G93/C	MFI	113@6000	116@4500	3.19x3.50	9.5:1	41@2000
	Eclipse	2.0 (1996)	420A/Y	MFI	140@6000	130@4800	3.44x3.27	9.6:1	11@idle
	Eclipse	2.0 (1997)	4G63/F	MFI	113@6000	②	3.35x3.46	8.5:1	11@idle
	Eclipse Spyder	2.0 (1997)	4G63/F	MFI	①	②	3.35x3.46	8.5:1	11@idle
	Eclipse Spyder	2.4 (2350)	4G64/G	MFI	③	148@3000	3.41x3.94	9.5:1	41@2000
	Galant	2.4 (2350)	4G64/G	MFI	③	148@3000	3.41x3.94	9.5:1	41@2000
	Galant	3.0 (2972)	6G72/L	MFI	175@5500	185@3000	3.59x2.99	8.9:1	30-80@2000
	3000GT	3.0 (2972)	6G72/H	MFI	175@5500	185@3000	3.59x2.99	8.9:1	30-80@2000
	3000GT	3.0 (2972)	6G72/J	MFI	202@6000	201@3500	3.59x2.99	10.0:1	30-80@2000
	3000GT	3.0 (2972)	6G72/K	MFI	320@6000	315@2500	3.59x2.99	8.0:1	30-80@2000
	Diamante	3.5 (3497)	6G74/P	MFI	214@5000	228@3000	3.66x3.38	9.5:1	30-80@2000
1999	Mirage	1.5 (1468)	4G15/A	MFI	92@6000	93@3000	2.97x3.23	9.2:1	54@2000
	Mirage	1.8 (1834)	4G93/C	MFI	113@6000	116@4500	3.19x3.50	9.5:1	41@2000
	Eclipse	2.4 (2350)	4G64/G	MFI	③	148@3000	3.41x3.94	9.5:1	41@2000
	Eclipse	3.0 (2972)	6G72/L	MFI	175@5500	185@3000	3.59x2.99	8.9:1	30-80@2000
	Eclipse Spyder	2.4 (2350)	4G64/G	MFI	③	148@3000	3.41x3.94	9.5:1	41@2000
	Eclipse Spyder	3.0 (2972)	6G72/L	MFI	175@5500	185@3000	3.59x2.99	8.9:1	30-80@2000
	3000GT	3.0 (2972)	6G72/H	MFI	175@5500	185@3000	3.59x2.99	8.9:1	30-80@2000
	3000GT	3.0 (2972)	6G72/J	MFI	202@6000	201@3500	3.59x2.99	10.0:1	30-80@2000
	3000GT	3.0 (2972)	6G72/K	MFI	320@6000	315@2500	3.59x2.99	8.0:1	30-80@2000
	Galant	2.4 (2350)	4G64/G	MFI	③	148@3000	3.41x3.94	9.5:1	41@2000
	Galant	3.0 (2972)	6G72/L	MFI	175@5500	185@3000	3.59x2.99	8.9:1	30-80@2000
	Diamante	3.5 (3497)	6G74/P	MFI	214@5000	228@3000	3.66x3.38	9.5:1	30-80@2000
2000	Mirage	1.5 (1468)	4G15/A	MFI	92@6000	93@3000	2.97x3.23	9.2:1	54@2000
	Mirage	1.8 (1834)	4G93/C	MFI	113@6000	116@4500	3.19x3.50	9.5:1	41@2000
	Eclipse	2.4 (2350)	4G64/G	MFI	③	148@3000	3.41x3.94	9.5:1	41@2000
	Eclipse	3.0 (2972)	6G72/L	MFI	175@5500	185@3000	3.59x2.99	8.9:1	30-80@2000
	Eclipse Spyder	2.4 (2350)	4G64/G	MFI	③	148@3000	3.41x3.94	9.5:1	41@2000
	Eclipse Spyder	3.0 (2972)	6G72/L	MFI	175@5500	185@3000	3.59x2.99	8.9:1	30-80@2000
	Galant	2.4 (2350)	4G64/G	MFI	③	148@3000	3.41x3.94	9.5:1	41@2000
	Galant	3.0 (2972)	6G72/L	MFI	175@5500	185@3000	3.59x2.99	8.9:1	30-80@2000
	Diamante	3.5 (3497)	6G74/P	MFI	214@5000	228@3000	3.66x3.38	9.5:1	30-80@2000
2001	Mirage	1.5 (1468)	4G15/A	MFI	92@6000	93@3000	2.97x3.23	9.2:1	54@2000
	Mirage	1.8 (1834)	4G93/C	MFI	113@6000	116@4500	3.19x3.50	9.5:1	41@2000
	Eclipse	2.4 (2350)	4G64/G	MFI	③	148@3000	3.41x3.94	9.5:1	41@2000
	Eclipse	3.0 (2972)	6G72/L	MFI	175@5500	185@3000	3.59x2.99	8.9:1	30-80@2000
	Eclipse Spyder	2.4 (2350)	4G64/G	MFI	③	148@3000	3.41x3.94	9.5:1	41@2000
	Eclipse Spyder	3.0 (2972)	6G72/L	MFI	175@5500	185@3000	3.59x2.99	8.9:1	30-80@2000
	Galant	2.4 (2350)	4G64/G	MFI	③	148@3000	3.41x3.94	9.5:1	41@2000
	Galant	3.0 (2972)	6G72/L	MFI	175@5500	185@3000	3.59x2.99	8.9:1	30-80@2000
	Diamante	3.5 (3497)	6G74/P	MFI	214@5000	228@3000	3.66x3.38	9.5:1	30-80@2000

MFI: Multiport fuel injection

① Manual transaxle: 210@6000
   Automatic transaxle: 205@6000

② Manual transaxle: 214@3000
   Automatic transaxle: 220@3000

③ California: 138@5500
   Except California: 141@5500

93471CH3

ENGINE TUNE-UP SPECIFICATIONS

Year	Engine Displacement Liters (cc)	Engine ID/VIN	Spark Plugs Gap (in.)	Ignition Timing (deg.) MT	Ignition Timing (deg.) AT	Fuel Pump (psi)	Idle Speed (rpm) MT	Idle Speed (rpm) AT	Valve Clearance In.	Valve Clearance Ex.
1998	1.5 (1468)	4G15/A	0.039-0.043	2-8B	2-8B	38	600-800	600-800	HYD	HYD
	1.8 (1834)	4G93/C	0.039-0.043	2-8B	2-8B	38	600-800	600-800	HYD	HYD
	2.0 (1996)	420A/Y	0.048-0.053	2-8B	2-8B	47-50	700-900	700-900	HYD	HYD
	2.0 (1997)	4G63/F	0.028-0.031	2-8B	2-8B	33	650-850	650-850	HYD	HYD
	2.4 (2350)	4G64/G	0.039-0.043	2-8B	2-8B	38	650-850	650-850	HYD	HYD
	3.0 (2972)	6G72/H	0.039-0.043	5B	5B	38	600-800	600-800	HYD	HYD
	3.0 (2972)	6G72/J	0.039-0.043	5B	5B	38	600-800	600-800	HYD	HYD
	3.0 (2972)	6G72/K	0.039-0.043	5B	5B	34	600-800	600-800	HYD	HYD
	3.0 (2972)	6G72/L	0.039-0.043	5B	5B	38	600-800	600-800	HYD	HYD
	3.5 (3497)	6G74/P	0.039-0.043	2-8B	2-8B	38	600-800	600-800	HYD	HYD
1999	1.5 (1468)	4G15/A	0.039-0.043	2-8B	2-8B	38	600-800	600-800	HYD	HYD
	1.8 (1834)	4G93/C	0.039-0.043	2-8B	2-8B	38	600-800	600-800	HYD	HYD
	2.4 (2350)	4G64/G	0.039-0.043	2-8B	2-8B	38	650-850	650-850	HYD	HYD
	3.0 (2972)	6G72/H	0.039-0.043	5B	5B	38	600-800	600-800	HYD	HYD
	3.0 (2972)	6G72/J	0.039-0.043	5B	5B	38	600-800	600-800	HYD	HYD
	3.0 (2972)	6G72/K	0.039-0.043	5B	5B	34	600-800	600-800	HYD	HYD
	3.0 (2972)	6G72/L	0.039-0.043	5B	5B	38	600-800	600-800	HYD	HYD
	3.5 (3497)	6G74/P	0.039-0.043	2-8B	2-8B	38	600-800	600-800	HYD	HYD
2000	1.5 (1468)	4G15/A	0.039-0.043	2-8B	2-8B	38	600-800	600-800	HYD	HYD
	1.8 (1834)	4G93/C	0.039-0.043	2-8B	2-8B	38	600-800	600-800	HYD	HYD
	2.4 (2350)	4G64/G	0.039-0.043	2-8B	2-8B	38	650-850	650-850	HYD	HYD
	3.0 (2972)	6G72/L	0.039-0.043	5B	5B	38	600-800	600-800	HYD	HYD
	3.5 (3497)	6G74/P	0.039-0.043	2-8B	2-8B	38	600-800	600-800	HYD	HYD
2001	1.5 (1468)	4G15/A	0.039-0.043	2-8B	2-8B	38	600-800	600-800	HYD	HYD
	1.8 (1834)	4G93/C	0.039-0.043	2-8B	2-8B	38	600-800	600-800	HYD	HYD
	2.4 (2350)	4G64/G	0.039-0.043	2-8B	2-8B	38	650-850	650-850	HYD	HYD
	3.0 (2972)	6G72/L	0.039-0.043	5B	5B	38	600-800	600-800	HYD	HYD
	3.5 (3497)	6G74/P	0.039-0.043	—	2-8B	38	600-800	600-800	HYD	HYD

NOTE: The Vehicle Emission Control Information label often reflects specification changes made during production. The label figures must be used if they differ from those in this chart.

B: Before top dead center

HYD: Hydraulic

93471CH4

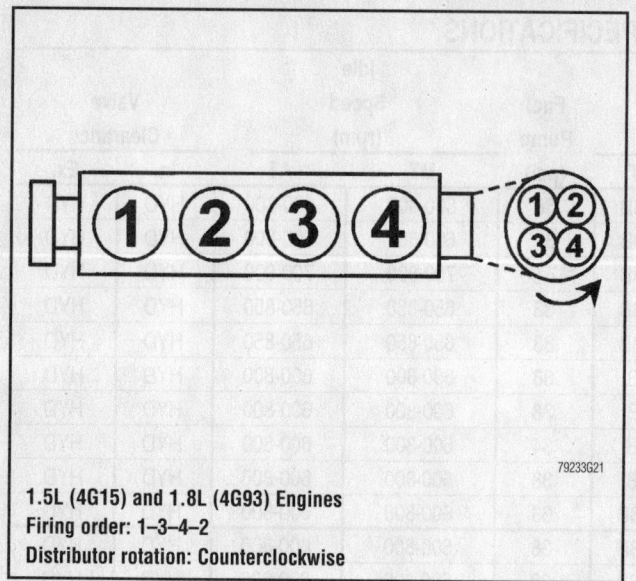

1.5L (4G15) and 1.8L (4G93) Engines
Firing order: 1–3–4–2
Distributor rotation: Counterclockwise

79233G21

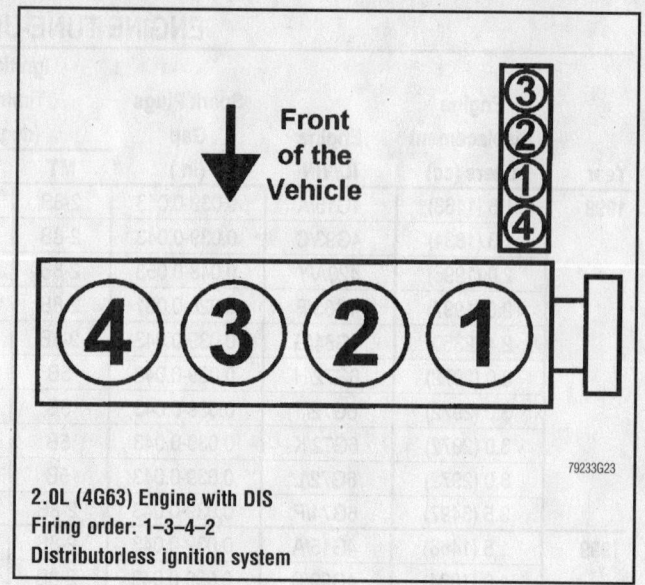

2.0L (4G63) Engine with DIS
Firing order: 1–3–4–2
Distributorless ignition system

79233G23

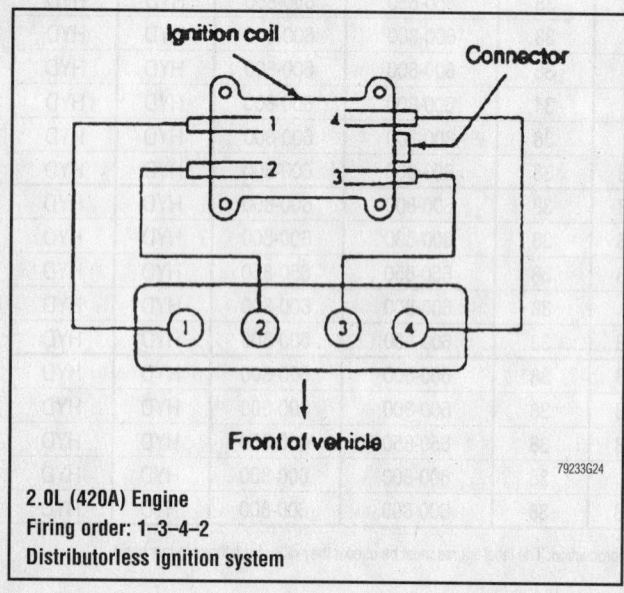

2.0L (420A) Engine
Firing order: 1–3–4–2
Distributorless ignition system

79233G24

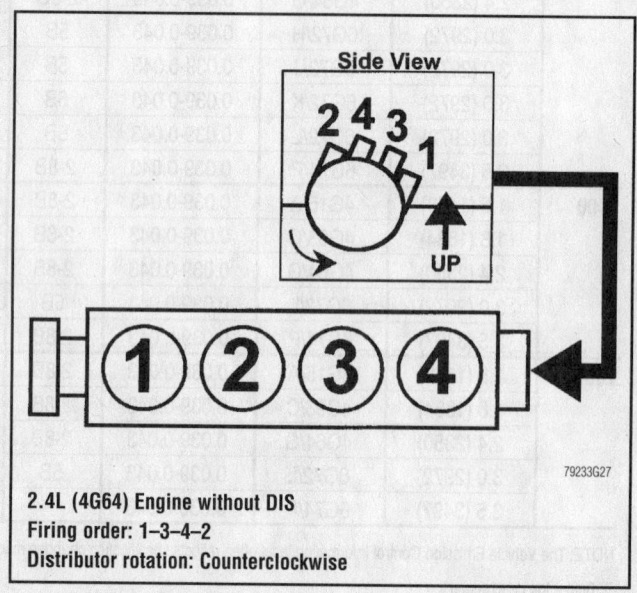

2.4L (4G64) Engine without DIS
Firing order: 1–3–4–2
Distributor rotation: Counterclockwise

79233G27

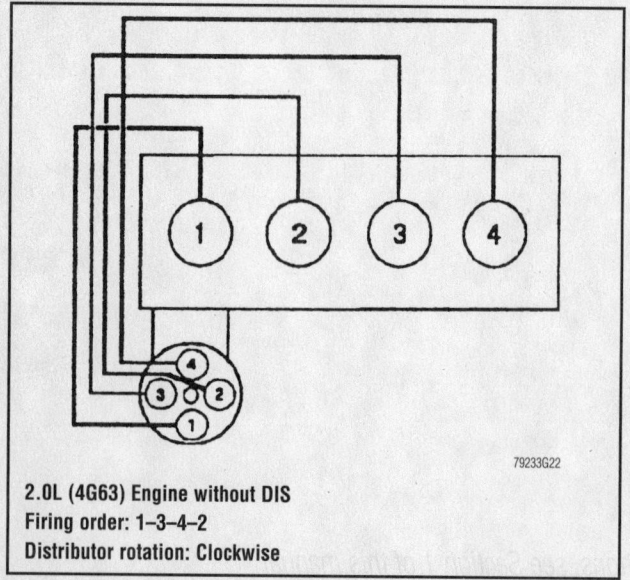

2.0L (4G63) Engine without DIS
Firing order: 1–3–4–2
Distributor rotation: Clockwise

79233G22

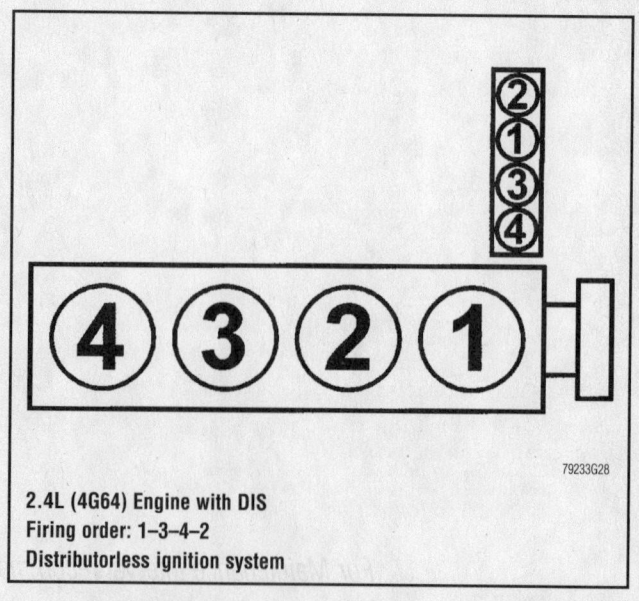

2.4L (4G64) Engine with DIS
Firing order: 1–3–4–2
Distributorless ignition system

79233G28

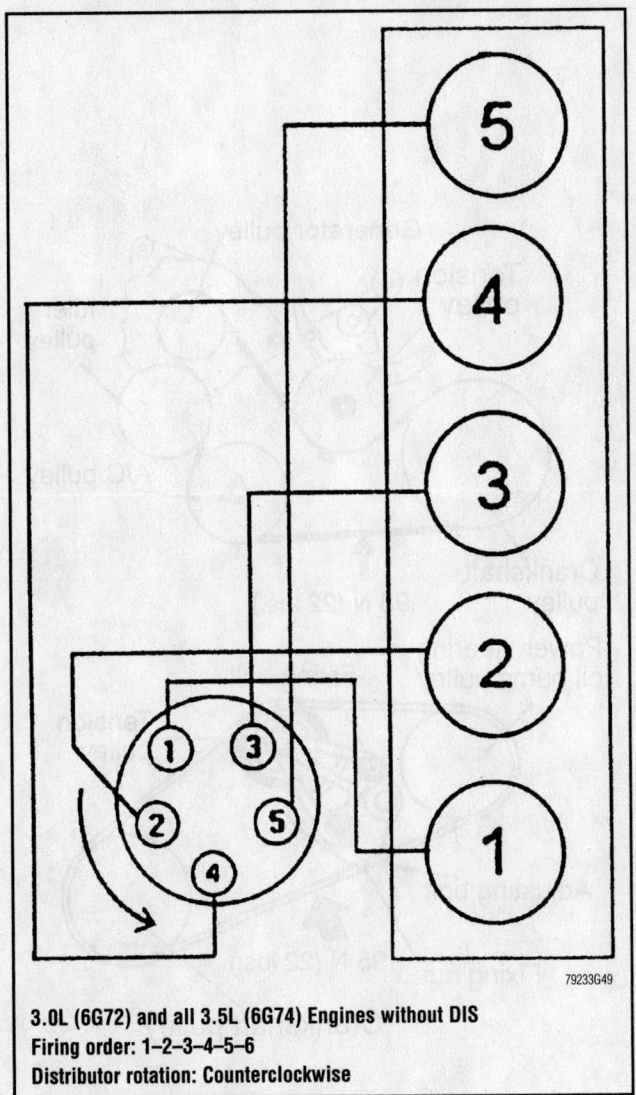

3.0L (6G72) and all 3.5L (6G74) Engines without DIS
Firing order: 1–2–3–4–5–6
Distributor rotation: Counterclockwise

79233G49

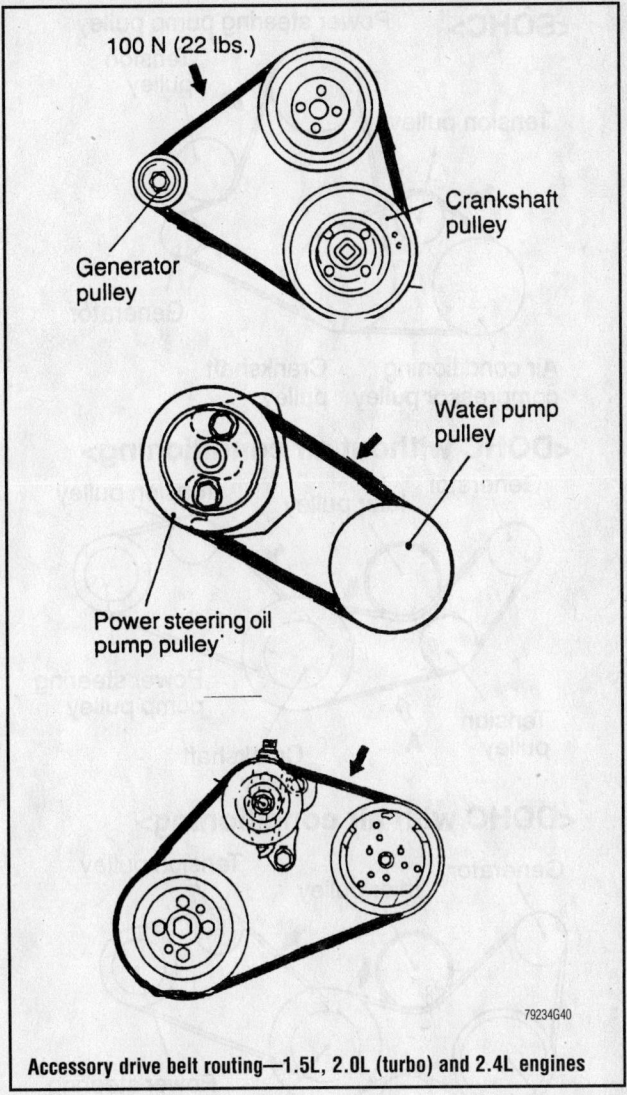

Accessory drive belt routing—1.5L, 2.0L (turbo) and 2.4L engines

79234G40

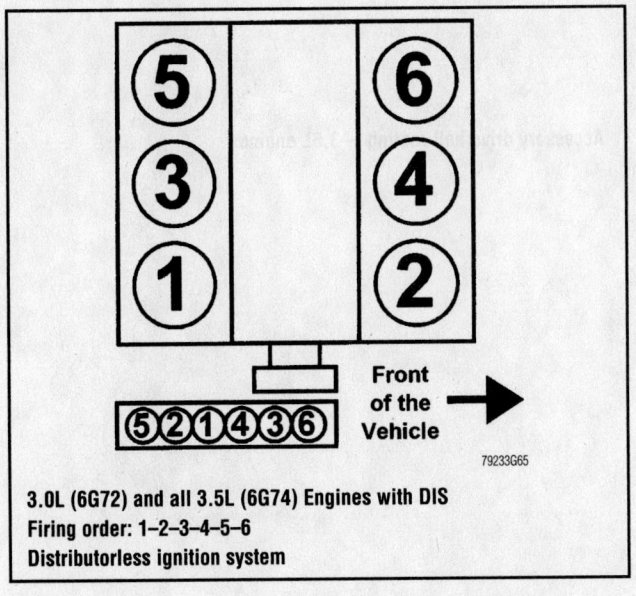

Front
of the
Vehicle

3.0L (6G72) and all 3.5L (6G74) Engines with DIS
Firing order: 1–2–3–4–5–6
Distributorless ignition system

79233G65

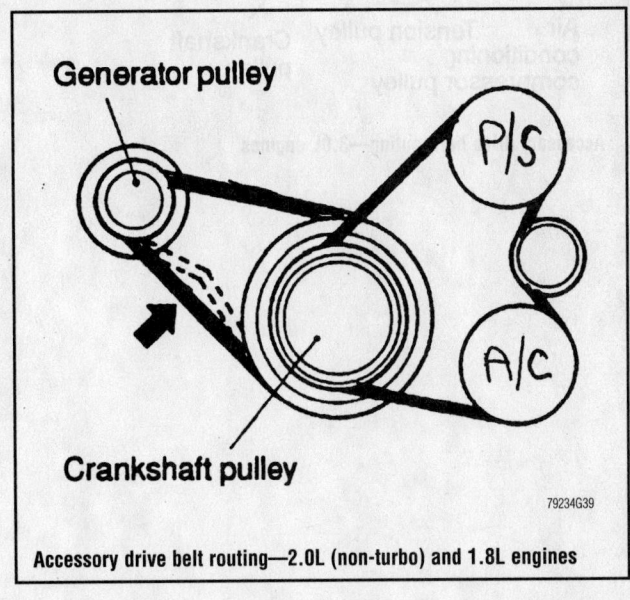

Accessory drive belt routing—2.0L (non-turbo) and 1.8L engines

79234G39

For Tune-up, Capacities and Firing orders, see Section 1 of this manual

<SOHC>

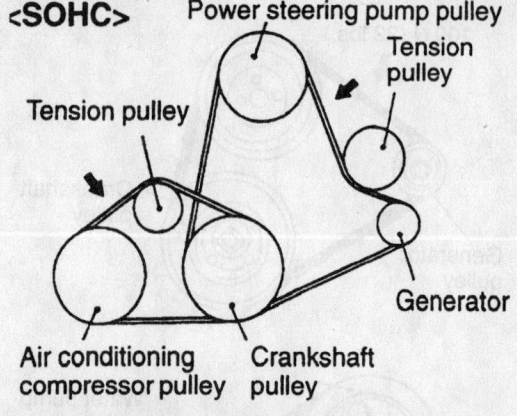

Power steering pump pulley
Tension pulley
Tension pulley
Generator
Air conditioning compressor pulley
Crankshaft pulley

<DOHC without air conditioning>

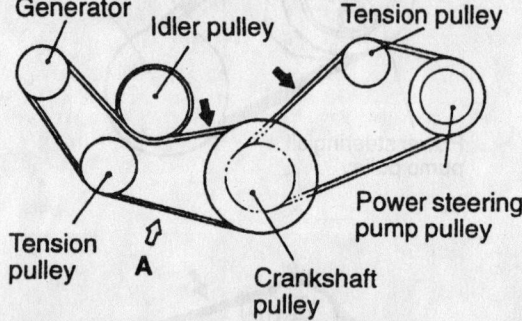

Generator
Idler pulley
Tension pulley
Tension pulley
A
Crankshaft pulley
Power steering pump pulley

<DOHC with air conditioning>

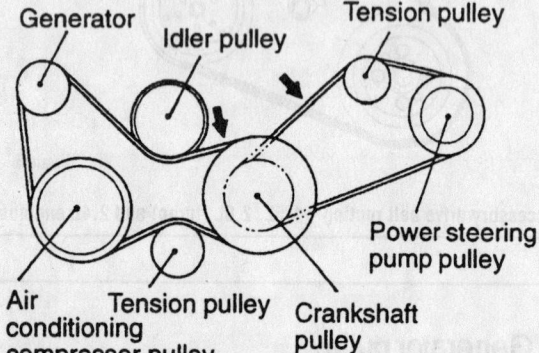

Generator
Idler pulley
Tension pulley
Air conditioning compressor pulley
Tension pulley
Crankshaft pulley
Power steering pump pulley

79234G37

Accessory drive belt routing—3.0L engines

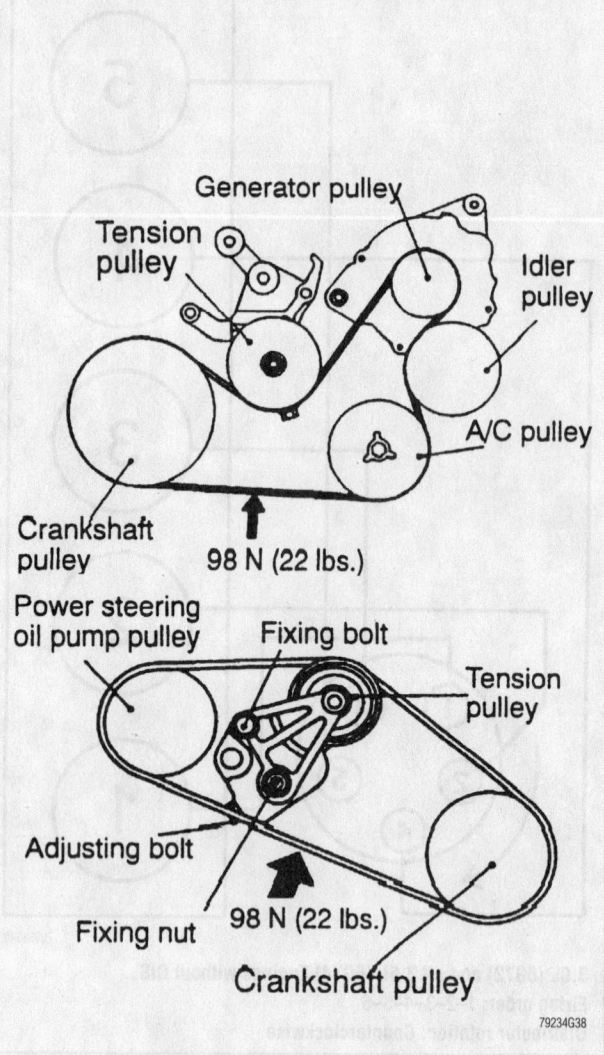

Generator pulley
Tension pulley
Idler pulley
A/C pulley
Crankshaft pulley
98 N (22 lbs.)

Power steering oil pump pulley
Fixing bolt
Tension pulley
Adjusting bolt
Fixing nut
98 N (22 lbs.)
Crankshaft pulley

79234G38

Accessory drive belt routing —3.5L engines

CAPACITIES

Year	Model	Engine Displacement Liters (cc)	Engine ID/VIN	Engine Oil with Filter	Transmission (pts.) 5 or 6-Spd	Auto.	Transfer Case (pts.)	Drive Axle Front (pts.)	Rear (pts.)	Fuel Tank (gal.)	Cooling System (qts.)
1998	Mirage	1.5 (1468)	4G15/A	3.7	3.8	12.6	—	—	—	13.2	5.3
	Mirage	1.8 (1834)	4G93/C	4.2	3.8	12.6	—	—	—	13.2	6.3
	Eclipse	2.0 (1996)	420A/Y	4.5	4.2	18.2	—	—	—	17.0	7.4
	Eclipse	2.0 (1997)	4G63/F	4.5	①	14.2	1.20	—	1.48	17.0	7.4
	Eclipse Spyder	2.0 (1997)	4G63/F	4.5	①	14.2	1.20	—	1.48	17.0	7.4
	Eclipse Spyder	2.4 (2350)	4G64/G	4.5	4.2	12.8	—	—	—	17.0	14.8
	Galant	2.4 (2350)	4G64/G	4.5	4.6	12.8	—	—	—	17.0	14.8
	Galant	3.0 (2972)	6G72/L	4.5	4.6	15.8	—	—	—	19.8	8.5
	3000GT	3.0 (2972)	6G72/H	4.5	②	15.8	0.64	—	—	19.8	8.5
	3000GT	3.0 (2972)	6G72/J	4.5	②	15.8	0.64	—	—	19.8	8.5
	3000GT	3.0 (2972)	6G72/K	4.5	②	15.8	0.64	—	2.3	19.8	8.5
	Diamante	3.5 (3497)	6G74/P	4.7	—	18.0	—	—	—	18.7	10.0
1999	Mirage	1.5 (1468)	4G15/A	3.7	3.8	12.6	—	—	—	13.2	5.3
	Mirage	1.8 (1834)	4G93/C	4.2	3.8	12.6	—	—	—	13.2	6.3
	Eclipse	2.0 (1996)	420A/Y	4.5	4.2	18.2	—	—	—	17.0	7.4
	Eclipse	2.0 (1997)	4G63/F	4.5	①	14.2	1.20	—	1.48	17.0	7.4
	Eclipse Spyder	2.0 (1997)	4G63/F	4.5	①	14.2	1.20	—	1.48	17.0	7.4
	Eclipse Spyder	2.4 (2350)	4G64/G	4.5	4.2	12.8	—	—	—	17.0	14.8
	Galant	2.4 (2350)	4G64/G	4.5	4.6	12.8	—	—	—	17.0	14.8
	Galant	3.0 (2972)	6G72/L	4.5	4.6	15.8	—	—	—	19.8	8.5
	3000GT	3.0 (2972)	6G72/H	4.5	②	15.8	0.64	—	—	19.8	8.5
	3000GT	3.0 (2972)	6G72/J	4.5	②	15.8	0.64	—	—	19.8	8.5
	3000GT	3.0 (2972)	6G72/K	4.5	②	15.8	0.64	—	2.3	19.8	8.5
	Diamante	3.5 (3497)	6G74/P	4.7	—	18.0	—	—	—	18.7	10.0
2000	Mirage	1.5 (1468)	4G15/A	3.7	3.8	12.6	—	—	—	13.2	5.3
	Mirage	1.8 (1834)	4G93/C	4.2	3.8	12.6	—	—	—	13.2	6.3
	Eclipse	2.4 (2350)	4G64/G	4.5	4.6	16.4	—	—	—	16.3	7.4
	Eclipse	3.0 (2972)	6G72/L	4.5	6.0	17.8	—	—	—	16.3	8.5
	Eclipse Spyder	2.4 (2350)	4G64/G	4.5	4.6	16.4	—	—	—	16.3	7.4
	Eclipse Spyder	3.0 (2972)	6G72/L	4.5	6.0	17.8	—	—	—	16.3	8.5
	Galant	2.4 (2350)	4G64/G	4.5	4.6	12.8	—	—	—	17.0	14.8
	Galant	3.0 (2972)	6G72/L	4.5	4.6	15.8	—	—	—	19.8	8.5
	Diamante	3.5 (3497)	6G74/P	4.7	—	18.0	—	—	—	18.7	10.0

93471CH5

For complete service labor times, order the Chilton Labor Guide

CAPACITIES

Year	Model	Engine Displacement Liters (cc)	Engine ID/VIN	Engine Oil with Filter	Transmission (pts.)		Transfer Case (pts.)	Drive Axle		Fuel Tank (gal.)	Cooling System (qts.)
					5 or 6-Spd	Auto.		Front (pts.)	Rear (pts.)		
2001	Mirage	1.5 (1468)	4G15/A	3.3	4.4	16.2	—	—	—	12.4	5.3
	Mirage	1.8 (1834)	4G93/C	3.9	4.6	16.2	—	—	—	12.4	6.3
	Eclipse	2.4 (2350)	4G64/G	4.5	4.6	16.4	—	—	—	16.4	7.4
	Eclipse	3.0 (2972)	6G72/L	4.5	6.0	17.8	—	—	—	16.4	8.5
	Eclipse Spyder	2.4 (2350)	4G64/G	4.5	4.6	16.4	—	—	—	16.4	7.4
	Eclipse Spyder	3.0 (2972)	6G72/L	4.5	6.0	17.8	—	—	—	16.4	8.5
	Galant	2.4 (2350)	4G64/G	4.5	—	16.2	—	—	—	16.3	14.8
	Galant	3.0 (2972)	6G72/L	4.5	—	17.8	—	—	—	16.3	8.5
	Diamante	3.5 (3497)	6G74/P	4.5	—	17.8	—	—	—	18.7	10.0

NOTE: All capacities are approximate. Add fluid gradually and ensure a proper fluid level is obtained.

① FWD: 4.2 pts.
   AWD: 4.6 pts.
② FWD: 4.8 pts.
   AWD: 5.0 pts.

93471CH6

VALVE SPECIFICATIONS

Year	Engine Displacement Liters (cc)	Engine ID/VIN	Seat Angle (deg.)	Face Angle (deg.)	Spring Test Pressure (lbs. @ in.)	Spring Installed Height (in.)	Stem-to-Guide Clearance (in.)		Stem Diameter (in.)	
							Intake	Exhaust	Intake	Exhaust
1998	1.5 (1468)	4G15/A	44.5-45	45-45.5	①	1.570	0.0008-0.0020	0.0020-0.0035	0.260	0.260
	1.8 (1834)	4G93/C	44.5-45	45-45.5	59@1.740	1.740	0.0008-0.0020	0.0020 0.0035	0.234	0.234
	2.0 (1996)	420A/Y	45	45-45.5	55-60@1.496	1.496	0.0019-0.0025	0.0020-0.0037	0.234	0.233
	2.0 (1997)	4G63/F	44.5-45	45-45.5	54@1.570	1.570	0.0008-0.0020	0.0020-0.0035	0.260	0.256
	2.4 (2350)	4G64/G	44.5-45	45-45.5	60@1.740	1.740	0.0008-0.0020	0.0008-0.0028	0.236	0.232
	3.0 (2972)	6G72/H	44.5-45	45-45.5	40.4@1.591	1.591	0.0012-0.0024	0.0020-0.0035	0.315	0.311
	3.0 (2972)	6G72/J	44.5-45	45-45.5	37.9@1.492	1.492	0.0008-0.0020	0.0020-0.0035	0.260	0.256
	3.0 (2972)	6G72/K	44.5-45	45-45.5	37.9@1.492	1.492	0.0008-0.0020	0.0020-0.0035	0.260	0.256
	3.0 (2972)	6G72/L	44.5-45	45-45.5	40.4@1.591	1.591	0.0012-0.0024	0.0020-0.0035	0.315	0.311
	3.5 (3497)	6G74/P	44-44.5	45-45.5	60@1.740	1.740	0.0008-0.0020	0.0016-0.0028	0.236	0.236
1999	2.0 (1996)	420A/Y	45	45-45.5	55-60@1.496	1.496	0.0019-0.0025	0.0020-0.0037	0.234	0.233
	1.5 (1468)	4G15/A	44.5-45	45-45.5	①	1.570	0.0008-0.0020	0.0020-0.0035	0.260	0.260
	2.0 (1997)	4G63/F	44.5-45	45-45.5	54@1.570	1.570	0.0008-0.0020	0.0020-0.0035	0.260	0.256
	2.4 (2350)	4G64/G	44.5-45	45-45.5	60@1.740	1.740	0.0008-0.0020	0.0008-0.0028	0.236	0.232
	1.8 (1834)	4G93/C	44.5-45	45-45.5	59@1.740	1.740	0.0008-0.0020	0.0020 0.0035	0.234	0.234
	3.0 (2972)	6G72/H	44.5-45	45-45.5	40.4@1.591	1.591	0.0012-0.0024	0.0020-0.0035	0.315	0.311
	3.0 (2972)	6G72/J	44.5-45	45-45.5	37.9@1.492	1.492	0.0008-0.0020	0.0020-0.0035	0.260	0.256
	3.0 (2972)	6G72/K	44.5-45	45-45.5	37.9@1.492	1.492	0.0008-0.0020	0.0020-0.0035	0.260	0.256
	3.0 (2972)	6G72/L	44.5-45	45-45.5	40.4@1.591	1.591	0.0012-0.0024	0.0020-0.0035	0.315	0.311
	3.5 (3497)	6G74/P	44-44.5	45-45.5	60@1.740	1.740	0.0008-0.0020	0.0016-0.0028	0.236	0.236

93471CH7

VALVE SPECIFICATIONS

Year	Engine Displacement Liters (cc)	Engine ID/VIN	Seat Angle (deg.)	Face Angle (deg.)	Spring Test Pressure (lbs. @ in.)	Spring Installed Height (in.)	Stem-to-Guide Clearance (in.)		Stem Diameter (in.)	
							Intake	Exhaust	Intake	Exhaust
2000	1.5 (1468)	4G15/A	44.5-45	45-45.5	①	1.570	0.0008-0.0020	0.0020-0.0035	0.260	0.260
	2.4 (2350)	4G64/G	44.5-45	45-45.5	60@1.740	1.740	0.0008-0.0020	0.0008-0.0028	0.236	0.232
	1.8 (1834)	4G93/C	44.5-45	45-45.5	59@1.740	1.740	0.0008-0.0020	0.0020-0.0035	0.234	0.234
	3.0 (2972)	6G72/L	44.5-45	45-45.5	40.4@1.591	1.591	0.0012-0.0024	0.0020-0.0035	0.315	0.311
	3.5 (3497)	6G74/P	44-44.5	45-45.5	60@1.740	1.740	0.0008-0.0020	0.0016-0.0028	0.236	0.236
2001	1.5 (1468)	4G15/A	44.5-45	45-45.5	①	1.570	0.0008-0.0020	0.0020-0.0035	0.260	0.260
	2.4 (2350)	4G64/G	44.5-45	45-45.5	60@1.740	1.740	0.0008-0.0020	0.0008-0.0028	0.236	0.232
	1.8 (1834)	4G93/C	44.5-45	45-45.5	59@1.740	1.740	0.0008-0.0020	0.0020-0.0035	0.234	0.234
	3.0 (2972)	6G72/L	44.5-45	45-45.5	40.4@1.591	1.591	0.0012-0.0024	0.0020-0.0035	0.315	0.311
	3.5 (3497)	6G74/P	44-44.5	45-45.5	60@1.740	1.740	0.0008-0.0020	0.0016-0.0028	0.236	0.236

① Intake: 51@1.57
  Exhaust: 64@1.57

93471CH8

CRANKSHAFT AND CONNECTING ROD SPECIFICATIONS
All measurements are given in inches.

Year	Engine Displacement Liters (cc)	Engine ID/VIN	Crankshaft				Connecting Rod		
			Main Brg. Journal Dia.	Main Brg. Oil Clearance	Shaft End-play	Thrust on No.	Journal Diameter	Oil Clearance	Side Clearance
1998	1.5 (1468)	4G15/A	1.8900	0.0008-0.0040	0.0020-0.0120	3	1.6500	0.0008-0.0040	0.0039-0.0160
	1.8 (1834)	4G93/C	1.9678-1.9685	0.0008-0.0040	0.0020-0.0098	3	1.7709-1.7717	0.0008-0.0040	0.0039-0.0160
	2.0 (1996)	420A/Y	2.0469-2.0475	0.0008-0.0040	0.0035-0.0094	3	1.8894-1.8900	0.0010-0.0030	0.0051-0.0150
	2.0 (1997)	4G63/F	2.2436-2.2441	0.0008-0.0040	0.0020-0.0098	3	1.7709-1.7717	0.0008-0.0040	0.0039-0.0160
	2.4 (2351)	4G64/G	2.2436-2.2441	0.0008-0.0040	0.0020-0.0098	3	1.7709-1.7717	0.0008-0.0040	0.0039-0.0160
	3.0 (2972)	6G72/H	2.3614-2.3622	0.0008-0.0040	0.0020-0.0120	3	2.1646-2.1654	0.0008-0.0040	0.0039-0.0160
	3.0 (2972)	6G72/J	2.3613-2.3620	0.0008-0.0040	0.0020-0.0120	3	2.1646-2.1654	0.0008-0.0040	0.0039-0.0160
	3.0 (2972)	6G72/K	2.3613-2.3620	0.0008-0.0040	0.0020-0.0120	3	2.1646-2.1654	0.0008-0.0040	0.0039-0.0160
	3.0 (2972)	6G72/L	2.3614-2.3622	0.0008-0.0040	0.0020-0.0120	3	2.1646-2.1654	0.0008-0.0040	0.0039-0.0160
	3.5 (3497)	6G74/P	2.3614-2.3622	0.0008-0.0040	0.0020-0.0120	3	1.9700	0.0008-0.0040	0.0039-0.0160
1999	1.5 (1468)	4G15/A	1.8900	0.0008-0.0040	0.0020-0.0120	3	1.6500	0.0008-0.0040	0.0039-0.0160
	1.8 (1834)	4G93/C	1.9678-1.9685	0.0008-0.0040	0.0020-0.0098	3	1.7709-1.7717	0.0008-0.0040	0.0039-0.0160
	2.0 (1996)	420A/Y	2.0469-2.0475	0.0008-0.0040	0.0035-0.0094	3	1.8894-1.8900	0.0010-0.0030	0.0051-0.0150
	2.0 (1997)	4G63/F	2.2436-2.2441	0.0008-0.0040	0.0020-0.0098	3	1.7709-1.7717	0.0008-0.0040	0.0039-0.0160
	2.4 (2351)	4G64/G	2.2436-2.2441	0.0008-0.0040	0.0020-0.0098	3	1.7709-1.7717	0.0008-0.0040	0.0039-0.0160
	3.0 (2972)	6G72/H	2.3614-2.3622	0.0008-0.0040	0.0020-0.0120	3	2.1646-2.1654	0.0008-0.0040	0.0039-0.0160
	3.0 (2972)	6G72/J	2.3613-2.3620	0.0008-0.0040	0.0020-0.0120	3	2.1646-2.1654	0.0008-0.0040	0.0039-0.0160
	3.0 (2972)	6G72/K	2.3613-2.3620	0.0008-0.0040	0.0020-0.0120	3	2.1646-2.1654	0.0008-0.0040	0.0039-0.0160
	3.0 (2972)	6G72/L	2.3614-2.3622	0.0008-0.0040	0.0020-0.0120	3	2.1646-2.1654	0.0008-0.0040	0.0039-0.0160
	3.5 (3497)	6G74/P	2.3614-2.3622	0.0008-0.0040	0.0020-0.0120	3	1.9700	0.0008-0.0040	0.0039-0.0160

93471CH9

Timing belt service is covered in Section 3 of this manual

CRANKSHAFT AND CONNECTING ROD SPECIFICATIONS

All measurements are given in inches.

Year	Engine Displacement Liters (cc)	Engine ID/VIN	Crankshaft				Connecting Rod		
			Main Brg. Journal Dia.	Main Brg. Oil Clearance	Shaft End-play	Thrust on No.	Journal Diameter	Oil Clearance	Side Clearance
2000	1.5 (1468)	4G15/A	1.8900	0.0008-0.0040	0.0020-0.0120	3	1.6500	0.0008-0.0040	0.0039-0.0160
	1.8 (1834)	4G93/C	1.9678-1.9685	0.0008-0.0040	0.0020-0.0098	3	1.7709-1.7717	0.0008-0.0040	0.0039-0.0160
	2.4 (2351)	4G64/G	2.2436-2.2441	0.0008-0.0040	0.0020-0.0098	3	1.7709-1.7717	0.0008-0.0040	0.0039-0.0160
	3.0 (2972)	6G72/L	2.3614-2.3622	0.0008-0.0040	0.0020-0.0120	3	2.1646-2.1654	0.0008-0.0040	0.0039-0.0160
	3.5 (3497)	6G74/P	2.3614-2.3622	0.0008-0.0040	0.0020-0.0120	3	1.9700	0.0008-0.0040	0.0039-0.0160
2001	1.5 (1468)	4G15/A	1.8900	0.0008-0.0040	0.0020-0.0120	3	1.6500	0.0008-0.0040	0.0039-0.0160
	1.8 (1834)	4G93/C	1.9678-1.9685	0.0008-0.0040	0.0020-0.0098	3	1.7709-1.7717	0.0008-0.0040	0.0039-0.0160
	2.4 (2351)	4G64/G	2.2436-2.2441	0.0008-0.0040	0.0020-0.0098	3	1.7709-1.7717	0.0008-0.0040	0.0039-0.0160
	3.0 (2972)	6G72/L	2.3614-2.3622	0.0008-0.0040	0.0020-0.0120	3	2.1646-2.1654	0.0008-0.0040	0.0039-0.0160
	3.5 (3497)	6G74/P	2.3614-2.3622	0.0008-0.0040	0.0020-0.0120	3	1.9700	0.0008-0.0040	0.0039-0.0160

93471CH0

PISTON AND RING SPECIFICATIONS

All measurements are given in inches.

Year	Engine Displacement Liters (cc)	Engine ID/VIN	Piston Clearance	Ring Gap			Ring Side Clearance		
				Top Compression	Bottom Compression	Oil Control	Top Compression	Bottom Compression	Oil Control
1998	1.5 (1468)	4G15/A	0.0008-0.0016	0.0079-0.0310	0.0079-0.0310	0.0079-0.0390	0.0012-0.0040	0.0008-0.0040	NA
	1.8 (1834)	4G93/C	0.0008-0.0016	0.0098-0.0310	0.0157-0.0310	0.0078-0.0390	0.0012-0.0039	0.0008-0.0039	NA
	2.0 (1996)	420A/Y	0.0008-0.0016	0.0098-0.0310	0.0157-0.0310	0.0078-0.0390	0.0016-0.0039	0.0008-0.0039	NA
	2.0 (1997)	4G63/F	0.0005-0.0017	0.0090-0.0310	0.0190-0.0390	0.0090-0.0390	0.0010-0.0040	0.0010-0.0040	0.0002-0.0070
	2.4 (2351)	4G64/G	0.0008-0.0016	0.0098-0.0310	0.0157-0.0310	0.0039-0.0390	0.0012-0.0040	0.0012-0.0040	NA
	3.0 (2972)	6G72/H	0.0008-0.0020	0.0118-0.0310	0.0177-0.0310	0.0079-0.0390	0.0012-0.0040	0.0008-0.0040	NA
	3.0 (2972)	6G72/J	0.0008-0.0020	0.0118-0.0310	0.0177-0.0310	0.0079-0.0390	0.0012-0.0040	0.0008-0.0040	NA
	3.0 (2972)	6G72/K	0.0008-0.0020	0.0118-0.0310	0.0177-0.0310	0.0079-0.0390	0.0012-0.0040	0.0008-0.0040	NA
	3.0 (2972)	6G72/L	0.0008-0.0020	0.0118-0.0310	0.0177-0.0310	0.0079-0.0390	0.0012-0.0040	0.0008-0.0040	NA
	3.5 (3497)	6G74/P	0.0008-0.0020	0.0118-0.0310	0.0177-0.0310	0.0079-0.0390	0.0012-0.0040	0.0008-0.0040	NA
1999	1.5 (1468)	4G15/A	0.0008-0.0016	0.0079-0.0310	0.0079-0.0310	0.0079-0.0390	0.0012-0.0040	0.0008-0.0040	NA
	1.8 (1834)	4G93/C	0.0008-0.0016	0.0098-0.0310	0.0157-0.0310	0.0078-0.0390	0.0012-0.0039	0.0008-0.0039	NA
	2.0 (1996)	420A/Y	0.0008-0.0016	0.0098-0.0310	0.0157-0.0310	0.0078-0.0390	0.0016-0.0039	0.0008-0.0039	NA
	2.0 (1997)	4G63/F	0.0005-0.0017	0.0090-0.0310	0.0190-0.0390	0.0090-0.0390	0.0010-0.0040	0.0010-0.0040	0.0002-0.0070
	2.4 (2351)	4G64/G	0.0008-0.0016	0.0098-0.0310	0.0157-0.0310	0.0039-0.0390	0.0012-0.0040	0.0012-0.0040	NA
	3.0 (2972)	6G72/H	0.0008-0.0020	0.0118-0.0310	0.0177-0.0310	0.0079-0.0390	0.0012-0.0040	0.0008-0.0040	NA
	3.0 (2972)	6G72/J	0.0008-0.0020	0.0118-0.0310	0.0177-0.0310	0.0079-0.0390	0.0012-0.0040	0.0008-0.0040	NA
	3.0 (2972)	6G72/K	0.0008-0.0020	0.0118-0.0310	0.0177-0.0310	0.0079-0.0390	0.0012-0.0040	0.0008-0.0040	NA
	3.0 (2972)	6G72/L	0.0008-0.0020	0.0118-0.0310	0.0177-0.0310	0.0079-0.0390	0.0012-0.0040	0.0008-0.0040	NA
	3.5 (3497)	6G74/P	0.0008-0.0020	0.0118-0.0310	0.0177-0.0310	0.0079-0.0390	0.0012-0.0040	0.0008-0.0040	NA

93471CI1

Heater Core replacement is covered in Section 2 of this manual

PISTON AND RING SPECIFICATIONS

All measurements are given in inches.

Year	Engine Displacement Liters (cc)	Engine ID/VIN	Piston Clearance	Ring Gap			Ring Side Clearance		
				Top Compression	Bottom Compression	Oil Control	Top Compression	Bottom Compression	Oil Control
2000	1.5 (1468)	4G15/A	0.0008-0.0016	0.0079-0.0310	0.0079-0.0310	0.0079-0.0390	0.0012-0.0040	0.0008-0.0040	NA
	1.8 (1834)	4G93/C	0.0008-0.0016	0.0098-0.0310	0.0157-0.0310	0.0078-0.0390	0.0012-0.0039	0.0008-0.0039	NA
	2.4 (2351)	4G64/G	0.0008-0.0016	0.0098-0.0310	0.0157-0.0310	0.0039-0.0390	0.0012-0.0040	0.0012-0.0040	NA
	3.0 (2972)	6G72/L	0.0008-0.0020	0.0118-0.0310	0.0177-0.0310	0.0079-0.0390	0.0012-0.0040	0.0008-0.0040	NA
	3.5 (3497)	6G74/P	0.0008-0.0020	0.0118-0.0310	0.0177-0.0310	0.0079-0.0390	0.0012-0.0040	0.0008-0.0040	NA
2001	1.5 (1468)	4G15/A	0.0008-0.0016	0.0079-0.0310	0.0079-0.0310	0.0079-0.0390	0.0012-0.0040	0.0008-0.0040	NA
	1.8 (1834)	4G93/C	0.0008-0.0016	0.0098-0.0310	0.0157-0.0310	0.0078-0.0390	0.0012-0.0039	0.0008-0.0039	NA
	2.4 (2351)	4G64/G	0.0008-0.0016	0.0098-0.0310	0.0157-0.0310	0.0039-0.0390	0.0012-0.0040	0.0012-0.0040	NA
	3.0 (2972)	6G72/L	0.0008-0.0020	0.0118-0.0310	0.0177-0.0310	0.0079-0.0390	0.0012-0.0040	0.0008-0.0040	NA
	3.5 (3497)	6G74/P	0.0008-0.0020	0.0118-0.0310	0.0177-0.0310	0.0079-0.0390	0.0012-0.0040	0.0008-0.0040	NA

93471CI2

TORQUE SPECIFICATIONS

All readings in ft. lbs.

Year	Engine Displacement Liters (cc)	Engine ID/VIN	Cylinder Head Bolts	Main Bearing Bolts	Rod Bearing Bolts	Crankshaft Damper Bolts	Flywheel Bolts	Manifold		Spark Plugs	Lug Nut
								Intake	Exhaust		
1998	1.5 (1468)	4G15/A	①	37	12 ②	93	95	12	12	18	65-80
	1.8 (1834)	4G93/C	③	18 ②	15 ②	131	71	15	④	18	65-80
	2.0 (1996)	420A/Y	⑤	55	20 ②	45	94-101	17	17	18	65-80
	2.0 (1997)	4G63/F	⑥	18 ②	14.5 ②	87	98	⑦	⑧	18	65-80
	2.4 (2350)	4G64/G	⑥	14.5 ②	14.5 ②	87	98	13	⑧	18	⑨
	3.0 (2972)	6G72/H	⑩	57	38	136	54	13	14	18	⑨
	3.0 (2972)	6G72/J	⑪	67	38	136	54	16	22	18	⑨
	3.0 (2972)	6G72/K	⑪	54	38	136	54	16	⑫	18	⑨
	3.0 (2972)	6G72/L	⑩	57	38	136	54	13	14	18	⑨
	3.5 (3497)	6G74/P	80	67	38	134	54	16	36	18	65-80
1999	1.5 (1468)	4G15/A	①	37	12 ②	93	95	12	12	18	65-80
	1.8 (1834)	4G93/C	③	18 ②	15 ②	131	71	15	④	18	65-80
	2.0 (1996)	420A/Y	⑥	55	20 ②	45	94-101	17	17	18	65-80
	2.0 (1997)	4G63/F	⑥	18 ②	14.5 ②	87	98	⑦	⑧	18	65-80
	2.4 (2350)	4G64/G	⑥	14.5 ②	14.5 ②	87	98	13	⑧	18	⑨
	3.0 (2972)	6G72/H	⑩	57	38	136	54	13	14	18	⑨
	3.0 (2972)	6G72/J	⑪	67	38	136	54	16	22	18	⑨
	3.0 (2972)	6G72/K	⑪	54	38	136	54	16	⑫	18	⑨
	3.0 (2972)	6G72/L	⑩	57	38	136	54	13	14	18	⑨
	3.5 (3497)	6G74/P	80	67	38	134	54	16	36	18	65-80
2000	1.5 (1468)	4G15/A	①	37	12 ②	93	95	12	12	18	65-80
	1.8 (1834)	4G93/C	③	18 ②	15 ②	131	71	15	④	18	65-80
	2.4 (2350)	4G64/G	⑥	14.5 ②	14.5 ②	87	98	13	⑧	18	⑨
	3.0 (2972)	6G72/L	⑩	57	38	136	54	13	14	18	⑨
	3.5 (3497)	6G74/P	80	67	38	134	54	16	36	18	65-80
2001	1.5 (1468)	4G15/A	①	37	12 ②	93	95	12	12	18	65-80
	1.8 (1834)	4G93/C	③	18 ②	15 ②	131	71	15	④	18	65-80
	2.4 (2350)	4G64/G	⑥	14.5 ②	14.5 ②	87	98	13	⑧	18	⑨
	3.0 (2972)	6G72/L	⑩	57	38	136	54	13	14	18	⑨
	3.5 (3497)	6G74/P	80	67	38	134	54	16	36	18	65-80

① Step 1: Tighten all bolts to 35 ft. lbs.

Step 2: Loosen all bolts to 0 ft. lbs.

Step 3: Tighten all bolts to 15 ft. lbs.

Step 4: Tighten all bolts 90 degrees.

Step 5: Tighten all bolts an additional 90 degrees.

② Torque to specification plus

an additional 90 degrees.

③ Step 1: Tighten all bolts to 54 ft. lbs.

Step 2: Loosen all bolts to 0 ft. lbs.

Step 3: Tighten all bolts to 15 ft. lbs.

Step 4: Tighten all bolts 90 degrees.

Step 5: Tighten all bolts an additional 90 degrees.

④ 8mm: 13 ft. lbs.

10mm: 21 ft. lbs

⑤ Step 1:

Tighten bolts 1-6: 24 ft. lbs.

Tighten bolts 7-10: 20 ft. lbs.

Step 2:

Tighten bolts 1-6: 48 ft. lbs.

Tighten bolts 7-10: 20 ft. lbs.

Step 3: Repeat Step 2

Step 4: Tighten all bolts an additional 90 degrees.

⑥ Step 1: Tighten all bolts to 58 ft. lbs.

Step 2: Loosen all bolts to 0 ft. lbs.

Step 3: Tighten all bolts to 14.5 ft. lbs.

Step 4: Tighten all bolts 90 degrees.

Step 5: Tighten all bolts an additional 90 degrees.

⑦ Bolts: 14 ft. lbs.

Nuts: 26 ft. lbs.

⑧ 8mm: 20 ft. lbs.

10mm: 21 ft. lbs.

⑨ Diamante/Galant: 65-80 ft. lbs.

Eclipse/3000GT: 85-100 ft. lbs.

⑩ Step 1: Tighten all bolts in 3 steps to 80 ft. lbs

Step 2: Loosen all bolts to 0 ft. lbs.

Step 3: Tighen all bolts in 3 steps to 80 ft. lbs

⑪ Step 1: Tighten all bolts in 3 steps to 91 ft. lbs

Step 2: Loosen all bolts to 0 ft. lbs.

Step 3: Tighen all bolts in 3 steps to 91 ft. lbs

⑫ Step 1: Tighten 5 inner nuts to 22 ft. lbs.

Step 2: Tighten 2 outer nuts to 36 ft. lbs.

Step 3: Loosen 2 outer nuts to 7 ft. lbs.

Step 4: Tighten 2 outer nuts to 22 ft. lbs.

93471CI3

Brake service is covered in Section 4 of this manual

BRAKE SPECIFICATIONS
All measurements in inches unless noted

Year	Model		Brake Disc Original Thickness	Brake Disc Minimum Thickness	Brake Disc Maximum Runout	Brake Drum Diameter Original Inside Diameter	Brake Drum Diameter Max. Wear Limit	Brake Drum Diameter Maximum Machine Diameter	Minimum Lining Thickness Front	Minimum Lining Thickness Rear	Brake Caliper Bracket Bolts (ft. lbs.)	Brake Caliper Mounting Bolts (ft. lbs.)
1998	3000GT	F	0.940	0.880	0.002	—	—	—	0.080	—	54	65
	FWD	R	0.710	0.650	0.0031	6.60	6.70	6.70	—	0.040	20	36-43
	3000GT	F	1.180	1.120	0.002	—	—	—	0.080	—	54	65
	AWD	R	0.790	0.720	0.0031	6.60	6.70	6.70	—	0.040	20	36-43
	Diamante	F	0.940	0.880	0.002	—	—	—	0.080	—	54	65
		R	0.410	0.330	0.0023	—	—	—	—	0.039	24	36-43
	Galant	F	0.940	0.880	0.0031	—	—	—	0.080	—	54	65
		R	—	—	—	8.976	9.078	9.078	—	0.040	—	—
	Mirage	F	0.710	0.650	0.0024	—	—	—	0.080	—	36	67-81
		R	—	—	—	8.00	8.10	8.10	—	0.039	—	—
	Eclipse	F	0.940	0.882	0.003	—	—	—	0.080	—	58-72	46-62
		R	—	—	—	9.000	—	9.100	—	0.039	—	—
	Eclipse w/rear disc	F	0.940	0.882	0.003	—	—	—	0.080	—	58-72	46-62
		R	0.390	0.331	0.003	—	—	—	—	0.080	36-43	54
1999	3000GT	F	0.940	0.880	0.002	—	—	—	0.080	—	54	65
	FWD	R	0.710	0.650	0.0031	6.60	6.70	6.70	—	0.040	20	36-43
	3000GT	F	1.180	1.120	0.002	—	—	—	0.080	—	54	65
	AWD	R	0.790	0.720	0.0031	6.60	6.70	6.70	—	0.040	20	36-43
	Diamante	F	0.940	0.880	0.002	—	—	—	0.080	—	54	65
		R	0.410	0.330	0.0023	—	—	—	—	0.039	24	36-43
	Galant	F	0.940	0.880	0.0031	—	—	—	0.080	—	54	65
		R	—	—	—	8.976	9.078	9.078	—	0.040	—	—
	Mirage	F	0.710	0.650	0.0024	—	—	—	0.080	—	36	67-81
		R	—	—	—	8.00	8.10	8.10	—	0.039	—	—
	Eclipse	F	0.940	0.882	0.003	—	—	—	0.080	—	58-72	46-62
		R	—	—	—	9.000	—	9.100	—	0.039	—	—
	Eclipse w/rear disc	F	0.940	0.882	0.003	—	—	—	0.080	—	58-72	46-62
		R	0.390	0.331	0.003	—	—	—	—	0.080	36-43	54
2000	Diamante	F	0.940	0.880	0.002	—	—	—	0.080	—	54	65
		R	0.410	0.330	0.0023	—	—	—	—	0.039	24	36-43
	Galant	F	0.940	0.880	0.0031	—	—	—	0.080	—	54	65
		R	—	—	—	8.976	9.078	9.078	—	0.040	—	—
	Mirage	F	0.710	0.650	0.0024	—	—	—	0.080	—	36	67-81
		R	—	—	—	8.00	8.10	8.10	—	0.039	—	—
	Eclipse	F	0.940	0.882	R.003	—	—	—	0.080	—	58-72	46-62
		R	—	—	—	9.000	—	9.100	—	0.039	—	—
	Eclipse w/rear disc	F	0.940	0.882	0.003	—	—	—	0.080	—	58-72	46-62
		R	0.390	0.331	0.003	—	—	—	—	0.080	36-43	54

93471CI4

BRAKE SPECIFICATIONS

All measurements in inches unless noted

Year	Model		Brake Disc Original Thickness	Brake Disc Minimum Thickness	Brake Disc Maximum Runout	Brake Drum Diameter Original Inside Diameter	Brake Drum Diameter Max. Wear Limit	Brake Drum Diameter Maximum Machine Diameter	Minimum Lining Thickness Front	Minimum Lining Thickness Rear	Brake Caliper Bracket Bolts (ft. lbs.)	Brake Caliper Mounting Bolts (ft. lbs.)
1998	3000GT	F	0.940	0.880	0.002	—	—	—	0.080	—	54	65
	FWD	R	0.710	0.650	0.0031	6.60	6.70	6.70	—	0.040	20	36-43
	3000GT	F	1.180	1.120	0.002	—	—	—	0.080	—	54	65
	AWD	R	0.790	0.720	0.0031	6.60	6.70	6.70	—	0.040	20	36-43
	Diamante	F	0.940	0.880	0.002	—	—	—	0.080	—	54	65
		R	0.410	0.330	0.0023	—	—	—	—	0.039	24	36-43
	Galant	F	0.940	0.880	0.0031	—	—	—	0.080	—	54	65
		R	—	—	—	8.976	9.078	9.078	—	0.040	—	—
	Mirage	F	0.710	0.650	0.0024	—	—	—	0.080	—	36	67-81
		R	—	—	—	8.00	8.10	8.10	—	0.039	—	—
	Eclipse	F	0.940	0.882	0.003	—	—	—	0.080	—	58-72	46-62
		R	—	—	—	9.000	—	9.100	—	0.039	—	—
	Eclipse	F	0.940	0.882	0.003	—	—	—	0.080	—	58-72	46-62
	w/rear disc	R	0.390	0.331	0.003	—	—	—	—	0.080	36-43	54
1999	3000GT	F	0.940	0.880	0.002	—	—	—	0.080	—	54	65
	FWD	R	0.710	0.650	0.0031	6.60	6.70	6.70	—	0.040	20	36-43
	3000GT	F	1.180	1.120	0.002	—	—	—	0.080	—	54	65
	AWD	R	0.790	0.720	0.0031	6.60	6.70	6.70	—	0.040	20	36-43
	Diamante	F	0.940	0.880	0.002	—	—	—	0.080	—	54	65
		R	0.410	0.330	0.0023	—	—	—	—	0.039	24	36-43
	Galant	F	0.940	0.880	0.0031	—	—	—	0.080	—	54	65
		R	—	—	—	8.976	9.078	9.078	—	0.040	—	—
	Mirage	F	0.710	0.650	0.0024	—	—	—	0.080	—	36	67-81
		R	—	—	—	8.00	8.10	8.10	—	0.039	—	—
	Eclipse	F	0.940	0.882	0.003	—	—	—	0.080	—	58-72	46-62
		R	—	—	—	9.000	—	9.100	—	0.039	—	—
	Eclipse	F	0.940	0.882	0.003	—	—	—	0.080	—	58-72	46-62
	w/rear disc	R	0.390	0.331	0.003	—	—	—	—	0.080	36-43	54
2000	Diamante	F	0.940	0.880	0.002	—	—	—	0.080	—	54	65
		R	0.410	0.330	0.0023	—	—	—	—	0.039	24	36-43
	Galant	F	0.940	0.880	0.0031	—	—	—	0.080	—	54	65
		R	—	—	—	8.976	9.078	9.078	—	0.040	—	—
	Mirage	F	0.710	0.650	0.0024	—	—	—	0.080	—	36	67-81
		R	—	—	—	8.00	8.10	8.10	—	0.039	—	—
	Eclipse	F	0.940	0.882	0.003	—	—	—	0.080	—	58-72	46-62
		R	—	—	—	9.000	—	9.100	—	0.039	—	—
	Eclipse	F	0.940	0.882	0.003	—	—	—	0.080	—	58-72	46-62
	w/rear disc	R	0.390	0.331	0.003	—	—	—	—	0.080	36-43	54

93471CI5

For complete Engine Mechanical specifications, see Section 1 of this manual

BRAKE SPECIFICATIONS
All measurements in inches unless noted

| Year | Model | | Brake Disc | | | Brake Drum Diameter | | | Minimum Lining Thickness | | Brake Caliper | |
			Original Thickness	Minimum Thickness	Maximum Runout	Original Inside Diameter	Max. Wear Limit	Maximum Machine Diameter	Front	Rear	Bracket Bolts (ft. lbs.)	Mounting Bolts (ft. lbs.)
2001	Diamante	F	0.940	0.880	0.002	—	—	—	0.080	—	54	65
		R	0.410	0.330	0.0023	—	—	—	—	0.039	24	36-43
	Galant	F	②	③	0.002	—	—	—	0.080	—	①	65
		R	0.390	0.331	0.002	9.000	—	9.100	—	0.040	36-43	54
	Mirage	F	0.940	0.882	0.0012	—	—	—	0.080	—	36	67-81
		R	—	—	—	8.00	8.10	8.10	—	0.039	—	—
	Eclipse	F	0.940	0.882	0.002	—	—	—	0.080	—	58-72	46-62
		R	—	—	—	9.000	—	9.100	—	0.039	—	—
	Eclipse	F	0.940	0.882	0.003	—	—	—	0.080	—	①	46-62
	w/rear disc	R	0.390	0.331	0.003	—	—	—	—	0.080	36-43	54

NA: Not Available

F: Front

R: Rear

① Lock pin (2.4L): 55 ft. lbs.
   Lock bolt (3.0L): 28 ft. lbs.

② 2.4L: 0.940
   3.0L: 1.020

③ 2.4L: 0.88
   3.0L: 0.96

93471CI6

WHEEL ALIGNMENT

Year	Model		Caster Range (+/-Deg.)	Caster Preferred Setting (Deg.)	Camber Range (+/-Deg.)	Camber Preferred Setting (Deg.)	Toe-in (in.)	Steering Axis Inclination (Deg.)
1998	3000 GT	F	0.50	+3.92	0.50	0	0 +/- 0.13	—
		R	—	—	0.50	0	0.01 +/- 0.09	—
	Diamante ①	F	0.50	+3.00	0.50	0	0 +/- 0.13	—
		R	—	—	0.50	-0.69	0.13 +/- 0.13	—
	Diamante ②	F	0.50	+3.00	0.50	0	0 +/- 0.13	—
		R	—	—	0.50	-0.81	0.13 +/- 0.13	—
	Eclipse AWD	F	1.50	+4.69	0.50	-0.09	0 +/- 0.13	7.19
		R	—	—	0.50	-1.31	0.13 +/- 0.13	—
	Eclipse FWD ③	F	1.50	+4.69	0.50	-0.09	0 +/- 0.13	7.19
		R	—	—	0.50	-1.31	0.13 +/- 0.13	—
	Eclipse FWD ②	F	1.50	+4.69	0.50	-0.31	0 +/- 0.13	7.19
		R	—	—	0.50	-1.69	0.13 +/- 0.13	—
	Galant	F	0.50	+3.00	0.50	0	0 +/- 0.13	—
		R	—	—	0.50	-1.00	0 +/- 0.13	—
	Mirage	F	0.50	+2.84	0.50	0	0 +/- 0.13	—
		R	—	—	0.50	-0.69	0.13 +/- 0.10	—
1999	3000 GT	F	0.50	+3.92	0.50	0	0 +/- 0.13	—
		R	—	—	0.50	0	0.01 +/- 0.09	—
	Diamante ①	F	0.50	+3.00	0.50	0	0 +/- 0.13	—
		R	—	—	0.50	-0.69	0.13 +/- 0.13	—
	Diamante ②	F	0.50	+3.00	0.50	0	0 +/- 0.13	—
		R	—	—	0.50	-0.81	0.13 +/- 0.13	—
	Eclipse AWD	F	1.50	+4.69	0.50	-0.09	0 +/- 0.13	7.19
		R	—	—	0.50	-1.31	0.13 +/- 0.13	—
	Eclipse FWD ③	F	1.50	+4.69	0.50	-0.09	0 +/- 0.13	7.19
		R	—	—	0.50	-1.31	0.13 +/- 0.13	—
	Eclipse FWD ②	F	1.50	+4.69	0.50	-0.31	0 +/- 0.13	7.19
		R	—	—	0.50	-1.69	0.13 +/- 0.13	—
	Galant	F	0.50	+3.00	0.50	0	0 +/- 0.13	—
		R	—	—	0.50	-1.00	0 +/- 0.13	—
	Mirage	F	0.50	+2.84	0.50	0	0 +/- 0.13	—
		R	—	—	0.50	-0.69	0.13 +/- 0.10	—
2000	Diamante ①	F	0.50	+3.00	0.50	0	0 +/- 0.13	—
		R	—	—	0.50	-0.69	0.13 +/- 0.13	—
	Diamante ②	F	0.50	+3.00	0.50	0	0 +/- 0.13	—
		R	—	—	0.50	-0.81	0.13 +/- 0.13	—
	Eclipse ③	F	1.50	+4.69	0.50	-0.09	0 +/- 0.13	7.19
		R	—	—	0.50	-1.31	0.13 +/- 0.13	—
	Eclipse ②	F	1.50	+4.69	0.50	-0.31	0 +/- 0.13	7.19
		R	—	—	0.50	-1.69	0.13 +/- 0.13	—
	Galant	F	0.50	+3.00	0.50	0	0 +/- 0.13	—
		R	—	—	0.50	-1.00	0 +/- 0.13	—
	Mirage	F	0.50	+2.84	0.50	0	0 +/- 0.13	—
		R	—	—	0.50	-0.69	0.13 +/- 0.10	—

93471CI7

For Accessory Drive Belt illustrations, see Section 1 of this manual

WHEEL ALIGNMENT

Year	Model		Caster Range (+/-Deg.)	Caster Preferred Setting (Deg.)	Camber Range (+/-Deg.)	Camber Preferred Setting (Deg.)	Toe-in (in.)	Steering Axis Inclination (Deg.)
2001	Diamante ①	F	0.50	+3.00	0.50	0	0 +/- 0.13	—
		R	—	—	0.50	-0.69	0.13 +/- 0.13	—
	Diamante ②	F	0.50	+3.00	0.50	0	0 +/- 0.13	—
		R	—	—	0.50	-0.81	0.13 +/- 0.13	—
	Eclipse ③	F	1.50	+4.69	0.50	-0.09	0 +/- 0.13	7.19
		R	—	—	0.50	-1.31	0.13 +/- 0.13	—
	Eclipse ②	F	1.50	+4.69	0.50	-0.31	0 +/- 0.13	7.19
		R	—	—	0.50	-1.69	0.13 +/- 0.13	—
	Galant	F	0.50	+3.00	0.50	0	0 +/- 0.13	—
		R	—	—	0.50	-1.00	0 +/- 0.13	—
	Mirage	F	0.50	+2.84	0.50	0	0 +/- 0.13	—
		R	—	—	0.50	-0.69	0.13 +/- 0.10	—

① With 15 in. wheels

② With 16 in. wheels

③ With 14 in. wheels

93471CI8

TIRE, WHEEL AND BALL JOINT SPECIFICATIONS

Year	Model	OEM Tires		Tire Pressures (psi)		Wheel Size	Ball Joint Inspection
		Standard	Optional	Front	Rear		
1998	Mirage DE	P175/70R13	None	31	31	5-J	9-56 in. ①
	Mirage LS	P185/65R14	None	31	31	5.5-JJ	9-56 in. ①
	Mirage GS Spyder	P215/50VR17	None	32	30	6.5-JJ	9-56 in. ①
	Eclipse RS	P195/70HR14	None	32	30	5.5-JJ	U: 3-13 in. ① / L: 13 in.
	Eclipse GS	P195/70HR14	None	32	30	5.5-JJ	U: 3-13 in. ① / L: 13 in.
	Eclipse GS-T	P205/55HR16	P215/50VR17	32	30	6-JJ	U: 3-13 in. ① / L: 13 in.
	Eclipse GS-X	P205/55HR16	P215/50VR17	32	30	6-JJ	U: 3-13 in. ① / L: 13 in.
	3000 GT, base	255/55VR16	None	32	30	8-JJ	84-108 in. ①
	3000 GT SL	245/45ZR17	None	32	30	8.5-JJ	84-108 in. ①
	3000 GT VR-4	245/40ZR18	None	32	30	8.5-JJ	84-108 in. ①
	Galant, exc. LS	185/70HR14	None	29	26	5.5-JJ	3-13 in. ①
	Galant LS	195/60HR15	None	30	26	6-JJ	3-13 in. ①
	Diamante	205/65VR15	None	32	30	6-JJ	87-109 in. ①
	Diamante Wagon	P205/65HR15	None	32	30	6-JJ	87-109 in. ①
1999	Diamante	205/65VR15	None	32	30	6-JJ	87-109 in. ①
	3000 GT, base	255/55VR16	None	32	30	8-JJ	87-109 in. ①
	3000 GT SL	225/55VR17	None	32	30	8.5-JJ	84-108 in. ①
	3000 GT VR-4	245/40ZR18	None	32	30	8.5-JJ	84-108 in. ①
	Galant LS/GT-Z/ES V6	205/55R16	None	32	29	6-JJ	3-13 in. ①
	Galant DE/ES 4-cyl	195/65R15	None	32	29	6-JJ	3-13 in. ①
	Mirage DE	P175/70R13	None	31	31	5-J	9-56 in. ①
	Mirage LS	P185/65R14	None	31	31	5.5-JJ	9-56 in. ①
	Mirage GS Spyder	P215/50VR17	None	32	30	6.5-JJ	9-56 in. ①
	Eclipse RS	P195/70HR14	None	32	30	5.5-JJ	U: 3-13 in. ① / L: 13 in.
	Eclipse GS	P195/70HR14	None	32	30	5.5-JJ	U: 3-13 in. ① / L: 13 in.
	Eclipse GS-T	P205/55HR16	P215/50VR17	32	30	6-JJ	U: 3-13 in. ① / L: 13 in.
	Eclipse GS-X	P205/55HR16	P215/50VR17	32	30	6-JJ	U: 3-13 in. ① / L: 13 in.
2000	Diamante	205/65VR15	None	32	30	6-JJ	87-109 in. ①
	Galant LS/GT-Z/ES V6	205/55R16	None	32	29	6-JJ	3-13 in. ①
	Galant DE/ES 4-cyl	195/65R15	None	32	29	6-JJ	3-13 in. ①
	Mirage DE	P175/65R14	None	31	31	5.5-JJ	9-56 in. ①
	Mirage LS	P185/65R14	None	31	31	5.5-JJ	9-56 in. ①
	Eclipse RS	195/65R15	None	32	29	6-JJ	U: 3-13 in. ① / L: 13 in.
	Eclipse GS	205/55R16	None	32	29	6-JJ	U: 3-13 in. ① / L: 13 in.
	Eclipse GS-T	215/50R17	None	32	29	6.5-JJ	U: 3-13 in. ① / L: 13 in.
	Eclipse GS-X	P205/55HR16	P215/50VR17	32	30	6-JJ	U: 3-13 in. ① / L: 13 in.

93471CI9

For Tire, Wheel and Ball Joint specifications, see Section 1 of this manual

TIRE, WHEEL AND BALL JOINT SPECIFICATIONS

| Year | Model | OEM Tires | | Tire Pressures (psi) | | Wheel Size | Ball Joint Inspection |
		Standard	Optional	Front	Rear		
2001	Diamante	205/65VR15	None	32	30	6-JJ	87-109 in. ①
	Galant LS/GT-Z/ES V6	205/55R16	None	32	29	6-JJ	3-13 in. ①
	Galant DE/ES 4-cyl	185/70HR14	None	29	26	5.5-JJ	3-13 in. ①
	Mirage DE	P175/70R13	None	31	31	5-J	9-56 in. ①
	Mirage LS	P185/65R14	None	31	31	5.5-JJ	9-56 in. ①
	Mirage GS Spyder	P215/50VR17	None	32	30	6.5-JJ	9-56 in. ①
	Eclipse RS	P195/70HR14	None	32	30	5.5-JJ	U: 3-13 in. ① L: 13 in.
	Eclipse GS	P195/70HR14	None	32	30	5.5-JJ	U: 3-13 in. ① L: 13 in.
	Eclipse GS-T	P205/55HR16	P215/50VR17	32	30	6-JJ	U: 3-13 in. ① L: 13 in.

OEM: Original Equipment Manufacturer

PSI: Pounds Per Square Inch

STD: Standard

OPT: Optional

L: Lower

U: Upper

① Torque required in inch lbs. to rotate ball joint when removed from the knuckle

93471CI0

SCHEDULED MAINTENANCE INTERVALS
Mitsubishi—Diamante, Eclipse, Galant, Mirage, 3000GT

TO BE SERVICED	TYPE OF SERVICE	VEHICLE MILEAGE INTERVAL (x1000)												
		7.5	15	22.5	30	37.5	45	52.5	60	67.5	75	82.5	90	97.5
Engine oil & filter ①	R	✓	✓	✓	✓	✓	✓	✓	✓	✓	✓	✓	✓	✓
Automatic transaxle fluid & filter	S/I		✓		✓		✓		✓		✓		✓	
Brake hoses	S/I		✓		✓		✓		✓		✓		✓	
Disc brake pads	S/I		✓		✓		✓		✓		✓		✓	
Driveshaft boots	S/I		✓		✓		✓		✓		✓		✓	
Valve clearance (Mirage)	S/I		✓		✓		✓		✓		✓		✓	
Air cleaner element	R				✓				✓				✓	
Engine coolant	R				✓				✓				✓	
Spark plugs (except Diamante & 3000GT w/platinum tip)	R				✓				✓				✓	
Spark plugs (Diamante & 3000GT w/platinum tip)	R								✓					
Ball joints & steering linkage seals	S/I				✓				✓				✓	
Drive belt(s)	S/I				✓				✓				✓	
Exhaust system	S/I				✓				✓				✓	
Fuel hoses	S/I				✓				✓				✓	
Manual transaxle fluid (Mirage)	S/I				✓				✓				✓	
Manual transaxle fluid (including transfer) (Eclipse & 3000GT)	S/I				✓				✓				✓	
Manual transaxle fluid (Galant)	S/I				✓				✓				✓	
Rear axle oil (Eclipse & 3000GT AWD)	S/I				✓				✓				✓	
Rear drum brake linings & rear wheel cylinders (Eclipse, Galant & Mirage)	S/I				✓				✓				✓	
Ignition cables	R								✓					
Timing belt(s)	R								✓					
Distributor cap & rotor (except 3000GT)	S/I								✓					
EVAP system (except canister)	S/I								✓					

93471CJ1

For Wheel Alignment specifications, see Section 1 of this manual

SCHEDULED MAINTENANCE INTERVALS
Mitsubishi—Diamante, Eclipse, Galant, Mirage, 3000GT

TO BE SERVICED	TYPE OF SERVICE	VEHICLE MILEAGE INTERVAL (x1000)												
		7.5	15	22.5	30	37.5	45	52.5	60	67.5	75	82.5	90	97.5
Fuel system (tank, pipe line, connection & fuel tank filler tube cap)	S/I								✓					

R: Replace   S/I: Service or Inspect

① 3000GT turbo: replace every 5000 miles.

FREQUENT OPERATION MAINTENANCE (SEVERE SERVICE)

If a vehicle is operated under any of the following conditions it is considered severe service:

- Extremely dusty areas.

- 50% or more of the vehicle operation is in 32°C (90°F) or higher temperatures, or constant operation in temperatures below 0°C (32°F).

- Prolonged idling (vehicle operation in stop and go traffic).

- Frequent short running periods (engine does not warm to normal operating temperatures).

- Police, taxi, delivery usage or trailer towing usage.

Oil & oil filter: change every 3000 miles.

Disc brake pads: service or inspect ever 6000 miles.

Air filter element: service or inspect every 15,000 miles.

Automatic transaxle fluid & filter: replace every 15,000 miles.

Spark plugs (except Diamante & 3000GT w/platinum tip): replace every 15,000 miles.

Manual transaxle oil (including transfer (Galant, Mirage & 3000GT): replace every 30,000 miles.

93471CJ2

SCHEDULED MAINTENANCE INTERVALS
MITSUBISHI
DIAMANTE, ECLIPSE, GALANT, MIRAGE, 3000GT

The following should be used as a guide when determining the amount of work required for a particular service. In estimating how long a particular Scheduled Maintenance Service should take, please observe the following:

- Labor Time is time based on field research and data supplied by the vehicle manufacturer.
- Labor time operations are given in hours and tenths of an hour.
- All labor operations are to be used as a guide.

Mechanic Skill Level Codes:
(A) PRECISION: Highly skilled with multiple certification.
(B) GENERAL: Normally skilled with certification.
(C) MAINTENANCE: Semi-skilled working on certification.

	LABOR TIME
7500 Mile Service (C)	
All Models	.4
15000 Mile Service (C)	
All Models	.9
w/AT add	.1
22500 Mile Service (C)	
All Models	.4
30000 Mile Service (B)	
All Models	3.1
Inspect rear brake lining & wheel cyls	
Eclipse, Galant & Mirage add	.2
w/AT add	.1
w/AWD add	.1
37500 Mile Service (C)	
All Models	.4

	LABOR TIME
45000 Mile Service (C)	
All Models	.9
w/AT add	.1
52500 Mile Service (C)	
All Models	.4
60000 Mile Service (B)	
All Models	6.8
Inspect rear brake lining & wheel cyls	
Eclipse, Galant & Mirage add	.2
Inspect dist. cap & rotor, add	.2
w/AT add	.1
w/AWD add	.1

	LABOR TIME
67500 Mile Service (C)	
All Models	.4
75000 Mile Service (C)	
All Models	.9
w/AT add	.1
82500 Mile Service (C)	
All Models	.4
90000 Mile Service (B)	
All Models	3.1
Inspect rear brake lining & wheel cyls	
Eclipse, Galant & Mirage add	.2
w/AT add	.1
w/AWD add	.1
97500 Mile Service (C)	
All Models	.4

93471CJ3

For Maintenance Interval recommendations, see Section 1 of this manual

NISSAN
200SX • 240SX • ALTIMA • MAXIMA • SENTRA

ENGINE AND VEHICLE IDENTIFICATION

| | Engine | | | | | | | Model Year | |
Code ①	Liters (cc)	Cu. In.	Cyl.	Fuel Sys.	Engine Type	Eng. Mfg.		Code ②	Year
GA16DE	1.6 (1597)	97	4	MFI	DOHC	Nissan		W	1998
SR20DE	2.0 (1998)	122	4	MFI	DOHC	Nissan		X	1999
KA24DE	2.4 (2389)	146	4	MFI	DOHC	Nissan		Y	2000
VQ30DE	3.0 (2988)	182	6	MFI	DOHC	Nissan		1	2001
								2	2002

MFI: Multi-port Fuel Injection

DOHC: Double Overhead Camshaft

① The Engine Code is stamped on the engine block near the starter.

② 10th position of the Vehicle Identification Number (VIN)

93471CJ4

GENERAL ENGINE SPECIFICATIONS

Year	Model	Engine Displacement Liters (cc)	Engine Series (ID/VIN)	Net Horsepower @ rpm	Net Torque @ rpm (ft. lbs.)	Bore x Stroke (in.)	Compression Ratio	Oil Pressure @ rpm
1998	200SX	1.6 (1597)	GA16DE	115@6000	108@4000	2.99x3.46	9.9:1	50@3000
	200SX	2.0 (1998)	SR20DE	140@6400	132@4800	3.39x3.39	9.5:1	46@3200
	240SX	2.4 (2389)	KA24DE	155@5600	160@4400	3.50x3.78	9.5:1	60@3000
	Altima	2.4 (2389)	KA24DE	150@5600	154@4400	3.50x3.78	9.2:1	60@3000
	Maxima	3.0 (2988)	VQ30DE	190@5600	205@4000	3.66x2.89	10.0:1	63@3000
	Sentra	1.6 (1597)	GA16DE	115@6000	108@4000	2.99x3.46	9.9:1	50@3000
1999	Altima	2.4 (2389)	KA24DE	150@5600	154@5600	3.50x3.78	9.2:1	60@3000
	Maxima	3.0 (2988)	VQ30DE	190@5600	205@4000	3.66x2.89	10.0:1	63@3000
	Sentra	1.6 (1597)	GA16DE	115@6000	108@4000	2.99x3.46	9.9:1	50@3000
	Sentra	2.0 (1998)	SR20DE	140@6400	132@4800	3.39x3.39	9.5:1	46@3200
2000	Altima	2.4 (2389)	KA24DE	150@5600	154@5600	3.50x3.78	9.2:1	60@3000
	Maxima	3.0 (2988)	VQ30DE	190@5600	205@4000	3.66x2.89	10.0:1	63@3000
	Sentra	1.8 (1769)	QG18DE	126@6000	129@2400	3.15x3.46	9.5:1	50@3000
	Sentra	2.0 (1998)	SR20DE	145@6400	136@4800	3.39x3.39	9.5:1	46@3200
2001	Altima	2.4 (2389)	KA24DE	150@5600	154@5600	3.50x3.78	9.2:1	60@3000
	Maxima	3.0 (2988)	VQ30DE	190@5600	205@4000	3.66x2.89	10.0:1	63@3000
	Sentra	1.8 (1769)	QG18DE	126@6000	129@2400	3.15x3.46	9.5:1	50@3000
	Sentra	2.0 (1998)	SR20DE	145@6400	136@4800	3.39x3.39	9.5:1	46@3200

93471CJ5

For Tune-up, Capacities and Firing orders, see Section 1 of this manual

ENGINE TUNE-UP SPECIFICATIONS

Year	Engine Displacement Liters (cc)	Engine ID/VIN	Spark Plug Gap (in.)	Ignition Timing (deg.) MT	AT	Fuel Pump (psi) ①	Idle Speed (rpm) MT	AT ②	Valve Clearance Intake ③	Exhaust ③
1998	1.6 (1597)	GA16DE	0.041	8B	8B	36	625 ⑤	725	0.015	0.016
	2.0 (1998)	SR20DE	0.033	15B	15B	36	800	800	HYD	HYD
	2.4 (2389)	KA24DE	0.041 ④	20B	20B	33	650	650	0.015	0.015
	3.0 (2988)	VQ30DE	0.041 ④	15B	15B	34	650	700	0.014	0.015
1999	1.6 (1597)	GA16DE	0.041	8B	8B	36	625 ⑤	725	0.015	0.016
	2.0 (1998)	SR20DE	0.033	15B	15B	36	800	800	HYD	HYD
	2.4 (2389)	KA24DE	0.041	20B	20B	33	650	650	0.015	0.016
	3.0 (2988)	VQ30DE	0.041	15B	15B	34	650	700	0.014	0.015
2000	1.8 (1769)	QG18DE	0.041	9B	9B	36	625 ⑤	725	0.015	0.016
	2.0 (1998)	SR20DE	0.033	15B	15B	36	800	800	HYD	HYD
	2.4 (2389)	KA24DE	0.041	20B	20B	33	650	650	0.015	0.016
	3.0 (2988)	VQ30DE	0.041	15B	15B	34	650	700	0.014	0.015
2001	1.8 (1769)	QG18DE	0.041	9B	9B	36	625 ⑤	725	0.015	0.016
	2.0 (1998)	SR20DE	0.033	15B	15B	36	800	800	HYD	HYD
	2.4 (2389)	KA24DE	0.041	20B	20B	33	650	650	0.015	0.016
	3.0 (2988)	VQ30DE	0.041	15B	15B	34	650	700	0.014	0.015

NOTE: The Vehicle Emission Control Information label often reflects specification changes made during production. The label figures must be used if they differ from those in this chart.

B: Before top dead center

HYD: Hydraulic

① System pressure at idle with vacuum hose connected; should increase to 43 psi when disconnected

② Automatic transmission in neutral

③ Engine warm

④ Do not check or adjust gap on platinum-tipped spark plugs

⑤ Canada: 750

93471CJ6

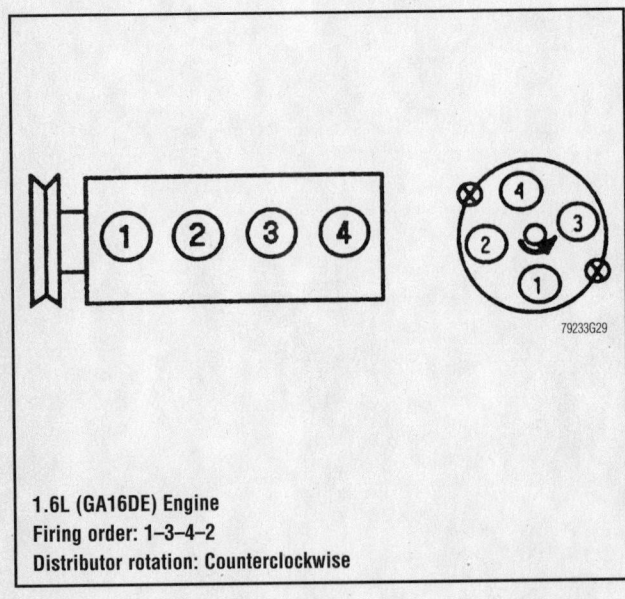

1.6L (GA16DE) Engine
Firing order: 1–3–4–2
Distributor rotation: Counterclockwise

79233G29

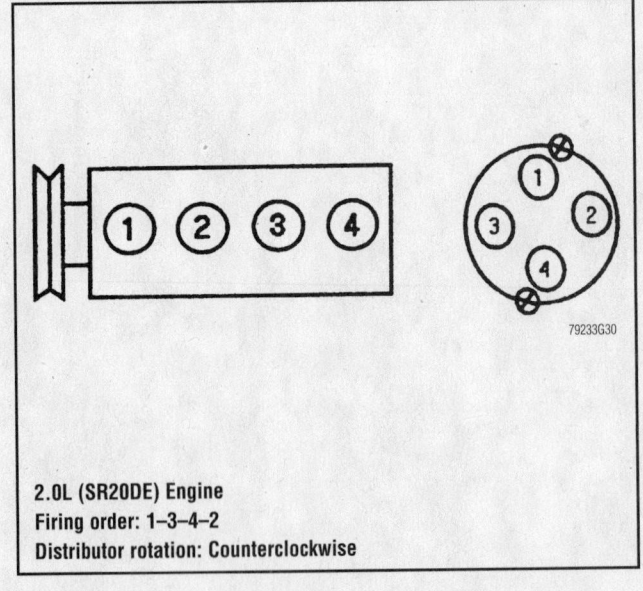

2.0L (SR20DE) Engine
Firing order: 1–3–4–2
Distributor rotation: Counterclockwise

79233G30

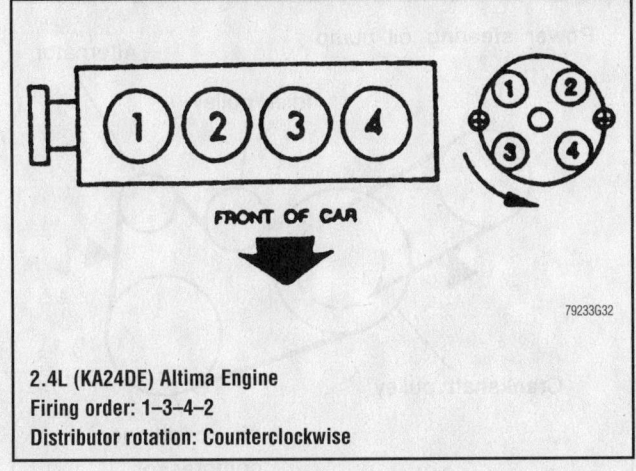

2.4L (KA24DE) Altima Engine
Firing order: 1–3–4–2
Distributor rotation: Counterclockwise

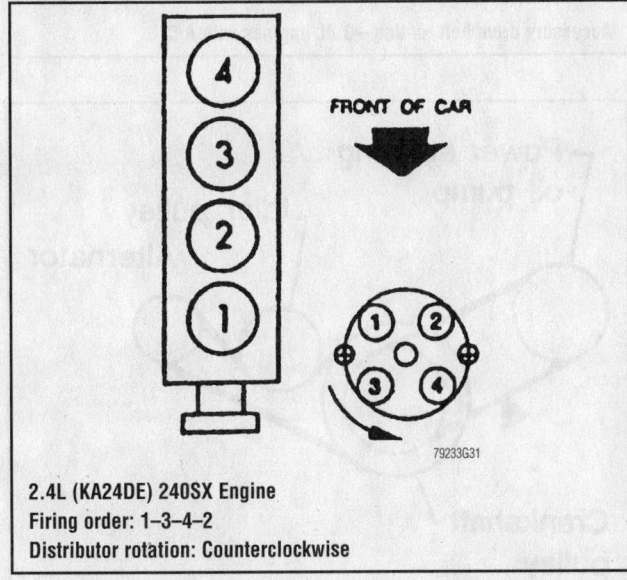

2.4L (KA24DE) 240SX Engine
Firing order: 1–3–4–2
Distributor rotation: Counterclockwise

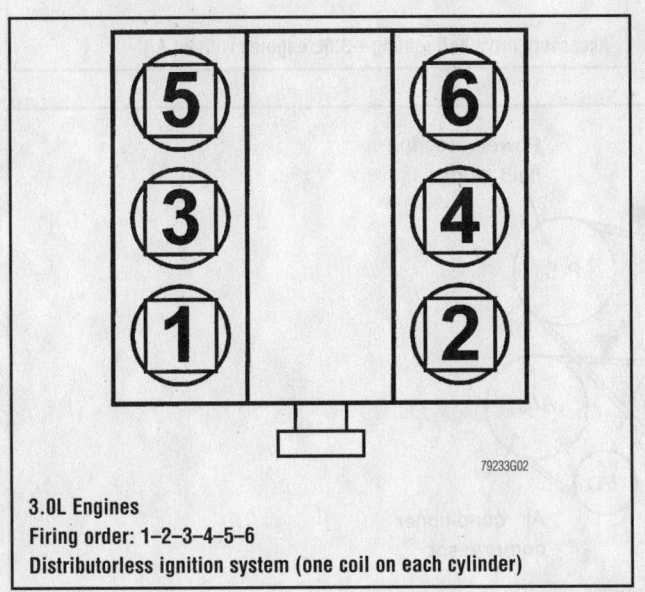

3.0L Engines
Firing order: 1–2–3–4–5–6
Distributorless ignition system (one coil on each cylinder)

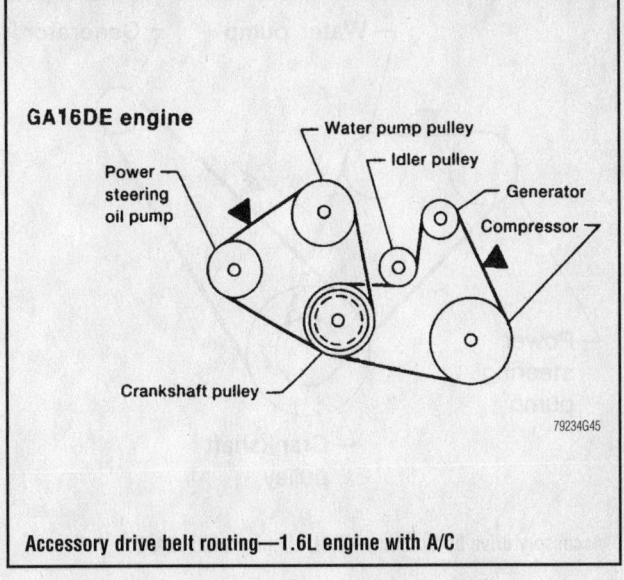

Accessory drive belt routing—1.6L engine with A/C

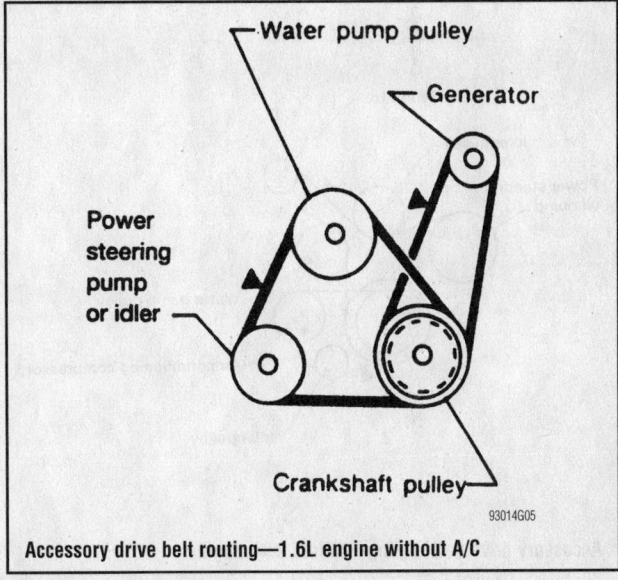

Accessory drive belt routing—1.6L engine without A/C

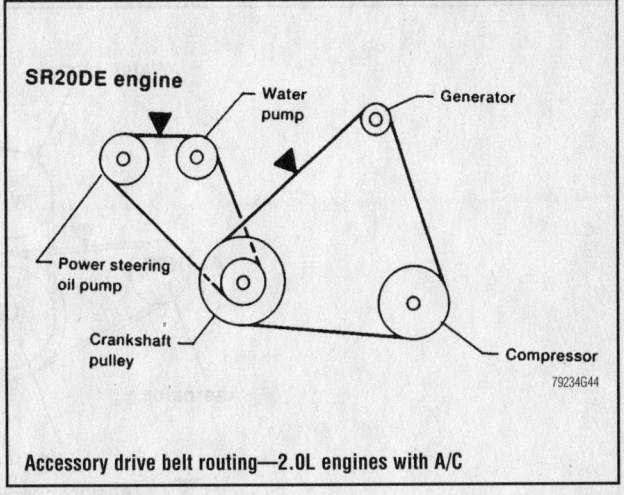

Accessory drive belt routing—2.0L engines with A/C

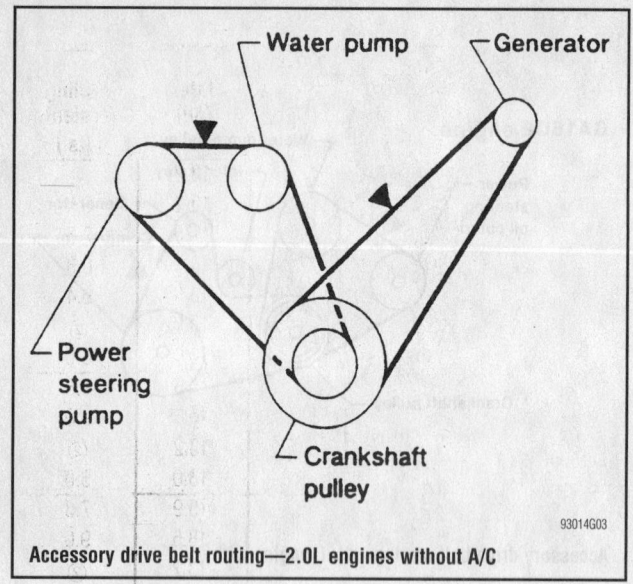

Accessory drive belt routing—2.0L engines without A/C

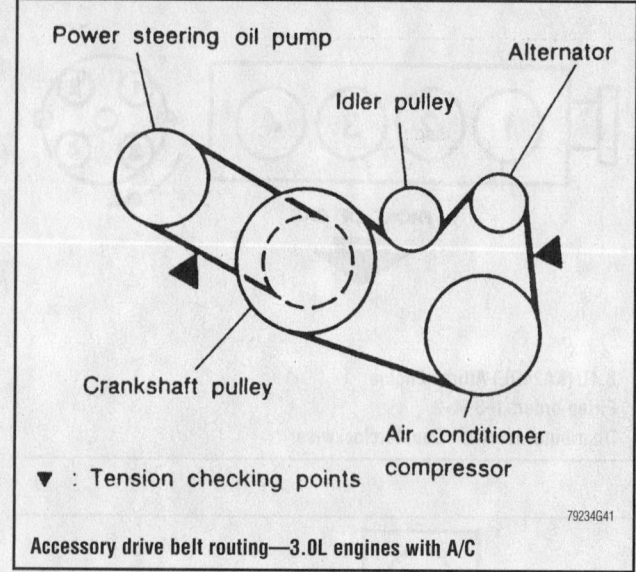

▼ : Tension checking points

Accessory drive belt routing—3.0L engines with A/C

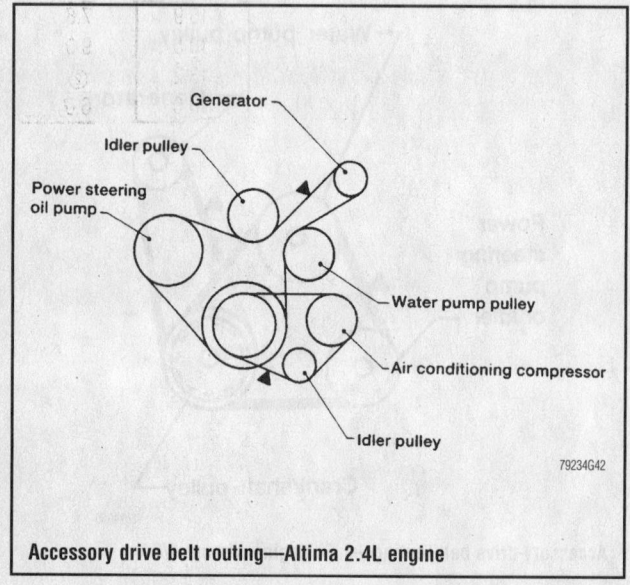

Accessory drive belt routing—Altima 2.4L engine

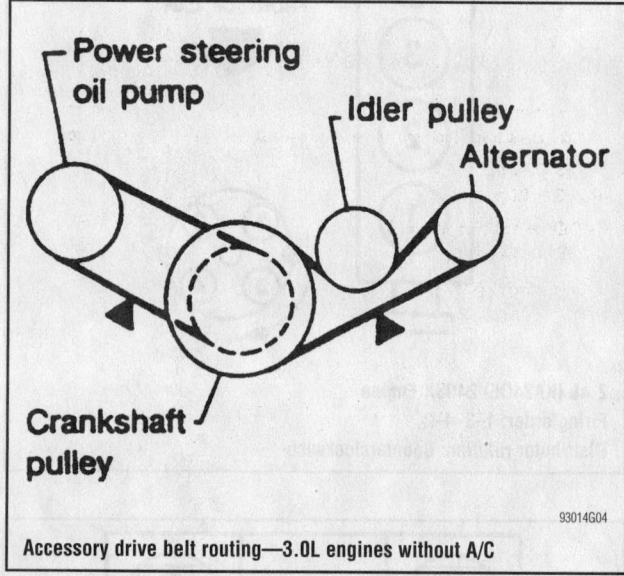

Accessory drive belt routing—3.0L engines without A/C

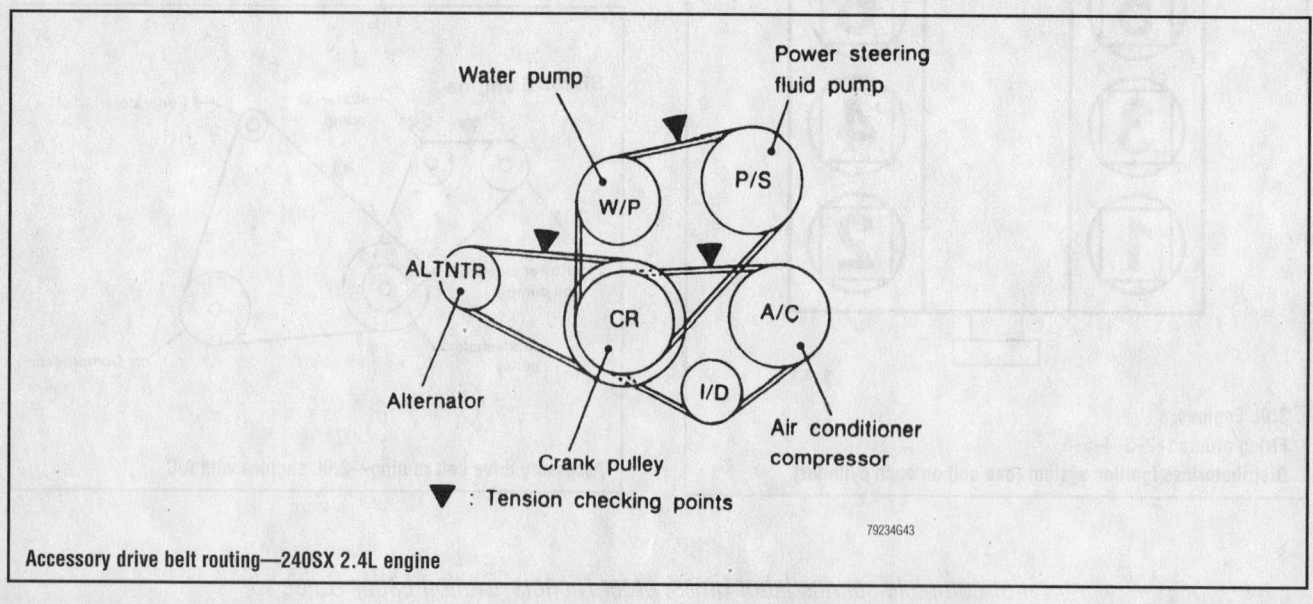

▼ : Tension checking points

Accessory drive belt routing—240SX 2.4L engine

CAPACITIES

Year	Model	Engine ID/VIN	Engine Displacement Liters (cc)	Engine Oil with Filter (qts.)	Transmission (pts.)		Drive Axle Rear (pts.)	Fuel Tank (gal.)	Cooling System (qts.)
					5-Spd	Auto.			
1998	200SX	GA16DE	1.6 (1597)	3.5	①	14.8	—	13.2	②
	200SX	SR20DE	2.0 (1998)	3.6	①	14.8	—	13.2	6.5
	240SX	KA24DE	2.4 (2389)	4.0	5.3	17.5	③	17.2	7.3
	Altima	KA24DE	2.4 (2389)	4.1	10.0	20.0	—	15.9	8.3
	Maxima	VQ30DE	3.0 (2988)	4.3	9.5	20.0	—	18.5	9.4
	Sentra	GA16DE	1.6 (1597)	3.5	①	14.8	—	13.2	②
	Sentra	SR20DE	2.0 (1998)	3.4	7.5	14.8	—	13.0	6.0
1999	Altima	KA24DE	2.4 (2389)	4.1	10.0	20.0	—	15.9	7.8
	Maxima	VQ30DE	3.0 (2988)	4.3	9.5	20.0	—	18.5	9.0
	Sentra	GA16DE	1.6 (1597)	3.5	①	14.8	—	13.2	②
	Sentra	SR20DE	2.0 (1998)	3.4	7.5	14.8	—	13.0	6.0
2000	Altima	KA24DE	2.4 (2389)	4.1	10.0	20.0	—	15.9	7.8
	Maxima	VQ30DE	3.0 (2988)	4.3	9.5	20.0	—	18.5	9.0
	Sentra	QG18DE	1.8 (1769)	3.5	①	14.8	—	13.2	②
	Sentra	SR20DE	2.0 (1998)	3.4	7.5	14.8	—	13.0	6.0
2001	Altima	KA24DE	2.4 (2389)	4.1	10.0	20.0	—	15.9	7.8
	Maxima	VQ30DE	3.0 (2988)	4.3	9.5	20.0	—	18.5	9.0
	Sentra	QG18DE	1.8 (1769)	3.5	①	14.8	—	13.2	②
	Sentra	SR20DE	2.0 (1998)	3.4	7.5	14.8	—	13.0	6.0

NOTE: All capacities are approximate. Add fluid gradually and check to be sure a proper fluid level is obtained.

① RS5F31A: 6.5 pts.
RS5F32V: 8.0 pts.
② GA16DE with MT: 5.5 qts.
GA16DE with AT: 6.0 qts.
③ With limited slip: 3.1 pts.
Standard: 2.8 pts.

93471CJ7

VALVE SPECIFICATIONS

Year	Engine ID/VIN	Engine Displacement Liters (cc)	Seat Angle (deg.)	Face Angle (deg.)	Spring Test Pressure (lbs. @ in.)	Spring Installed Height (in.)	Stem-to-Guide Clearance (in.)		Stem Diameter (in.)	
							Intake	Exhaust	Intake	Exhaust
1998	GA16DE	1.6 (1597)	45	45.25-45.75	77@0.995	NA	0.0008-0.0020	0.0016-0.0028	0.2152-0.2157	0.2144-0.2150
	SR20DE	2.0 (1998)	45	45.25-45.75	137@1.181	NA	0.0008-0.0021	0.0016-0.0029	0.2348-0.2354	0.2341-0.2346
	KA24DE	2.4 (2389)	45	45.25-45.75	123@1.024	NA	0.0008-0.0021	0.0016-0.0029	0.2742-0.2748	0.2734-0.2740
	VQ30DE	3.0 (2988)	45	45.25-45.75	102@1.085	NA	0.0008-0.0021	0.0016-0.0029	0.2348-0.2354	0.2341-0.2346
1999	GA16DE	1.6 (1597)	45	45.25-45.75	77@0.995	NA	0.0008-0.0020	0.0016-0.0028	0.2152-0.2157	0.2144-0.2150
	SR20DE	2.0 (1998)	45	45.25-45.75	137@1.181	NA	0.0008-0.0021	0.0016-0.0029	0.2348-0.2354	0.2341-0.2346
	KA24DE	2.4 (2389)	45	45.25-45.75	123@1.024	NA	0.0008-0.0021	0.0016-0.0029	0.2742-0.2748	0.2734-0.2740
	VQ30DE	3.0 (2988)	45	45.25-45.75	102@1.085	NA	0.0008-0.0021	0.0016-0.0029	0.2348-0.2354	0.2341-0.2346
2000	QG18DE	1.8 (1769)	45	45.25-45.75	83@0.931	NA	0.0008-0.0020	0.0016-0.0028	0.2152-0.2157	0.2144-0.2150
	SR20DE	2.0 (1998)	45	45.25-45.75	137@1.181	NA	0.0008-0.0021	0.0016-0.0029	0.2348-0.2354	0.2341-0.2346
	KA24DE	2.4 (2389)	45	45.25-45.75	123@1.024	NA	0.0008-0.0021	0.0016-0.0029	0.2742-0.2748	0.2734-0.2740
	VQ30DE	3.0 (2988)	45	45.25-45.75	102@1.085	NA	0.0008-0.0021	0.0016-0.0029	0.2348-0.2354	0.2341-0.2346
2001	QG18DE	1.8 (1769)	45	45.25-45.75	83@0.931	NA	0.0008-0.0020	0.0016-0.0028	0.2152-0.2157	0.2144-0.2150
	SR20DE	2.0 (1998)	45	45.25-45.75	137@1.181	NA	0.0008-0.0021	0.0016-0.0029	0.2348-0.2354	0.2341-0.2346
	KA24DE	2.4 (2389)	45	45.25-45.75	123@1.024	NA	0.0008-0.0021	0.0016-0.0029	0.2742-0.2748	0.2734-0.2740
	VQ30DE	3.0 (2988)	45	45.25-45.75	102@1.085	NA	0.0008-0.0021	0.0016-0.0029	0.2348-0.2354	0.2341-0.2346

NA: Not Available

93471CJ8

CRANKSHAFT AND CONNECTING ROD SPECIFICATIONS

All measurements are given in inches.

Year	Engine Displacement Liters (cc)	Engine ID/VIN	Crankshaft				Connecting Rod		
			Main Brg. Journal Dia.	Main Brg. Oil Clearance	Shaft End-play	Thrust on No.	Journal Diameter	Oil Clearance	Side Clearance
1998	1.6 (1597)	GA16DE	1.9668-1.9671	0.0007-0.0017	0.0024-0.0071	3	1.5735-1.5738	0.0141-0.0390	0.0079-0.0185
	2.0 (1998)	SR20DE	2.1643-2.1646	0.0002-0.0009	0.0039-0.0102	3	1.8885-1.8887	0.0008-0.0018	0.0079-0.0138
	2.4 (2389)	KA24DE	2.3609-2.3612	0.0008-0.0019	0.0020-0.0070	3	1.9672-1.9675	0.0004-0.0014	0.0080-0.0160
	3.0 (2988)	VQ30DE	2.3610-2.3612	0.0014-0.0021	0.0039-0.0098	3	1.7704-1.7706	0.0013-0.0023	0.0079-0.0138
1999	1.6 (1597)	GA16DE	1.9668-1.9671	0.0007-0.0017	0.0024-0.0071	3	1.5735-1.5738	0.0141-0.0390	0.0079-0.0185
	2.0 (1998)	SR20DE	2.1643-2.1646	0.0002-0.0009	0.0039-0.0102	3	1.8885-1.8887	0.0008-0.0018	0.0079-0.0138
	2.4 (2389)	KA24DE	2.3609-2.3612	0.0008-0.0019	0.0020-0.0070	3	1.9672-1.9675	0.0004-0.0014	0.0080-0.0160
	3.0 (2988)	VQ30DE	2.3610-2.3612	0.0014-0.0021	0.0039-0.0098	3	1.7704-1.7706	0.0013-0.0023	0.0079-0.0138
2000	1.8 (1769)	QG18DE	1.9668-1.9671	0.0007-0.0017	0.0024-0.0071	3	1.6929-1.6934	0.0006-0.0015	0.0079-0.0185
	2.0 (1998)	SR20DE	2.1643-2.1646	0.0002-0.0009	0.0039-0.0102	3	1.8885-1.8887	0.0008-0.0018	0.0079-0.0138
	2.4 (2389)	KA24DE	2.3609-2.3612	0.0008-0.0019	0.0020-0.0070	3	1.9672-1.9675	0.0004-0.0014	0.0080-0.0160
	3.0 (2988)	VQ30DE	2.3610-2.3612	0.0014-0.0021	0.0039-0.0098	3	1.7704-1.7706	0.0013-0.0023	0.0079-0.0138
2001	1.8 (1769)	QG18DE	1.9668-1.9671	0.0007-0.0017	0.0024-0.0071	3	1.6929-1.6934	0.0006-0.0015	0.0079-0.0185
	2.0 (1998)	SR20DE	2.1643-2.1646	0.0002-0.0009	0.0039-0.0102	3	1.8885-1.8887	0.0008-0.0018	0.0079-0.0138
	2.4 (2389)	KA24DE	2.3609-2.3612	0.0008-0.0019	0.0020-0.0070	3	1.9672-1.9675	0.0004-0.0014	0.0080-0.0160
	3.0 (2988)	VQ30DE	2.3610-2.3612	0.0014-0.0021	0.0039-0.0098	3	1.7704-1.7706	0.0013-0.0023	0.0079-0.0138

93471CJ9

Timing belt service is covered in Section 3 of this manual

PISTON AND RING SPECIFICATIONS

All measurements are given in inches.

Year	Engine Displacement Liters (cc)	Engine ID/VIN	Piston Clearance	Ring Gap			Ring Side Clearance		
				Top Compression	Bottom Compression	Oil Control	Top Compression	Bottom Compression	Oil Control
1998	1.6 (1597)	GA16DE	0.0006-0.0014	0.0079-0.0173	—	0.0079-0.0272	0.0016-0.0031	—	SNUG
	2.0 (1998)	SR20DE	0.0006-0.0014	0.0079-0.0118	0.0138-0.0197	0.0079-0.0236	0.0018-0.0031	0.0012-0.0026	SNUG
	2.4 (2389)	KA24DE	0.0006-0.0014	0.0110-0.0205	0.0079-0.0272	0.0100-1.0000	0.0016-0.0031	0.0012-0.0028	SNUG
	3.0 (2988)	VQ30DE	0.0006-0.0014	0.0087-0.0126	0.0126-0.0185	0.0079-0.0236	0.0016-0.0031	0.0012-0.0028	SNUG
1999	1.6 (1597)	GA16DE	0.0006-0.0014	0.0079-0.0173	—	0.0079-0.0272	0.0016-0.0031	—	SNUG
	2.0 (1998)	SR20DE	0.0006-0.0014	0.0079-0.0118	0.0138-0.0197	0.0079-0.0236	0.0018-0.0031	0.0012-0.0026	SNUG
	2.4 (2389)	KA24DE	0.0006-0.0014	0.0110-0.0205	0.0079-0.0272	0.0100-1.0000	0.0016-0.0031	0.0012-0.0028	SNUG
	3.0 (2988)	VQ30DE	0.0006-0.0014	0.0087-0.0126	0.0126-0.0185	0.0079-0.0236	0.0016-0.0031	0.0012-0.0028	SNUG
2000	1.8 (1769)	QG18DE	0.0010-0.0018	0.0079-0.0154	0.0126-0.0220	0.0079-0.0272	0.0018-0.0031	0.0012-0.0028	0.0026-0.0053
	2.0 (1998)	SR20DE	0.0006-0.0014	0.0079-0.0118	0.0138-0.0197	0.0079-0.0236	0.0018-0.0031	0.0012-0.0026	SNUG
	2.4 (2389)	KA24DE	0.0006-0.0014	0.0110-0.0205	0.0079-0.0272	0.0100-1.0000	0.0016-0.0031	0.0012-0.0028	SNUG
	3.0 (2988)	VQ30DE	0.0006-0.0014	0.0087-0.0126	0.0126-0.0185	0.0079-0.0236	0.0016-0.0031	0.0012-0.0028	SNUG
2001	1.8 (1769)	QG18DE	0.0010-0.0018	0.0079-0.0154	0.0126-0.0220	0.0079-0.0272	0.0018-0.0031	0.0012-0.0028	0.0026-0.0053
	2.0 (1998)	SR20DE	0.0006-0.0014	0.0079-0.0118	0.0138-0.0197	0.0079-0.0236	0.0018-0.0031	0.0012-0.0026	SNUG
	2.4 (2389)	KA24DE	0.0006-0.0014	0.0110-0.0205	0.0079-0.0272	0.0100-1.0000	0.0016-0.0031	0.0012-0.0028	SNUG
	3.0 (2988)	VQ30DE	0.0006-0.0014	0.0087-0.0126	0.0126-0.0185	0.0079-0.0236	0.0016-0.0031	0.0012-0.0028	SNUG

93471CJ0

TORQUE SPECIFICATIONS

All readings in ft. lbs.

Year	Engine Displacement Liters (cc)	Engine ID/VIN	Cylinder Head Bolts	Main Bearing Bolts	Rod Bearing Bolts	Crankshaft Damper Bolts	Flywheel Bolts	Manifold		Spark Plugs	Lug Nuts
								Intake	Exhaust		
1998	1.6 (1597)	GA16DE	①	34-38	②	98-112	③	14	19	18	79
	2.0 (1998)	SR20DE	④	⑤	⑥	105-112	61-69	14	30	18	79
	2.4 (2389)	KA24DE	⑦	34-41	⑥	105-112	105-112	14	32	18	80
	3.0 (2988)	VQ30DE	⑧	⑨	⑩	⑪	61-69	⑫	23	18	80
1999	1.6 (1597)	GA16DE	①	34-38	②	98-112	③	14	19	18	79
	2.0 (1998)	SR20DE	④	⑤	⑥	105-112	61-69	14	30	18	79
	2.4 (2389)	KA24DE	⑦	34-41	⑥	105-112	105-112	14	32	18	80
	3.0 (2988)	VQ30DE	⑧	⑨	⑩	⑪	61-69	⑫	23	18	80
2000	1.8 (1769)	QG18DE	⑭	34-38	②	98-112	③	14	19	18	79
	2.0 (1998)	SR20DE	④	⑤	⑥	105-112	61-69	14	30	18	79
	2.4 (2389)	KA24DE	⑦	34-41	⑥	105-112	105-112	14	32	18	80
	3.0 (2988)	VQ30DE	⑧	⑨	⑩	⑪	61-69	⑫	23	18	80
2001	1.8 (1769)	QG18DE	⑭	34-38	②	98-112	③	14	19	18	79
	2.0 (1998)	SR20DE	④	⑤	⑥	105-112	61-69	14	30	18	79
	2.4 (2389)	KA24DE	⑦	34-41	⑥	105-112	105-112	14	32	18	80
	3.0 (2988)	VQ30DE	⑧	⑨	⑩	⑪	61-69	⑫	23	18	80

① Bolt Nos. 1-10:
Step 1: 22 ft. lbs.
Step 2: 43 ft. lbs.
Step 3: Loosen completely then retorque to 22 ft. lbs.
Step 4: 43 ft. lbs. or an additional 50-55 degrees
Bolt Nos. 11-15: Torque last, to 72 inch lbs.

② Step 1: 12 ft. lbs.
Step 2: 19 ft. lbs. or an additional 35-40 degrees

③ Manual transmission: 61-69 ft. lbs.
Automatic transmission: 69-76 ft. lbs.

④ Step 1: 29 ft. lbs.
Step 2: 58 ft. lbs.
Step 3: Loosen completely then retorque to 30 ft. lbs.
Step 4: Turn each bolt, in sequence,
an additional 90-100 degrees
Step 5: Repeat Step 4

⑤ Step 1: 20-24 ft. lbs.
Step 2: 75-80 degrees
Step 3: Loosen completely and retorque to 24-28 ft. lbs.
Step 4: 45-50 degree turn

⑥ 12 ft. lbs. plus an additional 60-65 degrees

⑦ Step 1: 22 ft. lbs.
Step 2: 58 ft. lbs.
Step 3: Loosen completely then retorque to 22 ft. lbs.
Step 4: 58 ft. lbs. or an additional 80-85 degrees

⑧ Step 1: 29-36 ft. lbs.
Step 2: Plus 60-65 degrees

⑨ Step 1: 3.6-7.2 ft. lbs.
Step 2: 20-23 ft. lbs.

⑩ Step 1: 29 ft. lbs.
Step 2: 90 ft. lbs.
Step 3: Loosen completely and retorque to 25-33 ft. lbs.
Step 4: Plus 90 ft. lbs. or 70 degrees
Step 5: Tighten two bolts marked with an "X" to 7-9 ft. lbs.

⑪ Step 1: 29-36 ft. lbs.
Step 2: 60-66 degrees

⑫ Step 1: 10-12 ft. lbs.
Step 2: 43-48 ft. lbs. or an additional 60-65 degrees

⑬ Bolt Nos. 1-10:
Step 1: 22 ft. lbs.
Step 2: 43 ft. lbs.
Step 3: Loosen completely then retorque to 22 ft. lbs.
Step 4: 43 ft. lbs. or an additional 50-55 degrees
Bolt Nos. 11-14: Torque last, to 72 inch lbs.

93471CK1

Heater Core replacement is covered in Section 2 of this manual

BRAKE SPECIFICATIONS
All measurements in inches unless noted

Year	Model		Brake Disc Original Thickness	Brake Disc Minimum Thickness	Brake Disc Maximum Run-out	Brake Drum Diameter Original Inside Diameter	Brake Drum Diameter Max. Wear Limit	Brake Drum Diameter Maximum Machine Diameter	Minimum Lining Thickness Front	Minimum Lining Thickness Rear	Brake Caliper Bracket Bolts (ft. lbs.)	Brake Caliper Mounting Bolts (ft. lbs.)
1998	240SX	F	0.790	0.710	0.003	—	—	—	0.079	—	40-47	—
		R	0.350	0.310	0.003	—	—	—	—	0.079	40-47	—
	Altima	F	0.870	0.787	0.003	—	—	—	0.079	—	53-72	16-23
		R	0.390	0.315	0.003	9.00	NA	9.06	—	0.059	—	—
	Maxima	F	0.870	0.787	0.003	—	—	—	0.079	—	53-72	16-23
		R	0.350	0.315	0.003	—	—	—	—	0.059	—	—
	Sentra/200SX	F	0.710	0.630	0.003	—	—	—	0.079	—	40-47	—
		R	0.280	0.236	0.003	7.09	7.13	7.13	—	0.059	—	—
1999	Altima	F	0.870	0.787	0.003	—	—	—	0.079	—	53-72	16-23
		R	0.390	0.315	0.003	9.00	NA	9.06	—	0.059	—	—
	Maxima	F	0.870	0.787	0.003	—	—	—	0.079	—	53-72	16-23
		R	0.350	0.315	0.003	—	—	—	—	0.059	—	—
	Sentra	F	0.710	0.630	0.003	—	—	—	0.079	—	40-47	12-14
		R	0.280	0.236	0.003	7.09	7.13	7.13	—	0.059	—	—
2000	Altima	F	0.870	0.787	0.003	—	—	—	0.079	—	53-72	16-23
		R	0.390	0.315	0.003	9.00	NA	9.06	—	0.059	—	—
	Maxima	F	0.870	0.787	0.003	—	—	—	0.079	—	53-72	16-23
		R	0.350	0.315	0.003	—	—	—	—	0.059	—	—
	Sentra	F	0.710	0.630	0.003	—	—	—	0.079	—	40-47	12-14
		R	0.280	0.236	0.003	7.09	7.13	7.13	—	0.059	—	—
2001	Altima	F	0.870	0.787	0.003	—	—	—	0.079	—	53-72	16-23
		R	0.390	0.315	0.003	9.00	NA	9.06	—	0.059	—	—
	Maxima	F	0.870	0.787	0.003	—	—	—	0.079	—	53-72	16-23
		R	0.350	0.315	0.003	—	—	—	—	0.059	—	—
	Sentra	F	0.710	0.630	0.003	—	—	—	0.079	—	40-47	12-14
		R	0.280	0.236	0.003	7.09	7.13	7.13	—	0.059	—	—

NA: Not Available

93471CK2

WHEEL ALIGNMENT

Year	Model		Caster Range (+/-Deg.)	Caster Preferred Setting (Deg.)	Camber Range (+/-Deg.)	Camber Preferred Setting (Deg.)	Toe-in (in.)	Steering Axis Inclination (Deg.)
1998	200SX	F	0.75	+1.44	0.75	-0.56	0.08 +/- 0.08	14.75
		R	—	—	0.75	-1.00	0.05 +/- 0.14	—
	240 SX	F	0.75	+6.75	0.75	0.75	0.09 +/- 0.04	13.66
		R	—	—	0.50	-1.16	0.09 +/- 0.09	—
	Altima	F	0.75	+2.66	0.75	-0.10	0.04 +/- 0.04	14.09
		R	—	—	0.75	-1.25	0.08 +/- 0.04	—
	Maxima ①	F	0.75	+2.75	0.75	-0.25	0.04 +/- 0.04	—
		R	—	—	0.75	-1.00	0.04 +/- 0.16	—
	Maxima ②	F	0.75	+2.75	0.75	-0.33	0.04 +/- 0.04	—
		R	—	—	0.75	-1.00	0.04 +/- 0.16	—
	Sentra	F	0.75	+1.42	0.75	-0.58	0.08 +/- 0.08	—
		R	—	—	0.75	+1.00	0.04 +/- 0.15	—
1999	Altima	F	0.75	+2.66	0.75	-0.10	0.04 +/- 0.04	14.09
		R	—	—	0.75	-1.25	0.08 +/- 0.04	—
	Maxima ①	F	0.75	+2.75	0.75	-0.25	0.04 +/- 0.04	—
		R	—	—	0.75	-1.00	0.04 +/- 0.16	—
	Maxima ②	F	0.75	+2.75	0.75	-0.33	0.04 +/- 0.04	—
		R	—	—	0.75	-1.00	0.04 +/- 0.16	—
	Sentra	F	0.75	+1.42	0.75	-0.58	0.08 +/- 0.08	—
		R	—	—	0.75	+1.00	0.04 +/- 0.15	—
2000	Altima	F	0.75	+2.66	0.75	-0.10	0.04 +/- 0.04	14.09
		R	—	—	0.75	-1.25	0.08 +/- 0.04	—
	Maxima ①	F	0.75	+2.75	0.75	-0.25	0.04 +/- 0.04	—
		R	—	—	0.75	-1.00	0.04 +/- 0.16	—
	Maxima ②	F	0.75	+2.75	0.75	-0.33	0.04 +/- 0.04	—
		R	—	—	0.75	-1.00	0.04 +/- 0.16	—
	Sentra	F	0.75	+1.42	0.75	-0.58	0.08 +/- 0.08	—
		R	—	—	0.75	+1.00	0.04 +/- 0.15	—
2001	Altima	F	0.75	+2.66	0.75	-0.10	0.04 +/- 0.04	14.09
		R	—	—	0.75	-1.25	0.08 +/- 0.04	—
	Maxima ①	F	0.75	+2.75	0.75	-0.25	0.04 +/- 0.04	—
		R	—	—	0.75	-1.00	0.04 +/- 0.16	—
	Maxima ②	F	0.75	+2.75	0.75	-0.33	0.04 +/- 0.04	—
		R	—	—	0.75	-1.00	0.04 +/- 0.16	—
	Sentra	F	0.75	+1.42	0.75	-0.58	0.08 +/- 0.08	—
		R	—	—	0.75	+1.00	0.04 +/- 0.15	—

① With P225/55R16, P215/55R16 tires
② With P205/65R15 tires

93471CK3

Brake service is covered in Section 4 of this manual

TIRE, WHEEL AND BALL JOINT SPECIFICATIONS

Year	Model	OEM Tires		Tire Pressures (psi)		Wheel Size	Ball Joint Inspection
		Standard	Optional	Front	Rear		
1998	Altima	P195/65R15	P205/60R15	30	30	6-JJ	①
	Maxima GLE	P205/65SR15	None	29	29	6.5-JJ	①
	Maxima GXE	P205/65SR15	None	29	29	6JJ	①
	Maxima SE	P215/55R16	None	29	29	6.5JJ	①
	Sentra, base	P155/80R13	None	26	26	5-J	①
	Sentra XE	P175/70R13	None	26	26	5-J	①
	Sentra GXE	P175/65R14	None	26	26	5.5-JJ	①
	Sentra GLE	P175/65R14	None	26	26	5.5-JJ	①
	Sentra SE	P195/55R15	None	30	30	6-JJ	①
	200SX, base	P175/70R13	None	26	26	5-J	①
	200SX SE	P175/65R14	None	26	26	5.5-JJ	①
	200SX SE-R	P195/55R15	None	26	26	6-JJ	①
	240SX	P195/60HR15	P205/60R15 P205/55VR16	26	26	6-JJ	①
1999	Altima	P195/65R15	P205/60R15	30	30	6-JJ	①
	Maxima GLE	P205/65SR15	None	29	29	6.5-JJ	①
	Maxima GXE	P205/65SR15	None	29	29	6JJ	①
	Maxima SE	P215/55R16	None	29	29	6.5JJ	①
	Sentra, base	P155/80R13	None	26	26	5-J	①
	Sentra XE	P175/70R13	None	26	26	5-J	①
	Sentra GXE	P175/65R14	None	26	26	5.5-JJ	①
	Sentra GLE	P175/65R14	None	26	26	5.5-JJ	①
	Sentra SE	P195/55R15	None	30	30	6-JJ	①
2000-01	Altima	P195/65R15	P205/60R15	30	30	6-JJ	①
	Maxima GLE	P215/55R16	None	29	29	6.5-JJ	①
	Maxima GXE	P205/65SR15	None	29	29	6JJ	①
	Maxima SE	P215/55R16	P225/50R17	29	29	6.5J/7J	①
	Sentra, base	P155/80R13	None	26	26	5-J	①
	Sentra XE	P175/70R13	None	26	26	5-J	①
	Sentra GXE	P175/65R14	None	26	26	5.5-JJ	①
	Sentra GLE	P175/65R14	None	26	26	5.5-JJ	①
	Sentra SE	P195/55R15	None	30	30	6-JJ	①

OEM: Original Equipment Manufacturer

PSI: Pounds Per Square Inch

STD: Standard

OPT: Optional

① Replace if any measurable movement is found.

93471CK4

SCHEDULED MAINTENANCE INTERVALS
Nissan Altima, Sentra, Maxima, 200 SX, 240 SX

TO BE SERVICED	TYPE OF SERVICE	VEHICLE MILEAGE INTERVAL (x1000)												
		7.5	15	22.5	30	37.5	45	52.5	60	67.5	75	82.5	90	97.5
Engine oil & filter	R	✓	✓	✓	✓	✓	✓	✓	✓	✓	✓	✓	✓	✓
Brake lines & cables	S/I		✓		✓		✓		✓		✓		✓	
Brake pads, discs, drums & linings	S/I		✓		✓		✓		✓		✓		✓	
Differential gear oil (240SX)	S/I		✓		✓		✓		✓		✓		✓	
Driveshaft boots (Altima, Maxima, Sentra & 200SX)	S/I		✓		✓		✓		✓		✓		✓	
Exhaust system	S/I			✓					✓				✓	
Transmission or transaxle fluid	S/I		✓		✓		✓		✓		✓		✓	
Air cleaner filter	R				✓				✓				✓	
Spark plugs (except below)	R				✓				✓				✓	
Spark plugs (platinum tip) (Sentra, 200SX SR20DE, Maxima & 240SX	R								✓					
Idle RPM (GA16DE)	S/I				✓				✓				✓	
Steering gear & linkage, axle & suspension parts	S/I				✓				✓				✓	
Engine coolant	R								✓					
Timing belt	R								✓					
Drive belts	S/I								✓					
Fuel lines	S/I								✓					
Vapor lines	S/I								✓					

R: Replace   S/I: Service or Inspec

FREQUENT OPERATION MAINTENANCE (SEVERE SERVICE)

If a vehicle is operated under any of the following conditions it is considered severe service:

- Extremely dusty areas.

- 50% or more of the vehicle operation is in 32°C (90°F) or higher temperatures, or constant operation in temperatures below 0°C (32°F).

- Prolonged idling (vehicle operation in stop and go traffic).

- Frequent short running periods (engine does not warm to normal operating temperatures).

- Police, taxi, delivery usage or trailer towing usage.

Oil & oil filter: change every 3750 miles.

Brake pads & discs: service or inspect every 7500 miles.

Driveshaft boots (Altima, Maxima, Sentra & 200SX): service or inspect every 7500 miles.

Exhaust system: service or inspect every 7500 miles.

Steering gear & linkage, axle & suspension parts: service or inspect every 7500 miles.

Steering linkage ball joints & front suspension ball joints: service or inspect every 7500 miles.

Air cleaner filter: service or inspect every 15,000 miles.

93471CK5

For complete Engine Mechanical specifications, see Section 1 of this manual

SCHEDULED MAINTENANCE INTERVALS
NISSAN
200SX, 240SX, ALTIMA, MAXIMA, SENTRA

The following should be used as a guide when determining the amount of work required for a particular service.
In estimating how long a particular Scheduled Maintenance Service should take, please observe the following:

- Labor Time is time based on field research and data supplied by the vehicle manufacturer.
- Labor time operations are given in hours and tenths of an hour.
- All labor operations are to be used as a guide.

Mechanic Skill Level Codes:
(A) PRECISION: Highly skilled with multiple certification.
(B) GENERAL: Normally skilled with certification.
(C) MAINTENANCE: Semi-skilled working on certification.

	LABOR TIME		LABOR TIME		LABOR TIME
7500 Mile Service (C)		**37500 Mile Service (C)**		**75000 Mile Service (C)**	
All Models	.4	All Models	.4	200SX, 240SX	1.0
15000 Mile Service (C)		**45000 Mile Service (C)**		300ZX	1.1
200SX, 240SX	1.0	200SX, 240SX	1.0	Altima, Maxima	.9
300ZX	1.1	300ZX	1.1	Sentra	.9
Altima, Maxima	.9	Altima, Maxima	.9	**82500 Mile Service (C)**	
Sentra	.9	Sentra	.9	All Models	.4
22500 Mile Service (C)		**52500 Mile Service (C)**		**90000 Mile Service (B)**	
All Models	.4	All Models	.4	200SX	2.1
30000 Mile Service (B)		**60000 Mile Service (B)**		240SX	1.9
200SX	2.1	200SX	2.7	300ZX	2.4
240SX	1.9	300ZX	3.1	Altima, Maxima	1.7
300ZX	2.4	Altima	2.9	Sentra	2.0
Altima, Maxima	1.7	Maxima	2.9	*w/GA16DE engine add*	.2
Sentra	2.0	Sentra	3.1	**97500 Mile Service (C)**	
w/GA16DE engine add	.2	*w/GA16DE engine add*	.2	All Models	.4
		67500 Mile Service (C)			
		All Models	.4		

93471CK6

SAAB
9-3 • 9-5 • 900 • 9000

ENGINE AND VEHICLE IDENTIFICATION CHART

Engine							Model Year	
Code ①	Liters (cc)	Cu. In.	Cyl.	Fuel Sys.	Engine Type	Eng. Mfg.	Code ②	Year
B	2.3 (2290)	140	I4	MFI	DOHC	Saab	W	1998
M	2.3 (2290)	140	I4	MFI-Turbo	DOHC	Saab	X	1999
N	2.0 (1985)	121	I4	MFI-Turbo	DOHC	Saab	Y	2000
R	2.3 (2290)	140	I4	MFI-Turbo	DOHC	Saab	1	2001
U	2.3 (2290)	140	I4	MFI-Turbo	DOHC	Saab	2	2002
W	3.0 (2962)	180	V6	MFI	DOHC	Saab		

MFI: Multiport Fuel Injection

① 8th position of VIN

② 10th position of VIN

93471CK7

For Accessory Drive Belt illustrations, see Section 1 of this manual

GENERAL ENGINE SPECIFICATIONS

Year	Model	Engine Displacement Liters (cc)	Engine ID/VIN	Fuel System	Net Horsepower @ rpm	Net Torque @ rpm (ft. lbs.)	Bore x Stroke (in.)	Compression Ratio	Oil Pressure @ rpm
1998	900	2.0 (1985)	B204L/N	MFI-Turbo	185@5500	194@2100	3.54x3.07	9.2:1	39@2000
	900	2.3 (2290)	B234I/B	MFI	150@5700	155@4300	3.54x3.54	10.5:1	39@2000
	9000	2.3 (2290)	B234R/R	MFI-Turbo	225@5500	153@1800	3.54x3.54	9.25:1	39@2000
	9-3	2.0 (1985)	B204L/N	MFI-Turbo	185@5500	194@2100	3.54x3.07	9.2:1	39@2000
	9-5	2.3 (2290)	B235R/R	MFI-Turbo	185@5500	207@1800	3.54x3.54	9.3:1	39@2000
1999	9-3	2.0 (1985)	B204L/N	MFI-Turbo	185@5500	194@2100	3.54x3.07	9.2:1	39@2000
	9-5	2.3 (2290)	B235R/R	MFI-Turbo	185@5500	207@1800	3.54x3.54	9.3:1	39@2000
2000	9-3	2.0 (1985)	B204L/N	MFI-Turbo	185@5500	194@2100	3.54x3.07	9.2:1	39@2000
	9-5	2.3 (2290)	B235R/R	MFI-Turbo	185@5500	207@1800	3.54x3.54	9.3:1	39@2000
2001	9-3	2.0 (1985)	B204L/N	MFI-Turbo	185@5500	194@2100	3.54x3.07	9.2:1	39@2000
	9-5	2.3 (2290)	B235R/R	MFI-Turbo	185@5500	207@1800	3.54x3.54	9.3:1	39@2000
	9-5	3.0 (2962)	B308E/E	MFI-Turbo	200@5000	229@2500	3.38x3.35	9.5:1	N/A

MFI: Multiport Fuel Injection

NA: Not Available

93471CK8

TUNE-UP SPECIFICATIONS

Year	Engine Displacement Liters (cc)	Engine ID/VIN	Spark Plugs Gap (in.)	Ignition Timing (deg.) MT	AT	Fuel Pump (psi)	Idle Speed (rpm) MT	AT	Valve Clearance In.	Ex.
1998	2.0 (1985)	B204L/N	0.039	①	①	43 ②	850	850	HYD	HYD
	2.3 (2290)	B234I/B	0.024	①	①	43 ②	850	850	HYD	HYD
	2.3 (2290)	B234R/R	0.040	①	①	43 ②	850	850	HYD	HYD
1999	2.0 (1985)	B204L/N	0.039	①	①	43 ②	850	850	HYD	HYD
	2.3 (2290)	B234I/B	0.024	①	①	43 ②	850	850	HYD	HYD
	2.3 (2290)	B234R/R	0.040	①	①	43 ②	850	850	HYD	HYD
2000	2.0 (1985)	B204L/N	0.039	①	①	43 ②	850	850	HYD	HYD
	2.3 (2290)	B234I/B	0.024	①	①	43 ②	850	850	HYD	HYD
	2.3 (2290)	B234R/R	0.040	①	①	43 ②	850	850	HYD	HYD
2001	2.0 (1985)	B204L/N	0.039	①	①	43 ②	850	850	HYD	HYD
	2.3 (2290)	B234I/B	0.024	①	①	43 ②	850	850	HYD	HYD
	2.3 (2290)	B234R/R	0.040	①	①	43 ②	850	850	HYD	HYD
	2.3 (2290)	B234I/B	0.038	①	①	43 ②	850	850	HYD	HYD
	3.0 (2962)	B234R/R	0.040	①	①	43 ②	N/A	N/A	HYD	HYD

NOTE: The Vehicle Emission Control Information label often reflects specification changes made during production. The label figures must be used if they differ from those in this chart.

B: Before top dead center

HYD: Hydraulic

① Pre-programmed in ECU and cannot be adjusted

② Fuel line pressure regulator before the control pressure regulator

93471CK9

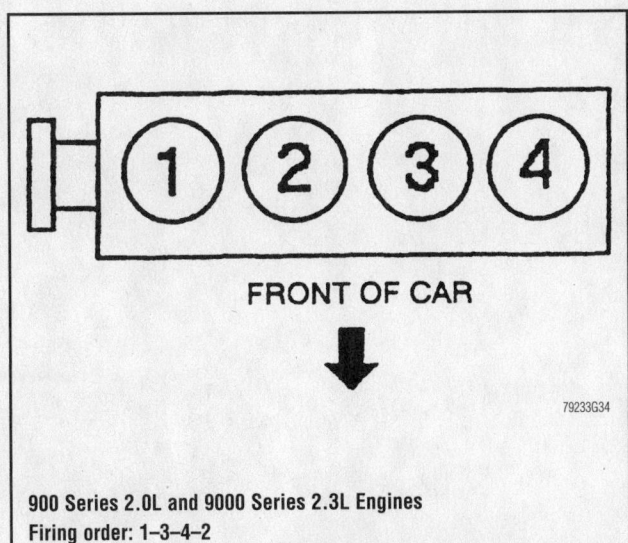

900 Series 2.0L and 9000 Series 2.3L Engines
Firing order: 1-3-4-2
Distributorless ignition system (one coil per cylinder)

79233G34

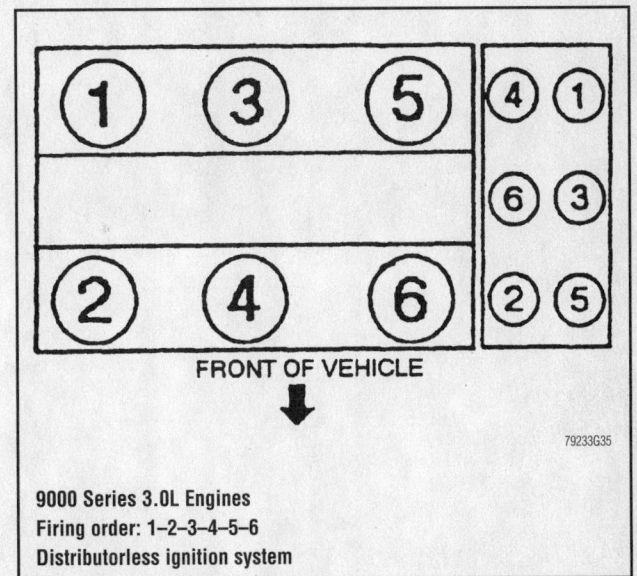

9000 Series 3.0L Engines
Firing order: 1-2-3-4-5-6
Distributorless ignition system

79233G35

For Tire, Wheel and Ball Joint specifications, see Section 1 of this manual

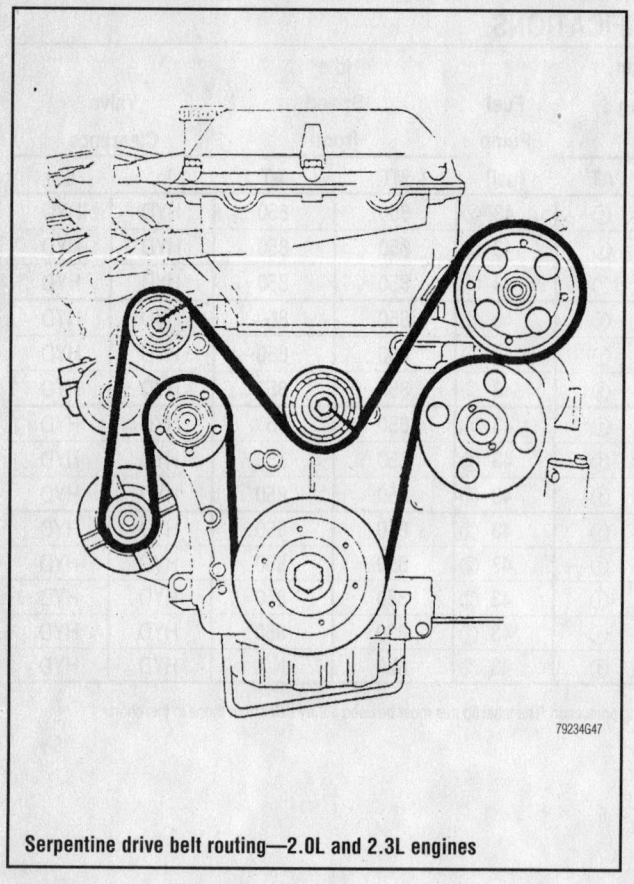

Serpentine drive belt routing—2.0L and 2.3L engines

79234G47

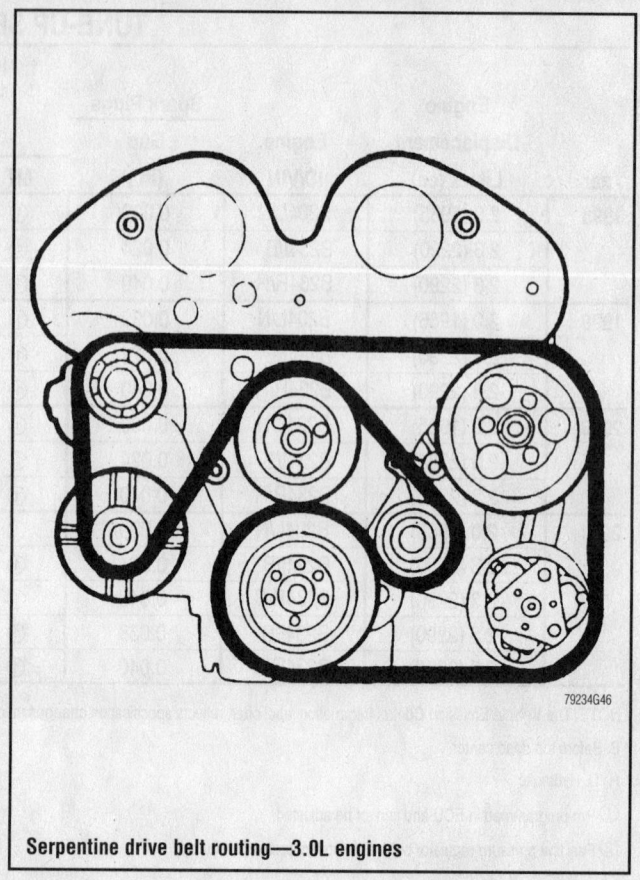

Serpentine drive belt routing—3.0L engines

79234G46

CAPACITIES

Year	Model	Engine Displacement Liters (cc)	Engine ID/VIN	Engine Oil with Filter (qts.)	Transmission (pts.)			Transfer Case (pts.)	Drive Axle		Fuel Tank (gal.)	Cooling System (qts.)
					4-Spd	5-Spd	Auto.		Front (pts.)	Rear (pts.)		
1998	900	2.0 (1985)	B204L/N	4.1	—	7.2	13.6	—	—	—	18.0	8.7
	900	2.3 (2290)	B234I/B	4.1	—	7.2	13.6	—	—	—	18.0	8.7
	9000	2.3 (2290)	B234R/R	5.1	—	7.2	—	—	—	—	17.4	9.5
	9-3	2.0 (1985)	B204L/N	4.1	—	7.2	13.6	—	—	—	18.0	8.7
	9-5	2.3 (2290)	B235R/R	4.2	—	7.2	—	—	—	—	18.5	7.8
1999	9-3	2.0 (1985)	B204L/N	4.1	—	7.2	13.6	—	—	—	18.0	8.7
	9-5	2.3 (2290)	B235R/R	4.2	—	7.2	—	—	—	—	18.5	7.8
2000	9-3	2.0 (1985)	B204L/N	4.1	—	7.2	13.6	—	—	—	18.0	8.7
	9-5	2.3 (2290)	B235R/R	4.2	—	7.2	—	—	—	—	18.5	7.8
2001	9-3	2.0 (1985)	B204L/N	4.1	—	7.2	13.6	—	—	—	18.0	8.7
	9-5	2.3 (2290)	B235R/R	4.2	—	7.2	—	—	—	—	18.5	7.8
	9-5	3.0 (2962)	B308E/E	4.7	—	7.2	14	—	—	—	18.5	7.6

NOTE: All capacities are approximate. Add fluid gradually and check to be sure a proper fluid level is obtained.

93471CK0

For Wheel Alignment specifications, see Section 1 of this manual

VALVE SPECIFICATIONS

Year	Engine Displacement Liters (cc)	Engine ID/VIN	Seat Angle (deg.)	Face Angle (deg.)	Spring Test Pressure (lbs. @ in.)	Spring Installed Height (in.)	Stem-to-Guide Clearance (in.)		Stem Diameter (in.)	
							Intake	Exhaust	Intake	Exhaust
1998	2.0 (1985)	B204L/N	45	44.5	138-150@ 1.12	1.46	0.020	0.020	0.2740-0.2746	0.2738-0.2748
	2.3 (2290)	B234I/B	45	44.5	138-150@ 1.12	1.46	0.020	0.020	0.2740-0.2746	0.2738-0.2748
	2.3 (2290)	B234R/R	45	44.5	138-141@ 1.12	1.46	0.020	0.020	0.2740-0.2746	0.2738-0.2748
1999	2.0 (1985)	B204L/N	45	44.5	138-150@ 1.12	1.46	0.020	0.020	0.2740-0.2746	0.2738-0.2748
	2.3 (2290)	B234I/B	45	44.5	138-150@ 1.12	1.46	0.020	0.020	0.2740-0.2746	0.2738-0.2748
	2.3 (2290)	B234R/R	45	44.5	138-141@ 1.12	1.46	0.020	0.020	0.2740-0.2746	0.2738-0.2748
2000	2.0 (1985)	B204L/N	45	44.5	138-150@ 1.12	1.46	0.020	0.020	0.2740-0.2746	0.2738-0.2748
	2.3 (2290)	B234I/B	45	44.5	138-150@ 1.12	1.46	0.020	0.020	0.2740-0.2746	0.2738-0.2748
	2.3 (2290)	B234R/R	45	44.5	138-141@ 1.12	1.46	0.020	0.020	0.2740-0.2746	0.2738-0.2748
2001	2.0 (1985)	B204L/N	45	44.5	138-150@ 1.12	1.46	0.020	0.020	0.2740-0.2746	0.2738-0.2748
	2.3 (2290)	B234I/B	45	44.5	138-150@ 1.12	1.46	0.020	0.020	0.2740-0.2746	0.2738-0.2748
	2.3 (2290)	B234R/R	45	44.5	138-141@ 1.12	1.46	0.020	0.020	0.2740-0.2746	0.2738-0.2748
	2.3 (2290)	B235R/R	45	44.5	138-150@ 1.24	1.46	0.020	0.020	0.2740-0.2746	0.2738-0.2748
	3.0 (2962)	B308E/E	45	45.3	138-141@ 1.33	1.64	0.020	0.020	0.2403-0.2409	0.2399-0.2405

NA: Not Available

93471CL1

CRANKSHAFT AND CONNECTING ROD SPECIFICATIONS
All measurements are given in inches.

Year	Engine Displacement Liters (cc)	Engine ID/VIN	Crankshaft Main Brg. Journal Dia.	Crankshaft Main Brg. Oil Clearance	Crankshaft Shaft End-play	Crankshaft Thrust on No.	Connecting Rod Journal Diameter	Connecting Rod Oil Clearance	Connecting Rod Side Clearance
1998	2.0 (1985)	B204L/N	2.283	0.0005-0.0025	0.002-0.012	NA	2.046	0.0008-0.0027	NA
	2.3 (2290)	B234I/B	2.283	0.0005-0.0025	0.002-0.012	NA	2.046	0.0008-0.0027	NA
	2.3 (2290)	B234R/R	2.283	0.0005-0.0025	0.002-0.012	NA	2.046	0.0008-0.0027	NA
1999	2.0 (1985)	B204L/N	2.283	0.0005-0.0025	0.002-0.012	NA	2.046	0.0008-0.0027	NA
	2.3 (2290)	B234I/B	2.283	0.0005-0.0025	0.002-0.012	NA	2.046	0.0008-0.0027	NA
	2.3 (2290)	B234R/R	2.283	0.0005-0.0025	0.002-0.012	NA	2.046	0.0008-0.0027	NA
2000	2.0 (1985)	B204L/N	2.283	0.0005-0.0025	0.002-0.012	NA	2.046	0.0008-0.0027	NA
	2.3 (2290)	B234I/B	2.283	0.0005-0.0025	0.002-0.012	NA	2.046	0.0008-0.0027	NA
	2.3 (2290)	B234R/R	2.283	0.0005-0.0025	0.002-0.012	NA	2.046	0.0008-0.0027	NA
2001	2.0 (1985)	B204L/N	2.283	0.0005-0.0025	0.002-0.012	NA	2.046	0.0008-0.0027	NA
	2.3 (2290)	B234I/B	2.283	0.0005-0.0025	0.002-0.012	NA	2.046	0.0008-0.0027	NA
	2.3 (2290)	B234R/R	2.283	0.0005-0.0025	0.002-0.012	NA	2.046	0.0008-0.0027	NA
	2.3 (2290)	B235R/R	2.283	0.0005-0.0025	0.002-0.012	NA	2.046	0.0008-0.0027	NA
	3.0 (2962)	B308E/E	2.676	0.0005-0.0150	0.004-0.030	NA	1.928	0.0005-0.0150	NA

NA: Not Available

93471CL2

For Maintenance Interval recommendations, see Section 1 of this manual

PISTON AND RING SPECIFICATIONS

All measurements are given in inches.

Year	Engine Displacement Liters (cc)	Engine ID/VIN	Piston Clearance	Ring Gap Top Compression	Ring Gap Bottom Compression	Ring Gap Oil Control	Ring Side Clearance Top Compression	Ring Side Clearance Bottom Compression	Ring Side Clearance Oil Control
1998	2.0 (1985)	B204LN	0.0002-0.0016	0.012-0.020	0.006-0.026	0.015-0.055	0.002-0.004	0.0016-0.0028	NA
	2.3 (2290)	B234I/B	0.0002-0.0016	0.012-0.020	0.006-0.026	0.015-0.055	0.002-0.004	0.0016-0.0028	NA
	2.3 (2290)	B234R/R	0.0002-0.0016	0.012-0.020	0.006-0.026	0.015-0.055	0.002-0.004	0.0016-0.0028	NA
1999	2.0 (1985)	B204LN	0.0002-0.0016	0.012-0.020	0.006-0.026	0.015-0.055	0.002-0.004	0.0016-0.0028	NA
	2.3 (2290)	B234I/B	0.0002-0.0016	0.012-0.020	0.006-0.026	0.015-0.055	0.002-0.004	0.0016-0.0028	NA
	2.3 (2290)	B234R/R	0.0002-0.0016	0.012-0.020	0.006-0.026	0.015-0.055	0.002-0.004	0.0016-0.0028	NA
2000	2.0 (1985)	B204LN	0.0002-0.0016	0.012-0.020	0.006-0.026	0.015-0.055	0.002-0.004	0.0016-0.0028	NA
	2.3 (2290)	B234I/B	0.0002-0.0016	0.012-0.020	0.006-0.026	0.015-0.055	0.002-0.004	0.0016-0.0028	NA
	2.3 (2290)	B234R/R	0.0002-0.0016	0.012-0.020	0.006-0.026	0.015-0.055	0.002-0.004	0.0016-0.0028	NA
2001	2.0 (1985)	B204LN	0.0002-0.0016	0.012-0.020	0.006-0.026	0.015-0.055	0.002-0.004	0.0016-0.0028	NA
	2.3 (2290)	B234I/B	0.0002-0.0016	0.012-0.020	0.006-0.026	0.015-0.055	0.002-0.004	0.0016-0.0028	NA
	2.3 (2290)	B234R/R	0.0002-0.0016	0.012-0.020	0.006-0.026	0.015-0.055	0.002-0.004	0.0016-0.0028	NA
	2.3 (2290)	B235R/R	0.0002-0.0016	0.012-0.020	0.006-0.026	0.015-0.055	0.002-0.004	0.0016-0.0028	NA
	3.0 (2962)	B308E/E	0.0010-0.0018	0.012-0.020	0.012-0.020	0.015-0.055	N/A	N/A	NA

NA: Not Available

93471CL3

TORQUE SPECIFICATIONS
All readings in ft. lbs.

Year	Engine Displacement Liters (cc)	Engine ID/VIN	Cylinder Head Bolts	Main Bearing Bolts	Rod Bearing Bolts	Crankshaft Damper Bolts	Flywheel Bolts	Manifold Intake	Manifold Exhaust	Spark Plugs	Lug Nut
1998	2.0 (1985)	B204L/N	①	81	35	130	59	16	19	20	80-90
	2.3 (2290)	B234I/B	①	81	35	130	59	16	13	20	80-90
	2.3 (2290)	B234R/R	①	81	35	130	59	16	19	20	80-90
1999	2.0 (1985)	B204L/N	①	81	35	130	59	16	19	20	80-90
	2.3 (2290)	B234I/B	①	81	35	130	59	16	13	20	80-90
	2.3 (2290)	B234R/R	①	81	35	130	59	16	19	20	80-90
2000	2.0 (1985)	B204L/N	①	81	35	130	59	16	19	20	80-90
	2.3 (2290)	B234I/B	①	81	35	130	59	16	13	20	80-90
	2.3 (2290)	B234R/R	①	81	35	130	59	16	19	20	80-90
2001	2.0 (1985)	B204L/N	①	81	35	130	59	16	19	20	80-90
	2.3 (2290)	B234I/B	①	81	35	130	59	16	13	20	80-90
	2.3 (2290)	B234R/R	①	81	35	130	59	16	19	20	80-90
	2.3 (2290)	B235R/R	①	81	35	130	59	16	13	20	80-90
	3.0 (2962)	B308E/E	②	③	④	⑤	N/A	15	15	18.5	80-90

NA: NA: Not Available

① Step 1: 44 ft. lbs.
   Step 2: 59 ft. lbs. If 17mm head bolts, torque to 70 ft. lbs.
   Step 3: Tighten each bolt an additional 90 degrees

② Step 1: 18 ft. lbs.
   Step 2: Tighten each bolt an additional 90 degrees
   Step 3: Tighten each bolt an additional 90 degrees

③ Step 1: 37 ft. lbs.
   Step 2: Tighten each bolt an additional 60 degrees
   Step 3: Tighten each bolt an additional 15 degrees

④ Step 1: 26 ft. lbs.
   Step 2: Tighten each bolt an additional 45 degrees
   Step 3: Tighten each bolt an additional 15 degrees

⑤ Step 1: 185 ft. lbs.
   Step 2: Tighten each bolt an additional 45 degrees

93471CL4

For Tune-up, Capacities and Firing orders, see Section 1 of this manual

BRAKE SPECIFICATIONS

All measurements in inches unless noted

Year	Model		Brake Disc			Minimum Lining Thickness		Brake Caliper	
			Original Thickness	Minimum Thickness	Maximum Runout	Front	Rear	Bracket Bolts (ft. lbs.)	Mounting Bolts (ft. lbs.)
1998	900	F	0.940	0.870	0.002	0.200	—	78	19
		R	0.390	0.310	0.003	—	0.200	59	—
	9000	F	0.980	0.900	0.003	0.160	—	62	21
		R	0.350	0.290	0.003	—	0.160	35	21
	9-3	F	0.940	0.870	0.002	0.200	—	78	19
		R	0.390	0.310	0.003	—	0.200	59	—
	9-5	F	0.980	0.870	0.003	0.910	—	86	21
		R	0.400	0.310	0.003	—	0.330	59	—
1999	9-3	F	0.940	0.870	0.002	0.200	—	78	19
		R	0.390	0.310	0.003	—	0.200	59	—
	9-5	F	0.980	0.870	0.003	0.910	—	86	21
		R	0.400	0.310	0.003	—	0.330	59	—
2000	9-3	F	0.940	0.870	0.002	0.200	—	78	19
		R	0.390	0.310	0.003	—	0.200	59	—
	9-5	F	0.980	0.870	0.003	0.910	—	86	21
		R	0.400	0.310	0.003	—	0.330	59	—
2001	9-3	F	0.940	0.870	0.002	0.200	—	78	19
		R	0.390	0.310	0.003	—	0.200	59	—
	9-5	F	0.980	0.870	0.003	0.910	—	86	21
		R	0.400	0.310	0.003	—	0.330	59	—

93471CL5

WHEEL ALIGNMENT

Year	Model		Caster Range (+/-Deg.)	Caster Preferred Setting (Deg.)	Camber Range (+/-Deg.)	Camber Preferred Setting (Deg.)	Toe-in (in.)	Steering Axis Inclination (Deg.)
1998	900	F	—	—	0.30	-1.7	0.39	—
		R	—	—	—	—	—	—
	9000	F	—	—	0.25	—	0.39	—
		R	—	—	—	—	—	—
	9-3	F	0.50	+2.90	—	-1.1	0.39	12.6
		R	—	—	—	—	—	—
	9-5	F	0.50	+2.90	0.50	-0.80	0.04+/-0.01	12.6
		R	—	—	0.25	-0.80	0.06+/-0.03	—
1999	9-3	F	0.50	+2.90	—	-1.1	0.39	12.6
		R	—	—	—	—	—	—
	9-5	F	0.50	+2.90	0.50	-0.80	0.04+/-0.01	12.6
		R	—	—	0.25	-0.80	0.06+/-0.03	—
2000	9-3	F	0.50	+2.90	—	-1.1	0.39	12.6
		R	—	—	—	—	—	—
	9-5	F	0.50	+2.90	0.50	-0.80	0.04+/-0.01	12.6
		R	—	—	0.25	-0.80	0.06+/-0.03	—
2001	9-3	F	0.50	+2.90	—	-1.1	0.39	12.6
		R	—	—	—	—	—	—
	9-5	F	0.50	+2.90	0.50	-0.80	0.04+/-0.01	12.6
		R	—	—	0.25	-0.80	0.06+/-0.03	—

93471CL6

TIRE, WHEEL AND BALL JOINT SPECIFICATIONS

Year	Model	OEM Tires		Tire Pressures (psi)		Wheel Size	Ball Joint Inspection
		Standard	Optional	Front	Rear		
1998	900S, SE	195/60VR15	None	33	33	6-J	0.040 in.
	900 SE Turbo	205/50VR16	None	33	33	6.5-J	0.040 in.
	9000 CS, CSE, CDE	195/65VR15	None	32	32	6-J	0.040 in.
	9000 CSE Turbo	205/60VR15	None	32	32	6-J	0.040 in.
1999	9-3	195/60VR15	205/50R16	32	32	6.5	0.040 in.
	9-5	215/55R16	205/65VR16	Std: 32	Std: 32	6.5	0.040 in.
				Opt: 30	Opt: 30	6.5	0.040 in.
2000	9-3 base	195/60VR15	None	32	32	6.5	0.040 in.
	9-3 SE	205/50ZR16	None	32	32	7.5J	0.040 in.
	9-3 Viggen	215/45ZR17	None	32	32	7.5J	0.040 in.
	9-5 2.3T	215/55VR15	None	32	32	6.5J	0.040 in.
	9-5 SE	215/55VR15	None	32	32	6.5J	0.040 in.
	9-5 Aero	225/45R17	None	32	32	7J	0.040 in.
2001	9-3 base	195/60VR15	None	32	32	6.5	0.040 in.
	9-3 SE	205/50ZR16	None	32	32	7.5J	0.040 in.
	9-3 Viggen	215/45ZR17	None	32	32	7.5J	0.040 in.
	9-5 2.3T	215/55VR15	None	32	32	6.5J	0.040 in.
	9-5 SE	215/55VR15	None	32	32	6.5J	0.040 in.
	9-5 Aero	225/45R17	None	32	32	7J	0.040 in.

OEM: Original Equipment Manufacturer

PSI: Pounds Per Square Inch

STD: Standard

OPT: Optional

93471CL7

SCHEDULED MAINTENANCE INTERVALS
900, 9000, 9-3 and 9-5

TO BE SERVICED	TYPE OF SERVICE	VEHICLE MILEAGE INTERVAL (x1000)												
		5	10	15	20	25	30	35	40	45	50	55	60	65
Engine oil and filter	R	✓	✓	✓	✓	✓	✓	✓	✓	✓	✓	✓	✓	✓
Battery electrolyte	S/I	✓		✓		✓		✓		✓		✓		✓
Brake fluid	S/I	✓		✓		✓		✓		✓		✓		✓
Brake lines and hoses	S/I	✓		✓		✓		✓		✓		✓		✓
Brake pads & discs	S/I	✓		✓		✓		✓		✓		✓		✓
Drive belts	S/I	✓		✓		✓		✓		✓		✓		✓
Engine coolant strength	S/I	✓		✓		✓		✓		✓		✓		✓
Exhaust system	S/I	✓		✓		✓		✓		✓		✓		✓
Final drive oil level (900 A/T)	S/I	✓		✓		✓		✓		✓		✓		✓
Gear oil	S/I	✓		✓		✓		✓		✓		✓		✓
Outer & inner drive joint boots	S/I	✓		✓		✓		✓		✓		✓		✓
Rotate tires (front to rear)	S/I	✓		✓		✓		✓		✓		✓		✓
Automatic transmission fluid & filter	R	✓						✓						✓
Air cleaner element	R							✓						✓
Engine coolant	R							✓						✓
Spark plugs	R							✓						✓
Ventilation air filter	R							✓						✓
Ball joint clearance	S/I							✓						✓
Engine cooling system, hoses & cap	S/I	✓						✓						✓
Front wheel alignment	S/I							✓						✓
Fuel lines	S/I							✓						✓
Shock absorbers & bushings	S/I							✓						✓
Fuel filter	R													✓
Power steering fluid	R							✓						✓
Crankcase ventilation & vacuum lines	S/I													✓
Distributor cap & rotor	S/I													✓
EVAP system	S/I													✓
Front suspension, rear axle mountings	S/I		✓											
Parking brake adjustment	S/I		✓											
Spart plug wires	S/I													✓

R: Replace   S/I: Inspect and service, if needed

FREQUENT OPERATION MAINTENANCE (SEVERE SERVICE)

If a vehicle is operated under any of the following conditions it is considered severe service:

- Extremely dusty areas.

- 50% or more of the vehicle operation is in 32°C (90°F) or higher temperatures, or constant operation in temperatures below 0°C (32°F).

- Prolonged idling (vehicle operation in stop and go traffic).

- Frequent short running periods (engine does not warm to normal operating temperatures).

- Police, taxi, delivery usage or trailer towing usage.

Oil and oil filter: change every 2500 miles.

Air cleaner element: service or inspect every 15,000 miles.

93471CL8

SCHEDULED MAINTENANCE INTERVALS
SAAB
900, 9000, 9-3, 9-5

The following should be used as a guide when determining the amount of work required for a particular service.
In estimating how long a particular Scheduled Maintenance Service should take, please observe the following:

- Labor Time is time based on field research and data supplied by the vehicle manufacturer.
- Labor time operations are given in hours and tenths of an hour.
- All labor operations are to be used as a guide.

Mechanic Skill Level Codes:
(A) PRECISION: Highly skilled with multiple certification.
(B) GENERAL: Normally skilled with certification.
(C) MAINTENANCE: Semi-skilled working on certification.

	LABOR TIME		LABOR TIME		LABOR TIME
5000 Mile Service (B)		**25000 Mile Service (B)**		**50000 Mile Service (C)**	
All Models	2.2	All Models	1.9	All models	.4
w/AT add	.6	*w/AT add*	.1	**55000 Mile Service (B)**	
10000 Mile Service (C)		**30000 Mile Service (C)**		All Models	1.9
All models	.4	All models	.4	*w/AT add*	.1
15000 Mile Service (B)		**35000 Mile Service (B)**		**60000 Mile Service (C)**	
All Models	1.9	All Models	5.5	All models	.4
w/AT add	.1	*w/AT add*	.6	**65000 Mile Service (B)**	
20000 Mile Service (C)		**40000 Mile Service (B)**		All Models	5.2
All models	.4	All models	.4	*w/AT add*	.6
		45000 Mile Service (B)			
		All Models	1.9		
		w/AT add	.1		

93471CL9

SUBARU
IMPREZA • LEGACY • SVX

ENGINE AND VEHICLE IDENTIFICATION CHART

Code ①	Liters (cc)	Cu. In.	Cyl.	Fuel Sys.	Type	Eng. Mfg.
4	2.2 (2212)	135	4	MFI	SOHC	Subaru
6	2.5 (2457)	150	4	MFI	DOHC	Subaru
6	2.5 (2457)	150	4	MFI	SOHC	Subaru

MFI: Multiport Fuel Injection

SOHC: Single Overhead Camshaft

DOHC: Double Overhead Camshaft

① 6th digit of the VIN

② 10th digit of the VIN

Code ②	Year
W	1998
X	1999
Y	2000
1	2001
2	2002

93471CL0

Timing belt service is covered in Section 3 of this manual

GENERAL ENGINE SPECIFICATIONS

Year		Engine Displacement Liters (cc)	Engine ID/VIN	Fuel System Type	Net Horsepower @ rpm	Net Torque @ rpm (ft. lbs.)	Bore x Stroke (in.)	Compression Ratio	Oil Pressure @ rpm
1998	Impreza ①	2.2 (2212)	4	MFI	137@5400	145@4000	3.82x2.95	9.7:1	14@600
	Impreza RS	2.5 (2457)	6	MFI	165@5600	162@4000	3.92x3.11	9.7:1	14@600
	Legacy	2.2 (2212)	4	MFI	137@5400	145@4000	3.82x2.95	9.7:1	14@600
	Legacy ①	2.5 (2457)	6	MFI	165@5600	162@4000	3.92x3.11	9.7:1	14@600
1999	Impreza ①	2.2 (2212)	4	MFI	137@5400	145@4000	3.82x2.95	9.7:1	14@600
	Impreza RS	2.5 (2457)	6	MFI	165@5600	162@4000	3.92x3.11	9.7:1	14@600
	Legacy	2.2 (2212)	4	MFI	137@5400	145@4000	3.82x2.95	9.7:1	14@600
	Legacy ②	2.5 (2457)	6	MFI	165@5600	162@4000	3.92x3.11	9.7:1	14@600
2000	Impreza ①	2.2 (2212)	4	MFI	142@5600	149@3600	3.82x2.95	10.0:1	14@600
	Impreza RS	2.5 (2457)	6	MFI	165@5600	166@4000	3.92x3.11	10.0:1	14@600
	Legacy ②	2.5 (2457)	6	MFI	165@5600	166@4000	3.92x3.11	10.0:1	14@600
2001	Impreza ①	2.2 (2212)	4	MFI	142@5600	149@3600	3.82x2.95	10.0:1	14@600
	Impreza RS	2.5 (2457)	6	MFI	165@5600	166@4000	3.92x3.11	10.0:1	14@600
	Legacy ②	2.5 (2457)	6	MFI	165@5600	166@4000	3.92x3.11	10.0:1	14@600

MFI: Multi-port Fuel Injection

Note: All capacities are approximate. Add fluid gradually and check to be sure a proper fluid level is obtained.

① Includes Outback

② Includes Outback and Sport Utility Sedan

93471CM1

ENGINE TUNE-UP SPECIFICATIONS

Year	Model	Engine Displacement Liters (cc)	Engine ID/VIN	Spark Plug Gap (in.)	Ignition Timing (deg.) ①		Fuel Pump (psi)	Idle Speed (rpm)		Valve Clearance ②	
					MT	AT		MT	AT	In.	Ex.
1998	Impreza/ Outback	2.2 (2212)	4	0.039-0.043	6-22	12-28	34-38	600-800	600-800	0.0071-0.0087	0.0090-0.0106
	Impreza RS	2.5 (2457)	6	0.039-0.043	7-23	7-23	34-38	600-800	600-800	0.0071-0.0087	0.0090-0.0106
	Legacy	2.2 (2212)	4	0.039-0.043	6-22	12-28	34-38	600-800	600-800	0.0071-0.0087	0.0090-0.0106
	Legacy/ Outback	2.5 (2457)	6	0.039-0.043	7-23	7-23	34-38	600-800	600-800	0.0071-0.0087	0.0090-0.0106
1999	Impreza/ Outback	2.2 (2212)	4	0.039-0.043	6-22	12-28	34-38	600-800	600-800	0.0071-0.0087	0.0090-0.0106
	Impreza RS	2.5 (2457)	6	0.039-0.043	7-23	7-23	34-38	600-800	600-800	0.0071-0.0087	0.0090-0.0106
	Legacy	2.2 (2212)	4	0.039-0.043	6-22	12-28	34-38	600-800	600-800	0.0071-0.0087	0.0090-0.0106
	Legacy/ Outback/SUS ③	2.5 (2457)	6	0.039-0.043	7-23	7-23	34-38	600-800	600-800	0.0071-0.0087	0.0090-0.0106
2000	Impreza/ Outback	2.2 (2212)	4	0.039-0.043	6-22	12-28	34-38	600-800	600-800	0.0071-0.0087	0.0090-0.0106
	Impreza RS	2.5 (2457)	6	0.039-0.043	7-23	7-23	34-38	600-800	600-800	0.0071-0.0087	0.0090-0.0106
	Legacy/ Outback/SUS ③	2.5 (2457)	6	0.039-0.043	7-23	7-23	34-38	600-800	600-800	0.0071-0.0087	0.0090-0.0106
2001	Impreza/ Outback	2.2 (2212)	4	0.039-0.043	6-22	12-28	34-38	600-800	600-800	0.0071-0.0087	0.0090-0.0106
	Impreza RS	2.5 (2457)	6	0.039-0.043	7-23	7-23	34-38	600-800	600-800	0.0071-0.0087	0.0090-0.0106
	Legacy/ Outback/SUS ③	2.5 (2457)	6	0.039-0.043	7-23	7-23	34-38	600-800	600-800	0.0071-0.0087	0.0090-0.0106

Note: The Vehicle Emission Control Information label often reflects specification changes made during production. The lable figures mudst be used if they differ from those in this chart.

① Before Top Dead Center

② Measured with engine cold

③ Sport Utility Sedan

93471CM2

Heater Core replacement is covered in Section 2 of this manual

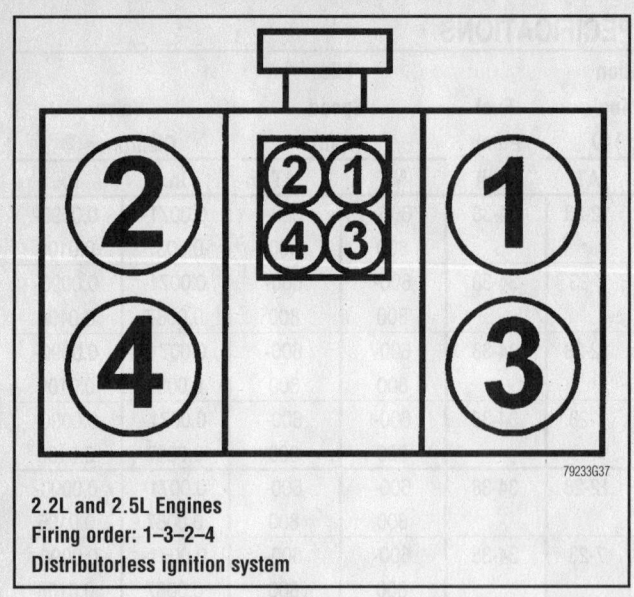

2.2L and 2.5L Engines
Firing order: 1–3–2–4
Distributorless ignition system

79233G37

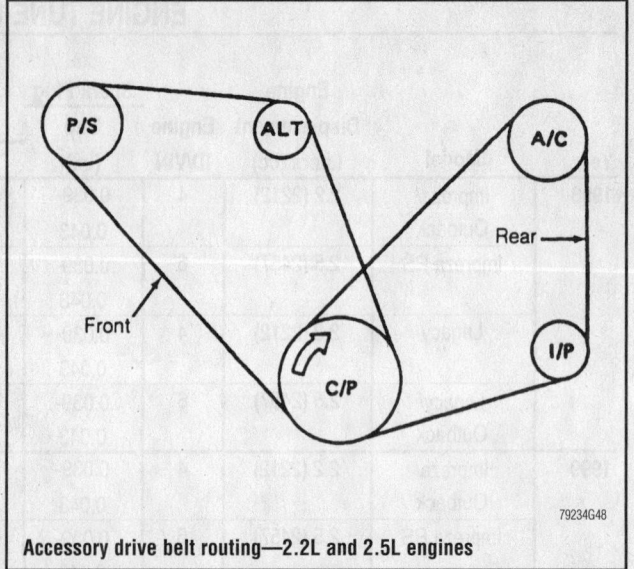

Accessory drive belt routing—2.2L and 2.5L engines

79234G48

CAPACITIES

Year	Model	Engine Displacement Liters (cc)	Engine ID/VIN	Engine Oil with Filter (qts.)	Transmission (pts.)		Transfer Case (pts.)	Drive Axle		Fuel Tank (gal.)	Cooling System (qts.)
					5-Spd	Auto.		Front ① (pts.)	Rear (pts.)		
1998	Impreza ②	2.2 (2212)	4	4.4	7.4	16.8	–	2.6	1.6	13.2	6.1
	Impreza RS	2.5 (2457)	6	4.7	7.4	20.0	–	2.6	1.6	13.2	6.3
	Legacy	2.2 (2212)	4	4.4	7.4	15.0	–	2.6	1.6	15.9	6.1
	Legacy ②	2.5 (2457)	6	4.7	7.4	20.0	–	2.6	1.6	15.9	6.3
1999	Impreza ②	2.2 (2212)	4	4.4	7.4	16.8	–	2.6	1.6	13.2	6.1
	Impreza RS	2.5 (2457)	6	4.7	7.4	20.0	–	2.6	1.6	13.2	6.3
	Legacy	2.2 (2212)	4	4.4	7.4	15.0	–	2.6	1.6	15.9	6.1
	Legacy ③	2.5 (2457)	6	4.7	7.4	20.0	–	2.6	1.6	15.9	6.3
2000	Impreza ②	2.2 (2212)	4	4.4	7.4	16.8	–	2.6	1.6	13.2	6.6
	Impreza RS	2.5 (2457)	6	4.7	7.4	20.0	–	2.6	1.6	13.2	6.6
	Legacy ③	2.5 (2457)	6	4.7	7.4	20.0	–	2.6	1.6	16.9	6.6
2001	Impreza ②	2.2 (2212)	4	4.4	7.4	16.8	–	2.6	1.6	13.2	6.6
	Impreza RS	2.5 (2457)	6	4.7	7.4	20.0	–	2.6	1.6	13.2	6.6
	Legacy ③	2.5 (2457)	6	4.7	7.4	20.0	–	2.6	1.6	16.9	6.6

Note: All capacities are approximate. Add fluid gradually and check to be sure a proper fluid level is obtained.

① A/T differential only

② Includes Outback

③ Includes Outback and Sport Utility Sedan

93471CM3

Brake service is covered in Section 4 of this manual

VALVE SPECIFICATIONS

Year	Engine Displacement Liters (cc)	Engine ID/VIN	Seat Angle (deg.)	Face Angle (deg.)	Spring Test Pressure (lbs. @ in.)	Spring Installed Height (in.)	Stem-to-Guide Clearance (in.)		Stem Diameter (in.)	
							Intake	Exhaust	Intake	Exhaust
1998	2.2 (2212)	4	45	45	91 - 103@ 1.110	①	0.0014-0.0059	0.0016-0.0059	0.2344-0.2350	0.2341-0.2346
	2.5 (2457)	6	45	45	102 - 118@ 1.315	②	0.0014-0.0059	0.0016-0.0059	0.2343-0.2348	0.2341-0.2346
1999	2.2 (2212)	4	45	45	91 - 103@ 1.110	①	0.0014-0.0059	0.0016-0.0059	0.2344-0.2350	0.2341-0.2346
	2.5 (2457)	6	45	45	102 - 118@ 1.315	②	0.0014-0.0059	0.0016-0.0059	0.2343-0.2348	0.2341-0.2346
2000	2.2 (2212)	4	45	45	91 - 103@ 1.110	①	0.0014-0.0059	0.0016-0.0059	0.2344-0.2350	0.2341-0.2346
	2.5 (2457)	6	45	45	102 - 118@ 1.315	②	0.0014-0.0059	0.0016-0.0059	0.2343-0.2348	0.2341-0.2346
2001	2.2 (2212)	4	45	45	91 - 103@ 1.110	①	0.0014-0.0059	0.0016-0.0059	0.2344-0.2350	0.2341-0.2346
	2.5 (2457)	6	45	45	102 - 118@ 1.315	②	0.0014-0.0059	0.0016-0.0059	0.2343-0.2348	0.2341-0.2346

① Free length: 1.7342 in.

② Free length: 1.8913 in.

93471CM4

CRANKSHAFT AND CONNECTING ROD SPECIFICATIONS

All measurements are given in inches.

Year	Engine Displacement Liters (cc)	Engine ID/VIN	Crankshaft				Connecting Rod		
			Main Brg. Journal Dia.	Main Brg. Oil Clearance	Shaft End-play	Thrust on No.	Journal Diameter	Oil Clearance	Side Clearance
1998	2.2 (2212)	4	2.3619-2.3625	①	0.0012-0.0098	3	2.0466-2.0472	0.0006-0.0020	0.0028-0.0160
	2.5 (2457)	6	2.3619-2.3625	②	0.0012-0.0098	3	1.8891-1.8898	0.0004-0.0020	0.0028-0.0160
1999	2.2 (2212)	4	2.3619-2.3625	①	0.0012-0.0098	3	2.0466-2.0472	0.0006-0.0020	0.0028-0.0160
	2.5 (2457)	6	2.3619-2.3625	②	0.0012-0.0098	3	1.8891-1.8898	0.0004-0.0020	0.0028-0.0160
2000	2.2 (2212)	4	2.3619-2.3625	①	0.0012-0.0098	3	2.0466-2.0472	0.0006-0.0020	0.0028-0.0160
	2.5 (2457)	6	2.3619-2.3625	②	0.0012-0.0098	3	1.8891-1.8898	0.0004-0.0020	0.0028-0.0160
2001	2.2 (2212)	4	2.3619-2.3625	①	0.0012-0.0098	3	2.0466-2.0472	0.0006-0.0020	0.0028-0.0160
	2.5 (2457)	6	2.3619-2.3625	②	0.0012-0.0098	3	1.8891-1.8898	0.0004-0.0020	0.0028-0.0160

① Journals 1 and 5: 0.0001 - 0.0016 in.
  Journals 2 and 4: 0.0004 - 0.0014 in.
  Journal 3: 0.0004 - 0.0014 in.

② Journals 1 and 5: 0.0001 - 0.0016 in.
  Journals 2 and 4: 0.0004 - 0.0018 in.
  Journal 3: 0.0004 - 0.0016 in.

93471CM5

For complete Engine Mechanical specifications, see Section 1 of this manual

PISTON AND RING SPECIFICATIONS

All measurements are given in inches.

Year	Engine Displacement Liters (cc)	Engine ID/VIN	Piston Clearance	Ring Gap			Ring Side Clearance		
				Top Compression	Bottom Compression	Oil Control	Top Compression	Bottom Compression	Oil Control
1998	2.2 (2212)	4	0.0004-0.0020	0.0079-0.0390	0.0079-0.0390	0.0079-0.0590	0.0016-0.0590	0.0012-0.0590	NA
	2.5 (2457)	6	0.0004-0.0020	0.0079-0.0390	0.0146-0.0390	0.0079-0.0590	0.0016-0.0059	0.0012-0.0059	NA
1999	2.2 (2212)	4	0.0004-0.0020	0.0079-0.0390	0.0079-0.0390	0.0079-0.0590	0.0016-0.0590	0.0012-0.0590	NA
	2.5 (2457)	6	0.0004-0.0020	0.0079-0.0390	0.0146-0.0390	0.0079-0.0590	0.0016-0.0059	0.0012-0.0059	NA
2000	2.2 (2212)	4	0.0004-0.0020	0.0079-0.0390	0.0079-0.0390	0.0079-0.0590	0.0016-0.0590	0.0012-0.0590	NA
	2.5 (2457)	6	0.0004-0.0020	0.0079-0.0390	0.0146-0.0390	0.0079-0.0590	0.0016-0.0059	0.0012-0.0059	NA
2001	2.2 (2212)	4	0.0004-0.0020	0.0079-0.0390	0.0079-0.0390	0.0079-0.0590	0.0016-0.0590	0.0012-0.0590	NA
	2.5 (2457)	6	0.0004-0.0020	0.0079-0.0390	0.0146-0.0390	0.0079-0.0590	0.0016-0.0059	0.0012-0.0059	NA

NA: Not Available

93471CM6

TORQUE SPECIFICATIONS
All readings in ft. lbs.

Year	Engine Displacement Liters (cc)	Engine ID/VIN	Cylinder Head Bolts	Main ① Bearing Bolts	Rod Bearing Bolts	Crankshaft Damper Bolts	Flywheel Bolts	Manifold		Spark Plugs	Lug Nut
								Intake	Exhaust		
1998	2.2 (2212)	4	②	③	31 - 44	87 - 101	51 - 55	17 - 20	19 - 26	13 - 17	58 - 72
	2.5 (2457)	6	②	③	31 - 44	123 - 137	51 - 55	17 - 20	19 - 26	13 - 17	58 - 72
1999	2.2 (2212)	4	②	③	31 - 44	87 - 101	51 - 55	17 - 20	19 - 26	13 - 17	58 - 72
	2.5 (2457)	6	②	③	31 - 44	123 - 137	51 - 55	17 - 20	19 - 26	13 - 17	58 - 72
2000	2.2 (2212)	4	②	③	31 - 44	87 - 101	51 - 55	17 - 20	19 - 26	13 - 17	58 - 72
	2.5 (2457)	6	②	③	31 - 44	123 - 137	51 - 55	17 - 20	19 - 26	13 - 17	58 - 72
2001	2.2 (2212)	4	②	③	31 - 44	87 - 101	51 - 55	17 - 20	19 - 26	13 - 17	58 - 72
	2.5 (2457)	6	②	③	31 - 44	123 - 137	51 - 55	17 - 20	19 - 26	13 - 17	58 - 72

① Engine block connecting bolts

② Step 1: Tighten all bolts to 22 ft. lbs.

Step 2: Tighten all bolts to 51 ft. lbs.

Step 3: Loosen all botls 180 degrees.

Step 4: Repeat Step 3.

Step 5: Tighten bolts 1 and 2 to 25 ft. lbs.

Step 6: Tighten bolts 3, 4, 5 and 6 to 11 ft. lbs.

Step 7: Tighten all bolts 80 to 90 degrees.

Step 8: Repeat Step 7. Do not exceed 180 degrees total tightening.

③ Split engine case connecting bolts:

Short bolts: 17-20 ft. lbs.

Long bolts: 33-37 ft. lbs.

Smaller short bolts (if used) 5 ft. lbs.

93471CM7

For Accessory Drive Belt illustrations, see Section 1 of this manual

BRAKE SPECIFICATIONS
All measurements in inches unless noted

Year	Model		Brake Disc Original Thickness	Brake Disc Minimum Thickness	Brake Disc Maximum Runout	Brake Drum Diameter Original Inside Diameter	Brake Drum Diameter Max. Wear Limit	Brake Drum Diameter Maximum Machine Diameter	Minimum Lining Thickness Front	Minimum Lining Thickness Rear	Brake Caliper Bracket Bolts (ft. lbs.)	Brake Caliper Mounting Bolts (ft. lbs.)
1998	Impreza	F	0.940	0.870	0.003	—	—	—	0.059	—	51-65	23-30
		R	0.390	0.335	0.004	9.000 ①	9.079 ②	NA	—	0.059	—	23-30
	Legacy	F	0.940	0.870	0.003	—	—	—	0.059	—	51-65	23-30
		R	0.390	0.335	0.004	9.000 ①	9.079 ②	NA	—	0.059	—	23-30
1999	Impreza	F	0.940	0.870	0.003	—	—	—	0.059	—	51-65	23-30
		R	0.390	0.335	0.004	9.000 ①	9.079 ②	NA	—	0.059	—	23-30
	Legacy	F	0.940	0.870	0.003	—	—	—	0.059	—	51-65	23-30
		R	0.390	0.335	0.004	9.000 ①	9.079 ②	NA	—	0.059	—	23-30
2000	Impreza	F	0.940	0.870	0.003	—	—	—	0.059	—	51-65	23-30
		R	0.390	0.335	0.004	9.000 ①	9.079 ②	NA	—	0.059	—	23-30
	Legacy	F	0.940	0.870	0.003	—	—	—	0.059	—	51-65	23-30
		R	0.390	0.335	0.004	9.000 ①	9.079 ②	NA	—	0.059	—	23-30
2001	Impreza	F	0.940	0.870	0.003	—	—	—	0.059	—	51-65	23-30
		R	0.390	0.335	0.004	9.000 ①	9.079 ②	NA	—	0.059	—	23-30
	Legacy	F	0.940	0.870	0.003	—	—	—	0.059	—	51-65	23-30
		R	0.390	0.335	0.004	9.000 ①	9.079 ②	NA	—	0.059	—	23-30

NA: Not Available

① Parking brake drum on vehicles with rear disc brakes: 6.69 in.

② Specification is for the parking brake drum.

93471CM8

WHEEL ALIGNMENT

Year	Model		Caster Range (+/-Deg.)	Caster Preferred Setting (Deg.)	Camber Range (+/-Deg.)	Camber Preferred Setting (Deg.)	Toe-in (in.)	Steering Axis Inclination (Deg.)
1998	Impreza 2.2L	F	1.00	+3.00	0.50	0	0+/-0.12	—
		R	—	—	0.75	-0.92	0+/-0.12	—
	Impreza 2.5L	F	1.00	+3.05	0.50	-0.42	0+/-0.12	—
		R	—	—	0.75	-1.17	0+/-0.12	—
	Legacy Sedan	F	0.75	+3.05	0.50	-0.05	0+/-0.12	—
		R	—		0.75	-0.50	0+/-0.12	—
	Legacy Wagon	F	0.75	+2.05	0.50	-0.05	0+/-0.12	—
		R	—		0.75	-0.20	0+/-0.12	—
	Outback	F	0.75	+2.40	0.50	+0.33	0+/-0.12	—
		R	—		0.75	-0.17	0+/-0.12	—
1999	Impreza 2.2L	F	1.00	+3.00	0.50	0	0+/-0.12	—
		R	—	—	0.75	-0.92	0+/-0.12	—
	Impreza 2.5L	F	1.00	+3.05	0.50	-0.42	0+/-0.12	—
		R	—	—	0.75	-1.17	0+/-0.12	—
	Legacy Sedan	F	0.75	+3.05	0.50	-0.05	0+/-0.12	—
		R	—		0.75	-0.50	0+/-0.12	—
	Legacy Wagon	F	0.75	+2.05	0.50	-0.05	0+/-0.12	—
		R	—		0.75	-0.20	0+/-0.12	—
	Outback	F	0.75	+2.40	0.50	+0.33	0+/-0.12	—
		R	—		0.75	-0.17	0+/-0.12	—
2000	Impreza 2.2L	F	1.00	+3.00	0.50	0	0+/-0.12	—
		R	—	—	0.75	-0.92	0+/-0.12	—
	Impreza 2.5L	F	1.00	+3.05	0.50	-0.42	0+/-0.12	—
		R	—	—	0.75	-1.17	0+/-0.12	—
	Legacy Sedan	F	0.75	+3.05	0.50	-0.05	0+/-0.12	—
		R	—		0.75	-0.50	0+/-0.12	—
	Legacy Wagon	F	0.75	+2.05	0.50	-0.05	0+/-0.12	—
		R	—		0.75	-0.20	0+/-0.12	—
	Outback	F	0.75	+2.40	0.50	+0.33	0+/-0.12	—
		R	—		0.75	-0.17	0+/-0.12	—
2001	Impreza 2.2L	F	1.00	+3.00	0.50	0	0+/-0.12	—
		R	—	—	0.75	-0.92	0+/-0.12	—
	Impreza 2.5L	F	1.00	+3.05	0.50	-0.42	0+/-0.12	—
		R	—	—	0.75	-1.17	0+/-0.12	—
	Legacy Sedan	F	0.75	+3.05	0.50	-0.05	0+/-0.12	—
		R	—		0.75	-0.50	0+/-0.12	—
	Legacy Wagon	F	0.75	+2.05	0.50	-0.05	0+/-0.12	—
		R	—		0.75	-0.20	0+/-0.12	—
	Outback	F	0.75	+2.40	0.50	+0.33	0+/-0.12	—
		R	—		0.75	-0.17	0+/-0.12	—

93471CM9

For Tire, Wheel and Ball Joint specifications, see Section 1 of this manual

TIRE, WHEEL AND BALL JOINT SPECIFICATIONS

| Year | Model | OEM Tires | | Tire Pressures (psi) | | Wheel Size | Ball Joint Inspection |
		Standard	Optional	Front	Rear		
1998	Impreza	P195/60R15	None	32	29	6-JJ	0.012 in.
	Impreza RS	P205/55R16	None	32	29	7-JJ	0.012 in.
	Impreza Outback	P205/60R15	None	32	29	7-JJ	0.012 in.
	Legacy	P185/70R14	P195/60R15 P205/55R16 P205/70R15	32	30	6.5-JJ	0.012 in.
1999	Impreza	P195/60R15	None	32	29	6-JJ	0.012 in.
	Impreza RS	P205/55R16	None	32	29	7-JJ	0.012 in.
	Impreza Outback	P205/60R15	None	32	29	7-JJ	0.012 in.
	Legacy	P185/70R14	P195/60R15 P205/55R16 P205/70R15	32	30	6.5-JJ	0.012 in.
2000	Impreza	P195/60R15	None	32	29	6-JJ	0.012 in.
	Impreza RS	P205/55R16	None	32	29	7-JJ	0.012 in.
	Impreza Outback	P205/60R15	None	32	29	7-JJ	0.012 in.
	Legacy	P185/70R14	P195/60R15 P205/55R16 P205/70R15	32	30	6.5-JJ	0.012 in.
2001	Impreza	P195/60R15	None	32	29	6-JJ	0.012 in.
	Impreza RS	P205/55R16	None	32	29	7-JJ	0.012 in.
	Impreza Outback	P205/60R15	None	32	29	7-JJ	0.012 in.
	Legacy	P185/70R14	P195/60R15 P205/55R16 P205/70R15	32	30	6.5-JJ	0.012 in.

OEM: Original Equipment Manufacturer

PSI: Pounds Per Square Inch

93471CM0

SCHEDULED MAINTENANCE INTERVALS

Subaru Impreza OutBack, Impreza OutBack Sport, Legacy and Legacy OutBack

TO BE SERVICED	TYPE OF SERVICE	VEHICLE MILEAGE INTERVAL (x1000)												
		7.5	15	22.5	30	37.5	45	52.5	60	67.5	75	82.5	90	97.5
Engine oil & filter	R	✓	✓	✓	✓	✓	✓	✓	✓	✓	✓	✓	✓	✓
Brake lines	S/I		✓		✓		✓		✓		✓		✓	
Clutch & hill holder system	S/I		✓		✓		✓		✓		✓		✓	
Disc brake pads & discs, front & rear axle boots & axle shaft joint portions	S/I		✓		✓		✓		✓		✓		✓	
Parking brake	S/I		✓		✓		✓		✓		✓		✓	
Steering & suspension	S/I		✓		✓		✓		✓		✓		✓	
Air filter element	R					✓			✓				✓	
Engine coolant	R					✓			✓				✓	
Fuel filter	R					✓			✓				✓	
Spark plugs	R								✓					
Automatic transmission fluid & filter	S/I				✓				✓				✓	
Brake fluid	S/I				✓				✓				✓	
Brake linings & drums	S/I				✓				✓				✓	
Camshaft drive belt ①	S/I				✓				✓				✓	
Coolant level, hoses & clamps	S/I				✓				✓				✓	
Drive belts	S/I				✓				✓				✓	
Fuel system, hoses & connections	S/I				✓				✓				✓	
Transmission and/or differential gear fluid	S/I				✓								✓	
Front & rear wheel bearing repack	S/I								✓					

R: Replace   S/I: Service or Inspect

① Non-California vehicles: replace every 60,000 miles.

FREQUENT OPERATION MAINTENANCE (SEVERE SERVICE)

If a vehicle is operated under any of the following conditions it is considered severe service:

- Extremely dusty areas.

- 50% or more of the vehicle operation is in 32°C (90°F) or higher temperatures, or constant operation in temperatures below 0°C (32°F).

- Prolonged idling (vehicle operation in stop and go traffic).

- Frequent short running periods (engine does not warm to normal operating temperatures).

- Police, taxi, delivery usage or trailer towing usage.

Oil & oil filter change: change every 3750 miles.

Clutch & hill holder system: service or inspect every 7500 miles.

Disc brake pads & discs, front & rear axle boots & axle shaft joint portions: service or inspect every 7500 miles.

Steering & suspension: service or inspect every 7500 miles.

Air filter element: service or inspect every 15,000 miles.

Automatic transmission fluid: service or inspect every 15,000 miles.

Brake linings & drums: service or inspect every 15,000 miles.

Coolant level, hoses & clamps: service or inspect every 15,000 miles.

Drive belts: service or inspect every 15,000 miles.

Transmission/differential gear oil (except SVX): service or inspect every 15,000 miles.

Front & rear wheel bearing repack: service or inspect every 30,000 miles.

93471CN1

For Wheel Alignment specifications, see Section 1 of this manual

SCHEDULED MAINTENANCE INTERVALS
SUBARU
IMPREZA OUTBACK, IMPREZA OUTBACK SPORT,
LEGACY, LEGACY OUTBACK

The following should be used as a guide when determining the amount of work required for a particular service. In estimating how long a particular Scheduled Maintenance Service should take, please observe the following:

- Labor Time is time based on field research and data supplied by the vehicle manufacturer.
- Labor time operations are given in hours and tenths of an hour.
- All labor operations are to be used as a guide.

Mechanic Skill Level Codes:
(A) PRECISION: Highly skilled with multiple certification.
(B) GENERAL: Normally skilled with certification.
(C) MAINTENANCE: Semi-skilled working on certification.

	LABOR TIME		LABOR TIME		LABOR TIME
7500 Mile Service (C)		**37500 Mile Service (C)**		**75000 Mile Service (C)**	
All Models	.4	All Models	.4	All Models	1.0
15000 Mile Service (C)		**45000 Mile Service (C)**		**82500 Mile Service (C)**	
All Models	1.0	All Models	1.0	All Models	.4
22500 Mile Service (C)		**52500 Mile Service (C)**		**90000 Mile Service (B)**	
All Models	.4	All Models	.4	All Models	6.7
30000 Mile Service (B)		**60000 Mile Service (B)**		**97500 Mile Service (C)**	
All Models	6.7	All Models	6.1	All Models	.4
		67500 Mile Service (C)			
		All Models	.4		

93471CN2

SUZUKI
ESTEEM • SWIFT

VEHICLE AND ENGINE IDENTIFICATION

Engine								Model Year	
Code	Liters (cc)	Cu. in.	Cyl.	Fuel Sys.	Engine Type	Eng. Mfg.		Code	Year
2	1.3 (1298)	79.3	4	TFI	SOHC	Suzuki		W	1998
3	1.6 (1590)	97.7	4	MFI	SOHC	Suzuki		X	1999
MFI: Multi-port Fuel Injection								Y	2000
TFI: Throttle body Fuel Injection								1	2001
SOHC: Single Overhead Camshaft								2	2002

93471CN3

For Maintenance Interval recommendations, see Section 1 of this manual

GENERAL ENGINE SPECIFICATIONS

Year	Engine ID/VIN	Engine Displacement Liters (cc)	Fuel System Type	Net Horsepower @ rpm	Net Torque @ rpm (ft. lbs.)	Bore x Stroke (in.)	Com-pression Ratio	Oil Pressure @ rpm
1998	2	1.3 (1298)	TFI	70@5500	74@3000	2.91x2.97	9.5:1	47-61@3000
	3	1.6 (1590)	MFI	98@6000	94@3200	2.95x3.54	9.5:1	47-61@4000
1999	2	1.3 (1298)	TFI	70@5500	74@3000	2.91x2.97	9.5:1	47-61@3000
	3	1.6 (1590)	MFI	98@6000	94@3200	2.95x3.54	9.5:1	47-61@4000
2000	2	1.3 (1298)	TFI	70@5500	74@3000	2.91x2.97	9.5:1	47-61@3000
	3	1.6 (1590)	MFI	98@6000	94@3200	2.95x3.54	9.5:1	47-61@4000
2001	2	1.3 (1298)	TFI	70@5500	74@3000	2.91x2.97	9.5:1	47-61@3000
	3	1.6 (1590)	MFI	98@6000	94@3200	2.95x3.54	9.5:1	47-61@4000

MFI: Multiport Fuel Injection

TFI: Throttle body Fuel Injection

93471CN4

GASOLINE ENGINE TUNE-UP SPECIFICATIONS

Year	Engine Displacement Liters (cc)	Engine ID/VIN	Spark Plugs Gap (in.)	Ignition Timing (deg.)		Fuel Pump (psi)		Idle Speed (rpm)		Valve Clearance	
				MT	AT			MT	AT	In.	Ex.
1998	1.3 (1298)	2	0.029	5B	5B	13-20	①	750	850	HYD	HYD
	1.6 (1590)	3	0.029	5B	5B	28-34	①	750-800	750-800	②	②
1999	1.3 (1298)	2	0.029	5B	5B	13-20	①	750	850	HYD	HYD
	1.6 (1590)	3	0.029	5B	5B	28-34	①	750-800	750-800	②	②
2000	1.3 (1298)	2	0.029	5B	5B	13-20	①	750	850	HYD	HYD
	1.6 (1590)	3	0.029	5B	5B	28-34	①	750-800	750-800	②	②
2001	1.3 (1298)	2	0.029	5B	5B	13-20	①	750	850	HYD	HYD
	1.6 (1590)	3	0.029	5B	5B	28-34	①	750-800	750-800	②	②

Note: The Vehicle Emission Control Information label often reflects specification changes made during production. The label figures must be used if they differ from those in this chart.

HYD: Hydraulic

B: Before top dead center

① At idle

② When cold: 0.005-0.007
   When hot: 0.007-0.008

93471CN5

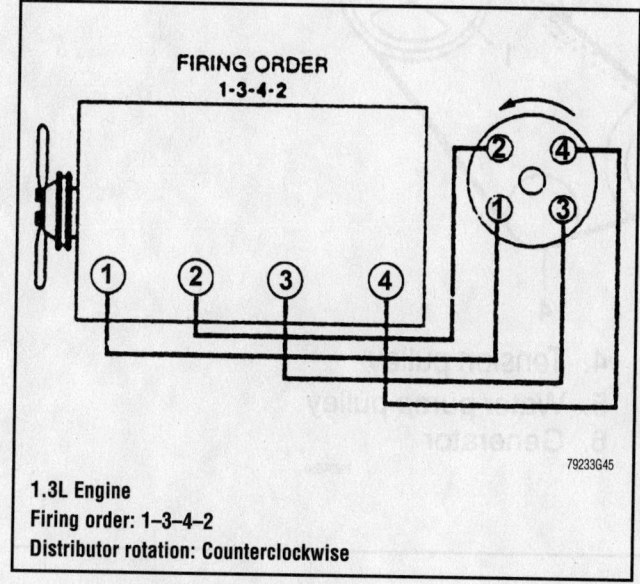

FIRING ORDER
1-3-4-2

1.3L Engine
Firing order: 1–3–4–2
Distributor rotation: Counterclockwise

79233G45

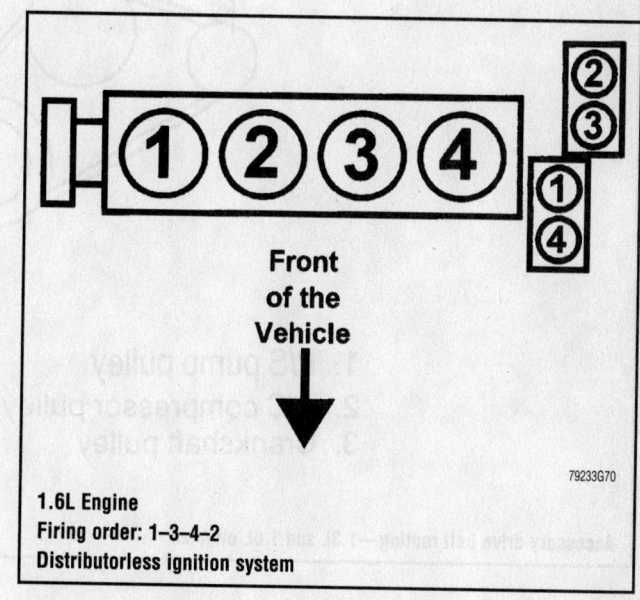

Front of the Vehicle

1.6L Engine
Firing order: 1–3–4–2
Distributorless ignition system

79233G70

For Tune-up, Capacities and Firing orders, see Section 1 of this manual

For vehicle equipped with A/C

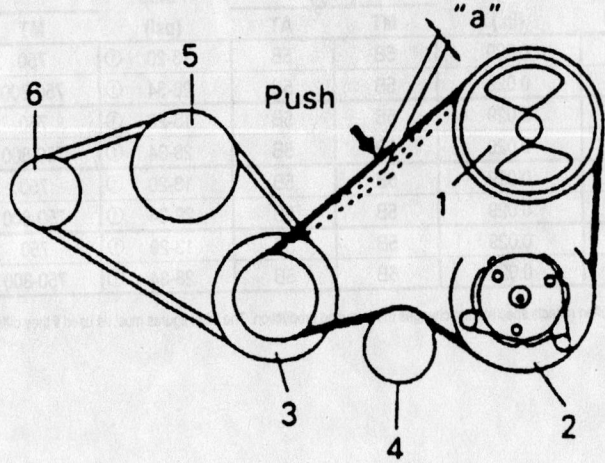

For vehicle not equipped with A/C

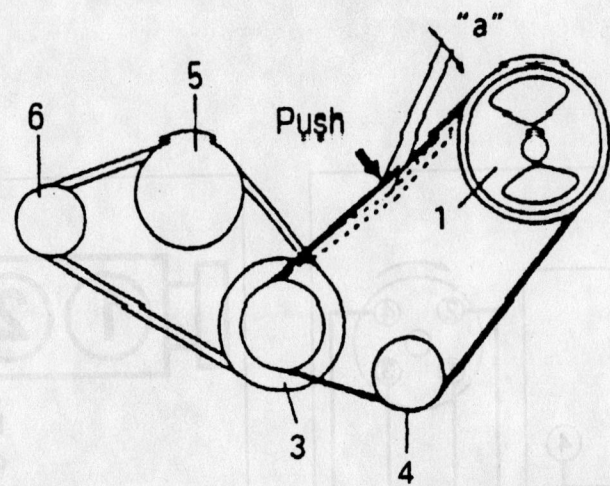

1. P/S pump pulley
2. A/C compressor pulley
3. Crankshaft pulley
4. Tension pulley
5. Water pump pulley
6. Generator

79234G50

Accessory drive belt routing—1.3L and 1.6L engines

CAPACITIES

Year	Model	Engine Displacement Liters (cc)	Engine ID/VIN	Engine Oil with Filter (qts.)	Transmission (pts.)		Fuel Tank (gal.)	Cooling System (qts.)
					5-Spd	Auto.		
1998	Esteem	1.6 (1590)	3	3.3	5.0	10.4 ①	13.5	③
	Swift	1.3 (1298)	2	3.3	5.0	10.4 ①	10.6	②
1999	Esteem	1.6 (1590)	3	3.3	5.0	10.4 ①	13.5	③
	Swift	1.3 (1298)	2	3.3	5.0	10.4 ①	10.6	②
2000	Esteem	1.6 (1590)	3	3.3	5.0	10.4 ①	13.5	③
	Swift	1.3 (1298)	2	3.3	5.0	10.4 ①	10.6	②
2001	Esteem	1.6 (1590)	3	3.3	5.0	10.4 ①	13.5	③
	Swift	1.3 (1298)	2	3.3	5.0	10.4 ①	10.6	②

Note: All capacities are approximate. Add fluid gradualy and check to be sure a proper fluid level is obtained.

① Specification for automatic transaxle is after complete overhaul. Drain and fill will be less

② Manual transmission: 4.8 qts.
   Automatic transmission: 4.9 qts.

③ Manual transmission: 4.8 qts.
   Automatic transmission: 4.7 qts.

93471CN6

VALVE SPECIFICATIONS

Year	Engine ID/VIN	Engine Displacement Liters (cc)	Seat Angle (deg.)	Face Angle (deg.)	Spring Test Pressure (lbs. @ in.)	Spring Installed Height (in.)	Stem-to-Guide Clearance (in.)		Stem Diameter (in.)	
							Intake	Exhaust	Intake	Exhaust
1998	2	1.3 (1298)	45	45	55-64 @ 1.63	1.941	0.0008-0.0019	0.0014-0.0025	0.2742-0.2748	0.2737-0.2742
	3	1.6 (1590)	45	45	24-28 @ 1.24	1.450	0.0008-0.0018	0.0018-0.0028	0.2152-0.2157	0.2142-0.2148
1999	2	1.3 (1298)	45	45	55-64 @ 1.63	1.941	0.0008-0.0019	0.0014-0.0025	0.2742-0.2748	0.2737-0.2742
	3	1.6 (1590)	45	45	24-28 @ 1.24	1.450	0.0008-0.0018	0.0018-0.0028	0.2152-0.2157	0.2142-0.2148
2000	2	1.3 (1298)	45	45	55-64 @ 1.63	1.941	0.0008-0.0019	0.0014-0.0025	0.2742-0.2748	0.2737-0.2742
	3	1.6 (1590)	45	45	24-28 @ 1.24	1.450	0.0008-0.0018	0.0018-0.0028	0.2152-0.2157	0.2142-0.2148
2001	2	1.3 (1298)	45	45	55-64 @ 1.63	1.941	0.0008-0.0019	0.0014-0.0025	0.2742-0.2748	0.2737-0.2742
	3	1.6 (1590)	45	45	24-28 @ 1.24	1.450	0.0008-0.0018	0.0018-0.0028	0.2152-0.2157	0.2142-0.2148

93471CN7

CRANKSHAFT AND CONNECTING ROD SPECIFICATIONS

All measurements are given in inches.

Year	Engine ID/VIN	Engine Displacement Liters (cc)	Crankshaft				Connecting Rod		
			Main Brg. Journal Dia.	Main Brg. Oil Clearance	Shaft End-play	Thrust on No.	Journal Diameter	Oil Clearance	Side Clearance
1998	3	1.3 (1298)	①	0.0008-0.0023	0.0044-0.0149	3	1.6529-1.6535	0.0008-0.0031	0.0039-0.0137
	6	1.6 (1590)	②	0.0008-0.0016	0.0044-0.0122	3	1.7316-1.7322	0.0008-0.0019	0.0039-0.0078
1999	3	1.3 (1298)	①	0.0008-0.0023	0.0044-0.0149	3	1.6529-1.6535	0.0008-0.0031	0.0039-0.0137
	6	1.6 (1590)	②	0.0008-0.0016	0.0044-0.0122	3	1.7316-1.7322	0.0008-0.0019	0.0039-0.0078
2000	3	1.3 (1298)	①	0.0008-0.0023	0.0044-0.0149	3	1.6529-1.6535	0.0008-0.0031	0.0039-0.0137
	6	1.6 (1590)	②	0.0008-0.0016	0.0044-0.0122	3	1.7316-1.7322	0.0008-0.0019	0.0039-0.0078
2001	3	1.3 (1298)	①	0.0008-0.0023	0.0044-0.0149	3	1.6529-1.6535	0.0008-0.0031	0.0039-0.0137
	6	1.6 (1590)	②	0.0008-0.0016	0.0044-0.0122	3	1.7316-1.7322	0.0008-0.0019	0.0039-0.0078

① No. 1: 1.7714-1.7716
No. 2: 1.7712-1.7714
No. 3: 1.7710-1.7712

② No. 1: 2.0470-2.0472
No. 2: 2.0468-2.0470
No. 3: 2.0465-2.0468

93471CN8

PISTON AND RING SPECIFICATIONS
All measurements are given in inches.

| Year | Engine ID/VIN | Engine Displacement Liters (cc) | Piston Clearance | Ring Gap | | | Ring Side Clearance | | |
				Top Compression	Bottom Compression	Oil Control	Top Compression	Bottom Compression	Oil Control
1998	2	1.3 (1298)	0.0008-0.0015	0.0079-0.0118	0.0079-0.0118	0.0079-0.0275	0.0012-0.0027	0.0008-0.0023	snug
	3	1.6 (1590)	0.0008-0.0015	0.0079-0.0137	0.0079-0.0137	0.0079-0.0275	0.0012-0.0027	0.0008-0.0023	snug
1999	2	1.3 (1298)	0.0008-0.0015	0.0079-0.0118	0.0079-0.0118	0.0079-0.0275	0.0012-0.0027	0.0008-0.0023	snug
	3	1.6 (1590)	0.0008-0.0015	0.0079-0.0137	0.0079-0.0137	0.0079-0.0275	0.0012-0.0027	0.0008-0.0023	snug
2000	2	1.3 (1298)	0.0008-0.0015	0.0079-0.0118	0.0079-0.0118	0.0079-0.0275	0.0012-0.0027	0.0008-0.0023	snug
	3	1.6 (1590)	0.0008-0.0015	0.0079-0.0137	0.0079-0.0137	0.0079-0.0275	0.0012-0.0027	0.0008-0.0023	snug
2001	2	1.3 (1298)	0.0008-0.0015	0.0079-0.0118	0.0079-0.0118	0.0079-0.0275	0.0012-0.0027	0.0008-0.0023	snug
	3	1.6 (1590)	0.0008-0.0015	0.0079-0.0137	0.0079-0.0137	0.0079-0.0275	0.0012-0.0027	0.0008-0.0023	snug

93471CN9

TORQUE SPECIFICATIONS
All readings in ft. lbs.

Year	Engine ID/VIN	Engine Displacement Liters (cc)	Cylinder Head Bolts	Main Bearing Bolts	Rod Bearing Bolts	Crankshaft Damper Bolts	Flywheel Bolts	Manifold		Spark Plugs	Lug Nut
								Intake	Exhaust		
1998	2	1.3 (1298)	①	36-41	24-26	76-83 ②	41-47	13-20	13-20	14-21	36-57
	3	1.6 (1590)	①	36-41	24-26	76-83 ②	57	13-20	13-20	14-21	58-80
1999	2	1.3 (1298)	①	36-41	24-26	76-83 ②	41-47	13-20	13-20	14-21	36-57
	3	1.6 (1590)	①	36-41	24-26	76-83 ②	57	13-20	13-20	14-21	58-80
2000	2	1.3 (1298)	①	36-41	24-26	76-83 ②	41-47	13-20	13-20	14-21	36-57
	3	1.6 (1590)	①	36-41	24-26	76-83 ②	57	13-20	13-20	14-21	58-80
2001	2	1.3 (1298)	①	36-41	24-26	76-83 ②	41-47	13-20	13-20	14-21	36-57
	3	1.6 (1590)	①	36-41	24-26	76-83 ②	57	13-20	13-20	14-21	58-80

① Step 1: 26 ft. lbs. (35 Nm)
　Step 2: 41 ft. lbs. (55 Nm)
　Step 3: 49 ft. lbs. (68 Nm)

② Specification shown is for crankshaft timing sprocket nut

93471CN0

Timing belt service is covered in Section 3 of this manual

BRAKE SPECIFICATIONS
All measurements in inches unless noted

| Year | Model | Brake Disc | | | Brake Drum Diameter | | | Minimum Lining Thickness | | Brake Caliper | |
		Original Thickness	Minimum Thickness	Maximum Runout	Original Inside Diameter	Max. Wear Limit	Maximum Machine Diameter	Front	Rear	Bracket bolts ft lbs.	Mounting bolts ft lbs.
1998	Swift ①	0.670	0.620	0.004	7.09	7.87	7.87	0.236 ②	0.110 ②	36	27
	Swift	0.670	0.620	0.004	7.16	7.95	7.95	0.236 ②	0.110 ②	36	27
	Esteem	0.790	0.710	0.004	7.87	7.95	7.95	0.240 ②	0.110 ②	62	16
1999	Swift ①	0.670	0.620	0.004	7.09	7.87	7.87	0.236 ②	0.110 ②	36	27
	Swift	0.670	0.620	0.004	7.16	7.95	7.95	0.236 ②	0.110 ②	36	27
	Esteem	0.790	0.710	0.004	7.87	7.95	7.95	0.240 ②	0.110 ②	62	16
2000	Swift ①	0.670	0.620	0.004	7.09	7.87	7.87	0.236 ②	0.110 ②	36	27
	Swift	0.670	0.620	0.004	7.16	7.95	7.95	0.236 ②	0.110 ②	36	27
	Esteem	0.790	0.710	0.004	7.87	7.95	7.95	0.240 ②	0.110 ②	62	16
2001	Swift ①	0.670	0.620	0.004	7.09	7.87	7.87	0.236 ②	0.110 ②	36	27
	Swift	0.670	0.620	0.004	7.16	7.95	7.95	0.236 ②	0.110 ②	36	27
	Esteem	0.790	0.710	0.004	7.87	7.95	7.95	0.240 ②	0.110 ②	62	16

① Hatchback

② Measurement is for lining and backing together.

93471C01

WHEEL ALIGNMENT

Year	Model		Caster Range (+/-Deg.)	Caster Preferred Setting (Deg.)	Camber Range (+/-Deg.)	Camber Preferred Setting (Deg.)	Toe-in (in.)	Steering Axis Inclination (Deg.)
1998	Swift	F	2.00	+3.00	1.00	+0.50	0.08+/-0.08	—
		R	—	—	—	—	0.18+/-0.06	—
	Esteem	F	2.00	+2.70	1.00	0	0+/-0.08	—
		R	—	—	—	0	0.08+/-0.08	—
1999	Swift	F	2.00	+3.00	1.00	+0.50	0.08+/-0.08	—
		R	—	—	—	—	0.18+/-0.06	—
	Esteem	F	2.00	+2.70	1.00	0	0+/-0.08	—
		R	—	—	—	0	0.08+/-0.08	—
2000	Swift	F	2.00	+3.00	1.00	+0.50	0.08+/-0.08	—
		R	—	—	—	—	0.18+/-0.06	—
	Esteem	F	2.00	+2.70	1.00	0	0+/-0.08	—
		R	—	—	—	0	0.08+/-0.08	—
2001	Swift	F	2.00	+3.00	1.00	+0.50	0.08+/-0.08	—
		R	—	—	—	—	0.18+/-0.06	—
	Esteem	F	2.00	+2.70	1.00	0	0+/-0.08	—
		R	—	—	—	0	0.08+/-0.08	—

93471C02

Heater Core replacement is covered in Section 2 of this manual

TIRE, WHEEL AND BALL JOINT SPECIFICATIONS

| Year | Model | OEM Tires | | Tire Pressures (psi) | | Wheel | Ball Joint |
		Standard	Optional	Front	Rear	Size	Inspection
1998	Swift	P155/80R13	None	32	32	4.5J	①
	Esteem GL	P175/70R13	None	30	30	5J	①
	Esteem GL Wagon	P175/70R13	None	30	30	5JJ	①
	Esteem GLX	P185/60R14	None	30	30	5.5JJ	①
	Esteem GLX Wagon	P185/55R14	None	30	30	5.5JJ	①
1999	Swift	P155/80R13	None	32	32	4.5J	①
	Esteem GL	P175/70R13	None	30	30	5J	①
	Esteem GL Wagon	P175/70R13	None	30	30	5JJ	①
	Esteem GLX	P185/60R14	None	30	30	5.5JJ	①
	Esteem GLX Wagon	P185/55R14	None	30	30	5.5JJ	①
2000	Swift	P155/80R13	None	32	32	4.5J	①
	Esteem GL	P175/70R13	None	30	30	5J	①
	Esteem GL Wagon	P185/60R14	None	30	30	5.5JJ	①
	Esteem GLX	P185/60R14	None	30	30	5.5JJ	①
	Esteem GLX Wagon	P195/55R15	None	29	29	5.5JJ	①
2001	Swift	P155/80R13	None	32	32	4.5J	①
	Esteem GL	P175/70R13	None	30	30	5J	①
	Esteem GL Wagon	P185/60R14	None	30	30	5.5JJ	①
	Esteem GLX	P185/60R14	None	30	30	5.5JJ	①
	Esteem GLX Wagon	P195/55R15	None	29	29	5.5JJ	①

OEM: Original Equipment Manufacturer

PSI: Pounds Per Square Inch

STD: Standard

OPT: Optional

① Replace if any measurable movement is found.

93471C03

SCHEDULED MAINTENANCE INTERVALS
SUZUKI—ESTEEM & SWIFT

TO BE SERVICED	TYPE OF SERVICE	VEHICLE MILEAGE INTERVAL (x1000)												
		7.5	15	22.5	30	37.5	45	52.5	60	67.5	75	82.5	90	97.5
Engine oil & filter	R	✓	✓	✓	✓	✓	✓	✓	✓	✓	✓	✓	✓	✓
Automatic transmission fluid & filter ①	S/I	✓	✓	✓	✓	✓	✓	✓	✓	✓	✓	✓	✓	✓
Clutch pedal free travel	S/I	✓	✓	✓	✓	✓	✓	✓	✓	✓	✓	✓	✓	✓
Drive axle boots	S/I	✓	✓	✓	✓	✓	✓	✓	✓	✓	✓	✓	✓	✓
Gear shift control lever/shift operation	S/I	✓	✓	✓	✓	✓	✓	✓	✓	✓	✓	✓	✓	✓
Inspect & rotate tires	S/I	✓	✓	✓	✓	✓	✓	✓	✓	✓	✓	✓	✓	✓
Manual transmission oil ②	S/I	✓	✓	✓	✓	✓	✓	✓	✓	✓	✓	✓	✓	✓
Power steering system	S/I	✓	✓	✓	✓	✓	✓	✓	✓	✓	✓	✓	✓	✓
Suspension system	S/I	✓	✓	✓	✓	✓	✓	✓	✓	✓	✓	✓	✓	✓
Brake discs, pads, drums & shoes	S/I	✓			✓		✓		✓		✓		✓	✓
Brake hoses, pipes, brake lever & cable	S/I	✓			✓		✓		✓		✓		✓	✓
Brake fluid ③	S/I		✓		✓		✓		✓		✓		✓	
Brake pedal	S/I		✓		✓		✓		✓		✓		✓	
Cooling system, hoses & connections	S/I		✓		✓		✓		✓		✓		✓	
Fuel tank, cap & lines	S/I		✓		✓		✓		✓		✓		✓	
Valve lash (clearance)	S/I		✓		✓		✓		✓		✓		✓	
Air cleaner filter element	R				✓				✓				✓	
Engine coolant	R				✓				✓				✓	
Spark plugs	R				✓				✓				✓	
Drive belts	S/I				✓				✓				✓	
Exhaust system	S/I				✓				✓				✓	

93471C04

Brake service is covered in Section 4 of this manual

SCHEDULED MAINTENANCE INTERVALS
SUZUKI—ESTEEM & SWIFT

TO BE SERVICED	TYPE OF SERVICE	VEHICLE MILEAGE INTERVAL (x1000)												
		7.5	15	22.5	30	37.5	45	52.5	60	67.5	75	82.5	90	97.5
Automatic transmission fluid hose	R						✓							
Camshaft timing belt	R								✓					
Ignition wiring	S/I								✓					

R: Replace   S/I: Service or Inspect

① Replace every 100,000 miles.

② Replace every 15,000 miles.

③ Replace every 60,000 miles.

FREQUENT OPERATION MAINTENANCE (SEVERE SERVICE)

If a vehicle is operated under any of the following conditions it is considered severe service:

- Extremely dusty areas.

- 50% or more of the vehicle operation is in 32°C (90°F) or higher temperatures, or constant operation in temperatures below 0°C (32°F).

- Prolonged idling (vehicle operation in stop and go traffic).

- Frequent short running periods (engine does not warm to normal operating temperatures).

- Police, taxi, delivery usage or trailer towing usage.

Oil & oil filter: change every 3000 miles.

Brake discs, pads, drums & shoes: service or inspect initially at 3000 miles, 6000 miles, & every 12,000 miles thereafter.

Brake hoses & pipes: service or inspect initially at 3000 miles, 6000 miles & every 12,000 miles thereafter.

Air cleaner filter element: service or inspect ever 3000 miles & replace every 30,000 miles (if not replaced previously).

Automatic transmission fluid & filter: service or inspect every 6000 miles & replace every 15,000 miles (if not replaced previously).

Clutch pedal free travel: service or inspect every 6000 miles.

Inspect & rotate tires: service or inspect every 6000 miles.

Manual transmission oil: service or inspect every 6000 miles & replace every 12,000 miles (if not replaced previously).

Power steering system: service or inspect every 6000 miles.

Steering system: service or inspect every 6000 miles.

Suspension system: service or inspect every 6000 miles.

Drive belts: service or inspect every 15,000 miles.

Exhaust system: service or inspect every 15,000 miles.

93471C05

SCHEDULED MAINTENANCE INTERVALS
SUZUKI
SWIFT, ESTEEM

The following should be used as a guide when determining the amount of work required for a particular service. In estimating how long a particular Scheduled Maintenance Service should take, please observe the following:

● Labor Time is time based on field research and data supplied by the vehicle manufacturer.
● Labor time operations are given in hours and tenths of an hour.
● All labor operations are to be used as a guide.

Mechanic Skill Level Codes:
(A) PRECISION: Highly skilled with multiple certification.
(B) GENERAL: Normally skilled with certification.
(C) MAINTENANCE: Semi-skilled working on certification.

	LABOR TIME			LABOR TIME			LABOR TIME
7500 Mile Service (C)			**37500 Mile Service (C)**			**75000 Mile Service (B)**	
All Models	1.8		All Models	1.8		All Models	2.2
w/AT add	.1		w/AT add	.1		w/AT add	.1
15000 Mile Service (C)			**45000 Mile Service (B)**			**82500 Mile Service (C)**	
All Models	2.0		All Models	2.5		All Models	1.8
w/AT add	.1		w/AT add	.1		w/AT add	.1
22500 Mile Service (C)			**52500 Mile Service (C)**			**90000 Mile Service (B)**	
All Models	1.8		All Models	1.8		All Models	4.0
w/AT add	.1		w/AT add	.1		w/AT add	.1
30000 Mile Service (B)			**60000 Mile Service (B)**			**97500 Mile Service (C)**	
All Models	3.7		All Models	4.1		All Models	1.8
w/AT add	.1		w/AT add	.1		w/AT add	.1
			67500 Mile Service (C)				
			All Models	1.8			
			w/AT add	.1			

93471C06

For complete Engine Mechanical specifications, see Section 1 of this manual

TOYOTA
AVALON • CAMRY • CELICA • COROLLA • ECHO • SUPRA • TERCEL

ENGINE AND VEHICLE IDENTIFICATION

Engine							Model Year	
Code ①	Liters (cc)	Cu. In.	Cyl.	Fuel Sys.	Engine Type	Eng. Mfg.	Code ②	Year
1MZ-FE	3.0 (2952)	180	6	EFI	DOHC	Toyota	W	1998
2JZ-GE	3.0 (2997)	183	6	EFI	DOHC	Toyota	X	1999
5S-FE	2.2 (2264)	138	4	EFI	DOHC	Toyota	Y	2000
2JZ-GTE	3.0 (2997)	183	6	EFI	DOHC	Toyota	1	2001
5E-FE	1.5 (1495)	91	4	EFI	DOHC	Toyota	2	2002
1ZZ-FE	1.8 (1794)	109	4	EFI	DOHC	Toyota		
2ZZ-GE	1.8 (1796)	109.5	4	EFI	DOHC	Toyota		
1NZ-FE	1.5 (1496)	91	4	EFI	DOHC	Toyota		
1MZ-FE	3.0 (2995)	183	6	EFI	DOHC	Toyota		

EFI: Electronic Fuel Injection

DOHC: Double Overhead Camshaft

① 8th digit of VIN

② 10th digit of VIN

93471C07

GENERAL ENGINE SPECIFICATIONS

Year	Model	Engine Displacement Liters (cc)	Engine Series (ID/VIN)	Fuel System	Net Horsepower @ rpm	Net Torque @ rpm (ft. lbs.)	Bore x Stroke (in.)	Com-pression Ratio	Oil Pressure @ rpm
1998	Avalon	3.0 (2995)	1MZ-FE	EFI	192@5200	210@4400	3.44x3.27	10.5:1	4.3
	Camry	2.2 (2164)	5S-FE	EFI	130@5400	145@4400	3.43x3.58	9.5:1	4.3
	Camry	3.0 (2995)	1MZ-FE	EFI	192@5200	210@4400	3.44x3.27	10.5:1	4.3
	Celica	2.2 (2164)	5S-FE	EFI	135@5400	145@4400	3.43x3.58	9.5:1	4.3
	Corolla	1.8 (1794)	1ZZ-FE	EFI	120@5600	122@4400	3.11x3.60	10.0:1	4.3
	Supra	3.0 (2997)	2JZ-GE	EFI	220@5800	210@4800	3.39x3.39	10.0:1	7
	Supra	3.0 (2997) ①	2JZ-GTE	EFI	320@5600	315@4000	3.39x3.39	8.5:1	7
	Tercel	1.5 (1497)	5E-FE	EFI	100@6400	91@3200	2.91x3.43	9.4:1	4.3
1999	Avalon	3.0 (2995)	1MZ-FE	EFI	200@5200	214@5200	3.44x3.27	10.5:1	4.3
	Camry	2.2 (2164)	5S-FE	EFI	133@5200	147@4400	3.43x3.58	9.5:1	4.3
	Camry	3.0 (2995)	1MZ-FE	EFI	194@5200	209@4400	3.44x3.27	10.5:1	4.3
	Camry Solara	2.2 (2164)	5S-FE	EFI	135@5200	147@4400	3.43x3.58	9.5:1	4.3
	Camry Solara	3.0 (2995)	1MZ-FE	EFI	200@5200	214@4400	3.44x3.27	10.5:1	4.3
	Celica	2.2 (2164)	5S-FE	EFI	130@5400	145@4400	3.43x3.58	9.5:1	4.3
	Corolla	1.8 (1794)	1ZZ-FE	EFI	120@5600	122@4400	3.11x3.60	10.0:1	4.3
	Tercel	1.5 (1497)	5E-FE	EFI	100@6400	91@3200	2.91x3.43	9.4:1	4.3
2000	Avalon	3.0 (2995)	1MZ-FE	EFI	200@5200	214@5200	3.44x3.27	10.5:1	4.3
	Camry	2.2 (2164)	5S-FE	EFI	133@5200	147@4400	3.43x3.58	9.5:1	4.3
	Camry	3.0 (2995)	1MZ-FE	EFI	194@5200	209@4400	3.44x3.27	10.5:1	4.3
	Camry Solara	2.2 (2164)	5S-FE	EFI	135@5200	147@4400	3.43x3.58	9.5:1	4.3
	Camry Solara	3.0 (2995)	1MZ-FE	EFI	200@5200	214@4400	3.44x3.27	10.5:1	4.3
	Corolla	1.8 (1794)	1ZZ-FE	EFI	120@5600	122@4400	3.11x3.60	10.0:1	4.3
	Celica	1.8 (1794)	1ZZ-FE	EFI	140@6400	125@4200	3.11x3.60	10.0:1	4.3
	Celica GT-S	1.8 (1796)	2ZZ-GE	EFI	180@7600	130@6800	3.23x3.35	11.5:1	4.3
	Echo	1.5 (1496)	1NZ-FE	EFI	108@6000	105@4000	2.95x3.33	10.5:1	4.3
	Tercel	1.5 (1497)	5E-FE	EFI	100@6400	91@3200	2.91x3.43	9.4:1	4.3

EFI: Electronic Fuel Injection

① Twin Turbo Charged

93471C09

For Accessory Drive Belt illustrations, see Section 1 of this manual

ENGINE TUNE-UP SPECIFICATIONS

Year	Engine Displacement Liters (cc)	Engine ID/VIN	Spark Plug Gap (in.)	Ignition Timing (deg.) ①	Fuel Pump (psi)	Idle Speed (rpm)		Valve Clearance	
						MT	AT	In.	Ex.
1998	1.5 (1497)	5E-FE	0.043	8-12 BTDC	41-42	700-800	700-800	0.006-0.010	0.012-0.016
	1.8 (1794)	1ZZ-FE	0.043	8-12 BTDC	44-50	650-750	650-750	0.006-0.010	0.010-0.014
	2.2 (2164)	5S-FE	0.043	8-12 BTDC	44-50	700-800	700-800	0.007-0.011	0.011-0.015
	3.0 (2952)	1MZ-FE	0.043	8-12 BTDC	44-50	650-750	650-750	0.006-0.010	0.010-0.014
	3.0 (2997)	2JZ-GE	0.043	8-12 BTDC	44-50	650-750	650-750	0.006-0.010	0.010-0.014
	3.0 (2997)	2JZ-GTE	0.043	8-12 BTDC	33-40	600-700	600-700	0.006-0.010	0.010-0.014
1999	1.5 (1497)	5E-FE	0.043	8-12 BTDC	41-42	700-800	700-800	0.006-0.010	0.012-0.016
	1.8 (1794)	1ZZ-FE	0.043	8-12 BTDC	44-50	650-750	650-750	0.006-0.010	0.010-0.014
	2.2 (2164)	5S-FE	0.043	8-12 BTDC	44-50	700-800	700-800	0.007-0.011	0.011-0.015
	3.0 (2952)	1MZ-FE	0.043	8-12 BTDC	44-50	650-750	650-750	0.006-0.010	0.010-0.014
2000	1.5 (1497)	5E-FE	0.043	8-12 BTDC	41-42	700-800	700-800	0.006-0.010	0.012-0.016
	1.8 (1794)	1ZZ-FE	0.043	10-18 BTDC	44-50	650-750	700-800	0.006-0.010	0.010-0.014
	1.8 (1796)	2ZZ-GE	0.043	8-12 BTDC	44-50	750-850	700-800	0.006-0.010	0.014-0.018
	2.2 (2164)	5S-FE	0.043	8-12 BTDC	44-50	700-800	700-800	0.007-0.011	0.011-0.015
	1.5 (1496)	1NZ-FE	0.043	8-12 BTDC	44-50	700-800	700-800	0.006-0.010	0.011-0.014
	3.0 (2952)	1MZ-FE	0.043	8-12 BTDC	44-50	650-750	650-750	0.006-0.010	0.010-0.014

Note: The Vehicle Emission Control Information label often reflects specification changes made during production. The label figures must be used if they differ from those in this chart.

① With terminal TE1 and E1 connected of DLC1

93471C00

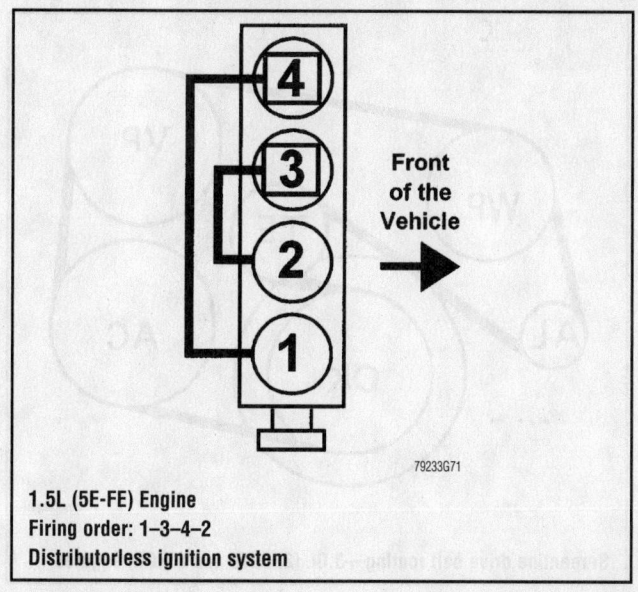

1.5L (5E-FE) Engine
Firing order: 1–3–4–2
Distributorless ignition system

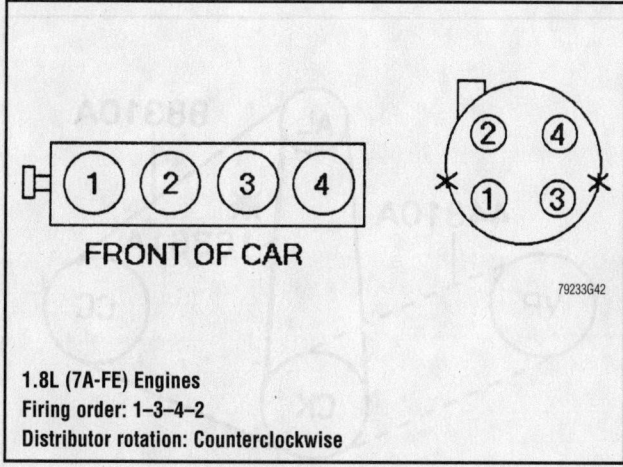

1.8L (7A-FE) Engines
Firing order: 1–3–4–2
Distributor rotation: Counterclockwise

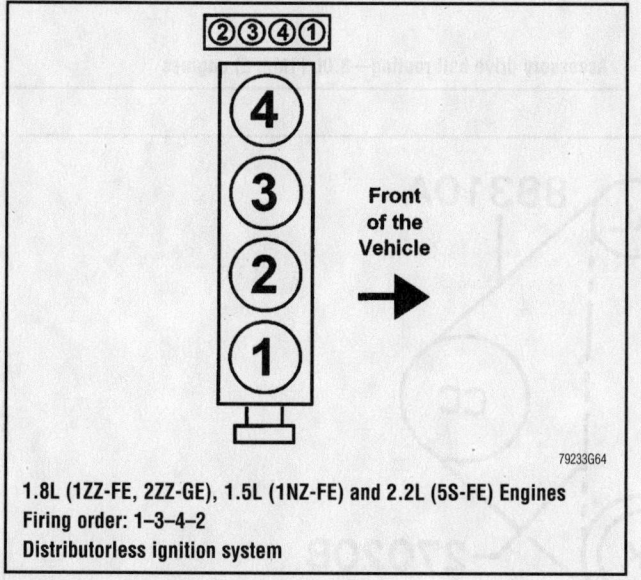

1.8L (1ZZ-FE, 2ZZ-GE), 1.5L (1NZ-FE) and 2.2L (5S-FE) Engines
Firing order: 1–3–4–2
Distributorless ignition system

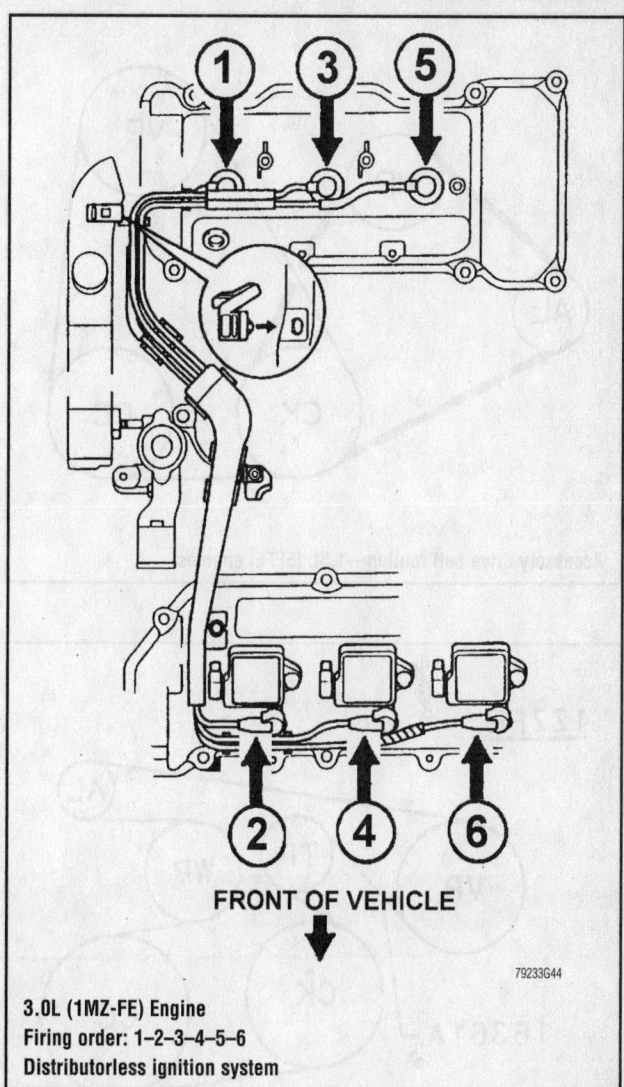

3.0L (1MZ-FE) Engine
Firing order: 1–2–3–4–5–6
Distributorless ignition system

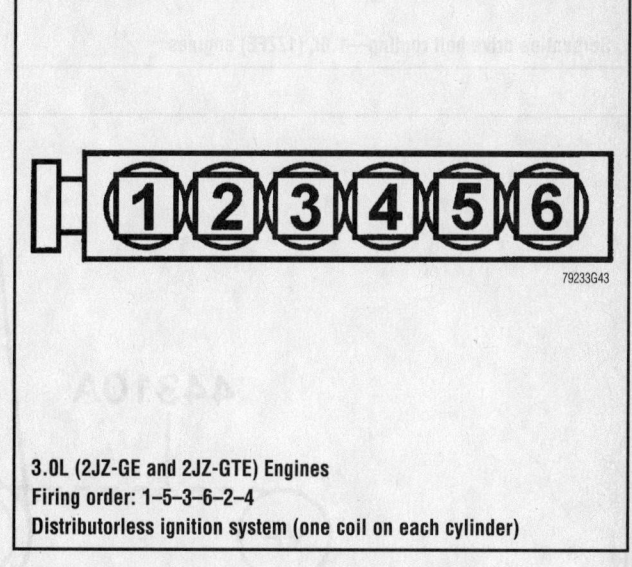

3.0L (2JZ-GE and 2JZ-GTE) Engines
Firing order: 1–5–3–6–2–4
Distributorless ignition system (one coil on each cylinder)

For Tire, Wheel and Ball Joint specifications, see Section 1 of this manual

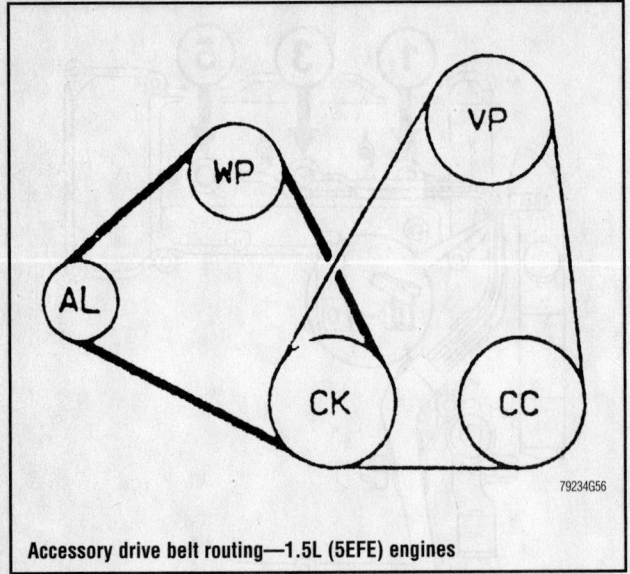

Accessory drive belt routing—1.5L (5EFE) engines

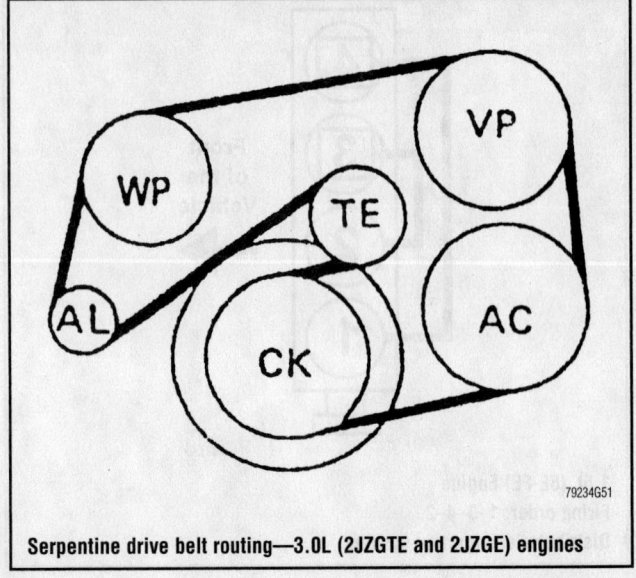

Serpentine drive belt routing—3.0L (2JZGTE and 2JZGE) engines

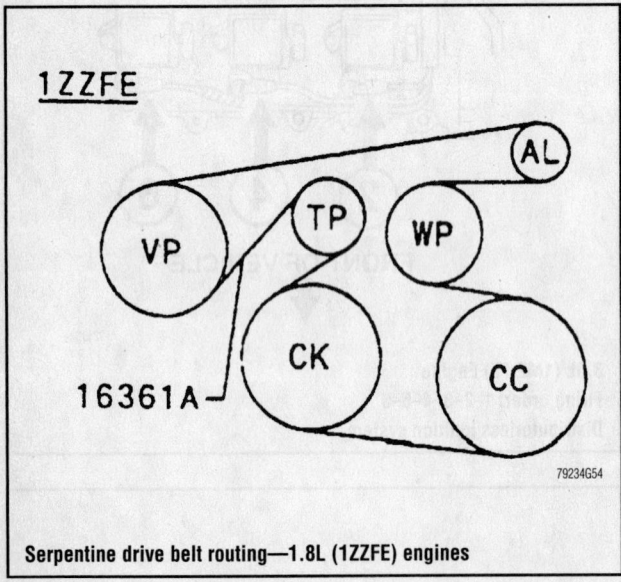

Serpentine drive belt routing—1.8L (1ZZFE) engines

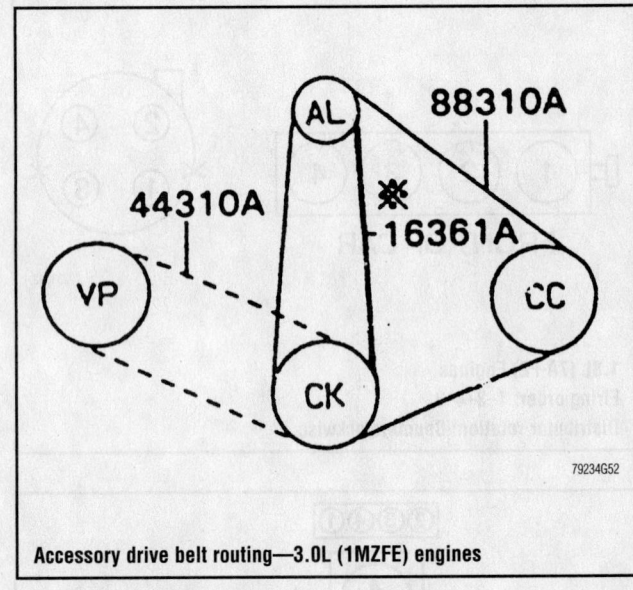

Accessory drive belt routing—3.0L (1MZFE) engines

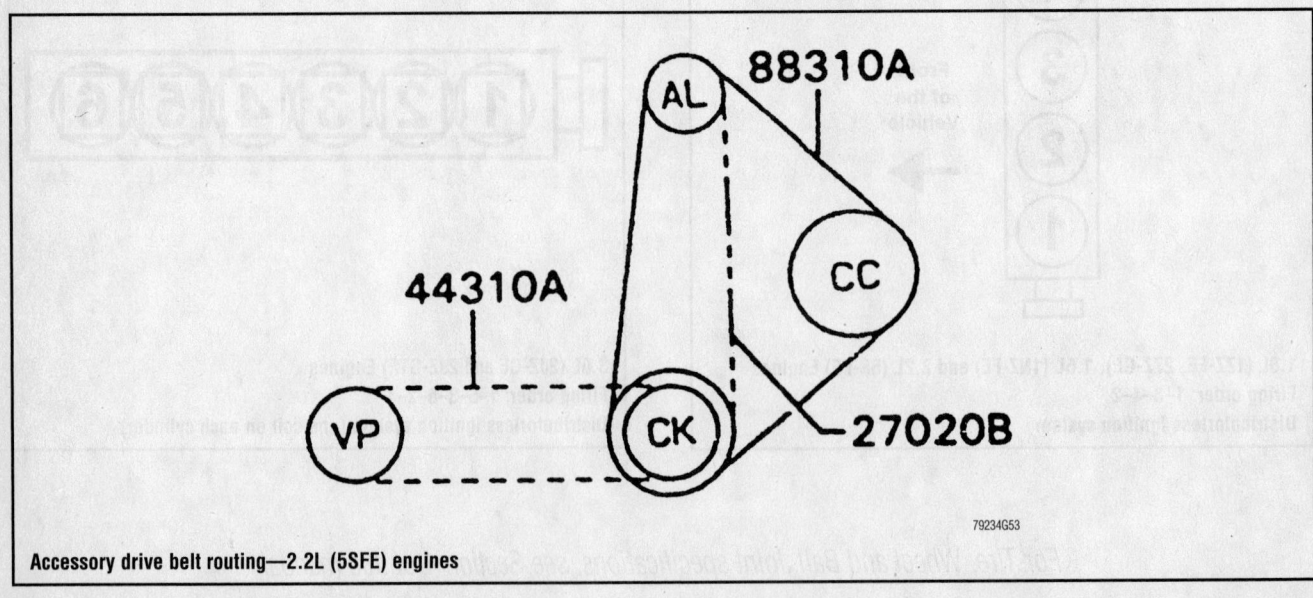

Accessory drive belt routing—2.2L (5SFE) engines

CAPACITIES

Year	Model	Engine Displacement Liters (cc)	Engine ID/VIN	Engine Oil with Filter	Transmission (pts.)			Drive Axle		Fuel Tank (gal.)	Cooling System (qts.)
					4-Spd	5-Spd	Auto.	Front (pts.)	Rear (pts.)		
1998	Avalon	3.0 (2995)	1MZ-FE	5.0	—	—	7.4	1.8	—	18.5	9.8
	Camry	3.0 (2995)	1MZ-FE	5.0	—	9.8	7.4	1.8	—	18.5	9.6
	Camry	2.2 (2164)	5S-FE	3.8	—	4.6	5.2	3.4	—	18.5	7.3
	Celica	2.2 (2164)	5S-FE	4.1	—	5.4	5.2	3.4	—	15.9	①
	Corolla	1.8 (1794)	1ZZ-FE	3.2	—	4.0	5.2	3.0	—	13.2	②
	Supra	3.0 (2997)	2JZ-GE	5.5	—	—	3.4	—	2.9	18.5	8.5
	Supra	3.0 (2997)	2JZ-GTE	5.3	—	3.8	4.0	—	2.9	18.5	③
	Tercel	1.5 (1497)	5E-FE	3.0	5.0	5.0	④	3.0 ⑤	—	11.9	⑥
1999	Avalon	3.0 (2995)	1MZ-FE	5.0	—	—	7.4	1.8	—	18.5	9.8
	Camry	2.2 (2164)	5S-FE	3.8	—	4.6	5.2	3.4	—	18.5	7.3
	Camry	3.0 (2995)	1MZ-FE	5.0	—	9.8	7.4	1.8	—	18.5	9.6
	Camry Solara	2.2 (2164)	5S-FE	3.8	—	4.6	5.2	3.4	—	18.5	7.3
	Camry Solara	3.0 (2995)	1MZ-FE	5.0	—	9.8	7.4	1.8	—	18.5	9.6
	Celica	2.2 (2164)	5S-FE	4.1	—	5.4	5.2	3.4	—	15.9	①
	Corolla	1.8 (1794)	1ZZ-FE	3.2	—	4.0	5.2	3.0	—	13.2	②
	Tercel	1.5 (1497)	5E-FE	3.0	5.0	5.0	④	3.0 ⑤	—	11.9	⑥
2000	Avalon	3.0 (2995)	1MZ-FE	5.0	—	—	7.4	1.8	—	18.5	9.8
	Camry	2.2 (2164)	5S-FE	3.8	—	4.6	5.2	3.4	—	18.5	7.3
	Camry	3.0 (2995)	1MZ-FE	5.0	—	9.8	7.4	1.8	—	18.5	9.6
	Camry Solara	2.2 (2164)	5S-FE	3.8	—	4.6	5.2	3.4	—	18.5	7.3
	Camry Solara	3.0 (2995)	1MZ-FE	5.0	—	9.8	7.4	1.8	—	18.5	9.6
	Celica	1.8 (1794)	1ZZ-FE	⑦	—	4.0	8.6	3.0	—	13.2	6.0
	Celica GT-S	1.8 (1796)	2ZZ-GE	⑧	—	4.8	6.1	⑨	—	14.5	6.0
	Corolla	1.8 (1794)	1ZZ-FE	3.2	—	4.0	5.2	3.0	—	13.2	②
	Echo	1.5 (1496)	1NZ-FE	4.0	—	4.0	6.2	⑨	—	11.9	⑩
	Tercel	1.5 (1497)	5E-FE	3.0	5.0	5.0	④	3.0 ⑤	—	11.9	⑥

Note: All capacities are approximate. Add fluid gradually and check to be sure a proper fluid level is obtained.

① Manual trans.: 6.4
Automatic trans.: 7.0

② M/T with Nippodenso radiator: 5.6
A/T with Nippodenso radiator: 6.2
M/T with Harrison radiator: 6.3
A/T with Harrison radiator: 6.2

③ Manual tra
Automatic transmission: 9.3

④ A132L transmission: 5.2
A242L transmission: 6.6

⑤ M/T with Nippodenso radiator: 5.8
A/T with Nippodenso radiator: 6.2
M/T with Harrison radiator: 6.6
A/T with Harrison radiator: 6.4

⑥ Manual transmission: 5.2
Automatic transmission: 5.7

⑦ w/oil cooler: 4.0
wo/oil cooler: 3.7

⑧ w/oil cooler: 4.8
wo/oil cooler: 4.6

⑨ Included in transaxle capacity

⑩ w/MT: 4.7
w/AT: 4.5

93471CP1

For Wheel Alignment specifications, see Section 1 of this manual

VALVE SPECIFICATIONS

Year	Engine Displacement Liters (cc)	Engine ID/VIN	Seat Angle (deg.)	Face Angle (deg.)	Spring Test Pressure (lbs. @ in.)	Spring Installed Height (in.)	Stem-to-Guide Clearance (in.)		Stem Diameter (in.)	
							Intake	Exhaust	Intake	Exhaust
1998	1.5 (1497)	5E-FE	45	44.5	33.3-36.8@ 1.252	1.252	0.0010-0.0024	0.0012-0.0026	0.2350-0.2356	0.2348-0.2354
	1.8 (1794)	1ZZ-FE	45	44.5	31.3-34.8@ 1.252	1.323	0.0010-0.0024	0.0012-0.0025	0.2154-0.2352	0.2152-0.2350
	2.2 (2164)	5S-FE	45	44.5	36.8-42.5@ 1.366	1.366	0.0010-0.0024	0.0012-0.0026	0.2350-0.2356	0.2348-0.2354
	3.0 (2995)	1MZ-FE	45	44.5	41.9-46.3@ 1.331	1.331	0.0010-0.0024	0.0012-0.0026	0.2154-0.2159	0.2152-0.2157
	3.0 (2997)	2JZ-GE	45	44.5	42.0-46.0@ 1.358	1.358	0.0010-0.0024	0.0012-0.0026	0.2350-0.2356	0.2348-0.2358
	3.0 (2997)	2JZ-GTE	45	44.5	42.0-46.0@ 1.358	1.358	0.0010-0.0024	0.0012-0.0026	0.2350-0.2356	0.2348-0.2358
1999	1.5 (1497)	5E-FE	45	44.5	33.3-36.8@ 1.252	1.252	0.0010-0.0024	0.0012-0.0026	0.2350-0.2356	0.2348-0.2354
	1.8 (1794)	1ZZ-FE	45	44.5	31.3-34.8@ 1.252	1.323	0.0010-0.0024	0.0012-0.0025	0.2154-0.2352	0.2152-0.2350
	2.2 (2164)	5S-FE	45	44.5	36.8-42.5@ 1.366	1.366	0.0010-0.0024	0.0012-0.0026	0.2350-0.2356	0.2348-0.2354
	3.0 (2995)	1MZ-FE	45	44.5	41.9-46.3@ 1.331	1.331	0.0010-0.0024	0.0012-0.0026	0.2154-0.2159	0.2152-0.2157
2000	1.5 (1497)	5E-FE	45	44.5	33.3-36.8@ 1.252	1.252	0.0010-0.0024	0.0012-0.0026	0.2350-0.2356	0.2348-0.2354
	1.8 (1794)	1ZZ-FE	45	44.5	31.3-34.8@ 1.252	1.323	0.0010-0.0024	0.0012-0.0025	0.2154-0.2352	0.2152-0.2350
	1.8 (1796)	2ZZ-GE	45	44.5	①	1.516	0.0010-0.0023	0.0012-0.0025	0.2150-0.2156	0.2144-0.2154
	2.2 (2164)	5S-FE	45	44.5	36.8-42.5@ 1.366	1.366	0.0010-0.0024	0.0012-0.0026	0.2350-0.2356	0.2348-0.2354
	1.5 (1496)	1NZ-FE	45	44.5	33.5-37@ 1.280	1.280	0.0010-0.0024	0.0012-0.0026	0.1957-0.1963	0.1955-0.1961
	3.0 (2995)	1MZ-FE	45	44.5	41.9-46.3@ 1.331	1.331	0.0010-0.0024	0.0012-0.0026	0.2154-0.2159	0.2152-0.2157

① Intake: 49.6-55.5@1.516
Exhaust: 47.6-52.6@1.516

93471CP2

CRANKSHAFT AND CONNECTING ROD SPECIFICATIONS
All measurements are given in inches.

Year	Engine Displacement Liters (cc)	Engine ID/VIN	Crankshaft				Connecting Rod		
			Main Brg. Journal Dia.	Main Brg. Oil Clearance	Shaft End-play	Thrust on No.	Journal Diameter	Oil Clearance	Side Clearance
1998	1.5 (1497)	5E-FE	1.9679-1.9685	0.0006-0.0014	0.0008-0.0078	3	1.6923-1.6929	0.0006-0.0019	0.0059-0.0138
	1.8 (1762)	1ZZ-FE	1.8891-1.8898	0.0006-0.0013	0.0008-0.0087	3	1.5742-1.5748	0.0008-0.0017	0.0059-0.0098
	2.2 (2164)	5S-FE	2.1653-2.6550	0.0010-0.0017	0.0008-0.0087	3	2.0466-2.0472	0.0009-0.0022	0.0063-0.0123
	3.0 (2995)	1MZ-FE	2.4011-2.4016	0.0010-0.0018	0.0016-0.0095	2	2.0863-2.8660	0.0015-0.0025	0.0059-0.0118
	3.0 (2997)	2JZ-GE	2.4403-2.4409	0.0010-0.0016	0.0008-0.0087	4	2.0465-2.0472	0.0009-0.0016	0.0098-0.0158
	3.0 (2997)	2JZ-GE	2.4403-2.4409	0.0010-0.0016	0.0008-0.0087	4	2.0465-2.0472	0.0009-0.0016	0.0098-0.0158
1999	1.5 (1497)	5E-FE	1.9679-1.9685	0.0006-0.0014	0.0008-0.0078	3	1.6923-1.6929	0.0006-0.0019	0.0059-0.0138
	1.8 (1762)	1ZZ-FE	1.8891-1.8898	0.0006-0.0013	0.0008-0.0087	3	1.5742-1.5748	0.0008-0.0017	0.0059-0.0098
	2.2 (2164)	5S-FE	2.1653-2.6550	0.0010-0.0017	0.0008-0.0087	3	2.0466-2.0472	0.0009-0.0022	0.0063-0.0123
	3.0 (2995)	1MZ-FE	2.4011-2.4016	0.0010-0.0018	0.0016-0.0095	2	2.0863-2.8660	0.0015-0.0025	0.0059-0.0118
2000	1.5 (1497)	5E-FE	1.9679-1.9685	0.0006-0.0014	0.0008-0.0078	3	1.6923-1.6929	0.0006-0.0019	0.0059-0.0138
	1.5 (1496)	1NZ-FE	①	0.0004-0.0009	0.0035-0.0075	3	1.5745-1.5748	0.0006-0.0016	0.0063-0.0142
	1.8 (1762)	1ZZ-FE	②	0.0006-0.0013	0.0008-0.0087	3	1.7320-1.7323	0.0011-0.0024	0.0063-0.0135
	1.8 (1796)	2ZZ-GE	②	0.0006-0.0013	0.0016-0.0094	3	1.7713-1.7717	0.0011-0.0020	0.0063-0.0135
	2.2 (2164)	5S-FE	2.1653-2.6550	0.0010-0.0017	0.0008-0.0087	3	2.0466-2.0472	0.0009-0.0022	0.0063-0.0123
	3.0 (2995)	1MZ-FE	2.4011-2.4016	0.0010-0.0018	0.0016-0.0095	2	2.0863-2.8660	0.0015-0.0025	0.0059-0.0118

① Reference mark:
0: 1.81102-1.81110
1: 1.81110-1.81118
2: 1.81118-1.81126
3: 1.81126-1.81133
4: 1.81133-1.81141
5: 1.81141-1.81149

② Reference mark:
0: 1.8897-1.8898
1: 1.8896-1.8897
2: 1.8895-1.8896
3: 1.8894-1.8895
4: 1.8893-1.8894
5: 1.8892-1.8893

93471CP3

PISTON AND RING SPECIFICATIONS
All measurements are given in inches.

Year	Engine Displacement Liters (cc)	Engine ID/VIN	Piston Clearance	Ring Gap			Ring Side Clearance		
				Top Compression	Bottom Compression	Oil Control	Top Compression	Bottom Compression	Oil Control
1998	1.5 (1497)	5E-FE	0.0035-0.0043	0.0102-0.0189	0.0118-0.0224	0.0059-0.0197	0.016-0.0031	0.0012-0.0028	SNUG
	1.8 (1762)	1ZZ-FE	0.0033-0.0041	0.0098-0.0138	0.0138-0.0197	0.0039-0.0157	0.0018-0.0033	0.0012-0.0028	SNUG
	2.2 (2164)	5S-FE	0.0055-0.0063	0.0106-0.0197	0.0138-0.0236	0.0079-0.0217	0.0016-0.0031	0.0012-0.0028	SNUG
	3.0 (2995)	1MZ-FE	0.0033-0.0042	0.0098-0.0138	0.0138-0.0177	0.0059-0.0157	0.0008-0.0028	0.0008-0.0024	SNUG
	3.0 (2997)	2JZ-GE	0.0022-0.0031	0.0118-0.0185	0.0138-0.0205	0.0051-0.0177	0.0004-0.0028	0.0012-0.0028	SNUG
	3.0 (2997)	2JZ-GTE	0.0029-0.0038	0.0118-0.0157	0.0138-0.0178	0.0051-0.0150	0.0016-0.0031	0.0012-0.0028	SNUG
1999	1.5 (1497)	5E-FE	0.0035-0.0043	0.0102-0.0189	0.0118-0.0224	0.0059-0.0197	0.016-0.0031	0.0012-0.0028	SNUG
	1.8 (1762)	1ZZ-FE	0.0033-0.0041	0.0098-0.0138	0.0138-0.0197	0.0039-0.0157	0.0018-0.0033	0.0012-0.0028	SNUG
	2.2 (2164)	5S-FE	0.0055-0.0063	0.0106-0.0197	0.0138-0.0236	0.0079-0.0217	0.0016-0.0031	0.0012-0.0028	SNUG
	3.0 (2995)	1MZ-FE	0.0033-0.0042	0.0098-0.0138	0.0138-0.0177	0.0059-0.0157	0.0008-0.0028	0.0008-0.0024	SNUG
2000	1.5 (1497)	5E-FE	0.0035-0.0043	0.0102-0.0189	0.0118-0.0224	0.0059-0.0197	0.016-0.0031	0.0012-0.0028	SNUG
	1.8 (1762)	1ZZ-FE	0.0033-0.0041	0.0098-0.0138	0.0138-0.0197	0.0039-0.0157	0.0012-0.0028	0.0012-0.0028	SNUG
	1.8 (1796)	2ZZ-GE	0.0003-0.0015	0.0098-0.0138	0.0138-0.0197	0.0059-0.0157	0.0012-0.0028	0.0012-0.0028	SNUG
	2.2 (2164)	5S-FE	0.0055-0.0063	0.0106-0.0197	0.0138-0.0236	0.0079-0.0217	0.0016-0.0031	0.0012-0.0028	SNUG
	1.5 (1496)	1NZ-FE	0.0022-0.0023	0.0098-0.0138	0.0138-0.0197	0.0039-0.0138	0.0012-0.0028	0.0012-0.0028	SNUG
	3.0 (2995)	1MZ-FE	0.0033-0.0042	0.0098-0.0138	0.0138-0.0177	0.0059-0.0157	0.0008-0.0028	0.0008-0.0024	SNUG

93471CP4

TORQUE SPECIFICATIONS
All readings in ft. lbs.

Year	Engine Displacement Liters (cc)	Engine ID/VIN	Cylinder Head Bolts	Main Bearing Bolts	Rod Bearing Bolts	Crankshaft Damper Bolts	Flywheel Bolts	Manifold		Spark Plugs	Lug Nuts
								Intake	Exhaust		
1998	1.5 (1497)	5E-FE	①	42	29	112	65	14	35	13	76
	1.8 (1794)	1ZZ-FE	②	③	④	102	②	14	27	13	76
	2.2 (2164)	5S-FE	②	43	⑤	80	⑥	14	36	13	76
	3.0 (2995)	1MZ-FE	⑦	⑤	⑧	159	61	11	36	13	76
	3.0 (2997)	2JZ-GE	⑨	①	⑩	243	②	21	30	13	76
	3.0 (2997)	2JZ-GTE	⑨	①	⑩	243	②	21	30	13	76
1999	1.5 (1497)	5E-FE	①	42	29	112	65	14	35	13	76
	1.8 (1794)	1ZZ-FE	②	③	④	102	②	14	27	13	76
	2.2 (2164)	5S-FE	②	43	⑧	80	④	14	36	13	76
	3.0 (2995)	1MZ-FE	⑦	⑤	⑧	159	61	11	36	13	76
2000	1.5 (1497)	5E-FE	①	42	29	112	65	14	35	13	76
	1.8 (1794)	1ZZ-FE	②	⑫	④	102	②	14	27	13	76
	1.8 (1796)	2ZZ-GE	⑪	⑫	⑩	87	②	⑬	37	13	76
	2.2 (2164)	5S-FE	②	43	⑧	80	④	14	36	13	76
	1.5 (1496)	1NZ-FE	⑭	⑮	⑯	94	②	22	20	13	76
	3.0 (2995)	1MZ-FE	⑦	⑤	⑧	159	61	11	36	13	76

① Step 1: 33 ft. lbs.
Step 2: 90 degree turn

② Step 1: 36 ft. lbs.
Step 2: 90 degree turn

③ Step 1: 16 ft. lbs.
Step 2: 32 ft. lbs.
Steps 3 and 4: 45 degree turn

④ Step 1: 15 ft. lbs.
Step 2: 90 degree turn

⑤ 12 pt.: 16 ft. lbs. + 90 degrees
6 pt.: 20 ft. lbs.

⑥ Manual transmission: 65 ft. lbs.

⑦ Step 1: 40 ft. lbs.
Step 2: 90 degree turn

⑧ Step 1: 18 ft. lbs.
Step 2: 90 degree turn

⑨ Step 1: 25 ft. lbs.
Step 2: 90 degree turn
Step 3: 90 degree turn
Recessed head bolt: 13 ft. lbs.

⑩ Step 1: 22 ft. lbs.
Step 2: 90 degree turn
Recessed head bolt: 13 ft. lbs.

⑪ Step 1: 26 ft. lbs.
Step 2: 180 degree turn

⑫ 12 pointed bolts:
Step 1: 16 ft. lbs.
Step 2: 32 ft. lbs.
Step 3: 45 degree turn
Step 4: 45 degree turn
Hex head bolts: 14 ft. lbs.

⑬ 4 upper bolts and 1 nut: 20 ft. lbs.
1 lower bolt: 34 ft. lbs.

⑭ Step 1: 22 ft. lbs.
Step 2: 90 degree turn
Step 3: 90 degree turn

⑮ Step 1: 16 ft. lbs.
Step 2: 90 degree turn

⑯ Step 1: 11 ft. lbs.
Step 2: 90 degree turn

93471CP5

For Tune-up, Capacities and Firing orders, see Section 1 of this manual

BRAKE SPECIFICATIONS
All measurements in inches unless noted

Year	Model		Brake Disc Original Thickness	Brake Disc Minimum Thickness	Brake Disc Maximum Runout	Brake Drum Diameter Original Inside Diameter	Brake Drum Diameter Max. Wear Limit	Brake Drum Diameter Maximum Machine Diameter	Minimum Lining Thickness	Brake Caliper Bracket Bolts (ft. lbs.)	Brake Caliper Mounting Bolts (ft. lbs.)
1998	Avalon	F	1.102	1.024	0.0020	—	—	—	0.039	25	79
		R	0.354	0.315	0.0059	—	—	—	0.039	25	34
	Camry	F	1.102	1.024	0.0020	—	—	—	0.039	25	79
		R	0.394	0.354	0.0059	9.00	—	9.08	0.039	14	20
	Celica	F	①	②	0.0020	—	—	—	0.039	25	79
		R	0.394	0.354	0.0059	7.87	—	7.91	0.039	19	34
	Corolla		0.866	0.787	0.0020	7.87	—	7.91	0.039	25	65
	Supra	F	③	④	0.0020	—	—	—	0.039	25	87
		R	0.630	0.591	—	7.48	—	7.52	0.039	25	77
	Tercel		0.709	0.669	0.0035	7.09	—	7.13	0.039	18	65
1999	Avalon	F	1.102	1.024	0.0020	—	—	—	0.039	25	79
		R	0.354	0.315	0.0059	—	—	—	0.039	25	34
	Camry	F	1.102	1.024	0.0020	—	—	—	0.039	25	79
		R	0.394	0.354	0.0059	9.00	—	9.08	0.039	14	20
	Camry Solara	F	1.102	1.024	0.0020	—	—	—	0.039	25	79
		R	0.394	0.354	0.0059	9.00	—	9.08	0.039	14	20
	Celica	F	1.102	1.024	0.0020	—	—	—	0.039	25	79
		R	0.354	0.314	0.0059	7.87	—	7.91	0.039	19	34
	Corolla		0.866	0.787	0.0020	7.87	—	7.91	0.039	25	65
	Tercel		0.709	0.630	0.0020	7.09	—	7.13	0.039	18	65
2000	Avalon	F	1.102	1.024	0.0020	—	—	—	0.039	25	79
		R	0.354	0.315	0.0059	—	—	—	0.039	25	34
	Camry	F	1.102	1.024	0.0020	—	—	—	0.039	25	79
		R	0.394	0.354	0.0059	9.00	—	9.08	0.039	14	20
	Camry Solara	F	1.102	1.024	0.0020	—	—	—	0.039	25	79
		R	0.394	0.354	0.0059	9.00	—	9.08	0.039	14	20
	Celica GT-S	F	0.984	0.906	0.002	—	—	—	0.039	25	79
		R	0.354	0.295	0.006	—	—	—	0.039	—	34
	Celica	F	1.102	1.024	0.0020	—	—	—	0.039	25	79
		R	0.354	0.314	0.0059	7.87	—	7.91	0.039	—	34
	Corolla		0.866	0.787	0.0020	7.87	—	7.91	0.039	25	65
	Echo		0.709	0.630	0.0020	7.09	—	7.13	0.039	25	65
	Tercel		0.709	0.630	0.0020	7.09	—	7.13	0.039	18	65

F: Front
R: Rear

① 7A-FE engine: 0.906
  5S-FE engine: 1.024

② 3S-GTE engine: 1.181
  5S-FE engine: 0.984

③ 2JZ-GTE engine: 1.102
  2JZ-GE engine: 1.181

④ 7A-FE engine: 0.984
  5S-FE engine: 1.102

93471CP6

WHEEL ALIGNMENT

Year	Model		Caster Range (+/-Deg.)	Caster Preferred Setting (Deg.)	Camber Range (+/-Deg.)	Camber Preferred Setting (Deg.)	Toe-in (in.)	Steering Axis Inclination (Deg.)
1998	Avalon	F	0.75	+2.17	0.75	-0.62	0+/-0.08	13.07+/-0.75
		R	—	—	0.75	-0.72	0.16+/-0.08	—
	Camry 4-cyl.	F	0.75	+2.18	0.75	-0.60	0+/-0.08	13.08+/-0.75
		R	—	—	0.75	-0.70	0.16+/-0.08	—
	Camry 6-cyl.	F	0.75	+2.09	0.75	-0.60	0+/-0.08	13.08+/-0.75
		R	—	—	0.75	-0.75	0.16+/-0.08	—
	Celica	F	0.75	+2.05	0.75	-0.77	0+/-0.08	14.97+/-0.75
		R	—	—	0.75	-1.17	0.14+/-0.08	—
	Corolla	F	0.75	+1.19	0.75	-0.18	0.04+/-0.08	12.38+/-0.75
		R	—	—	0.75	-0.92	0.16+/-0.08	—
	Supra 2JZ-GE	F	0.75	+3.33	0.75	-0.33	0+/-0.08	9.58+/-0.75
	Supra 2JZ-GTE	F	0.75	+3.50	0.75	-0.50	0+/-0.08	9.75+/-0.75
		R	—	—	0.75	-1.50	0.12+/-0.08	—
	Tercel	F	0.75	+1.33	0.75	-0.33	0.04+/-0.08	12.16+/-0.75
		R	—	—	0.75	-0.50	0.12+/-0.12	—
1999	Avalon	F	0.75	+2.17	0.75	-0.62	0+/-0.08	13.07+/-0.75
		R	—	—	0.75	-0.72	0.16+/-0.08	—
	Camry 4-cyl.	F	0.75	+2.18	0.75	-0.60	0+/-0.08	13.08+/-0.75
		R	—	—	0.75	-0.70	0.16+/-0.08	—
	Camry 6-cyl.	F	0.75	+2.09	0.75	-0.60	0+/-0.08	13.08+/-0.75
		R	—	—	0.75	-0.75	0.16+/-0.08	—
	Celica	F	0.75	+2.05	0.75	-0.77	0+/-0.08	14.97+/-0.75
		R	—	—	0.75	-1.17	0.14+/-0.08	—
	Corolla	F	0.75	+1.19	0.75	-0.18	0.04+/-0.08	12.38+/-0.75
		R	—	—	0.75	-0.92	0.16+/-0.08	—
	Solara	F	0.75	+2.08	0.75	-0.52	0+/-0.08	12.09+/-0.75
		R	—	—	0.75	-0.65	0.16+/-0.08	—
	Tercel	F	0.75	+1.33	0.75	-0.33	0.04+/-0.08	12.16+/-0.75
		R	—	—	0.75	-0.50	0.12+/-0.12	—
2000	Avalon	F	0.75	+2.17	0.75	-0.62	0+/-0.08	13.07+/-0.75
		R	—	—	0.75	-0.72	0.16+/-0.08	—
	Camry 4-cyl.	F	0.75	+2.18	0.75	-0.60	0+/-0.08	13.08+/-0.75
		R	—	—	0.75	-0.70	0.16+/-0.08	—
	Camry 6-cyl.	F	0.75	+2.09	0.75	-0.60	0+/-0.08	13.08+/-0.75
		R	—	—	0.75	-0.75	0.16+/-0.08	—
	Celica	F	0.75	+2.05	0.75	-0.77	0+/-0.08	14.97+/-0.75
		R	—	—	0.75	-1.17	0.14+/-0.08	—
	Corolla	F	0.75	+1.19	0.75	-0.18	0.04+/-0.08	12.38+/-0.75
		R	—	—	0.75	-0.92	0.16+/-0.08	—
	Echo	F	0.75	+1.60	0.75	-0.58	0+/-0.08	10.08+/-0.75
		R	—	—	0.75	-0.93	0.11+/-0.12	—
	Solara	F	0.75	+2.08	0.75	-0.52	0+/-0.08	12.09+/-0.75
		R	—	—	0.75	-0.65	0.16+/-0.08	—

93471CP7

For complete service labor times, order the Chilton Labor Guide

TIRE, WHEEL AND BALL JOINT SPECIFICATIONS

| Year | Model | OEM Tires | | Tire Pressures (psi) | | Wheel Size | Ball Joint Inspection |
		Standard	Optional	Front	Rear		
1998	Avalon	P205/65HR15	None	32	32	6-JJ	9-30 in. ①
	Camry 4-cyl.	P195/70SR14	None	30	30	5.5-JJ	9-26 in. ①
	Camry 6-cyl.	P205/65HR15	None	32	35	6-JJ	9-26 in. ①
	Celica	205/55VR15	P205/55VR15	33	33	6.5-JJ	9-26 in. ①
	Corolla VE, CE	P175/65R14	None	30	30	5.5-JJ	9-26 in. ①
	Corolla LE	P185/65R14	None	30	30	5.5-JJ	9-26 in. ①
	Supra, exc. Turbo	Fr: 225/50VR16	None	33	33	Fr: 8-JJ	9-26 in. ①
		Rr: 245/45VR16	None			Rr: 9.5-JJ	9-26 in. ①
	Supra Turbo	Fr: 235/45ZR17	None	36	36	Fr: 8-JJ	9-26 in. ①
		Rr: 255/40ZR17				Rr: 9.5-JJ	9-26 in. ①
	Tercel	P155/80SR13	P175/65R14	28	28	6-JJ	9-26 in. ①
			P185/60R14				
	Paseo	P185/60R14	None	26	26	5.5-JJ	9-26 in. ①
1999	Avalon	P205/65HR15	None	32	32	6-JJ	9-30 in. ①
	Camry 4-cyl.	P195/70SR14	None	30	30	5.5-JJ	9-26 in. ①
	Camry 6-cyl.	P205/65HR15	None	32	35	6-JJ	9-26 in. ①
	Celica	205/55VR15	P205/55VR15	33	33	6.5-JJ	9-26 in. ①
	Corolla VE, CE	P175/65R14	None	30	30	5.5-JJ	9-26 in. ①
	Corolla LE	P185/65R14	None	30	30	5.5-JJ	9-26 in. ①
	Paseo	P185/60R14	None	26	26	5.5-JJ	9-26 in. ①
2000	Avalon	P205/65HR15	None	32	32	6-JJ	9-30 in. ①
	Camry 4-cyl.	P195/70SR14	None	30	30	5.5-JJ	9-26 in. ①
	Camry 6-cyl.	P205/65HR15	None	32	35	6-JJ	9-26 in. ①
	Celica	205/55VR15	P205/55VR15	33	33	6.5-JJ	9-26 in. ①
	Corolla VE, CE	P175/65R14	None	30	30	5.5-JJ	9-26 in. ①
	Corolla LE	P185/65R14	None	30	30	5.5-JJ	9-26 in. ①
	Echo	P175/65R14	None	32	32	5.5-JJ	9-26 in. ①
	MR 2 Spider	Fr: 205/50R16	None	33	33	6-JJ	9-26 in. ①
		Rr: 215/50R16					
	Paseo	P185/60R14	None	26	26	5.5-JJ	9-26 in. ①

OEM: Original Equipment Manufacturer

PSI: Pounds Per Square Inch

STD: Standard

OPT: Optional

① Torque required in inch lbs. to rotate ball joint when removed from the knuckle

93471CP8

SCHEDULED MAINTENANCE INTERVALS
TOYOTA AVALON, CAMRY, CELICA, COROLLA, PASEO, SUPRA & TERCEL

TO BE SERVICED	TYPE OF SERVICE	VEHICLE MILEAGE INTERVAL (x1000)												
		7.5	15	22.5	30	37.5	45	52.5	60	67.5	75	82.5	90	97.5
Engine oil & filter ①	R	✓	✓	✓	✓	✓	✓	✓	✓	✓	✓	✓	✓	✓
Drive belts	S/I								✓	✓	✓	✓	✓	✓
Automatic transaxle fluid & filter	S/I		✓		✓		✓		✓		✓		✓	
Ball joints & dust covers	S/I		✓		✓		✓		✓		✓		✓	
Bolts & nuts on body & chassis	S/I		✓		✓		✓		✓		✓		✓	
Brake line pipes & hoses	S/I		✓		✓		✓		✓		✓		✓	
Brake linings & drums (except MR2)	S/I		✓		✓		✓		✓		✓		✓	
Brake pads & discs (front & rear if equipped)	S/I		✓		✓		✓		✓		✓		✓	
Differential oil (Camry, Celica, Corolla & Supra) ②	S/I		✓		✓		✓		✓		✓		✓	
Drive shaft boots (except Supra)	S/I		✓		✓		✓		✓		✓		✓	
Manual transaxle oil	S/I		✓		✓		✓		✓		✓		✓	
Steering gear housing oil	S/I		✓		✓		✓		✓		✓		✓	
Steering linkage	S/I		✓		✓		✓		✓		✓		✓	
Air filter	R				✓				✓				✓	
Rear wheel bearings (Tercel)	R				✓				✓				✓	
Spark plugs (Corolla & Tercel)	R				✓				✓				✓	
Spark plugs (platinum tip) (Avalon, Camry, Celica & Supra)	R								✓					
Exhaust system	S/I				✓				✓				✓	
Fuel lines & connections	S/I				✓				✓				✓	
Valve clearance	S/I				✓				✓				✓	
Engine coolant	R						✓				✓			
Fuel tank cap gasket	R								✓					
Charcoal canister	S/I								✓					

R: Replace   S/I: Service or Inspect
① Supra 2JZ-GTE: change every 5000 miles.
② Supra w/LSD: replace every 30,000 miles.

FREQUENT OPERATION MAINTENANCE (SEVERE SERVICE)
If a vehicle is operated under any of the following conditions it is considered severe service:

- Extremely dusty areas.

- 50% or more of the vehicle operation is in 32°C (90°F) or higher temperatures, or constant operation in temperatures below 0°C (32°F).

- Prolonged idling (vehicle operation in stop and go traffic).

- Frequent short running periods (engine does not warm to normal operating temperatures).

- Police, taxi, delivery usage or trailer towing usage.

Oil & oil filter: change every 6000 miles.

Oil & oil filter change (Supra 2JZ-GTE): change every 2500 miles.

Bolts & nuts on chassis & body: tighten every 7500 miles.

Ball joints & dust covers: service or inspect every 12,000 miles.

Brake linings & drums: service or inspect ever 12,000 miles.

Brake pads & discs (front & rear if equipped): service or inspect every 12,000 miles.

Drive shaft boots & except Supra): service or inspect every 12,000 miles.

Steering linkage: service or inspect every 12,000 miles.

Air filter: service or inspect every 15,000 miles.

Exhaust system: service or inspect every 15,000 miles.

Timing belt: replace every 60,000 miles.

93471CP9

SCHEDULED MAINTENANCE INTERVALS
TOYOTA
AVALON, CAMRY, CELICA, COROLLA, PASEO, SUPRA, TERCEL

The following should be used as a guide when determining the amount of work required for a particular service. In estimating how long a particular Scheduled Maintenance Service should take, please observe the following:

- Labor Time is time based on field research and data supplied by the vehicle manufacturer.
- Labor time operations are given in hours and tenths of an hour.
- All labor operations are to be used as a guide.

Mechanic Skill Level Codes:
(A) PRECISION: Highly skilled with multiple certification.
(B) GENERAL: Normally skilled with certification.
(C) MAINTENANCE: Semi-skilled working on certification.

	LABOR TIME		LABOR TIME		LABOR TIME
7500 Mile Service (C)		**37500 Mile Service (C)**		**67500 Mile Service (C)**	
All Models	.4	All Models	.4	All Models	.5
Paseo add	.2	*Paseo add*	.2	*Paseo add*	.2
15000 Mile Service (C)		**45000 Mile Service (C)**		**75000 Mile Service (C)**	
All Models	1.2	All Models	1.2	All Models	1.3
Inspect differential oil		*Inspect differential oil*		*Inspect differential oil*	
Camry, Celica, Corolla & Supra add	.1	*Camry, Celica, Corolla & Supra add*	.1	*Camry, Celica, Corolla & Supra add*	.1
Inspect drive shaft boots		*Inspect drive shaft boots*		*Inspect drive shaft boots*	
all models except Supra add	.1	*all models except Supra add*	.1	*all models except Supra add*	.1
w/AT add	.1	*w/AT add*	.1	**82500 Mile Service (C)**	
22500 Mile Service (C)		**52500 Mile Service (C)**		All Models	.5
All Models	.4	All Models	.4	*Paseo add*	.2
Paseo add	.2	*Paseo add*	.2	**90000 Mile Service (B)**	
30000 Mile Service (B)		**60000 Mile Service (B)**		Avalon	2.3
Avalon	2.2	Avalon	3.4	Camry, Celica	2.3
Camry	2.2	Camry, Celica	3.4	Corolla	2.8
Celica	2.1	Corolla	3.2	Paseo, Tercel	2.8
Corolla	2.7	Paseo	3.1	Supra	2.2
Paseo	3.2	Supra	3.3	*Supra w/LSD add*	.2
Tercel	3.2	Tercel	3.1	*w/AT add*	.1
Supra	2.1	*Supra w/LSD add*	.2	**97500 Mile Service (C)**	
Supra w/LSD add	.2	*w/AT add*	.1	All Models	.5
w/AT add	.1			*Paseo add*	.2

93471CP0

VOLKSWAGEN
BEETLE • CABRIO • GOLF • GTI • JETTA • PASSAT

ENGINE AND VEHICLE IDENTIFICATION

Engine							Model Year	
Code ①	Liters (cc)	Cu. In.	Cyl.	Fuel Sys.	Engine Type	Eng. Mfg.	Code ②	Year
AAA	2.8 (2792)	170	6	Motronic	DOHC	Volkswagen	W	1998
AAZ	1.9 (1896)	116	4	Diesel	SOHC	Volkswagen	X	1999
ABA	2.0 (1984)	121	4	Motronic	SOHC	Volkswagen	Y	2000
AEB	1.8 (1781)	109	4	Motronic	DOHC	Volkswagen	1	2001
AEG	2.0 (1984)	121	4	Motronic	SOHC	Volkswagen	2	2002
AHA	2.8 (2792)	170	6	Motronic	DOHC	Volkswagen		
AHH	1.9 (1896)	116	4	Diesel	SOHC	Volkswagen		
ALH	1.9 (1896)	116	4	Diesel	SOHC	Volkswagen		

DOHC: Double Overhead Camshafts

SOHC: Single Overhead Camshaft

① Located on the vehicle data plate

② 10th digit of VIN

93471CQ1

Timing belt service is covered in Section 3 of this manual

GENERAL ENGINE SPECIFICATIONS

Year	Model	Engine Displacement Liters (cc)	Engine ID/VIN	Fuel System Type	Net Horsepower @ rpm	Net Torque@rpm (ft. lbs.)	Bore x Stroke (in.)	Compression Ratio	Oil Pressure @ rpm
1998	Cabrio	2.0 (1984)	ABA	Motronic	115@5400	122@3200	3.25x3.65	10.0:1	29@2000
	Golf	1.9 (1896)	AAZ	DSL	75@4200	111@3400	3.13x3.76	22.5:1	29@2000
	Golf	2.0 (1984)	ABA	Motronic	115@5400	122@3200	3.25x3.65	10.0:1	29@2000
	GTI	2.0 (1984)	ABA	Motronic	115@5400	122@3200	3.25x3.65	10.0:1	29@2000
	GTI	2.8 (2792)	AAA	Motronic	172@5700	173@4200	3.19x3.56	10.0:1	29@2000
	Jetta	1.9 (1896)	AAZ	DSL	75@4200	111@3400	3.13x3.76	22.5:1	29@2000
	Jetta	2.0 (1984)	ABA	Motronic	115@5400	122@3200	3.25x3.65	10.0:1	29@2000
	Jetta	2.8 (2792)	AAA	Motronic	172@5700	173@4200	3.19x3.56	10.0:1	29@2000
	New Beetle	1.9 (1896)	ALH	DSL	90@3750	154@1900	3.13x3.76	19.5:1	29@2000
	New Beetle	2.0 (1984)	AEG	Motronic	115@5400	125@2400	3.25x3.65	10.0:1	29@2000
	Passat	1.8 (1781)	AEB	Motronic	150@5700	155@4600	3.19x3.40	9.5:1	29@2000
	Passat	1.9 (1896)	AHH	DSL	90@3750	154@1900	3.13x3.76	19.5:1	29@2000
	Passat	2.8 (2792)	AHA	Motronic	190@6000	206@3200	3.25x3.40	10.6:1	29@2000
1999	Cabrio	2.0 (1984)	ABA	Motronic	115@5400	122@3200	3.25x3.65	10.0:1	29@2000
	Golf	1.9 (1896)	AAZ	DSL	90@3750	155@1900	3.13x3.76	19.5:1	29@2000
	Golf	2.0 (1984)	ABA	Motronic	115@5200	122@2600	3.25x3.65	10.0:1	29@2000
	GTI	2.0 (1984)	ABA	Motronic	115@5200	122@2600	3.25x3.65	10.0:1	29@2000
	GTI	2.8 (2792)	AAA	Motronic	174@5800	181@3200	3.19x3.56	10.0:1	29@2000
	Jetta	1.9 (1896)	AAZ	DSL	90@3750	155@1900	3.13x3.76	19.5:1	29@2000
	Jetta	2.0 (1984)	ABA	Motronic	115@5500	122@2600	3.25x3.65	10.5:1	29@2000
	Jetta	2.8 (2792)	AAA	Motronic	174@5800	181@3200	3.19x3.56	10.0:1	29@2000
	New Beetle	1.9 (1896)	ALH	DSL	90@3750	154@1900	3.13x3.76	19.5:1	29@2000
	New Beetle	2.0 (1984)	AEG	Motronic	115@5400	125@2400	3.25x3.65	10.0:1	29@2000
	Passat	1.8 (1781)	AEB	Motronic	150@5700	155@1750-4600	3.19x3.40	9.5:1	29@2000
	Passat	1.9 (1896)	AHH	DSL	90@3750	154@1900	3.13x3.76	19.5:1	29@2000
	Passat	2.8 (2771)	AHA	Motronic	190@6000	206@3200	3.25x3.40	10.6:1	29@2000
2000	Cabrio	2.0 (1984)	ABA	Motronic	115@5400	122@3200	3.25x3.65	10.0:1	29@2000
	Golf	1.9 (1896)	AAZ	DSL	90@3750	155@1900	3.13x3.76	19.5:1	29@2000
	Golf	2.0 (1984)	ABA	Motronic	115@5200	122@2600	3.25x3.65	10.0:1	29@2000
	GTI	2.0 (1984)	ABA	Motronic	115@5200	122@2600	3.25x3.65	10.0:1	29@2000
	GTI	2.8 (2792)	AAA	Motronic	174@5800	181@3200	3.19x3.56	10.0:1	29@2000
	Jetta	1.9 (1896)	AAZ	DSL	90@3750	155@1900	3.13x3.76	19.5:1	29@2000
	Jetta	2.0 (1984)	ABA	Motronic	115@5500	122@2600	3.25x3.65	10.5:1	29@2000
	Jetta	2.8 (2792)	AAA	Motronic	174@5800	181@3200	3.19x3.56	10.0:1	29@2000
	New Beetle	1.9 (1896)	ALH	DSL	90@3750	154@1900	3.13x3.76	19.5:1	29@2000
	New Beetle	2.0 (1984)	AEG	Motronic	115@5400	125@2400	3.25x3.65	10.0:1	29@2000
	Passat	1.8 (1781)	AEB	Motronic	150@5700	155@1750-4600	3.19x3.40	9.5:1	29@2000
	Passat	1.9 (1896)	AHH	DSL	90@3750	154@1900	3.13x3.76	19.5:1	29@2000
	Passat	2.8 (2771)	AHA	Motronic	190@6000	206@3200	3.25x3.40	10.6:1	29@2000

DSL: Diesel

93471CQ2

GASOLINE ENGINE TUNE-UP SPECIFICATIONS

Year	Engine Displacement Liters (cc)	Engine ID/VIN	Spark Plug Gap (in.)	Ignition Timing (deg.)		Fuel Pump (psi)	Idle Speed (rpm)		Valve Clearance	
				MT	AT		MT	AT	Intake	Exhaust
1998	1.8 (1781)	AEB	0.035–0.043	5-7B ①	5-7B ①	52 ②	820-920	820-920	HYD	HYD
	2.0 (1984)	ABA	0.024	5-7B ①	5-7B ①	44 ②	800-880	800-880	HYD	HYD
	2.0 (1984)	AEG	0.035–0.043	5-7B ①	5-7B ①	52 ②	760-880	760-880	HYD	HYD
	2.8 (2792)	AAA	0.028	5-7B ①	5-7B ①	58 ②	650-750	650-750	HYD	HYD
	2.8 (2792)	AHA	0.063	5-7B ①	5-7B ①	52 ②	620-740	620-740	HYD	HYD
1999	1.8 (1781)	AEB	0.035–0.043	5-7B ①	5-7B ①	52 ②	820-920	820-920	HYD	HYD
	2.0 (1984)	ABA	0.024	5-7B ①	5-7B ①	44 ②	800-880	800-880	HYD	HYD
	2.0 (1984)	AEG	0.035–0.043	5-7B ①	5-7B ①	52 ②	760-880	760-880	HYD	HYD
	2.8 (2792)	AAA	0.028	5-7B ①	5-7B ①	58 ②	650-750	650-750	HYD	HYD
	2.8 (2792)	AHA	0.063	5-7B ①	5-7B ①	52 ②	620-740	620-740	HYD	HYD
2000	1.8 (1781)	AEB	0.035–0.043	5-7B ①	5-7B ①	52 ②	820-920	820-920	HYD	HYD
	2.0 (1984)	ABA	0.024	5-7B ①	5-7B ①	44 ②	800-880	800-880	HYD	HYD
	2.0 (1984)	AEG	0.035–0.043	5-7B ①	5-7B ①	52 ②	760-880	760-880	HYD	HYD
	2.8 (2792)	AAA	0.028	5-7B ①	5-7B ①	58 ②	650-750	650-750	HYD	HYD
	2.8 (2792)	AHA	0.063	5-7B ①	5-7B ①	52 ②	620-740	620-740	HYD	HYD

Note: The Vehicle Emission Control Information label often reflects specification changes made during production. The label figures must be used if they differ from those in this chart.

B: Before top dead center.

HYD: Hydraulic

① Specifications for reference only. The ignition timing is controlled bt the ECM and is not adjustable.

② System pressure at idle.

93471CQ3

Heater Core replacement is covered in Section 2 of this manual

DIESEL ENGINE TUNE-UP SPECIFICATIONS

| Year | Engine Displacement cu. in. (cc) | Engine ID/VIN | Valve Clearance | | Intake Valve Opens (deg.) | Start of Injection Stroke @ TDC (cyl 1) (in.) | Injection Nozzle Pressure (psi) | | Idle Speed (rpm) | Cranking Compression Pressure (psi) |
			Intake (in.)	Exhaust (in.)			New	Used		
1998	1.9 (1896)	AAZ	HYD	HYD	6	①	2175-2291	2030	870-930	NA
	1.9 (1896)	AHH	HYD	HYD	8-14	NA	2175-2291	2030	870-930	NA
	1.9 (1896)	ALH	HYD	HYD	8-14	NA	2175-2291	2030	870-930	NA
1999	1.9 (1896)	AAZ	HYD	HYD	6	①	2175-2291	2030	870-930	NA
	1.9 (1896)	AHH	HYD	HYD	8-14	NA	2175-2291	2030	870-930	NA
	1.9 (1896)	ALH	HYD	HYD	8-14	NA	2175-2291	2030	870-930	NA
2000	1.9 (1896)	AAZ	HYD	HYD	6	①	2175-2291	2030	870-930	NA
	1.9 (1896)	AHH	HYD	HYD	8-14	NA	2175-2291	2030	870-930	NA
	1.9 (1896)	ALH	HYD	HYD	8-14	NA	2175-2291	2030	870-930	NA

Note: The Vehicle Emission Control Information label often reflects specification changes made during production. The label figures must be used if they differ from those in this chart

HYD: Hydraulic

B: Before top dead center

NA: Not Available

① Checking: 0.0287 - 0.0342 in.
Adjusting: 0.0307 - 0.0323 in.

93471CQ4

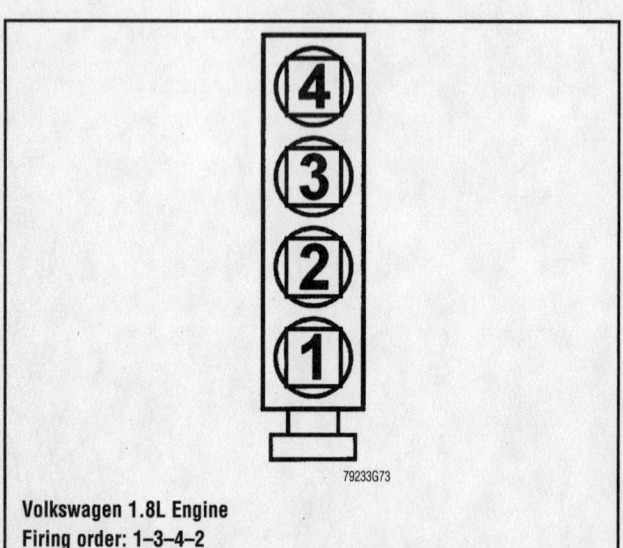

Volkswagen 1.8L Engine
Firing order: 1–3–4–2
Distributorless ignition system (one coil on each cylinder)

79233G73

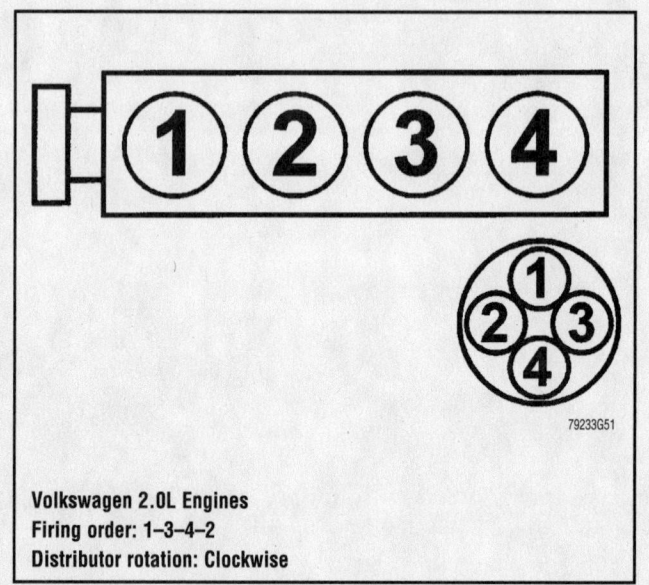

Volkswagen 2.0L Engines
Firing order: 1–3–4–2
Distributor rotation: Clockwise

79233G51

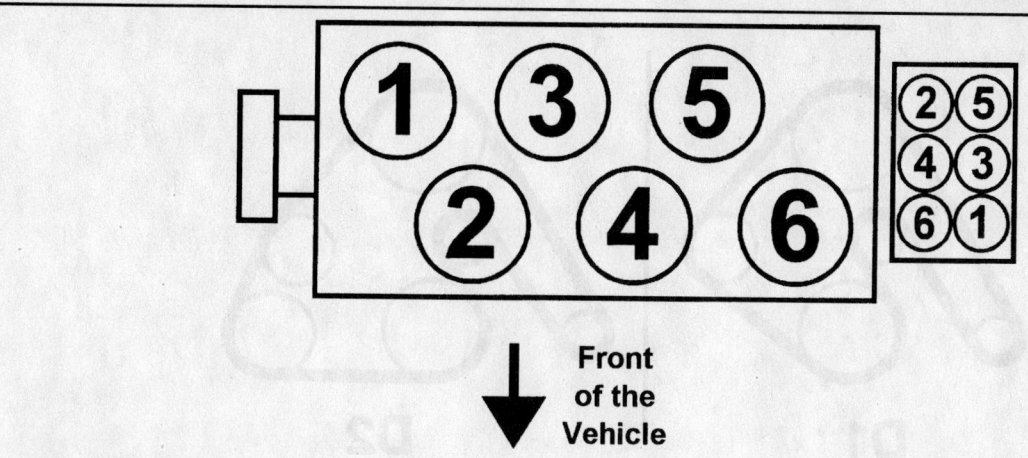

Front
of the
Vehicle

Volkswagen 2.8L Engine
Firing order: 1–5–3–6–2–4
Distributorless ignition system

79233G52

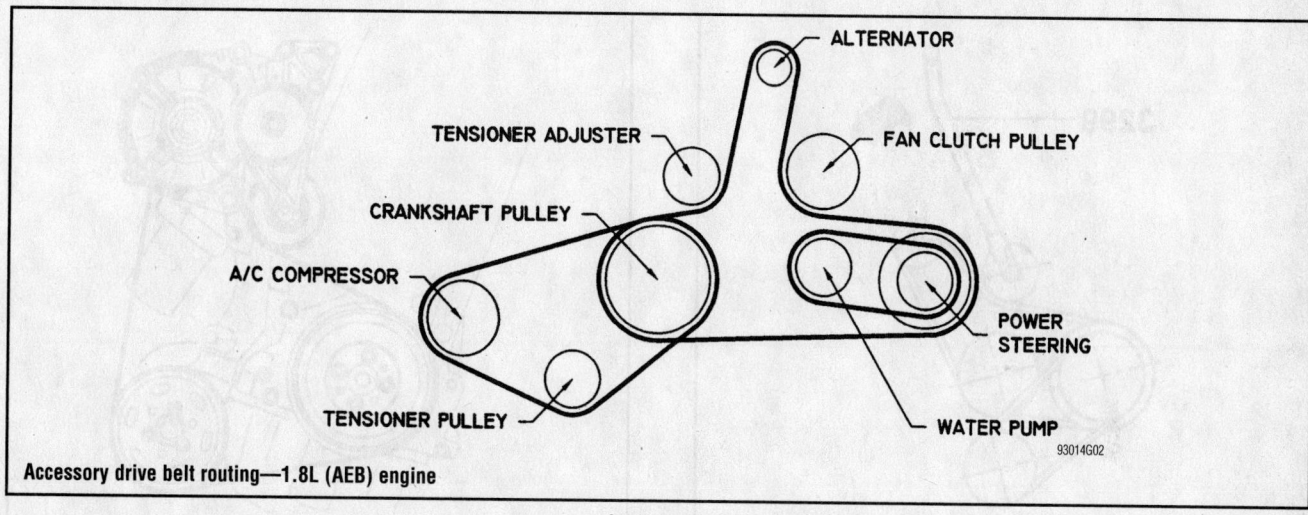

Accessory drive belt routing—1.8L (AEB) engine

93014G02

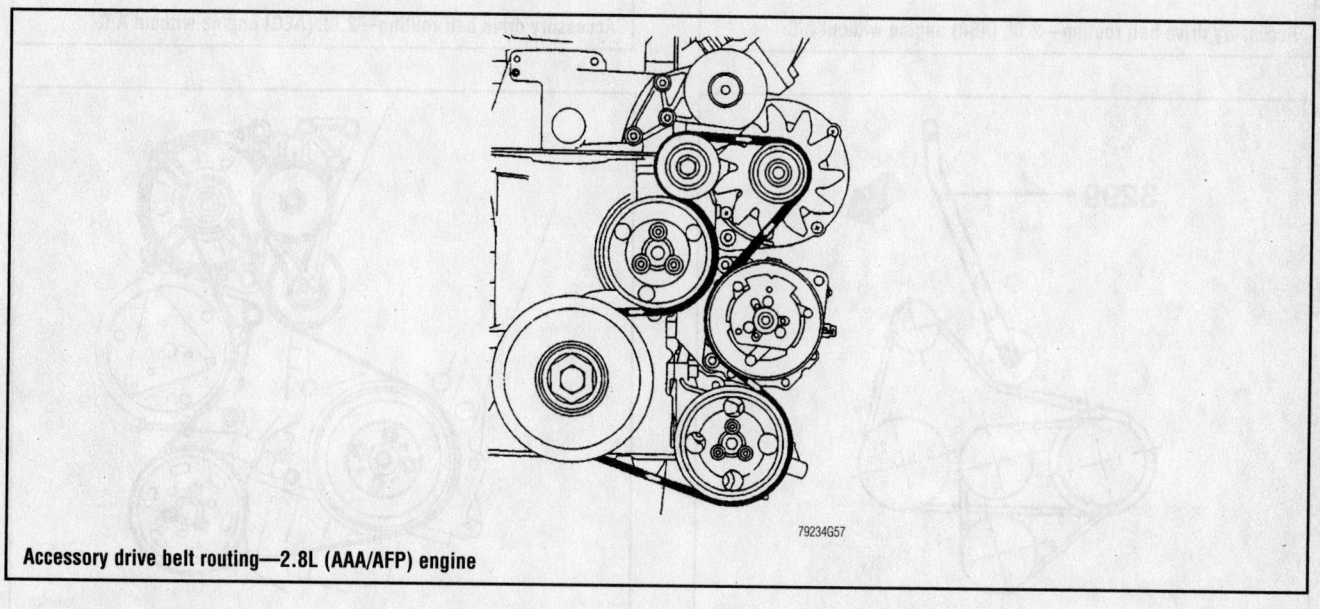

Accessory drive belt routing—2.8L (AAA/AFP) engine

79234G57

Brake service is covered in Section 4 of this manual

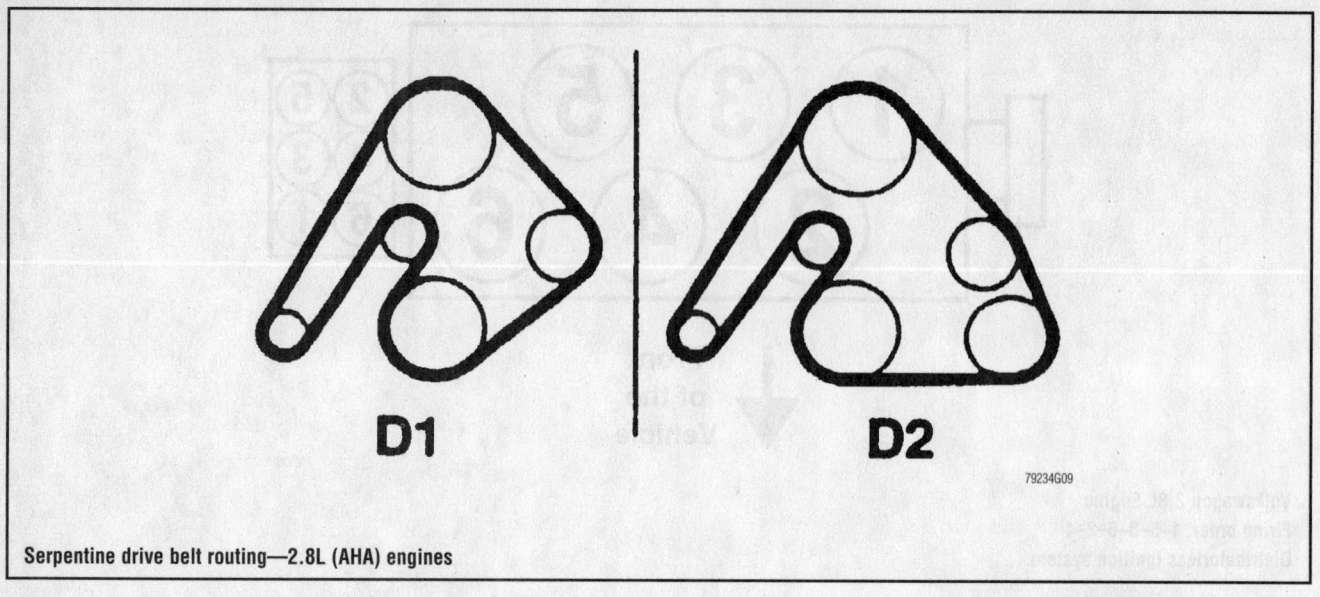

Serpentine drive belt routing—2.8L (AHA) engines

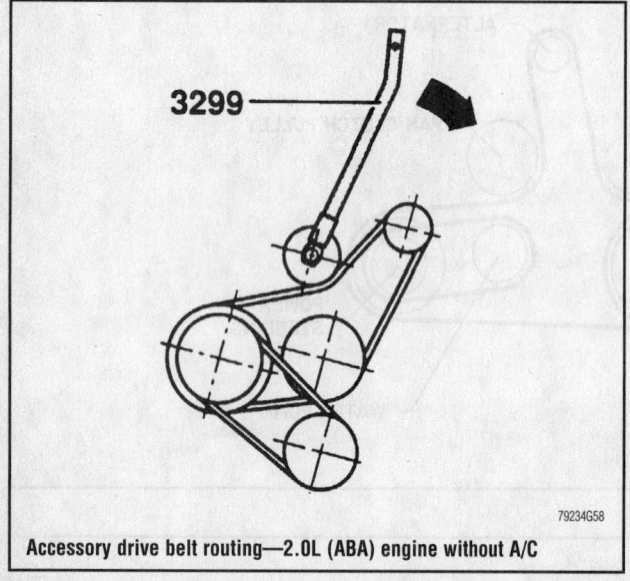

Accessory drive belt routing—2.0L (ABA) engine without A/C

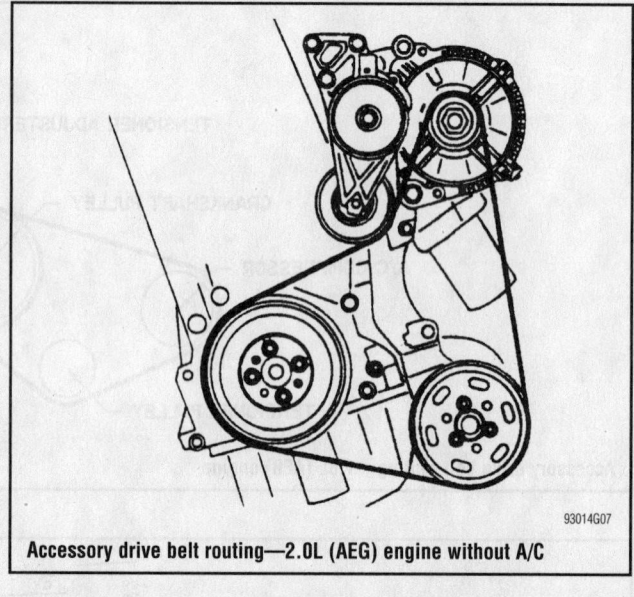

Accessory drive belt routing—2.0L (AEG) engine without A/C

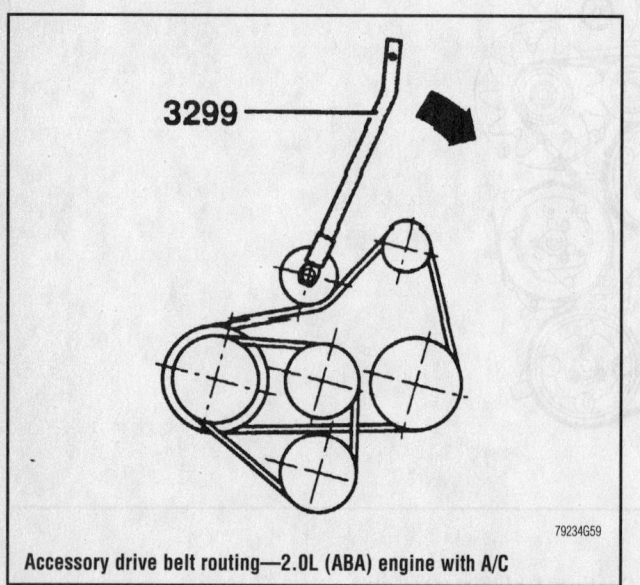

Accessory drive belt routing—2.0L (ABA) engine with A/C

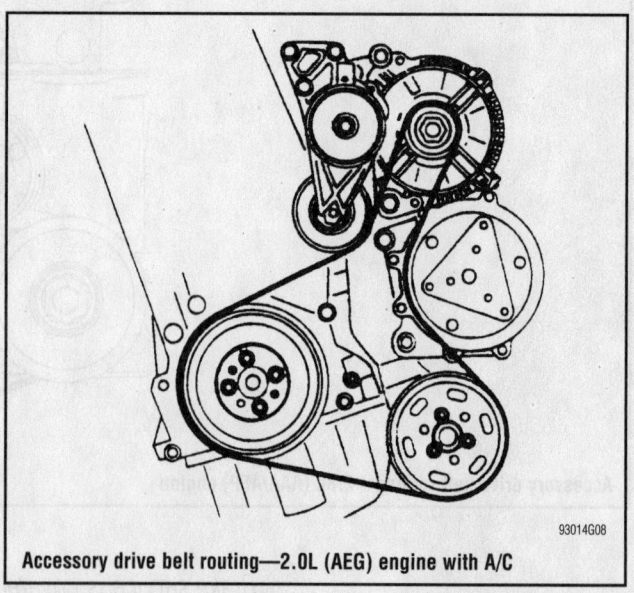

Accessory drive belt routing—2.0L (AEG) engine with A/C

CAPACITIES

Year	Model	Engine Displacement Liters (cc)	Engine ID/VIN	Engine Oil with Filter	Transmission (pts)		Final Drive (pts.)	Fuel Tank (gal.)	Cooling System (qts.)
					Manual	Auto.			
1998	Cabrio	2.0 (1984)	ABA	4.8	4.2	6.4	—	14.5	6.7
	Golf	1.9 (1896)	AAZ	4.4	4.2	6.4	—	14.5	6.4
	Golf	2.0 (1984)	ABA	4.8	4.2	6.4	—	14.5	6.7
	GTI	2.0 (1984)	ABA	4.8	4.2	6.4	—	14.5	6.7
	GTI	2.8 (2792)	AAA	5.8	4.2	6.4	—	14.5	8.5
	Jetta	1.9 (1896)	AAZ	4.4	4.2	6.4	—	14.5	6.4
	Jetta	2.0 (1984)	ABA	4.8	4.2	6.4	—	14.5	6.7
	Jetta	2.8 (2792)	AAA	5.8	4.2	6.4	—	14.5	8.5
	New Beetle	1.9 (1896)	ALH	4.8	4.2	6.8	—	14.5	6.7
	New Beetle	2.0 (1984)	AEG	4.2	4.2	6.8	—	14.5	6.7
	Passat	1.8 (1781)	AEB	4.2	4.8	7.4	—	18.5	6.9
	Passat	1.9 (1896)	AHH	3.7	4.8	7.4	—	N/A	N/A
	Passat	2.8 (2792)	AHA	5.8	4.8	6.4	—	16.4	8.5
1999	Cabrio	2.0 (1984)	ABA	4.8	4.2	6.4	—	14.5	6.7
	Golf	1.9 (1896)	AAZ	4.8	4.2	6.4	—	14.6	5.5
	Golf	2.0 (1984)	ABA	4.2	4.2	6.4	—	14.6	5.8
	GTI	2.0 (1984)	ABA	4.8	4.2	6.4	—	14.5	6.7
	GTI	2.8 (2792)	AAA	5.8	4.2	—	—	14.5	8.5
	Jetta	1.9 (1896)	AAZ	4.8	4.2	6.4	—	14.5	5.5
	Jetta	2.0 (1984)	ABA	4.2	4.2	6.4	—	14.5	5.8
	Jetta	2.8 (2792)	AAA	5.8	4.2	6.4	—	14.5	8.5
	New Beetle	1.9 (1896)	ALH	4.8	4.2	6.8	①	14.5	6.7
	New Beetle	2.0 (1984)	AEG	4.2	4.2	6.8	①	14.5	6.7
	Passat	1.8 (1781)	AEB	4.3	4.8	5.5	①	16.4	7.4
	Passat	1.9 (1896)	AHH	3.7	4.8	7.4	—	N/A	N/A
	Passat	2.8 (2792)	AHA	5.8	4.8	5.5	①	16.4	8.5
2000	Cabrio	2.0 (1984)	ABA	4.8	4.2	6.4	—	14.5	6.7
	Golf	1.9 (1896)	AAZ	4.8	4.2	6.4	—	14.6	5.5
	Golf	2.0 (1984)	ABA	4.2	4.2	6.4	—	14.6	5.8
	GTI	2.0 (1984)	ABA	4.8	4.2	6.4	—	14.5	6.7
	GTI	2.8 (2792)	AAA	5.8	4.2	—	—	14.5	8.5
	Jetta	1.9 (1896)	AAZ	4.8	4.2	6.4	—	14.5	5.5
	Jetta	2.0 (1984)	ABA	4.2	4.2	6.4	—	14.5	5.8
	Jetta	2.8 (2792)	AAA	5.8	4.2	6.4	—	14.5	8.5
	New Beetle	1.9 (1896)	ALH	4.8	4.2	6.8	①	14.5	6.7
	New Beetle	2.0 (1984)	AEG	4.2	4.2	6.8	①	14.5	6.7
	Passat	1.8 (1781)	AEB	4.3	4.8	5.5	①	16.4	7.4
	Passat	1.9 (1896)	AHH	3.7	4.8	7.4	—	N/A	N/A
	Passat	2.8 (2792)	AHA	5.8	4.8	5.5	①	16.4	8.5

Note: All capacities are approximate. Add fluid gradually and check often to avoid overfilling.

① For models equipped with an automatic transmission: 1.6 pints

93471CQ5

For complete Engine Mechanical specifications, see Section 1 of this manual

VALVE SPECIFICATIONS

Year	Engine Displacement Liters (cc)	Engine ID/VIN	Seat Angle (deg.)	Face Angle (deg.)	Spring Test Pressure (lbs. @ in.)	Spring Installed Height (in.)	Stem-to-Guide Clearance (in.)		Stem Diameter (in.)	
							Intake	Exhaust	Intake	Exhaust
1998	1.8 (1781)	AEB	45	45	NA	NA	0.039	0.051	0.2348	0.2340
	1.9 (1896)	AAZ	45	45	NA	NA	0.039	0.051	0.2741	0.2734
	1.9 (1896)	AHH	45	45	NA	NA	0.039	0.051	0.2741	0.2734
	1.9 (1896)	ALH	45	45	NA	NA	0.039	0.051	0.2741	0.2734
	2.0 (1984)	ABA	45	45	NA	NA	0.039	0.051	0.2744	0.2736
	2.0 (1984)	AEG	45	45	NA	NA	0.039	0.051	0.2724	0.2724
	2.8 (2792)	AAA	45	45	NA	NA	0.039	0.051	0.2744	0.2736
	2.8 (2792)	AHA	45	45	NA	NA	0.039	0.051	0.2350	0.2343
1999	1.8 (1781)	AEB	45	45	NA	NA	0.039	0.051	0.2348	0.2340
	1.9 (1896)	AAZ	45	45	NA	NA	0.039	0.051	0.2741	0.2734
	1.9 (1896)	AHH	45	45	NA	NA	0.039	0.051	0.2741	0.2734
	1.9 (1896)	ALH	45	45	NA	NA	0.039	0.051	0.2741	0.2734
	2.0 (1984)	ABA	45	45	NA	NA	0.039	0.051	0.2744	0.2736
	2.0 (1984)	AEG	45	45	NA	NA	0.039	0.051	0.2724	0.2724
	2.8 (2792)	AAA	45	45	NA	NA	0.039	0.051	0.2744	0.2736
	2.8 (2792)	AHA	45	45	NA	NA	0.039	0.051	0.2350	0.2343
2000	1.8 (1781)	AEB	45	45	NA	NA	0.039	0.051	0.2348	0.2340
	1.9 (1896)	AAZ	45	45	NA	NA	0.039	0.051	0.2741	0.2734
	1.9 (1896)	AHH	45	45	NA	NA	0.039	0.051	0.2741	0.2734
	1.9 (1896)	ALH	45	45	NA	NA	0.039	0.051	0.2741	0.2734
	2.0 (1984)	ABA	45	45	NA	NA	0.039	0.051	0.2744	0.2736
	2.0 (1984)	AEG	45	45	NA	NA	0.039	0.051	0.2724	0.2724
	2.8 (2792)	AAA	45	45	NA	NA	0.039	0.051	0.2744	0.2736
	2.8 (2792)	AHA	45	45	NA	NA	0.039	0.051	0.2350	0.2343

NA: Not Available

93471CQ6

CRANKSHAFT AND CONNECTING ROD SPECIFICATIONS
All measurements are given in inches.

Year	Engine Displacement Liters (cc)	Engine ID/VIN	Crankshaft				Connecting Rod		
			Main Brg. Journal Dia.	Main Brg. Oil Clearance	Shaft End-play	Thrust on No.	Journal Diameter	Oil Clearance	Side Clearance
1998	1.8 (1781)	AEB	2.1260	0.0008-0.0024	0.0028-0.0066	3	1.8819	0.0004-0.0024	0.0020-0.0122
	1.9 (1896)	AAZ	2.1600	0.0012-0.0032	0.0028-0.0066	3	1.8819	0.0012-0.0032	0.0028-0.0066
	1.9 (1896)	AHH	2.1260	0.0012-0.0032	0.0028-0.0066	3	1.8819	0.0012-0.0032	0.0028-0.0066
	1.9 (1896)	ALH	2.1260	0.0012-0.0032	0.0028-0.0066	3	1.8819	0.0012-0.0032	0.0028-0.0066
	2.0 (1984)	ABA	2.1260	0.0008-0.0024	0.0028-0.0066	3	1.8819	0.0004-0.0024	0.0020-0.0122
	2.0 (1984)	AEG	2.1260	0.0008-0.0024	0.0028-0.0066	3	1.8819	0.0004-0.0024	0.0020-0.0122
	2.8 (2782)	AAA	2.3606-2.3613	0.0008-0.0024	0.0028-0.0091	5	2.1243-2.1251	0.0004-0.0024	0.0020-0.0122
	2.8 (2792)	AHA	2.5590	0.0007-0.0018	0.0028-0.0091	3	2.1260	0.0006-0.0024	0.0020-0.0122
1999	1.8 (1781)	AEB	2.1260	0.0008-0.0024	0.0028-0.0066	3	1.8819	0.0004-0.0024	0.0020-0.0122
	1.9 (1896)	AAZ	2.1600	0.0012-0.0032	0.0028-0.0066	3	1.8819	0.0012-0.0032	0.0028-0.0066
	1.9 (1896)	AHH	2.1260	0.0012-0.0032	0.0028-0.0066	3	1.8819	0.0012-0.0032	0.0028-0.0066
	1.9 (1896)	ALH	2.1260	0.0012-0.0032	0.0028-0.0066	3	1.8819	0.0012-0.0032	0.0028-0.0066
	2.0 (1984)	ABA	2.1260	0.0008-0.0024	0.0028-0.0066	3	1.8819	0.0004-0.0024	0.0020-0.0122
	2.0 (1984)	AEG	2.1260	0.0008-0.0024	0.0028-0.0066	3	1.8819	0.0004-0.0024	0.0020-0.0122
	2.8 (2782)	AAA	2.3606-2.3613	0.0008-0.0024	0.0028-0.0091	5	2.1243-2.1251	0.0004-0.0024	0.0020-0.0122
	2.8 (2792)	AHA	2.5590	0.0007-0.0018	0.0028-0.0091	3	2.1260	0.0006-0.0024	0.0020-0.0122
2000	1.8 (1781)	AEB	2.1260	0.0008-0.0024	0.0028-0.0066	3	1.8819	0.0004-0.0024	0.0020-0.0122
	1.9 (1896)	AAZ	2.1600	0.0012-0.0032	0.0028-0.0066	3	1.8819	0.0012-0.0032	0.0028-0.0066
	1.9 (1896)	AHH	2.1260	0.0012-0.0032	0.0028-0.0066	3	1.8819	0.0012-0.0032	0.0028-0.0066
	1.9 (1896)	ALH	2.1260	0.0012-0.0032	0.0028-0.0066	3	1.8819	0.0012-0.0032	0.0028-0.0066
	2.0 (1984)	ABA	2.1260	0.0008-0.0024	0.0028-0.0066	3	1.8819	0.0004-0.0024	0.0020-0.0122
	2.0 (1984)	AEG	2.1260	0.0008-0.0024	0.0028-0.0066	3	1.8819	0.0004-0.0024	0.0020-0.0122
	2.8 (2782)	AAA	2.3606-2.3613	0.0008-0.0024	0.0028-0.0091	5	2.1243-2.1251	0.0004-0.0024	0.0020-0.0122
	2.8 (2792)	AHA	2.5590	0.0007-0.0018	0.0028-0.0091	3	2.1260	0.0006-0.0024	0.0020-0.0122

93471CQ7

For Accessory Drive Belt illustrations, see Section 1 of this manual

PISTON AND RING SPECIFICATIONS

All measurements are given in inches.

Year	Engine Displacement Liters (cc)	Engine ID/VIN	Piston Clearance	Ring Gap			Ring Side Clearance		
				Top Compression	Bottom Compression	Oil Control	Top Compression	Bottom Compression	Oil Control
1998	1.8 (1781)	AEB	0.0450	0.008-0.016	0.008-0.016	0.010-0.020	0.0035-0.0047	0.0020-0.0031	0.0012-0.0024
	1.9 (1896)	AAZ	0.0012	0.008-0.016	0.008-0.016	0.010-0.020	0.0035-0.0047	0.0020-0.0031	0.0012-0.0024
	1.9 (1896)	AHH	0.0400	0.008-0.016	0.008-0.016	0.010-0.020	0.0035-0.0047	0.0020-0.0031	0.0012-0.0024
	1.9 (1896)	ALH	0.0400	0.008-0.016	0.008-0.016	0.010-0.020	0.0035-0.0047	0.0020-0.0031	0.0012-0.0024
	2.0 (1984)	ABA	0.0010	0.008-0.016	0.008-0.016	0.010-0.020	0.0008-0.0020	0.0008-0.0020	0.0008-0.0020
	2.0 (1984)	AEG	0.0450	0.008-0.016	0.008-0.016	0.010-0.020	0.0035-0.0047	0.0020-0.0031	0.0012-0.0024
	2.8 (2782)	AAA	0.0010	0.008-0.016	0.008-0.016	0.010-0.020	0.0008-0.0020	0.0008-0.0020	0.0008-0.0019
	2.8 (2792)	AHA	0.0012	0.014-0.020	0.020-0.028	0.010-0.020	0.0008-0.0031	0.0008-0.0031	0.0008-0.0031
1999	1.8 (1781)	AEB	0.0450	0.008-0.016	0.008-0.016	0.010-0.020	0.0035-0.0047	0.0020-0.0031	0.0012-0.0024
	1.9 (1896)	AAZ	0.0012	0.008-0.016	0.008-0.016	0.010-0.020	0.0035-0.0047	0.0020-0.0031	0.0012-0.0024
	1.9 (1896)	AHH	0.0400	0.008-0.016	0.008-0.016	0.010-0.020	0.0035-0.0047	0.0020-0.0031	0.0012-0.0024
	1.9 (1896)	ALH	0.0400	0.008-0.016	0.008-0.016	0.010-0.020	0.0035-0.0047	0.0020-0.0031	0.0012-0.0024
	2.0 (1984)	ABA	0.0010	0.008-0.016	0.008-0.016	0.010-0.020	0.0008-0.0020	0.0008-0.0020	0.0008-0.0020
	2.0 (1984)	AEG	0.0450	0.008-0.016	0.008-0.016	0.010-0.020	0.0035-0.0047	0.0020-0.0031	0.0012-0.0024
	2.8 (2782)	AAA	0.0010	0.008-0.016	0.008-0.016	0.010-0.020	0.0008-0.0020	0.0008-0.0020	0.0008-0.0019
	2.8 (2792)	AHA	0.0012	0.014-0.020	0.020-0.028	0.010-0.020	0.0008-0.0031	0.0008-0.0031	0.0008-0.0031
2000	1.8 (1781)	AEB	0.0450	0.008-0.016	0.008-0.016	0.010-0.020	0.0035-0.0047	0.0020-0.0031	0.0012-0.0024
	1.9 (1896)	AAZ	0.0012	0.008-0.016	0.008-0.016	0.010-0.020	0.0035-0.0047	0.0020-0.0031	0.0012-0.0024
	1.9 (1896)	AHH	0.0400	0.008-0.016	0.008-0.016	0.010-0.020	0.0035-0.0047	0.0020-0.0031	0.0012-0.0024
	1.9 (1896)	ALH	0.0400	0.008-0.016	0.008-0.016	0.010-0.020	0.0035-0.0047	0.0020-0.0031	0.0012-0.0024
	2.0 (1984)	ABA	0.0010	0.008-0.016	0.008-0.016	0.010-0.020	0.0008-0.0020	0.0008-0.0020	0.0008-0.0020
	2.0 (1984)	AEG	0.0450	0.008-0.016	0.008-0.016	0.010-0.020	0.0035-0.0047	0.0020-0.0031	0.0012-0.0024
	2.8 (2782)	AAA	0.0010	0.008-0.016	0.008-0.016	0.010-0.020	0.0008-0.0020	0.0008-0.0020	0.0008-0.0019
	2.8 (2792)	AHA	0.0012	0.014-0.020	0.020-0.028	0.010-0.020	0.0008-0.0031	0.0008-0.0031	0.0008-0.0031

93471CQ8

TORQUE SPECIFICATIONS
All readings in ft. lbs.

Year	Engine Displacement Liters (cc)	Engine ID/VIN	Cylinder Head Bolts	Main Bearing Bolts	Rod Bearing Bolts	Crankshaft Damper Bolt	Flywheel Bolts	Manifold Intake	Manifold Exhaust	Spark Plugs	Lug Nuts
1998	1.8 (1781)	AEB	①	②	③	④	⑤	18	18	22	89
	1.9 (1896)	AAZ	①	②	③	④	⑤	18	18	—	89
	1.9 (1896)	AHH	①	②	③	④	⑤	18	18	—	89
	1.9 (1896)	ALH	①	②	③	⑥	⑤	18	18	—	89
	2.0 (1984)	ABA	①	②	⑥	④	⑤	15	15	22	89
	2.0 (1984)	AEG	⑧	②	③	④	⑨	18	18	22	89
	2.8 (2792)	AAA	①	⑩	⑥	④	⑤	15	15	22	89
	2.8 (2792)	AHA	①	⑪	③	⑫	⑬	18	18	22	89
1999	1.8 (1781)	AEB	①	②	③	④	⑤	18	18	22	89
	1.9 (1896)	AAZ	①	②	③	④	⑤	18	18	—	89
	1.9 (1896)	AHH	①	②	③	④	⑤	18	18	—	89
	1.9 (1896)	ALH	①	②	③	⑥	⑤	18	18	—	89
	2.0 (1984)	ABA	①	②	⑥	④	⑤	15	15	22	89
	2.0 (1984)	AEG	⑧	②	③	④	⑨	18	18	22	89
	2.8 (2792)	AAA	①	⑩	⑥	④	⑤	15	15	22	89
	2.8 (2792)	AHA	①	⑪	③	⑫	⑬	18	18	22	89
2000	1.8 (1781)	AEB	①	②	③	④	⑤	18	18	22	89
	1.9 (1896)	AAZ	①	②	③	④	⑤	18	18	—	89
	1.9 (1896)	AHH	①	②	③	④	⑤	18	18	—	89
	1.9 (1896)	ALH	①	②	③	⑥	⑤	18	18	—	89
	2.0 (1984)	ABA	①	②	⑥	④	⑤	15	15	22	89
	2.0 (1984)	AEG	⑧	②	③	④	⑨	18	18	22	89
	2.8 (2792)	AAA	①	⑩	⑥	④	⑤	15	15	22	89
	2.8 (2792)	AHA	①	⑪	③	⑫	⑬	18	18	22	89

① Torque in four steps: (use new bolts on all engines).
  Step 1: 30 ft. lbs.
  Step 2: 44 ft. lbs.
  Step 3: plus 90 degrees
  Step 4: plus 90 degrees

② 48 ft. lbs., plus an additional 90 degrees. Always replace bolt.

③ Torque to 22 ft. lbs. plus an additional 90 degrees

④ Step 1: 66 ft. lbs.
  Step 2: plus 90 degrees

⑤ 44 ft. lbs. plus an additional 90 degrees;
  except Passat GLX: 52 ft. lbs. plus 90 degrees.

⑥ 88 ft. lbs. Plus 90 degrees. Always replace bolt.

⑦ 22 ft. lbs. plus an additional 1/4 turn. Use new bolts.

⑧ Torque in three steps. Use new bolts.
  Step 1: 30 ft. lbs.
  Step 2: plus 90 degrees
  Step 3: plus 90 degrees

⑨ 30 ft. lbs. plus 90 degrees. Always replace bolt.

⑩ 22 ft. lbs. plus an additional 1/2 turn (180 degrees). Use new bolts.

⑪ 44 ft.lbs. plus an additional 1/2 turn (180 degrees)

⑫ Step 1: 148 ft. lbs.
  Step 2: 1/2 turn (180 degrees)

⑬ Driveplate: 44 ft.lbs. plus 1/4 turn (90 degrees)
  Flywheel: 44 ft. lbs. plus 1/2 turn (180 degrees)

93471CQ9

For Tire, Wheel and Ball Joint specifications, see Section 1 of this manual

BRAKE SPECIFICATIONS
All measurements in inches unless noted

Year	Model		Brake Disc Original Thickness	Brake Disc Minimum Thickness	Maximum Run-out	Drum Diameter Original Inside Diameter	Drum Diameter Maximum Machine Diameter	Minimum Lining Thickness	Brake Caliper Bracket Bolts (ft. lbs.)	Brake Caliper Mounting Bolts (ft. lbs.)
1998	Cabrio	F	0.790	0.709	0.002	—	—	0.28	18-26	92
		R	0.390	0.315	0.002	7.87	7.91	0.27 ①	22	41
	Golf ②	F	0.790	0.709	0.002	—	—	0.28	18-26	92
		R	0.390	0.315	0.002	7.87	7.91	0.27 ①	22	41
	Golf ③	F	0.870	0.787	0.002	—	—	0.28	18-26	92
		R	0.390	0.315	0.002	7.87	7.91	0.28	22	41
	GTI ④	F	0.790	0.709	0.002	—	—	0.28	18-26	92
		R	0.390	0.315	0.002	7.87	7.91	0.28	22	41
	GTI ⑤	F	0.870	0.787	0.002	—	—	0.28	18-26	92
		R	0.390	0.315	0.002	7.87	7.91	0.28	22	41
	Jetta ⑥	F	0.790	0.709	0.002	—	—	0.28	18-26	92
		R	0.390	0.315	0.002	7.87	7.91	0.27 ①	22	41
	Jetta ⑦	F	0.870	0.787	0.002	—	—	0.28	18-26	92
		R	0.390	0.315	0.002	7.87	7.91	0.28	22	41
	New Beetle	F	0.790	0.950	0.002	—	—	0.27	18	92
		R	0.390	0.315	0.002	9.06	9.09	0.27 ①	26	48
	Passat	F	0.980 ⑧	0.900 ⑨	0.002	—	—	0.28	22	89
		R	0.393	0.314	0.002	NA	NA	0.27	22	70
1999	Cabrio	F	0.790	0.709	0.002	—	—	0.28	18-26	92
		R	0.390	0.315	0.002	7.87	7.91	0.27 ①	22	41
	Golf ②	F	0.790	0.709	0.002	—	—	0.28	18-26	92
		R	0.390	0.315	0.002	7.87	7.91	0.27 ①	22	41
	Golf ③	F	0.870	0.787	0.002	—	—	0.28	18-26	92
		R	0.390	0.315	0.002	7.87	7.91	0.28	22	41
	GTI ④	F	0.790	0.709	0.002	—	—	0.28	18-26	92
		R	0.390	0.315	0.002	7.87	7.91	0.28	22	41
	GTI ⑤	F	0.870	0.787	0.002	—	—	0.28	18-26	92
		R	0.390	0.315	0.002	7.87	7.91	0.28	22	41
	Jetta ⑥	F	0.790	0.709	0.002	—	—	0.28	18-26	92
		R	0.390	0.315	0.002	7.87	7.91	0.27 ①	22	41
	Jetta ⑦	F	0.870	0.787	0.002	—	—	0.28	18-26	92
		R	0.390	0.315	0.002	7.87	7.91	0.28	22	41
	New Beetle	F	0.790	0.950	0.002	—	—	0.27	18	92
		R	0.390	0.315	0.002	9.06	9.09	0.27 ①	26	48
	Passat	F	0.980 ⑧	0.900 ⑨	0.002	—	—	0.28	22	89
		R	0.393	0.314	0.002	NA	NA	0.27	22	70
2000	Cabrio	F	0.790	0.709	0.002	—	—	0.28	18-26	92
		R	0.390	0.315	0.002	7.87	7.91	0.27 ①	22	41
	Golf ②	F	0.790	0.709	0.002	—	—	0.28	18-26	92
		R	0.390	0.315	0.002	7.87	7.91	0.27 ①	22	41
	Golf ③	F	0.870	0.787	0.002	—	—	0.28	18-26	92
		R	0.390	0.315	0.002	7.87	7.91	0.28	22	41
	GTI ④	F	0.790	0.709	0.002	—	—	0.28	18-26	92
		R	0.390	0.315	0.002	7.87	7.91	0.28	22	41
	GTI ⑤	F	0.870	0.787	0.002	—	—	0.28	18-26	92
		R	0.390	0.315	0.002	7.87	7.91	0.28	22	41
	Jetta ⑥	F	0.790	0.709	0.002	—	—	0.28	18-26	92
		R	0.390	0.315	0.002	7.87	7.91	0.27 ①	22	41
	Jetta ⑦	F	0.870	0.787	0.002	—	—	0.28	18-26	92
		R	0.390	0.315	0.002	7.87	7.91	0.28	22	41

93471CQ0

BRAKE SPECIFICATIONS

All measurements in inches unless noted

Year	Model		Brake Disc			Drum Diameter		Minimum Lining Thickness	Brake Caliper	
			Original Thickness	Minimum Thickness	Maximum Run-out	Original Inside Diameter	Maximum Machine Diameter		Bracket Bolts (ft. lbs.)	Mounting Bolts (ft. lbs.)
2000 (cont.)	New Beetle	F	0.790	0.950	0.002	—	—	0.27	18	92
		R	0.390	0.315	0.002	9.06	9.09	0.27 ①	26	48
	Passat	F	0.980 ⑧	0.900 ⑨	0.002	—	—	0.28	22	89
		R	0.393	0.314	0.002	NA	NA	0.27	22	70

F: Front
R: Rear
NA: Not Available
① VR6 Model
② GLX Model
③ GL model
④ 2.0 L Engine
⑤ 2.8 L Engine

⑥ ABS equipped models have a diameter of 0.937 in.
⑦ Models equipped with drum brakes have a lining limit of 0.098 in.
⑧ Lucas caliper: 0.87 in.
⑨ Lucas caliper: 0.78 in.

93471CR1

For Wheel Alignment specifications, see Section 1 of this manual

WHEEL ALIGNMENT

Year	Model		Caster Range (+/-Deg.)	Caster Preferred Setting (Deg.)	Camber Range (+/-Deg.)	Camber Preferred Setting (Deg.)	Toe-in (in.)	Steering Axis Inclination (Deg.)
1998	Cabrio							
	1.8L, 1.9L	F	0.50	+1.75	0.33	-0.50	0+/-0.16	—
	2.0L	F	0.50	+1.50	0.33	-0.58	0+/-0.16	
	2.8L	F	0.50	+3.27	0.33	-0.50	0+/-0.16	—
		R	—	—	0.16	-1.50	0.33+/-0.16	—
	Golf							
	Standard	F	0.50	+7.67	0.50	-0.50	0+/-0.16	—
		R	—	—	0.16	-1.45	0.33+/-0.16	
	Sport	F	0.50	+7.83	0.50	-0.58	0+/-0.16	
		R	—	—	0.16	-1.58	0.42+/-0.16	—
	Jetta							
	Standard	F	0.50	+7.67	0.50	-0.50	0+/-0.16	—
		R	—	—	0.16	-1.45	0.33+/-0.16	
	Sport	F	0.50	+7.83	0.50	-0.58	0+/-0.16	
		R	—	—	0.16	-1.45	0.42+/-0.16	—
	New Beetle							
	Standard	F	0.50	+7.67	0.50	-0.50	0+/-0.16	—
		R	—	—	0.16	-1.45	0.33+/-0.16	—
	Sport	F	0.50	+7.83	0.50	-0.55	0+/-0.16	—
		R	—	—	0.16	-1.45	0.42+/-0.16	—
	Passat FWD							
	Standard	F	NA	NA	0.42	-0.42	0.16+/-0.08	—
		R	—	—	0.33	-1.50	0.33+/-0.16	
	Sport	F	NA	NA	0.42	-0.67	0.16+/-0.08	
		R	—	—	0.33	-1.50	0.47+/-0.16	—
	Heavy Duty	F	NA	NA	0.42	-0.25	0.16+/-0.08	—
		R	—	—	0.33	-1.50	0.18+/-0.16	—
	Passat AWD							
	Standard	F	NA	NA	0.42	-0.42	0.16+/-0.08	—
		R	—	—	0.50	-0.67	0.27+/-0.16	—
	Sport	F	NA	NA	0.42	-0.67	0.16+/-0.08	—
		R	—	—	0.50	-0.67	0.27+/-0.16	
	Heavy Duty	F	NA	NA	0.42	-0.25	0.16+/-0.08	—
		R	—	—	0.50	-0.67	0.27+/-0.16	—
1999	Cabrio							
	1.8L, 1.9L	F	0.50	+1.75	0.33	-0.50	0+/-0.16	—
	2.0L	F	0.50	+1.50	0.33	-0.58	0+/-0.16	—
	2.8L	F	0.50	+3.27	0.33	-0.50	0+/-0.16	—
		R	—	—	0.16	-1.50	0.33+/-0.16	—
	Golf							
	Standard	F	0.50	+7.67	0.50	-0.50	0+/-0.16	—
		R	—	—	0.16	-1.45	0.33+/-0.16	—
	Sport	F	0.50	+7.83	0.50	-0.58	0+/-0.16	—
		R	—	—	0.16	-1.58	0.42+/-0.16	—
	Jetta							
	Standard	F	0.50	+7.67	0.50	-0.50	0+/-0.16	—
		R	—	—	0.16	-1.45	0.33+/-0.16	—

93471CR2

WHEEL ALIGNMENT

Year	Model		Caster Range (+/-Deg.)	Caster Preferred Setting (Deg.)	Camber Range (+/-Deg.)	Camber Preferred Setting (Deg.)	Toe-in (in.)	Steering Axis Inclination (Deg.)
1999 (cont.)	Sport	F	0.50	+7.83	0.50	-0.58	0+/-0.16	—
		R	—	—	0.16	-1.45	0.42+/-0.16	—
	New Beetle							
	Standard	F	0.50	+7.67	0.50	-0.50	0+/-0.16	—
		R	—	—	0.16	-1.45	0.33+/-0.16	—
	Sport	F	0.50	+7.83	0.50	-0.55	0+/-0.16	—
		R	—	—	0.16	-1.45	0.42+/-0.16	—
	Passat FWD							
	Standard	F	NA	NA	0.42	-0.42	0.16+/-0.08	—
		R	—	—	0.33	-1.50	0.33+/-0.16	—
	Sport	F	NA	NA	0.42	-0.67	0.16+/-0.08	—
		R	—	—	0.33	-1.50	0.47+/-0.16	—
	Heavy Duty	F	NA	NA	0.42	-0.25	0.16+/-0.08	—
		R	—	—	0.33	-1.50	0.18+/-0.16	—
	Passat AWD							
	Standard	F	NA	NA	0.42	-0.42	0.16+/-0.08	—
		R	—	—	0.50	-0.67	0.27+/-0.16	—
	Sport	F	NA	NA	0.42	-0.67	0.16+/-0.08	—
		R	—	—	0.50	-0.67	0.27+/-0.16	—
	Heavy Duty	F	NA	NA	0.42	-0.25	0.16+/-0.08	—
		R	—	—	0.50	-0.67	0.27+/-0.16	—
2000	Cabrio							
	1.8L, 1.9L	F	0.50	+1.75	0.33	-0.50	0+/-0.16	—
	2.0L	F	0.50	+1.50	0.33	-0.58	0+/-0.16	—
	2.8L	F	0.50	+3.27	0.33	-0.50	0+/-0.16	—
		R	—	—	0.16	-1.50	0.33+/-0.16	—
	Golf							
	Standard	F	0.50	+7.67	0.50	-0.50	0+/-0.16	—
		R	—	—	0.16	-1.45	0.33+/-0.16	—
	Sport	F	0.50	+7.83	0.50	-0.58	0+/-0.16	—
		R	—	—	0.16	-1.58	0.42+/-0.16	—
	Jetta							
	Standard	F	0.50	+7.67	0.50	-0.50	0+/-0.16	—
		R	—	—	0.16	-1.45	0.33+/-0.16	—
	Sport	F	0.50	+7.83	0.50	-0.58	0+/-0.16	—
		R	—	—	0.16	-1.45	0.42+/-0.16	—
	New Beetle							
	Standard	F	0.50	+7.67	0.50	-0.50	0+/-0.16	—
		R	—	—	0.16	-1.45	0.33+/-0.16	—
	Sport	F	0.50	+7.83	0.50	-0.55	0+/-0.16	—
		R	—	—	0.16	-1.45	0.42+/-0.16	—
	Passat FWD							
	Standard	F	NA	NA	0.42	-0.42	0.16+/-0.08	—
		R	—	—	0.33	-1.50	0.33+/-0.16	—
	Sport	F	NA	NA	0.42	-0.67	0.16+/-0.08	—
		R	—	—	0.33	-1.50	0.47+/-0.16	—
	Heavy Duty	F	NA	NA	0.42	-0.25	0.16+/-0.08	—
		R	—	—	0.33	-1.50	0.18+/-0.16	—
	Passat AWD							
	Standard	F	NA	NA	0.42	-0.42	0.16+/-0.08	—

93471CR3

For Maintenance Interval recommendations, see Section 1 of this manual

WHEEL ALIGNMENT

Year	Model		Caster Range (+/-Deg.)	Caster Preferred Setting (Deg.)	Camber Range (+/-Deg.)	Camber Preferred Setting (Deg.)	Toe-in (in.)	Steering Axis Inclination (Deg.)
2000 (cont.)	Sport	R	—	—	0.50	-0.67	0.27+/-0.16	—
		F	NA	NA	0.42	-0.67	0.16+/-0.08	—
		R	—	—	0.50	-0.67	0.27+/-0.16	—
	Heavy Duty	F	NA	NA	0.42	-0.25	0.16+/-0.08	—
		R	—	—	0.50	-0.67	0.27+/-0.16	—

93471CR4

TIRE, WHEEL AND BALL JOINT SPECIFICATIONS

Year	Model	OEM Tires		Tire Pressures (psi)		Wheel Size	Ball Joint Inspection
		Standard	Optional	Front	Rear		
1998	New Beetle	205/55R16	None	28	28	6.5-J	①
	Passat	195/65R15	None	32	32	6.5-J	①
	Cabrio	P195/60HR14	None	30	30	6-J	①
	GTI	205/50HR15	None	32	32	6.5-J	①
	Jetta GLX	205/50VR15	None	28	26	6.5-J	①
	Jetta GL, GLS	185/60HR14	195/60HR14	33	30	6-J	①
1999	New Beetle	205/55R16	None	28	28	6.5-J	①
	Passat	195/65R15	None	32	32	6.5-J	①
	Cabrio	P195/60HR14	None	30	30	6-J	①
	GTI	205/50HR15	None	32	32	6.5-J	①
	Jetta GLX	205/50VR15	None	28	26	6.5-J	①
	Jetta GL, GLS	185/60HR14	195/60HR14	33	30	6-J	①
2000	New Beetle	205/55R16	None	28	28	6.5-J	①
	Passat	195/65R15	None	32	32	6.5-J	①
	Cabrio	P195/60HR14	None	30	30	6-J	①
	GTI	205/50HR15	None	32	32	6.5-J	①
	Jetta GLX	205/50VR15	None	28	26	6.5-J	①
	Jetta GL, GLS	185/60HR14	195/60HR14	33	30	6-J	①

OEM: Original Equipment Manufacturer

PSI: Pounds Per Square Inch

STD: Standard

OPT: Optional

① Replace if any measurable movement is found.

93471CR5

For Tune-up, Capacities and Firing orders, see Section 1 of this manual

SCHEDULED MAINTENANCE INTERVALS
Volkswagen

TO BE SERVICED	TYPE OF SERVICE	VEHICLE MILEAGE INTERVAL (x1000)												
		7.5	15	22.5	30	37.5	45	52.5	60	67.5	75	82.5	90	97.5
Engine oil & filter	R	✓	✓	✓	✓	✓	✓	✓	✓	✓	✓	✓	✓	✓
Brake pad thickness	R	✓	✓	✓	✓	✓	✓	✓	✓	✓	✓	✓	✓	✓
A/T final drive fluid level	S/I		✓		✓		✓		✓		✓		✓	
Battery	S/I		✓		✓		✓		✓		✓		✓	
Brake system	S/I		✓		✓		✓		✓		✓		✓	
Cooling system	S/I		✓		✓		✓		✓		✓		✓	
Driveshaft boots	S/I		✓		✓		✓		✓		✓		✓	
Engine (check for leaks)	S/I		✓		✓		✓		✓		✓		✓	
Engine coolant level	S/I		✓		✓		✓		✓		✓		✓	
Engine exhaust	S/I		✓		✓		✓		✓		✓		✓	
Fuel system	S/I		✓		✓		✓		✓		✓		✓	
Idle speed (gasoline)	S/I		✓		✓		✓		✓		✓		✓	
Idle speed (diesel)	S/I				✓				✓				✓	
Intake air system	S/I		✓		✓		✓		✓		✓		✓	
OBD system - check for codes	S/I		✓		✓		✓		✓		✓		✓	
Power steering fluid level	S/I		✓		✓		✓		✓		✓		✓	
Steering system	S/I		✓		✓		✓		✓		✓		✓	
Timing belt (diesel)	S/I				✓				✓				✓	
Transaxle fluid level	S/I		✓		✓		✓		✓		✓		✓	
Water separator (diesel)	S/I		✓		✓		✓		✓		✓		✓	
Air filter element	R				✓				✓				✓	
Engine coolant	R				✓				✓				✓	
Fuel filter (diesel)	R				✓				✓				✓	
Spark plugs (w/o supercharger)	R				✓				✓				✓	
Spark plugs (w supercharger)	R								✓					
Passenger compartment air filter	R				✓				✓				✓	
Drive belts	S/I				✓				✓				✓	
Dust seals on ball joints, tie rod ends & tie rods	S/I				✓				✓				✓	
Brake fluid ①	R													

R: Replace   S/I: Service or Inspect

① Replace every two years regardless of mileage.

FREQUENT OPERATION MAINTENANCE (SEVERE SERVICE)

If a vehicle is operated under any of the following conditions it is considered severe service:

- Extremely dusty areas.

- 50% or more of the vehicle operation is in 32°C (90°F) or higher temperatures, or constant operation in temperatures below 0°C (32°F).

- Prolonged idling (vehicle operation in stop and go traffic).

- Frequent short running periods (engine does not warm to normal operating temperatures).

- Police, taxi, delivery usage or trailer towing usage.

Oil & oil filter change: change every 3750 miles.

Air filter element: service or inspect every 15,000 miles.

Automatic transaxle fluid & filter: replace every 30,000 miles.

93471CR6

SCHEDULED MAINTENANCE INTERVALS
VOLKSWAGEN
CABRIO, GOLF, JETTA, NEW BEETLE, PASSAT

The following should be used as a guide when determining the amount of work required for a particular service. In estimating how long a particular Scheduled Maintenance Service should take, please observe the following:

- Labor Time is time based on field research and data supplied by the vehicle manufacturer.
- Labor time operations are given in hours and tenths of an hour.
- All labor operations are to be used as a guide.

Mechanic Skill Level Codes:
(A) PRECISION: Highly skilled with multiple certification.
(B) GENERAL: Normally skilled with certification.
(C) MAINTENANCE: Semi-skilled working on certification.

	LABOR TIME
7500 Mile Service (C)	
All Models	.4
w/diesel add	.2
15000 Mile Service (B)	
All Models	2.1
w/AT add	.1
w/diesel add	.2
w/16V engine add	.2
22500 Mile Service (C)	
All Models	.4
w/diesel add	.2
30000 Mile Service (B)	
All Models	3.2
w/AT add	.1
w/diesel add	.7
w/16V engine add	.2

	LABOR TIME
37500 Mile Service (C)	
All Models	.4
w/diesel add	.2
45000 Mile Service (B)	
All Models	2.1
w/AT add	.1
w/diesel add	.2
w/16V engine add	.2
52500 Mile Service (C)	
All Models	.4
w/diesel add	.2
60000 Mile Service (B)	
All Models	3.2
w/AT add	.1
w/diesel add	.7
w/Cabrio add	.1
w/supercharger add	.3
67500 Mile Service (C)	
All Models	.4
w/diesel add	.2

	LABOR TIME
75000 Mile Service (B)	
All Models	2.1
w/AT add	.1
w/diesel add	.2
w/16V engine add	.2
82500 Mile Service (C)	
All Models	.4
w/diesel add	.2
90000 Mile Service (B)	
All Models	3.2
w/AT add	.1
w/diesel add	.7
w/16V engine add	.2
97500 Mile Service (C)	
All Models	.4
w/diesel add	.2

93471CR7

VOLVO
850 • 960 • C70 • S70T • S90 • V70

ENGINE AND VEHICLE IDENTIFICATION CHART

Code ①	Liters	Cu. in. (cc)	Cyl.	Fuel Sys.	Type	Eng. Mfg.
B5234T3/53	2.3 (2319)	144	5	EFI	DOHC	Volvo
B5254S/55	2.4 (2435)	151	5	EFI	DOHC	Volvo
B5254T/56	2.4 (2435)	151	5	EFI	DOHC	Volvo
B6304S3/97	2.9 (2922)	181	6	EFI	DOHC	Volvo

Engine

Model Year

Code ②	Year
W	1998
X	1999
Y	2000
1	2001
2	2002

EFI: Electronic Fuel Injection

DOHC: Double Overhead Camshafts

① Engine ID / 6th and 7th digits of the VIN

② 10th digit of the VIN

93471CR8

GENERAL ENGINE SPECIFICATIONS

Year	Model	Engine Displacement Liters (cc)	Engine ID/VIN	Fuel System Type	Net Horsepower @ rpm	Net Torque @ rpm (ft. lbs.)	Bore x Stroke (in.)	Compression Ratio	Oil Pressure @ rpm
1998	S70T-5	2.3 (2319)	B5234T3/53	EFI	236@5100	243@2100	3.19 x 3.54	8.5:1	49.8@4000
	C70	2.3 (2319)	B5234T3/53	EFI	236@5100	243@2700	3.19 x 3.54	8.5:1	49.8@4000
	V70T-5	2.3 (2319)	B5234T3/53	EFI	236@5100	243@2100	3.19 x 3.54	8.5:1	49.8@4000
	V70R AWD	2.3 (2319)	B5234T3/53	EFI	236@5100	243@2100	3.19 x 3.54	8.5:1	49.8@4000
	S70	2.4 (2435)	B5254S/55	EFI	168@6100	162@4700	3.27 x 3.54	10.5:1	49.8@4000
	S70GLT	2.4 (2435)	B5254T/56	EFI	190@5200	199@1800	3.27 x 3.54	10.5:1	49.8@4000
	V70	2.4 (2435)	B5254S/55	EFI	168@6100	162@4700	3.27 x 3.54	10.5:1	49.8@4000
	V70GLT	2.4 (2435)	B5254T/56	EFI	190@5200	199@1800	3.27 x 3.54	10.5:1	49.8@4000
	V70 AWD	2.4 (2435)	B5254T/56	EFI	190@5200	199@1800	3.27 x 3.54	10.5:1	49.8@4000
	V70XC AWD	2.4 (2435)	B5254T/56	EFI	190@5200	199@1800	3.27 x 3.54	10.5:1	49.8@4000
	S90	2.9 (2922)	B6304S3/97	EFI	181@5200	199@4100	3.27 x 3.54	10.7:1	36@2000
	V90	2.9 (2922)	B6304S3/97	EFI	181@5200	199@4100	3.27 x 3.54	10.7:1	36@2000
1999	S70T-5	2.3 (2319)	B5234T3/53	EFI	236@5100	243@2100	3.19 x 3.54	8.5:1	49.8@4000
	C70	2.3 (2319)	B5234T3/53	EFI	236@5100	243@2700	3.19 x 3.54	8.5:1	49.8@4000
	V70T-5	2.3 (2319)	B5234T3/53	EFI	236@5100	243@2100	3.19 x 3.54	8.5:1	49.8@4000
	V70R AWD	2.3 (2319)	B5234T3/53	EFI	236@5100	243@2100	3.19 x 3.54	8.5:1	49.8@4000
	S70	2.4 (2435)	B5254S/55	EFI	168@6100	162@4700	3.27 x 3.54	10.5:1	49.8@4000
	S70GLT	2.4 (2435)	B5254T/56	EFI	190@5200	199@1800	3.27 x 3.54	10.5:1	49.8@4000
	V70	2.4 (2435)	B5254S/55	EFI	168@6100	162@4700	3.27 x 3.54	10.5:1	49.8@4000
	V70GLT	2.4 (2435)	B5254T/56	EFI	190@5200	199@1800	3.27 x 3.54	10.5:1	49.8@4000
	V70 AWD	2.4 (2435)	B5254T/56	EFI	190@5200	199@1800	3.27 x 3.54	10.5:1	49.8@4000
	V70XC AWD	2.4 (2435)	B5254T/56	EFI	190@5200	199@1800	3.27 x 3.54	10.5:1	49.8@4000
2000	S70T-5	2.3 (2319)	B5234T3/53	EFI	236@5100	243@2100	3.19 x 3.54	8.5:1	49.8@4000
	C70	2.3 (2319)	B5234T3/53	EFI	236@5100	243@2700	3.19 x 3.54	8.5:1	49.8@4000
	V70T-5	2.3 (2319)	B5234T3/53	EFI	236@5100	243@2100	3.19 x 3.54	8.5:1	49.8@4000
	V70R AWD	2.3 (2319)	B5234T3/53	EFI	236@5100	243@2100	3.19 x 3.54	8.5:1	49.8@4000
	S70	2.4 (2435)	B5254S/55	EFI	168@6100	162@4700	3.27 x 3.54	10.5:1	49.8@4000
	S70GLT	2.4 (2435)	B5254T/56	EFI	190@5200	199@1800	3.27 x 3.54	10.5:1	49.8@4000
	V70	2.4 (2435)	B5254S/55	EFI	168@6100	162@4700	3.27 x 3.54	10.5:1	49.8@4000
	V70GLT	2.4 (2435)	B5254T/56	EFI	190@5200	199@1800	3.27 x 3.54	10.5:1	49.8@4000
	V70 AWD	2.4 (2435)	B5254T/56	EFI	190@5200	199@1800	3.27 x 3.54	10.5:1	49.8@4000
	V70XC AWD	2.4 (2435)	B5254T/56	EFI	190@5200	199@1800	3.27 x 3.54	10.5:1	49.8@4000

EFI: Electronic Fuel Injection

93471CR9

ENGINE TUNE-UP SPECIFICATIONS

Engine Displacement Liters (cc)	Engine ID/VIN	Spark Plug Gap (in.)	Ignition Timing (deg.) MT	Ignition Timing (deg.) AT	Fuel Pump (psi)	Idle Speed (rpm) MT	Idle Speed (rpm) AT	Valve Clearance In.	Valve Clearance Ex.
1998									
2.3 (2319)	B5234T3/53	0.030	6B	6B	58	800-900	800-900	HYD	HYD
2.3 (2319)	B5234T3/53	0.030	6B	6B	58	800-900	800-900	HYD	HYD
2.4 (2435)	B5254S/55	0.020	10B	10B	43	800-900	800-900	HYD	HYD
2.4 (2435)	B5254T/56	0.030	10B	10B	58	800-900	800-900	HYD	HYD
2.9 (2922)	B6304S3/97	0.020	10B	10B	43	800-900	800-900	HYD	HYD
1999									
2.3 (2319)	B5234T3/53	0.030	6B	6B	58	800-900	800-900	HYD	HYD
2.3 (2319)	B5234T3/53	0.030	6B	6B	58	800-900	800-900	HYD	HYD
2.4 (2435)	B5254S/55	0.020	10B	10B	43	800-900	800-900	HYD	HYD
2.4 (2435)	B5254T/56	0.030	10B	10B	58	800-900	800-900	HYD	HYD
2000									
2.3 (2319)	B5234T3/53	0.030	6B	6B	58	800-900	800-900	HYD	HYD
2.3 (2319)	B5234T3/53	0.030	6B	6B	58	800-900	800-900	HYD	HYD
2.4 (2435)	B5254S/55	0.020	10B	10B	43	800-900	800-900	HYD	HYD
2.4 (2435)	B5254T/56	0.030	10B	10B	58	800-900	800-900	HYD	HYD

HYD: Hydraulic lash adjusters

B: Before top dead center

93471CR0

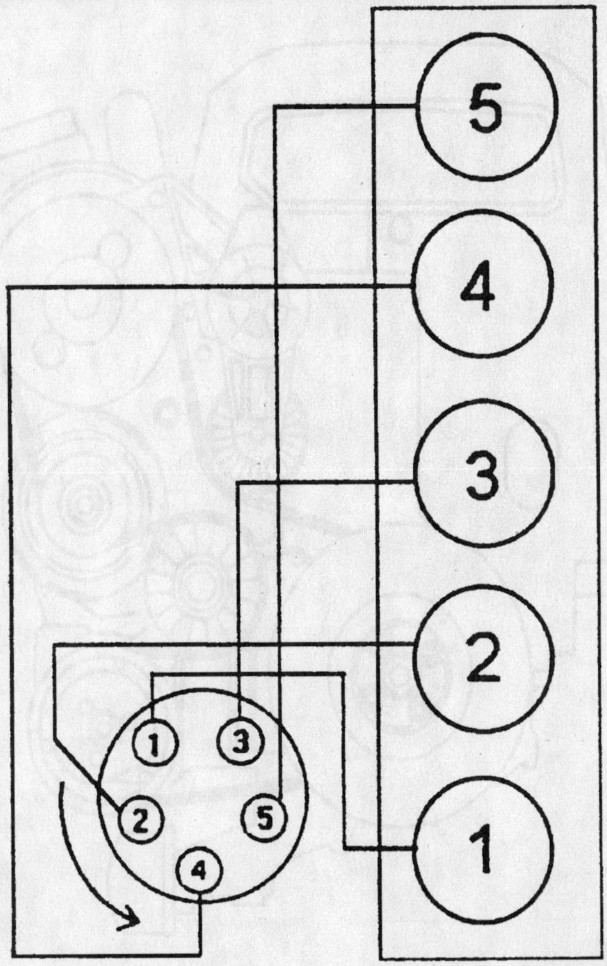

5 cylinder Engine
Firing order: 1–2–4–5–3
Distributor Rotation: Counterclockwise

90952G10

2.9L Engine
Firing order: 1–5–3–6–2–4
Distributorless ignition system (one coil on each cylinder)

79233G47

Timing belt service is covered in Section 3 of this manual

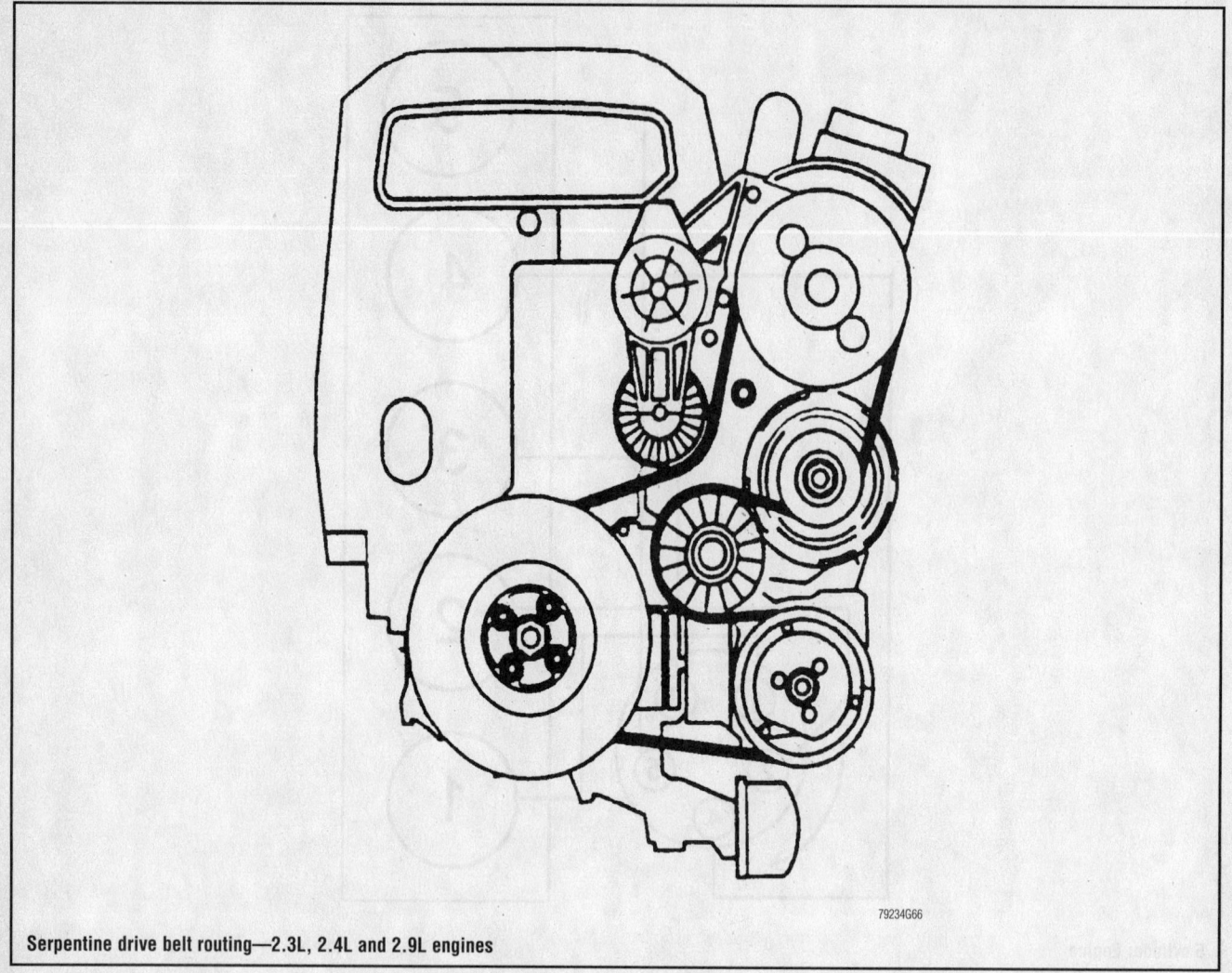

79234G66

Serpentine drive belt routing—2.3L, 2.4L and 2.9L engines

CAPACITIES

Year	Model	Engine Displacement Liters (cc)	Engine ID/VIN	Engine Oil with Filter (qts.) [1]	Transmission (pts.)		Transfer Case (pts.)	Drive Axle		Fuel Tank (gal.)	Cooling System (qts.)
					5-Spd	Auto.		Front (pts.)	Rear (pts.)		
1998	S70T-5	2.3 (2319)	B5234T3/53	6.1	2.2	8.4	—	—	—	18.5	7.6
	C70	2.3 (2319)	B5234T3/53	6.1	2.2	8.4	—	—	—	18.5	7.6
	V70T-5	2.3 (2319)	B5234T3/53	6.1	2.2	8.4	—	—	—	18.5	7.6
	V70R AWD	2.3 (2319)	B5234T3/53	6.1	—	8.4	1.7	—	2.9	18.5	7.6
	S70	2.4 (2435)	B5254S/55	6.1	2.2	8.4	—	—	—	18.5	7.6
	S70GLT	2.4 (2435)	B5254T/56	6.1	—	8.4	—	—	—	18.5	7.6
	V70	2.4 (2435)	B5254S/55	6.1	2.2	8.4	—	—	—	18.5	7.6
	V70GLT	2.4 (2435)	B5254T/56	6.1	—	8.4	—	—	—	18.5	7.6
	V70 AWD	2.4 (2435)	B5254T/56	6.1	—	8.4	1.7	—	2.9	18.5	7.6
	V70XC AWD	2.4 (2435)	B5254T/56	6.1	—	8.4	1.7	—	2.9	18.5	7.6
	S90	2.9 (2922)	B6304S3/97	7.3	2.2	16.4	—	—	[2]	21.8	10
	V90	2.9 (2922)	B6304S3/97	7.3	2.2	16.4	—	—	[2]	21.8	10
1999	S70T-5	2.3 (2319)	B5234T3/53	6.1	2.2	8.4	—	—	—	18.5	7.6
	C70	2.3 (2319)	B5234T3/53	6.1	2.2	8.4	—	—	—	18.5	7.6
	V70T-5	2.3 (2319)	B5234T3/53	6.1	2.2	8.4	—	—	—	18.5	7.6
	V70R AWD	2.3 (2319)	B5234T3/53	6.1	—	8.4	1.7	—	2.9	18.5	7.6
	S70	2.4 (2435)	B5254S/55	6.1	2.2	8.4	—	—	—	18.5	7.6
	S70GLT	2.4 (2435)	B5254T/56	6.1	—	8.4	—	—	—	18.5	7.6
	V70	2.4 (2435)	B5254S/55	6.1	2.2	8.4	—	—	—	18.5	7.6
	V70GLT	2.4 (2435)	B5254T/56	6.1	—	8.4	—	—	—	18.5	7.6
	V70 AWD	2.4 (2435)	B5254T/56	6.1	—	8.4	1.7	—	2.9	18.5	7.6
	V70XC AWD	2.4 (2435)	B5254T/56	6.1	—	8.4	1.7	—	2.9	18.5	7.6
2000	S70T-5	2.3 (2319)	B5234T3/53	6.1	2.2	8.4	—	—	—	18.5	7.6
	C70	2.3 (2319)	B5234T3/53	6.1	2.2	8.4	—	—	—	18.5	7.6
	V70T-5	2.3 (2319)	B5234T3/53	6.1	2.2	8.4	—	—	—	18.5	7.6
	V70R AWD	2.3 (2319)	B5234T3/53	6.1	—	8.4	1.7	—	2.9	18.5	7.6
	S70	2.4 (2435)	B5254S/55	6.1	2.2	8.4	—	—	—	18.5	7.6
	S70GLT	2.4 (2435)	B5254T/56	6.1	—	8.4	—	—	—	18.5	7.6
	V70	2.4 (2435)	B5254S/55	6.1	2.2	8.4	—	—	—	18.5	7.6
	V70GLT	2.4 (2435)	B5254T/56	6.1	—	8.4	—	—	—	18.5	7.6
	V70 AWD	2.4 (2435)	B5254T/56	6.1	—	8.4	1.7	—	2.9	18.5	7.6
	V70XC AWD	2.4 (2435)	B5254T/56	6.1	—	8.4	1.7	—	2.9	18.5	7.6

NOTE: All capacities are approximate. Add fluid gradualy and check to be sure a proper fluid level is obtained.

[1] On turbocharged engines, add 0.7 US qts. if the cooler is drained

[2] 1030 axle: 2.8
1031 axle: 3.4
1035 axle: 2.9
1041 axle: 3.1
1045 axle: 2.8
1055 axle: 3.2
1065 axle: 2.9

93471CS1

Heater Core replacement is covered in Section 2 of this manual

VALVE SPECIFICATIONS

Year	Engine Displacement Liters (cc)	Engine ID/VIN	Seat Angle (deg.)	Face Angle (deg.)	Spring Test Pressure (lbs. @ in.)	Spring Installed Height (in.)	Stem-to-Guide Clearance (in.)		Stem Diameter (in.)	
							Intake	Exhaust	Intake	Exhaust
1998	2.3 (2319)	B5234T3/53	45	44.5	①	NA	0.0012-0.0024	0.0015-0.0028	0.2738-0.2750	0.2734-0.2746
	2.4 (2435)	B5254S/55	45	44.5	①	NA	0.0012-0.0024	0.0012-0.0024	0.2738-0.2750	0.2734-0.2746
	2.4 (2435)	B5254T/56	45	44.5	①	NA	0.0012-0.0024	0.0015-0.0028	0.2738-0.2750	0.2734-0.2746
	2.9 (2922)	B6304S3/97	45	44.5	①	NA	0.0012-0.0024	0.0012-0.0024	0.2738-0.2744	0.2738-0.2744
1999	2.3 (2319)	B5234T3/53	45	44.5	①	NA	0.0012-0.0024	0.0015-0.0028	0.2738-0.2750	0.2734-0.2746
	2.4 (2435)	B5254S/55	45	44.5	①	NA	0.0012-0.0024	0.0012-0.0024	0.2738-0.2750	0.2734-0.2746
	2.4 (2435)	B5254T/56	45	44.5	①	NA	0.0012-0.0024	0.0015-0.0028	0.2738-0.2750	0.2734-0.2746
2000	2.3 (2319)	B5234T3/53	45	44.5	①	NA	0.0012-0.0024	0.0015-0.0028	0.2738-0.2750	0.2734-0.2746
	2.4 (2435)	B5254S/55	45	44.5	①	NA	0.0012-0.0024	0.0012-0.0024	0.2738-0.2750	0.2734-0.2746
	2.4 (2435)	B5254T/56	45	44.5	①	NA	0.0012-0.0024	0.0015-0.0028	0.2738-0.2750	0.2734-0.2746

Note: Exhaust valves for turbocharged engines are stellite-coated and must not be machined. They may be ground against the valve seat.

NA: Not Available

① Intake valve: 150 @ 1.00

Exhaust valve: 61 @ 1.34

93471CS2

CRANKSHAFT AND CONNECTING ROD SPECIFICATIONS
All measurements are given in inches.

Year	Engine Displacement Liters (cc)	Engine ID/VIN	Crankshaft				Connecting Rod		
			Main Brg. Journal Dia.	Main Brg. Oil Clearance	Shaft End-play	Thrust on No.	Journal Diameter	Oil Clearance	Side Clearance
1998	2.3 (2319)	B5234T3/53	2.5584-2.5592	0.0007-0.0017	0.003-0.007	5	1.9679-1.9685	NA	0.006-0.018
	2.4 (2435)	B5254S/55	2.5584-2.5592	0.0007-0.0017	0.003-0.007	5	1.9679-1.9685	NA	0.006-0.018
	2.4 (2435)	B5254T/56	2.5584-2.5592	0.0007-0.0017	0.003-0.007	5	1.9679-1.9685	NA	0.006-0.018
	2.9 (2922)	B6304S3/97	2.5584-2.5592	0.0007-0.0017	0.003-0.007	5	1.9679-1.9685	NA	0.006-0.018
1999	2.3 (2319)	B5234T3/53	2.5584-2.5592	0.0007-0.0017	0.003-0.007	5	1.9679-1.9685	NA	0.006-0.018
	2.4 (2435)	B5254S/55	2.5584-2.5592	0.0007-0.0017	0.003-0.007	5	1.9679-1.9685	NA	0.006-0.018
	2.4 (2435)	B5254T/56	2.5584-2.5592	0.0007-0.0017	0.003-0.007	5	1.9679-1.9685	NA	0.006-0.018
2000	2.3 (2319)	B5234T3/53	2.5584-2.5592	0.0007-0.0017	0.003-0.007	5	1.9679-1.9685	NA	0.006-0.018
	2.4 (2435)	B5254S/55	2.5584-2.5592	0.0007-0.0017	0.003-0.007	5	1.9679-1.9685	NA	0.006-0.018
	2.4 (2435)	B5254T/56	2.5584-2.5592	0.0007-0.0017	0.003-0.007	5	1.9679-1.9685	NA	0.006-0.018

NA: Not Available

93471CS3

Brake service is covered in Section 4 of this manual

PISTON AND RING SPECIFICATIONS
All measurements are given in inches.

Year	Engine Displacement Liters (cc)	Engine ID/VIN	Piston Clearance	Ring Gap			Ring Side Clearance		
				Top Compression	Bottom Compression	Oil Control	Top Compression	Bottom Compression	Oil Control
1998	2.3 (2319)	B5234T3/53	0.0004-0.0012	0.047	0.069	0.118	0.0020-0.0033	0.0011-0.0026	0.0008-0.0022
	2.4 (2435)	B5254S/55	0.0004-0.0012	0.047	0.069	0.118	0.0020-0.0033	0.0011-0.0026	0.0008-0.0022
	2.4 (2435)	B5254T/56	0.0004-0.0012	0.047	0.069	0.118	0.0020-0.0033	0.0011-0.0026	0.0008-0.0022
	2.9 (2922)	B6304S3/97	0.0004-0.0012	0.047	0.069	0.118	0.0020-0.0033	0.0011-0.0026	0.0008-0.0022
1999	2.3 (2319)	B5234T3/53	0.0004-0.0012	0.047	0.069	0.118	0.0020-0.0033	0.0011-0.0026	0.0008-0.0022
	2.4 (2435)	B5254S/55	0.0004-0.0012	0.047	0.069	0.118	0.0020-0.0033	0.0011-0.0026	0.0008-0.0022
	2.4 (2435)	B5254T/56	0.0004-0.0012	0.047	0.069	0.118	0.0020-0.0033	0.0011-0.0026	0.0008-0.0022
2000	2.3 (2319)	B5234T3/53	0.0004-0.0012	0.047	0.069	0.118	0.0020-0.0033	0.0011-0.0026	0.0008-0.0022
	2.4 (2435)	B5254S/55	0.0004-0.0012	0.047	0.069	0.118	0.0020-0.0033	0.0011-0.0026	0.0008-0.0022
	2.4 (2435)	B5254T/56	0.0004-0.0012	0.047	0.069	0.118	0.0020-0.0033	0.0011-0.0026	0.0008-0.0022

93471CS4

TORQUE SPECIFICATIONS

All readings in ft. lbs.

Year	Engine Displacement Liters (cc)	Engine ID/VIN	Cylinder Head Bolts	Main Bearing Bolts	Rod Bearing Bolts	Crankshaft Damper Bolts	Flywheel Bolts	Manifold		Spark Plugs	Lug Nut
								Intake	Exhaust		
1998	2.3 (2319)	B5234T3/53	①	②	③	133	④	15	18	18	81
	2.4 (2435)	B5254S/55	①	②	③	133	④	15	18	18	81
	2.4 (2435)	B5254T/56	①	②	③	133	④	15	18	18	81
	2.9 (2922)	B6304S3/97	①	②	③	221	④	15	18	18	81
1999	2.3 (2319)	B5234T3/53	①	②	③	133	④	15	18	18	81
	2.4 (2435)	B5254S/55	①	②	③	133	④	15	18	18	81
	2.4 (2435)	B5254T/56	①	②	③	133	④	15	18	18	81
2000	2.3 (2319)	B5234T3/53	①	②	③	133	④	15	18	18	81
	2.4 (2435)	B5254S/55	①	②	③	133	④	15	18	18	81
	2.4 (2435)	B5254T/56	①	②	③	133	④	15	18	18	81

① Step 1: 14 ft. lbs.
   Step 2: 43 ft. lbs.
   Step 3: Plus 90 degrees

② Tighten cylinder block, intermediate section, in stages:
   Step 1: M10 bolts: 15 ft. lbs. (20mm)
   Step 2: M10 bolts: 33 ft. lbs. (45mm)
   Step 3: M8 bolts: 18 ft. lbs. (25mm)
   Step 4: M7 bolts: 13 ft. lbs. (17mm)
   Step 5: M10 bolts: Plus 90 degrees

③ Step 1: 15 ft. lbs.
   Step 2: Plus 90 degrees

④ Step 1: 33 ft. lbs.
   Step 2: Plus 50 degrees

93471CS5

For complete Engine Mechanical specifications, see Section 1 of this manual

BRAKE SPECIFICATIONS
All measurements in inches unless noted

| Year | Model | | Brake Disc | | | Minimum Lining Thickness | Brake Caliper | |
			Original Thickness	Minimum Thickness	Maximum Runout		Bracket bolts (ft. lbs.)	Mounting bolts (ft. lbs.)
1998	C70	F	1.024	0.906	0.001	0.120	74	22
		R	0.378	0.330	0.002	0.075	37	22
	S70	F	1.024	0.906	0.001	0.120	74	22
		R	0.378	0.330	0.002	0.075	37	22
	S90	F	0.870	0.787	0.002	0.120	78	22
	Standard	R	0.378	0.330	0.004	0.078	44	22
	S90	F	1.024	0.910	0.002	0.120	78	22
	HD	R	0.393	0.314	0.003	0.078	44	22
	V70	F	1.024	0.906	0.001	0.120	74	22
		R	0.378	0.330	0.002	0.075	37	22
	V90	F	0.870	0.787	0.002	0.120	78	22
	Standard	R	0.378	0.330	0.004	0.078	44	22
	V90	F	1.024	0.910	0.002	0.120	78	22
	HD	R	0.393	0.314	0.003	0.078	44	22
1999	C70	F	1.024	0.906	0.001	0.120	74	22
		R	0.378	0.330	0.002	0.075	37	22
	S70	F	1.024	0.906	0.001	0.120	74	22
		R	0.378	0.330	0.002	0.075	37	22
	V70	F	1.024	0.906	0.001	0.120	74	22
		R	0.378	0.330	0.002	0.075	37	22
2000	C70	F	1.024	0.906	0.001	0.120	74	22
		R	0.378	0.330	0.002	0.075	37	22
	S70	F	1.024	0.906	0.001	0.120	74	22
		R	0.378	0.330	0.002	0.075	37	22
	V70	F	1.024	0.906	0.001	0.120	74	22
		R	0.378	0.330	0.002	0.075	37	22

NA: Not Available
F: Front
R: Rear

93471CS6

WHEEL ALIGNMENT

Year	Model		Caster Range (+/-Deg.)	Caster Preferred Setting (Deg.)	Camber Range (+/-Deg.)	Camber Preferred Setting (Deg.)	Toe-in (in.)	Steering Axis Inclination (Deg.)
1998	S70, V70	F	1.00	+3.35	1.00	0	0.33+/-0.08	—
		R	—	—	0.50	-1.00	0.10+/-0.20	—
	C70	F	1.00	+3.35	1.00	-0.50	0.33+/-0.08	—
		R	—	—	0.50	-1.00	0.10+/-0.20	—
1999	S70, V70	F	1.00	+3.35	1.00	0	0.33+/-0.08	—
		R	—	—	0.50	-1.00	0.10+/-0.20	—
	C70	F	1.00	+3.35	1.00	-0.50	0.33+/-0.08	—
		R	—	—	0.50	-1.00	0.10+/-0.20	—
	S80	F	1.00	+4.00	0.50	+0.10	0.30+/-0.10	—
		R	—	—	0.75	-0.20	0.10+/-0.20	—
2000	S40, V40	F	1.00	+3.20	1.00	0	0.15+/-0.05	—
		R	—	—	0.50	-0.67	—	—
	S70, V70	F	1.00	+3.35	1.00	0	0.33+/-0.08	—
		R	—	—	0.50	-1.00	0.10+/-0.20	—
	C70	F	1.00	+3.35	1.00	-0.50	0.33+/-0.08	—
		R	—	—	0.50	-1.00	0.10+/-0.20	—
	S80	F	1.00	+4.00	0.50	+0.10	0.30+/-0.10	—
		R	—	—	0.75	-0.20	0.10+/-0.20	—

93471CS7

For Accessory Drive Belt illustrations, see Section 1 of this manual

TIRE, WHEEL AND BALL JOINT SPECIFICATIONS

| Year | Model | OEM Tires | | Tire Pressures (psi) | | Wheel | Ball Joint |
		Standard	Optional	Front	Rear	Size	Inspection
1998	S70	185/65HR15	195/60VR15	Std: 36	Std: 36	6.5-J	0.12 in.
			205/55ZR15	VR: 38	VR: 41		
				ZR 42	ZR: 42		
	V70	195/65VR15	205/55WR15	Std: 36	Std: 41	Exc. ZR17: 6.5-J	0.12 in.
			205/55VR15	WR: 41	WR: 46	ZR17: 7-J	
			205/50ZR16	VR: 35	VR: 41		
			205/45ZR17	ZR16: 38	ZR16: 45		
				ZR17: 48	ZR17: 48		
1999	S70	185/65HR15	195/60VR15	Std: 36	Std: 36	6.5-J	0.12 in.
			205/55ZR15	VR: 38	VR: 41		
				ZR 42	ZR: 42		
	V70	195/65VR15	205/55WR15	Std: 36	Std: 41	Exc. ZR17: 6.5-J	0.12 in.
			205/55VR15	WR: 41	WR: 46	ZR17: 7-J	
			205/50ZR16	VR: 35	VR: 41		
			205/45ZR17	ZR16: 38	ZR16: 45		
				ZR17: 48	ZR17: 48		
2000	C70	225/50R16	None	33	30	7J	0.12 in.
	S70	185/65HR15	195/60VR15	Std: 36	Std: 36	6.5-J	0.12 in.
			205/55ZR15	VR: 38	VR: 41		
				ZR 42	ZR: 42		
	S80	215/55R15	225/50R17	29	29	7J	0.12 in.
	V70	195/65VR15	205/55WR15	Std: 36	Std: 41	Exc. ZR17: 6.5-J	0.12 in.
			205/55VR15	WR: 41	WR: 46	ZR17: 7-J	
			205/50ZR16	VR: 35	VR: 41		
			205/45ZR17	ZR16: 38	ZR16: 45		
				ZR17: 48	ZR17: 48		

OEM: Original Equipment Manufacturer

PSI: Pounds Per Square Inch

STD: Standard

OPT: Optional

93471CS8

SCHEDULED MAINTENANCE INTERVALS
(VOLVO 800 SERIES, 900 SERIES, C70, S90, V70 & V90)

TO BE SERVICED	TYPE OF SERVICE	VEHICLE MILEAGE INTERVAL (x1000)												
		5	10	20	30	40	50	60	70	80	90	100	110	120
Engine oil & filter ①	R	✓	✓	✓	✓	✓	✓	✓	✓	✓	✓	✓	✓	✓
Automatic transmission fluid	S/I	✓	✓	✓	✓	✓	✓	✓	✓	✓	✓	✓	✓	✓
Fluid levels (all)	S/I	✓	✓	✓	✓	✓	✓	✓	✓	✓	✓	✓	✓	✓
Rotate tires	S/I		✓	✓	✓	✓	✓	✓	✓	✓	✓	✓	✓	✓
Automatic transmission shift control	S/I		✓	✓	✓	✓	✓	✓	✓	✓	✓	✓	✓	✓
Brake pads & parking brake	S/I		✓	✓	✓	✓	✓	✓	✓	✓	✓	✓	✓	✓
Engine & transmission (check for leaks)	S/I		✓	✓	✓	✓	✓	✓	✓	✓	✓	✓	✓	✓
Reset service reminder ①	R		✓	✓	✓	✓	✓	✓	✓	✓	✓	✓	✓	✓
Driveshaft, U-joints	S/I				✓	✓	✓	✓	✓	✓	✓	✓	✓	✓
Clutch	S/I			✓		✓		✓		✓		✓		✓
Brake & fuel lines & hoses	S/I				✓			✓			✓			✓
Steering & suspension	S/I				✓			✓			✓			✓
Air cleaner filter	R					✓		✓			✓			✓
Spark plugs	R					✓		✓			✓			✓
Drive belt	S/I					✓		✓			✓			✓
Fuel line filter	R											✓		
EGR system	S/I							✓				✓		
PCV nipple orifice/hoses	S/I							✓				✓		
Check suspension	S/I		✓	✓	✓	✓	✓	✓	✓	✓	✓	✓	✓	✓
Brake fluid ②	R													

R: Replace   S/I: Service or Inspect

① Perform operation every 5000 miles on turbocharged models.

② Replace every 2 years or 30,000 miles, whichever comes first under normal conditions, more frequently in mountainous areas or moist climates.

FREQUENT OPERATION MAINTENANCE (SEVERE SERVICE)

If a vehicle is operated under any of the following conditions it is considered severe service:

- Extremely dusty areas.
- 50% or more of the vehicle operation is in 32°C (90°F) or higher temperatures, or constant operation in temperatures below 0°C (32°F).
- Prolonged idling (vehicle operation in stop and go traffic).
- Frequent short running periods (engine does not warm to normal operating temperatures).
- Police, taxi, delivery usage or trailer towing usage.

Oil & oil filter: change every 5000 miles.

Air filter element: service or inspect every 15,000 miles.

93471CS9

For Tire, Wheel and Ball Joint specifications, see Section 1 of this manual

SCHEDULED MAINTENANCE INTERVALS
VOLVO
800 SERIES, 900 SERIES, C70, S90, V70, V90

The following should be used as a guide when determining the amount of work required for a particular service. In estimating how long a particular Scheduled Maintenance Service should take, please observe the following:

- Labor Time is time based on field research and data supplied by the vehicle manufacturer.
- Labor time operations are given in hours and tenths of an hour.
- All labor operations are to be used as a guide.

Mechanic Skill Level Codes:
(A) PRECISION: Highly skilled with multiple certification.
(B) GENERAL: Normally skilled with certification.
(C) MAINTENANCE: Semi-skilled working on certification.

	LABOR TIME
5000 Mile Service (C)	
All Models	.5
10000 Mile Service (C)	
All Models	1.7
20000 Mile Service (C)	
S90, V90	1.8
C70, S70, V70	1.8
850	1.8
940	1.9
960	1.8
30000 Mile Service (B)	
S90, V90	2.8
C70, S70, V70	3.5
850	3.5
940	2.9
960	2.8
40000 Mile Service (B)	
S90, V90	2.1
C70, S70, V70	2.3
850	2.3
940	2.2
960	2.1

	LABOR TIME
50000 Mile Service (B)	
C70, S70, V70	1.8
S90, V90	3.8
850	1.8
940	1.8
960	3.8
Replace timing belt add	2.0
60000 Mile Service (B)	
S90, V90	4.2
C70, S70, V70	4.8
850	4.8
940	3.8
960	4.2
Inspect timing belt add	.2
70000 Mile Service (B)	
S90, V90	3.8
C70, S70, V70	4.1
850	4.1
940	1.8
960	3.8
80000 Mile Service (B)	
S90, V90	2.1
C70, S70, V70	2.4
850	2.4
940	2.2
960	2.1

	LABOR TIME
90000 Mile Service (B)	
S90, V90	3.2
C70, S70, V70	3.9
850	3.9
940	3.1
960	3.2
100000 Mile Service (B)	
S90, V90	2.3
C70, S70, V70	2.6
850	2.6
940	2.4
960	2.3
Replace timing belt add	2.0
110000 Mile Service (C)	
S90, V90	1.8
C70, S70, V70	2.1
850	2.1
940	1.8
960	1.8
Inspect timing belt add	.2
120000 Mile Service (B)	
S90, V90	3.8
C70, S70, V70	4.4
850	4.4
940	3.7
960	3.8

93471CT1

1998–01 ACURA

Integra

REMOVAL & INSTALLATION

➡**Be sure to acquire the anti-theft code for the radio; then, write down the frequencies for the preset buttons.**

1. Disconnect the negative battery cable.

✷✷ CAUTION

After disconnecting the negative battery cable, wait for at least 3 minutes for the SRS module to deplete its energy.

2. Drain the cooling system into a clean container for reuse.
3. Remove or disconnect the following:
   - Heater hoses from the heater core
   - Heater housing-to-cowl nut, located in the engine compartment
4. Remove the instrument panel by removing or disconnecting the following:
   - Front seats
   - Front and rear consoles
   - Lower dashboard-to-instrument panel screws and the lower dashboard, located on the driver's side
   - Knee bolster
   - Glove box
   - 4 glove box frame-to-instrument panel bolts and the frame
   - Clock
   - Moon roof switch
   - Stereo radio/cassette
   - Lower steering column cover clamps and the cover
   - Wrap a shop towel around the steering column to prevent damage to the column.
   - Steering column-to-instrument panel nuts/bolts and lower the steering column
   - SRS module-to-instrument panel nuts (located on the driver's side); then, carefully, remove the SRS module and disconnect the electrical connector
   - Air mix control cable and electrical connectors, located at the center of the instrument panel
   - Antenna lead
   - Electrical connectors, located at the under-dash fuse/relay box
   - Pry out the access panels, from both sides of the instrument panel
   - Instrument panel-to-chassis bolts and remove the instrument panel
   - Wiring harness clips; then, remove the 4 heater housing-to-blower motor duct screws and the duct, (models not equipped with air conditioning only)
5. If equipped with air conditioning,

▲ : **Bolt, nut locations**

A▲ : **Bolt, 2**

8 x 1.25 mm
22 N·m (2.2 kgf·m, 16 lbf·ft)

B▲ : **Nut, 2**

8 x 1.25 mm
13 N·m (1.3 kgf·m, 9 lbf·ft)
Replace.

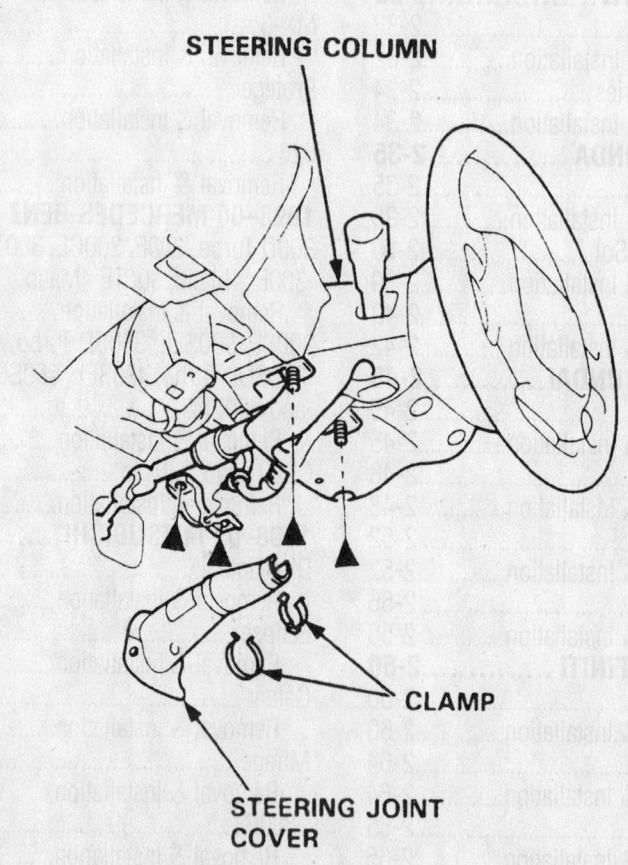

STEERING COLUMN

CLAMP

STEERING JOINT COVER

93112GP5

View of the steering column and related components—Integra

remove the evaporator housing removing or disconnecting the following:

- Discharge and recover the air conditioning system refrigerant.
- Refrigerant lines from the evaporator core, discard the O-rings and plug the openings to prevent contamination
- Cut the insulation pad (at the firewall) at the lower evaporator housing-to-chassis location
- Thermostat electrical connector, located at the evaporator housing
- Wiring harness clips from the evaporator housing
- Evaporator housing-to-chassis screws, the nut and bolt
- Drain hose
- Evaporator housing, carefully
- 2 SRS support beam bolts, nut and the SRS beam, located at the passenger's side
- Mode control motor connector and the wiring harness clip from the heater housing
- Heater housing-to-chassis nuts and the heater housing

6. Remove the damper arm cover-to-heater housing screw and the cover.

7. Remove or disconnect the following:

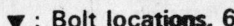
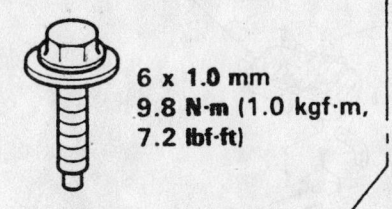

▼ : Bolt locations, 6

6 x 1.0 mm
9.8 N·m (1.0 kgf·m, 7.2 lbf·ft)

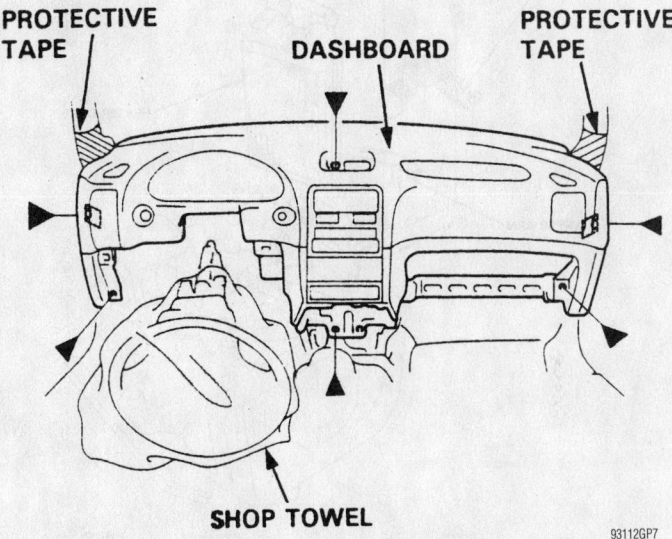

PROTECTIVE TAPE   DASHBOARD   PROTECTIVE TAPE

SHOP TOWEL

93112GP7

View of the instrument panel fastener locations—Integra

▲ : Nut locations, 4

6 x 1.0 mm
9.8 N·m
(1.0 kgf·m,
7.2 lbf·ft)

FRONT PASSENGER'S AIRBAG

FRONT PASSENGER'S AIRBAG BRACKET

SRS MAIN HARNESS

93112GP6

View of the passenger's side SRS module and related components—Integra

- Damper arm link; then, remove the damper arm-to-heater housing screw and the arm.
- 2 heater core cover-to-heater housing screws and the cover.
- Pipe clamp screw and the clamp.
- Heater core from the heater housing.

To install:

8. Install or connect the following:

- Heater core to the heater housing
- Pipe clamp screw and the clamp
- Heater core cover and the 2 cover-to-heater housing screws
- Damper arm-to-heater housing screw and the arm; then, connect the damper arm link
- Damper arm cover and the cover-to-heater housing screw
- Heater housing and the heater housing-to-chassis nuts
- Mode control motor connector and the wiring harness clip to the heater housing.
- SRS support beam, nut and the 2 SRS beam bolts on the passengers side

9. If equipped with air conditioning,

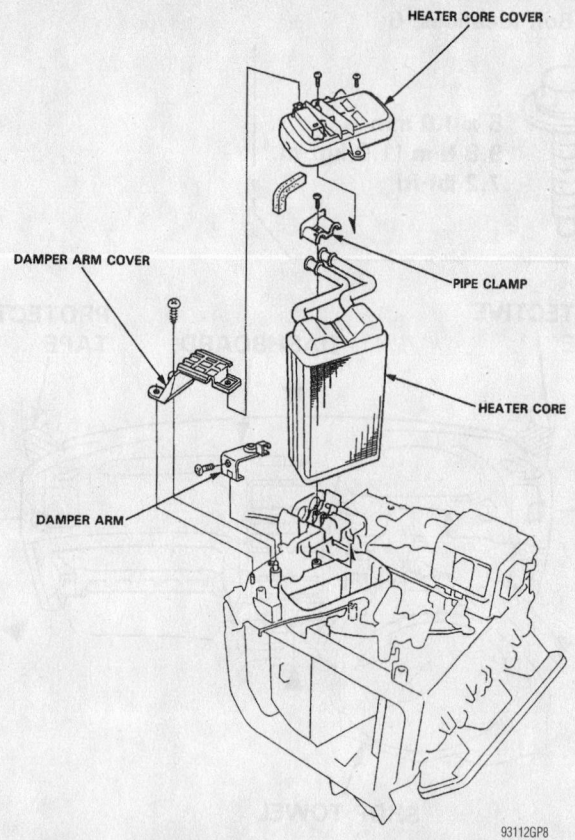

HEATER CORE COVER

DAMPER ARM COVER

PIPE CLAMP

HEATER CORE

DAMPER ARM

93112GP8

Exploded view of the heater core, heater housing and related components—Integra

install the evaporator housing by installing or connecting the following:
- Evaporator housing
- Drain hose
- 4 evaporator housing-to-chassis screws, the nut and bolt
- Wiring harness clips to the evaporator housing
- Thermostat electrical connector at the evaporator housing
- Refrigerant lines to the evaporator core making sure to use new O-rings
- Heater housing-to-blower motor duct and the 4 duct screws; then, connect the wiring harness clips (models not equipped with air conditioning only)

10. Install the instrument panel by installing or connecting the following:
- instrument panel and the instrument panel-to-chassis bolts
- access panels on both sides of the instrument panel
- Electrical connectors at the under-dash fuse/relay box
- Antenna lead.
- Air mix control cable and electrical connectors at the center of the instrument panel

- SRS module electrical connector (located on the passenger's side) and torque the SRS module-to-instrument panel nuts to 84 inch lbs. (9.8 Nm)
- Steering column and torque the steering column-to-instrument panel nuts to 108 ft. lbs. (13 Nm) and the bolts to 16 ft. lbs. (22 Nm)
- Lower steering column cover and the cover clamps
- Stereo radio/cassette
- Moon roof switch
- Clock
- Glove box frame and the 4 frame-to-instrument panel bolts
- Glove box
- Knee bolster
- Lower dashboard and the lower dashboard-to-instrument panel screws on the drivers side
- Front and rear consoles
- Front seats
- Heater housing-to-cowl nut (located in the engine compartment) and torque to 16 ft. lbs. (22 Nm)
- Heater hoses to the heater core.

11. Refill the cooling system.

12. Connect the negative battery cable.

13. Evacuate, charge and leak test the air conditioning system refrigerant.

14. Operate the engine to normal operating temperatures; then, check the climate control operation and check for leaks.

NSX

REMOVAL & INSTALLATION

✳✳ CAUTION

Before starting service procedures on components, especially under the instrument panel and/or near the steering column, disable the SRS system. In addition, the vehicle may be equipped with a radio anti-theft code (5 digits). Get this code from the customer before disconnecting the battery.

1. If equipped with SRS or an anti-theft radio, perform the following procedures:
- Disconnect the negative battery cable, then disconnect the positive battery cable.
- Remove the access panel beneath the air bag on the under-side of the steering wheel center housing.
- Disconnect the connector between the air bag and the cable reel and install the short connector (red) that is provided. Install this connector on the air bag side of the connector.
- The system is now disabled. When repairs are complete, unplug the red short connector, reconnect the air bag and cable reel connector and replace the access panel.
- Reconnect the battery. When the word "CODE" appears, re-enter the 5-digit code in the radio, if equipped.
- When the ignition is turned to the **II** position, the SRS light on the instrument cluster should come ON for about 6 seconds, then go out. If so, the system is okay.

✳✳ WARNING

Do not use electrically powered test equipment on this system or related circuits. All SRS wiring is covered with a special yellow cable housing for identification.

All SRS wiring harnesses are covered with yellow outer insulation. Before disconnecting any part of the SRS wire harness, install short connection to prevent accidental discharge of the air bag system. If the SRS harness assembly should have an open circuit or damaged wiring, replace the entire affected harness.

2. Disconnect the negative battery cable.
3. Properly drain the cooling system. Properly discharge the air conditioning system using an approved refrigerant recover/recycling system.
4. Remove or disconnect the following:
   • Blower assembly
   • Refrigerant lines from their connection at the firewall. Cap all opening immediately to keep dirt and moisture from contaminating the system

5. Remove the dashboard by removing or disconnecting the following:
   • Front seats
   • Knee bolster, pad, dashboard stay and center armrest
   • Clock, center air vent and center console panel
   • Heater/air conditioning control panel and the radio assembly
   • Right dashboard lower panel. Remove the glove box

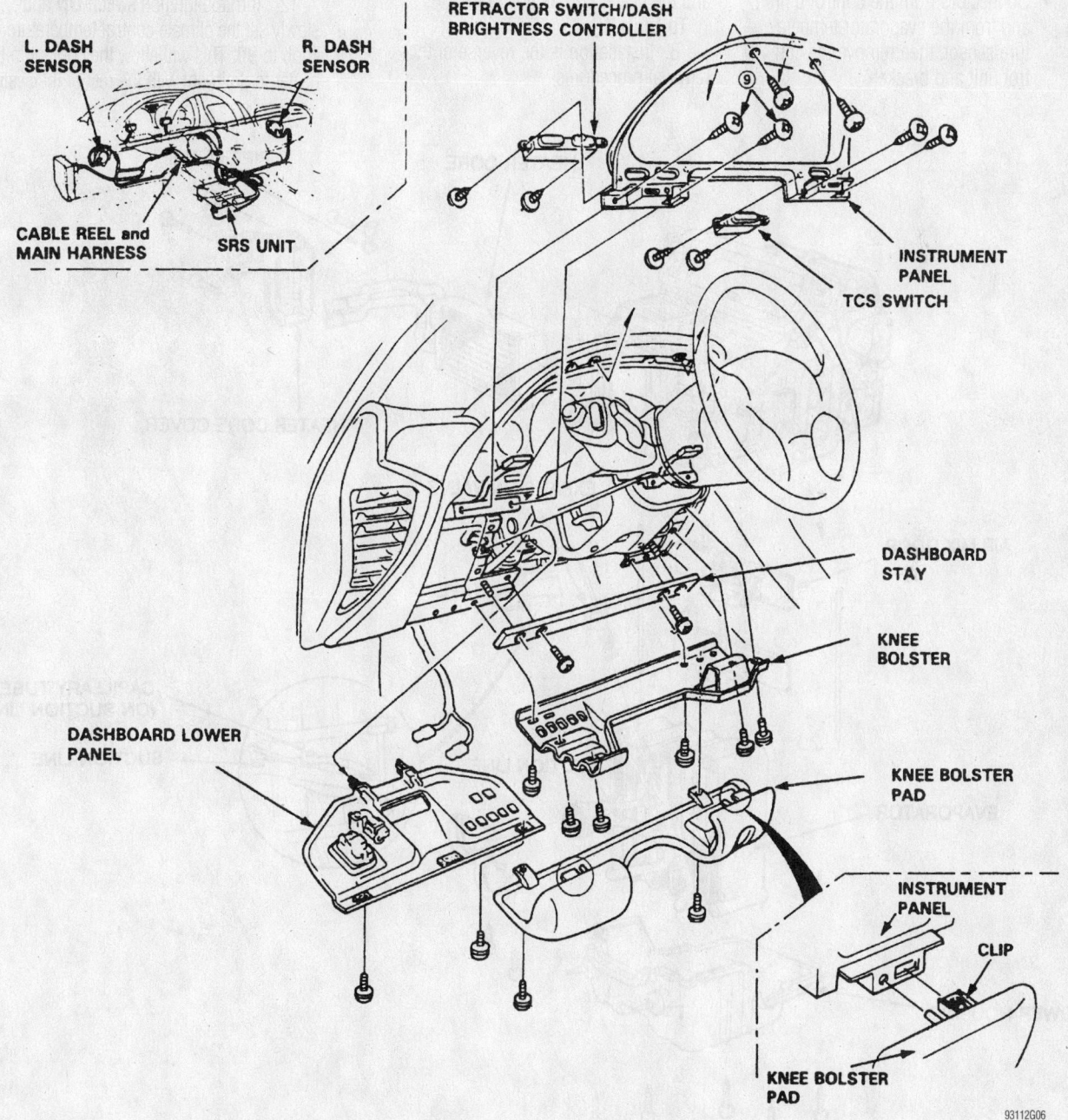

Exploded view of the dashboard and related components—NSX

93112G06

- Lower the steering column
- Instrument cluster bezel and the instrument cluster
- Air vents from both ends of the dashboard
- U-plate at the foot of the center console, rearward of the parking brake
- Defrost outlet grille, remove the mounting bolts and lift out the dashboard. Note the position of the center guide pin for reinstallation

6. Remove or disconnect the following:
- Heater duct
- Connectors from the control u nit and from the evaporator temperature sensor, then remove the control unit and bracket

- Sound system speaker
- Connectors from the control unit and from the evaporator temperature sensor, then remove the control unit and bracket
- Sound system speaker
- Actuator connectors and sensor connectors from the heater-evaporator unit
- 2 under-dash mounting bolts and remove the heater-evaporator unit through the passenger's door

7. The unit can now be disassembled and the heater core removed.

To install:
8. Installation is the reverse of the removal procedures.

9. Add 0.3 oz. of extra refrigerant oil if the evaporator core was replaced.

10. Adjust the heater valve cable, if needed.

11. Refill the cooling system and bleed the system by performing the following procedure:

❉❉ WARNING

Failure to properly bleed the air from the cooling system could cause engine damage.

12. Turn the ignition switch **ON** and slowly set the climate control temperature knob to **90**. This will allow the coolant in the heater to drain out with the rest of the system.

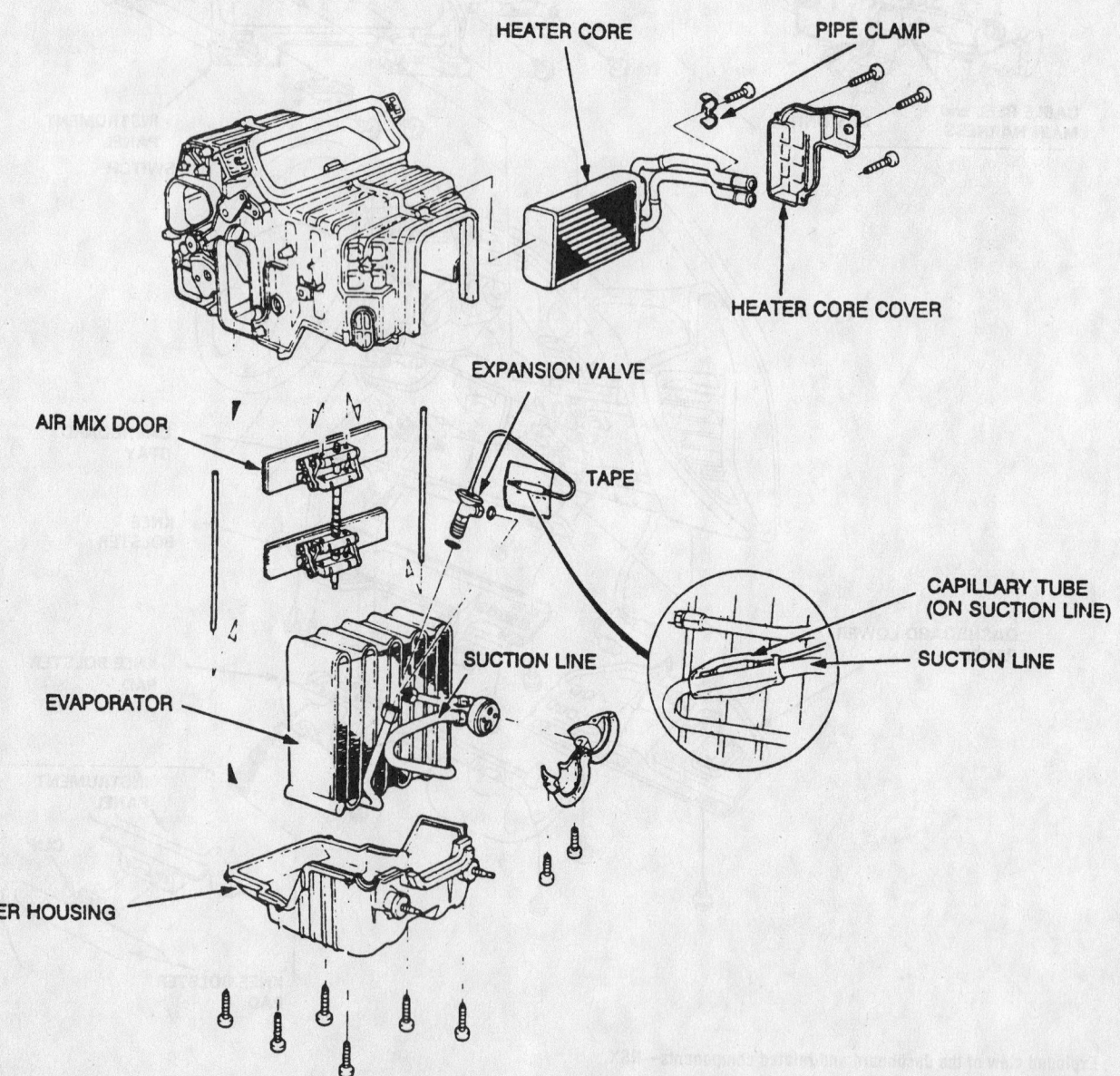

Exploded view of the heater core/evaporator assembly—NSX

93112G07

13. Open the hood, rear hatch and the engine cover.

14. Remove the cover protecting the water pipes and shift cables on the underside of the car.

15. Carefully loosen the coolant reserve tank cap. Loosen the drain plug at the bottom of the radiator. Remove the 2 drain bolts from the water pipes. Install rubber hoses to the drain bolts at the front and rear of the engine under the cylinder bank and loosen the drain bolts to drain the coolant.

16. Coolant will drain more quickly if all the air bleed bolts, plug and cap are opened. Be sure the coolant reserve tank has drained completely before opening the air bleed bolts.

17. Using new washers on the water pipe drain bolts, install the drain bolts and radiator drain plug.

18. Open all 4 air bleeder bolts (radiator, heater pipe, water pipe and engine thermostat cover).

19. Using approved coolant in a 50/50 mixture, fill the coolant reserve tank. Tighten the bleeders in sequence; thermostat cover bleed bolt, radiator bleed plug, heater pipe bleed cap, and water pipe bleed bolt as coolant runs out in a steady stream with no bubbles.

20. After tightening all the bleed bolts, fill the coolant tank to the MAX line. Loosen the thermostat bleed bolt to remove any remaining air.

21. When bleeding is complete, tighten the thermostat bolt and fill the coolant reserve tank to the MAX line again. Install the tank cap to the 1st detent.

22. Start the engine and run it to normal operating temperatures (thermostat opens and radiator cooling fan runs).

23. Turn the engine **OFF**, check and adjust the coolant to the MAX line if needed.

24. Install the coolant tank cap securely and install the car's undercover.

25. Evacuate, charge and leak test the air conditioning system.

26. Connect the negative battery cable.

2.3CL and 3.0CL

REMOVAL & INSTALLATION

➡**Be sure to acquire the anti-theft code for the radio; then, write down the frequencies for the preset buttons.**

1. Disconnect the negative battery cable.

✳✳ CAUTION

After disconnecting the negative battery cable, wait for at least 3 minutes for the SRS module to deplete its energy.

2. Drain the cooling system into a clean container for reuse.

3. Remove or disconnect the following:
- Heater hoses from the heater core
- 2 heater housing-to-cowl nuts, located in the engine compartment

4. Place the front wheel in the straight-ahead position.

5. Remove the SRS module and the steering wheel by removing or disconnecting the following:
- Access panel-to-steering wheel screws and the panel
- SRS module electrical connector
- SRS module-to-steering wheel covers from both sides of the steering wheel
- Both SRS module-to-steering wheel bolts, using a T30 Torx bit
- SRS module from the steering wheel
- Electrical connectors from the steering wheel
- Steering wheel from the steering column

6. Remove the steering column by removing or disconnecting the following:
- Coin pocket, the lower instrument panel-to-dash screw and the lower instrument panel from the driver's side
- Electrical connector and the air hose at the knee bolster; then, remove the knee bolster-to-dash bolts and the knee bolster
- Steering column cover screws and the covers
- Combination switch-to-steering column screws, disconnect the electrical connectors and remove the combination switch
- Ignition switch connectors
- Clamps, clips and the steering joint cover
- Steering joint from the steering column shaft
- Steering column-to-instrument panel nuts/bolts and the steering column

7. Remove the passenger's side SRS module by removing or disconnecting the following:

- SRS module electrical connector
- 5 SRS module-to-instrument panel nuts and carefully remove the SRS module

8. Remove the instrument panel by removing or disconnecting the following:
- Front and rear consoles
- Cruise control master switch and the panel brightness controller; then, disconnect the electrical connectors
- Instrument cluster-to-instrument panel screws and pry out the instrument cluster and disconnect the electrical connectors
- Glove box-to-damper screw
- 2 glove box-to-dash screws and the glove box
- Audio unit-to-instrument panel fasteners and the audio unit
- 3 passenger's dashboard panel screws (located on the passengers side); then, carefully pry out the dashboard panel and the climate control unit as an assembly
- Side defogger trim from both sides of the instrument panel
- Instrument panel electrical connectors, the recirculation control motor connector and the resistor connector
- Instrument panel-to-chassis bolts and the instrument panel
- Steering hanger beam-to-chassis nuts, bolts and the beam

9. Discharge and recover the air conditioning system refrigerant.

10. Remove the evaporator housing by removing or disconnecting the following:
- Refrigerant lines from the evaporator core. Discard the O-rings and plug the openings to prevent contamination.
- Thermostat electrical connector at the evaporator housing
- Wiring harness clips from the evaporator housing
- Evaporator housing-to-chassis screws, the nut and bolt
- Drain hose
- Evaporator housing

11. Remove or disconnect the following:
- Wiring harness clip; then, disconnect the mode control motor and the air mix control motor connectors
- Heater housing-to-chassis bolt and the heater housing
- Vent/defroster duct-to-heater housing screws and the duct

Timing belt service is covered in Section 3 of this manual

▶: Bolt, screw locations
A ▶, 2

B ▶, 8   C ▶, 1

▷: Clip locations
A ▷, 8   B ▷, 2   C ▷, 4

6 x 1.0mm
9.8 N·m, (1.0 kgf·m,
7.2 lbf·ft)

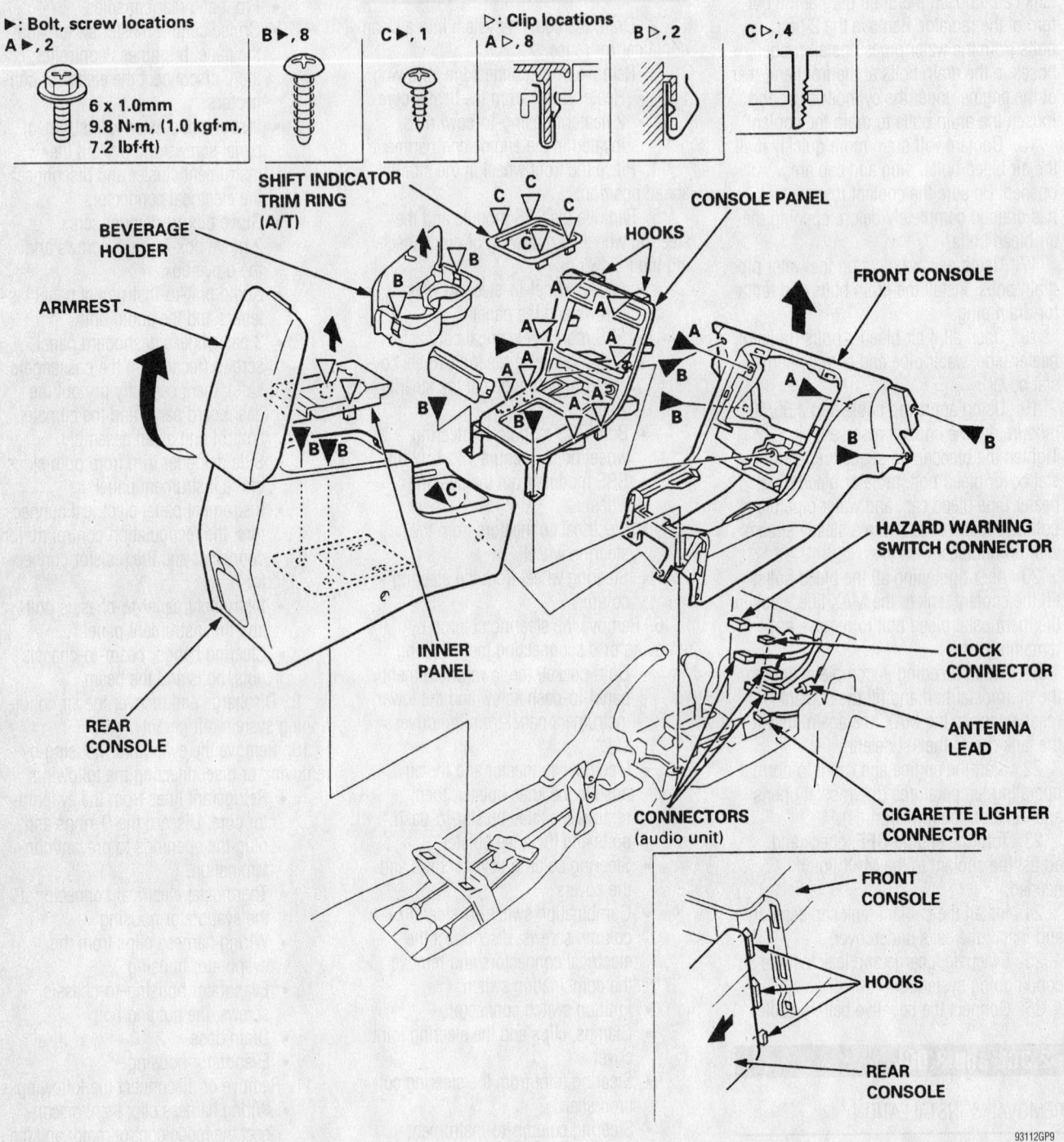

SHIFT INDICATOR
TRIM RING
(A/T)

CONSOLE PANEL

HOOKS

BEVERAGE
HOLDER

FRONT CONSOLE

ARMREST

HAZARD WARNING
SWITCH CONNECTOR

CLOCK
CONNECTOR

ANTENNA
LEAD

INNER
PANEL

REAR
CONSOLE

CONNECTORS
(audio unit)

CIGARETTE LIGHTER
CONNECTOR

FRONT
CONSOLE

HOOKS

REAR
CONSOLE

93112GP9

View of the front and rear consoles and related components—2.3CL and 3.0CL

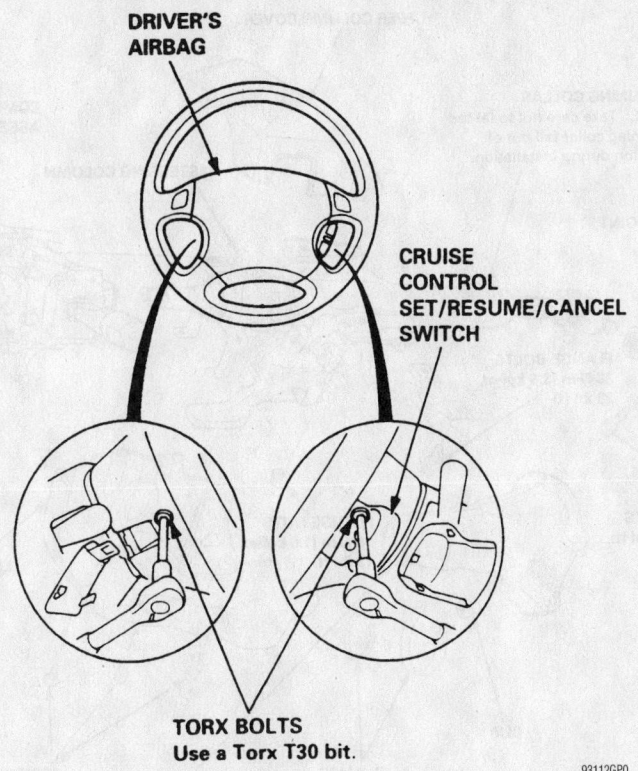

DRIVER'S
AIRBAG

CRUISE
CONTROL
SET/RESUME/CANCEL
SWITCH

TORX BOLTS
Use a Torx T30 bit.

93112GP0

Removing the driver's side SRS module-to-steering wheel bolts—2.3CL and 3.0CL

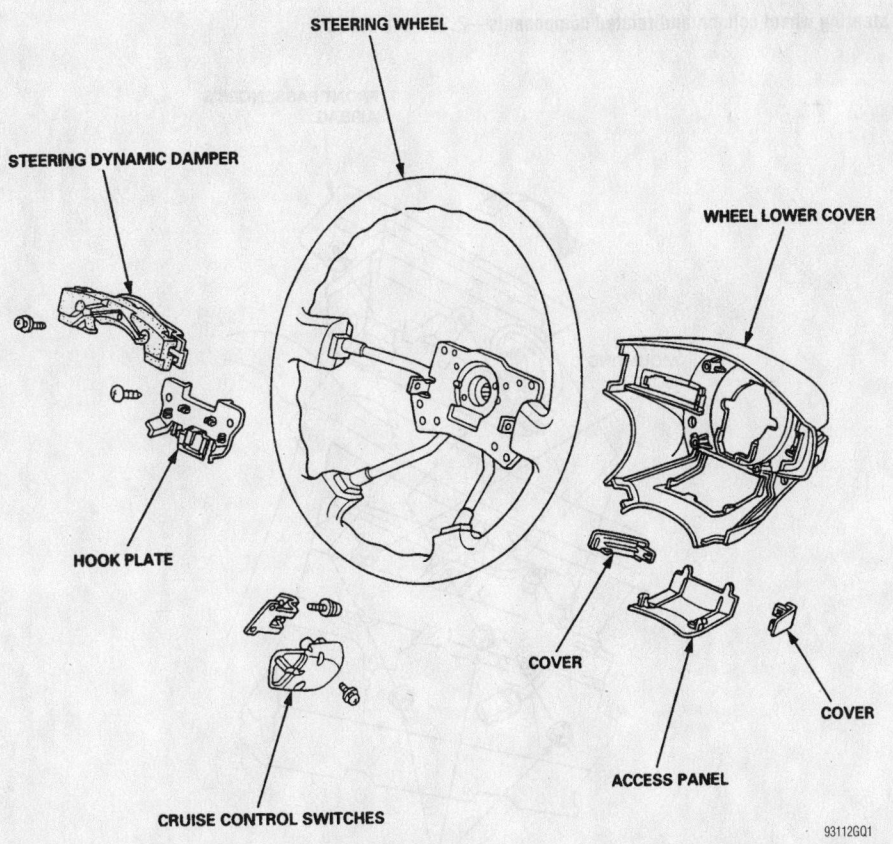

STEERING WHEEL

STEERING DYNAMIC DAMPER

WHEEL LOWER COVER

HOOK PLATE

COVER

COVER

ACCESS PANEL

CRUISE CONTROL SWITCHES

93112GQ1

Exploded view of the steering wheel and related components—2.3CL and 3.0CL

Heater Core replacement is covered in Section 2 of this manual

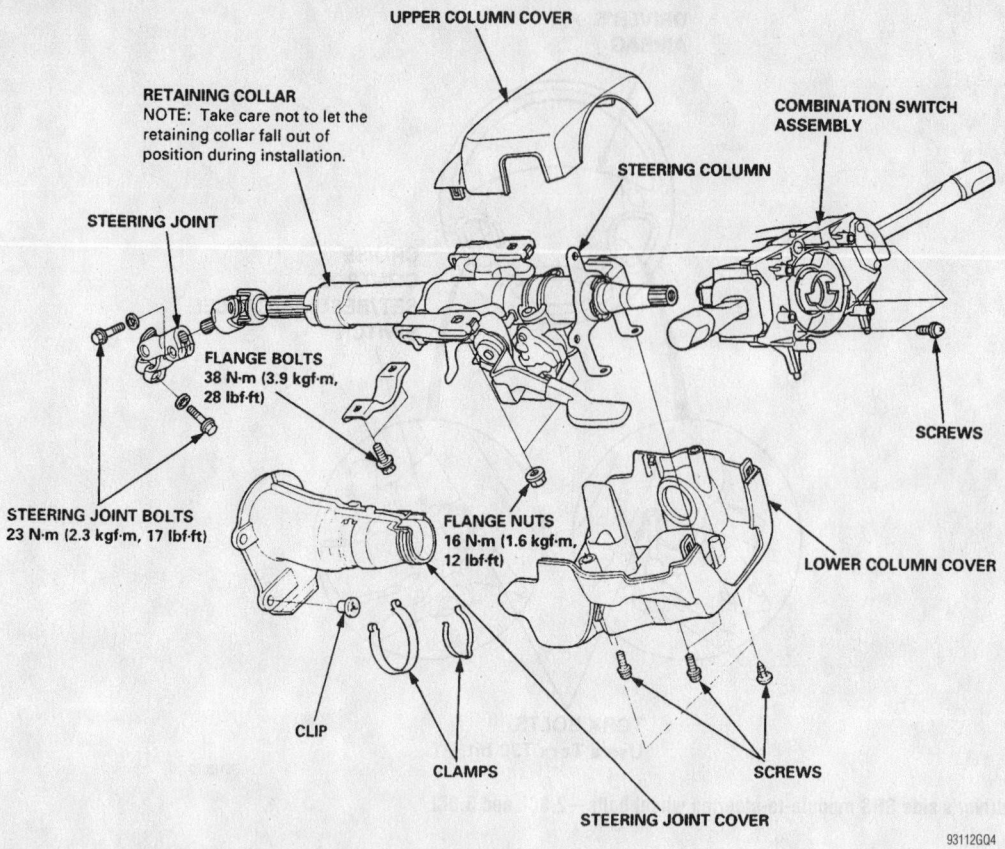

UPPER COLUMN COVER

RETAINING COLLAR
NOTE: Take care not to let the retaining collar fall out of position during installation.

COMBINATION SWITCH ASSEMBLY

STEERING COLUMN

STEERING JOINT

FLANGE BOLTS
38 N·m (3.9 kgf·m, 28 lbf·ft)

SCREWS

STEERING JOINT BOLTS
23 N·m (2.3 kgf·m, 17 lbf·ft)

FLANGE NUTS
16 N·m (1.6 kgf·m, 12 lbf·ft)

LOWER COLUMN COVER

CLIP

CLAMPS

SCREWS

STEERING JOINT COVER

93112GQ4

Exploded view of the steering wheel column and related components—2.3CL and 3.0CL

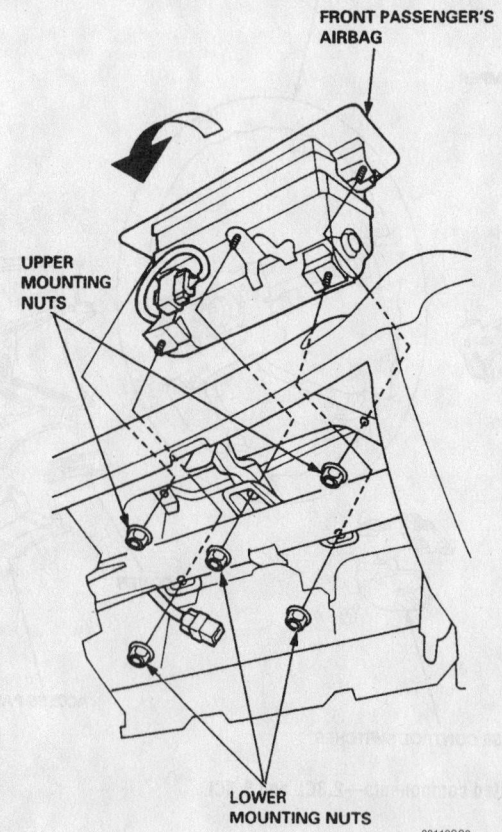

FRONT PASSENGER'S AIRBAG

UPPER MOUNTING NUTS

LOWER MOUNTING NUTS

93112GQ2

Exploded view of the passenger's side SRS module—2.3CL and 3.0CL

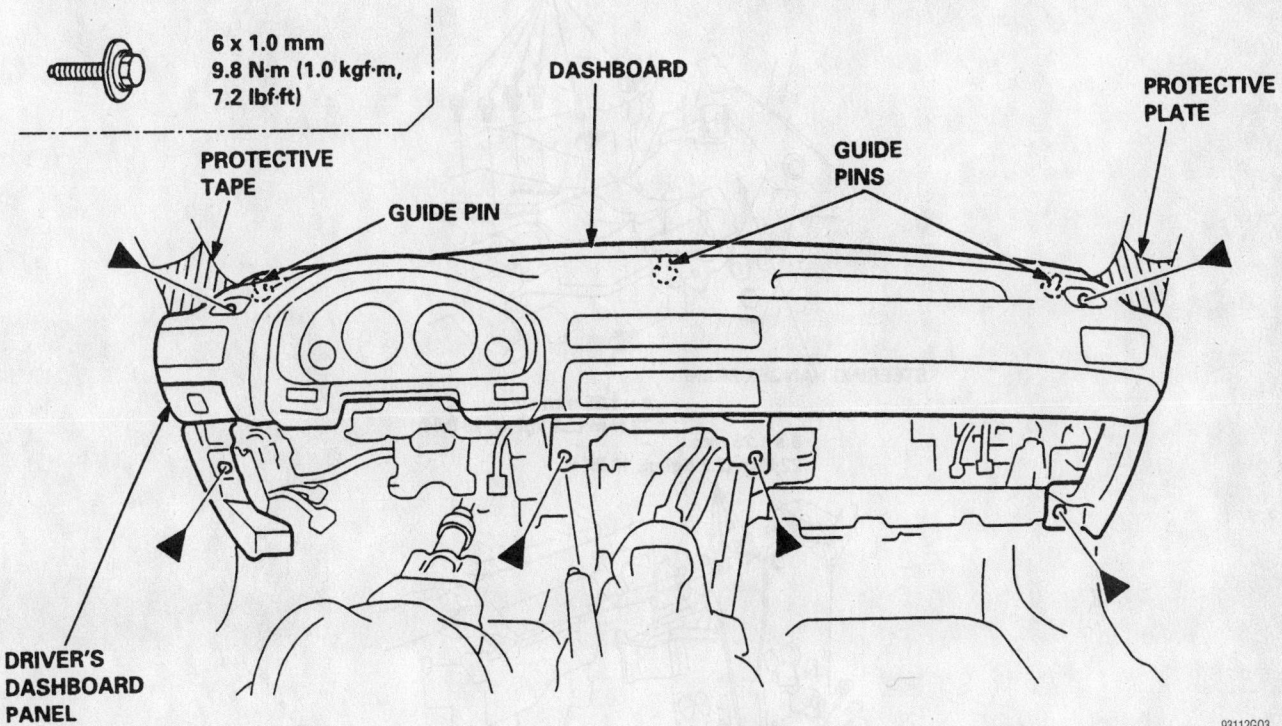

►: Bolt locations, 6

6 x 1.0 mm
9.8 N·m (1.0 kgf·m, 7.2 lbf·ft)

PROTECTIVE TAPE

GUIDE PIN

DASHBOARD

GUIDE PINS

PROTECTIVE PLATE

DRIVER'S DASHBOARD PANEL

93112GQ3

View of the dashboard bolt locations—2.3CL and 3.0CL

- Heater core pipe clamp-to-heater housing screw and the clamp
- Heater core clamp-to-heater housing screw and the heater core

To install:

12. Install or connect the following:
- Heater core clamp and the heater core-to-heater housing screw
- Heater core pipe clamp and the clamp-to-heater housing screw
- Vent/defroster duct and the duct-to-heater housing screws
- Heater housing and the heater housing-to-chassis bolt
- Wiring harness clip; then, connect the mode control motor and the air mix control motor connectors

13. Install the evaporator housing by installing or connecting the following:
- Evaporator housing
- Drain hose
- Evaporator housing-to-chassis screws, the nut and bolt
- Wiring harness clips to the evaporator housing
- Thermostat electrical connector, at the evaporator housing

- Refrigerant lines to the evaporator core, make sure to use new O-rings
- Steering hanger beam and the beam-to-chassis nuts and bolts

14. Install the instrument panel by installing or connecting the following:
- Instrument panel and the instrument panel-to-chassis bolts
- Instrument panel electrical connectors, the recirculation control motor connector and the resistor connector
- Side defogger trim at both sides of the instrument panel
- Dashboard panel and the climate control unit as an assembly; then, install the 3 passenger's dashboard panel screws at the passenger's side
- Audio unit and the audio unit-to-instrument panel fasteners
- Glove box and the 2 glove box-to-dash screws
- Glove box-to-damper screw
- Instrument cluster and the instrument cluster-to-instrument panel screws; then, connect the electrical connectors

- Cruise control master switch and the panel brightness controller; then, connect the electrical connectors
- Front and rear consoles

15. Install the passenger's side SRS module by installing or connecting the following:
- SRS module and the 5 SRS module-to-instrument panel nuts and torque to 84 inch lbs. (9.8 Nm)
- SRS module electrical connector

16. Install the steering column by installing or connecting the following:
- Steering column and the steering column-to-instrument panel nuts/bolts
- Steering joint to the steering column shaft
- Clamps, clips and the steering joint cover
- Ignition switch connectors
- Combination switch, connect the electrical connectors and install the combination switch-to-steering column screws
- Steering column covers and the cover screws.

Brake service is covered in Section 4 of this manual

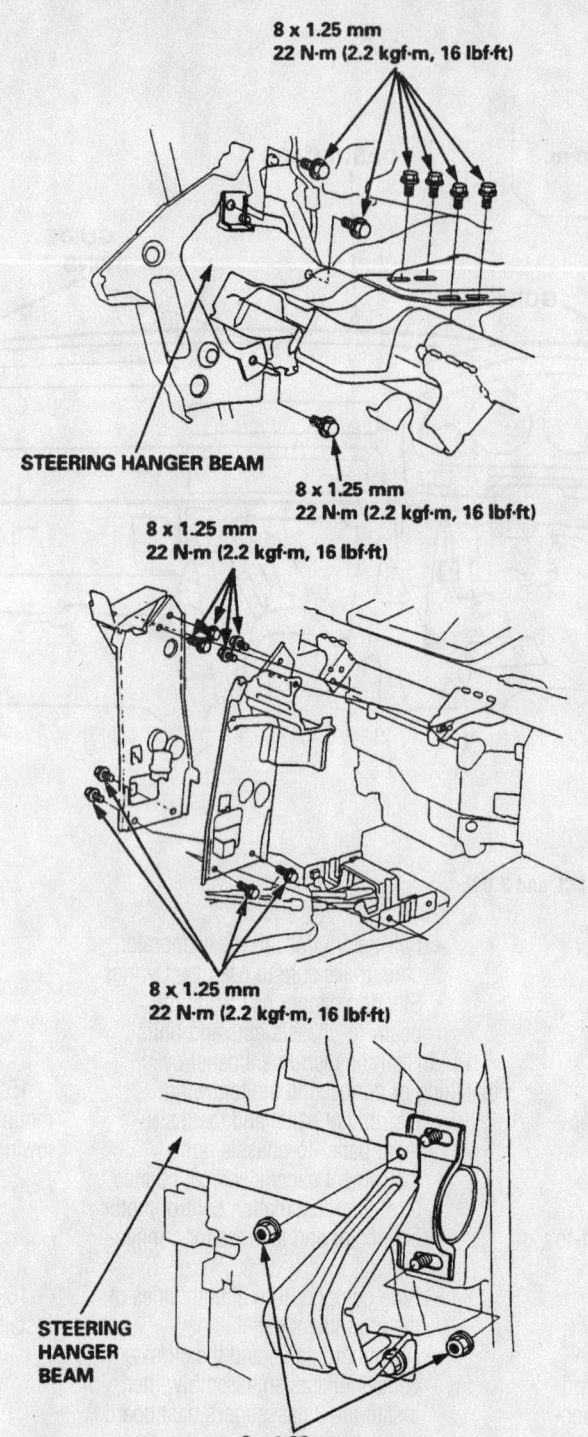

8 x 1.25 mm
22 N·m (2.2 kgf·m, 16 lbf·ft)

STEERING HANGER BEAM

8 x 1.25 mm
22 N·m (2.2 kgf·m, 16 lbf·ft)

8 x 1.25 mm
22 N·m (2.2 kgf·m, 16 lbf·ft)

8 x 1.25 mm
22 N·m (2.2 kgf·m, 16 lbf·ft)

STEERING HANGER BEAM

8 x 1.25 mm
22 N·m (2.2 kgf·m, 16 lbf·ft)

93112GQ5

View of the steering hanger beam and related components—2.3CL and 3.0CL

- Knee bolster and the knee bolster-to-dash bolts; then, connect the electrical connector and the air hose.
- Lower instrument panel, the lower instrument panel-to-dash screw and the coin pocket
17. Install the SRS module and the

steering wheel by installing or connecting the following:
- Steering wheel to the steering column and torque the nut to 28 ft. lbs. (38 Nm)
- Electrical connectors to the steering wheel
- SRS module to the steering wheel

- Both SRS module-to-steering wheel bolts and torque to 84 inch lbs. (9.8 Nm) using a T30 Torx bit
- SRS module-to-steering wheel covers.
- SRS module electrical connector.
- Access panel-to-steering wheel screws and the panel.

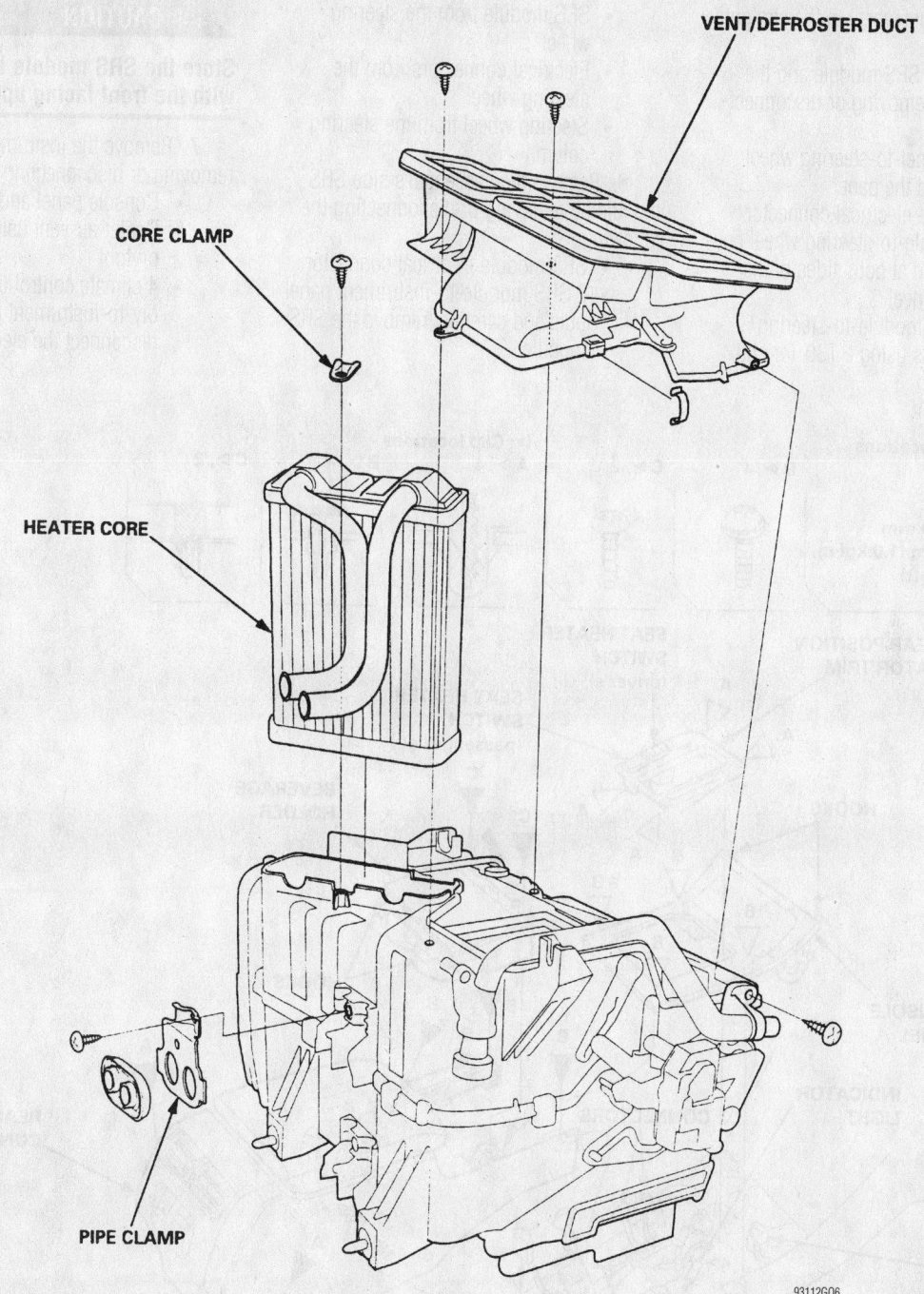

VENT/DEFROSTER DUCT

CORE CLAMP

HEATER CORE

PIPE CLAMP

93112GQ6

Exploded view of the heater core, heater housing and related components—2.3CL and 3.0CL

- 2 heater housing-to-cowl nuts located in the engine compartment
- Heater hoses to the heater core
18. Refill the cooling system.
19. Connect the negative battery cable.
20. Evacuate, charge and leak test the air conditioning system.
21. Operate the engine to normal operating temperatures; then, check the climate control operation and check for leaks.

3.5RL

REMOVAL & INSTALLATION

➡**Be sure to acquire the anti-theft code for the radio; then, write down the frequencies for the preset buttons.**

1. Disconnect the negative battery cable.

⁕⁕ CAUTION

After disconnecting the negative battery cable, wait for at least 3 minutes for the SRS module to deplete its energy.

2. Drain the cooling system into a clean container for reuse.
3. Disconnect the heater hoses from the heater core.

For complete Engine Mechanical specifications, see Section 1 of this manual

4. Place the front wheel in the straight-ahead position.

5. Remove the SRS module and the steering wheel by removing or disconnecting the following:

- Access panel-to-steering wheel screws and the panel
- RS module electrical connector
- SRS module-to-steering wheel covers, located at both sides of the steering wheel
- Both SRS module-to-steering wheel bolts using a T30 Torx bit

- SRS module from the steering wheel
- Electrical connectors from the steering wheel
- Steering wheel from the steering column

6. Remove the passenger's side SRS module by removing or disconnecting the following:

- SRS module electrical connector
- 2 SRS module-to-instrument panel nuts and carefully remove the SRS module

☀☀ CAUTION

Store the SRS module in a safe place with the front facing upward.

7. Remove the instrument panel by removing or disconnecting the following:

- Console panel and the rear console
- Center air vent using a suitable prytool
- 4 climate control unit/audio assembly-to-instrument panel bolts; then, disconnect the electrical connectors

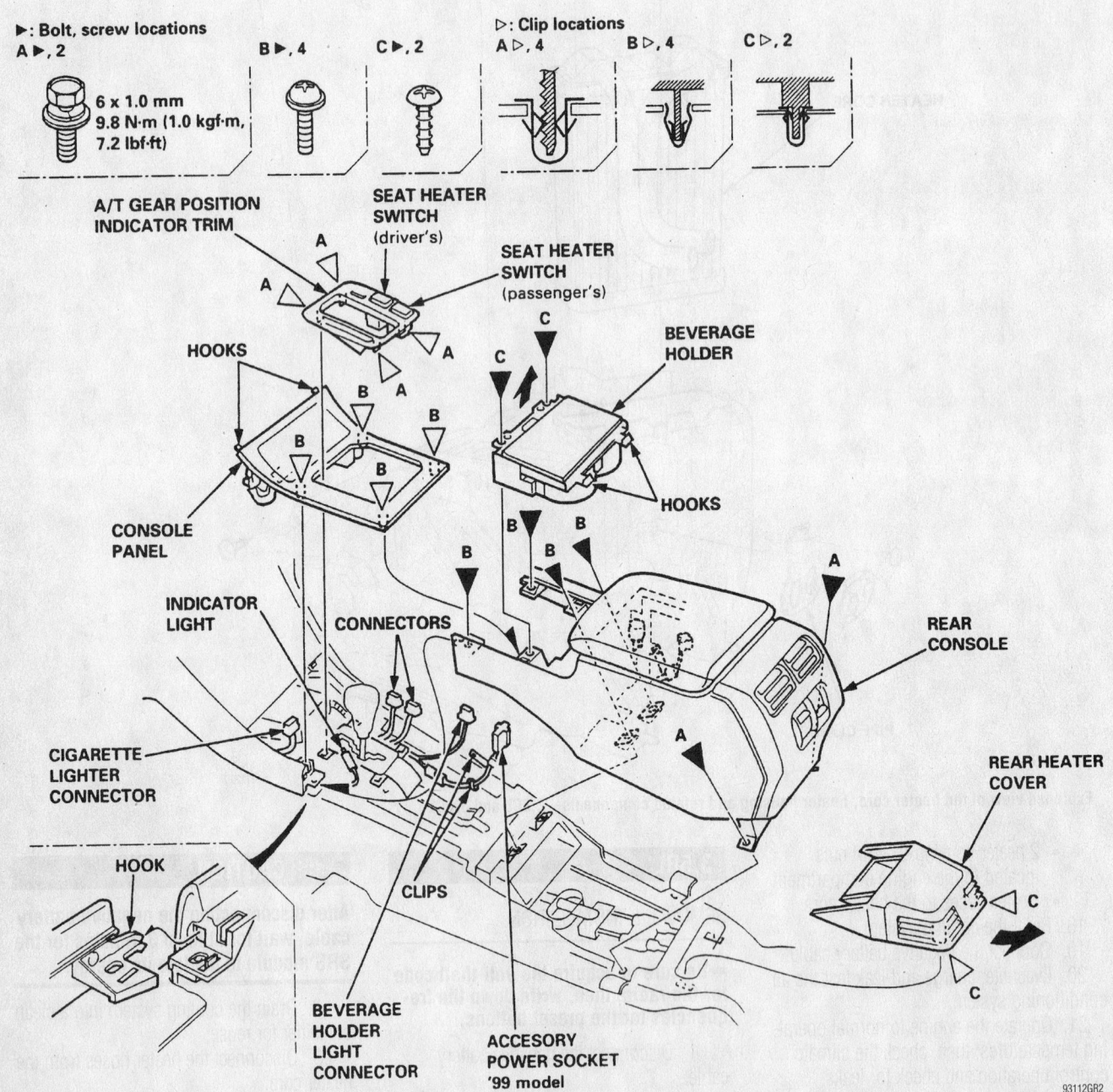

Exploded view of the console panel, rear console and related components—3.5RL

93112GR2

'96 – 98 models:

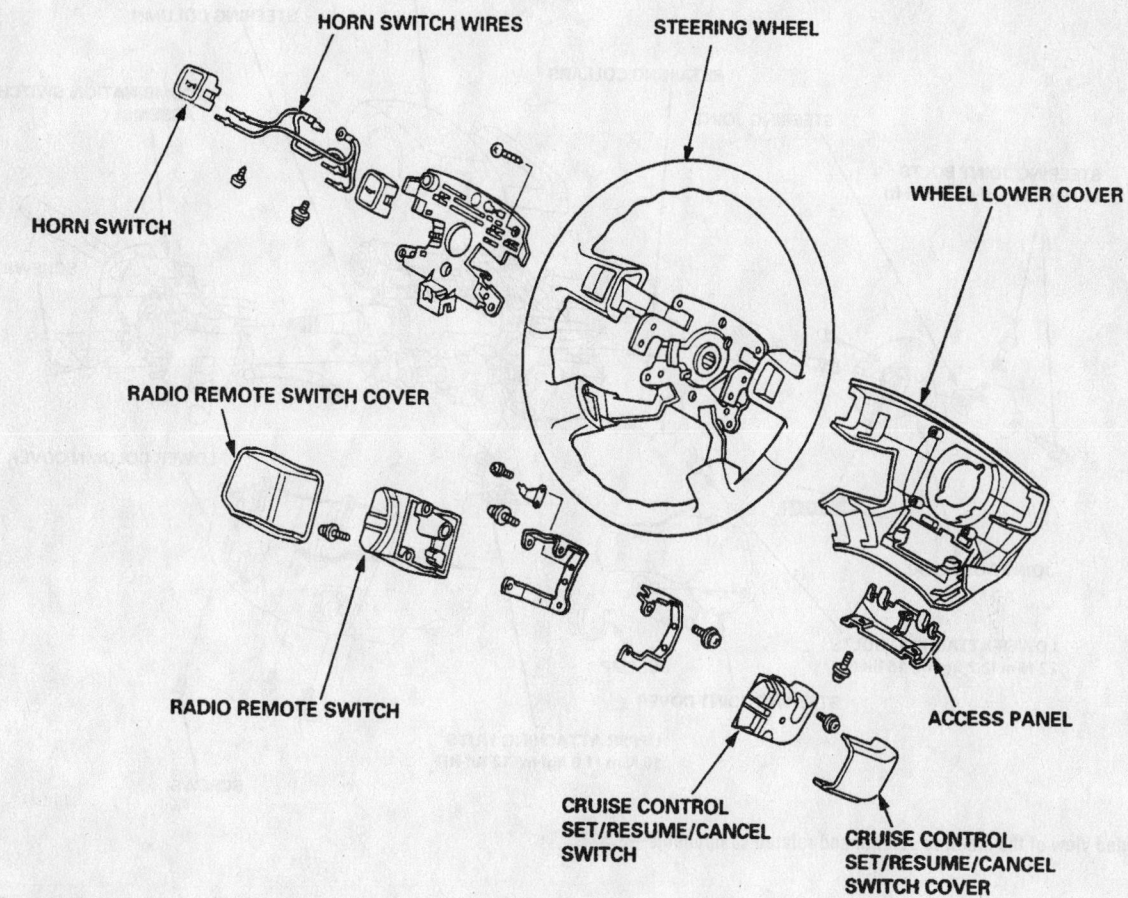

HORN SWITCH WIRES

STEERING WHEEL

HORN SWITCH

WHEEL LOWER COVER

RADIO REMOTE SWITCH COVER

RADIO REMOTE SWITCH

ACCESS PANEL

CRUISE CONTROL
SET/RESUME/CANCEL
SWITCH

CRUISE CONTROL
SET/RESUME/CANCEL
SWITCH COVER

'99 models:

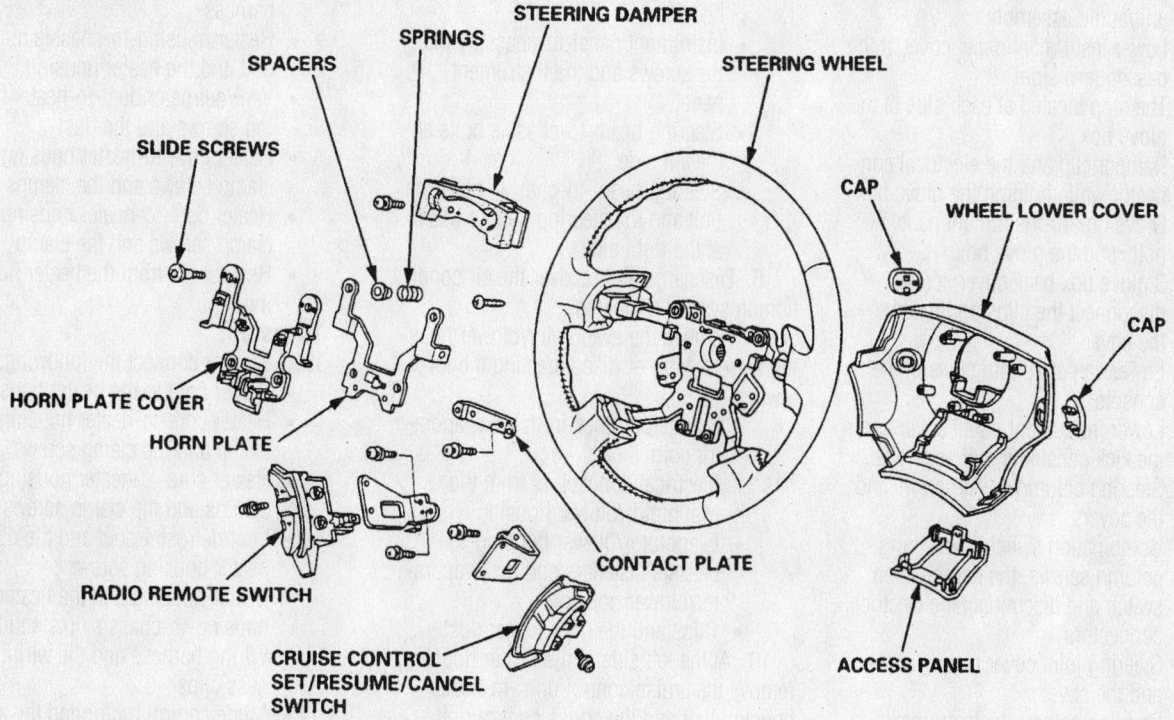

STEERING DAMPER

SPRINGS

SPACERS

STEERING WHEEL

SLIDE SCREWS

CAP

WHEEL LOWER COVER

CAP

HORN PLATE COVER

HORN PLATE

CONTACT PLATE

RADIO REMOTE SWITCH

CRUISE CONTROL
SET/RESUME/CANCEL
SWITCH

ACCESS PANEL

93112GR3

Exploded view of the SRS module, the steering wheel and related components—3.5RL

For Accessory Drive Belt illustrations, see Section 1 of this manual

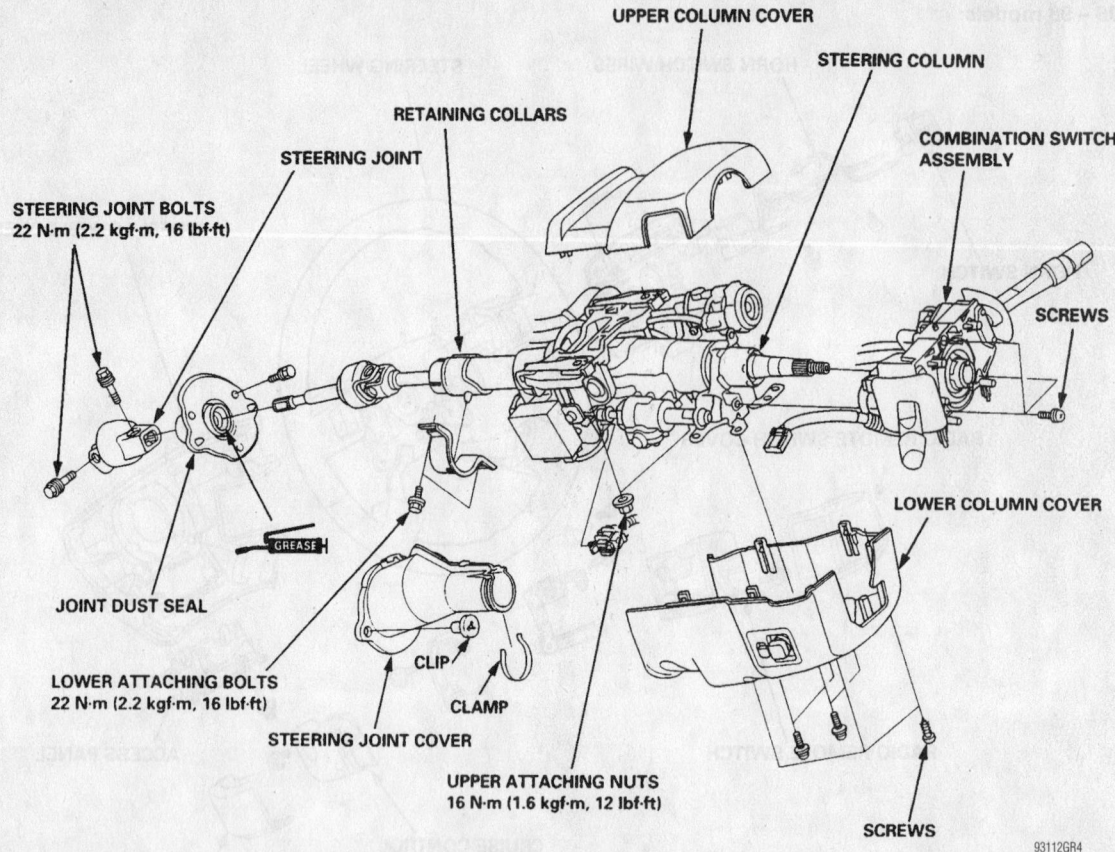

STEERING JOINT BOLTS 22 N·m (2.2 kgf·m, 16 lbf·ft)

STEERING JOINT

RETAINING COLLARS

UPPER COLUMN COVER

STEERING COLUMN

COMBINATION SWITCH ASSEMBLY

SCREWS

JOINT DUST SEAL

LOWER COLUMN COVER

GREASE

LOWER ATTACHING BOLTS 22 N·m (2.2 kgf·m, 16 lbf·ft)

CLIP

CLAMP

STEERING JOINT COVER

UPPER ATTACHING NUTS 16 N·m (1.6 kgf·m, 12 lbf·ft)

SCREWS

93112GR4

Exploded view of the steering column and related components—3.5RL

and remove the climate control unit/audio assembly
- Lower instrument panel cover at the passenger's side
- The stop located at each side of the glove box
- Damper clip and the electrical connector while holding the glove box
- Glove box-to-instrument panel bolts and the glove box
- 3 glove box back cover screws, disconnect the clips and remove the cover
- Lower carpet at both sides of the console
- Lower instrument panel cover and the kick panel at the driver's side
- Steering column cover screws and the covers
- Combination switch-to-steering column screws, the combination switch and disconnect the electrical connectors
- Steering joint cover clamp, screws and the cover
- Steering column-to-instrument panel nuts and bolts; then, lower the steering column
- Instrument panel wiring harness

connectors, the clip and the air hose
- Instrument panel-to-chassis bolts, the screws and the instrument panel
- Steering beam-to-chassis bolts at the left side
- Steering beam-to-chassis nuts, the bolt and the steering hanger beam at the right side

8. Discharge and recover the air conditioning system refrigerant.

9. Remove the evaporator/blower housing by removing or disconnecting the following:
- Refrigerant lines from the evaporator core
- Electrical connectors from the evaporator/blower housing
- Evaporator/blower housing-to-chassis fasteners and the evaporator/blower housing
- Clips and the floor heater duct

10. At the left side of the heater housing, remove the cruise control unit -to-heater housing bolt and the cruise control unit.

11. Remove or disconnect the following:
- Mode control motor and the air mix control motor connectors

- Wiring harness clips and the wiring harness
- Heater housing-to-chassis nuts, the bolt and the heater housing
- Vent/defroster duct-to-heater housing screws and the duct
- Heater core-to-heater housing pipe clamp screws and the clamps
- Heater core-to-heater housing clamp screws and the clamp
- Heater core from the heater housing

To install:

12. Install or connect the following:
- Heater core to the heater housing
- Heater core-to-heater housing clamp and the clamp screws
- Heater core-to-heater housing pipe clamps and the clamp screws
- Vent/defroster duct and the duct-to-heater housing screws
- Heater housing and the heater housing-to-chassis nuts and bolt
- Wiring harness and the wiring harness clips
- Mode control motor and the air mix control motor connectors
- Cruise control unit and the cruise control unit-to-heater housing bolt

at the left side of the heater housing
- Floor heater duct and the clips

13. Install the evaporator/blower housing by installing or connecting the following:

- Evaporator/blower housing and the evaporator/blower housing-to-chassis fasteners

- Electrical connectors to the evaporator/blower housing
- Refrigerant lines to the evaporator core using new O-rings
- Steering beam and the steering hanger beam-to-chassis nuts
- Steering beam-to-chassis bolts

14. Install the instrument panel by installing or connecting the following:

- Instrument panel and the instrument panel-to-chassis bolts and the screws
- Instrument panel wiring harness connectors, the clip and the air hose
- Lower the steering column and torque the steering column-to-instrument panel nuts to 12 ft. lbs.

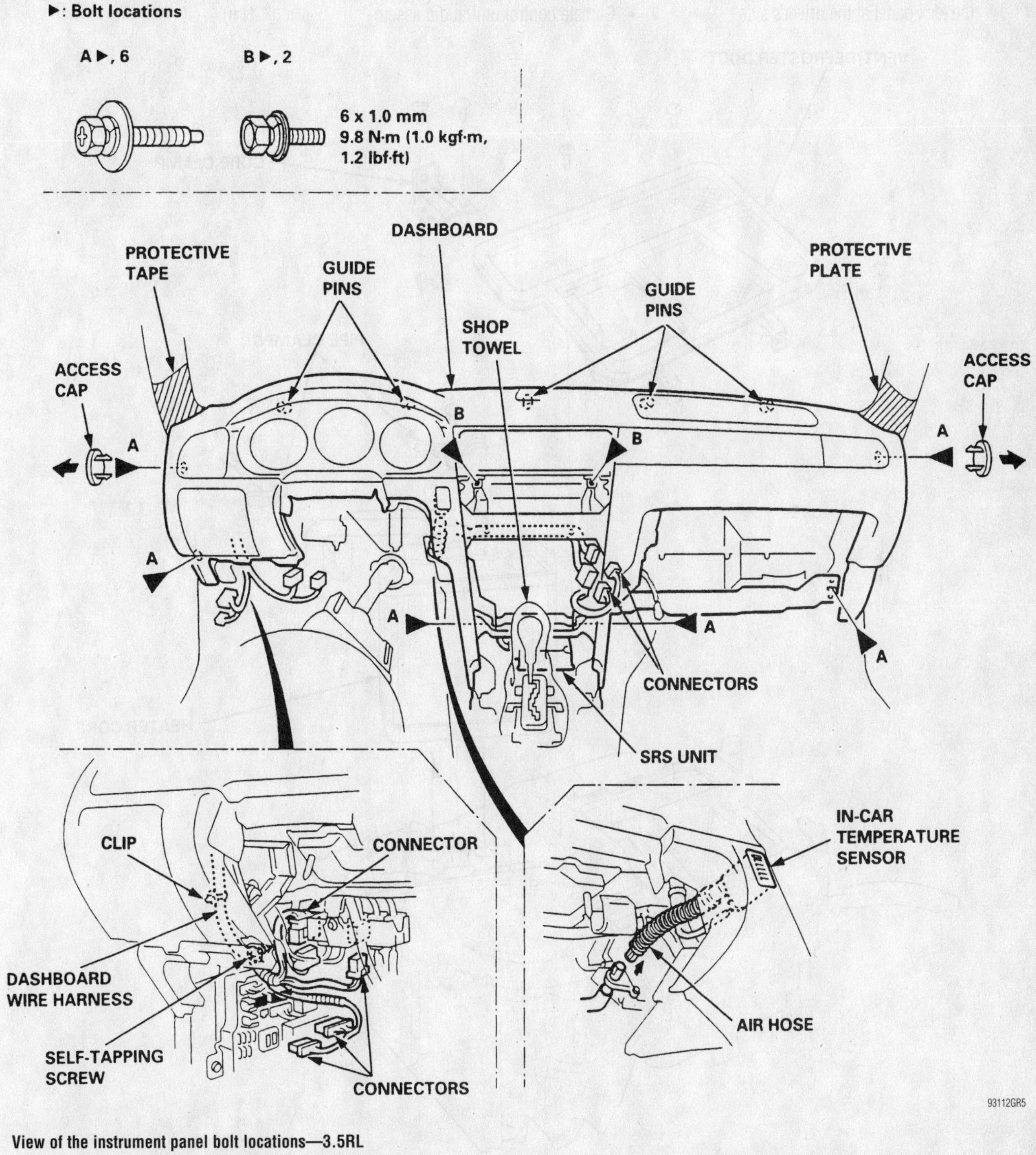

▶: Bolt locations

A ▶, 6   B ▶, 2

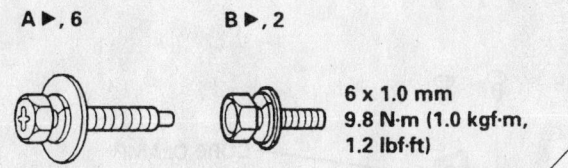

6 x 1.0 mm
9.8 N·m (1.0 kgf·m,
1.2 lbf·ft)

View of the instrument panel bolt locations—3.5RL

93112GR5

For Tire, Wheel and Ball Joint specifications, see Section 1 of this manual

(16 Nm) and bolts to 16 ft. lbs. (22 Nm)
- Steering joint cover and the cover clamp and screws
- Combination switch, the combination switch-to-steering column screws and connect the electrical connectors
- Steering column covers and the cover screws
- Lower instrument panel cover and the kick panel at the driver's side

- Lower carpet at both sides of the console
- Glove box back cover, connect the clips and install the 3 cover screws
- Glove box and the glove box-to-instrument panel bolts
- Damper clip and the electrical connector while holding the glove box
- The stop at each side of the glove box
- Lower instrument panel cover.
- Climate control unit/audio assem-

bly; then, connect the electrical connectors and install the 4 climate control unit/audio assembly-to-instrument panel bolts
- Center air vent.
- Console panel and the rear console
15. Install the passenger's side SRS module by installing or connecting the following:
- SRS module and the 2 SRS module-to-instrument panel nuts; then, torque the nuts to 84 inch lbs. (9.8 Nm)

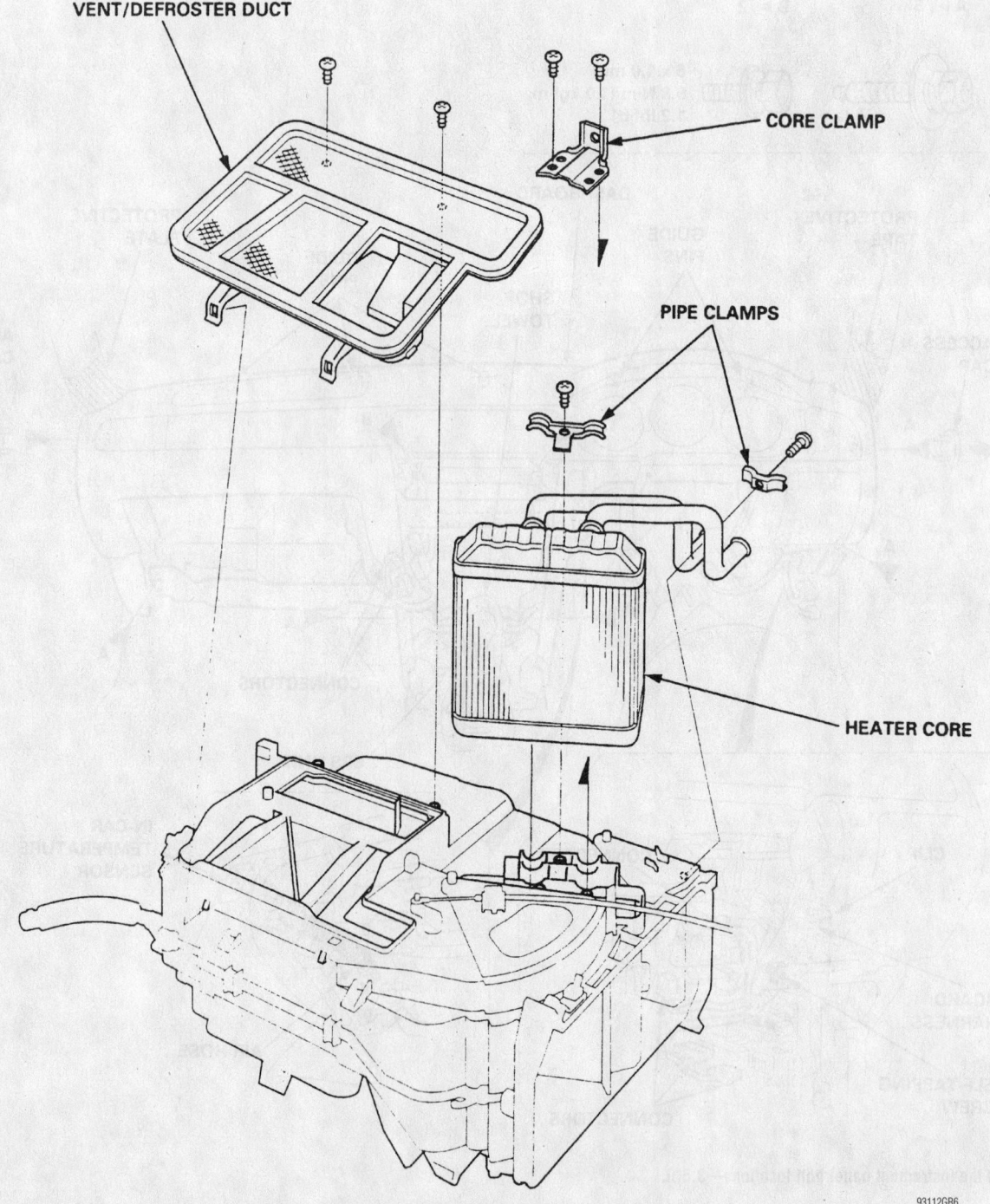

Exploded view of the heater core, heater housing and related components—3.5RL

93112GR6

- SRS module electrical connector.

16. Install the SRS module and the steering wheel by installing or connecting the following:

- Steering wheel to the steering column and torque the nut to 36 ft. lbs. (49 Nm)
- Electrical connectors to the steering wheel
- SRS module to the steering wheel
- Both SRS module-to-steering wheel bolts and torque to 84 inch lbs. (9.8 Nm) using a T30 Torx bit
- SRS module-to-steering wheel covers
- SRS module electrical connector.
- Access panel-to-steering wheel screws and the panel

17. Connect the heater hoses to the heater core.

18. Refill the cooling system.

19. Connect the negative battery cable.

20. Evacuate, charge and leak test the air conditioning system.

21. Operate the engine to normal operating temperatures; then, check the climate control operation and check for leaks.

3.2TL

REMOVAL & INSTALLATION

➡**Be sure to acquire the anti-theft code for the radio; then, write down the frequencies for the preset buttons.**

Fastener Locations

▶ : Screw, 4   ▷ : Clip, 10

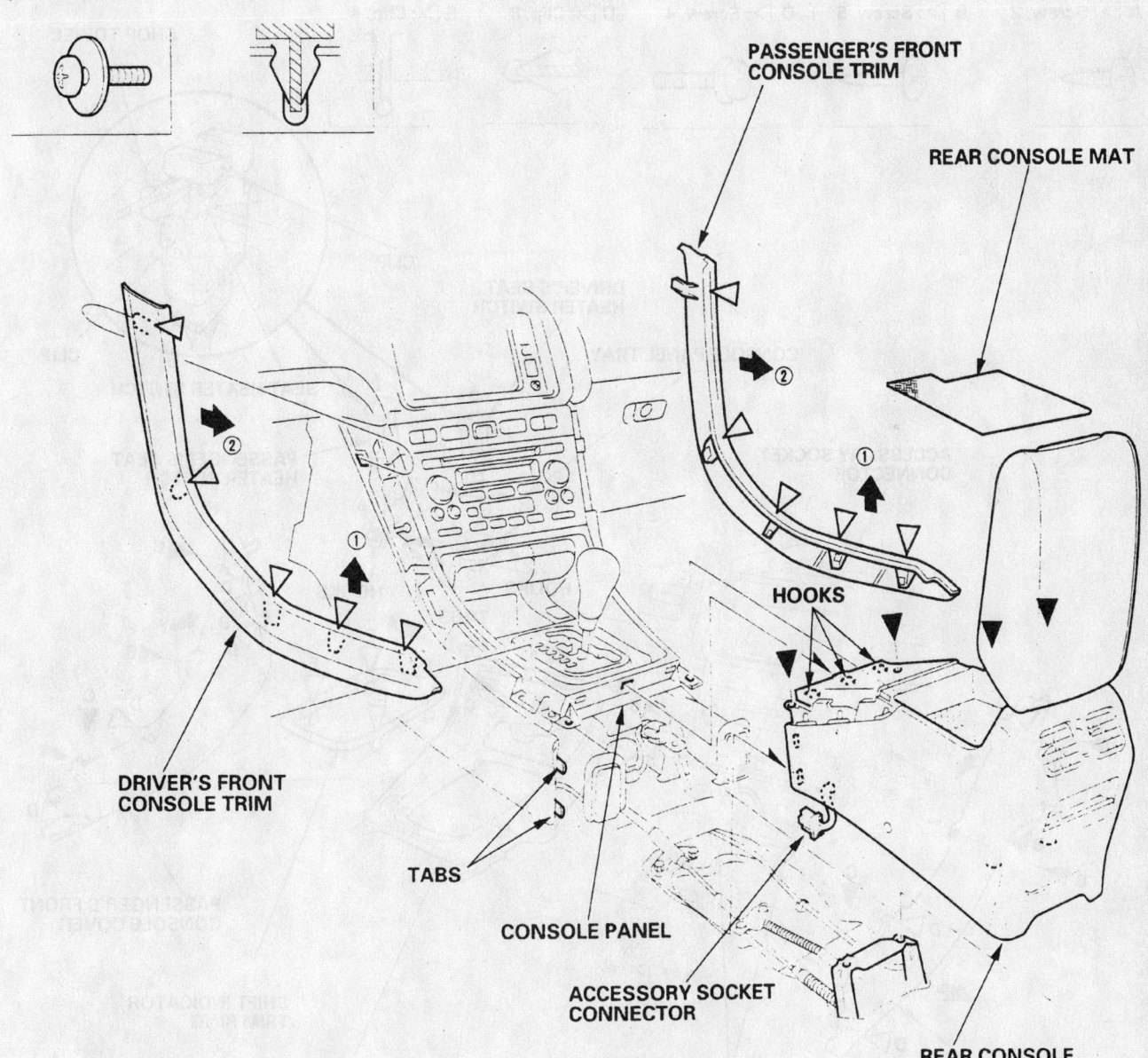

DRIVER'S FRONT CONSOLE TRIM

TABS

CONSOLE PANEL

ACCESSORY SOCKET CONNECTOR

PASSENGER'S FRONT CONSOLE TRIM

REAR CONSOLE MAT

HOOKS

REAR CONSOLE

93112GQ7

Exploded view of the rear console and related components—3.2TL

For Wheel Alignment specifications, see Section 1 of this manual

1. Disconnect the negative battery cable.

✳✳ CAUTION

After disconnecting the negative battery cable, wait for at least 3 minutes for the SRS module to deplete its energy.

2. Drain the cooling system into a clean container for reuse.

3. Disconnect the heater hoses from the heater core.

4. In the engine compartment, remove the heater housing-to-cowl nut.

5. Remove the instrument panel by removing or disconnecting the following:

- Clips and remove the dashboard side cover, located at the driver's side
- Lower dashboard cover screw, detach the clips and remove the lower dashboard cover
- Console panel screws and the console panel
- Glove box-to-instrument panel screw; then, remove the damper from the glove box
- Glove box stop located at each side while holding the glove box
- Both glove box-to-instrument panel screws and the glove box
- Clips and remove the dashboard side cover located at the passenger's side
- Glove box cover-to-instrument panel screws, disconnect the electrical connectors and remove the glove box cover

Fastener Locations

A ▷ : Screw, 2   B ▷ : Screw, 5   C ▷ : Screw, 4   D ▷ : Clip, 8   E ▷ : Clip, 4

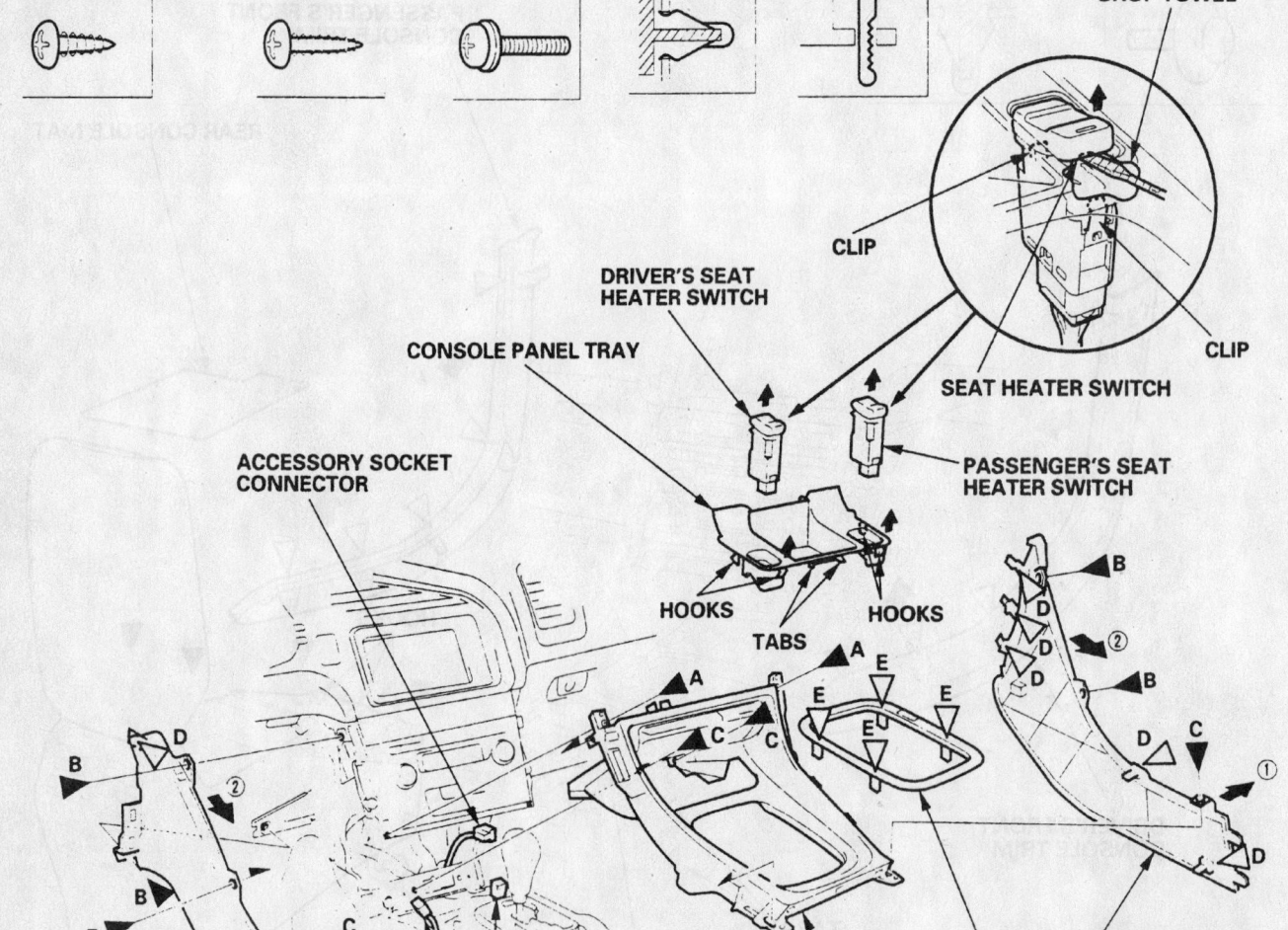

Exploded view of the console panel and related components—3.2TL

93112GQ8

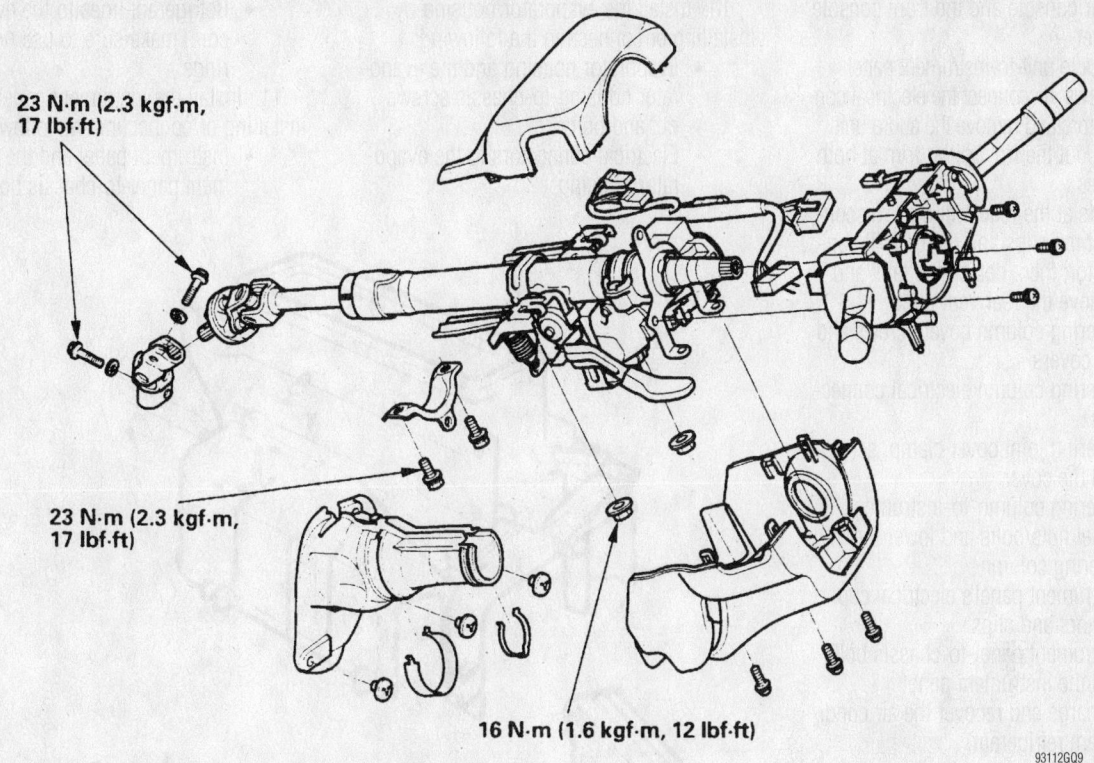

23 N·m (2.3 kgf·m, 17 lbf·ft)

23 N·m (2.3 kgf·m, 17 lbf·ft)

16 N·m (1.6 kgf·m, 12 lbf·ft)

93112GQ9

Exploded view of the steering column and related components—3.2TL

Fastener Locations

B ▶ : Bolt, 6   C ▶ : Bolt, 3   D ▶ : Bolt, 1   E ▶ : Bolt, 3   F ▶ : Bolt, 1

8 x 1.25 mm
22 N·m (2.2 kgf·m, 16 lbf·ft)

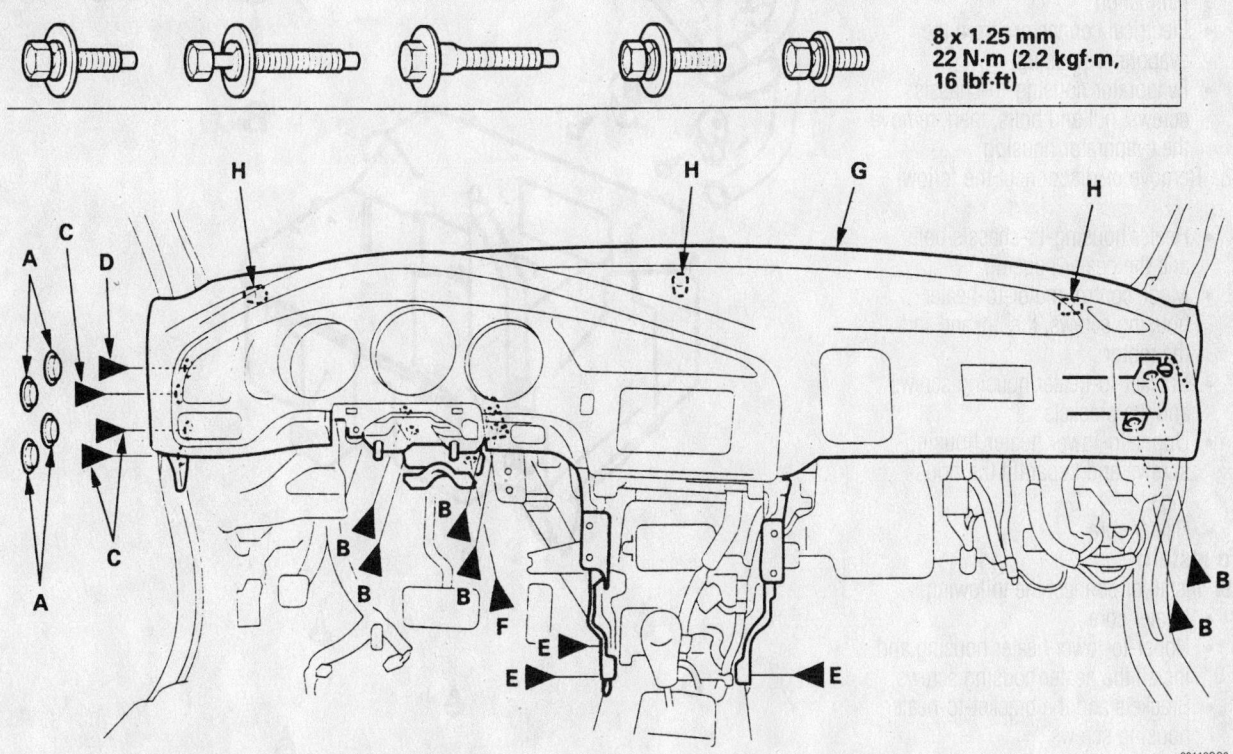

93112GQ0

View of the instrument panel and bolt locations—3.2TL

For Maintenance Interval recommendations, see Section 1 of this manual

- Rear console and the front console cover
- 4 audio unit-to-instrument panel screws, disconnect the electrical connectors and remove the audio unit
- Pry out the front pillar trim at both sides
- Bolts at the rear vent duct, disconnect the clips and the OBD-II connector; then, detach the tabs and remove the rear vent duct
- Steering column cover screws and the covers
- Steering column electrical connectors
- Steering joint cover clamp, screws and the cover
- Steering column-to-instrument panel nuts/bolts and lower the steering column
- Instrument panel's electrical connectors and clips
- Instrument panel-to-chassis bolts and the instrument panel

6. Discharge and recover the air conditioning system refrigerant.

7. Remove the evaporator housing by removing or disconnecting the following:
- Refrigerant lines from the evaporator core. Discard the O-rings and plug the openings to prevent contamination
- Electrical connectors from the evaporator housing
- Evaporator housing-to-chassis screws, nut and bolts; then, remove the evaporator housing

8. Remove or disconnect the following:
- Heater housing-to-chassis bolts and the heater housing
- Mode control motor-to-heater housing screws, the linkage and the motor
- Bracket-to-heater housing screws and the brackets
- Upper-to-lower heater housing screws and separate the housings
- Heater core

To install:

9. Install or connect the following:
- Heater core
- Upper-to-lower heater housing and install the heater housing screws
- Brackets and the bracket-to-heater housing screws
- Mode control motor, the linkage and the motor-to-heater housing screws
- Heater housing and the heater housing-to-chassis bolts

10. Install the evaporator housing by installing or connecting the following:
- Evaporator housing and the evaporator housing-to-chassis screws, nut and bolts
- Electrical connectors to the evaporator housing

- Refrigerant lines to the evaporator core, make sure to use new O-rings

11. Install the instrument panel by installing or connecting the following:
- Instrument panel and the instrument panel-to-chassis bolts.

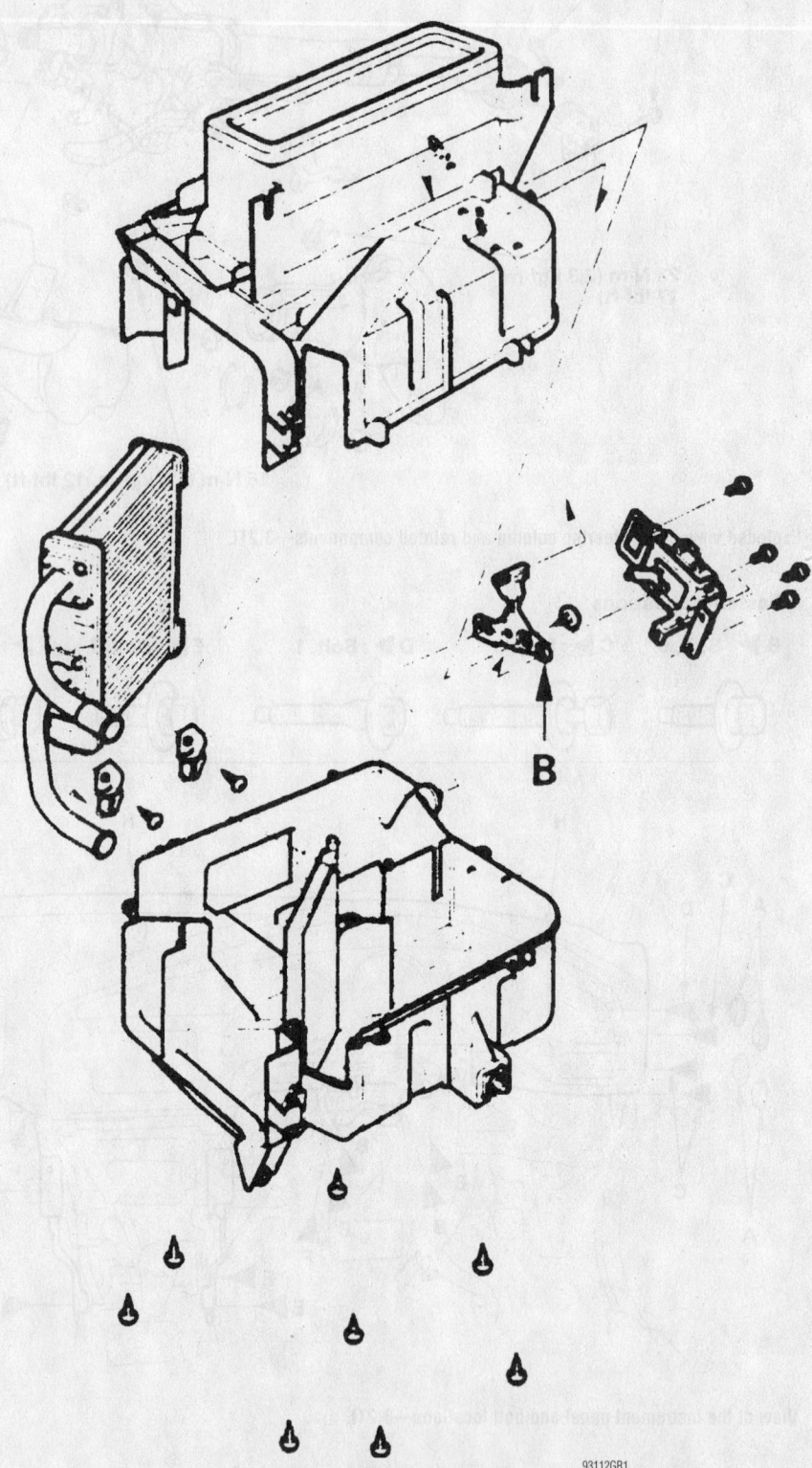

B

Exploded view of the heater core, heater housing and related components—3.2TL

93112GR1

- Instrument panel's electrical connectors and clips
- Steering column and torque the steering column-to-instrument panel nuts to 12 ft. lbs. (16 Nm) and bolts to 17 ft. lbs. (23 Nm)
- Steering joint cover and the cover clamp and screws
- Steering column electrical connectors
- Steering column covers and the cover screws
- Rear vent duct, connect the clips and the OBD-II connector, attach the tabs and install the bolts
- Front pillar trim. On both sides
- Audio unit, connect the electrical

connectors and install the 4 audio unit-to-instrument panel screws
- Rear console and the front console cover
- Glove box cover, connect the electrical connectors and install the glove box cover-to-instrument panel screws
- Dashboard side cover and attach the clips at the passenger's side
- Both glove box and the glove box-to-instrument panel screws
- Glove box stop (located at each side), while holding the glove box
- Damper to the glove box and the glove box-to-instrument panel screw

- Console panel and the console panel screws
- Lower dashboard cover, attach the clips and install the lower dashboard cover screw
- Dashboard side cover and attach the clips at the driver's side
- Heater housing-to-cowl nut located in the engine compartment
- Heater hoses to the heater core
12. Refill the cooling system.
13. Connect the negative battery cable.
14. Evacuate, charge and leak test the air conditioning system.
15. Operate the engine to normal operating temperatures; then, check the climate control operation and check for leaks.

1998–01 AUDI

A4

REMOVAL & INSTALLATION

1. Perform the Output Diagnostic Test Mode (DTM) using the VAG 1551 function 03 by performing the following procedure:
- Initiate the Output DTM and wait until the air flow flap (located on top) closes by way of the air flow flap motor.
- Once the air flow flap closes, cancel the Output DTM by pressing the "C" button.

➡**If equipped with power seats, move the seats as far rearward as possible. Also, obtain the anti-theft radio coding and the preset radio stations from the owner.**

2. Discharge and recover the air conditioning system refrigerant.
3. Remove or disconnect the following:
- Air plenum cover, the water guide and the dust/pollen filter, located at the right side
- Negative battery cable
- Positive battery cable and remove the battery
- Coolant recovery tank cap to relieve the pressure from the system
4. Drain the cooling system into a clean container for reuse.
5. Label and disconnect or clamp off the heater hoses from (at) the heater core.

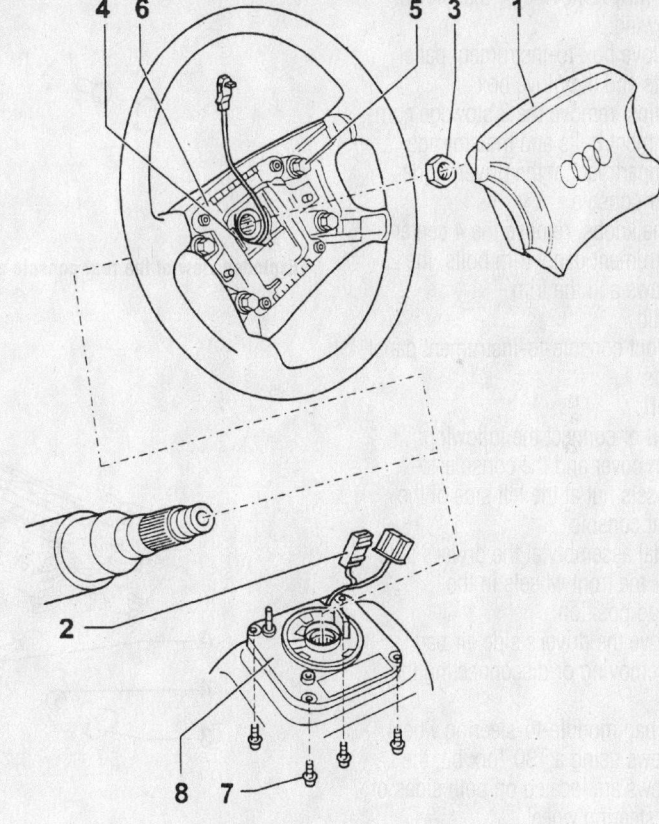

1 - Airbag unit

2 - Airbag and horn connectors

3 - Retaining nut for steering wheel

4 - Horn connector

93112GN6

Exploded view of the air bag module, steering wheel and related components—A4

For Tune-up, Capacities and Firing orders, see Section 1 of this manual

6. Place a container under the right heater core tube, induce pressurized air to the left tube and blow compressed into the tube to drain excess coolant.

7. Remove or disconnect the following:
- Air conditioning system vacuum supply hose and attach it at the heater core inlet/outlet
- Heater core-to-chassis boot
- Refrigerant lines-to-evaporator core clamp bolt, disconnect the refrigerant line clamp and discard the O-rings. Plug the openings to prevent contamination.
- Evaporator core-to-chassis boot
- Low pressure switch electrical connector and secure it to the evaporator fixture

8. Remove the glove box, the driver's side lower shelf and the instruments panel center section by removing or disconnecting the following:
- 5 glove box-to-instrument panel bolts and the glove box
- 3 clips, remove the 3 stowage compartment bolts and the stowage compartment at the driver's side
- Rear console
- 3 the knobs, remove the 4 center instrument panel trim bolts, the 2 screws and the trim
- Radio
- 4 front console-to-instrument panel bolts

To install:

9. Install or connect the following:
- Trim cover and the console-to-chassis nut at the left side of the front console
- Pedal assembly at the driver's side

10. Place the front wheels in the straight-ahead position.

11. Remove the driver's side air bag module by removing or disconnecting the following:
- Air bag module-to-steering wheel screws using a T30 Torx bit; the screws are located on both sides of the steering wheel
- Air bag module and disconnect the electrical connector
- Place the air bag module in a safe place with the front facing upward.

12. Remove the steering wheel by removing or disconnecting the following:
- Steering wheel nut
- Horn and air bag electrical connectors
- 4 carrier unit-to-steering wheel bolts and the carrier unit
- Steering wheel

13. At the steering column-to-instrument

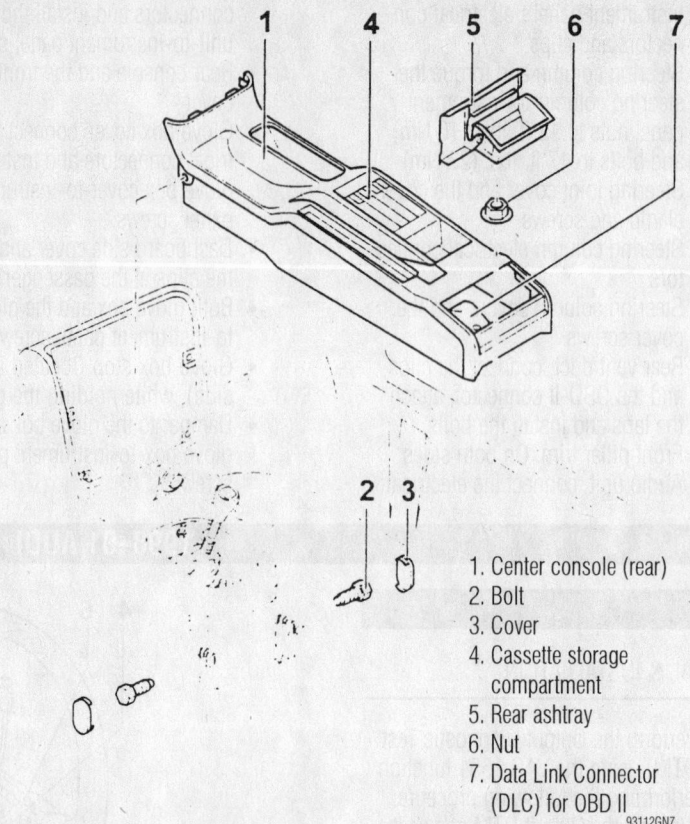

1. Center console (rear)
2. Bolt
3. Cover
4. Cassette storage compartment
5. Rear ashtray
6. Nut
7. Data Link Connector (DLC) for OBD II

93112GN7

Exploded view of the rear console assembly—A4

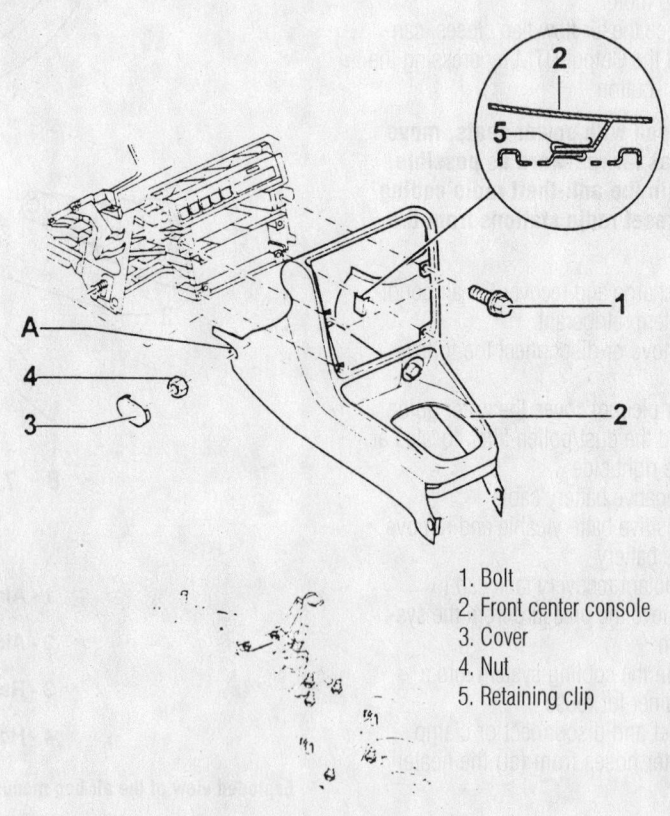

1. Bolt
2. Front center console
3. Cover
4. Nut
5. Retaining clip

93112GN8

Exploded view of the front console assembly—A4

panel connection, secure the steering column wire to keep the steering column from sliding apart.

14. Remove or disconnect the following:
   - Steering column-to-steering gear bolt
   - Electronic box electrical connectors located in the left air plenum chamber
   - All instrument panel-to-chassis electrical connectors

15. Remove the instrument panel by removing or disconnecting the following:
   - Instrument panel-to-chassis bolts
   - Any necessary electrical connectors
   - Instrument panel

16. Remove or disconnect the following:
   - Ducts, heater/air conditioning housing assembly-to-chassis bolts and the assembly
   - Heater core-to-heater/air conditioning housing screws
   - Press the heater core-to-heater housing catches and remove the heater core from the heater/air conditioning housing

To install:

17. Install or connect the following:
   - Heater core to the heater/air conditioning housing and press the heater core into heater housing until the latches catch
   - Heater core-to-heater/air conditioning housing screws
   - Heater/air conditioning housing assembly and connect the ducts

18. Install the instrument panel by installing or connecting the following:
   - Instrument panel
   - Any necessary electrical connectors
   - Instrument panel-to-chassis bolts and torque the bolts to 44 inch lbs. (5 Nm)
   - All instrument panel-to-chassis electrical connectors
   - Electronic box electrical connectors located in the left air plenum chamber
   - Steering column-to-steering gear bolt

19. Install the steering wheel by installing or connecting the following:
   - Steering wheel
   - Carrier unit and the 4 carrier unit-to-steering wheel bolts
   - Horn and air bag electrical connectors
   - Steering wheel nut

20. Install the driver's side air bag mod-

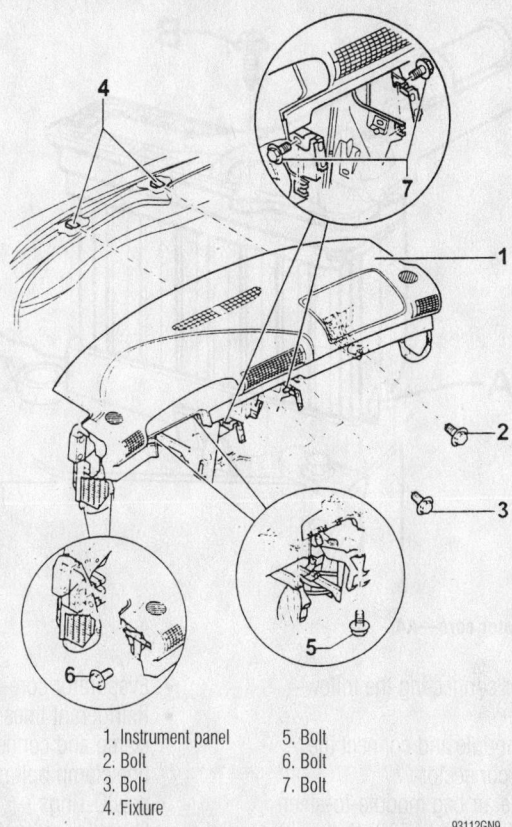

1. Instrument panel   5. Bolt
2. Bolt   6. Bolt
3. Bolt   7. Bolt
4. Fixture

93112GN9

Exploded view of the instrument panel assembly—A4

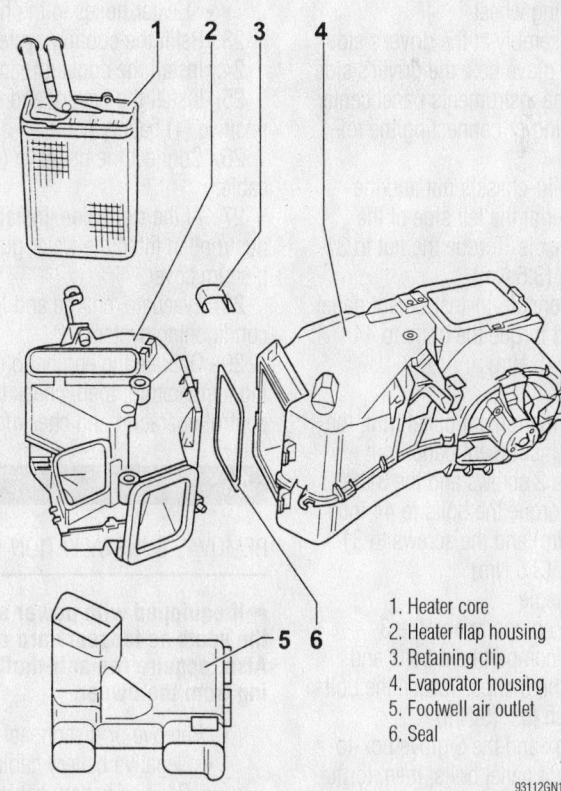

1. Heater core
2. Heater flap housing
3. Retaining clip
4. Evaporator housing
5. Footwell air outlet
6. Seal

93112GN1

Exploded view of the heater core, heater/air conditioning housing and related components—A4

For complete service labor times, order the Chilton Labor Guide

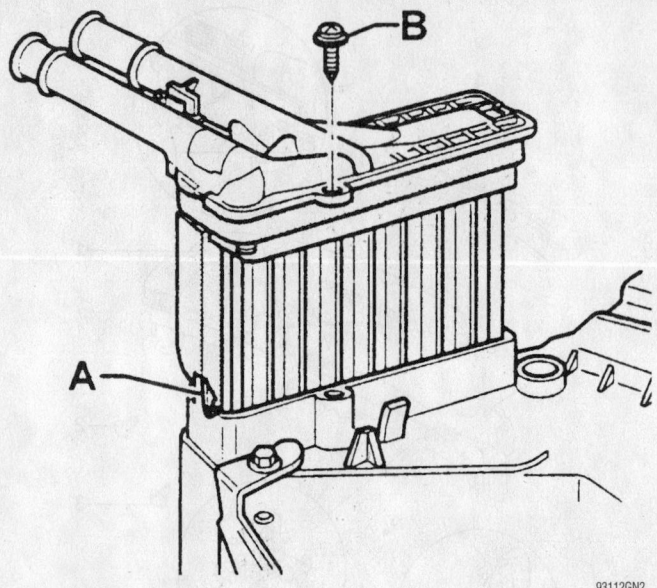

93112GN2

Removing the heater core—A4

ule by installing or connecting the following:

- Air bag module and connect the electrical connector
- Torque the air bag module-to-steering wheel screws to 53 inch lbs. (6 Nm) using a T30 Torx bit; the screws are located on both sides of the steering wheel
- Pedal assembly at the driver's side

21. Install the glove box, the driver's side lower shelf and the instruments panel center section by installing or connecting the following:

- Console-to-chassis nut and the trim cover at the left side of the front console. Torque the nut to 31 inch lbs. (3.5 Nm)
- 4 front console-to-instrument panel bolts and torque the bolts to 44 inch lbs. (5 Nm)
- Radio
- 4 center instrument panel trim, the 4 center instrument panel trim bolts, the 2 screws and the 3 the knobs. Torque the bolts to 44 inch lbs. (5 Nm) and the screws to 31 inch lbs. (3.5 Nm)
- Rear console
- Stowage compartment, the 3 stowage compartment bolts and engage the 3 clips. Torque the bolts to 44 inch lbs. (5 Nm)
- Glove box and the 5 glove box-to-instrument panel bolts; then, torque the bolts to 44 inch lbs. (5 Nm)

22. Install or connect the following:

- Low pressure switch electrical connector

- Evaporator core-to-chassis boot.
- Refrigerant lines-to-evaporator core clamp and connect the refrigerant line clamp bolt making sure to use new O-rings
- Heater core-to-chassis boot
- Air conditioning system vacuum supply hose
- Heater hoses to the heater core

23. Refill the cooling system.

24. Install the coolant recovery tank cap.

25. Install the battery and connect the positive (+) battery cable.

26. Connect the negative (() battery cable.

27. At the right side, install the dust/pollen filter, the water guide and the air plenum cover.

28. Evacuate, charge and leak test the air conditioning system.

29. Operate the engine to normal operating temperature; then, check the climate control operation and check for leaks.

A6

REMOVAL & INSTALLATION

➡**If equipped with power seats, move the seats as far rearward as possible. Also, acquire the anti-theft radio coding from the owner.**

1. Remove or disconnect the following:

- Negative battery cable
- Positive battery cable and remove the battery
- Coolant recovery tank cap to relieve the pressure from the system

- Coolant
- Heater hoses from (at) the heater core
- Heater core tubes-to-chassis grommet

2. Remove the glove box by removing or disconnecting the following:

- Open the glove box door and remove the 3 upper glove compartment-to-chassis bolts
- 2 lower glove compartment bolts; 1 located at each side
- Glove box and disconnect the electrical connector

3. At the driver's side, remove the storage compartment by removing or disconnecting the following:

- Instrument panel molding
- Instrument panel end trim
- 5 storage compartment-to-instrument panel bolts
- Lower the storage compartment; then, disconnect the footwell light connector and the DLC
- Storage compartment

4. Remove the instrument panel's center section by removing or disconnecting the following:

- Center console
- Open the ashtray and disconnect the trim section
- Radio
- Air conditioning control head
- Front ashtray
- Switch panel trim
- 4 front console-to-instrument panel bolts
- Trim cap and the console-to-instrument panel nut at the left side of the front console
- 6 center section-to-instrument panel bolts
- Center section and disconnect and wiring
- Right side instrument panel brace-to-chassis bolts and the brace
- Left side instrument panel brace-to-chassis bolts
- Coolant hose cover screws and the cover at the left side of the heater housing
- Pedal bracket-to-instrument panel cross member bolt and detach the bracket from the instrument panel
- Push the adjustable steering column inward as far as it will go
- Left side instrument panel cover
- Instrument panel trim
- 2 instrument panel-to-chassis bolts (at the left side of the instrument panel); then, pull the instrument

panel rearward about 0.394 in. (10mm)
- Central tube-to-chassis bolts at the driver's side
- Pull the left side of the instrument panel rearward far enough for the heater core to be removed and secure the panel in that position

5. Remove or disconnect the following:
- Heater tube bracket-to-heater housing screw and the bracket

- Heater tube clamp screws and the clamps
- Slightly, pull the heater core from the heater housing
- Heater tubes and discard the O-rings
- Heater core

To install:
6. Install or connect the following:
- Heater core
- Heater tubes using new O-rings
- Slightly, push the heater core into the heater housing

- Heater tube clamps and the clamp screws.
- Heater tube bracket and the bracket-to-heater housing screw.
- Push the left side of the instrument panel forward and secure it
- Central tube-to-chassis nuts and torque to 30 ft. lbs. (40 Nm)
- 2 instrument panel-to-chassis bolts at the left side of the instrument panel

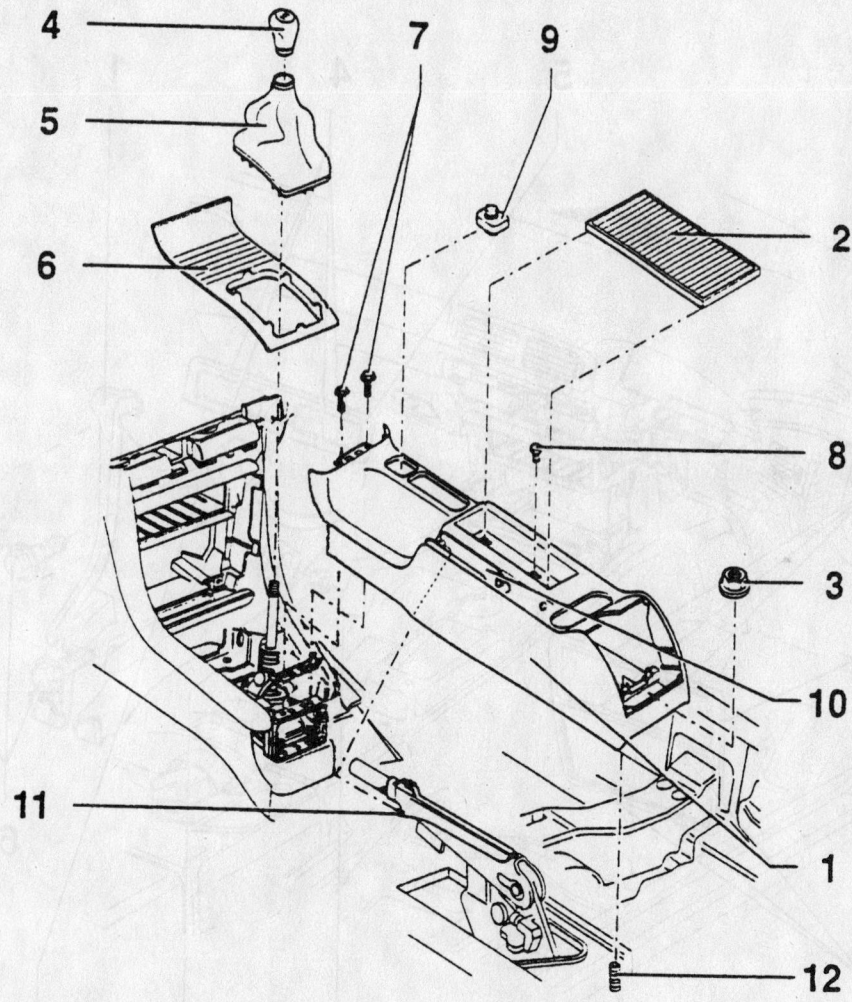

1. Rear section of center console
2. Mat lining
3. Nut
4. Shift lever knob
5. Protective boot
6. Center console insert
7. Bolts
8. Bolt
9. Mirror adjustment switch
10. Parking brake lever trim
11. Retaining tab
12. Stud

93112GN3

Exploded view of the rear console assembly—A6

- Instrument panel trim
- Left side instrument panel cover
- Pull the adjustable steering column outward
- Bracket to the instrument panel and install the pedal bracket-to-instrument panel cross member bolt.
- Coolant hose cover and the cover screws at the left side of the heater housing
- Left side instrument panel brace-to-chassis bolts and torque to 108 ft. lbs. (12 Nm)

- Right side instrument panel brace and torque the brace-to-chassis bolts to 108 ft. lbs. (12 Nm)

7. Install the instrument panel's center section by installing or connecting the following:

- Center section and connect and wiring
- 6 center section-to-instrument panel bolts and torque to 35 inch lbs. (4 Nm)
- Console-to-instrument panel nut and the trim cap at the left side of the front console

- 4 front console-to-instrument panel bolts and torque to 44 inch lbs. (5 Nm)
- Switch panel trim
- Front ashtray
- Air conditioning control head
- Radio
- Trim section
- Center console

8. At the driver's side, install the storage compartment by installing or connecting the following:

- Footwell light connector and the DLC to the storage compartment

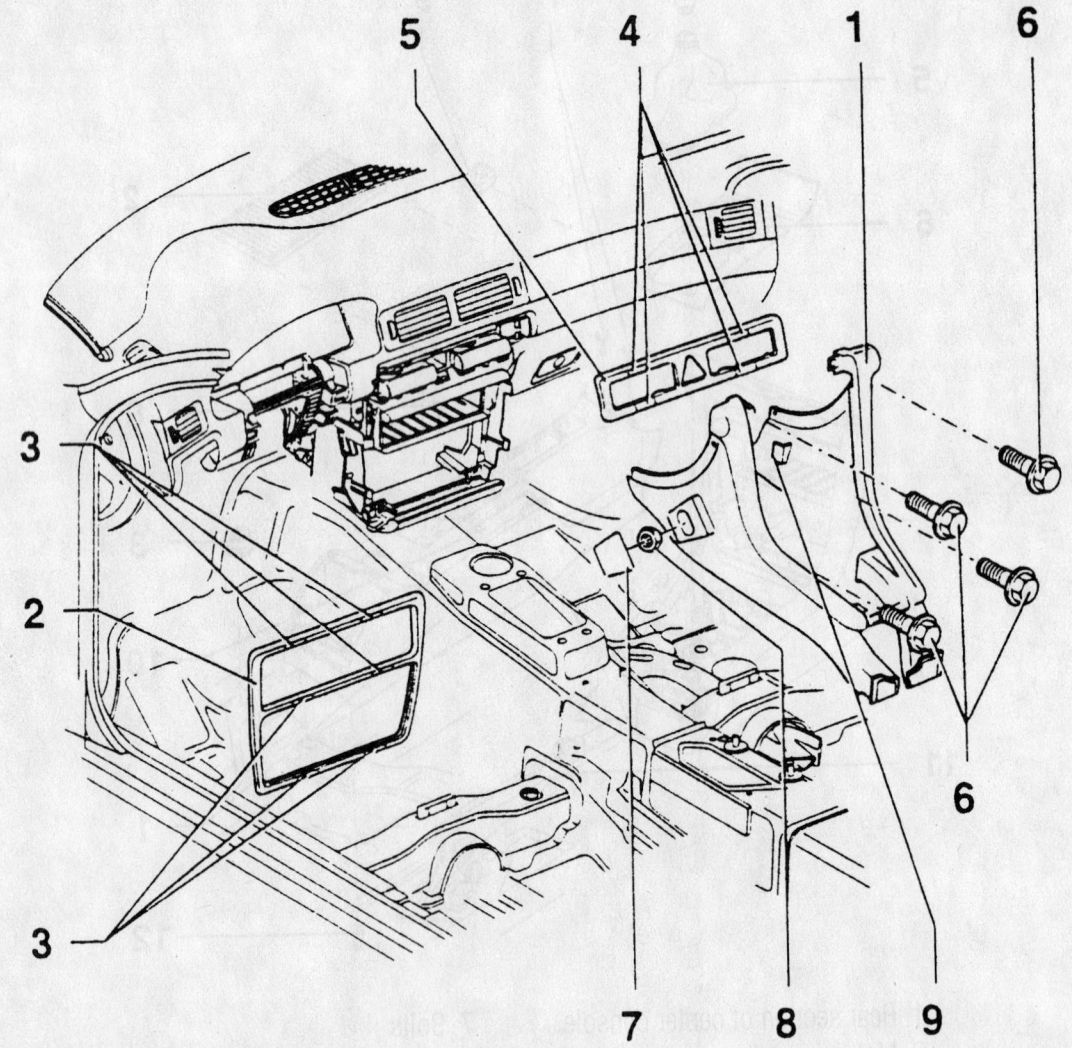

1. Front section of center console
2. Trim section
3. Fasteners
4. Fasteners
5. Switch panel trim
6. Bolts
7. Trim cap
8. Nut
9. Retainer

93112GN4

Exploded view of the front console and instrument panel center assemblies—A6

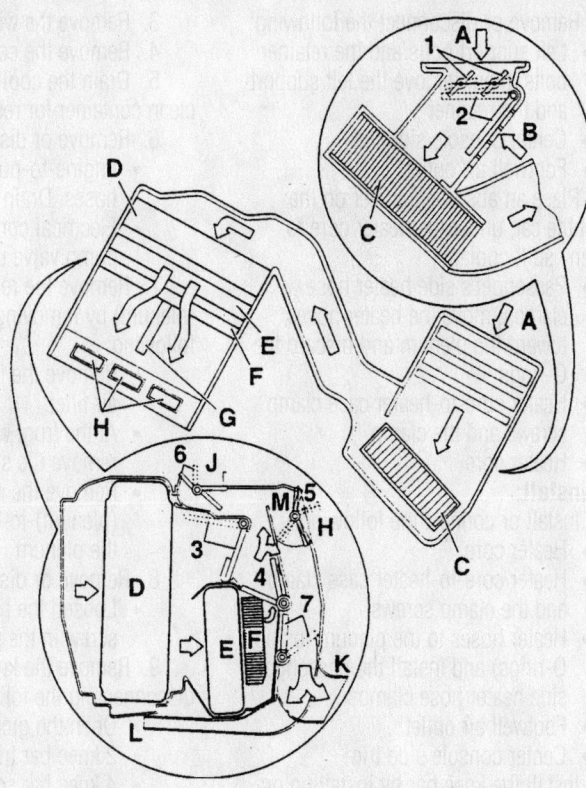

A. Plenum chamber/dust and
   pollen filter
B. Front passengers footwell
C. Fresh air blower
D. Evaporator
E. Heater core
F. Not applicable
G. Division between driver's side
   and passengers side
H. To instrument panel vents
J. To windshield vents
K. To rear footwell vents
L. To footwell vent
M. Wire mesh
1. Air flow flaps
2. Fresh air/recirculated air flap
3. Temperature flap 1
4. Temperature flap 2
5. Central flap
6. Defroster flap
7. Footwell flap

93112GM0

View of the air distribution system—A6

1. Bolt
2. Clips
3. Foam sealing pad
4. Bottom section of air distributor
   housing
5. Drip pan
6. Evaporator
7. Bolt
8. Refrigerant lines
9. Bracket
10. Bolt
11. O-rings
12. Central air flap motor
13. Bolt
14. Actuator for temperature flap right
15. Defroster flap motor
16. Bolt
17. Bracket for defroster flap motor
18. Air distributor housing
19. Coolant tube
20. Coolant tube
21. Bracket for coolant tubes
22. Cover
23. Bracket for coolant tubes
24. Actuator for temperature flap left
25. Clamp
26. O-ring
27. Heater core
28. Heater element for additional
    heater
29. Cap
30. Rubber mounting

93112GM9

Exploded view of the heater core, heater/air conditioning housing and related components—A6

Timing belt service is covered in Section 3 of this manual

- 5 storage compartment-to-instrument panel bolts
- Instrument panel end trim
- Instrument panel molding

9. Install the glove box by installing or connecting the following:
- Electrical connector and install the glove box
- 2 lower glove compartment bolts; 1 located at each side
- 3 upper glove compartment-to-chassis bolts
- Heater core tubes-to-chassis grommet
- Heater hoses to the heater core

10. Refill the cooling system.

11. Install the coolant recovery tank cap.

12. Install the battery and connect the positive (+) battery cable.

13. Connect the negative (()) battery cable.

14. Operate the engine to normal operating temperature; then, check the climate control operation and check for leaks.

A8

REMOVAL & INSTALLATION

Passenger's Side

1. Disconnect the negative battery cable.

2. Turn the ignition switch OFF.

3. Remove the cowl panel.

4. Drain the cooling system into the clean container for reuse.

5. Remove or disconnect the following:
- Engine-to-pump valve unit heater hoses. Drain and plug the openings
- Electrical connectors from the pump valve unit

6. Remove the reinforcement plate (plenum) by removing or disconnecting the following:
- Intake hose from the air filter
- Sound proofing mat at the front wall of the plenum
- Reinforcement plate (plenum)-to-chassis nuts/bolts and the plenum
- In the plenum, loosen the heater hose holder screw about 2 turns

7. Remove the knee bar by removing or disconnecting the following:
- Open the glove box
- 2 knee bar trim screws
- 4 knee bar screws
- Pull out the knee bar and disengage the latch
- Electrical connector and remove the knee bar

8. Remove or disconnect the following:
- Left support bolts and the retainer bolts; then, remove the left support and the retainer
- Center console side trim.
- Footwell air outlet

9. Place an absorbent cover on the floor of the car, under the heater core to catch any spilt coolant.
- Passenger's side heater hose clamps, move the heater hoses toward the plenum and discard the O-rings
- Heater core-to-heater case clamp screws and the clamp
- Heater core

To install:

10. Install or connect the following:
- Heater core
- Heater core-to-heater case clamp and the clamp screws
- Heater hoses to the plenum (using O-rings) and install the passenger's side heater hose clamps
- Footwell air outlet
- Center console side trim

11. Install the knee bar by installing or connecting the following:
- Left support and the retainer and the left support bolts and the retainer bolts; then, torque the bolts to 16 ft. lbs. (22 Nm)
- Electrical connector and install the knee bar
- Knee bar and engage the latch
- 4 knee bar screws and torque to 22 inch lbs. (2.5 Nm)
- 2 knee bar trim screws and torque to 22 inch lbs. (2.5 Nm)
- Tighten the heater hose holder screw in the plenum

12. Install the reinforcement plate (plenum) by installing or connecting the following:
- Reinforcement plate (plenum) and the plenum-to-chassis nuts/bolts
- Sound proofing mat at the front wall of the plenum
- Intake hose to the air filter
- Electrical connectors to the pump valve unit
- Engine-to-pump valve unit heater hoses

13. Refill the cooling system.

14. Install the cowl panel.

15. Connect the negative battery cable.

Driver's Side

1. Disconnect the negative battery cable.

2. Turn the ignition switch OFF.

3. Remove the windshield.

4. Remove the cowl panel.

5. Drain the cooling system into the clean container for reuse.

6. Remove or disconnect the following:
- Engine-to-pump valve unit heater hoses. Drain and plug the openings
- Electrical connectors from the pump valve unit

7. Remove the reinforcement plate (plenum) by removing or disconnecting the following:
- Remove the intake hose from the air filter.
- At the front wall of the plenum, remove the sound proofing mat.
- Remove the reinforcement plate (plenum)-to-chassis nuts/bolts and the plenum

8. Remove or disconnect the following:
- Loosen the heater hose holder screw in the plenum about 2 turns

9. Remove the knee bar by removing or disconnecting the following:
- Open the glove box
- 2 knee bar trim screws
- 4 knee bar screws
- Pull out the knee bar and disengage the latch
- Electrical connector and remove the knee bar

10. Remove or disconnect the following:
- Left support bolts and the retainer bolts; then, remove the left support and the retainer
- Center console side trim
- Passenger's footwell air outlet
- Passenger's side heater hose clamps, move the heater hoses toward the plenum and discard the O-rings
- Passenger's side heater core-to-heater case clamp screws and the clamp
- Passengers side heater core
- Driver's side shelf and center console's side trim
- Loosen the heater hose holder screws about 2 turns at the driver's side heater core
- Driver's side footwell air outlet
- Driver's side heater hose clamps, move the heater hoses toward the plenum and discard the O-rings
- Driver's side heater core through the passenger's side of the heater housing

To install:

11. Install or connect the following:
- Driver's side heater core through the passenger's side of the heater housing

- Move the heater hoses away from the plenum and install the driver's side heater hose clamps using new O-rings
- Driver's side footwell air outlet
- Heater hose holder screws at the driver's side heater core
- Driver's side shelf and center console's side trim
- Passenger's side heater core
- Heater core-to-heater case clamp and the clamp screws

- Heater hoses to the plenum using O-rings and install the passenger's side heater hose clamps
- Footwell air outlet
- Center console side trim

12. Install the knee bar by installing or connecting the following:

- Left support and the retainer and the left support bolts and the retainer bolts; then, torque the bolts to 16 ft. lbs. (22 Nm)

- Electrical connector and install the knee bar
- Knee bar and engage the latch
- 4 knee bar screws and torque to 22 inch lbs. (2.5 Nm)
- 2 knee bar trim screws and torque to 22 inch lbs. (2.5 Nm)

13. In the plenum, tighten the heater hose holder screw.

14. Install the reinforcement plate (plenum) by installing or connecting the following:

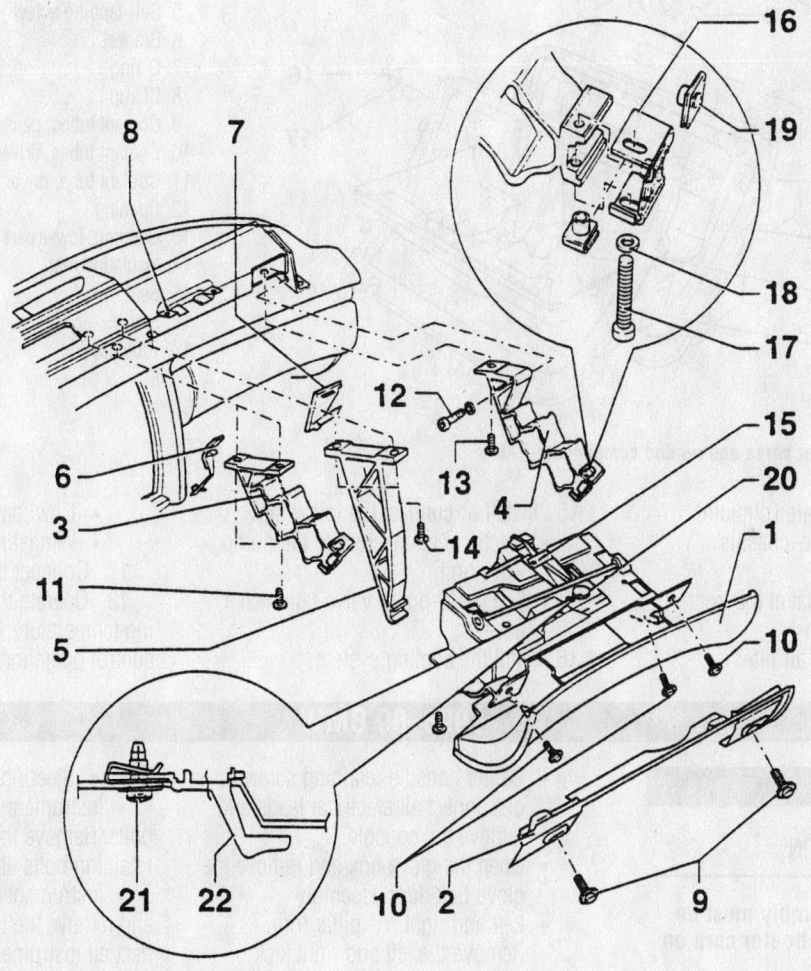

1. Knee bar	9. Screws	17. Screw
2. Knee bar trim	10. Screws	18. Washer
3. Support, left	11. Bolts	19. Clip
4. Support, right	12. Bolt	20. Adjusting eccentric
5. Retainer	13. Bolt	21. Screw
6. Knee bar bracket, left	14. Bolts	22. Eccentric arm
7. Knee bar bracket, right	15. Knee bar support	
8. Latch	16. Mounting bracket	

93112GN0

Exploded view of the glove box, knee bar assembly and related components—A8

Heater Core replacement is covered in Section 2 of this manual

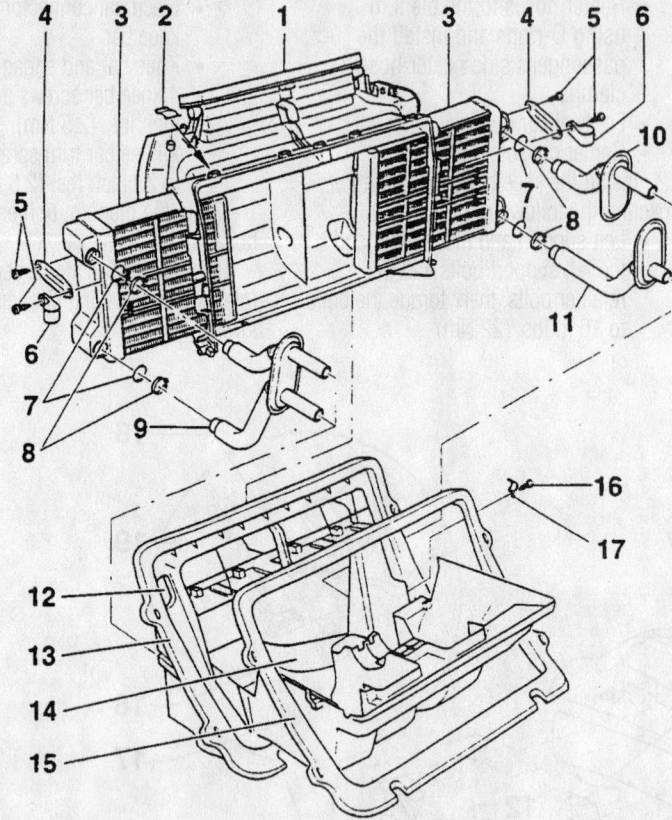

1. Air distribution housing
2. Clip
3. Heater core
4. Bracket
5. Self-tapping screw
6. Bracket
7. O-ring
8. Clamp
9. Coolant tubes, passenger side
10. Coolant tubes, driver side
11. Coolant tube, driver side lower
12. Opening
13. Housing, lower part
14. Insulation mat
15. Seal
16. Screw
17. Tube bracket

93112GM8

Exploded view of the heater cores and related components—A8

- Reinforcement plate (plenum) and the plenum-to-chassis nuts/bolts
- Sound proofing mat at the front wall of the plenum
- Intake hose to the air filter.

15. Install or connect the following:
 - Electrical connectors to the pump valve unit
 - Engine-to-pump valve unit heater hoses
16. Refill the cooling system.

- Cowl panel
- Windshield
17. Connect the negative battery cable.
18. Operate the engine to normal operating temperature; then, check the climate control operation and check for leaks.

1998–00 BMW

3-Series

REMOVAL & INSTALLATION

➡The heater case assembly must be removed to remove the heater core on all 3-Series vehicles.

1. Disconnect the negative battery cable.
2. Drain the cooling system into a clean container for reuse.
3. Discharge and recover the air conditioning system refrigerant.
4. If equipped with an automatic transmission, remove the screw retaining the shift lever T-handle. On manual transmission equipped vehicles, remove the shifter knob.
5. Remove or disconnect the following:
 - Shift lever cover and pull the window switches out of the console

- Center console retaining screws, disconnect all electrical leads and remove the console
- Open the glove box and remove the glove box door assembly
- Left and right "A" pillar trim. Remove the left and right kick panel trim
- Lower instrument panel trim pads. Remove the steering wheel

❊❊ CAUTION

When removing the steering wheel on airbag equipped vehicles, use extreme caution handling the airbag assembly. Store the airbag with the airbag facing up, to avid injury in case of accidental deployment. Replace damaged airbag assemblies. Do not try to repair or reuse deployed airbag assemblies.

- Steering column trim
- Instrument panel "A" pillar retaining bolts. Remove the lower instrument panel retaining bolts at the kick panels
- Instrument cluster retaining screws and remove the instrument cluster. Disconnect all instrument cluster electrical leads
- Radio assembly and the radio trim
- Ventilation control head
- Upper instrument panel retaining screws and pull the instrument panel away from the firewall
- All electrical multi-plugs and pull the instrument panel out of the vehicle
- Refrigerant lines from the evaporator
- Cowl cover from in the engine compartment. Disconnect the coolant lines from the heater core
- Electrical leads from the blower
- Air ducts from the heater case
- Heater case retaining bolts from the

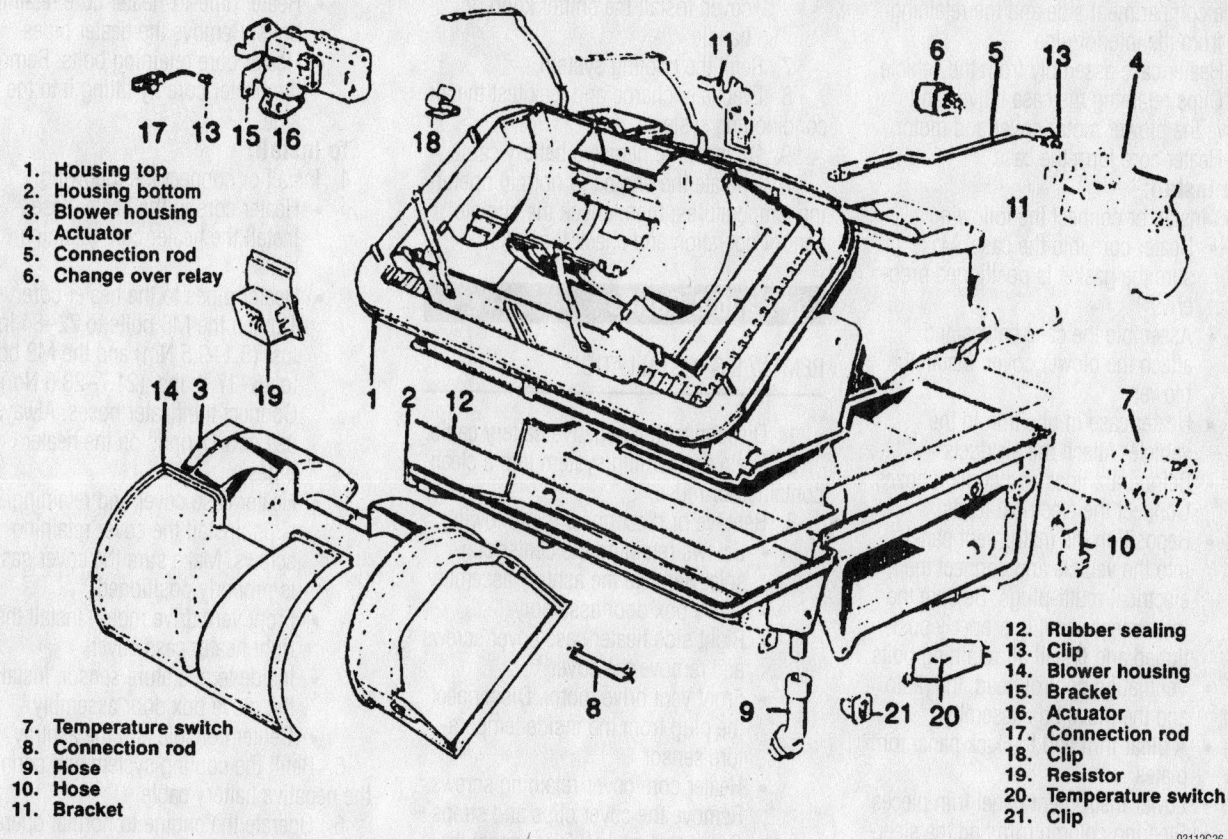

1. Housing top
2. Housing bottom
3. Blower housing
4. Actuator
5. Connection rod
6. Change over relay

7. Temperature switch
8. Connection rod
9. Hose
10. Hose
11. Bracket

12. Rubber sealing
13. Clip
14. Blower housing
15. Bracket
16. Actuator
17. Connection rod
18. Clip
19. Resistor
20. Temperature switch
21. Clip

93112G36

Exploded view of the heater/blower housing assembly—3-Series

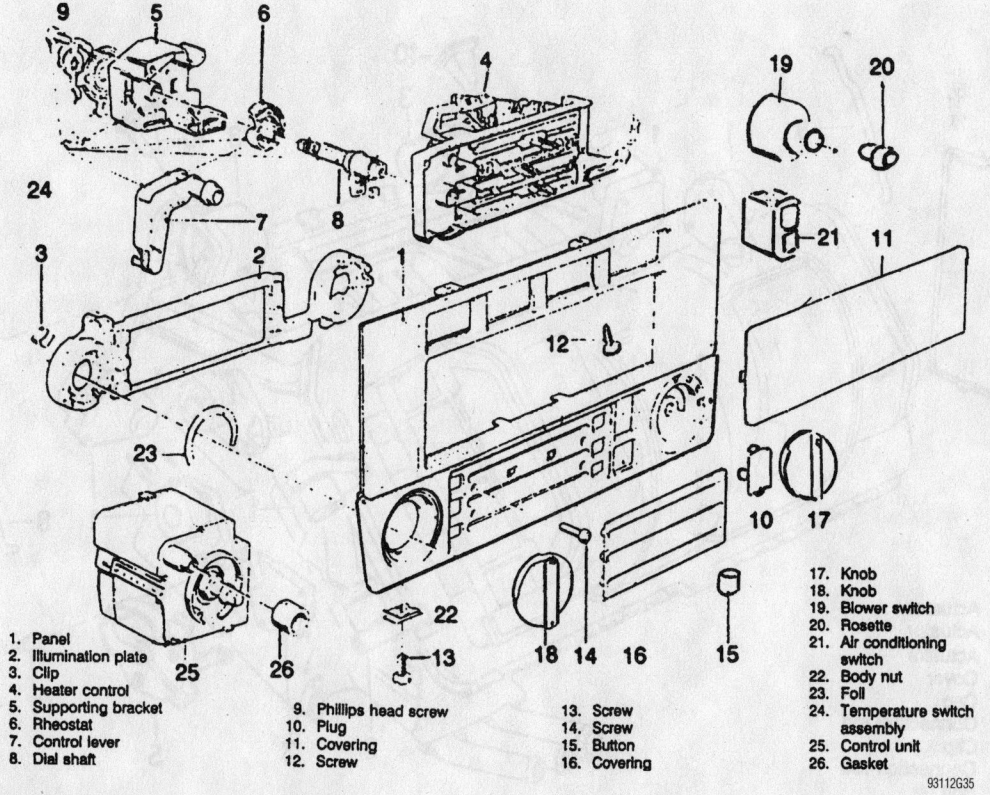

1. Panel
2. Illumination plate
3. Clip
4. Heater control
5. Supporting bracket
6. Rheostat
7. Control lever
8. Dial shaft

9. Phillips head screw
10. Plug
11. Covering
12. Screw

13. Screw
14. Screw
15. Button
16. Covering

17. Knob
18. Knob
19. Blower switch
20. Rosette
21. Air conditioning switch
22. Body nut
23. Foil
24. Temperature switch assembly
25. Control unit
26. Gasket

93112G35

Exploded view of the control head components—3-Series

Brake service is covered in Section 4 of this manual

engine compartment side and the retaining bolts from the interior side

- Heater case assembly from the vehicle
- Clips retaining the case halves and remove the blower motor cover and motor
- Heater core form the case

To install:

6. Install or connect the following:

- Heater core into the case. Make sure the gasket is positioned properly.
- Assemble the case halves and attach the blower cover. Install the blower
- Heater case in position in the vehicle. Attach the air ducts
- Refrigerant lines and coolant hoses. Connect the electrical leads
- Reposition the instrument panel into the vehicle and connect the electrical multi-plugs. Be sure the instrument panel is properly positioned and install all retaining bolts
- Ventilation control head, the radio and the glove box assembly
- A pillar trim and the kick panel trim plates
- Lower instrument panel trim pieces
- Steering column trim and the steering wheel
- Center console and the shift lever

cover. Install the shifter knob or handle

7. Refill the cooling system.
8. Evacuate, charge and leak test the air conditioning system.
9. Connect the negative battery cable.
10. Operate the engine to normal operating temperatures; then, check the climate control operation and check for leaks.

5, 7 and 8-Series

REMOVAL & INSTALLATION

1. Disconnect the negative battery cable.
2. Drain the cooling system into a clean container for reuse.
3. Remove or disconnect the following:

- Screws retaining the center console. Remove the ashtray assembly
- Glove box door assembly
- Right side heater case cover screws and remove the cover
- Front vent drive motor. Disconnect the plug from the inside temperature sensor
- Heater core cover retaining screws. Remove the cover clips and straps
- Cover and gasket. Disconnect the coolant hoses from the heater core

- Heater pipe-to-heater core retaining bolts. Remove the heater pipes
- Heater core retaining bolts. Remove the heater core by tilting it to the side

To install:

4. Install or connect the following:

- Heater core in the heater case. Install the heater core retaining bolts
- Heater pipes to the heater core. Tighten the M6 bolts to 72–84 inch lbs. (8.1–9.5 Nm) and the M8 bolts to 16–17 ft. lbs. (21.7–23.0 Nm). Connect the heater hoses. Always use new O-rings on the heater pipes.
- Heater case cover and retaining clips. Install the cover retaining screws. Make sure the cover gasket is properly positioned
- Front vent drive motor. Install the right heater case cover
- Inside temperature sensor. Install the glove box door assembly
- Center console and the ashtray

5. Refill the cooling system and connect the negative battery cable.

6. Operate the engine to normal operating temperatures; then, check the climate control operation and check for leaks.

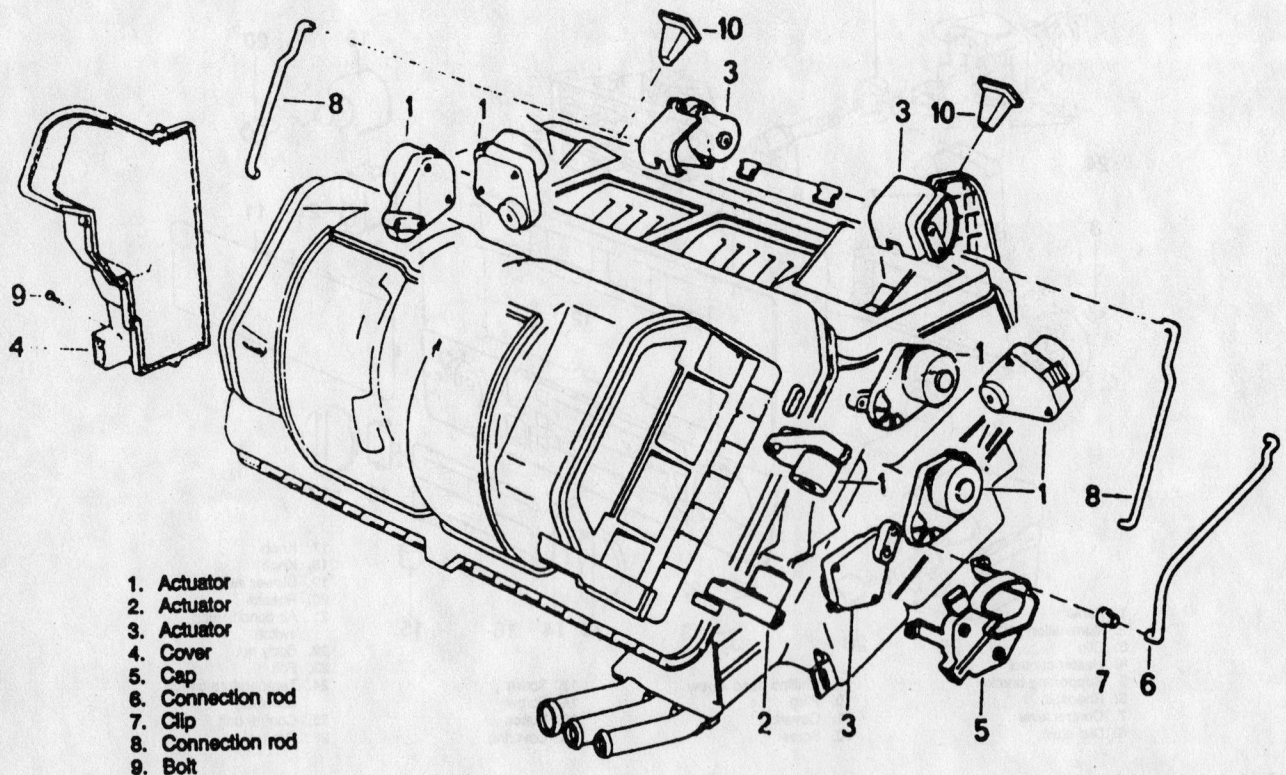

1. Actuator
2. Actuator
3. Actuator
4. Cover
5. Cap
6. Connection rod
7. Clip
8. Connection rod
9. Bolt
10. Covering

93112G39

View of the heater/blower housing assembly– 5, 7 and 8-Series

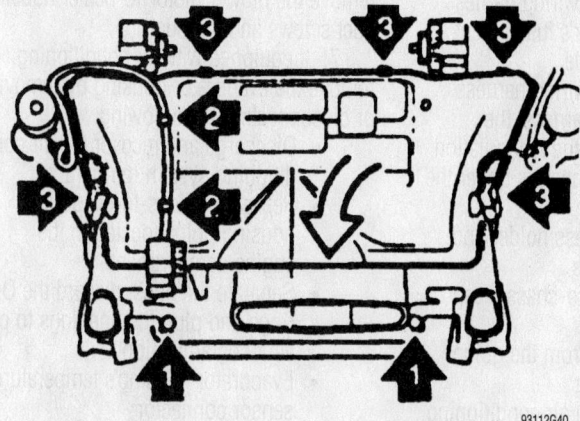

Removing the heater core cover– 5, 7 and 8-Series

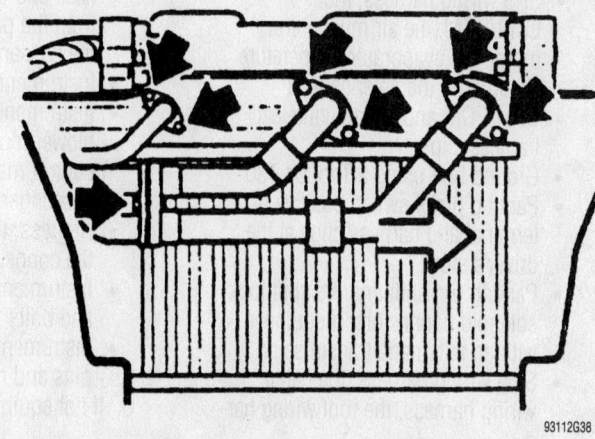

Removing the heater water pipes– 5, 7 and 8-Series

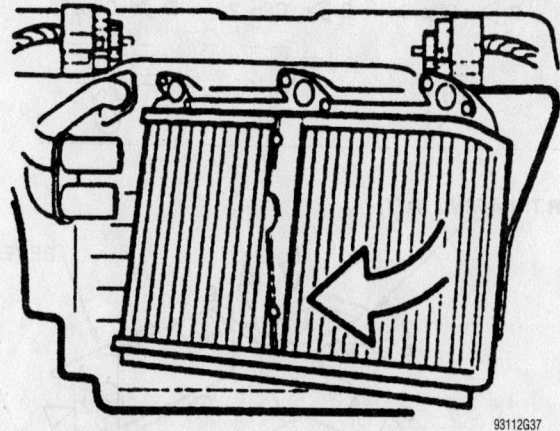

Remove the heater by pulling to the right–5, 7 and 8-Series

1998–01 HONDA

Accord

REMOVAL & INSTALLATION

1998–01 Models

➡ **Make sure to acquire the anti-theft code form the radio and write down the frequencies for the radio's preset buttons.**

1. Disconnect the negative battery cable.

❊❊ CAUTION

After disconnecting the negative battery cable, wait for at least 3 minutes for the air bag module to deplete its energy before working the on the instrument panel or steering wheel.

2. Drain the cooling system into a clean container for reuse.

3. In the engine compartment, open the heater valve cable clamp and disconnect the cable from the heater valve arm. Then, turn the heater valve to the fully opened position.

4. Remove or disconnect the following:
- Heater hoses from the heater core
- Heater housing-to-chassis nut

➡ **When removing the heater housing nut, be careful not to damage or bend the fuel lines, the brake lines or etc.**

- Center console

5. Remove the instrument panel by removing or disconnecting the following:
- Center lower cover
- Passenger's side lower cover
- Lower glove box screws
- Hooks at the top inner side of the glove box by placing a flat tipped screwdriver in the cap notch and pry out the cap

- 4 glove box screws, release the hooks and remove the glove box
- Front door opening trim and the front door pillar trim at both sides
- Combination switch, the ignition switch and the air bag connectors
- Steering column-to-instrument panel nuts and lower the steering column
- Side wiring harness connectors, cabin wire harness and door wire harness from the fuse box at the driver's side
- Instrument panel wiring harness connector
- Wiring harness, the steering hanger beam wiring harness and the brake switch connectors under the dash
- Clutch switch connectors, if equipped with a manual transmission
- Pull back the carpet at the passenger's side

For complete Engine Mechanical specifications, see Section 1 of this manual

- SRS wiring harness, the ECM/PCM, the air mix control motor, the evaporator temperature sensor and the antenna lead
- ECM/PCM and the antenna lead harness clips
- Ground bolt using a Torx bit T30
- Parking brake switch positive (+) terminal and harness clips at the driver's side
- Parking pin shift and the shift lock solenoid connectors, if equipped with an automatic transmission
- Side wiring harness, the cabin wiring harness, the roof wiring har-

ness and the door wiring harness from the passenger's fuse box at the passenger's side
- Instrument panel wiring harness
- Instrument pane harness, the blower motor and the recirculation control motor connectors under the passenger's side
- Harness, the harness holder and the connector clips
- Instrument panel-to-chassis cap and bolts
- Instrument panel from the guide pins and remove it
6. If not equipped with air conditioning,

remove the blower motor-to-heater housing duct screws and the duct.
7. If equipped with air conditioning, remove the evaporator housing by removing or disconnecting the following:

- Discharge and recover the air conditioning system refrigerant
- Refrigerant lines-to-evaporator housing bolts, located in the engine compartment
- Separate the lines, discard the O-rings and plug the openings to prevent contamination
- Evaporator housing's temperature sensor connector

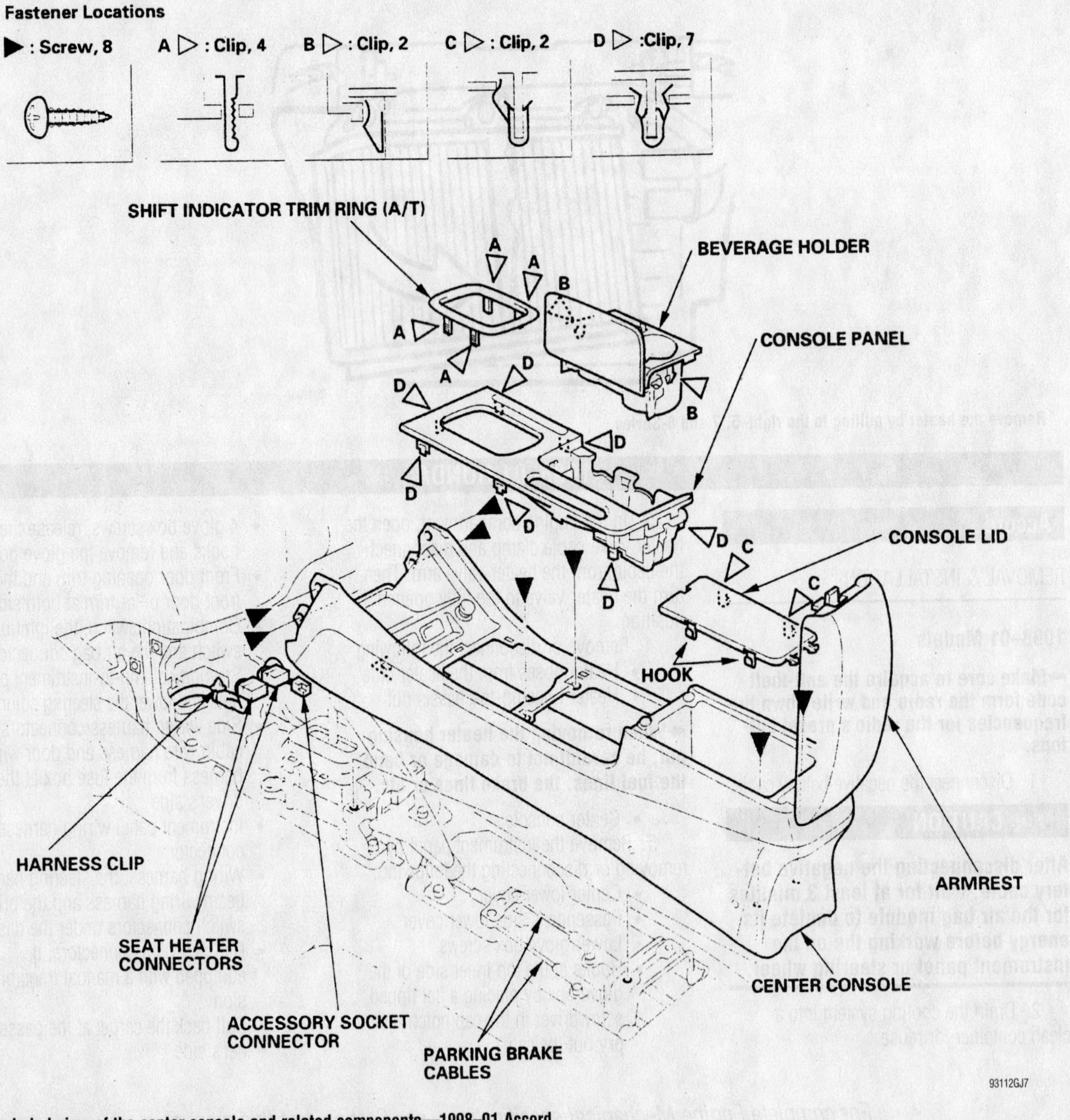

Fastener Locations

▶ : Screw, 8   A ▷ : Clip, 4   B ▷ : Clip, 2   C ▷ : Clip, 2   D ▷ : Clip, 7

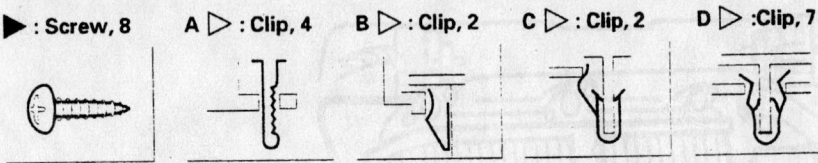

SHIFT INDICATOR TRIM RING (A/T)

BEVERAGE HOLDER

CONSOLE PANEL

CONSOLE LID

HOOK

ARMREST

CENTER CONSOLE

HARNESS CLIP

SEAT HEATER CONNECTORS

ACCESSORY SOCKET CONNECTOR

PARKING BRAKE CABLES

Exploded view of the center console and related components—1998–01 Accord

93112GJ7

Fastener Locations

B ▶ : Bolt, 6   C ▶ : Bolt, 4   D ▶ : Bolt, 3   E ▶ : Bolt, 1

8 x 1.25 mm
22 N·m (2.2 kgf·m,
16 lbf·ft)

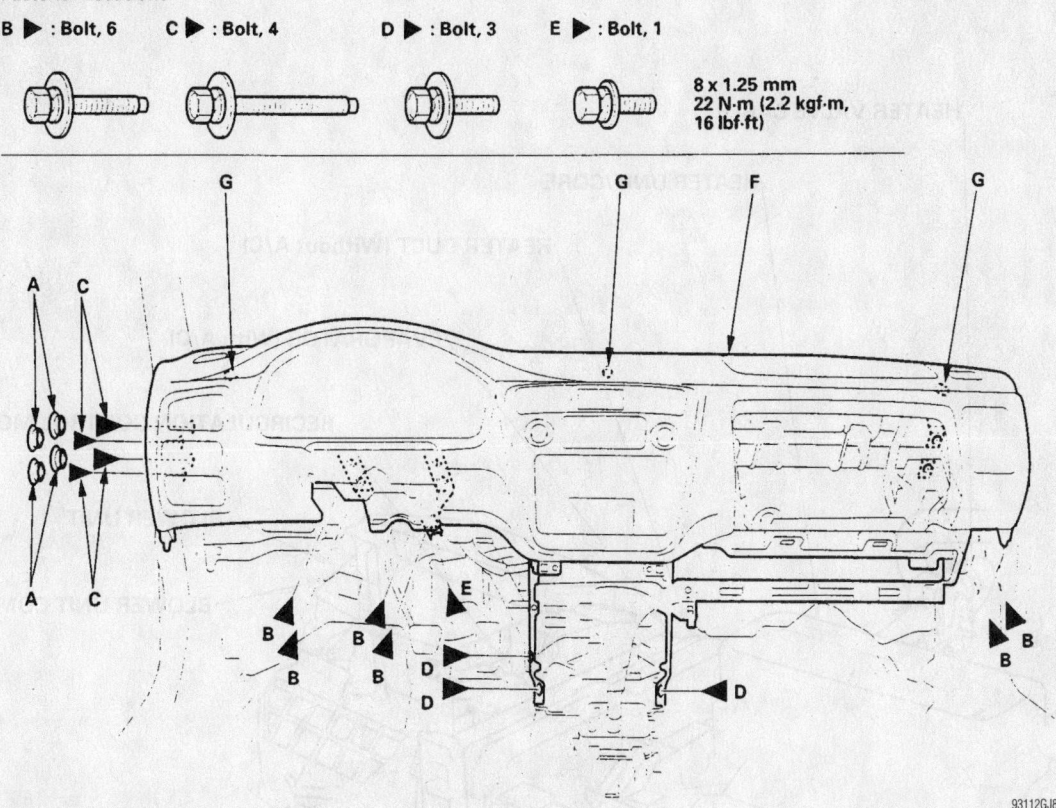

93112GJ8

View of the instrument panel bolt locations—1998–01 Accord

- Evaporator housing-to-chassis nut/bolts and remove the evaporator housing
8. Remove or disconnect the following:
  - Heater housing-to-chassis bolts and the heater housing
  - Heater core-to-heater housing bracket screws and the brackets
  - Heater core from the heater housing

To install:
9. Install or connect the following:
  - Heater core to the heater housing
  - Heater core-to-heater housing bracket and the bracket screws
  - Heater housing and the heater housing-to-chassis bolts
10. If not equipped with air conditioning, install the blower motor-to-heater housing duct and the duct screws.
11. If equipped with air conditioning, install the evaporator housing by installing or connecting the following:
  - Evaporator housing and the evaporator housing-to-chassis nut/bolts
  - Evaporator housing's temperature sensor connector
  - Refrigerant lines using new O-rings

- Refrigerant lines-to-evaporator housing bolts, located in the engine compartment
12. Install the instrument panel by installing or connecting the following:
  - Instrument panel to the guide pins
  - Instrument panel-to-chassis cap and bolts
  - Harness holder and the connector clips
  - Instrument pane harness, the blower motor and the recirculation control motor connectors under the passenger's side
  - Instrument panel wiring harness
  - Side wiring harness, the cabin wiring harness, the roof wiring harness and the door wiring harness to the passenger's fuse box at the passenger's side
  - Parking pin shift and the shift lock solenoid connectors, if equipped with an automatic transmission
  - Parking brake switch positive (+) terminal and harness clips a the driver's side
  - Ground bolt using a Torx bit T30

- ECM/PCM and the antenna lead harness clips
- SRS wiring harness, the ECM/PCM, the air mix control motor, the evaporator temperature sensor and the antenna lead
- Move back the carpet.
- Clutch switch connectors, if equipped with a manual transmission
- Wiring harness, the steering hanger beam wiring harness and the brake switch connectors, located under the dash
- Instrument panel wiring harness connector
- Side wiring harness connectors, cabin wire harness and door wire harness from the fuse box, located at the driver's side
- Steering column and the steering column-to-instrument panel nuts
- Combination switch, the ignition switch and the air bag connectors
- Front door pillar trim and the front door opening trim on both sides
- Glove box and the 4 glove box screws

For Accessory Drive Belt illustrations, see Section 1 of this manual

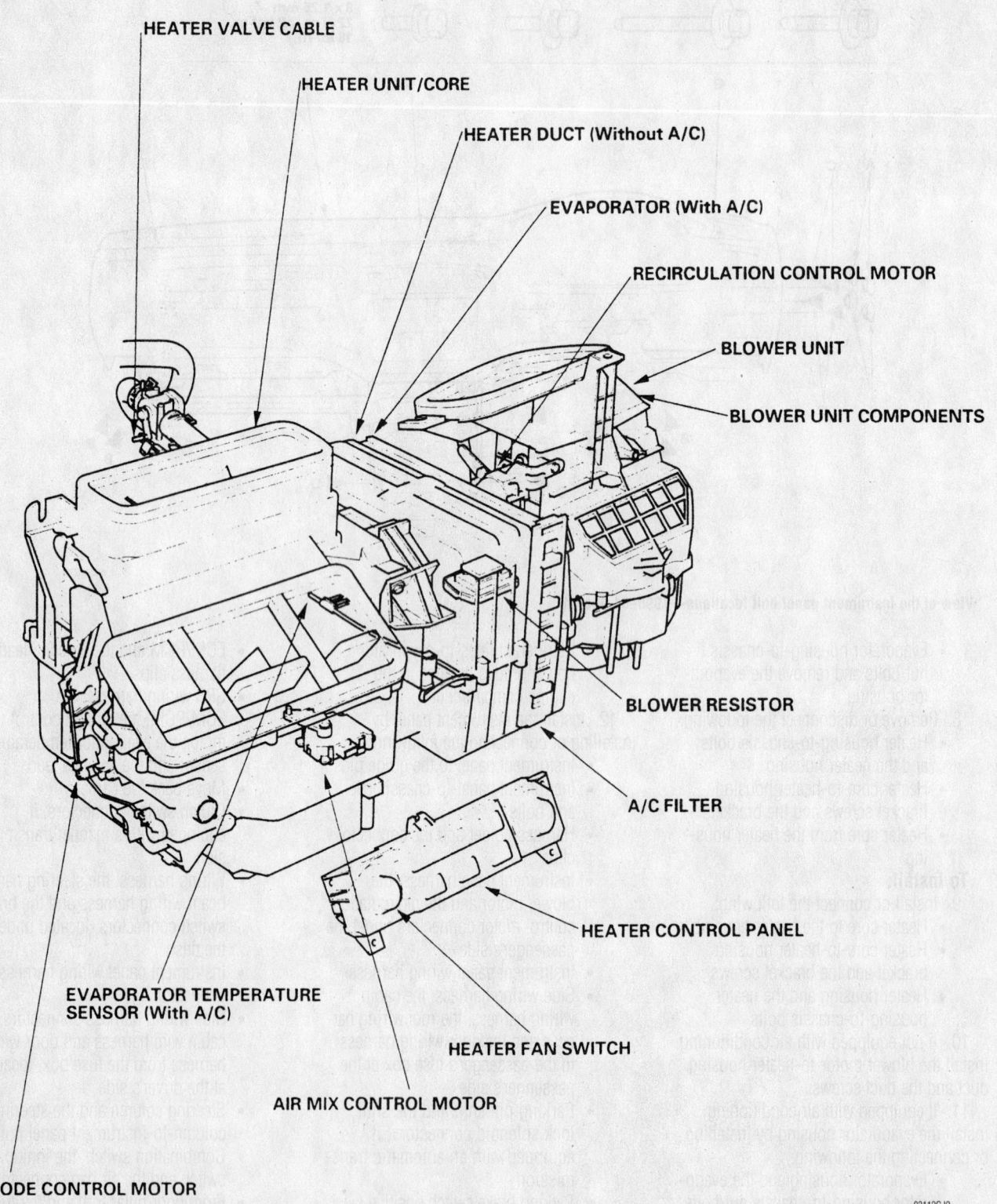

HEATER VALVE CABLE

HEATER UNIT/CORE

HEATER DUCT (Without A/C)

EVAPORATOR (With A/C)

RECIRCULATION CONTROL MOTOR

BLOWER UNIT

BLOWER UNIT COMPONENTS

BLOWER RESISTOR

A/C FILTER

HEATER CONTROL PANEL

EVAPORATOR TEMPERATURE
SENSOR (With A/C)

HEATER FAN SWITCH

AIR MIX CONTROL MOTOR

MODE CONTROL MOTOR

93112GJ0

View of the heater housing, evaporator housing and related components—1998–01 Accord

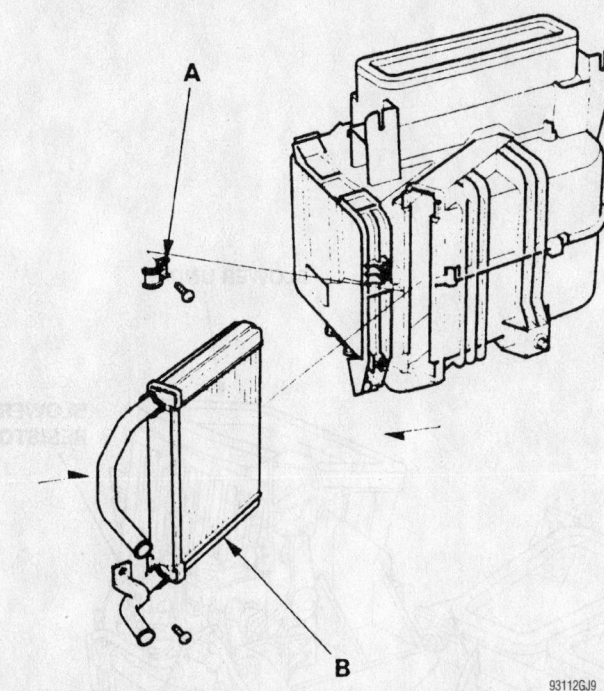

A

B

93112GJ9

Exploded view of the heater core and housing—1998–01 Accord

- Cap at the top inner side of the glove box
- Lower glove box screws
- Passenger's side lower cover
- Center lower cover
- Center console

➡️**When installing the heater housing nut, be careful not to damage or bend the fuel lines, the brake lines or etc.**

13. Install or connect the following:
    - Heater housing-to-chassis nut
    - Heater hoses to the heater core
14. Refill the cooling system.
15. Connect the negative battery cable.
16. Evacuate, charge and leak test the air conditioning system refrigerant.
17. Operate the engine to normal operating temperatures; then, check the climate control operation and check for leaks.

Civic and del Sol

REMOVAL & INSTALLATION

1. Disconnect the negative battery cable.

❋❋ CAUTION

Wait at least 3 minutes for the SRS to deplete its energy before working on the steering wheel or instrument panel.

2. In the engine compartment, remove the heater valve cable clamp; then, disconnect the heater valve cable and rotate the heater valve to the fully open position.

3. Drain the engine coolant into a clean container for reuse.

4. Remove or disconnect the following:
   - Heater hoses from the heater unit
   - Heater housing-to-chassis nut
   - Instrument panel.

5. If not equipped with air conditioning, remove the wiring harness from the heater duct; then, remove the 2 screws and the heater duct.

6. If equipped with air conditioning, remove or disconnect the following:
   - Discharge and recover the air conditioning system refrigerant
   - Refrigerant lines-to-evaporator bolts. Disconnect the lines. Discard the O-rings. Plug the openings to prevent contamination
   - Thermostat electrical connector and the wiring harness from the evaporator
   - 4 evaporator housing screws, the bolt and nut
   - Drain hose and remove the evaporator housing

7. On 198 models, disconnect the electrical connectors from the blower motor, the

blower resistor and the recirculation control motor.

8. On 1999–01 models, disconnect the electrical connectors from the blower motor, the power transistor, the blower motor high relay and the recirculation control motor.

9. Remove or disconnect the following:
   - Electrical connectors from the mode control motor and the air mix control motor (for 1999–01); then, remove the electrical harness clips and wiring harness from the heater housing
   - Heater duct clip
   - 2 heater housing-to-chassis nuts and the heater housing
   - Heater core cover-to-heater housing screws, the cover, the clamp and the heater core

To install:

10. Install or connect the following:
    - Heater core, the clamp, the cover and the heater core cover-to-heater housing screws
    - Heater housing and the 2 heater housing-to-chassis nuts
    - Heater duct clip
    - Electrical connectors to the mode control motor and the air mix control motor (for 1999–01); then, install the electrical harness clips and wiring harness to the heater housing

11. On 1999–01 models, connect the electrical connectors to the blower motor, the power transistor, the blower motor high relay and the recirculation control motor.

12. On 1998 models, connect the electrical connectors to the blower motor, the blower resistor and the recirculation control motor.

13. If not equipped with air conditioning, install the heater duct, the 2 screws and the wiring harness to the heater duct.

14. If equipped with air conditioning, install or connect the following:
    - Drain hose and install the evaporator housing
    - 4 evaporator housing screws, the bolt and nut
    - Thermostat electrical connector and the wiring harness to the evaporator
    - Refrigerant lines-to-evaporator bolts, connect the lines using new O-rings
    - Evacuate, charge and leak test the air conditioning system refrigerant.

15. Install or connect the following:

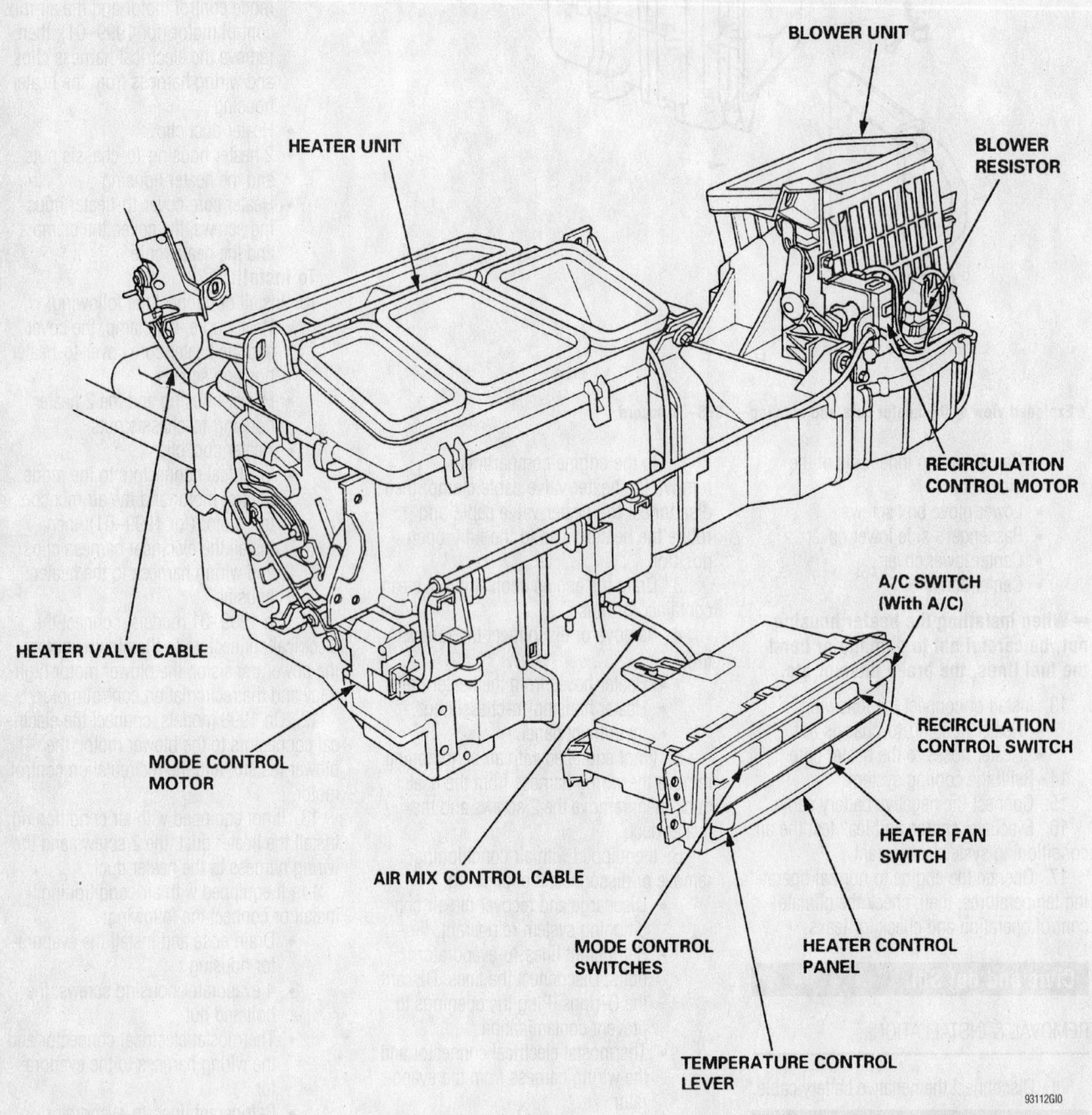

HEATER UNIT

BLOWER UNIT

BLOWER RESISTOR

RECIRCULATION CONTROL MOTOR

A/C SWITCH (With A/C)

RECIRCULATION CONTROL SWITCH

HEATER FAN SWITCH

HEATER CONTROL PANEL

TEMPERATURE CONTROL LEVER

MODE CONTROL SWITCHES

AIR MIX CONTROL CABLE

MODE CONTROL MOTOR

HEATER VALVE CABLE

93112GI0

View of the heater housing, evaporator housing and related components—1998 Civic

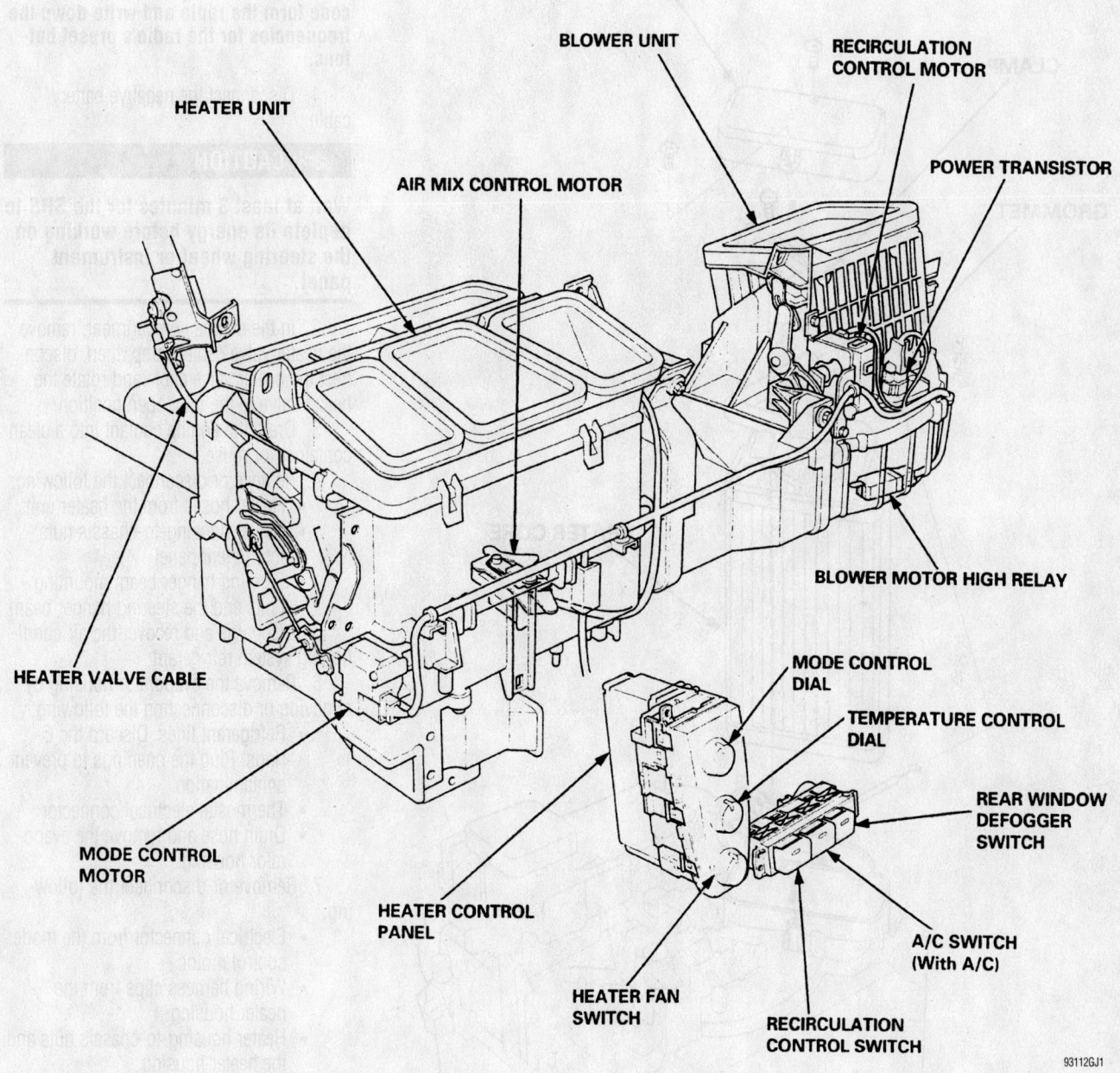

BLOWER UNIT

RECIRCULATION
CONTROL MOTOR

POWER TRANSISTOR

HEATER UNIT

AIR MIX CONTROL MOTOR

BLOWER MOTOR HIGH RELAY

HEATER VALVE CABLE

MODE CONTROL
DIAL

TEMPERATURE CONTROL
DIAL

REAR WINDOW
DEFOGGER
SWITCH

MODE CONTROL
MOTOR

HEATER CONTROL
PANEL

A/C SWITCH
(With A/C)

HEATER FAN
SWITCH

RECIRCULATION
CONTROL SWITCH

93112GJ1

View of the heater housing, evaporator housing and related components—1999–01 Civic

For Wheel Alignment specifications, see Section 1 of this manual

- Instrument panel
- Heater housing-to-chassis nut
- Heater hoses to the heater unit
16. Refill the cooling system.
17. In the engine compartment, install

the heater valve cable clamp and connect the heater valve cable.

18. Connect the negative battery cable.

19. Operate the engine to normal operating temperatures; then, check the cli-

mate control operation and check for leaks.

Prelude

REMOVAL & INSTALLATION

➡ **Make sure to acquire the anti-theft code form the radio and write down the frequencies for the radio's preset buttons.**

1. Disconnect the negative battery cable.

✳✳ CAUTION

Wait at least 3 minutes for the SRS to deplete its energy before working on the steering wheel or instrument panel.

2. In the engine compartment, remove the heater valve cable clamp; then, disconnect the heater valve cable and rotate the heater valve to the fully open position.

3. Drain the engine coolant into a clean container for reuse.

4. Remove or disconnect the following:
- Heater hoses from the heater unit
- Heater housing-to-chassis nuts
- Instrument panel
- Steering hanger beam mounting bolts and the steering hanger beam

5. Discharge and recover the air conditioning system refrigerant

6. Remove the evaporator housing by removing or disconnecting the following:
- Refrigerant lines. Discard the O-rings. Plug the openings to prevent contamination.
- Thermostat electrical connector
- Drain hose and remove the evaporator housing.

7. Remove or disconnect the following:
- Electrical connector from the mode control motor
- Wiring harness clips from the heater housing
- Heater housing-to-chassis nuts and the heater housing
- Heater housing screws and separate the housings
- Heater core from the heater housing

To install:

8. Install or connect the following:
- Heater core to the heater housing
- Assemble the housings and install the heater housing screws
- Heater housing and the heater housing-to-chassis nuts

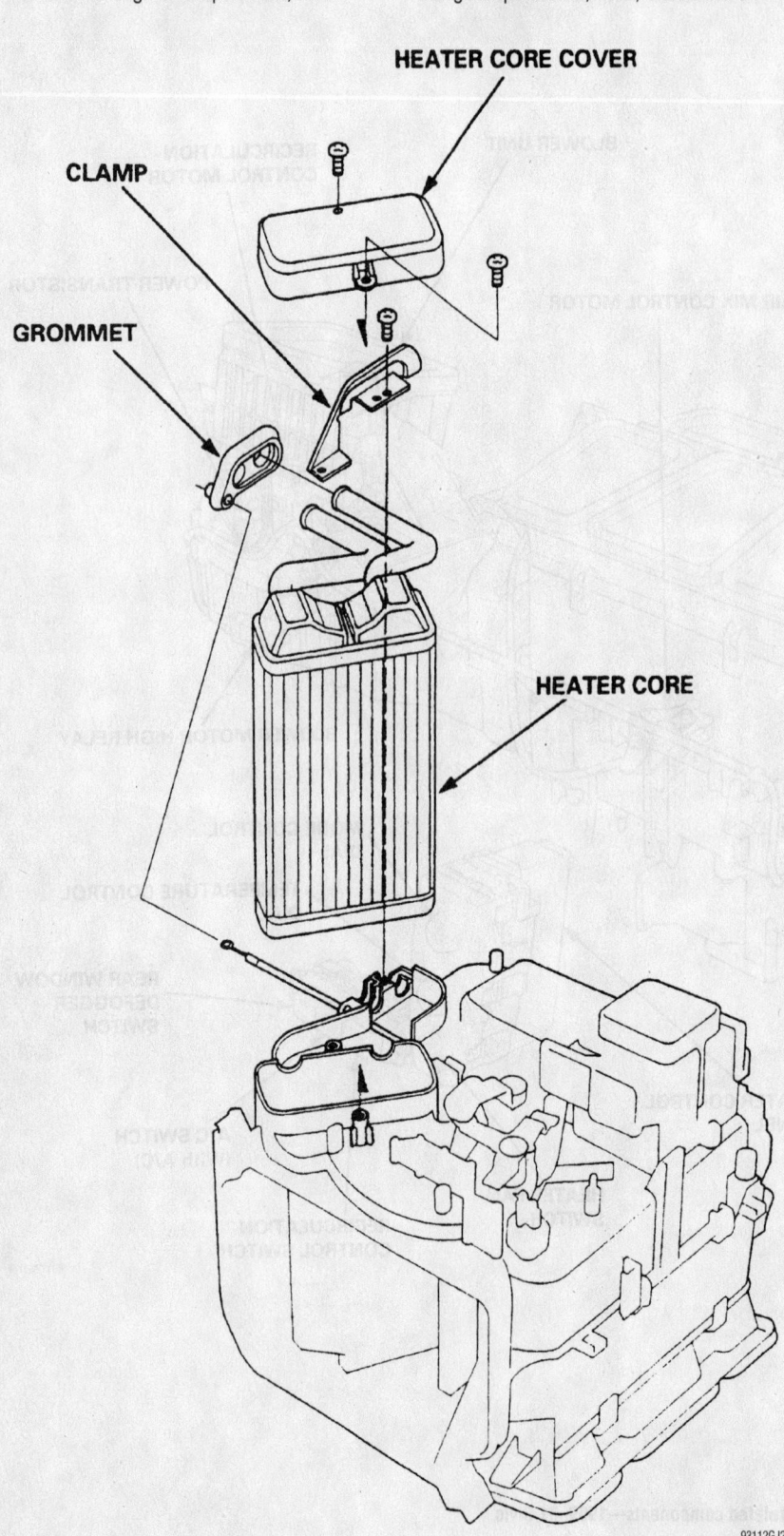

HEATER CORE COVER

CLAMP

GROMMET

HEATER CORE

93112GJ2

Exploded view of the heater core—Civic

- Wiring harness clips to the heater housing
- Electrical connector to the mode control motor

9. Install the evaporator housing:
- Evaporator housing and connect the drain hose
- Evaporator housing-to-chassis screws, the bolt and nuts
- Thermostat electrical connector and

the wiring harness to the evaporator
- Refrigerant lines using new O-rings
- Evacuate, charge and leak test the air conditioning system refrigerant

10. Install or connect the following:
- Steering hanger beam and the steering hanger beam mounting bolts
- Instrument panel

- Heater housing-to-chassis nuts
- Heater hoses to the heater unit

11. Refill the cooling system.

12. In the engine compartment, Install the heater valve cable clamp; then, connect the heater valve cable.

13. Connect the negative battery cable.

14. Operate the engine to normal operating temperatures; then, check the climate control operation and check for leaks.

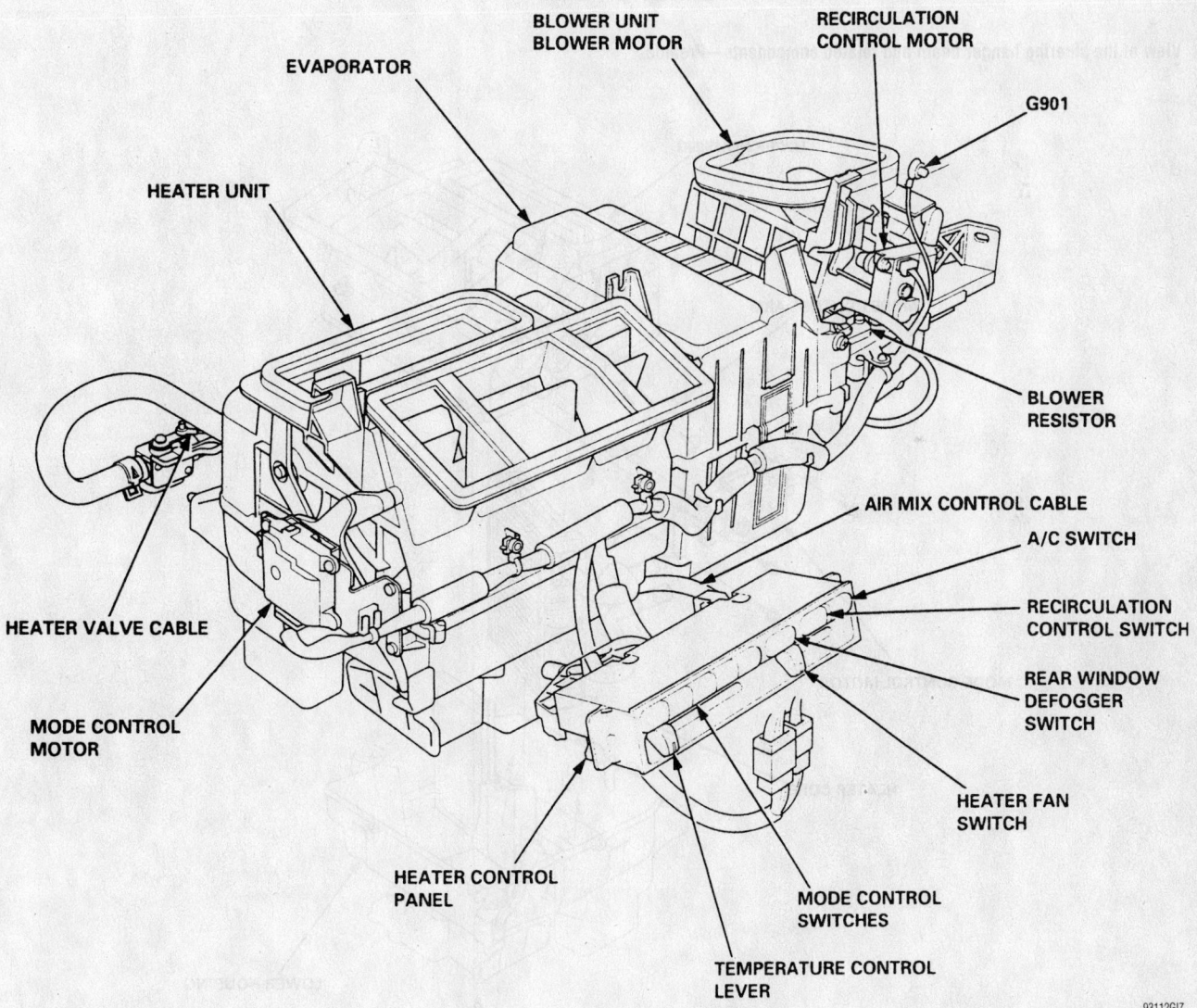

View of the heater housing, evaporator housing and related components—Prelude

For Maintenance Interval recommendations, see Section 1 of this manual

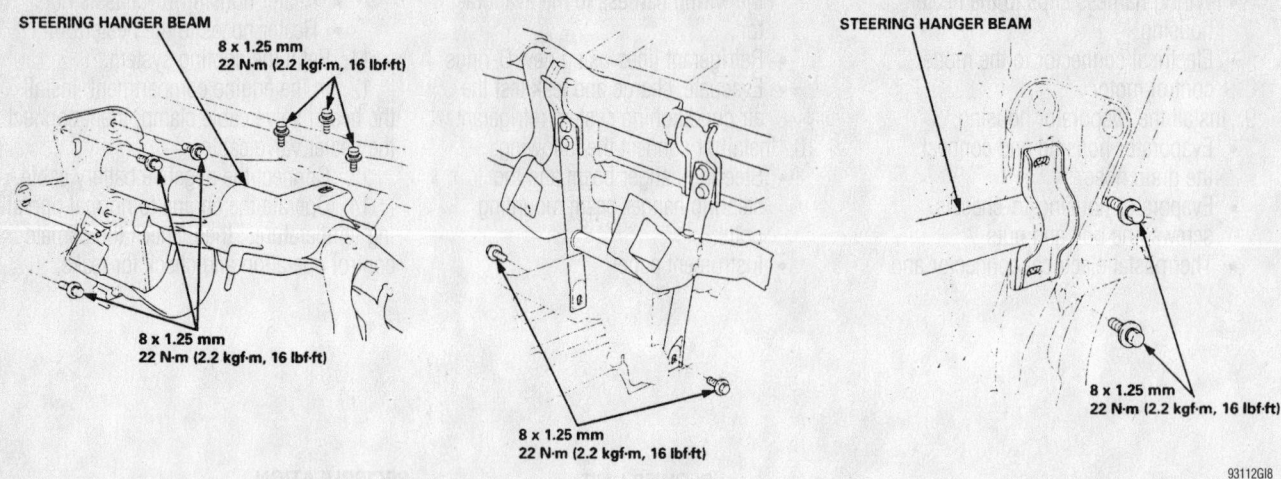

STEERING HANGER BEAM

8 x 1.25 mm
22 N·m (2.2 kgf·m, 16 lbf·ft)

8 x 1.25 mm
22 N·m (2.2 kgf·m, 16 lbf·ft)

STEERING HANGER BEAM

8 x 1.25 mm
22 N·m (2.2 kgf·m, 16 lbf·ft)

8 x 1.25 mm
22 N·m (2.2 kgf·m, 16 lbf·ft)

93112GI8

View of the steering hanger beam and related components—Prelude

UPPER HOUSING

MODE CONTROL ARM

MODE CONTROL MOTOR

HEATER CORE

LOWER HOUSING

93112GI9

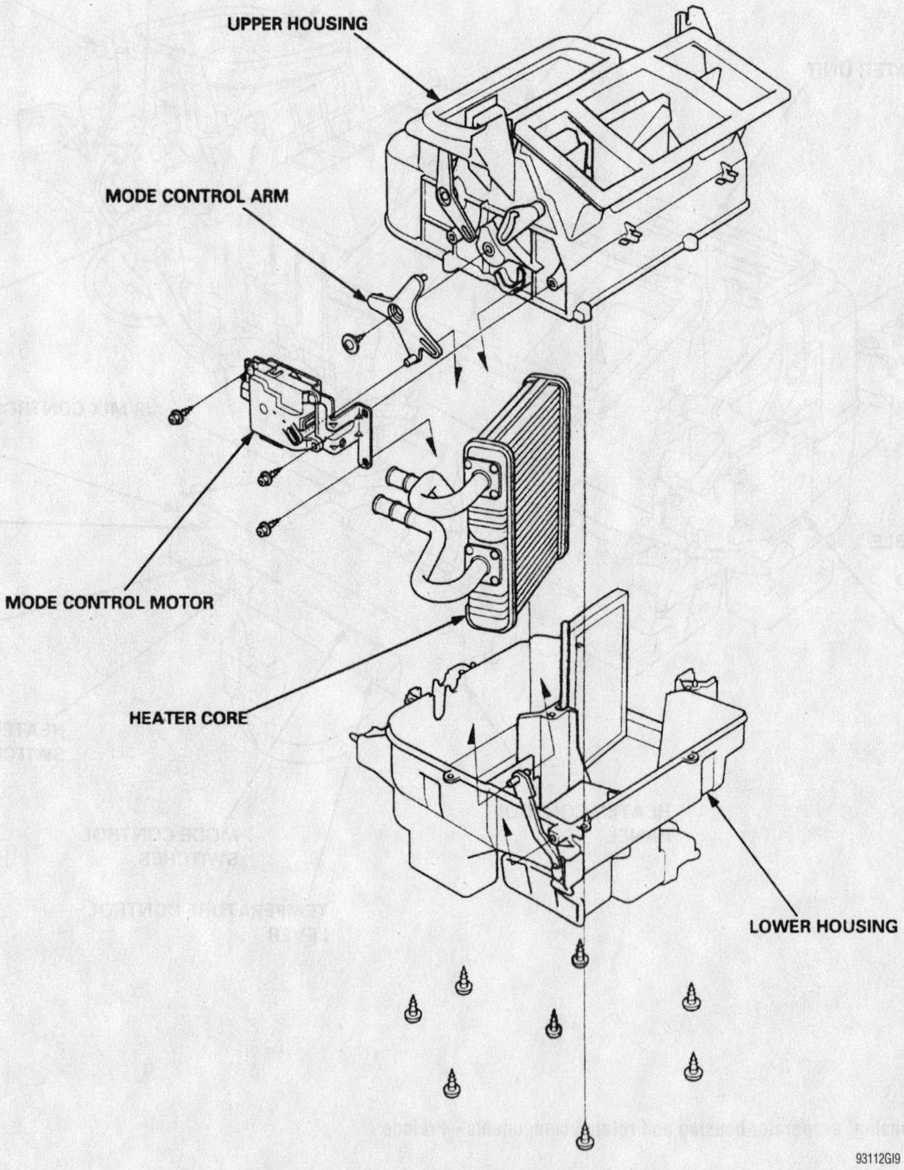

Exploded view of the heater core, the heater housing and related components—Prelude

1998-01 HYUNDAI

Accent

REMOVAL & INSTALLATION

1. Disconnect the negative battery cable and wait 90 seconds for the SRS memory battery to drain.

☀☀ CAUTION

After disconnecting the negative battery cable, wait for at least 30 seconds for the SRS module to deplete its stored energy.

2. Drain the cooling system into a clean container for reuse.
3. Remove or disconnect the following:
   - Heater hoses with the vacuum hose from the heater housing
   - Discharge and recover the air conditioning system refrigerant
   - Suction and discharge hoses from the evaporator assembly
4. Remove the SRS module and the steering wheel:

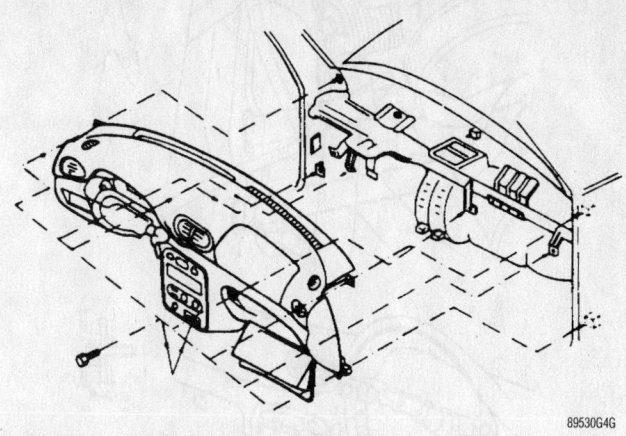

89530G4G

Instrument panel screw locations—Accent

- Steering wheel-to-SRS module nuts
- SRS module from the steering wheel and disconnect the electrical connector
- Steering wheel-to-steering column nut
- Steering wheel from the steering column

5. Remove or disconnect the following:
   - Multi-function switch assembly
   - Front and rear console assemblies
   - Lower left side crash pad
   - Center fascia panel and disconnect the connectors and vacuum connector from the heater control assembly

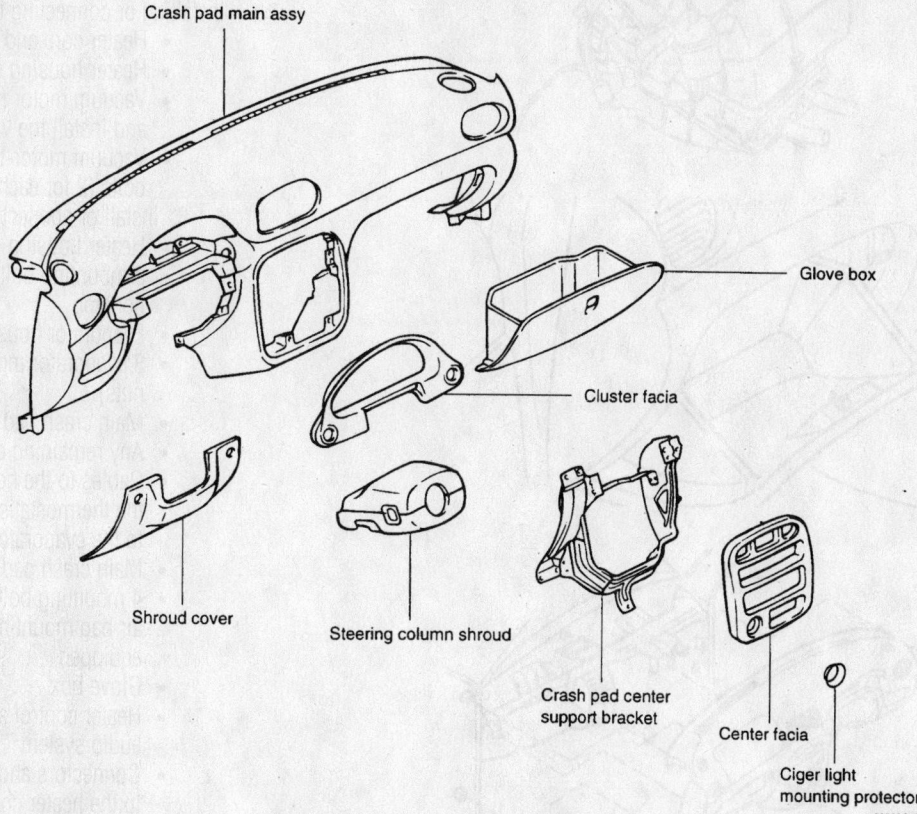

Crash pad main assy

Glove box

Cluster facia

Shroud cover

Steering column shroud

Crash pad center support bracket

Center facia

Ciger light mounting protector

89530G4F

Instrument panel assembly—Accent

For Tune-up, Capacities and Firing orders, see Section 1 of this manual

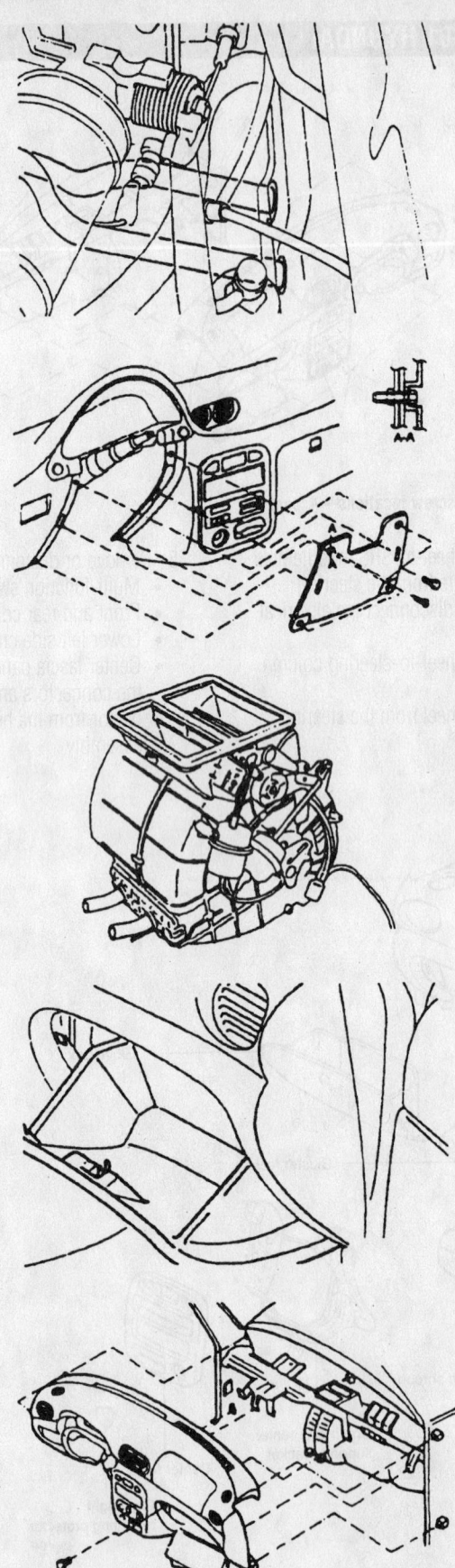

View of the heater housing assembly and related components—Accent

93112G30

- Heater control assembly and the audio system
- Glove box
- 4 mounting bolts from the passenger air bag mounting bracket, if equipped
- Main crash pad assembly
- Cables from the heater housing and the thermostatic switch connector from the evaporator housing
- Any remaining connectors
- Main crash pad assembly
- 3 evaporator mounting bolts (or nuts)
- Evaporator housing
- 3 mounting bolts from the heater housing
- Heater housing

6. Disassemble the heater housing by removing or disconnecting the following:

- Vacuum motor-to-heater housing bolts (2 for each vacuum motor)
- Vacuum motor rod end connection and remove the vacuum motors
- Heater housing cover clips
- Cover and the heater core

To install:

7. Assemble the heater housing by installing or connecting the following:

- Heater core and the cover
- Heater housing cover clips
- Vacuum motor rod end connection and install the vacuum motors
- Vacuum motor-to-heater housing bolts (2 for each vacuum motor)

8. Install or connect the following:

- Heater housing
- 3 mounting bolts to the heater housing
- Evaporator housing
- 3 evaporator mounting bolts (or nuts)
- Main crash pad assembly
- Any remaining connectors
- Cables to the heater housing and the thermostatic switch connector to the evaporator housing
- Main crash pad assembly
- 4 mounting bolts to the passenger air bag mounting bracket, if equipped
- Glove box
- Heater control assembly and the audio system
- Connectors and vacuum connector to the heater control assembly. Install the center fascia panel.
- Lower left side crash pad
- Front and rear console assemblies

9. Install the SRS module and the

* Heater Assembly

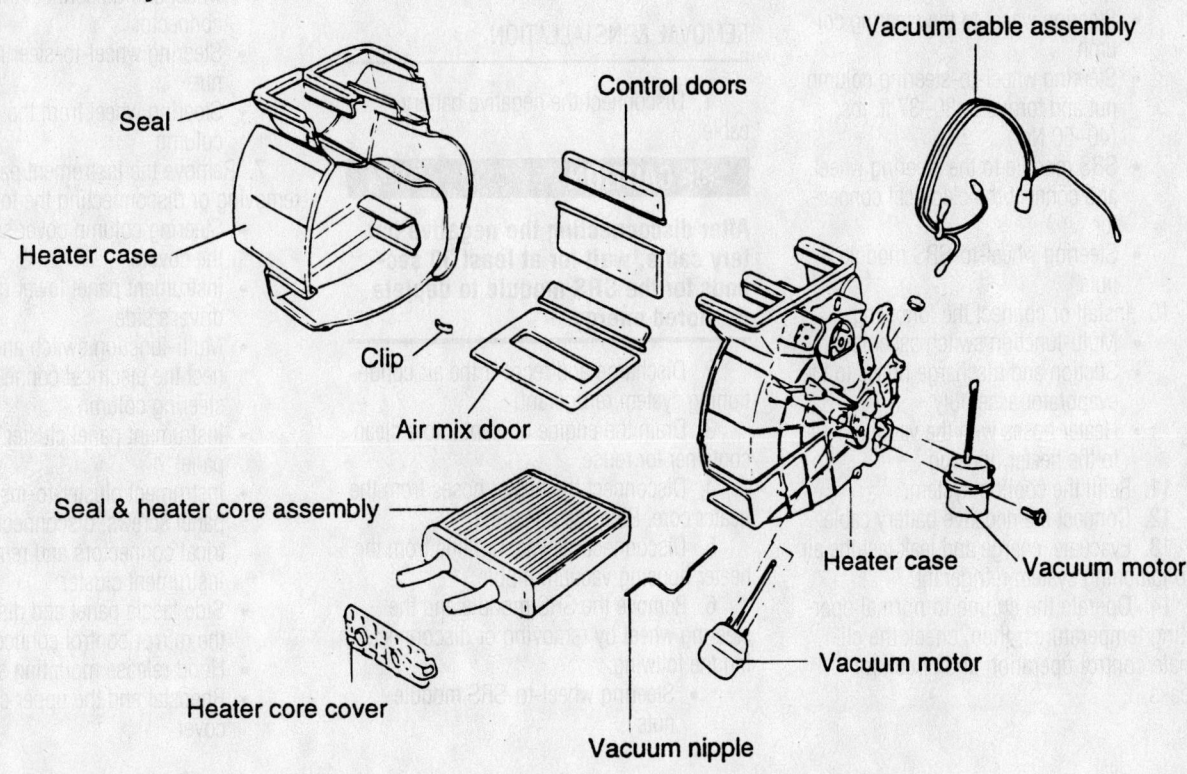

- Seal
- Heater case
- Clip
- Air mix door
- Control doors
- Vacuum cable assembly
- Heater case
- Vacuum motor
- Seal & heater core assembly
- Heater core cover
- Vacuum motor
- Vacuum nipple

* Vacuum Source Lines

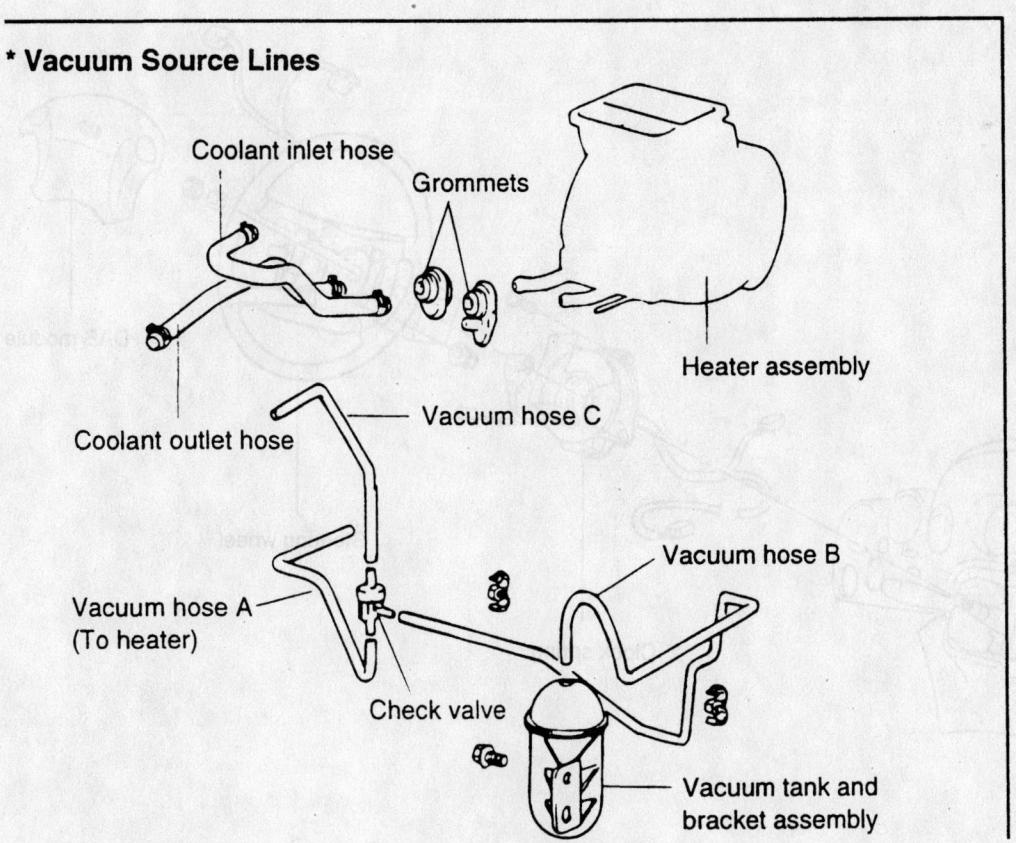

- Coolant inlet hose
- Grommets
- Coolant outlet hose
- Vacuum hose C
- Heater assembly
- Vacuum hose B
- Vacuum hose A (To heater)
- Check valve
- Vacuum tank and bracket assembly

93112GP4

Exploded view of the heater core, heater housing and related components—Accent

For complete service labor times, order the Chilton Labor Guide

steering wheel by installing or connecting the following:

- Steering wheel to the steering column
- Steering wheel-to-steering column nut and torque to 30–37 ft. lbs. (40–50 Nm)
- SRS module to the steering wheel and connect the electrical connector
- Steering wheel-to-SRS module nuts

10. Install or connect the following:
- Multi-function switch assembly
- Suction and discharge hoses to the evaporator assembly
- Heater hoses with the vacuum hose to the heater housing

11. Refill the cooling system.

12. Connect the negative battery cable.

13. Evacuate, charge and leak test the air conditioning system refrigerant.

14. Operate the engine to normal operating temperatures; then, check the climate control operation and check for leaks.

Elantra

REMOVAL & INSTALLATION

1. Disconnect the negative battery cable.

✳✳ CAUTION

After disconnecting the negative battery cable, wait for at least 30 seconds for the SRS module to deplete its stored energy.

2. Discharge and recover the air conditioning system refrigerant.

3. Drain the engine coolant into a clean container for reuse.

4. Disconnect the heater hoses from the heater core. Plug the openings.

5. Disconnect the vacuum line from the heater housing vacuum nipple.

6. Remove the SRS module and the steering wheel by removing or disconnecting the follwing:
- Steering wheel-to-SRS module nuts
- SRS module from the steering wheel and disconnect the electrical connector
- Steering wheel-to-steering column nut
- Steering wheel from the steering column

7. Remove the instrument panel by removing or disconnecting the following:
- Steering column cover screws and the covers
- Instrument panel lower cover at the driver's side
- Multi-function switch and disconnect the electrical connector at the steering column
- Instrument panel cluster fascia panel
- Instrument cluster-to-instrument panel screws, disconnect the electrical connectors and remove the instrument cluster
- Side fascia panel and disconnect the mirror control connector
- Hood release mounting screws
- Rheostat and the upper console cover

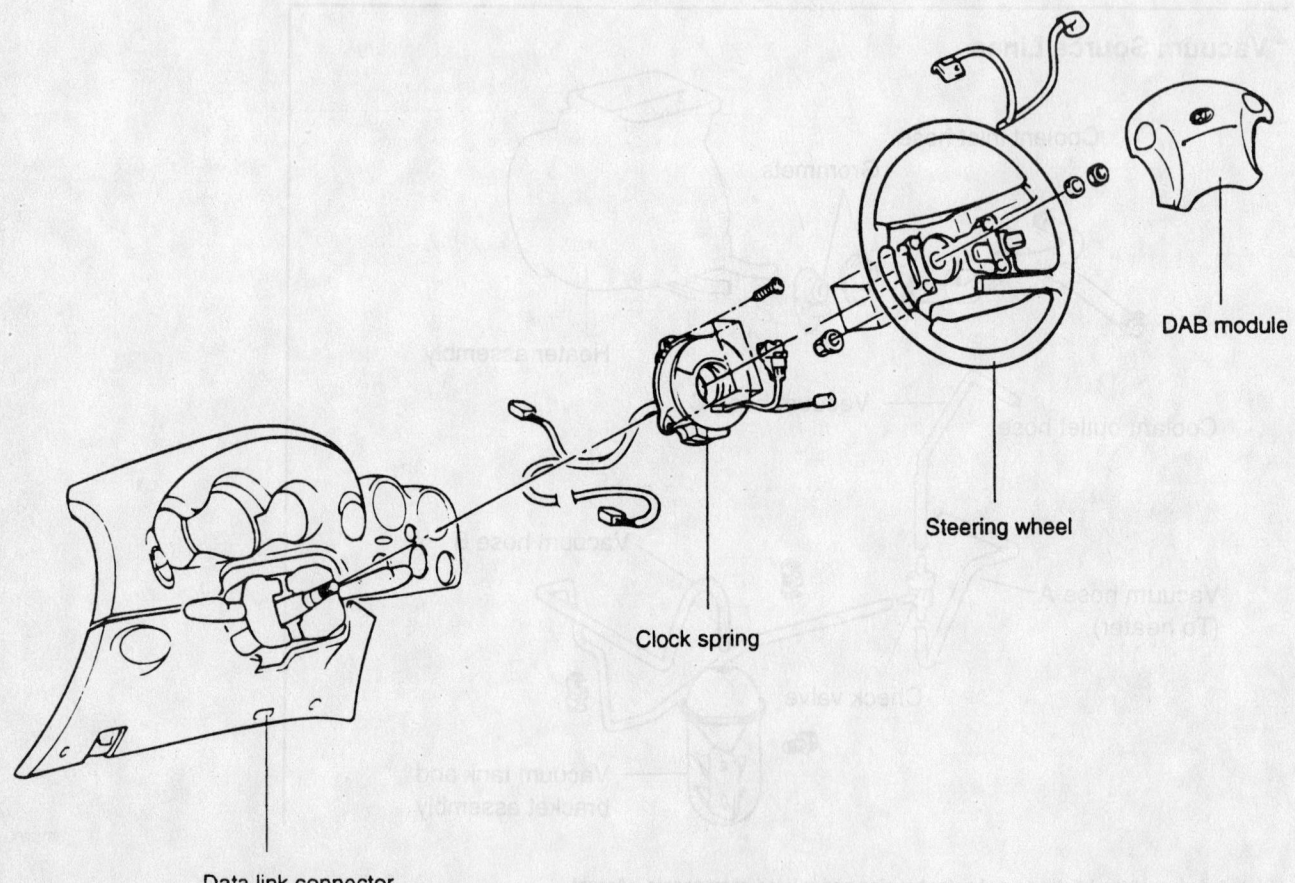

DAB module

Steering wheel

Clock spring

Data link connector

93112G07

Exploded view of the SRS module, steering wheel and related components—Elantra

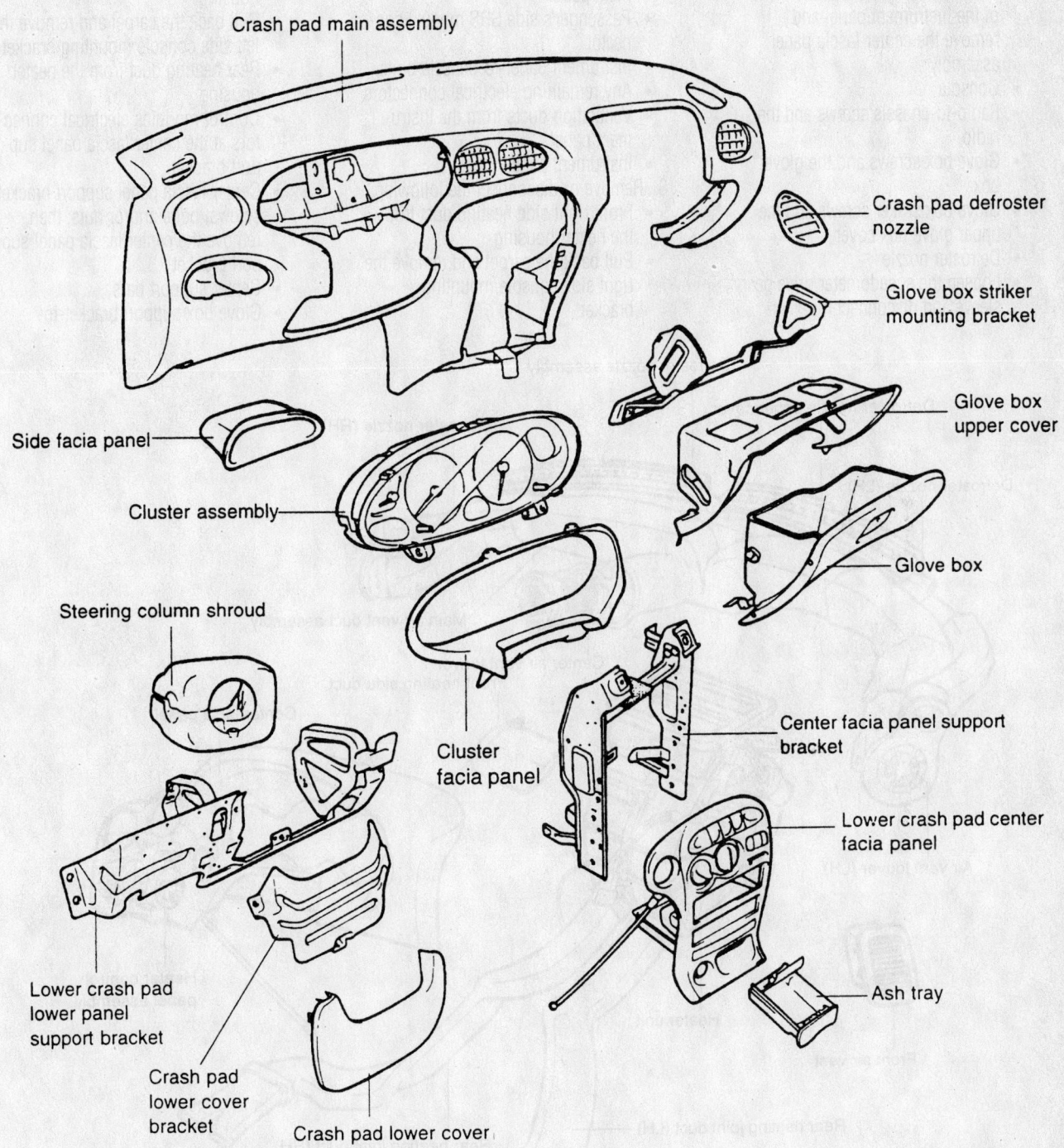

Crash pad main assembly

Crash pad defroster nozzle

Glove box striker mounting bracket

Glove box upper cover

Side facia panel

Cluster assembly

Glove box

Steering column shroud

Cluster facia panel

Center facia panel support bracket

Lower crash pad center facia panel

Lower crash pad lower panel support bracket

Crash pad lower cover bracket

Crash pad lower cover

Ash tray

TORQUE : Nm (kg·cm, lb·ft)

93112G08

Exploded view of the instrument panel and related components—Elantra

- Heater control cable
- Electrical connectors at the center of the instrument panel and remove the center fascia panel assembly
- Console
- Radio-to-chassis screws and the radio
- Glove box screws and the glove box
- Glove box striker screws and the upper glove box cover
- Defroster nozzle
- Loosen the speedometer drive gear sleeve and disconnect the speedometer cable from the instrument panel
- Passenger's side SRS module connector
- Instrument panel-to-chassis bolts
- Any remaining electrical connectors
- Ventilation ducts from the instrument panel
- Instrument panel
8. Remove or disconnect the following:
- Front right side heating duct from the heater housing
- Pull back the carpet and remove the right side console mounting bracket
- Front left side duct from the heater housing
- Pull back the carpet and remove the left side console mounting bracket
- Rear heating duct from the heater housing
- Control modules electrical connectors at the center fascia panel support bracket
- Center fascia panel support bracket screws, bolts and/or nuts; then, remove the center fascia panel support bracket
- Center support bars
- Glove box support bracket-to-

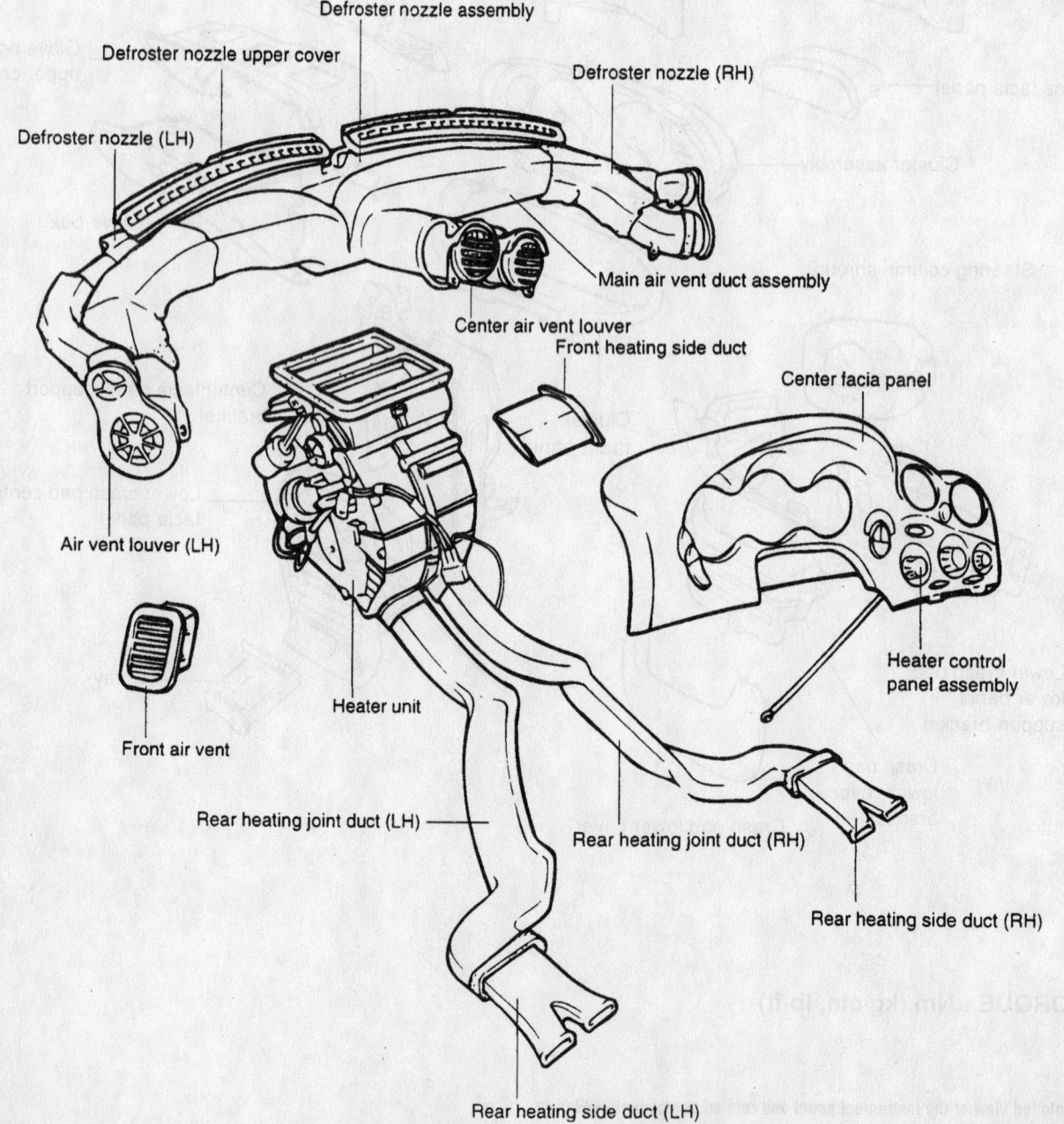

Exploded view of the heater housing, center fascia, distribution ducts and related components—Elantra Coupe

93112G00

instrument panel bolts and the bracket

9. Remove the evaporator housing by removing or disconnecting the following:

- Refrigerant lines from the evaporator housing and discard the O-rings
- Thermostatic switch connector
- Evaporator housing upper and lower bolts
- Evaporator housing

10. Remove or disconnect the following:

- Heater housing-to-chassis bolts and the housing
- Vacuum motor-to-heater housing bolts (2 for each vacuum motor)
- Vacuum motor rod end connection and remove the vacuum motors
- Heater housing cover clips
- Cover and the heater core

To install:

11. Assemble the heater housing by installing or connecting the following:

- Heater core and the cover

- Heater housing cover clips
- Vacuum motor rod end connection and install the vacuum motors
- Vacuum motor-to-heater housing bolts (2 for each vacuum motor)

12. Install or connect the following:

- Heater housing and the housing-to-chassis bolts
- Evaporator housing
- Evaporator housing upper and lower bolts
- Thermostatic switch connector

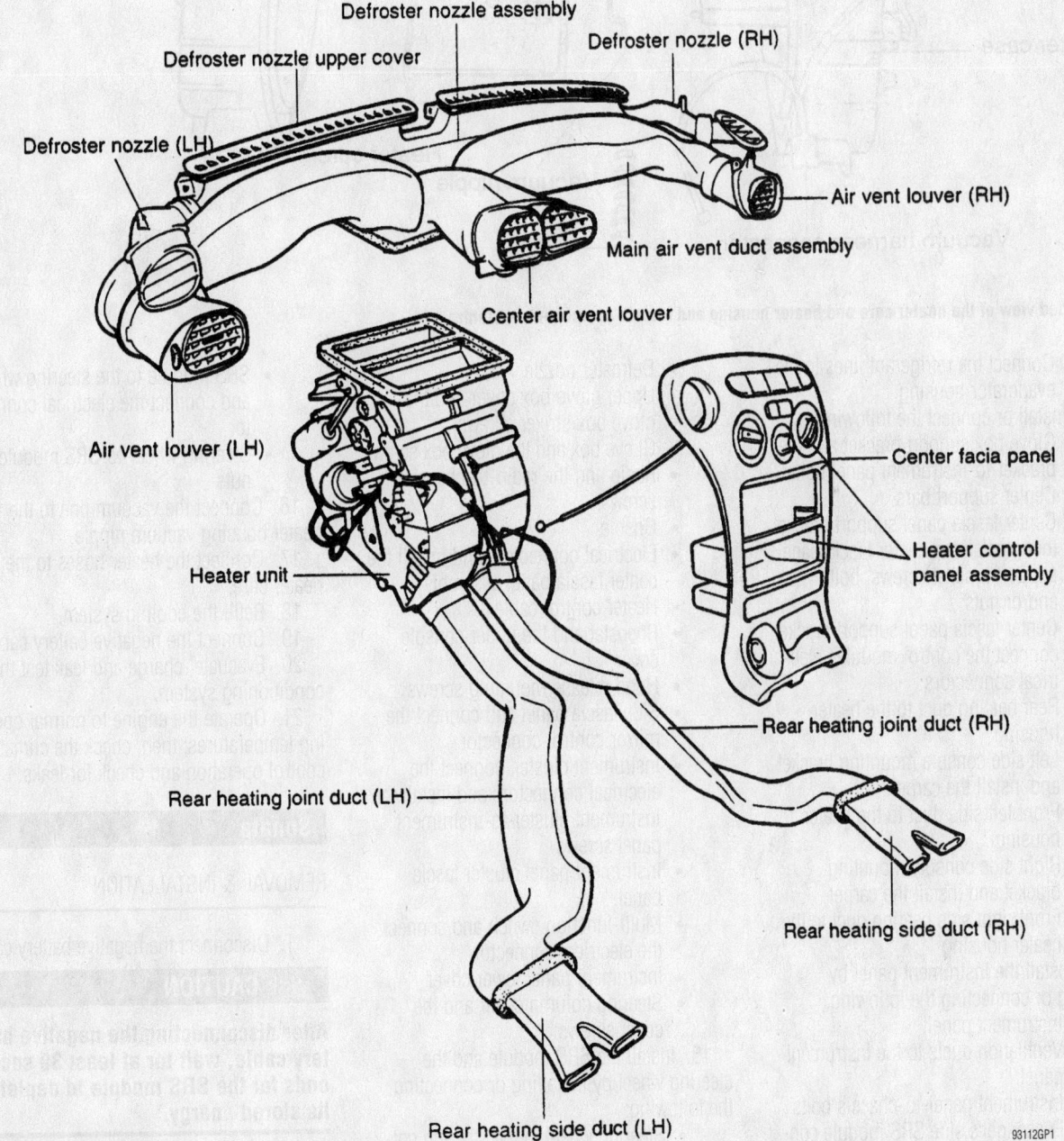

Exploded view of the heater housing, center fascia, distribution ducts and related components—Elantra Sedan and Wagon

Timing belt service is covered in Section 3 of this manual

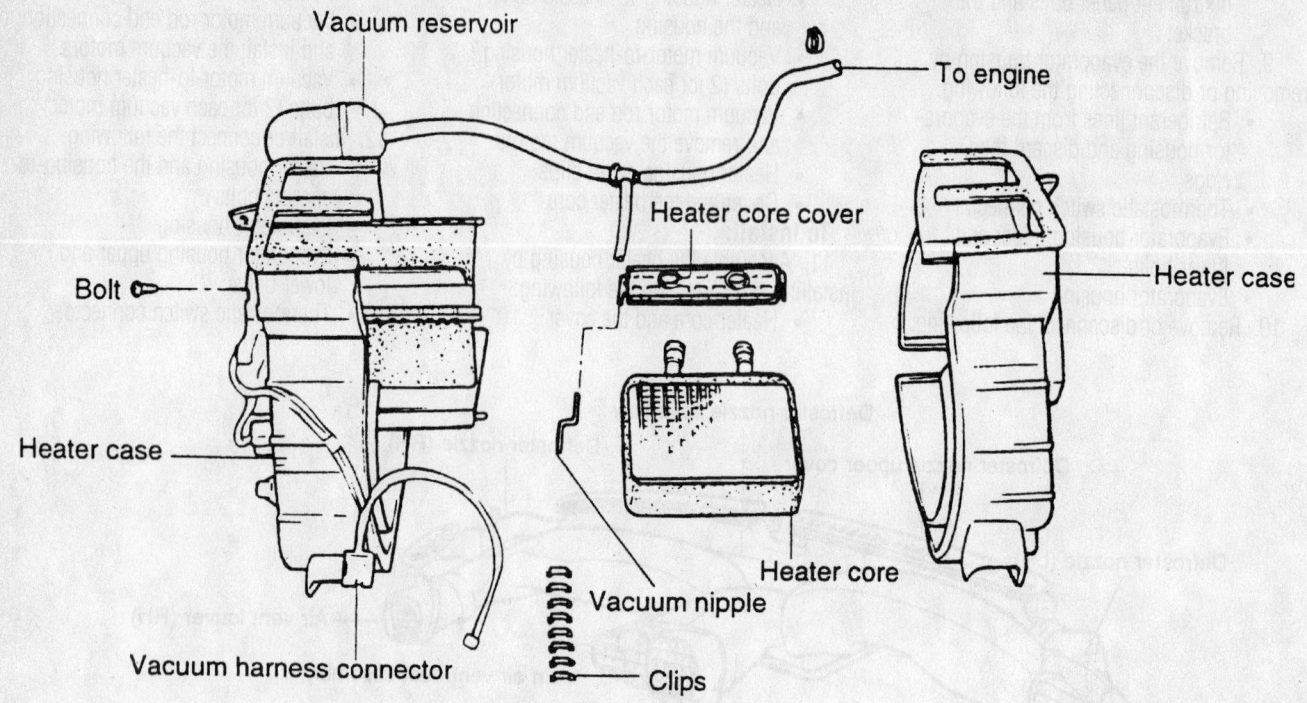

Vacuum reservoir

To engine

Heater core cover

Bolt

Heater case

Heater case

Heater core

Vacuum nipple

Vacuum harness connector

Clips

93112GP2

Exploded view of the heater core and heater housing and related components—Elantra

- Connect the refrigerant lines to the evaporator housing
13. Install or connect the following:
  - Glove box support bracket and the bracket-to-instrument panel bolts
  - Center support bars
  - Center fascia panel support bracket; then, install the center fascia panel support bracket screws, bolts and/or nuts
  - Center fascia panel support bracket, connect the control modules electrical connectors
  - Rear heating duct to the heater housing
  - Left side console mounting bracket and install the carpet
  - Front left side duct to the heater housing
  - Right side console mounting bracket and install the carpet
  - Front right side heating duct to the heater housing
14. Install the instrument panel by installing or connecting the following:
  - Instrument panel
  - Ventilation ducts to the instrument panel
  - Instrument panel-to-chassis bolts
  - Passenger's side SRS module connector
  - Speedometer cable to the instrument panel and tighten the speedometer drive gear sleeve

- Defroster nozzle
- Upper glove box cover and the glove box striker screws
- GLove box and the glove box screws
- Radio and the radio-to-chassis screws
- Cnsole
- Electrical connectors and install the center fascia panel assembly
- Heater control cable
- Rheostat and the upper console cover
- Hood release mounting screws
- Side fascia panel and connect the mirror control connector
- Instrument cluster, connect the electrical connectors and install the instrument cluster-to-instrument panel screws
- Instrument panel cluster fascia panel
- Multi-function switch and connect the electrical connector
- Instrument panel lower cover
- Steering column cover and the cover screws
15. Install the SRS module and the steering wheel by installing or connecting the following:
  - Steering wheel to the steering column
  - Steering wheel-to-steering column nut and torque to 30–37 ft. lbs. (40–50 Nm)

- SRS module to the steering wheel and connect the electrical connector
- Steering wheel-to-SRS module nuts
16. Connect the vacuum line to the heater housing vacuum nipple.
17. Connect the heater hoses to the heater core.
18. Refill the cooling system.
19. Connect the negative battery cable.
20. Evacuate, charge and leak test the air conditioning system.
21. Operate the engine to normal operating temperatures; then, check the climate control operation and check for leaks.

Sonata

REMOVAL & INSTALLATION

1. Disconnect the negative battery cable.

✳✳ CAUTION

After disconnecting the negative battery cable, wait for at least 30 seconds for the SRS module to deplete its stored energy.

2. Drain the cooling system into a clean container for reuse.
3. Remove the heater hoses from the heater housing.

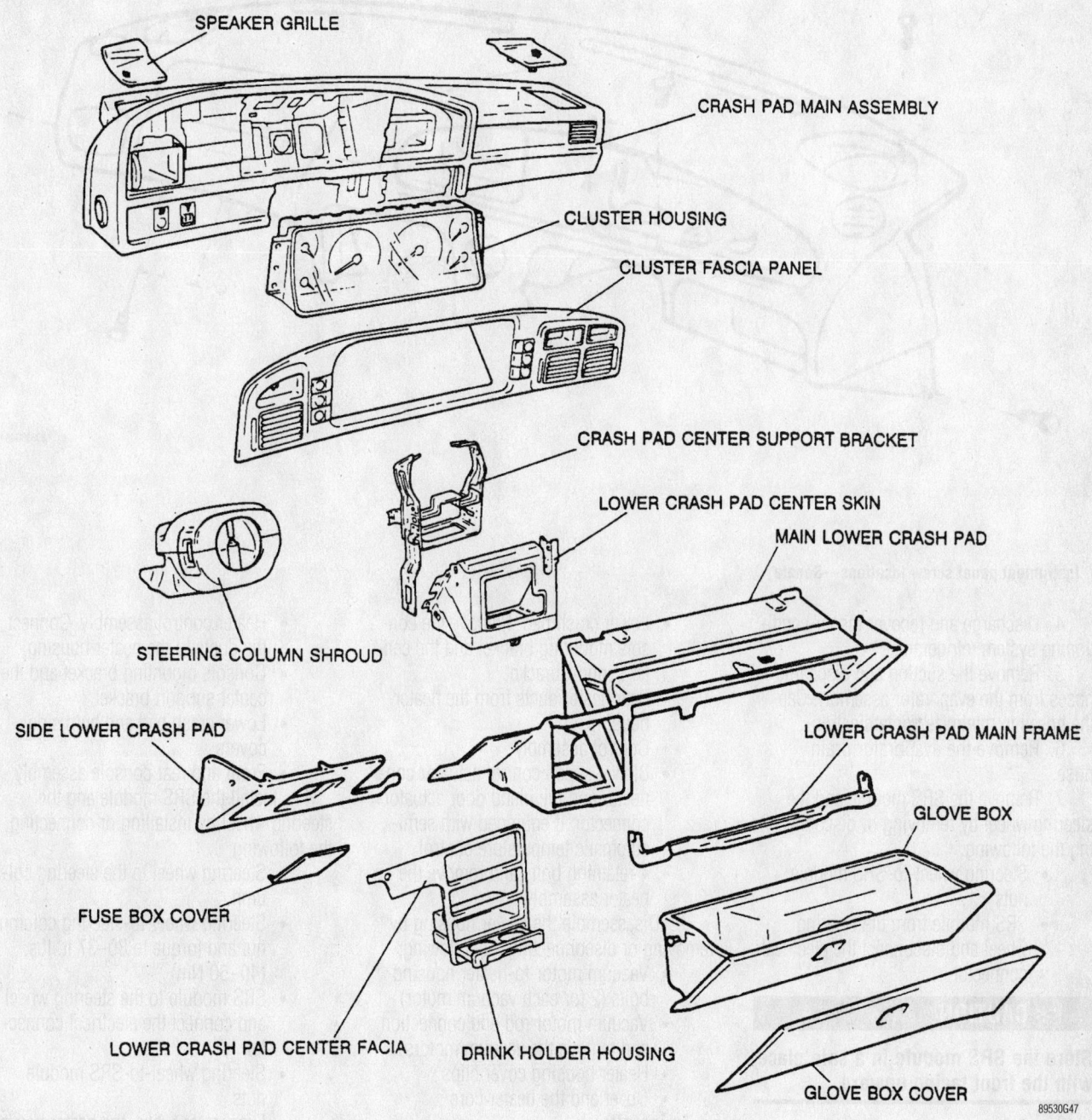

SPEAKER GRILLE

CRASH PAD MAIN ASSEMBLY

CLUSTER HOUSING

CLUSTER FASCIA PANEL

CRASH PAD CENTER SUPPORT BRACKET

LOWER CRASH PAD CENTER SKIN

MAIN LOWER CRASH PAD

STEERING COLUMN SHROUD

SIDE LOWER CRASH PAD

LOWER CRASH PAD MAIN FRAME

GLOVE BOX

FUSE BOX COVER

LOWER CRASH PAD CENTER FACIA

DRINK HOLDER HOUSING

GLOVE BOX COVER

89530G47

Instrument panel assembly—Sonata

Heater Core replacement is covered in Section 2 of this manual

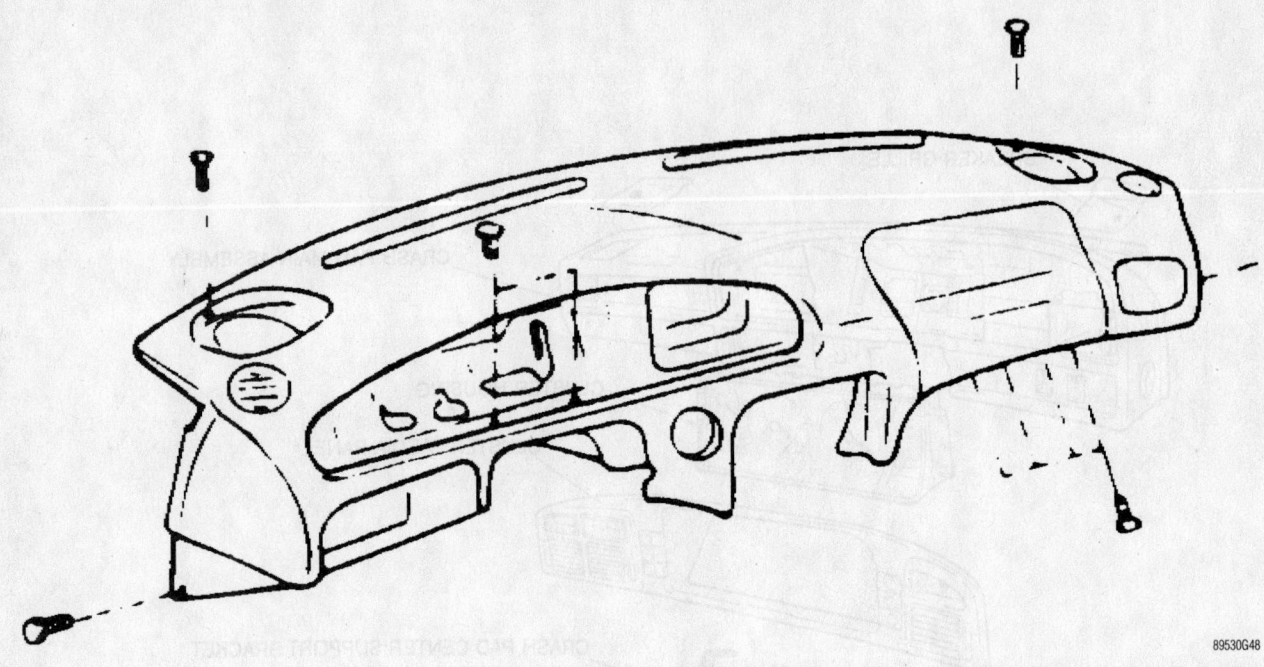

89530G48

Instrument panel screw locations—Sonata

4. Discharge and recover the air conditioning system refrigerant.

5. Remove the suction and discharge hoses from the evaporator assembly. Cap the hoses to minimize contamination.

6. Remove the evaporator drain hose.

7. Remove the SRS module and the steering wheel by removing or disconnecting the following:
- Steering wheel-to-SRS module nuts
- SRS module from the steering wheel and disconnect the electrical connector

✳✳ CAUTION

Store the SRS module in a safe place with the front facing upward.

- Steering wheel-to-steering column nut
- Steering wheel from the steering column

8. Remove or disconnect the following:
- Front and rear console assembly and remove both side covers
- Glove box, the center pad cover, the center crash pad and the cassette assembly

- Lower crash pad. Remove the console mounting bracket and the center support bracket
- Rear heater ducts from the heater housing
- Control assembly
- Blower speed control actuator connector and the blend door actuator connector, if equipped with semi-automatic temperature control
- 4 retaining bolts and remove the heater assembly

9. Disassemble the heater housing by removing or disconnecting the following:
- Vacuum motor-to-heater housing bolts (2 for each vacuum motor)
- Vacuum motor rod end connection and remove the vacuum motors
- Heater housing cover clips
- Cover and the heater core

To install:

10. Install or connect the following:
- Heater core and the cover
- Heater housing cover clips
- Vacuum motor rod end connection and install the vacuum motors
- Vacuum motor-to-heater housing bolts (2 for each vacuum motor)
- Heater assembly and attach it to the dash panel with the mounting bolts

- Heater control assembly. Connect the ducts to the heater housing
- Console mounting bracket and the center support bracket
- Lower crash pad and both side covers
- Front and rear console assembly

11. Install the SRS module and the steering wheel by installing or connecting the following:
- Steering wheel to the steering column
- Steering wheel-to-steering column nut and torque to 30–37 ft. lbs. (40–50 Nm)
- SRS module to the steering wheel and connect the electrical connector
- Steering wheel-to-SRS module nuts
- Evaporator tubes, the heater hoses and the drain hose

12. Refill the cooling system.

13. Connect the negative battery cable.

14. Evacuate, charge and leak test the air conditioning system.

15. Operate the engine to normal operating temperatures; then, check the climate control operation and check for leaks.

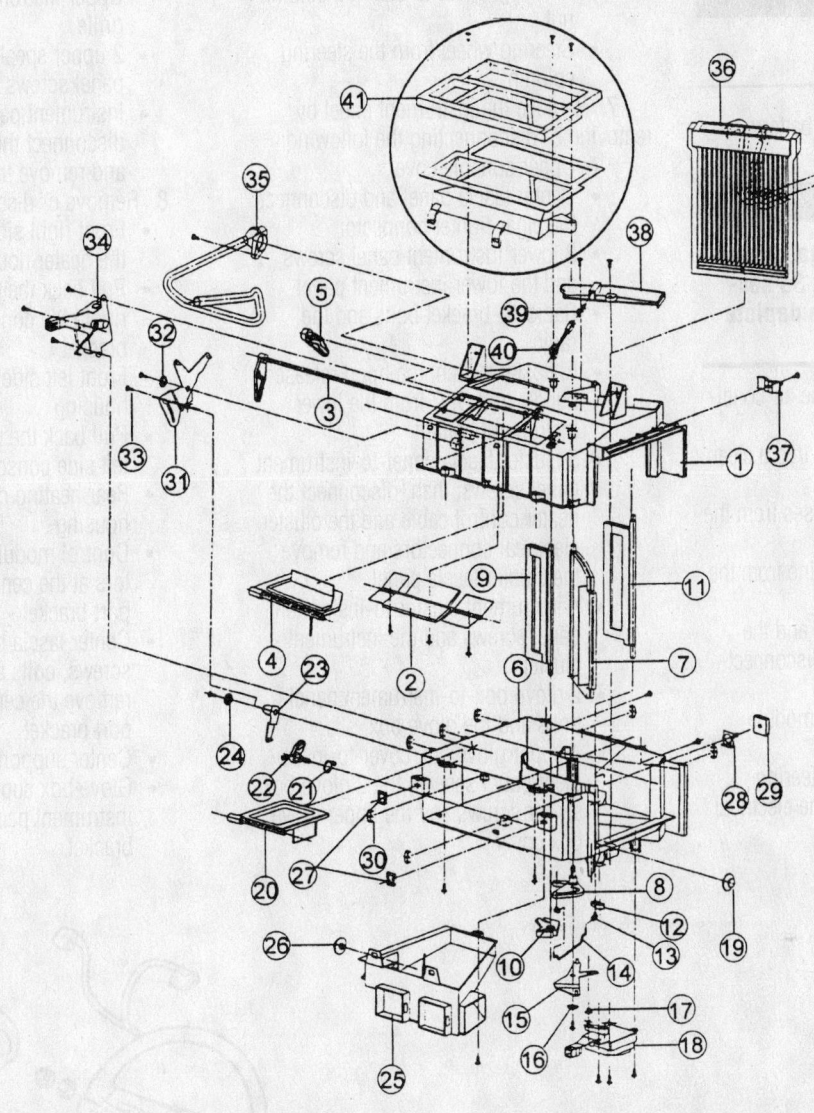

1	Case-heater upper	11	Door ass'y-by pass
2	Door ass'y-vent	12	Arm-By pass door
3	Shaft ass'y-vent door	13	Holder-rod link
4	Door ass'y-defrost	14	Link
5	Arm defrost door	15	Lever-temp. door
6	Case-heater lower	16	Spring washer
7	Door ass'y-temp.	17	Guide bush
8	Arm-temp. door	18	Blend door actuator
9	Door ass'y (A)-temp. door		(For AUTO A/C only)
10	Arm (A)-temp. door (A)	19	Guide bush
		20	Door ass'y-floor

21	Spring	31	Cam-mode
22	Arm-floor door	32	Spring-washer
23	Lever-floor door	33	Holder-rod link
24	Spring washer	34	Mode actuator
25	Duct-floor	35	Aspirator & hose ass'y
26	Guide bush	36	Heater core
27	U-nut	37	Clip
28	Clip & Bolt ass'y	38	Cover-heater core
29	Seal (A)-heater to D/panel	39	Stopper
30	Clip	40	Sensor
		41	Plenum duct ass'y

93112GP3

Exploded view of the heater housing assembly—Sonata

Brake service is covered in Section 4 of this manual

Tiburon

REMOVAL & INSTALLATION

1. Disconnect the negative battery cable.

※※ CAUTION

After disconnecting the negative battery cable, wait for at least 30 seconds for the SRS module to deplete its stored energy.

2. Discharge and recover the air conditioning system refrigerant.

3. Drain the engine coolant into a clean container for reuse.

4. Disconnect the heater hoses from the heater core. Plug the openings.

5. Disconnect the vacuum line from the heater housing vacuum nipple.

6. Remove the SRS module and the steering wheel by removing or disconnecting the following:
- Steering wheel-to-SRS module nuts
- SRS module from the steering wheel and disconnect the electrical connector

- Steering wheel-to-steering column nut
- Steering wheel from the steering column

7. Remove the instrument panel by removing or disconnecting the following:
- Upper console cover
- Center fascia panel and disconnect the cigar lighter connector
- 3 lower instrument panel screws and the lower instrument panel
- Radio-to-bracket bolts and the radio
- Rheostat switch, the hood release handle and DLC from the lower instrument panel
- 5 cluster fascia panel-to-instrument panel screws; then, disconnect the heater control cable and the cluster electrical connectors and remove the cluster fascia panel
- 4 instrument cluster-to-instrument panel screws and the instrument cluster
- 2 glove box-to-instrument panel bolts and the glove box
- 4 upper glove box cover-to-instrument panel screws, the 2 glove box striker screws and the upper glove box cover

- Upper instrument panel speaker grille
- 2 upper speaker-to-instrument panel screws
- Instrument panel-to-chassis bolts, disconnect the electrical connectors and remove the instrument panel

8. Remove or disconnect the following:
- Front right side heating duct from the heater housing
- Pull back the carpet and remove the right side console mounting bracket
- Front left side duct from the heater housing
- Pull back the carpet and remove the left side console mounting bracket
- Rear heating duct from the heater housing
- Control modules electrical connectors at the center fascia panel support bracket
- Center fascia panel support bracket screws, bolts and/or nuts; then, remove the center fascia panel support bracket
- Center support bars
- Glove box support bracket-to-instrument panel bolts and the bracket

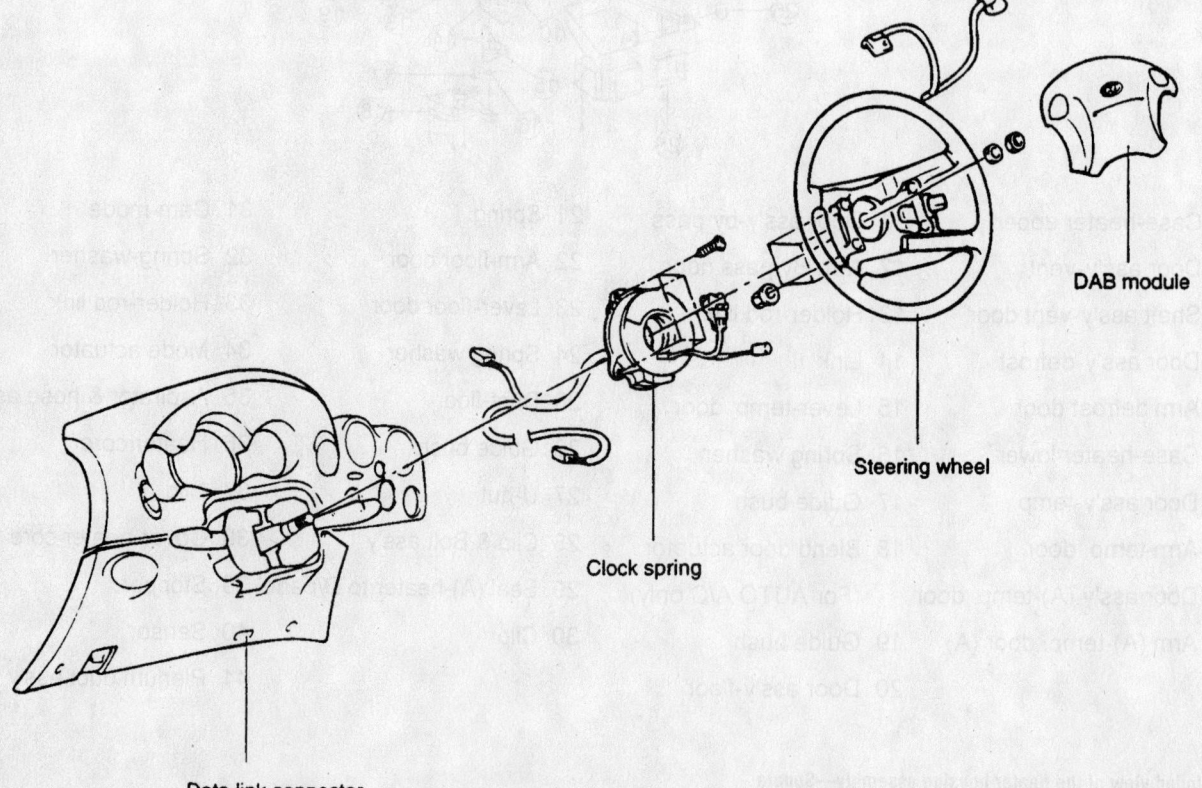

Data link connector

Clock spring

Steering wheel

DAB module

Exploded view of the SRS module, steering wheel and related components—Tiburon

93112G07

9. Remove the evaporator housing by removing or disconnecting the following:
- Refrigerant lines (located in the engine compartment), from the evaporator housing and discard the O-rings
- Thermostatic switch connector
- Evaporator housing upper and lower bolts
- Evaporator housing
- Heater housing-to-chassis bolts and the housing

10. Disassemble the heater housing by removing or disconnecting the following:
- Vacuum motor-to-heater housing bolts (2 for each vacuum motor)
- Vacuum motor rod end connection and remove the vacuum motors
- Heater housing cover clips
- Cover and the heater core

To install:

11. Assemble the heater housing by installing or connecting the following:
- Heater core and the cover

- Heater housing cover clips
- Vacuum motor rod end connection and install the vacuum motors
- Vacuum motor-to-heater housing bolts (2 for each vacuum motor)
- Heater housing and the housing-to-chassis bolts

12. Install the evaporator housing by installing or connecting the following:
- Evaporator housing
- Evaporator housing upper and lower bolts

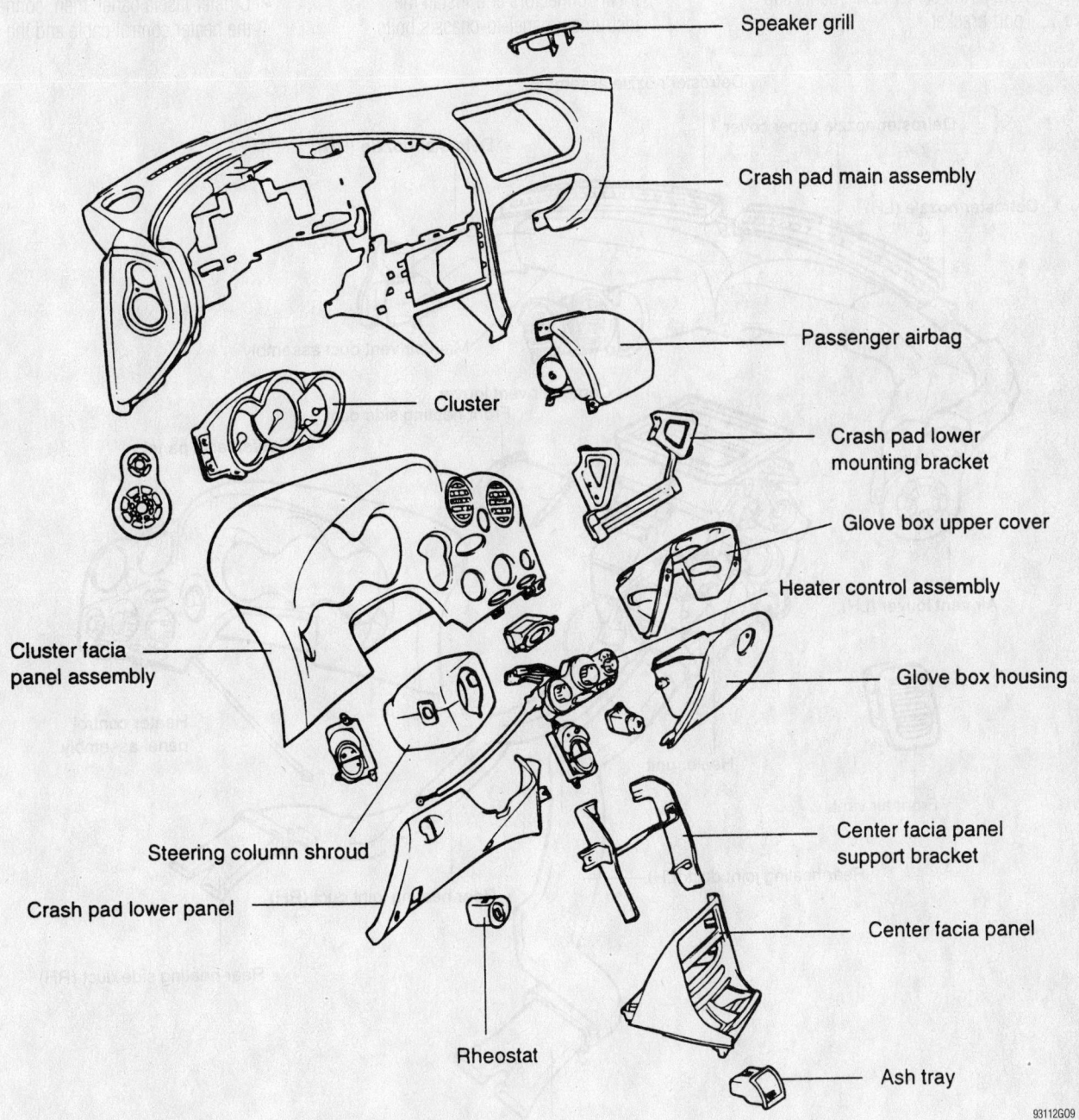

Speaker grill

Crash pad main assembly

Passenger airbag

Cluster

Crash pad lower mounting bracket

Glove box upper cover

Heater control assembly

Cluster facia panel assembly

Glove box housing

Steering column shroud

Center facia panel support bracket

Crash pad lower panel

Center facia panel

Rheostat

Ash tray

93112G09

Exploded view of the instrument panel and related components—Tiburon

For complete Engine Mechanical specifications, see Section 1 of this manual

- Thermostatic switch connector
- Connect the refrigerant lines to the evaporator housing using new O-rings

13. Install or connect the following:
- Glove box support bracket and the bracket-to-instrument panel bolts
- Center support bars
- Center fascia panel support bracket; then, install the center fascia panel support bracket screws, bolts and/or nuts
- Control modules electrical connectors at the center fascia panel support bracket

- Rear heating duct to the heater housing
- Left side console mounting bracket and install the carpet
- Front left side duct to the heater housing
- Right side console mounting bracket and install the carpet
- Front right side heating duct to the heater housing

14. Install the instrument panel by installing or connecting the following:
- Instrument panel, connect the electrical connectors and install the instrument panel-to-chassis bolts

- 2 upper speaker-to-instrument panel screws
- Upper instrument panel speaker grille
- Upper glove box cover, the 2 glove box striker screws and the 4 upper glove box cover-to-instrument panel screws
- Glove box and the 2 glove box-to-instrument panel bolts
- Instrument cluster and the 4 instrument cluster-to-instrument panel screws
- Cluster fascia panel; then, connect the heater control cable and the

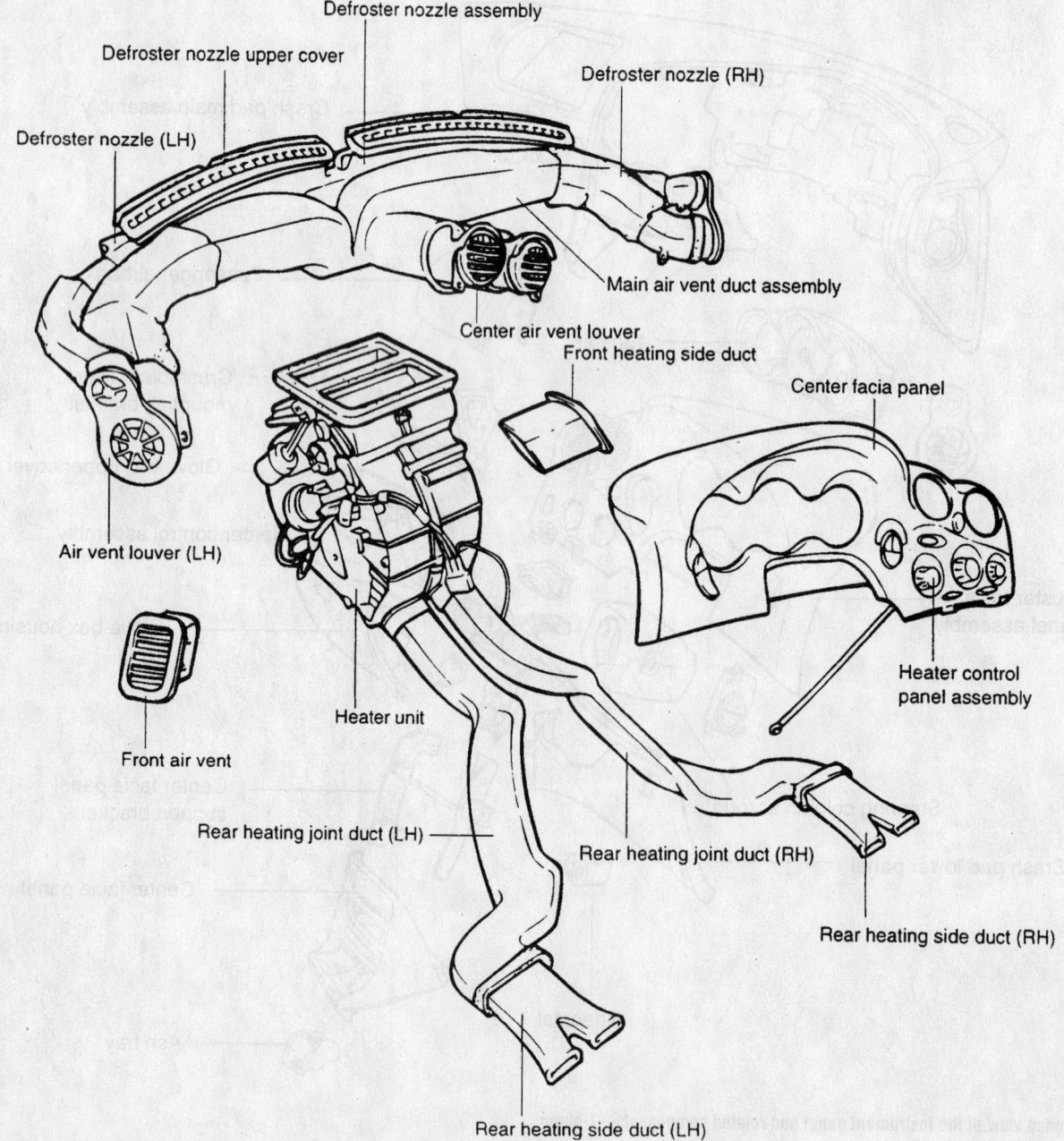

93112G00

Exploded view of the heater housing, center fascia, distribution ducts and related components—Tiburon Coupe

cluster electrical connectors and install the 5 cluster fascia panel-to-instrument panel screws

- Rheostat switch, the hood release handle and DLC to the lower instrument panel
- Radio and the radio-to-bracket bolts
- Lower instrument panel and the 3 lower instrument panel screws
- Cigar lighter connector and install the center fascia panel

- Upper console cover
15. Install the SRS module and the steering wheel by installing or connecting the following:
  - Steering wheel to the steering column
  - Steering wheel-to-steering column nut and torque to 30–37 ft. lbs. (40–50 Nm)
  - SRS module to the steering wheel and connect the electrical connector

- Steering wheel-to-SRS module nuts
16. Install or connect the following:
  - Vacuum line to the heater housing vacuum nipple
  - Heater hoses to the heater core
17. Refill the cooling system.
18. Connect the negative battery cable.
19. Evacuate, charge and leak test the air conditioning system.
20. Operate the engine to normal operating temperatures; then, check the climate control operation and check for leaks.

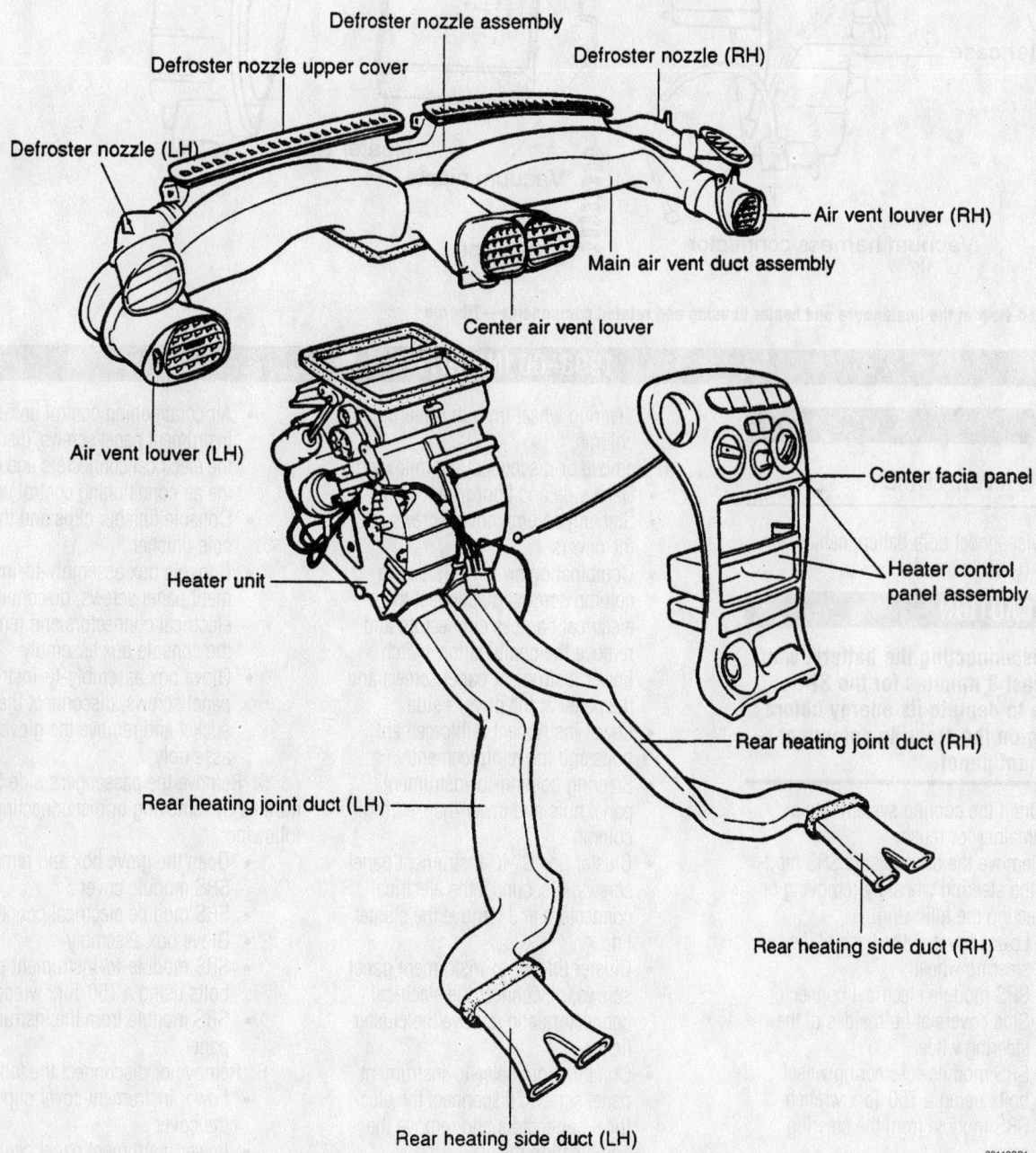

Exploded view of the heater housing, center fascia, distribution ducts and related components—Tiburon Sedan and Wagon

For Accessory Drive Belt illustrations, see Section 1 of this manual

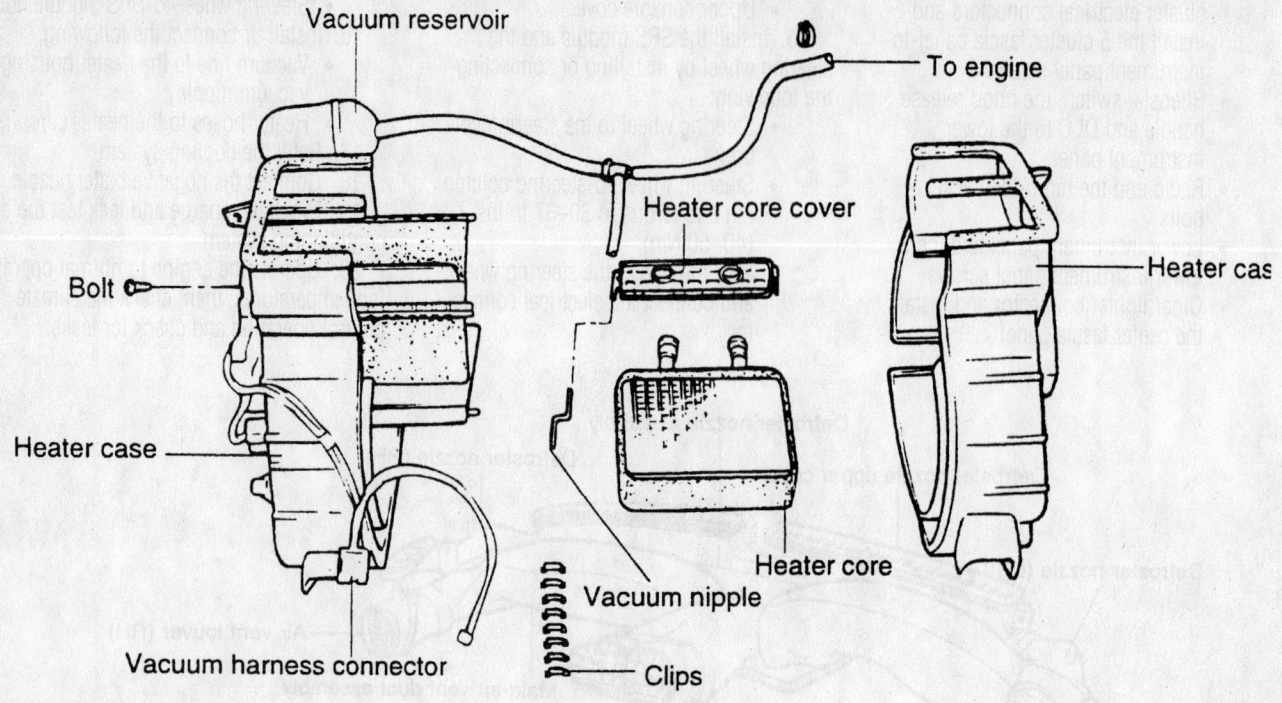

Vacuum reservoir

To engine

Heater core cover

Heater case

Bolt

Heater case

Heater core

Vacuum nipple

Vacuum harness connector

Clips

93112GP2

Exploded view of the heater core and heater housing and related components—Tiburon

1998–00 INFINITI

G20

REMOVAL & INSTALLATION

1. Disconnect both battery cables, the negative (()) cable first.

❊❊ CAUTION

After disconnecting the battery, wait for a least 3 minutes for the SRS module to deplete its energy before working on the steering column or instrument panel.

2. Drain the cooling system into a clean container for reuse.

3. Remove the driver's side SRS module and the steering wheel by removing or disconnecting the following:
  - Lower cover at the base of the steering wheel
  - SRS module electrical connector
  - Side covers at both sides of the steering wheel
  - SRS module-to-steering wheel bolts using a T50 Torx wrench
  - SRS module from the steering wheel
  - Place the front wheel in the straight-ahead position
  - Horn connector and remove the steering wheel nut

  - Steering wheel from the steering column

4. Remove or disconnect the following:
  - Dash side and floor trim
  - Steering column cover screws and the covers
  - Combination switch-to-steering column screws, disconnect the electrical harness connectors and remove the combination switch
  - Lower instrument panel screws and the panel at the driver's side
  - Lower instrument reinforcement bolts and the reinforcement
  - Steering column-to-instrument panel nuts and lower the steering column
  - Cluster lid "C"-to-instrument panel screws, disconnect the electrical connectors and remove the cluster lid
  - Cluster lid "A"-to-instrument panel screws, disconnect the electrical connectors and remove the cluster lid
  - Combination meter-to-instrument panel screws, disconnect the electrical connectors and remove the combination meter
  - Audio center-to-instrument panel bolts, disconnect the electrical connectors and remove the audio center

  - Air conditioning control unit-to-instrument panel screws, disconnect the electrical connectors and remove the air conditioning control unit
  - Console finisher clips and the console finisher
  - Console box assembly-to-instrument panel screws, disconnect the electrical connectors and remove the console box assembly
  - Glove box assembly-to-instrument panel screws, disconnect the lamp socket and remove the glove box assembly

5. Remove the passenger's side SRS module by removing or disconnecting the following:
  - Open the glove box and remove the SRS module cover
  - SRS module electrical connector
  - Glove box assembly
  - SRS module-to-instrument panel bolts using a T50 Torx wrench
  - SRS module from the instrument panel

6. Remove or disconnect the following:
  - Lower instrument cover clip and the cover
  - Lower instrument panel center-to-instrument panel screws and the panel
  - Connectors and remove the defroster grilles

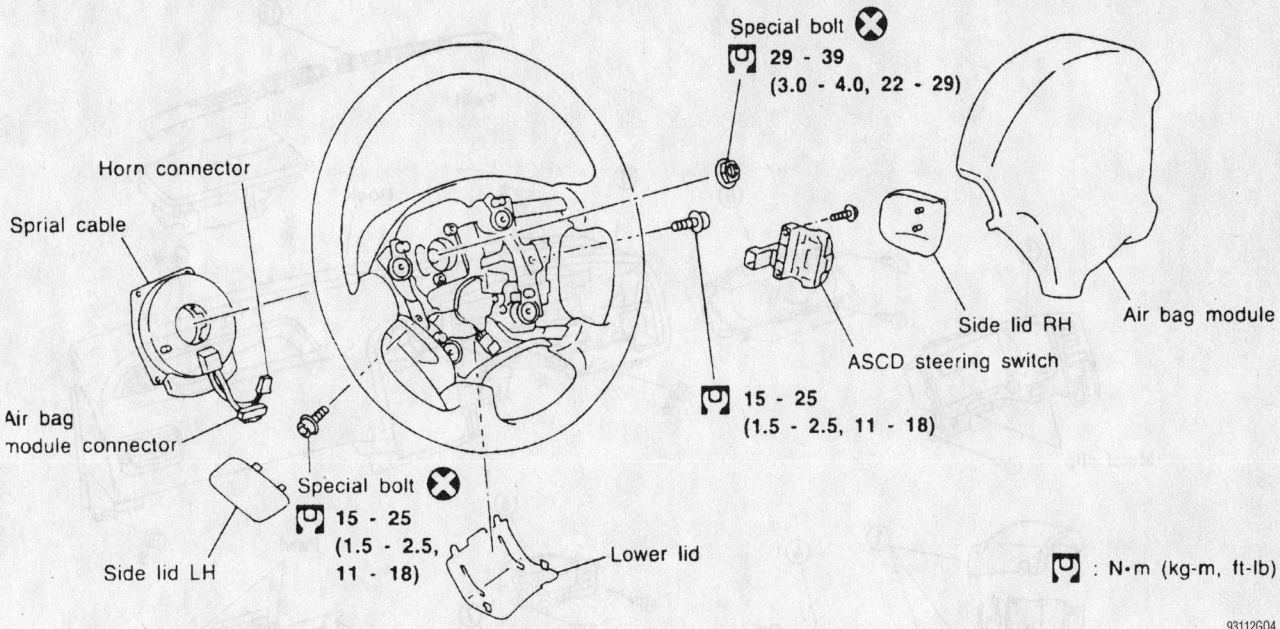

Special bolt ⊗
🔧 29 - 39
(3.0 - 4.0, 22 - 29)

Horn connector

Sprial cable

Air bag
module connector

Side lid LH

Special bolt ⊗
🔧 15 - 25
(1.5 - 2.5,
11 - 18)

Side lid RH

Air bag module

ASCD steering switch

🔧 15 - 25
(1.5 - 2.5, 11 - 18)

Lower lid

🔧 : N·m (kg-m, ft-lb)

93112G04

Exploded view of the air bag module, the steering wheel and related components—G20

- Front pillar garnish
- Instrument panel-to-chassis bolts/nuts and the instrument panel
- Heater hoses from the heater core
- Rear duct from the heater unit
- Air conditioning housing-to-heater housing fasteners
- Heater housing-to-chassis fasteners
- Heater unit from the vehicle

7. Disassemble and remove the heater core from the heater housing.

To install:

8. Assemble and install the heater core to the heater housing.

9. Install or connect the following:
- Heater unit to the vehicle
- Heater housing-to-chassis fasteners
- Air conditioning housing-to-heater housing fasteners
- Rear duct to the heater unit
- Heater hoses to the heater core
- Instrument panel and the instrument panel-to-chassis bolts/nuts
- Front pillar garnish
- Defroster grilles and connect the connectors
- Lower instrument panel center and the panel-to-instrument panel screws
- Lower instrument cover and the cover clip

10. Install the passenger's side SRS module by installing or connecting the following:

- SRS module to the instrument panel
- SRS module-to-instrument panel bolts and torque the bolts using a T50 Torx wrench to 11–18 ft. lbs. (15–25 Nm)
- Glove box assembly
- SRS module electrical connector
- SRS module cover

11. Install or connect the following:
- Glove box assembly, connect the lamp socket and install the glove box assembly-to-instrument panel screws
- Console box assembly, connect the electrical connectors and install the console box assembly-to-instrument panel screws
- Console finisher and engage the clips
- Air conditioning control unit, connect the electrical connectors and Install the air conditioning control unit-to-instrument panel screws
- Audio center, connect the electrical connectors and Install the audio center-to-instrument panel bolts
- Combination meter, connect the electrical connectors and Install the combination meter-to-instrument panel screws
- Cluster lid "A", connect the electrical connectors and install the cluster lid-to-instrument panel screws

- Cluster lid "C", connect the electrical connectors and install the cluster lid-to-instrument panel screws
- Steering column and torque the steering column-to-instrument panel nuts to 11–14 ft. lbs. (15–19 Nm)
- Lower instrument reinforcement and the reinforcement bolts
- Lower instrument panel and the panel screws at the driver's side
- Combination switch and the combination switch-to-steering column screws and connect the electrical harness connectors
- Steering column covers and the cover screws
- Dash side and floor trim

12. Install the driver's side SRS module and the steering wheel by installing or connecting the following:
- Steering wheel to the steering column
- Steering wheel nut and torque to 22–29 ft. lbs. (29–39 Nm). Connect the horn connector
- SRS module to the steering wheel
- Torque the SRS module-to-steering wheel bolts to 11–18 ft. lbs. (15–25 Nm) using a T50 Torx wrench
- Side covers at both sides of the steering wheel
- SRS module electrical connector

For Tire, Wheel and Ball Joint specifications, see Section 1 of this manual

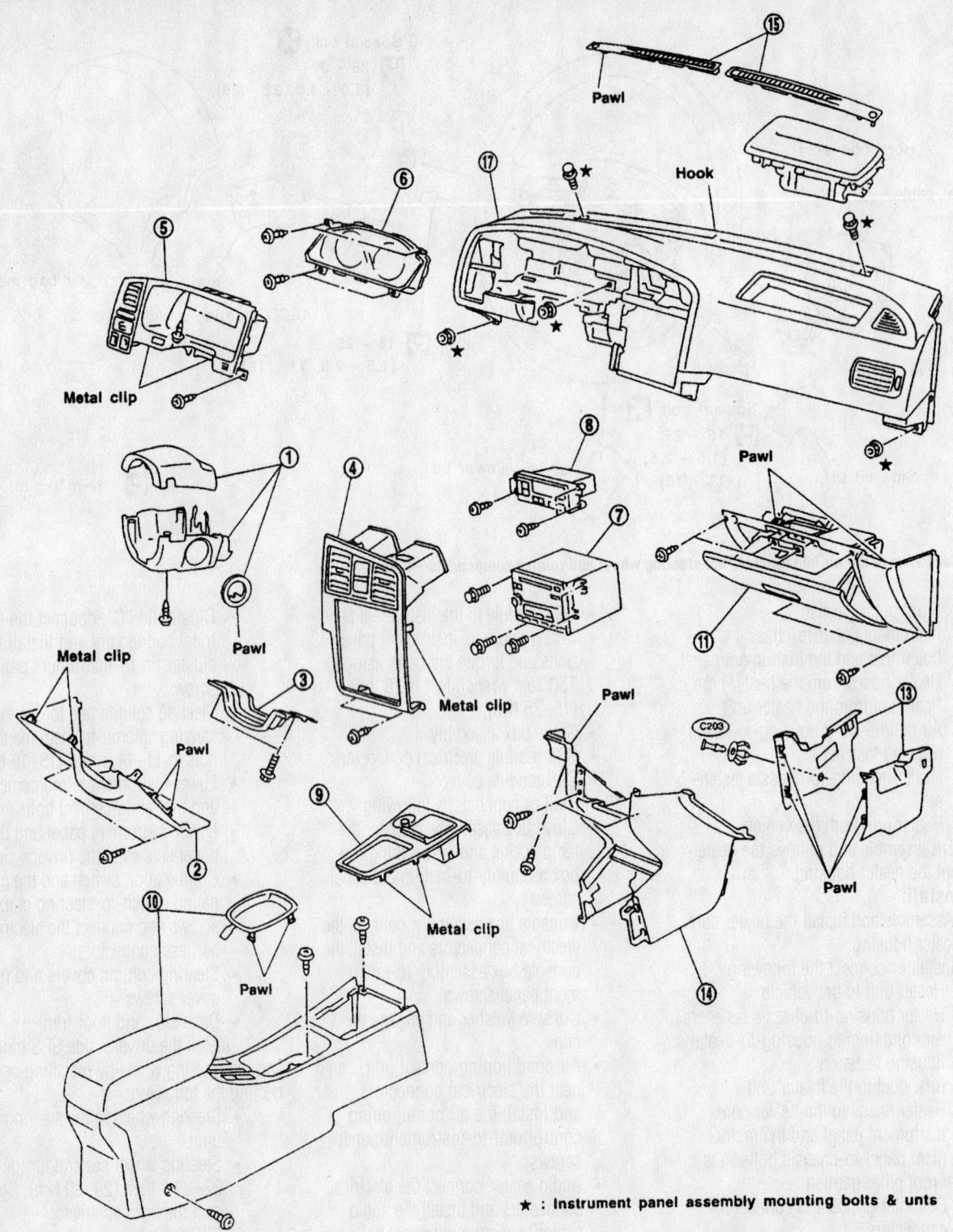

★ : Instrument panel assembly mounting bolts & unts

1. Steering column cover
   and combination switch
2. Instrument lower panel
   on driver side
3. Instrument lower reinforcement
4. Cluster lid C
5. Cluster lid A

6. Combination meter
7. Audio
8. A/C control unit
9. Console M/T or A/T finisher
10. Console box assembly
11. Glove box assembly

12. Passenger air bag module
13. Lower instrument cover
14. Lower instrument panel center
15. Defroster grille
16. Front pillar garnish
17. Instrument panel and pads

93112G05

Exploded view of the instrument panel—G20

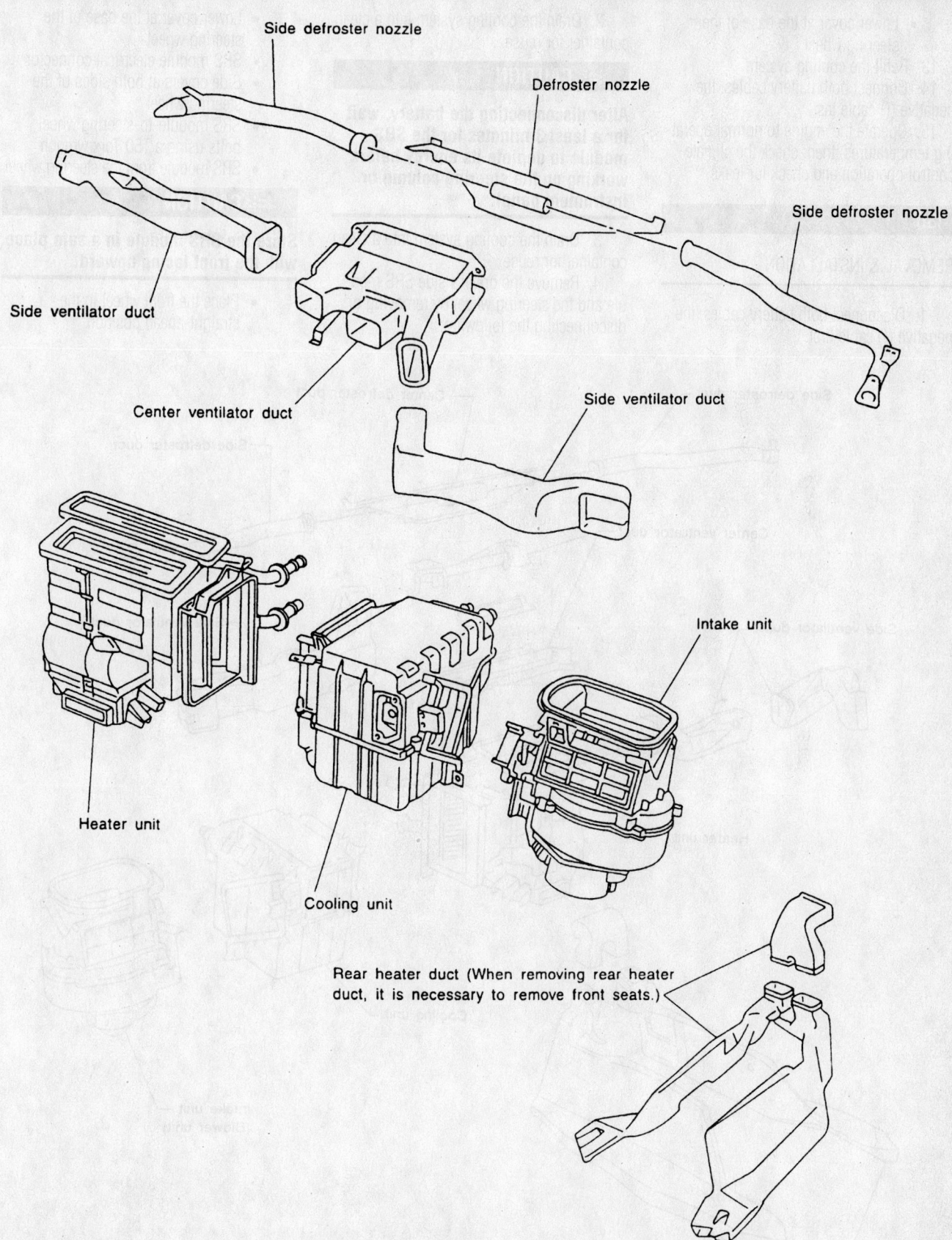

Side defroster nozzle

Defroster nozzle

Side defroster nozzle

Side ventilator duct

Center ventilator duct

Side ventilator duct

Intake unit

Heater unit

Cooling unit

Rear heater duct (When removing rear heater duct, it is necessary to remove front seats.)

93112G06

Exploded view of the heater housing, air conditioning housing, blower motor and ventilation ducts—G20

For Wheel Alignment specifications, see Section 1 of this manual

- Lower cover at the base of the steering wheel
13. Refill the cooling system.
14. Connect both battery cables, the negative (()) cable last.
15. Operate the engine to normal operating temperatures; then, check the climate control operation and check for leaks.

I30

REMOVAL & INSTALLATION

1. Disconnect both battery cables, the negative (()) cable first.

2. Drain the cooling system into a clean container for reuse.

✳✳ CAUTION

After disconnecting the battery, wait for a least 3 minutes for the SRS module to deplete its energy before working on the steering column or instrument panel.

3. Drain the cooling system into a clean container for reuse.
4. Remove the driver's side SRS module and the steering wheel by removing or disconnecting the following:

- Lower cover at the base of the steering wheel
- SRS module electrical connector
- Side covers at both sides of the steering wheel
- SRS module-to-steering wheel bolts using a T50 Torx wrench
- SRS module from the steering wheel

✳✳ CAUTION

Store the SRS module in a safe place with the front facing upward.

- Place the front wheel in the straight-ahead position

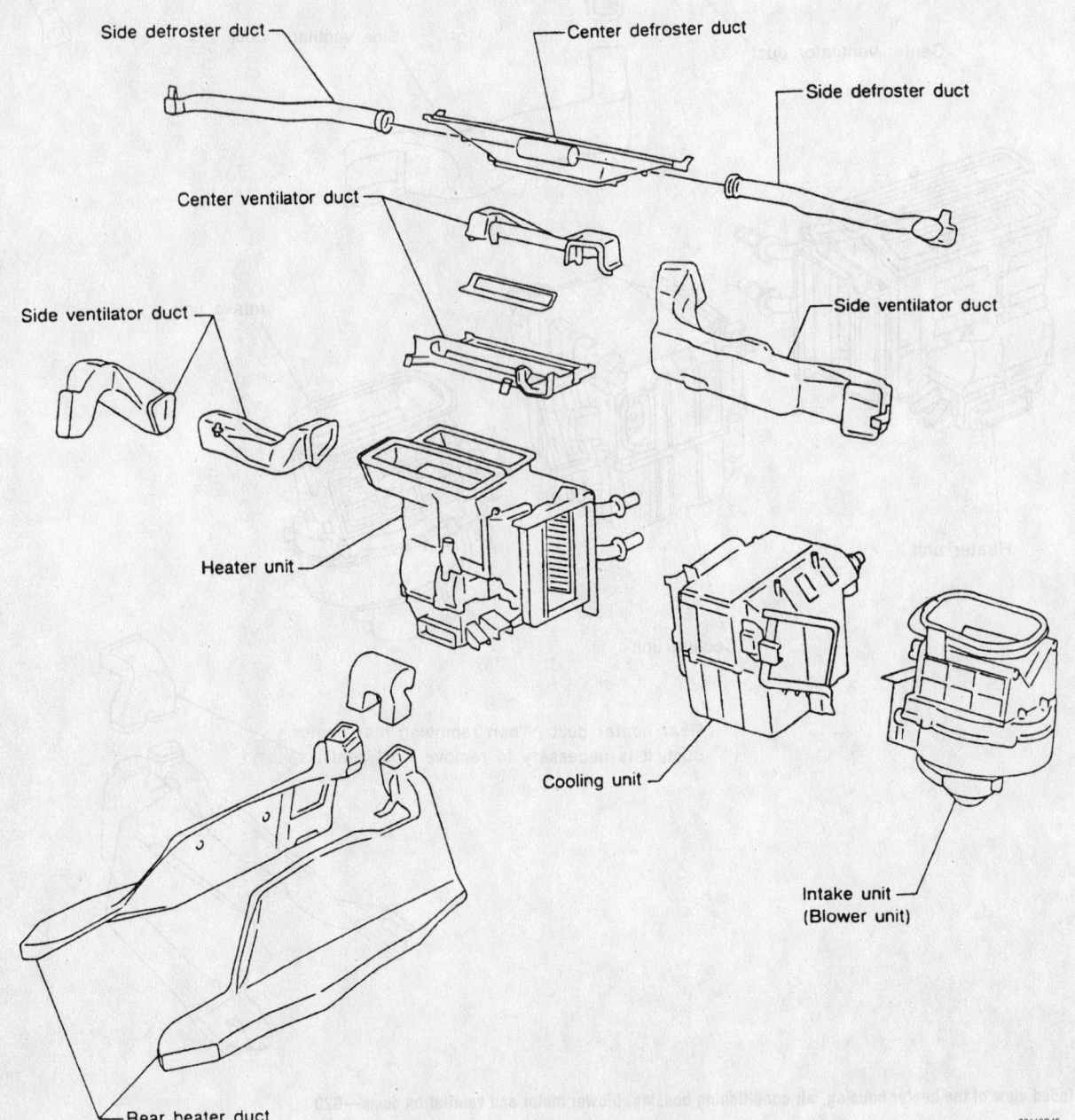

Exploded view of the heating-air conditioning system assemblies and related components—I30

Side defroster duct
Center defroster duct
Side defroster duct
Center ventilator duct
Side ventilator duct
Side ventilator duct
Heater unit
Cooling unit
Intake unit (Blower unit)
Rear heater duct

93112G45

- Horn connector and remove the steering wheel nut
- Steering wheel from the steering column

5. Remove or disconnect the following:
- Instrument panel assembly
- Front seats
- Defroster ducts, the ventilator ducts and the floor ducts from the heater unit
- Vacuum hoses and electrical connectors leading to the heater unit
- Heater unit from the cooling unit. Take care not to damage the air conditioning tubes
- Heater unit attaching bolts. Remove the heater unit from the passenger compartment

6. Disassemble the heater unit and remove the heater core.

To install:

7. Install the heater core and assemble the heater unit.

8. Install or connect the following:
- Heater unit in the passenger compartment and tighten the attaching bolts securely
- All vacuum hoses and electrical connectors
- Defroster ducts, the ventilator ducts and the floor ducts to the heater unit
- Front seats
- Instrument panel assembly
- Heater hoses

9. Install the driver's side SRS module and the steering wheel by installing or connecting the following:
- Steering wheel to the steering column
- Steering wheel nut and torque to 22–29 ft. lbs. (29–39 Nm). Connect the horn connector
- SRS module to the steering wheel
- SRS module-to-steering wheel bolts to 11–18 ft. lbs. (15–25 Nm) using a T50 Torx wrench
- Side covers at both sides of the steering wheel
- SRS module electrical connector
- Lower cover at the base of the steering wheel

10. Refill the cooling system.

11. Connect both battery cables, the negative (() cable last.

12. Operate the engine to normal operating temperatures; then, check the climate control operation and check for leaks.

J30

REMOVAL & INSTALLATION

1. Disconnect the negative battery cable.
2. Drain the cooling system into a clean container for reuse.
3. Remove or disconnect the following:
- Instrument panel assembly
- Front seats
- Defroster ducts, the ventilator ducts and the floor ducts from the heater unit
- All vacuum hoses and electrical connectors leading to the heater unit
- Heater unit from the cooling unit. Take care not to damage the air conditioning tubes
- Heater unit attaching bolts. Remove the heater unit from the passenger compartment

4. Disassemble the heater unit and remove the heater core.

To install:

5. Install the heater core and assemble the heater unit.

6. Install or connect the following:
- Heater unit in the passenger compartment and tighten the attaching bolts securely
- All previously disconnected vacuum hoses and electrical connectors
- Defroster ducts, the ventilator ducts and the floor ducts to the heater unit
- Front seats
- Instrument panel assembly
- Heater hoses

7. Refill the cooling system.
8. Connect the negative battery cable.
9. Operate the engine to normal operating temperatures; then, check the climate control operation and check for leaks.

Q45

REMOVAL & INSTALLATION

1. Disconnect both battery cables, the negative (() cable first.

✳✳ CAUTION

After disconnecting the battery, wait for a least 3 minutes for the SRS module to deplete its energy before working on the steering column or instrument panel.

2. Drain the cooling system into a clean container for reuse.
3. Remove the driver's side SRS module and the steering wheel by removing or disconnecting the following:
- Lower cover at the base of the steering wheel
- SRS module electrical connector
- Side covers at both sides of the steering wheel
- SRS module-to-steering wheel bolts using a T50 Torx wrench
- SRS module from the steering wheel
- Place the front wheel in the straight-ahead position
- Horn connector and remove the steering wheel nut
- Steering wheel from the steering column

4. Remove or disconnect the following:
- Dash side lower finishers
- Steering column cover screws and the covers
- Combination switch-to-steering column screws, disconnect the electrical harness connectors and remove the combination switch
- Lower instrument panel screws/bolts and the panel. Disconnect the electrical harness connectors and the in-vehicle sensor at the driver's side
- Lower instrument reinforcement bolts and the reinforcement
- Steering column-to-instrument panel nuts and lower the steering column
- Cluster lid "A"-to-instrument panel screws and remove the cluster lid
- Steering lock escutcheon screws and the escutcheon
- Cluster lid "D"
- Combination meter-to-instrument panel screws, disconnect the electrical connectors and remove the combination meter
- Glove box assembly-to-instrument panel screws, disconnect the lamp socket and remove the glove box assembly

5. Remove the passenger's side SRS module by removing or disconnecting the following:
- Lower instrument panel cover
- SRS module electrical connector
- SRS module-to-instrument panel bolts using a T50 Torx wrench
- SRS module from the instrument panel

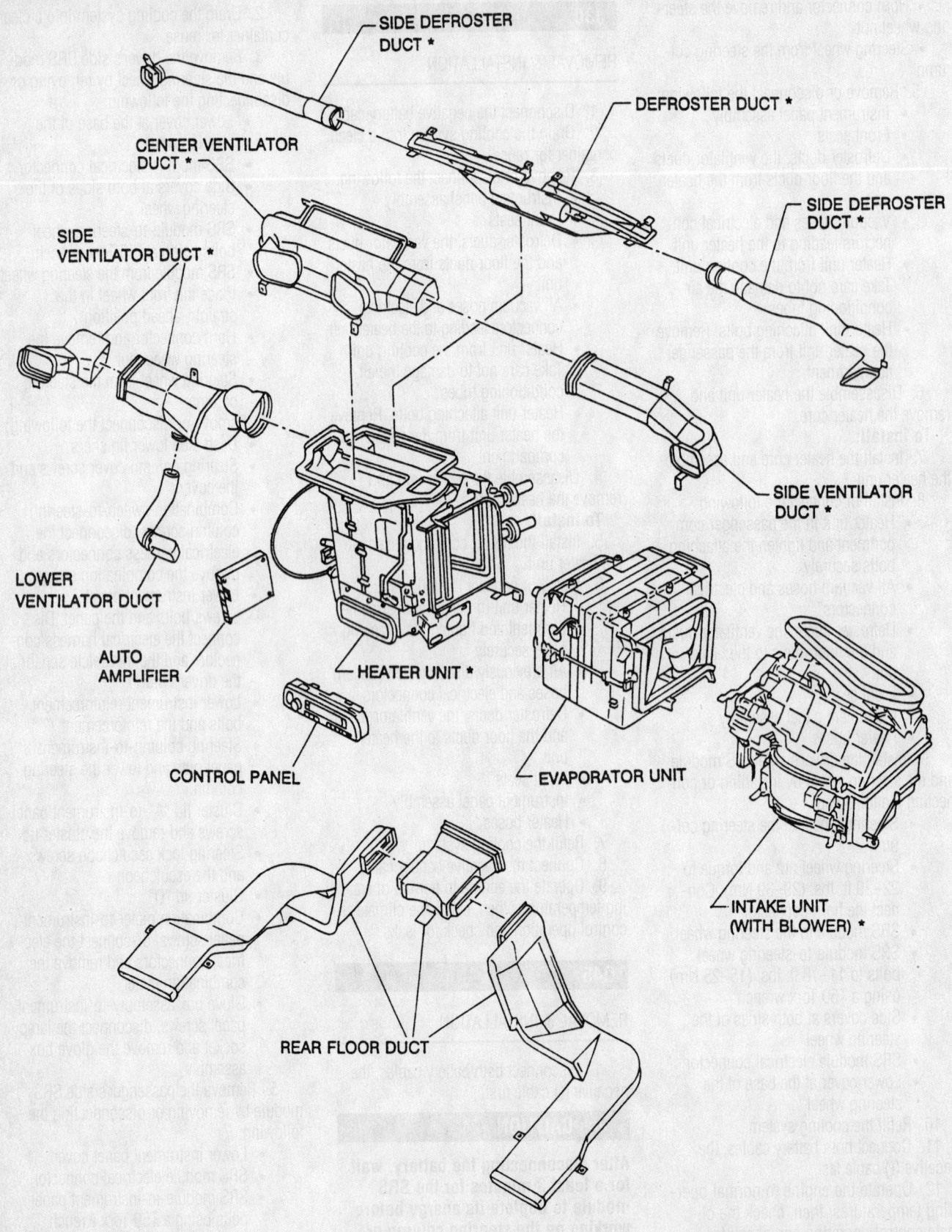

SIDE DEFROSTER DUCT *

CENTER VENTILATOR DUCT *

DEFROSTER DUCT *

SIDE DEFROSTER DUCT *

SIDE VENTILATOR DUCT *

SIDE VENTILATOR DUCT *

LOWER VENTILATOR DUCT

AUTO AMPLIFIER

HEATER UNIT *

CONTROL PANEL

EVAPORATOR UNIT

INTAKE UNIT (WITH BLOWER) *

REAR FLOOR DUCT

ITEMS MARKED WITH "*" REQUIRE INSTRUMENT PANEL REMOVAL

93112G11

Exploded view of the heating-air conditioning system assemblies and related components—J30

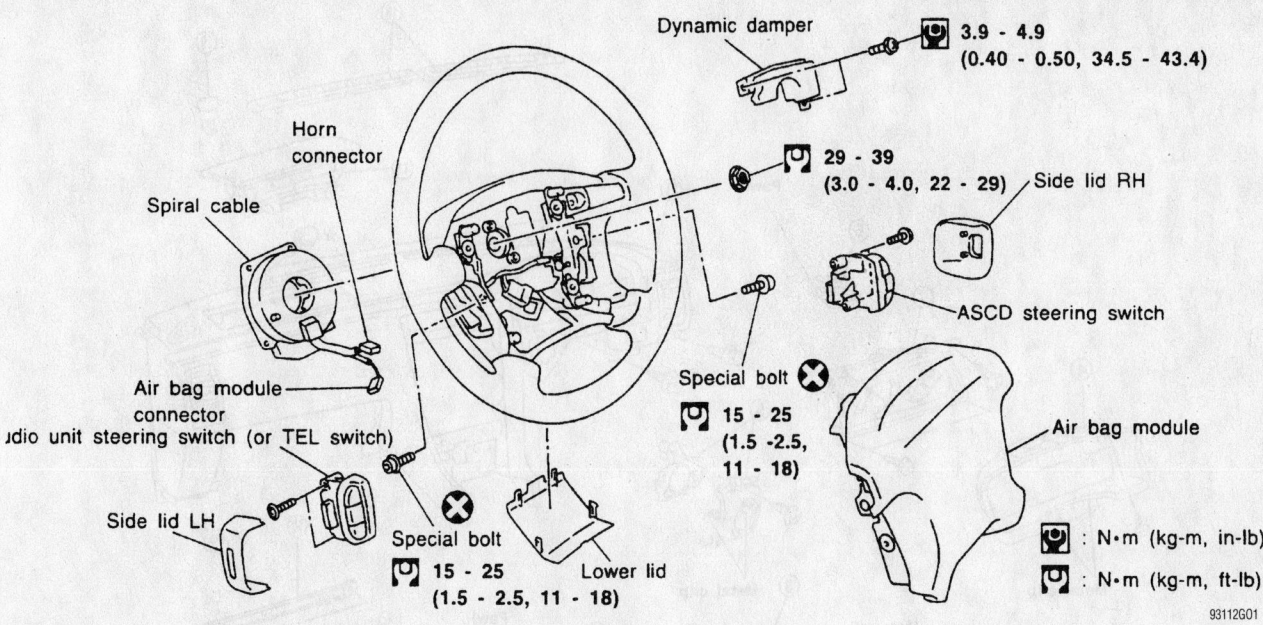

Dynamic damper

3.9 - 4.9
(0.40 - 0.50, 34.5 - 43.4)

29 - 39
(3.0 - 4.0, 22 - 29)

Side lid RH

ASCD steering switch

Horn connector

Spiral cable

Special bolt

15 - 25
(1.5 -2.5, 11 - 18)

Air bag module

Air bag module connector

udio unit steering switch (or TEL switch)

Side lid LH

Special bolt

15 - 25
(1.5 - 2.5, 11 - 18)

Lower lid

: N•m (kg-m, in-lb)

: N•m (kg-m, ft-lb)

93112G01

Exploded view of the air bag module, the steering wheel and related components—Q45

✳✳ CAUTION

Store the SRS module in a safe place with the front facing upward.

6. Remove or disconnect the following:
- Center ventilation grille using a suitable prytool
- Lock console
- Card pocket assembly screws and the assembly
- Console finisher clips and the console finisher
- Audio center, the cluster lid "C" and the air conditioning control unit-to-instrument panel screws, disconnect the electrical connectors and remove the panel
- Console box assembly-to-instrument panel screws, disconnect the electrical connectors and remove the console box assembly
- Connectors and remove the defroster grilles
- Connectors and remove the sunload sensor
- Front pillar garnish
- Instrument panel-to-chassis bolts/nuts and the instrument panel
- Heater hoses from the heater core
- Rear duct from the heater unit
- Air conditioning housing-to-heater housing fasteners
- Heater unit from the vehicle

7. Disassemble and remove the heater core from the heater housing.

To install:

8. Assemble and install the heater core to the heater housing.

9. Install or connect the following:
- Heater unit to the vehicle
- Heater housing-to-chassis fasteners
- Air conditioning housing-to-heater housing fasteners
- Rear duct to the heater unit
- Heater hoses to the heater core
- Instrument panel and the instrument panel-to-chassis bolts/nuts
- Front pillar garnish
- Defroster grilles and connect the connectors
- Lower instrument panel center and the panel-to-instrument panel screws
- Lower instrument cover and the cover clip
- Center ventilation grille
- Lock console
- Card pocket assembly and the assembly screws
- Console finisher and engage the clips
- Audio center, the cluster lid "C" and the air conditioning control unit panel, connect the electrical connectors and install the panel-to-instrument panel screws
- Console box assembly, connect the

electrical connectors and install the console box assembly-to-instrument panel screws
- Defroster grilles and connect the connector
- Sunload sensor and connectors

10. Install the passenger's side SRS module by installing or connecting the following:
- SRS module to the instrument panel
- SRS module-to-instrument panel bolts and torque the bolts to 11–18 ft. lbs. (15–25 Nm) using a T50 Torx wrench
- SRS module electrical connector
- Lower instrument panel cover
- Glove box assembly, connect the lamp socket and install the glove box assembly-to-instrument panel screws

11. Install or connect the following:
- Combination meter, connect the electrical connectors and install the combination meter-to-instrument panel screws
- Cluster lid "D"
- Steering lock escutcheon and the escutcheon screws
- Cluster lid "A" and the cluster lid-to-instrument panel screws
- Steering column and torque the steering column-to-instrument panel nuts to 11–14 ft. lbs. (15–19 Nm)

For Tune-up, Capacities and Firing orders, see Section 1 of this manual

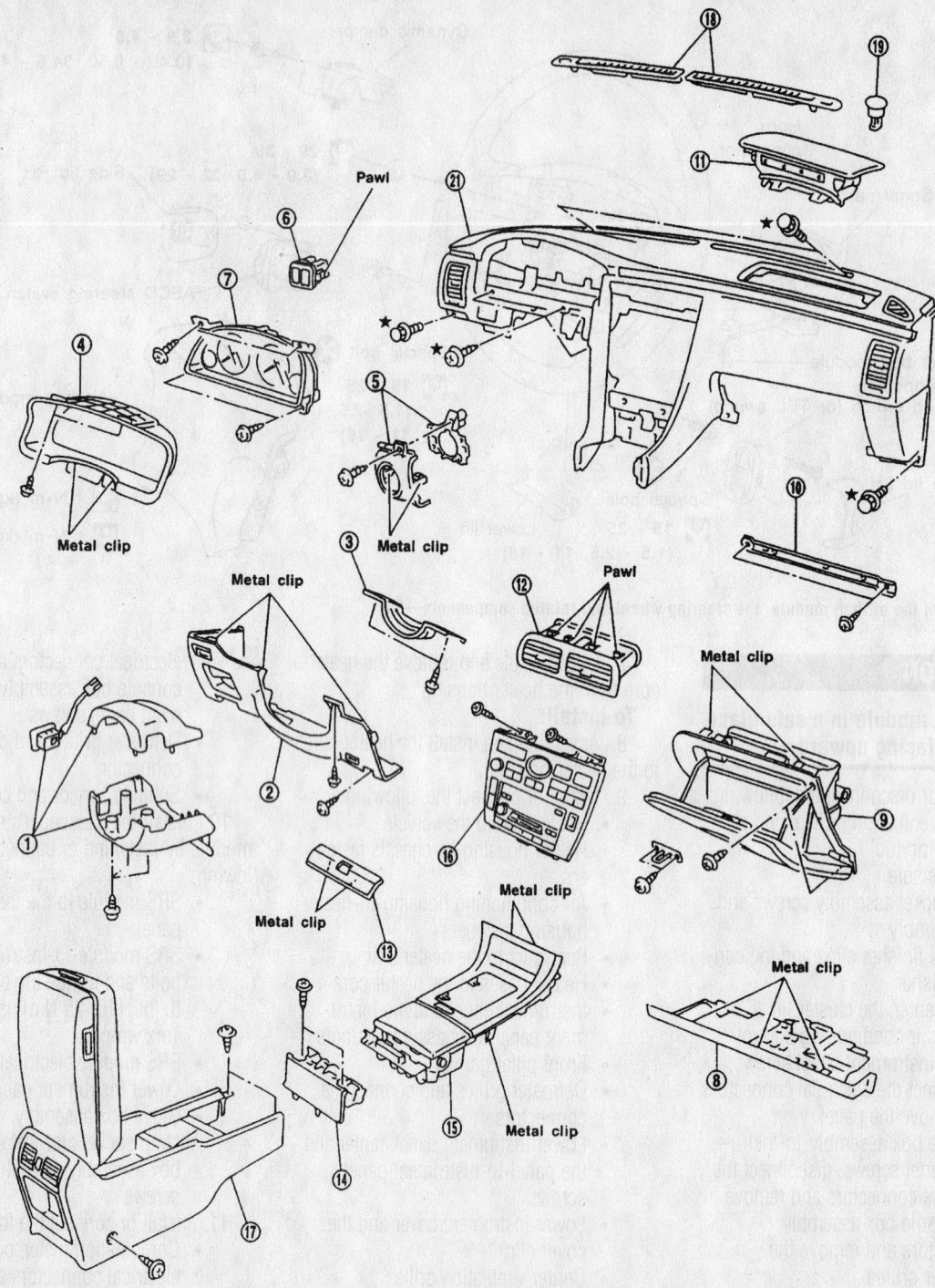

Pawl

★ : Instrument panel assembly mounting bolts and screws

1. Steering column cover and combination switch
2. Instrument lower panel on driver side
3. Instrument lower reinforcement
4. Cluster lid A
5. Steering lock escutcheon
6. Cluster lid D
7. Combination meter
8. Instrument lower cover on passenger side
9. Glove box assembly
10. Instrument panel reinforcement
11. Passenger air bag module

12. Center ventilation grille
13. Lock console
14. Card pocket assembly
15. Console A/T finisher
16. Audio, cluster lid C and A/C control unit
17. Console box assembly
18. Defroster grille
19. Sunload sensor
20. Front pillar garnish
21. Instrument panel and pads

93112G02

Exploded view of the instrument panel—Q45

- Lower instrument reinforcement and the reinforcement bolts.
- Electrical harness connectors and the in-vehicle sensor; then, install the lower instrument panel and the panel screws/bolts at the driver's side
- Combination switch and the combination switch-to-steering column screws and connect the electrical harness connectors
- Steering column covers and the cover screws

- Dash side lower finishers

12. Install the driver's side SRS module and the steering wheel by installing or connecting the following:
- Steering wheel to the steering column
- Steering wheel nut and torque to 22–29 ft. lbs. (29–39 Nm). Connect the horn connector
- SRS module to the steering wheel
- Torque the SRS module-to-steering wheel bolts to 11–18 ft. lbs.

(15–25 Nm) using a T50 Torx wrench
- Side covers at both sides of the steering wheel
- SRS module electrical connector
- Lower cover at the base of the steering wheel

13. Refill the cooling system.

14. Connect both battery cables, the negative (()) cable last.

15. Operate the engine to normal operating temperatures; then, check the climate control operation and check for leaks.

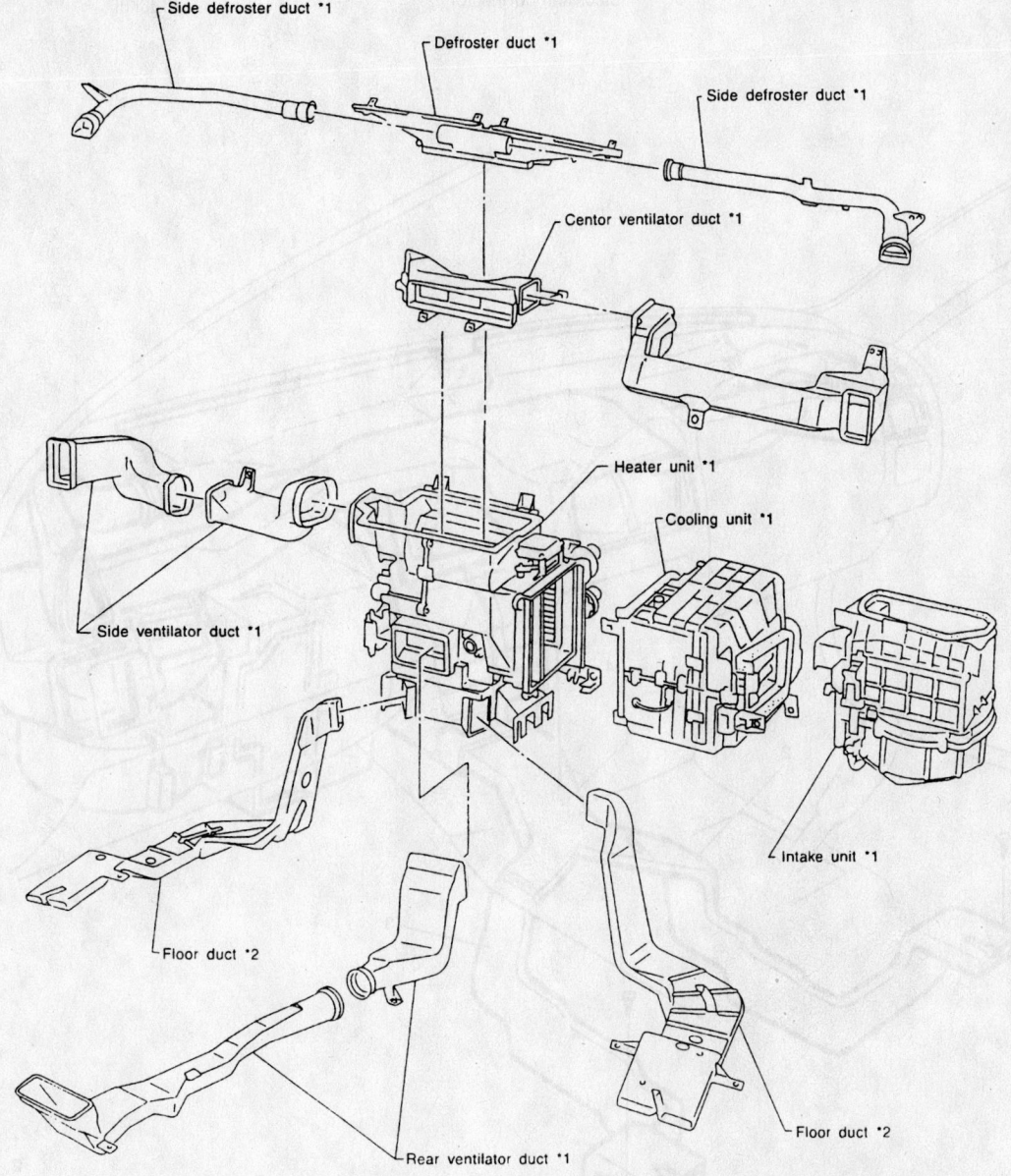

*1 : For removal, it is necessary to remove instrument assembly.

*2 : For removal, it is necessary to remove front seat.

93112G03

Exploded view of the heater housing, air conditioning housing, blower motor and ventilation ducts—Q45

For complete service labor times, order the Chilton Labor Guide

1998–00 KIA

Sephia

REMOVAL & INSTALLATION

1. Disconnect the negative battery cable.

※※ CAUTION

After disconnecting the negative battery cable, wait for at least 10 min-utes for the SRS module to deplete its stored energy.

2. Remove the driver's side air bag and steering wheel by removing or disconnecting the following:
- Position the front wheels in the straight-ahead position
- 4 steering wheel-to-air bag module bolts
- Air bag module and disconnect the electrical connector

- Steering wheel-to-steering column nut
- Steering wheel from the steering column

3. Remove the passenger's side air bag module by removing or disconnecting the following:
- 2 screws and the glove box at the bottom of the glove box
- Side cover and pull the connector from the "T" bar side bracket

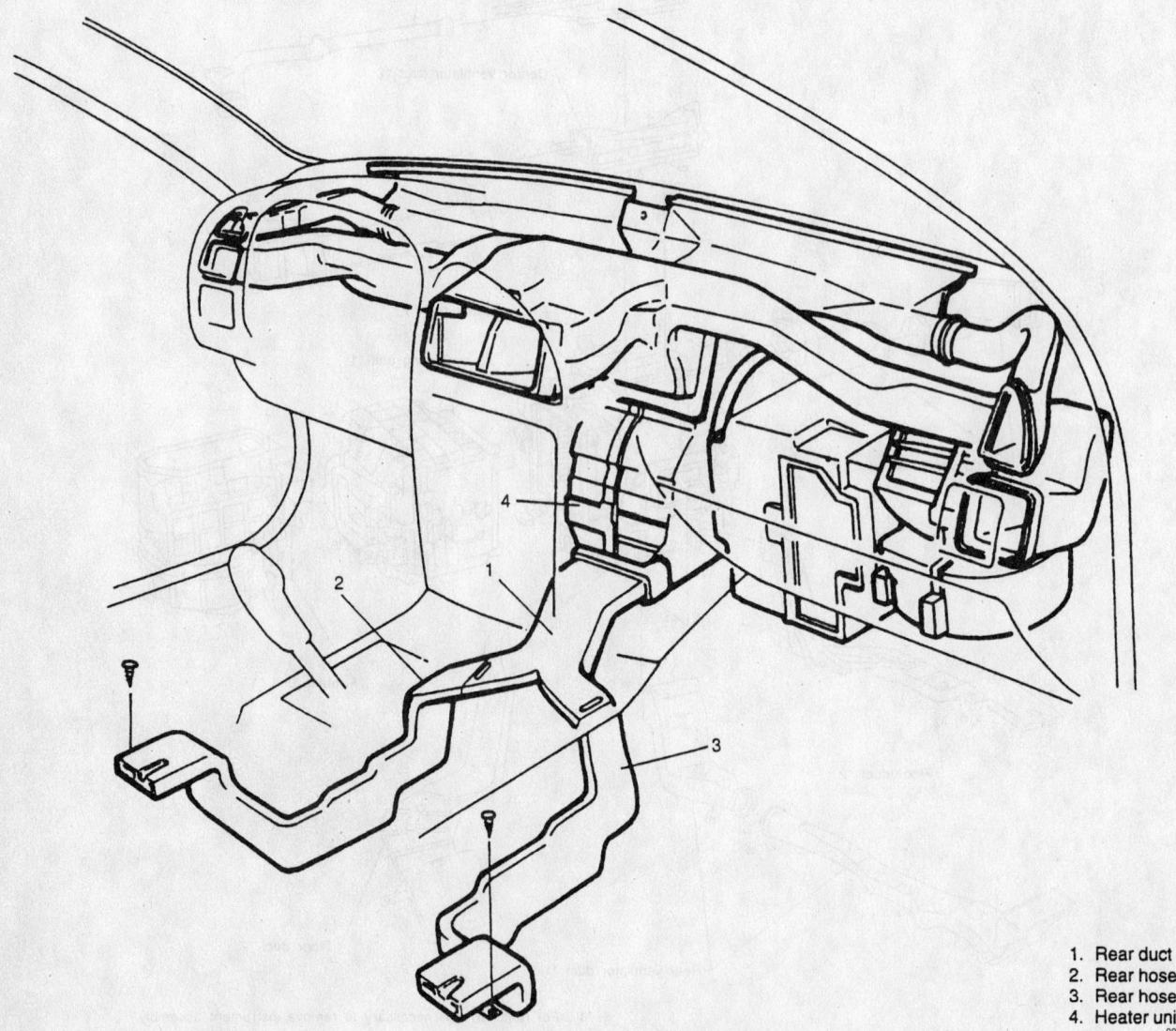

1. Rear duct
2. Rear hose LH
3. Rear hose RH
4. Heater unit

93112GI4

Exploded view of the steering wheel and air bag module assembly—Sephia

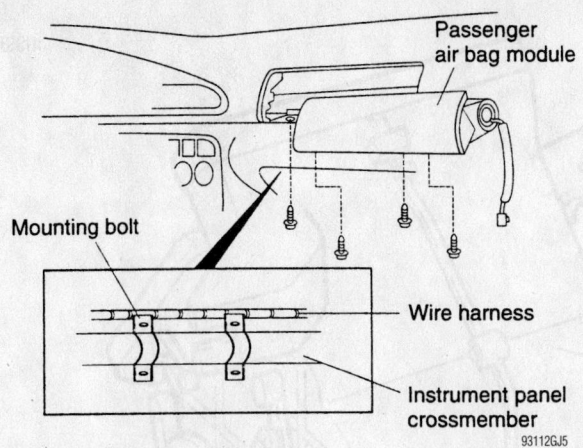

Passenger
air bag module

Mounting bolt

Wire harness

Instrument panel
crossmember

93112GJ5

Exploded view of the passenger's side air bag module assembly—Sephia

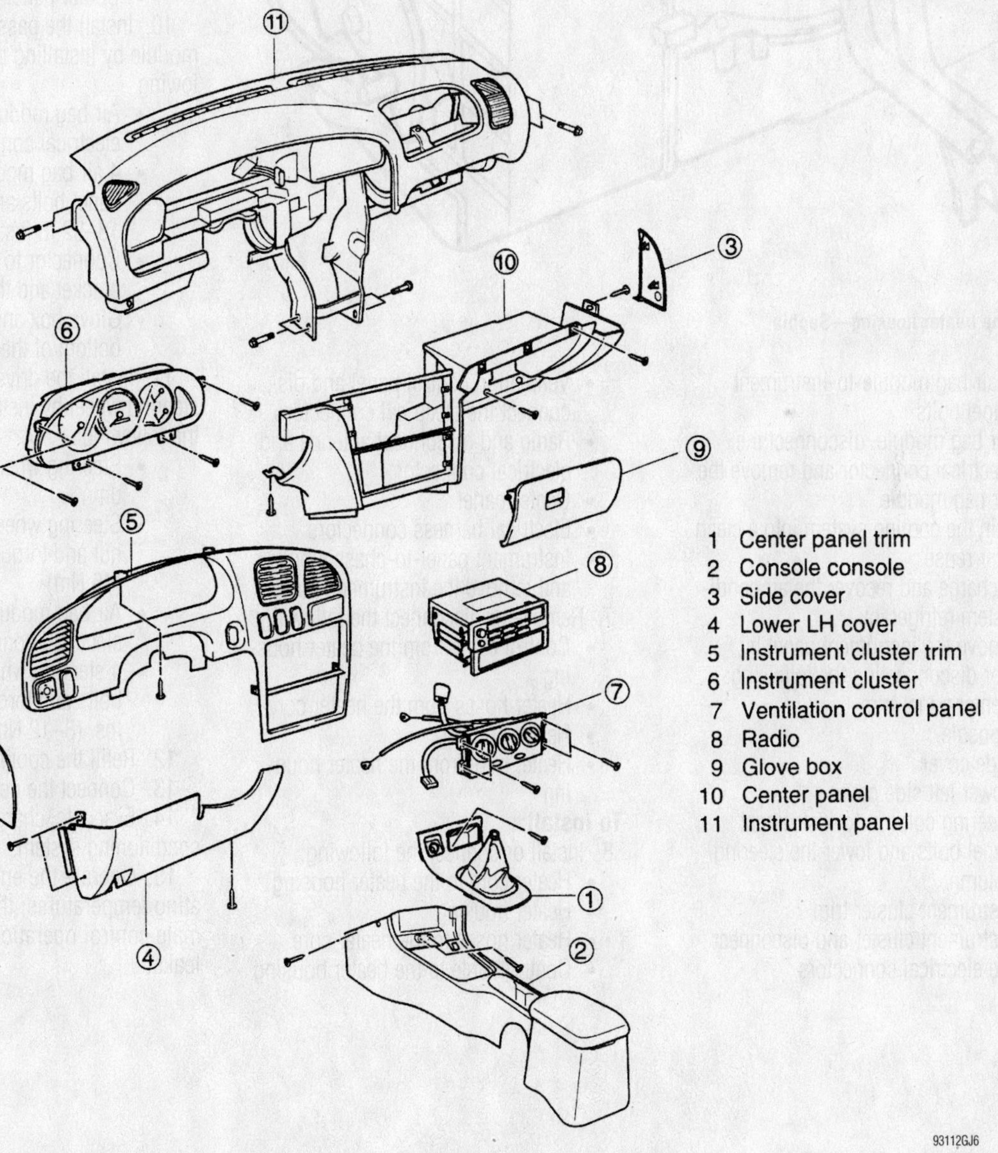

1 Center panel trim
2 Console console
3 Side cover
4 Lower LH cover
5 Instrument cluster trim
6 Instrument cluster
7 Ventilation control panel
8 Radio
9 Glove box
10 Center panel
11 Instrument panel

93112GJ6

Exploded view of the instrument panel assembly—Sephia

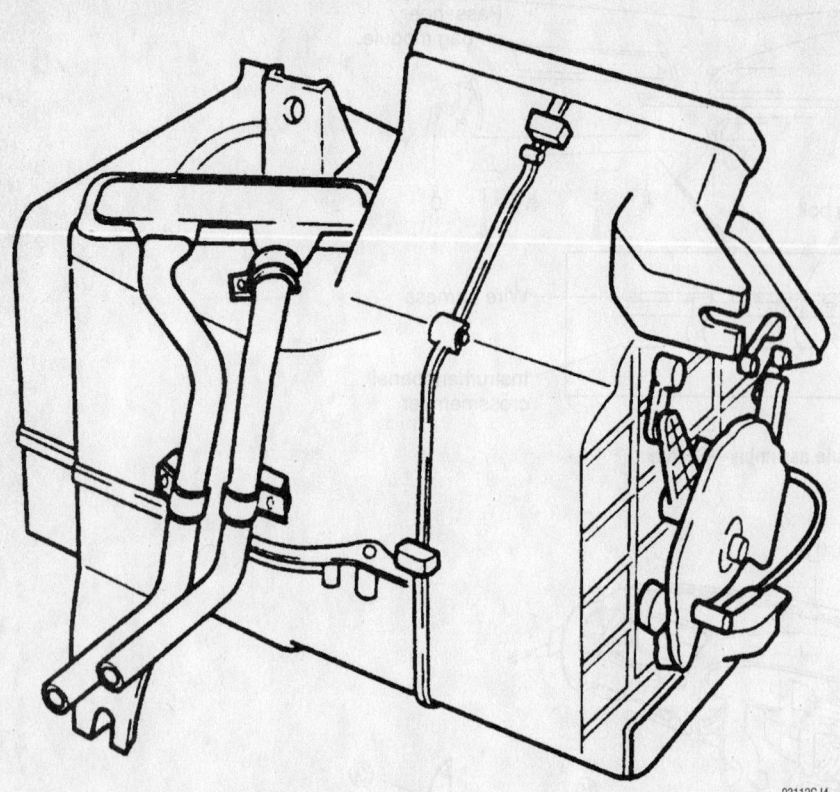

93112GJ4

View of the heater housing—Sephia

- 4 air bag module-to-instrument panel bolts
- Air bag module, disconnect the electrical connector and remove the air bag module

4. Drain the cooling system into a clean container for reuse.

5. Discharge and recover the air conditioning system refrigerant.

6. Remove the instrument panel by removing or disconnecting the following:
- Center panel trim
- Console
- Side cover
- Lower left side cover
- Steering column-to-instrument panel bolts and lower the steering column
- Instrument cluster trim
- Instrument cluster and disconnect the electrical connectors
- Ventilation control panel and disconnect the electrical connectors
- Radio and disconnect antenna and electrical connector
- Center panel
- Electrical harness connectors
- Instrument panel-to-chassis bolts and remove the instrument panel

7. Remove or disconnect the following:
- Control cable from the heater housing
- Heater hoses from the heater core
- Heater housing
- Heater core from the heater housing

To install:

8. Install or connect the following:
- Heater core to the heater housing
- Heater housing
- Heater hoses to the heater core
- Control cable to the heater housing

9. Install the instrument panel by installing or connecting the following:
- Instrument panel and the instrument panel-to-chassis bolts
- Electrical harness connectors
- Center panel
- Antenna and electrical connector and install the radio
- Electrical connectors and install the ventilation control panel
- Electrical connectors and install the instrument cluster
- Instrument cluster trim
- Steering column-to-instrument panel bolts
- Lower left side cover
- Side cover
- Console
- Center panel trim

10. Install the passenger's side air bag module by installing or connecting the following
- Air bag module and connect the electrical connector
- 4 air bag module-to-instrument panel bolts and torque the bolts to 18–32 ft. lbs. (24–43 Nm)
- Connector to the "T" bar side bracket and the side cover
- Glove box and the 2 screws at the bottom of the glove box

11. Install the driver's side air bag and steering wheel by installing or connecting the following:
- Steering wheel to the steering column
- Steering wheel-to-steering column nut and torque the nut to 33 ft. lbs. (45 Nm)
- Air bag module and connect the electrical connector
- 4 steering wheel-to-air bag module bolts and torque to 72–106 inch lbs. (8–12 Nm)

12. Refill the cooling system.

13. Connect the negative battery cable.

14. Evacuate, charge and leak test the air conditioning system.

15. Operate the engine to normal operating temperatures; then, check the climate control operation and check for leaks.

1998–01 LEXUS

ES 300

REMOVAL & INSTALLATION

1. Disconnect the negative battery cable. Wait 90 seconds before doing any further work while the airbag system de-energizes.

2. Drain the cooling system into a clean container for reuse.

3. Remove or disconnect the following:

- 2 hood release lever screws and the lever
- No. 1 lower panel-to-instrument panel bolt/screw, discon-

nect the electrical connectors and remove the No. 1 lower panel

- Lower left hand panel
- 3 heater protector clips and remove the heater protector
- 2 screws and the 2 clamps holding the heater core in place

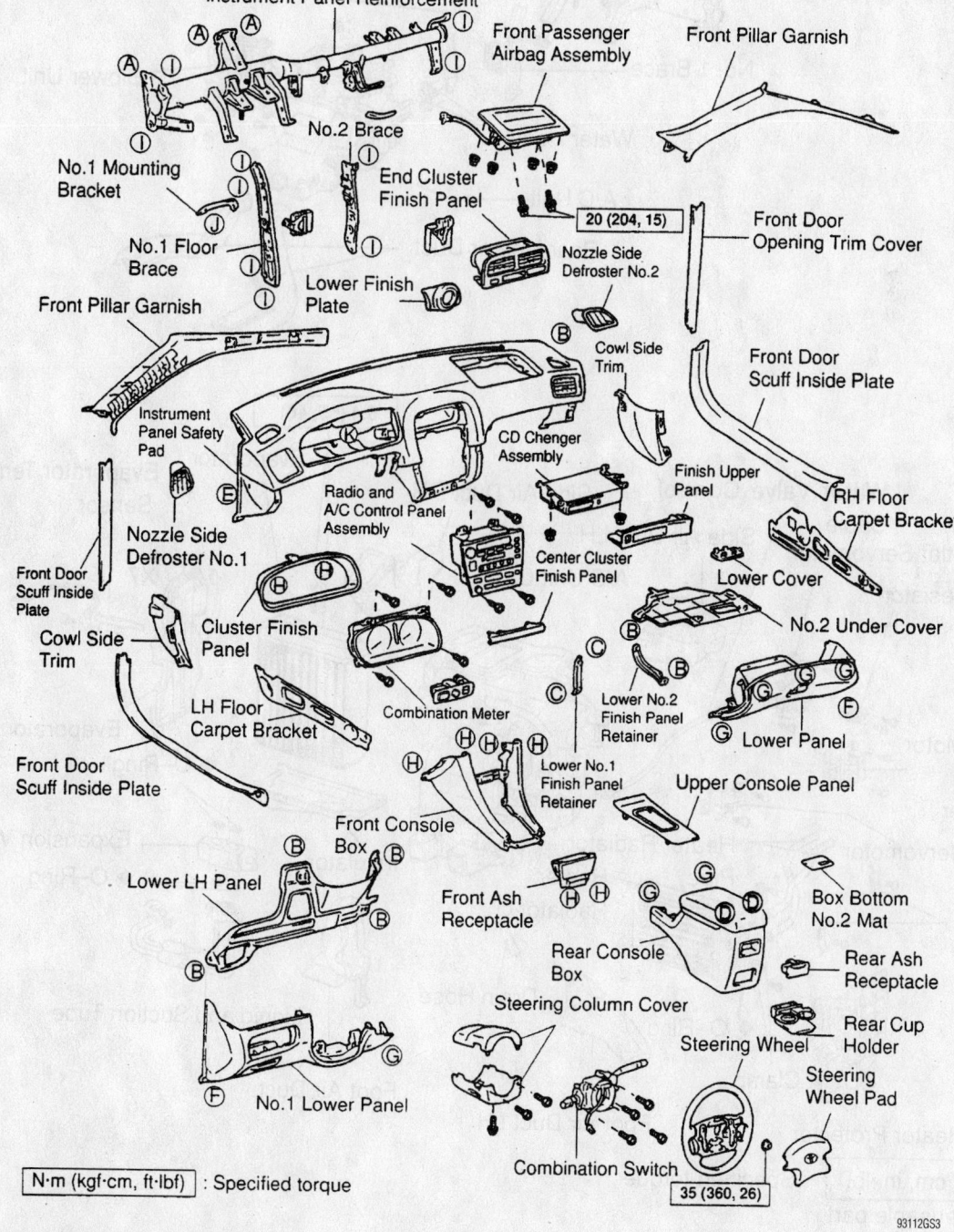

Instrument Panel Reinforcement

Front Passenger Airbag Assembly

Front Pillar Garnish

No.2 Brace

No.1 Mounting Bracket

End Cluster Finish Panel

Nozzle Side Defroster No.2

20 (204, 15)

Front Door Opening Trim Cover

No.1 Floor Brace

Lower Finish Plate

Cowl Side Trim

Front Pillar Garnish

Front Door Scuff Inside Plate

Instrument Panel Safety Pad

CD Chenger Assembly

Finish Upper Panel

RH Floor Carpet Bracket

Front Door Scuff Inside Plate

Nozzle Side Defroster No.1

Radio and A/C Control Panel Assembly

Lower Cover

Cowl Side Trim

Cluster Finish Panel

Center Cluster Finish Panel

No.2 Under Cover

LH Floor Carpet Bracket

Combination Meter

Lower No.2 Finish Panel Retainer

Lower Panel

Front Door Scuff Inside Plate

Lower No.1 Finish Panel Retainer

Upper Console Panel

Front Console Box

Lower LH Panel

Front Ash Receptacle

Box Bottom No.2 Mat

Rear Console Box

Rear Ash Receptacle

Steering Column Cover

Steering Wheel

Rear Cup Holder

Steering Wheel Pad

No.1 Lower Panel

Combination Switch

35 (360, 26)

N·m (kgf·cm, ft·lbf) : Specified torque

Exploded view of the instrument assembly—ES 300

93112GS3

Timing belt service is covered in Section 3 of this manual

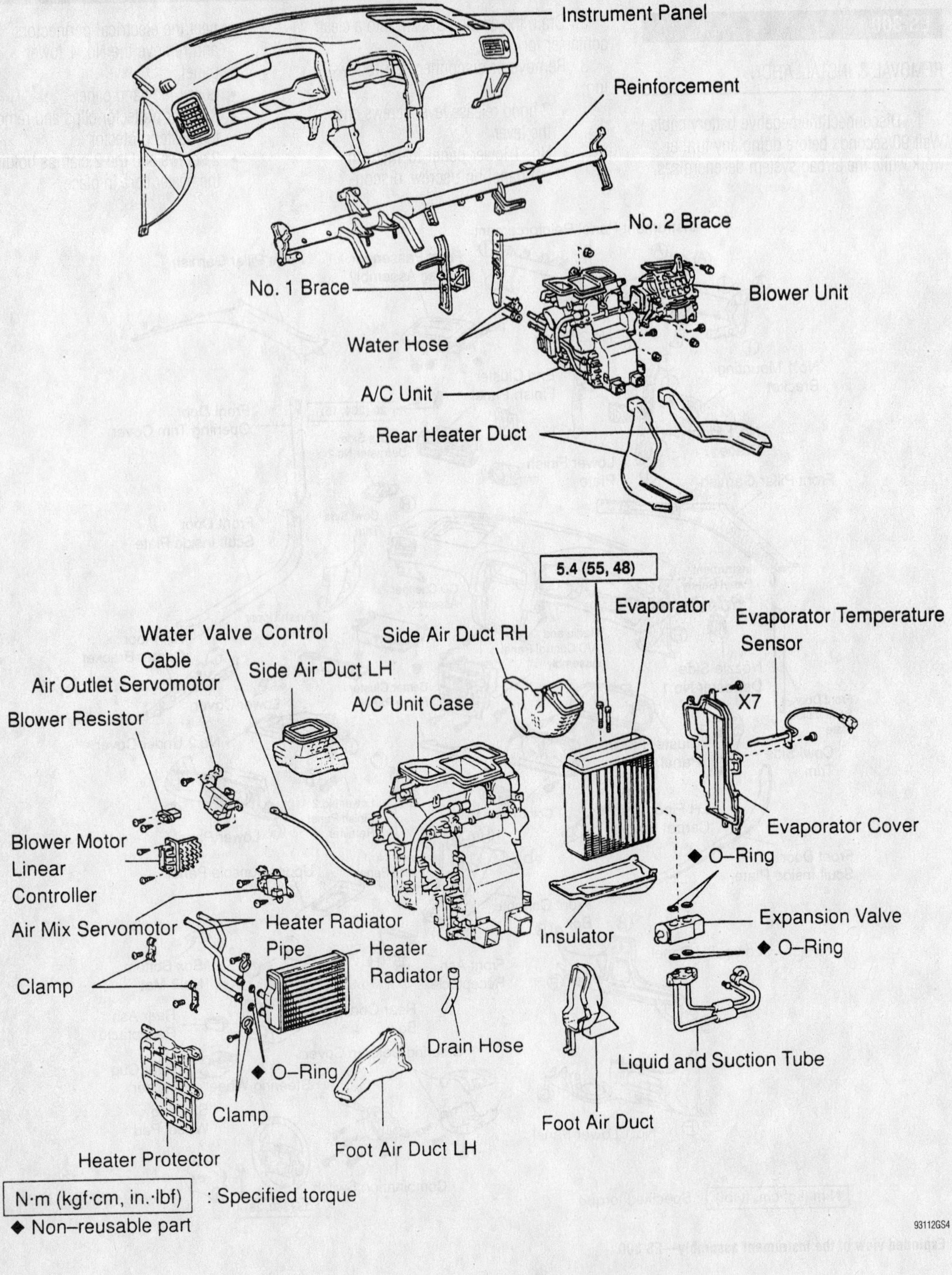

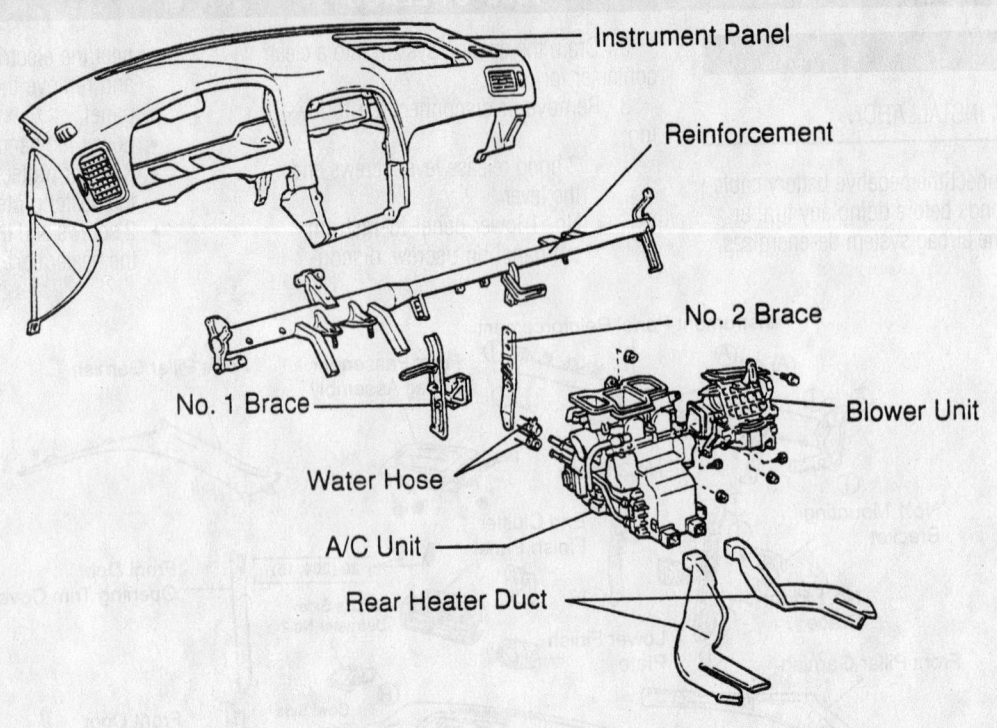

Instrument Panel

Reinforcement

No. 2 Brace

No. 1 Brace

Water Hose

A/C Unit

Blower Unit

Rear Heater Duct

5.4 (55, 48)

Water Valve Control
Cable

Air Outlet Servomotor

Blower Resistor

Blower Motor
Linear
Controller

Air Mix Servomotor

Clamp

Heater Protector

Side Air Duct LH

A/C Unit Case

Heater Radiator
Pipe

Clamp

◆ O–Ring

Foot Air Duct LH

Side Air Duct RH

Heater
Radiator

Drain Hose

Evaporator

Insulator

Foot Air Duct

Evaporator Temperature
Sensor

X7

Evaporator Cover

◆ O–Ring

Expansion Valve

◆ O–Ring

Liquid and Suction Tube

| N·m (kgf·cm, in.·lbf) | : Specified torque

◆ Non–reusable part

93112GS4

Exploded view of the heater core, heater/air conditioning housing and related components—ES 300

- Heater core hoses and discard the O-rings
- Pull out heater core from the heater housing

To install:

4. Install or connect the following:
- Heater core to the heater housing
- Heater core hoses using new O-rings
- 2 clamps and the 2 screws holding the heater core in place
- Heater protector and connect the 3 heater protector clips
- Lower left hand panel
  No. 1 lower panel, connect the electrical connectors and install the No. 1 lower panel-to-instrument panel bolt/screw
- Hood release lever and the 2 lever screws

5. Refill the cooling system.
6. Connect the negative battery cable.
7. Operate the engine to normal operating temperatures; then, check the climate control operation and check for leaks.

GS 300/GS400

REMOVAL & INSTALLATION

1998–01 Models

1. Disconnect the negative battery cable. Wait 90 seconds before doing any further work while the airbag system de-energizes.
2. Drain the cooling system into a clean container for reuse.
3. Discharge and recover the air conditioning system refrigerant.
4. Remove or disconnect the following:

- Refrigerant lines-to-evaporator bolt, slide the plate clockwise, disconnect both lines and discard the O-rings. Plug the openings to prevent contamination
- Heater hoses from the heater core
- No. 1 grommet, the heater pipe grommet and the drain hose grommet

5. Remove the steering wheel by removing or disconnecting the following:
- Place the front wheels in the straight-ahead position.
- Torx bolt covers at both sides of the steering wheel
- Loosen the Torx bolts (using a Torx wrench), until the circumference

ring on the bolt catches on the screw case
- Lift the air bag, disconnect the electrical connector and remove it
- Steering wheel nut and press the steering wheel from the steering column

6. Remove the instrument by removing or disconnecting the following:
- Front pillar garnishes and the front door scuff plates
- Steering column cover screws and the covers
- Electrical connectors and remove the combination switch
- End pad
- 2 No. 1 undercover-to-instrument panel screws and the undercover
- 2 hood lock release screws and the release
- 4 No. 1 safety pad-to-instrument panel bolts, screw and the safety pad
- Parking brake handle and the No. 1 switch hole base
- 4 steering column-to-instrument panel nuts, disconnect the spring from the brake pedal and remove the steering column
- Instrument cluster finish panel using a suitable prytool
- 4 instrument cluster-to-instrument panel screws, disconnect the electrical connectors and remove the instrument cluster
- No. 2 undercover using a suitable prytool
- Plate and disconnect the air bag electrical connector inside the glove box
- Glove box-to-instrument panel 2 bolts, 3 screws, and the glove box
- 3 CD changer-to-instrument panel nuts, disconnect the electrical connectors and remove the CD changer
- Ashtray
- No. 2 register using a suitable prytool and disconnect the connector
- Audio unit-to-instrument panel 2 bolts, 2 screws and the audio unit
- Cluster finish panel using a suitable prytool, disconnect the connectors and remove the panel.
- Console box carpet
- Lower rear console box
- Rear console armrest
- No. 3 console box mounting bracket

- Console box
- No. 1 console box duct
- No. 7 heater-to-register
- 5 instrument panel-to-chassis bolts, the nut, the screw and the instrument panel
- No. 1 and No. 2 brace
- Reinforcement-to-chassis 5 nuts, 4 bolts and the reinforcement
- Ventilation nozzles from the heater/air conditioning housing

7. Remove the blower unit by removing or disconnecting the following:
- Connector clamp
- 3 air duct-to-blower housing screws and the air duct
- Electrical connector bracket, the wiring harness clamps and the wiring harness
- Blower housing connectors
- 2 blower housing-to-bracket bolts and the bracket
- Blower housing-to-chassis bolt, screw, nut and the blower housing

8. Remove or disconnect the following:
- 2 center air duct-to-heater/air conditioning housing screws and the air duct
- Move the floor carpet rearward
- Wiring harness clamps
- 2 air duct bolts and the ducts at both sides

9. Remove the heater/air conditioning housing by removing or disconnecting the following:
- Electrical connector
- Wiring harness set nut
- Wiring harness clamp
- Heater/air conditioning housing 2 nuts and bolt
- Heater/air conditioning housing

10. Remove the heater core-to-heater/air conditioning housing clamp screw and the clamp.
11. Pull the heater core from the heater/air conditioning housing.

To install:

12. Install the heater core to the heater/air conditioning housing.
13. Install the heater core clamp and the clamp-to-heater/air conditioning housing screw.
14. Install the heater/air conditioning housing by installing or connecting the following:
- Heater/air conditioning housing
- Heater/air conditioning housing 2 nuts and bolt
- Wiring harness clamp

Heater Core replacement is covered in Section 2 of this manual

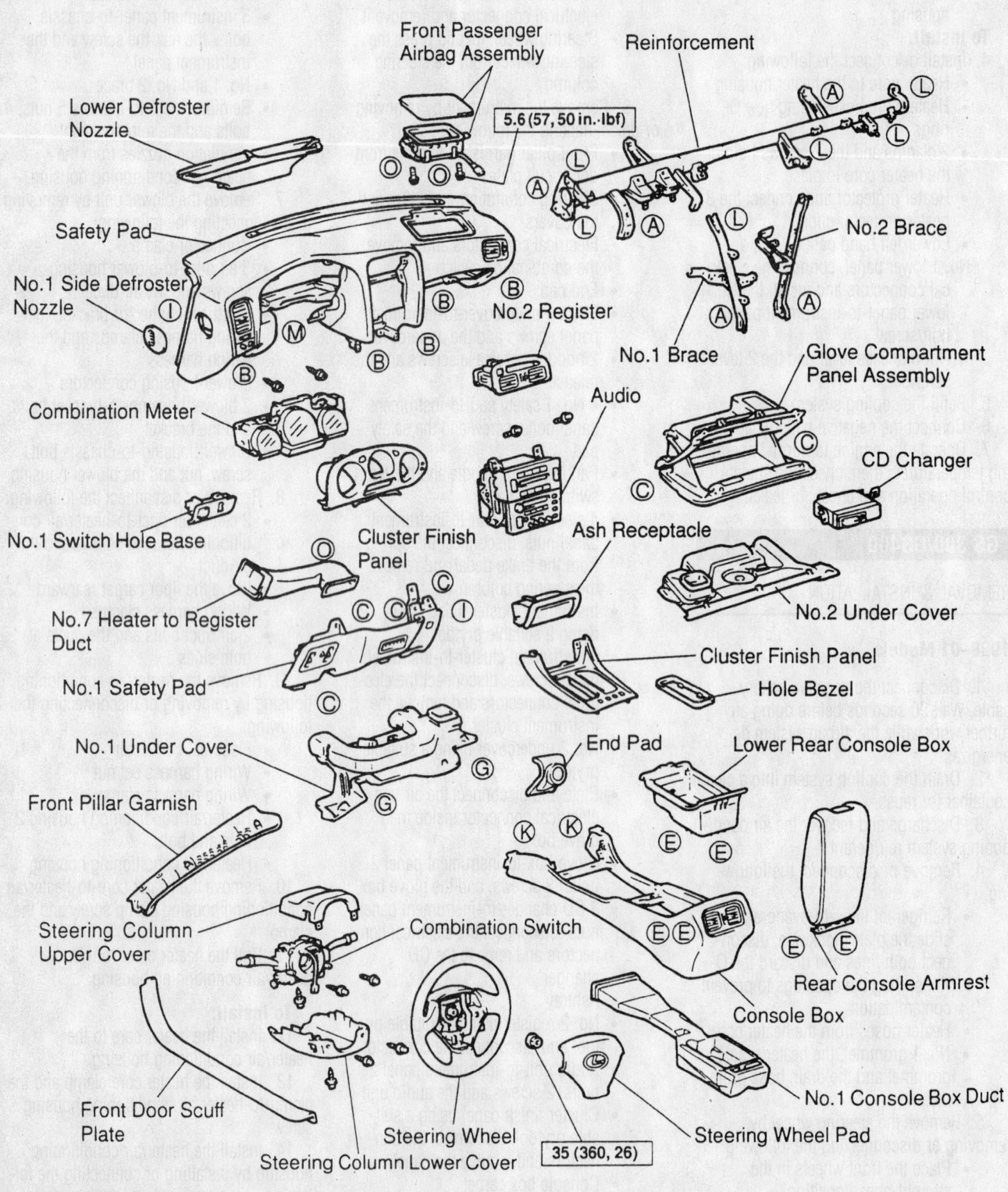

Front Passenger Airbag Assembly

Reinforcement

5.6 (57, 50 in.·lbf)

Lower Defroster Nozzle

Safety Pad

No.2 Brace

No.1 Side Defroster Nozzle

No.1 Brace

No.2 Register

Glove Compartment Panel Assembly

Audio

CD Changer

Combination Meter

Ash Receptacle

No.2 Under Cover

No.1 Switch Hole Base

Cluster Finish Panel

Cluster Finish Panel

No.7 Heater to Register Duct

Hole Bezel

No.1 Safety Pad

End Pad

Lower Rear Console Box

No.1 Under Cover

Front Pillar Garnish

Steering Column Upper Cover

Combination Switch

Rear Console Armrest

Console Box

Front Door Scuff Plate

Steering Wheel

No.1 Console Box Duct

Steering Column Lower Cover

Steering Wheel Pad

35 (360, 26)

N·m (kgf·cm, ft·lbf) : Specified torque

93112GT4

Exploded view of the instrument panel and related components—GS 300 and 1998–01 GS 400

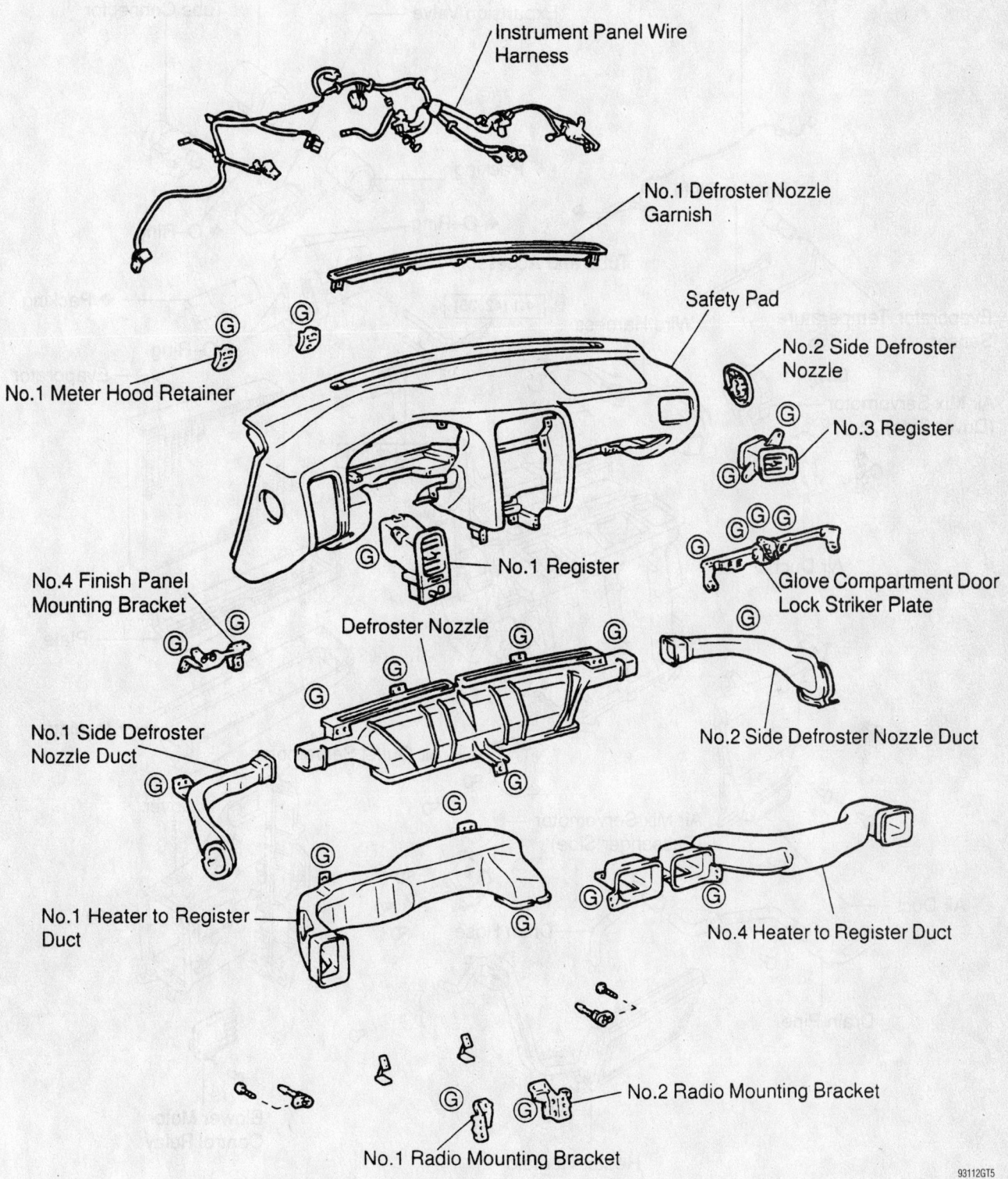

Instrument Panel Wire Harness

No.1 Defroster Nozzle Garnish

Safety Pad

No.2 Side Defroster Nozzle

No.1 Meter Hood Retainer

No.3 Register

No.1 Register

No.4 Finish Panel Mounting Bracket

Glove Compartment Door Lock Striker Plate

Defroster Nozzle

No.2 Side Defroster Nozzle Duct

No.1 Side Defroster Nozzle Duct

No.1 Heater to Register Duct

No.4 Heater to Register Duct

No.2 Radio Mounting Bracket

No.1 Radio Mounting Bracket

93112GT5

Exploded view of the instrument panel, ventilation ducts and related components—GS 300 and 1998–01 GS 400

Brake service is covered in Section 4 of this manual

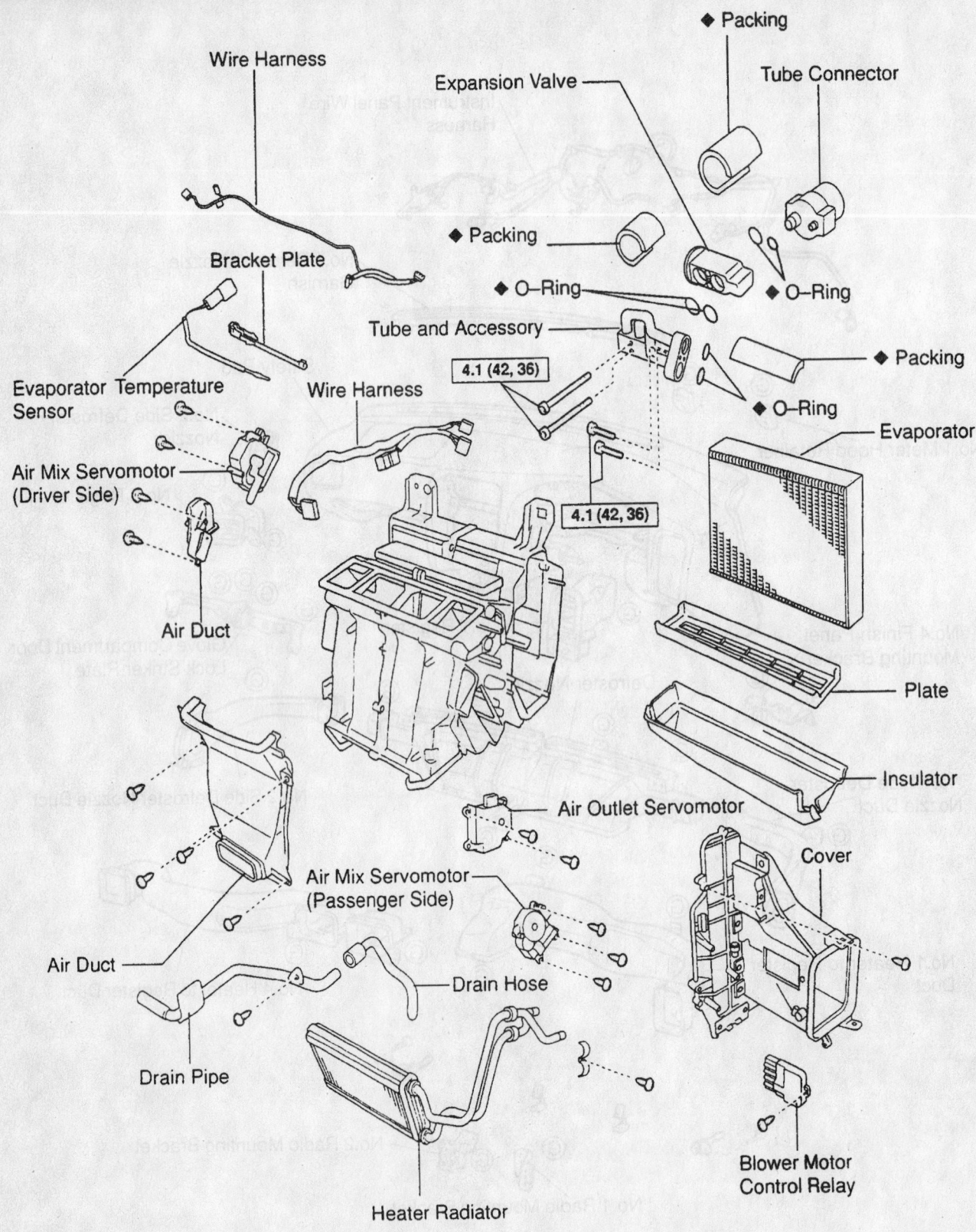

Wire Harness

Expansion Valve

◆ Packing

Tube Connector

◆ Packing

◆ O-Ring

◆ O-Ring

◆ Packing

◆ O-Ring

Bracket Plate

Tube and Accessory

4.1 (42, 36)

Evaporator

Evaporator Temperature Sensor

Wire Harness

Air Mix Servomotor (Driver Side)

4.1 (42, 36)

Air Duct

Plate

Insulator

Air Outlet Servomotor

Cover

Air Mix Servomotor (Passenger Side)

Air Duct

Drain Hose

Drain Pipe

Heater Radiator

Blower Motor Control Relay

N·m (kgf·cm, in.·lbf) : Specified torque

◆ Non-reusable part

93112GT6

Exploded view of the heater/air conditioning housing and related components—GS 300 and 1998–01 GS 400

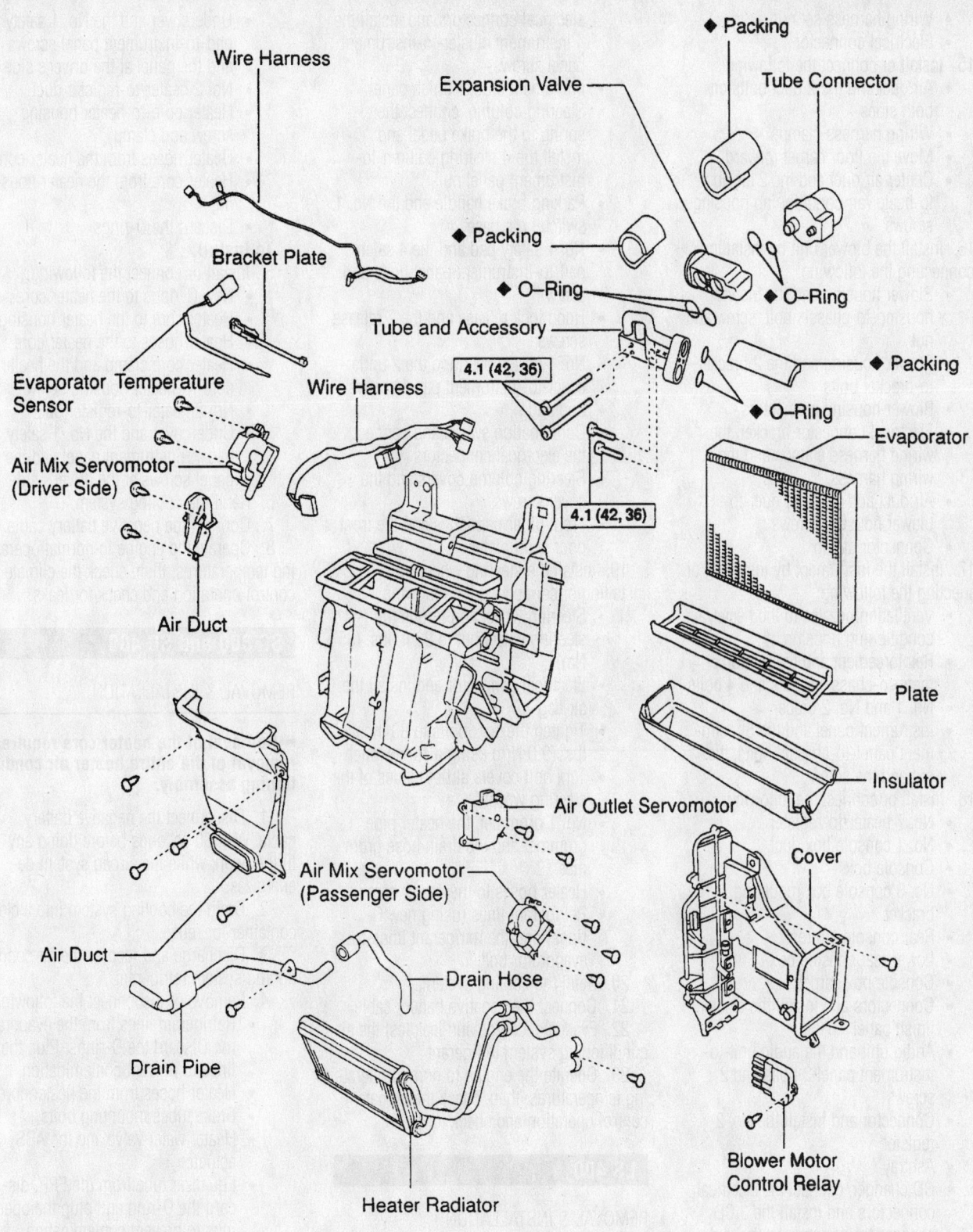

◆ Packing

Tube Connector

Wire Harness

Expansion Valve

◆ Packing

◆ O–Ring

◆ O–Ring

Bracket Plate

Tube and Accessory

◆ Packing

4.1 (42, 36)

◆ O–Ring

Evaporator Temperature Sensor

Wire Harness

Evaporator

Air Mix Servomotor (Driver Side)

4.1 (42, 36)

Air Duct

Plate

Insulator

Air Outlet Servomotor

Cover

Air Mix Servomotor (Passenger Side)

Air Duct

Drain Hose

Drain Pipe

Blower Motor Control Relay

Heater Radiator

N·m (kgf·cm, in.·lbf) : Specified torque

◆ Non–reusable part

93112GT7

Exploded view of the heater core, heater housing and related components—GS 300 and 1998–01 GS 400

For complete Engine Mechanical specifications, see Section 1 of this manual

- Wiring harness set nut
- Electrical connector

15. Install or connect the following:
- Air duct and the 2 duct bolts on both sides
- Wiring harness clamps
- Move the floor carpet forward
- Center air duct and the 2 air duct-to-heater/air conditioning housing screws

16. Install the blower unit by installing or connecting the following:
- Blower housing and the blower housing-to-chassis bolt, screw and nut
- Blower housing and the 2 bracket-to-bracket bolts
- Blower housing connectors
- Electrical connector bracket, the wiring harness clamps and the wiring harness
- Air duct and the 3 air duct-to-blower housing screws
- Connector clamp

17. Install the instrument by installing or connecting the following:
- Ventilation nozzles to the heater/air conditioning housing
- Reinforcement and the reinforcement-to-chassis 5 nuts and 4 bolts
- No. 1 and No. 2 brace
- Instrument panel and the 5 instrument panel-to-chassis bolts, the nut and the screw

18. Install or connect the following:
- No. 7 heater-to-register
- No. 1 console box duct
- Console box
- No. 3 console box mounting bracket
- Rear console armrest
- Lower rear console box
- Console box carpet
- Connectors and install the cluster finish panel
- Audio unit and the audio unit-to-instrument panel 2 bolts and 2 screws
- Connector and install the No. 2 register
- Ashtray
- CD changer, connect the electrical connectors and install the 3 CD changer-to-instrument panel nuts
- Glove box and the glove box-to-instrument panel 2 bolts and 3 screws
- Air bag electrical connector and install the plate inside the glove box
- No. 2 undercover
- Instrument cluster, connect the

electrical connectors and install the 4 instrument cluster-to-instrument panel screws
- Instrument cluster finish panel
- Steering column, connect the spring to the brake pedal and install the 4 steering column-to-instrument panel nuts
- Parking brake handle and the No. 1 switch hole base
- No. 1 safety pad and the 4 safety pad-to-instrument panel bolts and screw
- Hood lock release and the 2 release screws
- No. 1 undercover and the 2 under-cover-to-instrument panel screws
- End pad
- Combination switch and connect the electrical connectors
- Steering column covers and the cover screws
- Front pillar garnishes and the front door scuff plates

19. Install the steering wheel by installing or connecting the following:
- Steering wheel and torque the steering wheel nut to 26 ft. lbs. (35 Nm)
- Electrical connector and install the air bag
- Tighten the Torx bolts to 80 inch lbs. (9.0 Nm) using a Torx wrench
- Torx bolt covers at both sides of the steering wheel
- No. 1 grommet, the heater pipe grommet and the drain hose grommet
- Heater hoses to the heater core
- Refrigerant lines (using new O-rings) and the refrigerant lines-to-evaporator bolt

20. Refill the cooling system.
21. Connect the negative battery cable.
22. Evacuate, charge and leak test the air conditioning system refrigerant.
23. Operate the engine to normal operating temperatures; then, check the climate control operation and check for leaks.

LS 400

REMOVAL & INSTALLATION

1. Disconnect the negative battery cable. Wait 90 seconds before doing any further work while the airbag system de-energizes.
2. Disconnect the negative battery cable.
3. Drain the cooling system into a clean container for reuse.
4. Remove or disconnect the following:

- Undercover and the No. 1 safety pad-to-instrument panel screws and the panel at the driver's side
- No. 2 heater-to-register duct
- Heater core-to-heater housing screw and clamp
- Heater hoses from the heater core
- Heater core from the heater housing
- Discard the O-rings

To install:
5. Install or connect the following:
- New O-rings to the heater core
- Heater core to the heater housing
- Heater hoses to the heater core
- Heater core clamp and the heater core-to-heater housing screw
- No. 2 heater-to-register duct
- Undercover and the No. 1 safety pad-to-instrument panel and the panel screws at the driver's side

6. Refill the cooling system.
7. Connect the negative battery cable.
8. Operate the engine to normal operating temperatures; then, check the climate control operation and check for leaks.

SC 300 and SC 400

REMOVAL & INSTALLATION

➡**Removal of the heater core requires removal of the entire heater air conditioning assembly.**

1. Disconnect the negative battery cable. Wait 90 seconds before doing any further work while the airbag system de-energizes.
2. Drain the cooling system into a clean container for reuse.
3. Discharge and recover the air conditioning system refrigerant.
4. Remove or disconnect the following:
- Refrigerant lines from the evaporator. Discard the O-rings. Plug the lines to prevent contamination
- Heater hoses from the heater core
- Brake tubes mounting bolts
- Heater water valve and the ABS actuator
- Equalizer tube from the EPR, discard the O-ring and plug the openings to prevent contamination
- 2 insulator retainer bolts and the retainer at the heater core

5. Remove the steering wheel by removing or disconnecting the following:
- Place the front wheels in the straight-ahead position
- Torx bolt covers at both sides of the steering wheel

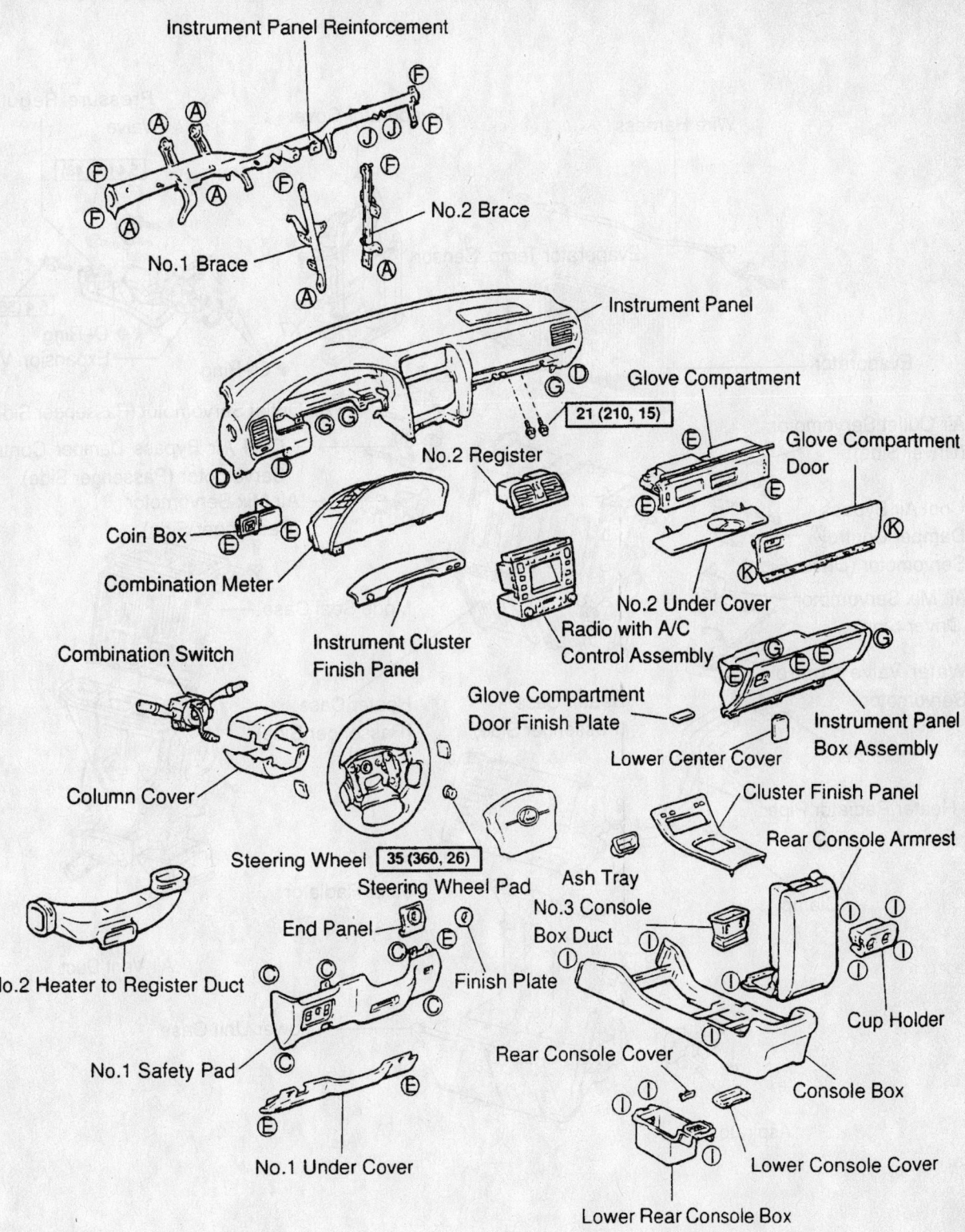

Instrument Panel Reinforcement

No.2 Brace

No.1 Brace

Instrument Panel

Glove Compartment

Glove Compartment Door

21 (210, 15)

No.2 Register

Coin Box

Combination Meter

No.2 Under Cover

Radio with A/C Control Assembly

Glove Compartment Door Finish Plate

Lower Center Cover

Instrument Panel Box Assembly

Combination Switch

Instrument Cluster Finish Panel

Cluster Finish Panel

Rear Console Armrest

Column Cover

Steering Wheel   35 (360, 26)

Steering Wheel Pad

Ash Tray

No.3 Console Box Duct

Cup Holder

No.2 Heater to Register Duct

End Panel

Finish Plate

No.1 Safety Pad

Rear Console Cover

Console Box

Lower Console Cover

No.1 Under Cover

Lower Rear Console Box

N·m (kgf·cm, ft·lbf) : Specified torque

Exploded view of the instrument panel—LS 400

93112GS1

For Accessory Drive Belt illustrations, see Section 1 of this manual

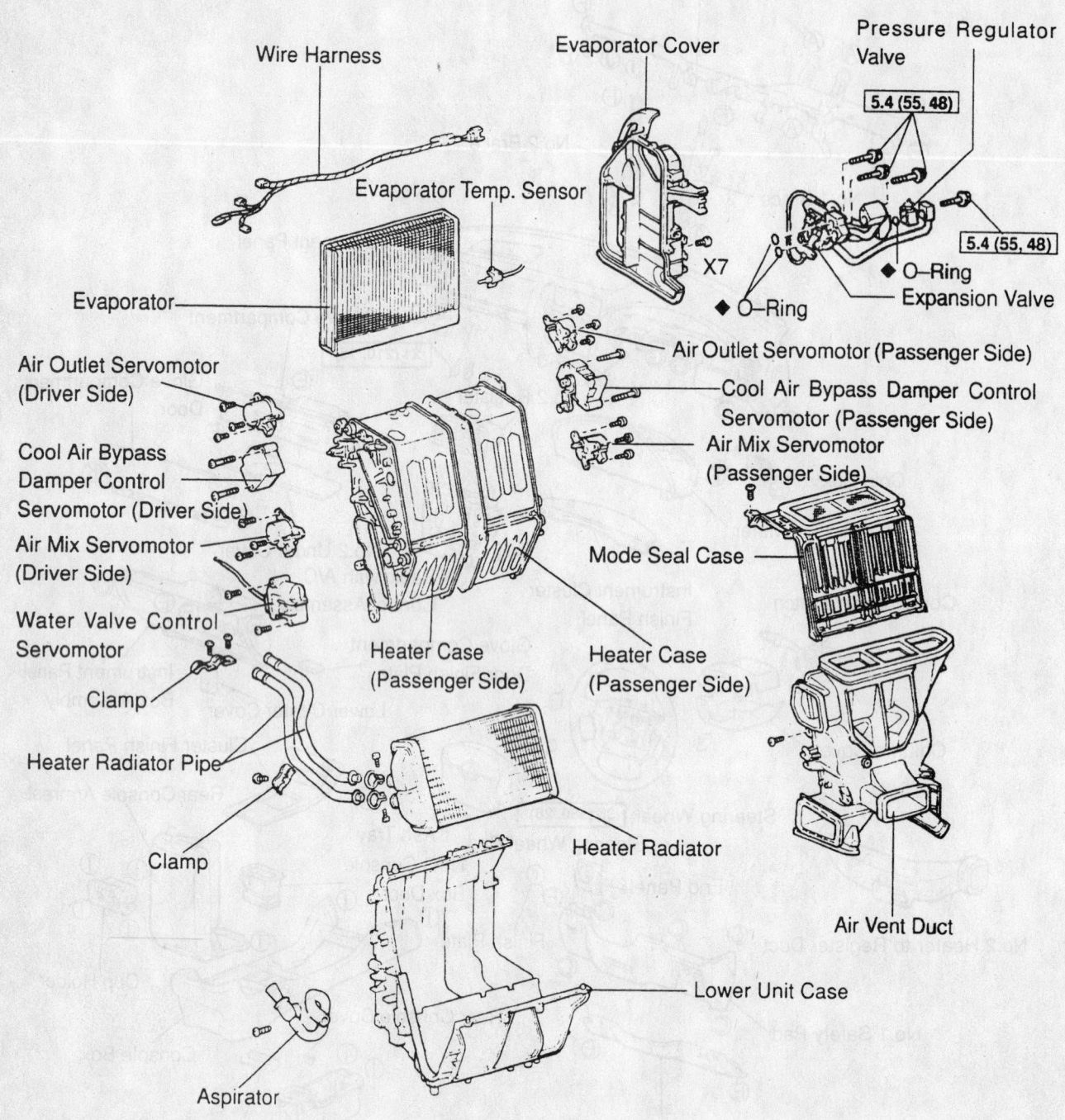

Wire Harness

Evaporator Cover

Pressure Regulator Valve

5.4 (55, 48)

5.4 (55, 48)

Evaporator Temp. Sensor

X7

◆ O—Ring

◆ O—Ring

Expansion Valve

Evaporator

Air Outlet Servomotor (Passenger Side)

Air Outlet Servomotor (Driver Side)

Cool Air Bypass Damper Control Servomotor (Passenger Side)

Cool Air Bypass Damper Control Servomotor (Driver Side)

Air Mix Servomotor (Passenger Side)

Air Mix Servomotor (Driver Side)

Mode Seal Case

Water Valve Control Servomotor

Clamp

Heater Case (Passenger Side)

Heater Case (Passenger Side)

Heater Radiator Pipe

Clamp

Heater Radiator

Air Vent Duct

Lower Unit Case

Aspirator

N·m (kgf·cm, in.·lbf) : Specified torque

◆ Non—reusable part

93112GS2

Exploded view of the heater core, heater/air conditioning housing and related components—LS 400

- Loosen the Torx bolts (using a Torx wrench) until the circumference ring on the bolt catches on the screw case
- Lift the air bag, disconnect the electrical connector and remove it
- Steering wheel nut and press the steering wheel from the steering column

6. Remove the instrument panel and reinforcement by removing or disconnecting the following:

- Tilt the steering column down and pull out the steering wheel
- Front assist grips
- Front pillar garnishes and the front scuff plates
- Steering column cover
- Shift lever knob
- Upper console panel rear
- Upper console panel
- Radio with the air conditioning control assembly
- No. 2 undercover
- Finish plate and disconnect the air bag electrical connector located inside the glove box
- Glove compartment assembly
- 4 lower right side finish panel bolts and the panel
- No. 4 heater-to-register duct screw and the duct
- Passenger's side air bag module-to-instrument panel 3 bolts and 2 nuts; then, remove the air bag
- No. 4 undercover-to-instrument panel cover screws and the cover
- Console box
- End pad at the left side of the instrument panel
- key center plate and the center pad
- 2 hood release lever screws and the lever
- 3 No. 1 lower finish panel screws, disconnect the electrical connectors and remove the panel
- 4 lower left side finish panel bolts and the panel
- No. 2 heater-to-register duct
- Combination switch
- Cluster finish panel
- Combination meter
- Steering column assembly
- Instrument panel-to-chassis fasteners and the instrument panel
- Instrument panel reinforcement fasteners and the reinforcement

7. Remove or disconnect the following:
- PPS ECU
- Cooling fan ECU
- ABS ECU
- Move the floor carpet rearward
- Heater-to-No. 3 duct register screw and the duct
- 2 connector bracket screws and the bracket at the under side of the blower motor
- Electrical connector; then, remove the 6 heater/air conditioning housing-to-chassis bolts and the housing
- 2 heater core plate-to-heater housing screws and the plate
- 2 heater core clamp screws and the clamps
- Heater core

To install:
8. Install or connect the following:
- Heater core
- Heater core clamps and the 2 clamp screws
- Heater core plate and the 2 plate-to-heater housing screws
- Electrical connector; then, install the heater/air conditioning housing and the 6 housing-to-chassis bolts
- Connector bracket and the 2 bracket screw at the under side of the blower motor
- Heater-to-No. 3 duct register and the duct screw
- Floor carpet forward
- ABS ECU
- Cooling fan ECU
- PPS ECU

9. Install the instrument panel and reinforcement by installing or connecting the following:

- Instrument panel reinforcement and the reinforcement fasteners
- Instrument panel and the instrument panel-to-chassis fasteners
- Steering column assembly
- Combination meter
- Cluster finish panel
- Combination switch
- No. 2 heater-to-register duct
- Lower left side finish panel and the 4 panel bolts
- No. 1 lower finish panel, connect the electrical connectors and install the 3 panel screws
- Hood release lever and the 2 lever screws

- key center plate and the center pad
- End pad at the left side of the instrument panel
- Console box
- No. 4 undercover-to-instrument panel cover and the cover screws
- Passenger's side air bag module and torque the air bag-to-instrument panel 3 bolts and 2 nuts to 15 ft. lbs. (21 Nm)
- No. 4 heater-to-register duct and the duct screw
- Lower right side finish panel and the 4 panel bolts
- Glove compartment assemble
- Air bag electrical connector and install the finish plate inside the glove box
- No. 2 undercover
- Radio with the air conditioning control assembly
- Upper console panel
- Upper console panel rear
- Shift lever knob
- Steering column cover
- Front pillar garnishes and the front scuff plates
- Front assist grips

10. Install the steering wheel by installing or connecting the following:
- Steering wheel and torque the steering wheel nut to 26 ft. lbs. (35 Nm)
- Electrical connector and install the air bag
- Torx bolts to 80 inch lbs. (9.0 Nm) using a Torx wrench
- Torx bolt covers at both sides of the steering wheel

11. Install or connect the following:
- Insulator retainer and the 2 retainer bolts at the heater core
- Equalizer tube to the EPR using a new O-ring
- Heater water valve and the ABS actuator
- Brake tubes mounting bolts
- Heater hoses to the heater core
- Refrigerant lines to the evaporator using new O-rings

12. Refill the cooling system.
13. Connect the negative battery cable.
14. Evacuate, charge and leak test the air conditioning system refrigerant
15. Operate the engine to normal operating temperatures; then, check the climate control operation and check for leaks.

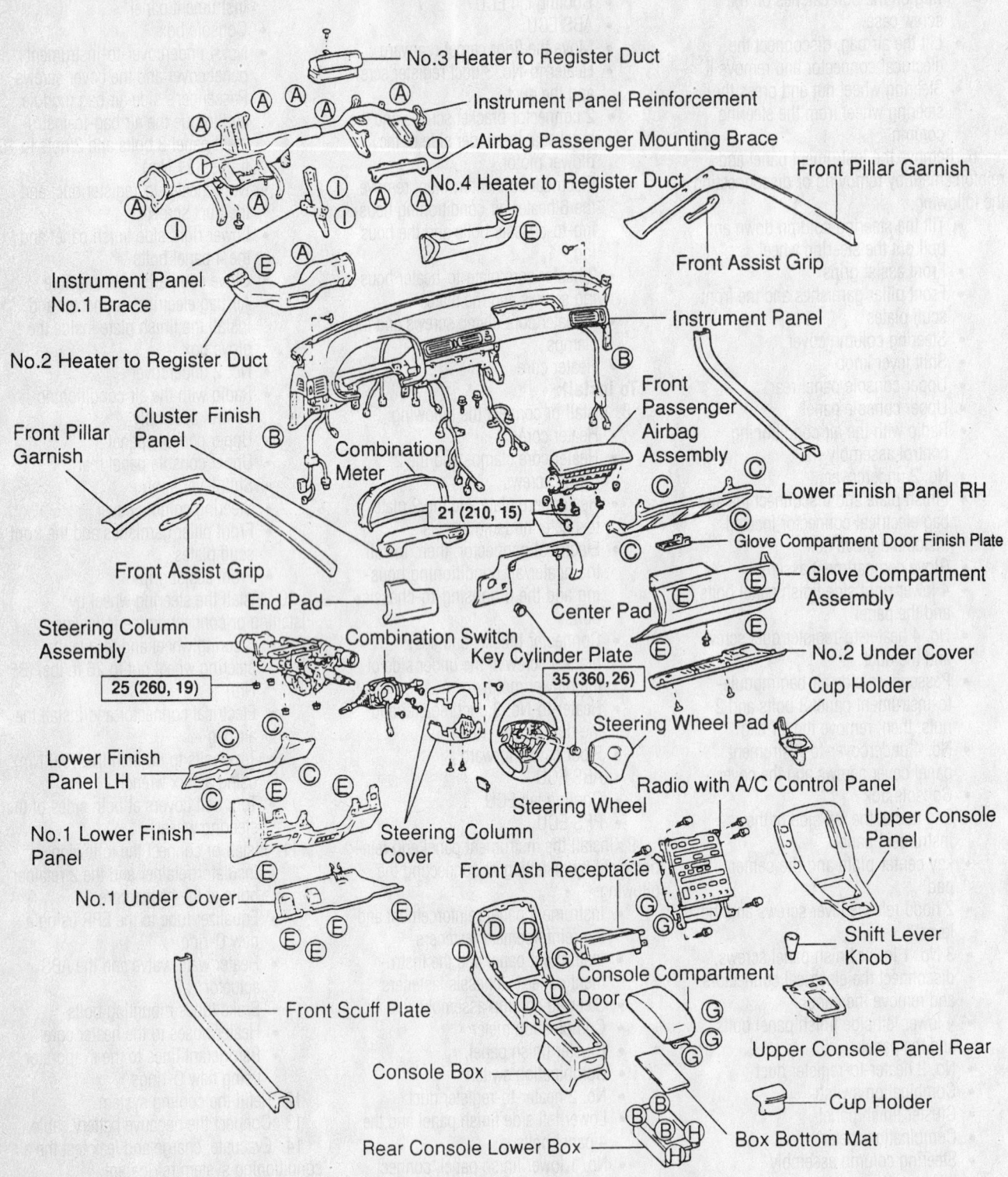

No.3 Heater to Register Duct

Instrument Panel Reinforcement

Airbag Passenger Mounting Brace

No.4 Heater to Register Duct

Front Pillar Garnish

Front Assist Grip

Instrument Panel No.1 Brace

Instrument Panel

No.2 Heater to Register Duct

Front Passenger Airbag Assembly

Front Pillar Garnish

Cluster Finish Panel

Combination Meter

21 (210, 15)

Front Assist Grip

Lower Finish Panel RH

Glove Compartment Door Finish Plate

Glove Compartment Assembly

End Pad

Center Pad

Steering Column Assembly

Combination Switch

Key Cylinder Plate

35 (360, 26)

No.2 Under Cover

Cup Holder

25 (260, 19)

Steering Wheel Pad

Lower Finish Panel LH

Radio with A/C Control Panel

Upper Console Panel

No.1 Lower Finish Panel

Steering Column Cover

Steering Wheel

Front Ash Receptacle

Shift Lever Knob

No.1 Under Cover

Front Scuff Plate

Console Compartment Door

Upper Console Panel Rear

Console Box

Cup Holder

Rear Console Lower Box

Box Bottom Mat

N·m (kgf·cm, ft·lbf) : Specified torque

93112GS5

Exploded view of the instrument panel and related components—SC 300 and SC 400

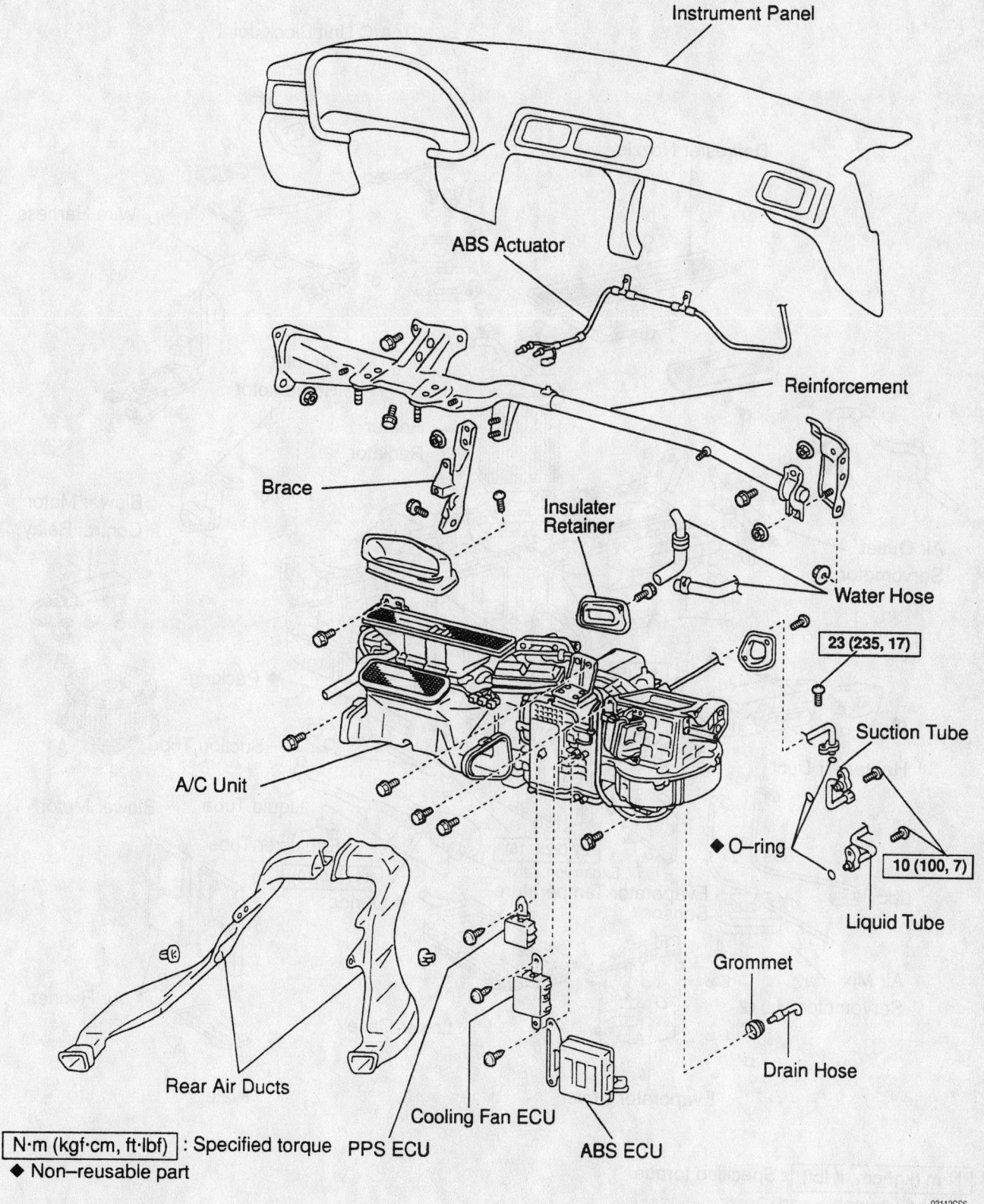

Instrument Panel

ABS Actuator

Reinforcement

Brace

Insulater Retainer

Water Hose

23 (235, 17)

Suction Tube

A/C Unit

◆ O–ring

10 (100, 7)

Liquid Tube

Grommet

Drain Hose

Rear Air Ducts

Cooling Fan ECU

PPS ECU

ABS ECU

N·m (kgf·cm, ft·lbf) : Specified torque

◆ Non–reusable part

93112GS6

Exploded view of the heater/air conditioning housing, reinforcement, ventilation ducts and related components—SC 300 and SC 400

For Wheel Alignment specifications, see Section 1 of this manual

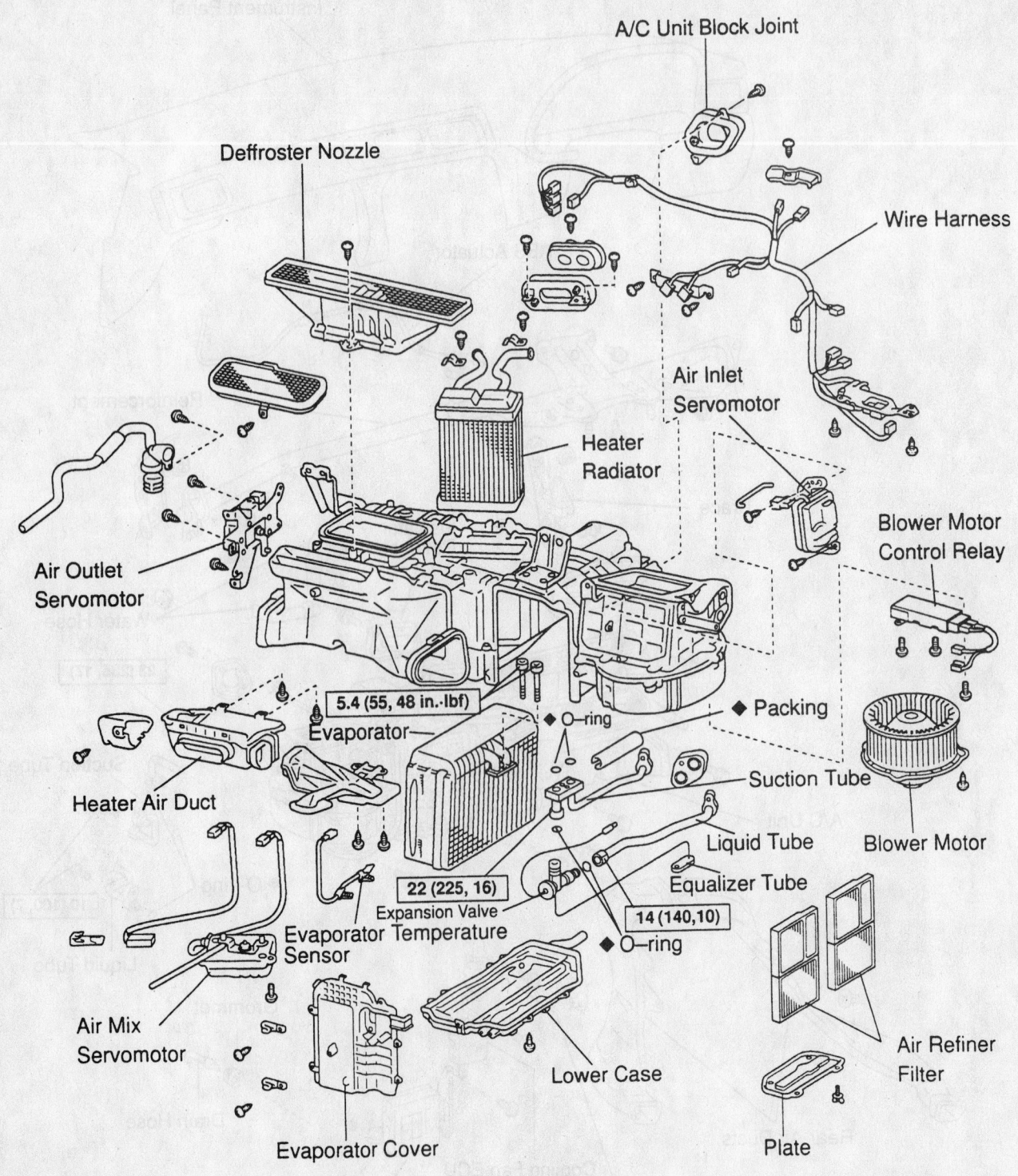

A/C Unit Block Joint

Deffroster Nozzle

Wire Harness

Air Inlet Servomotor

Heater Radiator

Blower Motor Control Relay

Air Outlet Servomotor

5.4 (55, 48 in.·lbf)

Evaporator

◆ Packing

O–ring

Suction Tube

Liquid Tube

Equalizer Tube

Blower Motor

Heater Air Duct

22 (225, 16)

Expansion Valve

14 (140,10)

◆ O–ring

Evaporator Temperature Sensor

Air Mix Servomotor

Air Refiner Filter

Evaporator Cover

Lower Case

Plate

N·m (kgf·cm, ft·lbf) : Specified torque

◆ Non–reusable part

93112GS7

Exploded view of the heater core, heater/air conditioning housing and related components—SC 300 and SC 400

1998-00 MAZDA

Miata

REMOVAL & INSTALLATION

1998–00 Models

1. Disconnect the negative battery cable.

✳✳ CAUTION

After disconnecting the battery, wait for more than 1 minute for the SAS to deplete its stored energy.

2. Drain the cooling system into a clean container for reuse.
3. Disconnect the heater hoses from the heater core.
4. Discharge and recover the air conditioning system refrigerant.
5. At the driver's side, remove the SAS module and the steering wheel by removing or disconnecting the following:
   - Place the wheel in the straight-ahead position and turn the ignition switch to LOCK
   - Cover clips at both sides of the steering wheel
   - Steering wheel-to-SAS module bolts
   - SAS module from the steering wheel and disconnect the electrical connector
   - Steering wheel-to-column nut
   - Steering wheel from the steering column using a suitable puller
6. At the passenger's side, remove the SAS module by removing or disconnecting the following:
   - Glove compartment and the glove compartment cover
   - SAS module-to-dash bolts
   - SAS module and disconnect the electrical connector
   - Console
7. Remove the instrument cluster by removing or disconnecting the following:
   - A-pillar trim at both sides
   - Lower panel
   - Instrument cluster hood
   - Instrument cluster-to-dash screws and the instrument cluster
   - Hood release lever
   - Control wire from the heater unit and the blower unit
   - Steering column-to-instrument

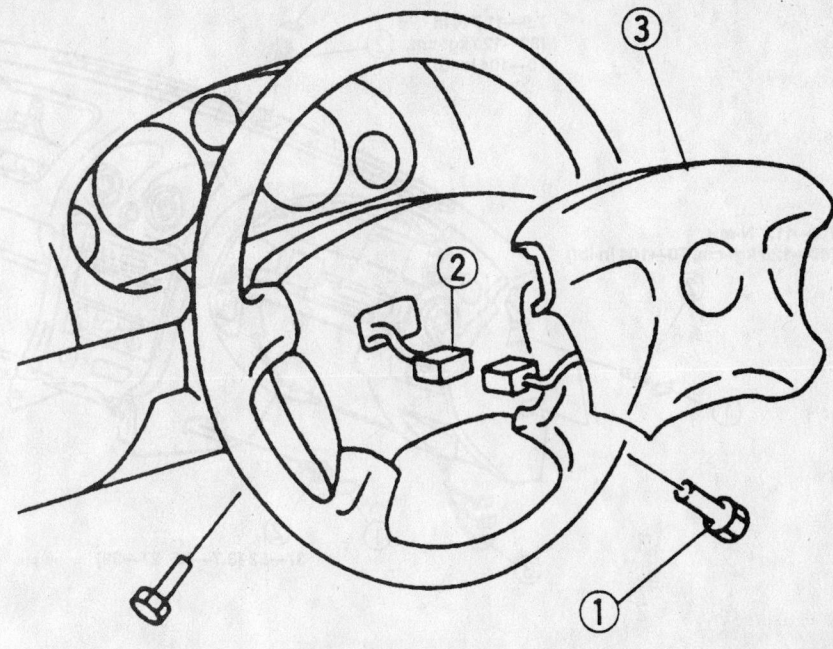

7.9—11.7 N·m {80—120 kgf·cm, 70—104 in·lbf}
93112GG0

View of the SAS module and the steering wheel—1998–00 Miata

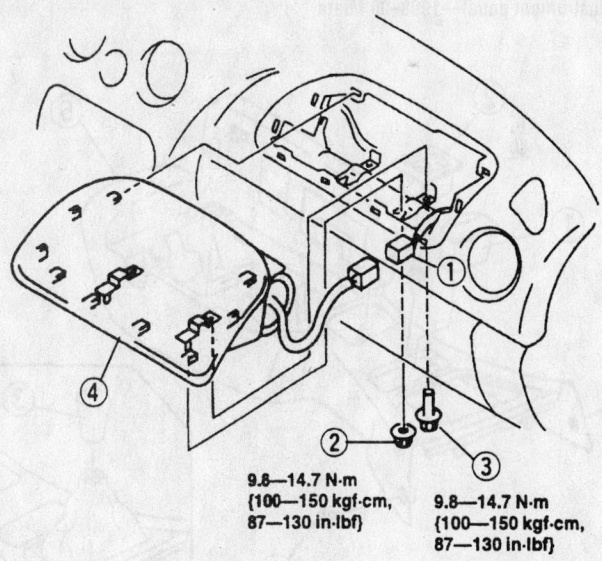

9.8—14.7 N·m {100—150 kgf·cm, 87—130 in·lbf}

9.8—14.7 N·m {100—150 kgf·cm, 87—130 in·lbf}

1	Connector
2	Nut
3	Bolt
4	Passenger-side air bag module

93112GH1

View of the passenger's side SAS module—1998–00 Miata

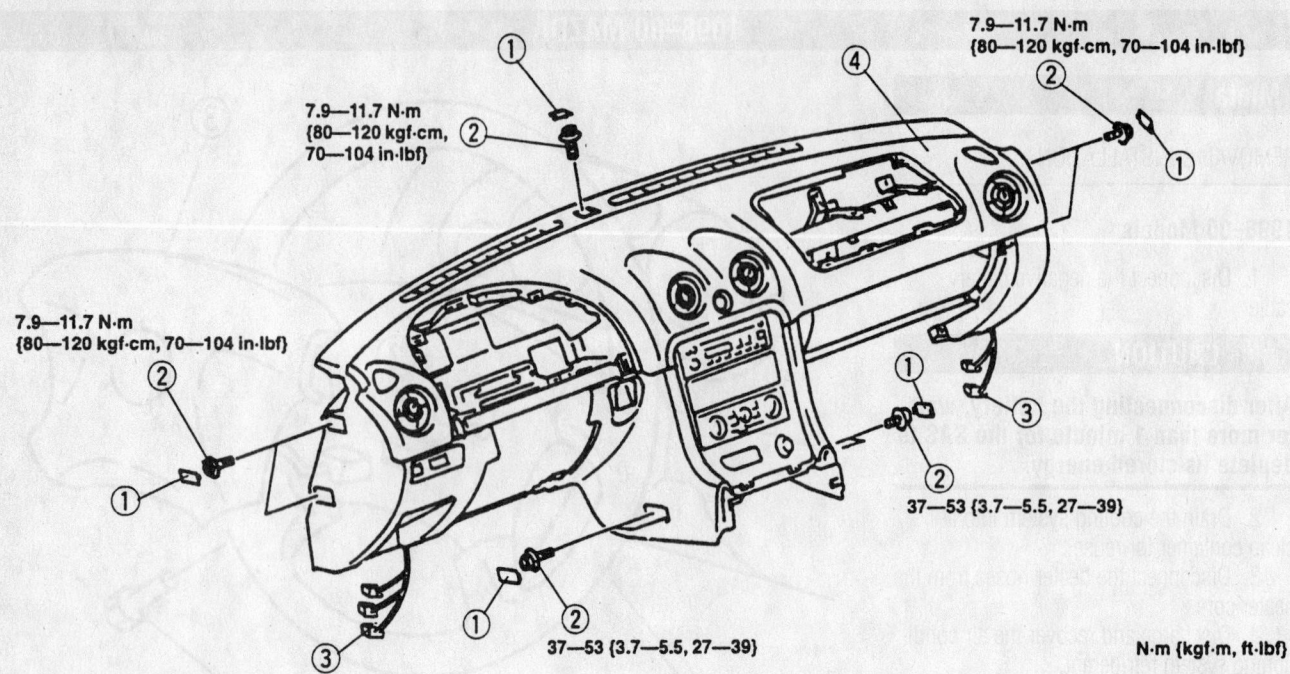

7.9—11.7 N·m
{80—120 kgf·cm,
70—104 in·lbf}

7.9—11.7 N·m
{80—120 kgf·cm, 70—104 in·lbf}

7.9—11.7 N·m {80—120 kgf·cm, 70—104 in·lbf}

37—53 {3.7—5.5, 27—39}

37—53 {3.7—5.5, 27—39}

N·m {kgf·m, ft·lbf}

93112GH2

1	Cover	3	Connector
2	Bolt	4	Dashboard

View of the instrument panel—1998-00 Miata

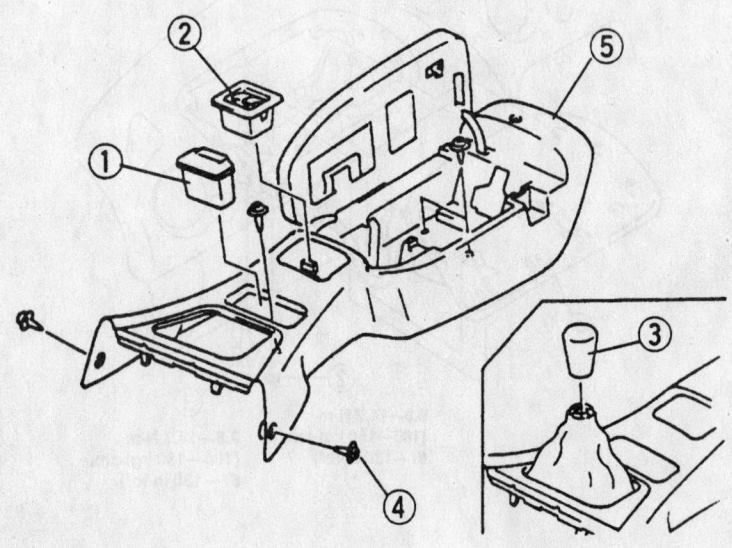

1   Ashtray

2   Power window switch

3   Shift lever knob (MT)

4   Screw

5   Console

93112GH3

View of the console—1998-00 Miata

panel bolts and lower the steering column
- Instrument panel-to-chassis bolt covers and the bolts
- Instrument panel with the help of an assistant

8. Remove or disconnect the following:
- Heater unit-to-evaporator housing seal plate
- Heater unit-to-chassis nuts and the heater unit

9. Separate the heater unit cases and remove the heater core.

To install:

10. Install the heater core and assemble the heater unit cases.

11. Install or connect the following:
- Heater unit-to-chassis nuts and the heater unit
- Heater unit-to-evaporator housing seal plate

12. Install the instrument cluster by installing or connecting the following:
- Instrument panel with the help of an assistant
- Instrument panel-to-chassis bolt covers and the bolts
- Steering column-to-instrument panel bolts

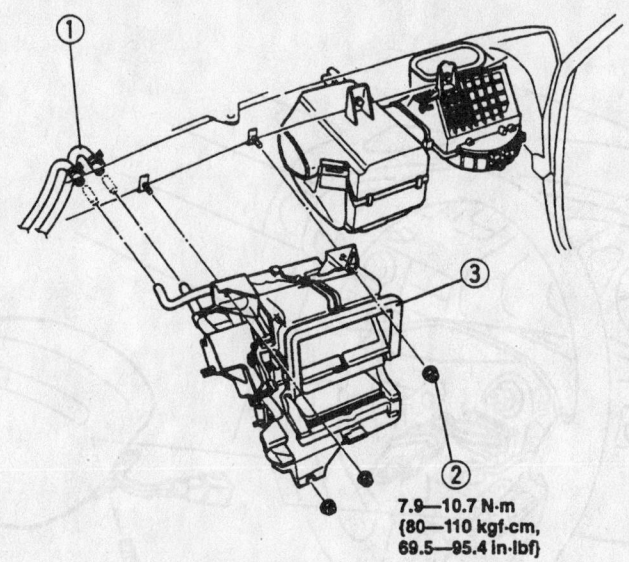

7.9—10.7 N·m
(80—110 kgf-cm,
69.5—95.4 in-lbf)

1 Heater hose
2 Nut
3 Heater unit

93112GH4

View of the heater unit and the heater unit-to-evaporator unit seal plate—1998–00 Miata

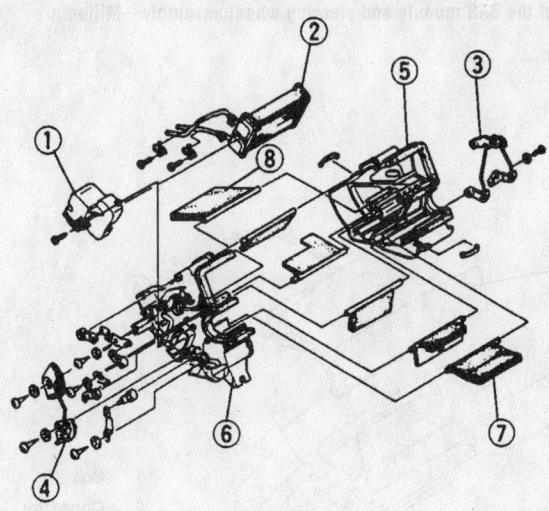

1 Cover
2 Heater core
3 Air mix link
4 Airflow mode link
5 Case (RH)
6 Case (LH)
7 Air mix door
8 Airflow mode door

93112GH5

Exploded view of the heater core and heater unit assembly—1998–00 Miata

- Control wire to the heater unit and the blower unit
- Hood release lever
- Instrument cluster and the instrument cluster-to-dash screws
- Instrument cluster hood
- Lower panel
- A-pillar trim at both sides
- Console

13. At the passenger's side, install the SAS module by installing or connecting the following:
- SAS module and connect the electrical connector
- SAS module-to-dash bolts
- Glove compartment cover and the glove compartment

14. At the driver's side, install the SAS module and the steering wheel by installing or connecting the following:
- Steering wheel-to-column nut. Torque the steering wheel nut to 29–36 ft. lbs. (40–49 Nm)
- SAS module to the steering wheel and connect the electrical connector
- Steering wheel-to-SAS module bolts. Torque the steering column-to-SAS module bolts to 70–104 inch lbs. (8–12 Nm)
- Cover clips at both sides of the steering wheel

15. Connect the heater hoses to the heater core.

16. Refill the cooling system.

17. Connect the negative battery cable.

18. Evacuate, charge and leak test the air conditioning system refrigerant.

19. Operate the engine to normal operating temperatures; then, check the climate control operation and check for leaks.

Millenia

REMOVAL & INSTALLATION

1. Disconnect the negative battery cable.

✳✳ CAUTION

After disconnecting the battery, wait for more than 1 minute for the SAS to deplete its stored energy.

2. Drain the cooling system into a clean container for reuse.

3. Discharge and recover the air conditioning system refrigerant.

4. At the driver's side, remove the SAS

module and the steering wheel by removing or disconnecting the following:

- Place the wheel in the straight-ahead position and turn the ignition switch to LOCK
- Cover clips at both sides of the steering wheel
- Steering wheel-to-SAS module clips
- SAS module from the steering wheel and disconnect the electrical connector
- Steering wheel-to-column nut
- Steering wheel from the steering column using a suitable puller

5. At the passenger's side, remove the SAS module by removing or disconnecting the following:

- Glove compartment and the glove compartment cover
- SAS module-to-dash bolts
- SAS module and disconnect the electrical connector

6. Remove the rear console by removing or disconnecting the following:

- Rear console's box
- Bake boot
- Center panel
- Rear console

7. Remove or disconnect the following:

- A-pillar trim at both sides
- Undercover at the passenger's side
- Upper and lower steering column covers
- Electrical connectors and remove the combination switch from the steering column
- Meter hood
- Electrical connectors and remove the instrument cluster
- Steering column-to-chassis bolts and the steering column
- Hood release lever
- Both side panels
- Instrument panel with the help of an assistant

8. Remove the evaporator housing by removing or disconnecting the following:

- Air conditioning system refrigerant lines and discard the gaskets
- Aspirator hose
- Power transistor connector
- MAX-HI connector
- Evaporator temperature connector
- Evaporator housing assembly

9. Disconnect and remove the heater unit assembly.

10. Separate the heater unit cases and remove the heater core.

To install:

11. Install the heater core and assemble the heater unit cases.

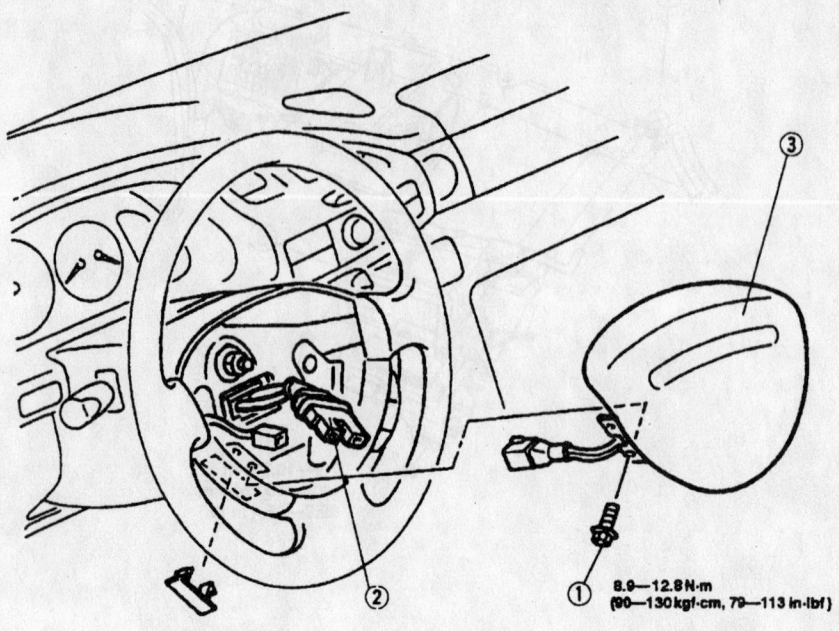

8.9—12.8 N·m
(90—130 kgf·cm, 79—113 in·lbf)

1	Bolt
2	Connector
3	Driver-side air bag module

93112GE5

Exploded view of the SAS module and steering wheel assembly—Millenia

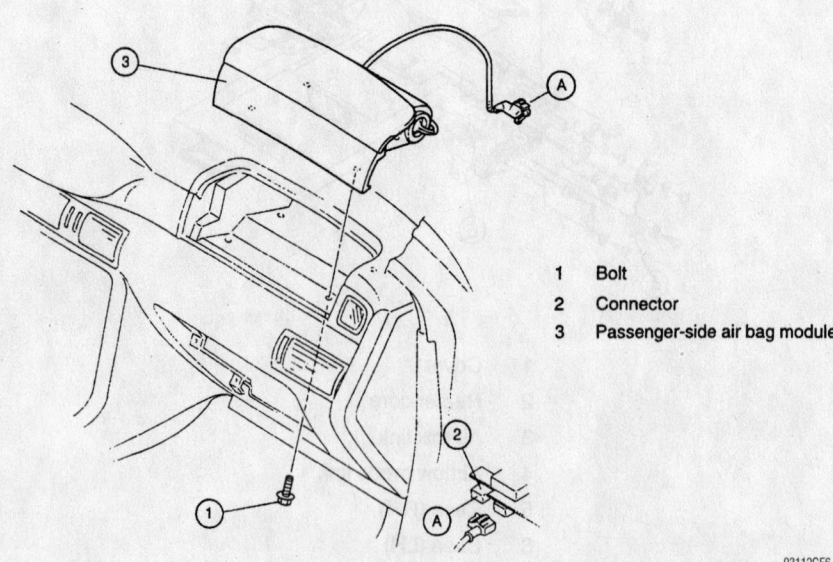

1	Bolt
2	Connector
3	Passenger-side air bag module

93112GE6

Exploded view of the passenger's side SAS module—Millenia

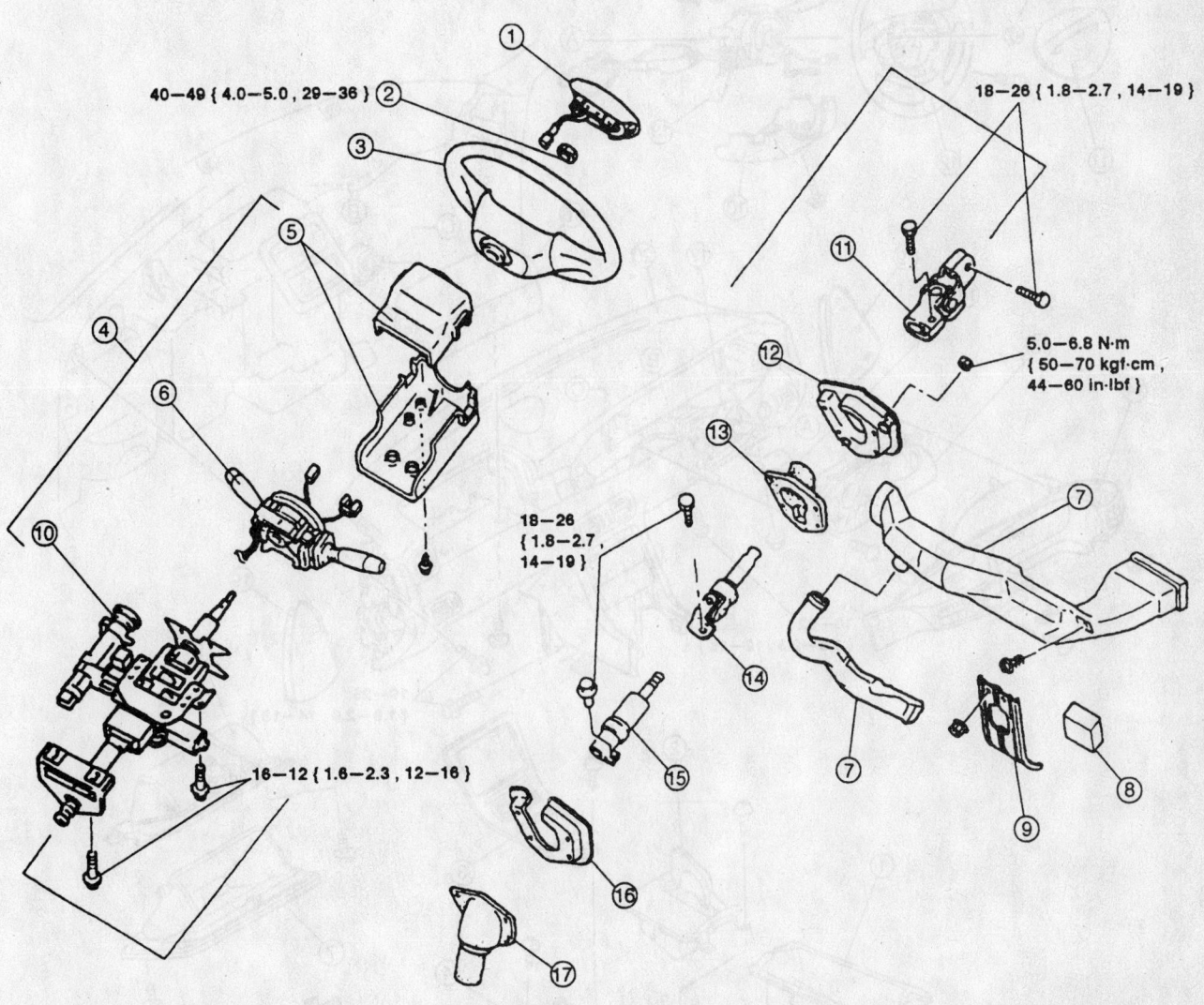

40—49 { 4.0—5.0 , 29—36 }

18—26 { 1.8—2.7 , 14—19 }

5.0—6.8 N·m
{ 50—70 kgf·cm ,
44—60 in·lbf }

18—26
{ 1.8—2.7 ,
14—19 }

16—12 { 1.6—2.3 , 12—16 }

N·m { kgf·m , ft·lbf }

1	Air bag module	10	Steering shaft component
2	Locknut	11	Universal joint (intermediate shaft)
3	Steering wheel	12	Cover
4	Dashboard, console, and steering shaft component	13	Shaft seal
5	Column cover	14	Intermediate shaft
6	Combination switch	15	Collapsible shaft
7	Air duct	16	Set plate
8	Flasher unit	17	Dust cover
9	Bracket		

93112GE9

Exploded view of the steering wheel and steering column assembly—Millenia

For complete service labor times, order the Chilton Labor Guide

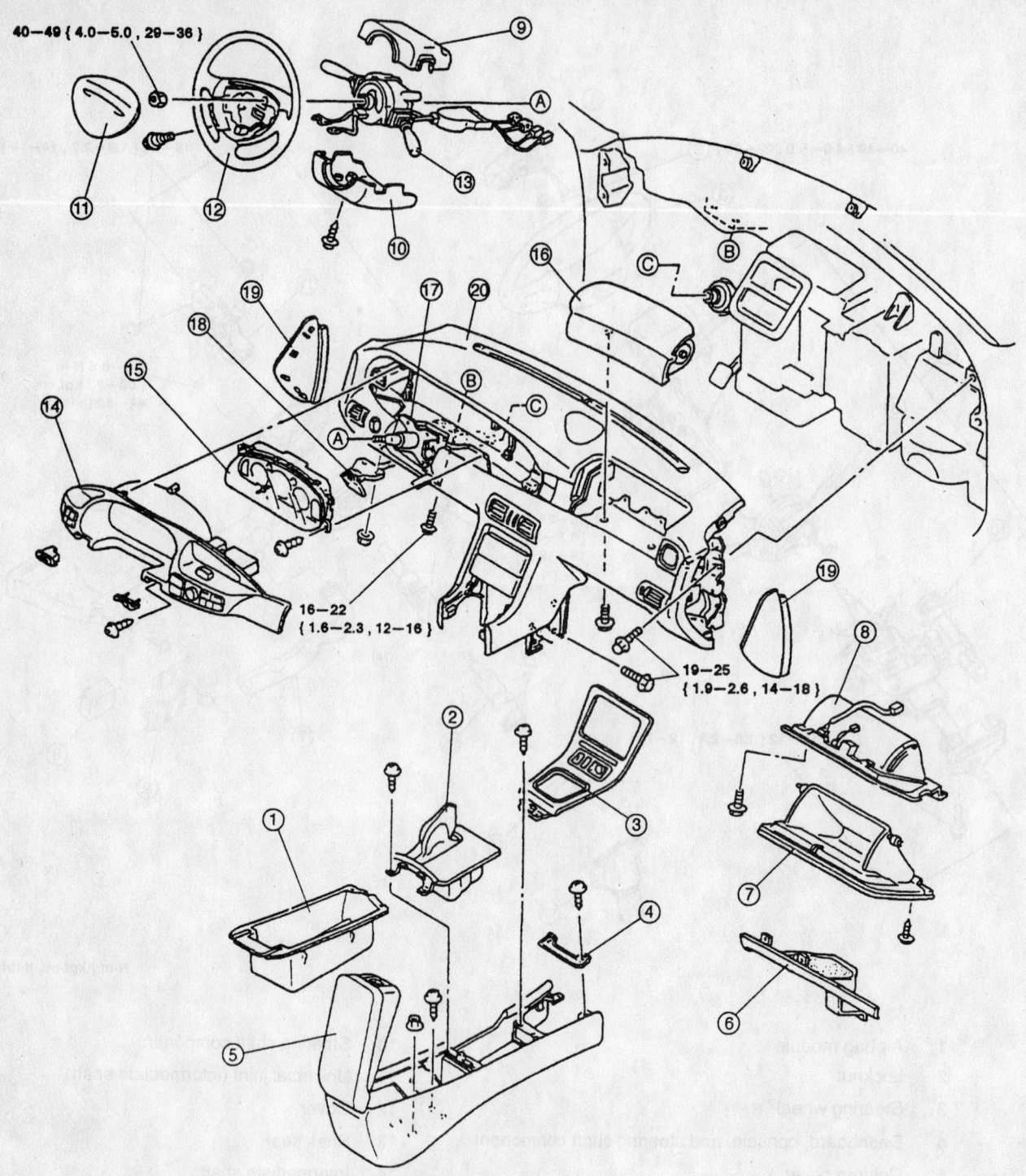

40—49 { 4.0—5.0 , 29—36 }

16—22
{ 1.6—2.3 , 12—16 }

19—25
{ 1.9—2.6 , 14—18 }

1	Rear console box	8	Glove compartment cover	15	Instrument cluster
2	Brake boot	9	Upper column cover	16	Passenger-side air bag module
3	Center panel	10	Lower column cover	17	Steering shaft
4	Bracket	11	Driver-side air bag module	18	Hood release lever
5	Rear console	12	Steering wheel	19	Side panel
6	Under cover	13	Combination switch	20	Dashboard
7	Glove compartment	14	Meter hood		

Exploded view of the instrument panel and rear console assemblies—Millenia

93112GE8

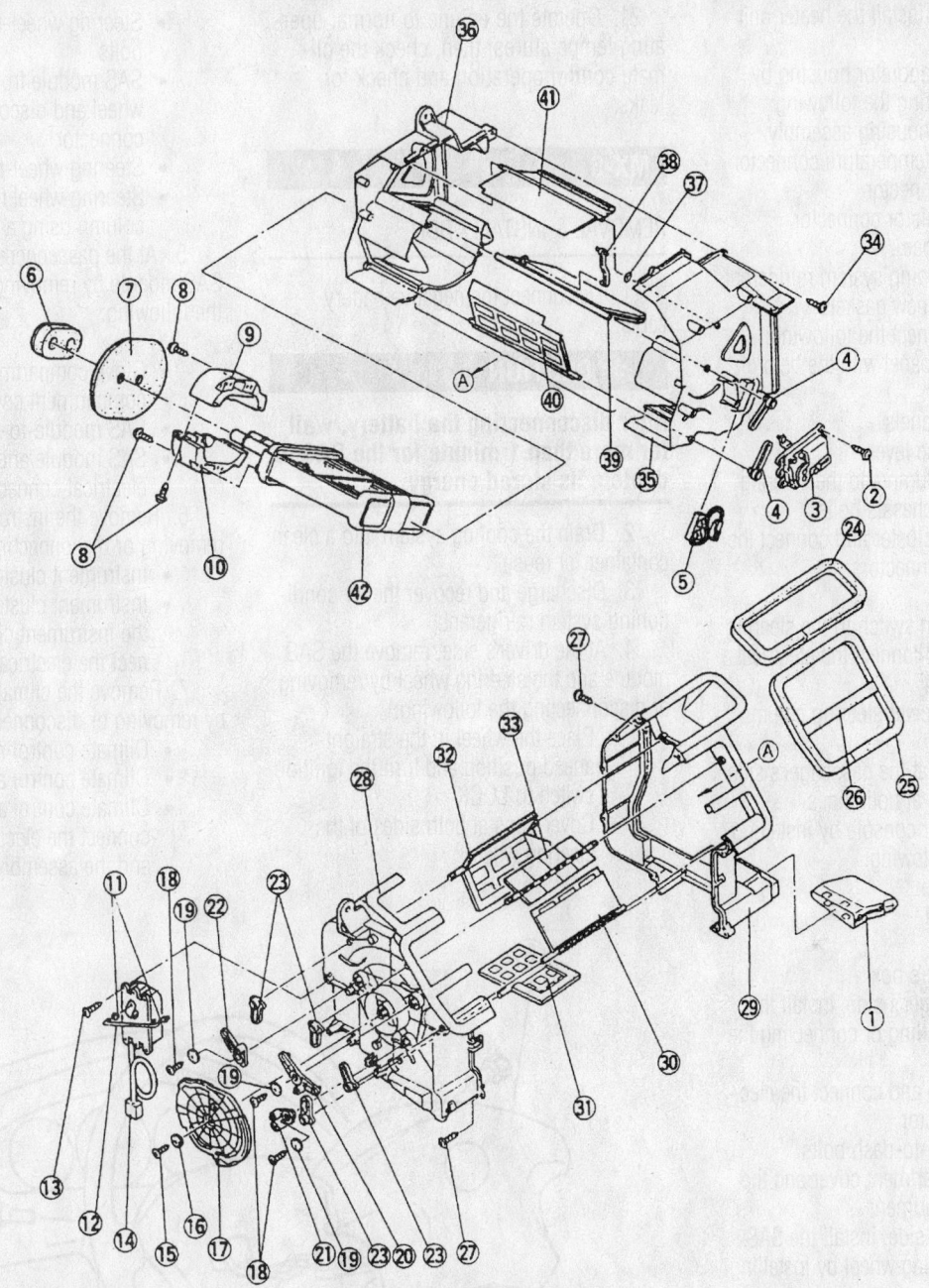

1	Heat duct	16	Link collar	30	Vent door
2	Tapping screw	17	Airflow mode main link	31	Heat door
3	Air mix actuator		☞ Assembly Note	32	Defroster door
4	Air mix crank	18	Tapping screw	33	Side vent door
5	Water temperature sensor	19	Link collar	34	Tapping screw
6	Polyurethane foam (thick)	20	Airflow mode sub link (VENT)	35	Heater case (1)
7	Polyurethane foam (thin)	21	Airflow mode sub link (HEAT)	36	Heater case (2)
8	Tapping screw	22	Airflow mode sub link (DEFROSTER)	37	Collar
9	Heater core bracket (1)	23	Airflow mode crank	38	Air mix rod
10	Heater core bracket (2)	24	Polyurethane protector (DEFROSTER)	39	Air mix main door
11	Rod stopper	25	Polyurethane protector (VENT)	40	Air mix sub door
12	Airflow mode rod	26	Polyurethane protector (SIDE VENT)	41	Air mix guide door
13	Tapping screw	27	Tapping screw	42	Heater core
14	Airflow mode actuator	28	Heater case (4)		
15	Tapping screw	29	Heater case (3)		

93112GE7

Exploded view of the heater core and heater case assembly—Millenia

12. Connect and install the heater unit assembly.

13. Install the evaporator housing by installing or connecting the following:
- Evaporator housing assembly
- Evaporator temperature connector
- MAX-HI connector
- Power transistor connector
- Aspirator hose
- Air conditioning system refrigerant lines using new gaskets

14. Install or connect the following:
- Instrument panel with the help of an assistant
- Both side panels
- Hood release lever
- Steering column and the steering column-to-chassis bolts
- Instrument cluster and connect the electrical connectors
- Meter hood
- Combination switch to the steering column and connect the electrical connectors
- Upper and lower steering column covers
- Undercover at the passenger's side
- A-pillar trim at both sides

15. Install the rear console by installing or connecting the following:
- Rear console
- Center panel
- Brake boot
- Rear console's box

16. At the passenger's side, install the SAS module by installing or connecting the following:
- SAS module and connect the electrical connector
- SAS module-to-dash bolts
- Glove compartment cover and the glove compartment

17. At the driver's side, install the SAS module and the steering wheel by installing or connecting the following:
- Steering wheel to the steering column
- Steering wheel-to-column nut. Torque the nut to 29–36 ft. lbs. (40–49 Nm)
- SAS module to the steering wheel and connect the electrical connector. Torque the bolts to 79–113 inch lbs. (9–13 Nm)
- Steering wheel-to-SAS module clips
- Cover clips at both sides of the steering wheel

18. Refill the cooling system.
19. Connect the negative battery cable.
20. Evacuate, charge and leak test the air conditioning system refrigerant.

21. Operate the engine to normal operating temperatures; then, check the climate control operation and check for leaks.

MX-6

REMOVAL & INSTALLATION

1. Disconnect the negative battery cable.

✳✳ CAUTION

After disconnecting the battery, wait for more than 1 minute for the SAS to deplete its stored energy.

2. Drain the cooling system into a clean container for reuse.
3. Discharge and recover the air conditioning system refrigerant.
4. At the driver's side, remove the SAS module and the steering wheel by removing or disconnecting the following:
- Place the wheel in the straight-ahead position and turn the ignition switch to LOCK
- Cover clips at both sides of the steering wheel
- Steering wheel-to-SAS module bolts
- SAS module from the steering wheel and disconnect the electrical connector
- Steering wheel-to-column nut
- Steering wheel from the steering column using a suitable puller

5. At the passenger's side, remove the SAS module by removing or disconnecting the following:
- Glove compartment and the glove compartment cover
- SAS module-to-dash bolts
- SAS module and disconnect the electrical connector

6. Remove the instrument cluster by removing or disconnecting the following:
- Instrument cluster meter hood
- Instrument cluster-to-dash screws, the instrument cluster and disconnect the electrical connectors

7. Remove the climate control assembly by removing or disconnecting the following:
- Climate control meter hood
- Climate control assembly screws
- Climate control assembly and disconnect the electrical connector and the assembly

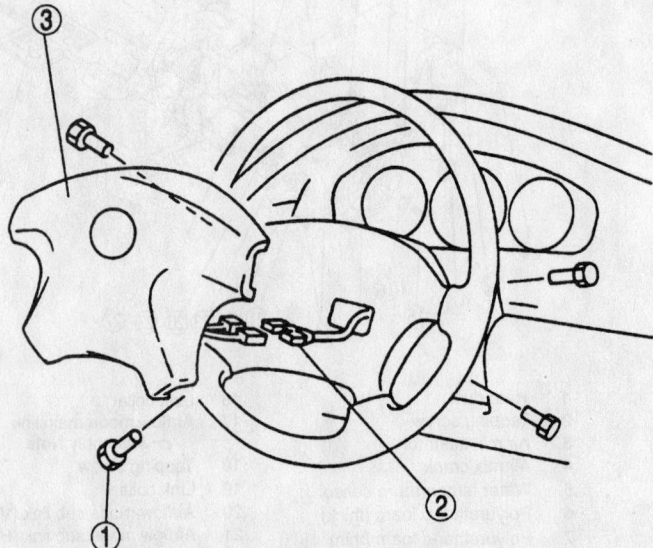

7.9—11.7 N·m {80—120 kgf·cm, 70—104 in·lbf}

1	Bolt
2	Connector
3	Driver-side air bag module

93112GH6

Exploded view of the steering wheel and SAS module—1998–00 MX-6

8. Remove the audio unit by removing or disconnecting the following:
- Hole covers by inserting a small tape-wrapped flathead screwdriver into the slot and carefully pry off the hole covers
- Using 2 removal tools 49 UN01 050 or equivalent, insert them into sides of the audio unit
- Slide audio unit outward and forward
- Electrical connectors and the antenna jack

9. Remove the instrument panel by removing or disconnecting the following:
- Upper center panel cover
- Dash panel-to-chassis bolts
- Dash panel with the help of an assistant

10. Remove the evaporator housing by removing or disconnecting the following:
- Center lower panel
- Refrigerant lines from the air conditioning evaporator and discard the gaskets.
- Blower motor assembly, if necessary
- Evaporator assembly fasteners and remove the evaporator assembly

11. Remove or disconnect the following:
- Rear heater duct
- Heater housing fasteners and the heater housing
- Airflow mode actuator

12. Separate the heater housing and remove the heater core.

To install:

13. Install the heater core and assemble the heater housing.

14. Install or connect the following:
- Airflow mode actuator
- Heater housing and the heater housing fasteners
- Rear heater duct

15. Install the evaporator housing by installing or connecting the following:
- Center lower panel
- Refrigerant lines to the air conditioning evaporator using new gaskets
- Blower motor assembly, if necessary
- Evaporator assembly and the evaporator assembly fasteners

16. Install the instrument panel by installing or connecting the following:
- Upper center panel cover
- Dash panel-to-chassis bolts

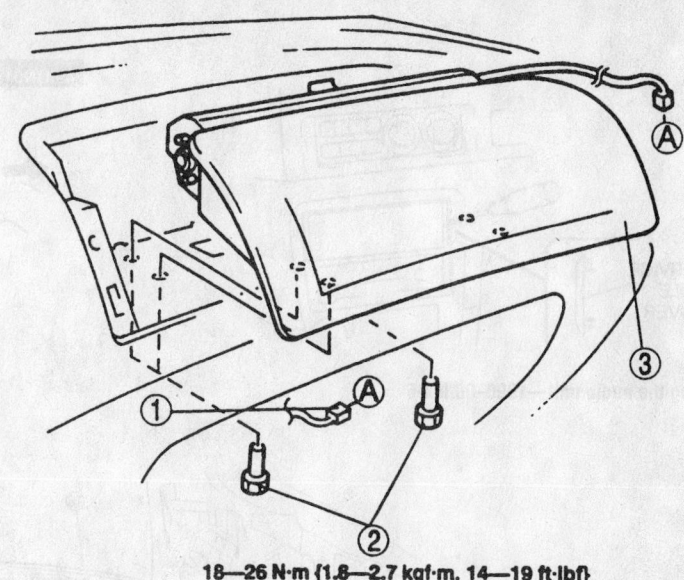

18—26 N·m {1.8—2.7 kgf·m, 14—19 ft·lbf}

1 **Connector**

2 **Bolt**

3 **Passenger-side air bag module**

93112GH7

Exploded view of the passenger's side SAS module—1998–00 MX-6

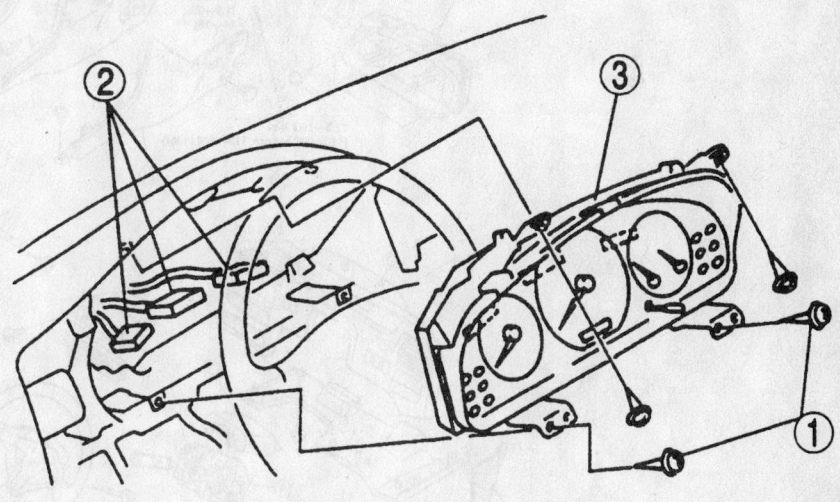

1 **Screw**

2 **Connector**

3 **Instrument cluster**

93112GH8

View of the instrument cluster assembly—1998–00 MX-6

Timing belt service is covered in Section 3 of this manual

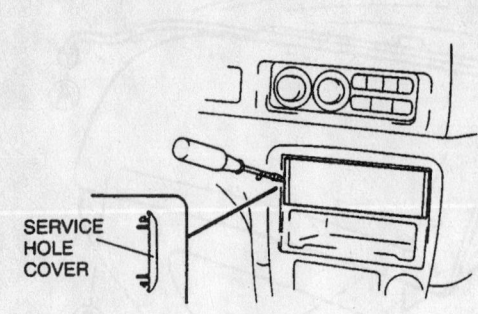

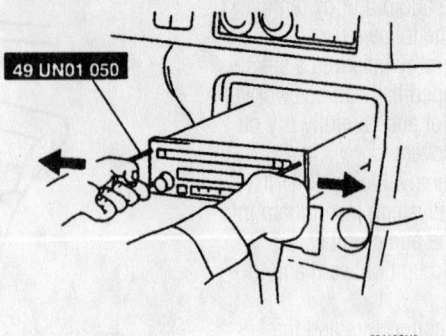

SERVICE
HOLE
COVER

49 UN01 050

93112GH9

Removing the audio unit—1998–00 MX-6

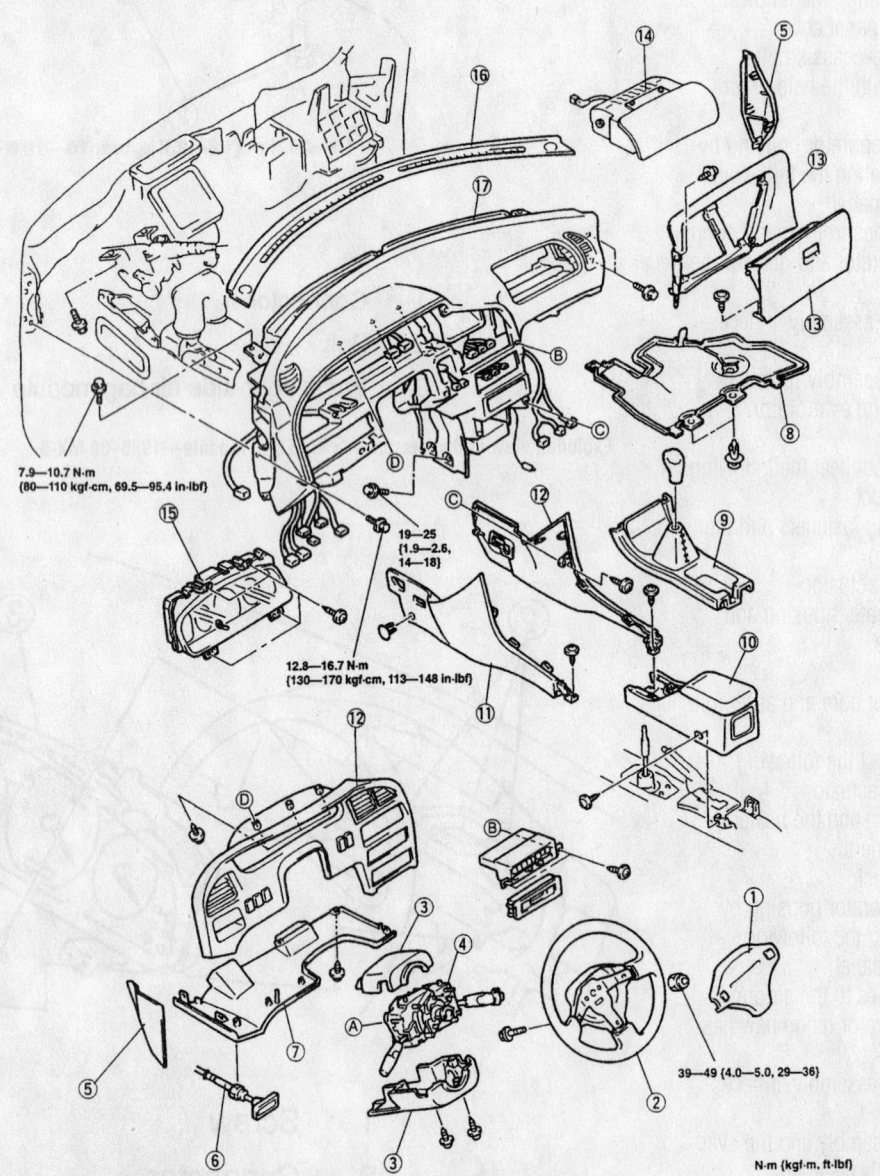

7.9—10.7 N·m
{80—110 kgf-cm, 69.5—95.4 in-lbf}

19—25
{1.9—2.6,
14—18}

12.8—16.7 N·m
{130—170 kgf-cm, 113—148 in-lbf}

39—49 {4.0—5.0, 29—36}

N·m {kgf-m, ft-lbf}

1. Driver-side air bag module
2. Steering wheel
3. Column cover
4. Combination switch
5. Side panel
6. Hood release knob
7. Lower panel
8. Undercover
9. Front console
10. Rear console
11. Side wall
12. Meter hood
13. Glove compartment
14. Passenger-side air bag module
15. Instrument cluster
16. Upper garnish
17. Dashboard

93112GJ3

View of the instrument panel assembly—1998–00 MX-6

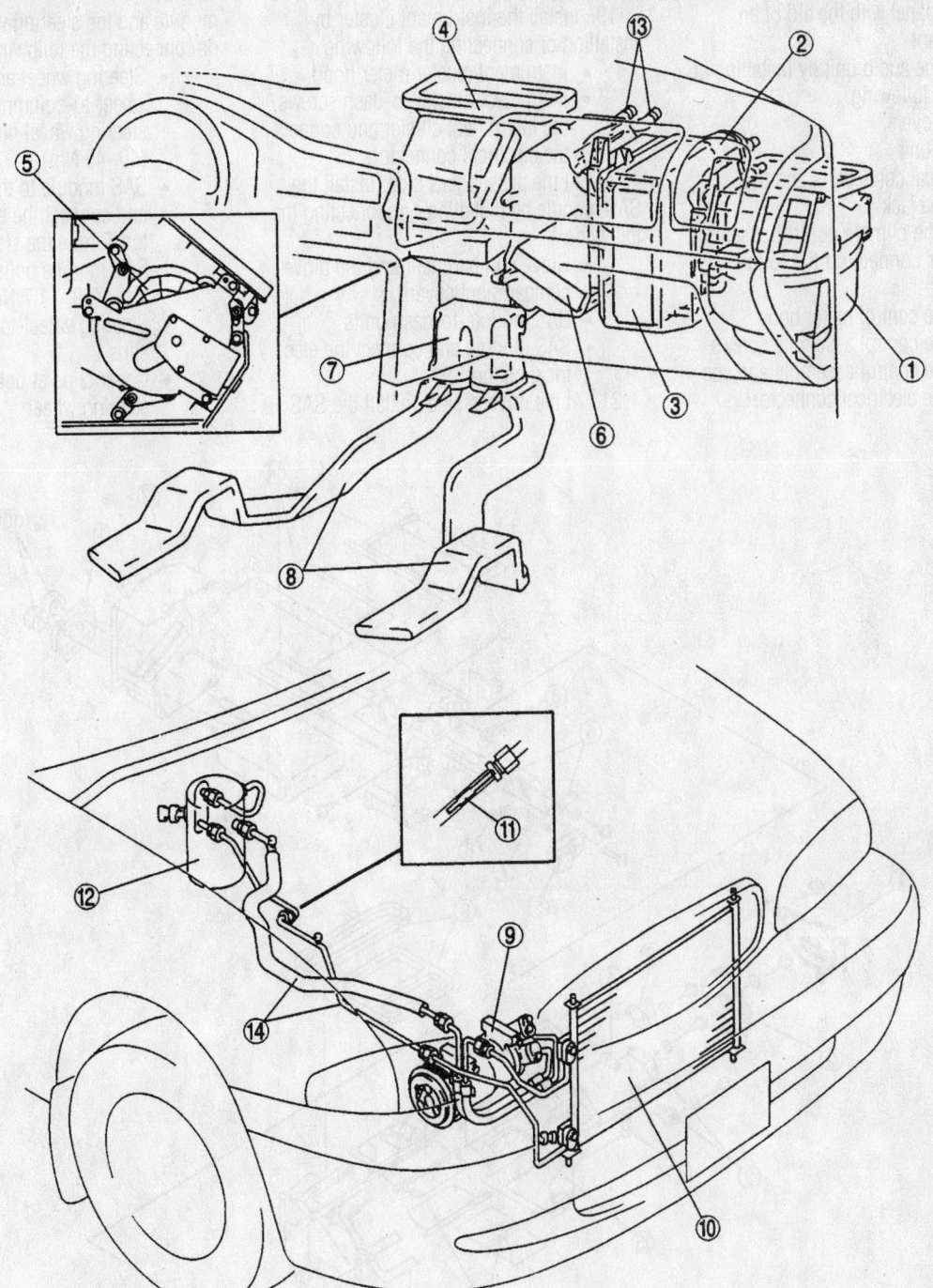

1	Blower unit	8	Rear heat duct (CANADA only)
2	Cooling unit	9	A/C compressor
3	Evaporator	10	Condenser
4	Heater unit	11	Orifice tube
5	Airflow mode main link	12	Accumulator tank
6	Heater core	13	Heater hose
7	Rear duct (CANADA only)	14	Refrigerant lines

93112GI1

View of the heater and air conditioning housing assemblies—1998–00 MX-6

Heater Core replacement is covered in Section 2 of this manual

- Dash panel with the aid of an assistant

17. Install the audio unit by installing or connecting the following:
- Hole covers
- Audio unit
- Electrical connectors and the antenna jack

18. Install the climate control assembly by installing or connecting the following:
- Climate control meter hood
- Climate control assembly screws
- Climate control assembly and connect the electrical connector

19. Install the instrument cluster by installing or connecting the following:
- Instrument cluster meter hood
- Instrument cluster-to-dash screws, the instrument cluster and connect the electrical connectors

20. At the passenger's side, install the SAS module by installing or connecting the following:
- Glove compartment and the glove compartment cover
- SAS module-to-dash bolts
- SAS module and connect the electrical connector

21. At the driver's side, install the SAS module and the steering wheel by installing or connecting the following:
- Steering wheel and the steering wheel-to-column nut. Torque the steering wheel nut to 29–36 ft. lbs. (40–49 Nm)
- SAS module to the steering wheel and connect the electrical connector. Torque the steering column-to-SAS module bolts to 70–104 inch lbs. (7. 9–11.7 Nm)
- Steering wheel-to-SAS module clips
- Cover clips at both sides of the steering wheel

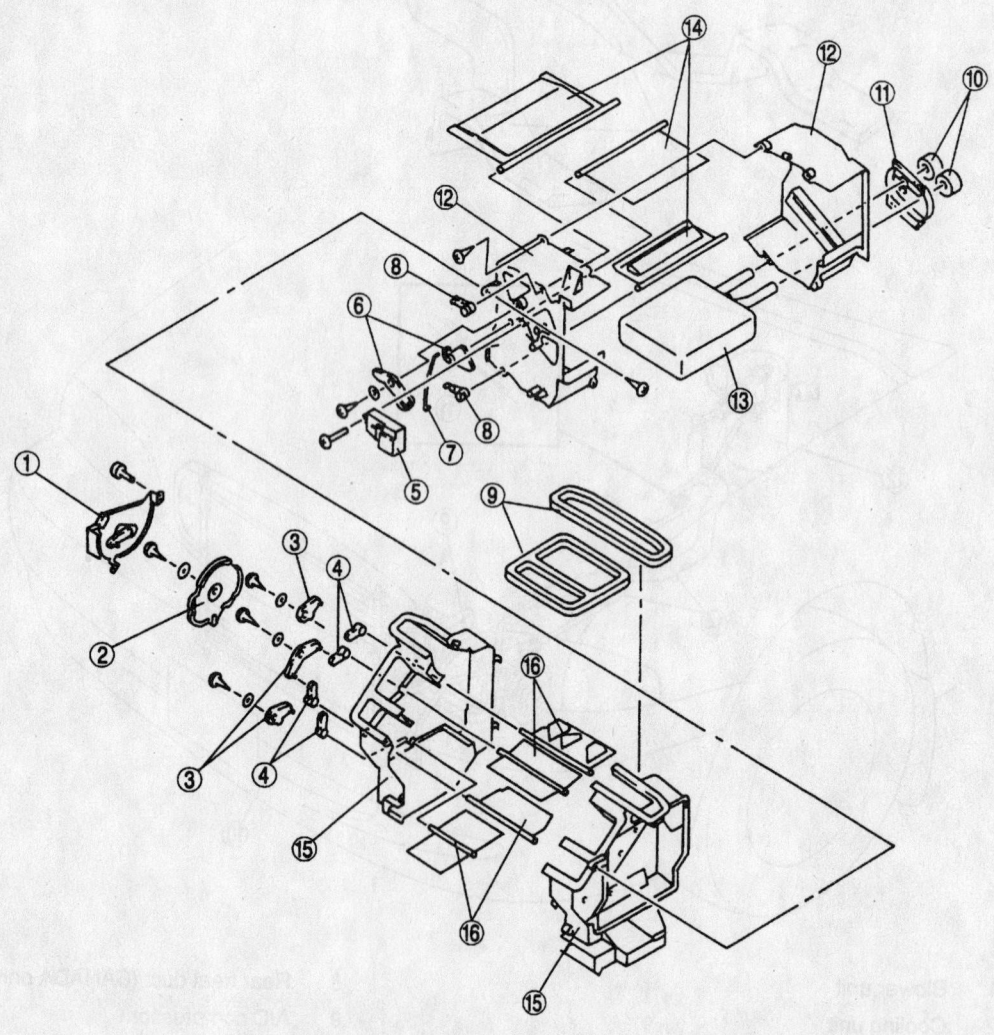

1	Airflow mode actuator	9	Polyurethane protector
2	Airflow mode main link	10	Seal
3	Airflow mode sub link	11	Cover
4	Airflow mode crank	12	Case
5	Air mix actuator	13	Heater core
6	Air mix link	14	Air mix door
7	Air mix rod	15	Case
8	Air mix crank	16	Airflow mode door

93112GI2

Exploded view of the heater core and heater housing assembly—1998–00 MX-6

22. Refill the cooling system.

23. Connect the negative battery cable.

24. Evacuate, charge and leak test the air conditioning system refrigerant.

25. Operate the engine to normal operating temperatures; then, check the climate control system and check for leaks.

Protege

REMOVAL & INSTALLATION

1998 Models

➡**If equipped with an air bag or anti-theft coded radio, disable the air bag system and obtain the anti-theft codes before proceeding.**

1. Disconnect the negative battery cable.

2. Drain the engine coolant into a clean container for reuse.

3. Remove or disconnect the following:
- Heater hoses from the heater core
- Transmission selector lever or knob
- Center console
- Steering wheel
- Upper and lower steering column covers
- Combination switch
- Instrument meter hood
- Instrument cluster assembly
- Speedometer cable
- Heater ducts
- Instrument housing lower panels
- Glove box assembly
- Heater control switch and cables
- Header and the side trim, if required
- Center cap, the side covers and the center bracket bolts on the instrument panel
- Steering shaft bolts
- Any necessary wire harness connectors
- Instrument panel
- Seal plate
- Attaching nuts
- Heater unit
- Attaching clips on the heater unit and separate the assembly
- Heater core

To install:

4. Install or connect the following:
- Heater core into the heater unit
- Heater case halves with the clips
- Heater unit
- Seal plates
- Instrument panel
- Wiring harness connectors

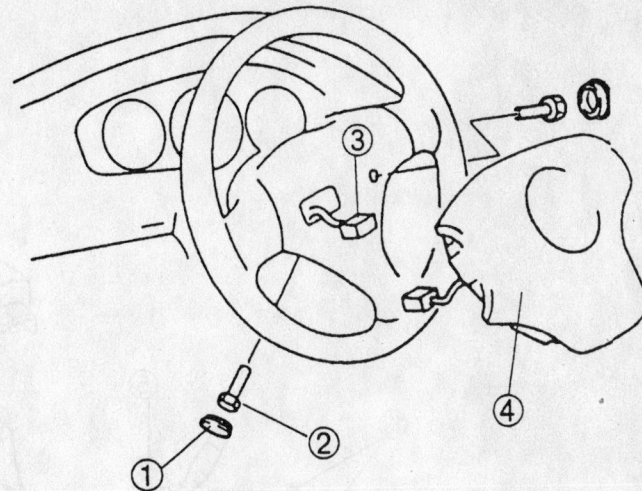

7.9—11.7 N·m {80—120 kgf·cm, 70—104 in·lbf}

1	Cap
2	Bolt
3	Connector
4	Driver-side air bag module

93112GG4

Exploded view of the SAS module and the steering wheel assembly—1999–00 Protege

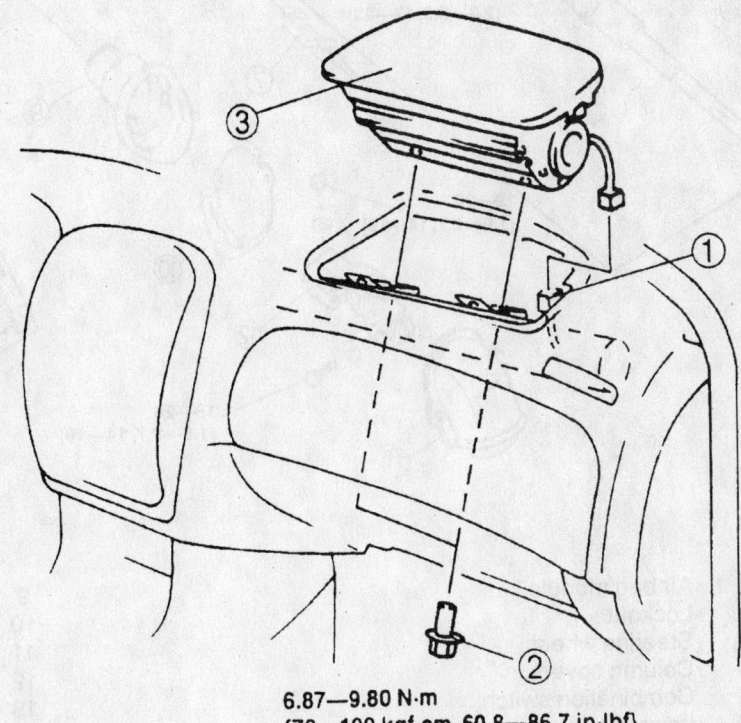

6.87—9.80 N·m
{70—100 kgf·cm, 60.8—86.7 in·lbf}

93112GG5

Exploded view of the passenger's side SAS module—1999–00 Protege

Brake service is covered in Section 4 of this manual

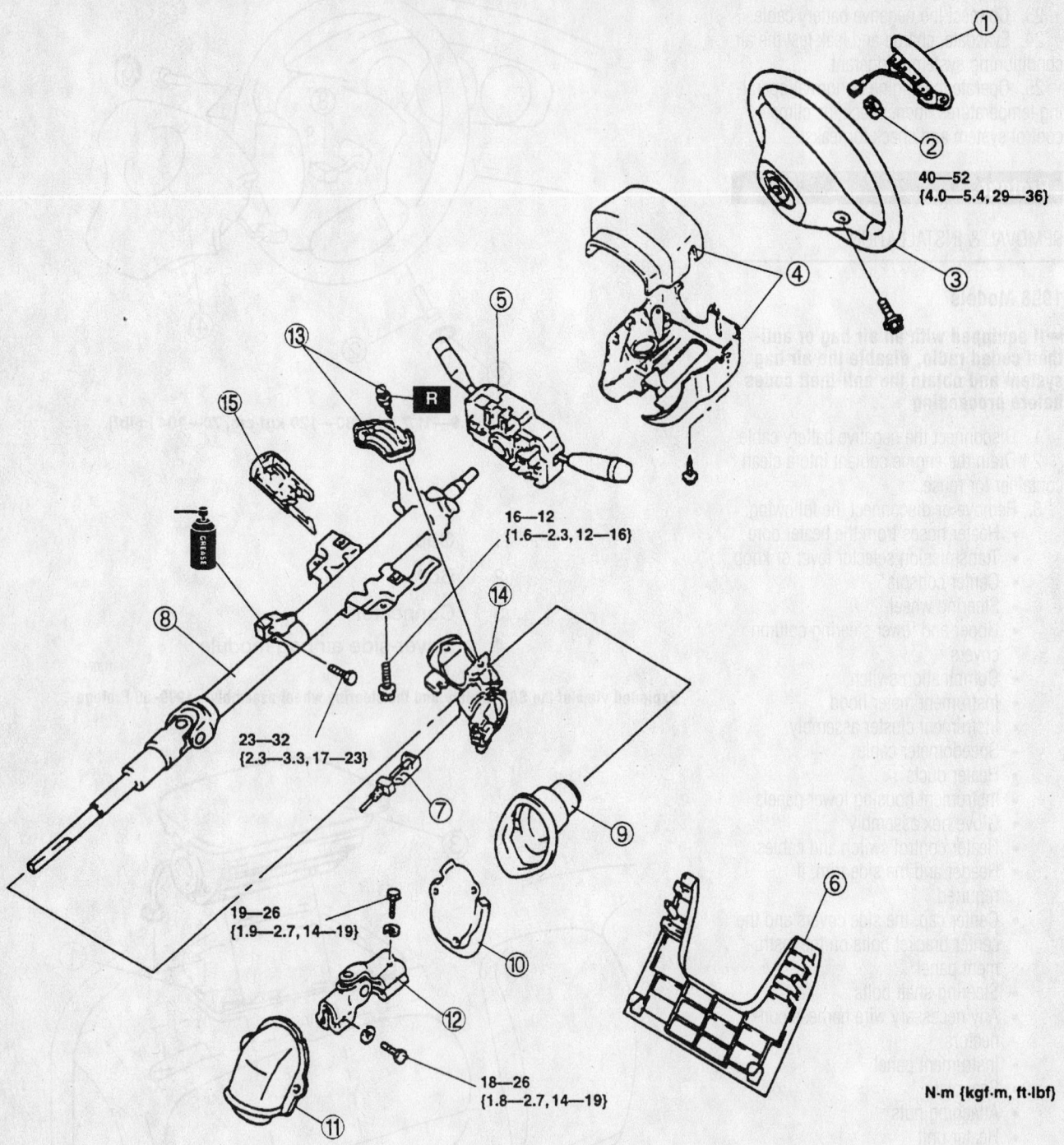

40—52
{4.0—5.4, 29—36}

R

16—12
{1.6—2.3, 12—16}

23—32
{2.3—3.3, 17—23}

19—26
{1.9—2.7, 14—19}

18—26
{1.8—2.7, 14—19}

N·m {kgf·m, ft·lbf}

1	Air bag module	9	Shaft seal
2	Locknut	10	Set plate
3	Steering wheel	11	Dust cover
4	Column cover	12	Universal joint
5	Combination switch	13	Steering lock mounting bolts and bracket
6	Lower panel	14	Steering lock component
7	Key interlock cable	15	Cylinder outer component
8	Steering shaft		

Exploded view of the steering column assembly—1999–00 Protege

93112GG6

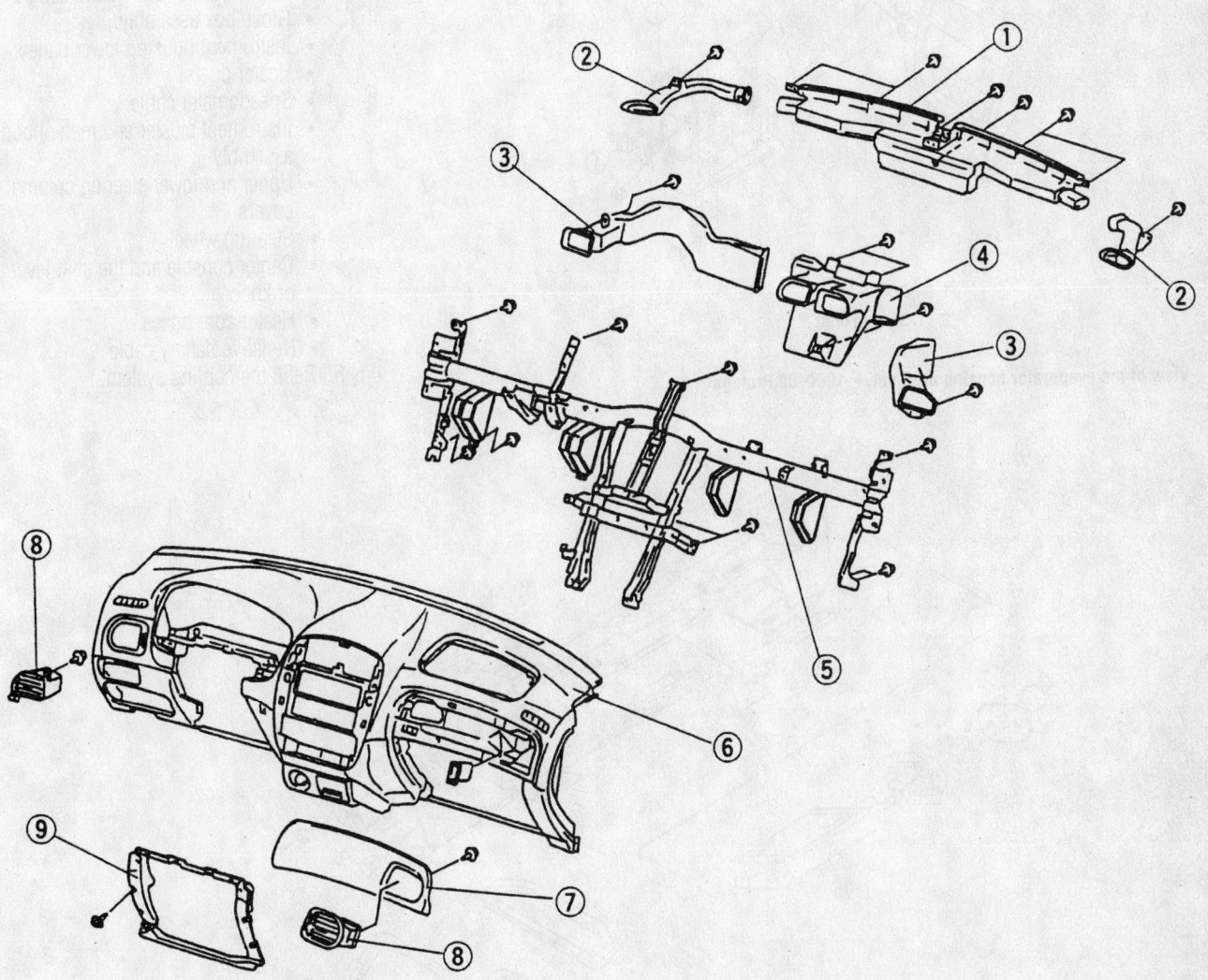

1   Defroster nozzle

2   Side demister duct

3   Duct

4   Center duct

5   Dashboard member

6   Crush pad

7   Pad

8   Ventilator grille

9   Passenger-side lower panel

93112GG7

Exploded view of the dashboard assembly—1999–00 Protege

For complete Engine Mechanical specifications, see Section 1 of this manual

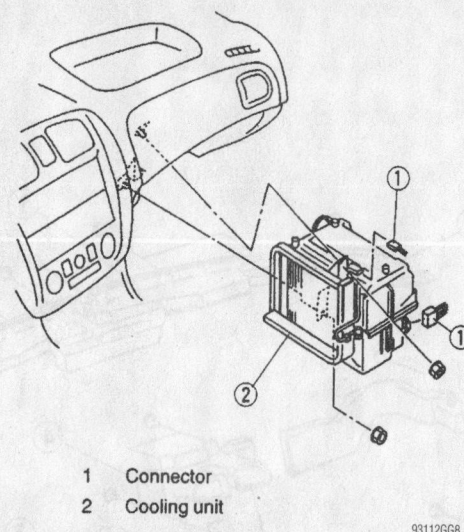

1   Connector
2   Cooling unit

93112GG8

View of the evaporator housing assembly—1999–00 Protege

- Steering shaft bolts
- Center cap, the side covers and the center bracket bolts on the instrument panel
- Header and side trim, if removed.
- Heater control switch and cables
- Glove box assembly
- Instrument housing lower panels
- Heater ducts
- Speedometer cable
- Instrument cluster and meter hood assembly
- Upper and lower steering column covers
- Steering wheel
- Center console and the shift lever or knob
- Heater core hoses
- Negative battery cable

5. Refill the cooling system.

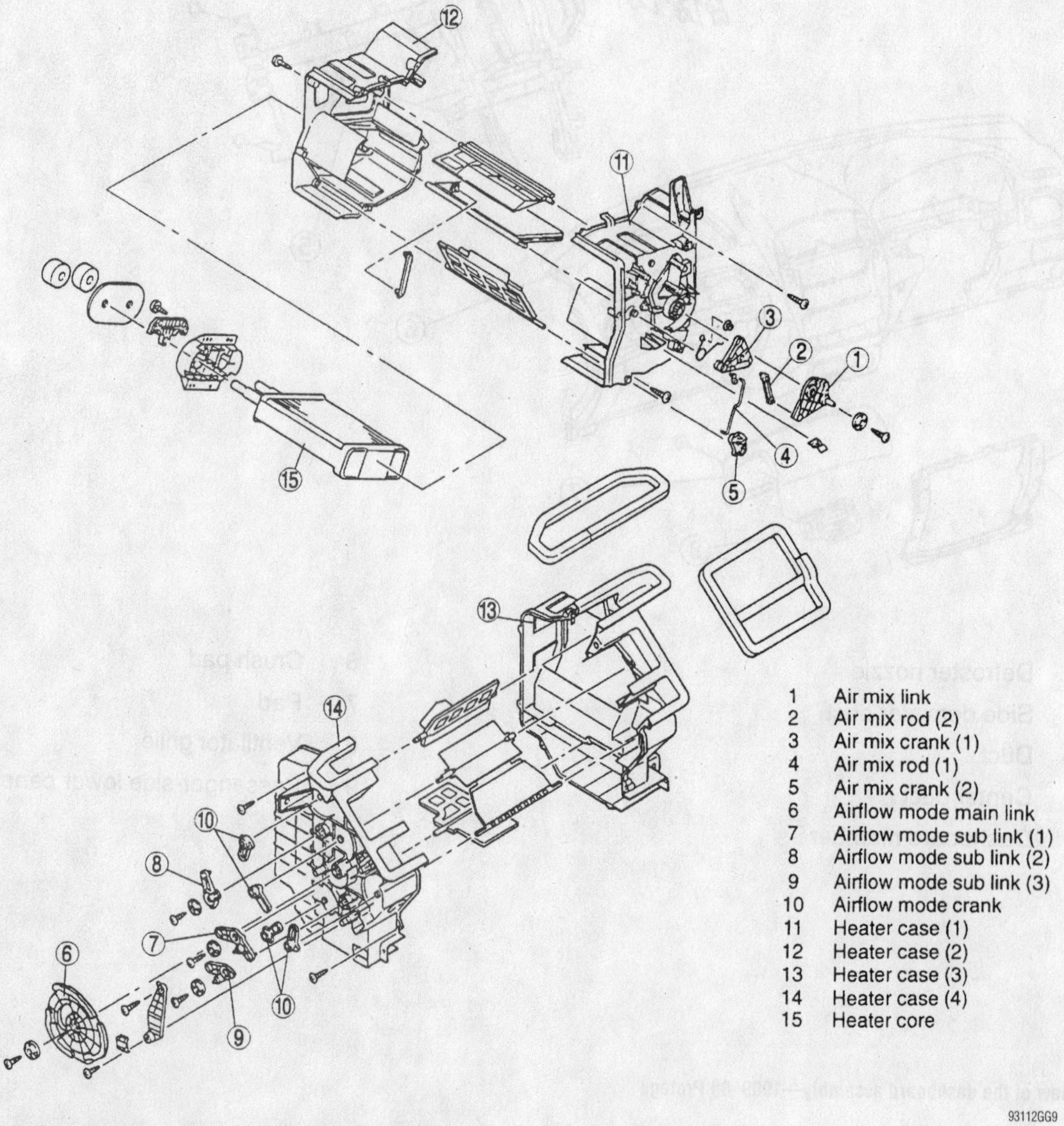

1   Air mix link
2   Air mix rod (2)
3   Air mix crank (1)
4   Air mix rod (1)
5   Air mix crank (2)
6   Airflow mode main link
7   Airflow mode sub link (1)
8   Airflow mode sub link (2)
9   Airflow mode sub link (3)
10   Airflow mode crank
11   Heater case (1)
12   Heater case (2)
13   Heater case (3)
14   Heater case (4)
15   Heater core

93112GG9

Exploded view of the heater housing assembly—1999–00 Protege

6. Operate the engine to normal operating temperatures; then, check the climate control operation and check for leaks.

1999–00 Models

1. Disconnect the negative battery cable.

After disconnecting the battery, wait for more than 1 minute for the SAS to deplete its stored energy.

2. Drain the cooling system into a clean container for reuse.

3. Disconnect the heater hoses from the heater core.

4. Discharge and recover the air conditioning system refrigerant.

5. Place the wheel in the straight-ahead position and turn the ignition switch to LOCK.

6. At the driver's side, remove the SAS module and the steering wheel by removing or disconnecting the following:
- Cover clips at both sides of the steering wheel
- Steering wheel-to-SAS module bolts
- SAS module from the steering wheel and disconnect the electrical connector
- Steering wheel-to-column nut
- Steering wheel from the steering column using a suitable puller

7. At the passenger's side, remove the SAS module by removing or disconnecting the following:
- Glove compartment and the glove compartment cover
- SAS module-to-dash bolts
- SAS module and disconnect the electrical connector

8. Remove the console by removing or disconnecting the following:
- Shift lever knob, if equipped with a manual transmission
- Console's cover
- Console-to-chassis screws and console

9. Remove the combination switch by removing or disconnecting the following:
- Steering column cover
- Electrical connectors and remove the combination switch-to-steering column bolts and the combination switch

10. Remove the instrument cluster by:
- Meter hood

- Instrument cluster-to-dash panel screws
- Electrical connectors and remove the instrument cluster

11. Remove or disconnect the following:
- Lower panel
- Hood release cable installation nut
- Side wall trim
- "A" pillar trim at both sides
- Side panel
- Antenna connector
- Blower motor and heater unit electrical connectors, if equipped with the wire-type climate control unit
- Electrical connectors and the bolts
- Dashboard-to-chassis bolts
- Dashboard from the vehicle with the help of an assistant
- Passenger's side lower panel
- Air intake wire from the climate control unit
- Air conditioning refrigerant lines from the evaporator. Discard the O-rings
- Evaporator electrical connector(s)
- Evaporator housing

12. Disassemble the heater housing and remove the heater core.

To install:

13. Install the heater core and assemble the heater housing.

14. Install or connect the following:
- Evaporator housing
- Evaporator electrical connector(s)
- Air conditioning refrigerant lines to the evaporator using new O-rings
- Air intake wire to the climate control unit
- Passenger's side lower panel
- Dashboard to the vehicle with the help of an assistant
- Dashboard-to-chassis bolts
- Electrical connectors and the bolts
- Blower motor and heater unit electrical connectors, if equipped with the wire-type climate control unit
- Antenna connector
- Side panel
- A-pillar trim at both sides
- Side wall trim
- Hood release cable installation nut
- Lower panel

15. Install the instrument cluster by installing or connecting the following:
- Instrument cluster and connect the electrical connectors
- Instrument cluster-to-dash panel screws
- Meter hood

16. Install the combination switch by installing or connecting the following:
- Electrical connectors and install the combination switch-to-steering column bolts and the combination switch
- Steering column cover

17. Install the console by installing or connecting the following:
- Console and the console-to-chassis screws
- Console's cover
- Shift lever knob, if equipped with a manual transmission

18. At the passenger's side, install the SAS module by installing or connecting the following:
- SAS module and connect the electrical connector
- SAS module-to-dash bolts
- Glove compartment and the glove compartment cover

19. At the driver's side, install the SAS module and the steering wheel by installing or connecting the following:
- Steering wheel-to-column nut
- SAS module from the steering wheel and connect the electrical connector
- Steering wheel-to-SAS module bolts
- Cover clips at both sides of the steering wheel

20. Connect the heater hoses to the heater core.

21. Refill the cooling system.

22. Connect the negative battery cable.

23. Evacuate, charge and leak test the air conditioning system.

24. Operate the engine to normal operating temperatures; then, check the climate control operation and check for leaks.

626

REMOVAL & INSTALLATION

1. Disconnect the negative battery cable.

After disconnecting the battery, wait for more than 1 minute for the SAS to deplete its stored energy.

2. Drain the cooling system into a clean container for reuse.

3. Discharge and recover the air conditioning system refrigerant.

For Accessory Drive Belt illustrations, see Section 1 of this manual

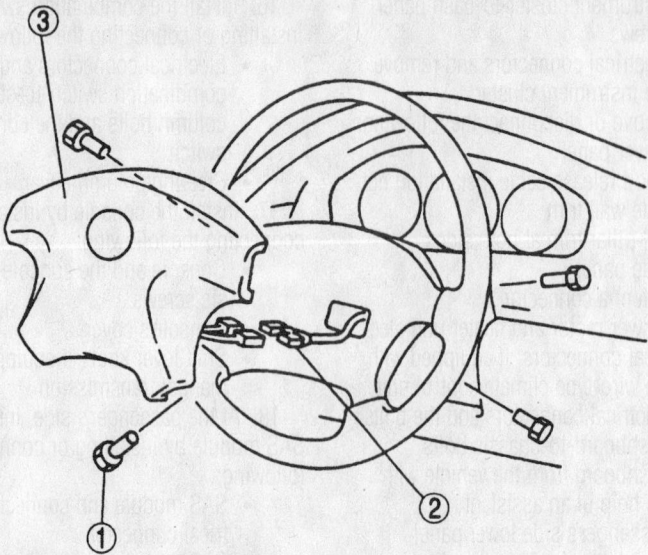

7.9—11.7 N·m {80—120 kgf·cm, 70—104 in·lbf}

1 Bolt

2 Connector

3 Driver-side air bag module

93112GH6

Exploded view of the steering wheel and SAS module—1998–00 626

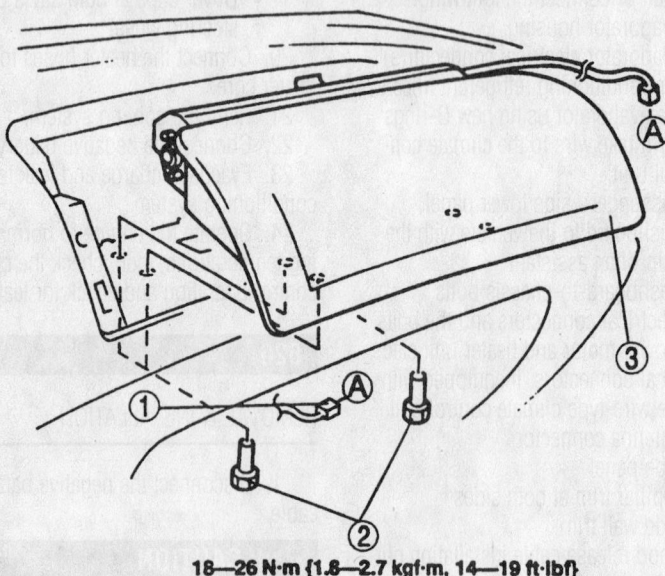

18—26 N·m {1.8—2.7 kgf·m, 14—19 ft·lbf}

1 Connector

2 Bolt

3 Passenger-side air bag module

93112GH7

Exploded view of the passenger's side SAS module—1998–00 626

4. Place the wheel in the straight-ahead position and turn the ignition switch to LOCK.

5. At the driver's side, remove the SAS module and the steering wheel removing or disconnecting the following:
- Cover clips at both sides of the steering wheel
- Steering wheel-to-SAS module bolts
- SAS module from the steering wheel and disconnect the electrical connector
- Steering wheel-to-column nut
- Steering wheel from the steering column using a suitable puller

6. At the passenger's side, remove the SAS module by removing or disconnecting the following:
- Glove compartment and the glove compartment cover
- SAS module-to-dash bolts
- SAS module and disconnect the electrical connector

7. Remove the instrument cluster by removing or disconnecting the following:
- Instrument cluster meter hood
- Instrument cluster-to-dash screws, the instrument cluster and disconnect the electrical connectors

8. Remove the climate control assembly by removing or disconnecting the following:
- Climate control meter hood
- Climate control assembly screws
- Climate control assembly and disconnect the electrical connector and the assembly

9. Remove the audio unit by removing or disconnecting the following:
- Hole covers by inserting a small tape-wrapped flathead screwdriver into the slot and carefully pry off the hole covers
- Using 2 removal tools 49 UN01 050 or equivalent, insert them into sides of the audio unit.
- Slide audio unit outward and forward
- Electrical connectors and the antenna jack

10. Remove the instrument panel by removing or disconnecting the following:
- Upper center panel cover
- Dash panel-to-chassis bolts
- Dash panel with the help of an assistant

11. Remove the evaporator housing by removing or disconnecting the following:
- Center lower panel
- Refrigerant lines from the air conditioning evaporator and discard the gaskets

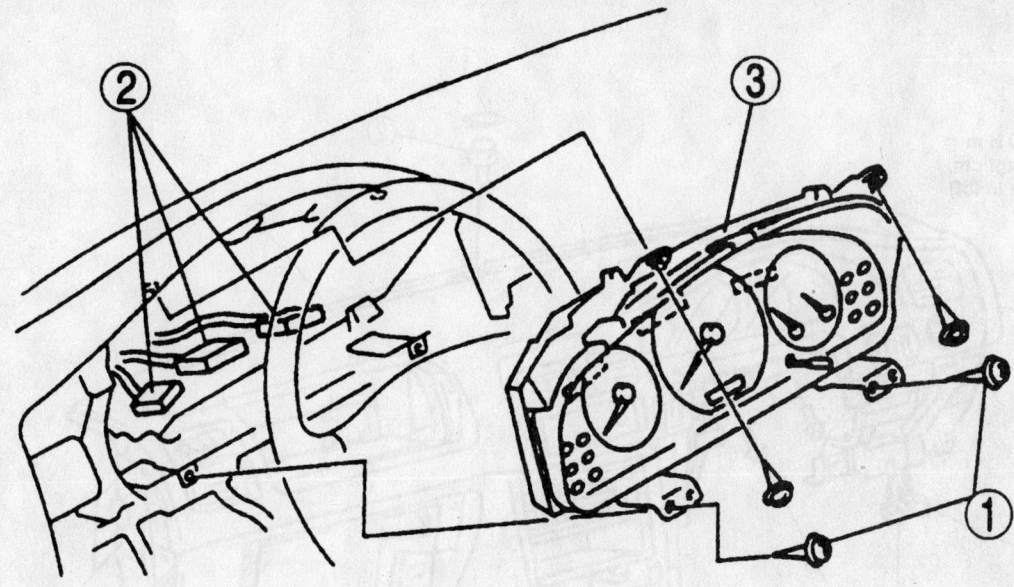

1 Screw

2 Connector

3 Instrument cluster

93112GH8

View of the instrument cluster assembly—1998–00 626

- Blower motor assembly, if necessary
- Evaporator assembly fasteners and remove the evaporator assembly
12. Remove or disconnect the following:
  - Rear heater duct
  - Heater housing fasteners and the heater housing
  - Airflow mode actuator

13. Separate the heater housing and remove the heater core.
  To install:
14. Install the heater core and assemble the heater housing.
15. Install or connect the following:
  - Airflow mode actuator
  - Heater housing and the heater housing fasteners

- Rear heater duct
16. Install the evaporator housing by installing or connecting the following:
  - Center lower panel
  - Refrigerant lines to the air conditioning evaporator using new gaskets
  - Blower motor assembly, if necessary
  - Evaporator assembly and the evap-

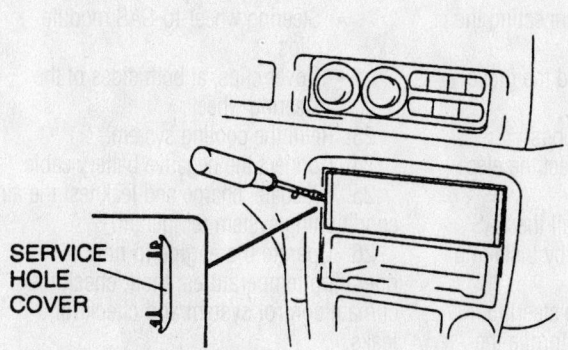

SERVICE HOLE COVER

Removing the audio unit—1998–00 626

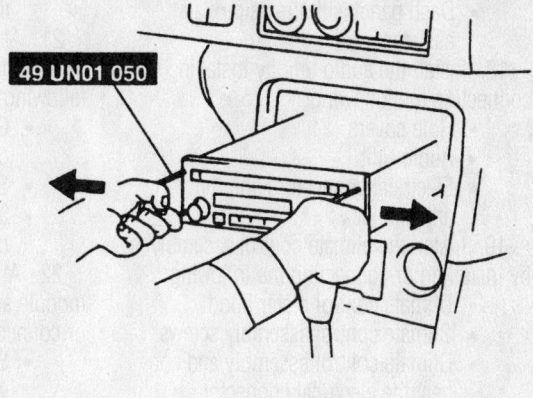

49 UN01 050

93112GH9

For Tire, Wheel and Ball Joint specifications, see Section 1 of this manual

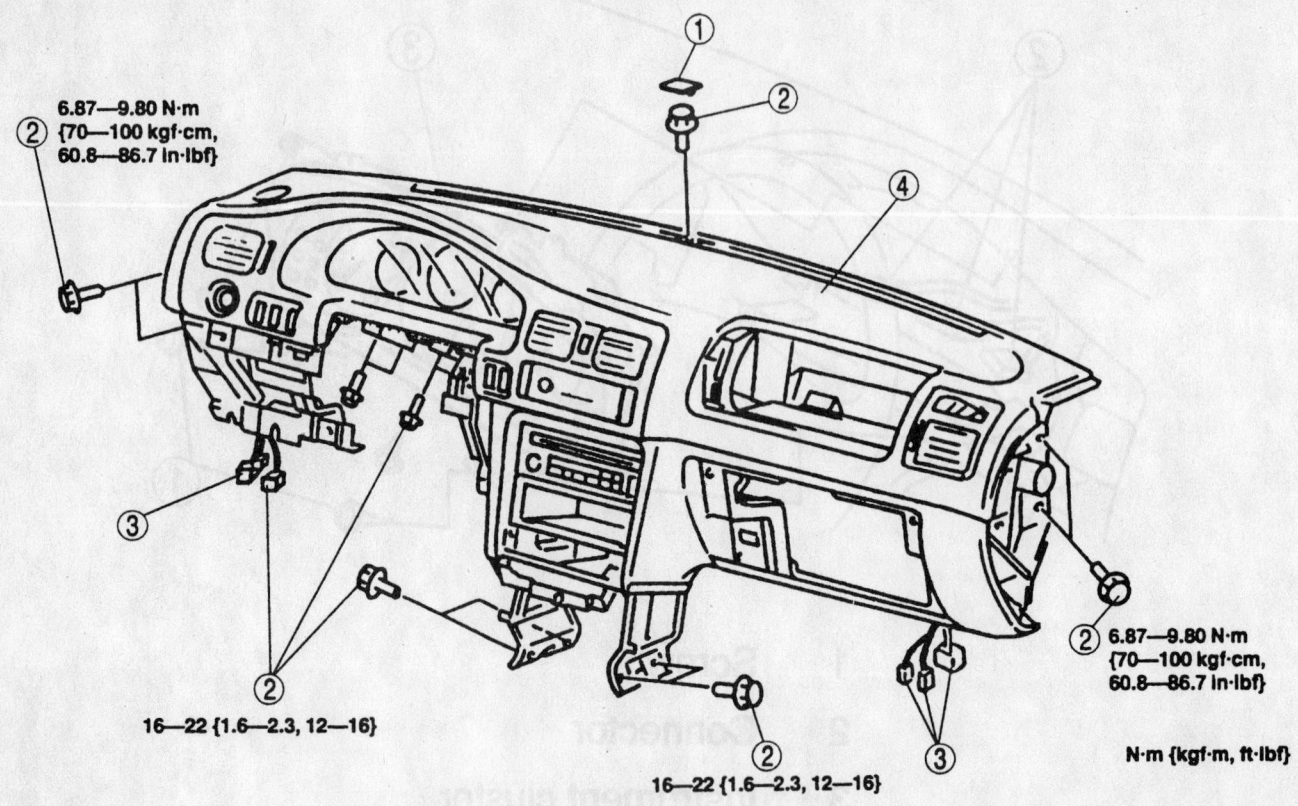

6.87—9.80 N·m
{70—100 kgf·cm,
60.8—86.7 in·lbf}

16—22 {1.6—2.3, 12—16}

16—22 {1.6—2.3, 12—16}

6.87—9.80 N·m
{70—100 kgf·cm,
60.8—86.7 in·lbf}

N·m {kgf·m, ft·lbf}

1	Cover	3	Connector
2	Bolt	4	Dashboard

93112GH0

View of the instrument panel assembly—1998–00 626

orator assembly fasteners

17. Install the instrument panel by installing or connecting the following:
- Upper center panel cover
- Dash panel-to-chassis bolts
- Dash panel with the help of an assistant

18. Install the audio unit by installing or connecting the following:
- Hole covers
- Audio unit
- Electrical connectors and the antenna jack

19. Install the climate control assembly by installing or connecting the following:
- Climate control meter hood
- Climate control assembly screws
- Climate control assembly and connect the electrical connector

20. Install the instrument cluster by installing or connecting the following:
- Instrument cluster meter hood
- Instrument cluster-to-dash screws, the instrument cluster and connect the electrical connectors

21. At the passenger's side, install the SAS module by installing or connecting the following:
- Glove compartment and the glove compartment cover
- SAS module-to-dash bolts
- SAS module and connect the electrical connector

22. At the driver's side, install the SAS module and the steering wheel by installing or connecting the following:
- Steering wheel and the steering wheel-to-column nut. Torque the

steering wheel nut to 29–36 ft. lbs. (40–49 Nm)
- SAS module to the steering wheel and connect the electrical connector. Torque the steering column-to-SAS module bolts to 70–104 inch lbs. (8–12 Nm)
- Steering wheel-to-SAS module clips
- Cover clips, at both sides of the steering wheel

23. Refill the cooling system.

24. Connect the negative battery cable.

25. Evacuate, charge and leak test the air conditioning system refrigerant.

26. Operate the engine to normal operating temperatures; then, check the climate control system and check for leaks.

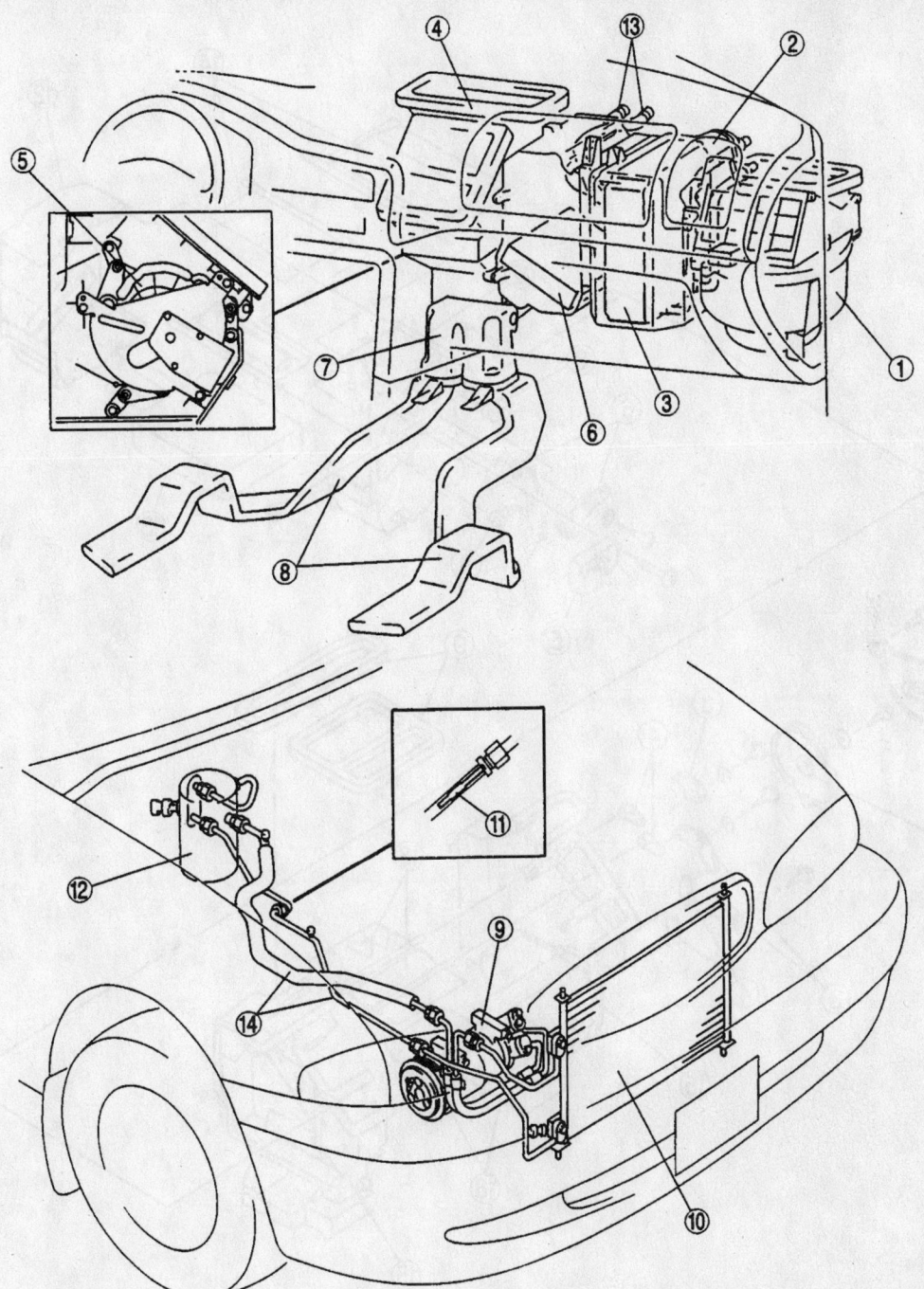

1	Blower unit	8	Rear heat duct (CANADA only)
2	Cooling unit	9	A/C compressor
3	Evaporator	10	Condenser
4	Heater unit	11	Orifice tube
5	Airflow mode main link	12	Accumulator tank
6	Heater core	13	Heater hose
7	Rear duct (CANADA only)	14	Refrigerant lines

93112GI1

View of the heater and air conditioning housing assemblies—1998–00 626

For Wheel Alignment specifications, see Section 1 of this manual

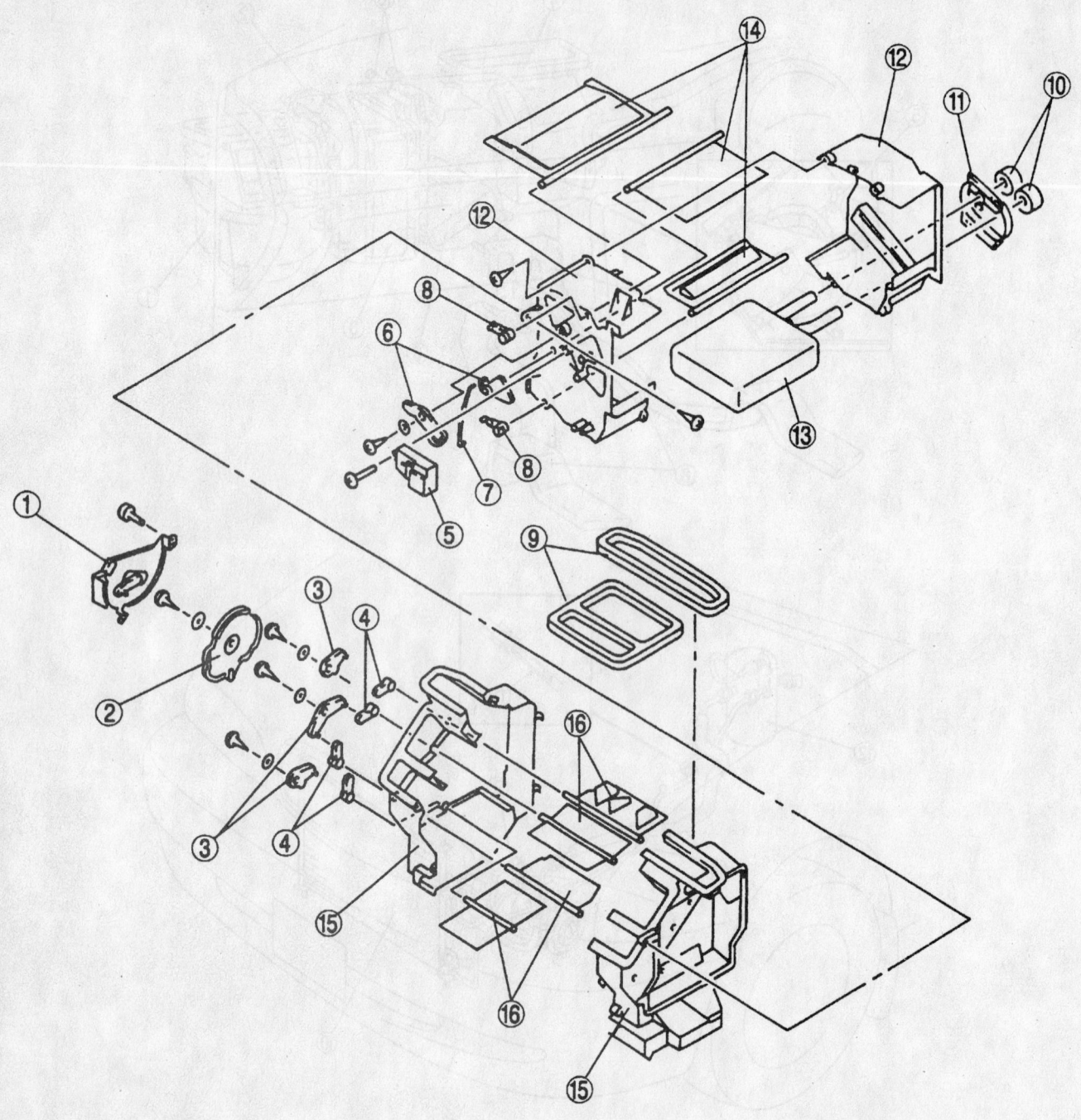

1	Airflow mode actuator	9	Polyurethane protector
2	Airflow mode main link	10	Seal
3	Airflow mode sub link	11	Cover
4	Airflow mode crank	12	Case
5	Air mix actuator	13	Heater core
6	Air mix link	14	Air mix door
7	Air mix rod	15	Case
8	Air mix crank	16	Airflow mode door

93112GI2

Exploded view of the heater core and heater housing assembly—1998–00 626

1998–00 MERCEDES-BENZ

300D Turbo, 300E, 300CE, 300TE, 300E, 4Matic, 300TE 4Matic

REMOVAL & INSTALLATION

Behr Style

1. Disconnect the negative battery cable.
2. Drain the cooling system into a clean container for reuse.
3. Remove or disconnect the following:
   - Hoses from the heater supply and return pipes. Using compressed air, blow the residual coolant from the heater core
   - Instrument panel
   - Upper heater box cover-to-heater box screws, the 6 clips and the cover
   - Heater supply pipe-to-chassis clamp in the engine compartment
   - Heater supply and return pipes-to-heater core screws from the top of

the heater box and swing the pipes away from the heater box
   - Pull the heater core from the top of the heater box
4. Clean any spilled coolant from inside the heater box.

To install:

5. Install or connect the following:
   - Heater core into the heater box
   - Heater supply and return pipes to the heater box using 3 new sealing rings
   - Heater supply pipe-to-chassis clamp in the engine compartment
   - Upper cover to the heater box, the 6 clips and the screws
   - Instrument panel
   - Hoses to the heater supply and return pipes
6. Refill the cooling system.
7. Connect the negative battery cable.
8. Operate the engine to normal operating temperatures; then, check the climate control operation and check for leaks.

Valeo Style

1. Disconnect the negative battery cable.
2. Drain the cooling system into a clean container for reuse.
3. Remove or disconnect the following:
   - Hoses from the heater supply and return pipes. Using compressed air, blow the residual coolant from the heater core
   - Instrument panel and the center console
   - Pull off the temperature sensor heat exchanger, the left and right 2-pole couplings from the lower front of the heater box
   - Pull the air ducts from the bottom of the heater box
   - Crossmember and strut from the front of the heater box
   - Bracket from under the heater box
   - Loosen the ignition switch housing-to-steering column bolt. Turn

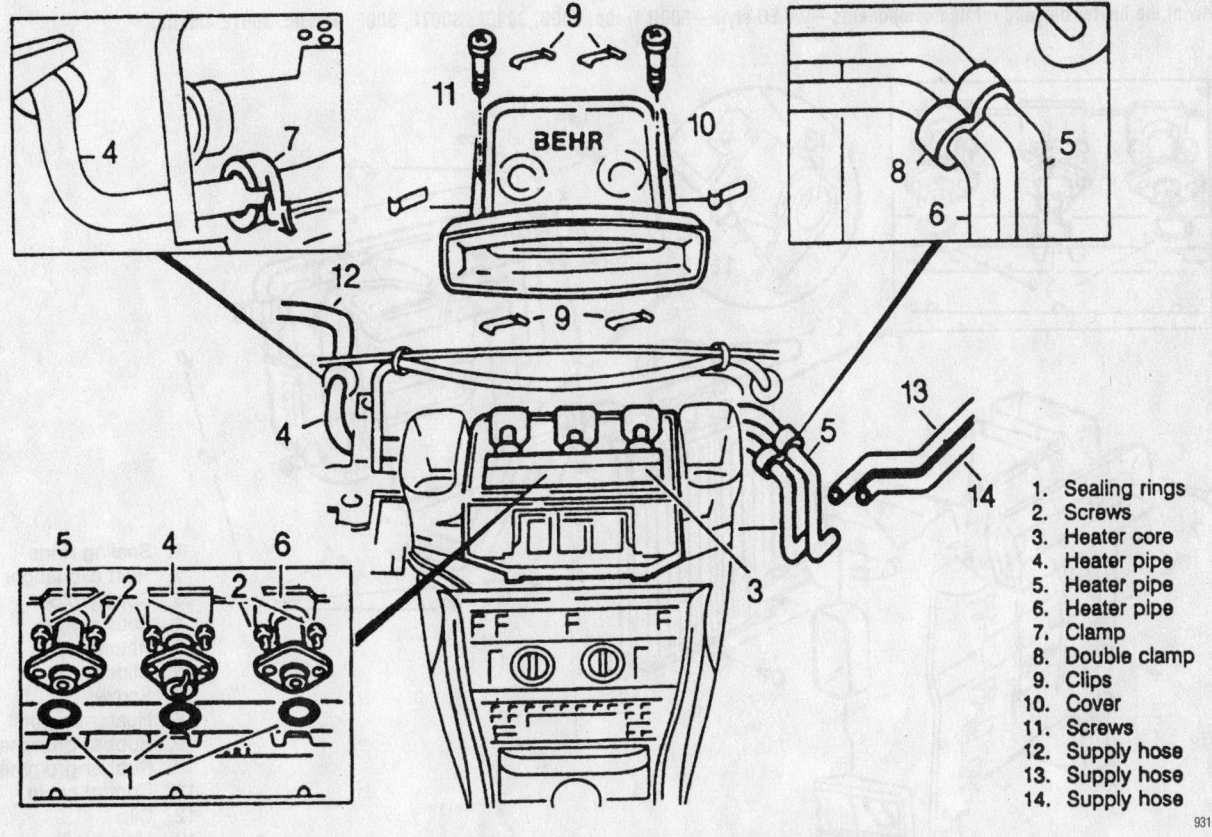

1. Sealing rings
2. Screws
3. Heater core
4. Heater pipe
5. Heater pipe
6. Heater pipe
7. Clamp
8. Double clamp
9. Clips
10. Cover
11. Screws
12. Supply hose
13. Supply hose
14. Supply hose

93112G58

View of the heater box and related components—BEHR style—300D Turbo, 300E, 300CE, 300TE, 300E, 4Matic, 300TE 4Matic

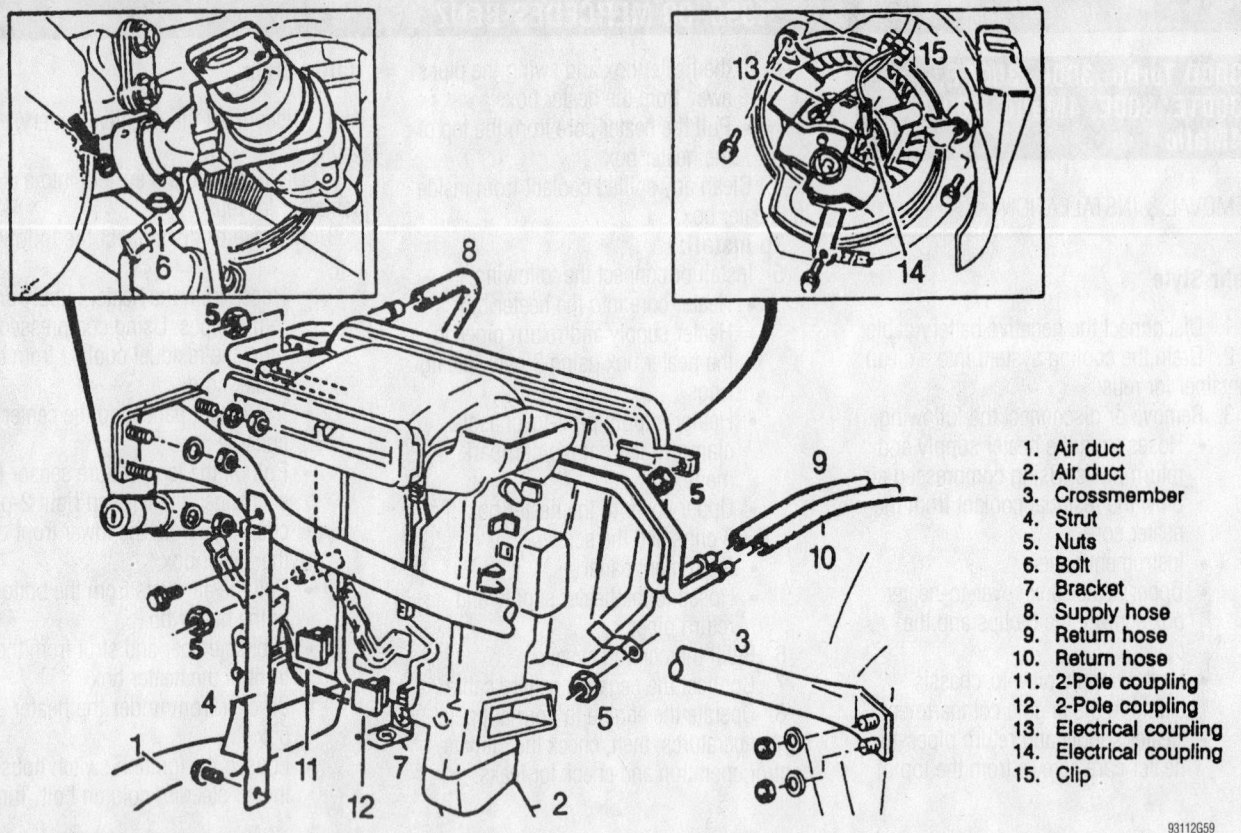

1. Air duct
2. Air duct
3. Crossmember
4. Strut
5. Nuts
6. Bolt
7. Bracket
8. Supply hose
9. Return hose
10. Return hose
11. 2-Pole coupling
12. 2-Pole coupling
13. Electrical coupling
14. Electrical coupling
15. Clip

93112G59

View of the heater box and related components—VALEO style—300D Turbo, 300E, 300CE, 300TE, 300E, 4Matic, 300TE 4Matic

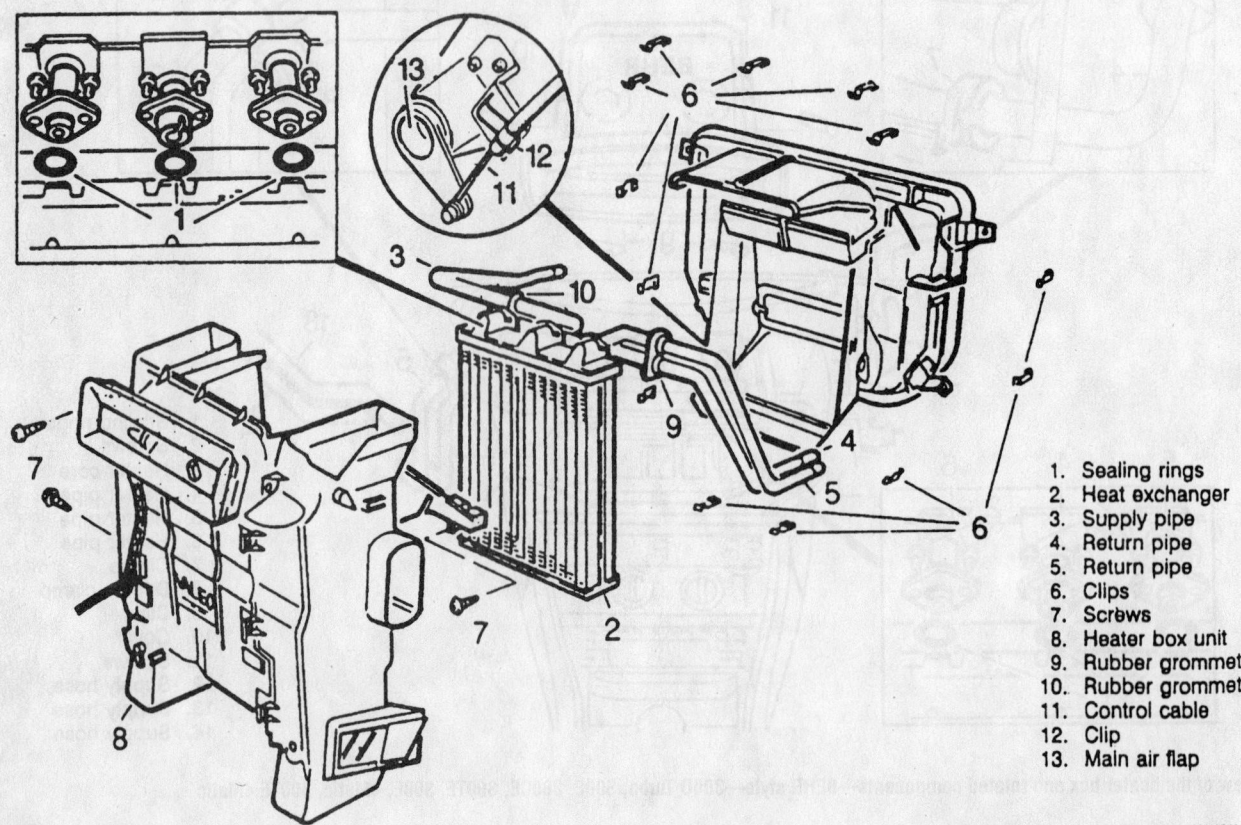

1. Sealing rings
2. Heat exchanger
3. Supply pipe
4. Return pipe
5. Return pipe
6. Clips
7. Screws
8. Heater box unit
9. Rubber grommet
10. Rubber grommet
11. Control cable
12. Clip
13. Main air flap

93112G60

Exploded view of the heater box and related components—VALEO style—300D Turbo, 300E, 300CE, 300TE, 300E, 4Matic, 300TE 4Matic

the ignition key to position **1**, press the ignition switch housing locking button and pull the steering lock from the steering column
- Pull the air ducts from both sides of the heater box
- Heater box-to-chassis nuts and pull the heater box from the firewall
- Electrical connectors and clip from the blower motor. Disconnect the main air flap control cable from the heater box
- Heater box retaining clips and pull the front case from the rear case. Remove the heater core from the heater case
- Heater supply and return pipes-to-heater core screws and the pipes from the heater core

4. Clean any spilled coolant from inside the heater box.

To install:

5. Install or connect the following:
- Heater supply and return pipes to the heater box using 3 new sealing rings
- Heater core into the heater box; make sure the rubber pipe grommets are seated properly
- Front case onto the heater box and secure with the clips
- Main air flap control cable to the heater box. Connect the electrical connectors and clip to the blower motor
- Position the heater box against the firewall (using a new seal) and install the heater box-to-chassis nuts.
- Air ducts to both sides of the heater box
- Slide the steering lock into the steering column and tighten the bolt
- Bracket under the heater box
- Crossmember and strut at the front of the heater box
- Air ducts to the bottom of the heater box
- Temperature sensor heat exchanger, the left and right 2-pole couplings at the lower front of the heater box

6. Refill the cooling system.
7. Connect the negative battery cable.
8. Operate the engine to normal operating temperatures; then, check the climate control operation and check for leaks.

300SE, 300SEL, 350SD Turbo, 350SDL Turbo, 420SEL, 560SEC and 560SEL

REMOVAL & INSTALLATION

1. Disconnect the negative battery cable.
2. Discharge and recover the air conditioning system refrigerant.
3. Drain the cooling system into a clean container for reuse.
4. Detach and plug the heater hoses from the firewall tube connections.
5. Raise and safely support the vehicle.
6. Remove or disconnect the following:
- Panel behind the front right side member
- Move both front seats rearward and cover
- Floor mats from both sides of the vehicle
- Instrument panel and the center console
- Control unit, if equipped with ABS
- Left and right air ducts, from the rear passenger compartment, and the air ducts from the transmission tunnel, on the driver's floor
- 12-pole electrical connector for the temperature control from the right side of the heater/air conditioning box
- 5-pole and 6-pole electrical connectors from the temperature dial and the 2-pole electrical connector from the temperature sensor, air volume and the air distributing switch
- Vacuum lines from the air volume switch
- Heater/air conditioning box-to-stiffening strut screws
- Cable straps from the blower motor housing. Loosen the main cable harness straps from the heater/air conditioning box
- Pull the right and left air ducts from the fresh air nozzles on the heater/air conditioning box

7. From the heater/air conditioning box, removing or disconnecting the following:
- 2-pole electrical connectors from both temperature sensors
- Electric lines cable connector screw
- Separate the electrical plug connector and unclip the coupling member from the holder

8. Remove or disconnect the following:

- Expansion valve; be sure to plug the openings
- Condensate drain hoses from both sides of the evaporator housing
- 2-pole electrical connector from the switch-over valve

9. From the heater/air conditioning box, removing or disconnecting the following:
- Lower heater/air conditioning box-to-holding angle screws
- Angle bracket-to-blower housing nut
- Right side upper angle-to-chassis nut
- Left side upper angle-to-chassis nut

10. Remove or disconnect the following:
- Pull the heater/air conditioning box rearward to disengage the heater pipes from the front wall.
- Lift the heater/air conditioning box above the front passenger's leg room and remove from the vehicle; be sure to keep the heater pipes vertical so coolant will not drain out
- Evaporator housing and pull the heater core from the heater/air conditioning box

To install:

➡ **When assembling the cases, be sure to seal them especially well along the horizontal separating joint. Pay particular attention to the areas behind the expansion valve and temperature or blower control.**

11. Install or connect the following:
- Heater core and the evaporator housing onto the heater/air conditioning box
- Heater/air conditioning box against the front wall and position the heater pipes through the front wall
- Heater/air conditioning box to the left side upper angle bracket and screw
- Heater/air conditioning box to the right side upper angle bracket and nuts
- Lower angle bracket with screws
- 2-pole electrical connector to the switch-over valve
- Condensate drain hoses to the heater/air conditioning box

12. Using refrigerant oil, lubricate the O-rings and threads; then, connect the refrigerant lines to the expansion valve.

13. Install or connect the following:
- Electrical lines on the cable connector
- 2-pole electrical connectors to the temperature sensors

- Both fresh air duct nozzles onto the heater/air conditioning box
- Main cable harness onto the heater/air conditioning box and secure with the cable straps
- Electric lines onto the blower housing and secure with the cable straps
- Stiffening strut and screws
- Vacuum lines to the air volume switch
- 2-pole, 5-pole and 6-pole electrical connectors onto the temperature dials
- 2-pole connector onto the flow sensor switch and the air distributing switch
- 12-pole connector onto the electronic switch gear
- Both heater air ducts for the rear passenger compartment
- Heater boxes
- ABS control unit, if equipped with ABS

14. Refill the cooling system.
15. Install or connect the following:
   - Center console and the instrument panel
   - Both foot mats into the leg room
   - Side member panel (located under the vehicle), if removed
16. Connect the negative battery cable.
17. Evacuate, charge and leak test the air conditioning system.
18. Operate the engine to normal operating temperatures; then, check the climate control operation and check for leaks.

C220 and C280

REMOVAL & INSTALLATION

1. Disconnect the negative battery.
2. Drain the cooling system into a clean container for reuse.

Diamante

REMOVAL & INSTALLATION

1. Disconnect the negative battery cable.
2. Drain the cooling system into a clean container for reuse.
3. Discharge and recover the air conditioning system refrigerant.
4. Remove or disconnect the following:
   - Heater hoses from the heater core

---

3. Discharge and recover the air conditioning system refrigerant.
4. Remove or disconnect the following:
   - Air inlet cover
   - Water collector
   - Series resistor for the heater blower
   - Loosen the clamps on the hot water supply and return hoses and pull off the hoses. Seal off the openings
   - Instrument panel
   - Center console
   - Left and right floor covering
   - Front passenger airbag unit, if equipped
   - Jacket tube
   - Cable duct screws and loosen the cable duct
   - Bracket screws and bracket
   - Traverse pipe attaching the bolts and nuts and remove the traverse pipe
   - Electrical cables at the ground point
   - Open the cable tie and remove the left and right rear air ducts
   - Vacuum lines at the cockpit separation point
   - Right ground strap
   - Plug connection for the blower motor
   - Air ducts to the left and right side outlets
   - Heater box
   - Screws and clips and remove the cover
   - Screws and loosen the guide for the hot water supply pipe
   - Screws and loosen the hot water return pipe
   - Retaining clip for the hot water supply pipe
   - Heat core together with the hot water pipes
   - Hot water pipe attaching screws and remove the pipes

1998–01 MITSUBISHI

- Refrigerant lines from the evaporator core and discard the O-rings

❋❋ CAUTION

After disconnecting the negative battery cable, wait at least 60 seconds before working on the SRS module or instrument panel.

5. Remove the passenger's side air bag by removing or disconnecting the following:

   - Dash undercover

To install:

5. Install or connect the following:
   - Heater core
   - Hot water pipe and screws
   - Heat core together with the hot water pipes
   - Retaining clip for the hot water supply pipe
   - Hot water return pipe and the screws
   - Guide and the screws for the hot water supply pipe
   - Cover and the screws and clips
   - Heater box
   - Air ducts to the left and right side outlets
   - Plug connection for the blower motor
   - Right ground strap
   - Vacuum lines at the cockpit separation point
   - Left and right rear air ducts
   - Electrical cables at the ground point
   - Traverse pipe, the bolts and nuts
   - Bracket and screws
   - Cable duct and the cable duct screws
   - Jacket tube
   - Front passenger air bag unit, if equipped
   - Left and right floor covering
   - Center console
   - Instrument panel
   - Return hoses and the hoses to the hot water supply
   - Series resistor for the heater blower
   - Water collector
   - Air inlet cover
6. Evacuate, charge and leak test the air conditioning system.
7. Refill the cooling system.
8. Connect the negative battery.
9. Operate the engine to normal operating temperatures; then, check the climate control operation and check for leaks.

- Glove box assembly
- Glove box case
- Air bag-to-dash bolts and the air bag; then, disconnect the electrical connector
6. Remove or disconnect the following:

   - Floor console
   - Front pillar trim at both sides
7. Remove the instrument panel by removing or disconnecting the following:
   - Steering column covers
   - Hood lock release handle

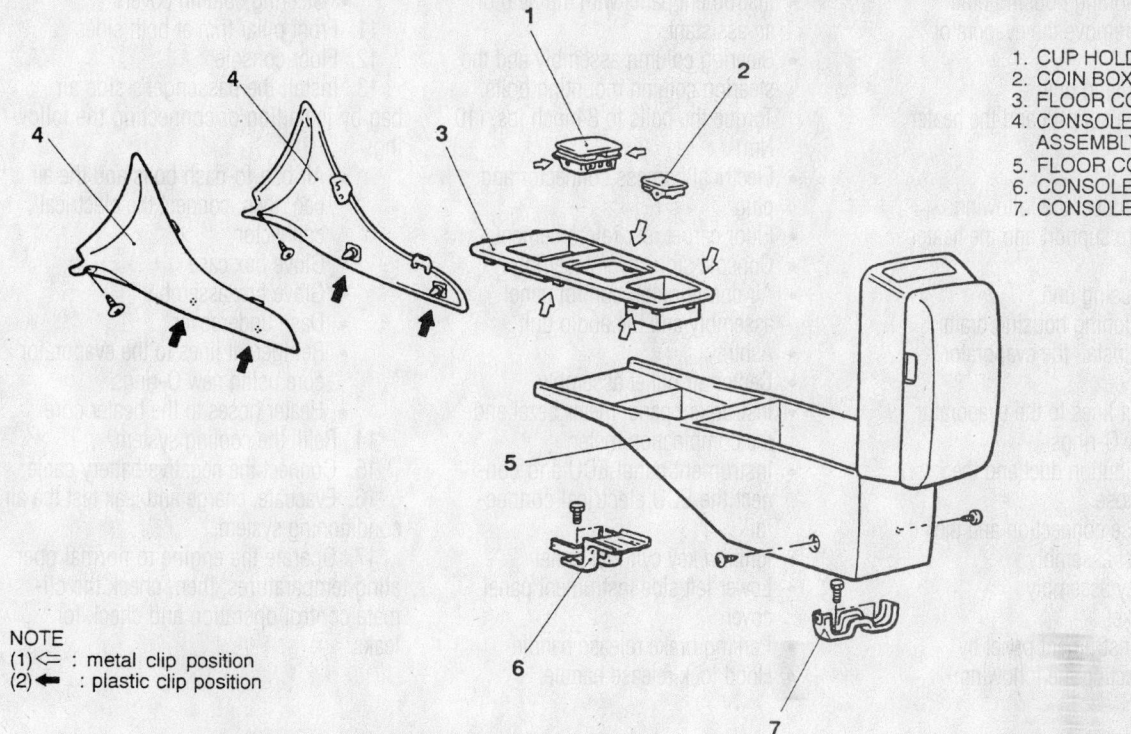

1. CUP HOLDER ASSEMBLY
2. COIN BOX ASSEMBLY
3. FLOOR CONSOLE PANEL
4. CONSOLE SIDE COVER
   ASSEMBLY
5. FLOOR CONSOLE BOX
6. CONSOLE BRACKET A
7. CONSOLE BRACKET C

NOTE
(1) ⇐ : metal clip position
(2) ◄ : plastic clip position

93112GF0

Exploded view of the floor console and related components—Diamante

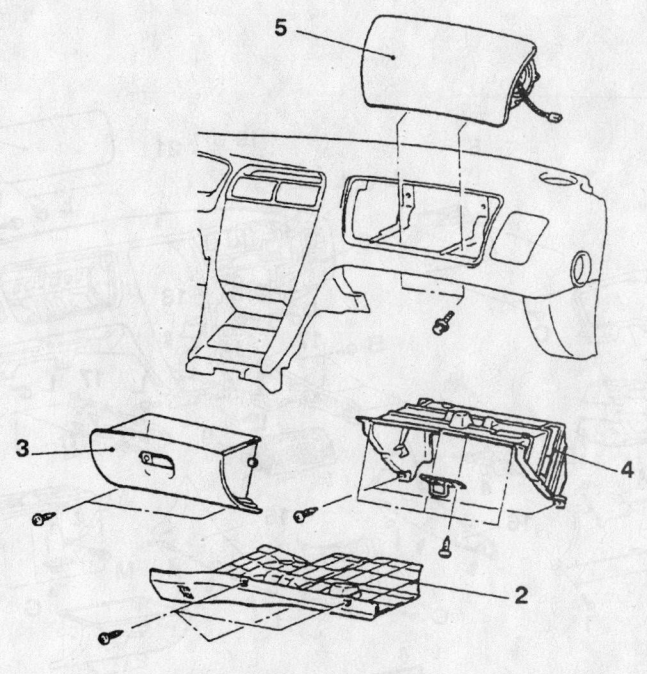

2. UNDERCOVER
3. GLOVE BOX ASSEMBLY
4. GLOVE BOX CASE
5. AIR BAG MODULE

93112GG1

Exploded view of the passenger's air bag module—Diamante

- Parking brake release handle
- Lower left side instrument panel cover
- Ignition key cylinder panel
- Instrument panel ECU and remove the ECU
- Instrument panel meter bezel and the combination meter
- Center air outlet assembly
- Ashtray
- Air conditioning control panel assembly and the audio unit
- Console side cover assembly
- Floor carpet rear reinforcement
- Electrical harness connector and plug
- Steering column mounting bolts and lower the steering column assembly
- Instrument panel with the help of an assistant

8. Remove or disconnect the following:

- ECU bracket
- Center stay assembly
- Heater hose connection and the center duct assembly
- Foot distribution duct and the breather hose
- Refrigerant lines from the evaporator and discard the O-rings

For complete service labor times, order the Chilton Labor Guide

- Air conditioning housing drain hose and remove the evaporator housing
- Heater housing unit
- Heater core support and the heater core

To install:

9. Install or connect the following:
- Heater core support and the heater core
- Heater housing unit
- Air conditioning housing drain hose and Install the evaporator housing
- Refrigerant lines to the evaporator using new O-rings
- Foot distribution duct and the breather hose
- Heater hose connection and the center duct assembly
- Center stay assembly
- ECU bracket

10. Install the instrument panel by installing or connecting the following:

- Instrument panel with the help of an assistant
- Steering column assembly and the steering column mounting bolts. Torque the bolts to 84 inch lbs. (10 Nm)
- Electrical harness connector and plug
- Floor carpet rear reinforcement
- Console side cover assembly
- Air conditioning control panel assembly and the audio unit
- Ashtray
- Center air outlet assembly
- Instrument panel meter bezel and the combination meter
- Instrument panel ECU and connect the ECU electrical connector
- Ignition key cylinder panel
- Lower left side instrument panel cover
- Parking brake release handle
- Hood lock release handle

- Steering column covers
11. Front pillar trim at both sides
12. Floor console
13. Install the passenger's side air bag by installing or connecting the following:

- Air bag-to-dash bolts and the air bag; then, connect the electrical connector
- Glove box case
- Glove box assembly
- Dash undercover
- Refrigerant lines to the evaporator core using new O-rings
- Heater hoses to the heater core

14. Refill the cooling system.
15. Connect the negative battery cable.
16. Evacuate, charge and leak test the air conditioning system.
17. Operate the engine to normal operating temperatures; then, check the climate control operation and check for leaks.

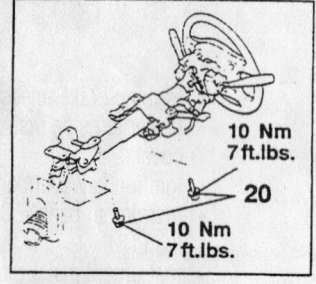

1. COLUMN COVER
2. HOOD LOCK RELEASE HANDLE
3. PARKING BRAKE RELEASE HANDLE
4. INSTRUMENT PANEL LOWER COVER ASSEMBLY (LH)
5. KEY CYLINDER PANEL
6. INSTRUMENT PANEL ECU
7. METER BEZEL
8. COMBINATION METER
9. CENTER AIR OUTLET ASSEMBLY
10. ASHTRAY
11. AIR CONTROL PANEL ASSEMBLY & AUDIO UNIT
12. UNDERCOVER ASSEMBLY
13. GLOVEBOX ASSEMBLY
14. GLOVEBOX OUTER CASE
15. PASSENGER SIDE AIRBAG MODULE
16. CONSOLE SIDE COVER ASSEMBLY
17. FLOOR CARPET REAR REINFORCEMENT
18. HARNESS CONNECTOR
19. PLUG
20. STEERING COLUMN MOUNTIN BOLT
21. INSTRUMENT PANEL

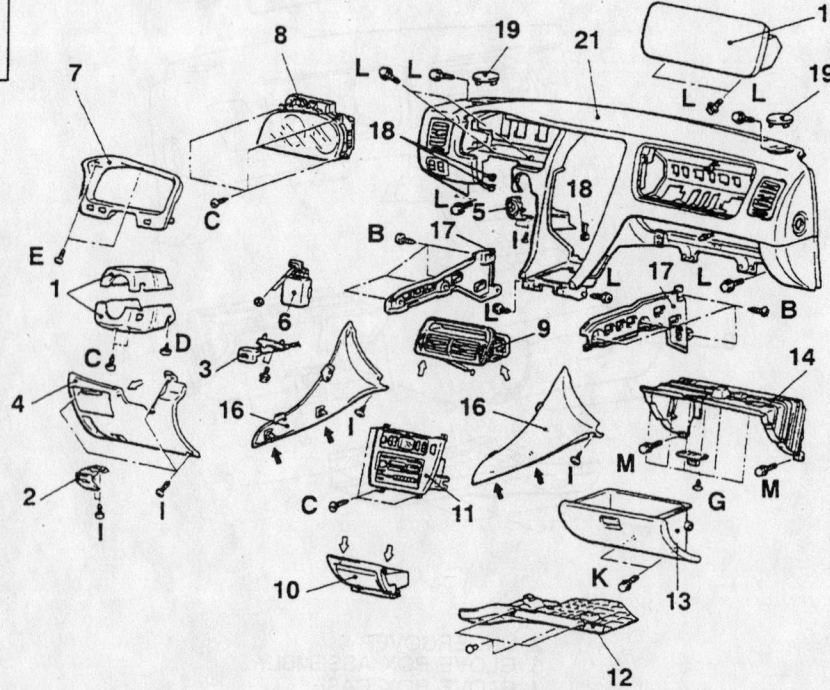

NOTE
(1) ⇦ : metal clip position
(2) ⬅ : plastic clip position

93112GG2

Exploded view of the instrument panel and steering column assembly—Diamante

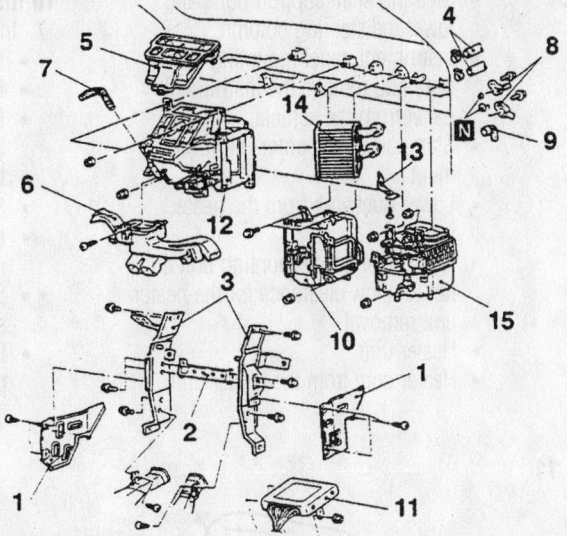

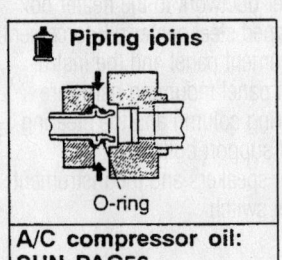

Piping joins

O-ring

A/C compressor oil:
SUN PAG56

1. FLOOR CARPET FRONT REINFORCEMENT
3. ECU BRACKET
4. CENTER STAY ASSEMBLY
5. HEATER HOSE CONNECTION
6. CENTER DUCT ASSEMBLY
7. FOOT DISTRIBUTION DUCT
8. BREATHER HOSE
9. SUCTION PIPE, LIQUID PIPE B AND COOLING UNIT CONNECTION
10. DRAIN HOSE
11. EVAPORATOR
12. ENGINE CONTROL MODULE
13. HEATER UNIT
14. HEATER CORE SUPPORT
15. HEATER CORE

93112GG3

Exploded view of the heater core, heater housing and related components—Diamante

Eclipse

REMOVAL & INSTALLATION

✳✳ CAUTION

If equipped with an air bag, wait for 1 minute after disconnecting the negative battery cable before working inside the vehicle. The air bag system is set to deploy for a short period of time after the battery is disconnected.

1. Disconnect the negative battery cable.
2. Drain the cooling system into a clean container for reuse.
3. Disconnect the heater hoses from the heater core tubes at the firewall. Do not allow the coolant to damage the vehicle speed sensor located below the heater hoses on the non-turbo manual transmission vehicles.

✳✳ WARNING

To prevent damage to the air bag control unit during removal or installation of the floor console, avoid shocks or impact. Do not drop.

4. Remove the floor console by removing or disconnecting the following:
   - Center console trim panel
   - Ashtray and cup holder assembly
   - Shift lever knob on manual transmission

   - Retaining screws
   - Floor console assembly
5. Locate the rectangular plugs in the knee protector on either side of the steering column. Pry these plugs out and remove the screws.
6. Remove or disconnect the following:
   - Driver's side air bag assembly, the

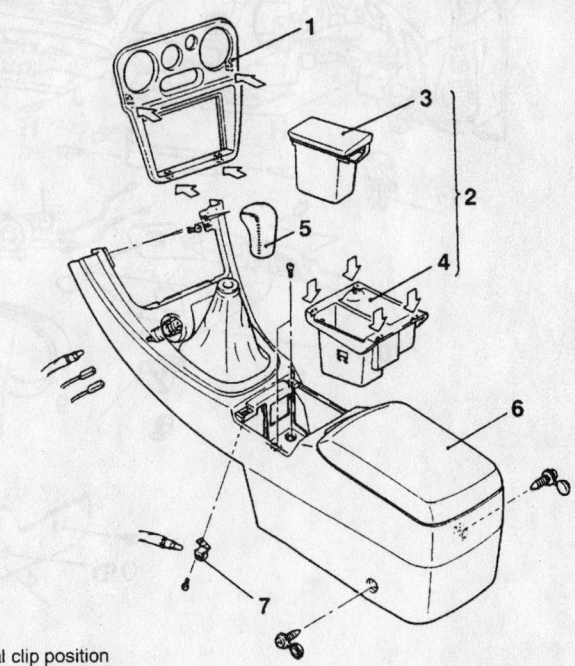

NOTE
◁ : Metal clip position

1. Center console panel
2. Ashtray and cupholder assembly
3. Ashtray
4. Cup holder
5. Shift lever knob <M/T>
6. Floor console assembly
7. Ashtray illumination light bracket

93112G56

Exploded view of the floor console and related components—Eclipse

steering wheel and the passenger's side air bag assembly
- Lap cooler duct and steering column covers
- Instrument cluster bezel and then the instrument cluster
- Radio
- Glove box
- Center air outlet assembly
- Hood release handle and the lower cover
- Heater control assembly
- Front speakers and the instrument panel switch

- Steering shaft support bolts and lower the steering column
- Instrument panel mounting hardware and remove the instrument panel from the vehicle
- Stamped steel center reinforcement
- Lower ductwork from the heater box
- Evaporator case mounting bolt and nut to allow clearance for the heater unit removal
- Heater unit
- Heater core from the heater unit

To install:
7. Install or connect the following:
- Heater core to the heater unit
- Heater unit
- Evaporator case mounting bolt and nut
- Lower ductwork to the heater box
- Stamped steel center reinforcement
- Instrument panel and the instrument panel mounting hardware
- Steering column and the steering shaft support bolts
- Front speakers and the instrument panel switch

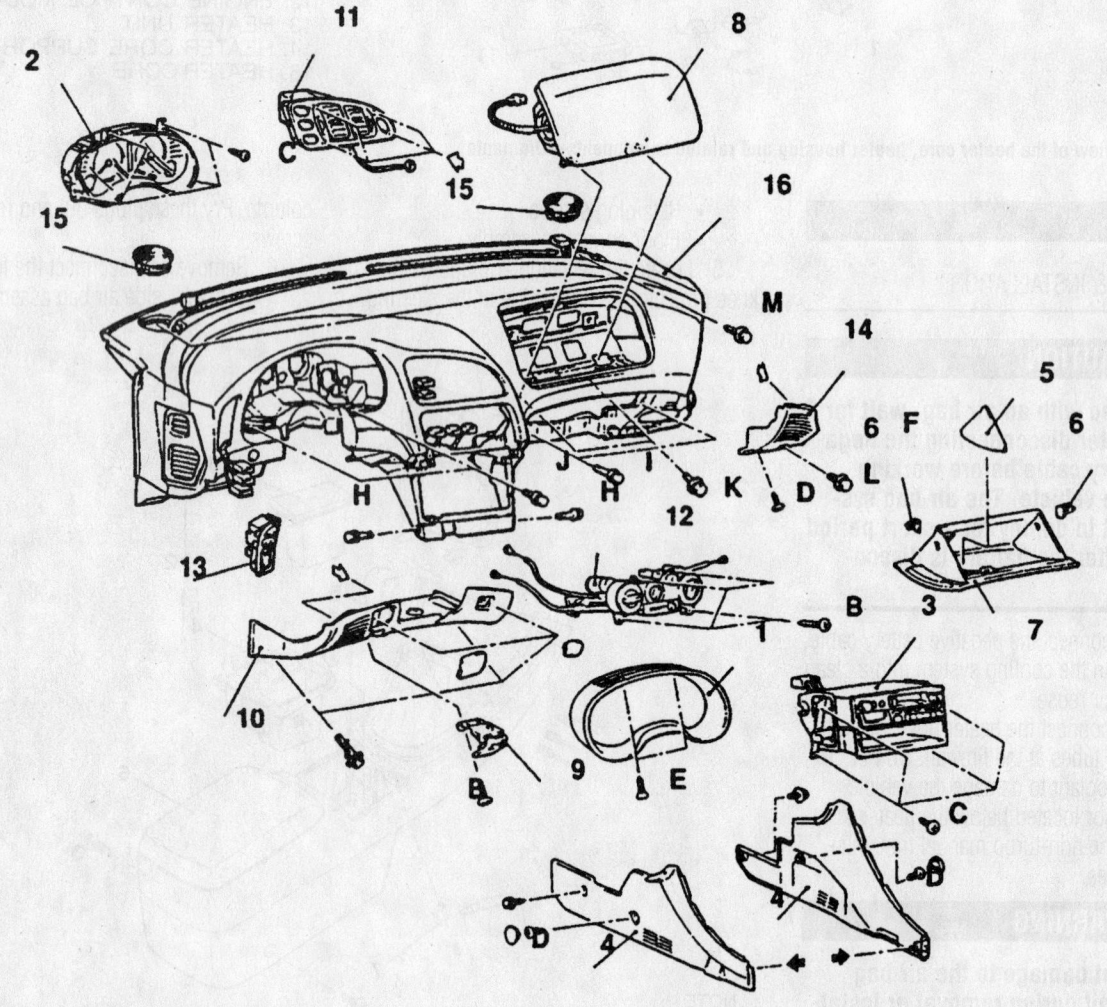

1. Meter bezel
2. Combination meter
3. Radio and tape player, and box
4. Console side cover
5. Sunglasses holder
6. Stopper
7. Glove box
8. Passenger's side air bag module assembly
9. Hood lock release handle
10. Instrument under cover L.H.
11. Center air outlet assembly
12. Heater control assembly
13. Instrument panel switch
14. Instrument under cover R.H.
15. Front speaker
16. Instrument panel assembly

93112G74

Exploded view of the instrument panel and related components—Eclipse

- Heater control assembly
- Hood release handle and the lower cover
- Center air outlet assembly
- Glove box
- Radio
- Instrument cluster and the instrument cluster bezel
- Steering column covers and the lap cooler duct
- Steering wheel, the driver's side air

bag assembly and the passenger's side air bag assembly
- Screws and the rectangular plugs in the knee protector on either side of the steering column

8. Install the floor console by installing or connecting the following:
- Floor console assembly
- Retaining screws
- Shift lever knob on manual transmission

- Ashtray and cup holder assembly
- Center console trim panel
- Center console trim panel

9. Connect the heater hoses to the heater core tubes at the firewall.

10. Refill the cooling system.

11. Connect the negative battery cable.

12. Operate the engine to normal operating temperatures; then, check the climate control operation and check for leaks.

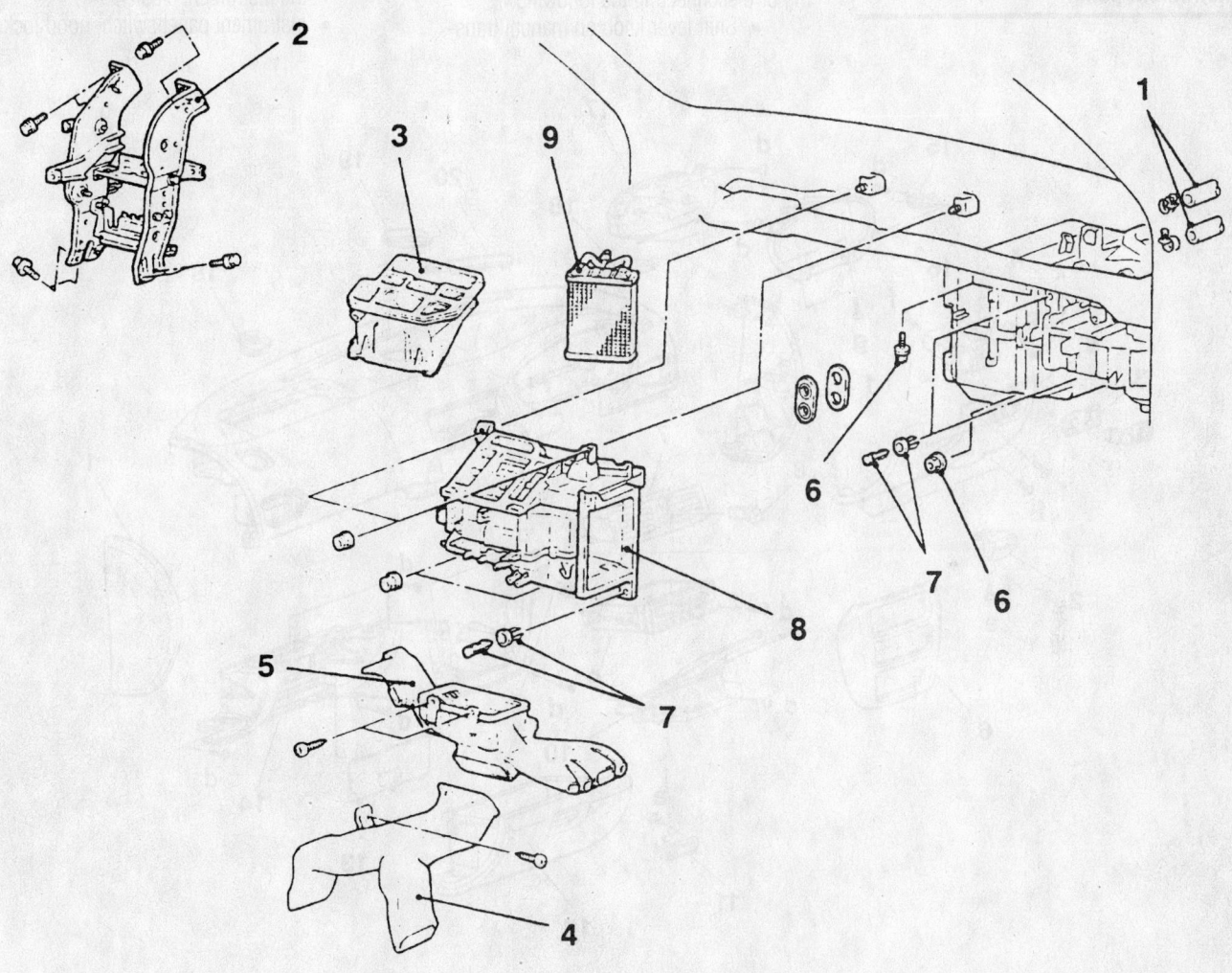

1. Heater hose connection
2. Center stay
3. Center duct
4. Semi rear heater duct
5. Foot distribution duct
6. Cooling unit installation bolt and nut
7. Clip
8. Heater unit
9. Heater core

93112GE0

Exploded view of the heater core, heater case and related components—Eclipse

Timing belt service is covered in Section 3 of this manual

Galant

REMOVAL & INSTALLATION

1. Disconnect the negative battery cable.

✳✳ CAUTION

After disconnecting the negative battery cable, wait at least 60 seconds before working on the SRS module or instrument panel.

2. Drain the cooling system into a clean container for reuse.

3. Disconnect the heater hoses from the heater core at the firewall.

✳✳ WARNING

To prevent damage to the air bag control unit during removal or installation of the floor console, avoid shocks or impact. Do not drop.

4. Remove the floor console by removing or disconnecting the following:
- Shift lever knob on manual trans-

mission vehicles or the shift indicator plate on automatic transmissions
- Coin holder behind the shifter, then the center console trim cover in front of the shifter
- Center console retaining bolt cover plugs, then remove the bolts
- Console assembly, then the brackets
5. Remove or disconnect the following:
- Steering column covers
- Instrument cluster bezel and then the instrument cluster
- Instrument panel switch, hood lock

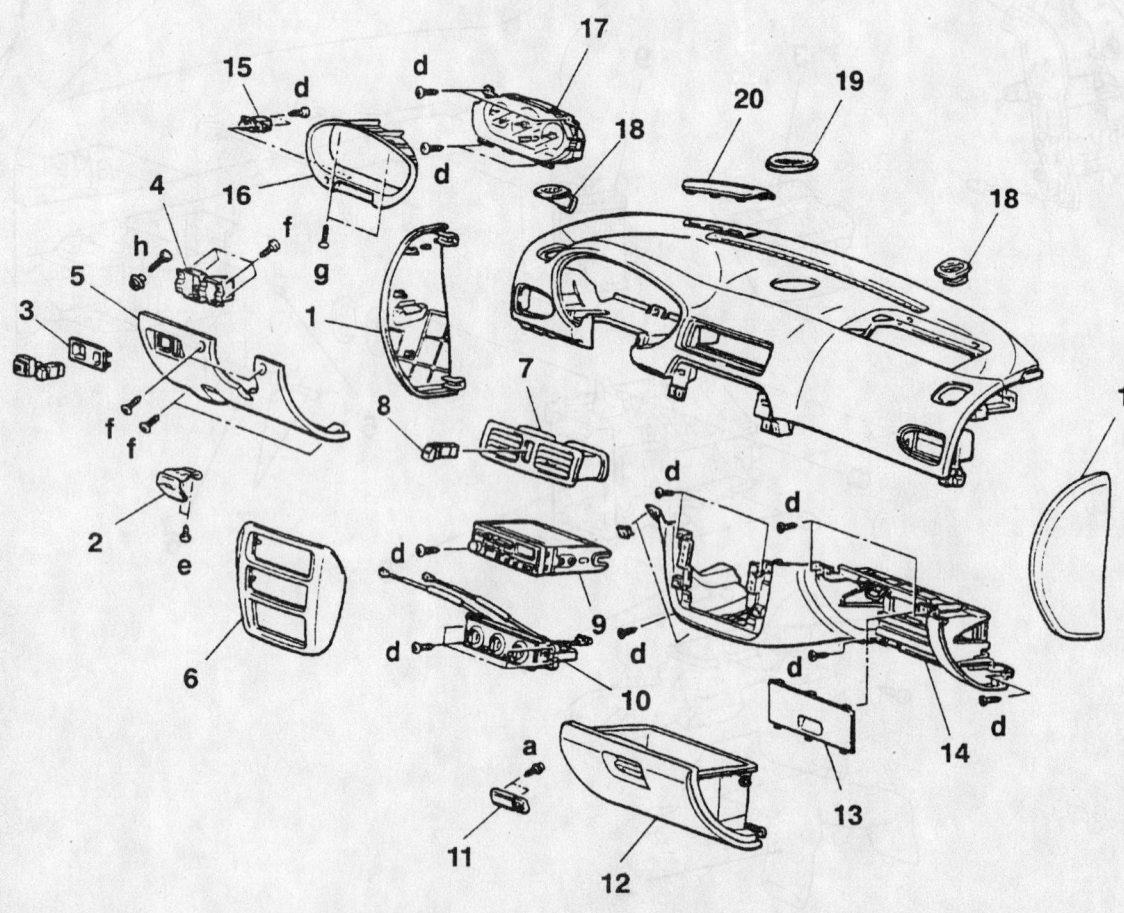

1. INSTRUMENT PANEL SIDE COVER
2. HOOD LOCK RELEASE HANDLE
3. SWITCH PANEL ASSEMBLY
4. CONNECTOR HOLDER
5. FRONT DRIVER'S SIDE UNDER COVER
6. CENTER PANEL ASSEMBLY
7. CENTER AIR OUTLET ASSEMBLY
8. HAZARD WARNING LIGHT SWITCH
9. RADIO AND TAPE PLAYER
10. HEATER CONTROL ASSEMBLY
11. GLOVE BOX STRIKER

12. GLOVE BOX
13. FRONT PASSENGER'S UNDER COVER PLUG
14. FRONT PASSENGER'S SIDE UNDER COVER
15. RHEOSTAT
16. METER BEZEL
17. COMBINATION METER
18. SIDE DEFROSTER GRILLE
19. SPEAKER GRILLE
20. INSTRUMENT PANEL UPPER PLUG

93112GF1

Exploded view of the instrument panel assembly—Galant

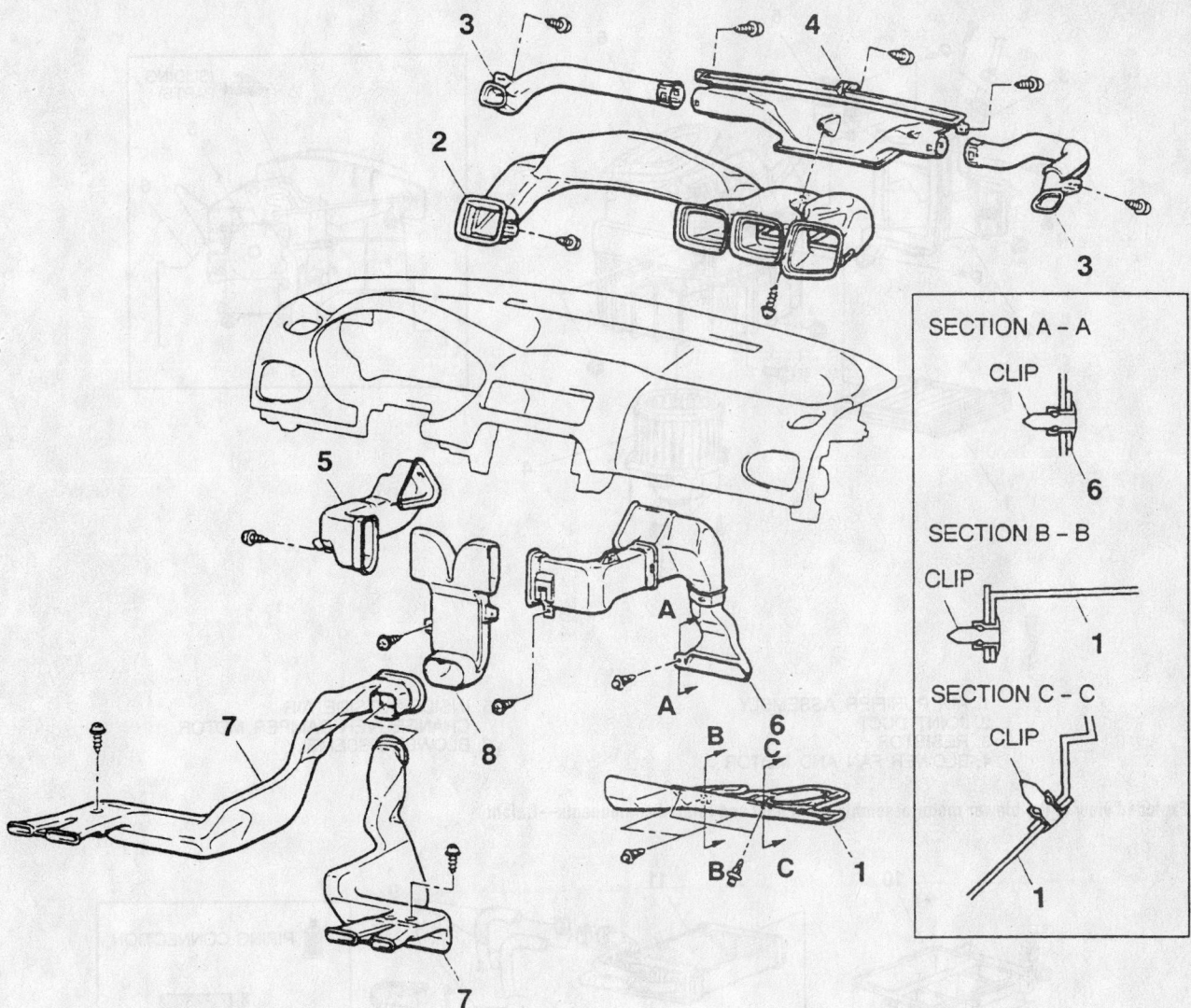

SECTION A – A
CLIP
6

SECTION B – B
CLIP
1

SECTION C – C
CLIP
1

1. UNDER COVER
2. DISTRIBUTION DUCT
3. SIDE DEFROSTER DUCT
4. DEFROSTER NOZZLE ASSEMBLY

5. FOOT DUCT (LH)
6. FOOT DUCT (RH)
7. REAR HEATER DUCT
8. FOOT CENTER DUCT

93112GF2

Exploded view of the ventilator assembly—Galant

release handle and the lower duct work
- Driver's knee protector and the left side air outlet cover
- Center panel assembly
- Glove box undercover, then the glove box and the right side panel cover
- Radio and cup holder
- Cables from the heater assembly and the blower, then pull out the heater control panel assembly, not-

ing the location of the boss in the center reinforcement
- Cool air bypass damper lever cable connection
- Passenger's side air bag module and disconnect the harness connector, if equipped
- Steering column bolts and lower the column
- Harness connector at the lower left side of the instrument panel
- Instrument panel mounting hard-

ware and remove the instrument panel from the vehicle
- Joint duct between the heater case and the blower assembly (on models without air conditioning)
- Both stamped steel center reinforcement piece
- ECM bracket
- Evaporator retaining nut and remove the heater case assembly, if equipped with air conditioning
- Heater core from the case

Heater Core replacement is covered in Section 2 of this manual

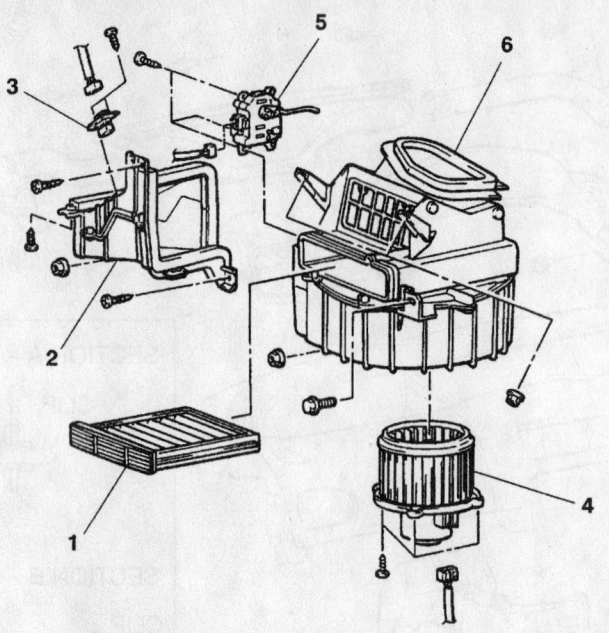

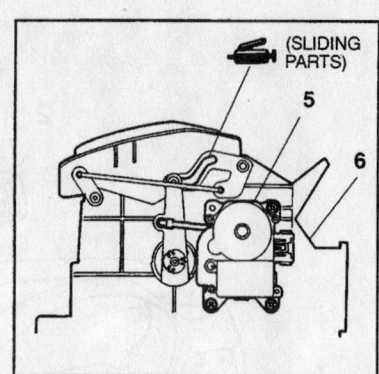

1. AIR PURIFIER ASSEMBLY
2. JOINT DUCT
3. RESISTOR
4. BLOWER FAN AND MOTOR

5. INSIDE/OUTSIDE AIR
   CHANGEOVER DAMPER MOTOR
6. BLOWER ASSEMBLY

93112GF3

Exploded view of the blower motor assembly, joint duct and related components—Galant

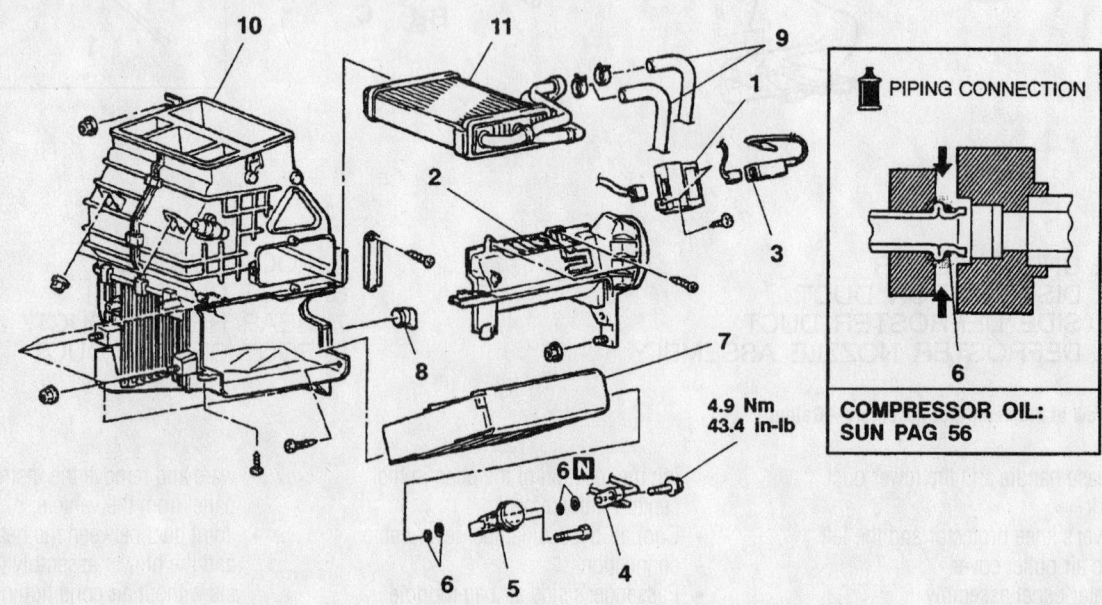

1. BELT LOCK CONTROLLER
2. COVER
3. AUTOMATIC COMPRESSOR
   CONTROLLER
4. A/C PIPE
5. EXPANSION VALVE

6. O-RING
7. EVAPORATOR
8. DRAIN HOSE
9. HEATER HOSE
10. HEATER/COOLER UNIT
11. HEATER CORE

4.9 Nm
43.4 in-lb

PIPING CONNECTION

COMPRESSOR OIL:
SUN PAG 56

93112GF4

Exploded view of the heater core, heater housing and related components—Galant

To install:

6. Install or connect the following:
- Heater core to the case
- Evaporator retaining nut and Install the heater case assembly, if equipped with air conditioning
- ECM bracket
- Both stamped steel center reinforcement pieces
- Joint duct between the heater case and the blower assembly (on models without air conditioning)
- Instrument panel and install the instrument panel mounting hardware
- Harness connector at the lower left side of the instrument panel
- Steering column bolts
- Passenger's side air bag module and connect the harness connector, if equipped
- Cool air bypass damper lever cable connection
- Cables to the heater assembly and the blower and install the heater control panel assembly

- Radio and cup holder
- Glove box undercover, then the glove box and the right side panel cover
- Center panel assembly
- Left side air outlet cover and the driver's knee protector
- Lower duct work, the instrument panel switch and hood lock release handle
- Instrument cluster and the instrument cluster bezel
- Steering column covers

7. Install the floor console installing or connecting the following:
- Console assembly brackets and the console
- Center console retaining bolts, then the bolt cover plugs
- Coin holder behind the shifter, then the center console trim cover in front of the shifter
- Shift lever knob on manual transmission vehicles or the shift indicator plate on automatic transmissions

- Heater hoses to the heater core at the firewall
8. Refill the cooling system.
9. Connect the negative battery cable.
10. Operate the engine to normal operating temperatures; then, check the climate control operation and check for leaks.

Mirage

REMOVAL & INSTALLATION

1. Disconnect the negative battery cable.
2. Drain the cooling system into a clean container for reuse.
3. Remove the air cleaner cover and the air intake hose.
4. Disconnect the heater hoses from the heater core.
5. If equipped with air conditioning, discharge and recover the air conditioning system refrigerant.
6. If equipped with air conditioning, disconnect the refrigerant lines from the evaporator core and discard the O-rings.

NOTE
⟵ : metal clip position

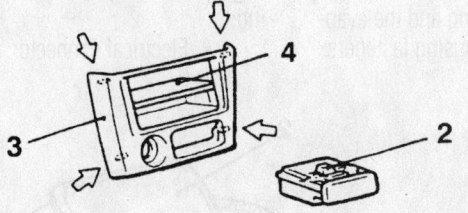

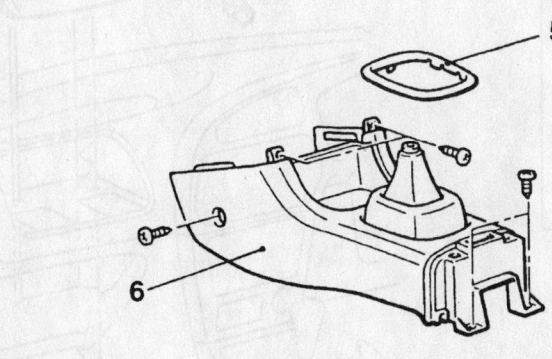

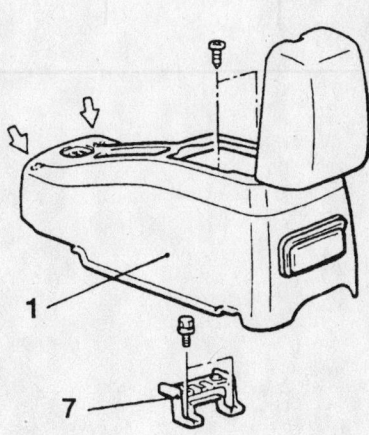

1. Rear floor console assembly
2. Ashtray
3. Audio panel
4. Box
- Shift lever knob
5. A/T panel
6. Front floor console assembly
7. Rear console bracket

93112GF5

Exploded view of the floor console and related components—Mirage

Brake service is covered in Section 4 of this manual

✳✳ CAUTION

After disconnecting the negative battery cable, wait at least 60 seconds before working on the SRS module or instrument panel.

7. Remove the passenger's side air bag by removing or disconnecting the following:
- Glove box assembly
- Air bag-to-dash bolts and the air bag; then, disconnect the electrical connector

8. Remove the floor console.

9. Remove the instrument panel by removing or disconnecting the following:
- Rheostat
- Hood release handle
- Knee protector plug and the knee protector assembly
- Steering column cover
- Meter bezel and the combination meter
- Mirror control switch or plug, if equipped
- Auto-cruise main switch, fog light switch or plug, if equipped
- Side air outlet assembly
- Radio and tape player
- Cup holder

- Heater control panel
- Heater control assembly
- Steering column bolts and lower the steering column
- Instrument panel assembly with the help of an assistant

10. If not equipped with air conditioning, remove the blower motor-to-heater housing joint duct.

11. If equipped with air conditioning, remove the evaporator housing-to-heater housing fasteners and remove the evaporator housing.

12. Remove or disconnect the following:
- Center reinforcement
- Center ventilation duct
- Foot distribution duct
- Heater housing
- Heater core from the heater housing

To install:

13. Install or connect the following:
- Heater core to the heater housing
- Heater housing
- Foot distribution duct
- Center ventilation duct
- Center reinforcement

14. If equipped with air conditioning, install the evaporator housing and the evaporator housing-to-heater housing fasteners.

15. If not equipped with air conditioning, install the blower motor-to-heater housing joint duct.

16. Install the instrument panel by installing or connecting the following:
- Instrument panel assembly with the help of an assistant
- Steering column and install the steering column bolts
- Heater control assembly
- Heater control panel
- Cup holder
- Radio and tape player
- Side air outlet assembly
- Auto-cruise main switch, fog light switch or plug, if equipped
- Mirror control switch or plug, if equipped
- Combination meter and the meter bezel
- Steering column cover
- Knee protector assembly and the knee protector plug
- Hood release handle
- Rheostat
- Floor console

17. Install the passenger's side air bag by installing or connecting the following:
- Electrical connector; then, install

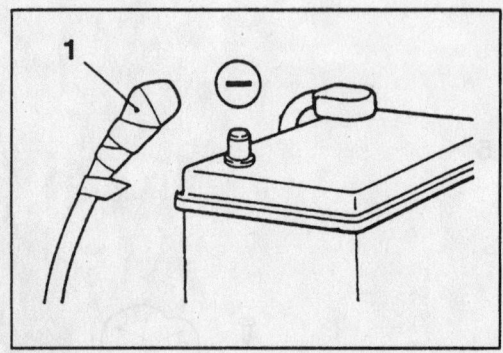

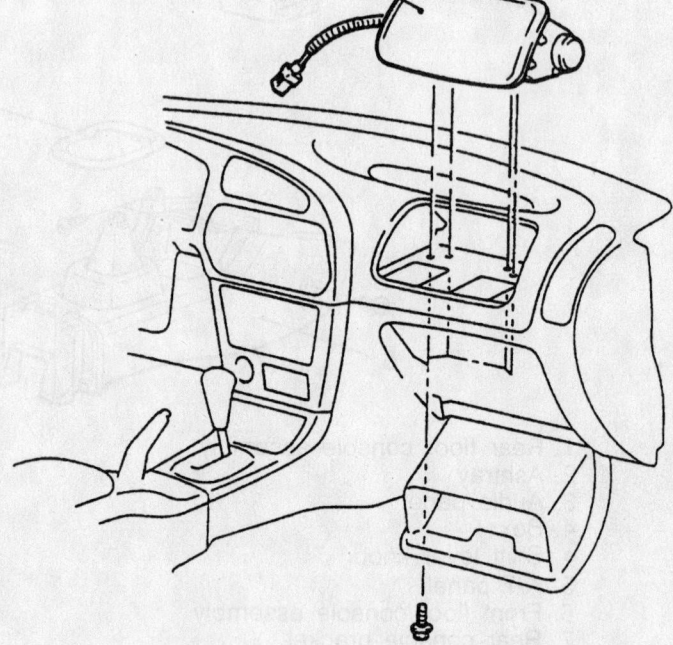

1. Negative (–) battery cable connection
2. Air bag module

93112GF6

Exploded view of the passenger's air bag module—Mirage

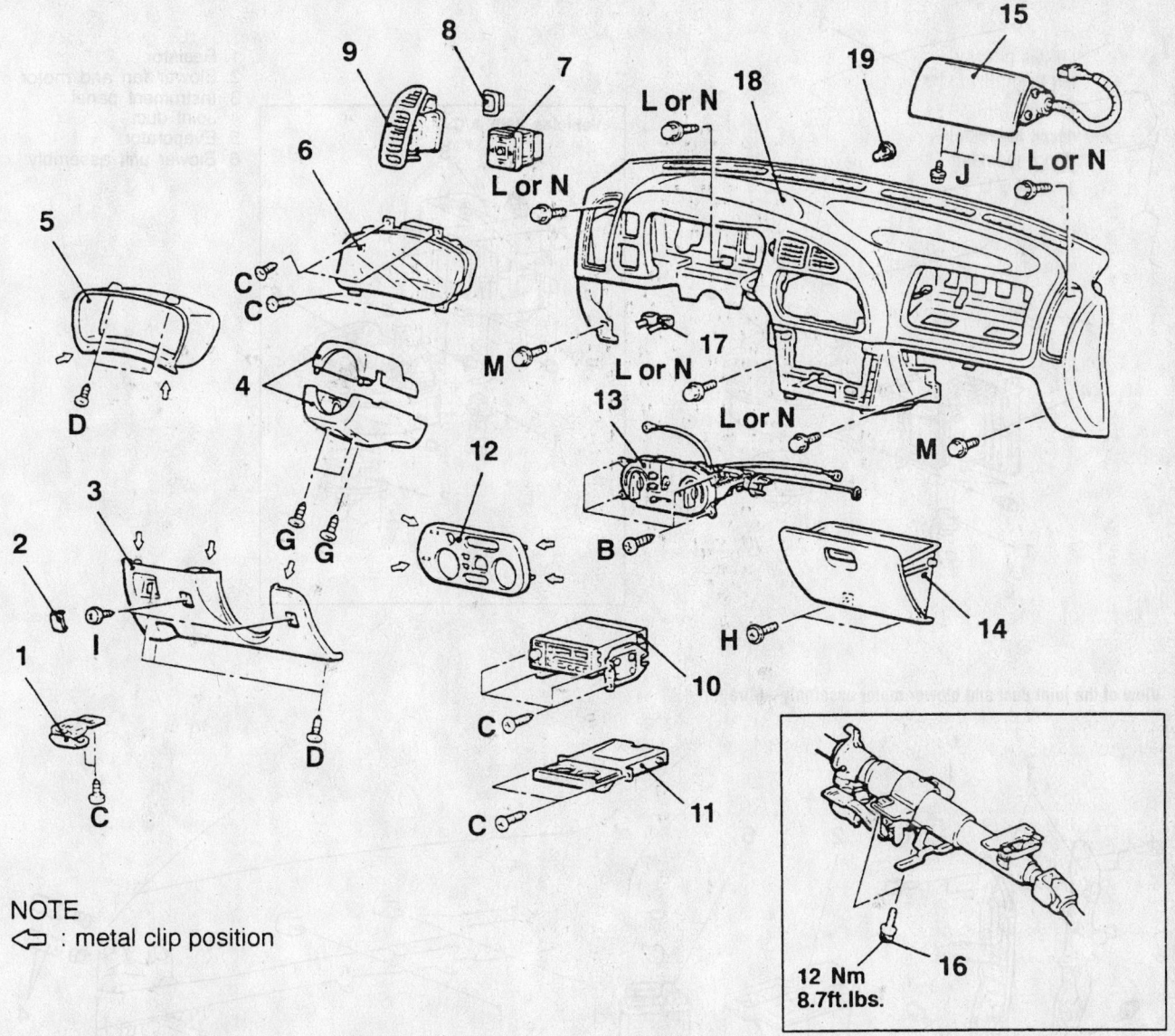

NOTE
◁ : metal clip position

1. Hood lock release handle
2. Knee protector plug
3. Knee protector assembly
4. Column cover
5. Meter bezel
6. Combination meter
7. Door mirror control switch or plug
8. Auto-cruise control main switch, fog light switch or plug
9. Side air outlet assembly
10. Radio and tape player
11. Cup holder
12. Heater control panel
13. Heater control assembly
14. Glove box
15. Front passenger's air bag module assembly
16. Steering column assembly installation bolt
17. Harness connector
18. Instrument panel assembly
19. Grommet

93112GF7

Exploded view of the instrument panel and steering column assembly—Mirage

For complete Engine Mechanical specifications, see Section 1 of this manual

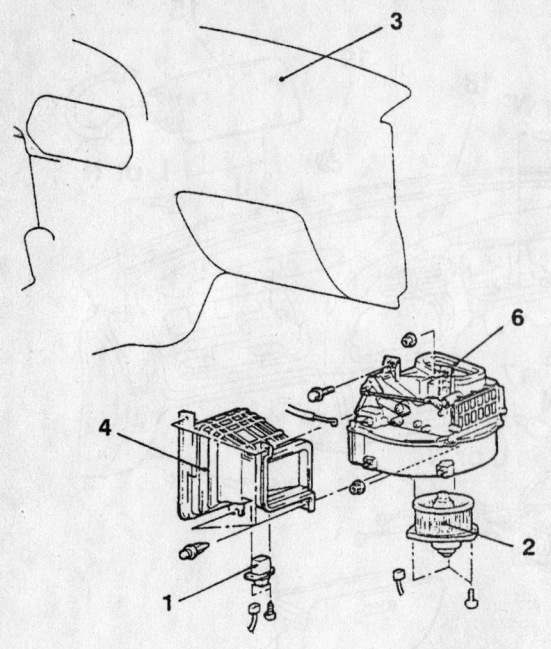

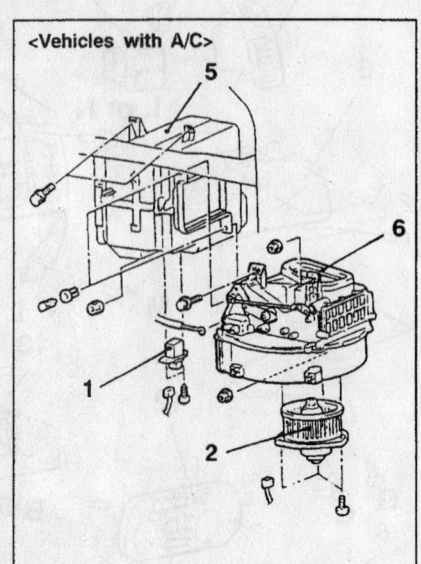

<Vehicles with A/C>

1. Resistor
2. Blower fan and motor
3. Instrument panel
4. Joint duct
5. Evaporator
6. Blower unit assembly

93112GF9

View of the joint duct and blower motor assembly—Mirage

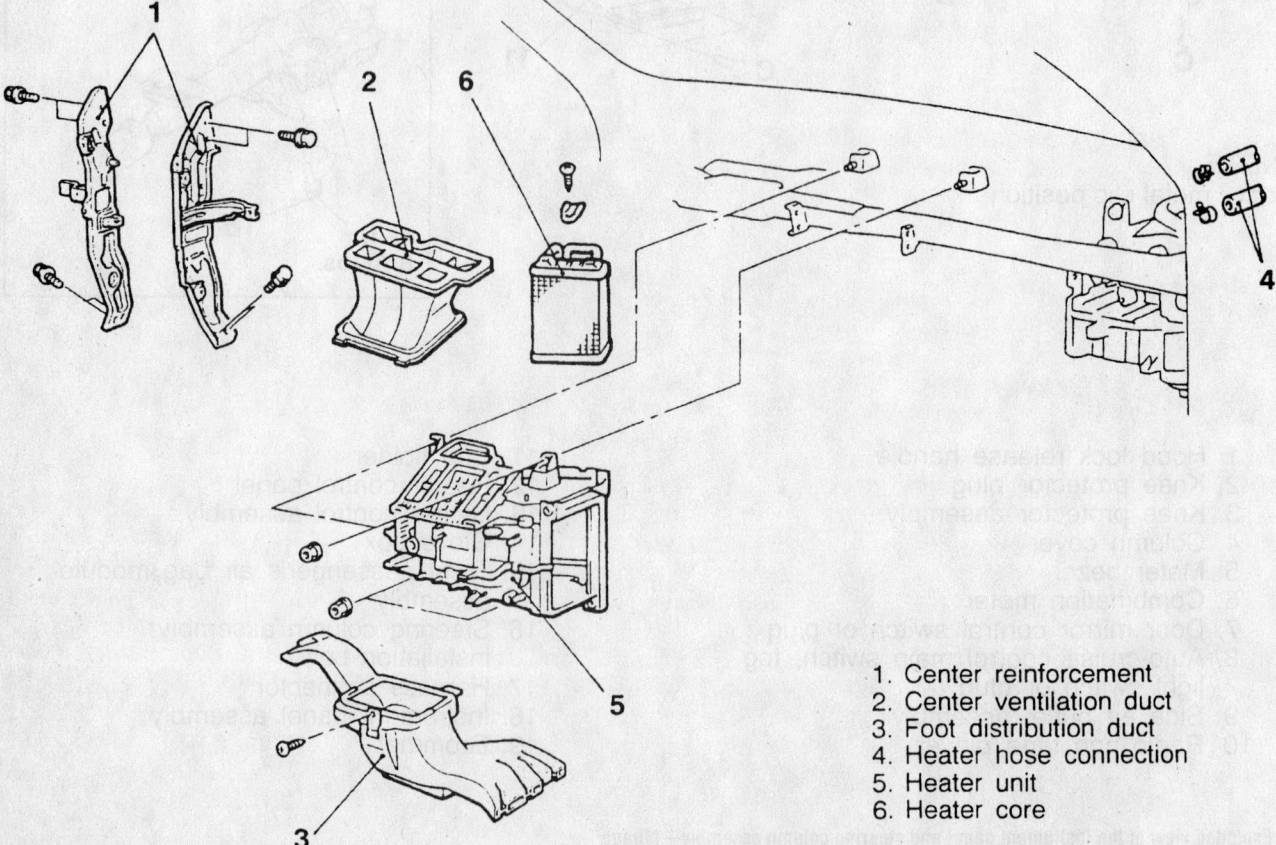

1. Center reinforcement
2. Center ventilation duct
3. Foot distribution duct
4. Heater hose connection
5. Heater unit
6. Heater core

93112GF8

Exploded view of the heater core, heater housing and related components—Mirage

the air bag and the air bag-to-dash
bolts
- Glove box assembly

18. If equipped with air conditioning,
use new O-rings and connect the refrigerant
lines to the evaporator core.
19. Install or connect the following:
- Heater hoses to the heater core
- Air cleaner cover and the air intake
hose

20. Refill the cooling system.
21. Connect the negative battery
cable.
22. If equipped with air conditioning,

evacuate, charge and leak test the air conditioning system refrigerant.
23. Operate the engine to normal operating temperatures; then, check the climate
control operation and check for leaks.

3000GT

REMOVAL & INSTALLATION

1. Disconnect the negative battery cable.
2. Drain the cooling system into a clean
container for reuse.

3. Disconnect the heater hoses from the
heater core.
4. Disconnect the refrigerant lines from
the evaporator core and discard the O-rings.

✲✲ CAUTION

**After disconnecting the negative bat-
tery cable, wait at least 60 seconds
before working on the SRS module or
instrument panel.**

5. Remove the passenger's side air bag
by removing or disconnecting the following:
- Glove box assembly

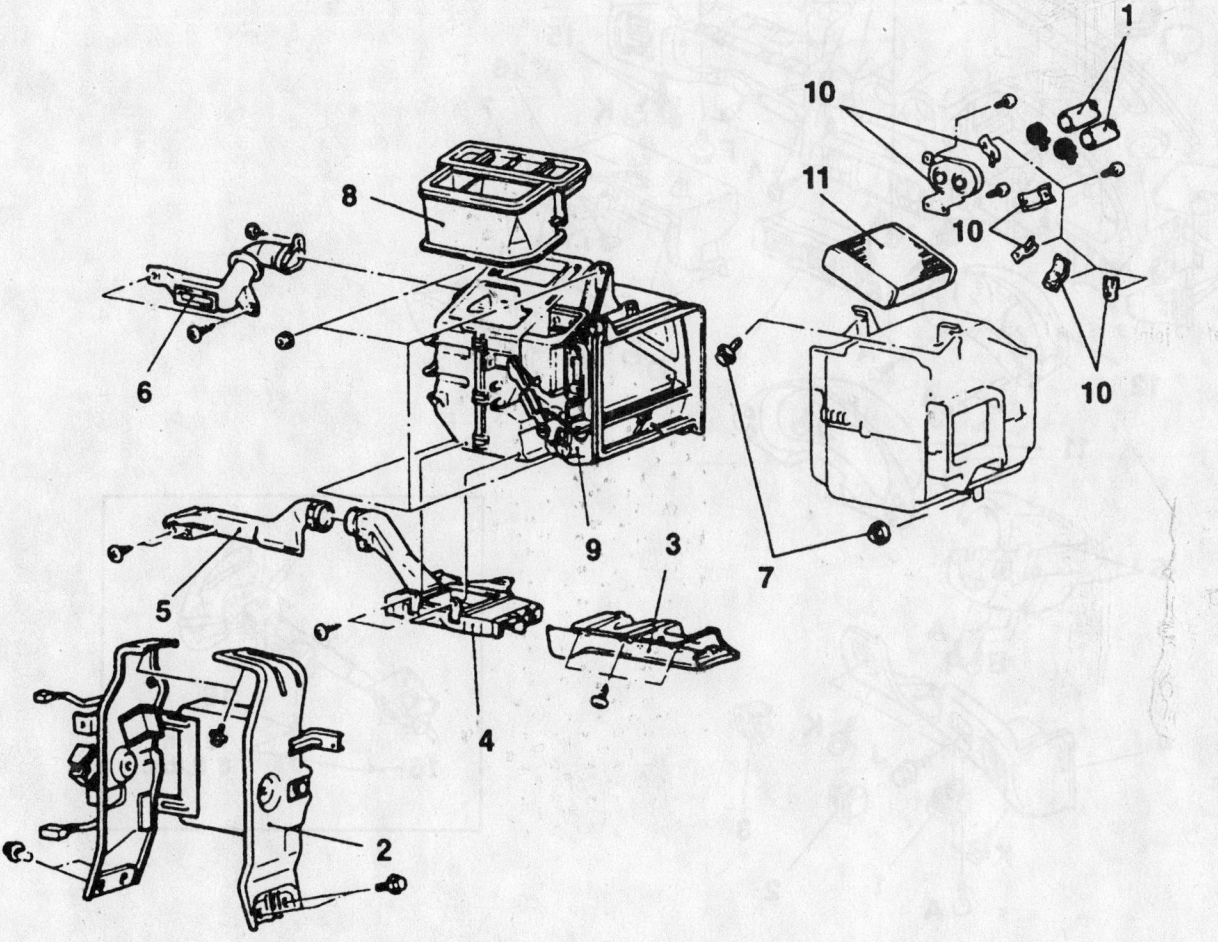

1. Water hoses connection
2. Center reinforcement
3. Under cover
4. Distribution duct (foot)
5. Foot shower duct
6. Lap cooler duct

7. Evaporator mounting bolt and nut
   <Vehicles with air conditioning>
8. Center duct
9. Heater unit
10. Plate
11. Heater core

93112G71

Exploded view of the heater case and related components—Mitsubishi 3000GT

For Accessory Drive Belt illustrations, see Section 1 of this manual

- Cross pipe cover
- Air bag-to-dash bolts and the air bag; then, disconnect the electrical connector
6. Remove the floor console.

7. Remove the instrument panel by removing or disconnecting the following:
- Hood lock release handle
- Rheostat
- Switch garnish "B"

- Knee protector assembly
- Steering column cover
- Center outlet assembly
- Heater control assembly-to-dash screws

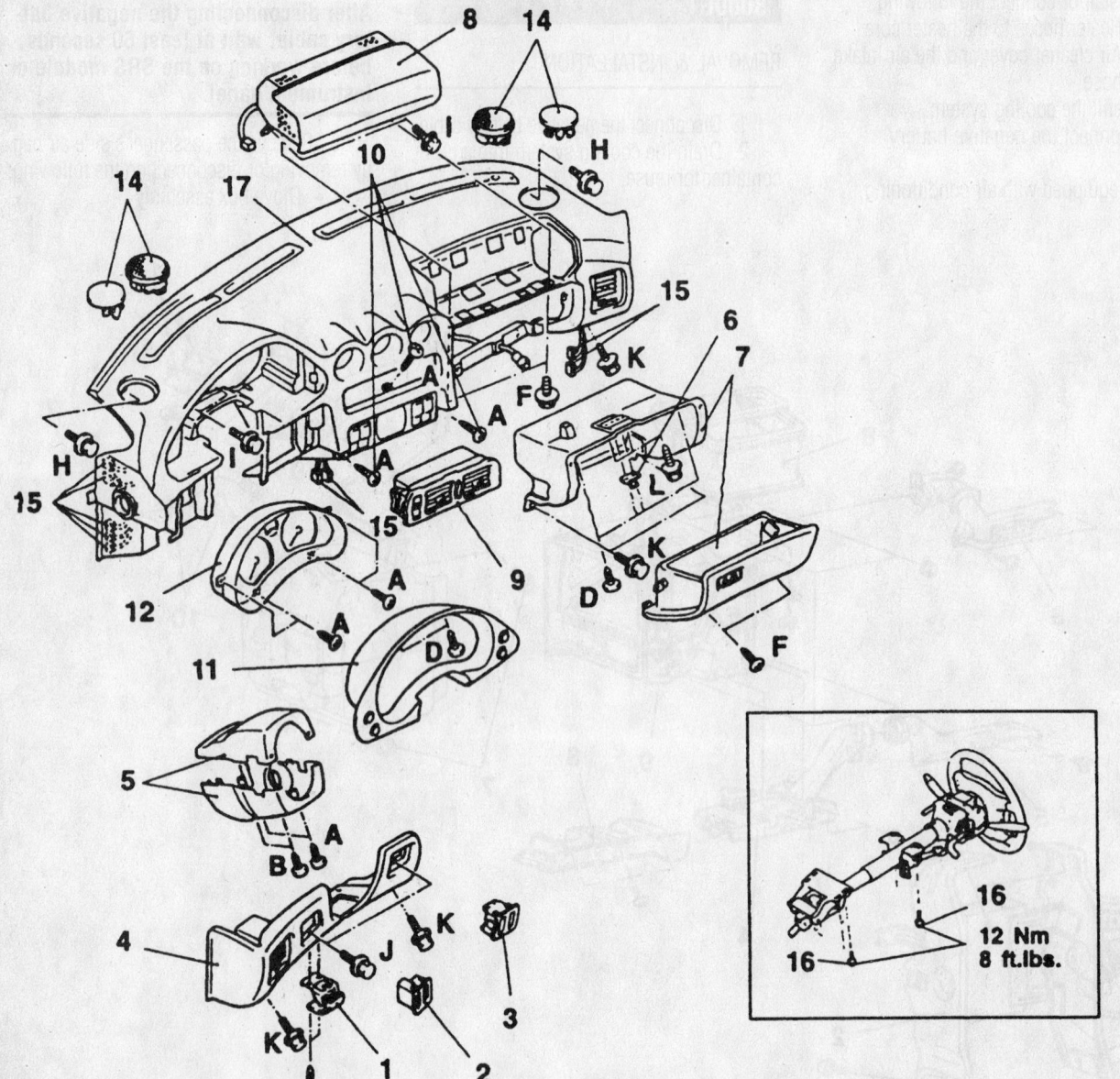

Removal steps

1. Hood lock release handle
2. Rheostat
3. Switch garnish B
4. Knee protector assembly
5. Column cover
6. Glove box striker
7. Glove box and cross pipe cover
8. Passenger seat air bag module

9. Center air outlet assembly
10. Heater control assembly installation screws
11. Meter bezel
12. Combination meter
14. Speaker or plug
15. Harness connector
16. Steering shaft mounting bolts
17. Instrument panel assembly

Exploded view of the instrument panel and related components—Mitsubishi 3000GT

93112G72

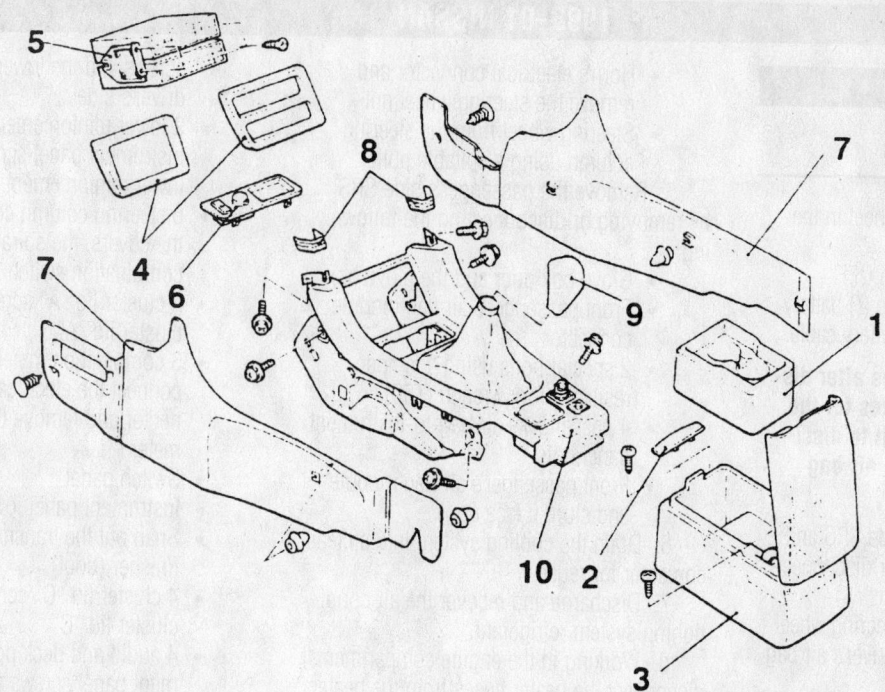

1. Cup holder
2. Console plug
3. Rear console assembly
4. Radio panel
5. Radio
6. Switch garnish C
7. Console side cover
8. Front console garnish
9. Manual transaxle shift lever knob
10. Front console assembly

93112GE4

Exploded view of the floor console and related components—Mitsubishi 3000GT

- Meter bezel and the combination bezel
- Speaker or plug and disconnect the harness connector
- Steering column-to-dash bolts and lower the steering column
- Instrument panel-to-dash bolts
- Instrument panel from the vehicle with the help of an assistant

8. Install or connect the following:
  - Center reinforcement
  - Under cover
  - Foot distribution duct
  - Foot shower duct
  - Evaporator housing-to-heater housing nut and bolt
  - Center duct
  - Heater housing unit
  - Heater housing plate and the heater core from the housing

To install:
9. Install or connect the following:

- Heater housing plate and the heater core to the housing
- Heater housing unit
- Center duct
- Evaporator housing-to-heater housing nut and bolt
- Foot shower duct
- Foot distribution duct
- Under cover
- Center reinforcement

10. Install the instrument panel by installing or connecting the following:
  - Instrument panel to the vehicle with the help of an assistant
  - Instrument panel-to-dash bolts
  - Steering column-to-dash and the steering column bolts
  - Speaker or plug and connect the harness connector
  - Meter bezel and the combination bezel

- Heater control assembly-to-dash screws
- Center outlet assembly
- Steering column cover
- Knee protector assembly
- Switch garnish "B"
- Rheostat
- Hood lock release handle

11. Install the floor console.
12. Install the passenger's side air bag by installing or connecting the following:
  - Air bag-to-dash bolts and the air bag; then, connect the electrical connector
  - Cross pipe cover
  - Glove box assembly

13. Using new O-rings, connect the refrigerant lines to the evaporator core.
14. Connect the heater hoses to the heater core.
15. Refill the cooling system.
16. Connect the negative battery cable.

1998–01 NISSAN

Altima

REMOVAL & INSTALLATION

1. Position the steering wheel in the straight-ahead position.

2. Turn the ignition switch OFF.

3. Disconnect the negative (()) battery cable; then, the positive (+) battery cable.

➡ **Wait for a least 3 minutes after disconnecting the battery cables for the charge in the air bag circuit to dissipate before working on the air bag module(s).**

4. Remove the driver's side SRS and steering wheel by removing or disconnecting the following:

- Lower lid from the steering wheel and disconnect the driver's air bag module connector
- Left and right side lids from the steering wheel
- Special bolts from both side of the steering wheel using a tamper resistant Torx wrench (T50)
- Air bag module and store it face up

- Horn's electrical connector and remove the steering wheel nut
- Steering wheel from the steering column using a suitable puller

5. Remove the passenger's side SRS by removing or disconnecting the following:

- Glove box door and the glove box
- Front passenger's air bag module connector
- 2 special bolts using a tamper resistant Torx wrench (T50)
- 4 passenger's air bag-to-instrument panel nuts
- Front passenger's air bag module and store it face up.

6. Drain the cooling system into a clean container for reuse.

7. Discharge and recover the air conditioning system refrigerant.

8. Working in the engine compartment, disconnect the heater hoses from the heater core tubes.

9. Remove the instrument panel by removing or disconnecting the following:

- Kick plate and dash side finisher on the driver's side
- 2 lower panel-to-instrument panel

screws and the lower panel on the driver's side
- 2 lower reinforcement panel-to-instrument panel screws and the lower reinforcement panel
- 6 steering column cover screws, the covers, the spiral cable and combination switch
- 2 cluster lid "A" screws and the cluster lid "A"
- 3 combination meter screws, disconnect the electrical harness connector and remove the combination meter
- Switch panel
- Instrument panel lower covers
- Snap out the transmission shifter finisher (boot)
- 4 cluster lid "C" screws and the cluster lid "C"
- 4 audio and deck pocket-to-instrument panel screws and the audio and deck pocket
- 5 center console screws and the center console
- 2 center instrument panel screws and the center panel
- front defroster grilles

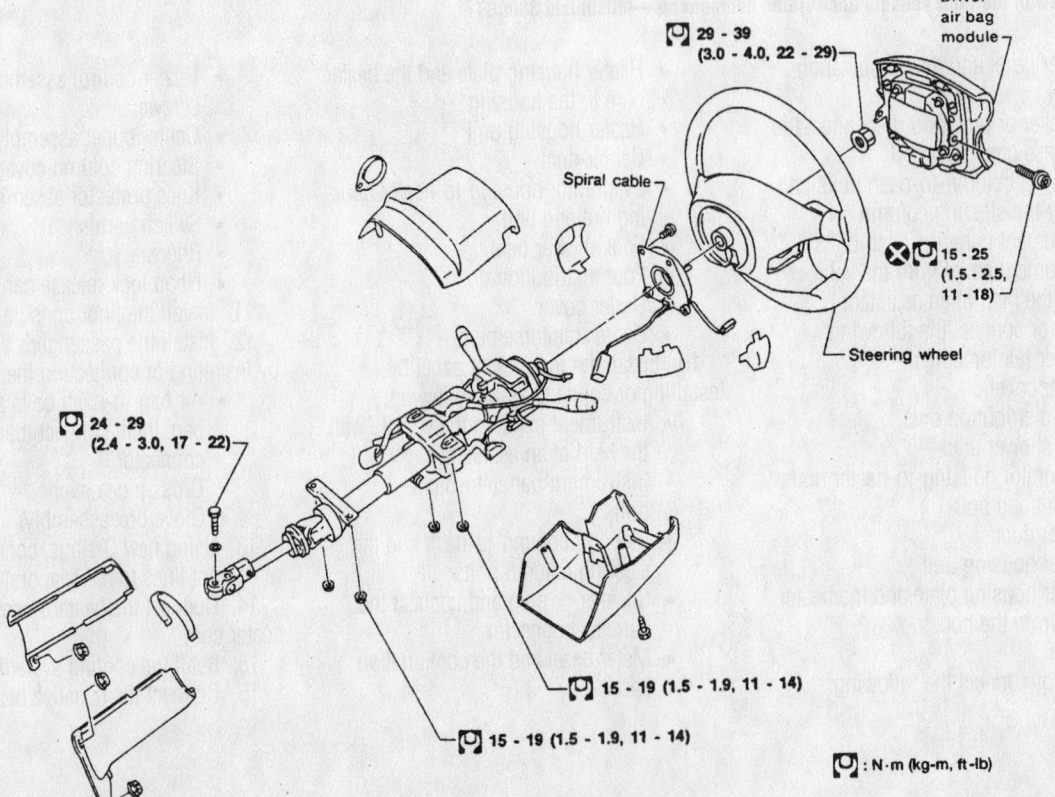

Exploded view of the steering wheel and air bag module

29 - 39 (3.0 - 4.0, 22 - 29)

Driver air bag module

Spiral cable

⊗ 15 - 25 (1.5 - 2.5, 11 - 18)

Steering wheel

24 - 29 (2.4 - 3.0, 17 - 22)

15 - 19 (1.5 - 1.9, 11 - 14)

15 - 19 (1.5 - 1.9, 11 - 14)

: N·m (kg-m, ft-lb)

93112GD9

- Front pillar garnish
- Instrument panel 3 nuts/4 screws and the instrument panel
- 8 instrument stay assembly nuts and the stay
- Steering member assembly 5 nuts/1 bolt and the steering member

10. Remove the air conditioning housing assembly by removing or disconnecting the following:
- Refrigerant lines from the air conditioning housing assembly

- Thermo control amp
- Air conditioning housing assembly

11. Remove or disconnect the following:
- Heater unit
- Heater core from the heater unit

To install:

12. Install or connect the following:
- Heater core to the heater unit
- Heater unit

13. Install the air conditioning housing assembly by installing or connecting the following:

- Air conditioning housing assembly
- Thermo control amp
- Refrigerant lines to the air conditioning housing assembly

14. Install the instrument panel by installing or connecting the following:
- Steering member assembly and the steering member 5 nuts/1 bolt
- Instrument stay assembly and the 8 stay nuts
- Instrument panel and the instrument panel 3 nuts/4 screws

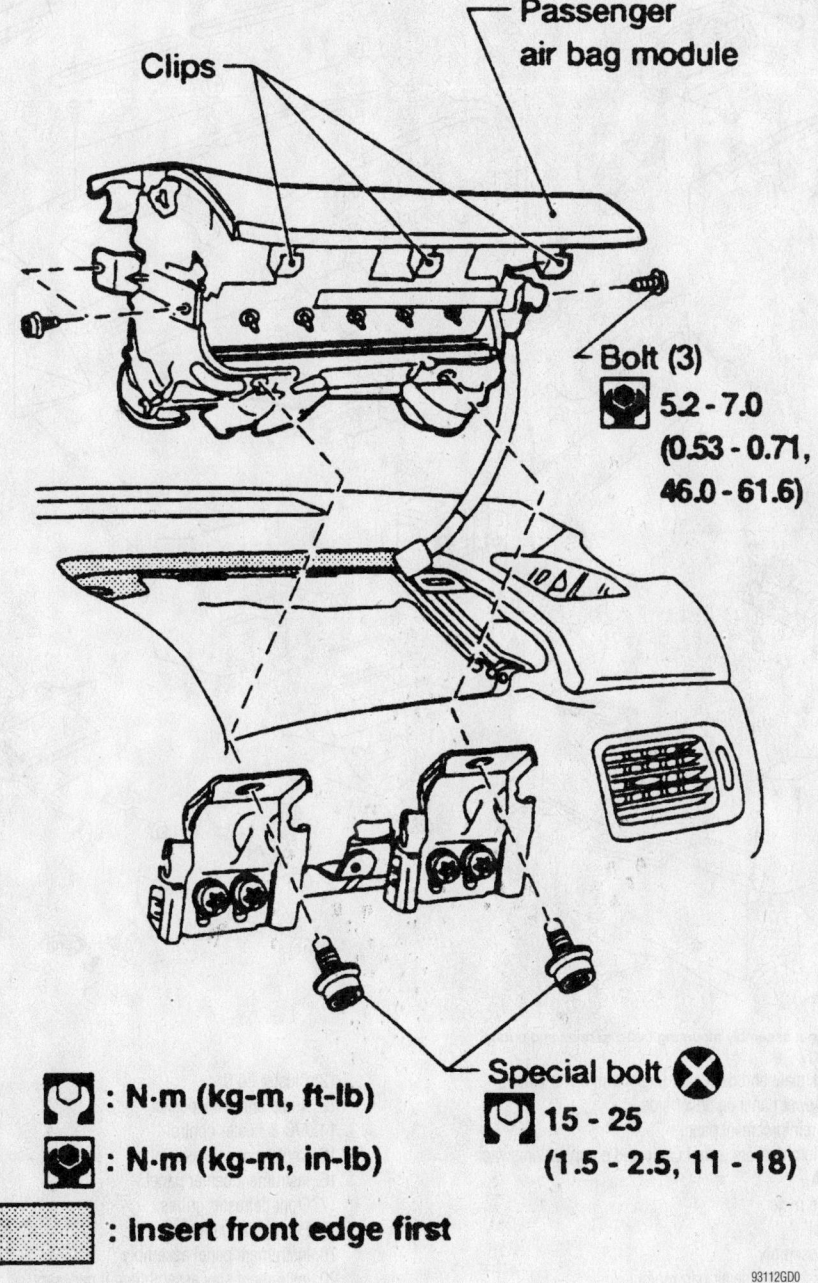

Exploded view of the passenger's side air bag module

93112GD0

For Wheel Alignment specifications, see Section 1 of this manual

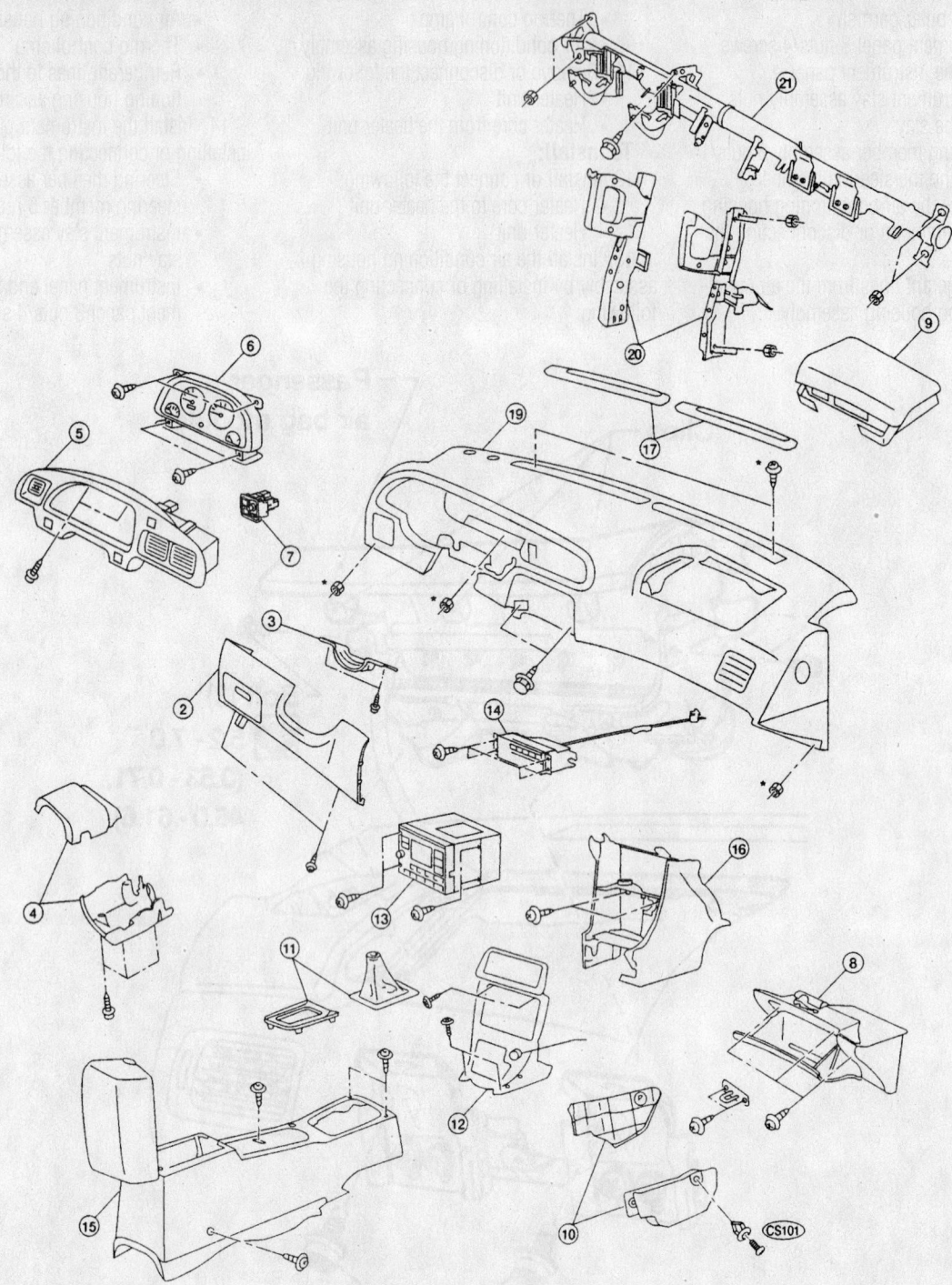

*: Instrument panel assembly mounting bolts, screws and nuts.

1. Remove kick plate and dash side finisher on driver side
2. Instrument lower panel on driver side
3. Dash lower reinforcement panel
4. Steering column covers, spiral cable and combination switch
5. Cluster lid A
6. Combination meter
7. Switch panel
8. Glove box assembly
9. Remove passenger side air bag moldule
10. Instrument lower covers
11. A/T finisher or M/T boot

12. Cluster lid C
13. Audio and deck pocket
14. A/C & heater control
15. Center console assembly
16. Instrument center panel
17. Front defroster grilles
18. Front pillar garnish
19. Instrument panel assembly
20. Instrument stay assemblies, if necesary
21. Steering member assembly, if necessary

93112GE1

Exploded view of the instrument panel assembly

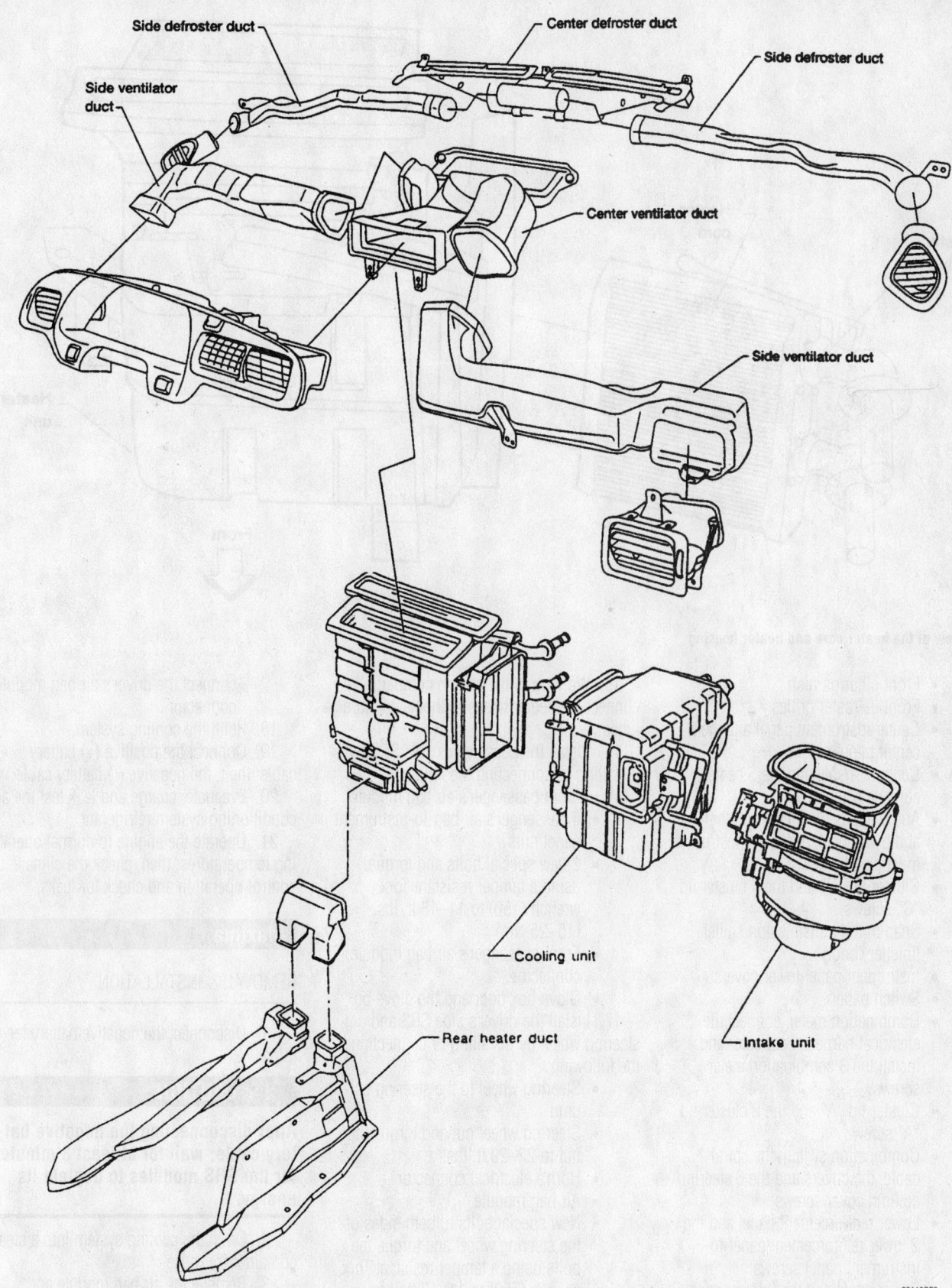

Side defroster duct

Center defroster duct

Side defroster duct

Side ventilator duct

Center ventilator duct

Side ventilator duct

Cooling unit

Intake unit

Rear heater duct

93112GE2

Exploded view of the heater housing assembly and related components

For Maintenance Interval recommendations, see Section 1 of this manual

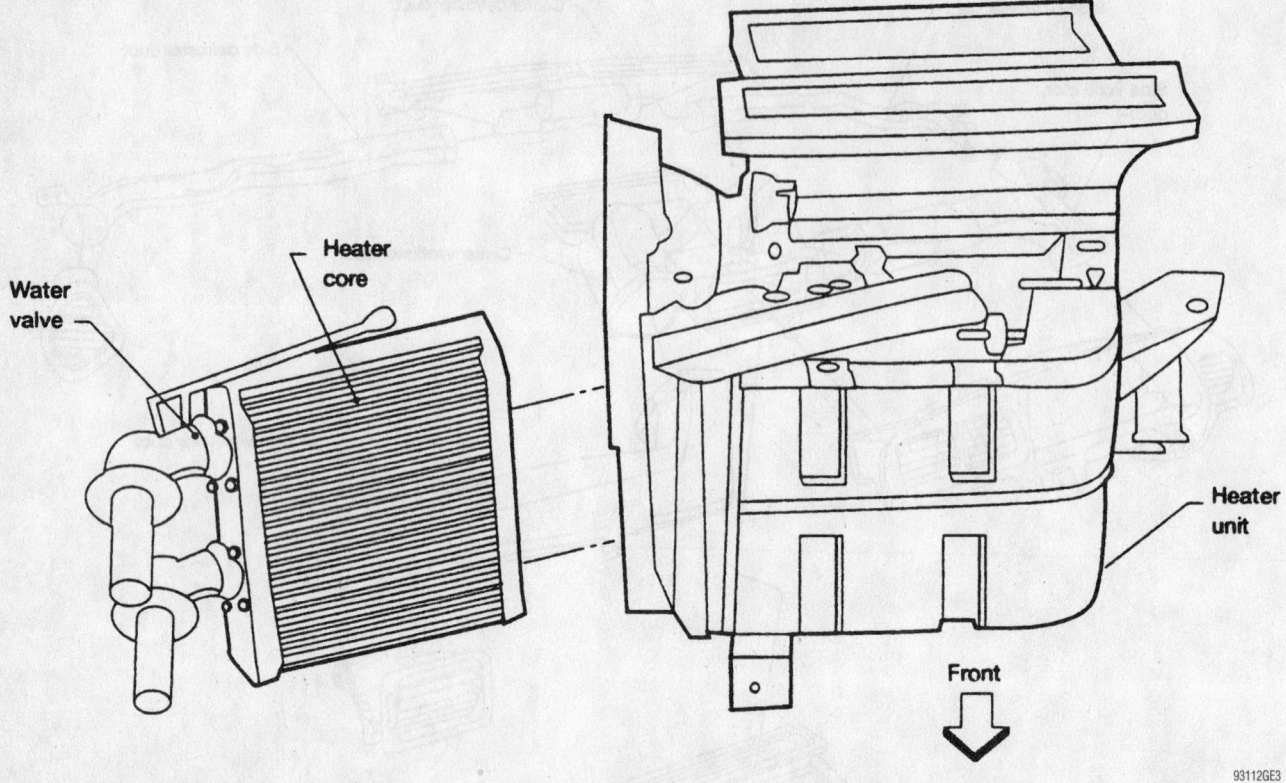

View of the heater core and heater housing

- Front pillar garnish
- Front defroster grilles
- Center instrument panel and the 2 center panel screws
- Center console and the 5 center console screws
- Audio and deck pocket and the 4 audio and deck pocket-to-instrument panel screws
- Cluster lid "C" and the 4 cluster lid "C" screws
- Snap in the transmission shifter finisher (boot)
- Instrument panel lower covers
- Switch panel
- Combination meter, connect the electrical harness connector and install the 3 combination meter screws
- Cluster lid "A" and the 2 cluster lid "A" screw
- Combination switch, the spiral cable, the covers and the 6 steering column cover screws
- Lower reinforcement panel and the 2 lower reinforcement panel-to-instrument panel screws
- Lower panel and the 2 lower panel-to-instrument panel screws, on the driver's side
- Kick plate and dash side finisher, on the driver's side

15. Working in the engine compartment, connect the heater hoses to the heater core tubes.

16. Install the passenger's side SRS by installing or connecting the following:

- Front passenger's air bag module
- 4 passenger's air bag-to-instrument panel nuts
- 2 new special bolts and torque using a tamper resistant Torx wrench (T50) to 11–18 ft. lbs. (15–25 Nm)
- Front passenger's air bag module connector
- Glove box door and the glove box

17. Install the driver's side SRS and steering wheel by installing or connecting the following:

- Steering wheel to the steering column
- Steering wheel nut and torque the nut to 22–29 ft. lbs.
- Horn's electrical connector
- Air bag module
- New special bolts to both sides of the steering wheel and torque the bolts using a tamper resistant Torx wrench (T50) to 11–18 ft. lbs. (15–25 Nm).
- Both the left and right side lids to the steering wheel
- Lower lid to the steering wheel and

connect the driver's air bag module connector

18. Refill the cooling system.

19. Connect the positive (+) battery cable; then, the negative (() battery cable.

20. Evacuate, charge and leak test the air conditioning system refrigerant.

21. Operate the engine to normal operating temperatures; then, check the climate control operation and check for leaks.

Maxima

REMOVAL & INSTALLATION

1. Disconnect the negative battery terminal.

✳✳ CAUTION

After disconnecting the negative battery cable, wait for at least 3 minutes for the SRS modules to deplete its energy.

2. Drain the cooling system into a clean container for reuse.

3. Remove the air bag module and steering wheel by removing or disconnecting the following:

- Place the front wheels in the straight-ahead position

- Lower lid and disconnect the air bag electrical connector at the bottom of the steering wheel
- Side lids from both sides of the steering wheel
- Torx bolts using a Torx wrench T50 from both side of the steering wheel; then, discard the bolts
- Air bag module from the steering wheel
- Steering wheel nut
- Steering wheel from the steering column using a suitable puller

4. Disarm the passenger's side air bag by removing or disconnecting the following:

- Glove box lid
- Passenger's air bag electrical connector

5. Remove the instrument panel by removing or disconnecting the following:

- Upper and lower glove box screws and remove the glove box
- Lower instrument panel screws and the panel at the driver's side
- Knee protector screws and the knee protector
- Steering column cover screws and the cover
- Combination switch-to-steering column screws, disconnect the electrical connector and the combination switch

- Cluster lid "A" screws and the lid
- Combination meter screws, disconnect the electrical connectors and the combination meter
- Center ventilator with the switch panel using a suitable prytool
- Cover plate (automatic transmission) or the shifter cover plate (manual transmission)
- Ashtray
- Upper and lower audio/air conditioning control unit assembly screws and the assembly
- Console box screws and the console box (under the shifter cover plate); be sure to remove the rear screws
- Front pillar garnish
- Left and right lower cover and the center lower cover at the instrument panel dash
- Defroster grille
- Instrument panel-to-chassis nuts/bolts and the instrument panel

6. Remove or disconnect the following:

- Rear heater ducts
- Side ventilator ducts
- Center defroster duct and the center ventilator duct
- Heater housing-to-chassis fasteners and remove the heater housing
- Heater core from the heater housing

To install:

7. Install or connect the following:

- Heater core to the heater housing
- Heater housing and the heater housing-to-chassis fasteners
- Center ventilator duct and the center defroster duct
- Side ventilator ducts
- Rear heater ducts

8. Install the instrument panel by installing or connecting the following:

- Instrument panel and the instrument panel-to-chassis nuts/bolts
- Defroster grille
- Left and right lower cover and the center lower cover
- Front pillar garnish
- Console box and the console box screws under the shifter cover plate; be sure to install the rear screws
- Audio/air conditioning control unit assembly and the upper and lower assembly screws
- Ashtray
- Cover plate (automatic transmission) or the shifter cover plate (manual transmission)
- Center ventilator with the switch panel
- Combination meter, connect the electrical connectors and the combination meter screws

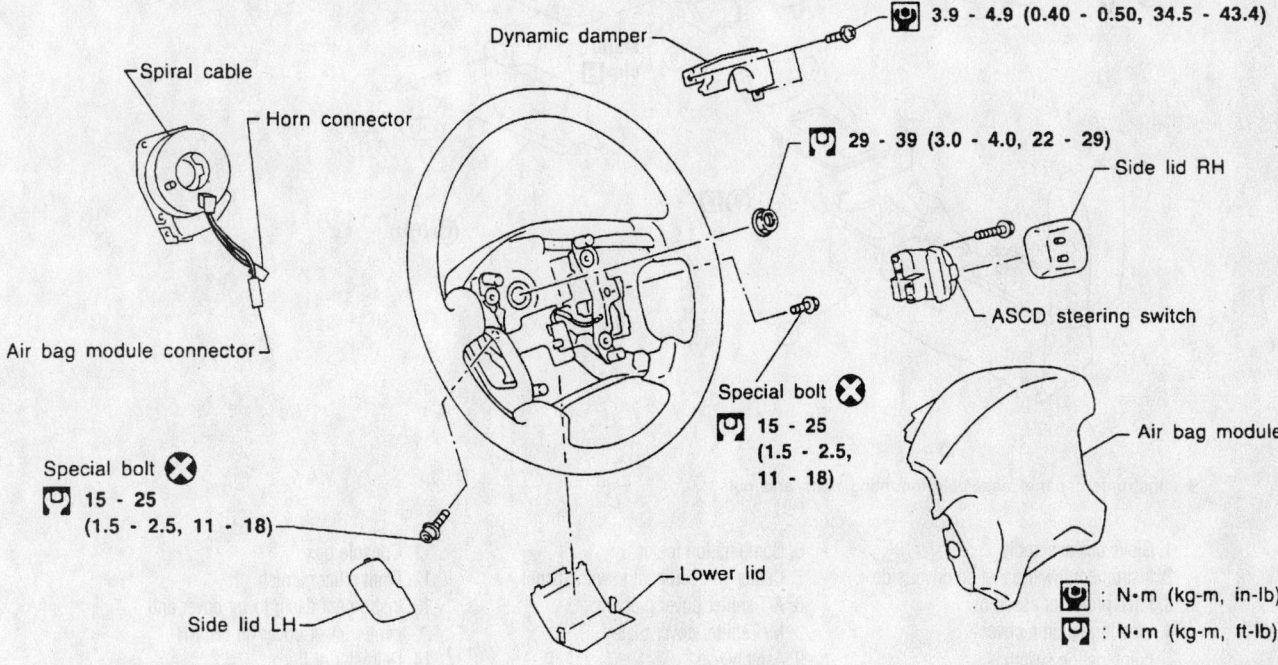

Exploded view of the air bag module and steering wheel—1999–01 Maxima

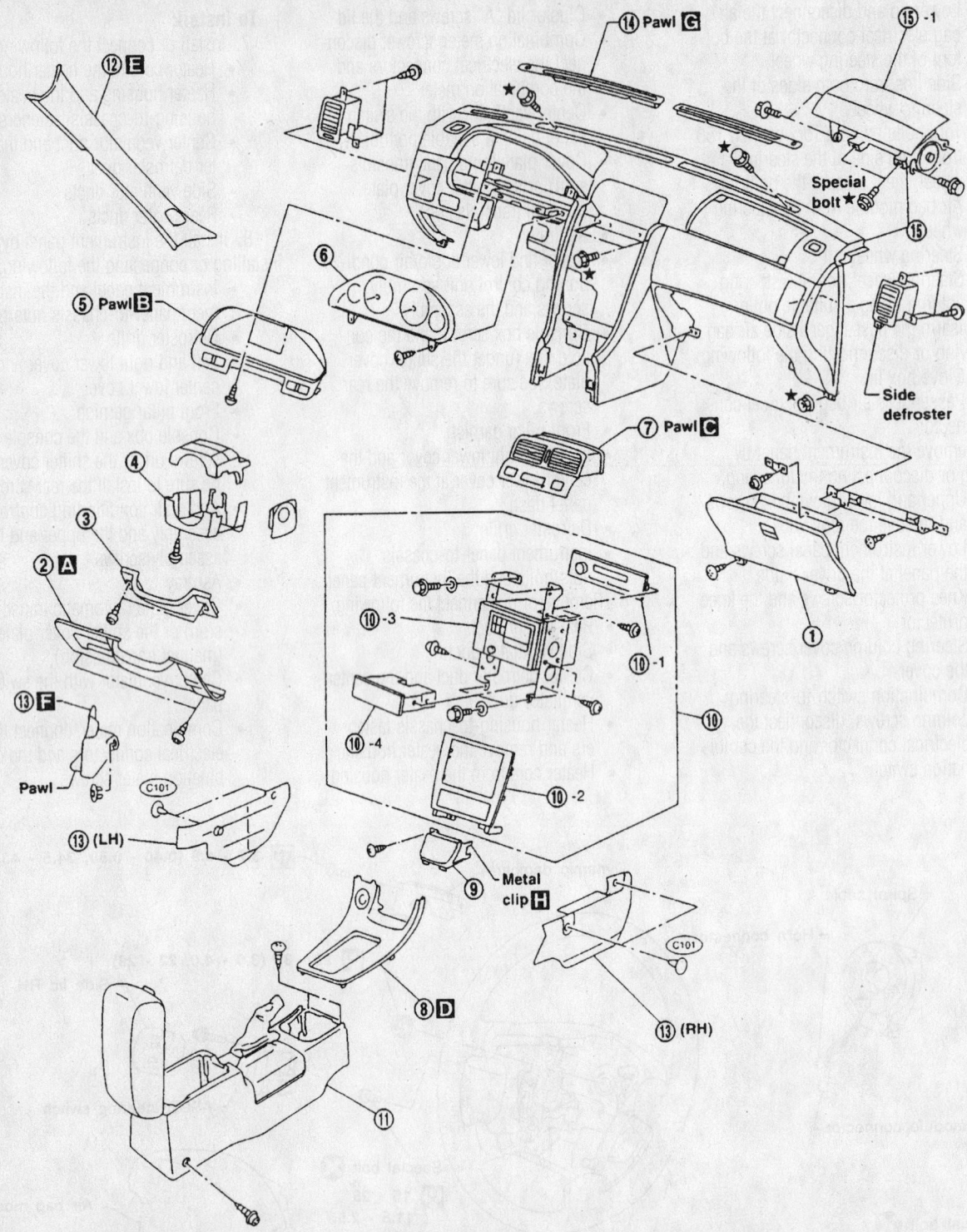

★: Instrument panel assembly mounting bolts and nuts

1. Glove box assembly
2. Instrument lower panel on driver side
3. Knee protector assembly
4. Steering column cover & combination switch
5. Cluster lid A

6. Combination meter
7. Center ventilator with switch panel
8. A/T shifter cover plate or M/T shifter cover plate
9. Ashtray
10. Audio & A/C control unit assembly

11. Console box
12. Front pillar garnish
13. Instrument dash: lower cover and center lower cover on LH, RH
14. Defroster grille
15. Instrument panel assembly
15.-1 Passenger air bag module

93112GK2

Exploded view of the instrument panel, console and related components—1999–01 Maxima

- Cluster lid "A" and the lid screws
- Combination switch, connect the electrical connector and the combination switch-to-steering column screws
- Steering column cover and the cover screws
- Knee protector and the knee protector screws
- Lower instrument panel and the panel screws on the driver's side

- Glove box and the upper and lower glove box screws

9. Arm the passenger's side air bag by installing or connecting the following:
  - Passenger's air bag electrical connector
  - Glove box lid
10. Install the air bag module and steering wheel by installing or connecting the following:
  - Steering wheel to the steering column

- Steering wheel nut and torque it to 22–29 ft. lbs. (29–39 Nm)
- Air bag module to the steering wheel
- Torque the new Torx bolts (using a Torx wrench T50), at both side of the steering wheel to 11–18 ft. lbs. (15–25 Nm).
- Side lids to both sides of the steering wheel
- Air bag electrical connector and install the lower lid at the bottom of the steering wheel

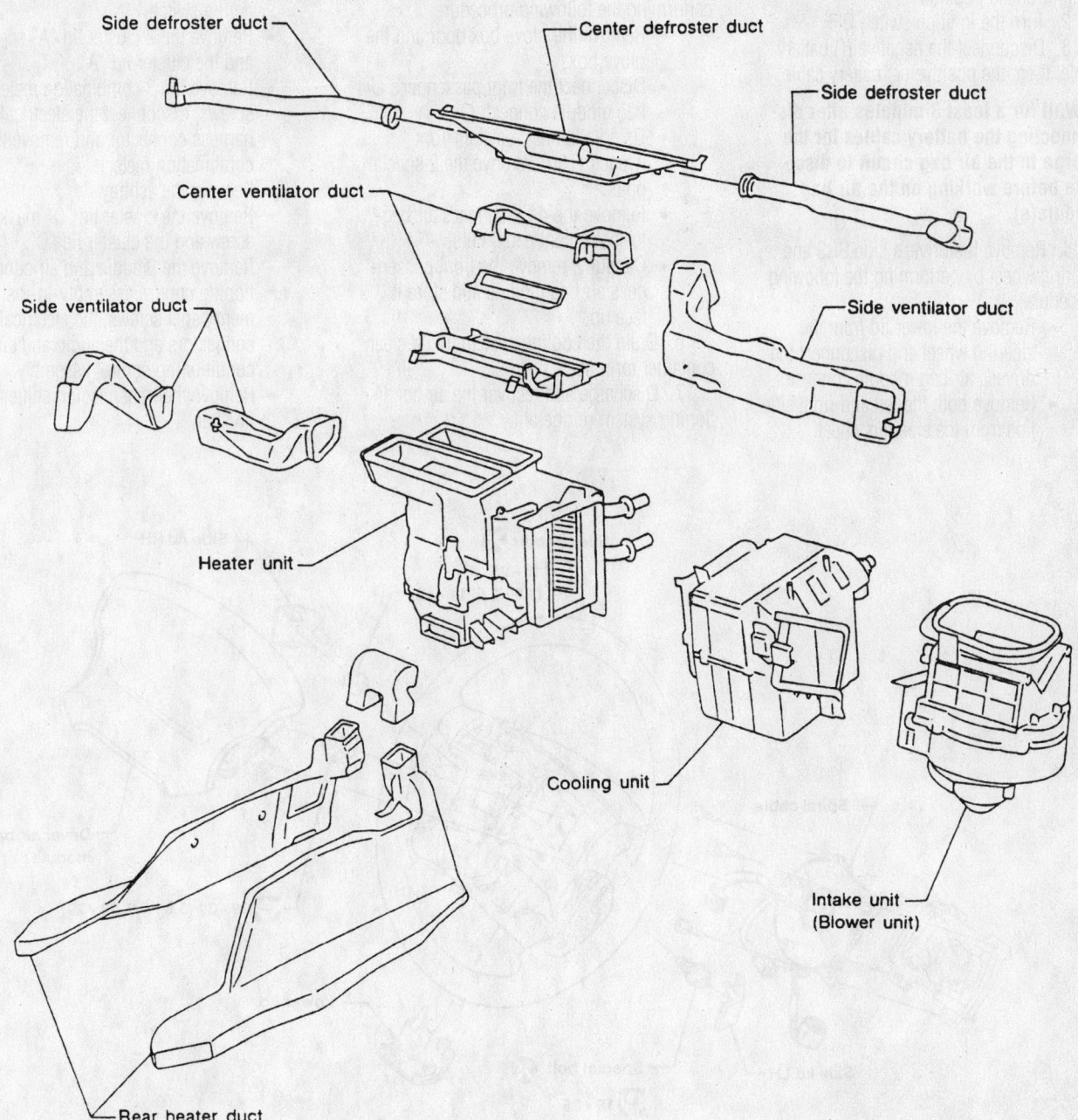

Side defroster duct — Center defroster duct — Side defroster duct — Center ventilator duct — Side ventilator duct — Side ventilator duct — Heater unit — Cooling unit — Intake unit (Blower unit) — Rear heater duct

93112GK3

Exploded view of the heater housing, evaporator housing, ventilator system and related components—1999–01 Maxima

For complete service labor times, order the Chilton Labor Guide

11. Refill the cooling system.

12. Connect the negative battery terminal.

13. Operate the engine to normal operating temperatures; then, check the climate control operation and check for leaks.

200SX and Sentra

REMOVAL & INSTALLATION

1. Position the steering wheel in the straight-ahead position.

2. Turn the ignition switch OFF.

3. Disconnect the negative (()) battery cable; then, the positive (+) battery cable.

➡ **Wait for a least 3 minutes after disconnecting the battery cables for the charge in the air bag circuit to dissipate before working on the air bag module(s).**

4. Remove the driver's side SRS and steering wheel by performing the following procedure:
- Remove the lower lid from the steering wheel and disconnect the driver's air bag module connector.
- Remove both the left and right side lids from the steering wheel.

- Using a tamper resistant Torx wrench (T50), remove the special bolts from both side of the steering wheel.
- Carefully, remove the air bag module and store it face up.
- Disconnect the horn's electrical connector and remove the steering wheel nut.
- Using a steering wheel puller, press the steering wheel from the steering column.

5. Remove the passenger's side SRS by performing the following procedure:
- Remove the glove box door and the glove box.
- Disconnect the front passenger's air bag module connector.
- Using a tamper resistant Torx wrench (T50), remove the 2 special bolts.
- Remove the 4 passenger's air bag-to-instrument panel nuts.
- Carefully, remove the front passenger's air bag module and store it face up.

6. Drain the cooling system into a clean container for reuse.

7. Discharge and recover the air conditioning system refrigerant.

8. Working in the engine compartment, disconnect the heater hoses from the heater core tubes.

9. Remove the instrument panel by performing the following procedures:
- On the driver's side, remove the 2 lower panel-to-instrument panel screws and the lower panel.
- Remove the 2 lower reinforcement panel-to-instrument panel screws and the lower reinforcement panel.
- Remove the 6 steering column cover screws, the cover and combination switch.
- Remove the 2 cluster lid "A" screws and the cluster lid "A".
- Remove the 3 combination meter screws, disconnect the electrical harness connector and remove the combination meter.
- Remove the ashtray.
- Remove the cluster lid "C" mask, screw and the cluster lid "C".
- Remove the 8 audio and air conditioning control assembly-to-instrument panel screws, the electrical connectors and the audio and air conditioning control assembly.
- Remove the transmission shifter finisher.

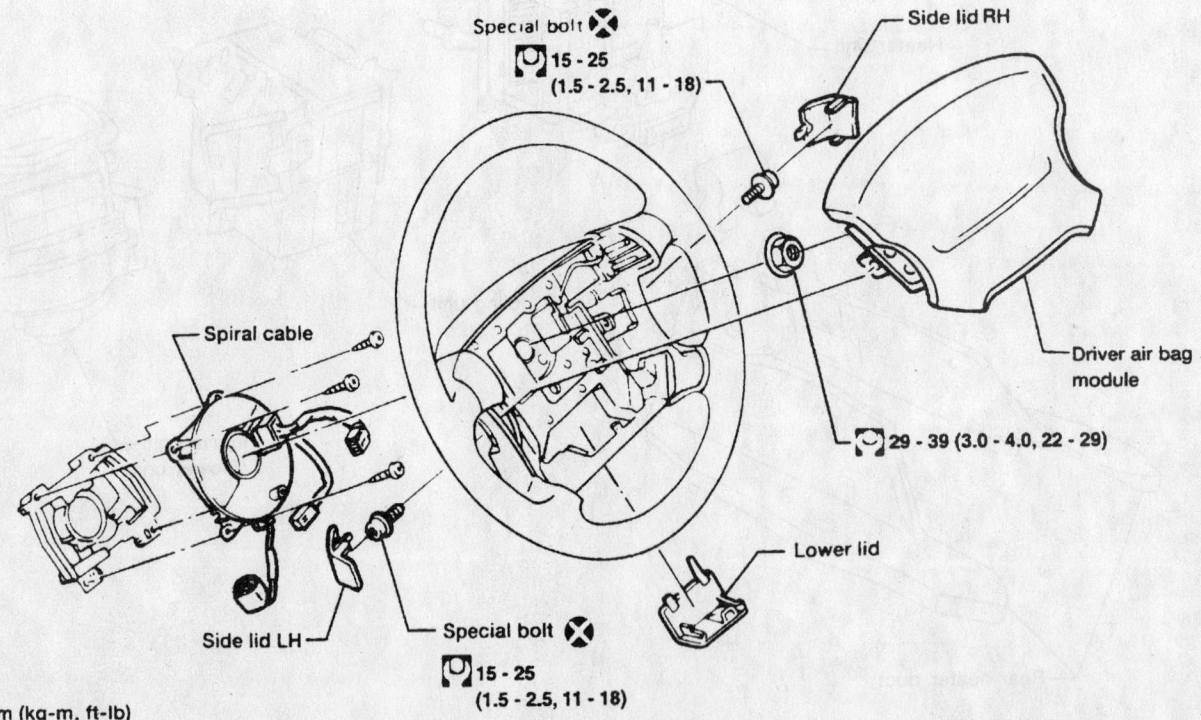

Special bolt ✕
15 - 25
(1.5 - 2.5, 11 - 18)

Side lid RH

Spiral cable

Driver air bag module

29 - 39 (3.0 - 4.0, 22 - 29)

Side lid LH

Special bolt ✕
15 - 25
(1.5 - 2.5, 11 - 18)

Lower lid

N·m (kg-m, ft-lb)

93112GD4

Exploded view of the steering wheel and air bag module—1998 200SX and Sentra

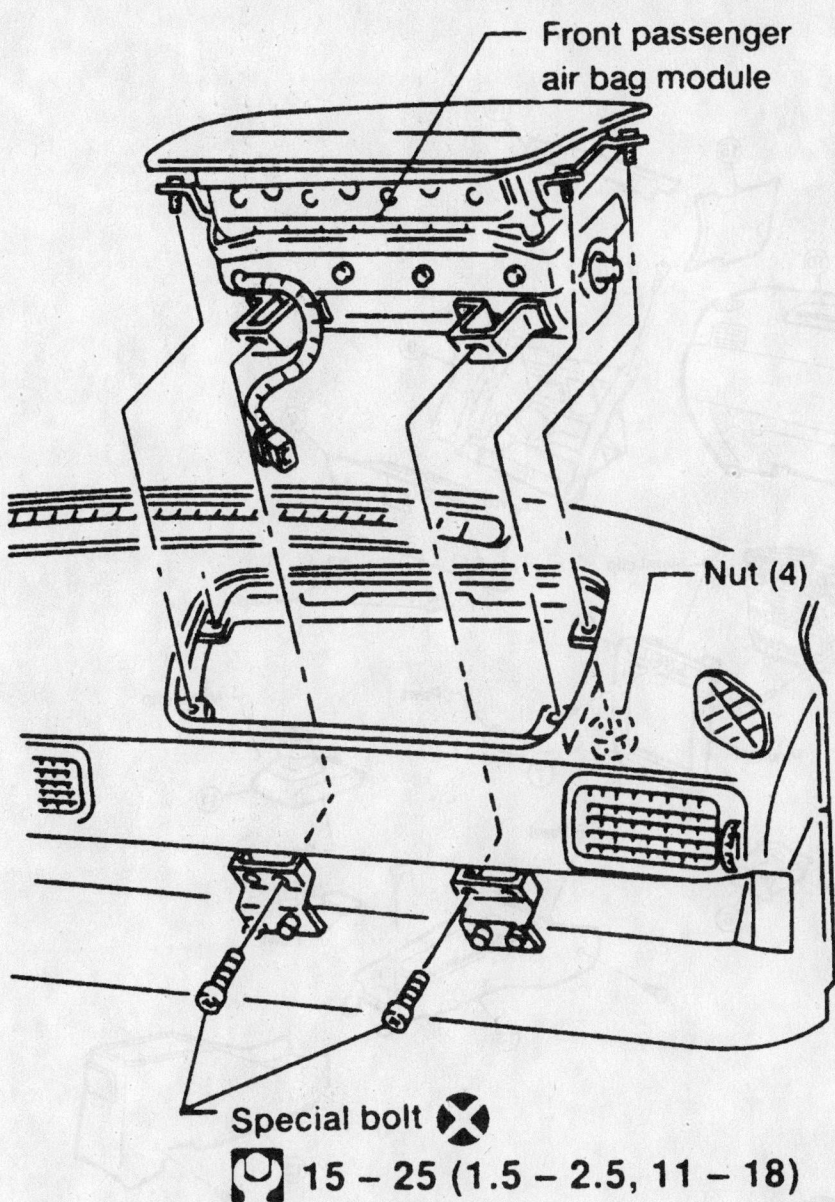

Front passenger air bag module

Nut (4)

Special bolt ✖

🔲 15 – 25 (1.5 – 2.5, 11 – 18)

🔲 : N·m (kg-m, ft-lb)

93112GD5

Exploded view of the passenger's side air bag module—1998 200SX and Sentra

- Remove the rear console mask, the 4 screws and the rear console.
- Remove the 4 front console screws and the front console.
- Remove the front pillar garnish.
- Remove the lower dash side garnish.
- Remove the instrument panel mask.
- Remove the instrument panel-to-chassis nuts/bolts and the instrument panel.

10. Remove the air conditioning housing assembly by performing the following procedure:
  - Disconnect the refrigerant lines from the air conditioning housing assembly.
  - Disconnect the thermo control amp.
  - Remove the air conditioning housing assembly.
11. Remove the heater unit.
12. Remove the heater core from the heater unit.

To install:
13. Install the heater core to the heater unit.
14. Install the heater unit.
15. Install the air conditioning housing assembly by performing the following procedure:
  - Install the air conditioning housing assembly.
  - Connect the thermo control amp.
  - Connect the refrigerant lines to the air conditioning housing assembly.
16. Install the instrument panel by performing the following procedures:
  - Install the instrument panel and the instrument panel-to-chassis nuts/bolts.
  - Install the instrument panel mask.
  - Install the lower dash side garnish.
  - Install the front pillar garnish.
  - Install the front console and the 4 front console screws.
  - Install the rear console, the 4 screws and the rear console mask.
  - Install the transmission shifter finisher.
  - Install the air conditioning control assembly, the electrical connectors and the audio and the 8 audio and air conditioning control assembly-to-instrument panel screws.
  - Install the cluster lid "C", the screw and the cluster lid "C" mask.
  - Install the ashtray.
  - Install the combination meter, connect the electrical harness connector and install the 3 combination meter screws.
  - Install the cluster lid "A" and the 2 cluster lid "A" screws.
  - Install the combination switch, the cover and the 6 steering column cover screws.
  - Install the lower reinforcement panel and the 2 lower reinforcement panel-to-instrument panel screws.
  - On the driver's side, install the lower panel and the 2 lower panel-to-instrument panel screws.
17. Working in the engine compartment, connect the heater hoses to the heater core tubes.
18. Install the passenger's side SRS by performing the following procedure:
  - Carefully, install the front passenger's air bag module.

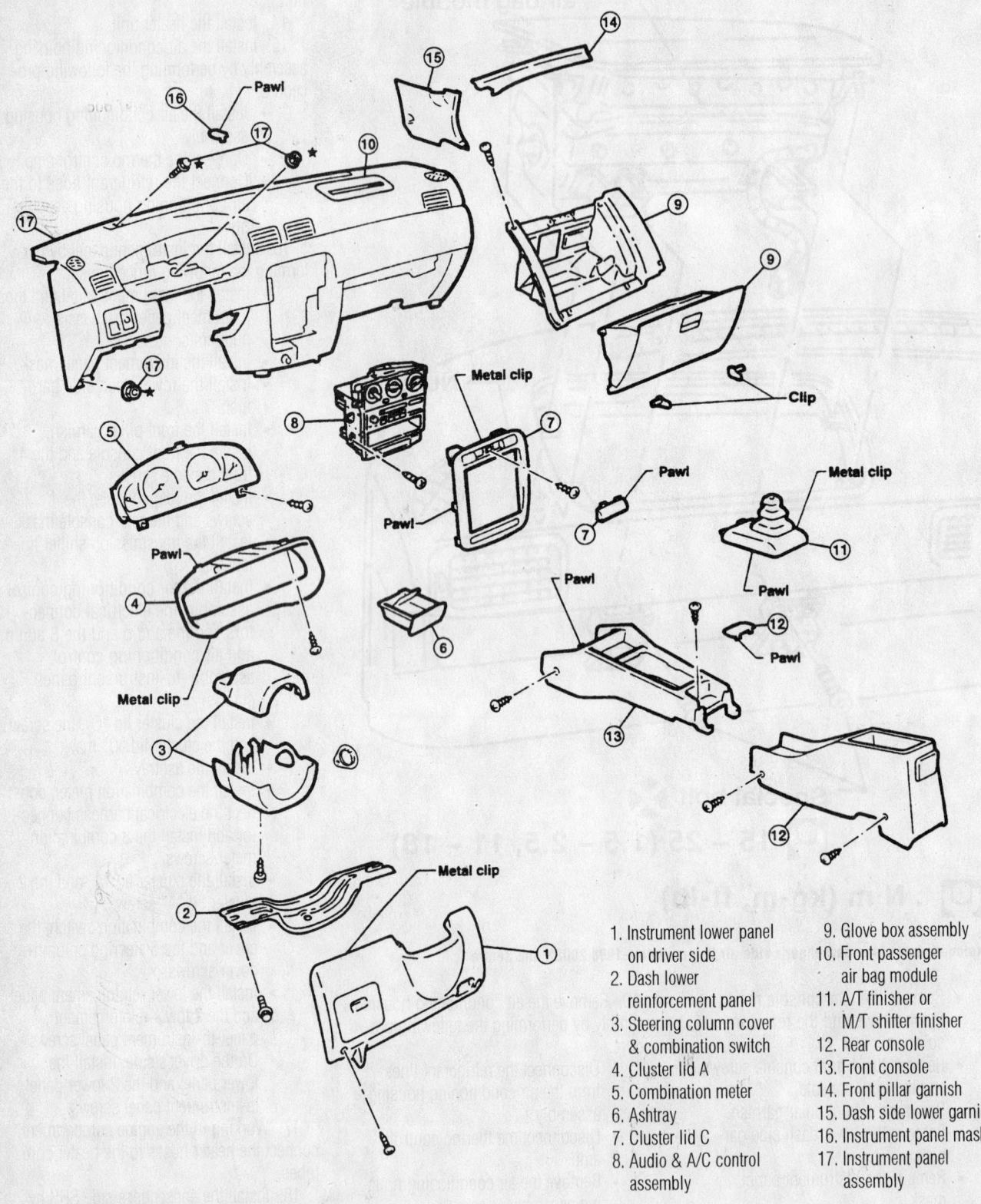

1. Instrument lower panel on driver side
2. Dash lower reinforcement panel
3. Steering column cover & combination switch
4. Cluster lid A
5. Combination meter
6. Ashtray
7. Cluster lid C
8. Audio & A/C control assembly
9. Glove box assembly
10. Front passenger air bag module
11. A/T finisher or M/T shifter finisher
12. Rear console
13. Front console
14. Front pillar garnish
15. Dash side lower garnish
16. Instrument panel mask
17. Instrument panel assembly

★ : Instrument panel assembly mounting bolts and nuts.

93112GD6

Exploded view of the instrument panel assembly—1998 200SX and Sentra

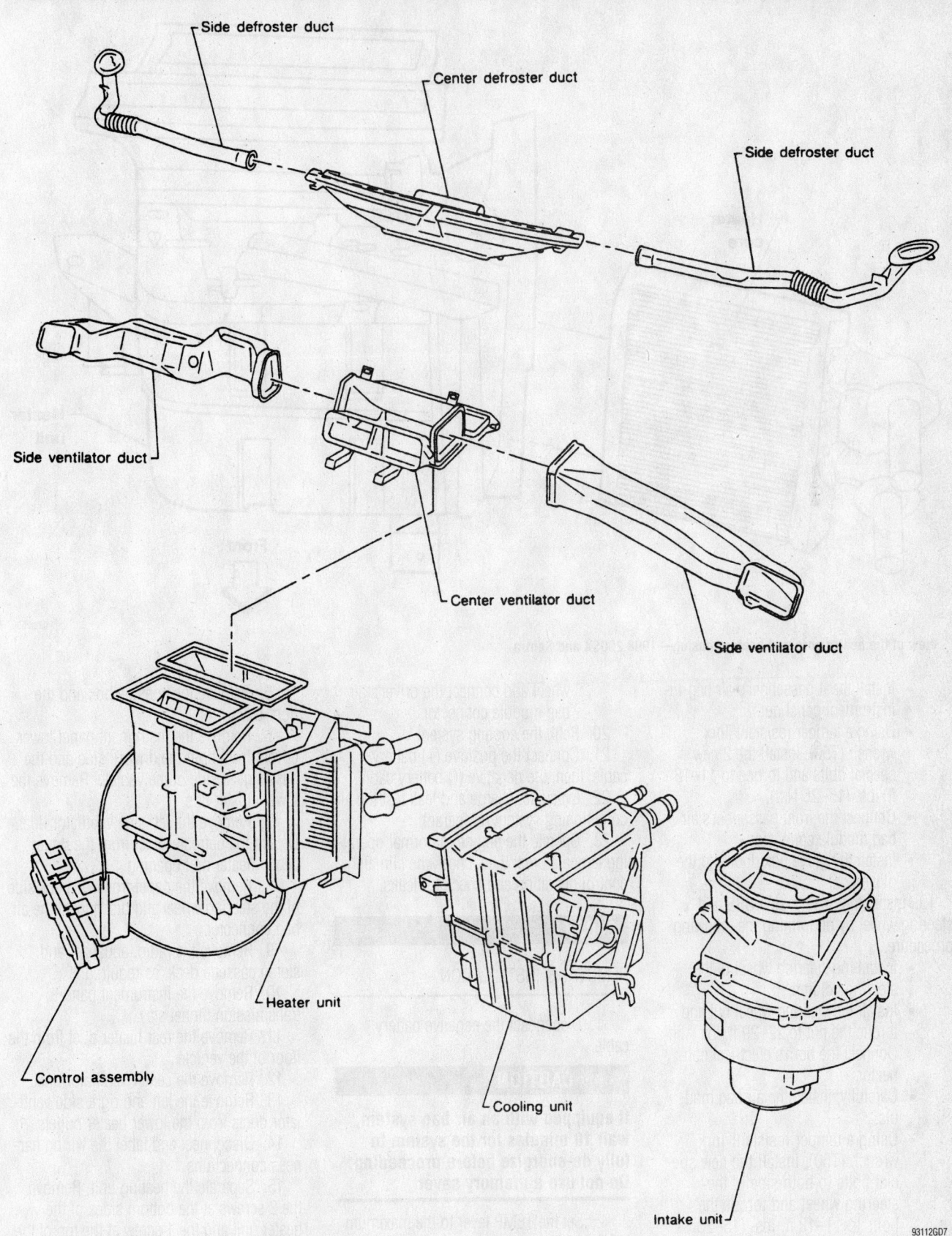

Side defroster duct

Center defroster duct

Side defroster duct

Side ventilator duct

Center ventilator duct

Side ventilator duct

Heater unit

Cooling unit

Control assembly

Intake unit

93112GD7

Exploded view of the heater housing assembly and related components—1998 200SX and Sentra

Timing belt service is covered in Section 3 of this manual

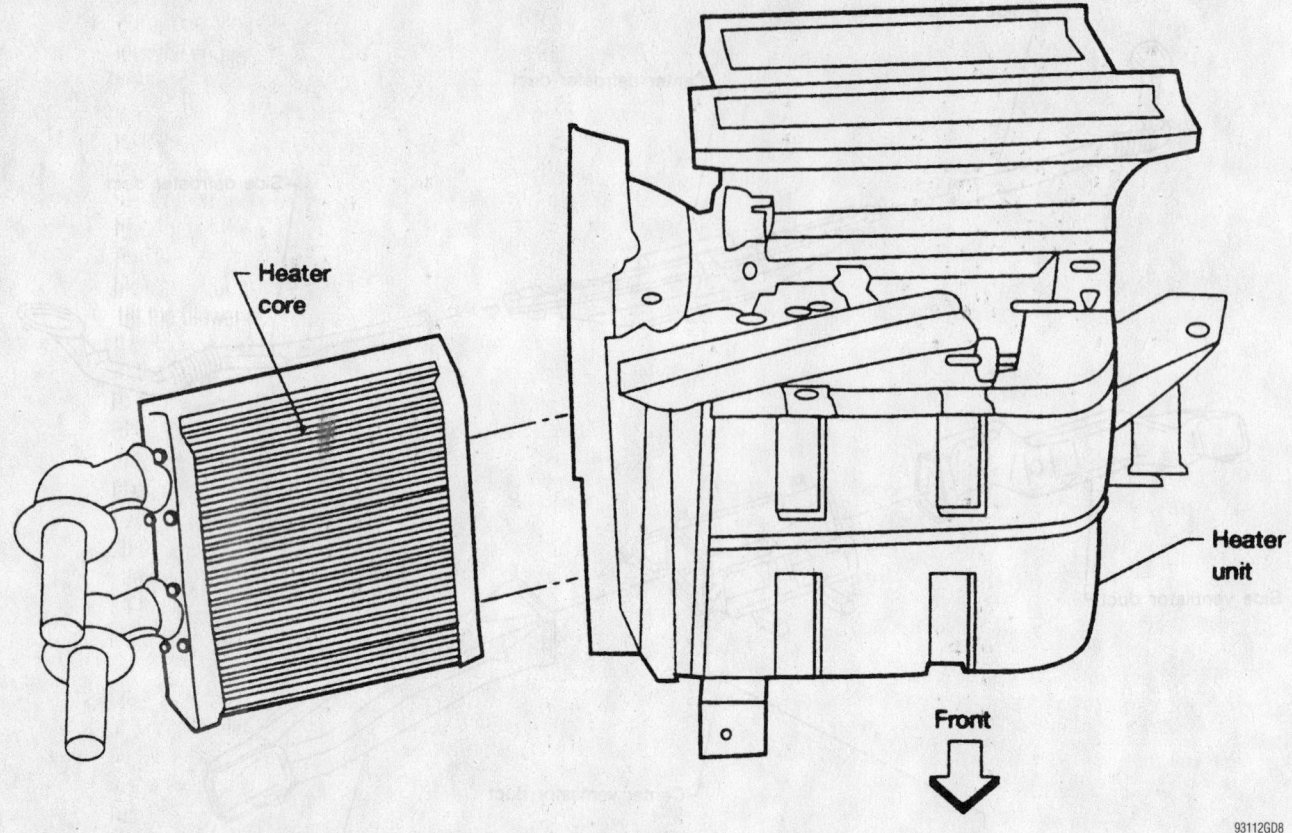

View of the heater core and heater housing—1998 200SX and Sentra

- Install the 4 passenger's air bag-to-instrument panel nuts.
- Using a tamper resistant Torx wrench (T50), install the 2 new special bolts and torque to 11–18 ft. lbs. (15–25 Nm).
- Connect the front passenger's air bag module connector.
- Install the glove box door and the glove box.

19. Install the driver's side SRS and steering wheel by performing the following procedure:

- Install the steering wheel to the steering column.
- Install the steering wheel nut and torque the nut to 22–29 ft. lbs. Connect the horn's electrical connector.
- Carefully, install the air bag module.
- Using a tamper resistant Torx wrench (T50), install the new special bolts to both side of the steering wheel and torque the bolts to 11–18 ft. lbs. (15–25 Nm).
- Install both the left and right side lids to the steering wheel.
- Install the lower lid to the steering

wheel and connect the driver's air bag module connector.

20. Refill the cooling system.

21. Connect the positive (+) battery cable; then, the negative (()) battery cable.

22. Evacuate, charge and leak test the air conditioning system refrigerant.

23. Operate the engine to normal operating temperatures; then, check the climate control operation and check for leaks.

240SX

REMOVAL & INSTALLATION

1. Disconnect the negative battery cable.

✴ CAUTION

If equipped with an air bag system, wait 10 minutes for the system to fully de-energize before proceeding. Do not use a memory saver.

2. Set the TEMP lever to the maximum **HOT** position.

3. Drain the cooling system into a clean container for reuse.

4. Disconnect the heater hoses from the driver's side of the heater unit.

5. Remove the console box and the floor mats.

6. Remove the instrument panel lower covers from both the driver's side and the passenger's side of the vehicle. Remove the lower cluster lids.

7. Remove the left side ventilator duct. Detach the defroster duct from the upper center heater unit opening.

8. Remove the panel from the back side of the steering wheel and disconnect the air bag connector.

9. Remove the radio, equalizer and stereo cassette deck, as required.

10. Remove the instrument pane-to-transmission tunnel stay.

11. Remove the rear heater duct from the floor of the vehicle.

12. Remove the center ventilator duct.

13. Remove the left and right side ventilator ducts from the lower heater outlets.

14. Disconnect and label the wiring harness connections.

15. Separate the heating unit. Remove the 2 screws at the bottom sides of the heater unit and the 1 screw at the top of the unit and remove the unit together with heater control assembly.

16. Separate the heater case halves and slide the core from the case.

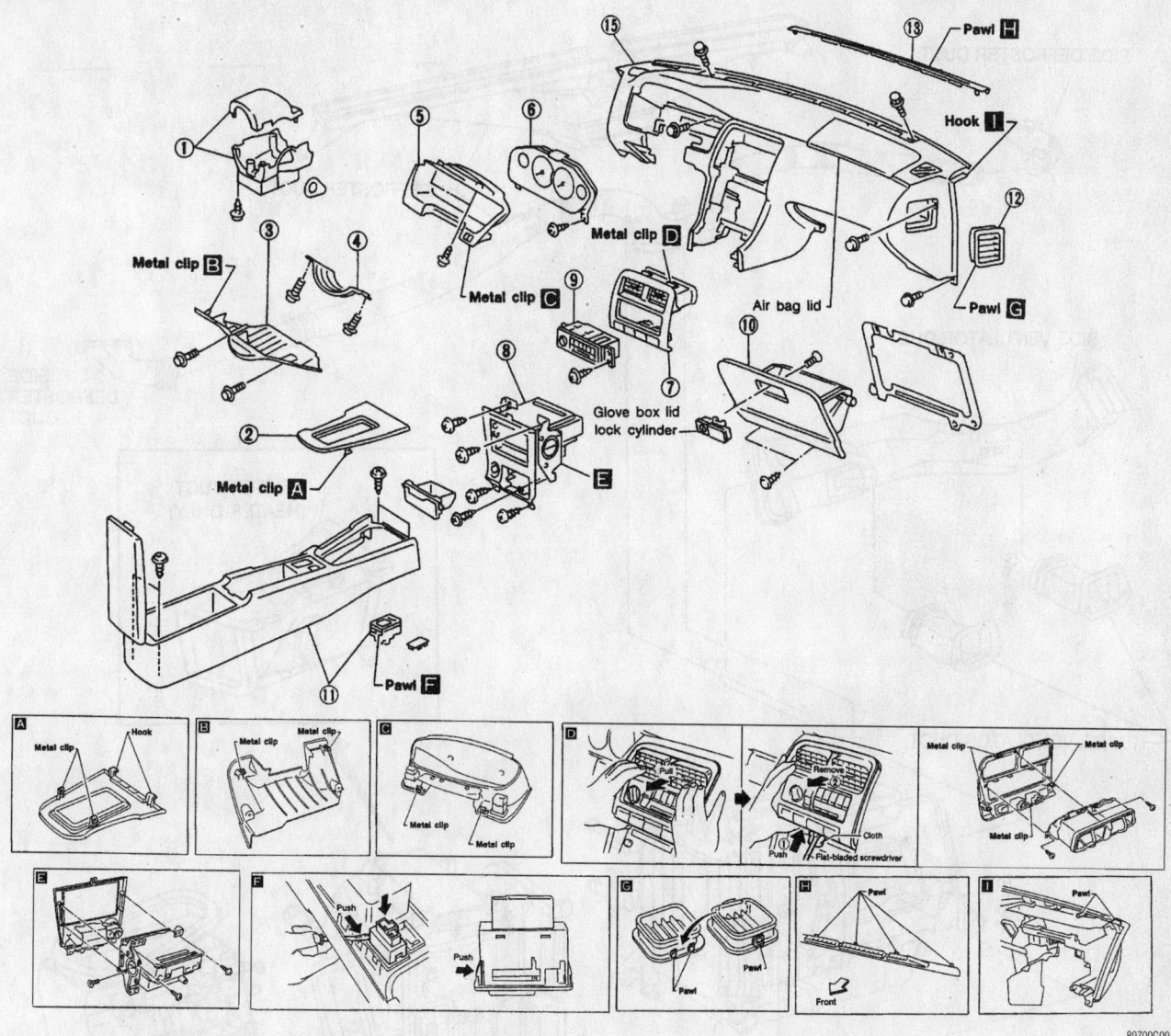

Exploded view of the instrument panel—240SX

To install:

17. Install the heater core and assemble the heater case halves. Use new gaskets and seals, as required.

18. Mount the heater unit/control assembly and install the upper and lower attaching screws.

19. Plug in the wiring harness connectors.

20. Connect the left and right side ducts to the lower heater outlets.

21. Connect the center ventilator duct.

22. Connect the rear heater duct.

23. Attach the instrument panel-to-transmission stay.

24. Install the cassette deck, equalizer and radio.

25. Connect the upper defroster duct to the upper center heater opening. Connect the left side ventilator duct.

26. Install the lower cluster lids and lower instrument panel covers.

27. Install the floor mats and console box.

28. Install the front seats. Torque the seat bolts to 32–41 ft. lbs. (43–55 Nm).

29. Connect the heater hoses. Use new grommets, as required.

30. Refill the cooling system.

31. If equipped with an air bag, connect the module connector at the steering wheel.

32. Connect the negative battery cable.

33. Operate the engine to normal operating temperatures; then, check the climate control operation and check for leaks.

Heater Core replacement is covered in Section 2 of this manual

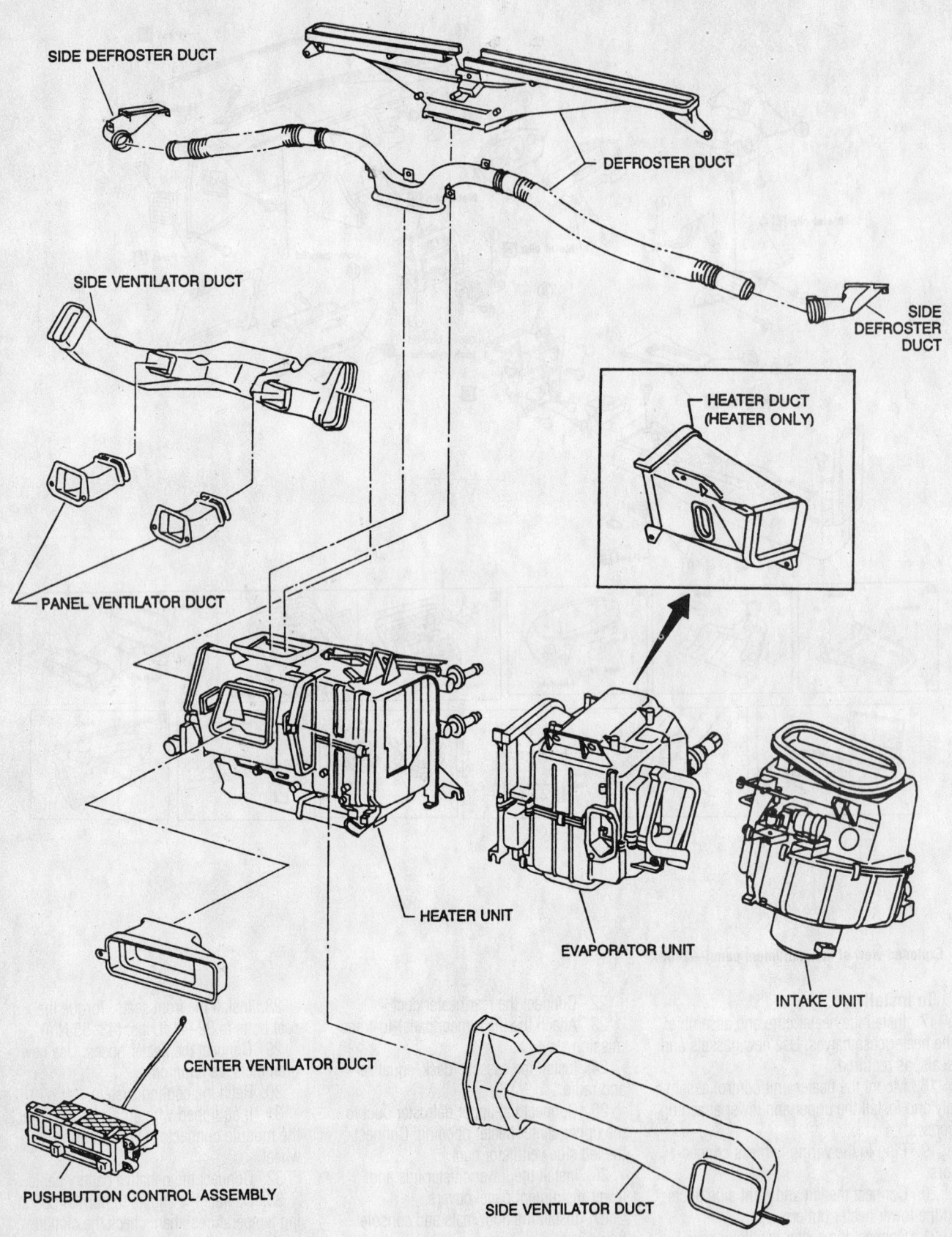

SIDE DEFROSTER DUCT

DEFROSTER DUCT

SIDE DEFROSTER DUCT

SIDE VENTILATOR DUCT

HEATER DUCT (HEATER ONLY)

PANEL VENTILATOR DUCT

HEATER UNIT

EVAPORATOR UNIT

INTAKE UNIT

CENTER VENTILATOR DUCT

PUSHBUTTON CONTROL ASSEMBLY

SIDE VENTILATOR DUCT

93112G90

Exploded view of the heater/air conditioning assemblies and related components—1998 240SX

1998–01 SAAB

900 and 9–3 Models

REMOVAL & INSTALLATION

1. Disconnect the negative battery cable.

2. Drain the cooling system into a clean container for reuse.

3. Disconnect the heater hoses at the firewall in the engine compartment. Plug the ends of the fittings on the valve.

4. Blow any remaining coolant from the heater core with compressed air.

5. Remove the glove compartment retaining screws, bolt, quick-release pin and bracket catch. Pull the glove compartment out partway to disconnect the lamp; the remove the glove compartment.

6. Remove the floor console by performing the following procedure:

   • Apply the handbrake.
   • Remove the ignition switch cover plate by loosening the left front corner first and working around the plate counterclockwise. Disconnect the ignition switch lighting connector.
   • Remove the rear ashtray. Remove the rear trim panel retaining screws and trim panel.
   • Remove the floor console retaining screws and nuts.
   • Pull the floor console to the rear and lift to gain access to the window lift and interior light switch connectors.
   • Remove and disconnect the window lift and interior light switches.
   • Lift and remove the floor console.

7. Remove the center console side panel screws and remove both side panels.

8. If equipped with a manual transmission, release the shift lever boot clips and remove the boot.

9. Reach in through the side panel openings and push out the heating and air conditioning control panel.

10. Disconnect the electrical and mechanical connectors (as applicable) and remove the heating and air conditioning control panel.

11. Disconnect the center console electrical connectors.

12. Remove the center console-to-instrument panel quick-release pins.

13. Remove the center console.

14. Cut the tie-wraps and remove the rear air ducts.

15. Open the heater core case.

16. Disconnect the hoses from the heater core.

17. Remove the toggle clips at the sides of the heater core case.

18. Pull the heater hoses down and lift out the heater core.

To install:

19. Install the heater core and connect the heater hoses.

20. Install the toggle clips at the sides of the heater core case.

21. Connect the hoses to the heater core.

22. Close the heater core case.

23. Install the rear air ducts.

24. Install the center console.

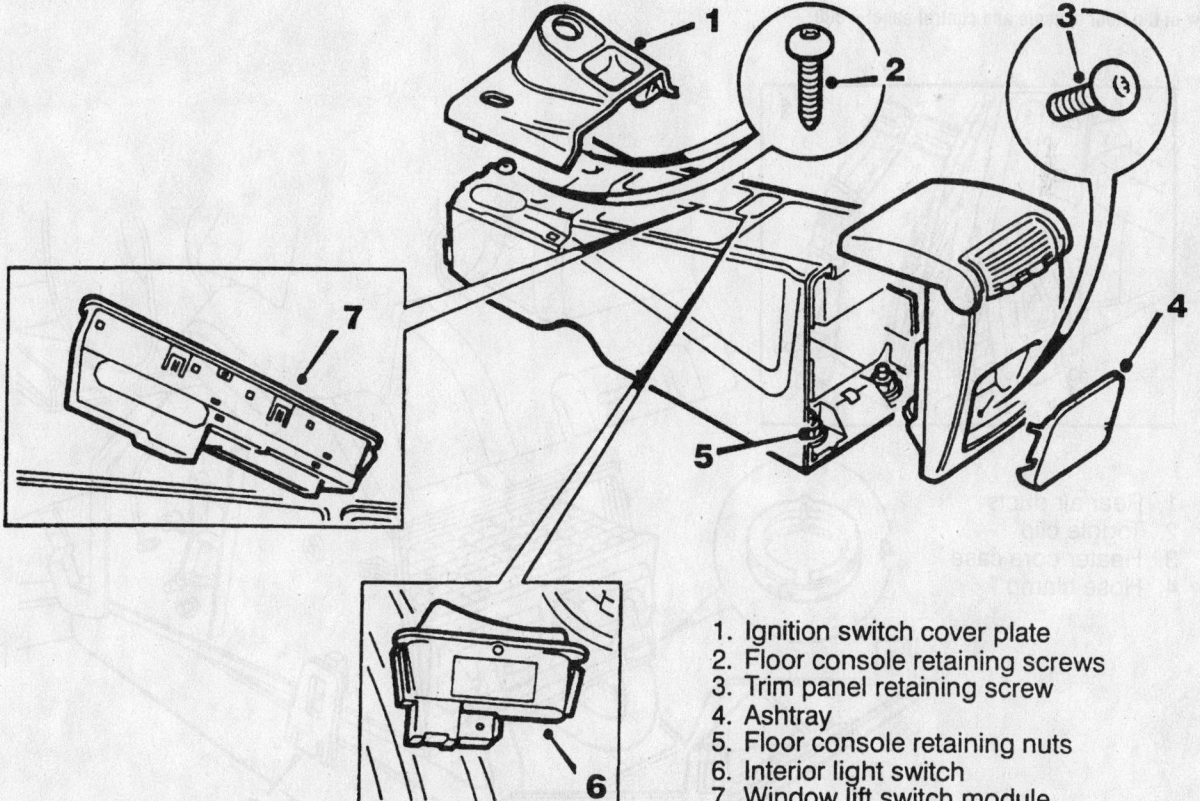

1. Ignition switch cover plate
2. Floor console retaining screws
3. Trim panel retaining screw
4. Ashtray
5. Floor console retaining nuts
6. Interior light switch
7. Window lift switch module

93112G84

Exploded view of the floor console and related components—900

Brake service is covered in Section 4 of this manual

1. Center console side panels
2. Manual shift lever boot
3. Heater/air conditioner connectors
4. Heat control cable
5. Heater/air conditioner control panel

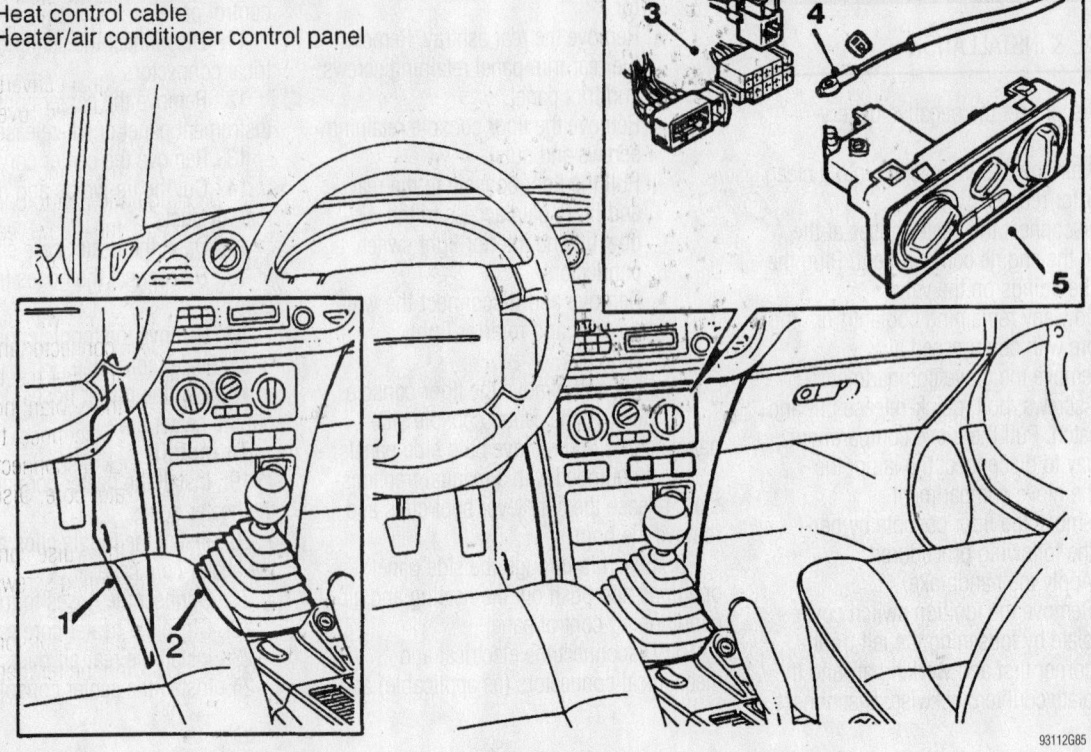

View of the floor console and control panel—900

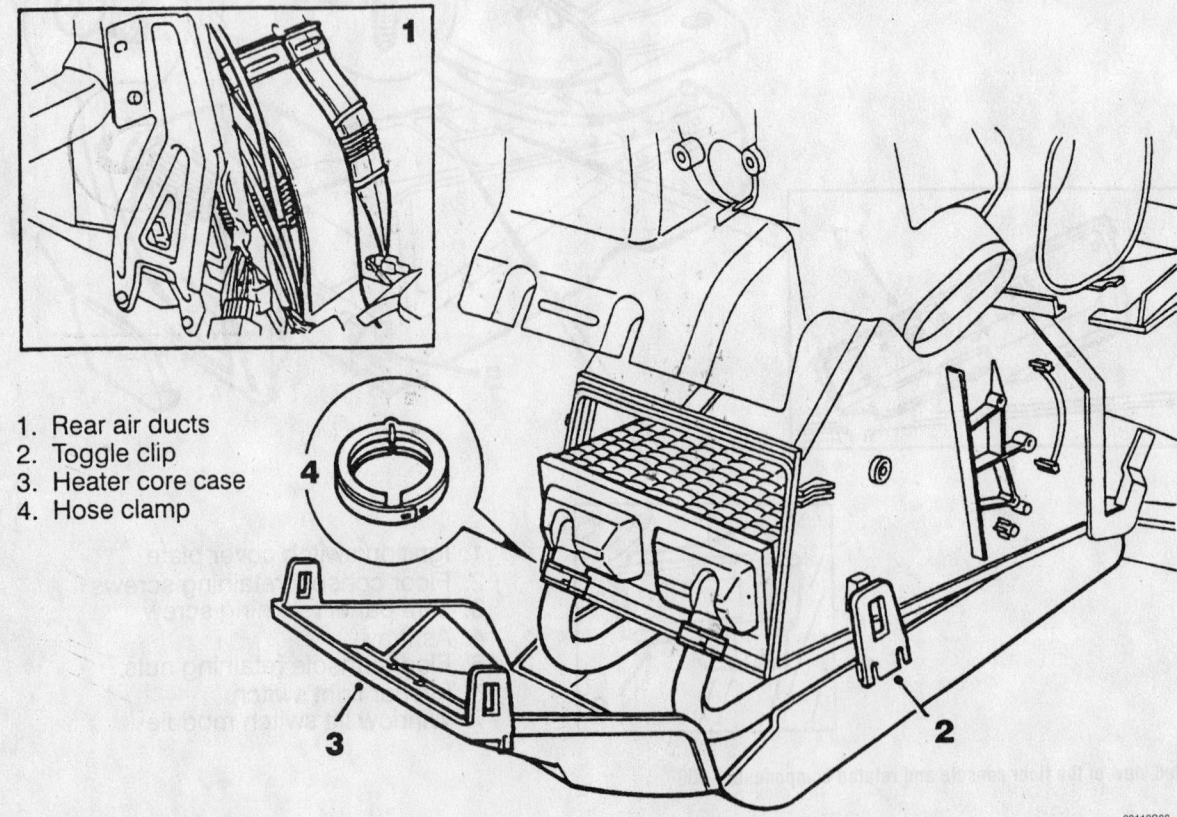

1. Rear air ducts
2. Toggle clip
3. Heater core case
4. Hose clamp

View of the heater core removal—900

25. Install the center console-to-instrument panel quick-release pins.

26. Connect the center console electrical connectors.

27. Install the heating and air conditioning control panel and connect the electrical and mechanical connectors (as applicable).

28. Install the heating and air conditioning control panel.

29. If equipped with a manual transmission, install the boot and the shift lever boot clips.

30. Install the center console side panels and screws.

31. Install the floor console by performing the following procedure:
- Install the floor console.
- Connect and install the window lift and interior light switches.
- Install the floor console retaining screws and nuts.
- Install the rear trim panel and retaining screws. Install the rear ashtray.
- Connect the ignition switch lighting connector. Install the ignition switch cover plate.
- Release the handbrake.

32. Install the glove compartment, the bracket catch, the glove compartment retaining screws, the bolt and the quick-release pin.

33. Connect the heater hoses at the firewall in the engine compartment.

34. Refill the cooling system.

➡**On 6-cylinder models, bleed the cooling system.**

35. Connect the negative battery cable.

36. Operate the engine to normal operating temperatures; then, check the climate control operation and check for leaks.

9000

REMOVAL & INSTALLATION

1. Disconnect the negative battery cable.

2. Drain the cooling system into a clean container for reuse.

3. Remove the blower motor assembly by performing the following procedures:
- Remove the hood.
- Remove the windshield wiper arms.
- Remove the windshield wiper motor

and air conditioning evaporator cover panels.
- Disconnect the blower motor control unit electrical connector.
- Remove the cowl cover panel seal and the signal converter.
- Remove the cowl cover panel.
- Remove the electronic ignition control unit mounting bolts and move the ignition unit to one side.
- Remove the 4 cowl lead-through panel screws and remove the lead-through panel.
- Disconnect the windshield wiper electrical connector and remove the windshield wiper mechanism.
- Remove the coolant hose grommets and disconnect the coolant hose quick-disconnect couplings at the heater core. Discard the O-rings.
- Remove the cruise control vacuum pump mounting screws and push the pump aside.
- Remove the evaporator retaining screws and the refrigerant hose retaining clips.

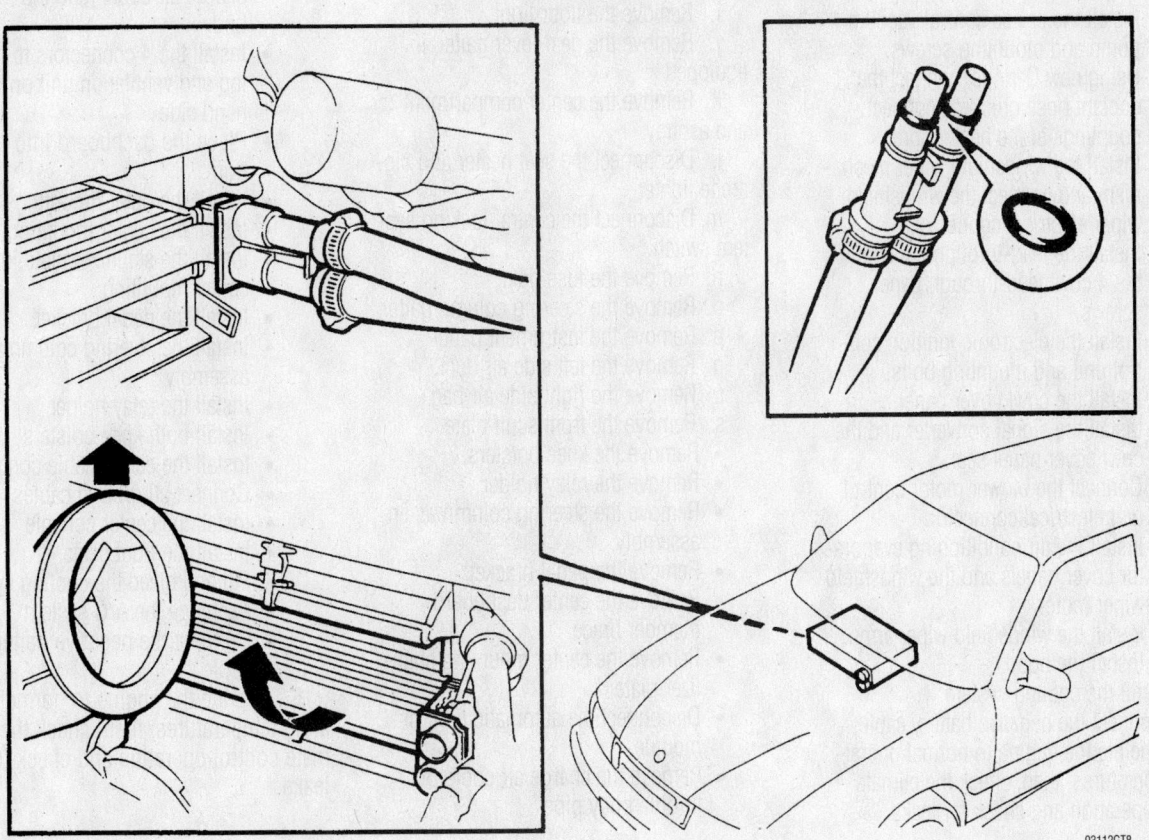

Exploded view of the heater core location—9000

93112GT8

For complete Engine Mechanical specifications, see Section 1 of this manual

- Disconnect the temperature control cable from the temperature valve.
- Remove the engine bracket from the right rear corner of the engine compartment.
- Remove the rear motor-mount nut. Using an engine sling attached to the rear engine-lift hook, carefully tilt the engine forward.
- Carefully lift the evaporator, release the blower motor mounting clips and tilt the blower motor assembly to lift out.

4. Remove the heater core.

To install:

5. Install the heater core.

6. Install the blower motor assembly by performing the following procedures:
- Install the blower motor assem-bly.
- Install the rear motor-mount nut.
- Install the engine bracket to the right rear corner of the engine compartment.
- Connect the temperature control cable to the temperature valve.
- Install the refrigerant hose retaining clips and the evaporator retaining screws.
- Install the cruise control vacuum pump and mounting screws.
- Using new O-rings, connect the coolant hose quick-disconnect couplings at the heater core.
- Install the windshield wiper mechanism and connect the windshield wiper electrical connector.
- Install the lead-through panel and the 4 cowl lead-through panel screws.
- Install the electronic ignition control unit and mounting bolts.
- Install the cowl cover panel.
- Install the signal converter and the cowl cover panel seal.
- Connect the blower motor control unit electrical connector.
- Install the air conditioning evaporator cover panels and the windshield wiper motor.
- Install the windshield wiper arms.
- Install the hood.

7. Refill the cooling system.
8. Connect the negative battery cable.
9. Operate the engine to normal operating temperatures; then, check the climate control operation and check for leaks.

9-5

REMOVAL & INSTALLATION

1. Disconnect the negative battery cable.
2. Drain the cooling system into a clean container for reuse.
3. Evacuate the A/C system.
- Remove the center console.
- Remove the front seats.
- Fold away the floor carpeting and remove the air ducts from under the seats.

4. Remove the dashboard as follows:
   a. Remove the speaker grille and remove the speakers.
   b. Remove the center dashboard cover.
   c. Remove the glove box.
   d. Remove the air duct.
   e. Remove the left side air bag and horn.
   f. Remove the steering wheel nut and remove the steering wheel.
   g. Data Link Connector (DLC).
   h. Remove the lower dashboard panel and air duct.
   i. Remove the floor light.
   j. Remove the gear lever gaiter, if equipped
   k. Remove the center compartment and ashtray.
   l. Disconnect the seat heater and cigarette lighter
   m. Disconnect the central locking system switch.
   n. Remove the fuse box.
   o. Remove the steering column gaiter.
   p. Remove the instrument panel.
   q. Remove the left side air duct.
   r. Remove the right side air bag.
   s. Remove the front scuff plate.
- Remove the knee bolsters.
- Remove the relay holder
- Remove the steering column as an assembly.
- Remove the pedal bracket.
- Remove the center dashboard member brace.
- Remove the center mounting stabilizer plate
- Disconnect the automatic control module.
- Remove the charge air cooler to throttle body pipe.

- Remove the charge air bypass valve solenoid.
- Remove the A/C pipes from the expansion valve.
- Remove the expansion valve.
- Disconnect the drain hoses from the climate unit.
- Remove the heating and ventilation unit from the bulkhead.

To install:
- Install the heating and ventilation unit to the bulkhead
- Connect the drain hoses on both sides of the unit
- Install the air ducts under the front seats
- Install the screws that hold the heating and ventilation unit to the bulkhead
- Install the expansion valve
- Install the A/C pipes to the expansion valve with new O-rings
- Connect the heat exchanger hoses at the bulkhead
- Solenoid valve and holder to the bulkhead
- Solenoid valve cover
- Connect the pipe between the charge air cooler and the throttle body
- Install the 4 connectors to the heating and ventilation unit on the right hand side
- Place the dashboard into position
- Install the outer mountings
- Install the center mounting
- Install the stabilizer plate in the center mounting
- Install the pedal bracket
- Install the steering column as an assembly
- Install the relay holder
- Install both knee bolsters
- Install the center cable conduit
- Connect all ground cables
- Install the center console
- Install the front seats
- Fill and bleed the cooling system
- Recharge the A/C system
- Connect the negative battery cable

5. Operate the engine to normal operating temperatures; then, check the climate control operation and check for leaks.

1998–01 SUBARU

Impreza

REMOVAL & INSTALLATION

1. Disconnect the negative battery cable.

✳✳ CAUTION

If equipped with an air bag system, wait 10 minutes after disconnecting the negative battery cable before performing any further work while the system fully de-energizes in order to avoid accidental deployment. All air bag system wiring is yellow. Do not use electrically powered test equipment on these circuits.

2. Drain the cooling system into a clean container for reuse.

3. Disconnect the heater hoses from the heater core. Plug the heater core and heater hoses.

4. Remove the radio box or console.

5. Remove the instrument panel by:
   - Remove the rear console box.
   - Pull the cup holder.
   - Turn over the shift lever boot (manual transaxle models) or remove

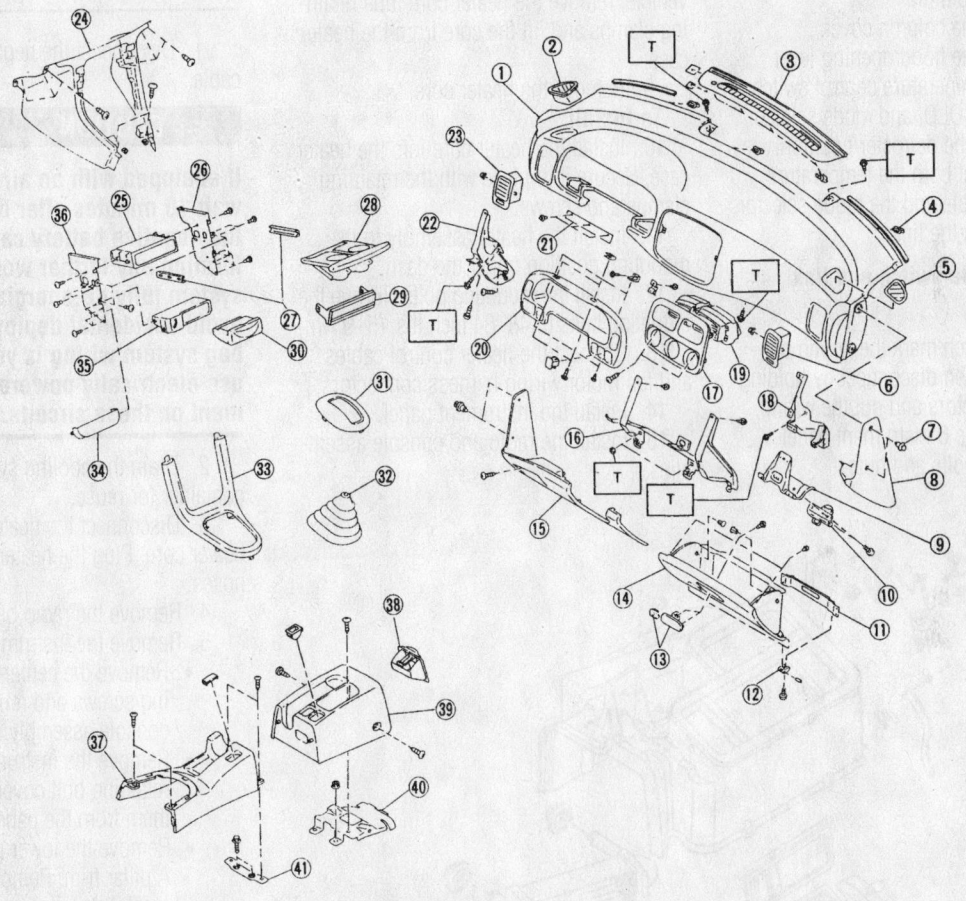

① Pad & frame	⑯ Reinf. CTR	㉛ Panel (AT) ASSY
② Grille SD def. (D)	⑰ Panel CTR (A)	㉜ Shift boot
③ Front def. grille	⑱ Reinf. (P)	㉝ Console cover
④ Grommet	⑲ Grille CTR def.	㉞ Panel (Airbag)
⑤ Grille SD def. (P)	⑳ Meter visor	㉟ Housing (Ash tray)
⑥ Grille vent (P)	㉑ Cover	㊱ BRKT (Radio) LH
⑦ Clip	㉒ Reinf. (D)	㊲ Center console
⑧ SD panel (P)	㉓ Grille vent (D)	㊳ Ash tray
⑨ Reinforcement striker	㉔ Instrument panel console	㊴ Rear console box
⑩ Striker	㉕ Pocket CTR	㊵ Rear console BRKT
⑪ Frame pocket	㉖ BRKT (Radio) RH	㊶ Center console BRKT
⑫ Hinge	㉗ Rail (Cup holder)	
⑬ Lock ASSY	㉘ Cup holder	**Tightening torque: N·m (kg-cm, in-lb)**
⑭ Pocket ASSY	㉙ Panel (Radio)	**T: 6.9 ± 1.0 (70 ± 10, 60.8 ± 8.7)**
⑮ Lower cover ASSY	㉚ Ash tray	

87970G59

Exploded view of the instrument panel and center console—Impreza

For Accessory Drive Belt illustrations, see Section 1 of this manual

select lever cover (automatic transaxle models).
- Remove the console cover.
- Remove the audio assembly and disconnect the antenna cable and connectors.
- Remove the lower cover and then disconnect the seat belt timer connector.
- Remove the glove box.
- Remove the instrument panel console.
- Remove the 2 bolts and lower the steering column.
- Remove the column cover.
- Remove the hood opening lever.
- Set the temperature control switch to MAX. COLD, and mode selector switch to the defroster position.
- Disconnect both the temperature control cable and the mode selector cable from the link.

➡**Do not move the switch and link when installing.**

- Tag or match mark the wiring connectors, then disconnect by holding the connectors and not the wiring.
- Remove the 6 instrument panel retaining bolts and nuts.

- Remove the front defroster grille and 2 bolts.
- Carefully remove the instrument panel from the body and then disconnect the speedometer cable from the back of the combination meter.

6. Disconnect the heater control cables and the fan motor wiring harness.

7. Disconnect the duct between the heater unit and the blower heater unit. Lift up and out on the heater unit and remove it.

8. With the heater assembly out of the vehicle, remove the heater core tube retaining clamps and lift the core from the heater case.

9. Remove the heater core.

To install:

10. Install the heater core into the heater case. Secure it in place with the retaining clamps and screws.

11. Install the heater assembly to its mounting position under the dash.

12. Install the mounting bolts. Torque the mounting bolts to 48–84 inch lbs. (5–9 Nm).

13. Connect the heater control cables and fan motor wiring harness connectors.

14. Install the instrument panel.

15. Install the radio and console assemblies.

16. Connect the heater hoses in the engine compartment.

17. Refill the cooling system.

18. Connect the negative battery cable.

19. Evacuate, charge and leak test the air conditioning system.

20. Operate the engine to normal operating temperatures; then, check the climate control operation and check for leaks.

Legacy

REMOVAL & INSTALLATION

1. Disconnect the negative battery cable.

✳✳ CAUTION

If equipped with an air bag system, wait 10 minutes after disconnecting the negative battery cable before performing any further work while the system fully de-energizes in order to avoid accidental deployment. All air bag system wiring is yellow. Do not use electrically powered test equipment on these circuits.

2. Drain the cooling system into a clean container for reuse.

3. Disconnect the heater hoses from the heater core. Plug the heater core and heater hoses.

4. Remove the radio box or console.

5. Remove the instrument panel by:
- Remove the center console retaining screws and remove the center console assembly.
- Remove the instrument panel retaining bolt covers by prying them from the panel.
- Remove the lower part of the front A pillar trim. Remove the instrument panel under covers from the driver's and passenger's sides.
- Remove the hood release cable from the hood release lever.
- Disconnect the wiring harness connectors under the instrument panel.
- Remove the instrument cluster assembly. Remove the glove box assembly
- Disconnect the ventilation control cables and electrical connectors at the heater unit. Disconnect the vacuum line at the blower housing.
- Disconnect the radio antenna feeder wire. Disconnect the main harness connector at the fuse box.
- Remove the lower steering column covers. Remove the steering col-

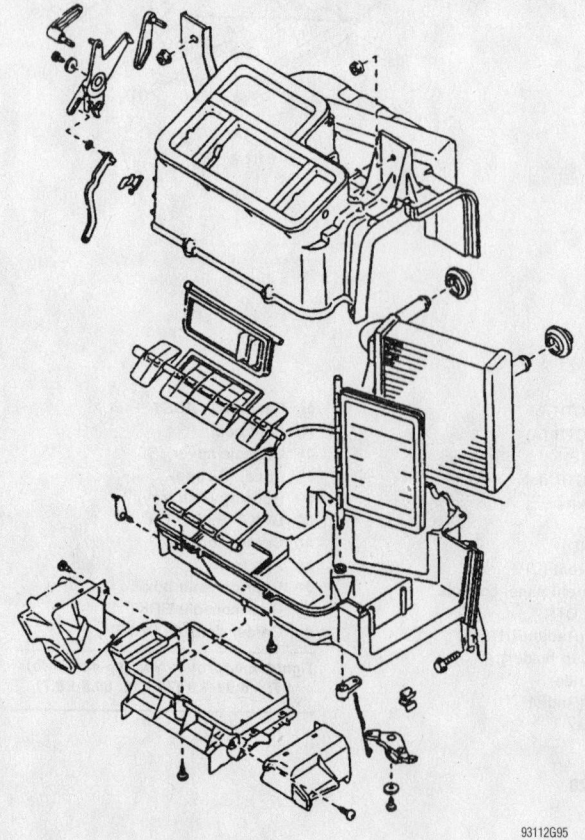

93112G95

Exploded view of the heater assembly—Impreza

umn retaining bolts and allow the column to hang down.

- Remove the instrument panel retaining bolts.

➡ **When removing the instrument panel, check that all wiring and cables are disconnected before pulling it completely away from the firewall.**

- With the help of an assistant, lift and remove the instrument panel from the vehicle.

6. Disconnect the heater control cables and the fan motor wiring harness.

7. Disconnect the duct between the heater unit and the blower heater unit. Lift up and out on the heater unit and remove it.

8. Remove the evaporator case assembly by performing the following procedure:

- Discharge and recover the air conditioning system refrigerant.
- Disconnect the low and high pressure line from the evaporator outlet and cap the fittings.
- Remove the inlet and outlet pipe grommets.
- Remove the glove box and support bracket.
- Disconnect the air conditioning wiring harness from the evaporator.

Disconnect the drain hose from the evaporator.

- Remove the evaporator mounting nut and bolt.
- Remove the evaporator case assembly from the vehicle.

9. With the heater assembly out of the vehicle, remove the heater core tube retaining clamps and lift the core from the heater case.

10. Remove the heater core.

To install:

11. Install the heater core into the heater case. Secure it in place with the retaining clamps and screws.

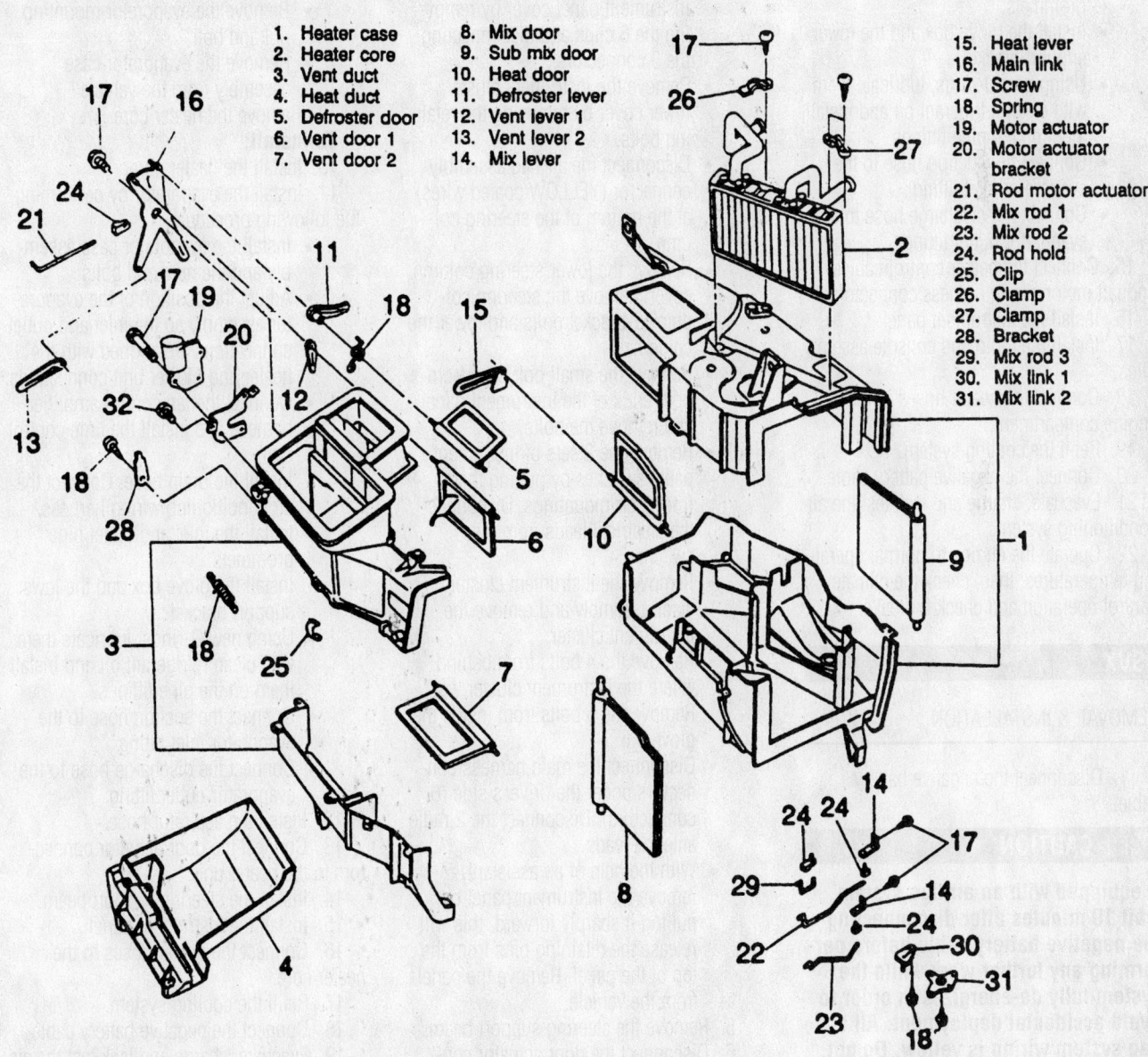

1. Heater case	8. Mix door	15. Heat lever
2. Heater core	9. Sub mix door	16. Main link
3. Vent duct	10. Heat door	17. Screw
4. Heat duct	11. Defroster lever	18. Spring
5. Defroster door	12. Vent lever 1	19. Motor actuator
6. Vent door 1	13. Vent lever 2	20. Motor actuator bracket
7. Vent door 2	14. Mix lever	21. Rod motor actuator
		22. Mix rod 1
		23. Mix rod 2
		24. Rod hold
		25. Clip
		26. Clamp
		27. Clamp
		28. Bracket
		29. Mix rod 3
		30. Mix link 1
		31. Mix link 2

Exploded view of the heater assembly—Legacy

93112G93

12. Install the heater assembly to its mounting position under the dash.

13. Install the mounting bolts. Torque the mounting bolts to 48–84 inch lbs. (5–9 Nm).

14. Install the evaporator by performing the following procedure:

- Install the evaporator case assembly and the nuts and bolts.
- Adjust the position of the evaporator assembly so the inlet and outlet connections are aligned with the heater and blower unit connections.
- Install the drain hose. Connect the air conditioning wiring harness.
- Install the inlet and outlet pipe grommets.
- Install the glove box and the lower support bracket.
- Using new O-rings, lubricate them with clean refrigerant oil and install them on the pipe fittings.
- Connect the suction hose to the evaporator inlet fitting.
- Connect the discharge hose to the evaporator outlet fitting.

15. Connect the heater control cables and fan motor wiring harness connectors.

16. Install the instrument panel.

17. Install the radio and console assemblies.

18. Connect the heater hoses in the engine compartment.

19. Refill the cooling system.

20. Connect the negative battery cable.

21. Evacuate, charge and leak test the air conditioning system.

22. Operate the engine to normal operating temperatures; then, check the climate control operation and check for leaks.

SVX

REMOVAL & INSTALLATION

1. Disconnect the negative battery cable.

✳✳ CAUTION

If equipped with an air bag system, wait 10 minutes after disconnecting the negative battery cable before performing any further work while the system fully de-energizes in order to avoid accidental deployment. All air bag system wiring is yellow. Do not use electrically powered test equipment on these circuits.

2. Drain the cooling system into a clean container for reuse.

3. Disconnect the heater hoses from the heater core. Plug the heater core and heater hoses.

4. Remove the instrument panel by:

- Remove the center console assembly retaining screws and remove the center console.
- Remove the front A pillar upper trim pieces.
- Remove the radio grounding wire, which is screwed to the floor just behind the shifter assembly.
- Remove the power mirror switch and remove the bolt located behind it.
- Remove the lower driver's side instrument panel cover by removing the 6 clips and disconnecting the 3 connectors.
- Remove the instrument cluster lower cover by removing the retaining bolts.
- Disconnect the air bag assembly connector (YELLOW coated wires) at the bottom of the steering column.
- Remove the lower steering column cover. Remove the steering column-to-bracket bolts and lower the column.
- Remove the small bolt caps from both ends of the instrument panel and remove the bolts.
- Remove the 2 sets of instrument panel switches by pulling them from their mountings. Disconnect the electrical leads from the switches.
- Remove the instrument cluster visor assembly and remove the instrument cluster.
- Remove the 4 bolts from behind where the instrument cluster was. Remove the 5 bolts from inside the glove box.
- Disconnect the main harness connectors under the driver's side (6 connectors). Disconnect the 2 radio antenna leads.
- With the help of an assistant, remove the instrument panel by pulling it sharply forward, this will release the retaining pins from the top of the panel. Remove the panel from the vehicle.

5. Remove the steering support beam.

6. Disconnect the door actuator connectors from the heater unit.

7. Remove the aspirator hose.

8. Remove the evaporator case assembly by performing the following procedure:

- Discharge and recover the air conditioning system refrigerant.
- Disconnect the low and high pressure line from the evaporator outlet and cap the fittings.
- Remove the inlet and outlet pipe grommets.
- Remove the glove box and support bracket.
- Remove the time control unit and detach the fan control amplifier harness connector.
- Disconnect the air conditioning wiring harness from the evaporator. Disconnect the drain hose from the evaporator.
- Remove the evaporator mounting nut and bolt.
- Remove the evaporator case assembly from the vehicle.

9. Remove the heater core.

To install:

10. Install the heater core.

11. Install the evaporator by performing the following procedure:

- Install the evaporator case assembly and the nuts and bolts.
- Adjust the position of the evaporator assembly so the inlet and outlet connections are aligned with the heater and blower unit connections.
- Connect the fan control amplifier harness and install the time control unit.
- Install the drain hose. Connect the air conditioning wiring harness.
- Install the inlet and outlet pipe grommets.
- Install the glove box and the lower support bracket.
- Using new O-rings, lubricate them with clean refrigerant oil and install them on the pipe fittings.
- Connect the suction hose to the evaporator inlet fitting.
- Connect the discharge hose to the evaporator outlet fitting.

12. Install the aspirator hose.

13. Connect the door actuator connectors to the heater unit.

14. Install the steering support beam.

15. Install the instrument panel.

16. Connect the heater hoses to the heater core.

17. Refill the cooling system.

18. Connect the negative battery cable.

19. Evacuate, charge and leak test the air conditioning system.

20. Operate the engine to normal operating temperatures; then, check the climate control operation and check for leaks.

Esteem

REMOVAL & INSTALLATION

1. Disconnect the negative battery cable.
2. Wait for a least 1 minute for the air bag system to deplete its energy before working on any part the instrument panel.
3. Drain the cooling system into a clean container for reuse.
4. Remove the instrument panel.
5. Remove the mode actuator by performing the following procedure:
   - Disconnect the mode actuator coupler.
   - At the heater housing, remove the mode actuator rod.
   - Remove the mode actuator from the heater housing.
6. Remove the rear air duct.
7. Remove the heater housing.
8. Remove the heater housing clips and screws; then, separate the heater housings.
9. Remove the heater core from the heater housing.

To install:

10. Install the heater core to the heater housing.
11. Assemble the heater housings and install the heater housing clips and screws.
12. Install the heater housing.
13. Install the rear air duct.
14. Install the mode actuator by performing the following procedure:
   - Install the mode actuator to the heater housing.
   - At the heater housing, install the mode actuator rod.
   - Connect the mode actuator coupler.
15. Install the instrument panel.
16. Refill the cooling system.
17. Connect the negative battery cable.
18. Operate the engine to normal operating temperatures; then, check the climate control operation and check for leaks.

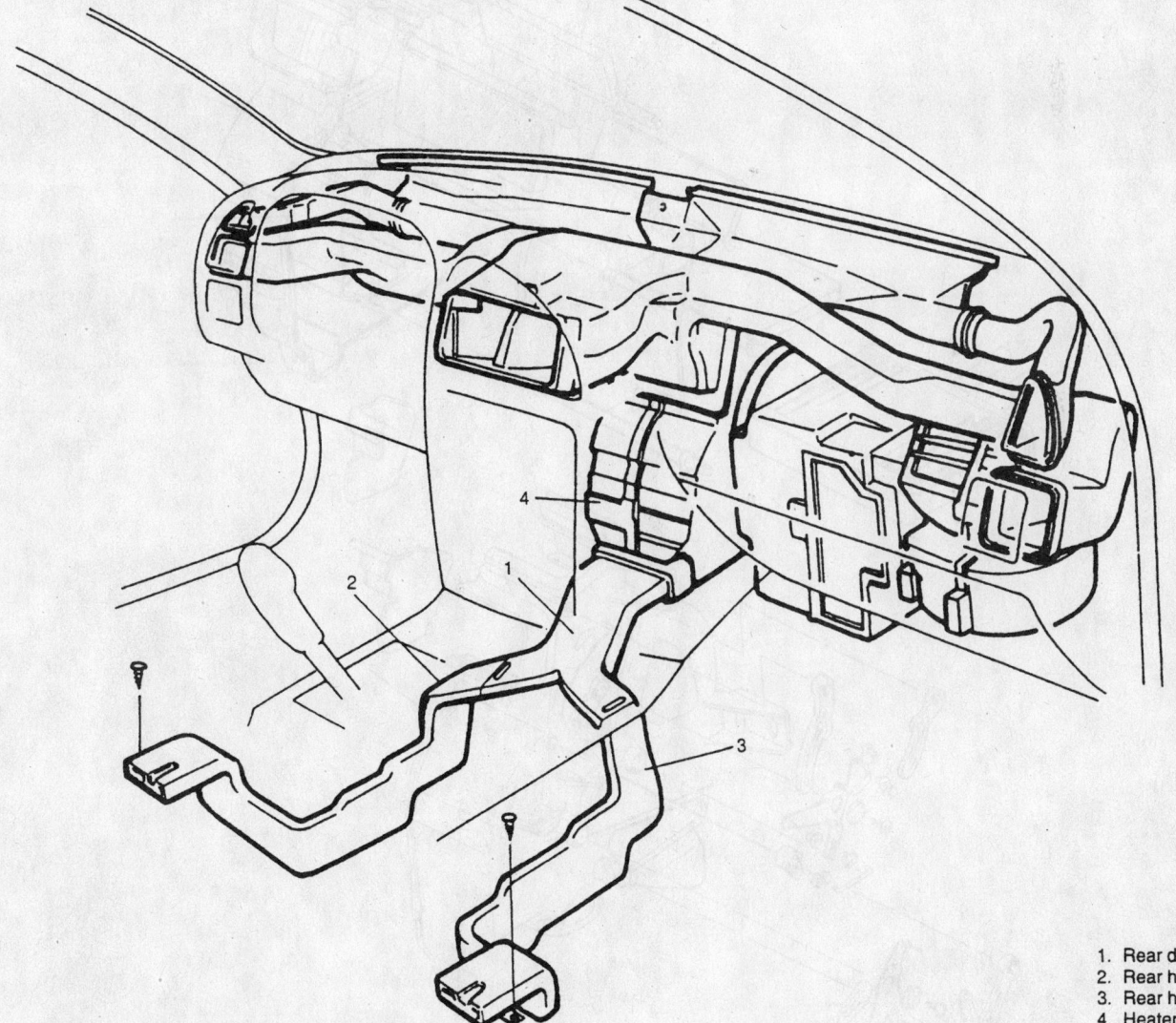

1. Rear duct
2. Rear hose LH
3. Rear hose RH
4. Heater unit

93112GI4

Assembled view of the heater housing and related components—Esteem

For Wheel Alignment specifications, see Section 1 of this manual

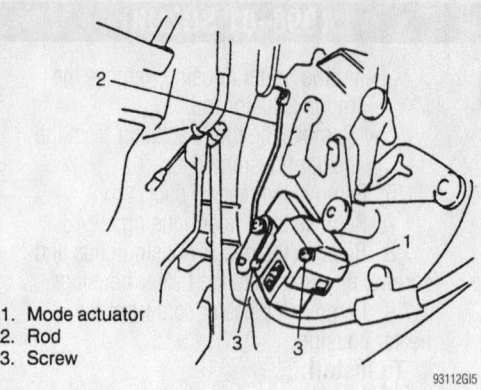

1. Mode actuator
2. Rod
3. Screw
2
1
3
3
93112GI5

View of the heater housing mode actuator—Esteem

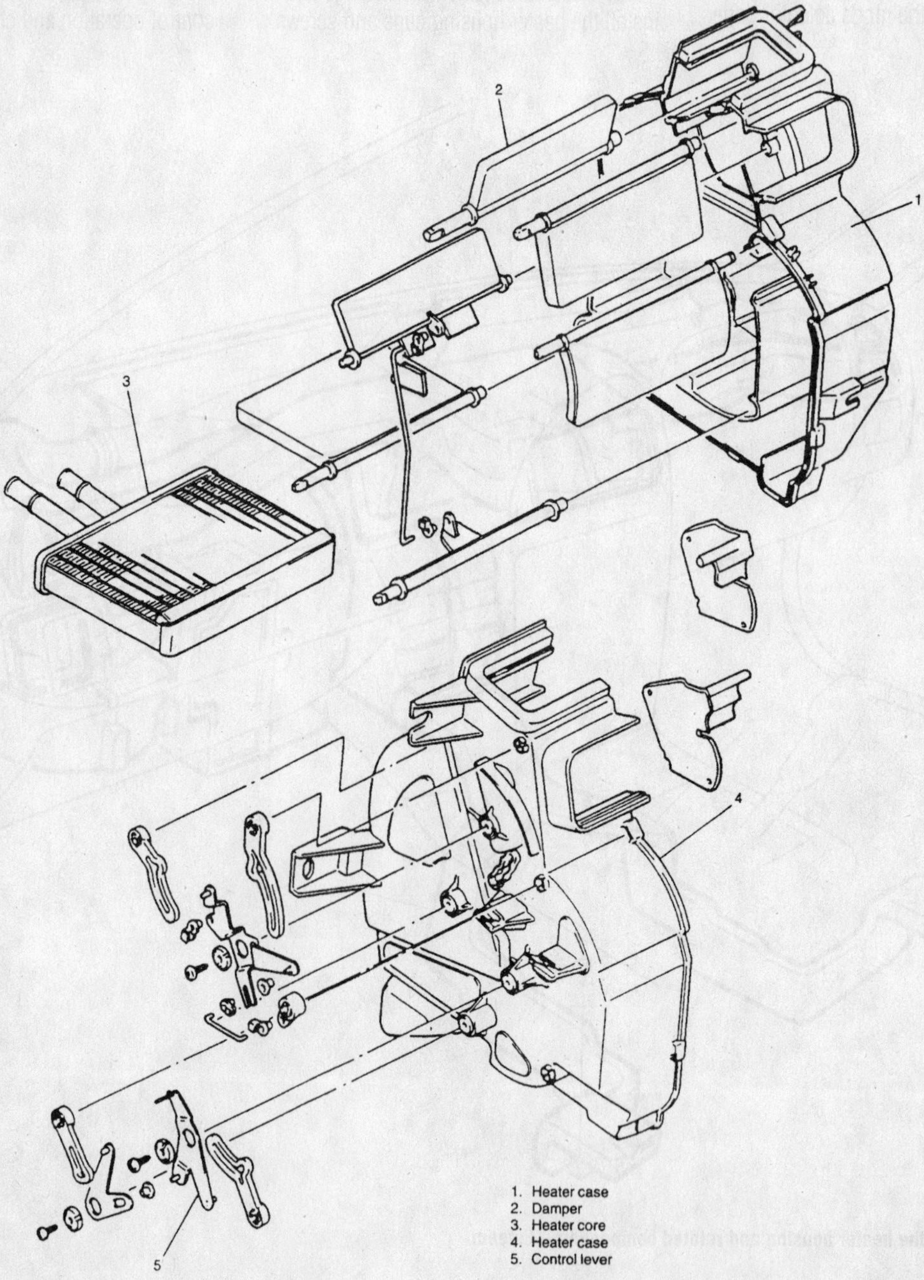

1
2
3
4
5
1. Heater case
2. Damper
3. Heater core
4. Heater case
5. Control lever
93112GI6

Exploded view of the heater core, heater housing and related components—Esteem

Swift

REMOVAL & INSTALLATION

1. Disconnect the negative battery cable.

2. Drain the cooling system into a clean container for reuse.

3. Disconnect the heater hoses at the heater core. Plug the heater hoses.

4. Remove the console box.

5. Disconnect the electrical wires and control cables from the heater housing.

6. Disconnect the steering joint upper bolt and remove the steering column unit from the vehicle.

7. Remove the speedometer assembly, the left and right speaker covers, the hood opening cable and the center garnish from the dash.

8. Remove the dashboard mounting bolts and the dashboard being careful to protect it from damage.

9. Remove the heater housing mounting bolts and remove it from the vehicle.

10. Remove the attaching screws and clips which hold the heater housing together and separate the housing to gain access to the heater core.

11. Remove the heater core from the case.

To install:

12. Install the heater core to the case.

13. Install the heater housing and the attaching screws and clips.

14. Install the heater housing and mounting bolts.

15. Install the dashboard and mounting bolts.

16. Install the speedometer assembly, the left and right speaker covers, the hood opening cable and the center garnish to the dash.

17. Install the steering column unit and connect the steering joint upper bolt.

18. Connect the electrical wires and control cables to the heater housing.

19. Install the console box.

20. Connect the heater hoses at the heater core.

21. Refill the cooling system.

22. Connect the negative battery cable.

23. Operate the engine to normal operating temperatures; then, check the climate control operation and check for leaks.

1998–00 TOYOTA

Avalon

REMOVAL & INSTALLATION

1. Disconnect the negative battery cable.

✳✳ CAUTION

After disconnecting the negative battery cable, wait for at least 1½ minutes for the SRS to deplete its energy.

2. Drain the cooling system into a clean container for reuse.

3. Remove the air bag module and the steering wheel by performing the following procedure:
- Place the front wheels in the straight-ahead position.
- At both sides of the steering wheel, remove the screw covers.
- Using a Torx socket, loosen the 2 air bag module-to-steering wheel Torx screws until the circumference ring catches on the screw case.
- Carefully, remove the air bag module and disconnect the electrical connector.
- Remove the steering wheel nut.
- Using a steering wheel puller, press the steering wheel from the steering column.

4. Remove the instrument panel by performing the following procedure:
- Remove the front pillar garnishes and the door scuff plates.
- Remove the hood lock release lever and the cowl side trims.
- Remove the steering column covers and the combination switch.
- Remove the lower finish panel assembly and the instrument panel finish lower left side panel.
- Remove the fuse box bolt and the No. 2 heater-to-register duct.
- Remove the parking brake release lever and the No. 2 undercover.
- Remove the lower No. 2 finish panel.
- If equipped with a column shifter, disconnect the shift control cable from the shift lever housing; then, disconnect the shift control cable from the steering column cable bracket.
- Matchmark the steering column shaft and the control valve shaft.
- Remove the steering column shaft-to-intermediate shaft bolt.
- Remove the steering column-to-instrument panel nuts and remove the steering column assembly.
- Inside the glove compartment, pry out the glove compartment door finish plate.
- Pull out the air bag electrical connector and disconnect it.
- Remove the 3 glove compartment door-to-instrument panel nuts and the door.
- Remove the 4 glove compartment-to-instrument panel screws and the glove compartment; then, disconnect the glove box light connector.
- Remove the passenger's side air bag module-to-instrument panel 2 bolts and 4 nuts. Carefully, remove the air bag module from the instrument panel.
- Remove the center cluster finish panel and the radio.
- Remove the heater control assembly.
- If equipped with a floor shifter, remove the upper console panel, the rear console box and the front console box.
- If equipped with a column shifter, remove the finish panel.
- Pry out the cluster finish panel.
- Remove the 6 cluster finish panel screws, the cluster finish panel assembly.
- Remove the 4 combination meter screws and the combination meter.
- Disconnect instrument panel electrical connectors.
- Remove the instrument panel-to-chassis nuts/bolts and the remove the instrument panel.

5. Remove the instrument panels No. 2 brace.

6. Disconnect the heater hoses from the heater core.

7. Remove the 2 heater pipes-to-heater core screws and clips; then, disconnect the heater pipes from the heater core.

8. Remove the heater core O-rings and discard them.

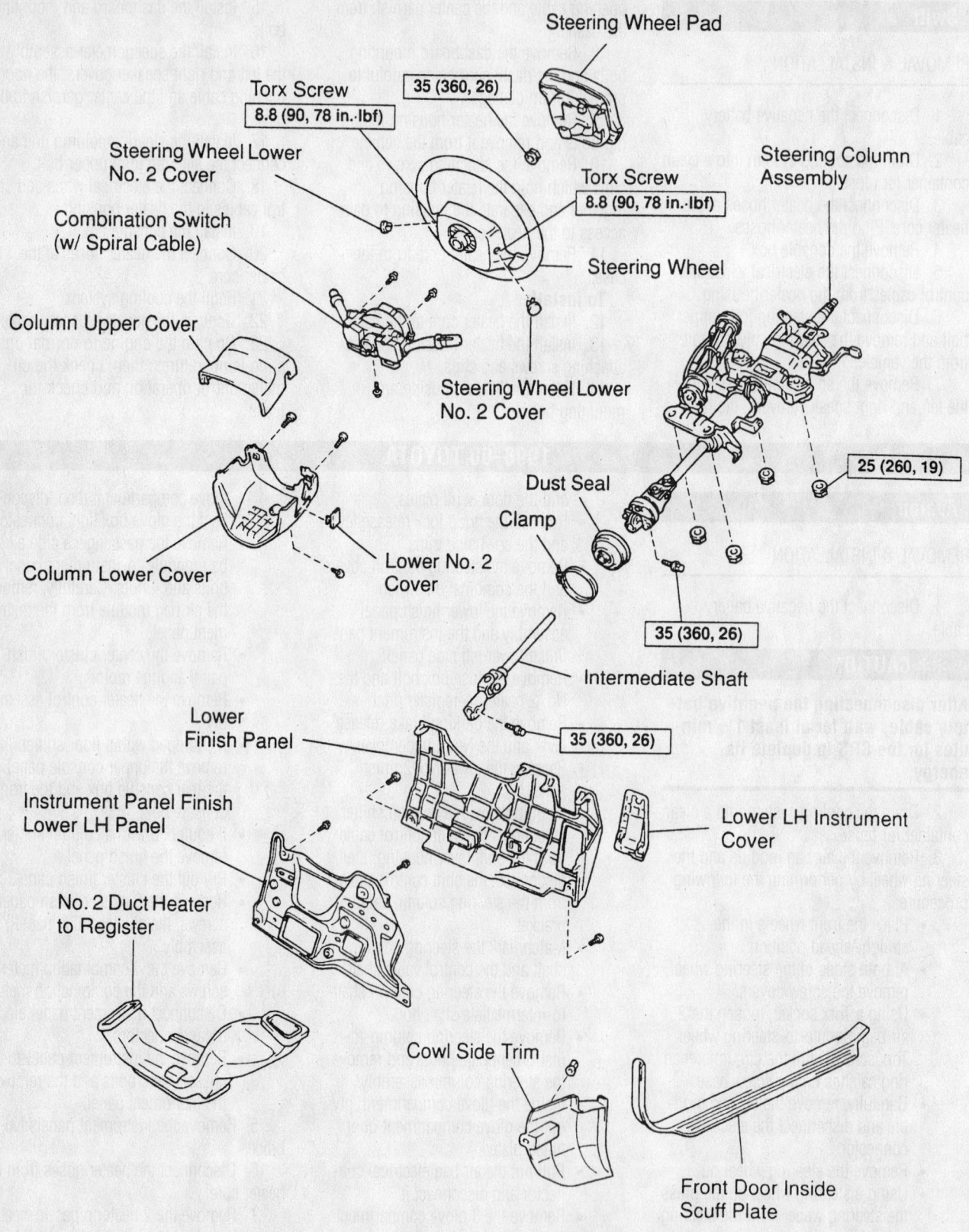

Steering Wheel Pad

Torx Screw
8.8 (90, 78 in.·lbf)

35 (360, 26)

Steering Wheel Lower
No. 2 Cover

Steering Column
Assembly

Combination Switch
(w/ Spiral Cable)

Torx Screw
8.8 (90, 78 in.·lbf)

Steering Wheel

Column Upper Cover

Steering Wheel Lower
No. 2 Cover

Column Lower Cover

Dust Seal

Clamp

Lower No. 2
Cover

25 (260, 19)

35 (360, 26)

Intermediate Shaft

Lower
Finish Panel

Instrument Panel Finish
Lower LH Panel

35 (360, 26)

Lower LH Instrument
Cover

No. 2 Duct Heater
to Register

Cowl Side Trim

Front Door Inside
Scuff Plate

N·m (kgf·cm, ft·lbf) : Specified torque

93112GL2

Exploded view of the air bag module, the steering wheel, the floor shift steering column and related components—Avalon

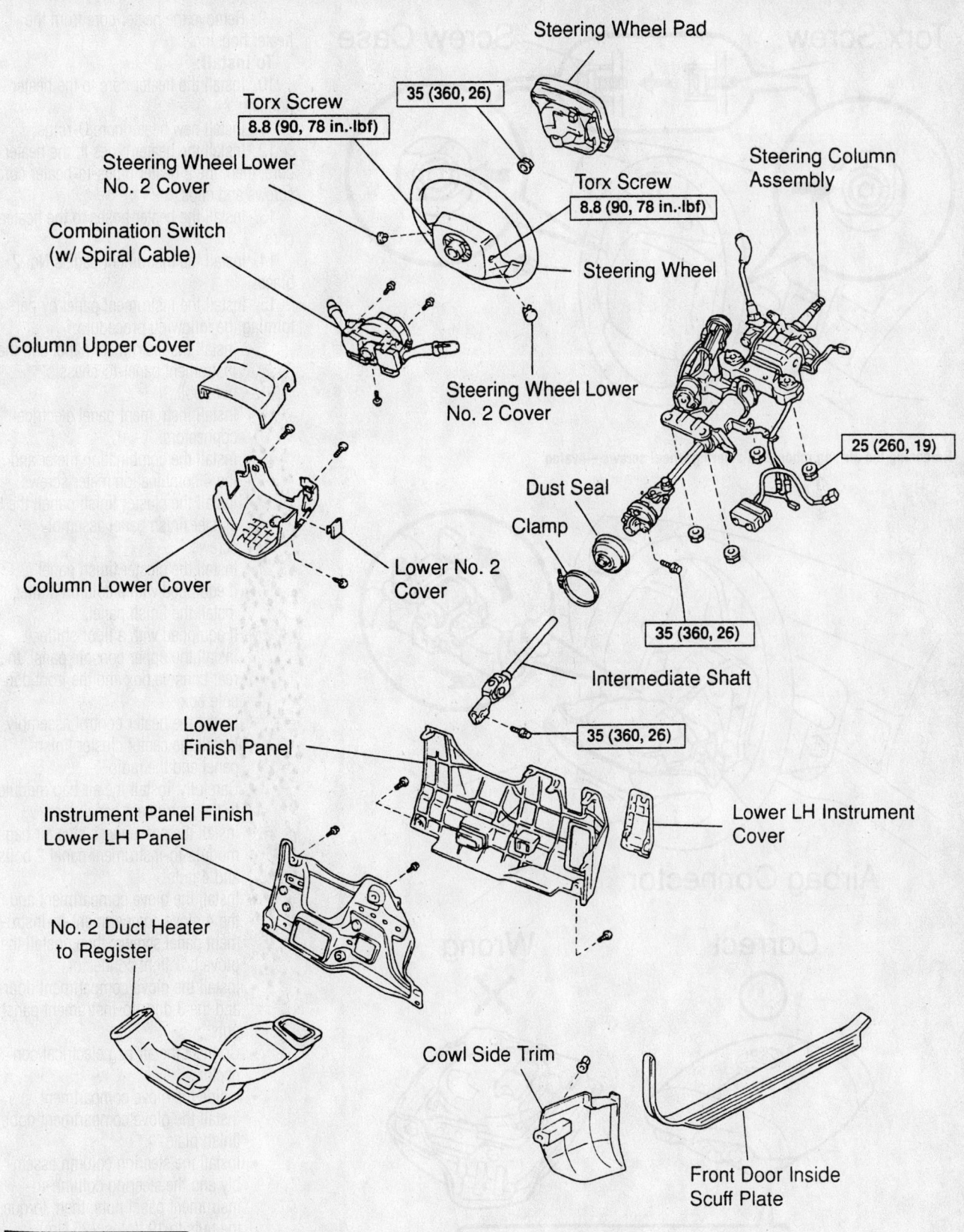

Steering Wheel Pad

Torx Screw
8.8 (90, 78 in.·lbf)

35 (360, 26)

Steering Wheel Lower
No. 2 Cover

Steering Column
Assembly

Combination Switch
(w/ Spiral Cable)

Torx Screw
8.8 (90, 78 in.·lbf)

Steering Wheel

Column Upper Cover

Steering Wheel Lower
No. 2 Cover

Column Lower Cover

Dust Seal

Clamp

Lower No. 2
Cover

25 (260, 19)

35 (360, 26)

Intermediate Shaft

Lower
Finish Panel

35 (360, 26)

Instrument Panel Finish
Lower LH Panel

Lower LH Instrument
Cover

No. 2 Duct Heater
to Register

Cowl Side Trim

Front Door Inside
Scuff Plate

N·m (kgf·cm, ft·lbf) : Specified torque

93112GL3

Exploded view of the air bag module, the steering wheel, the column shift steering column and related components—Avalon

For Tune-up, Capacities and Firing orders, see Section 1 of this manual

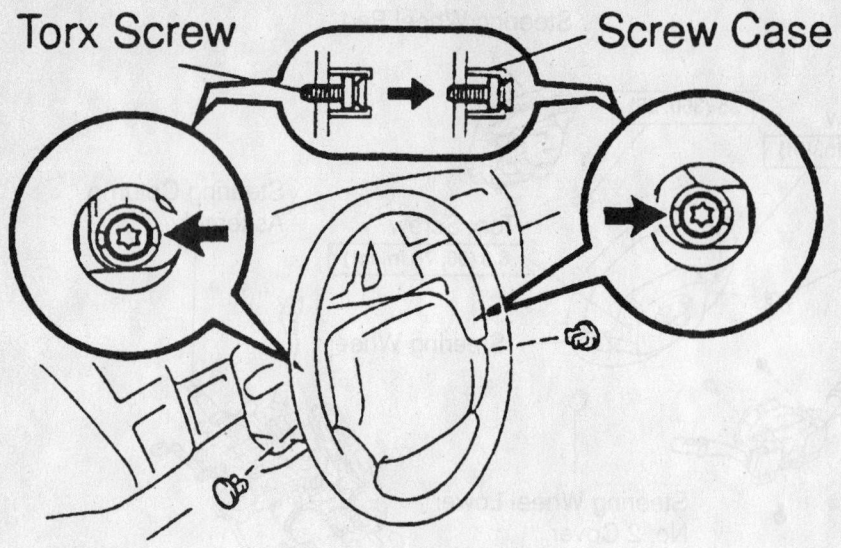

Releasing the air bag module-to-steering wheel screws—Avalon

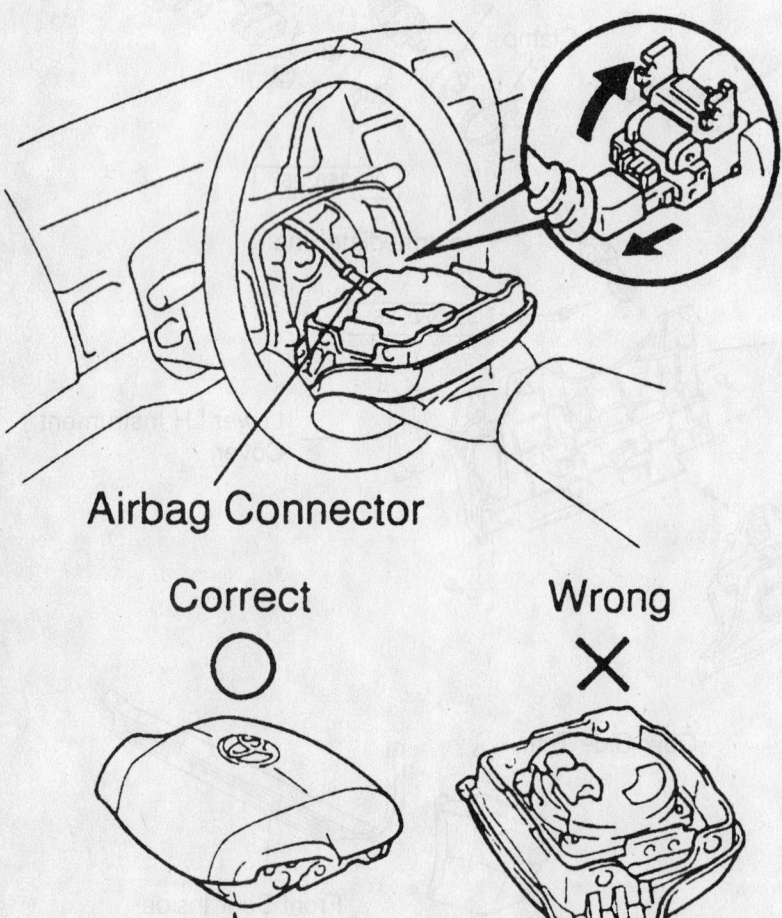

Disconnecting and positioning the air bag module—Avalon

9. Remove the heater core from the heater housing.

To install:

10. Install the heater core to the heater housing.

11. Install new heater core O-rings.

12. Install the heater pipes to the heater core; then, the 2 heater pipes-to-heater core screws and clips.

13. Install the heater hoses to the heater core.

14. Install the instrument panels No. 2 brace.

15. Install the instrument panel by performing the following procedure:

- Install the instrument panel and the instrument panel-to-chassis nuts/bolts.
- Install instrument panel electrical connectors.
- Install the combination meter and the 4 combination meter screws.
- Install the cluster finish panel, the 6 cluster finish panel assembly screws.
- Install the cluster finish panel.
- If equipped with a column shifter, install the finish panel.
- If equipped with a floor shifter, install the upper console panel, the rear console box and the front console box.
- Install the heater control assembly.
- Install the center cluster finish panel and the radio.
- Carefully, install the air bag module to the instrument panel; then, install the passenger's side air bag module-to-instrument panel 2 bolts and 4 nuts.
- Install the glove compartment and the 4 glove compartment-to-instrument panel screws; then, install the glove box light connector.
- Install the glove compartment door and the 3 door-to-instrument panel nuts.
- Connect the air bag electrical connector.
- Inside the glove compartment, install the glove compartment door finish plate.
- Install the steering column assembly and the steering column-to-instrument panel nuts; then, torque the nuts to 19 ft. lbs. (25 Nm).
- Align the matchmarks and install the steering column shaft-to-intermediate shaft bolt.
- If equipped with a column shifter, install the shift control cable to the shift lever housing; then, install the

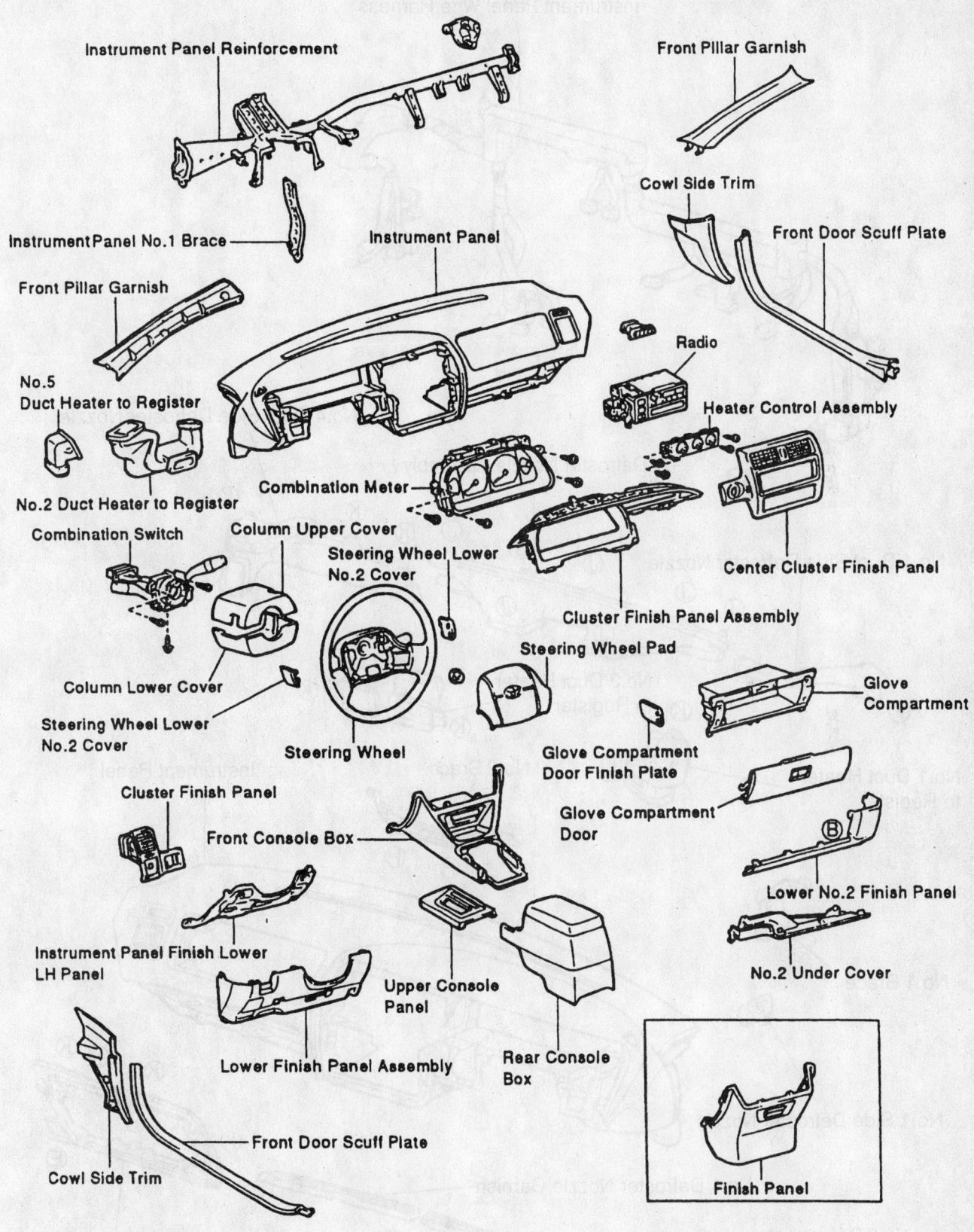

Instrument Panel Reinforcement

Front Pillar Garnish

Cowl Side Trim

InstrumentPanel No.1 Brace

Instrument Panel

Front Door Scuff Plate

Front Pillar Garnish

Radio

No.5
Duct Heater to Register

Heater Control Assembly

No.2 Duct Heater to Register

Combination Meter

Center Cluster Finish Panel

Combination Switch

Column Upper Cover

Steering Wheel Lower
No.2 Cover

Cluster Finish Panel Assembly

Column Lower Cover

Steering Wheel Pad

Glove
Compartment

Steering Wheel Lower
No.2 Cover

Steering Wheel

Glove Compartment
Door Finish Plate

Glove Compartment
Door

Lower No.2 Finish Panel

Cluster Finish Panel

Front Console Box

No.2 Under Cover

Instrument Panel Finish Lower
LH Panel

Upper Console
Panel

Lower Finish Panel Assembly

Rear Console
Box

Cowl Side Trim

Front Door Scuff Plate

Finish Panel

93112GA5

Exploded view of the instrument panel and related components—Avalon

For complete service labor times, order the Chilton Labor Guide

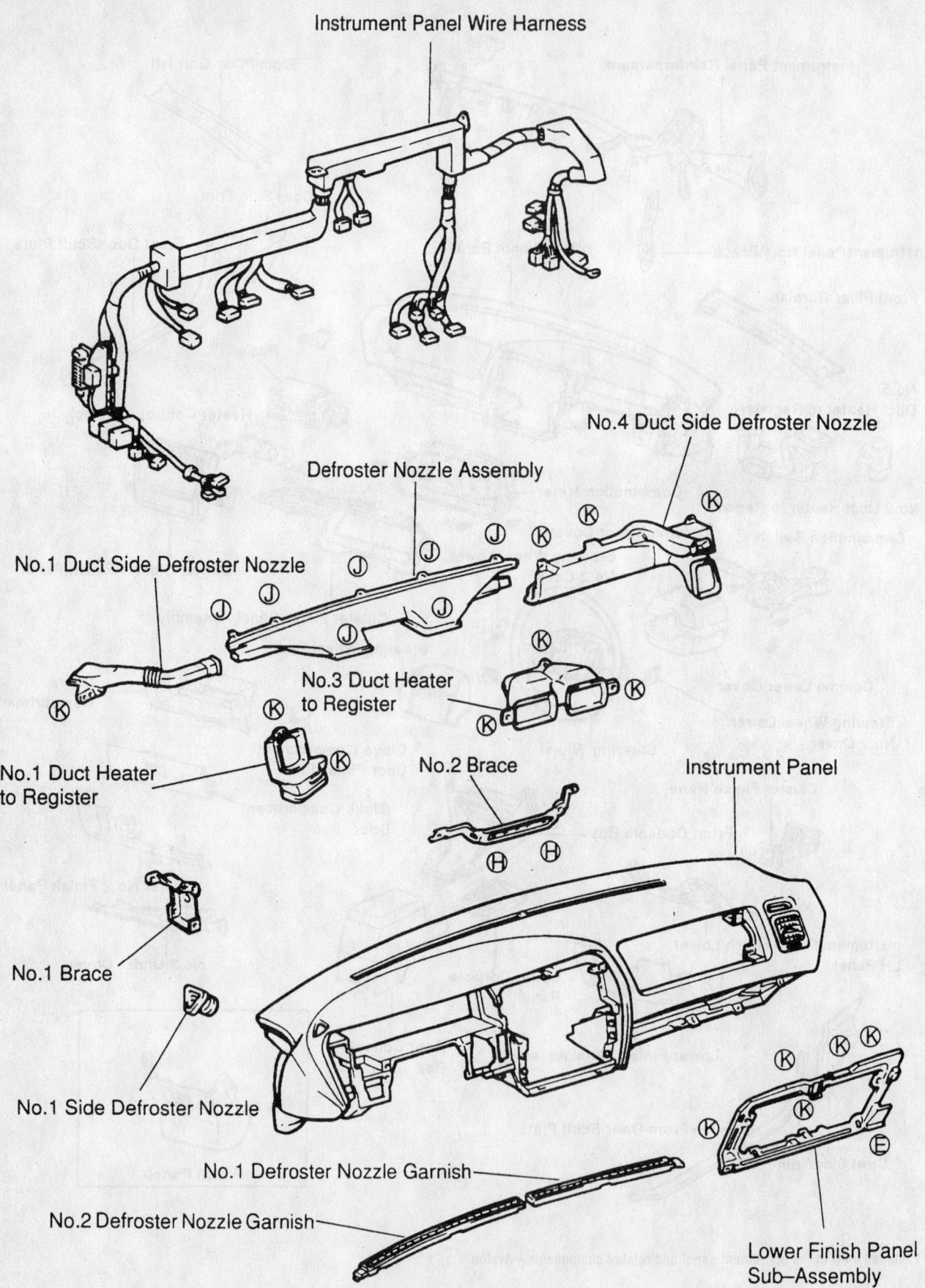

Instrument Panel Wire Harness

No.4 Duct Side Defroster Nozzle

Defroster Nozzle Assembly

No.1 Duct Side Defroster Nozzle

No.3 Duct Heater to Register

No.1 Duct Heater to Register

No.2 Brace

Instrument Panel

No.1 Brace

No.1 Side Defroster Nozzle

No.1 Defroster Nozzle Garnish

No.2 Defroster Nozzle Garnish

Lower Finish Panel Sub–Assembly

93112GL4

Exploded view of the wiring harness, ventilation system and related components—Avalon

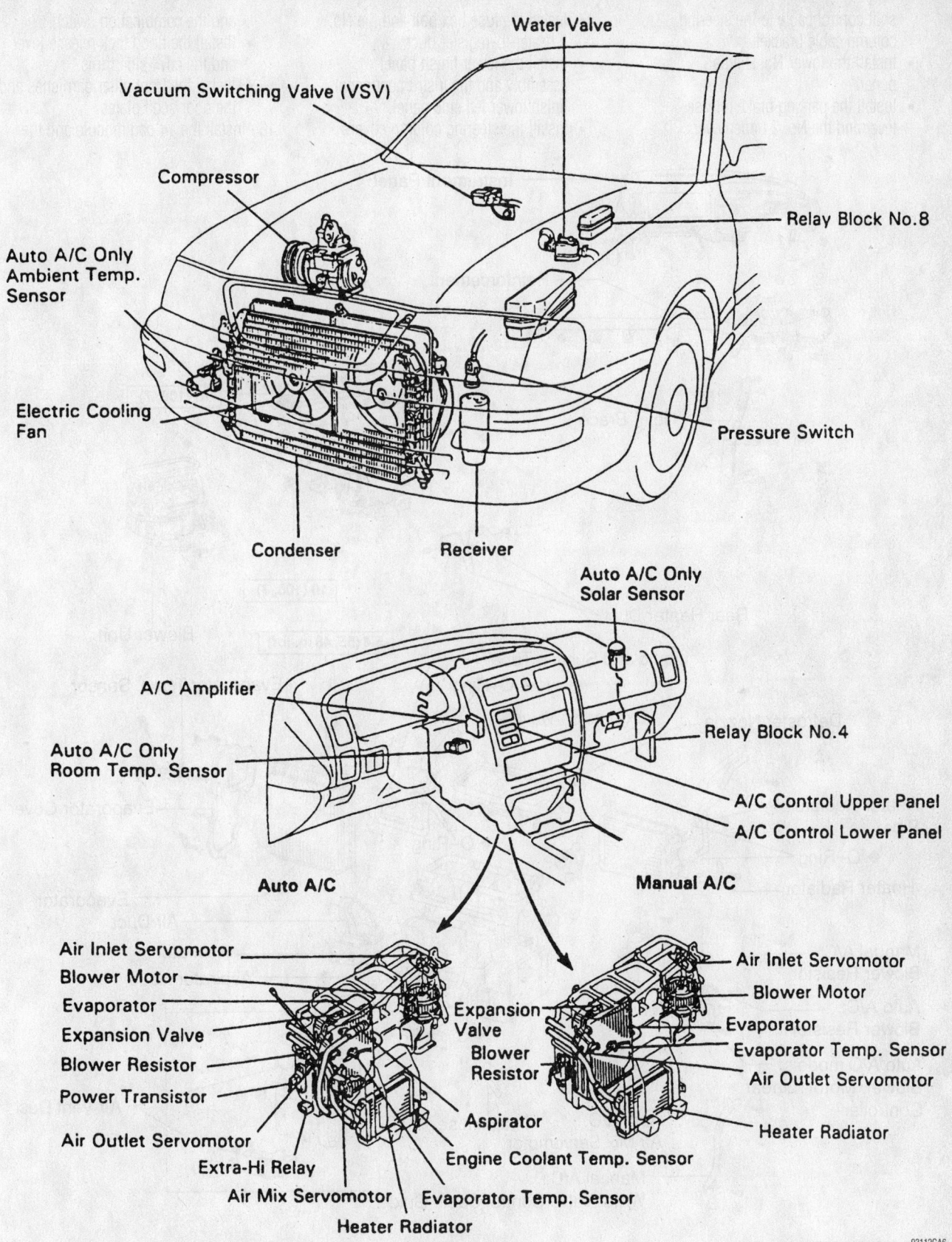

Water Valve

Vacuum Switching Valve (VSV)

Compressor

Relay Block No.8

Auto A/C Only
Ambient Temp.
Sensor

Electric Cooling
Fan

Pressure Switch

Condenser

Receiver

Auto A/C Only
Solar Sensor

A/C Amplifier

Relay Block No.4

Auto A/C Only
Room Temp. Sensor

A/C Control Upper Panel

A/C Control Lower Panel

Auto A/C

Manual A/C

Air Inlet Servomotor
Blower Motor
Evaporator
Expansion Valve
Blower Resistor
Power Transistor

Expansion
Valve

Blower
Resistor

Air Inlet Servomotor
Blower Motor
Evaporator
Evaporator Temp. Sensor
Air Outlet Servomotor

Air Outlet Servomotor

Aspirator

Heater Radiator

Extra-Hi Relay

Engine Coolant Temp. Sensor

Air Mix Servomotor

Evaporator Temp. Sensor

Heater Radiator

93112GA6

View of the heater/air conditioning assembly and related components—Avalon

Please visit our web site at www.chiltononline.com

shift control cable to the steering column cable bracket.
- Install the lower No. 2 finish panel.
- Install the parking brake release lever and the No. 2 undercover.

- Install the fuse box bolt and the No. 2 heater-to-register duct.
- Install the lower finish panel assembly and the instrument panel finish lower left side panel.
- Install the steering column covers

and the combination switch.
- Install the hood lock release lever and the cowl side trims.
- Install the front pillar garnishes and the door scuff plates.
16. Install the air bag module and the

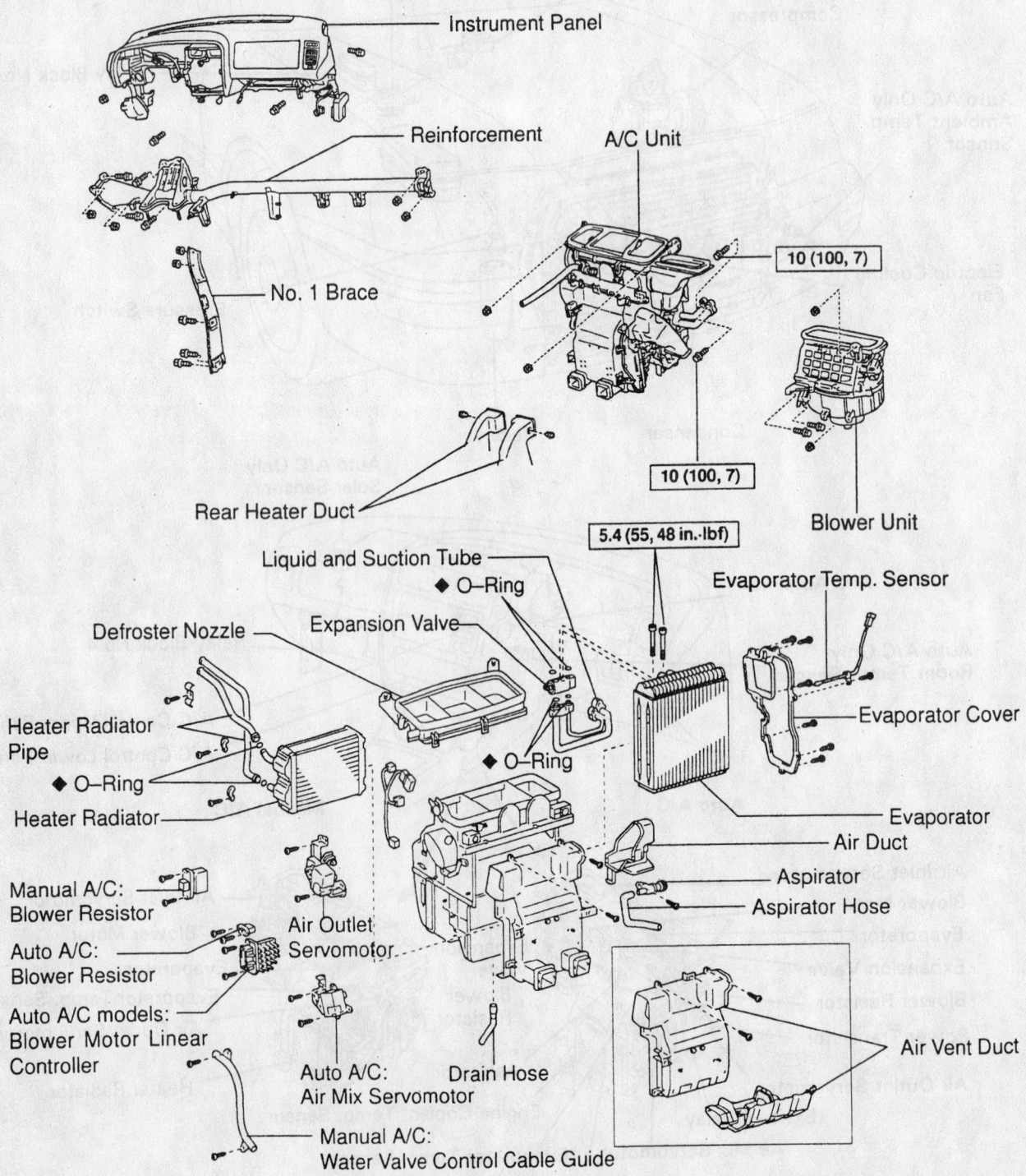

Instrument Panel

Reinforcement

A/C Unit

10 (100, 7)

No. 1 Brace

10 (100, 7)

Blower Unit

Rear Heater Duct

5.4 (55, 48 in.·lbf)

Liquid and Suction Tube

Evaporator Temp. Sensor

◆ O–Ring

Expansion Valve

Defroster Nozzle

Heater Radiator Pipe

Evaporator Cover

◆ O–Ring

◆ O–Ring

Heater Radiator

Evaporator

Air Duct

Manual A/C: Blower Resistor

Aspirator

Auto A/C: Blower Resistor

Aspirator Hose

Air Outlet Servomotor

Auto A/C models: Blower Motor Linear Controller

Air Vent Duct

Auto A/C: Air Mix Servomotor

Drain Hose

Manual A/C: Water Valve Control Cable Guide

N·m (kgf·cm, ft·lbf) : Specified torque

◆ Non–reusable part

93112GL5

Exploded view of the evaporator housing, heater housing, heater core and related components—Avalon

steering wheel by performing the following procedure:

- Align the matchmarks and install the steering wheel to the steering column.
- Install the steering wheel nut and torque to 26 ft. lbs. (35 Nm).
- Carefully, connect the electrical connector and install the air bag module.
- Using a Torx socket, tighten the 2 air bag module-to-steering wheel Torx screws to 78 inch lbs. (8.8 Nm).
- At both sides of the steering wheel, install the screw covers.

17. Refill the cooling system.
18. Install the negative battery cable.
19. Operate the engine to normal operating temperatures; then, check the climate control operation and check for leaks.

Camry

REMOVAL & INSTALLATION

1. Disconnect the negative battery cable.
2. Drain the cooling system into a clean container for reuse.

3. Disconnect the heater hoses from the heater core.
4. At the driver's side, remove the lower instrument panel.
5. Remove the left hand instrument lower panel.
6. At the heater/air conditioning housing, disengage the 3 heater protector-to-heater/air conditioning housing clips and remove the heater protector.
7. Remove the 3 heater core pipe clamp screws and the clamps.
8. Remove the 2 heater core pipe clamp

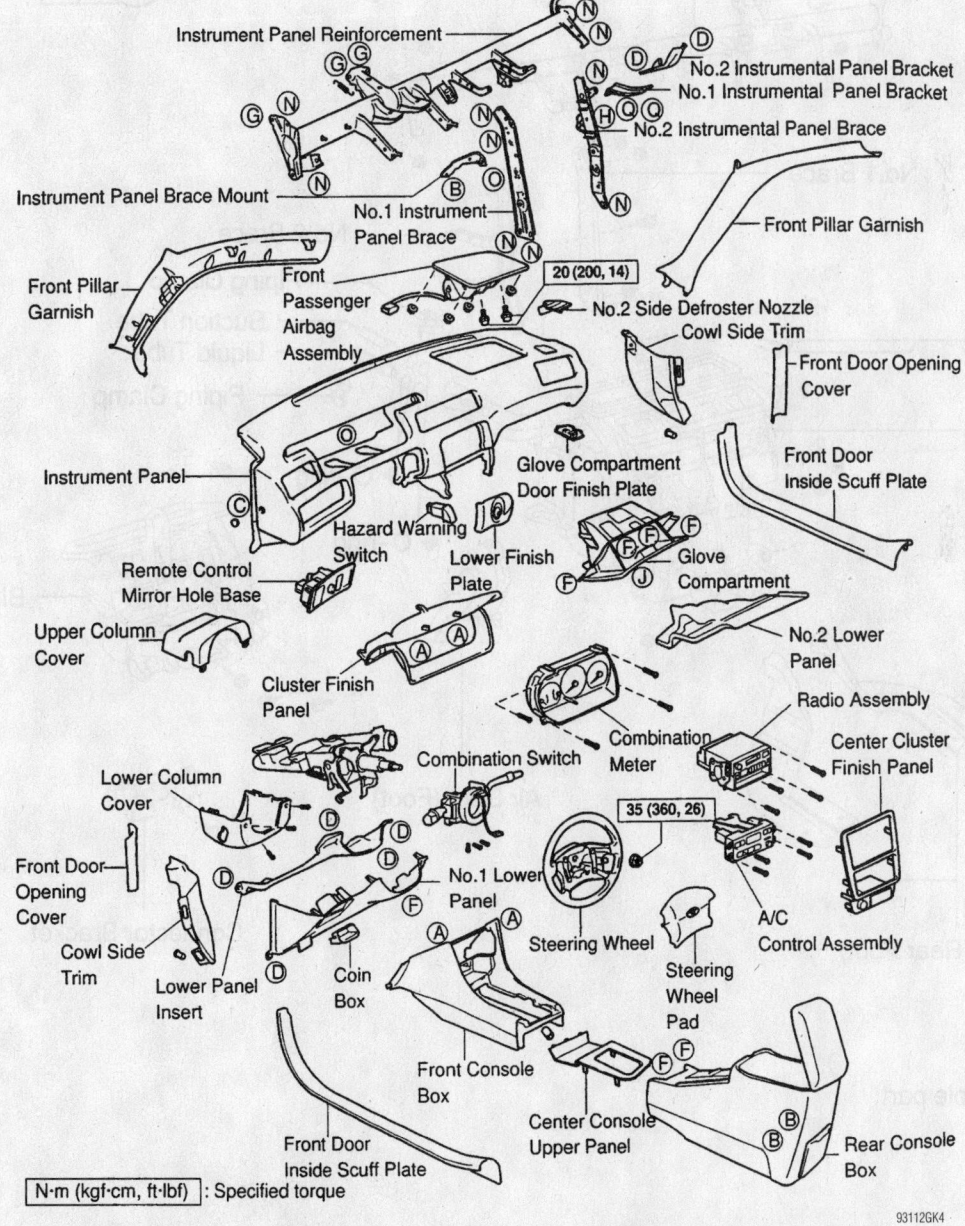

Exploded view of the instrument panel, heater/air conditioning housing and related components—Camry

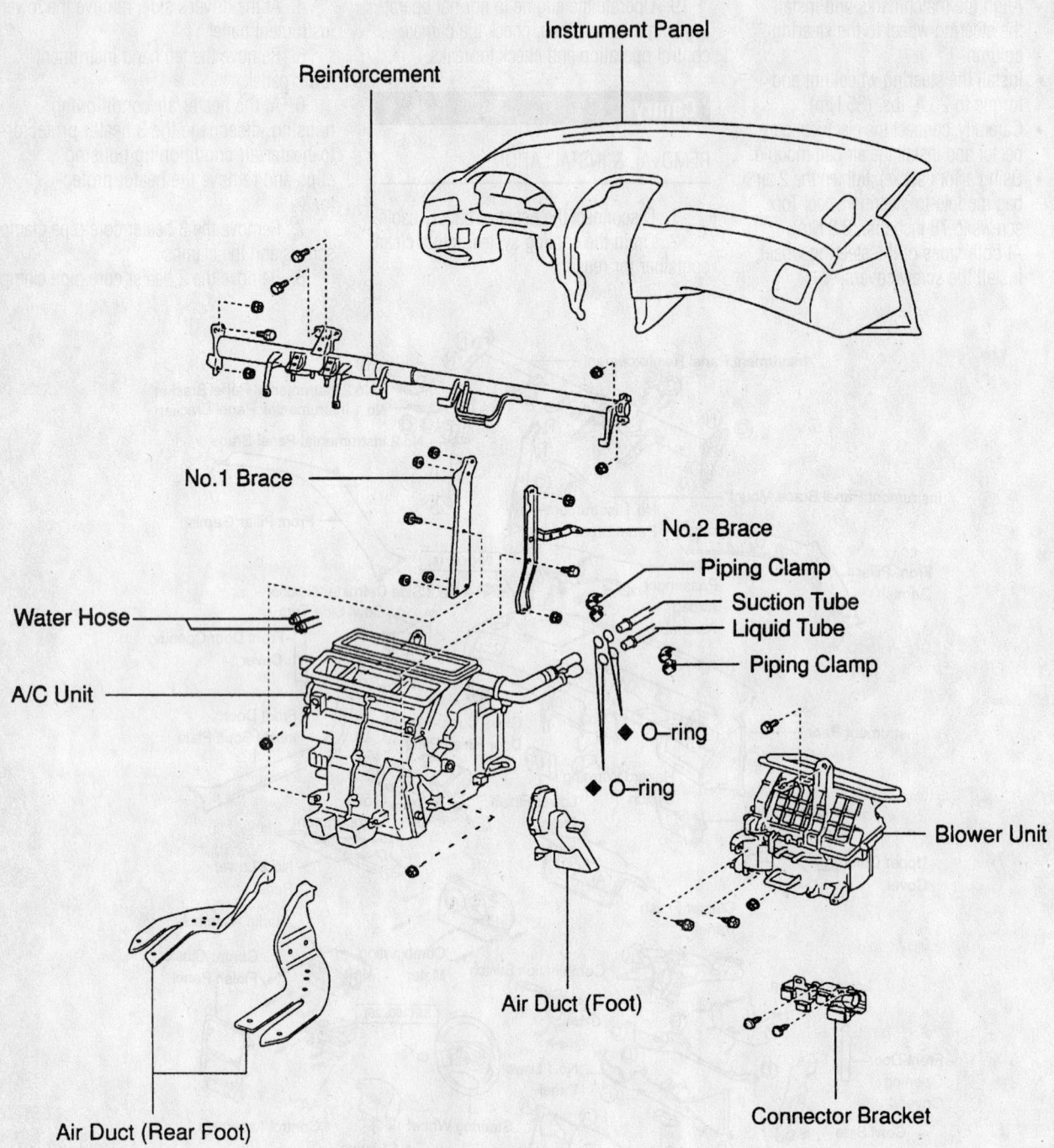

Instrument Panel

Reinforcement

No.1 Brace

No.2 Brace

Piping Clamp

Suction Tube

Liquid Tube

Piping Clamp

Water Hose

A/C Unit

◆ O-ring

◆ O-ring

Blower Unit

Air Duct (Foot)

Air Duct (Rear Foot)

Connector Bracket

◆ Non-reusable part

93112GK5

Exploded view of the instrument panel, heater/air conditioning housing and related components—Camry

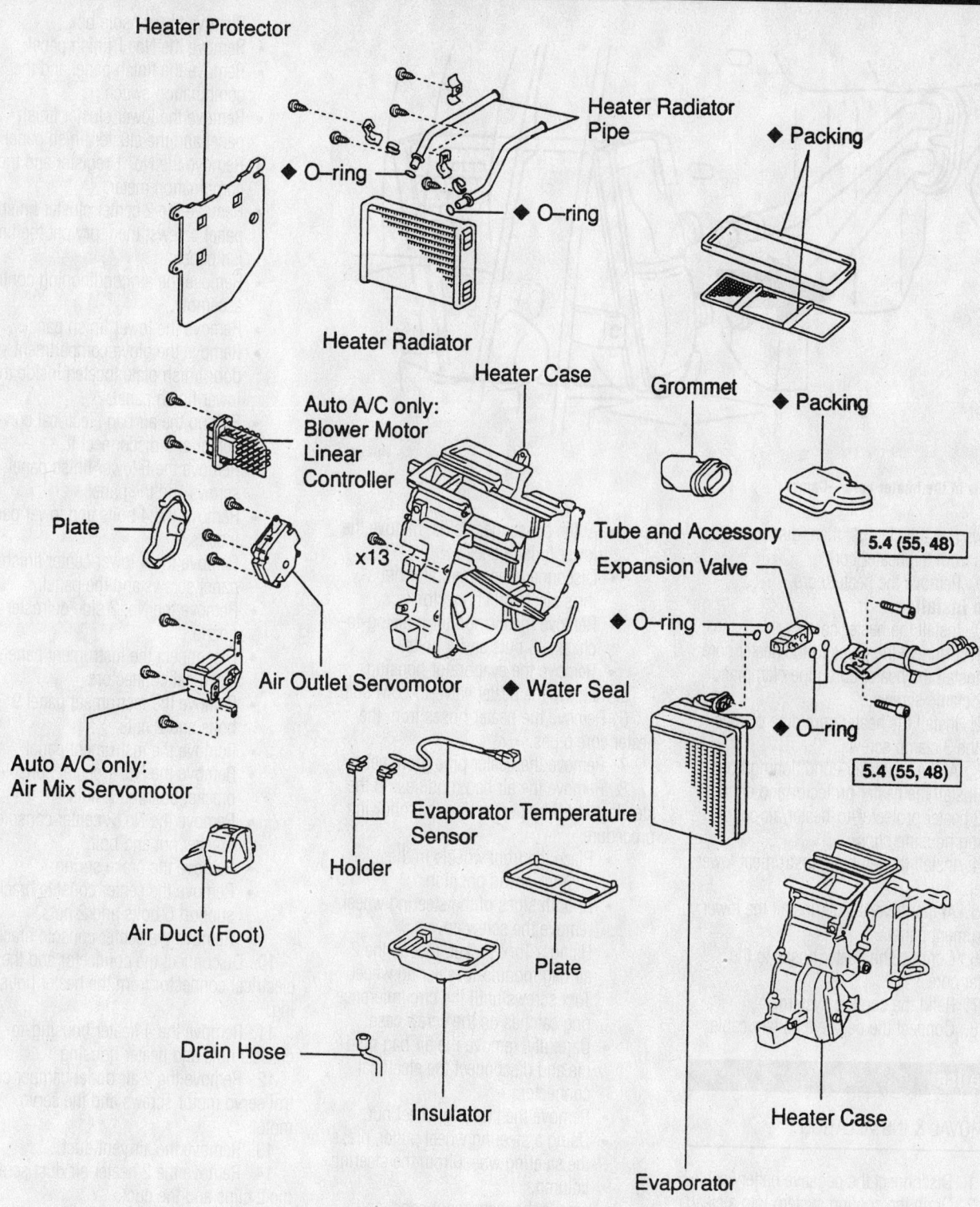

Heater Protector

Heater Radiator Pipe

◆ Packing

◆ O-ring

◆ O-ring

Heater Radiator

Heater Case

Grommet

◆ Packing

Auto A/C only:
Blower Motor Linear Controller

Tube and Accessory

Expansion Valve

5.4 (55, 48)

Plate

x13

◆ O-ring

Air Outlet Servomotor

◆ Water Seal

◆ O-ring

Auto A/C only:
Air Mix Servomotor

5.4 (55, 48)

Air Duct (Foot)

Evaporator Temperature Sensor

Holder

Plate

Drain Hose

Insulator

Heater Case

Evaporator

N·m (kgf·cm, in.·lbf) : Specified torque

◆ Non-reusable part

93112GK6

Exploded view of the heater/air conditioning housing, heater core, evaporator core and related components—Camry

Heater Core replacement is covered in Section 2 of this manual

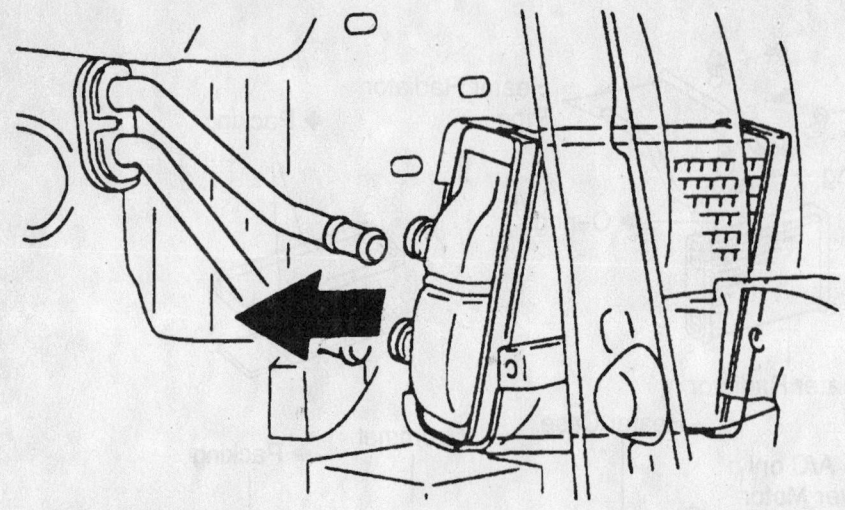

93112GK7

View of the heater core—Camry

screws and the clamps; then, disconnect the pipes from the heater core.

9. Remove the heater core.

To install:

10. Install the heater core.

11. Connect the pipes to the heater core and install the heater core pipe clamp and the 2 clamp screws.

12. Install the heater core pipe clamps and the 3 clamp screws.

13. At the heater/air conditioning housing, install the heater protector and engage the 3 heater protector-to-heater/air conditioning housing clips.

14. Install the left hand instrument lower panel.

15. At the driver's side, install the lower instrument panel.

16. Connect the heater hoses to the heater core.

17. Refill the cooling system.

18. Connect the negative battery cable.

Celica

REMOVAL & INSTALLATION

1. Disconnect the negative battery cable.

2. Drain the cooling system into a clean container for reuse.

3. Discharge and recover the air conditioning system refrigerant.

4. Remove the evaporator housing by performing the following procedure:

- Disconnect the refrigerant lines from the evaporator core. Discard the O-rings. Plug the openings to prevent contamination.
- Remove the 2 grommets and the drain pipe grommet.

- At the passenger's side, remove the lower finish panel.
- Disconnect the evaporator housing's electrical connector.
- Remove the evaporator housing-to-chassis 3 nuts and 4 bolts.
- Remove the evaporator housing.

5. Remove the heater valve-to-cowl bolt.

6. Remove the heater hoses from the heater core pipes.

7. Remove the heater pipe grommets.

8. Remove the air bag module and the steering wheel by performing the following procedure:

- Place the front wheels in the straight-ahead position.
- At both sides of the steering wheel, remove the screw covers.
- Using a Torx socket, loosen the 2 air bag module-to-steering wheel Torx screws until the circumference ring catches on the screw case.
- Carefully, remove the air bag module and disconnect the electrical connector.
- Remove the steering wheel nut.
- Using a steering wheel puller, press the steering wheel from the steering column.

9. Remove the instrument panel and reinforcement by performing the following procedure:

- Remove the front pillar lower garnishes and the pillar garnishes.
- Remove the front door scuff plates.
- Remove the cowl side trim boards.
- Remove the steering column covers.
- Pry out the upper console panel by disengaging the 4 clips.

- Remove the console box.
- Remove the No. 1 finish panel.
- Remove the finish panel and the combination switch.
- Remove the lower cluster finish panel and the cluster finish panel.
- Remove the No. 1 register and the combination meter.
- Remove the 2 center cluster finish panel screws; then, pry out the finish panel.
- Remove the air conditioning control assembly.
- Remove the lower finish panel.
- Remove the glove compartment door finish plate located inside the lower finish panel.
- Pull up the air bag electrical connector and disconnect it.
- Remove the 5 lower finish panel screws and the panel.
- Remove the 4 bolts and lower pad inserts.
- Remove the 4 lower center finish panel screws and the panel.
- Remove the No. 2 side defroster nozzle.
- Disconnect the instrument panel electrical connectors.
- Remove the instrument panel 9 bolts and 2 nuts.
- Remove the instrument panel.
- Remove the No. 1 center console bracket bolt and 2 nuts.
- Remove the No. 2 center console bracket nut and bolt.
- Remove the brake spring.
- Remove the center console bracket support 6 bolts and 2 nuts.
- Remove the center console bracket.

10. Disconnect the connector and the electrical connector from the heater housing.

11. Remove the 4 heater housing-to-chassis nuts and heater housing.

12. Remove the 2 air outlet damper control servo motor screws and the servo motor.

13. Remove the air vent duct.

14. Remove the 2 heater air duct screws, the 2 clips and the duct.

15. Remove the 3 heater pipe clamp screws and the clamps.

16. Remove the heater core from the heater housing.

To install:

17. Install the heater core to the heater housing.

18. Install the heater pipe clamp and the 3 clamp screws.

19. Install the heater air duct, the 2 clips and the 2 duct screws.

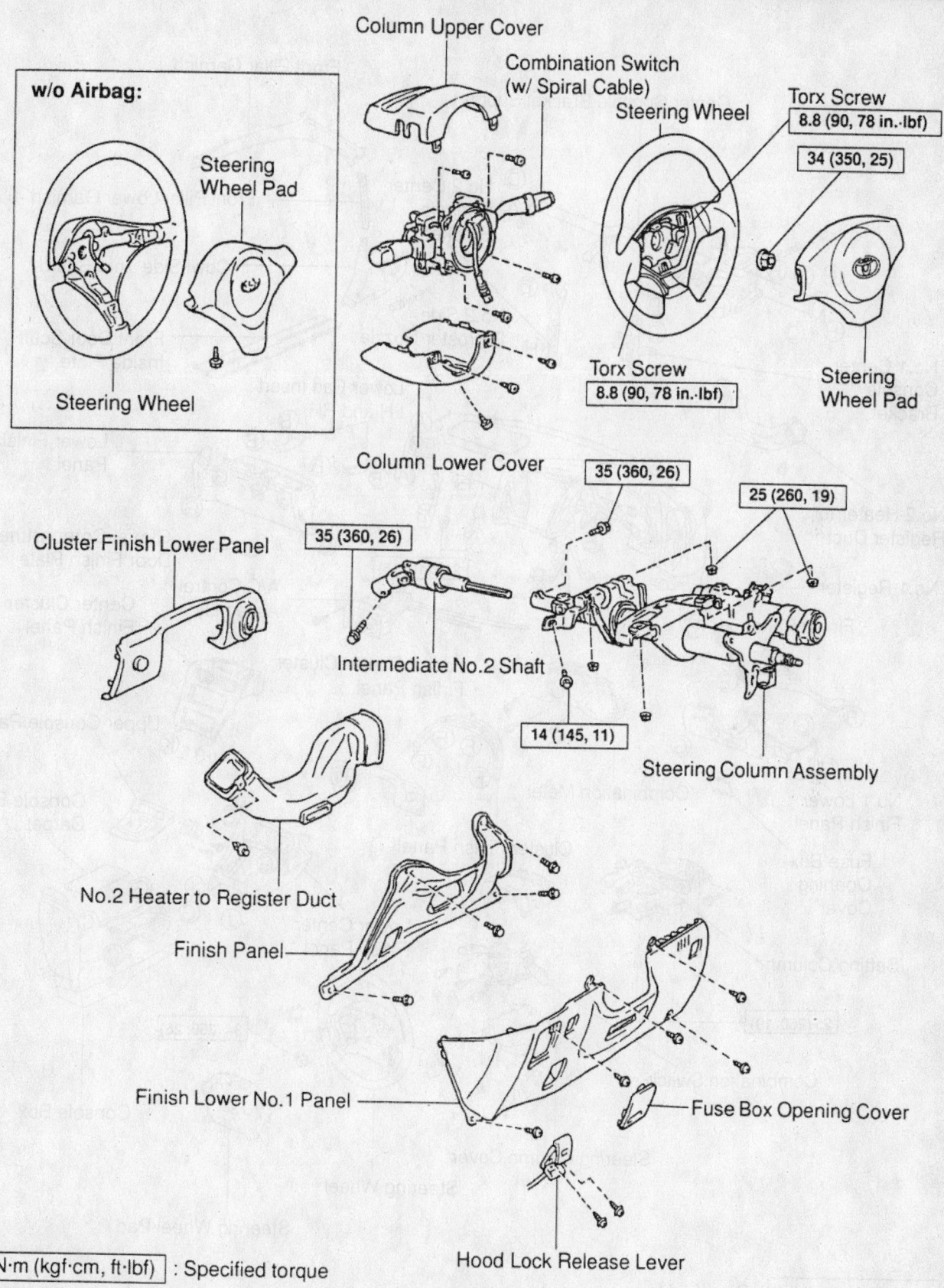

w/o Airbag:

Steering
Wheel Pad

Steering Wheel

Column Upper Cover

Combination Switch
(w/ Spiral Cable)

Steering Wheel

Torx Screw
8.8 (90, 78 in.·lbf)

34 (350, 25)

Torx Screw
8.8 (90, 78 in.·lbf)

Steering
Wheel Pad

Column Lower Cover

Cluster Finish Lower Panel

35 (360, 26)

35 (360, 26)

25 (260, 19)

Intermediate No.2 Shaft

14 (145, 11)

Steering Column Assembly

No.2 Heater to Register Duct

Finish Panel

Finish Lower No.1 Panel

Fuse Box Opening Cover

Hood Lock Release Lever

N·m (kgf·cm, ft·lbf) : Specified torque

93112GK8

Exploded view of the air bag module, steering wheel and related components (typical)—Celica

Brake service is covered in Section 4 of this manual

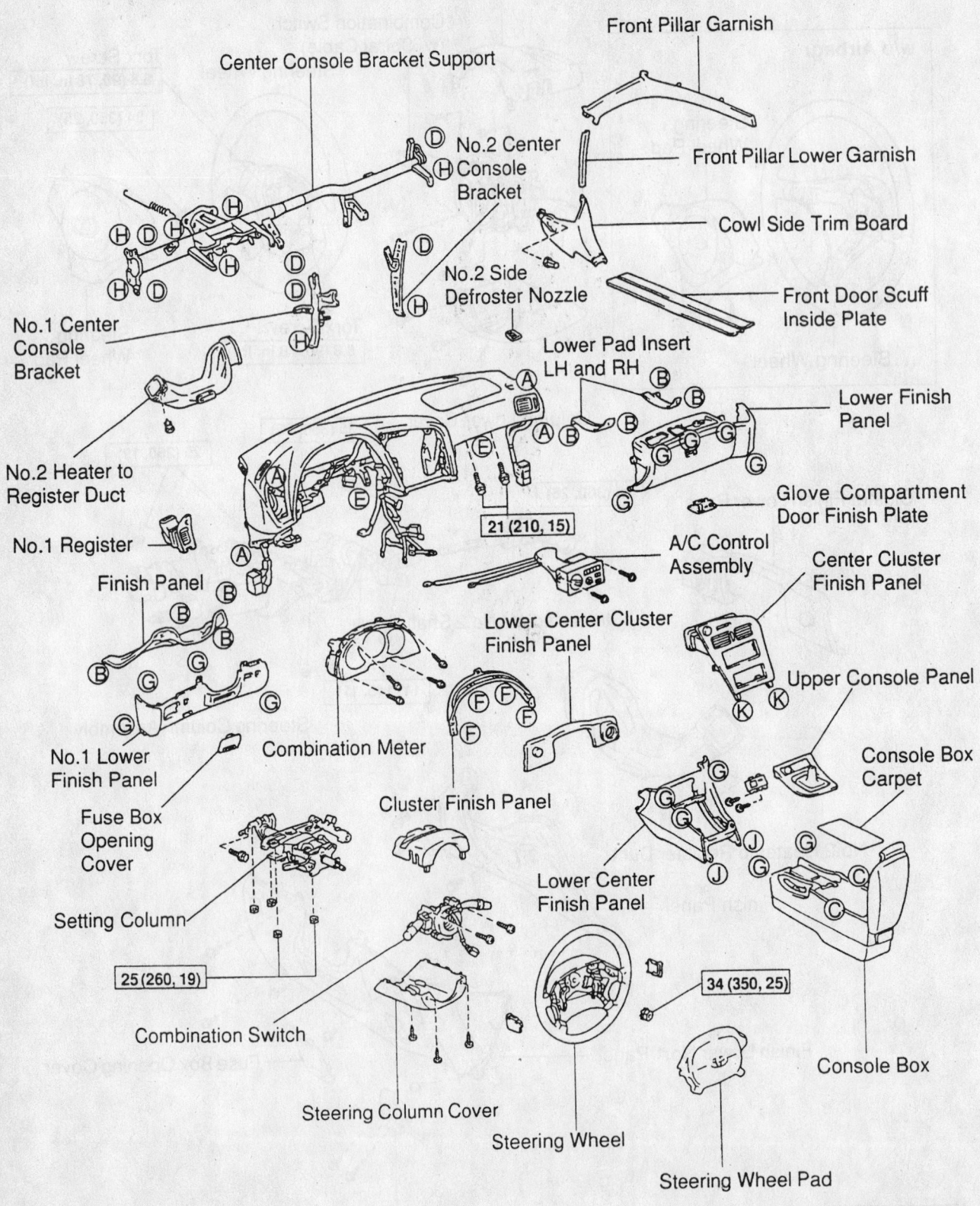

Center Console Bracket Support

No.2 Center Console Bracket

No.2 Side Defroster Nozzle

No.1 Center Console Bracket

No.2 Heater to Register Duct

No.1 Register

Finish Panel

No.1 Lower Finish Panel

Fuse Box Opening Cover

Setting Column

25 (260, 19)

Combination Switch

Steering Column Cover

Front Pillar Garnish

Front Pillar Lower Garnish

Cowl Side Trim Board

Front Door Scuff Inside Plate

Lower Pad Insert LH and RH

Lower Finish Panel

21 (210, 15)

A/C Control Assembly

Glove Compartment Door Finish Plate

Center Cluster Finish Panel

Upper Console Panel

Console Box Carpet

Lower Center Cluster Finish Panel

Combination Meter

Cluster Finish Panel

Lower Center Finish Panel

Steering Wheel

34 (350, 25)

Console Box

Steering Wheel Pad

N·m (kgf·cm, ft·lbf) : Specified torque

93112GK9

Exploded view of the instrument panel and related components—Celica

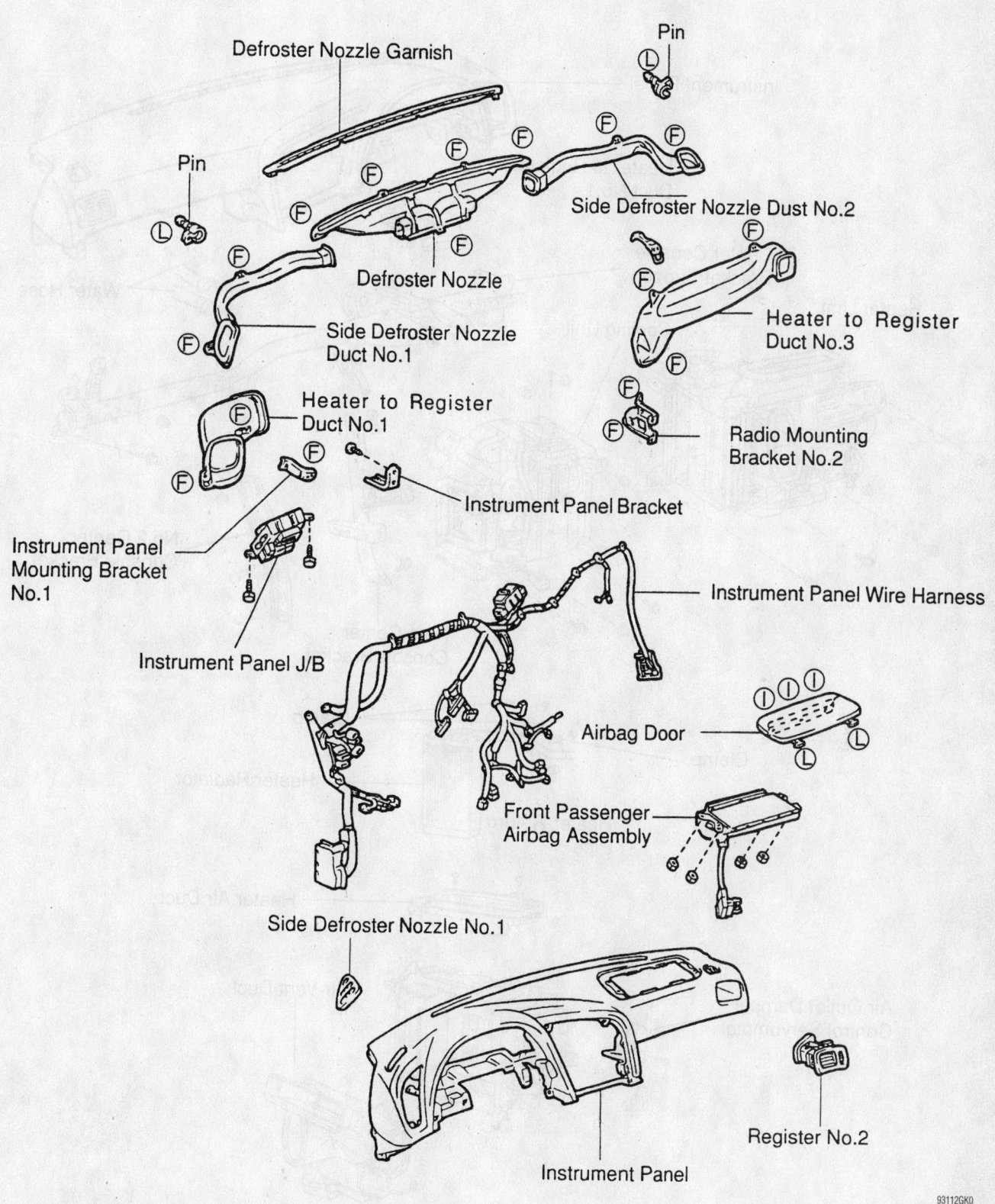

Defroster Nozzle Garnish

Pin

Pin

Side Defroster Nozzle Dust No.2

Defroster Nozzle

Heater to Register Duct No.3

Side Defroster Nozzle Duct No.1

Heater to Register Duct No.1

Radio Mounting Bracket No.2

Instrument Panel Bracket

Instrument Panel Mounting Bracket No.1

Instrument Panel Wire Harness

Instrument Panel J/B

Airbag Door

Front Passenger Airbag Assembly

Side Defroster Nozzle No.1

Register No.2

Instrument Panel

93112GK0

Exploded view of the ventilation system, wiring harness and related components—Celica

For complete Engine Mechanical specifications, see Section 1 of this manual

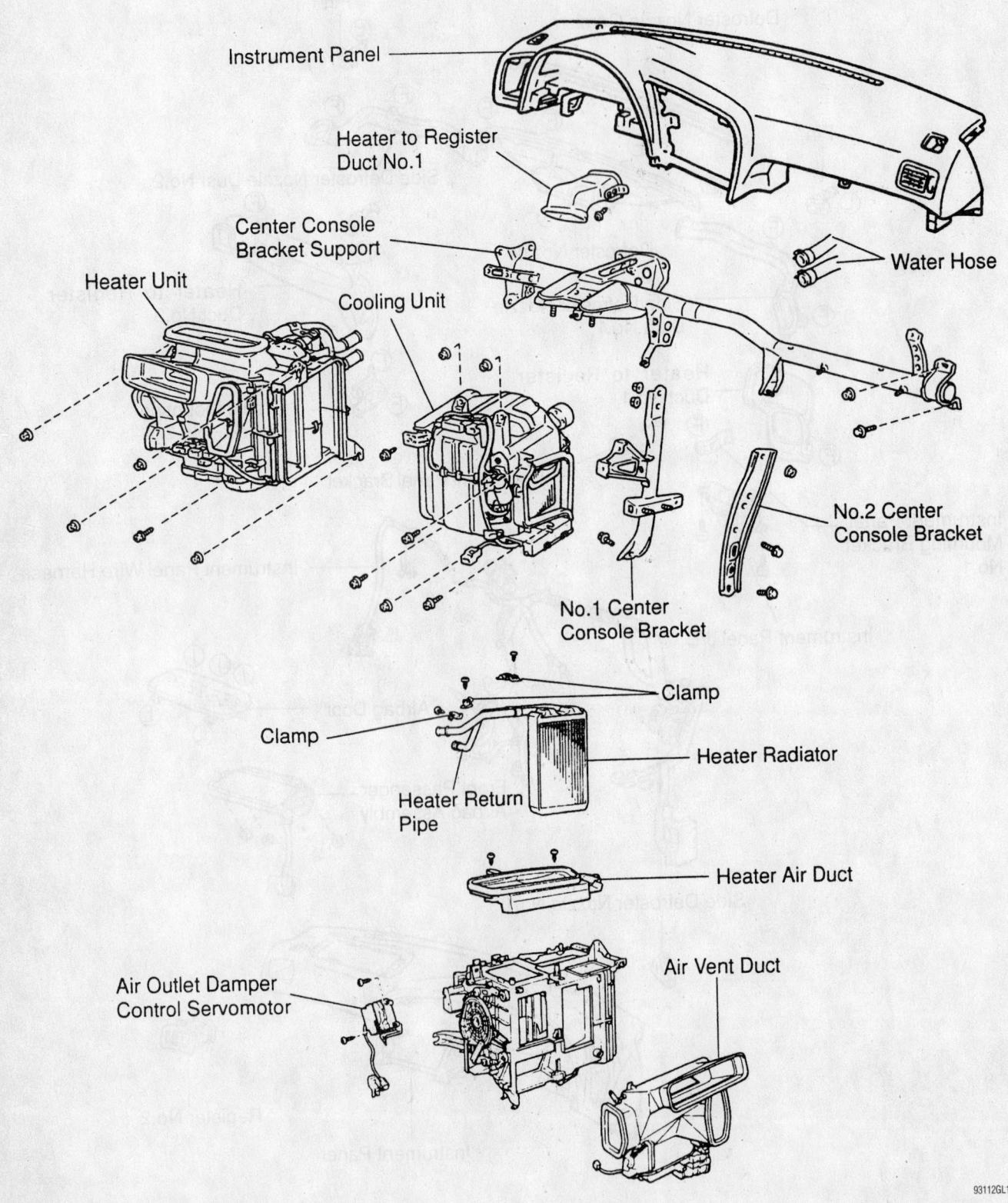

Instrument Panel

Heater to Register Duct No.1

Center Console Bracket Support

Heater Unit

Cooling Unit

Water Hose

No.2 Center Console Bracket

No.1 Center Console Bracket

Clamp

Clamp

Heater Radiator

Heater Return Pipe

Heater Air Duct

Air Outlet Damper Control Servomotor

Air Vent Duct

93112GL1

Exploded view of the heater core, heater housing, evaporator housing and related components—Celica

20. Install the air vent duct.

21. Install the air outlet damper control servo motor and the 2 servo motor screws.

22. Install the heater housing and 4 heater housing-to-chassis nuts.

23. Connect the connector and the electrical connector to the heater housing.

24. Install the air bag module and the steering wheel by performing the following procedure:

- Align the matchmarks and install the steering wheel to the steering column.
- Install the steering wheel nut and torque the nut to 26 ft. lbs. (35 Nm).
- Carefully, connect the electrical connector and install the air bag module.
- Using a Torx socket, tighten the 2 air bag module-to-steering wheel Torx screws to 78 inch lbs. (8.8 Nm).
- At both sides of the steering wheel, install the screw covers.

25. Install the instrument panel and reinforcement by performing the following procedure:

- Install the center console bracket.
- Install the center console bracket support 6 bolts and 2 nuts.
- Install the brake spring.
- Install the No. 2 center console bracket nut and bolt.
- Install the No. 1 center console bracket bolt and 2 nuts.
- Install the instrument panel.
- Install the instrument panel 9 bolts and 2 nuts.
- Connect the instrument panel electrical connectors.
- Install the No. 2 side defroster nozzle.
- Install the lower center finish panel and the 4 panel screws.
- Install the 4 bolts and lower pad inserts.
- Install the lower finish panel and the 5 panel screws.
- Connect the air bag electrical connector.
- Install the glove compartment door finish plate located inside the lower finish panel.
- Install the lower finish panel.
- Install the air conditioning control assembly.
- Install the center cluster finish panel and the 2 finish panel screws.

- Install the No. 1 register and the combination meter.
- Install the lower cluster finish panel and the cluster finish panel.
- Install the finish panel and the combination switch.
- Install the No. 1 finish panel.
- Install the console box.
- Pry out the upper console panel by engaging the 4 clips.
- Install the steering column covers.
- Install the cowl side trim boards.
- Install the front door scuff plates.
- Install the front pillar lower garnishes and the pillar garnishes.

26. Install the heater pipe grommets.

27. Install the heater hoses to the heater core pipes.

28. Install the heater valve-to-cowl bolt.

29. Install the evaporator housing by performing the following procedure:

- Install the evaporator housing.
- Install the evaporator housing-to-chassis 3 nuts and 4 bolts.
- Connect the evaporator housing's electrical connector.
- At the passenger's side, install the lower finish panel.
- Install the 2 grommets and the drain pipe grommet.
- Using new O-rings, connect the refrigerant lines to the evaporator core.

30. Refill the cooling system.

31. Connect the negative battery cable.

32. Evacuate, charge and leak test the air conditioning system.

33. Operate the engine to normal operating temperatures; then, check the climate control operation and check for leaks.

Corolla

REMOVAL & INSTALLATION

1. Disconnect the negative battery cable.

✲✲ CAUTION

After Connecting the negative battery cable, wait for at least 1½ minutes for the SRS to deplete its energy.

2. Drain the cooling system into a clean container for reuse.

3. Disconnect the heater hoses from the heater core.

4. Discharge and recover the air conditioning system refrigerant.

5. Remove the air bag module and the steering wheel by performing the following procedure:

- Place the front wheels in the straight-ahead position.
- At both sides of the steering wheel, remove the screw covers.
- Using a Torx socket, loosen the 2 air bag module-to-steering wheel Torx screws until the circumference ring catches on the screw case.
- Carefully, remove the air bag module and disconnect the electrical connector.
- Remove the steering wheel nut.
- Using a steering wheel puller, press the steering wheel from the steering column.

6. Remove the passenger's side air bag module by performing the following procedure:

- Remove the glove box-to-instrument panel 2 bolts and 3 screws; then, pull out the glove box and remove it.
- Disconnect the passenger's side air bag module electrical connector.
- Remove the 2 passenger's air bag module-to-instrument panel bolts and nuts. Carefully, remove the air bag module.

7. Remove the following items in the following order:

- The front door scuff plates
- The cowl side trims
- The front pillar garnishes
- Remove the 2 lower finish panel bolts and the panel.
- Disconnect the hood lock control cable.
- Remove the 2 lower insert bolts and the lower insert.
- Remove the 2 cluster finish panel screws; then, pry out the panel.
- Remove the steering column cover.
- Remove the steering column's combination switch.
- Remove the No. 2 heater-to-register duct.
- Disconnect and remove the combination meter.
- Remove the steering column-to-instrument panel fasteners, the steering column shaft pinch bolt and the steering column.

8. Remove the following items in the following order:

- The lower left side panel
- The lower right side panel

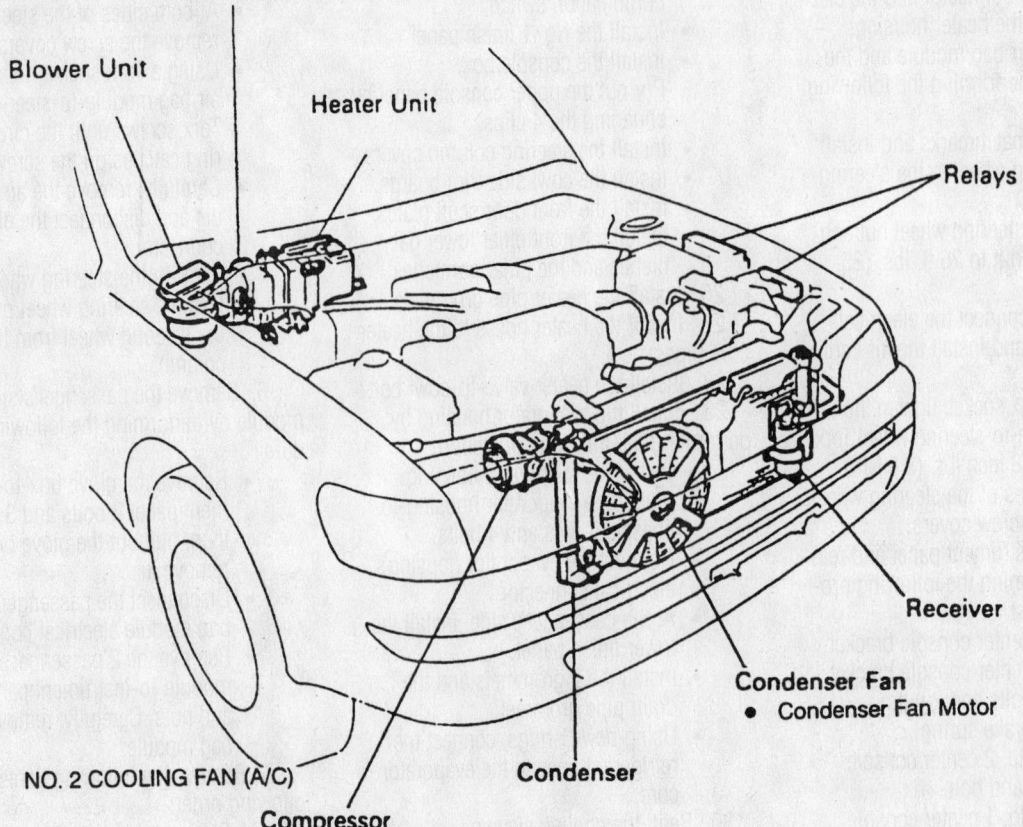

Cooling Unit
- Expansion Valve
- Evaporator
- Blower Resistor
- Thermistor

Blower Unit

Heater Unit

Relays

Receiver

Condenser Fan
- Condenser Fan Motor

Condenser

NO. 2 COOLING FAN (A/C)

Compressor
- Magnetic Clutch
- Refrigerant Temperature Switch

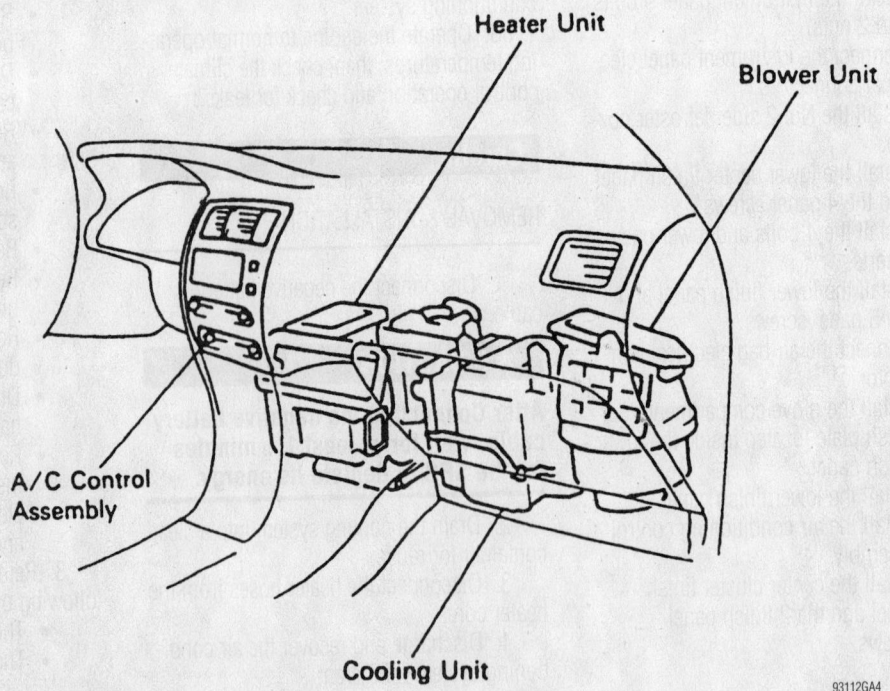

Heater Unit

Blower Unit

A/C Control
Assembly

Cooling Unit

93112GA4

View of the heater/air conditioning assembly and related components—Corolla

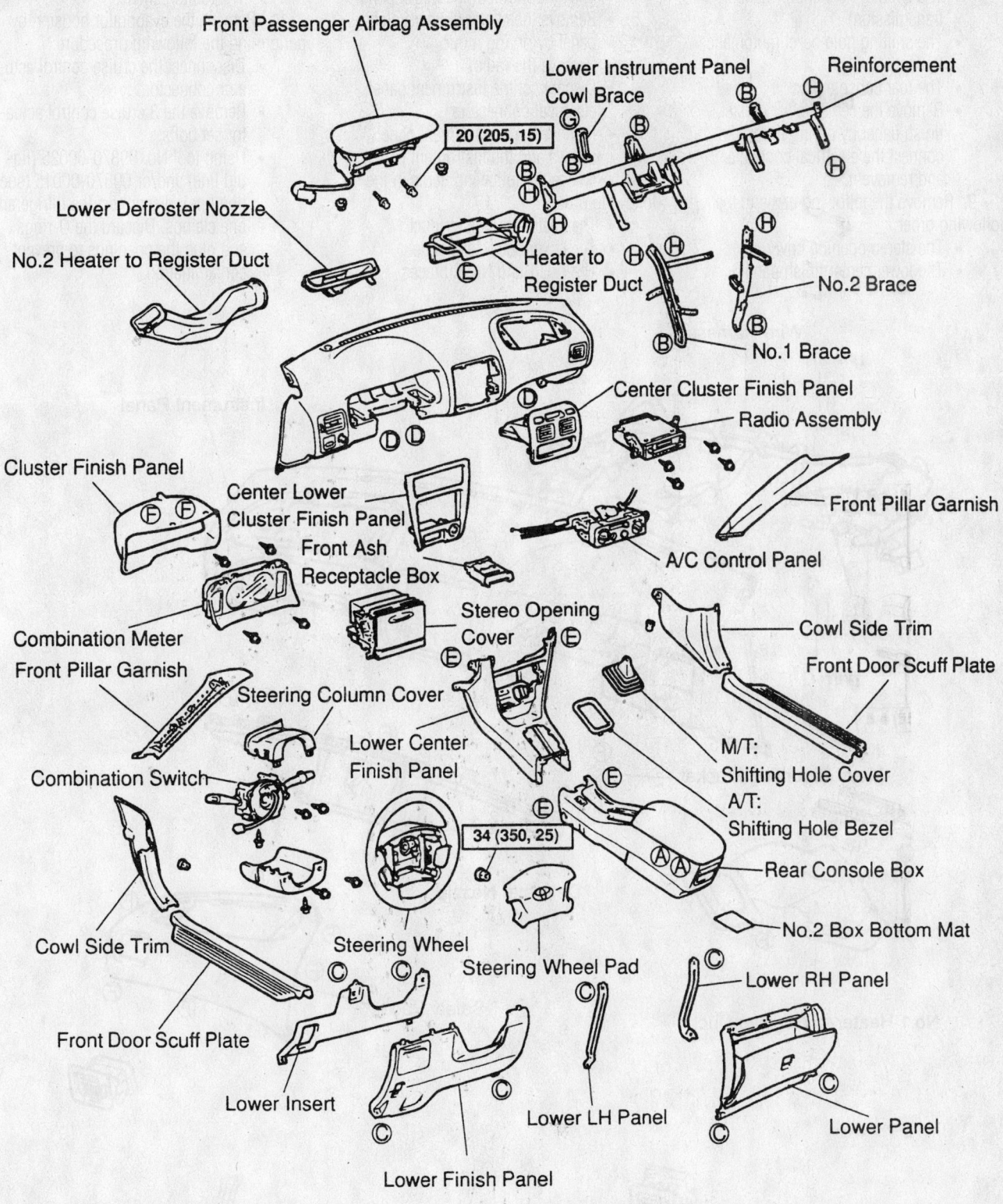

Front Passenger Airbag Assembly

Lower Instrument Panel Cowl Brace

Reinforcement

20 (205, 15)

Lower Defroster Nozzle

No.2 Heater to Register Duct

Heater to Register Duct

No.2 Brace

No.1 Brace

Center Cluster Finish Panel

Radio Assembly

Cluster Finish Panel

Center Lower Cluster Finish Panel

Front Ash Receptacle Box

Front Pillar Garnish

A/C Control Panel

Stereo Opening Cover

Cowl Side Trim

Front Door Scuff Plate

Combination Meter

Front Pillar Garnish

Steering Column Cover

Lower Center Finish Panel

M/T: Shifting Hole Cover

A/T: Shifting Hole Bezel

Rear Console Box

Combination Switch

Cowl Side Trim

Steering Wheel

34 (350, 25)

No.2 Box Bottom Mat

Lower RH Panel

Front Door Scuff Plate

Lower Insert

Steering Wheel Pad

Lower LH Panel

Lower Panel

Lower Finish Panel

N·m (kgf·cm, ft·lbf) : Specified torque

93112GM4

Exploded view of the instrument panel and related components—Corolla

For Tire, Wheel and Ball Joint specifications, see Section 1 of this manual

- The shifting hole knob (manual transmission)
- The shifting hole bezel (automatic transmission)
- The rear console box
- Remove the center lower cluster finish panel by prying it out, disconnect the electrical connector and remove it.

9. Remove the following items in the following order:
- The stereo opening cover
- The lower center finish panel

- The air conditioning control panel
- Remove the center cluster finish panel by prying it out.
- Remove the radio.
- Disconnect the instrument panel electrical connectors.
- Remove the 4 instrument panel screws and the instrument panel.

10. Remove the following items in the following order:
- The heater-to-register duct
- The lower defroster nozzle
- The No. 1 and No. 2 braces

- The reinforcement

11. Remove the evaporator housing by performing the following procedure:
- Disconnect the cruise control actuator connector.
- Remove the 3 cruise control actuator set bolts.
- Using tool No. 09870-00025 (liquid line) and/or 09870-00015 (suction line), disconnect the refrigerant line clamps. Discard the O-rings and plug the openings to prevent contamination.

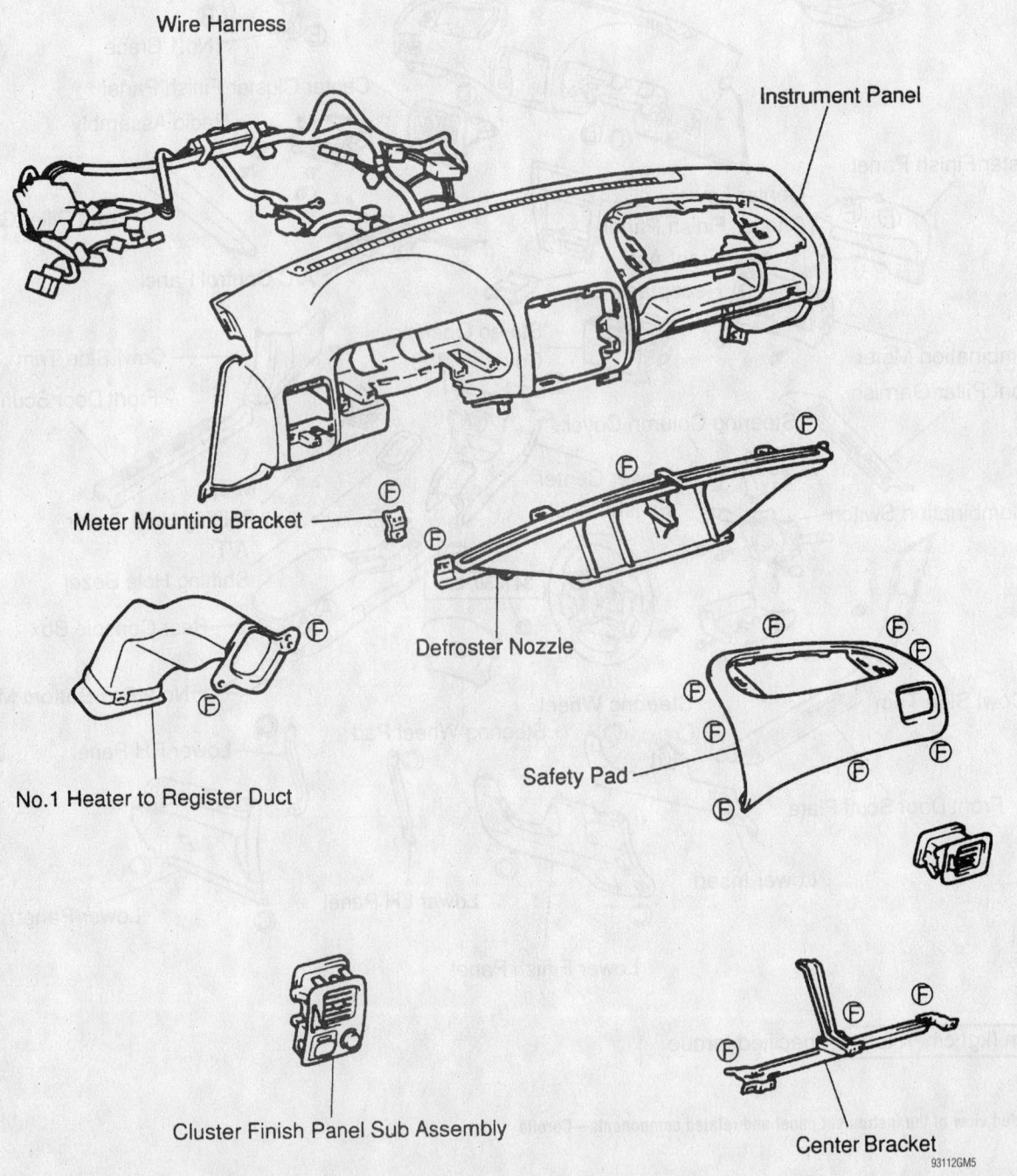

Exploded view of the ventilation system, wiring harness and related components—Corolla

93112GM5

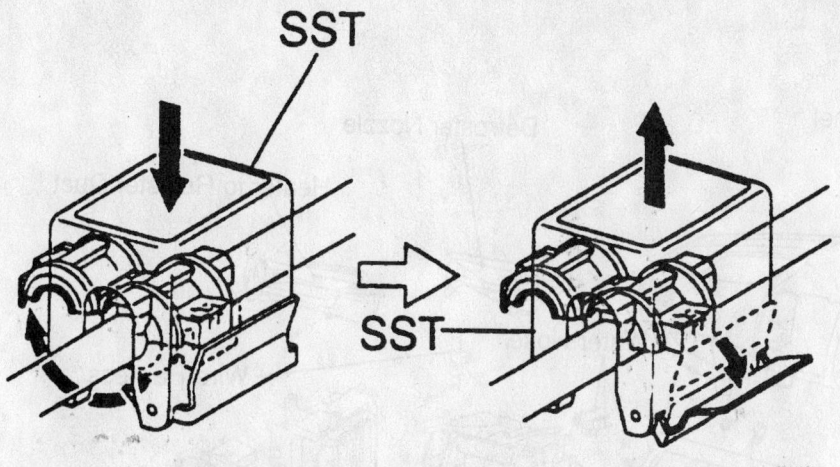

Using the special tool to remove the air conditioning refrigerant line clamps—Corolla

93112GM6

- Remove the tube grommet and the drain hose grommet.
- Disconnect the air conditioning amplifier electrical connector.
- Remove the 2 air conditioning amplifier nuts and the amplifier.
- Disconnect the evaporator housing's electrical connectors.
- Remove the evaporator housing-to-chassis 3 screws, nut and the housing.

12. Pull back the carpet and remove the rear heater ducts.

13. Disconnect the wiring harness from the heater housing.

14. Remove the 3 heater housing-to-chassis nuts and the heater housing.

15. Remove the air duct-to-heater housing screw, the 5 clips and the duct.

16. Release the 2 heater core cover clips and the cover.

17. Remove the heater core-to-heater housing screw, clamp and the heater core.

To install:

18. Install the heater core, clamp and the heater core-to-heater housing screw.

19. Install the heater core cover and the 2 cover clips.

20. Install the air duct, the 5 clips and the duct-to-heater housing screw.

21. Install the heater housing and the 3 heater housing-to-chassis nuts.

22. Connect the wiring harness to the heater housing.

23. Install the rear heater ducts and the carpet.

24. Install the evaporator housing by performing the following procedure:
- Install the evaporator housing-to-

chassis 3 screws, nut and the housing.
- Connect the evaporator housing's electrical connectors.
- Install the 2 air conditioning amplifier nuts and the amplifier.
- Connect the air conditioning amplifier electrical connector.
- Install the tube grommet and the drain hose grommet.
- Using new O-rings, connect the refrigerant line clamps.
- Install the 3 cruise control actuator set bolts.
- Connect the cruise control actuator connector.

25. Install the following items in the following order:
- The reinforcement
- The No. 1 and No. 2 braces
- The lower defroster nozzle
- The heater-to-register duct
- Install the instrument panel and the 4 instrument panel screws.
- Connect the instrument panel electrical connectors.
- Install the radio.
- Install the center cluster finish panel.

26. Install the following items in the following order:
- The air conditioning control panel
- The lower center finish panel
- The stereo opening cover
- Connect the electrical connector and install the center lower cluster finish panel.

27. Install the following items in the following order:
- The rear console box

- The shifting hole bezel (automatic transmission)
- The shifting hole knob (manual transmission)
- The lower right side panel
- The lower left side panel
- Install the steering column, the steering column shaft pinch bolt and the steering column-to-instrument panel fasteners.
- Connect and install the combination meter.
- Install the No. 2 heater-to-register duct.
- Install the steering column's combination switch.
- Install the steering column cover.
- Install the cluster finish panel and the 2 panel screws.
- Install the lower insert and the 2 lower insert bolts.
- Connect the hood lock control cable.
- Install the lower finish panel and the 2 panel bolts.

28. Install the following items in the following order:
- The front pillar garnishes
- The cowl side trims
- The front door scuff plates

29. Install the passenger's side air bag module by performing the following procedure:
- Carefully, Install the air bag module and the 2 passenger's air bag module-to-instrument panel bolts and nuts.
- Connect the passenger's side air bag module electrical connector.
- Install the glove box and the glove box-to-instrument panel 2 bolts and 3 screws.

30. Install the air bag module and the steering wheel by performing the following procedure:
- Align the matchmarks and install the steering wheel to the steering column.
- Install the steering wheel nut and torque the nut to 26 ft. lbs. (35 Nm).
- Carefully, connect the electrical connector and install the air bag module.
- Using a Torx socket, tighten the 2 air bag module-to-steering wheel Torx screws to 78 inch lbs. (8.8 Nm).
- At both sides of the steering wheel, install the screw covers.

For Wheel Alignment specifications, see Section 1 of this manual

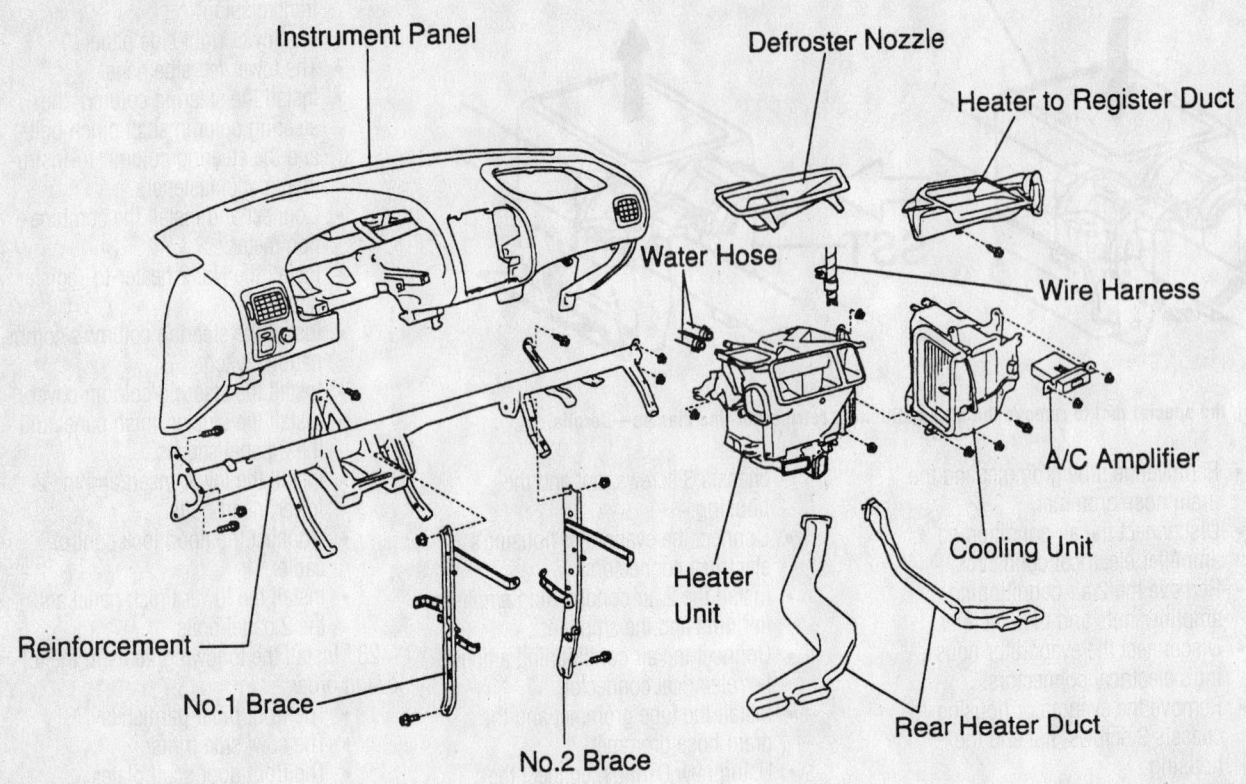

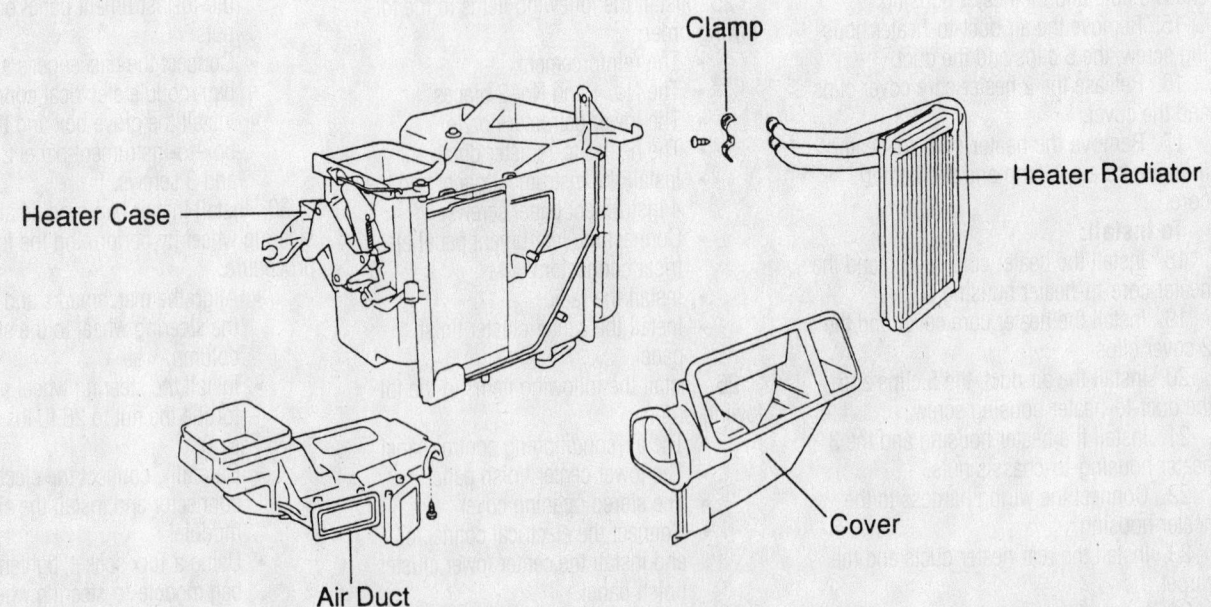

Exploded view of the heater core, heater housing, evaporator housing and related components—Corolla

31. Connect the heater hoses to the heater core.

32. Refill the cooling system.

33. Connect the negative battery cable.

34. Evacuate, charge and leak test the air conditioning system refrigerant.

Paseo

REMOVAL & INSTALLATION

1. Disconnect the negative battery cable.

2. Drain the cooling system into a clean container for reuse.

✱✱ CAUTION

If the equipped with an SRS system, when the ignition is turned to LOCK and the negative battery cable is disconnected, do not work inside the passenger compartment for at least 90 seconds to give the SRS a chance to become fully disarmed.

3. Discharge and recover the air conditioning system refrigerant.

4. Remove the instrument panel by performing the following procedure:

- Remove the pillar trim and the door sill plate.
- Remove the steering wheel.
- Remove the steering column cover, the center console, the hood release handle, the panel trim below the steering column, the combination switch and the glove box door.
- Remove the instrument cluster bezel and the instrument cluster.
- Remove the center panel bezel, the radio, the console over the shift

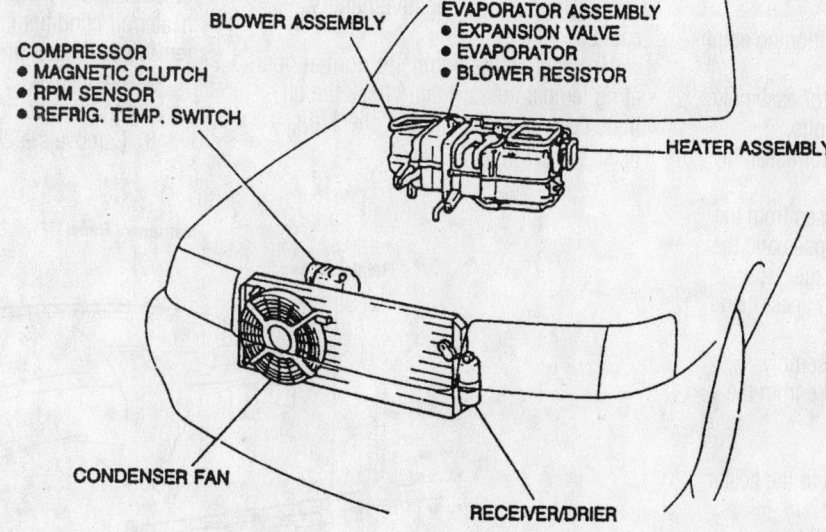

COMPRESSOR
- MAGNETIC CLUTCH
- RPM SENSOR
- REFRIG. TEMP. SWITCH

BLOWER ASSEMBLY

EVAPORATOR ASSEMBLY
- EXPANSION VALVE
- EVAPORATOR
- BLOWER RESISTOR

HEATER ASSEMBLY

CONDENSER FAN

RECEIVER/DRIER

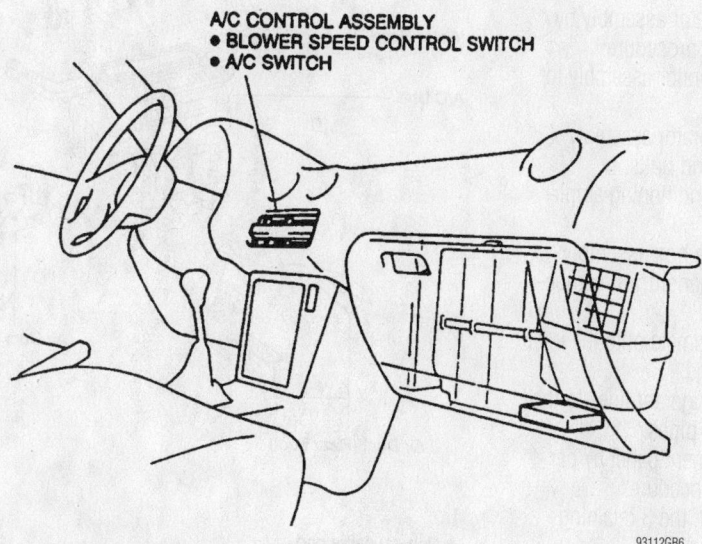

A/C CONTROL ASSEMBLY
- BLOWER SPEED CONTROL SWITCH
- A/C SWITCH

93112GB6

View of the heater/air conditioning assembly locations—1998 Paseo

For Maintenance Interval recommendations, see Section 1 of this manual

lever, the clock and the heater/air conditioning control panel.
- Remove the heater duct and the junction block from the left side.
- Remove the 3 retaining nuts and 1 bolt and remove the panel.

5. Remove the evaporator assembly by performing the following procedure:
- Disconnect the refrigerant lines from the evaporator core pipes. Cap all the openings.
- Remove the grommets from the inlet and outlet fittings.
- Remove the glove box assembly and the lower panel trim, as necessary.
- Remove the air conditioning amplifier.
- Remove the evaporator assembly attaching nuts and bolts.
- Remove the evaporator assembly from the vehicle.

6. Remove the heater hoses from the heater core tubes. Plug the hoses and the tubes to prevent coolant leakage.

7. Remove the 2 center braces, then remove the heater assembly.

8. Remove the heater assembly.

9. Remove the heater core from the heater assembly.

To install:

10. Install the heater core to the heater assembly.

11. Install the heater assembly.

12. Install the heater assembly and the 2 center braces.

13. Install the heater hoses to the heater core tubes.

14. Install the evaporator assembly by performing the following procedure:
- Install the evaporator assembly to the vehicle.
- Install the evaporator assembly attaching nuts and bolts.
- Install the air conditioning amplifier.
- Install the glove box assembly and the lower panel trim, as necessary.
- Install new grommets onto the inlet and outlet fittings.
- Connect the refrigerant lines to the evaporator core pipes.

15. Install the instrument panel by performing the following procedure:
- Install the panel, the 3 retaining nuts and 1 bolt.
- Install the heater duct and the junction block to the left side.
- Install the center panel bezel, the radio, the console over the shift

lever, the clock and the heater/air conditioning control panel.
- Install the instrument cluster bezel and the instrument cluster.
- Install the glove box door, the combination switch, the panel trim below the steering column, the hood release handle, the center console and the steering column cover.
- Install the steering wheel.
- Install the door sill plate and the pillar trim.

16. Evacuate, charge and leak test the air conditioning system.

17. Refill the cooling system.

18. Connect the negative battery cable.

19. Operate the engine to normal operating temperatures; then, check the climate control operation and check for leaks.

Solara

REMOVAL & INSTALLATION

1. Disconnect the negative battery cable.

2. Drain the cooling system into a clean container for reuse.

3. Disconnect the heater hoses from the heater core.

4. At the driver's side, remove the lower instrument panel.

5. Remove the left hand instrument lower panel.

6. At the heater/air conditioning housing, disengage the 3 heater protector-to-heater/air conditioning housing clips and remove the heater protector.

7. Remove the 3 heater core pipe clamp screws and the clamps.

8. Remove the 2 heater core pipe clamp

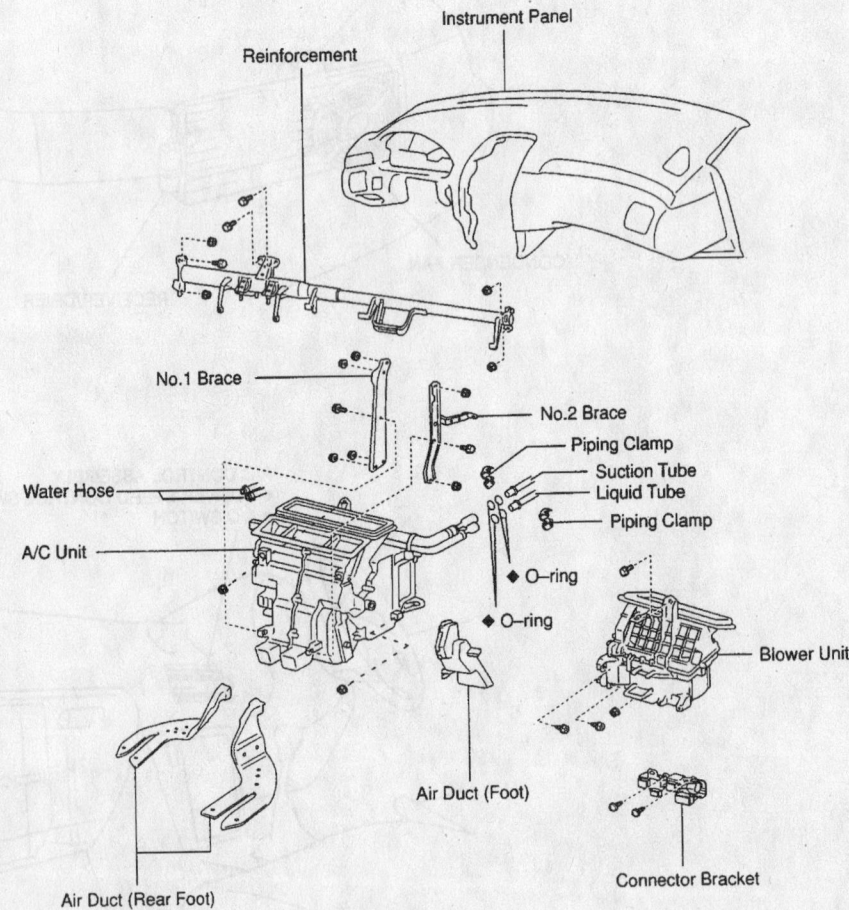

Instrument Panel

Reinforcement

No.1 Brace

No.2 Brace

Piping Clamp

Suction Tube

Liquid Tube

Piping Clamp

Water Hose

A/C Unit

O–ring

O–ring

Blower Unit

Air Duct (Foot)

Connector Bracket

Air Duct (Rear Foot)

◆ Non–reusable part

93112GK5

Exploded view of the instrument panel, heater/air conditioning housing and related components—1999–00 Solara

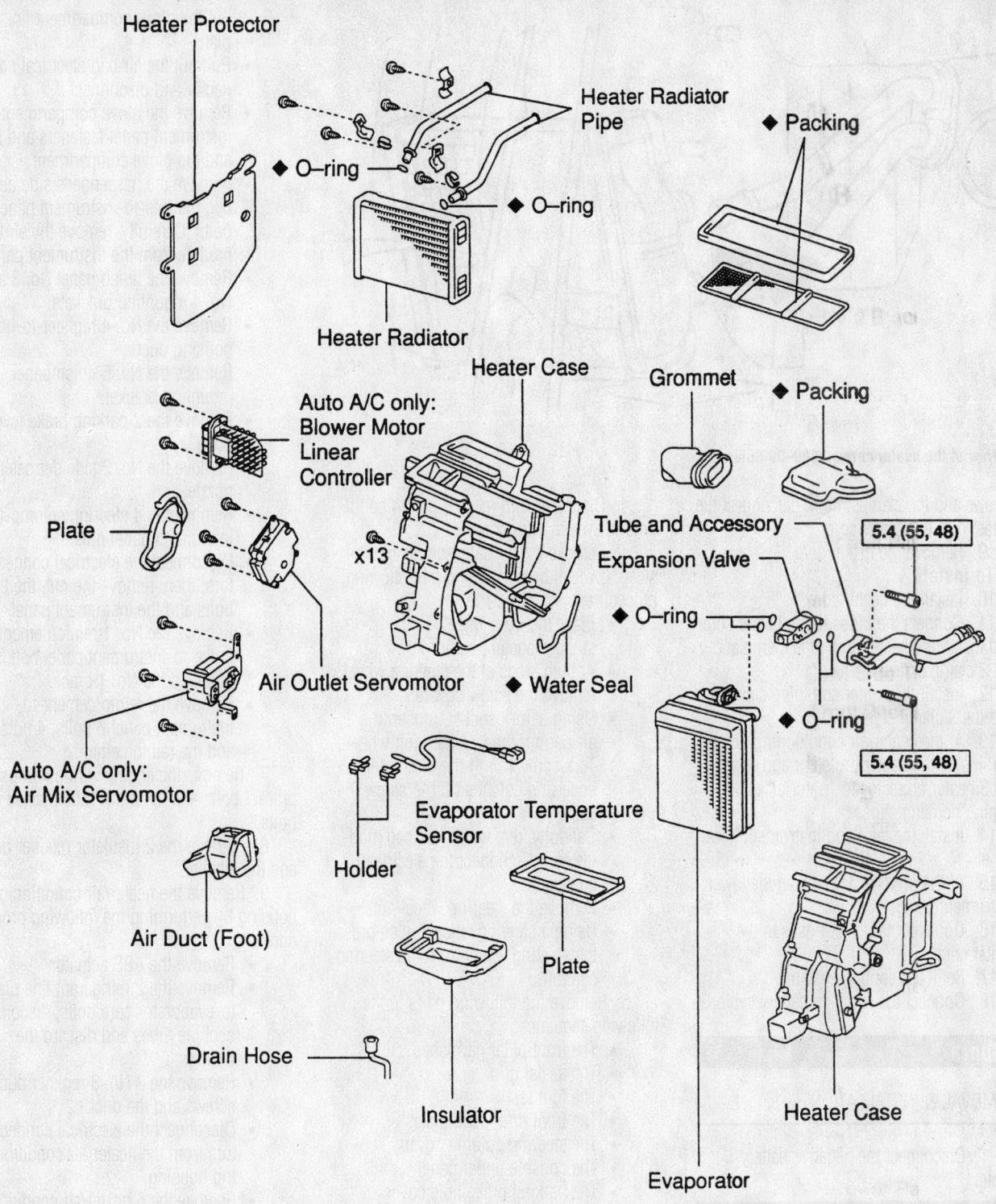

Heater Protector

Heater Radiator Pipe

◆ Packing

◆ O-ring

◆ O-ring

Heater Radiator

Heater Case

Auto A/C only:
Blower Motor
Linear
Controller

Grommet

◆ Packing

Plate

x13

Tube and Accessory

Expansion Valve

5.4 (55, 48)

◆ O-ring

Air Outlet Servomotor

◆ Water Seal

◆ O-ring

Auto A/C only:
Air Mix Servomotor

5.4 (55, 48)

Evaporator Temperature
Sensor

Air Duct (Foot)

Holder

Plate

Drain Hose

Insulator

Heater Case

Evaporator

N·m (kgf·cm, in.·lbf) : Specified torque

◆ Non-reusable part

93112GK6

Exploded view of the heater/air conditioning housing, heater core, evaporator core and related components—1999–00 Solara

For Tune-up, Capacities and Firing orders, see Section 1 of this manual

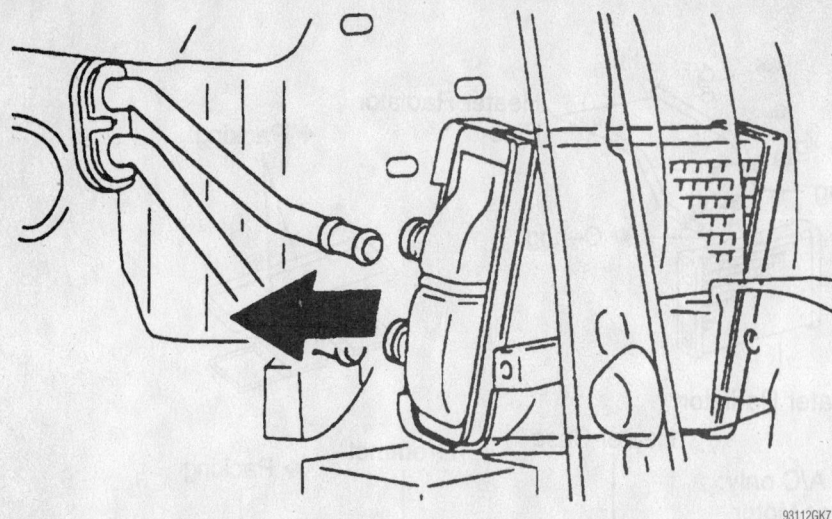

View of the heater core—1999–00 Solara

93112GK7

screws and the clamps; then, disconnect the pipes from the heater core.

9. Remove the heater core.

To install:

10. Install the heater core.

11. Connect the pipes to the heater core and install the heater core pipe clamp and the 2 clamp screws.

12. Install the heater core pipe clamps and the 3 clamp screws.

13. At the heater/air conditioning housing, install the heater protector and engage the 3 heater protector-to-heater/air conditioning housing clips.

14. Install the left hand instrument lower panel.

15. At the driver's side, install the lower instrument panel.

16. Connect the heater hoses to the heater core.

17. Refill the cooling system.

18. Connect the negative battery cable.

Supra

REMOVAL & INSTALLATION

1. Disconnect the negative battery cable.

❊❊ CAUTION

After disconnecting the negative battery cable, wait for at least 1½ minutes for the SRS to deplete it's energy.

2. Drain the cooling system into a clean container for reuse.

3. Disconnect the heater hoses from the heater core.

4. Discharge and recover the air conditioning system refrigerant.

5. Remove the air bag module and the steering wheel by performing the following procedure:

- Place the front wheels in the straight-ahead position.
- At both sides of the steering wheel, remove the screw covers.
- Using a Torx socket, loosen the 2 air bag module-to-steering wheel Torx screws until the circumference ring catches on the screw case.
- Carefully, remove the air bag module and disconnect the electrical connector.
- Remove the steering wheel nut.
- Using a steering wheel puller, press the steering wheel from the steering column.

6. Remove the following parts in the following sequence:

- The front pillar garnishes
- The assist grip
- The foot rest
- The front door scuff plates
- The steering column cover
- The console upper panel
- The parking brake hole cover
- The console box
- The finish panel
- The lower left side finish panel
- The cluster finish panel
- The center cluster finish panel
- The left side finish panel
- The right side finish panel
- The combination meter
- The audio receiver assembly
- The computer cover
- Inside the glove compartment, pry

out the glove compartment finish plate.

- Pull out the air bag electrical connector and disconnect it.
- Remove the glove compartment-to-instrument panel fasteners and the and the glove compartment.
- Remove the passenger's side air bag module-to-instrument panel 4 bolts. Carefully, remove the air bag module from the instrument panel.
- Remove the finish panel No. 3 and No. 4 mounting brackets.
- Remove the No. 4 register-to-heater housing duct.
- Remove the No. 5 finish panel mounting bracket.
- Remove the 2 parking brake lever bolts.
- Remove the No. 2 side defroster nozzle.
- Remove the 4 steering column-to-instrument panel nuts.
- Disconnect the electrical connectors; then, remove the nut, the 8 bolts and the instrument panel.
- Remove the No. 1 reinforcement brace-to-instrument panel bolt, 2 nuts and the No. 1 brace.
- Remove the reinforcement-to-instrument panel 6 bolts, 4 nuts and the reinforcement.

7. Remove the engine wiring harness bracket bolts and the brake tube bracket bolts.

8. Remove the 2 insulator retainer bolts and the retainer.

9. Remove the heater/air conditioning housing by performing the following procedure:

- Remove the ABS actuator.
- Remove the 2 refrigerant line plate-to-evaporator core bolts, disconnect the tubes and discard the O-rings.
- Remove the 3 No. 3 register duct screws and the duct.
- Disconnect the electrical connectors from the heater/air conditioning housing.
- Remove the 6 heater/air conditioning housing-to-chassis bolts and the heater/air conditioning housing assembly.

10. Remove the 2 heater core-to-heater/air conditioning housing plate screws and the plate.

11. Remove the 2 heater core-to-housing clamp screws and the clamps.

12. Remove the 3 heater valve-to-housing screws and the heater core from the housing.

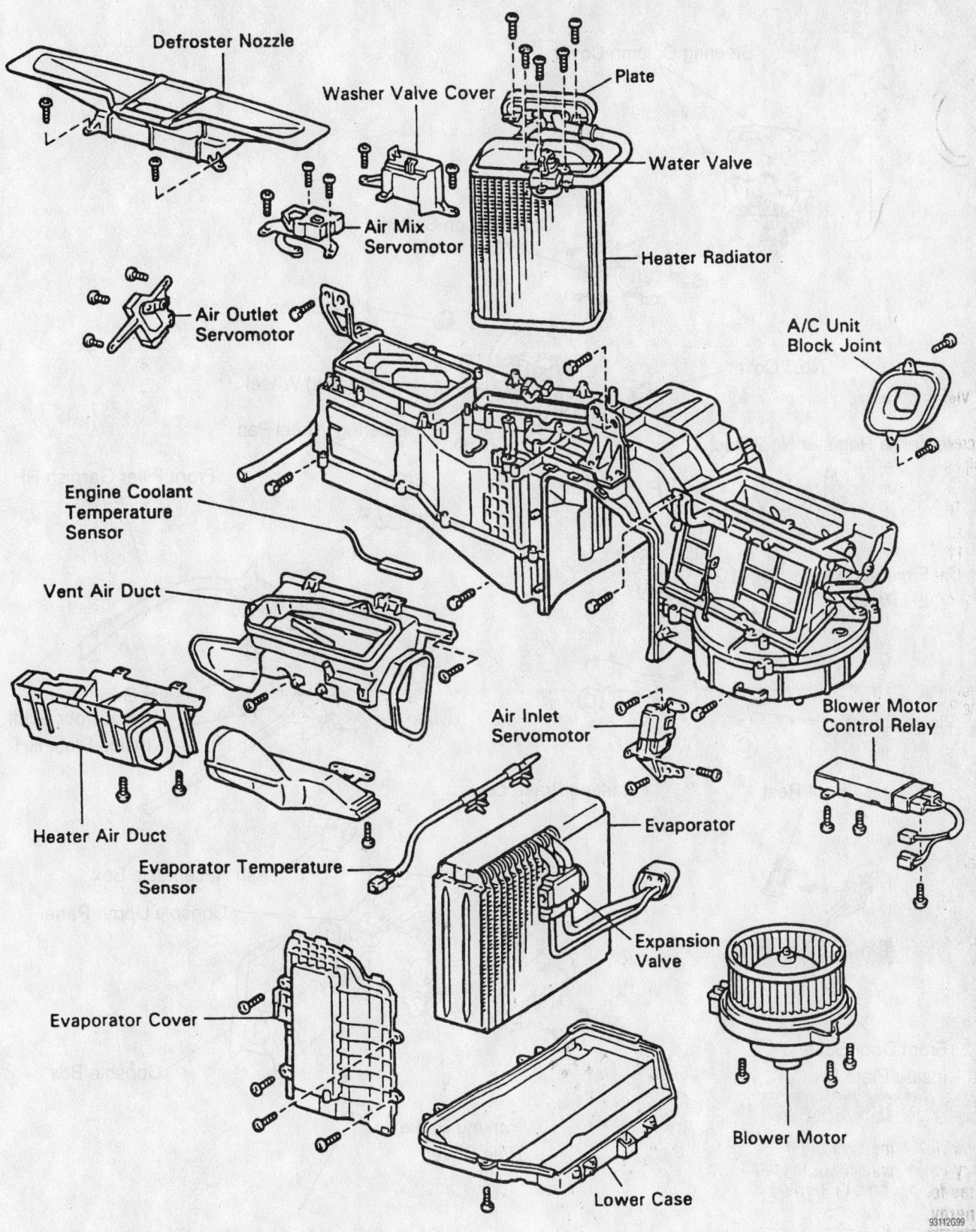

Defroster Nozzle

Washer Valve Cover

Plate

Water Valve

Air Mix Servomotor

Heater Radiator

Air Outlet Servomotor

A/C Unit Block Joint

Engine Coolant Temperature Sensor

Vent Air Duct

Heater Air Duct

Air Inlet Servomotor

Blower Motor Control Relay

Evaporator Temperature Sensor

Evaporator

Expansion Valve

Evaporator Cover

Blower Motor

Lower Case

93112G99

View of the heater/air conditioning assemblies location—Supra

For complete service labor times, order the Chilton Labor Guide

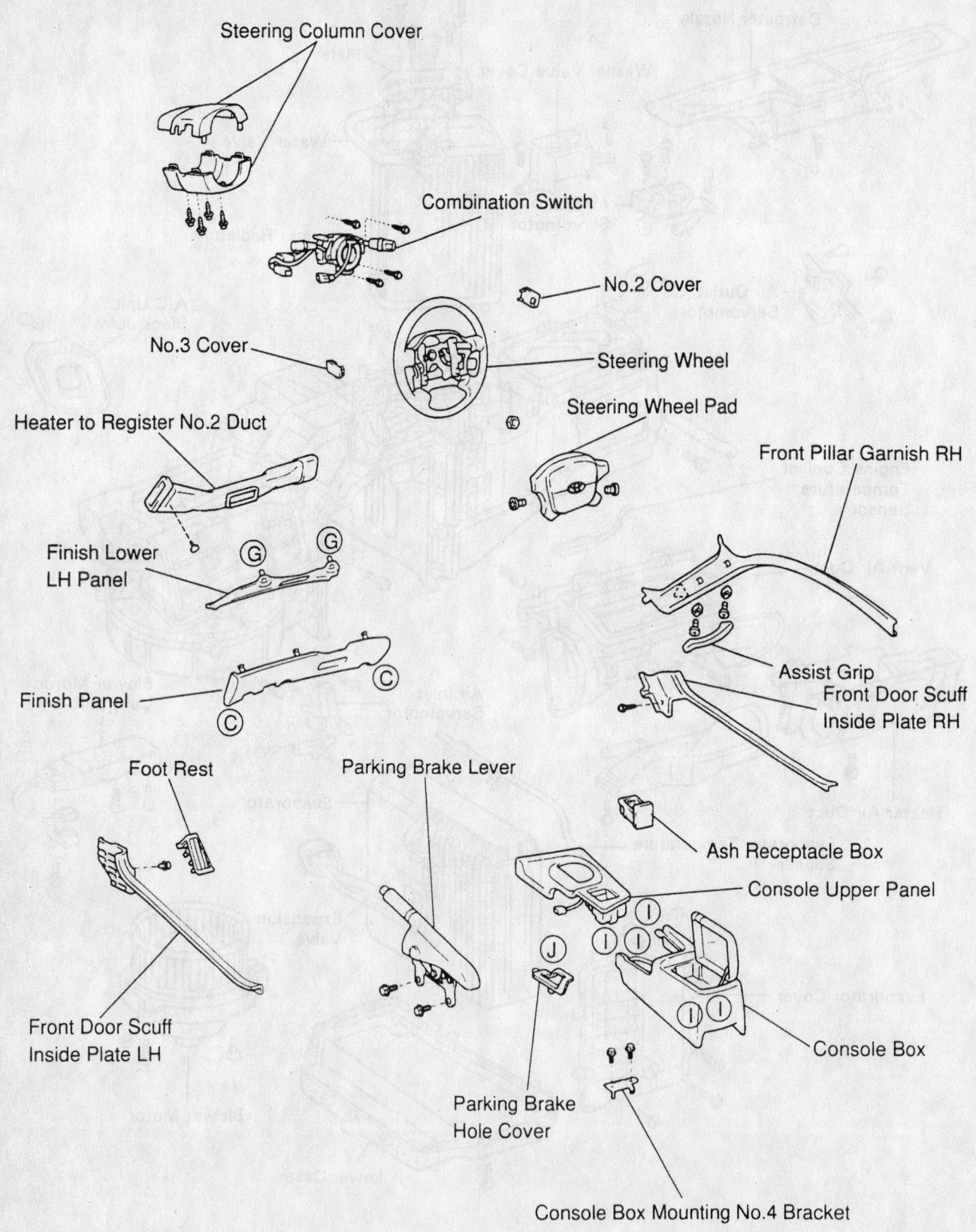

Steering Column Cover

Combination Switch

No.2 Cover

No.3 Cover

Steering Wheel

Steering Wheel Pad

Heater to Register No.2 Duct

Front Pillar Garnish RH

Finish Lower LH Panel

Assist Grip
Front Door Scuff
Inside Plate RH

Finish Panel

Foot Rest

Parking Brake Lever

Ash Receptacle Box

Console Upper Panel

Front Door Scuff
Inside Plate LH

Parking Brake
Hole Cover

Console Box

Console Box Mounting No.4 Bracket

93112GL8

Exploded view of the air bag module, steering wheel, console and related components—Supra

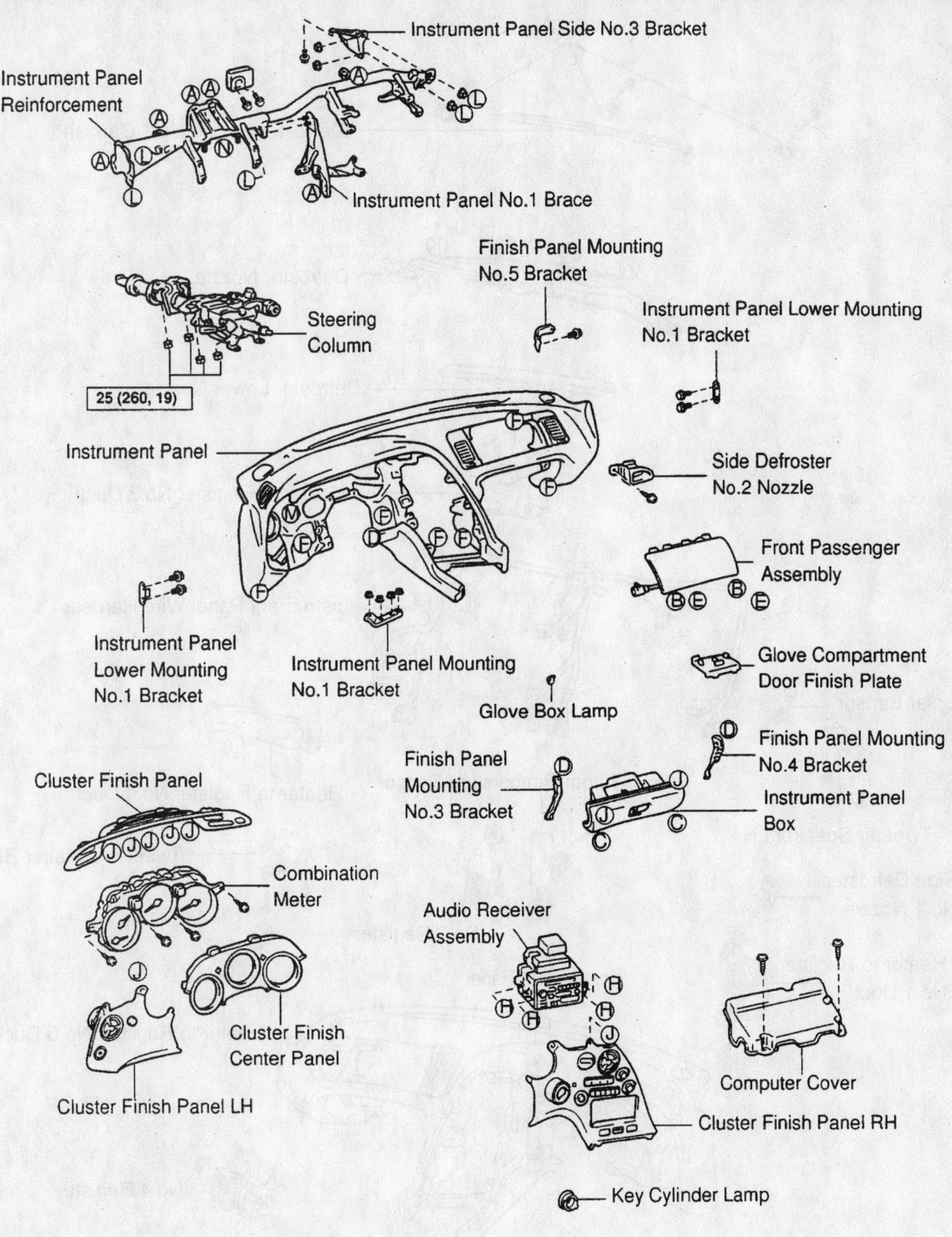

Instrument Panel Side No.3 Bracket

Instrument Panel Reinforcement

Instrument Panel No.1 Brace

Finish Panel Mounting No.5 Bracket

Instrument Panel Lower Mounting No.1 Bracket

Steering Column

25 (260, 19)

Instrument Panel

Side Defroster No.2 Nozzle

Instrument Panel Lower Mounting No.1 Bracket

Instrument Panel Mounting No.1 Bracket

Glove Box Lamp

Front Passenger Assembly

Glove Compartment Door Finish Plate

Finish Panel Mounting No.3 Bracket

Finish Panel Mounting No.4 Bracket

Instrument Panel Box

Cluster Finish Panel

Combination Meter

Audio Receiver Assembly

Computer Cover

Cluster Finish Center Panel

Cluster Finish Panel LH

Cluster Finish Panel RH

Key Cylinder Lamp

N·m (kgf·cm, ft·lbf) : Specified torque

93112GL9

Exploded view of the instrument panel, reinforcement and related components—Supra

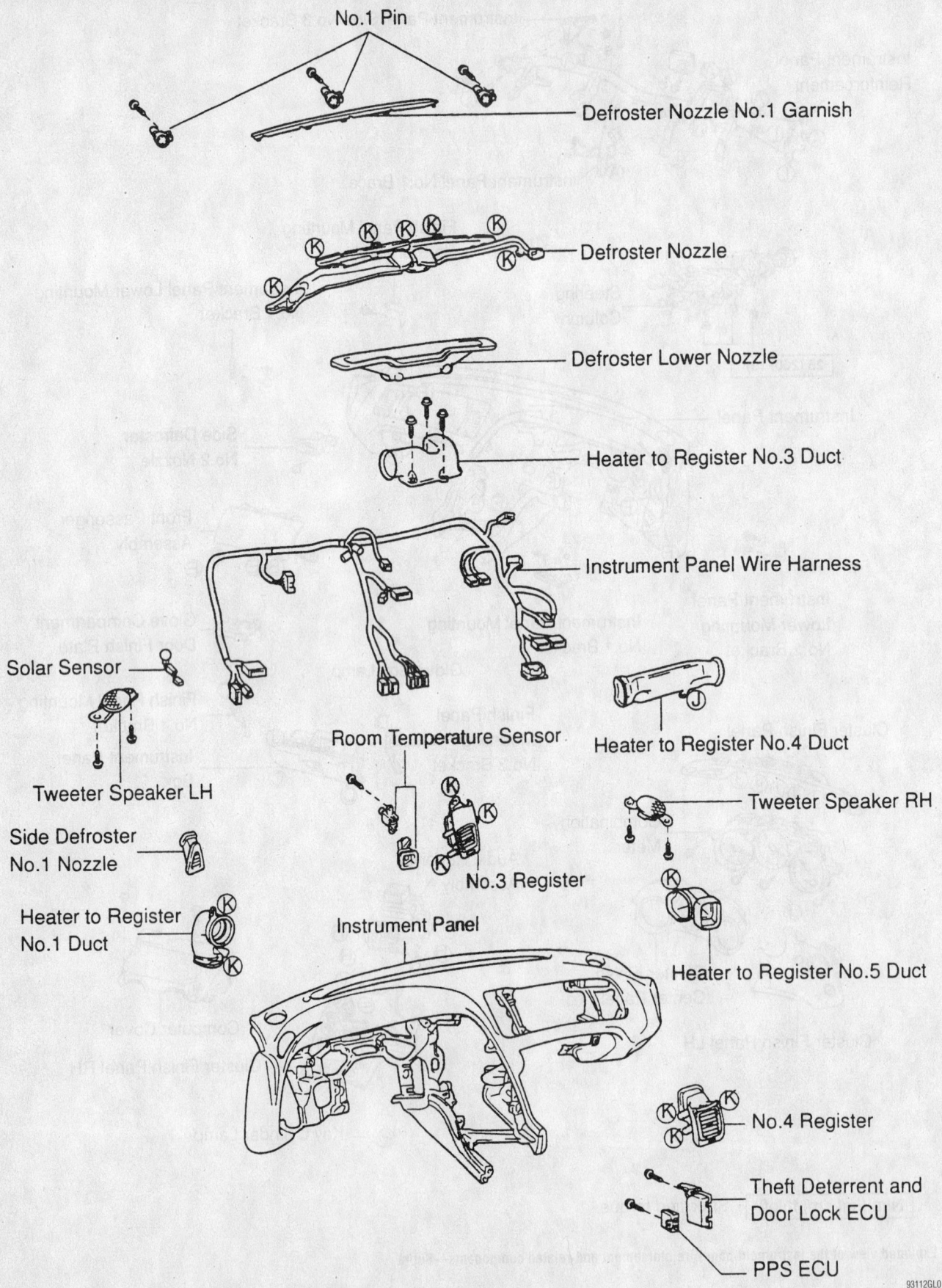

No.1 Pin

Defroster Nozzle No.1 Garnish

Defroster Nozzle

Defroster Lower Nozzle

Heater to Register No.3 Duct

Instrument Panel Wire Harness

Solar Sensor

Room Temperature Sensor

Heater to Register No.4 Duct

Tweeter Speaker LH

Tweeter Speaker RH

Side Defroster No.1 Nozzle

No.3 Register

Heater to Register No.1 Duct

Instrument Panel

Heater to Register No.5 Duct

No.4 Register

Theft Deterrent and Door Lock ECU

PPS ECU

93112GL0

Exploded view of the ventilation system and wiring harness and related components—Supra

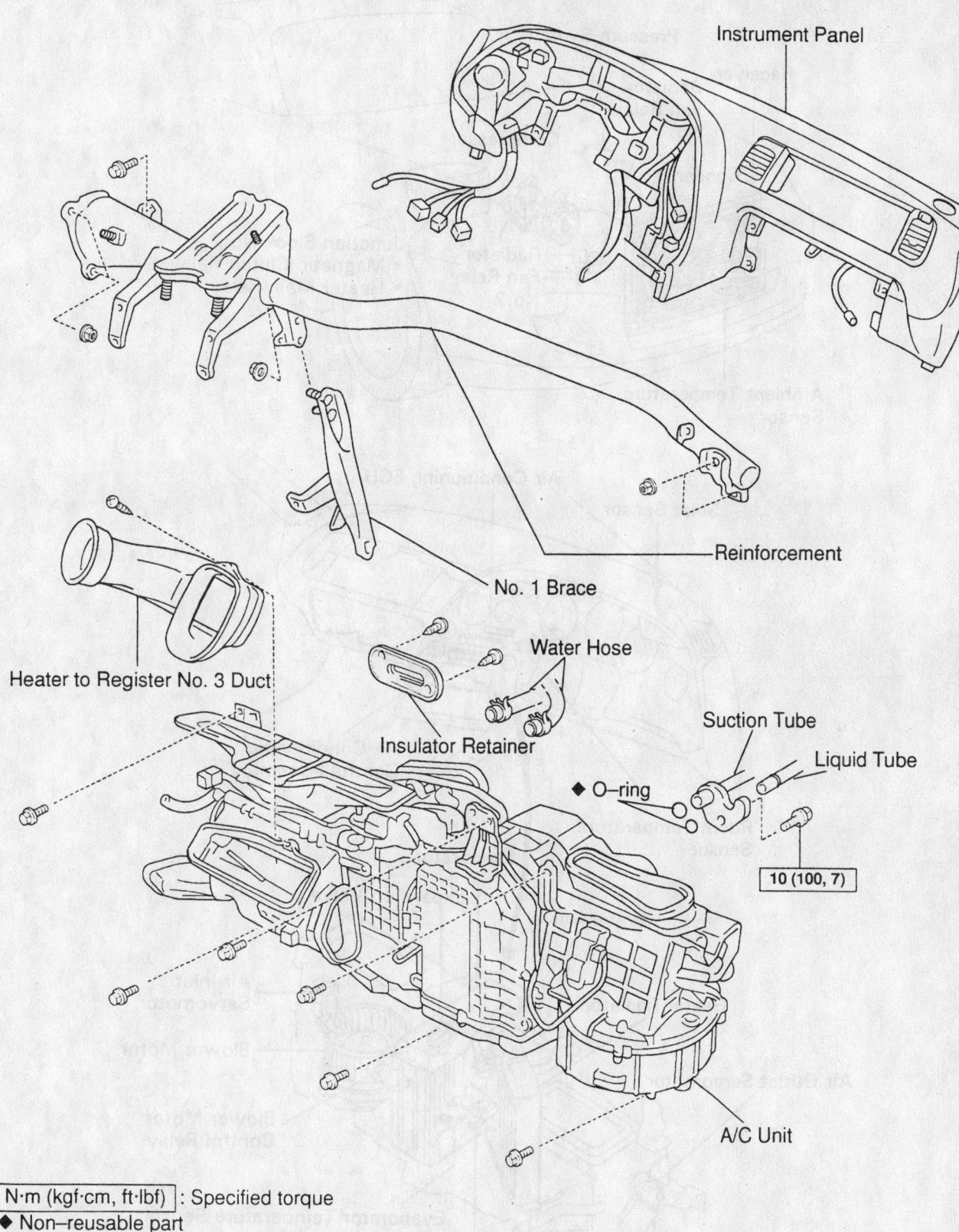

Instrument Panel

Reinforcement

No. 1 Brace

Water Hose

Heater to Register No. 3 Duct

Insulator Retainer

Suction Tube

Liquid Tube

◆ O–ring

10 (100, 7)

A/C Unit

N·m (kgf·cm, ft·lbf) : Specified torque
◆ Non–reusable part

93112GM1

Exploded view of the reinforcement, register, heater/air conditioning housing assembly and related components—Supra

Timing belt service is covered in Section 3 of this manual

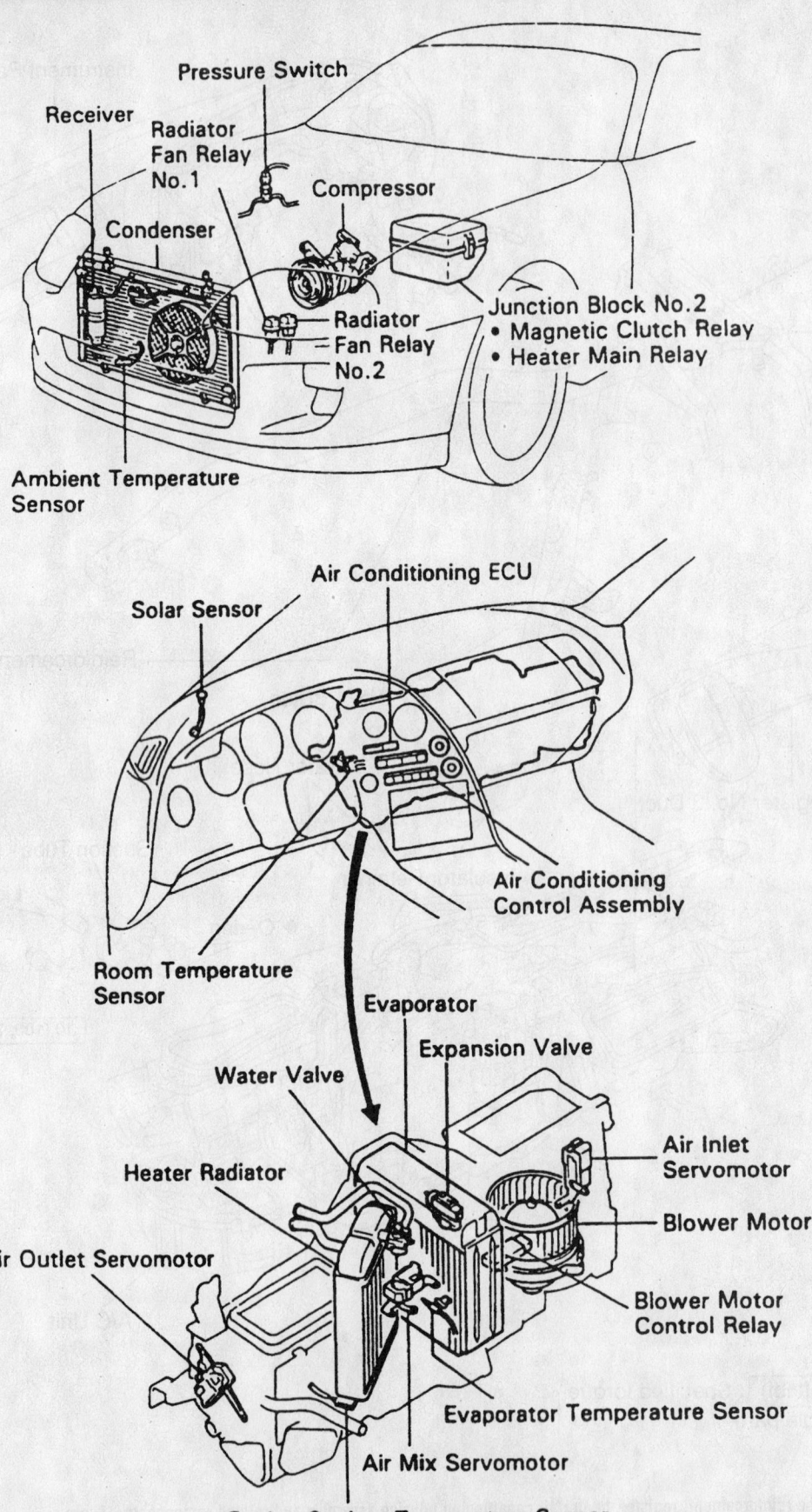

Exploded view of the heater/air conditioning assembly and related components—1998 Supra

To install:

13. Install the heater core and the 3 heater valve-to-housing screws.

14. Install the heater core-to-housing clamp and the 2 clamp screws.

15. Install the heater core-to-heater/air conditioning housing plate and the 2 plate screws.

16. Install the heater/air conditioning housing by performing the following procedure:

- Install the heater/air conditioning housing and the 6 heater/air conditioning housing assembly-to-chassis bolts.
- Connect the electrical connectors to the heater/air conditioning housing.
- Install the No. 3 register duct and the 3 duct screws.
- Using new O-rings, install the 2 refrigerant line plate, connect the tubes and install the plate-to-evaporator core bolts.
- Install the ABS actuator.

17. Install the insulator retainer and the 2 retainer bolts.

18. Install the engine wiring harness bracket bolts and the brake tube bracket bolts.

19. Install the instrument panel by performing the following procedure:

- Install the reinforcement and the reinforcement-to-instrument panel 6 bolts, 4 nuts.
- Install the No. 1 reinforcement brace and the No. 1 brace-to-instrument panel bolt, 2 nuts.
- Connect the electrical connectors; then, install the nut, the 8 bolts and the instrument panel.
- Install the 4 steering column-to-instrument panel nuts.
- Install the No. 2 side defroster nozzle.
- Install the 2 parking brake lever bolts.
- Install the No. 5 finish panel mounting bracket.
- Install the No. 4 register-to-heater housing duct.
- Install the finish panel No. 3 and No. 4 mounting brackets.
- Carefully, install the air bag module to the instrument panel and the 4 passenger's side air bag module-to-instrument panel bolts.
- Install the glove compartment-to-instrument panel and the glove compartment fasteners.
- Connect the air bag electrical connector.

- Install the glove compartment finish plate.

20. Install the following parts in the following sequence:
- The computer cover
- The audio receiver assembly
- The combination meter
- The right side finish panel
- The left side finish panel
- The center cluster finish panel
- The cluster finish panel
- The lower left side finish panel
- The finish panel
- The console box
- The parking brake hole cover
- The console upper panel
- The steering column cover
- The front door scuff plates
- The foot rest
- The assist grip
- The front pillar garnishes

21. Install the air bag module and the steering wheel by performing the following procedure:
- Align the matchmarks and install the steering wheel to the steering column.
- Install the steering wheel nut and torque to 26 ft. lbs. (34 Nm).
- Carefully, install the air bag module and connect the electrical connector.
- Using a Torx socket, install the 2 air bag module-to-steering wheel Torx screws and torque to 78 inch lbs. (8.8 Nm).
- At both sides of the steering wheel, install the screw covers.

22. Connect the heater hoses to the heater core.

23. Refill the cooling system.

24. Connect the negative battery cable.

25. Evacuate, charge and leak test the air conditioning system refrigerant.

26. Operate the engine to normal operating temperatures; then, check the climate control operation and check for leaks.

Tercel

REMOVAL & INSTALLATION

1. Disconnect the negative battery cable.

✳✳ CAUTION

After Connecting the negative battery cable, wait for at least 1½ minutes for the SRS to deplete it's energy.

2. Drain the cooling system into a clean container for reuse.

3. Disconnect the heater hoses from the heater core.

4. Discharge and recover the air conditioning system refrigerant.

5. Remove the glove box-to-instrument panel 4 bolts and 3 screws; then, pull out the glove box and remove it.

6. Disconnect the passenger's side air bag module electrical connector.

7. Remove the 2 passenger's air bag module-to-instrument panel bolts and nuts. Carefully, remove the air bag module.

8. Remove the evaporator housing by performing the following procedure:
- Disconnect the air conditioning amplifier electrical connector.
- Remove the 2 air conditioning amplifier screws and the amplifier.
- Disconnect the evaporator housing's electrical connector.
- Remove the evaporator housing-to-chassis 3 screws, 2 nuts and the housing.

9. Remove the 3 heater housing-to-chassis nuts and the heater housing.

10. Remove the defroster nozzle-to-heater housing screw and the nozzle.

11. Disconnect the air vent control link, the 2 air vent duct screws and the duct.

12. Remove the 3 heater core cover screws and the cover.

13. Remove the heater core-to-heater housing 3 screws, 2 clamps and the heater core.

To install:

14. Install the heater core and the heater core-to-heater housing 3 screws and 2 clamps.

15. Install the heater core cover and the 3 cover screws.

16. Connect the air vent control link, the air vent duct and the 2 duct screws.

17. Install the defroster nozzle and the nozzle-to-heater housing screw.

18. Install the heater housing and the 3 heater housing-to-chassis nuts.

19. Install the evaporator housing by performing the following procedure:
- Install the evaporator housing and the housing-to-chassis 3 screws, 2 nuts.
- Connect the evaporator housing's electrical connector.
- Install the air conditioning amplifier and the 2 amplifier screws.
- Connect the air conditioning amplifier electrical connector.

Heater Core replacement is covered in Section 2 of this manual

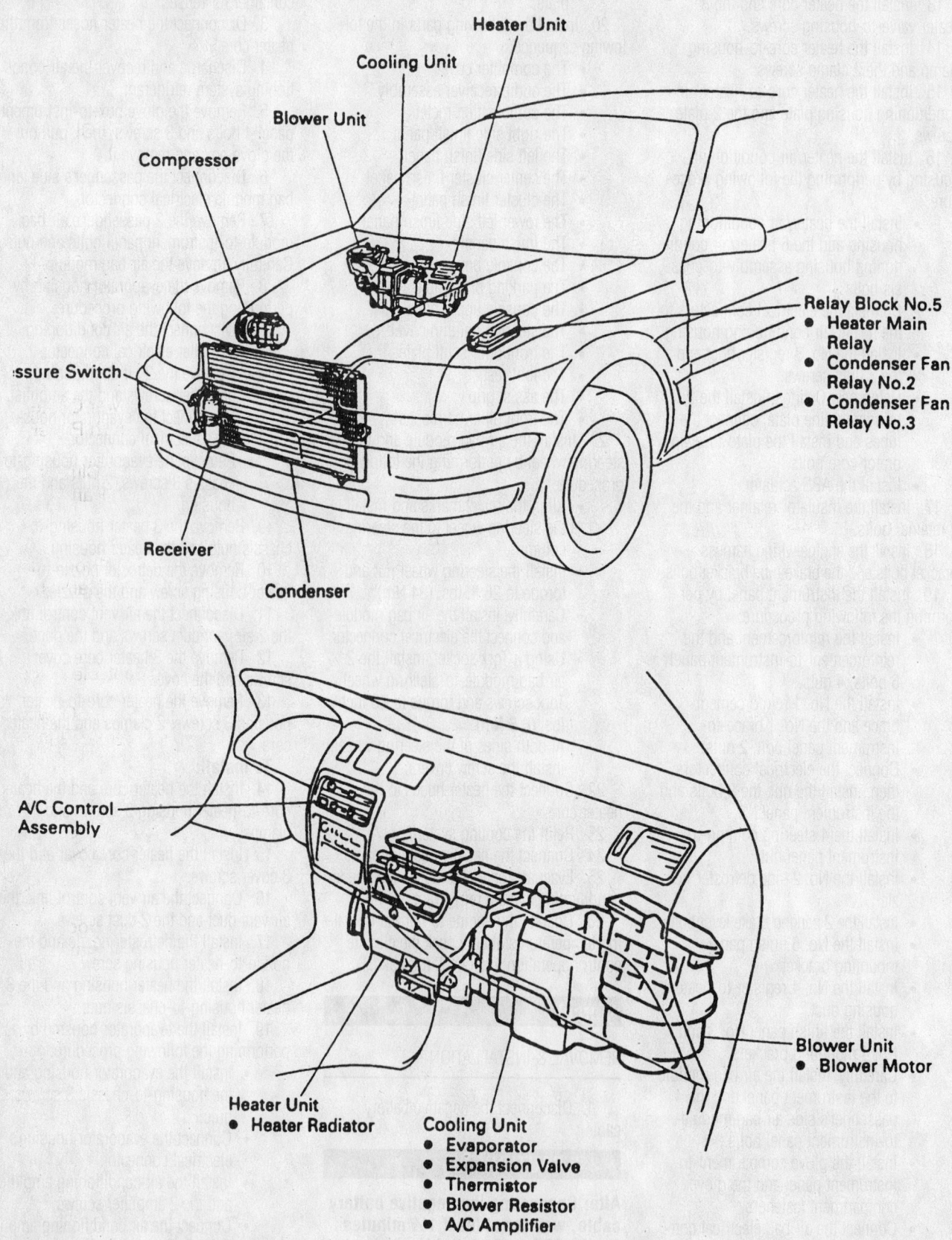

Heater Unit

Cooling Unit

Blower Unit

Compressor

Relay Block No.5
- Heater Main Relay
- Condenser Fan Relay No.2
- Condenser Fan Relay No.3

ssure Switch

Receiver

Condenser

A/C Control Assembly

Blower Unit
- Blower Motor

Heater Unit
- Heater Radiator

Cooling Unit
- Evaporator
- Expansion Valve
- Thermistor
- Blower Resistor
- A/C Amplifier

93112G98

View of the heater/air conditioning assembly and related components—Tercel

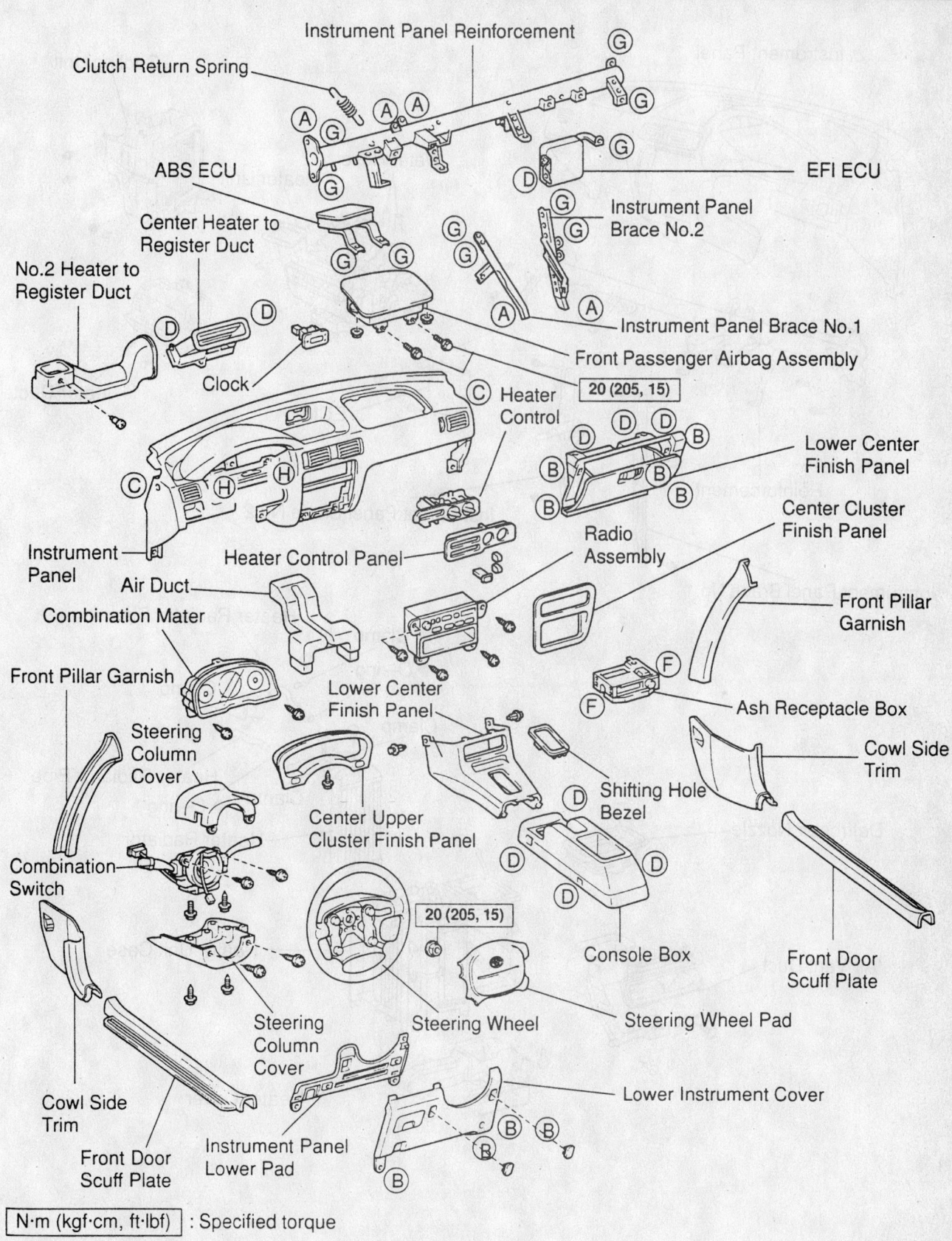

Instrument Panel Reinforcement

Clutch Return Spring

ABS ECU

EFI ECU

Center Heater to Register Duct

Instrument Panel Brace No.2

No.2 Heater to Register Duct

Instrument Panel Brace No.1

Front Passenger Airbag Assembly

Clock

20 (205, 15)

Heater Control

Lower Center Finish Panel

Instrument Panel

Center Cluster Finish Panel

Heater Control Panel

Radio Assembly

Air Duct

Front Pillar Garnish

Combination Mater

Front Pillar Garnish

Ash Receptacle Box

Steering Column Cover

Cowl Side Trim

Combination Switch

Lower Center Finish Panel

Center Upper Cluster Finish Panel

Shifting Hole Bezel

Console Box

Front Door Scuff Plate

Steering Column Cover

Steering Wheel

Steering Wheel Pad

Cowl Side Trim

Lower Instrument Cover

Front Door Scuff Plate

Instrument Panel Lower Pad

20 (205, 15)

N·m (kgf·cm, ft·lbf) : Specified torque

93112GM2

Exploded view of the instrument panel and related components—Tercel

Brake service is covered in Section 4 of this manual

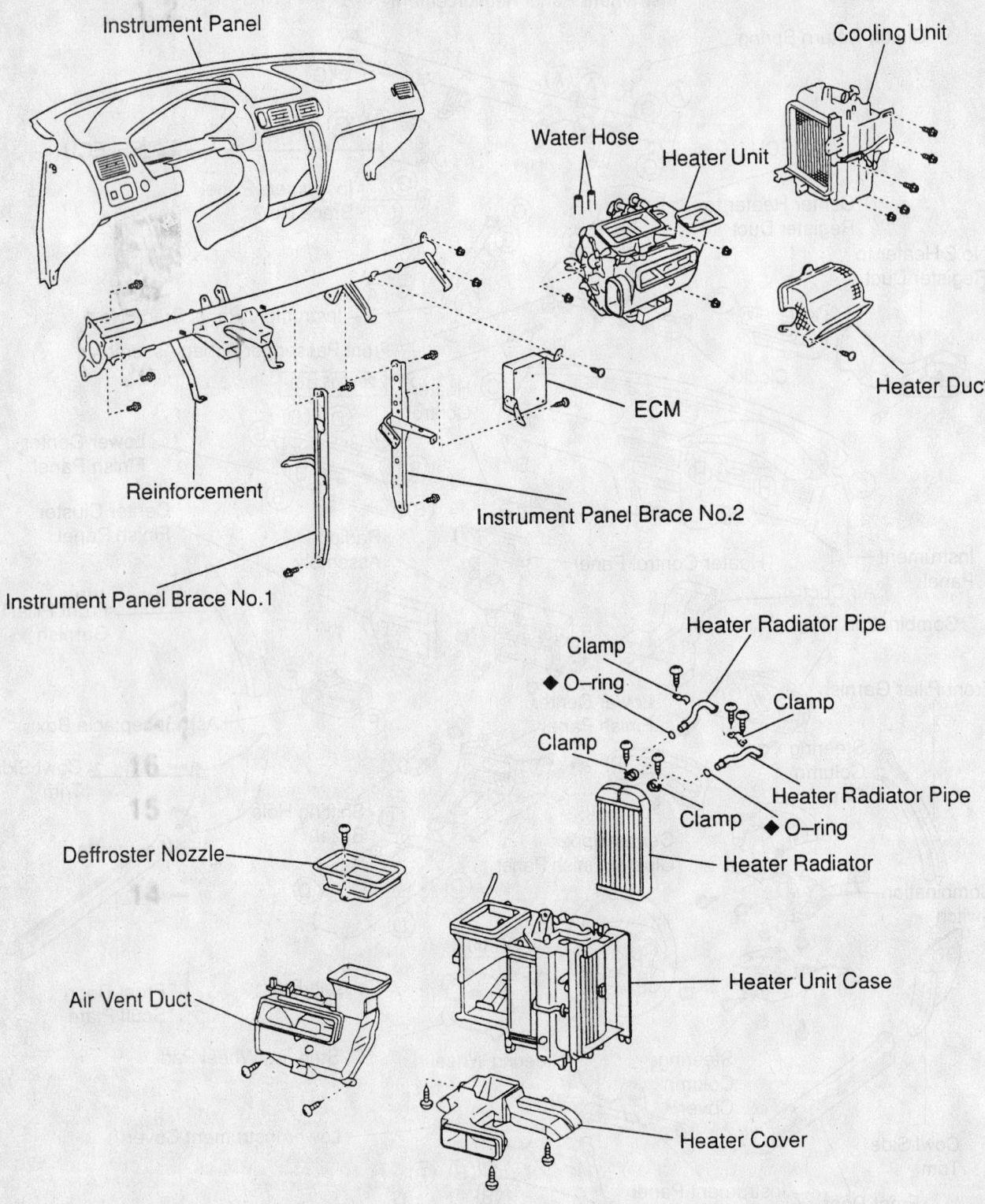

Instrument Panel

Cooling Unit

Water Hose

Heater Unit

Heater Duct

ECM

Reinforcement

Instrument Panel Brace No.2

Instrument Panel Brace No.1

Heater Radiator Pipe

Clamp

◆ O–ring

Clamp

Clamp

Heater Radiator Pipe

Clamp

◆ O–ring

Deffroster Nozzle

Heater Radiator

Heater Unit Case

Air Vent Duct

Heater Cover

◆ Non–reusable part

93112GM3

Exploded view of the heater core, heater housing and related components—Tercel

20. Carefully, install the passenger's air bag module and the 2 air bag module-to-instrument panel bolts and nuts.

21. Connect the passenger's side air bag module electrical connector.

22. Install the glove box and the glove box-to-instrument panel 4 bolts and 3 screws.

23. Connect the heater hoses to the heater core.

24. Refill the cooling system.

25. Connect the negative battery cable.

26. Evacuate, charge and leak test the air conditioning system refrigerant.

27. Operate the engine to normal operating temperatures; then, check the climate control operation and check for leaks.

1998–00 VOLKSWAGEN

New Beetle

REMOVAL & INSTALLATION

➡**Be sure to obtain the anti-theft code for the radio.**

1. Disconnect the negative battery cable.

2. Drain the cooling system into a clean container for reuse.

3. Remove the driver's side air bag by performing the following procedure:

- Release the steering column adjustment.
- Rotate the steering wheel until the spokes are vertical (90 degrees off center).
- Adjust the steering column to the completely out down position and lock the adjustment.
- At the rear of the steering wheel, insert a 7 in. (175mm) screwdriver into the hole in the hub approximately 1¾ in. (45mm) deep.
- Twist the screwdriver counterclockwise (as viewed from the driver's seat) to release the air bag clip.
- Rotate the steering wheel 180 degrees in the opposite direction and release the clip on the other side.
- Center the steering wheel with the front wheels facing straight-ahead.
- Disconnect the electrical connector and remove the air bag.

4. Remove the steering wheel by performing the following procedure:

- Place the front wheels in the straight-ahead position.
- Disconnect the horn's electrical connector.
- Remove the steering wheel-to-steering column bolt and discard it.
- Remove the steering wheel from the steering column.

5. Remove the instrument panel by performing the following procedure:

- Remove the upper steering column cover screws and the cover.
- Remove the lower steering column cover screws, release the steering wheel height adjustment and remove the lower steering column switch trim.
- Remove the combination switch-to-steering column bolt, disconnect the electrical connectors and remove the combination switch.
- At the driver's side, remove the lower instrument panel cover screws and the cover.
- At the passenger's side of instrument panel, remove the side cover.
- At the center of the instrument panel, remove both switch assembly covers, the 4 switch assembly-to-panel screws and the switch assembly.
- Remove the 4 heater/ventilator control unit-to-instrument panel screws and move the control unit under the instrument panel with the cables attached.
- Remove the 6 glove box-to-instrument panel screws, disconnect the glove box light connector and remove the glove box.
- At the driver's side, remove the instrument panel's end cover and remove the 2 light switch screws.
- Rotate the light switch knob to **0**, press the knob inward, rotate it clockwise; then, pull the switch outward and disconnect the electrical connector.

93112GR7

Releasing the driver's side air bag—1998–00 New Beetle

For complete Engine Mechanical specifications, see Section 1 of this manual

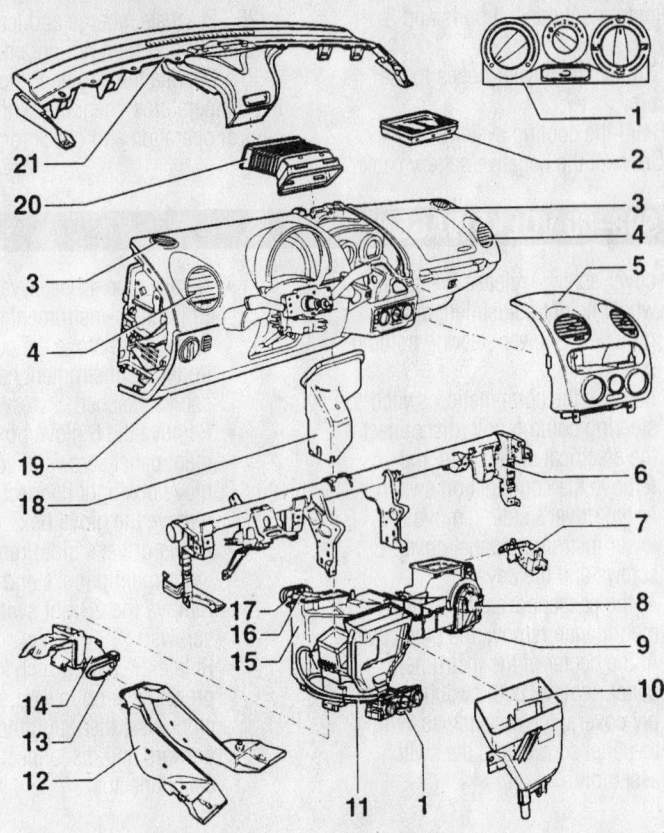

1. Heating and ventilation controls
2. Dust and pollen filter
3. Side window air outlet
4. Side air outlets
5. Center air outlet console
6. Instrument panel cross member
7. Servo motor for fresh/recirculating air door
8. Fresh air blower
9. Fresh air blower series resistance with fuse
10. Intermediate duct
11. Cables
12. Rear footwell air outlet duct
13. Gasket
14. Connecting duct
15. Heating and ventilation unit
16. Heater core
17. Heater core/bulkhead seal
18. Defroster duct
19. Instrument panel
20. Center air outlet duct
21. Defroster air outlet panel

93112GR8

Exploded view of the instrument panel, crossmember and related components—1998–00 New Beetle

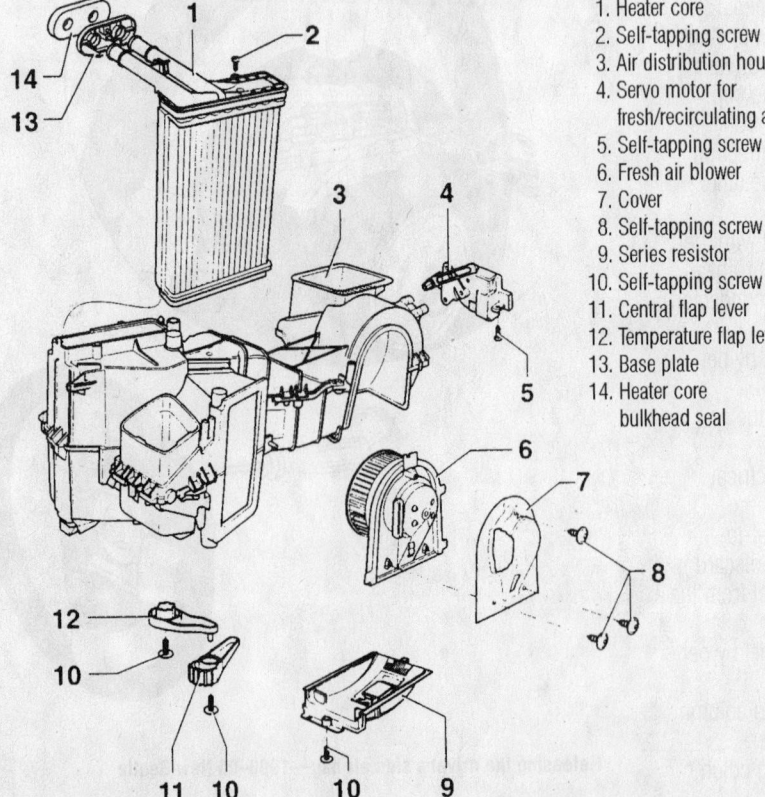

1. Heater core
2. Self-tapping screw
3. Air distribution housing
4. Servo motor for fresh/recirculating air door
5. Self-tapping screw
6. Fresh air blower
7. Cover
8. Self-tapping screw
9. Series resistor
10. Self-tapping screw
11. Central flap lever
12. Temperature flap lever
13. Base plate
14. Heater core bulkhead seal

93112GR9

Exploded view of the heater core, heater/ventilation housing and related components—1998–00 New Beetle without air conditioning

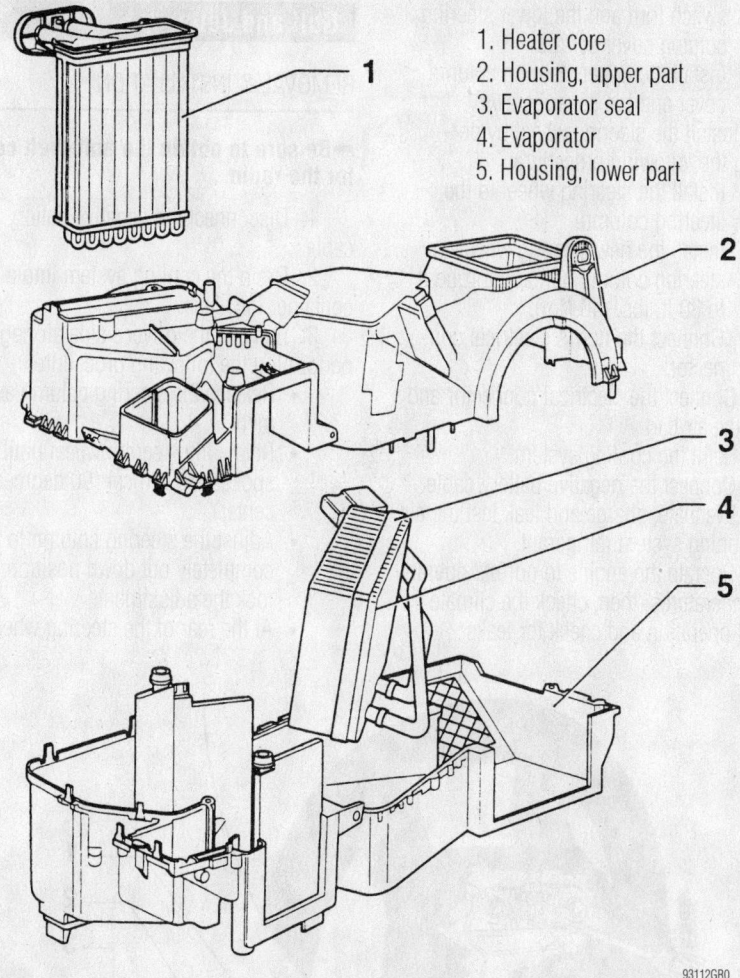

1. Heater core
2. Housing, upper part
3. Evaporator seal
4. Evaporator
5. Housing, lower part

93112GR0

Exploded view of the heater core, heater/air conditioning housing and related components—1998–00 New Beetle with air conditioning

- Unclip the illumination control switch and disconnect the electrical connector.
- At the steering column, remove the instrument panel trim-to-instrument panel screws and the trim.
- Remove the knee bar-to-instrument panel screws and the knee bar.
- Slide the center instrument panel cover forward out of the retainer and remove it; then, remove the 3 instrument panel-to-chassis screws.
- At the upper center of the instrument panel, lift both (left and right) plenum panel covers from the front and rear edge clips; then, move the covers toward the center and out of the A-pillar retainers.
- Unclip and remove the instrument cluster trim.
- Remove the 2 instrument cluster-to-instrument panel screws, dis-

connect the multi-pin electrical connector and remove the instrument cluster.
- Slide the Radio Release tool No. 3344 into the slots at both sides of the radio until they engage; then, pull the radio out of the instrument panel and disconnect the electrical connectors.

➡**To remove the Radio Release tool(s) No. 3344, press the locating tabs (located on both sides of the radio) inward and remove them.**

- If not equipped with a radio, remove the center instrument panel's trim cover.
- Remove the 2 center instrument panel-to-instrument panel screws and the center panel.
- At the center of the instrument panel, remove the air duct.

- At the glove box location, remove the 2 frame-to-instrument panel screws and the frame.
- Remove the instrument panel-to-chassis screws and the instrument panel from the crossmember.
6. Loosen the instrument panel crossmember-to-chassis bolts and lift the crossmember upward.
7. Disconnect the heater hoses from the heater core.
8. If not equipped with air conditioning, perform the following procedure:
- Partially remove the heater/ventilation housing assembly.
- Depress the retainer clips and remove the heater core from the heater housing.
9. If equipped with air conditioning, perform the following procedure:
- Discharge and recover the air conditioning system refrigerant.
- Disconnect the refrigerant lines from the evaporator core, discard the O-rings and plug the openings to prevent contamination.
- Remove the ventilation ducts from the heater/air conditioning housing assembly.
- Remove the heater core from the heater/air conditioning housing assembly.

To install:
10. If equipped with air conditioning, perform the following procedure:
- Install the heater core to the heater/air conditioning housing assembly.
- Install the ventilation ducts to the heater/air conditioning housing assembly.
- Using new O-rings, connect the refrigerant lines to the evaporator core.
11. If not equipped with air conditioning, perform the following procedure:
- Install the heater core to the heater housing.
- Install the heater/ventilation housing assembly.
12. Connect the heater hoses to the heater core.
13. Lower the crossmember, install the instrument panel crossmember-to-chassis bolts and torque the bolts to 18 ft. lbs. (25 Nm).
14. Install the instrument panel by performing the following procedure:
- Install the instrument panel and the

instrument panel-to-chassis screws.

- At the glove box location, install the frame and the frame-to-instrument panel screws.
- At the center of the instrument panel, install the air duct.
- Install the center instrument panel and the 2 center panel-to-instrument panel screws.
- If not equipped with a radio, install the center instrument panel's trim cover.
- Connect the electrical connectors and slide the radio into the instrument panel.
- Install the instrument cluster, connect the multi-pin electrical connector and install the 2 instrument cluster-to-instrument panel screws.
- Install the instrument cluster trim.
- Install both (left and right) plenum panel covers.
- Install the 3 instrument panel-to-chassis screws and the center instrument panel cover.
- Install the knee bar and the knee bar-to-instrument panel screws.
- At the steering column, install the instrument panel trim and the trim-to-instrument panel screws.
- Connect the electrical connector and install the illumination control switch.
- Connect the light switch electrical connector and install the switch.
- At the driver's side, install the 2 light switch screws and the instrument panel's end cover.
- Install the glove box, connect the glove box light connector and install the 6 glove box-to-instrument panel screws.
- Install the heater/ventilator control unit and the 4 heater/ventilator control unit-to-instrument panel screws.
- At the center of the instrument panel, install the switch assembly, the 4 switch assembly-to-panel screws and both switch assembly covers.
- At the passenger's side of instrument panel, install the side cover.
- At the driver's side, install the lower instrument panel cover and the cover screws.
- Install the combination switch, connect the electrical connectors and install the combination switch-to-steering column bolt.
- Install the lower steering column

switch trim and the lower steering column cover screws.
- Install the upper steering column cover and the cover screws.

15. Install the steering wheel by performing the following procedure:
- Install the steering wheel to the steering column.
- Install the new steering wheel-to-steering column bolt and torque it to 30 ft. lbs. (40 Nm).
- Connect the horn's electrical connector.

16. Connect the electrical connector and install the air bag.

17. Refill the cooling system.

18. Connect the negative battery cable.

19. Evacuate, charge and leak test the air conditioning system refrigerant.

20. Operate the engine to normal operating temperatures; then, check the climate control operation and check for leaks.

Golf and Jetta

REMOVAL & INSTALLATION

➡ **Be sure to obtain the anti-theft code for the radio.**

1. Disconnect the negative battery cable.

2. Drain the cooling system into a clean container for reuse.

3. Remove the driver's side air bag by performing the following procedure:
- Release the steering column adjustment.
- Rotate the steering wheel until the spokes are vertical (90 degrees off center).
- Adjust the steering column to the completely out down position and lock the adjustment.
- At the rear of the steering wheel,

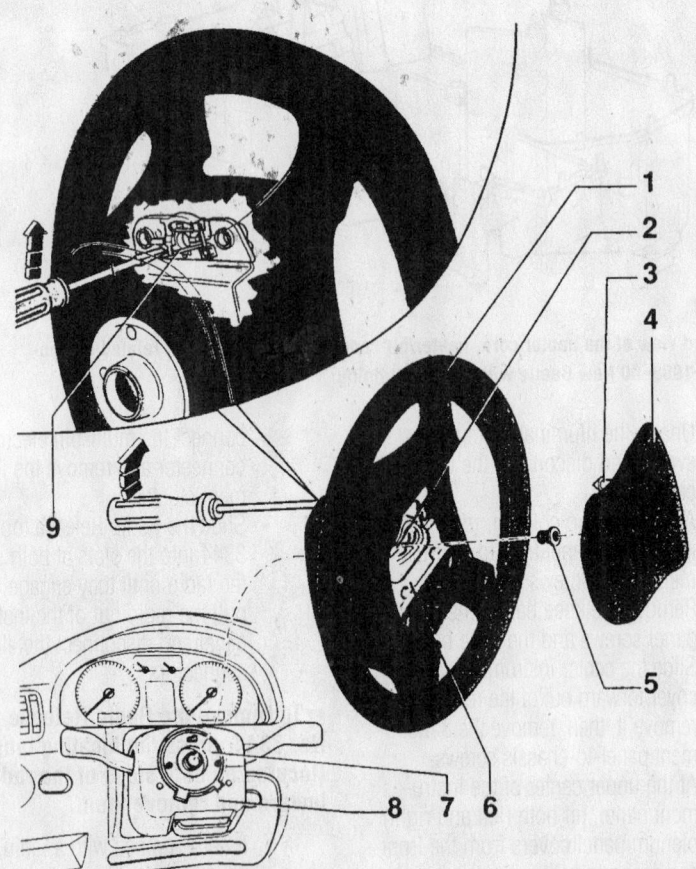

1. Steering wheel
2. Connector
3. Locking lug
4. Airbag unit (with airbag igniter, driver's side)
5. Multi-point socket head bolt
6. Securing plate
7. Coil spring for airbag/return spring with slip ring
8. Trim
9. Clip

93112GS8

Releasing the driver's side air bag—Golf and Jetta

insert a 7 in. (175mm) screwdriver into the hole in the hub approximately 1¾ in. (45mm) deep.

- Raise the screwdriver (as viewed from the driver's seat) to release the air bag clip.
- Rotate the steering wheel 180 degrees in the opposite direction and release the clip on the other side.
- Center the steering wheel with the front wheels facing straight-ahead.
- Disconnect the electrical connector and remove the air bag.

4. Remove the steering wheel by performing the following procedure:
- Place the front wheels in the straight-ahead position.
- Remove the steering wheel-to-steering column bolt.

➡**The Multi-point socket head bolt can be used up to 5 times; be sure to place a punch mark on it each time it is reused.**

- Remove the steering wheel from the steering column.

5. Remove the center console-to-chassis bolts and the console.

6. Remove the instrument panel by performing the following procedure:
- Remove the upper steering column cover screws and the cover.
- Remove the 3 lower steering column cover screws, release the steering wheel height adjustment and remove the lower steering column switch trim.
- Remove the combination switch-to-steering column bolt, disconnect the electrical connectors and remove the combination switch.
- At the passenger's side of instrument panel, remove the side cover.

- Remove the 7 glove box-to-instrument panel screws, disconnect the glove box light connector and remove the glove box.
- At the driver's side, remove the both lower instrument panel cover screws and the covers.
- Remove the DLC-to-bracket screws.
- At the driver's side, remove the 6 sound damper panel-to-instrument panel screws and the panel.
- Slide the Radio Release tool No. 3316 into the slots at both sides of the radio until they engage; then, pull the radio out of the instrument panel and disconnect the electrical connectors.

➡**To remove the Radio Release tool(s) No. 3316, press the locating tabs (located on both sides of the radio) inward and remove them.**

- At the center of the instrument pane (above the radio), unclip and remove the switches and disconnect the electrical connectors.
- Unclip and remove the Climatronic control (heater control) trim.
- Remove the 4 Climatronic control (heater control)-to-instrument panel screws and the control; then, disconnect the electrical connector.
- Remove the 5 center reinforcement-to-chassis screws and the reinforcement.
- At the driver's side, remove the 3 footwell cover screws and the cover.
- Remove the 2 instrument cluster-to-instrument panel screws, disconnect the multi-pin electrical connector and remove the instrument cluster.
- At the upper center of the instrument panel, unclip the photo sensor and disconnect the electrical connector.
- Rotate the light switch knob to **0**, press the knob inward, rotate it clockwise; then, pull the switch outward and disconnect the electrical connector.
- Unclip the illumination control switch and disconnect the electrical connector.
- Remove the instrument panel-to-

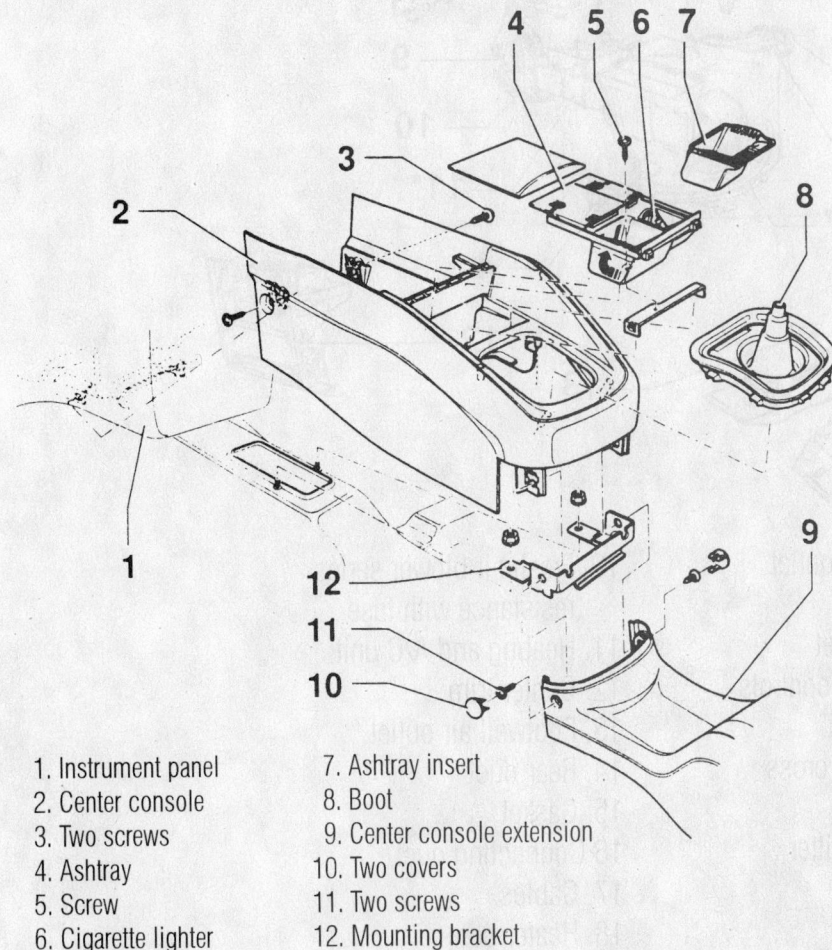

1. Instrument panel
2. Center console
3. Two screws
4. Ashtray
5. Screw
6. Cigarette lighter
7. Ashtray insert
8. Boot
9. Center console extension
10. Two covers
11. Two screws
12. Mounting bracket

93112GS9

Exploded view of the center console and related components—Golf and Jetta

For Tire, Wheel and Ball Joint specifications, see Section 1 of this manual

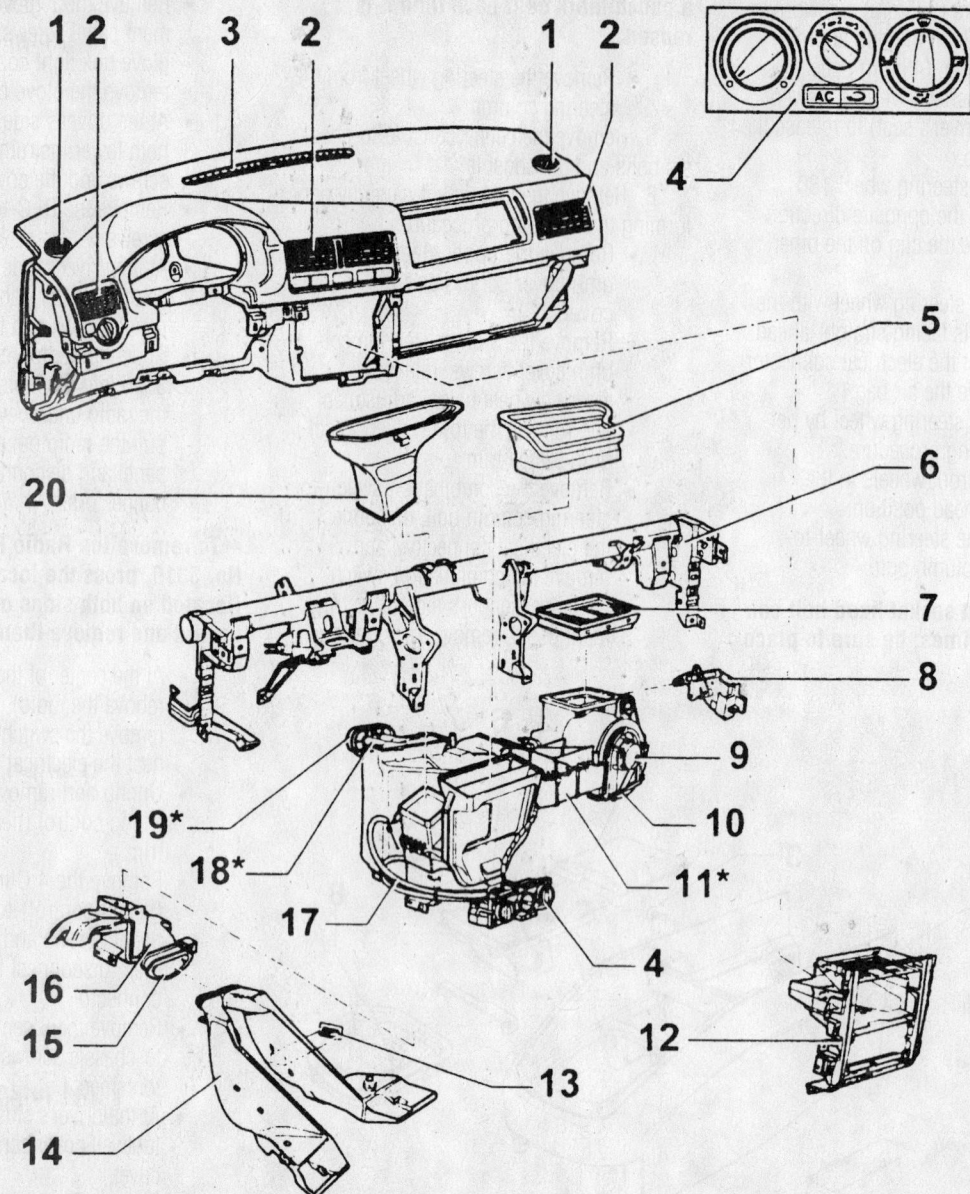

1. Side window air outlet
2. Side air outlets
3. Defroster air outlet
4. Heating and A/C controls
5. Intermediate duct
6. Instrument panel cross member
7. Dust and pollen filter
8. Servo motor for fresh/recirculated air door
9. Fresh air blower
10. Fresh air blower series resistance with fuse
11. Heating and A/C unit
12. Center trim
13. Footwell air outlet
14. Rear duct
15. Gasket
16 Connecting duct
17. Cables
18. Heater core
19. Heater core/bulkhead seal
20. Defroster duct

93112GS0

Exploded view of the instrument panel, crossmember and related components—Golf and Jetta

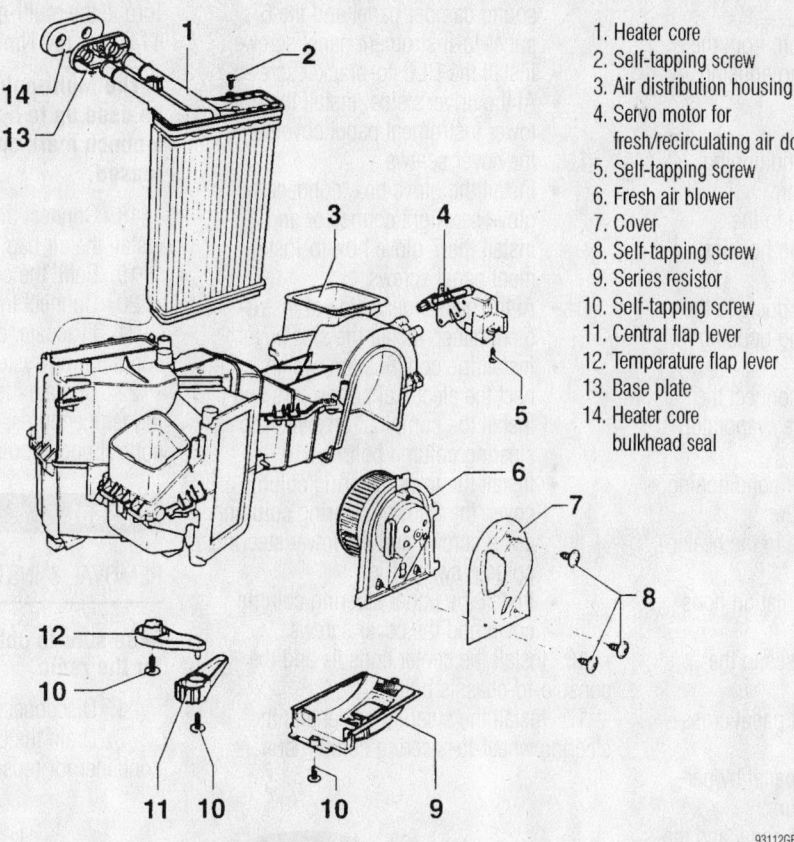

1. Heater core
2. Self-tapping screw
3. Air distribution housing
4. Servo motor for fresh/recirculating air door
5. Self-tapping screw
6. Fresh air blower
7. Cover
8. Self-tapping screw
9. Series resistor
10. Self-tapping screw
11. Central flap lever
12. Temperature flap lever
13. Base plate
14. Heater core bulkhead seal

93112GR9

Exploded view of the heater core, heater/ventilation housing and related components—Golf and Jetta without air conditioning

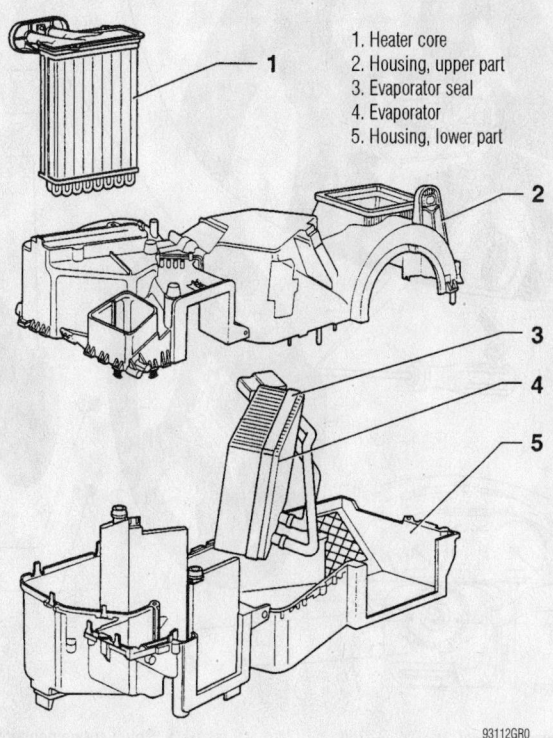

1. Heater core
2. Housing, upper part
3. Evaporator seal
4. Evaporator
5. Housing, lower part

93112GR0

Exploded view of the heater core, heater/air conditioning housing and related components—Golf and Jetta with air conditioning

chassis screws and the instrument panel from the crossmember.

7. Loosen the instrument panel cross-member-to-chassis bolts and lift the cross-member upward.

8. Disconnect the heater hoses from the heater core.

9. If not equipped with air conditioning, perform the following procedure:

- Partially remove the heater/ventilation housing assembly.
- Depress the retainer clips and remove the heater core from the heater housing.

➡️**If the retainer clips break, use a self-tapping screw to hold the heater core in place.**

10. If equipped with air conditioning, perform the following procedure:

- Discharge and recover the air conditioning system refrigerant.
- Disconnect the refrigerant lines from the evaporator core, discard the O-rings and plug the openings to prevent contamination.
- Remove the ventilation ducts from the heater/air conditioning housing

For Wheel Alignment specifications, see Section 1 of this manual

assembly.

• Remove the heater core from the heater/air conditioning housing assembly.

To install:

11. If equipped with air conditioning, perform the following procedure:

• Install the heater core to the heater/air conditioning housing assembly.

• Install the ventilation ducts to the heater/air conditioning housing assembly.

• Using new O-rings, connect the refrigerant lines to the evaporator core.

12. If not equipped with air conditioning, perform the following procedure:

• Install the heater core to the heater housing.

• Install the heater/ventilation housing assembly.

13. Connect the heater hoses to the heater core.

14. Tighten the instrument panel cross-member-to-chassis bolts.

15. Install the instrument panel by performing the following procedure:

• Install the instrument panel and the instrument panel-to-chassis screws.

• Connect the electrical connectors; then, install the light switch knob and the illumination control switch.

• At the upper center of the instrument panel, connect the electrical connector and install the photo sensor.

• Install the instrument cluster, connect the multi-pin electrical connector and install the 2 instrument cluster-to-instrument panel screws.

• At the driver's side, install the footwell cover and the 3 cover screws.

• Install the center reinforcement and the 5 reinforcement-to-chassis screws.

• Connect the electrical connector; then, install the Climatronic control (heater control) and the 4 control-to-instrument panel screws.

• Install the Climatronic control (heater control) trim.

• At the center of the instrument pane (above the radio), Install the switches and connect the electrical connectors.

• Connect the electrical connectors and install the radio.

• At the driver's side, install the

sound damper panel and the 6 panel-to-instrument panel screws.

• Install the DLC-to-bracket screws.

• At the driver's side, install the both lower instrument panel covers and the cover screws.

• Install the glove box, connect the glove box light connector and install the 7 glove box-to-instrument panel screws.

• At the passenger's side of instrument panel, install the side cover.

• Install the combination switch, connect the electrical connectors and install the combination switch-to-steering column bolt.

• Install the lower steering column cover, the 3 lower steering column cover screws and the lower steering column switch trim.

• Install the upper steering column cover and the cover screws.

16. Install the center console and the console-to-chassis bolts.

17. Install the steering wheel and the steering wheel-to-steering column bolt.

Torque the multi-point socket head bolt to 44 ft. lbs. (60 Nm).

➡**The Multi-point socket head bolt can be used up to 5 times; be sure to place a punch mark on it each time it is reused.**

18. Connect the electrical connector and install the air bag.

19. Refill the cooling system.

20. Connect the negative battery cable.

21. Evacuate, charge and leak test the air conditioning system.

22. Operate the engine to normal operating temperatures; then, check the climate control operation and check for leaks.

Passat

REMOVAL & INSTALLATION

➡**Be sure to obtain the anti-theft code for the radio.**

1. Disconnect the negative battery cable.

2. Drain the cooling system into a clean container for reuse.

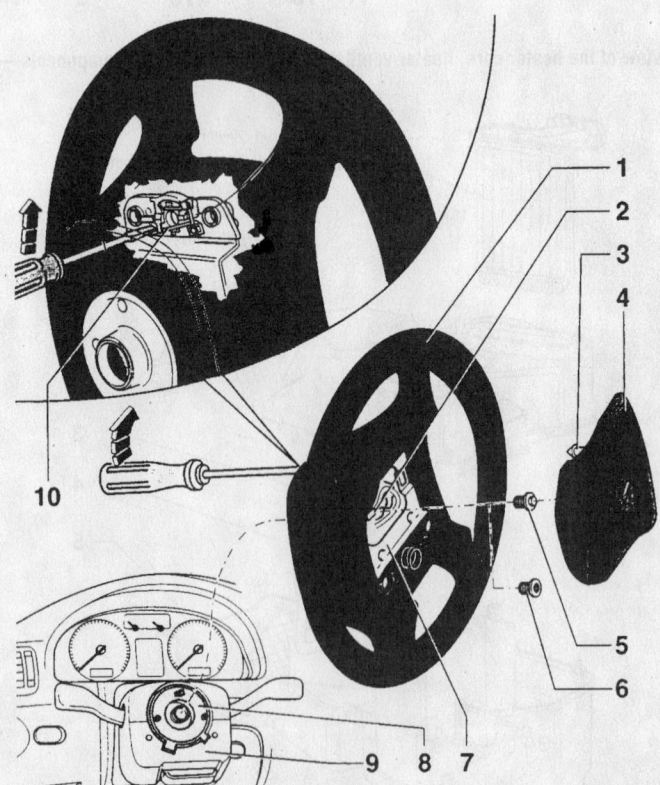

1. Steering wheel	5. Bolt	8. Spiral spring connector with slip ring
2. Connector	6. Multi-point socket head bolt	9. Trim panel
3. Locking lug	7. Securing plate	10. Clip
4. Airbag unit		

93112GT1

Releasing the driver's side air bag—1998–00 Passat

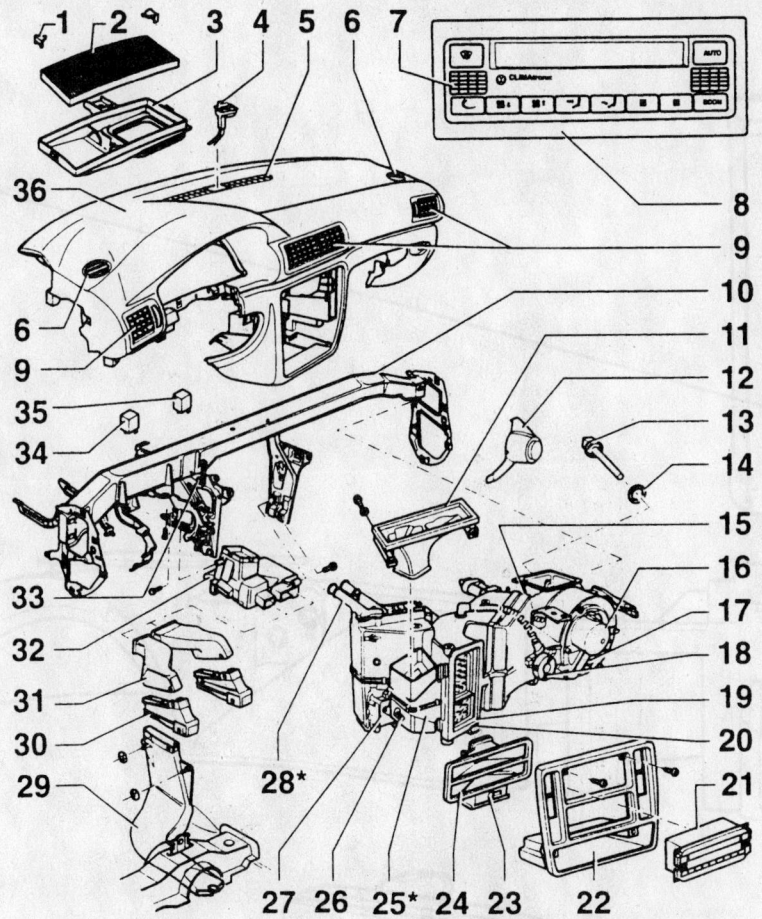

1. Clip
2. Dust and pollen filter
3. Air intake housing
4. Sunlight photo sensor
5. Defroster air outlet
6. Side window air outlet
7. Instrument panel interior temperature sensor
8. A/C control head
9. Center air outlet
10. Instrument panel cross member
11. Defroster duct
12. Evaporator water drain guide
13. Fresh air intake duct temperature sensor
14. Seal
15. Air flow flap motor
16. Fresh air blower
17. Control module for fresh air blower
18. Air outlet for glove box
19. Temperature regulator flap motor
20. Central air flap motor
21. Climatronic control unit
22. Center trim
23. Sender for outlet temperature, center
24. Center air outlet duct
25. Heater/evaporator unit
26. Sender for outlet temperature, floor outlet
27. Footwell/defroster flap motor
28. Heater core
29. Left rear duct
30. Lower duct
31. Upper duct
32. Footwell air outlet
33. Instrument panel cross member and left side support securing bolt
34. Relay
35. A/C clutch relay
36. Instrument panel

93112GT2

Exploded view of the instrument panel, crossmember and related components—1998–00 Passat

For Maintenance Interval recommendations, see Section 1 of this manual

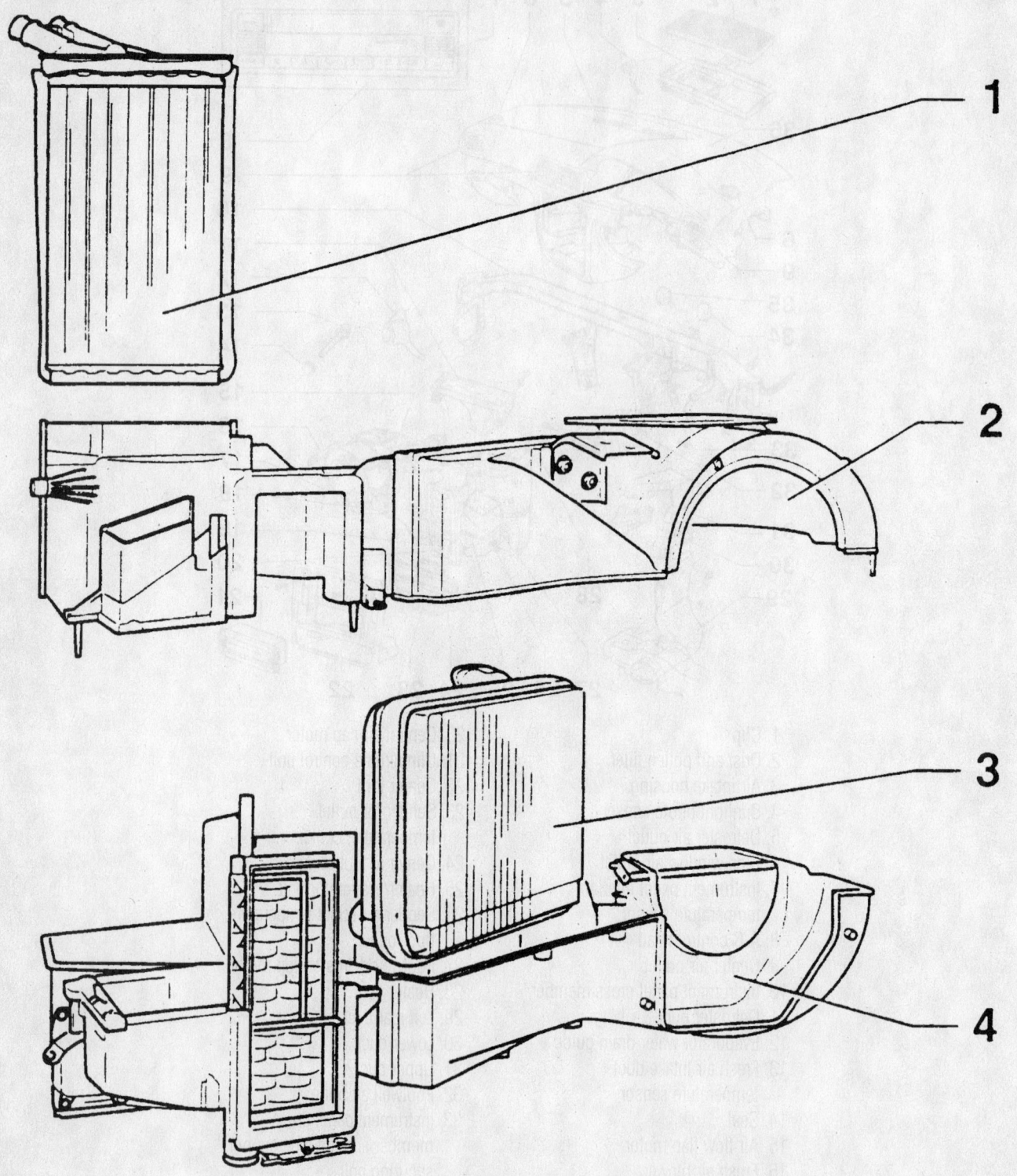

1. Heater core
2. Housing, upper part
3. Evaporator
4. Housing, lower part

Exploded view of the heater core, heater/air conditioning housing and related components—1998–00 Passat

93112GT3

3. Disconnect the heater hoses from the heater core.

4. Discharge and recover the air conditioning system refrigerant.

5. Remove the driver's side air bag by performing the following procedure:

- Release the steering column adjustment.
- Rotate the steering wheel until the spokes are vertical (90 degrees off center).
- Adjust the steering column to the completely out down position and lock the adjustment.
- At the rear of the steering wheel, insert a 7 in. (175mm) screwdriver into the hole in the hub approximately 1¾ in. (45mm) deep.
- Raise the screwdriver (as viewed from the driver's seat) to release the air bag clip.
- Rotate the steering wheel 180 degrees in the opposite direction and release the clip on the other side.
- Center the steering wheel with the front wheels facing straight-ahead.
- Disconnect the electrical connector and remove the air bag.

6. Remove the steering wheel by performing the following procedure:

- Place the front wheels in the straight-ahead position.
- Remove the steering wheel-to-steering column bolt.

➡**Two kinds of bolts are used: Standard (disconnected shortly after production) and Multi-point socket head. If equipped with a standard bolt, discard it. If equipped with a multi-point socket head bolt, it can be used up to 5 times; be sure to place a punch mark on it each time it is reused.**

- Remove the steering wheel from the steering column.

7. Remove the instrument panel by performing the following procedure:

- Remove the 2 upper steering column cover screws and the cover.
- Remove the 4 lower steering column cover screws/bolt, release the steering wheel height adjustment and remove the lower steering column switch trim.
- Remove the combination switch-to-steering column bolt, disconnect the electrical connectors and remove the combination switch.

- At the driver's side, remove the lower "A" pillar trim bolt cover, the "A" pillar trim-to-chassis bolts and the trim.
- At the driver's side, remove the instrument panel's end cover.
- Remove the 2 stowage compartment-to-instrument panel bolts, the 2 instrument panel-to-frame bolts and the stowage compartment.
- Disconnect the electrical connectors from the light switch and the headlight brightness control.
- At the steering column, slide the cover upward and place a screwdriver handle between the cover the steering column to secure it.
- Remove the 4 cover-to-instrument panel screws and the cover.
- Between the steering column and the instrument cluster, remove the 2 instrument panel insert screws and the panel.
- Disconnect the electrical connectors.
- At both sides of the steering column, remove the 2 instrument panel-to-frame bolts.
- Slide the Radio Release tool No. 3316 into the slots at both sides of the radio until they engage; then, pull the radio out of the instrument panel and disconnect the electrical connectors.

➡**To remove the Radio Release tool(s) No. 3316, press the locating tabs (located on both sides of the radio) inward and remove them.**

- At the center of the instrument panel, unclip and remove the control cluster trim.
- Remove the 8 center instrument panel cover-to-instrument panel bolts and the center panel; then, disconnect the electrical connector.
- At the top of the center instrument panel cover opening, remove the 5 instrument panel-to-bracket bolts.
- Under the left side of the glove box, remove the small piece of trim bolt and the trim.
- Under the right side of the glove box, remove the lower "A" pillar trim bolt cover, the "A" pillar trim-to-chassis bolts and the trim.
- Remove the 2 lower glove box bolts and open the glove box.
- Remove the 5 glove box-to-instru-

ment panel screws, disconnect the glove box light connector and remove the glove box.
- At the passenger's side of instrument panel, remove the side cover.
- Remove the 4 instrument panel-to-bracket bolts and disconnect the electrical connector.
- Remove the instrument cluster.

8. Remove the ventilation ducts from the heater/air conditioning housing assembly.

9. Remove the heater/air conditioning housing assembly-to-chassis fasteners and the assembly.

10. Remove the heater core.

To install:

11. Install the heater core.

12. Install the heater/air conditioning housing assembly and the assembly-to-chassis fasteners.

13. Install the ventilation ducts to the heater/air conditioning housing assembly.

14. Install the instrument panel by performing the following procedure:

- Install the instrument cluster.
- Install the 4 instrument panel-to-bracket bolts and connect the electrical connector.
- At the passenger's side of instrument panel, install the side cover.
- Install the glove box, connect the glove box light connector and install the 5 glove box-to-instrument panel screws.
- Install the glove box and open the 2 lower glove box bolts.
- Under the right side of the glove box, install the lower "A" pillar trim, the "A" pillar trim-to-chassis bolts and the trim bolt cover.
- Under the left side of the glove box, install the small piece of trim and the bolt.
- At the top of the center instrument panel cover opening, install the 5 instrument panel-to-bracket bolts.
- Install the center instrument panel cover and the 8 center panel cover-to-instrument panel bolts; then, connect the electrical connector.
- At the center of the instrument panel, install the control cluster trim.
- Connect the electrical connectors and install the radio.

- At both sides of the steering column, install the 2 instrument panel-to-frame bolts.
- Connect the electrical connectors.
- Between the steering column and the instrument cluster, install instrument panel insert and the 2 panel screws.
- At the steering column, install the cover and the 4 cover-to-instrument panel screws.
- Connect the electrical connectors to the light switch and the headlight brightness control.
- Install the stowage compartment, the 2 instrument panel-to-frame bolts and the 2 stowage compartment-to-instrument panel bolts.
- At the driver's side, install the instrument panel's end cover.
- At the driver's side, install the lower "A" pillar trim, the "A" pillar trim-to-chassis bolts and the trim bolt cover.
- Install the combination switch, connect the electrical connectors and install the combination switch-to-steering column bolt.
- Install the lower steering column cover trim and the lower steering column switch trim screws/bolt.
- Install the upper steering column cover and the 2 cover screws.

15. Install the steering wheel and the steering wheel-to-steering column bolt. Torque the standard bolt to 55 ft. lbs. (75 Nm) or the multi-point socket head bolt to 44 ft. lbs. (60 Nm).
16. Connect the electrical connector and install the air bag.
17. Connect the heater hoses to the heater core.
18. Refill the cooling system.
19. Connect the negative battery cable.
20. Evacuate, charge and leak test the air conditioning system refrigerant.
21. Operate the engine to normal operating temperatures; then, check the climate control operation and check for leaks.

1998–00 VOLVO

960 Series

REMOVAL & INSTALLATION

With Automatic Climate Control (ACC) Or Manual Climate Control (MCC)

➥On some vehicles equipped with Automatic Climate Control, a thermal switch is located on the outlet hose from the heater core. It switches ON and starts the fan motor only when the water temperature exceeds approximately 95°F (35°C). This prevents cold air from being blown into the passenger compartment during winter. The thermal switch is bypassed in the defrost position.

1. Disconnect the negative battery cable.
2. Drain the cooling system into a clean container for reuse.
3. Disconnect the heater hoses from the heater core assembly.
4. Remove the ashtray, the ashtray holder, the cigarette lighter and the console's storage compartment.
5. Remove the console assembly from the gearshift lever and the parking brake.
6. Disconnect the electrical connector. Remove the rear ashtray, the console and light.
7. Remove the screws beneath the plastic cover in the bottom of the storage compartment and the parking brake console.
8. From the left side of the passenger compartment, remove the panel from under the dashboard.
9. Pull down the floor mat on the left side and remove the side panel screws, the front and rear edge.
10. From the right side of the passenger compartment, remove the panel from under the glove compartment and the glove compartment box with the lighting.
11. Pull down the floor mat on the right side and remove the side panel screws, the front and rear edge.
12. Remove the radio compartment assembly screws.
13. Remove the screws from the heater control, the radio compartment assembly console and the control panel.
14. Loosen the heater control head assembly retaining screws and remove the assembly and mount from the dash.
15. Remove the center dash panel, the distribution duct screw and the air duct-to-panel vents/distribution duct screws.
16. Remove the screws holding the air ducts top-to-rear seats and the air distribution duct section-to-rear seat ducts.
17. Remove the vacuum hoses from the vacuum motors and the hose from the aspirator, if equipped with an ACC unit.
18. Remove the distribution unit housing from the vehicle.
19. Remove the retaining clips and the heater core assembly.
20. If the vacuum motors must be replaced, remove the panel from the distribution unit and replace the vacuum motor.

To install:
21. Clean the heater core housing of all dirt, leaves etc. before installation. Install the heater core assembly and the retaining clips.
22. Install the distribution unit into the vehicle.
23. Connect the vacuum hoses to the vacuum motors and the hose to the aspirator, if equipped with an ACC unit.
24. Install the air ducts top-to-rear seats and the air distribution duct section-to-rear seat ducts.

25. Install the center dash panel, the distribution duct screw and the air duct-to-panel vents/distribution duct screws.
26. Install the heater control head assembly unit and the mount to the dash.
27. Install the heater control, the radio compartment console and the control panel.
28. Install the radio compartment screws.
29. At the right side of the passenger compartment, install the panel under the glove compartment and the glove compartment box with lighting.
30. Install the side panel screws, the front and rear edge.
31. At the left side of the passenger's compartment, install the panel under the dashboard.
32. Install the plastic cover in the bottom of the storage compartment and the parking brake console.
33. Connect the electrical connector. Install the rear ashtray, the console and light.
34. Install the console assembly to the gearshift lever and the parking brake.
35. Install the ashtray holder, the ashtray, the cigarette lighter and the console's storage compartment.
36. Reconnect the heater core hoses.
37. Refill the cooling system.
38. Connect the negative battery cable.
39. Operate the engine to normal operating temperatures; then, check the climate control operation and check for leaks.

With Electronic Climate Control (ECC)

➥On some vehicles equipped with Automatic Climate Control, a thermal switch is located on the outlet hose from the heater core. It switches ON and starts the fan motor only when the water temperature exceeds approximately 95°F (35°C). **This prevents cold air from being blown into the passenger compartment during winter. The thermal switch is bypassed in the defrost position.**

1. Disconnect the negative battery cable.
2. Drain the cooling system into a clean container for reuse.
3. Disconnect the heater hoses from the heater core assembly.
4. Remove the heater core cover plate.
5. Remove the dashboard by performing the following procedures:
   - From the right side, remove the lower glove box panel, the glove box, the footwell panel and the "A" post panel. Disconnect the solar sensor electrical connector and cut the cable ties.
   - From the left side, remove the lower steering wheel sound-proofing, the knee bolster (leave the bracket attached to the bolster), the footwell panel and the "A" post panel.
   - From the left side, remove the defroster grille, the plastic fuse box screws, the ashtray, the dashboard-to-center console screws, the parking brake-to-console screws (move the console rearward) and the lower center console screws (located below the ashtray).

➡**Before performing the next procedure, be sure the front wheels are in the straight-ahead position.**

6. If equipped with an SRS, remove:
   - steering column adjuster (Allen)
   - steering column covers
   - air bag assembly (Torx)
   - steering wheel center bolt
   - plastic tape label screw from the steering wheel hub (use the lockscrew, label attached, to lock the contact reel through the steering wheel hub hole)
   - steering wheel
7. Remove the contact reel and the steering column combination switch assembly.

➡**After securing the contact reel, do not turn the steering wheel for it will shear off the contact reel pin.**

8. From the left side of the steering column, push out the light switch panel. Remove the small trim moldings and the light switch.

9. From the right side of the steering column, push out the switch panel. Remove the ECC control panel, the radio console and the small trim molding.
10. Remove the outer air vent grille by lifting it upwards, grasp it at the bottom and pull it upwards to release it. Remove the instrument panel cover-to-dash screws and the cover.
11. Remove the combined instrument assembly-to-dash screws and the assembly; disconnect any electrical connectors and/or the vacuum hoses.
12. From the rear of the dashboard, cut the cable ties.
   At the dashboard-to-firewall area, turn the retaining clip ⅓ turn (to release), pull the dash out slightly and pass the fuse box through the opening. Disconnect the cable harnesses from the dashboard and carefully lift it from the vehicle.
13. From the left side of the heater housing, remove the lower duct. Disconnect the vacuum hoses from the diaphragms and the electrical connector. Remove the heater core cover-to-housing screws and the cover.
14. Remove the heater core-to-housing bracket and carefully remove the heater core.

To install:
15. Clean the heater housing of al dirt and debris before installation. Install the heater core and the bracket.
16. Install the heater core cover to the housing. Connect the electrical connector and the vacuum hoses. Install the lower duct to the housing assembly.
17. Install the dashboard by performing the following procedures:
   - Install the dashboard, connect the cable harnesses, pass the fuse box through the opening. Secure the dash clips by turning them ⅓ turn.
   - Install the small trim molding, the radio console, the ECC control panel and the right side switch panel.
   - At the left side of the steering column, install the light switch, the small trim moldings and the light switch panel.
   - If equipped with an SRS, install the steering column combination switch assembly and the contact reel. Install the steering wheel and remove the lockscrew. Install the steering wheel center bolt, the air bag assembly and the steering column adjuster.

   - If not equipped with an SRS, install the steering column combination switch assembly, the steering column covers, the steering wheel adjustment assembly and the steering wheel.
   - At the left side, install the lower center console screw, the parking brake-to-console screws, the dashboard-to-center console screws, the ashtray, the plastic fuse box screws and the defroster grille.
   - At the left side, install the "A" post panel, the footwell panel, the knee bolster (with bracket) and the lower steering wheel sound-proofing.
   - At the right side, connect the solar sensor electrical connector, install the "A" post panel, the footwell panel, the glove box and the lower glove box panel.
18. Install the heater core cover plate and connect the heater hoses to the heater core.
19. Refill the cooling system.
20. Connect the negative battery cable.
21. Operate the engine to normal operating temperatures; then, check the climate control operation and check for leaks.

C70, S70 and V70

REMOVAL & INSTALLATION

1. Disconnect the negative battery cable.
2. Drain the cooling system into a clean container for reuse.
3. Using tools No. 155 8957, disconnect the heater hoses from the heater core by pressing in the hose connections, squeezing the locking catches and pulling out the hoses. Discard the O-rings.
4. At the driver's side, remove the soundproofing panel-to-dashboard Torx 20 screw and the soundproofing panel.
5. Open the glove box door; then, remove the 4 glove compartment-to-dashboard Torx 20 screws and the glove compartment.
6. At the passenger's side, remove the 2 soundproofing panel-to-dashboard screws and the soundproofing panel.
7. At both the driver's side and passenger's side, remove the carpet supports.
8. If equipped, remove the Road Traffic Information (RTI) control module bracket or booster bracket.

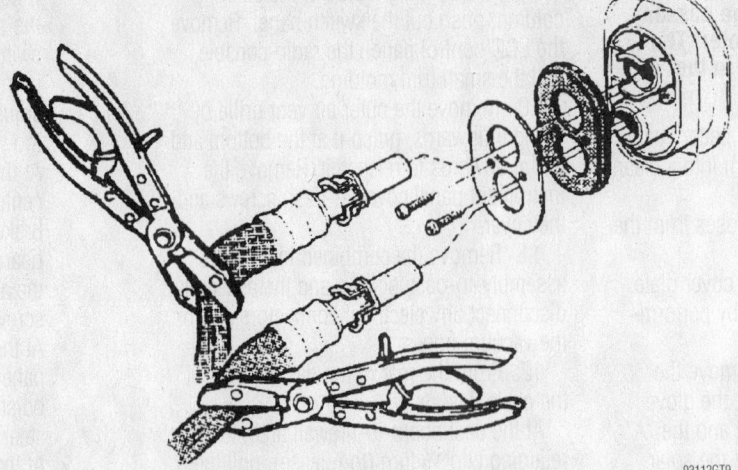

93112GT9

Exploded view of the heater hose connections—1998–00 C70, S70 and V70

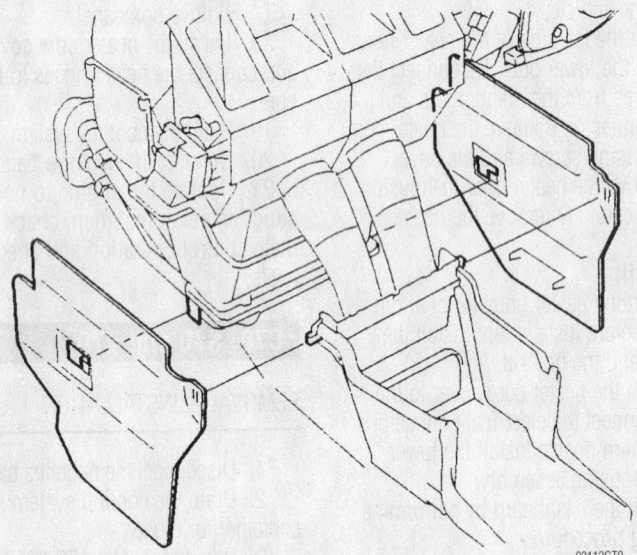

93112GT0

Exploded view of the carpet supports—1998–00 C70, S70 and V70

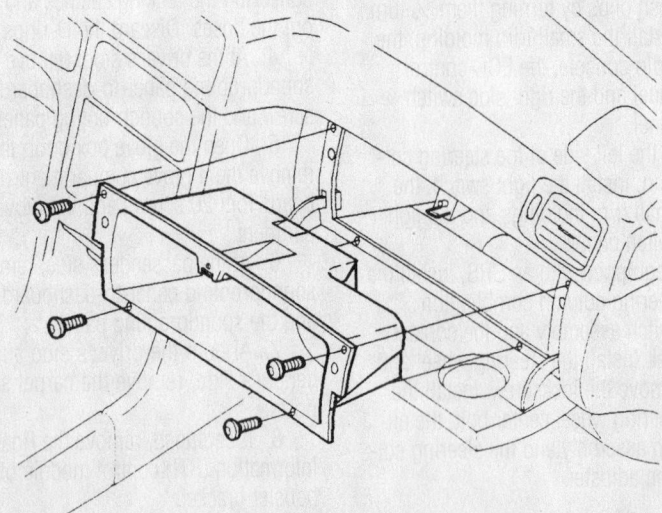

93112GU1

Exploded view of the glove box assembly—1998–00 C70, S70 and V70

9. If equipped, remove the knee bolster.

10. Disconnect the drain hose and fold it out of the way.

11. At both sides of the heater/air conditioning housing assembly, remove the heater core cover screws.

12. Remove the heater housing pipe flange screw.

13. Carefully, remove the heater core with the heater core cover.

14. Remove the heater core from the cover.

To install:

15. Install the heater core to the cover.

16. Carefully, install the heater core with the heater core cover.

17. Install the heater housing pipe flange screw.

18. At both sides of the heater/air conditioning housing assembly, install the heater core cover screws.

19. Connect the drain hose.

20. If equipped, install the knee bolster.

21. If equipped, install the Road Traffic Information (RTI) control module bracket or booster bracket.

22. At both the driver's side and passenger's side, install the carpet supports.

23. At the passenger's side, install the soundproofing panel and the 2 soundproofing panel-to-dashboard screws.

24. Install the glove compartment and the 4 glove compartment-to-dashboard Torx 20 screws.

25. At the driver's side, install the soundproofing panel and the soundproofing panel-to-dashboard Torx 20 screw.

26. Connect the heater hoses to the heater core by inserting the hoses, squeezing the locking catches and pressing in the hose connections.

27. Refill the cooling system.

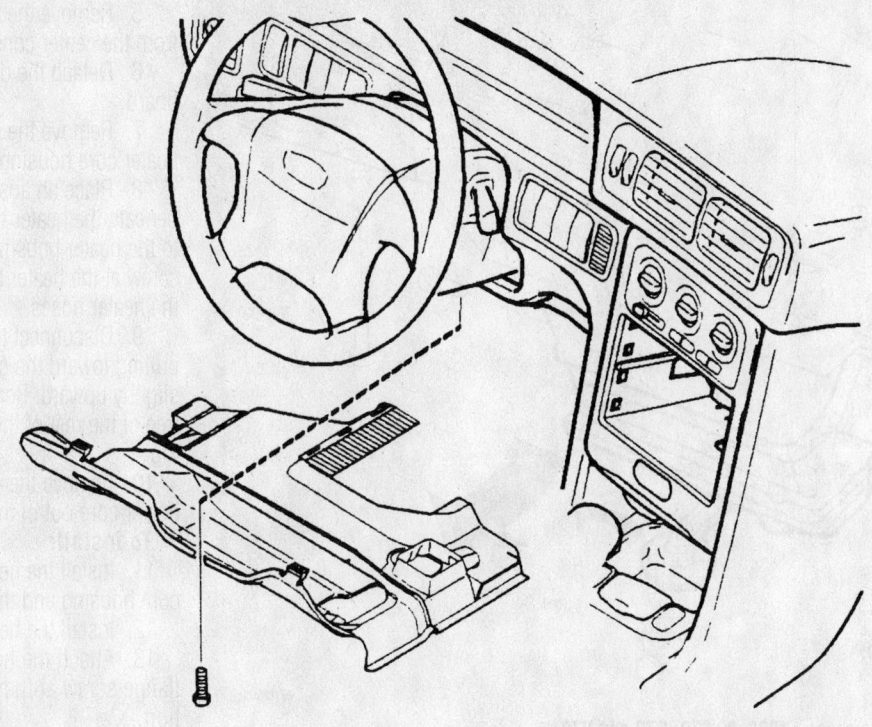

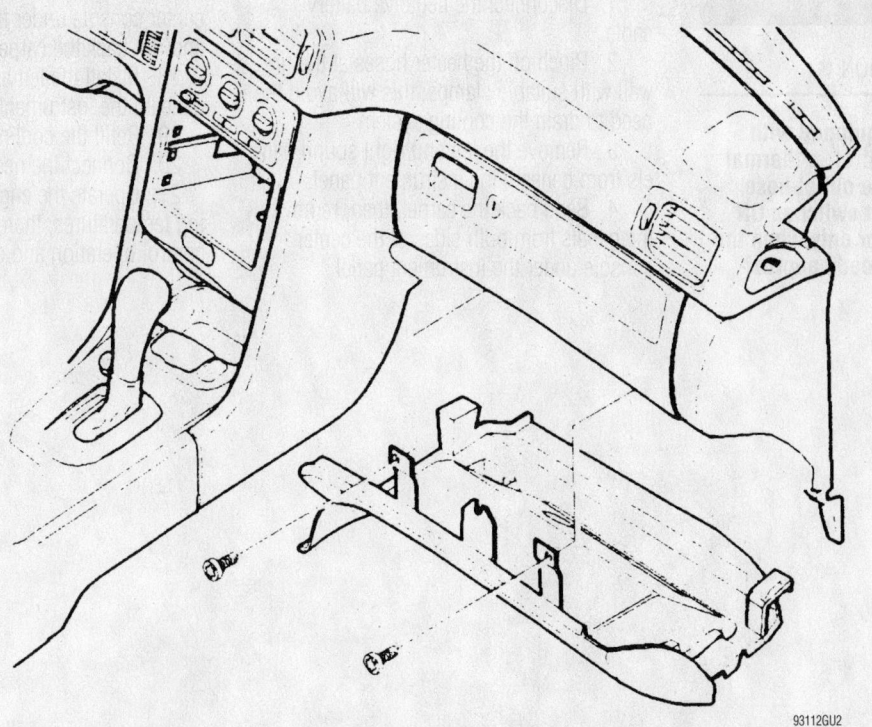

93112GU2

Exploded view of the driver's side and passenger's side soundproofing panels—1998–00 C70, S70 and V70

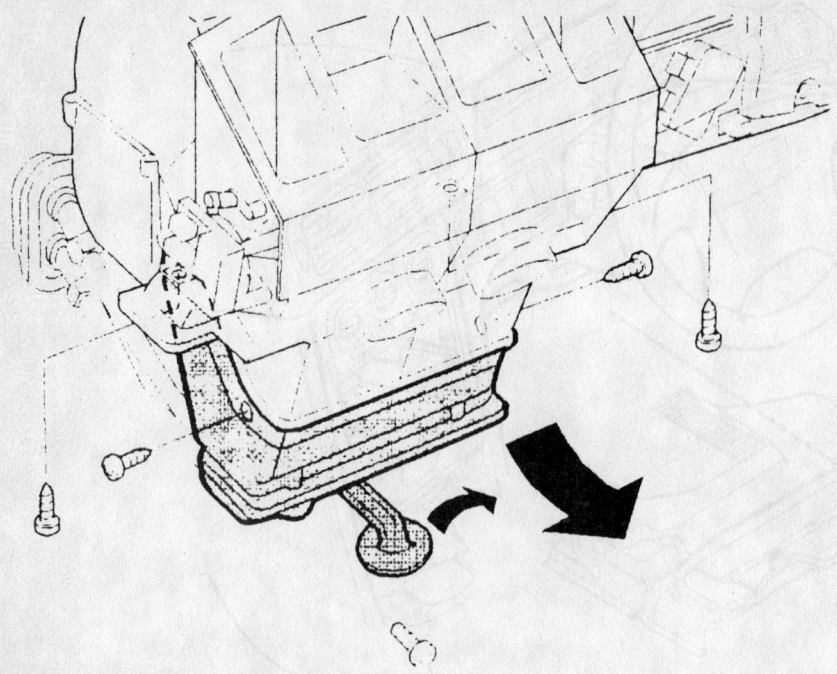

View of the heater core cover—1998–00 C70, S70 and V70

93112GU3

28. Connect the negative battery cable.

29. Operate the engine to normal operating temperatures; then, check the climate control operation and check for leaks.

850 Series

REMOVAL & INSTALLATION

➡️**On some vehicles equipped with Automatic Climate Control, a thermal switch is located on the outlet hose from the heater core. It switches ON and starts the fan motor only when the water temperature exceeds approximately 95°F (35°C). This prevents cold air from being blown into the passenger compartment during winter. The thermal switch is bypassed in the defrost position.**

1. Disconnect the negative battery cable.

2. Pinch off the heater hoses at the firewall with suitable clamps; this will avoid the need to drain the cooling system.

3. Remove the left and right sound panels from beneath the instrument panel.

4. Bend back the carpet, then, remove the panels from both sides of the center console under the instrument panel.

5. Remove the amplifier and bracket from the center console, if equipped.

6. Detach the drain hose from the floor board.

7. Remove the 4 screws holding the heater core housing to the air duct.

8. Place an absorbent cloth or paper beneath the heater hoses at the connection to the heater housing. Remove the flange screw at the heater hose connection. Detach the heater hoses.

9. Disconnect the heater housing by pulling toward the gearshift while twisting slightly upward. Remove it to one side when free of the rest of the climate control housing.

10. Remove the 4 screws and take the heater core out of the heater core housing.

To install:

11. Install the heater core to the heater core housing and the 4 screws.

12. Install the heater housing.

13. Attach the heater hoses. Install the flange screw at the heater hose connection.

14. Install the 4 screws holding the heater core housing to the air duct.

15. Attach the drain hose to the floor board.

16. Install the amplifier and bracket to the center console, if equipped.

17. Install the panels to both sides of the center console under the instrument panel and roll back the carpet.

18. Install the left and right sound panels beneath the instrument panel.

19. Refill the cooling system.

20. Connect the negative battery cable.

21. Operate the engine to normal operating temperatures; then, check the climate control operation and check for leaks.

TIMING BELTS

3

GENERAL INFORMATION

Timing belts are typically only used on overhead camshaft engines. Timing belts are used to synchronize the crankshaft with the camshaft, similar to a timing chain on an overhead valve (pushrod) engine. Unlike a timing belt, a timing chain will normally last the life of the engine without needing service or replacement. Timing belts use raised teeth to mesh with sprockets to operate the valvetrain of an overhead camshaft engine.

Whenever a vehicle with an unknown service history comes into your repair facility or is recently purchased, here are some points that should be asked to help prevent costly engine damage:
- Does the owner know if, or when the belt was replaced?
- If the vehicle purchased is used, or the

condition and mileage of the last timing belt replacement are unknown, it is recommended to inspect, replace or at least inform the owner that the vehicle is equipped with a timing belt.
- Note the mileage of the vehicle. The average replacement interval for a timing belt is approximately 60,000 miles (96,000 km).

Interference Engines

Engines, chain-or belt-driven, can be classified as either free-running or interference, depending on what would happen if the piston-to-valve timing were disrupted. A free-running engine is designed with enough clearance between the pistons and valves to allow the crankshaft to rotate (pistons still moving) while the camshaft stays

in one position (several valves fully open). If this condition occurs normally, no internal engine damage will result. In an interference engine, there is not enough clearance between the pistons and valves to allow the crankshaft to turn without the camshaft being in time.

An interference engine can suffer extensive internal damage if a timing belt fails. The piston design does not allow clearance for the valve to be fully open and the piston to be at the top of its stroke. If the belt fails, the piston will collide with the valve and will bend or break the valve, damage the piston, and/or bend a connecting rod. When this type of failure occurs, the engine will need to be replaced or disassembled for further internal inspection; either choice costing many times that of replacing the timing belt.

TIMING BELT SERVICE

Inspection

➡**For manufacturers recommended service interval, refer to the maintenance interval chart located in this manual.**

The average replacement interval for a timing belt is approximately 60,000 miles (96,000km). If, however, the timing belt is inspected earlier or more frequently than suggested, and shows signs of wear or defects, the belt should be replaced at that time.

✳✳ WARNING

Never allow antifreeze, oil or solvents to come into with a timing belt. If this occurs immediately wash the solution from the timing belt. Also, never excessive bend or twist the timing belt; this can damage the belt so that its lifetime is severely shortened.

Inspect both sides of the timing belt. Replace the belt with a new one if any of the following conditions exist:

• Hardening of the rubber—back side is glossy without resilience and leaves no indentation when pressed with a fingernail
• Cracks on the rubber backing
• Cracks or peeling of the canvas backing
• Cracks on rib root
• Cracks on belt sides
• Missing teeth or chunks of teeth
• Abnormal wear of belt sides—the sides are normal if they are sharp, as if cut by a knife

If none of these conditions exist, the

TCCS1242

Never bend or twist a timing belt excessively, and do not allow solvents, antifreeze, gasoline, acid or oil to come into contact with the belt

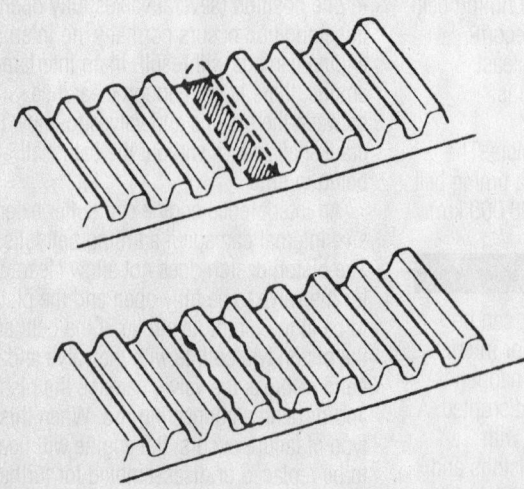

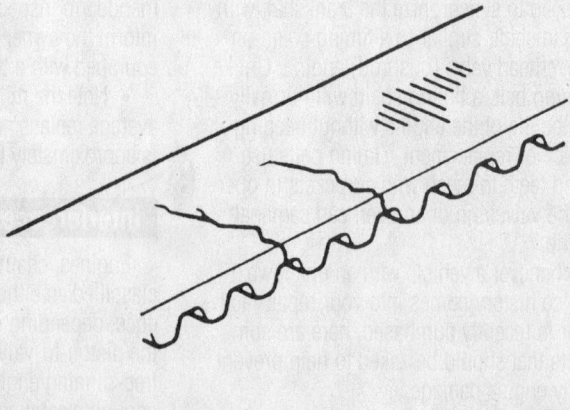

TCCS1244

Broken tooth may be due to a damaged pulley

TCCS1245

Back surface worn or cracked from a possible overheated engine or interference with the belt cover

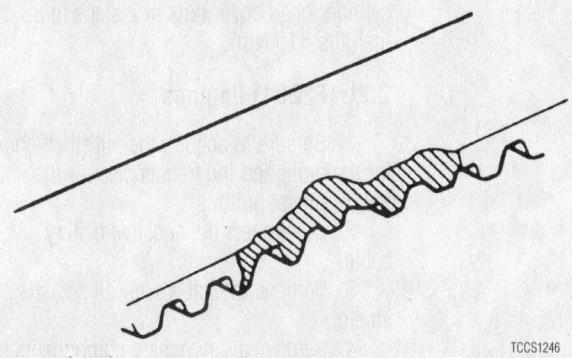

TCCS1246

Side wear from improper installation

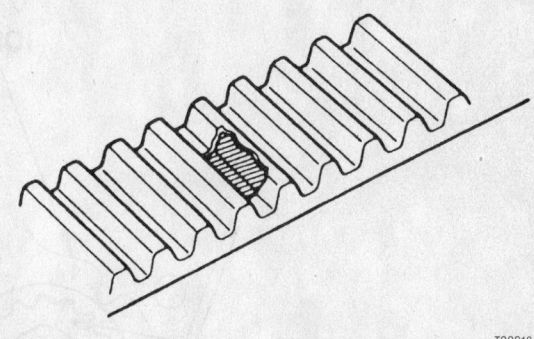

TCCS1247

Worn teeth from excessive belt tension, camshaft or distributor not turning properly, or fluid leaking on the belt

belt does not need replacement unless it is at the recommended interval. The belt MUST be replaced at the recommended interval.

✳✳ WARNING

On interference engines, it is very important to replace the timing belt at the recommended intervals, otherwise expensive engine damage will likely result if the belt fails.

Removal & Installation

ACURA MODELS

➡**The radio may have a coded theft protection circuit. Obtain the code before disconnecting the battery, removing the radio fuse, or removing the radio.**

1.8L (B18B1 and B18C1) Engines

1. Turn the crankshaft pulley until cylinder No. 1 is set to Top Dead Center (TDC) on the compression stroke. The white crankshaft pulley mark should be aligned with the pointer on the lower timing belt cover.

2. Remove all necessary components to gain access to the cylinder head and timing belt covers.

3. Remove the cylinder head and timing belt covers.

4. Remove the crankshaft pulley bolt and the crankshaft pulley. To remove the crankshaft pulley bolt a pulley holder (holder attachment tool part No. 07MAB-PY3010A and holder handle tool part No. 07JAB-001020A) will be needed to keep the crankshaft from turning.

✳✳ WARNING

Do not use the timing belt covers to store small parts. Grease or oil can transfer from the parts to the cover, then to the belt. Clean the covers thoroughly before installation.

5. Recheck that the No. 1 piston is at TDC on its compression stroke. Align the groove on the toothed side of the crankshaft timing belt drive sprocket to the arrow pointer on the oil pump.

6. To set the camshafts to top dead center for the No. 1 cylinder, align the hole in each camshaft with the holes in the No. 1 camshaft holders, then push 5.0mm pin

punches into the holes. Be sure that the **UP** arrows are pointing up and that the TDC marks on the intake and exhaust sprockets are aligned.

7. Loosen the tensioner adjusting bolt 180 degrees (½ turn). Push on the tensioner to remove the tension from the timing belt, then retighten the bolt. If the timing belt is to be reinstalled, mark the direction of rotation on the belt with a crayon or white paint.

8. Remove the timing belt from the sprockets.

➡**Be sure the water pump pulley turns counterclockwise freely. Check for signs of seal leakage; a small amount of weeping from the bleed hole is normal.**

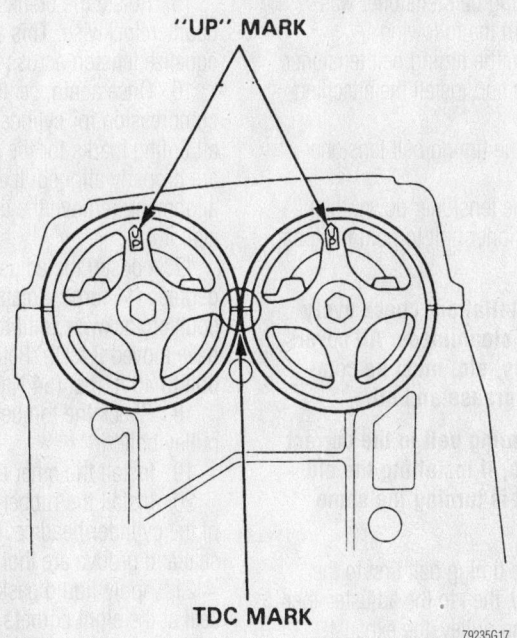

Camshaft timing belt sprocket alignment marks for TDC—Acura 1.8L (B18B1 and B18C1) engines

For complete service labor times, order the Chilton Labor Guide

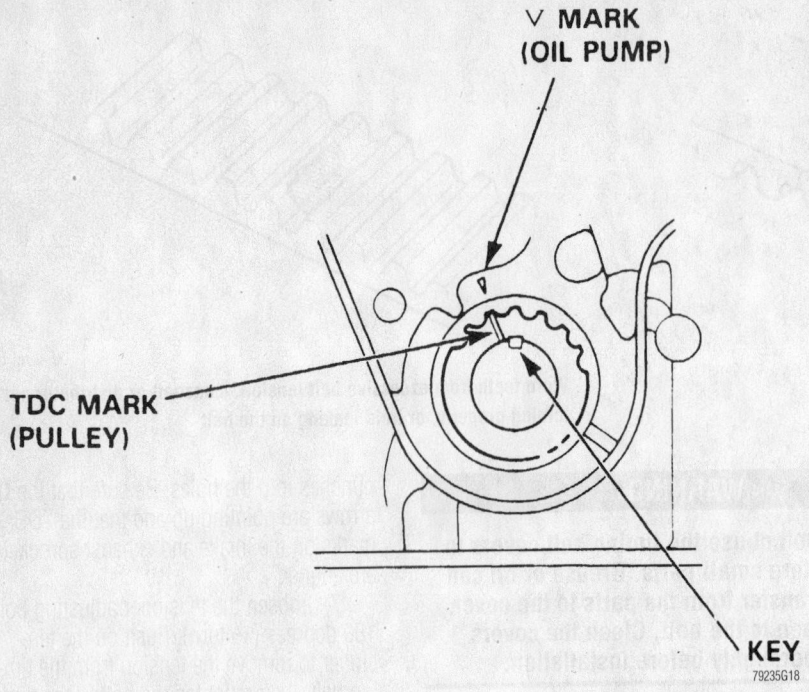

∨ **MARK**
(OIL PUMP)

TDC MARK
(PULLEY)

KEY

79235G18

Crankshaft timing belt sprocket alignment marks for TDC—Acura 1.8L (B18B1 and B18C1) engines

9. If necessary, remove the timing belt tensioner by performing the following:

a. Remove the timing belt tensioner spring.

b. Remove the bolt from the timing belt tensioner and remove the tensioner.

To install:

10. If the timing belt tensioner was removed, perform the following:

a. Position the timing belt tensioner on the engine and install the attaching bolt loosely.

b. Install the timing belt tensioner spring.

c. Push the tensioner down, then snug the tensioner bolt to hold this position.

➡**Before reinstallation, check every component for cleanliness. All covers, pulleys, shields, etc. must be completely free of grease and oil.**

➡**Install the timing belt in the correct sequence. Also, if installing the old belt, be sure it is turning the same direction.**

11. Install the timing belt first to the crankshaft pulley, then to the adjuster, then to the water pump pulley, the exhaust camshaft and finally to the intake camshaft pulley.

12. Install the lower belt cover. Install the crankshaft pulley, tightening the bolt to 130 ft. lbs. (177 Nm). Lubricate the threads and the flange of the bolt with engine oil before installation.

13. Loosen the adjusting bolt, allowing the adjuster to tension the belt. Retighten the bolt to 40 ft. lbs. (54 Nm).

14. Remove the pin punches from the camshafts.

15. Rotate the crankshaft 4–6 turns counterclockwise. This allows the belt to equalize tension across all of the pulleys.

16. Once again, set the engine to TDC compression for cylinder No. 1. Check that all timing marks for the cam and crankshaft are properly aligned. If any mark is out of alignment, remove the timing belt and reinstall it.

17. Loosen the adjusting bolt 180 degrees (½ turn). Rotate the crankshaft counterclockwise until the camshaft pulleys have moved 3 teeth. Retighten the adjusting bolt to 40 ft. lbs. (54 Nm).

18. Check the torque of the crankshaft pulley bolt.

19. Install the other timing belt covers.

20. Install the rubber seal in the groove of the cylinder head cover. Be sure that the seal and groove are thoroughly clean first.

21. Apply liquid gasket to the rubber seal at the eight corners of the recesses. Do not install the parts if 20 minutes or more have elapsed since applying the liquid gasket. Instead, reapply liquid gasket after removing the old residue.

22. Install the cylinder head cover and all other applicable components. Tighten the cylinder head cover nuts in 2 steps to 88 inch lbs. (10 Nm).

2.2L (F22B1) Engines

1. Be sure to acquire the anti-theft code for the radio and the frequencies for the radio's preset buttons.

2. Disconnect the negative battery cable.

3. Remove the left wheelwell splash shield.

4. Remove all necessary components to gain access to the cylinder head (valve) and timing belt covers.

5. Remove the cylinder head and timing belt covers.

6. Turn the engine to align the timing marks and set cylinder No. 1 to Top Dead Center (TDC). The white mark on the crankshaft sprocket should align with the pointer on the timing belt cover. The words **UP** embossed on the camshaft sprocket should be aligned in the upward position. The marks on the edge of the sprocket should be aligned with the cylinder head or the back cover upper edge. Once in this position, the engine must **NOT** be turned or disturbed.

7. There are two belts in this system; the belt running to the camshaft sprocket is the timing belt, the other, shorter belt drives the balance shaft and is referred to as the balancer shaft belt or timing balancer belt. Lock the timing belt adjuster in position, by installing one of the lower timing belt cover bolts in the adjuster arm.

8. Loosen the timing belt and balancer shaft tensioner adjuster nut, do not loosen the nut more than one revolution. Push the tensioner for the balancer belt away from the belt to relieve the tension. Hold the tensioner and tighten the adjusting nut to hold the tensioner in place.

9. Carefully remove the balancer belt. Do not crimp or bend the belt; protect it from contact with oil or coolant.

10. Remove the balancer belt sprocket from the crankshaft.

11. Loosen the lockbolt, installed in the timing belt adjuster and the adjusting nut. Push on the timing belt adjuster to remove the tension on the timing belt, then tighten the adjuster nut.

12. Remove the timing belt by sliding it off the sprockets. Do not crimp or bend the belt; protect it from contact with oil or coolant.

13. If necessary, remove the belt tensioners by performing the following:

a. Remove the springs from the balancer belt and the timing belt tensioners.

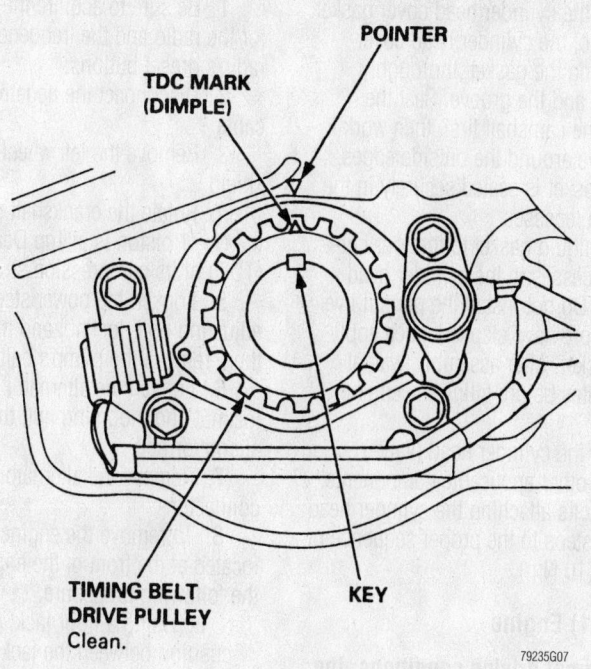

Crankshaft timing belt sprocket alignment mark locations—Acura 2.2L (F22B1) and 2.3L (F23A1) engines

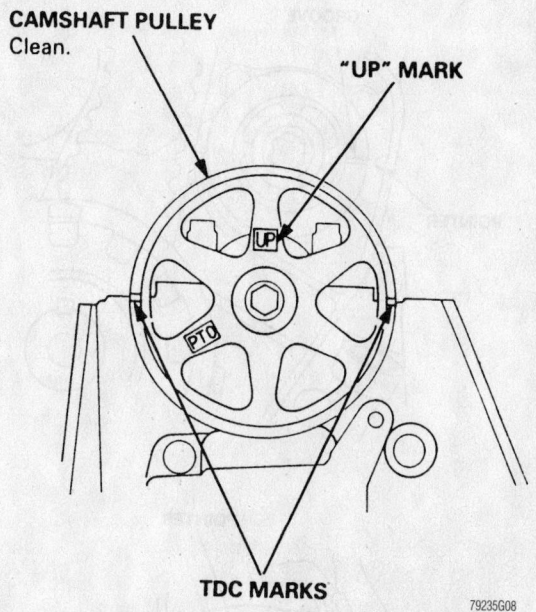

Camshaft timing belt sprocket alignment mark locations—Acura 2.2L (F22B1) and 2.3L (F23A1) engines

b. Remove the adjusting nut from the belt tensioners.

c. Remove the bolt from the balancer belt adjuster lever, then remove the lever and the tensioner pulley.

d. Remove the lockbolt from the tim-

ing belt tensioner lever, then remove the tensioner pulley and lever from the engine.

14. This is an excellent time to check or replace the water pump. Even if the timing belt is only being replaced as part of a good

maintenance schedule, consider replacing the pump at the same time.

To install:

15. If the water pump is to be replaced, install a new O-ring on the pump and make certain it is properly seated. Install the water pump and tighten the mounting bolts to 106 inch lbs. (12 Nm).

16. If the tensioners were removed, perform the following to install them:

a. Install the timing belt tensioner lever and the tensioner pulley.

b. Install the balancer belt pulley and adjuster lever.

c. Install the adjusting nut and bolt to the balancer belt adjuster lever.

d. Install the springs to the tensioners.

e. Install the lockbolt to the timing belt tensioner, then move the tensioner it's full deflection and tighten the lockbolt.

f. Move the balancer belt tensioner it's full deflection and tighten the adjusting nut to hold its position.

17. The pointer on the crankshaft sprocket should be aligned with the pointer on the oil pump; the camshaft sprocket must be aligned so that the word **UP** is at the top of the sprocket and the marks on the edge of the sprocket are aligned with the surfaces of the head or the back cover upper edge.

18. Install the timing belt in the following sequence:

a. Start the belt on the crankshaft sprocket.

b. Then, around the tensioner sprocket.

c. On the water pump sprocket.

d. Finally, around the camshaft sprocket.

19. Check the timing marks to be sure that they did not move.

20. Loosen, then retighten the timing belt adjusting nut.

21. Install the timing/balancer belt drive sprocket and the lower timing belt cover.

22. Install the crankshaft pulley and bolt, tighten the bolt to 181 ft. lbs. (245 Nm). Rotate the crankshaft sprocket five or six revolutions to properly position the timing belt on the sprockets.

23. Set the No. 1 cylinder to TDC and loosen the timing belt adjusting nut one turn. Turn the crankshaft counterclockwise until the cam sprocket has moved 3 teeth; this creates the proper tension on the timing belt.

24. Tighten the timing belt adjusting nut.

25. Set the crankshaft sprocket and the camshaft sprocket to TDC. If the sprockets do not align, remove the belt to realign the marks, then install the belt.

26. Remove the crankshaft pulley and the lower cover.

27. With the timing marks aligned, lock the timing belt adjuster in place with one of the lower cover mounting bolts.

28. Loosen the adjusting nut and ensure the timing balancer belt adjuster moves freely.

29. Align the rear timing balancer sprocket using a 6x100mm bolt or rod. Mark the bolt or rod at a point 2.9 in. (74mm) from the end. Remove the bolt from the maintenance hole on the side of the block; insert the bolt/rod into the hole and align the 2.9 in. (74mm) mark with the face of the hole. This will hold the shaft in place during installation.

30. Align the groove on the front balancer shaft sprocket with the pointer on the oil pump.

31. Install the balancer belt. Once the belts are in place, be sure that all the engine alignment marks are still correct. If not, remove the belts, realign the engine and reinstall the belts. Once the belts are properly installed, slowly loosen the adjusting nut, allowing the tensioner to move against the belt. Remove the bolt from the maintenance hole and reinstall the bolt and washer.

32. Install the crankshaft pulley, then turn the crankshaft sprocket one turn counterclockwise and tighten the timing belt adjusting nut to 33 ft. lbs. (45 Nm).

33. Remove the crankshaft pulley and the bolt locking the timing belt adjuster in place.

34. Install the lower timing belt cover and tighten the bolts to 106 inch lbs. (12 Nm).

35. Install a new seal around the adjusting nut. Do not loosen the adjusting nut.

36. Install the crankshaft pulley. Coat the threads and seating face of the pulley bolt with engine oil, then install and tighten the bolt to 181 ft. lbs. (250 Nm).

37. Install the dipstick tube, then install the side engine mount. Tighten the bolt and nut attaching the mount to the engine to 40 ft. lbs. (55 Nm). Tighten the through-bolt and nut to 47 ft. lbs. (65 Nm), then remove the jack from under the engine.

38. Install the upper timing belt cover. Tighten the bolt toward the exhaust manifold to 89 inch lbs. (10 Nm) and the bolt toward the intake manifold to 106 inch lbs. (12 Nm).

39. Install the cylinder head cover gasket in the groove of the cylinder head cover. Before installing the gasket, thoroughly clean the seal and the groove. Seat the recesses for the camshaft first, then work it into the groove around the outside edges. Be sure the gasket is seated securely in the corners of the recesses.

40. Apply liquid gasket to the four corners of the recesses in the cylinder head cover gasket. Do not install the parts if five minutes or more have elapsed since applying liquid gasket. After assembly, wait at least 20 minutes before filling the engine with oil.

41. Install the cylinder head (valve) cover and all other applicable components. Tighten the bolts attaching the cylinder head cover in two steps to the proper sequence of 89 inch lbs. (10 Nm).

2.3L (F23A1) Engine

➡Under normal driving conditions, the timing belt and timing balancer belt are to be replaced at 105,000 miles (168,000 km).

1. Be sure to acquire the anti-theft code for the radio and the frequencies for the radio's preset buttons.

2. Disconnect the negative battery cable.

3. Remove the left wheel-well splash shield.

4. Rotate the crankshaft pulley so that the No. 1 piston is at Top Dead Center (TDC) of its compression stroke.

5. Loosen the power steering pump's adjusting bolt, locknut and mounting nut; then, remove the pump's belt.

6. Loosen the alternator adjusting bolt, locknut and mounting nut; then, remove the alternator belt.

7. Remove the alternator terminal and connector.

8. To remove the engine's side mount, located at the front of the engine, perform the following procedure:

   a. Using a floor jack, position a cushion between the jack and the oil pan.

   b. Raise the engine slightly to take the weight off of the side mount.

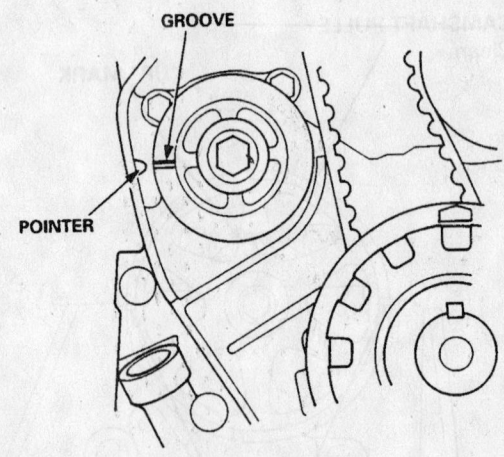

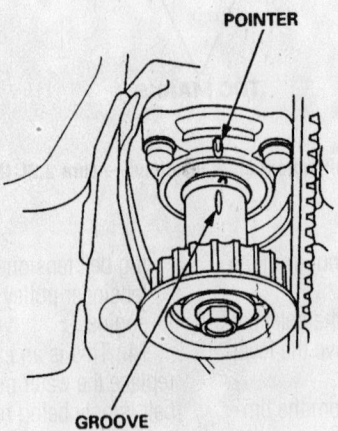

Balancer shaft alignment mark locations—Acura 2.3L (F23A1) engine

93015G05

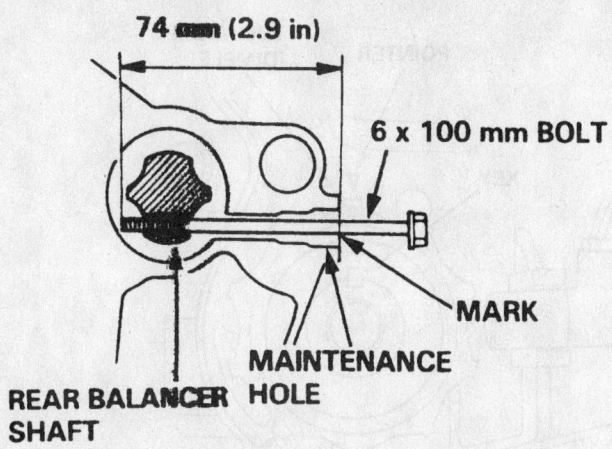

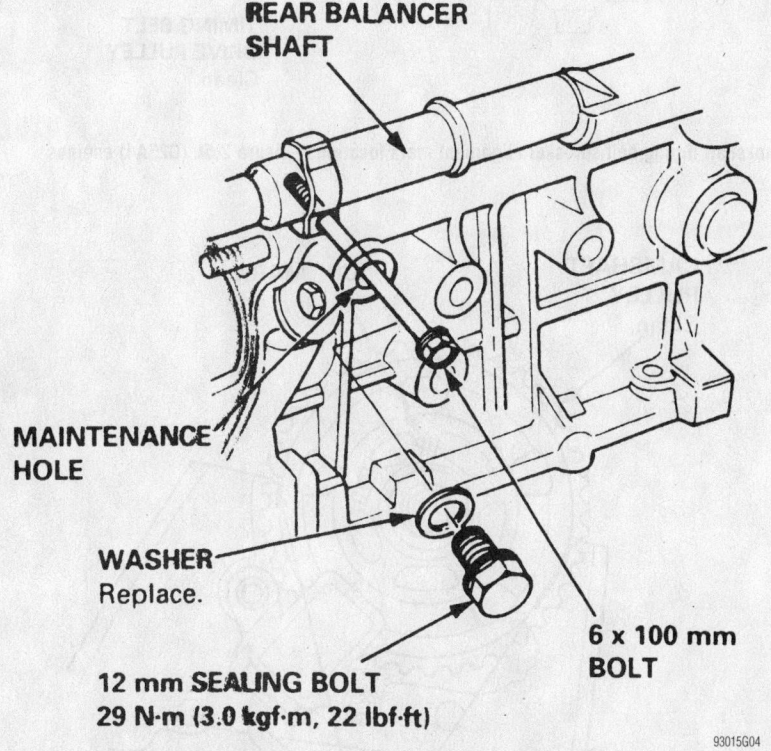

Securing the rear balancer shaft—Acura 2.3L (F23A1) engine

c. Remove the side mount-to-engine bolts, the side mount-to-chassis bolts and the side mount.

9. Remove the dipstick and the dipstick tube-to-engine bolt; then, pull the tube from its O-ring mount.

10. Remove the cylinder head cover.

11. Remove the crankshaft pulley by performing the following procedure:

a. Using the Holder Handle and the 50mm Offset Holder Attachment tool 07MAB-PY3010A and a 19mm socket, secure the crankshaft pulley and remove the crankshaft pulley bolt.

b. Remove the washer and pull the crankshaft pulley from the engine.

12. From the lower timing belt cover, remove the rubber seal concealing the adjusting nut. Remove the upper timing belt cover-to-engine bolts and the cover.

13. Loosen the adjusting nut ⅔–1 turn. Push the tensioner to relieve the tension from the timing belt and the balancer belt; then, retighten the adjusting nut.

14. Remove the balancer and the timing belts.

To install:

15. Clean the upper and lower timing belt covers.

16. Remove the balancer belt drive pulley from the crankshaft.

17. Position the timing belt pulley so the No. 1 piston is at TDC of its compression stroke. Align the mark on the pulley (near keyway) with the pointer mark on the oil pump.

18. Adjust the camshaft pulley so that the No. 1 piston is the TDC of the compression stroke. Align the TDC marks on the pulley with the upper surface of the back cover and the "UP" mark should be facing upward.

19. Install the timing belt in the following sequence:

a. Crankshaft pulley sprocket.
b. Adjusting pulley.
c. Water pump pulley.
d. Camshaft pulley.

✳✳ WARNING

Make sure that the camshaft and crankshaft pulleys are at TDC.

20. Loosen and retighten the adjusting nut to tension the timing belt.

21. Install the balancer belt drive pulley and the lower timing belt cover.

22. Install the crankshaft pulley and finger-tighten the bolt and washer. Using the Holder Handle and the 50mm Offset Holder Attachment tool 07MAB-PY3010A and a 19mm socket with a torque wrench, tighten the crankshaft pulley bolt to 181 ft. lbs. (245 Nm).

23. Rotate the crankshaft pulley about 5–6 turns counterclockwise to position the timing belt on the pulleys.

24. To adjust the timing belt tension, perform the following procedure:

a. Make sure that the No. 1 piston is at TDC of its compression stroke.

b. Loosen the adjusting nut ⅔–1 turn. Rotate the crankshaft counterclockwise three teeth on the camshaft pulley.

c. Tighten the adjusting nut to 33 ft. lbs. (44 Nm).

25. Retighten the crankshaft pulley bolt to 181 ft. lbs. (245 Nm).

26. Make sure the crankshaft and camshaft pulleys are at TDC.

➡ **If the camshaft or crankshaft pulley is not at TDC, remove the timing belt and re-perform the adjustment procedure.**

Timing belt service is covered in Section 3 of this manual

27. Remove the crankshaft pulley and the timing belt lower cover.

28. Install a 6 x 1.0mm bolt to lock the timing belt adjuster arm in place.

29. Loosen the adjusting nut ⅔–1 turn and verify that the balancer belt adjuster moves freely.

30. Position the tensioner to remove tension from the balancer belt, then tighten the adjusting nut.

31. Align the rear balancer shaft pulley by performing the following procedure:

    a. Using a 6 x 100mm bolt, scribe a line 2.9 in. (74mm) from the end of the bolt.

    b. Insert the bolt into the maintenance hole to the scribed line.

    c. Align the groove on the front balancer shaft pulley with the pointer on the oil pump housing.

32. Install the balancer belt. Loosen the adjuster nut ⅔–1 turn to place tension on the balancer belt.

33. Remove the 6 x 100mm bolt from the rear balancer shaft and install the 12mm sealing bolt.

34. Install the crankshaft pulley and tighten the bolt to 181 ft. lbs. (245 Nm).

35. Rotate the crankshaft pulley 1 turn counterclockwise and tighten the adjusting nut to 33 ft. lbs. (44 Nm).

36. Remove the 6 x 1.0mm bolt from the timing belt adjuster arm.

37. Remove the crankshaft pulley and install the lower timing belt cover.

38. Install the rubber seal over the adjusting nut.

➡**DO NOT loosen the adjusting nut.**

39. Install the crankshaft pulley and tighten the bolt to 181 ft. lbs. (245 Nm).

40. To complete the installation, reverse the removal procedures. Adjust the tension of the drive belts.

2.5L (G25A4) Engines

1. Set the No. 1 piston at Top dead Center (TDC) on the compression stroke. The white TDC mark on the crankshaft pulley must align with the pointers on the lower belt cover.

2. Remove all necessary components to gain access to the timing belt covers.

3. Remove the timing belt upper cover.

4. Be sure the **UP** mark and the TDC marks on the camshaft sprocket are correctly positioned.

5. Rotate the crankshaft to align the white timing mark on the crankshaft pulley with the pointer on the lower cover. There

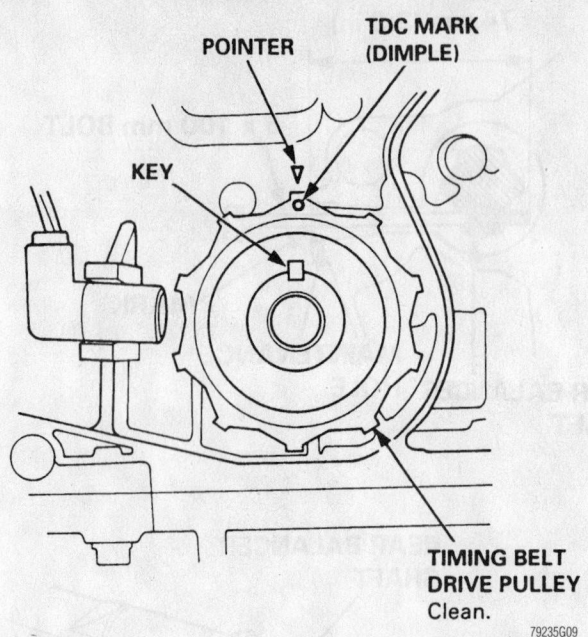

Crankshaft timing belt sprocket alignment mark locations—Acura 2.5L (G25A4) engines

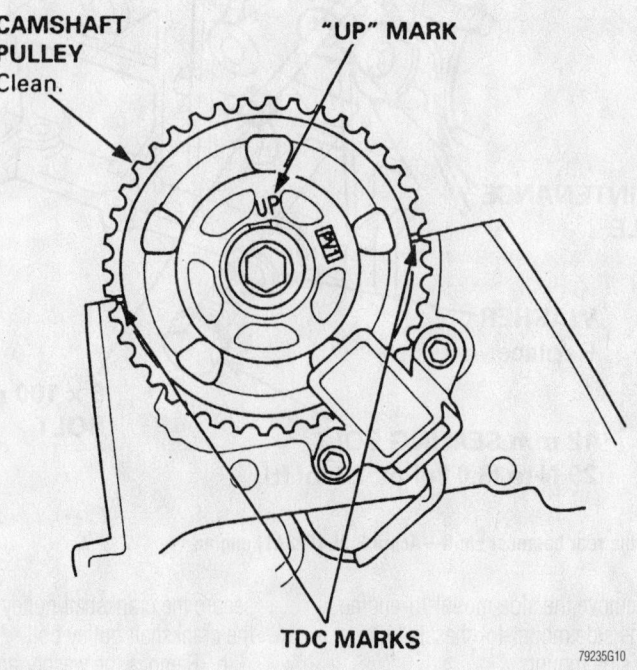

Camshaft timing belt sprocket alignment mark locations—Acura 2.5L (G25A4) engines

are similar alignment marks on the crankshaft sprocket and oil pump housing

6. Remove the timing belt lower cover.

7. Loosen the adjusting bolt 180 degrees (½ turn), then push the tensioner down to relieve the belt tension. Retighten the adjusting bolt to 33 ft. lbs. (44 Nm).

➡**Do not remove the adjusting bolt and tensioner pulley unless they are to be replaced. The bolt is only loosened and tightened in this procedure to tension the timing belt.**

8. Remove the timing belt.

9. Inspect the timing belt tensioner pulley and tension spring for signs of wear.

Remove and replace the tensioner assembly as necessary.

To install:

➡**Replace the timing belt if it shows any signs of wear, damage or contamination from oil or coolant. The source of any oil or coolant contamination must be determined and corrected before the new timing belt may be installed.**

10. Verify that the crankshaft and camshaft sprocket matchmarks are properly aligned.

11. Install the timing belt in the following order: first onto the crankshaft sprocket, then the tensioner pulley, the water pump sprocket and finally the camshaft sprocket.

12. Loosen the tensioner adjusting bolt to allow the spring to set the tension. Then, tighten the bolt to 33 ft. lbs. (44 Nm). Rotate the crankshaft six full turns to seat the belt and verify that the timing marks align properly.

13. Install the lower and upper timing belt covers, and tighten the bolts to 106 inch lbs. (12 Nm).

14. Oil only the threads on the crankshaft pulley bolt. Install the crankshaft pulley and tighten the bolt to 181 ft. lbs. (245–250 Nm).

15. Install the dipstick tube.

16. Install the cylinder head cover with new gaskets and washers. Tighten the cap to 106 inch lbs. (12 Nm.

3.2L (C32A6) Engines—3.2TL

➡**Under normal driving conditions, the timing belt and timing balancer belt are to be replaced at 105,000 miles (168,000 km).**

1. Be sure to acquire the anti-theft code for the radio and the frequencies for the radio's preset buttons.

2. Disconnect the negative battery cable.

3. Remove the left wheelwell splash shield.

4. Loosen the power steering pump's adjusting bolt, locknut and mounting nut; then, remove the pump's belt.

5. Loosen the alternator adjusting bolt, locknut and mounting nut; then, remove the alternator belt.

6. Remove the alternator terminal and connector.

7. To remove the engine-to-chassis center bracket, located at the front of the engine, perform the following procedure:

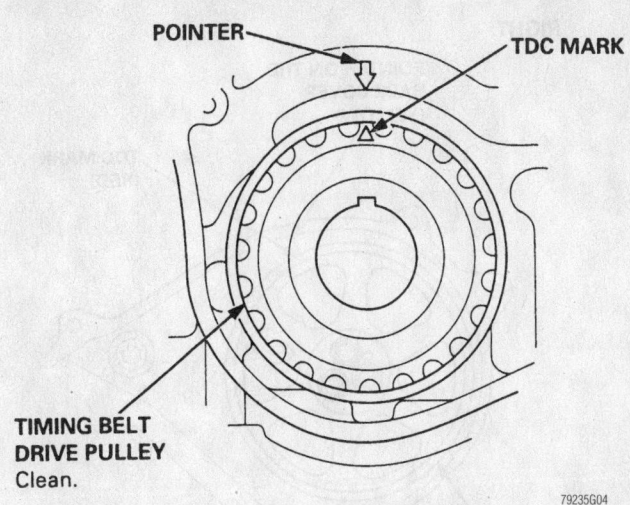

Crankshaft timing belt sprocket alignment mark locations—Acura 3.2TL 3.2L (C32A6) engines

LEFT:

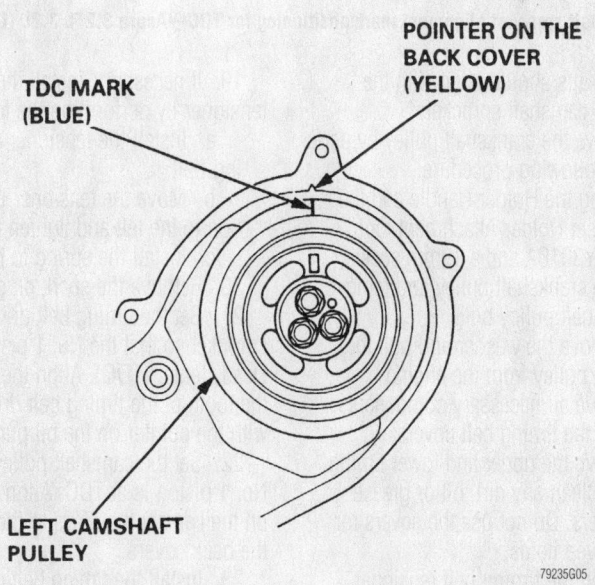

Left camshaft sprocket alignment mark positioning for TDC—Acura 3.2TL 3.2L (C32A6) engines

a. Using a floor jack, position a cushion between the jack and the oil pan.

b. Raise the engine slightly to take the weight off of the center bracket.

c. Remove the center bracket-to-engine bolts, the center bracket-to-chassis bolts and the center bracket.

8. Remove the TCS upper and lower brackets.

9. Disconnect the TCS throttle sensor and actuator connectors and remove the TCS control valve assembly.

10. From the front of the engine, remove the oil pressure switch connector, the engine ground cable and the engine wire harness cover.

11. Remove the dipstick and the dipstick tube-to-engine bolt; then, pull the tube from its O-ring mount. Discard the O-ring.

12. Turn the engine to align the timing marks and set cylinder No. 1 to Top Dead Center (TDC) on the compression stroke. The white mark on the crankshaft pulley should align with the pointer on the timing belt cover. Remove the inspection caps on the upper timing belt covers to check the alignment of the timing marks. The pointers

Heater Core replacement is covered in Section 2 of this manual

RIGHT:

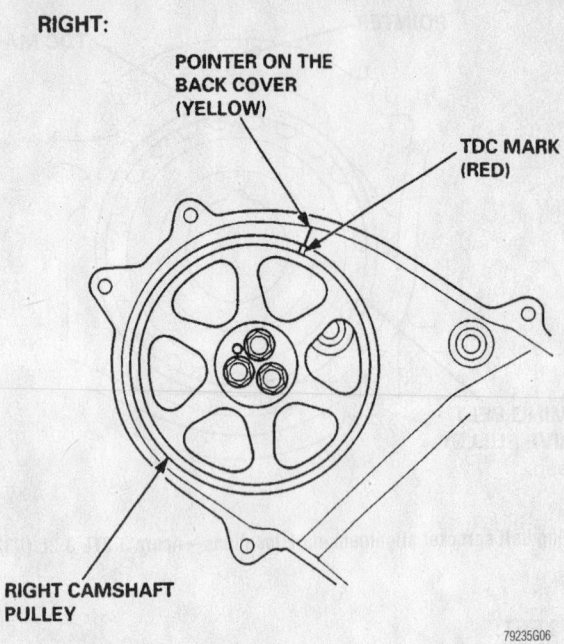

POINTER ON THE
BACK COVER
(YELLOW)

TDC MARK
(RED)

RIGHT CAMSHAFT
PULLEY

79235G06

Right camshaft sprocket alignment mark positioning for TDC—Acura 3.2TL 3.2L (C32A6)

for the camshafts should align with the marks on the camshaft sprockets.

13. Remove the crankshaft pulley by performing the following procedure:

   a. Using the Holder Handle and the 50mm Offset Holder Attachment tool 07MAB-PY3010A and a 19mm socket, secure the crankshaft pulley and remove the crankshaft pulley bolt.

   b. Remove the washer and pull the crankshaft pulley from the engine.

14. Remove all necessary components for access to the timing belt covers.

15. Remove the upper and lower timing belt covers. Clean any dirt, oil or grease from the covers. Do not use the covers for storing removed items.

16. Loosen the timing belt tensioner adjusting bolt 180 degrees (½ turn). Push on the tensioner to remove tension from the timing belt, then tighten the adjusting bolt.

17. Remove the timing belt.

18. If necessary, remove the timing belt tensioner by performing the following:

   a. Remove the spring from the tensioner.

   b. Remove the bolt mounting the tensioner, then remove the tensioner.

   To install:

✳✳ CAUTION

Do not rotate the crankshaft pulley or camshaft pulleys with the timing belt removed. The pistons may hit the valves and cause damage.

19. If necessary, install the timing belt tensioner by performing the following:

   a. Install the tensioner and the attaching bolt.

   b. Move the tensioner its full deflection to the left and tighten the bolt.

   c. Install the spring to the tensioner.

20. Remove the spark plugs.

21. Set the timing belt drive (crankshaft) sprocket so that the No. 1 piston is at Top Dead Center (TDC). Align the TDC mark on the tooth of the timing belt drive sprocket with the pointer on the oil pump.

22. Set the camshaft pulleys so that the No. 1 piston is at TDC. Align the TDC mark on the camshaft pulleys to the pointers on the back covers.

23. Install the timing belt on the sprockets in the following sequence: drive sprocket (crankshaft), tensioner pulley, left camshaft sprocket, water pump pulley, right camshaft sprocket.

24. Loosen, then retighten the timing belt adjuster bolt to tension the timing belt.

25. Install the lower timing belt cover.

26. Install the crankshaft pulley and finger-tighten the bolt and washer. Using the Holder Handle and the 50mm Offset Holder Attachment tool 07MAB-PY3010A and a 19mm socket with a torque wrench, tighten the crankshaft pulley bolt to 181 ft. lbs. (245 Nm).

27. Rotate the crankshaft 5–6 turns clockwise so that the timing belt positions itself properly on the sprockets.

28. Set cylinder No. 1 to TDC by aligning the timing marks. If the timing marks do not align, remove the timing belt, then adjust the components and reinstall the timing belt.

29. Rotate the crankshaft clockwise enough to move the camshaft pulley nine teeth (the blue mark on the crankshaft pulley should align with the pointer on the lower cover).

30. Loosen the timing belt adjusting bolt 180 degrees (½ turn), then tighten the bolt to 31 ft. lbs. (42 Nm).

31. Install the upper timing belt covers, then install all applicable components. When installing the center bracket, tighten the bolts attaching the brackets to 40 ft. lbs. (54 Nm), then the mount through-bolt to 40 ft. lbs. (54 Nm).

32. To complete the installation, reverse the removal procedures. Adjust the tension of the drive belts.

3.0L (J30A1) and 3.2L (J32A1) Engines

1. Disconnect the negative battery cable.

2. Remove the ignition coil cover.

3. Remove the front tire/wheel assemblies. Remove the splash shield from under the vehicle.

4. Move the alternator drive belt tensioner to relieve the tension; then, remove the alternator belt.

5. Loosen the power steering pump's adjusting bolt, locknut and mounting nut; then, remove the pump's belt.

6. To remove the engine-to-chassis side mount, located at the front of the engine, perform the following procedure:

   a. Using a floor jack, position a cushion between the jack and the oil pan.

   b. Raise the engine slightly to take the weight off of the side mount.

   c. Remove the side mount-to-engine bracket bolt, the side mount-to-chassis bolts and the side mount.

7. Remove the dipstick and the dipstick tube-to-engine bolt; then, pull the tube from its O-ring mount. Discard the O-ring.

8. Turn the engine to align the timing marks and set cylinder No. 1 to Top Dead Center (TDC) on the compression stroke. The white mark on the crankshaft pulley should align with the pointer on the timing belt cover. Remove the inspection caps on the upper timing belt covers to check the alignment of the timing marks. The pointers for the camshaft pulleys should align with the marks on rear upper cover mark.

9. Remove the crankshaft pulley by performing the following procedure:

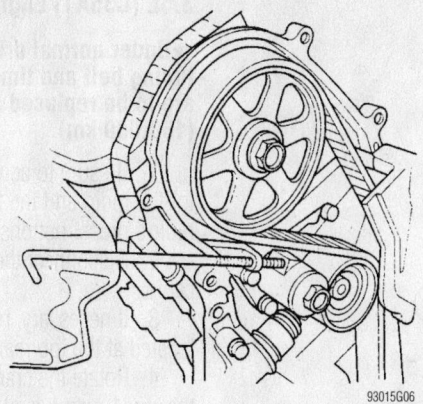

Using battery clamp bolt to hold timing belt adjuster in position—Acura 3.0L (J30A1) and 3.2L (J32A1) engines

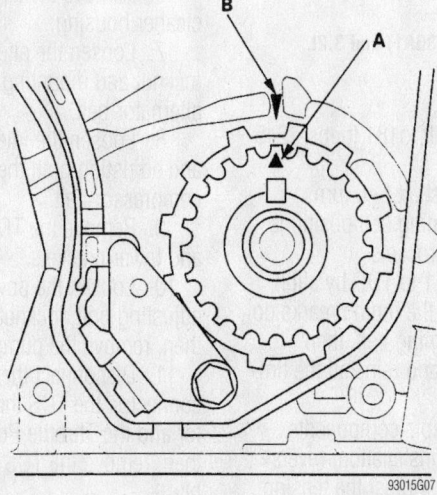

Crankshaft sprocket alignment mark positioning for TDC—Acura 3.0L (J30A1) and 3.2L (J32A1) engines

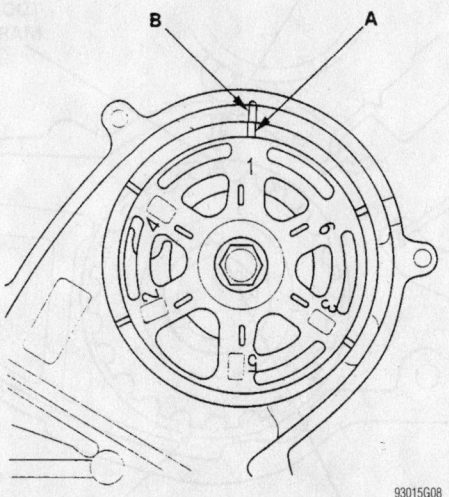

Left camshaft sprocket alignment mark positioning for TDC—Acura 3.0L (J30A1) and 3.2L (J32A1) engines

a. Using the Holder Handle and the 50mm Offset Holder Attachment tool 07MAB-PY3010A and a 19mm socket, secure the crankshaft pulley and remove the crankshaft pulley bolt.

b. Remove the washer and pull the crankshaft pulley from the engine.

10. Remove all necessary components for access to the timing belt covers.

11. Remove the upper and lower timing belt covers. Clean any dirt, oil or grease from the covers. Do not use the covers for storing removed items.

12. Remove one of the battery clamp bolts and grind a 45° bevel on the threaded end. Screw the battery clamp bolt into hole provided at the base of the right cylinder head to hold the timing belt adjuster in it's current position; tighten the bolt by hand, DO NOT use a wrench.

13. Remove the engine mount bracket.

14. At the base of the left cylinder head, loosen the idler pulley bolt about 5–6 turns; then, remove the timing belt.

To install:

✷✷ CAUTION

Do not rotate the crankshaft pulley or camshaft pulleys with the timing belt removed. The pistons may hit the valves and cause damage.

15. Remove the spark plugs.

16. Set the timing belt drive (crankshaft) sprocket so that the No. 1 piston is at Top Dead Center (TDC). Align the TDC mark on the tooth of the timing belt drive sprocket with the pointer on the oil pump.

17. Set the camshaft pulleys so that the No. 1 piston is at TDC. Align the TDC mark on the camshaft pulleys to the pointers on the back covers.

18. Remove the battery clamp bolt from the back cover. Remove the auto-tensioner.

19. Service the auto-tensioner by performing the following procedure:

a. Position the auto-tensioner in a soft jawed vise with the maintenance bolt facing upward. DO NOT grip the body of the auto-tensioner.

b. Remove the maintenance bolt.

c. Be careful not to spill the oil from inside the tensioner. If oil is spilled, replenish it; the total capacity is 0.22 fl. oz. (6.5 ml).

d. Using Stopper tool 14540-P8A-A01, position it on the auto-tensioner while turning the internal screw.

e. Insert a flat-blade screwdriver into

Brake service is covered in Section 4 of this manual

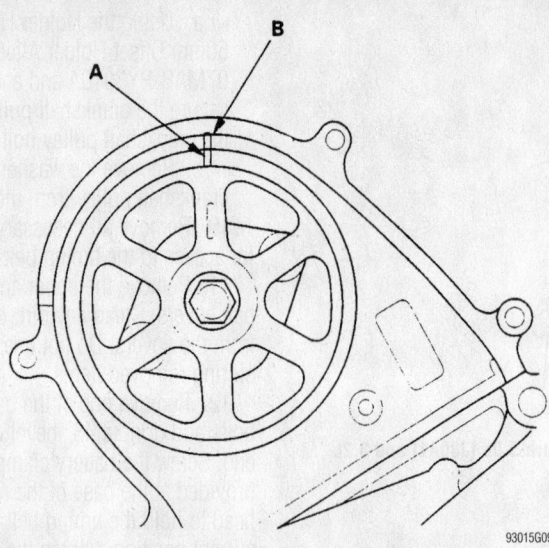

Right camshaft sprocket alignment mark positioning for TDC—Acura 3.0L (J30A1) and 3.2L (J32A1) engines

the maintenance hole and turn it clockwise to compress the bottom.

✳✳ WARNING

Be careful not to damage the threads or the gasket contact surface with the screwdriver.

f. Using a new gasket, reinstall the maintenance bolt and torque it to 6 ft. lbs. (8 Nm).

g. Make sure that no oil is leaking around the maintenance bolt and install the auto-tensioner; torque the bolts 33 ft. lbs. (44 Nm).

➡**Make sure that the Stopper tool 14540-P8A-A01 stays in place.**

20. Install the timing belt on the sprockets in the following sequence: drive sprocket (crankshaft), idler pulley, left camshaft sprocket, water pump pulley, right camshaft sprocket and adjusting pulley.

21. Torque the idler pulley bolt to 33 ft. lbs. (44 Nm).

22. Remove the Stopper tool from the auto-tensioner.

23. Install the engine mount-to-engine and torque the No. 10 bolts to 33 ft. lbs. (44 Nm) and the No. 6 bolt to 8.7 ft. lbs. (12 Nm).

24. Install the lower and upper timing belt covers.

25. Install the crankshaft pulley and finger-tighten the bolt and washer. Using the Holder Handle and the 50mm Offset Holder Attachment tool 07MAB-PY3010A and a 19mm socket with a torque wrench, tighten

the crankshaft pulley bolt to181 ft. lbs. (245 Nm).

26. Rotate the crankshaft 5–6 turns clockwise so that the timing belt positions itself properly on the sprockets.

27. Set cylinder No. 1 to TDC by aligning the timing marks. If the timing marks do not align, remove the timing belt, then adjust the components and reinstall the timing belt.

28. Install all applicable components.

29. To complete the installation, reverse the removal procedures. Adjust the tension of the drive belts.

3.5L (C35A1) Engines—3.5RL

➡**Under normal driving conditions, the timing belt and timing balancer belt are to be replaced at 105,000 miles (168,000 km).**

1. Be sure to acquire the anti-theft code for the radio and the frequencies for the radio's preset buttons.

2. Disconnect the negative battery cable.

3. If necessary, remove the strut brace located at the top rear of the engine.

4. Rotate the crankshaft pulley so that the No. 1 piston is at Top Dead Center (TDC) of its compression stroke.

5. Remove the top engine cover-to-engine bolts and the cover.

6. Remove the air intake duct and air cleaner housing.

7. Loosen the alternator adjusting rod, locknut and mounting bolt; then, remove the alternator belt.

8. Loosen the idler pulley center nut and adjusting bolt; then, remove the A/C compressor belt.

9. Remove the TCS control valve upper and lower brackets.

10. Loosen the power steering pump's adjusting bolt, locknut and mounting bolt; then, remove the pump's belt.

11. Disconnect the TCS throttle sensor connector, the TCS throttle actuator connector and the Throttle Position (TP) sensor; then, remove the TCS control valve assembly.

12. Disconnect the Vehicle Speed Sen-

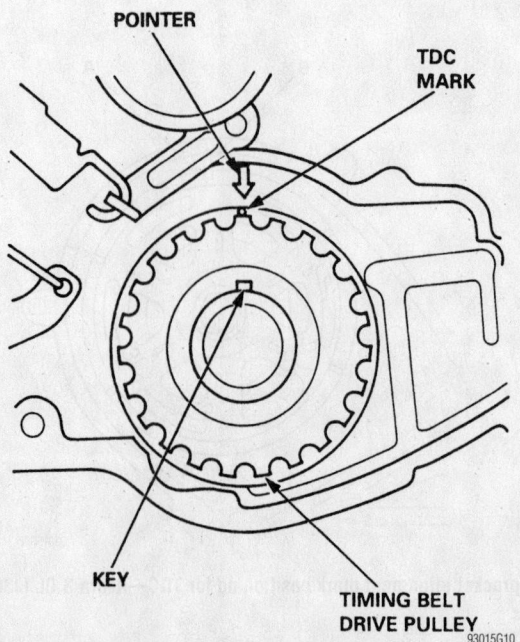

Crankshaft timing belt pulley alignment mark locations—Acura 3.5L (C35A1) engine

sor (VSS) harness connector and remove the wire harness holder.

13. Remove the breather and vacuum hoses.

14. From the right timing belt cover, remove the Ignition Control Module (ICM) bracket.

15. From the left side of the crankshaft pulley, remove the idler pulley bracket.

16. Remove the dipstick and the dipstick tube-to-engine bolt; then, pull the tube from its O-ring mount. Discard the O-ring.

17. Remove the crankshaft pulley by performing the following procedure:

   a. Using the Holder Handle and the 50mm Offset Holder Attachment tool 07MAB-PY3010A and a 19mm socket, secure the crankshaft pulley and remove the crankshaft pulley bolt.

   b. Remove the washer and pull the crankshaft pulley from the engine.

18. From the upper and lower timing belt cover.

19. Loosen the balancer belt adjusting nut 180° (½) turn. Push the tensioner to relieve the tension from the balancer belt; then, retighten the adjusting bolt. Remove the balancer belt.

20. Loosen the timing belt adjusting nut 180° (½) turn. Push the tensioner to relieve the tension from the timing belt; then, retighten the adjusting bolt. Remove the timing belt.

To install:

21. Remove the spark plugs.

22. Remove the balancer belt drive pulley and the timing belt guide plate from the crankshaft.

23. Clean the upper and lower timing belt covers.

24. Position the timing belt pulley so the No. 1 piston is at TDC of its compression stroke. Align the mark on the pulley (near keyway) with the pointer mark on the oil pump.

25. Adjust the camshaft pulley so that the No. 1 piston is the TDC of the compression stroke. Align the TDC marks on the pulley with the upper surface of the back cover; the arrow mark on the left camshaft pulley and the "1" on the right camshaft pulley should be facing the back cover pointer.

26. Install the timing belt in the following sequence:

   a. Crankshaft timing belt pulley sprocket.

   b. Adjusting pulley.

   c. Left camshaft pulley.

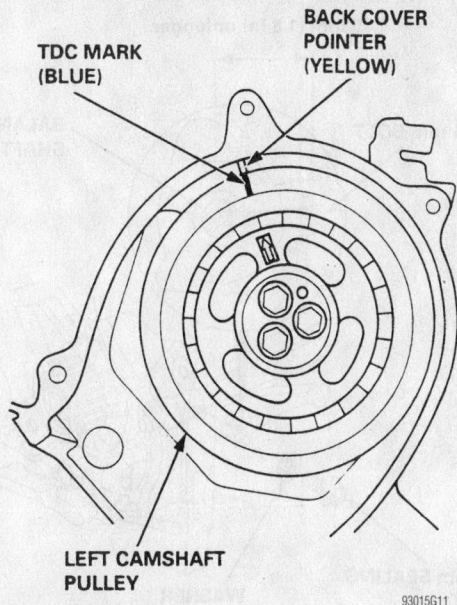

Left camshaft timing belt pulley alignment mark locations—Acura 3.5L (C35A1) engine

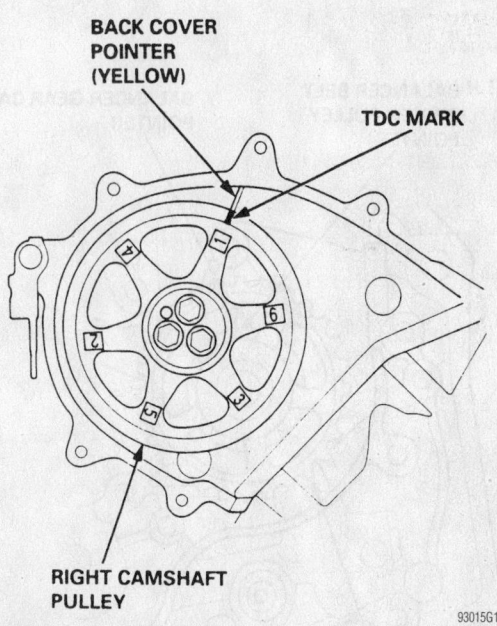

Right camshaft timing belt pulley alignment mark locations—Acura 3.5L (C35A1) engine

   d. Water pump pulley.

   e. Right camshaft pulley.

✳✳ WARNING

Make sure that the camshaft and crankshaft pulleys are at TDC.

➡For easier installation, turn the right camshaft pulley about ½ tooth from TDC.

27. Loosen and retighten the timing belt adjusting bolt to tension the timing belt.

28. Install the lower cover and the crankshaft pulley.

29. Rotate the crankshaft pulley about 5–6 turns clockwise to position the timing belt on the pulleys.

30. To adjust the timing belt tension, perform the following procedure:

   a. Make sure that the No. 1 piston is at TDC of its compression stroke.

For complete Engine Mechanical specifications, see Section 1 of this manual

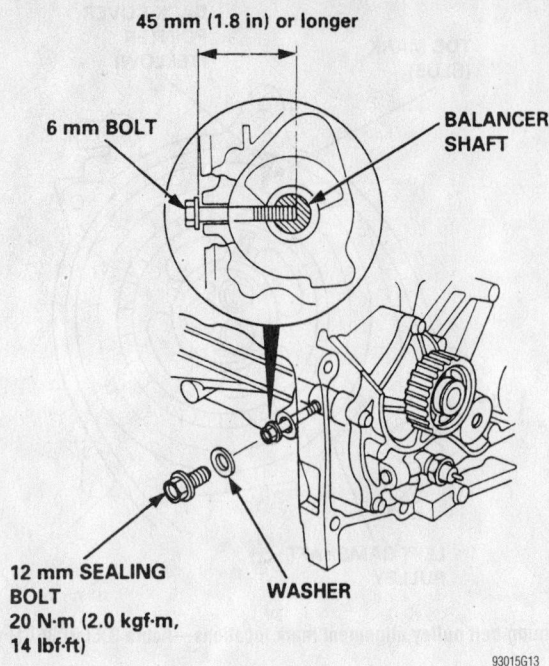

Securing the balancer shaft—Acura 3.5L (C35A1) engine

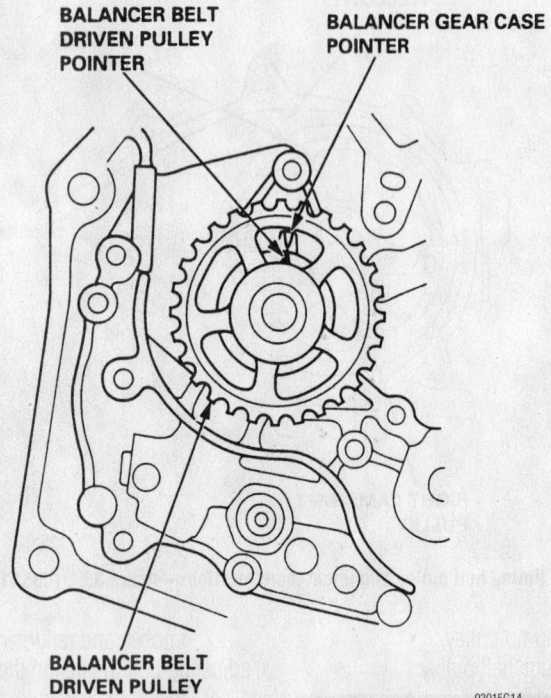

Balancer shaft alignment mark locations—Acura 3.5L (C35A1) engine

b. Rotate the crankshaft clockwise ten teeth on the camshaft pulley; the blue mark on the crankshaft pulley should align with the lower cover pointer.

c. Loosen the adjusting nut 180° (½ turn).

d. Tighten the adjusting nut to 31 ft. lbs. (42 Nm).

31. Remove the crankshaft pulley and the lower cover; then, install the timing belt guide plate and the balancer belt drive pulley.

32. Align the balancer shaft pulley by performing the following procedure:

a. Using a 6 x 45mm bolt, insert it into the maintenance hole and the balancer shaft.

b. Align the pointer on the balancer belt pulley with the pointer on the balancer gear case.

33. Adjust the timing belt drive pulley so that the No. 1 piston is at TDC of the compression stroke.

34. Install the balancer belt drive pulley and the balancer belt.

35. Loosen and retighten the balancer adjuster bolt to place tension on the balancer belt.

36. Remove the 6mm bolt and install the sealing bolt in the maintenance hole using a new washer.

37. Install the crankshaft pulley. Rotate the crankshaft pulley about 5–6 turns clockwise to position the timing belt on the pulleys.

38. Loosen the balancer belt adjuster bolt 180° (½ turn) and retighten the bolt to 33 ft. lbs. (44 Nm).

39. Remove the crankshaft pulley.

40. Install the upper and lower timing belt covers and the crankshaft pulley.

41. Install the crankshaft pulley and finger-tighten the bolt and washer. Using the Holder Handle and the 50mm Offset Holder Attachment tool 07MAB-PY3010A and a 19mm socket with a torque wrench, tighten the crankshaft pulley bolt to 181 ft. lbs. (245 Nm).

42. Make sure the crankshaft and camshaft pulleys are at TDC.

➡**If the camshaft or crankshaft pulley is not at TDC, remove the timing belt and re-perform the adjustment procedure.**

43. To complete the installation, reverse the removal procedures. Adjust the tension of the drive belts.

AUDI MODELS

1.8L (AEB) 4-Cylinder Engines

1. Disconnect the negative battery cable.

2. Raise and safely support the vehicle.

3. From under the vehicle, remove the splash shield.

4. Remove the front bumper.

5. Remove the intake air duct between the grille/front end assembly and the air cleaner housing.

6. Remove the grille/front end assembly-to-chassis bolts and the grille/front end assembly-to-vehicle fasteners.

7. If installed, remove the wiring harness retaining clamps from the left side of the radiator frame.

8. Install Support tools 3369 bolts into

the grille/front end assembly-to-chassis holes; then, pull the grille/front end assembly forward until it hits the stops.

➡**If necessary to secure the grille/front end assembly, install M6 bolts into the rear bored holes of the grille/front end assembly and the fender.**

9. Loosen the A/C compressor's serpentine belt tensioner bolts; release the belt tension and remove the belt.

10. Place an open-end wrench on the alternator belt tensioner and rotate it clockwise toward the alternator to release the belt's tension. Remove the alternator serpentine drive belt and release the tensioner.

➡**If necessary to lock the alternator tensioner in position, align the housing holes and insert an Allen wrench into the holes to secure its movement.**

11. Using a 5 x 60mm bolt, secure the viscous fan pulley. Using a hex wrench, remove the viscous fan-to-pulley bolts. Remove the viscous fan assembly.

12. Remove the upper timing belt cover.

➡**If reusing the timing belt, mark its rotational direction so it may be installed in its original position.**

13. Using the center bolt, rotate the crankshaft in the direction of engine rotation to position the No. 1 cylinder at Top Dead Center (TDC) of its compression stroke.

14. Remove the damper pulley-to-crankshaft bolts and the damper.

15. Remove the lower timing belt cover.

16. Using a Torx Wrench T45, loosen the timing belt tensioner, push the tensioner downward and remove the timing belt.

To install:

17. Align the camshaft sprocket timing mark with the cylinder head cover mark.

18. Install the timing belt on the crankshaft sprocket with the arrow facing the rotational direction.

19. Install the lower timing belt cover.

20. Using a bolt, secure the damper/belt pulley on the crankshaft.

21. Align the crankshaft damper/belt pulley with the housing timing mark so that the No. 1 cylinder is at TDC of its compression stroke.

22. Install the timing belt on the camshaft sprocket and belt tensioner.

23. Using a 2-pin Spanner Matra V159 Wrench, lift (turn clockwise) the timing belt tensioner cylinder No. 1 until it is fully extended and tensioner cylinder No. 2 is

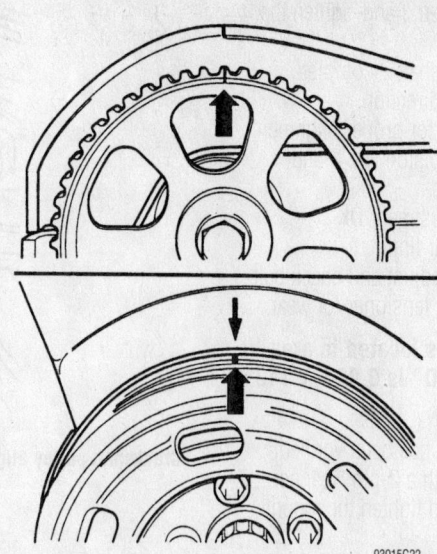

93015G22

Crankshaft pulley and camshaft sprocket alignment locations—Audi 1.8L (AEB) 4-Cyl engine

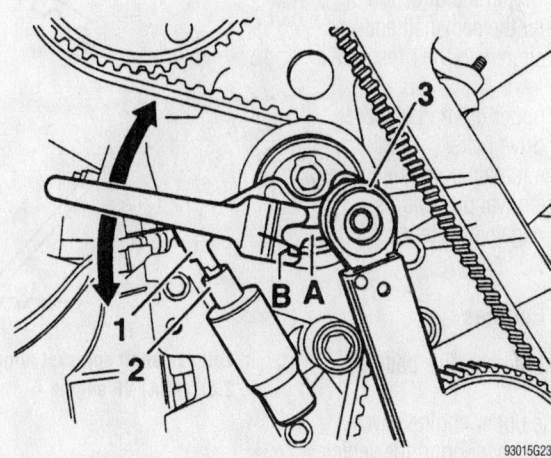

93015G23

Timing belt tension adjustment—Audi 1.8L (AEB) 4-Cyl engine

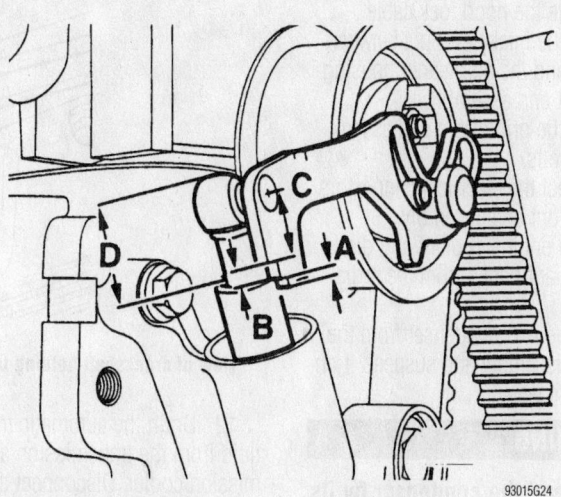

93015G24

Timing belt tension wear limits—Audi 1.8L (AEB) 4-Cyl engine

For Accessory Drive Belt illustrations, see Section 1 of this manual

raised approx. 1mm; then, hand-tighten the mounting bolt.

24. Rotate the crankshaft 2 complete rotation in the running direction.

25. Inspect area "A" for proper alignment with the upper edge of piston No. 2 and adjust if necessary.
- Area "A"—adjustment OK
- Area "B"—wear limit
- Area "C"—re-adjust and check belt drive including tensioner for wear.

➡️If the piston edge is located in area "A", measurement "D" is 0.984–1.142 in. (25–29mm).

26. After adjustment has been verified, secure the tensioner with a 2-pin Spanner Matra V159 Wrench and tighten the mounting bolt.

27. Complete the damper to crankshaft installation.

28. Using the center bolt, rotate the crankshaft 2 rotations in the direction of engine rotation until the camshaft and crankshaft marks align with their respective reference points.

29. Install the upper timing belt cover.

30. Install the drive belts.

31. Replace the remaining components by reversing the removal procedures.

32. Install the negative battery cable last.

33. Test drive the vehicle.

2.8L (AHA) V6 Engines

1. Disconnect the negative battery cable.

2. Remove the upper engine cover.

3. Raise and safely support the vehicle.

4. From under the vehicle, remove the splash shield.

5. Remove the front bumper.

6. Disengage the hood lock cable

7. Remove the intake air duct between the lock carrier and the air cleaner housing at the grille/front end assembly.

8. Remove the grille/front end assembly-to-chassis bolts.

9. Disconnect the electrical connectors from the grille/front end assembly.

10. Drain the engine coolant and disconnect the coolant hoses from the radiator.

11. Detach the A/C condenser from the grille/front end assembly and suspend it on a wire at the front wheel.

⁑ WARNING

DO NOT suspend the condenser by its lines. The condenser lines must not be bent or kinked.

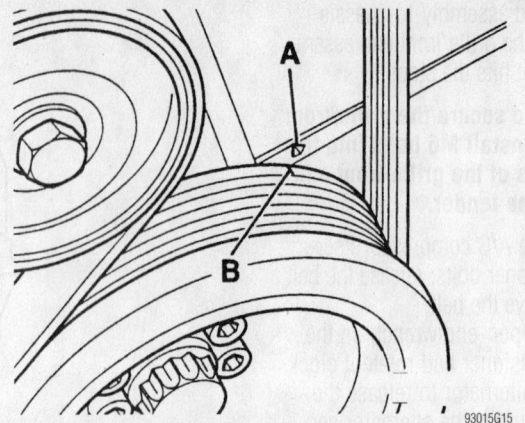

Crankshaft pulley alignment location for TDC—Audi 2.8L (AHA) V6 engine

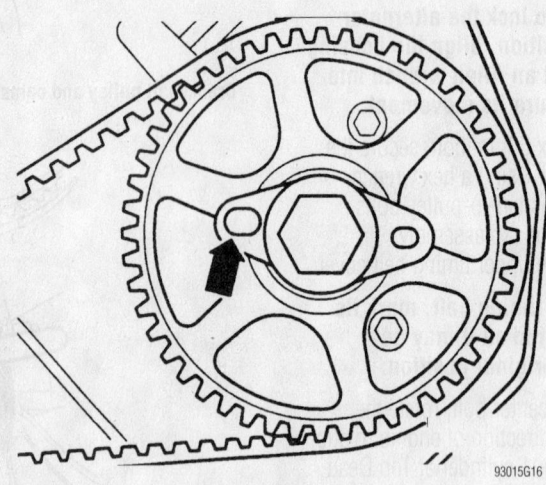

Left camshaft sprocket alignment position for TDC; right camshaft position is similar—Audi 2.8L (AHA) V6 engine

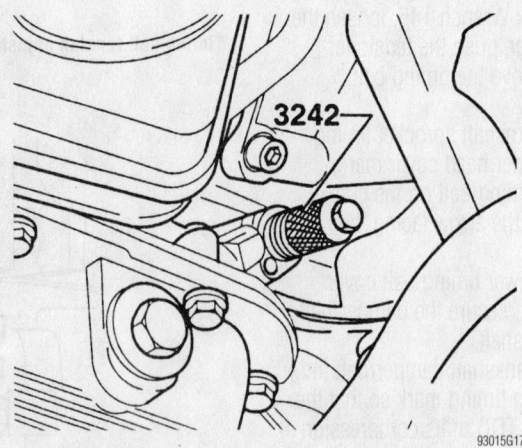

View of crankshaft holding tool installed—Audi 2.8L (AHA) V6 engine

12. Drain the automatic transmission fluid from the transmission and the transmission cooler. Disconnect the hydraulic lines from the transmission cooler.

13. If equipped, remove the charge air cooler.

14. Remove the grille/front end assembly-to-vehicle fasteners and the grille/front end assembly from the vehicle.

15. Remove the serpentine drive belt by performing the following procedure:

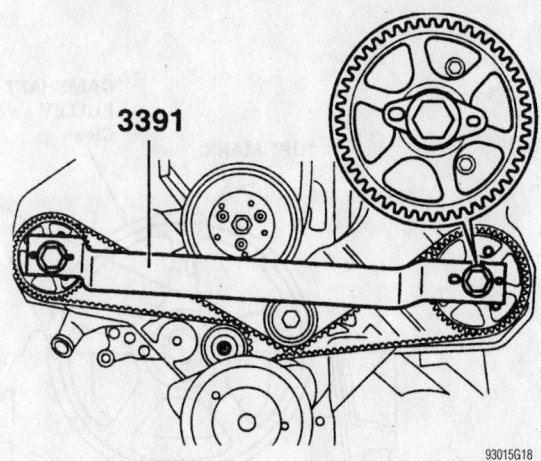

3391

View of camshaft locator bar installed—Audi 2.8L (AHA) V6 engine

93015G18

a. Using Spanner Wrench No. 3212, secure the viscous fan pulley. Using an Open-end Spanner Wrench 3212, remove the viscous fan bearing housing by turning it clockwise.

➡**The viscous fan is mounted with a left-handed thread; turn it clockwise to loosen it.**

b. Place a 17mm box wrench on the serpentine drive belt tensioner and turn the tensioner clockwise until the 2 holes align; insert drift 3204 into the holes to secure the tensioner in place.

c. Mark the running direction of the serpentine drive belt and remove it from the pulleys.

16. Rotate the crankshaft by hand to align the crankshaft pulley mark with the arrow on the engine housing and the large hole in each camshaft sprocket must face inward and must align; this should be Top Dead Center (TDC) of the No. 1 cylinder's compression stroke. If these conditions are not correct, rotate the crankshaft one complete revolution and realign.

17. On the left side of the cylinder block near the crankshaft, remove the sealing plug.

18. Insert Crankshaft Holder tool No. 3242 into the sealing plug hole to secure the crankshaft.

19. Using a 8mm Allen® wrench, rotate the timing belt tensioner roller clockwise until the tensioner is compressed; then, insert a 2mm spring pin through the tensioner housing and tensioner plunger to secure it in place. When the plunger is secure, release the wrench tension.

20. Remove the damper-to-crankshaft bolts and the damper.

➡**It is not necessary to remove the center bolt when removing the crankshaft damper.**

21. Remove the serpentine belt idler and the crankshaft damper guard.

22. Mark the running direction of the timing belt and remove it from the pulleys.

To install:

23. Make sure that the camshaft pulleys and the crankshaft pulley are in alignment with TDC of the No. 1 cylinder's compression stroke.

24. Install the timing belt; make sure the timing belt is installed in the correct running direction from which it was removed.

25. Using a 8mm Allen® wrench, rotate the timing belt tensioner roller clockwise until the tensioner is compressed; then, remove the 2mm spring pin from the tensioner housing. Slowly, release the tensioner's spring pressure to put pressure on the timing belt.

26. Install the crankshaft damper guard and the serpentine belt idler pulley; torque the idler pulley bolts to 33 ft. lbs. (45 Nm).

27. Install the crankshaft damper and torque the damper-to-crankshaft bolts to 15 ft. lbs. If the damper-to-crankshaft center bolt was removed, torque it to 147 ft. lbs. (200 Nm) plus 180° ½ turn).

28. Remove the Crankshaft Holder tool No. 3242 and install the sealing plug.

29. Replace the remaining components by reversing the removal procedures.

30. Refill the cooling system and the automatic transaxle. Connect the electrical connectors. Install the negative battery cable last.

31. Test drive the vehicle.

HONDA MODELS

1.6L and 1.7L Engines

1. Rotate the crankshaft to set the engine at Top Dead Center (TDC) on the compression stroke for the No. 1 piston. The white mark on the crankshaft pulley should align with the pointers on the timing cover. Once the engine is in this position, it must not be disturbed.

2. Remove all necessary components for access to the cylinder head and timing belt covers. Cover the rocker arm and shaft assemblies with a towel or sheet of plastic to keep out dust and foreign objects.

3. Remove the timing belt covers.

4. Loosen the timing belt adjusting bolt 180 degrees (½ turn). Push the tensioner pulley down to release the belt tension. After releasing the tension, retighten the tensioner pulley bolt until snug.

➡**Do not remove the tensioner pulley unless it is to be replaced.**

5. Remove the timing belt. Mark the direction of the belt's rotation if it is to be reinstalled.

To install:

➡**Inspect the water pump when replacing the timing belt; the manufacturer recommends replacing the water pump at the timing belt's service interval. Replace the timing belt if it shows any signs of wear, or if it is contaminated with oil or coolant.**

6. Verify that the timing is set at TDC on the compression stroke for the No. 1 cylinder as follows:

a. The groove in the crankshaft sprocket must align with the pointer on the oil pump.

b. The TDC marks on the camshaft sprockets must align with the pointer located between the sprockets. The TDC marks will also be in line with the upper surface of the head.

c. On other engines, the TDC mark on the camshaft sprocket must align with the pointer on the back cover.

d. The **UP** mark on the camshaft sprocket must point up.

7. Install the timing belt onto the crankshaft sprocket, then around the adjusting pulley and water pump sprocket, and finally over the camshaft sprocket.

8. Loosen the adjusting pulley bolt 180

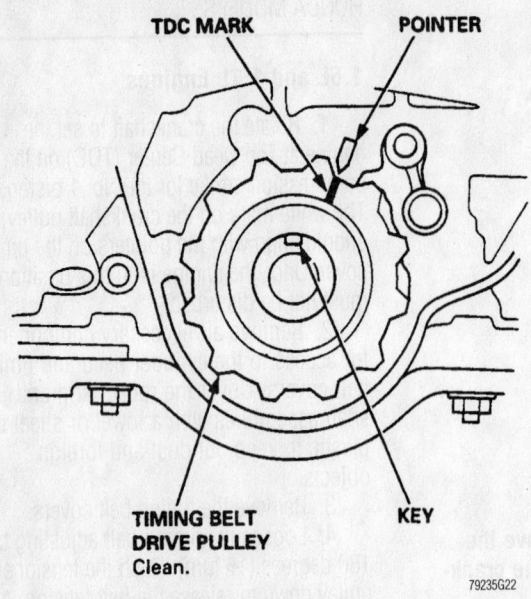

TDC MARK   POINTER

TIMING BELT
DRIVE PULLEY
Clean.   KEY

79235G22

TDC alignment mark locations for the crankshaft sprocket—Honda 1.6L and 1.7L SOHC engines

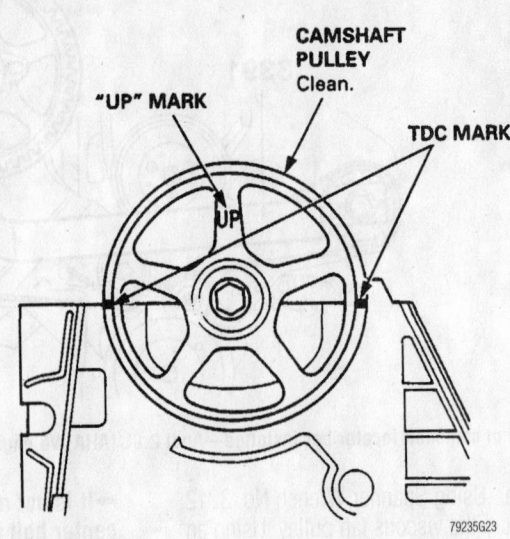

CAMSHAFT
PULLEY
Clean.

"UP" MARK   TDC MARK

79235G23

Single camshaft timing belt sprocket TDC mark positioning for timing belt installation—Honda 1.6L and 1.7L SOHC engines

degrees (½ turn). Then, tighten the adjusting bolt to 40 ft. lbs. (55 Nm).

9. Install the lower timing belt cover and the crankshaft pulley. Apply a light coat of fresh oil to the pulley bolt threads, then tighten it to 134 ft. lbs. (181 Nm).

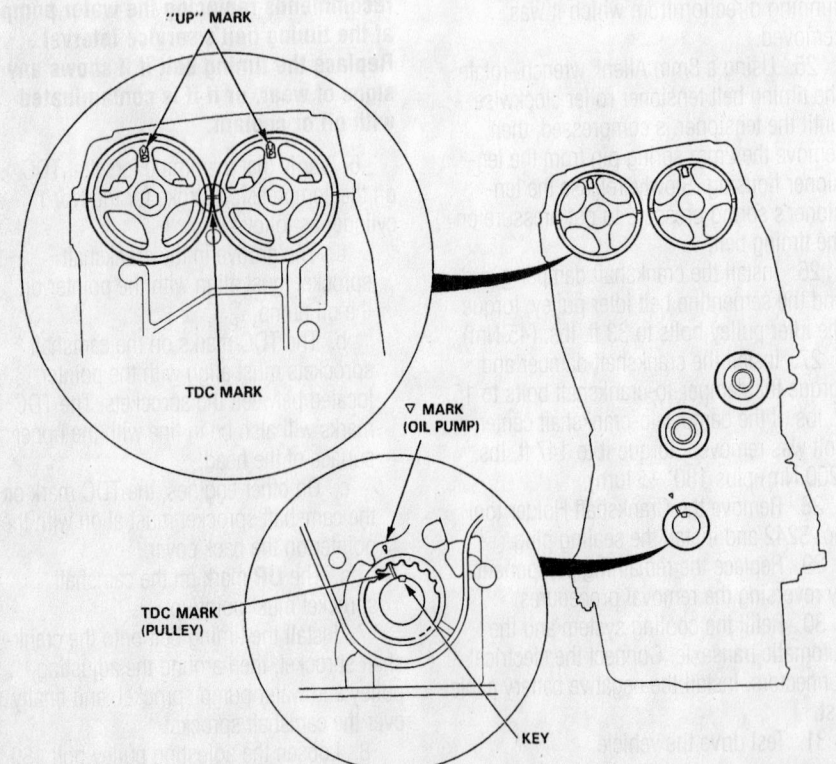

"UP" MARK

TDC MARK

▽ MARK
(OIL PUMP)

TDC MARK
(PULLEY)

KEY

79235G32

Twin camshaft timing belt alignment marks—Honda 1.6L DOHC engines

10. Rotate the crankshaft 5–6 turns counterclockwise to position the belt on the sprockets.

11. Adjust the timing belt tension, as follows:

a. Set the No. 1 piston at TDC on the compression stroke for the No. 1 cylinder.

b. Loosen the adjusting pulley bolt 180 degrees (½ turn).

c. Rotate the crankshaft counterclockwise so that the camshaft sprocket moves 3 teeth from the TDC/compression position.

d. Tighten the adjusting bolt to 33 ft. lbs. (45 Nm).

e. Tighten the crankshaft pulley to 134 ft. lbs. (181 Nm).

12. Verify that the crankshaft and camshaft sprockets will align properly at the TDC/compression position. If the camshaft pulley is not at TDC/compression, remove the timing belt, adjust the sprocket positions and reinstall the belt.

13. Install the upper timing and cylinder head covers, and all other applicable components. When reattaching the side engine mount, tighten the support nuts to 54 ft. lbs. (75 Nm).

2.2L and 2.3L Engines

1. Remove the cylinder head (valve) and upper timing belt covers.

2. Turn the engine to align the timing marks and set cylinder No. 1 to Top Dead Center (TDC). The white mark on the crankshaft sprocket should align with the pointer on the timing belt cover. The words **UP** embossed on the camshaft sprocket should be aligned in the upward position. The marks on the edge of the sprocket should be aligned with the cylinder head or the back cover upper edge. Once in this posi-

tion, the engine must NOT be turned or disturbed.

3. Remove all necessary components for access to the lower timing belt cover, then remove the cover.

4. There are two belts in this system; the one running to the camshaft sprocket is the timing belt. The other, shorter one drives the balance shaft and is referred to as the balancer shaft belt or timing balancer belt. Lock the timing belt adjuster in position, by installing one of the lower timing belt cover bolts to the adjuster arm.

5. Loosen the timing belt and balancer shafts tensioner adjuster nut, do not loosen the nut more than one turn. Push the tensioner for the balancer belt away from the belt to relieve the tension. Hold the tensioner and tighten the adjusting nut to hold the tensioner in place.

6. Carefully remove the balancer belt. Do not crimp or bend the belt; protect it from contact with oil or coolant.

7. Remove the balancer belt sprocket from the crankshaft.

8. Loosen the lockbolt installed to the timing belt adjuster and loosen the adjusting nut. Push the timing belt adjuster to remove the tension on the timing belt, then tighten the adjuster nut.

9. Remove the timing belt by sliding it off the sprockets. Do not crimp or bend the belt; protect it from contact with oil or coolant.

10. If defective, remove the belt tensioners by performing the following:

    a. Remove the springs from the balancer belt and the timing belt tensioners.

    b. Remove the adjusting nut from the belt tensioners.

    c. Remove the bolt from the balancer belt adjuster lever, then remove the lever and the tensioner pulley.

    d. Remove the lockbolt from the timing belt tensioner lever, then remove the tensioner pulley and lever from the engine.

11. This is an excellent time to check or replace the water pump. Even if the timing belt is only being replaced as part of a good maintenance schedule, consider replacing the pump at the same time.

To install:

12. If the water pump is to be replaced, install a new O-ring and make certain it is properly seated. Install the water pump and tighten the mounting bolts to 106 inch lbs. (12 Nm).

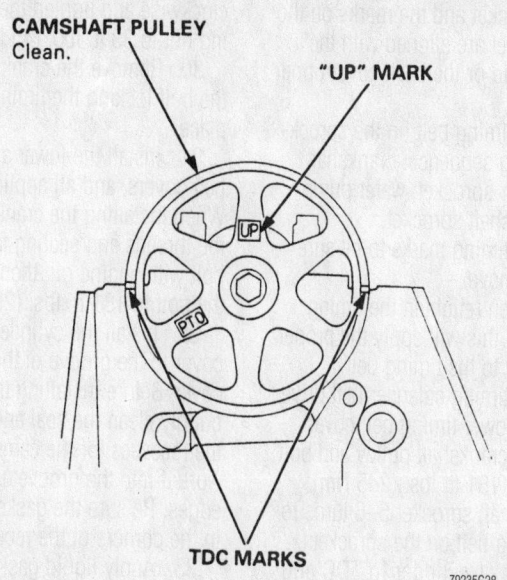

Position the camshaft sprocket as indicated for timing belt installation—Honda 2.2L and 2.3L engines

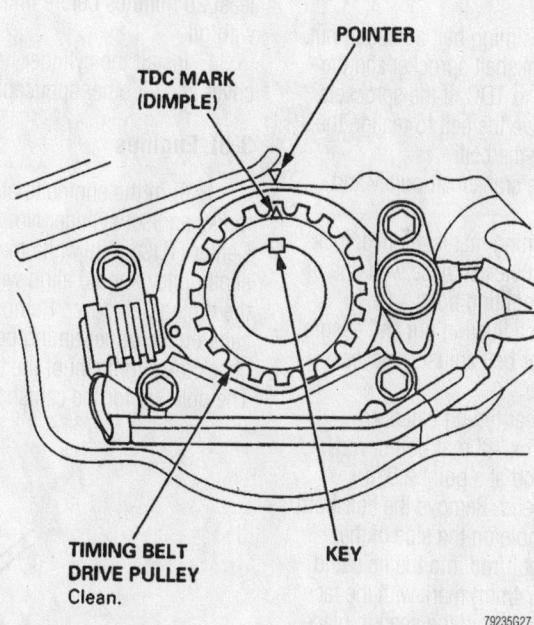

Before installing the timing belt, ensure the crankshaft sprocket marks are properly aligned—Honda 2.2L and 2.3L engines

13. If the tensioners were removed, perform the following procedures:

    a. Install the timing belt tensioner lever and the tensioner pulley.

    b. Install the balancer belt pulley and adjuster lever.

    c. Install the adjusting nut and the bolt to the balancer belt adjuster lever.

    d. Install the springs to the tensioners.

    e. Install the lockbolt to the timing belt tensioner, then move it its full deflection and tighten the lockbolt.

    f. Move the balancer it's full deflection and tighten the adjusting nut to hold its position.

14. The pointer on the crankshaft sprocket should be aligned with the pointer on the oil pump; the camshaft sprocket must be aligned so that the word **UP** is at

the top of the sprocket and the marks on the edge of the sprocket are aligned with the surfaces of the head or the back cover upper edge.

15. Install the timing belt on the sprockets in the following sequence: crankshaft sprocket, tensioner sprocket, water pump sprocket and camshaft sprocket.

16. Check the timing marks to be sure that they did not move.

17. Loosen, then retighten the timing belt adjusting nut; this will apply the proper amount of tension to the timing belt.

18. Install the timing balancer belt drive sprocket and the lower timing belt cover.

19. Install the crankshaft pulley and bolt, tighten the bolt to 181 ft. lbs. (245 Nm). Rotate the crankshaft sprocket 5–6 turns to position the timing belt on the sprockets.

20. Set the No. 1 cylinder to TDC and loosen the timing belt adjusting nut one turn. Turn the crankshaft counterclockwise until the cam sprocket has moved 3 teeth; this creates tension on the timing belt.

21. Tighten the timing belt adjusting nut.

22. Set the crankshaft sprocket and the camshaft sprocket to TDC. If the sprockets do not align, remove the belt to realign the marks, then install the belt.

23. Remove the crankshaft pulley and the lower cover.

24. With the timing marks aligned, lock the timing belt adjuster in place with one of the lower cover mounting bolts.

25. Loosen the adjusting nut and ensure the timing balancer belt adjuster moves freely.

26. Align the rear timing balancer sprocket using a 6 x 100mm bolt or rod. Mark the bolt or rod at a point 2.9 in. (74mm) from the end. Remove the bolt from the maintenance hole on the side of the block; insert the bolt/rod into the hole and align the 2.9 in. (74mm) mark with the face of the hole. This will hold the shaft in place during installation.

27. Align the groove on the front balancer shaft sprocket with the pointer on the oil pump.

28. Install the balancer belt. Once the belts are in place, be sure that all the engine alignment marks are still correct. If not, remove the belts, realign the engine and reinstall the belts. Once the belts are properly installed, slowly loosen the adjusting nut, allowing the tensioner to move against the belt. Remove the bolt from the maintenance hole and reinstall the bolt and washer.

29. Install the crankshaft pulley, then turn the crankshaft sprocket 1 turn counter-clockwise and tighten the timing belt adjusting nut to 33 ft. lbs. (45 Nm).

30. Remove the crankshaft pulley and the bolt locking the timing belt adjuster in place.

31. Install the lower and upper timing belt covers, and all applicable components. When installing the crankshaft pulley, coat the threads and seating face of the pulley bolt with engine oil, then install and tighten the bolt to 181 ft. lbs. (250 Nm).

32. Install the cylinder head cover gasket cover to the groove of the cylinder head cover. Before installing the gasket thoroughly clean the seal and the groove. Seat the recesses for the camshaft first, then work it into the groove around the outside edges. Be sure the gasket is seated securely in the corners of the recesses.

33. Apply liquid gasket to the four corners of the recesses of the cylinder head cover gasket. Do not install the parts if 5 minutes or more have elapsed since applying liquid gasket. After assembly, wait at least 20 minutes before filling the engine with oil.

34. Install the cylinder head (valve) cover and all other applicable components.

3.0L Engines

1. Turn the engine to align the timing marks and set cylinder No. 1 to Top Dead Center (TDC). The white mark on the crankshaft pulley should align with the pointer on the timing belt cover. Remove the inspection caps on the upper timing belt covers to check the alignment of the timing marks. The pointers for the camshafts should align with the green marks on the camshaft sprockets.

2. Remove all necessary components for access to the timing belt covers, then remove the covers.

➡️**Do not use the covers to store removed items.**

3. Loosen the timing belt adjuster bolt 180 degrees (½ turn). Push the tensioner to remove the tension from the timing belt, then retighten the adjusting bolt.

4. Remove the timing belt. Do not crimp or bend the belt; protect it from contact with oil or coolant. Slide the belt off the sprockets.

5. Remove the bolts attaching the camshaft sprockets to the camshafts, then remove the sprockets.

6. If the timing belt tensioner is defective, remove the spring from the timing belt tensioner. Remove the tensioner pulley adjusting bolt and the adjuster assembly from the engine.

➡️**This is an excellent time to check or replace the water pump. Even if the timing belt is only being replaced as part of a good maintenance schedule, consider replacing the pump at the same time.**

To install:

7. If the water pump is to be replaced, install a new O-ring and make certain it is properly seated. Install the water pump and retaining bolts. Tighten the mounting bolts to 16 ft. lbs. (22 Nm).

8. If removed, install the tensioner pulley and the adjusting bolt, be sure the ten-

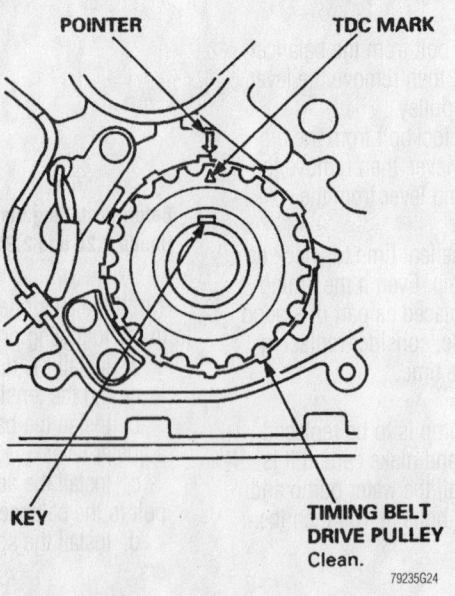

POINTER **TDC MARK**

KEY **TIMING BELT DRIVE PULLEY** Clean.

79235G24

Crankshaft timing belt sprocket alignment mark locations—Honda 3.0L engine

FRONT:

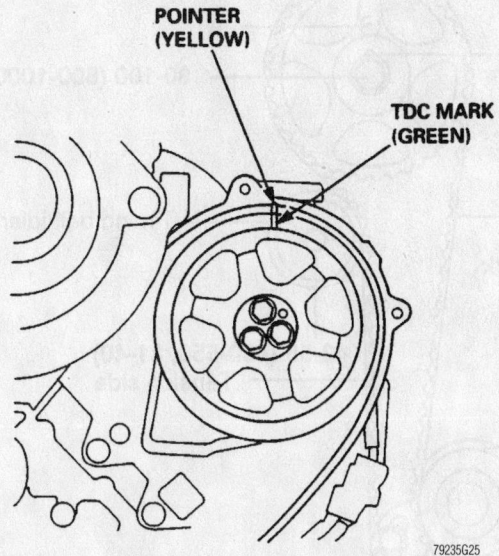

Left camshaft timing belt sprocket alignment mark location—Honda 3.0L engine

REAR:

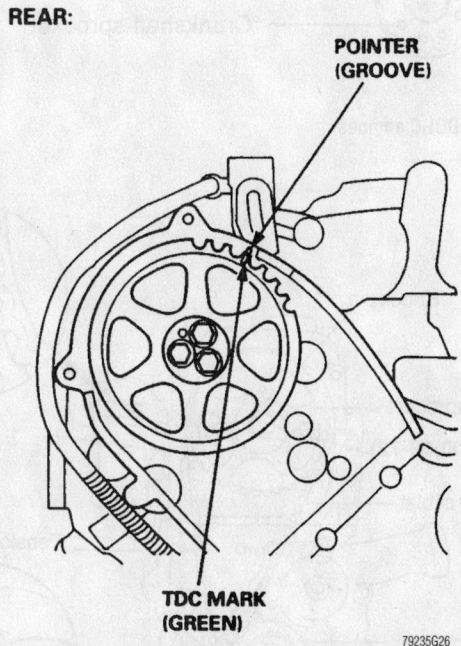

Rear camshaft timing belt sprocket alignment mark location—Honda 3.0L engine

sioner is properly positioned on its pivot pin. Install the spring to the tensioner, then push the tensioner to its full deflection and tighten the adjusting bolt.

9. Set the timing belt drive sprocket so that the No. 1 piston is at TDC. Align the TDC mark on the tooth of the timing belt drive sprocket with the pointer on the oil pump.

10. Set the camshaft sprockets so that the No. 1 piston is at TDC. Align the TDC marks (green mark) on the camshaft sprockets to the pointers on the back covers.

11. Install the timing belt onto the sprockets in the following sequence: crankshaft sprocket, tensioner pulley, front camshaft sprocket, water pump pulley and rear camshaft sprocket.

12. Loosen, then retighten the timing belt adjuster bolt to tension the timing belt.

13. Install the lower timing belt cover.

14. Install the crankshaft sprocket and the crankshaft pulley bolt. Tighten the bolt to 181 ft. lbs. (245 Nm) with the aid of the crank pulley holder.

15. Rotate the crankshaft five or six turns clockwise so that the timing belt positions on the sprockets.

16. Set cylinder No. 1 to TDC by aligning the timing marks. If the timing marks do not align, remove the timing belt, then adjust the components and reinstall the timing belt.

17. Loosen the timing belt adjusting bolt 180 degrees (½ turn) and retighten the adjusting bolt. Tighten the adjusting bolt to 31 ft. lbs. (42 Nm).

18. Install the upper timing belt cover and all other applicable components. When installing the side engine mount to the engine, use 3 new attaching bolts. Tighten the new bolts to 40 ft. lbs. (54 Nm).

HYUNDAI MODELS

❋❋ CAUTION

Timing belt maintenance is extremely important. All Hyundai models use interference-type non-freewheeling engines. Should the timing belt break in these engines, the valves in the cylinder head will come in contact with the pistons, causing major engine damage. The recommended replacement interval for timing belts is 60,000 miles.

1.5L and 1.6L Engines

1. Remove all necessary components for access to the timing belt cover, then remove the cover.

2. Rotate the crankshaft clockwise and align the timing marks so No. 1 piston will be at Top Dead Center (TDC) of the compression stroke.

3. Loosen the tensioning bolt and the pivot bolt on the timing belt tensioner. Move the tensioner as far as it will go toward the water pump. Tighten the adjusting bolt.

4. Mark the timing belt with an arrow showing direction of rotation.

5. Remove the timing belt.

6. If defective, remove the timing belt tensioner.

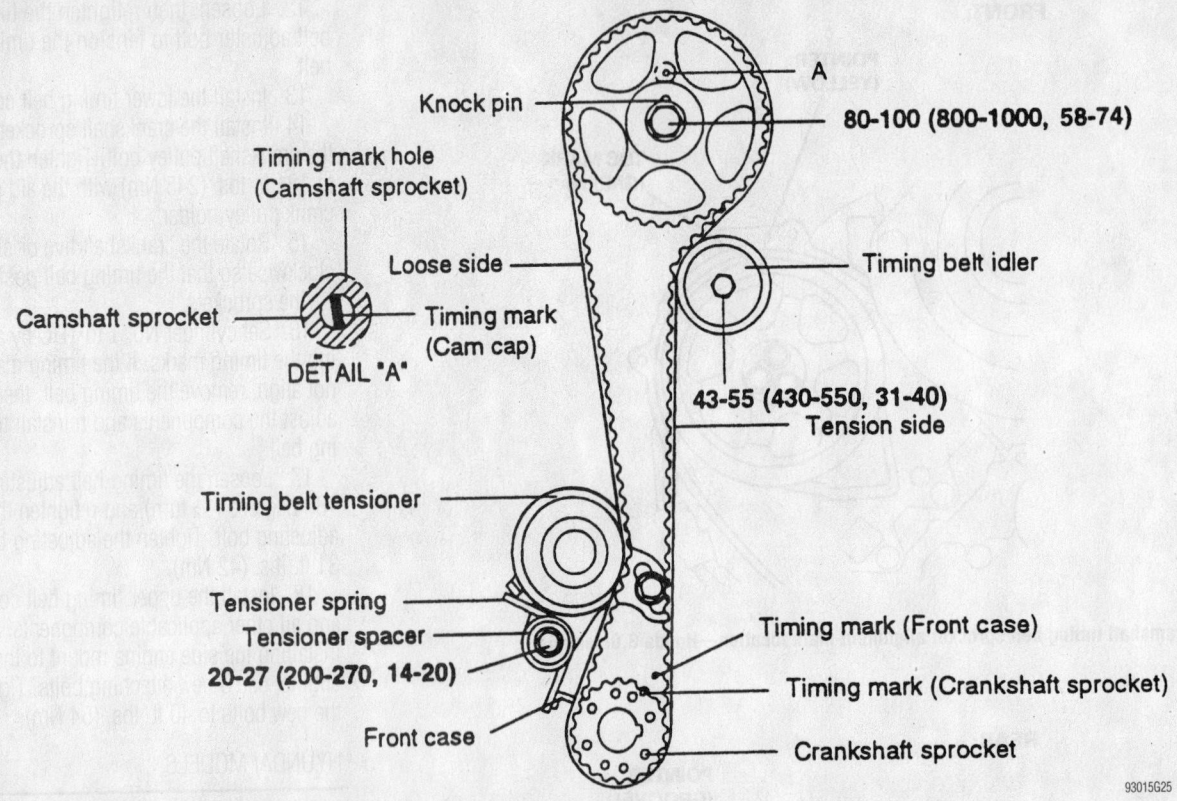

Knock pin

A

80-100 (800-1000, 58-74)

Timing mark hole
(Camshaft sprocket)

Loose side

Timing belt idler

Camshaft sprocket

Timing mark
(Cam cap)

DETAIL "A"

43-55 (430-550, 31-40)
Tension side

Timing belt tensioner

Tensioner spring
Tensioner spacer

Timing mark (Front case)

20-27 (200-270, 14-20)

Timing mark (Crankshaft sprocket)

Front case

Crankshaft sprocket

93015G25

Proper pulley alignment for timing belt installation at TDC—Hyundai 1.5L DOHC engines

To install:

7. Align the timing marks of the camshaft sprocket and check that the crankshaft timing marks are still in alignment.

8. If removed, install the timing belt tensioner, spring and spacer with the bottom end of the spring free. Tighten the adjusting bolt slightly with the tensioner moved as far as possible away from the water pump.

9. Install the free end of the spring into the locating tang on the front case.

10. Position the timing belt over the crankshaft sprocket, then over the camshaft sprocket. Slip the back of the belt over the tensioner wheel.

11. Turn the camshaft sprocket in the opposite of its normal direction of rotation until the straight side of the belt is tight and be sure the timing marks align.

➡If the timing marks are not properly aligned, shift the belt 1 tooth at a time in the appropriate direction until they are aligned.

12. Loosen the tensioner mounting bolts so the tensioner works, without the interference of any friction, under spring pressure. Be sure the belt follows the curve of the camshaft pulley so the teeth are engaged all

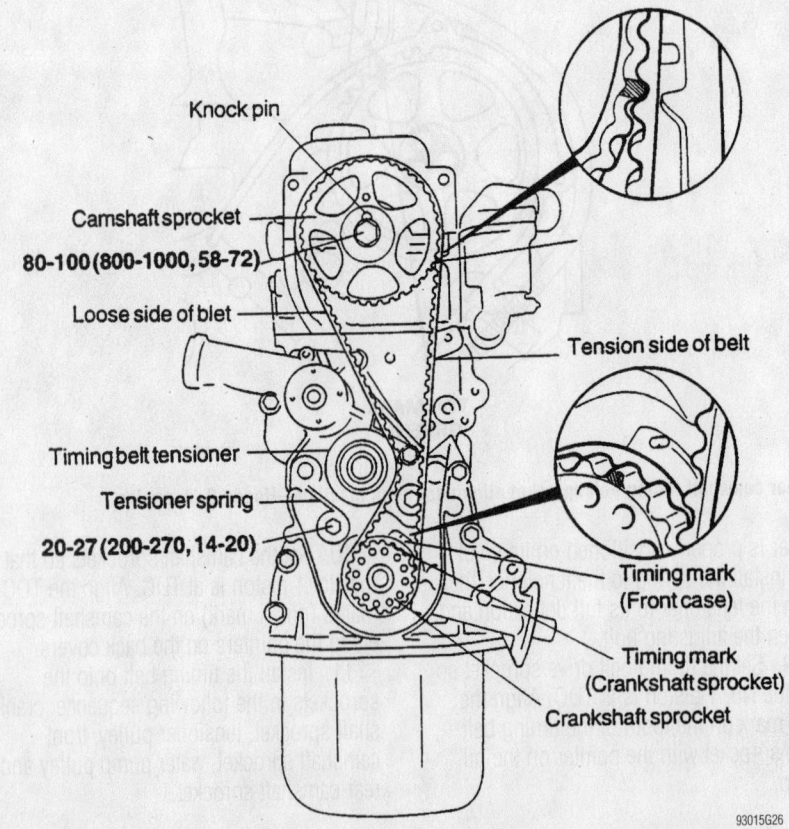

Knock pin

Camshaft sprocket

80-100 (800-1000, 58-72)

Loose side of blet

Tension side of belt

Timing belt tensioner

Tensioner spring

20-27 (200-270, 14-20)

Timing mark
(Front case)

Timing mark
(Crankshaft sprocket)

Crankshaft sprocket

93015G26

Proper pulley alignment for timing belt installation at TDC—Hyundai 1.5L SOHC engines

the way around. Correct the path of the belt, if necessary.

13. Tighten the tensioner adjusting bolt, then the tensioner pivot bolt to 15–18 ft. lbs. (20–26 Nm).

➡**Bolts must be tightened in the stated order or tension won't be correct.**

14. Turn the crankshaft 1 turn clockwise until timing marks again align to seat the belt.

15. Loosen both tensioner attaching bolts and let the tensioner position itself under spring tension. Retighten the bolts.

16. Check belt tension by putting a finger on the water pump side of the tensioner wheel and pull the belt toward the water pump. The belt should move toward the pump until the teeth are approximately ½ of the way across the head of the tensioner adjusting bolt. Re-tension the belt, if necessary.

17. Install the timing belt covers and all other related components.

2.0L (VIN P) Engines

1. Remove all necessary components for access to the timing belt covers, then remove the covers.

➡**Always rotate the crankshaft in a clockwise direction.**

2. Rotate the crankshaft clockwise and align the timing marks so the No. 1 piston will be at Top Dead Center (TDC) of the compression stroke. At this time the timing marks on the camshaft sprocket and the upper surface of the cylinder head should coincide, and the dowel pin of the camshaft sprocket should be at the upper side.

3. Remove the outer timing belt tensioner.

4. Mark the timing belts, indicating the direction of rotation.

5. Remove the outer timing belt.

6. Remove the camshaft sprockets.

7. Insert a prytool with a 0.32 in. (8mm) diameter shaft into the left side cylinder block plug hole. The prytool will hold the counterbalance shaft stable while removing the oil pump sprocket retaining nut.

8. Remove the oil pump sprocket.

9. Loosen the right counterbalance shaft sprocket bolt.

10. Remove the inner timing belt tensioner.

11. Remove the inner timing belt.

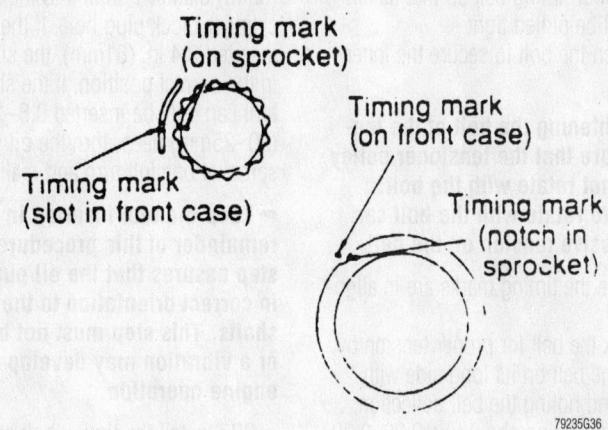

Oil pump and crankshaft inner timing belt sprocket alignment marks—Hyundai 2.0L (VIN P) engines

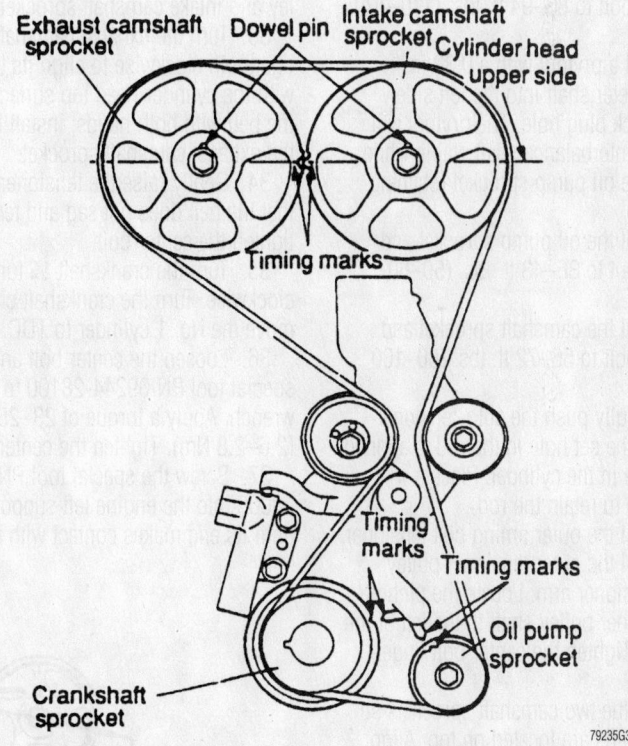

Timing belt sprocket alignment mark locations and positioning for belt removal and installation—Hyundai 2.0L (VIN P) engines

To install:

12. Install the counterbalance shaft sprocket and tighten the flange bolt finger-tight.

13. Align the timing mark on each sprocket with the corresponding timing mark on the front case.

14. Install the inner timing belt.

➡**When installing the inner timing belt, ensure that the tension side has no slack.**

15. Install the inner timing belt tensioner with the center of the pulley on the left side of the mounting bolt and with the pulley flange facing the front of the engine.

16. Lift the inner timing belt tensioner to

tighten the inner timing belt so that its tension side will be pulled tight.

17. Tighten the bolt to secure the inner tensioner.

➡ **When tightening the bolt of the tensioner, ensure that the tensioner pulley shaft does not rotate with the bolt. Allowing it to rotate with the bolt can cause excessive tension on the belt.**

18. Ensure the timing marks are in alignment.

19. Check the belt for proper tension by depressing the belt on its long side with your finger and noting the belt deflection. The desired deflection should be 0.20–0.28 in. (5–7mm).

20. Install the flange, crankshaft sprocket and washer on the crankshaft. The flange on the crankshaft sprocket must be installed towards the inner timing belt sprocket. Tighten the bolt to 80–94 ft. lbs. (110–130 Nm).

21. Insert a prytool with a 0.32 in. (8mm) diameter shaft into the left side cylinder block plug hole. The prytool will hold the counterbalance shaft stable while removing the oil pump sprocket retaining nut.

22. Install the oil pump sprocket and tighten the nut to 36–43 ft. lbs. (50–60 Nm).

23. Install the camshaft sprocket and tighten the bolt to 56–72 ft. lbs. (80–100 Nm).

24. Carefully push the auto-tensioner rod in until the set hole in the rod is aligned with the hole in the cylinder. Place a wire into the hole to retain the rod.

25. Install the outer timing belt tensioner.

26. Install the outer tensioner pulley onto the tensioner arm. Locate the pinhole in the tensioner pulley shaft to the left of the center bolt. Tighten the center bolt finger-tight.

27. Turn the two camshaft sprockets so their dowel pins are located on top. Align the timing marks facing each other with the top surface of the cylinder head.

➡ **Both camshaft sprockets are used for the intake and exhaust camshafts and are provided with two timing marks. When the sprocket is mounted on the exhaust camshaft, use the timing mark on the right with the dowel pin hole on top. For the intake camshaft sprocket, use the mark on the left with the dowel pin hole on top.**

28. Align the crankshaft sprocket and oil pump sprocket timing marks.

29. Insert a prytool with a 0.32 in.

(8mm) diameter shaft into the left side cylinder block plug hole. If the shaft can be inserted 2.4 in. (61mm), the silent shaft is in the correct position. If the shaft of the tool can only be inserted 0.8–1.0 in. (20–25mm) deep, turn the oil pump sprocket one full turn and realign the marks.

➡ **Keep the tool inserted in hole for the remainder of this procedure. The above step assures that the oil pump socket is in correct orientation to the silent shafts. This step must not be skipped or a vibration may develop during engine operation.**

30. Install the timing belt around the tensioner pulley and crankshaft sprocket. Hold the belt with your left-hand.

31. Pulling the belt with your right-hand, install it around the oil pump sprocket.

32. Install the belt around the idler pulley and intake camshaft sprocket.

33. Turn the exhaust camshaft sprocket one tooth clockwise to align its timing mark with the cylinder head top surface. Pulling the belt with both hands, install it around the exhaust camshaft sprocket.

34. Gently raise the tensioner pulley so that the belt does not sag and temporarily tighten the center bolt.

35. Turn the crankshaft ¼ turn counterclockwise. Turn the crankshaft clockwise to move the No. 1 cylinder to TDC.

36. Loosen the center bolt and attach special tool PN 09244-28100 to a torque wrench. Apply a torque of 23–25 inch lbs. (2.6–2.8 Nm). Tighten the center bolt.

37. Screw the special tool PN 09244-28000 into the engine left support bracket until its end makes contact with the ten-

sioner arm. At this point, screw the special tool in some more and remove the set wire attached to the auto-tensioner, if the wire was not previously removed. Remove the special tool.

38. Rotate the crankshaft 2 complete turns clockwise and let it sit for approximately 15 minutes. Then, measure the auto-tensioner protrusion (the distance between the tensioner arm and auto-tensioner body) to ensure that it is within 0.15–0.18 in. (3.8–4.5mm).

39. If the timing belt tension adjustment is being performed with the engine mounted in the vehicle, and clearance between the tensioner arm and the auto-tensioner body cannot be measured, the following alternative method can be used:

   a. Screw in special tool PN 09244-28000 until its end makes contact with the tensioner arm.

   b. After the tool makes contact with the arm, screw it in some more to retract the auto-tensioner pushrod while counting the number of turns the tool makes until the tensioner arm is brought into contact with the auto-tensioner body. Be sure the number of turns the special tool makes conforms with the standard value of 2½–3 turns.

   c. Install the rubber plug to the timing belt rear cover.

40. Install the timing belt covers.

1.8L (VIN M) and 2.0L (VIN F) Engines

1. Remove all necessary components for access to the timing belt cover, then remove the cover.

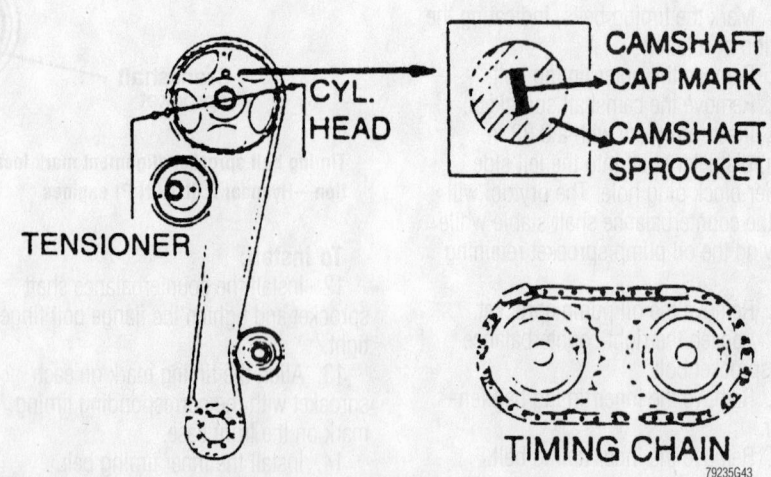

Proper alignment of the timing belt alignment marks for belt removal and installation—Hyundai 1.8L and 2.0L (VIN F) engines

2. Rotate the crankshaft clockwise and align the timing marks so No. 1 piston will be at Top Dead Center (TDC) of the compression stroke.

3. Remove the timing belt tensioner and idler pulley.

4. Mark the timing belt with an arrow showing direction of rotation.

5. Remove the timing belt.

To install:

6. Align the timing marks of the camshaft sprocket and check that the crankshaft timing marks are still in alignment.

7. Install the timing belt tensioner.

8. Install the idler pulley, if equipped. Tighten bolt to 32–41 ft. lbs. (43–55 Nm).

9. Position the timing belt over the camshaft sprocket, then over the crankshaft sprocket.

10. Tension the timing belt and tighten the tensioner pulley bolt to 32–41 ft. lbs. (43–55 Nm). When properly tensioned, the timing belt should deflect 0.16–0.24 in. (4–6mm) when a force of 5 lbs. (2.2kg) is placed on the longest span of the belt.

11. Turn the crankshaft sprocket one turn clockwise and realign the crankshaft sprocket timing mark.

12. Recheck the belt tension and adjust as necessary.

13. Install the timing belt cover and all other applicable components.

3.0L Engine

1. Remove all necessary components for access to the timing belt covers, then remove the covers.

2. Turn the crankshaft until the timing marks on the camshaft sprocket and cylinder head are aligned.

3. Loosen the timing belt tensioner bolt and turn the tensioner counterclockwise as far as it will go. Tighten the adjusting bolt.

4. Mark the timing belt with an arrow showing direction of rotation.

5. Remove the timing belt.

6. If defective, remove the timing belt tensioner.

To install:

7. If necessary, install the timing belt tensioner.

8. Attach the top of the tensioner spring on the engine coolant pump pin. Ensure the hook on the pin is facing down and the hook on the tensioner is facing away from the engine

9. Rotate the timing belt tensioner to the extreme counterclockwise position. Temporarily lock the tensioner in place.

10. Align the timing marks of the camshaft and crankshaft sprockets.

11. Install the timing belt on the crankshaft sprocket, then onto the rear camshaft sprocket.

12. Route the belt to the coolant pump pulley, the front camshaft sprocket and the timing belt tensioner.

13. Apply force counterclockwise to the rear camshaft sprocket with tension on the tight side of the belt and check that timing marks are aligned.

14. Loosen the tensioner bolt 1–2 turns and tighten the timing belt to a tension of 57–84 lbs. (260–380 N).

15. Turn the crankshaft two turns clockwise.

16. Readjust the sprocket timing marks and tighten the tensioner bolts.

17. Install the timing belt covers.

18. Install the crankshaft pulley and tighten to 108-116 ft. lbs. (150-160 Nm).

19. Install all applicable components.

INFINITI MODELS

3.0L (VG30DE) Engine

1. Remove all necessary components for access to the front timing covers, then remove the covers.

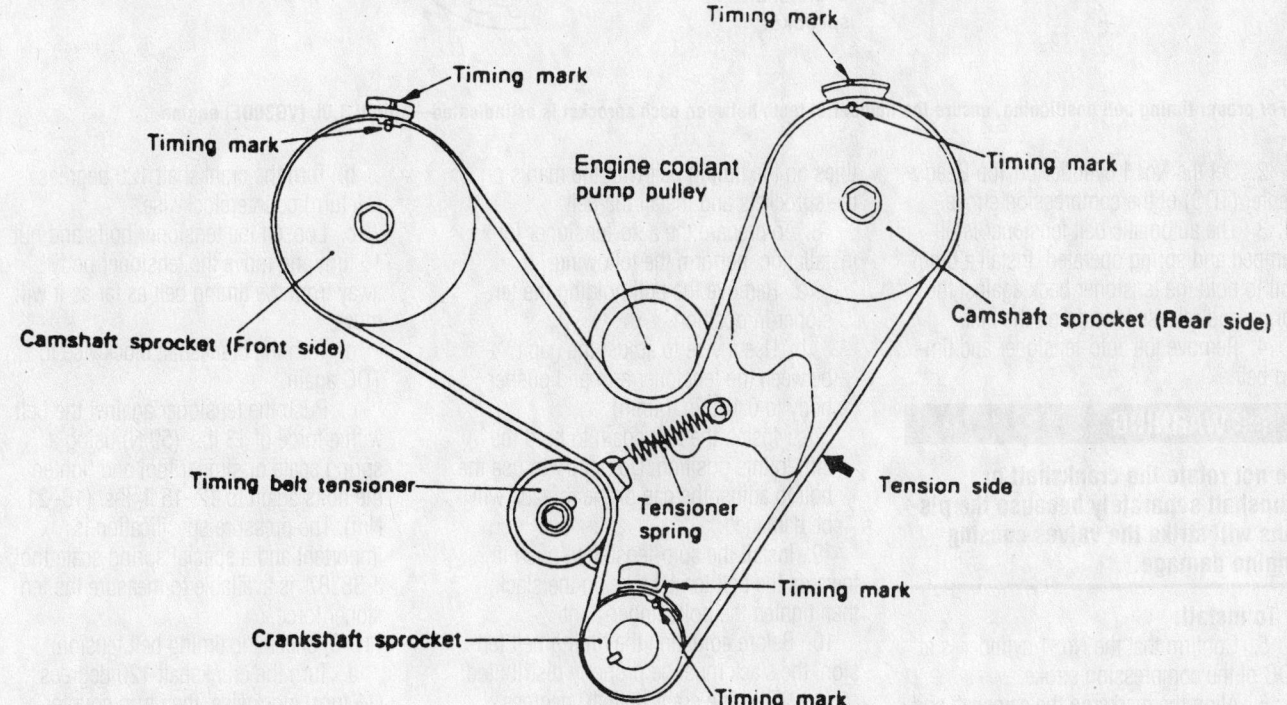

Timing belt sprocket alignment mark positioning for belt removal and installation—Hyundai 3.0L engine

79235G34

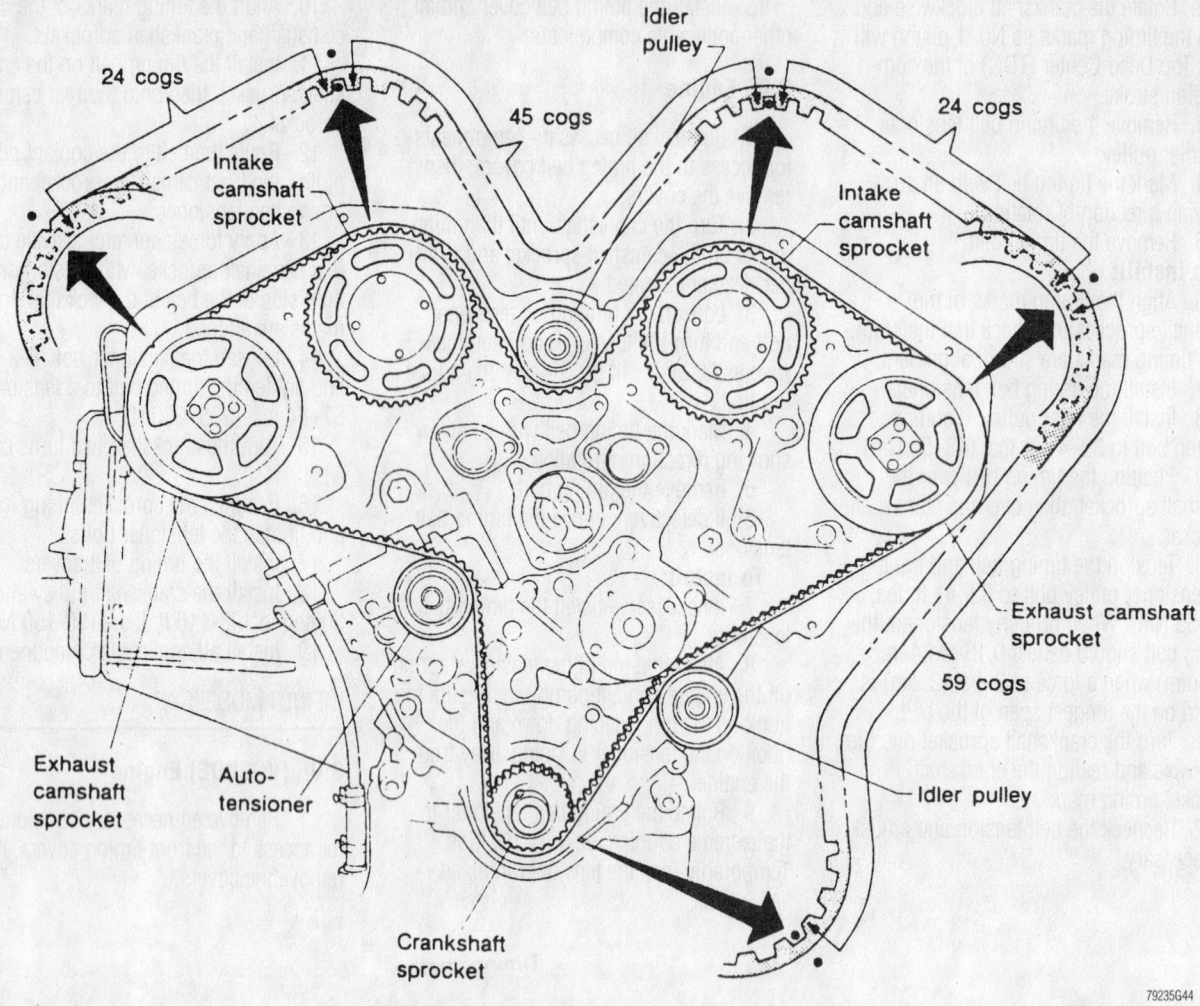

For proper timing belt positioning, ensure the number of teeth between each sprocket is as indicated—Infiniti 3.0L (VG30DE) engine

2. Set the No. 1 cylinder on Top Dead Center (TDC) of the compression stroke.

3. The automatic belt tensioner is oil damped and spring operated. Install a 6mm bolt to hold the tensioner back against the spring and release tension on the belt.

4. Remove the auto-tensioner and timing belt.

✶✶ WARNING

Do not rotate the crankshaft or camshaft separately because the pistons will strike the valves causing engine damage.

To install:

5. Confirm that the No. 1 cylinder is at TDC of the compression stroke.

6. Align the marks on the camshaft and crankshaft sprockets with the marks on the rear belt cover and oil pump housing.

7. With the arrows on the timing belt pointing towards the front, align the white

lines on the timing belt with the marks on the sprockets and install the belt.

8. To prepare the auto-tensioner for installation, perform the following:

a. Remove the bolt holding the tensioner in position.

b. Use a vise to adjust the gap between the tensioner arm and pusher body to 0.160 in. (4mm).

c. Install the bolt again to hold the arm in this position. Do not try to use the bolt to adjust the gap or the threads will be damaged.

9. Install the auto-tensioner, push it towards the belt to just take up the slack, then tighten the bolts finger-tight.

10. Before adjusting the timing belt tension, the slack must be properly distributed:

a. Turn the crankshaft 10 degrees clockwise and tighten the tensioner bolts and nut to 12–15 ft. lbs. (16–21 Nm). Do not push the auto-tensioner hard or the belt will be adjusted too tight.

b. Turn the crankshaft 120 degrees (⅓ turn) counterclockwise.

c. Loosen the tensioner bolts and nut ½ turn and move the tensioner body away from the timing belt as far as it will move.

d. Turn the crankshaft clockwise to TDC again.

e. Push the tensioner against the belt with a force of 13 lbs. (59 N) using a spring scale or similar tool and tighten the bolts again to 12–15 ft. lbs. (16–21 Nm). The pressure specification is important and a special spring scale tool, J-38387, is available to measure the tensioner force.

11. To check the timing belt tension:

a. Turn the crankshaft 120 degrees (⅓ turn) clockwise, then turn counterclockwise and return the engine to TDC.

b. Prepare a steel plate that is approximately ⅜ in. (10mm) wide and longer than the width of the belt.

c. Set the plate on the timing belt between two camshaft sprockets and push against the plate with a force of 11 lbs. (49 N). Note the belt deflection.

d. Repeat the procedure between the other camshaft sprockets and between the exhaust sprockets and idler/tensioner pulleys. There will be a total of four measurements.

e. Add the deflection measurements and divide by four. The average deflection must be 0.240–0.280 in. (6–7mm).

If belt tension is not correct, start the entire adjustment procedure again.

12. Confirm the auto-tensioner mounting nuts are tightened to 12–15 ft. lbs. (16–21 Nm) and remove the auto-tensioner stopper bolt.

13. After 5 minutes, measure the clearance between the tensioner arm and the pusher. It should be 0.138–0.205 in. (3.5–5.2mm).

14. Be sure all the sprocket timing marks are correctly aligned. Install the timing belt

covers and tighten the bolts to 24–38 inch lbs. (3–5 Nm).

15. Install all applicable components.

KIA SEPHIA

1. Disconnect the negative battery cable.

2. Remove the accessory drive belts.

3. Remove the generator.

4. Remove the water pump and crankshaft pulleys.

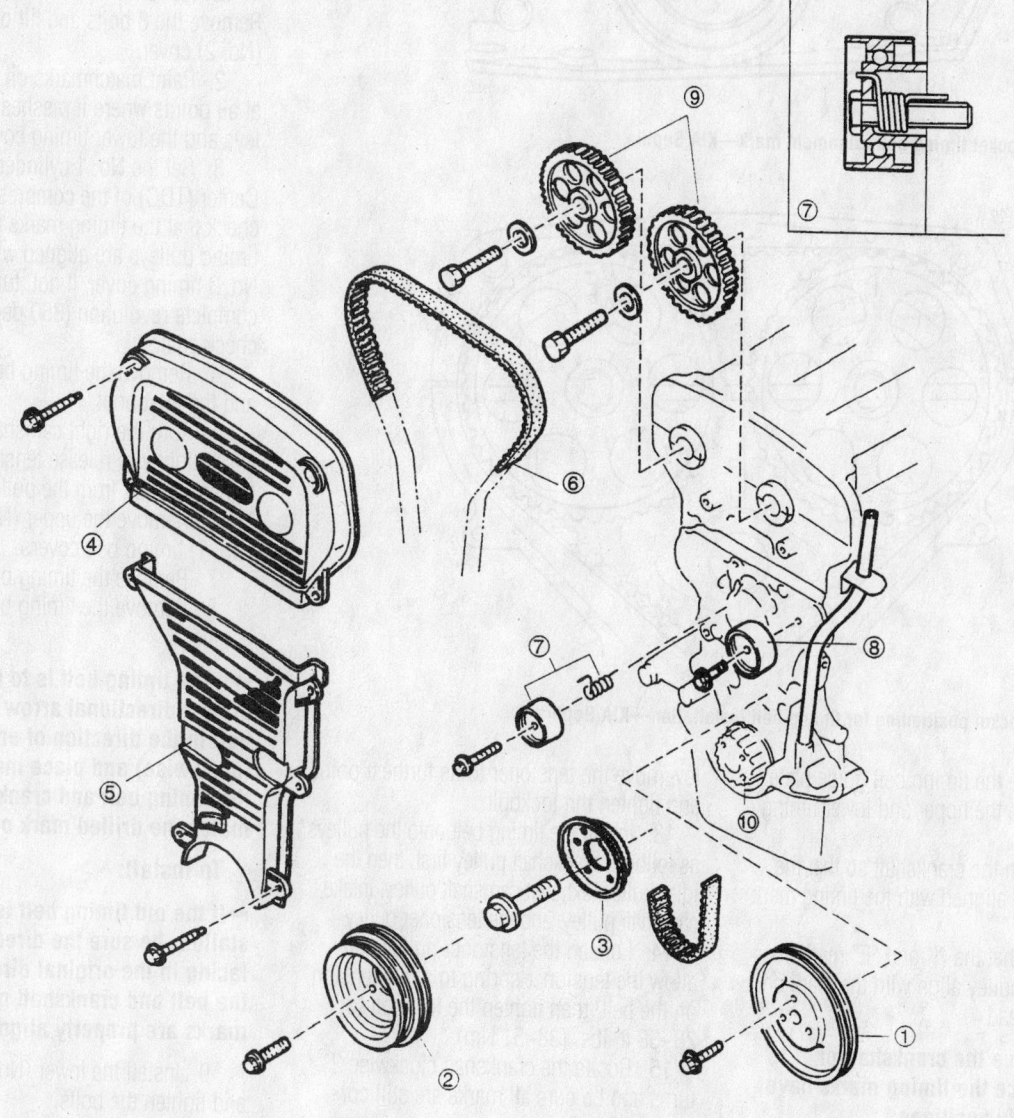

(1) Water pump pulley
(2) Crankshaft pulley
(3) Timing belt guide plate
(4) Timing belt cover (Upper)
(5) Timing belt cover (Lower)
(6) Timing belt
(7) Timing belt tensioner pulley & spring
(8) Idler pulley
(9) Camshaft pulley
(10) Timing belt pulley

93015G01

Exploded view of the timing belt cover mounting and related components–Sephia

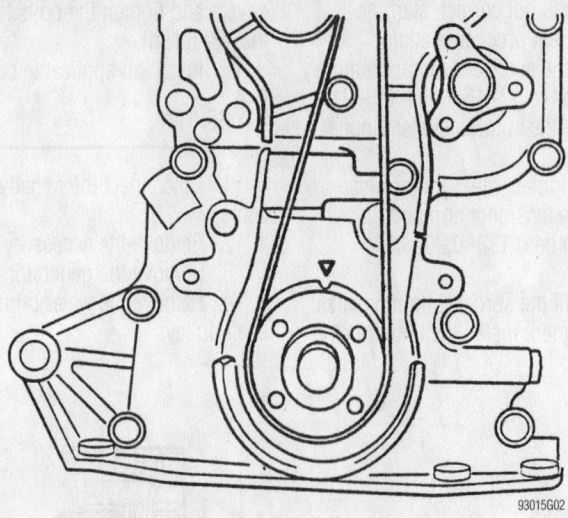

93015G02

Crankshaft sprocket timing belt alignment mark—KIA Sephia

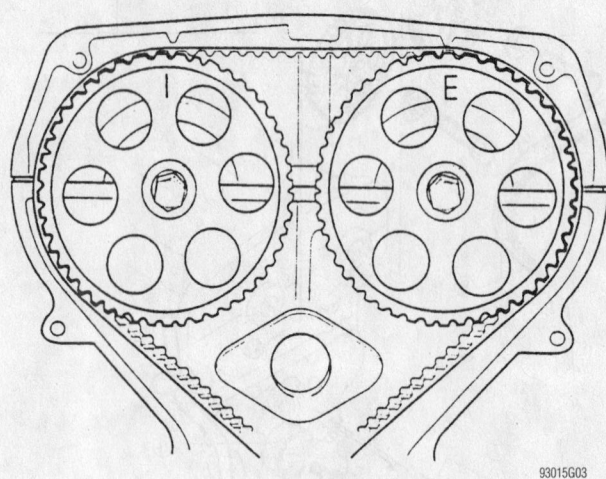

93015G03

Camshaft sprocket positioning for timing belt installation—KIA Sephia

5. Remove the timing belt guide plate.

6. Remove the upper and lower timing belt covers.

7. Position the crankshaft so that the timing mark is aligned with the timing mark on the engine.

8. Verify that the "I" and "E" mark on the camshaft pulley align with the mark on the cylinder head.

➡**Do not move the crankshaft or camshaft once the timing marks have been correctly positioned.**

9. If the timing belt is to be reused, mark the direction of rotation on the timing belt.

10. Remove the timing belt tensioner pulley.

11. Remove the timing belt.

To install:

12. Install the timing belt tensioner pul-

ley, move the tensioner to its furthest point and tighten the lockbolt.

13. Install the timing belt onto the pulleys, as follows, crankshaft pulley first, then the idler pulley, exhaust camshaft pulley, intake camshaft pulley, and the tensioner pulley.

14. Loosen the tensioner pulley and allow the tensioner spring to apply tension on the belt, then tighten the lockbolt to 28–38 ft. lbs. (38–51 Nm).

15. Rotate the crankshaft clockwise 2 turns and be sure all marks are still correctly aligned.

16. Install the remaining components in the revere order of the removal noting the following torque specifications:

- Crankshaft pulley: 9–13 ft. lbs. (12–17 Nm)
- Water pump pulley: 9–13 ft. lbs. (12–17 Nm)

17. Connect the negative battery cable.

LEXUS MODELS

> ✸✸ **CAUTION**
>
> **On models with an air bag, wait at least 90 seconds from the time that the ignition switch is turned to the LOCK position and the battery is disconnected before performing any further work.**

3.0L (1MZ-FE) Engine

1. Remove all necessary components for access to the upper timing belt cover. Remove the 8 bolts and lift off the upper (No. 2) cover.

2. Paint matchmarks on the timing belt at all points where it meshes with the pulleys and the lower timing cover.

3. Set the No. 1 cylinder to Top Dead Center (TDC) of the compression stroke and check that the timing marks on the camshaft timing pulleys are aligned with those on the No. 3 timing cover. If not, turn the engine 1 complete revolution (360 degrees) and check again.

4. Remove the timing belt tensioner and the dust boot.

5. Turn the right camshaft pulley clockwise slightly to release tension, then remove the timing belt from the pulleys.

6. Remove the upper (No. 3) and lower (No. 1) timing belt covers.

7. Remove the timing belt guide.

8. Remove the timing belt from the engine.

➡If the timing belt is to be reused, draw a directional arrow on the timing belt in the direction of engine rotation (clockwise) and place matchmarks on the timing belt and crankshaft gear to match the drilled mark on the pulley.

To install:

➡If the old timing belt is being reinstalled, be sure the directional arrow is facing in the original direction and that the belt and crankshaft gear matchmarks are properly aligned.

9. Install the lower (No. 1) timing cover and tighten the bolts.

10. Set the No. 1 cylinder to TDC again. Turn the right camshaft until the knock pin hole is aligned with the timing mark on the No. 3 belt cover. Turn the left pulley until the marks on the pulley are aligned with the mark on the No. 3 timing cover.

11. Check that the mark on the belt matches with the edge of the lower cover. If not, shift it on the crank pulley until it does.

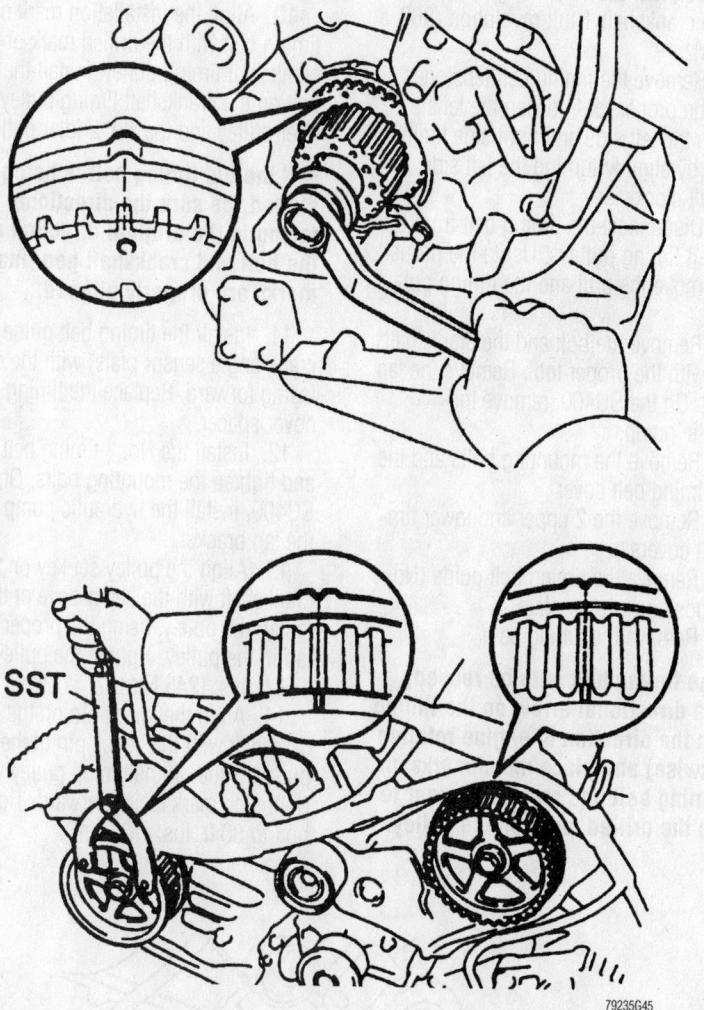

Camshaft and crankshaft pulley positioning for timing belt installation—Lexus 3.0L (1MZ-FE) engine

Turn the left pulley clockwise a bit and align the mark on the timing belt with the timing mark on the pulley. Slide the belt over the left pulley. Now move the pulley until the marks on it align with the one on the No. 3 cover. There should be tension on the belt between the crankshaft pulley and the left camshaft pulley.

12. Align the installation mark on the timing belt with the mark on the right side camshaft pulley. Hang the belt over the pulley with the flange facing inward. Align the timing marks on the right pulley with the one on the No. 3 cover and slide the pulley onto the end of the camshaft. Move the pulley until the camshaft knock pin hole is aligned with the groove in the pulley, then install the knock pin. Tighten the bolt to 55 ft. lbs. (75 Nm).

13. Position a plate washer between the timing belt tensioner and the block, then press in the pushrod until the holes are aligned between it and the housing. Slide a 0.05 in. Allen wrench through the hole to keep the pushrod set. Install the dust boot, then install the tensioner. Tighten the bolts to 20 ft. lbs. (26 Nm). Don't forget to pull out the Allen wrench.

14. Turn the crankshaft clockwise 2 complete revolutions and check that all marks are still in alignment. If they aren't, remove the timing belt and start over again.

15. Install the remaining components.

3.0L (2JZ-GE) Engine

1. Remove all necessary components for access to the upper timing belt covers. Using a 5mm Allen wrench, remove the 9 bolts and lift off the two upper (No. 2 and No. 3) timing belt covers.

2. Rotate the crankshaft pulley clockwise so its groove is aligned with the **0** mark in the No. 1 (lower) timing cover. Check that the timing marks on the camshaft timing sprockets are aligned with the marks on the No. 4 (inner) cover. If not, rotate the crankshaft 1 complete revolution (360 degrees).

3. Alternately loosen the 2 tensioner mounting bolts and remove them, the tensioner and the dust boot. Slide the timing belt off of the 2 camshaft sprockets. Its a good idea to matchmark the belt to the pulleys.

4. Ensuring the timing belt is securely supported, hold the crankshaft pulley with a spanner wrench and loosen the mounting bolt. Remove the bolt and the pulley.

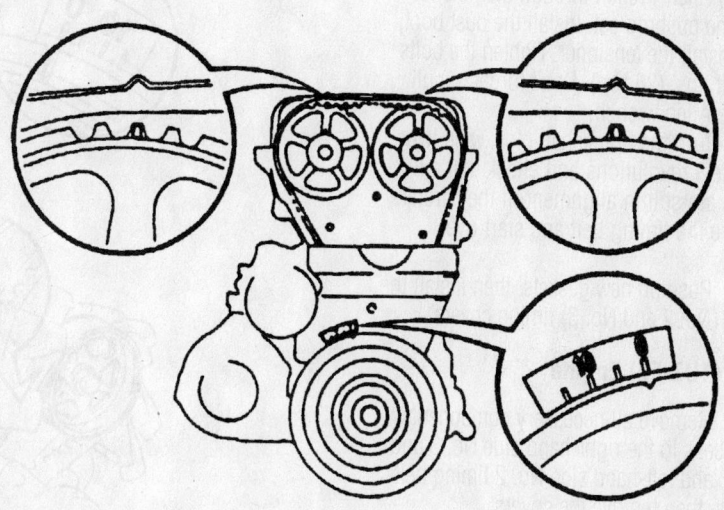

Set the engine to TDC by aligning the marks before removing the lower timing cover—Lexus 3.0L (2JZ-GE) engine

Timing belt service is covered in Section 3 of this manual

5. Remove the 5 bolts, then lift off the lower No. 1 timing belt cover.

6. Remove the timing belt guide.

7. Remove the timing belt.

➡ If the timing belt is to be reused, draw a directional arrow on the timing belt in the direction of engine rotation (clockwise) and place matchmarks on the timing belt and crankshaft gear to match the drilled mark on the pulley.

To install:

8. Install the timing belt on the crankshaft timing pulley and the idler pulleys.

➡ If the old timing belt is being reinstalled, be sure the directional arrow is facing in the original direction and that the belt and crankshaft gear matchmarks are properly aligned.

9. Install the timing belt guide. Install the lower (No. 1) timing cover and tighten the bolts.

10. Align the crankshaft pulley set key with the key groove on the pulley and slide the pulley on. Tighten the bolt to 239 ft. lbs. (324 Nm).

11. Set the No. 1 cylinder to TDC again. Turn the camshaft until the sprocket timing marks are aligned with the timing marks on the No. 4 belt cover.

12. Check that the marks on the belt matches with those on the sprockets, then slide it over the sprockets. If not, shift it on the crank pulley until it does.

13. Position a plate washer between the timing belt tensioner and the block, then press in the pushrod until the holes are aligned between it and the housing. Slide a 1.5mm Allen wrench through the hole to keep the pushrod set. Install the dust boot, then install the tensioner. Tighten the bolts to 20 ft. lbs. (26 Nm). Don't forget to pull out the Allen wrench.

14. Turn the crankshaft clockwise two complete revolutions and check that all marks are still in alignment. If they aren't, remove the timing belt and start over again.

15. Position new gaskets, then install the upper (No. 2 and No. 3) timing covers.

4.0L (1UZ-FE) Engine

1. Remove all necessary components for access to the right-hand side No. 3 and No. 2, and left-hand side No. 2 timing belt covers, then remove the covers.

2. Turn the crankshaft pulley and align it's groove with the timing mark **0** of the No. 1 timing cover. Check that the timing marks of the camshaft timing pulleys and

timing belt rear plates are aligned. If not, turn the crankshaft 1 full revolution (360 degrees).

3. Remove the timing belt tensioner. Using the proper tool, loosen the tension between the left side and right side timing pulleys by slightly turning the left side camshaft clockwise.

4. Disconnect the timing belt from the camshaft timing pulleys. Using the proper tool, remove the bolt and the timing pulleys.

5. Remove the bolt and the crankshaft pulley with the proper tool. Remove the fan bracket. On the SC400, remove the hydraulic pump.

6. Remove the mounting bolts and the No. 1 timing belt cover.

7. Remove the 2 upper and lower timing belt covers.

8. Remove the timing belt guide (No. 1 crank position sensor plate).

9. Remove the timing belt.

➡ If the timing belt is to be reused, draw a directional arrow on the timing belt in the direction of engine rotation (clockwise) and place matchmarks on the timing belt and crankshaft gear to match the drilled mark on the pulley.

To install:

10. Align the installation mark on the timing belt with the drilled mark of the crankshaft timing pulley. Install the timing belt on the crankshaft timing pulley, No. 1 idler pulley and the No. 2 idler pulley.

➡ If the old timing belt is being reinstalled, be sure the directional arrow is facing in the original direction and that the belt and crankshaft gear matchmarks are properly aligned.

11. Install the timing belt guide (No. 1 crank angle sensor plate) with the cup side facing forward. Replace the timing belt cover spacer.

12. Install the No. 1 timing belt cover and tighten the mounting bolts. On the SC400, install the hydraulic pump. Install the fan bracket.

13. Align the pulley set key on the crankshaft with the key groove of the pulley. Install the pulley, using the proper tool to tap in the pulley. Tighten the pulley bolt to 181 ft. lbs. (245 Nm).

14. Align the knock pin on the right side camshaft with the knock pin of the timing pulley. Slide on the timing pulley with the right side mark facing forward. Tighten the bolt to 80 ft. lbs. (108 Nm).

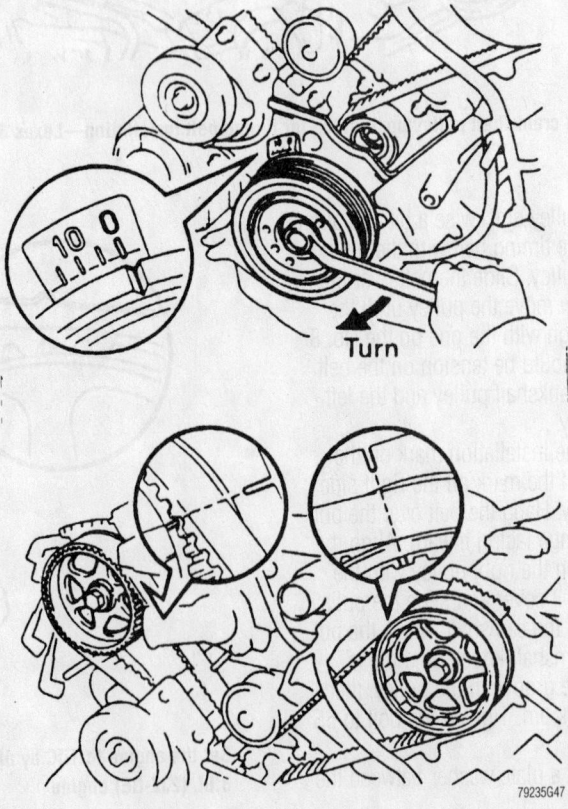

79235G47

Timing belt sprocket mark alignment for belt installation—Lexus 4.0L (1UZ-FE) engine

15. Align the knock pin on the left side camshaft with the knock pin of the timing pulley. Slide on the timing pulley with the left side mark facing forward. Tighten the bolt to 80 ft. lbs. (108 Nm).

16. Turn the crankshaft pulley and align its groove with the **0** timing mark on the No. 1 timing belt cover. Using the proper tool, turn the crankshaft timing pulley and align the timing marks of the camshaft timing pulley and the timing belt rear plate.

17. Install the timing belt to the left side camshaft timing pulley by:

a. Using the proper tool, slightly turn the left side timing pulley clockwise. Align the installation mark of the timing belt with the timing mark of the camshaft timing pulley and hang the timing belt on the left side camshaft pulley.

b. Using the proper tool, align the timing marks of the left side camshaft pulley and the timing belt rear plate.

c. Check that the timing belt has tension between crankshaft timing pulley and the left side camshaft pulley.

18. Install the timing belt to the right side camshaft timing pulley by:

a. Using the proper tool, slightly turn the right side timing pulley clockwise. Align the installation mark of the timing belt with the timing mark of the camshaft timing pulley and hang the timing belt on the right side camshaft pulley.

b. Using the proper tool, align the timing marks of the right side camshaft pulley and the timing belt rear plate.

c. Check that the timing belt has tension between the crankshaft timing pulley and the right side camshaft pulley.

19. The timing belt tensioner must be set prior to installation. The tensioner can be set as follows:

a. Place a plate washer between the tensioner and a block. Using a suitable press, press in the pushrod using 220–2205 lbs. (100–1000kg) of pressure.

b. Align the holes of the pushrod and housing, pass the proper tool (0.05 in. Allen wrench) through the holes to keep the setting position of the pushrod.

c. Release the press and install the dust boot on the tensioner.

20. Install the tensioner and tighten the bolts to 20 ft. lbs. (26 Nm). Remove the tool from the tensioner.

21. Turn the crankshaft pulley two complete revolutions from TDC-to-TDC. Always turn the crankshaft clockwise. Check that each pulley aligns with the timing marks.

22. Install all remaining components in the reverse order of removal.

MAZDA MODELS

Protégé 1.5L (Z5D) and 1.8L (BPD) Engines

1. Remove all necessary components for access to the timing belt covers, then remove the covers.

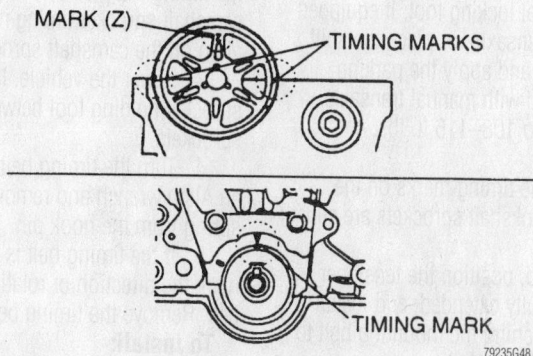

Proper timing belt sprocket mark alignment for belt installation—Mazda 1.5L (Z5D) engine

2. Turn the crankshaft until the timing mark on the crankshaft sprocket aligns with the timing mark on the oil pump and the camshaft sprocket timing marks align on the camshaft sprockets.

3. Remove the crankshaft pulley lock-bolt and pulley boss.

4. Lower the vehicle. Insert a camshaft sprocket holding tool between the camshaft sprockets.

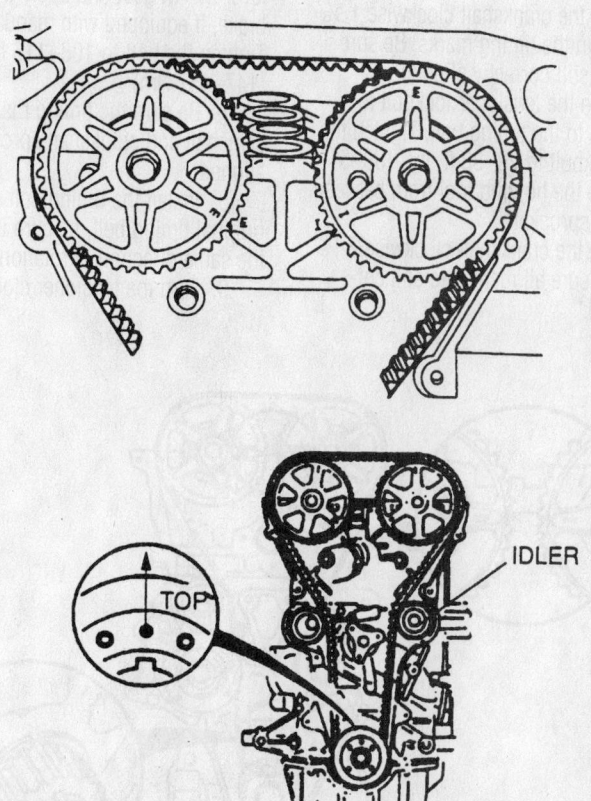

When properly positioning the timing belt sprocket marks, be sure the two I's are aligned and the two E's are aligned as indicated—Mazda Protégé and Miata 1.8L (BPD) engines

Heater Core replacement is covered in Section 2 of this manual

5. Loosen the tensioner pulley lockbolt. Pull the tensioner pulley away from the center of the engine to reduce the tension on the timing belt.

6. If the timing belt is to be reused, mark the direction of rotation on the timing belt. Remove the timing belt.

7. To remove the tensioner, unhook the tensioner spring, and remove the pulley lockbolt and tensioner.

To install:

8. Install the crankshaft sprocket bolt. Install the flywheel locking tool, if equipped with automatic transaxle, or place the shift lever in **4th** gear and apply the parking brake, if equipped with manual transaxle. Tighten the bolt to 108–116 ft. lbs. (147–157 Nm).

9. Be sure the timing marks on the camshaft and crankshaft sprockets are still aligned.

10. If removed, position the tensioner with the spring fully extended, and install the lockbolt tightening the mounting bolt to 28–38 ft. lbs. (38–51 Nm).

11. Install the timing belt. If reusing the original timing belt, be sure it is installed in the same direction of rotation.

12. Rotate the crankshaft clockwise 1 5/6 turns and align the timing marks. Be sure all marks are still correctly aligned.

13. Loosen the tensioner lockbolt to apply tension to the timing belt. Tighten the tensioner lockbolt to 28–38 ft. lbs. (38–51 Nm). Remove the holding tool from between the camshaft sprockets.

14. Rotate the crankshaft clockwise 2 1/6 turns and be sure all marks are still correctly aligned.

15. Raise and safely support the vehicle. Install the crankshaft pulley lockbolt and boss. Tighten the bolt to 116–122 ft. lbs. (157–166 Nm).

16. Install the timing belt covers.

2.0L (FS) Engine

1. Remove the timing belt covers. Temporarily reinstall the crankshaft pulley bolt.

2. Turn the crankshaft until the timing mark on the crankshaft sprocket aligns with the timing mark on the oil pump and the camshaft sprocket timing marks **E** and **I** align on the camshaft sprockets.

3. Lower the vehicle. Insert a camshaft sprocket holding tool between the camshaft sprockets.

4. Turn the timing belt tensioner with an Allen wrench and remove the tensioner spring from the hook pin.

5. If the timing belt is to be reused, mark the direction of rotation on the timing belt. Remove the timing belt.

To install:

6. Install the crankshaft sprocket bolt. Install the flywheel locking tool, if equipped with automatic transaxle, or place the shift lever in **4th** gear and apply the parking brake, if equipped with manual transaxle. Tighten the bolt to 108–116 ft. lbs. (147–157 Nm).

7. Be sure the timing marks on the camshaft and crankshaft sprockets are still aligned.

8. Install the timing belt. If reusing the original timing belt, be sure it is installed in the same direction of rotation.

9. Turn the tensioner clockwise with an

Allen wrench and install the tensioner spring. Remove the holding tool from between the camshaft sprockets.

10. Rotate the crankshaft 2 turns in the normal direction of rotation and align the timing marks. Be sure all marks are still correctly aligned.

11. Raise and safely support the vehicle. Remove the crankshaft pulley bolt and install the timing belt covers.

Miata 1.8L (BPD) Engine

1. Remove all necessary components for access to the valve cover, then remove the cover.

2. Remove the spark plugs.

➡**Spark plugs are removed to make it easier to rotate the engine.**

3. Remove the upper, middle and lower timing belt covers.

4. Turn the crankshaft until the timing marks on the crankshaft and camshaft sprockets are aligned. The pin on the pulley boss must face upward. Hold the crankshaft pulley boss with a suitable tool and remove the pulley lockbolt, being careful not to rotate the crankshaft. Remove the crankshaft pulley boss.

5. Mark the direction of rotation on the timing belt. Loosen the tensioner lockbolt and pry the tensioner outward. Tighten the lockbolt with the tensioner spring fully extended. Remove the timing belt.

➡**Protect the tensioner with a shop towel before prying on it. Do not rotate the crankshaft after the timing belt has been removed.**

6. Remove the tensioner and spring. If necessary, remove the idler pulley.

7. Inspect the belt for wear, peeling, cracking, hardening or signs of oil contamination. Inspect the tensioner for free and smooth rotation. Check the tensioner spring free length; it should not exceed 2.315 in. (58.8mm). Inspect the sprocket teeth for wear or damage. Replace parts, as necessary.

To install:

8. Install the crankshaft sprocket bolt. Install the flywheel locking tool, if equipped with automatic transaxle, or place the shift lever in **4th** gear and apply the parking brake, if equipped with manual transaxle. Tighten the bolt to 108–116 ft. lbs. (147–157 Nm).

9. If removed, install the idler pulley and tighten the bolt to 38 ft. lbs. (52 Nm).

10. Install the tensioner and tensioner spring. Pry the tensioner outward and tem-

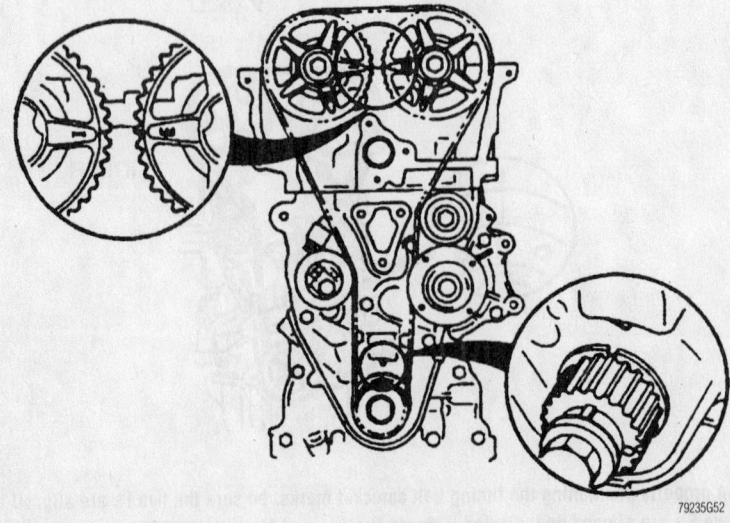

When properly aligned for belt removal, the cam gear marks should face each other—Mazda 2.0L (FS) engines

porarily tighten the tensioner lockbolt with the tensioner spring fully extended.

11. Be sure the crankshaft sprocket timing mark is aligned with the mark on the oil pump housing. Be sure the camshaft sprocket timing marks are aligned with the marks on the seal plate.

12. Install the timing belt so there is no looseness at the idler pulley side or between the camshaft sprockets. If reusing the old belt, be sure it is installed in the same direction of rotation.

13. Temporarily install the pulley boss and lockbolt.

14. Turn the crankshaft 2 turns clockwise and align the crankshaft sprocket timing mark. Face the pin on the pulley boss upright. Be sure the camshaft sprocket timing marks are aligned. If they are not, repeat the alignment steps.

15. Turn the crankshaft 1 ⅝ turns clockwise and align the crankshaft sprocket timing mark with the tension set mark for proper belt tension adjustment. Remove the lockbolt and pulley boss.

16. Be sure the crankshaft sprocket timing mark is aligned with the tension set mark. Loosen the tensioner lockbolt, and allow the spring to apply tension to the belt. Tighten the tensioner lockbolt to 28–38 ft. lbs. (38–52 Nm).

17. Install the pulley boss and lockbolt.

18. Turn the crankshaft 2 ⅙ turns clockwise and be sure the timing marks are correctly aligned.

19. Apply approximately 22 lbs. (10kg) pressure to the timing belt at a point midway between the camshaft sprockets. The belt should deflect 0.35–0.45 in. (9.0–11.5mm). If the deflection is not correct, repeat the alignment and tensioning procedure.

20. Hold the pulley boss with a suitable tool, and tighten the lockbolt to 123 ft. lbs. (167 Nm).

21. Install the timing belt covers and tighten the bolts to 95 inch lbs. (11 Nm).

22. Install the valve cover and spark plugs, along with all other applicable components.

2.5L (KL) Engines

1. Remove the timing belt covers. Temporarily reinstall the crankshaft pulley bolt.

2. On Millenia models, support the engine, and remove the nuts and through-bolt from the right side (No. 3) engine mount sub bracket. Remove the sub bracket.

3. Turn the crankshaft until the timing

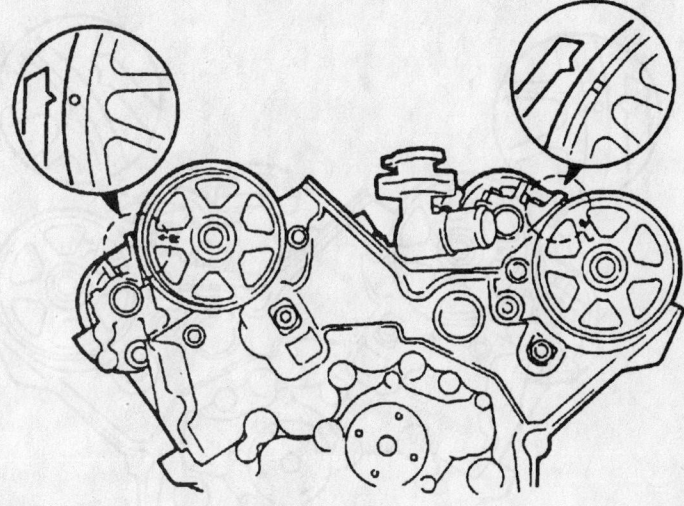

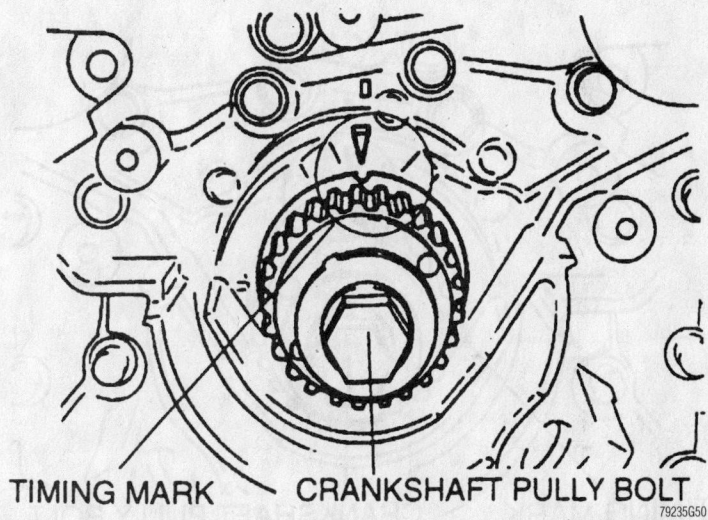

TIMING MARK — CRANKSHAFT PULLY BOLT

79235G50

Timing belt sprocket positioning for proper timing belt installation—Mazda 626/MX6 2.5L (KL) engines

mark on the crankshaft sprocket aligns with the timing mark on the oil pump and the camshaft sprocket timing marks align with the marks on the cylinder head. The No. 1 piston should be at Top Dead Center (TDC) of the compression stroke.

4. Remove the two bolts from the automatic tensioner, removing the lower one first. Keep the bolt holes aligned by holding the tensioner to reduce the chance of stripping the threads on the bolts.

5. If the timing belt is to be reused, mark the direction of rotation on the timing belt.

6. Remove the number one idler pulley. Remove the timing belt.

To install:

7. Install the crankshaft sprocket bolt. Install the flywheel locking tool. Tighten the bolt to 116–122 ft. lbs. (157–166 Nm). remove the flywheel locking tool.

8. Position the automatic tensioner in a suitable press. Set a flat washer under the tensioner body to prevent damage to the body plug.

9. Compress the tensioner until the hole in the piston is aligned with the 2nd hole in the tensioner case. Insert a 0.060 in. (1.6mm) diameter wire or pin through the 2nd hole to keep the piston compressed.

10. Be sure the camshaft sprocket timing marks are still aligned. Turn the crankshaft counterclockwise until the timing sprocket is aligned.

Brake service is covered in Section 4 of this manual

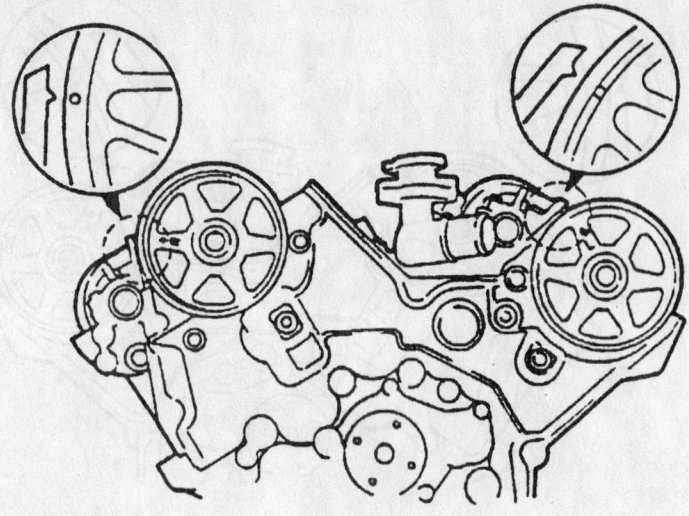

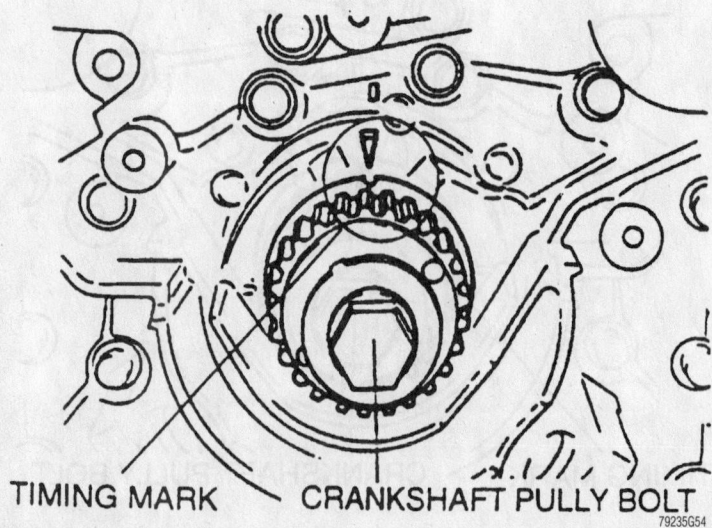

TIMING MARK   CRANKSHAFT PULLY BOLT

Position the crankshaft and camshaft sprockets as shown for belt installation—Mazda Millennia 2.5L (KL) engine

11. With the No. 1 idler pulley removed, install the timing belt. If the original belt is being reused, be sure it is installed in the same direction of rotation. The order of installation is: timing belt (crankshaft) sprocket, No. 2 idler pulley, left-hand camshaft sprocket, tensioner pulley and right-hand camshaft sprocket.

12. Install the No. 1 idler pulley while applying pressure on the timing belt. Tighten the bolt to 28–38 ft. lbs. (38–51 Nm).

13. Install the automatic belt tensioner and tighten the bolts to 14–18 ft. lbs. (19–25 Nm). Remove the wire or pin from the tensioner.

14. Turn the crankshaft clockwise, until the crankshaft sprocket timing mark is again at TDC. This should place all of the belt slack in the automatic tensioner portion of the belt.

15. Rotate the crankshaft 2 turns in the normal direction of rotation and align the timing marks. Be sure all marks are still correctly aligned.

16. Inspect timing belt deflection, 0.24–0.31 in. (6–8mm), between the crankshaft sprocket and the tensioner pulley. If it is out of specification, replace the auto-tensioner.

17. On Millenia models, install the right side (No. 3) engine mount sub-bracket. Tighten the nuts to 55–77 ft. lbs. (75–104 Nm) and the through-bolt to 63–86 ft. lbs. (86–116 Nm). Remove the engine support.

18. Remove the crankshaft damper bolt and install the timing belt covers.

2.3L (KJ) Engine

1. Remove the timing belt covers. Temporarily reinstall the crankshaft pulley bolt.

2. Remove the power steering auto-tensioner and pulley.

3. Turn the crankshaft until the timing mark on the crankshaft sprocket aligns with the timing mark on the oil pump and the camshaft sprocket timing marks align with the marks on the cylinder head. The No. 1 piston should be at Top Dead Center (TDC) of the compression stroke.

4. Remove the two bolts from the automatic tensioner, removing the lower one first. Keep the bolt holes aligned by holding the tensioner to reduce the chance of stripping the threads on the bolts.

5. If the timing belt is to be reused, mark the direction of rotation on the timing belt.

6. Remove the timing belt.

To install:

7. Install the crankshaft sprocket bolt. Install the flywheel locking tool. Tighten the bolt to 116–122 ft. lbs. (157–166 Nm). Remove the flywheel locking tool.

8. Position the automatic tensioner in a press. Set a flat washer under the tensioner body to prevent damage to the body plug.

9. Compress the tensioner until the hole in the piston is aligned with the 2nd hole in the tensioner case. Insert a 0.063 in. (1.6mm) diameter wire or pin through the 2nd hole to keep the piston compressed.

10. Be sure the camshaft sprocket timing marks are still aligned. Turn the crankshaft clockwise until the timing sprocket is aligned.

11. Install the timing belt. If the original belt is being reused, be sure it is installed in the same direction of rotation. The order of installation is: timing belt (crankshaft) sprocket, No. 2 idler pulley, left-hand camshaft sprocket, both No. 1 idler pulleys, right-hand camshaft sprocket and the tensioner pulley.

12. Install the automatic belt tensioner and tighten the bolts to 14–18 ft. lbs. (19–25 Nm). Remove the wire or pin from the tensioner.

13. Turn the crankshaft clockwise, until the crankshaft sprocket timing mark is again at TDC. This should place all of the belt slack in the automatic tensioner portion of the belt.

14. Rotate the crankshaft two turns in the

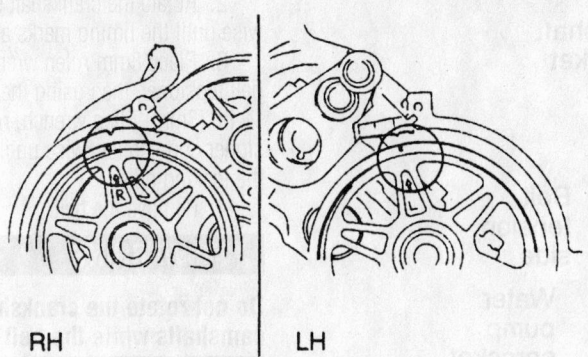

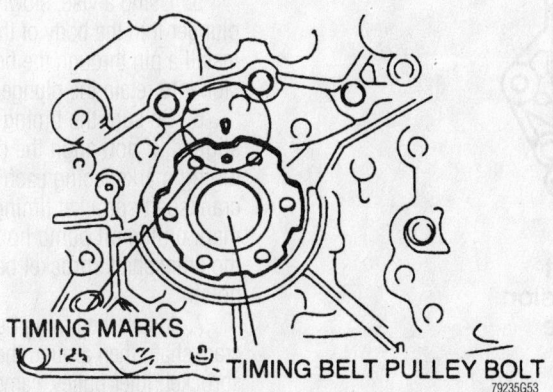

Proper crankshaft and camshaft timing belt sprocket alignment mark positioning—Mazda 2.3L (KJ) engines

normal direction of rotation and align the timing marks. Be sure all marks are still correctly aligned.

15. Inspect timing belt deflection, 0.24–0.31 in. (6–8mm), between the crankshaft sprocket and the tensioner pulley. If it is out of specification, replace the auto-tensioner.

16. Install the power steering auto-tensioner and tighten the bolts to 14–18 ft. lbs. (19–25 Nm). Install the pulley, and tighten the bolt to 29–34 ft. lbs. (40–47 Nm).

17. Remove the crankshaft damper bolt and install the timing belt covers.

MITSUBISHI

1.5L and 1.8L Engines

1. Remove the timing belt upper and lower covers.

2. Make a mark on the back of the timing belt indicating the direction of rotation so it may be reassembled in the same direction if it is to be reused. Loosen the timing belt tensioner and move the tensioner to provide slack to the

timing belt. Tighten the tensioner in this position.

3. Remove the timing belt.

✳✳ WARNING

Coolant and engine oil will damage the rubber in the timing belt, drastically reducing its life. Do not allow engine oil or coolant to contact the timing belt, the sprockets or tensioner assembly.

4. If defective, replace the tensioner spacer, tensioner spring and tensioner assembly.

To install:

5. Position the tensioner, tensioner spring and tensioner spacer on engine block.

6. Align the timing marks on the camshaft sprocket and crankshaft sprocket. This will position No. 1 piston on Top Dead Center (TDC) on the compression stroke.

7. Position the timing belt on the crankshaft sprocket and keeping the tension side of the belt tight, set it on the camshaft sprocket, then the tensioner.

8. Apply slight counterclockwise force to the camshaft sprocket to give tension to the belt and be sure all timing marks are aligned.

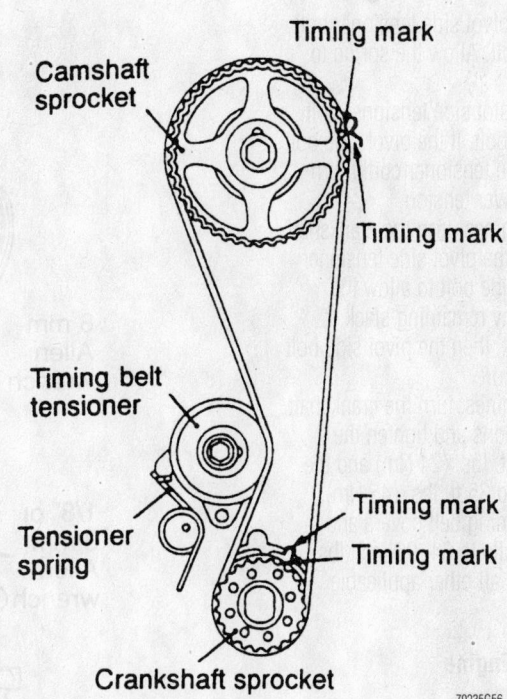

Timing mark locations with engine at Top Dead Center (TDC) of compression stroke—Mitsubishi 1.5L (4G15) engine

For complete Engine Mechanical specifications, see Section 1 of this manual

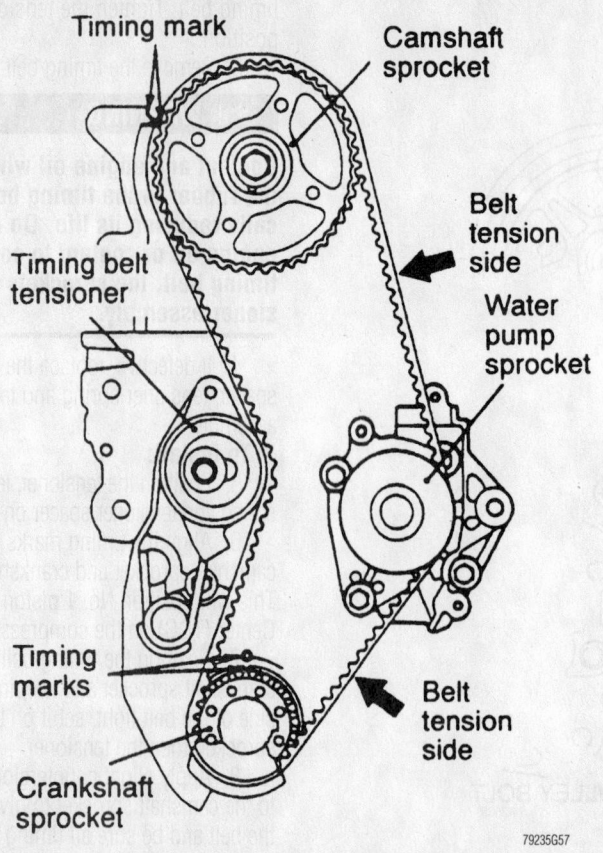

Timing mark locations with engine at Top Dead Center (TDC) of compression stroke—Mitsubishi 1.8L (4G93) engine

2. Rotate the crankshaft sprocket clockwise until the timing marks are aligned.

3. Place 8mm Allen wrench into the belt tensioner, then using the long end of a ⅛ in. (3mm) Allen wrench, rotate the tensioner counterclockwise until it slides into the locking hole.

4. Remove the belt.

✳✳ WARNING

Do not rotate the crankshaft or the camshafts while the belt is removed.

To install:

5. Using a vise, slowly compress the plunger into the body of the tensioner and install a pin through the body of the tensioner to retain the plunger.

6. Be sure the timing marks are still aligned, if not, align the camshaft sprocket timing marks facing each other. Align the crankshaft sprocket timing mark with the mark on the oil pump housing, then turn the crankshaft sprocket backward ½ notch.

7. Install the timing belt, starting at the crankshaft, then around the water pump sprocket, idler pulley, camshaft sprockets and the tensioner pulley.

8. Turn the crankshaft sprocket ½ notch to Top Dead Center (TDC) to take up the slack in the belt.

9. Loosen the pivot side tensioner bolt and the slot side bolt. Allow the spring to remove the slack.

10. Tighten the slot side tensioner bolt, then the pivot side bolt. If the pivot side bolt is tightened first, the tensioner could turn with bolt, causing over tension.

11. For 1.5L engines, turn the crankshaft clockwise. Loosen the pivot side tensioner bolt, then the slot side bolt to allow the spring to take up any remaining slack. Tighten the slot bolt, then the pivot side bolt to 17 ft. lbs. (24 Nm).

12. For 1.8L engines, turn the crankshaft clockwise two rotations and tighten the adjuster bolt to 18 ft. lbs. (24 Nm) and the pivot (spring) bolt to 35 ft. lbs. (45 Nm).

13. Install the timing belt covers and tighten the cover bolts to 84–96 inch lbs. (10–11 Nm). Install all other applicable components.

2.0L Non-Turbo Engine

1. Remove the front timing belt cover.

➡ If the timing belt is going to be reused, mark the direction of rotation on the belt with an arrow. Install the belt in the same direction.

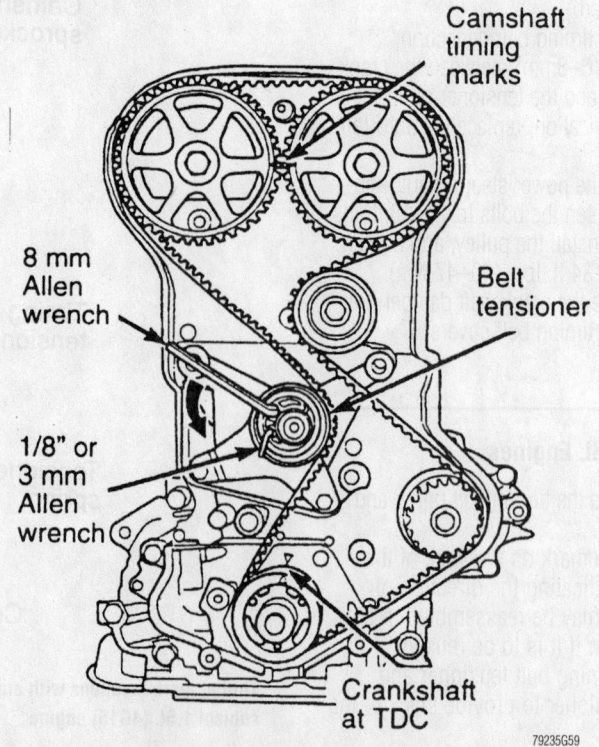

Timing belt sprocket mark alignment for belt service—Mitsubishi 2.0L non-turbo engines

9. Install the tensioner on the engine, but do not tighten the bolts.

10. Place a torque wrench on the tensioner pulley and apply 21 ft. lbs. (28 Nm) of torque in the direction of the water pump. Push the tensioner up against the tensioner pulley and tighten the mounting bolts to 23 ft. lbs. (31 Nm).

11. Pull the pin out of the tensioner. Belt tension is correct when the pin can be removed and installed.

12. Rotate the crankshaft two revolutions and check the timing marks for alignment. Repeat the previous steps, if necessary.

13. Install the timing belt cover and all other applicable components.

2.0L Turbo Engine

1. Remove all necessary components for access to the timing belt covers.

2. Remove the stud bolt from the engine support bracket and remove the timing belt covers.

3. Rotate the crankshaft clockwise to align the camshaft timing marks. Always turn the crankshaft in the normal direction of rotation only.

4. Loosen the tension pulley center bolt.

➡ **If the timing belt is to be reused, mark the direction of rotation on the flat side of the belt with an arrow.**

5. Move the tension pulley towards the water pump and remove the timing belt.

6. Remove the crankshaft sprocket center bolt using special tool MB990767 to hold the crankshaft sprocket while removing the center bolt. Then, use MB998778 puller to remove the sprocket.

7. Mark the direction of rotation on the timing belt "B" with an arrow.

8. Loosen the center bolt on the tensioner and remove the belt.

❄❄ WARNING

Do not rotate the camshafts or the crankshaft while the timing belt is removed.

To install:

9. Place the crankshaft sprocket on the crankshaft. Use tool MB990767 to hold the crankshaft sprocket while tightening the center bolt. Tighten the center bolt to 80–94 ft. lbs. (108–127 Nm).

10. Align the timing marks on the crankshaft sprocket "B" and the balance shaft.

11. Install timing belt "B" on the sprock-

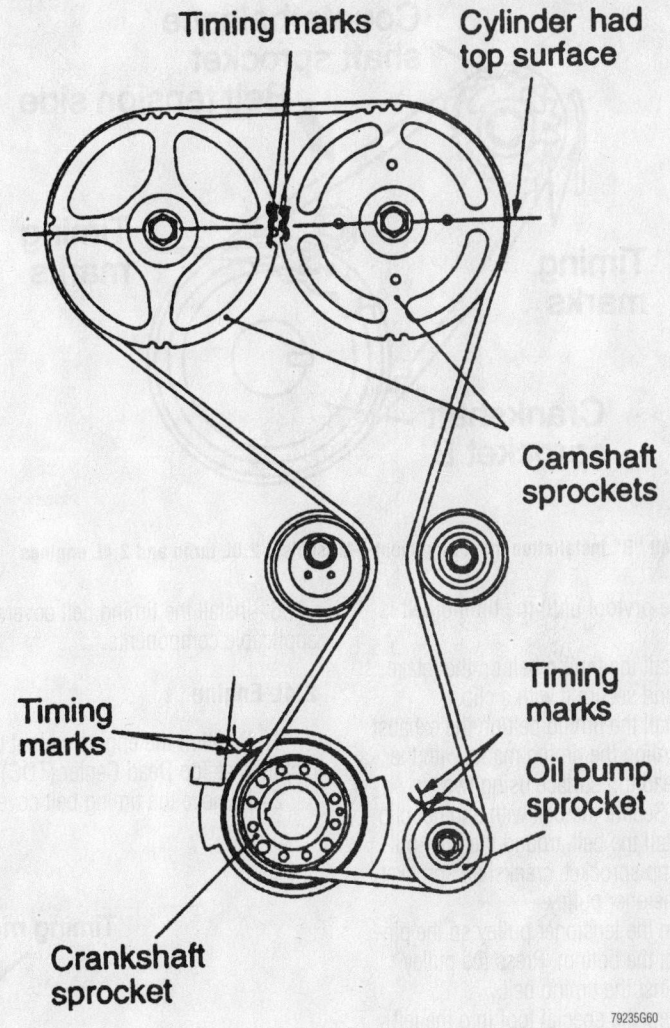

Camshaft and crankshaft timing belt sprocket TDC alignment mark positioning for timing belt removal and installation—Mitsubishi 2.0L turbo engine

ets. Position the center of the tensioner pulley to the left and above the center of the mounting bolt.

12. Push the pulley clockwise toward the crankshaft to apply tension to the belt and tighten the mounting bolt to 14 ft. lbs. (19 Nm). Do not let the pulley turn when tightening the bolt because it will cause excessive tension on the belt. The belt should deflect 0.20–0.28 in. (5–7mm) when finger pressure is applied between the pulleys.

13. Install the crankshaft sensing blade and the crankshaft sprocket. Apply engine oil to the mounting bolt and tighten the bolt to 80–94 ft. lbs. (108–127 Nm).

14. Use a press or vise to compress the auto-tensioner pushrod. Insert a set pin when the holes are lined up.

❄❄ WARNING

Do not compress the pushrod too quickly, damage to the pushrod can occur.

15. Install the auto-tensioner on the engine.

16. Align the timing marks on the camshaft sprocket, crankshaft sprocket and the oil pump sprocket.

17. After aligning the mark on the oil pump sprocket, remove the cylinder block plug and insert a prytool in the hole to check the position of the counterbalance shaft. The prytool should go in at least 2.36 in. (60mm) or more, if not, rotate the oil pump sprocket once and realign the timing mark so the prytool goes in. Do not

For Accessory Drive Belt illustrations, see Section 1 of this manual

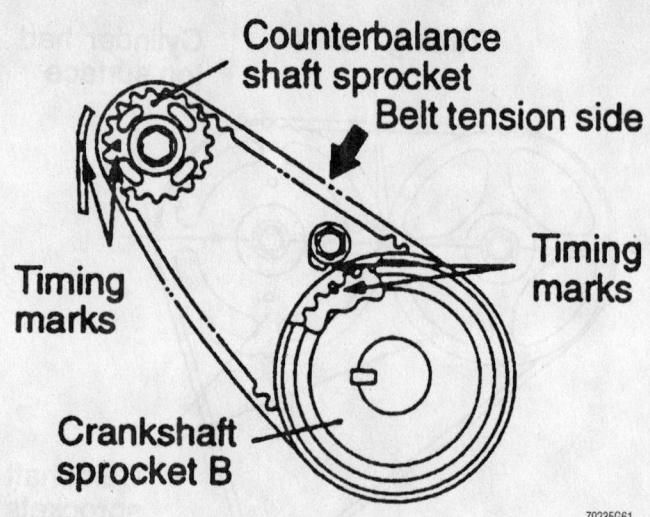

Counterbalance shaft sprocket

Belt tension side

Timing marks

Timing marks

Crankshaft sprocket B

79235G61

Timing belt "B" installation mark alignment—Mitsubishi 2.0L turbo and 2.4L engines

remove the prytool until the timing belt is installed.

18. Install the timing belt on the intake camshaft and secure it with a clip.

19. Install the timing belt on the exhaust camshaft. Align the timing marks with the cylinder head top surface using two wrenches. Secure the belt with another clip.

20. Install the belt around the idler pulley, oil pump sprocket, crankshaft sprocket and the tensioner pulley.

21. Turn the tensioner pulley so the pinholes are at the bottom. Press the pulley lightly against the timing belt.

22. Screw the special tool into the left engine support bracket until it contacts the tensioner arm, then screw the tool in a little more and remove the pushrod pin from the auto-tensioner. Remove the special tool and tighten the center bolt to 35 ft. lbs. (48 Nm).

23. Turn the crankshaft ¼ turn counterclockwise, then clockwise until the timing marks are aligned.

24. Loosen the center bolt. Install Mitsubishi Special tool MD998767 on the tensioner pulley. Turn the tensioner pulley counterclockwise with a torque of 2.6 ft. lbs. (3.5 Nm) and tighten the center bolt to 35 ft. lbs. (48 Nm). Do not let the tensioner pulley turn when tightening the bolt.

25. Turn the crankshaft clockwise two revolutions and align the timing marks. After 15 minutes, measure the protrusion of the pushrod on the auto-tensioner. The standard measurement is 0.150–0.177 in (3.8–4.5mm). If the protrusion is out of specification, loosen the tensioner pulley, apply the proper torque to the belt and retighten the center bolt.

26. Install the timing belt covers and all applicable components.

2.4L Engine

1. Position the engine so that the No. 1 piston is at Top Dead Center (TDC).

2. Remove the timing belt covers.

→If the timing belts are going to be reused, mark the direction of rotation on the belt. This will ensure the belt is reinstalled in same direction, extending belt life.

3. To loosen the timing (outer) belt tensioner, install Mitsubishi Special tool MD998738 to the slot and screw inward to move the tensioner toward the water pump. Once the tension has been relieved, remove the outer timing belt.

4. If tensioner replacement is required, align the pin hole in the tensioner rod to the hole in the tensioner cylinder. Insert a 0.055 in. (1.4mm) wire in the hole and remove the special tool from the slot. With the cylinder tension relieved, remove the auto-tensioner cylinder assembly two mounting bolts.

5. Remove the outer crankshaft sprocket and flange.

6. Loosen the silent shaft (inner) belt tensioner and remove the belt.

To install:

❄❄ WARNING

Do not spray or immerse the sprockets or tensioners in cleaning solvent. The sprocket may absorb the solvent

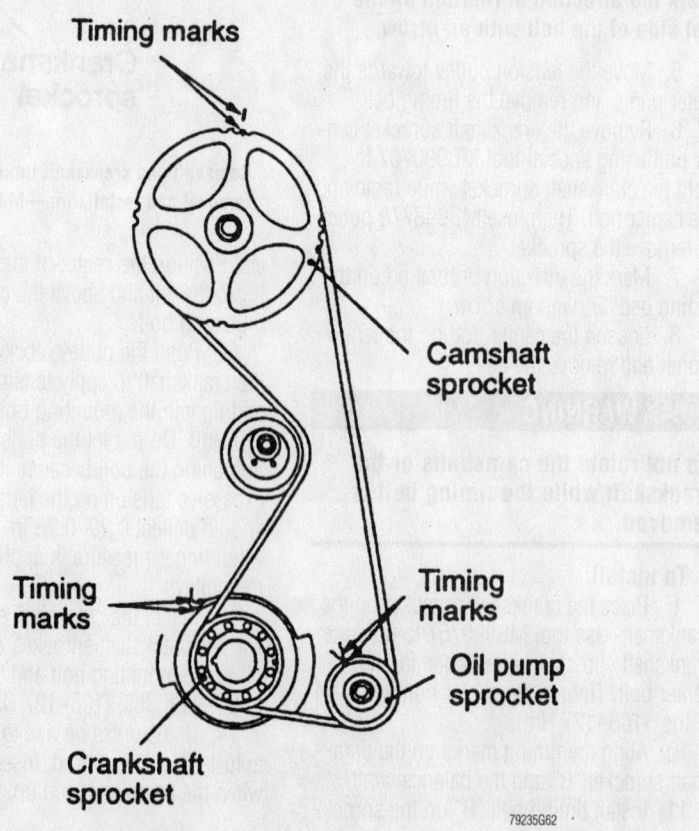

Timing marks

Camshaft sprocket

Timing marks

Timing marks

Oil pump sprocket

Crankshaft sprocket

79235G62

Proper alignment of the timing belt sprocket marks for belt service—Mitsubishi 2.4L engines

and transfer it to the belt. The tensioners are internally lubricated and the solvent will dilute or dissolve the lubricant.

7. Align the timing marks of the silent shaft sprockets and the crankshaft sprocket with the timing marks on the front case. Route the timing belt around the sprockets so there is no slack in the upper span of the belt and the timing marks are still aligned.

8. Install the tensioner pulley and move the pulley by hand so the long side of the belt deflects approximately ¼ in. (6mm).

9. Hold the pulley tightly so the pulley cannot rotate when the bolt is tightened. Tighten the bolt to 14 ft. lbs. (19 Nm) and recheck the deflection.

10. Align the timing marks of the camshaft, crankshaft and oil pump sprockets with their corresponding marks on the front case or rear cover.

➡There is a possibility to align all timing marks and have the oil pump sprocket and silent shaft out of time, causing an engine vibration during operation. If the following step is not followed exactly, there is a 50 percent chance that the silent shaft alignment will be 180 degrees (½ turn) off.

11. Before installing the timing belt, ensure that the left side (rear) silent shaft (oil pump sprocket) is in the correct position as follows:

a. Remove the plug from the rear side of the block and insert a tool with shaft diameter of 0.31 in. (8mm) into the hole.

b. With the timing marks still aligned, the shaft of the tool must be able to go in at least 2 ½ in. (63.5mm). If the tool can only go in approximately 1 in. (25mm), the shaft is not in the correct orientation and will cause a vibration during engine operation. Remove the tool from the hole and turn the oil pump sprocket one complete revolution. Realign the timing marks and insert the tool. The shaft of the tool must go in at least 2 ¼ in. (63.5mm).

c. Recheck and realign the timing marks.

d. Leave the tool in place to hold the silent shaft while continuing.

12. If the camshaft belt tensioner was removed, use a vise to carefully push the auto-tensioner rod in until the set hole in

the rod is aligned with the hole in the cylinder. Place a wire into the hole to retain the rod. Mount the tensioner to the engine block and tighten the mounting bolt to 17 ft. lbs. (23 Nm).

13. Install the belt to the crankshaft sprocket, oil pump sprocket, then camshaft sprocket, in that order. While doing so, be sure there is no slack between the sprocket except where the tensioner is installed.

14. To adjust the timing (outer) belt perform the following steps:

a. Turn the crankshaft ¼ turn counterclockwise, then turn it clockwise to move No. 1 cylinder to TDC.

b. Loosen the center bolt. Using tool MD998752 and a torque wrench, apply a torque of 2.6 ft. lbs. (3.6 Nm) to the tensioner. Tighten the center bolt.

c. Screw the special tool into the engine left support bracket until its end makes contact with the tensioner arm. At this point, screw the special tool in some more and remove the set wire attached to the auto-tensioner, if the wire was not previously removed. Then, remove the special tool.

d. Rotate the crankshaft two complete turns clockwise and let it sit for approximately 15 minutes. Then, measure the auto-tensioner protrusion (the distance between the tensioner arm and auto-tensioner body) to ensure that it is within 0.15–0.18 in. (3.8–4.5mm). If out of specification, repeat substeps **a** through **d** until the specified value is obtained.

➡Do not manually overtighten the belt or it will howl.

15. Install the upper and lower timing belt covers.

3.0L (6G72) SOHC Engine

1. Position the engine so the No. 1 cylinder is at Top Dead Center (TDC) of its compression stroke.

❋❋ CAUTION

Wait at least 90 seconds after the negative battery cable is disconnected to prevent possible deployment of the air bag.

2. Remove all necessary components for access to the timing belt covers, then remove the covers from the engine.

3. If the same timing belt will be

reused, mark the direction of the timing belt's rotation for installation in the same direction. Be sure the engine is positioned so the No. 1 cylinder is at the TDC of its compression stroke and the timing marks are aligned with the engine's timing mark indicators.

4. Loosen the timing belt tensioner bolt and remove the belt. If the tensioner is not being removed, position it as far away from the center of the engine as possible and tighten the bolt.

5. If the tensioner is being removed, mark the outside of the spring to ensure that it is not installed backwards. Unbolt the tensioner and remove it along with the spring.

❋❋ WARNING

Do not rotate the camshafts when the timing belt is removed from the engine. Turning the camshaft when the timing belt is removed could cause the valves to interfere with the pistons thus causing severe internal engine damage.

To install:

6. Install the tensioner, if removed, and hook the upper end of the spring to the water pump pin and the lower end to the tensioner in exactly the same position as originally installed.

7. Ensure both camshafts are still positioned so the timing marks align with those on the rear timing covers. Rotate the crankshaft so the timing mark aligns with the mark on the front cover.

8. Install the timing belt on the crankshaft sprocket and while keeping the belt tight on the tension side, install the belt on the front (left) camshaft sprocket.

9. Install the belt on the water pump pulley, then the rear (right) camshaft sprocket and the tensioner.

10. Loosen the bolt that secures the adjustment of the tensioner and lightly press the tensioner against the timing belt.

11. Check that the timing marks are in alignment.

12. Rotate the crankshaft 2 full turns in the clockwise direction only, then realign the timing marks.

13. Tighten the bolt that secures the tensioner to 19 ft. lbs. (26 Nm).

14. Install the lower and the upper timing belt covers, along with all other applicable components.

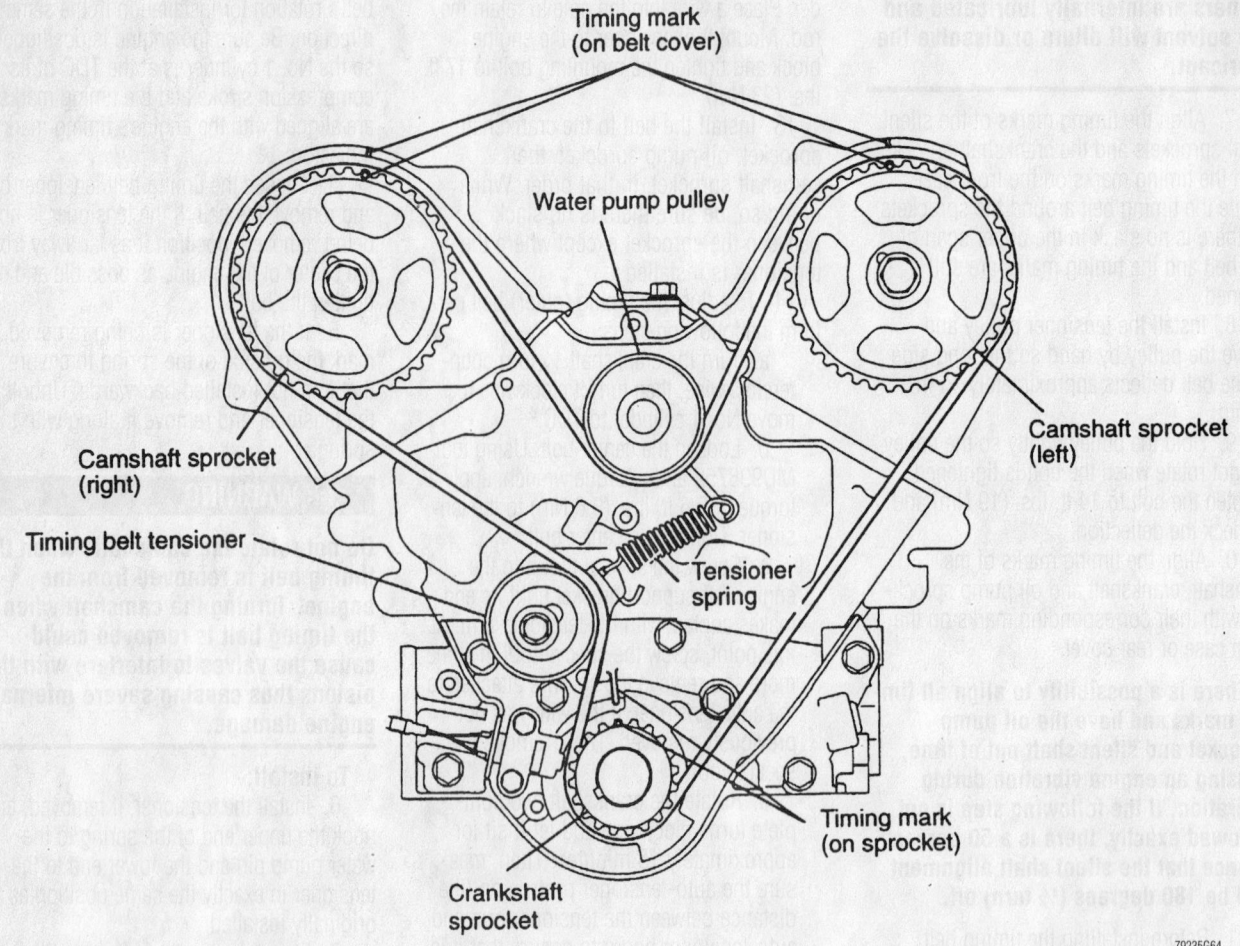

Align the sprockets properly before removing or installing the timing belt—Mitsubishi 3000 GT and Diamante with the 3.0L (6G72) SOHC engine

3.0L (6G72) DOHC Engine

1. Position the engine so the No. 1 cylinder is at Top Dead Center (TDC) of its compression stroke.

2. Remove all necessary components for access to the timing belt covers, then remove the covers from the engine.

✳✳ CAUTION

Be sure to disconnect the negative battery cable. Wait at least 90 seconds after the negative battery cable is disconnected to prevent possible deployment of the air bag.

3. If the same timing belt will be reused, mark the direction of the timing belt's rotation for installation in the same direction. Be sure the engine is positioned so the No. 1 cylinder is at the TDC of its compression stroke and the timing marks are aligned with the engine's timing mark indicators on the rear timing covers.

4. Remove the timing belt.

✳✳ WARNING

Turning the camshaft sprocket when the timing belt is removed could cause the valves to contact with the pistons, resulting in severe engine damage.

5. Remove the bolts that secure the auto-tensioner to the engine block and remove the tensioner.

To install:

➡**The auto-tensioner assembly must be reset to correctly adjust belt tension.**

6. Loosen the center bolt of tensioner pulley to provide timing belt slack. Remove the timing belt assembly.

7. Position the auto-tensioner into a vise with soft jaws. The plug at the rear of tensioner protrudes, be sure to use a washer as a spacer to protect the plug from contacting vise jaws.

8. Slowly push the rod into the ten-

sioner until the set hole in rod is aligned with set hole in the auto-tensioner.

9. Insert a 0.055 in. (1.4mm) wire into the aligned set holes. Unclamp the tensioner from the vise and install it on the engine. Tighten tensioner mounting bolts to 17 ft. lbs. (24 Nm).

✳✳ WARNING

DO NOT rotate or turn the camshafts when removing the sprockets or severe engine damage will result from internal component interference.

10. Align the mark on the crankshaft sprocket with the mark on the front case. Then, move the crankshaft sprocket 1 tooth counterclockwise.

11. Align the timing marks of the camshafts with the marks on the rear covers.

12. Using large paper clips to secure the timing belt to the sprockets, install the timing belt in the following order. Be sure

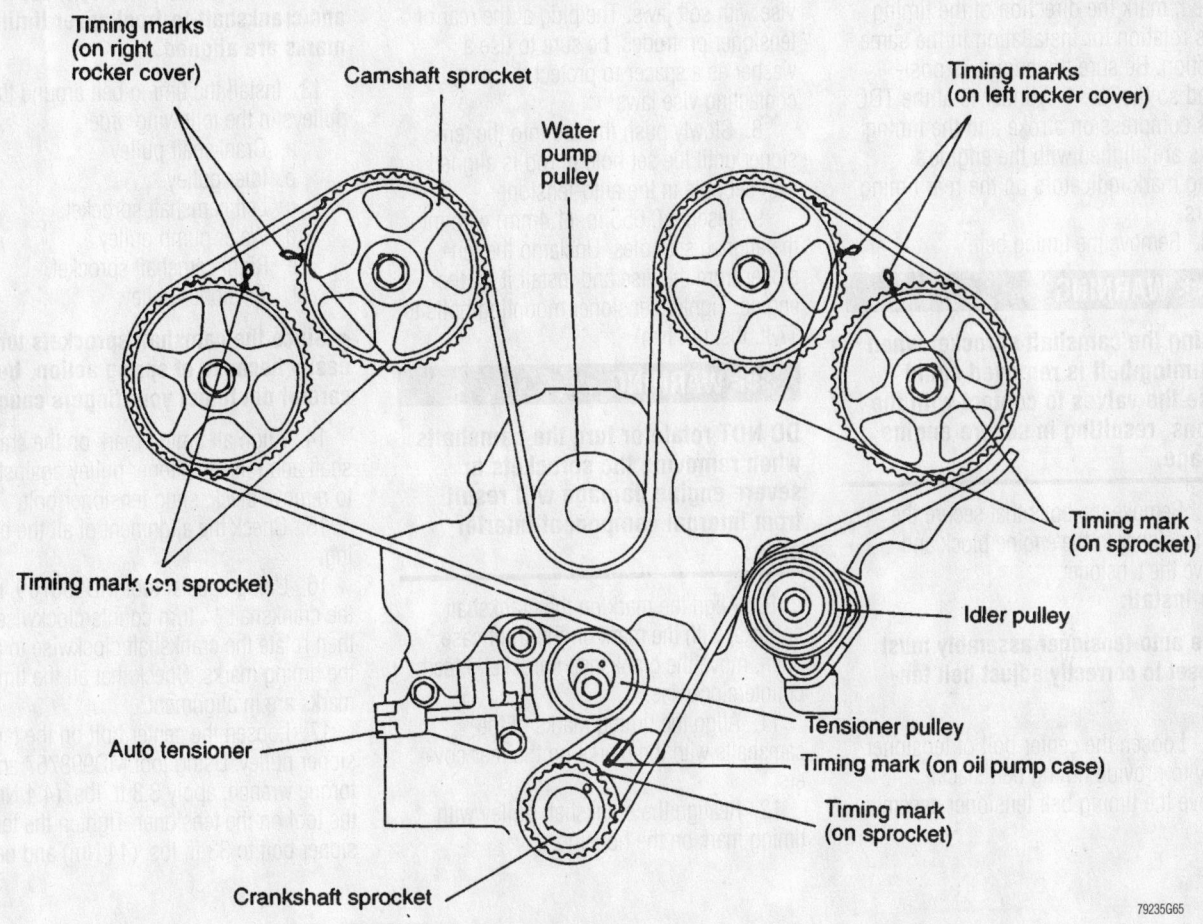

Timing marks (on right rocker cover)

Camshaft sprocket

Water pump pulley

Timing marks (on left rocker cover)

Timing mark (on sprocket)

Timing mark (on sprocket)

Idler pulley

Auto tensioner

Tensioner pulley

Timing mark (on oil pump case)

Timing mark (on sprocket)

Crankshaft sprocket

79235G65

Sprocket alignment for timing belt installation—Mitsubishi 3000 GT and Diamante with the 3.0L (6G72) DOHC engine

camshafts-to-cylinder heads and crankshaft-to-front cover timing marks are aligned. Install the timing belt around the pulleys in the following order:

   a. Exhaust camshaft sprocket (front bank).

   b. Intake camshaft sprocket (front bank).

   c. Water pump pulley.

   d. Intake camshaft sprocket (rear bank).

   e. Exhaust camshaft sprocket (rear bank).

   f. Tensioner pulley.

   g. Crankshaft pulley.

   h. Idler pulley.

➡**Since the camshaft sprockets turn easily, secure them with box wrenches when installing the timing belt.**

13. Align all timing mark on the crankshaft and raise tensioner pulley against belt to remove slack, snug tensioner bolt.

14. Check the alignment of all the timing marks and remove the clips that secure the timing belt to the camshaft sprockets.

15. Rotate the engine ¼ turn counterclockwise, then rotate the engine clockwise to align the timing marks. Check that all the timing marks are in alignment.

16. Loosen the center bolt on the tensioner pulley. Using tool MD998752 and a torque wrench, apply 84 inch lbs. (10 Nm) to the tool on the tensioner. Tighten the tensioner bolt to 35 ft. lbs. (49 Nm) and be sure the tensioner does not rotate with the bolt.

17. Rotate the crankshaft two complete turns clockwise and let it sit for approximately five minutes. Then, check that the set pin can easily be inserted and removed from the hole in the auto-tensioner.

18. Remove the set wire attached to the auto-tensioner.

19. Measure the auto-tensioner protrusion (the distance between the tensioner arm and auto-tensioner body) to ensure that

it is within 0.15–0.18 in. (3.8–4.5mm). If out of specification, repeat adjustment procedure until the specified value is obtained.

20. Check again that the timing marks on all sprockets are in proper alignment.

21. Install the timing belt covers and all other applicable components.

3.5L (6G74) SOHC Engine

1. Position the engine so the No. 1 cylinder is at Top Dead Center (TDC) of its compression stroke.

2. Remove all necessary components for access to the timing belt covers, then remove the covers from the engine.

✹✹ CAUTION

Be sure to disconnect the negative battery cable. Wait at least 90 seconds after the negative battery cable is disconnected to prevent possible deployment of the air bag.

For Wheel Alignment specifications, see Section 1 of this manual

3. If the same timing belt will be reused, mark the direction of the timing belt's rotation for installation in the same direction. Be sure the engine is positioned so the No. 1 cylinder is at the TDC of its compression stroke and the timing marks are aligned with the engine's timing mark indicators on the rear timing covers.

4. Remove the timing belt.

※※ WARNING

Turning the camshaft sprocket when the timing belt is removed could cause the valves to contact with the pistons, resulting in severe engine damage.

5. Remove the bolts that secure the auto-tensioner to the engine block and remove the tensioner.

To install:

➡**The auto-tensioner assembly must be reset to correctly adjust belt tension.**

6. Loosen the center bolt of tensioner pulley to provide timing belt slack. Remove the timing belt tensioner assembly.

7. Position the auto-tensioner into a vise with soft jaws. The plug at the rear of tensioner protrudes, be sure to use a washer as a spacer to protect the plug from contacting vise jaws.

8. Slowly push the rod into the tensioner until the set hole in rod is aligned with set hole in the auto-tensioner.

9. Insert a 0.055 in. (1.4mm) wire into the aligned set holes. Unclamp the tensioner from the vise and install it on the engine. Tighten tensioner mounting bolts to 17 ft. lbs. (24 Nm).

※※ WARNING

DO NOT rotate or turn the camshafts when removing the sprockets or severe engine damage will result from internal component interference.

10. Align the mark on the crankshaft sprocket with the mark on the front case. Then, move the crankshaft sprocket 3 teeth counterclockwise.

11. Align the timing marks of the camshafts with the marks on the rear covers.

12. Realign the crankshaft pulley with timing mark on the housing.

➡**Be sure camshafts-to-cylinder heads and crankshaft-to-front cover timing marks are aligned.**

13. Install the timing belt around the pulleys in the following order:
   a. Crankshaft pulley.
   b. Idler pulley.
   c. Left camshaft sprocket.
   d. Water pump pulley.
   e. Right camshaft sprocket.
   f. Tensioner pulley.

➡**Since the camshaft sprockets turn easily because of spring action, be careful not to get your fingers caught.**

14. Align all timing mark on the crankshaft and raise tensioner pulley against belt to remove slack, snug tensioner bolt.

15. Check the alignment of all the timing.

16. Using special tool MD998769, rotate the crankshaft ¼ turn counterclockwise, then rotate the crankshaft clockwise to align the timing marks. Check that all the timing marks are in alignment.

17. Loosen the center bolt on the tensioner pulley. Using tool MD998767 and a torque wrench, apply 3.3 ft. lbs. (4.4 Nm) to the tool on the tensioner. Tighten the tensioner bolt to 33 ft. lbs. (44 Nm) and be

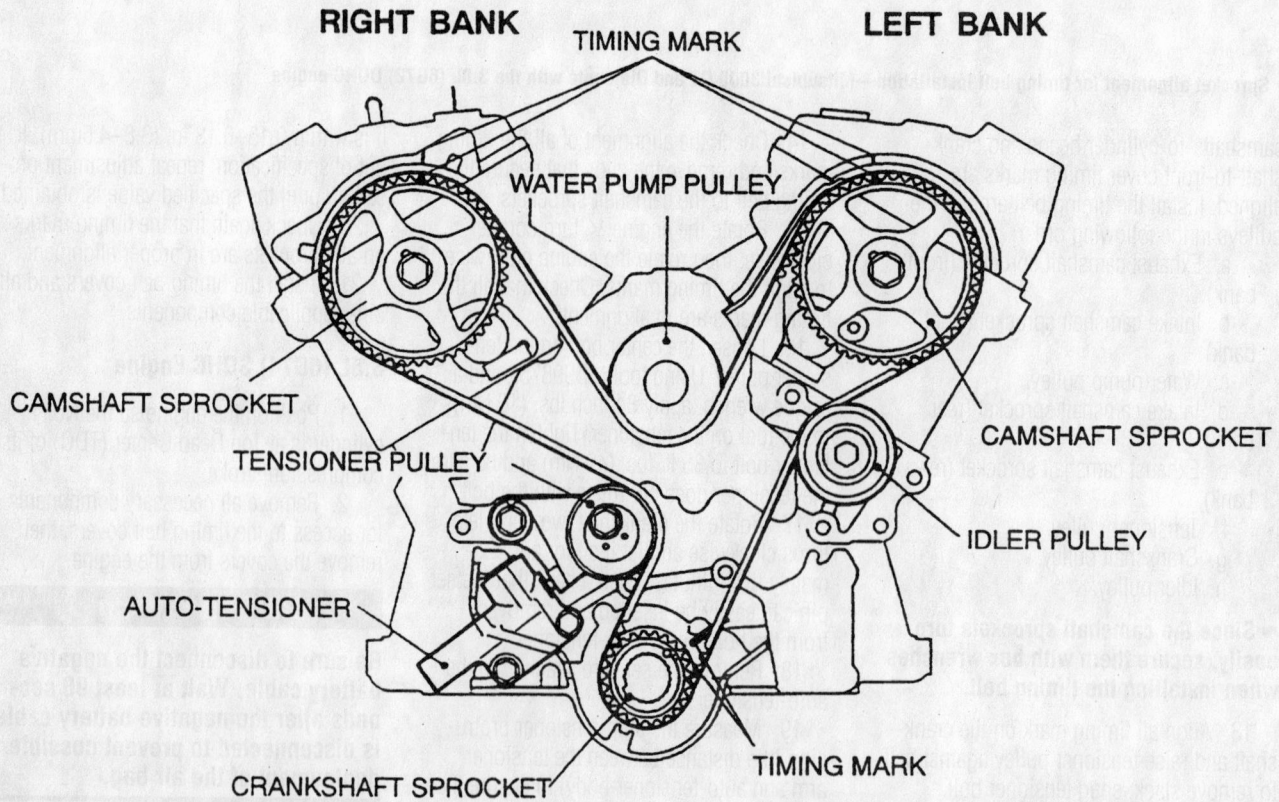

RIGHT BANK   TIMING MARK   **LEFT BANK**

WATER PUMP PULLEY

CAMSHAFT SPROCKET

CAMSHAFT SPROCKET

TENSIONER PULLEY

IDLER PULLEY

AUTO-TENSIONER

TIMING MARK

CRANKSHAFT SPROCKET

93015G27

Sprocket alignment for timing belt installation—Mitsubishi Diamante with 3.5L (6G74) SOHC engine

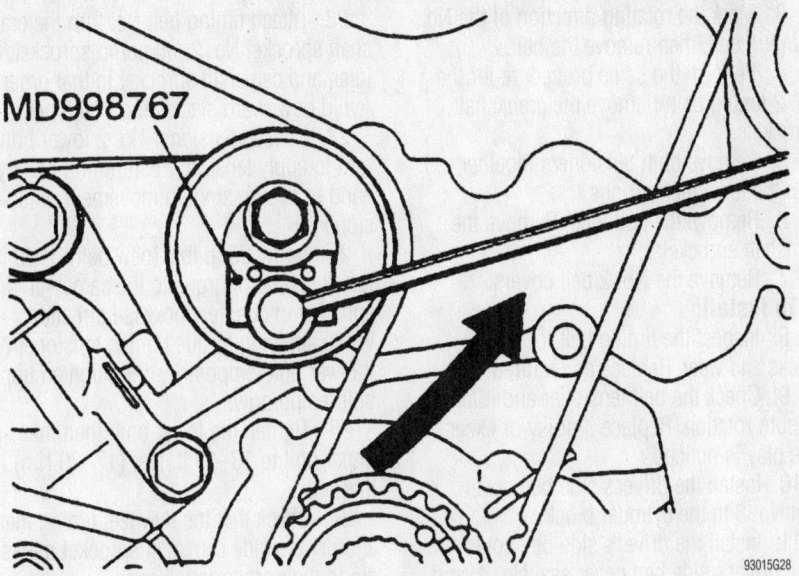

MD998767

Special tool used for tightening timing belt—Mitsubishi Diamante with 3.5L (6G74) SOHC engine

93015G28

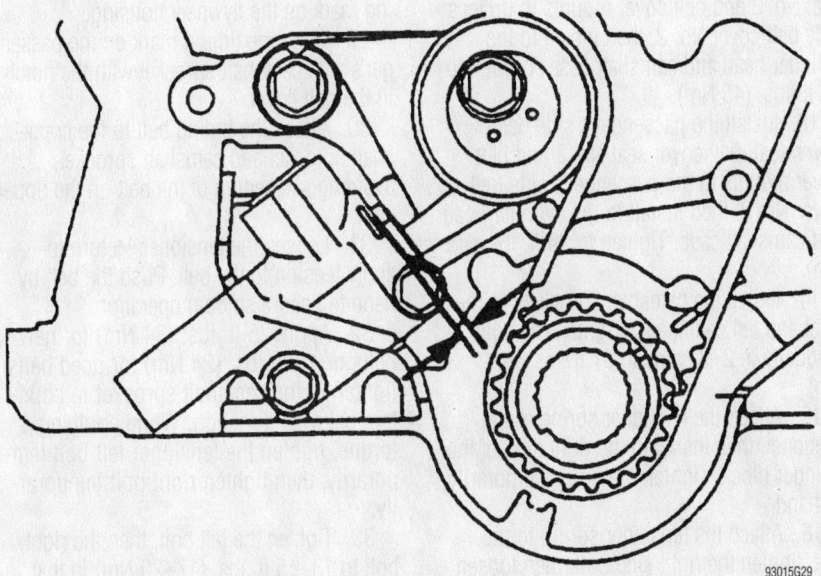

Measuring the standard value of the timing belt tensioner—Mitsubishi Diamante with 3.5L (6G74) SOHC engine

93015G29

sure the tensioner does not rotate with the bolt.

18. Rotate the crankshaft two complete turns clockwise and let it sit for approximately five minutes. Then, check that the set pin can easily be inserted and removed from the hole in the auto-tensioner.

19. Remove the set wire attached to the auto-tensioner.

20. Measure the auto-tensioner protru-sion (the distance between the tensioner arm and auto-tensioner body) to ensure that it is within 0.150–0.196 in. (3.8–5.0mm). If out of specification, repeat adjustment procedure until the specified value is obtained.

21. Check again that the timing marks on all sprockets are in proper align-ment.

22. Install the timing belt covers and all other applicable components.

SAAB MODELS

3.0L Engine

> ✱✱ **WARNING**

To avoid damage to the valves, DO NOT rotate the camshafts once the timing belt is removed from the engine. The crankshaft may only be turned between 0–60 degrees Before Top Dead Center (BTDC) when the camshafts are locked in position with the appropriate locking tool.

1. Remove all necessary components for access to the timing cover, then remove the cover.

➡ **When removing the crankshaft pul-ley, remove the 6 outer bolts only, DO NOT remove the center bolt.**

2. Remove the crankshaft pulley.
3. Rotate the crankshaft to Top Dead Center (TDC) of No. 1 cylinder.
4. The timing marks on the crankshaft and camshafts should be in alignment with their respective marks on the engine. Install camshaft locking tools (such as Saab tools KM-800-1 for camshaft sprockets No. 1 and 2 and KM-800-2 for sprockets No. 3 and 4) and a flywheel locking tool (such as Saab tool 83-94-868).
5. Mark the direction of rotation of the timing belt for reassembly.
6. Release tension from and remove the timing belt. Loosen the timing belt adjuster bolts.
7. Rotate the crankshaft back to 60 degrees BTDC, to prevent damage to the valves.
8. Remove the bracket with the upper timing belt adjuster and tensioner rollers.

To install:

9. Remove the flywheel locking tool and install the flywheel inspection cover.
10. Install the bracket with the upper timing belt adjuster and tensioner pulleys.
11. Install both camshaft locking tools.
12. Rotate the crankshaft forward to just before 0 degrees TDC and install the crank-shaft locking tool on the crankshaft. Care-fully rotate the engine until the arm of the tool is against the water pump flange. Be sure the crankshaft is at 0 degrees TDC and all timing marks are aligned. Remove the locking tool.
13. If reusing the belt, fit the timing belt according to its marked direction of rotation and timing marks. Adjust the tensioning

roller loosely by hand to prevent the belt from slipping out of the cogs. Always adjust counterclockwise.

14. Measure the belt tension with Saab tool 83-93-985.

15. Snug the center bolts of the adjusting rollers. Turn the lower adjusting roller counterclockwise, until a belt tension of 202–220 ft. lbs. (275–300 Nm) is reached. Tighten the adjusting roller center bolts to 30 ft. lbs. (40 Nm).

➡ **This is a preliminary adjustment of the belt tension and must not be used as a check when the belt is finally adjusted.**

16. Continue to carry out the adjustment by means of the tensioning roller, mark against mark. Remove the locking tool for camshaft sprockets No. 1 and 2. Carry out the final adjustment with the upper center adjusting roller until camshaft sprocket No. 2 moves 0.04–0.08 in. (1–2mm) forward.

17. Remove the locking tool for camshaft sprockets No. 3 and 4 and also remove the crankshaft locking tool.

18. Tighten the tensioning roller to 15 ft. lbs. (20 Nm). Tighten the upper adjusting roller to 30 ft. lbs. (40 Nm) and tighten the lower adjusting roller to 15 ft. lbs. (20 Nm).

19. Rotate the engine 2 complete revolutions to just before 0 degrees TDC and install the locking tool on the crankshaft. Carefully turn the crankshaft until the arm of the locking tool is against the water pump flange and tighten the locking tool. Set Saab tool KM-800-20 into position. Be sure that the timing marks on the camshaft sprockets are aligned with the marks on the tool and that the edge of the timing belt is flush with the edge of the camshaft sprockets.

➡ **Check that the alignment marks on the tensioner pulley are still aligned.**

20. If necessary, install the crankshaft pulley and tighten the retaining bolts to 15 ft. lbs. (20 Nm).

21. Install the timing belt cover, and tighten the bolts to 6 ft. lbs. (8 Nm). Install all of the remaining components in the reverse order of the removal procedure.

SUBARU MODELS

1.8L Engine

1. Remove all necessary components for access to the timing belt covers, then remove the covers.

2. Loosen the timing belt tensioner mounting bolts ½ turn and slacken the timing belt. Tighten the mounting bolts.

3. Mark the rotating direction of the No. 1 timing belt, then remove the belt.

4. Perform the same procedure for the No. 2 timing belt. Remove the crankshaft sprockets.

5. Remove both tensioners together with the tensioner springs.

6. Remove the belt idler. Remove the camshaft sprockets.

7. Remove the No. 2 belt covers.

To install:

8. Inspect the timing belt for breaks, cracks and wear. Replace as required.

9. Check the belt tensioner and idler for smooth rotation. Replace if noisy or excessive play is noticed.

10. Install the driver's side belt cover seal No. 3 to the cylinder block.

11. Install the driver's side belt cover seal, driver's side belt cover seal No. 4, and belt cover mount to the right rear belt cover, then install the assembly on the cylinder block. Tighten to 34 ft. lbs. (45 Nm).

12. Install the driver's side belt cover seal No. 2 and belt cover mounts to driver's side belt cover No. 2, then install to the cylinder head and camshaft case. Tighten to 34 ft. lbs. (45 Nm).

13. Install the passenger's side belt cover seal, belt cover seal No. 2 and belt cover mounts to the passenger's side belt cover No. 2, then install to the cylinder head and camshaft case. Tighten to 34 ft. lbs. (45 Nm).

14. Install the camshaft sprockets to the right and left camshafts. Tighten the bolts gradually in 2–3 steps to 67 ft. lbs. (91 Nm).

15. Attach the tensioner spring to the tensioner, then install to the right side of the cylinder block. Tighten the bolts temporarily by hand.

16. Attach the tensioner spring to the bolt, tighten the right side bolt, then loosen it ½ turn.

17. Push down the tensioner until it stops, then temporarily tighten the left bolt.

18. Install the left side tensioner in the same manner.

19. Install the belt idler to the cylinder block using care not to turn the seal. Tighten to 29–35 ft. lbs. (39–47 Nm).

20. Install the sprockets on the crankshaft. Install the crankshaft pulley and tighten the bolt temporarily.

21. Align the center of the three lines scribed on the flywheel with the timing mark on the flywheel housing.

22. Align the timing mark on the driver's side camshaft sprocket with the notch on the belt cover.

23. Attach timing belt No. 2 to the crankshaft sprocket No. 2, oil pump sprocket, belt idler, and camshaft sprocket in that order. Avoid downward slackening of the belt.

24. Loosen tensioner No. 2 lower bolt ½ turn to apply tension. Push timing belt by hand to ensure smooth movement of tensioner.

25. Apply 25 ft. lbs. (new belt) or 18 ft. lbs. (used belt) torque to the camshaft sprocket in counterclockwise direction. While applying torque tighten tensioner No. 2 lower bolt temporarily, then tighten upper bolt temporarily.

26. Tighten the lower bolt, then the upper bolt to 13–15 ft. lbs. (17–20 Nm) in that order.

27. Check that the flywheel timing mark and drivers side camshaft sprocket marks are in their proper positions.

28. Turn the crankshaft one turn clockwise from the position where timing belt No. 2 was installed, and align the center of the three lines on the flywheel with the timing mark on the flywheel housing.

29. Align the timing mark on the passenger's side camshaft sprocket with the notch in the belt cover.

30. Attach the timing belt to the crankshaft sprocket and camshaft sprocket, avoiding slackening of the belt on the upper side.

31. Loosen the tensioner ½ turn to apply tension to the belt. Push the belt by hand to ensure smooth operation.

32. Apply 25 ft. lbs. (34 Nm) for new belts or 18 ft. lbs. (24 Nm) for used belts, tighten to the camshaft sprocket in counterclockwise direction. While applying torque, tighten the tensioner left bolt temporarily, then tighten right bolt temporarily.

33. Tighten the left bolt, then the right bolt to 13–15 ft. lbs. (17–20 Nm) in that order.

34. Check that the flywheel timing mark and drivers side camshaft sprocket marks are in their proper positions.

35. Remove the crankshaft pulley.

36. Install the right front belt cover seals and belt cover plug. Install the belt covers to the cylinder block.

37. On turbo-charged engines, install the belt cover plate.

38. Install the crankshaft pulley and tighten to 66–79 ft. lbs. (89–107 Nm).

39. Install the water pump pulley and tighten to 67 ft. lbs. (91 Nm). Install the pulley cover, oil level guide and gauge and oil pressure switch connector.

40. Install and properly tension the accessory drive belt.

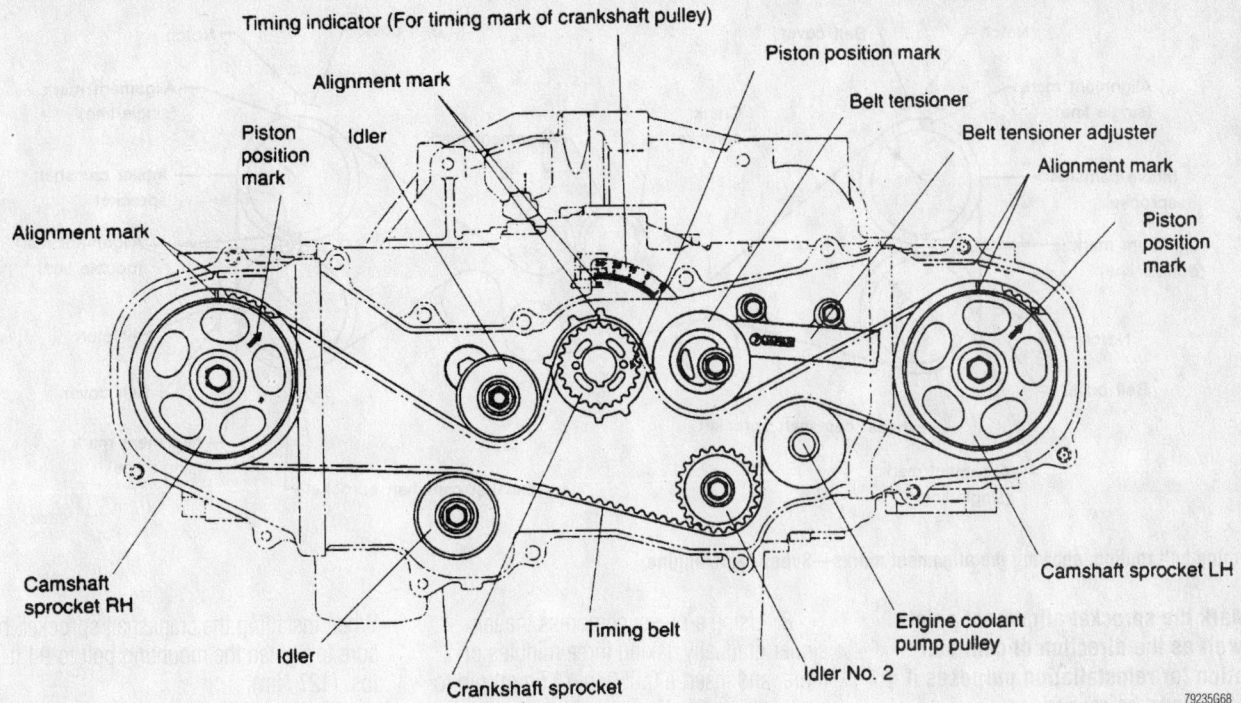

Timing indicator (For timing mark of crankshaft pulley)

Timing belt components and alignment mark locations—Subaru 1.8L and 2.2L engines

2.2L Engine

The engine uses a single cam belt drive system with a serpentine type belt. The left side of the engine uses a hydraulic cam belt tensioner, which is self-adjusting.

➡**It is recommended that the timing belt be replaced at least every 60,000 miles (96,618 km).**

1. Disconnect the negative battery cable.

2. Position the No. 1 piston to Top Dead Center (TDC) of its compression stroke.

3. Remove the engine drive belts.

4. Remove the timing belt covers.

5. Align the camshaft sprockets so each sprocket notch aligns with the cam cover notches. Align the crankshaft sprocket top tooth notch, located at the rear of the tooth, with the notch on the crank angle sensor boss. Mark the three alignment points as well as the direction of cam belt rotation.

6. Loosen the tensioner adjusting bolts. Remove the bottom three idlers, the cam belt and the cam belt tensioner. The cam sprockets can, then be removed with a modified camshaft sprocket wrench tool.

7. Remove the sprockets, if necessary. Note the reference sensor at the rear of the left cam sprocket.

To install:

8. Install the sprockets, if removed and tighten the retaining bolts to 47–54 ft. lbs. (64–74 Nm).

9. Install the crankshaft sprocket and the non-adjustable right side idler. Do not install the tensioner idler at this time.

10. Compress the hydraulic tensioner in a vise slowly and temporarily secure the plunger with a pin or suitable Allen wrench. Install the tensioner and the pulley with the adjustable idler pulley. Temporarily tighten the tensioner while the tensioner is pushed to the right.

11. Align the crankshaft sprocket notch on the rear sprocket tooth with the crank angle sensor boss. This places the sprocket notch in the 12 o'clock position.

12. Align the camshaft sprockets with the notches in the cam rear belt cover. This places the sprocket notch in the 12 o'clock position for each camshaft.

13. Install the timing belt with the directional mark and alignment marks properly positioned (if the belt is reused).

14. Loosen the tensioner retaining bolts and slide the tensioner to the left. Tighten the mounting bolts.

15. After verifying the timing marks are correct, remove the stopper pin from the tensioner.

16. Verify the correctness of the timing by noting that the notches on the 2 cam pulleys and the notch on the crankshaft pulley all point to the 12 o'clock position when the belt is properly installed.

17. Complete the engine component assembly by installing the cam belt covers, the crankshaft pulley bolt and pulley and the remaining components.

18. Connect the negative battery cable.

2.5L Engine

The engine uses a single cam belt drive system with a serpentine type belt. The left side of the engine uses a hydraulic cam belt tensioner, which is self-adjusting.

➡**It is recommended that the timing belt be replaced at least every 60,000 miles (96,618 km).**

1. Remove all necessary components for access to the left, right and center timing belt covers, then remove the covers.

2. Align the camshaft sprockets so each sprocket notch aligns with the rear cover notches. Align the crankshaft sprocket top tooth notch, located at the rear of the tooth with the notch. The crankshaft notch will be at 12 o'clock and the keyway will be at 6 o'clock.

For Tune-up, Capacities and Firing orders, see Section 1 of this manual

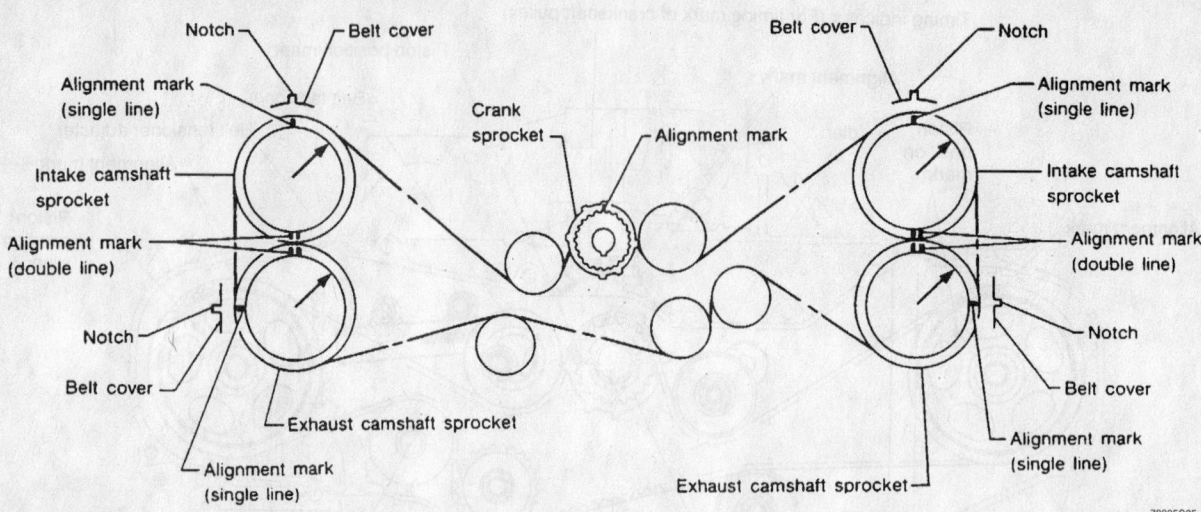

Timing belt routing, showing the alignment marks—Subaru 2.5L engine

➡Mark the sprocket alignment points as well as the direction of cam belt rotation for reinstallation purposes if the belt is to be reused.

3. Loosen the tensioner adjusting bolts.
4. Remove the lower timing belt idler.
5. Remove the timing belt from the pulleys.

❋❋ WARNING

After the timing belt is removed, DO NOT rotate the camshaft sprockets or the crankshaft. Severe internal damage will result from the valve and/or piston contact.

6. Remove the timing belt tensioner and the timing belt tension adjuster.

To install:

➡Inspect the timing belt and tensioner for wear or damage and replace as necessary.

7. Inspect the timing belt tensioner as follows:

a. When compressing the pushrod of the tensioner with a force of 33 lbs. (147 N), the tensioner should not sink.

b. When compressing the pushrod of the tensioner with a force of 33–110 lbs. (147–490 N), the tensioner should not sink within 8.5 seconds.

c. Measure the extension of the rod beyond the body of the tensioner for a length of 0.606–0.646 in. (15.4–16.4mm). If not within specifications, replace the tensioner.

➡Check the idler sprockets for smooth operation. Replace as necessary.

8. Using a press, compress the tensioner gradually, taking three minutes or more, and insert a 0.059 in. (1.5mm) pin to secure the rod.

9. Install the tensioner and the pulley with the adjustable idler pulley. Temporarily tighten the tensioner while the tensioner is pushed to the right.

10. Align the crankshaft sprocket notch on the rear sprocket tooth with the crank angle sensor boss. This places the sprocket notch in the 12 o'clock position and the keyway at the 6 o'clock position.

11. Align the camshaft sprockets with the notches in the cam rear belt cover. This places the sprocket notch in the 12 o'clock position for each camshaft.

12. Install the timing belt in a clockwise direction starting at the crankshaft with the directional mark and alignment marks properly positioned (if the belt was reused).

13. Install the lower timing belt idler and tighten the mounting bolt to 29 ft. lbs. (39 Nm).

❋❋ WARNING

Be sure all the timing marks are properly aligned.

14. Loosen the tensioner retaining bolts and slide the tensioner to the left. Tighten the mounting bolts to 18 ft. lbs. (25 Nm).

15. After verifying the timing marks are correct, remove the stopper pin from the tensioner and recheck the timing marks.

16. Install the center, right, then the center timing belt covers. Tighten the bolts to 44 inch lbs. (5 Nm).

17. Install the remaining components in the reverse order of the removal procedure.

When installing the crankshaft sprocket, be sure to tighten the mounting bolt to 94 ft. lbs. (127 Nm).

3.3L Engine

1. Disconnect the negative battery cable.

2. Remove the timing belt covers.

3. Matchmark the timing belt to the sprocket, cover and block marks as follows:

a. Turn the crankshaft to align the timing marks on the crankshaft sprocket with the mark on the block.

b. With the crankshaft marks aligned be sure the left and right camshaft sprocket marks are aligned with marks on the timing covers.

c. If all the marks align, use white paint to mark the direction of rotation of the belt as well as mark the spots on the belt where it crosses over the timing marks on the pulleys.

4. Loosen the belt tensioner bolts.
5. Remove belt idler pulley No. 1.
6. Remove belt idler pulley No. 2.
7. Remove the timing belt.
8. Remove the tensioner pulley bolt and remove the tensioner pulley.
9. Remove the two bolts and the tensioner assembly.

To install:

10. Insert a 0.059 in. (1.5mm) diameter stopper pin into place while pushing the tension adjuster rod into the tensioner body.

11. Install the tensioner and tighten the bolts to 18 ft. lbs. (24 Nm), while the tensioner is pushed all the way to the right.

12. Install the tensioner pulley and mounting bolt. DO NOT tighten the idler pulley bolt completely.

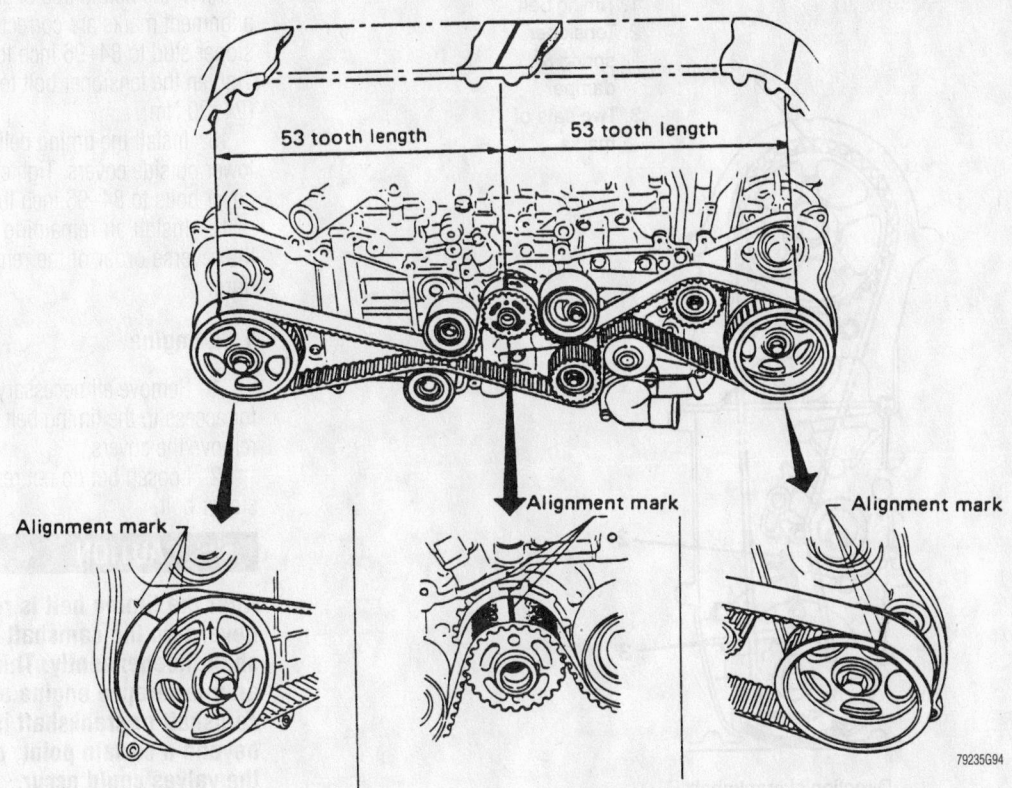

To ensure proper installation of the timing belt, be sure the proper number of belt teeth are between each sprockets, as indicated—Subaru 3.3L engine

13. Be sure the crankshaft and both camshaft sprockets are still aligned with their respective timing marks.

14. Install the timing belt onto the sprockets with the direction of rotation arrow in the correct direction and the timing marks on the belt align with the marks on the sprockets.

15. Install No. 1 and 2 idler pulleys and tighten the mounting bolts to 29 ft. lbs. (39 Nm).

16. Loosen the tensioner pulley bolt and the tensioner assembly mounting bolts. Slide the tensioner assembly all the way to the left and tighten the bolts to 18 ft. lbs. (24 Nm).

17. Check again that all the timing marks are still in alignment. If they are remove the stopper pin from the tensioner assembly.

18. Install the timing belt covers.

19. Connect the negative battery cable.

SUZUKI MODELS

1.3L Engines

1. Remove all necessary components for access to the upper and lower timing belt outside covers, then remove the covers.

2. Align the camshaft timing belt pulley with its timing marks. The crankshaft and camshaft marks are straight up.

3. Remove the resonator and the timing belt outside cover.

4. Remove the tensioner stud and loosen the tensioner bolt.

5. Remove the tensioner spring and damper, then remove the timing belt.

✳✳ WARNING

After the timing belt is removed never turn the camshaft or the crankshaft. Interference may occur between the pistons and the valves causing component damage.

6. Remove the tensioner and the tensioner plate.

To install:

7. Install the timing belt tensioner plate and tensioner. Only hand-tighten the tensioner bolt.

➡**Be sure that the lug on the tensioner plate is inserted into the hole on the tensioner.**

8. Be sure the tensioner plate and the tensioner move uniformly. If they do not move together remove the tensioner and the tensioner plate and reinsert the plate lug into the tensioner hole.

9. Check the camshaft sprocket to verify that it has not moved.

10. Check the crankshaft alignment by verifying that the punch mark on the timing belt pulley is aligned with the arrow on the oil pump case.

11. Remove the cylinder head cover.

➡**This is to permit the free rotation of the camshaft. When installing the timing belt on the pulleys, the tensioner spring force should correctly tension the belt. If the camshaft does not rotate freely the belt will not be correctly tensioned.**

12. With the timing marks aligned, hold the tensioner plate up by hand and install the timing belt on the pulleys so there is no slack on the drive side of the belt.

13. Turn the crankshaft 2 rotations clockwise. Confirm that the timing marks are still properly aligned.

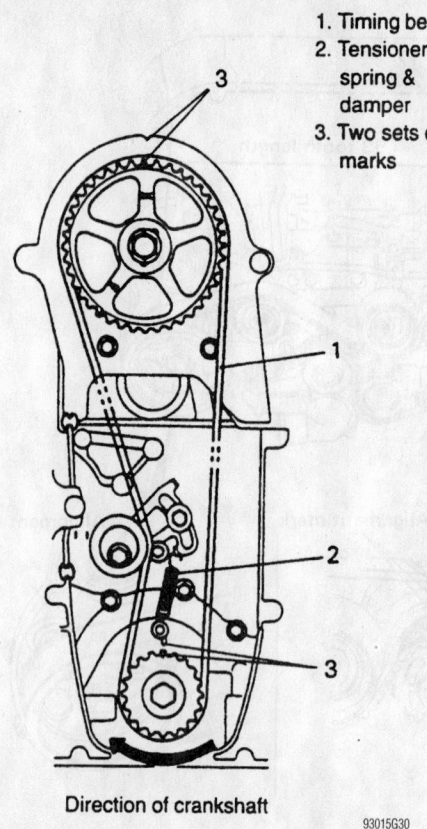

1. Timing belt
2. Tensioner spring & damper
3. Two sets of marks

Direction of crankshaft

93015G30

View of the timing belt and timing marks—Suzuki 1.0L engine

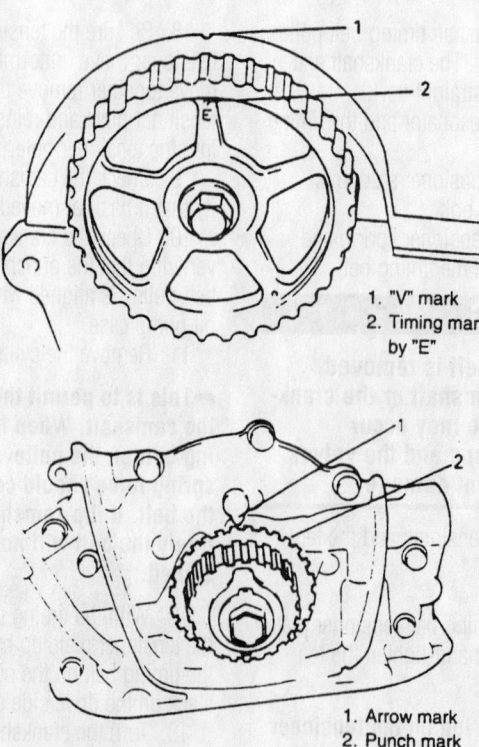

1. "V" mark
2. Timing mark by "E"

1. Arrow mark
2. Punch mark

79235G69

Match the "V" notch to the "E" mark on the camshaft, and the punch and arrow on the crankshaft to properly position the engine for belt service—Suzuki 1.3L and 1.6L engines

14. If the belt is free of slack and the alignment marks are correct tighten the tensioner stud to 84–96 inch lbs. (9–12 Nm). Tighten the tensioner bolt to 17–21 ft. lbs. (24–30 Nm).

15. Install the timing belt upper and lower outside covers. Tighten the timing cover bolts to 84–96 inch lbs. (9–12 Nm).

16. Install all remaining components in the reverse order of the removal procedure.

1.6L Engine

1. Remove all necessary components for access to the timing belt covers, then remove the covers.

2. Loosen but do not remove the tensioner bolt.

❊❊ CAUTION

After the timing belt is removed, never turn the camshaft and crankshaft independently. This engine is an interference engine and if the camshaft or crankshaft is turned beyond a certain point, damage to the valves could occur.

3. Loosen the timing belt tensioner adjusting bolt and pivot nut. Apply pressure to the tensioner to loosen the timing belt, and remove the timing belt from the camshaft and crankshaft sprockets.

4. Remove the timing belt tensioner, tensioner plate and tensioner spring.

To install:

5. Install the timing belt tensioner, plate and spring. Hand-tighten the tensioner bolt and stud only at this time.

6. Turn the camshaft sprocket clockwise and align the timing marks.

7. Turn the crankshaft clockwise, using a 17mm wrench to crank the timing belt sprocket bolt.

8. Align the punch mark on the timing belt sprocket with the arrow mark on the oil pump.

9. With the timing marks aligned, remove any slack from the drive side of the belt. Tighten the tensioner bolt to 16–20 ft. lbs. (22–28 Nm).

10. To allow the belt to be free of any slack, turn the crankshaft clockwise 2 full rotations. Confirm that the timing marks are aligned.

11. Install the timing cover and tighten the bolts to 84–96 inch lbs. (9–12 Nm).

12. Install all remaining components in the reverse order of the removal procedure.

TOYOTA MODELS

1.5L (5E-FE) Engine

1. Remove all necessary components for access to the timing belt covers.

> **✻✻ CAUTION**
>
> **If equipped with an air bag, be sure to disconnect the negative battery cable and wait at least 90 seconds before proceeding.**

2. Remove the No. 2 timing belt cover.

3. Rotate the engine clockwise until the crankshaft pulley is aligned with the **0** mark on the No. 1 timing belt cover. Verify the hole in the camshaft timing pulley is aligned with the timing mark on the No. 1 bearing cap. If not as specified, rotate the crankshaft an additional 360 degrees.

4. For vehicles with A/C and/or power steering, remove the four bolts to the No. 2 crankshaft pulley. Then, remove the No. 2 crankshaft pulley.

5. Using Toyota tools SST 09213-

14010 and SST 09330-00021, remove the No. 1 crankshaft pulley bolt.

6. Using a crankshaft pulley/damper puller (such as Toyota tool SST 09950-50010), remove the No. 1 crankshaft pulley from the crankshaft.

7. Remove the No. 3 timing belt cover (plug).

8. Remove the No. 1 timing belt cover and timing belt guide.

9. Place matchmarks on the timing belt on both sides of the cam and crankshaft gear timing marks. Also, place an arrow on the top surface of the belt to indicate the direction of travel.

10. Using pliers, remove the tension spring.

11. Loosen the No. 1 idler pulley bolt and push the pulley to the left as far as it will go, then temporarily tighten the bolt.

12. Remove the belt.

To install:

13. For vehicles with a distributor (distributor ignition), use the crankshaft bolt to turn the crankshaft until the timing marks on the sprocket and oil pump body align. This is method is used to set the piston at Top

Dead Center (TDC) before the marks on the belt cover can be seen.

14. For vehicles with a crankshaft position sensor (distributorless ignition), use the crankshaft bolt to turn the crankshaft until the rotor side of the crankshaft position sensor faces inward.

15. Turn the camshaft and align the hole of the camshaft timing pulley with the timing mark of the bearing cap. The matchmarks, if using the old belt should align. Place the belt over the crankshaft pulley and the idler pulleys.

16. Install the belt on the crankshaft gear (using the matchmarks if reinstalling the old belt) and install the timing belt guide with flange out.

17. Loosen the No. 1 idler pulley bolt until the pulley is moved slightly by the spring tension.

18. Turn the crankshaft pulley two revolutions from TDC-to-TDC.

➡**Always rotate the crankshaft clockwise.**

19. Check that the pulleys align with the reference marks. If not, reinstall the belt.

20. When the timing is verified, tighten the adjuster pulley (No. 1 idler pulley) to 13 ft. lbs. (18 Nm).

21. Install No. 1 and No. 3 lower timing belt covers.

22. Install the No. 1 crankshaft pulley and tighten the pulley bolt to 112 ft. lbs. (152 Nm).

23. Tighten the four No. 2 crankshaft pulley bolts to 14 ft. lbs. (19 Nm).

24. Install the No. 2 timing belt cover with the four bolts.

25. Install the remaining components in the reverse order of the removal procedure. When installing the right-hand engine mounting insulator, tighten the bracket bolt to 47 ft. lbs. (64 Nm) and the through-bolt to 54 ft. lbs. (73 Nm).

1.6L (4A-FE) and 1.8L (7A-FE) Corolla engines

1. Remove all necessary components for access to the timing belt covers.

> **✻✻ CAUTION**
>
> **If equipped with an air bag, be sure to disconnect the negative battery cable and wait at least 90 seconds before proceeding.**

2. Turn the crankshaft to align the timing mark on crankshaft pulley at **0**, setting

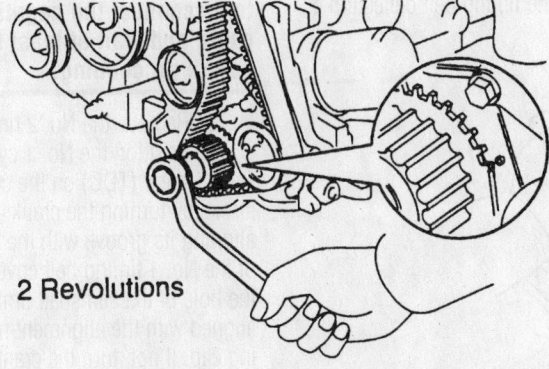

2 Revolutions

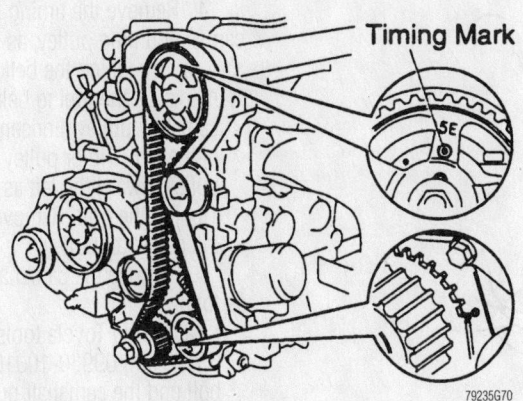

Timing Mark

79235G70

Turn the engine two revolutions, then ensure the timing marks are still aligned—Toyota 1.5L (5E-FE) engine the piston in No. 1 cylinder at Top Dead Center (TDC) on the compression stroke. Check that the valve lifters on the No. 1 cylinder are loose. If not, turn crankshaft pulley 1 complete revolution (360 degrees).

3. Remove the nine bolts and timing belt covers from the engine.

4. Slide the timing belt guide from crankshaft.

5. Set the camshaft and crankshaft timing sprockets to align the marks.

✹✹ WARNING

Do not turn crankshaft or camshaft independently after removal of timing belt. Binding or damage to engine components could result. If the timing belt is to be reused, mark timing belt with arrow showing direction of engine revolution. Place matchmarks where timing belt meets with crankshaft timing sprocket and camshaft timing sprocket to ensure installation in the same position.

6. Remove the timing belt tensioner bolt, tensioner and the tension spring.

7. Remove the timing belt from camshaft and crankshaft timing sprockets.

8. Remove the timing belt.

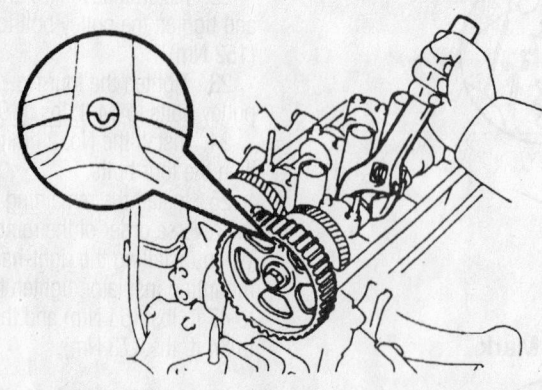

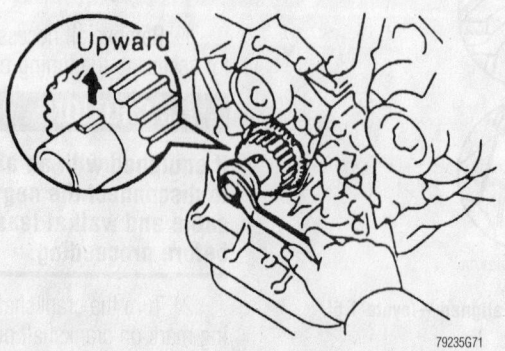

Sprocket alignment for timing belt replacement—Toyota 1.6L (4A-FE) and 1.8L (7A-FE) engines

✹✹ WARNING

Do not bend, twist or turn the timing belt inside out. Do not allow the belt to come in contact with oil, coolant or steam.

To install:

➡Inspect the camshaft timing sprocket to ensure mark is still aligned as indicated.

9. Reinstall the timing belt tensioner and the tension spring. Pry the tensioner to the left as far as it will go and temporarily tighten the retaining bolt.

10. Install the timing belt. If reinstalling the old belt, observe the matchmarks made during removal. Be sure the belt is fully and squarely seated on the sprockets.

11. Loosen the retaining bolt for the timing belt tensioner and allow it to tension the belt.

12. Temporarily install the crankshaft pulley bolt and turn the crank clockwise 2 full revolutions from TDC to TDC. Insure that each timing mark realigns exactly.

13. Tighten the timing belt tensioner bolt to 27 ft. lbs. (37 Nm).

14. Measure the timing belt deflection at the **SIDE** point, looking for ¼ in. (5–6mm) of deflection at 4.4 lbs. (2 kg) of pressure. If the deflection is not correct, adjust with the timing belt tensioner.

15. Install the timing belt guide, with the cup side facing outward, onto the crankshaft and install the timing belt covers from the lowest to the highest. Tighten the nine cover bolts to 62 inch lbs. (7 Nm).

16. Install all applicable remaining components. During assembly, be sure to tighten the crankshaft pulley bolt to 87 ft. lbs. (118 Nm), the mounting bracket-to-engine mount bolt to 47 ft. lbs. (64 Nm), the mounting bracket-to-engine mount nuts to 38 ft. lbs. (52 Nm), the engine mount-to-body bolt **A** to 19 ft. lbs. (25 Nm), the engine mount-to-body bolts **B** to 19 ft. lbs. (25 Nm), the engine mount-to-body bolt **C** to 19 ft. lbs. (25 Nm), if equipped with cruise control.

2.2L (5S-FE) Engines

1. Remove all necessary components for access to the timing belt covers.

✹✹ CAUTION

If equipped with an air bag, be sure to disconnect the negative battery cable and wait at least 90 seconds before proceeding.

2. Remove the No. 2 timing cover.

3. Position the No. 1 cylinder to Top Dead Center (TDC) on the compression stroke by turning the crankshaft pulley and aligning its groove with the timing mark **0** of the No. 1 timing belt cover. Check that the hole of the camshaft timing pulley is aligned with the alignment mark of the bearing cap. If not, turn the crankshaft one revolution (360 degrees).

4. Remove the timing belt from the camshaft timing pulley, as follows:

a. If reusing the belt, place matchmarks on the timing belt and the camshaft pulley. Loosen the mount bolt of the No. 1 idler pulley and position the pulley toward the left as far as it will go. Tighten the bolt. Remove the belt from the camshaft pulley.

5. Remove the camshaft timing pulley as follows:

a. Using Toyota tools Nos. 09249-63010 and 09960-10010, remove the bolt and the camshaft pulley.

6. Remove the crankshaft pulley as follows:

a. Use Toyota tools Nos. 09213-54015 and 09330-00021 to hold the crankshaft pulley. Remove the pulley set

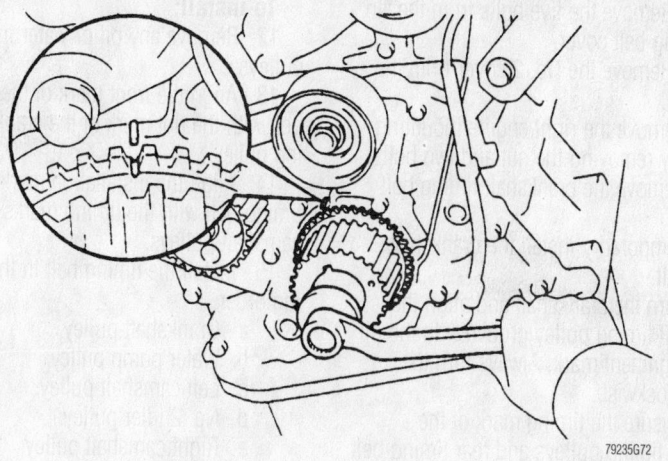

Crankshaft positioning for timing belt removal and installation—Toyota 2.2L (5S-FE) engine

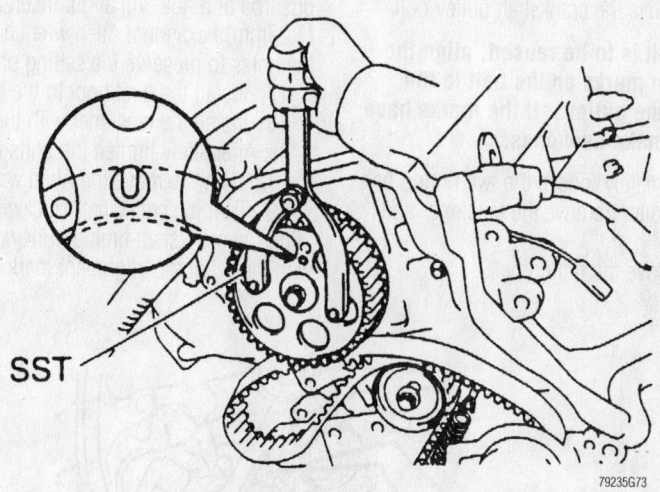

Using a spanner wrench, turn the camshaft into position so that the alignment mark is visible through the hole in the sprocket—Toyota 2.2L (5S-FE) engine

Sprocket alignment for timing belt replacement—Toyota 2.2L (5S-FE) engine

bolt and remove the pulley using a puller.

7. Remove the No. 1 timing belt cover.

8. Remove the timing belt and the belt guide. If reusing the belt mark the belt and the crankshaft pulley in the direction of engine rotation and matchmark for correct installation.

To install:

9. Install the crankshaft timing pulley, as follows:

   a. Align the timing pulley set key with the key groove of the pulley.

   b. Slide on the timing pulley with the flange side facing inward.

10. Install the No. 2 idler pulley and tighten the bolt to 31 ft. lbs. (42 Nm). Be sure that the pulley moves freely.

11. Temporarily install the No. 1 idler pulley and tension spring. Pry the pulley toward the left as far as it will go. Tighten the bolt. Be sure that the pulley rotates freely.

12. Temporarily install the timing belt, as follows:

   a. Using the crankshaft pulley bolt, turn the crankshaft and align the timing marks of the crankshaft timing pulley and the oil pump body.

   b. If reusing the old belt, align the marks made during removal, and install the belt with the arrow pointing in the direction of the engine revolution.

13. Install the timing belt guide with the cup side facing outward.

14. Install the No. 1 timing belt cover.

15. Install the crankshaft pulley. Align the pulley set key with the key groove of the pulley and slide on the pulley. Tighten the bolt to 80 ft. lbs. (108 Nm).

16. Install the camshaft timing pulley as follows:

   a. Align the camshaft knock pin with the knock pin groove of the pulley and slide on the timing pulley. Tighten the bolt to 40 ft. lbs. (54 Nm).

17. With the No. 1 cylinder set at TDC on the compression stroke, install the timing belt (all timing marks aligned). If reusing the belt, align with the marks made during the removal procedure:

   a. Turn the crankshaft pulley and align its groove with the timing mark **0** of the No. 1 timing belt cover. Be sure the camshaft sprocket hole is aligned with the mark on the bearing cap.

18. Connect the timing belt to the camshaft timing pulley.

19. Check that the matchmark on the

Timing belt service is covered in Section 3 of this manual

timing belt matches the end of the No. 1 timing belt cover.

20. Once the belt is installed, be sure that there is tension between the crankshaft timing pulley and the camshaft pulley.

21. Check the valve timing as follows:

a. Loosen the No. 1 idler pulley mount bolt ½ turn. Turn the crankshaft pulley two revolutions from TDC in the clockwise direction. Always turn the crankshaft pulley clockwise.

b. Be sure that the all the timing marks are aligned.

c. Slowly turn the crankshaft pulley 1 ⅞ revolutions. Align its groove with the mark at 45° Before Top Dead Center (BTDC) on the No. 1 timing belt cover for the No. 1 cylinder.

d. Tighten the No. 1 idler pulley mount bolt to 31 ft. lbs. (42 Nm).

22. Install the No. 2 timing belt cover as follows:

a. Install the upper gasket to the No. 1 timing belt cover.

b. Disconnect the engine wire protector between the cylinder head cover and the No. 3 timing belt cover.

c. Install the gasket to the timing belt cover.

d. Install the belt covers and all remaining components. During assembly, tighten the right engine mount bracket bolts to 38 ft. lbs. (52 Nm), the engine mount insulator bolt to 47 ft. lbs. (64 Nm), the through-bolt to 54 ft. lbs. (78 Nm), the power steering reservoir bracket bolt to 21 ft. lbs. (28 Nm), the power steering reservoir-to-bracket bolt to 27 ft. lbs. (37 Nm) and the nut to 38 ft. lbs. (52 Nm).

3.0L (1MZ-FE) Engine

1. Remove all necessary components for access to the timing belt covers.

✴✴ CAUTION

If equipped with an air bag, be sure to disconnect the negative battery cable and wait at least 90 seconds before proceeding.

2. Remove the lower timing belt cover by removing the four bolts.

3. Remove the No. 2 timing belt cover as follows:

a. Remove the bolt and disconnect the engine wire protector from the No. 3 (rear) timing belt cover.

b. Disconnect the engine wire protector clamp from the No. 3 timing belt cover.

c. Remove the five bolts from the No. 2 timing belt cover.

d. Remove the No. 2 cover from the engine.

4. Remove the right engine mounting bracket by removing the nut and two bolts.

5. Remove the crankshaft timing belt guide.

6. Temporarily install the crankshaft pulley bolt.

7. Turn the crankshaft and align the crankshaft timing pulley groove with the oil pump alignment mark. Always turn the engine clockwise.

8. Ensure the timing mark of the camshaft timing pulleys and rear timing belt covers are aligned. If not, turn the engine over an additional 360 degrees (one revolution).

9. Remove the crankshaft pulley bolt.

➡**If the belt is to be reused, align the installation marks on the belt to the marks on the pulleys. If the marks have worn off, make new ones.**

10. Alternately loosen the two timing belt tensioner bolts. Remove the tensioner and dust boot.

11. Remove the timing belt.

To install:

12. Remove any oil or water from the pulleys.

13. Align the front mark of the timing belt with the dot mark of the crankshaft timing pulley.

14. Align the installation marks on the timing belt with the timing marks of the camshaft pulleys.

15. Install the timing belt in the following order:

a. Crankshaft pulley.
b. Water pump pulley.
c. Left camshaft pulley.
d. No. 2 idler pulley.
e. Right camshaft pulley.
f. No. 1 idler pulley.

16. Using a press, slowly press the timing belt tensioner until the holes of the pushrod and housing align. Insert a 0.05 in. (1.27mm) hexagonal Allen wrench through the holes to preserve the setting position.

17. Install the dust boot to the tensioner.

18. Install the tensioner with the two bolts. Alternately tighten the bolts to 20 ft. lbs. (27 Nm). Remove the Allen wrench.

19. Turn the crankshaft clockwise and align the crankshaft timing pulley groove with the oil pump alignment mark.

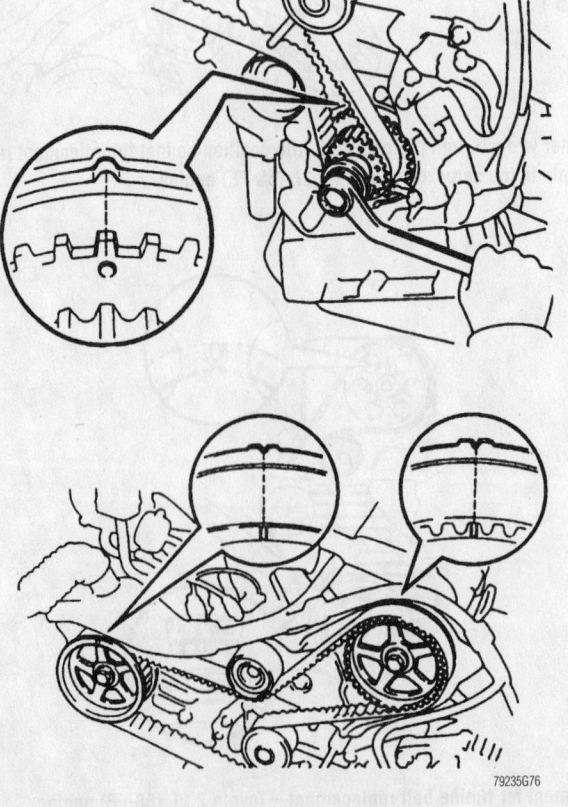

79235G76

Camshaft and crankshaft timing belt sprocket alignment mark positioning for belt service—Toyota 3.0L (1MZ-FE) engine

20. Ensure the camshaft timing marks align with the timing marks on the rear timing belt cover.

21. Install the timing belt guide.

22. Install the right engine mounting bracket and tighten the bolts to 21 ft. lbs. (28 Nm).

23. Install the upper timing belt cover with the five bolts. Tighten the bolts to 74 inch lbs. (8 Nm).

24. Install the engine wire protector clamp to the No. 3 timing belt cover.

25. Install the engine wire protector to the No. 3 timing belt cover with the bolt.

26. Install the lower timing belt cover by installing the four bolts. Tighten the bolts to 74 inch lbs. (8 Nm).

27. Install the remaining components. During installation be sure to tighten the crankshaft pulley bolt to 159 ft. lbs. (215 Nm) and the No. 2 alternator bracket nut to 21 ft. lbs. (28 Nm).

3.0L (2JZ-GTE and 2JZ-GE) Engines

1. Remove all necessary components for access to the timing belt covers.

✴✴ CAUTION

If equipped with an air bag, be sure to disconnect the negative battery cable and wait at least 90 seconds before proceeding.

2. Remove the upper two timing belt covers (Nos. 2 and 3).

3. Remove the drive belt tensioner.

4. Set the No. 1 cylinder to Top Dead Center (TDC) on the compression stroke. Turn the crankshaft pulley clockwise to align the groove with the **0** mark on the lower (No. 1) timing belt cover. Check that the timing marks on the camshaft pulleys are aligned with the marks on the rear belt cover. If the camshaft marks do not align, turn the crankshaft another 360 degrees.

5. Alternately loosen the two bolts holding the timing belt tensioner. Remove the bolts and remove the tensioner.

6. Remove the timing belt from the camshaft pulleys. If the belt is to be reused, place matchmarks on the belt and gears before removing the belt. Mark the belt with an arrow to show direction of rotation.

7. Using Toyota tool SST 09960-10010, remove the bolts for the camshaft timing gears.

8. Remove the camshaft gears from the engine.

9. If necessary, disconnect the oil cooler tubes from the front of the engine by removing the two bolts and hose clamps.

10. Remove the crankshaft pulley by using Toyota tool 09330-0021 to hold the pulley and using tool 09213-70010 to remove the pulley bolt.

11. Remove the lower (No. 1) timing belt cover and the timing belt guide.

12. Remove the timing belt. If the belt is to be reused, protect it from contact with oil, grease or fluids.

To install:

13. Use the crankshaft pulley bolt to turn the crankshaft (clockwise) until the mark on the gear aligns with the oil pump body.

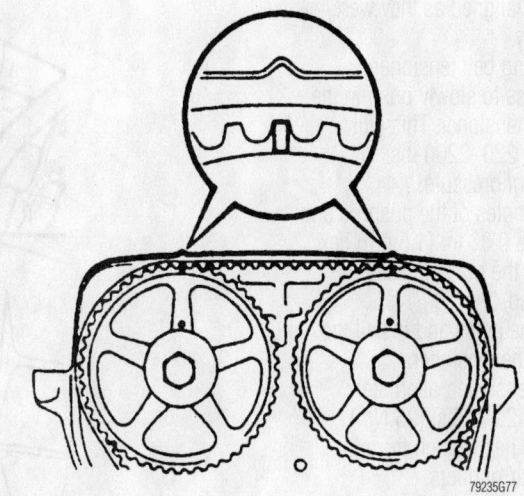

Camshaft timing mark alignment—Toyota 3.0L (2JZ-GTE and 2JZ-GE) engines

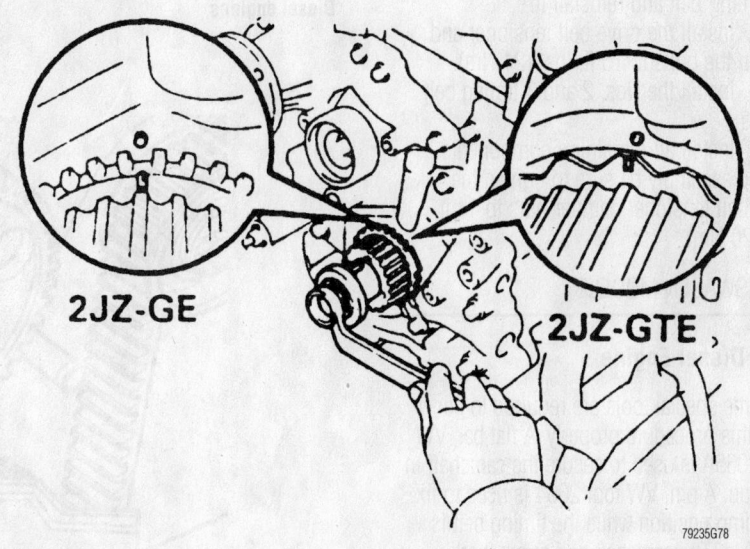

Crankshaft timing marks; notice the timing mark difference between the engines—Toyota 3.0L (2JZ-GTE and 2JZ-GE) engines

Check all the pulleys and gears for cleanliness; remove any grease, oil or coolant. Install the timing belt onto the crankshaft gear and idler pulleys.

14. Install the timing belt guide with the cupped side facing outward.

15. Install the No. 1 timing belt cover.

16. Install the crankshaft pulley. Align the set key with the groove. Hold the pulley with the proper tool and tighten the pulley bolt to 239 ft. lbs. (324 Nm).

17. If equipped with automatic transmission, connect the oil cooler tubes with the clamps and two bolts.

18. Install the camshaft gears as follows:

Heater Core replacement is covered in Section 2 of this manual

a. Align the camshaft knock pin with the groove on the gear and slide on the timing gear.

b. Temporarily install the timing gear bolt.

c. Using the same tools as removal, tighten the camshaft gear bolts to 59 ft. lbs. (79 Nm).

d. Turn the crankshaft pulley and align its groove with the timing mark, **0** on the No. 1 timing belt cover.

e. Align the timing marks on the camshaft timing gears and the No. 4 timing belt cover.

19. Finish installing the timing belt.

20. Double check that all the timing marks for the crankshaft pulley and the camshaft gears are aligned as they were during disassembly.

21. Set the timing belt tensioner:

a. Use a press to slowly push in the pushrod on the tensioner. This will require between 220–2200 lbs. (100–1000 kg) of pressure.

b. Align the holes of the pushrod and housing. Place a 0.06 in. (1.5mm) hex wrench through the holes to keep the pushrod retracted.

c. Release the press and install the dust boot onto the tensioner.

22. Install the tensioner; alternately tighten the bolts to 20 ft. lbs. (26 Nm).

23. Remove the hex wrench from the tensioner with a pair of pliers.

24. Turn the crankshaft pulley two full turns clockwise. Check that each pulley's timing marks align correctly after the two turns. If any mark does not align, remove the timing belt and reinstall it.

25. Install the drive belt tensioner and tighten the bolts to 15 ft. lbs. (21 Nm).

26. Install the Nos. 2 and 3 timing belt covers.

27. Install all remaining components. During assembly be sure to tighten the drive belt tensioner damper nuts to 14 ft. lbs. (20 Nm).

VOLKSWAGEN MODELS

1.9L Diesel Engine

Some special tools are required to perform this procedure properly. A flat bar, VW tool 2065A is used to secure the camshaft in position. A pin, VW tool 2064 is used to fix the pump position while the timing belt is removed. The camshaft and pump work against spring pressure and will move out of position when the timing belt is removed. It is not difficult to find substitutes but do not remove the timing belt without these tools.

✳✳ WARNING

Do not turn the engine or camshaft with the timing belt removed. The pistons will contact the valves and cause internal engine damage.

1. Disconnect the negative battery cable and remove the accessory drive belts, crankshaft pulley and the timing belt cover(s). Remove the camshaft cover and rubber plug at the back end of the camshaft.

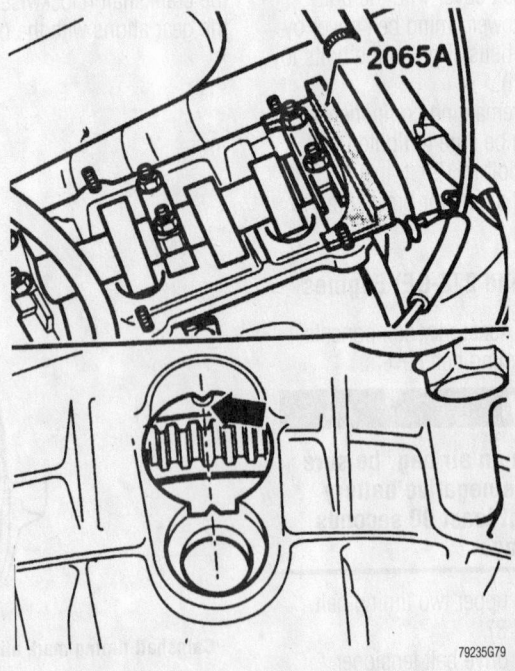

Use the VW tool to lock the camshaft at TDC for timing belt replacement—Volkswagen 1.9L Diesel engines

2. Temporarily reinstall the crankshaft pulley bolt and turn the crankshaft to Top Dead Center (TDC) of No. 1 piston. The mark on the camshaft sprocket should be aligned with the mark on the inner timing belt cover or the edge of the cylinder head.

3. With the engine at TDC, insert the bar into the slot at the back of the camshaft. The bar rests on the cylinder head to will hold the camshaft in position.

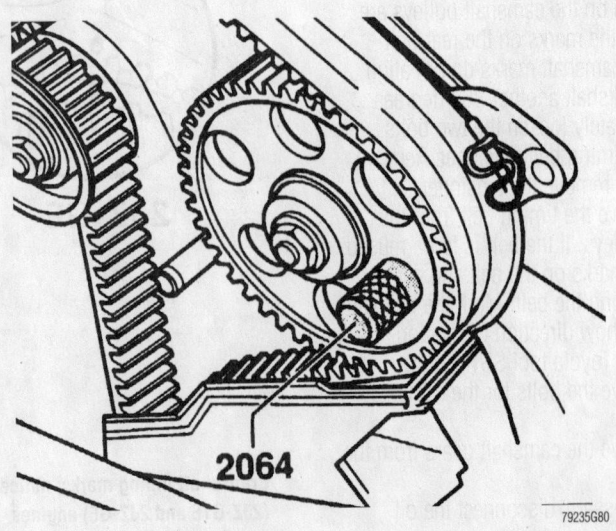

Locking the injection pump with the VW tool as shown—Volkswagen 1.9L Diesel engines

4. Insert the pin into the injection pump drive sprocket to hold the pump in position.

5. Loosen the locknut on the tensioner pulley and turn the tensioner counterclockwise to relieve the tension on the timing belt. Slide the timing belt from the sprockets.

To install:

6. Install the new timing belt and adjust the tension so the belt can be twisted 45 degrees at the halfway point between the camshaft and pump sprockets. Tighten the tensioner nut to 33 ft. lbs. (45 Nm).

7. Remove the holding tools.

8. Turn the engine 2 full revolutions to return to TDC for the No. 1 cylinder. Recheck belt tension and timing mark alignment, readjust as required.

9. Install the belt cover and accessory drive belts.

➡**If the belt is too tight, there will be a growling noise that rises and falls with engine speed.**

1.8L (ACC) and 2.0L (ABA) Engines

➡**Do not turn the engine or camshaft with the camshaft drive belt removed. The pistons will contact the valves and cause internal engine damage.**

1. Disconnect the negative battery cable and remove the accessory drive belts, crankshaft pulley and the timing belt cover(s).

2. Temporarily reinstall the crankshaft pulley bolt, if removed, and turn the crankshaft to Top Dead Center (TDC) of No. 1 piston. The mark on the camshaft sprocket should be aligned with the mark on the inner drive belt cover, if equipped, or the edge of the cylinder head.

3. On 8-valve engines, the notch on the crankshaft pulley should align with the dot on the intermediate shaft sprocket. With the distributor cap removed, the rotor should be pointing toward the No. 1 mark on the rim of the distributor housing.

4. Loosen the locknut on the tensioner pulley and turn the tensioner counterclockwise to relieve the tension on the timing belt.

5. Slide the timing belt off the sprockets.

To install:

6. Install the new timing belt and tension the belt so that it can be twisted 90 degrees at the middle of its longest section, between the camshaft and intermediate sprockets.

7. Recheck the alignment of the timing

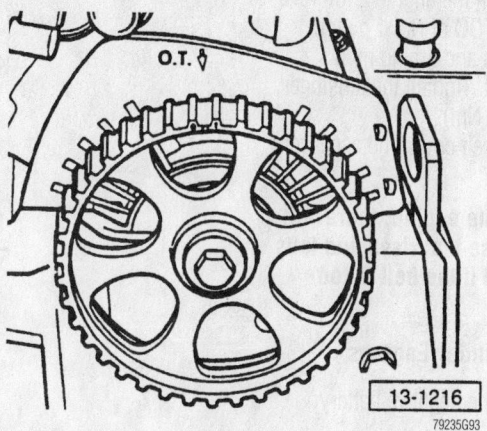

Camshaft timing belt sprocket TDC alignment mark—Volkswagen 1.8L (ACC) and 2.0L (ABA) engines

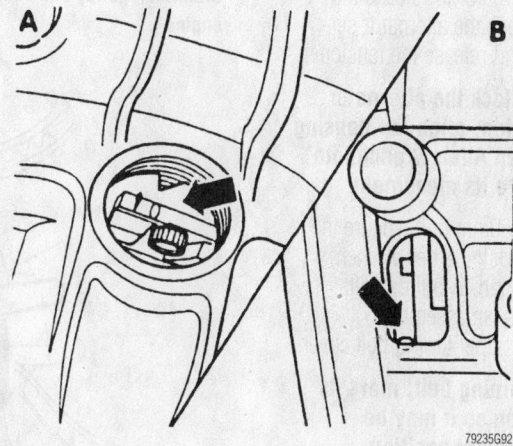

Align the flywheel (A) or driveplate (B) as shown for TDC alignment for cylinder No. 1—Volkswagen 1.8L (ACC) and 2.0L (ABA) engines

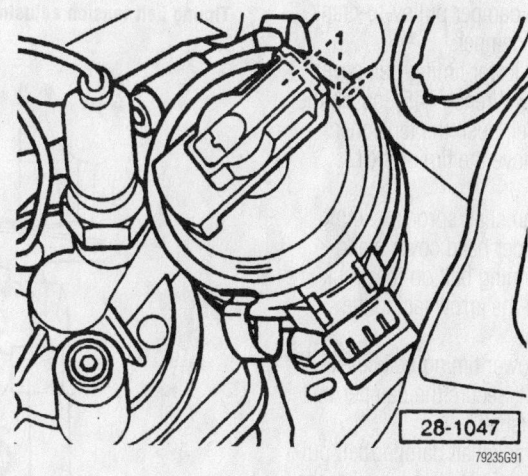

When the No. 1 cylinder is at TDC, the ignition rotor should face the notch in the distributor housing—Volkswagen 2.0L (ABA) engine

Brake service is covered in Section 4 of this manual

marks, if correct, turn the engine 2 full revolutions to return to TDC of No. 1 piston. Recheck belt tension and timing marks. Readjust as required. Tighten the tensioner nut to 33 ft. lbs. (45 Nm).

8. Reinstall the belt cover and accessory drive belts.

➡ **When running the engine, there will be a growling noise that rises and falls with engine speed if the belt is too tight.**

1.8L (AEB) 4-Cylinder Engines

1. Disconnect the negative battery cable.

2. Remove the necessary components to gain access to the front of the engine.

3. Place an open-end wrench on the alternator belt tensioner and rotate it clockwise toward the alternator to release the belt's tension. Remove the alternator serpentine drive belt and release the tensioner.

➡ **If necessary to lock the alternator tensioner in position, align the housing holes and insert an Allen wrench into the holes to secure its movement.**

4. Using a 5 x 60mm bolt, secure the viscous fan pulley. Using a hex wrench, remove the viscous fan-to-pulley bolts. Remove the viscous fan assembly.

5. Remove the upper timing belt cover.

➡ **If reusing the timing belt, mark its rotational direction so it may be installed in its original position.**

6. Using the center bolt, rotate the crankshaft in the direction of engine rotation to position the No. 1 cylinder at Top Dead Center (TDC) of its compression stroke.

7. Remove the damper pulley-to-crankshaft bolts and the damper.

8. Remove the lower timing belt cover.

9. Using a Torx Wrench T45, loosen the timing belt tensioner, push the tensioner downward and remove the timing belt.

To install:

10. Align the camshaft sprocket timing mark with the cylinder head cover mark.

11. Install the timing belt on the crankshaft sprocket with the arrow facing the rotational direction.

12. Install the lower timing belt cover.

13. Using a bolt, secure the damper/belt pulley on the crankshaft.

14. Align the crankshaft damper/belt pulley with the housing timing mark so that the No. 1 cylinder is at TDC of its compression stroke.

15. Install the timing belt on the camshaft sprocket and belt tensioner.

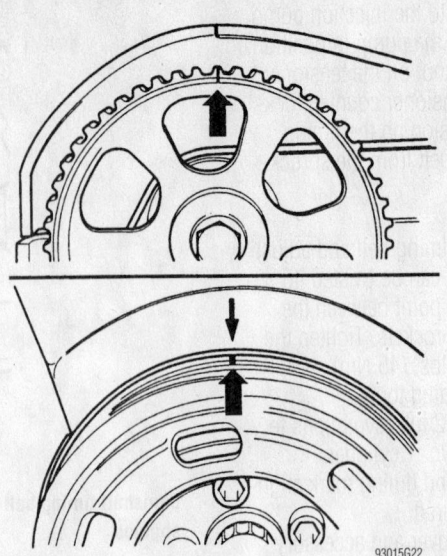

Crankshaft pulley and camshaft sprocket alignment locations—Volkswagen 1.8L (AEB) 4-Cyl engine

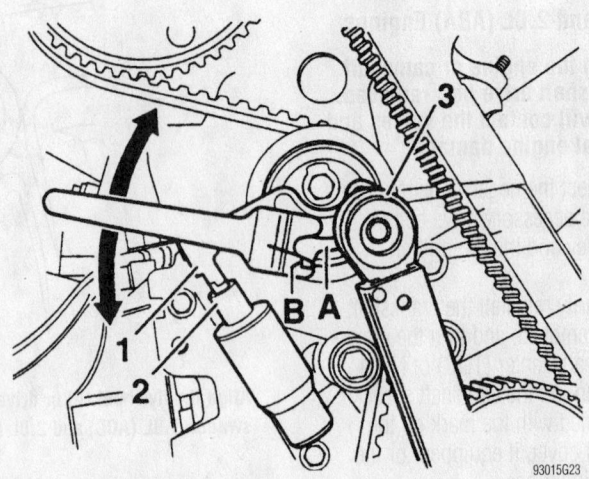

Timing belt tension adjustment— Volkswagen 1.8L (AEB) 4-Cyl engine

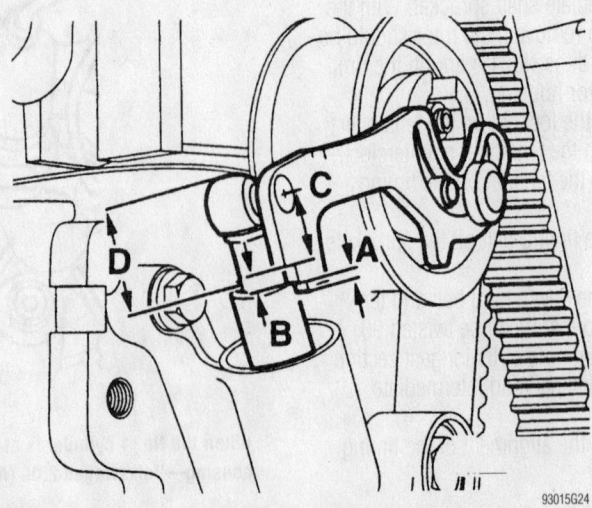

Timing belt tension wear limits— Volkswagen 1.8L (AEB) 4-Cyl engine

16. Using a 2-pin Spanner Matra V159 Wrench, lift (turn clockwise) the timing belt tensioner cylinder No. 1 until it is fully extended and tensioner cylinder No. 2 is raised approx. 1mm; then, hand-tighten the mounting bolt.

17. Rotate the crankshaft 2 complete rotation in the running direction.

18. Inspect area "A" for proper alignment with the upper edge of piston No. 2 and adjust if necessary.

- Area "A"—adjustment OK
- Area "B"—wear limit
- Area "C"—re-adjust and check belt drive including tensioner for wear.

➡️**If the piston edge is located in area "A", measurement "D" is 0.984–1.142 in. (25–29mm).**

19. After adjustment has been verified, secure the tensioner with a 2-pin Spanner Matra V159 Wrench and tighten the mounting bolt.

20. Complete the damper to crankshaft installation.

21. Using the center bolt, rotate the crankshaft 2 rotations in the direction of engine rotation until the camshaft and crankshaft marks align with their respective reference points.

22. Install the upper timing belt cover.

23. Install the drive belts.

24. Replace the remaining components by reversing the removal procedures.

25. Install the negative battery cable last.

26. Test drive the vehicle.

2.0L (AEG) 4-Cylinder Engine—Beetle

1. Disconnect the negative battery cable.

2. Remove the necessary components to gain access to the front of the engine.

3. Remove the serpentine drive belt by performing the following procedure:

a. Using an open-end wrench, rotate the serpentine drive belt tensioner clockwise to relieve the belt tension. Using tool 3090 or a drift awl, lock the tensioner in place.

b. Remove the serpentine drive belt.

4. Remove the drive tensioner from the engine.

5. Rotate the crankshaft to position the No. 1 cylinder on Top Dead Center (TDC) of its compression stroke.

6. Remove the upper timing belt cover.

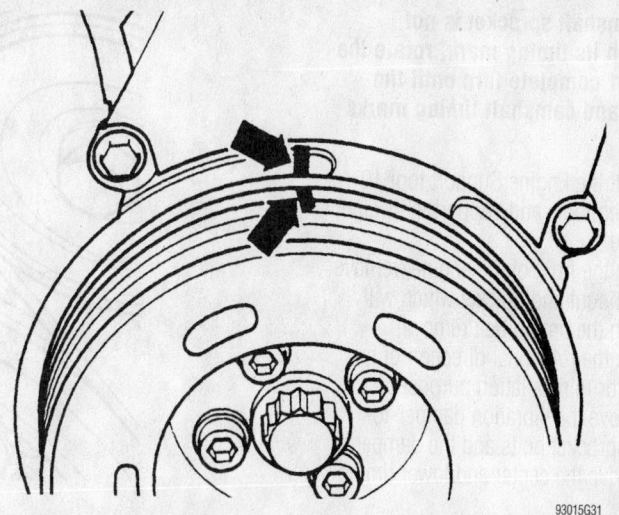

View of crankshaft damper aligned with the timing mark—Volkswagen 2.0L (AEG) 4-Cyl. engine—Beetle

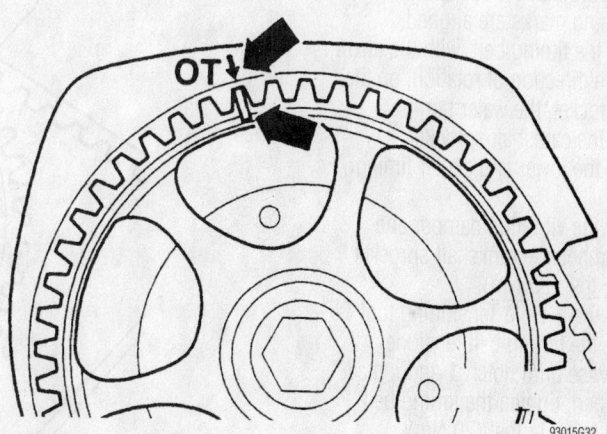

View of camshaft damper aligned with the timing mark—Volkswagen 2.0L (AEG) 4-Cyl. engine—Beetle

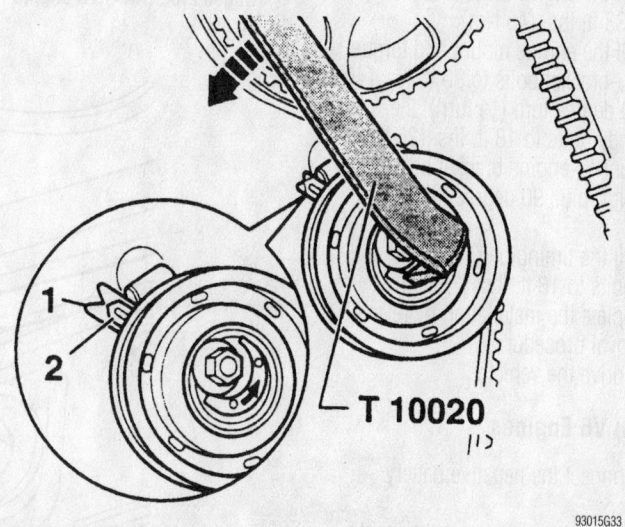

Adjusting the timing belt tensioner—Volkswagen 2.0L (AEG) 4-Cyl. engine—Beetle

For complete Engine Mechanical specifications, see Section 1 of this manual

➡️If the camshaft sprocket is not aligned with its timing mark, rotate the crankshaft 1 complete turn until the crankshaft and camshaft timing marks align.

7. Install the Engine Support tool 10-222A with Leg tools and support the weight of the engine.

8. From the front of the engine, remove the engine mount and bracket which will interfere with the timing belt removal.

9. Mark the rotational direction of the timing belt for reinstallation purposes.

10. Remove the vibration damper-to-crankshaft sprocket bolts and the damper.

11. Remove the center and lower timing belt covers.

12. Release the timing belt tensioner and remove the timing belt.

To install:

13. Make sure that the camshaft and crankshaft timing marks are aligned.

14. Install the timing belt, with the arrow pointing in the direction of rotation, on the crankshaft sprocket, the water pump sprocket and the camshaft sprocket.

15. Install the lower and center timing belt covers.

16. Install the vibration damper and torque the damper-to-crankshaft sprocket bolts to 18 ft. lbs. (25 Nm).

17. Using the 2-Hole Tensioning tool T-10020, rotate the timing belt tensioner counterclockwise until notch **1** and indicator **2** align. Then, tighten the timing belt tensioner nut to 15 ft. lbs. (20 Nm).

18. Rotate the crankshaft 2 complete revolutions and recheck the timing marks.

19. Install the upper timing belt cover.

20. Install the engine bracket and torque the bolts to 33 ft. lbs. (45 Nm).

21. Install the engine mount and torque the mount-to-bracket bolts to 30 ft. lbs. (40 Nm) plus 90 degree turn (¼ turn), the mount-to-body bolts to 18 ft. lbs. (25 Nm) and the mount-to-engine bracket bolts to 44 ft. lbs. (60 Nm) plus 90 degree turn (¼ turn).

22. Install the timing belt tensioner and torque the bolts to 18 ft. lbs. (25 Nm).

23. Complete the installation by reversing the removal procedures.

24. Test drive the vehicle.

2.8L (AHA) V6 Engines

1. Disconnect the negative battery cable.

2. Remove the necessary components to gain access to the front of the engine.

3. Remove the serpentine drive belt by performing the following procedure:

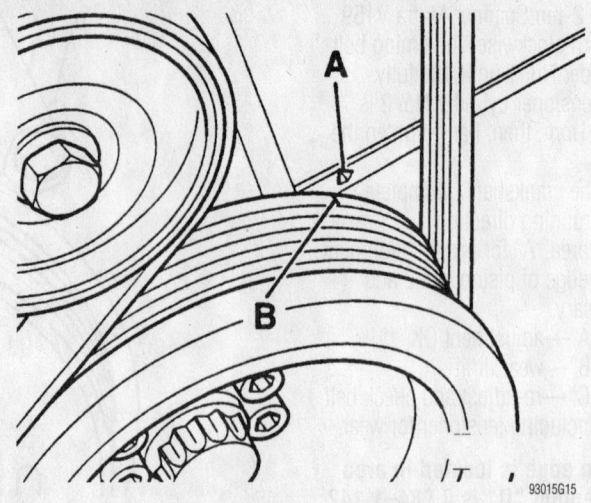

Crankshaft pulley alignment location for TDC—Volkswagen 2.8L (AHA) V6 engine

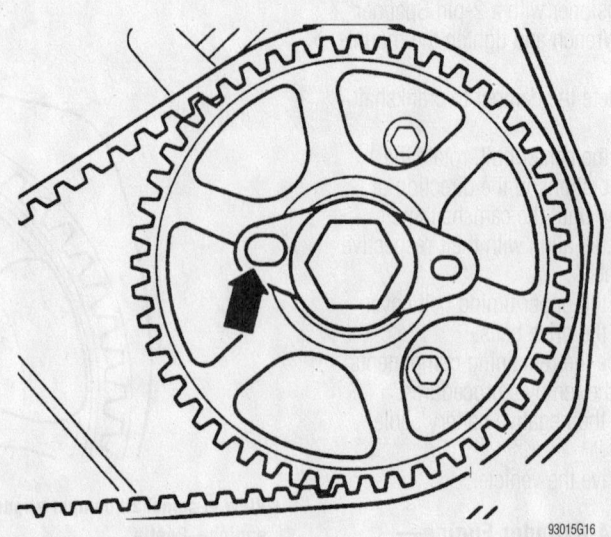

Left camshaft sprocket alignment position for TDC; right camshaft position is similar—Volkswagen 2.8L (AHA) V6 engine

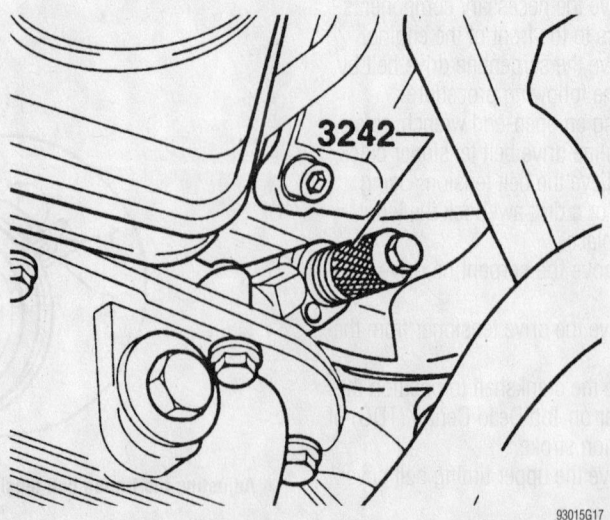

View of crankshaft holding tool installed—Volkswagen 2.8L (AHA) V6 engine

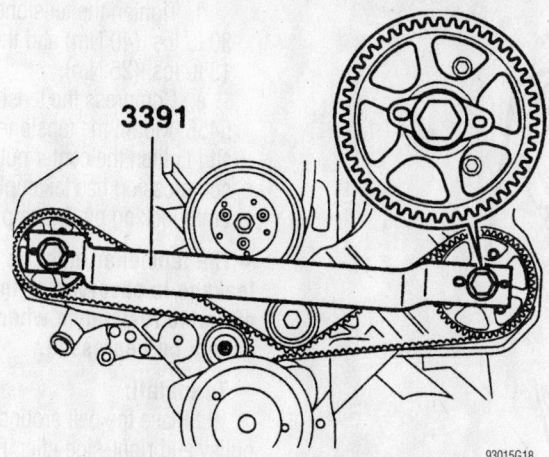

3391

93015G18

View of camshaft locator bar installed—Volkswagen 2.8L (AHA) V6 engine

a. Using Spanner Wrench No. 3212, secure the viscous fan pulley. Using an Open-end Spanner Wrench 3212, remove the viscous fan bearing housing by turning it clockwise.

➡**The viscous fan is mounted with a left-handed thread; turn it clockwise to loosen it.**

b. Place a 17mm box wrench on the serpentine drive belt tensioner and turn the tensioner clockwise until the 2 holes align; insert drift 3204 into the holes to secure the tensioner in place.

c. Mark the running direction of the serpentine drive belt and remove it from the pulleys.

4. Rotate the crankshaft by hand to align the crankshaft pulley mark with the arrow on the engine housing and the large hole in each camshaft sprocket must face inward and must align; this should be Top Dead Center (TDC) of the No. 1 cylinder's compression stroke. If these conditions are not correct, rotate the crankshaft one complete revolution and realign.

5. On the left side of the cylinder block near the crankshaft, remove the sealing plug.

6. Insert Crankshaft Holder tool No. 3242 into the sealing plug hole to secure the crankshaft.

7. Using a 8mm Allen® wrench, rotate the timing belt tensioner roller clockwise until the tensioner is compressed; then, insert a 2mm spring pin through the tensioner housing and tensioner plunger to secure it in place. When the plunger is secure, release the wrench tension.

8. Remove the dampener-to-crankshaft bolts and the damper.

➡**It is not necessary to remove the center bolt when removing the crankshaft damper.**

9. Remove the serpentine belt idler and the crankshaft damper guard.

10. Mark the running direction of the timing belt and remove it from the pulleys.

To install:

11. Make sure that the camshaft pulleys and the crankshaft pulley are in alignment with TDC of the No. 1 cylinder's compression stroke.

12. Install the timing belt; make sure the timing belt is installed in the correct running direction from which it was removed.

13. Using a 8mm Allen® wrench, rotate the timing belt tensioner roller clockwise until the tensioner is compressed; then, remove the 2mm spring pin from the tensioner housing. Slowly, release the tensioner's spring pressure to put pressure on the timing belt.

14. Install the crankshaft damper guard and the serpentine belt idler pulley; torque the idler pulley bolts to 33 ft. lbs. (45 Nm).

15. Install the crankshaft damper and torque the damper-to-crankshaft bolts to 15 ft. lbs. If the damper-to-crankshaft center bolt was removed, torque it to 147 ft. lbs. (200 Nm) plus 180° (½ turn).

16. Remove the Crankshaft Holder tool No. 3242 and install the sealing plug.

17. Replace the remaining components by reversing the removal procedures.

18. Refill the cooling system and the automatic transaxle. Connect the electrical connectors. Install the negative battery cable last.

19. Test drive the vehicle.

VOLVO MODELS

2.3L and 2.4L 5-Cylinder Engines

1. Disconnect the negative battery cable.

2. Remove the coolant expansion tank and place it on top of the engine.

3. Remove the spark plug cover and drive belts.

4. Remove the timing belt cover.

5. Wait five minutes after aligning marks, then install Volvo Gauge 998-8500 between the exhaust camshaft and water pump. Read the gauge using a mirror, while still installed. For 23mm belts, the tension should be 2.7–4.0 units.

➡**If the belt tension is incorrect, the tensioner must be replaced.**

6. Remove the upper tensioner bolt and loosen the lower bolt, turning the tensioner to free up the pulley.

7. Remove the lower bolt and the tensioner. Remove the timing belt.

To install:

8. Turn all the pulleys listening for bearing noise. Check to see that the contact surfaces are clean and smooth. Remove the tensioner pulley lever and idler pulley, lubricate the contact surfaces and bearing with grease. If the tensioner pulley lever or idler is seized, replace it.

9. Install the tensioner pulley lever and idler pulley and tighten to 18 ft. lbs. (25 Nm).

10. Compress the tensioner with Volvo tool 999-5456 and insert a 0.079 in. (2.0mm) lockpin in the piston. If the tensioner leaks, has no resistance, or will not compress, replace it. Install the tensioner and tighten to 18 ft. lbs. (25 Nm).

11. Install the timing belt in the following order:

a. Around the crankshaft sprocket.

b. Around the right idler pulley

c. Around the camshaft sprockets

d. Around the water pump

e. Onto the tensioner pulley

12. Pull the lockpin out from the tensioner and install the upper timing cover. Turn the crankshaft two complete revolutions and check to see that the timing marks on the crankshaft and camshaft pulleys are aligned.

13. Install the timing belt covers and the fuel line clips.

14. Install the accessory belts.

For Accessory Drive Belt illustrations, see Section 1 of this manual

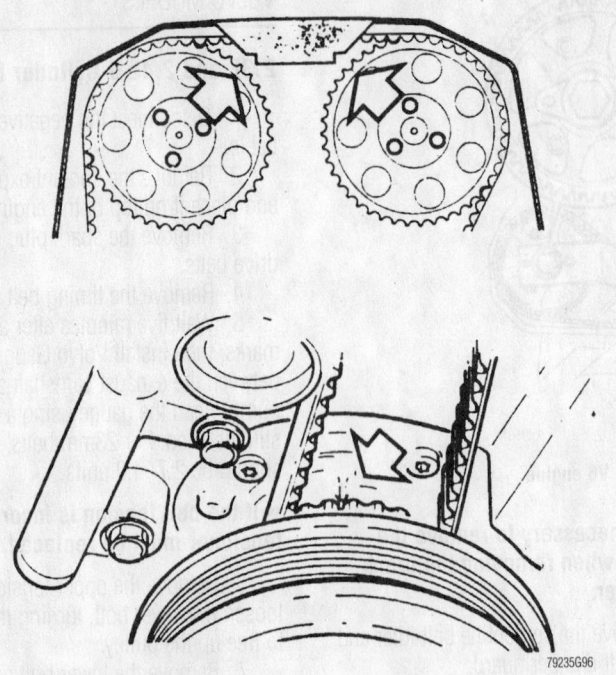

79235G96

Before removing or installing the timing belt, align the timing marks as indicated—Volvo 2.3L 5 cylinder, 2.4L and 2.9L engines

15. Install the vibration damper guard and the inner fender well.

16. Install the spark plug cover and any remaining components.

2.9L 6-Cylinder Engines

1. Disconnect the negative battery cable.

2. Remove the drive belts.

3. Remove the timing belt cover.

4. Remove the splash guard, vibration damper guard and ignition coil cover.

5. Rotate the crankshaft clockwise, until the timing marks on the camshaft pulleys and timing belt mounting plate and crankshaft pulley/oil pump housing are aligned.

6. Remove the tensioner upper mounting bolts. Loosen the tensioner lower mounting bolt and twist the tensioner to free the plunger. Remove the lower mounting bolt and remove the tensioner.

7. Remove the timing belt.

➡**Do not rotate the crankshaft while the timing belt is removed.**

8. Check the tensioner and idler pulleys, as follows:

　a. Spin the pulleys and listen for bearing noise.

　b. Check that the pulley surfaces in contact with the belt are clean and smooth.

　c. Check the tensioner pulley arm and idler pulley mountings.

　d. Tighten the tensioner pulley arm to 30 ft. lbs. (40 Nm) and the idler pulley to 18 ft. lbs. (25 Nm).

　e. Compress the tensioner using tool 5456. Mount the tensioner in the tool and tighten the center nut fully. Wait until compression has taken place and insert a 2mm locking pin in the plunger.

➡**The tensioner must be replaced if leakage is observed or the plunger offers no resistance when depressed or cannot be depressed.**

To install:

9. Place the belt around the crankshaft pulley and right-side idler. Place the belt over the camshaft pulleys. Position the belt around the water pump and press over the tensioner pulley.

10. Insert the tensioner mounting bolts. Tighten to 18 ft. lbs. (25 Nm).

➡**The lever bushing must be greased every time the belt is replace or the pulley is removed. Service the bushing, using the following procedure:**

11. Remove the locking pin from the tensioner. Install the front timing belt cover as follows:

　a. Remove the lever mounting bolt, tensioner pulley and sleeve.

　b. Grease the surfaces of the bushing, bolt and sleeve, using Volvo Part No. 1161246-2.

　c. Install the sleeve, tensioner pulley and lever mounting bolt.

　d. Tighten the bolt to 30 ft. lbs. (40 Nm).

12. Turn the crankshaft two revolutions and check that the timing marks on the crankshaft and camshaft pulleys are correctly aligned.

13. Install the remaining components.

14. Connect the negative battery lead, start and check the engine operation.

BRAKES

4

PRECAUTIONS

Many of the vehicles covered in this section are equipped with anti-lock brakes. Before servicing any vehicle, please be sure to read all of the following precautions, which deal with personal safety, prevention of component damage, and important points to take into consideration when servicing a motor vehicle:

• When servicing anti-lock brakes, do not force brake fluid through the anti-lock solenoids when compressing the calipers. Instead, open a bleeder screw and allow the brake fluid to escape when compressing calipers.

• Some anti-lock systems store brake fluid under high pressure. Pump the brake pedal with the engine off to bleed off residual pressure before servicing any part of the braking system.

• Brake fluid often contains polyglycol ethers and polyglycols. Avoid contact with the eyes and wash your hands thoroughly after handling brake fluid. If you do get brake fluid in your eyes, flush your eyes with clean, running water for 15 minutes. If eye irritation persists, or if you have taken brake fluid internally, IMMEDIATELY seek medical assistance.

• Clean, high quality brake fluid from a sealed container is essential to the safe and proper operation of the brake system. You should always buy the correct type of brake fluid for your vehicle. If the brake fluid becomes contaminated, completely flush the system with new fluid. Never reuse any brake fluid. Any brake fluid that is removed from the system should be discarded. Also, do not allow any brake fluid to come in contact with a painted surface; it will damage the paint.

• When servicing drum brakes, only disassemble and assemble one side at a time, leaving the remaining side intact for reference.

1998–01 ACURA

Brake Caliper

REMOVAL & INSTALLATION

All Models

FRONT

1. Before servicing the vehicle, refer to the precautions in the beginning of this section.
2. Remove the wheels.
3. Remove the banjo bolt and disconnect the brake line from the caliper.
4. Remove the caliper mounting bolts and remove the caliper.
5. Remove the brake pads and shims.
6. If there is a pad spring, remove it from the caliper body.
7. Remove the caliper bracket mounting bolts and remove the bracket.

To install:

8. Install the bracket and torque the bolts to 80 ft. lbs. (109 Nm)
9. Install the pad spring, brake pads, shims, caliper and slide mounting bolts.
10. Caliper slide mounting bolt torque:
   • Integra: 23 ft. lbs. (31 Nm)
   • TL and RL: 36 ft. lbs. (49 Nm)
   • CL: 54 ft. lbs. (74 Nm)
11. Connect the brake line and the banjo bolt. Replace the crush washers and torque the banjo bolt to 25 ft. lbs. (35 Nm).
12. Bleed the brake system. Torque the bleed screws to 84 inch lbs. (9 Nm).
13. Install the front wheels and tighten the wheel nuts to 80 ft. lbs. (109 Nm).

REAR

1. Before servicing the vehicle, refer to the precautions in the beginning of this section.
2. Remove the wheels.
3. Remove the caliper dust shield. Disconnect the parking brake cable from the caliper arm, if equipped.
4. Disconnect the brake line from the caliper.
5. Remove the caliper mounting bolts and pull the caliper off the bracket.
6. Remove the pads, shim, and pad retainer spring.
7. Remove the caliper bracket mounting bolts. Remove the bracket from the rotor.

To install:

8. Install the caliper bracket. Torque the mounting bolts to 28 ft. lbs. (39 Nm).
9. Install the pads, shims, and pad retainer springs.
10. Install the caliper. Torque the mounting bolts to 17 ft. lbs. (23 Nm).
11. Connect the brake hose with new crush washers and torque the banjo bolt to 25 ft. lbs. (35 Nm). Install the parking brake cable, if equipped.
12. Bleed the brake system.
13. Install the caliper dust shield and tighten the bolts to 84 inch lbs. (10 Nm).

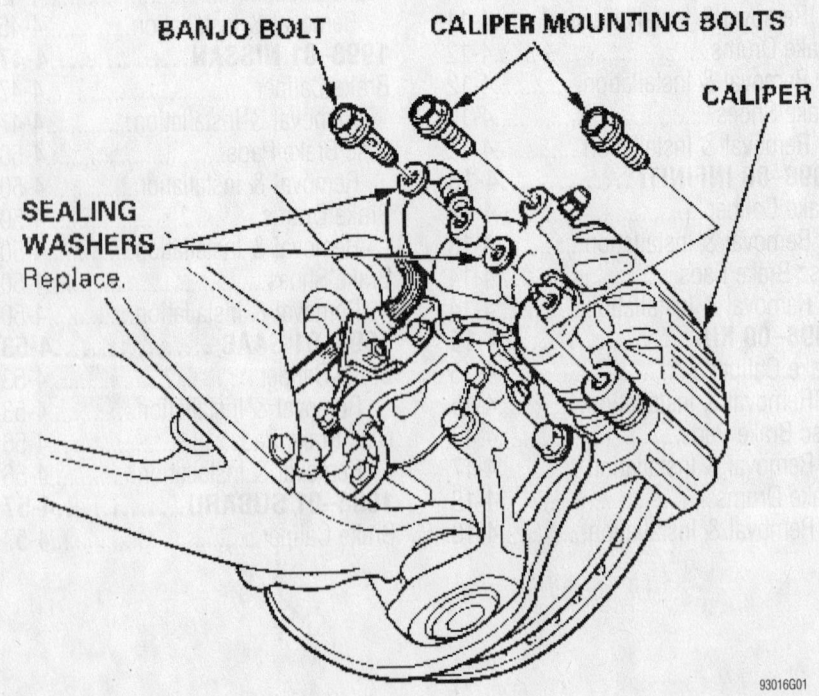

Front caliper mounting—Acura

93016G01

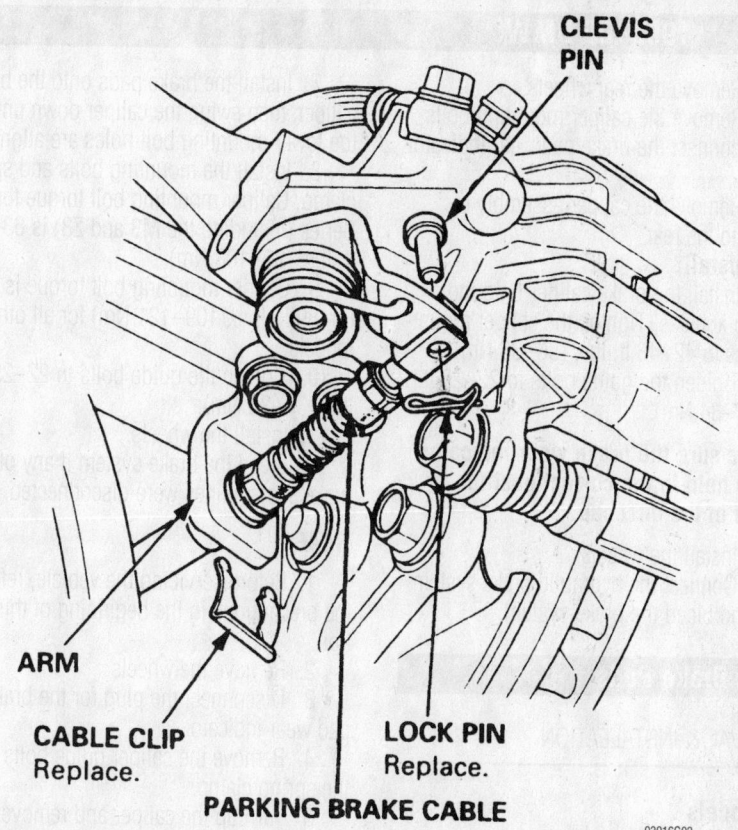

CLEVIS PIN

ARM

CABLE CLIP
Replace.

LOCK PIN
Replace.

PARKING BRAKE CABLE

93016G02

Rear caliper mounting and parking brake cable attachment—Acura

14. Install the rear wheels and torque the wheel nuts to 80 ft. lbs. (109 Nm).

Disc Brake Pads

REMOVAL & INSTALLATION

All Models

FRONT

1. Before servicing the vehicle, refer to the precautions in the beginning of this section.
2. Remove the wheels.
3. Remove the lower caliper bolt and pivot the caliper up and away from the rotor.
4. Remove the pads, shims, and pad retainer springs.

To install:

5. Clean all points where the pads and shims touch the caliper and mount. Apply a thin film of silicone grease to the cleaned areas.
6. Place the pad retainers in position on the caliper bracket.
7. Apply high temperature brake grease to the back side of the pads and both sides of shims and wipe off the excess. Install the pads and shims.
8. Install the inner brake pad with the wear indicator facing upward.
9. Loosen the bleed screw slightly and push in the caliper piston to allow mounting of the caliper over the rotor. Torque the bleed screw to 84 inch lbs. (9 Nm).
10. Pivot the caliper down over the rotor

and install the caliper bolts. Torque the bolts to the following:

- Integra: 23 ft. lbs. (31 Nm)
- TL and RL: 36 ft. lbs. (50 Nm)
- CL: 54 ft. lbs. (74Nm)

11. If disconnected, install the brake pad wear indicator connector. Install the wheels.

REAR

1. Before servicing the vehicle, refer to the precautions in the beginning of this section.
2. Remove the wheels.
3. Remove the caliper dust shield, if equipped.
4. Remove both caliper mounting bolts and pull the caliper off the bracket. Be sure to hang the caliper with a piece of wire so no tension is on the brake line.
5. Remove the pads, shim and pad retainer spring. Clean all points where the pads and shims touch the caliper and mount. Apply a thin film of silicone grease to the cleaned areas.

To install:

6. Apply high temperature brake grease to the pads and shims. Install the pads and shims, making sure the inner pad has the wear indicator facing down.
7. Rotate the brake caliper piston clockwise into the cylinder, using a locknut wrench (part # 07916-6390001). Align the cutout in the piston with the tab on the inner pad by turning the piston back.

✳✳ WARNING

Lubricate the piston boot with grease to avoid twisting the piston boot. If the piston boot is twisted, back the piston out so it sits properly.

8. Install the caliper on the bracket and torque the bolts to 17 ft. lbs. (23 Nm)
9. Install the parking brake cable and the dust shield, if removed.
10. Install the wheels and lower the vehicle.

1998–00 BMW

Brake Caliper

REMOVAL & INSTALLATION

All Models

FRONT

1. Before servicing the vehicle, refer to the precautions in the beginning of this section.
2. Draw off brake fluid with a syringe.
3. Disconnect the hydraulic brake lines.
4. Remove the wheels.
5. Remove the caliper mounting bolts and disconnect the brake pad wear indicator plug.
6. Remove the caliper assembly.

To install:

7. Install the brake caliper onto the steering knuckle. Caliper mounting bolt torque for 3 Series (including the M3 and Z3) is 63–79 ft. lbs. (85–108 Nm).
8. Caliper mounting bolt torque is 80–89 ft. lbs. (109–121 Nm) for all other models.
9. Tighten the guide bolts to 22–25 ft. lbs. (30–34 Nm).

➡ **Make sure the brake wear indicator wire is held in the correct position by the tab of the dust cap.**

10. Install the front wheels.
11. Connect the hydraulic brake system lines and bleed the brake system.

REAR

1. Before servicing the vehicle, refer to the precautions in the beginning of this section.
2. Draw off brake fluid with a syringe.
3. Disconnect the hydraulic brake lines.
4. Remove the rear wheels.
5. Remove the caliper mounting bolts and disconnect the brake pad wear indicator plug.
6. Remove the caliper assembly by pulling to the rear.

To install:

7. Install the brake caliper onto the steering knuckle. Tighten the caliper mounting bolts to 42–48 ft. lbs. (56–66 Nm).
8. Tighten the guide bolts to 22–25 ft. lbs. (30–34 Nm).

➡ **Make sure the brake wear indicator wire is held in the correct position by the tab of the dust cap.**

9. Install the wheels.
10. Connect the hydraulic brake system lines and bleed the brake system.

Disc Brake Pads

REMOVAL & INSTALLATION

All Models

FRONT

1. Before servicing the vehicle, refer to the precautions in the beginning of this section.
2. Remove the wheels.
3. Disconnect the brake pad wear indicator connector.
4. Remove the caliper guide bolts and the spring clamp.
5. Turn up the caliper and remove the brake pads. The inner pad is located with a spring in the piston.

To install:

6. Lubricate the mounting pads with suitable grease.

7. Install the brake pads onto the brake caliper, then swing the caliper down until the lower mounting bolt holes are aligned.
8. Install the mounting bolts and spring clamp. Caliper mounting bolt torque for 3 Series (including the M3 and Z3) is 63–79 ft. lbs. (85–108 Nm).
9. Caliper mounting bolt torque is 80–89 ft. lbs. (109–121 Nm) for all other models.
10. Tighten the guide bolts to 22–25 ft. lbs. (30–34 Nm).
11. Install the wheels.
12. Bleed the brake system if any of the brake system lines were disconnected.

REAR

1. Before servicing the vehicle, refer to the precautions in the beginning of this section.
2. Remove the wheels.
3. Disconnect the plug for the brake pad wear indicator
4. Remove the caliper guide bolts and the spring clamp.
5. Turn up the caliper and remove the brake pads. The inner pad is located with a spring in the piston.

To install:

6. Lubricate the mounting pads with suitable grease.
7. Install the brake pads onto the brake caliper, then swing the caliper down until the lower mounting bolt holes are aligned.
8. Install the mounting bolts and spring clamp. Tighten the caliper mounting bolts to 42–48 ft. lbs. (56–66 Nm), and the guide bolts to 22–25 ft. lbs. (30–34 Nm).
9. Install the wheels.
10. Bleed the brake system if any of the brake system lines were disconnected.

1998–01 HONDA

Brake Caliper

REMOVAL & INSTALLATION

Civic and del Sol

FRONT

➡ **Two distinct types of front calipers are used on these vehicles. The caliper body will be marked either 5410 or 2056 depending on type. Servicing procedures are similar, but the different calipers use different pads.**

1. Before servicing the vehicle, refer to the precautions in the beginning of this section.
2. Remove the front wheels.
3. Remove the banjo bolt and disconnect the brake hose from the caliper.
4. Remove the mounting bolts, and then remove the caliper from its mounting bracket.
5. If necessary for servicing, remove the caliper mounting bracket from the steering knuckle.

To install:

6. If the caliper mounting bracket was removed, install it. Apply brake seal grease to the caliper pins and install them with new pin boots. Apply anti-seize paste to the caliper mounting bolts and tighten them to 80 ft. lbs. (108 Nm).
7. Install the brake pads.
8. Fit the caliper over the pads and onto its mounting bracket.
9. On vehicles equipped with type 2056 calipers, torque the top caliper bolt to 25 ft. lbs. (35 Nm). Torque the lower bolt to 20–24 ft. lbs. (27–32 Nm).
10. On vehicles equipped with type 5410 calipers, torque both caliper bolts to 24 ft. lbs. (33 Nm).

- Coat the piston, piston seal, and caliper bore with clean brake fluid.
- Replace all rubber parts with new ones whenever disassembled.

GREASE : Use recommended rubber grease in the caliper seal set.

GREASE : Use recommended seal grease in the caliper seal set.

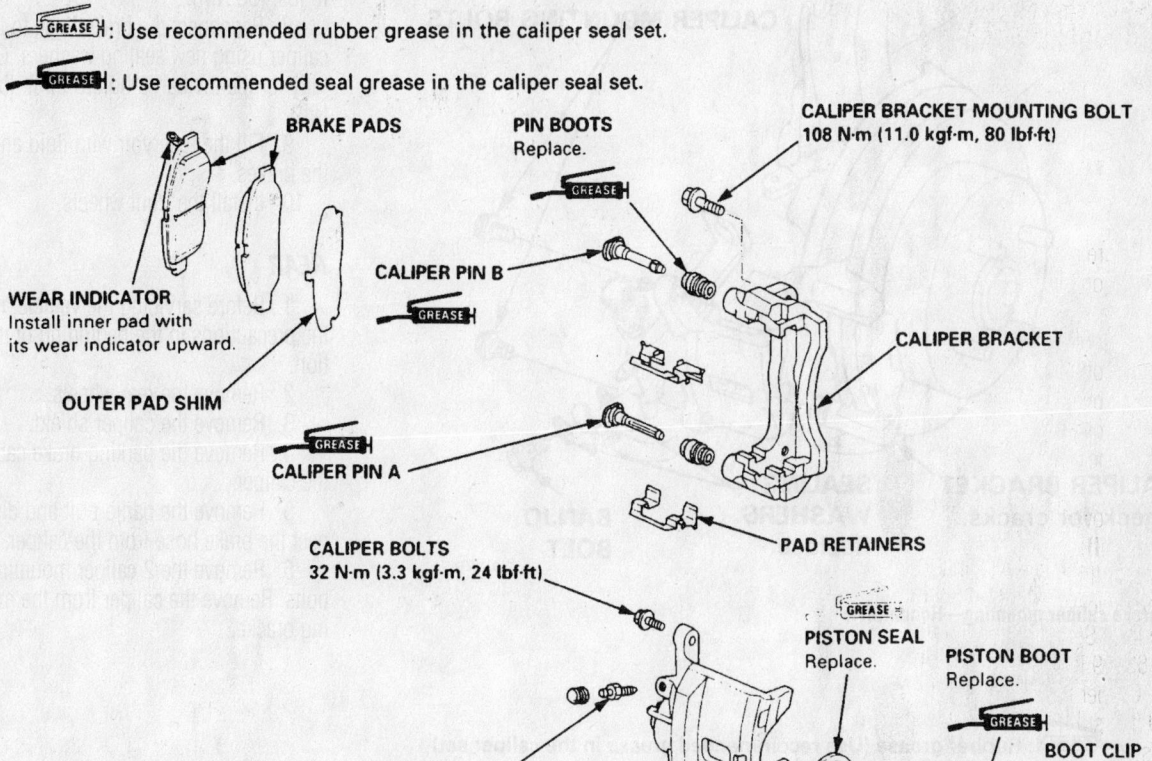

Exploded view of the front brakes—Honda 5410 Type shown

11. Reconnect the brake hose to the caliper using new sealing washers. Carefully torque the banjo bolt to 25 ft. lbs. (35 Nm).

12. Fill the reservoir with fresh brake fluid and bleed the brake system.

13. Install the front wheels.

REAR

1. Before servicing the vehicle, refer to the precautions in the beginning of this section.

2. Remove the rear wheels.

3. Remove the caliper shield.

4. Remove the lock pin and clevis pin from the parking brake cable. Remove the cable securing clip and disconnect the parking brake cable from the caliper.

5. Remove the banjo bolt and disconnect the brake hose from the caliper.

6. Remove the 2 caliper mounting bolts. Remove the caliper from its mounting bracket.

7. If necessary for servicing, remove the caliper mounting bracket from the trailing arm.

To install:

8. If the caliper mounting bracket was removed, install it. Apply brake seal grease to the caliper pins and install them with new pin boots. Apply anti-seize paste to the caliper bracket mounting bolts and tighten them to 28 ft. lbs. (39 Nm).

9. Install the brake pads.

10. Rotate the caliper piston clockwise into place in the cylinder and then align the groove in the piston with the tab on inner pad. Fit the caliper over the pads and onto its mounting bracket.

11. Tighten the caliper bolts to 17 ft. lbs. (23 Nm).

12. Grease the parking brake linkage and connect the parking brake cable.

13. Install the caliper shield.

14. Fill the reservoir with fresh brake

fluid and bleed the brake system. Adjust the parking brake if necessary.

15. Install the rear wheels.

Accord and Prelude

FRONT

1. Before servicing the vehicle, refer to the precautions in the beginning of this section.

2. Remove the front wheels.

3. Remove the banjo bolt and disconnect the brake hose from the caliper.

4. Remove the mounting bolts and remove the caliper from its mounting bracket.

To install:

5. Fit the caliper over the pads and onto its mounting bracket.

6. On vehicles equipped with long caliper pins, torque the top caliper bolts to 54 ft. lbs. (74 Nm).

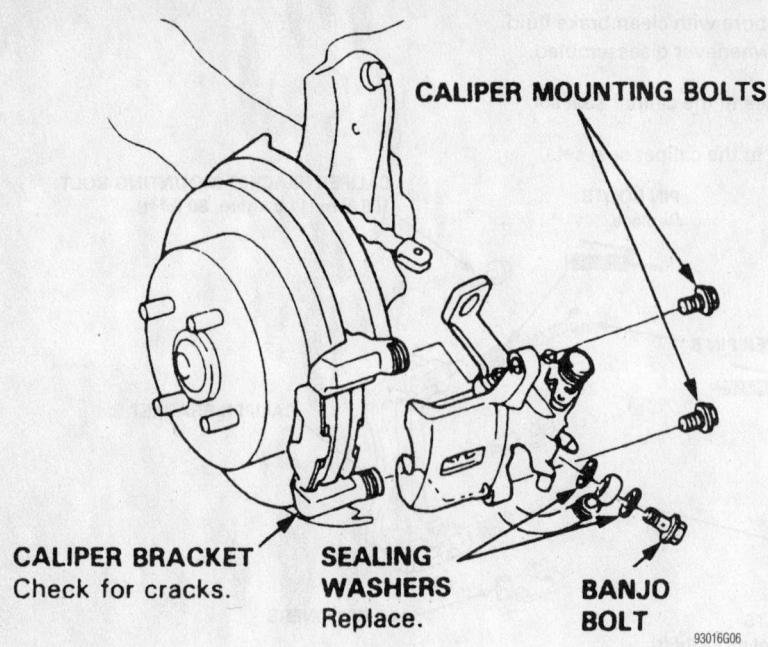

CALIPER MOUNTING BOLTS

CALIPER BRACKET
Check for cracks.

SEALING WASHERS
Replace.

BANJO BOLT

93016G06

Rear brake caliper mounting—Honda Civic

7. On vehicles equipped with short caliper bolts, torque the caliper bolts to 36 ft. lbs. (50 Nm).

8. Reconnect the brake hose to the caliper using new sealing washers. Carefully torque the banjo bolt to 25 ft. lbs. (35 Nm).

9. Fill the reservoir with fluid and bleed the brakes.

10. Install the front wheels.

REAR

1. Before servicing the vehicle, refer to the precautions in the beginning of this section.

2. Remove the rear wheels.

3. Remove the caliper shield.

4. Remove the parking brake cable from the caliper.

5. Remove the banjo bolt and disconnect the brake hose from the caliper.

6. Remove the 2 caliper mounting bolts. Remove the caliper from the mounting bracket.

: Rubber grease (Use recommended grease in the caliper set)

: Silicone grease (Use recommended seal grease and pin grease in the caliper set)

INNER SHIM B

INNER SHIM A

WEAR INDICATOR
Install inner pad with its wear indicator upward.

OUTER PAD SHIM

12 mm FLANGE BOLTS
108 N·m (11.0 kgf·m, 79.6 lbf·ft)

PIN BOOT

PIN B

BANJO BOLT
34 N·m (3.5 kgf·m, 25 lbf·ft)

BRAKE PADS

PIN BOOTS
Replace.

SEALING WASHERS
Replace.

CALIPER BOLTS
49 N·m (5.0 kgf·m, 36 lbf·ft)

PIN A

BRAKE HOSE

PAD SPRING

PAD RETAINERS

CALIPER BRACKET

BLEED SCREW
9 N·m (0.9 kgf·m, 6.5 lbf·ft)

CALIPER BODY

PISTON SEAL
Replace.

PISTON

PISTON BOOT
Replace.

93016G07

Exploded view of the front brakes—Accord V6 shown

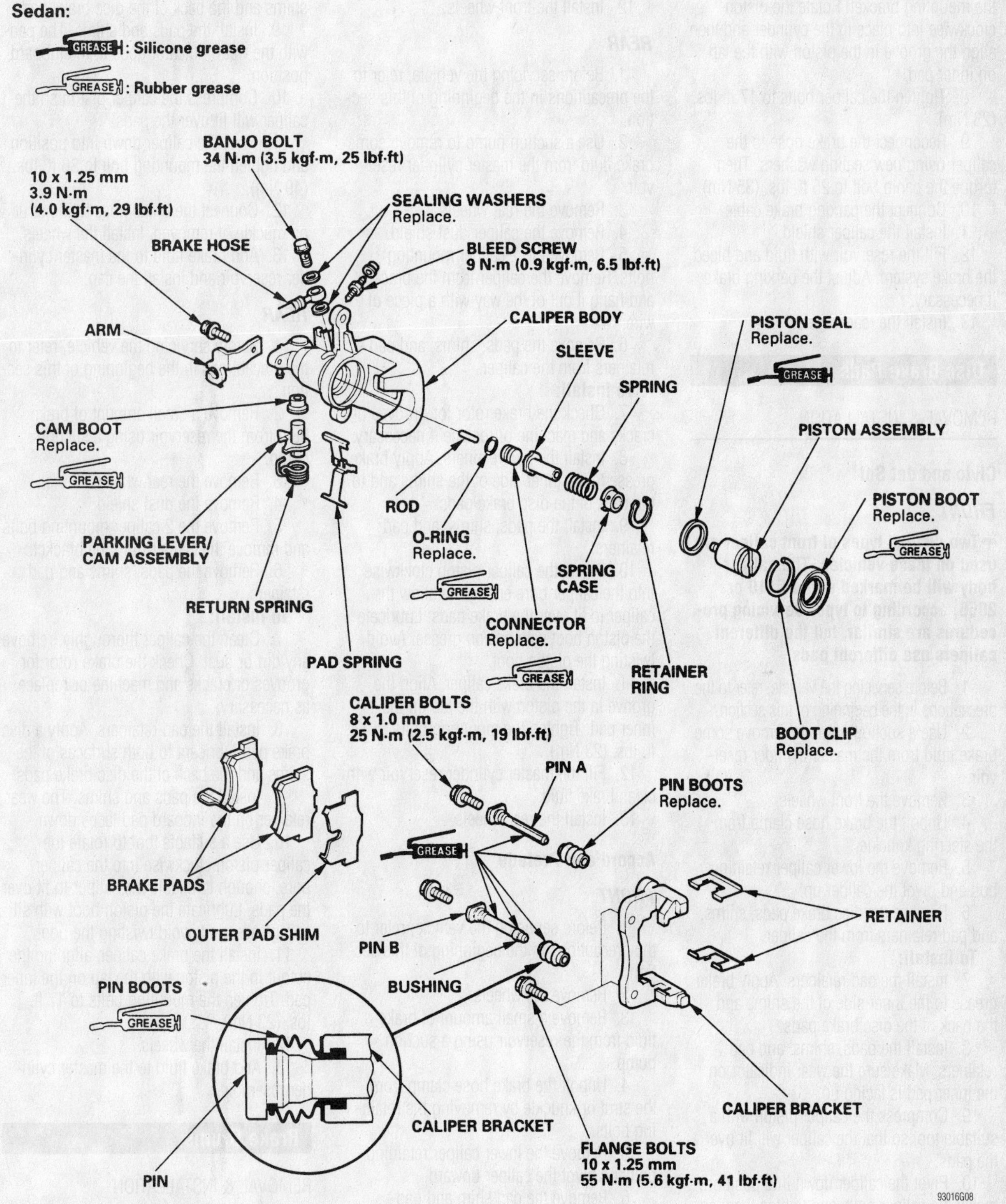

Sedan:

[GREASE]: Silicone grease

[GREASE]: Rubber grease

BANJO BOLT
34 N·m (3.5 kgf·m, 25 lbf·ft)

10 x 1.25 mm
3.9 N·m
(4.0 kgf·m, 29 lbf·ft)

SEALING WASHERS
Replace.

BRAKE HOSE

BLEED SCREW
9 N·m (0.9 kgf·m, 6.5 lbf·ft)

ARM

CALIPER BODY

SLEEVE

SPRING

PISTON SEAL
Replace.

[GREASE]

CAM BOOT
Replace.

[GREASE]

PISTON ASSEMBLY

PARKING LEVER/
CAM ASSEMBLY

ROD

O-RING
Replace.

[GREASE]

SPRING
CASE

PISTON BOOT
Replace.

[GREASE]

RETURN SPRING

CONNECTOR
Replace.

[GREASE]

RETAINER
RING

BOOT CLIP
Replace.

PAD SPRING

CALIPER BOLTS
8 x 1.0 mm
25 N·m (2.5 kgf·m, 19 lbf·ft)

PIN A

PIN BOOTS
Replace.

[GREASE]

BRAKE PADS

RETAINER

OUTER PAD SHIM

PIN B

PIN BOOTS

[GREASE]

BUSHING

CALIPER BRACKET

PIN

CALIPER BRACKET

FLANGE BOLTS
10 x 1.25 mm
55 N·m (5.6 kgf·m, 41 lbf·ft)

93016G08

Rear disc brakes—Accord Sedan shown

Timing belt service is covered in Section 3 of this manual

To install:

7. Fit the caliper over the pads and onto the mounting bracket. Rotate the piston clockwise into place in the cylinder and then align the groove in the piston with the tab on inner pad.

8. Tighten the caliper bolts to 17 ft. lbs. (23 Nm).

9. Reconnect the brake hose to the caliper using new sealing washers. Then torque the banjo bolt to 25 ft. lbs. (35 Nm).

10. Connect the parking brake cable.

11. Install the caliper shield.

12. Fill the reservoir with fluid and bleed the brake system. Adjust the parking brake if necessary.

13. Install the rear wheels.

Disc Brake Pads

REMOVAL & INSTALLATION

Civic and del Sol

FRONT

➡ **Two distinct types of front caliper are used on these vehicles. The caliper body will be marked either 5410 or 2056, according to type. Servicing procedures are similar, but the different calipers use different pads.**

1. Before servicing the vehicle, refer to the precautions in the beginning of this section.

2. Use a suction pump to remove some brake fluid from the master cylinder reservoir.

3. Remove the front wheels.

4. Unbolt the brake hose clamp from the steering knuckle.

5. Remove the lower caliper retaining bolt and pivot the caliper up.

6. Remove the disc brake pads, shims, and pad retainers from the caliper.

To install:

7. Install the pad retainers. Apply brake grease to the inner side of the shims and the back of the disc brake pads.

8. Install the pads, shims, and pad retainers. Make sure the wear indicator on the inner pad is facing up.

9. Compress the caliper piston with a suitable tool so that the caliper will fit over the pads.

10. Pivot the caliper down into position. Install caliper bolts and tighten them as follows:

- 5410 calipers: 24 ft. lbs. (33 Nm)
- 2056 calipers: 25 ft. lbs. (35 Nm) (top) and 20 ft. lbs. (27 Nm) (bottom)

11. Fill the reservoir with clean brake fluid.

12. Install the front wheels.

REAR

1. Before servicing the vehicle, refer to the precautions in the beginning of this section.

2. Use a suction pump to remove some brake fluid from the master cylinder reservoir.

3. Remove the rear wheels.

4. Remove the caliper dust shield.

5. Remove the 2 caliper mounting bolts. Remove the caliper from the bracket and hang it out of the way with a piece of wire.

6. Remove the pads, shims, and pad retainers from the caliper.

To install:

7. Check the brake rotor for grooves or cracks and machine or replace if necessary.

8. Install the pad retainers. Apply brake grease to the inner side of the shims and to the back of the disc brake pads.

9. Install the pads, shims, and pad retainers.

10. Rotate the caliper piston clockwise into the caliper bore enough to allow the caliper to fit over the brake pads. Lubricate the piston boot with silicon grease. Avoid twisting the piston boot.

11. Install the brake caliper. Align the groove in the piston with the tab on the inner pad. Tighten the mounting bolts to 17 ft. lbs. (23 Nm).

12. Fill the master cylinder reservoir with clean brake fluid.

13. Install the rear wheels.

Accord and Prelude

FRONT

1. Before servicing the vehicle, refer to the precautions in the beginning of this section.

2. Remove the wheels.

3. Remove a small amount of brake fluid from the reservoir using a suction pump.

4. Unbolt the brake hose clamp from the strut or knuckle by removing the retaining bolts.

5. Remove the lower caliper retaining bolt and pivot the caliper upward.

6. Remove the pad shim and pad retainers. Remove the disc brake pads from the caliper.

To install:

7. Check the brake rotor for grooves or cracks. If any heavy scoring is present, the rotor must be replaced.

8. Install the pad retainers. Apply a disc brake pad lubricant to both surfaces of the shims and the back of the disc brake pads.

9. Install the pads and shims. The pad with the wear indicator goes in the inboard position.

10. Compress the caliper piston so the caliper will fit over the pads.

11. Pivot the caliper down into position and tighten the mounting bolt to 36 ft. lbs. (49 Nm).

12. Connect the brake hose to the strut or knuckle, if removed. Install the wheels.

13. Add brake fluid to the master cylinder reservoir and install the cap.

REAR

1. Before servicing the vehicle, refer to the precautions in the beginning of this section.

2. Remove a small amount of brake fluid from the reservoir using a suction pump.

3. Remove the rear wheels.

4. Remove the dust shield.

5. Remove the 2 caliper mounting bolts and remove the caliper from the bracket.

6. Remove the pads, shims and pad retainers.

To install:

7. Clean the caliper thoroughly; remove any dirt or dust. Check the brake rotor for grooves or cracks and machine or replace, as necessary.

8. Install the pad retainers. Apply a disc brake pad lubricant to both surfaces of the shims and the back of the disc brake pads.

9. Install the pads and shims. The wear retainer on the inboard pad faces down.

10. Use a suitable tool to rotate the caliper piston clockwise into the caliper bore, enough to enable the caliper to fit over the pads. Lubricate the piston boot with silicon grease, and avoid twisting the boot.

11. Install the brake caliper, aligning the cutout in the piston with the tab on the inner pad. Tighten the mounting bolts to 17 ft. lbs. (23 Nm).

12. Install the wheels.

13. Add brake fluid to the master cylinder reservoir.

Brake Drums

REMOVAL & INSTALLATION

Accord, Civic and del Sol

1. Before servicing the vehicle, refer to the precautions in the beginning of this section.

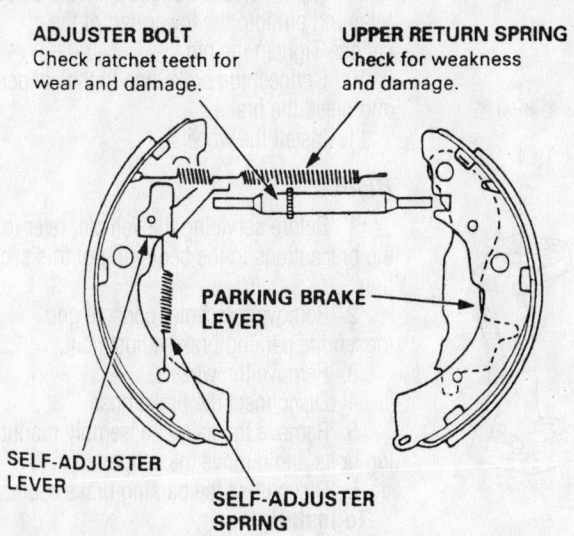

ADJUSTER BOLT
Check ratchet teeth for wear and damage.

UPPER RETURN SPRING
Check for weakness and damage.

PARKING BRAKE LEVER

SELF-ADJUSTER LEVER

SELF-ADJUSTER SPRING

Rear drum brakes—Civic showm

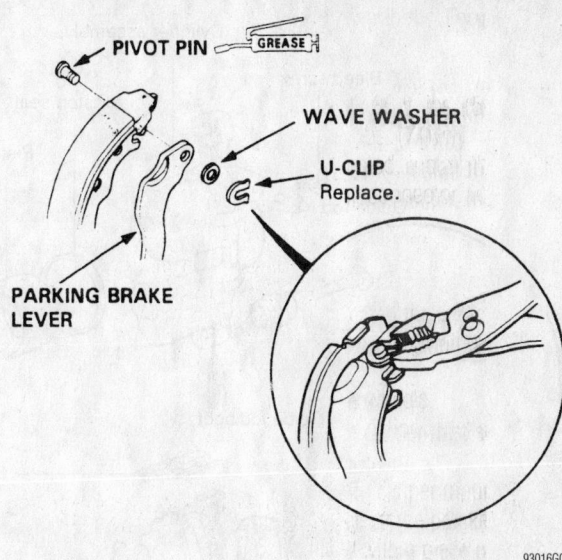

PIVOT PIN — GREASE

WAVE WASHER

U-CLIP Replace.

PARKING BRAKE LEVER

93016G09

2. Remove the rear wheels.
3. Remove the rear brake drum.
To install:
4. Make certain the brake shoes are adjusted to allow the drum clearance during installation. Fit the drum into position.
5. Install the rear wheels.

Brake Shoes

REMOVAL & INSTALLATION

Accord, Civic and del Sol

1. Before servicing the vehicle, refer to the precautions in the beginning of this section.
2. Remove the rear wheels and brake drums.
3. Disconnect the upper return spring from the brake shoes.
4. Push the retainer springs and turn

the tension pins to release the shoes from the backing plate.
5. Lower the brake shoe assembly and remove the lower return spring.
6. Remove the brake shoe assembly from the backing plate.
7. Disconnect the parking brake cable from the brake shoe lever.
8. Remove the upper return spring, self-adjuster lever, and self-adjuster spring. Separate the brake shoes.
9. Separate the wave washer, parking brake lever, and pivot pin from the brake shoe by removing the U-clip.
To install:
10. Apply brake cylinder grease to the sliding surface of the pivot pin and insert the pin into the brake shoe.
11. Install the parking brake lever and wave washer on the pivot pin and pinch the U-clip with a pair of pliers to secure the pivot pin to the shoe.

12. Connect the parking brake cable to the parking brake lever.
13. Hook the adjuster spring to the adjuster lever first, then to the brake shoe.
14. Install the adjuster bolt/clevis assembly and the upper return spring.
15. Install the brake shoes to the backing plate.
16. Install the lower return spring, the tension pins and retaining springs.
17. Connect the upper return spring.
18. Turn the adjuster bolt to force the brake shoes out until the brake drum will not easily go on. Back off the adjuster bolt just enough that the brake drum will go on and turn easily.
19. Install the wheels.
20. Depress the brake pedal several times to set the self-adjusting brake. Adjust the parking brake.

1998–01 HYUNDAI

Brake Caliper

REMOVAL & INSTALLATION

Accent

1. Before servicing the vehicle, refer to the precautions in the beginning of this section.
2. Remove the front wheels.
3. Loosen the brake line at the caliper and disconnect it.
4. Remove the brake pads.

5. Remove the pin and sleeve boots.
6. Remove the lower caliper bolt and raise the caliper up and out to remove it.
To install:
7. Position the caliper onto its mounting and install the lower mounting bolt. Torque the bolt to 16–24 ft. lbs. (22–32 Nm).
8. Install the pin boots, sleeve boots and brake pads.
9. Connect the brake line to the caliper with 2 new metal gaskets. Torque the brake

line union bolt to 18–22 ft. lbs. (25–30 Nm).
10. Bleed the system.
11. Mount the front wheels.

Elantra

FRONT

1. Before servicing the vehicle, refer to the precautions in the beginning of this section.
2. Remove the front wheels.

Heater Core replacement is covered in Section 2 of this manual

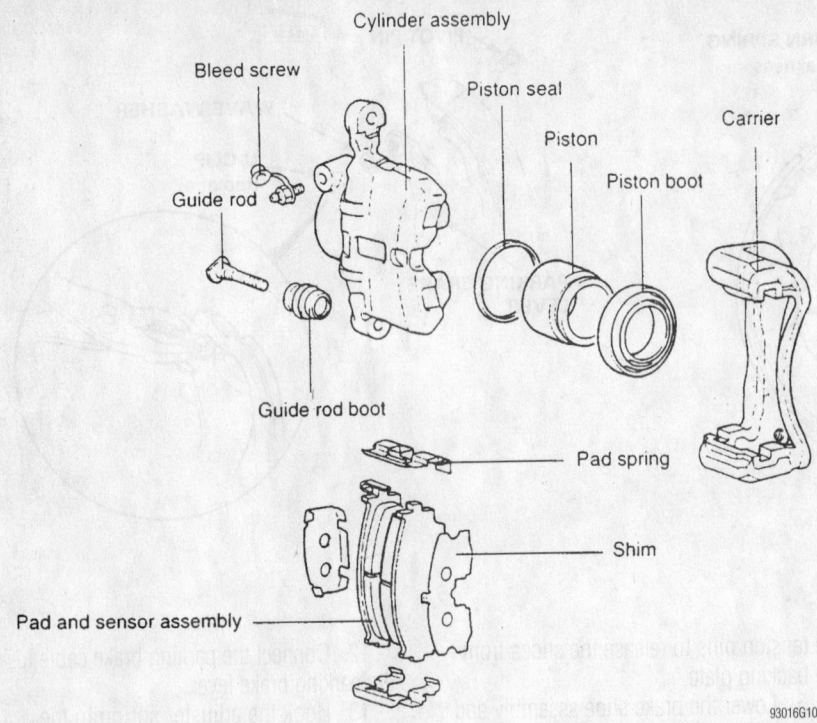

Cylinder assembly

Bleed screw

Piston seal

Piston

Carrier

Guide rod

Piston boot

Guide rod boot

Pad spring

Shim

Pad and sensor assembly

93016G10

Front caliper—Sonata shown

3. Loosen the brake line at the caliper and disconnect it.

4. Remove the brake pads.

5. Remove the pin and sleeve boots.

6. Remove the lower caliper bolt and raise the caliper up and out to remove it.

To install:

7. Position the caliper onto its mounting and install the lower mounting bolt. Torque the bolt to 16–24 ft. lbs. (22–32 Nm).

8. Install the pin boots, sleeve boots and brake pads.

9. Connect the brake line to the caliper with 2 new metal gaskets. Torque the brake line union bolt to 18–22 ft. lbs. (24–30 Nm).

10. Bleed the system.

11. Mount the front wheels.

REAR

1. Before servicing the vehicle, refer to the precautions in the beginning of this section.

2. Remove the center console and loosen the parking brake adjustment.

3. Remove the wheels.

4. Disconnect the brake hose.

5. Remove the caliper assembly mounting bolts and remove the caliper.

6. Disconnect the parking brake cable.

To install:

7. Connect the parking brake cable and

install the caliper. Tighten the mounting bolts to 16–23 ft. lbs. (22–32 Nm).

8. Connect the brake hose.

9. Fill the master cylinder with clean fluid and bleed the hydraulic system.

10. Install the wheel.

11. Adjust the parking brake.

12. Install the center console.

Sonata

FRONT

1. Before servicing the vehicle, refer to the precautions in the beginning of this section.

2. Remove the front wheels.

3. Disconnect the brake tube from the brake hose. Release the brake hose clip and remove the brake hose from the strut.

4. Loosen and disconnect the brake line from the caliper.

5. Remove the small retaining pin from the lower part of the caliper.

6. Swing the caliper up until it clears the rotor and pads.

7. Slide the caliper inboard until the locating pin disengages from its groove in the caliper. Pull the caliper from the locating pin.

To install:

8. Lubricate the locating pin bore with white silicone compound and mount the caliper onto the locating pin.

9. Lower the caliper until the small

retaining pin holes are aligned. Install a new retaining pin into the lower part of the caliper. Tighten the pin.

10. Connect the brake line to the caliper and bleed the brakes.

11. Install the wheels.

REAR

1. Before servicing the vehicle, refer to the precautions in the beginning of this section.

2. Remove the center console and loosen the parking brake adjustment.

3. Remove the wheels.

4. Disconnect the brake hose.

5. Remove the caliper assembly mounting bolts and remove the caliper.

6. Disconnect the parking brake cable.

To install:

7. Connect the parking brake cable and install the caliper. Tighten the mounting bolts to 16–23 ft. lbs. (22–32 Nm).

8. Connect the brake hose.

9. Fill the master cylinder with clean fluid and bleed the hydraulic system.

10. Install the wheel.

11. Adjust the parking brake.

12. Install the center console.

Tiburon

FRONT

1. Before servicing the vehicle, refer to the precautions in the beginning of this section.

2. Remove the wheels.

3. Disconnect the brake hose.

4. Unbolt and remove the caliper.

5. Unbolt and remove the caliper support.

To install:

6. Install the caliper support and tighten the bolts to 44–63 ft. lbs. (69–85 Nm).

7. Install the caliper. Tighten the guide rod bolts to 16–24 ft. lbs. (22–32 Nm).

8. Connect the brake hose and tighten to 18–22 ft. lbs. (25–30 Nm).

9. Install the wheels.

10. Fill the master cylinder with clean brake fluid and bleed the hydraulic system.

REAR

1. Before servicing the vehicle, refer to the precautions in the beginning of this section.

2. Remove the wheels.

3. Disconnect the brake hose.

4. Unbolt and remove the caliper.

5. Disconnect the parking brake cable.

To install:

6. Connect the parking brake cable and

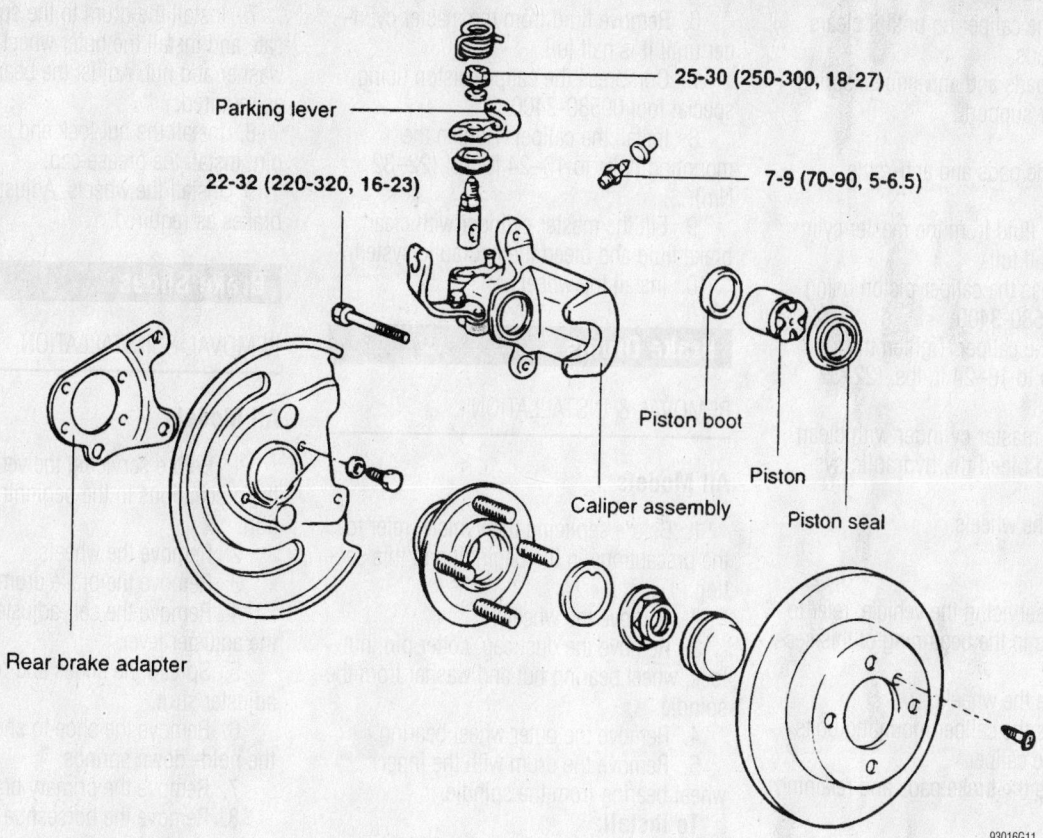

25-30 (250-300, 18-27)

Parking lever

22-32 (220-320, 16-23)

7-9 (70-90, 5-6.5)

Piston boot

Piston

Caliper assembly

Piston seal

Rear brake adapter

93016G11

Rear caliper—Elantra shown

install the caliper. Tighten the mounting bolts to 16–24 ft. lbs. (22–32 Nm).

7. Connect the brake hose.

8. Fill the master cylinder with clean brake fluid and bleed the hydraulic system.

9. Install the wheels.

Disc Brake Pads

REMOVAL & INSTALLATION

Accent

1. Before servicing the vehicle, refer to the precautions in the beginning of this section.

2. Remove the front wheels.

3. Remove the lower caliper mounting bolt and rotate the caliper upward.

4. Remove the pads from the caliper support.

To install:

5. Install the pad clips.

6. Install the pads onto the pad clips.

7. Compress the caliper piston using a C-clamp.

8. Rotate the caliper downward and install the mounting bolt.

9. Install the wheels.

Elantra

FRONT

1. Before servicing the vehicle, refer to the precautions in the beginning of this section.

2. Remove the front wheels.

3. Remove the lower caliper mounting bolt and rotate the caliper upward.

4. Remove the pads from the caliper support.

To install:

5. Install the pad clips.

6. Install the pads onto the pad clips.

7. Compress the caliper piston using a C-clamp.

8. Rotate the caliper downward and install the mounting bolt.

9. Install the wheels.

REAR

1. Before servicing the vehicle, refer to the precautions in the beginning of this section.

2. Remove the wheels.

3. Remove the caliper mounting bolts and remove the caliper.

4. Remove the brake pads and retaining clips.

To install:

5. Install new pads and retainers.

6. Remove fluid from the master cylinder until it is half full.

7. Compress the caliper piston using special tool 09580-3400.

8. Install the caliper. Tighten the mounting bolts to 16–24 ft. lbs. (22–32 Nm).

9. Fill the master cylinder with clean brake fluid and bleed the hydraulic system.

10. Install the wheels.

Sonata

FRONT

1. Before servicing the vehicle, refer to the precautions in the beginning of this section.

2. Remove the front wheels.

3. Remove the small retaining pin from the lower part of the caliper.

4. Swing the caliper up until it clears the rotor and pads.

5. Lift the pads and anti-squeal spring from the caliper support.

To install:

6. Install the pads and anti-rattle spring.

7. Remove fluid from the master cylinder until it is half full.

8. Compress the caliper piston using special tool 09580-3400.

9. Install the caliper. Tighten the mounting bolts to 16–24 ft. lbs. (22–32 Nm).

10. Fill the master cylinder with clean brake fluid and bleed the hydraulic system.

11. Install the wheels.

REAR

1. Before servicing the vehicle, refer to the precautions in the beginning of this section.

2. Remove the wheels.

3. Remove the caliper mounting bolts and remove the caliper.

4. Remove the brake pads and retaining clips.

To install:

5. Install new pads and retainers.

---

6. Remove fluid from the master cylinder until it is half full.

7. Compress the caliper piston using special tool 09580-3400.

8. Install the caliper. Tighten the mounting bolts to 16–24 ft. lbs. (22–32 Nm).

9. Fill the master cylinder with clean brake fluid and bleed the hydraulic system.

10. Install the wheels.

Brake Drums

REMOVAL & INSTALLATION

All Models

1. Before servicing the vehicle, refer to the precautions in the beginning of this section.

2. Remove the wheels.

3. Remove the dust cap, cotter pin, nut lock, wheel bearing nut and washer from the spindle.

4. Remove the outer wheel bearing.

5. Remove the drum with the inner wheel bearing from the spindle.

To install:

6. Lubricate and install the inner wheel bearing. Install a new grease seal.

---

7. Install the drum to the spindle. Lubricate and install the outer wheel bearing, washer and nut. Adjust the bearing preload as required.

8. Install the nut lock and a new cotter pin. Install the grease cap.

9. Install the wheels. Adjust the rear brakes as required.

Brake Shoes

REMOVAL & INSTALLATION

All Models

1. Before servicing the vehicle, refer to the precautions in the beginning of this section.

2. Remove the wheels.

3. Remove the brake drum.

4. Remove the self-adjuster spring and the adjuster lever.

5. Spread the shoes and remove the adjuster strut.

6. Remove the shoe to shoe spring and the hold-down springs.

7. Remove the primary brake shoe.

8. Remove the horseshoe clip and the parking brake lever from the secondary brake shoe.

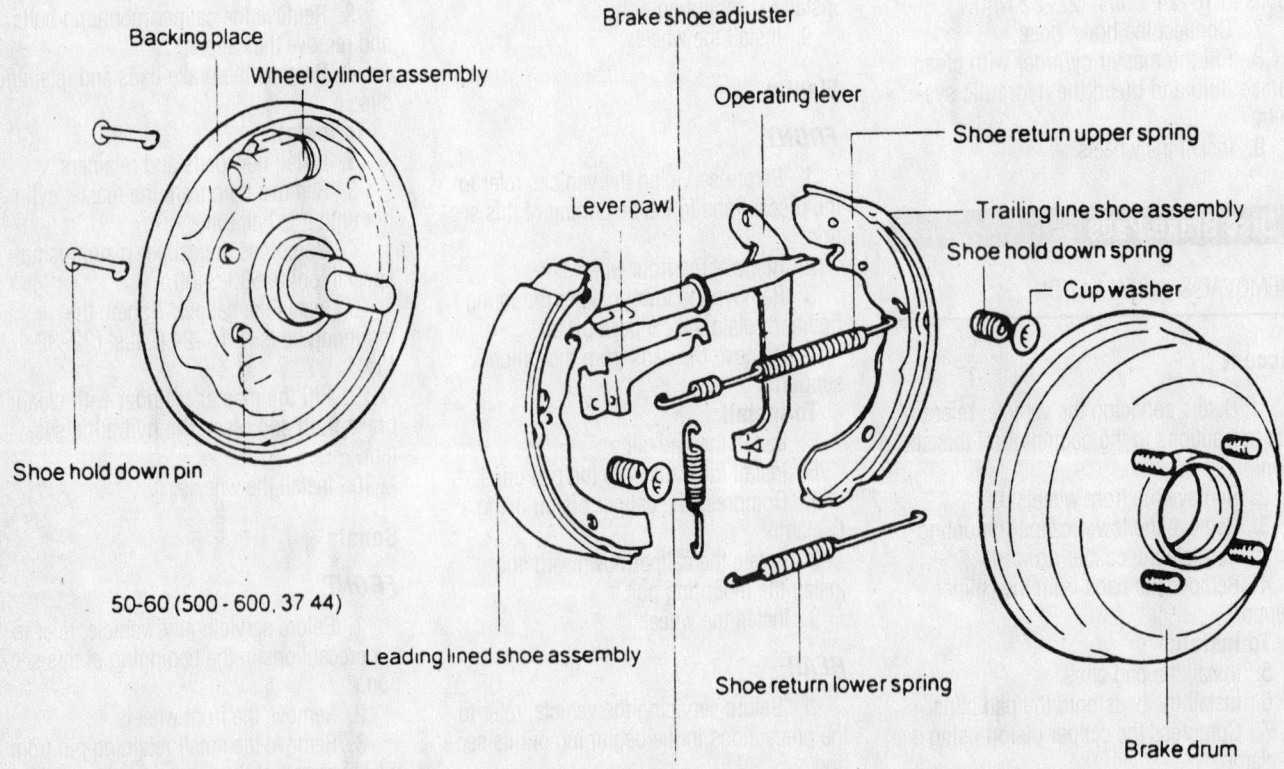

50-60 (500 - 600, 37 44)

Rear drum brakes—Accent shown

93016G12

To install:

9. Clean the backing plate with brake cleaning solvent.

10. Apply a light coating of lithium grease to the friction points on the backing plate.

11. Position the primary shoe on the backing plate and install the hold down spring and pin.

12. Install the parking brake lever to the secondary shoe and install the shoe to the backing plate.

13. Install the adjuster strut assembly and the adjuster lever and spring.

14. Install the lower shoe to shoe spring.

15. Install the brake drum and the wheel.

16. Adjust the brake shoes.

1998–00 INFINITI

Brake Caliper

REMOVAL & INSTALLATION

G20 Models

FRONT

1. Before servicing the vehicle, refer to the precautions in the beginning of this section.

2. Remove the wheels.

3. Loosen the brake hose connecting bolt.

4. Remove the bolts connecting the caliper to the torque member.

5. Slide the caliper out from the rotor.

6. Remove the brake hose connecting bolt from the caliper.

7. Remove the caliper from the vehicle.

To install:

8. Fit the caliper onto the torque member and torque the bolts to 16–23 ft. lbs. (22–31 Nm).

9. Using new copper washers, connect the hydraulic hose to the caliper. Torque the union bolt to 12–14 ft. lbs. (17–19 Nm).

10. Bleed the air from the system.

REAR

1. Before servicing the vehicle, refer to the precautions in the beginning of this section.

2. Remove the rear wheels.

3. Remove the brake cable mounting bolt and lock spring.

4. Disconnect the parking brake cable from the caliper.

5. Disconnect the brake fluid hose.

6. Remove the torque member mounting bolts and remove the caliper assembly.

To install:

7. Fit the caliper onto the torque member and torque the bolts to 16–23 ft. lbs. (22–31 Nm).

8. Using new copper washers, connect the hydraulic hose to the caliper. Torque the union bolt to 12–14 ft. lbs. (17–19 Nm).

9. Connect the parking brake cable to the rear caliper.

10. Bleed the brake system.

11. Install the rear wheels.

J30 Models

FRONT

1. Before servicing the vehicle, refer to the precautions in the beginning of this section.

2. Remove the wheels.

3. Place a drain pan under the caliper and loosen the brake hose connecting bolt.

4. Remove the torque member mounting bolts and disconnect the brake fluid hose from the caliper.

5. Slide the caliper off of the rotor.

6. Remove the caliper from the vehicle.

To install:

7. Position the torque member on the knuckle assembly and install the mounting bolts. Torque the bolts to 53–72 ft. lbs. (72–97 Nm).

8. Using new copper washers, connect the hydraulic hose to the caliper. Torque the union bolt to 12–14 ft. lbs. (17–19 Nm).

9. Bleed the air from the system.

REAR

1. Before servicing the vehicle, refer to the precautions in the beginning of this section.

2. Remove the rear wheels.

3. Disconnect the brake fluid hose.

4. Remove the torque member mounting bolts and remove the caliper assembly.

To install:

5. Fit the caliper over the rotor. Install the mounting bolts. Torque the bolts to 28–38 ft. lbs. (38–52 Nm).

6. Using new copper washers, connect the hydraulic hose to the caliper. Torque the union bolt to 12–14 ft. lbs. (17–19 Nm).

7. Bleed the brake system.

8. Install the rear wheels.

I30 Models

FRONT

1. Before servicing the vehicle, refer to the precautions in the beginning of this section.

2. Remove the front wheels.

3. Remove both guide pin bolts securing the caliper to the steering knuckle.

4. Loosen and remove the brake hose connector from the caliper.

5. Remove the caliper assembly from the vehicle.

To install:

6. Using new copper washers, install the brake line to the brake caliper and torque the connecting bolt to 12–14 ft. lbs. (17–20 Nm).

7. Install the caliper to the steering knuckle using the guide pins bolts.

8. Install the wheels and tighten the lug nuts to the proper specification.

9. Bleed the brake system and top off the master cylinder as necessary.

REAR

1. Before servicing the vehicle, refer to the precautions in the beginning of this section.

2. Remove the rear wheels.

3. Remove the parking brake cable stay fixing bolt and the lock spring.

4. Remove the brake fluid hose from the caliper.

5. Remove the guide pin bolts and remove the caliper.

To install:

6. Install the caliper body into position and torque the caliper-to-torque member pin bolts to 16–23 ft. lbs. (22–31 Nm).

7. Reconnect the brake fluid hose and tighten the flare nut to 12–14 ft. lbs. (17–20 Nm).

8. Install the lock spring and the parking brake stay attaching bolt.

9. Bleed the brake system and top off the master cylinder as necessary.

10. Install the wheels.

Q45 Models

FRONT

1. Before servicing the vehicle, refer to the precautions in the beginning of this section.

2. Remove the wheels.

3. Remove the brake hose connecting bolt.

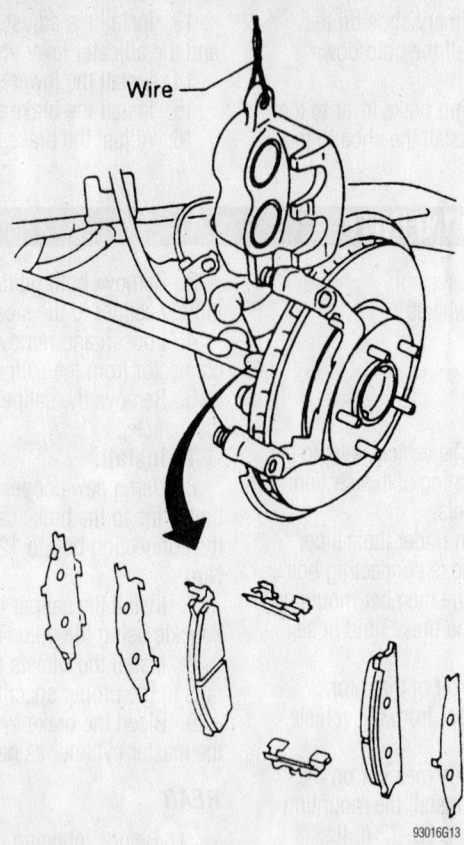

Front brake caliper—Q45 shown

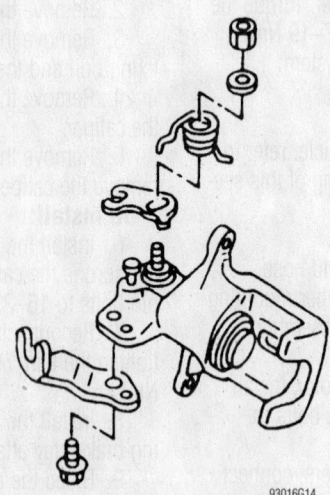

Rear caliper parking brake lever—G20 shown

4. Remove the torque member mounting bolts and disconnect the brake fluid hose from the caliper.

5. Slide the caliper off of the rotor.

6. Remove the caliper from the vehicle.

To install:

7. Position the torque member on the knuckle assembly and install the mounting bolts. Torque the bolts to 118–137 ft. lbs. (160–186 Nm).

8. Using new copper washers, connect

the hydraulic hose to the caliper. Torque the union bolt to 12–14 ft. lbs. (17–19 Nm).

9. Bleed the air from the system and fill the master cylinder with clean brake fluid.

REAR

1. Before servicing the vehicle, refer to the precautions in the beginning of this section.

2. Remove the rear wheels.

3. Disconnect the brake fluid hose.

4. Remove the torque member mounting bolts and remove the caliper assembly.

To install:

5. Fit the caliper over the rotor and install the mounting bolts. Torque the bolts to 28–38 ft. lbs. (38–52 Nm).

6. Using new copper washers, connect the hydraulic hose to the caliper. Torque the union bolt to 12–14 ft. lbs. (17–19 Nm).

7. Bleed the brake system.

8. Install the rear wheels and lower the vehicle to the floor.

Disc Brake Pads

REMOVAL & INSTALLATION

G20 Models

FRONT

1. Before servicing the vehicle, refer to the precautions in the beginning of this section.

2. Remove the cap from the master cylinder reservoir and extract about ⅓ of the brake fluid from the reservoir to prevent overflow when the caliper piston is compressed.

3. Remove the wheels.

4. Remove the lower pin bolt.

5. Pivot the caliper body upward and secure it with a length of wire. Remove the retainers and inner and outer shims and pads.

To install:

6. Place an old pad over the caliper piston. Use a C-clamp to compress the piston.

7. Install the new pads and shims and rotate caliper down onto rotor. Install the pin bolt and torque it to 16–23 ft. lbs. (22–31 Nm).

8. Install the wheels and lower the vehicle to the floor.

9. Check and then refill the master cylinder if needed.

REAR

1. Before servicing the vehicle, refer to the precautions in the beginning of this section.

2. Remove the cap from the master cylinder reservoir and extract about ⅓ of the brake fluid from the reservoir to prevent overflow when the caliper piston is compressed.

3. Remove the wheels.

4. Remove the brake cable mounting bracket bolt and lock spring.

5. Disconnect the parking brake cable.

6. Remove the lower pin bolt.

7. Pivot the caliper body upward and secure it with a length of wire. Remove the retainers and inner and outer shims and pads.

To install:

8. Push the piston into the cylinder body by turning the piston clockwise.

9. Install the new pads and shims and rotate the caliper down onto rotor. Install the pin bolt and torque it to 16–23 ft. lbs. (22–31 Nm).

10. Connect the parking brake cable and install the bracket.

11. Install the wheels.

12. Check and then refill the master cylinder if needed.

J30 Models

FRONT AND REAR

1. Before servicing the vehicle, refer to the precautions in the beginning of this section.

2. Remove the cap from the master cylinder reservoir and extract about ⅓ of the brake fluid from the reservoir to prevent overflow when the caliper piston is compressed.

3. Remove the wheels.

4. Remove the lower pin bolt.

5. Pivot the caliper body upward and secure it with a length of wire. Remove the retainers and inner and outer shims and pads.

To install:

6. Place an old pad over the caliper piston. Use a C-clamp to compress the piston.

7. Install the new pads and shims and rotate caliper down onto rotor. Install the pin bolt and torque it to 16–23 ft. lbs. (22–31 Nm) for the front caliper, and 28–38 ft. lbs. (38–52 Nm) for the rear caliper.

8. Install the wheels.

9. Check and then refill the master cylinder if needed.

I30 Models

FRONT

1. Before servicing the vehicle, refer to the precautions in the beginning of this section.

2. Remove the wheels.

3. Remove the bottom guide pin from the caliper and swing the caliper cylinder body upward. Support the caliper with a wire.

4. Remove the brake pad retainers and the pads.

To install:

5. Compress the piston of the disc brake caliper.

6. Install the brake pads and caliper assembly. Torque the guide pin to 16–23 ft. lbs. (22–31 Nm).

7. Install the wheels.

8. Check the master cylinder and add fluid if necessary.

REAR

1. Before servicing the vehicle, refer to the precautions in the beginning of this section.

2. Remove the rear wheels.

3. Remove the parking brake cable mounting bolt and lock spring.

4. Disconnect the cable from the caliper.

5. Remove the upper pin bolt.

6. Pivot the caliper body downward.

7. Pull out the pad springs and then remove the pads and shims.

To install:

8. Turn the piston clockwise back into the caliper body. Take care not to damage the piston boot.

9. Coat the pad contact area on the mounting support with grease.

10. Install the pads, shims, and the pad springs.

11. Position the caliper body in the mounting support and tighten the pin bolts to 16–23 ft. lbs. (22–31 Nm).

12. Install the wheels.

13. Check the master cylinder and add fluid if necessary.

Q45 Models

FRONT AND REAR

1. Before servicing the vehicle, refer to the precautions in the beginning of this section.

2. Remove the cap from the master cylinder reservoir and extract about ⅓ of the brake fluid from the reservoir to prevent overflow when the caliper piston is compressed.

3. Remove the wheels.

4. Remove the lower pin bolt.

5. Pivot the caliper body upward and secure it with a length of wire. Remove the retainers and inner and outer shims and pads.

To install:

6. Place an old pad over the caliper piston. Use a C-clamp to compress the piston.

7. Install the new pads and shims and rotate caliper down onto rotor. Install the pin bolt and torque it to 61–69 ft. lbs. (83–93 Nm) for the front caliper and 23–30 ft. lbs. (31–41 Nm) for the rear caliper.

8. Install the wheels.

9. Check and refill master cylinder if needed.

1998–00 KIA

Brake Caliper

REMOVAL & INSTALLATION

Front

1. Before servicing the vehicle, refer to the precautions in the beginning of this section.

2. Remove the wheels.

3. Disconnect the brake hose from the caliper.

4. Remove the caliper guide pin bolts and remove the caliper.

To install:

5. Install the caliper and tighten the guide pin bolts to 19–21 ft. lbs. (26–28 Nm).

6. Install the brake hose and tighten to 9–13 ft. lbs. (13–18 Nm).

7. Bleed the brakes and fill the master cylinder with clean brake fluid.

8. Install the wheels.

Rear

1. Before servicing the vehicle, refer to the precautions in the beginning of this section.

2. Remove the wheels.

3. Remove the parking brake cable and clip.

4. Remove the brake hose banjo bolt.

5. Remove the caliper lock bolts and remove the caliper.

To install:

6. Install the caliper and tighten the lock bolts to 22–29 ft. lbs. (29–39 Nm).

7. Install the brake hose and tighten the banjo bolt to 16–22 ft. lbs. (22–29 Nm).

8. Install the park brake cable and clip.

9. Bleed the brakes and fill the master cylinder with clean brake fluid.

10. Install the wheels.

For Accessory Drive Belt illustrations, see Section 1 of this manual

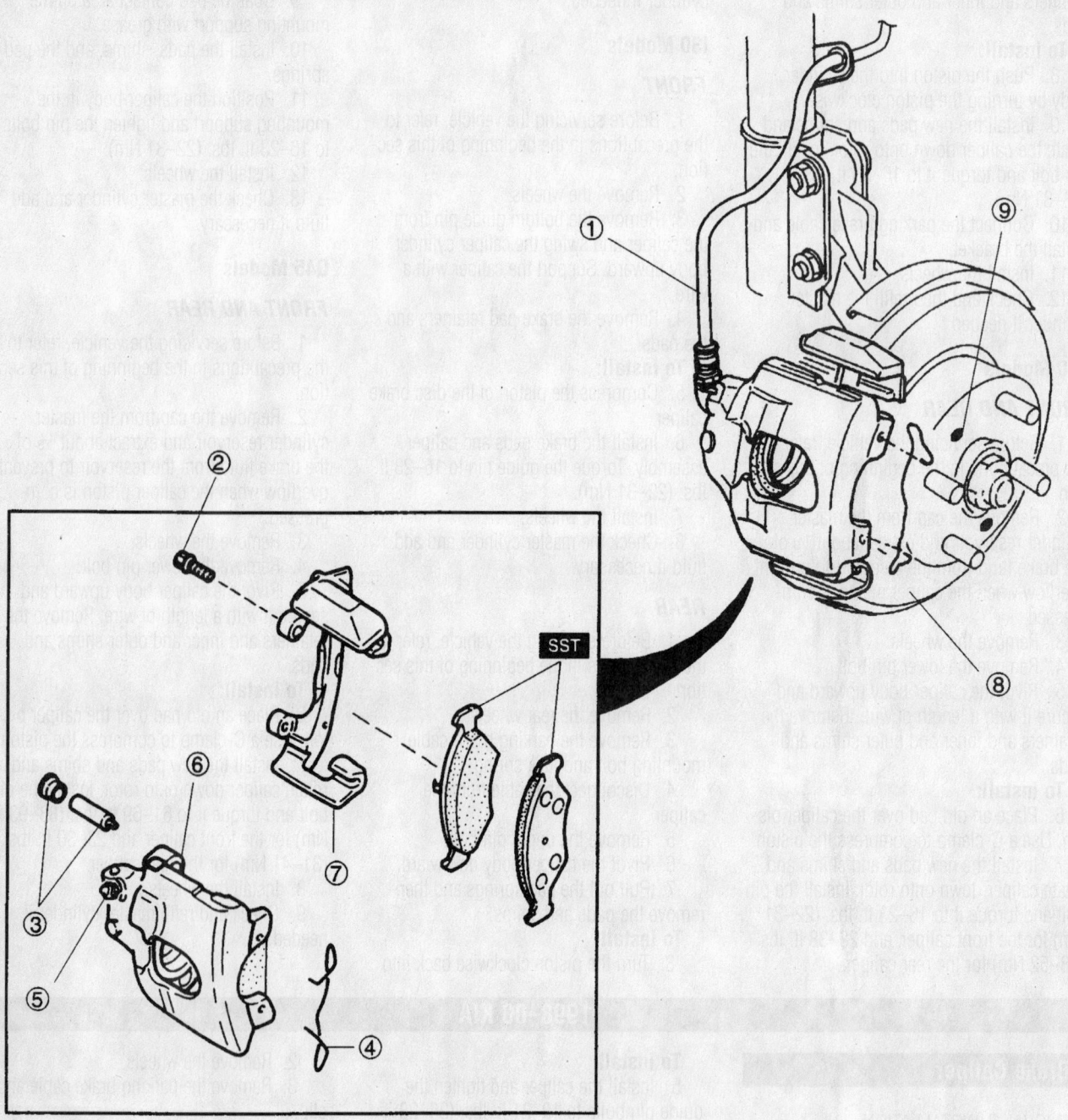

(1) Flexible hose
(2) Bolt
(3) Cap, bolt (square head) and bushing
(4) Spring
(5) Cap and bleeder screw
(6) Caliper
(7) Supporting plate
(8) Brake rotor (Disc)
(9) Mounting screws

93016G15

Exploded view of the front brakes—Kia

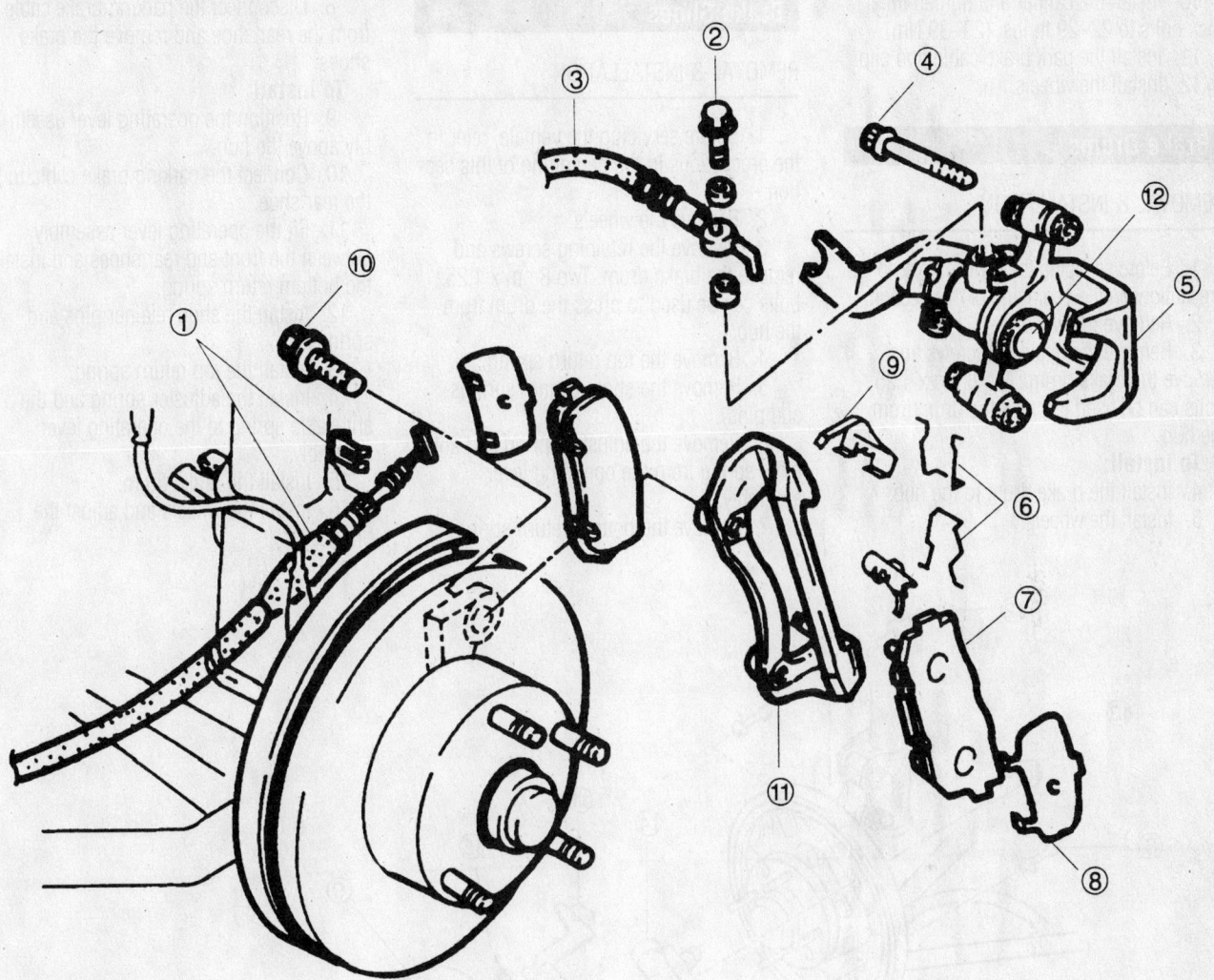

(1) Parking cable, clip
(2) Connecting bolt
(3) Brake hose
(4) Lock bolt

(5) Caliper
(6) V-spring
(7) Disc pad
(8) Shim

(9) Guide plate
(10) Bolt
(11) Mounting support
(12) Caliper piston

93016G16

Exploded view of the rear disc brakes—Kia

Disc Brake Pads

REMOVAL & INSTALLATION

Front

1. Before servicing the vehicle, refer to the precautions in the beginning of this section.
2. Remove the wheels.
3. Remove the caliper guide pin bolts and lift the caliper away from the rotor.
4. Remove the brake pads and retainer spring from the caliper.

To install:
5. Compress the caliper piston into the bore.
6. Install the brake pads and retainer spring to the caliper.
7. Position the caliper on the caliper support bracket and install the guide pin bolts.
8. Install the wheels.

Rear

1. Before servicing the vehicle, refer to the precautions in the beginning of this section.
2. Remove the wheels.
3. Remove the parking brake cable and clip.
4. Remove the caliper lock bolts and remove the caliper.
5. Remove the V-springs from the brake pads.
6. Remove the brake pads and shims.
To install:
7. Compress the caliper piston into the bore by rotating the piston with special tool OK9A4 263 001.
8. Install the new pads and shims.
9. Install the V-springs.

For Tire, Wheel and Ball Joint specifications, see Section 1 of this manual

10. Install the caliper and tighten the lock bolts to 22–29 ft. lbs. (29–39 Nm).

11. Install the park brake cable and clip.

12. Install the wheels.

Brake Drums

REMOVAL & INSTALLATION

1. Before servicing the vehicle, refer to the precautions in the beginning of this section.

2. Remove the wheels.

3. Remove the retaining screws and remove the brake drum. Two 8mm x 1.25 bolts can be used to press the drum from the hub.

To install:

4. Install the brake drum to the hub.

5. Install the wheels.

Brake Shoes

REMOVAL & INSTALLATION

1. Before servicing the vehicle, refer to the precautions in the beginning of this section.

2. Remove the wheels.

3. Remove the retaining screws and remove the brake drum. Two 8mm x 1.25 bolts can be used to press the drum from the hub.

4. Remove the top return spring.

5. Remove the shoe retainer springs and pins.

6. Remove the adjuster spring and anti-rattle spring from the operating lever assembly.

7. Remove the bottom return spring.

8. Disconnect the parking brake cable from the rear shoe and remove the brake shoes.

To install:

9. Position the operating lever assembly above the hub.

10. Connect the parking brake cable to the rear shoe.

11. Fit the operating lever assembly between the front and rear shoes and install the bottom return spring.

12. Install the shoe retainer pins and springs.

13. Install the top return spring.

14. Install the adjuster spring and the anti-rattle spring to the operating lever assembly.

15. Install the brake drum.

16. Install the wheels and adjust the brakes.

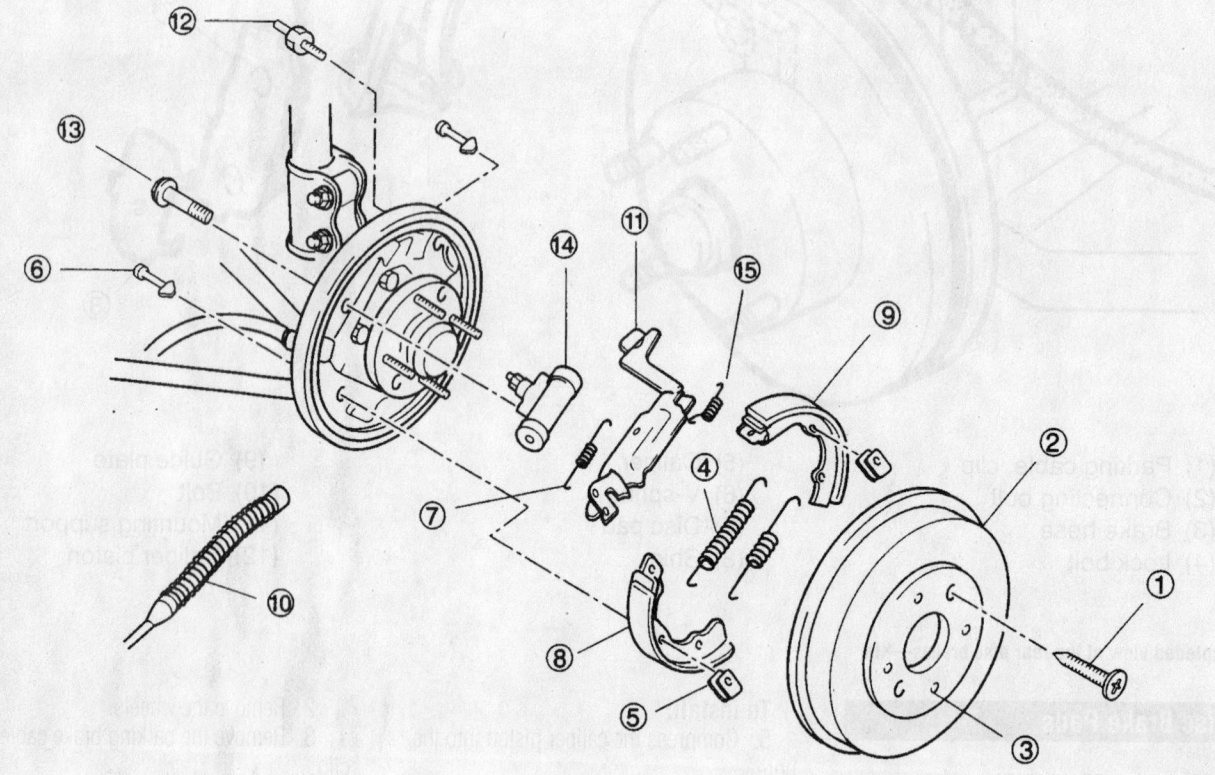

(1) Mounting screws
(2) Brake drum
(3) Drum pulling threads
(4) Return springs
(5) Spring clips
(6) Hold down pins
(7) Adjuster spring
(8) Brake shoe-leading

(9) Brake shoe-trailing
(10) Parking brake cable
(11) Operating lever assembly
(12) Brake line
(13) Bolts
(14) Wheel cylinder assembly
(15) Anti-rattle spring

93016G17

Exploded view of the rear drum brakes—Kia

Brake Caliper

REMOVAL & INSTALLATION

ES300

FRONT AND REAR

1. Before servicing the vehicle, refer to the precautions in the beginning of this section.
2. Remove the wheels.

3. Disconnect the brake hose from the caliper.
4. Remove the bolts that attach the caliper to the torque plate.
5. Lift the bottom of the caliper up and remove the caliper assembly.

To install:

6. Grease the caliper slides and bolts with lithium grease. Install the caliper. Torque the bolts to 25 ft. lbs. (34 Nm).
7. Reconnect the brake hose to the

caliper using 2 new washers. Torque the union bolt to 21 ft. lbs. (29 Nm).
8. Fill the master cylinder to the proper level and bleed the brake system.

GS300 and GS400

FRONT AND REAR

1. Before servicing the vehicle, refer to the precautions in the beginning of this section.

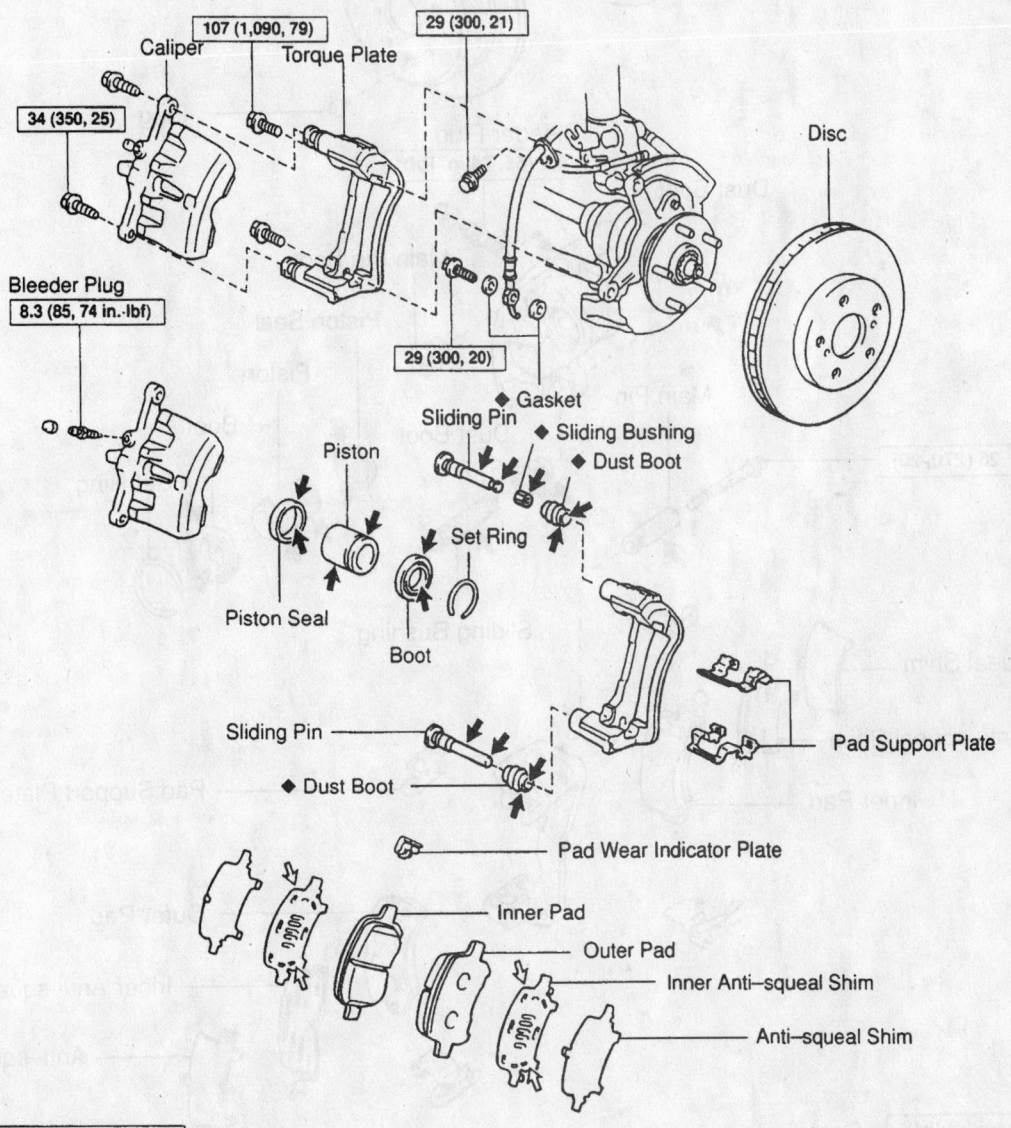

N·m (kgf·cm, ft·lbf) : Specified torque
◆ Non–reusable part
← Lithium soap base glycol grease
⇐ Disc brake grease

93016G18

Front disc brakes—ES300

For Wheel Alignment specifications, see Section 1 of this manual

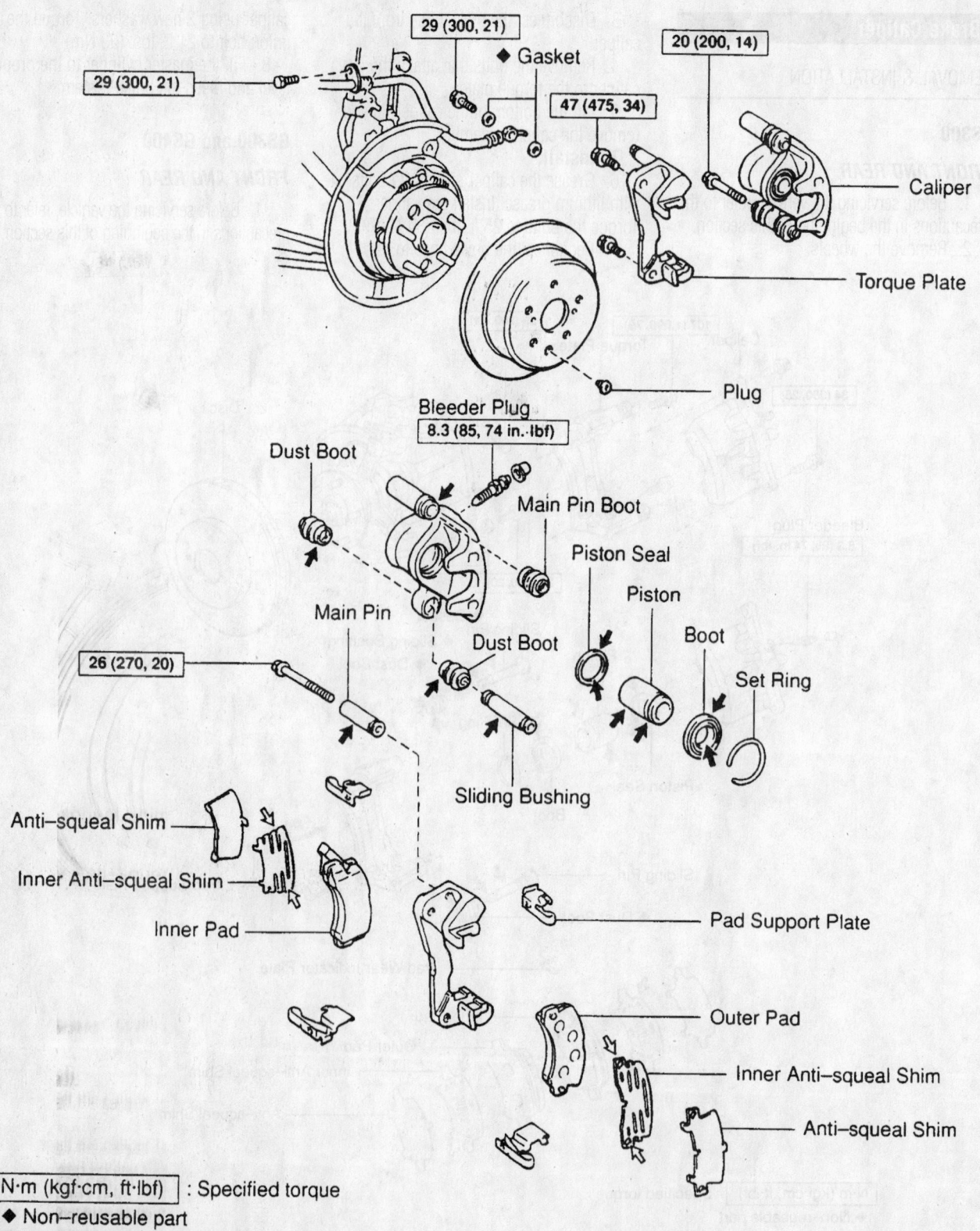

29 (300, 21)

29 (300, 21)

◆ Gasket

20 (200, 14)

47 (475, 34)

Caliper

Torque Plate

Plug

Bleeder Plug
8.3 (85, 74 in.·lbf)

Dust Boot

Main Pin Boot

Piston Seal

Piston

Boot

Main Pin

Set Ring

26 (270, 20)

Dust Boot

Sliding Bushing

Anti–squeal Shim

Inner Anti–squeal Shim

Inner Pad

Pad Support Plate

Outer Pad

Inner Anti–squeal Shim

Anti–squeal Shim

N·m (kgf·cm, ft·lbf) : Specified torque

◆ Non–reusable part

◀ Lithium soap base glycol grease

◁ Disc brake grease

93016G19

Rear disc brakes—ES300

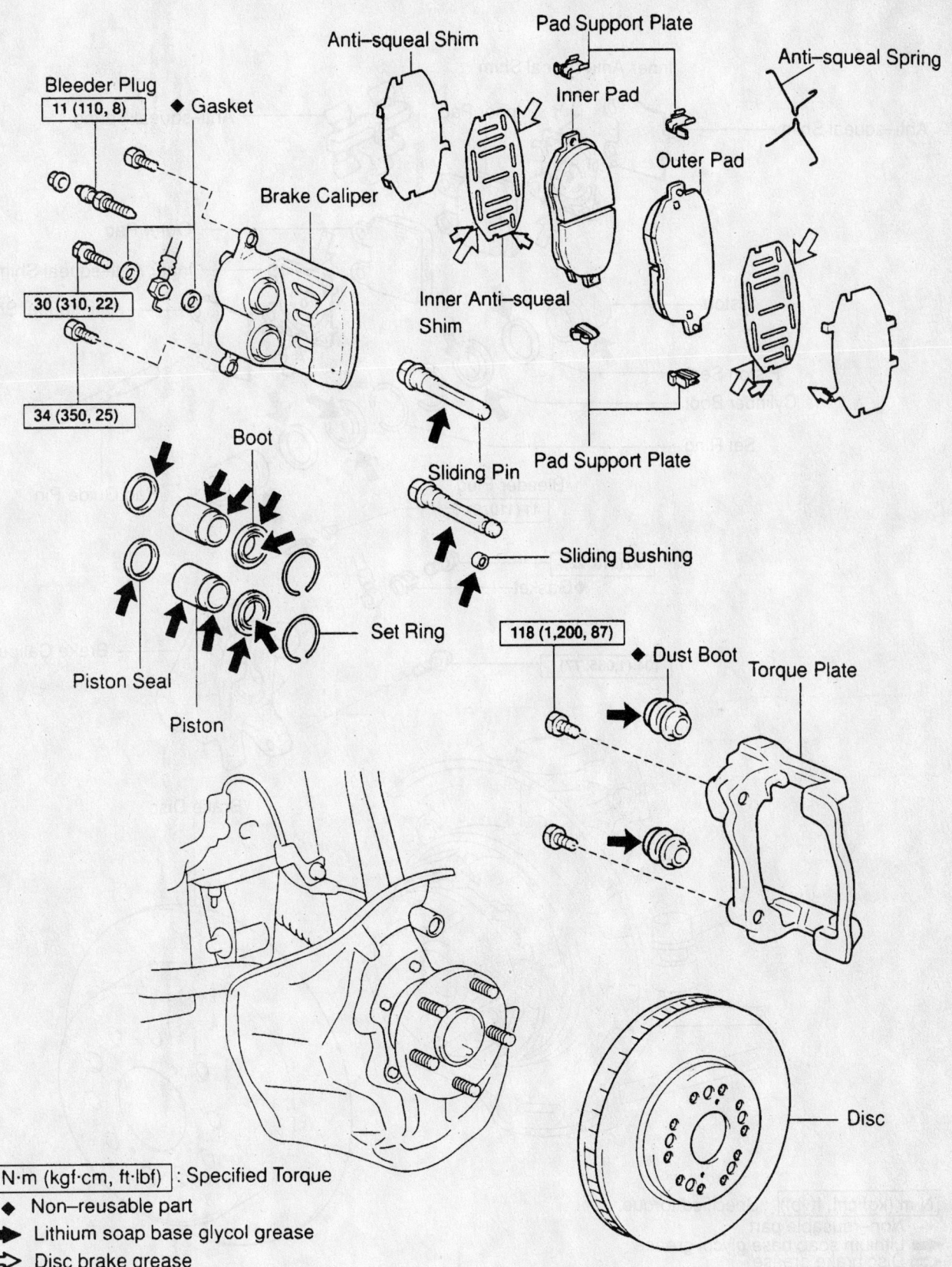

Bleeder Plug
11 (110, 8)
◆ Gasket

Anti-squeal Shim

Pad Support Plate

Inner Pad

Anti-squeal Spring

Outer Pad

Brake Caliper

30 (310, 22)

34 (350, 25)

Inner Anti-squeal Shim

Pad Support Plate

Boot

Sliding Pin

Sliding Bushing

Piston Seal

Set Ring

118 (1,200, 87)

◆ Dust Boot

Torque Plate

Piston

Disc

N·m (kgf·cm, ft·lbf) : Specified Torque

◆ Non-reusable part

➡ Lithium soap base glycol grease

⇨ Disc brake grease

93016G20

Front disc brakes—GS300 and GS400

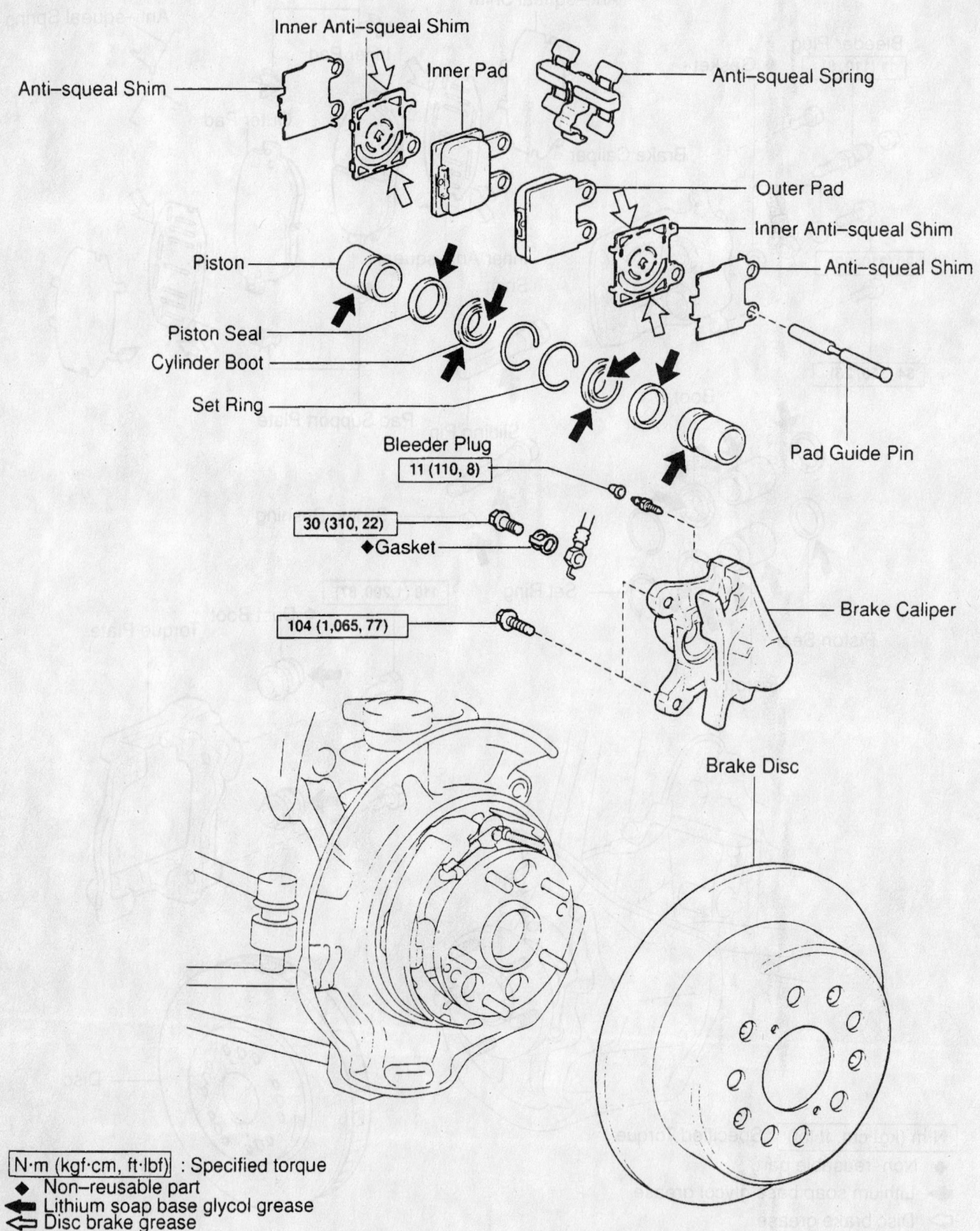

Anti-squeal Shim

Inner Anti-squeal Shim

Inner Pad

Anti-squeal Spring

Outer Pad

Inner Anti-squeal Shim

Anti-squeal Shim

Piston

Piston Seal

Cylinder Boot

Set Ring

Pad Guide Pin

Bleeder Plug

11 (110, 8)

30 (310, 22)

◆Gasket

104 (1,065, 77)

Brake Caliper

Brake Disc

N·m (kgf·cm, ft·lbf) : Specified torque
◆ Non-reusable part
◀ Lithium soap base glycol grease
◁ Disc brake grease

93016G21

Rear disc brakes—GS300 and GS400

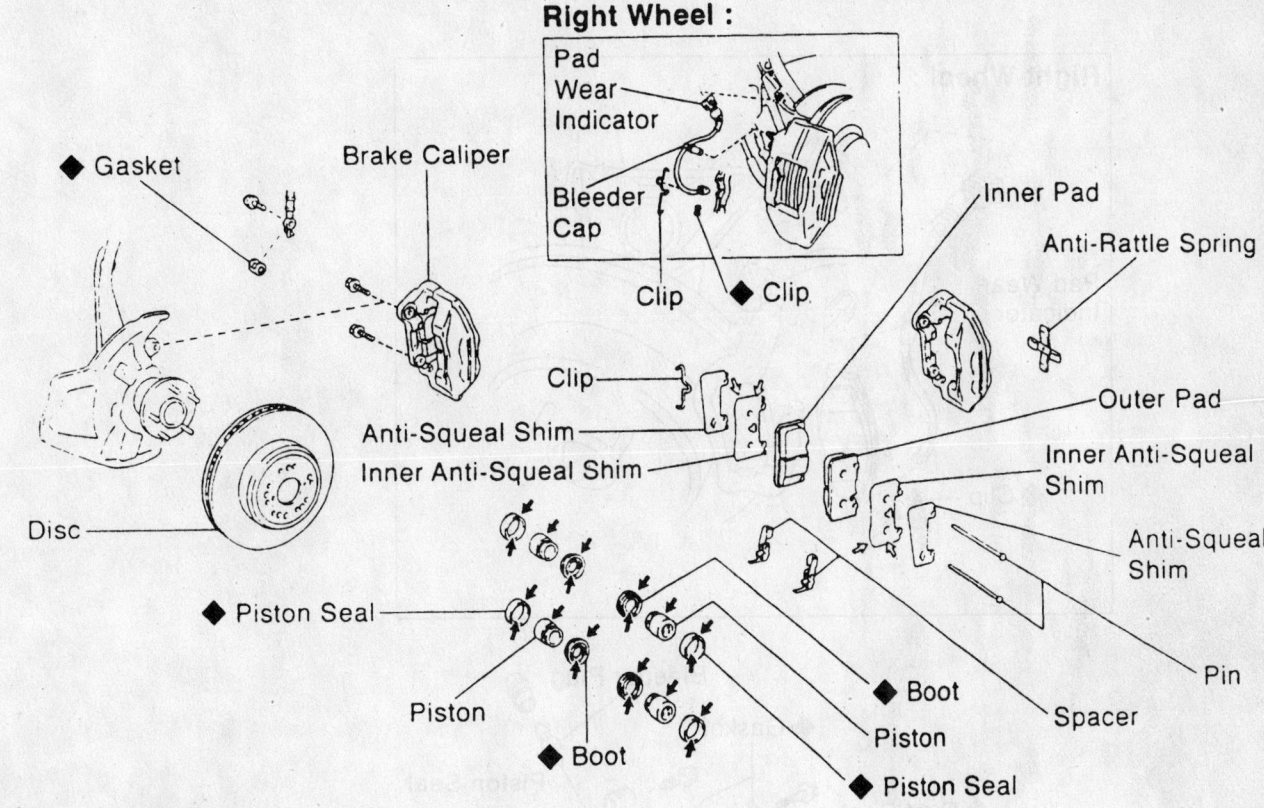

Right Wheel :

Gasket · Brake Caliper

Pad Wear Indicator

Bleeder Cap

Clip · Clip

Inner Pad

Anti-Rattle Spring

Clip

Anti-Squeal Shim

Inner Anti-Squeal Shim

Outer Pad

Inner Anti-Squeal Shim

Anti-Squeal Shim

Disc

Piston Seal

Piston

Boot

Boot

Piston

Piston Seal

Pin

Spacer

93016G23

◆ Non-reusable part
➔ Lithium soap base glycol grease
⇨ Disc brake grease

Rear disc brakes—LS400

2. Remove the wheels.
3. Disconnect the brake line at the caliper.
4. Hold the sliding pin with a wrench and remove the mounting bolts.
5. Remove the caliper assembly.

To install:
6. Install the caliper. Hold the sliding pin and tighten the mounting bolts to 25 ft. lbs. (34 Nm).
7. Connect the brake line with 2 new gaskets and tighten the union bolt to 22 ft. lbs. (30 Nm).
8. Bleed the brake system.
9. Install the wheels.
10. Check and if necessary fill the master cylinder reservoir.

LS400

FRONT

1. Before servicing the vehicle, refer to the precautions in the beginning of this section.

2. Remove the front wheels.
3. Disconnect the brake line at the caliper.
4. Remove the 2 bolts to the holding the caliper to the steering knuckle and remove the caliper assembly.

To install:
5. Install the caliper. Install the 2 bolts and torque to 87 ft. lbs. (118 Nm).
6. Connect the brake line with 2 new gaskets and tighten the union to 29 ft. lbs. (39 Nm).
7. Refill the reservoir as necessary and bleed the brake system.

REAR

1. Before servicing the vehicle, refer to the precautions in the beginning of this section.
2. Remove the rear wheels.
3. Disconnect and plug the brake line at the caliper.
4. Remove the mounting bolts and the caliper assembly.

To install:
5. Temporarily install the caliper on the torque plate with the 2 installation bolts.
6. Hold the sliding pin and tighten the mounting bolts to 25 ft. lbs. (34 Nm).
7. Connect the brake line with 2 new gaskets and tighten the union to 29 ft. lbs. (39 Nm).
8. Refill the reservoir as necessary and bleed the brake system.

SC300 and SC400

FRONT AND REAR

1. Before servicing the vehicle, refer to the precautions in the beginning of this section.
2. Remove the wheels.
3. Disconnect the brake line at the caliper by removing the union bolt and 2 gaskets.
4. Hold the sliding pin with a wrench and then remove the mounting bolts.
5. Remove the caliper from the caliper support.

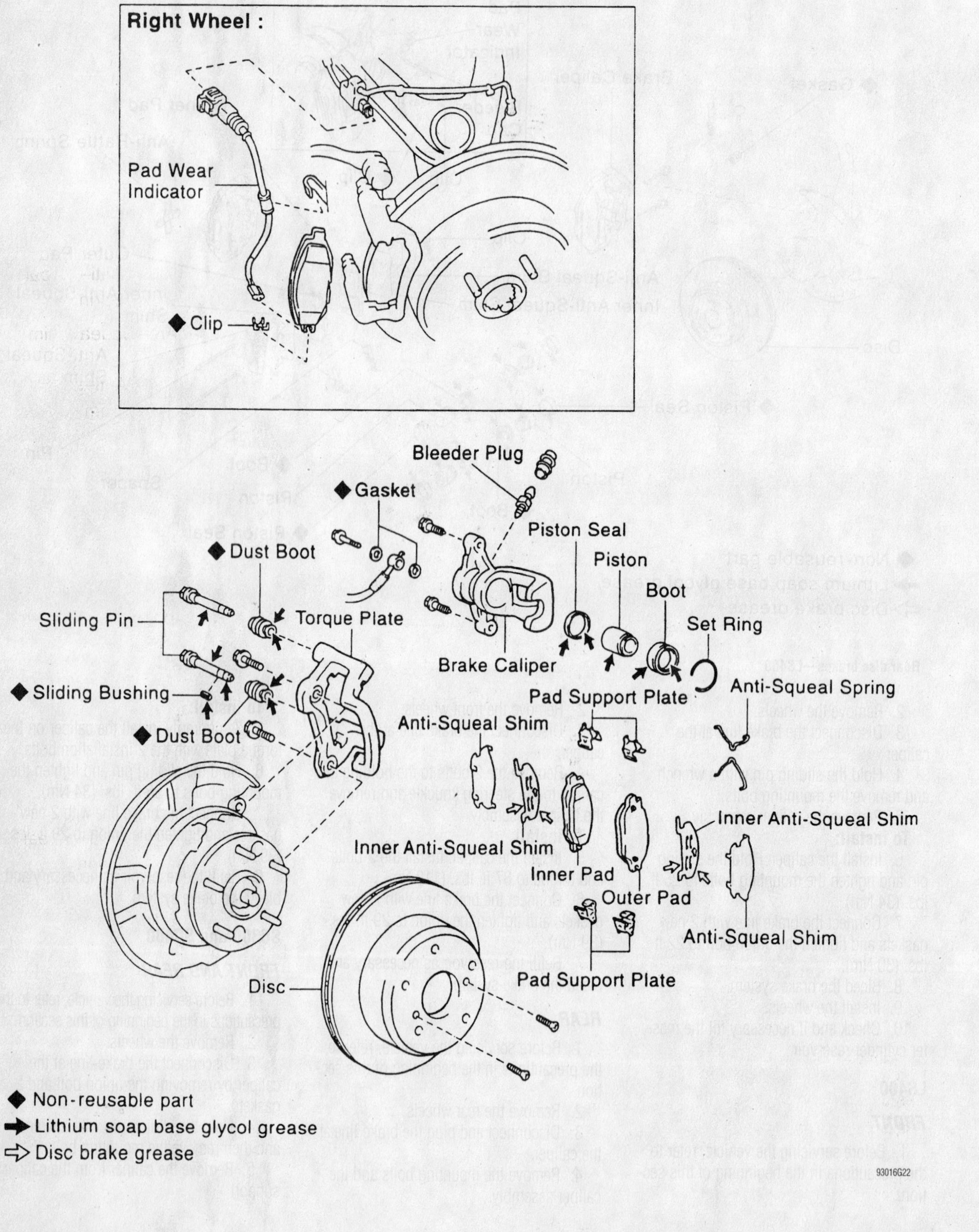

Right Wheel :

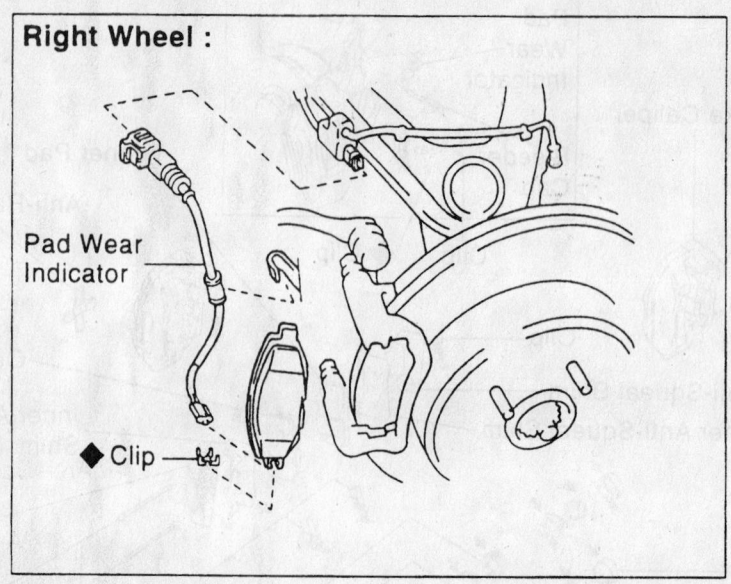

Pad Wear Indicator

◆ Clip

Bleeder Plug

◆ Gasket

Piston Seal

Piston

Boot

◆ Dust Boot

Set Ring

Sliding Pin

Torque Plate

Brake Caliper

Anti-Squeal Spring

◆ Sliding Bushing

Pad Support Plate

◆ Dust Boot

Anti-Squeal Shim

Inner Anti-Squeal Shim

Inner Anti-Squeal Shim

Inner Pad

Outer Pad

Anti-Squeal Shim

Disc

Pad Support Plate

◆ Non-reusable part

➡ Lithium soap base glycol grease

⇨ Disc brake grease

93016G22

Front disc brakes—LS400

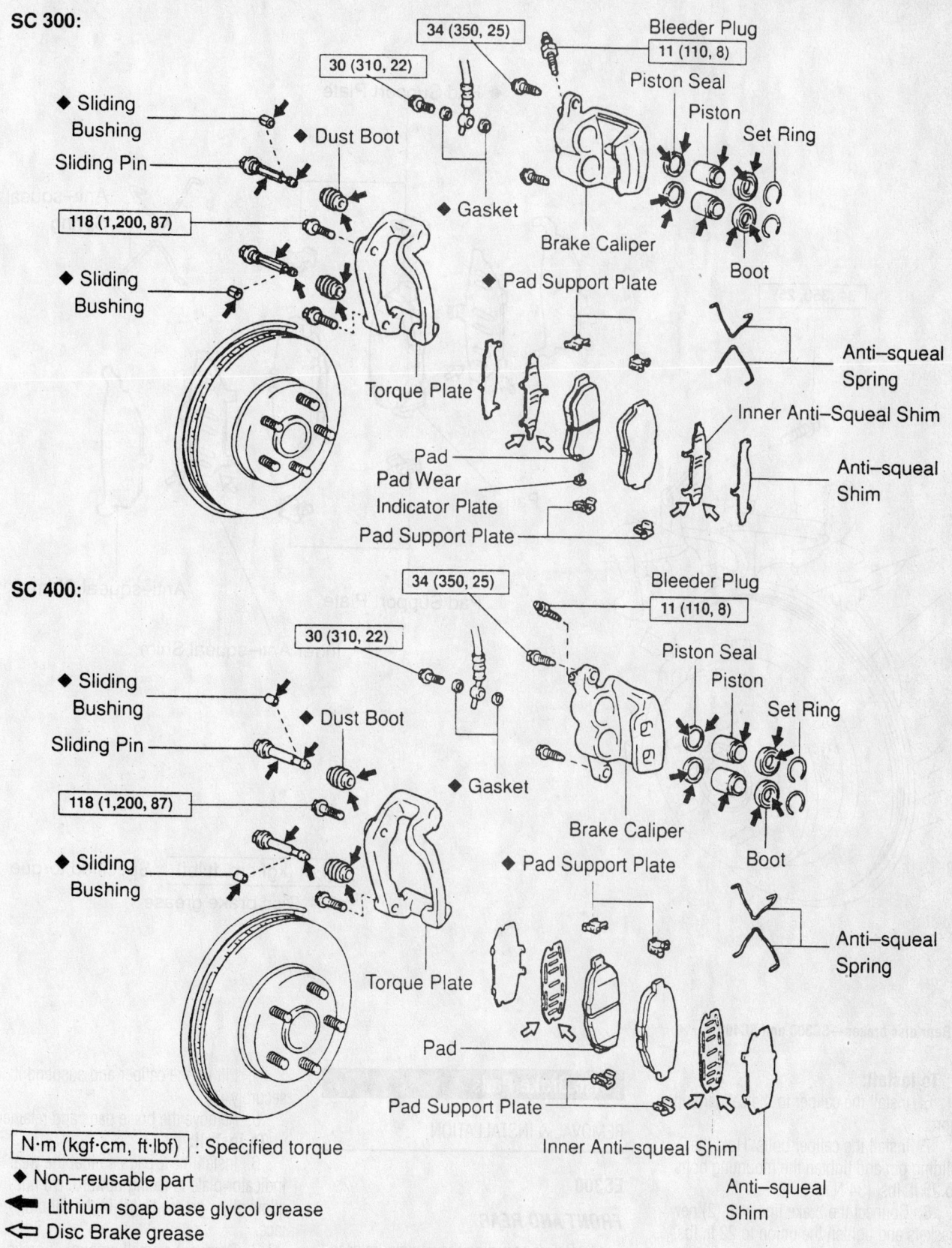

SC 300:

◆ Sliding Bushing

Sliding Pin

118 (1,200, 87)

◆ Sliding Bushing

◆ Dust Boot

34 (350, 25)

30 (310, 22)

◆ Gasket

Bleeder Plug
11 (110, 8)

Piston Seal

Piston

Set Ring

Brake Caliper

Boot

◆ Pad Support Plate

Torque Plate

Anti–squeal Spring

Inner Anti–Squeal Shim

Anti–squeal Shim

Pad
Pad Wear Indicator Plate
Pad Support Plate

SC 400:

◆ Sliding Bushing

Sliding Pin

118 (1,200, 87)

◆ Sliding Bushing

◆ Dust Boot

34 (350, 25)

30 (310, 22)

◆ Gasket

Bleeder Plug
11 (110, 8)

Piston Seal

Piston

Set Ring

Brake Caliper

Boot

◆ Pad Support Plate

Torque Plate

Anti–squeal Spring

Pad

Pad Support Plate

Inner Anti–squeal Shim

Anti–squeal Shim

N·m (kgf·cm, ft·lbf) : Specified torque

◆ Non–reusable part

◀ Lithium soap base glycol grease

◁ Disc Brake grease

93016G24

Front disc brakes—SC300 and SC400

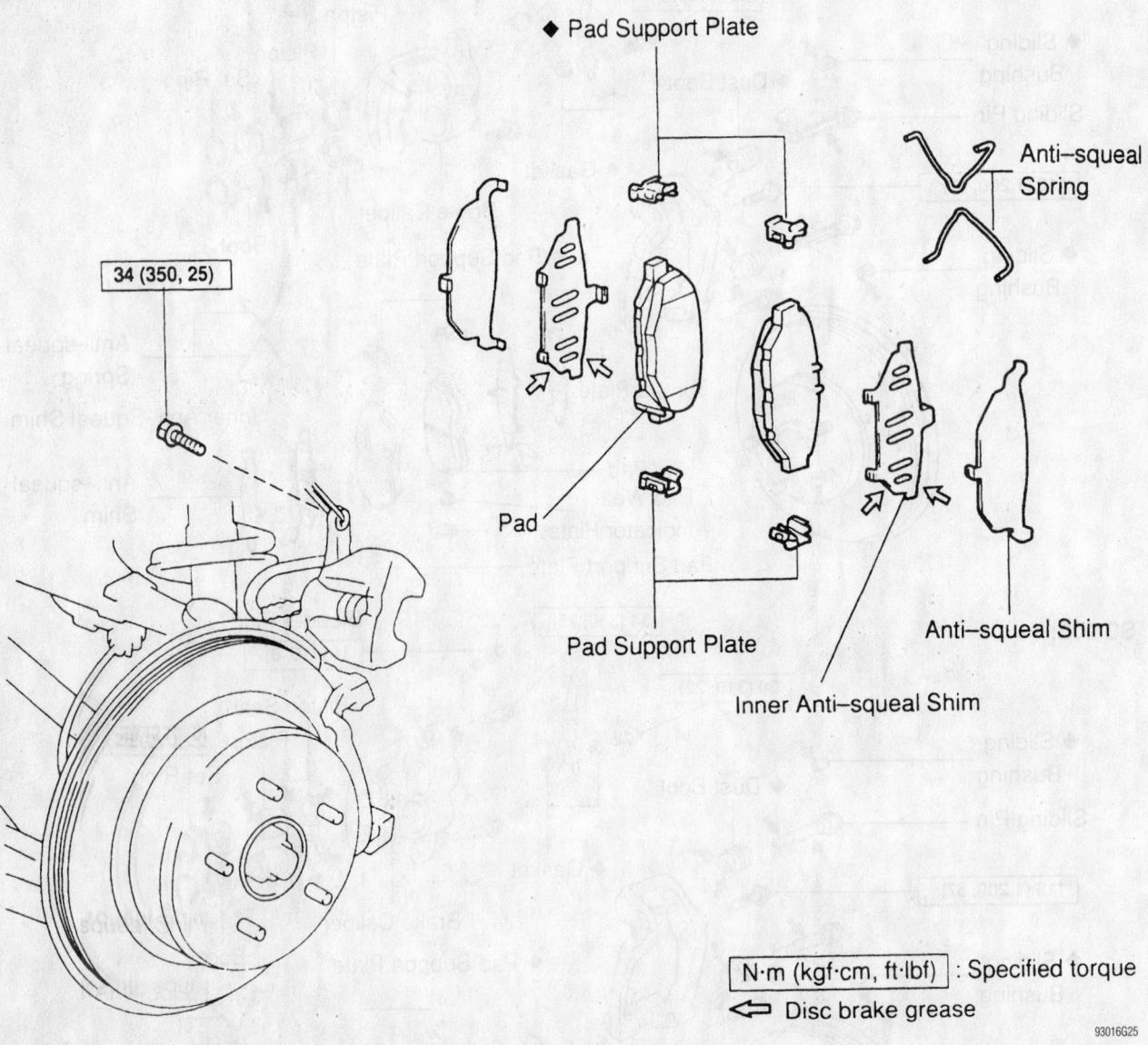

34 (350, 25)

◆ Pad Support Plate

Anti–squeal Spring

Pad

Pad Support Plate

Anti–squeal Shim

Inner Anti–squeal Shim

N·m (kgf·cm, ft·lbf) : Specified torque
⇐ Disc brake grease

93016G25

Rear disc brakes—SC300 and SC400

To install:

6. Install the caliper to the caliper support.

7. Install the caliper bolts. Hold the sliding pin and tighten the mounting bolts to 25 ft. lbs. (34 Nm).

8. Connect the brake line with (2) new gaskets and tighten the union to 22 ft. lbs. (30 Nm).

9. Bleed the brake system.

10. Install the wheels.

11. Check and fill the master cylinder reservoir, if needed.

Disc Brake Pads

REMOVAL & INSTALLATION

ES300

FRONT AND REAR

1. Before servicing the vehicle, refer to the precautions in the beginning of this section.

2. Remove the wheels.

3. Remove the lower installation bolt.

4. Lift up the caliper and suspend it securely.

5. Remove the brake pads and retainers.

To install:

6. Install the 2 pads so that the wear indicator plate is facing upward. Do not allow oil or grease to get in the rubbing face.

7. Draw out a small amount of brake fluid from the brake reservoir. Press in the caliper piston with a suitable tool.

8. Lower and install the caliper. Torque the sliding main pin to 25 ft. lbs. (34 Nm).

9. Install the wheels.

10. Check the fluid level in the master cylinder and add as necessary.

GS300, SC300 and SC400

FRONT AND REAR

1. Before servicing the vehicle, refer to the precautions in the beginning of this section.

2. Remove the wheels.

3. Hold the sliding pin on the lower mounting bolt and remove the bolt. Swivel the caliper upward and out of the way.

4. Remove the brake pads and retainers.

To install:

5. Install the pad support plates and the pad wear indicator plate on the inside pad.

6. Install both pads with the wear indicator plates facing downward.

7. Compress the caliper pistons and install the caliper.

8. Hold the sliding pin and tighten the mounting bolts to 25 ft. lbs. (34 Nm).

9. Install the wheels.

10. Check the brake fluid level in the reservoir.

LS400

FRONT

1. Before servicing the vehicle, refer to the precautions in the beginning of this section.

2. Remove the front wheels.

3. Remove the 2 bolts holding the caliper to the steering knuckle and remove the caliper.

4. Remove the brake pads and retainers.

To install:

5. Install the pad support plates and the pad wear indicator plate on the inside pad.

6. Install both pads with the wear indicator plates.

7. Compress the caliper pistons and install the caliper.

8. Install the 2 bolts to hold the caliper to the steering knuckle. Torque the bolts to 87 ft. lbs. (118 Nm).

9. Install the front wheels.

10. Check the fluid level in the reservoir.

REAR

1. Before servicing the vehicle, refer to the precautions in the beginning of this section.

2. Remove the rear wheels.

3. Hold the sliding pin on the lower mounting bolt and remove the bolt. Swivel the caliper upward and out of the way.

4. Remove the brake pads and retainers.

To install:

5. Install the pad support plates and the pad wear indicator plate on the inside pad.

6. Install both pads with the wear indicator plates facing downward.

7. Compress the caliper pistons and install the caliper.

8. Hold the sliding pin and tighten the mounting bolts to 25 ft. lbs. (34 Nm).

9. Install the rear wheels.

10. Check the brake fluid level in the reservoir.

1998–00 MAZDA

Brake Caliper

REMOVAL & INSTALLATION

Protege

FRONT

1. Before servicing the vehicle, refer to the precautions in the beginning of this section.

2. Remove the wheels.

3. Disconnect the flexible brake hose from the caliper.

4. Remove the disc pad retaining pins. Remove the brake pads.

5. Remove the upper and lower caliper bolts. Remove the caliper from the vehicle.

To install:

6. Position the caliper on the brake disc. Install the caliper mounting bolts and tighten the bolts to 29–36 ft. lbs. (40–49 Nm).

7. Install the disc pads and retaining pins.

8. Replace the washers for the brake line. Connect the brake hose to the caliper and tighten the hose nut to 16–21 ft. lbs. (22–29 Nm).

9. Bleed the brake system.

10. Install the wheels.

REAR

1. Before servicing the vehicle, refer to the precautions in the beginning of this section.

2. Remove the wheels.

3. Disconnect the parking brake cable from the cable bracket and the operating lever.

4. Disconnect the flexible brake line from the caliper assembly.

5. Turn the manual adjustment gear counterclockwise with an Allen wrench to pull the caliper piston inward (turn until it stops).

6. Remove the caliper mounting bolts. Remove the caliper from the vehicle.

To install:

7. Install the caliper. Torque the caliper mount bolts to 34–44 ft. lbs. (46–60 Nm).

8. Install the brake hose. Torque the line bolt to 16–22 ft. lbs. (22–30 Nm).

9. Connect the parking brake cable.

10. Install the wheels and bleed the brake system.

626 and MX6

FRONT

1. Before servicing the vehicle, refer to the precautions in the beginning of this section.

2. Remove the wheels.

3. Disconnect the flexible brake hose from the caliper.

4. Remove the lower caliper bolt and pivot the caliper upward. Slide the top of the caliper off of the top pin and remove it from the vehicle.

To install:

5. Lubricate the caliper pin and slide the caliper onto the guide pin. Pivot the caliper over the brake pads.

6. Connect the brake hose to the caliper and tighten the hose nut to 16–21 ft. lbs. (22–29 Nm).

7. Install the caliper mounting bolt and tighten the bolt to 22–29 ft. lbs. (30–40 Nm).

8. Bleed the brake system and inspect the brake system for proper operation.

9. Install the wheels.

REAR

1. Before servicing the vehicle, refer to the precautions in the beginning of this section.

2. Remove the wheels. Loosen the parking brake cable adjustment from inside the vehicle.

3. Disconnect the parking brake cable from the cable bracket and the operating lever.

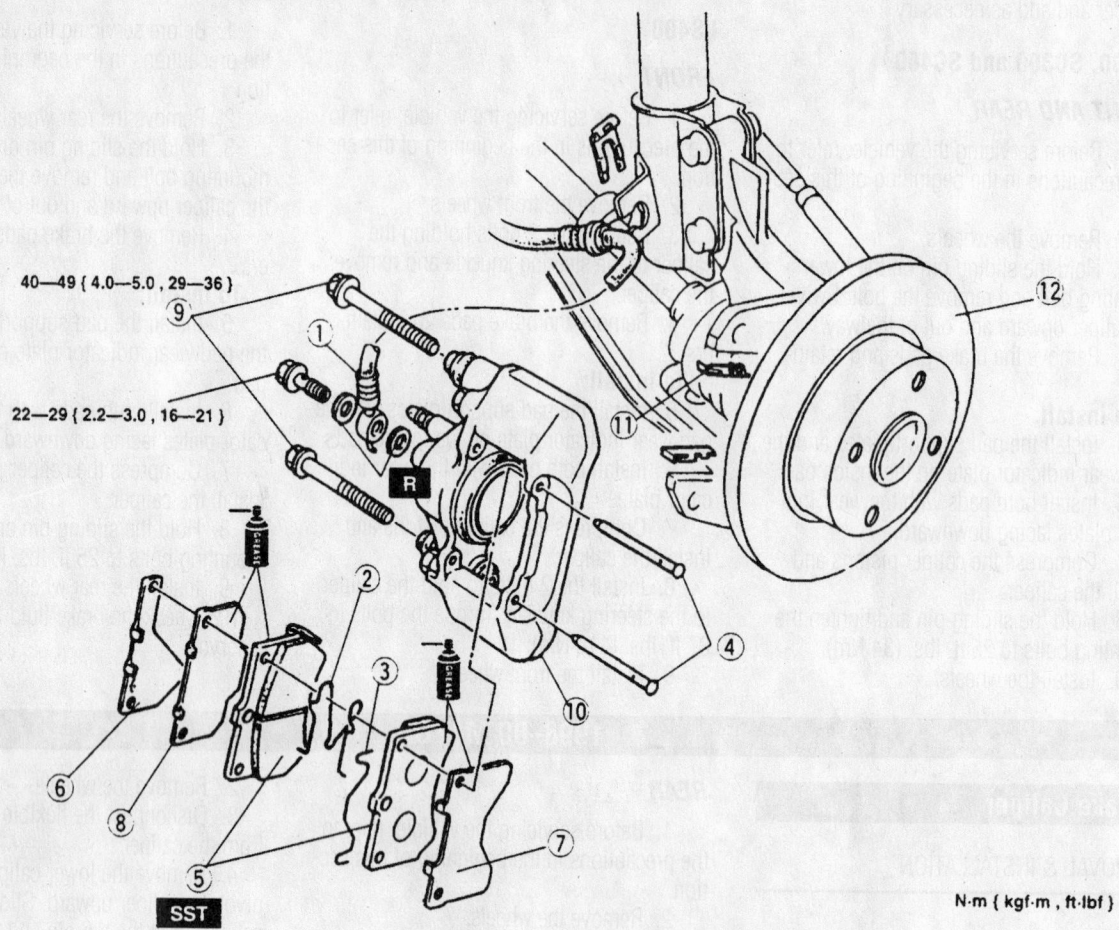

40—49 { 4.0—5.0 , 29—36 }

22—29 { 2.2—3.0 , 16—21 }

⑨ ① ⑪ ⑫

⑥ ② ③ ④ ⑩

⑧ ⑤ ⑦

R

SST

N·m { kgf·m , ft·lbf }

1	Flexible hose		8	Inner shim
2	M-spring		9	Bolt
3	W-clip		10	Caliper
4	Pad pin		11	Guide plate
5	Disc pad ☞ Installation Note		12	Disc plate ☞ Removal Note ☞ Installation Note
6	Anti-squeak shim			
7	Outer shim			

Disc Plate Removal Note

- Mark the wheel hub bolt and disc plate before removal for reference during installation.

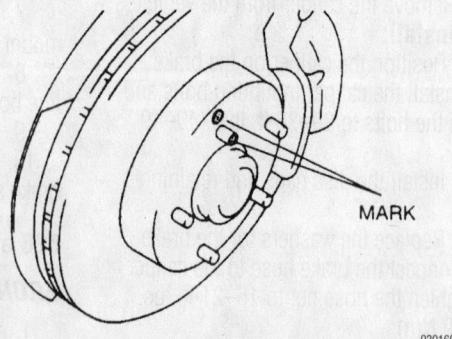

MARK

93016G26

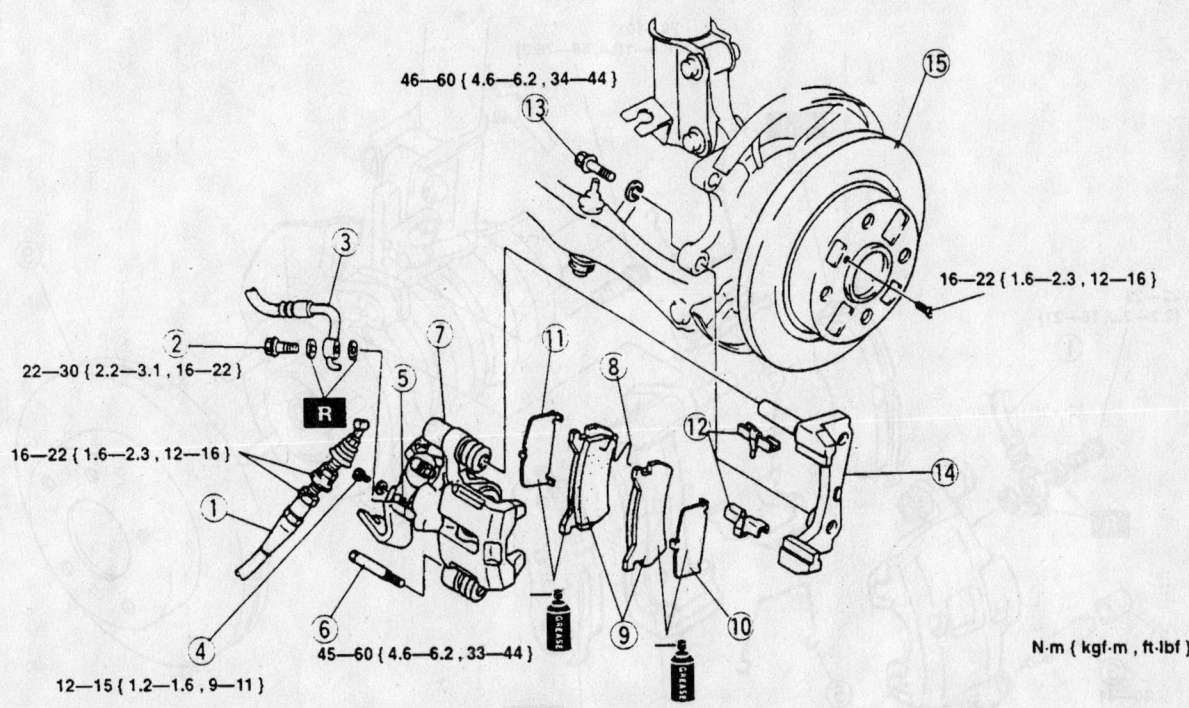

1	Parking cable, clip ☞ 04-12 PARKING BRAKE (LEVER TYPE) RE-MOVAL / INSTALLATION		8	M-spring
2	Connecting bolt		9	Disc pad ☞ Installation Note
3	Brake hose		10	Outer shim
4	Screw plug		11	Inner shim
5	Manual adjustment gear ☞ Removal Note ☞ Installation Note		12	Guide plate
			13	Bolt
6	Lock bolt		14	Mounting support
7	Caliper		15	Disc plate

Manual Adjustment Gear Removal Note
- Turn the manual adjustment gear counterclockwise by using an Allen wrench to pull the brake caliper piston inward. (Turn until it stops.)

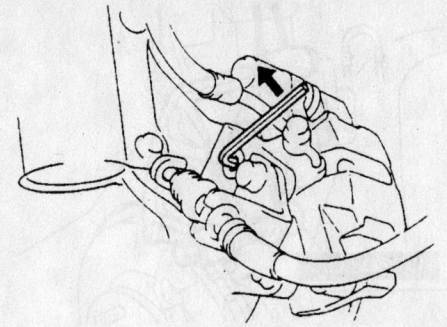

Disc Plate Removal Note
- Mark the wheel hub bolt and disc plate before removal for reference during installation.

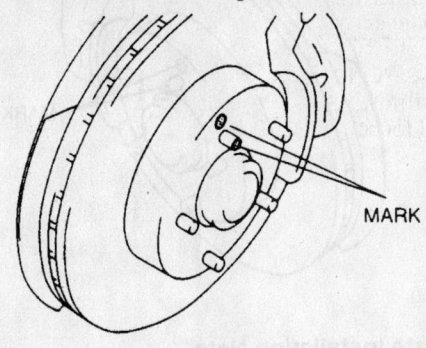

Rear disc brakes—Protégé

Timing belt service is covered in Section 3 of this manual

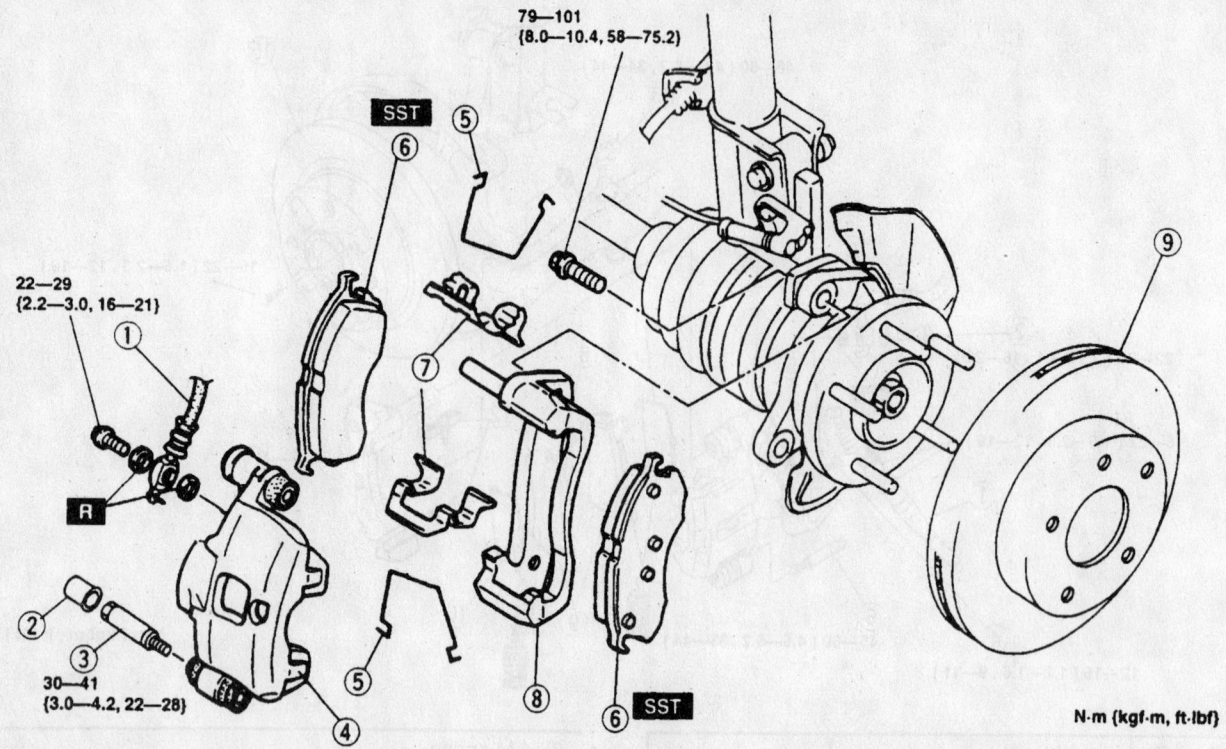

79—101
{8.0—10.4, 58—75.2}

22—29
{2.2—3.0, 16—21}

30—41
{3.0—4.2, 22—28}

N·m {kgf·m, ft·lbf}

1	Flexible hose	
2	Cap	
3	Lock bolt	
4	Caliper	
5	M-spring	
6	Disc pad ☞ Installation Note	

7	Guide plate	
8	Mounting support	
9	Disc plate ☞ Removal Note ☞ Installation Note	

Disc Plate Removal Note

● Mark the wheel hub bolt and disc plate for guide at installation before removal.

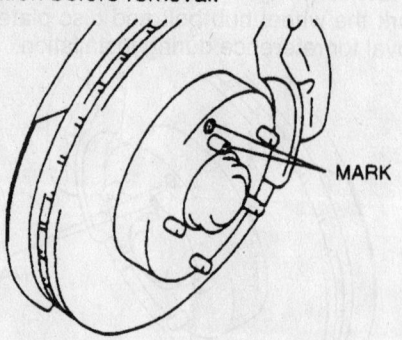

MARK

Disc Plate Installation Note

1. Remove any rust or grime on the contact face of the disc plate and wheel hub.

2. Install the disc plate and align the marks put before removal.

Disc Pad Installation Note

1. Push the piston fully inward by using the **SST**.
2. Install the disc pad.

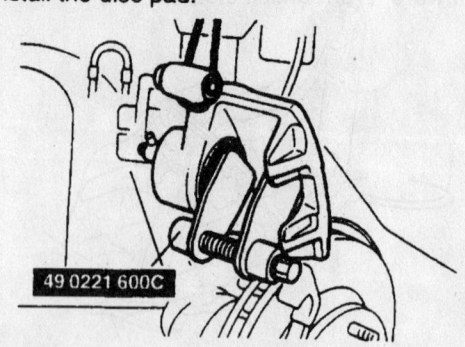

49 0221 600C

93016G28

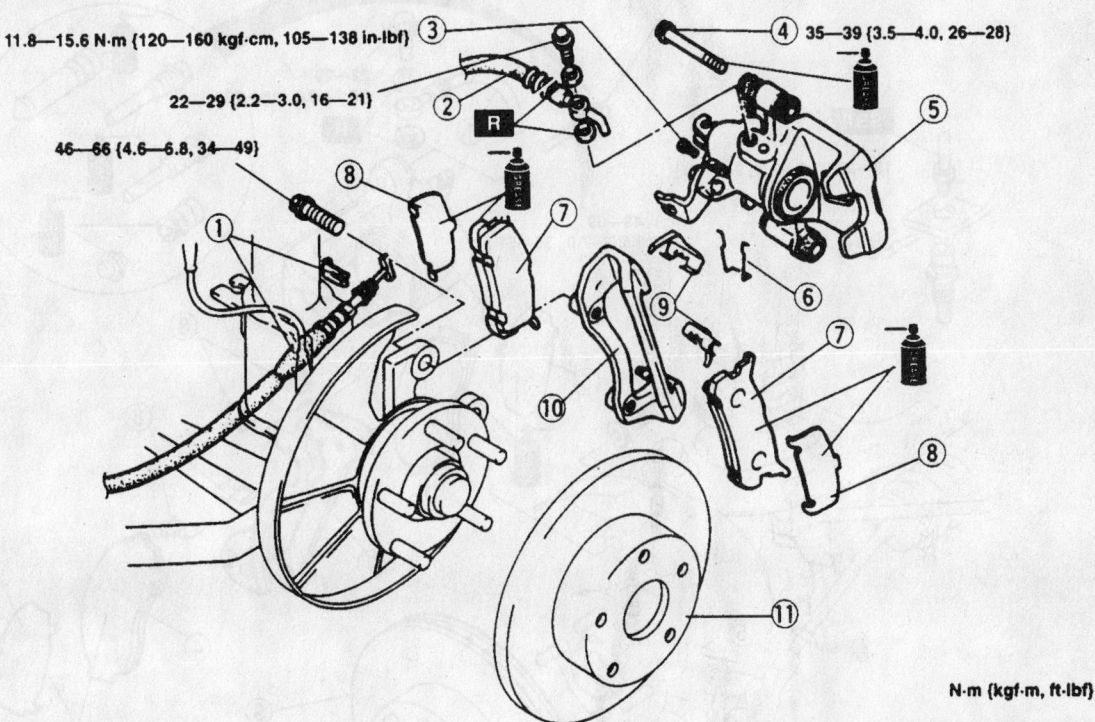

11.8—15.6 N·m {120—160 kgf·cm, 105—138 in·lbf}

22—29 {2.2—3.0, 16—21}

46—66 {4.6—6.8, 34—49}

35—39 {3.5—4.0, 26—28}

R

N·m {kgf·m, ft·lbf}

1	Parking brake cable, clip
2	Flexible hose
3	Screw plug
4	Lock bolt
5	Caliper
6	Spring
7	Disc pad ☞ Installation Note

8	Shim
9	Guide plate
10	Mounting support
11	Disc plate ☞ FRONT BRAKE (DISC) REMOVAL/INSTALLATION, Disc Plate Removal Note ☞ FRONT BRAKE (DISC) REMOVAL/INSTALLATION, Disc Plate Installation Note

Disc Pad Installation Note
1. Turn the manual adjustment gear counterclockwise with an Allen wrench to pull the brake caliper piston inward. (Turn until it stops.)
2. Install the disc pads.
3. Turn the manual adjustment gear clockwise until the brake pads just touch the disc plate. Turn the manual adjustment gear back 1/3-turn.

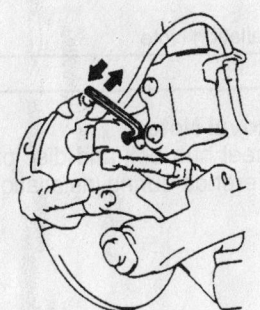

93016G29

Heater Core replacement is covered in Section 2 of this manual

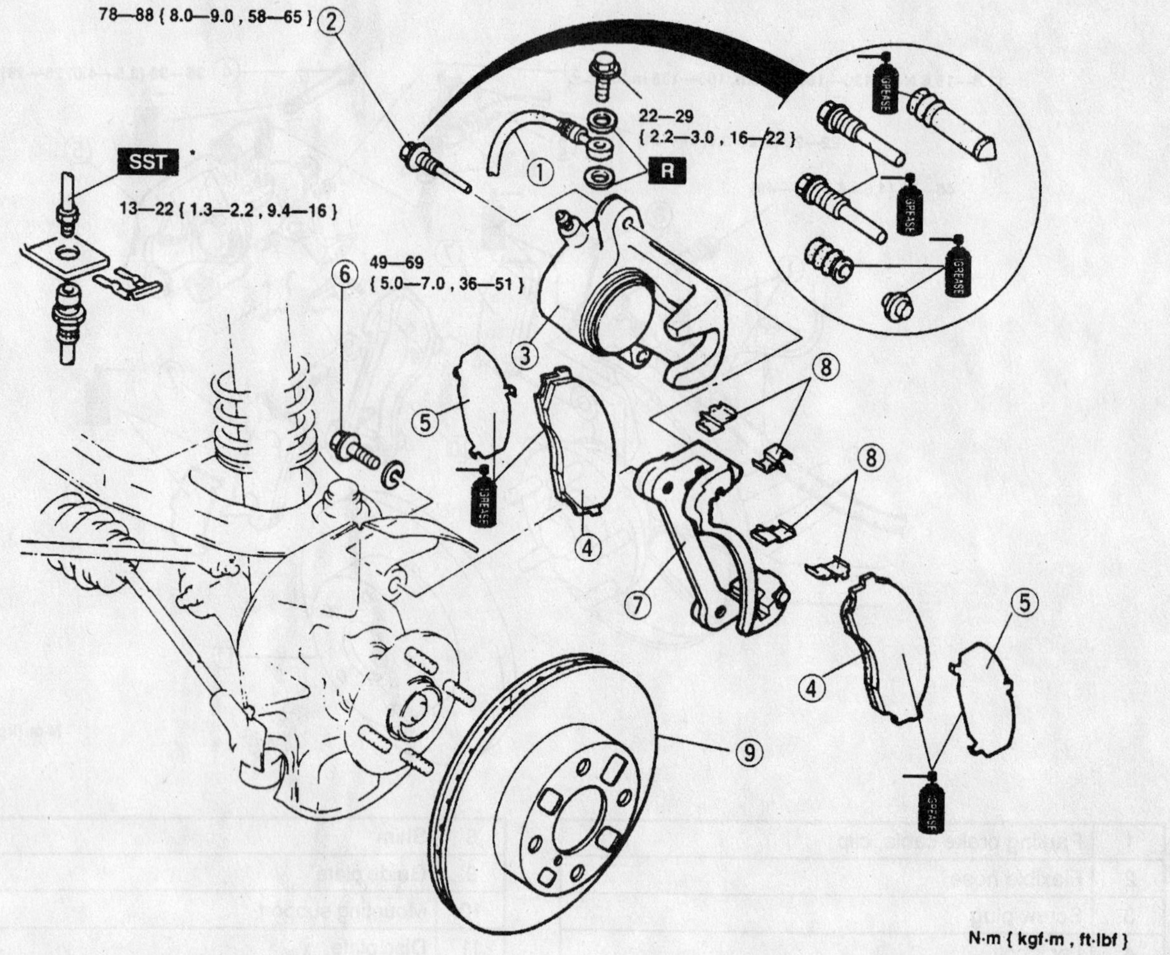

78—88 { 8.0—9.0 , 58—65 } ②

22—29
{ 2.2—3.0 , 16—22 }

SST

13—22 { 1.3—2.2 , 9.4—16 }

49—69
{ 5.0—7.0 , 36—51 } ⑥

N·m { kgf·m , ft·lbf }

1	Brake hose
2	Connecting bolt
3	Caliper
4	Disc pad ☞ Installation Note
5	Shim

6	Bolt
7	Mounting support
8	Guide plate
9	Disc plate ☞ Removal Note ☞ Installation Note

Disc Plate Removal Note
- Mark the wheel hub bolt and disc plate before removal for reference during installation.

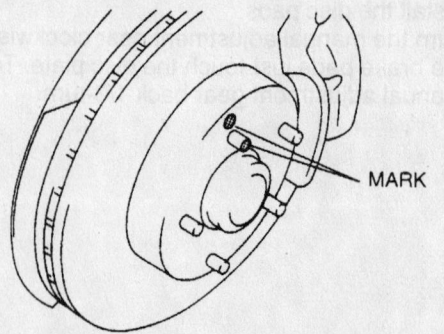

MARK

93016G30

Front disc brakes—Miata

4. Disconnect the flexible brake line from the caliper assembly.

5. Remove the caliper upper mounting bolt and pivot the caliper downward. Slide the caliper off of the guide pin. Remove the caliper from the vehicle.

To install:

6. Lubricate the caliper pin and slide the caliper onto the guide pin. Pivot the caliper over the brake pads.

7. Connect the brake hose to the caliper and tighten the hose nut to 16–21 ft. lbs. (22–29 Nm).

8. Install the upper caliper mounting bolt and tighten the bolt to 26–28 ft. lbs. (35–39 Nm).

9. Connect the parking brake cable to the cable bracket and the operating lever.

10. Bleed the brake system.

11. Install the wheels.

Miata

FRONT

1. Before servicing the vehicle, refer to the precautions in the beginning of this section.

2. Remove the wheels.

3. Disconnect the flexible brake hose from the caliper.

4. Remove the upper and lower caliper bolts. Remove the caliper from the vehicle.

To install:

5. Position the caliper on the brake disc. Install the caliper mounting bolts and tighten the bolts to 36–51 ft. lbs. (49–69 Nm).

6. Replace the washers for the brake line. Connect the brake hose to the caliper and tighten the hose nut to 16–22 ft. lbs. (22–29 Nm).

7. Bleed the brake system and inspect the brake system for proper operation.

8. Install the wheels.

REAR

1. Before servicing the vehicle, refer to the precautions in the beginning of this section.

2. Loosen the parking brake cable adjustment from inside the vehicle.

3. Remove the wheels.

4. Disconnect the parking brake cable from the cable bracket and the operating lever.

5. Disconnect the flexible brake line from the caliper assembly.

6. Remove the cover for the manual adjustment gear. Insert an Allen wrench and turn counterclockwise to retract the caliper piston.

7. Remove the caliper mounting bolts and remove the caliper from the vehicle.

To install:

8. Position the caliper and install the mounting bolts. Tighten the upper bolt to 33–36 ft. lbs. (45–49 Nm), and the lower bolt to 25–29 ft. lbs. (34–39 Nm).

9. Turn the manual adjustment gear clockwise to return the caliper until the brake pads just touch the disc, then turn counterclockwise ⅓ of a turn. Replace the cover.

10. After replacing the washers, connect the brake hose to the caliper. Tighten the hose nut to 16–21 ft. lbs. (22–29 Nm).

11. Connect the parking brake cable to the cable bracket and the operating lever.

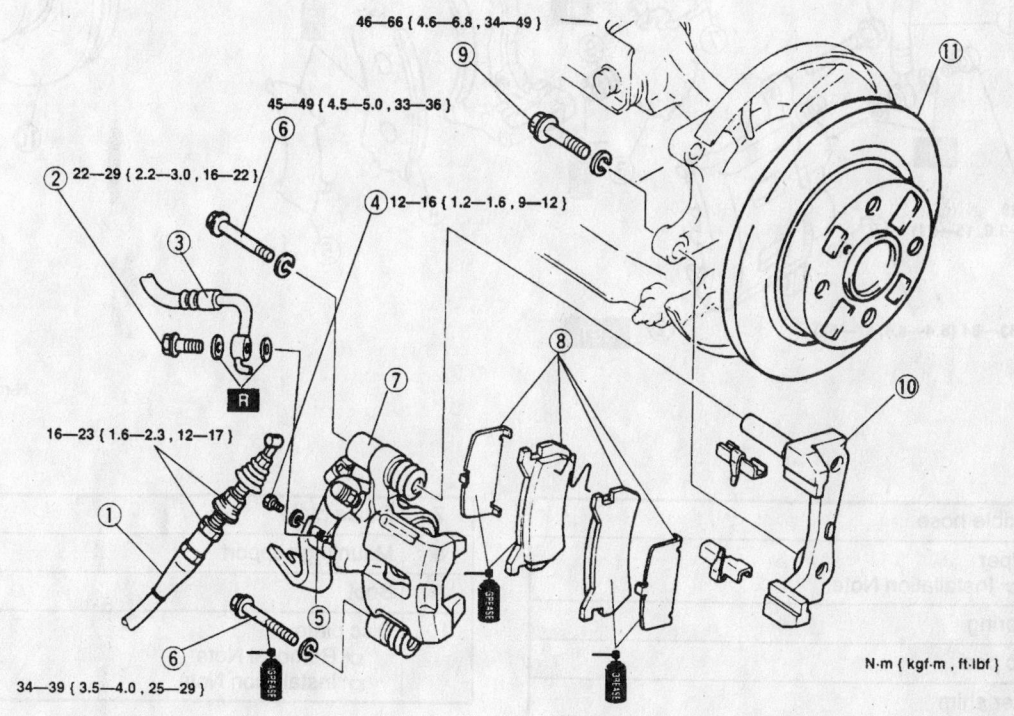

46–66 { 4.6–6.8 , 34–49 }
45–49 { 4.5–5.0 , 33–36 }
22–29 { 2.2–3.0 , 16–22 }
12–16 { 1.2–1.6 , 9–12 }
16–23 { 1.6–2.3 , 12–17 }
34–39 { 3.5–4.0 , 25–29 }

N·m { kgf·m , ft·lbf }

1	Parking brake cable
2	Connecting bolt
3	Brake hose
4	Plug

5	Manual adjustment gear ☞ Removal Note ☞ Installation Note
6	Lock bolt

93016G31

Rear disc brakes—Miata

Brake service is covered in Section 4 of this manual

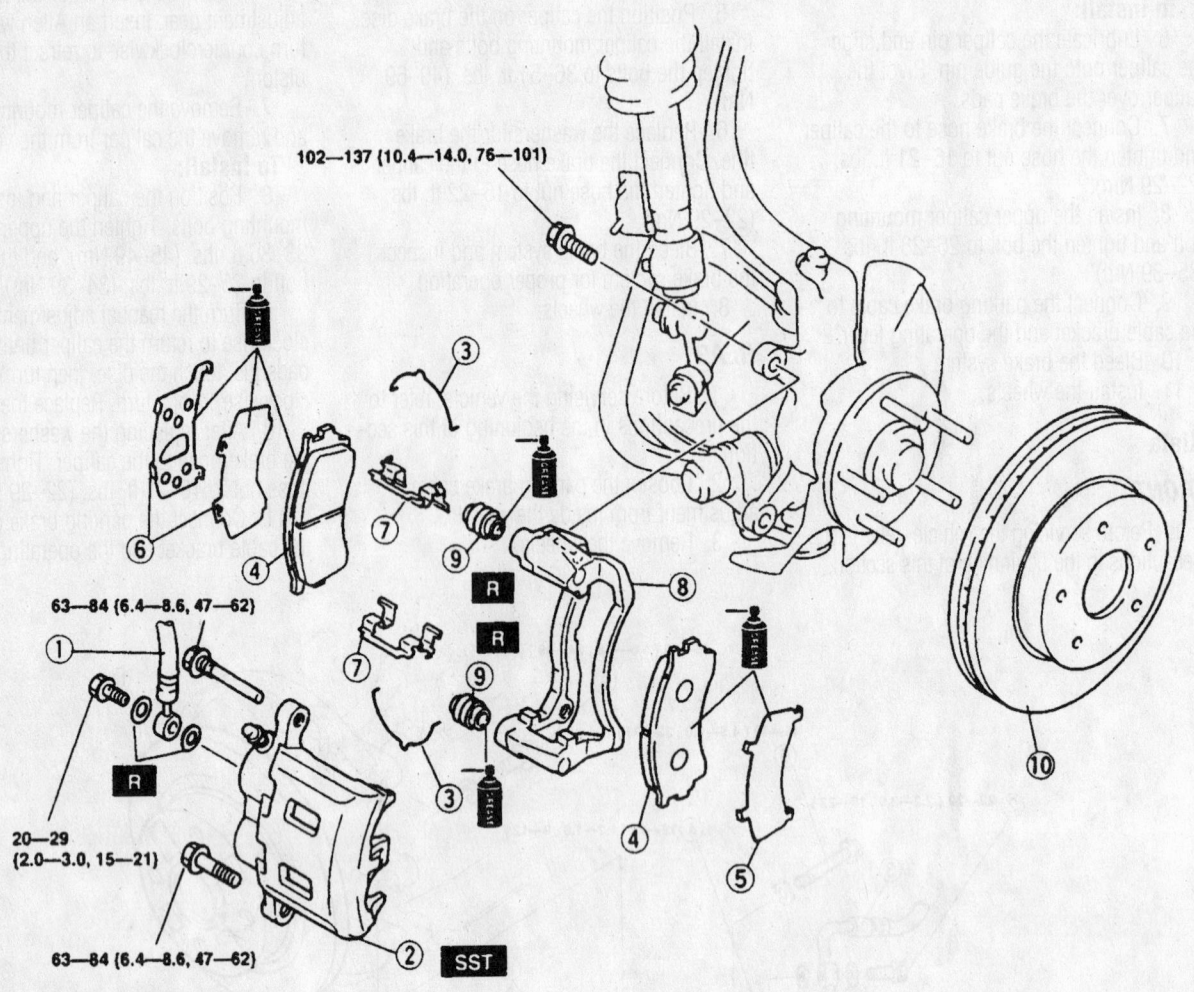

102—137 (10.4—14.0, 76—101)

63—84 (6.4—8.6, 47—62)

20—29
(2.0—3.0, 15—21)

63—84 (6.4—8.6, 47—62)

N·m (kgf·m, ft·lbf)

1	Flexible hose		7	Guide plate
2	Caliper ☞ Installation Note		8	Mounting support
3	V-spring		9	Boot
4	Disc pad		10	Disc plate ☞ Removal Note ☞ Installation Note
5	Outer shim			
6	Inner shim			

93016G32

Front disc brakes—Millenia

12. Bleed the brake system and adjust the parking brake.

13. Install the wheels.

Millenia

FRONT

1. Before servicing the vehicle, refer to the precautions in the beginning of this section.

2. Remove the wheels.

3. Disconnect the brake hose and brake pipe.

4. Remove the caliper mounting bolts and remove the caliper.

To install:

5. Install the caliper. Torque the caliper mounting bolts to 47–62 ft. lbs. (63–84 Nm).

6. Connect the brake hose and pipe. Fill the master cylinder with clean brake fluid and bleed the hydraulic system.

7. Install the wheels.

REAR

1. Before servicing the vehicle, refer to the precautions in the beginning of this section.

2. Remove the wheels.

3. Disconnect the brake hose.

4. Remove the caliper bracket mounting bolts and remove the caliper.

To install:

5. Install the caliper. Torque the caliper mounting bolts to 12–17 ft. lbs. (16–23 Nm).

6. Connect the brake hose and pipe. Fill the master cylinder with clean brake fluid and bleed the hydraulic system.

7. Install the wheels.

REMOVAL & INSTALLATION

Protege

FRONT

1. Before servicing the vehicle, refer to the precautions in the beginning of this section.

2. Remove some of the brake fluid from the master cylinder reservoir. The reservoir should be no more than ½ full.

3. Remove the wheels.

4. Remove the pad pins and retaining springs from the brake pads.

5. Remove the brake pads.

6. Push the caliper piston into the caliper bore.

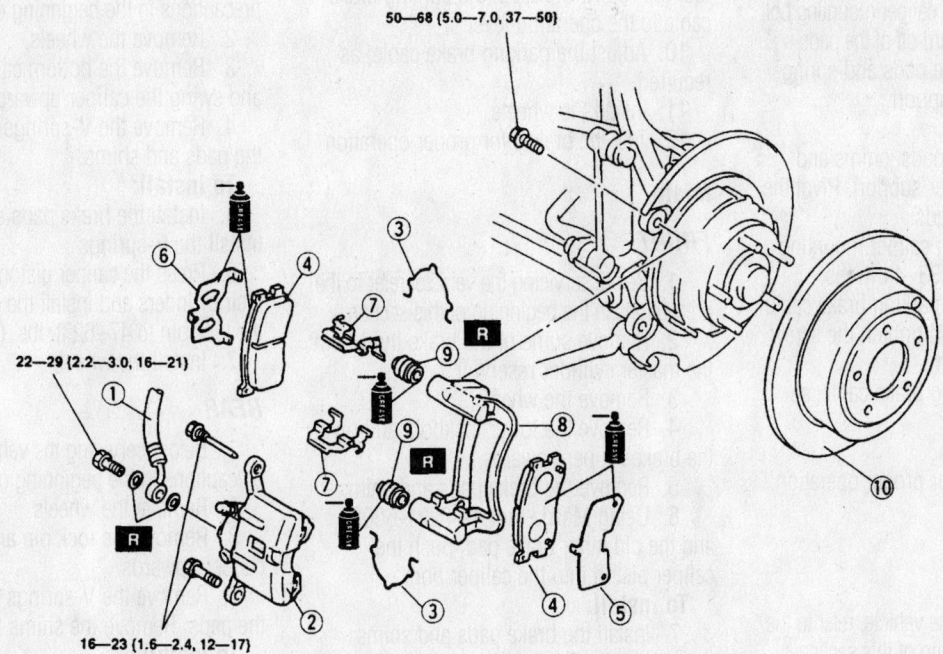

50—68 (5.0—7.0, 37—50)

22—29 (2.2—3.0, 16—21)

16—23 (1.6—2.4, 12—17)

N·m {kgf·m, ft·lbf}

1	Flexible hose
2	Caliper ☞ Installation Note
3	V-spring
4	Disc pad
5	Outer shim
6	Inner shim
7	Guide plate

8	Mounting support
9	Boot
10	Disc plate ☞ 04–12 PARKING BRAKE SHOE REMOVAL/INSTALLATION, Disc Plate Removal Note ☞ 04–12 PARKING BRAKE SHOE REMOVAL/INSTALLATION, Disc Plate Installation Note

93016G33

Rear disc brakes—Millenia

To install:

7. Install the brake pads and shims to the caliper.

8. Insert the pad pins and retaining springs.

9. Install the wheels.

10. Test the brakes for proper operation.

REAR

1. Before servicing the vehicle, refer to the precautions in the beginning of this section.

2. Remove some of the brake fluid from the master cylinder reservoir.

3. Loosen the parking brake cable adjustment from inside the vehicle.

4. Remove the wheels.

5. Disconnect the parking brake cable from the cable bracket and the operating lever. Remove the screw plug. Turn the manual adjustment gear counterclockwise with an Allen wrench to pull the brake caliper piston inward (turn until it stops).

6. Remove the upper caliper mounting bolt and pivot the caliper upward off of the pads.

7. Remove the brake pads and spring clips from the caliper support.

To install:

8. Install the brake pads, shims and spring clips to the caliper support. Pivot the caliper over the brake pads.

9. Install the bottom caliper mounting bolt. Tighten the bolt to 33–44 ft. lbs. (46–60 Nm). Attach the parking brake cable to the operating lever and tighten the screw plug to 10 ft. lbs. (14 Nm).

10. Adjust the parking brake cable, as required.

11. Install the wheels.

12. Test the brakes for proper operation.

626 and MX6

FRONT

1. Before servicing the vehicle, refer to the precautions in the beginning of this section.

2. Remove some of the brake fluid from the master cylinder reservoir.

3. Remove the wheels.

4. Remove the caliper lower mounting bolt and pivot the caliper up and support it.

5. Remove the brake pads, shims and pin.

6. Using Mazda tool 49-0221-600C and the old inner brake pad, push the caliper piston into the caliper bore.

To install:

7. Install the brake pads and shims to the caliper support. Install the caliper over the brake pads.

8. Install the caliper mounting bolt and torque to 22–29 ft. lbs. (30–40 Nm).

9. Install the wheels.

10. Test the brakes for proper operation.

REAR

1. Before servicing the vehicle, refer to the precautions in the beginning of this section.

2. Remove some of the brake fluid from the master cylinder reservoir.

3. Loosen the parking brake cable adjustment from inside the vehicle.

4. Remove the wheels.

5. Disconnect the parking brake cable from the cable bracket and the operating lever.

6. Remove the upper caliper mounting bolt and pivot the caliper downward off of the pads.

7. Remove the brake pads and spring clips from the caliper support.

To install:

8. Install the brake pads, shims and spring clips to the caliper support. Pivot the caliper over the brake pads.

9. Lubricate and install the top caliper mounting bolt. Tighten the bolt to 26–28 ft. lbs. (35–39 Nm). Attach the parking brake cable to the operating lever.

10. Adjust the parking brake cable, as required.

11. Install the wheels.

12. Test the brakes for proper operation.

Miata

FRONT

1. Before servicing the vehicle, refer to the precautions in the beginning of this section.

2. Remove some of the brake fluid from the master cylinder reservoir.

3. Remove the wheels.

4. Remove the lower lockbolt, and pivot the brake caliper upwards.

5. Remove the brake pads and shim.

6. Using Mazda tool 49-0221-600C and the old inner brake pad, push the caliper piston into the caliper bore.

To install:

7. Install the brake pads and shims onto the caliper.

8. Pivot the caliper into position. Install the lockbolt and tighten to 58–65 ft. lbs. (78–88 Nm).

9. Install the wheels.

10. Test the brakes for proper operation.

REAR

1. Before servicing the vehicle, refer to the precautions in the beginning of this section.

2. Remove some of the brake fluid from the master cylinder reservoir.

3. Remove the wheels.

4. Remove the plug for the manual adjustment gear. Using an Allen wrench turn the gear counterclockwise to retract the caliper piston.

5. Remove the lower caliper mounting bolt and pivot the caliper upward off of the pads.

6. Remove the brake pads and spring clips from the caliper support.

To install:

7. Install the brake pads, shims and spring clips to the caliper support. Pivot the caliper over the brake pads.

8. Lubricate and install the lower caliper mounting bolt. Tighten the bolt to 25–29 ft. lbs. (34–39 Nm).

9. Turn the manual adjusting gear clockwise until the piston contacts the brake disc, then turn it clockwise ⅓ turn. Replace the plug.

10. Install the wheels.

11. Test the brakes for proper operation.

Millenia

FRONT

1. Before servicing the vehicle, refer to the precautions in the beginning of this section.

2. Remove the wheels.

3. Remove the bottom caliper lock pin and swing the caliper upwards.

4. Remove the V-springs and remove the pads and shims.

To install:

5. Install the brake pads and shims. Install the V-springs

6. Press the caliper pistons back into their cylinders and install the calipers. Torque the lock pin to 47–62 ft. lbs. (63–84 Nm).

7. Install the wheels.

REAR

1. Before servicing the vehicle, refer to the precautions in the beginning of this section.

2. Remove the wheels.

3. Remove the lock pin and rotate the caliper upwards.

4. Remove the V-springs and remove the pads. Remove the shims from the pads.

To install:

5. Press the caliper piston back into the cylinder.

6. Install the pads, shims, and V-springs.

7. Install the caliper and torque the lock pin to 37–50 ft. lbs. (50–68 Nm).

8. Install the wheels.

Brake Drums

REMOVAL & INSTALLATION

All Models

1. Before servicing the vehicle, refer to the precautions in the beginning of this section.

2. Remove the rear wheel.

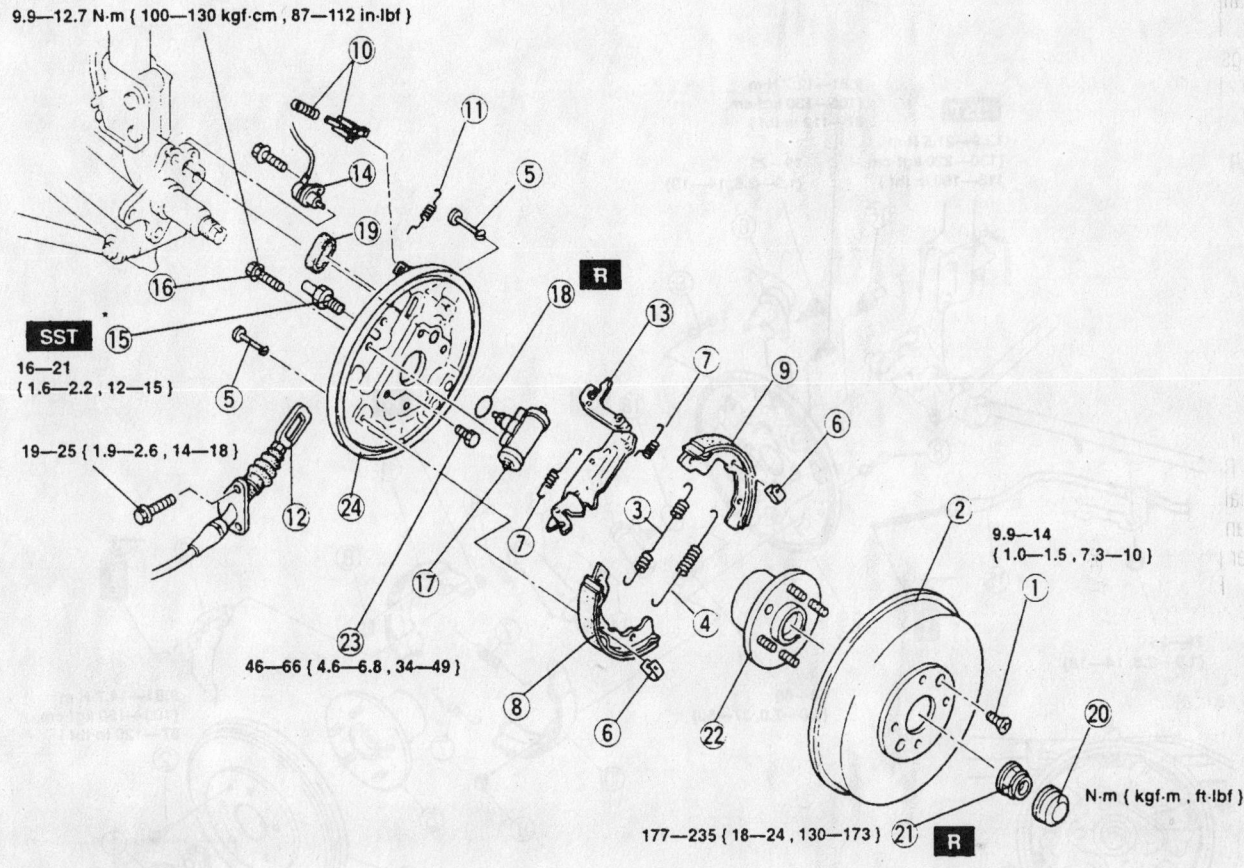

9.9—12.7 N·m { 100—130 kgf·cm , 87—112 in·lbf }

SST

16—21
{ 1.6—2.2 , 12—15 }

19—25 { 1.9—2.6 , 14—18 }

46—66 { 4.6—6.8 , 34—49 }

9.9—14
{ 1.0—1.5 , 7.3—10 }

N·m { kgf·m , ft·lbf }

177—235 { 18—24 , 130—173 }

1	Screw		17	Wheel cylinder
2	Brake drum ☞ Removal Note ☞ Installation Note		18	Wheel cylinder gasket
			19	Dust boot
			20	Hub cap
3	Return spring (upper)		21	Locknut ☞ 03–10 WHEEL HUB, STEERING KNUCKLE REMOVAL / INSTALLATION, Locknut Removal Note ☞ 03–10 WHEEL HUB, STEERING KNUCKLE REMOVAL / INSTALLATION, Locknut Installation Note
4	Return spring (lower)			
5	Hold pin			
6	Hold spring			
7	Anti-rattle spring			
8	Brake shoe (leading side)			
9	Brake shoe (trailing side)		22	Wheel hub component ☞ 03–10 WHEEL HUB, STEERING KNUCKLE REMOVAL / INSTALLATION, Wheel Hub Component Removal Note ☞ 03–10 WHEEL HUB, STEERING KNUCKLE REMOVAL / INSTALLATION, Wheel hub Component Installation Note
10	Stop spring and clip			
11	Spring			
12	Parking cable			
13	Operating lever component			
14	ABS wheel-speed sensor		23	Bolt
15	Brake pipe		24	Backing plate ☞ Installation Note
16	Bolt			

93016G34

Rear drum brakes—Protégé

For Accessory Drive Belt illustrations, see Section 1 of this manual

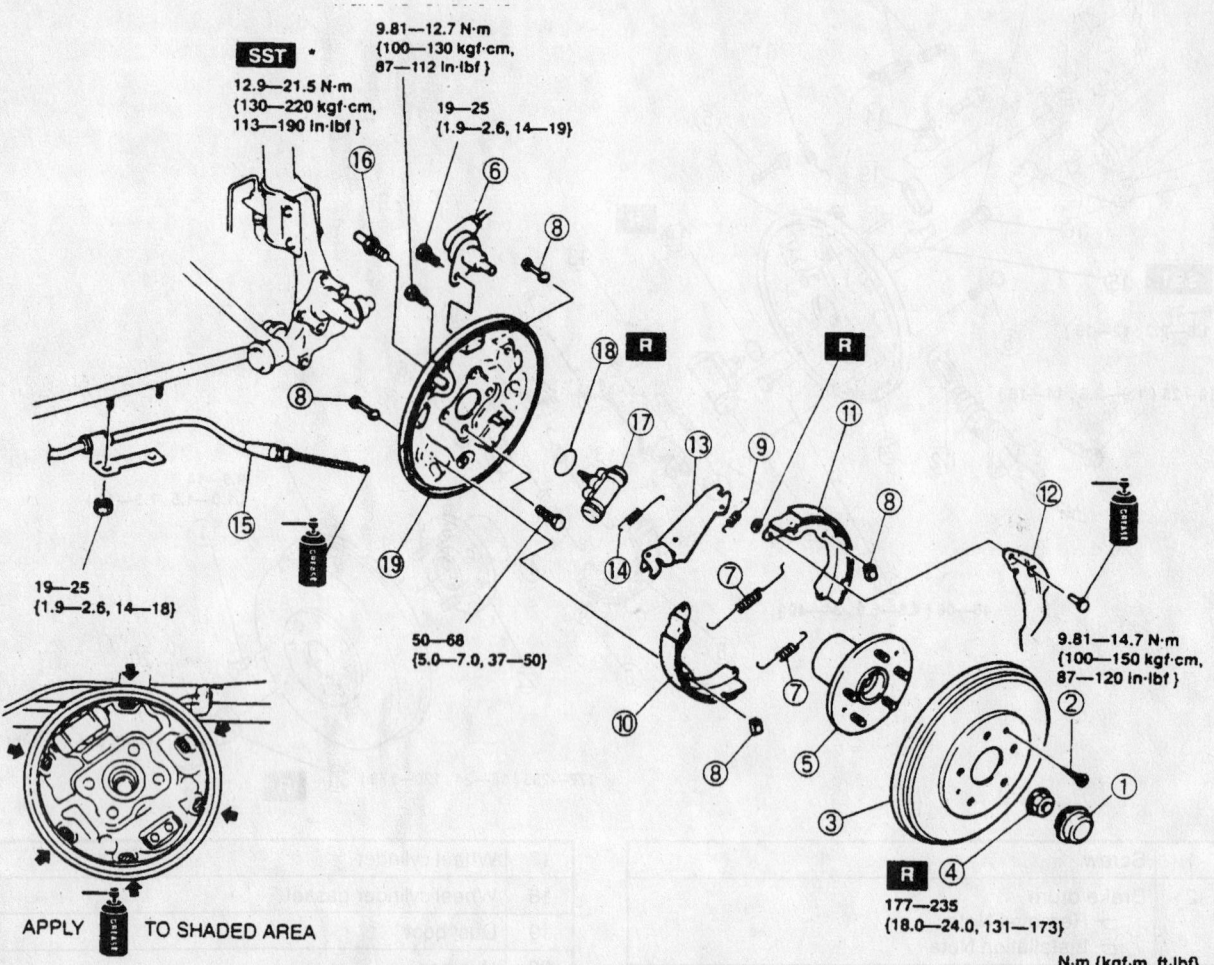

SST

12.9—21.5 N·m
{130—220 kgf·cm,
113—190 in·lbf }

9.81—12.7 N·m
{100—130 kgf·cm,
87—112 in·lbf }

19—25
{1.9—2.6, 14—19}

19—25
{1.9—2.6, 14—18}

50—68
{5.0—7.0, 37—50}

9.81—14.7 N·m
{100—150 kgf·cm,
87—120 in·lbf }

177—235
{18.0—24.0, 131—173}

N·m {kgf·m, ft·lbf}

APPLY [GREASE] TO SHADED AREA

1	Hub cap
2	Screw
3	Brake drum ☞ Removal Note ☞ Installation Note
4	Locknut ☞ 03–11 WHEEL HUB, STEERING KNUCKLE REMOVAL/INSTALLATION, Locknut Removal Note ☞ 03–11 WHEEL HUB, STEERING KNUCKLE REMOVAL/INSTALLATION, Locknut Installation Note
5	Wheel hub
6	ABS wheel-speed sensor (if equipped)
7	Return spring

8	Hold pin and hold spring
9	Anti-rattle spring
10	Leading shoe
11	Trailing shoe
12	Operating lever
13	Quadrant
14	Quadrant spring
15	Parking brake cable
16	Brake pipe
17	Wheel cylinder
18	O-ring
19	Backing plate

93016G35

Rear drum brakes—626

3. Remove the screw securing the rear brake drum, and pull the brake drum outward to remove.

To install:

4. Install the drum and retaining screw.

5. Install the wheels and adjust the rear brakes as necessary.

Brake Shoes

REMOVAL & INSTALLATION

All Models

1. Before servicing the vehicle, refer to the precautions in the beginning of this section.

2. Remove the wheel and remove the brake drum.

3. Disconnect the parking brake cable from the backside of the brake backing plate.

4. Remove the upper return spring.

5. From the lower (leading side) brake shoe, remove the hold pin and the spring.

6. Remove the lower (leading side) brake shoe and lower return spring and the anti-rattle spring.

7. Remove the upper (trailing side) brake shoe hold pin and spring and remove the upper brake shoe.

To install:

8. Install the upper (trailing side) brake shoe to the operating lever and then to the wheel cylinder and backing plate. Install the brake shoe hold spring and hold pin.

9. Install the anti-rattle spring.

10. Install the lower return spring to both brake shoes.

11. Install the leading side brake shoe to the operating lever and then to the wheel cylinder and anchor plate.

12. Install the hold spring and hold pin to the leading side brake shoe.

13. Install the upper return spring.

14. Install the brake drum.

15. Install the wheels.

1998–01 MITSUBISHI

Brake Caliper

REMOVAL & INSTALLATION

Mirage

FRONT

1. Before servicing the vehicle, refer to the precautions in the beginning of this section.

2. Remove the wheels.

3. Disconnect the brake hose.

4. Remove the caliper guide and lock pins. Remove the caliper assembly from the caliper support.

To install

5. Position the brake caliper onto the caliper support. Install and tighten the guide and lock pins.

6. Reconnect the brake hose.

7. Bleed the brake system.

8. Install the wheels.

Eclipse

FRONT

1. Before servicing the vehicle, refer to the precautions in the beginning of this section.

2. Remove the wheels.

3. Disconnect the brake hose.

4. Remove the caliper guide and lock pins. Remove the caliper assembly from the caliper support.

To install

5. Position the brake caliper onto the caliper support. Install and tighten the guide and lock pins.

6. Reconnect the brake hose.

7. Bleed the brake system.

8. Install the wheels.

REAR

1. Before servicing the vehicle, refer to the precautions in the beginning of this section.

2. Loosen the parking brake cable adjustment from inside the vehicle.

3. Remove the wheels.

4. Disconnect the brake hose.

5. Remove the caliper lock pin. Pivot the caliper upward, and slide the caliper assembly from the caliper support.

To install:

6. Install the caliper over the brake pads.

7. Lubricate and install the lock pin and tighten to 23 ft. lbs. (32 Nm).

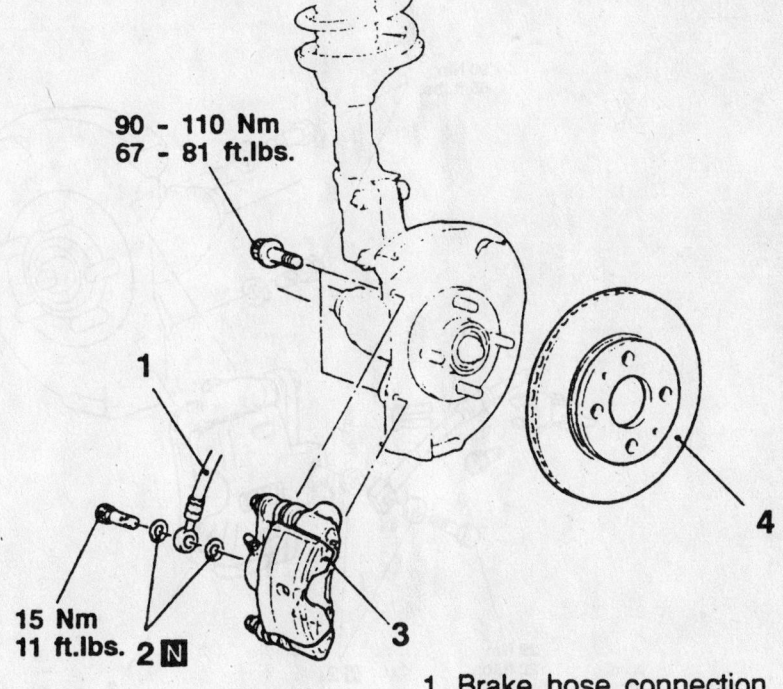

90 - 110 Nm
67 - 81 ft.lbs.

15 Nm
11 ft.lbs. 2 **N**

1. Brake hose connection
2. Gasket
3. Disc brake assembly
4. Brake disc

93016G36

Front disc brakes—Mirage

For Tire, Wheel and Ball Joint specifications, see Section 1 of this manual

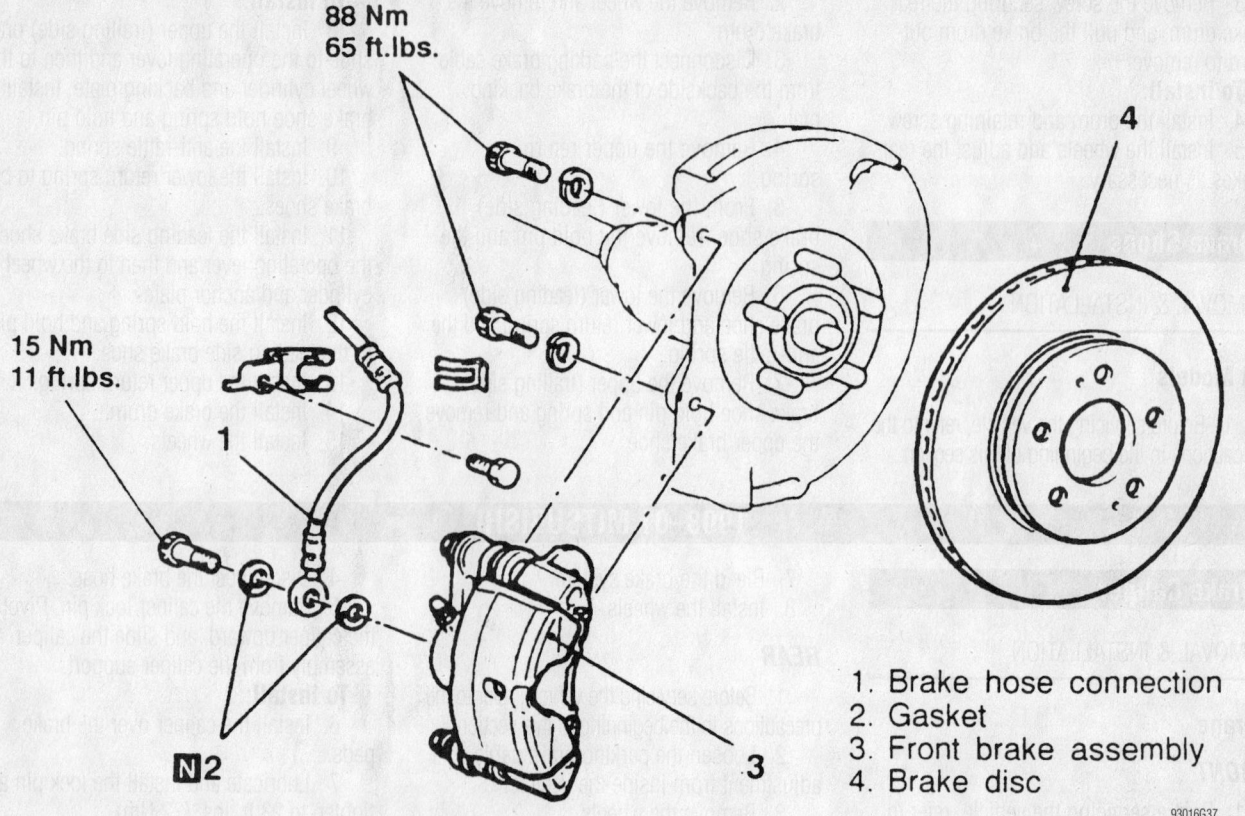

88 Nm
65 ft.lbs.

15 Nm
11 ft.lbs.

1

N 2

3

1. Brake hose connection
2. Gasket
3. Front brake assembly
4. Brake disc

4

93016G37

Front and rear disc brakes—Eclipse

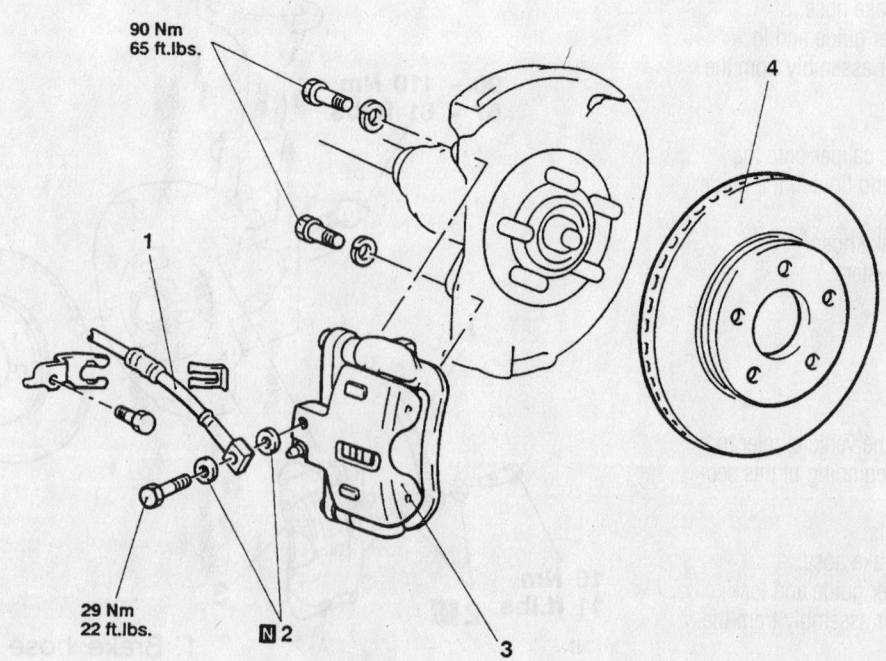

90 Nm
65 ft.lbs.

1

4

29 Nm
22 ft.lbs.

N 2

3

1. CONNECTION FOR THE BRAKE
   HOSE
2. GASKET
3. FRONT BRAKE ASSEMBLY
4. BRAKE DISC

93016G39

Front disc brakes—Diamante

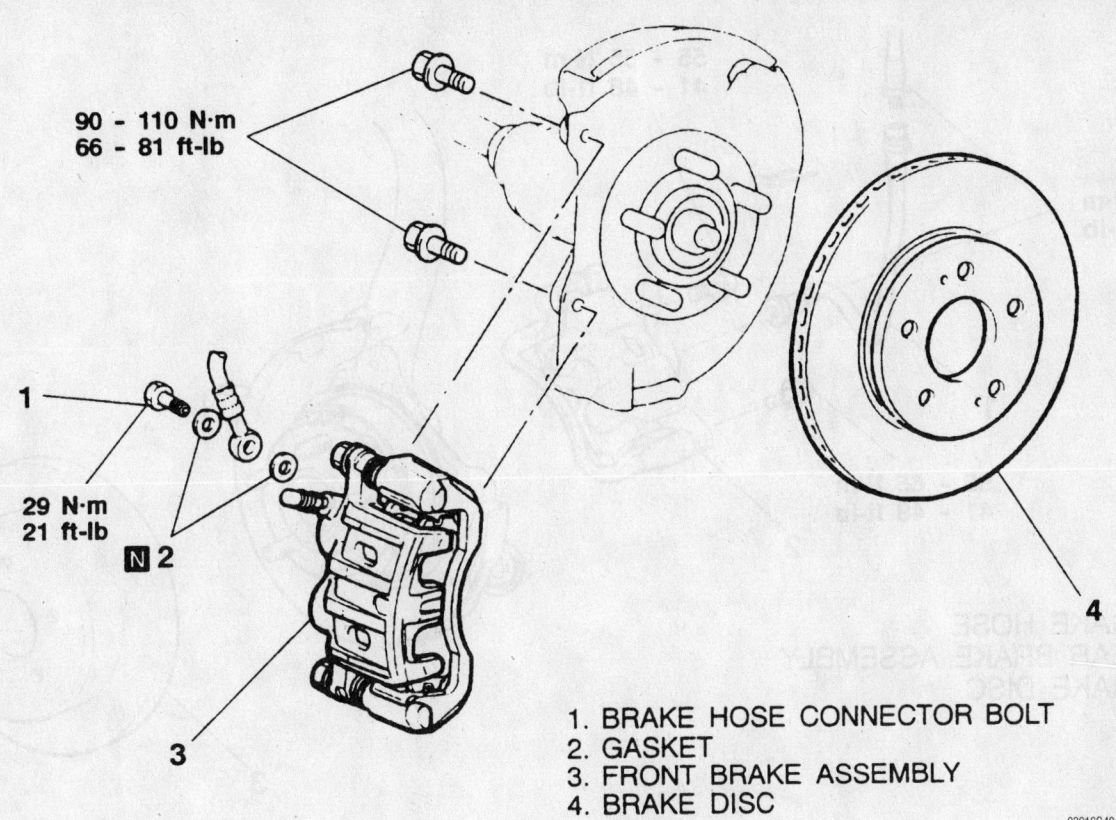

90 – 110 N·m
66 – 81 ft-lb

29 N·m
21 ft-lb

2

1. BRAKE HOSE CONNECTOR BOLT
2. GASKET
3. FRONT BRAKE ASSEMBLY
4. BRAKE DISC

93016G40

Front disc brakes—Galant

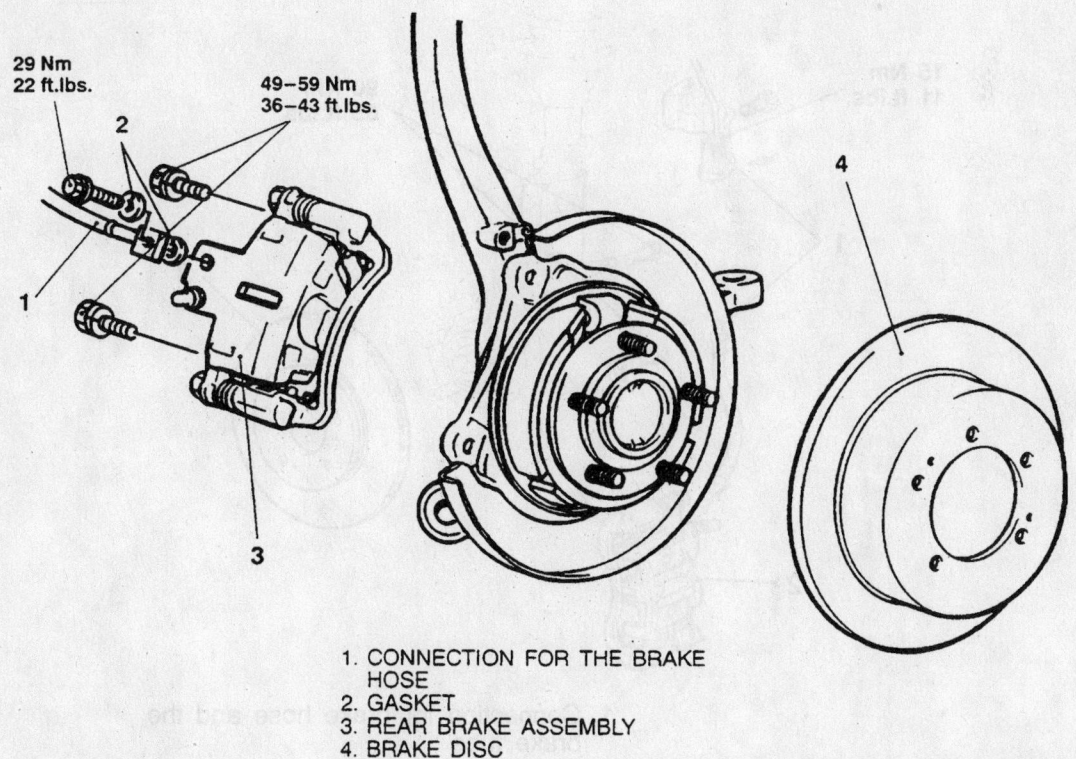

29 Nm
22 ft.lbs.

49–59 Nm
36–43 ft.lbs.

1. CONNECTION FOR THE BRAKE
   HOSE
2. GASKET
3. REAR BRAKE ASSEMBLY
4. BRAKE DISC

93016G41

Rear disc brakes—Diamante

For Wheel Alignment specifications, see Section 1 of this manual

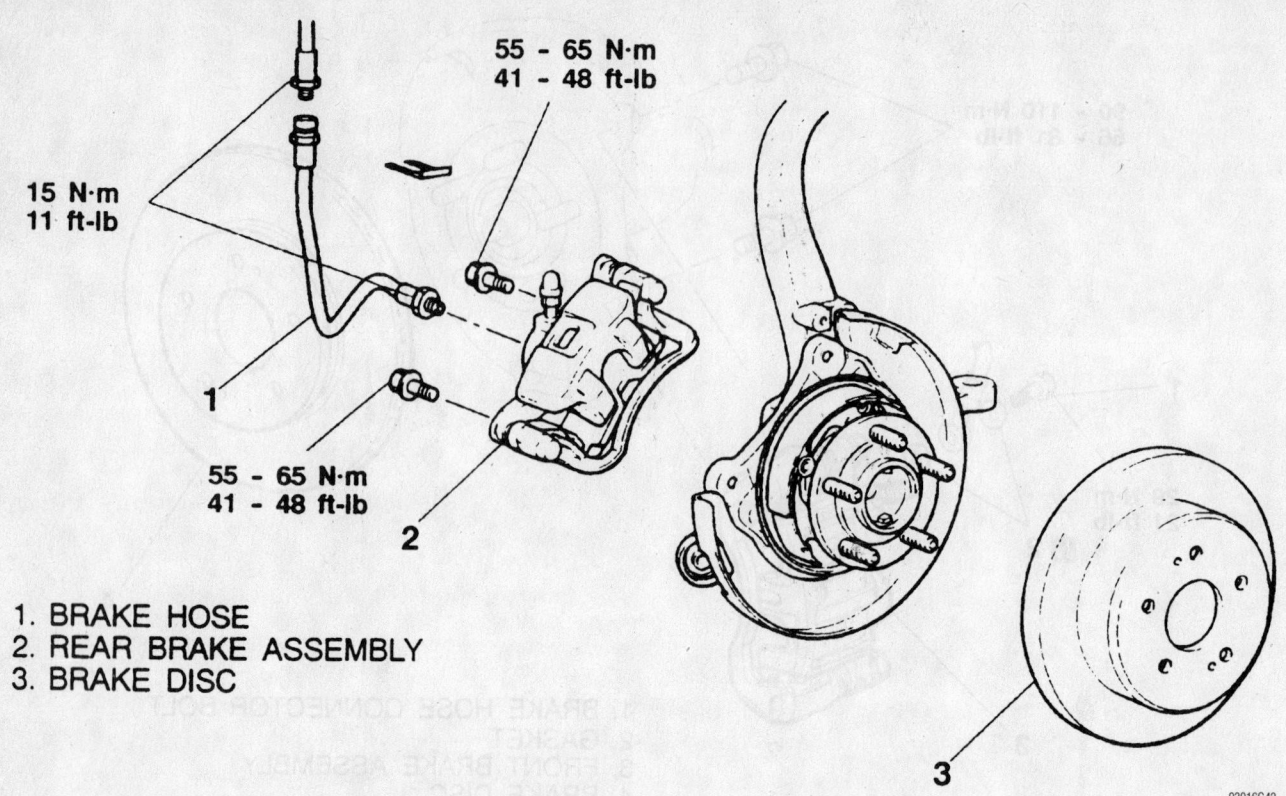

55 - 65 N·m
41 - 48 ft-lb

15 N·m
11 ft-lb

1

55 - 65 N·m
41 - 48 ft-lb

2

1. BRAKE HOSE
2. REAR BRAKE ASSEMBLY
3. BRAKE DISC

3

93016G42

Rear disc brakes—Galant

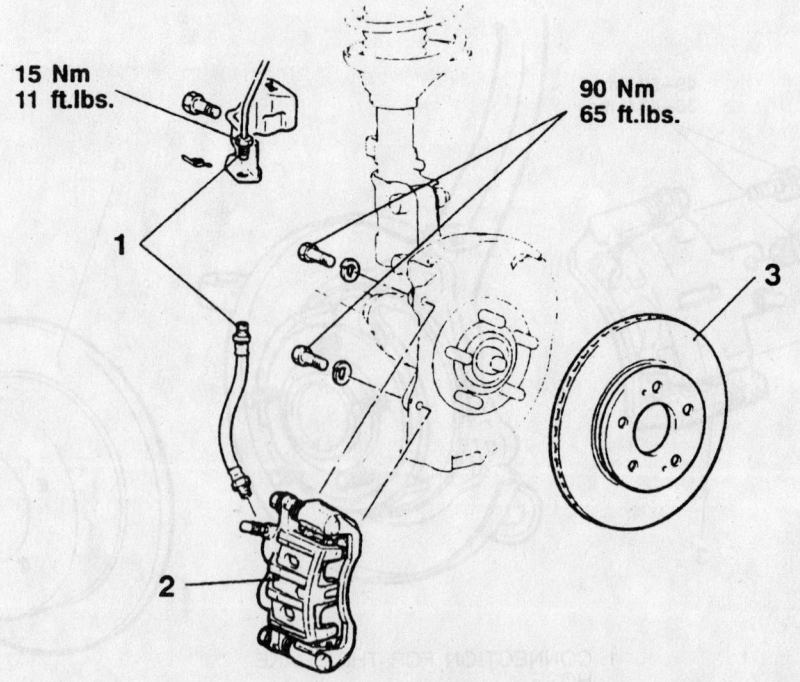

15 Nm
11 ft.lbs.

90 Nm
65 ft.lbs.

1

3

2

1. Connection for brake hose and the brake tube
2. Front brake assembly
3. Brake disc

93016G43

Front disc brakes—3000GT

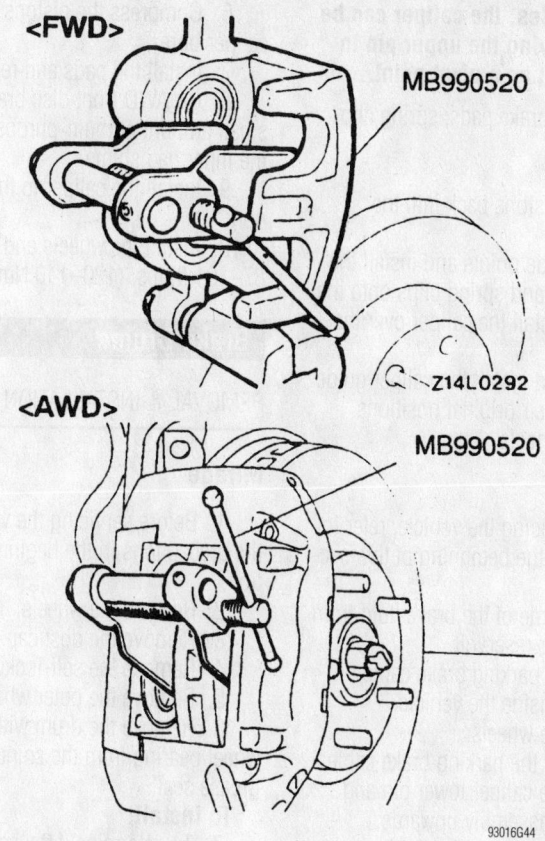

Compressing the caliper pistons—3000GT

8. Install the brake hose to the caliper.
9. Bleed the brake system.
10. Install the wheels.

Diamante and Galant

FRONT AND REAR

1. Before servicing the vehicle, refer to the precautions in the beginning of this section.
2. Remove the wheels.
3. Disconnect the brake hose from the caliper.
4. Remove the caliper guide and lock pins and lift the caliper assembly from the caliper support.

To install

5. Position the caliper onto the caliper support. Install the guide pin and lock pin. Tighten to specification.
6. Reconnect the brake hose or install the banjo bolt with new washers.
7. Bleed the brake system.
8. Install the wheels.

3000GT

FRONT AND REAR

1. Before servicing the vehicle, refer to the precautions in the beginning of this section.

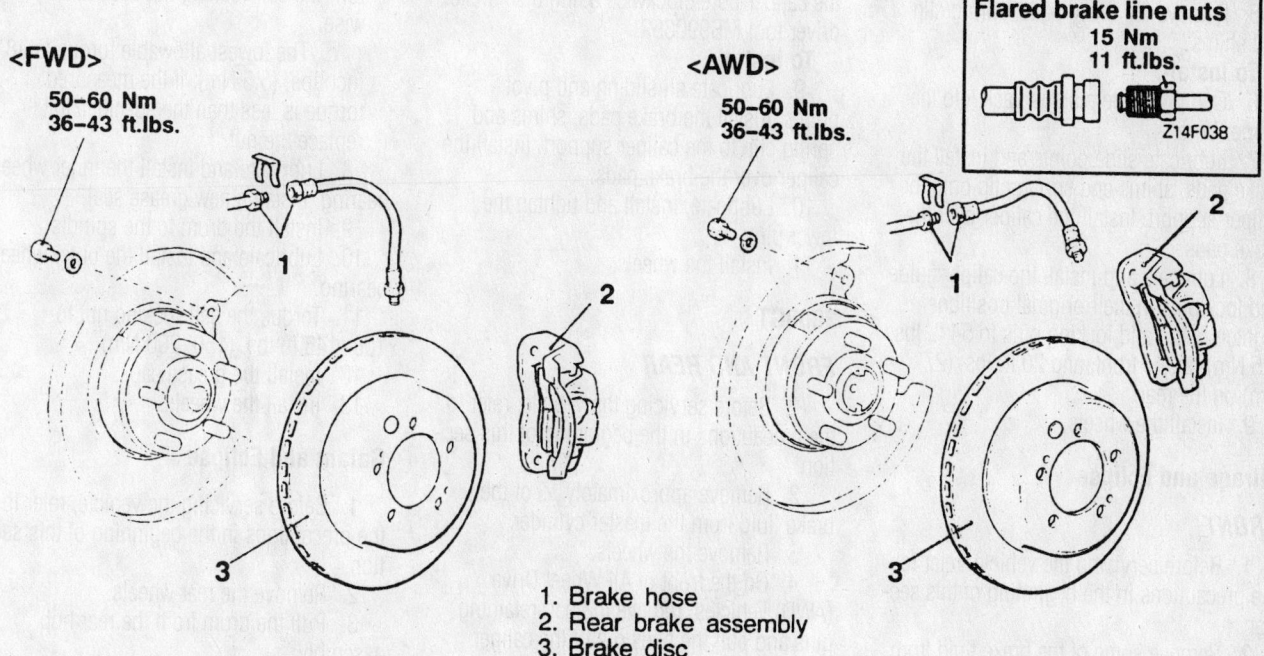

**Flared brake line nuts
15 Nm
11 ft.lbs.**

Z14F038

1. Brake hose
2. Rear brake assembly
3. Brake disc

Rear disc brakes—3000GT

For Maintenance Interval recommendations, see Section 1 of this manual

2. Remove the wheels.
3. Disconnect the brake hose.
4. Remove the caliper lock pins and remove the caliper.

To install:

5. Install the brake caliper to the caliper support and torque the bolts to 65 ft. lbs. (90 Nm).

6. Connect the brake hose and torque to 11 ft. lbs. (15 Nm).

7. Refill the brake fluid as required and bleed the brakes.

Disc Brake Pads

REMOVAL & INSTALLATION

Diamante and Galant

FRONT AND REAR

1. Before servicing the vehicle, refer to the precautions in the beginning of this section.

2. Remove some of the brake fluid from the master cylinder reservoir.

3. Remove the wheels.

4. Remove the caliper guide and lock pins and lift the caliper assembly from the caliper support.

➡**On some vehicles, the caliper can be flipped up by leaving the upper pin in place and using it as a pivot point.**

5. Remove the brake pads, spring clip and shims.

To install:

6. Compress the pistons back into the caliper bore.

7. Lubricate slide points and install the brake pads, shims and spring clip onto the caliper support. Install the caliper over the brake pads.

8. Lubricate and install the caliper guide and lock pins in their original positions. Tighten guide and locking pins to 54 ft. lbs. (75 Nm) on the front, and 20 ft. lbs. (27 Nm) on the rear.

9. Install the wheels.

Mirage and Eclipse

FRONT

1. Before servicing the vehicle, refer to the precautions in the beginning of this section.

2. Remove some of the brake fluid from the master cylinder reservoir.

3. Remove the wheels.

4. Remove the caliper guide and lock pins and lift the caliper assembly from the caliper support.

➡**On some vehicles, the caliper can be flipped up by leaving the upper pin in place and using it as a pivot point.**

5. Remove the brake pads, spring clip and shims.

To install:

6. Compress pistons back into the caliper bore.

7. Lubricate slide points and install the brake pads, shims and spring clips onto the caliper support. Install the caliper over the brake pads.

8. Lubricate and install the caliper guide and lock pins in their original positions.

9. Install the wheels.

REAR

1. Before servicing the vehicle, refer to the precautions in the beginning of this section.

2. Remove some of the brake fluid from the master cylinder reservoir.

3. Loosen the parking brake cable adjustment from inside the vehicle.

4. Remove the wheels.

5. Disconnect the parking brake cable.

6. Remove the caliper lower pin and swing the caliper assembly upwards.

7. Remove the outer shim, brake pads and spring clips from the caliper support.

8. On the Eclipse model, compress the piston into the caliper bore. On the Mirage and Galant models, thread the piston into the caliper bore clockwise using disc brake driver tool MB990652.

To install:

9. Lubricate all sliding and pivot points. Install the brake pads, shims and spring clip to the caliper support. Install the caliper over the brake pads.

10. Lubricate, install and tighten the lower pin.

11. Install the wheels.

3000GT

FRONT AND REAR

1. Before servicing the vehicle, refer to the precautions in the beginning of this section.

2. Remove approximately ⅓ of the brake fluid from the master cylinder.

3. Remove the wheels.

4. On the front of All Wheel Drive (AWD) vehicles, remove the pad retaining pins and pull the pads out of the caliper body.

5. On others, remove the lock pin and swing the caliper upward and support it with a wire. Remove the pads and retainers.

To install:

6. Compress the pistons back into the caliper bore.

7. Install the pads and retainers.

8. On AWD front disc brakes, apply a small amount of multi-purpose grease to the inner pad shims.

9. Install the caliper to the caliper support.

10. Install the wheels and torque to 87–101 ft. lbs. (120–140 Nm).

Brake Drums

REMOVAL & INSTALLATION

Mirage

1. Before servicing the vehicle, refer to the precautions in the beginning of this section.

2. Remove the wheels.

3. Remove the dust cap.

4. Remove the self-locking nut.

5. Remove the outer wheel bearing.

6. Remove the drum with the inner wheel bearing from the spindle. Remove the grease seal.

To install:

7. To determine if the self-locking nut is reusable:

a. Screw in the self-locking nut until about ⅛ in. of the spindle is showing.

b. Measure the torque required to turn the self-locking nut counterclockwise.

c. The lowest allowable torque is 48 inch lbs. (5.5 Nm). If the measured torque is less than the specification, replace the nut.

8. Lubricate and install the inner wheel bearing. Install a new grease seal.

9. Install the drum to the spindle.

10. Lubricate and install the outer wheel bearing.

11. Torque the self-locking nut to 108–145 ft. lbs. (150–200 Nm).

12. Install the grease cap.

13. Install the wheels.

Galant and Eclipse

1. Before servicing the vehicle, refer to the precautions in the beginning of this section.

2. Remove the rear wheels.

3. Pull the drum from the rear hub assembly.

To install:

4. Install the drum on the rear hub assembly.

5. Install the wheels.

Brake Shoes

REMOVAL & INSTALLATION

Mirage

1. Before servicing the vehicle, refer to the precautions in the beginning of this section.
2. Remove the wheels.
3. Remove the brake drum. Remove the shoe-to-shoe spring.
4. Remove the shoe-to-lever spring and remove the adjuster assembly.
5. Remove the shoe hold-down clips and the brake shoes.
6. Disconnect the parking brake cable from the rear shoes by spreading the horse-shoe clip apart.

To install

7. Lubricate the backing plate bosses, anchor pin, and parking brake actuating mechanism with lithium-based grease.
8. Connect the parking brake arm to the appropriate brake shoe.
9. Install the brake shoes and the shoe hold-down clips.
10. Install the adjuster assembly and the shoe-to-lever spring.
11. Install the shoe-to-shoe spring.
12. Preadjust the shoes so the drum slides on with a light drag and install brake drum.
13. Install a new wheel bearing self-locking nut and torque to 130 ft. lbs. (180 Nm).
14. Install the wheel bearing dust cap and adjust the rear brake shoes.

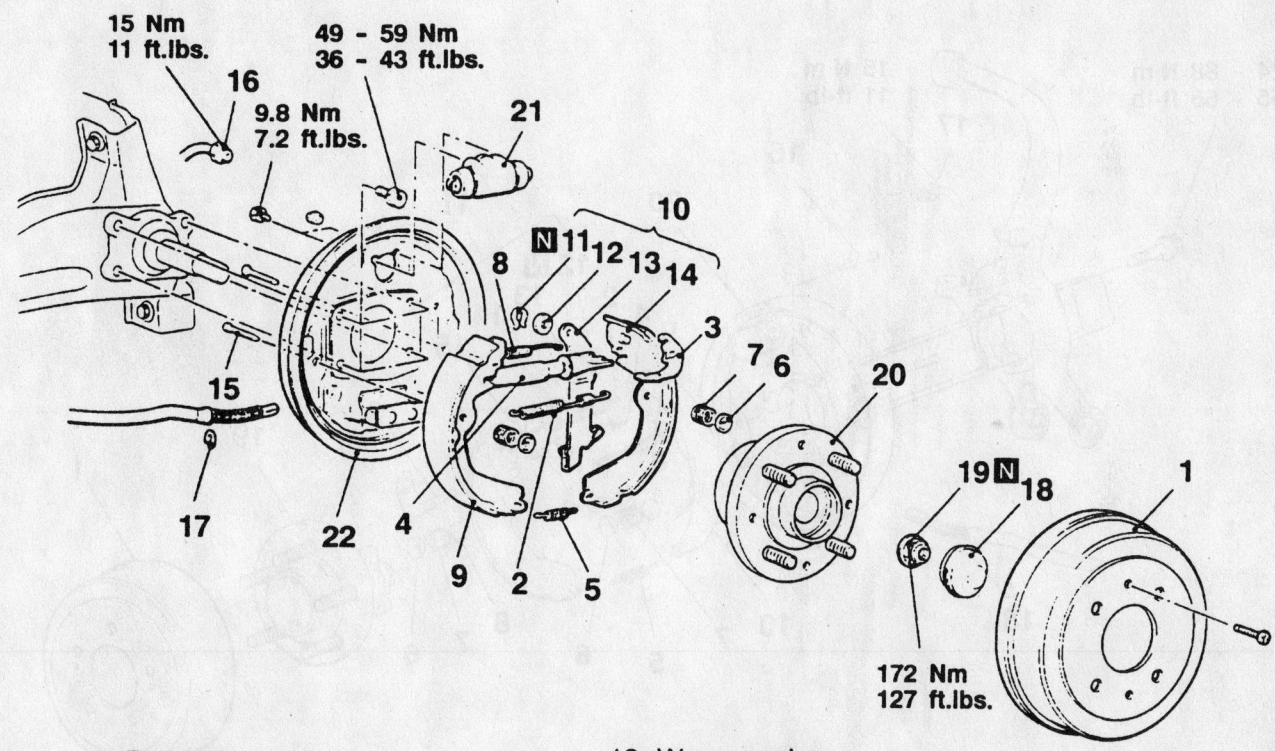

1. Brake drum
2. Shoe-to-lever spring
3. Adjuster lever
4. Auto adjuster assembly
5. Retainer spring
6. Shoe hold-down cup
7. Shoe hold-down spring
8. Shoe-to-shoe spring
9. Shoe and lining assembly
10. Shoe, lining and lever assembly
11. Retainer
12. Wave washer
13. Parking lever
14. Shoe and lining assembly
15. Shoe hold-down pin
16. Brake pipe connection
17. Snap ring
18. Hub cap
19. Flange nut
20. Rear hub assembly
21. Wheel cylinder
22. Backing plate

93016G46

Rear drum brakes—Mirage

Galant and Eclipse

1. Before servicing the vehicle, refer to the precautions in the beginning of this section.
2. Remove the rear wheels and drums.
3. Remove the lever return spring.
4. Remove the shoe-to-lever spring.
5. Remove the adjuster lever.
6. Remove the auto-adjuster assembly.
7. Remove the retainer spring.
8. Remove the brake shoe hold-down springs and spring cups.
9. Remove the shoe-to-shoe spring.

10. Remove the brake shoes.
11. Disconnect the parking brake cable from the lever on the rear shoe.

To install:

12. Remove the parking brake lever from the used shoe and install it on the new brake shoe. Make sure the wave washer is installed in the proper direction.
13. Clean the backing plate and lightly apply brake grease to the 6 shoe support pads.
14. Clean the adjuster assembly and apply brake grease to the threads.
15. Connect the parking brake cable to the lever on the rear shoe. Position the rear

shoe on the backing plate and install the hold-down spring and pin.
16. Position the front shoe on the backing plate and install the hold-down spring and pin.
17. Position the adjuster assembly between the 2 shoes.
18. Install the shoe-to-shoe spring.
19. Install the retainer spring.
20. Install the adjuster lever.
21. Install the shoe-to-lever spring.
22. Install the lever return spring.
23. Adjust the brake shoes and install the drum.
24. Install the wheels.

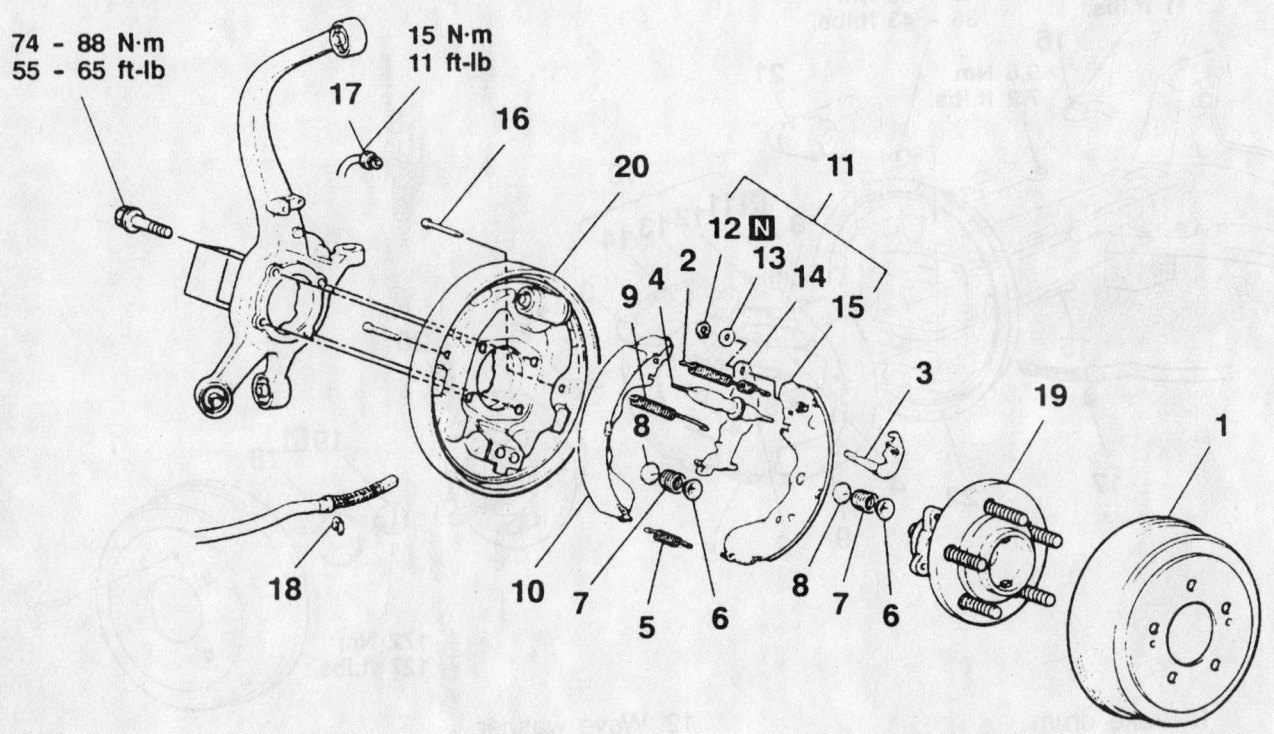

1. BRAKE DRUM
2. SHOE-TO-LEVER SPRING
3. ADJUSTER LEVER
4. AUTO ADJUSTER ASSEMBLY
5. RETAINER SPRING
6. SHOE HOLD-DOWN CUP
7. SHOE HOLD-DOWN SPRING
8. SHOE HOLD-DOWN CUP
9. SHOE-TO-SHOE SPRING
10. SHOE AND LINING ASSEMBLY
11. SHOE AND LEVER ASSEMBLY
12. RETAINER
13. WAVE WASHER
14. PARKING LEVER
15. SHOE AND LINING ASSEMBLY
16. SHOE HOLD-DOWN PIN
17. BRAKE TUBE CONNECTION
18. SNAP RING
19. REAR HUB ASSEMBLY
20. BACKING PLATE

93016G47

Rear drum brakes—Galant and Eclipse

1998–01 NISSAN

Brake Caliper

REMOVAL & INSTALLATION

Altima, Maxima, 240SX, Sentra and 200SX

FRONT

1. Before servicing the vehicle, refer to the precautions in the beginning of this section.

2. Remove the front wheels.
3. Disconnect the brake fluid hose.
4. Remove the pin bolts.
5. Remove the caliper assembly from the vehicle.

To install:

6. Use a large C-clamp to press the caliper piston back into the caliper.
7. Install the new pads, new shims and pad retainers.

8. Install the brake caliper and torque the pin bolts to 23 ft. lbs. (31 Nm).
9. Using new copper washers, install the brake line to the brake caliper and torque the connecting bolt to 12–14 ft. lbs. (17–20 Nm).
10. Install the wheels.
11. Bleed the brake system and top off the master cylinder as necessary.

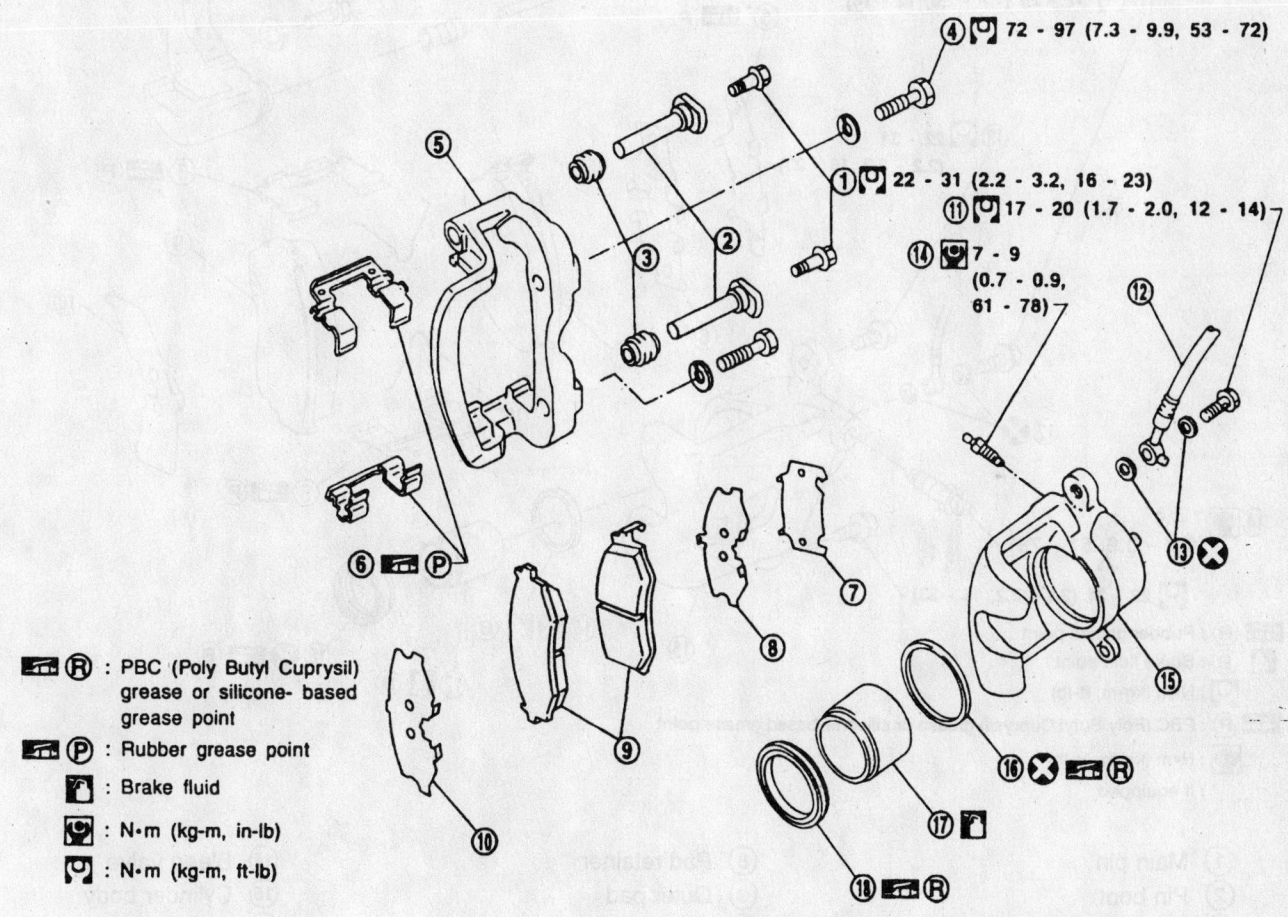

⊞ⓇR : PBC (Poly Butyl Cuprysil) grease or silicone- based grease point

⊞Ⓟ : Rubber grease point

▮ : Brake fluid

🔧 : N•m (kg-m, in-lb)

🔧 : N•m (kg-m, ft-lb)

① Main pin bolt	⑦ Shim cover
② Pin	⑧ Inner shim
③ Pin boot	⑨ Pad
④ Torque member fixing bolt	⑩ Outer shim
⑤ Torque member	⑪ Connecting bolt
⑥ Pad retainer	⑫ Brake hose

⑬ Copper washer
⑭ Air bleeder
⑮ Cylinder body
⑯ Piston seal
⑰ Piston
⑱ Piston boot

93016G50

Front brake caliper—240 SX

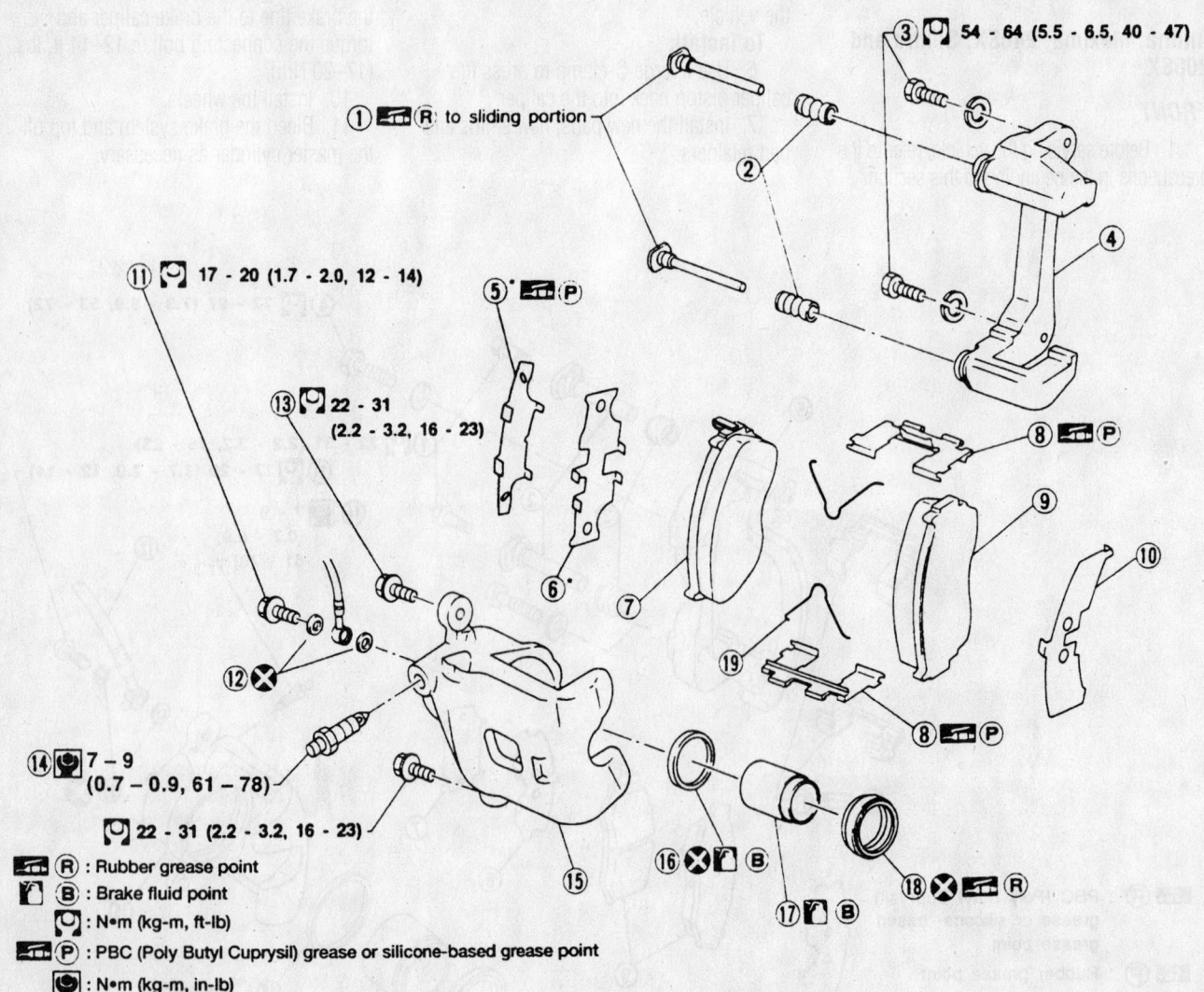

① 🔧 ℝ to sliding portion

③ 🔧 54 - 64 (5.5 - 6.5, 40 - 47)

④

⑪ 🔧 17 - 20 (1.7 - 2.0, 12 - 14)

⑤ • 🔧 ℙ

⑬ 🔧 22 - 31 (2.2 - 3.2, 16 - 23)

⑧ 🔧 ℙ

⑨

⑩

⑦

⑥

⑲

⑧ 🔧 ℙ

⑫ ⊗

⑭ 🔧 7 - 9 (0.7 - 0.9, 61 - 78)

🔧 22 - 31 (2.2 - 3.2, 16 - 23)

⑮

⑯ ⊗ 🔧 B

⑰ 🔧 B

⑱ ⊗ 🔧 ℝ

🔧 ℝ : Rubber grease point
🔧 B : Brake fluid point
🔧 : N•m (kg-m, ft-lb)
🔧 ℙ : PBC (Poly Butyl Cuprysil) grease or silicone-based grease point
🔧 : N•m (kg-m, in-lb)
  * : If equipped

① Main pin
② Pin boot
③ Torque member fixing bolt
④ Torque member
⑤ Shim cover*
⑥ Inner shim*
⑦ Inner pad

⑧ Pad retainer
⑨ Outer pad
⑩ Outer shim
⑪ Connecting bolt
⑫ Copper washer
⑬ Main pin bolt

⑭ Bleed valve
⑮ Cylinder body
⑯ Piston seal
⑰ Piston
⑱ Piston boot
⑲ Pad return spring

93016G51

Front brake caliper—Sentra

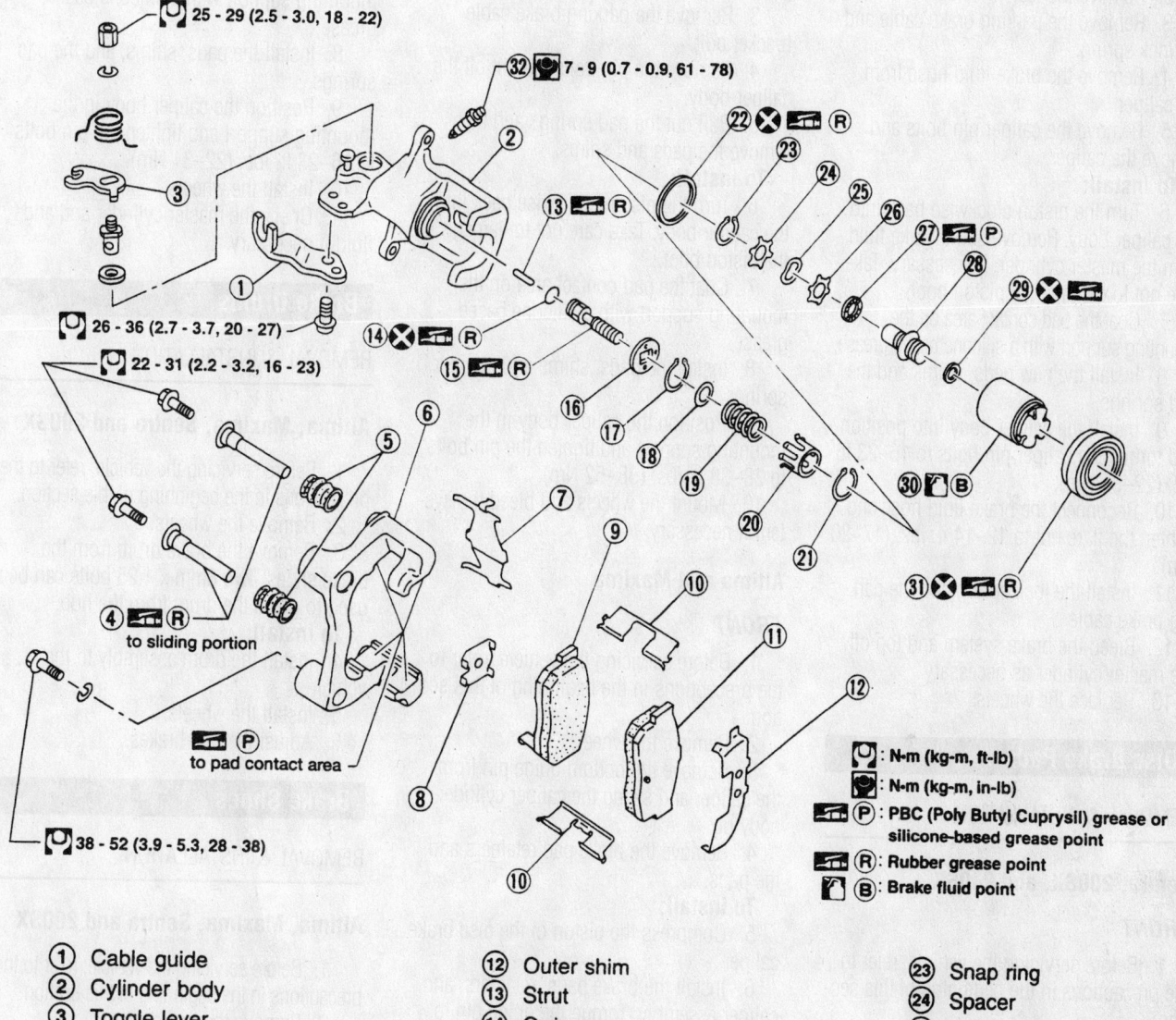

25 - 29 (2.5 - 3.0, 18 - 22)

32 7 - 9 (0.7 - 0.9, 61 - 78)

13

26 - 36 (2.7 - 3.7, 20 - 27)

22 - 31 (2.2 - 3.2, 16 - 23)

14

15

4 to sliding portion

to pad contact area

38 - 52 (3.9 - 5.3, 28 - 38)

: N-m (kg-m, ft-lb)

: N-m (kg-m, in-lb)

: PBC (Poly Butyl Cuprysil) grease or
silicone-based grease point

: Rubber grease point

: Brake fluid point

① Cable guide	⑫ Outer shim	㉓ Snap ring
② Cylinder body	⑬ Strut	㉔ Spacer
③ Toggle lever	⑭ O-ring	㉕ Wave washer
④ Pin	⑮ Push rod	㉖ Spacer
⑤ Pin boot	⑯ Key plate	㉗ Bearing
⑥ Torque member	⑰ Snap ring	㉘ Adjuster
⑦ Retainer	⑱ Seat	㉙ Cup
⑧ Inner shim	⑲ Spring	㉚ Piston
⑨ Inner pad	⑳ Spring cover	㉛ Piston boot
⑩ Pad retainer	㉑ Snap ring	㉜ Air bleeder
⑪ Outer pad	㉒ Piston seal	

93016G52

Rear disc brakes—Sentra shown

REAR

1. Before servicing the vehicle, refer to the precautions in the beginning of this section.
2. Remove the rear wheels.
3. Remove the parking brake cable and the lock spring.
4. Remove the brake fluid hose from the caliper.
5. Remove the caliper pin bolts and remove the caliper.

To install:
6. Turn the piston clockwise back into the caliper body. Remove some brake fluid from the master cylinder, if necessary. Take care not to damage the piston boot.
7. Coat the pad contact area on the mounting support with a silicone based grease.
8. Install the new pads, shims and the pad springs.
9. Install the caliper body into position and torque the caliper pin bolts to 16–23 ft. lbs. (22–31 Nm).
10. Reconnect the brake fluid hose and tighten the flare nut to 12–14 ft. lbs. (17–20 Nm).
11. Install the lock spring and the parking brake cable.
12. Bleed the brake system and top off the master cylinder as necessary.
13. Replace the wheels.

Disc Brake Pads

REMOVAL & INSTALLATION

Sentra, 200SX, and 240SX

FRONT

1. Before servicing the vehicle, refer to the precautions in the beginning of this section.
2. Remove the wheels.
3. Remove the bottom guide pin from the caliper and swing the caliper cylinder body up.
4. Remove the brake pad retainers and the pads.

To install:
5. Compress the piston of the disc brake caliper.
6. Install the brake pads, shims, and retainers.
7. Install the caliper assembly. Torque the guide pin to 23–30 ft. lbs. (31–41 Nm).
8. Install the wheels.
9. Check the master cylinder and add fluid if necessary.

REAR

1. Before servicing the vehicle, refer to the precautions in the beginning of this section.
2. Remove the wheels.
3. Remove the parking brake cable bracket bolt.
4. Remove the pin bolts and lift off the caliper body.
5. Pull out the pad springs and then remove the pads and shims.

To install:
6. Turn the piston clockwise back into the caliper body. Take care not to damage the piston boot.
7. Coat the pad contact area on the mounting support with a silicone based grease.
8. Install the pads, shims and retainer springs.
9. Position the caliper body in the mounting support and tighten the pin bolts to 28–38 ft. lbs. (38–52 Nm).
10. Mount the wheels and bleed the system if necessary.

Altima and Maxima

FRONT

1. Before servicing the vehicle, refer to the precautions in the beginning of this section.
2. Remove the wheels.
3. Remove the bottom guide pin from the caliper and swing the caliper cylinder body up.
4. Remove the brake pad retainers and the pads.

To install:
5. Compress the piston of the disc brake caliper.
6. Install the brake pads, retainers, and caliper assembly. Torque the guide pin to 16–23 ft. lbs. (22–31 Nm).
7. Install the wheels.
8. Check the master cylinder and add fluid if necessary.

REAR

1. Before servicing the vehicle, refer to the precautions in the beginning of this section.
2. Remove the rear wheels.
3. Remove the parking brake cable bracket bolt.
4. Remove the pin bolts and lift off the caliper body.
5. Pull out the pad springs and then remove the pads and shims.

To install:
6. Turn the piston clockwise back into the caliper body. Take care not to damage the piston boot.
7. Coat the pad contact area on the mounting support with a silicone based grease.
8. Install the pads, shims, and the pad springs.
9. Position the caliper body in the mounting support and tighten the pin bolts to 16–23 ft. lbs. (22–31 Nm).
10. Install the wheels.
11. Check the master cylinder and add fluid if necessary.

Brake Drums

REMOVAL & INSTALLATION

Altima, Maxima, Sentra and 200SX

1. Before servicing the vehicle, refer to the precautions in the beginning of this section.
2. Remove the wheels.
3. Remove the brake drum from the brake shoes. Two 8mm x 1.25 bolts can be used to press the drum from the hub.

To install:
4. Install the drum assembly to the vehicle.
5. Install the wheels.
6. Adjust the rear brakes.

Brake Shoes

REMOVAL & INSTALLATION

Altima, Maxima, Sentra and 200SX

1. Before servicing the vehicle, refer to the precautions in the beginning of this section.
2. Remove the brake drum.
3. Remove the return springs, adjuster assembly, hold-down springs, and brake shoes.
4. Disconnect the parking brake cable from the toggle lever.

To install:
5. Reconnect the parking brake cable.
6. Install the shoes with the hold-down springs.
7. Hook the return springs into the new shoes.
8. Install the adjuster assembly.
9. Install the drums and wheels. Adjust the brakes and bleed the hydraulic system, if necessary.
10. Check the parking brake adjustment.

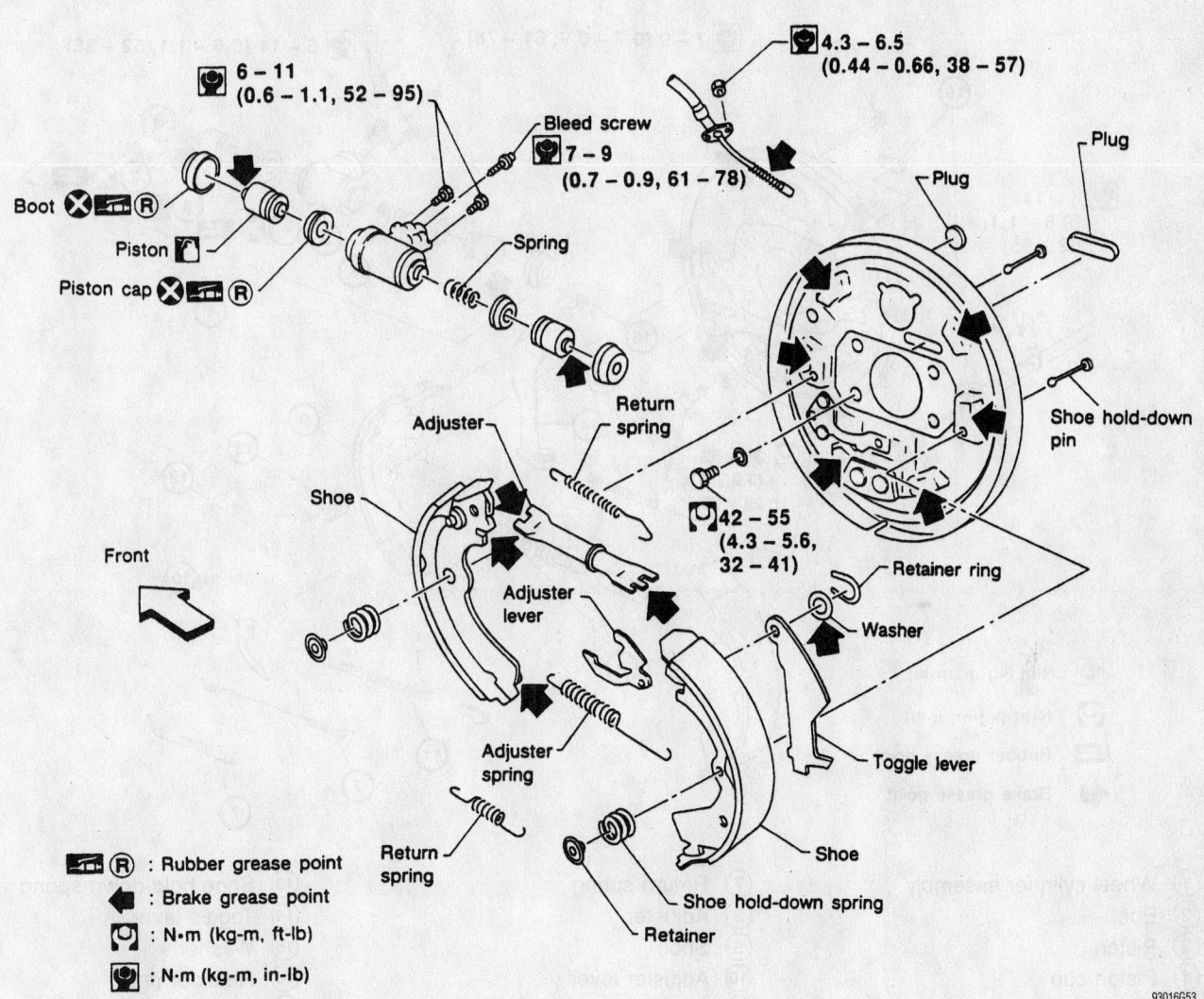

6 – 11
(0.6 – 1.1, 52 – 95)

Bleed screw
7 – 9
(0.7 – 0.9, 61 – 78)

4.3 – 6.5
(0.44 – 0.66, 38 – 57)

Plug

Plug

Boot

Piston

Piston cap

Spring

Return
spring

Adjuster

Shoe hold-down
pin

Shoe

Front

Adjuster
lever

42 – 55
(4.3 – 5.6,
32 – 41)

Retainer ring

Washer

Adjuster
spring

Toggle lever

Return
spring

Shoe

Shoe hold-down spring

Retainer

: Rubber grease point

: Brake grease point

: N·m (kg-m, ft-lb)

: N·m (kg-m, in-lb)

93016G53

Exploded view of the rear drum brakes—Sentra

Timing belt service is covered in Section 3 of this manual

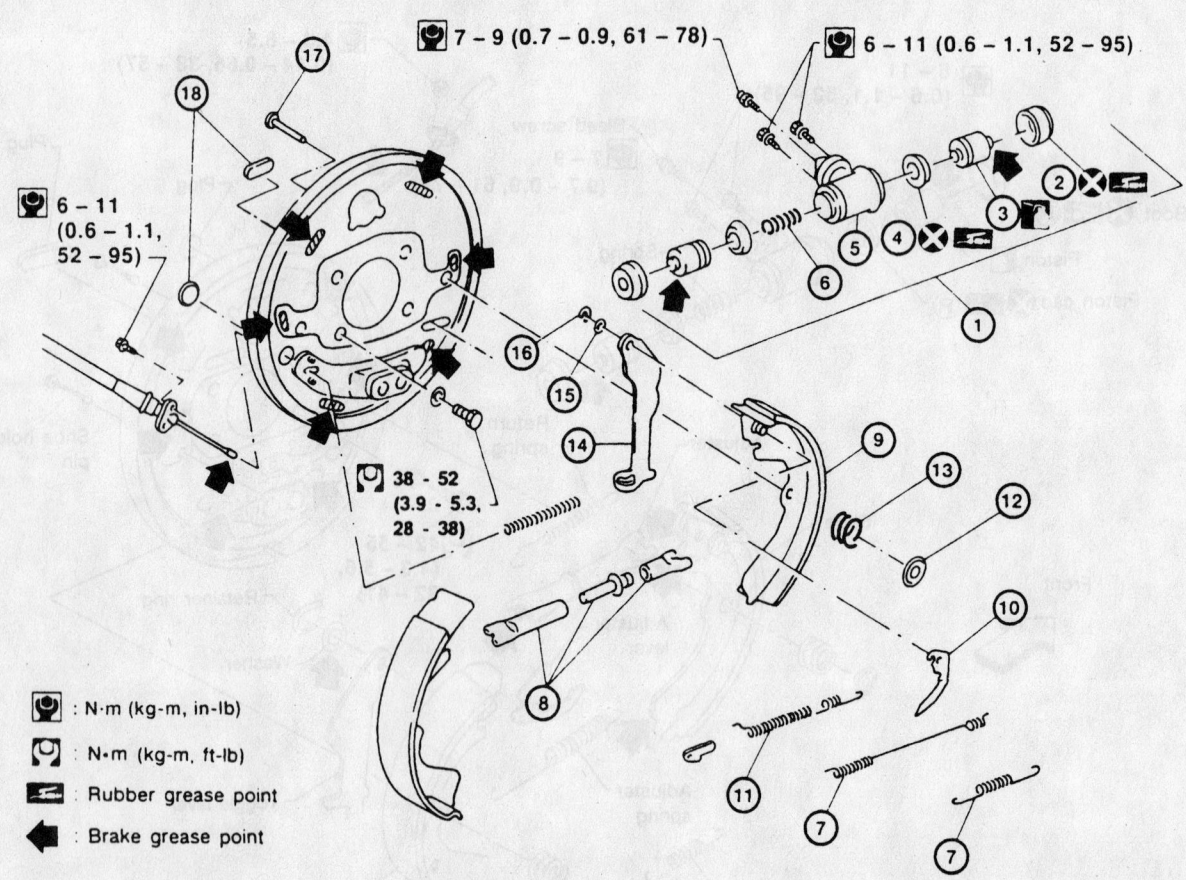

7 – 9 (0.7 – 0.9, 61 – 78)

6 – 11 (0.6 – 1.1, 52 – 95)

6 – 11
(0.6 – 1.1,
52 – 95)

38 – 52
(3.9 – 5.3,
28 – 38)

: N·m (kg-m, in-lb)

: N·m (kg-m, ft-lb)

: Rubber grease point

: Brake grease point

① Wheel cylinder assembly	⑦ Return spring	⑬ Shoe hold-down spring
② Boot	⑧ Adjuster	⑭ Toggle lever
③ Piston	⑨ Shoe	⑮ Washer
④ Piston cup	⑩ Adjuster lever	⑯ Retainer ring
⑤ Cylinder body	⑪ Adjuster spring	⑰ Shoe hold-down pin
⑥ Spring	⑫ Retainer	⑱ Plug

93016G54

Rear drum brakes—1998 Altima

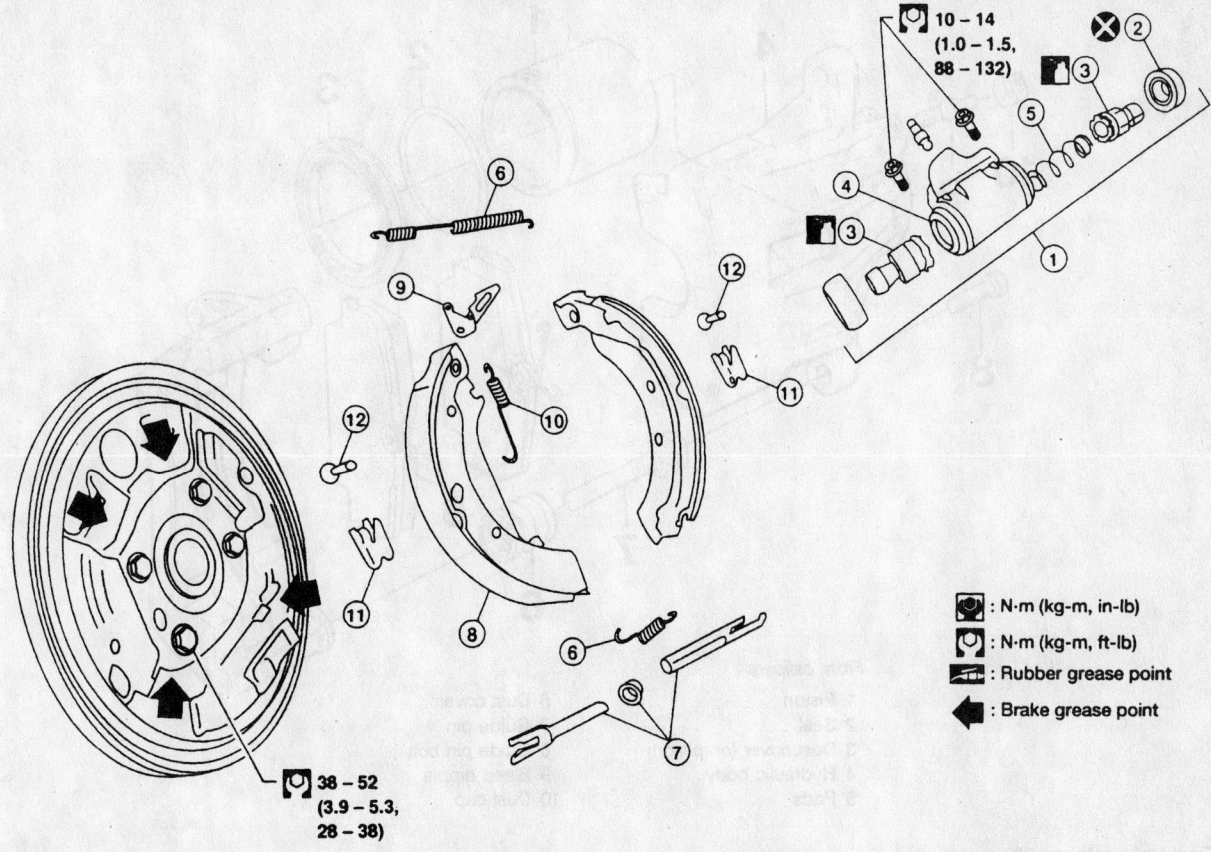

10 – 14
(1.0 – 1.5,
88 – 132)

38 – 52
(3.9 – 5.3,
28 – 38)

: N·m (kg-m, in-lb)

: N·m (kg-m, ft-lb)

: Rubber grease point

: Brake grease point

① Wheel cylinder assembly
② Boot
③ Piston
④ Cylinder body
⑤ Spring
⑥ Return spring
⑦ Adjuster
⑧ Shoe
⑨ Adjuster lever
⑩ Adjuster spring
⑪ Retainer
⑫ Shoe hold-down pin

93016G55

Rear drum brakes—1999 Altima

1998–01 SAAB

Brake Caliper

REMOVAL & INSTALLATION

900 and 9–3 Series

FRONT

1. Before servicing the vehicle, refer to the precautions in the beginning of this section.
2. Remove the wheels.
3. Disconnect the brake hose from the brake caliper.
4. Remove the brake caliper mounting bolts and remove the brake caliper.

To install:

5. Fit the brake caliper onto the strut.

Apply thread-locking compound to the mounting bolts and tighten the caliper mounting bolts to 59 ft. lbs. (80 Nm).

6. Install the brake hose onto the caliper. Be careful to install the brake hose in its original position.
7. Bleed the brake system and fill the fluid to the proper level.
8. Install the wheels.

REAR

1. Before servicing the vehicle, refer to the precautions in the beginning of this section.
2. Remove the wheels.
3. Remove the locking spring and caliper pins and remove the caliper.
4. Remove the brake pads.

To install:

5. Compress the caliper piston.
6. Install the brake pads, caliper pins and locking spring.
7. Install the wheels and tighten the lug nuts.

9000 Series

FRONT

1. Before servicing the vehicle, refer to the precautions in the beginning of this section.
2. Remove the front wheels.
3. Remove the guide pins and the retaining clip.
4. Disconnect the brake hose from the caliper. Remove the caliper assembly.

Heater Core replacement is covered in Section 2 of this manual

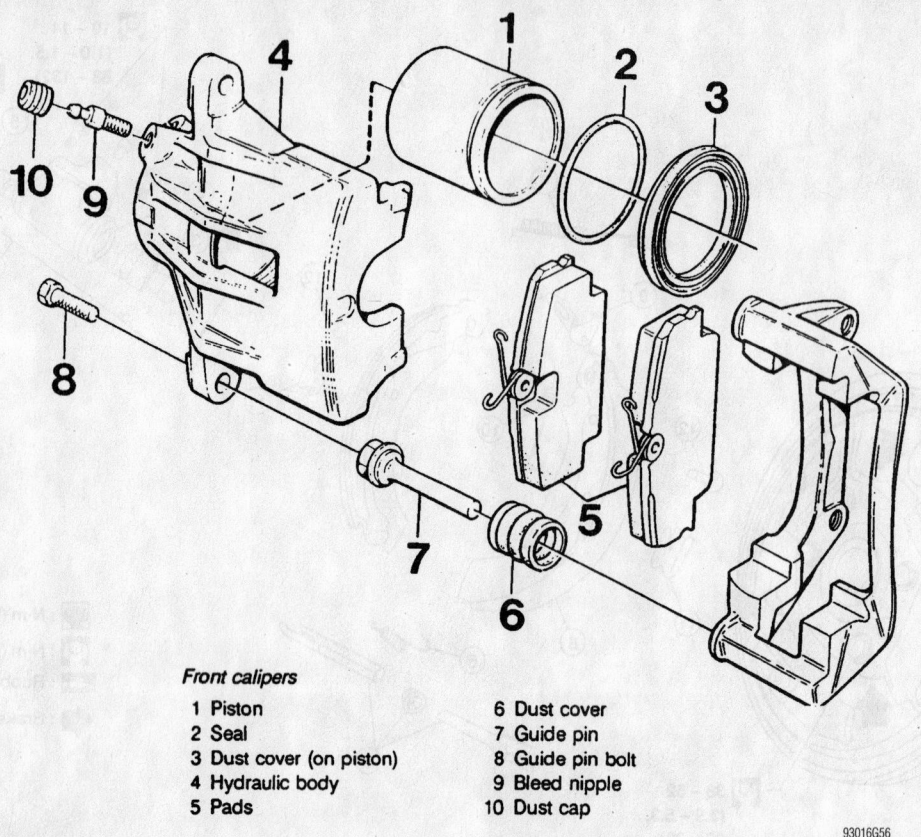

Front calipers

1 Piston
2 Seal
3 Dust cover (on piston)
4 Hydraulic body
5 Pads
6 Dust cover
7 Guide pin
8 Guide pin bolt
9 Bleed nipple
10 Dust cap

93016G56

Front brake caliper—900

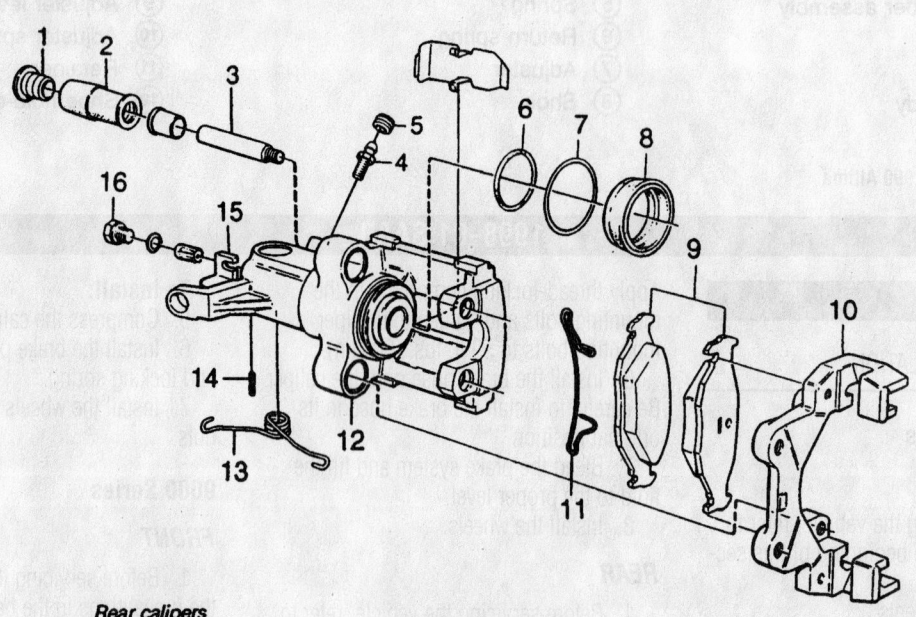

Rear calipers

1 Dust cap
2 Spacer sleeve
3 Guide pin
4 Bleed nipple
5 Dust cap
6 Piston seal
7 Retaining ring
8 Dust cover
9 Brake pads
10 Carrier
11 Pad retaining clip
12 Hydraulic body
13 Return spring
14 Stop pin
15 Lever
16 Screw plug (over adjusting screw)

93016G57

Rear caliper—900

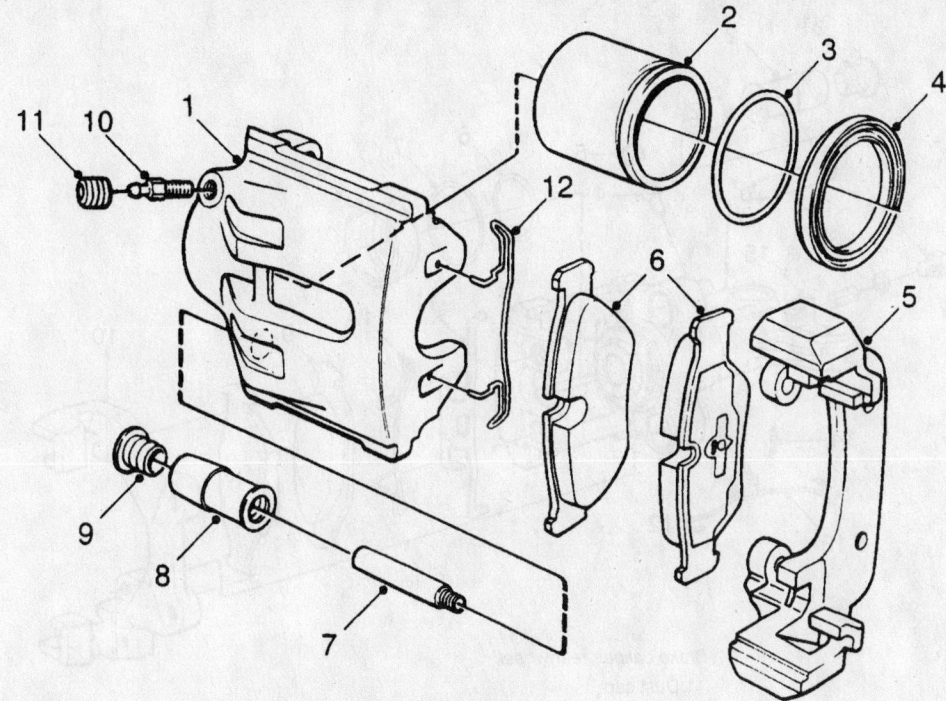

Brake caliper, front wheel

1 Hydraulic body
2 Brake piston
3 Piston sealing ring
4 Dust excluder
5 Carrier
6 Brake pad
7 Guide pin
8 Spacer sleeve
9 Dust cap
10 Bleed nipple
11 Dust cap
12 Retaining clip

93016G58

Front brake caliper—9000

To install:

5. Install the caliper. Install the retaining clip and guide pins.

6. Reconnect the brake hose to the caliper. Bleed the hydraulic system.

7. Install the wheels and tighten the wheel lug nuts. Torque the wheel lug nuts to 80–90 ft. lbs. (108–122 Nm).

REAR

1. Before servicing the vehicle, refer to the precautions in the beginning of this section.

2. Remove the rear wheels.

3. Disconnect the parking brake cable from the back of the caliper assembly.

4. Remove the disc brake pad retaining clip.

5. Remove the screw plug from the

adjusting screw and unscrew the adjusting screw using a 4mm hex key tool.

6. Remove the caliper guide pins.

7. Remove the caliper assembly. Disconnect the brake line from the brake caliper.

To install:

8. Reconnect the brake line to the brake caliper.

9. Install the brake caliper into position. Install and tighten the caliper guide pins.

10. Install the brake pad retaining clip into place.

11. Reconnect the parking brake cable to the parking brake lever.

12. Adjust the parking brake to proper specifications. Bleed the brake system. Top off the brake fluid, if needed.

13. Install the wheels.

9–5 Series

FRONT

1. Before servicing the vehicle, refer to the precautions in the beginning of this section.

2. Remove the front wheels.

3. Press in the brake piston with slip joint pliers.

4. Depress the brake pedal slightly with a brake pedal clamp.

5. Remove fuse No. 1.

6. Remove the brake hose from the caliper.

7. Remove the caliper retaining bolts and remove the caliper.

To install:

8. Install the brake caliper and torque the retaining bolts to 86 ft. lbs. (117 Nm).

Brake service is covered in Section 4 of this manual

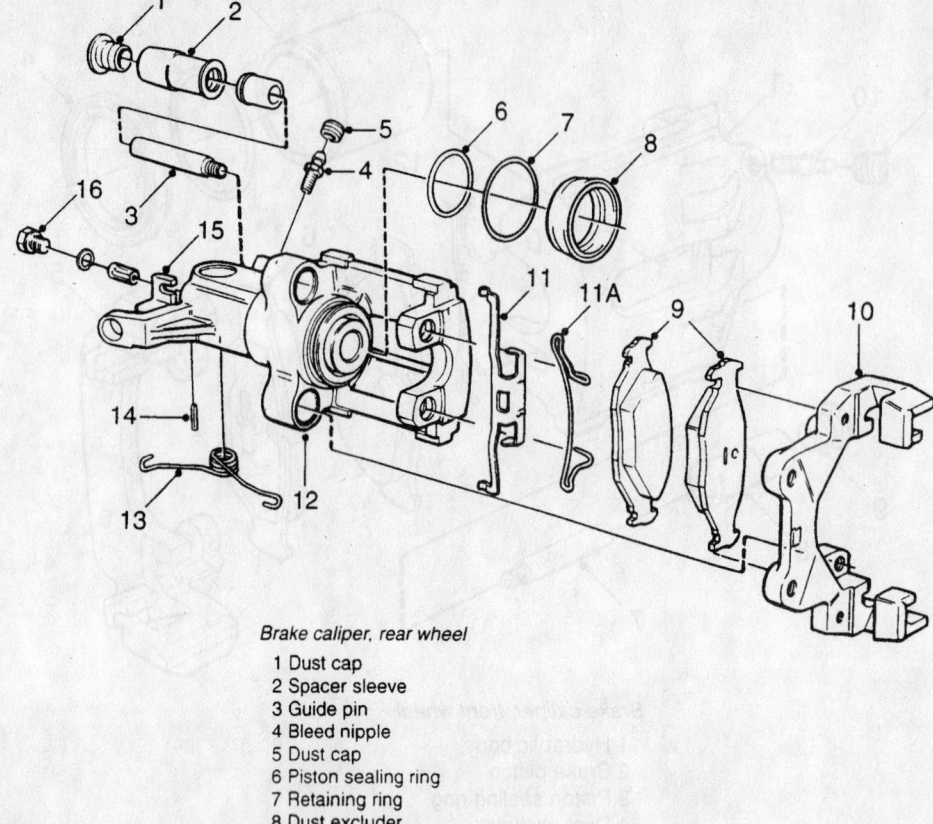

Brake caliper, rear wheel

1 Dust cap
2 Spacer sleeve
3 Guide pin
4 Bleed nipple
5 Dust cap
6 Piston sealing ring
7 Retaining ring
8 Dust excluder
9 Brake pad
10 Carrier
11 Retaining clip (early version),
   11A Retaining clip (later version)
12 Hydraulic body
13 Return spring
14 Stop pin
15 Lever
16 Threaded plug (for adjusting screw)

93016G59

Rear caliper—9000

9. Install the brake hose to the caliper and torque it to 29 ft. lbs. (40 Nm).

10. Remove the brake pedal clamp and replace the fuse

11. Bleed the brake system and top off the fluid, if necessary.

12. Install the front wheels.

REAR

1. Before servicing the vehicle, refer to the precautions in the beginning of this section.

2. Remove the rear wheels.

3. Remove the brake pads.

4. Depress the brake pedal slightly with a brake pedal clamp.

5. Remove fuse No. 1.

6. Remove the clip from the brake pedal pipe, swinging caliper only.

7. Remove the hose from the brake pedal pipe.

8. Remove the caliper guide pins and remove the hydraulic body, swinging caliper only.

9. Remove the brake hose from the caliper.

10. Remove the caliper.

To install:

11. Install the brake lines to the hydraulic body, swinging caliper only. Torque the line to 32 ft. lbs. (43 Nm).

12. Install the hydraulic body and torque the bolts to 21 ft. lbs. (28 Nm).

13. Install the caliper and torque the bolts to 59 ft. lbs. (80 Nm), fixed caliper only.

14. Install the brakes lines and torque them to 11 ft. lbs. (16 Nm), fixed caliper only.

15. Install the brake hose to the pipe and install the clip, swinging caliper only.

16. Install the brake pads.

17. Remove the brake pedal clamp and replace the fuse.

18. Bleed the brake system and top off the fluid, if necessary.

19. Install the rear wheels.

Disc Brake Pads

REMOVAL & INSTALLATION

900 and 9–3 Series

FRONT

1. Before servicing the vehicle, refer to the precautions in the beginning of this section.

2. Remove the wheels.

3. Remove the retaining clip from the brake caliper.

4. Remove the caliper guide pins.

5. Lift off the brake caliper and remove the pads.

To install:

6. Install the new brake pads in the caliper and place the brake caliper back into its original position.

7. Install the caliper guide pins. Torque the guide pin bolts to 78 ft. lbs. (105 Nm).

8. Install the caliper retaining clip.

9. Install the wheels and tighten the wheel lug nuts.

REAR

1. Before servicing the vehicle, refer to the precautions in the beginning of this section.

2. Remove the rear wheels.

3. Compress the caliper piston into the caliper bore.

4. Remove the brake pad retaining pins and lock spring.

5. Remove the brake pads.

To install:

6. Install the brake pads into the brake caliper.

7. Install the lock spring and the brake pad retaining pins.

8. Install the wheels and tighten the lug nuts. Torque the lug nuts to 80–90 ft. lbs. (108–122 Nm).

9000 Series

FRONT

1. Before servicing the vehicle, refer to the precautions in the beginning of this section.

2. Remove the front wheels.

3. Remove the caliper guide pins.

4. Remove the brake pad retaining clip.

5. Lift off the caliper assembly and remove the brake pads.

To install:

6. Compress the brake pad and piston into the caliper.

---

7. Install the new brake pads onto the caliper assembly along with the anti-rattle spring.

8. Install the caliper assembly and tighten the guide pins. Install the brake pad retaining clip.

9. Install the wheels and wheel lug nuts. Torque the lug nuts to 80–90 ft. lbs. (108–122 Nm).

REAR

1. Before servicing the vehicle, refer to the precautions in the beginning of this section.

2. Remove the rear wheels.

3. Disconnect the parking brake cable from the parking brake lever on the caliper.

4. Remove the screw plug to access the parking brake adjusting screw.

5. Move the brake caliper piston back against its stop by turning in the adjusting screw.

6. Remove the caliper guide pins.

7. Remove the brake pad retaining clip.

8. Move the brake caliper out of the way and remove the brake pads.

To install:

9. Install the new brake pads, fitting the brake pad with the clip against the caliper piston.

10. Install the caliper into position and install the caliper guide pins. Torque the guide pins to 18–22 ft. lbs. (25–30 Nm).

11. Install the guide pin dust caps. Install the brake pad retaining clips.

12. Rotate the adjusting screw all the way in and then back it out ¼–½ turn. Check that the brake rotor freely rotates. Install the adjusting screw plug.

13. Reconnect the parking brake cable to the parking brake lever on the caliper and adjust the parking brake.

14. Install the wheels and lug nuts.

9-5 Series

FRONT

1. Before servicing the vehicle, refer to the precautions in the beginning of this section.

2. Remove the front wheels.

3. Press back the caliper piston and remove the clip.

4. Remove the caliper guide pins.

5. Remove the caliper and suspend it from the strut.

6. Remove the brake pads.

To install:

7. Install the brake pads in the caliper.

8. Install the brake caliper and torque the bolts to 21 ft. lbs. (28 Nm).

9. Install the brake clip depress the brake pedal to force out the pistons.

10. Install the front wheels.

11. Check and top off the brake fluid, if necessary.

REAR

1. Before servicing the vehicle, refer to the precautions in the beginning of this section.

2. Remove the rear wheels.

3. Press back the piston and remove the caliper clip.

4. Remove the caliper guide pins.

5. Remove the brake caliper.

6. Remove the brake pads.

To install:

7. Install the brake pads.

8. Install the caliper and torque the pins to 21 ft. lbs. (28 Nm).

9. Install the caliper clip and depress the brake pedal to force out the pistons.

10. Install the rrear wheels.

11. Check and top off the brake fluid, if necessary.

1998–01 SUBARU

Brake Caliper

REMOVAL & INSTALLATION

Front

1. Before servicing the vehicle, refer to the precautions in the beginning of this section.

2. Remove the front wheels.

3. Remove the brake hose from the caliper body.

4. Remove the caliper bracket retainer bolts.

---

5. Slide the caliper and bracket assembly off the spindle and rotor.

To install:

6. Compress the piston assembly into the cylinder bore.

7. Install the caliper bracket to the spindle assembly, and secure in place with the retainer bolts. Tighten the retainer bolts to 25–33 ft. lbs. (32–43 Nm).

8. Connect the brake hose using new sealing washers, and tighten the fitting to 11–15 ft. lbs. (15–21 Nm).

9. Bleed the brake system. Install the

---

wheels. Check the fluid level in the master cylinder.

Rear

1. Before servicing the vehicle, refer to the precautions in the beginning of this section.

2. Remove the rear wheels.

3. Remove the brake hose from the caliper body.

4. Remove the caliper bracket retainer bolts.

5. Slide the caliper and bracket assembly off the rotor.

To install:

6. Compress the piston assembly into the cylinder bore.

7. Install the caliper bracket to the spindle assembly, and secure in place with the retainer bolts. Tighten the retainer bolts to 12–17 ft. lbs. (16–22 Nm).

8. Connect the brake hose using new sealing washers, and tighten the fitting to 12–14 ft. lbs. (16–20 Nm).

9. Bleed the brake system. Install the wheels. Check the fluid level in the master cylinder.

Disc Brake Pads

REMOVAL & INSTALLATION

Front

1. Before servicing the vehicle, refer to the precautions in the beginning of this section.

2. Remove the wheels.

3. Disconnect the cable from the caliper lever, if equipped.

4. Remove the lock pin bolts from the lower portion of the caliper.

5. Swing the caliper upward to access the pads.

6. Remove the disc brake pads.

To install:

➡ **If equipped with a parking brake, use a suitable tool to rotate the piston back into the caliper bore.**

7. Compress the caliper pistons.

8. Install the new pads into the caliper brackets, being sure all shims and clips are in their original positions.

9. Swing the calipers down into posi-

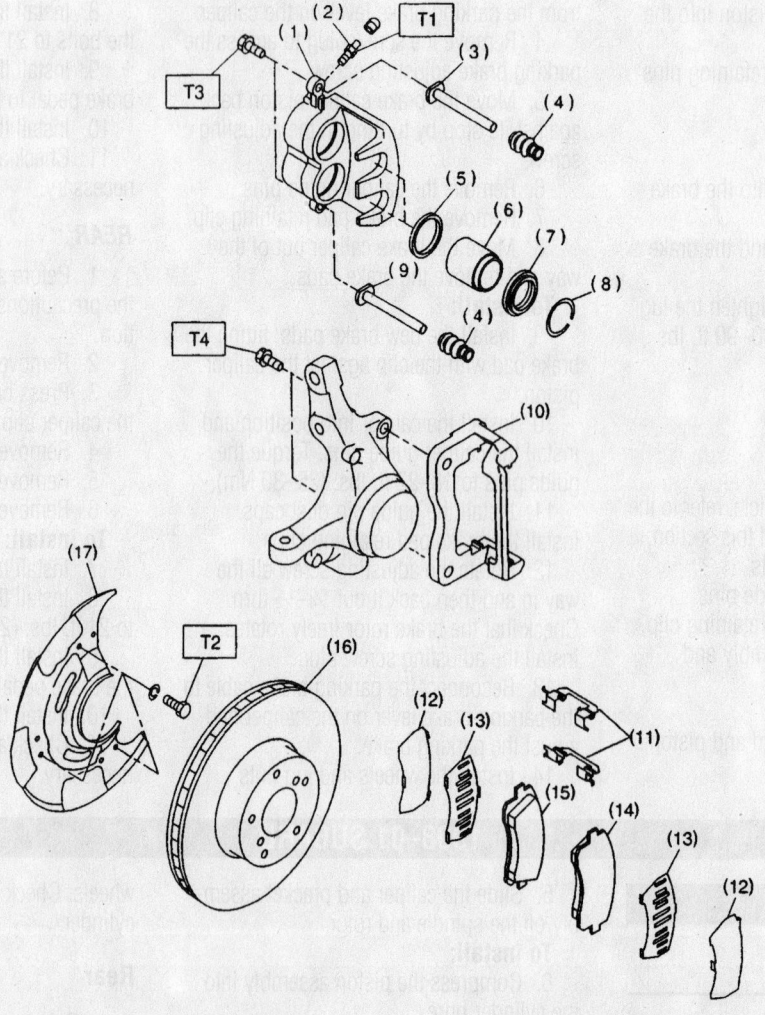

(1)	Caliper body
(2)	Air bleeder screw
(3)	Guide pin (Green)
(4)	Pin boot
(5)	Piston seal
(6)	Piston
(7)	Piston boot
(8)	Boot ring
(9)	Lock pin (Yellow)
(10)	Support
(11)	Pad clip
(12)	Outer shim
(13)	Inner shim
(14)	Pad (Outside)
(15)	Pad (Inside)
(16)	Disc rotor
(17)	Disc cover

Tightening torque: N·m (kg-m, ft-lb)

T1: 8±1 (0.8±0.1, 5.8±0.7)
T2: 18±5 (1.8±0.5, 13.0±3.6)
T3: 37±5 (3.8±0.5, 27.5±3.6)
T4: 78±10 (8.0±1.0, 58±7)

93016G60

Exploded view of the front disc brakes—Subaru

tion and install the lock pin bolts. Tighten the lock pin bolts to
25–33 ft. lbs. (34–44 Nm).
10. Reconnect the parking brake cable, if equipped, and fill the master cylinder reservoir.
11. Install the wheels.

Rear

1. Before servicing the vehicle, refer to the precautions in the beginning of this section.

2. Remove a small portion of brake fluid from the master cylinder reservoir.
3. Remove the wheels.
4. Disconnect the parking brake cable from the caliper lever, if equipped.
5. Remove the lock pin bolts from the lower portion of the caliper.
6. Swing the caliper upward to access the pads.
7. Remove the disc brake pads.

To install:

➡️**If equipped with a parking brake, use a suitable tool to rotate the piston back into the caliper bore.**

8. Compress the caliper pistons.
9. Install the new pads into the caliper bracket, being sure all shims and clips are in their original positions.
10. Swing the caliper down into position and install the lock pin bolts. Tighten the

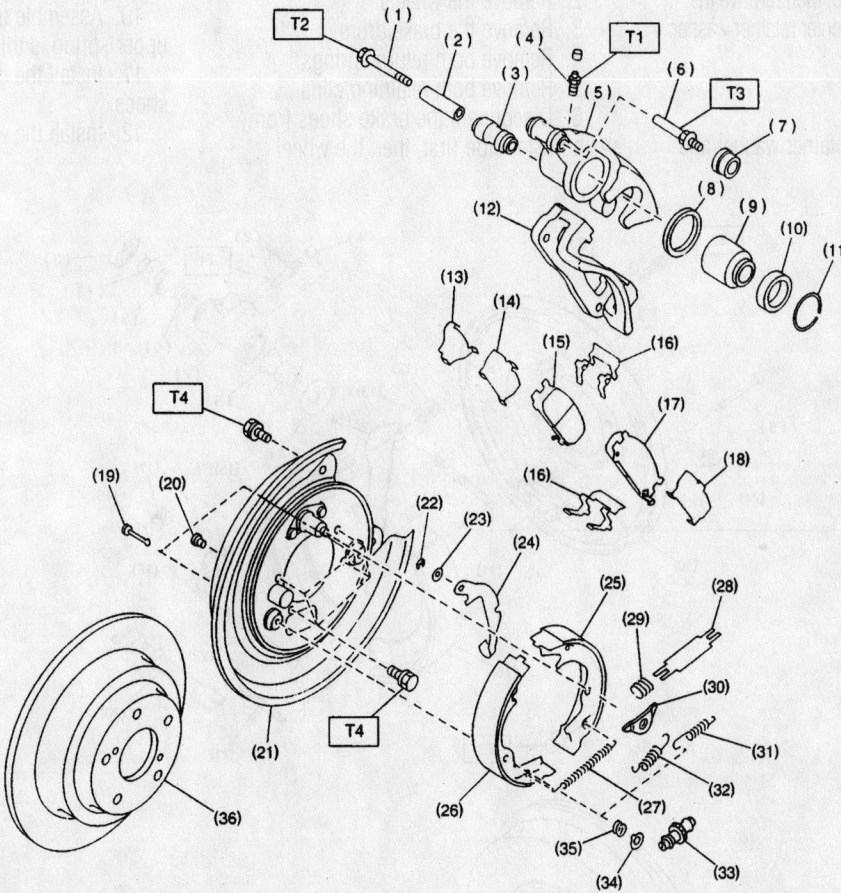

(1) Lock pin	(15) Inner pad	(29) Strut shoe spring
(2) Lock pin sleeve	(16) Pad clip	(30) Shoe guide plate
(3) Lock pin boot	(17) Outer pad	(31) Secondary shoe return spring
(4) Air bleeder screw	(18) Outer shim	(32) Primary shoe return spring
(5) Caliper body	(19) Shoe hold-down pin	(33) Adjuster
(6) Guide pin	(20) Cover	(34) Shoe hold-down cup
(7) Guide pin boot	(21) Back plate	(35) Shoe hold-down spring
(8) Piston seal	(22) Retainer	(36) Disc rotor
(9) Piston	(23) Spring washer	
(10) Piston boot	(24) Parking brake lever	
(11) Boot ring	(25) Parking brake shoe (Secondary)	
(12) Support	(26) Parking brake shoe (Primary)	
(13) Shim	(27) Adjusting spring	
(14) Inner shim	(28) Strut	

Tightening torque: N·m (kg-m, ft-lb)
T1: 8 ± 1 (0.8 ± 0.1, 5.8 ± 0.7)
T2: 20 ± 4 (2.0 ± 0.4, 14.5 ± 2.9)
T3: 26 ± 5 (2.7 ± 0.5, 19.5 ± 3.6)
T4: 52 ± 6 (5.3 ± 0.6, 38.3 ± 4.3)

93016G61

Exploded view of the rear disc brakes—Subaru

For Accessory Drive Belt illustrations, see Section 1 of this manual

lock pin bolt to 12–17 ft. lbs. (16–24 Nm).

11. Reconnect the parking brake cable, and install the wheels.

Brake Drums

REMOVAL & INSTALLATION

1. Before servicing the vehicle, refer to the precautions in the beginning of this section.
2. Remove the rear wheels.
3. Pry off the center cap, then remove the cotter pin, castle nut, and center retainer washer.
4. Remove the drum.

To install:

5. Install the drum.
6. Install the center retainer washer and

castle nut. Tighten the castle nut to 108 ft. lbs. (147 Nm). Install a new cotter pin.

7. Install the center cap.
8. Adjust the brake shoes.
9. Install the rear wheels.

Brake Shoes

REMOVAL & INSTALLATION

1. Before servicing the vehicle, refer to the precautions in the beginning of this section.
2. Remove the wheels.
3. Remove the brake drum.
4. Remove both return springs.
5. Remove both retaining clips.
6. Disconnect the brake shoes from the adjuster side first, then the wheel

cylinder side, and pull them off the backing plate.

7. If equipped with rear drum parking brakes, unfasten the parking brake cable from the parking lever on the trailing brake shoe.

To install:

8. Apply brake grease to the backing plate where the brake shoes contact it.
9. Install the brake shoes to the wheel cylinder, then to the adjuster. Secure in place with the 2 pins and retaining clips.
10. Assemble the return springs. The upper spring is thinner.
11. Install the drum and adjust the brake shoes.
12. Install the wheels.

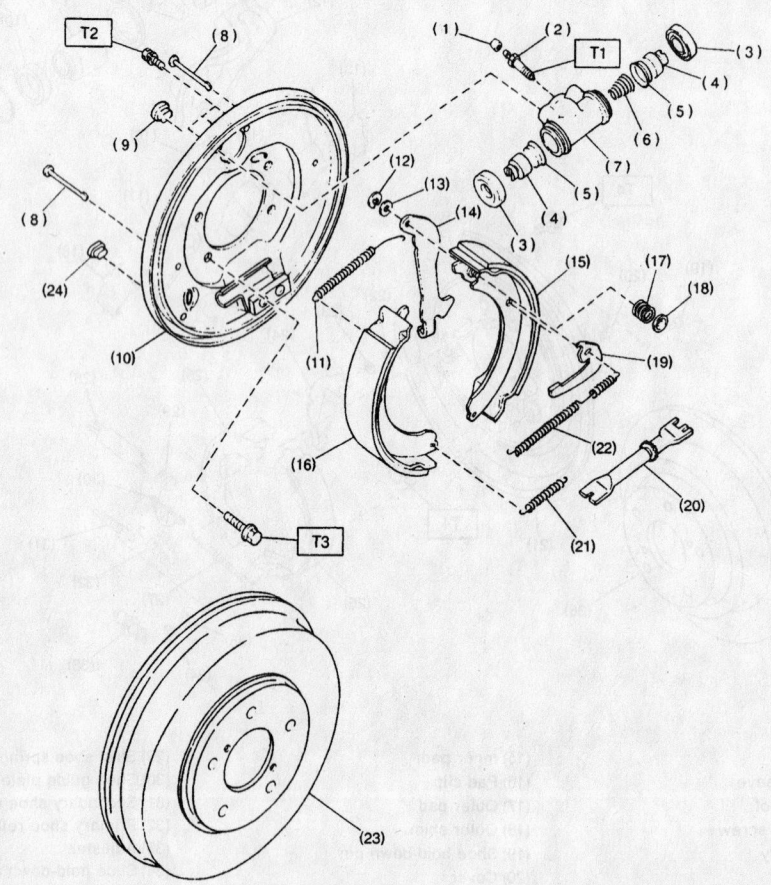

(1) Air bleeder cap	(11) Upper shoe return spring	(21) Lower shoe return spring
(2) Air bleeder screw	(12) Retainer	(22) Adjusting spring
(3) Boot	(13) Washer	(23) Drum
(4) Piston	(14) Parking brake lever	(24) Plug
(5) Cup	(15) Brake shoe (Trailing)	
(6) Spring	(16) Brake shoe (Leading)	**Tightening torque: N·m (kg-m, ft-lb)**
(7) Wheel cylinder body	(17) Shoe hold-down spring	T1: 8±1 (0.8±0.1, 5.8±0.7)
(8) Pin	(18) Cup	T2: 10±2 (1.0±0.2, 7.2±1.4)
(9) Plug	(19) Adjusting lever	T3: 52±6 (5.3±0.6, 38.3±4.3)
(10) Back plate	(20) Adjuster	

93016G62

Exploded view of the rear drum brakes—Subaru

1998–01 SUZUKI

Brake Caliper

REMOVAL & INSTALLATION

Swift and Esteem

FRONT

1. Before servicing the vehicle, refer to the precautions in the beginning of this section.

2. Remove the wheels.
3. Remove the brake hose mounting bolt from the brake caliper.
4. Remove the 2 caliper mounting bolts, then remove the caliper from the mounting carrier.

To install:

5. Install the caliper assembly to the mounting carrier. Tighten the mounting bolts to 16–23 ft. lbs. (22–32 Nm).

6. Install the brake hose to the caliper. Always use new washers and tighten the mounting bolt to 14–18 ft. lbs. (20–25 Nm).
7. Bleed the brake system and install the front wheels.
8. Check the level of the brake fluid in the master cylinder reservoir.

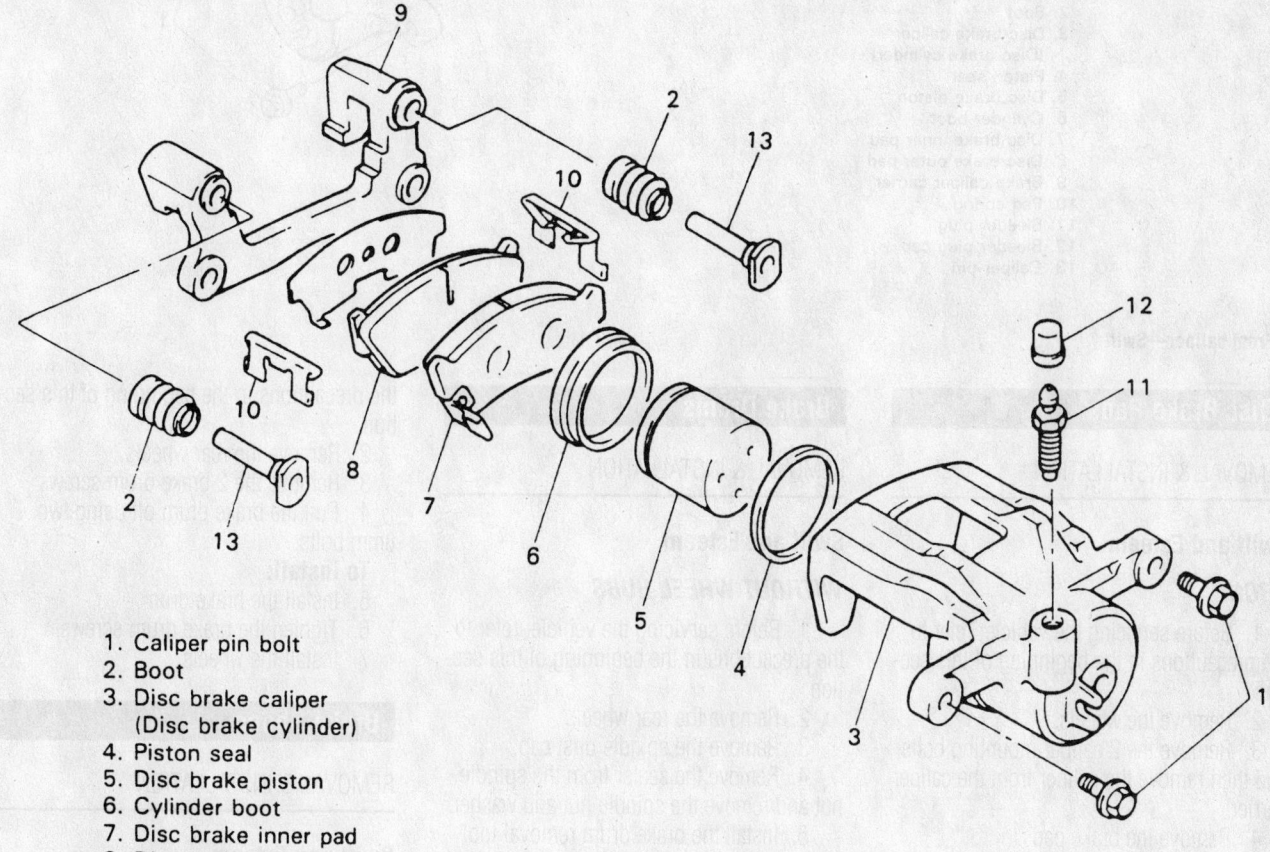

1. Caliper pin bolt
2. Boot
3. Disc brake caliper
   (Disc brake cylinder)
4. Piston seal
5. Disc brake piston
6. Cylinder boot
7. Disc brake inner pad
8. Disc brake outer pad
9. Brake caliper carrier
10. Pad spring
11. Bleeder plug
12. Bleeder plug cap
13. Caliper pin

93016G63

Front caliper—Esteem

For Tire, Wheel and Ball Joint specifications, see Section 1 of this manual

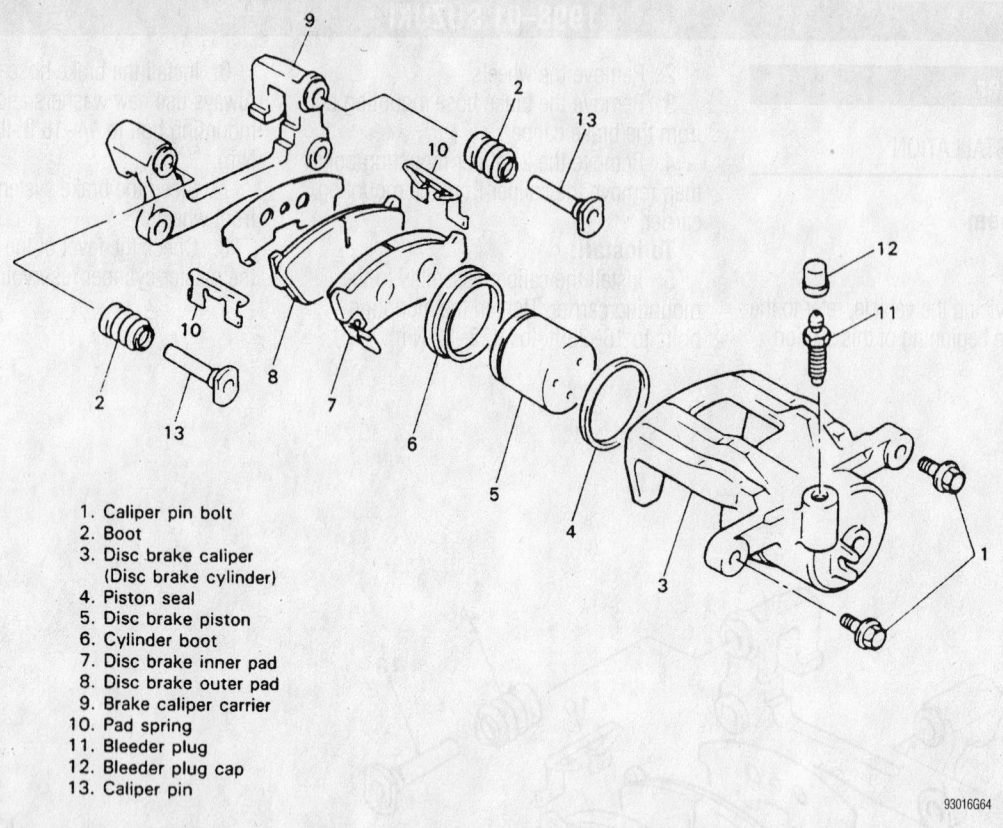

1. Caliper pin bolt
2. Boot
3. Disc brake caliper
   (Disc brake cylinder)
4. Piston seal
5. Disc brake piston
6. Cylinder boot
7. Disc brake inner pad
8. Disc brake outer pad
9. Brake caliper carrier
10. Pad spring
11. Bleeder plug
12. Bleeder plug cap
13. Caliper pin

93016G64

Front caliper—Swift

Disc Brake Pads

REMOVAL & INSTALLATION

Swift and Esteem

FRONT

1. Before servicing the vehicle, refer to the precautions in the beginning of this section.
2. Remove the wheels.
3. Remove the 2 caliper mounting bolts and then remove the caliper from the caliper carrier.
4. Remove the brake pads.

To install:

5. Install the pads into the caliper carrier.
6. On front calipers, press the caliper piston back into the caliper.
7. On rear calipers, use a piston installer 09945-16030, to rotate the caliper piston clockwise into the caliper bore. Lubricate the piston boot with silicone grease to avoid twisting the piston boot.
8. Install the caliper assembly to the caliper carrier. Tighten the bolts to 16–23 ft. lbs. (22–32 Nm).
9. Install the wheels.

Brake Drums

REMOVAL & INSTALLATION

Swift and Esteem

WITHOUT WHEEL HUBS

1. Before servicing the vehicle, refer to the precautions in the beginning of this section.
2. Remove the rear wheels.
3. Remove the spindle dust cap.
4. Remove the sealer from the spindle nut and remove the spindle nut and washer.
5. Install the brake drum removal tool 09943-17911, to the brake drum, then attach a slide hammer to the tool and remove the brake drum.

To install:

6. Install the brake drum.
7. Install the spindle washer and a new spindle nut. Torque the spindle nut to 58–86 ft. lbs. (80–120 Nm).
8. Apply sealer to the spindle nut.
9. Install the spindle dust cap.
10. Install the wheels.

WITH WHEEL HUBS

1. Before servicing the vehicle, refer to

the precautions in the beginning of this section.
2. Remove the rear wheels.
3. Remove the 2 brake drum screws.
4. Pull the brake drum off using two 8mm bolts.

To install:

5. Install the brake drum.
6. Tighten the brake drum screws.
7. Install the wheels.

Brake Shoes

REMOVAL & INSTALLATION

Swift and Esteem

1. Before servicing the vehicle, refer to the precautions in the beginning of this section.
2. Remove the rear wheels.
3. Remove the rear brake drums,
4. Remove the upper and lower springs from the brake shoes.
5. Remove the anti-rattle spring and brake shoe adjustment strut.
6. Remove the primary and secondary brake shoe hold-down springs and remove the shoes from vehicle.
7. Remove the clip securing the parking brake shoe lever to the secondary shoe.

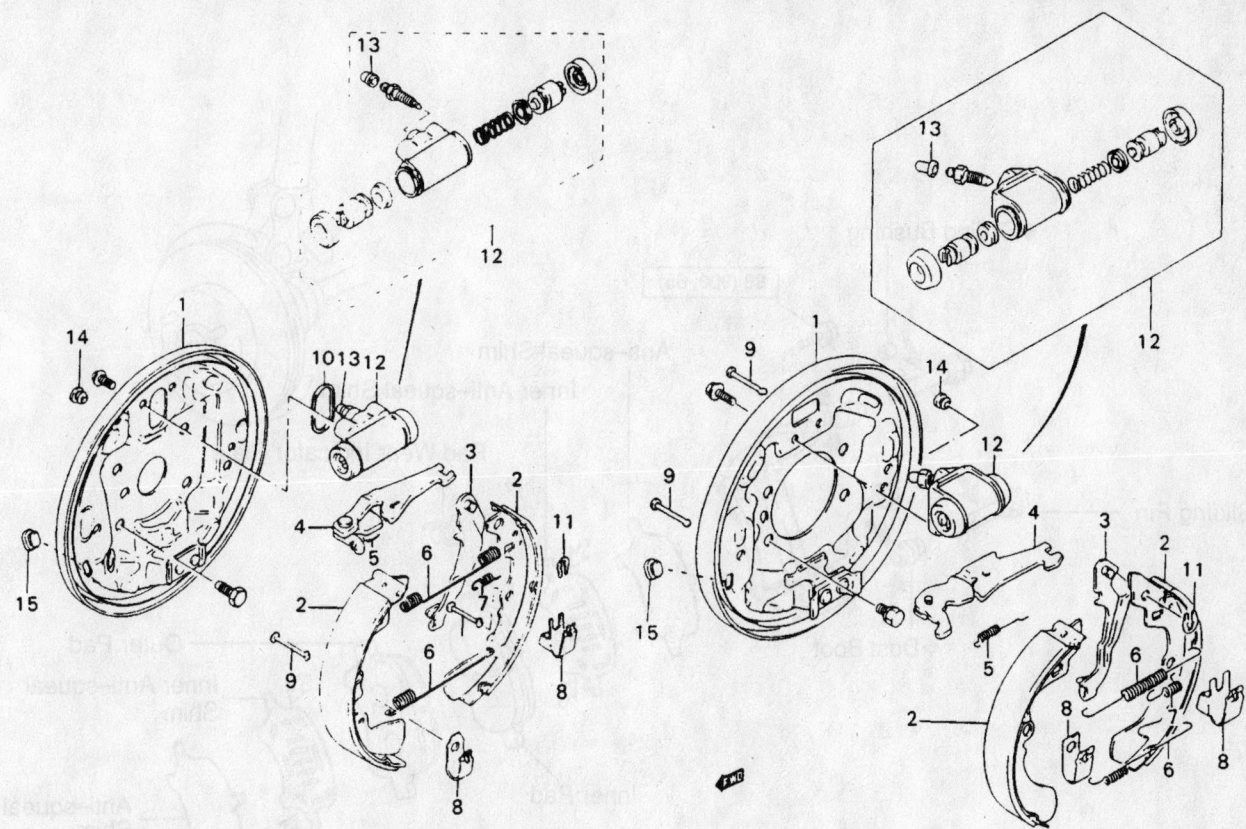

1. Brake back plate
2. Brake shoe
3. Parking brake shoe lever
4. Brake strut
5. Quadrant spring
6. Shoe return spring
7. Antirattle spring
8. Shoe hold down spring
9. Shoe hold down pin
10. Packing
11. Parking lever retainer
12. Wheel cylinder
13. Bleeder plug cap
14. Rubber plug
15. Rubber plug

93016G65

Rear drum brakes—Swift and Esteem

To install:

8. Install the clip securing the parking brake shoe lever to the secondary shoe.

9. Install the primary and secondary brake shoes to the vehicle and secure with the hold-down springs.

10. Install the brake adjustment strut and anti-rattle spring.

11. Install the upper and lower return springs to the primary and secondary brake shoes.

12. Install the brake drum

13. Install the wheels.

14. Press the brake pedal 3–5 times to adjust the brake shoe clearance.

15. Adjust the parking brake cable.

1998–00 TOYOTA

Brake Caliper

REMOVAL & INSTALLATION

All Models

FRONT AND REAR

1. Before servicing the vehicle, refer to the precautions in the beginning of this section.

2. Remove the wheels.

3. Disconnect the brake hose from the caliper.

4. Remove the bolts that attach the caliper to the torque plate. If applicable, hold the flats of the sliding pin with a wrench while loosening the caliper attaching bolts.

5. Lift up and remove the caliper assembly.

To install:

6. Install the caliper and loosely install the bolts.

7. Hold the flats of the sliding pin with

a wrench, then tighten the bolts. On front calipers, tighten single piston types to 18 ft. lbs. (25 Nm) and dual piston types to 25 ft. lbs. (34 Nm). Tighten the bolts on rear calipers to 14 ft. lbs. (20 Nm).

8. Connect the brake hose to the caliper, using 2 new washers.

9. Fill the brake system to the proper level and bleed the brake system.

10. Add brake fluid to the reservoir to fill to the correct level.

11. Lower the vehicle to the ground.

For Wheel Alignment specifications, see Section 1 of this manual

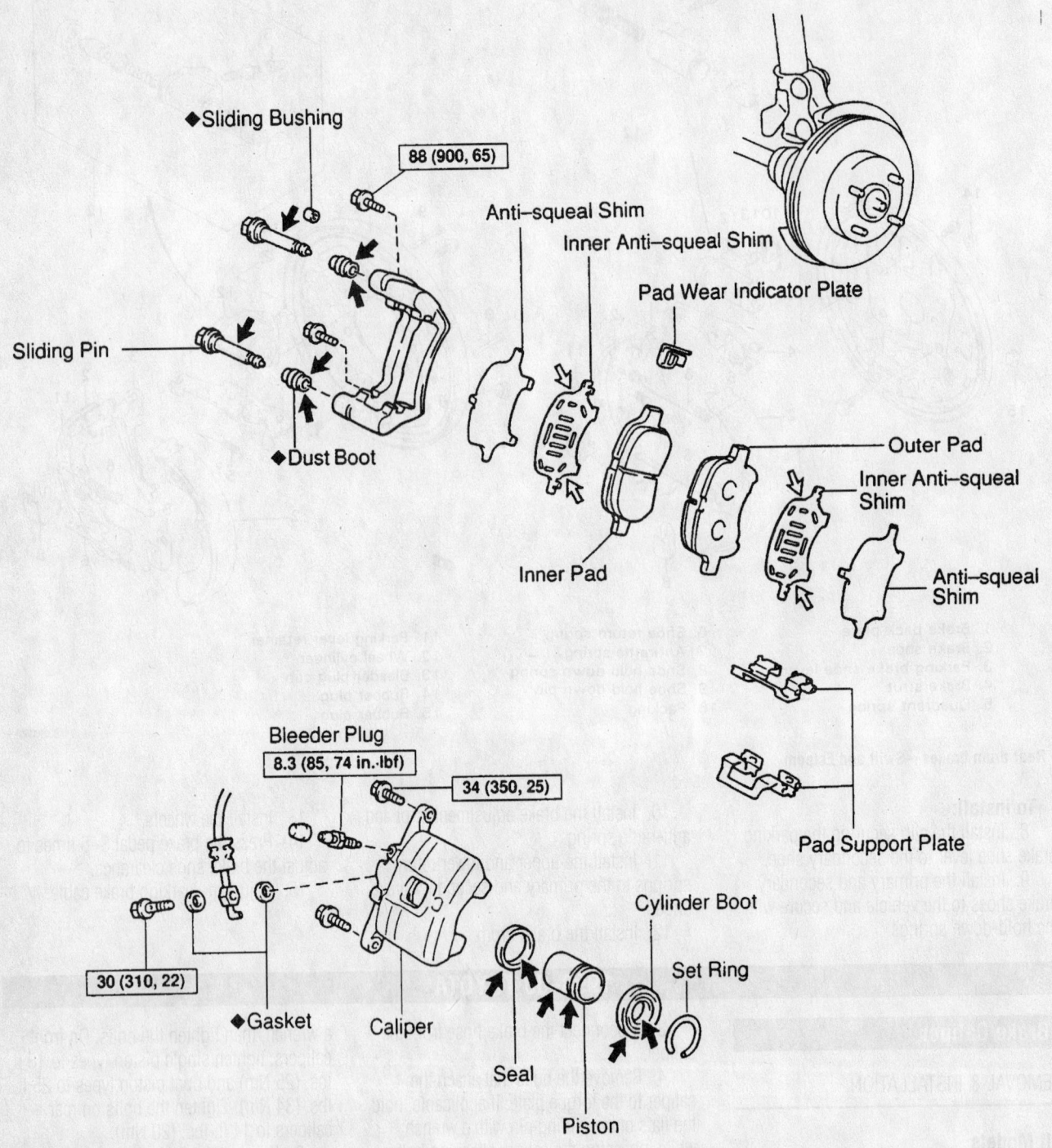

◆Sliding Bushing

88 (900, 65)

Anti–squeal Shim

Inner Anti–squeal Shim

Pad Wear Indicator Plate

Sliding Pin

◆Dust Boot

Outer Pad

Inner Anti–squeal Shim

Inner Pad

Anti–squeal Shim

Bleeder Plug

8.3 (85, 74 in.·lbf)

34 (350, 25)

Pad Support Plate

Cylinder Boot

Set Ring

30 (310, 22)

◆Gasket   Caliper   Seal   Piston

N·m (kgf·cm, ft·lbf) : Specified torque

◆ Non–reusable part

➡ Lithium soap base glycol grease

⇨ Disc brake grease

93016G66

Front caliper—Corolla

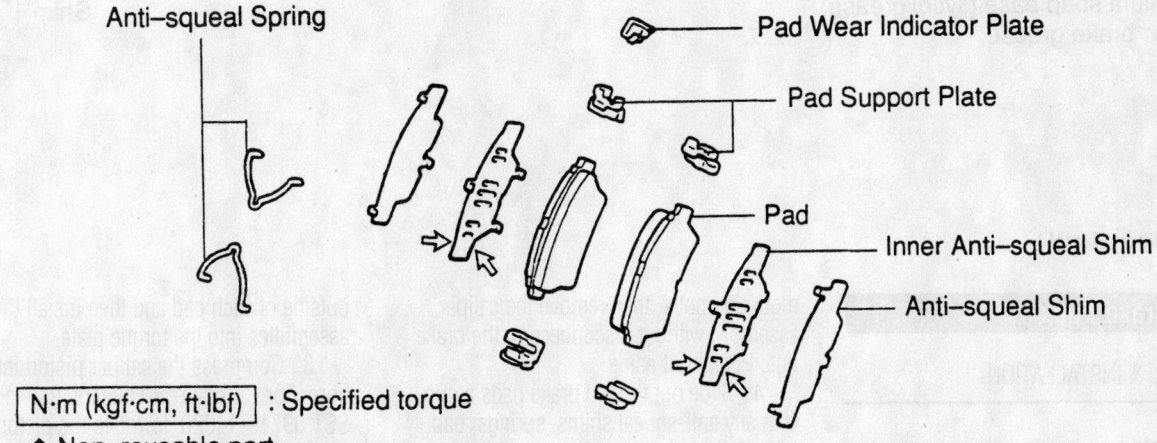

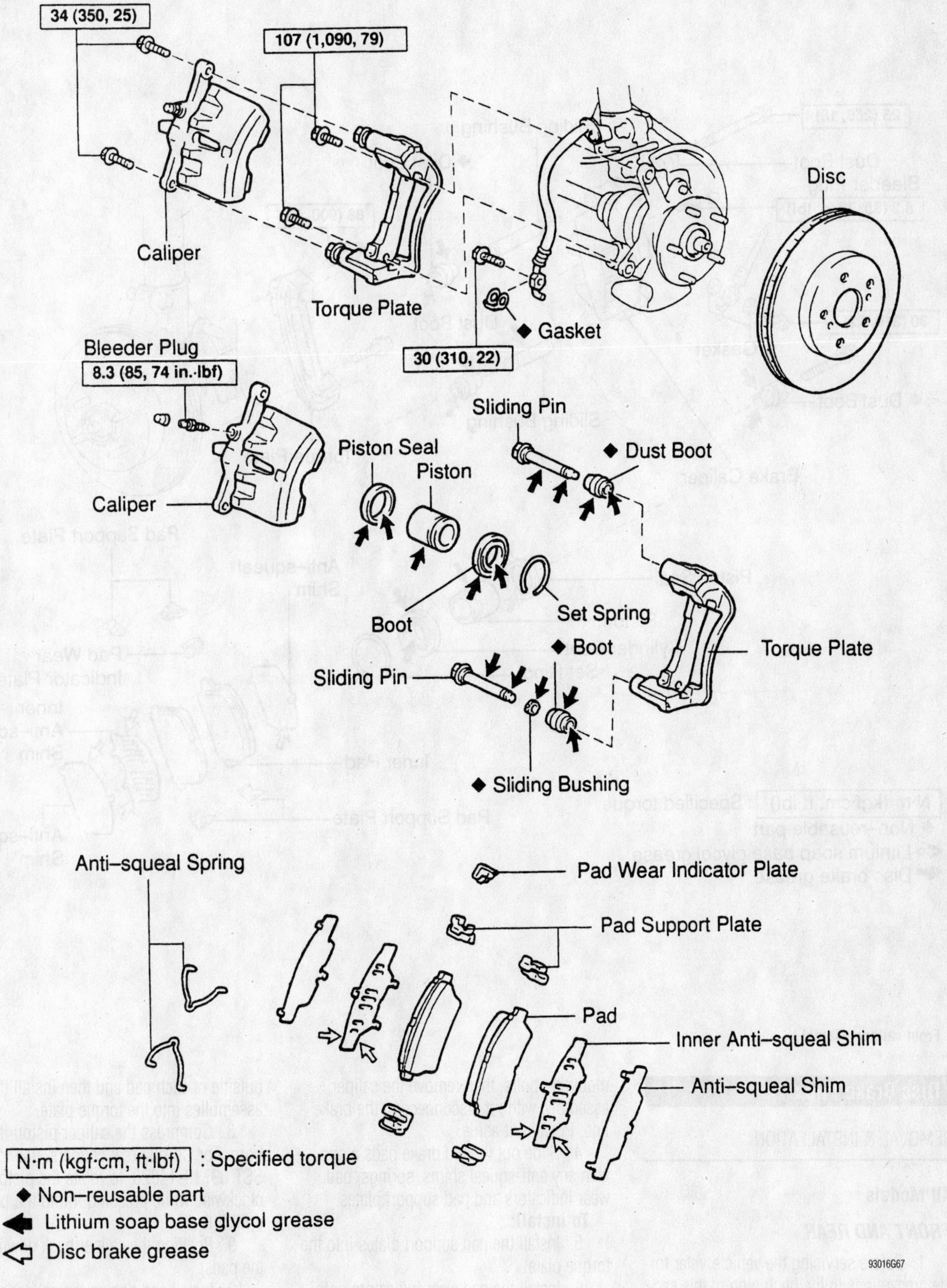

34 (350, 25)

107 (1,090, 79)

Disc

Caliper

Torque Plate

◆ Gasket

30 (310, 22)

Bleeder Plug
8.3 (85, 74 in.·lbf)

Sliding Pin

◆ Dust Boot

Caliper

Piston Seal
Piston

Boot

Set Spring

◆ Boot

Torque Plate

Sliding Pin

◆ Sliding Bushing

Anti–squeal Spring

Pad Wear Indicator Plate

Pad Support Plate

Pad

Inner Anti–squeal Shim

Anti–squeal Shim

N·m (kgf·cm, ft·lbf) : Specified torque

◆ Non–reusable part

◀ Lithium soap base glycol grease

◁ Disc brake grease

93016G67

Front caliper—Celica, Camry, Avalon

For Maintenance Interval recommendations, see Section 1 of this manual

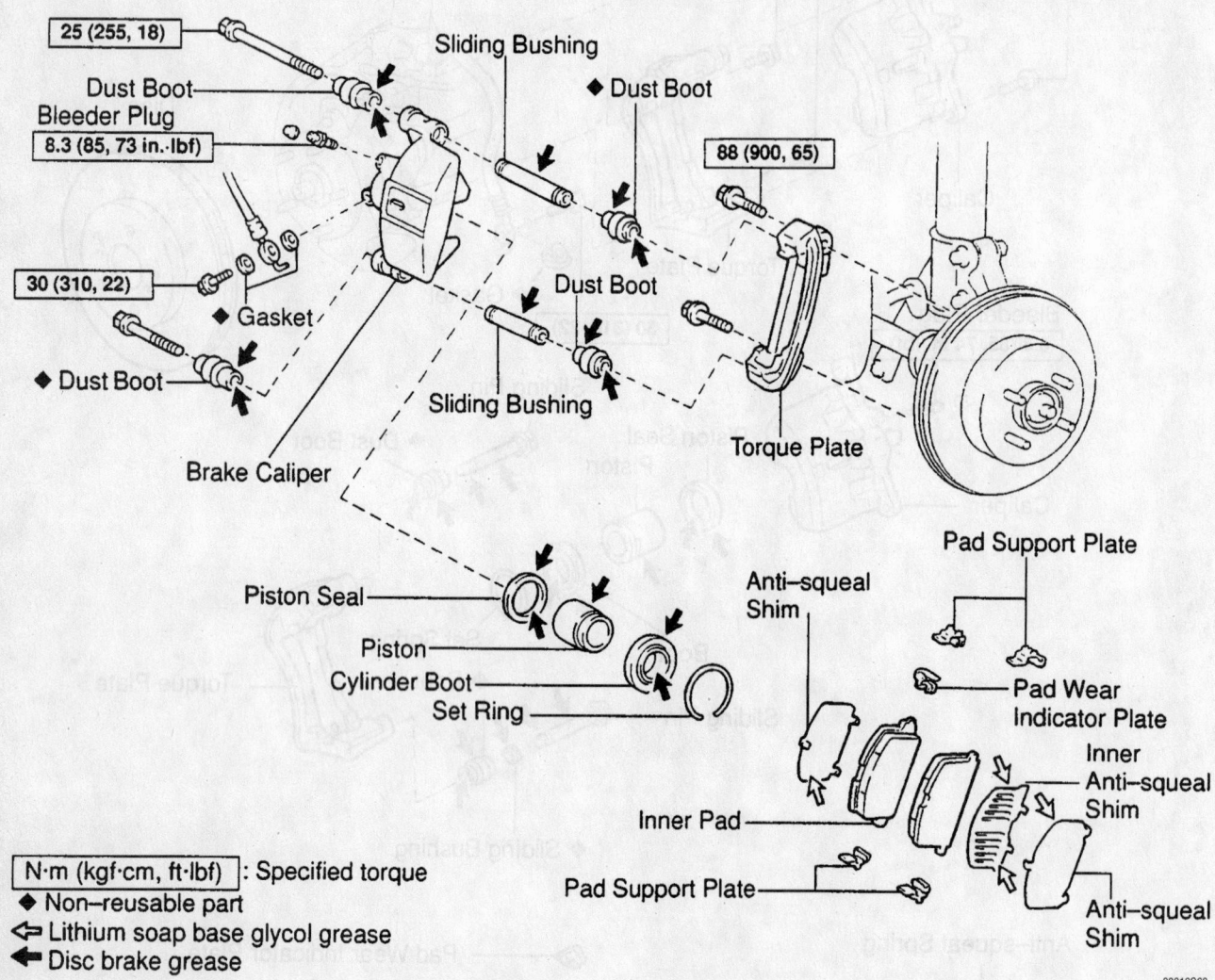

25 (255, 18)

Dust Boot

Bleeder Plug

8.3 (85, 73 in.·lbf)

Sliding Bushing

◆ Dust Boot

88 (900, 65)

30 (310, 22)

◆ Gasket

◆ Dust Boot

Brake Caliper

Dust Boot

Sliding Bushing

Torque Plate

Piston Seal

Piston

Cylinder Boot

Set Ring

Anti-squeal Shim

Pad Support Plate

Pad Wear Indicator Plate

Inner Anti-squeal Shim

Inner Pad

Pad Support Plate

Anti-squeal Shim

N·m (kgf·cm, ft·lbf) : Specified torque
◆ Non–reusable part
⇦ Lithium soap base glycol grease
⬅ Disc brake grease

93016G68

Front caliper—Tercel

Disc Brake Pads

REMOVAL & INSTALLATION

All Models

FRONT AND REAR

1. Before servicing the vehicle, refer to the precautions in the beginning of this section.
2. Remove the wheels.
3. Loosen and remove the caliper mounting bolts, then remove the caliper assembly, without disconnecting the brake line. Position it aside.
4. Slide out the old brake pads along with any anti-squeal shims, springs, pad wear indicators and pad support plates.

To install:

5. Install the pad support plates into the torque plate.
6. Install the pad wear indicators onto the pads. Be sure the arrow on the indicator plate is pointing in the direction of rotation.
7. Install the anti-squeal shims on the outside of each pad and then install the pad assemblies into the torque plate.
8. Compress the caliper piston into the bore. For Corolla rear calipers, use tool SST 09719-14020, to rotate the piston clockwise while pressing it into the bore until it locks.
9. Position the caliper back down over the pads.
10. Install and tighten the caliper mounting bolts.
11. Install the wheels. Check the brake fluid level.

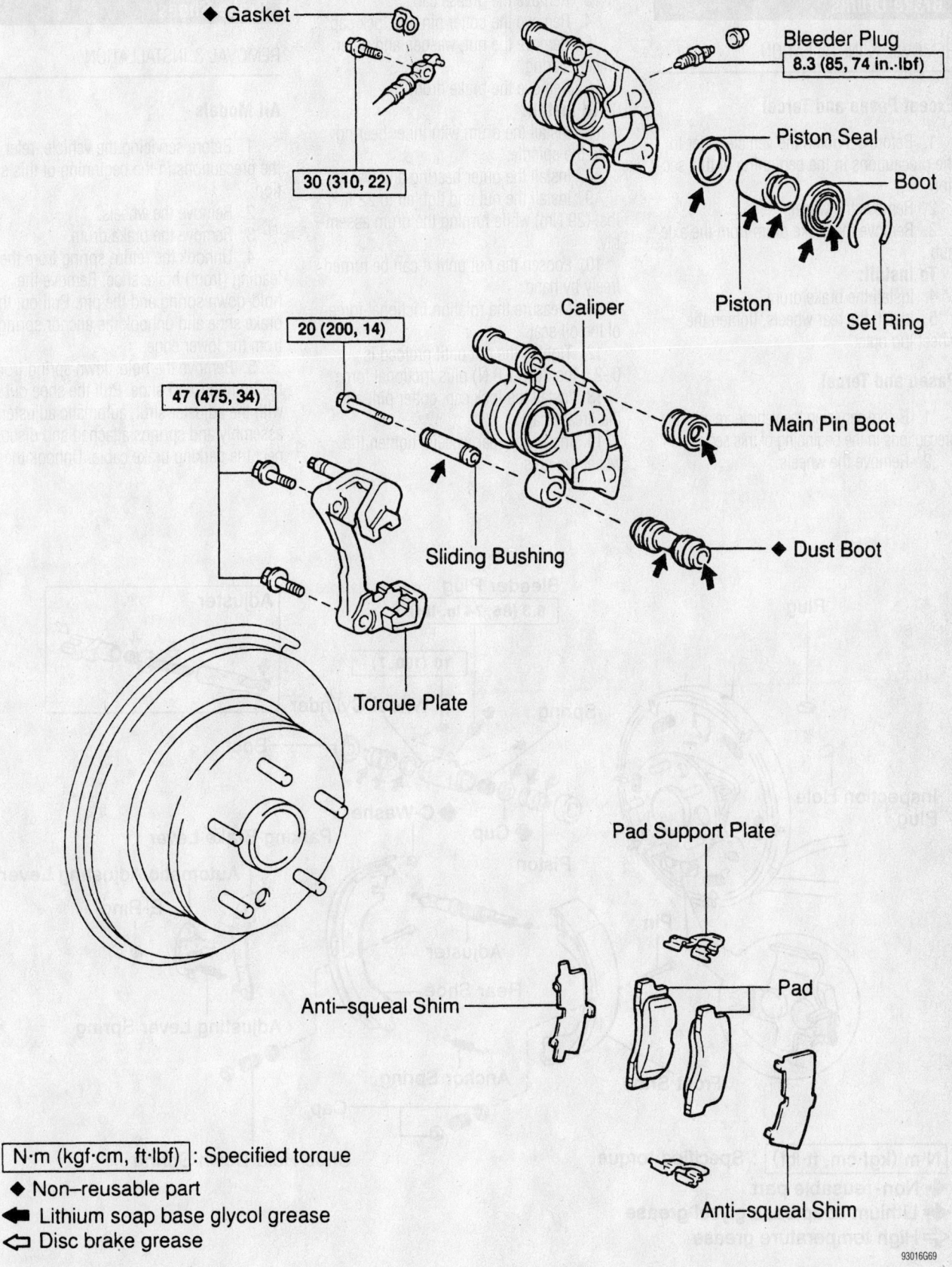

◆ Gasket

Bleeder Plug
8.3 (85, 74 in.·lbf)

Piston Seal

Boot

30 (310, 22)

Piston

Set Ring

Caliper

20 (200, 14)

Main Pin Boot

47 (475, 34)

Sliding Bushing

◆ Dust Boot

Torque Plate

Pad Support Plate

Anti–squeal Shim

Pad

Anti–squeal Shim

93016G69

N·m (kgf·cm, ft·lbf) : Specified torque

◆ Non–reusable part

◀ Lithium soap base glycol grease

◁ Disc brake grease

Rear caliper— Celica, Camry, Avalon

For Tune-up, Capacities and Firing orders, see Section 1 of this manual

Brake Drums

REMOVAL & INSTALLATION

Except Paseo and Tercel

1. Before servicing the vehicle, refer to the precautions in the beginning of this section.
2. Remove the wheels.
3. Remove the brake drum from the axle hub.

To install:

4. Install the brake drum.
5. Install the rear wheels, tighten the wheel lug nuts.

Paseo and Tercel

1. Before servicing the vehicle, refer to the precautions in the beginning of this section.
2. Remove the wheels.

3. Remove the grease cap.
4. Remove the cotter pin and lock cap.
5. Remove the nut, washer, and outer wheel bearing.
6. Remove the brake drum.

To install:

7. Install the drum with inner bearing onto the spindle.
8. Install the outer bearing and washer.
9. Install the nut and tighten to 22 ft. lbs. (29 Nm) while turning the drum assembly.
10. Loosen the nut until it can be turned freely by hand.
11. Measure the rotation frictional force of the oil seal.
12. Tighten the nut until preload is 0–2.6 lbs. (0–11.8 N) plus frictional force.
13. Install the lock cap, cotter pin, and the grease cap.
14. Install the rear wheels, tighten the wheel lug nuts.

Brake Shoes

REMOVAL & INSTALLATION

All Models

1. Before servicing the vehicle, refer to the precautions in the beginning of this section.
2. Remove the wheels.
3. Remove the brake drum.
4. Unhook the return spring from the leading (front) brake shoe. Remove the hold-down spring and the pin. Pull out the brake shoe and unhook the anchor spring from the lower edge.
5. Remove the hold-down spring from the trailing (rear) shoe. Pull the shoe out with the adjuster strut, automatic adjuster assembly and springs attached and disconnect the parking brake cable. Unhook the

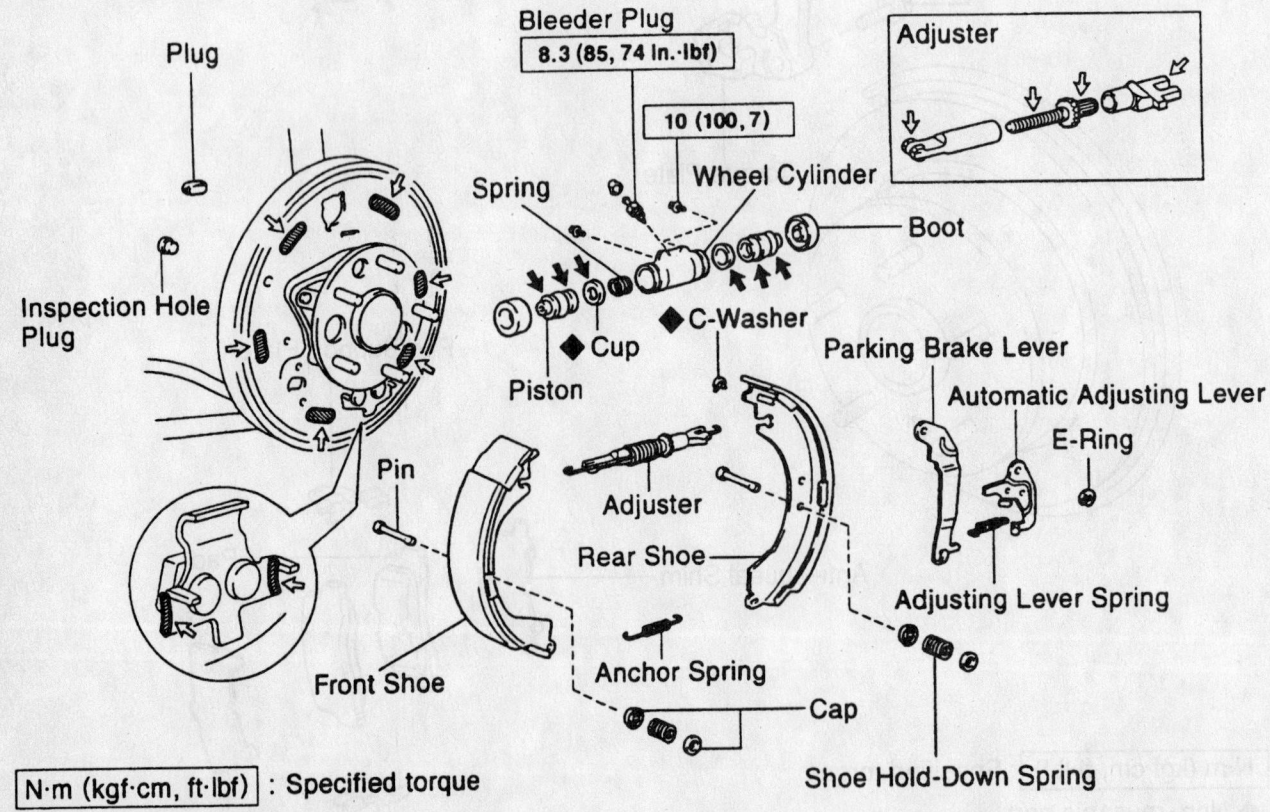

N·m (kgf·cm, ft·lbf) : Specified torque
◆ Non-reusable part
← Lithium soap base glycol grease
⇐ High temperature grease

93016G70

Rear drum brakes—Camry

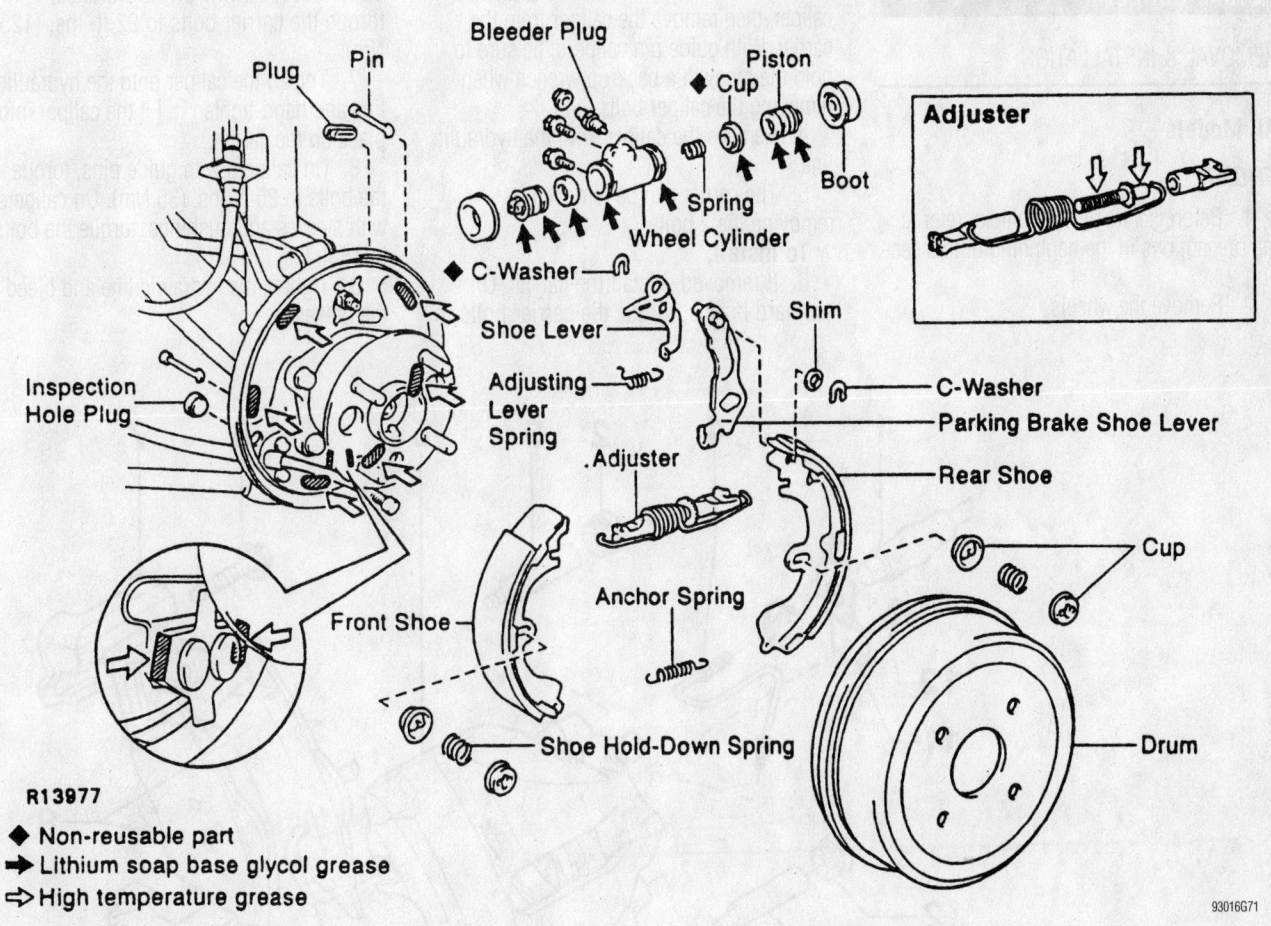

Plug   Pin

Bleeder Plug

Piston
◆ Cup
Boot

Spring

Wheel Cylinder

◆ C-Washer

Shoe Lever

Adjusting
Lever
Spring

Adjuster

Anchor Spring

Inspection
Hole Plug

Front Shoe

Shoe Hold-Down Spring

Shim

C-Washer

Parking Brake Shoe Lever

Rear Shoe

Cup

Drum

Adjuster

R13977

◆ **Non-reusable part**
➡ **Lithium soap base glycol grease**
⇨ **High temperature grease**

93016G71

Rear drum brakes—Corolla and Tercel

return spring and then remove the adjusting strut. Remove the anchor spring.

6. Remove the adjusting strut. Unhook the adjusting lever spring from the rear shoe and then remove the automatic adjuster assembly by popping out the C-clip.

To install:

7. Mount the automatic adjuster assembly onto a new rear brake shoe. Make sure

the C-clip fits properly. Connect the adjusting strut/return spring and then install the adjusting spring.

8. Connect the parking brake cable to the rear shoe and then position the shoe so the lower end rides in the anchor plate and the upper end is against the boot in the wheel cylinder. Install the pin and the hold-down spring.

9. Install the anchor spring between the front and rear shoes. Install the hold-down spring and pin.

10. Connect the return spring/adjusting strut between the 2 shoes so it rides freely.

11. Install the drum.

12. Install the wheel.

Brake Caliper

REMOVAL & INSTALLATION

All Models

FRONT

1. Before servicing the vehicle, refer to the precautions in the beginning of this section.

2. Remove the wheels.

3. Loosen the hydraulic line at the caliper, then remove the caliper from the carrier. With guide pin calipers, be sure to hold the pin with a back-up wrench when removing the caliper bolts.

4. Remove the caliper from the hydraulic line.

5. The carrier can be removed by removing the 2 bolts.

To install:

6. If removed, install the carrier. On standard brakes, torque the carrier bolts to 52 ft. lbs. (70 Nm). On ABS brakes, torque the carrier bolts to 92 ft. lbs. (125 Nm).

7. Thread the caliper onto the hydraulic line and hand-tighten it. Fit the caliper into place on the carrier.

8. On calipers with guide pins, torque the bolts to 25 ft. lbs. (35 Nm). On calipers with sleeves and bushings, torque the bolts to 18 ft. lbs. (25 Nm).

9. Tighten the hydraulic line and bleed the brakes.

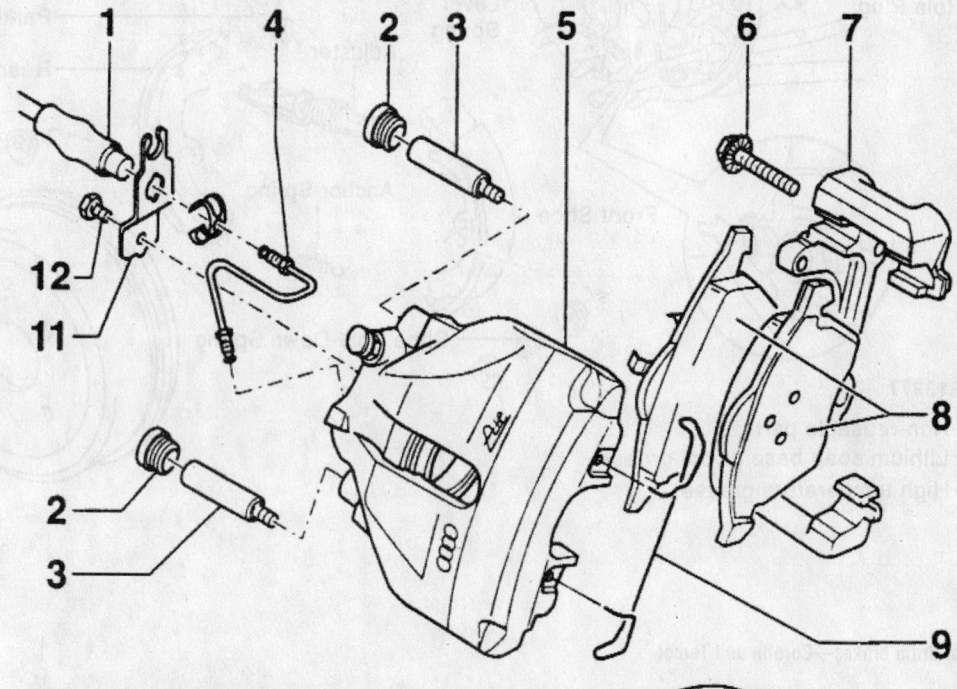

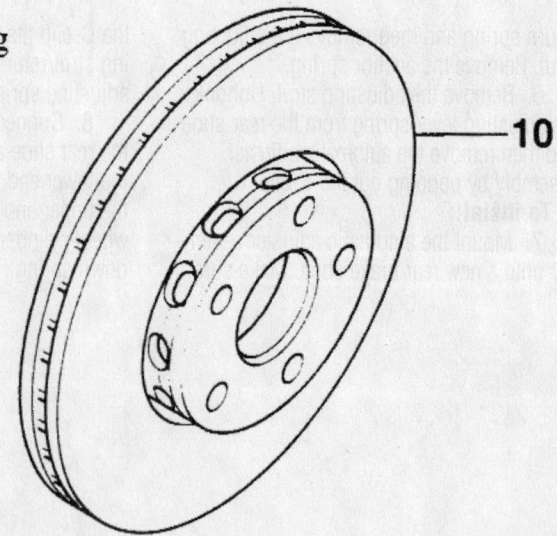

1. Brake hose
2. Bolt cap
3. Guide pins
4. Brake line
5. Caliper housing
6. Ribbed combination bolt
7. Brake carrier
8. Brake pads
9. Retaining spring
10. Brake disc
11. Retainer
12. Retainer bolt

93016G72

Front brake caliper—Passat shown

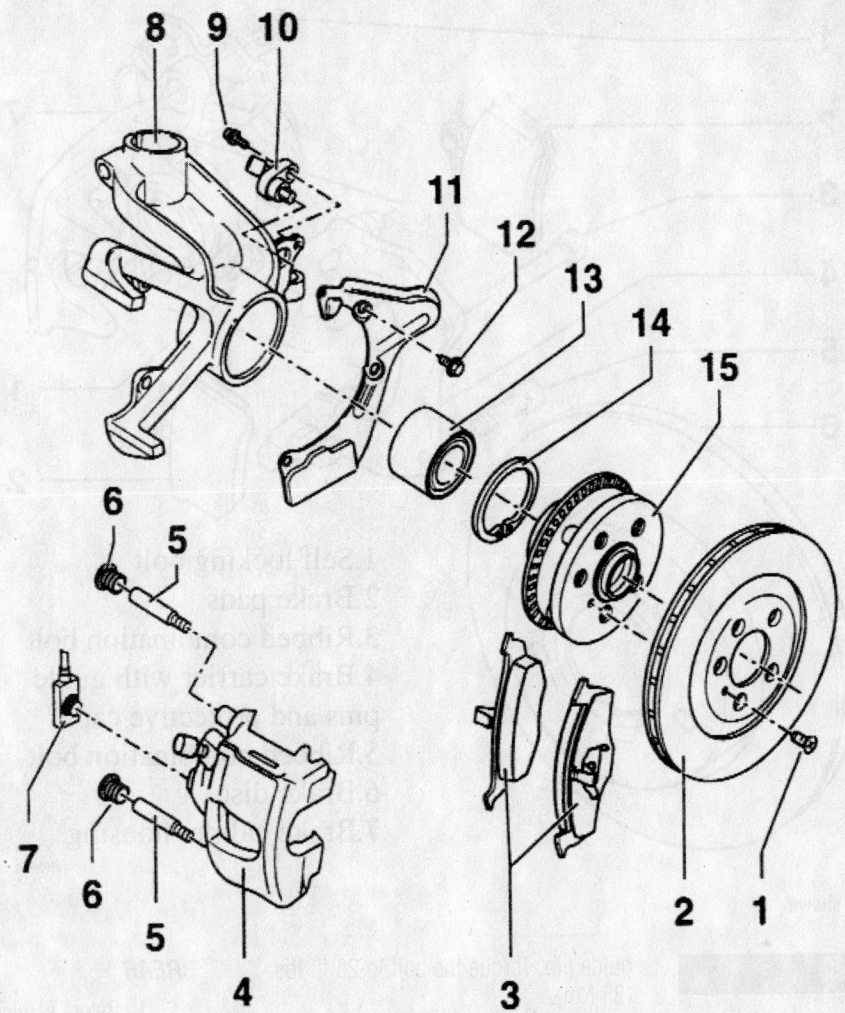

1. Screw
2. Brake disc
3. Brake pads
4. Brake caliper
5. Guide pin
6. Protective cap
7. Brake hose and banjo bolt
8. Wheel bearing housing

9. Socket head bolt
10. ABS sensor
11. Splash shield
12. Bolt
13. Wheel bearing
14. Circlip
15. Wheel hub and rotor

93016G73

Front caliper—New Beetle

REAR

1. Before servicing the vehicle, refer to the precautions in the beginning of this section.

2. If equipped with ABS, make sure the ignition switch stays **OFF** and pump the brake pedal 25–35 times to relieve the system pressure.

3. Remove the wheels.

4. Disconnect the parking brake cable.

5. Loosen the hydraulic line.

6. Use a back-up wrench to hold the guide pins and remove the caliper bolts.

7. Lift the caliper off the carrier and unscrew it from the hydraulic line.

To install:

8. Thread the caliper onto the hydraulic line and hand-tighten it. Fit the caliper into place on the carrier. Torque the bolts to 26 ft. lbs. (35 Nm).

9. Bleed the brakes.

10. Install the wheels.

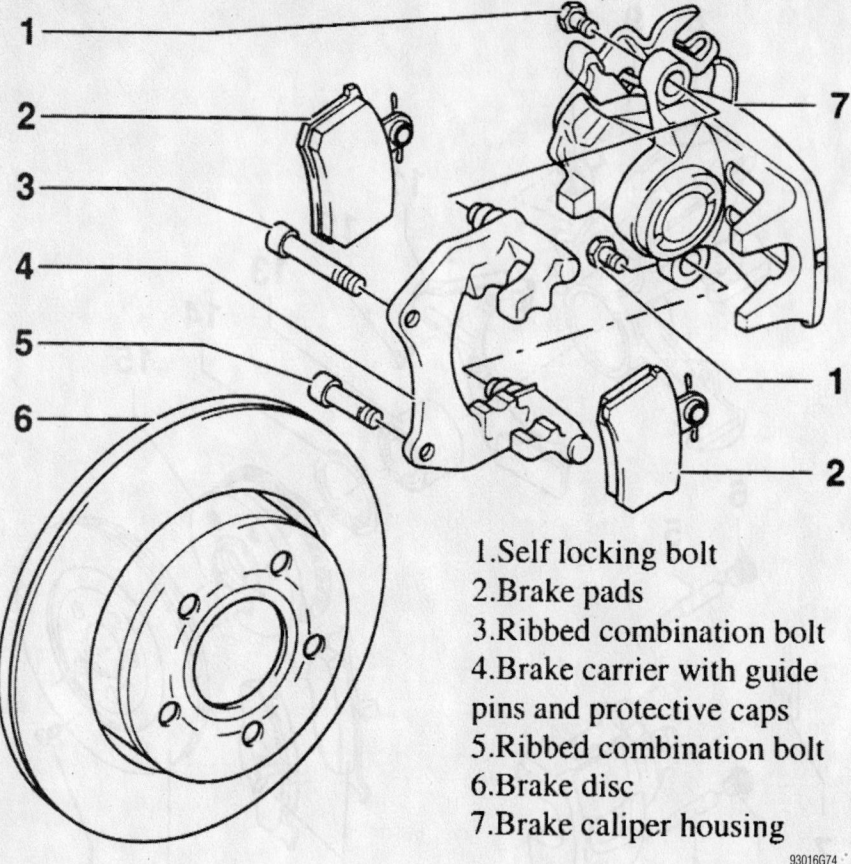

1. Self locking bolt
2. Brake pads
3. Ribbed combination bolt
4. Brake carrier with guide pins and protective caps
5. Ribbed combination bolt
6. Brake disc
7. Brake caliper housing

93016G74

Rear disc brakes—Passat shown

Brake Pads

REMOVAL & INSTALLATION

All Models

FRONT CALIPER WITH GUIDE PINS

1. Before servicing the vehicle, refer to the precautions in the beginning of this section.
2. Remove the front wheels.
3. Hold the lower guide pin with an open wrench and remove the bolt securing the caliper to the guide pin.
4. Pivot the caliper up on the upper guide pin and slide the pads straight out to remove them.

To install:

5. Compress the caliper piston into the bore.
6. Fit the new pads into the carrier and pivot the caliper into place.
7. The original bolts are micro-encapsulated with a thread locking compound. Install a new bolt or clean the old bolt and apply a thread-locking compound.
8. When tightening the bolt, be sure to use a back-up wrench to hold the

guide pin. Torque the bolt to 26 ft. lbs. (35 Nm).
9. Install the wheels.

FRONT CALIPER WITH SLEEVES AND BUSHINGS

1. Before servicing the vehicle, refer to the precautions in the beginning of this section.
2. Remove the front wheels.
3. Remove the 2 bolts holding the caliper to the carrier. Push the caliper up and pivot the bottom of the caliper out of the carrier.
4. Remove the anti-rattle springs and the pads from the carrier and note their location.

To install:

5. Fit the anti-rattle springs into place and slide the new pads onto the carrier.
6. Fit the caliper into place at the top and push up so it can be pivoted into place at the bottom. The tabs on the anti-rattle springs should be pushing against the inside of the caliper.
7. Make sure the caliper mounting bolts are clean. Install the bolts and torque them to 18 ft. lbs. (25 Nm).
8. Install the wheels.

REAR

1. Before servicing the vehicle, refer to the precautions in the beginning of this section.
2. Remove the rear wheels.
3. Remove the parking brake cable clip from the caliper. Disconnect the parking brake cable.
4. Hold the guide pin with a back-up wrench and remove the upper mounting bolt from the brake caliper.
5. Swing the caliper downward and remove the brake pads.

To install:

6. Retract the piston into the housing by rotating the piston clockwise.
7. Install the new brake pads onto the pad carrier.
8. Install the caliper to the pad carrier using a new self locking bolt or a thread locking compound and torque to 26 ft. lbs. (35 Nm).
9. Attach the hand brake cable to the caliper.
10. Check the parking brake operation and adjust the cable if necessary.
11. Install the wheels.

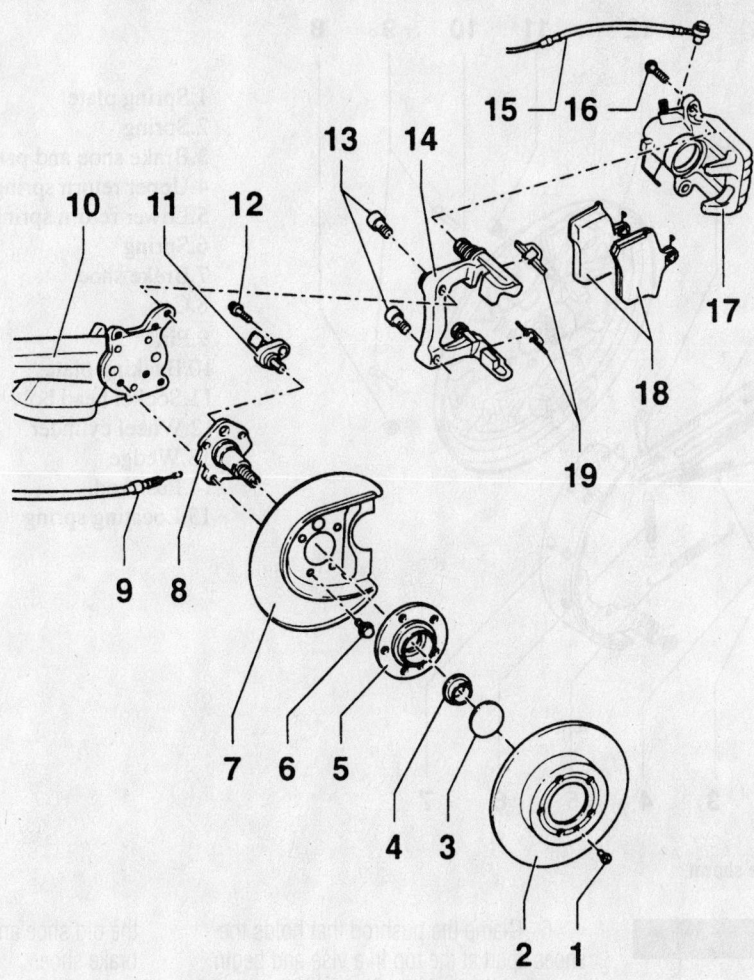

1. Screw
2. Brake disc
3. Cap
4. Self locking nut
5. Wheel hub
6. Bolt
7. Splash shield
8. Stub axle
9. Parking brake cable
10. Axle beam
11. ABS sensor
12. Socket head bolt
13. Socket head bolt
14. Brake carrier
15. Brake hose
16. Self locking nut
17. Brake caliper
18. Brake pads
19. Pad retaining springs

93016G75

Rear disc brakes—New Beetle

Brake Drums

REMOVAL & INSTALLATION

All Models

1. Before servicing the vehicle, refer to the precautions in the beginning of this section.

2. Remove the rear wheels.

3. Insert a small pry tool through a wheel bolt hole and push up on the adjusting wedge to slacken the rear brake adjustment.

4. Remove the grease cap, cotter pin, locking ring, axle nut, thrust washer, and outer bearing.

5. Remove the drum.
To install:

6. Install the drum. Install the outer bearing, washer, and nut.

7. Install the lock ring and cotter pin. Install the grease cap.

8. Install the wheels.

Timing belt service is covered in Section 3 of this manual

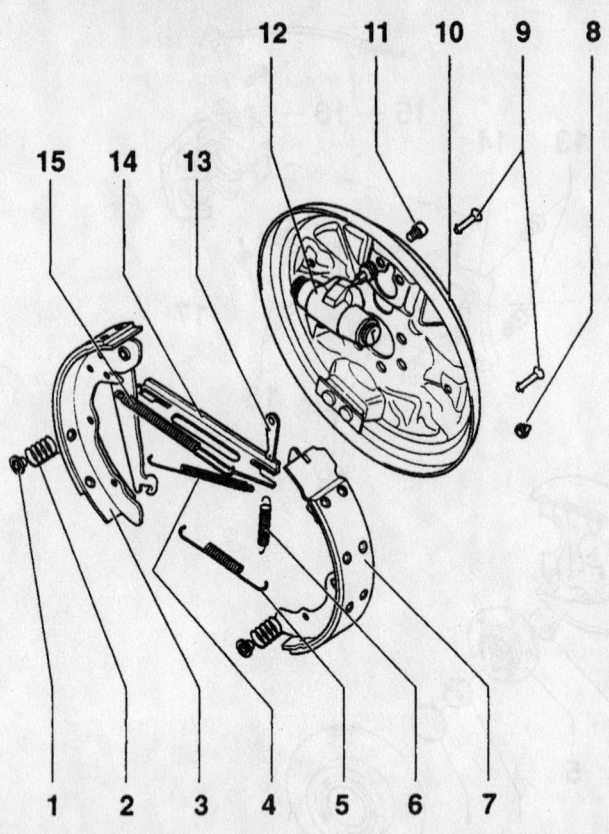

1. Spring plate
2. Spring
3. Brake shoe and park brake lever
4. Upper return spring
5. Lower return spring
6. Spring
7. Brake shoe
8. Cap
9. Pin
10. Backing plate
11. Socket head bolt
12. Wheel cylinder
13. Wedge
14. Push rod
15. Locating spring

93016G76

Rear drum brakes—New Beetle shown

Brake Shoes

REMOVAL & INSTALLATION

All Models

1. Before servicing the vehicle, refer to the precautions in the beginning of this section.
2. Remove the rear wheels.
3. Remove the rear brake drum.
4. Remove the brake shoe hold-down spring retainers.
5. Remove the shoes from the back plate by pulling first 1 shoe, then the other against the upper spring and from its wheel cylinder slot. Detach the parking brake cable from the brake lever. Remove the brake shoe assembly from the vehicle.

6. Clamp the pushrod that holds the shoes apart at the top in a vise and begin removing the springs. Start with the lower return spring, adjusting wedge spring, upper return spring and then the tensioning spring and adjusting wedge.
7. On most vehicles, the parking brake lever must be removed from the old shoes and reused.

To install:

8. Clean the back plate and lubricate the shoe contact points with a suitable brake lubricant.
9. With the push rod clamped in a vise, attach the front brake shoe and tensioning spring.
10. Insert the adjusting wedge between the front shoe and pushrod so its lug is pointing toward the backing plate.
11. Remove the parking brake lever from

the old shoe and attach it onto the new rear brake shoe.
12. Put the rear brake shoe and parking brake lever assembly onto the pushrod and hook up the spring.
13. Connect the parking brake cable to the lever and place the whole assembly onto the backing plate.
14. Install the hold-down springs.
15. Install the upper and lower return springs.
16. Install the adjusting wedge spring.
17. Center the brake shoes on the backing plate, making sure the adjusting wedge is fully released (all the way up) before installing the drum.
18. Install the drum and wheel assembly.
19. Apply the brake pedal a few times to bring the brake shoe into adjustment.

Brake Caliper

REMOVAL & INSTALLATION

700 Series, 900 Series, S90 and V90 Models

FRONT

1. Before servicing the vehicle, refer to the precautions in the beginning of this section.
2. Remove the wheels.
3. Disconnect the ABS lead and brake hose from their clips.
4. Disconnect the hose from the line.
5. Disconnect the hose from the caliper.
6. Remove the caliper guide pin bolts and separate the caliper from the mounting bracket.

To install:

7. Lubricate the guide pins with silicone grease.
8. Install the caliper and tighten the guide pin bolts to 20 ft. lbs. (27 Nm).
9. Install the brake hose.

10. Reconnect brake hose to line and the ABS lead to the hose.
11. Bleed the brake system.
12. Install the wheels.

REAR

1. Before servicing the vehicle, refer to the precautions in the beginning of this section.
2. Remove the wheels.
3. Disconnect the ABS wires.
4. Disconnect the brake hose from the line. Disconnect the brake hose from the caliper.
5. Disconnect the parking brake cable from the bracket.
6. For models with independent rear suspension, remove the caliper guide pin bolts and remove the caliper.
7. For models with a solid rear axle, unbolt and remove the caliper with the brake pads.

To install:

8. For independent rear suspension, install the caliper and tighten the guide pin retaining bolts to 25 ft. lbs. (34 Nm).

9. For solid rear axle, install the caliper and tighten the mounting bolts to 43 ft. lbs. (58 Nm).
10. Connect the brake hose to the caliper and brake line.
11. Connect the parking brake cable to the bracket.
12. Connect the ABS wires.
13. Bleed the brake system.
14. Install the wheels.

850, S70, C70 and V70 Series

FRONT

1. Before servicing the vehicle, refer to the precautions in the beginning of this section.
2. Remove the wheels.
3. Disconnect the ABS wires.
4. Loosen the brake hose a half turn.
5. Remove the caliper bolts, lift the caliper off and unscrew the caliper from the hose.

To install:

6. Grease the caliper bolts with lithium grease and insert them into the sleeves.
7. Screw the caliper onto the brake hose.
8. Install the caliper and tighten the caliper bolts to 22 ft. lbs. (30 Nm).
9. Tighten the brake hose to 13 ft. lbs. (18 Nm).
10. Fill the master cylinder and bleed the brake system.
11. Connect the ABS wires.
12. Install the wheels.

REAR

1. Before servicing the vehicle, refer to the precautions in the beginning of this section.
2. Remove the wheels.
3. Disconnect the ABS wires.
4. Remove the caliper mounting bolts and lift the caliper off.

To install:

5. Install the caliper and new mounting bolts. Tighten the bolts to 44 ft. lbs. (60 Nm).
6. Connect the brake line to the caliper and tighten to 10 ft. lbs. (14 Nm).
7. Fill the brake master cylinder and bleed the brake system.
8. Connect the ABS wires.
9. Install the wheels.

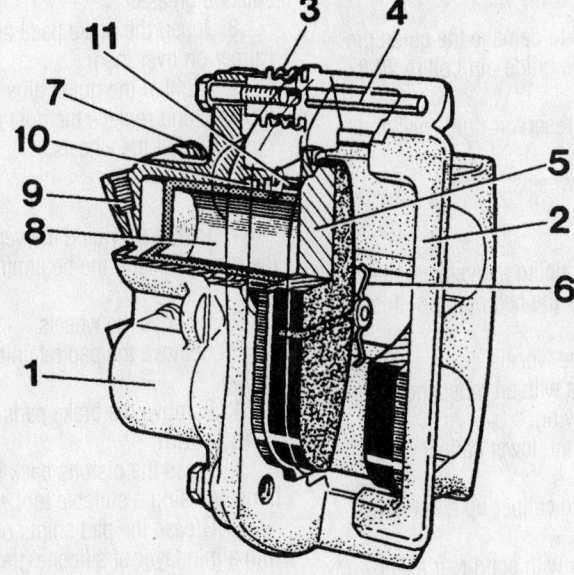

1. Caliper	7. Bleeding nipple
2. Retainer	8. Piston
3. Rubber gaiter	9. Brake fluid inlet
4. Guide pin	10. Fluid seal
5. Brake pad	11. Dust seal
6. Spring	

93016G77

Front caliper—Volvo dual piston type shown

Heater Core replacement is covered in Section 2 of this manual

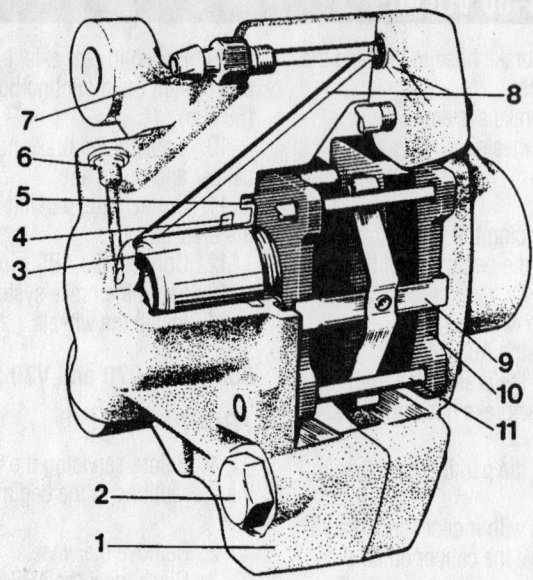

1. Housing	7. Bleeding nipple
2. Bolt	8. Passage
3. Piston	9. Damping spring
4. Fluid seal	10. Brake pad
5. Dust seal	11. Guide pin
6. Brake fluid inlet	

93016G78

Rear caliper—Volvo

Brake Pads

REMOVAL & INSTALLATION

700 Series, 900 Series, S90 and V90 Models

FRONT

1. Before servicing the vehicle, refer to the precautions in the beginning of this section.
2. Remove the wheels.
3. Remove the lower caliper guide pin bolt and swing the caliper upwards.
4. Remove the brake pads.
To install:
5. Press the piston back into the caliper.
6. Check to see that the metal guide plates are in position and install the pads. Check the guide pin boots for damage and replace them if necessary.
7. Swing the caliper down into position,

being careful not to damage the guide pin boots. Tighten the guide pin bolt to 20 ft. lbs. (27 Nm).
8. Check the reservoir fluid level and add as necessary.
9. Install the wheels.

REAR

1. Before servicing the vehicle, refer to the precautions in the beginning of this section.
2. Remove the wheels.
3. On models with an independent rear axle, do the following:
    a. Remove the lower caliper guide bolt.
    b. Swing the caliper up and remove the brake pads.
4. On models with solid rear axles, remove the brake pads in this sequence:
    a. Using a 3mm punch, drive out the guide pins.
    b. Remove the spring plate.

c. Remove the brake pads and shims.
5. Press the piston back into the caliper, taking care not to damage the dust boot.
6. Inspect the rubber guide pin boots and replace if necessary.
To install:
7. On models with an independent rear axle, install the new pads, lower the caliper and tighten the lower guide pin bolt to 25 ft. lbs. (34 Nm).
8. On models with a solid rear axle, install the guide pins with a new spring plate.
9. Install the wheels.
10. Check the brake fluid reservoir level.

850, S70, C70 and V70 Series

FRONT

1. Before servicing the vehicle, refer to the precautions in the beginning of this section.
2. Remove the front wheels.
3. Remove the caliper guide pins.
4. Remove caliper from the carrier.
5. Remove the brake pads.
To install:
6. Press the piston back into the caliper cylinder using a suitable tool.
7. Lubricate the caliper guide pins with silicone grease.
8. Insert the brake pads and slide the caliper on over them.
9. Tighten the guide pins to 22 ft. lbs. (30 Nm) and replace the dust caps.
10. Install the wheels.

REAR

1. Before servicing the vehicle, refer to the precautions in the beginning of this section.
2. Remove the wheels.
3. Remove the pad retaining pins and spring.
4. Remove the brake pads and shims.
To install:
5. Press the pistons back into their housing using a suitable tool.
6. Grease the pad shims on both sides with a thin layer of silicone grease.
7. Install the shims on the pads.
8. Install the pads in the caliper.
9. Install the retaining pins and spring.
10. Install the wheels.

ACURA

1998–01

2.3CL • 2.5TL • 3.0CL • 3.2TL • 3.2CL • 3.5RL • Integra • NSX

5

PRECAUTIONS

Before servicing any vehicle, please be sure to read all of the following precautions, which deal with personal safety, prevention of component damage and important points to take into consideration when servicing a motor vehicle:

• Never open, service or drain the radiator or cooling system when the engine is hot; serious burns can occur from the steam and hot coolant.

• Observe all applicable safety precautions when working around fuel. Whenever servicing the fuel system, always work in a well-ventilated area. Do not allow fuel spray or vapors to come in contact with a spark, open flame, or excessive heat (a hot drop light, for example). Keep a dry chemical fire extinguisher near the work area. Always keep fuel in a container specifically designed for fuel storage; also, always properly seal fuel containers to avoid the possibility of fire or explosion. Refer to the additional fuel system precautions later in this section.

• Fuel injection systems often remain pressurized, even after the engine has been turned **OFF**. The fuel system pressure must be relieved before disconnecting any fuel lines. Failure to do so may result in fire and/or personal injury.

• Brake fluid often contains polyglycol ethers and polyglycols. Avoid contact with the eyes and wash your hands thoroughly after handling brake fluid. If you do get brake fluid in your eyes, flush your eyes with clean, running water for 15 minutes. If eye irritation persists, or if you have taken brake fluid internally, seek medical assistance IMMEDIATELY.

• The EPA warns that prolonged contact with used engine oil may cause a number of skin disorders, including cancer! You should make every effort to minimize your exposure to used engine oil. Protective gloves should be worn when changing oil. Wash your hands and any other exposed skin areas as soon as possible after exposure to used engine oil. Soap and water, or waterless hand cleaner should be used.

• All new vehicles are now equipped with an air bag system. The system must be disabled before performing service on or around system components, steering column, instrument panel components, wiring and sensors. Failure to follow safety and disabling procedures could result in accidental air bag deployment, possible personal injury and unnecessary system repairs.

• Always wear safety goggles when working with, or around, the air bag system. When carrying a non-deployed air bag, be sure the bag and trim cover are pointed away from your body. When placing a non-deployed air bag on a work surface, always face the bag and trim cover upward, away from the surface. This will reduce the motion of the module if it is accidentally deployed. Refer to the additional air bag system precautions later in this section.

• Clean, high quality brake fluid from a sealed container is essential to the safe and proper operation of the brake system. You should always buy the correct type of brake fluid for your vehicle. If the brake fluid

becomes contaminated, completely flush the system with new fluid. Never reuse any brake fluid. Any brake fluid that is removed from the system should be discarded. Also, do not allow any brake fluid to come in contact with a painted surface; it will damage the paint.

• Never operate the engine without the proper amount and type of engine oil; doing so WILL result in severe engine damage.

• Timing belt maintenance is extremely important! Many models utilize an interference type, non-freewheeling engine. If the timing belt breaks, the valves in the cylinder head may strike the pistons, causing potentially serious (also time consuming and expensive) engine damage. Refer to the maintenance interval charts in the front of this manual for the recommended replacement interval for the timing belt and to the timing belt section for belt replacement and inspection.

• Disconnecting the negative battery cable on some vehicles may interfere with the functions of the on-board computer system(s) and may require the computer to undergo a relearning process once the negative battery cable is reconnected.

• When servicing drum brakes, only disassemble and assemble one side at a time, leaving the remaining side intact for reference.

• Only an MVAC-trained, EPA-certified automotive technician should service the air conditioning system or its components.

• The radio may contain a coded theft protection circuit. Always obtain the code number before disconnecting the battery.

ENGINE REPAIR

➥Disconnecting the negative battery cable on some vehicles may interfere with the functions of the on board computer system. The computer may undergo a relearning process once the negative battery cable is reconnected.

Distributor

REMOVAL & INSTALLATION

1.8L and 2.3L Engines

1. Before servicing the vehicle, refer to the precautions in the beginning of this section.
2. Remove or disconnect the following:
   • Negative battery cable

• Engine wiring harness and connectors from the distributor
• Spark plug wires from the distributor cap

3. If removing the ignition coil, remove the distributor cap, rotor, and cap seal, then remove the leak cover.

4. Remove or disconnect the following:
   • 2 screws to disconnect the wires from the coil
   • Ignition coil screws and slide the ignition coil out of the distributor housing
   • Distributor hold-down bolts
   • Distributor from the cylinder head

To install:

5. Use a new O-ring coated with engine oil, on the distributor housing.

6. Slip the distributor into position.

➥Lugs on the end of the distributor and the matching grooves in the camshaft end are offset to eliminate any possibility of installing the distributor 180 degrees out of time.

7. Install the hold-down bolts and hand tighten them.

8. Slide the ignition coil into the distributor housing and install the 2 mounting screws.

9. Install or connect the following:
   • 2 wires to the coil and install the 2 screws.
   • Distributor leak cover, rotor, cap seal, and cap
   • Engine wiring harness and connector to distributor

- Spark plug wires
- Negative battery cable

10. Set the timing, then tighten the hold down bolts to 16 ft. lbs. (22 Nm).

2.5L and 3.0L Engines

1. Before servicing the vehicle, refer to the precautions in the beginning of this section.
2. Remove or disconnect the following:
   - The negative battery cable
   - The spark plug and coil wires from the distributor cap
   - Harness connector(s) from the distributor
   - Distributor mounting bolts
   - Distributor from the cylinder head

To install:
3. Install or connect the following:
   - New O-ring coated with engine oil, on the distributor housing
   - Distributor
   - Mounting bolts and tighten them to 13 ft. lbs. (18 Nm)
   - Spark plug and coil wires
   - Negative battery cable
4. Check the ignition timing with a timing light. The timing marks are located on the crankshaft pulley and lower timing cover. If the timing is not within specification replace the PCM module.

Alternator

REMOVAL & INSTALLATION

1.8L, 2.3L, 2.5L, 3.5L Engines

1. Remove or disconnect the following:
   - Both battery cables
   - 4 prong connector and the black wire from the rear of the alternator
   - Alternator adjusting bolt(s)
   - Mounting bolt(s)
   - Alternator belt
   - Alternator assembly

To install:
2. Install or connect the following:
   - Alternator assembly
   - Mounting bolts and tighten them to 33 ft. lbs. (44Nm)
   - Alternator adjusting bolt, hand tight
   - Alternator belt
3. Adjust alternator belt to tension.
4. Tighten the 8 x 1.25mm locknut/lock bolt(s) to 16 ft. lbs. (22 Nm).
5. Install or connect the following:

- 4 prong connector and the black wire to the rear of the alternator.
- Both battery cables.

※※ **WARNING**

Be sure to adjust the alternator belt to the proper tension or alternator bearing failure may occur.

➡**The Powertrain Control Module (PCM) idle memory must be reset after reconnecting the battery. Start the engine and hold it at 3000 rpm until the cooling fan comes on. Then allow the engine to idle for about 5 minutes with all accessories OFF and with the transmission in Park or Neutral.**

3.0L Engine

1. Remove or disconnect the following:
   - Battery cover
   - Both battery cables
   - Engine cover
   - Accessory drive belt
   - Fan motor connector and A/C compressor clutch wiring connector from the fan shroud
   - A/C condenser fan shroud assembly
   - Ground cable
   - 4 prong connector
   - Alternator and bracket assembly

To install:
2. Install or connect the following:
   - Alternator and bracket assembly and tighten the 10 x 1.25 mounting bolts to 33 ft. lbs. (44Nm) and the 8 x 1.25mm locknut/lock bolt(s) to 16 ft. lbs. (22 Nm)
   - 4 prong connector
   - Ground cable
   - A/C condenser fan shroud assembly
   - Accessory drive belt
   - Front engine cover

➡**There is no belt tension adjustment due to the use of an automatic tensioner.**

- Both battery cables
- Battery cover

3. The Powertrain Control Module (PCM) idle memory must be reset after reconnecting the battery. Start the engine and hold it at 3000 rpm until the cooling fan comes on. Then allow the engine to idle for about 5 minutes with all accessories OFF and with the transmission in Park or Neutral.

3.2L Engine

1. Remove or disconnect the following:
   - Both battery cables
   - Adjusting bolts
   - Accessory drive belt
   - Mounting bolts
2. Turn the alternator 90° in a counterclockwise direction.
   - Alternator
   - 4 prong connector
   - Harness clip and bracket assembly
   - Black wire from the terminal
   - Alternator from the vehicle

To install:
3. Install or connect the following:
   - Alternator
   - Black wire
   - Harness clip and bracket assembly and t tighten the bolt to 8.7lbf. ft. (12Nm)
   - 4 prong connector
   - Mounting bolts and tighten the 10 x 1.25mm bolts to 33 ft. lbs. (44Nm), the 8 x 1.25mm locknut/lock bolt(s) to 16 ft. lbs. (22 Nm) and the 6 x 1.0mm 104 inch lbs. (12Nm)
   - Accessory drive belt

➡**There is no belt tension adjustment due to the use of an automatic tensioner.**

4. The Powertrain Control Module (PCM) idle memory must be reset after reconnecting the battery. Start the engine and hold it at 3000 rpm until the cooling fan comes on. Then allow the engine to idle for about 5 minutes with all accessories OFF and with the transmission in Park or Neutral.

5. Connect the positive, then the negative battery cable.

Ignition Timing

ADJUSTMENT

1.8L and 2.3L Engines

1. Before servicing the vehicle, refer to the precautions in the beginning of this section.
2. Start the engine and hold the engine speed at 3000 rpm, until the radiator fan comes on. The engine should be at idle speed and at normal operating temperature. Be sure all electrical devices (radio, air conditioning, lights, etc.) are turned OFF.

3. Locate the Service Check (SCS) connector:

- 1.8L engines: behind the right kick panel
- 2.3L engines: centrally located under the dash

4. Connect the SCS service connector (part number 07PAZ-0010100) to the service check connector.

5. Connect a timing light to No. 1 ignition wire and point the light toward the pointer on the timing belt cover.

6. Check the idle speed and adjust if necessary.

7. The red mark on the crankshaft pulley should be aligned with the pointer on the timing belt cover.

➡**The white mark on the crank pulley is TDC.**

8. Adjust the ignition timing by loosening the distributor mounting bolts and rotating the distributor housing to adjust the timing. Set as follows:

- 1.8L (Except Type R): 16 degrees BTDC at 700–800 rpm
- 1.8L (Type R): 16 degrees BTDC at 750–850 rpm
- 2.3L: 12 degrees BTDC at 650–750 rpm

9. Tighten the distributor bolts to 17 ft. lbs. (24 Nm) and recheck the timing.

10. Remove the SCS service connector from the service check connector.

2.5L and 3.0L Engines

➡**The ignition timing is controlled by the PCM and can be checked for diagnostic purposes. If the timing is out of specification, all mechanical and electrical systems should be checked for proper operation before replacing the PCM.**

1. Before servicing the vehicle, refer to the precautions in the beginning of this section.

2. To check the ignition timing, start the engine and allow it to fast idle at 3000 rpm with all electrical accessories off and the transmission in **N** or **P**. Allow the engine to warm up and reach normal operating temperature. The engine cooling fan should cycle at least 1 time.

3. Locate the Service Check (SCS) connector under the glove box. Connect the service connector tool part number 07PAZ-0010100 to the SCS terminals.

4. Check the idle speed and adjust if necessary.

5. Connect a timing light to the No. 1 plug wire. While engine idles, point the light toward the pointer on the timing belt cover.

6. Inspect the ignition timing at idle. The specifications are as follows:

- 2.5L: 13–17 degrees BTDC at 650–750 rpm
- 3.0L: 8–12 degrees BTDC at 700–800

➡**All mechanical and electrical systems should checked for proper operation before replacing the PCM.**

7. If the ignition timing is incorrect, replace the PCM.

8. Remove the service connector.

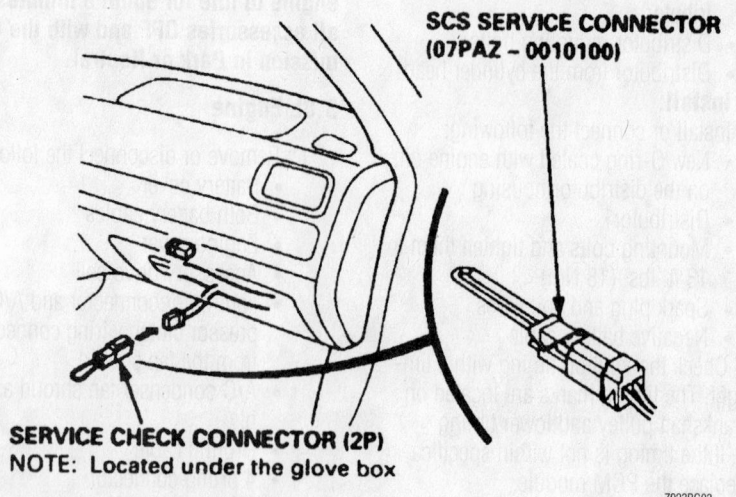

Service check connector—3.2TL and 3.5RL

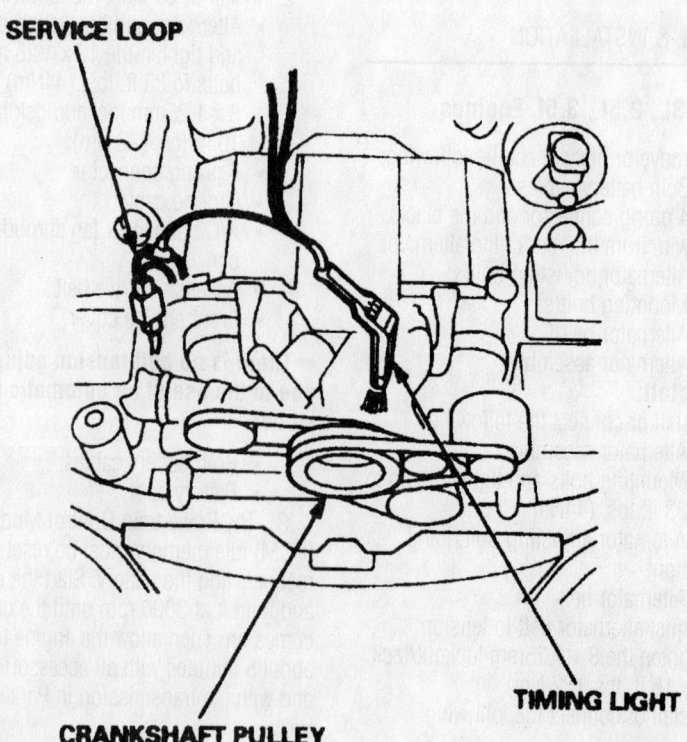

Timing light attachment—3.2TL

3.2L and 3.5L Engines

➡**The ignition timing is controlled by the PCM and can be checked for diagnostic purposes. If the timing is out of specification, all mechanical and electrical systems should be checked for proper operation before replacing the PCM.**

1. Before servicing the vehicle, refer to the precautions in the beginning of this section.

2. To check the ignition timing, start the engine and allow it to fast idle at 3000 rpm

with all electrical accessories off and the transmission in **N** or **P**. Allow the engine to warm up and reach normal operating temperature. The engine cooling fan should cycle at least 1 time.

3. Locate the Service Check (SCS) connector under the glove box and connect the service connector tool part number 07PAZ-0010100 to it.

4. Check the idle speed and adjust if necessary.

5. Connect a timing light to the No. 1 plug wire. With the engine idling at normal operating temperature point the timing light toward the pointer on the timing belt cover.

6. Inspect the ignition timing. The specifications are as follows:
- 3.2L: 13–17 degrees BTDC at 590–690 rpm
- 3.5L: 13–17 degrees BTDC at 700–800 rpm

➡️**All mechanical and electrical systems should checked for proper operation before replacing the PCM.**

7. If the ignition timing is incorrect, replace the PCM.
   Only replace the PCM as a last resort.
8. Remove the timing light.
9. Disconnect the special tool (SCS service connector) from the service check connector.

Engine Assembly

REMOVAL & INSTALLATION

1.8L Engines

1. Before servicing the vehicle, refer to the precautions in the beginning of this section.
2. Relieve the fuel system pressure.
3. Remove or disconnect the following:
- Both battery cables
- Hood
- Strut brace (if equipped)
- Battery cables from the under-hood fuse/relay box and under-hood Antilock Brake (ABS) System fuse/relay box
- Air cleaner assembly and mounting bracket
- Evaporative emission (EVAP) control canister hose and vacuum hose from the intake manifold
- Brake booster vacuum hose
- Fuel return hose

- Engine wiring harness connectors on the right side of the engine compartment
- Fuel feed hose
- Throttle cable

➡️**Be careful not to bend the cable when removing it. Replace the cable if it gets kinked.**

- Engine wiring harness connectors on the left side of the engine compartment
- Cruise control actuator
- Engine ground cable
- Power steering belt
- Power steering pump without disconnecting the hoses
- Air conditioning compressor belt
- Clutch slave cylinder and pipe/hose assembly (if equipped). Do not disconnect the pipe/hose assembly.
- Transmission ground cable and hose clamp
- Front wheels
- Lower splash shield

4. Drain the engine coolant, engine oil, and transmission fluid into sealable containers. Reinstall the drain plugs using new washers. Be careful not to overtighten the drain plugs.

5. Remove or disconnect the following:
- Upper and lower radiator hoses and the heater hoses from the engine
- Transmission oil cooler hoses (if equipped)
- Radiator assembly
- Air conditioning compressor without disconnecting the hoses
- Heated Oxygen (HO$_2$S) sensor connector
- Front exhaust pipe from the exhaust manifold
- Shift rod and extension rod from the transaxle (if equipped)
- Shift cable cover, then disconnect the shift cable from the transaxle (if equipped)
- Damper fork
- Lower ball joints
- Halfshafts from the transaxle

6. Attach a hoist to the engine.
- Left and right front engine mounts and brackets
- Rear engine mount and bracket
- Side engine mount
- Transmission mount

7. Check that the engine is completely clear of all vacuum, fuel and coolant hoses, and electrical wiring.

8. Raise the engine and transaxle assembly all the way and remove it from the vehicle.

9. Separate the engine and transaxle.

To install:

10. Installation is the reverse of the removal procedure, while using the following torque values:
- Transaxle bolts: 47 ft. lbs. (64 Nm) (if equipped with a manual transaxle), or 43 ft. lbs. (59 Nm) (if equipped with an automatic transaxle)
- Rear mounting bracket: 87 ft. lbs. (118 Nm)
- Upper mounting bolts: 54 ft. lbs. (74 Nm).
- Torque converter bolts: 104 inch lbs. (12 Nm) (if equipped)
- Transmission mount bolts: 47 ft. lbs. (64 Nm)
- Engine side mount bolt: 47 ft. lbs. (64 Nm)
- Rear mount bracket bolts: 40 ft. lbs. (54 Nm)
- Right front mount/bracket bolts: 47 ft. lbs. (64 Nm)
- Damper fork bolts: 47 ft. lbs. (64 Nm)
- Lower ball joints nuts: 40 ft. lbs. (54 Nm)
- Shift cable bolt: 10 ft. lbs. (14 Nm)
- Shift rod and extension rod bolt: 16 ft. lbs. (22 Nm)
- Front exhaust pipe nuts: 40 ft. lbs. (54 Nm)
- A/C compressor bolts: 17 ft. lbs. (24 Nm)
- Power steering pump bolts: 17 ft. lbs. (24 Nm)
- Cruise control actuator bolts: 17 ft. lbs. (24 Nm)

2.3L and 3.0L Engines

1. Before servicing the vehicle, refer to the precautions in the beginning of this section.
2. Obtain the anti-theft code for the radio.
3. Drain the engine oil, coolant, and transmission oil (or fluid) into sealable containers and carefully reinstall the drain plugs using new sealing washers.
4. Properly relieve the fuel system pressure.
5. Remove or disconnect the following:
- Air cleaner assembly
- Hood support struts and support the hood in a vertical position

- Negative battery cable and then the positive cable
- Strut brace
- Battery, battery tray and the engine ground cable
- Accelerator and cruise control cables from the throttle body and bracket
- Battery cables from the underhood Antilock Brake System (ABS) fuse/relay box and underhood fuse/relay box assemblies
- Engine wiring harness on the right side of the engine
- Evaporative emission (EVAP) control canister hose
- Brake booster vacuum hose
- Fuel feed and return hoses
- Engine wiring harness on the left side of the engine
- Accessory drive belt(s)
- Power steering pump leaving the hoses attached
- Shift cable (if equipped with a manual transmission)
- Clutch slave cylinder leaving (if equipped) the hydraulic line attached
- Reverse light switch connector (if equipped with a manual transmission)
- Cruise control vacuum tank (if equipped)
- Front wheels
- Lower splash shield
- Center support beam
- Heated Oxygen (HO$_2$S) sensor connector
- Front exhaust pipe from the manifold
- Shift selector cable and cover (if equipped with an automatic transmission)
- Damper fork
- Lower ball joints
- Driveshafts from the transaxle
- Crankshaft pulley (if equipped with 3.0L engine)
- VTEC/oil filter housing (if equipped with 3.0L engine)
- Upper and lower radiator hoses
- Transmission oil cooler hoses (if equipped)
- Radiator
- Distributor (if equipped with 2.3L engine)
- Intake Air Control (IAC) valve (if equipped with 2.3L engine)
- Air conditioning compressor leaving the hoses attached
- Heater hoses

6. Attach a suitable engine lifting hoist

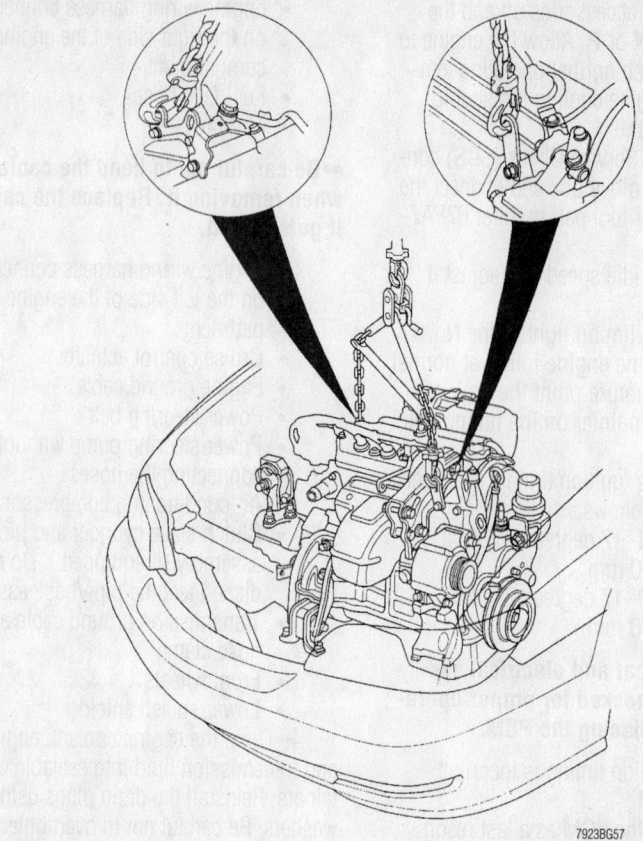

Engine lifting points—2.3CL

7923BG57

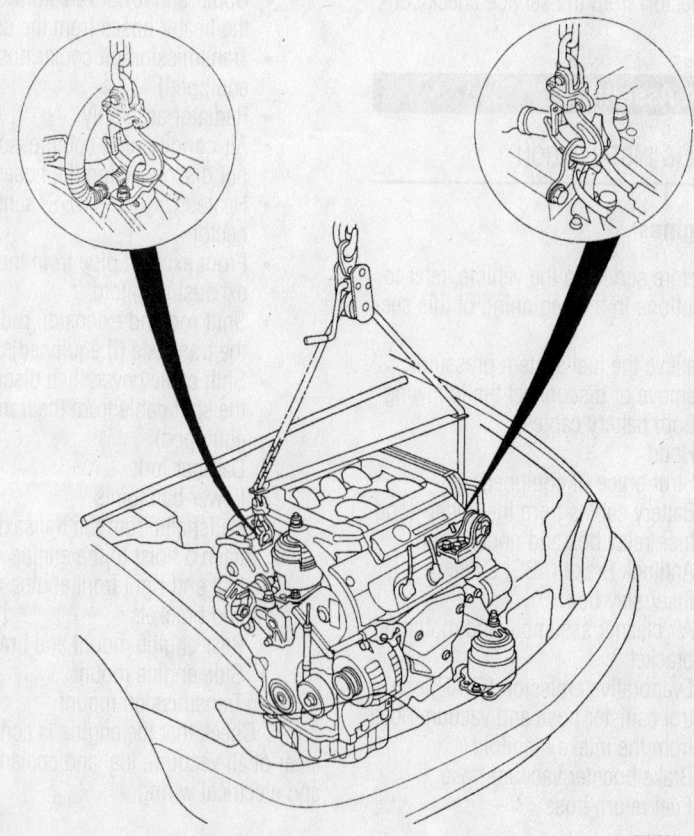

Engine lifting points—3.0CL

7923BG58

to the engine lifting hooks and secure the engine.

- Front engine mount
- Rear engine mount
- Side engine mount
- Transmission mount

7. Lift the engine slightly and check that all hoses, cables and wires have been properly disconnected.

8. Carefully raise the engine/transmission from the vehicle.

To install:

9. Installation is the reverse of the removal procedure, while using the following torque values:

- Transmission mount bolts: 47 ft. lbs. (64 Nm)
- Front engine mount bolts: 47 ft. lbs. (64 Nm)
- Side engine mount bolts: 47 ft. lbs. (64 Nm)
- Rear engine mount bolts: 47 ft. lbs. (64 Nm)
- A/C compressor bolts: 16 ft. lbs. (22 Nm)
- VTEC/oil filter housing bolts (if equipped with 3.0L engine): 16 ft. lbs. (22 Nm)
- Crankshaft pulley bolt (if equipped with 3.0L engine): 181 ft. lbs. (245 Nm)
- Lower ball joint nuts: 40 ft. lbs. (54 Nm)
- Damper fork bolts: 47 ft. lbs. (64 Nm)
- Front exhaust pipe nuts: 40 ft. lbs. (54 Nm)
- Center support beam bolts: 37 ft. lbs. (50 Nm)
- Splash shield bolts: 84 inch lbs. (9.5 Nm)
- Power steering pump bolts: 16 ft. lbs. (22 Nm)
- Strut brace bolts (if equipped): 16 ft. lbs. (22 Nm)

2.5L Engine

1. Before servicing the vehicle, refer to the precautions in the beginning of this section.

2. Relieve the fuel system pressure.

3. Drain the engine oil, coolant, differential oil and transmission fluid.

4. Open the hood and secure it in its fully opened position (vertical). The hood may be removed if more clearance and working room is desired.

5. Remove or disconnect the following:

- Negative and positive battery cables
- Battery and battery tray
- Engine ground cables and ignition coil wire
- Battery cables from the under-hood fuse/relay box
- Right side engine wiring harness connectors
- Battery cable from the Antilock Brake System (ABS) fuse/relay box
- Air cleaner assembly
- Accessory drive belts
- Power steering pump without disconnecting the hoses
- Air conditioning compressor without disconnecting the refrigerant lines
- Throttle cable
- Left side engine wiring harness connectors
- Brake booster vacuum hose
- Evaporative Emissions (EVAP) canister hoses
- Fuel feed and return hoses
- Vehicle Speed Sensor (VSS)
- Torque converter cover
- Torque converter bolts
- Upper transmission bolts and the 26mm differential shim—Note the location of the shim for installation.
- Transmission sub-harness connector
- Splash shield
- Front tires
- Damper fork
- Ball joints
- Halfshafts from the transmission
- Heated Oxygen (HO2S) sensor connector
- Front exhaust pipe from the manifold
- Transmission mount and bracket

6. Be sure the transmission is in **P** park. Remove the 33mm extension shaft sealing bolt on the lower left side of transmission housing.

7. Use an extension shaft puller tool to remove the extension shaft from the differential.

8. Lower the vehicle and install a chain hoist onto the engine lifting hooks.

9. Remove or disconnect the following:

- Heater hoses
- Upper and lower radiator hoses
- Transmission oil cooler hoses
- Cooling fan electrical connector
- Radiator assembly

- Front engine mounts
- Mid mounts

10. Support the transmission and remove the transmission mounting bolts.

11. Install the transmission mid-mounts and spacer to hold the transmission in the vehicle. Be sure the engine will separate from the transmission, and, that the transmission will be supported by the mounts, after the engine has been removed.

12. Unbolt the front engine mounts.

13. Slowly raise the engine slightly to separate it from the transmission. Verify that all wiring harnesses, fuel and coolant lines, and vacuum hoses are disconnected.

14. Lift the engine out of the vehicle. Be sure the engine clears the mounts, the transmission case, and the differential extension shaft.

To install:

15. Install or connect the following:

- Engine in the vehicle
- Transmission mounting bolts, Do not tighten the bolts at this time

16. Remove the mid-mounts and torque the transmission mounting bolts to 47 ft. lbs. (64 Nm).

- Mid-mounts and torque the bolts to 28 ft. lbs. (38 Nm)
- Left front engine mount and torque the nut to 54 ft. lbs. (75 Nm)
- Right front engine mount and torque the nut to 54 ft. lbs. (75 Nm)
- Extension shaft
- Transmission mount and torque the bolts to 47 ft. lbs. (64 Nm)

17. Torque the mid mount nuts to 32 ft. lbs. (43 Nm) and the bolts to 28 ft. lbs. (38 Nm).

18. The remainder of the installation procedure is the reverse of the removal procedure, while using the following torque values:

- Extension shaft sealing bolt: 58 ft. lbs. (80 Nm)
- Left front strut bracket bolt: 40 ft. lbs. (55 Nm).
- Front exhaust pipe to the manifold nuts: 40 ft. lbs. (54 Nm)
- Torque converter bolts: 108 inch lbs. (12 Nm).
- Ball joint nuts: 40 ft. lbs. (54 Nm)
- Damper fork bolts: 32 ft. lbs. (43 Nm)
- Splash shield bolts: 84 inch lbs. (9.5 Nm)
- A/C compressor bolts: 16 ft. lbs. (22 Nm)
- Power steering pump bolt: 36 ft.

lbs. (49 Nm). Nut: 16 ft. lbs. (22 Nm)

3.2L Engine

1. Before servicing the vehicle, refer to the precautions in the beginning of this section.

2. Do not remove the hood. Disconnect the hood support strut and reconnect it to hold the hood in a vertical position.

3. Properly relieve the fuel pressure.

4. Drain the engine oil, coolant, transmission fluid and the differential fluid.

5. Remove or disconnect the following:

- Negative battery cable, then the positive battery cable
- battery and battery tray
- Air cleaner assembly
- Water bypass hoses (on 3.2CL)
- Traction Control System (TCS) control valve actuator connector (on 3.2CL)
- TCS control valve angel sensor connector (on 3.2CL)
- Throttle cable and cruise control cable from the throttle and bracket
- Left side engine wiring harness connectors
- Fuel feed and return hoses
- Brake booster vacuum hose
- Battery cables from the under-hood fuse/relay box
- Under-hood fuse/relay box
- Powertrain Control Module (PCM) electrical connectors
- Accessory drive belts
- Power steering pump without disconnecting the hoses
- Vehicle Speed sensor (VSS) connector, then remove the VSS/power steering sensor leaving the fluid hoses attached
- Front wheels
- Lower splash shield
- Heated Oxygen (HO2S) sensor connector
- Front exhaust pipe from the manifold
- Front damper forks
- Lower ball joints from the steering knuckles
- Halfshafts from the differential and the intermediate shaft
- Shift cable cover
- Shift cable with the control lever from the transaxle
- Power steering hose clamps and the engine mount control vacuum hose
- Upper and lower radiator hoses
- Heater hoses

- Transmission fluid cooler hoses
- Ground cable
- Power steering hose clamp from the rear beam assembly

6. Attach a suitable chain hoist to the engine lifting hooks and support the engine.

➡**The engine and transmission assembly is removed by lowering it from the vehicle. Be sure the vehicle is in a position that will allow the engine and transmission assembly enough clearance to be moved from the vehicle once it is lowered away from the vehicle.**

7. Remove or disconnect the following:

- Side, rear and front engine mount support fasteners
- Front suspension radius rod bolts

8. Make alignment marks on the front beam and remove the front beam.

9. Remove or disconnect the following:

- Air conditioning compressor leaving the hoses attached
- Rear mounts from the engine and transmission

10. Check that all hoses, cables and wires have been properly disconnected and slowly lower the engine about 6 inches (150 mm). Recheck that all hoses, cables and wires have been properly disconnected.

11. Carefully lower the engine/transmission assembly from the vehicle.

To install:

12. Installation is the reverse of the removal procedure, while using the following torque values:

- Transmission rear mount bolts: 28 ft. lbs. (38 Nm)
- Air conditioning compressor bolts: 16 ft. lbs. (22 Nm)
- Front beam nuts: 28 ft. lbs. (38 Nm)
- Front beam bolts: 76 ft. lbs. (103 Nm)
- Rear bolts: 28 ft. lbs. (38 Nm)
- Front suspension radius rod bolts: 119 ft. lbs. (162 Nm)
- Front engine mount nut: 40 ft. lbs. (54 Nm)
- Rear engine mount nut: 40 ft. lbs. (54 Nm)
- Rear engine mount bolt: 28 ft. lbs. (38 Nm)
- Side engine mount bracket bolts: 33 ft. lbs. (44 Nm)
- Side engine mount through bolt: 40 ft. lbs. (54 Nm)
- Lower ball joint nuts: 40 ft. lbs. (54 Nm)
- Damper fork bolts: 47 ft. lbs. (64 Nm)

- Front exhaust pipe nuts: 40 ft. lbs. (54 Nm)
- Shift cable and control lever bolts: 10 ft. lbs. (14 Nm)
- Power steering pump bolts: 16 ft. lbs. (22 Nm)

3.5L Engine

1. Before servicing the vehicle, refer to the precautions in the beginning of this section.

2. Move the front passenger's seat forward.

3. Relieve the fuel system pressure.

4. Drain the engine oil, coolant, transmission fluid and differential fluid.

5. Remove or disconnect the following:

- Hood
- Negative battery cable, then the positive battery cable
- Strut brace
- Engine cover
- Air cleaner assembly and intake duct
- Throttle cover
- Throttle cable and cruise control cable from the throttle and bracket
- Battery
- Battery tray
- Relay box
- Ground cable and wiring harness clips from the firewall
- Alternator and battery cables from the under-hood fuse/relay box
- Underhood fuse/relay box
- Left side engine wiring harness connectors
- Fuel feed and return hoses
- Brake booster vacuum hose
- Evaporative emissions (EVAP) canister hose
- Transmission sub-harness connector
- Control box
- Right side engine wiring harness connectors
- Spark plug voltage detection module
- Engine ground cables
- Accessory drive belts
- Power Steering Pressure (PSP) switch connector
- Power steering pump leaving the hoses attached

6. Pull the carpet back under the front passengers seat and detach the secondary Heated Oxygen (HO2S) sensor connector.

7. Remove or disconnect the following:

- Front wheels
- Splash shield
- Front suspension damper forks
- Lower ball joints from the steering knuckles

Bracket Bolts Torque Specifications:

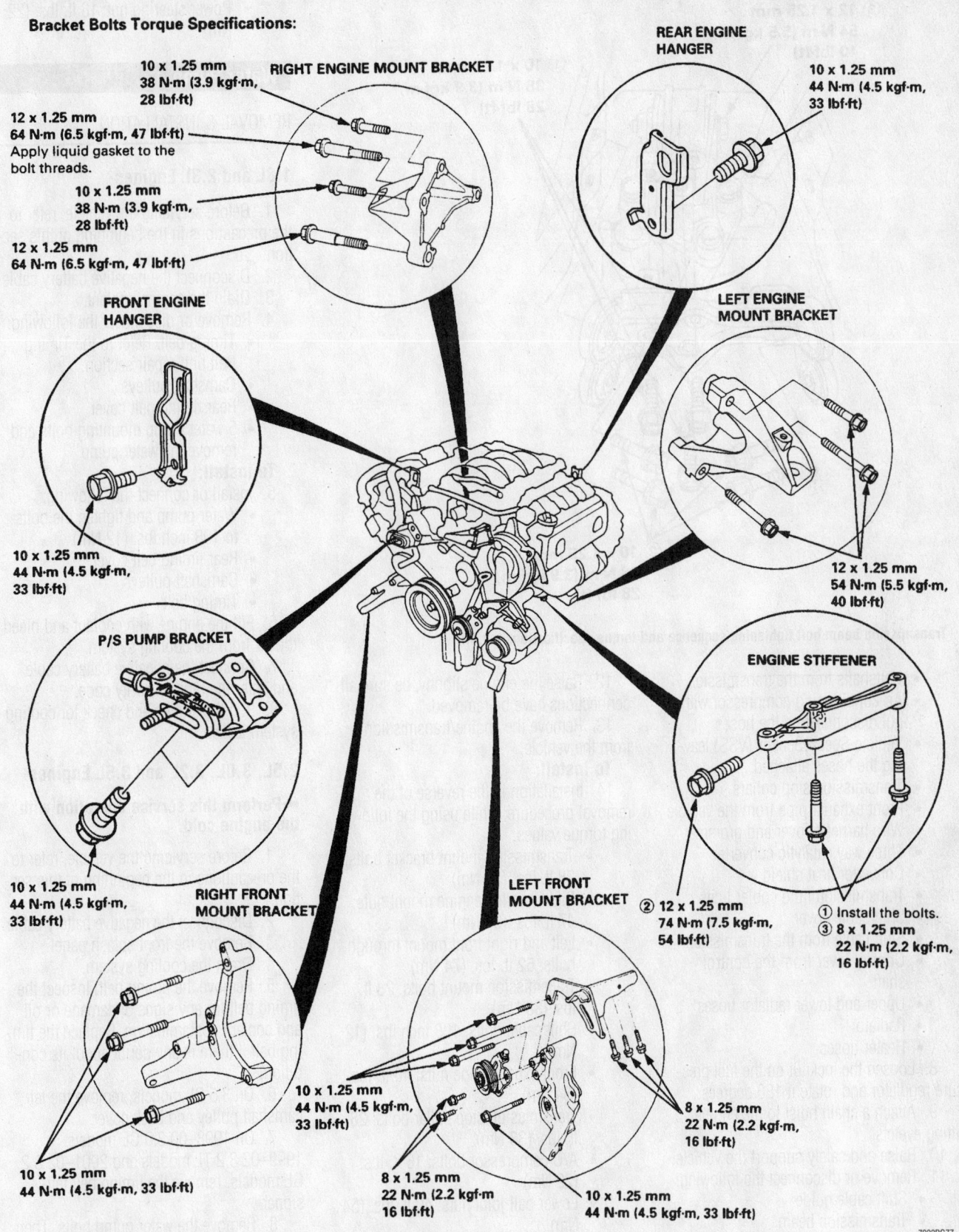

10 x 1.25 mm
38 N·m (3.9 kgf·m, 28 lbf·ft)

RIGHT ENGINE MOUNT BRACKET

REAR ENGINE HANGER

10 x 1.25 mm
44 N·m (4.5 kgf·m, 33 lbf·ft)

12 x 1.25 mm
64 N·m (6.5 kgf·m, 47 lbf·ft)
Apply liquid gasket to the bolt threads.

10 x 1.25 mm
38 N·m (3.9 kgf·m, 28 lbf·ft)

12 x 1.25 mm
64 N·m (6.5 kgf·m, 47 lbf·ft)

FRONT ENGINE HANGER

LEFT ENGINE MOUNT BRACKET

10 x 1.25 mm
44 N·m (4.5 kgf·m, 33 lbf·ft)

12 x 1.25 mm
54 N·m (5.5 kgf·m, 40 lbf·ft)

P/S PUMP BRACKET

ENGINE STIFFENER

10 x 1.25 mm
44 N·m (4.5 kgf·m, 33 lbf·ft)

RIGHT FRONT MOUNT BRACKET

LEFT FRONT MOUNT BRACKET

② 12 x 1.25 mm
74 N·m (7.5 kgf·m, 54 lbf·ft)

① Install the bolts.
③ 8 x 1.25 mm
22 N·m (2.2 kgf·m, 16 lbf·ft)

10 x 1.25 mm
44 N·m (4.5 kgf·m, 33 lbf·ft)

8 x 1.25 mm
22 N·m (2.2 kgf·m, 16 lbf·ft)

10 x 1.25 mm
44 N·m (4.5 kgf·m, 33 lbf·ft)

8 x 1.25 mm
22 N·m (2.2 kgf·m, 16 lbf·ft)

10 x 1.25 mm
44 N·m (4.5 kgf·m, 33 lbf·ft)

View of the engine mounting bracket showing torque specifications—3.5RL

7923BG77

Heater Core replacement is covered in Section 2 of this manual

③ 12 x 1.25 mm
54 N·m (5.5 kgf·m,
40 lbf·ft)

① 10 x 1.25 mm
38 N·m (3.9 kgf·m,
28 lbf·ft)

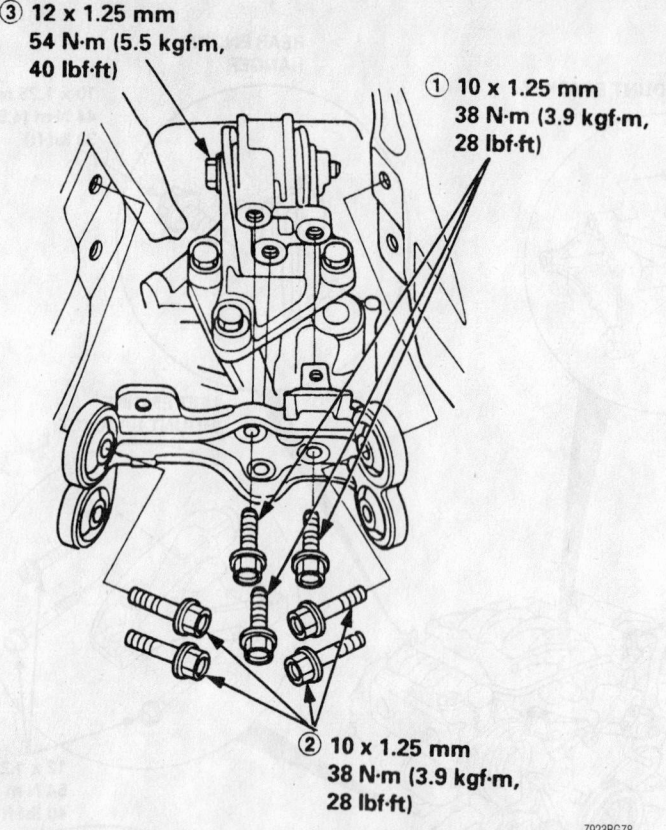

② 10 x 1.25 mm
38 N·m (3.9 kgf·m,
28 lbf·ft)

7923BG78

Transmission beam bolt tightening sequence and torque specifications—3.5RL

- Halfshafts from the transmission
- Air conditioning compressor without disconnecting the hoses
- Vehicle Speed Sensor (VSS) leaving the hoses attached
- Transmission stop collars
- Front exhaust pipe from the vehicle
- Wire harness cover and grommet
- Three way catalytic converter
- Converter heat shield
- Transmission fluid cooler lines
- Shift cable cover
- Shift cable from the transmission
- Control lever from the control shaft
- Upper and lower radiator hoses
- Radiator
- Heater hoses

8. Loosen the locknut on the fuel pressure regulator and rotate it 180 degrees.

9. Attach a chain hoist to the engine lifting eyelets.

10. Raise and safely support the vehicle.

11. Remove or disconnect the following:
- Shift cable guide
- Transmission beam
- Transmission mount and bracket
- Left and right front mount brackets from the mounts
- Right and left engine mounts

12. Raise the engine slightly, be sure all connections have be removed.

13. Remove the engine/transmission from the vehicle.

To install:

14. Installation is the reverse of the removal procedure, while using the following torque values:
- Transmission mount bracket bolts: 28 ft. lbs. (38 Nm)
- Right and left engine mount nuts: 47 ft. lbs. (64 Nm)
- Left and right front mount through bolts: 52 ft. lbs. (74 Nm)
- Transmission mount bolts: 28 ft. lbs. (38 Nm)
- Shift cable bolts: 108 inch lbs. (12 Nm)
- Front exhaust pipe nuts: 40 ft. lbs. (54 Nm)
- Transmission stop collar bolts: 28 ft. lbs. (38 Nm)
- A/C compressor bolts: 16 ft. lbs. (22 Nm)
- Lower ball joint nuts: 40 ft. lbs. (54 Nm)
- Front suspension damper fork bolts: 41 ft. lbs. (69 Nm)
- Power steering pump bolt: 33 ft. lbs. (44 Nm)
- Power steering nut: 16 ft. lbs. (22 Nm)

Water Pump

REMOVAL & INSTALLATION

1.8L and 2.3L Engines

1. Before servicing the vehicle, refer to the precautions in the beginning of this section.

2. Disconnect the negative battery cable.

3. Drain the engine coolant.

4. Remove or disconnect the following:
- Timing belt. Refer to the Timing Belt unit repair section.
- Camshaft pulleys
- Rear timing belt cover
- 5 water pump mounting bolts and remove the water pump

To install:

5. Install or connect the following:
- Water pump and tighten the bolts to 108 inch lbs. (12 Nm)
- Rear timing belt cover
- Camshaft pulleys
- Timing belt

6. Fill the engine with coolant and bleed the air from the cooling system.

7. Connect the negative battery cable and enter the radio security code.

8. Run the engine and check for cooling system leaks.

2.5L, 3.0L, 3.2L and 3.5L Engines

➡**Perform this service operation with the engine cold.**

1. Before servicing the vehicle, refer to the precautions in the beginning of this section.

2. Disconnect the negative battery cable.

3. Remove the front splash panel.

4. Drain the cooling system.

5. Remove the timing belt. Inspect the timing belt for any signs of damage or oil and coolant contamination. Replace the timing belt if there is any doubt about its condition.

6. On 3.5 RL models, remove the left camshaft pulley and back cover.

7. On 1998–00 3.0 CL models, 1999–02 3.2 TL models and 2001–02 3.2 CL models, remove the timing belt tensioner.

8. Remove the water pump bolts. Then, remove the water pump and sprocket assembly from the engine block.

To install:

9. Install the water pump with a new O-

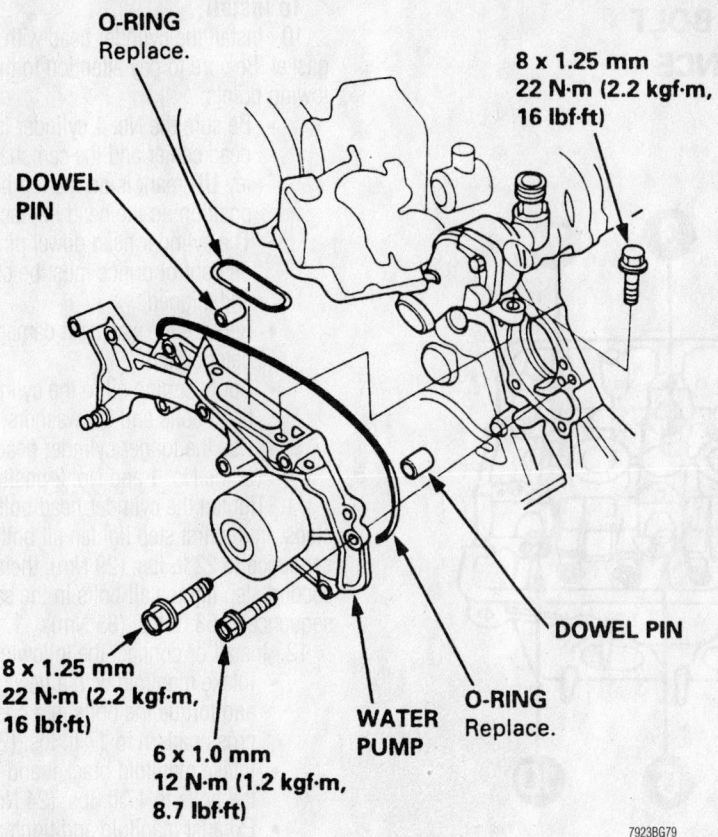

Water pump mounting and bolt torque specifications—3.5L engine

O-RING
Replace.

DOWEL
PIN

8 x 1.25 mm
22 N·m (2.2 kgf·m,
16 lbf·ft)

DOWEL PIN

8 x 1.25 mm
22 N·m (2.2 kgf·m,
16 lbf·ft)

6 x 1.0 mm
12 N·m (1.2 kgf·m,
8.7 lbf·ft)

WATER
PUMP

O-RING
Replace.

be sure the engine temperature is below 100°F (38°C); a fully cooled engine is best.

3. Disconnect the negative battery cable.

4. Be sure the crankshaft is at TDC on No. 1 cylinder by aligning the white mark on the crankshaft pulley with the pointer on the lower timing belt cover.

5. Drain the engine coolant.

6. Properly relieve the fuel system pressure.

7. Remove or disconnect the following:
- Cylinder head cover
- Crankshaft pulley
- Middle and lower timing belt covers
- Timing belt
- Camshaft pulleys
- Back cover
- Exhaust manifold cover
- Exhaust manifold and bracket

8. Loosen the locknuts and adjusting screws, then remove the camshaft holder bolts. Remove the camshaft holders, camshafts and rocker arms.

9. Remove the cylinder head bolts, then remove the cylinder head. To prevent warpage, unscrew the bolts in the reverse of the torque sequence, ⅓ turn at a time. Repeat the sequence until all bolts are loosened.

ring. Use new bolts and tighten the 6mm mounting bolts evenly to 104 inch lbs. (12 Nm) and the 8mm bolts to 16 ft. lbs. (22 Nm).

10. If removed, install the timing belt rear cover and camshaft pulley.

11. If removed, install the timing belt tensioner.

12. Install or connect the following:
- Timing belt and timing belt covers
- Accessory drive belts

13. Close the cooling system drain plug. Refill and bleed the cooling system.

14. Connect the negative battery cable.

15. Start the engine, allow it to reach normal operating temperature, check for leaks, and top off as necessary.

Cylinder Head

REMOVAL & INSTALLATION

1.8L Engines

1. Before servicing the vehicle, refer to the precautions in the beginning of this section.

2. Before removing the cylinder head,

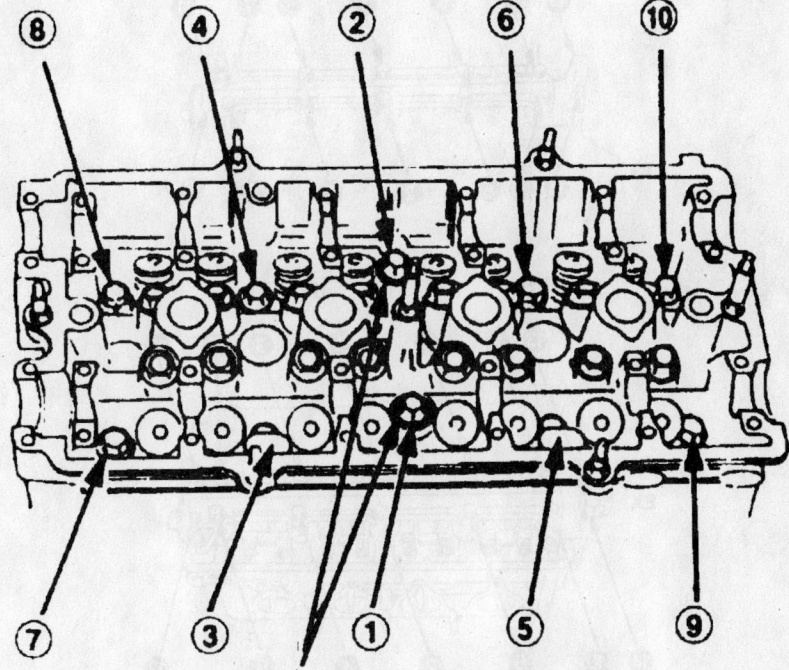

NOTE: Put longer bolts here.

Cylinder head torque sequence—1.8L (B18B1) engine

CYLINDER HEAD BOLT TORQUE SEQUENCE

11 x 1.5 mm
81 N·m (8.3 kgf·m, 60 lbf·ft)

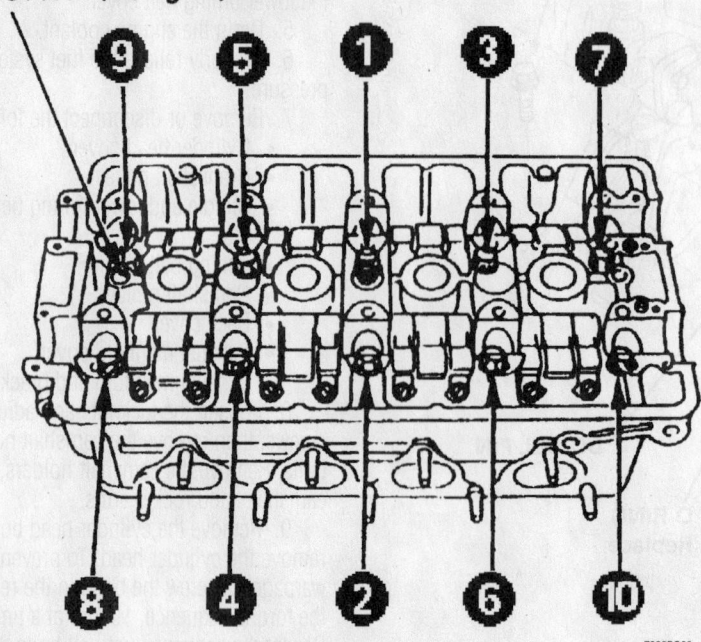

7923BG08

Cylinder head torque sequence—1.8L (B18C1, B18C5) engines

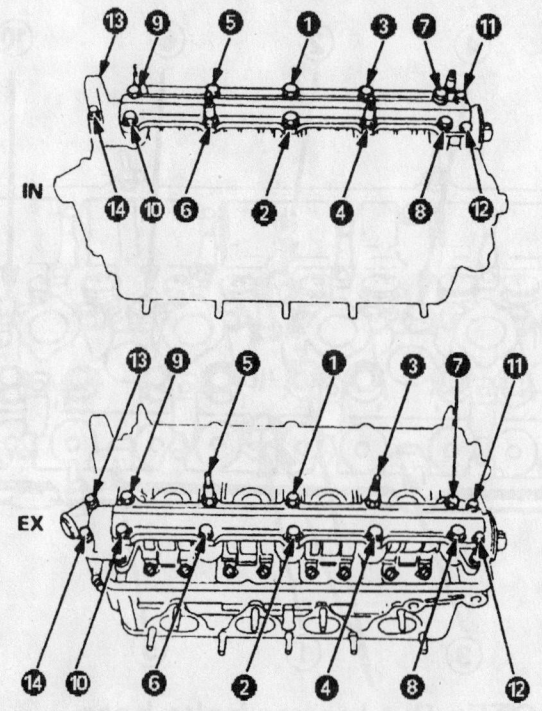

❶ – ❿: 8 x 1.25 mm 27 N·m (2.8 kgf·m, 20 lbf·ft)
⓫ – ⓮: 6 x 1.0 mm 9.8 N·m (1.0 kgf·m, 7.2 lbf·ft)

7923BG09

Camshaft plate torque sequence—1.8L engines

To install:

10. Install the cylinder head with a new gasket. Be sure to pay attention to the following points:
   - Be sure the No. 1 cylinder is at top dead center and the camshaft pulley **UP** mark is on the top before positioning the head in place.
   - The cylinder head dowel pins and oil control orifice must be cleaned and aligned.
   - Replace the washer if damaged or deteriorated.
   - Apply engine oil to the cylinder head bolts and the washers.
   - Use the longer cylinder head bolts at the No. 1 and No. 2 positions.

11. Tighten the cylinder head bolts in 2 steps. In the first step tighten all bolts in sequence to 22 ft. lbs. (29 Nm), then in the second step tighten all bolts in the same sequence to 63 ft. lbs. (85 Nm).

12. Install or connect the following:
   - Intake manifold with a new gasket and torque the bolts in a criss-cross pattern to 17 ft. lbs. (24 Nm)
   - Intake manifold bracket and torque the bolts to 17 ft. lbs. (24 Nm)
   - Exhaust manifold and tighten the new self-locking nuts in a criss-cross pattern in to 23 ft. lbs. (31 Nm)
   - Exhaust pipe with a new gasket and tighten the new nuts to 40 ft. lbs. (54 Nm)

13. Be sure that the keyways on the camshafts are facing up and that the rocker arms are in their original position. The valve locknuts should be loosened and the adjusting screw backed off before installation.

14. Place the rocker arms on the pivot bolts and the valve stems.

15. Install the camshafts, then install the camshaft seals with the open side facing in. Install the rubber cap with liquid gasket applied. If the rubber cap has 2 horizontal marks, align the marks with the cylinder head upper surface.

16. Apply liquid gasket to the cylinder head mating surfaces of the No. 1 and No. 6 intake and exhaust camshaft holders and install them, along with No. 2, 3, 4 and 5. Be sure to pay attention to the following points:
   - **I** or **E** marks are stamped on the camshaft holders.
   - Do not apply oil to the holder mating surface of camshaft seals.
   - The arrows marked on the camshaft holders should point to the timing belt.

17. Tighten the camshaft holders temporarily. Be sure that the rocker arms are properly positioned on the valve stems.

18. Tighten each bolt in 2 steps to ensure that the rockers do not bind on the valves. Tighten the 6mm bolts to 86 inch lbs. (9.8 Nm) and the 8mm bolts to 20 ft. lbs. (27 Nm) working from the middle outward.

19. Install the keys into the camshaft grooves. To set the No. 1 piston at TDC, align the holes on the camshaft with the holes in the No. 1 camshaft holders and insert 5.0mm pin punches into the holes.

20. Install or connect the following:
   • Rear timing belt cover and torque the bolts to 84 inch lbs. (9.5 Nm)
   • Camshaft pulleys and tighten the retaining bolts to 27 ft. lbs. (37 Nm) (B18B1 engine) or 41 ft. lbs. (56 Nm) (B18C1 and B18C5 engines)
   • Timing belt and adjust the tension
   • Timing belt cover(s) and torque the bolts to 84 inch lbs. (9.5 Nm)

21. Adjust the valve clearance.
   • Cylinder head cover and torque the nuts to 86 inch lbs. (9.8 Nm)
   • Engine side mount and torque the mounting bolts to 38 ft. lbs. (52 Nm) and the through bolt to 54 ft. lbs. (74 Nm)
   • Crankshaft pulley and torque the bolt to 130 ft. lbs. (177 Nm)

22. Connect the negative battery cable and enter the radio security code.

23. After installation, check to see that all hoses and wires are installed correctly.

24. Fill and bleed the air from the cooling system.

25. Attach the negative battery cable.

26. Enter the radio security code.

2.3L Engine

1. Before servicing the vehicle, refer to the precautions in the beginning of this section.

2. Disconnect the negative battery cable.

3. Turn the crankshaft so the No. 1 piston is at Top Dead Center (TDC).

➡**The No. 1 piston is at TDC when the pointer on the block aligns with the white painted mark on the flywheel (manual transaxle) or driveplate (automatic transaxle).**

4. Drain the engine coolant into a sealable container.

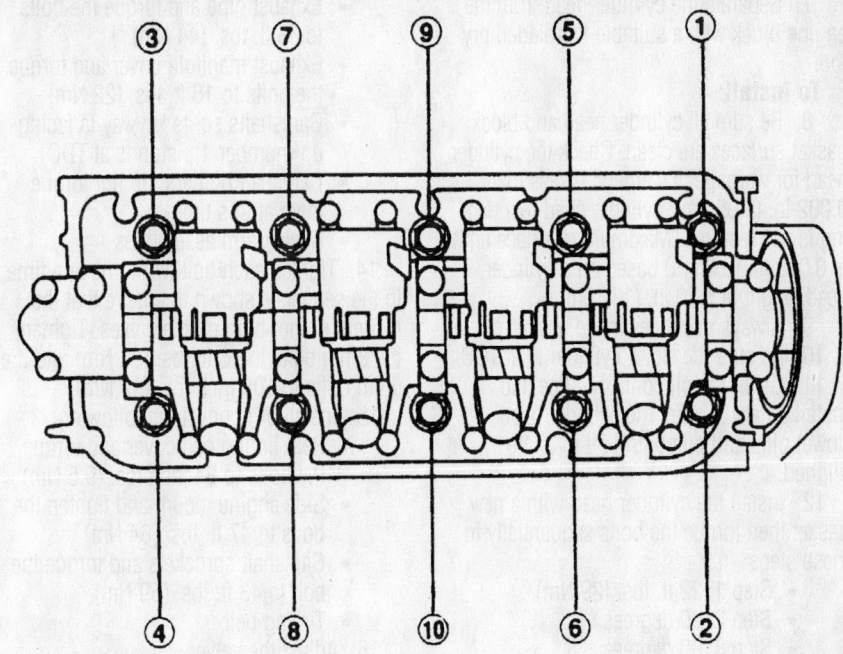

Cylinder head loosening sequence—2.3L engine

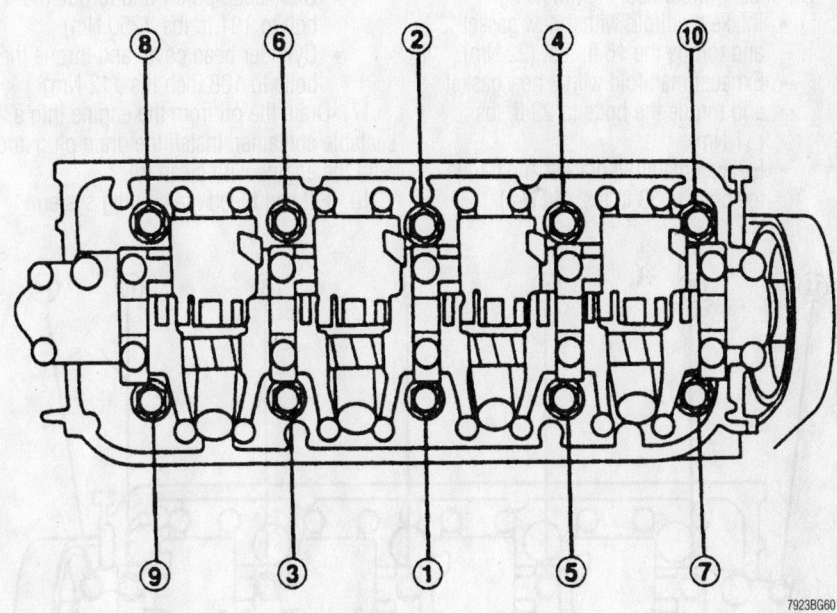

Cylinder head torque sequence—2.3L engine

5. Properly relieve the fuel system pressure.

6. Remove or disconnect the following:
   • Cylinder head cover
   • Crankshaft pulley
   • Timing belt cover
   • Timing belt
   • Camshaft pulley
   • Rear timing belt cover
   • Lower splash shield
   • Intake manifold
   • Exhaust pipe from the exhaust manifold
   • Exhaust manifold

➡**To prevent warpage, unscrew the bolts in sequence ⅓ turn at a time. Repeat the sequence until all bolts are loosened.**

7. Separate the cylinder head from the engine block with a suitable flat bladed pry-tool.

To install:

8. Be sure all cylinder head and block gasket surfaces are clean. Check the cylinder head for warpage. If warpage is less than 0.002 in. (0.05mm), cylinder head resurfacing is not required. Maximum resurface limit is 0.008 in. (0.2mm) based on a cylinder head height of 5.20 in. (132.0mm).

9. Always use a new head gasket.

10. Be sure the No. 1 cylinder is at TDC.

11. Clean the oil control orifice and install a new O-ring. The cylinder head dowel pins and oil control jet must be aligned.

12. Install the cylinder head with a new gasket then torque the bolts sequentially in these steps:

- Step 1: 22 ft. lbs. (29 Nm)
- Step 2: 90 degrees
- Step 3: 90 degrees
- Step 4 (Only if using new bolts): 90 degrees

13. Install or connect the following:

- Intake manifold with a new gasket and torque the 16 ft. lbs. (22 Nm)
- Exhaust manifold with a new gasket and torque the bolts to 23 ft. lbs. (31 Nm)
- Exhaust manifold bracket and torque the bolts to 33 ft. lbs. (44 Nm)

- Exhaust pipe and torque the bolts to 33 ft. lbs. (44 Nm)
- Exhaust manifold cover and torque the bolts to 16 ft. lbs. (22 Nm)
- Camshafts so its keyway is facing up (number 1 piston is at TDC)
- Camshaft holders, do not torque them at this time
- Rocker arm assemblies

14. Tighten each bolt two turns at a time in the sequence shown to ensure that the rockers do not bind on the valves. Tighten the 8mm bolts to 16 ft. lbs. (22 Nm) and the 6mm bolts to 108 inch lbs. (12 Nm).

15. Install or connect the following:

- Rear timing belt cover and torque the bolts to 84 inch lbs. (9.5 Nm)
- Side engine mount and tighten the bolts to 47 ft. lbs. (64 Nm)
- Camshaft sprockets and torque the bolt to 43 ft. lbs. (59 Nm)
- Timing belt

16. Adjust the valves.

- Timing belt cover and torque the bolts to 108 inch lbs. (12 Nm)
- Crankshaft pulley and torque the bolt to 181 ft. lbs. (250 Nm)
- Cylinder head cover and torque the bolts to 108 inch lbs. (12 Nm)

17. Drain the oil from the engine into a sealable container. Install the drain plug and refill the engine with clean oil.

18. Fill and bleed the cooling system.

19. Connect the negative battery cable and enter the radio security code.

20. Start the engine, checking carefully for any leaks.

2.5L Engine

1. Before servicing the vehicle, refer to the precautions in the beginning of this section.

2. Drain the engine coolant.

3. Properly relieve the fuel system pressure.

4. Remove or disconnect the following:

- Negative battery cable
- Cylinder head cover
- Distributor
- Heated Oxygen (HO_2S) sensor connector
- Exhaust manifold heat shields
- Exhaust pipe from the manifold
- Exhaust manifold and bracket
- Intake manifold
- Crankshaft pulley
- Upper and lower timing belt covers
- Timing belt

➡**Replace the belt if it shows any signs of stress or damage.**

- Camshaft position (CMP) sensor
- Camshaft sprocket
- Crankshaft Speed Fluctuation (CKF) sensor
- Rear timing belt cover

5. Loosen each cylinder head bolt about ⅓ turn at a time. Follow the reverse of the installation sequence to prevent warping the head. Repeat until all bolts are loose and can be removed.

6. If the cylinder head is stuck to the block, there are pry points at each end of the cylinder head. Do not pry against the gasket surfaces.

7. Carefully remove the cylinder head from the vehicle.

To install:

8. Verify that the crankshaft and camshaft are both at TDC for the number 1 piston.

9. Install the cylinder head with a new gasket. Be sure the oil control orifice is properly aligned.

10. Lightly oil the threads and washer surfaces of the cylinder head bolts and install them. Tighten the bolts in sequence as follows:

- Step 1: 29 ft. lbs. (39 Nm)
- Step 2: 51 ft. lbs. (69 Nm)
- Step 3: 72 ft. lbs. (981 Nm)

11. Install or connect the following:

- Intake manifold with a new gasket and tighten the nuts in a crisscross

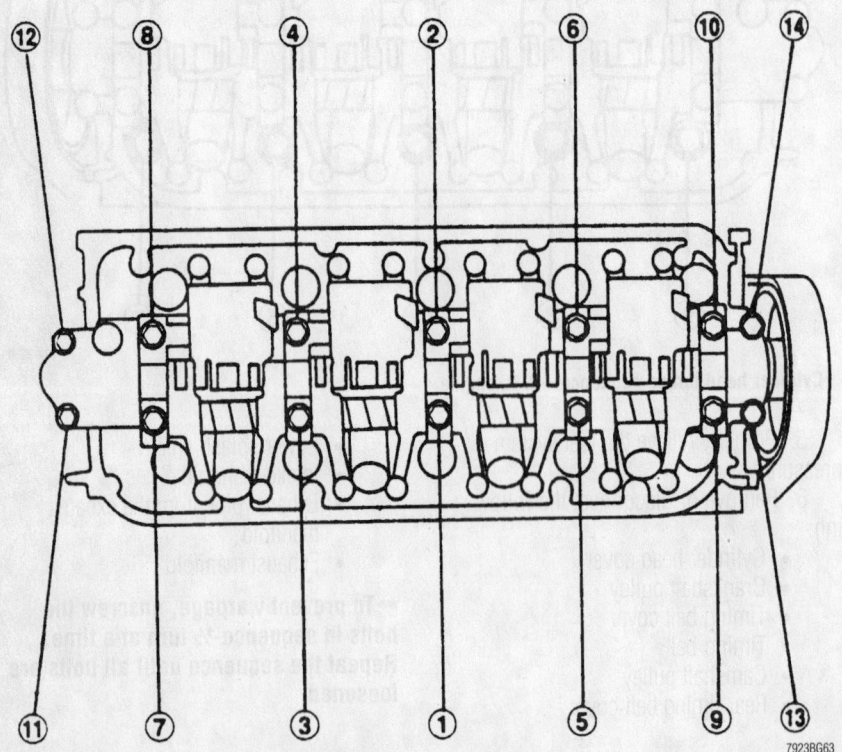

Camshaft holder torque sequence

7923BG63

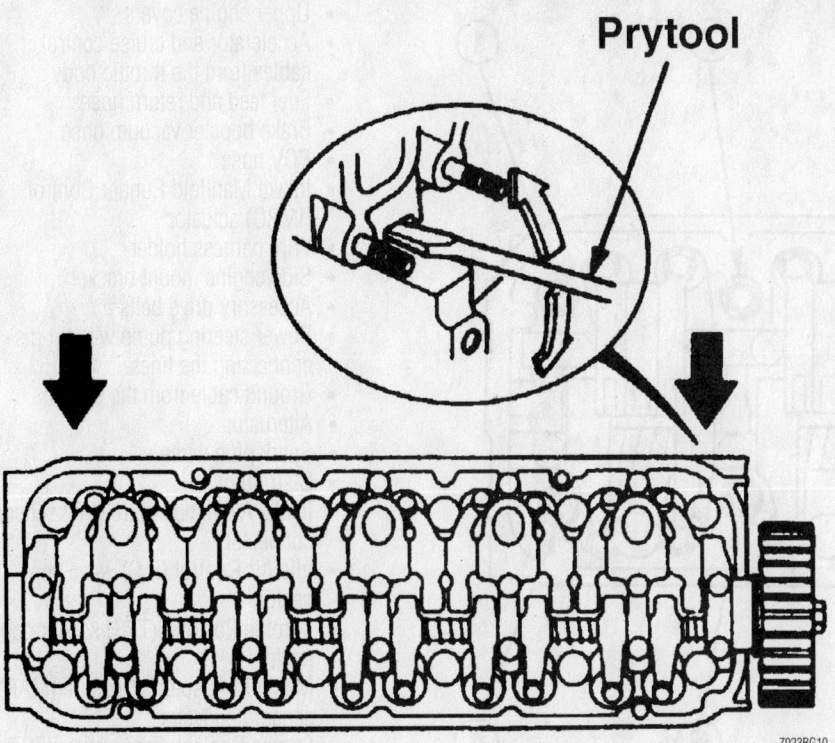

Prytool

7923BG10

Cylinder head prying points—2.5L engine

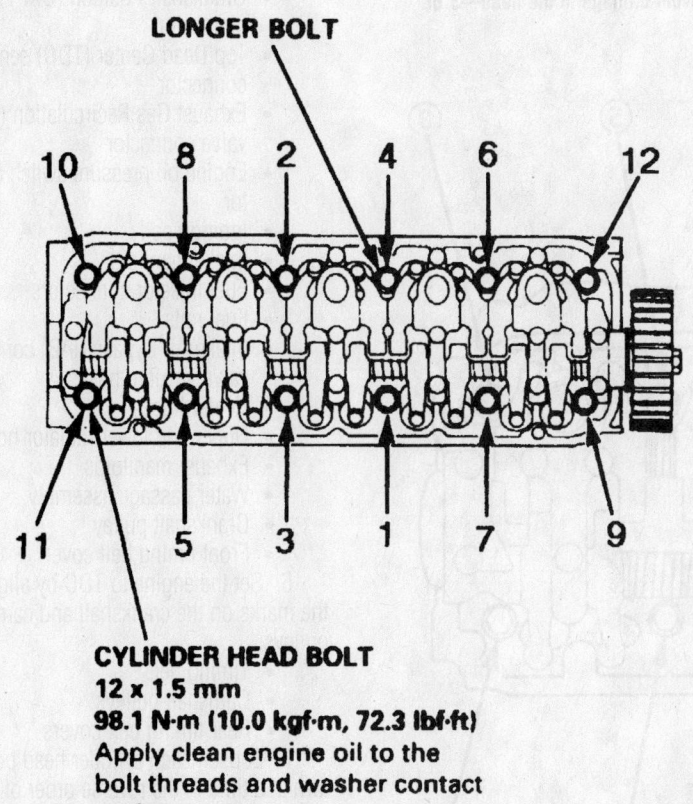

LONGER BOLT

10 8 2 4 6 12

11 5 3 1 7 9

CYLINDER HEAD BOLT
12 x 1.5 mm
98.1 N·m (10.0 kgf·m, 72.3 lbf·ft)
Apply clean engine oil to the
bolt threads and washer contact
surfaces.

7923BG11

Cylinder head torque sequence—2.5L engine

pattern in 2 steps to 16 ft. lbs. (22 Nm)

• Exhaust manifold with a new gasket and torque the nuts in a criss-cross pattern to 23 ft. lbs. (31 Nm)
• Exhaust manifold brackets and torque the nuts to 22 ft. lbs. (29 Nm)
• Exhaust pipe and torque the nuts to 40 ft. lbs. (54 Nm)
• Exhaust manifold shields and torque the 22 ft. lbs. (29 Nm)
• HO_2S sensor
• Rear timing belt cover and torque the bolts to 108 inch lbs. (12 Nm)
• Camshaft pulley and torque the bolt to 51 ft. lbs. (69 Nm)
• CKF sensor and torque the bolts to 108 inch lbs. (12 Nm)
• Timing belt
• Front timing belt cover and torque the bolts to 108 inch lbs. (12 Nm)
• Crankshaft pulley and torque the bolt to 181 ft. lbs. (245 Nm)

12. Adjust the valves.

13. Apply sealant to the ends of the cylinder head near the camshaft holders. Install the cylinder head cover with new rubber seals as required and torque the nuts to 108 inch lbs. (12 Nm)

14. Replace the engine oil and filter.

15. Fill and bleed the cooling system.

16. Verify that all wiring, grounds, hoses, and cables are properly connected.

17. Connect the negative battery cable. Run the engine and check for leaks.

18. Enter the radio security code.

3.0L, 3.2 and 3.5L Engines

1. Before servicing the vehicle, refer to the precautions in the beginning of this section.

2. Disconnect the negative battery cable.

3. Drain the coolant.

4. Relieve the fuel system pressure.

5. Remove or disconnect the following:

• Engine covers
• Strut brace
• Water bypass hose
• Traction Control System (TCS) control valve from the throttle body (if equipped)
• Evaporative Emissions (EVAP) canister hose from the throttle body
• Intake air duct

For Accessory Drive Belt illustrations, see Section 1 of this manual

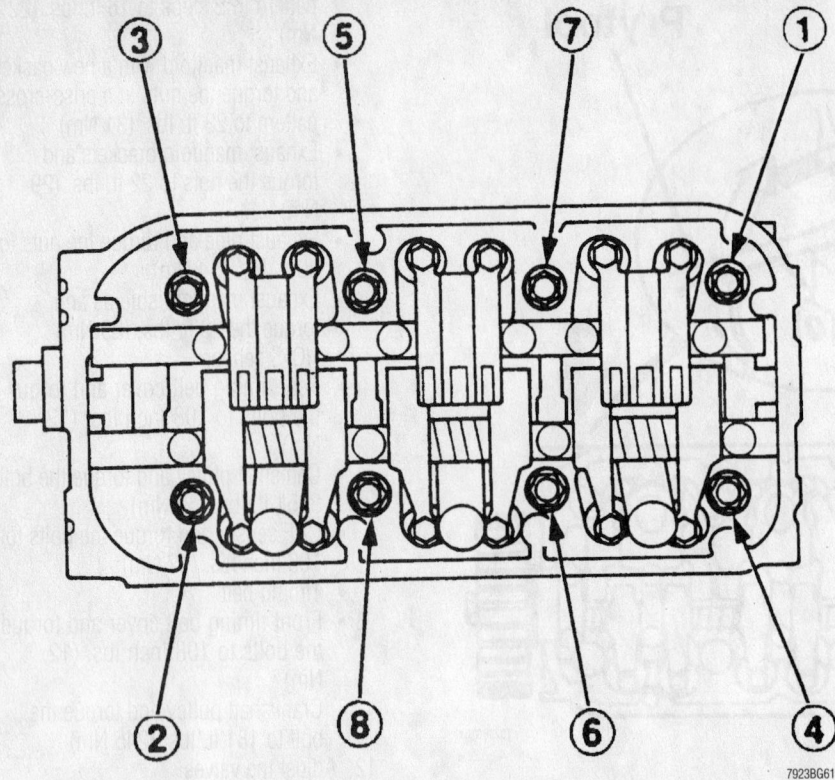

Loosen the cylinder head bolts in the sequence shown to prevent damage to the head—3.0L engine

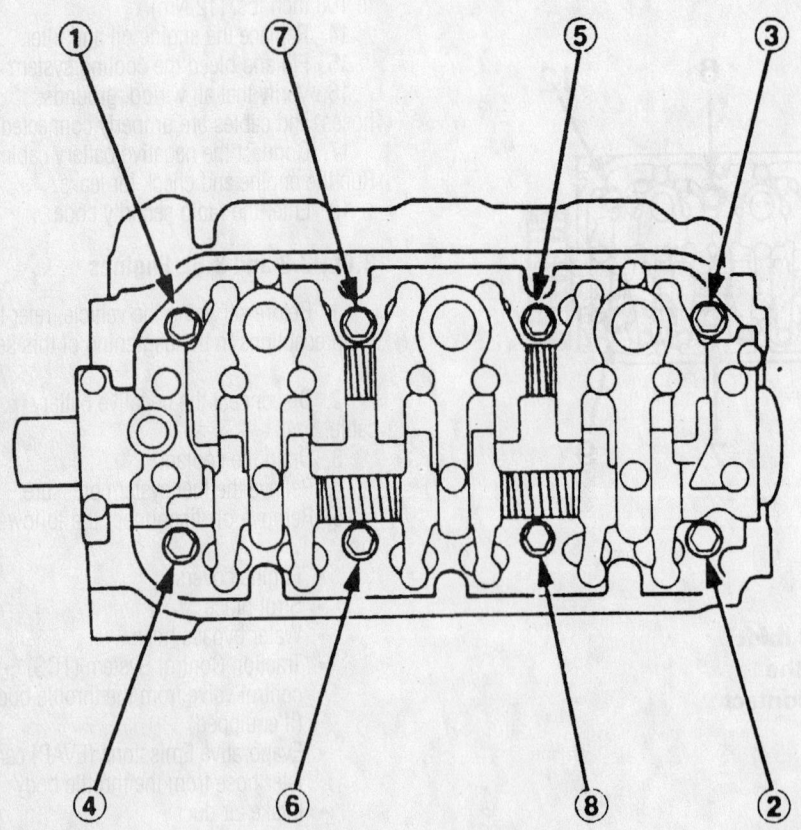

Loosen the cylinder head bolts in the sequence shown—3.5L engine

- Upper engine covers
- Accelerator and cruise control cables from the throttle body
- Fuel feed and return hoses
- Brake booster vacuum hose
- PCV hose
- Intake Manifold Runner Control (IMRC) actuator
- Wire harness holder
- Side engine mount bracket
- Accessory drive belts
- Power steering pump without disconnecting the lines
- Ground cable from the engine
- Alternator
- Spark plug wires
- Distributor
- Intake Air Temperature (IAT) sensor connector
- Idle Air Control (IAC) valve connector
- Throttle Position (TPS) sensor connector
- Manifold Absolute Pressure (MAP) sensor connector
- Engine Coolant Temperature (ECT) sensor connector
- Radiator fan switch connectors
- Crankshaft Position (CKP) sensor connector
- Top Dead Center (TDC) sensor connector
- Exhaust Gas Recirculation (EGR) valve connector
- Engine oil pressure switch connector
- Ignition coils
- Intake manifold
- Fuel injector connectors
- Fuel rails
- Intake Air Bypass (IAB) control valve vacuum hoses
- Heater hoses
- Upper and lower radiator hoses
- Exhaust manifolds
- Water passage assembly
- Crankshaft pulley
- Front timing belt cover

6. Set the engine to TDC by aligning the marks on the crankshaft and camshaft pulleys.

- Timing belt
- Camshaft pulleys
- Rear timing belt covers

7. Loosen each cylinder head bolt ⅓ turn at a time in the reverse order of the tightening sequence.

8. Remove the cylinder heads and the oil control orifices.

To install:

9. Install the oil control orifices (A) using new O-rings (B).

10. If removed, install the dowel pins (C).

11. Position new cylinder head gaskets (D) on the cylinder block.

12. If moved, set the crankshaft and camshaft pulleys to TDC by aligning the marks on the pulley and oil pump.

13. Carefully position the cylinder heads on the engine.

14. Lubricate the cylinder head bolts with clean engine oil.

15. Tighten the cylinder head bolts in 3 steps. Be sure to follow the tightening torque sequence.
* Step 1: 29 ft. lbs. (39 Nm)
* Step 2: 51 ft. lbs. (69 Nm)
* Step 3: 72 ft. lbs. (98 Nm)

16. Install or connect the following:
* Exhaust manifolds with new gaskets and torque the nuts to 23 ft. lbs. (31 Nm)
* Rear timing belt covers and torque the bolts to 16 ft. lbs. (22 Nm)

17. Install the camshaft pulleys and torque to:
* 3.0L and 3.2L engines: 67 ft. lbs. (90 Nm)
* 3.5 L engines: 23 ft. lbs. (31 Nm)

18. Install or connect the following:
* Timing belt
* Front timing belt cover and torque the bolts to 108 inch lbs. (12 Nm)

19. Check and adjust the valve clearance.
* Crankshaft pulley and torque the bolt to 181 ft. lbs. (245 Nm)
* Water passage assembly and torque the bolts to 16 ft. lbs. (22 Nm)
* Intake manifold with new gaskets and O-rings and tighten the bolts to 16 ft. lbs. (22 Nm)
* Cylinder head cover and tighten the bolts to 108 inch lbs. (12 Nm)
* Ignition coils and torque the bolts to 108 inch lbs. (12 Nm)
* Exhaust manifolds and torque the bolts to 23 ft. lbs. (31 Nm)
* Upper and lower radiator hoses
* Heater hoses
* IAB vacuum hoses
* Fuel rails and torque the bolts to 84 inch lbs. (9.5 Nm)
* Fuel injector connectors
* Intake manifold and torque the bolts to 16 ft. lbs. (22 Nm)
* Engine oil pressure switch connector

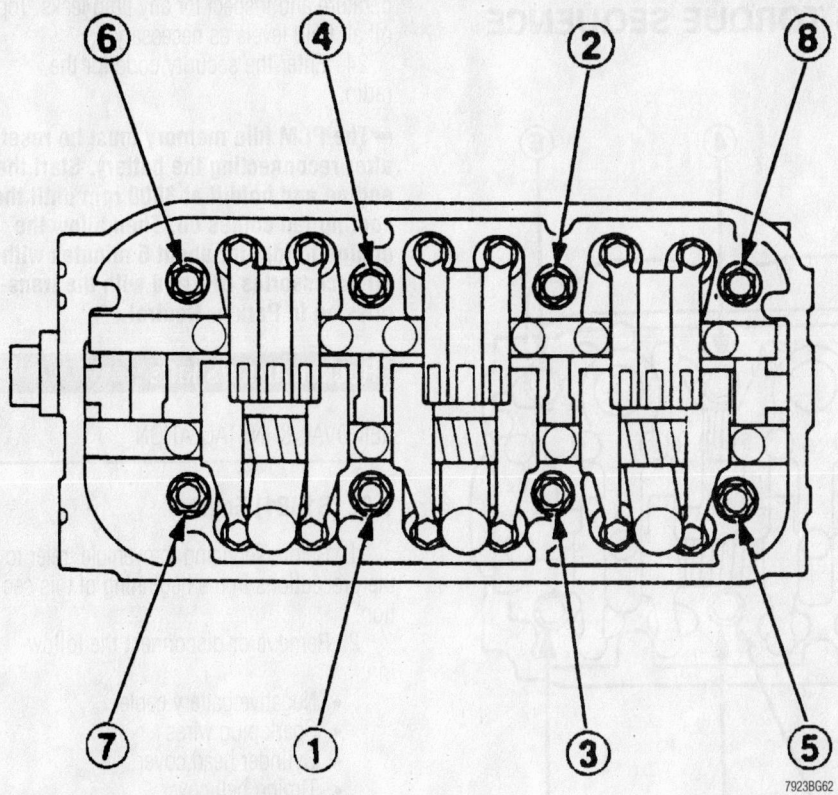

Cylinder head torque sequence—3.0L engine

7923BG62

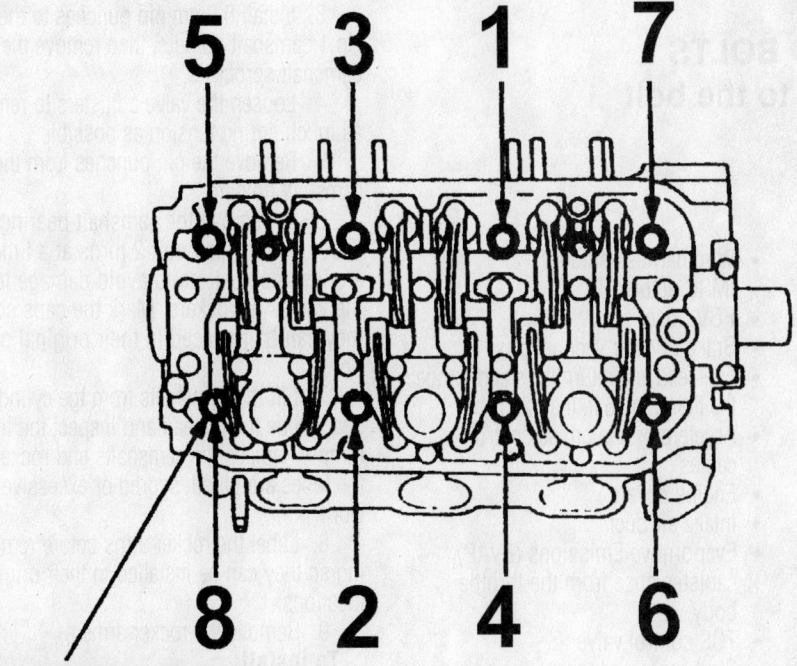

CYLINDER HEAD BOLTS
Apply engine oil to the bolt threads.

Cylinder head torque sequence—3.2L engine

7923BG12

For Tire, Wheel and Ball Joint specifications, see Section 1 of this manual

CYLINDER HEAD BOLTS TORQUE SEQUENCE

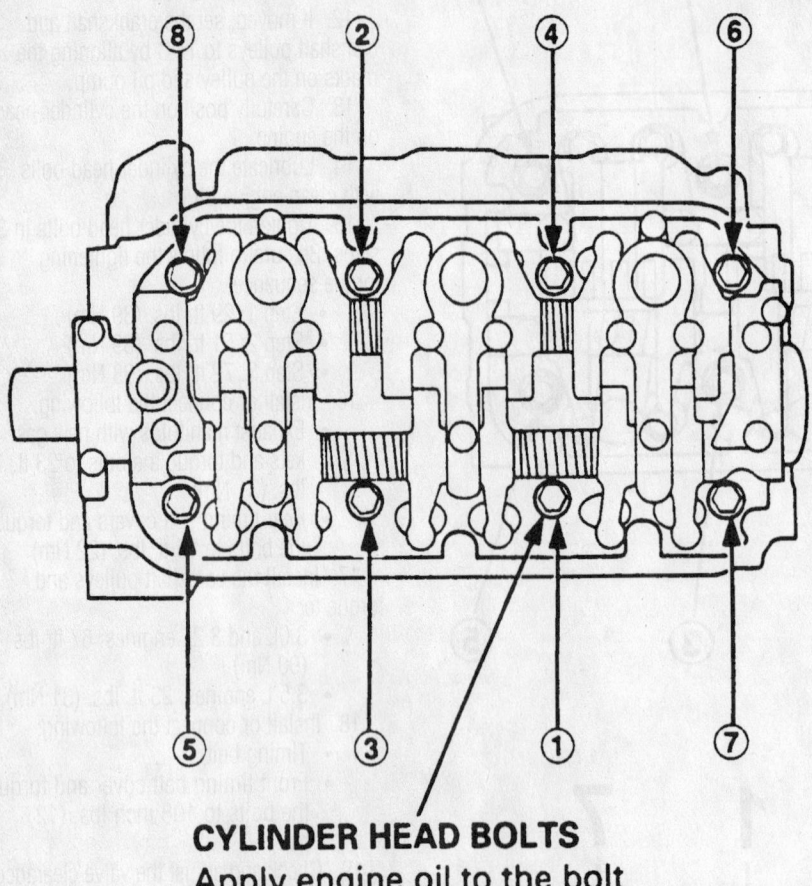

CYLINDER HEAD BOLTS
Apply engine oil to the bolt threads.

7923BG82

Cylinder head torque sequence—3.5L engine

- EGR valve connector
- TDC sensor connector
- CKP sensor connector
- Radiator fan switch connectors
- MAP sensor connector
- ECT sensor connector
- TPS sensor connector
- IAC sensor connector
- IAT sensor connector
- Spark plug wires
- Distributor
- Alternator and torque the upper bolt to 16 ft. lbs. (22 Nm) and the lower bolt to 33 ft. lbs. (44 Nm)
- Ground cable
- Power steering pump and torque the bolts to 17 ft. lbs. (24 Nm)
- Accessory drive belts
- Side engine mount and torque the mounting bolts to 33 ft. lbs. (44 Nm) and the through bolt to 40 ft. lbs. (54 Nm)

- Wire harness holder
- IMRC actuator
- PCV hose
- Brake booster vacuum hose
- Fuel feed and return lines and torque the fitting to 16 ft. lbs. (22 Nm)
- Accelerator and cruise control cables
- Engine covers
- Intake air duct
- Evaporative Emissions (EVAP) canister hose from the throttle body
- TCS control valve
- Strut brace and torque the bolts to 16 ft. lbs. (22 Nm)
- Engine covers and torque the bolts to 108 inch lbs. (12 Nm)
20. Change the engine oil and filter.
21. Fill and bleed the cooling system.
22. Connect the negative battery cable.
23. Bring the engine to operating tem-

perature and inspect for any fluid leaks. Top off all fluid levels as necessary.

24. Enter the security code for the radio.

➡**The PCM idle memory must be reset after reconnecting the battery. Start the engine and hold it at 3000 rpm until the cooling fan comes on. Then allow the engine to idle for about 5 minutes with all accessories OFF and with the transmission in Park or Neutral.**

Rocker Arms/Shafts

REMOVAL & INSTALLATION

1.8L (B18B1) Engine

1. Before servicing the vehicle, refer to the precautions in the beginning of this section.
2. Remove or disconnect the following:

- Negative battery cable
- Spark plug wires
- Cylinder head cover
- Timing belt cover
- Timing belt. Refer to the Timing Belt unit repair section.
- Distributor

3. Install 5.0mm pin punches to the No.1 camshaft holders, then remove the camshaft sprockets.
4. Loosen the valve adjusters to remove as much spring tension as possible.
5. Remove the pin punches from the camshaft holders.
6. To remove the camshaft bearing caps, loosen each bolt 2 turns at a time in a crisscross pattern to avoid damage to the valves or rockers. Mark the caps so they can be replaced in their original position.
7. Lift the camshafts from the cylinder head, wipe them clean and inspect the lift ramps. Replace the camshafts and rockers if the lobes are pitted, scored or excessively worn.
8. Label the rocker arms before removing so they can be installed in their original locations.
9. Remove the rocker arms.
To install:
10. Check the following before installing the camshafts:

   a. Be certain the keyways on the camshafts are facing UP (No. 1 cylinder at TDC).
   b. The valve adjuster locknuts should be loosened and the adjusting screws backed off before installation.

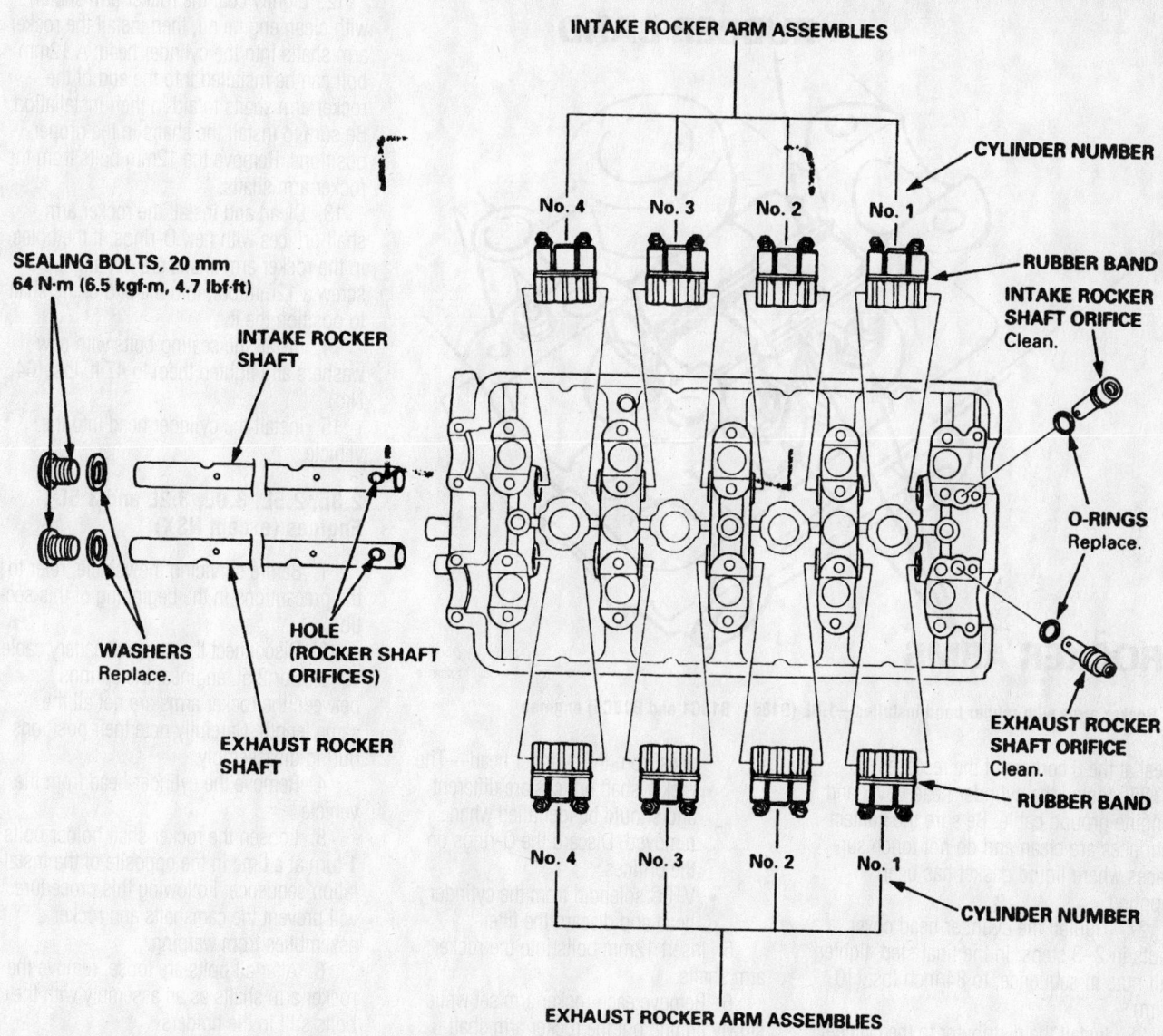

INTAKE ROCKER ARM ASSEMBLIES

CYLINDER NUMBER

No. 4   No. 3   No. 2   No. 1

RUBBER BAND
INTAKE ROCKER SHAFT ORIFICE
Clean.

SEALING BOLTS, 20 mm
64 N·m (6.5 kgf·m, 4.7 lbf·ft)

INTAKE ROCKER SHAFT

O-RINGS
Replace.

WASHERS
Replace.

HOLE (ROCKER SHAFT ORIFICES)

EXHAUST ROCKER SHAFT

EXHAUST ROCKER SHAFT ORIFICE
Clean.

RUBBER BAND

No. 4   No. 3   No. 2   No. 1

CYLINDER NUMBER

EXHAUST ROCKER ARM ASSEMBLIES

7923BG13

Rocker arms and shafts—1.8L (B18B1, B18C1, and B18C5) engines

11. Lubricate the rocker arms and camshafts with clean engine oil.

12. Place the rocker arms on the pivot bolts and the valve stems, making sure that the rocker arms are in their original positions.

13. Install the camshaft seals with the open side (spring) facing in. Lubricate the lip of the seal.

14. Be sure the keyways on the camshafts are facing up and install the camshafts to the cylinder head.

15. Apply liquid gasket to the head mating surfaces of the No. 1 and No. 6 camshaft holders, then install them along with Nos. 2, 3, 4 and 5 camshaft holders.

The arrows stamped on the holders should point toward the timing belt. Do not apply oil to the holder mating surface where the camshaft seals are housed.

16. Tighten the camshaft holders temporarily and be sure that the rocker arms are properly positioned.

17. Press the oil seals into the No.1 camshaft holders with a seal driver.

18. Tighten the bolts in a crisscross pattern to 84 inch lbs. (10 Nm). Check that the rockers do not bind on the valves.

19. Install the cylinder head plug to the end of the cylinder head. If the plug has alignment marks, align the marks.

20. Install the rear timing belt cover and tighten the bolts to 108 inch lbs. (12 Nm).

21. Install 5.0mm pin punches to the No.1 camshaft holders, then install the camshaft pulley keys onto the grooves in the camshafts.

22. Push the camshaft pulleys onto the camshafts, then tighten the retaining bolts to 27 ft. lbs. (38 Nm).

23. Install the timing belt and timing belt covers. Remove the pin punches from the camshaft holders.

24. Adjust the valves and pour oil over the camshafts and rocker arms.

25. Apply liquid gasket to the rubber

RUBBER BAND

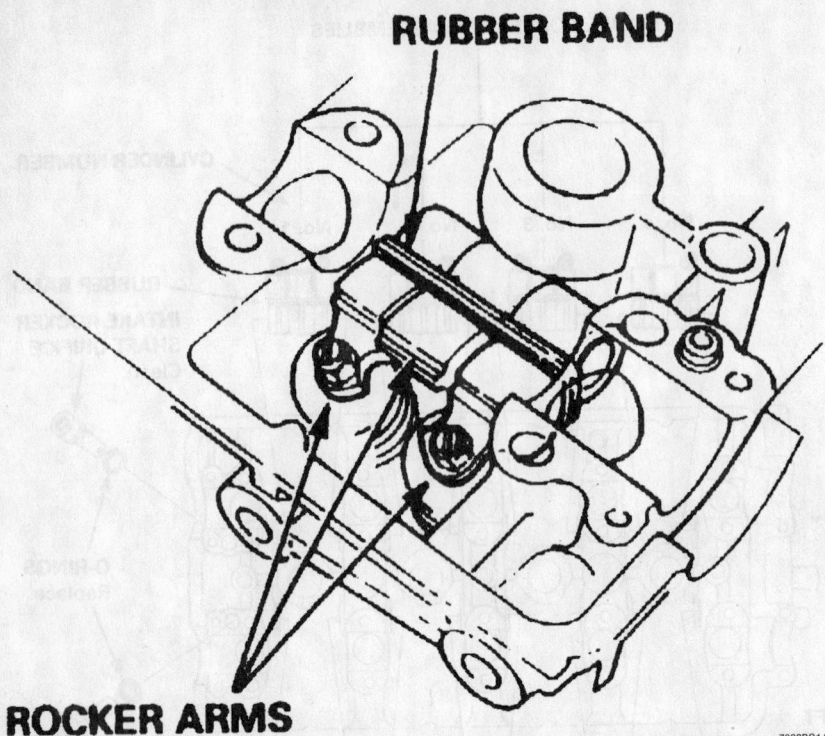

ROCKER ARMS

Rocker arms with rubber band installed—1.8L (B18B1, B18C1 and B18C5) engines

7923BG14

seal at the 8 corners of the recesses.

26. Install the cylinder head cover and engine ground cable. Be sure the contact surfaces are clean and do not touch surfaces where liquid gasket has been applied.

27. Tighten the cylinder head cover nuts in 2–3 steps. In the final step, tighten all nuts in sequence, to 84 inch lbs. (10 Nm).

28. Install the distributor to the cylinder head and reconnect the spark plug wires to the spark plugs.

29. Connect the negative battery cable and enter the radio security code.

30. Change the engine oil. Wait at least 20 minutes for the sealant to cure before filling the engine with oil.

1.8L (B18C1, B18C5) Engine and NSX

1. Before servicing the vehicle, refer to the precautions in the beginning of this section.

2. Remove or disconnect the following:
   • Negative battery cable
   • Cylinder head from the vehicle

3. Hold each rocker arm assembly together with a rubber band to prevent them from separating.

4. Remove or disconnect the following:
   • Intake and exhaust rocker shaft ori-

fices from the cylinder head—The rocker shaft orifices are different and should be identified when removed. Discard the O-rings on the orifices.
   • VTEC solenoid from the cylinder head and discard the filter

5. Insert 12mm bolts into the rocker arm shafts.

6. Remove each rocker arm set while slowly pulling out the rocker arm shaft.

➡️**Tag each rocker arm set to assure installation in their original locations.**

7. Inspect the rocker arm pistons. If they do not move smoothly, replace the rocker arm assembly.

8. Remove the lost motion assembly from the cylinder head. Inspect the lost motion assembly by pushing the plunger with your finger. Replace the lost motion assembly if it does not move smoothly.

To install:

9. Install the lost motion assembly to the cylinder head.

10. Apply engine oil to the rocker arm pistons, then bundle the rocker arms with a rubber band. Apply a light coat of clean engine oil to the rocker arms.

11. Position the rocker arms in their original locations, if they are being reused. If new assembles are being used place them in the cylinder head.

12. Lightly coat the rocker arm shafts with clean engine oil, then install the rocker arm shafts into the cylinder head. A 12mm bolt can be installed into the end of the rocker arm shafts to aid in their installation. Be sure to install the shafts in the proper positions. Remove the 12mm bolts from the rocker arm shafts.

13. Clean and install the rocker arm shaft orifices with new O-rings. If the holes in the rocker arm shafts are not aligned screw a 12mm bolt into the end of the shaft to position the it.

14. Install the sealing bolts with new washers and tighten them to 47 ft. lbs. (64 Nm).

15. Install the cylinder head into the vehicle.

2.3L, 2.5L, 3.0L, 3.2L and 3.5L Engines (except NSX)

1. Before servicing the vehicle, refer to the precautions in the beginning of this section.

2. Disconnect the negative battery cable.

3. For 2.5L engines, the springs between the rocker arms are not all the same length. Carefully note their positions during disassembly.

4. Remove the cylinder head from the vehicle.

5. Loosen the rocker shaft holder bolts 1 turn at a time in the opposite of the installation sequence. Following this procedure will prevent the camshafts and rocker assemblies from warping.

6. After all bolts are loose, remove the rocker arm shafts as an assembly with the bolts still in the holders.

7. If the rocker shafts are to be disassembled, note that each rocker arm has a letter **A** or **B** stamped into the side. Before disassembling the rocker arms, make a note of the position of each letter so the arms can be reassembled the same way.

8. For 3.2L and 3.5L engines, do not remove the hydraulic tappets from the rocker arms unless they are to be replaced. Handle the rocker arms carefully so the oil does not drain out of the tappets.

9. Lift the camshafts from the cylinder head, wipe them clean and inspect the lift ramps. Replace the camshafts and rockers if the lobes are pitted, scored, or excessively worn.

To install:

10. Lubricate the camshaft and its journals with fresh engine oil.

11. Place a new camshaft seal on the end of the camshaft. The spring side of the seal must face in. Lubricate the jour-

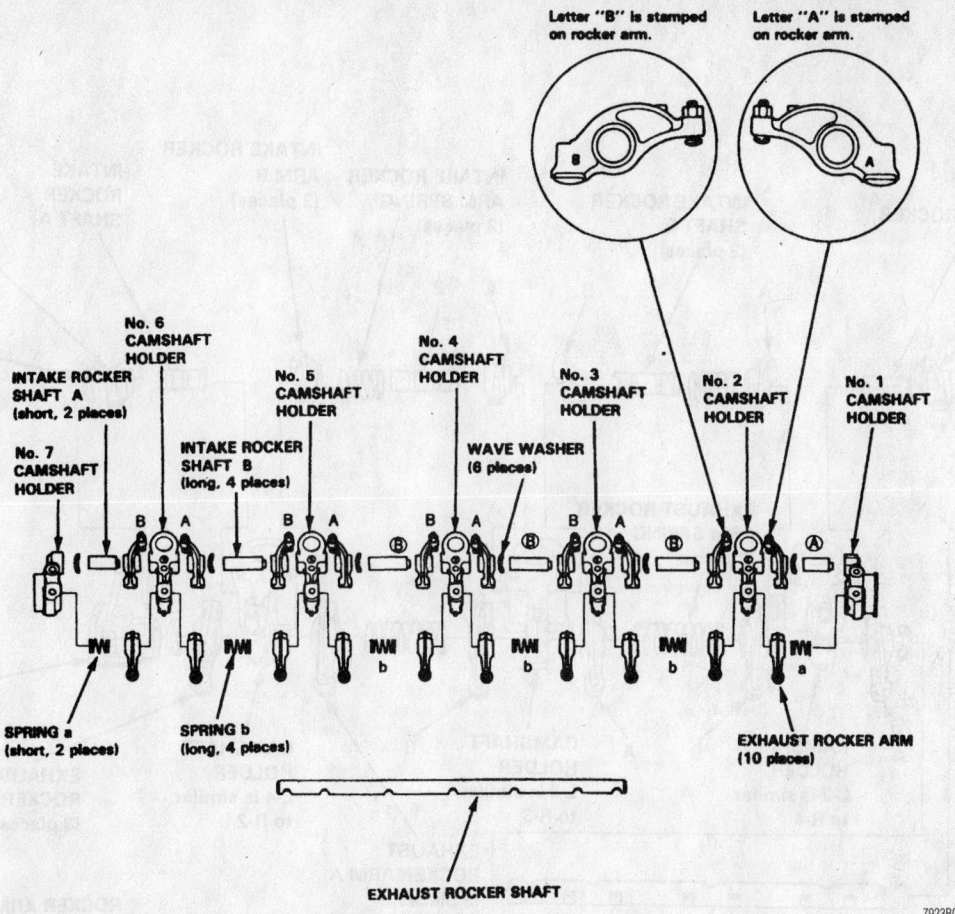

Letter "B" is stamped on rocker arm.

Letter "A" is stamped on rocker arm.

Rocker arm and shaft assembly—2.5L engine

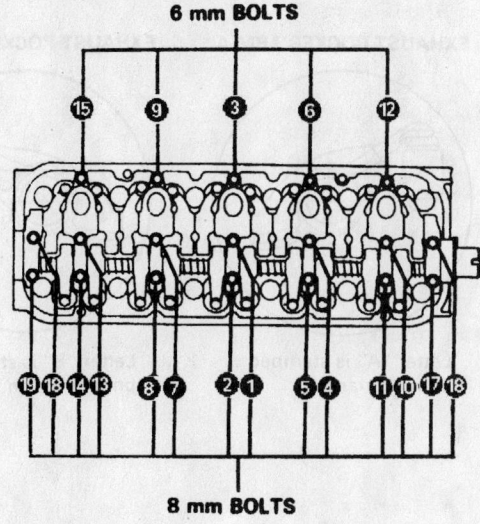

Specified torque:
8 mm bolts: 22 N·m (2.2 kg-m, 16 lb-ft)
6 mm bolts: 12 N·m (1.2 kg-m, 9 lb-ft)

6 mm BOLTS

8 mm BOLTS

Rocker arm assembly holder bolt torque sequence—2.5L engine

nals and set the camshaft in place on the head.

12. Install the camshaft onto the cylinder head with the keyway pointed up.

13. Apply liquid gasket to the mounting surfaces of the camshaft end holders.

14. Set the rocker arm assemblies in place and start all the cam holder bolts. Be sure the rocker arms are properly positioned and turn each bolt in sequence 2 turns at a time until the holders are seated on the head. Follow this procedure to avoid damaging the camshaft and rocker assemblies.

15. When all the camshaft and rocker holders are seated, tighten the bolts in the same sequence. Tighten the 8mm bolts to 16 ft. lbs. (22 Nm) and the 6mm bolts to 104 inch lbs. (12 Nm).

16. Install or connect the following:
   • Cylinder head
   • Distributor (2.5L engines)
   • Negative battery cable

17. Check for proper engine and valve train operation.

For Maintenance Interval recommendations, see Section 1 of this manual

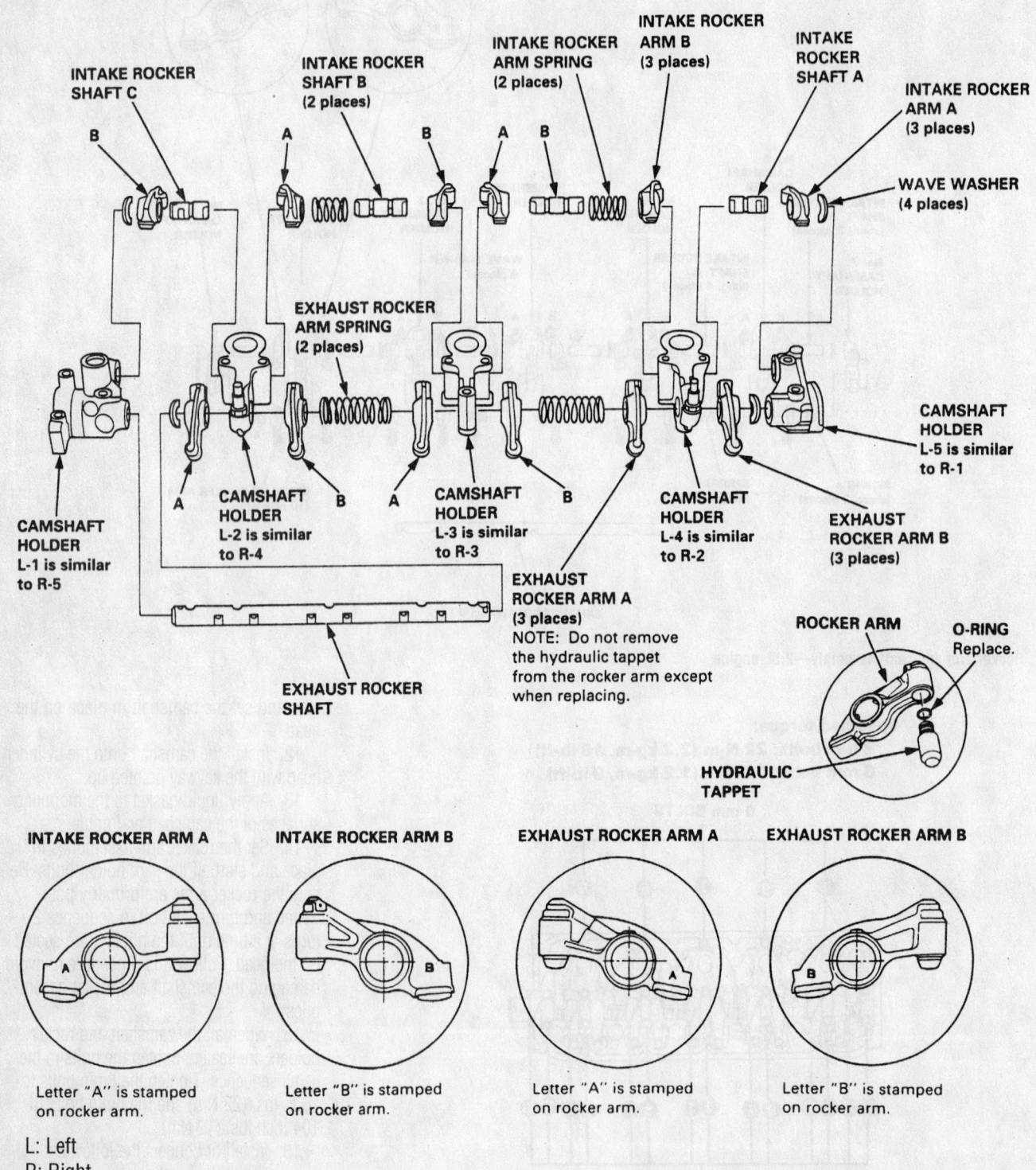

INTAKE ROCKER
SHAFT C

INTAKE ROCKER
SHAFT B
(2 places)

INTAKE ROCKER
ARM SPRING
(2 places)

INTAKE ROCKER
ARM B
(3 places)

INTAKE
ROCKER
SHAFT A

INTAKE ROCKER
ARM A
(3 places)

WAVE WASHER
(4 places)

EXHAUST ROCKER
ARM SPRING
(2 places)

CAMSHAFT
HOLDER
L-5 is similar
to R-1

CAMSHAFT
HOLDER
L-1 is similar
to R-5

CAMSHAFT
HOLDER
L-2 is similar
to R-4

CAMSHAFT
HOLDER
L-3 is similar
to R-3

CAMSHAFT
HOLDER
L-4 is similar
to R-2

EXHAUST
ROCKER ARM B
(3 places)

EXHAUST
ROCKER ARM A
(3 places)
NOTE: Do not remove
the hydraulic tappet
from the rocker arm except
when replacing.

EXHAUST ROCKER
SHAFT

ROCKER ARM

O-RING
Replace.

HYDRAULIC
TAPPET

INTAKE ROCKER ARM A

Letter "A" is stamped
on rocker arm.

INTAKE ROCKER ARM B

Letter "B" is stamped
on rocker arm.

EXHAUST ROCKER ARM A

Letter "A" is stamped
on rocker arm.

EXHAUST ROCKER ARM B

Letter "B" is stamped
on rocker arm.

L: Left
R: Right

7923BG83

Exploded view of the rocker arms and related components—3.5L engine

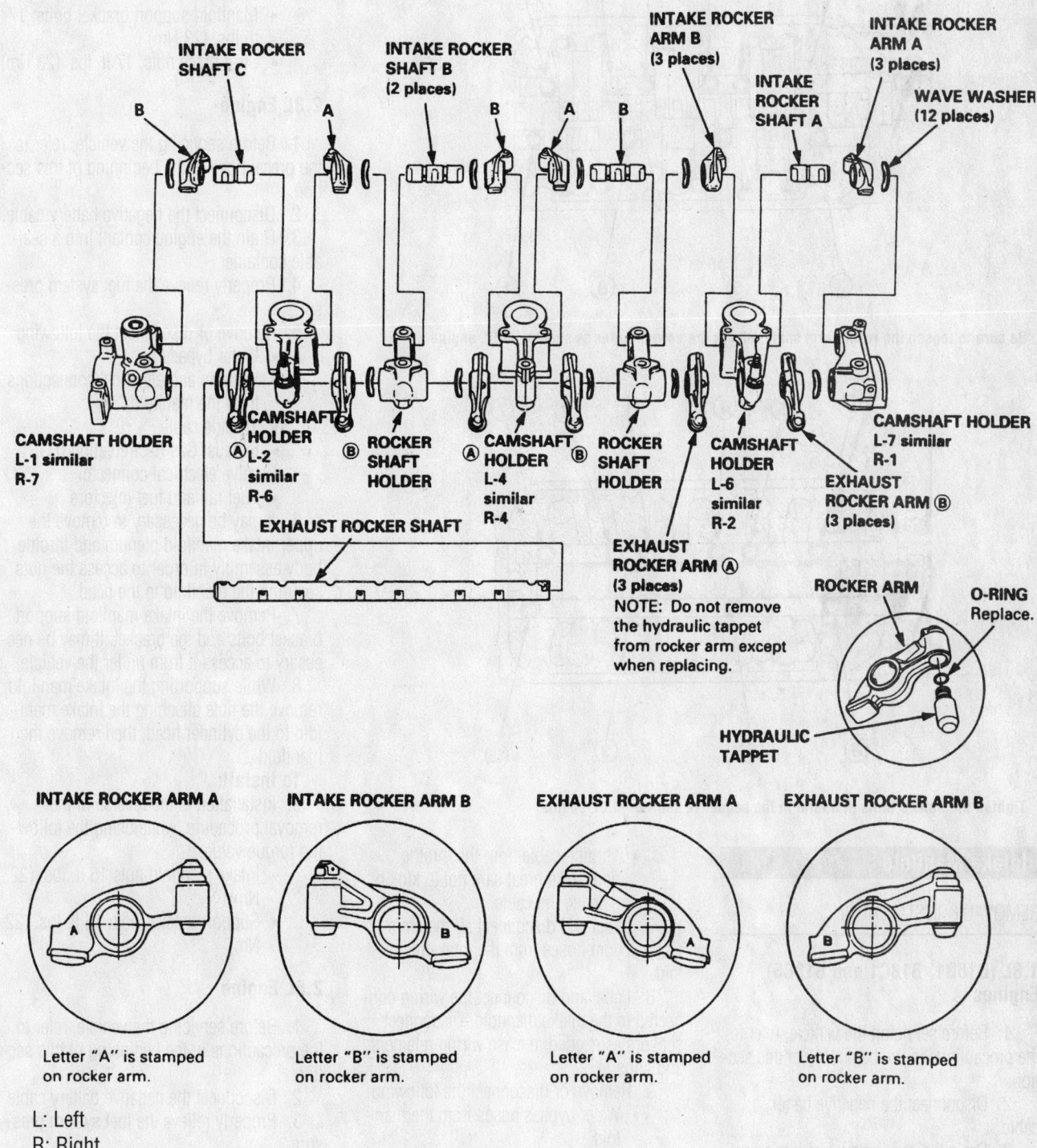

INTAKE ROCKER SHAFT C

INTAKE ROCKER SHAFT B (2 places)

INTAKE ROCKER ARM B (3 places)

INTAKE ROCKER ARM A (3 places)

INTAKE ROCKER SHAFT A

WAVE WASHER (12 places)

CAMSHAFT HOLDER L-1 similar R-7

CAMSHAFT HOLDER Ⓐ L-2 similar R-6

Ⓑ ROCKER SHAFT HOLDER

CAMSHAFT HOLDER L-4 similar R-4

Ⓑ ROCKER SHAFT HOLDER

CAMSHAFT HOLDER L-6 similar R-2

CAMSHAFT HOLDER L-7 similar R-1

EXHAUST ROCKER ARM Ⓑ (3 places)

EXHAUST ROCKER SHAFT

EXHAUST ROCKER ARM Ⓐ (3 places)
NOTE: Do not remove the hydraulic tappet from rocker arm except when replacing.

ROCKER ARM

O-RING Replace.

HYDRAULIC TAPPET

INTAKE ROCKER ARM A

Letter "A" is stamped on rocker arm.

INTAKE ROCKER ARM B

Letter "B" is stamped on rocker arm.

EXHAUST ROCKER ARM A

Letter "A" is stamped on rocker arm.

EXHAUST ROCKER ARM B

Letter "B" is stamped on rocker arm.

L: Left
R: Right

7923BG84

Exploded view of the rocker arms and related components—3.2L engine

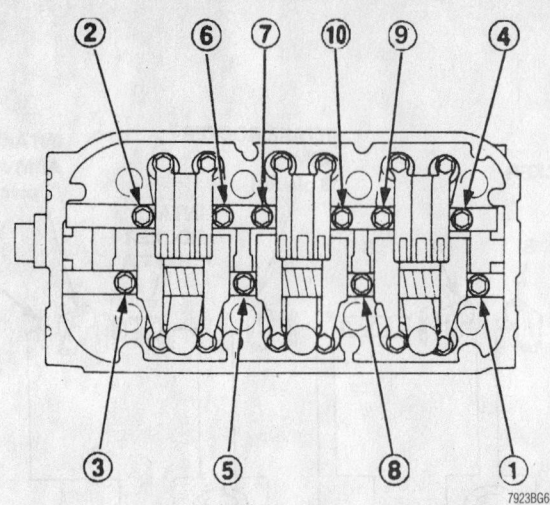

Be sure to loosen the rocker arm shaft bolts in the correct order as shown—3.0L engine

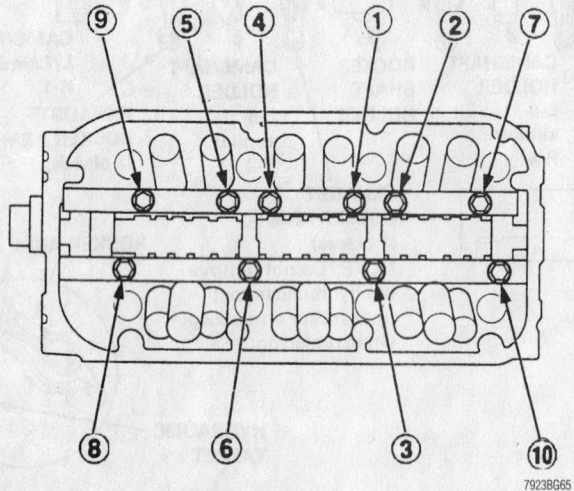

Tighten the bolts 2 turns at a time in the sequence shown—3.0L engine

Intake Manifold

REMOVAL & INSTALLATION

1.8L (B18B1, B18C1 and B18C5) Engines

1. Before servicing the vehicle, refer to the precautions in the beginning of this section.

2. Disconnect the negative battery cable.

3. Drain the cooling system into a sealable container.

4. Remove the strut brace (if equipped).

5. Properly relieve the fuel pressure.

6. Remove or disconnect the following:
- Intake air duct
- Fuel feed and return hoses
- PCV hose
- Brake booster vacuum hose

- Throttle cable from the throttle body, take great care not to kink or damage the cable

7. Label and disconnect all the emission vacuum hoses from the intake manifold.

8. Label and disconnect the wiring connected to the intake manifold. Disconnect sensors as needed; release wiring retainers and clips.

9. Remove or disconnect the following:
- Water bypass hoses from the manifold
- Attaching the intake manifold to the support bracket
- Nuts attaching the intake manifold to the cylinder head in a crisscross pattern, beginning from the center and moving out to both ends
- Intake manifold

To install:

10. Installation is the reverse of the

removal procedure, while using the following torque values:
- Intake manifold nuts:17 ft. lbs. (23 Nm)
- Manifold support bracket bolts: 17 ft. lbs. (23 Nm)
- Strut brace nuts: 17 ft. lbs. (23 Nm)

2.3L Engine

1. Before servicing the vehicle, refer to the precautions in the beginning of this section.

2. Disconnect the negative battery cable.

3. Drain the engine coolant into a sealable container.

4. Properly relieve the fuel system pressure.

5. Remove or disconnect the following:
- Water bypass hose
- Vacuum and electrical connections from the manifold
- Throttle cable
- Exhaust Gas Recirculation (EGR) valve electrical connector
- Fuel rail and fuel injectors

6. It may be necessary to remove the upper intake manifold plenum and throttle body assembly in order to access the nuts securing the manifold to the head.

7. Remove the intake manifold support bracket bolts and the bracket. It may be necessary to access it from under the vehicle.

8. While supporting the intake manifold, remove the nuts attaching the intake manifold to the cylinder head, then remove the manifold.

To install:

9. Installation is the reverse of the removal procedure, while using the following torque values:
- Intake manifold nuts:16 ft. lbs. (22 Nm)
- Support bracket bolt: 16 ft. lbs. (22 Nm)

2.5L Engine

1. Before servicing the vehicle, refer to the precautions in the beginning of this section.

2. Disconnect the negative battery cable.

3. Properly relieve the fuel system pressure.

4. Remove or disconnect the following:
- Fuel feed and return hoses
- Throttle cable
- Engine harness cover
- All vacuum hoses and wiring from the intake manifold

5. To avoid having to drain the cooling system, remove the fast idle valve and the Idle Air Control (IAC) valve without discon-

necting the coolant hoses. Move these components out of the work area so that they will not be damaged.

6. Remove or disconnect the following:
   - EGR pipe and the vacuum pipe
   - Fuel rail
   - Fuel injectors from the manifold
   - Intake manifold brackets
   - Nuts that secure the manifold to the head. Remove the intake manifold from the engine

To install:

7. Installation is the reverse of the removal procedure, while using the following torque values:
   - Intake manifold bolts: 16 ft. lbs. (22 Nm)
   - Manifold bracket bolts: 16 ft. lbs. (22 Nm)

3.0L Engines

1. Before servicing the vehicle, refer to the precautions in the beginning of this section.

2. Remove or disconnect the following:
   - Negative battery cable
   - Engine covers
   - Strut brace
   - Intake air duct
   - Throttle body and intake manifold covers
   - Throttle and cruise control cables
   - Fuel feed and return hoses
   - Brake booster vacuum hose
   - Electrical connectors from the manifold
   - Intake manifold

To install:

3. Installation is the reverse of the removal procedure, while using the following torque values:
   - Intake manifold bolts:16 ft. lbs. (22 Nm)
   - Strut brace bolts: 16 ft. lbs. (22 Nm)

3.2L and 3.5L Engines

1. Before servicing the vehicle, refer to the precautions in the beginning of this section.

2. Disconnect the negative battery cable.
3. Drain the cooling system.
4. Properly relieve the fuel system pressure.
5. Remove or disconnect the following:
   - Intake air duct
   - Strut brace (if equipped)
   - Intake manifold cover

- Water bypass hose
- Traction Control System (TCS) actuator (if equipped)
- Evaporative Emission (EVAP) canister hose
- Throttle and cruise control cables
- Brake booster vacuum hose
- Upper intake manifold cover
- Intake manifold nuts and bolts in a crisscross pattern, beginning from the center and moving out

6. Verify that all vacuum lines are disconnected and remove the intake manifold and throttle body as a unit.

7. Inspect the manifold for cracks, flatness, or other damage; replace any damaged parts. If the intake manifold is to be replaced, transfer all the necessary components to the new manifold.

To install:
- Intake fasteners:16 ft. lbs. (22 Nm)
- Upper intake manifold cover fasteners: 108 inch lbs. (12 Nm)
- TCS actuator bolts: 16 ft. lbs. (22 Nm)
- Strut brace bolts: 16 ft. lbs. (22 Nm)

Exhaust Manifold

REMOVAL & INSTALLATION

1.8L Engines

1. Before servicing the vehicle, refer to the precautions in the beginning of this section.

2. Remove or disconnect the following:
   - Negative battery cable
   - Exhaust manifold cover
   - Front exhaust pipe from the manifold
   - Exhaust manifold support bracket
   - Exhaust manifold from the engine
   - Rear cover from the exhaust manifold (if necessary)

To install:

➡**Use new exhaust manifold nuts for installation.**

- Rear heat shield bolts: 17 ft. lbs. (24 Nm)
- Exhaust manifold nuts: 23 ft. lbs. (31 Nm)
- Bracket bolts: 33 ft. lbs. (44 Nm)

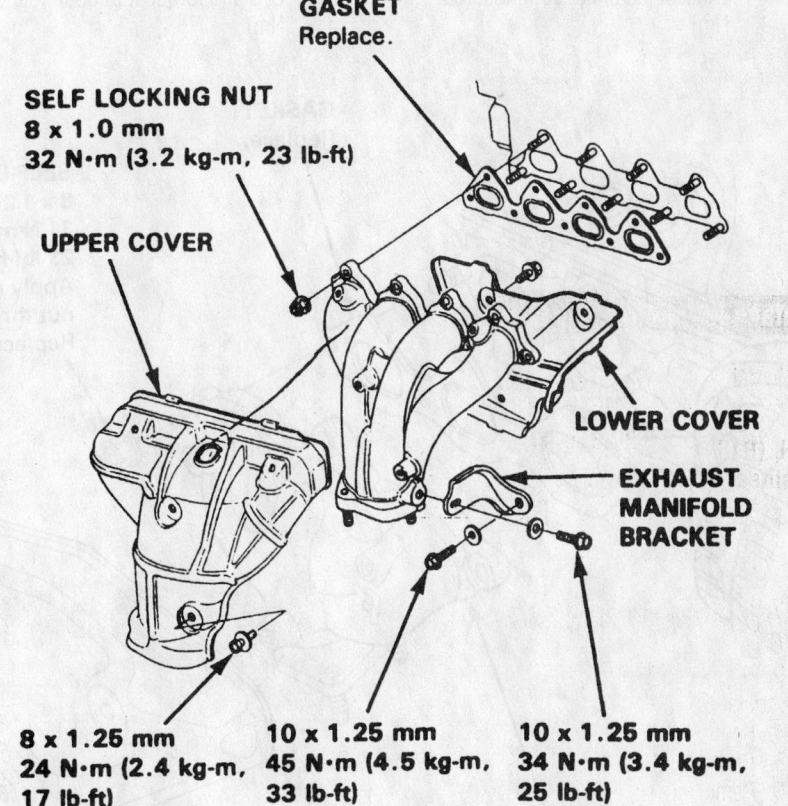

GASKET Replace.

SELF LOCKING NUT 8 x 1.0 mm 32 N·m (3.2 kg-m, 23 lb-ft)

UPPER COVER

LOWER COVER

EXHAUST MANIFOLD BRACKET

8 x 1.25 mm 24 N·m (2.4 kg-m, 17 lb-ft)

10 x 1.25 mm 45 N·m (4.5 kg-m, 33 lb-ft)

10 x 1.25 mm 34 N·m (3.4 kg-m, 25 lb-ft)

7923BG18

Exhaust manifold—1.8L engines

- Front exhaust pipe nuts: 40 ft. lbs. (54 Nm)
- Exhaust manifold heat shield bolts: 17 ft. lbs. (24 Nm)

2.3L Engine

1. Before servicing the vehicle, refer to the precautions in the beginning of this section.
2. Disconnect the negative battery cable.
3. Remove or disconnect the following:
   - Exhaust manifold cover
   - Heat insulator from the manifold, if equipped with air conditioning
   - Oxygen (O2S) sensor electrical connector
   - Front exhaust pipe from the manifold
   - Exhaust manifold bracket
   - Exhaust manifold attaching nuts, using a crisscross pattern (starting from the center)
   - Exhaust manifold

To install:

4. Installation is the reverse of the removal procedure, while using the following torque values:
   - Exhaust manifold: 23 ft. lbs. (32 Nm)

- Bracket mounting bolts: 33 ft. lbs. (44 Nm)
- Heat shield bolts: 16 ft. lbs. (22 Nm)
- Front exhaust pipe nuts: 40 ft. lbs. (55 Nm)

2.5L Engine

1. Before servicing the vehicle, refer to the precautions in the beginning of this section.
2. Remove or disconnect the following:
   - Negative battery cable
   - Outer manifold heat shields
   - Oxygen (O2S) sensor from the manifold
   - Front exhaust pipe from the manifold
   - Manifold support bracket
   - Exhaust manifold

To install:

3. Install or connect the following:
   - Exhaust manifold with a new gasket
   - New manifold nuts and torque them to 23 ft. lbs. (31 Nm)
   - Manifold support bracket and torque the bolts to 22 ft. lbs. (29 Nm)
   - O2S and torque it to 33 ft. lbs. 944 Nm)

- Front exhaust pipe to the manifold with a new gasket and torque the nuts to 40 ft. lbs. (54 Nm)
- Manifold heat shields and tighten the bolts to 22 ft. lbs. (29 Nm)

4. Connect the negative battery cable. Start the engine and check for exhaust leaks.

3.0L Engine

1. Before servicing the vehicle, refer to the precautions in the beginning of this section.
2. Remove or disconnect the following:
   - Manifold cover
   - Exhaust pipe from the manifold to be removed
   - Mounting nuts and the exhaust manifold

To install:

3. Install or connect the following:
   - Exhaust manifold with a new gasket and tighten the nuts to 23 ft. lbs. (31 Nm)
   - Exhaust pipe to the manifold using a new gasket and tighten the nuts to 40 ft. lbs. (54 Nm)
   - Manifold cover and tighten the bolts to 16 ft. lbs. (22 Nm)

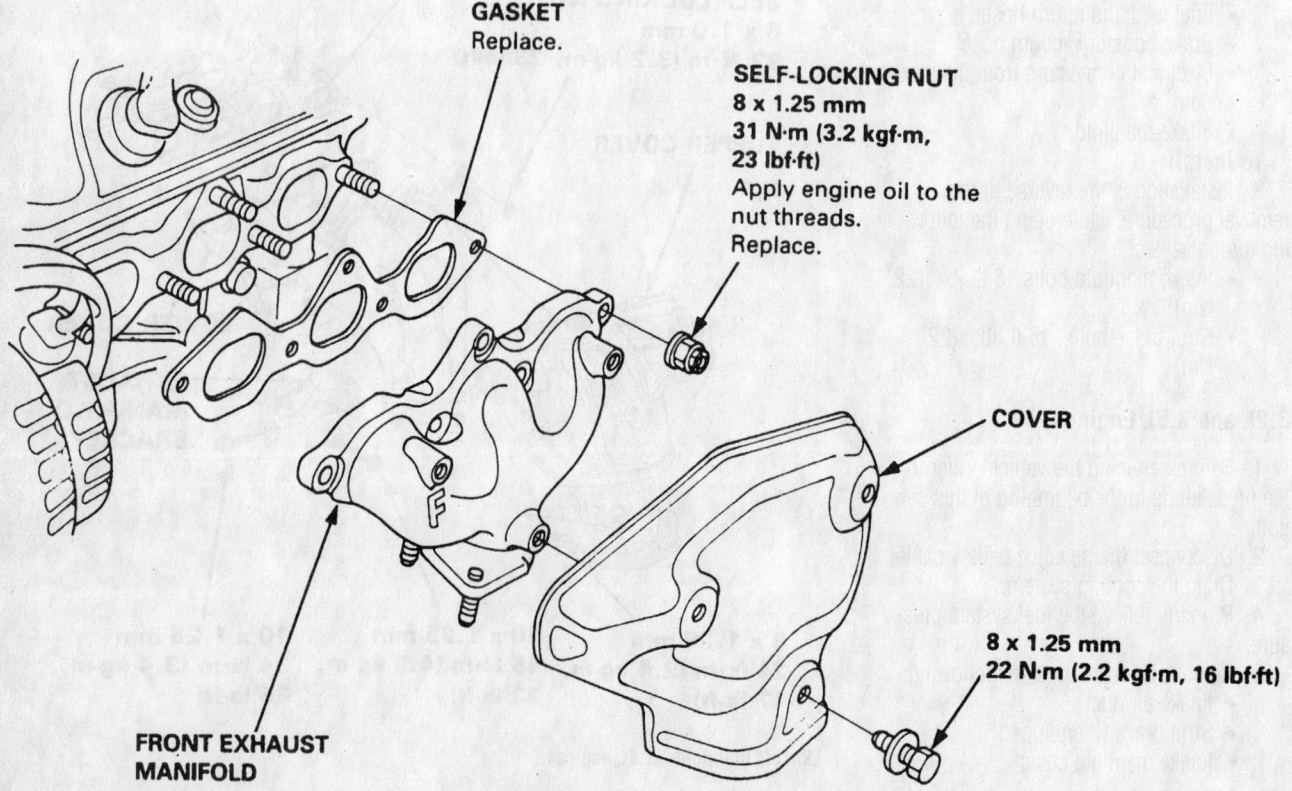

GASKET
Replace.

SELF-LOCKING NUT
8 x 1.25 mm
31 N·m (3.2 kgf·m, 23 lbf·ft)
Apply engine oil to the nut threads.
Replace.

COVER

8 x 1.25 mm
22 N·m (2.2 kgf·m, 16 lbf·ft)

FRONT EXHAUST MANIFOLD

Exploded view of the front exhaust manifold mounting—3.0L engine

7923BG68

3.2L and 3.5L Engines

1. Before servicing the vehicle, refer to the precautions in the beginning of this section.

2. Remove or disconnect the following:
- Negative battery cable
- Exhaust manifold covers
- Small heat shields from the cylinder heads (if equipped)
- Exhaust pipe from the manifold
- Heated Oxygen (HO$_2$S) sensors
- Exhaust manifold nuts in a crisscross pattern starting from the center of the manifold
- Exhaust manifold

To install:

3. Install or connect the following:
- Exhaust manifold with a new gasket and new nuts and tighten the nuts in a crisscross pattern starting from the center to 22 ft. lbs. (30 Nm)
- Small heat shields and tighten the attaching bolts to 16 ft. lbs. (22 Nm) (if equipped)
- Exhaust pipe to the manifold with a new gasket and tighten the nuts to 40 ft. lbs. (55 Nm)
- HO$_2$S sensor and tighten it to 33 ft. lbs. (45 Nm)
- Manifold covers and tighten the bolts to 16 ft. lbs. (22 Nm)

4. Verify that all vacuum lines and wiring are properly connected.

5. Reconnect the negative battery cable.

6. Start the engine and check for leaks.

Front Crankshaft Seal

REMOVAL & INSTALLATION

1. Before servicing the vehicle, refer to the precautions in the beginning of this section.

2. Disconnect negative cable at the battery.

3. Raise and safely support the vehicle. Drain the engine oil and properly dispose of it.

4. Be sure the crankshaft is at TDC on No. 1 cylinder by aligning the white mark on the crankshaft pulley with the pointer on the lower timing belt cover.

5. Remove or disconnect the following:
- Crankshaft pulley
- Cylinder head cover
- Timing belt cover
- Timing belt

- Crankshaft Speed Fluctuation (CKF) sensor (if equipped)
- Timing belt drive gear from the crankshaft

6. Using a suitable pry tool, carefully remove the seal.

To install:

7. Apply a light coat of oil to the seal lip.

8. Position the seal, then using a seal driver, install the seal into the housing.

9. Install or connect the following:
- Timing belt drive gear
- Timing belt
- Timing belt cover
- Cylinder head cover
- CKF sensor and tighten the attaching bolts to 96 inch lbs. (11 Nm) (if equipped)
- Crankshaft pulley

10. Lower the vehicle and check and fill the engine with oil as necessary.

11. Connect the negative battery cable and enter the radio security code.

12. Run the engine and check for leaks.

13. Turn off engine and check the oil level. Top off the oil level if necessary.

Camshaft and Valve Lifters

REMOVAL & INSTALLATION

➡The radio may have a coded theft protection circuit. Obtain the code from the owner before disconnecting the battery, removing the radio fuse, or removing the radio.

1.8L (B18B1) Engines

1. Before servicing the vehicle, refer to the precautions in the beginning of this section.

2. Remove or disconnect the following:
- Negative battery cable
- Spark plug wires
- Cylinder head cover and timing belt cover

3. Rotate the crankshaft to TDC, compression of No. 1 piston and remove the timing belt.

4. Remove the distributor from the cylinder head.

5. Install 5.0mm pin punches to the No.1 camshaft holders, then remove the camshaft sprockets.

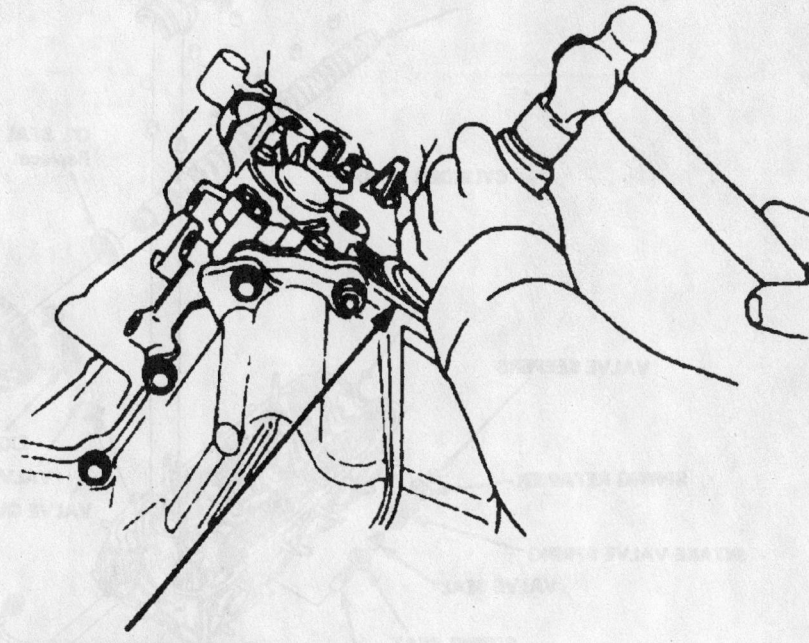

SEAL DRIVER
Install seal with the
part number side
facing out.

7923BG19

Installing the seal

6. Loosen the valve adjusters to remove as much spring tension as possible.

7. Remove the pin punches from the camshaft holders.

To install:

8. Check the following before installing the camshafts:

a. Be certain the keyways on the camshafts are facing UP (No. 1 cylinder at TDC).

b. The valve adjuster locknuts should be loosened and the adjusting screws backed off before installation.

9. Lubricate the rocker arms and camshafts with clean oil.

10. Place the rocker arms on the pivot bolts and the valve stems, making sure that the rocker arms are in their original positions.

11. Install the camshaft seals with the open side (spring) facing in. Lubricate the lip of the seal.

12. Be sure the keyways on the camshafts are facing up and install the camshafts to the cylinder head.

13. Apply liquid gasket to the head mating surfaces of the No. 1 and No. 6 camshaft holders, then install them along with No. 2, 3, 4 and 5 camshaft holders. The arrows stamped on the holders should

point toward the timing belt. Do not apply oil to the holder mating surface where the camshaft seals are housed.

14. Tighten the camshaft holders temporarily and be sure that the rocker arms are properly positioned.

15. Press the oil seals into the No.1 camshaft holders with a seal driver.

16. Tighten the bolts in a crisscross pattern to 84 inch lbs. (10 Nm). Check that the rockers do not bind on the valves.

17. Install the cylinder head plug to the end of the cylinder head. If the plug has alignment marks, align the marks with the cylinder head upper surface.

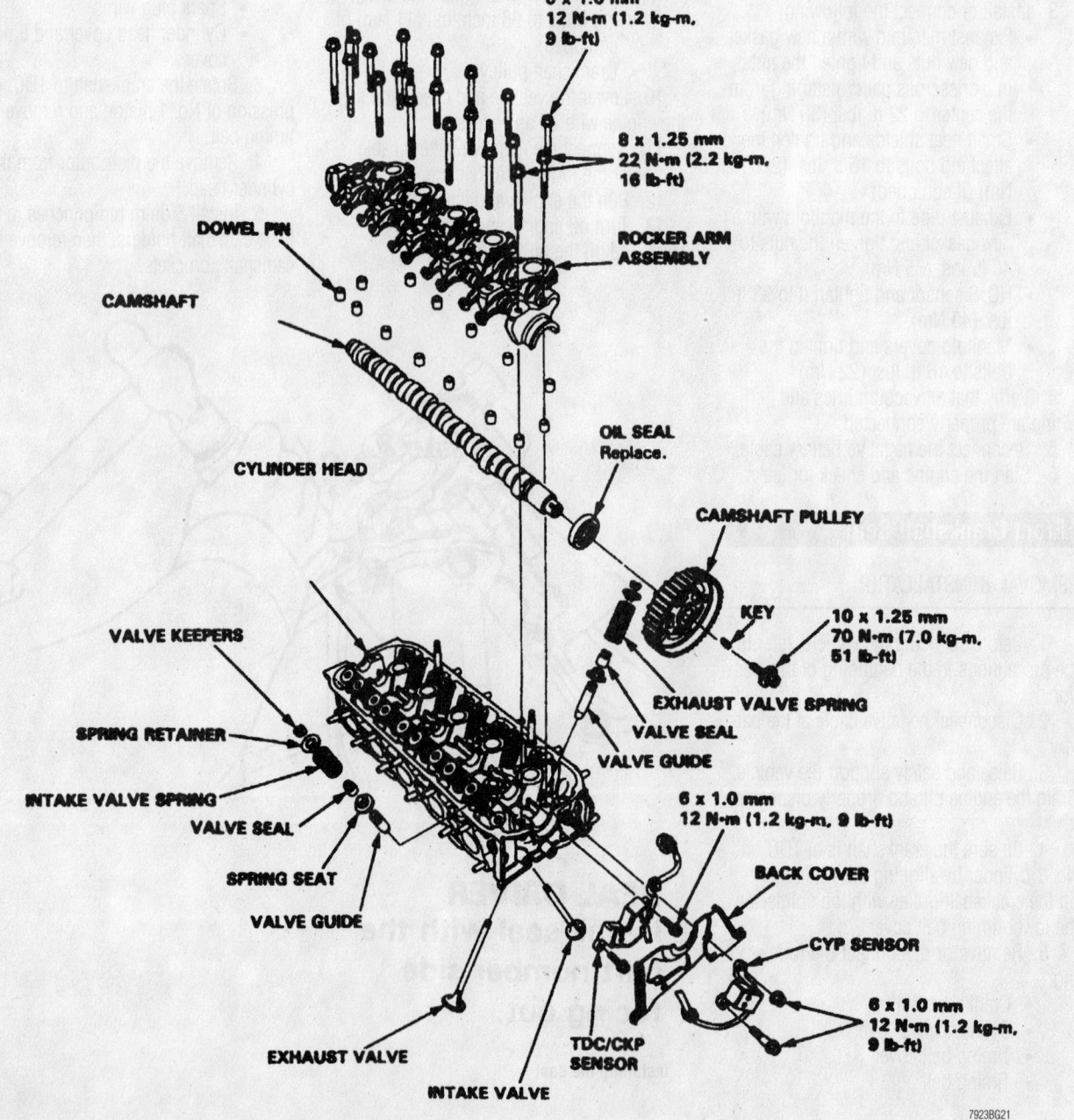

Exploded view of the cylinder head—1.8L (B18B1) engine

18. If equipped with a timing belt back cover, install the cover and tighten the bolts to 84 inch lbs. (10 Nm).

19. Install 5.0mm pin punches to the No.1 camshaft holders, then install the camshaft pulley keys onto the grooves in the camshafts.

20. Push the camshaft pulleys onto the camshafts, then tighten the retaining bolts to 27 ft. lbs. (38 Nm).

21. Install the timing belt and timing belt covers. Remove the pin punches from the camshaft holders.

22. Adjust the valves and pour oil over the camshafts and rocker arms.

23. Install or connect the following:
- Cylinder head cover and engine ground cable
- Distributor to the cylinder head and reconnect the spark plug wires to the spark plugs
- Negative battery cable and enter the radio security code

24. Change the engine oil. Wait at least 20 minutes for the sealant to cure before filling the engine with oil.

1.8L (B18C1, B18C5) Engines

1. Before servicing the vehicle, refer to the precautions in the beginning of this section.

2. Disconnect the negative battery cable.

3. Be sure the crankshaft is at TDC/compression on No. 1 cylinder by aligning the white mark on the crankshaft pulley with the pointer on the lower timing belt cover.

4. Remove or disconnect the following:
- Strut brace
- Cylinder head cover, timing belt cover, and timing belt
- Camshaft pulleys and back cover

5. Loosen the rocker arm locknuts and adjusting screws.

6. Remove the camshaft holder bolts, then, remove the camshaft holder plates, the camshaft holders, and camshafts.

To install:

7. Be sure that the keyways on the camshafts are facing up and that the rocker arms are in their original position. The valve locknuts should be loosened and the adjusting screw backed off before installation

8. Install or connect the following:
- Camshafts
- Camshaft seals with the open side facing in

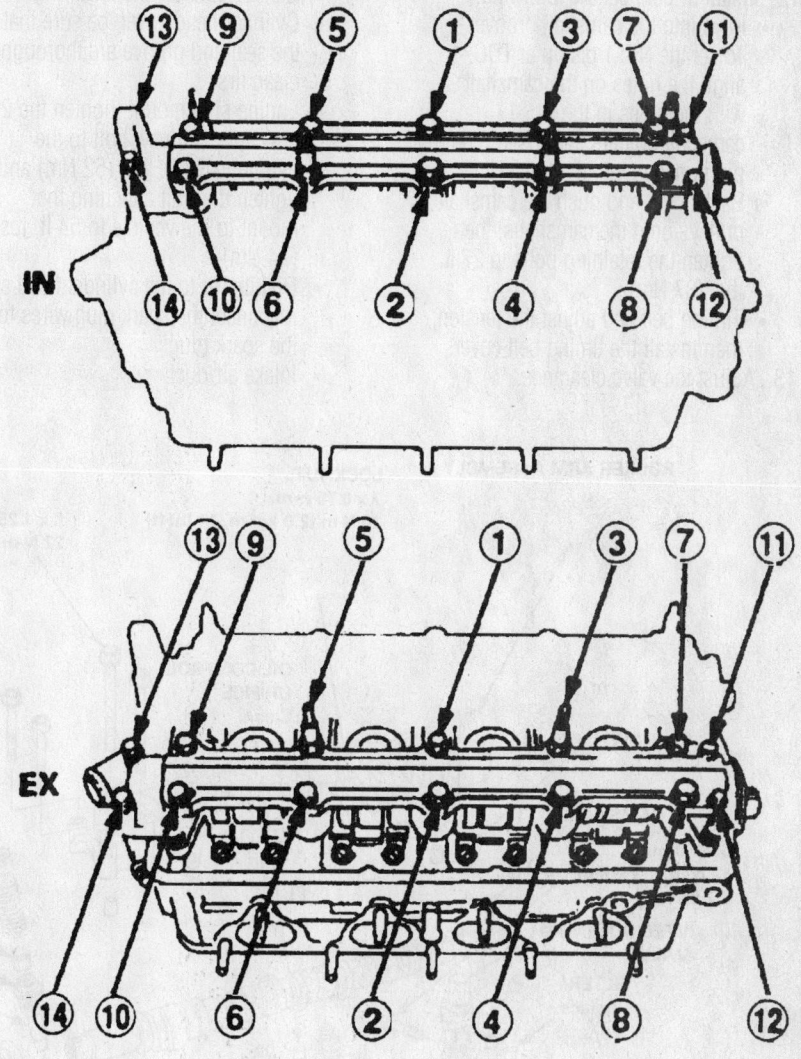

$$1\text{-}10 : 8 \times 1.25 \text{ mm } 27 \text{ N·m (2.8 kgf·m, 20 lbf·ft)}$$
$$11\text{-}14 : 6 \times 1.0 \text{ mm } 9.8 \text{ N·m (1.0 kgf·m, 7.2 lbf·ft)}$$

7923BG20

Camshaft holder plates torque sequence—1.8L (B18C1, B18C5) engines

- Rubber cap with liquid gasket applied
- O-ring and the dowel pin to the oil passage of the No. 3 camshaft holder

9. Apply liquid gasket to the head of the mating surfaces of the No. 1 and No. 5 camshaft holders, then install them, along with No. 2, 3, and 4. Be sure to pay attention to the following points:
- Do not apply oil to the holder mating surface of camshaft seals.

- The arrows marked on the camshaft holders should point to the timing belt.

10. Tighten the camshaft holders temporarily. Be sure that the rocker arms are properly positioned on the valve stems.

11. Tighten the camshaft holder bolts in 2 steps, following the proper sequence, to ensure that the rockers do not bind on the valves. Tighten the 8 x 1.25mm bolts to 20 ft. lbs. (27 Nm). Tighten the 6 x 1.0mm bolts to 84 inch lbs. (10 Nm).

Timing belt service is covered in Section 3 of this manual

12. Install or connect the following:
- Keys into the camshaft grooves—To set the No. 1 piston at TDC, align the holes on the camshaft with the holes in the No. 1 camshaft holders and insert 5.0mm pin punches into the holes.
- Back cover and push the camshaft pulleys onto the camshafts, then tighten the retaining bolts to 27 ft. lbs. (37 Nm)
- Timing belt and adjust the tension, then install the timing belt covers

13. Adjust the valve clearance.

14. Install or connect the following:
- Cylinder head cover, be sure that the seal and groove are thoroughly clean first
- Engine side mount, tighten the 2 new nuts and new bolt to the engine to 38 ft. lbs. (52 Nm) and tighten the bolt attaching the mount to the vehicle to 54 ft. lbs. (74 Nm)
- Distributor to the cylinder head and reconnect the spark plug wires to the spark plugs
- Intake air duct

- Strut brace, tighten the nuts to 17 ft. lbs. (24 Nm)
- Negative battery cable and enter the radio security code

15. Drain the engine oil. Wait at least 20 minutes before filling the engine with oil; the time delay allows the sealant to cure.

2.3L Engine

1. Before servicing the vehicle, refer to the precautions in the beginning of this section.
2. Disconnect the negative battery cable.

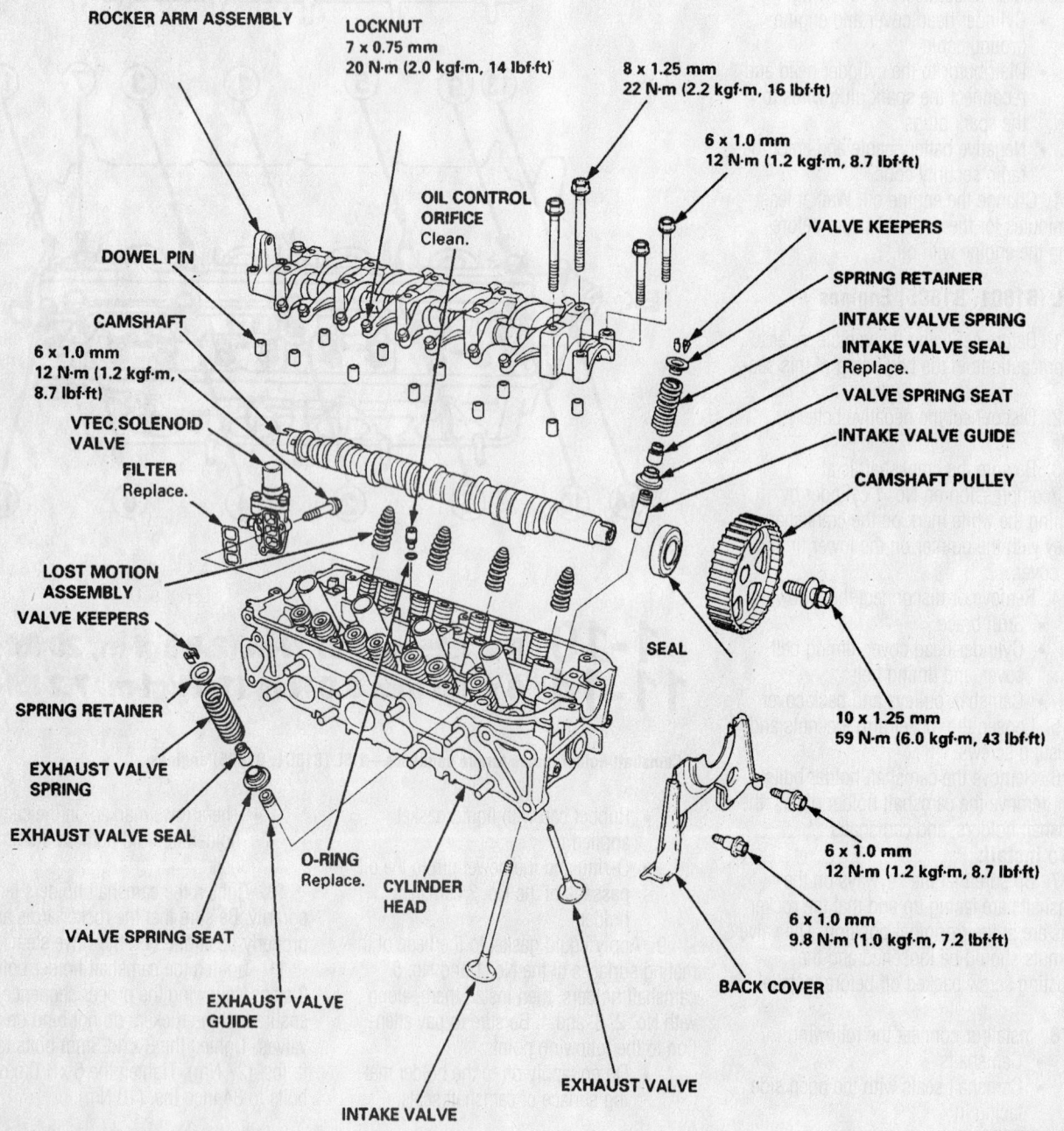

ROCKER ARM ASSEMBLY

LOCKNUT
7 x 0.75 mm
20 N·m (2.0 kgf·m, 14 lbf·ft)

8 x 1.25 mm
22 N·m (2.2 kgf·m, 16 lbf·ft)

6 x 1.0 mm
12 N·m (1.2 kgf·m, 8.7 lbf·ft)

OIL CONTROL ORIFICE
Clean.

VALVE KEEPERS

SPRING RETAINER

INTAKE VALVE SPRING

INTAKE VALVE SEAL
Replace.

VALVE SPRING SEAT

INTAKE VALVE GUIDE

CAMSHAFT PULLEY

DOWEL PIN

CAMSHAFT

6 x 1.0 mm
12 N·m (1.2 kgf·m, 8.7 lbf·ft)

VTEC SOLENOID VALVE

FILTER
Replace.

LOST MOTION ASSEMBLY

VALVE KEEPERS

SPRING RETAINER

EXHAUST VALVE SPRING

EXHAUST VALVE SEAL

VALVE SPRING SEAT

EXHAUST VALVE GUIDE

SEAL

10 x 1.25 mm
59 N·m (6.0 kgf·m, 43 lbf·ft)

6 x 1.0 mm
12 N·m (1.2 kgf·m, 8.7 lbf·ft)

6 x 1.0 mm
9.8 N·m (1.0 kgf·m, 7.2 lbf·ft)

BACK COVER

O-RING
Replace.

CYLINDER HEAD

EXHAUST VALVE

INTAKE VALVE

Exploded view of the cylinder head and related components—2.3L engine

7923BG70

3. Turn the crankshaft so the No. 1 piston is at TDC.

➡**The No. 1 piston is at top dead center when the pointer on the block aligns with the white painted mark on the flywheel (manual transaxle) or driveplate (automatic transaxle).**

4. Remove or disconnect the following:
- Air intake duct
- Engine ground cable from the cylinder head cover
- Connector and the terminal from the alternator, then remove the engine wiring harness from the valve cover
- Ignition coil

5. Label, then disconnect the electrical connectors from the distributor and the spark plug wires from the spark plugs. Mark the position of the distributor and remove it from the cylinder head. Disconnect the ignition coil wire from the distributor.

6. Remove or disconnect the following:
- PCV hose, then remove the cylinder head cover, replace the rubber seals if damaged or deteriorated
- Timing belt middle cover

7. Ensure the words **UP** embossed on the camshaft pulleys are aligned in the upward position.

8. Mark the rotation of the timing belt if it is to be used again. Loosen the timing belt adjusting nut 1/2 turn, then release the tension on the timing belt. Push the tensioner to release tension from the belt, then tighten the adjusting nut.

9. Remove the timing belt from the camshaft sprockets.

10. Remove the side engine mount bracket, then the timing belt back cover from behind the camshaft sprockets.

11. Loosen all of the rocker arm adjusting screws, then remove the pin punches from the camshaft caps.

12. Remove or disconnect the following:
- Camshaft holders, note the holders locations for ease of installation, loosen the bolts in the reverse order of the installation
- Camshafts from the cylinder head, then discard the camshaft seals
- Rubber cap from the head, located at the end of the intake camshaft
- Rocker arms from the cylinder head, note the locations of the rocker arms

➡**The rocker arms have to be installed to their original locations if being reused.**

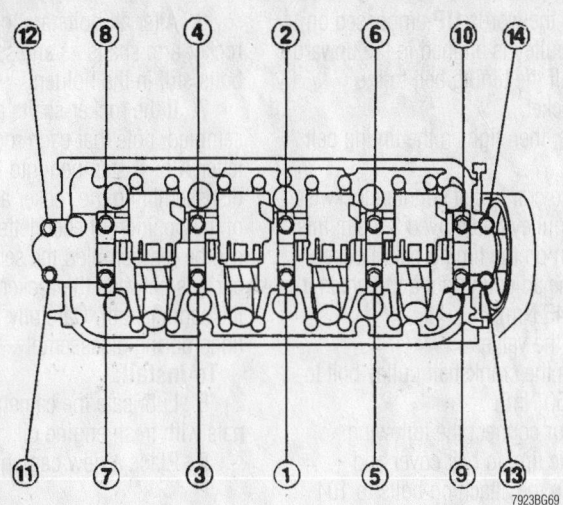

Camshaft holder torque sequence—2.3L engine

Number 1 Piston at TDC:

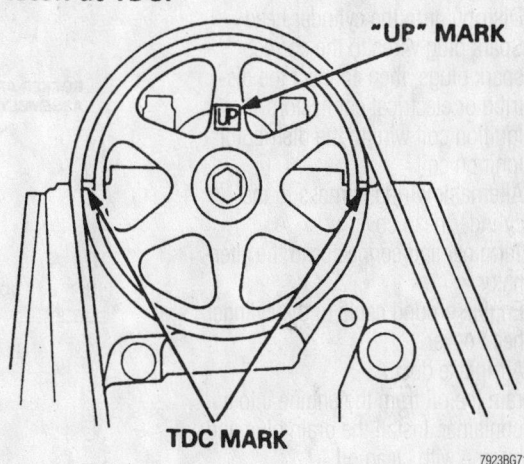

Camshaft sprocket alignment—2.3L engine

To install:

13. Lubricate the rocker arms with clean oil, then install the rocker arms on the pivot bolts and the valve stems. If the rocker arms are being reused, install them to their original locations. The locknuts and adjustment screws should be loosened before installing the rocker arms.

14. Lubricate the camshafts with clean oil.

15. Install the camshaft seals to the end of the camshafts that the timing belt sprockets attach to. The open side (spring) should be facing into the cylinder head when installed.

16. Be sure the keyways on the camshaft is facing up and install the camshaft to the cylinder head.

17. Apply liquid gasket to the head mating surfaces of the No. 1 and No. 5 camshaft holders, then install them along with No. 2, 3 and 4.

18. Snug the camshaft holders in place.

19. Press the camshaft seals securely into place.

20. Tighten the camshaft holder bolts in 2 steps, following the proper sequence, to ensure that the rockers do not bind on the valves. Tighten all the 6 mm bolts to 104 inch lbs. (12 Nm). Tighten the 8 mm bolts to 16 ft. lbs. (22 Nm).

21. Install or connect the following:
- Timing belt back cover
- Side engine mount bracket B— Tighten the bolt attaching the bracket to the cylinder head to 33 ft. lbs. (45 Nm). Tighten the bolts attaching the bracket to the side engine mount to 16 ft. lbs. (22 Nm).

22. Push the camshaft sprocket onto the camshaft, then tighten the retaining bolt to 43 ft. lbs. (59 Nm).

Heater Core replacement is covered in Section 2 of this manual

23. Ensure the words **UP** embossed on the camshaft pulley is aligned in the upward position. Install the timing belt to the camshaft sprocket.

24. Loosen, then tighten the timing belt adjuster nut.

25. Turn the crankshaft counterclockwise until the cam pulley has moved 3 teeth; this creates tension on the timing belt. Loosen, then tighten the adjusting nut and tighten it to 33 ft. lbs. (45 Nm).

26. Adjust the valves.

27. Tighten the crankshaft pulley bolt to 181 ft. lbs. (250 Nm).

28. Install or connect the following:
- Middle timing belt cover and tighten the attaching bolts to 104 inch lbs. (12 Nm)
- Cylinder head cover and tighten the cap nuts to 104 inch lbs. (12 Nm)
- PCV hose to the cylinder head cover
- Distributor to the cylinder head
- Spark plug wires to the correct spark plugs, then connect the distributor electrical connectors
- Ignition coil wire to the distributor
- Ignition coil
- Alternator wiring harness to the cylinder head cover
- Terminal and connector to the alternator
- Engine ground cable to the cylinder head cover
- Air intake duct

29. Drain the oil from the engine into a sealable container. Install the drain plug and refill the engine with clean oil.

30. Connect the negative battery cable and enter the radio security code.

31. Start the engine, checking carefully for any leaks.

32. Enter the radio security code.

2.5L Engine

1. Before servicing the vehicle, refer to the precautions in the beginning of this section.

2. Disconnect the negative battery cable. Remove the timing belt covers and cylinder head covers.

3. Rotate the crankshaft to TDC compression of No. 1 piston and remove the timing belt.

4. Remove or disconnect the following:
- Camshaft sprocket
- Cylinder head from the vehicle

5. Loosen the rocker shaft holder bolts 1 turn at a time in the opposite of the installation sequence. Following this procedure will prevent the camshafts and rocker assemblies from warping.

6. After all bolts are loose, remove the rocker arm shafts as an assembly with the bolts still in the holders.

7. If the rocker shafts are to be disassembled, note that each rocker arm has a letter **A** or **B** stamped into the side. Before disassembling the rocker arms, make a note of the position of each letter so the arms can be reassembled the same way. The springs between the rocker arms are not all the same length. Carefully note their positions during disassembly.

To install:

8. Lubricate the camshaft and its journals with fresh engine oil.

9. Place a new camshaft seal on the end of the camshaft. The spring side of the seal must face in. Lubricate the journals and set the camshaft in place on the head.

10. Install the camshaft onto the cylinder head with the keyway pointed up.

11. Apply liquid gasket to the mounting surfaces of the camshaft end holders.

12. Set the rocker arm assemblies in place and start all the cam holder bolts. Be sure the rocker arms are properly positioned and turn each bolt in sequence 2 turns at a time until the holders are seated on the head. Follow this procedure to avoid damaging the camshaft and rocker assemblies.

Prior to reassembling, clean all the parts in solvent, dry them and apply lubricant to any contact parts.

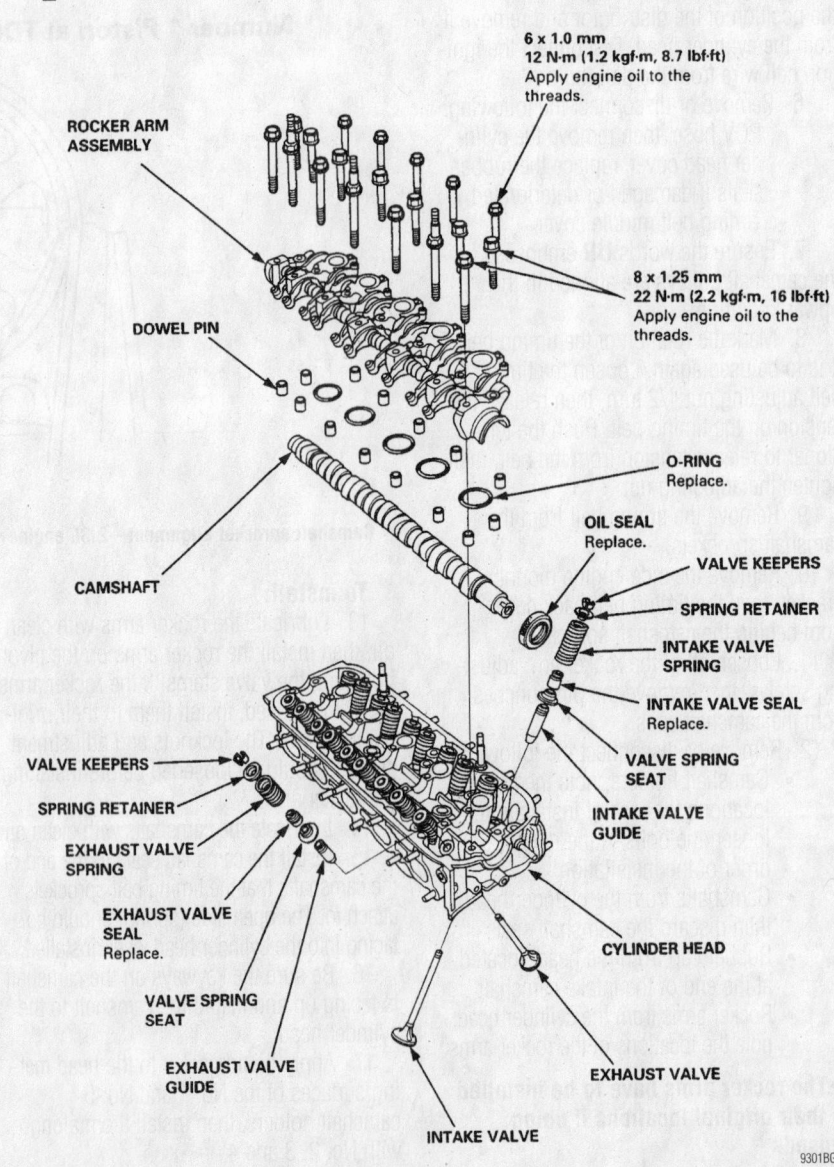

6 x 1.0 mm
12 N·m (1.2 kgf·m, 8.7 lbf·ft)
Apply engine oil to the threads.

8 x 1.25 mm
22 N·m (2.2 kgf·m, 16 lbf·ft)
Apply engine oil to the threads.

ROCKER ARM ASSEMBLY

DOWEL PIN

CAMSHAFT

O-RING
Replace.

OIL SEAL
Replace.

VALVE KEEPERS

SPRING RETAINER

INTAKE VALVE SPRING

INTAKE VALVE SEAL
Replace.

VALVE SPRING SEAT

INTAKE VALVE GUIDE

VALVE KEEPERS

SPRING RETAINER

EXHAUST VALVE SPRING

EXHAUST VALVE SEAL
Replace.

VALVE SPRING SEAT

EXHAUST VALVE GUIDE

CYLINDER HEAD

EXHAUST VALVE

INTAKE VALVE

9301BGA5

Camshaft and rocker arm assembly—2.5L engine

13. When all of the camshaft and rocker holders are seated, tighten the bolts in the same sequence. Tighten the 8mm bolts to 16 ft. lbs. (22 Nm) and the 6mm bolts to 104 inch lbs. (12 Nm).

14. Install or connect the following:
- Cylinder head
- Camshaft sprocket and tighten the bolts to 51 ft. lbs. (70 Nm)
- Timing belt, adjust the valves and oil the camshaft before completing the assembly
- Cylinder head cover and timing cover
- Distributor
- Negative battery cable

15. Check for proper engine and valve train operation.

3.0L Engine

1. Before servicing the vehicle, refer to the precautions in the beginning of this section.

2. Remove or disconnect the following:
- Negative battery cable
- Timing belt
- Cylinder head
- Camshaft sprocket and rear cover
- Rocker arm/shaft assembly
- Camshaft thrust cover and O-ring

3. Pull out the camshaft.

To install:

4. Lubricate the camshaft with clean engine oil.

5. Slide the camshaft into position.

6. Install or connect the following:
- Thrust plate using a new O-ring, tighten the bolts to 16 ft. lbs. (22 Nm)
- Rocker arm/shaft assembly
- Cylinder head
- Rear cover and camshaft sprocket, tighten the bolt to 67 ft. lbs. (90 Nm)
- Timing belt

7. Adjust the valves.

3.2L and 3.5L Engines

1. Before servicing the vehicle, refer to the precautions in the beginning of this section.

2. Remove or disconnect the following:
- Negative battery cable
- Timing belt covers and cylinder head covers

3. Rotate the crankshaft to TDC for the No. 1 piston and remove the timing belt.

4. Remove the camshaft sprockets.

5. Loosen the rocker shaft holder bolts 1 turn at a time in the reverse of the torque

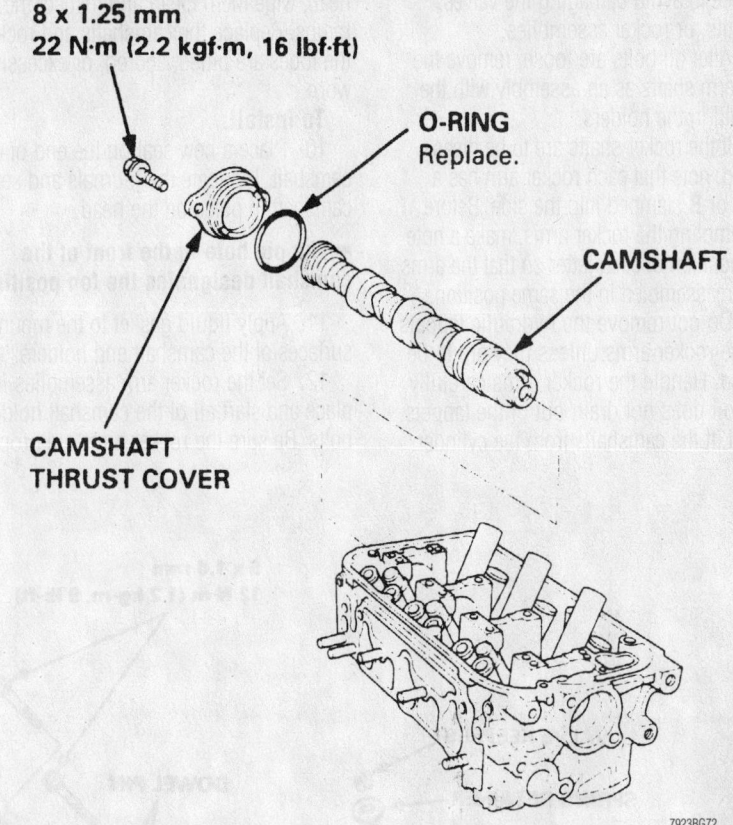

8 x 1.25 mm
22 N·m (2.2 kgf·m, 16 lbf·ft)

O-RING
Replace.

CAMSHAFT

CAMSHAFT THRUST COVER

7923BG72

Camshaft installation—3.0L engine

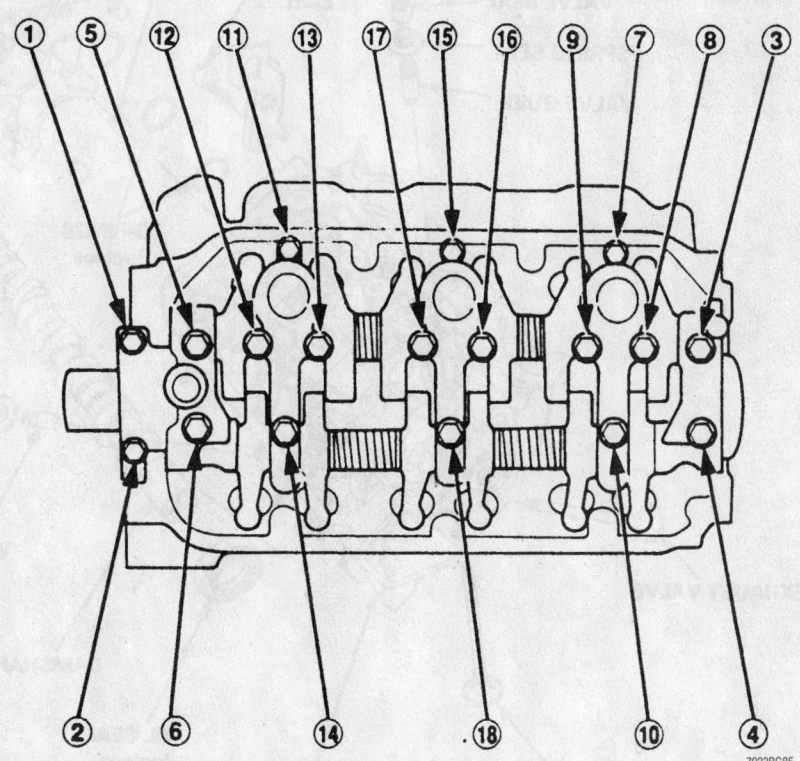

7923BG85

Loosen the camshaft holder bolts in the specified sequence—3.5L engine

Brake service is covered in Section 4 of this manual

sequence to avoid damaging the valves, camshafts, or rocker assemblies.

6. After all bolts are loose, remove the rocker arm shafts as an assembly with the bolts still in the holders.

7. If the rocker shafts are to be disassembled, note that each rocker arm has a letter **A** or **B** stamped into the side. Before disassembling the rocker arms, make a note of the position of each letter so that the arms can be reassembled in the same position.

8. Do not remove the hydraulic tappets from the rocker arms unless they are to be replaced. Handle the rocker arms carefully so the oil does not drain out of the tappets.

9. Lift the camshafts from the cylinder head, wipe them clean and inspect the lift ramps. Replace the camshafts and rockers if the lobes are pitted, scored, or excessively worn.

To install:

10. Place a new seal on the end of the camshaft, lubricate the journals and set the camshaft in place on the head.

➡ **The pin hole in the front of the camshaft designates the top position.**

11. Apply liquid gasket to the mounting surfaces of the camshaft end holders.

12. Set the rocker arm assemblies in place and start all of the camshaft holder bolts. Be sure the rocker arms are properly positioned and turn each bolt in sequence 2 turns at a time until the holders are seated on the head to avoid damaging the valves or rocker assemblies.

13. When all the camshaft and rocker holders are seated, tighten the bolts in the same sequence. Tighten the 8mm bolts to 16 ft. lbs. (22 Nm) and the 6mm bolts to 104 inch lbs. (12 Nm).

14. Install or connect the following:
- Camshaft pulleys and tighten the bolts to 23 ft. lbs. (32 Nm)
- Timing belt and pour oil over the camshafts
- Cylinder head cover and reassemble accessory components

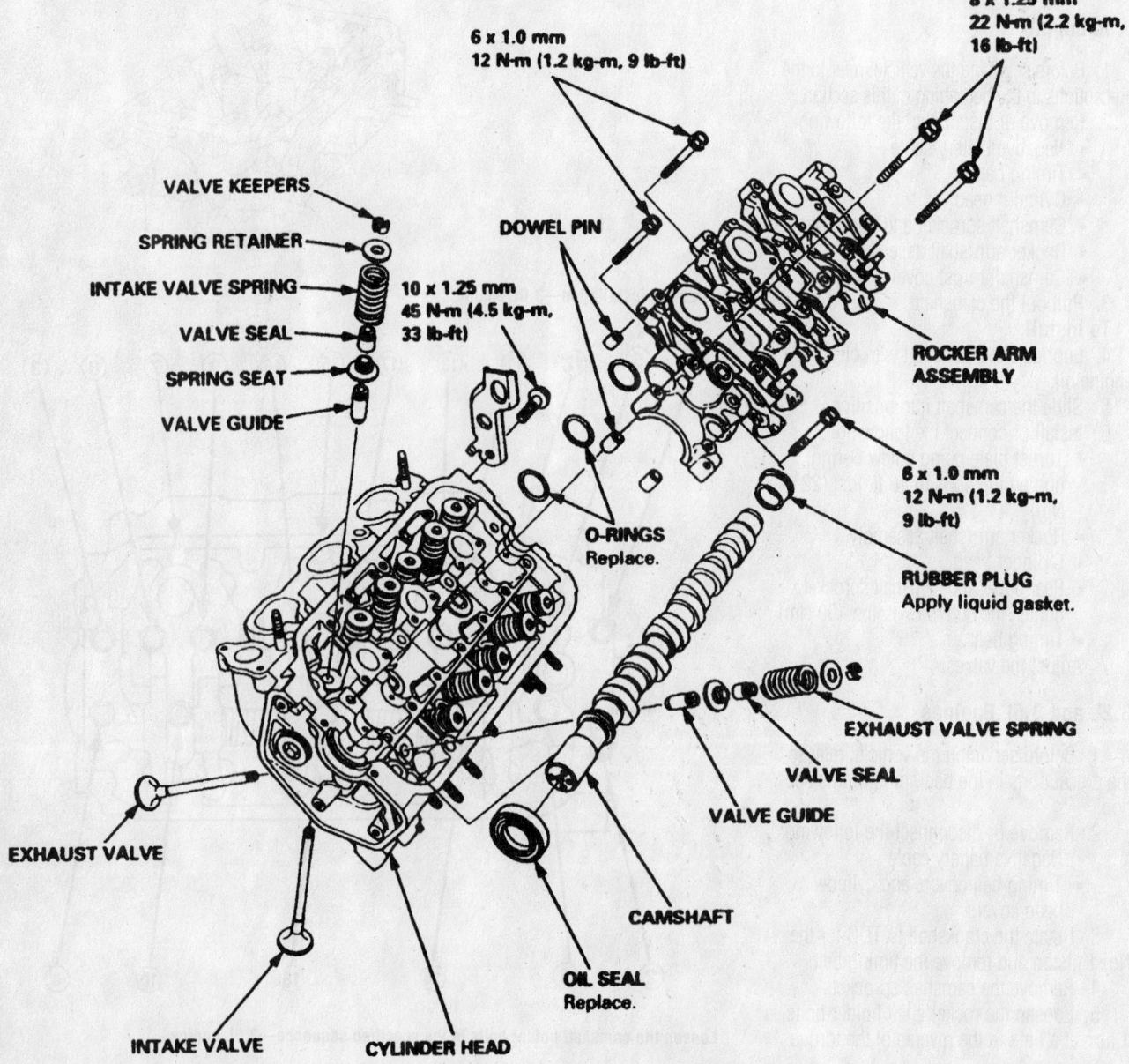

Camshaft and rocker arm assembly—3.2L engines

7923BG22

Specified torque:
8 mm bolts: 22 N·m (2.2 kg-m, 16 lb-ft)
6 mm bolts: 12 N·m (1.2 kg-m, 9 lb-ft)

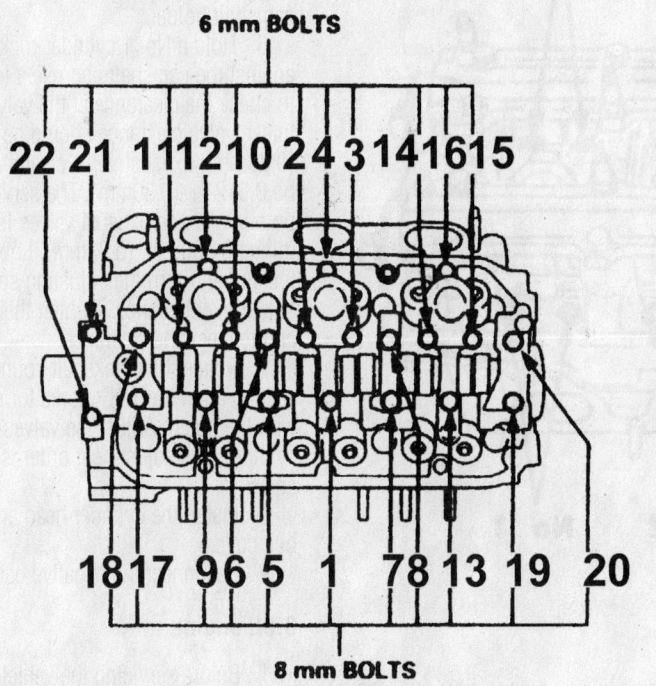

Camshaft holder bolt tightening sequence—3.2L engine

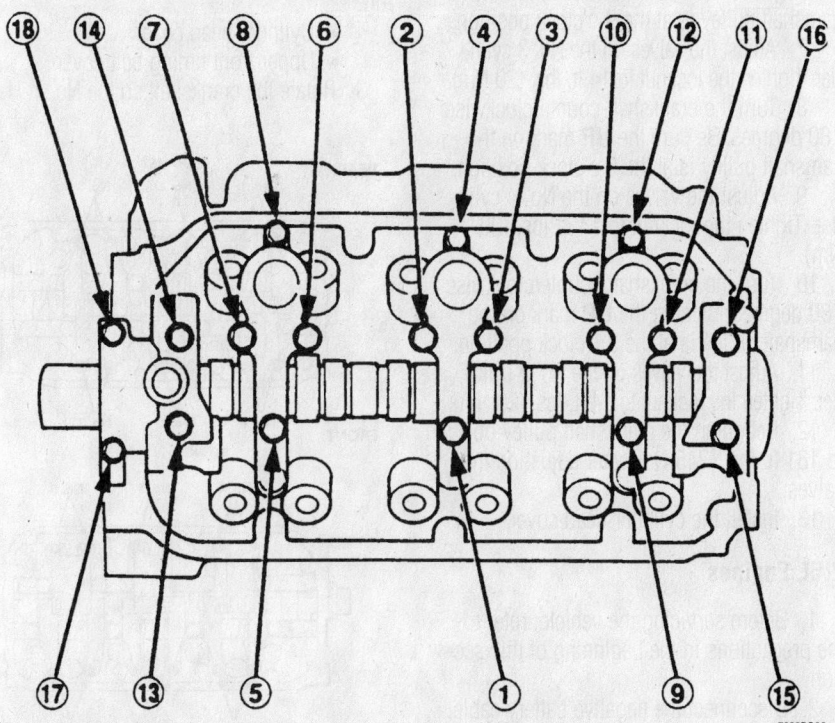

Tighten the camshaft holder bolts in the specified sequence—3.5L engine

15. Verify that all electrical connections and vacuum lines are connected.

16. Reconnect the negative battery cable.

17. Run the engine and check for leaks and proper operation..

Valve Lash

ADJUSTMENT

1.8L (B18B1, B18C1 and B18C5) Engines

➡ **While all valve adjustments must be as accurate as possible, it is better to have the valve adjustment slightly loose rather than too tight. Burned valves may result from overly tight adjustments. Perform the valve adjustment for each cylinder in the same sequence as the firing order: 1–3–4–2.**

1. Before servicing the vehicle, refer to the precautions in the beginning of this section.

2. Be sure the engine is cold; cylinder head temperature must be below 100° F (38° C). Overnight cold is best.

3. Remove the cylinder head cover and the upper timing belt cover.

4. Set the No. 1 cylinder to TDC. The word **UP** should appear at the top and the TDC grooves on the pulley should align with the cylinder head surface or the mark on the rear belt cover.

5. Valve clearances are:
   - B18B1 engine: Intake—0.003–0.005 in. (0.08–0.12mm) Exhaust—0.006–0.008 in. (0.16–0.20mm)
   - B18C1 and B18C5 VTEC engine: Intake—0.006–0.007 in. (0.15–0.19mm) Exhaust—0.007–0.008 in. (0.17–0.20mm)

6. With the No. 1 cylinder at TDC, adjust the valves of the No. 1 cylinder by performing the following procedures:

   a. Hold the rocker arm against the valve and place the feeler gauge between the rocker arm and the camshaft lobe. There should be a slight drag on the feeler gauge.

   b. If adjustment is required, loosen the valve adjusting the screw locknut.

   c. Turn the adjusting screw to obtain the proper clearance.

   d. Hold the adjusting screw and tighten the locknut(s) to 18 ft. lbs. (25 Nm).

   e. Recheck the clearance.

7. Turn the crankshaft 180 degrees

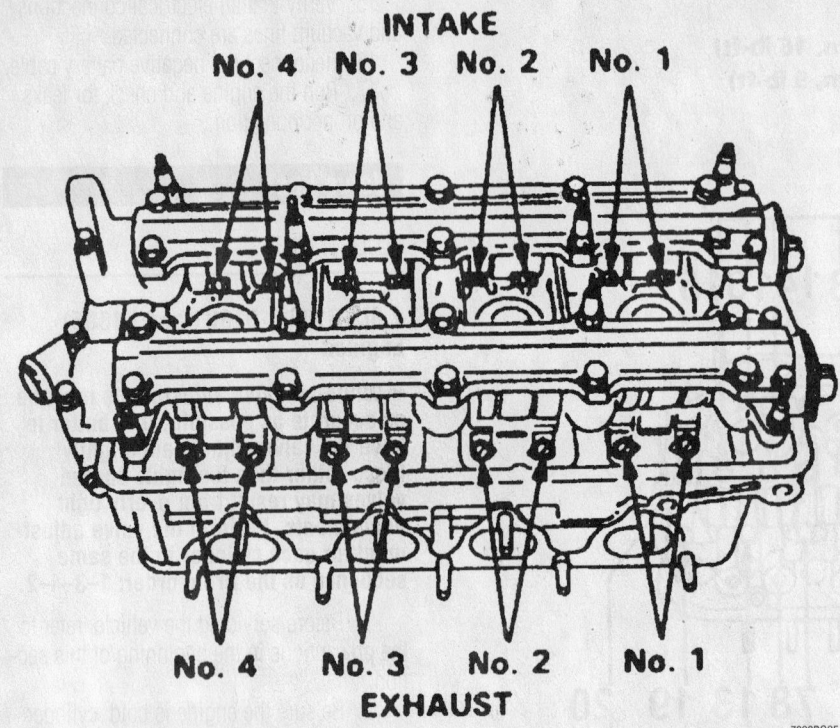

INTAKE

No. 4 No. 3 No. 2 No. 1

No. 4 No. 3 No. 2 No. 1

EXHAUST

Valve arrangement—1.8L (B18B1, B18C1, and B18C5) engines

counterclockwise; the cam pulley will turn 90 degrees. With the No. 3 cylinder at TDC, the **UP** marks should be at the exhaust side. Adjust the valves on the No. 3 cylinder.

8. Turn the crankshaft 180 degrees counterclockwise; the cam pulley will turn 90 degrees. With the No. 4 cylinder at TDC, both **UP** marks should be at the bottom. Adjust the valves on the No. 4 cylinder.

9. Turn the crankshaft 180 degrees counterclockwise. The No. 2 cylinder will now be on TDC and the **UP** marks should be at the intake side. Adjust the valves on the No. 2 cylinder.

10. Install the cylinder head cover and upper timing belt cover.

2.3L Engine

➡**The valves should be adjusted only when the engine temperature is below 100°F (38°C).**

1. Before servicing the vehicle, refer to the precautions in the beginning of this section.

2. Remove the cylinder head cover.

3. Turn the crankshaft so the No. 1 piston is at TDC. Be sure the **UP** mark on the camshaft pulley is at the 12 o'clock position.

4. Adjust the valves on the No. 1 cylinder. To the following specifications:
   - Intake: 0.010 in. (0.26mm)
   - Exhaust: 0.012 in. (0.30mm)

5. Tighten the locknut to 14 ft. lbs. (20 Nm).

6. Turn the crankshaft counterclockwise 180degrees. Be sure the **UP** mark on the camshaft pulley is at the 9 o'clock position.

7. Adjust the valves on the No. 3 cylinder. Tighten the locknut to 14 ft. lbs. (20 Nm).

8. Turn the crankshaft counterclockwise 180 degrees. Be sure the **UP** mark on the camshaft pulley is at the 6 o'clock position.

9. Adjust the valves on the No. 4 cylinder. Tighten the locknut to 14 ft. lbs. (20 Nm).

10. Turn the crankshaft counterclockwise 180 degrees. Be sure the **UP** mark on the camshaft pulley is at the 3 o'clock position.

11. Adjust the valves on the No. 2 cylinder. Tighten the locknut to 14 ft. lbs. (20 Nm).

12. Retighten the crankshaft pulley bolt to 181 ft. lbs. (245 Nm) after adjusting the valves.

13. Install the cylinder head cover.

2.5L Engines

1. Before servicing the vehicle, refer to the precautions in the beginning of this section.

2. Disconnect the negative battery cable.

3. Remove the cylinder head cover and the upper timing belt cover.

4. Rotate the crankshaft to align the white TDC on the crankshaft pulley with the pointer on the cover. Be sure the **UP** mark

on the camshaft sprocket is up and the TDC marks align with the edge of the cylinder head.

5. Align the No. 1 mark on the back of the camshaft sprocket with the notch in the camshaft holder.

6. Hold a No. 1 cylinder rocker arm against the camshaft and use a feeler gauge to check the clearance at the valve stem. Intake valve clearance should be 0.010 in. (0.26mm), exhaust valve clearance should be 0.012 in. (0.30mm). The service limit for both intake and exhaust valves is plus or minus 0.0008 in. (0.02mm). Loosen the locknut and turn the adjusting screw to adjust the clearance. Tighten the locknut and recheck the clearance.

7. Rotate the crankshaft counterclockwise to align the TDC marks for each piston with the notch. Adjust the valves of each cylinder. The adjustment order is 1, 2, 4, 5 and 3.

8. Install the cylinder head and timing belt covers.

9. Reconnect the negative battery cable.

3.0L engine

1. Before servicing the vehicle, refer to the precautions in the beginning of this section.

2. Remove or disconnect the following:
   - Cylinder head cover
   - Upper front timing belt cover

3. Rotate the crankshaft so the No. 1

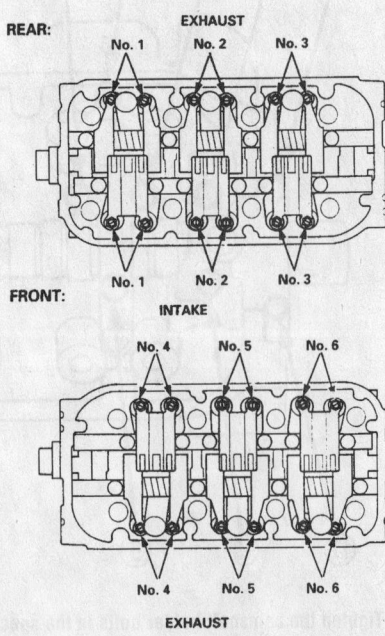

REAR:

EXHAUST

No. 1 No. 2 No. 3

No. 1 No. 2 No. 3

FRONT:

INTAKE

No. 4 No. 5 No. 6

No. 4 No. 5 No. 6

EXHAUST

Adjusting screw locations for valve lash adjustment—3.0L engine

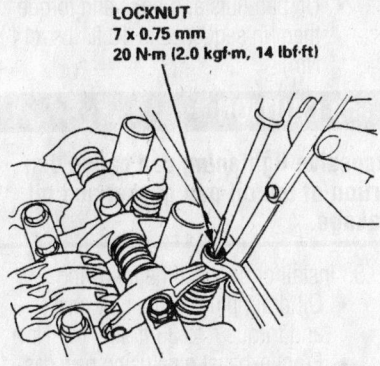

LOCKNUT
7 x 0.75 mm
20 N·m (2.0 kgf·m, 14 lbf·ft)

7923BG74

Slide the feeler gauge between the valve and rocker arm while turning the adjusting screw—3.0L engine

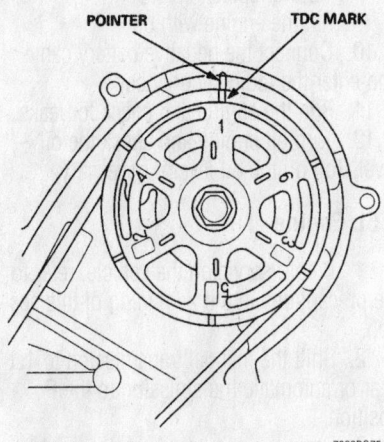

POINTER   TDC MARK

7923BG75

Camshaft sprocket position when No. 1 piston is at TDC—3.0L engine

piston is at TDC on compression to adjust the valves for the No. 1 cylinder.

4. Loosen the locknuts and adjust the screws until a slight drag can be felt with the feeler gage when the gage is placed between the valve and rocker arm tip as shown. The specifications are as follows:
- Intake: 0.006–0.007 in. (0.15–0.18mm)
- Exhaust: 0.007–0.008 in. (0.18–0.20mm)

5. Rotate the crankshaft clockwise until the No. 4 on the camshaft sprocket is near the pointer on the rear cover. This is the No. 4 cylinder firing position.

6. Adjust the valves for the No. 4 cylinder while the sprocket is in this position. Tighten the locknuts to 14 ft. lbs. (20 Nm).

7. Continue to rotate the crankshaft and adjust the valves for cylinders 3 and 2.

8. Install the timing belt and cylinder head covers.

3.2L and 3.5L Engines

These engines are equipped with hydraulic valve lash adjusters on the rocker arms. No valve clearance adjustments are possible or necessary.

Starter Motor

REMOVAL & INSTALLATION

Except 3.5L Engine

1. Disconnect the negative battery cable.
2. On the 1.8L engine, remove the intake air duct.
3. On 2.3L engine, remove the radiator lower hose from the bracket on the starter
4. On 2.5L engines, remove the intake manifold rear bracket assembly.
5. On 3.0L engines, remove the automatic transmission cooler hose from the bracket on the starter motor.
6. Remove or disconnect the following:
- Starter electrical connectors
- 2 starter mounting bolts
- Starter motor

To install:

7. Installation is the reverse of the removal procedure. Tighten the starter motor bolts to 33 ft. lbs. (44 Nm).

➡**When installing the starter cable, be sure to place the closed loop connector over the stud on the starter with the crimped side of the connector facing up. This is to ensure a proper fit against the stud.**

3.5L Engine

➡**This procedure requires the use of an engine hoist to lift the engine slightly.**

1. Before servicing the vehicle, refer to the precautions in the beginning of this section.
2. Remove or disconnect the following:
- Both battery cables
- Battery and tray
- Alternator and belt
- Left exhaust manifold cover
- Left damper fork
- Left lower ball joint from the suspension
- Left drive shaft
- Transmission stop collar
- Exhaust system Y-pipe
- Front mounting bolts
3. Attach a suitable engine hoist and slightly lift the engine.
4. Remove or disconnect the following:

- Left engine mount bracket
- Starter electrical connectors
- Starter motor

To install:

5. Install or connect the following:
- Starter motor and tighten the bolts to 33 ft. lbs. (44 Nm)
- Starter electrical connectors

➡**Upon installation of the starter cable and the black/white wire, make sure that the crimped side of the connector is facing up.**

6. Lower the engine onto the motor mount and tighten the nut to 47 ft. lbs. (64 Nm) and the bolts to 40 ft. lbs. (54Nm)

7. Install or connect the following:
- Exhaust system Y-pipe with new gaskets and tighten the 10mm bolts and nuts to 40 ft. lbs. (54Nm) and the 8mm nuts to 16 ft. lbs. (22Nm)
- Transmission stop collar and tighten the bolts to 28 ft. lbs. (38Nm)
- Left drive shaft
- Damper fork
- Left lower ball joint
- Left exhaust manifold cover
- Alternator and belt
- Battery tray
- Battery

Oil Pan

REMOVAL & INSTALLATION

1.8L and 2.3L Engines

1. Before servicing the vehicle, refer to the precautions in the beginning of this section.
2. Raise and safely support the vehicle.
3. Drain the engine oil.
4. Remove or disconnect the following:
- Negative battery cable
- Splash shield
- Heated Oxygen (HO2S) sensor connector
- Front exhaust pipe from the vehicle
- Center beam from the subframe
5. Loosen the oil pan bolts in a criss-cross pattern. To remove the oil pan, lightly tap the corners of the oil pan with a rubber or plastic faced mallet.

To install:

6. Apply liquid gasket to the oil pan mating surface where the oil pump and the right side cover meet the engine block.
7. Install or connect the following:
- Oil pan with a new gasket

For Accessory Drive Belt illustrations, see Section 1 of this manual

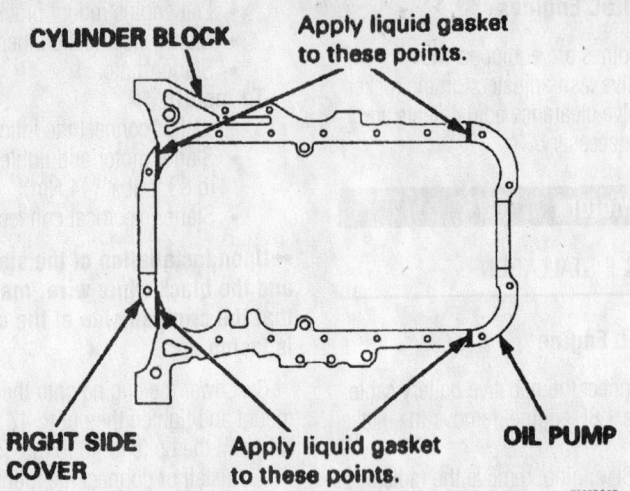

Apply liquid gasket to the oil pan as shown—1.8L (B18B1, B18C1 and B18C5) engines

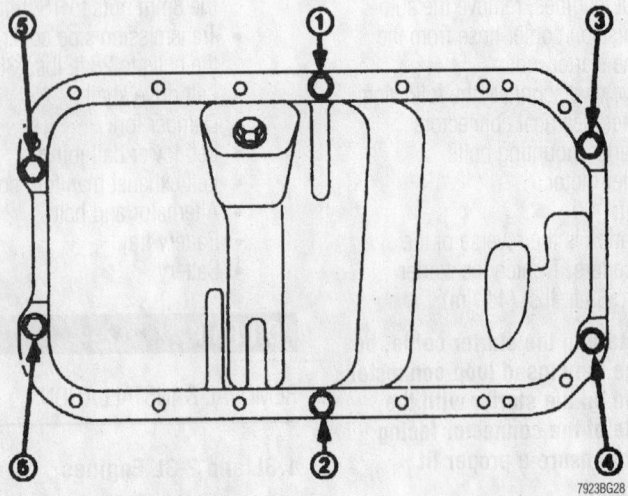

Oil pan bolt tightening sequence—1.8L (B18B1, B18C1 and B18C5) engines

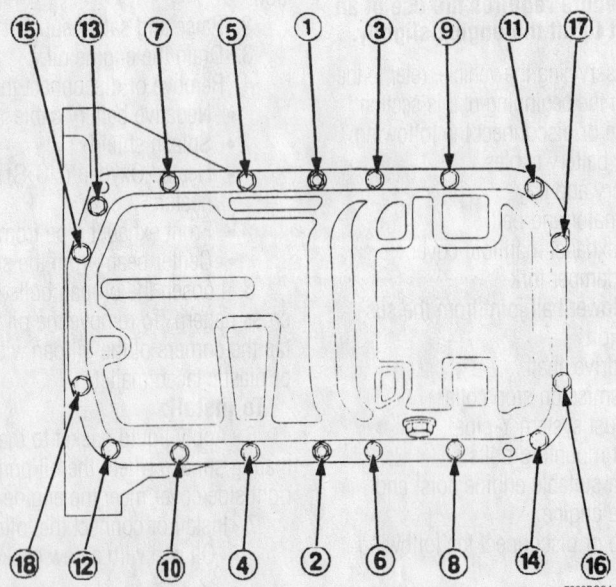

Oil pan bolt tightening sequence—2.3L engine

• Oil pan nuts and bolts and torque them in sequence to 10 ft. lbs. (14 Nm)

✳✳ CAUTION

Excessive tightening can cause distortion of the oil pan gasket and oil leakage.

8. Install or connect the following:
   • Oil drain plug with a new gasket and torque it to 33 ft. lbs. (44 Nm)
   • Front exhaust pipe using new gaskets and locknuts and tighten the manifold nuts to 40 ft. lbs. (54 Nm) and the others to 16 ft. lbs. (22 Nm)
   • HO2S sensor connector
   • Lower splash shield
9. Fill the engine with oil.
10. Connect the negative battery cable and enter the radio security code.
11. Run the engine and check for leaks.
12. Turn off engine and check the oil level. Top off the oil level if necessary.

2.5L Engine

1. Before servicing the vehicle, refer to the precautions in the beginning of this section.
2. Shift the manual transmission to 1st gear or automatic transmission to the **P** position.
3. Drain the engine oil, engine coolant and differential oil.
4. Remove or disconnect the following:
   • Negative battery cable
   • Intake air duct
   • A/C compressor without disconnecting the hoses
   • Front wheels
   • Strut forks
   • Lower ball joints
5. Attach a chain hoist to the engine.
   • Transmission mount and bracket
   • Extension shaft from the differential using a puller
   • Side splash shield
   • Left front engine mount bracket
   • Vehicle speed/power steering speed sensor without disconnecting the hoses
   • Differential oil cooler hoses
   • Differential mounting bolts and the 26mm shim
   • Differential from the vehicle
   • Intermediate shaft from the vehicle
   • A/C compressor mounting bracket
   • Set plate and oil pan inner pipe
   • Oil pan

To install:
6. Apply an even bead of liquid gasket

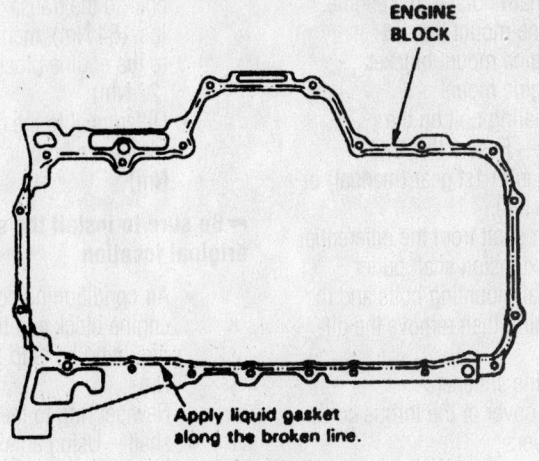

ENGINE BLOCK

Apply liquid gasket along the broken line.

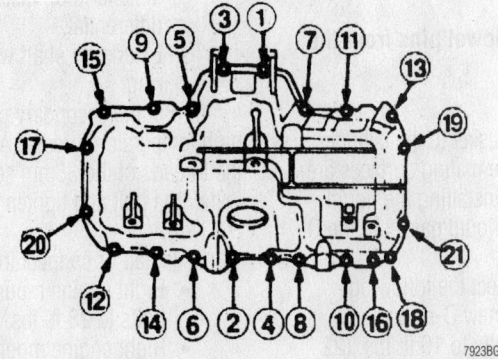

Apply liquid gasket and tighten the oil pan bolts as shown—2.5L engine

to the engine block sealing surface. Apply some liquid gasket to the inner threads of the bolt holes.

7. Install or connect the following:
  • Oil pan and torque the bolts to 16 ft. lbs. (22 Nm)
  • Oil pan inner pipe with new O-rings
  • Inner pipe set plate and torque the bolts to 108 inch lbs. (12 Nm)
  • A/C compressor mounting bracket and torque the bolts to 36 ft. lbs. (49 Nm)
  • Intermediate shaft and tighten the bolts to 16 ft. lbs. (22 Nm)
  • Differential, making sure the shim is in the proper position and tighten the bolts to 54 ft. lbs. (75 Nm)

8. Pack the extension shaft cavity with high temperature grease and install the 33mm sealing bolt. Tighten the bolt to 58 ft. lbs. (80 Nm) and install the secondary cover.

9. Install or connect the following:
  • Differential oil cooler hoses
  • Vehicle speed/power steering speed sensor and tighten the mounting bolt to 89 inch lbs. (10 Nm)
  • Left front engine mount bracket and tighten the bolts to 47 ft. lbs. (64 Nm)

  • Side splash shield
  • New set ring on the extension shaft—Coat the splines and their mating surfaces with high temperature grease

  • Extension shaft
  • Transmission mount bracket and torque the bolts to 40 ft. lbs. (54 Nm)
  • Transmission mount and torque the bolts to 47 ft. lbs. (64 Nm)
  • Lower ball joints and torque the nuts to 36–43 ft. lbs. (49–59 Nm)
  • Strut forks and torque the bolts to 47 ft. lbs. (64 Nm)
  • Front wheels
  • A/C compressor onto the mount and tighten the bolts to 16 ft. lbs. (22 Nm)
  • Intake air duct
  • Negative battery cable

10. Refill the differential.
11. Remove the chain hoist.
12. Refill the engine oil and cooling system.
13. Install the air cleaner and intake duct.
14. Bleed the cooling system by opening the bleeder on the upper radiator hose inlet when filling the system.
15. Connect the negative battery cable.

3.0L Engine

1. Before servicing the vehicle, refer to the precautions in the beginning of this section.
2. Raise and safely support the vehicle.
3. Drain the engine oil.
4. Remove or disconnect the following:
  • Front exhaust pipe
  • Oil pan mounting bolts
  • Oil pan

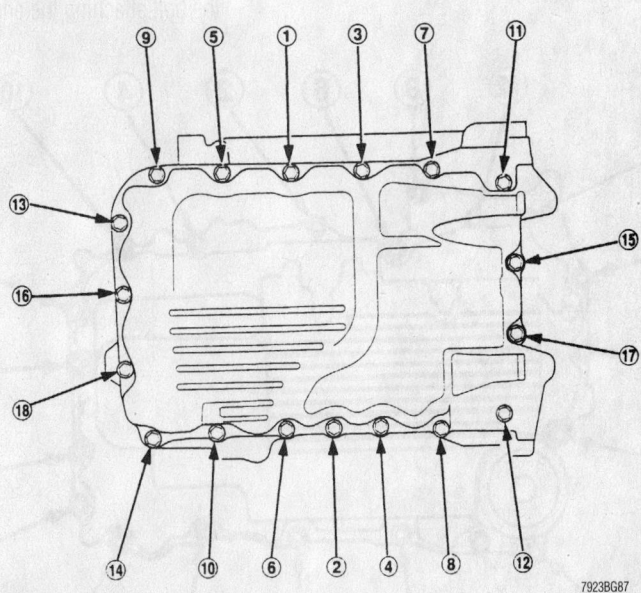

Oil pan bolt tightening sequence—3.0L engine

For Tire, Wheel and Ball Joint specifications, see Section 1 of this manual

To install:

5. Apply a bead of sealant to the oil pan flange and install the oil pan. Tighten the bolts in the sequence shown to 108 inch lbs. (12 Nm).

6. Install the front exhaust pipe with new gaskets and torque the manifold nuts to 40 ft. lbs. (54 Nm) and the catalytic converter nuts to 25 ft. lbs. (33 Nm).

7. Add the correct amount of engine oil to the crankcase.

8. Start the engine and check for leaks.

3.2L and 3.5L Engines

1. Before servicing the vehicle, refer to the precautions in the beginning of this section.

2. Drain the engine oil.

3. Drain the differential oil.

4. Remove or disconnect the following:
- Negative battery cable
- Accessory drive belts
- Power steering pump without disconnecting the lines
- Exhaust manifold covers
- Front wheels
- Splash shield
- Strut forks
- Lower ball joints
- Halfshafts from the differential
- Intermediate shaft
- Vehicle speed/power steering speed sensor without disconnecting the hoses
- Lower plate from the rear beam
- A/C compressor without disconnecting the lines

5. Attach a chain hoist to the engine.
- Left engine mount bracket
- Right engine mount bracket
- Right engine mount
- 36mm sealing bolt on the transaxle—Ensure that the transaxle is in 1st gear (manual) or **P** (automatic).
- Extension shaft from the differential with an extension shaft puller
- Differential mounting bolts and the 26mm shim, then remove the differential
- Rear engine stiffener
- Flywheel cover or the torque converter covers
- Oil pan

➡**Do not lose the dowel pins from the oil pan**

To install:

6. Apply liquid gasket to the cylinder block. Be sure that the mating surfaces are clean and dry before installing the liquid gasket. Do not apply liquid gasket to the O-ring grooves.

7. Install or connect the following:
- Oil pan with new O-rings and torque the bolts to 16 ft. lbs. (22 Nm)

➡**Be sure the dowel pins are still in place.**

- Flywheel or torque converter cover and torque the bolts to 108 inch lbs. (12 Nm)
- Rear engine stiffener and tighten the bolt attaching the engine stiffener to the transaxle first, to 47 ft. lbs. (64 Nm), then tighten the bolts to the engine block to 16 ft. lbs. (22 Nm)
- Differential to the engine and torque the bolts to 47 ft. lbs. (64 Nm)

➡**Be sure to install the shim in the original location**

- Air conditioning compressor to the engine block and tighten the mounting bolts to 16 ft. lbs. (22 Nm)
- New set ring to the extension shaft—Using an extension shaft installer tool, install the shaft to the differential.
- Extension shaft with a new set ring

8. Fill the secondary gear with super high temperature grease. Applying sealer to the threads of the 36mm sealing bolt, then install the bolt and tighten it to 58 ft. lbs. (78 Nm).

9. Install or connect the following:
- Right engine mount and torque the bolts to 28 ft. lbs. (38 Nm)
- Right engine mount bracket and torque the nut and bolts to 47 ft. lbs. (64 Nm)
- Left engine mount bracket and torque the bolts to 40 ft. lbs. (54 Nm)
- A/C compressor and torque the bolts to 16 ft. lbs. (22 Nm)
- Lower plate and torque the bolts to 28 ft. lbs. (38 Nm)
- Vehicle speed/power steering speed sensor and torque the bolts to 108 inch lbs. (12 Nm)
- Intermediate shaft and the halfshafts
- Lower ball joints and tighten the nuts to 40 ft. lbs. (54 Nm)
- Strut forks and torque the bolts to 51 ft. lbs. (69 Nm)
- Engine splash shield and tighten the bolts to 84 inch lbs. (9.5 Nm)
- Front wheels
- Exhaust manifold covers and torque the bolts to 16 ft. lbs. (22 Nm)
- Power steering pump and torque the bolt to 33 ft. lbs. (44 Nm) and the nut to 16 ft. lbs. (22 Nm)
- Accessory drive belts
- Negative battery cable

10. Fill the engine with oil.

11. Fill the differential with oil.

12. Run the engine and check for leaks.

13. Check the front wheel alignment.

Be sure to tighten the oil pan bolts in the sequence shown—3.2L engines

7923BG30

Oil Pump

REMOVAL & INSTALLATION

All Models Except NSX

1. Before servicing the vehicle, refer to the precautions in the beginning of this section.

2. Drain the engine oil.

3. Be sure the crankshaft is at Top Dead Center (TDC) on the No. 1 cylinder.

4. Remove or disconnect the following:
- Negative battery cable
- Cylinder head cover
- Timing belt cover
- Timing belt. Refer to the Timing Belt unit repair section.
- Crankshaft Position (CKP) sensor (if necessary)
- Crankshaft timing belt gear
- Oil pan
- Pickup screen
- Oil pump from the front of the engine

➡Any time the oil pump is removed, the front oil seal should be replaced.

To install:

5. Install a new oil seal in the oil pump.

6. Apply liquid gasket to the mounting surface of the oil pump.

7. Install the oil pump, using new O-

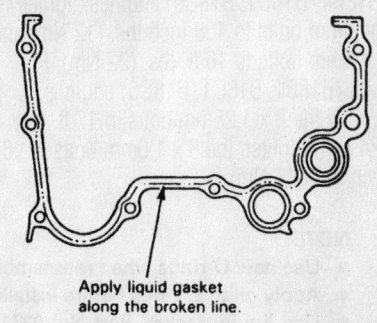

OIL PUMP HOUSING

Apply liquid gasket along the broken line.

7923BG32

Apply sealant to the oil pump sealing surface as shown—2.5L engine

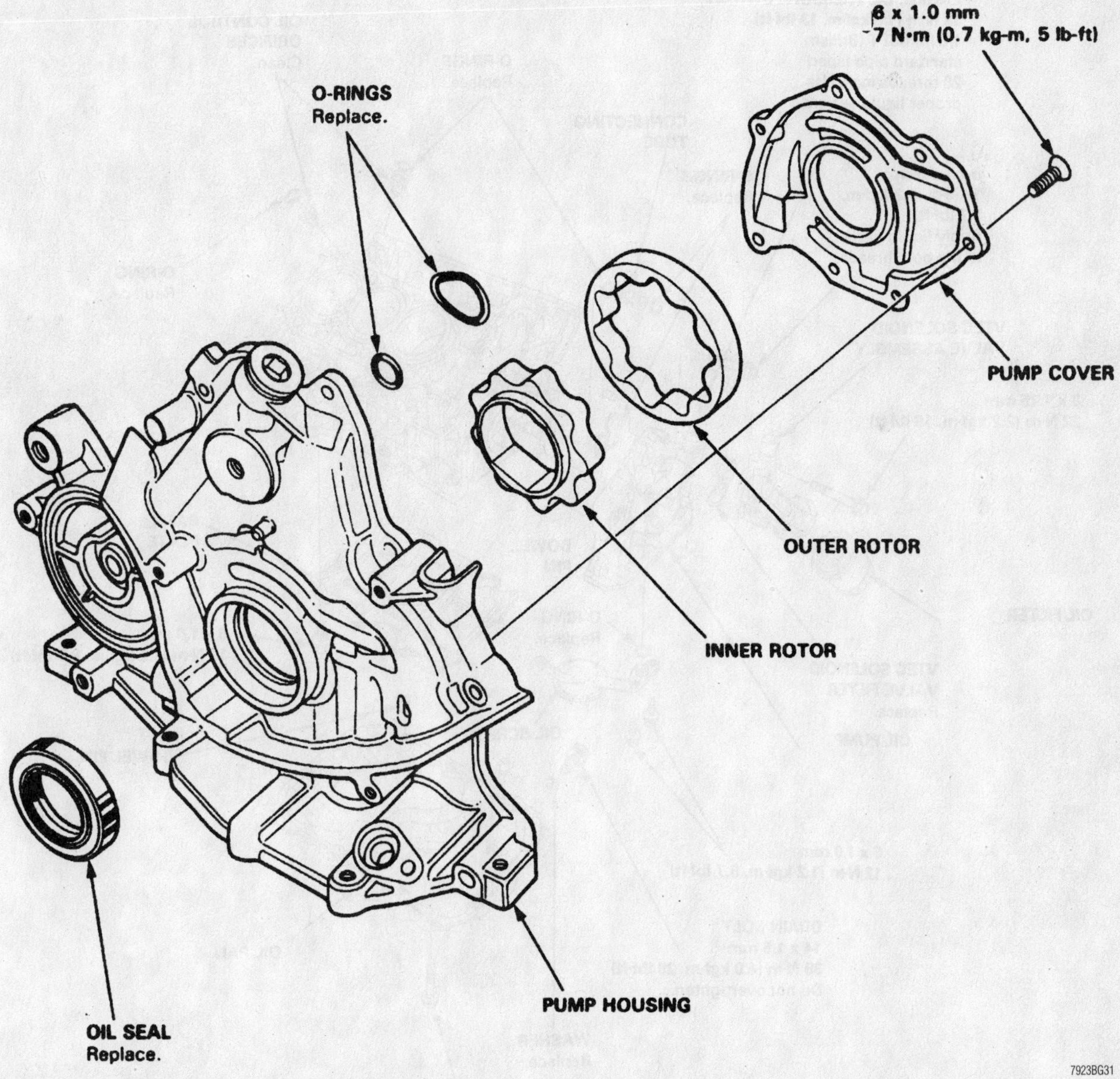

6 x 1.0 mm
7 N·m (0.7 kg-m, 5 lb-ft)

O-RINGS
Replace.

PUMP COVER

OUTER ROTOR

INNER ROTOR

PUMP HOUSING

OIL SEAL
Replace.

7923BG31

Exploded view of the oil pump—1.8L (B18B1) engines

rings. For all engines, except the 1.8L (B18B1, B18C1, B18C5) engines, tighten the 6mm bolts to 108 inch lbs. (12 Nm) and the 8mm bolts to 16 ft. lbs. (22 Nm). For 1.8L (B18B1, B18C1, B18C5) engines, tighten the 8 x 1.25mm bolts to 17 ft. lbs. (24 Nm), tighten the 6 x 1.0mm bolts to 96 inch lbs. (11 Nm).

NOTE:
- Use new O-rings when reassembling.
- Apply oil to O-rings before installation.
- Use liquid gasket, Part No. 08718 – 0001 or 08718 – 0003.
- Clean the oil control orifice before installing.

✸✸ WARNING

The B18B1, B18C1 and B18C5 engines use different oil pumps, be sure that you have the correct oil pump. Match the crankshaft timing mark on the new oil pump with the timing mark on the old oil pump, because the timing marks are in different locations. If an oil pump is used with the timing mark in the wrong position the pistons may contact the valves.

8. Install or connect the following:
- Oil pump pickup screen

CAUTION: Do not overtighten the drain bolt.

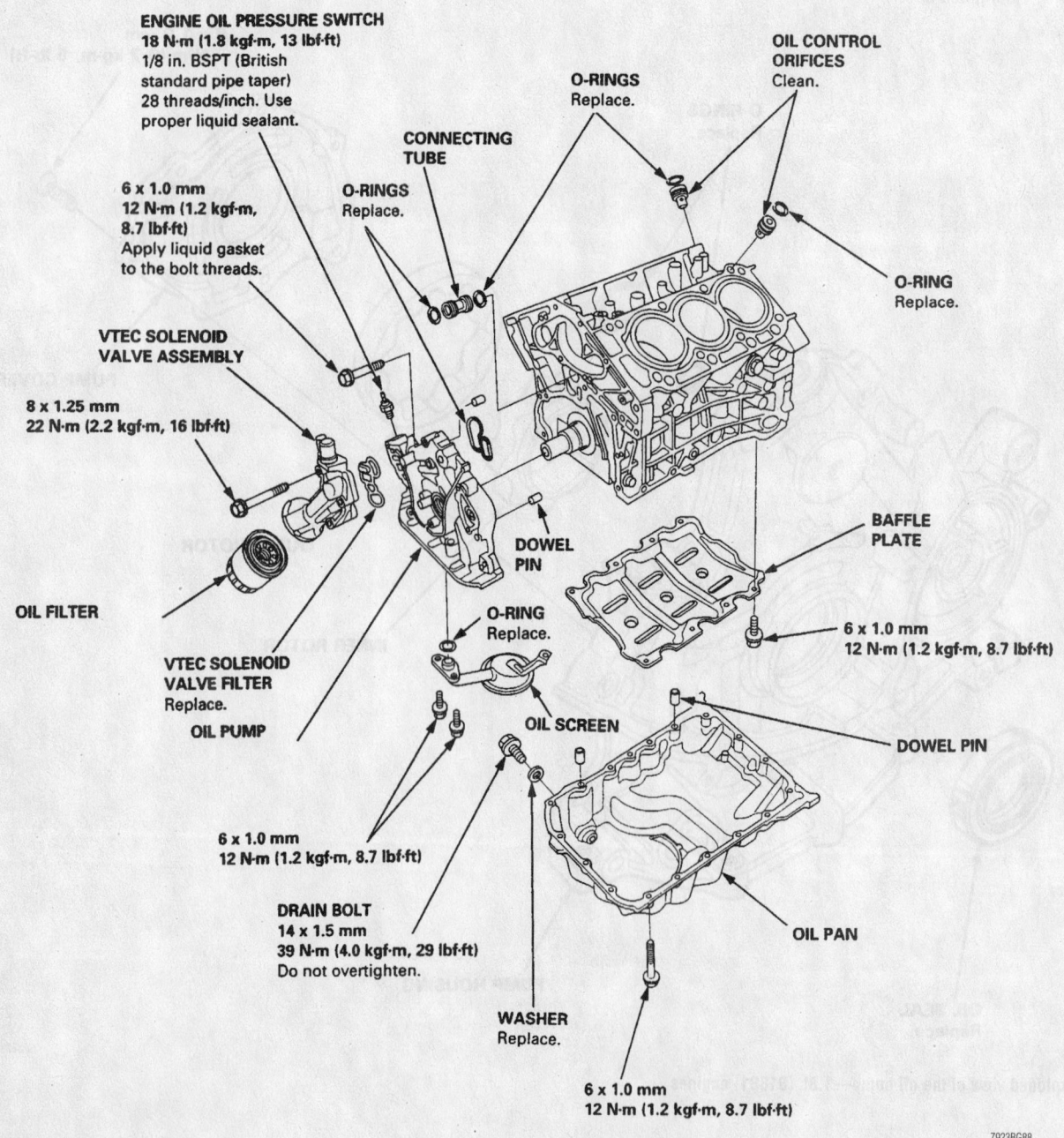

ENGINE OIL PRESSURE SWITCH
18 N·m (1.8 kgf·m, 13 lbf·ft)
1/8 in. BSPT (British standard pipe taper)
28 threads/inch. Use proper liquid sealant.

6 x 1.0 mm
12 N·m (1.2 kgf·m, 8.7 lbf·ft)
Apply liquid gasket to the bolt threads.

VTEC SOLENOID VALVE ASSEMBLY

8 x 1.25 mm
22 N·m (2.2 kgf·m, 16 lbf·ft)

OIL FILTER

VTEC SOLENOID VALVE FILTER
Replace.
OIL PUMP

O-RINGS
Replace.

CONNECTING TUBE

O-RINGS
Replace.

OIL CONTROL ORIFICES
Clean.

O-RING
Replace.

DOWEL PIN

O-RING
Replace.

OIL SCREEN

BAFFLE PLATE

6 x 1.0 mm
12 N·m (1.2 kgf·m, 8.7 lbf·ft)

DOWEL PIN

6 x 1.0 mm
12 N·m (1.2 kgf·m, 8.7 lbf·ft)

DRAIN BOLT
14 x 1.5 mm
39 N·m (4.0 kgf·m, 29 lbf·ft)
Do not overtighten.

WASHER
Replace.

OIL PAN

6 x 1.0 mm
12 N·m (1.2 kgf·m, 8.7 lbf·ft)

Lubrication system—3.0L engine

7923BG88

NOTE:
- Use new O-rings when reassembling.
- Apply oil to O-rings before installation.
- Use liquid gasket, Part No. 08718 – 0001 or 08718 – 0003.
- Clean the oil control orifice before installing.
- Remove the balancer shaft

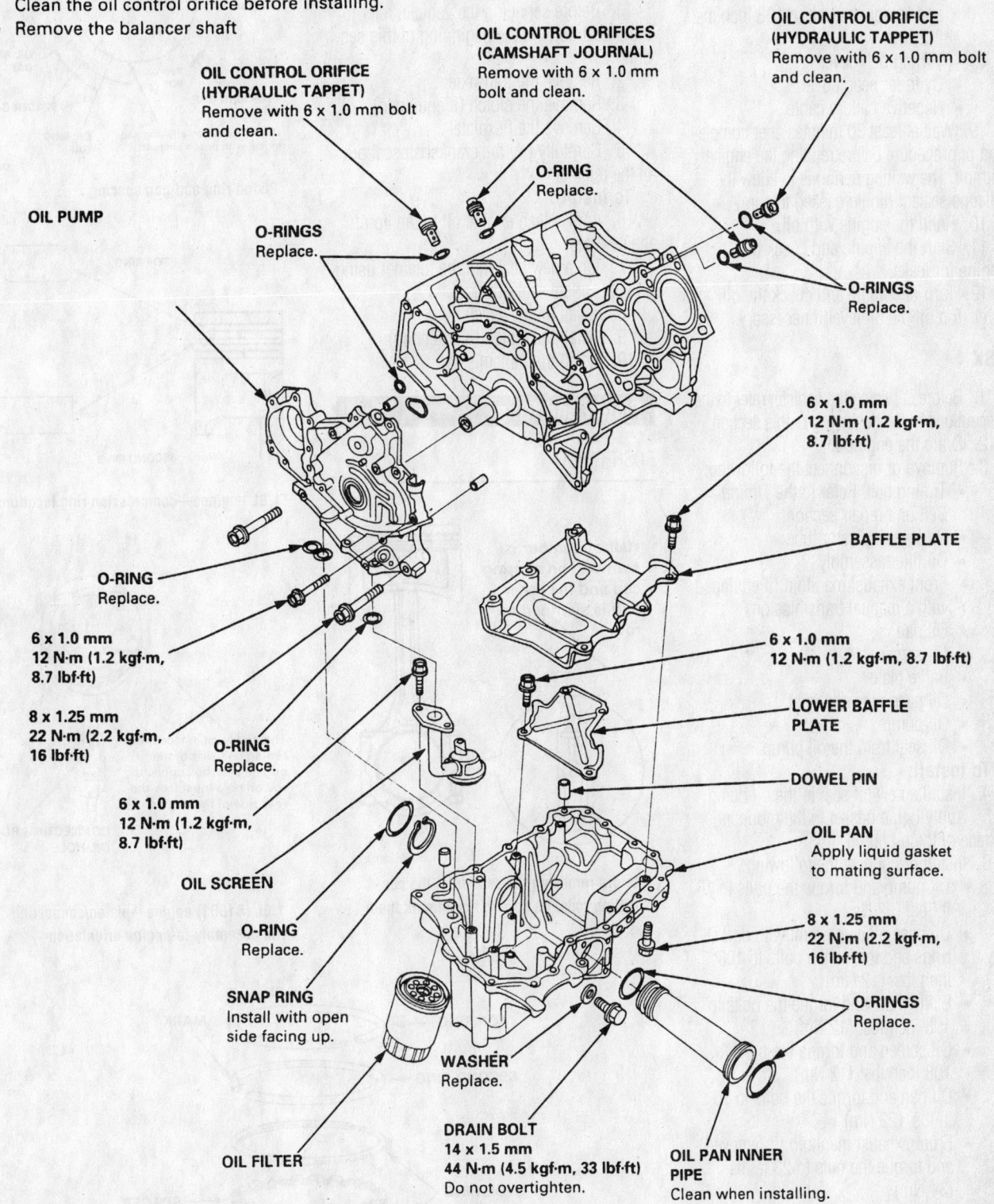

OIL CONTROL ORIFICE (HYDRAULIC TAPPET)
Remove with 6 x 1.0 mm bolt and clean.

OIL CONTROL ORIFICES (CAMSHAFT JOURNAL)
Remove with 6 x 1.0 mm bolt and clean.

OIL CONTROL ORIFICE (HYDRAULIC TAPPET)
Remove with 6 x 1.0 mm bolt and clean.

O-RING
Replace.

O-RINGS
Replace.

OIL PUMP

O-RINGS
Replace.

O-RING
Replace.

6 x 1.0 mm
12 N·m (1.2 kgf·m, 8.7 lbf·ft)

BAFFLE PLATE

6 x 1.0 mm
12 N·m (1.2 kgf·m, 8.7 lbf·ft)

8 x 1.25 mm
22 N·m (2.2 kgf·m, 16 lbf·ft)

O-RING
Replace.

6 x 1.0 mm
12 N·m (1.2 kgf·m, 8.7 lbf·ft)

LOWER BAFFLE PLATE

DOWEL PIN

OIL PAN
Apply liquid gasket to mating surface.

OIL SCREEN

O-RING
Replace.

SNAP RING
Install with open side facing up.

8 x 1.25 mm
22 N·m (2.2 kgf·m, 16 lbf·ft)

O-RINGS
Replace.

WASHER
Replace.

OIL FILTER

DRAIN BOLT
14 x 1.5 mm
44 N·m (4.5 kgf·m, 33 lbf·ft)
Do not overtighten.

OIL PAN INNER PIPE
Clean when installing.

7923BG89

Exploded view of the lubrication system—3.5L engine

- Oil pan and tighten the bolts to 108 inch lbs. (12 Nm)
- Crankshaft timing belt gear
- Timing belt
- Crankshaft Position (CKP) sensor and torque the bolt to 108 inch lbs. (12 Nm)
- Timing belt cover
- Cylinder head cover
- Negative battery cable

9. Wait at least 30 minutes after completion of procedure before refilling the engine with oil. The waiting period is to allow the silicone sealant (liquid gasket) to cure.

10. Refill the engine with oil.

11. Start the engine and check the engine for leaks.

12. Turn off engine and check the oil level. Top off the oil level if necessary.

NSX

1. Before servicing the vehicle, refer to the precautions in the beginning of this section.

2. Drain the engine oil.

3. Remove or disconnect the following:
- Timing belt. Refer to the Timing Belt unit repair section.
- Oil level indicator tube
- Oil filter assembly
- Front exhaust manifold (if equipped with a manual transmission)
- Oil pan
- Oil screen
- Baffle plate
- Oil pass pipe and joint
- Oil pump
- Oil seal from the oil pump

To install:

4. Install a new oil seal in the oil pump

5. Apply liquid gasket to the mounting surface of the oil pump.

6. Install or connect the following:
- Oil pump and torque the bolts to 16 ft. lbs. (22 Nm)
- Oil pass pipe and joint with new O-rings and torque the bolts to 108 inch lbs. (12 Nm)
- Baffle plate and torque the bolts to 108 inch lbs. (12 Nm)
- Oil screen and torque the bolts to 108 inch lbs. (12 Nm)
- Oil pan and torque the bolts to 16 ft. lbs. (22 Nm)
- Front exhaust manifold (if removed) and torque the nuts to 23 ft. lbs. (31 Nm)
- Oil filter assembly
- Oil level indicator tube
- Timing belt

7. Refill the engine oil.

8. Start the engine and check for leaks.

Rear Main Seal

REMOVAL & INSTALLATION

1. Before servicing the vehicle, refer to the precautions in the beginning of this section.

2. Remove the transaxle.

3. Remove the clutch (if equipped)

4. Remove the flexplate.

5. Carefully pry the crankshaft seal out of the retainer.

To install:

6. Apply clean engine oil to the lip of the new seal.

7. Tap a new seal into the retainer using an appropriate seal driver.

8. Install the flywheel.

9. Install the clutch (if equipped)

10. Install the transmission.

Piston and Ring

POSITIONING

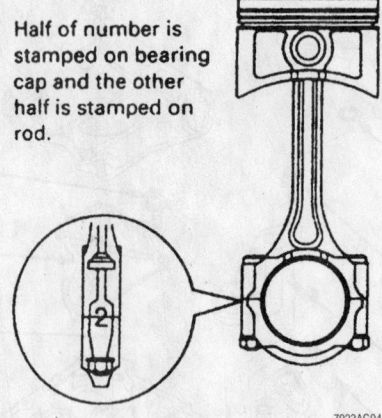

Before removing the caps from the connecting rods, be sure to matchmark them as shown

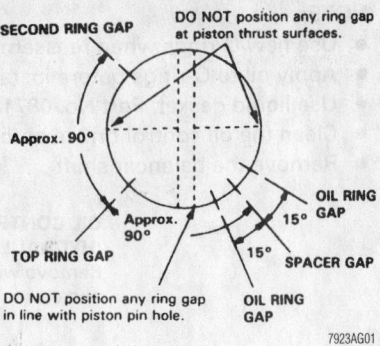

Piston ring end-gap spacing

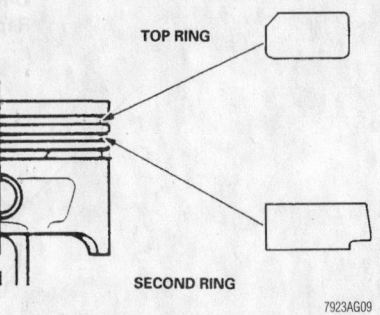

1.8L engines—compression ring locations

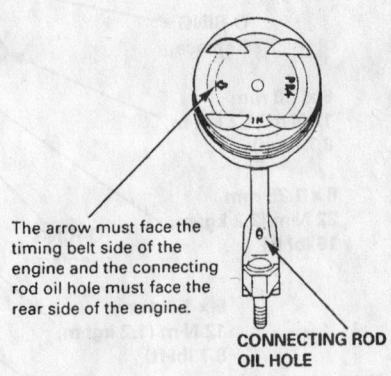

1.8L (B18B1) engine—piston/connecting rod assembly-to-engine orientation

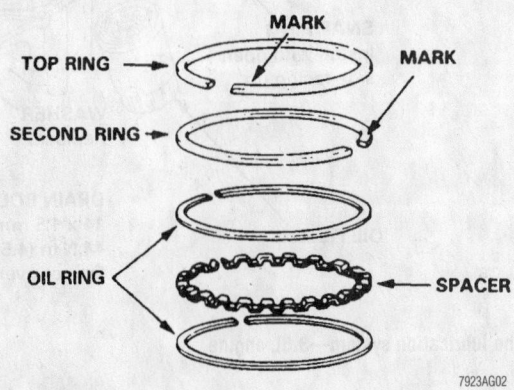

Piston ring positioning

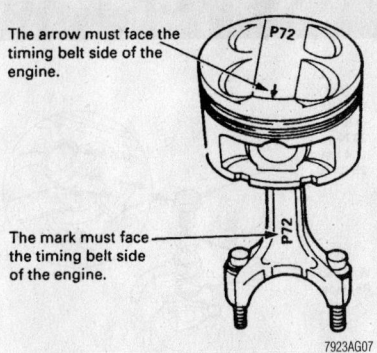

1.8L (B18C1) engine—piston/connecting rod assembly-to-engine orientation

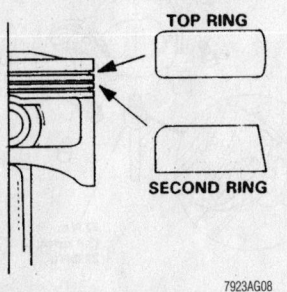

2.3L, 2.5L, 3.0L (J30A1) and 3.2L (C32A1) engines—compression ring locations

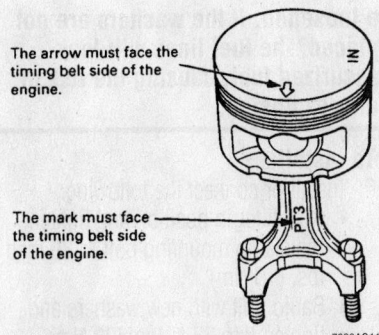

2.3L (F23A1) engines—piston/connecting rod assembly-to-engine orientation

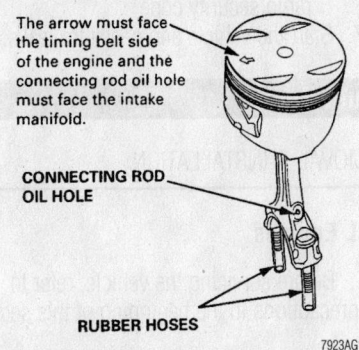

2.5L engine—piston/connecting rod assembly-to-engine orientation

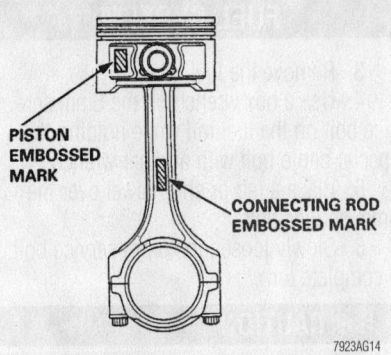

3.0L (J30A1) engine—piston-to-connecting rod assembly

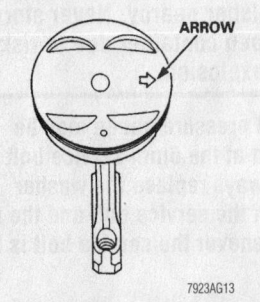

3.0L (J30A1) engine—piston directional arrow location

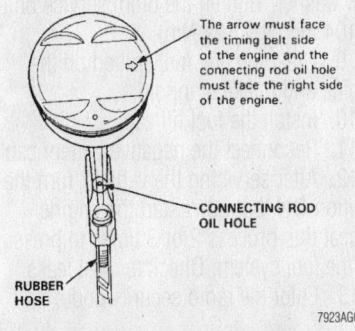

3.2L (C32A6) and (J32A1) engines—piston/connecting rod assembly-to-engine orientation

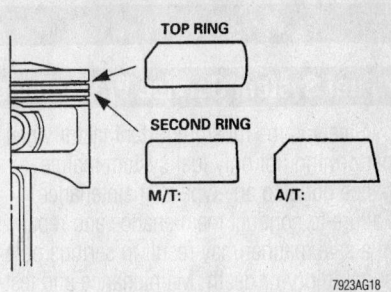

3.0L (C30A1), (C32A6), (J32A1) and 3.2L (C32B1) engines—compression ring locations

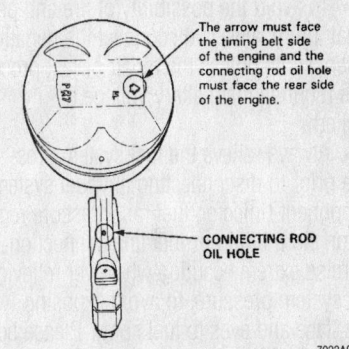

3.0L (C30A1) and 3.2L (C32B1) engines—piston/connecting rod assembly-to-engine orientation

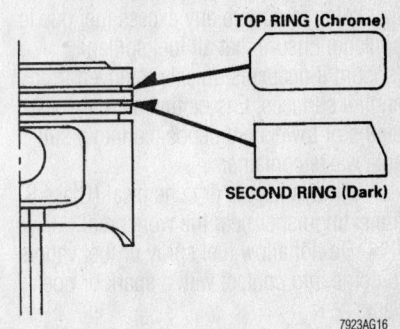

3.5L engine—compression ring locations

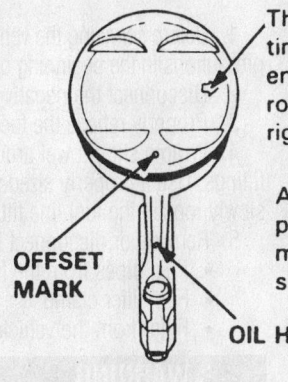

The arrow must face the timing belt side of the engine and the connecting rod oil hole must face the right side of the engine.

Assemble the rod and the piston with the off-set mark and oil hole on the same side.

3.5L engine—piston/connecting rod assembly-to-engine orientation

For Tune-up, Capacities and Firing orders, see Section 1 of this manual

FUEL SYSTEM

Fuel System Service Precautions

Safety is the most important factor when performing not only fuel system maintenance but also any type of maintenance. Failure to conduct maintenance and repairs in a safe manner may result in serious personal injury or death. Maintenance and testing of the vehicle's fuel system components can be accomplished safely and effectively by adhering to the following rules and guidelines.

• To avoid the possibility of fire and personal injury, always disconnect the negative battery cable unless the repair or test procedure requires that battery voltage be applied.

• Always relieve the fuel system pressure prior to disconnecting any fuel system component (injector, fuel rail, pressure regulator, etc.), fitting or fuel line connection. Exercise extreme caution whenever relieving fuel system pressure, to avoid exposing skin, face and eyes to fuel spray. Please be advised that fuel under pressure may penetrate the skin or any part of the body that it contacts.

• Always place a shop towel or cloth around the fitting or connection prior to loosening to absorb any excess fuel due to spillage. Ensure that all fuel spillage (should it occur) is quickly removed from engine surfaces. Ensure that all fuel soaked cloths or towels are deposited into a suitable waste container.

• Always keep a dry chemical (Class B) fire extinguisher near the work area.

• Do not allow fuel spray or fuel vapors to come into contact with a spark or open flame.

• Always use a back-up wrench when loosening and tightening fuel line connection fittings. This will prevent unnecessary stress and torsion to fuel line piping. Always follow the proper torque specifications.

• Always replace worn fuel fitting O-rings with new. Do not substitute fuel hose or equivalent, where fuel pipe is installed.

Fuel System Pressure

RELIEVING

1. Before servicing the vehicle, refer to the precautions in the beginning of this section.
2. Disconnect the negative battery cable.

3. Remove the fuel fill cap.
4. Use a box wrench on the 6mm service bolt on the fuel rail while holding the special banjo bolt with another wrench.
5. Place a rag or shop towel over the 6mm service bolt.
6. Slowly loosen the 6mm service bolt 1 complete turn.

✳✳ CAUTION

Do not allow fuel spray or fuel vapors to come in contact with a spark or open flame. Keep a dry chemical fire extinguisher nearby. Never store fuel in an open container due to risk of fire or explosion.

➡**A fuel pressure gauge may be attached at the 6mm service bolt location. Always replace the washer between the service bolt and the banjo bolt whenever the service bolt is loosened.**

7. Properly dispose of the rag or shop towel.
8. Remove the service bolt and install a new washer. Tighten the 6mm service bolt to 104 inch lbs. (12 Nm).
9. Clean up any fuel spilled on the engine and intake manifold.
10. Install the fuel fill cap.
11. Reconnect the negative battery cable.
12. After servicing the vehicle, turn the ignition **ON**, but don't start the engine. Repeat this process 2 or 3 times to pressurize the fuel system. Check for fuel leaks.
13. Enter the radio security code.

Fuel Filter

REMOVAL & INSTALLATION

1. Before servicing the vehicle, refer to the precautions in the beginning of this section.
2. Disconnect the negative battery cable.
3. Properly relieve the fuel pressure.
4. Wrap a shop towel around the filter fittings. Use a properly sized wrench to slowly loosen the fuel line fittings.
5. Remove or disconnect the following:
• Fuel pipes from the fuel filter
• Fuel filter clamp
• Filter from the vehicle

✳✳ WARNING

It is very important that ALL of the fuel line banjo bolt washers be replaced every time the banjo bolts

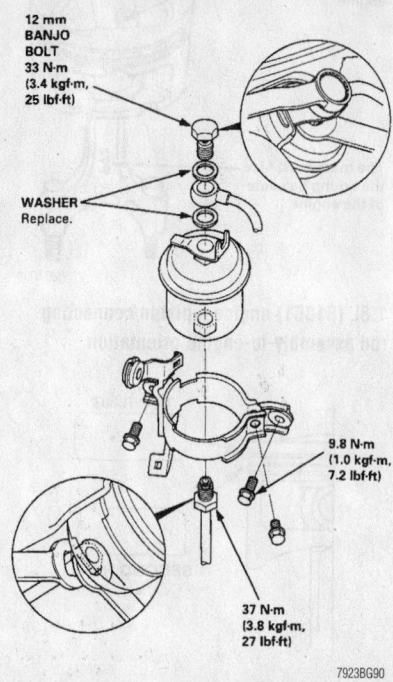

12 mm BANJO BOLT
33 N·m (3.4 kgf·m, 25 lbf·ft)

WASHER
Replace.

9.8 N·m (1.0 kgf·m, 7.2 lbf·ft)

37 N·m (3.8 kgf·m, 27 lbf·ft)

7923BG90

Fuel filter assembly—3.0CL shown, others are similar

are loosened. If the washers are not replaced, the fuel lines will leak pressurized fuel, causing the risk of fire or explosion.

To install:

6. Install or connect the following:
• New filter in position and tighten the clamp mounting bolt to 89 inch lbs. (10 Nm)
• Banjo bolt with new washers and tighten it to 25 ft. lbs. (33 Nm)
• Fuel feed line and tighten the fitting to 27 ft. lbs. (37 Nm)
• Negative battery cable and enter the radio security code
7. Start the vehicle and check for leaks.

Fuel Pump

REMOVAL & INSTALLATION

1.8L Engines

1. Before servicing the vehicle, refer to the precautions in the beginning of this section.
2. Disconnect the negative battery cable.
3. Properly relieve the fuel system pressure.
4. Remove the rear seat to gain access to the fuel pump access panel.

5. Remove the maintenance access cover.

6. Disconnect the electrical connector from the fuel pump.

7. If equipped with quick connect fittings, hold the fuel line connector with one hand and press down the retainer tabs with the other hand, and then pull the connector off. Check the contact area of the pipe for dirt or damage, clean or replace the pipe or pump as required. Remove the old retainer from the pipe and discard. Cover the connector and pipe with plastic bags to prevent damage and keep foreign material out.

8. Remove the fuel pump mounting nuts, then remove the fuel pump from the fuel tank.

To install:

9. Installation is the reverse of the removal procedure. Tighten the fuel pump mounting nuts to 53 inch lbs. (6 Nm).

2.3L and 3.0L Engines

1. Before servicing the vehicle, refer to the precautions in the beginning of this section.

2. Properly relieve the fuel system pressure.

3. Lower the fuel tank and detach the electrical connector and fuel lines from the pump assembly.

4. Remove the nuts and the fuel pump from the tank.

To install:

5. Installation is the reverse of the removal procedure. Tighten the fuel pump mounting nuts to 53 inch lbs. (6 Nm).

2.5L and 3.2L Engines

1. Before servicing the vehicle, refer to the precautions in the beginning of this section.

2. Remove or disconnect the following:
   - Negative battery cable
   - Left rear wheel
   - Tank drain bolt and drain the fuel into an approved container
   - Pump and float wiring connectors located under the trunk floor
   - Fuel hose and pipe covers from the inside of the quarter panel

3. Support the tank with a transmission jack, remove the straps and lower the tank out of the vehicle. If it sticks on the undercoating, carefully pry it free using a blunt or wooden instrument as a lever.

4. Disconnect the fuel line by removing the banjo bolt or uncoupling the quick-connect fittings.

5. Remove the fuel pump mounting nuts. Remove the fuel pump from the fuel tank.

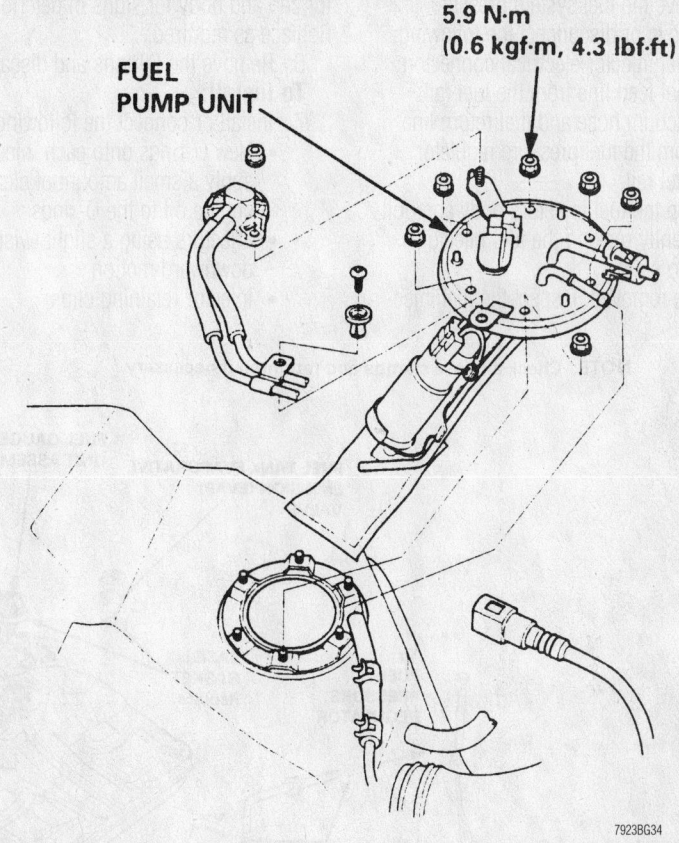

**5.9 N·m
(0.6 kgf·m, 4.3 lbf·ft)**

**FUEL
PUMP UNIT**

7923BG34

Fuel pump mounting—2.5L engine

To install:

6. Installation is the reverse of the removal procedure, while using the following torque values:
   - Fuel pump mounting nuts: 48 inch lbs. (6 Nm)
   - Fuel tank strap bolts: 28 ft. lbs. (38 Nm)
   - Fuel tank drain bolt: 36 ft. lbs. (49–50 Nm)

3.5L Engine

1. Before servicing the vehicle, refer to the precautions in the beginning of this section.

2. Properly relieve the fuel system pressure.

3. Remove or disconnect the following:
   - Rear seat cushion
   - Access panel from the floor
   - Fuel line and wiring from the fuel pump assembly
   - Mounting nuts and the fuel pump from the fuel tank

To install:

4. Installation is the reverse of the removal procedure. Tighten the fuel pump mounting nuts to 53 inch lbs. (6 Nm).

Fuel Injector

REMOVAL & INSTALLATION

❋❋ CAUTION

Fuel injection systems remain under pressure, even after the engine has been turned OFF. The fuel system pressure must be relieved before disconnecting any fuel lines. Failure to do so may result in fire and/or personal injury. Observe all applicable safety precautions when working around fuel. Whenever servicing the fuel system, always work in a well-ventilated area. Do not allow fuel spray or vapors to come in contact with a spark or open flame. Keep a dry chemical fire extinguisher near the work area. Always keep fuel in a container specifically designed for fuel storage; also, always properly seal fuel containers to avoid the possibility of fire or explosion.

1. Disconnect the negative battery cable.

2. Relieve the fuel system pressure.
3. Remove or disconnect the following:
   - Fuel injector electrical connectors
   - Fuel feed line from the fuel rail
   - Vacuum hose and fuel return line from the fuel pressure regulator
   - Fuel rail
4. Grasp the fuel injectors body and pull up while gently rocking the fuel injector from side to side.
5. Once removed, inspect the fuel injec-

tor cap and body for signs of deterioration. Replace as required.
6. Remove the O-rings and discard.

To install:
7. Install or connect the following:
   - New O-rings onto each injector and apply a small amount of clean engine oil to the O-rings
   - Injectors using a slight twisting downward motion
   - Injector retaining clips

- Fuel rail
- Fuel feed line
- Vacuum hose and fuel return line to the fuel pressure regulator
- Fuel injector electrical connectors
- Negative battery cable
8. Run the engine at idle for 2 minutes, then turn the engine **OFF** and check for fuel leaks and proper operation.

NOTE: Check all hose clamps and retighten if necessary.

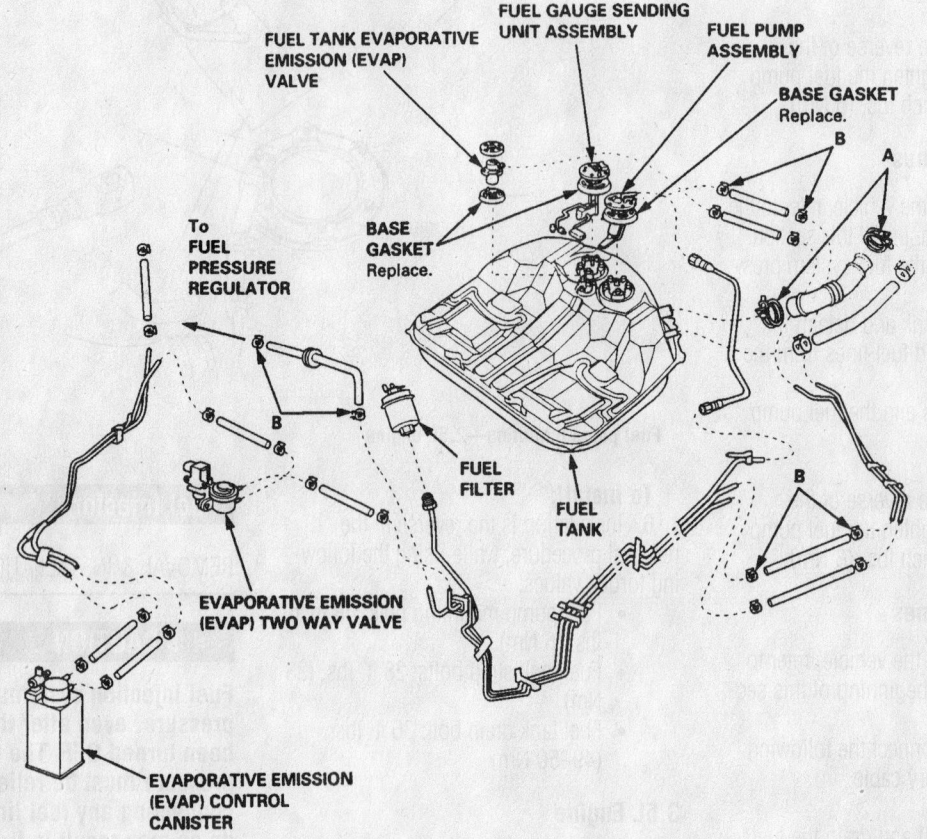

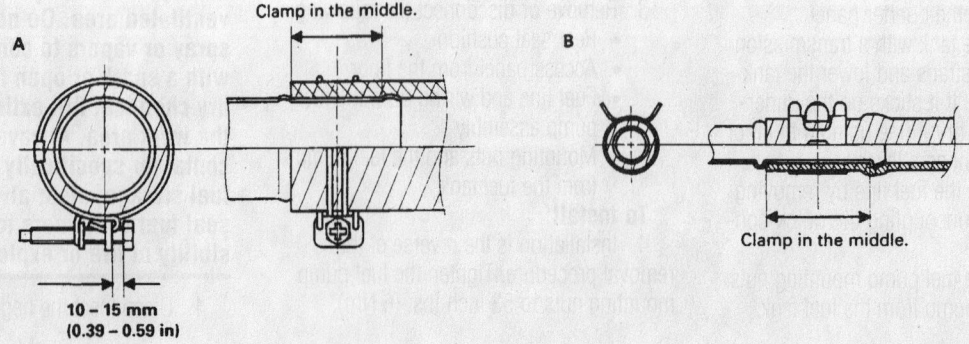

Exploded view of the fuel line routing—3.5L engine

DRIVE TRAIN

Transaxle Assembly

REMOVAL & INSTALLATION

Manual

INTEGRA

1. Before servicing the vehicle, refer to the precautions in the beginning of this section.

2. Disconnect the negative battery cable, then the positive battery cable.

3. Drain the transaxle oil. Install the drain plug with a new washer.

4. Remove or disconnect the following:

- Air cleaner housing and the intake air tube
- Backup light switch connector
- Transaxle ground wire
- Lower radiator hose clamp from the transaxle hanger
- Wiring harness clips
- Starter motor electrical connectors
- Vehicle Speed (VSS) sensor electrical connector
- Clutch pipe bracket and slave cylinder

✳✳ CAUTION

Do not operate the clutch pedal once the slave cylinder has been removed.

- 3 upper transaxle mounting bolts and the lower starter mounting bolt
- Engine splash shield
- Heated Oxygen (HO2S) sensor connector
- Front exhaust pipe from the vehicle
- Ball joints from the lower control arms
- Right strut fork
- Both halfshafts
- Intermediate shaft
- Extension rod
- Shift rod
- Front and rear engine stiffeners
- Clutch cover
- Right front mount/bracket

5. Place a transmission jack under the transaxle and a jackstand under the engine.

- Transaxle mount
- Rear mounting bracket
- Transaxle mounting bolts
- Transaxle from the vehicle

To install:

6. Installation is the reverse of the removal procedure, while using the following torque values:

- Transaxle mounting bolts: 47 ft. lbs. (64 Nm)
- Rear mounting bracket bolts: 87 ft. lbs. (118 Nm)
- Transaxle mount fasteners: 47 ft. lbs. (64 Nm)
- Transaxle mount through bolt: 54 ft. lbs. (74 Nm)
- Starter motor bolts: 33 ft. lbs. (44 Nm)
- Right front mount/bracket self-locking bolt: 61 ft. lbs. (83 Nm)
- Right front bracket and mount bolts, except self-locking bolt: 33 ft. lbs. (44 Nm)
- Clutch cover 6mm bolts: 108 inch lbs. (12 Nm)
- Clutch cover 8mm bolt: 17 ft. lbs. (24 Nm)
- Clutch cover 12mm bolt:42 ft. lbs. (57 Nm)
- Front and rear engine stiffener transaxle bolts: 42 ft. lbs. (57 Nm)
- Front and rear engine stiffener engine bolts: 17 ft. lbs. (24 Nm)
- Extension rod bolt: 16 ft. lbs. (22 Nm)
- Intermediate shaft mounting bolts: 29 ft. lbs. (39 Nm)
- Halfshafts
- Ball joint nuts:36–43 ft. lbs. (49–59 Nm)
- Right strut fork pinch bolt: 32 ft. lbs. (43 Nm)
- Right strut fork lower nut and bolt: 47 ft. lbs. (64 Nm)
- Exhaust pipe nuts: 40 ft. lbs. (54 Nm)
- Catalytic converter nuts: 25 ft. lbs. (33 Nm)
- Slave cylinder bolts: 16 ft. lbs. (22 Nm)

2.3CL

1. Before servicing the vehicle, refer to the precautions in the beginning of this section.

2. Drain the transaxle fluid.

3. Place the transmission in reverse.

4. Remove or disconnect the following:
- Both battery cables
- Battery and tray
- Air cleaner assembly

- Intake air resonator
- Starter motor
- Back up light switch electrical connector
- Ground cable from the transaxle
- Vehicle Speed (VSS) sensor electrical connector
- Wire harness clamp
- Shift cable, select cable and bracket
- Clutch damper bracket
- Slave cylinder and clutch line
- 2 upper transmission mounting bolts
- Engine splash shield
- Ball joints from the control arms
- Right Strut fork
- Right radius rod
- Halfshafts and intermediate shaft
- Center beam from the sub-frame
- Engine stiffener
- Intake manifold bracket
- 3 bolts from the rear mount bracket

5. Raise the transaxle slightly to take the weight off the mounts.

- Transaxle mount and bracket
- 3 lower transaxle-to-engine mounting bolts
- Transaxle from the engine

✳✳ CAUTION

Do not damage the clutch hydraulic lines.

To install:

6. Installation is the reverse of the removal procedure, while using the following torque values:

- Transaxle flange bolts: 47 ft. lbs. (64 Nm)
- Rear transaxle mount bolts: 40 ft. lbs. (54 Nm)
- Intake manifold bracket bolts: 16 ft. lbs. (22 Nm)
- Engine stiffener bolts: 33 ft. lbs. (44 Nm)
- Center beam bolts: 37 ft. lbs. (50 Nm)
- Right radius rod bolts: 76 ft. lbs. (103 Nm)
- Right radius arm nut: 32 ft. lbs. (43 Nm)
- Right strut fork bolt: 47 ft. lbs. (64 Nm)
- Ball joint nuts: 36-43 ft. lbs. (49-59 Nm)
- Slave cylinder and clutch line bolts: 16 ft. lbs. (22 Nm)

- Clutch damper bracket bolts: 16 ft. lbs. (22 Nm)
- Starter motor bolts: 33 ft. lbs. (44 Nm)

NSX

1. Before servicing the vehicle, refer to the precautions in the beginning of this section.
2. Drain the transaxle fluid.
3. Remove or disconnect the following:
   - Negative battery cable, then the positive battery cable
   - Strut bar
   - Air cleaner assembly
   - Control box
   - Transmission ground cable
   - Back up light switch electrical connector
   - Neutral position switch electrical connector
   - Differential speed sensor electrical connector
   - Reverse lockout solenoid electrical connector
   - Vehicle Speed (VSS) sensor electrical connector
   - Starter motor
   - Transmission mount
   - Parking brake cable holders from the rear beam rod
   - Rear beam rod
   - Parking brake cable holder from the rear sub frame
   - Wheel sensor wire clamps from the lower control arms
   - Toe control arms from the side beams
   - Strut forks
   - Lower control arm from the side beam
   - Half shafts from the differential
   - Intermediate shaft from the differential
   - Lower cover
   - Change wire bracket
   - Upper cover
   - Shift and select cables
   - Slave cylinder
   - Release fork from the clutch release hanger
4. Attach a chain hoist to the transmission hangers
   - Front engine mounting bolts on the transmission side
   - Transmission mounting bolts and stiffener
   - Transmission housing mounting bolts
   - Transmission from the vehicle

To install:
5. Installation is the reverse of the

removal procedure, while using the following torque values:
- Transmission and engine stiffener bolts: 47 ft. lbs. (64 Nm)
- Front engine mounting bolts: 43 ft. lbs. (60 Nm)
- Slave cylinder bolts: 16 ft. lbs. (22 Nm)
- Upper cover bolts: 108 inch lbs. (12 Nm)
- Change wire bracket bolts: 19 ft. lbs. (25 Nm)
- Lower cover bolts: 108 inch lbs. (12 Nm)
- Lower control arm bolts: 90 ft. lbs. (123 Nm)
- Strut fork bolts: 69 ft. lbs. (93 Nm)
- Toe control arms and torque the bolts to 69 ft. lbs. (93 Nm)
- Parking brake cable holder bolts: 16 ft. lbs. (22 Nm)
- Rear beam rod bolts: 43 ft. lbs. (59 Nm)
- Transmission mount bolts: 43 ft. lbs. (59 Nm)
- Starter motor bolts: 54 ft. lbs. (74 Nm)
- Control box bolt: 16 ft. lbs. (22 Nm)
- Strut bar bolts: 16 ft. lbs. (22 Nm)

Automatic

INTEGRA

1. Before servicing the vehicle, refer to the precautions in the beginning of this section.
2. Drain the transaxle fluid.
3. Remove or disconnect the following:
   - Negative battery cable, then the positive battery cable
   - Air cleaner housing assembly with intake air duct
   - Starter motor cables and holder
   - Transaxle ground cable from the transaxle hanger
   - Lockup control solenoid valve connector
   - Shift control solenoid valve connector
   - Harness clamp on the lockup control solenoid harness from the harness stay
   - Vehicle Speed (VSS) sensor electrical connector
   - Main shaft speed sensor electrical connector
   - Counter shaft speed sensor electrical connector
   - Upper transaxle mounting bolts
   - Splash shield
   - Front wheels

- Ball joints from the lower control arms
- Right strut fork
- Both halfshafts from the vehicle
- Heated Oxygen (HO2S) sensor connector
- Front exhaust pipe from the vehicle
- Intermediate shaft
- Shift cable cover
- Shift cable by removing the control lever

✳✳ WARNING

Do not bend the shift control cable when removing it.

- Right front mount/bracket
- End of the throttle control cable from the throttle control drum
- Transmission oil cooler hoses from the joint pipes
- Engine stiffener
- Torque converter cover
- 8 drive plate bolts
4. Support the transaxle.
   - Transaxle mounting bolts and rear engine mounting bolts
   - Transaxle from the engine
   - Starter motor from the transaxle (if necessary)

To install:
5. Installation is the reverse of the removal procedure, while using the following torque values:
- Starter motor bolts: 33 ft. lbs. (45 Nm)
- Transaxle mounting bolts: 47 ft. lbs. (64 Nm)
- Rear engine mount bolts: 87 ft. lbs. (118 Nm)
- Transmission mount bolt: 54 ft. lbs. (74 Nm)
- Transmission mount nuts: 47 ft. lbs. (64 Nm)
- Drive plate bolts: 108 inch lbs. (12 Nm)
- Torque converter cover 6mm bolts: 108 inch lbs. (12 Nm)
- Torque converter cover 10mm bolt: 33 ft. lbs. (44 Nm)
- Engine stiffener transaxle bolt: 32 ft. lbs. (43 Nm)
- Engine stiffener engine bolts: 17 ft. lbs. (24 Nm)
- Right front mount/bracket 12mm bolts: 47 ft. lbs. (64 Nm)
- Right front mount/bracket 10mm bolts: 33 ft. lbs. (44 Nm)
- Control lever bolt: 10 ft. lbs. (14 Nm)
- Intermediate shaft mounting bolts: 29 ft. lbs. (39 Nm)

- Front exhaust pipe manifold nuts: 40 ft. lbs. (54 Nm)
- Catalytic converter nuts: 16 ft. lbs. (22 Nm)
- Strut fork pinch bolt: 32 ft. lbs. (43 Nm)
- Strut fork lower bolt: 47 ft. lbs. (64 Nm)
- Splash shield bolts: 108 inch lbs. (12 Nm)

2.3CL AND 3.0CL

1. Before servicing the vehicle, refer to the precautions in the beginning of this section.
2. Drain the transaxle fluid into a suitable container.
3. Remove or disconnect the following:
    - Negative battery cable, then the positive cable
    - Battery and tray
    - Air cleaner assembly with the intake air duct
    - Starter cables
    - Transmission ground cable
    - Clutch pressure control solenoid valve electrical connector
    - Mainshaft speed sensor electrical connector
    - Clutch pressure switch electrical connector
    - Lock-up control solenoid valve electrical connector
    - Transmission oil cooler hoses
    - Shift control solenoid electrical connectors
    - Transaxle mounting bolts
    - Splash shield
    - Front mount bracket bolts
    - Center beam
    - Ball joints from the lower control arms
    - Halfshafts from the differential
    - Right damper fork from the damper
    - Both radius rods
    - Engine stiffener
    - Shift cable cover
    - Shift cable with the control lever
    - Drive plate bolts
4. Use a suitable jack to raise the transaxle slightly.
    - Transmission mount
    - Intake manifold bracket
    - Rear mount bracket
    - Transaxle from the engine
    - Torque converter from the transaxle (if necessary)
    - Starter motor from the transaxle (if necessary)

To install:

5. Installation is the reverse of the removal procedure, while using the following torque values:
    - Starter motor bolts: 33 ft. lbs. (44 Nm)
    - Rear mount bracket bolts: 40 ft. lbs. (54 Nm)
    - Intake manifold bracket bolts: 16 ft. lbs. (22 Nm)
    - Transaxle mounting bolts: 47 ft. lbs. (64 Nm)
    - Drive plate bolts: 108 inch lbs. (12 Nm)
    - Shift cable/control lever bolt: 10 ft. lbs. (14 Nm)
    - Engine stiffener bolts: 33 ft. lbs. (44 Nm)
    - Shift cable cover bolts: 20 ft. lbs. (26 Nm)
    - Right damper fork pinch bolt: 32 ft. lbs. (43 Nm)
    - Radius rod bolts: 76 ft. lbs. (103 Nm)
    - Right radius rod nuts: 40 ft. lbs. (54 Nm)
    - Ball joint nuts: 36-43 ft. lbs. (49-59 Nm)
    - Center beam bolts: 37 ft. lbs. (50 Nm)

2.5TL

1. Before servicing the vehicle, refer to the precautions in the beginning of this section.
2. Shift the transmission into **P**.
3. Drain the transmission fluid.
4. Remove or disconnect the following:
    - Negative then the positive battery cable
    - Battery and tray
    - Heat shield

➡**The distributor may be removed for better access to the transmission case bolts.**

- Emission control equipment box from the firewall without disconnecting the vacuum hoses
- Torque converter cover
- Drive plate bolts
- Transmission wiring sub-harness connector
- Transmission ground cable
- Upper transmission bolts and the 26mm shim
- Transmission left side mount and bracket
- Oil cooler hoses

- Sealing bolt from the differential
- Extension shaft from the differential
- Front exhaust pipe and its mounting brackets
- Shift cable cover
- Shift cable holder from the transmission
- Control lever from the control shaft
- Gear position switch electrical connector
- Mid mount nuts

5. Raise the transmission with a jack just enough to take the weight off of the mounts.
    - Mid mounts and spacer
    - Transmission mounting bolts
    - Transmission from the engine and carefully lower it from the vehicle

To install:

6. Installation is the reverse of the removal procedure, while using the following torque values:
    - Transmission flange bolts: 47 ft. lbs. (64 Nm)
    - Mid-mount and bracket bolts: 28 ft. lbs. (39 Nm)
    - Mid-mount bracket nuts: 32 ft. lbs. (43 Nm)
    - Shaft sealing bolt: 58 ft. lbs. (78 Nm)
    - Transmission mount bracket bolts: 40 ft. lbs. (54 Nm)
    - Transmission left side mount bolts: 47 ft. lbs. (64 Nm)
    - Shift cable holder bolts: 108 inch lbs. 912 Nm)
    - Control lever bolt: 10 ft. lbs. (14 Nm)
    - Exhaust manifold bracket bolts: 20 ft. lbs. (27 Nm)
    - Exhaust pipe manifold nuts: 40 ft. lbs. (54 Nm)
    - Catalytic converter nuts: 25 ft. lbs. (33 Nm)
    - Drive plate bolts: 108 inch lbs. (12 Nm)
    - Torque converter cover bolts: 108 inch lbs. (12 Nm)
    - Heat shield bolts: 84 inch lbs. (9.5 Nm)
    - Battery tray bolts: 16 ft. lbs. (22 Nm)

3.2CL AND 3.2TL

1. Before servicing the vehicle, refer to the precautions in the beginning of this section.
2. Shift the transmission into **P**.
3. Drain the transmission fluid.

Timing belt service is covered in Section 3 of this manual

4. Remove or disconnect the following:
- Both battery cables
- Battery and the tray
- Intake air duct
- Transmission oil cooler hoses
- Starter motor
- Transmission ground cable
- Shift control solenoid valve connectors
- Clutch pressure switch electrical connector
- Mainshaft speed sensor electrical connector
- Clutch pressure control solenoid valve electrical connector
- Lock-up control solenoid valve electrical connector
- Countershaft speed sensor electrical connector
- Gear position switch connector
- Vehicle Speed (VSS) sensor without disconnecting the hoses
- Front mount nut
- Engine cover
- Splash shield
- Strut forks
- Ball joints from the control arms
- Radius rods from the control arms
- Both halfshafts
- Front beam

5. Raise the transmission with a jack to take the pressure off of the mounts.
- Lower rear mount
- Intermediate shaft
- Shift cable holder
- Shift cable cover
- Shift cable with the control lever
- Torque converter cover
- Drive plate bolts
- Engine stiffener
- Front mount bracket
- Transmission-to-engine bolts
- Transmission from the vehicle

To install:
6. Installation is the reverse of the removal procedure, while using the following torque values:
- Transmission mounting bolts: 47 ft. lbs. (64 Nm)
- Front mount bracket bolts: 28 ft. lbs. (38 Nm)
- Engine stiffener bolts: 28 ft. lbs. (38 Nm)
- Drive plate bolts: 20 ft. lbs. (26 Nm)
- Torque converter cover bolts: 108 inch lbs. (12 Nm)
- Control lever bolt: 10 ft. lbs. (14 Nm)
- Cable cover bolts: 16 ft. lbs. (22 Nm)

- Shift cable holder bolts: 84 inch lbs. (9.5 Nm)
- Intermediate shaft bolts: 30 ft. lbs. (39 Nm)
- Lower transmission mount bolts: 28 ft. lbs. (38 Nm)
- Front beam 10mm bolts: 28 ft. lbs. (38 Nm)
- Front beam 12mm bolts: 47 ft. lbs. (64 Nm)
- Front beam 14mm bolts: 76 ft. lbs. (103 Nm)
- Lower transmission mount nuts: 28 ft. lbs. (38 Nm)
- Strut fork pinch bolts: 32 ft. lbs. (43 Nm)
- Strut fork through bolts to 47 ft. lbs. (64 Nm)
- Ball joint nuts: 36–43 ft. lbs. (49–59 Nm)
- Radius rod bolts: 132 ft. lbs. (179 Nm)
- Front mount nut: 40 ft. lbs. (54 Nm)
- VSS sensor bolts: 20 ft. lbs. (26 Nm)
- Transmission ground cable bolt: 20 ft. lbs. (26 Nm)
- Starter motor bolts: 33 ft. lbs. (44 Nm)

3.5RL

1. Disconnect the negative, then the positive battery cables.
2. Shift the transaxle into **P**.
3. Drain the transmission.
4. Remove or disconnect the following:
- Strut brace
- Control box
- Transaxle sub-harness connectors, and remove the sub-harness clamp
- 3 bolts securing the transaxle dipstick pipe bracket
- Upper transaxle mounting bolts

5. Pull the carpet back under the passenger seat to expose the secondary Heated Oxygen (HO2S) sensor connector. Detach the connector and push it out from the inside of the vehicle.
6. Remove or disconnect the following:
- Transmission stop collars
- Front exhaust pipe from the vehicle
- HO2S sensor wiring harness cover and grommet
- Catalytic converter
- Exhaust heat shield from the floor of the vehicle
- Transaxle oil cooler hoses
- Shift solenoid valve electrical connector
- Shift cable cover from the transaxle
- Shift cable holder from the holder base

- Control lever from the control shaft
- Transaxle dipstick pipe from the torque converter housing
- Range switch connector
- Lower plate from under the steering gear, then re-install the 2 steering gear mounting bolts
- Shift cable guide bracket from the transaxle beam

7. Raise the transmission slightly to take the weight off of the mounts.
- Transaxle beam
- Rear transaxle mount bracket and the mount
- Exhaust pipe bracket
- Sealing bolt from the differential
- Extension shaft from the differential
- Transmission-to-engine bolts
- Engine stiffener
- Torque converter covers
- Drive plate bolts
- Transmission from the vehicle

To install:
8. Installation is the reverse of the removal procedure, while using the following torque values:
- Drive plate bolts: Step 1: 108 inch lbs. (12 Nm). Step 2: 20 ft. lbs. (26 Nm)
- Torque converter cover bolts: 108 inch lbs. (12 Nm)
- Transaxle housing 8mm bolts: 16 ft. lbs. (22 Nm)
- Transaxle housing 12mm bolts: 47 ft. lbs. (64 Nm)
- Engine stiffener bolts: 16 ft. lbs. (22 Nm)
- Transaxle beam bolts: 28 ft. lbs. (38 Nm)
- Rear transaxle mount bracket bolts: 28 ft. lbs. (38 Nm)
- Rear transaxle mount bolts: 40 ft. lbs. (54 Nm).
- Shift cable guide bolt: 84 inch lbs. (9.5 Nm)
- Differential sealing bolt: 58 ft. lbs. (78 Nm)
- Lower plate bolts: 28 ft. lbs. (38 Nm)
- Steering gear bolts: 43 ft. lbs. (59 Nm)
- Shift control lever nut: 12 ft. lbs. (16 Nm)
- Shift cable holder bolts: 108 inch lbs. (12 Nm)
- Shift cable cover bolts: 108 inch lbs. (12 Nm)
- Exhaust heat shield bolts: 84 inch lbs. (9.5 Nm)
- Catalytic converter nuts: 25 ft. lbs. (33 Nm)

- Front exhaust pipe manifold nuts: 40 ft. lbs. (54 Nm)
- Transmission stop collar bolts: 28 ft. lbs. (38 Nm)
- Control box mounting bolts: 108 inch lbs. (12 Nm)
- Strut brace bolts: 16 ft. lbs. (22 Nm)

NSX

1. Before servicing the vehicle, refer to the precautions in the beginning of this section.
2. Drain the transmission.
3. Remove or disconnect the following:

- Both battery cables
- Strut brace
- Air cleaner assembly
- Control box
- Vehicle Speed (VSS) sensor electrical connector
- Transmission ground cable
- Starter cables
- Lock-up solenoid valve electrical connector
- Shift control solenoid valve electrical connector
- Transaxle oil cooler
- Starter motor
- Upper transmission housing and mounting bolts
- Parking brake cable holders from the rear beam
- Rear beam
- Front exhaust pipe
- Toe control arms from the side beams
- Damper forks
- Lower control arms from the side beams
- Both halfshafts and the intermediate shaft
- Shift cable cover and holder
- Shift cable from the control lever
- Torque converter cover
- Drive plate bolts

4. Place a jack under the transmission and raise it just enough to take the weight off of the mounts.

- Front engine mount bolts on the transaxle side
- Rear transaxle mount bolts
- Transaxle-to-engine bolts
- Transmission from the vehicle

To install:

5. Installation is the reverse of the removal procedure, while using the following torque values:

- Transaxle mounting bolts: 54 ft. lbs. (74 Nm)
- Rear transaxle mount bolts: 43 ft. lbs. (59 Nm)
- Front engine mount bolts: 43 ft. lbs. (59 Nm)
- Drive plate bolts: 108 inch lbs. (12 Nm)
- Torque converter cover bolts: 108 inch lbs. (12 Nm)
- Shift cable cover bolts: 6 ft. lbs. (8 Nm)
- Intermediate shaft bolts: 16 ft. lbs. (22 Nm)
- Intermediate shaft heat shield bolts: 84 inch lbs. (9.5 Nm)
- Lower control arm bolts: 90 ft. lbs. (123 Nm)
- Damper fork bolts: 69 ft. lbs. (93 Nm)
- Toe control arm bolts: 69 ft. lbs. (93 Nm)
- Front exhaust pipe manifold nuts: 40 ft. lbs. (54 Nm)
- Catalytic converter nuts: 25 ft. lbs. (33 Nm)
- Rear beam 10mm bolts: 43 ft. lbs. (59 Nm)

- Rear beam nuts and 12mm bolts: 69 ft. lbs. (93 Nm)
- Parking brake cable holder bolts: 16 ft. lbs. (22 Nm)
- Starter motor bolts: 33 ft. lbs. (44 Nm)
- Transaxle oil cooler bolts: 13 ft. lbs. (18 Nm)
- Strut bar bolts: 28 ft. lbs. (38 Nm)

Clutch

REMOVAL & INSTALLATION

1. Before servicing the vehicle, refer to the precautions in the beginning of this section.
2. Disconnect the negative battery cable.
3. Remove the transaxle assembly from the vehicle.
4. Insert a clutch alignment tool. Use a feeler gauge and measure the clearance between the pressure plate spring fingers and the clutch alignment disc. There should be a maximum of 0.02 in. (0.6mm) of clearance for a new pressure plate with 0.03 in. (0.8mm) limit for a used pressure plate. If

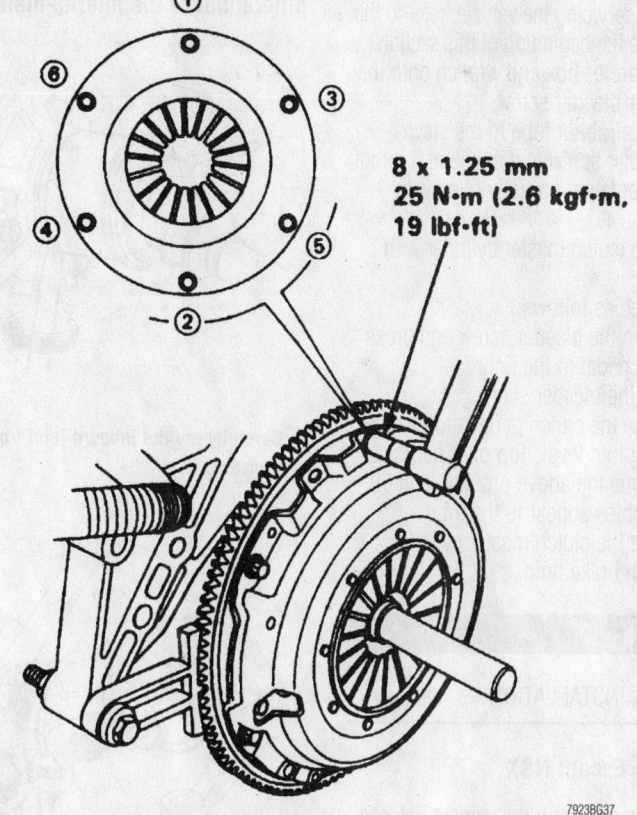

8 x 1.25 mm
25 N·m (2.6 kgf·m, 19 lbf·ft)

Pressure plate bolt torque sequence—Integra

7923BG37

Heater Core replacement is covered in Section 2 of this manual

the height is more than the service limit, replace the pressure plate.

5. Remove the clutch alignment disc.

6. Install a flywheel holder to aid in the removal of the pressure plate and clutch disc.

7. Matchmark the flywheel and pressure plate for easy reassembly. Remove the pressure plate bolts in a crisscross pattern 2 turns at a time to prevent warping the plate.

8. Remove the pressure plate, then the clutch disc with the alignment shaft.

To install:

9. Installation is the reverse of the removal procedure, while using the following torque values:

- Flywheel mounting bolts: 76 ft. lbs. (103Nm)
- Pressure plate bolts: 19 ft. lbs. (25 Nm)

Hydraulic Clutch System

BLEEDING

➡**Use DOT 3 brake fluid in the clutch master and slave cylinders. Brake fluid will damage the vehicle's paint. Immediately clean up any spills.**

1. Before servicing the vehicle, refer to the precautions in the beginning of this section.

2. Fit a flare or box end wrench onto the slave cylinder bleeder screw.

3. Attach a rubber tube to the slave cylinder bleeder screw and suspend it into a clear drain container partially filled with brake fluid.

4. Fill the clutch master cylinder with brake fluid.

5. Proceed as follows:

   a. Open the bleeder screw and press the clutch pedal to the floor.

6. Close the bleeder screw.

7. Release the clutch pedal and recheck the reservoir fluid level. Top off if necessary.

8. Continue the above procedure until no more bubbles appear in the tube.

9. Top off the clutch master cylinder reservoir with brake fluid.

Halfshaft

REMOVAL & INSTALLATION

All Models Except NSX

1. Before servicing the vehicle, refer to the precautions in the beginning of this section.

2. Drain the differential or transmission lubricant.

3. Remove or disconnect the following:
- Negative battery cable
- Wheel(s)
- Axle nut
- Strut fork
- Lower ball joint from the control arm

4. Pull the knuckle outward and remove the halfshaft outboard CV-joint from the knuckle using a plastic mallet.

5. Using a small pry bar carefully pry out the inboard CV-joint approximately ½ in. (13mm) in order to force the spring clip out of the groove in the differential side gears.

➡**Be careful not to damage the oil seal. Do not pull on the inboard CV-joint, it may come apart.**

6. Pull the halfshaft out of the differential or the intermediate shaft.

7. Remove the halfshaft from the wheel hub.

To install:

➡**Always use a new set ring whenever the driveshaft is being installed. Be sure the driveshaft locks in the differential side gear groove and that the CV-joint stub-axle bottoms in the differential or the intermediate shaft.**

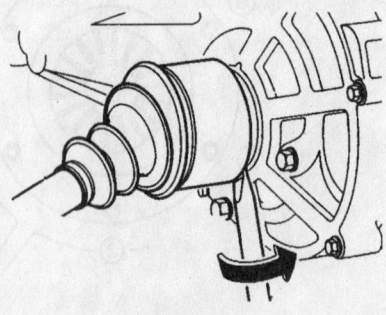

Carefully pry the inboard joint from the transaxle

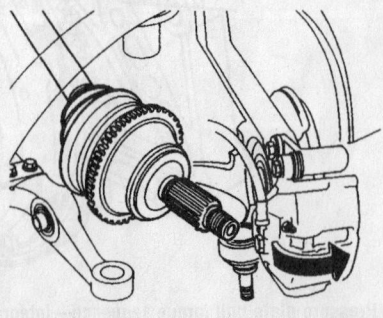

Pull the hub assembly from the outboard joint

8. Install or connect the following:
- Outboard joint to the wheel hub
- Inboard joint with a new set ring into the differential or intermediate shaft until the set ring locks in the groove
- Ball joint to the control arm and torque the nut to 36-43 ft. lbs. (49-59 Nm)
- Strut fork and torque the pinch bolt to 32 ft. lbs. (43 Nm) and the lower bolt to 47 ft. lbs. (64 Nm)
- Axle nut and torque it to 181 ft. lbs. (245 Nm)
- Wheel(s)

9. Refill the transmission or differential with the correct amount and type of fluid.

10. Reconnect the battery cable and enter the radio security code.

11. Measure and adjust the wheel alignment.

NSX

1. Before servicing the vehicle, refer to the precautions in the beginning of this section.

2. Drain the differential or transmission lubricant.

3. Remove or disconnect the following:
- Negative battery cable
- Rear wheel(s)
- Axle nut
- Brake hose from the caliper and from the knuckle
- Wheel sensor from the knuckle and lower control arm
- Parking brake cable from the control arm
- Toe control arm from the knuckle
- Damper mounting nut
- Stabilizer link
- Lower control arm from the sub-frame
- Outboard joint from the knuckle
- Inboard joint from the differential or intermediate shaft using a suitable pry tool

To install:

4. Install or connect the following:
- Inboard joint with a new set ring to the differential or intermediate shaft
- Outboard joint to the knuckle
- Lower control arm to the sub-frame and torque the bolts to 90 ft. lbs. (123 Nm)
- Stabilizer link and torque the nut to 61 ft. lbs. (83 Nm)
- Damper mounting nut and torque it to 69 ft. lbs. (93 Nm)
- Toe control arm to the knuckle and torque the bolt to 69 ft. lbs. (93 Nm)

- Parking brake cable and torque the bolt to 16 ft. lbs. (22 Nm)
- Wheel sensor and torque the bolts to 16 ft. lbs. (22 Nm)
- Brake hose and torque the bracket bolts to 84 inch lbs. (10 Nm) and the banjo bolt to 25 ft. lbs. (34 Nm)
- Axle nut and torque it to 242 ft. lbs. (329 Nm)
- Rear wheel(s)
- Negative battery cable

5. Fill the transaxle with the proper amount and type of fluid.

6. Fill and bleed the brake system.

7. Measure and adjust the wheel alignment.

CV-Joints

OVERHAUL

Integra, 2.3CL, 2.5TL, 3.2TL and 3.5RL

➡The outer joint is not serviceable, if wear or excessive play is found the joint must be replace.

1. Before servicing the vehicle, refer to the precautions in the beginning of this section.

2. Remove or disconnect the following:
- Negative battery cable
- Halfshaft from the vehicle
- Set ring from the inner joint
- Inner joint boot bands
- Inboard joint after marking relationship of joint-to-rollers for later installation
- Rollers from the spider
- Circlip from the shaft
- Spider using a bearing remover
- Stop ring

- Inboard boot
- Dynamic damper band and damper (if equipped)
- Outer joint boot bands and boot (if necessary)

To install:

3. Wrap the splines with vinyl tape to prevent damage to the boots..

4. Install or connect the following:
- Outer joint boot
- Dynamic damper
- Inner joint boot then remove the tape

5. Pack the outer joint boot with grease included with the kit (approx. 4.7 oz.)
- Stop ring into the halfshaft groove
- Spider on the halfshaft
- Circlip into the halfshaft groove
- Rollers onto the spider

6. Pack the inner joint with the grease included with the kit (approx. 4.7 oz.)
- Inner joint on the halfshaft matching the marks made earlier

7. Adjust the length of the halfshaft to the specification shown in the illustration.

8. Adjust the boots to halfway between full compression and full extension.

9. Position the dynamic damper (if equipped) as shown in the illustration.

10. Install a new dynamic damper band and bend down both locking tabs.

11. Install new inner and outer joint boot bands.

12. Install a new set ring in the halfshaft groove.

13. Install the halfshaft in the vehicle.

NSX, 3.0CL and 3.2CL

1. Before servicing the vehicle, refer to the precautions in the beginning of this section.

2. Remove or disconnect the following:

- Negative battery cable
- Halfshaft from the vehicle
- Boot bands
- Circlip from the shaft (outer joint only)
- Joint from the shaft after match-marking the rollers and the joint for re-installation

➡**There is a spring located under the outer joint.**

- Rollers from the spider
- Set ring
- Spider using a bearing puller
- Stop ring

To install:

3. Wrap the splines of the halfshaft with vinyl tape to protect the boots from damage.

4. Install or connect the following:
- CV boot, then remove the tape
- Stop ring
- Spider
- Set ring
- Rollers onto the spider
- Pack the outer joint with the grease provided (approx. 6.2 oz. for outer joint or 4.4 oz. for the inner)
- Spring and cap (outer joint only)
- CV joint onto the shaft, matching the marks made earlier
- Circlip into the outer joint inner groove

5. Adjust the length of the halfshaft as shown in the illustration.

6. Adjust the boots to halfway between full compression and full extension.

7. Install new boot bands on the boots and bend both sets of locking tabs down.

8. Install a new set ring in the halfshaft groove.

9. Install the halfshaft in the vehicle.

STEERING AND SUSPENSION

Air Bag

❊❊ CAUTION

Some vehicles are equipped with an air bag system, also known as the Supplemental Restraint System (SRS). The system must be disabled before performing service on or around system components, steering column, instrument panel components, wiring and sensors. Failure to follow safety and disabling procedures could result in accidental air bag deployment, possible personal injury and unnecessary system repairs.

PRECAUTIONS

Several precautions must be observed when handling the inflator module to avoid accidental deployment and possible personal injury.

- Never carry the inflator module by the wires or connector on the underside of the module.
- When carrying a live inflator module, hold securely with both hands, and ensure that the bag and trim cover are pointed away.
- Place the inflator module on a bench or other surface with the bag and trim cover facing up.
- With the inflator module on the bench, never place anything on or close to the module which may be thrown in the event of an accidental deployment.

Brake service is covered in Section 4 of this manual

DISARMING

Integra and NSX

❊❊ CAUTION

The Supplemental Restraint System (SRS, air bag system) must be disarmed before any of its components are disconnected or removed. Failing to disable the SRS before servicing its components may cause accidental deployment of the air bag, resulting in unnecessary repairs and possible personal injury.

1. Before servicing the vehicle, refer to the precautions in the beginning of this section.
2. Turn the ignition switch **OFF**.
3. Wait 3 minutes to let the capacitor in the backup circuit discharge.
4. Disconnect the negative battery cable, then disconnect the positive battery cable.
5. For the driver air bag:
   a. Remove the access panel lid below the air bag assembly on the steering wheel and remove the red shorting connector.
   b. Disconnect the connector between the air bag and cable reel.
   c. Connect the red shorting connector to the air bag side of the connector.
6. For the passenger air bag:
   a. If necessary, remove the glove box, then remove the red shorting connector from its holder.
   b. Disconnect the 3 pin connector between the passenger air bag and the main harness.
   c. Connect the shorting connector to the air bag side of the connector.
7. After installing the shorting connectors on the air bags, connect shorting connector 07-MAZ-SP0020A, or equivalent, on

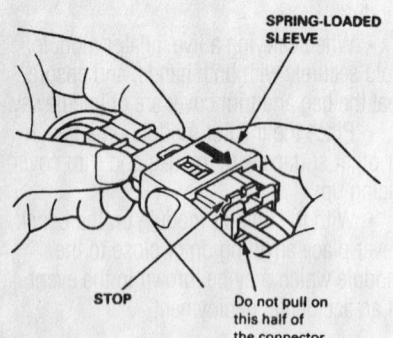

SPRING-LOADED SLEEVE

STOP

Do not pull on this half of the connector.

7923BG42

Spring-sleeve connectors

the cable reel connector and another on the main harness connector of the passenger's side air bag. This will prevent static electricity from setting off the seat belt pre-tensioners before you disconnect them.
8. For the seat belt pre-tensioners, disarm them one side at a time:
   a. Remove the B-pillar trim panels.
   b. Remove the red shorting connector from the short connector holder.
   c. Disconnect the pre-tensioner 3-pin connector, then install the shorting connector to the pre-tensioner side of the connector.

To enable:
9. Enable the seat belt pre-tensioners:
   a. Disconnect the shorting connector from the 3-pin connector. Then, reconnect the 3-pin connector.
   b. Fit the shorting connector into its holder and reinstall the B-pillar trim panels.
10. Enable the passenger air bag:
   a. Disconnect the shorting connectors from the air bag and main harness connectors.
   b. Reconnect the air bag connector to the main harness.
   c. Fit the short connector into its holder.
   d. If removed, install the glove box.
11. Disconnect the shorting connector from the cable reel connection.
12. Enable the driver's air bag:
   a. Disconnect the shorting connector from the air bag connector.
   b. Reconnect the air bag and cable reel connectors.
   c. Fit the shorting connector back into its holder.
   d. Install the steering wheel access cover.
13. Reconnect the positive and negative battery cables.
14. Turn the ignition switch to the **ON** position, but don't start the engine. The SRS indicator light should turn on for 6 seconds, then turn off. If the SRS indicator light doesn't come on, or stays on longer than 6 seconds, the system fault must be diagnosed.
15. Enter the radio security code.

2.5TL, 3.2CL and 3.2TL

❊❊ CAUTION

The SRS must be disarmed before any of its components are disconnected or removed. Failing to disable the SRS before servicing its components may cause accidental deploy-

ment of the air bag, resulting in unnecessary repairs and possible personal injury.

1. Before servicing the vehicle, refer to the precautions in the beginning of this section.
2. Turn the ignition switch to the **LOCK** position. Remove the key.
3. Disconnect the negative and positive battery cables.
4. Always wait at least 3 minutes after disconnecting the battery before working around the air bags.
5. Remove or disconnect the following:
   • Steering wheel lower access cover
   • Clip securing the air bag module/cable reel connection to the steering column
   • Air bag and cable reel connection—Immediately install the red shorting connector onto the air bag module connector.

➡ **The driver's side air bag connection contains a spring-contact self-disabling device. A shorting connector doesn't need to be installed on the driver's air bag connector.**

6. After servicing has been completed, couple the air bag and cable reel connectors.
7. Install or connect the following:
   • Clip securing the air bag/cable reel connection to the steering column
   • Access cover

To enable:
8. Reconnect the positive and negative battery cables.
9. Turn the ignition switch to the **ON** position, but don't start the engine. The SRS indicator light should turn on for 6 seconds, then turn off. If the SRS indicator light doesn't come on, or stays on longer than 6 seconds, the system fault must be diagnosed.
10. Enter the radio security code.

2.3CL, 3.0CL and 3.5RL

1. Before servicing the vehicle, refer to the precautions in the beginning of this section.
2. Disconnect the negative battery cable, then the positive cable.
3. Wait 3 minutes for the air bag reserve power to discharge before preceding with work.

To enable:
4. Reconnect the positive and negative battery cables.
5. Turn the ignition switch to the **ON**

position, but don't start the engine. The SRS indicator light should turn on for 6 seconds, then turn off. If the SRS indicator light doesn't come on, or stays on longer than 6 seconds, the system fault must be diagnosed.

6. Enter the radio security code.

Rack and Pinion Steering Gear

REMOVAL & INSTALLATION

Power

2.3CL, 3.0CL, 3.2CL AND INTEGRA

1. Before servicing the vehicle, refer to the precautions in the beginning of this section.
2. Drain the power steering fluid.
3. Remove or disconnect the following:
   - Both battery cables
   - Wheels
   - Steering wheel lower access panel
   - Supplemental Inflatable Restraint (SIR) electrical connector
   - Steering wheel side panels

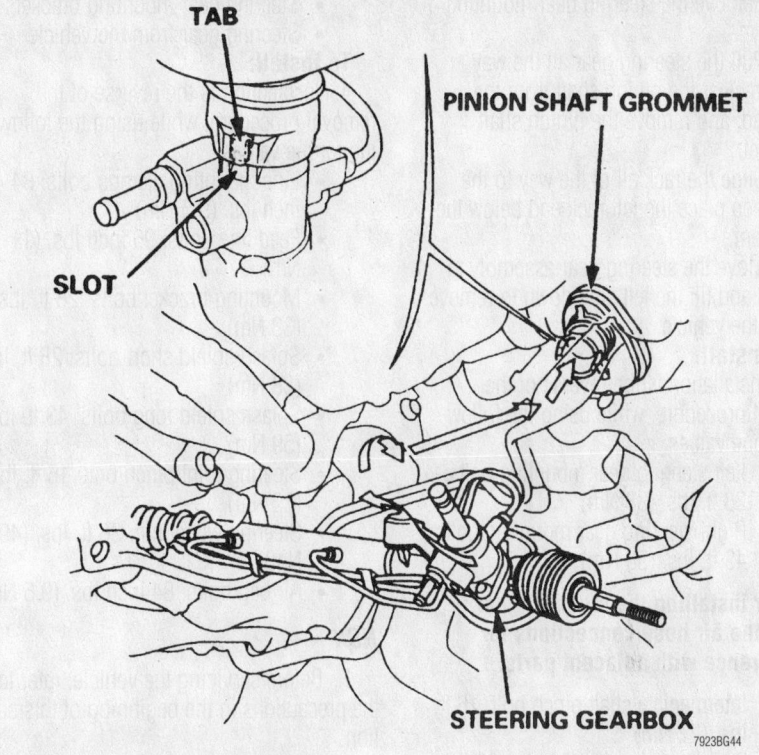

Installing the steering gear—Integra

- Air bag
- Horn electrical connector
- Cruise control electrical connector
- Steering wheel
- Steering joint cover
- Steering joint lower bolt and pull the joint toward the column
- Tie rods from the steering knuckles
- Shift linkage (if equipped with a manual transmission)
- Heated Oxygen (HO2S) sensor electrical connector
- Catalytic converter from the vehicle
- Return line from the steering gear
- Rear beam brace rod
- Left tie rod end, then slide the steering gear all of the way to the right
- 2 lines from the valve body unit on the steering gear

※※ CAUTION

After disconnecting the hose and pipe, plug or cap the hose and pipe to prevent foreign materials from entering the valve body unit.

➡ Do not loosen the cylinder pipes between the valve body unit and the cylinder.

8 x 1.25 mm
22 N·m (2.2 kgf·m, 16 lbf·ft)

SHIFT CABLE HOLDER

SHIFT CABLE

SHIFT CABLE COVER

CONTROL SHAFT

CONTROL LEVER

LOCK WASHER
Replace.

6 x 1.0 mm
14 N·m (1.4 kgf·m, 10 lbf·ft)

6 x 1.0 mm
12 N·m (1.2 kgf·m, 8.7 lbf·ft)

Automatic transaxle shift cable attachment—Integra

4. Remove the steering gear mounting bolts.

5. Pull the steering gear all the way down to clear the pinion shaft from the bulkhead, and remove the pinion shaft grommet.

6. Slide the rack all of the way to the right, then place the left rack end below the rear beam.

7. Move the steering gear assembly to the left, and tilt the left side down to remove it from the vehicle.

To install:

8. Installation is the reverse of the removal procedure, while using the following torque values:
- Left steering gear mounting bolts: 28 ft. lbs. (38 Nm)
- Right steering gear mounting bolts: 43 ft. lbs. (58 Nm)

➡ **After installing the steering gear, check the air hose connections for interference with adjacent parts.**

- Intermediate shaft pinch bolt: 16 ft. lbs. (22 Nm)
- Rear beam brace rod bolts: 28 ft. lbs. (38 Nm)
- Catalytic converter nuts: 25 ft. lbs. (33 Nm)
- Tie rod end nuts: 33 ft. lbs. (44 Nm)
- Steering wheel nut: 36 ft. lbs. (49 Nm)
- Air bag bolts: 84 inch lbs. (9.5 Nm)

2.5TL, 3.2TL AND 3.5RL

1. Before servicing the vehicle, refer to the precautions in the beginning of this section.

2. Drain the power steering fluid.

3. Remove or disconnect the following:
- Negative then the positive battery cables
- Steering wheel lower access panel
- Supplemental Inflatable Restraint (SIR) electrical connector
- Steering wheel side panels
- Air bag
- Horn electrical connector
- Cruise control electrical connector
- Steering wheel
- Steering joint bolts then move the joint toward the column
- Front wheels
- Tie rods from the steering knuckles
- Splash guard
- Feed line from the valve body
- Line mounting clamps
- Feed line from the line mounting cushions
- Sensor inlet line and both return lines from the hoses

- Steering gear mounting brackets
- Steering gear from the vehicle

To install:

4. Installation is the reverse of the removal procedure, while using the following torque values:
- Line mounting clamps bolts: 84 inch lbs. (9.5 Nm)
- Feed line bolts: 96 inch lbs. (11 Nm)
- Mounting bracket bolts: 28 ft. lbs. (38 Nm)
- Splash shield short bolts: 28 ft. lbs. (38 Nm)
- Splash shield long bolts: 43 ft. lbs. (59 Nm)
- Steering joint pinch bolt: 16 ft. lbs. (22 Nm)
- Steering wheel nut: 36 ft. lbs. (49 Nm)
- Air bag bolts: 84 inch lbs. (9.5 Nm)

NSX

1. Before servicing the vehicle, refer to the precautions in the beginning of this section.

2. Drain the power steering fluid.

3. Remove or disconnect the following:
- Negative then the positive battery cables
- Battery
- Steering joint cover
- Steering joint bolts then move the joint toward the column to separate it
- Wheels
- Tie rods from the steering knuckles
- Spare tire
- Spare tire holder plate
- Spare tire holder
- Floor under cover
- Terminal guard
- Ground cable
- Wires from the steering gear terminals
- Radiator pipe bracket at the front compartment bulkhead
- Radiator pipe bracket at the floor

4. Support the steering gear and the front crossbeam
- Steering gear and front crossbeam bolts and nuts
- Steering gear and crossbeam from the vehicle

To install:

5. Installation is the reverse of the removal procedure, while using the following torque values:
- Steering gear and crossbeam fasteners: 43 ft. lbs. (59 Nm)
- Radiator pipe bracket bolts: 16 ft. lbs. (22 Nm)

- Ground cable bolt: 84 inch lbs. (9.5 Nm)
- Terminal guard bolts: 84 inch lbs. (9.5 Nm)
- Floor under cover bolts: 84 inch lbs. (9.5 Nm)
- Spare tire holder bolt: 18 ft. lbs. (25 Nm)
- Steering joint bolts: 16 ft. lbs. (22 Nm)

Strut

REMOVAL & INSTALLATION

Front

2.3CL, 2.5TL, 3.0CL, 3.2CL, 3.2TL, 3.5RL AND INTEGRA

1. Before servicing the vehicle, refer to the precautions in the beginning of this section.

2. Support the lower suspension arm with a jack.

3. Remove or disconnect the following:
- Wheel(s)
- Brake hose from the strut
- Strut fork
- Upper strut mounting nuts
- Strut from the vehicle

To install:

4. Install the strut and loosely install the mounting nuts

5. Install the strut fork, do not tighten the bolts at this time

➡ **All suspension nuts and bolts should be tightened with the vehicle on the ground, or with a floor jack supporting the vehicle's weight.**

6. Tighten the pinch bolt to 32 ft. lbs. (43 Nm) (except Integra 16 ft. lbs. [22 Nm]).

7. Tighten the lower strut fork bolt to 47 ft. lbs. (64 Nm).

8. Install the brake hose to the strut fork and tighten the bolt to 16 ft. lbs. (22 Nm).

9. Install the wheels.

10. Tighten the upper strut nuts to 28 ft. lbs. (38 Nm) (except Integra and 3.2 models 37 ft. lbs. [50 Nm])

11. Measure and adjust the wheel alignment.

NSX

1. Before servicing the vehicle, refer to the precautions in the beginning of this section.

2. Remove or disconnect the following:
- Wheel(s)
- Brake hose from the strut

CAUTION:
- Replace the self-locking nuts after removal.
- The vehicle should be on the ground before any bolts or nuts connected to rubber mounts or bushings are tightened.
- Torque the castle nut to the lower torque specification, then tighten it only far enough to align the slot with the pin hole. Do not align the nut by loosening.

NOTE: Wipe off the grease before tightening the nut at the ball joint.

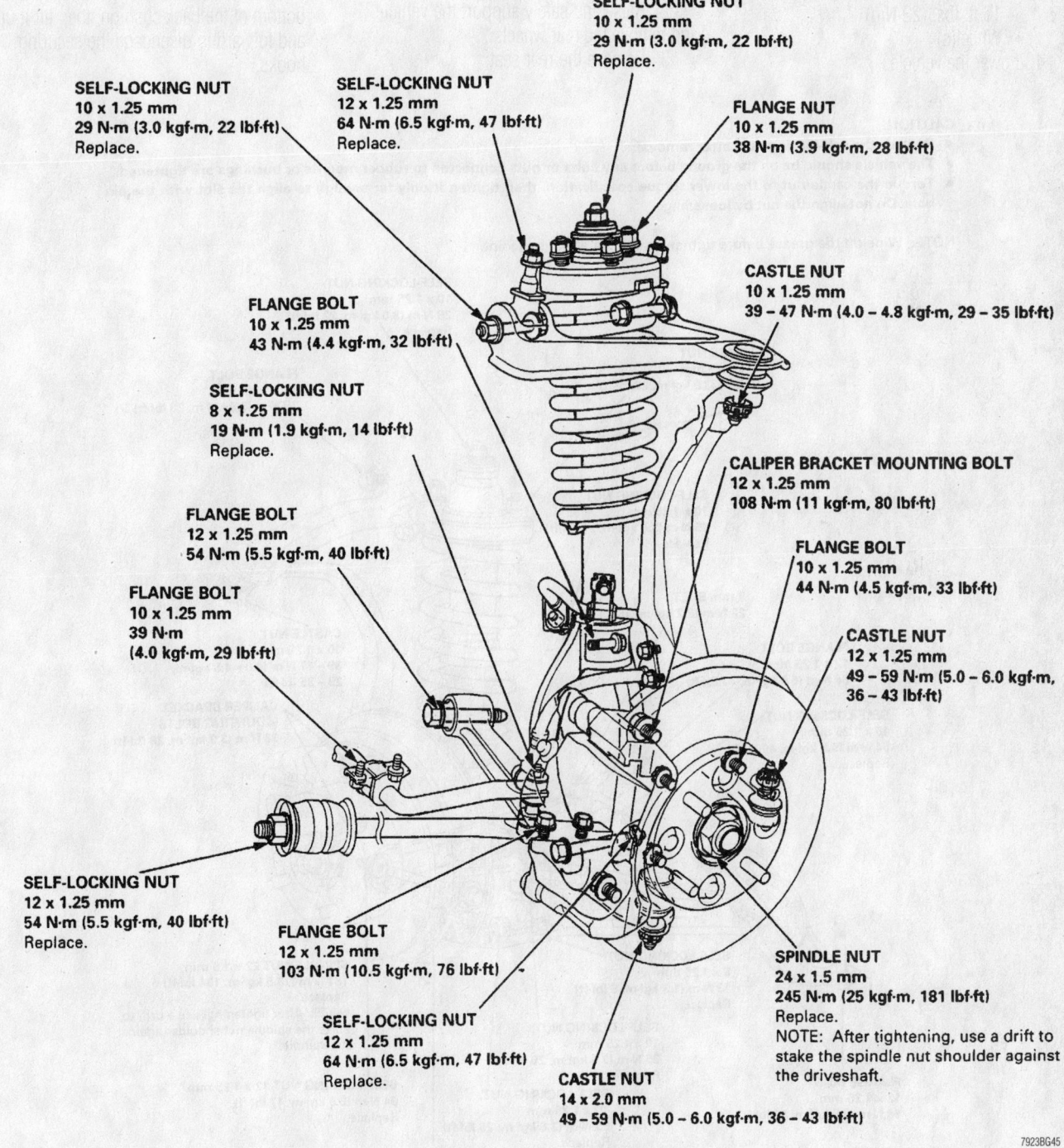

SELF-LOCKING NUT
10 x 1.25 mm
29 N·m (3.0 kgf·m, 22 lbf·ft)
Replace.

SELF-LOCKING NUT
12 x 1.25 mm
64 N·m (6.5 kgf·m, 47 lbf·ft)
Replace.

SELF-LOCKING NUT
10 x 1.25 mm
29 N·m (3.0 kgf·m, 22 lbf·ft)
Replace.

FLANGE NUT
10 x 1.25 mm
38 N·m (3.9 kgf·m, 28 lbf·ft)

FLANGE BOLT
10 x 1.25 mm
43 N·m (4.4 kgf·m, 32 lbf·ft)

CASTLE NUT
10 x 1.25 mm
39 – 47 N·m (4.0 – 4.8 kgf·m, 29 – 35 lbf·ft)

SELF-LOCKING NUT
8 x 1.25 mm
19 N·m (1.9 kgf·m, 14 lbf·ft)
Replace.

CALIPER BRACKET MOUNTING BOLT
12 x 1.25 mm
108 N·m (11 kgf·m, 80 lbf·ft)

FLANGE BOLT
12 x 1.25 mm
54 N·m (5.5 kgf·m, 40 lbf·ft)

FLANGE BOLT
10 x 1.25 mm
44 N·m (4.5 kgf·m, 33 lbf·ft)

FLANGE BOLT
10 x 1.25 mm
39 N·m
(4.0 kgf·m, 29 lbf·ft)

CASTLE NUT
12 x 1.25 mm
49 – 59 N·m (5.0 – 6.0 kgf·m, 36 – 43 lbf·ft)

SELF-LOCKING NUT
12 x 1.25 mm
54 N·m (5.5 kgf·m, 40 lbf·ft)
Replace.

FLANGE BOLT
12 x 1.25 mm
103 N·m (10.5 kgf·m, 76 lbf·ft)

SELF-LOCKING NUT
12 x 1.25 mm
64 N·m (6.5 kgf·m, 47 lbf·ft)
Replace.

SPINDLE NUT
24 x 1.5 mm
245 N·m (25 kgf·m, 181 lbf·ft)
Replace.
NOTE: After tightening, use a drift to stake the spindle nut shoulder against the driveshaft.

CASTLE NUT
14 x 2.0 mm
49 – 59 N·m (5.0 – 6.0 kgf·m, 36 – 43 lbf·ft)

7923BG45

Front suspension showing the torque specifications—2.5TL shown, other vehicles are similar

For Accessory Drive Belt illustrations, see Section 1 of this manual

- Lower strut mounting bolt
- Upper strut mounting nuts
- Strut from the vehicle

To install:

3. Install or connect the following:
- Strut to the vehicle and loosely install the lower mounting bolt
- New upper mounting nuts hand tight
- Brake hose and torque the bolt to 16 ft. lbs. (22 Nm)
- Wheel(s)
4. Lower the vehicle.

5. Torque the upper mounting nuts to 32 ft. lbs. (43 Nm) and the lower mounting bolt to 69 ft. lbs. (93 Nm).

Rear

2.3CL, 2.5TL, 3.0CL AND 3.2TL

1. Before servicing the vehicle, refer to the precautions in the beginning of this section.
2. Raise and safely support the vehicle and remove the rear wheels.
3. Remove the rear seat:

a. Remove the lower cushion bolts located under the armrest and near the floor.

b. Pull the rear of the lower cushion up and lift it forward to release it from the clips.

c. Pull down the trunk bulkhead trim and release the armrest lid clips.

d. Remove the bolts from the top and bottom of the back cushion, then lift it up and forward to disengage the securing hooks.

CAUTION:
- Replace the self-locking nuts after removal.
- The vehicle should be on the ground before any bolts or nuts connected to rubber mounts or bushings are tightened.
- Torque the castle nut to the lower torque specification, then tighten it only far enough to align the slot with the pin hole. Do not align the nut by loosening.

NOTE: Wipe off the grease before tightening the nut at the ball joint.

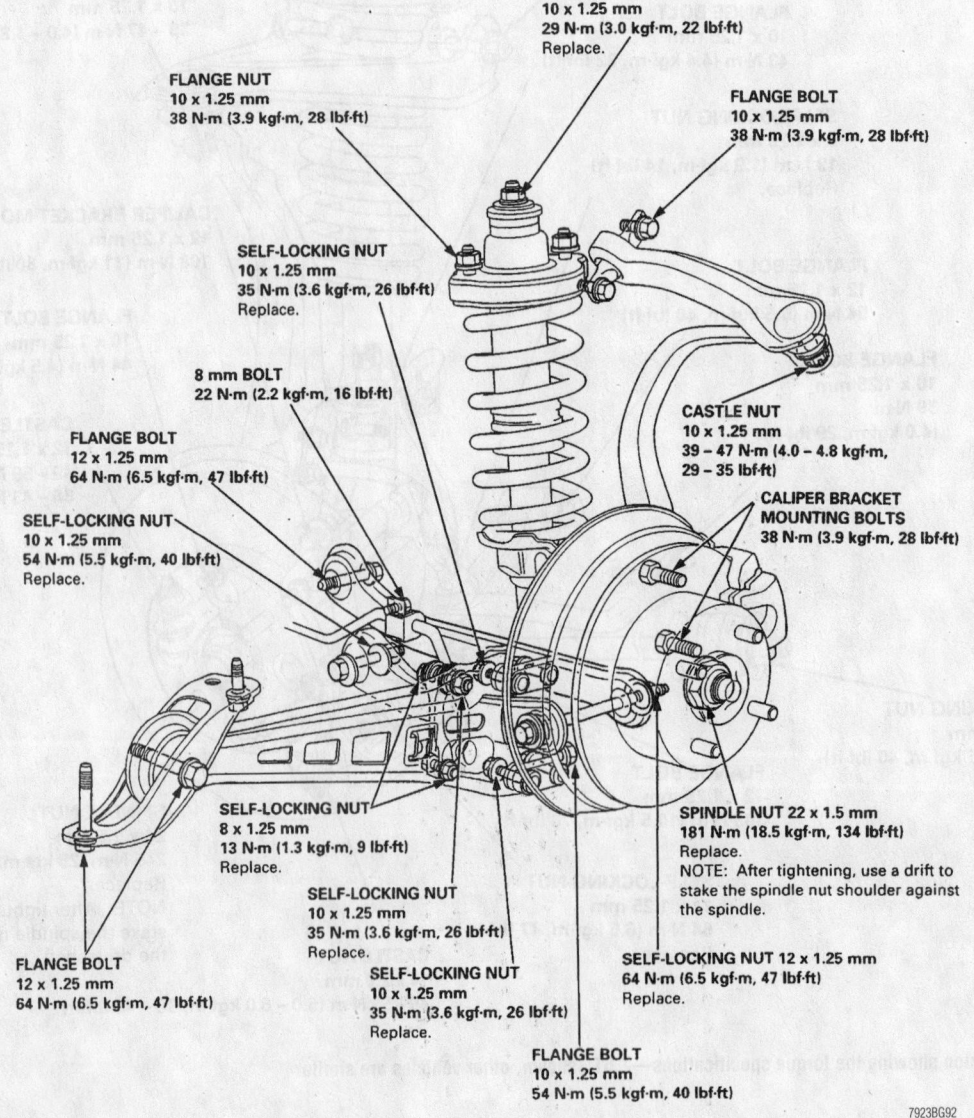

SELF-LOCKING NUT
10 x 1.25 mm
29 N·m (3.0 kgf·m, 22 lbf·ft)
Replace.

FLANGE NUT
10 x 1.25 mm
38 N·m (3.9 kgf·m, 28 lbf·ft)

FLANGE BOLT
10 x 1.25 mm
38 N·m (3.9 kgf·m, 28 lbf·ft)

SELF-LOCKING NUT
10 x 1.25 mm
35 N·m (3.6 kgf·m, 26 lbf·ft)
Replace.

8 mm BOLT
22 N·m (2.2 kgf·m, 16 lbf·ft)

FLANGE BOLT
12 x 1.25 mm
64 N·m (6.5 kgf·m, 47 lbf·ft)

SELF-LOCKING NUT
10 x 1.25 mm
54 N·m (5.5 kgf·m, 40 lbf·ft)
Replace.

FLANGE BOLT
12 x 1.25 mm
64 N·m (6.5 kgf·m, 47 lbf·ft)

CASTLE NUT
10 x 1.25 mm
39 – 47 N·m (4.0 – 4.8 kgf·m, 29 – 35 lbf·ft)

CALIPER BRACKET MOUNTING BOLTS
38 N·m (3.9 kgf·m, 28 lbf·ft)

SELF-LOCKING NUT
8 x 1.25 mm
13 N·m (1.3 kgf·m, 9 lbf·ft)
Replace.

SELF-LOCKING NUT
10 x 1.25 mm
35 N·m (3.6 kgf·m, 26 lbf·ft)
Replace.

SELF-LOCKING NUT
10 x 1.25 mm
35 N·m (3.6 kgf·m, 26 lbf·ft)
Replace.

SPINDLE NUT 22 x 1.5 mm
181 N·m (18.5 kgf·m, 134 lbf·ft)
Replace.
NOTE: After tightening, use a drift to stake the spindle nut shoulder against the spindle.

SELF-LOCKING NUT 12 x 1.25 mm
64 N·m (6.5 kgf·m, 47 lbf·ft)
Replace.

FLANGE BOLT
10 x 1.25 mm
54 N·m (5.5 kgf·m, 40 lbf·ft)

7923BG92

Rear suspension showing the torque specifications—2.5TL

4. Place a floor jack under the lower arm and slightly compress the spring.

5. Remove the upper mounting nuts and the lower flange bolt.

6. Lower the jack to remove the strut. Be sure and mark the right and left struts so they can be reinstalled on the proper sides.

To install:

7. Install the strut into the vehicle. Loosely install the mounting nuts and mounting bolt, but do not tighten them until the weight of the vehicle is on the suspension.

8. Raise the rear suspension with a floor jack until the weight of the vehicle is on the strut. Tighten the upper mounting nuts to 28 ft. lbs. (39 Nm), then tighten the lower mounting bolts to 40 ft. lbs. (55 Nm). Be careful not to pinch the ABS speed sensor wire between the strut and bracket.

9. Install the rear wheels and lower the vehicle.

10. Install the rear seat back and torque the bolts to 84 inch lbs. (9.5 Nm).

11. Install the armrest lid.

12. Install the lower seat cushion and torque the bolts to 84 inch lbs. (9.5 Nm).

3.5RL

1. Before servicing the vehicle, refer to the precautions in the beginning of this section.

2. Raise and safely support the vehicle and remove the rear wheels.

3. Remove or disconnect the following:
- Upper strut mount cover from the rear panel, just below the speaker—On sedans, remove the trunk side panel.
- Trim cover, then remove the upper mount nuts
- Wheel sensor wire brackets, on cars with ABS—do not disconnect the wheel sensor connector
- Lower strut mounting bolt

4. On the Integra, remove the flange bolt that connects the lower arm to the trailing arm.

5. Lower the rear suspension and remove the strut assembly from the vehicle.

6. If necessary, use a spring compressor to remove the spring from the strut assembly.

To install:

7. Reassemble the strut and coil spring assembly. Tighten the strut self-locking nut to 22 ft. lbs. (30 Nm).

8. Lower the rear suspension and position the strut assembly in the vehicle. The

nut welded to the lower strut mounting should face the front of the vehicle.

9. Loosely install the upper mounting nuts.

10. On the Integra, raise the rear suspension and install the bolt connecting the lower arm to the trailing arm.

11. Install the strut lower mounting bolt.

12. Raise the vehicle until the vehicle just lifts off the safety stand and tighten the lower strut bolt and lower control arm bolt. On the integra, tighten the lower strut bolt and the control arm bolt to 40 ft. lbs. (54 Nm). In the 3.5RL, tighten the lower strut mounting bolt to 76 ft. lbs. (103 Nm).

13. Install the wheel sensor wire bracket on cars with ABS.

14. Tighten the upper mounting nuts to 36 ft. lbs. (49 Nm).

15. Install the rear wheels, then lower the vehicle.

16. Install the trim panel, or the trunk side panel.

17. Check the vehicle's alignment.

INTEGRA

1. Before servicing the vehicle, refer to the precautions in the beginning of this section.

2. Raise and safely support the vehicle.

3. Support the control arm with a jack.

4. Remove or disconnect the following:
- Rear wheels
- Strut access panel
- Upper mounting nuts
- Wheel sensor wire brackets (if equipped)
- Lower mounting bolt

5. Lower the jack and remove the strut from the vehicle.

To install:

6. Install or connect the following:
- Strut and hand tighten the upper mounting nuts
- Wheel sensor wire bracket and torque the bolts to 84 inch lbs. (9.5 Nm)

7. Raise the control arm and install the lower mounting bolt and torque it to 40 ft. lbs. (54 Nm).

8. Torque the upper mounting nuts to 36 ft. lbs. (49 Nm).

9. Install the strut access panel.

10. Install the wheels.

NSX

1. Before servicing the vehicle, refer to the precautions in the beginning of this section.

2. Remove or disconnect the following:
- Rear wheels
- Lower rear hatch glass trim
- Upper strut mounting nuts
- Rear strut brace (NSX-T model only)
- Lower strut mounting nut
- Stabilizer link
- Strut from the vehicle

To install:

3. Install or connect the following:
- Strut
- Stabilizer link and torque the nut (at the stabilizer bar) to 61 ft. lbs. (83 Nm)
- Lower strut mounting bolt and torque it to 69 ft. lbs. (93 Nm)
- New upper strut mounting nuts and torque them to 39 ft. lbs. (53 Nm)
- Strut brace (if equipped) and torque the bolts to 16 ft. lbs. (22 Nm)
- Rear hatch glass trim
- Rear wheels

4. Measure and adjust the wheel alignment.

Coil Spring

REMOVAL & INSTALLATION

All Models

1. Before servicing the vehicle, refer to the precautions in the beginning of this section.

2. Raise and support the vehicle and remove the front wheels.

3. Remove the strut (damper).

4. Place the strut assembly in a coil spring compressor.

5. Compress the coil spring and remove the locking nut from the top of the strut.

6. Release the pressure from the spring compressor.

7. Remove the coil spring and related pieces from the strut.

To install:

➡**Use new self-locking nuts and bolts when assembling the strut.**

8. Install the strut, coil spring and related components on the spring compressor.

9. Compress the spring.

10. Install the mounting washer, and loosely install a new self-locking nut.

11. Hold the strut piston rod with a hex wrench and tighten the self-locking nut to 22 ft. lbs. (30 Nm).

For Tire, Wheel and Ball Joint specifications, see Section 1 of this manual

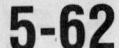

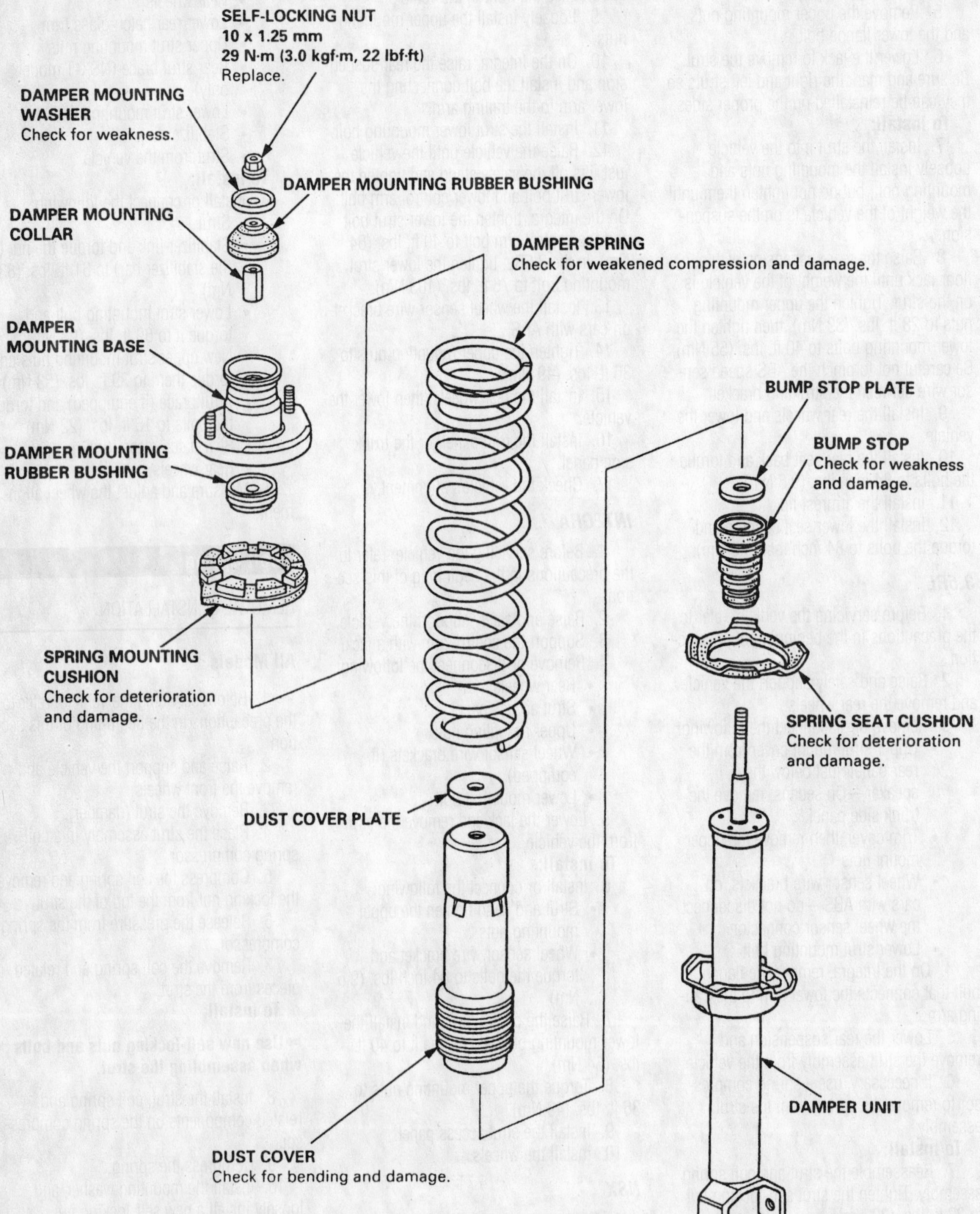

SELF-LOCKING NUT
10 x 1.25 mm
29 N·m (3.0 kgf·m, 22 lbf·ft)
Replace.

DAMPER MOUNTING WASHER
Check for weakness.

DAMPER MOUNTING RUBBER BUSHING

DAMPER MOUNTING COLLAR

DAMPER SPRING
Check for weakened compression and damage.

DAMPER MOUNTING BASE

BUMP STOP PLATE

DAMPER MOUNTING RUBBER BUSHING

BUMP STOP
Check for weakness and damage.

SPRING MOUNTING CUSHION
Check for deterioration and damage.

SPRING SEAT CUSHION
Check for deterioration and damage.

DUST COVER PLATE

DAMPER UNIT

DUST COVER
Check for bending and damage.

7923BG93

Exploded view of the rear strut (damper)—2.5TL shown, other vehicles are similar

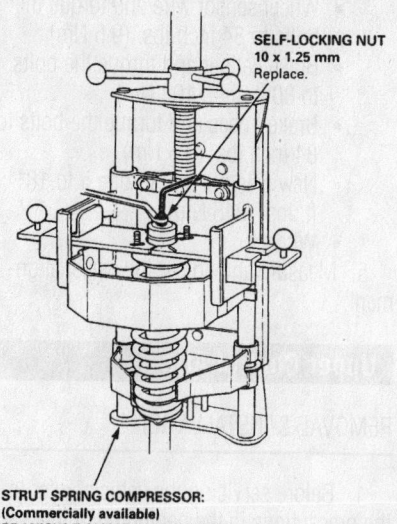

SELF-LOCKING NUT
10 x 1.25 mm
Replace.

STRUT SPRING COMPRESSOR:
(Commercially available)
BRANICK® T/N MST-580A, T/N7200,
or equivalent

7923BG94

Compress the coil spring until the spring moves away from the seat and use a hex wrench to hold the piston rod while removing the nut—all models

12. Install the strut in the vehicle.
13. Check and adjust the vehicle's front wheel alignment.

Upper Ball Joint

REMOVAL & INSTALLATION

All Models

➡The upper ball joint cannot be removed from the control arm. If the ball joint is damaged, the upper arm assembly must be replaced.

Lower Ball Joint

REMOVAL & INSTALLATION

Integra, 2.3CL

1. Before servicing the vehicle, refer to the precautions in the beginning of this section.
2. Raise and support the vehicle safely.
3. Remove or disconnect the following:
   • Front wheels
   • Axle nut
   • Brake hose from the knuckle
   • Brake caliper
   • Brake rotor
   • Wheel sensor wire bracket
   • Wheel sensor from the knuckle

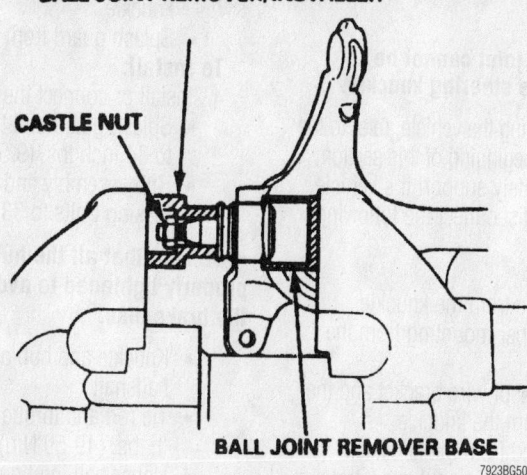

BALL JOINT REMOVER/INSTALLER

CASTLE NUT

BALL JOINT REMOVER BASE

7923BG50

Use a vise or press to remove the ball joint from the steering knuckle

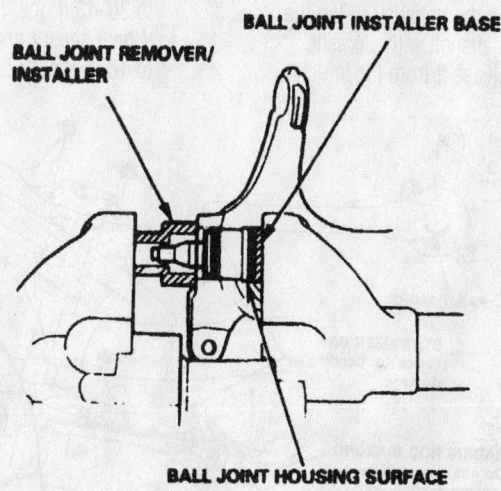

BALL JOINT REMOVER/INSTALLER

BALL JOINT INSTALLER BASE

BALL JOINT HOUSING SURFACE

7923BG51

Press the new ball joint into the steering knuckle

   • Tie rod from the knuckle
   • Lower ball joint from the knuckle
   • Upper ball joint from the knuckle
   • Steering knuckle
   • Boot by prying off the snapring
4. Press the ball joint from the knuckle.
   To install:
5. Place the ball joint in position by hand. Install the ball joint into the tool and press in the new ball joint.

❋❋ WARNING

After installing the boot, check the ball joint pin tapered section for grease contamination and wipe it if necessary.

6. Install or connect the following:
   • Ball joint boot and snapring

   • Steering knuckle to the vehicle
   • Lower ball joint and torque the nut to 40 ft. lbs. (54 Nm)
   • Tie rod and torque the nut to 32 ft. lbs. (43 Nm)
   • Upper ball joint and torque the nut to 32 ft. lbs. (43 Nm)
   • Wheel sensor and torque the bolts to 84 inch lbs. (9.5 Nm)
   • Brake rotor and torque the bolts to 84 inch lbs. (9.5 Nm)
   • Brake caliper and torque the bolts to 80 ft. lbs. (108 Nm)
   • Brake hose and torque the bolts to 84 inch lbs. (9.5 Nm)
   • New axle nut and torque it to 134 ft. lbs. (181 Nm)
   • Front wheels
7. Check the front wheel alignment and adjust if necessary.

For Wheel Alignment specifications, see Section 1 of this manual

2.5TL, 3.0CL, 3.2TL, 3.2CL, 3.5RL and NSX

➡️ **The lower ball joint cannot be removed from the steering knuckle.**

1. Before servicing the vehicle, refer to the precautions in the beginning of this section.
2. Raise and safely support the vehicle
3. Remove or disconnect the following:
   - Wheels
   - Axle nut
   - Brake hose from the knuckle
   - Brake caliper mounting from the knuckle
   - Wheel sensor wire bracket and the sensor from the knuckle
   - Tie rod end from the knuckle
   - Lower ball joint from the control arm
   - Upper ball joint from the knuckle
   - Knuckle and hub by sliding the assembly off the halfshaft—Tap the end of the halfshaft with a plastic mallet to release it from the knuckle.

- Hub and rotor assembly from the knuckle
- Splash guard from the knuckle

To install:

4. Install or connect the following:
   - Splash guard and torque the bolts to 84 inch lbs. (9.5 Nm)
   - Hub assembly and tighten the self-locking bolts to 33 ft. lbs. (45 Nm)

➡️ **Be sure that all the hub bolts are properly tightened to avoid warpage of the brake disc.**

- Knuckle and hub assembly onto the halfshaft
- Tie rod and torque the nut to 36-43 ft. lbs. (49-59 Nm)
- Upper ball joint and torque the nut to 29-35 ft. lbs. (39-47 Nm)
- Lower ball joint and torque the nut to 36-43 ft. lbs. (49-59 Nm)
- Wheel sensor and torque the bolts to 16 ft. lbs. (22 Nm)

- Wheel sensor wire and torque the bolts to 84 inch lbs. (9.5 Nm)
- Brake caliper and torque the bolts to 80 ft. lbs. (108 Nm)
- Brake hoses and torque the bolts to 84 inch lbs. (9.5 Nm)
- New axle nut and torque it to 181 ft. lbs. (245 Nm)
- Wheel

5. Measure and adjust the wheel alignment.

Upper Control Arm

REMOVAL & INSTALLATION

1. Before servicing the vehicle, refer to the precautions in the beginning of this section.
2. Raise and safely support the vehicle.
3. Remove or disconnect the following:
   - Front wheel

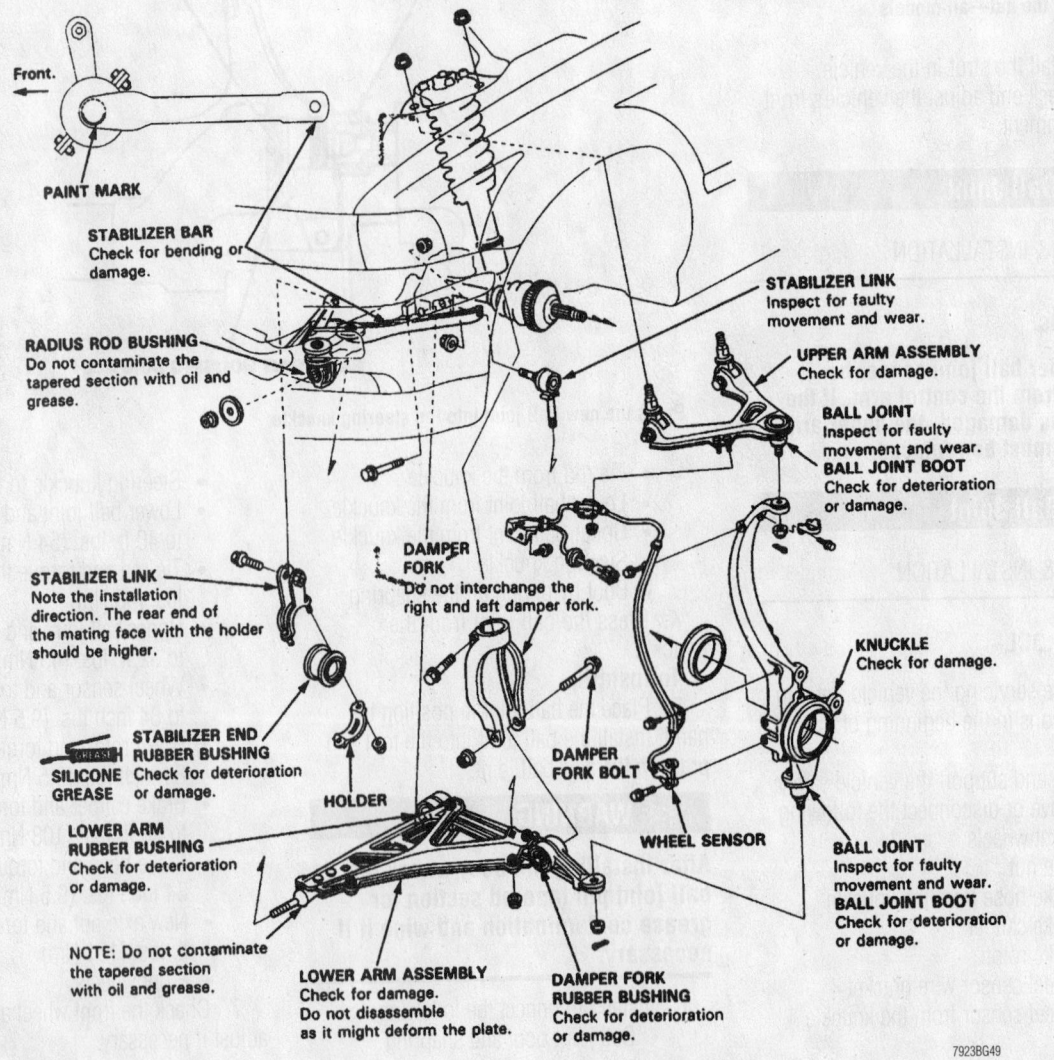

Front.

PAINT MARK

STABILIZER BAR
Check for bending or damage.

RADIUS ROD BUSHING
Do not contaminate the tapered section with oil and grease.

STABILIZER LINK
Note the installation direction. The rear end of the mating face with the holder should be higher.

STABILIZER END RUBBER BUSHING
Check for deterioration or damage.

SILICONE GREASE

HOLDER

LOWER ARM RUBBER BUSHING
Check for deterioration or damage.

NOTE: Do not contaminate the tapered section with oil and grease.

LOWER ARM ASSEMBLY
Check for damage.
Do not disassemble as it might deform the plate.

DAMPER FORK
Do not interchange the right and left damper fork.

DAMPER FORK BOLT

WHEEL SENSOR

DAMPER FORK RUBBER BUSHING
Check for deterioration or damage.

STABILIZER LINK
Inspect for faulty movement and wear.

UPPER ARM ASSEMBLY
Check for damage.

BALL JOINT
Inspect for faulty movement and wear.
BALL JOINT BOOT
Check for deterioration or damage.

KNUCKLE
Check for damage.

BALL JOINT
Inspect for faulty movement and wear.
BALL JOINT BOOT
Check for deterioration or damage.

7923BG49

A common upper control arm and ball joint assembly

- Upper ball joint from the knuckle
- Upper control arm nuts
- Upper control arm from the vehicle

To install:

4. Install or connect the following:
   - Upper control arm
   - Upper control arm-to-chassis nuts and torque them to 47 ft. lbs. (64 Nm)
   - Ball joint to the steering knuckle and torque the nut to 29-35 ft. lbs. (39-47 Nm)
   - Front wheel

5. Check the front wheel alignment and adjust if necessary.

CONTROL ARM BUSHING REPLACEMENT

➡The bushings are an integral part of the control arm and are not replaceable. If they are damaged the control arm should be replaced.

Lower Control Arm

REMOVAL & INSTALLATION

Front

ALL MODELS EXCEPT NSX

1. Before servicing the vehicle, refer to the precautions in the beginning of this section.
2. Raise and safely support the vehicle.
3. Remove or disconnect the following:
   - Front wheels
   - Lower damper fork bolt
   - Stabilizer bar from the arm
   - Lower ball joint from the steering knuckle
   - Radius rod from the lower control arm
   - Lower control arm mounting bolts
   - Lower arm from the vehicle

To install:

4. Install or connect the following:
   - Lower control arm and torque the bolts to 40 ft lbs. (54 Nm)
   - Lower ball joint to the steering knuckle and torque the nut to 36-43 ft. lbs. (49-59 Nm)
   - Stabilizer bar and torque the bolts to 16 ft. lbs. (22Nm)
   - Lower damper fork bolt and torque it to 47 ft. lbs. (64 Nm)
   - Radius rod and torque the bolts to 76 ft. lbs. (103 Nm)
   - Front wheels

5. Measure and adjust the wheel alignment.

NSX

1. Before servicing the vehicle, refer to the precautions in the beginning of this section.
2. Raise and safely support the vehicle.
3. Remove or disconnect the following:
   - Front wheels
   - Steering knuckle from the control arm
   - Lower strut mounting bolt
   - Stabilizer link from the control arm
   - Ball joint from the compliance pivot assembly
   - Control arm adjusting bolt
   - Control arm from the vehicle

To install:

4. Install or connect the following:
   - Lower control arm to the vehicle and torque the adjusting bolt to 90 ft. lbs. (123 Nm)
   - Ball joint to the compliance pivot assembly and torque the nut to 40-47 ft. lbs. (54-64 Nm)
   - Stabilizer link and torque the nut to 61 ft. lbs. (83 Nm)
   - Lower strut mounting bolt and torque it to 69 ft. lbs. (93 Nm)
   - Steering knuckle to the control arm and torque the nut to 40-47 ft. lbs. (54-64 Nm)
   - Front wheels

5. Measure and adjust the wheel alignment.

Rear

INTEGRA

1. Before servicing the vehicle, refer to the precautions in the beginning of this section.
2. Raise and safely support the vehicle.
3. Remove or disconnect the following:
   - Rear wheels
   - Hub/bearing assembly
   - Splash guard
   - Lower strut mounting bolt
   - Upper arm from the control arm
   - Lower arm from the control arm
   - Compensator arm
   - Control arm mounting bolts
   - Control arm from the vehicle

To install:

4. Install or connect the following:
   - Control arm and torque the bolts to 47 ft. lbs. (64 Nm)
   - Compensator arm and torque the bolt to 47 ft. lbs. (64 Nm)

- Lower arm and torque the bolt to 40 ft. lbs. (54 Nm)
- Upper arm and torque the bolt to 40 ft. lbs. (54 Nm)
- Lower strut mounting bolt and torque it to 40 ft. lbs. (54 Nm)
- Hub/bearing assembly and torque the nut to 134 ft. lbs. (181 Nm)

2.3CL, 2.5TL, 3.0CL AND 3.5RL

1. Before servicing the vehicle, refer to the precautions in the beginning of this section.
2. Raise and safely support the vehicle.
3. Remove or disconnect the following:
   - Rear wheels
   - Stabilizer link from the control arm
   - Control arm from the knuckle
   - Control arm from the bracket
   - Control arm from the vehicle

To install:

   - Control arm to the vehicle
   - Control arm bracket bolt and torque it to 47 ft. lbs. (64 Nm)
   - Control arm to the knuckle and torque the nuts to 47 ft. lbs. (64 Nm)
   - Stabilizer link and torque the nut to 22 ft. lbs. (29 Nm)
   - Rear wheels

4. Measure and adjust the wheel alignment.

3.2CL AND 3.2TL

1. Before servicing the vehicle, refer to the precautions in the beginning of this section.
2. Raise and safely support the vehicle.
3. Remove or disconnect the following:
   - Rear wheels
   - Control arm from the knuckle
   - Control arm from the subframe
   - Control arm from the vehicle

To install:

4. Install the control arm to the vehicle and torque the subframe nut to 40 ft. lbs. (54 Nm) and the knuckle nut to 43 ft. lbs. (59 Nm).

5. Measure and adjust the wheel alignment.

NSX

1. Before servicing the vehicle, refer to the precautions in the beginning of this section.
2. Raise and safely support the vehicle.
3. Remove or disconnect the following:
   - Rear wheels
   - Knuckle from the control arm
   - Control arm from the vehicle

To install:

4. Install the control arm to the vehicle and torque the mounting bolts to 90 ft. lbs. (123 Nm).

5. Install the knuckle to the control arm and torque the nut to 40-47 ft. lbs. (54-64 Nm).

6. Install the wheels.

7. Measure and adjust the wheel alignment.

CONTROL ARM BUSHING REPLACEMENT

➡ **The bushings are an integral part of the control arm and are not replaceable. If they are damaged the control arm should be replaced.**

Wheel Bearings

ADJUSTMENT

The front and rear wheel bearings are not adjustable or repairable and should be replaced if found defective.

REMOVAL & INSTALLATION

Front

INTEGRA

1. Before servicing the vehicle, refer to the precautions in the beginning of this section.

2. Raise and safely support the vehicle.

3. Remove or disconnect the following:
   - Front wheels
   - Axle nut
   - Brake hose mounting bolts
   - Brake caliper bolts and remove the caliper from the knuckle
   - Disc brake rotor
   - Wheel sensor wire bracket
   - Wheel sensor from the knuckle
   - Lower ball joint
   - Upper ball joint

4. Pull the knuckle outward and remove the halfshaft outboard joint from the knuckle using a plastic hammer, then remove the knuckle.

5. Place the knuckle on a base. Insert a disassembly tool into the hub, then using a press, remove the hub from the knuckle.

6. Remove the circlip and the splash guard from the knuckle.

7. Place the knuckle on the disassembly base and install a driver to the bearing. Using a press, remove the bearing from the knuckle.

8. Press the wheel bearing inner race from the hub using the hub disassembly tool and a bearing separator.

To install:

9. Press a new inner race and wheel bearing into the knuckle with a suitable driver.

10. Install or connect the following:
    - Splash guard and tighten the screws to 43 inch lbs. (5 Nm)
    - Circlip securely in the knuckle groove

11. Use a suitable driver to press the knuckle onto the hub.

12. Install or connect the following:
    - Knuckle/hub assembly onto the halfshaft
    - Lower ball joint to the knuckle and torque the nut to 36-43 ft. lbs. (49-59 Nm)
    - Tie rod and tighten the nut to 29-35 ft. lbs. (39-47 Nm)
    - Upper ball joint and tighten the nut to 29-35 ft. lbs. (39-47 Nm)
    - Wheel sensor and torque the bolts to 84 inch lbs. (9.5 Nm)
    - Wheel sensor wire bracket and torque the bolts to 84 inch lbs. (9.5 Nm)
    - Brake rotor and tighten the screws to 84 inch lbs. (10 Nm)
    - Brake caliper and torque the bolts to 80 ft. lbs. (108 Nm)
    - Brake hose mounting bolts and torque them to 84 inch lbs. (9.5 Nm)

- New axle nut and tighten it to 134 ft. lbs. (181 Nm)
- Wheels

13. Measure and adjust the wheel alignment.

2.3CL AND 3.0CL

1. Before servicing the vehicle, refer to the precautions in the beginning of this section.

2. Remove or disconnect the following:
   - Wheel
   - Axle nut
   - Brake hose from the knuckle
   - Brake caliper
   - Brake rotor from the hub
   - Wheel sensor wire bracket
   - Wheel sensor
   - Tie rod from the knuckle
   - Lower ball joint
   - Upper ball joint
   - Steering knuckle
   - Hub/bearing from the knuckle

3. Press the bearing out of the hub assembly.

4. Press the inner race out of the hub/flange.

To install:

5. Press a new inner race into the hub.

6. Press a new wheel bearing into the hub.

7. Install or connect the following:

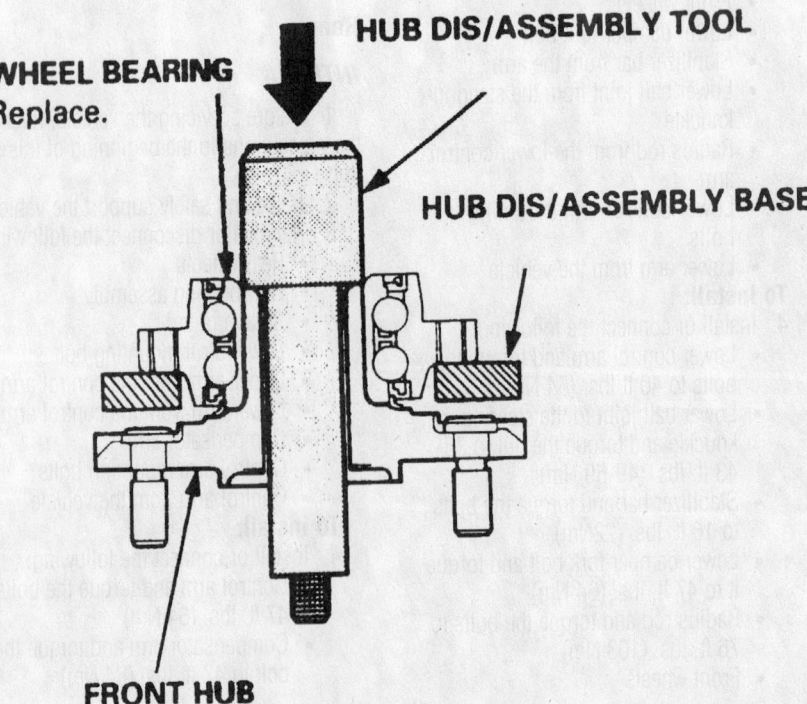

Press the hub/flange out of the bearing assembly—3.0CL

7923BGA0

HUB DIS/ASSEMBLY TOOL

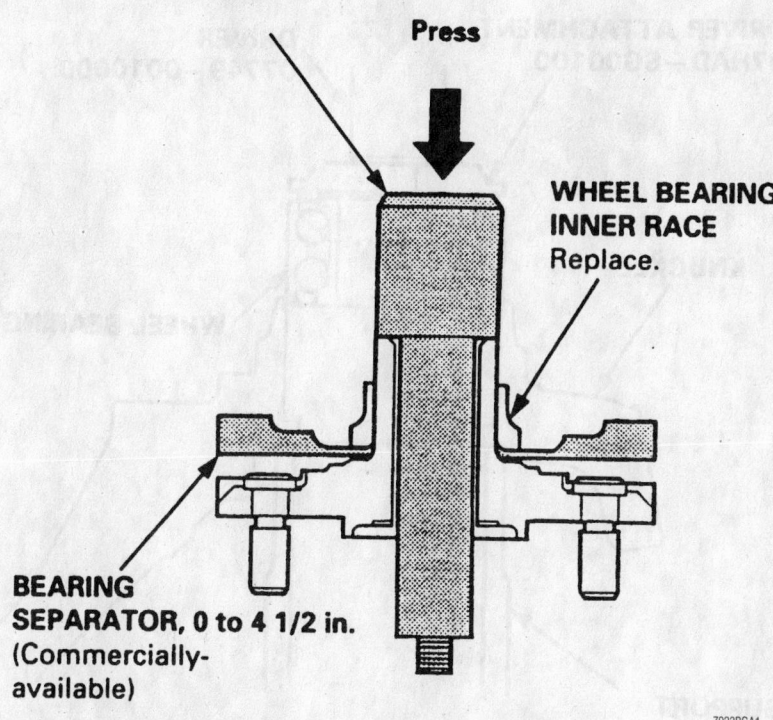

Use a press to remove the inner race from the hub/flange—3.0CL

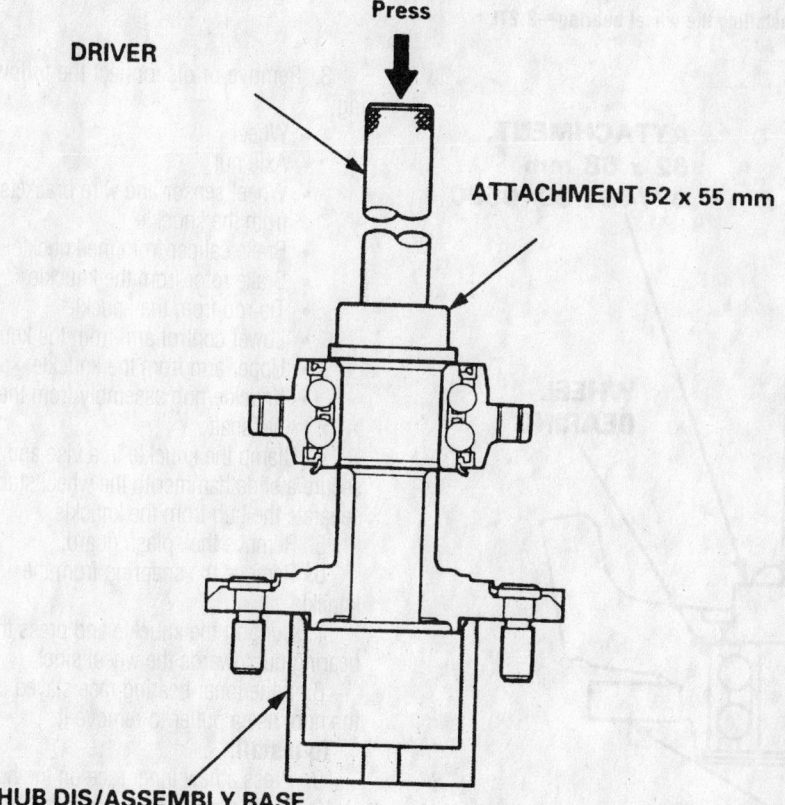

Press the new bearing onto the hub/flange—3.0CL

- Hub/bearing assembly on the knuckle and torque the 4 bolts to 33 ft. lbs. (44 Nm)
- Knuckle assembly onto the half-shaft
- Lower ball joint and torque the nut to 36-43 ft. lbs. (49-59 Nm)
- Upper ball joint and torque the nut to 29-35 ft. lbs. (39-47 Nm)
- Tie rod and torque the nut to 29-35 ft. lbs. (39-47 Nm)
- Wheel sensor and torque the bolts to 16 ft. lbs. (22 Nm)
- Wheel sensor wire brackets and torque the bolts to 84 inch lbs. (9.5 Nm)
- Brake rotor and torque the bolts to 84 inch lbs. (9.5 Nm)
- Brake caliper and torque the bolts to 80 ft. lbs. (108 Nm)
- Brake hose and torque the bolts to 84 inch lbs. (9.5 Nm)
- New axle nut and torque it to 181 ft. lbs. (245 Nm)
- Wheel

8. Measure and adjust the wheel alignment.

2.5TL

1. Before servicing the vehicle, refer to the precautions in the beginning of this section.
2. Raise and safely support the vehicle.
3. Remove or disconnect the following:
   - Wheel
   - Axle nut
   - Brake caliper from the knuckle
   - Wheel sensor from the knuckle
   - Tie rod from the knuckle
   - Lower ball joint from the knuckle
   - Upper ball joint from the knuckle
   - Knuckle and hub by sliding the assembly off of the halfshaft
   - Hub/rotor assembly from the knuckle
   - Hub assembly from the rotor
4. Press the bearing from the hub.
5. Press the inner race from the hub.

To install:

➡ **When pressing on a new bearing, be sure to press only on the inner race or the bearing will be damaged.**

6. Press a new inner race into the hub.
7. Press a new wheel bearing into the hub.
8. Install or connect the following:
   - Brake disc and tighten the bolts to 40 ft. lbs. (54 Nm)

- Knuckle to the hub and tighten the bolts to 33 ft. lbs. (45 Nm)
- Knuckle and hub assembly onto the halfshaft
- Knuckle on the tie rod and torque the nut to 36-43 ft. lbs. (49-59 Nm)
- Upper control arm and torque the nut to 29-35 ft. lbs. (39-47 Nm)
- Lower control arm and tighten the nut to 36-43 ft. lbs. (49-59 Nm)
- Wheel sensor and torque the bolts to 16 ft. lbs. (22 Nm)
- Wheel sensor wire bracket and torque the bolts to 84 inch lbs. (9.5 Nm)
- Brake caliper and torque the bolts to 80 ft. lbs. (108 Nm)
- Brake hoses and torque the bolts to 84 inch lbs. (9.5 Nm)
- New axle nut and torque it to 181 ft. lbs. (245 Nm)
- Wheel

9. Measure and adjust the wheel alignment.

3.2CL, 3.2TL AND 3.5RL

1. Before servicing the vehicle, refer to the precautions in the beginning of this section.
2. Raise and safely support the vehicle.

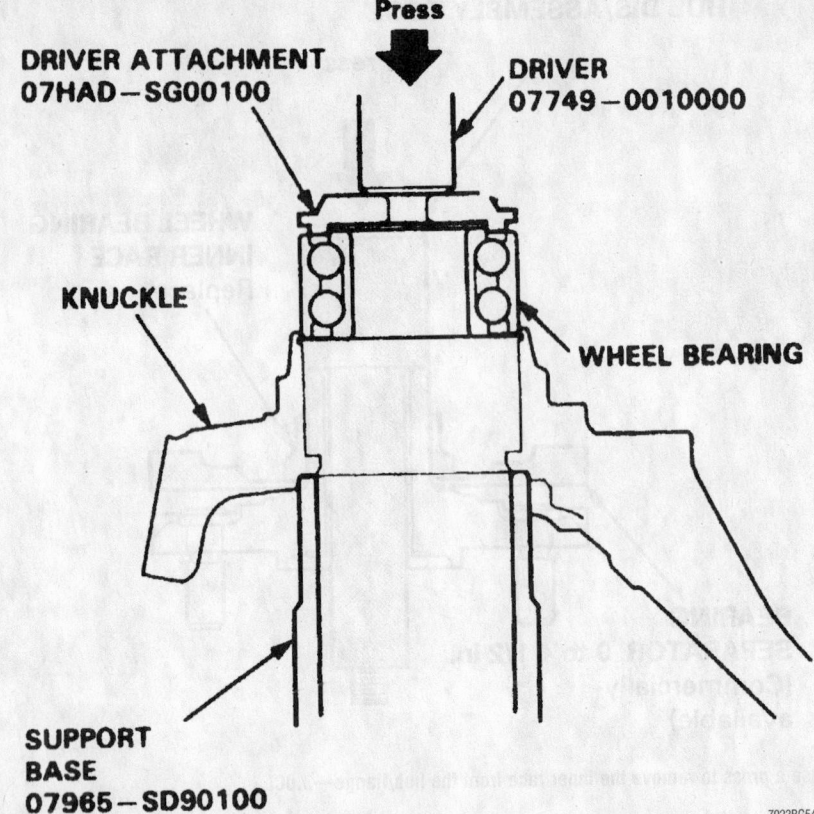

Installing the wheel bearing—3.2TL

3. Remove or disconnect the following:

- Wheel
- Axle nut
- Wheel sensor and wire brackets from the knuckle
- Brake caliper from the knuckle
- Brake rotor from the knuckle
- Tie rod from the knuckle
- Lower control arm from the knuckle
- Upper arm from the knuckle
- Knuckle/hub assembly from the halfshaft

4. Clamp the knuckle in a vise and secure a slide hammer to the wheel studs to separate the hub from the knuckle.
5. Remove the splash guard.
6. Remove the snapring from the knuckle.
7. Support the knuckle and press the bearing out towards the wheel side.
8. If the inner bearing race stayed on the hub, use a puller to remove it.

To install:

9. Press a new inner race on the hub.
10. Press a new bearing into the knuckle.
11. Install the outer snapring.
12. Install the splash guard and torque the bolts to 4 ft. lbs. (4.9 Nm).

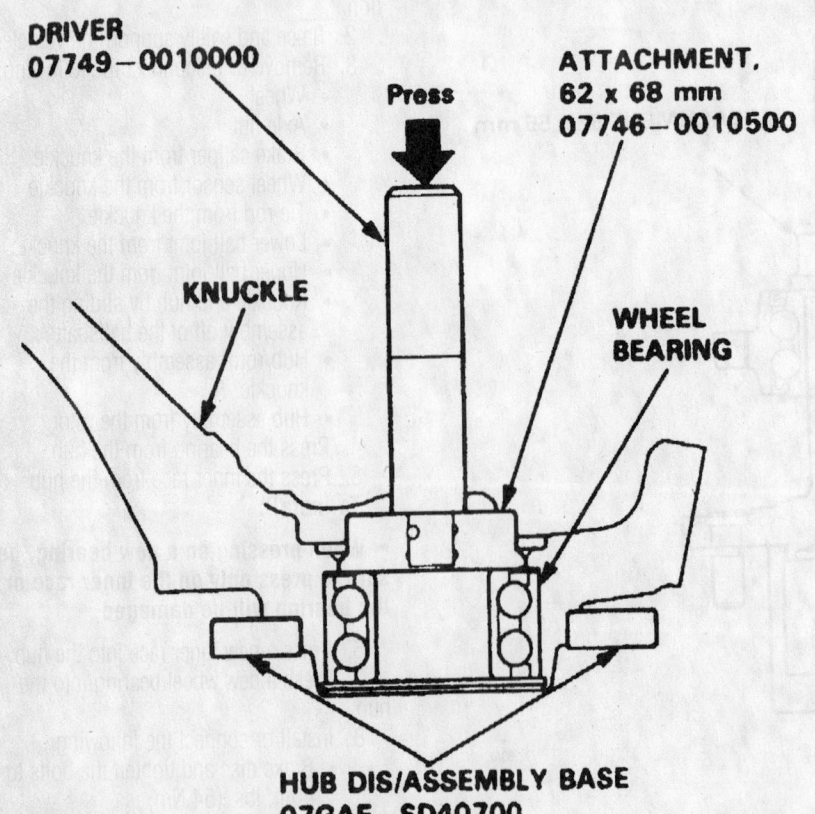

Pressing out the wheel bearing—3.2TL

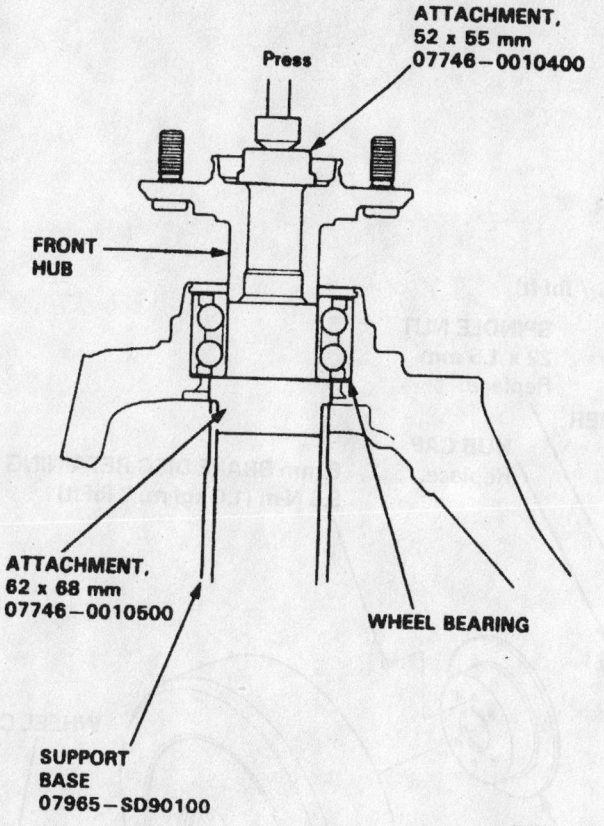

ATTACHMENT,
52 x 55 mm
07746–0010400

Press

FRONT
HUB

ATTACHMENT,
62 x 68 mm
07746–0010500

WHEEL BEARING

SUPPORT
BASE
07965–SD90100

7923BG55

Hub installation—3.2TL

13. Properly support the knuckle and press the hub into the bearing.

✳✳ CAUTION

Do not press on the wheel studs or they will press out of the hub.

14. Install or connect the following:
- Knuckle/hub assembly onto the halfshaft
- Lower ball joint and torque the nut to 51-58 ft. lbs. (69-78 Nm)
- Upper ball joint and torque the nut to 29-35 ft. lbs. (40-48 Nm)
- Tie rod end and torque the nut to 36-43 ft. lbs. (50-60 Nm)
- Brake rotor and torque the bolts to 84 inch lbs. (9.5 Nm)
- Brake caliper and torque the bolts to 80 ft. lbs. (108 Nm)
- Brake hose brackets and torque the bolts to 16 ft. lbs. (22 Nm)
- Wheel sensor and torque the bolts to 84 inch lbs. (9.5 Nm)
- Wheel sensor wire brackets and torque the bolts to 84 inch lbs. (9.5 Nm)

- New axle nut and torque it to 181 ft. lbs. (245 Nm)
- Wheel

15. Measure and adjust the wheel alignment

NSX

1. Before servicing the vehicle, refer to the precautions in the beginning of this section.
2. Raise and safely support the vehicle.
3. Remove or disconnect the following:
- Wheel
- Wheel sensor and wire brackets from the knuckle
- Brake caliper and hose brackets from the knuckle
- Brake rotor from the knuckle
- Hub assembly from the knuckle
- Spindle nut
- Pulser from the hub using a puller
4. Press the bearing from the hub.
5. Press the inner race from the hub.

To install:
6. Press a new inner race on the hub
7. Press a new bearing into the hub.
8. Install or connect the following:

- Pulser
- New spindle nut and torque it to 242 ft. lbs. (329 Nm)
- Hub assembly and torque the nuts to 47 ft. lbs. (64 Nm)
- Brake rotor and torque the bolts to 84 inch lbs. (9.5 Nm)
- Brake caliper and torque the bolts to 80 ft. lbs.(108 Nm)
- Brake hose brackets and torque the bolts to 16 ft. lbs. (22 Nm)
- Wheel sensor and torque the bolts to 16 ft. lbs. (22 Nm)
- Wheel sensor wire brackets and torque the bolts to 84 inch lbs. (9.5 Nm)
- Wheel

9. Measure and adjust the wheel alignment.

Rear

ALL MODELS EXCEPT NSX

➡The rear wheel bearings on these vehicles is part of the wheel hub and is not serviceable. If the bearing is bad the wheel hub must be replaced.

1. Before servicing the vehicle, refer to the precautions in the beginning of this section.
2. Be sure the emergency brake is disengaged.
3. Raise and safely support the vehicle.
4. Remove or disconnect the following:
- Wheel
- Spindle hub cap
- Spindle nut
- Caliper shield (if equipped)
- Brake hose mounting bolts from the knuckle
- Brake caliper
- Brake rotor
- Hub assembly from the knuckle

To install:
5. Install or connect the following:
- Hub/bearing assembly
- Brake disc and tighten the bolts to 84 inch lbs. (9.5 Nm)
- Brake caliper and torque the bolts to 41 ft. lbs. (56 Nm)
- Brake hose clamps and torque the bolts to 16 ft. lbs. (22 Nm)
- Brake caliper shield and torque the mounting bolts to 84 inch lbs. (9.5 Nm) (if equipped)
- New spindle nut and torque it to 134 ft. lbs. (181 Nm)
- Spindle hub cap
- Rear wheels

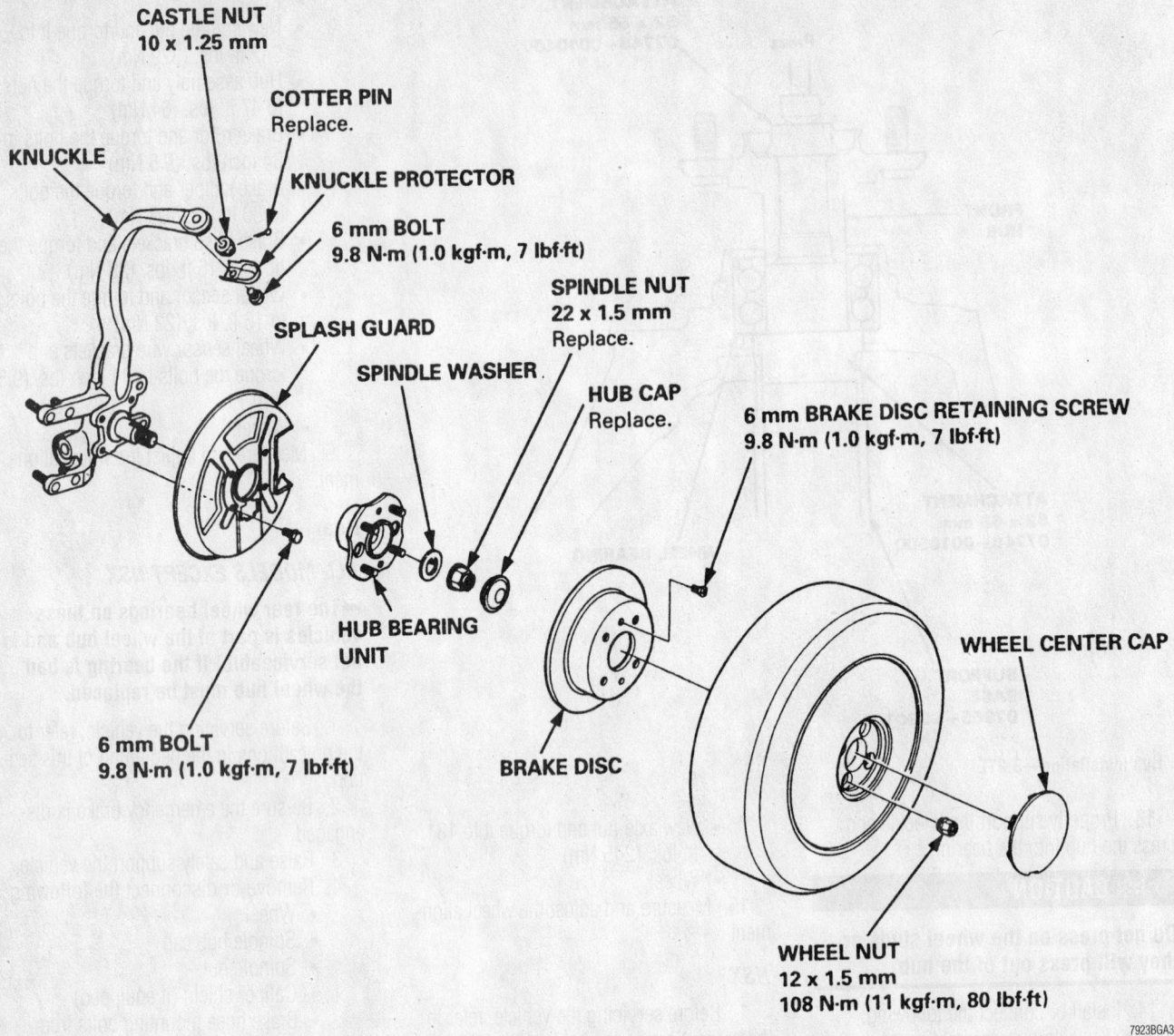

CASTLE NUT
10 x 1.25 mm

COTTER PIN
Replace.

KNUCKLE

KNUCKLE PROTECTOR

6 mm BOLT
9.8 N·m (1.0 kgf·m, 7 lbf·ft)

SPINDLE NUT
22 x 1.5 mm
Replace.

SPLASH GUARD

SPINDLE WASHER

HUB CAP
Replace.

6 mm BRAKE DISC RETAINING SCREW
9.8 N·m (1.0 kgf·m, 7 lbf·ft)

HUB BEARING UNIT

WHEEL CENTER CAP

6 mm BOLT
9.8 N·m (1.0 kgf·m, 7 lbf·ft)

BRAKE DISC

WHEEL NUT
12 x 1.5 mm
108 N·m (11 kgf·m, 80 lbf·ft)

7923BGA3

Exploded view of the rear bearing and related components—3.0CL

NSX

1. Before servicing the vehicle, refer to the precautions in the beginning of this section.

2. Raise and safely support the vehicle.

3. Remove or disconnect the following:
- Wheel
- Spindle nut
- Wheel sensor and wire brackets
- Brake hose from the caliper
- Brake caliper
- Brake rotor
- Hub/bearing from the knuckle

4. Press the bearing from the hub.

5. Press the inner race from the hub.

To install:

6. Press a new inner race onto the hub.

7. Press a new wheel bearing onto the hub.

8. Install or connect the following:
- Hub/bearing and torque the bolts to 47 ft. lbs. (64 Nm)
- Brake rotor and torque the bolts to 84 inch lbs. (9.5 Nm)
- Brake caliper and torque the bolts to 80 ft. lbs. (108 Nm)
- Brake caliper hose and torque the bolt to 25 ft. lbs. (34 Nm)
- Wheel sensor and torque the bolts to 16 ft. lbs. (22 Nm)
- Wheel sensor wire brackets and torque the bolts to 84 inch lbs. (9.5 Nm)
- New spindle nut and torque it to 242 ft. lbs. (329 Nm)
- Wheel

9. Fill and bleed the brake system.

AUDI

1998–01

A4 • A4 Avant • A6 • A6 Avant • A6 Avant Allroad • Cabriolet • TT

6

PRECAUTIONS

Before servicing any vehicle, please be sure to read all of the following precautions, which deal with personal safety, prevention of component damage and important points to take into consideration when servicing a motor vehicle:

• Never open, service or drain the radiator or cooling system when the engine is hot; serious burns can occur from the steam and hot coolant.

• Observe all applicable safety precautions when working around fuel. Whenever servicing the fuel system, always work in a well-ventilated area. Do not allow fuel spray or vapors to come in contact with a spark, open flame, or excessive heat (a hot drop light, for example). Keep a dry chemical fire extinguisher near the work area. Always keep fuel in a container specifically designed for fuel storage; also, always properly seal fuel containers to avoid the possibility of fire or explosion. Refer to the additional fuel system precautions later in this section.

• Fuel injection systems often remain pressurized, even after the engine has been turned **OFF**. The fuel system pressure must be relieved before disconnecting any fuel lines. Failure to do so may result in fire and/or personal injury.

• Brake fluid often contains polyglycol ethers and polyglycols. Avoid contact with the eyes and wash your hands thoroughly after handling brake fluid. If you do get brake fluid in your eyes, flush your eyes with clean, running water for 15 minutes. If eye irritation persists, or if you have taken brake fluid internally, seek medical assistance IMMEDIATELY.

• The EPA warns that prolonged contact with used engine oil may cause a number of skin disorders, including cancer! You should make every effort to minimize your exposure to used engine oil. Protective gloves should be worn when changing oil. Wash your hands and any other exposed skin areas as soon as possible after exposure to used engine oil. Soap and water, or waterless hand cleaner should be used.

• All new vehicles are now equipped with an air bag system. The system must be disabled before performing service on or around system components, steering column, instrument panel components, wiring and sensors. Failure to follow safety and disabling procedures could result in accidental air bag deployment, possible personal injury and unnecessary system repairs.

• Always wear safety goggles when working with, or around, the air bag system. When carrying a non-deployed air bag, be sure the bag and trim cover are pointed away from your body. When placing a non-deployed air bag on a work surface, always face the bag and trim cover upward, away from the surface. This will reduce the motion of the module if it is accidentally deployed. Refer to the additional air bag system precautions later in this section.

• Clean, high quality brake fluid from a sealed container is essential to the safe and proper operation of the brake system. You should always buy the correct type of brake fluid for your vehicle. If the brake fluid becomes contaminated, completely flush the system with new fluid. Never reuse any brake fluid. Any brake fluid that is removed from the system should be discarded. Also, do not allow any brake fluid to come in contact with a painted surface; it will damage the paint.

• Never operate the engine without the proper amount and type of engine oil; doing so WILL result in severe engine damage.

• Timing belt maintenance is extremely important! Many models utilize an interference-type, non-freewheeling engine. If the timing belt breaks, the valves in the cylinder head may strike the pistons, causing potentially serious (also time-consuming and expensive) engine damage. Refer to the maintenance interval charts in the front of this manual for the recommended replacement interval for the timing belt and to the timing belt section for belt replacement and inspection.

• Disconnecting the negative battery cable on some vehicles may interfere with the functions of the on-board computer system(s) and may require the computer to undergo a relearning process once the negative battery cable is reconnected.

• When servicing drum brakes, only disassemble and assemble one side at a time, leaving the remaining side intact for reference.

• Only an MVAC-trained, EPA-certified automotive technician should service the air conditioning system or its components.

ENGINE REPAIR

➡Disconnecting the negative battery cable on some vehicles may interfere with the functions of the on-board computer systems and may require the computer to undergo a relearning process, once the negative battery cable is disconnected.

Alternator

REMOVAL

1. Before servicing the vehicle, refer to the precautions in the beginning of this section.
2. Remove or disconnect the following:
   • Negative battery cable
   • Alternator drive belt

• Alternator wiring harness connectors
• Alternator

INSTALLATION

1. Install or connect the following:
   • Alternator. Torque the bolt to 18 ft. lbs. (25 Nm).
   • Alternator wiring harness connectors
   • Alternator drive belt
2. Adjust the alternator belt.

Ignition Timing

All vehicles in this section are equipped with distributorless ignition systems. No adjustments are possible.

Engine Assembly

REMOVAL & INSTALLATION

1.8L Engine

➡To allow clearance for removal of the engine assembly, the front bumper and the hood lock carrier assembly must be removed from the front of the vehicle.

1. Before servicing the vehicle, refer to the precautions in the beginning of this section.
2. Turn the ignition switch to the **OFF** position, then disconnect the negative battery cable.
3. Position the wipers to the vertical position.

4. Properly relieve the fuel system pressure.

5. Drain the engine coolant.

6. Remove or disconnect the following:

- Negative battery cable
- Lower engine slash shield
- Water pump housing drain plug, drain coolant and reinstall
- Left and right lower bumper air grilles
- Inner fender lining
- Front bumper
- Power steering cooling coil
- Transaxle oil cooling lines
- Electric cooling fan thermal switch
- Air intake duct/assembly
- Headlight height adjuster wiring harness
- Turn signal bulb sockets
- Power steering fluid reservoir cap/dipstick
- Anti-lock Brake System (ABS) hydraulic unit wiring harness
- Horn electrical connectors
- Air guides at the left and right sides of the radiator
- Wires or connectors that would inhibit hood lock carrier assembly removal
- Hood release cable
- Air guide between the lock carrier and air filter
- Outside temperature electrical connector
- Upper and lower radiator hoses from the radiator
- Outside air temperature sensor/cooling line
- Power steering hydraulic cooling line bracket without disconnecting the fluid lines
- Air conditioning condenser assembly without disconnecting the pressure hoses

✳✳ CAUTION

Use care to not kink the A/C lines. Do not allow the condenser to hang on the lines.

- Outside air temperature sensor/cooling line bracket
- Power steering hydraulic cooling line bracket from bottom of the radiator. Do not open the fluid lines
- Hood weather seal
- Front hood lock carrier
- Air conditioning low-pressure switch
- Green harness connector from the air conditioning compressor magnetic clutch
- Engine covers
- Wiring harness connectors for the wastegate bypass regulator valve
- EVAP canister purge regulator valve
- Power output stage
- Mass Air Flow (MAF) sensor
- Engine Coolant Level (ECL) warning switch
- Coolant hoses at the expansion tank
- Coolant tank
- Actuating rod from the throttle valve control module and the vacuum hose from the vacuum unit, if equipped with cruise control
- Accelerator pedal cable from the throttle valve control module
- Hose for the Leak Detection Pump (LDP)
- Fuel supply and return lines
- Brake booster vacuum hose
- Vacuum hose for the Evaporative Emissions (EVAP) canister purge regulator valve
- Uncover the E-box
- Engine Control Module (ECM) retaining bracket
- Wiring harness to the ECM
- Kickdown switch connector, if equipped with an automatic transaxle
- Heated Oxygen (HO2S) sensor wiring harness
- Ground connection at the plenum chamber
- Heater hoses from the heater core
- Vehicle Speed Sensor (VSS) from the transaxle
- Backup light switch connector from the transaxle, if equipped with a manual transaxle
- Cooling fan
- Accessory drive belts
- A/C compressor from the mounting bracket
- Power steering pump

➡The flex pipe at the front exhaust pipe must not be bent more than 10 degrees. Otherwise it may be damaged.

Bumper mounting fastener (1), energy absorbing strut fasteners (2), strut assembly (3), saddle (4), bumper cover (5), spoiler (6) and liner fastener (7)—A4 models

9301CG03

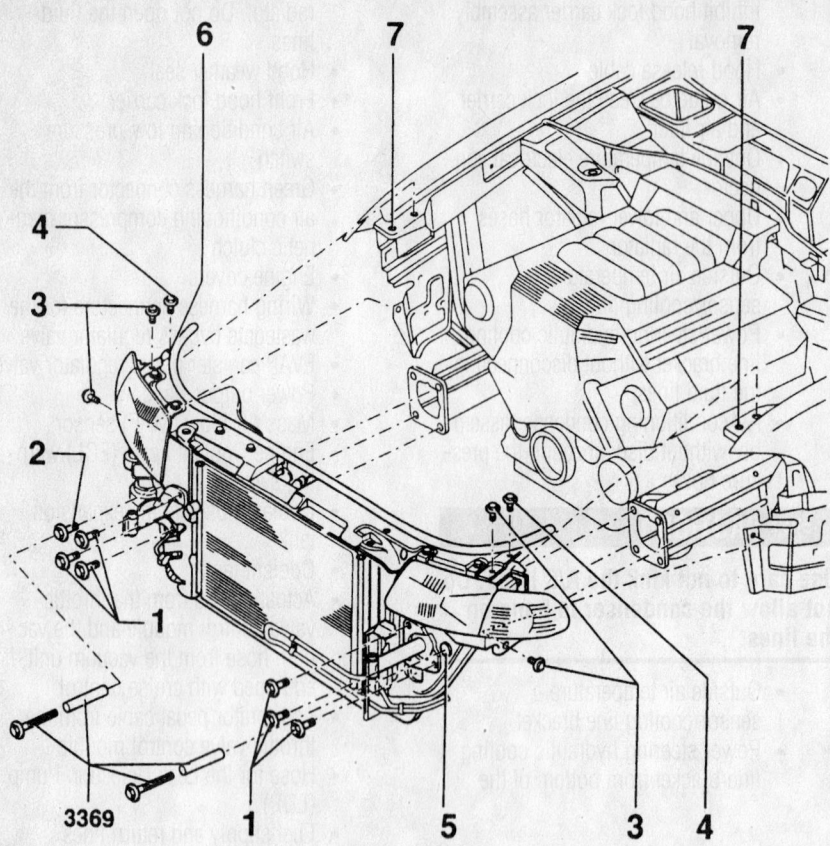

1. Bolts 33 ft. lbs. (45 Nm)
2. Bolts 33 ft. lbs. (45 Nm)
3. Bolts 7 ft. lbs. (10 Nm)
4. Bolts 7 ft. lbs. (10 Nm)
5. Bore for support tool
6. Lock carrier bore
7. Fender bore

Exploded view of the hood lock carrier assembly—A4 models

- Catalytic converter from the turbocharger
- Starter
- Ground strap at the right engine mount
- 3 torque converter-to-driveplate bolts through the starter opening, if equipped with an automatic transaxle

7. Loosen the upper nuts for the left and right engine mounts.

8. Place matchmarks on the threaded bolt and centering sleeves at the bottom of the left and right engine mounts, then remove the mounting nuts.

9. Remove or disconnect the following:
- Lower engine-to-transaxle mounting bolts

- Cooler line bracket form the left side of the engine, if equipped with an automatic transaxle
- Upper nuts from the engine mounts

10. Position an Engine Support Bridge tool 10-222A to the bolted flanges of the fenders with the spindle facing forward.

11. Attach the Support Adapter tool 3147 to the bolt hole above the starter area in the transaxle bell housing.

12. Connect the Engine Support Adapter tool 3147 to the Engine Support Bridge tool 10-222A using Adapter tool 2024A/1 and Extension tool 2024A/2 and support the transaxle.

13. Attach an engine sling between the engine and the hoist.

14. Remove or disconnect the following:

- Upper engine-to-transaxle bolts
- Engine from the transaxle, then out the front of the engine compartment

15. If equipped with an automatic transaxle, secure the torque converter to prevent it from falling out.

To install:

16. Installation is the reverse of the removal procedure, while using the following torque values:

- Transaxle flange M12 bolts: 48 ft. lbs. (65 Nm)
- Transaxle flange M10 bolts: 33 ft. lbs. (45 Nm)
- Engine mounting fasteners: 18 ft. lbs. (25 Nm)
- Torque converter bolts: 63 ft. lbs. (85 Nm), if equipped with an automatic transaxle
- Catalytic converter bolts: 22 ft. lbs. (30 Nm)

2.8L Engine

➡The engine is removed without the transaxle, through the top of the engine compartment. On all A4 and A6 models, the front bumper and the front lock carrier assembly must be removed.

1. Before servicing the vehicle, refer to the precautions in the beginning of this section.

2. Relieve the fuel system pressure.

3. Drain the engine coolant.

4. Remove or disconnect the following:
- Negative battery cable

➡On some vehicles the battery is located under the rear seat.

- Engine undercover
- Soundproofing material holder from the engine mount

5. Remove the front bumper and the hood lock carrier assembly, for A4 models and A6 models.

6. Remove the front bumper as follows:

7. Front wheels

8. Fasteners attaching the wheel housing lining to the front bumper

9. Bumper cover flanged nuts

10. Lower air grille assemblies

11. Front bumper mounting bolts

12. Front bumper

13. Remove or disconnect the following:
- Hood release cable
- Air intake duct for the air cleaner housing
- Wring connectors for the horns, headlights and electrical equipment mounted on the front hood lock carrier assembly

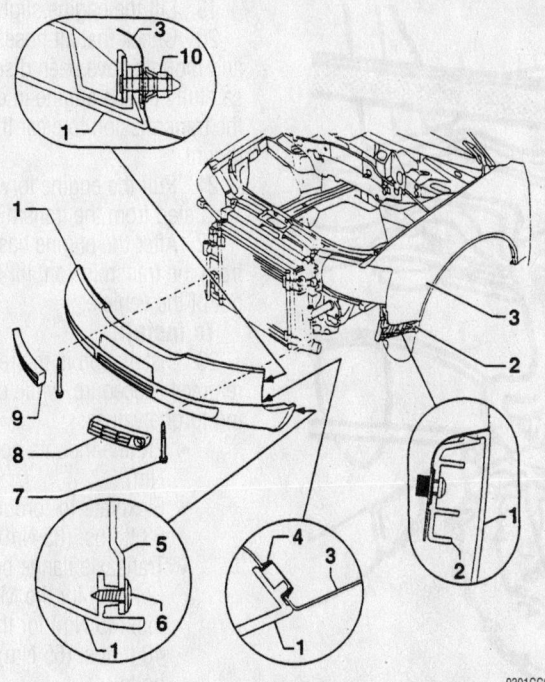

9301CG02

Bumper cover (1), saddle (2), fender (3), alignment tab (4), inner fender liner (5), liner fastener (6), bumper mounting fastener (7), air grilles (8, 9) and flanged nut (10)—1998–02 A6 models

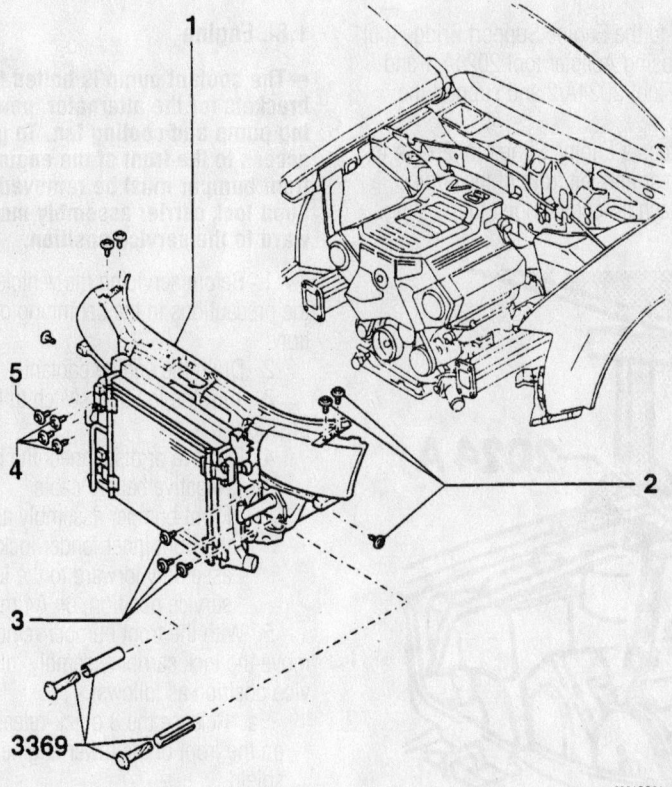

9301CG01

The hood lock carrier assembly (1), fender fasteners (2) and energy absorbing strut fasteners (3). Tool 3369 is used when placing the carrier in the service position—1998–02 A6 models

- Hood lock carrier assembly to the front fenders fasteners
- Radiator coolant hoses
- A/C condenser from the brackets and move it aside with the hoses attached

❊❊ CAUTION

The condenser lines must not be bent or kinked.

- Power steering cooling line brackets located on the engine support and the transmission and move them aside with the power steering hoses attached
- Cooling lines, if equipped with an automatic transmission
- Hood seal at the front fenders
- Energy absorbing bumper struts to the carrier fasteners
- Carrier assembly
- Stabilizer brace from the right rear of the engine compartment
- Wiper arms and water deflector trim, on A4 and A6 models
- Hoses
- Wiring harnesses
- Lines and cables for engine removal, as necessary
- Accessory drive belt guard
- Accessory drive belt
- Front engine mount at the cross-member
- Hydraulic lines that route above the cylinder head covers from the power steering pump
- Ground strap from the right side engine support
- Air intake for the alternator
- Alternator
- Air conditioning condenser brackets and move the condenser aside with the hoses attached
- Air conditioning compressor and move it aside with the hoses connected
- Exhaust system from the manifolds
- Crossmember to access the catalytic converters and the exhaust pipe
- Front exhaust pipes with the catalytic converters
- Starter
- Oil filter
- Oil cooler
- Flexplate to torque converter bolts through the starter opening, if equipped with an automatic transmission
- Engine-to-transmission bolts

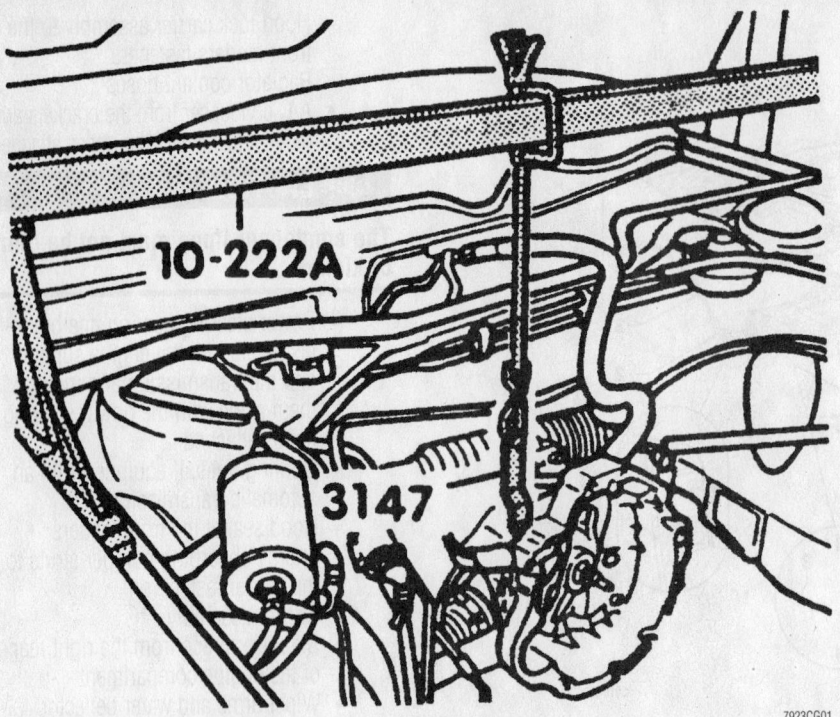

The engine support bridge and hook used to support the weight of the transmission for engine removal or weight of the engine for transmission removal—shown supporting the transmission with the engine removed

14. Position an Engine Support Bridge tool 10-222A to the bolted flanges of the fenders with the spindle facing forward.

15. Attach the Support Adapter tool 3147 to the bolt hole above the starter mounting area in the transaxle bell housing.

16. Connect the Engine Support Adapter tool 3147 to the Engine Support Bridge tool 10-222A using Adapter tool 2024A/1 and Extension tool 2024A/2 and support the transaxle.

17. Attach a Engine Sling tool 2024A to the right rear and left front of the engine.

18. Attach an engine hoist to the sling.

Be sure to attach the engine sling properly

19. Lift the engine slightly.

20. Check that all hoses, wires, cables and mounts have been disconnected and carefully lift the engine in conjunction with the transmission to clear the right engine mount.

21. Pull the engine forward until it is separated from the transmission.

22. After the engine has been separated from the transmission, lift the engine up and out of the vehicle.

To install:

23. Installation is the reverse of the removal procedure, while using the following torque values:

- Engine mounts bolts: 33 ft. lbs. (45 Nm).
- Flexplate-to-torque converter bolts: 63 ft. lbs. (85 Nm)
- Transaxle flange bolts: 18 ft. lbs. (24 Nm) for the M8 bolts, 33 ft. lbs. (45 Nm) for the M10 bolts and 48 ft. lbs. (65 Nm) for the M12 bolts

Water Pump

REMOVAL & INSTALLATION

1.8L Engine

➥**The coolant pump is bolted to the brackets for the alternator, power steering pump and cooling fan. To gain access to the front of the engine, the front bumper must be removed and the hood lock carrier assembly moved forward to the service position.**

1. Before servicing the vehicle, refer to the precautions in the beginning of this section.

2. Drain the engine coolant.

3. Turn the ignition switch to the **OFF** position.

4. Remove or disconnect the following:

- Negative battery cable
- Front bumper assembly and move the front inner fender lock carrier assembly forward to the lock carrier service position, on A4 models

5. With the front bumper removed, move the lock carrier assembly into the service position as follows:

  a. Release the 3 quick-release screws on the front of the lower engine splash shield.

  b. Unbolt the air guide between the lock carrier and the air filter assembly.

  c. Remove the 2 bolts that attach the lock carrier assembly to the side of the front fender and the 2 front bolts that mount the carrier to the top of the fender.

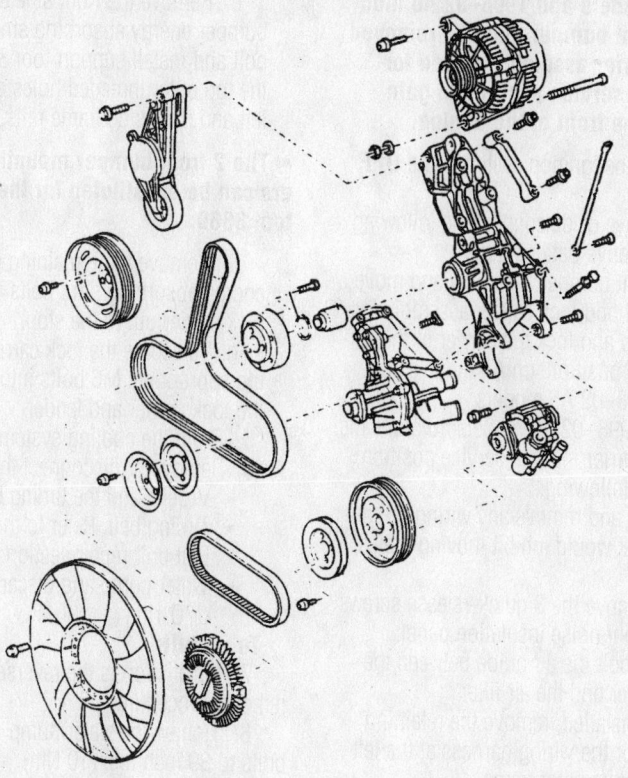

9301CG04

The water pump, alternator and power steering pump all mount to the same engine bracket—
1.8L engines

d. Remove the right side top outer bumper energy absorbing strut mounting bolt and install support tool 3369 into the top outer threaded holes on both the left and right sides of the bumper energy absorbing strut mounting surface.

➡ **The 2 front bumper-to-bumper energy absorbing strut fasteners can be substituted for support tool 3369.**

e. Remove the remaining bumper energy absorbing strut mounting bolts.

f. Remove the remaining 2 bolts that attach the lock carrier assembly to the top of the front fenders. Pull the lock carrier assembly forward until the rearmost bolt holes of the carrier align with the front-most threaded mounting points on the top of the front fender. Install the 2 carrier mounting bolts through the carrier into threaded mounting points of front fender to secure the lock carrier assembly in this position.

➡ **The 2 front bumper mounting fasteners can be substituted for Support tool 3369.**

g. Remove the remaining bumper

energy absorbing strut bolts and pull the lock carrier out to the stop.

h. To secure the lock carrier, install the appropriate M6 bolts into the rear of the lock carrier and fender.

6. Remove or disconnect the following:

- Accessory drive belt
- Cooling fan
- Lower engine slash shield

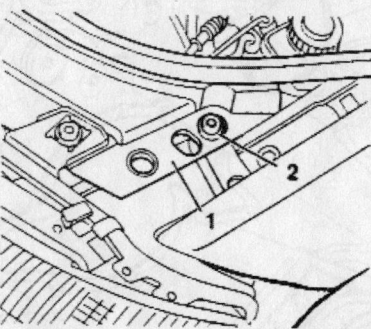

9301CG05

The hood lock carrier assembly (1) secured in the service position with the mounting fastener (2) installed—A4 models

7. Loosen the clamps for the coolant hoses at the water pump.

8. Remove or disconnect the following:

- Intake air duct between the intake manifold and the charge air cooler
- Alternator mounting bolts and slide it forward
- Alternator wiring

9. Unbolt the following supports and brackets for the alternator, power steering pump and engine cooling fan:

- Intake manifold support
- Support for the engine torque bracket
- Brace to the cylinder block
- Alternator brackets
- Power steering pump brackets
- Cooling fan brackets. Position the brackets for the alternator, power steering pump and engine cooling fan to the left side using a piece of wire
- Coolant hoses from the pump
- Coolant hoses from the thermostat housing
- Coolant pump housing from the timing belt cover
- Coolant pump
- Impeller housing from the pump housing

10. Clean all gasket and O-ring sealing surfaces.

To install:

11. Install or connect the following:

- Coolant pump to the pump housing, using a new gasket. Torque the bolts to 84 inch lbs. (10 Nm).
- Coolant pump, using a new gasket and O-ring. Torque the bolts in sequence to 18 ft. lbs. (25 Nm).
- Coolant pump housing to the timing belt cover. Torque it to 84 inch lbs. (10 Nm).

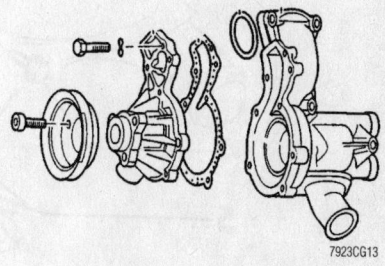

7923CG13

Exploded view of the water pump, housing and related components—1.8L Engine

Timing belt service is covered in Section 3 of this manual.

- Coolant hoses to the pump and thermostat housing
- Brackets. Torque the bolts to 18 ft. lbs. (25 Nm)
- Alternator and wire connectors
- Air intake duct between the intake manifold and the charge air cooler

12. The remaining steps are the reverse of the removal procedure noting the following items:

- Fill the engine with coolant
- Verify that the key is in the **OFF** position before connecting the battery
- Fully close all power windows to stop, operate all window switches for at least 1 second in the close direction to activate the one touch opening/closing function
- After installing the lock carrier, check the wiring for proper routing near the cooling fan

2.8L Engine

1. Before servicing the vehicle, refer to the precautions in the beginning of this section.

→On A4 models and 1998–02 A6 models, the front bumper must be removed and the carrier assembly moved forward to the service position to gain access to the front of the engine.

2. Turn the ignition switch to the **OFF** position.

3. Remove or disconnect the following:

- Negative battery cable
- Front bumper assembly and move the hood lock carrier assembly forward and lock the carrier in the service position, on A4 models and 1998–02 A6 models

4. On 1998–02 A6 models, to move the hood lock carrier into the service position, perform the following:

a. Tag and remove any wiring or connector that would inhibit moving the carrier.

b. Remove the 3 quick-release screws on the front noise insulation panel.

c. Unbolt the air guide between the lock carrier and the air filter.

d. If installed, remove the retaining clamps for the wiring harness at the left side of the radiator frame.

e. Remove the right side top outer bumper energy absorbing strut mounting bolt and install support tool 3369 into the top outer threaded holes on both the left and right side frame rails.

→The 2 front bumper mounting fasteners can be substituted for the Support tool 3369.

f. Remove the remaining bumper energy absorbing strut bolts and pull the lock carrier out to the stop.

g. To secure the lock carrier, install the appropriate M6 bolts into the rear of the lock carrier and fender.

5. Drain the cooling system.

6. Remove or disconnect the following

- V-belts and the timing belt covers
- Timing belt. Refer to the Timing Belt unit repair section.
- Water pump and discard the gasket or O-ring

To install:

7. Installation is the reverse of the removal procedure.

8. Tighten the water pump retaining bolts to 89 inch lbs. (10 Nm).

1. Thermostat	4. Bolt	7. Bolt
2. Seal	5. Gasket	
3. Thermostat housing	6. Coolant pump	

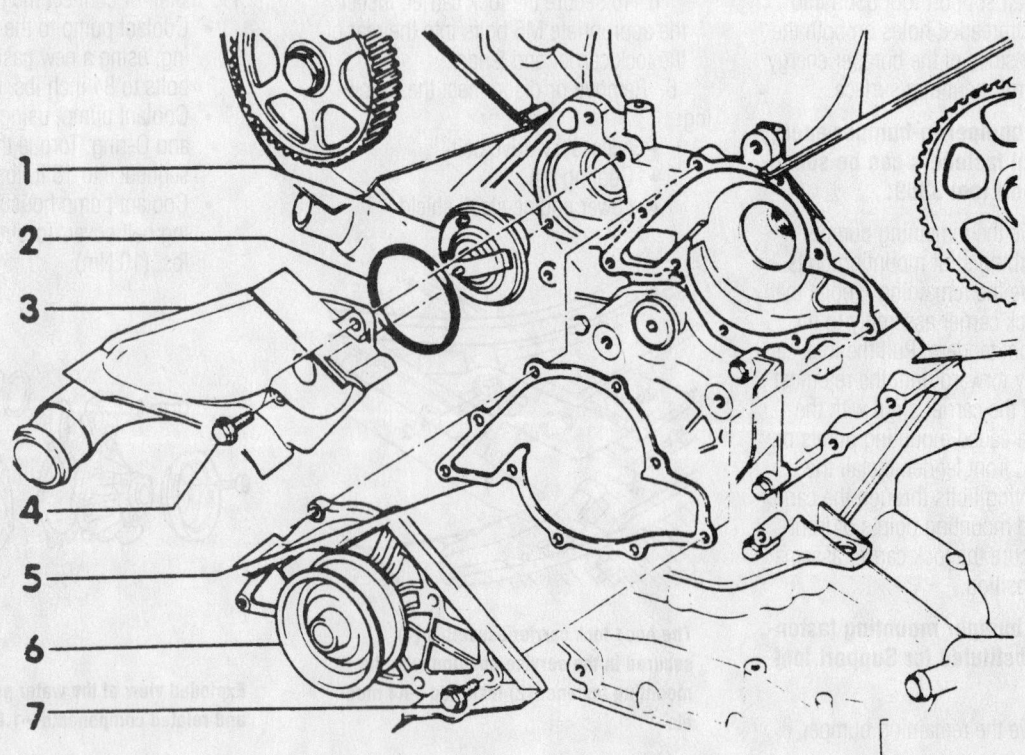

Exploded view of the water pump and related components—2.8L engine

7923CG03

9. Reinstall the timing belt.

10. Properly tension the belt with the water pump. Refer to the necessary service procedures.

11. Fill and bleed the cooling system.

Cylinder Head

➡Before removing or installing the cylinder head, align the engine timing marks at Top Dead Center (TDC). Rotate the crankshaft mark away about ¼ turn Before Top Dead Center (BTDC). This will prevent the valves from hitting the piston heads. Be sure to turn the crankshaft to the proper position after cylinder head installation.

REMOVAL & INSTALLATION

✳✳ CAUTION

Cylinder head removal should not be attempted unless the engine is cold.

1.8L Engine

1. Before servicing the vehicle, refer to the precautions in the beginning of this section.

2. Remove or disconnect the following:
   • Front bumper

3. Place the hood lock carrier into the service position.

4. Turn the ignition switch to the **OFF** position.

5. Remove or disconnect the following:
   • Negative battery cable
   • Accessory drive belt
   • Cooling fan

6. Drain the engine coolant.

7. Remove or disconnect the following:
   • Intake manifold
   • Accessory drive belts
   • Wastegate bypass regulator valve
   • Evaporative Emissions (EVAP) canister purge regulator valve
   • Power outage stage
   • Mass Air Flow (MAF) sensor
   • Air cleaner housing
   • Engine Temperature Control (ETC) and the temperature II sensor harness connector
   • All connections from the cylinder head
   • Crankcase breather line
   • Oil supply line at the cylinder head
   • Exhaust manifold heat shield
   • Turbocharger from the exhaust manifold

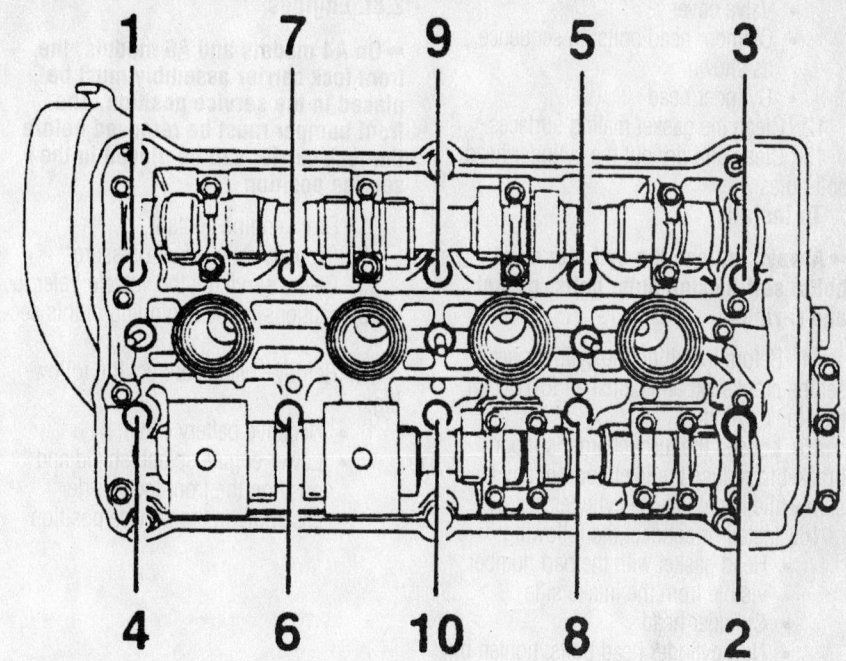

Cylinder head bolt removal sequence—1.8L engine

7923CG14

   • Coolant hose to the heat exchanger at the rear of the cylinder head
   • Upper timing belt cover

8. Turn the crankshaft, in the direction of rotation (clockwise), until the No. 1 cylinder is at TDC.

9. Using Torx® wrench T45, loosen the timing belt tensioner.

10. Push down on the tensioner and remove the belt from the camshaft gear.

11. Remove or disconnect the following:
   • Torx® bolt and swing the tensioner assembly bracket forward

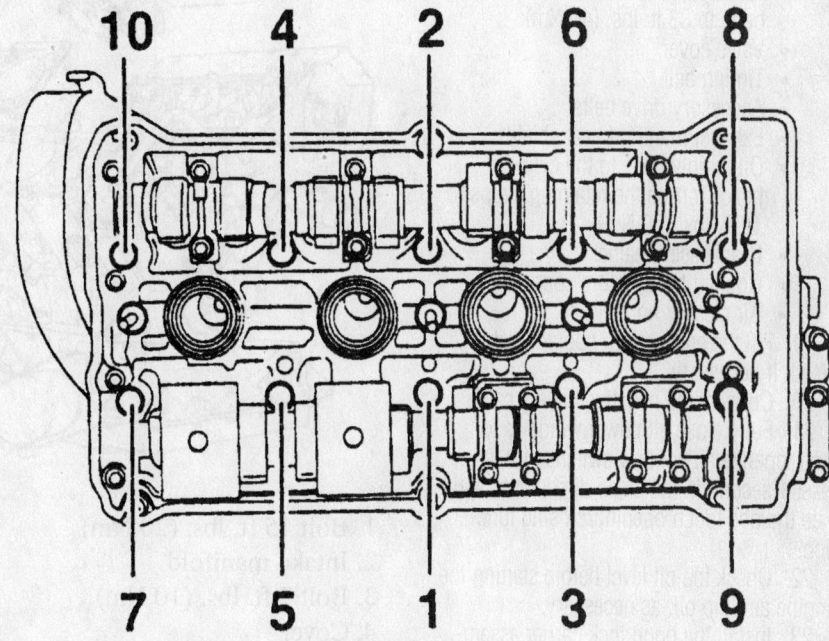

Cylinder head torque sequence—1.8L engine

7923CG15

Heater Core replacement is covered in Section 2 of this manual

- Valve cover
- Cylinder head bolts, in sequence, as shown
- Cylinder head

12. Clean the gasket mating surfaces.

13. Clean and dry out the cylinder head bolt holes.

To install:

➡ **Always replace the cylinder head bolts, self-locking nuts, bolts, gaskets and O-rings.**

14. Before installing the cylinder head, set the crankshaft and camshaft to TDC for the No. 1 cylinder.

15. Loosen the turbocharger support bracket to reduce the likelihood of any tension while installing the cylinder head.

16. Install or connect the following:
- Head gasket with the part number visible from the intake side
- Cylinder head
- New cylinder head bolts, tighten by hand

17. Tighten the new cylinder head bolts in sequence in 2 steps:
   a. Step 1: 44 ft. lbs. (60 Nm)
   b. Step 2: Plus 180 degrees

18. Install or connect the following:
- Turbocharger to the exhaust manifold using new gaskets and the bolts coated with Hot Bolt Paste G 052 112 A3. Torque the bolts to 26 ft. lbs. (35 Nm).
- Turbo support bracket. Torque the bolts to 33 ft. lbs. (40 Nm).
- Valve cover
- Timing belt
- Accessory drive belts
- Exhaust manifold heat shield
- Oil supply lines to the cylinder head. Torque the retaining straps to 15 ft. lbs. (20 Nm).
- Crankcase breather
- Coolant temperature sensors
- Air cleaner housing

19. Fill the engine with coolant and bleed, if necessary.

20. Connect the negative battery cable.

21. Fully close all power windows to stop, operate all window switches for at least 1 second in the close direction to activate the one touch opening/closing function.

22. Check the oil level before starting the engine and top off, as necessary.

23. Install the hood lock carrier assembly and front bumper.

24. Adjust the headlights.

25. Start the vehicle, check for leaks and repair if necessary.

2.8L Engines

➡ **On A4 models and A6 models, the front lock carrier assembly must be placed in the service position. The front bumper must be removed before the lock carrier can be placed in the service position**

1. Drain engine coolant.

2. Properly relieve fuel pressure.

3. Before servicing the vehicle, refer to the precautions in the beginning of this section.

4. Remove or disconnect the following:
- Negative battery cable
- Lower engine splash shield and position the front lock carrier assembly in the service position
- Accessory drive belt
- Timing belt. Refer to the Timing Belt unit repair section.
- Exhaust pipe from the manifold
- Exhaust Gas Recirculation (EGR) valve hose at manifold
- Air guide hose between air mass sensor and intake manifold
- Spark plug wires
- Injector connectors
- Crankcase breathers on the left and right cylinder head covers
- Fuel feed and return lines
- Left side cover for fuel line
- Throttle cable
- Vacuum hoses from vacuum pump and intake manifold
- Idling stabilization valve and throttle valve potentiometer

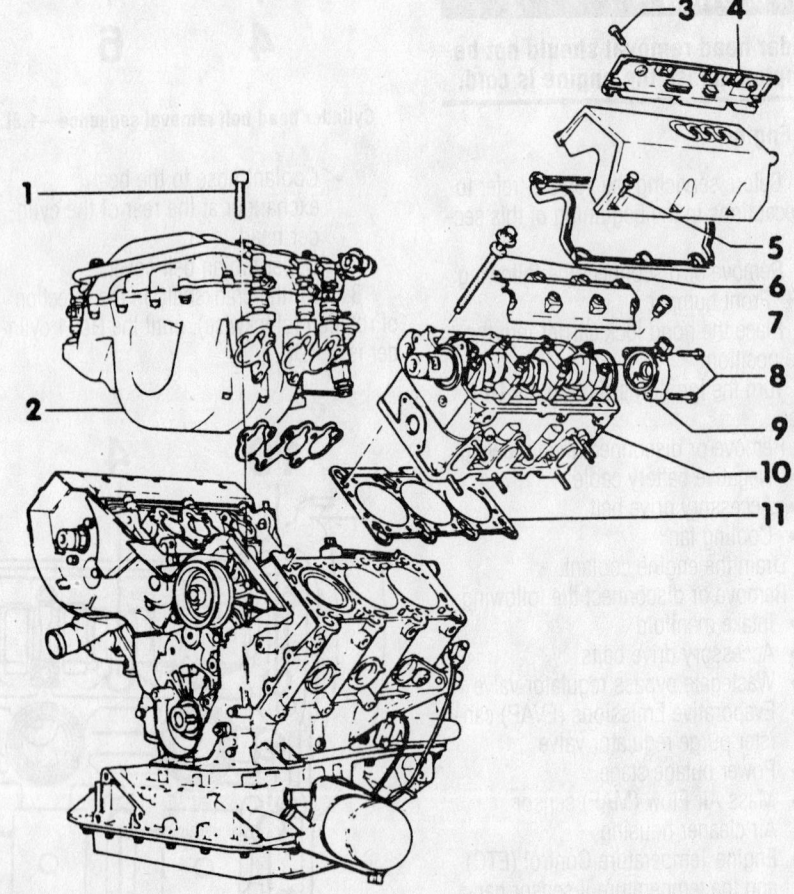

1. Bolt 15 ft. lbs. (20 Nm)
2. Intake manifold
3. Bolt 7 ft. lbs. (10 Nm)
4. Cover
5. Bolt 7 ft. lbs. (10 Nm)
6. Cylinder head bolts
7. Pressure relief valve
8. Bolt 7 ft. lbs. (10 Nm)
9. Camshaft Position (CMP) sensor
10. Cylinder head
11. Cylinder head gasket

Exploded view of the cylinder head and related components—2.8L SOHC engine

7923CG06

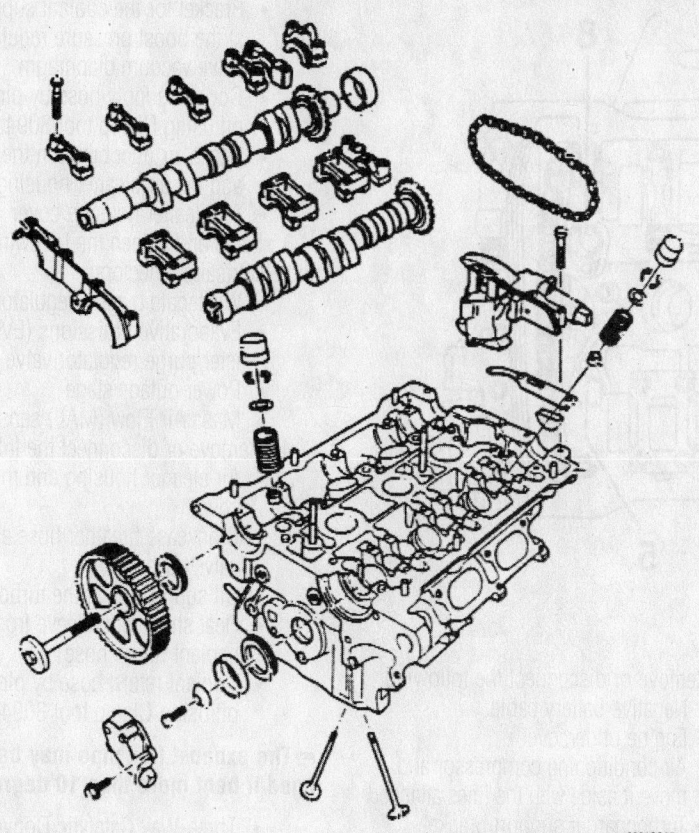

Exploded view of the cylinder head and related components—2.8L DOHC engine

- Vacuum hose on vacuum control unit
- Oil pressure sender
- Oil pressure switch
- Hall sender sensor

- EGR valve from the intake manifold
- Intake manifold assembly
- Coolant pipe at the rear of the cylinder head

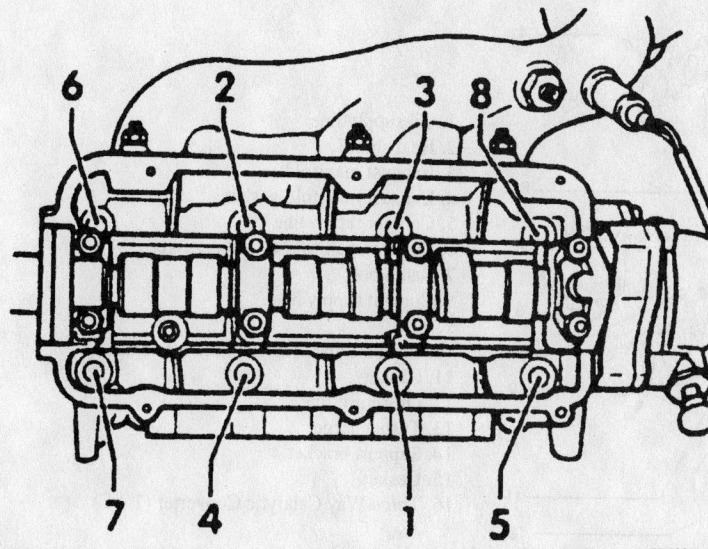

Cylinder head torque sequence—2.8L SOHC engine

- Oxygen (O_2S) sensor
- Heat shield on the exhaust manifold
- Cylinder head cover
- Timing belt rear belt guard
- Hydraulic line from reservoir to pump
- Cylinder head bolts by loosening them in the reverse of the torque sequence
- Cylinder head

To install:

5. Clean all sealing surfaces. Check cylinder head for distortion. Measure at several locations. The maximum permissible distortion is 0.0039 in. (0.1mm).

6. Install or connect the following:
- New cylinder head gasket with the lettering facing upwards
- Cylinder head

7. Tighten the cylinder head bolts following the torque tightening sequence in 2 steps as follows:
- a. Step 1: 44 ft. lbs. (59.8 Nm)
- b. Step 2: Plus 180 degrees

8. Install or connect the following:
- Timing belt rear belt guard
- Cylinder head cover
- O_2S sensor
- Heat shield on the exhaust manifold
- Intake manifold assembly
- EGR valve to the intake manifold
- Hall sender sensor
- Oil pressure sender
- Oil pressure switch
- Connectors on idling stabilization valve and throttle valve potentiometer
- Vacuum hoses to vacuum pump and intake manifold
- Throttle cable
- Fuel feed and return lines
- Crankcase breather on the cylinder head cover
- Spark plug wires and injector connectors
- Exhaust manifold
- Timing belt and accessory drive belt

9. Fill and bleed the cooling system, as necessary.

10. Connect the negative battery cable.

11. Check all fluid levels and top off as necessary.

12. Operate the engine and check for leaks.

Brake service is covered in Section 4 of this manual

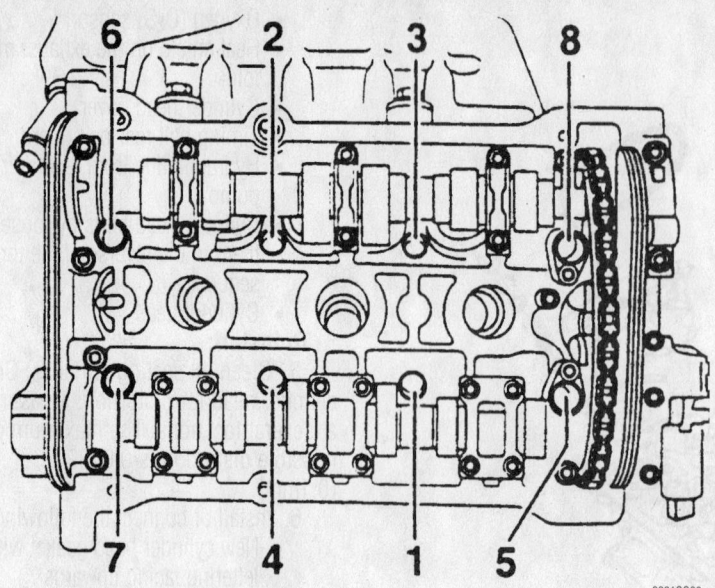

Cylinder head torque sequence—2.8L DOHC engine

13. Road test the vehicle for proper operation.

Turbocharger

REMOVAL & INSTALLATION

1.8L Engine

1. Before servicing the vehicle, refer to the precautions in the beginning of this section.

2. Remove or disconnect the following:
- Negative battery cable
- Engine undercover
- Air conditioning compressor and move it aside with the lines attached
- Turbocharger support bracket
- Oil return line at the turbocharger
- Air hoses from the turbocharger
- Oil feed line at the turbocharger
- Hose for the boost pressure regulation valve vacuum diaphragm

- Bracket for the coolant supply line at the boost pressure regulation valve vacuum diaphragm
- Coolant supply hose by pinching it off using Clamp tool 3094
- Intake air duct between the cowl and the air cleaner housing
- Air cleaner housing cover

3. Label and detach the following lines and electrical connectors:
- Wastegate bypass regulator valve
- Evaporative Emissions (EVAP) canister purge regulator valve
- Power outage stage
- Mass Air Flow (MAF) sensor

4. Remove or disconnect the following:
- Air cleaner housing and the engine cover
- Crankcase breather hose at the valve cover
- Oil supply line at the turbocharger
- Heat shield and sleeve from the coolant return hose
- Coolant return hose by pinching it off using Clamp tool 3094 first

➡**The exhaust flex pipe may be damaged if bent more than 10 degrees.**

- Three-Way Catalytic Converter (TWC) from the turbo
- Turbo from the exhaust manifold
- Coolant supply banjo fitting at the turbocharger
- Turbocharger

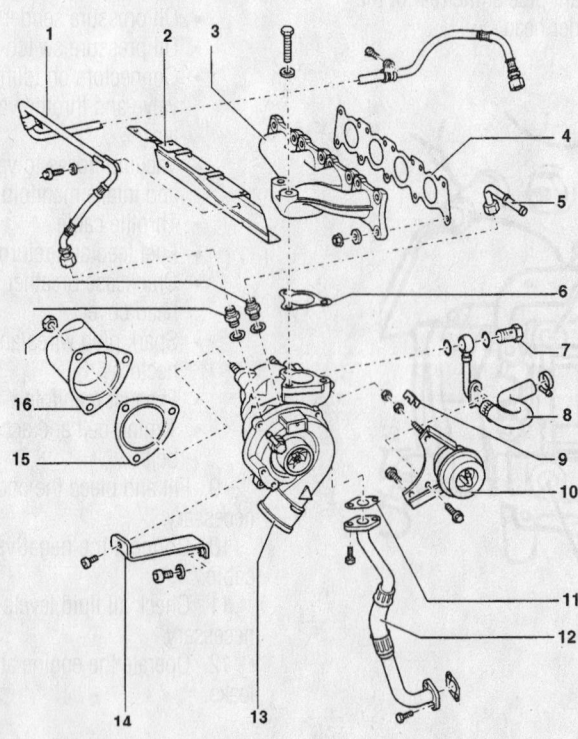

1. Oil supply line
2. Heat shield
3. Exhaust manifold
4. Exhaust manifold gasket
5. Coolant return line
6. Exhaust manifold-to-turbo gasket
7. Banjo bolt
8. Coolant supply hose
9. Fuse
10. Vacuum diaphragm for the wastegate
11. Gasket
12. Oil return line
13. Turbocharger
14. Support bracket
15. Gasket
16. Three Way Catalytic Converter (TWC)

Exploded view of the turbocharger and related components—1.8L engine

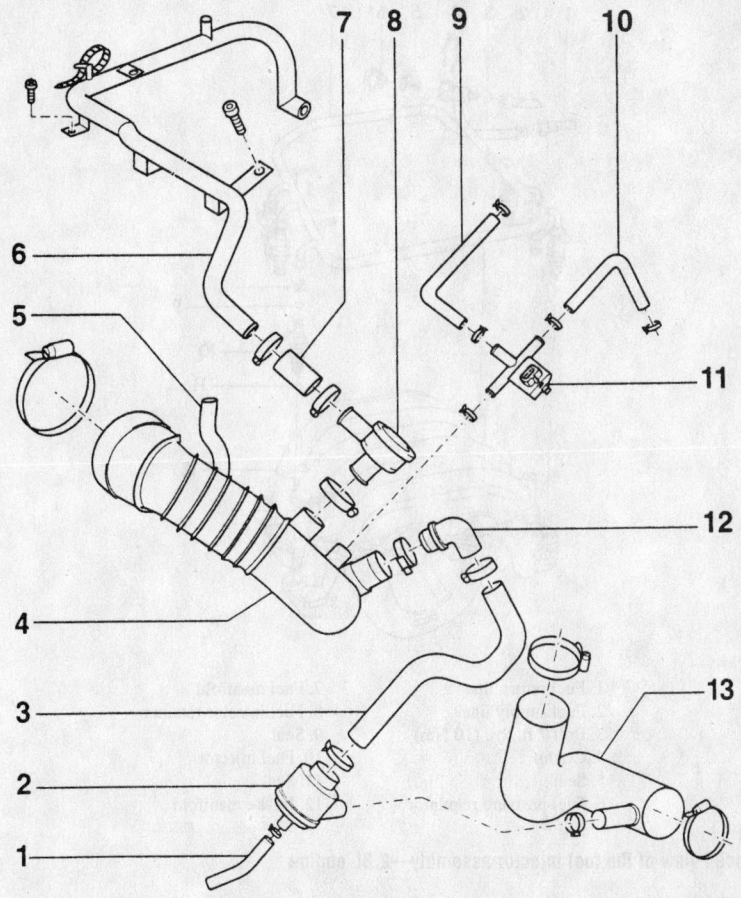

1. Vacuum hose
2. Boost pressure recirculation valve
3. Hose
4. Intake air duct
5. EVAP hose
6. Crankcase ventilation hose
7. Crankcase ventilation hose
8. PCV valve
9. Hose
10. Wastegate vacuum hose
11. Wastegate bypass regulator valve
12. Elbow
13. Hose to the turbocharger

7923CG18

Exploded view of the vacuum hoses related to the turbocharger—1.8L engine

To install:

5. Installation is the reverse of the removal procedure, while using the following torque values:

- Coolant supply banjo fitting: 18 ft. lbs. (25 Nm)
- Turbocharger exhaust manifold bolts: 26 ft. lbs. (35 Nm)
- Turbo support bracket bolts: 33 ft. lbs. (45 Nm)
- Turbo oil supply line: 18 ft. lbs. (25 Nm)

Intake Manifold

REMOVAL & INSTALLATION

1.8L Engine

1. Before servicing the vehicle, refer to the precautions in the beginning of this section.
2. Drain the engine coolant.
3. Properly relieve the fuel system pressure.

4. Turn the ignition switch to the **OFF**.
5. Remove or disconnect the following:

- Negative battery cable
- Engine covers
- Hose for the Leak Detection Pump (LDP)
- Accelerator pedal cable from the throttle valve control module
- Air guide hose from the throttle valve control module
- Vacuum line from the Evaporative Emissions (EVAP) canister
- Brake booster vacuum hose
- Intake Air Temperature (IAT) sensor and the throttle valve control module
- Camshaft Position (CMP) sensor wiring harness connector
- Fuel rail with the injectors
- Coolant hoses from to the intake manifold
- Crankcase breather hose at the intake manifold
- Intake manifold brace and the oil dipstick
- Manifold at the mounting flange

6. Clean all gasket surfaces.

To install:

7. Install or connect the following:

- Intake manifold using new gaskets. Torque the fasteners to 89 inch lbs. (10 Nm).
- Manifold brace. Torque the bolts to 15 ft. lbs. (20 Nm).
- Dipstick
- Crankcase breather and coolant hoses to the intake manifold
- Fuel injector sealing O-ring
- Fuel rail with the injectors. Torque the bolts to 89 inch lbs. (10 Nm).
- CMP and the IAT sensors to the throttle valve control module
- Brake booster and EVAP canister vacuum hoses
- Air guide hose and the accelerator pedal cable to the throttle valve control module
- Hose for the LDP

8. Top off the engine coolant and bleed, if necessary.

For complete Engine Mechanical specifications, see Section 1 of this manual

9. Connect the negative battery cable.

10. Fully close all power windows to stop, operate all window switches for at least 1 second in the close direction to activate the one-touch opening/closing function.

11. Check the oil level before starting the engine and top off, as necessary.

2.8L Engine

1. Before servicing the vehicle, refer to the precautions in the beginning of this section.

2. Disconnect the negative battery cable.

3. Properly relieve the fuel system pressure.

4. Remove or disconnect the following:
- Fuel supply and return lines
- Fuel injector electrical connectors and the retainers for the fuel injectors
- Fuel pressure regulator clamp and the regulator
- Fuel manifold
- Injectors
- Any hoses or connectors associated with the intake manifold
- Upper intake manifold mounting bolts
- Upper intake manifold and discard the gasket
- Idle Air Control (IAC) valve
- Exhaust Gas Recirculation (EGR) valve
- Accelerator cable form the throttle body
- Lower intake manifold mounting bolts
- Lower intake manifold

5. Clean all gasket mating surfaces.

To install:

6. Install or connect the following:
- Lower intake manifold using a new gasket. Torque the bolts to 15 ft. lbs. (20 Nm).
- EGR valve
- Accelerator cable to the throttle body
- IAC valve. Torque the bolts to 89 inch lbs. (10 Nm).
- Upper intake manifold. Torque the short bolts to 89 inch lbs. (10 Nm) and the long bolts to 15 ft. lbs. (20 Nm).
- All hoses or connectors that were removed
- Fuel injectors using new O-rings
- Fuel manifold. Torque the bolts to 89 inch lbs. (10 Nm).
- Pressure regulator and clamp
- Fuel injector retainers

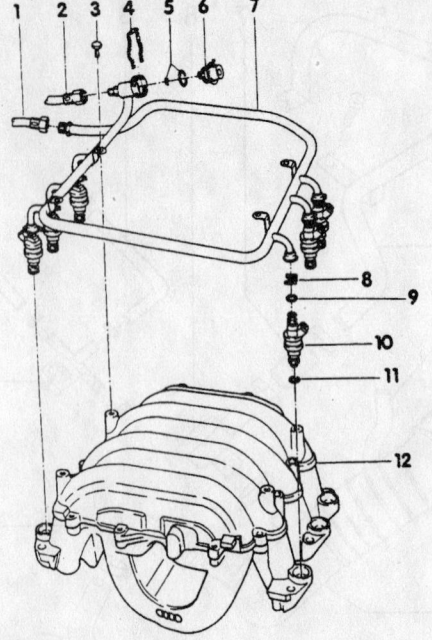

1. Fuel return line
2. Fuel supply line
3. Bolt 7 ft. lbs. (10 Nm)
4. Clamp
5. Seal
6. Fuel pressure regulator
7. Fuel manifold
8. Fuel injector retainer
9. Seal
10. Fuel injector
11. Seal
12. Intake manifold

7923CG07

Exploded view of the fuel injector assembly—2.8L engine

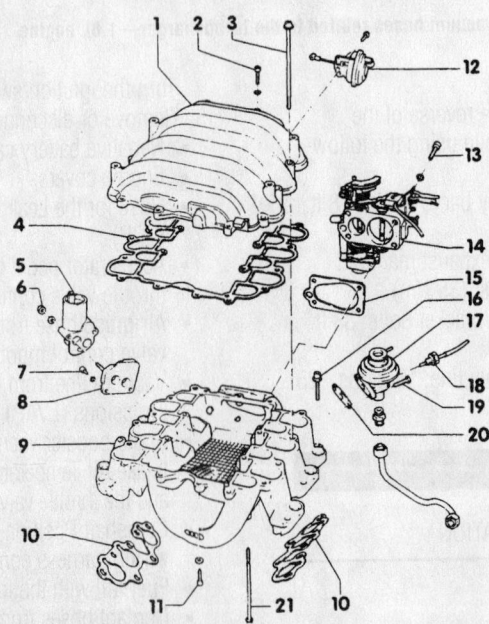

1. Upper intake manifold
2. Bolt 7 ft. lbs. (10 Nm)
3. Bolt 15 ft. lbs. (20 Nm)
4. Gasket
5. Idle Air Control (IAC) valve
6. Bolt 7 ft. lbs. (10 Nm)
7. Bolt 53 inch lbs. (6 Nm)
8. Flange
9. Lower intake manifold
10. Gasket
11. Bolt 7 ft. lbs. (10 Nm)
12. Vacuum unit
13. Bolt 15 ft. lbs. (20 Nm)
14. Throttle body
15. Gasket
16. Bolt 15 ft. lbs. (20 Nm)
17. EGR valve
18. EGR temp sensor
19. Bolt 7 ft. lbs. (10 Nm)
20. Gasket
21. Bolt 7 ft. lbs. (10 Nm)

7923CG08

Exploded view of the upper/lower intake manifold assembly and related components—2.8L engine

- Electrical connectors
- Fuel return and supply lines
- Negative battery cable

Exhaust Manifold

REMOVAL & INSTALLATION

1.8L Engine

1. Before servicing the vehicle, refer to the precautions in the beginning of this section.
2. Remove or disconnect the following:
   - Cylinder head cover
   - Intake manifold cover
   - Turbocharger to charcoal filter hose
   - Upper air pipe and heat shield
   - Turbocharger
   - Exhaust manifold nuts
   - Exhaust manifold
3. Clean the gasket mounting surfaces.

To install:

4. Install or connect the following:
   - Exhaust manifold using a new gasket. Torque the nuts to 18 ft. lbs. (25 Nm).
   - Turbocharger. Torque the bolts to 22 ft. lbs. (30 Nm).
   - Upper air pipe and heat shield. Torque the bolts to 89 inch lbs. (10 Nm).
   - Turbocharger to charcoal filter hose
   - Intake manifold cover
   - Cylinder head cover

2.8L Engine

RIGHT MANIFOLD

1. Before servicing the vehicle, refer to the precautions in the beginning of this section.
2. Remove or disconnect the following:
   - Negative battery cable
   - Heated Oxygen (HO$_2$S) sensor
   - Exhaust system from the manifold
   - Heat shield
   - Manifold nuts
   - Exhaust manifold
3. Clean all gasket mating surfaces.

To install:

4. Install or connect the following:
   - Exhaust manifold using a new gasket. Torque the nuts to 18 ft. lbs. (24 Nm)
   - Heat shield
   - Exhaust system to the manifold using a new gasket
   - HO$_2$S sensor

- Negative battery cable

5. Start the vehicle, check for leaks and repair if necessary.

LEFT MANIFOLD

1. Before servicing the vehicle, refer to the precautions in the beginning of this section.
2. Disconnect the negative battery cable.
3. Drain the engine coolant.
4. Remove or disconnect the following:
   - Heated Oxygen (O$_2$S) sensor
   - Exhaust system from the manifold
   - Coolant tube from the cylinder head
   - Heat shield
   - Exhaust Gas Recirculation (EGR) tube from the rear of the manifold
   - Exhaust manifold nuts
   - Exhaust manifold
5. Clean all gasket mating surfaces.

To install:

6. Install or connect the following:
   - Manifold using a new gasket. Torque the nuts to 18 ft. lbs. (24 Nm)
   - Exhaust system to the manifold using a new gasket
   - HO$_2$S sensor
   - EGR tube to the rear of the manifold
   - Heat shield
   - Coolant tube to the cylinder head
   - Negative battery cable
7. Fill the engine with coolant.
8. Start the vehicle, check for leaks and repair if necessary.

Front Crankshaft Seal

REMOVAL & INSTALLATION

1.8L Engine

1. Before servicing the vehicle, refer to the precautions in the beginning of this section.
2. Place the lock carrier into the service position.
3. Turn the ignition switch to the **OFF** position.
4. Remove or disconnect the following:
   - Negative battery cable
   - Accessory drive belts
   - Timing belt. Refer to the Timing Belt unit repair section.
   - Torque arm mounting bracket
5. Hold the crankshaft timing belt gear using tool 3099.

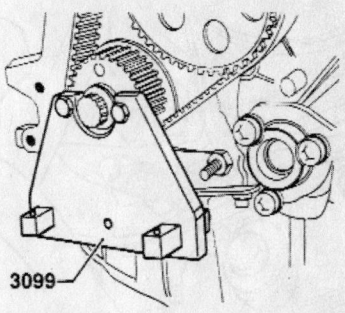

3099

9301CG08

Tool 3099 is used to hold the crankshaft toothed gear, allowing removal of the center bolt—1.8L engines

6. Remove or disconnect the following:
   - Crankshaft timing belt gear retaining bolt
   - Crankshaft timing belt gear
   - Crankshaft oil seal

To install:

7. Install or connect the following:
   - New oil seal lubricated with engine oil using a Seal Driver until it is flush
   - Timing belt gear. Torque the new bolt to 66 ft. lbs. (90 Nm) plus ¼ (90 degree) turn.
   - Torque arm mounting bracket. Torque the bolts to 18 ft. lbs. (25 Nm)
   - Timing belt
   - Accessory drive belts
   - Negative battery cable
8. Fully close all power windows to stop, operate all window switches for at least 1 second in the close direction to activate the one-touch opening/closing function.
9. Check the oil level before starting the engine.

2.8L Engine

1. Before servicing the vehicle, refer to the precautions in the beginning of this section.
2. On all A4 A6 models, place the lock carrier into the service position.
3. Remove the accessory drive belt.
4. Rotate the engine to Top Dead Center (TDC).
5. Remove or disconnect the following:
   - Sealing plug from the left side of the engine
   - Crankshaft by locking it in position using tool 3242
   - Timing belt. Refer to the Timing Belt unit repair section.

For Accessory Drive Belt illustrations, see Section 1 of this manual

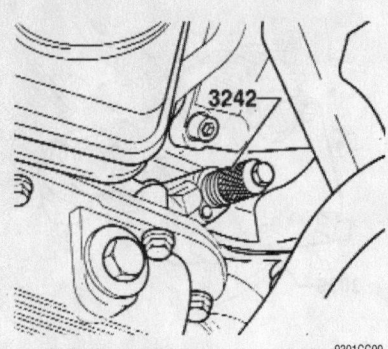

9301CG09

The crankshaft is locked in the TDC position by installing tool 3242 through the sealing plug on the left side of the engine block—2.8L engines

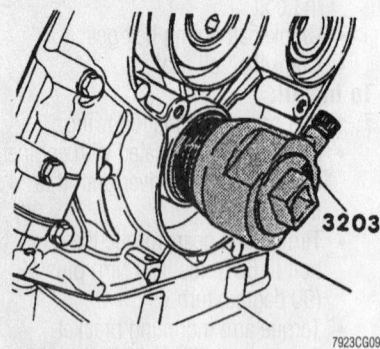

7923CG09

Removing the seal using the seal remover—2.8L engine

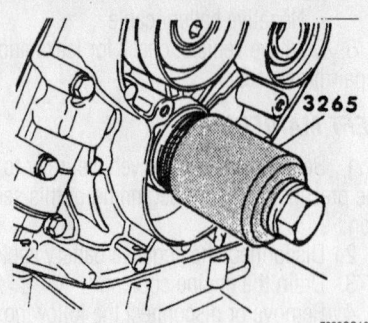

7923CG10

Installing the seal using the seal installer and the crankshaft center bolt—2.8L engine

- Crankshaft timing belt sprocket from the crankshaft
- Seal using a Seal remover

6. Clean the running and sealing surfaces.

To install:

7. Slide the new seal over the Installing Sleeve tool 3202.

8. Install or connect the following:
- New oil seal using a Seal Installer tool 3265 until it is flush
- Center crankshaft bolt
- Timing belt sprocket. Torque the center crankshaft bolt to 148 ft. lbs. (200 Nm), plus an additional ½ (180 degree) turn

- Timing belt
- Accessory drive belt

Camshaft

REMOVAL & INSTALLATION

1.8L Engine

1. Before servicing the vehicle, refer to the precautions in the beginning of this section.

2. Turn the ignition switch to the **OFF** position.

3. Disconnect the negative battery cable.

4. Place the lock carrier into the service position.

5. Remove or disconnect the following:
- Accessory drive belts
- Engine covers
- Timing belt upper cover

6. Turn the crankshaft, in the direction of rotation (clockwise), until the No. 1 cylinder is at Top Dead Center (TDC).

7. Remove or disconnect the following:
- Timing belt tensioner by loosening it using Torx® wrench T45
- Belt from the camshaft gear by pushing the tensioner downward
- Torx® bolt and swing the tensioner assembly bracket forward
- Valve cover

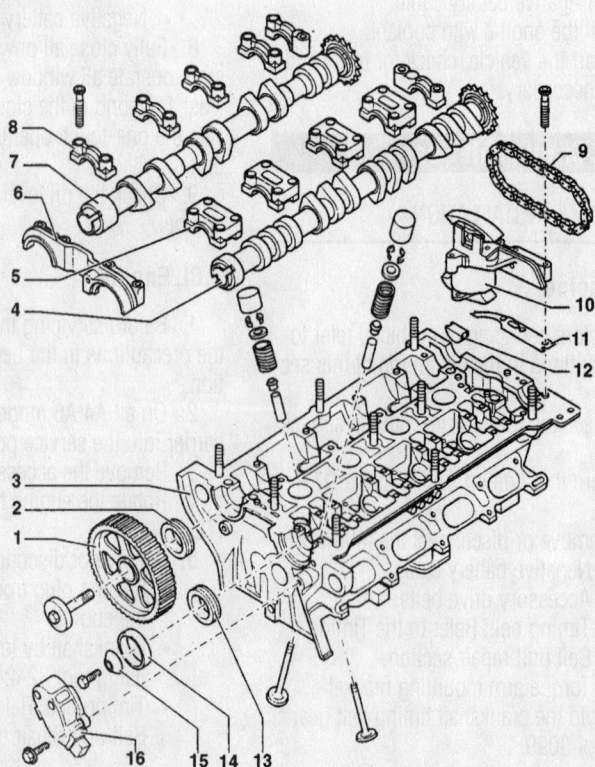

1. Camshaft gear
2. Oil seal
3. Cylinder head
4. Intake camshaft
5. Intake camshaft bearing cap
6. Double bearing cap
7. Exhaust camshaft
8. Exhaust camshaft bearing cap
9. Drive chain
10. Hydraulic chain tensioner
11. Rubber/metal seal
12. Gasket
13. Oil seal
14. Shutter wheel for the CMP
15. Washer
16. CMP sensor housing

7923CG20

Exploded view of the camshaft mounting and related components—1.8L engine

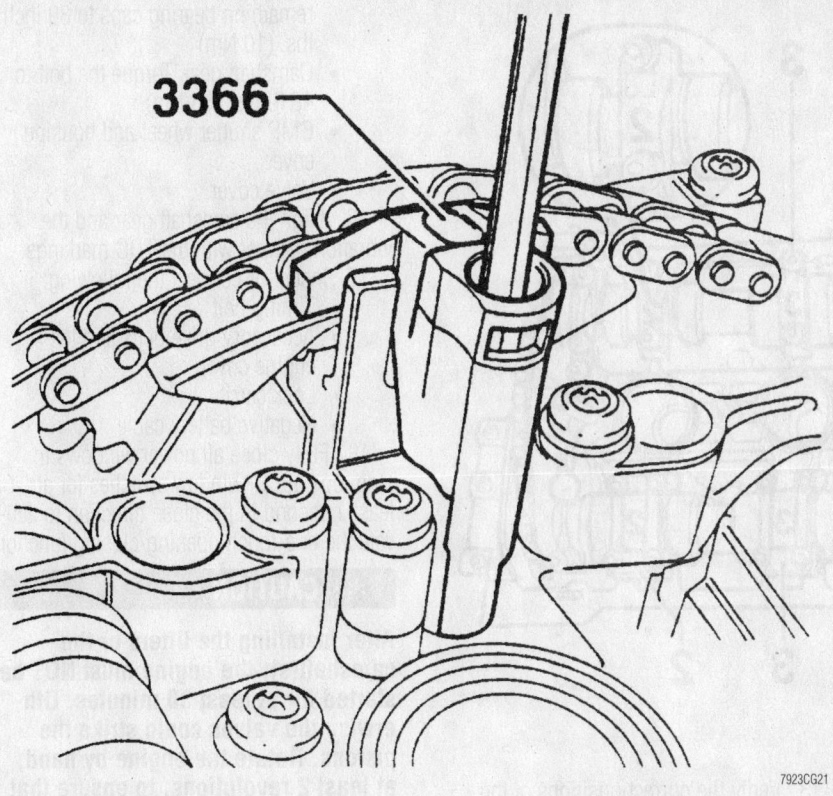

Do not overtighten the chain tensioner tool 3366, it can be damaged—1.8L and 2.8L DOHC engines

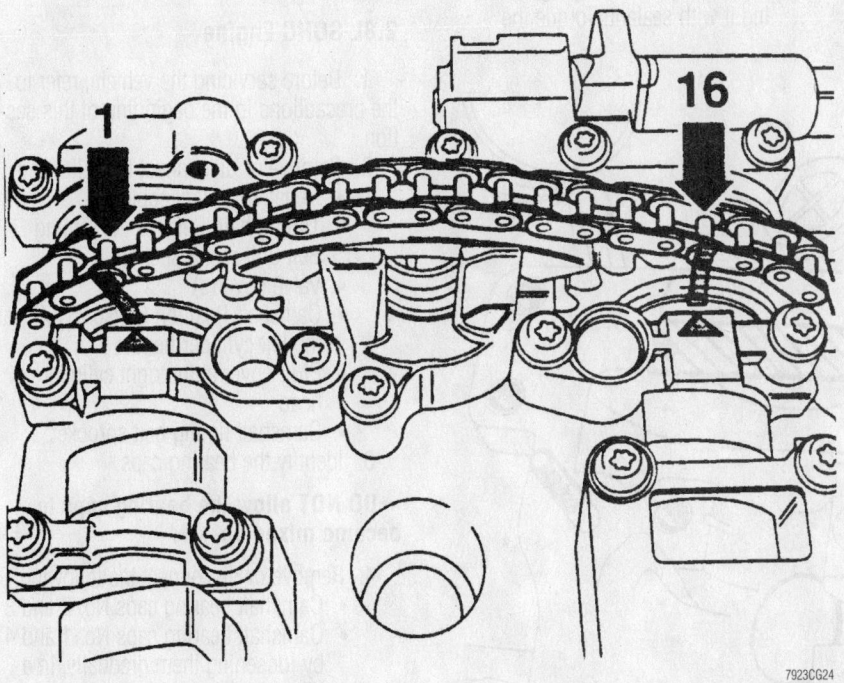

To ensure proper installation, matchmark the chain-to-camshaft position—1.8L and 2.8L DOHC engines

- Cam gear retaining bolt by loosening it using retainer tool 3036
- Camshaft gear
- Camshaft Position (CMP) housing sensor and shutter wheel
- Hydraulic chain tensioner by securing it with bracket tensioner tool 3366

8. Verify that the camshafts are at TDC for the No. 1 cylinder. Both camshaft markings must align with arrows on the bearing caps.

9. Clean the drive chain and the cam chain gears opposite both arrows on the bearing caps. Matchmark the installed position using paint.

➡ **The distance between the 2 arrows/paint marks is equivalent to 16 drive chain rollers and the notch on the exhaust camshaft is slightly offset inward toward the drive chain roller.**

10. Remove or disconnect the following:
- Bearing caps No. 3 and 5 from the intake and exhaust camshafts
- Double bearing cap
- Both bearing caps from the chain gears on the intake and exhaust camshafts
- Hydraulic chain tensioner retaining bolts
- Intake and exhaust manifold bearing caps No. 2 and 4 by loosening them in an alternating and diagonal sequence
- Camshafts with the hydraulic chain tensioner

To install:

11. Replace the rubber/metal chain tensioner gasket and apply sealant to the hatched area, as shown.

12. Install or connect the following:
- Drive chain on the camshaft

➡ **If installing the old chain, align the paint marks with the camshaft marks. If installing a new chain, the distance between the notches A and B on the camshafts must equal the distance between 16 drive chain rollers.**

- Hydraulic chain tensioner by sliding it between the drive chain
- Camshafts with the chain tensioner lubricated with engine oil into the cylinder head

➡ **When installing the bearing caps, verify the markings on the caps are readable from the intake side of the cylinder head.**

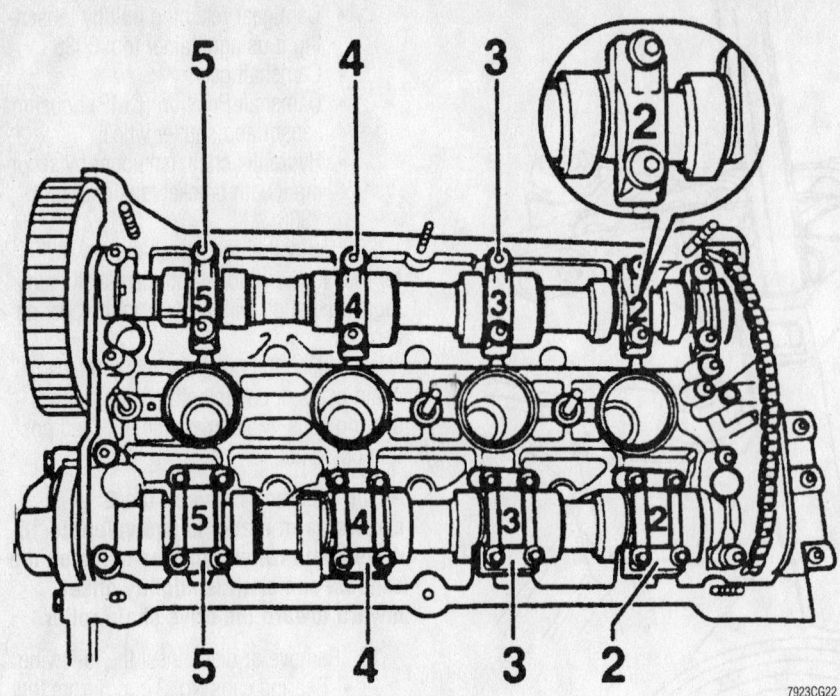

Camshaft bearing cap identification—1.8L engine

• Intake and exhaust camshafts bearing caps No. 2 and 4. Torque them in an alternating diagonal sequence to 89 inch lbs. (10 Nm).

• Both the intake and exhaust camshafts bearing caps on the chain sprockets. Torque the bolts to 89 inch lbs. (10 Nm).

13. Verify the correct positions of the camshafts.
14. Remove the bracket tensioner.
15. Install or connect the following:
   • Cylinder head-to-double bearing cap mating surface by lightly coating it with sealant. Torque the

remaining bearing caps to 89 inch lbs. (10 Nm).
   • Camshaft gear. Torque the bolt to 48 ft. lbs. (65 Nm)
   • CMP shutter wheel and housing cover
   • Valve cover

16. Align the camshaft gear and the vibration damper with the TDC markings.

17. Install or connect the following:
   • Timing belt
   • Accessory drive belts and the engine cover
   • Lock carrier
   • Negative battery cable

18. Fully close all power windows to stop, operate all window switches for at least 1 second in the close direction to activate the one-touch opening/closing function

✳✳ CAUTION

After installing the lifters or the camshaft(s), the engine must NOT be started for at least 30 minutes. Otherwise the valves could strike the pistons. Rotate the engine by hand, at least 2 revolutions, to ensure that the valves do not strike the pistons.

19. Check the oil level before starting the engine.
20. Adjust the headlights.

2.8L SOHC Engine

1. Before servicing the vehicle, refer to the precautions in the beginning of this section.

2. Remove or disconnect the following:
   • Negative battery cable
   • Timing belt. Refer to the Timing Belt unit repair section.
   • Valve cover(s)
   • Camshaft Position Sensor (CMP) at the left cylinder head
   • Plug/cover at the right cylinder head
   • Camshaft timing belt sprocket

3. Identify the bearing caps.

➡**DO NOT allow the bearing caps to become mixed up.**

4. Remove or disconnect the following:
   • Camshaft bearing caps No. 2 and 3
   • Camshaft bearing caps No. 1 and 4 by loosening them gradually, in a diagonal sequence
   • Camshaft
   • Valve lifter

➡**If the valve lifter is to be reused, it must go in the bore from which it was removed.**

To ensure a proper seal, be sure to apply sealant to the hatched area—1.8L and 2.8L DOHC Engines

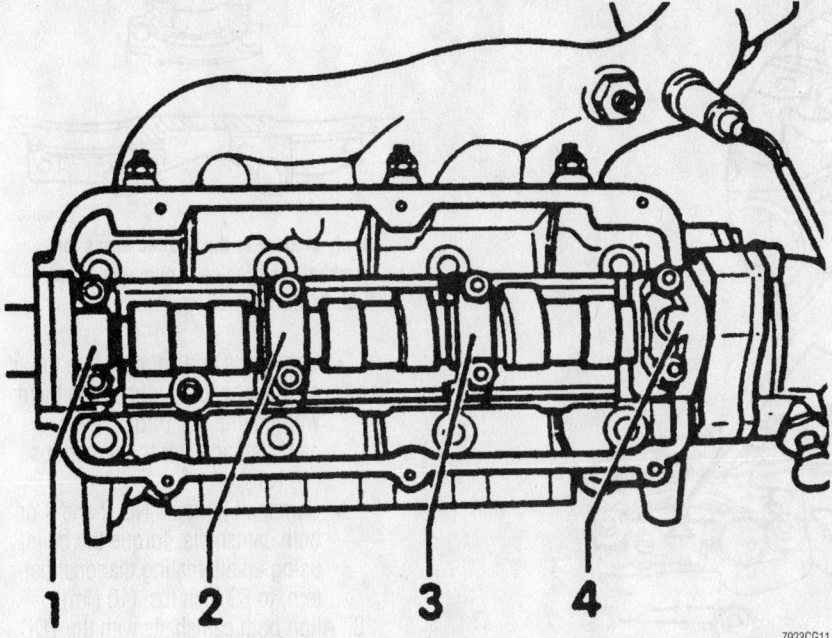

Camshaft bearing cap identification—2.8L SOHC engine

7923CG11

To install:

5. Install or connect the following:
- Lifters into their respective bore
- Bearing caps No. 1 and 4 in an alternating and diagonal sequence
- Bearing caps No. 2 and 3. Torque all bearing cap bolts to 15 ft. lbs. (20 Nm)

❄❄ CAUTION

After installing the lifters or the camshaft(s), the engine must NOT be started for at least 30 minutes. Otherwise the valves could strike the pistons. Rotate the engine by hand, at least 2 revolutions, to ensure that the valves do not strike the pistons.

- Camshaft timing belt sprocket. Torque the bolt to 52 ft. lbs. (71 Nm)
- Plug/cover, on the right cylinder head
- CMP on the left cylinder head. Torque the bolts to 89 inch lbs. (10 Nm)
- Valve cover
- Timing belt
- Negative battery cable

6. Fully close all power windows to stop, operate all window switches for at least 1 second in the close direction to activate the one-touch opening/closing function.

7. Check the oil level before starting engine.

8. Check and adjust the headlights.

2.8L DOHC Engine

1. Before servicing the vehicle, refer to the precautions in the beginning of this section.

2. Properly relieve the fuel system pressure.

3. On A4 and A6 models, place the hood lock carrier assembly in the service position.

4. Disconnect the battery negative cable.

5. Rotate the engine to Top Dead Center (TDC).

6. Remove or disconnect the following:
- Accessory drive belt
- Tooted cam belt
- Engine cosmetic covers

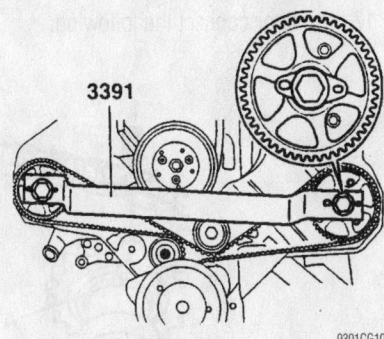

The camshaft sprockets are held using tool 3391—2.8L DOHC engine

9301CG10

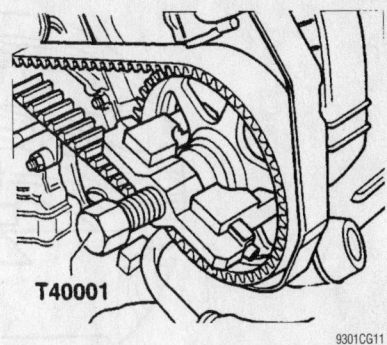

T40001

9301CG11

The camshaft gears are removed using tool T40001—2.8L DOHC engine

- Engine lower splash shield
- Crankshaft housing ventilation line from the cylinder head cover
- Intake air duct
- Coolant expansion tank and place aside
- Mass Air Flow (MAF) connector
- Evaporative Emissions (EVAP) purge regulator and valve

7. Rotate the secondary air injection pump duct 45 degrees clockwise.

8. Remove or disconnect the following:
- Air cleaner assembly
- Camshaft Position (CMP) sensor housing and shutter wheel
- Cylinder head covers

9. Install the Camshaft Locator Bar tool 3391 onto the camshaft locking plates.

10. Remove or disconnect the following:
- Camshaft bolts by loosening them 5 revolutions
- Camshaft locator bar
- Camshaft gears using tool T40001
- Oil supply lines for the camshaft bearings

➡**Use care to not damage the positioning clips.**

11. Secure the camshaft adjuster using the bracket tensioner tool 3366. Use care to not over tighten the bracket adjuster as the camshaft adjuster could be damaged.

12. Verify that both camshafts are at TDC. The notch on the camshaft behind the cam sprocket should align with the arrows on the camshaft bearing caps.

13. Make sure the camshafts are at TDC.

14. Clean the camshaft sprockets and cam drive chain and make alignment marks on the sprocket and chain in alignment with the top dead center marks.

15. Remove or disconnect the following:
- Camshaft adjuster retaining bolts, leaving the adjuster in place

For Wheel Alignment specifications, see Section 1 of this manual

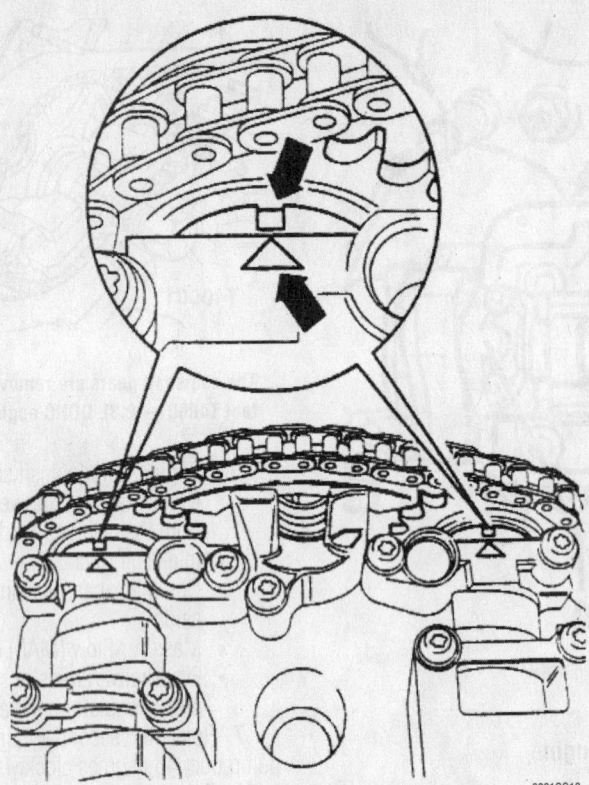

The notches in the camshaft align with the arrows on the cam journal caps when the camshafts are at TDC—2.8L DOHC engine

- Camshaft journals No. 1, 3 and 5 from both cams and journal No. 7 from the exhaust camshaft
- Camshaft journal caps No. 2 and 4 by working in a diagonal sequence
- Camshafts with the camshaft adjuster assembly

To install:

16. Install or connect the following:
    - Rubber/metal gasket for the camshaft adjuster assembly by applying a thin layer of sealant, as illustrated

Replace the rubber/metal gasket for the camshaft adjuster assembly, and apply a thin layer of sealant on the hatched surface as illustrated

- Drive chain on the camshaft sprockets by aligning the match-marks

➡**If a new chain is installed or the marks are no longer visible, install the chain on the camshaft sprockets such that the distance between the 2 camshaft sprockets is 16 drive chain rollers.**

➡**Use an assistant to place the camshaft adjuster between the camshaft drive chain.**

17. Install or connect the following:

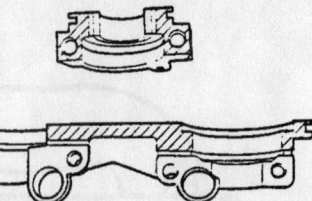

Apply sealant to the shaded areas for camshaft bearing caps numbers 1 and 7—2.8L DOHC engines

- Camshafts and adjuster into the cylinder head by lubricating them with engine oil. Torque the camshaft adjuster to 89 inch lbs. (10 Nm)
- Camshaft journals No. 2 and 4 on both camshafts. Torque the bolts, using an alternating diagonal pattern, to 89 inch lbs. (10 Nm)

18. Align both camshafts with the TDC marks.

19. Apply sealant to the double camshaft bearing cap (No. 1) and the outer bearing cap (No. 7) at their mating surfaces. Refer to the shaded area in the illustration.

➡**The notch of the exhaust camshaft is slightly offset inward from the drive chain roller.**

20. Install or connect the following:
    - Remaining camshaft bearing caps. Torque the bolts to 89 inch lbs. (10 Nm)
    - Camshaft bearing oil lines

21. Remove the cam chain bracket tensioner tool 3366.

22. Apply sealant to the corners of the camshaft journal caps and the gasket surface for the head covers.

23. Install or connect the following:
    - New camshaft oil seal
    - Camshaft cam belt sprocket. Torque the bolt to 41 ft. lbs. (55 Nm)

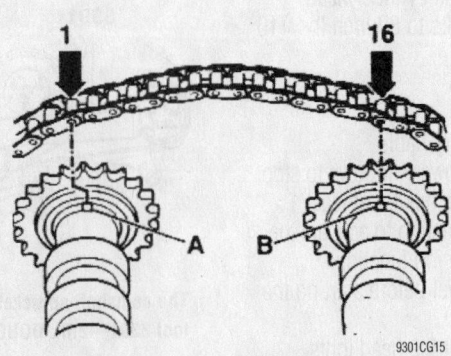

The camshaft drive chain is installed on the camshaft sprockets 16 drive chain rollers apart—1.8L and 2.8L DOHC engines

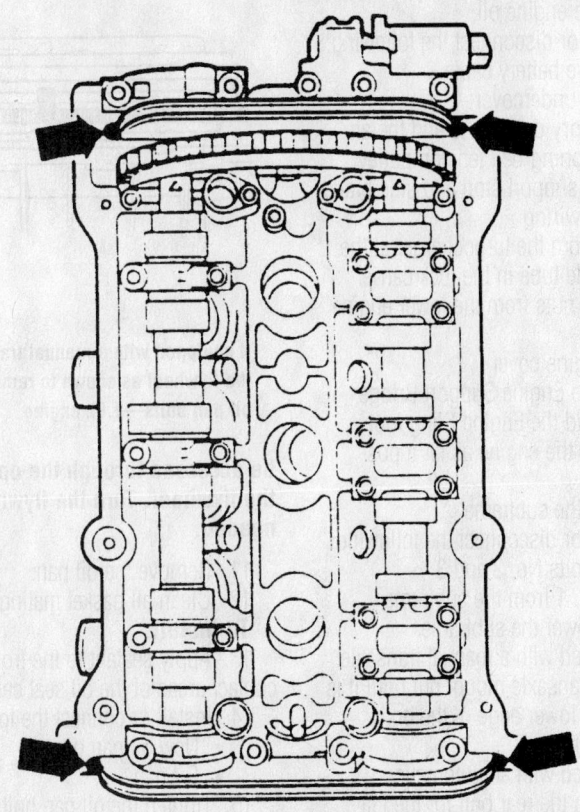

Apply a thin bead of sealant at the corners of the outer camshaft journals—2.8L DOHC engines

9301CG17

• Cylinder head covers. Torque the fasteners to 89 inch lbs. (10 Nm)
• Camshaft toothed belt
• Accessory drive belt
• Remaining equipment in the reverse order of removal
• Negative battery cable

24. Fully close all power windows to stop, operate all window switches for at least 1 second in the close direction to activate the one-touch opening/closing function

25. Check the oil level before starting the engine.

26. Adjust the headlights if necessary.

Valve Lash

ADJUSTMENT

Audi engines are equipped with hydraulic lash adjusters. No adjustment is possible.

Starter Motor

REMOVAL & INSTALLATION

1.8L Engine

1. Before servicing the vehicle, refer to the precautions in the beginning of this section.

2. Remove or disconnect the following:
• Negative battery cable
• Noise insulation panel
• Front bumper and move the lock carrier into the service position
• Loosen the A/C compressor belt tensioner
• A/C compressor and move it aside
• Starter electrical connectors
• Starter

To install:

3. Install or connect the following:
• Starter motor. Torque the mounting bolts to 48 ft. lbs. (65 Nm)
• Starter electrical connectors
• A/C compressor and torque the bolts to 18 ft. lbs. (25 Nm)
• A/C belt and torque the tensioner bolt to 15 ft. lbs. (20 Nm)
• Front bumper
• Noise insulation panel
• Negative battery cable

4. Enter the radio code and the preset frequencies.

2.8L Engine

➡️ **It may be necessary to remove the alternator prior to removal of the starter.**

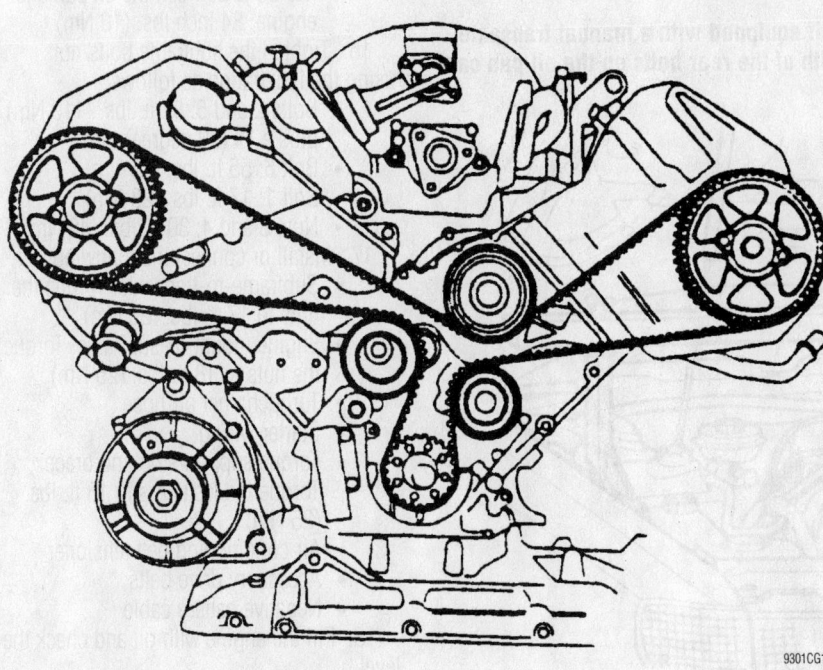

Camshaft belt routing—2.8L DOHC engines

9301CG12

For Maintenance Interval recommendations, see Section 1 of this manual

1. Before servicing the vehicle, refer to the precautions in the beginning of this section.
2. Remove or disconnect the following:
   • Negative battery cable
   • Air duct
   • Starter wiring connectors
   • Rear noise insulation panel, if equipped with a manual transmission
   • Right front wheel , if equipped with an automatic transmission
   • Starter

To install:

3. Install or connect the following:
   • Starter and torque the bolts to 48 ft. lbs. (65 Nm)
   • Right front wheel, if equipped with an automatic transmission
   • Rear noise insulation panel, if equipped with a manual transmission
   • Starter wiring connectors
   • Air duct
   • Negative battery cable
   • Alternator, if removed

Oil Pan

REMOVAL & INSTALLATION

1.8L Engine

1. Before servicing the vehicle, refer to the precautions in the beginning of this section.

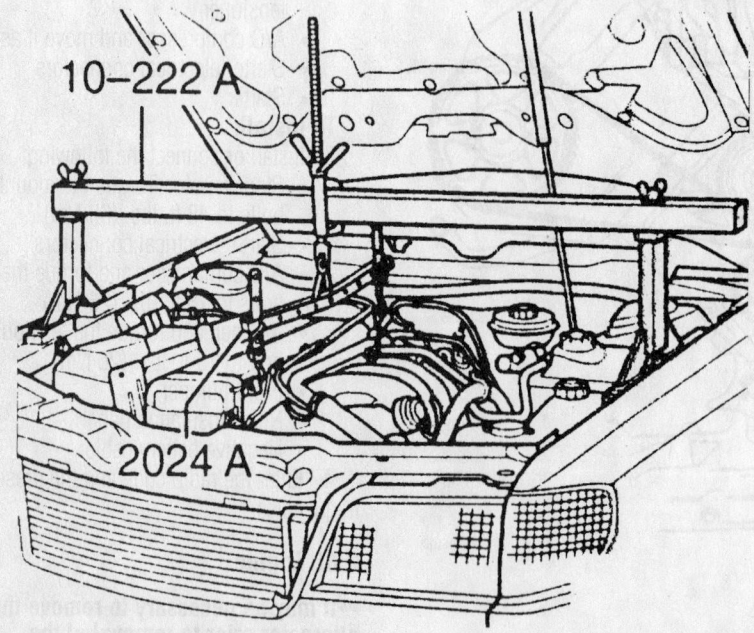

The engine must be supported, because the subframe mounting bolts must be loosened—1.8L engine

2. Drain the engine oil.
3. Remove or disconnect the following:
   • Negative battery cable
   • Engine undercover
   • Accessory drive belts and the air conditioning belt tension pulley
   • Torque support stop and side brace
   • Starter wiring
   • Hose from the turbocharger at the air guide tube in the lock carrier
   • Bottom nuts from the lower engine mount
   • Top engine cover
4. Install the Engine Support Bridge tool 10-222A and the Engine Sling tool 2024A; then, lift the engine as far a possible.
5. Support the subframe.
6. Remove or disconnect the following:
   • Front bolts No. 2 and 3
   • Bolt No. 1 from the subframe
7. Slowly lower the subframe.
8. If equipped with a manual transaxle, loosen the left transaxle mount nut until it is aligned with the lower edge of the bolt (approx. 4 turns).
9. If equipped with an automatic transaxle, loosen the rear bolt for the left transaxle mount several turns, then remove the front bolt for the transaxle mount.
10. At the right transaxle mount, loosen the rear bolt mount several turns and remove the front bolt.

➡**If equipped with a manual transaxle, both of the rear bolts on the oil pan can**

If equipped with a manual transaxle, align the flywheel as shown to remove the rear oil pan bolts—1.8L engine

be accessed through the opening on the flywheel. Turn the flywheel as needed.

11. Remove the oil pan.
12. Clean all gasket mating surfaces.

To install:

13. Apply sealant to the front and rear contact areas of the oil seal carriers.
14. Install or connect the following:
   • New oil pan gasket
   • Oil pan
15. Tighten the oil pan bolts as follows:
   • Oil pan-to-engine bolts: 44 inch lbs. (5 Nm)
   • M10 bolts between the oil pan and engine: 33 ft. lbs. (45 Nm)
   • M6 bolts between the oil pan and engine: 84 inch lbs. (10 Nm)
16. Tighten the subframe bolts/nuts using the illustration as follows:
   • Bolts 2 and 5: 81 ft. lbs. (110 Nm) plus a ¼ (90 degree) turn
   • Bolt 6: 55 ft. lbs. (75 Nm)
   • Bolt 1: 17 ft. lbs. (23 Nm)
   • Nuts 3 and 4: 30 ft. lbs. (40 Nm)
17. Install or connect the following:
   • Subframe-to-transaxle. Torque the nuts to 17 ft. lbs. (23 Nm)
   • Engine mount-to-subframe. Torque the nuts to 18 ft. lbs. (25 Nm)
   • Turbocharger air hose
   • Starter wiring
   • Torque support stop and brace. Torque the fasteners to 18 ft. lbs. (25 Nm)
   • Air conditioning belt tensioner
   • Accessory drive belts
   • Negative battery cable
18. Fill the engine with oil and check the level.
19. Start the vehicle and check for leaks, then recheck the engine oil level.
20. Install or connect the following:
   • Engine cover
   • Undercover

2.8L Engine

1. Before servicing the vehicle, refer to the precautions in the beginning of this section.
2. Drain the engine oil.
3. Remove or disconnect the following:
   - Oil pan bolts
   - Oil pan from the engine

To install:

4. Be sure the gasket surface is flat.
5. Install or connect the following:
   - New oil pan gasket
   - Oil pan. Torque the bolts in a criss-cross pattern to 11 ft. lbs. (15 Nm)
6. Fill the engine with oil.
7. Start the engine, check for leaks and repair if necessary.

Oil Pump

REMOVAL & INSTALLATION

1.8L Engine

1. Before servicing the vehicle, refer to the precautions in the beginning of this section.

2. Drain the engine oil.
3. Remove or disconnect the following:
   - Oil pan
   - Baffle plate
   - Oil pump-to-engine bolts
   - Oil pump by pressing down on the subframe

To install:

4. Install or connect the following:
   - Oil pump by pressing down on the subframe. Torque the Oil pump-to-engine bolts to 18 ft. lbs. (25 Nm)
   - Baffle plate
   - Oil pan
5. Fill the engine with clean oil.
6. Start the vehicle, check for leaks and repair if necessary.

2.8L Engine

The oil pump is part of the engine front cover, and the cooling system does not have to be opened during this procedure.

1. Before servicing the vehicle, refer to the precautions in the beginning of this section.
2. Drain the engine oil.
3. Remove or disconnect the following:

- Negative battery cable
- Timing belt. Refer to the Timing Belt unit repair section.
- Engine under cover
- Engine oil dipstick tube
- Engine oil tube from the crankcase
- Crankshaft vibration damper and sprocket
- Timing belt idler and tensioner pulleys
- Any wiring, hoses, lines and cables that interfere with oil pump removal
- Oil filter
- Oil cooler line from the filter housing
- Engine support front bolts

✳✳ CAUTION

Be prepared for the engine support to drop ³⁄₈ inch.

- Starter bolts
- Upper and lower oil pans
- Oil suction tube
- Oil pump

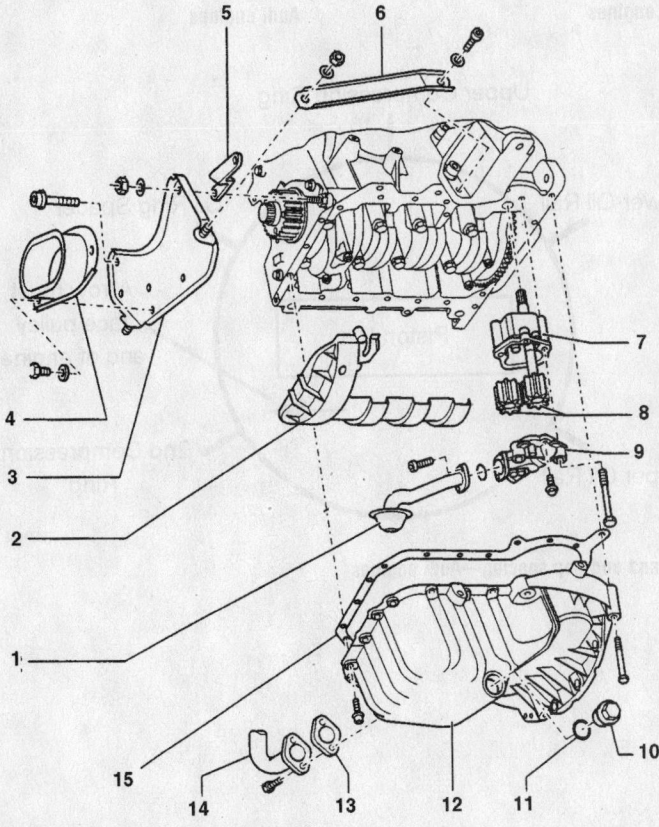

1. Suction pipe
2. Baffle plate
3. Bracket
4. Stop for torque support
5. Brace
6. Side brace
7. Oil pump housing
8. Gears
9. Oil pump cover with pressure relief valve
10. Oil drain plug
11. Sealing washer
12. Oil pan
13. Gasket
14. Oil return line
15. Gasket

7923CG27

Exploded view of the oil pan and pump—1.8L engine

To install:

4. Clean the gasket mating surfaces.
5. Install or connect the following:
   - Oil pump and verify the oil pump drive is properly engaged
   - Upper and lower oil pans
   - Starter bolts
   - Engine support bolts
   - Oil cooler line to the oil filter housing
   - New oil filter
   - Wiring, hoses, lines and cables that were removed
   - Timing belt idler pulley. Torque the bolt to 15 ft. lbs. (20 Nm)
   - Tensioner pulleys. Tighten after adjusting the timing belt.
   - Crankshaft sprocket and vibration damper. Torque the center bolt to 148 ft. lbs. (200 Nm) plus an additional ½ (180 degree) turn
   - Dipstick tube and dipstick
   - Engine under cover
   - Timing belt
   - Negative battery cable
6. Fully close all power windows to stop, operate all window switches for at least 1 second in the close direction to activate the one-touch opening/closing function
7. Fill the engine with clean oil.
8. Start the vehicle, check for leaks and repair if necessary.

Rear Main Seal

REMOVAL & INSTALLATION

1. Before servicing the vehicle, refer to the precautions in the beginning of this section.
2. Remove or disconnect the following:
   - Negative battery cable
   - Transaxle
   - Flywheel/flexplate assembly
   - Oil seal by prying it out of the housing.

To install:

3. Install or connect the following:
   - New oil seal by coating it with engine oil and press it into place

➡ **Be careful not to damage the seal or score the crankshaft.**

   - Flywheel/flexplate
   - Transaxle
   - Negative battery cable

Piston and Ring

POSITIONING

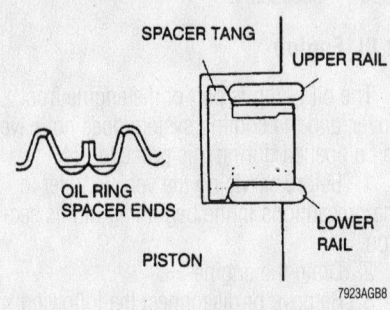

Piston ring positioning mark and location—Audi engines

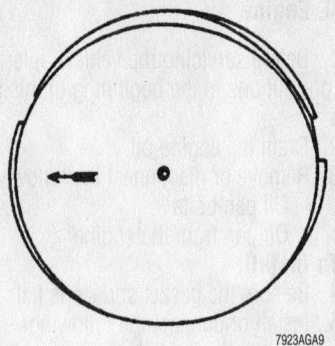

Arrow on the piston crown must face the front of the engine—Audi engines

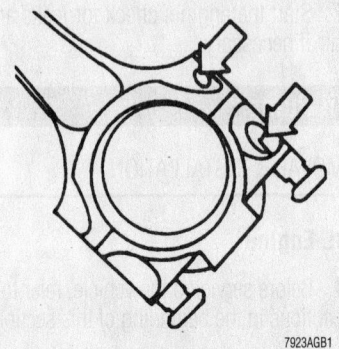

Connecting rod to bearing cap assembly—Audi engines

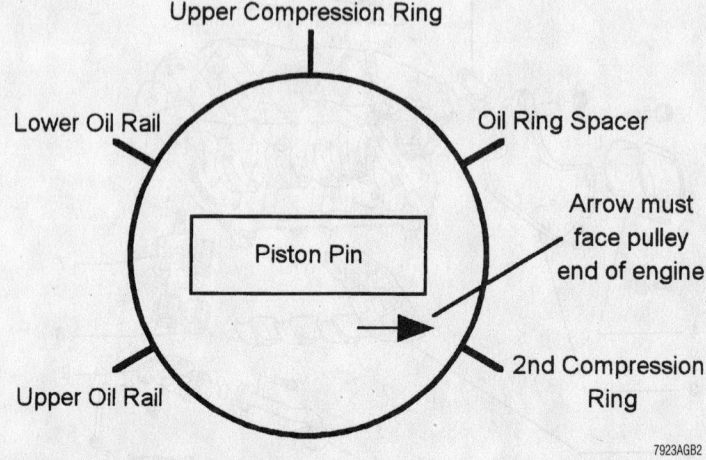

Piston ring and end-gap spacing—Audi engines

FUEL SYSTEM

Fuel System Service Precautions

Safety is the most important factor when performing not only fuel system maintenance but also any type of maintenance. Failure to conduct maintenance and repairs in a safe manner may result in serious personal injury or death. Maintenance and testing of the vehicle's fuel system components can be accomplished safely and effectively by adhering to the following rules and guidelines.

- To avoid the possibility of fire and personal injury, always disconnect the negative battery cable unless the repair or test procedure requires that battery voltage be applied.
- Always relieve the fuel system pressure prior to disconnecting any fuel system component (injector, fuel rail, pressure regulator, etc.), fitting or fuel line connection. Exercise extreme caution whenever relieving fuel system pressure, to avoid exposing skin, face and eyes to fuel spray. Please be advised that fuel under pressure may penetrate the skin or any part of the body that it contacts.
- Always place a shop towel or cloth around the fitting or connection prior to loosening to absorb any excess fuel due to spillage. Ensure that all fuel spillage is quickly removed from engine surfaces. Ensure that all fuel soaked cloths or towels are deposited into a suitable waste container.
- Always keep a dry chemical (Class B) fire extinguisher near the work area.
- Do not allow fuel spray or fuel vapors to come into contact with a spark or open flame.
- Always use a back-up wrench when loosening and tightening fuel line connection fittings. This will prevent unnecessary stress and torsion to fuel line piping. Always follow the proper torque specifications.
- Always replace worn fuel fitting O-rings with new. Do not substitute fuel hose where fuel pipe is installed.

Fuel System Pressure

RELIEVING

The fuel injection system operates under high pressure. This makes it necessary to first relieve the system of pressure before servicing. The pressurized fuel, when released, may ignite or cause personal injury.

✳✳ CAUTION

The fuel injection system remains under pressure after the engine has been turned OFF. Properly relieve fuel pressure before disconnecting any fuel lines. Failure to do so may result in fire or personal injury.

1. Before servicing the vehicle, refer to the precautions in the beginning of this section.
2. Remove or disconnect the following:
   - Power to the fuel pump by removing the relay or the fuel pump fuse. The fuse can be removed to stop the fuel pump from running. With the engine operating at idle, wait until the engine stalls from fuel starvation.
3. Switch the ignition **OFF** and remove the negative battery cable.
4. Slowly and carefully open the fuel tank filler cap for a brief moment and reinstall.
5. Carefully loosen the fuel line on the control pressure regulator or component to be serviced.
6. Wrap a clean rag around the connection, while loosening, to catch any fuel.
7. After service is complete, discard the fuel soaked rag in the proper manner and reconnect negative battery cable, relay or fuses.

Fuel Filter

REMOVAL & INSTALLATION

Most vehicles use a fuel filter mounted under the vehicle, below the fuel tank or in the engine compartment. An arrow on the filter indicates fuel flow direction. Use care not to mix up fuel supply or return lines. Fuel pressure applied to the return side of the system will cause damage.

1. Before servicing the vehicle, refer to the precautions in the beginning of this section.
2. Properly relieve the residual fuel pressure.
3. Remove or disconnect the following:
   - Negative battery cable

- Fuel lines leading into and out of the fuel filter.
- Filter retaining bracket
- Fuel filter

To install:

4. Install or connect the following:
   - New fuel filter in the bracket
   - Fuel filter bracket

➡ **Be sure the arrows are pointing in the direction of the fuel flow**

- Fuel lines
- Negative battery cable

5. Start the engine, check for leaks and repair if necessary.

Fuel Pump

REMOVAL & INSTALLATION

The fuel pump is located in the fuel tank. It is recommended that the fuel tank not be filled more than ⅓ full. If necessary the fuel must be drained using an approved fuel cart.

1. Before servicing the vehicle, refer to the precautions in the beginning of this section.
2. Properly relieve fuel pressure.
3. Remove or disconnect the following:
   - Negative battery cable
   - Inspection cover by lifting the cargo area trim
4. Mark the fuel pump/gauge sending unit assembly supply and return lines for reassembly.
5. Remove or disconnect the following:
   - Electrical connector
   - Union nut by loosening it using tool 3217
   - Fuel pump/gauge sending unit assembly by matchmarking it

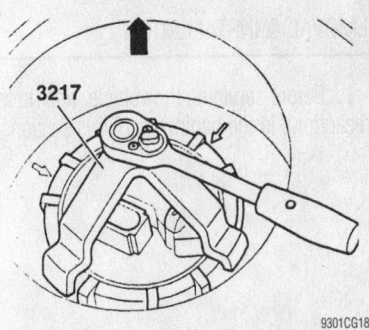
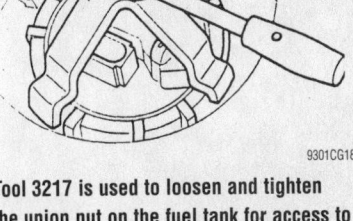

Tool 3217 is used to loosen and tighten the union nut on the fuel tank for access to the fuel pump and fuel level sending unit

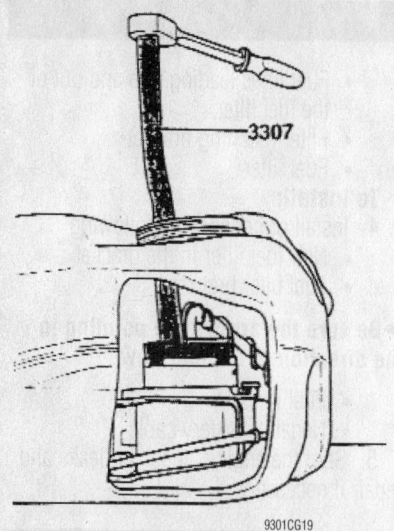

9301CG19

Tool 3307 is used to remove and install the fuel pump in the fuel tank

- Level connector and the fuel return line
6. Using tool 3307, turn the fuel pump module to the left (counterclockwise) about 15 degrees.
7. Remove or disconnect the following:
- Fuel pump supply hose
- Electrical connectors
- Fuel pump

To install:

8. The installation is reverse of the removal procedure.
9. Install or connect the following:
- Flange cover using a new O-ring lubricated the O-ring with fuel
- Union nut using tool 3217. Torque it to 59 ft. lbs. (80 Nm)
- Negative battery cable
10. Start the vehicle, check for leaks and repair if necessary.

Fuel Injector

REMOVAL & INSTALLATION

1. Before servicing the vehicle, refer to the precautions in the beginning of this section.

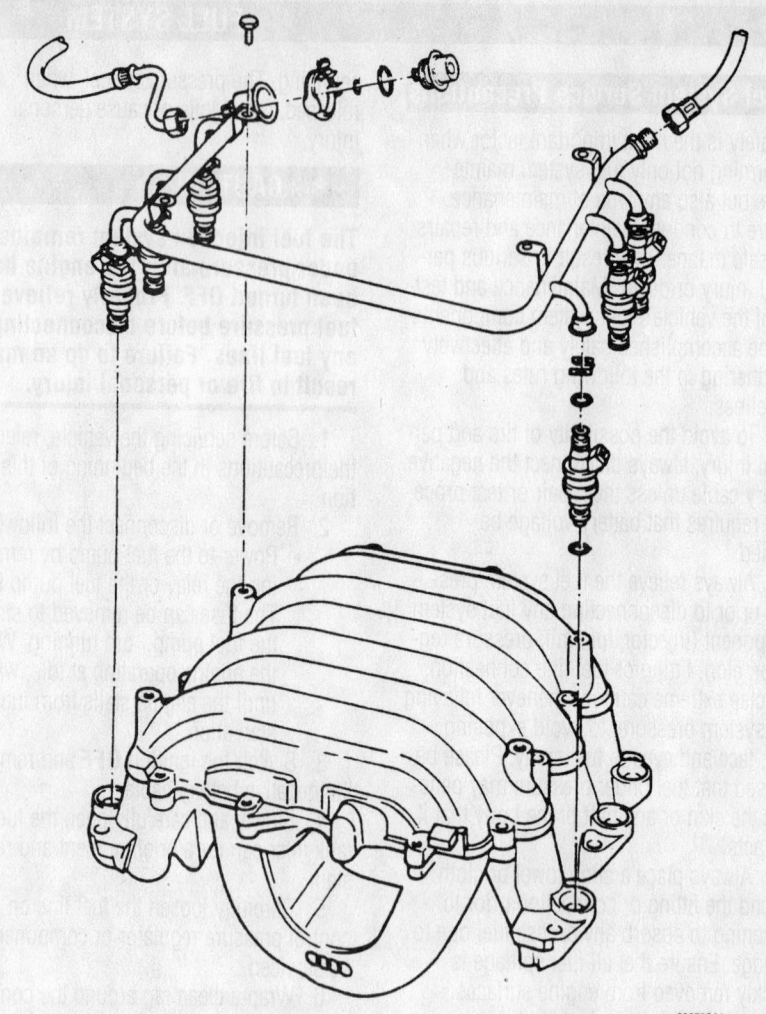

93070G01

Fuel injectors and rail—2.8L V6 engine

2. Relieve the fuel system pressure.
3. Remove or disconnect the following:
- Negative battery cable
- Engine under cover, if applicable
- Fuel lines
- Fuel injector connectors
- Fuel pressure line
- Fuel supply rail with injectors attached
- Fuel injectors from the supply rail

To install:

4. Install or connect the following:

- Fuel injectors to the fuel supply rail using new O-rings
- Fuel supply rail with injectors attached. Tighten the bolts to 89 inch lbs. (10 Nm)
- Fuel injector connectors
- Fuel pressure line
- Fuel lines
- Engine under cover, if applicable
- Negative battery cable
5. Start the engine, check for leaks and repair if necessary.

DRIVE TRAIN

Transaxle Assembly

REMOVAL & INSTALLATION

Manual

A4, A6 AND CABRIOLET MODELS

1. Before servicing the vehicle, refer to the precautions in the beginning of this section.
2. Remove or disconnect the following:
   - Negative battery cable
   - Engine undercover
   - Front exhaust pipe with the catalytic converter
   - Driveshaft for All Wheel Drive (AWD) models only
   - Starter
   - Shift rod and joint bolt at the transaxle and separate from the rear of the shift rod
   - Shift rod
   - Vehicle Speed Sensor (VSS) connector
   - Backup light switch connector
3. Support the transaxle, using a transmission/transaxle jack.
4. Remove or disconnect the following:
   - Transaxle mount heat shield
   - Right mount at the transaxle
   - Left mount with the bushings
   - Left and right halfshafts
   - Axle heat shield
   - Remaining engine-to-transaxle mounting bolts
5. Pry the transaxle off the dowel sleeves and carefully lower the transaxle until the slave cylinder is accessible, approx. 6 inches. (15cm).
6. Remove or disconnect the following:
   - Clutch slave cylinder from the transaxle with the hydraulic line attached
   - Transaxle

To install:

7. Installation is the reverse of the removal procedure, while using the following torque values:
   - Transaxle flange M12 bolts: 48 ft. lbs. (65 Nm)
   - Transaxle flange M10 bolts: 33 ft. lbs. (45 Nm)
   - Transaxle flange M8 bolts: 18 ft. lbs. (25 Nm)
   - Slave cylinder bolts: 18 ft. lbs. (25 Nm)
   - Transaxle mount bolts: 30 ft. lbs. (40 Nm)
   - Shift rod bolts: 15 ft. lbs. (20 Nm)

Automatic

1. Before servicing the vehicle, refer to the precautions in the beginning of this section.
2. Remove or disconnect the following:
   - Negative battery cable
   - Upper engine-to-transaxle bolts
3. Using an engine support tool, secure it to the engine and the vehicle.
4. Remove or disconnect the following:
   - Both top bolts at the front of the engine
   - Starter
   - Torque converter-to-driveplate bolts through the starter opening
   - Torque converter cover plate
   - Coolant hoses at the transmission cooler by clamping them off
5. Support the halfshafts
6. Remove or disconnect the following:
   - Inner halfshaft-to-transaxle bolts
   - Remove the ball joint and the support
   - Oil filler tube from the oil pan and drain the fluid
   - Exhaust pipe-to-transaxle bracket
   - Selector cable bracket from the transaxle
   - Selector cable circlip and the cable at the transaxle shift lever
   - Accelerator cable bracket and the cable from the operating lever
   - Center bolt, from the transaxle mount.
7. Using the engine support tool, lift the engine slightly.
8. Remove the throttle cable bracket bolts and the bracket.
9. Support the transaxle and lift it slightly.
10. Remove or disconnect the following:
   - Lower transaxle-to-engine bolts
   - Transaxle from the engine

➡**Be sure to secure the torque converter.**

To install:

11. Installation is the reverse of the removal procedure, while using the following torque values:
   - Transaxle flange bolts: 41 ft. lbs. (56 Nm)
   - Subframe bolts: 52 ft. lbs. (71 Nm)
   - Transaxle mount center bolt: 30 ft. lbs. (40 Nm)
   - Torque converter bolts: 63 ft. lbs. (85 Nm)
   - Halfshaft bolts: 33 ft. lbs. (45 Nm)
   - Ball joint bolts: 48 ft. lbs. (65 Nm)

Clutch

ADJUSTMENT

All vehicles use a hydraulic clutch release mechanism. No free-play adjustment is required or possible.

REMOVAL & INSTALLATION

1. Before servicing the vehicle, refer to the precautions in the beginning of this section.
2. Remove or disconnect the following:
   - Negative battery cable
   - Transaxle

➡**If the pressure plate is to be reused, matchmark its relationship to the flywheel.**

3. Using a Flywheel Locking tool, lock the flywheel.
4. Remove or disconnect the following:
   - Pressure plate from the flywheel by loosening the bolts alternately, a little at a time, to prevent warpage
   - Clutch disc

To install:

5. Install or connect the following:
   - Clutch with the driven plate on the pressure plate so the spring cage is facing the pressure plate.
6. Hold the clutch assembly against the flywheel, aligning the matchmarks and the dowel pins on the flywheel with the pressure plate. Insert an alignment shaft tool through the pressure plate and the driven plate into the crankshaft pilot bearing.
7. Install the pressure plate to the flywheel. Torque the bolts evenly to 18 ft. lbs. (24 Nm), in a diagonal pattern, to avoid distortion.
8. Remove the alignment shaft.

✳✳ WARNING

The clutch release bearing in the front of the transaxle should be checked before reassembly. It is retained by 2 springs.

9. Install or connect the following:
   • Transaxle
   • Negative battery cable

Hydraulic Clutch System

BLEEDING

The clutch system can be bled using a pressure bleeder. Follow the instructions that come with the pressure bleeder for the proper pressure bleeding procedure. The maximum line pressure while pressure bleeding must not exceed 36 psi (248 kPa).

1. Before servicing the vehicle, refer to the precautions in the beginning of this section.

2. To bleed the system perform the following:

   a. Top off the hydraulic fluid reservoir using a fluid that meets the standards of the vehicle's hydraulic system.

   b. Open the clutch slave cylinder bleed screw and press the clutch pedal to the floor and hold the pedal down.

   c. Close the clutch slave cylinder bleed screw.

   d. Release the clutch pedal.

   e. Check the hydraulic fluid level and top off as necessary.

3. Repeat the above steps until the discharged fluid is clean and no air bubbles appear during the bleeding process.

Halfshaft

REMOVAL & INSTALLATION

Front

A6 AND CABRIOLET MODELS

1. Before servicing the vehicle, refer to the precautions in the beginning of this section.

2. Remove or disconnect the following:
   • Halfshaft nut
   • Front wheels
   • Anti-lock Brake System (ABS) speed sensor by sliding it partly out of its mount
   • Halfshaft from the transaxle flange

3. Press the halfshaft upward toward the front of the vehicle.

4. Turn the steering to full lock.

5. Remove the halfshaft.

To install:

➡**If equipped, replace the inner CV-joint gasket.**

   • Halfshaft into the wheel hub
   • Halfshaft-to-transaxle flange.

Torque the bolts to 33 ft. lbs. (45 Nm) for the M8 bolts and to 59 ft. lbs. (80 Nm) for the M10 bolts
   • ABS speed sensor
   • Halfshaft nut. Torque it to 148 ft. lbs. (200 Nm) plus an additional ¼ (90 degree) turn
   • Front wheels

A4 MODELS

1. Before servicing the vehicle, refer to the precautions in the beginning of this section.

2. Remove or disconnect the following:
   • Hub cap or center cap
   • Hex collar bolt by loosening it
   • Front wheels
   • Halfshaft-to-transaxle flange bolts
   • Hex collar bolt
   • Anti-lock Brake System (ABS) wheel speed sensor cable from the brake caliper bracket
   • ABS speed sensor by sliding it partly out of its mount
   • Remove nut/bolt No. 1, as shown

3. Pull both arms up and out of the swing arm

✳ WARNING

The slots in the swing arm must not be widened. Do not loosen the bolts No. 3 and 4, otherwise the axle geometry must be checked.

4. Tilt the swing arm out and to the rear of the vehicle

5. Remove the halfshaft

To install:

6. Install or connect the following:
   • Halfshaft into the wheel hub
   • Swing arm bolt. Torque it to 30 ft. lbs. (40 Nm)
   • Halfshaft-to-transaxle flange. Torque the bolts to 30 ft. lbs. (40 Nm) for the M8 bolts and to 57 ft. lbs. (77 Nm) for the M10 bolts
   • ABS wheel speed sensor
   • Sensor cable into the caliper bracket
   • Front wheels

7. Tighten the axle bolt as follows:
   • M14 bolt: 85 ft. lbs. (115 Nm) plus an additional ¼ (90 degree) turn
   • M16 bolt: 140 ft. lbs. (190 Nm) plus an additional ¼ (90degree) turn

Rear

A6 QUATTRO MODELS

1. Before servicing the vehicle, refer to the precautions in the beginning of this section.

2. Remove or disconnect the following:
   • Halfshaft bolt
   • Rear wheel

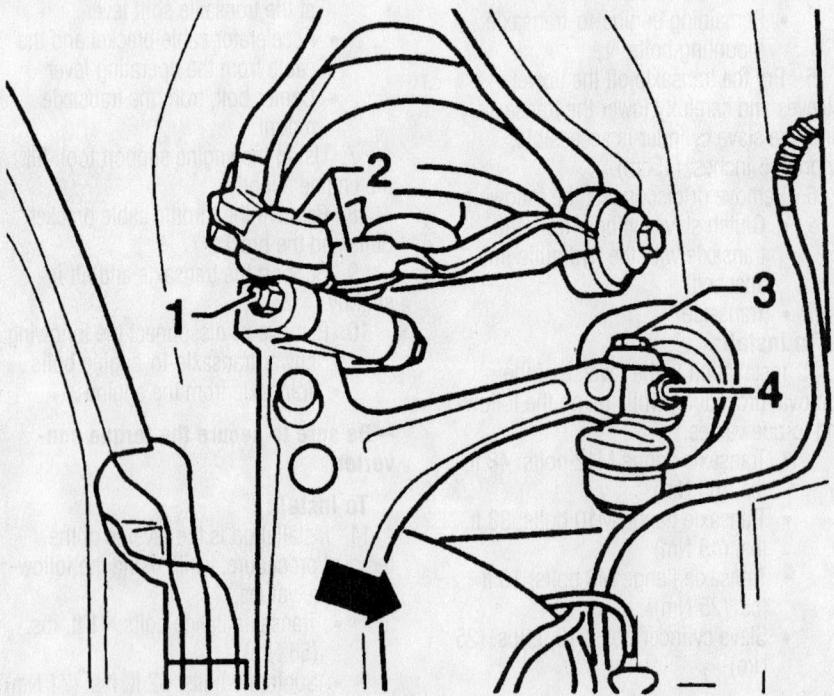

Loosen nut (1), remove the hex bolt and pull both arms (2) upward and out—A4 models

7923CG28

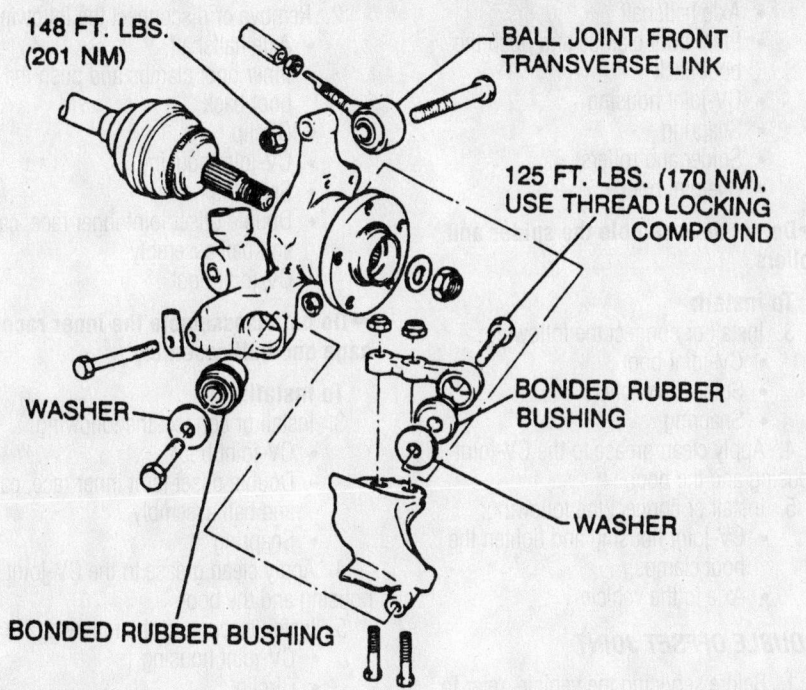

Exploded view of the rear suspension and halfshaft—A6 Quattro models

- Brake caliper without disconnecting the hydraulic line and support it on a wire
- Brake rotor
- Inner halfshaft flange bolts

3. Support the halfshaft.

4. Remove or disconnect the following:

- Fuel tank cover plate and/or inner CV-joint heat shield
- Anti-lock Brake System (ABS) speed sensor by sliding it partly out of its mount
- Lower strut bolt
- Transverse link from the wheel bearing housing
- Halfshaft by pressing down on wheel bearing housing

5. Clean the halfshaft splines of any grease, dirt or locking compound.

To install:

6. Use a new inner flange gasket and reverse the removal procedures.

7. Torque the halfshaft flange bolts as follows:

- M8 bolts: 33 ft. lbs. (45 Nm)
- M10 bolts: 59 ft. lbs. (80 Nm)

8. Install or connect the following:

- Brake caliper. Torque the bolts to 48 ft. lbs. (65 Nm)

➡**Adjustment of parking brake may be necessary.**

- Halfshaft bolt and washer assembly. Torque the bolt to 148 ft. lbs. (200 Nm) plus an additional ¼ (90 degree) turn
- ABS speed sensor
- Rear wheels

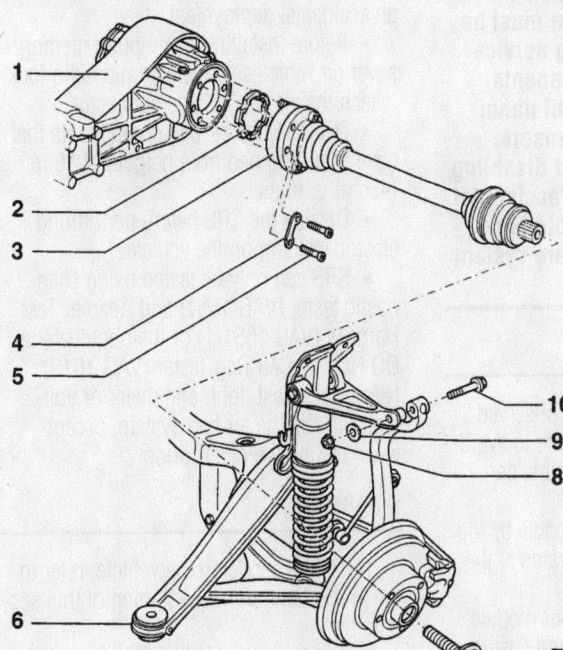

1. Rear final drive
2. Gasket
3. Halfshaft
4. Spacer plate
5. Halfshaft retaining bolts
6. Subframe
7. Collar bolt
8. Self-locking nut
9. Washer
10. Halfshaft retaining bolt

Exploded view of the rear halfshaft and related component mounting—A4 Quattro models

A4 QUATTRO MODELS

1. Before servicing the vehicle, refer to the precautions in the beginning of this section.

2. Remove or disconnect the following:

- Halfshaft bolt
- Rear wheel
- Anti-lock Brake System (ABS) speed sensor by sliding it partly out of its mount
- CV-joint from the final drive

3. Loosen the sway bar link mounting bolt at the wheel bearing housing.

4. Loosen the upper control arm mounting bolt at the wheel bearing housing.

5. Remove or disconnect the following:

- The center and rear mufflers, if servicing the left halfshaft
- Halfshaft from the final drive and the wheel bearing housing

To install:

6. Install or connect the following:

- Halfshaft into the wheel bearing and the final drive
- Upper control arm. Torque the bolt to 52 ft. lbs. (70 Nm) plus a ¼ (90 degree) turn
- Sway bar link. Torque the bolt to 37 ft. lbs. (50 Nm)
- Center and rear mufflers, if removed

- CV-joint to the final drive. Torque the bolts to 30 ft. lbs. (40 Nm)
- ABS wheel speed sensor
- Halfshaft bolt. Torque the bolt to 85 ft. lbs. (115 Nm) plus ¼ (90 degree) turn
- Rear wheel

CV-Joints

OVERHAUL

Outer CV-Joint

The outer CV-joint is serviced with the axle halfshaft as an assembly. The outer CV-joint boot may be serviceable by first removing the inner joint.

Inner CV-Joint

TRI-POD JOINT

1. Before servicing the vehicle, refer to the precautions in the beginning of this section.
2. Remove or disconnect the following:

- Axle halfshaft
- Inner boot clamps and push the boot back
- CV-joint housing
- Snapring
- Spider and rollers
- CV-joint boot

➡Do not disassemble the spider and rollers.

To install:

3. Install or connect the following:
   - CV-joint boot
   - Spider and rollers
   - Snapring
4. Apply clean grease to the CV-joint housing and the boot.
5. Install or connect the following:
   - CV-joint housing and tighten the boot clamps
   - Axle to the vehicle

DOUBLE OFFSET JOINT

1. Before servicing the vehicle, refer to the precautions in the beginning of this section.

2. Remove or disconnect the following:
   - Axle halfshaft
   - Inner boot clamps and push the boot back
   - Circlip
   - CV-joint housing
   - Snapring
   - Double offset joint inner race, cage and ball assembly
   - CV-joint boot

➡Do not disassemble the inner race, cage and ball assembly.

To install:

3. Install or connect the following:
   - CV-joint boot
   - Double offset joint inner race, cage and ball assembly
   - Snapring
4. Apply clean grease to the CV-joint housing and the boot.
5. Install or connect the following:
   - CV-joint housing
   - Circlip
   - Boot clamps by tightening them
   - Axle to the vehicle

STEERING AND SUSPENSION

Air Bag

✳✳ CAUTION

Some vehicles are equipped with an air bag system. The system must be disabled before performing service on or around system components, steering column, instrument panel components, wiring and sensors. Failure to follow safety and disabling procedures could result in accidental air bag deployment, possible personal injury and unnecessary system repairs.

PRECAUTIONS

Several precautions must be observed when handling the inflator module to avoid accidental deployment and possible personal injury.

- Never carry the inflator module by the wires or connector on the underside of the module.
- When carrying a live inflator module, hold securely with both hands and ensure that the bag and trim cover are pointed away from you.
- Place the inflator module on a bench

or other surface with the bag and trim cover facing up.

- With the inflator module on the bench, never place anything on or close to the module that may be thrown in the event of an accidental deployment.
- Before installing a computer memory saver on vehicles with electronic radio lock, detach the air bag voltage connector.
- DO NOT use air bag components that have been dropped from heights of 18 in. (45cm) or higher.
- Disable the SRS before performing electric welding on the vehicle.
- SRS can only be tested using Diagnostic tester (VAG 1551) and Adapter Test Harness (VAG 1551/1) or their equivalents. DO NOT use Air Bag Tester (VAG 1619). Never use a test light, ohmmeter or voltmeter to test the air bag system, except when testing the clockspring.

DISARMING

1. Before servicing the vehicle, refer to the precautions in the beginning of this section.
2. Disconnect and shield the negative battery cable.
3. Wait at least 5 minutes before servicing the vehicle.

ARMING

1. Before servicing the vehicle, refer to the precautions in the beginning of this section.
2. Remove the shield.
3. Connect the negative battery cable.

Power Rack and Pinion Steering Gear

REMOVAL & INSTALLATION

A6 and Cabriolet Models

1. Before servicing the vehicle, refer to the precautions in the beginning of this section.
2. Remove or disconnect the following:
   - Negative battery cable
   - Sound cover by prying it off
   - Crankcase breather hose
   - Vacuum hose to the sound cover
3. Drain the brake fluid from the reservoir.
4. Remove or disconnect the following:
   - Hoses and lines from the brake fluid reservoir and master cylinder
   - Check valve from the brake booster
   - Vacuum unit from the left valve cover

- Left side storage shelf and footwell under the instrument panel
- Center electrical panel and/or relay panel
- Brake booster clevis pin
- Pedal bracket mounting nuts
- Fuel line by unclipping it from the left side of the instrument panel bracket
- Brake booster
- Steering gear pinion U-joint-to-steering column bolt
- U-joint

5. Clamp the pressure and return hydraulic hoses, using Camp 3094.

6. Remove or disconnect the following:

- Hoses from the steering gear
- Both tie rods with the carrier
- Cable tie from the right wheel housing wiring harness
- Right side steering gear mounting bolts
- Anti-lock Brake System (ABS) wheel speed sensor connector from the bracket
- Front wheels

- Steering gear bolts on the left and right wheel housing

7. Pull the steering gear toward the front of the vehicle, to clear the instrument panel seal.

8. With the aid of an assistant, pull the steering gear to the left in the left wheel housing hole so that the wiring harness can be pushed toward the back on the right side of the steering gear. The steering gear can be removed from the opening in the right wheel housing.

To install:

9. Install or connect the following:

- Steering gear into the vehicle through the right wheel housing. Torque the bolts to 37 ft. lbs. (50 Nm)

➡**Ensure the proper routing of the wiring harness around the steering gear.**

- Tie rods. Torque the bolts to 44 ft. lbs. (60 Nm) with the vehicle on its wheels
- Hydraulic hoses to the steering gear. Torque the banjo bolts to 30 ft. lbs. (40 Nm)

- Left front wheel ABS wheel speed sensor bracket

10. The completion of installation is the reverse of the removal, with following these additional points:

a. Remove the hose clamps and check the fluid levels.

b. Ensure the brake lines are connected and bled properly.

c. Torque the U-joint bolts to 18 ft. lbs. (25 Nm).

d. Torque the pedal bracket bolts to 18 ft. lbs. (25 Nm).

e. Install the brake booster clevis pin.

❊❊ WARNING

If the power steering fluid needs to be topped off, use only an approved fluid otherwise internal damage may occur.

A4 Models

1. Before servicing the vehicle, refer to the precautions in the beginning of this section.

2. Remove or disconnect the following:
- Battery

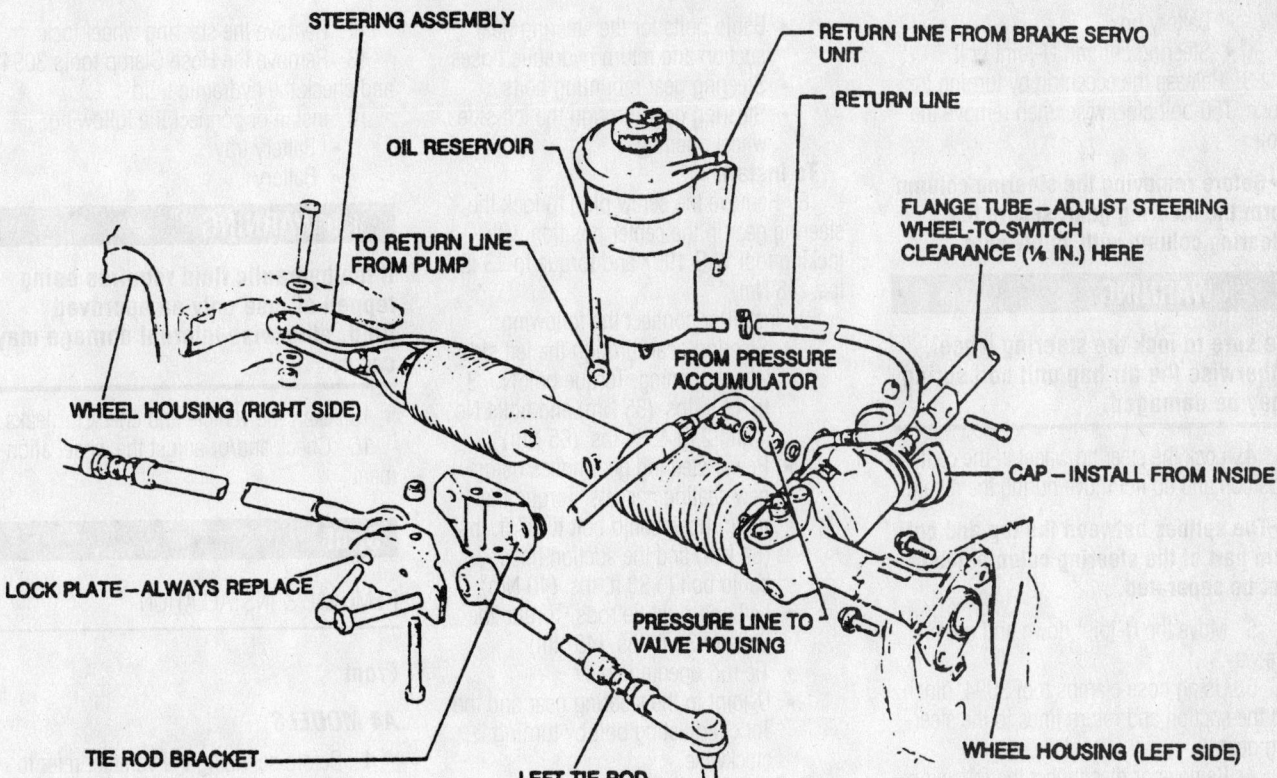

STEERING ASSEMBLY

RETURN LINE FROM BRAKE SERVO UNIT

RETURN LINE

OIL RESERVOIR

TO RETURN LINE FROM PUMP

FLANGE TUBE—ADJUST STEERING WHEEL-TO-SWITCH CLEARANCE (⅛ IN.) HERE

FROM PRESSURE ACCUMULATOR

WHEEL HOUSING (RIGHT SIDE)

CAP—INSTALL FROM INSIDE

LOCK PLATE—ALWAYS REPLACE

PRESSURE LINE TO VALVE HOUSING

TIE ROD BRACKET

LEFT TIE ROD

WHEEL HOUSING (LEFT SIDE)

7923CG32

Exploded view of the steering rack assembly—A6 and cabriolet models

Heater Core replacement is covered in Section 2 of this manual

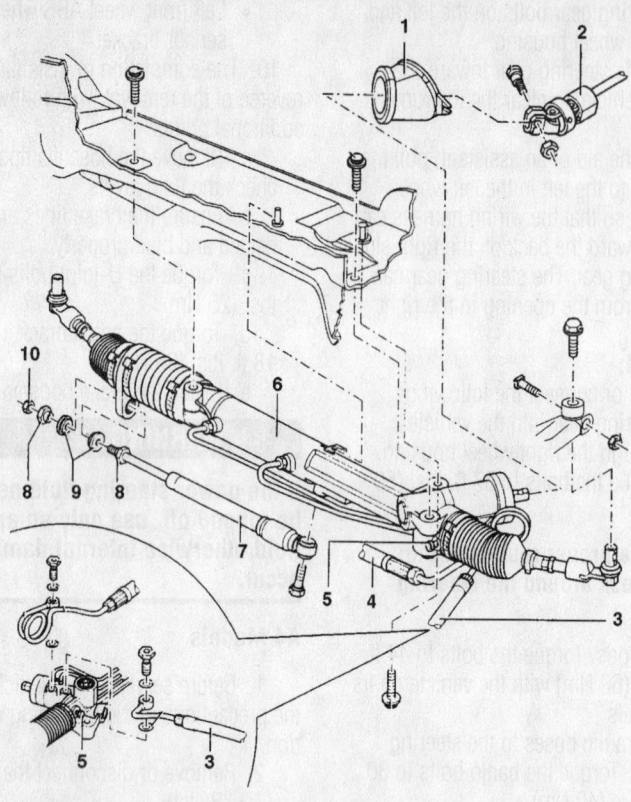

1. Boot seal
2. Steering column
3. Return hose
4. Flexible hose
5. Screw plug for centering the steering wheel
6. Rack and pinion steering gear
7. Steering damper
8. Bushing
9. Two-piece rubber bushing
10. Nut

7923CG33

Exploded view of the steering gear mounting—A4 models

- Battery box
- Steering column U-joint bolt

3. Release the eccentric by turning the Torx® T50 bolt clockwise, then remove the bolt.

➥**Before removing the steering column form the steering gear, secure the steering column with safety wire.**

✳✳ WARNING

Be sure to lock the steering wheel, otherwise the air bag unit coil spring may be damaged.

4. Lock the steering wheel in the center position and do not move during the repairs.

➥**The splines between the top and bottom part of the steering column must not be separated.**

5. Move the U-joint down and out of the way.

6. Using hose clamps tool 3094, pinch off the suction and return lines to the steering gear.

7. Remove or disconnect the following:
- Front wheels
- Left and right tie rods
- Tie rod opening cover

➥**Place a drip tray under the vehicle to catch any residual power steering fluid.**

- Banjo bolts for the steering gear suction and return hydraulic hoses
- Steering gear mounting bolts
- Steering gear through the left side wheel opening

To install:

8. Remove the screw plug to lock the steering gear in the center position with locking tool VAG 1907 and torque to 13 ft. lbs. (18 Nm)

9. Install or connect the following:
- Steering gear through the left side wheel opening. Torque bolt No. 3 to 48 ft. lbs. (65 Nm) and bolts No. 1 and 2 to 48 ft. lbs. (65 Nm)
- Power steering gear hoses using new sealing gaskets. Torque the return hose banjo bolt to 37 ft. lbs. (50 Nm) and the suction hose banjo bolt to 30 ft. lbs. (40 Nm)
- Left and right tie rods. Torque the bolts to 33 ft. lbs. (45 Nm)
- Tie rod opening cover
- U-joint to the steering gear and the Torx® adjusting bolt by turning it clockwise

10. Remove the locking tool VAG 1907.
11. Install or connect the following:
- Screw plug. Torque it to 13 ft. lbs. (18 Nm)
- Adjusting bolt. Torque the nut to 30 ft. lbs. (40 Nm)

12. Remove the steering wheel lock
13. Remove the Hose Clamp tools 3094 and check the hydraulic fluid
14. Install or connect the following:
- Battery tray
- Battery

✳✳ WARNING

If the hydraulic fluid requires being topped of, use only an approved fluid, otherwise internal damage may occur.

15. Start the vehicle and check for leaks.
16. Check and/or adjust the wheel alignment.

Strut

REMOVAL & INSTALLATION

Front

A4 MODELS

1. Before servicing the vehicle, refer to the precautions in the beginning of this section.

2. Remove or disconnect the following:
- Front wheels
- Rubber grommets from the plenum chamber

- Upper strut-to-body mounting nuts
- Anti-lock Brake System (ABS) wheel speed sensor wire from the brake caliper bracket
- Upper control arm pinch bolt and both upper control links
- Guide link ball joint, by swiveling the wheel bearing housing aside
- Lower strut mounting bolt

➡**When removing the strut, be sure not to damage the CV-joint boot.**

- Strut

To install:

➡**The bonded rubber bushing can only turned to a limited extent. The bolted connections between the suspension strut and the lower track control links should therefore only be tightened when the vehicle is standing on the ground.**

3. Install or connect the following:
   - Strut by positioning it so that the spring hole plate faces the middle of the vehicle. Torque the bolt to 66 ft. lbs. (90 Nm)
   - Upper control links to the wheel bearing housing. Torque the pinch bolt to 30 ft. lbs. (40 Nm)

➡**It may be necessary to hold the ball joint stud with a 4mm hex wrench.**

- Ball joint. Torque the nut to 74 ft. lbs. (100 Nm)
- ABS wheel speed sensor wire into the brake caliper holder
- Upper strut-to-body nuts. Torque them to 15 ft. lbs. (20 Nm)
- Rubber grommets into the plenum chamber
- Front wheels

4. Test drive the vehicle.
5. Check and/or adjust the front alignment.

A6 AND CABRIOLET MODELS

1. Before servicing the vehicle, refer to the precautions in the beginning of this section.
2. Remove or disconnect the following:
   - Negative battery cable
   - Halfshaft nut or bolt
   - Wheel assembly
   - Brake caliper without disconnecting the brake line and support it aside
   - Speed sensor, if equipped
   - Disc brake rotor
   - Ball joint clamp bolt and nut
   - Tie rod end from the strut

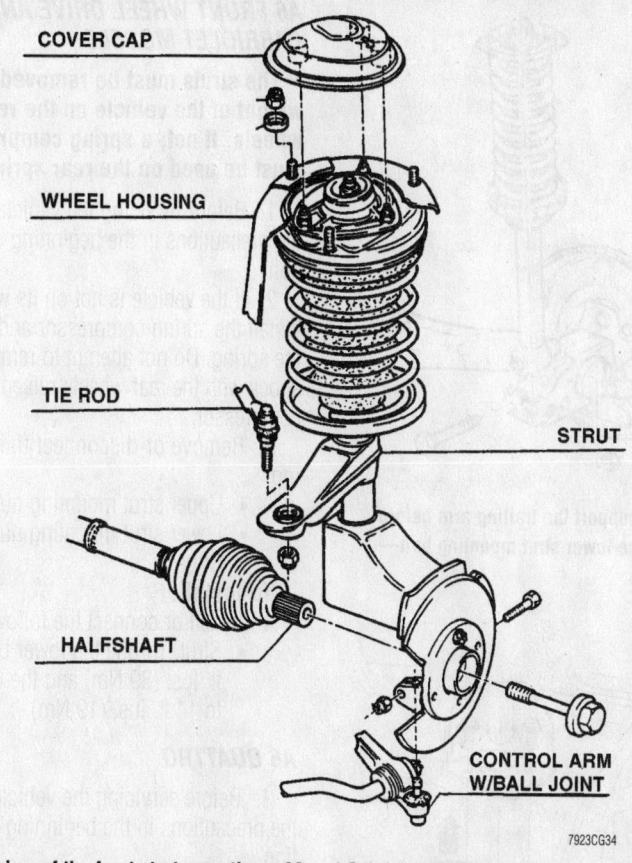

Exploded view of the front strut mounting—A6 and Cabriolet models

Labels: COVER CAP, WHEEL HOUSING, TIE ROD, HALFSHAFT, STRUT, CONTROL ARM W/BALL JOINT

7923CG34

- Stabilizer bar end clamps, if equipped, and pivot the bar downward
- 2 center stabilizer bar clamps
- Stabilizer bar from the lower control arm
- Steering knuckle pinch bolt
- Lower ball joint from the knuckle
- Halfshaft by pressing it from the hub
- Upper strut cover
- 3 strut retaining nuts
- Strut assembly

To install:

3. Installation is the reverse of the removal procedures.
4. Install or connect the following:
   - Strut. Torque the 3 upper strut retaining nuts to 22 ft. lbs. (30 Nm)
   - Stabilizer bar

➡**When installing the stabilizer bar, the position is correct if the clamps are difficult to install in the rubber bushings. Attach the clamps loosely.**

- Seat the speed sensor

➡**When installing the axle shaft, apply a bead of thread locking compound to the splines.**

5. When the vehicle is on the ground, torque the center nut or bolt to 148 ft. lbs. (200 Nm) plus ¼ (90 degree) turn.
6. After test driving to seat stabilizer bushings in correct position, torque the clamps to 18 ft. lbs. (24 Nm).

Rear

A4 MODELS

1. Before servicing the vehicle, refer to the precautions in the beginning of this section.
2. Support the trailing arms.
3. Remove or disconnect the following:
   - Lower strut mounting bolt
   - Rear seat backrest side bolster cover or backrest to access the upper mounting
   - Upper strut-to-body mounting nuts

➡**In addition to the bolted connection, the strut is also attached to the body by 4 retaining lugs.**

4. Turn the strut until the retaining lugs are positioned above the recesses, then pull the strut downward out of its mount.

Brake service is covered in Section 4 of this manual

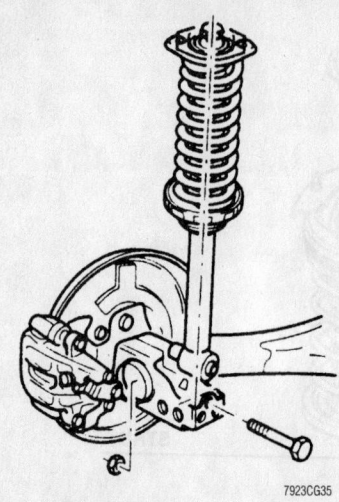

Be sure to support the trailing arm before removing the lower strut mounting bolt—a4 models

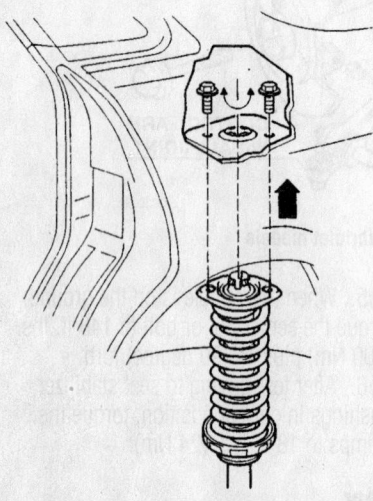

After loosening the 2 attaching bolts, rotate the upper strut mount to disengage the strut from the vehicle—A4 models

To install:

➡**The bonded rubber bushing can only turned to a limited extent. The bolted connections between the suspension strut and the rear axle should therefore only be tightened when the vehicle is standing on the ground.**

5. Install or connect the following:
- Strut by engaging the retaining lugs
- Upper strut-to-body. Torque the nuts to 18 ft. lbs. (25 Nm)
- Lower strut. Torque the bolt to 37 ft. lbs. (50 Nm) plus a ¼ (90 degree) turn with the suspension loaded
- Rear seat backrest side bolster cover or backrest

A6 FRONT WHEEL DRIVE AND CABRIOLET MODELS

➡**The struts must be removed with the weight of the vehicle on the rear wheels. If not, a spring compressor must be used on the rear springs.**

1. Before servicing the vehicle, refer to the precautions in the beginning of this section.

2. If the vehicle is not on its wheels, install the spring compressor and compress the spring. Do not attempt to remove the shock with the rear wheels raised without a compressor.

3. Remove or disconnect the following:

- Upper strut mounting nut
- Lower strut mounting nut
- Strut

To install:

4. Install or connect the following:
- Strut. Torque the lower bolts to 66 ft. lbs. (89 Nm) and the upper bolts to 14 ft. lbs. (19 Nm)

A6 QUATTRO

1. Before servicing the vehicle, refer to the precautions in the beginning of this section.

2. Remove or disconnect the following:
- Rear wheel
- Strut covers, the remove the strut-to-body nuts/bolts.
- Strut-to-rear wheel knuckle assembly. Remove the strut from the vehicle

To install:

3. Install or connect the following:
- Strut. Torque the strut to body bolts to 15 ft. lbs. (20 Nm) and the knuckle bolt to 66 ft. lbs. (89 Nm)
- Strut covers
- Rear wheel

Coil Spring

REMOVAL & INSTALLATION

1. Before servicing the vehicle, refer to the precautions in the beginning of this section.

2. Remove the strut from the vehicle.

3. Clamp the spring compressor tool VAG 1752/2, in a vise.

4. Install the strut into the spring compressor.

5. Pry off the mounting bolt cap.

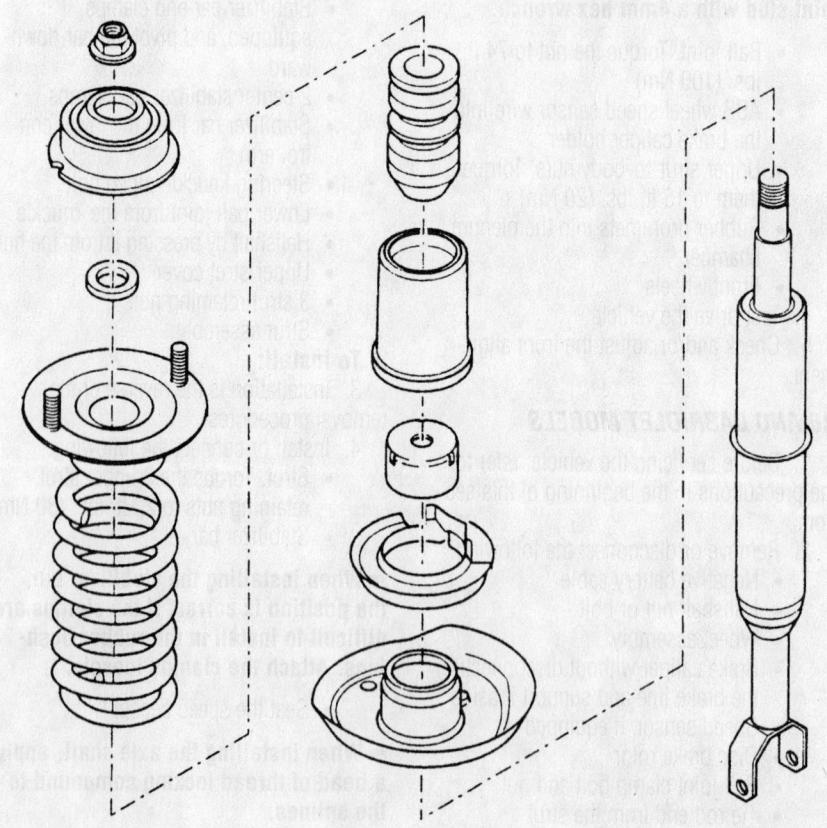

Exploded view of the front strut—A4 model

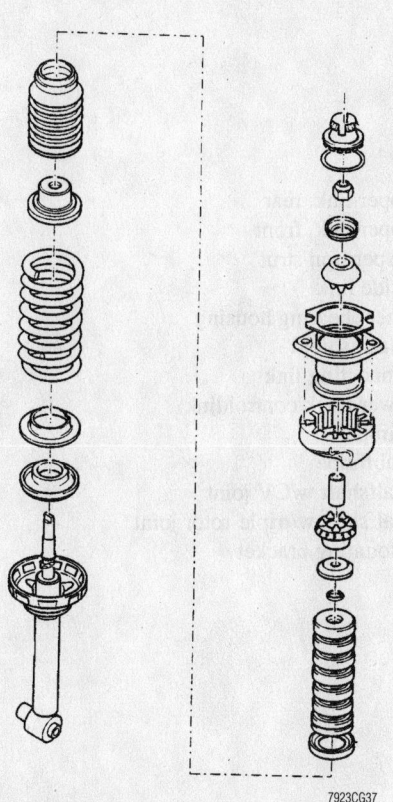

Exploded view of the rear strut—A4 model

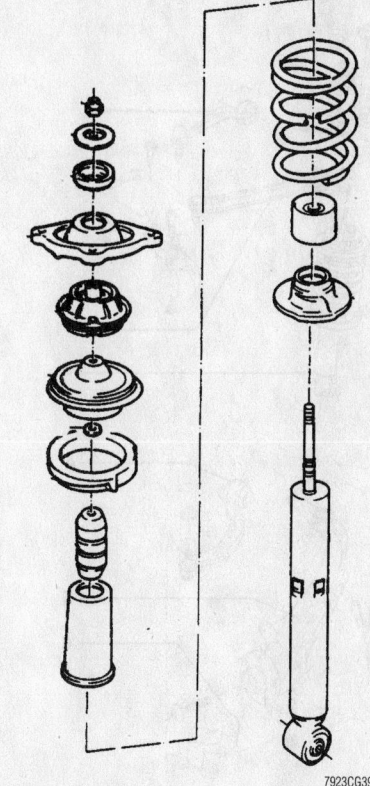

Exploded view of the rear strut—except A4 model

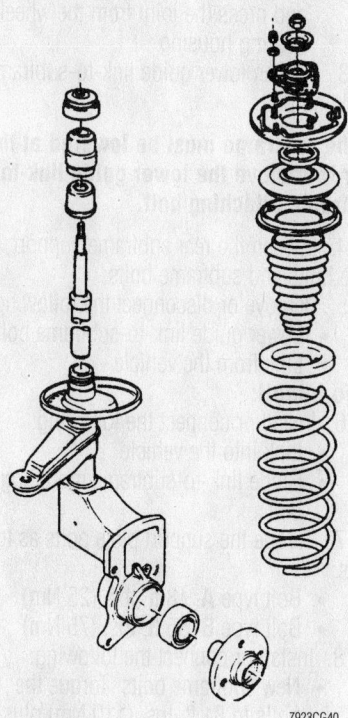

Exploded view of the front strut—except A4 model

6. Compress the coil spring and remove the self-locking nut from the piston rod.

7. Matchmark the position of the spring retainer and spring mount.

8. Remove or disconnect the following:
   - Spring seat and related components
   - Strut from the spring compressor

9. Release the tension on the coil spring.

10. Remove the spring out of the compressor.

To install:

11. Install the spring into the compressor.

12. Compress the spring and insert the strut through the spring.

13. Install or connect the following:
   - Spring seat and related components in the reverse order as they were removed by aligning the matchmarks
   - New self-locking nut
   - Mounting bolt cap

14. Release the spring compressor.

15. Install the strut into the vehicle.

Upper Ball Joint

REMOVAL & INSTALLATION

The Audi A4 front suspension is equipped with 2 separate upper ball joints that are not replaceable, therefore the upper link (front or rear) must be replaced. To remove this link, perform the following procedures.

1. Before servicing the vehicle, refer to the precautions in the beginning of this section.

2. Remove or disconnect the following:
   - Front wheels
   - Pinch bolt and pull both control arms upward and out

3. Cover the steering gear boot.

4. Remove or disconnect the following:
   - Guide link ball joint and press off the joint
   - Anti-lock Brake System (ABS) wheel speed sensor wire from the brake caliper bracket

5. Support the suspension from excessive rebound travel.

6. Remove or disconnect the following:
   - Lower strut bolt and swing the wheel bearing housing aside
   - Rubber grommets from the plenum chamber
   - Upper strut-to-body nuts
   - Strut together with the mounting bracket

7. Clamp the strut in a vise with the protective jaw covers.

8. Remove or disconnect the following:
   - Upper link bolts and detach both of the links
   - Bracket-to-strut nuts, then separate

To install:

9. Install or connect the following:
   - Brackets and links, as shown. Torque the bracket-to-strut mounting nuts to 15 ft. lbs. (20 Nm)
   - Links by aligning them, as shown. Torque to 37 ft. lbs. (50 Nm) plus a ¼ (90 degree) turn
   - Strut with mounting bracket. Torque the upper strut-to-body nuts to 48 ft. lbs. (75 Nm)
   - Lower strut bolt. Torque it to 66 ft. lbs. (90 Nm)
   - Nut on the ball joint. Torque to 74 ft. lbs. (100 Nm)
   - Upper links to the wheel bearing housing. Torque the pinch bolt to 30 ft. lbs. (40 Nm)

For complete Engine Mechanical specifications, see Section 1 of this manual

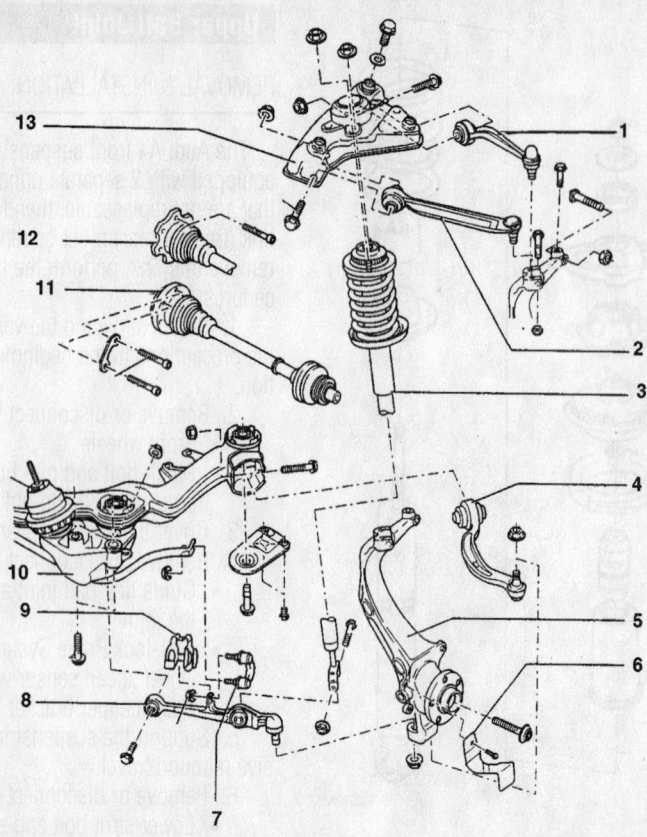

1. Upper link, rear
2. Upper link, front
3. Suspension strut
4. Guide link
5. Wheel bearing housing
6. Splash shield
7. Connecting link
8. Lower track control link
9. Clamp
10. Subframe
11. Halfshaft w/CV joint
12. Halfshaft w/triple-rotor joint
13. Mounting bracket

7923CG41

Exploded view of the front suspension—A4 models

- ABS wiring to the brake caliper bracket
- Wheels

10. Check the front suspension alignment.

Lower Ball Joint

REMOVAL & INSTALLATION

➡The A4 A6 models are equipped with 2 lower ball joints that are not serviceable. The control arms must be replaced if a joint is worn. The lower track control link ball joint stud faces down, and the guide link ball joint stud faces up.

LOWER TRACK CONTROL LINK

1. Before servicing the vehicle, refer to the precautions in the beginning of this section.

2. Remove or disconnect the following:
- Front wheels
- Nut from the lower track control link

3. Press the ball joint out of the tapered seat.

4. Support the wheel bearing housing to prevent excessive rebound travel in the suspension.

5. Remove or disconnect the following:
- Stabilizer link and lower strut mounting bolt
- Lower track control link-to-subframe attaching bolt
- Lower track control link

To install:

6. Install or connect the following:
- Lower track control link
- Subframe attaching bolt. Torque the bolt to 74 ft. lbs. (100 Nm)

7. Install or connect the following:
- Stabilizer link. Torque the upper bolt to 30 ft. lbs. (40 Nm) plus ¼ (90 degree) turn and the lower bolt to 74 ft. lbs. (100 Nm)

8. Load the suspension and torque the subframe bolt to 59 ft. lbs. (80 Nm) plus ¼ (90 degree) turn

9. Front wheels

10. Check and/or adjust the front suspension alignment.

LOWER GUIDE LINK

1. Before servicing the vehicle, refer to the precautions in the beginning of this section.

2. Remove or disconnect the following:
- Front wheels
- Nut from the lower guide link joint

and press the joint from the wheel bearing housing

3. Loosen lower guide link-to-subframe attaching bolt

➡The subframe must be lowered at the rear to remove the lower guide link-to-subframe attaching bolt.

4. Loosen the rear subframe support plate bolts and subframe bolts.

5. Remove or disconnect the following:
- Lower guide link-to-subframe bolt
- Link from the vehicle

To install:

6. Install or connect the following:
- Link into the vehicle
- Guide link-to-subframe mounting bolt

7. Torque the support plate bolts as follows:
- Bolt type **A**: 18 ft. lbs. (25 Nm)
- Bolt type **B**: 55 ft. lbs. (75 Nm)

8. Install or connect the following:
- New subframe bolts. Torque the bolts to 81 ft. lbs. (110 Nm) plus a ¼ (90 degree) turn
- Joint end into the wheel bearing housing. Torque the nut to 74 ft. lbs. (100 Nm)

9. Load the suspension. Torque the

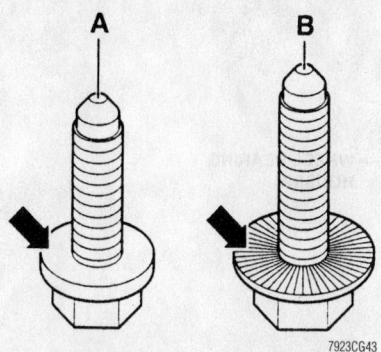

Subframe support bracket bolt identification—A4 model

lower guide link-to-subframe attaching bolt to 66 ft. lbs. (90 Nm) plus ¼ (90 degree) turn.

10. Install the front wheels.

11. Check and/or adjust the front suspension alignment.

Upper Control Arm

REMOVAL & INSTALLATION

1. Before servicing the vehicle, refer to the precautions in the beginning of this section.

2. Remove or disconnect the following:
   - Negative battery cable
   - Wheel

3. Loosen the upper strut mounting nuts.

4. Loosen, but do not remove, the upper strut rod nut.

✴✴ CAUTION

DO NOT completely remove the upper strut nut at this time.

5. Remove or disconnect the following:
   - Brake caliper, leaving the line attached and secure it out of the way
   - Anti-lock Brake System (ABS) speed sensor and harness, if applicable
   - Cotter pin and nut from the upper control arm
   - Upper control arm from the steering knuckle
   - Stabilizer bar from the link, if applicable
   - Cotter pin and nut from the lower control arm
   - Strut
   - Upper strut mounting nuts
   - Strut
   - Upper control arm

To install:

6. Install or connect the following:
   - Upper suspension arm
   - Strut. Torque the upper nuts to 42 ft. lbs. (56 Nm)
   - Strut to the lower arm
   - Stabilizer bar bracket
   - Stabilizer bar to the link
   - Upper suspension arm to the steering knuckle. Torque the nut to 64 ft. lbs. (87 Nm)
   - New cotter pin
   - ABS speed sensor. Torque the bolt to 69 inch lbs. (8 Nm)
   - Brake caliper
   - Front wheel

7. Bounce the vehicle several times to stabilize the suspension.

8. Tighten the lower strut bolt

9. Check and/or adjust the front wheel alignment.

CONTROL ARM BUSHING REPLACEMENT

The upper control arm bushings are serviced with the control arm as an assembly.

Lower Control Arm

REMOVAL & INSTALLATION

1. Remove or disconnect the following:
   - Front wheels
   - Cotter pin and castle nut from the tie rod end
   - Tie rod end from the steering knuckle
   - Cotter pin and castle nut from the ball joint stud
   - Knuckle from the ball joint
   - Stabilizer bar link from the lower arm
   - Nuts and bolts that connect the lower control arm
   - Lower control arm from the vehicle

To install:

➡**Tightening of the suspension component fasteners should be performed with the full weight of the vehicle resting on the suspension.**

2. Install or connect the following:
   - Control arm to the crossmember and install the through bolt, lockwasher and nut
   - Lower control arm to the tension rod nuts
   - Stabilizer bar transverse link to the lower arm and tighten the link

- Knuckle to the lower ball joint.
- Tie rod end to the knuckle. Torque the castle nut to 22–36 ft. lbs. (29–49 Nm)
- New cotter pin
- Front wheels

3. Tighten all nuts and bolts to specification
   - Check and/or adjust the wheel alignment.

CONTROL ARM BUSHING REPLACEMENT

Front Bushings

1. Before servicing the vehicle, refer to the precautions in the beginning of this section.

2. Remove the lower control arm from the vehicle.

3. Press the front bushing out of the control arm.

To install:

4. Lubricate the front bushing with soap and press into the control arm.

5. Install the control arm to the vehicle.

6. Check and/or adjust the wheel alignment.

Rear Bushings

1. Before servicing the vehicle, refer to the precautions in the beginning of this section.

2. Remove or disconnect the following:
   - Rear wheel
   - Rear control arm
   - Press the rear control arm bushing out

To install:

3. Lube the rear bushing with soap

4. Install or connect the following:
   - Rear bushing
   - Rear control arm
   - Front wheel

5. Check and/or adjust the wheel alignment.

Wheel Bearings

ADJUSTMENT

Front

The front wheel bearings are sealed and no adjustment is necessary or possible. If the bearings are found to be loose or noisy, they must be replaced.

For Accessory Drive Belt illustrations, see Section 1 of this manual

Rear

CABRIOLET AND NON-QUATTRO MODELS

1. Before servicing the vehicle, refer to the precautions in the beginning of this section.

2. Remove or disconnect the following:
- Grease cap
- Cotter pin and the locking nut

3. While turning the wheel, so the wheel bearing does not jam, tighten the adjusting nut firmly.

4. Back the nut off slightly. The nut is properly adjusted when it is possible to pry the thrust washer side to side with some drag but using light pressure on the tool.

5. Install the locking nut and a new cotter pin.

6. When installing the cap, be sure it is securely in place.

QUATTRO MODELS

The wheel bearings are sealed; no adjustment is necessary or possible. If the bearings are found to be loose or noisy, they must be replaced.

REMOVAL & INSTALLATION

Front

A4 AND A6 MODELS

1. Before servicing the vehicle, refer to the precautions in the beginning of this section.

2. Loosen the halfshaft retaining bolt.

3. Remove or disconnect the following:
- Front wheel
- Anti-lock Brake System (ABS) wheel speed sensor
- Caliper bracket
- Rotor
- Brake splash guard

4. Loosen the mounting nuts for the lower guide and track links.

5. Remove or disconnect the following:
- Tie rod end from the wheel bearing housing
- Mounting nuts for the lower guide and track links and press out the joints
- Upper control arm pinch bolt and the arms
- Wheel bearing housing

6. Place the wheel bearing housing on a press.

7. Drive out the hub with the wheel bearing.

8. Using a bearing separator and press, drive hub out of the bearing.

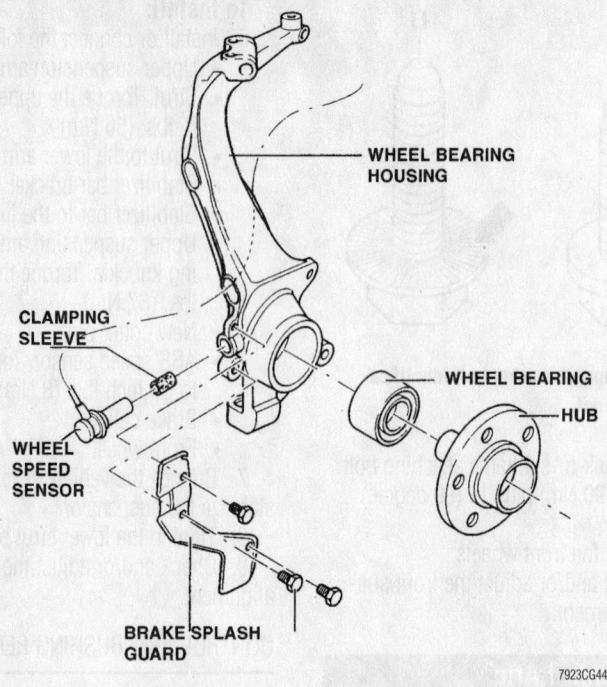

Exploded view of the front wheel bearing housing—A4 models

(labels: WHEEL BEARING HOUSING, CLAMPING SLEEVE, WHEEL SPEED SENSOR, BRAKE SPLASH GUARD, WHEEL BEARING, HUB)

7923CG44

To install:

9. Press the new wheel bearing into the bearing housing using the appropriate bearing driver.

10. Press the hub into the wheel bearing using the appropriate bearing driver.

11. Install or connect the following:
- Wheel bearing housing
- CV-joint by sliding it through the wheel hub and hand-tighten the new nut
- Lower track control and guide link. Torque the new self-locking nut to 74 ft. lbs. (100 Nm)
- Both of the upper link ball joints into the wheel bearing. Torque the pinch bolt to 30 ft. lbs. (40 Nm)
- Tie rod end. Torque the new self-locking nut to 37 ft. lbs. (50 Nm) and the bolt to 44 inch lbs. (5 Nm)
- ABS wheel speed sensor
- Brake splash guard. Torque the bolts to 84 inch lbs. (10 Nm)
- Brake rotor
- Bake caliper. Torque the bolt to 89 ft. lbs. (120 Nm)
- Front wheel

12. Tighten the halfshaft retaining bolt as follows:
- M14 bolt: 85 ft. lbs. (115 Nm) plus ½ (180 degree) turn
- M16 bolt: 140 ft. lbs. (190 Nm) plus ½ (180 degree) turn

13. Check and/or adjust the front alignment, if necessary.

Rear

NON-QUATTRO MODELS

1. Before servicing the vehicle, refer to the precautions in the beginning of this section.
- Wheel
- Brake caliper, without disconnecting the hydraulic line and suspend it on a wire.
- Grease cap
- Cotter pin, nut and washer
- Outer bearing
- Brake rotor
- Bearing inner bearing and seal from the rotor hub, using a soft drift or press
- Bearing inner and outer race(s) from the rotor, using a soft drift or press

To install:

2. Clean and inspect mating surfaces for bearing races.

3. Install or connect the following:
- New races using soft drift or press
- New bearing packed with grease and set it into the inner race
- Seal, making sure it is square in the rotor hub
- Rotor, outer bearing, washer, and nut and adjust bearing play
- Cotter pin and dust cap
- Brake caliper

4. If hydraulic lines were removed, install and bleed brakes.

5. If parking brake cable has been remove, install and adjust as necessary.

6. Install the wheel .

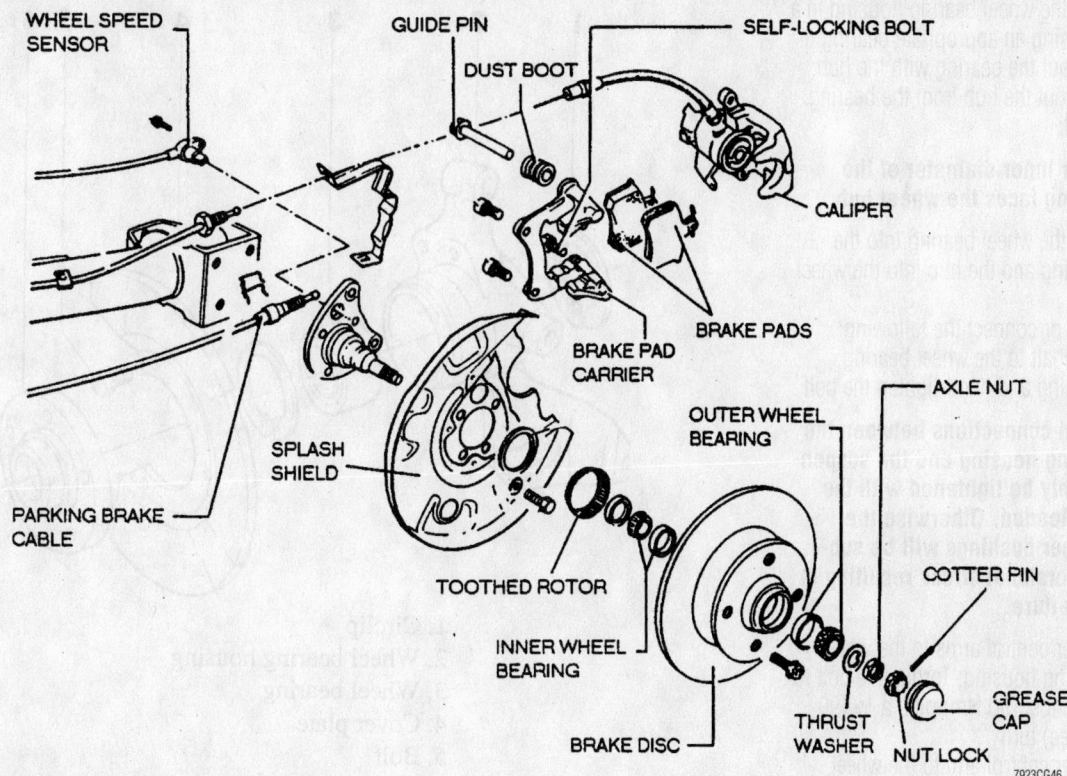

Exploded view of the rear wheel bearing—front wheel drive vehicles

7. Check the brakes for proper operation.

A4 QUATTRO MODELS

1. Before servicing the vehicle, refer to the precautions in the beginning of this section.
2. Remove or disconnect the following:
   - Wheel
   - Halfshaft bolt
   - Anti-lock Brake System (ABS) wheel speed sensor from the wheel bearing housing
   - Stabilizer bar link to the wheel bearing housing nut
   - Track rod from the wheel bearing housing
   - Brake caliper
   - Brake rotor
3. Matchmark the position of the eccentric washer for the lower control arm-to-wheel bearing housing bolt and remove it.
4. Remove or disconnect the following:
   - Wheel bearing housing-to-upper control arm bolt
   - Halfshaft from the wheel bearing housing
5. Clean any dirt and debris from around the machined area for the wheel bearing path.

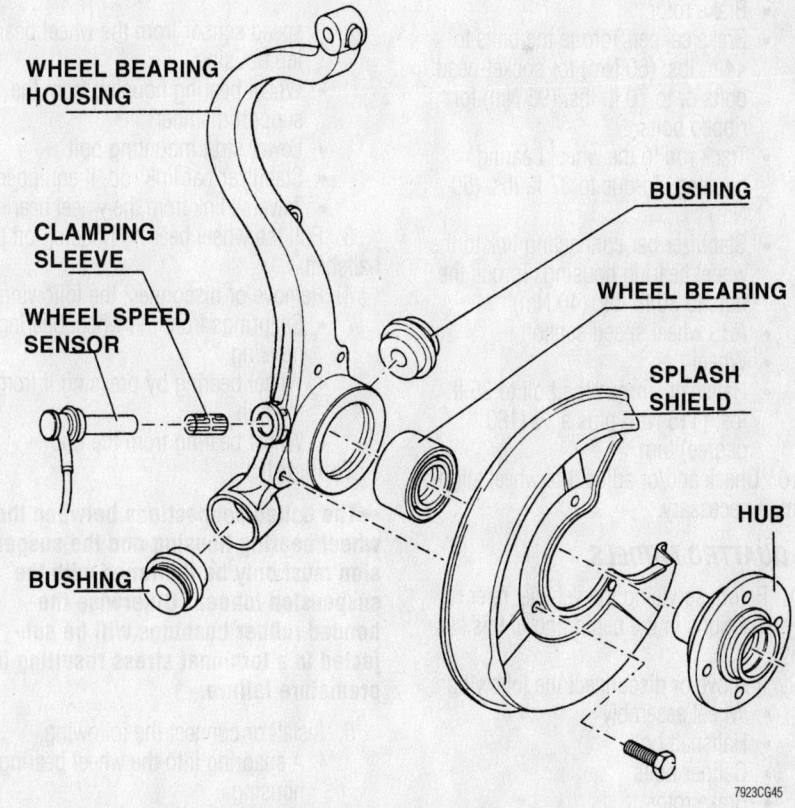

Exploded view of the rear wheel bearing housing—A4 Quattro models

For Tire, Wheel and Ball Joint specifications, see Section 1 of this manual

6. Place the wheel bearing housing in a press, then using an appropriate bearing driver, press out the bearing with the hub.

7. Press out the hub from the bearing.

To install:

➡ **The larger inner diameter of the wheel bearing faces the wheel hub.**

8. Press the wheel bearing into the bearing housing and the hub into the wheel bearing.

9. Install or connect the following:
- Halfshaft to the wheel bearing housing and hand-tighten the bolt

➡ **The bolted connections between the wheel bearing housing and the suspension must only be tightened with the suspension loaded. Otherwise the bonded rubber bushings will be subjected to a torsional stress resulting in premature failure.**

- Upper control arms to the wheel bearing housing. Torque the bolt to 37 ft. lbs. (50 Nm) plus a ¼ (90 degree) turn
- Lower control arm to the wheel bearing housing by aligning the matchmarks. Torque the bolt to 70 ft. lbs. (95 Nm)
- Brake rotor
- Brake caliper. Torque the bolts to 44 ft. lbs. (60 Nm) for socket-head bolts or to 70 ft. lbs. (95 Nm) for ribbed bolts
- Track rod to the wheel bearing housing. Torque to 37 ft. lbs. (50 Nm)
- Stabilizer bar connecting link to the wheel bearing housing. Torque the nuts to 30 ft. lbs. (40 Nm)
- ABS wheel speed sensor
- Wheel
- Halfshaft. Torque the bolt to 85 ft. lbs. (115 Nm) plus a ½ (180 degree) turn

10. Check and/or adjust the wheel alignment, if necessary.

A6 QUATTRO MODELS

1. Before servicing the vehicle, refer to the precautions in the beginning of this section.

2. Remove or disconnect the following:
- Wheel assembly
- Halfshaft bolt
- Caliper bolts
- Brake rotor
- Trapezoidal arm from the wheel bearing housing
- Anti-lick Brake System (ABS) wheel

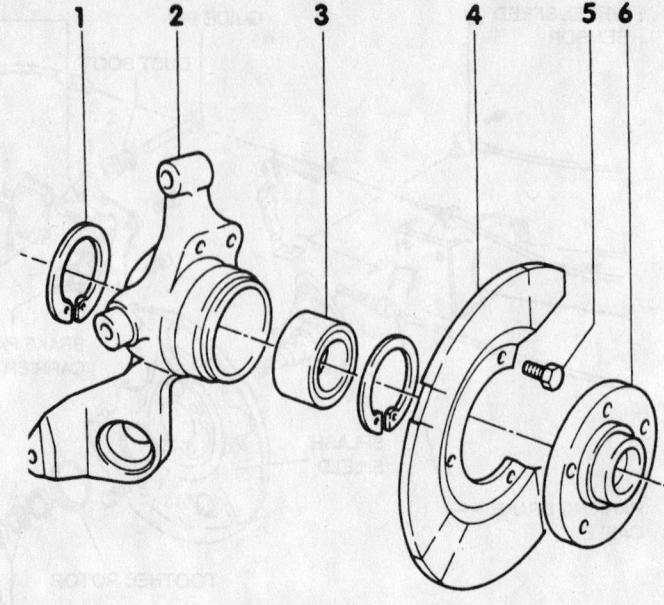

1. Circlip
2. Wheel bearing housing
3. Wheel bearing
4. Cover plate
5. Bolt
6. Wheel hub

7923CG47

Exploded view of the wheel bearing and related components—A6 Quattro models

speed sensor from the wheel bearing housing
- Wheel bearing housing from the support member
- Lower strut mounting bolt
- Stabilizer bar link rod, if equipped
- Traverse link from the wheel bearing

3. Pull the wheel bearing housing off the halfshaft.

4. Remove or disconnect the following:
- Snaprings from the wheel bearing housing
- Wheel bearing by pressing it from the hub
- Wheel bearing from the hub

To install:

➡ **The bolted connections between the wheel bearing housing and the suspension must only be tightened with the suspension loaded. Otherwise the bonded rubber bushings will be subjected to a torsional stress resulting in premature failure.**

5. Install or connect the following:
- A snapring into the wheel bearing housing
- New wheel bearing by pressing it in from the opposite side of the wheel bearing housing

- Outer snapring after the bearing is seated against the snapring
- Both snaprings, position them so that the gap s points downward
- Hub by pressing it into the wheel bearing
- Wheel bearing housing to the halfshaft and hand-tighten the bolt
- Transverse link to the wheel bearing housing. Torque to 148 ft. lbs. (200 Nm)
- Stabilizer link. Torque the self-locking nuts to 33 ft. lbs. (45 Nm)
- Lower strut mounting bolt. Torque the bolt to 66 ft. lbs. (90 Nm)
- Wheel bearing housing to the support and trapezoidal arms. Torque the bolts to 104 ft. lbs. (170 Nm) plus a ¾ (270 degree) turn
- ABS wheel speed sensor
- Brake rotor
- Brake caliper. Torque the bolts to 48 ft. lbs. (65 Nm)
- Wheel assembly
- Halfshaft and torque the bolt to 148 ft. lbs. (200 Nm) plus a ¼ (90 degree) turn

6. Check and/or adjust the wheel alignment, if necessary turn.

BMW

1998–00

M3 • Z3 • Z8 • 3 • 5 • 7 • 8 Series

7

PRECAUTIONS

Before servicing any vehicle, please be sure to read all of the following precautions, which deal with personal safety, prevention of component damage and important points to take into consideration when servicing a motor vehicle:

• Never open, service or drain the radiator or cooling system when the engine is hot; serious burns can occur from the steam and hot coolant.

• Observe all applicable safety precautions when working around fuel. Whenever servicing the fuel system, always work in a well-ventilated area. Do not allow fuel spray or vapors to come in contact with a spark, open flame, or excessive heat (a hot drop light, for example). Keep a dry chemical fire extinguisher near the work area. Always keep fuel in a container specifically designed for fuel storage; also, always properly seal fuel containers to avoid the possibility of fire or explosion. Refer to the additional fuel system precautions later in this section.

• Fuel injection systems often remain pressurized, even after the engine has been turned **OFF**. The fuel system pressure must be relieved before disconnecting any fuel lines. Failure to do so may result in fire and/or personal injury.

• Brake fluid often contains polyglycol ethers and polyglycols. Avoid contact with the eyes and wash your hands thoroughly after handling brake fluid. If you do get brake fluid in your eyes, flush your eyes with clean, running water for 15 minutes. If eye irritation persists, or if you have taken

brake fluid internally, seek medical assistance IMMEDIATELY.

• The EPA warns that prolonged contact with used engine oil may cause a number of skin disorders, including cancer. You should make every effort to minimize your exposure to used engine oil. Protective gloves should be worn when changing oil. Wash your hands and any other exposed skin areas as soon as possible after exposure to used engine oil. Soap and water, or waterless hand cleaner should be used.

• All new vehicles are now equipped with an air bag system. The system must be disabled before performing service on or around system components, steering column, instrument panel components, wiring and sensors. Failure to follow safety and disabling procedures could result in accidental air bag deployment, possible personal injury and unnecessary system repairs.

• Always wear safety goggles when working with, or around, the air bag system. When carrying a non-deployed air bag, be sure the bag and trim cover are pointed away from your body. When placing a non-deployed air bag on a work surface, always face the bag and trim cover upward, away from the surface. This will reduce the motion of the module if it is accidentally deployed. Refer to the additional air bag system precautions later in this section.

• Clean, high quality brake fluid from a sealed container is essential to the safe and proper operation of the brake system. You should always buy the correct type of brake

fluid for your vehicle. If the brake fluid becomes contaminated, completely flush the system with new fluid. Never reuse any brake fluid. Any brake fluid that is removed from the system should be discarded. Also, do not allow any brake fluid to come in contact with a painted surface; it will damage the paint.

• Never operate the engine without the proper amount and type of engine oil; doing so WILL result in severe engine damage.

• Timing belt maintenance is extremely important. Many models utilize an interference-type, non-freewheeling engine. If the timing belt breaks, the valves in the cylinder head may strike the pistons, causing potentially serious (also time-consuming and expensive) engine damage. Refer to the maintenance interval charts in the front of this manual for the recommended replacement interval for the timing belt and to the timing belt section for belt replacement and inspection.

• Disconnecting the negative battery cable on some vehicles may interfere with the functions of the on-board computer system(s) and may require the computer to undergo a relearning process once the negative battery cable is reconnected.

• When servicing drum brakes, only disassemble and assemble one side at a time, leaving the remaining side intact for reference.

• Only an MVAC-trained, EPA-certified automotive technician should service the air conditioning system or its components.

ENGINE REPAIR

➥Disconnecting the negative battery cable on some vehicles may interfere with the operation of the on-board computer system. The computer to undergo a relearning process once the negative battery cable is reconnected.

Alternator

REMOVAL & INSTALLATION

During removal and installation of the alternator, refer to the following information for proper tightening of the fasteners.
Alternator fastener torque specifications for all models:

• M6 fasteners: 61 inch lbs. (7 Nm)
• M8 fasteners: 115 inch lbs. (13 Nm)

• Alternator mounting fasteners: 31 ft. lbs. (43 Nm)
• Alternator pulley nut (V-belt type): 33 ft. lbs. (45 Nm)
• Alternator pulley nut (ribbed drive belt type): 51 ft. lbs. (70 Nm)
  Alternator brush minimum length:
• Distance from housing to end of brush: 0.20 inches (5 mm)

E30 Models

1. Before servicing the vehicle, refer to the precautions in the beginning of this section.

2. Set the ignition switch to the **OFF** position.

3. Disconnect the negative battery cable. Remove the air cleaner assembly.

4. Disconnect the wires from the rear of the alternator, marking them for installation. Note the presence of a ground wire on some vehicles. On the 325i and M3 models, it may be easier to remove the alternator mounting bolts first, then turn it, and remove the wires.

5. If equipped with an alternator cooling duct, loosen the hose clamp on the alternator and remove the cooling duct.

6. Loosen the lock bolt, turn the tensioning bolt to loosen the belt tension and remove the belt. Remove the mounting bolts and remove the alternator.

To install:

7. Install the alternator in position and install the retaining bolts.

8. The tensioning bolt on the front of the alternator must be turned so as to ten-

sion the belt, using a torque wrench, until the torque is approximately 60 inch lbs. (7 Nm). Then, hold the adjustment nut with one wrench while tightening the locknut at the rear of the unit. Make sure, if the unit has a ground wire on the alternator, it has been reconnected as removed.

9. Install the fan and cowl.

10. Install the air cleaner and connect the hoses as necessary.

11. Install the alternator cooling duct, then reconnect the wires to the alternator.

12. The balance of the installation is in reverse order of removal.

E36 Four Cylinder Engines

1. Before servicing the vehicle, refer to the precautions in the beginning of this section.

2. Set the ignition switch to the **OFF** position.

3. Disconnect the negative battery cable.

4. Rotate the electrical connector at the mass air flow sensor counter clockwise and carefully pull off connector.

5. Remove the intake air filter upper housing complete with the air-flow sensor.

6. On M42 engines:

   a. Loosen the alternator belt adjustment fasteners and remove the drive belt.

   b. Disconnect the alternator-to-body ground strap at the body.

7. On M44 engines:

   a. Remove the alternator drive belt.

   b. Remove the upper belt roller to access the upper alternator mounting bolt.

8. Remove the wire cover cap and disconnect the wires at the alternator.

9. Remove the alternator mounting bolts and remove the alternator assembly.

10. Installation is in reverse order of removal making sure to re-code the radio as necessary.

E36 Six Cylinder Engines

1. Before servicing the vehicle, refer to the precautions in the beginning of this section.

2. Set the ignition switch to the **OFF** position.

3. Disconnect the negative battery cable.

4. Remove the intake air filter housing as follows:

   a. Rotate the electrical connector at the mass air flow sensor counter clockwise and carefully pull off connector.

   b. Loosen the hose clamp at the intake manifold side of the 90° air intake bellows and remove the hose from the throttle plate.

   c. Loosen the air filter mounting fasteners at their support brackets.

   d. Remove the air filter housing complete with the mass air flow sensor and the 90° air intake bellows. If necessary, release the clips between the mass air flow sensor and the air filter housing and remove separately.

➡**The fan clutch assembly has left hand threads. Loosen by turning clockwise.**

5. Remove the radiator cooling fan and fan clutch assembly by supporting the water pump pulley by using Tool No. 11 5 030 or its equivalent and loosen the fan clutch by turning it clockwise with Tool No. 11 5 040 or its equivalent.

6. If equipped with air conditioning, and additional space is desired, remove the air conditioner drive belt.

7. Remove the alternator belt.

8. If equipped, remove the vent hose from the rear of the alternator.

9. Remove the alternator wire terminal protective cap from the back of the alternator.

10. Label and remove the wires from the alternator.

11. If equipped, remove the belt guide roller from the top of the alternator.

12. While supporting the alternator, remove the alternator mounting fasteners, then remove the alternator.

13. Installation is in reverse order of

A top side view of the alternator installed on M42 engines

84273001

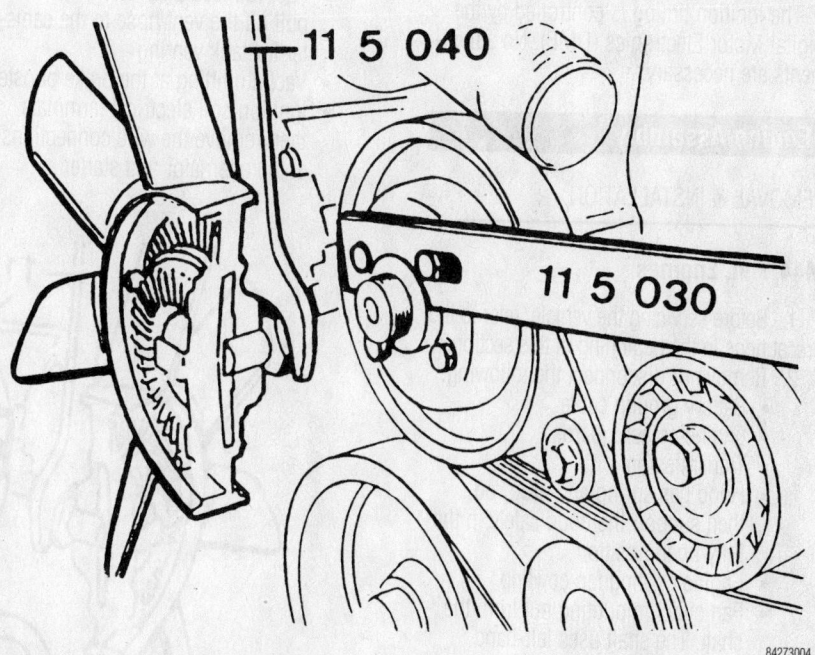

11 5 040

11 5 030

84273004

The engine cooling fan is removed by holding the drive pulley and rotating the fan/fan clutch assembly clockwise

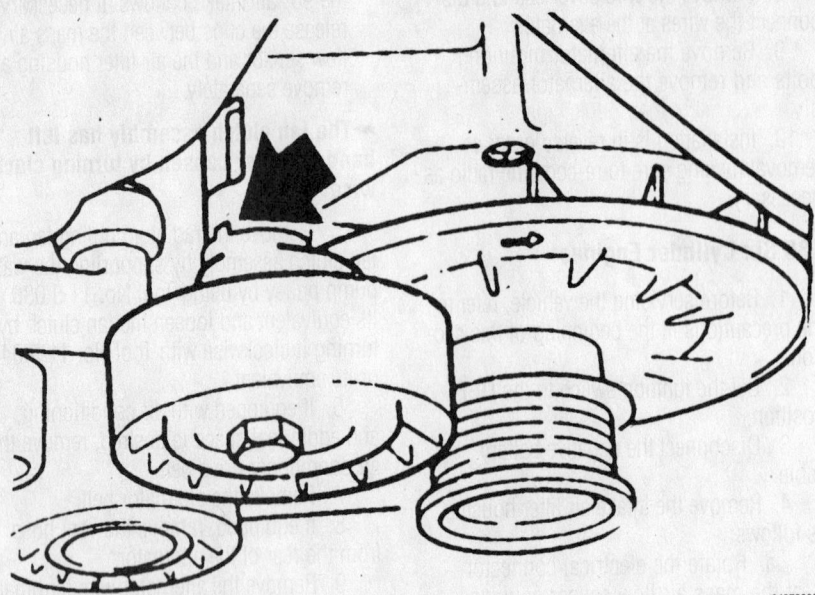

When installing the belt idler tensioner roller, make sure the alignment tab is seated into the slot—M50 engine shown

84273003

removal making sure to re-code the radio as necessary.

Ignition Timing

ADJUSTMENT

The ignition timing is controlled by the Digital Motor Electronics (DME). No adjustments are necessary.

Engine Assembly

REMOVAL & INSTALLATION

M44 1.9L Engines

1. Before servicing the vehicle, refer to the precautions in the beginning of this section.
2. Remove or disconnect the following:
   - Battery ground cable
   - Engine splash guards
   - Transmission
   - Hood gas spring and prop rod, then support the hood safely in the fully open position
   - Engine cooling fan cowling
   - Fan clutch mounting nut from the shaft. The shaft uses left-hand threads; thus the nut is turned clockwise to remove.
   - Coolant
   - Lower radiator expansion tank hose, the engine coolant hoses and the heater hoses from the firewall

- Air flow meter electrical terminal, and then lift the air sensor with the air cleaner housing up and out of the engine compartment
- Unclip the throttle cable and release the cable with the rubber holder
- Relieve the residual fuel line pressure
- Fuel rail feed and return lines, then pull off the vent hose to the canister for tank venting
- Vacuum fitting at the brake booster
- Ignition coil electrical terminals, then remove the wire connections at the alternator and starter

- Wire connections from the throttle valve potentiometer, then remove the tank venting valve plug located next to the air cleaner
- Fuel injector connector located near to the fuel pipes, then disconnect the idle speed control connector at the rear of the intake manifold and the oil pressure switch electrical connection
- Front and rear intake manifold supports
- Coolant temperature senders for the gauge and the Digital Motor Electronics (DME)
- Remaining electrical wiring harnesses

3. Support the weight of the engine and remove the engine mounting bolts and the engine ground strap.

4. Lift the engine and remove the engine from the vehicle.

To install:

5. Installation is the reverse of the removal procedure, while using the following torque values:
- 6mm fasteners: 61 inch lbs. (7 Nm)
- 7mm fasteners: 10 ft. lbs. (13 Nm)
- 8mm fasteners: 16 ft. lbs. (22 Nm)
- 10mm fasteners: 31 ft. lbs. (42 Nm)

M52 2.5L, M52 2.8L, and S52 3.2L Engines

1998–02 318, Z3, 328, M3, AND 1998–02 323 MODELS

1. Before servicing the vehicle, refer to the precautions in the beginning of this section.

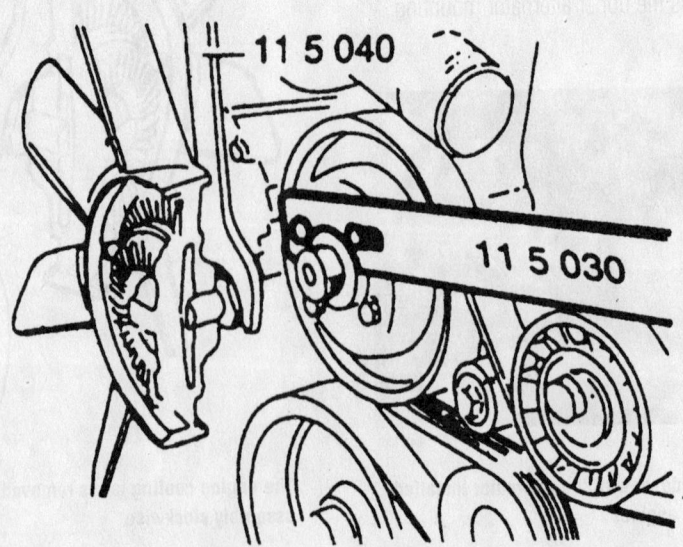

Cooling fan removal tools

7923DG01

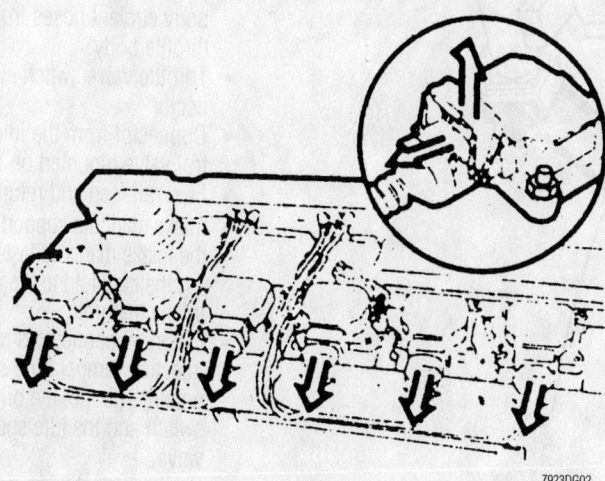

Ignition leads on coil—3 Series with M52/S52 engines

7923DG02

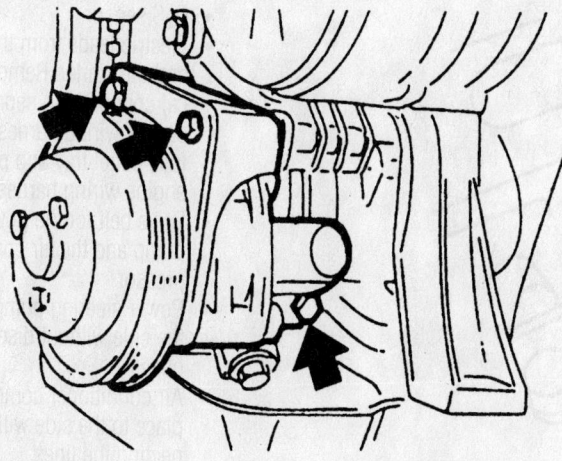

Power steering pump mounting bolts—3 Series with M52/S52 engines

7923DG03

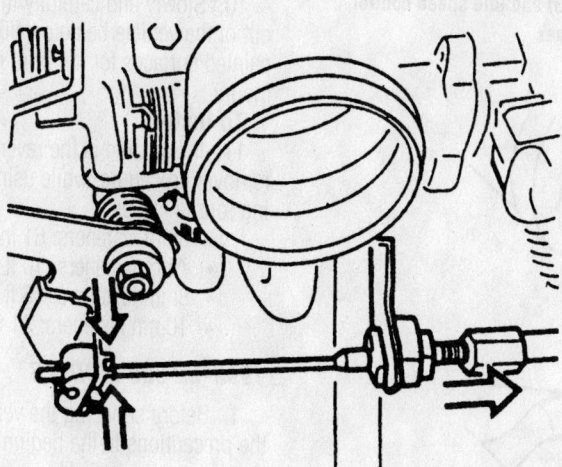

Throttle cable—3 Series with M52/S52 engines

7923DG04

2. Drain the cooling system.

3. Relieve the fuel system pressure.

4. Remove or disconnect the following:
   • Negative battery terminal
   • Engine splash guard
   • Transmission assembly

5. Press the hood hinge so it goes over center and support hood safely in the fully open position.

6. Remove or disconnect the following:
   • Air mass sensor connector and loosen the air intake duct hose clamp, then remove the air cleaner housing assembly
   • Hoses for the idle speed control valve and the crankcase breather
   • Air ducting for the alternator. Pull out the fan cowl expansion rivets and remove the cowl upwards.
   • Fan and fan clutch assembly, keeping it upright
   • Upper and lower radiator hoses
   • Coolant level switch and plug the cooler lines for the automatic transmission
   • Right side coolant hose and the temperature sensor
   • Radiator
   • Heater coolant hoses from the heater and heater valve
   • Grill from the air intake cowl at the base of the windshield

7. To remove the cowl:
   a. Remove the electrical lead tray.
   b. Remove the fasteners on the right side cowl holder bracket and the screw on the left side.
   c. Remove the cowl from the engine compartment.

8. Remove or disconnect the following:
   • Throttle cable cover. Unclip the throttle cable and pull the cable out along with the rubber holder.
   • Vacuum fitting from the brake booster and plug the openings
   • Engine and intake manifold covers. Remove the bolt holding the ground strap on the front engine lifting bracket. Reinstall the bolt before lifting the engine.
   • Bolts holding the ignition coil wire harness plug plate and remove the plug plate
   • Ignition coil electrical plugs. Remove the plug plate complete with the electrical leads.
   • Cylinder head vent hose and the intake air temperature sensor connector

Front and rear intake manifold supports—3 Series with M52/S52 engines

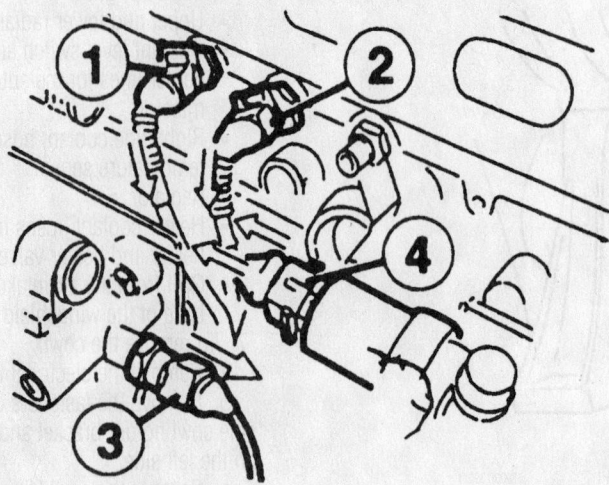

Temperature sensor (1), temperature gauge (2), oil pressure switch (3) and idle speed control valve (4) electrical connector locations—3 Series with M52/S52 engines

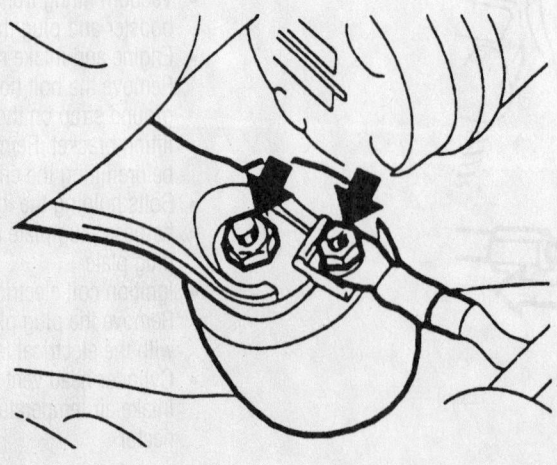

Right engine mount, showing the ground cable attachment—3 Series with M52/S52 engines

- Fuel tank vent hose and the throttle body coolant hoses from the throttle body
- Throttle valve switch electrical connector
- Connector from the idle speed control valve mounted on the manifold
- Fuel rail feed and return hoses
- Intake manifold support brackets and the intake manifold fasteners mounting the manifold to the intake ports
- Intake manifold
- Electrical connectors from the fuel injection temperature sensor, temperature gauge, the oil pressure switch and the idle speed control valve
- Cylinder identifying sender plug (black) and the pulse sender plug (gray) for the Digital Motor Electronics (DME)
- O_2S sensor
- Electric leads from the alternator and the starter. Remove the electrical connector to separate the engine wiring harness from the fuse panel tray and place the engine wiring harness to the side.
- Drive belt for the power steering pump and the air conditioner compressor
- Power steering pump and place to the side without disconnecting the hoses
- Air conditioner compressor and place to the side without disconnecting the lines

9. Carefully support the engine and remove the engine mounts and ground strap.

10. Slowly and carefully lift the engine out of the vehicle being careful of the painted surfaces for the front radiator mount.

To install:

11. Installation is the reverse of the removal procedure, while using the following torque values:

- 6mm fasteners: 61 inch lbs. (7 Nm)
- 7mm fasteners: 10 ft. lbs. (13 Nm)
- 8mm fasteners: 16 ft. lbs. (22 Nm)
- 10mm fasteners: 31 ft. lbs. (42 Nm)

1998–02 528 MODELS

1. Before servicing the vehicle, refer to the precautions in the beginning of this section.

2. Drain the cooling system.

3. Relieve the fuel system pressure.

4. Remove or disconnect the following:

- Negative battery cable, battery, and battery tray

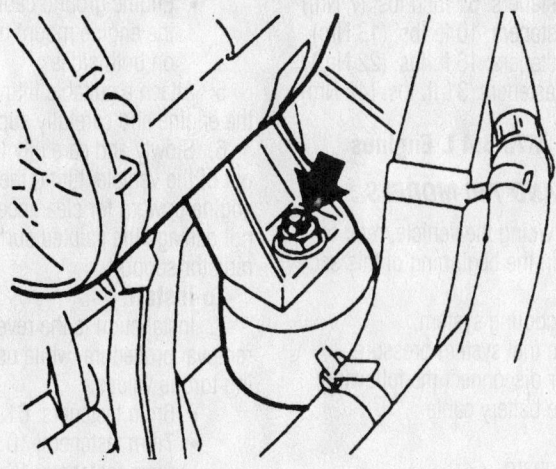

7923DG08

Left engine mount—3 Series with M52/S52 engines

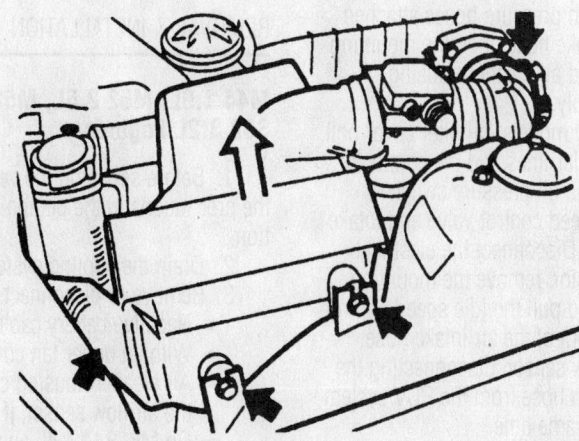

7923DG09

Air cleaner and air mass sensor—3 Series with M52/S52 engines

- Transmission assembly
- Alternator air cooling duct
- Plug to the intake air flow meter and air cleaner housing assembly
- Fan cowl
- Fan pulley
- Coolant hoses from the radiator and coolant level switch plug
- With automatic transmission, oil lines to the radiator and plug the lines
- Lower radiator hose and trim panel from the right side of the engine compartment
- Electrical connector for the air conditioner temperature switch
- Radiator
- Heater hoses from the heater valve and the heater core

- Throttle cable cover and throttle cable cover
- Vacuum fitting from the brake booster and plug the openings
- Engine and intake manifold covers. Remove the bolt holding the ground strap on the front engine lifting bracket. Reinstall the bolt before lifting the engine.
- Bolts holding the plug plate for the ignition coil wiring harness and remove the plug plate
- Ignition coil electrical connectors. Remove the plug plate complete with the electrical terminals.
- Cylinder head vent hose and air temperature sensor plug
- Tank venting hose and the throttle body coolant hoses from the throttle body

- Throttle valve switch plug. Unclip the idle speed control valve mounted on the manifold.
- Fuel feed and return hoses from the fuel rail
- Intake manifold
- Electrical connectors from the temperature sensor, temperature gauge, the oil pressure switch and the idle speed control valve. Disconnect the cylinder identifying sender connector (black) and the pulse sender connector (gray) for the Digital Motor Electronics (DME), then disconnect the Oxygen (O_2S) sensor connector from the holder.
- Electric wire harness from the alternator and the starter. Remove the connector for the main engine wiring harness and place the engine wiring harness to the side.
- Drive belt for the power steering pump and the air conditioner compressor
- Power steering pump and place to the side without disconnecting the hoses
- Air conditioner compressor and place to the side without disconnecting the high pressure lines
- Engine mounts and ground strap. Slowly lift the engine out of the vehicle using care to not damage the painted surfaces of the front radiator mount.

To install:

5. Installation is the reverse of the removal procedure, while using the following torque values:
- 6mm fasteners: 61 inch lbs. (7 Nm)
- 7mm fasteners: 10 ft. lbs. (13 Nm)
- 8mm fasteners: 16 ft. lbs. (22 Nm)
- 10mm fasteners: 31 ft. lbs. (42 Nm)

M62 4.4L Engines

1998–02 540 MODELS

1. Before servicing the vehicle, refer to the precautions in the beginning of this section.
2. Drain the cooling system.
3. Relieve the fuel system pressure.
4. Remove or disconnect the following:
- Negative battery cable
- Engine splash shield
- Radiator
- If equipped, nut holding the transmission oil cooler lines to the engine oil pan

Timing belt service is covered in Section 3 of this manual

- Transmission assembly
- Drive belts for the power steering pump and the air conditioner compressor, then remove the bolts holding the pump and compressor to the engine and remove them from the engine, keeping the high pressure hoses connected. Secure the pump and compressor out of the way and without any tension or binding on the hoses.
- Hoses from the coolant reservoir expansion tank, then remove the fasteners on the side of the expansion tank and remove the expansion tank from the engine compartment
- Heater hoses from the heater control valve and the heater core inlet pipe
- Electrical connections to the ignition coil, then the air cleaner housing assembly
- Idle speed control valve from the intake duct
- Wire harness connectors for the air flow meter, then the ducting to the air flow meter and crankcase breather vacuum line
- With a cable operated throttle, cruise control cable and the throttle cable at the throttle, then remove the cable mounting bracket
- With an electronically controlled throttle, electrical connector to the throttle control unit
- Wiring harness from the starter, then the electrical connectors located near the starter motor assembly
- Oil level sender connector and the wiring harness from the alternator
- Air duct to the alternator
- Fuel tank venting valve and the hose to the canister
- Fuels lines for the fuel rail
- Vacuum line from the brake booster and plug the opening
- Wiring harness connections to the temperature sensors and the Digital Motor Electronics (DME)
- Engine ground cable and verify that all remaining fluid lines or electrical leads have been disconnected and properly placed aside
- Engine mount nuts and bolts and carefully lift the engine from the vehicle

To install:

5. Installation is the reverse of the removal procedure, while using the following torque values:

- 6mm fasteners: 61 inch lbs. (7 Nm)
- 7mm fasteners: 10 ft. lbs. (13 Nm)
- 8mm fasteners: 16 ft. lbs. (22 Nm)
- 10mm fasteners: 31 ft. lbs. (42 Nm)

M62 4.4L and M73 5.4 L Engines

1998–02 740 AND 750 MODELS

1. Before servicing the vehicle, refer to the precautions in the beginning of this section.
2. Drain the cooling system.
3. Relieve the fuel system pressure.
4. Remove or disconnect the following:

- Negative battery cable
- Hood
- Splash guard
- Transmission assembly
- Power steering pump with the hoses attached
- Air conditioner compressor, leaving the high pressure hoses attached
- Air intake hose, then the mounting nut, and air cleaner housing assembly
- On 750 models, oil filter cover bolt, oil cooler lines and the connector from the oil pressure switch
- Idle speed control valve and intake hoses. Disconnect the electrical connector, remove the mounting nut, and pull the idle speed control valve out of the air intake hose.
- Air flow sensor, disconnecting the vacuum hose from the PCV system at the same time
- Coolant reservoir expansion tank
- Engine coolant hoses at both the control valve and at the heater core
- Throttle and cruise control cables at the throttle lever, then remove the cable housing retainer and remove the housing and cables
- Light gauge wires from the starter motor, and with the positive battery cable disconnected from the battery, remove the starter motor with the battery cable attached
- Fresh air hose from the alternator
- Electrical connection for the O$_2$S sensor, and disconnect the surrounding electrical connections and place the wire harness safely aside
- Fuel rail supply and return lines
- Fuel pipe at the injector supply manifold, then disconnect the harness electrical connectors. Disconnect the electrical connector at the throttle body, lift off the protective caps, remove the attaching nuts for the protective cover for the fuel injector wiring harness and remove it.

- Engine ground cable, then remove the engine mount nut from the top on both sides

5. Attach a suitable lifting attachment to the engine and carefully support the engine.

6. Slowly and carefully lift the engine out of the vehicle, tilting the front of the engine upward for clearance. Use care to not damage the painted surfaces of the front radiator support.

To install:

7. Installation is the reverse of the removal procedure, while using the following torque values:

- 6mm fasteners: 61 inch lbs. (7 Nm)
- 7mm fasteners: 10 ft. lbs. (13 Nm)
- 8mm fasteners: 16 ft. lbs. (22 Nm)
- 10mm fasteners: 31 ft. lbs. (42 Nm)

Water Pump

REMOVAL & INSTALLATION

M44 1.9L, M52 2.5L, M52 2.8L, and S52 3.2L Engines

1. Before servicing the vehicle, refer to the precautions in the beginning of this section.
2. Drain the cooling system.
3. Remove or disconnect the following:

- Negative battery cable
- With an upper fan cowl, fan cowl
- Air cleaner housing complete with the air flow sensor, if necessary
- On M62B44 4.4L and M73B54 5.4L engines, coolant hoses from the upper water pump housing
- Fan clutch mounting nut from the water pump shaft. The shaft uses left-hand threads; thus the nut is turned clockwise to remove.
- Accessory drive belt
- Fasteners securing the pulley to the water pump
- Water pump mounting bolts
- Water pump assembly

➡The water pump can be separated from the engine by using 2 M6 bolts threaded into the tapped bores of the water pump housing. To press the water pump away from the engine thread the bolts evenly.

To install:

4. Clean and remove any residual debris or gasket material from the engine mounting surface for the water pump.

5. Lubricate and install a new O-ring.

6. Install the water pump and tighten the bolts as follows:

- M6 bolts: 78 inch lbs. (9 Nm)
- M8 bolts:16 ft. lbs. (22 Nm)

7. Install or disconnect the following:
- Water pump pulley onto the water pump
- Cooling fan and fan clutch assembly onto the threaded water pump spindle

➡**The water pump spindle threads for the cooling fan and fan clutch assembly are left hand threads hence the fan clutch assembly must be rotated counterclockwise to install. As there is a minimum amount of clearance between the water pump and the radiator assembly, rotating the fan clutch assembly to install it onto the water pump spindle may be difficult. To assist with the installation of the fan clutch assembly, use a piece of thin, sturdy string about 3 feet (1 meter) long. Wind the string around the fan clutch mounting nut in a clockwise direction 15-20 revolutions. Position the fan clutch assembly on the water pump spindle squarely and while holding the fan, slowly pull the string to spin the fan clutch mounting nut in a counterclockwise direction. Once the nut begins to thread onto the water pump spindle, remove the remainder of the string.**

- Fan clutch assembly using tool No. 11-5-040. Nut: 29 ft. lbs. (40 Nm). If using the fan tool, torque the nut to 22 ft. lbs. (30 Nm); as the additional length of the tool multiplies the torque to achieve 29 ft. lbs. (40 Nm) at the nut.

8. The remaining components are installed in the reverse order from which they were removed.
9. Fill the cooling system.
10. Start the engine and check for leaks.

➡**It may be necessary to bleed the cooling system a second time once the engine has been started.**

M62 4.4L Engines

1. Before servicing the vehicle, refer to the precautions in the beginning of this section.
2. Drain the cooling system.
3. Remove or disconnect the following:
- Negative battery cable
- With an upper fan cowl, fan cowl
- Air cleaner housing complete with the air flow sensor, if needed

- Accessory drive belts
- 8 bolts that secure the crankshaft pulley vibration damper. Do not remove the center bolt.
- Coolant hoses from the upper water pump housing
- Fan clutch mounting nut from the water pump shaft. The shaft uses left-hand threads; thus the nut is turned clockwise to remove.
- Fasteners securing the pulley to the water pump
- Thermostat housing from the water pump assembly
- Water pump mounting bolts
- Water pump assembly

To install:

4. Clean and remove any residual debris or gasket material from the engine mounting surface for the water pump.
5. Install or connect the following:
- Water pump with a new gasket. Tighten the M6 bolts to 78 inch lbs. (9 Nm), and the M8 bolts to 16 ft. lbs. (22 Nm).
- Thermostat and thermostat housing onto the water pump assembly using a new housing gasket sealing ring
- Crankshaft pulley vibration damper using new fasteners. Be sure to align the damper with the dowel.
- Water pump pulley onto the water pump
- Accessory drive belts
- Cooling fan and fan clutch assembly onto the threaded water pump spindle

➡**The water pump spindle threads for the cooling fan and fan clutch assembly are left hand threads hence the fan clutch assembly must be rotated counterclockwise to install. As there is a minimum amount of clearance between the water pump and the radiator assembly, rotating the fan clutch assembly to install it onto the water pump spindle may be difficult. To assist with the installation of the fan clutch assembly, use a piece of thin, sturdy string about 3 feet (1 meter) long. Wind the string around the fan clutch mounting nut in a clockwise direction 15-20 revolutions. Position the fan clutch assembly on the water pump spindle squarely and while holding the fan, slowly pull the string to spin the fan clutch mounting nut in a counterclockwise direction. Once the**

nut begins to thread onto the water pump spindle, remove the remainder of the string.

- Fan clutch assembly using tool No. 11-5-040. Nut: 29 ft. lbs. (40 Nm). If using the fan tool, torque the nut to 22 ft. lbs. (30 Nm); as the additional length of the tool multiplies the torque to achieve 29 ft. lbs. (40 Nm) at the nut.

6. The remaining components are installed in the reverse order from which they were removed.
7. Fill the cooling system.
8. Start the engine and check for leaks.

➡**It may be necessary to bleed the cooling system a second time once the engine has been started.**

M73 5.4 L Engines

1. Before servicing the vehicle, refer to the precautions in the beginning of this section.
2. Remove or disconnect the following:
- Negative battery cable
- Air filter housing electrical connector
- Air inlet hose
- Mass Air Flow (MAF) sensor
- Cooling fan and fan clutch
- Oil filler cap drip tray
- Non-return valve pressure hoses
- Coolent vent hose bracket
- Cylinder 1-6 non-return valve
- Pressure pipe
- Left side heat baffle from the front axle support
- Coolant from the radiator and the engine block drain
- Cooling system reservoir expansion tank
- Fan shroud assembly
- Air conditioning drive belt
- Coolant hoses from the upper water pump housing
- Engine coolant expansion tank hose bracket
- Electrical connector for the coolant temperature sensor
- Fasteners securing the pulley to the water pump
- Pump mounting bolts
- Water pump assembly

➡**The water pump can be separated from the engine by using 3 M6 bolts threaded into the tapped bores of the water pump housing. To press the water pump away from the engine thread the bolts evenly.**

Heater Core replacement is covered in Section 2 of this manual

3. If replacing the water pump, secure the water pump in a vice. Take care not to damage the housing or gasket mating surfaces. Then remove the coolant temperature sensor, the thermostat housing cover and the thermostat assembly and transfer these parts to the replacement pump using a new thermostat housing gasket sealing ring. It is also suggested to renew the thermostat at this time.

To install:

4. Clean and remove any residual debris or gasket material from the engine mounting surface for the water pump.

5. Install or connect the following:

- Lubricate and install new O-rings
- If replacing the water pump, coolant temperature sensor, thermostat, and thermostat housing using a new gasket
- Water pump. Tighten the M6 bolts to 78 inch lbs. (9 Nm) and the M8 bolts to 16 ft. lbs. (22 Nm).
- Coolant temperature sensor electrical connector
- Water pump pulley onto the water pump
- Cooling fan and fan clutch assembly onto the threaded water pump spindle

➡ **The water pump spindle threads for the cooling fan and fan clutch assembly are left hand threads hence the fan clutch assembly must be rotated counterclockwise to install. As there is a minimum amount of clearance between the water pump and the radiator assembly, rotating the fan clutch assembly to install it onto the water pump spindle may be difficult. To assist with the installation of the fan clutch assembly, use a piece of thin, sturdy string about 3 feet (1 meter) long. Wind the string around the fan clutch mounting nut in a clockwise direction 15-20 revolutions. Position the fan clutch assembly on the water pump spindle squarely and while holding the fan, slowly pull the string to spin the fan clutch mounting nut in a counterclockwise direction. Once the nut begins to thread onto the water pump spindle, remove the remainder of the string.**

- Fan clutch assembly using tool No. 11-5-040. Nut: 29 ft. lbs. (40 Nm). If using the fan tool, torque the nut to 22 ft. lbs. (30 Nm); as the additional length of the tool multiplies the torque to achieve 29 ft. lbs. (40 Nm) at the nut.

6. The remaining components in the reverse order from which they were removed.

7. Fill the cooling system.

8. Start the engine and check for leaks.

➡ **It may be necessary to bleed the cooling system a second time once the engine has been started.**

Cylinder Head

REMOVAL & INSTALLATION

M44 1.9L Engines

1. Before servicing the vehicle, refer to the precautions in the beginning of this section.

2. Drain the cooling system.

3. Relieve the fuel system pressure.

4. Remove or disconnect the following:

- Negative battery cable
- Fuel feed and return lines
- Intake manifold
- Ignition coil cover and remove the spark plug connectors
- All the fasteners securing the brackets for the ignition coil wire harness, then move the harness aside to access the cylinder head cover
- Cylinder head cover
- Coolant hoses and temperature sensor
- Thermostat housing and thermostat
- Upper timing case cover

5. Rotate the engine in the direction of the rotation until the camshaft peaks of the intake and exhaust camshafts for cylinder No. 1 face each other. The arrows on the cam sprocket should face up.

6. Remove or disconnect the following:

- Chain tensioner, the upper chain guide, the chain guide bolt on the right side and the camshaft sprockets
- Cylinder head bolts in the reverse order of the tightening sequence (from the outside to the inside) in at least 3 steps
- Cylinder head

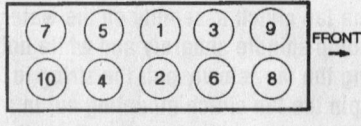

Cylinder head torque sequence—M44B19 1.9L engine

To install:

7. Install the cylinder head with a new gasket and bolts. Tighten the bolts in sequence as follows:

- Step 1: 22 ft. lbs. (30 Nm)
- Step 2: +90 degrees
- Step 3: +90 degrees

8. The balance of installation is the reverse of the removal procedure.

9. Fill the cooling system.

10. Start the engine and check for leaks.

➡ **It may be necessary to bleed the cooling system a second time after the engine has been started.**

M52 2.5L, 2.8L, and S52 3.2L Engines

1. Before servicing the vehicle, refer to the precautions in the beginning of this section.

2. Drain the cooling system.

3. Relieve the fuel system pressure.

4. Remove or disconnect the following:

- Negative battery cable
- Lower engine splash shield
- Exhaust manifolds from the exhaust pipes
- Rear hood seal, then remove the fasteners securing the sealed plastic housing for the engine wiring harness from the passenger compartment fresh air intake shroud
- Fresh air intake shroud
- If equipped, plastic trim covers from the top of the engine
- MAF sensor and remove intake manifold air duct along with the air filter housing assembly
- Throttle cable from the throttle body linkage, coolant hoses, the TP sensor electrical connector, the throttle body mounting fasteners, and remove the throttle body assembly
- Intake manifold support bracket fasteners from the intake manifold and loosen the support bracket fasteners on the engine block
- Fuel feed and return lines
- Vacuum hoses from the intake manifold, and loosen the clamps for the idle speed stabilizer underneath the manifold, and disconnect the hoses
- Intake manifold intake port fasteners and carefully lift the manifold away from the engine
- Ground cable at the front of the cylinder head, then remove the 2 radiator hoses from the thermostat housing, and remove the housing
- Exhaust manifolds

- Wire connectors for the ignition coils and cylinder head sensors and remove the coils, then remove the cylinder head cover

5. Rotate the engine in the direction of rotation to TDC for cylinder number one. Cylinder number 1 will be at TDC when the intake and exhaust camshaft peaks for cylinder number 1 face each other

6. Lock the engine in the TDC position by placing the holding dowel tool No. 11-2-300 through the machined hole in the engine block, just inside of the transmission bell housing mounting tab located on the left lower portion of the engine block. Slide the locating dowel through the machined hole in the block and into the machined hole in the flywheel to prevent movement of the crankshaft.

7. The camshafts are held in the TDC position by placing tool No. 11-3-240 on the valve cover mating surface at the back of the cylinder head and onto the squared ends of the camshafts. Secure the camshafts so that 2 sides of the squared ends are parallel with the cam cover gasket mating surface. With the camshafts in this position, the arrows on the sprockets will be facing up.

8. Remove or disconnect the following:
- Valve cover mounting studs
- The 2 hex plugs at the front of the cylinder head to access the exhaust camshaft sprocket mounting bolts, then loosen the exhaust cam sprocket bolts 2 turns

9. Press down on the secondary cam chain tensioner between the 2 camshaft sprockets and install tool No. 11-3-292 through the back side of the tensioner housing to hold the tensioner down. A similar sized and suitably hardened drill bit can be substituted for tool No. 11-3-292.

10. Remove the fasteners from the front of the cylinder head securing the hydraulic variable camshaft control (VANOS) unit to the cylinder head, and inspect to make sure any hydraulic or sensor connectors have been removed or disconnected.

11. On engines with a spring plate installed on the intake camshaft, place tool No. 11-5-490, onto the exhaust camshaft sprocket. Carefully rotate the sprocket clockwise to allow the helical gear of the VANOS unit to release the intake camshaft and to allow the VANOS unit to be pulled away from the front of the cylinder head.

12. If tool No. 11-5-490 is not available, move the camshaft sprockets to release the VANOS by using a suitable drift and soft faced mallet and lightly tapping on a sprocket tooth of the intake cam sprocket to rotate both cam sprockets clockwise, while alternately pulling on the VANOS unit to release it. This procedure may need to be repeated several times to fully release the VANOS unit, and must be performed very carefully, in such a manner to not distort or damage the teeth of the cam sprocket.

13. With the VANOS assembly removed, remove the intake and exhaust camshaft sprockets, the hydraulic cam chain tensioner, and the cam chain guide.

14. From the side of the right front area of the engine, remove the cap nut for the cam chain tensioner for the cam chain that runs between the crankshaft and the exhaust camshaft sprocket. Use care when removing the tensioner cap nut as the cam tensioner spring applies pressure to the cap nut.

15. Remove the exhaust cam tensioner, and then release the sprocket from the cam chain.

16. Attach a wire tie or mechanic's wire to the cam chain and temporarily secure the chain fully extended.

17. Remove the dowel from the engine and flywheel locking the engine in the TDC position.

18. While holding the crankshaft to exhaust camshaft cam chain, rotate the engine 30 degrees counterclockwise to avoid damaging the valves during cylinder head removal and reinstallation.

19. With the attached wire tie or mechanic's wire secured to the cam chain, and carefully lower the chain downward making sure there is enough exposed wire to retrieve the chain for reinstallation.

20. Remove the fasteners located in a recessed area near the front of the camshafts securing the front of the cylinder head to the engine block.

21. In the reverse order of the tightening sequence, using a proper sized Torx® bit or tool No. 11-2-250, loosen the cylinder head mounting bolts in 3 steps as follows:
- Step 1: 90 degrees
- Step 2: 90 degrees
- Step 3: Completely remove the bolt

22. With all of the cylinder head bolts removed, verify that all electrical connectors and fluid lines have been removed and lift the cylinder head off the engine block.

➡ It is not recommended to mill the cylinder head surface of the S52 engine. If milling the cylinder head of an M52 engine, a thicker head gasket is available for reassembly.

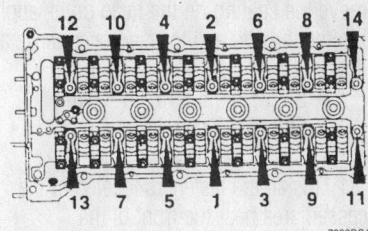

Cylinder head torque sequence—M52 2.5L, 2.8L, S52 3.2L engines

To install:

23. Thoroughly clean the deck surface of the engine cylinder block.

24. If the camshafts have been removed and reinstalled a waiting period dependent on the ambient temperature is necessary before mounting the cylinder head on the engine. At room temperature wait 4 minutes to allow the lifters to compress fully. At temperatures down to 50°F (10°C), wait 11 minutes. At temperatures lower than 50°F (10°C) wait 30 minutes. This is to prevent contact between the valves and the piston tops. The engine may not be cranked under the same condition for a period of 10 minutes at room temperature; 30 minutes for temperatures down to 50°F (10°C); 75 minutes for temperatures below 50°F (10°C).

25. Make sure the cylinder head has been checked for warpage and coolant leakage.

26. Do not to drop any pieces of gasket or debris into the oil or coolant passages. Check the condition of the head locating dowel sleeves.

27. Place a new head gasket on the engine block over the locating dowels and gently place the head on the engine with the dowel sleeves properly aligned and check that the head sits flat on the engine.

➡ **The cylinder head bolts may not be reused. The bolts are a different length and have a different torque specification for aluminum engine blocks than those for the cast iron engine blocks.**

28. Install and torque the new cylinder head bolts in 3 steps following the tightening sequence.

29. On cast iron engine blocks, wash the bolts in a cleaning solvent, then apply oil to the threads of the bolts and torque in sequence as follows:
- Step 1: 22.l ft. lbs. (30 Nm)
- Step 2: 90 degrees
- Step 3: 90 degrees

30. On aluminum engine blocks do not

Brake service is covered in Section 4 of this manual

remove the coating on the head bolts, apply oil to the threads and torque in sequence as follows:

- Step 1: 29.5 ft. lbs. (40 Nm)
- Step 2: 90 degrees
- Step 3: 90 degrees

31. Install the fasteners located in the recessed area near the front of the camshafts securing the front of the cylinder head to the engine block and torque to 96 inch lbs. (10 Nm).

32. With the attached wire secured to the crankshaft cam chain, carefully raise the chain upward and apply a light tension to the chain, then while holding the chain, carefully rotate the engine clockwise to TDC on cylinder number one.

33. Lock the engine in the TDC position by placing the holding dowel tool No. 11-2-300 through the machined hole in the engine block, just inside of the transmission bell housing mounting tab located on the left lower portion of the engine block. Slide the locating dowel through the machined hole in the block and into the machined hole in the flywheel to prevent movement of the crankshaft.

34. Hold the camshafts in the TDC position by placing tool No. 11-3-240, on the valve cover mating surface at the back of the cylinder head and onto the squared ends of the camshafts, securing the camshafts so that 2 sides of the squared ends are parallel with the cam cover gasket mating surface. With the camshafts in this position, the arrows on the sprockets will be facing up.

35. Position the crankshaft cam chain over the exhaust cam chain sprocket and install the sprocket on the exhaust cam such that the slotted holes are centered with the fastener bores in the camshaft.

36. Install the hydraulic cam chain tensioner, and the cam chain guide, then the intake and exhaust camshaft sprockets along with the cam chain and remove the tool used to hold the hydraulic cam chain tensioner collapsed.

37. Install the crankshaft cam chain tensioner assembly, spring and cap nut.

38. Apply sealant to the upper corners of the cylinder head where the VANOS housing mounts and install a new VANOS housing gasket.

39. Press the helical gear of the VANOS assembly toward the housing and install the VANOS assembly. To do so on engines with a spring plate installed on the intake camshaft, place tool No. 11-5-490 onto the exhaust camshaft sprocket. Carefully rotate the sprocket counterclockwise to allow the helical gear of the VANOS unit to thread into the intake camshaft and to allow the VANOS

unit to be pulled into the front of the cylinder head.

40. If tool No. 11-5-490 is not available, move the camshaft sprockets to install the VANOS by using a suitable drift and soft faced mallet and lightly tapping on a sprocket tooth of the exhaust cam sprocket to rotate both cam sprockets counterclockwise, while alternately pressing on the VANOS unit to install it. This procedure may need to be repeated several times to fully install the VANOS unit, and must be performed very carefully, in such a manner to not distort or damage the teeth of the cam sprocket.

➡**Make sure when assembling the VANOS unit is able to rest on the front of the cylinder head without being forced or without binding. If the VANOS unit does not fully seat it may be necessary to reposition the camshaft sprockets such that the slots in the camshaft sprockets allow enough movement of the sprockets for the helical gear of the VANOS assembly to be fully seated during assembly.**

41. Tighten the VANOS unit fastener and then torque the camshaft sprocket bolts of both the intake and exhaust cams to 16 ft. lbs. (22 Nm).

42. Remove the crankshaft TDC holding dowel tool No. 11-3-240 from the engine, and remove the camshaft TDC positioning tool No. 11-3-240 from the valve cover mating surface at the back of the cylinder head.

43. Slowly and carefully rotate the engine clockwise 4 complete revolutions bringing cylinder number 1 to TDC. If the engine binds for any reason stop immediately to evaluate and rectify the cause of the binding.

44. With cylinder number 1 at TDC slide the camshaft TDC positioning tool No. 11-3-240 over the ends of the cams and onto the valve cover mating surface. If the tool slides over the cams and is flush with the mating surface, the camshafts are properly timed. If the tool does not slide easily over the ends of the cams, or if the tool is not flush with the valve cover mating surface, the camshaft timing must be repeated until the tool fits squarely.

45. The balance of installation is the reverse of the removal procedure.

46. Top off the engine cooling system and bleed as necessary.

➡**It may be necessary to bleed the cooling system a second time after the engine has been started.**

47. Change the engine oil and filter.

48. Connect the negative battery cable, start the engine and check for any leaks.

M62 4.4L Engines

1. Before servicing the vehicle, refer to the precautions in the beginning of this section.

2. Drain the cooling system.

3. Relieve the fuel system pressure.

4. Remove or disconnect the following:

- Negative battery cable
- Both exhaust manifolds from each side of the engine, then remove the heat shields from the front axle carrier
- Coolant expansion tank
- Upper timing case cover
- Oil pipes from the cylinder head
- Intake manifold
- Cylinder head cover
- Engine vent pipe together with the O-ring, disconnect the coolant hoses on the coolant collector, remove the coolant collector mounting bolts, and then remove the coolant collector
- Spark plugs
- Camshaft sprockets, and the timing chain tensioner
- Bolts retaining the guide rail on the cylinder head's left-hand side
- Cylinder head bolts in the reverse order of the tightening sequence
- Cylinder head

➡**The cylinder head bolts must be replaced.**

To install:

5. Thoroughly clean all mounting surfaces and check the head for warpage. Take care not to drop any pieces of gasket or debris into the oil or coolant passages. Check the condition of the head locating dowel sleeves and clean out the bolt threads with a tap.

6. Mount the cylinder head on the block and use new bolts. Do not remove the coating on the head bolts. Apply oil to the threads and torque following the tightening sequence in 3 steps as follows:

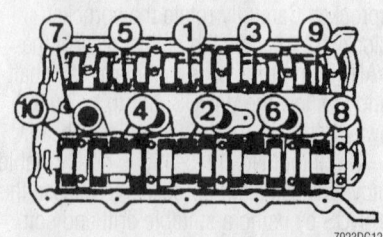

Cylinder head torque sequence—M62 4.4L engine

- Step 1: 22 ft. lbs. (30 Nm)
- Step 2: 80 degrees
- Step 3: 80 degrees

7. The balance of installation is the reverse of the removal procedure.

8. Fill the cooling system.

9. Start the engine and check for leaks.

➡ **It may be necessary to bleed the cooling system a second time after the engine has been started.**

M73 5.4L Engines

1. Before servicing the vehicle, refer to the precautions in the beginning of this section.

2. Drain the cooling system.

3. Relieve the fuel system pressure.

4. Remove or disconnect the following:
- Negative battery cable
- Exhaust pipe connections at the manifold and at the transmission pipe clamp
- Splash shield from under the engine
- Engine oil
- Fan shroud and engine cooling fan
- Fresh air inlet hose and air cleaner housing
- Idle speed control valve. To do so, loosen the hose clamps and remove the hoses. Detach the electrical connector. Remove the mounting nut, then pull the idle speed control out of the air intake hose.
- Retainers for the air flow sensor, then pull the unit off the mounts, and disconnect the vacuum hose from the PCV system
- Electrical connector from the coolant expansion tank. Remove the nuts on both sides. Loosen the clamps, then disconnect all hoses and remove the tank.
- Heater hoses at both the control valve and at the heater core and remove the valve
- Throttle and cruise control cables at the throttle lever. Unbolt the cable housing retainer and remove the housing and cables.
- Plugs near the thermostat housing. Loosen the hose clamps and remove the coolant hoses.
- Connector for the O2S sensor. Disconnect the surrounding electrical connectors.
- Fuel supply and return lines
- Fuel pipe running along the cylin-

der head, near the manifold, then remove the electrical connector at the throttle body
- Bracket covers, then remove the attaching bolts, wiring harness and carrier for the fuel injectors
- Ignition coil high tension lead, the high tension wires at the spark plugs, then, remove the mounting nuts and the carrier for the high tension wires from the head
- Camshaft cover

5. Turn the engine until the timing marks are at TDC with the cylinder number 6 valves at overlap. Both valves should be slightly open.

6. Remove or disconnect the following:
- Upper timing case cover
- Timing chain tensioner piston
- Upper timing chain sprocket bolts and pull the sprocket off, and while holding it upward, support it securely so the relationship between the chain and sprockets both top and bottom will not be lost
- Upper radiator hose at the thermostat housing, then remove the bolts and remove the support for the intake manifold
- Cylinder head bolts in reverse of the tightening sequence
- Cylinder head

➡ **The cylinder head bolts must be replaced.**

To install:

7. Thoroughly clean all mounting surfaces and check the head for warpage. Check the condition of the head locating dowel sleeves and clean out the bolt threads with a tap.

8. Check the cylinder head and block deck surface to be sure they are true and install a new head gasket.

➡ **Use a thicker head gasket if the head has been machined.**

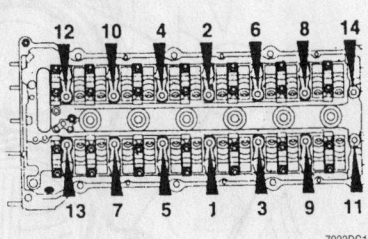

Cylinder head torque sequence—M73 5.4L engines

9. Apply sealant to the joints between the engine block and the upper and lower timing covers.

10. Mount the cylinder head on the block and use new bolts. Do not remove the coating on the head bolts, apply oil to the threads and torque following the tightening sequence in 3 steps as follows:
- Step 1: 22 ft. lbs. (30 Nm)
- Step 2: 60 degrees
- Step 3: 60 degrees

11. Reinstall the timing sprocket to the camshaft. Be sure the camshaft is in proper time and use new lockplates. The camshaft for cylinder number 1 is at TDC firing position when the recess of the camshaft flange is flush with the bolt holding the oil baffle.

12. The balance of installation is the reverse of the removal procedure.

13. Fill the cooling system.

14. Start the engine and check for leaks.

➡ **It may be necessary to bleed the cooling system a second time after the engine has been started.**

Rocker Arms/Shafts

REMOVAL & INSTALLATION

1. Before servicing the vehicle, refer to the precautions in the beginning of this section.

2. Remove or disconnect the following:
- Negative battery cable
- Cylinder head cover
- Spark plugs
- Engine cooling fan

3. To remove a rocker, rotate the engine until the lobe of the camshaft for the rocker to be removed is pointing up.

4. Insert Tool No. 11 5 130 between the camshaft and the valve, then press the valve down and remove the rocker.

➡ **The Hydraulic Valve Adjuster (HVA) can also be replaced once the rocker is removed.**

To install:

➡ **The rockers must be installed in the exact position from which they were removed. The rocker has a slight bend away from the camshaft journal.**

5. Make sure the camshaft lobe is pointing upward.

6. Coat the valve, camshaft and HVA adjuster with fresh engine oil.

7. Press the valve down using tool No. 11 5 130 and install the rocker.

For complete Engine Mechanical specifications, see Section 1 of this manual

8. The remaining components are installed in the reverse order from which they were removed.

9. Check all fluid levels and top off as necessary.

10. Reconnect the negative battery cable.

11. Start the engine and inspect for any fluid leaks.

Intake Manifold

REMOVAL & INSTALLATION

M44 1.9L Engines

1. Before servicing the vehicle, refer to the precautions in the beginning of this section.

2. Remove or disconnect the following:
   - Fuel system pressure
   - Negative battery cable
   - Upper manifold section
   - Rear mounting bracket and remove the coolant hose
   - Front mounting bracket and disconnect the holder for the preheater
   - Mounting bolts and lift off the upper manifold section, then the hose from the fuel pressure regulator
   - Plug plate off the fuel injectors and the wire holding clamp
   - Injection pipe with the fuel injectors attached, then remove the lower manifold section

   To install:

3. Installation is the reverse of the removal procedure, while using the following torque values:
   - 6mm fasteners: 88 inch lbs. (10 Nm)
   - 7mm fasteners: 11 ft. lbs. (15 Nm)
   - 8mm fasteners: 16 ft. lbs. (22 Nm)

M52 2.5L, 2.8L, and S52 3.2L Engines

1. Before servicing the vehicle, refer to the precautions in the beginning of this section.

2. Remove or disconnect the following:
   - Fuel system pressure
   - Negative battery cable
   - Coolant
   - Throttle cable cover and throttle cable with the rubber holder
   - Vacuum fitting from the brake booster and plug the openings
   - Engine and intake manifold covers, remove the bolt holding the ground strap on the front lifting eye and reinstall the bolt

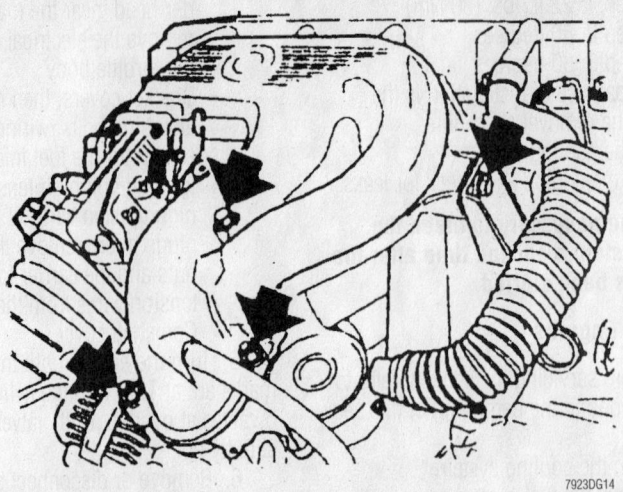

7923DG14

Upper intake manifold and support bracket mounting bolt locations—M44 1.9L engines

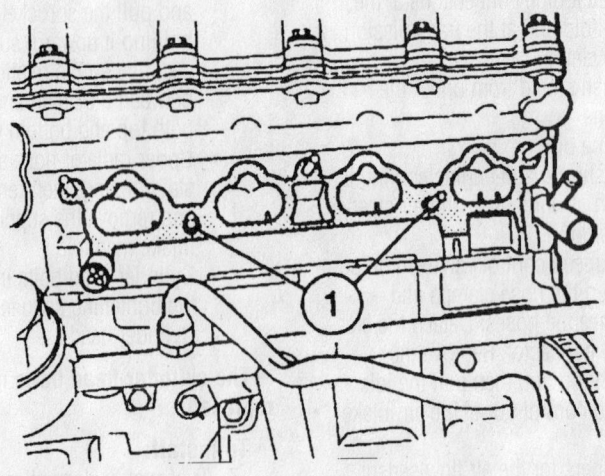

7923DG16

Be sure the hollow bushings (1) are properly installed—M44 1.9L engines

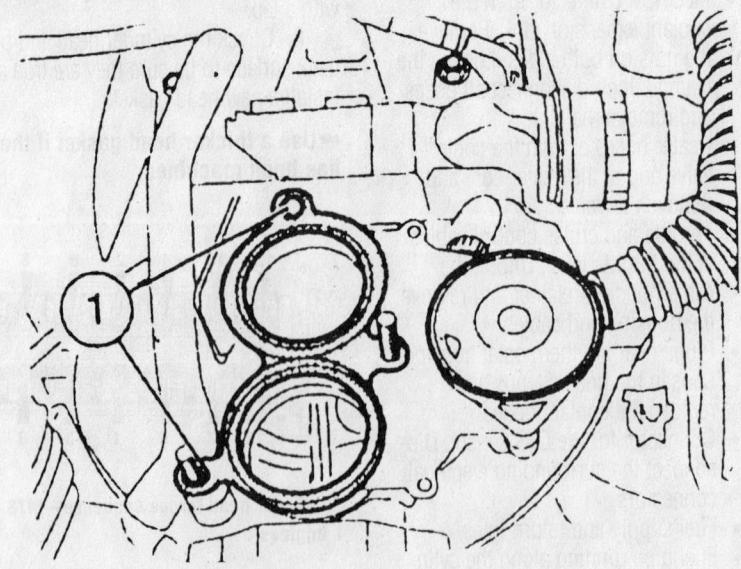

7923DG15

Check the dowel sleeves (1) for damage and correct installation position—M44 1.9L engines

- Bolts holding the ignition coil wire harness plug plate. Be careful not to damage the rubber seals. Take off the ignition coil electrical plugs, then remove the plug plate complete with the electrical leads.
- Cylinder head vent hose and remove the air temperature sensor plug. Remove the tank venting hose and the coolant hoses from the throttle body. Remove the throttle valve switch connector.
- Idle speed control valve mounted on the manifold, then disconnect the fuel hoses from the feed and return lines
- Hardware holding the intake manifold to the cylinder head. Remove the intake manifold taking care not to drop anything into the exposed ports.

To install:

3. Installation is the reverse of the removal procedure, while using the following torque values:
- 6mm fasteners: 88 inch lbs. (10 Nm)
- 7mm fasteners: 11 ft. lbs. (15 Nm)
- 8mm fasteners: 16 ft. lbs. (22 Nm)

M62 4.4L Engines

1. Before servicing the vehicle, refer to the precautions in the beginning of this section.
2. Remove or disconnect the following:
- Fuel system pressure
- Negative battery cable
- Center cover from the cylinder head cover

- Hose clamps on the idle speed control and the throttle valve assembly
- Plug on the mass air flow sensor
- Upper section of the air cleaner assembly along with the MAF sensor
- Right-hand cover from the cylinder head cover
- Plug for the oil level switch
- Connectors for the ignition coils
- Both KS, for cylinders 1 and 2 along with 3 and 4, and the pulse sensor
- IAT sensor, TVS and the ISC valve
- Diagnostic connector from the mounting bracket and disconnect the engine wiring connector
- Ignition coil ground wire, located near the rear engine lifting eye, then the temperature sensor (black) for the temperature gauge, and the temperature sensor (white) for the Digital Motor Electronics (DME)
- 4 bolts for the intake manifold cover, and remove the holder
- Throttle cable
- Left-hand cover from the cylinder head cover
- Ignition coil electrical connectors
- Both KS and the camshaft reference sender
- Coolant expansion tank plug and overflow hose, then remove the 2 mounting bolts, and move the tank aside
- Oil pressure switch electrical connector and remove the wiring
- Fasteners for the wiring ducts on the cylinder heads
- Vacuum hoses on the radiator, and

loosen the hose clamp. Pull the vacuum hose off of the brake booster.
- Tank vapor venting hose from the throttle valve assembly
- Fuel feed and return lines
- Hose off of the end cover on the back of the manifold
- Mounting bolts, and remove the end cover together with the pressure regulating valve straight back to prevent damaging the vent pipe
- Intake manifold

To install:

3. Installation is the reverse of the removal procedure, while using the following torque values:
- 6mm fasteners: 88 inch lbs. (10 Nm)
- 7mm fasteners: 11 ft. lbs. (15 Nm)
- 8mm fasteners: 16 ft. lbs. (22 Nm)

M73 5.4L Engines

1. Before servicing the vehicle, refer to the precautions in the beginning of this section.
2. Remove or disconnect the following:
- Fuel system pressure
- Negative battery cable
- Vacuum hoses for the pressure regulators. Lift out the injection pipes with the injectors attached.
- Distributor caps and the throttle valve necks on the manifolds
- Spark plug wires and the ignition lead pipes
- Crankcase breather hose and loosen the manifold support nuts
- Nose guard
- Intake manifold

To install:

3. Installation is the reverse of the removal procedure, while using the following torque values:
- 6mm fasteners: 88 inch lbs. (10 Nm)
- 7mm fasteners: 11 ft. lbs. (15 Nm)
- 8mm fasteners: 16 ft. lbs. (22 Nm)

Exhaust Manifold

REMOVAL & INSTALLATION

M44 1.9L Engine

1. Before servicing the vehicle, refer to the precautions in the beginning of this section.
2. Remove or disconnect the following:
- Negative battery terminal
- Exhaust pipe from the manifold.

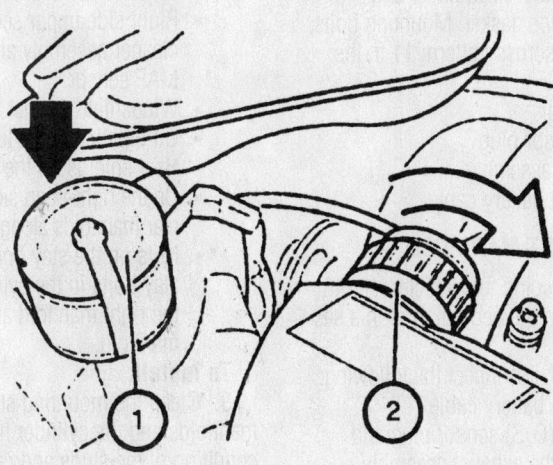

7923DG17

Diagnosis connector (1) and engine connector (2) locations—M62 4.4L engines

For Accessory Drive Belt illustrations, see Section 1 of this manual

Remove the nuts on the flange connection and lower the exhaust pipes. Support the exhaust system. Be sure the Oxygen (O_2S) sensor wire is not being stretched.

• Nuts securing the manifold to the cylinder head, then remove the manifolds

To install:

3. Clean the mounting surfaces on the manifolds and the cylinder head. Check the condition of the studs and replace if necessary.

4. Coat the exhaust studs with an anti-locking compound.

5. Install or connect the following:
• New exhaust manifold gaskets with the graphite side towards the cylinder head and install the manifolds. Nuts: 11ft. lbs. (15 Nm).
• Exhaust pipe to the manifolds
• Negative battery terminal

➡**After 1200 miles, loosen, then tighten each nut to 10 ft. lbs. (12 Nm).**

M52 2.5L, 2.8L, and S52 3.2L Engines

1. Before servicing the vehicle, refer to the precautions in the beginning of this section.

2. Remove or disconnect the following:
• Negative battery terminal
• Mounting nuts on each flange connection and separate the exhaust pipes from the manifold. Support the exhaust system. Be sure the O_2S sensor wire is not being stretched.
• Nuts securing the manifold to the cylinder head
• Manifolds

To install:

3. Clean the mounting surfaces on the manifolds and the cylinder head. Check the condition of the studs and replace if necessary.
• New exhaust manifold gaskets with the graphite side towards the cylinder head and install the manifolds. Nuts, with exhaust studs coated with an anti-locking compound: 14 ft. lbs. (19 Nm).
• Exhaust pipe to the manifolds using new mounting nuts
• Negative battery terminal

M62 4.4L Engine

LEFT EXHAUST MANIFOLD

1. Before servicing the vehicle, refer to the precautions in the beginning of this section.

2. Remove or disconnect the following:
• Negative battery cable
• O_2S sensor plug and the exhaust assembly
• Alternator
• Left cylinder head trim cover
• Complete air cleaner upper section along with the MAF sensor
• Bolts from the left and right engine mounts at the bottom
• Rear engine splash guard
• Bolts of the center of gravity mount to front axle carrier
• Left heat shields on the front axle carrier

3. Support the engine and ensure clearance between the engine and the firewall.

4. Remove the manifold fasteners and remove the manifolds downwards.

To install:

5. Remove the old gasket off of the cylinder head and exhaust manifold and replace the gasket. The gasket beads face the exhaust manifolds.

6. Install or connect the following:
• Exhaust manifold, with upper row of the exhaust fasteners bolts coated with a locking agent. Mounting fasteners: 16 ft. lbs. (22 Nm).
• Left heat shields and the center of gravity mount-to-front axle carrier bolts
• Rear engine splash guard
• Bolts for the left and right engine mounts at the bottom. 10mm bolts: 31 ft. lbs. (42 Nm). 8mm bolts: 16 ft. lbs. (22 Nm).
• Complete air cleaner upper section along with the mass air flow sensor
• Left cylinder head cover and replace the gasket. Mounting bolts, in a crisscross pattern: 11 ft. lbs. (15 Nm)
• Alternator
• O_2S sensor plug
• Exhaust assembly
• Negative battery cable

RIGHT EXHAUST MANIFOLD

1. Before servicing the vehicle, refer to the precautions in the beginning of this section.

2. Remove or disconnect the following:
• Negative battery cable
• Oxygen (O_2S) sensor plug and remove the exhaust assembly
• Right heat shields on the front axle carrier
• Rear engine splash guard
• Windshield washer fluid tank

➡**Remove the manifold for cylinders number 2 and 4 first.**

• Manifold fasteners and remove the manifolds upwards

To install:

3. Remove the old gasket from the cylinder head and exhaust manifold and replace the gasket. The gasket beads face the exhaust manifolds.

4. Install or connect the following:
• Exhaust manifold, with upper row of the exhaust fasteners coated with a locking agent. Mounting fasteners: 16 ft. lbs. (22 Nm).
• Washer fluid tank
• Rear engine splash guard
• Right heat shields
• O_2S sensor plug and install the exhaust assembly
• Negative battery cable

M73 5.4L Engine

1. Before servicing the vehicle, refer to the precautions in the beginning of this section.

2. Remove or disconnect the following:
• Negative battery terminal
• Left side upper section of the air cleaner assembly along with the MAF sensor
• Clamp on the left and right split pipes
• Heat shields on the left manifold and on the steering gear
• Manifold and split pipe bolts on the left-hand side
• On the left-hand side, the front and rear manifolds along with the gaskets
• Nuts on the stay bolts. Remove the stay bolts in the cylinder head for the left manifold.
• Right side upper section of the air cleaner assembly along with the MAF sensor
• Windshield washer fluid tank
• Oil dipstick guide tube
• Heat shields on the right manifold
• On the right-hand side, the front and rear manifolds along with the gaskets
• Nuts on the stay bolts. Remove the stay bolts in the cylinder head for the right manifold and remove the manifold.

To install:

3. Clean the mounting surfaces on the manifolds and the cylinder head. Check the condition of the studs and replace if necessary.

4. Install or connect the following:
• Stay bolts in the cylinder head for the right manifold. Install the nuts onto the stay bolts.

- On the right-hand side, new exhaust manifold heat shield gaskets and the manifolds. Nuts: 16–18 ft. lbs. (22–25 Nm). Use new self-locking nuts.
- Heat shields on the right manifold
- Oil dipstick guide tube
- Windshield washer fluid tank
- Right side upper section of the air cleaner assembly along with the air mass sensor
- Stay bolts in the cylinder head for the left manifold. Install the nuts onto the stay bolts.
- On the left-hand side, new exhaust manifold heat shield gaskets and the manifolds. Nuts: 16–18 ft. lbs. (22–25 Nm). Use new self-locking nuts.
- Manifold and split pipe bolts on the left-hand side
- Heat shields on the left manifold and on the steering gear
- Clamp on the left and right split pipes
- Left side upper section of the air cleaner assembly along with the MAF sensor
- Negative battery terminal

Camshaft and Valve Lifters

REMOVAL & INSTALLATION

E 36 M44 Engines

1. Before servicing the vehicle, refer to the precautions in the beginning of this section.

➡**The lifters can be lifted out of the cylinder head once the rockers are removed.**

2. Properly relieve the residual fuel system pressure.

➡**The rockers must be kept in exact order when removed and reinstalled.**

3. Remove or disconnect the following:
- Negative battery cable
- Valve cover
- Spark plugs
- Camshaft rocker fingers

➡**The Hydraulic Valve Adjuster (HVA) can also be replaced once the rocker is removed.**

4. Remove the front, upper timing chain cover.

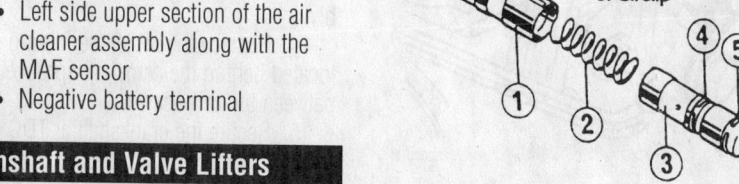

The camshaft sprockets should be positioned as shown when the engine is at TDC on the No. 1 cylinder—M44 1.9L engines

1. Sleeve
2. Spring
3. Hydr. piston
4. Circlip
5. Circlip

Hydraulic lash adjuster components—M44 1.9L engines

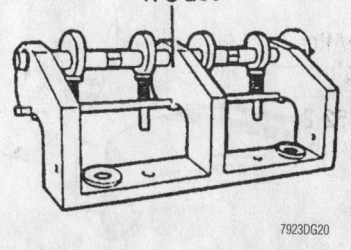

11 3 260

Camshaft removal tool No. 11-3-260 used to hold the camshaft in place during camshaft journal removal—M44 1.9L engines

Bearing cap bolt locations—M44 1.9L engines

Bearing caps are marked with A1 through A5 for exhaust side and E1 through E5 for intake side—M44 1.9L engines

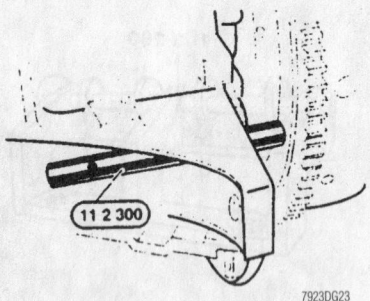

Hold the crankshaft in TDC position with tool No. 11-2-300—M44 1.9L engines

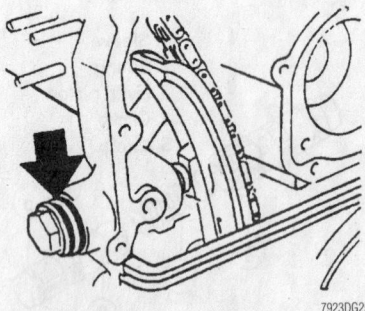

Chain tensioner location—M44 1.9L engines

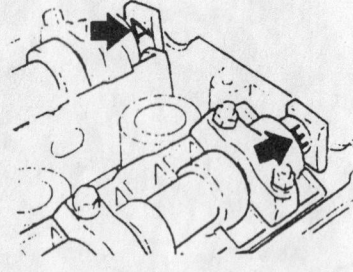

Camshafts are marked with "A" for exhaust and "E" for intake—M44 1.9L engines

Upper chain guide location—M44 1.9L engines

5. Rotate the engine clockwise to Top Dead Center (TDC) of the firing stroke for cylinder number 1. In this position, the camshaft peaks of the intake and exhaust camshafts for cylinder No. 1 face each other. The arrows on the cam sprocket should face up and the dimples on the machined surface at the rear of the camshafts should be facing up.

6. Remove the plug for the crankshaft TDC alignment hole located in the engine block just below the starter motor.

7. Secure the crankshaft at TDC by inserting Tool No. 11 2 300 or its equivalent through the engine block and into the flywheel.

✳✳ CAUTION

Do not forget to remove the crankshaft alignment tool before starting the engine.

8. Install Tool No. 11 3 240 or its equivalent to secure the camshafts at TDC on the compression stroke for cylinder number 1.

9. Remove the timing chain hydraulic tensioner, then remove the camshaft sprocket bolts and sprockets

10. Remove the upper camshaft chain guide.

11. Make note of the camshaft bearing cap locations, label if necessary and remove. The intake cam bearing caps should be stamped from the factory from the front to the rear starting with E1 through E5, and the exhaust cam bearing caps from the front to the rear starting with A1 through A5.

12. Once the cam journals are removed, the camshafts can be lifted away from the cylinder head.

➡ **The intake camshaft is labeled with the letter E, the exhaust camshaft is labeled with the letter A.**

13. Thoroughly clean and inspect all of the removed components prior to reassembly.

14. Replace the oil jet feed fitting seal located behind the camshaft sprockets between the camshafts.

15. Secure the crankshaft at TDC by inserting Tool No. 11 2 300 or its equivalent through the engine block and into the flywheel.

✳✳ CAUTION

Do not forget to remove the tool before starting the engine.

16. Apply fresh engine oil to the camshaft journals in the cylinder head, then install the camshafts with the dimples on the machined surface at the rear of the camshafts facing up.

17. Apply fresh engine oil to the camshaft journal and the cam cap fasteners, then install the camshaft journals in the proper order. Tighten the caps evenly in at least two steps and torque as follows:
- M6 fasteners: 88.5 inch lbs. (10 Nm)
- M7 fasteners: 10 ft. lbs. (14 Nm)
- M8 fasteners: 14.7 ft. lbs. (20 Nm)

18. Install the upper camshaft chain guide.

19. Install Tool No. 11 3 240 or its equivalent to secure the camshafts at TDC on the compression stroke for cylinder number 1. Make sure the dimples on the machined surface at the rear of the camshafts are facing up.

20. Install the camshaft sprockets onto

the camshafts with the timing chain attached making sure the sprocket and sensor plate timing marks are facing up. Install the camshaft sprocket bolts until they lightly bottom, then back off enough such that there is no free play, but the sprockets can rotate when the cam chain is tensioned.

21. Collapse the hydraulic chain tensioner before installing as follows:

    a. Drain the oil chamber of the tensioner.

    b. Place the tensioner in a vice using soft protective jaws or wooden paint stirrers to avoid damaging the machined surface.

    c. Compress the tensioner slowly, and carefully compress it until the snap ring is almost even with the outer housing. Repeat this compression procedure a second time.

22. Install the tensioner and using a new sealing washer, install and tighten the access plug to 29.5 ft. lbs. (40 Nm).

23. Tighten down the camshaft sprocket bolts to 88.5 inch lbs. (10 Nm).

24. Remove Tool Nos. 11 2 300 and 11 3 240 or their equivalents.

25. Apply fresh engine oil to the valve tips, valve stem and the HVA hydraulic valve adjuster, then apply an assembly lubricant to the camshaft lobes and install the rockers.

26. Install and reseal the front cover as follows:

    a. Obtain the upper front cover-to-timing chain cover rubber replacement seal.

    b. Cut the rubber seal in the timing chain cover where it meets the cylinder head.

    c. Thoroughly clean the groove in the timing chain cover with a brake cleaner or rubbing alcohol.

    d. Apply a gasket sealant such as Drei Bond® 1209 to the lower corners of the cylinder head-to-timing cover and the upper corner of the cylinder head-to-upper timing cover-to-valve cover area.

    e. Install four long guide studs, Tool No. 11 4 110 or their equivalent into the cylinder head in place of the four upper timing cover through bolts.

    f. Coat the new timing cover seals and the groove end pockets in the timing chain cover with a gasket sealant such as Drei Bond® 1209.

    g. Apply a light, thin coat of grease to the upper portion of the seal.

    h. Use tool No. 11 2 330 or a thin rectangular section of sheet metal larger than the size of the timing cover, seal and guide studs, and apply a light coating of grease to both flat surfaces.

    i. Place the tool onto the seal on top of the timing chain cover.

    j. Install the upper timing chain cover onto the guide studs and carefully press the cover into place.

    k. Install the upper timing chain cover mounting fasteners in the unoccupied bolt holes, but only tighten until resistance is felt, making sure the cover is fully seated.

    l. Carefully slide the tool away from between the timing covers.

    m. Temporarily remove the oil jet feed seal at the front of the cylinder head.

    n. Install the valve cover without the gasket using two M6 x 1.0 mm bolts and fender washers threaded into the first two threaded holes for the valve cover in the cylinder head. Tighten the bolts evenly until the valve cover presses down the upper timing chain cover such that the upper edge of the timing cover is level with the valve cover gasket surface of the cylinder head.

    o. Tighten the installed upper timing cover fasteners, then remove the guide tool and install and tighten the remaining cover fasteners. Recheck the tightness of all the upper timing cover fasteners. Timing cover tightening specifications:
    • M6: 88.5 inch lbs. (10 Nm)
    • M7 11 ft. lbs. (15 Nm)
    • M8 16.2 ft. lbs. (22 Nm)
    • M10 34.6 ft. lbs. (47 Nm)

    p. Remove the valve cover and install the oil jet feed seal.

✳✳ CAUTION

Failure to install the oil jet feed seal will result in engine damage.

27. The balance of installation is the reverse of the removal procedure noting the following.

✳✳ WARNING

Remove the crankshaft holding tool before operating the engine.

28. Fill the engine cooling system with the recommended mixture of coolant and bleed as necessary.

➡️**It may be necessary to bleed the cooling system a second time after the engine has been started and cooled.**

29. Change the engine oil and filter assembly.

30. Check and top off all fluid levels as necessary.

31. Connect the negative battery cable.

32. Start the engine and check for proper operation.

➡️**A new or reinstalled timing chain tensioner is without oil. During initial start up the engine must be operated at 3500 rpm for approximately 20 seconds.**

M52 2.5L, 2.8L, and S52 3.2L Engines

1. Before servicing the vehicle, refer to the precautions in the beginning of this section.

2. Remove or disconnect the following:
    • Fuel pressure
    • Negative battery cable
    • Cylinder head, if necessary. If the fresh air intake shroud on the firewall is removable, removing the shroud should allow enough room to remove the camshafts. If the body of the vehicle interferes with the camshaft removal, removing the cylinder head is necessary

➡️**There are several tools recommended for removing the camshafts. Without these tools there is a risk the camshaft could break during removal, or damage to a valve could occur.**

    • Rear hood seal, then remove the fasteners securing the sealed plastic housing for the engine wiring harness from the passenger compartment fresh air intake shroud
    • Fasteners securing the fresh air intake shroud to the firewall, and remove the shroud

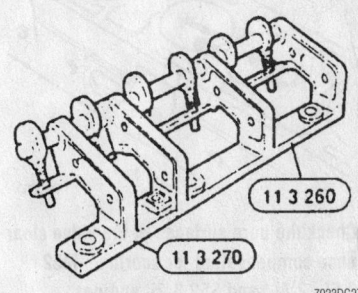

7923DG27

Camshaft removal tools 11-3-260 and 11-3-270 used to hold the camshaft journals in place during camshaft removal—M52 2.5L, 2.8L, S52 3.2L engines

7923DG28

Camshaft bearing journal bolt locations—M52 2.5L, 2.8L, S52 3.2L engines

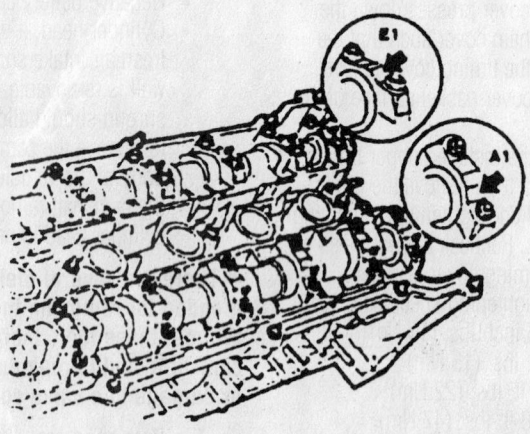

7923DG29

Bearing cap ID markings. The intake camshaft journals are indicated by the letter E, the exhaust journals are indicated by the letter A—M52 2.5L, 2.8L, and S52 3.2L engines

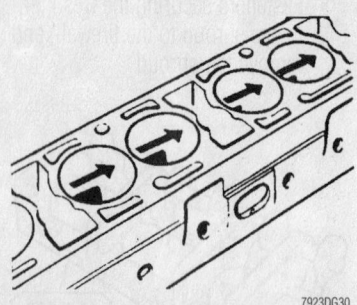

7923DG30

Check the bore surfaces of the valve clearance compensators for scoring—M52 2.5L, 2.8L, and S52 3.2L engines

7923DG31

The lower camshaft journal and tappet bore assemblies are marked A for exhaust and E for intake—M52 2.5L, 2.8L, S52 3.2L engines

- If equipped, plastic trim covers from the top of the engine
- Wire connectors for the ignition coils and cylinder head sensors and remove the coils, then remove the cylinder head cover

3. Rotate the engine in the direction of rotation to TDC for cylinder number one. Cylinder number 1 will be at TDC when the intake and exhaust camshaft peaks for cylinder number 1 face each other

4. Lock the engine in the TDC position by placing the holding dowel tool No. 11-2-300 through the machined hole in the engine block, just inside of the transmission bell housing mounting tab located on the left lower portion of the engine block. Slide the locating dowel through the machined hole in the block and into the machined hole in the flywheel to prevent movement of the crankshaft.

5. The camshafts are held in the TDC position by placing tool No. 11-3-240 on the valve cover mating surface at the back of the cylinder head and onto the squared ends of the camshafts, securing the camshafts such that 2 sides of the squared ends are parallel with the cam cover gasket mating surface. With the camshafts in this position, the arrows on the sprockets will be facing up.

6. Remove or disconnect the following:
- Valve cover mounting studs
- 2 hex plugs at the front of the cylinder head to access the exhaust camshaft sprocket mounting bolts, then loosen the exhaust cam sprocket bolts 2 turns

7. Press down on the secondary cam chain tensioner between the 2 camshaft sprockets and install tool No. 11-3-292 through the back side of the tensioner housing to hold the tensioner down. A similar sized and suitably hardened drill bit can be substituted for tool No. 11-3-292.

8. Remove the fasteners from the front of the cylinder head securing the hydraulic variable camshaft control (VANOS) unit to the cylinder head, and inspect to make sure any hydraulic or sensor connectors have been removed or disconnected.

9. On engines with a spring plate installed on the intake camshaft, place tool No. 11-5-490 onto the exhaust camshaft sprocket and carefully rotate the sprocket clockwise to allow the helical gear of the VANOS unit to release the intake camshaft and to allow the VANOS unit to be pulled away from the front of the cylinder head.

10. If tool No. 11-5-490 is not available, move the camshaft sprockets to release the VANOS by using a suitable drift and soft faced mallet and lightly tapping on a sprocket tooth of the intake cam sprocket to rotate both cam sprockets clockwise, while alternately pulling on the VANOS unit to release it. This procedure may need to be repeated several times to fully release the VANOS unit, and must be performed very carefully, in such a manner to not distort or damage the teeth of the cam sprocket.

11. With the VANOS assembly removed, remove the intake and exhaust camshaft sprockets, the hydraulic cam chain tensioner, and the cam chain guide.

12. From the side of the right front area of the engine, remove the cap nut for the cam chain tensioner for the cam chain that runs between the crankshaft and the exhaust camshaft sprocket. Use care when removing the tensioner cap nut as the cam tensioner spring applies pressure to the cap nut.

13. Remove the exhaust cam tensioner, and then release the sprocket from the cam chain.

14. Attach a wire tie or mechanic's wire to the cam chain and temporarily secure the chain fully extended.

15. Remove the dowel from the engine and flywheel locking the engine in the TDC position.

16. While holding the crankshaft to exhaust camshaft cam chain, rotate the engine 30 degrees counterclockwise to avoid damaging the valves during camshaft removal and reinstallation.

17. With the attached wire tie or mechanic's wire secured to the cam chain, and carefully lower the chain downward making sure there is enough exposed wire to retrieve the chain for reinstallation.

18. Remove the spark plugs and install the upper camshaft journal holding device, tool No. 11-3-260/270/250. Tighten the hold down bolts in the spark plug bores to 17 ft. lbs. (23 Nm).

19. Apply a load to the bearing caps by rotating the eccentric shaft. This relieves the tension on the bearing cap bolts. Loosen and remove the bearing cap bolts.

20. Slowly and carefully rotate the eccentric of the cam cap holding fixture to release the camshaft.

21. Remove and rotate the camshaft fixture tool 180 degrees and repeat the above procedures to remove the other camshaft.

22. Remove the camshafts and the bearing caps. Note that the intake camshaft is marked **E** and the exhaust camshaft is marked **A**. The camshaft bearing are consecutively numbered and lettered with **A** or **E** to designate intake or exhaust side.

23. Hold the valve lash compensators in place using tool 11-3-250 and remove the bearing plate along with the valve plungers.

To install:

24. Apply fresh engine oil to the camshaft journals and lobes.

25. Place the camshaft in cam journals of the cylinder head.

26. Place the upper camshaft journal bearing caps on the camshaft in the correct order. Note that the intake camshaft is marked **E** and the exhaust camshaft is

marked **A**. The camshaft bearing are consecutively numbered and lettered with **A** or **E** to designate intake or exhaust side.

27. With the spark plugs removed install the upper camshaft journal holding device, tool No. 11-3-260/270/250. Tighten the hold down bolts in the spark plug bores to 17 ft. lbs. (23 Nm).

28. Rotate the eccentric shaft of camshaft journal holding fixture to seat the camshaft and journals into the cylinder head.

29. Use the holding fixture to secure the camshaft journals and camshaft into the cylinder head Install the camshaft journal bolts and torque as follows:
- M6 fasteners: 89 inch lbs. (10 Nm)
- M7 fasteners: 10 ft. lbs. (14 Nm)
- M8 fasteners: 14 ft. lbs. (19 Nm)

30. Repeat the above procedures to install the second camshaft.

31. Remove the camshaft journal fixture tool.

32. Align the camshafts so that lobes of the intake and exhaust cams face each other for the No. 1 cylinder. The camshafts can be turned on the hexagon casting using a 1 1/16 inch or 27mm open end wrench.

➡**Once the camshafts have been removed and reinstalled a waiting period dependent on the ambient temperature is necessary before rotating the engine. At room temperature wait 4 minutes to allow the lifters to compress fully. At temperatures down to 50˚F (10˚C) wait 11 minutes. At temperatures lower than 50˚F (10˚C) wait 30 minutes. This is to prevent contact between the valves and the piston tops. The engine may not be cranked under the same condition for a period of 10 minutes at room temperature; 30 minutes for temperatures down to 50˚F (10˚C); 75 minutes for temperatures below 50˚F (10˚C).**

33. With the attached wire secured to the crankshaft cam chain, carefully raise the chain upward and apply a light tension to the chain, then while holding the chain, carefully rotate the engine clockwise to TDC on cylinder number one.

34. Lock the engine in the TDC position by placing the holding dowel tool No. 11-2-300 through the machined hole in the engine block, just inside of the transmission bell housing mounting tab located on the left lower portion of the engine block. Slide the locating dowel through the machined

hole in the block and into the machined hole in the flywheel to prevent movement of the crankshaft.

35. Hold the camshafts TDC position by placing tool No. 11-3-240 on the valve cover mating surface at the back of the cylinder head and onto the squared ends of the camshafts, securing the camshafts such that 2 sides of the squared ends are parallel with the cam cover gasket mating surface. With the camshafts in this position, the arrows on the sprockets will be facing up.

36. Position the crankshaft cam chain over the exhaust cam chain sprocket and install the sprocket on the exhaust cam such that the slotted holes are centered with the fastener bores in the camshaft.

37. Install the hydraulic cam chain tensioner, and the cam chain guide, then the intake and exhaust camshaft sprockets along with the cam chain and remove the tool used to hold the hydraulic cam chain tensioner collapsed.

38. Install the crankshaft cam chain tensioner assembly, spring and cap nut.

39. Apply sealant to the upper corners of the cylinder head where the VANOS housing mounts and install a new VANOS housing gasket.

40. Press the helical gear of the VANOS assembly toward the housing and install the VANOS assembly. To do so on engines with a spring plate installed on the intake camshaft, place tool No. 11-5-490 onto the exhaust camshaft sprocket and carefully rotate the sprocket counterclockwise to allow the helical gear of the VANOS unit to thread into the intake camshaft and to allow the VANOS unit to be pulled into the front of the cylinder head.

41. If tool No. 11-5-490 is not available, move the camshaft sprockets to install the VANOS by using a suitable drift and soft faced mallet and lightly tapping on a sprocket tooth of the exhaust cam sprocket to rotate both cam sprockets counterclockwise, while alternately pressing on the VANOS unit to install it. This procedure may need to be repeated several times to fully install the VANOS unit, and must be performed very carefully, in such a manner to not distort or damage the teeth of the cam sprocket.

➡**Make sure when assembling the VANOS unit is able to rest on the front of the cylinder head without being forced or without binding. If the VANOS unit does not fully seat it may be necessary to reposition the camshaft** sprockets such that the slots in the camshaft sprockets allow enough movement of the sprockets for the helical gear of the VANOS assembly to be fully seated during assembly.

42. Tighten the VANOS unit fastener and then torque the camshaft sprocket bolts of both the intake and exhaust cams to 16 ft. lbs. (22 Nm).

43. Remove the crankshaft TDC holding dowel tool No. 11-3-240 from the engine, and remove the camshaft TDC positioning tool No. 11-3-240 from the valve cover mating surface at the back of the cylinder head.

44. Slowly and carefully rotate the engine clockwise 4 complete revolutions bringing cylinder number 1 to TDC. If the engine binds for any reason stop immediately to evaluate and rectify the cause of the binding.

45. With cylinder number 1 at TDC slide the camshaft TDC positioning tool No. 11-3-240 over the ends of the cams and onto the valve cover mating surface. If the tool slides over the cams and is flush with the mating surface, the camshafts are properly timed. If the tool does not slide easily over the ends of the cams, or if the tool is not flush with the valve cover mating surface, the camshaft timing must be repeated until the tool fits squarely.

46. The balance of the assembly is in reverse order of disassembly.

47. Check all fluid levels as necessary.

48. Connect the negative battery cable.

M62 4.4L Engines

LEFT CAMSHAFT (CYLINDER BANK 5–8)

1. Before servicing the vehicle, refer to the precautions in the beginning of this section.

2. Remove or disconnect the following:
   • Negative battery cable
   • Left and right cylinder head covers
   • Spark plugs
   • Top left timing case cover
   • Splash guard
   • Oil lines to the left and right cylinder head

3. Rotate the crankshaft in direction of rotation until cylinder number 1 is in TDC firing position.

4. Brace the camshaft on the hex head with a suitable open-end wrench and loosen the 3 accessible fasteners on the right sprocket approximately ½ a turn.

   a. Turn the engine over once and loosen the remaining 3 fasteners on each right sprocket approximately ½ a turn.

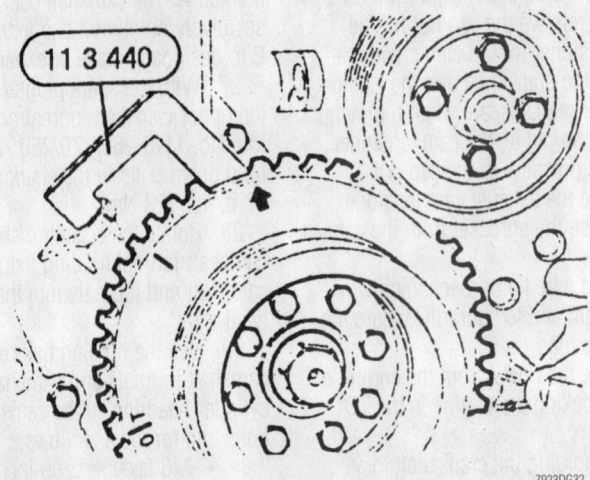

Gap in increment gear must fit in tool No. 11-3-440—M62 4.4L engines

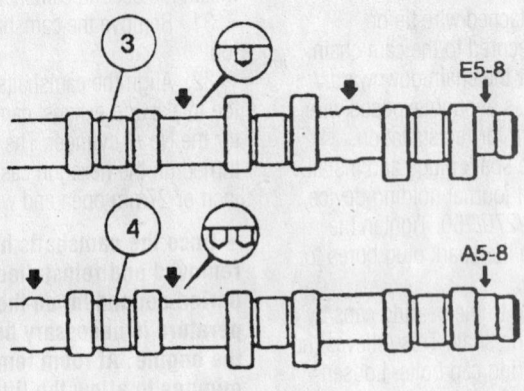

Left side camshaft identification (cylinder bank 5 to 8): hex head (3) on intake camshaft between cylinders 7 and 8, hex head (4) on exhaust camshaft between cylinders 5 and 6—M62 4.4L engines

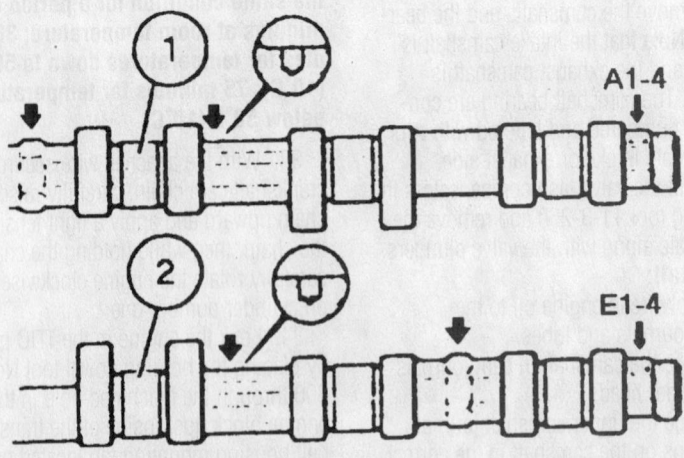

Right side camshaft identification (cylinder bank 1 to 4): hex-head (2) on intake camshaft between cylinders 3 and 4, hex-head (1) on exhaust camshaft between cylinders 1 and 2—M62 4.4L engines

Camshaft positioning—M62 4.4L engines

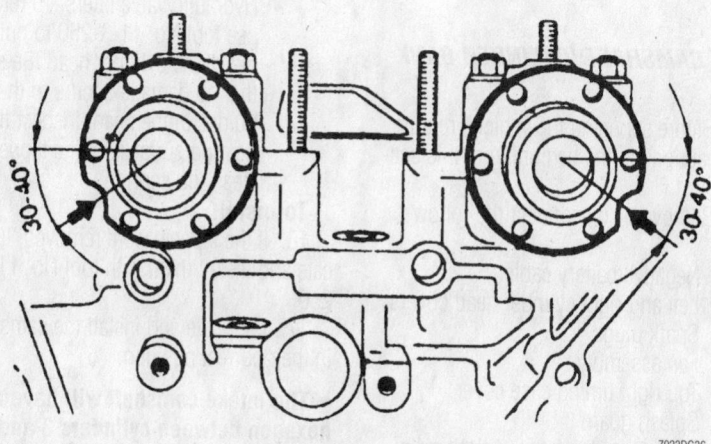

Install left-hand camshafts: Recesses in camshaft point downwards approximately 30–40 degrees from plane of cylinder head—M62 4.4L engines

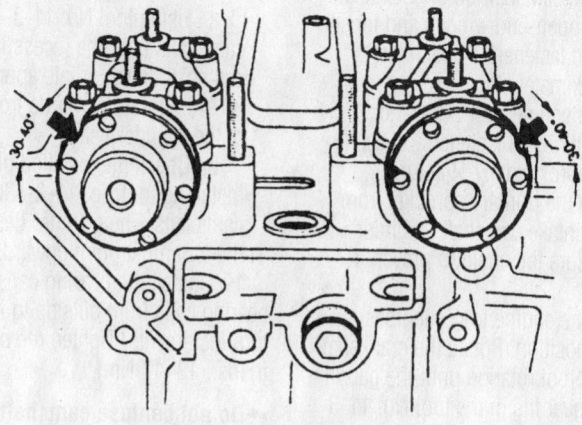

Install right-hand camshafts: Recesses in camshaft point upward approximately 30–40 degrees from plane of cylinder head—M62 4.4L engines

b. Remove the primary sprocket from the left-hand intake camshaft (cylinder bank 5–8). Secure the chain to prevent it from dropping into the engine.

5. Rotate the engine to 45 degrees BTDC. Rotate the crankshaft against direction of rotation until the gap in the increment gear fits in tool No. 11-3-440.

6. Remove or disconnect the following:
- Fasteners on the exhaust camshaft
- Fasteners on the exhaust camshaft sprocket. Do not remove the sprocket.

7. Compress the chain tensioner and install tool No. 11-3-420 to lock the tensioner in place.

8. Lift off both secondary camshaft sprockets together with the chain.

9. Rotate the intake and exhaust camshafts to the installed position as follows:

a. Using tool No. 11-3-430, rotate the camshafts until the recess in both camshaft flange points approximately 30–40 degrees downwards from the plane of the cylinder head.

b. Check the installed position by installing tool No. 11-2-430 to the camshafts. The cylinder designation of the tool must point upwards.

10. Loosen the both camshaft bearing caps uniformly from outside to inside ½ turn.

11. Remove or disconnect the following:
- All bearing caps. Label each bearing cap to facilitate reassembly and place aside.
- Camshafts noting their locations
- Hydraulic valve lifters. To remove, use tool No. 11-3-250 to pull them out of the cylinder head. Be sure that no damage occurs to the guides in the head. Inspect the bearing surfaces of the tappets for wear and scoring.

To install:

12. If the lifters were removed, install them with tool No. 11-32-250.

13. Lubricate and install the camshafts in their correct position.

➡ **The intake camshaft will have a hexagon between cylinders 7 and 8. The exhaust camshaft will have the hexagon between cylinders 5 and 6.**

14. Rotate the intake and exhaust camshafts to the installed position as follows:

a. Using tool No. 11-3-430, rotate the camshafts until the recess in both camshaft flange points approximately 30–40 degrees downwards from the plane of the cylinder head.

b. Check the installed position by installing tool No. 11-2-430 to the camshafts. The cylinder designation of the tool must point upwards.

15. Install the bearing caps, tightened from outside to inside in 1/2 turn increments to 9–13 ft. lbs. (12–17 Nm).

➡**Do not confuse camshaft bearing caps of cylinders No. 1–4 and 5–8. The exhaust camshaft bearing caps are marked A1 through A5. The intake camshaft bearing caps are marked E1 through E5 from intake end.**

16. Fit tool No. 11-3-430 to the camshaft. Rotate the camshaft until the marker bores face upwards.

17. Install tool Nos. 11-2-442/446 to the camshaft on cylinder bank 5–8.

18. Install tool Nos. 11-2-441/445 to the camshaft on cylinder bank 1–4.

19. Using a suitable open-end wrench, align all camshafts in such a way that the tools fit on the cylinder heads without any gaps.

20. Fit tool Nos. 11-2-443 to tool Nos. 11-2-441/442/445/446 and secure them with tool No. 11-2-444 using the spark plug threads.

21. Install:
- Secondary sprockets together with chain to the camshafts on cylinder bank 5–8
- Fasteners on the exhaust camshaft sprocket and tighten

22. Remove the tool used to lock the chain tensioner in position.

23. Rotate the engine from 45 degrees BTDC in direction of rotation as far as TDC. Install tool No. 11-2-300 at the flywheel to lock the crankshaft in TDC position.

24. Assemble the primary sprocket and chain to the intake camshaft with the arrow pointing upwards (in cylinder axis) and the long bores centrally aligned. Install the fasteners snuggly.

25. Install tool No. 11-3-390 in the right timing case cover and with a suitable torque wrench, tension the tool to 11 inch lbs. (1.2 Nm).

26. Tighten the sprockets to 11 ft. lbs. (15 Nm) in the following order:
- All fasteners on the left exhaust camshaft
- The 3 fasteners in the right exhaust camshaft
- All fasteners on the left intake camshaft

- The 3 fasteners in the right intake camshaft

27. Remove:
- Tool Nos. 11-2-444/443/441/445 or their equivalents
- Tool Nos. 11-2-444/443/442/446 or their equivalents
- Tool No. 11-2-300 or the equivalent used to locked the crankshaft in TDC position

28. Turn the engine over once.

29. Tighten the remaining 3 fasteners on right exhaust camshaft and remaining 3 fasteners on the right intake camshaft to 11 ft. lbs. (15 Nm).

30. Relieve the load and remove tool No. 11-3-390 from the right timing case cover.

31. The balance of installation is the reverse of the removal procedure.

32. Check and top off all fluid levels as necessary.

RIGHT CAMSHAFT (CYLINDER BANK 1–4)

1. Before servicing the vehicle, refer to the precautions in the beginning of this section.

2. Remove or disconnect the following:
- Negative battery cable
- Left and right cylinder head covers
- Spark plugs
- Fan assembly
- Top right timing case cover
- Splash guard
- Oil lines to the left and right cylinder head

3. Rotate the crankshaft in direction of rotation until the first cylinder is in TDC firing position.

4. Brace the camshaft on the hex head with a suitable open-end wrench and loosen the 3 accessible fasteners on each left sprocket approximately ½ a turn.

5. Turn the engine over once and loosen the remaining 3 fasteners on each left sprocket approximately ½ a turn.

6. Remove the primary sprocket from the right-hand intake camshaft (cylinder bank 1–4). Secure the chain to prevent it from dropping.

7. Rotate the engine to 45 degrees BTDC setting position. Rotate the crankshaft against direction of rotation until the gap in the increment gear fits in the tool No. 11-3-440.

8. Remove the fasteners on the exhaust camshaft sprocket. Do not remove the sprocket.

9. Compress the chain tensioner and install tool No. 11-3-420 to lock the tensioner in place.

10. Lift off both secondary camshaft sprockets together with the chain.

11. Rotate the intake and exhaust camshafts to the installed position.

12. Using tool No. 11-3-430, rotate the camshafts until the recess in both camshaft flange points approximately 30–40 degrees upwards from the plane of the cylinder head.

13. Check the installed position by installing tool No. 11-2-430 to the camshafts. The cylinder designation of the tool must point upwards.

14. Loosen the both camshaft bearing caps uniformly from outside to inside ½ turn.

15. Remove or disconnect the following:
- Bearing journal caps. Label each bearing cap to facilitate reassembly and position aside.
- Camshafts noting their locations
- Hydraulic valve lifters. To remove, use tool No. 11-3-250 to pull them out of the cylinder head. Be sure that no damage occurs to the guides in the head. Inspect the bearing surfaces of the tappets for wear and scoring.

To install:

16. If the tappets were removed, lubricate and install them with tool No. 11-32-250.

17. Lubricate and install the camshafts in their correct position.

➡**The intake camshaft will have a hexagon between cylinders 3 and 4. The exhaust camshaft will have the hexagon between cylinders 1 and 2.**

18. Rotate the intake and exhaust camshafts to the installed position and perform the following:

a. Using tool No. 11-3-430, rotate the camshafts until the recess in both camshaft flange points approximately 30–40 degrees upwards from the plane of the cylinder head.

b. Check the installed position by installing tool No. 11-2-430 to the camshafts. The cylinder designation of the tool must point upwards.

19. Install the bearing caps. Tighten the bearing caps from outside to inside in ½ turn increments. Tighten the bolts to 10–13 ft. lbs. (13–17 Nm).

➡**Do not confuse camshaft bearing caps of cylinders No. 1–4 and 5–8. The exhaust camshaft bearing caps are marked A1 through A5 counting from intake side. The intake camshaft bearing caps are marked E1 through E5 counting from the intake end.**

20. Fit tool No. 11-3-430 to the camshaft. Rotate the camshaft until the marker bores face upwards, then proceed as follows:

　　a. Install tool Nos. 11-2-442/446 to the camshaft on cylinder bank 5–8.

　　b. Install tool Nos. 11-2-441/445 to the camshaft on cylinder bank 1–4.

　　c. Using a suitable open-end wrench, align all camshafts in such a way that the tools fit on the cylinder heads without any gaps.

　　d. Fit tool Nos. 11-2-443 to tool Nos. 11-2-441/442/445/446 and secure them with tool No. 11-2-444 using the spark plug threads.

21. Install:

- Secondary sprockets together with chain to the camshafts on cylinder bank 1–4
- Fasteners on the exhaust camshaft sprocket and tighten

22. Remove the tool used to lock the chain tensioner in position.

23. Rotate the engine from 45 degrees BTDC in direction of rotation as far as TDC setting. Install tool No. 11-2-300 ore equivalent at the flywheel to lock the crankshaft in TDC position.

24. Assemble the primary sprocket and chain with sensor pin to the intake camshaft with the arrow pointing upwards (in cylinder axis) and the long bores centrally aligned. Install the fasteners.

25. Install tool No. 11-2-400 to the right cylinder head (cylinder bank 1–4). Install tool No. 11-3-390 to tool No. 11-2-400. Using a suitable torque wrench, tension the tool to 11.5 inch lbs. (1.3 Nm).

26. Tighten the sprockets to 11 ft. lbs. (15 Nm) in the following order:

- 3 fasteners on the left exhaust camshaft
- All fasteners on the right exhaust camshaft
- The 3 fasteners on the left intake camshaft
- All fasteners in the right intake camshaft

27. Remove:

- Tool Nos. 11-2-444/443/441/445 or the equivalent
- Tool Nos. 11-2-444/443/442/446 or the equivalent
- Tool No. 11-2-300 used to locked the crankshaft in TDC position

28. Turn the engine over once.

29. Tighten the remaining 3 fasteners on left exhaust camshaft and remaining 3 fas-teners on left intake camshaft to 11 ft. lbs. (15 Nm).

30. Relieve the load and remove the tool Nos. 11-3-390 and 11-2-400 or their equiv-alents.

31. Install the remaining components in the reverse order of removal.

32. Check and top off the fluid levels as necessary.

M73 5.4L Engines

1. Before servicing the vehicle, refer to the precautions in the beginning of this sec-tion.

2. Remove or disconnect the following:

- Negative battery cable
- Coolant
- Fan assembly

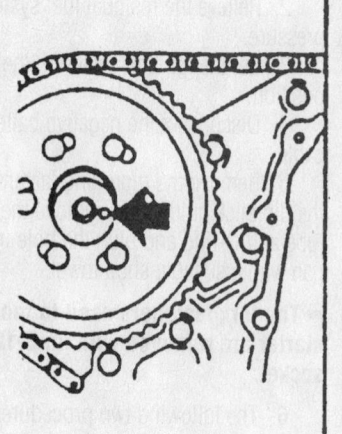
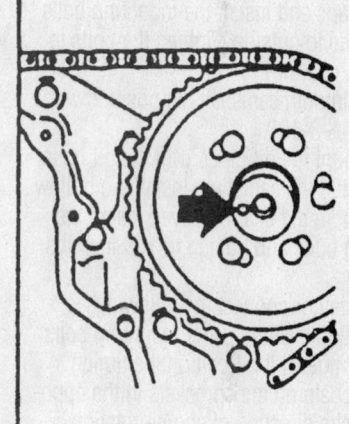

At TDC, the dowel pins in the camshaft sprockets should face each other—M73 5.4L engines

- Both intake manifolds and distribu-tor housings
- Cylinder head covers
- Mounting bolts and lift out the upper timing cover

3. Set the engine to TDC. Install a holder in the crankshaft. The valves of cylinders 1 and 7 should be closed and the dowel pins in the camshafts should face in.

4. Press off the anti-tamper lock for the chain tensioner with a screwdriver. Loosen the nut, then loosen the adjusting screw several turns, and then remove the plug. Remove the timing chain tensioning piston, using care not to lose the spring that is between the plug and the piston.

5. Remove or disconnect the following:

- Mounting bolts for the timing

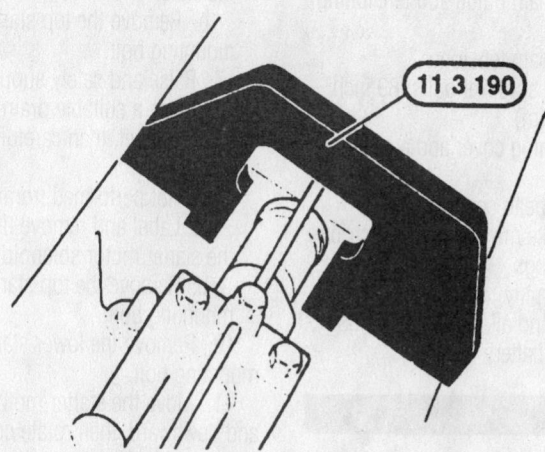

Camshaft alignment gauge—M73 5.4L engines

chain guide, and remove the guide, then remove the tensioning rail
- Mounting bolts on the camshaft sprockets, and carefully remove the sprockets. Do not allow the timing chain to fall into the engine.
- Oil pipe mounting bolts from the top of the camshaft bearings
- Camshaft

✳✳ WARNING

The bearing caps are matched with the bearings, do not mix the order of the caps.

To install:

6. With the crankshaft positioned at TDC, install the camshaft with the dowel pin facing the center of the engine. Position the bearing caps and install the mounting bolts from inside to outside. Tighten the bolts to 11 ft. lbs. (15 Nm).

7. Hold both camshafts in position with tool No. 11-3-190.

8. Mount the oil pipes with the oil outlet bores facing the camshaft. Install the hollow union bolt in the bearing cover. Install the mounting bolts and tighten to 108 inch lbs. (12 Nm).

9. Install or connect the following:
- Camshaft sprocket mounting bolts finger-tight. Position the timing chain on the sprockets in the opposite direction of engine rotation, beginning at the crankshaft. Verify that the timing chain is correctly aligned on all the sprockets, and remove the crankshaft holder.
- Timing chain guide and tensioning rail
- Timing chain tensioner
- Camshaft sprocket bolts: 89 inch lbs. (10 Nm)
- Upper timing cover and a new gasket
- Cylinder head covers
- Both intake manifolds and distributor housings
- Fan assembly
- Coolant and all remaining fluids
- Negative battery cable

Valve Lash

ADJUSTMENT

All engines are equipped with hydraulic valve lash adjusters. No adjustments are possible.

Starter Motor

REMOVAL & INSTALLATION

Starter motor electrical terminal fastener torque specifications for all models:
- M5 terminal fasteners: 44 inch lbs. (5 Nm)
- M6 terminal fasteners: 53 inch lbs. (6 Nm)
- M8 terminal fasteners: 115 inch lbs. (13 Nm)

Starter motor mounting fasteners:
- Hex head bolts: 36 ft. lbs. (49 Nm)
- Torx bolts: 32 ft. lbs. (43 Nm)

M42 And M 44 Engines

1. Before servicing the vehicle, refer to the precautions in the beginning of this section.

2. Relieve the residual fuel system pressure.

3. Set the ignition switch to the **OFF** position.

4. Disconnect the negative battery cable.

5. Remove the mounting fastener for the dip stick guide tube, remove the guide tube and O-ring and plug the hole in the oil pan with a suitable shop towel.

➡ **The torx fasteners used to mount the starter are removed with an E-12 sized socket.**

6. The following two procedures can be performed working from above or from underneath the vehicle after the vehicle is raised and safely supported:
   a. Label and remove the wires from the starter motor solenoid.
   b. Remove the top starter motor mounting bolt.

7. Raise and safely support the vehicle.

8. Place a suitable drain pan below the fuel lines. Label, then carefully remove the fuel lines

9. If not performed from above:
   a. Label and remove the wires from the starter motor solenoid.
   b. Remove the top starter motor mounting bolt.

10. Remove the lower starter motor mounting bolt.

11. Move the starter motor assembly out and downward, then rotate counterclockwise to remove, taking care not to damage the electrical leads.

12. Visually inspect the starter pinion and ring gears for any signs of damage.

13. Installation is in reverse order of removal.

E36 Six Cylinder Engines

AUTOMATIC TRANSMISSION MODELS

1. Before servicing the vehicle, refer to the precautions in the beginning of this section.

2. Set the ignition switch to the **OFF** position.

3. Disconnect the negative battery cable.

4. Remove the climate control air intake duct at the rear of the engine compartment as follows:
   a. Release the grille clips at the rear and lift the grille upward from the rubber seal, then remove the seal.
   b. Remove the two self tapping Phillips screws from inside the duct to release the electrical harness housing.
   c. Press the electrical housing downward to expose and remove the two hex head self tapping screws on each side of the duct and lift the duct up and out of the engine compartment.

5. Remove the intake manifold trim and valve cover trim covers.

6. Disconnect the ignition coil electrical connectors and ground straps, label for reinstallation and place aside.

7. Remove the two fasteners that secure the fuel injection electrical connector harness to the rail. Remove the harness and carefully place aside toward the back of the engine compartment behind the right side strut tower.

8. Remove the intake manifold.

9. Remove the electrical leads from the starter motor and place aside.

10. Remove the starter motor mounting fasteners at the transmission bell housing, then remove the two mounting nuts at the L-shaped mounting bracket.

11. Loosen the L-shaped bracket to engine block mounting bolt slightly, then remove the bolt while supporting the starter motor.

12. Remove the bracket, then lift the starter motor upward away from the engine.

13. Visually inspect the starter pinion and ring gears for any signs of damage.

14. Installation is in reverse order of removal.

MANUAL TRANSMISSION MODELS

1. Before servicing the vehicle, refer to the precautions in the beginning of this section.

2. Set the ignition switch to the **OFF** position.

3. Disconnect the negative battery cable.

4. Raise and safely support the vehicle.

5. Remove the fasteners for the reinforcement mounted between the frame rails just behind the rear lower control arm bushing brackets. Mark the bar's direction for proper reinstallation and remove the bar.

6. Remove the fuel line cover located near the left side rear lower control arm bushing bracket.

7. Release the fuel lines and reverse light wiring from their routing brackets.

8. Note the routing of the starter motor electrical wires, remove the wire mounting nuts, and place the wires aside.

9. Remove the top and bottom starter motor to engine-transmission mounting fasteners.

10. Remove the two mounting nuts at the L-shaped mounting bracket.

11. Loosen the L-shaped bracket to engine block mounting bolt slightly, then remove the bolt while supporting the starter motor.

12. Remove the bracket and carefully lower the starter motor downward away from the engine.

13. Visually inspect the starter pinion and ring gears for any signs of damage.

14. Installation is in reverse order of removal.

Oil Pan

REMOVAL & INSTALLATION

M44 1.9L Engines

1. Before servicing the vehicle, refer to the precautions in the beginning of this section.
2. Remove or disconnect the following:
   • Negative battery cable
   • Engine oil
   • Exhaust pipe, if necessary
   • Lower oil pan mounting bolts and lower oil pan
   • Upper section oil pan bolts and upper oil pan

To install:

3. Clean the mounting surfaces and install new gaskets.
4. Install or connect the following:
   • Oil pan and mounting bolts. Bolts: 72 inch lbs. (9 Nm).
   • Exhaust pipe, if removed
   • Oil pan drain plug, then fill engine with oil
   • Negative battery cable
5. Start the engine and check that oil pressure is present; if the oil pressure lamp

does not extinguish within 5–7 seconds of starting the engine, turn the engine **OFF**.

M52 2.5L, 2.8L, and S52 3.2L Engines

1998–02 3 SERIES MODELS

1. Before servicing the vehicle, refer to the precautions in the beginning of this section.
2. Remove or disconnect the following:
   • Negative battery cable
   • Front lower splash guard, if necessary
   • Oil
   • Holding bolt for the oil dipstick guide pipe and remove the clamp. Pull the guide tube free of the pan.
   • Electrical terminal from the oil level sending unit
   • Power steering box, leaving the hoses attached from the front axle carrier
3. Support the engine, then remove the engine mount fasteners.
4. Support the crossmember and remove the fasteners securing the front cross member to the vehicle's chassis. Carefully lower the jack.
5. Lift the engine slightly.
6. Remove or disconnect the following:
   • Flywheel cover
   • Oil pan bolts and lower the oil pan away from the engine and over the cross member

To install:

7. Before installing the oil pan, clean the gasket surfaces and install a new gasket on the oil pan.
8. Coat the joints on the ends of the front engine cover with a universal sealing compound.
9. Install or connect the following:
   • Oil pan and tighten the fasteners
   • Flywheel cover
10. Raise the crossmember into position and install the mounting bolts.
11. Slowly lower then engine onto the engine mounts and install the engine mount fasteners.
12. Install or connect the following:
   • Power steering box
   • Dipstick guide tube using a new base seal and tighten the holding bolt
   • Electrical wiring harness to oil sending unit
   • Front lower splash guard, if removed

   • Oil pan drain plug, then fill the engine with oil
   • Negative battery cable
13. Start the engine and check that oil pressure is present; if the oil pressure lamp does not turn off within 5–7 seconds of starting the engine, turn the engine **OFF**. Check for any oil leaks.

1998–02 5 SERIES MODELS

1. Before servicing the vehicle, refer to the precautions in the beginning of this section.
2. Install or connect the following:
   • Negative battery terminal
   • Engine oil
   • Holding bolt for the oil dipstick guide pipe and remove the clamp. Pull the guide tube free of the pan.
   • All oil pan bolts and remove the pan. Raise the engine slightly if needed for clearance.

To install:

3. Apply sealer to the joint between the pan, front cover and block.
4. Install or connect the following:
   • Oil pan, with new gaskets. Mounting bolts: 78–96 inch lbs. (9–11 Nm).
   • Dipstick guide tube using a new base seal and tighten the holding bolt
   • Engine oil
   • All remaining fluids as necessary

M62 4.4L Engines

1. Before servicing the vehicle, refer to the precautions in the beginning of this section.
2. Remove or disconnect the following:
   • Negative battery cable
   • Intake manifold cover. Remove the top clips on the radiator.
   • Cooling fan
   • Guide tube for the oil dipstick
   • Engine splash guards
   • Cover for the oil filter so the oil will run back to the pan
   • Engine oil, then disconnect the plug for the level switch
   • Lower oil pan. Remove the gasket and clean the mounting surfaces.
   • Left and right engine mounts at the bottom
   • Unbolt the power steering pump at the holder. On automatic transmissions, remove the oil pipes at the power steering pump.
   • Banjo bolt for the oil return pipe from the oil filter at the oil pan

3. Attach a suitable engine hoist and lift the engine by the front eye hook. Observe the distance between the engine and the firewall while lifting the engine.

4. Remove the oil pan bolts and remove the oil pan.

To install

5. Clean the mounting surfaces and install a new gasket.

6. Install upper oil pan. Bolts: 84–96 inch lbs. (9–11 Nm). Lower the engine.

7. Check the seals on the oil pipes and replace it if necessary. Lubricate the seals with oil.

8. Install or connect the following:
- Banjo bolt for the oil return pipe from the oil filter at the oil pan
- Power steering pump and connect the oil lines, if equipped
- Ground strap, then connect the left and right engine mounts at the bottom and tighten to 32 ft. lbs. (43 Nm).

➡If replacing the lower oil pan, remove the level switch from the old pan and install it in the new pan with a new O-ring.

- Lower oil pan with a new gasket. Bolts, beginning in the middle and working to the outside: 84–96 inch lbs. (9–11 Nm).
- Plug for the level switch, making sure to replace the O-ring
- Engine splash guards
- Oil dipstick guide tube, making sure to replace the O-ring
- Cooling fan
- Intake manifold cover. Install the top clips on the radiator.
- Engine oil
- Negative battery cable

M73 5.4L Engines
7 SERIES

1. Before servicing the vehicle, refer to the precautions in the beginning of this section.

2. Remove or disconnect the following:
- Negative battery cable
- Transmission and the oil pump assembly
- Windshield washer tank and the coolant expansion tank
- Guide tube for the oil dipstick. Disconnect the oil pipe on the tandem pump. Remove the mounting bracket.
- Belt tensioner and the oil drain hose
- Flywheel

- Left and right engine mounts at the bottom. Remove the pipe adapter.
- Oil pump consoles. Remove the oil pan bolts and remove the oil pan.

To install

3. Clean the mounting surfaces, and install a new gasket.

4. Install or connect the following:
- Oil pan. Mounting bolts: 84 inch lbs. (11 Nm).
- Left and right engine mounts at the bottom, and tighten to 32 ft. lbs. (43 Nm).
- Oil consoles and tighten to 25 ft. lbs. (34 Nm).
- Flywheel. Bolts: 72 ft. lbs. (97 Nm).
- The remainder of the installation is the reverse of the removal procedure
- Engine oil

Oil Pump

REMOVAL & INSTALLATION

M44 1.9L Engines

➡The engine oil pump is located in the front engine housing.

1. Before servicing the vehicle, refer to the precautions in the beginning of this section.

2. Remove or disconnect the following:
- Negative battery cable
- Engine oil
- Timing case cover
- Oil pump cover mounting bolts and oil pump assembly

To install:

3. Clean the oil pump mounting surfaces

4. Install or connect the following:
- Oil pump on the engine. Install the oil pump cover mounting bolts.
- Timing case cover
- Oil pan drain plug, then fill the engine with oil
- Negative battery cable

5. Start the engine and check that oil pressure is present; if the oil pressure lamp does not turn off within 5–7 seconds of starting the engine, turn the engine **OFF**. Check for any oil leaks.

M52 2.5L, 2.8L, and S52 3.2L Engines

1. Before servicing the vehicle, refer to the precautions in the beginning of this section.

2. Remove or disconnect the following:
- Negative battery cable
- Oil
- Oil pan, to access the oil pump drive sprocket
- Oil pump drive sprocket nut. Note that it is a left-hand thread. Remove the oil pump drive sprocket from the oil pump shaft. Check the shaft splines.
- Oil pump body. Check the condition of the dowel sleeves.

To install:

3. Clean the oil pump mounting surfaces.

4. Install or connect the following:
- Oil pump on the engine. Mounting bolts: 16 ft. lbs. (22 Nm).
- Oil pump drive sprocket onto the oil pump shaft. Oil pump drive sprocket nut: 18 ft. lbs. (25 Nm). The sprocket nut must be tightened in a counterclockwise direction; it has a reverse, or left hand thread.
- Oil pan
- Oil pan drain plug, then fill the engine with oil
- Negative battery cable

5. Start the engine and check that oil pressure is present; if the oil pressure lamp does not extinguish within 5–7 seconds of starting the engine, turn the engine **OFF**. Check for any oil leaks.

M62 4.4L Engines

1. Before servicing the vehicle, refer to the precautions in the beginning of this section.

2. Remove or disconnect the following:
- Negative battery cable
- Lower oil pan
- Left and right engine mounts at the bottom
- Power steering pump at the holder. On automatic transmissions, remove the oil pipes at the power steering pump.
- Banjo bolt for the oil return pipe from the oil filter at the oil pan
- Mounting bolt on the sprocket for the oil pump and remove the sprocket along with the chain
- 3 oil pump mounting bolts and remove the oil pump. Remove the oil pipes out of the crankcase.

To install

3. Check the seals on the oil pipes and replace it if necessary. Lubricate the seals with oil and the oil pipes.

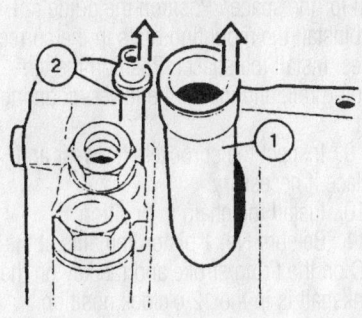

Fresh oil pipe (1), and pure oil pipe (2) locations—M62 4.4L engines

4. Check the seal in the oil pump and replace it if necessary. Screw the hexagon adapter back into the oil pump until it stops.

5. Install or connect the following:
- Oil pump and 2 right side oil pump mounting bolts. Bolts: 14–17 ft. lbs. (20–24Nm).

6. Position the chain on the pump and the sprocket and install the sprocket. Bolt: 35 ft. lbs. (47 Nm). Verify that the chain is positioned correctly.

7. Adjust the chain sag to 0.315–0.472 inches (8–12mm) by turning the hexagon adapter in the oil pump, then install the left side mounting bolt.

8. Install or connect the following:
- Banjo bolt for the oil return pipe from the oil filter at the oil pan
- Power steering pump and connect the oil lines, if equipped
- Lower oil pan
- Engine oil
- Negative battery cable

M73 5.4L Engines

1. Before servicing the vehicle, refer to the precautions in the beginning of this section.

2. Remove or disconnect the following:
- Negative battery cable
- Oil pan
- Bolts retaining the sprocket to the oil pump shaft and remove the sprocket
- Oil pump retaining bolts and lower the oil pump from the engine block. There are 3 bolts at the front and 2 bolts attaching the rear of the oil pick-up to the lower end of a support bracket. It is necessary to remove all 5 bolts.

3. Do not loosen the chain adjusting shims from the 2 mounting locations.

To install:

4. Add or subtract shims between the oil pump body and the engine block to obtain a slight movement of the chain under light thumb pressure.

5. Install oil pump in position.

➡ **When used, the 2 shim thicknesses must be the same. Tighten the pump holder at the pick-up end after shimming is completed to avoid stress on the pump.**

6. After the main pump mounting bolts are tightened, loosen the bolts at the bracket on the rear of the pick-up, allowing the pick-up to assume its most natural position. This will relieve tension on the bracket.

7. The balance of installation is the reverse of the removal procedure.

8. Install or connect the following:
- Oil pan
- Oil pan drain plug, then fill the engine with oil
- Negative battery cable

➡ **Start the engine and check that oil pressure is present; if the oil pressure lamp does not extinguish within 5–7 seconds of starting the engine, turn the engine OFF.**

Rear Main Seal

REMOVAL & INSTALLATION

The rear main bearing oil seal can be replaced after the transmission and flywheel have been removed from the engine.

1. Before servicing the vehicle, refer to the precautions in the beginning of this section.

2. Remove or disconnect the following:
- Transmission
- Flywheel assembly
- Oil seal, using a suitable seal removal tool

To install:

3. Coat the sealing lips of the new seal with oil and install the new seal into the end cover housing with a suitable seal installation tool. On the 6-cylinder engines, press the seal in until it is about 0.039–0.079 inches (0.991–2.070mm) deeper than the standard seal, which was installed flush with the housing.

4. Install the remaining components in the opposite order from which they were removed.

5. Connect the negative battery cable.

6. Start the engine and check that oil pressure is present; if the oil pressure lamp does not extinguish within 5–7 seconds of starting the engine, turn the engine **OFF**.

7. Check and top off all fluid levels.

Timing Chain, Sprockets, Front Cover and Seal

REMOVAL & INSTALLATION

M44 1.9L, M52 2.5L, 2.8L, and S52 3.2L Engines

1. Before servicing the vehicle, refer to the precautions in the beginning of this section.

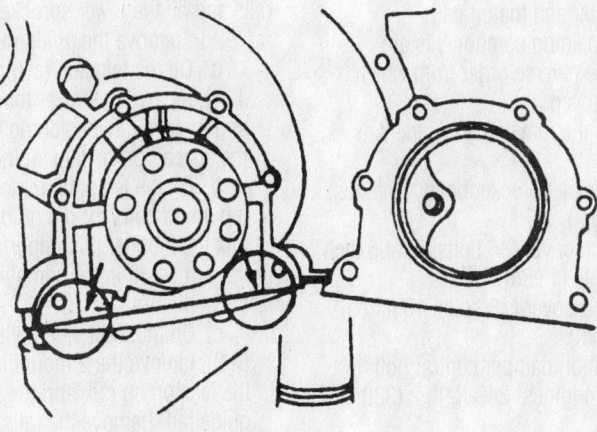

When installing, apply sealer to the joints (as marked) of the rear main seal housing if it has been removed

2. Drain the cooling system.

3. Remove or disconnect the following:

- Negative battery cable
- Radiator and fan assembly
- Drive belts and any accessories that block access to the timing cover. Remove the engine splash shield.
- Vibration damper using a suitable tool, then remove the central bolt and remove the vibration damper hub
- Timing case cover bolts and timing cover

➡**The timing case cover can be removed without removing the water pump.**

- Upper chain guide and top bolt on the right chain guide
- Timing chain sprockets and the lift out the chain. Remove the timing chain guide.
- Tensioning rail, if necessary. Remove the crankshaft sprocket with a suitable tool and lift out the Woodruff key.
- Reversing roller, if needed

➡**The reversing roller can only be replaced complete with bearings.**

To install:

4. Install the Woodruff key into the channel in the crankshaft. Slide the crankshaft sprocket over the end of the crankshaft with the Woodruff key aligning with the channel in the crankshaft. Use the central mounting bolt to draw the sprocket entirely into position.

5. Apply sealant at the intersections of the timing cover and the oil pan.

6. The remaining components are installed in the reverse order from which they were removed.

7. Tighten the remaining fasteners as follows:

- Camshaft sprocket bolts: 16 ft. lbs. (22 Nm)
- Timing cover M6 bolts: 78–96 inch lbs. (9–11 Nm)
- Timing cover M8 bolts: 16 ft. lbs. (22 Nm)
- Vibration damper central bolt: M44 engines: 243 ft. lbs. (330 Nm)
- Vibration damper central bolt: M52 and S52 engines: 302 ft. lbs. (410 Nm)
- Vibration damper pulley bolts: 17 ft. lbs. (23 Nm)

8. Connect the negative battery cable.

M62 4.4L Engines

1. Before servicing the vehicle, refer to the precautions in the beginning of this section.

2. Remove or disconnect the following:

- Negative battery cable
- Accessory drive belt
- front engine splash shield
- Left cylinder head cover
- Alternator
- Oil filter housing
- Left upper timing cover
- Right cylinder head cover
- Mass Air Flow (MAF) sensor
- Timing chain tensioner and mount
- Camshaft Position (CMP) sensor
- Oil dipstick tube
- Right timing cover
- Engine cooling fan
- Water pump pulley
- Crankshaft damper pulley and hub
- Front crankshaft seal using tool No. 11-2-380 and 11-2-383
- Lower timing cover

3. Turn the engine in the direction of rotation and set cylinder number 1 to TDC firing. The arrows on the sprockets should face up in the cylinder axis. Use a crankshaft holder to keep the TDC position.

4. Loosen and remove the camshaft sprocket bolts from both banks of cylinders. Compress the hydraulic tensioning element to loosen the timing chain. Lock the element with tool No. 11-3-420, or equivalent and remove the sprockets with the chain. Do not rotate the engine with the timing chain removed.

5. Guide the chain out of the tensioner rails and off the lower sprocket.

6. To remove the guide rail:

a. On the left side (cylinder bank 1–4), remove the lower mounting bolt and remove the tensioning rail. Remove the spacer for the tensioning rail.

b. On the left side, remove the 2 mounting bolts for the guide rail. Do not mix the 2 bolts, it is important to install the same bolt in the same hole. Remove the sliding rail.

c. On the right side (cylinder bank 5–8), remove the 2 mounting bolts on the tensioning rail, and the 2 bolts on the guide rail. Remove the rails.

To install:

7. On the right cylinder bank, position the guide rail and install the mounting bolts. Position the tensioning rail, and install the mounting bolts.

8. On the left cylinder bank, check the seal for the spacer. Position the guide rail, and install the mounting bolts in the correct holes. Install the spacer. Position the tensioning rail, and install the lower mounting bolt.

9. Inspect the sprockets for wear and replace if necessary.

10. Install the chain in position.

11. Be sure No. 1 piston remains at the TDC on the firing stroke and the key on the crankshaft is in the 12 o'clock position.

12. Position the chain on the guide rail and swing the chain inward and to the left.

13. Engage the chain on the crankshaft gear and install the camshaft sprockets into the chain.

14. The sprocket, on the intake camshaft for cylinder bank 1–4, has a sender pin. With the arrow pointing up, align the pin in the middle of the slots, then install the camshaft sprockets. Remove tool No. 11-3-420 and remove the crankshaft holder.

15. Install the chain tensioner piston, spring and cap plug, but do not tighten.

16. To bleed the chain tensioner, fill the oil pocket located on the upper timing housing cover with engine oil and move the tensioner back and forth with a suitable pry-tool until oil is expelled at the cap plug, then tighten the cap plug securely.

17. Check for the correct seating of the dowel sleeves. Clean the sealing surfaces thoroughly, then place a new gasket on the lower cover.

18. Trim the protruding ends of the gasket, making sure the cutting tool is level. Do not allow the pieces to fall into the engine.

19. Position the lower cover and install the mounting bolts with an even distribution of pressure. Tighten the 6mm bolts to 84 inch lbs. (10 Nm), 8mm bolts to 16 ft. lbs. (22 Nm) and 10mm bolts to 35 ft. lbs. (47 Nm)

20. Install the oil seal in the timing case cover using tool No. 11 1 220 or an equivalent seal driver. Make sure the seal is flush with the cover.

21. Install the vibration damper hub and install the mounting bolt. Tighten the hub bolt as follows:

- Step 1: 74 ft. lbs. (100 Nm)
- Step 2: Plus 60 degrees
- Step 3: Plus 60 degrees
- Step 4: Plus 30 degrees

22. Raise and safely support the vehicle. Install the cooling air guide for the alternator, located on the engine carrier. Install the front engine splash shield and lower the vehicle.

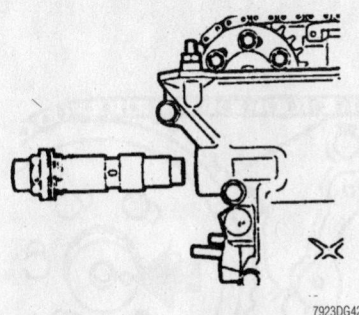

The timing chain tensioner slides out of the side of the front cover—M62 4.4L engines

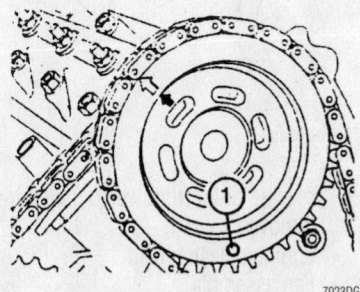

Camshaft sprocket sender pin—M62 4.4L engines

23. Position the vibration damper pulleys and install the mounting bolts for the damper.

24. Install the pulley on the water pump.

25. Install the drive belt and the cooling fan. Rotate the fan counterclockwise to install.

26. Install the intake hose between the throttle body and the air volume meter, then install the manifold cover.

27. Replace the hydraulic tensioner oil seal in the timing case cover.

28. Check for the correct seating of the dowel sleeves. Clean the sealing surfaces to remove any oil or old gasket material and install a new gasket.

29. Mount the timing case cover. Screw in the vertically mounted bolts until the cover contacts the cylinder head. Do not fully tighten the bolts.

30. Install the horizontally mounted bolts, then tighten the vertically mounted bolts in 2 steps. After the vertically mounted bolts have been tightened, torque the horizontally mounted bolts in 2 steps to a final torque as follows:
- M6 bolts: 84 inch lbs. (10 Nm)
- M8 bolts: 16 ft. lbs. (22 Nm)
- M10 bolts: 35 ft. lbs. (47 Nm)

31. Install the oil dipstick guide tube, making sure to replace and lubricate the O-ring.

32. Install the camshaft sender fastener and the chain tensioner.

33. Install the air cleaner upper section along with the MAF sensor.

34. Install the right cylinder head cover.

35. Check for the correct seating of the dowel sleeves. Clean the sealing surfaces to remove any oil or old gasket material and install a new gasket.

36. Mount the timing case cover together with the inserted bolt. This bolt cannot be installed with the cover in place. Install the rest of the mounting bolts and fasten the vertically mounted bolts until the cover just contacts the cylinder head. Do not fully tighten the bolts.

37. Install the horizontally mounted bolts, then tighten the vertically mounted bolts in 2 steps. After the vertically mounted bolts have been tightened, torque the horizontally mounted bolts in 2 steps to a final torque as follows: M6 bolts: 84 inch lbs. (10 Nm), M8 bolts: 16 ft. lbs. (22 Nm), M10 bolts: 35 ft. lbs. (47 Nm)

38. Install or connect the following:
- Battery positive cable for the alternator and install the protective tube mounting fasteners and connect the remaining wires to the alternator
- Oil filter housing and the return pipe and replace the housing cover
- Alternator and cylinder head cover
- Negative battery cable

M73 5.4L Engines

1. Before servicing the vehicle, refer to the precautions in the beginning of this section.

2. Remove or disconnect the following:
- Negative battery cable
- Coolant
- Fan assembly by rotating it clockwise
- Drive belts and engine splash shield. Remove the tensioning bolt.
- Both intake manifolds and distributor housings
- Round rubber mounts, bolts and nuts. Remove both cylinder head covers.
- Mounting bolts and lift out the timing cover
- Bolts but do not remove the vibration damper. Remove the central hub bolt with a suitable tool.
- Vibration damper using a suitable tool to pull the vibration damper hub from the crankshaft
- Engine oil, and then the lower section of the oil pan
- Bottom mounting fasteners from the timing case cover and loosen

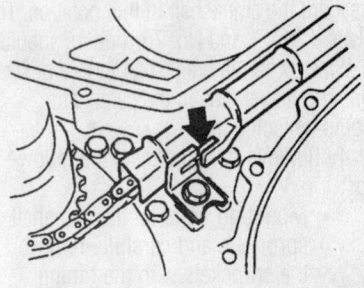

With the timing case cover fitted correctly, the retaining tab is not visible—M73 5.4L engines

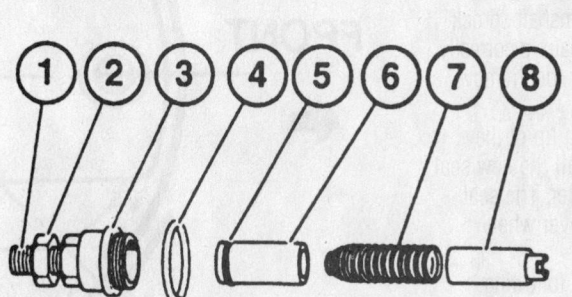

Timing chain tensioner components—M73 5.4L engines

1. Adjusting screw
2. Lock nut
3. Screw plug
4. Replace sealing ring
5. Replace o-ring
6. Dowel sleeve
7. Compression spring
8. Chain tensioning piston

Heater Core replacement is covered in Section 2 of this manual

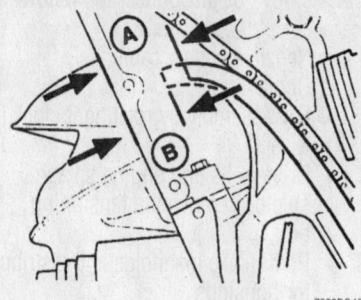

7923DG46

Timing chain tensioner adjustment: dimension "B" from "A" should be 0.216–0.256 inches (5.48–6.5mm)—M73 5.4L engines

the adjacent oil pan bolts on both sides
- Timing chain tensioner and reference mark sender
- Timing case mounting fasteners and remove the timing case cover
- Front cover oil seal using tool No. 11-1-210

3. Set the engine to TDC firing for cylinder number one. Install suitable holder to secure the crankshaft in this position. The valves of No. 1 and No. 7 cylinders should be closed with the crankshaft in this position. The dowel pins in the camshafts should be facing in.

4. Remove or disconnect the following:

- Mounting bolts on the camshaft sprockets and carefully remove the sprockets with the timing chain
- Mounting bolts for the timing chain guide and remove the guide

To install:

5. Position and install the timing chain guide.

6. Position the timing chain on the sprockets and install the camshaft sprockets. Verify that the timing chain is correctly aligned on all the sprockets, then remove the crankshaft holder.

7. Lubricate the sealing lip of the shaft seal with oil and install the new seal using a suitable seal installer. The seal should be flush with the cover when installed.

8. Install or connect the following:
- Timing chain cover and mounting bolts
- Reference mark sender and the timing chain tensioner
- Remaining components in the reverse order of removal

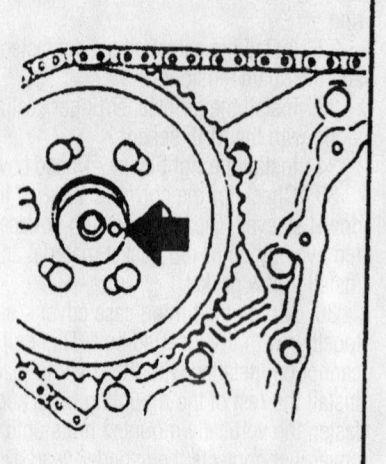
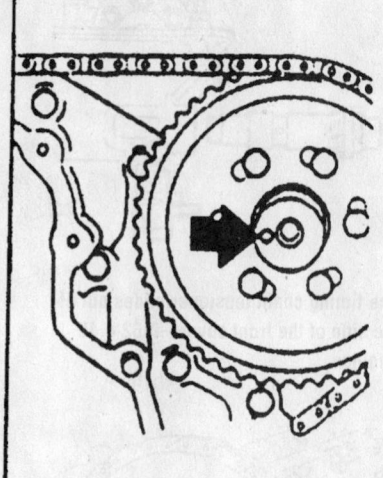

7923DG47

Camshaft dowel pins with the engine at TDC—M73 5.4L engines

9. Torque the central vibration damper hub bolt as follows:
- Step 1: 74 ft. lbs. (100 Nm)
- Step 2: turn an additional 60 degrees
- Step 3: turn an additional 60 degrees

10. Tighten the vibration damper pulleys to 17 ft. lbs. (25 Nm).

11. Install or connect the following:
- Both cylinder head covers
- The remaining components in the reverse order from which they were removed
- Engine splash shield
- Fan assembly
- Negative battery cable

12. Fill the engine with clean oil.
13. Fill and bleed the cooling system.

Piston and Ring

POSITIONING

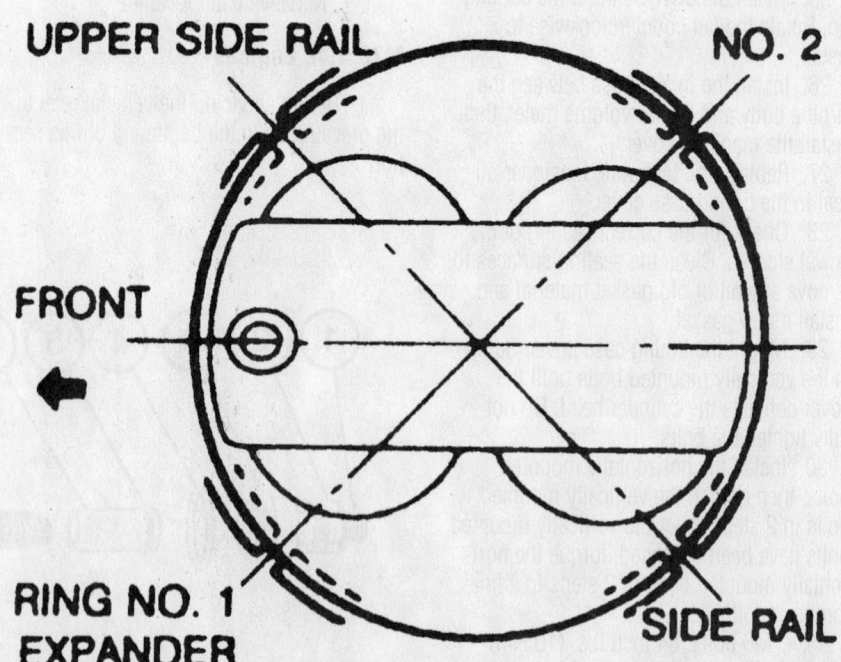

UPPER SIDE RAIL

NO. 2

FRONT

RING NO. 1 EXPANDER

SIDE RAIL

7923AGB3

Piston ring end-gap spacing—All engines

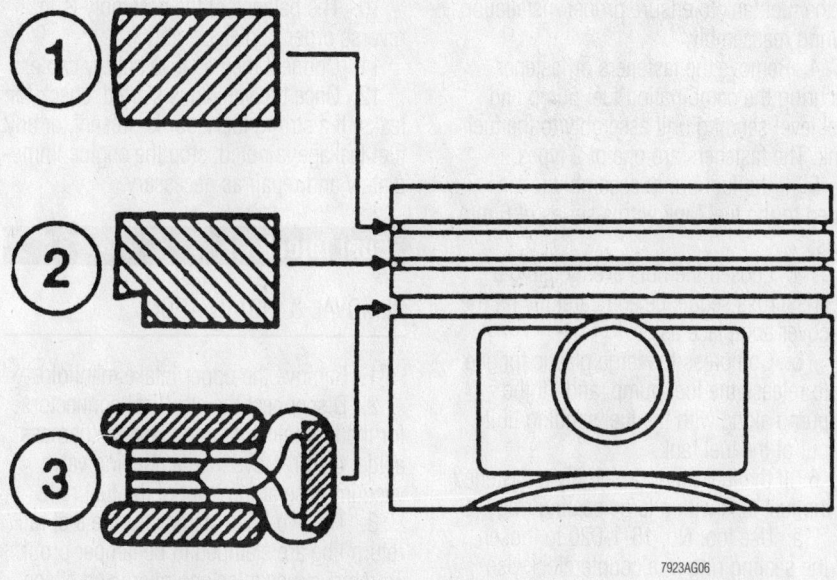

Compression and oil control ring locations—All engines

7923AG06

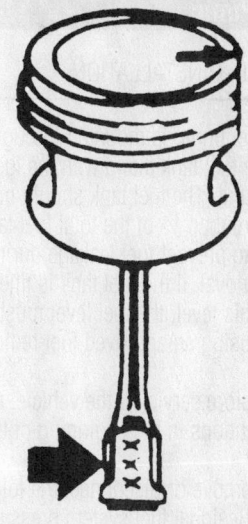

7923AG05

Connecting rod-to-piston positioning—All engines

FUEL SYSTEM

Fuel System Service Precautions

Safety is the most important factor when performing not only fuel system maintenance but also any type of maintenance. Failure to conduct maintenance and repairs in a safe manner may result in serious personal injury or death. Maintenance and testing of the vehicle's fuel system components can be accomplished safely and effectively by adhering to the following rules and guidelines.

• To avoid the possibility of fire and personal injury, always disconnect the negative battery cable unless the repair or test procedure requires that battery voltage be applied.

• Always relieve the fuel system pressure prior to disconnecting any fuel system component (injector, fuel rail, pressure regulator, etc.), fitting or fuel line connection. Exercise extreme caution whenever relieving fuel system pressure, to avoid exposing skin, face and eyes to fuel spray. Fuel under pressure may penetrate the skin or any part of the body that it contacts.

• Always place a shop towel or cloth around the fitting or connection prior to loosening to absorb any excess fuel due to spillage. Ensure that all fuel spillage (should it occur) is quickly removed from engine surfaces. Ensure that all fuel soaked cloths or towels are deposited into a suitable waste container.

• Always keep a dry chemical (Class B) fire extinguisher near the work area.

• Do not allow fuel spray or fuel vapors to come into contact with a spark or open flame.

• Always use a back-up wrench when loosening and tightening fuel line connection fittings. This will prevent unnecessary stress and torsion to fuel line piping. Always follow the proper torque specifications.

• Always replace worn fuel fitting O-rings with new. Do not substitute fuel hose where fuel pipe is installed.

Fuel System Pressure

RELIEVING

To relieve the pressure in the system, locate fuel pump relay located on the cowl. The relay can sometimes be distinguished by the orange color of the housing. Unplug and remove the relay, and place it in a safe location. With the fuel pump relay removed, start the engine and operate it until it stalls. Crank the engine for 10 seconds after it stalls to remove any residual pressure.

Fuel Filter

REMOVAL & INSTALLATION

On filters that are located near the fuel tank, it is necessary to clamp the fuel lines closed before disconnecting them, or fuel will run out continuously.

1. Before servicing the vehicle, refer to the precautions in the beginning of this section.

2. Remove or disconnect the following:
   • Fuel system pressure
   • Negative battery cable

3. Clamp the fuel lines closed if the filter is mounted low, near the fuel tank or underneath the vehicle. Then, loosen the clamps and disconnect the inlet and outlet hoses. Remove the hose clamps or slide them back, well off the connections to make it easier to remove the hoses, if necessary.

4. The filters are usually attached to a bracket on the frame, floor pan or wheel well. Loosen the bracket and remove the filter. Note the direction of flow for reinstallation or replacement of the filter.

To install:

5. Position the new fuel filter and install the fuel lines onto the correct fuel filter fittings. Tighten the fuel line clamps until tight, but not to the point where the fuel lines become excessively pinched or damaged, then tighten the mounting bracket until snug.

6. Connect negative battery cable. Cycle the ignition **ON** and **OFF** several times to build fuel pressure.

7. Inspect the fuel filter and fuel lines for any fuel leaks.

Brake service is covered in Section 4 of this manual

Fuel Pump

REMOVAL & INSTALLATION

The fuel pump is mounted through the top of the fuel tank along with the fuel level sending unit. The fuel tank should not be filled more than 1/3 of the total fuel tank capacity to prevent fuel leakage during fuel pump removal. If the fuel tank is filled beyond this level, the fuel level must be reduced using an approved fuel removal device.

1. Before servicing the vehicle, refer to the precautions in the beginning of this section.
2. Remove or disconnect the following:
   - Residual fuel system pressure
   - Negative battery cable
   - Fuel, if filled beyond 1/3 of the capacity of the fuel tank. Drain the fuel tank enough to prevent spillage when removing the pump using an approved fuel removal device.

➡ **The fuel pump must be removed through the top of the fuel tank, thus the location of the fuel tank determines whether the fuel pump is accessed by removal of the rear seat, or removal of the trim panels in the trunk.**

   - Rear seat, or the trim panels in the trunk, depending on fuel tank location to access the top of the fuel tank

➡ **On models which require removal of the rear seat, the insulation mat under the seat must be cut in a "U" shape to allow the insulation to be folded up to access the top of the fuel tank.**

   - Fasteners securing the metal cover located above the fuel tank, and remove the cover
   - Electrical connector at the top of the combination fuel pump and fuel level sending unit assembly
   - Fuel feed and return lines

3. Match mark the combination fuel pump and fuel level sending unit assembly

to the fuel tank to ensure proper installation during reassembly.

4. Remove the fasteners or fastener securing the combination fuel pump and fuel level sending unit assembly to the fuel tank. The fasteners are one of 2 types.

5. If the fuel pump assembly is fastened to the fuel tank with a series of 6 mm nuts:

   a. Loosen the nuts evenly using a crisscross sequence and carefully lift the cover and place aside.

   b. Compress the large plastic tongue to release the fuel pump, and lift the pump along with the fuel sending unit out of the fuel tank.

6. If the fuel pump assembly is fastened to the fuel tank with a large sealing ring:

   a. Use tool No. 16-1-020 to loosen the sealing ring in a counterclockwise direction.

   b. With the seal ring removed, lift the fuel pump assembly out of the fuel tank.

To install:

➡ **Always use a new seal or gasket when installing the fuel pump or fuel level gauge sending unit assembly.**

7. Install the fuel pump into the fuel tank taking care not to bend or damage the fuel sending unit assembly.

8. If the fuel pump is held in place by a plastic bracket in the fuel tank perform the following:

   a. Make sure the fuel pump is fully snapped in place.

   b. Install the fuel tank cover plate with a new gasket and torque the fasteners using a crisscross pattern to 57 inch lbs. (6.5 Nm).

9. If the fuel pump is held in place with a sealing ring perform the following:

   a. Ensure the pump is properly aligned with the fuel tank matchmarks made during disassembly.

   b. Install a new seal and torque the sealing ring using tool No. 16-1-020 as follows:

   - Metal sealing rings: 26 ft. lbs. (35 Nm)
   - Plastic sealing rings: 41 ft. lbs. (55 Nm)

10. The balance of the assembly is in reverse order of disassembly.
11. Connect the negative battery cable.
12. Once the vehicle is started, check for leaks. If a strong fuel odor is present, or any fuel leakage is noted, stop the engine immediately and repair as necessary.

Fuel Injector

REMOVAL & INSTALLATION

1. Remove the upper intake manifold.
2. Disconnect the electrical connectors for the fuel injectors and place the harness aside, then remove the dampening valve vacuum hose at the front of the fuel rail.
3. The two hose clamps for the fuel rail return line are stamped to be tamper proof. To remove, use a locking pliers and attach then onto the clamp's adjustment screw head securely. Turn the adjustment screw counterclockwise until the clamp is loosened. Loosen both clamps and slide to the middle of the fuel line.
4. Using a 12 mm line wrench, loosen the fuel rail feel line flare nut and once loose, slide the flare nut away from the fuel rail.
5. To avoid debris from entering the injector ports, If compressed air is available, using an air nozzle, and wearing eye protection, carefully spray compressed air around the base of the injectors. If excessive debris exists, spray the base of the injectors with an evaporating brake cleaner and spray again with compressed air. If compressed air is not available, spray the brake cleaner to clean as best as possible.
6. Using a 10 mm socket, short extension and a ratchet, remove the two fuel rail mounting bolts.
7. Use a clean shop towel folded over several times to form a small pad, and place on the corner of the valve cover. Then using a suitable and sturdy prytool, slowly and carefully lift the fuel rail until the injectors are just removed from the manifold, then remove the fuel feed and fuel return lines away from the fuel rail and remove the entire assembly with the large vent hoses still attached and set aside.

DRIVE TRAIN

Transmission Assembly

REMOVAL & INSTALLATION

Manual

3 SERIES

1. Before servicing the vehicle, refer to the precautions in the beginning of this section.

2. Remove or disconnect the following:
   - Negative battery cable
   - Exhaust system. Remove the cross brace and heat shield.
   - Flexible coupling at the front of the driveshaft. Some vehicles may also have a vibration damper located at this point in the drive train. This damper is mounted on the transmission output flange with bolts that are pressed into the damper. On these vehicles, remove the nuts located behind the damper.

3. Loosen the threaded sleeve on the driveshaft. Use a suitable tool to hold the splined portion of the shaft while turning the sleeve.

4. Remove or disconnect the following:
   - Mounting bolts and center driveshaft mount. Then, bend the driveshaft downward at the center and pull it off the transmission output flange. Keep the sections of the driveshaft from pulling apart and suspend it from the vehicle.
   - Retainer and washer and pull out the shift selector rod
   - Self-locking bolts that retain the shift rod bracket at the rear of the transmission, then remove the bracket. If equipped with a shift arm, use a suitable prytool to pry the spring clip up off the boss on the transmission case and swing it upward. Then, pull out the shift shaft pin.
   - Clutch slave cylinder and support it so the hydraulic line can remain connected

5. The transmission incorporates sending units for flywheel rotating speed and position. Remove the heat shield that protects these from the exhaust heat, then remove the retaining bolt for each sending unit. Note that the speed sending unit, which has no identifying ring is located on the right, and that the

reference mark sending unit, which has a marking ring, is located on the left. If the sending units are installed in reverse positions, the engine will not run. Remove both units from the flywheel housing.

6. Detach the wiring connector going to the back-up light switch and place the wires aside.

7. Support the transmission. Remove the mounting bolts and crossmember holding the rear of the transmission to the body. Then, lower the transmission onto the front axle carrier.

8. Using the proper tool, remove the bolts holding the transmission flywheel housing to the engine. Be sure to retain the washers with the bolts. Pull the transmission rearward to slide the input shaft out of the clutch disc, then lower the transmission and carefully remove it from the vehicle.

To install:

9. Installation is the reverse of the removal procedure, while using the following torque values:
   - Front mounting bolts: 46–58 ft. lbs. (62–80 Nm)
   - Shift rod bracket bolts: 16 ft. lbs. (22 Nm)
   - Driveshaft center bearing bolts: 16–17 ft. lbs. (21–23 Nm)
   - Flexible coupling bolts: 83–94 ft. lbs. (114–129 Nm)

1998–02 5 SERIES MODELS

1. Before servicing the vehicle, refer to the precautions in the beginning of this section.

2. Remove or disconnect the following:
   - Negative battery cable
   - Exhaust system, to provide clearance for transmission removal. Remove the heat shield brace and transmission heat shield.
   - Driveshaft coupling at the rear of the transmission
   - Screw-on ring type connection that attaches the driveshaft to the center bearing. Then, unbolt the center bearing mount. Bend the driveshaft downward and pull it off the centering pin. If equipped with a vibration damper, turn it and pull it back over the output flange before pulling the driveshaft off the guide pin. Suspend it from the vehicle.
   - The wires for the back-up light switch. Remove the passenger

compartment console to disconnect it from the top of the transmission by removing the self-locking bolts. Discard and replace the bolts.
   - Shift rod at the rear of the transmission, by pulling out the locking clip

3. If the transmission is linked to the shift lever with an arm, use a small prytool to lift the spring out of the holder on the bracket, then raise the arm, then remove the shift shaft bolt.

4. Remove or disconnect the following:
   - With a flywheel housing cover (semi-circular in shape), the mounting bolts and cover
   - The speed sensor and reference mark sensor. Note their locations. The speed sensor is located in the upper bore, marked **D**. The reference mark sensor, which has a ring, is located in the lower bore, marked **B**. Check the O-rings for the sensors and install new ones if they are damaged.
   - Rear transmission crossmember, with transmission supported
   - Upper and lower attaching nuts and remove the clutch slave cylinder, supporting it so the hydraulic line need not be disconnected. Disconnect the reverse gear back-up light switch and pull the wires out of the holders and place aside.
   - The bolts fastening the transmission to the bell housing, using the proper tool. Pull the transmission rearward until the input shaft has disengaged from the clutch disc, then lower and remove it.

To install:

5. Installation is the reverse of the removal procedure, while using the following torque values:
   - Transmission-to-bell housing: 52–58 ft. lbs. (70–80 Nm)
   - Rear/top transmission Torx® bolts: 46–58 ft. lbs. (62–80 Nm)
   - Center mount-to-body: 16–17 ft. lbs. (21–23 Nm)
   - Front joint-to-transmission: 83–94 ft. lbs. (114–129 Nm)

7 SERIES

1. Before servicing the vehicle, refer to the precautions in the beginning of this section.

2. Remove or disconnect the following:
- Negative battery cable
- Exhaust system. Remove the attaching bolts and remove the heat shield mounted just to the rear of the transmission on the floorpan.
- Crossmember that supports the transmission at the rear from the body by removing the mounting bolts on both sides
- Bolts, passing through the vibration damper and front universal joint at the front of the driveshaft
- Mounting bolts, then center driveshaft mount. Bend the driveshaft downward at the center and pull it off the transmission output flange. Keep the sections of the driveshaft from pulling apart and suspend it from the vehicle.

3. Pull out the circlip, slide off the washer, and then pull the shift selector rod off the transmission shift shaft. Disconnect the back-up light switch.

4. Lower the transmission slightly for access. Then, use a small prytool to lift the spring out of the holder on the bracket, then raise the arm. Pull out the shift shaft bolt.

5. Remove or disconnect the following:
- Upper and lower attaching nuts and remove the clutch slave cylinder, supporting it so the hydraulic line need not be disconnected
- Bolts fastening the transmission to the bell housing. Be sure to retain the washer with each bolt to ensure that they can be readily removed later, if necessary. Pull the transmission rearward until the input shaft has disengaged from the clutch disc, then lower and remove the transmission.

To install:

6. Install the transmission in position under the vehicle. Align the input shaft and install the transmission.

7. Preload the center bearing mount forward of its most natural position 0.157–0.236 inches (3.98–5.99mm).

8. When reconnecting the nuts and bolt at the transmission coupling, replace the nuts with new ones and turn only the nut, holding the bolts stationary. Tighten the center mount-to-body nut to 16–17 ft. lbs. (21–23 Nm). Tighten the front joint-to-transmission nut to 58 ft. lbs. (80 Nm).

9. Install or connect the following:
- Remaining components in the reverse order of removal

- Shift arm, if equipped, and lubricate the bolt with a light layer of a suitable lubricant, then check the O-ring for crushing, cracks or cuts, replacing it, if damaged
- Exhaust system
- Negative battery cable
- All remaining fluids

Automatic Transmission

3 SERIES

➡ To perform this operation, a support for the transmission, tool Nos. 24-0-120 and 00-2-020 or the equivalents and a tool for tightening the driveshaft locking ring, tool No. 26-1-040, are recommended.

1. Before servicing the vehicle, refer to the precautions in the beginning of this section.

2. Remove or disconnect the following:

- Negative battery cable
- Throttle cable adjusting nuts, release the cable tension and disconnect the cable at the throttle lever. Then, remove and retain the nuts and pull the cable housing out of the bracket.
- Exhaust system at the manifold and hangers and lower it aside. Remove the hanger that runs across under the driveshaft. Remove the exhaust heat shield from under the center of the vehicle.
- Transmission oil. Remove the oil filler neck. Disconnect the oil cooler lines at the transmission by removing the flare nuts and plug the open connections.

3. Support the transmission. Separate the torque converter housing from the transmission by removing the Torx® bolts with the proper tool, then remove the bolts from underneath. Retain the washers used with the Torx® bolts.

4. Remove or disconnect the following:

- Bolts attaching the torque converter housing to the engine, making sure to retain the spacer used behind one of the bolts. Then, loosen the mounting bolts for the oil level switch just enough so the plate can be removed while pushing the switch mounting bracket to one side.
- Bolts attaching the torque converter to the driveplate. Turn the flywheel as necessary to gain

access to each of the bolts. Be sure to re-use the same bolts and retain the washers.
- Speed and reference mark sensors. Remove the attaching bolt for each and remove each sensor.
- Bayonet type electrical connector, turning it counterclockwise, then pull the plug out of the socket. Then, lift the wiring harness out of the harness bails
- Crossmember that supports the transmission at the rear
- Transmission shift rod. Then, remove the nuts, then the through-bolts from the damper-type U-joint at the front of the transmission.
- Transmission locking ring at the center mount, if equipped, using a suitable tool. Then, remove the bolts and remove the center mount. Bend the driveshaft downward and pull it off the centering pin. Suspend it from the underside of the vehicle.

5. Lower the transmission as far as possible. Then, remove all the Torx® or standard type bolts attaching the transmission to the engine.

6. Remove the small grill from the bottom of the transmission. Then, press the converter off with a large prytool passing through this opening while sliding the transmission out.

To install:

7. Position the transmission under the vehicle and install the torque converter onto the transmission.

8. Be sure the converter is fully seated onto the transmission, such that the ring on the front is inside the edge of the case. Install the small air grille onto the bottom of the transmission.

9. Install or connect the following:
- Engine-to-transmission attaching bolts
- The remaining components in the reverse order from which they were removed

10. Make note of the following points during reassembly:
- When reinstalling the driveshaft, tighten the lockring with the proper tool. Be sure to replace the self-locking nuts on the driveshaft flexible joint and to hold the bolts while tightening the nuts to keep from distorting the joint.
- When installing the center mount, preload it forward from its most natural position 0.157–0.236

inches (3.98–5.99mm).

- When reconnecting the bayonet type electrical connector, be sure the alignment marks are aligned after the plug it twisted into its final position.
- When reinstalling the speed and reference mark sensors, inspect the O-rings used on the sensors and install new ones, if necessary. Be sure to install the speed sensor into the bore marked **D** and the reference mark sensor, which is marked with a ring, into the bore marked **B**.
- Tighten the crossmember mounting bolts to 16–17 ft. lbs. (21–23 Nm).
- If O-rings are used with the transmission oil cooler connections, replace them

11. Install or connect the following:
- Negative battery cable, then check and adjust the throttle cables as necessary
- Transmission fluid
- All remaining fluids

5 AND 7 SERIES

➡ **To perform this operation, the following tools are recommended. Transmission support tool No. 24-0-120 and 00-2-020 and driveshaft locking ring tool No. 26-1-040, or the equivalents.**

1. Before servicing the vehicle, refer to the precautions in the beginning of this section.
2. Remove or disconnect the following:
- Negative battery cable
- Throttle cable adjusting nuts, release the cable tension and disconnect the cable at the throttle lever. Then, remove the nuts and pull the cable housing out of the bracket.
- Exhaust system at the manifold and hangers and lower it out of the way. Remove the hanger that runs across under the driveshaft. Remove the exhaust heat shield from under the center of the vehicle.
- Crossmember that supports the transmission at the rear, with transmission supported
- Driveshaft coupling through-bolts and nuts or the CV-joint through bolts and nuts. Either type is located right at the rear of the

transmission. Discard and replace the self-locking coupling nuts. Keep the CV-joint clean and replace its gasket upon reassembly.
- Transmission locking ring at the center mount, if equipped. Then, remove the bolts and remove the center mount. Bend the driveshaft downward and pull it off the centering pin. Suspend it from the underside of the vehicle.
- Transmission oil. Remove the oil filler neck. Disconnect the oil cooler lines at the transmission by removing the flare nuts and plug the open connections.
- If equipped, converter cover by removing the Torx® bolts from behind and the regular bolts from underneath.
- Bolts fastening the torque converter to the driveplate, turning the flywheel as necessary to gain access from below
- If equipped, the guard for the speed and reference mark sensors. Remove the attaching bolt for each and remove each sensor. Keep the sensors clean.
- Shift cable by loosening the locknut fastening it to the shift lever and disconnecting the cable at the cable housing bracket

3. If the transmission has an electrical connection, turn the bayonet fastener to the left to release the connection, disconnect it and pull the wire out of the brackets and place aside.
4. Lower the transmission as far as possible. Then, remove all the fasteners attaching the transmission to the engine.
5. Remove the small grill from the bottom of the transmission. Then, press the converter off with a large prytool through this opening while sliding the transmission out.

To install:

6. Place the transmission under the vehicle and raise it into position. Slide the torque converter and the transmission together before installing the transmission completely to the engine. Make sure the converter is fully seated onto the transmission, such that the ring on the front is inside the edge of the case.
7. Install or disconnect the following:
- Small grill onto the transmission
- Transmission to the engine
- The remaining components in the

reverse order from which they were removed
8. Make note of the following points during reassembly:
- Driveshaft, as follows: Tighten the lockring with a suitable tool. If the driveshaft has a simple coupling, rather than a CV-joint, be sure to replace the self-locking nuts and to hold the bolts still while tightening the nuts to keep from distorting the coupling.
- Center mount. Preload it forward from its most natural position 0.157–0.236 inches (3.98–5.99mm).
- Transmission fluid
- All remaining fluids as necessary
- Negative battery cable, then check and adjust the throttle cables as necessary

SHIFT LINKAGE ADJUSTMENT

1. Before servicing the vehicle, refer to the precautions in the beginning of this section.
2. Move the selector lever to **P** position. Loosen the nut until the cable is free.
3. Push the transmission lever to the **D** or **P** position. Then, push the cable rod in the opposite direction.
4. Clamp down the cable rod without tension.
5. Tighten the nut to 84–102 inch lbs. (9–11 Nm).

➡ **Do not bend the cable.**

THROTTLE LINKAGE ADJUSTMENT

1. Before servicing the vehicle, refer to the precautions in the beginning of this section.
2. On the injection system throttle body, loosen the 2 locknuts at the end of the throttle cable and adjust the cable until there is a play of 0.010–0.030 inches (0.254–0.762mm).
3. Loosen the locknut and lower the kickdown stop under the accelerator pedal. Have someone depress the accelerator pedal until the transmission detent can be felt. Then, back the kickdown stop back out until it just touches the pedal.
4. Check that the distance from the seal at the throttle body end of the cable housing is at least 1.732 inches (43.9mm) from the rear end of the threaded sleeve and tighten the locknuts.

For Accessory Drive Belt illustrations, see Section 1 of this manual

Clutch

ADJUSTMENT

These vehicles are equipped with a hydraulic clutch actuating system. No adjustment is possible.

REMOVAL & INSTALLATION

1. Before servicing the vehicle, refer to the precautions in the beginning of this section.
2. Remove or disconnect the following:
   - Negative battery cable
   - Heat shield, then the mounting bolts. Disconnect the speed and reference mark sensors at the flywheel housing. Mark the plugs for reinstallation.
   - Transmission and clutch housing
   - With a 265/6 transmission (without an integral clutch housing), clutch housing. On vehicles with 6-cylinder engines, a Torx® socket is required.
3. Prevent the flywheel from turning, using a suitable locking tool.
4. Loosen the mounting bolts one after another gradually, 1–1 ½ turns at a time, to relieve tension from the clutch.
5. Remove the mounting bolts, clutch and driveplate. Coat the splines of the transmission input shaft with Molykote® Long-term 2, Microlube® GL 2611. Be sure the clutch pilot bearing, located in the center of the crankshaft, turns easily.
6. Check the clutch driven disc for excess wear or cracks. Check the integral torsional damping springs, used with lighter flywheels only, for tight fit. Inspect the rivets to be sure they are all tight. Inspect the flywheel to be sure it is not scored, cracked, or burned. Use a straightedge to verify the contact surface is true. Replace any defective parts.

To install:

➠On vehicles with 6-cylinder engines, the clutch pressure plate must fit over dowel pins.

7. To install, fit the new clutch disk on a suitable clutch alignment tool and center the tool with the flywheel by centering it with the clutch pilot bearing. Mount the clutch pressure plate and install the mounting bolts. Clutch mounting bolts: 16–19 ft. lbs. (21–26 Nm).
8. When installing the clutch retaining bolts turn them in gradually to evenly tighten the clutch pressure plate to prevent warpage.
9. Install or connect the following:
   - Clutch housing, if equipped with separate housing, then the transmission. If the clutch housing is part of the transmission, install the transmission.
   - If equipped, speed and reference mark sensors, then the heat shield
   - The remaining components in the reverse order from which they were removed
   - All remaining fluids as necessary
   - Negative battery cable

Hydraulic Clutch System

BLEEDING

The clutch system can be bled using a pressure bleeder. Follow the instructions that come with the pressure bleeder for the proper pressure bleeding procedure. The maximum line pressure while pressure bleeding must not exceed 36 pi (248 pa).

1. Before servicing the vehicle, refer to the precautions in the beginning of this section.
2. To bleed a clutch manually requires the following:
   - The assistance of a second person
   - A section of hose that is compatible with brake fluid (preferably clear) and fits the slave cylinder bleeder valve snugly
   - A container to catch the fluid that is bled through the system.

➠Cleanliness is of the utmost importance. Always be sure the fluid reservoir is as clean as possible to avoid bleeding debris or contaminated fluid into the system check valves and seals. As brake hydraulic fluid easily absorbs moisture, always use fresh fluid when bleeding a clutch hydraulic system. If the clutch hydraulic system has been run dry, or if the hydraulic fluid line, clutch slave or master cylinder has just been replaced, it may be necessary to repeat the bleeding procedure a number of times to remove all of the air. Make sure the reservoir does not run dry during bleeding. If the reservoir runs dry the bleeding process must be repeated until no air is present.

3. To bleed the system perform the following:
   a. Top off the clutch hydraulic fluid reservoir using a fluid that meets the standards of the vehicle's clutch hydraulic system.
   b. With one end of a section of hose attached to the slave cylinder bleed valve and the other end of the hose submerged into a container of fresh brake hydraulic fluid, open the clutch slave cylinder bleed valve. Slowly press the clutch pedal to the floor and hold the pedal down.
   c. Close the clutch slave cylinder bleed valve.
   d. Slowly release the clutch pedal.
   e. Check the hydraulic fluid level and top off as necessary.
4. Repeat the above steps until the discharged fluid is clean and no air bubbles appear during the bleeding process.

Halfshafts

REMOVAL & INSTALLATION

3 Series

1. Before servicing the vehicle, refer to the precautions in the beginning of this section.
2. Remove or disconnect the following:
   - Rear tire and wheel assembly
   - Output shaft at the outer flange and suspend it
   - Caliper, and suspend it with the brake line connected. Unbolt and remove the rear disc.
   - Large nut and lock plate. If equipped with ABS, disconnect, then remove the ABS speed sensor.
   - Collar nut. Then, remove the drive flange with a suitable press tool or install the collar nut until it is just flush with the end of the shaft and use a suitable soft faced hammer to knock out the shaft.
   - Snapring. Pull out the wheel bearings, using a suitable tool.
3. Pull the inner bearing race off the axle shaft with tool No. 00-7-500.

To install:

4. Install or connect the following:
   - New bearing assembly, using suitable bearing driver tools. Install the snapring.
   - Rear axle shaft with tool Nos.: 23-

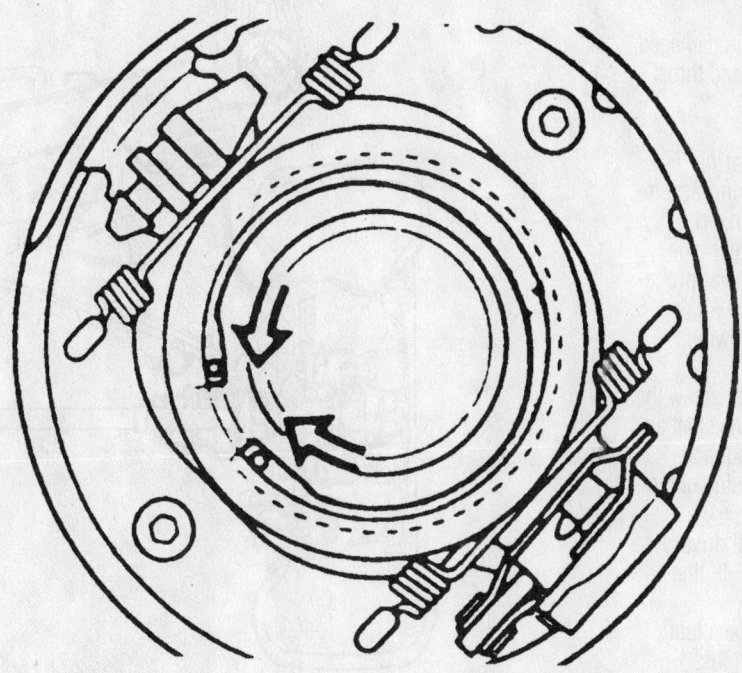

Remove the snapring from the hub—3 and 5 series

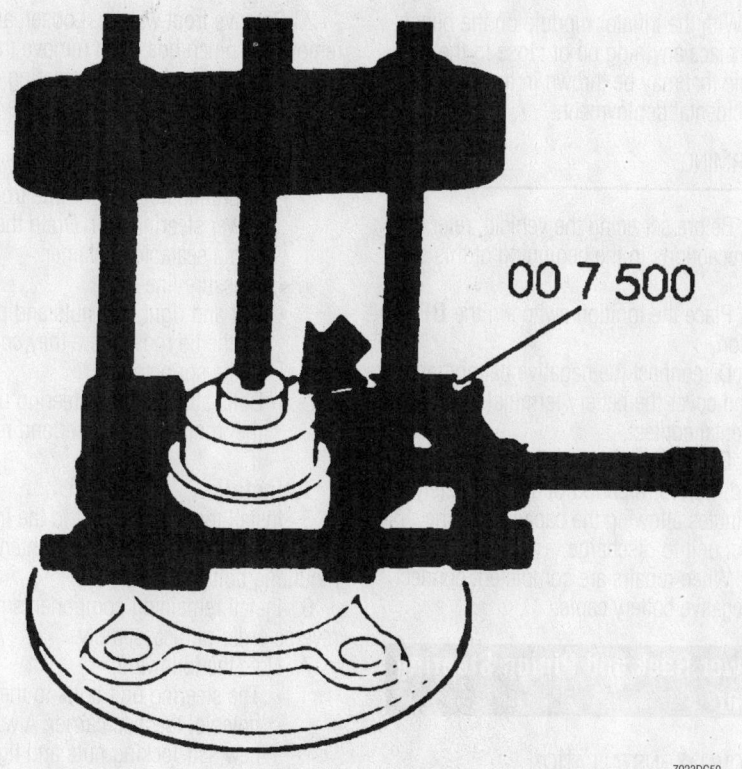

00 7 500

Tool No. 00-7-500 bearing puller used to remove the inner bearing race—3 and 5 series

1-300, 33-4-080 and 33-4-020 or their equivalents.
- Collar nut, after lubricating, and drive in the lock plate using a suitable tool. Collar nut: 148 ft. lbs. (200 Nm).
- ABS speed sensor
- Brake disc and caliper
- Output shaft
- Rear tire and wheel assembly

5 and 7 Series

1. Before servicing the vehicle, refer to the precautions in the beginning of this section.

2. Remove or disconnect the following:

- Rear tire and wheel assembly
- Lock plate and if equipped, the ABS sensor
- Retaining nut from the output flange. Remove the flange.
- Output half shaft from the final drive (differential carrier) by pressing out with a suitable tool and suspend it
- Output shaft from the drive flange hub using a suitable tool
- Rear axle shaft with a suitable tool
- Snapring. Then, pull out the wheel bearings, using a suitable tool.

For Tire, Wheel and Ball Joint specifications, see Section 1 of this manual

3. Pull out the seal with a suitable tool.

4. If the inner bearing shell is damaged, pull it off with a suitable puller and thrust pad.

To install:

5. Using an appropriate bearing installer, install the wheel bearing assembly, then install the seal, and insert the snapring, then install the rear axle shaft, all in reverse of the removal procedure.

6. Install or connect the following:
- Axle shaft seal
- Output shaft, as follows: screw the threaded spindle into the shaft all the way, then use the nut and washer against the outside of the bridge
- Output shaft to the final drive. Mounting bolts: 42–46 ft. lbs. (58–63 Nm).
- Outer nut, bearing surface lubricated. Nut: 169–188 ft. lbs. (234–260 Nm).
- ABS sensor, if removed
- Rear tire and wheel assembly

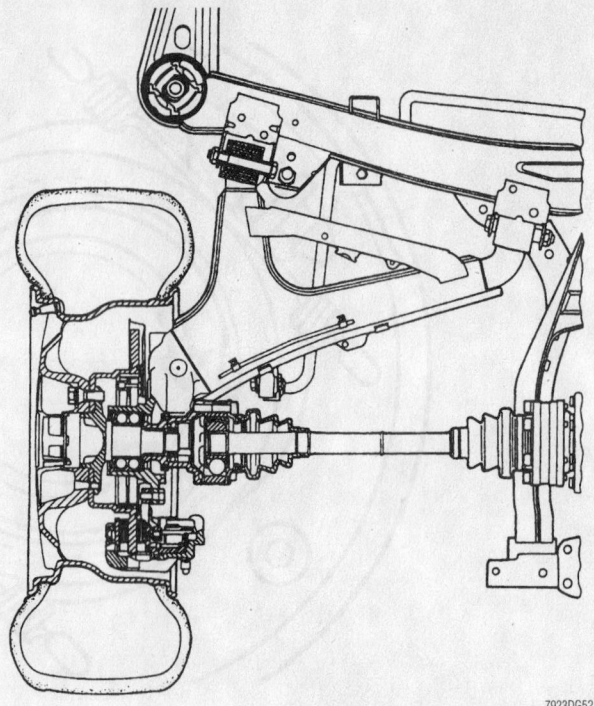

7923DG52

Cut-away schematic of the rear halfshaft and suspension system—5 Series

STEERING AND SUSPENSION

Air Bag

✳✳ CAUTION

The vehicles are equipped with an air bag system. The system must be disarmed before performing service on, or around, system components, the steering column, instrument panel components, wiring and sensors. Failure to follow the safety precautions and the disarming procedure could result in accidental air bag deployment, possible personal injury and unnecessary system repairs.

PRECAUTIONS

Several precautions must be observed when handling the inflator module to avoid accidental deployment and possible personal injury.

- Never carry the inflator module by the wires or connector on the underside of the module.
- When carrying a live inflator module, hold securely with both hands, and ensure that the bag and trim cover are pointed away.
- Place the inflator module on a bench or other surface with the bag and trim cover facing up.

- With the inflator module on the bench, never place anything on or close to the module that may be thrown in the event of an accidental deployment.

DISARMING

1. Before servicing the vehicle, refer to the precautions in the beginning of this section.

2. Place the ignition switch in the **OFF** position.

3. Disconnect the negative battery terminal and cover the battery terminal to prevent accidental contact.

4. Once the battery has been disconnected, wait for a period of approximately 10 minutes allowing the capacitor in the control unit to discharge.

5. When repairs are completed, connect the negative battery cable.

Power Rack and Pinion Steering Gear

REMOVAL & INSTALLATION

3 Series

1. Before servicing the vehicle, refer to the precautions in the beginning of this section.

2. Remove front wheels. Loosen and remove the pinch bolt, then remove the steering shaft spindle off the steering gear.

3. Use a siphon tool to empty the power steering fluid reservoir.

4. Remove or disconnect the following:
- Hydraulic fluid return line from the power steering unit. Drain the fluid into a sealable container.
- Pressure line
- Left and right side nuts and press off the tie rods where they connect to the spring struts
- Bolts attaching the steering unit to the front crossmember and remove it

To install:

5. Install the steering unit to the front crossmember, then install and tighten the mounting bolts.

6. Install remaining components in reverse order of disassembly.

7. Note the following:
- The steering unit bolts to the rear holes of the axle carrier. Always use new self-locking nuts and tighten them to 29–34 ft. lbs. (40–46 Nm).
- When reconnecting tie rods to the spring struts, be sure tie rod pins and strut bores are clean. Replace self-locking nut and tighten to 40–48 ft. lbs. (54–66 Nm).

- Replace the seals on the power steering pump connection and tighten the bolt to 29–32 ft. lbs. (40–43 Nm).

8. Connect the negative battery cable.

9. Refill the fluid reservoir with specified fluid. Idle the engine and turn the steering wheel back and forth until it has reached full lock both right and left twice in each direction. Then, turn the engine **OFF** and refill the reservoir as necessary.

Power Recirculating Ball Steering Gear

REMOVAL & INSTALLATION

5 and 7 Series

1. Before servicing the vehicle, refer to the precautions in the beginning of this section.

2. Remove or disconnect the following:
   - Negative battery cable
   - Steering wheel

3. Discharge the pressure reservoir by pushing in on the brake pedal about 10 times. Draw off hydraulic fluid in the supply tank.

4. Remove or disconnect the following:
   - Bolt and press the tie rod off the steering drop arm with the proper tool
   - Heat shield on the steering gear and disconnect the ride level height control pipes on 750iL models
   - U-joint from the steering gear. Disconnect and plug the hydraulic lines.
   - Steering gear mounting bolts and remove the steering gear

➡ **If necessary, move the steering drop arm by turning the steering stub to enable the removal of the gear assembly.**

To install:

5. Install or connect the following:
   - Steering gear and tighten the mounting bolts
   - Hydraulic lines, using new seals

6. Turn the steering wheel counterclockwise or clockwise against the stop, then back about 1¾ turn until the marks are aligned.

7. Install or connect the following:
   - U-joint to the steering gear making sure the bolt is in the locking groove of the steering stub

- Tie rod to the steering drop arm and replace the self-locking nut
- Heat shield on the steering gear and connect the ride level height control pipes on 750iL models
- Hydraulic fluid
- Steering wheel
- Negative battery cable

Strut

REMOVAL & INSTALLATION

Front

3 SERIES

1. Before servicing the vehicle, refer to the precautions in the beginning of this section.

2. Remove or disconnect the following:
   - Negative battery cable
   - Tire and wheel assembly
   - Brake pad wear indicator plug and ground wire. Pull the wires out of the holder on the strut. Remove the ABS pulse sender, if equipped.
   - Caliper and pull it away from the strut, suspending it from the body. Do not disconnect the brake line.
   - Attaching nut, then detach the pushrod on the stabilizer bar at the strut
   - Attaching nut and press off the guide joint with the proper tool
   - Nut and press off the tie rod joint

3. Press the bottom of the strut outward and push it over the guide joint pin, using the proper tool. Support the bottom of the strut.

4. Remove the nuts at the top of the strut, from inside the engine compartment, then remove the strut.

To install:

5. Position the strut in the vehicle. Upper strut mounting nuts: 16–17 ft. lbs. (21–23 Nm). The upper strut mounting nuts must be replaced with new self-locking nuts.

6. The remaining components are installed in the reverse order from which they were removed. Tie rod and guide joints must have both pins and bores clean for reassembly. Replace both self-locking nuts. Tighten the control arm to spring strut attaching nut to 43–51 ft. lbs. (59–69 Nm).

7. Install the front tire and wheel assemblies.

8. Connect the negative battery cable.

5 AND 7 SERIES

1. Before servicing the vehicle, refer to the precautions in the beginning of this section.

2. Remove or disconnect the following:
   - Negative battery cable
   - Tire and wheel assembly
   - Brake pad wear indicator plug and ground wire. Pull the wires out of the holder on the strut. Remove the ABS pulse sender, if equipped.
   - Stabilizer pushrod with the proper tool
   - Lower strut bolts at the control arm

3. Support the bottom of the strut and remove the nuts at the top of the strut, from inside the engine compartment. Remove the strut.

To install:

4. The installation is the reverse of the removal procedure.

Shock Absorber

REMOVAL & INSTALLATION

Rear

3 SERIES

1. Before servicing the vehicle, refer to the precautions in the beginning of this section.

2. Remove the trunk trim panel to expose the upper shock mounts.

3. Support the trailing arm and remove the lower mounting bolt.

✳✳ WARNING

The support must not be removed until the new shock absorber is installed, and the vehicle must not be raised since this could damage the halfshafts.

4. Remove the cap and remove the upper mounting nuts and remove the shock from the vehicle.

To install:

5. Place the shock into position with new seals fitted between the shock absorber and body. Renew the upper self-locking nuts: 11 ft. lbs. (15 Nm) for the Z3 and 318ti, 16 ft. lbs. (22 Nm) for all other 3 Series vehicles.

6. Install the trunk trim panel.

7. Install the lower shock mounting to the rear axle assembly. The thrust washer

For Wheel Alignment specifications, see Section 1 of this manual

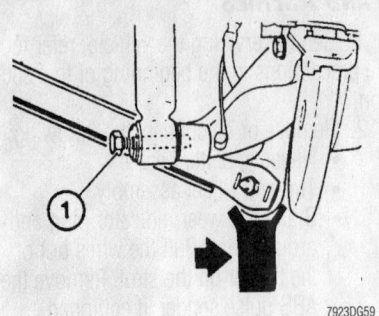

Support the trailing arm and remove the bolt (1)—3 Series

7923DG59

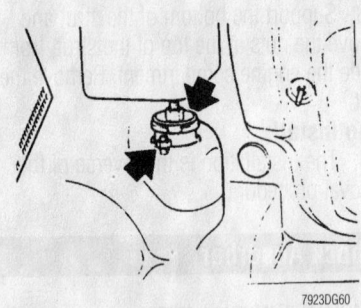

Upper mounting nut locations—3 Series

7923DG60

on the rubber mount must face the screw head.

8. With the vehicle resting at standard ride height, tighten the mounting bolt to 63 ft. lbs. (87 Nm), or to 94 ft. lbs. (130 Nm) if marked with 10.9, for the Z3 and 318ti models, or to 74 ft. lbs. (100 Nm) for all other 3 Series models.

5 AND 7 SERIES WITH STANDARD SUSPENSION

1. Before servicing the vehicle, refer to the precautions in the beginning of this section.
2. Remove or disconnect the following:
   • Rear seat cushion and back rest

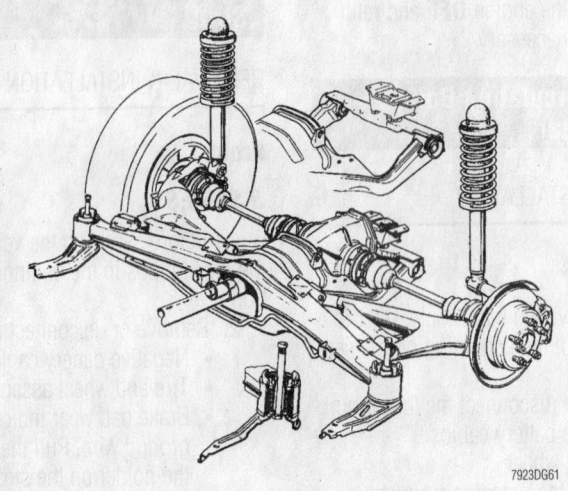

Rear suspension system—5 Series

7923DG61

REAR SPRING STRUT LAYOUT DRAWING

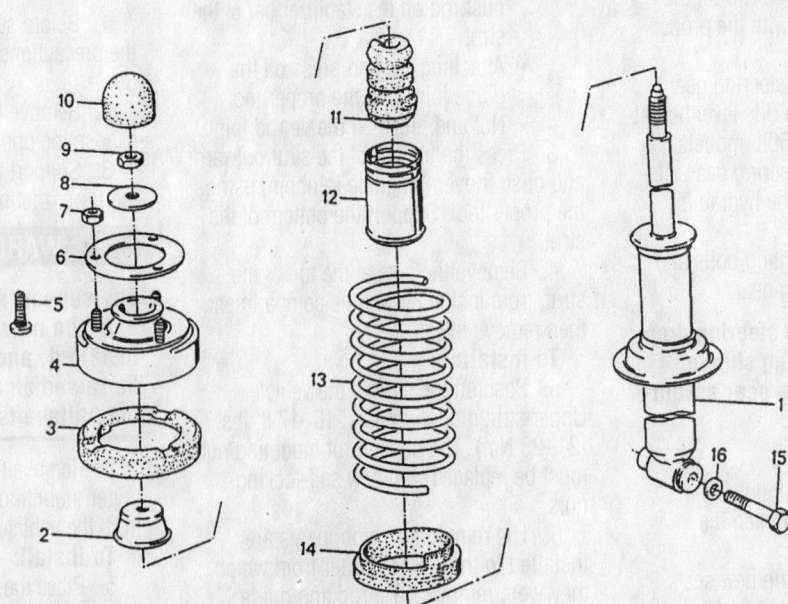

3	Upper spring ring	12	Protective tube
4	Mount	13	Coil spring
5	Bolt	14	Lower spring ring
6	Insulator	15	Bolt M 14 x 1.5 x 85
7	Collar nut M 8	16	Washer
8	Disc		
9	Hexagon nut M 10 x 1.8 ZN		

7923DG62

Exploded view of the shock and coil spring assembly—5 and 7 Series

- Trim panel over the strut mount
- With the control arm supported, rubber cap and remove the nuts at the top of the strut mount
- Lower mounting bolt and lower the spring/shock assembly. Remove the assembly from the vehicle.

3. Use a spring compressor and compress the spring. Remove the top nut and pull the top mount off. Remove the spring.

To install:

4. Compress the new spring or replace the old spring on the shock. Install the mount and washers. Use a new locknut and tighten to 18 ft. lbs. (25 Nm). Release the spring.

5. Install the shock. Upper mount nuts: 16 ft. lbs. (21.5 Nm). Loosely install the lower mounting bolt.

6. With the vehicle lowered to the ground and at standard riding height, tighten the lower mount to 94 ft. lbs. (130 Nm).

7. Install the trim and seat cushions.

5 AND 7 SERIES WITH RIDE LEVEL HEIGHT CONTROL SUSPENSION

1. Before servicing the vehicle, refer to the precautions in the beginning of this section.

2. Remove or disconnect the following:
- Rear seat cushion and back rest
- Trim panel over the strut mount

➡**The coil spring, shock absorber assembly acts as a strap so the control arm should always be supported.**

- Low pressure switch electrical connection and turn on the ignition
- Control rod nut, holding the collar with an 8mm wrench against torque. Don't disconnect the rod at the ball joint.

3. Operate the lever on the control switch in the "discharge" direction for about 20 seconds to discharge fluid from the lines.

4. Remove or disconnect the following:
- Hydraulic line on the strut and turn off the ignition
- With the control arm supported, rubber cap and remove the nuts at the top of the strut mount
- Lower mounting bolt and lower the spring strut assembly. Remove the assembly from the vehicle.

5. Use a spring compressor and compress the spring. Remove the top nut and pull the top mount off. Remove the spring.

To install:

6. Compress the new spring or replace the old spring on the strut. Install the mount and washers. Use a new locknut and tighten to 18 ft. lbs. (25 Nm). Release the spring.

7. Install or connect the following:
- Spring strut. Upper mount nuts: 16 ft. lbs. (21.5 Nm). Loosely install the lower mounting bolt.
- Hydraulic line on the strut
- Control rod nut, holding the collar with an 8mm wrench against torque
- Low pressure switch electrical connection

8. With the vehicle lowered to the ground and at standard riding height, tighten the lower mount to 94 ft. lbs. (130 Nm).

9. Install the trim and seat cushions.

Coil Spring

REMOVAL & INSTALLATION

Front

✳✳ CAUTION

This procedure calls for the spring to be compressed. A compressed spring has high potential energy and if released suddenly can cause severe damage and personal injury.

1. Before servicing the vehicle, refer to the precautions in the beginning of this section.

2. Remove the strut from the vehicle and mount in a vise using a strut holder. This will prevent damage to the strut tube.

3. Using a proper spring compressor, compress the spring and lock into place.

4. Remove the top nut of the strut mount. Counterhold the strut rod during removal.

5. Pull the strut mount off the strut rod. Note the positioning of the spacers and washer for replacement.

6. Pull the spring off the strut and place aside in a safe area.

7. slowly release the compression of the spring.

To install:

8. Install the spring in the compressor and compress.

9. Install the spring and strut mount with all the spacers and washers in their original positions. New strut rod nut: 47 ft. lbs. (65 Nm).

10. Release the spring slowly and check

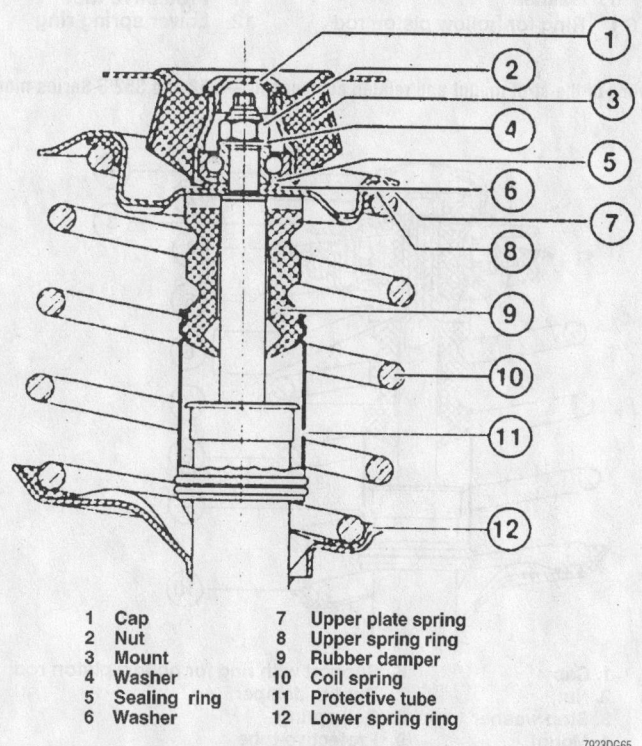

1	Cap	7	Upper plate spring
2	Nut	8	Upper spring ring
3	Mount	9	Rubber damper
4	Washer	10	Coil spring
5	Sealing ring	11	Protective tube
6	Washer	12	Lower spring ring

7923DG65

Cut away view of the strut mount and related components—M44 1.9L 3 Series models

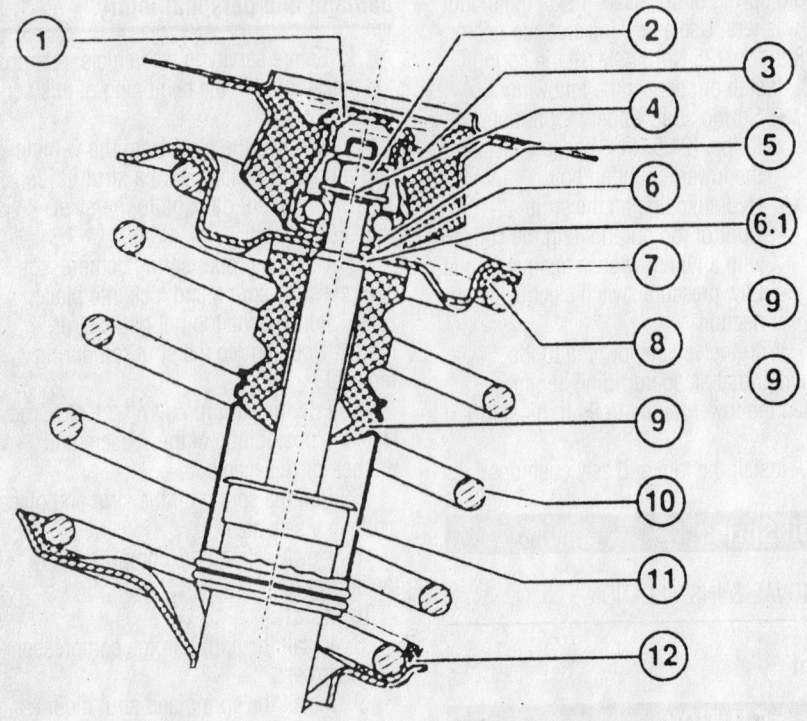

1	Cap	7	Upper plate spring
2	Nut	8	Upper spring ring
3	Mount	9	Rubber damper
4	Washer	10	Coil spring
5	Sealing ring	11	Protective tube
6	Washer	12	Lower spring ring
6.1	Ring for hollow piston rod		

7923DG66

Cut away view of the strut mount and related components—M52 and S52 3 Series models

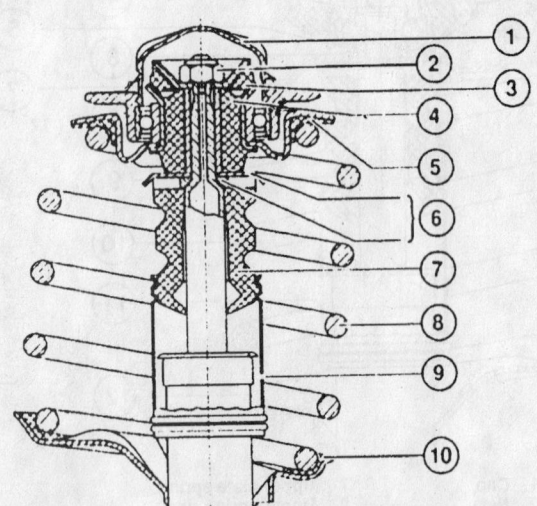

1.	Cap	6.	Support with ring for hollow piston rod
2.	Nut	7.	Rubber damper
3.	Stop washer	8.	Coil spring
4.	Mount	9.	Protective tube
5.	Upper spring ring	10.	Lower spring ring

7923DG67

Cut away view of the strut mounting with a separate support bearing and related components— M52 and S52 3 Series models

that it seats in the spring holders. Install the strut in the vehicle.

Rear

3 SERIES—EXCEPT Z3 AND 318TI

1. Before servicing the vehicle, refer to the precautions in the beginning of this section.

2. Remove the tire and wheel assembly.

3. Support the lower trailing arm at the hub with a suitable jack and disconnect the stabilizer bar at the control arm and the subframe.

4. Remove the shock absorber lower mounting bolt.

5. Lower the trailing arm slowly and remove the spring to the side.

To install:

6. Install the spring with the bushing in place and the top of the upper spring ring lubricated.

7. Raise the trailing arm to a level where the bolt can be replaced in the lower shock mount. Connect the stabilizer bar. Do not fully tighten the bolts at this time.

8. Install the tire and wheel assembly.

9. Tighten the stabilizer bolt to 16 ft. lbs. (21.5 Nm) and the shock bolt to 63 ft. lbs. (87 Nm) with the control arm in the normal ride position.

Z3 AND 318TI

1. Before servicing the vehicle, refer to the precautions in the beginning of this section.

2. Remove or disconnect the following:

- Rear portion of the exhaust system and support it from the body
- Final drive rubber mount, push it downward, and hold it down with a suitable wedge
- Bolt that connects the rear stabilizer bar to the strut on the side being worked on. Be careful not to damage the brake line.

➡**Support the lower control arm securely with a suitable jack or other device that will permit it to be lowered gradually, while maintaining a secure support.**

3. Then, to prevent damage to the output shaft joints, lower the control arm only enough to slip the coil spring off the retainer.

To install:

4. Be sure, when replacing the spring, that the same part number, color code, and proper rubber ring are used. Install the spring, making sure that the spring is in proper position.

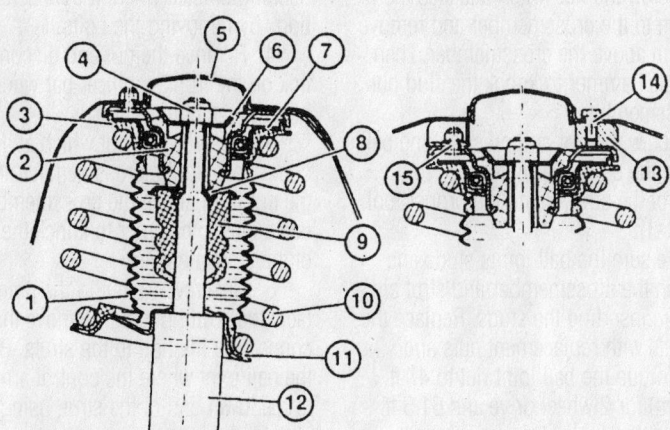

1. Rubber Gaiter
2. Thrust Bearing
3. Top spring support
4. Nut
5. Cap
6. Joint seat
7. Top spring plate
8. Support disc
9. Auxiliary spring
10. Coil spring
11. Bottom spring support
12. Spring strut shock absorber

7923DG68

Cut away view of the strut mount and related components—5, 7 and 8 Series

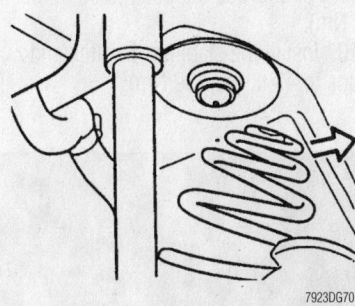

7923DG70

Remove the coil spring—3 Series, except Z3 and 318ti

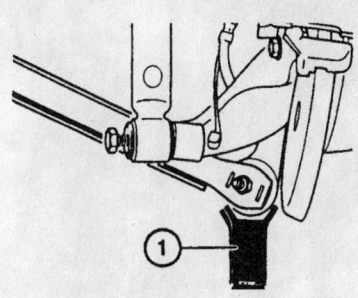

7923DG69

Support the trailing arm (1)—3 Series, except Z3 and 318ti

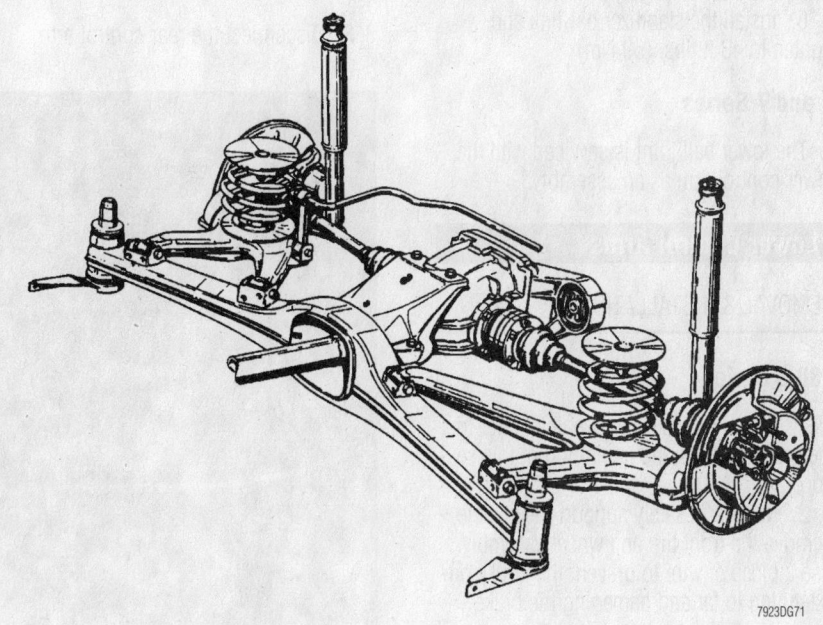

7923DG71

Rear suspension setup on Z3 and 318ti models

5. Keep the control arm securely supported while raising and replace the shock bolt. Install the bolts in the final drive rubber mount and tighten to 69 ft. lbs. (95 Nm).

6. Tighten the stabilizer bolt to 16 ft. lbs. (21.5 Nm), and the shock bolt to 63 ft. lbs. (87 Nm) with the control arm in the normal ride position. Install the exhaust system.

5 AND 7 SERIES

The coil spring is removed along with the shock absorber. The 5 and 7 Series use a "coil over" type shock absorber where the spring is mounted to the shock in one compact unit. Once the shock is removed from the vehicle, the spring can be compressed and separated from the shock absorber.

Lower Ball Joint

REMOVAL & INSTALLATION

3 Series

1. Before servicing the vehicle, refer to the precautions in the beginning of this section.

2. Remove or disconnect the following:
- Front tire and wheel assembly
- Rear control arm bushing bracket where it connects to the body by removing the bolts
- Nut and disconnect the link on the front stabilizer bar where it connects to the control arm
- Nut that attaches the control arm to the crossmember, and remove the nut from above the crossmember. Then, use a soft faced hammer to knock the stud out of the cross-member.
- Nut to the point it contacts the strut housing. Remove the bolts connecting the hub to the struts. Press off the ball joint where the control arm attaches to the lower end of the strut, using a suitable tool.

To install:

3. Clean the threaded holes in the hub. New micro-encapsulated hub mounting bolts to the strut: 58 ft. lbs. (80 Nm).

4. Be sure the ball joints studs and the bores in the crossmember and strut are clean before inserting the studs. Replace the original nuts with replacement nuts and washers. Ball joint nut: 47 ft. lbs. (65 Nm). Control arm to subframe nut: 61 ft. lbs. (85 Nm).

5. Install the control arm bushing bracket. Bolts: 30 ft. lbs. (42 Nm).

6. Install the stabilizer bar link and tighten to 43 ft. lbs. (59 Nm).

5 and 7 Series

The lower ball joint is serviced with the lower control arm as an assembly.

Lower Control Arms

REMOVAL & INSTALLATION

E30 3 Series

1. Before servicing the vehicle, refer to the precautions in the beginning of this section.

2. Raise and safely support the vehicle. Remove the front tire and wheel assembly. Use a piece of wire to prevent the strut from extending to far and damaging the brake hose.

3. Disconnect the rear control arm bushing bracket where it connects to the body by removing the bolts.

4. Remove the nut and disconnect the link on the front stabilizer bar where it connects to the stabilizer bar.

5. Unscrew the nut which attaches the control arm to the crossmember and remove the nut from above the crossmember. Then, use a plastic hammer to knock the stud out of the crossmember.

6. Unscrew the nut and press off the ball joint where the control arm attaches to the lower end of the strut, using the proper tool.

To install:

7. Make sure the ball joints studs and the bores in the crossmember and strut are clean before inserting the studs. Replace the original nuts with replacement nuts and washers. Torque the ball joint nut to 47 ft. lbs. (65 Nm) for 2 wheel drive and 61.5 ft. lbs. (93 Nm) for 4 wheel drive. Torque the control arm to subframe nut to 61 ft. lbs. (85 Nm) for 2 wheel drive and 72 ft. lbs. (100 Nm) for 4 wheel drive.

8. Install the control arm bushing bracket and torque the bolts to 30 ft. lbs. (42 Nm).

9. Install the stabilizer bar link and torque to 43 ft. lbs. (59 Nm).

E36 3 Series

1. Before servicing the vehicle, refer to the precautions in the beginning of this section.

2. Raise and safely support the vehicle. Remove the front tire and wheel assembly. Use a piece of wire to prevent the strut from extending to far and damaging the brake hose.

3. Disconnect the rear control arm

bushing bracket where it connects to the body by removing the bolts.

4. Remove the nut and disconnect the link on the front stabilizer bar where it connects to the control arm.

5. Unscrew the nut which attaches the control arm to the crossmember and remove the nut from above the crossmember. Then, use a plastic hammer to knock the stud out of the crossmember.

6. Unscrew the nut to the point it contacts the strut housing. Remove the bolts connecting the hub to the struts. Press off the ball joint where the control arm attaches to the lower end of the strut, using proper tool.

To install:

7. Clean the threaded holes in the hub. Install new micro-encapsulated hub mounting bolts to the strut and torque to 58 ft. lbs. (80 Nm).

8. Make sure the ball joints studs and the bores in the crossmember and strut are clean before inserting the studs. Replace the original nuts with replacement nuts and washers. Torque the ball joint nut to 47 ft. lbs. (65 Nm) for 2 wheel drive. Torque the control arm to subframe nut to 61 ft. lbs. (85 Nm) for 2 wheel drive.

9. Install the control arm bushing bracket and torque the bolts to 30 ft. lbs. (42 Nm).

10. Install the stabilizer bar link and torque to 43 ft. lbs. (59 Nm).

84278004

Control arm bracket mounting. Matchmark the control arm position before pressing the bushing off during replacement—E36 3 Series shown, E30 3 Series uses the same design

Control arm to subframe mount ball joint—E36 3 Series shown, E30 3 Series uses the same design

84278005

BUSHING REPLACEMENT

The bushings must be pressed out of the housing bores. BMW bushings are notoriously hard to press out of the housings. Use a high capacity hydraulic press, pene-

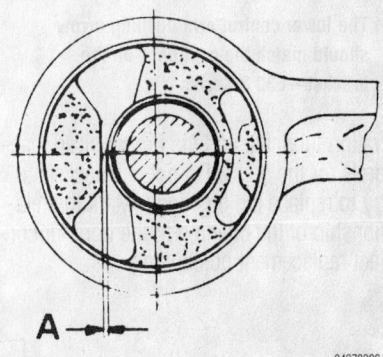

84278006

With the vehicle loaded and at normal ride height, the gap at A in the lower control arm bushing should be 0.028–0.067 in. (0.7–1.7 mm)—E30 3 Series

5 and 7 Series

1. Before servicing the vehicle, refer to the precautions in the beginning of this section.

2. Raise and safely support the front end. Do not place the jackstands under any suspension parts. Remove the wheel.

3. Remove the 3 bolts holding the steering knuckle to the bottom of the strut.

4. Remove the ball joint nut and press the stud out of the steering knuckle with a ball joint remover tool.

5. Remove the nut and bolt at the subframe end of the control arm. Remove the control arm.

To install:

6. Install the control arm to the subframe using a new nut and washer on both sides. Do not torque at this point.

7. Clean the grease and dirt off of the ball joint stud and bore. Install the ball joint stud into the steering knuckle and torque the new nut to 67 ft. lbs. (93 Nm).

8. Clean the threads and bores of the steering knuckle mounting bolts and the strut housing. Install the bolts, using threadlocker, and torque to 80 ft. lbs. (110 Nm). There is a groove that will align the strut and knuckle.

9. Install the wheel and lower the vehicle to the ground. Load 150 lbs. into each of the front seats and in the center of the rear seat. Torque the control arm to subframe bolt to 56 ft. lbs. (77.5 Nm).

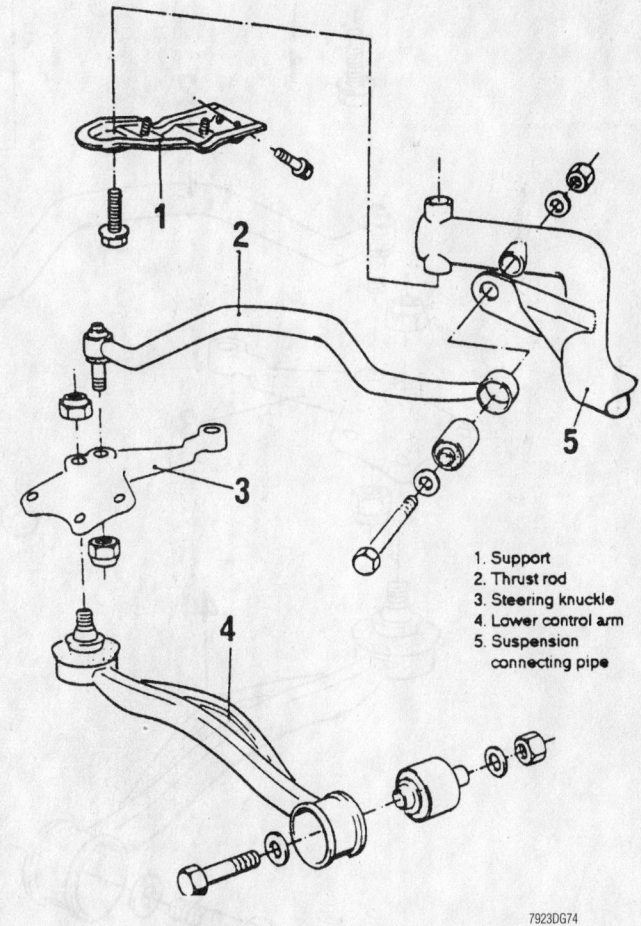

1. Support
2. Thrust rod
3. Steering knuckle
4. Lower control arm
5. Suspension connecting pipe

7923DG74

Exploded view of the front suspension—5 and 7 Series

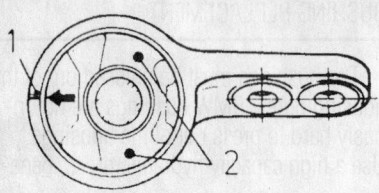

84278007

The lower control arm bushing arrow should match the cast boss on the bracket—E30 3 Series

trating lubricant and the proper sized mandrels for the press. Do not use sockets to try to replace the bushings. Mark the relationship of the bushing to the bore for correct replacement positioning.

Thrust Rod

REMOVAL & INSTALLATION

5 and 7 Series

The E34 5 Series uses a multi-link type suspension. There is a lower control arm to support the strut housing and a thrust rod to control fore and aft motion. The thrust rod is not used on 3 Series vehicles.

Always replace the strut rods in pairs. If the strut rods are not replaced in pains, uneven driving response may result.

1. Before servicing the vehicle, refer to the precautions in the beginning of this section.

2. Raise and safely support the front end. Do not place the jackstands under any suspension parts. Remove the wheel.

3. Remove the thrust rod ball joint nut and press the stud out of the steering knuckle with a ball joint remover tool.

4. Remove the nut and bolt at the subframe end of the strut rod. Remove the strut rod.

To install:

5. Install the strut to the subframe using a new nut and washer on both sides. Do not torque at this point.

6. Clean the grease and dirt off of the ball joint stud and bore. Install the ball joint

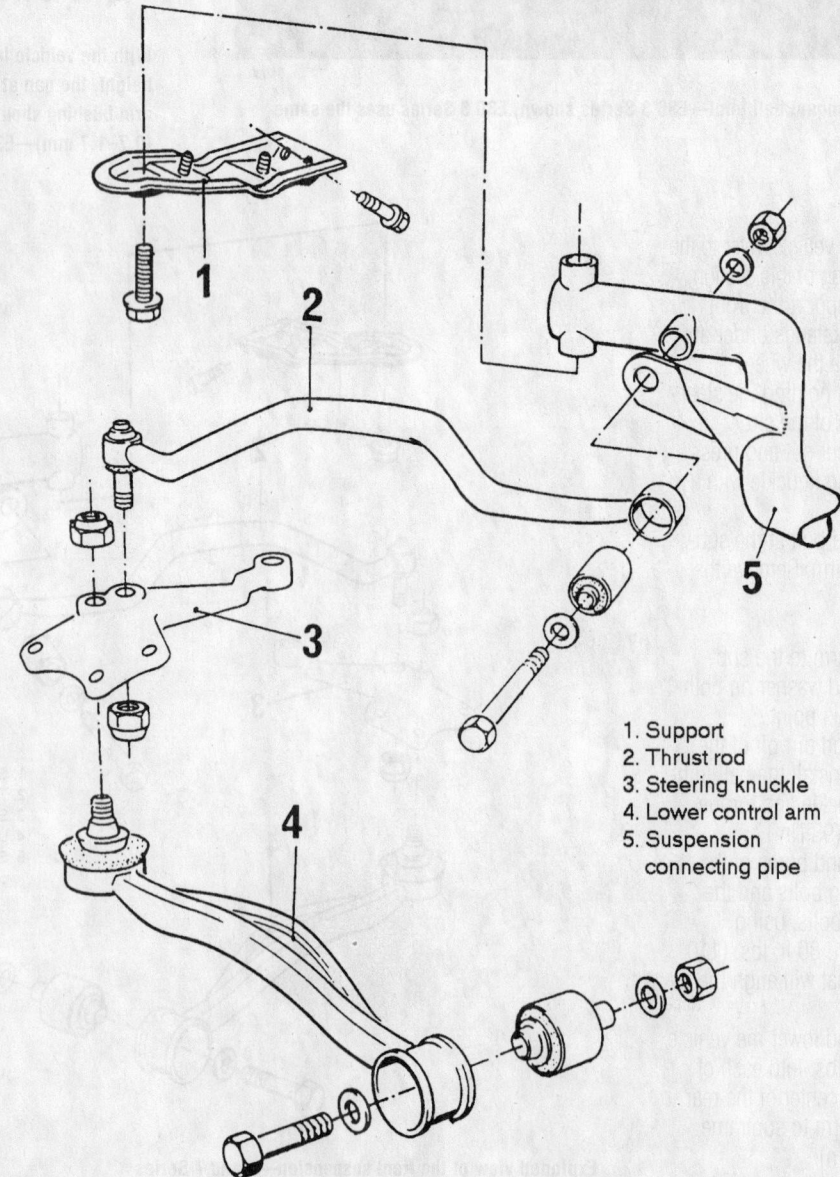

1. Support
2. Thrust rod
3. Steering knuckle
4. Lower control arm
5. Suspension connecting pipe

84278008

Front suspension components—E34 5 Series shown

stud into the steering knuckle and torque the new nut to 67 ft. lbs. (93 Nm).

7. Install the wheel and lower the vehicle to the ground. Load 150 lbs. into each of the front seats and in the center of the rear seat. Torque the control arm to subframe bolt to 92 ft. lbs. (127 Nm).

BUSHING REPLACEMENT

The bushings must be pressed out of the housing bores. BMW bushings are notoriously hard to press out of the housings. Use a high capacity hydraulic press, penetrating lubricant and the proper sized mandrels for the press. Do not use sockets to try to replace the bushings. Mark the relationship of the bushing to the bore for correct replacement positioning.

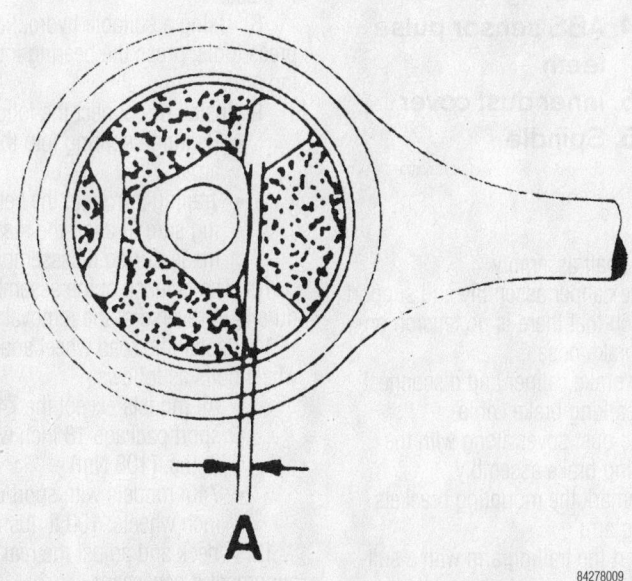

With the vehicle loaded and at normal ride height, the gap at A in the lower control arm bushing should be 0.039–0.079 in. (1.0–2.0 mm)

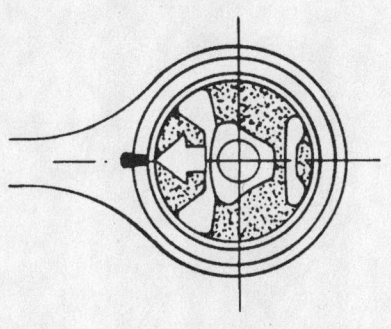

The lower control arm bushing arrow should match the cast boss on the bracket

Wheel Bearings

ADJUSTMENT

Wheel bearings can not be adjusted. The rear bearings must be replaced as a unit and never be reused once removed. The front wheel bearings are pressed into the hub and are not available separately. If a front wheel bearing is in need of replacement, it is replaces as a unit with the hub.

REMOVAL & INSTALLATION

Front

➡The wheel bearings are only removed if they are worn. They cannot be removed without destroying them (due to side thrust created by the bear-

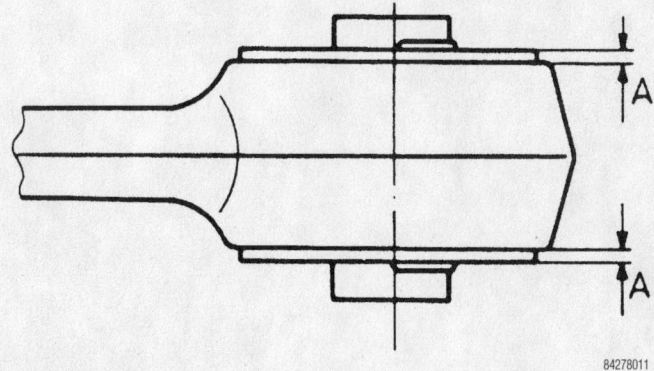

The bushing should protrude evenly when installing a replacement

ing puller). They cannot be disassembled, repacked or adjusted.

1. Before servicing the vehicle, refer to the precautions in the beginning of this section.

2. Remove or disconnect the following:
   • Front wheel. Remove the attaching bolts and remove, then suspend the brake caliper to avoid putting stress on the brake line.
   • Setscrew, with a suitable hex wrench. Remove the brake disc, then remove the dust cover suing a suitable prytool or chisel to remove the pressed on cover.

3. Using a suitable chisel, knock the tab on the collared nut away from the shaft. Remove and discard the nut.

4. Remove the bearing with a puller set tool Nos. 31-2-101/102/104 or their equivalents and discard it. On M3 models, use a puller set tool No. 31-2-102/105/106 or their equivalents. On M3 models, install the main bracket of the puller with 3 wheel bolts.

5. If the inside bearing inner race remains on the stub axle, unbolt and remove the dust guard. Bend back the inner dust guard and pull the inner race off using tool No. 00-7-500 and 33-1-309 or their equivalents, which are capable of getting under the race to remove it. Reinstall the dust guard.

To install:

6. If the dust guard has been removed, install a new one. Install tool No. 31-2-120; except the M3 models, which require tool No. 31-2-110. Place the tool over the stub axle and screw it in the entire length of the guide sleeve's threads, then press the bearing on.

7. Reverse the remaining removal procedures to install the disc and caliper. Using a

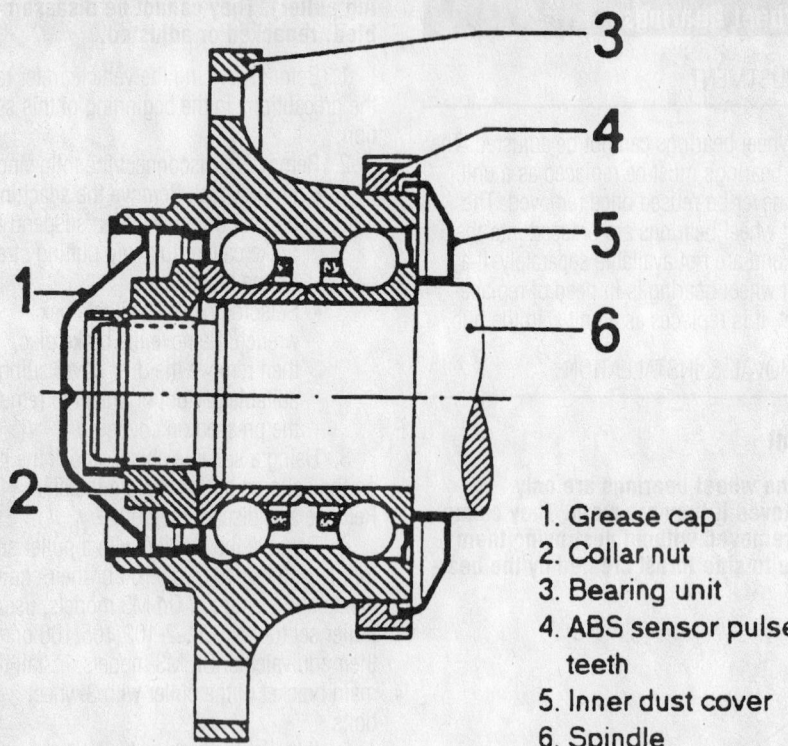

1. Grease cap
2. Collar nut
3. Bearing unit
4. ABS sensor pulse teeth
5. Inner dust cover
6. Spindle

7923DG73

Cut away view of the front wheel bearing

new collared mounting nut, tighten the wheel hub collar nut to 210 ft. lbs. (290 Nm). Lock the collar nut by bending over the tab into the groove of the spindle.

Rear

1. Before servicing the vehicle, refer to the precautions in the beginning of this section.

2. Remove or disconnect the following:
   • Negative battery cable
   • Wheel assembly
   • Half shaft assembly
   • Brake caliper assembly and support it such that there is no tension on the brake hose
   • Rear brake caliper and disconnect the parking brake cable
   • Brake dust cover along with the parking brake assembly

3. Matchmark the mounting brackets for the trailing arm

4. Support the trailing arm with a suitable jack and remove the fasteners that attach the rear shock assembly to the trailing arm.

5. Remove or disconnect the following:
   • Stabilizer bar, and if installed, the upper control arm
   • Trailing arm assembly
   • Large snapring in front of the wheel bearing

6. Using suitable press tools, support the trailing arm in a suitable hydraulic press and press the bearing from the trailing arm assembly.

To install:

7. Inspect the trailing arm for damage and thoroughly clean the area where the wheel bearing is installed. Make sure there are no sharp edges on the entrance of the bore where the bearing is to be installed. If necessary, chamfer the edges of the housing to avoid binding when pressing the bearing in place.

8. Using a suitable hydraulic press and press tools, press the bearing into the trailing arm.

9. Install or connect the following:
   • Bearing snapring into the trailing arm
   • Trailing arm onto the vehicle, making sure to align the match marks made during disassembly

10. The balance of the assembly procedure is in reverse of the removal procedure.

11. Install the road wheel and torque the wheel bolts as follows:
    • All models except the 740i with sport package 18 inch wheels: 80 ft. lbs. (108 Nm)
    • 740i models with sport package 18 inch wheels: 100 ft. lbs. (136 Nm)

12. Check and adjust the rear wheel alignment if necessary.

DAEWOO

1998–01

Lanos • Leganza • Nubira

8

PRECAUTIONS

Before servicing any vehicle, please be sure to read all of the following precautions, which deal with personal safety, prevention of component damage and important points to take into consideration when servicing a motor vehicle:

• Never open, service or drain the radiator or cooling system when the engine is hot; serious burns can occur from the steam and hot coolant.

• Observe all applicable safety precautions when working around fuel. Whenever servicing the fuel system, always work in a well ventilated area. Do not allow fuel spray or vapors to come in contact with a spark, open flame, or excessive heat (a hot drop light, for example). Keep a dry chemical fire extinguisher near the work area. Always keep fuel in a container specifically designed for fuel storage; also, always properly seal fuel containers to avoid the possibility of fire or explosion. Refer to the additional fuel system precautions later in this section.

• Fuel injection systems often remain pressurized, even after the engine has been turned **OFF**. The fuel system pressure must be relieved before disconnecting any fuel lines. Failure to do so may result in fire and/or personal injury.

• Brake fluid often contains polyglycol ethers and polyglycols. Avoid contact with the eyes and wash your hands thoroughly after handling brake fluid. If you do get brake fluid in your eyes, flush your eyes with clean, running water for 15 minutes. If eye irritation persists, or if you have taken brake fluid internally, seek medical assistance IMMEDIATELY.

• The EPA warns that prolonged contact with used engine oil may cause a number of skin disorders, including cancer! You should make every effort to minimize your exposure to used engine oil. Protective gloves should be worn when changing oil. Wash your hands and any other exposed skin areas as soon as possible after exposure to used engine oil. Soap and water, or waterless hand cleaner should be used.

• All new vehicles are now equipped with an air bag system, often referred to as a Supplemental Restraint System (SRS) or Supplemental Inflatable Restraint (SIR) system. The system must be disabled before performing service on or around system components, steering column, instrument panel components, wiring and sensors. Failure to follow safety and disabling procedures could result in accidental air bag deployment, possible personal injury and unnecessary system repairs.

• Always wear safety goggles when working with, or around, the air bag system. When carrying a non-deployed air bag, be sure the bag and trim cover are pointed away from your body. When placing a non-deployed air bag on a work surface, always face the bag and trim cover upward, away from the surface. This will reduce the motion of the module if it is accidentally deployed. Refer to the additional air bag system precautions later in this section.

• Clean, high quality brake fluid from a sealed container is essential to the safe and proper operation of the brake system. You should always buy the correct type of brake fluid for your vehicle. If the brake fluid becomes contaminated, completely flush the system with new fluid. Never reuse any brake fluid. Any brake fluid that is removed from the system should be discarded. Also, do not allow any brake fluid to come in contact with a painted surface; it will damage the paint.

• Never operate the engine without the proper amount and type of engine oil; doing so WILL result in severe engine damage.

• Timing belt maintenance is extremely important! Many models utilize an interference-type, non-freewheeling engine. If the timing belt breaks, the valves in the cylinder head may strike the pistons, causing potentially serious (also time-consuming and expensive) engine damage. Refer to the maintenance interval charts in the front of this manual for the recommended replacement interval for the timing belt and to the timing belt section for belt replacement and inspection.

• Disconnecting the negative battery cable on some vehicles may interfere with the functions of the on-board computer system(s) and may require the computer to undergo a relearning process once the negative battery cable is reconnected.

• When servicing drum brakes, only dissemble and assemble one side at a time, leaving the remaining side intact for reference.

• Only an MVAC-trained, EPA-certified automotive technician should service the air conditioning system or its components.

ENGINE REPAIR

➡**Disconnecting the negative battery cable on some vehicles may interfere with the operation of the on-board computer system. The computer may undergo a relearning process once the negative battery cable is reconnected.**

Distributor

All models utilize a Distributorless Ignition System (DIS). This system uses a signal from the Crankshaft Position (CKP) sensor to the Powertrain Control Module (PCM) to determine proper Electronic Spark Timing (EST). The PCM then triggers the ignition coil to fire.

Alternator

REMOVAL

1.6L Engine

1. Before servicing the vehicle, refer to the precautions in the beginning of this section.

2. Remove or disconnect the following:
   • Negative battery cable
   • Intake Air Temperature (IAT) sensor electrical connector from the intake tube
   • Breather tube clamp and all other

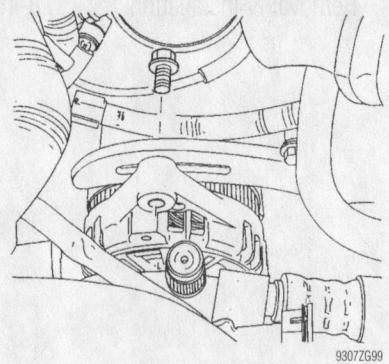

93072G99

Alternator mounting—1.6L engines

clamps to facilitate air intake tube removal
- Air intake tube
- Battery harness connector nut from the alternator
- Batter harness connector from the alternator
- Alternator shackle bracket bolt and washer
- Drive belt
- Bolt and retaining clamp of the harness
- Nuts and washers that retain the alternator to the lower bracket
- Alternator

2.0L Engine

1. Before servicing the vehicle, refer to the precautions in the beginning of this section.

2. Remove or disconnect the following:
- Negative battery cable
- Intake Air Temperature (IAT) sensor electrical connector from the intake tube
- All clamps from the air intake tube hose and set the tube aside

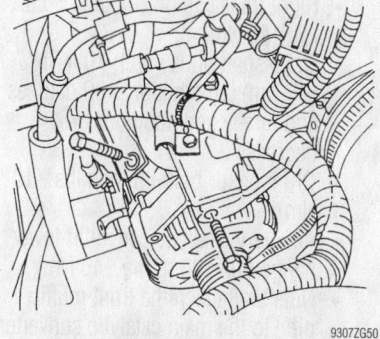

9307ZG50

Alternator upper mounting bolts—2.0L engine

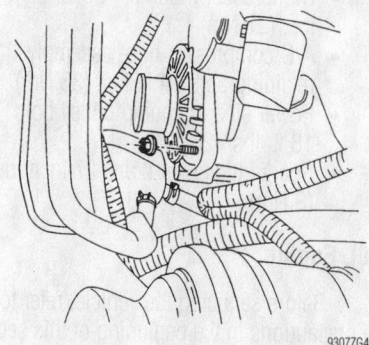

9307ZG49

Alternator lower mounting bolts—2.0L engine

- Harness connector from the back of the alternator
- Accessory drive belt
- Push up on the power steering reservoir and set it aside
- Alternator upper mounting bolts
- Nuts and washer that retain the alternator lower bracket-to-alternator bolt
- Lower bracket-to-alternator bolt
- Alternator with the lower bracket
- Lower bracket nut, bolt, washer and the bracket

2.2L Engine

1. Before servicing the vehicle, refer to the precautions in the beginning of this section.

2. Remove or disconnect the following:
- Negative battery cable
- Intake Air Temperature (IAT) sensor electrical connector from the intake tube
- Air intake tube
- Alternator lead nut and the lead from the back of the alternator
- Harness connector from the back of the alternator
- Accessory drive belt
- Alternator-to-intake manifold and cylinder head bracket bolts and the alternator-to-intake manifold strap bracket bolt
- Alternator lower bracket-to-engine bolts
- Alternator and bracket
- Alternator from the bracket

INSTALLATION

1.6L Engine

Install or connect the following:
- Alternator
- Nuts and washers that retain the alternator to the lower bracket. Tighten to 15 ft. lbs. (20 Nm).
- Drive belt
- Alternator shackle bracket bolt and washer. Tighten to 15 ft. lbs. (20 Nm).
- Batter harness connector to the alternator
- Battery harness connector nut to the alternator. Tighten to 11 ft. lbs. (15 Nm).
- Breather tube clamp and all other clamps to facilitate air intake tube removal

- Harness retaining clamp and bolt
- Air intake tube
- IAT sensor electrical connector to the intake tube
- Negative battery cable

2.0L Engine

1. Before servicing the vehicle, refer to the precautions in the beginning of this section.

2. Install or connect the following:
- Alternator lower bracket to the alternator and insert the alternator bolt
- Alternator and lower support bracket to the block. Tighten the lower bracket-to-engine bolt to 22 ft. lbs. (30 Nm)
- Alternator upper mounting bolts. Tighten the alternator-to-intake manifold and cylinder head bolts to 26 ft. lbs. (35 Nm) and the alternator-to-intake manifold strap bracket bolts and the intake manifold-to-cylinder body strap bracket bolts to 15 ft. lbs. (20 Nm).
- Harness connector to the back of the alternator
- Accessory drive belt
- Power steering reservoir
- Air intake tube hose and clamps
- IAT sensor electrical connector
- Negative battery cable

2.2L Engine

1. Before servicing the vehicle, refer to the precautions in the beginning of this section.

2. Remove or disconnect the following:
- Bracket to the alternator and tighten the nut to 15 ft. lbs. (20 Nm)
- Alternator and bracket
- Alternator lower bracket-to-engine bolts and tighten to 22 ft. lbs. (30 Nm)
- Alternator-to-intake manifold and cylinder head bracket bolts. Tighten to 26 ft. lbs. (35 Nm).
- Alternator-to-intake manifold strap bracket bolt and tighten to 15 ft. lbs. (20 Nm)
- Harness connector to the back of the alternator
- Alternator lead and nut to the back of the alternator and tighten the nut to 11 ft. lbs. (15 Nm)
- Accessory drive belt
- Air intake tube

- IAT sensor electrical connector
- Negative battery cable

Ignition Timing

ADJUSTMENT

Ignition timing is controlled by the Powertrain Control Module (PCM). The PCM receives signals from various sensors mounted on the engine. No ignition timing adjustment is possible.

Engine Assembly

REMOVAL & INSTALLATION

1.6L Engine

1. Before servicing the vehicle, refer to the precautions in the beginning of this section.
2. Relieve the fuel system pressure.
3. Remove the hood.
4. Drain the engine oil.
5. Remove or disconnect the following:
   - Battery cables
   - Negative battery cable from the vehicle frame
6. If equipped with air conditioning, discharge the system using an approved recovery/recycling machine.
7. Remove or disconnect the following:
   - Intake Air Temperature (IAT) sensor connector
   - Air intake tube from the throttle body and air filter housing
   - Breather tubes from the valve cover
   - Right front wheel
   - Right front splash shield
   - Air conditioning drive belt, if equipped
   - Alternator drive belt
   - Power steering pump pulley bolts and the pulley, if equipped
8. Drain the coolant.
9. Remove or disconnect the following:
   - Cooling fans and the radiator
   - Upper radiator hose from the thermostat housing
   - Power steering hose return and pressure hoses from the pump, if equipped
10. Disconnect the following electrical connections:
    - Ignition coil
    - powertrain Control Module (PCM)
    - Ground terminal at the intake manifold and the starter
    - Oxygen (O2S) sensor
    - Fuel injector harness connectors

- Idle Air Control (IAC) valve
- Throttle Position (TP) sensor
- Engine Coolant Temperature (ECT) sensor
- Alternator voltage regulator connection

11. Remove or disconnect the following:
    - Spark plug cover bolts and the cover
    - Camshaft Position (CMP) sensor electrical connection
    - All necessary vacuum lines, including those at the brake booster. Tag the lines for identification as necessary.
    - Fuel return line from the pressure regulator
    - Fuel feed line at the fuel rail
    - Throttle cable from the throttle body and intake manifold bracket
    - Surge tank coolant hose at the throttle body
    - Heater outlet hose at the coolant pipe
    - Heater inlet hose at the cylinder head
    - Surge tank coolant hose from the coolant pipe
    - Lower radiator hose from the coolant pipe
    - Starter solenoid **S** terminal wire
    - Air conditioning compressor hose assembly retaining bolt
    - Air conditioning compressor hose assembly from the compressor
    - Electrical connector from the air conditioning compressor coil
    - Compressor mounting bolts
    - Compressor
    - Compressor mounting bracket bolts and the bracket from the block
    - Auxiliary catalytic converter lower flange nuts from the exhaust manifold studs and the bolts at the bracket
    - Nuts that attach the front muffler pipe to the main catalytic converter
    - Catalytic converter and the exhaust pipe as an assembly
    - Crankshaft pulley bolt and the pulley
    - Vacuum lines from the Evaporative (EVAP) emissions canister solenoid, if not already done
    - Electrical connection from the EVAP canister solenoid
    - Electrical connection from the oil pressure switch
    - Electrical connections from the Crankshaft Position (CKP) sensor and the Knock Sensor (KS)

- CKP bolt and the sensor
- Right transaxle brace bolts from the transaxle
- Flywheel or flexplate inspection cover
- Transaxle torque converter bolts, if equipped with an automatic transaxle
- Transaxle bell housing bolts

12. Support the transaxle with a floor jack and attach a suitable engine lifting device
13. Remove or disconnect the following:
    - Right engine mount bracket-to-engine mount bolts
    - Right engine mount bracket from the block
    - Engine from the transaxle
    - Engine from the vehicle

To install:

14. Installation is the reverse of the removal procedure, while using the following torque values:
    - Transaxle bell housing bolts: 55 ft. lbs. (75 Nm)
    - Right engine mount bracket-to-engine mount bolts: 44 ft. lbs. (60 Nm)
    - Transaxle torque converter bolts, if equipped: 48 ft. lbs. (65 Nm)
    - Right transaxle brace bolts: 30 ft. lbs. (40 Nm)
    - Crankshaft pulley bolt: 70 ft. lbs. (95 Nm), an additional 30 degrees and finally and additional 15 degrees
    - CKP sensor bolt: 89 inch lbs. (10 Nm)
    - Auxiliary catalytic converter lower flange nuts: 30 ft. lbs. (40 Nm)
    - Nuts that attach the front muffler pipe to the main catalytic converter: 22 ft. lbs. (30 Nm).
    - Compressor mounting bracket bolts: 37 ft. lbs. (50 Nm)
    - Compressor mounting bolts: 20 ft. lbs. (27 Nm)
    - A/C compressor hose assembly retaining bolt: 24 ft. lbs. (33 Nm)
    - Power steering pump pulley bolt: 18 ft. lbs. (25 Nm)
    - Spark plug cover bolts: 27 inch lbs. (3 Nm)

2.0L Engine

1. Before servicing the vehicle, refer to the precautions in the beginning of this section.
2. Relieve the fuel system pressure.
3. Remove the hood.
4. Drain the engine oil.
5. Disconnect the battery cables.

6. If equipped with air conditioning, discharge the system using an approved recovery/recycling machine.

7. Remove or disconnect the following:
- Intake Air Temperature (IAT) sensor connector
- Air intake tube from the throttle body and air filter housing
- Breather tubes from the valve cover
- Right front wheel
- Right front splash shield
- Drive belt

8. Drain the coolant.

9. Remove or disconnect the following:
- Cooling fans and the radiator
- Upper radiator hose from the thermostat housing
- Power steering hose return and pressure hoses from the pump, if equipped

10. Disconnect the following electrical connections:
- Ignition coil
- powertrain Control Module (PCM)
- Pre-converter Oxygen (O2S) sensor
- Idle Air Control (IAC) valve
- Throttle Position (TP) sensor
- Engine Coolant Temperature (ECT) sensor
- Coolant Temperature (CT) sensor
- Alternator voltage regulator and power connections

11. Remove or disconnect the following:
- All necessary vacuum lines, including those at the brake booster. Tag the lines for identification as necessary.
- Fuel return line from the pressure regulator
- Fuel feed line at the fuel rail
- Fuel rail and injector channel cover as an assembly
- Throttle cable from the throttle body and intake manifold bracket
- Coolant hose from the throttle body
- Heater outlet hose at the coolant pipe
- Coolant bypass hose at the cylinder head
- Surge tank coolant hose at the coolant pipe
- Lower radiator hose from the coolant pipe
- Starter solenoid **S** terminal wire
- Air conditioning compressor
- Auxiliary catalytic converter upper flange nuts from the exhaust manifold studs and the bolts at the bracket

- Nuts that attach the front muffler pipe to the main catalytic converter
- Both catalytic converters
- Crankshaft pulley bolt and the pulley
- Vacuum lines from the Evaporative (EVAP) emissions canister solenoid, if not already done
- Electrical connection from the EVAP canister solenoid
- Electrical connection from the oil pressure switch
- Camshaft Position (CMP) sensor and Crankshaft Position (CKP) sensor electrical connections
- Knock Sensor (KS) electrical connection
- Transaxle torque converter bolts, if equipped with an automatic transaxle
- Transaxle bell housing bolts

12. Support the transaxle with a floor jack and attach a suitable engine lifting device

13. Remove or disconnect the following:
- Right engine mount bracket-to-engine mount bolts
- Right engine mount bracket from the block and frame mount
- Engine from the transaxle
- Engine from the vehicle

To install:

14. Installation is the reverse of the removal procedure, while using the following torque values:
- Transaxle bell housing bolts: 55 ft. lbs. (75 Nm)
- Oil pan flange-to-transaxle bolts: 30 ft. lbs. (40 Nm)
- Right engine mount bracket-to-engine mount bolts: 44 ft. lbs. (60 Nm).
- Transaxle torque converter bolts, if equipped with an automatic transaxle: 44 ft. lbs. (60 Nm).
- Crankshaft pulley bolt: 15 ft. lbs. (20 Nm)
- Nuts that attach the front muffler pipe to the main catalytic converter: 22 ft. lbs. (30 Nm)
- Support bracket bolts: 30 ft. lbs. (40 Nm)
- Auxiliary catalytic converter lower nuts: 30 ft. lbs. (40 Nm)

2.2L Engine

1. Before servicing the vehicle, refer to the precautions in the beginning of this section.

2. Relieve the fuel system pressure.

3. Remove the hood.

4. Drain the engine oil.

5. Disconnect the battery cables.

6. If equipped with air conditioning, discharge the system using an approved recovery/recycling machine.

7. Remove or disconnect the following:
- Intake Air Temperature (IAT) sensor connector
- Air intake tube
- Breather tubes from the valve cover
- Right front wheel
- Right front splash shield
- Drive belt

8. Drain the coolant.

9. Remove or disconnect the following:
- Cooling fans and the radiator
- Upper radiator hose from the thermostat housing
- Power steering hose return and pressure hoses from the pump, if equipped

10. Disconnect the following electrical connections:
- Ignition coil
- Powertrain Control Module (PCM)
- Oxygen (O2S) sensor
- Idle Air Control (IAC) valve
- Manifold Absolute Pressure (MAP) sensor
- Throttle Position (TP) sensor
- Engine Coolant Temperature (ECT) sensor
- Coolant Temperature (CT) sensor
- Alternator voltage regulator and power connections
- Camshaft Position (CMP) sensor electrical connection

11. Remove or disconnect the following:
- All necessary vacuum lines, including those at the brake booster. Tag the lines for identification as necessary.
- Fuel return line from the pressure regulator
- Fuel feed line at the fuel rail
- Fuel rail and injector channel cover as an assembly
- Throttle cable from the throttle body and intake manifold bracket
- Coolant hose from the throttle body
- Heater outlet hose at the coolant pipe
- Coolant bypass hose at the cylinder head
- Surge tank coolant hose at the coolant pipe
- Lower radiator hose from the coolant pipe

- Starter solenoid **S** terminal wire
- Air conditioning compressor
- Auxiliary catalytic converter lower flange nuts
- Nuts that attach the front muffler pipe to the main catalytic converter
- Rubber hangers that attach the connecting pipe to the vehicle
- Connecting pipe bracket bolts and the bracket
- Main catalytic converter and pipe
- Crankshaft pulley bolt and the pulley
- Vacuum lines from the Evaporative (EVAP) emissions canister solenoid, if not already done
- Electrical connection from the EVAP canister solenoid
- Electrical connection from the oil pressure switch
- Crankshaft Position (CKP) sensor electrical connections
- Knock Sensor (KS) electrical connection

12. Support the transaxle with a floor jack.

13. Remove the center member.

14. Attach a suitable engine lifting device

15. Remove or disconnect the following:
- Transaxle torque converter bolts, if equipped with an automatic transaxle
- Transaxle bell housing and oil pan flange bolts
- Right engine mount bracket
- Engine from the transaxle
- Resonator
- Air filter housing
- Engine from the vehicle

To install:

16. Installation is the reverse of the removal procedure, while using the following torque values:
- Transaxle bell housing bolts: 55 ft. lbs. (75 Nm)
- Oil pan flange-to-transaxle bolts: 30 ft. lbs. (40 Nm)
- Right engine mount bracket-to-engine mount bolts: 44 ft. lbs. (60 Nm)
- Transaxle torque converter bolts, if equipped with an automatic transaxle: 44 ft. lbs. (60 Nm)
- Crankshaft pulley bolt: 15 ft. lbs. (20 Nm)
- Nuts that attach the front muffler pipe to the main catalytic converter: 22 ft. lbs. (30 Nm)
- Support bracket bolts: 30 ft. lbs. (40 Nm)
- Auxiliary catalytic converter lower fasteners: 30 ft. lbs. (40 Nm)

- Connecting pipe bracket nuts: 22 ft. lbs. (30 Nm)

Water Pump

REMOVAL & INSTALLATION

1.6L Engine

1. Before servicing the vehicle, refer to the precautions in the beginning of this section.

2. Drain the cooling system to a level below the pump.

3. Remove or disconnect the following:
- Rear timing belt cover
- Water pump bolts
- Water pump
- Ring seal from the pump

4. Inspect the pump for damage and/or wear.

5. Clean the gasket mating surfaces.

To install:

6. Install or connect the following:
- New ring seal on the water pump. Coat the seal with Lubriplate®.
- Water pump and it's retaining bolts. Tighten the bolts to 89 inch lbs. (10 Nm).
- Rear timing belt cover

7. Refill the cooling system, start the vehicle and check for leaks.

2.0L and 2.2L Engines

1. Before servicing the vehicle, refer to the precautions in the beginning of this section.

2. Drain the cooling system to a level below the pump.

3. Remove or disconnect the following:
- Timing belt. Refer to the Timing Belt unit repair section.
- Timing belt tension roller retaining bolt and the roller

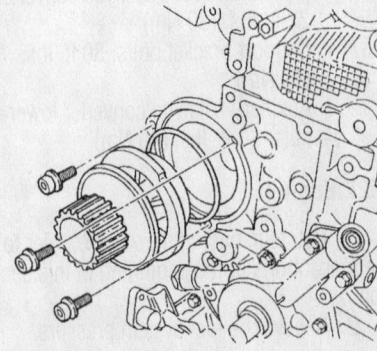

9307ZG98

Water pump mounting—1.6L engines

- Water pump bolts
- Water pump
- Ring seal from the pump

4. Inspect the pump for damage and/or wear.

5. Clean the gasket mating surfaces.

To install:

6. Install or connect the following:
- New ring seal on the water pump. Coat the seal with Lubriplate®.
- Water pump and it's retaining bolts. Tighten the bolts to 15 ft. lbs. (20 Nm).
- Timing belt tension roller to the oil pump with the flange inserted into the recess of the oil pump
- Roller bolt and finger tighten only
- Timing belt and tighten the roller bolt

7. Refill the cooling system, start the vehicle and check for leaks.

Cylinder Head

REMOVAL & INSTALLATION

1.6L Engine

1. Before servicing the vehicle, refer to the precautions in the beginning of this section.

2. Relieve the fuel system pressure.

3. Remove or disconnect the following:
- Negative battery cable
- Powertrain Control Module (PCM) ground terminal from the intake manifold

4. Drain the cooling system.

5. Remove or disconnect the following:
- Intake Air Temperature (IAT) sensor connector
- Breather tube from the valve cover
- Air intake tube from the throttle body
- Ignition coil electrical connector
- Pre-converter Oxygen (O_2S) sensor connector
- Fuel Injector harness connectors
- Idle Air Control (IAC) valve connector
- Throttle Position (TP) sensor connector
- Alternator drive belt
- Engine Coolant Temperature (ECT) sensor connector
- Coolant Temperature (CT) sensor connector
- Air filter housing bolts and the housing
- Right front wheel
- Right front splash shield

- Upper radiator hose from the thermostat housing
- Air Conditioner (A/C) compressor drive belt, if equipped
- Power steering pulley bolts, if equipped

➡ **Push the engine assembly towards the battery to remove the power steering pump pulley.**

- Power steering pump pulley, if equipped
- Crankshaft pulley bolt and the pulley
- Upper and lower front timing cover bolts and the covers
- Power steering pump, if equipped. Set the pump aside without disconnecting the lines.

6. If the power steering pump was removed, install the engine mount-to-engine bracket bolts and tighten to 44 ft. lbs. (60 Nm).

7. Align the camshaft gear timing marks

8. Loosen the water pump retaining bolts slightly.

9. Rotate the water pump counterclock-

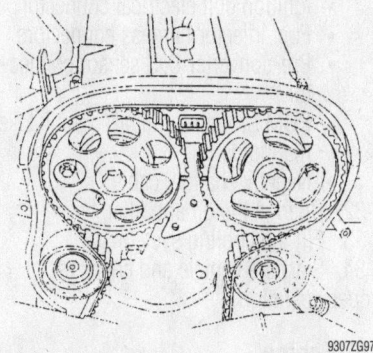

9307ZG97

Align the camshaft gear timing marks as shown—1.6L engine

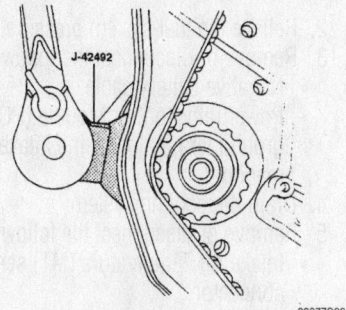

9307ZG96

Rotate the water pump counterclockwise using timing belt adjuster J 42492 to relieve timing belt tension —1.6L engine

wise using timing belt adjuster J 42492 to relieve timing belt tension.

10. Remove or disconnect the following:
- Timing belt. Refer to the Timing Belt unit repair section.
- Crankcase ventilation tube from the valve cover
- Spark plug cover bolts and the cover
- Camshaft Position (CMP) sensor connector
- Spark plug wires from the plugs
- Valve cover nuts, washers, cover and gasket

➡ **Be very careful not to nick or scratch the camshafts.**

11. Hold the flats of the intake camshaft in position using an open ended wrench and remove the camshaft gear bolt. Remove the gear.

12. Hold the flats of the exhaust camshaft in position using an open ended wrench and remove the camshaft gear bolt. Remove the gear.

13. Remove or disconnect the following:
- Rear timing belt cover
- Exhaust manifold heat shield bolts and the shield
- Auxiliary catalytic converter nuts from the exhaust manifold
- All necessary vacuum hoses
- Fuel return line from the pressure regulator

- Fuel feed line from the fuel rail
- Heater inlet hose from the cylinder head
- Surge tank coolant hose from the throttle body
- Throttle cable from the intake manifold and the throttle body
- Upper intake manifold support bracket bolts
- Cylinder head bolts in the sequence illustrated using several passes
- Cylinder head with the intake and exhaust manifolds as an assembly
- Cylinder head gasket

14. Clean the gasket mating surfaces.

15. Inspect the cylinder head for nicks, scratches and warpage.

To install:

16. Install or connect the following:
- Cylinder head gasket
- Cylinder head with the intake and exhaust manifolds as an assembly

17. Install the cylinder head bolts and tighten in the sequence illustrated as follows:

a. Step 1: in sequence to 18 ft. lbs. (25 Nm).

b. Step 2: in sequence an additional 60 degrees.

c. Step 3: in sequence an additional 60 degrees.

d. Step 4: in sequence an additional 60 degrees.

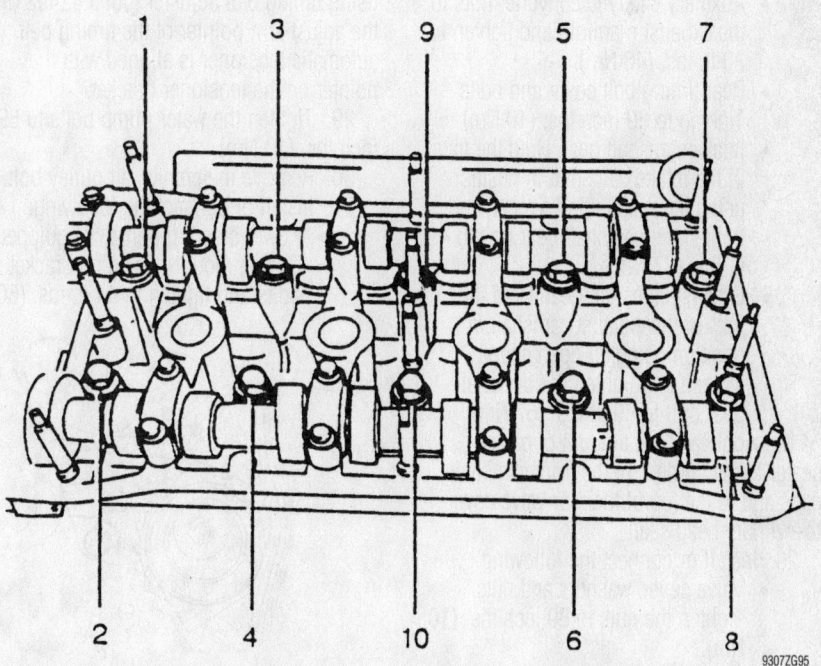

9307ZG95

Loosen the cylinder head bolts in this sequence using several passes—1.6L engine

Timing belt service is covered in Section 3 of this manual

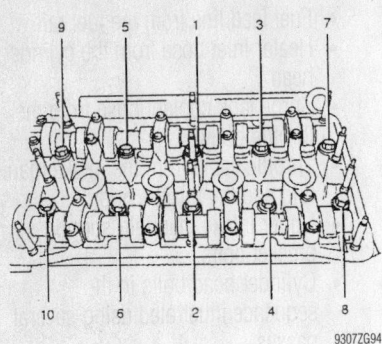

Cylinder head torque sequence—1.6L engine

e. Step 5: in sequence an additional 10 degrees.
18. Install or connect the following:
- Throttle cable to the intake manifold and the throttle body
- Upper intake manifold support bracket bolts and tighten to 18 ft. lbs. (25 Nm)
- Surge tank coolant hose to the throttle body
- Heater inlet hose to the cylinder head
- Fuel feed line to the fuel rail
- Fuel return line to the pressure regulator
- All vacuum hoses
- Exhaust manifold heat shield and bolts. Tighten the bolts to 11 ft. lbs. (15 Nm).
- Auxiliary catalytic converter nuts to the exhaust manifold and tighten to 30 ft. lbs. (40 Nm).
- Rear timing belt cover and bolts. Tighten to 89 inch lbs. (10 Nm).
- Intake camshaft gear. Hold the flats of the intake camshaft in position using an open ended wrench and tighten the camshaft gear bolt to 49 ft. lbs. (67 Nm).
- Exhaust camshaft gear. Hold the flats of the exhaust camshaft in position using an open ended wrench and tighten the camshaft gear bolt to 49 ft. lbs. (67 Nm).
19. Apply a small amount of gasket sealer to the corners of the front camshaft caps and to the top of the rear valve cover-to-cylinder head seal.
20. Install or connect the following:
- Valve cover, washers and nuts. Tighten the nuts to 89 inch lbs. (10 Nm).
- Spark plug wires to the plugs
- Spark plug cover and bolts. Tighten the bolts to 27 inch lbs. (3 Nm).
- CMP sensor connector

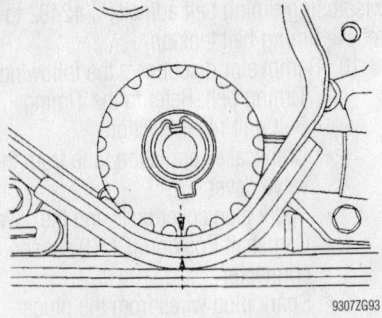

Align the mark on the crankshaft gear to the notch at the bottom of the rear timing belt cover—1.6L engine

- Crankcase ventilation tube to the valve cover
21. Align the camshaft gear timing marks
22. Align the mark on the crankshaft gear to the notch at the bottom of the rear timing belt cover.
23. Install the timing belt.
24. Rotate the water pump clockwise using timing belt adjuster tool J 42492 until the adjust arm pointer of the timing belt automatic tensioner is aligned with the notch in the tensioner bracket.
25. Tighten the water pump bolts.
26. Using the crankshaft pulley bolt, rotate the crankshaft two full turns clockwise.
27. Loosen the water pump bolts.
28. Rotate the water pump clockwise using timing belt adjuster tool J 42492 until the adjust arm pointer of the timing belt automatic tensioner is aligned with the pointer on the tensioner bracket.
29. Tighten the water pump bolts to 89 inch lbs. (10 Nm).
30. Remove the crankshaft pulley bolt.
31. Install or connect the following:
- Power steering pump, if equipped
- Engine mount-to-engine bracket bolts and tighten to 44 ft. lbs. (60

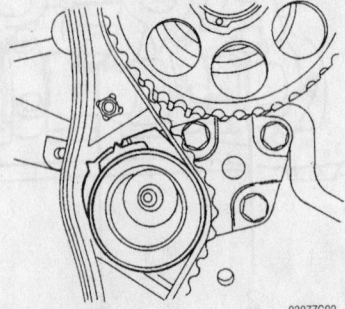

Using tool J 42492 rotate the water pump until the adjust arm pointer of the automatic tensioner is aligned with the notch in the tensioner bracket —1.6L engine

Nm), if the power steering pump was removed.
- Upper and lower front timing covers and their bolts. Tighten the bolts to 89 inch lbs. (10 Nm).
- Crankshaft pulley and bolt. Tighten the bolt to 70 ft. lbs. (95 Nm). Retighten the bolt an additional 30 degrees and finally an additional 15 degrees.
- Power steering pump pulley and bolts, if equipped. Tighten the bolts to 18 ft. lbs. (25 Nm).
- CT sensor connector
- ECT sensor connector
- Alternator drive belt
- A/C compressor drive belt, if equipped
- Upper radiator hose to the thermostat housing
- Right front splash shield
- Right front wheel
- Air filter housing and tighten the bolts to 106 inch lbs. (12 Nm)
- Air intake tube to the throttle body
- Breather tube to the valve cover
- IAT sensor connector
- IAC valve connector
- TP sensor connector
- Ignition coil electrical connector
- Fuel Injector harness connectors
- Pre-converter O_2S sensor connector
- PCM ground terminal to the intake manifold
- Negative battery cable
32. Change the oil and filter.
33. Fill the cooling system.
34. Start the vehicle and check for proper operation.

2.0L Engine

1. Before servicing the vehicle, refer to the precautions in the beginning of this section.
2. Relieve the fuel system pressure.
3. Remove or disconnect the following:
- Negative battery cable
- Powertrain Control Module (PCM) ground terminal from the intake manifold
4. Drain the cooling system.
5. Remove or disconnect the following:
- Intake Air Temperature (IAT) sensor connector
- Breather tube from the valve cover
- Air intake tube from the throttle body
- Ignition coil electrical connector
- Pre-converter Oxygen (O_2S) sensor connector

- Idle Air Control (IAC) valve connector
- Throttle Position (TP) sensor connector
- Engine Coolant Temperature (ECT) sensor connector
- Coolant Temperature (CT) sensor connector
- Camshaft Position (CMP) sensor connector
- Air filter housing bolts and the housing
- Right front wheel
- Right front splash shield

6. Install a suitable engine support fixture.

7. Remove or disconnect the following:
- Engine right mount bracket and bolts
- Upper radiator hose from the thermostat housing
- Drive belt
- Crankshaft pulley bolt and the pulley
- Front timing cover bolts and the cover
- Timing belt. Refer to the Timing Belt unit repair section.
- Crankcase ventilation tube from the valve cover
- Spark plug cover bolts and the cover
- Spark plug wires from the plugs
- Valve cover nuts, washers, cover and gasket

➡**Be very careful not to nick or scratch the camshafts.**

8. Hold the flats of the intake camshaft in position using an open ended wrench and remove the camshaft gear bolt. Remove the gear.

9. Hold the flats of the exhaust camshaft in position using an open ended wrench and remove the camshaft gear bolt. Remove the gear.

10. Remove or disconnect the following:
- Timing belt tensioner
- Timing belt idler pulleys
- Engine mount
- Rear timing belt cover
- Auxiliary catalytic converter nuts from the exhaust manifold
- All necessary vacuum hoses
- Fuel return line from the pressure regulator
- Fuel feed line from the fuel rail
- Alternator adjusting bracket
- Coolant hose from the rear cylinder

head and coil Exhaust Gas Recirculation (EGR) bracket
- Surge tank coolant hose from the throttle body
- Fuel rail assembly
- Alternator-to-intake manifold support bracket bolt at the cylinder head and alternator
- Intake manifold-to-alternator strap bracket bolt and loosen the bolt on the alternator
- Strap clear of the intake manifold
- EVAP canister solenoid bracket bolt and move the bracket aside
- Throttle cable from the intake manifold and the throttle body
- Cylinder head bolts in the sequence illustrated using several passes
- Cylinder head
- Cylinder head gasket

11. Clean the gasket mating surfaces.

12. Inspect the cylinder head for nicks, scratches and warpage.

To install:

13. Install or connect the following:
- Cylinder head gasket
- Cylinder head

14. Install the cylinder head bolts and tighten in the sequence illustrated as follows:

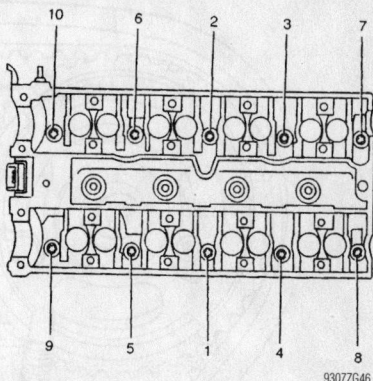

Cylinder head torque sequence—2.0L engine

a. Step 1: in sequence to 18 ft. lbs. (25 Nm).

b. Step 2: in sequence an additional 90 degrees.

c. Step 3: in sequence an additional 90 degrees.

d. Step 4: in sequence an additional 90 degrees.

15. Install or connect the following:
- Throttle cable to the intake manifold and the throttle body
- Alternator-to-intake manifold support bracket and tighten the bolts to 26 ft. lbs. (35 Nm)

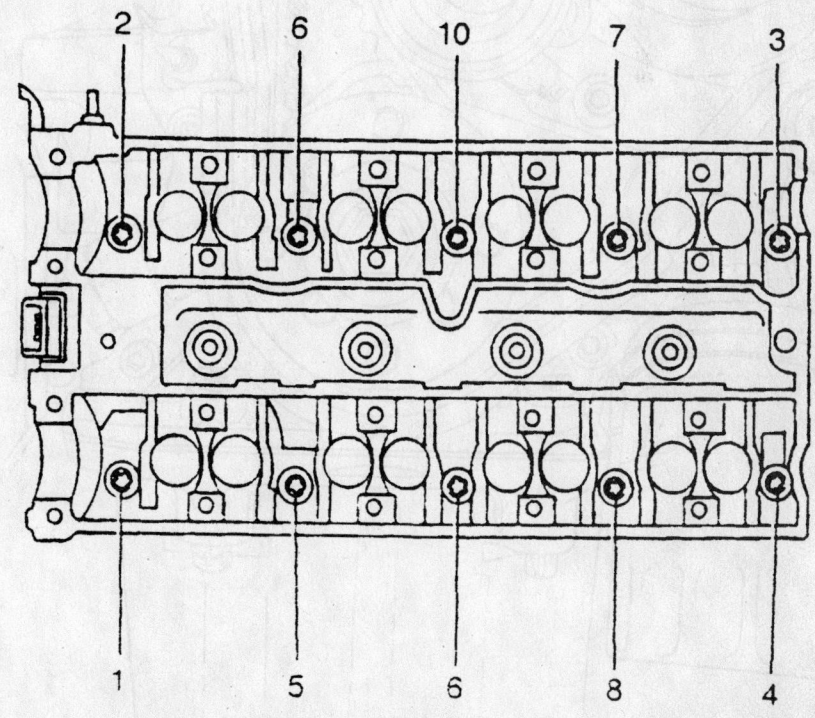

Loosen the cylinder head bolts in this sequence using several passes—2.0L engine

Heater Core replacement is covered in Section 2 of this manual

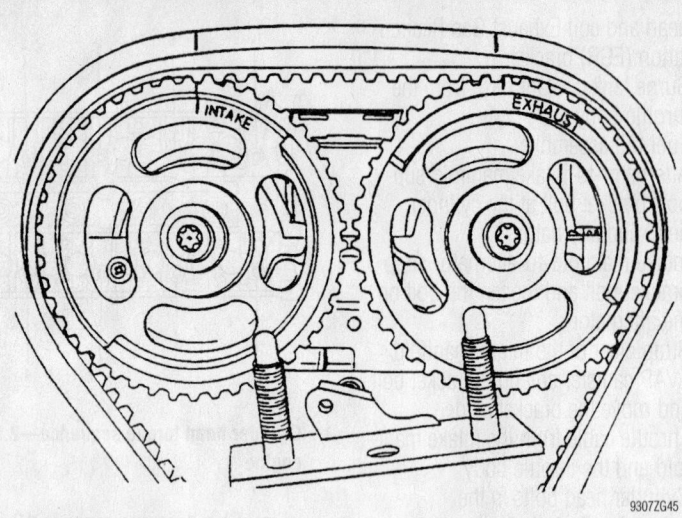

9307ZG45

Align the camshaft gear timing marks to the notches on the valve cover—2.0L engine

- Intake manifold-to-alternator strap bracket to the alternator and tighten to 15 ft. lbs. (20 Nm)
- Surge tank coolant hose to the throttle body
- Coolant hose to the rear cylinder head and coil EGR bracket
- Fuel feed line to the fuel rail
- Fuel return line to the pressure regulator
- All vacuum hoses
- Fuel rail assembly
- Auxiliary catalytic converter nuts to the exhaust manifold and tighten to 30 ft. lbs. (40 Nm)
- Rear timing belt cover and bolts. Tighten to 53 inch lbs. (6 Nm).
- Timing belt tensioner and tighten the bolt to 18 ft. lbs. (25 Nm)
- Timing belt idler pulleys and

9307ZG44

Align the mark on the crankshaft gear to the notch at the bottom of the rear timing belt cover— 2.0L engine

tighten the bolt and nut to 18 ft. lbs. (25 Nm)

16. Install the camshaft gears with the timing marks at the front. Insert the guide pin of the intake camshaft into the **IN** bore and the guide pin of the exhaust camshaft into the **EX** bore.

17. Hold the flats of the camshafts in position using an open ended wrench and tighten the camshaft gear bolt to 37 ft. lbs. (50 Nm), tighten the bolt an additional 60 degrees and finally an additional 15 degrees.

18. Apply a small amount of gasket sealer to the corners of the front camshaft caps and to the top of the rear valve cover-to-cylinder head seal.

19. Install or connect the following:
- Valve cover, washers and nuts. Tighten the nuts to 71 inch lbs. (8 Nm).
- Spark plug wires to the plugs
- Spark plug cover and bolts. Tighten the bolts to 27 inch lbs. (3 Nm).
- Crankcase ventilation tube to the valve cover

20. Align the camshaft gear timing marks to the notches on the valve cover.

21. Align the mark on the crankshaft gear to the notch at the bottom of the rear timing belt cover.

22. Install or connect the following:
- Timing belt
- Front timing cover and bolts. Tighten the bolts to 53 inch lbs. (6 Nm).
- Crankshaft pulley and bolt. Tighten the bolt to 15 ft. lbs. (20 Nm).
- Right engine mount bracket and tighten the bolts to 44 ft. (60 Nm)

23. Remove the engine support assembly.

24. Install or connect the following:
- Drive belt
- Upper radiator hose to the thermostat housing
- Right front splash shield
- Right front wheel
- Air filter housing and tighten the bolts to 71 inch lbs. (8 Nm)
- Air intake tube to the throttle body
- Breather tube to the valve cover
- IAT sensor connector
- CMP sensor connector
- CT sensor connector
- ECT sensor connector
- IAC valve connector
- TP sensor connector
- EVAP canister solenoid bracket bolt and tighten to 44 inch lbs. (5 Nm)
- Ignition coil electrical connector
- Pre-converter O$_2$S sensor connector

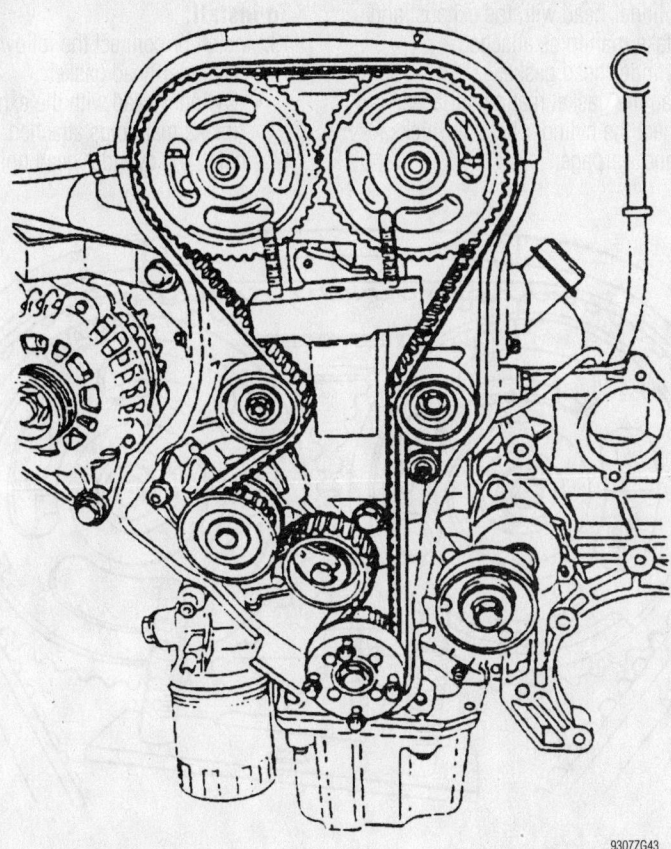

Correct timing belt routing—2.0L engine

- Oxygen (O$_2$S) sensor connector
- Idle Air Control (IAC) valve connector
- Throttle Position (TP) sensor connector
- Engine Coolant Temperature (ECT) sensor connector
- Coolant Temperature (CT) sensor connector
- Camshaft Position (CMP) sensor connector
- Air filter housing bolts and the housing
- Right front wheel
- Right front splash shield
6. Install a suitable engine support fixture.
7. Remove or disconnect the following:
- Engine right mount bracket and bolts
- Upper radiator hose from the thermostat housing
- Drive belt
- Crankshaft pulley bolt and the pulley
- Front timing cover bolts and the cover
- Timing belt. Refer to the Timing Belt unit repair section.
- Crankcase ventilation tube from the valve cover

- PCM ground terminal to the intake manifold
- Negative battery cable
25. Change the oil and filter.
26. Fill the cooling system.
27. Start the vehicle and check for proper operation.

2.2L Engine

1. Before servicing the vehicle, refer to the precautions in the beginning of this section.
2. Relieve the fuel system pressure.
3. Remove or disconnect the following:
- Negative battery cable
- Powertrain Control Module (PCM) ground terminal
4. Drain the cooling system.
5. Remove or disconnect the following:
- Intake Air Temperature (IAT) sensor connector
- Breather tube from the valve cover
- Resonator
- Air intake tube
- Evaporative Emission (EVAP) solenoid bracket bolt and move the solenoid aside
- Ignition coil electrical connector

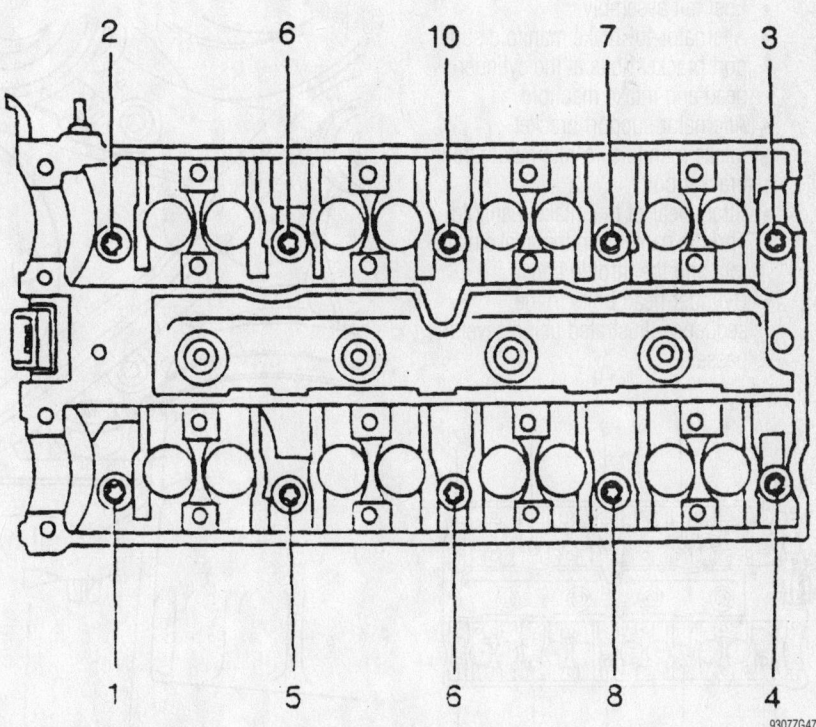

Loosen the cylinder head bolts in this sequence using several passes—2.0L and 2.2L engines

Brake service is covered in Section 4 of this manual

- Spark plug cover bolts and the cover
- Spark plug wires from the plugs
- Valve cover bolts, washers, cover and gasket

➡ **Be very careful not to nick or scratch the camshafts.**

8. Hold the flats of the intake camshaft in position using an open ended wrench and remove the camshaft gear bolt. Remove the gear.

9. Hold the flats of the exhaust camshaft in position using an open ended wrench and remove the camshaft gear bolt. Remove the gear.

10. Remove or disconnect the following:
- Timing belt tensioner
- Timing belt idler pulleys
- Engine mount
- Rear timing belt cover
- Exhaust manifold heat shield
- Auxiliary catalytic converter upper flange nuts
- All necessary vacuum hoses
- Fuel return line from the pressure regulator
- Fuel feed line from the fuel rail
- Coolant hose from the rear cylinder head and coil Exhaust Gas Recirculation (EGR) bracket
- Surge tank coolant hose from the throttle body
- Fuel rail assembly
- Alternator-to-intake manifold support bracket bolts at the cylinder head and intake manifold
- Alternator support bracket
- Intake manifold-to-alternator strap bracket bolt
- Strap clear of the intake manifold
- Throttle cable from the intake manifold and the throttle body
- Cylinder head bolts in the sequence illustrated using several passes

- Cylinder head with the exhaust and intake manifolds attached
- Cylinder head gasket

11. Clean the gasket mating surfaces.

12. Inspect the cylinder head for nicks, scratches and warpage.

To install:

13. Install or connect the following:
- Cylinder head gasket
- Cylinder head with the exhaust and intake manifolds attached

14. Install the cylinder head bolts and

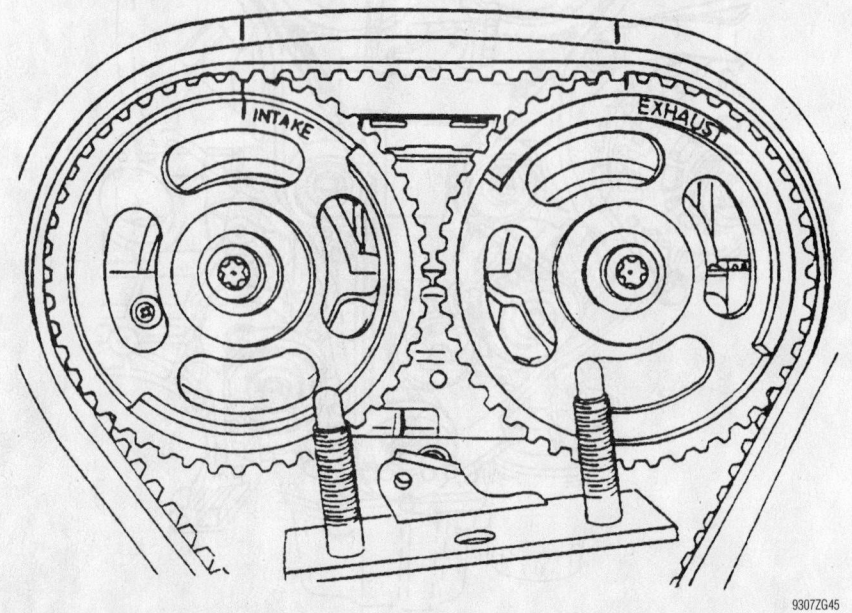

Align the camshaft gear timing marks to the notches on the valve cover—2.0L and 2.2L engines

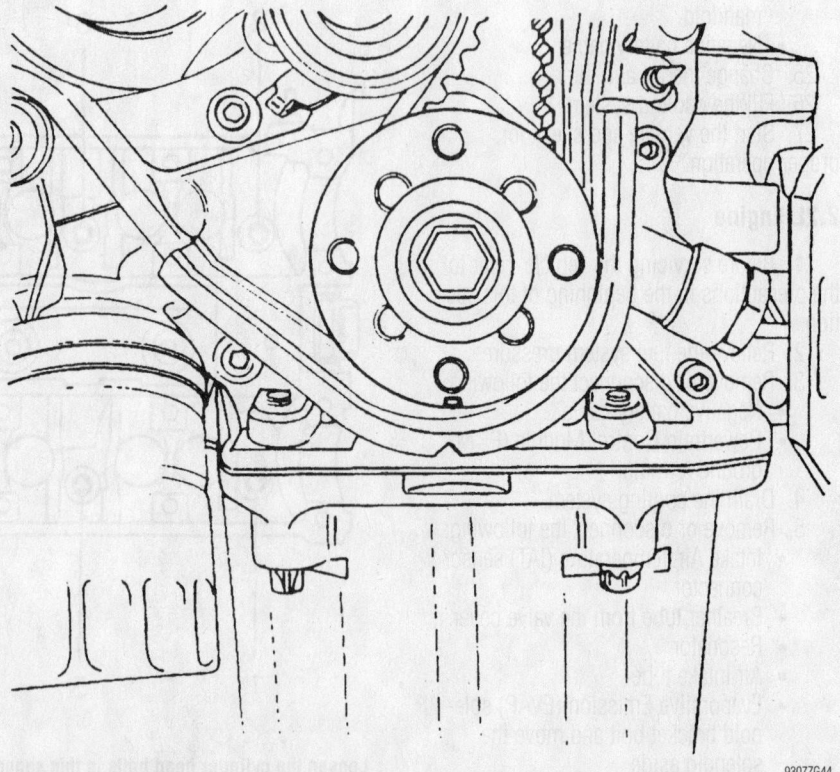

Align the mark on the crankshaft gear to the notch at the bottom of the rear timing belt cover—2.0L and 2.2L engines

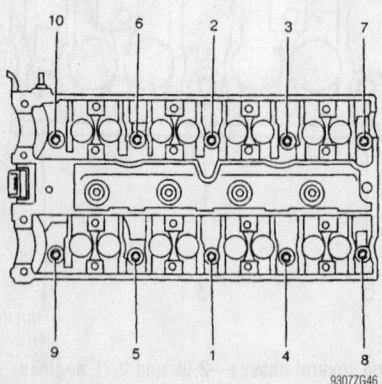

Cylinder head torque sequence—2.2L engine

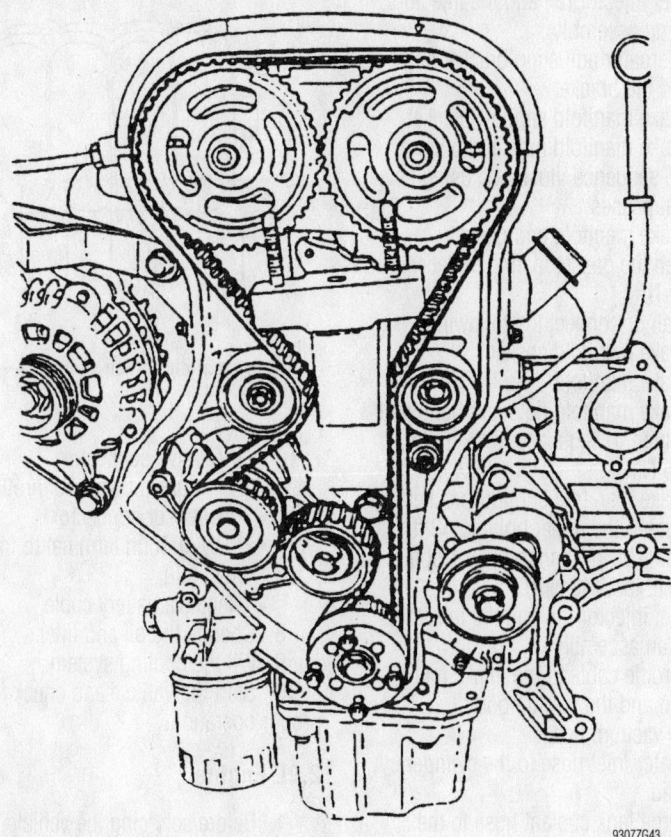

9307ZG43

Correct timing belt routing—2.0L and 2.2L engines

tighten in the sequence illustrated as follows:

    a. Step 1: in sequence to 18 ft. lbs. (25 Nm).

    b. Step 2: in sequence an additional 90 degrees.

    c. Step 3: in sequence an additional 90 degrees.

    d. Step 4: in sequence an additional 90 degrees.

15. Install or connect the following:
- Throttle cable to the intake manifold and the throttle body
- Alternator-to-intake manifold support bracket and tighten the bolts to 26 ft. lbs. (35 Nm)
- Intake manifold-to-alternator strap bracket to the alternator and tighten to 15 ft. lbs. (20 Nm)
- Surge tank coolant hose to the throttle body
- Coolant hose to the rear cylinder head and coil EGR bracket
- Fuel feed line to the fuel rail
- Fuel return line to the pressure regulator
- All vacuum hoses
- Fuel rail assembly

- Auxiliary catalytic converter nuts to the exhaust manifold and tighten to 30 ft. lbs. (40 Nm)
- Exhaust manifold heat shield and tighten the bolts to 71 inch lbs. (8 Nm)
- Rear timing belt cover and bolts. Tighten to 53 inch lbs. (6 Nm).
- Timing belt tensioner and tighten the bolt to 18 ft. lbs. (25 Nm)
- Timing belt idler pulleys and tighten the bolt and nut to 18 ft. lbs. (25 Nm)

16. Install the camshaft gears with the timing marks at the front. Insert the guide pin of the intake camshaft into the **IN** bore and the guide pin of the exhaust camshaft into the **EX** bore.

17. Hold the flats of the camshafts in position using an open ended wrench and tighten the camshaft gear bolt to 37 ft. lbs. (50 Nm), tighten the bolt an additional 60 degrees and finally an additional 15 degrees.

18. Apply a small amount of gasket sealer to the corners of the front camshaft caps and to the top of the rear valve cover-to-cylinder head seal.

19. Install or connect the following:

- Valve cover, washers and nuts. Tighten the nuts to 71 inch lbs. (8 Nm).
- Spark plug wires to the plugs
- Spark plug cover and bolts. Tighten the bolts to 27 inch lbs. (3 Nm).
- Crankcase ventilation tube to the valve cover

20. Align the camshaft gear timing marks to the notches on the valve cover.

21. Align the mark on the crankshaft gear to the notch at the bottom of the rear timing belt cover.

22. Install or connect the following:
- Timing belt
- Front timing cover and bolts. Tighten the bolts to 71 inch lbs. (8 Nm).
- Crankshaft pulley and bolt. Tighten the bolt to 15 ft. lbs. (20 Nm).
- Right engine mount bracket and tighten the bolts to 44 ft. (60 Nm)

23. Remove the engine support assembly.

24. Install or connect the following:
- Drive belt
- Upper radiator hose to the thermostat housing
- Right front splash shield
- Right front wheel
- Air filter housing and tighten the bolts to 71 inch lbs. (8 Nm)
- Air intake tube to the throttle body
- Breather tube to the valve cover
- IAT sensor connector
- Resonator and tighten the bolts to 71 inch lbs. (8 Nm)
- CMP sensor connector
- CT sensor connector
- ECT sensor connector
- IAC valve connector
- TP sensor connector
- EVAP canister solenoid bracket bolt and tighten to 44 inch lbs. (5 Nm)
- Ignition coil electrical connector
- O$_2$S sensor connection
- PCM ground terminal
- Negative battery cable

25. Change the oil and filter.
26. Fill the cooling system.
27. Start the vehicle and check for proper operation.

Rocker Arms/Shafts

REMOVAL & INSTALLATION

These engines are equipped with hydraulic lash adjusters. The valve lash is not adjustable.

Intake Manifold

REMOVAL & INSTALLATION

1.6L Engine

1. Before servicing the vehicle, refer to the precautions in the beginning of this section.
2. Relieve the fuel system pressure.
3. Remove or disconnect the following:
   - Negative battery cable
   - Powertrain Control Module (PCM) ground terminal from the intake manifold
4. Drain the cooling system.
5. Remove or disconnect the following:
   - Intake Air Temperature (IAT) sensor connector
   - Air intake tube from the throttle body
   - Idle Air Control (IAC) valve connector
   - Throttle Position (TP) sensor connector
   - Alternator drive belt
   - Engine Coolant Temperature (ECT) sensor connector
   - Coolant Temperature (CT) sensor connector
   - Heater inlet hose from the cylinder head
   - Surge tank coolant hose from the throttle body
   - All necessary vacuum hoses
   - Throttle cable from the intake manifold and the throttle body

- Fuel injector rail and the injectors as an assembly
- Alternator adjusting bracket bolt and the bracket
- Intake manifold support bracket
- Intake manifold nuts and bolts in the sequence illustrated using several passes
- Intake manifold and gasket
6. Clean the gasket mating surfaces.

To install:

7. Install or connect the following:
   - Intake manifold gasket
   - Intake manifold
   - Intake manifold nuts and bolts and tighten in sequence to 18 ft. lbs. (25 Nm)
   - Intake manifold support bracket. Tighten the upper bolts to 18 ft. lbs. (25 Nm) and the lower bolt to 30 ft. lbs. (40 Nm).
   - Fuel injector rail and the injectors as an assembly
   - Throttle cable to the intake manifold and the throttle body
   - All vacuum hoses
   - Heater inlet hose to the cylinder head
   - Surge tank coolant hose to the throttle body
   - Alternator adjusting bracket and tighten the bolts to 11 ft. lbs. (15 Nm)
   - Alternator drive belt
   - CT sensor connector
   - ECT sensor connector
   - TP sensor connector

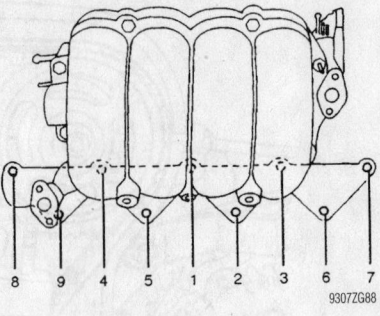

Intake manifold torque sequence—1.6L engine

- IAC valve connector
- Air intake tube to the throttle body
- IAT sensor connector
- PCM ground terminal to the intake manifold
- Negative battery cable
8. Change the oil and filter.
9. Fill the cooling system.
10. Start the vehicle and check for proper operation.

2.0L Engine

1. Before servicing the vehicle, refer to the precautions in the beginning of this section.
2. Relieve the fuel system pressure.
3. Remove or disconnect the following:
   - Negative battery cable
   - Evaporative (EVAP) canister purge solenoid from the intake manifold
4. Drain the cooling system.
5. Remove or disconnect the following:
   - Intake Air Temperature (IAT) sensor connector
   - Air intake tube from the throttle body
   - Idle Air Control (IAC) valve connector
   - Throttle Position (TP) sensor connector
   - Manifold Absolute Pressure (MAP) sensor
   - Coolant hoses from the throttle body
   - All necessary vacuum hoses
   - Throttle cable from the intake manifold and the throttle body
   - Throttle cable bracket from the intake manifold
   - Fuel rail and injector cover as an assembly
   - Alternator-to-intake manifold strap bracket
   - Alternator-to-intake manifold support bracket
   - Intake manifold support bracket

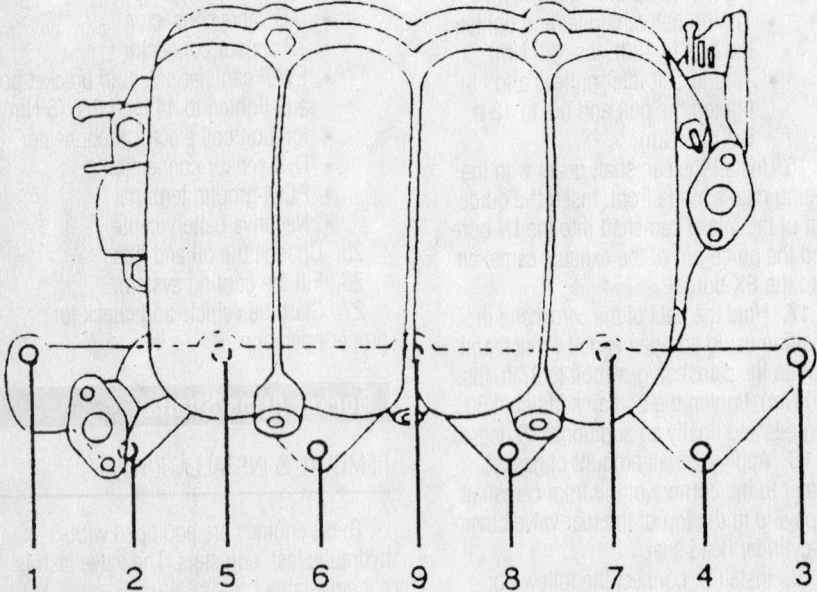

Loosen the intake manifold bolts in this sequence using several passes—1.6L engine

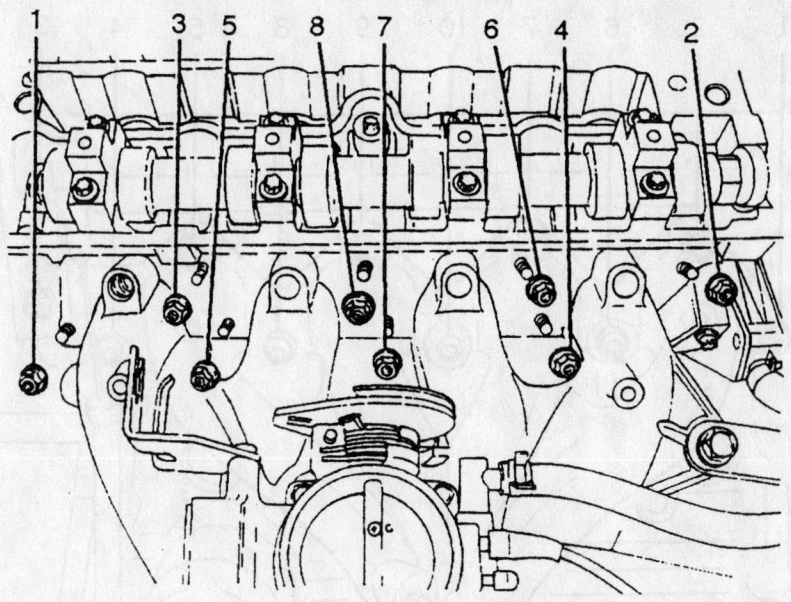

Loosen the intake manifold bolts in this sequence using several passes—2.0L and 2.2L engines

- Intake manifold nuts and bolts in the sequence illustrated using several passes
- Intake manifold and gasket

6. Clean the gasket mating surfaces.

To install:

7. Install or connect the following:
- Intake manifold gasket
- Intake manifold
- Intake manifold nuts and bolts and tighten in sequence to 16 ft. lbs. (22 Nm)
- Alternator-to-intake manifold strap bracket and tighten the bolts to 15 ft. lbs. (20 Nm)
- Intake manifold support bracket and tighten the bolts to 15 ft. lbs. (20 Nm)

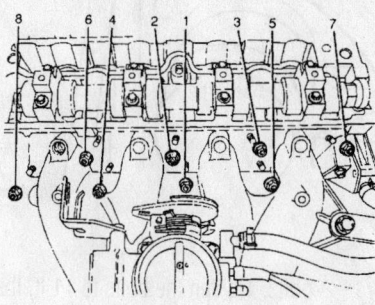

9307ZG41

Intake manifold torque sequence—2.0L engine

- Alternator-to-intake manifold support bracket. Tighten the lower bolt to 15 ft. lbs. (20 Nm) and the upper bolt to 26 ft. lbs. (35 (Nm).
- Fuel rail and injector cover as an assembly
- Throttle cable bracket to the intake manifold and tighten the bolts to 71 inch lbs. (8 Nm)
- Throttle cable to the intake manifold and the throttle body
- All vacuum hoses
- MAP sensor connector
- Coolant hoses to the throttle body
- IAC valve connector
- TP sensor connector
- Air intake tube to the throttle body
- IAT sensor connector
- EVAP canister purge solenoid to the manifold and tighten the bolt to 44 inch lbs. (5 Nm)
- Negative battery cable

8. Change the oil and filter.
9. Fill the cooling system.
10. Start the vehicle and check for proper operation.

2.2L Engine

1. Before servicing the vehicle, refer to the precautions in the beginning of this section.
2. Relieve the fuel system pressure.
3. Remove or disconnect the following:

- Negative battery cable
- Evaporative (EVAP) canister purge solenoid from the intake manifold and loosen the bolt

4. Drain the cooling system.
5. Remove or disconnect the following:

- Intake Air Temperature (IAT) sensor connector
- Air intake tube from the throttle body
- Idle Air Control (IAC) valve connector
- Throttle Position (TP) sensor connector
- Manifold Absolute Pressure (MAP) sensor
- Coolant hoses from the throttle body
- All necessary vacuum hoses
- Throttle cable from the intake manifold and the throttle body
- Throttle cable bracket from the intake manifold
- Alternator-to-intake manifold strap bracket
- Power steering hose clamp bolt and move the hose aside
- Fuel rail and injector cover as an assembly
- Alternator-to-intake manifold support bracket
- Intake manifold support bracket
- Intake manifold nuts and bolts in the sequence illustrated using several passes
- Intake manifold and gasket

6. Clean the gasket mating surfaces.

To install:

7. Install or connect the following:
- Intake manifold gasket
- Intake manifold
- Intake manifold nuts and bolts and

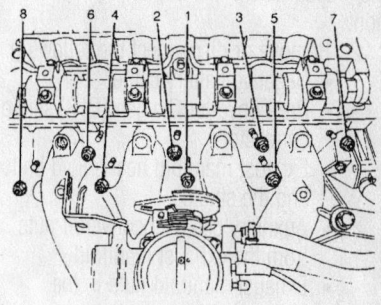

9307ZG41

Intake manifold torque sequence—2.2L engine

For Accessory Drive Belt illustrations, see Section 1 of this manual

tighten in sequence to 13 ft. lbs. (18 Nm)

- Alternator-to-intake manifold strap bracket and tighten the bolts to 15 ft. lbs. (20 Nm)
- Intake manifold support bracket and tighten the bolts to 15 ft. lbs. (20 Nm)
- Alternator-to-intake manifold support bracket. Tighten the bolt to 26 ft. lbs. (35 (Nm).
- Fuel rail and injector cover as an assembly
- Throttle cable bracket to the intake manifold and tighten the bolts to 71 inch lbs. (8 Nm)
- Throttle cable to the intake manifold and the throttle body
- All vacuum hoses
- MAP sensor connector
- Coolant hoses to the throttle body
- IAC valve connector
- TP sensor connector
- Power steering hose clamp and tighten the bolt to 71 inch lbs. (8 Nm)
- Air intake tube to the throttle body
- IAT sensor connector
- EVAP canister purge solenoid to the manifold and tighten the bolt to 44 inch lbs. (5 Nm)
- Negative battery cable

8. Change the oil and filter.
9. Fill the cooling system.
10. Start the vehicle and check for proper operation.

Exhaust Manifold

REMOVAL & INSTALLATION

1.6L Engine

1. Before servicing the vehicle, refer to the precautions in the beginning of this section.
2. Remove or disconnect the following:
   - Negative battery cable
   - Pre-converter Oxygen (O_2S) sensor connector
   - Exhaust manifold heat shield bolts and the shield
   - Auxiliary catalytic converter nuts from the exhaust manifold
   - Exhaust manifold nuts in the sequence illustrated using several passes
   - Exhaust manifold and gasket
3. Clean the gasket mating surfaces.

To install:
4. Install or connect the following:

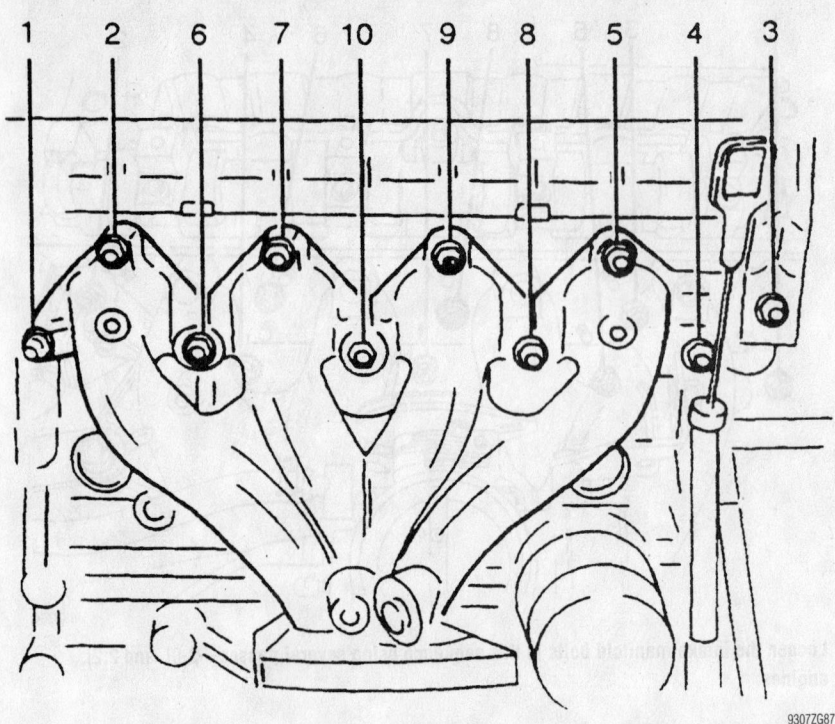

Loosen the exhaust manifold nuts in the sequence shown using several passes—1.6L engine

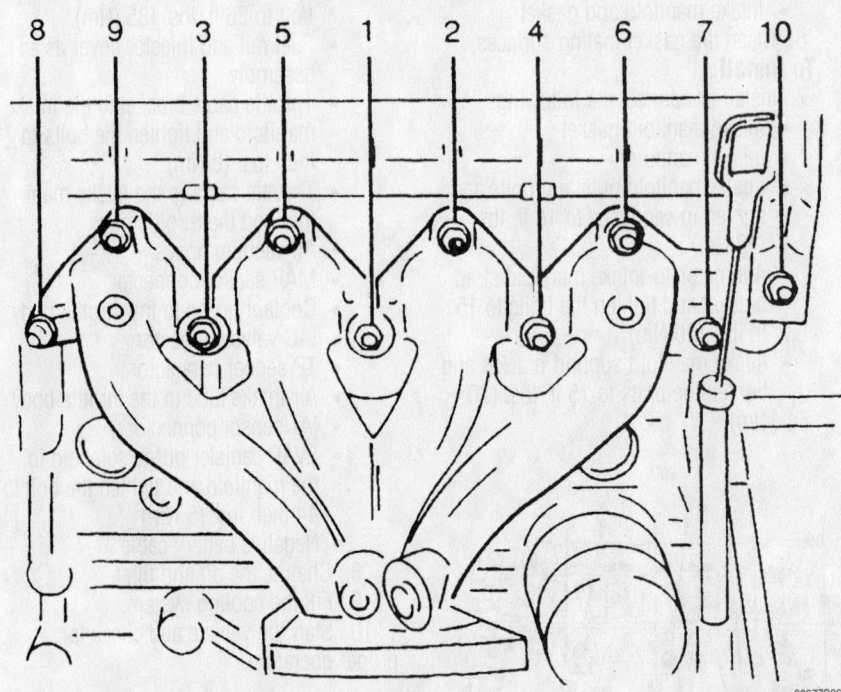

Exhaust manifold nut torque sequence—1.6L engine

- Exhaust manifold gasket and the manifold. Tighten the nuts in sequence to 18 ft. lbs. (25 Nm).
- Auxiliary catalytic converter nuts to the exhaust manifold and tighten to 30 ft. lbs. (40 Nm)
- Exhaust manifold heat shield and bolts. Tighten the bolts to 11 ft. lbs. (15 Nm).
- Pre-converter O_2S sensor connector
- Negative battery cable

5. Start the vehicle and check for proper operation.

2.0L Engine

1. Before servicing the vehicle, refer to the precautions in the beginning of this section.
2. Remove or disconnect the following:
   - Negative battery cable
   - Pre-converter Oxygen (O₂S) sensor connector
   - Exhaust manifold heat shield bolts and the shield
   - Auxiliary catalytic converter nuts from the exhaust manifold
   - Exhaust manifold nuts in the sequence illustrated using several passes
   - Exhaust manifold and gasket
3. Clean the gasket mating surfaces.

To install:
4. Install or connect the following:
   - Exhaust manifold gasket and the manifold. Tighten the nuts in sequence to 11 ft. lbs. (15 Nm)
   - Auxiliary catalytic converter nuts to the exhaust manifold and tighten to 30 ft. lbs. (40 Nm).
   - Exhaust manifold heat shield and bolts. Tighten the bolts to 71 inch lbs. (8 Nm).
   - Pre-converter O₂S sensor connector
   - Negative battery cable
5. Start the vehicle and check for proper operation.

2.2L Engine

1. Before servicing the vehicle, refer to the precautions in the beginning of this section.
2. Remove or disconnect the following:
   - Negative battery cable
   - Oxygen (O₂S) sensor connector
   - Exhaust manifold heat shield bolts and the shield
   - Auxiliary catalytic converter nuts from the exhaust manifold
   - Exhaust manifold nuts in the sequence illustrated using several passes
   - Exhaust manifold and gasket
3. Clean the gasket mating surfaces.

To install:
4. Install or connect the following:
   - Exhaust manifold gasket and the manifold. Tighten the nuts in sequence using four passes and retighten bolts 1. 2 and 3 to 11 ft. lbs. (15 Nm)
   - Auxiliary catalytic converter nuts to the exhaust manifold and tighten to 30 ft. lbs. (40 Nm).
   - Exhaust manifold heat shield and bolts. Tighten the bolts to 71 inch lbs. (8 Nm).
   - O₂S sensor connector
   - Negative battery cable
5. Start the vehicle and check for proper operation.

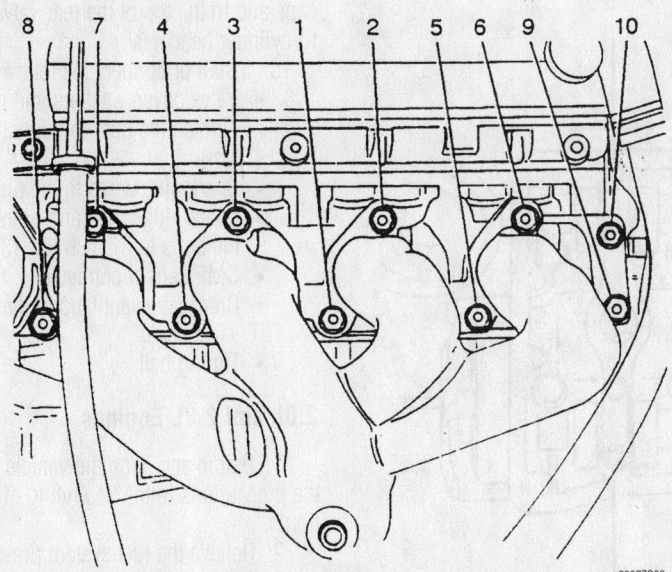

Loosen the exhaust manifold nuts in the sequence shown using several passes—2.0L and 2.2L engines

9307ZG40

Exhaust manifold nut torque sequence—2.0L and 2.2L engines

9307ZG39

Camshaft and Valve Lifters

REMOVAL & INSTALLATION

1.6L Engine

1. Before servicing the vehicle, refer to the precautions in the beginning of this section.

2. Relieve the fuel system pressure.
3. Remove or disconnect the following:
- Negative battery cable
- Timing belt. Refer to the Timing Belt unit repair section.
- Crankcase ventilation tube from the valve cover

- Spark plug cover bolts and the cover
- Camshaft Position (CMP) sensor connector
- Spark plug wires from the plugs
- Valve cover nuts, washers, cover and gasket

➡ **Be very careful not to nick or scratch the camshafts.**

4. Hold the flats of the intake camshaft in position using an open ended wrench and remove the camshaft gear bolt. Remove the gear.

5. Hold the flats of the exhaust camshaft in position using an open ended wrench and remove the camshaft gear bolt. Remove the gear.

6. Mark the camshaft cap locations so that they may be returned to their original positions.

7. Remove the camshaft cap bolts in the sequence illustrated.

8. Remove the caps and the camshafts.

To install:

9. Lubricate the camshafts and caps with engine oil.

10. Install the caps and the camshafts in their original positions.

11. Tighten the camshaft cap bolts in the sequence illustrated to 12 ft. lbs. (16 Nm).

12. Measure the camshaft endplay on both camshafts. The endplay should be 0.003–0.009 in. (0.10–0.25mm).

13. Install the intake camshaft gear. Hold the flats of the intake camshaft in position using an open ended wrench and tighten the camshaft gear bolt to 49 ft. lbs. (67 Nm).

14. Install the exhaust camshaft gear. Hold the flats of the exhaust camshaft in position using an open ended wrench and tighten the camshaft gear bolt to 49 ft. lbs. (67 Nm).

15. Apply a small amount of gasket sealer to the corners of the front camshaft caps and to the top of the rear valve cover-to-cylinder head seal.

16. Install or connect the following:
- Valve cover, washers and nuts. Tighten the nuts to 89 inch lbs. (10 Nm).
- Spark plug wires to the plugs
- Spark plug cover and bolts. Tighten the bolts to 27 inch lbs. (3 Nm).
- CMP sensor connector
- Crankcase ventilation tube to the valve cover
- Timing belt

2.0L and 2.2L Engines

1. Before servicing the vehicle, refer to the precautions in the beginning of this section.

2. Relieve the fuel system pressure.

3. Remove or disconnect the following:

- Negative battery cable
- Timing belt. Refer to the Timing Belt unit repair section.

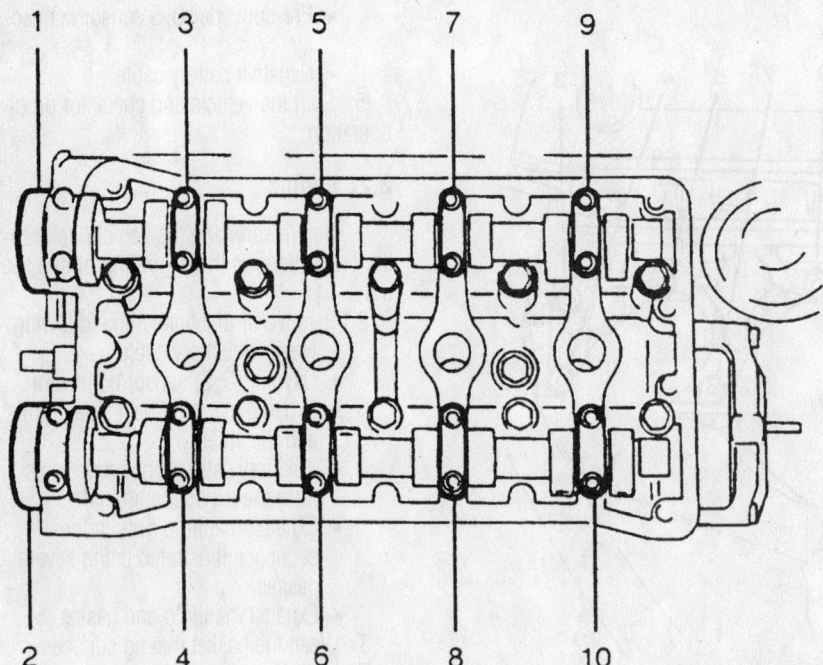

9307ZG85

Loosen the camshaft cap bolts in this sequence using several passes—1.6L engine

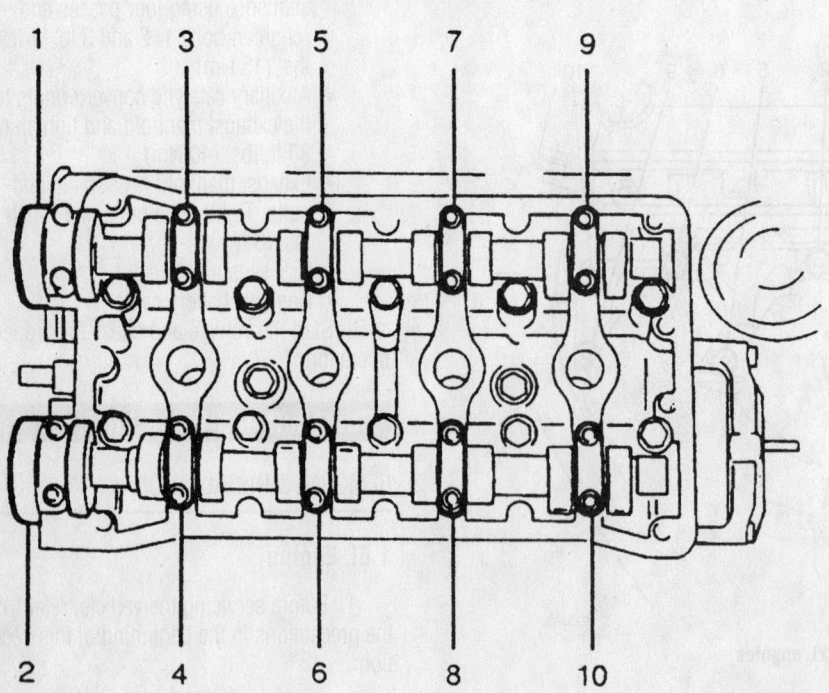

9307ZG84

Tighten the camshaft cap bolts in the sequence illustrated using several passes—1.6L engine

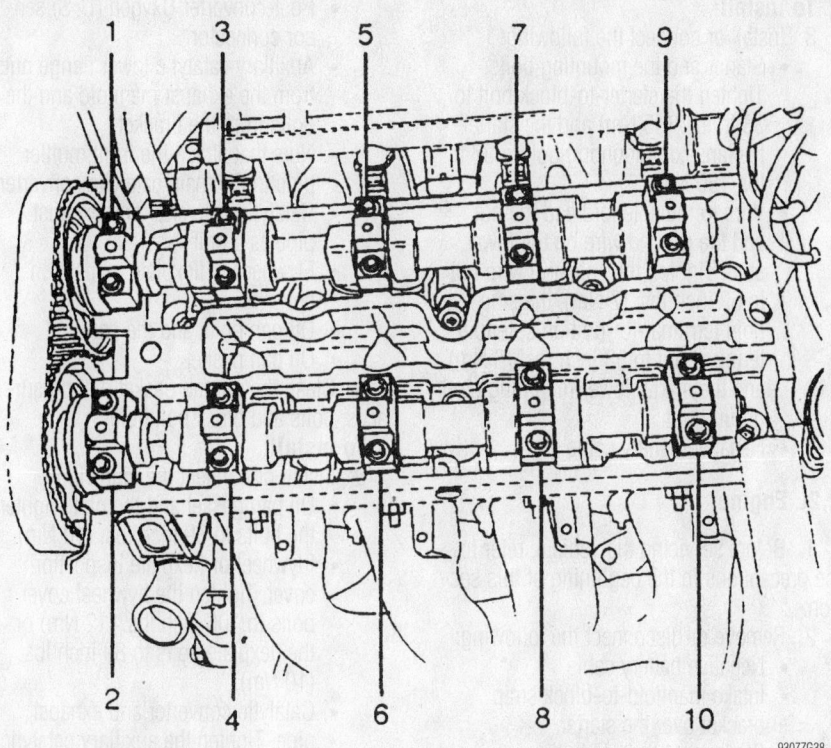

Remove the camshaft cap bolts in several passes of one half turn each in the sequence illustrated—2.0L and 2.2L engines

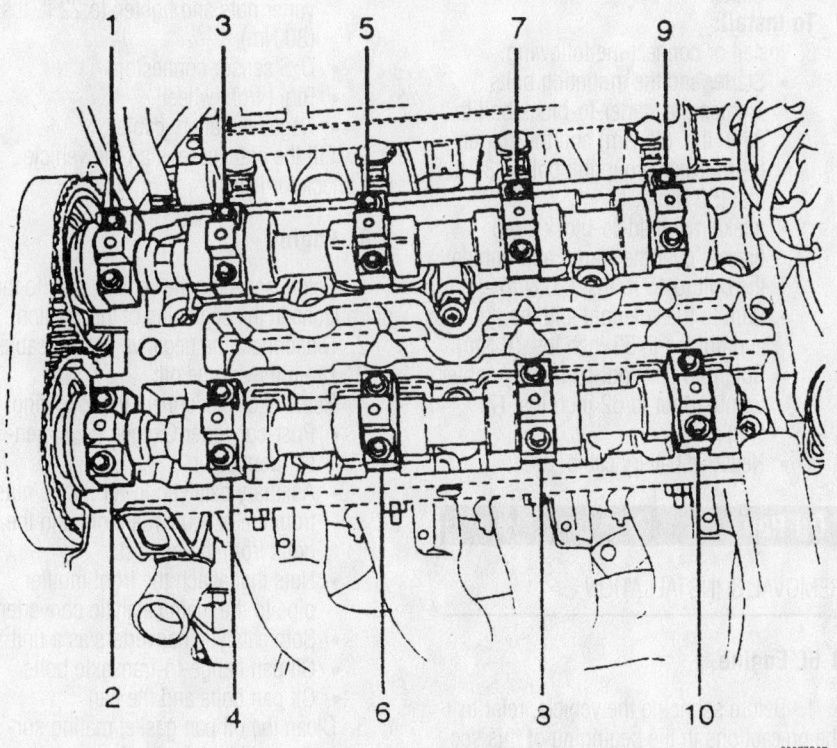

Tighten the camshaft cap bolts in the sequence illustrated using several passes—2.0L and 2.2L engines

- Crankcase ventilation tube from the valve cover
- Spark plug cover bolts and the cover
- Spark plug wires from the plugs
- Camshaft Position (CMP) sensor connector
- Valve cover nuts, washers, cover and gasket

➡ **Be very careful not to nick or scratch the camshafts.**

4. Hold the flats of the intake camshaft in position using an open ended wrench and remove the camshaft gear bolt. Remove the gear.

5. Hold the flats of the exhaust camshaft in position using an open ended wrench and remove the camshaft gear bolt. Remove the gear.

6. Mark the camshaft cap locations so that they may be returned to their original positions.

7. Remove the camshaft cap bolts in several passes of one half turn each in the sequence illustrated.

8. Remove the seal ring from the camshafts.

➡ **Make sure the camshafts detaches evenly from the bearing seats in the front guide bearing.**

9. Remove the caps and the camshafts.
   To install:
10. Lubricate the camshafts and caps with engine oil.

11. Install the caps and the camshafts in their original positions.

12. Tighten the camshaft cap bolts in the sequence illustrated to 71 inch lbs. (8 Nm).

13. Measure the camshaft endplay on both camshafts. The endplay should be 0.0015–0.0056 in. (0.040–0.144mm).

14. Install the intake camshaft gear. Hold the flats of the intake camshaft in position using an open ended wrench and tighten the camshaft gear bolt to 37 ft. lbs. (50 Nm), retighten the bolt an additional 60 degree turn and finally an additional 15 degree turn.

15. Install the exhaust camshaft gear. Hold the flats of the exhaust camshaft in position using an open ended wrench and tighten the camshaft gear bolt to 37 ft. lbs. (50 Nm), retighten the bolt an additional 60 degree turn and finally an additional 15 degree turn.

16. Install or connect the following:
- Valve cover, washers and nuts.

Tighten the nuts to 71 inch lbs. (8 Nm).
- CMP sensor connector
- Spark plug wires to the plugs
- Spark plug cover and bolts. Tighten the bolts to 27 inch lbs. (3 Nm).
- Crankcase ventilation tube to the valve cover
- Timing belt
- Negative battery cable

Valve Lash

ADJUSTMENT

No valve lash adjustment is possible on these engines because of the use of Hydraulic Lash Adjusters (HLA's). If valve noise occurs, look for excessive wear on the camshaft or hydraulic lash adjuster.

Starter Motor

REMOVAL & INSTALLATION

1.6L Engine

1. Before servicing the vehicle, refer to the precautions in the beginning of this section.
2. Remove or disconnect the following:
   - Negative battery cable
   - Upper and lower starter mounting bolts
   - Starter electrical connections
   - Starter

To install:
3. Install or connect the following:
   - Starter
   - Upper and lower starter mounting bolts and tighten to 32 ft. lbs. (43 Nm)
   - Starter electrical connections and tighten the nuts to 11 ft. lbs. (15 Nm)
   - Negative battery cable

2.0L Engine

1. Before servicing the vehicle, refer to the precautions in the beginning of this section.
2. Remove or disconnect the following:
   - Negative battery cable
   - Nut that attaches the starter ground wire to the lower mounting stud
   - Ground wire
   - Starter-to-block and starter-to-transaxle mounting bolts
   - Starter solenoid nuts and the cable
   - Starter

To install:
3. Install or connect the following:
   - Starter and the mounting bolts. Tighten the starter-to-block bolt to 33 ft. lbs. (45 Nm) and the starter-to-transaxle mounting bolt to 37 ft. lbs. (50 Nm).
   - Wire to the solenoid and the nut and the ground wire on the lower stud. Tighten the solenoid wire nut to 62 inch lbs. (7 Nm), the solenoid terminal-to-ignition solenoid terminal nut to 53 inch lbs. (6 Nm) and the starter lower mounting stud ground nut.
   - Negative battery cable

2.2L Engine

1. Before servicing the vehicle, refer to the precautions in the beginning of this section.
2. Remove or disconnect the following:
   - Negative battery cable
   - Intake manifold-to-block strap bracket over the starter
   - Starter-to-transaxle bolt
   - Starter-to-block bolt
   - Starter solenoid nuts and the cable
   - Starter

To install:
3. Install or connect the following:
   - Starter and the mounting bolts. Tighten the starter-to-block bolt to 33 ft. lbs. (45 Nm) and the starter-to-transaxle mounting bolt to 37 ft. lbs. (50 Nm).
   - Intake manifold-to-block strap bracket over the starter and tighten the bolt to 15 ft. lbs. (20 Nm)
   - Wire to the solenoid and the nut and tighten to 53 inch lbs. (6 Nm)
   - Solenoid terminal-to-battery cable terminal nut to 62 inch lbs. (7 Nm)
   - Negative battery cable

Oil Pan

REMOVAL & INSTALLATION

1.6L Engine

1. Before servicing the vehicle, refer to the precautions in the beginning of this section.
2. Remove or disconnect the following:
   - Negative battery cable
   - Right front wheel
3. Drain the engine oil.
4. Remove or disconnect the following:
   - Post-converter Oxygen (O_2S) sensor connector
   - Auxiliary catalytic lower flange nuts from the exhaust manifold and the bolts from the bracket
   - Nuts that attach the front muffler pipe to the main catalytic converter
   - Catalytic converter and exhaust pipe as a unit
   - Flywheel or flexplate inspection cover
   - Oil pan bolts and the pan
   - Oil pan gasket
5. Clean the oil pan gasket mating surfaces, bolts and block bolt holes.

To install:
6. Install or connect the following:
   - Oil pan gasket and the pan. Tighten the bolts to 89 inch lbs. (10 Nm).
   - Flywheel or flexplate inspection cover. Tighten the flywheel cover bolts to 106 inch lbs. (12 Nm) or the flexplate bolts to 89 inch lbs. (10 Nm).
   - Catalytic converter and exhaust pipe. Tighten the auxiliary catalytic converter-to-exhaust manifold nuts and the exhaust flexible pipe bracket bolts to 30 ft. lbs. (40 Nm).
   - Front muffler-to-main catalytic converter nuts and tighten to 22 ft. lbs. (30 Nm)
   - O_2S sensor connector
   - Right front wheel
   - Negative battery cable
7. Fill the crankcase, start the vehicle and check for leaks.

2.0L Engine

1. Before servicing the vehicle, refer to the precautions in the beginning of this section.
2. Disconnect the negative battery cable.
3. Drain the engine oil.
4. Remove or disconnect the following:
   - Post-converter Oxygen (O_2S) sensor connector
   - Auxiliary catalytic lower flange nuts from the exhaust manifold and the bolts from the bracket
   - Nuts that attach the front muffler pipe to the main catalytic converter
   - Both catalytic converters as a unit
   - Oil pan flange-to-transaxle bolts
   - Oil pan bolts and the pan
5. Clean the oil pan gasket mating surfaces, bolts and block bolt holes.

To install:
6. Coat the new oil pan gasket with sealant.

➡**Install the pan within 5 minutes of applying the sealant.**

7. Install or connect the following:
- Oil pan. Tighten the bolts to 89 inch lbs. (10 Nm).
- Oil pan flange-to-transaxle bolts and tighten to 30 ft. lbs. (40 Nm)
- Catalytic converters. Tighten the auxiliary catalytic converter-to-exhaust manifold nuts and the exhaust pipe support bracket bolts to 30 ft. lbs. (40 Nm).
- Front muffler-to-main catalytic converter nuts and tighten to 22 ft. lbs. (30 Nm)
- O2S sensor connector
- Negative battery cable

8. Fill the crankcase, start the vehicle and check for leaks.

2.2L Engine

1. Before servicing the vehicle, refer to the precautions in the beginning of this section.
2. Remove or disconnect the following:
- Negative battery cable
- Post-converter Oxygen (O2S) sensor connector
3. Drain the engine oil.
4. Remove or disconnect the following:
- Auxiliary catalytic lower flange nuts
- Nuts that attach the front muffler pipe to the main catalytic converter
- Rubber hangers that retain the connecting pipe
- Connecting pipe bracket
- Main catalytic converter and the connecting pipe
- Oil pan flange-to-transaxle bolts
- Oil pan bolts and the pan
5. Clean the oil pan gasket mating surfaces, bolts and block bolt holes.

To install:
6. Coat the new oil pan gasket with sealant.

➡**Install the pan within 5 minutes of applying the sealant.**

7. Install or connect the following:
- Oil pan. Tighten the bolts to 89 inch lbs. (10 Nm).
- Oil pan flange-to-transaxle bolts and tighten to 30 ft. lbs. (40 Nm)
- Main catalytic converter bolts in the front muffler flange
- Rubber hangers that retain the connecting pipe
- Front muffler-to-main catalytic converter nuts and tighten to 22 ft. lbs. (30 Nm)
- Auxiliary catalytic lower flange

bolts and tighten the nuts to 30 ft. lbs. (40 Nm)
- Connecting pipe bracket and tighten the nuts to 22 ft. lbs. (30 Nm)
- O2S sensor connector
- Negative battery cable

8. Fill the crankcase, start the vehicle and check for leaks.

Oil Pump

REMOVAL & INSTALLATION

1.6L Engine

1. Before servicing the vehicle, refer to the precautions in the beginning of this section.
2. Remove or disconnect the following:
- Negative battery cable
- Power steering pump, if equipped
- Timing belt. Refer to the Timing Belt unit repair section.
- Rear timing belt cover
- Oil pressure switch connector
- Crankshaft Position (CKP) sensor
- Oil pan
- Oil pump pick-up tube and support bracket bolts
- Pick-up tube
- Oil pump bolts, pump and gasket
3. Clean the pump gasket mating surfaces.
4. Remove the safety relief valve bolt, relief valve and the spring.
5. Remove the pump-to-crankshaft seal.
6. Remove the oil pump cover bolts and the cover.
7. Clean the pump housing and pump parts. Inspect for damage and replace as necessary.

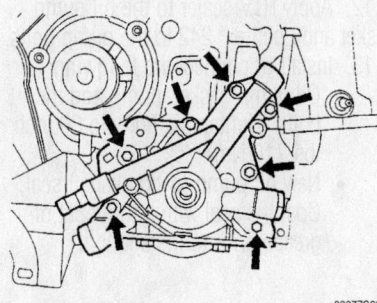

9307ZG80

Location of the oil pump bolts—1.6L engines

To install:
8. Coat the oil pump parts with oil and reinstall.
9. Pack the oil pump cavity with petroleum jelly to make sure the oil pump primes.
10. Apply Loctite ® 242 to the rear cover bolts and install the cover. Tighten the bolts to 53 inch lbs. (6 Nm).
11. Install the safety relief valve, spring, washer and bolt. Tighten the bolt to 22 ft. lbs. (30 Nm).
12. Apply RTV sealer to the oil pump gasket and Loctite ® 242 to the pump bolts.
13. Install or connect the following:
- Oil pump gasket, pump and the bolts. Tighten the bolts to 89 inch lbs. (10 Nm).
- New oil pump-to-crankshaft seal. Coat the seal with a thin coat of grease prior to installation.
14. Coat the threads of the pick-up tube and support bracket bolts with Loctite ® 242.
15. Install or connect the following:
- Pick-up tube and support bracket. Tighten the bolts to 89 inch lbs. (10 Nm).
- Oil pan
- CKP sensor and tighten the bolts to 89 inch lbs. (10 Nm)
- Oil pressure switch connector
- Timing belt
- Rear timing belt cover
- Power steering pump, if equipped
- Negative battery cable

2.0L Engine

1. Before servicing the vehicle, refer to the precautions in the beginning of this section.
2. Remove or disconnect the following:
- Negative battery cable
- Rear timing belt cover
- Oil pressure switch connector
- Oil pan
- Oil pump pick-up tube and support bracket bolts
- Pick-up tube
- Oil pump bolts, pump and gasket
3. Clean the pump gasket mating surfaces.
4. Remove the safety relief valve bolt, relief valve and the spring.
5. Remove the pump-to-crankshaft seal.
6. Remove the oil pump cover bolts and the cover.
7. Clean the pump housing and pump parts. Inspect for damage and replace as necessary.

9307ZG36

Location of the oil pump bolts—2.0L engines

9307ZG30

Location of the oil pump bolts—2.2L engines

To install:

8. Coat the oil pump parts with oil and reinstall.

9. Pack the oil pump cavity with petroleum jelly to make sure the oil pump primes.

10. Apply Loctite ® 242 to the rear cover bolts and install the cover. Tighten the bolts to 53 inch lbs. (6 Nm).

11. Install the safety relief valve, spring, washer and bolt. Tighten the bolt to 22 ft. lbs. (30 Nm).

12. Apply RTV sealer to the oil pump gasket and Loctite ® 242 to the pump bolts.

13. Install or connect the following:
- Oil pump gasket, pump and the bolts. Tighten the bolts to 89 inch lbs. (10 Nm).
- New oil pump-to-crankshaft seal. Coat the seal with a thin coat of grease prior to installation.

14. Coat the threads of the pick-up tube and support bracket bolts with Loctite ® 242.

15. Install or connect the following:
- Pick-up tube and support bracket. Tighten the bolts to 89 inch lbs. (10 Nm).
- Oil pan
- Oil pressure switch connector
- Rear timing belt cover
- Negative battery cable

2.2L Engine

1. Before servicing the vehicle, refer to the precautions in the beginning of this section.

2. Remove or disconnect the following:
- Negative battery cable

- Timing belt. Refer to the Timing Belt unit repair section.
- Rear timing belt cover
- Oil pressure switch connector
- Oil pan
- Oil pump pick-up tube and support bracket bolts
- Pick-up tube
- Oil pump bolts, pump and gasket

3. Clean the pump gasket mating surfaces.

4. Remove the safety relief valve bolt, relief valve and the spring.

5. Remove the pump-to-crankshaft seal.

6. Remove the oil pump cover bolts and the cover.

7. Clean the pump housing and pump parts. Inspect for damage and replace as necessary.

To install:

8. Coat the oil pump parts with oil and reinstall.

9. Pack the oil pump cavity with petroleum jelly to make sure the oil pump primes.

10. Apply Loctite ® 242 to the rear cover bolts and install the cover. Tighten the bolts to 53 inch lbs. (6 Nm).

11. Install the safety relief valve, spring, washer and bolt. Tighten the bolt to 22 ft. lbs. (30 Nm).

12. Apply RTV sealer to the oil pump gasket and Loctite ® 242 to the pump bolts.

13. Install or connect the following:
- Oil pump gasket, pump and the bolts. Tighten the bolts to 89 inch lbs. (10 Nm).
- New oil pump-to-crankshaft seal. Coat the seal with a thin coat of grease prior to installation.

14. Coat the threads of the pick-up tube and support bracket bolts with Loctite ® 242.

15. Install or connect the following:
- Pick-up tube and support bracket. Tighten the pick-up tube bolts to 71 inch lbs. (8 Nm) an the tube support bolts to 89 inch lbs. (10 Nm).
- Oil pan
- Oil pressure switch connector
- Rear timing belt cover
- Timing belt
- Negative battery cable

Rear Main Seal

REMOVAL & INSTALLATION

1. Before servicing the vehicle, refer to the precautions in the beginning of this section.

2. Remove or disconnect the following:
- Transaxle
- Flywheel or flexplate

3. Carefully pry the seal out of the retainer without damaging the crankshaft or the seal retainer.

To install:

4. Lubricate the seal with engine oil.

5. Install or connect the following:
- Seal into the retainer using the appropriate seal driver
- Flywheel or flexplate
- Transaxle

Piston and Ring

POSITIONING

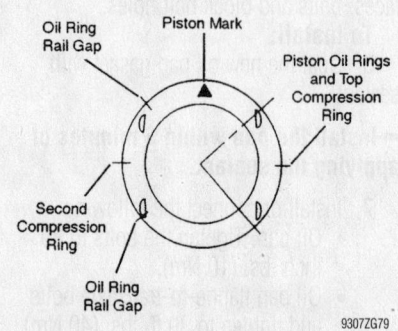

9307ZG79

Daewoo engines—piston ring end-gap spacing

FUEL SYSTEM

Fuel System Service Precautions

Safety is the most important factor when performing not only fuel system maintenance but any type of maintenance. Failure to conduct maintenance and repairs in a safe manner may result in serious personal injury or death. Maintenance and testing of the vehicle's fuel system components can be accomplished safely and effectively by adhering to the following rules and guidelines.

• To avoid the possibility of fire and personal injury, always disconnect the negative battery cable unless the repair or test procedure requires that battery voltage be applied.

• Always relieve the fuel system pressure prior to disconnecting any fuel system component (injector, fuel rail, pressure regulator, etc.), fitting or fuel line connection. Exercise extreme caution whenever relieving fuel system pressure, to avoid exposing skin, face and eyes to fuel spray. Please be advised that fuel under pressure may penetrate the skin or any part of the body that it contacts.

• Always place a shop towel or cloth around the fitting or connection prior to loosening to absorb any excess fuel due to spillage. Ensure that all fuel spillage (should it occur) is quickly removed from engine surfaces. Ensure that all fuel soaked cloths or towels are deposited into a suitable waste container.

• Always keep a dry chemical (Class B) fire extinguisher near the work area.

• Do not allow fuel spray or fuel vapors to come into contact with a spark or open flame.

• Always use a back-up wrench when loosening and tightening fuel line connection fittings. This will prevent unnecessary stress and torsion to fuel line piping. Always follow the proper torque specifications.

• Always replace worn fuel fitting O-rings with new. Do not substitute fuel hose or equivalent, where fuel pipe is installed.

Fuel System Pressure

RELIEVING

1. Before servicing the vehicle, refer to the precautions in the beginning of this section.

2. Remove or disconnect the following:
   • Fuel cap
   • Fuel pump fuse from the engine fuse block
3. Crank the engine and allow it to stall.
4. Crank the engine for an additional 10 seconds after it has stalled to completely deplete all fuel pressure.
5. After repairs have been made, install the fuse and the cap. Start the vehicle and check for leaks at the repaired component.

Fuel Filter

REMOVAL & INSTALLATION

1.6L Engine

1. Before servicing the vehicle, refer to the precautions in the beginning of this section.
2. Relieve the fuel system pressure.
3. Remove or disconnect the following:

   • Negative battery cable
   • Fuel inlet line by pushing the release sleeve towards the fuel line with Trim remover KM-475-B and pulling the hose off the filter tube
   • Fuel outlet line by moving the line connector lock forward and pulling the hose off the filler tube
   • Filter from the clamp by pulling it out

To install:
4. Install or connect the following:
   • Filter into the bracket making sure to note the flow direction
   • Inlet and outlet lines to the filter
   • Negative battery cable
5. Start the vehicle and check for leaks.

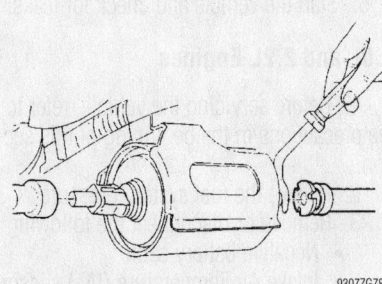

9307ZG78

Removing the fuel inlet line from the filter—1.6L engines

2.0L Engine

1. Before servicing the vehicle, refer to the precautions in the beginning of this section.
2. Relieve the fuel system pressure.
3. Remove or disconnect the following:
   • Negative battery cable
   • Fuel inlet/outlet lines by moving the line connector lock forward and pulling the hose off the filler tube
   • Filter mounting bracket both and the filter

To install:
4. Install or connect the following:
   • Filter into the bracket making sure to note the flow direction and tighten the bolt to 89 inch lbs. (10 Nm)
   • Inlet and outlet lines to the filter
   • Negative battery cable
5. Start the vehicle and check for leaks.

2.2L Engine

1. Before servicing the vehicle, refer to the precautions in the beginning of this section.
2. Relieve the fuel system pressure.
3. Remove or disconnect the following:
   • Negative battery cable
   • Fuel inlet line by moving the line connector lock forward and pulling the hose off the filler tube
   • Fuel outlet line by pushing the release sleeve towards the fuel line with Trim remover KM-475-B and pulling the hose off the filter tube
   • Filter from the clamp by pulling it out

To install:
4. Install or connect the following:

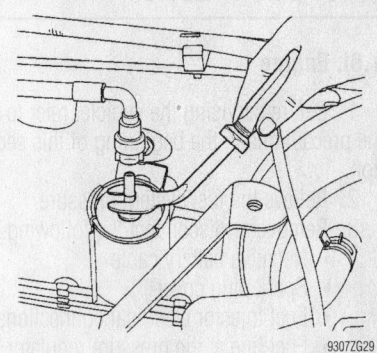

9307ZG29

Removing the fuel outlet line from the filter—2.2L engines

- Filter into the bracket making sure to note the flow direction
- Inlet and outlet lines to the filter
- Negative battery cable
5. Start the vehicle and check for leaks.

Fuel Pump

REMOVAL & INSTALLATION

1. Before servicing the vehicle, refer to the precautions in the beginning of this section.
2. Relieve the fuel system pressure.
3. Remove or disconnect the following:
   - Negative battery cable
   - Rear seat
   - Fuel pump access cover
   - Pump assembly electrical connections
   - Fuel outlet and return lines
   - Lock ring by turning it counterclockwise
   - Fuel pump assembly from the tank
   - Gasket and discard
4. Clean the gasket mating surface
To install:
5. Install or connect the following:
   - New gasket
   - Fuel pump assembly into the tank
   - Lock ring and turn it clockwise until it contacts the tank stop
   - Pump assembly electrical connections
   - Fuel outlet and return lines
   - Fuel pump access cover
   - Rear seat
   - Negative battery cable
6. Start the vehicle and check for proper operation.

Fuel Injector

REMOVAL & INSTALLATION

1.6L Engine

1. Before servicing the vehicle, refer to the precautions in the beginning of this section.
2. Relieve the fuel system pressure.
3. Remove or disconnect the following:
   - Negative battery cable
   - Spark plug cover
   - Fuel injector electrical connections
   - Fuel line at the pressure regulator using a back-up wrench to avoid damaging the line
   - Fuel inlet line
   - Fuel rail retaining bolts

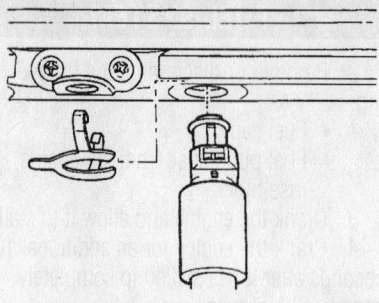

9307ZG77

Install the injectors to the fuel rail making sure the terminals face outward and secure them with the clips

- Fuel rail and the injectors as an assembly
- Injector clips
- Injectors by pulling them down and out
- O-rings and discard
To install:

➡**When ordering new injectors, make sure to order the same part numbers because other injectors may have a different calibrated flow rate.**

4. Lubricate the new O-rings with clean engine oil and install the O-rings on the injector(s).
5. Install or connect the following:
   - Fuel injectors to the fuel rail making sure the terminals face outward, then secure the injectors with the clips. Make sure the clip is parallel to the injector harness connector.
   - Fuel rail and injectors as an assembly. Tighten the mounting bolts to 18 ft. lbs. (25 Nm).
   - Fuel inlet line
   - Fuel line to the pressure regulator using a back-up wrench to avoid damaging the line
   - Fuel injector electrical connections
   - Spark plug cover
   - Negative battery cable
6. Start the vehicle and check for leaks.

2.0L and 2.2L Engines

1. Before servicing the vehicle, refer to the precautions in the beginning of this section.
2. Relieve the fuel system pressure.
3. Remove or disconnect the following:
   - Negative battery cable
   - Intake Air Temperature (IAT) sensor connector
   - Breather hose from the valve cover
   - Air intake tube and if necessary, the resonator

- Positive Crankcase Ventilation (PCV) valve from the valve cover
- Throttle cable from the throttle body and bracket
4. Remove the fuel pressure regulator by removing or disconnecting the following:
   - Vacuum hose from the regulator
   - Regulator retaining clamp
   - Regulator by turning it back and forth, then pull it out and discard the O-ring
5. Remove or disconnect the following:
   - Inlet and return lines from the fuel rail
   - Fuel rail retaining bolts
   - Throttle cable bracket, if necessary
   - Fuel rail and the injectors as an assembly
   - Fuel injector channel cover connectors
   - Injector clips
   - Injectors by pulling them down and out
   - O-rings and discard
To install:

➡**When ordering new injectors, make sure to order the same part numbers because other injectors may have a different calibrated flow rate.**

6. Lubricate the new O-rings with clean engine oil and install the O-rings on the injector(s).
7. Install or connect the following:
   - Fuel injectors to the fuel rail making sure the terminals face outward, then secure the injectors with new clips. Make sure the clip is parallel to the injector harness connector.
   - Fuel rail and injectors as an assembly. Tighten the mounting bolts to 18 ft. lbs. (25 Nm).
   - Fuel inlet and outlet lines line
8. Install the fuel pressure regulator by installing or connecting the following:
   - New O-ring lubricated with clean engine oil on the pressure regulator body
   - Regulator and tighten the retaining screw to 106 inch lbs. (12 Nm)
   - Vacuum hose to the regulator
9. Install or connect the following:
   - Throttle cable bracket, if necessary
   - PCV valve to the valve cover
   - Throttle cable to the throttle body and bracket
   - Air intake tube and resonator
   - Breather hose to the valve cover
   - IAT sensor connector
   - Negative battery cable
10. Start the vehicle and check for leaks.

DRIVE TRAIN

Transaxle Assembly

REMOVAL & INSTALLATION

Manual

LANOS AND NUBIRA

1. Before servicing the vehicle, refer to the precautions in the beginning of this section.
2. Remove or disconnect the following:
   - Negative battery cable
   - Back-up lamp switch electrical connection
   - Speedometer speed sensor electrical connection
   - Clip and bolt from the universal joint
   - Clutch release bracket bolts and the bracket
3. Install engine support fixture J 28467 B.
4. Remove or disconnect the following:
   - Upper transaxle-to-engine bolts
   - Both driveshafts
   - Flywheel inspection cover, if equipped
5. Support the transaxle with a suitable jack.
6. Remove or disconnect the following:
   - Left front transaxle support bracket bolts and the bracket
   - Left rear transaxle support bracket bolts and the bracket
   - Center rear transaxle support bracket bolts, on Nubira models
   - Lower transaxle-to-engine bolts
   - Transaxle

To install:

7. Install or connect the following:
   - Transaxle
   - Lower transaxle-to-engine bolts and tighten to 55 ft. lbs. (75 (Nm)
   - Left rear transaxle support bracket and tighten the bolts to 66 ft. lbs. (90 Nm)
   - Left front transaxle support bracket and tighten the bolts to 44 ft. lbs. (60 Nm)
   - Center rear transaxle support bracket and tighten the bolts to 44 ft. lbs. (60 Nm), on Nubira models
   - Both driveshafts
   - Upper transaxle-to-engine bolts and tighten to 55 ft. lbs. (75 (Nm)
8. Remove the engine support fixture.

9. Install or connect the following:
   - Clutch release bracket and tighten the bolts to 55 ft. lbs. (75 Nm)
   - Bolt and clip to the universal joint
   - Speedometer speed sensor electrical connection
   - Back-up lamp switch electrical connection
   - Negative battery cable
10. Check the transaxle fluid level and add as necessary.

LEGANZA

1. Before servicing the vehicle, refer to the precautions in the beginning of this section.
2. Remove or disconnect the following:
   - Negative battery cable
   - Back-up lamp switch electrical connection
   - Speedometer speed sensor electrical connection
   - Clip and bolt from the universal joint
   - Clutch release bracket bolts and the bracket
   - Transaxle upper brace bolts and the brace
3. Install engine support fixture J 28467 B.
4. Remove or disconnect the following:
   - Upper transaxle-to-engine bolts
   - Both driveshafts
   - Flywheel inspection cover, if equipped
5. Support the transaxle with a suitable jack.
6. Remove or disconnect the following:
   - Left front transaxle support bracket bolts and the bracket
   - Left rear transaxle support bracket bolts and the bracket
   - Center rear transaxle support bracket bolts
   - Lower transaxle-to-engine bolts
   - Transaxle

To install:

7. Install or connect the following:
   - Transaxle
   - Lower transaxle-to-engine bolts and tighten to 55 ft. lbs. (75 Nm)
   - Left rear transaxle support bracket and tighten the bolts to 44 ft. lbs. (60 Nm)
   - Left front transaxle support bracket and tighten the bolts to 44 ft. lbs. (60 Nm)

   - Center rear transaxle support bracket and tighten the bolts to 66 ft. lbs. (90 Nm)
   - Both driveshafts
   - Upper transaxle-to-engine bolts and tighten to 55 ft. lbs. (75 (Nm)
8. Remove the engine support fixture.
9. Install or connect the following:
   - Upper transaxle bracket and tighten the bolts to 55 ft. lbs. (75 Nm)
   - Clutch release bracket and tighten the bolts to 55 ft. lbs. (75 Nm)
   - Bolt and clip to the universal joint
   - Speedometer speed sensor electrical connection
   - Back-up lamp switch electrical connection
   - Negative battery cable
10. Check the transaxle fluid level and add as necessary.

Automatic

LANOS

1. Before servicing the vehicle, refer to the precautions in the beginning of this section.
2. Remove or disconnect the following:
   - Negative battery cable
   - Transaxle cooler lines from the transaxle
   - Transaxle wiring harness
   - Park/neutral switch electrical connection
   - Shift cable from the lever and mounting bracket
   - Vehicle Speed Sensor (VSS) electrical connection
   - Font exhaust pipe
   - Both driveshafts
   - Flywheel inspection cover
   - Flywheel bolts
3. Install engine support fixture J 28467 B.
4. Support the transaxle using a suitable jack.
5. Remove or disconnect the following:
   - Bell housing bolts
   - Right and left transaxle mount bolts
   - Rear transaxle mount bolts
   - Transaxle

To install:

6. Install or connect the following:
   - Transaxle
   - Rear transaxle mount bolts and tighten to 55 ft. lbs. (75 Nm)
   - Right and left transaxle mount bolts and tighten to 55 ft. lbs. (75 Nm)

- Bell housing bolts and tighten to 55 ft. lbs. (75 Nm)
- Flywheel bolts and tighten to 48 ft. lbs. (65 Nm)
- Flywheel inspection cover

7. Remove the transaxle jack and the engine support fixture.

8. Install or connect the following:
- Both driveshafts
- Font exhaust pipe
- VSS electrical connection
- Shift cable to mounting bracket and lever
- Park/neutral switch electrical connection
- Transaxle wiring harness
- Transaxle cooler lines to the transaxle
- Negative battery cable

9. Check the transaxle fluid level and add as necessary.

NUBIRA

1. Before servicing the vehicle, refer to the precautions in the beginning of this section.

2. Remove or disconnect the following:
- Negative battery cable
- Transaxle cooler lines from the transaxle
- Transaxle wiring harness
- Park/neutral switch electrical connection
- Shift cable from the lever and mounting bracket
- Vehicle Speed Sensor (VSS) electrical connection
- Font exhaust pipe
- Both driveshafts
- Flywheel inspection cover
- Flywheel bolts

3. Install engine support fixture J 28467 B.

4. Support the transaxle using a suitable jack.

5. Remove or disconnect the following:
- Bell housing bolts
- Transaxle brace mount bolts
- Transaxle mount bolts
- Transaxle mounting bracket bolts
- Transaxle

To install:

6. Install or connect the following:
- Transaxle
- Transaxle mounting bracket bolts and tighten to 45 ft. lbs. (61 Nm)
- Transaxle mount bolts and tighten to 60 ft. lbs. (81 Nm)
- Transaxle brace mount bolts and tighten to 55 ft. lbs. (75 Nm)
- Bell housing bolts and tighten to 55 ft. lbs. (75 Nm)

- Flywheel bolts and tighten to 48 ft. lbs. (65 Nm)
- Flywheel inspection cover

7. Remove the transaxle jack and the engine support fixture.

8. Install or connect the following:
- Both driveshafts
- Font exhaust pipe
- VSS electrical connection
- Shift cable to mounting bracket and lever
- Park/neutral switch electrical connection
- Transaxle wiring harness
- Transaxle cooler lines to the transaxle
- Negative battery cable

9. Check the transaxle fluid level and add as necessary.

LEGANZA

1. Before servicing the vehicle, refer to the precautions in the beginning of this section.

2. Install engine support fixture J 28467 B.

3. Remove or disconnect the following:
- Battery and tray
- Transaxle left mount
- Clip that attaches the shift control cable to the park/neutral switch
- Shift control cable
- Clip that attaches the shift control cable to the bracket and move the cable aside
- Park/neutral switch and transaxle sensor electrical connections
- Vehicle Speed Sensor (VSS) electrical connection
- Vent tube from the top of the transaxle
- Three upper transaxle-to-engine bolts
- Engine undercover

4. Drain the transaxle fluid.

5. Remove or disconnect the following:
- Transaxle cooler lines from the valve body, cap the lines and position them aside
- Both driveshafts
- Center member
- Power steering hose and position it aside
- Starter motor
- Two engine-to-transaxle bolts located next to the starter opening

6. Support the transaxle using a suitable jack.

7. Remove or disconnect the following:
- Front engine mount bracket
- Torque converter cover
- Torque converter-to-flex plate bolts

- Oil pan-to-transaxle bolts
- Transaxle

To install:

8. Install or connect the following:
- Transaxle
- Oil pan-to-transaxle bolts and finger tighten
- Two engine-to-transaxle bolts located next to the starter opening and finger tighten
- Torque converter-to-flex plate bolts and tighten to 44 ft. lbs. (60 Nm)
- Front engine mount bracket and tighten the bolts to 66 ft. lbs. (90 Nm)
- Tighten the oil pan-to-transaxle bolts to 30 ft. lbs. (40 Nm)
- Two engine-to-transaxle bolts located next to the starter opening to 55 ft. lbs. (75 Nm)
- Starter motor

9. Remove the transaxle jack.

10. Install or connect the following:
- Center member
- Power steering hose
- Both driveshafts
- Transaxle cooler lines to the valve body and tighten to 15–22 ft. lbs. (20–29 Nm)
- Engine undercover
- Three upper transaxle-to-engine bolts and tighten to 55 ft. lbs. (75 Nm)
- Vent tube to the top of the transaxle
- VSS electrical connection
- Park/neutral switch and transaxle sensor electrical connections
- Shift control cable to the bracket and attach the clip
- Shift control cable to the park/neutral switch and attach the clip
- Transaxle left mount
- Battery tray and battery

11. Remove the engine support fixture.

12. Fill the transaxle with the correct type and amount of fluid.

Clutch

ADJUSTMENTS

Clutch Pedal

1. Before servicing the vehicle, refer to the precautions in the beginning of this section.

2. To determine the clutch pedal play, depress the clutch pedal lightly with your hand and measure the distance when you feel resistance.

3. To adjust the pedal play, loosen the

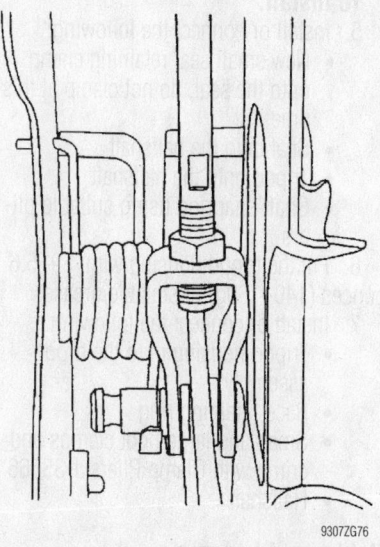

Loosen the locknut and turn the pushrod to adjust the pedal play

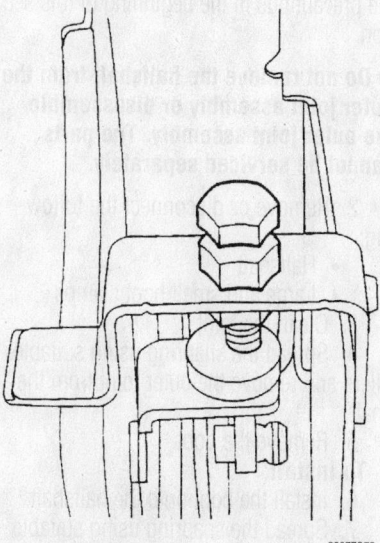

Loosen the locknut and turn the bolt to adjust the pedal travel

locknut and turn the pushrod. The pedal play should measure 0.2–0.5 in. (6–12mm). After the desired measurement has been reached, tighten the locknut.

4. To determine the clutch pedal travel, push the pedal to the floor; then measure from the starting point to the ending point.

5. To adjust the pedal travel, loosen the locknut and turn the bolt. The pedal travel should measure more than 5.1 in. (130mm) on Lanos and Nubira models and 5.5 in. (140mm) on Leganza models. After the desired measurement has been reached, tighten the locknut.

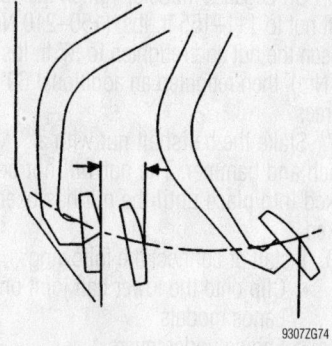

Measuring points to determine the clutch release point

6. Adjust the clutch release point as follows:
   a. Apply the parking brake.
   b. Start the vehicle and let it idle.
   c. While moving the shift lever into reverse, depress the clutch pedal slowly and measure the distance between the point when gear noise is not heard and the point where the clutch pedal is completely depressed. The distance should exceed 1.2 in. (30mm).
   d. If the distance is not as specified, check the clutch pedal height, clutch pedal play, air in the system and the clutch cover and disc.

REMOVAL & INSTALLATION

1. Before servicing the vehicle, refer to the precautions in the beginning of this section.
2. Disconnect the negative battery cable.
3. Remove or disconnect the following:
   • Left front wheel
   • Engine undercovers
   • Transaxle
4. Loosen the pressure plate bolts in small increments using a star-type pattern. Remove the pressure plate and clutch disc as an assembly

To install:
5. Coat the spline on the clutch plate with multi-purpose grease.

➡**Keep clutch disc and all clutch components clean during installation. Do not allow grease to contact the clutch disc.**

6. Insert tool J-42474 alignment tool into the clutch disc hub. Install the clutch disc and pressure plate on the tool and tighten the pressure plate bolts to 11 ft. lbs. (15 Nm) in 2–3 steps using a crisscross pattern. Remove the tool.

7. Install or connect the following:
   • Engine undercovers
   • Left front wheel
8. Lower vehicle, connect negative battery cable and road test vehicle.

Hydraulic Clutch System

BLEEDING

1. Before servicing the vehicle, refer to the precautions in the beginning of this section.

➡**Make sure the clutch/brake master cylinder is maintained at least at the MIN level during the bleeding procedure.**

2. Attach a transparent vinyl tube to the bleed screw, immersing the free end in a clean container of clean brake fluid.
3. Fill the clutch master cylinder with the proper fluid.
4. Slowly depress the clutch pedal all the way several times and hold it down.
5. Have an assistant open the bleeder valve about ¾ turn to release the air. Then, close the bleeder valve while the pedal is still depressed.
6. Repeat the above procedure until no more air bubbles are seen in the fluid container.
7. Remove the bleed tube.
8. Fill the master cylinder.

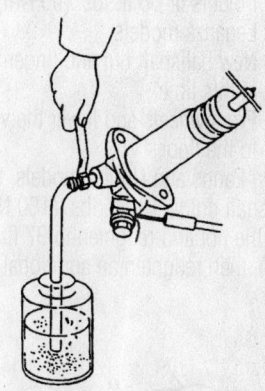

Bleeding the clutch hydraulic system

Halfshaft

REMOVAL & INSTALLATION

1. Before servicing the vehicle, refer to the precautions in the beginning of this section.

2. Remove or disconnect the following:
- Negative battery cable.
- Front wheels
- Engine undercovers
- Halfshaft nut and discard
- Lower ball joint clip and nut, Lanos models
- Lower ball joint nut and bolt, Nubira and Leganza models
- Steering knuckle from the lower ball joint using a suitable separator tool
- Tie rod nut
- Tie rod end using a suitable separator tool
- Halfshaft from the wheel hub. Tie the unfastened end of the halfshaft with wire to support it. Do not allow the halfshaft to hang freely as this may cause damage.
- Halfshaft from the transaxle using removal tool KM-406-A
- Halfshaft

To install:

3. Clean the hub and transaxle seals.
4. Install or connect the following:
- Halfshaft into the transaxle
- Wheel hub onto the halfshaft
- Steering knuckle onto the lower ball joint
- Tie rod into the knuckle/strut
- Tie rod nut and tighten to 44 ft. lbs. (60 Nm)
- Lower ball joint nut and tighten to 52 ft. lbs. (70 Nm) on Lanos models, 44 ft. lbs. (60 Nm) on Nubira models or 66 ft. lbs. (90 Nm) on Leganza models
- New halfshaft nut and finger tighten at this time
- Front wheels and lower the vehicle to the floor.

5. On Lanos and Nubira models, tighten the halfshaft nut to 134 ft. lbs. (180 Nm). Loosen the nut and retighten to 37 ft. lbs. (50 Nm), then retighten an additional 60 degrees.

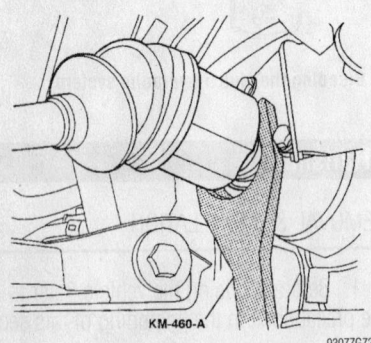

KM-460-A

9307ZG72

Separating the halfshaft from the transaxle

6. On Leganza models, tighten the halfshaft nut to 111–155 ft. lbs. (150–210 Nm). Loosen the nut and retighten to 37 ft. lbs. (50 Nm), then retighten an additional 60 degrees.

7. Stake the halfshaft nut with a punch and hammer. The nut will not be locked into place until the nut has been staked.

8. Install or connect the following:
- Clip onto the lower ball joint on Lanos models
- Engine undercovers

9. Refill the transaxle.

CV-Joints

OVERHAUL

Lanos

OUTER JOINT SEAL

1. Before servicing the vehicle, refer to the precautions in the beginning of this section.

➡**Do not remove the halfshaft from the outer joint assembly or disassemble the outer joint assembly. The parts cannot be serviced separately.**

2. Remove or disconnect the following:
- Halfshaft
- Inner tripod seal
- Large and small boot clamps

3. Clean the joint and remove the boot.

To install:

4. Install the boot onto the halfshaft and fill it with 2.6–3.4 ounces (75–95g) of grease.

5. Repack the joint with 2.6–3.4 ounces (75–95g) of grease.

6. Install or connect the following:
- Small and large boot clamps and crimp with Clamp Pliers J-35566
- Inner tripod seal
- Halfshaft

INNER TRIPOD SEAL

1. Before servicing the vehicle, refer to the precautions in the beginning of this section.

2. Remove or disconnect the following:
- Halfshaft
- Large and small boot clamps
- Race retaining ring
- Tripod housing from the seal

3. Clean the tripod assembly.

4. Remove or disconnect the following:
- Shaft snapring using suitable pliers
- Tripod from the halfshaft
- Tripod joint seal from the halfshaft

To install:

5. Install or connect the following:
- New small seal retaining clamp onto the seal, do not clamp at this time
- Seal onto the halfshaft
- Tripod onto the halfshaft
- Shaft snapring using suitable pliers

6. Fill the tripod housing with 4.9–5.6 ounces (140–160g) of suitable grease.

7. Install or connect the following:
- Tripod housing onto the tripod assembly
- Race retaining ring
- Small and large boot clamps and crimp with Clamp Pliers J-35566
- Halfshaft

Nubira and Leganza

OUTER JOINT SEAL

1. Before servicing the vehicle, refer to the precautions in the beginning of this section.

➡**Do not remove the halfshaft from the outer joint assembly or disassemble the outer joint assembly. The parts cannot be serviced separately.**

2. Remove or disconnect the following:
- Halfshaft
- Large and small boot clamps

3. Clean the joint.

4. Spread the snapring using suitable pliers and remove the outer joint from the shaft.

5. Remove the boot.

To install:

6. Install the boot onto the halfshaft.

7. Spread the snapring using suitable pliers and install the outer joint onto the shaft.

8. Fill the joint with 6.2–6.9 ounces (175–195g) of grease.

9. Repack the joint with 6.2–6.9 ounces (175–195g) of grease.

10. Install or connect the following:
- Small and large boot clamps and crimp with Clamp Pliers J-35566
- Inner tripod seal
- Halfshaft

INNER TRIPOD SEAL

1. Before servicing the vehicle, refer to the precautions in the beginning of this section.

2. Remove or disconnect the following:
- Halfshaft
- Large and small boot clamps

- Race retaining ring
- Tripod housing from the seal
3. Clean the tripod assembly.
4. Remove or disconnect the following:
  - Shaft snapring using suitable pliers
  - Tripod from the halfshaft
  - Tripod joint seal from the halfshaft

To install:
5. Install or connect the following:

- New small seal retaining clamp onto the seal, do not clamp at this time
- Seal onto the halfshaft
- Tripod onto the halfshaft
- Shaft snapring using suitable pliers
6. Fill the tripod housing with 6.9–7.6 ounces (195–215g) of suitable grease.

7. Repack the tripod with 6.9–7.6 ounces (195–215g) of grease.
8. Install or connect the following:
- Tripod housing onto the tripod assembly
- Race retaining ring
- Small and large boot clamps and crimp with Clamp Pliers J-35566
- Halfshaft

STEERING AND SUSPENSION

Air Bag

✳✳ CAUTION

Some vehicles are equipped with an air bag system. The system must be disarmed before performing service on, or around, system components, the steering column, instrument panel components, wiring and sensors. Failure to follow the safety precautions and the disarming procedure could result in accidental air bag deployment, possible personal injury and unnecessary system repairs.

PRECAUTIONS

Several precautions must be observed when handling the inflator module to avoid accidental deployment and possible personal injury.

• Never carry the inflator module by the wires or connector on the underside of the module.

• When carrying a live inflator module, hold securely with both hands, and ensure that the bag and trim cover are pointed away.

• Place the inflator module on a bench or other surface with the bag and trim cover facing up.

• With the inflator module on the bench, never place anything on or close to the module that may be thrown in the event of an accidental deployment.

DISARMING

The Supplemental Inflatable Restraint (SIR) system is disarmed by disconnecting the negative battery cable. Wait a few moments to allow system energy to deplete before working on the system.

Rack and Pinion Steering Gear

REMOVAL & INSTALLATION

Lanos

1. Before servicing the vehicle, refer to the precautions in the beginning of this section.
2. Remove or disconnect the following:
- Negative battery cable
- Inner tie rods from the rack by removing the lock plate, the inner tie rod bolts and the inner tie rod plate
3. Position the steering gear in the straight ahead position by turning the steering wheel until the spokes are centered diagonally and facing downward.
4. Remove or disconnect the following:
- Top pinch bolt on the coupling flange
- Coolant surge tank without disconnecting the hoses and move aside
- Inlet and outlet pipe fittings
- Left inner and outer tie rod from the vehicle
- Steering gear retaining bracket nuts from the bottom of each bracket

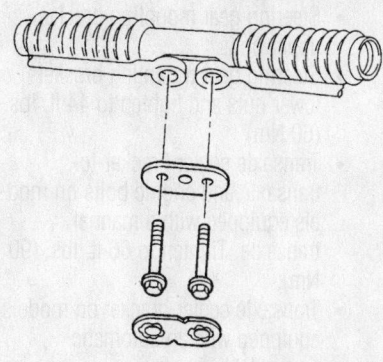

Inner tie rod plate—Lanos models

9307ZG71

- Steering gear bracket bolts from the top of each bracket
- Rack assembly through the wheel opening

To install:
5. Install or connect the following:
- Rack assembly through the wheel opening. Make sure the gear is in the straight ahead position.
- Steering gear bracket bolts and nuts, finger tighten only at this time
- Inlet and outlet pipe fittings and tighten to 20 ft. lbs. (27 Nm).

➡**Make sure the center housing cover washers are located between the tie rods and the steering gear. Always use a new lock plate.**

6. Install or connect the following:
- Both tie rods to the gear by threading the inner tie rod bolts though the inner tie rod plate, the tie rods and the center housing cover washers onto the rack.
- New lock plate and tighten the tie rod bolts to 66 ft. lbs. (90 Nm)
- Dash seal
- Coupling flange to the gear. Attach the flange with the pinch bolt and tighten to 16 ft. lbs. (22 Nm).
- Tighten all gear retaining bolts to 28 ft. lbs. (38 Nm)
- Coolant surge tank
- Negative battery cable
7. Fill the power steering reservoir.
8. Check the vehicle front end alignment and adjust as necessary.

Nubira

1. Before servicing the vehicle, refer to the precautions in the beginning of this section.
2. Remove or disconnect the following:
- Negative battery cable
- Power steering gear fluid inlet and outlet pipes

Timing belt service is covered in Section 3 of this manual

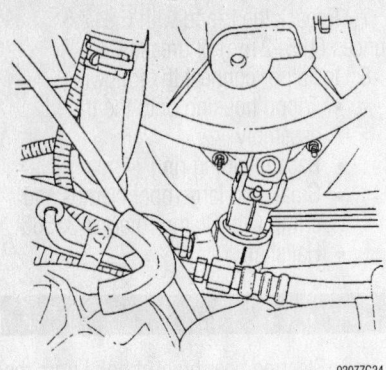

93072G34

Scribe a mark on the stub shaft housing that aligns with a mark on the intermediate shaft lower coupling—Nubira model

- Wheels
- Outer tie rods

3. Position the steering gear in the straight ahead position by turning the steering wheel until the spokes are vertical and pointing to the left.

4. Scribe a mark on the stub shaft housing that aligns with a mark on the intermediate shaft lower coupling.

5. Remove or disconnect the following:
- Intermediate shaft pinch bolt
- Crossmember
- Lower bolt and upper nut from the left side mounting bracket
- Upper bolt from the right side mounting bracket
- Rack assembly

To install:

6. Install or connect the following:
- Rack assembly. Make sure the gear is in the straight ahead position and the steering wheel spokes are vertical and pointing to the left. Make sure the marks are aligned on the shafts.
- Lower bolt and upper nut onto the left side mounting bracket and the upper bolt from the right side. Tighten the steering gear mounting bracket bolts and nuts to 44 ft. lbs. (60 Nm).
- Crossmember
- Intermediate shaft pinch bolt and tighten to 19 ft. lbs. (26 Nm)
- Outer tie rods, perform a wheel alignment and tighten the adjusting nut to 47 ft. lbs. (64 Nm)
- Wheels
- Power steering gear fluid inlet and outlet pipes and tighten the fitting to 21 ft. lbs. (28 Nm)
- Negative battery cable

7. Fill the power steering reservoir.

Leganza

1. Before servicing the vehicle, refer to the precautions in the beginning of this section.

2. Remove or disconnect the following:
- Negative battery cable
- Wheels
- Power steering gear fluid inlet and outlet pipes

3. Position the steering gear in the straight ahead position by turning the steering wheel until the spokes are vertical and pointing to the left.

4. Scribe a mark on the stub shaft housing that aligns with a mark on the intermediate shaft lower coupling.

5. Remove or disconnect the following:
- Intermediate shaft pinch bolt
- Outer tie rod nuts
- Outer tie rod ends from the strut assembly using a suitable separator tool
- Crossmember
- Transaxle center bracket on models equipped with an automatic transaxle
- Transaxle center bracket-to-transaxle and engine bolts on models equipped with a manual transaxle
- Steering gear mounting bracket nuts
- Steering gear mounting bracket bolts
- Rack assembly

To install:

6. Install or connect the following:
- Rack assembly. Make sure the gear is in the straight ahead position and the steering wheel spokes are vertical and pointing to the left. Make sure the marks are aligned on the shafts.
- Steering gear mounting bracket upper bolts
- Steering gear mounting bracket lower nuts and tighten to 44 ft. lbs. (60 Nm)
- Transaxle center bracket-to-transaxle and engine bolts on models equipped with a manual transaxle. Tighten to 66 ft. lbs. (90 Nm).
- Transaxle center bracket on models equipped with an automatic transaxle
- Crossmember
- Tie rod ends to the strut and tighten the nuts to 37 ft. lbs. (50 Nm)
- Intermediate shaft pinch bolt and tighten to 18 ft. lbs. (25 Nm)

- Power steering gear fluid inlet and outlet pipes and tighten the fitting to 21 ft. lbs. (28 Nm)
- Wheels

7. Fill the power steering reservoir.
8. Perform a wheel alignment.
9. Connect the negative battery cable.

Strut

REMOVAL & INSTALLATION

Lanos

FRONT

1. Before servicing the vehicle, refer to the precautions in the beginning of this section.

2. Remove or disconnect the following:
- Nuts that attach the top of the strut
- Halfshaft nut and discard

3. Place jackstands under the frame and lower the vehicle so that weight of the vehicle rests on the jackstands and not on the control arms.

4. Remove or disconnect the following:
- Brake caliper. Do not let the caliper hang by the brake hose, support it with wire
- Anti-Lock Brake Sensor (ABS) electrical connection
- Ball joint cotter pin
- Ball joint-to-knuckle strut nut
- Steering knuckle assembly from the ball joint using Ball Joint Removal Tool KM-507-B
- Outer tie rod from the steering knuckle assembly
- Halfshaft from the hub and tie the unfastened end of the halfshaft with wire to support it. Do not allow the halfshaft to hang freely as this may cause damage.
- Strut assembly-to-body nuts
- Strut assembly

To install:

5. Install or connect the following:
- Strut assembly
- Strut assembly-to-body nuts and tighten to 18 ft. lbs. (25 Nm)
- Halfshaft to the hub
- Outer tie rod to the steering knuckle assembly
- Ball joint-to-knuckle strut nut and tighten to 52 ft. lbs. (70 Nm)
- Cotter pin
- ABS electrical connection
- Brake caliper
- Wheel
- Halfshaft nut and tighten to 134 ft. lbs. (180 Nm). Loosen the nut and

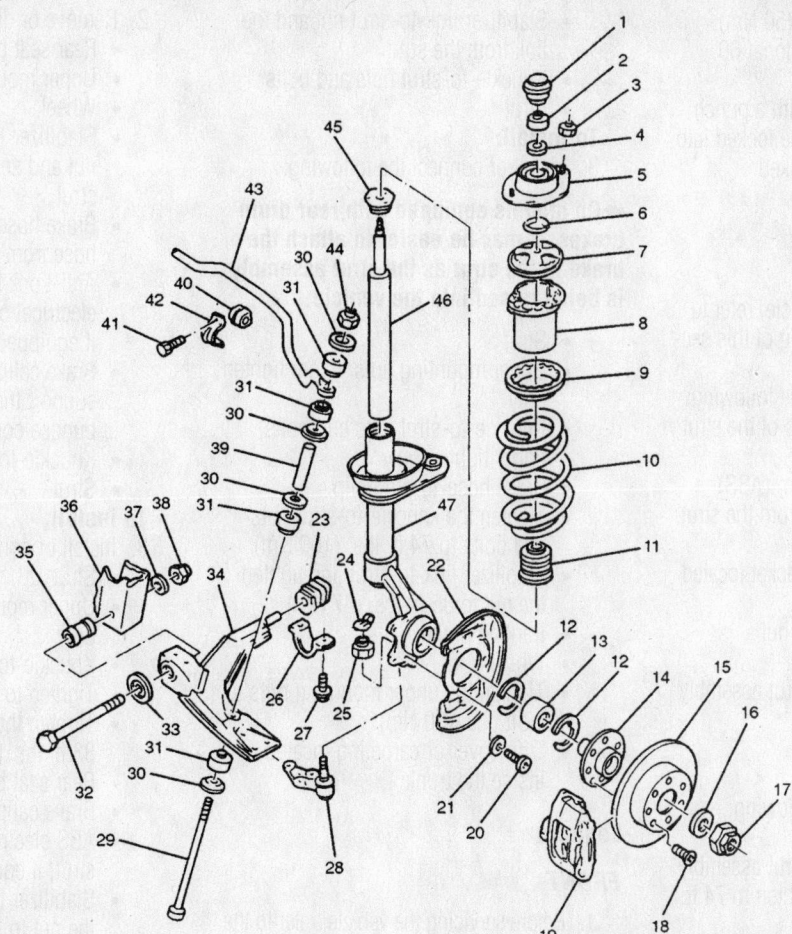

1. Upper Bearing Dust Cover
2. Piston Nut
3. Strut Assembly-to-Body Nut
4. Piston Nut Washer
5. Upper Strut Mount Bearing
6. Bearing Support Washer
7. Plastic Mount
8. Strut Shield
9. Upper Spring Insulator Ring
10. Coil Spring
11. Strut Bumper
12. Snap Ring
13. Front Wheel Bearing
14. Front Wheel Hub
15. Rotor
16. Drive Axle-to-Hub Nut Lock Washer
17. Drive Axle-to-Hub Caulking Nut
18. Detent Screw
19. Front Brake Caliper
20. Brake Shield Attachment Screw
21. Brake Shield Attachment Screw Washer
22. Brake Shield
23. Lower Control Arm Rear Bushing
24. Lower Ball Joint Cotter Pin
25. Lower Ball Joint Nut

26. Lower Control Arm Rear Bushing Mounting Bracket
27. Lower Control Arm Rear Bushing Mounting Bracket Bolt
28. Lower Ball Joint
29. Front Stabilizer Shaft Link Assembly Bolt
30. Front Stabilizer Shaft Link Grommet Washer
31. Front Stabilizer Shaft Link Grommet
32. Lower Control Arm Front Mounting Bracket Bolt
33. Lower Control Arm Front Mounting Bracket Bolt Washer
34. Lower Control Arm
35. Lower Control Arm Front Bushing
36. Vehicle Body
37. Lower Control Arm Front Mounting Bracket Nut Washer
38. Lower Control Arm Front Mounting Bracket Nut
39. Front Stabilizer Shaft Link Assembly Spacer
40. Front Stabilizer Shaft Insulator
41. Front Stabilizer Shaft Clamp Bolt
42. Front Stabilizer Shaft Clamp
43. Front Stabilizer Shaft
44. Front Stabilizer Shaft Link Assembly Nut
45. Strut Cartridge Closure Nut
46. Strut Cartridge
47. Knuckle and Support Tube

9307ZG70

Exploded view of the front strut and suspension assembly—Lanos models

Heater Core replacement is covered in Section 2 of this manual

retighten to 37 ft. lbs. (50 Nm), then retighten an additional 60 degrees.

6. Stake the halfshaft nut with a punch and hammer. The nut will not be locked into place until the nut has been staked.

Nubira

FRONT

1. Before servicing the vehicle, refer to the precautions in the beginning of this section.

2. Remove or disconnect the following:
   - Nuts that attach the top of the strut
   - Wheel
   - Anti-Lock Brake System (ABS) electrical connection from the strut, if equipped
   - Brake line from the bracket located on the strut
   - Stabilizer link-to-strut nut
   - Stabilizer shaft link
   - Steering knuckle-to-strut assembly nuts and bolts
   - Strut assembly

To install:

3. Install or connect the following:
   - Strut assembly
   - Steering knuckle-to-strut assembly nuts and bolts and tighten to 74 ft. lbs. (100 Nm)
   - Stabilizer shaft link
   - Stabilizer link-to-strut nut and tighten to 35 ft. lbs. (47 Nm)
   - ABS electrical connection to the strut, if equipped
   - Brake line to the bracket located on the strut
   - Wheel
   - Nuts that attach the top of the strut and tighten to 22 ft. lbs. (30 Nm)

REAR

1. Before servicing the vehicle, refer to the precautions in the beginning of this section.

2. Remove or disconnect the following:
   - Trim cover or carpeting from the upper mounting nuts located inside the trunk
   - Upper mounting nuts
   - Wheel
   - Parking brake

➡️**On models equipped with rear drum brakes, it may be easier to detach the brake hose from the strut once the strut assembly is being lowered from the vehicle.**

   - Brake hose-to-strut clip and the hose from the strut
   - Stabilizer link-to-strut nut and the link from the strut
   - Knuckle-to-strut nuts and bolts
   - Strut

To install:

3. Install or connect the following:

➡️**On models equipped with rear drum brakes, it may be easier to attach the brake to the strut as the strut assembly is being raised into the vehicle.**

   - Strut
   - Upper mounting nuts, finger tighten only
   - Knuckle-to-strut nuts and bolts, finger tighten only
   - Brake hose-to-strut clip
   - Tighten the knuckle-to-strut nuts and bolts to 74 ft. lbs. (100 Nm)
   - Stabilizer link-to-strut and tighten the nut to 35 ft. lbs. (47 Nm)
   - Parking brake
   - Wheel
   - Tighten the upper mounting nuts to 22 ft. lbs. (30 Nm)
   - Trim cover or carpeting located inside the trunk

Leganza

FRONT

1. Before servicing the vehicle, refer to the precautions in the beginning of this section.

2. Remove or disconnect the following:
   - Nuts that attach the top of the strut
   - Wheel
   - Anti-Lock Brake System (ABS) electrical connection from the strut, if equipped
   - Brake line from the bracket located on the strut
   - Steering knuckle-to-strut assembly nuts and bolts
   - Strut assembly

To install:

3. Install or connect the following:
   - Strut assembly
   - Steering knuckle-to-strut assembly nuts and bolts and tighten to 100 ft. lbs. (135 Nm)
   - Brake line to the bracket located on the strut
   - ABS electrical connection to the strut, if equipped
   - Wheel
   - Nuts that attach the top of the strut and tighten to 33 ft. lbs. (45 Nm)

REAR

1. Before servicing the vehicle, refer to the precautions in the beginning of this section.

2. Remove or disconnect the following:
   - Rear seat back
   - Upper mounting nuts
   - Wheel
   - Stabilizer link-to-strut assembly nut and separate link the from the strut
   - Brake hose-to-strut clip and the hose from the strut
   - Anti-Lock Brake System (ABS) electrical connection from the strut, if equipped
   - Brake caliper from the rotor and support the caliper with wire or a bungee cord
   - Knuckle-to-strut nuts and bolts
   - Strut

To install:

3. Install or connect the following:
   - Strut
   - Upper mounting nuts, finger tighten only
   - Knuckle-to-strut nuts and bolts. Tighten to 89 ft. lbs. (120 Nm).
   - Tighten the upper mounting nuts to 33 ft. lbs. (45 Nm)
   - Rear seat back
   - Brake caliper
   - ABS electrical connection to the strut, if equipped
   - Stabilizer link-to-strut and tighten the nut to 33 ft. lbs. (45 Nm)
   - Brake hose-to-strut
   - Wheel

Shock Absorber

REMOVAL & INSTALLATION

Lanos

REAR

1. Before servicing the vehicle, refer to the precautions in the beginning of this section.

2. Remove or disconnect the following:
   - Trim cover from the upper mounting nut located inside the trunk
   - Upper mounting nut by counterholding the threaded shock absorber shaft

3. Support the rear axle with jackstands.

4. Remove or disconnect the following:
   - Lower shock absorber-to-axle nut and bolt
   - Shock absorber

To install:

➡️**It is necessary to have the axle assembly at trim height before tightening the shock absorber bolts.**

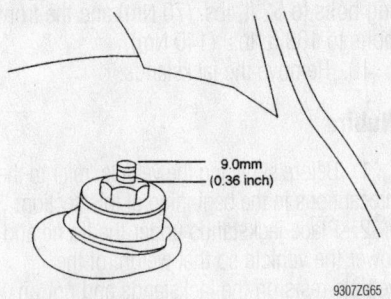

Tighten the upper nut until 0.36 in. (9mm) of the thread is visible—Lanos models

5. Install or connect the following:
   • Lower shock absorber-to-axle bolt and loosely attach the nut
   • Upper shock stud through the body opening and loosely install the nut
   • Tighten the lower shock absorber-to-axle-bolt to 52 ft. lbs. (70 Nm)
6. Lower the vehicle to the floor and tighten the upper nut until 0.36 in. (9mm) of the thread is visible.
7. Install the trim cover.

Coil Spring

REMOVAL & INSTALLATION

Lanos

FRONT

1. Before servicing the vehicle, refer to the precautions in the beginning of this section.
2. Mount spring compressor KM-465-A and spring compressor KM-329-A on a mounting trestle and a suitable workbench.
3. Remove the strut.
4. Attach the strut to the spring compressor making sure the hooks are seated on the spring properly.
5. Compress the front spring using the compressors.
6. Remove the dust cover from the support bearing.
7. Using an open ended wrench, hold the threaded portion piston rod while removing the piston rod nut with nut wrench J-42468.
8. Remove the washer.
9. Remove the support bearing, washer, plastic mount, shield and upper insulator.
10. Release the spring compressor.
11. Remove the spring and bumper.

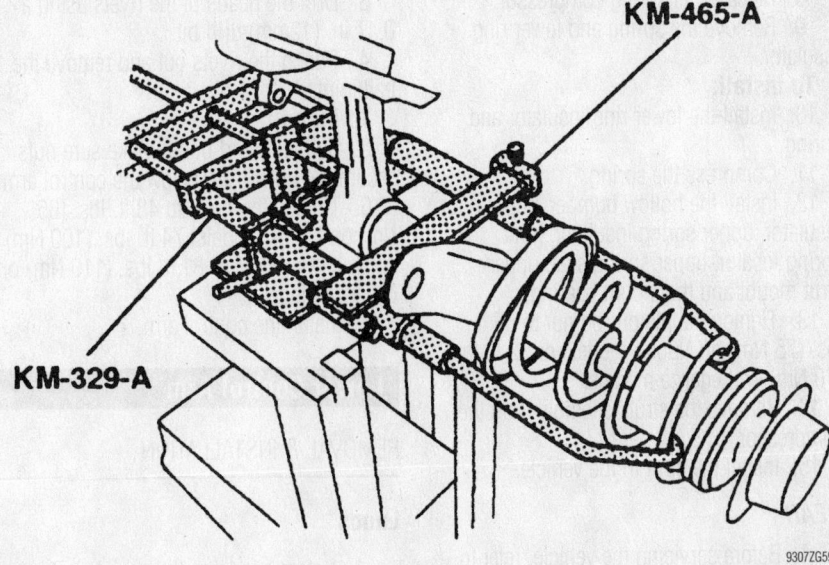

Spring compressors KM-465-A and KM-329-A are required to disassemble the strut assembly—Lanos

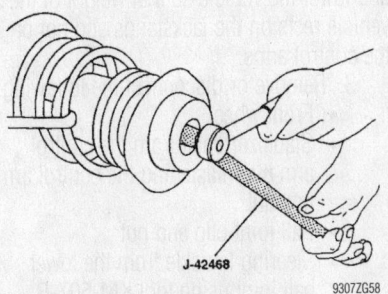

Removing the piston nut—Lanos

To install:
12. Lubricate the upper strut mount bearing with multi-purpose grease.
13. Install the bumper and spring.
14. Compress the spring compressors.
15. Install the upper insulator, shield, plastic mount, washer and the support bearing.
16. Tighten the piston rod nut to 41 ft. lbs. (55 Nm) using nut wrench J-42468.
17. Remove the strut assembly from the compressors.
18. Install the strut in the vehicle.

REAR

1. Before servicing the vehicle, refer to the precautions in the beginning of this section.
2. Support the rear control arms with jackstands.
3. Remove or disconnect the following:
   • Wheel
   • Right and left shock absorber bolts

4. Lower the rear axle and remove the springs and insulator.

To install:
5. Apply adhesive to the upper insulators and attach them to the body. The adhesive will hold them in place until the spring is installed.
6. Install or connect the following:
   • Upper insulator and lower bumper
   • Springs and raise the axle
   • Shock absorber bolts
   • Wheel
7. Remove the jackstands.

Nubira and Leganza Models

FRONT

1. Before servicing the vehicle, refer to the precautions in the beginning of this section.
2. Remove the strut.
3. Attach the strut to spring compressor KM-329-A making sure the hooks are seated on the spring properly.
4. Compress the front spring using the compressor.
5. Remove the dust cover from the support bearing.
6. Using an open ended wrench, hold the threaded portion piston rod while removing the piston rod nut with a double ring spanner that is sharply offset.
7. Remove the upper strut mount, mount bearing, upper spring seat, front spring locator, upper spring insulator, upper ring insulator and the hollow bumper.

Brake service is covered in Section 4 of this manual

8. Release the spring compressor.

9. Remove the spring and lower ring insulator.

To install:

10. Install the lower ring insulator and spring.

11. Compress the spring.

12. Install the hollow bumper, upper ring insulator, upper spring insulator, front spring locator, upper spring seat, upper strut mount and the mount bearing.

13. Tighten the piston rod nut to 55 ft. lbs. (75 Nm) on Nubira models or 52 ft. lbs. (70 Nm) on Leganza models.

14. Remove the strut assembly from the compressor.

15. Install the strut in the vehicle.

REAR

1. Before servicing the vehicle, refer to the precautions in the beginning of this section.

2. Remove the strut.

3. Attach the strut to spring compressor KM-329-A making sure the hooks are seated on the spring properly.

4. Compress the front spring using the compressor.

5. Remove the lock-nut from the dampener rod.

6. Remove the strut mount.

7. Remove the upper seat, dust cover and the hollow bumper.

8. Release the spring compressor.

9. Remove the spring and lower seat.

To install:

10. Install lower seat and spring.

11. Compress the spring.

12. Install the hollow bumper, dust cover and upper seat.

13. Strut mount and the lock-nut.

14. If equipped with an Anti-lock Brake system (ABS), align the flat side of the strut mount with the ABS sensor line bracket on the dampener so that both are parallel to each other.

15. Tighten the lock-nut to 55 ft. lbs. (75 Nm) on Nubira models and 59 ft. lbs. (80 Nm) on Leganza models.

16. Remove the strut assembly from the compressor.

17. Install the strut in the vehicle.

Lower Ball Joint

REMOVAL & INSTALLATION

1. Before servicing the vehicle, refer to the precautions in the beginning of this section.

2. Remove the control arm.

3. Drill the heads of the rivets using a 0.47 in. (12mm) drill bit.

4. Punch the rivets out and remove the ball joint.

To install:

5. Ball joint and bolts. Make sure nuts secure the bolts from below the control arm.

6. Tighten the bolts to 48 ft. lbs. (65 Nm) on Lanos models, 74 ft. lbs. (100 Nm) on Nubira models or 81 ft. lbs. (110 Nm) on Leganza models.

7. Install the control arm.

Lower Control Arm

REMOVAL & INSTALLATION

Lanos

1. Before servicing the vehicle, refer to the precautions in the beginning of this section.

2. Place jackstands under the frame and lower the vehicle so that weight of the vehicle rests on the jackstands and not on the control arms.

3. Remove or disconnect the following:
   - Front wheel
   - Stabilizer shaft from the control arm by unfastening the control arm link nut
   - Ball joint clip and nut
   - Steering knuckle from the lower ball joint using tool KM-507-B
   - Control arm front and rear mounting bolts
   - Control arm

To install:

4. Install or connect the following:
   - Control arm and install the front bolts and finger tighten

5. Apply thread sealer to the rear mounting bolts.

6. Install or connect the following:
   - Rear mounting bolts and finger tighten

➡**Use a new self-locking nut to install the control arm link bolt assembly.**

7. Install or connect the following:
   - Stabilizer shaft link bolt
   - Ball joint to the steering knuckle and tighten the bolt to 52 ft. lbs. (75 Nm)
   - Ball joint clip
   - Front wheel

8. Place jackstands under the frame and lower the vehicle so that weight of the vehicle rests on the jackstands and not on the control arms.

9. Tighten the control arm rear mount-

ing bolts to 52 ft. lbs. (70 Nm) and the front bolts to 103 ft. lbs. (140 Nm).

10. Remove the jackstands.

Nubira

1. Before servicing the vehicle, refer to the precautions in the beginning of this section.

2. Place jackstands under the frame and lower the vehicle so that weight of the vehicle rests on the jackstands and not on the control arms.

3. Remove or disconnect the following:
   - Front wheel
   - Ball joint bolt and nut
   - Steering knuckle from the lower ball joint using tool KM-507-B
   - Control arm-to-crossmember nut and bolt
   - Rear crossmember-to-body bolt
   - Control arm

To install:

4. Install or connect the following:
   - Control arm and install the rear crossmember-to-body bolt and finger tighten
   - Control arm-to-crossmember nut and bolt and tighten to 81 ft. lbs. (110 Nm)
   - Ball joint to the steering knuckle and tighten the bolt nut to 44 ft. lbs. (60 Nm)
   - Front wheel

5. Tighten the rear crossmember-to-body bolt to 145 ft. lbs. (196 Nm).

6. Remove the jackstands.

Leganza

1. Before servicing the vehicle, refer to the precautions in the beginning of this section.

2. Place jackstands under the frame and lower the vehicle so that weight of the vehicle rests on the jackstands and not on the control arms.

3. Remove or disconnect the following:
   - Front wheel
   - Stabilizer link-to-control arm nut and separate the link from the arm
   - Ball joint bolt and nut
   - Steering knuckle from the lower ball joint using tool KM-333
   - Control arm-to-crossmember nut and bolt
   - Control arm rear bushing clamp bolts and the clamp
   - Control arm

To install:

4. Install or connect the following:
   - Control arm and install the arm-to-crossmember bolt. Finger tighten only at this time.

- Control arm rear bushing clamp and bolts. Finger tighten only at this time.
- Ball joint to the steering knuckle and tighten the bolt nut to 66 ft. lbs. (90 Nm)
- Stabilizer link to the control arm and the nut. Lower the vehicle so that the arms are supported by jackstands.
- Tighten the control arm rear bushing clamp bolts and tighten to 55 ft. lbs. (75 Nm)
- Tighten the control arm front bushing nut to 103 ft. lbs. (140 Nm)
- Tighten the stabilizer link-to-control arm nut and tighten to 33 ft. lbs. (45 Nm)
- Front wheel

5. Remove the jackstands.

CONTROL ARM BUSHING REPLACEMENT

Lanos

FRONT

1. Before servicing the vehicle, refer to the precautions in the beginning of this section.

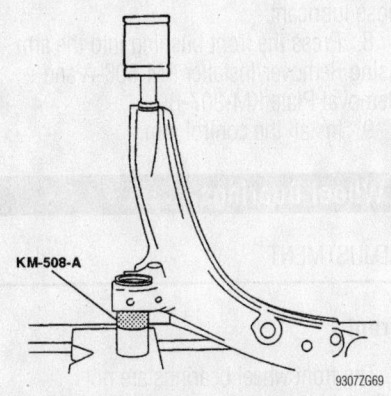

Installing the rear bushing—Lanos

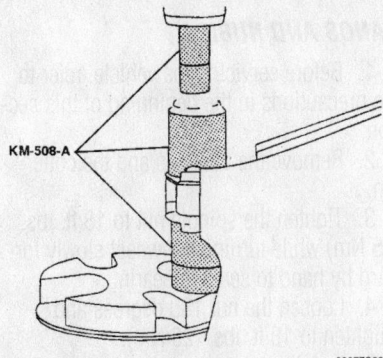

Installing the front bushing—Lanos

2. Remove the control arm.
3. Press the rear bushing from the arm using a press, Remover/Installer KM-158 and Removal Plate KM-307-B.
4. Press the front bushing from the arm using a press, Remover/Installer KM-508-A and Removal/Installer KM-158.

To install:

5. Coat the rear arm shaft with a multi-purpose lubricant.
6. Press the rear bushing onto the shaft using Remover/Installer KM-508-A to support the arm. Make sure the flat of the bushing is on the top side the same as the ball joint.
7. Coat the outside of the front bushing and the inside of the arm with a multi-purpose lubricant.
8. Press the front bushing into the arm from the back to the front using Remover/Installer KM-508-A.
9. Center the bushing.
10. Install the control arm.

REAR

1. Before servicing the vehicle, refer to the precautions in the beginning of this section.
2. Remove the rear axle and place it on a workbench.

3. Warm the rear axle bushing area to 122–158 degrees F (50–75 degrees C).
4. Place control arm bushing housing J-29376-A on the axle.
5. Slide the puller bolt/thrust washer J-21474-19 through the arm bushing remover/installer J-29376-6A, the arm bushing, the bushing plate J-29376-7 and into the nut J-21474-18.
6. Partially remove the bushing by turning nut J-21474-18 and counterholding puller bolt J-21474-19
7. Completely remove the bushing by striking the bushing remover/installer J-29376-6A with remover KM-66-A.

To install:

8. Place control arm bushing housing J-29376-A on the axle.
9. Slide the puller bolt/thrust washer J-21474-19, the arm bushing plate J-29376-7, rear bushing, arm bushing remover/installer J-29376-6A and into the nut J-21474-18.
10. Turn nut J-21474-18 and counterholding puller bolt J-21474-19.
11. Make sure the bushing angle is 40–50 degrees to the axis of the rear axle.
12. Install the rear axle.

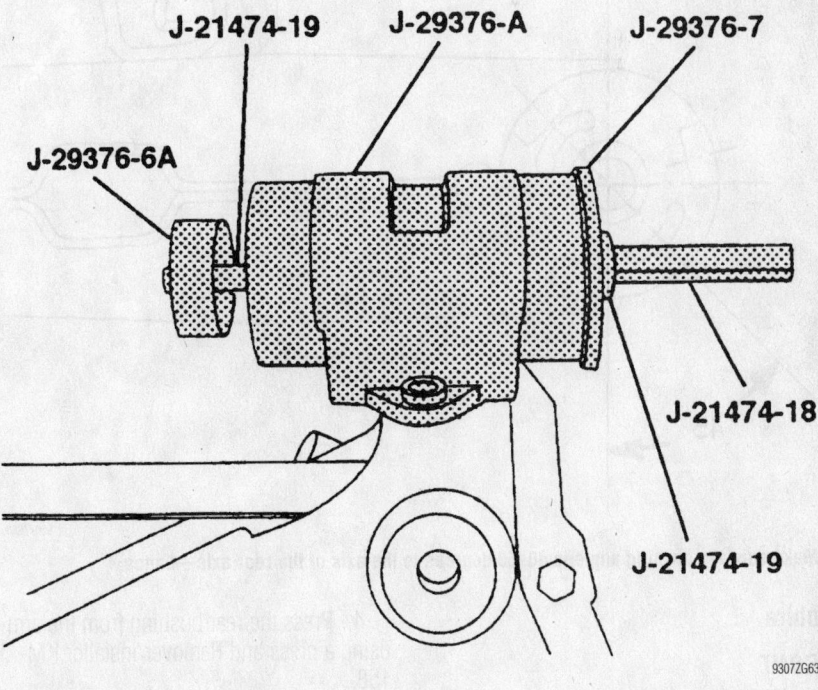

Assemble the tool as shown to remove the rear axle control arm bushing—Lanos

For complete Engine Mechanical specifications, see Section 1 of this manual

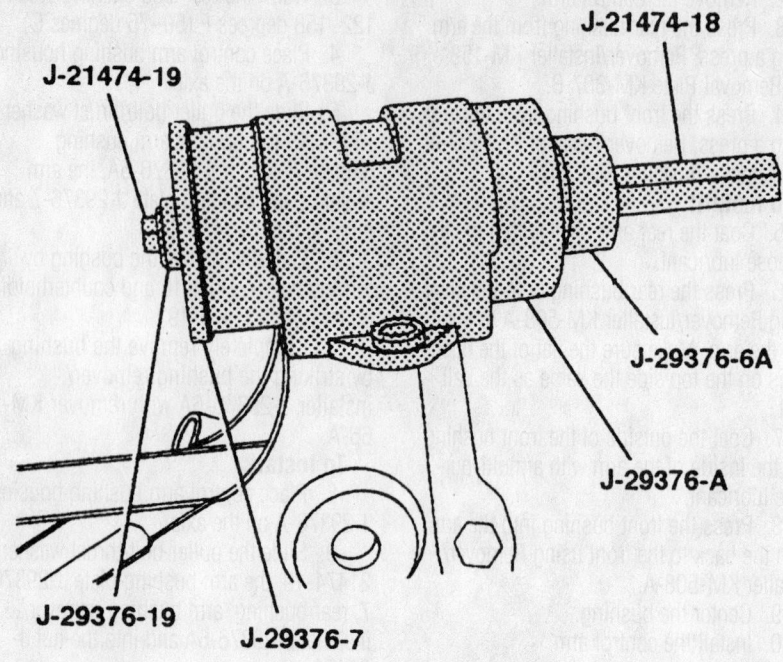

Assemble the tool as shown to install the rear axle control arm bushing—Lanos

J-21474-19
J-21474-18
J-29376-6A
J-29376-A
J-29376-19
J-29376-7

9307ZG62

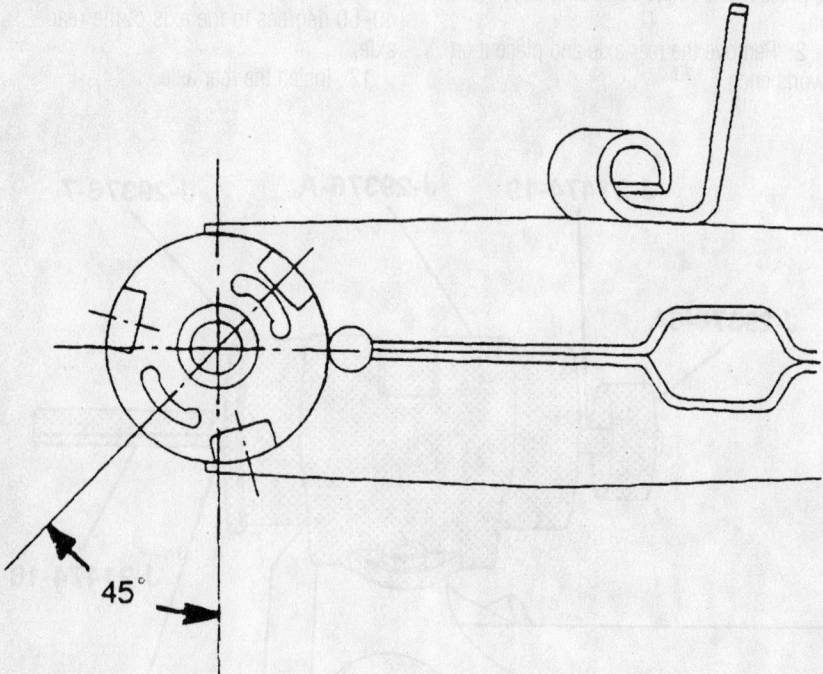

45°

Make sure the bushing angle is 40–50 degrees to the axis of the rear axle—Lanos

9307ZG61

Nubira

FRONT

1. Before servicing the vehicle, refer to the precautions in the beginning of this section.
2. Remove the control arm.
3. Remove the split sleeves from the rear bushing.

4. Press the rear bushing from the arm using a press and Remover/Installer KM-158.
5. Press the front bushing from the arm using a press and Remover/Installer KM-158.

To install:

6. Press the rear bushing into the arm using Remover/Installer KM-158.

7. Press the front bushing into the arm using Remover/Installer KM-158.
8. Install the split sleeves into the rear bushing.
9. Install the control arm.

Leganza

FRONT

1. Before servicing the vehicle, refer to the precautions in the beginning of this section.
2. Remove the control arm.
3. Press the rear bushing from the arm using a press, Remover/Installer KM-158 and Removal Plate KM-307-B.
4. Press the front bushing from the arm using a press, Remover/Installer KM-508-A and Removal Plate KM-307-B.

To install:

5. Coat the rear arm shaft with a multi-purpose lubricant.
6. Press the rear bushing onto the shaft using Remover/Installer KM-508-A to support the arm.
7. Coat the outside of the front bushing and the inside of the arm with a multi-purpose lubricant.
8. Press the front bushing into the arm using Remover/Installer KM-508-A and Removal Plate KM-307-B.
9. Install the control arm.

Wheel Bearings

ADJUSTMENT

Front

The front wheel bearings are not adjustable.

Rear

LANOS AND NUBIRA

1. Before servicing the vehicle, refer to the precautions in the beginning of this section.
2. Remove the dust cap and the cotter pin.
3. Tighten the spindle nut to 18 ft. lbs. (25 Nm) while turning the wheel slowly forward by hand to seat the bearing.
4. Loosen the nut 180 degrees and retighten to 18 ft. lbs. (25 Nm).
5. Install a new cotter pin.
6. Measure bearing end-play. The end-play should be 0.001–0.005 in. (0.03–0.13mm) when properly adjusted.
7. Install the dust cap.

LEGANZA

The rear bearings are a non-serviceable part of the rear hub assembly. If the bearings are found to be defective, replace the hub and bearing assembly.

REMOVAL & INSTALLATION

Front

1. Before servicing the vehicle, refer to the precautions in the beginning of this section.
2. Remove or disconnect the following:

- Halfshaft from the hub
- Inner snapring
- Wheel hub using Support Bridge (J-37105-B-1 on Lanos and Nubira models or J-37105-40 on Leganza models), hub adapter J-37105-B-3, hex nut 500-20 and forcing screw J-36661-2
- Brake shield
- Outer snapring
- Wheel bearing using Support Bridge (J-37105-B-1 on Lanos and Nubira models or J-37105-40 on Leganza models), bearing adapter J-37105-B-2, hex nut 500-20 and forcing screw J-36661-2

3. Clean the knuckle bore.

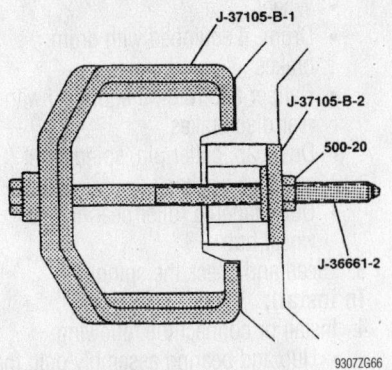

Removing the wheel bearing

To install:

4. Install or connect the following:

- Outer snapring and push the wheel bearing into position using support bridge (J-37105-B-1 on Lanos and Nubira models or J-37105-40 on Leganza models), bearing adapter J-37105-B-2, hex nut 500-20 and forcing screw J-36661-2
- Brake shield
- Inner snapring and push the wheel hub into position using hub adapter J-37105-B-3, bearing adapter J-37105-B-2, hex nut 500-20 and forcing screw J-36661-2
- Halfshaft into the hub

Rear

LANOS WITH ABS

Models equipped with Anti-Lock Brake Sensor (ABS) have a non-serviceable assembly. If the bearing is defective, the whole assembly must be replaced.

1. Before servicing the vehicle, refer to the precautions in the beginning of this section.
2. Remove or disconnect the following:

- Brake drum and detent screw
- Parking brake cable, loosen only
- Anti-Lock Brake Sensor (ABS) line
- Wheel hub assembly bolts and nuts
- Wheel hub assembly

3. Installation is the reverse of removal. Tighten the hub assembly nuts to 30 ft. lbs. (40 Nm), turn the nut an additional 60 degrees and another additional 15 degrees.

LANOS WITHOUT ABS

1. Before servicing the vehicle, refer to the precautions in the beginning of this section.
2. Remove or disconnect the following:

- Wheel
- Parking brake cable, loosen only
- Dust cap, cotter pin, spindle nut and lock washer
- Wheel hub and outer tapered roller bearing
- Seal ring from the hub using a suitable prytool
- Inner tapered roller bearing
- Races of the inner and outer bearings using a brass drift

3. Clean and check the spindle.

To install:

4. Install or connect the following:

- Outer race of the outer bearing into the hub using a press and installers J-36791 and KM-266-A
- Outer race of the inner bearing into

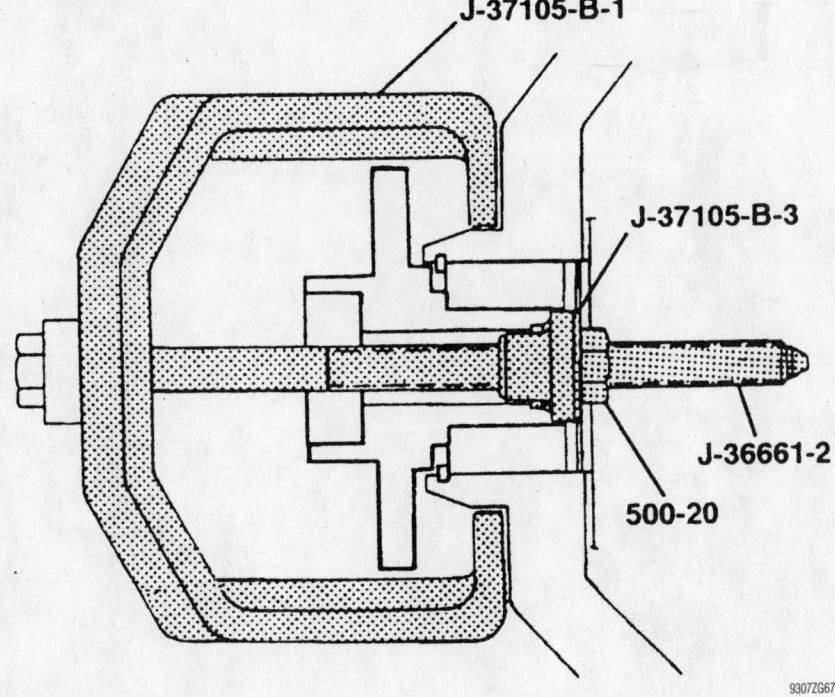

Removing the wheel hub

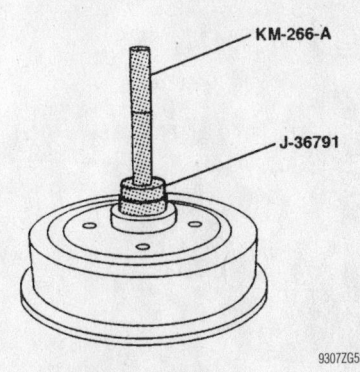

Installing the outer race of the outer bearing into the hub—Lanos

For Accessory Drive Belt illustrations, see Section 1 of this manual

the hub using a press and installers J-36791 and KM-266-A
- Inner tapered roller bearing

5. Coat the hollow spaces of both bearings, ring seal lip and the wheel hub with suitable grease

6. Install or connect the following:
- Seal into the hub using a press if necessary
- Hub and bearing assembly onto the spindle
- Lock washer and spindle nut. Hand-tighten the nut
- Wheel

7. Adjust the wheel bearing and the parking brake.

8. Install the dust cap.

NUBRIA

1. Before servicing the vehicle, refer to the precautions in the beginning of this section.

2. Remove or disconnect the following:

- Wheel
- Drum, if equipped with drum brakes
- Caliper and rotor, if equipped with rear disc brakes
- Dust cap, cotter pin, spindle nut and lock washer
- Outer tapered roller bearing and wheel hub

3. Clean and check the spindle.

To install:

4. Install or connect the following:
- Hub and bearing assembly onto the spindle
- Lock washer and spindle nut. Hand-tighten the nut.
- Drum, if equipped with drum brakes
- Rotor and caliper, if equipped with rear disc brakes
- Wheel

5. Adjust the wheel bearing.

6. Install the dust cap.

LEGANZA

1. Before servicing the vehicle, refer to the precautions in the beginning of this section.

2. Remove or disconnect the following:
- Wheel
- Caliper and rotor
- Dust cap and straighten the stake in the spindle nut with a hammer and a drift.
- Spindle nut
- Hub and bearing

3. Clean and check the spindle.

To install:

4. Install or connect the following:
- Hub and bearing assembly onto the spindle
- Spindle nut and tighten to 210 ft. lbs. (285 Nm). Stake the nut using a hammer and drift.
- Rotor and caliper
- Wheel

5. Install the dust cap.

HONDA

1998–01
Accord • Civic • Prelude • S2000

PRECAUTIONS

Before servicing any vehicle, please be sure to read all of the following precautions, which deal with personal safety, prevention of component damage, and important points to take into consideration when servicing a motor vehicle:

• Never open, service or drain the radiator or cooling system when the engine is hot; serious burns can occur from the steam and hot coolant.

• Observe all applicable safety precautions when working around fuel. Whenever servicing the fuel system, always work in a well-ventilated area. Do not allow fuel spray or vapors to come in contact with a spark, open flame, or excessive heat (a hot drop light, for example). Keep a dry chemical fire extinguisher near the work area. Always keep fuel in a container specifically designed for fuel storage; also, always properly seal fuel containers to avoid the possibility of fire or explosion. Refer to the additional fuel system precautions later in this section.

• Fuel injection systems often remain pressurized, even after the engine has been turned **OFF**. The fuel system pressure must be relieved before disconnecting any fuel lines. Failure to do so may result in fire and/or personal injury.

• Brake fluid often contains polyglycol ethers and polyglycols. Avoid contact with the eyes and wash your hands thoroughly after handling brake fluid. If you do get brake fluid in your eyes, flush your eyes with clean, running water for 15 minutes. If

eye irritation persists, or if you have taken brake fluid internally, IMMEDIATELY seek medical assistance.

• The EPA warns that prolonged contact with used engine oil may cause a number of skin disorders, including cancer! You should make every effort to minimize your exposure to used engine oil. Protective gloves should be worn when changing oil. Wash your hands and any other exposed skin areas as soon as possible after exposure to used engine oil. Soap and water, or waterless hand cleaner should be used.

• All new vehicles are now equipped with an air bag system. The system must be disabled before performing service on or around system components, steering column, instrument panel components, wiring and sensors. Failure to follow safety and disabling procedures could result in accidental air bag deployment, possible personal injury, and unnecessary system repairs.

• Always wear safety goggles when working with, or around, the air bag system. When carrying a non-deployed air bag, be sure the bag and trim cover are pointed away from your body. When placing a non-deployed air bag on a work surface, always face the bag and trim cover upward, away from the surface. This will reduce the motion of the module if it is accidentally deployed. Refer to the additional air bag system precautions later in this section.

• Clean, high quality brake fluid from a

sealed container is essential to the safe and proper operation of the brake system. You should always buy the correct type of brake fluid for your vehicle. If the brake fluid becomes contaminated, completely flush the system with new fluid. Never reuse any brake fluid. Any brake fluid that is removed from the system should be discarded. Also, do not allow any brake fluid to come in contact with a painted surface; it will damage the paint.

• Never operate the engine without the proper amount and type of engine oil; doing so WILL result in severe engine damage.

• Timing belt maintenance is extremely important! Many models utilize an interference-type, non-freewheeling engine. If the timing belt breaks, the valves in the cylinder head may strike the pistons, causing potentially serious (also time-consuming and expensive) engine damage. Refer to the maintenance interval charts in the front of this manual for the recommended replacement interval for the timing belt, and to the timing belt section for belt replacement and inspection.

• Disconnecting the negative battery cable on some vehicles may interfere with the functions of the on-board computer system(s) and may require the computer to undergo a relearning process once the negative battery cable is reconnected.

• When servicing drum brakes, only disassemble and assemble one side at a time, leaving the remaining side intact for reference.

ENGINE REPAIR

➡**Disconnecting the negative battery cable on some vehicles may interfere with the functions of the on board computer systems and may require the computer to undergo a relearning process, once the negative battery cable is reconnected.**

Distributor

REMOVAL

➡**The radio may contain a coded theft protection circuit. Always obtain the code number before disconnecting the battery. If the vehicle is equipped with 4WS, the steering control unit is shut down when the battery is disconnected. After connecting the battery, turn the steering wheel lock-to-lock to reset the steering control unit.**

1. Before servicing the vehicle, refer to the precautions in the beginning of this section.
2. Disconnect the negative battery cable.
3. Rotate the crankshaft to bring No. 1 cylinder to TDC, and align the white mark on the crankshaft pulley with the pointer on the timing belt cover.
4. Remove or disconnect the following:
   • Distributor cap with the ignition wires attached
   • Electrical connectors from the distributor
5. Mark the direction the ignition rotor is pointing on the distributor housing to aid in installation.
6. Matchmark the distributor housing with the cylinder head to aid in installation.
   • Distributor mounting bolts and remove the distributor

• O-ring from the distributor housing, then discard the O-ring

INSTALLATION

1. Coat a new O-ring with clean engine oil and install it to the distributor housing.
2. Align the ignition rotor with the mark made on the distributor housing. The drive lugs are offset so the distributor cannot be installed incorrectly. Fit the distributor into place and turn the rotor until the drive lugs engage and the distributor seats in the cylinder head.

➡**The lugs on the end of the distributor and their mating grooves in the camshaft end, are offset to eliminate the possibility of installing the distributor 180° out of time.**

3. Align the matchmark on the distribu-

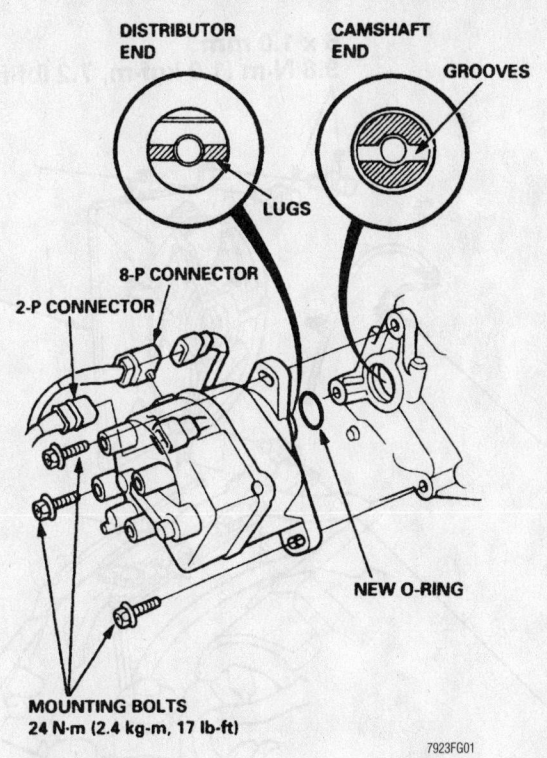

Distributor components—1.6L engines

tor housing and the cylinder head and install the mounting bolts snugly.

4. Install or connect the following:
- Distributor cap with the ignition wires
- Distributor electrical connectors
- Negative battery cable and enter the radio security code.

5. If equipped with 4-Wheel Steering (4WS), start the engine and turn the steering wheel lock-to-lock to reset the 4WS control unit.

6. Adjust the ignition timing.

7. Tighten the distributor mounting bolts to 16 ft. lbs. (22 Nm)

Alternator

1.6L and 1.7L Engines

1. Before servicing the vehicle, refer to the precautions in the beginning of this section.

2. Remove or disconnect the following:
- Negative battery cable
- Accessory drive belts
- Power steering pump
- 4P connector and battery terminal wire
- Alternator bolts
- Alternator

To install:
- Alternator and tighten the bolts to 33 ft. lbs. (44 Nm)
- 4P connector and battery terminal wire. Tighten the battery terminal wire nut to 108 inch lbs. (12 Nm).
- Power steering pump
- Accessory drive belts
- Negative battery cable

2.0L Engine

1. Before servicing the vehicle, refer to the precautions in the beginning of this section.

2. Remove or disconnect the following:
- Negative battery cable
- Accessory drive belt
- 4P connector and battery terminal wire
- Alternator bolts
- Alternator

To install:
- Alternator and tighten the bolts to 33 ft. lbs. (44 Nm)
- 4P connector and battery terminal wire. Tighten the battery terminal wire nut to 108 inch lbs. (12 Nm).

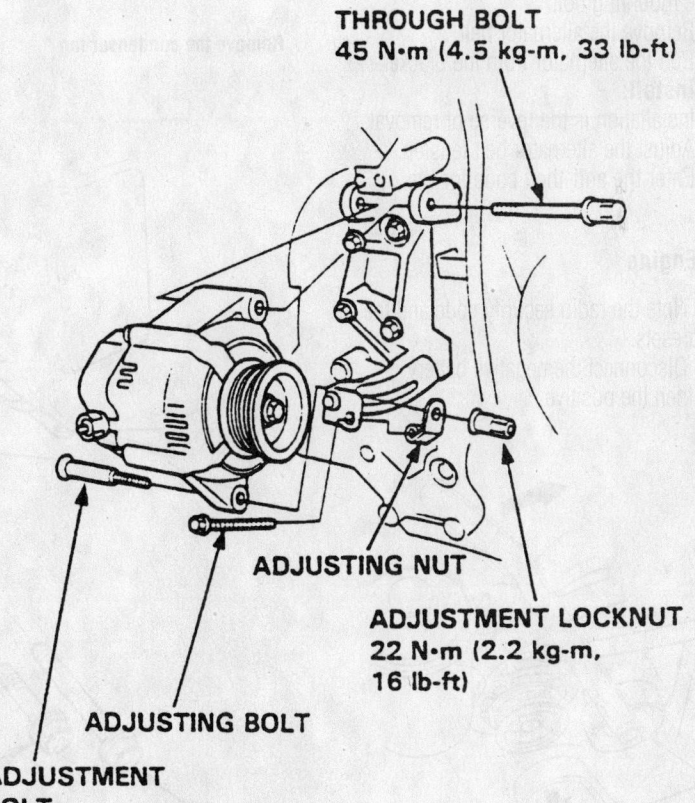

Alternator mounting bolt locations

- Accessory drive belt
- Negative battery cable

2.2L Engine

1. Disconnect the negative battery cable, then the positive.
2. Remove the power steering pump.
3. Detach the cruise control actuator but do not remove the cable.
4. Loosen the through bolt and then loosen adjusting bolt.
5. Remove the alternator belt.
6. Remove the adjusting bolt.
7. Remove the through bolt and then the alternator.

To install:

8. Installation is the reverse of removal.
9. Adjust the alternator belt tension.

2.3L Engines

1. Note the radio security code and the radio presets.
2. Disconnect the negative battery cable, then the positive.
3. Detach the 4P connector and battery terminal wire from the alternator.
4. Remove the adjusting bolt, locknut and the mounting bolt.
5. Remove the alternator belt.
6. Pull the alternator from the bracket.

To install:

7. Installation is the reverse of removal.
8. Adjust the alternator belt tension.
9. Enter the anti-theft code for the radio.

3.0L Engine

1. Note the radio security code and the radio presets.
2. Disconnect the negative battery cable, then the positive.

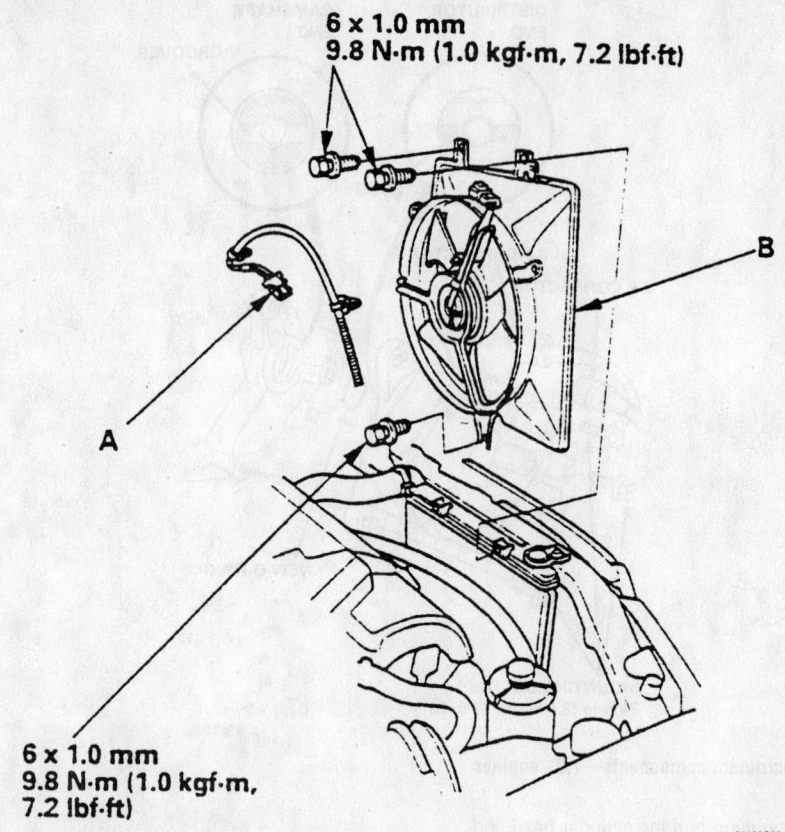

6 x 1.0 mm
9.8 N·m (1.0 kgf·m, 7.2 lbf·ft)

B

A

6 x 1.0 mm
9.8 N·m (1.0 kgf·m, 7.2 lbf·ft)

91182G22

Remove the condenser fan

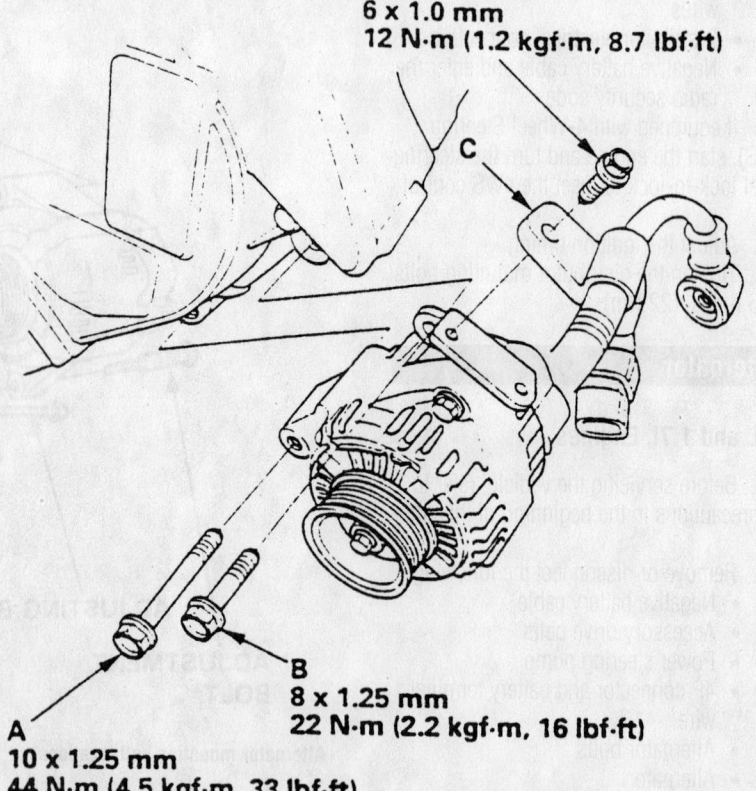

6 x 1.0 mm
12 N·m (1.2 kgf·m, 8.7 lbf·ft)

C

A
10 x 1.25 mm
44 N·m (4.5 kgf·m, 33 lbf·ft)

B
8 x 1.25 mm
22 N·m (2.2 kgf·m, 16 lbf·ft)

91182G23

Torque the alternator bolts to the specs shown

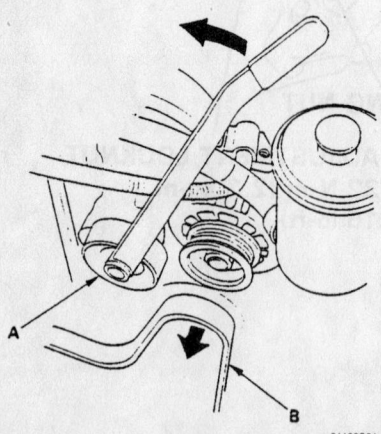

A

B

91182G21

Relieve the belt tension by pulling back on the tensioner

3. Relieve the alternator belt tension by pulling back on the adjuster and then remove the belt.

4. Detach the condenser fan motor connector from the shroud.

5. Remove the condenser fan assembly.

6. Disconnect the four prong connector from the rear of the alternator.

7. Remove the alternator mounting bolts.

8. Remove the wiring harness clamp.

9. Remove the alternator assembly.

To install:

10. Alternator installation is the reverse of the removal procedure.

11. Connect the positive battery cable, then the negative battery cable. Enter the radio security code and station presets.

❈❈ WARNING

Be sure to adjust the alternator belt to the proper tension or alternator bearing failure may occur.

Ignition Timing

ADJUSTMENT

1.6L Engines

1. Before servicing the vehicle, refer to the precautions in the beginning of this section.

2. Set the parking brake and block the front wheels.

3. Connect a timing light to the No. 1 spark plug wire.

4. Start the engine and allow it to warm up.

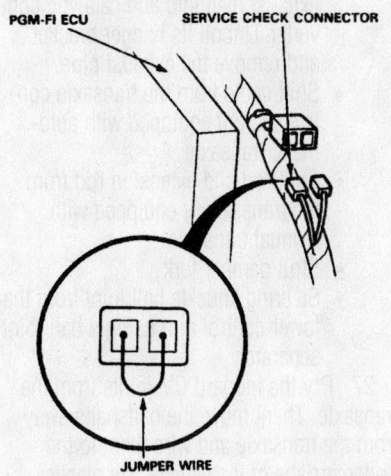

Service check connector—1.6L engines

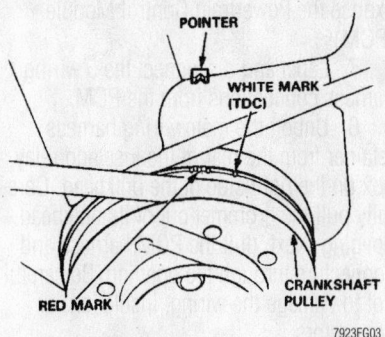

7923FG03

Typical crankshaft pulley timing mark location

5. Pull out the service check connector located behind the right kick panel. On the 2-P connector, connect the WHT/BGN or BRN and BLK terminals with service connector 07PAZ-0010100, or equivalent. Don't connect a jumper wire to the 3-P DLC.

6. Shift the transaxle to neutral. All electrical accessories must be off. If equipped with DRL's, turn them off by engaging the parking brake lever.

7. Connect a test tachometer to the test tachometer connector located on the left shock tower. Check the idle speed.

8. While the engine idles, point the timing light at the mark on the timing belt cover.

9. Timing specifications: B16A2 engines
  • manual transmission: 16° BTDC at 650–750 (USA) and 700–800 (Canada)

10. Timing specifications: D16Y5 engine
  • manual transmission: 10–14° BTDC at 620–720 rpm (USA only)
  • automatic transmission and CVT: 10–14° BTDC at 650–750 rpm (USA only)

11. Timing specifications: D16Y7 and D16Y8 engines
  • manual transmission: 10–14° BTDC at 620–730 rpm (USA) or 700–800 rpm (Canada)
  • automatic transmission: 10–14° BTDC at 650–750 rpm (USA) or 700–800 rpm (Canada)

12. If adjustment is needed, loosen the distributor adjusting bolts and turn the distributor counterclockwise to advance the timing or clockwise to retard the timing.

13. Tighten the distributor adjusting bolts to 16 ft. lbs. (22 Nm) and recheck the timing and the idle.

14. After everything has been rechecked, remove the service connector from the service check connector. Tuck the service check connector back behind the kick panel.

1.7L and 2.0L Engines

These engines are equipped with Distributorless Ignition Systems (DIS). No adjustment is necessary.

2.2L and 2.3L Engines

1. Before servicing the vehicle, refer to the precautions in the beginning of this section.

2. Connect a PGM tester (scan tool) to the Data Link Connector (DLC).

3. Connect a timing light to the No. 1 ignition cable.

4. Start the engine and allow it to warm up until the electric fan comes on.

5. Be sure to turn off all accessories.

6. Verify the idle speed is 650–750 rpm.

7. Point the light at the timing belt cover near the crankshaft pulley and read the timing. Correct timing is 10–14° BTDC for both automatic and manual transmissions. If necessary, loosen the distributor hold-

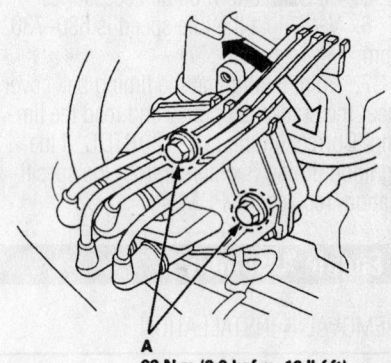

A
22 N·m (2.2 kgf·m, 16 lbf·ft)

7923FG04

Distributor hold-down bolt locations—2.3L (F23A1 and F23A4) engine

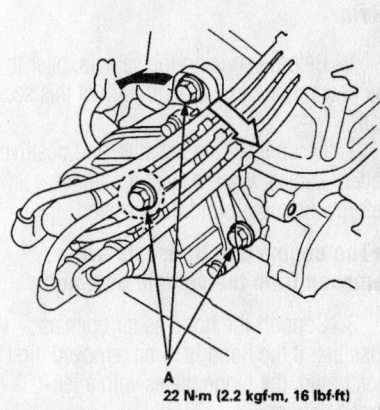

A
22 N·m (2.2 kgf·m, 16 lbf·ft)

7923FG05

Distributor hold-down bolt locations—2.3L (F23A5) engine

down bolt and rotate the distributor slightly to adjust the timing. Turn it counterclockwise to advance and clockwise to retard the timing.

8. Tighten the hold-down bolt to 16 ft. lbs. (22 Nm). Recheck the timing after the bolt is tight to confirm the correct timing.

9. Disconnect the PGM tester.

3.0L Engine

The ignition timing is only adjustable by the Powertrain Control Module (PCM), but the ignition base timing can be checked by performing the following:

1. Before servicing the vehicle, refer to the precautions in the beginning of this section.

2. Connect a PGM tester (scan tool) to the data link connector.

3. Connect a timing light to the No. 1 ignition cable.

4. Start the engine and allow it to warm up until the electric fan comes on.

5. Be sure to turn off all accessories.

6. Verify that the idle speed is 630–730 rpm.

7. Point the light at the timing belt cover near the crankshaft pulley and read the timing. Correct timing is 8–12° BTDC. If the ignition timing is different from the specification, replace the PCM.

Engine Assembly

REMOVAL & INSTALLATION

➡The original radio contains a coded anti-theft circuit. Obtain the security code number before disconnecting the battery cables.

Civic

1. Before servicing the vehicle, refer to the precautions in the beginning of this section.

2. Disconnect the negative and positive battery cables. Wait at least 3 minutes before working around the air bags.

➡The engine and transaxle are removed from the vehicle as 1 unit.

3. Support the hood as far open as possible. If the hood is to be removed, first matchmark the hinge plates with a felt-tipped marker.

4. Remove the battery from the vehicle. Unbolt and remove the battery tray.

5. Disconnect the battery and alternator cables from the underhood fuse and relay box on the right shock tower.

6. Remove the lower right kick panel to

expose the Powertrain Control Module (PCM).

7. Label and disconnect the 5 wiring harness connections from the PCM.

8. Unbolt the main wiring harness retainer from the rear of the fuse and relay box on the right side of the bulkhead. Carefully pull the grommet out of its bulkhead opening. Next, pull the PCM harness and connectors through the opening. Be careful not to damage the wiring, insulation, or connectors.

9. Relieve the fuel pressure:
   a. Loosen the fuel filler cap.
   b. Use a box-end wrench and a flare nut wrench to hold the fuel filter banjo fitting.
   c. Place a shop towel over the fuel filter to catch the fuel spray.
   d. Slowly loosen the fuel filter service bolt 1 full turn.
   e. Clean up any spilled fuel.

10. Remove or disconnect the following:
   • Intake air duct and air cleaner
   • Intake Air Temperature (IAT) sensor connector from the air cleaner case, if equipped
   • Fuel feed hose from the fuel filter
   • Fuel return hose from the fuel rail
   • Intake manifold/throttle body vacuum hoses
   • Brake booster vacuum hose
   • Evaporative emissions (EVAP) canister vacuum hose
   • Power Steering Pressure (PSP) switch and detach its clamp from the bracket below the brake booster
   • Transaxle ground cable
   • Radiator hose bracket

11. Loosen the throttle cable locknut, then disconnect the cable from the throttle body linkage. Move the cable aside without kinking it.

12. Loosen the power steering pump mounting bolts. Slip power steering belt off its pulleys. Unbolt the steering pump and move it out of the work area. Don't disconnect the hydraulic hoses.

➡Label the connectors before detaching them.

13. Remove or disconnect the following:
   • Engine wiring harness connectors at the left side of the engine compartment

14. Drain the coolant from the radiator and engine block.

15. Remove or disconnect the following:
   • Upper and lower radiator hoses
   • Heater hoses from the cylinder head

16. If equipped with a CVT transaxle,

loosen the shift cable locknut. Remove the spring clip and washers and disconnect the shift cable from its linkage. Be careful not to kink the cable or damage its boot.

17. Remove or disconnect the following:
   • Hydraulic line brackets from the top of the transaxle case, if equipped with a manual transaxle

18. Attach a chain hoist to the engine lifting brackets. Don't raise the hoist to lift the engine yet.

19. Raise the vehicle and support it safely. Remove the front wheels.

20. Remove or disconnect the following:
   • Engine splash shield

21. Drain the engine oil.

22. Drain the fluid from the transaxle.

23. Remove or disconnect the following:
   • Left front engine mount bracket from the shock tower, if equipped with air conditioning

24. Loosen the compressor idler pulley and adjusting bolt. Slip the belt around the engine mount stud to remove it.

25. Remove or disconnect the following:
   • Compressor mounting bolts to separate the compressor from its mounting plate. Move the compressor out of the work area. Do not disconnect the air conditioning refrigerant lines.

26. Remove or disconnect the following:
   • ATF cooler lines, if equipped. Plug the cooler lines to prevent fluid leakage and contamination
   • Slave cylinder from the transaxle case without disconnecting its hydraulic line, if equipped with manual transaxle
   • Front exhaust pipe from the exhaust manifold and catalytic converter. Unbolt its hanger bracket and remove the exhaust pipe.
   • Shift cable from the transaxle control shaft, if equipped with automatic transaxle
   • Shift rod and extension rod from the transaxle, if equipped with manual transaxle
   • Strut damper fork
   • Steering knuckle ball joint from the lower control arm using a ball joint separator

27. Pry the inboard CV-joints from the transaxle. Then, move the halfshafts away from the transaxle and wire them to the undercarriage of the vehicle. Tie plastic bags over the inboard CV-joints to prevent damage to the boots and splined shafts.

28. Raise the hoist slightly to take up the weight of the engine and transaxle assembly.

29. Disconnect the engine mounts in the following order:

   a. Unbolt and remove the left front engine mount.

   b. Unbolt and remove the right front engine mount and bracket assembly.

   c. Remove the rear engine mount through-bolt. Then, unbolt the rear mount bracket from the engine block.

30. If necessary, lower the vehicle slightly to gain access to the side engine and transaxle mounts. Do not release the tension on the chain hoist. The engine must be securely supported.

31. Unbolt the side engine mount bracket from the engine block bracket and mount damper.

32. Unbolt the transaxle mount bracket from the transaxle case. Then, unbolt the mount from the shock tower.

33. Raise the chain hoist to lift the engine a few inches off of its mounts.

34. Verify that all electrical, vacuum, and fuel lines have been disconnected.

35. Raise the engine and transaxle assembly and remove it from the vehicle.

To install:

➡ **Use new self-locking nuts and gaskets when installing the front exhaust pipe and when assembling the front suspension. Use new set rings on the inboard CV-joint splined shafts.**

36. Lower the engine and transaxle assembly into the vehicle.

37. Install and connect the engine and transaxle mounts and brackets. Use new self-locking nuts and color-coded bolts. At this point, only tighten the mounting nuts and bolts by hand.

38. Before installing the left front engine mount, fit the air conditioning compressor back into place and install the compressor belt. Tighten the compressor bolts to 17 ft. lbs. (24 Nm).

➡ **Failure to tighten the bolts in the proper sequence can cause excessive noise and vibration and reduce bushing life. Be sure to check that the bushings are not twisted or offset.**

39. The engine and transaxle mount and bracket fasteners must be tightened in the proper sequence with the weight of the engine resting upon them. This step is important for engine mount pre-loading. Tighten the engine mount bolts in the following sequence:

   a. Transaxle mount bolts: 47 ft. lbs.

(64 Nm); or 28 ft. lbs. (38 Nm) for CVT-equipped vehicles

   b. Side engine mount bracket nuts: 54 ft. lbs. (74 Nm)

   c. Rear mount bracket bolts: 61 ft. lbs. (83 Nm); or 43 ft. lbs. (59 Nm) for CVT-equipped vehicles

   d. Rear mount through-bolt: 43 ft. lbs. (59 Nm)

   e. Transaxle mount bracket nuts or bolts: 47 ft. lbs. (64 Nm).

   f. Transaxle mount through-bolt: 54 ft. lbs. (74 Nm).

   g. Right front mount bracket bolts: 33 ft. lbs. (44 Nm).

   h. Right front mount carrier bolts: 33 ft. lbs. (44 Nm).

   i. Left front mount: stud: 61 ft. lbs. (85 Nm); carrier bolts: 33 ft. lbs. (44 Nm); nut: 43 ft. lbs. (59 Nm).

40. Remove the chain hoist from the engine lifting hooks.

41. Install or connect the following:

- New set rings on the inboard splined shafts of each halfshaft. Check that the set ring on each inboard CV-joint clicks into place when the halfshafts are installed into the transaxle.

- Damper fork and reconnect the lower ball joint. When the weight of the vehicle is resting on its suspension, tighten the pinch bolt to 32 ft. lbs. (44 Nm) and the fork bolt to 47 ft. lbs. (65 Nm). Tighten the ball joint castle nut to 36–43 ft. lbs. (50–60 Nm). Next, tighten the castle nut only enough to install a new cotter pin.

- Slave cylinder, if equipped. Tighten the slave cylinder mounting bolts to 16 ft. lbs. (22 Nm). If the clutch hydraulic line was disconnected, air must be bled from the system.

- Transaxle shift and extension rods to the linkage at the transaxle case, if equipped. Install a new 8mm spring pin into the shift rod linkage. Then, install the retainer clip and boot. Tighten the extension rod bolt to 16 ft. lbs. (22 Nm).

- Shift cable to the control shaft, if equipped with an automatic transaxle. Use a new lockwasher and tighten the lockbolt to 10 ft. lbs. (14 Nm). Tighten the shift cable cover bolts to 16 ft. lbs. (22 Nm). Install the shift cable cover

and tighten its bolts to 16 ft. lbs. (22 Nm).

42. Install the front exhaust pipe using new self-locking nuts:

   a. If equipped with the D16Y8 engine, tighten the converter flange nuts to 16 ft. lbs. (22 Nm)

   b. Tighten the exhaust manifold nuts to 40 ft. lbs. (55 Nm).

   c. If equipped with the D16Y5 or D16Y7 engine, tighten the converter flange nuts to 25 ft. lbs. (33 Nm).

43. Install or connect the following:

- Tighten the exhaust flange bolts to 16 ft. lbs. (22 Nm).

- ATF cooler lines. If the rubber cooler lines are cracked or stressed, they must be replaced.

- Engine splash shield

44. Refill the engine with fresh oil.

45. Refill the transaxle with the proper fluid.

46. Lower the vehicle.

47. If equipped, fit the clutch hydraulic line brackets back into place. Tighten the 8mm bolts to 17 ft. lbs. (24 Nm). Tighten the 6mm bolts to 96 inch lbs. (11 Nm).

48. If equipped with a CVT transaxle, reconnect the shift cable to the linkage. Use new plastic washers and a new spring clip. Tighten the locknut to 22 ft. lbs. (29 Nm).

49. Adjust the alternator and air conditioning compressor belt tensions.

50. Install or connect the following:

- Upper and lower radiator hoses and the heater hoses

- Power steering pump into its mounts. Adjust the pump belt tension, then tighten the mounting bolts to 17 ft. lbs. (24 Nm).

- PSP switch connector and attach its harness clamp

- Intake manifold/throttle body vacuum hoses

- Brake booster vacuum hose

- EVAP canister vacuum hose

- Fuel line fittings to the fuel filter and fuel rail. Use new sealing washers. Tighten the banjo fittings to 25 ft. lbs. (33 Nm), and the service bolts to 11 ft. lbs. (15 Nm). Don't overtighten the fittings

- Throttle cable and adjust its deflection to 10–12mm (0.39–0.47 in.).

51. Feed the PCM harness through the hole in the bulkhead. Apply sealant to the grommet, then install the retainer.

52. Install or connect the following:

- Engine wiring harness and ground

Timing belt service is covered in Section 3 of this manual

cables that were disconnected during removal. Be sure the grounds are free of corrosion to ensure good contact.

- Fuse and relay box back into position
- Battery and alternator cables
- Air cleaner case and air intake duct
- IAT connector
- 5 PCM connectors and kick panel
- Battery tray and the battery

53. Verify that all wiring harnesses and grounds, vacuum lines, fuel lines have been reconnected.

54. Refill the radiator with fresh coolant.

55. If it was removed, install the hood. Reconnect the windshield washer tubing. After installation, check to be sure that the hood, fender, and grille panel gaps are equal.

56. Reconnect the positive and negative battery cables.

57. Turn the ignition switch to the **ON** position, but don't start the engine. Then, turn the ignition **OFF**. Repeat this procedure 2 or 3 times and check for fuel leaks.

58. Start the engine and allow it to warm up to its normal operating temperature.

59. Bleed the air from the cooling system with the heater valve open.

60. Check the throttle cable deflection and operation.

61. Check and adjust the ignition timing.

62. Shut the engine off and check the drive belt adjustments.

63. Check all fluid levels and top up as necessary.

64. Check and adjust the front wheel alignment.

65. Road test the vehicle.

Prelude

1. Before servicing the vehicle, refer to the precautions in the beginning of this section.

2. Secure the hood as far open as possible.

3. Remove or disconnect the following:
- Negative battery cable, then the positive battery cable
- Radiator cap

4. Raise and safely support the vehicle. Remove the front wheels and the engine splash shield.

5. Drain the engine coolant into a sealable container.

6. Drain the transaxle fluid into a sealable container. Install the drain plug with a new gasket.

7. Lower the vehicle to a working level.

8. Remove or disconnect the following:

- Air intake duct and the air cleaner case
- PAIR vacuum tank and bracket
- Battery and the battery base
- Battery cable and starter cable harnesses from the body

9. Relieve the pressure from the fuel system.

❊❊ CAUTION

The fuel injection system remains under pressure after the engine has been turned OFF. Properly relieve fuel pressure before disconnecting any fuel lines. Failure to do so may result in fire or personal injury.

10. Remove or disconnect the following:
- Fuel feed hose from the fuel rail and the fuel return line from the fuel pressure regulator
- Injector resistor connector on the left side of the engine compartment
- Throttle cable by loosening the locknut, then slip the cable end out of the throttle linkage. Take care not to bend the cable when removing it. Always replace any kinked cable with a new one.
- Engine wiring harness connectors, terminal, and clamps on the right side of the engine
- Power cable from the under-hood fuse/relay box
- Brake booster vacuum hose and emissions control vacuum tubes from the intake manifold
- Cruise control actuator electrical connector and vacuum tube, then the actuator
- Engine ground cable from the body side
- Power steering pump drive belt, then the pump
- Air conditioning condenser fan, then install a protector plate on the radiator
- Alternator mounting bolt, nut, and adjusting nut, then the alternator drive belt
- Air conditioning compressor electrical connector
- Compressor without disconnecting the air conditioning hoses. Support the compressor with a strong wire out of the way.
- Upper and lower radiator hoses, and the heater hoses from the engine
- Transaxle ground cable
- Cooler hoses, if equipped with an automatic transaxle

11. If equipped with a manual transaxle perform the following:

a. Disconnect the shift cable and the select cable from the transaxle. Do not bend the cables when removing them. Replace any kinked cable with a new one.

b. Remove the clutch slave cylinder and the pipe/hose assembly. Do not depress the clutch pedal once the slave cylinder has been removed.

c. Remove the clutch damper assembly.

12. Remove or disconnect the following:
- Vehicle Speed Sensor (VSS)/Power Steering speed sensor assembly. Do not disconnect the hoses
- Nuts attaching the exhaust pipe to the exhaust manifold and the catalytic converter
- Bolts from the exhaust pipe hanger, then the exhaust pipe and discard the gaskets

13. If equipped with an automatic transaxle, remove the shift cable cover, then disconnect the shift cable. Do not bend the cable and replace the cable if it becomes kinked.

14. Remove or disconnect the following:
- Left and the right side damper forks
- Lower ball joints from the lower control arms

15. Pry the halfshafts from the transaxle. Cover the inner CV-joints with plastic bags to protect them.

16. Swing the halfshafts under the fender out of the way.

17. Attach an engine hoist to the engine lifting points and raise the hoist to remove all slack from the chain.

18. Remove or disconnect the following:
- Rear engine mount bracket
- Front engine mount bracket
- Left side engine mount
- Transaxle mount and the mount bracket

19. Check that the engine is completely free of vacuum hoses, fuel and coolant hoses, and electrical wiring.

20. Slowly raise the engine approximately 6 in. (150mm). Check once again that all hoses and wires have been disconnected from the engine.

21. Raise the engine all the way and remove it from the vehicle.

22. Remove the transaxle.

23. If equipped with a manual transaxle, remove the clutch cover (pressure plate) and clutch disc.

24. Mount the engine on an engine stand, making sure the mounting bolts are tight. If an engine stand is not available,

support the engine in an upright position with blocks. Never leave an engine hanging from a lift or hoist.

To install:

25. Install or connect the following:
   - Clutch disc and pressure plate to the flywheel for manual transaxle vehicles
   - Transaxle
   - Engine into position and lower it into the car, aligning the mounts and bushings.

➡ **When installing the engine mounts and vibration dampers in the following steps, they must be tightened to the correct tension in the correct order if they are to damp vibration properly.**

   - Side engine mount and the through-bolt. Do not tighten the through-bolt at this time
   - Nut and bolt attaching the side mount to the engine: 40 ft. lbs. (55 Nm)
   - Transaxle mount and through-bolt. Do not tighten the through-bolt at this time
   - Rear engine mount and new bolts attaching the mount to the engine: 40 ft. lbs. (55 Nm)
   - New rear engine mount through-bolt: 47 ft. lbs. (65 Nm)
   - Front mount and the 3 bolts attaching the mount to the engine assembly, only snug the bolts in place
   - New through-bolt to the front mount: 47 ft. lbs. (65 Nm)
   - Nuts to the transaxle mount: 28 ft. lbs. (39 Nm)

26. Tighten the side engine mount through-bolt to 47 ft. lbs. (65 Nm).

27. Tighten the transaxle mount through-bolt to 47 ft. lbs. (65 Nm).

28. Tighten the 3 bolts attaching the front mount to the engine to 28 ft. lbs. (39 Nm).

29. Remove the hoist equipment from the engine.

30. Install or connect the following:
   - New spring clips to the inner CV-joints
   - Halfshafts into the transaxle. Be sure that the inner joint spring clips click into place
   - Lower ball joints to the lower control arms. Tighten the nuts to 36–43 ft. lbs. (50–60 Nm). Install a new cotter pin to the ball joint stud.

31. Using new self-locking bolts, attach the damper forks to the struts and tighten to

32 ft. lbs. (44 Nm). Tighten the new nut and bolt attaching the damper fork to the lower control arm to 47 ft. lbs. (65 Nm).

32. If equipped with an automatic transaxle, connect the shift cable to the transaxle. Install a new lockwasher and tighten the attaching bolt to 84 inch lbs. (10 Nm). Install the shift cable cover. Tighten the shift cable cover attaching bolts to 13 ft. lbs. (18 Nm).

33. Install or connect the following:
   - Exhaust pipe with new gaskets
   - New nuts attaching the exhaust pipe to the exhaust manifold: 40 ft. lbs. (55 Nm)
   - New nuts attaching the exhaust pipe to the catalytic converter: 25 ft. lbs. (34 Nm)
   - New attaching bolts to the exhaust pipe hanger: 13 ft. lbs. (18 Nm)
   - VSS, connect the electrical connector, and tighten the mounting bolt to 13 ft. lbs. (18 Nm)

34. If equipped with a manual transaxle perform the following:

   a. Install the clutch damper assembly and tighten the attaching bolts to 16 ft. lbs. (22 Nm).

   b. Install the clutch slave cylinder and the pipe/hose assembly and tighten the slave cylinder mounting bolts to 16 ft. lbs. (22 Nm).

   c. Connect the shift cable and the select cable to the transaxle. Adjust the shift cable and select cable.

35. Install or connect the following:
   - Cooler hoses, if equipped with an automatic transaxle
   - Transaxle ground cable
   - Upper and lower radiator hoses and heater hoses to the engine
   - Air conditioning compressor, connect the electrical connector. Tighten the mounting bolts to 16 ft. lbs. (22 Nm).
   - Alternator drive belt, then adjust

36. Remove the protector plate from the radiator and install the air conditioning condenser fan.

37. Install or connect the following:
   - Power steering pump and drive belt. Adjust the drive belt tension, then tighten the attaching nuts and bolts to 16 ft. lbs. (22 Nm).
   - Engine ground cable to the body
   - Cruise control actuator, electrical connector and vacuum tube. Tighten the mounting bolts to 84 inch lbs. (10 Nm)

   - Brake booster vacuum hose and the emissions control vacuum tubes to the intake manifold
   - Engine wiring harness connectors, terminal, and clamps
   - Throttle cable, then adjust
   - Injector resistor connector on the left side of the engine compartment
   - Fuel return hose to the regulator
   - Fuel feed hose to the fuel rail with new washers. Tighten the cap nut to 16 ft. lbs. (22 Nm)
   - Battery and starter cables to the body
   - Battery base and the battery. Tighten the battery base attaching bolts to 16 ft. lbs. (22 Nm)
   - PAIR vacuum tank and bracket. Tighten the mounting bolts to 96 inch lbs. (10 Nm)
   - Air cleaner duct and housing
   - Engine splash shield and the front wheels

38. Lower the vehicle.

39. Fill the engine with oil and the transaxle with fluid.

40. Fill and bleed the air from the cooling system.

41. Connect the positive, then the negative battery cable and enter the radio security code.

42. Switch the ignition **ON** but do not engage the starter. The fuel pump should run for approximately 2 seconds, building pressure within the lines. Switch the ignition **OFF**, then **ON** 2 or 3 more times to build full system pressure. Check for fuel leaks.

43. Start the engine, allowing it to idle. Check the hoses and lines carefully for any sign of leakage.

44. Check the timing and idle speed.

45. After the engine has warmed up fully and the fan(s) have come on at least once, recheck the engine for fluid leaks. Switch the engine OFF.

46. Adjust the belts and throttle cable as necessary.

47. If equipped with 4WS, start the engine and turn the steering wheel lock-to-lock to reset the 4WS control unit.

48. Road test the vehicle, then loosen and retighten the 3 bolts attaching the front engine mount to the engine. Tighten the bolts to 28 ft. lbs. (39 Nm).

Accord

1. Before servicing the vehicle, refer to the precautions in the beginning of this section.

Heater Core replacement is covered in Section 2 of this manual

2. Obtain the anti-theft code for the radio, then disconnect the battery cables. Be sure to disconnect the negative cable first.

3. Remove the air intake duct.

4. Secure the hood in the open position with a long prop rod such as P/N 74145-S84-A00.

5. Remove or disconnect the following:
- Both battery cables and the connector from the underhood relay box
- Battery and tray, on the 3.0L engine
- Bolt securing the relay box to the body
- Accelerator and cruise control cables from the throttle body and bracket

6. Properly relieve the fuel system pressure.

7. Remove or disconnect the following:
- Fuel hoses from the fuel rail
- Brake booster vacuum and evaporative emissions (EVAP) canister hoses

- Vacuum hose from the canister
- Hose securing the power steering hose on the engine
- Power steering pump belt, then remove the pump and position it out of the way. Use wire if necessary.
- Powertrain Control Module (PCM) connectors from the control module. Remove the grommet and pull the connectors through.
- Wiring harness connectors at the right side of the engine compartment for 2.3L engines and on the left side for the 3.0L engine.

8. On the 2.3L engine, remove the starter cable (A) and clamp (B). Remove the ground cable (C) and back-up light switch connectors (D). On the 3.0L engine, remove

the starter wiring from the engine compartment attaching points.

9. On vehicles with a manual transaxle, disconnect the shift and select cables from the transaxle. Remove the slave cylinder mounting bolts and position the cylinder out of the way. Be sure not to bend the line.

10. Remove or disconnect the following:
- Rear engine mount through-bolt and stiffener
- Front engine mount bracket mounting bolts and loosen the through-bolt
- Radiator cap

11. Raise and safely support the vehicle.

12. Remove or disconnect the following:
- Front tires
- Engine under cover

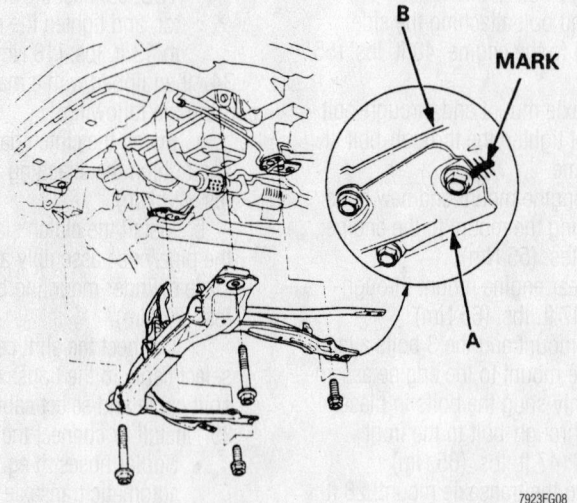

Be sure to mark the location of the front beams (A) on the rear beams (B) before removing the subframe—2.3L Accord

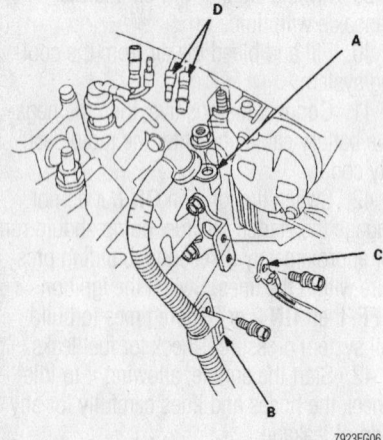

Starter cable, clamp, ground cable and back-up light switch connector locations— 2.3L Accord

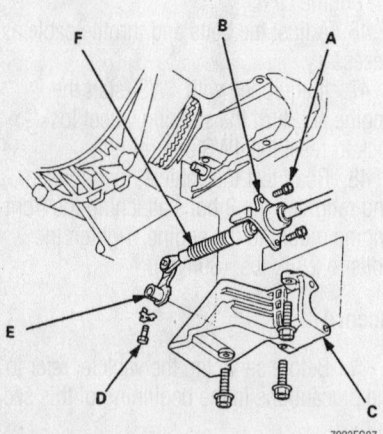

Automatic transaxle linkage components—2.3L Accord

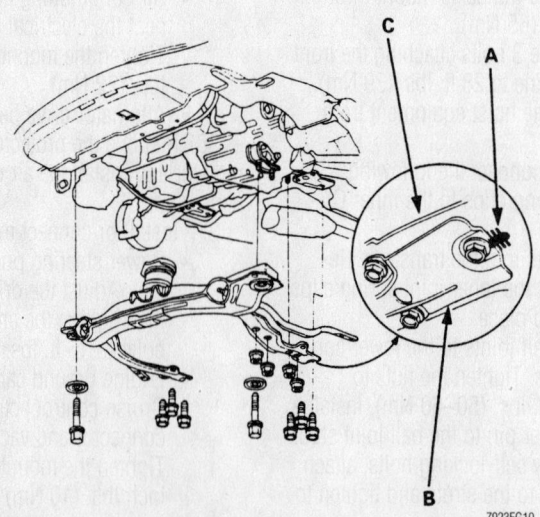

Mark the location of the front beams (A) on the rear beams (B) before removing the subframe—3.0L Accord

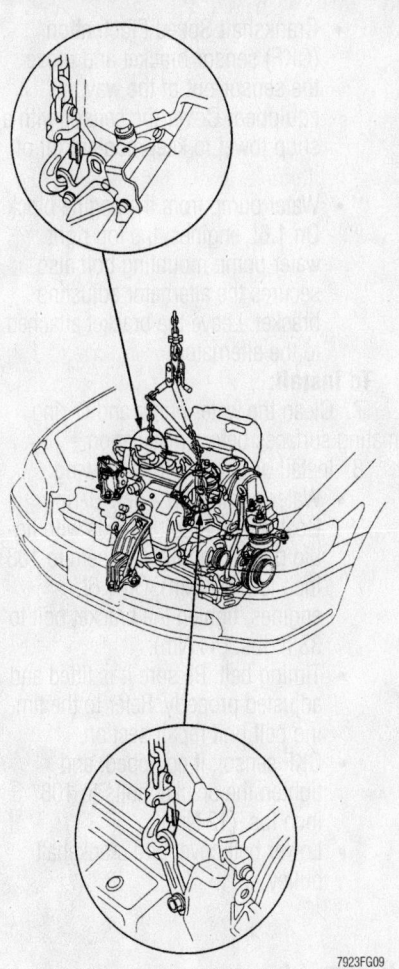

Engine lifting points—2.3L Accord

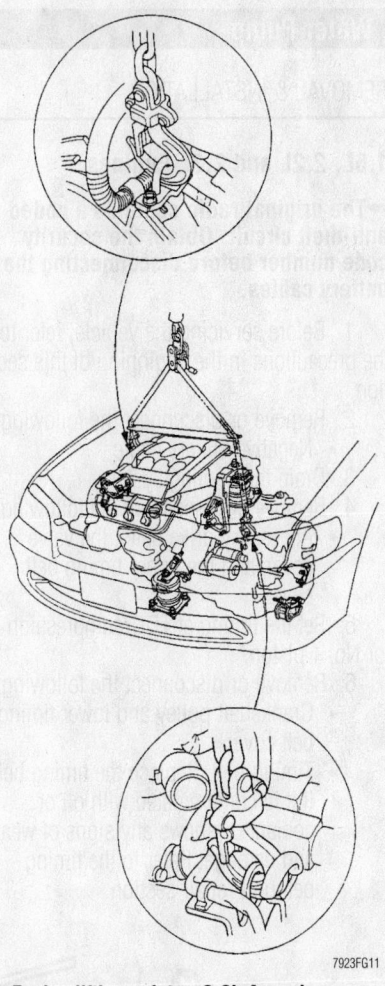

Engine lifting points—3.0L Accord

13. Loosen the radiator drain plug and drain the coolant.

14. Drain the transaxle oil or fluid, then reinstall the plug using a new washer.

15. Drain the engine oil, then reinstall the plug using a new washer.

16. Lower the vehicle and remove the upper and lower radiator hoses and heater hoses from the engine.

17. On vehicles with an automatic trans-axle, disconnect the ATF fluid cooler lines.

18. Remove the air conditioning com-pressor from the engine and position it to the side without disconnecting the hoses.

19. Raise the vehicle and remove the front exhaust pipe.

20. On vehicles with automatic transaxle, remove the 2 bolts (A) for the shift cable holder (B), then remove the shift cable cover (C). To prevent damage to the linkage, be sure to remove the shift cable holder before removing the bolts for the cover.

21. Remove or disconnect the following:
- Lockbolt (D) from the control lever

(E), then the shift cable (F) with the control lever
- Through-bolt securing the bottom of the shock absorber to the control arm
- Halfshafts
- Rear engine mounting bracket
- 2 flange bolts from each of the radius rods

22. Mark the location of the front beams (A) on the rear beams (B). Remove the 4 bolts and the subframe.

23. Lower the vehicle about half way and attach a chain hoist to the engine lifting points as shown. Apply slight upward pres-sure to the engine/transaxle assembly.

24. Remove the remaining engine and transaxle mounting brackets.

25. Lower the engine about 6 in. (150mm) and check that the engine/transaxle is free of any hoses, cables or wiring.

26. Lower the assembly completely and remove it from under the vehicle.

To install:

27. Lift the engine into position and install the engine mounting brackets. Tighten the retainers as follows:
   a. On the 2.3L, tighten the engine mounting bolts and nuts to 40 ft. lbs. (54 Nm).
   b. On the 3.0L, tighten the bolts to 28 ft. lbs. (38 Nm).

28. On the 3.0L engine, install the air conditioning compressor. Tighten the bolts to 16 ft. lbs. (22 Nm).

29. Install the transaxle mounting bracket, and tighten the retainers as follows:
   a. On the 2.3L engine, tighten the nuts to 28 ft. lbs. (38 Nm) and the through-bolt to 40 ft. lbs. (54 Nm).
   b. On the 3.0L engine, tighten the bolts to 28 ft. lbs. (38 Nm).

30. Install the sub-frame in its original position, and tighten the retainers as follows:
   a. On the 2.3L engine, tighten the rear bolts to 47 ft. lbs. (64 Nm) and the front bolts to 76 ft. lbs. (103 Nm).
   b. On the 3.0L engine, tighten the rear bolts to 40 ft. lbs. (54 Nm), front bolts to 76 ft. lbs. (103 Nm) and the nuts to 28 ft. lbs. (38 Nm).

31. On the 2.3L engine, install or con-nect the following:
- Radius rod bolts: 119 ft. lbs. (162 Nm)
- Rear mount bracket: 40 ft. lbs. (54 Nm)
- Stiffener. Tighten the through-bolt to 47 ft. lbs. (64 Nm) for manual transaxles or the nut and bolt to 28 ft. lbs. (38 Nm) for automatic transaxles.
- 3 front mounting bracket bolts: 28 ft. lbs. (38 Nm). Then, tighten the through-bolt to 47 ft. lbs. (64 Nm).
- Air conditioning compressor: 16 ft. lbs. (22 Nm)

32. On the 3.0L engine, install or con-nect the following:
- Radius rod bolts: 119 ft. lbs. (162 Nm)
- Front mounting bracket support nut: 40 ft. lbs. (54 Nm)
- Rear mounting bracket nut and bolt. Tighten the nut to 40 ft. lbs. (54 Nm) and the bolt to 28 ft. lbs. (38 Nm).
- Side mounting bracket. Tighten the bolts to 40 ft. lbs. (54 Nm) and the through-bolt to 40 ft. lbs. (54 Nm).
- Exhaust system

Brake service is covered in Section 4 of this manual

- Shift linkage, if equipped with an automatic transaxle

33. The remainder of the installation is the reverse of the removal.

34. Refill and bleed the cooling system.

✳✳ WARNING

Operating the engine without the proper amount and type of engine oil will result in severe engine damage.

35. Fill the engine with the correct amount of oil.

36. Install the battery if removed. Start the engine and check for leaks.

S2000

1. Before servicing the vehicle, refer to the precautions in the beginning of this section.

2. Drain the cooling system.

3. Relieve the fuel system pressure.

4. Remove or disconnect the following:
   - Battery
   - Front wheels
   - Transmission
   - Engine Control Module (ECM) connectors and the main wire harness connector. Pass the connectors through the cowl panel.
   - Vacuum tank
   - Throttle cable
   - Electrical Power Steering (EPS) control unit
   - Battery cable at the main underhood fuse/relay box
   - Battery cable at the auxiliary fuse box
   - Ground cable and harness clamps
   - Fuel lines
   - Brake booster vacuum line
   - Evaporative Emissions (EVAP) canister hose
   - Front motor mount and bracket
   - Heater hoses
   - Radiator hoses
   - Left motor mount
   - Right motor mount bracket

5. Carefully raise the engine out of the vehicle.

To install:

6. Installation is the reverse of the removal procedure, while using the following torque values:
   - Right motor mount bracket bolts: 28 ft. lbs. (38 Nm)
   - Motor mount nuts: 40 ft. lbs. (54 Nm)
   - Front motor mount bolts: 16 ft. lbs. (22 Nm)

Water Pump

REMOVAL & INSTALLATION

1.6L, 2.2L and 2.3L Engines

➡**The original radio contains a coded anti-theft circuit. Obtain the security code number before disconnecting the battery cables.**

1. Before servicing the vehicle, refer to the precautions in the beginning of this section.

2. Remove or disconnect the following:
   - Negative battery cable

3. Drain the cooling system.

4. Remove or disconnect the following:
   - Accessory drive belts, the valve cover, and the upper timing belt cover

5. Set the timing at TDC/compression for No. 1 piston.

6. Remove or disconnect the following:
   - Crankshaft pulley and lower timing belt cover
   - Timing belt. Replace the timing belt if it is contaminated with oil or coolant or shows any signs of wear and damage. Refer to the timing belt unit repair section

- Crankshaft Speed Fluctuation (CKF) sensor bracket and move the sensor out of the way, if equipped. Cover the sensor with a shop towel to keep coolant off of it.
- Water pump from the engine block. On 1.6L engines, the top right water pump mounting bolt also secures the alternator adjusting bracket. Leave the bracket attached to the alternator.

To install:

7. Clean the water pump and O-ring mating surfaces before installation.

8. Install or connect the following:
   - Water pump with a new O-ring. Coat only the bolt threads with liquid gasket and tighten them to 108 inch lbs. (12 Nm). On 1.6L engines, tighten the bracket bolt to 33 ft. lbs. (44 Nm).
   - Timing belt. Be sure it is fitted and adjusted properly. Refer to the timing belt unit repair section.
   - CKF sensor, if equipped, and tighten the bracket bolts to 108 inch lbs. (12 Nm).
   - Lower belt cover and crankshaft pulley

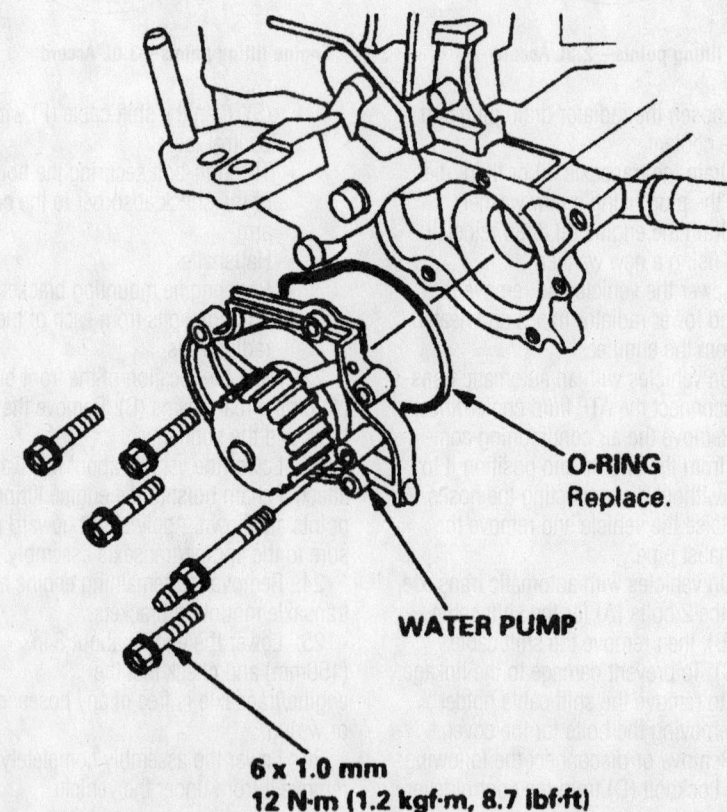

O-RING
Replace.

WATER PUMP

6 x 1.0 mm
12 N·m (1.2 kgf·m, 8.7 lbf·ft)

Water pump—2.2L and 2.3L engines

7923FG12

- Upper timing belt cover, the valve cover, and the accessory drive belts

9. Be sure the cooling system drain plug is closed. Refill and bleed the cooling system.

10. Connect the negative battery cable and enter the radio security code.

11. Start the engine, allow it to reach normal operating temperature, and check for coolant leaks.

12. If equipped with 4WS, and turn the steering wheel lock-to-lock to reset the 4WS control unit.

2.0L Engine

1. Before servicing the vehicle, refer to the precautions in the beginning of this section.

2. Drain the cooling system.

3. Remove or disconnect the following:
- Negative battery cable
- Accessory drive belt
- Water pump pulley
- Water pump

To install:

4. Install or connect the following:
- Water pump with a new O ring seal. Tighten the 8mm bolts to 16 ft. lbs. (22 Nm) and the 10mm bolt to 33 ft. lbs. (44 Nm).

- Water pump pulley and tighten the bolts to 10 ft. lbs. (14 Nm)
- Accessory drive belt
- Negative battery cable

5. Fill the cooling system.

6. Start the engine and check for leaks.

3.0L Engine

1. Before servicing the vehicle, refer to the precautions in the beginning of this section.

2. Remove or disconnect the following:
- Timing belt. Refer to the timing belt unit repair section
- Timing belt tensioner
- 5 water pump mounting bolts, then remove the pump and seal

To install:

3. Clean the seal groove and mating surfaces.

4. Install or connect the following:
- Water pump, with a new seal. Tighten the bolts to 104 inch lbs. (12 Nm).
- Timing belt tensioner
- Timing belt. Refer to the timing belt unit repair section

5. Refill the cooling system.

6. Start the engine and check for leaks.

7. Top off the cooling system if necessary after the engine has cooled.

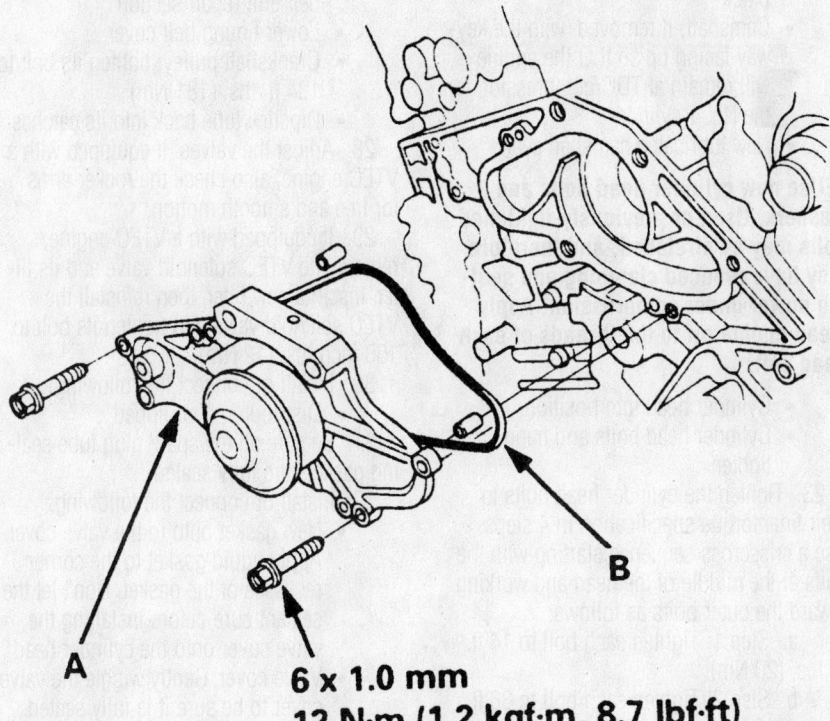

6 x 1.0 mm
12 N·m (1.2 kgf·m, 8.7 lbf·ft)

7923FG14

Exploded view of the water pump mounting—3.0L engine

Cylinder Head

REMOVAL & INSTALLATION

➡**The radio may contain a coded theft protection circuit. Always obtain the code number before disconnecting the battery. If the vehicle is equipped with 4WS, the steering control unit is shut down when the battery is disconnected. After connecting the battery, turn the steering wheel lock-to-lock to reset the steering control unit.**

1.6L (D16Y5, D16Y7, D16Y8) and 1.7L (D17A1, D17A2) Engines

1. Before servicing the vehicle, refer to the precautions in the beginning of this section.

2. Be sure the cylinder head is cool to the touch before beginning the removal procedure. The coolant temperature must be below 100°F (38°C).

3. Remove or disconnect the following:
- Negative battery cable

4. Drain the cooling system.

5. Remove or disconnect the following:

➡**Label the wires before disconnecting them.**

- Ignition wires
- Air intake duct and the air cleaner assembly

6. Relieve the fuel pressure.

7. Clean up any fuel that may have spilled on the engine or intake manifold.

8. Remove or disconnect the following:
- Upper radiator hose from the coolant inlet
- Coolant bypass hoses and the heater hose from the intake manifold
- Power steering pump belt
- Power steering pump from its mounting bracket and lift the power steering reservoir from its mount. Move the pump and reservoir out of the work area and secure them. Don't disconnect the hydraulic lines.

9. Place a block of wood on the pad of a floor jack. Place the floor jack under the engine for support.

10. Remove or disconnect the following:
- Left-front engine mount bracket, if equipped with air conditioning

➡**Slip the air conditioning compressor belt around the engine mount to remove it.**

For complete Engine Mechanical specifications, see Section 1 of this manual

- Air conditioning compressor belt
- Alternator belt

11. Be sure the engine is supported with the padded floor jack. Loosen the nuts from left side engine mount. Remove the engine mount bracket.

12. Remove or disconnect the following:
- Valve cover and the upper timing belt cover
- Crankshaft pulley and the lower timing belt cover
- Dipstick tube from its catches on the timing cover
- Timing belt. Refer to the timing belt unit repair section

13. With the timing belt removed, inspect the water pump and replace it if necessary.

14. Remove or disconnect the following:
- Distributor, if equipped
- Camshaft sprocket
- Fuel lines from the intake manifold fuel rail. Immediately plug the lines to prevent fuel leakage and contamination
- Throttle cable from the linkage by first loosening its locknut, then slipping it out of its holder.

➡**Label the electrical connectors, before disconnecting them.**

- Fuel injector wiring harness connectors
- VTEC solenoid valve and pressure switch connectors, if equipped
- Idle Air Control (IAC) valve connector
- Throttle Position (TP) sensor connector
- Exhaust Gas Recirculation (EGR) valve lift sensor connectors, if equipped
- Engine Coolant Temperature (ECT) sensor, switch, and gauge sender connectors
- Manifold Absolute Pressure (MAP) sensor connector
- Primary and secondary Heated Oxygen Sensor (HO2S) connectors
- Vacuum hoses and Positive Crankcase Ventilation (PCV) hose from the intake manifold and throttle body
- EVAP and breather hoses from the intake manifold
- Intake manifold together with the throttle body and plenum
- Exhaust manifold
- Power steering pump bracket

15. Loosen the cylinder head bolts in a 3-step crisscross pattern in the reverse order of the tightening sequence. Start with the outermost bolts and work toward the middle of the cylinder head. Loosen the bolts in the reverse order of installation.

16. Remove the cylinder head. If the head sticks to the engine block, tap it with a plastic or wooden mallet.

17. Inspect the cylinder head for warpage and cracking. Repair, machine, or replace as necessary. The warpage limit is 0.002 in. (0.05mm). Standard cylinder head height is 3.659–3.663 in. (92.95–93.05mm).

18. Remove the old cylinder head gasket and thoroughly clean the mating surfaces.

19. Cover the engine block with a sheet of plastic to keep out dust and foreign objects.

To install:

➡**Use new O-ring, seals, and gaskets when installing the cylinder head and its components.**

20. Be sure the cylinder head and the engine block surfaces are clean, level, and straight.

21. Be sure the cylinder head dowel pins and control orifice are aligned. Clean the oil control orifice and reinstall it with a new O-ring.

22. Install or connect the following:
- New head gasket onto the engine block
- Camshaft, if removed, with the keyway facing up so that the engine will remain at TDC/compression for the No. 1 cylinder
- New lubricated camshaft seal

➡**Use new cylinder head bolts and washers. Used or previously-tightened bolts may be stretched, and therefore they have reduced clamping and sealing power under compression. Apply clean engine oil to the threads of each head bolt.**

- Cylinder head into position
- Cylinder head bolts and hand-tighten

23. Tighten the cylinder head bolts to their final torque specification in 4 steps. Use a crisscross sequence starting with the bolts at the middle of the head and working toward the outer bolts as follows:
   a. Step 1: Tighten each bolt to 14 ft. lbs. (20 Nm)
   b. Step 2: Tighten each bolt to 36 ft. lbs. (49 Nm)
   c. Step 3: Tighten each bolt to 49 ft. lbs. (67 Nm)
   d. Step 4: Retighten only the 2 center bolts to 49 ft. lbs. (67 Nm)

24. Apply oil to the camshaft sprocket bolt. Install the sprocket with the UP mark and the keyway pointing straight up. Tighten the sprocket bolt to 27 ft. lbs. (37 Nm).

25. Install or connect the following:
- Intake manifold with a new gasket, and tighten the nuts in a crisscross pattern in 2 or 3 steps to 17 ft. lbs. (24 Nm) starting with the inner nuts
- Bolts that secure the intake manifold to its bracket: 17 ft. lbs. (24 Nm)
- Power steering pump bracket: 33 ft. lbs. (44 Nm)
- Exhaust manifold with a new gasket. Apply anti-seize paste to the studs, and tighten the nuts to 23 ft. lbs. (31 Nm) in a crisscross sequence
- Exhaust manifold to the front exhaust pipe. Tighten the self-locking nuts to 25 ft. lbs. (33 Nm). On vehicles with the D16Y8 engine, tighten the nuts to 40 ft. lbs. (55 Nm)

26. Verify that the engine is at TDC/compression for the No. 1 cylinder.

27. Install or connect the following:
- Timing belt. After the timing belt has been properly tensioned, tighten the adjusting bolt to 33 ft. lbs. (44 Nm). Refer to the timing belt unit repair section
- Lower timing belt cover
- Crankshaft pulley; tighten its bolt to 134 ft. lbs. (181 Nm)
- Dipstick tube back into its catches

28. Adjust the valves. If equipped with a VTEC engine, also check the rocker arms for free and smooth motion.

29. If equipped with a VTEC engine, remove the VTEC solenoid valve and its filter. Install a new filter, then reinstall the VTEC solenoid valve and tighten its bolt to 108 inch lbs. (12 Nm).

30. Install or connect the following:
- Distributor, if equipped

31. Be sure all the spark plug tube sealing gaskets are fully seated.

32. Install or connect the following:
- New gasket onto to the valve cover. Apply liquid gasket to the corner recesses of the gasket. Don't let the sealant cure before installing the valve cover onto the cylinder head
- Valve cover. Gently wiggle the valve cover to be sure it is fully seated. Tighten the valve cover bolts in a crisscross pattern to 84 inch lbs. (10 Nm)
- New spark plugs
- Ignition wires
- Upper radiator hose, heater hoses,

and intake manifold coolant bypass hoses

- Intake manifold vacuum lines, PCV, EVAP canister, and breather hoses
- Fuel lines to the fuel rail. Use new sealing washers on the banjo fitting. Carefully tighten the banjo fitting to 21 ft. lbs. (28 Nm) for the D16Y5 engine, or to 16 ft. lbs. (22 Nm) for all other engines. Tighten the service bolt to 10–11 ft. lbs. (13–15 Nm)
- Throttle cable. Adjust its tension so the cable has a deflection of 10–12mm (0.39–0.47 in.)

33. Installation of the remaining components is the reverse of removal.

1.6L (B16A2) Engine

1. Before servicing the vehicle, refer to the precautions in the beginning of this section.

2. Before beginning the cylinder head removal procedure, be sure the engine temperature is below 100°F (38°C). To prevent warping, the cylinder head should be removed when the engine is cold.

3. Remove or disconnect the following:
- Negative battery cable

➡**Label the wires before disconnecting them.**

- Ignition wires

4. Drain the engine coolant. Remove the radiator cap to speed draining.

5. Remove or disconnect the following:
- Strut brace
- Intake air duct and the breather hose

6. Relieve the fuel pressure as follows:

a. Loosen the fuel filler cap.

b. Hold the fuel filter banjo bolt with a back-up wrench. Hold the fuel filter service bolt with a box end wrench.

c. Place a shop rag over the fuel filter to absorb fuel spray.

d. Slowly loosen the fuel filter service bolt 1 complete turn.

7. Clean up any fuel that may have spilled on the engine or intake manifold.

8. Remove or disconnect the following:
- Upper radiator hose from the coolant inlet
- Coolant bypass hoses and the heater hose from the intake manifold
- Power steering pump belt
- Power steering pump from its mounting bracket and lift the power

steering reservoir from its mount Move the pump and reservoir out of the work area and secure them. Don't disconnect the hydraulic lines.

9. Place a block of wood on the pad of a floor jack. Place the floor jack under the engine for support.

10. Remove or disconnect the following:
- Left-front engine mount bracket, if equipped with air conditioning

➡**Slip the air conditioning compressor belt around the engine mount to remove it.**

- Air conditioning compressor drive belt
- Alternator belt

➡**Be sure the engine is supported with the padded floor jack. Loosen the left side engine mount nuts. Remove the engine mount bracket.**

- Valve cover and the upper timing belt cover
- Crankshaft pulley and the lower timing belt cover
- Timing belt. Refer to the timing belt unit repair section

11. With the timing belt removed, inspect the water pump and replace it if necessary.

12. Remove or disconnect the following:
- Distributor from the cylinder head as an assembly

- Fuel lines from the intake manifold fuel rail. Immediately plug the lines to prevent fuel leakage and contamination.
- Throttle cable from the linkage by first loosening its locknut, then slipping it out of its holder.

➡**Label all connectors and vacuum lines before disconnecting them.**

- Fuel injector wiring harness connectors
- VTEC solenoid valve and pressure switch connectors
- Idle Air Control (IAC) valve connector
- Throttle Position (TP) sensor connector
- Engine Coolant Temperature (ECT) sensor, switch, and gauge sender connectors
- Manifold Absolute Pressure (MAP) sensor connector
- Primary Heated Oxygen Sensor (HO₂S) connector
- Vacuum hoses and Positive Crankcase Ventilation (PCV) hose from the intake manifold and throttle body
- Evaporative emissions (EVAP) and breather hoses from the intake manifold

13. Loosen the intake manifold nuts in a crisscross sequence. Then, remove the

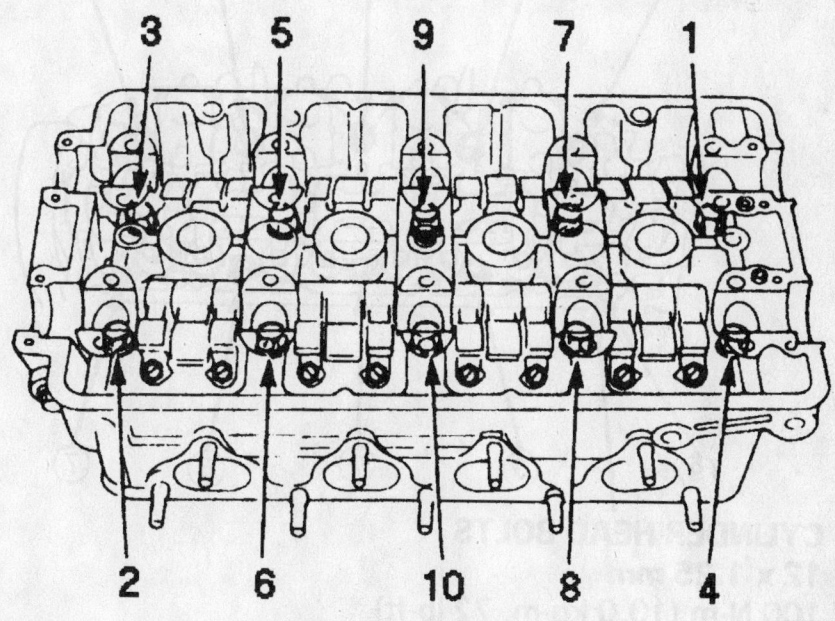

Cylinder head bolt loosening sequence—1.6L (B16A2) engines

7923FG16

For Accessory Drive Belt illustrations, see Section 1 of this manual

intake manifold together with the throttle body and plenum.

14. Remove or disconnect the following:
   • Exhaust manifold heat shield
   • Exhaust manifold nuts in a criss-cross sequence
   • Exhaust manifold

➡ Be careful not to damage the oxygen sensors when removing the manifold. Cover the front exhaust pipe flange with a shop towel to keep dirt out

   • Power steering pump bracket
   • Camshaft pulleys and back cover
   • Camshaft holder plate bolts in a crisscross sequence working toward the middle of the cylinder head.
15. Loosen the valve adjusting screws.
16. Remove or disconnect the following:
   • Camshaft holder plates and holders from the cylinder head. The holder bolts will keep the components together. Note the positions of each camshaft holder for reassembly.
   • Camshafts from the cylinder head. Mark the exhaust and intake camshafts so that they will not be confused.
   • Cylinder head bolts in a 3-step

crisscross pattern. Start with the outermost bolts and work toward the middle of the cylinder head.
   • Cylinder head. If the head sticks to the engine block, tap it with a plastic-faced or wooden mallet.
17. Inspect the cylinder head for warpage and cracking. Repair, machine, or replace as necessary. The warpage limit is 0.002 in. (0.05mm). Standard cylinder head height is 5.589–5.593 in. (141.95–142.05mm).

To install:

➡ Use new O-ring, seals, and gaskets when installing the cylinder head and its components.

18. Be sure the cylinder head and the engine block surfaces are clean, level, and straight.
19. Be sure the cylinder head dowel pins and oil control orifice are aligned. Clean the oil control orifice and reinstall it with a new O-ring.
20. Install or connect the following:
   • New head gasket onto the engine block

➡ Use new cylinder head bolts and washers. Used or previously-tightened bolts may be stretched; and therefore,

they have reduced clamping and sealing power under compression. Apply clean engine oil to the threads of each head bolt.

   • Cylinder head into position. Hand-tighten all the cylinder head bolts.
21. Tighten the cylinder head bolts to their final torque specification in 2 steps. Use a crisscross sequence starting with the bolts at the middle of the head and working toward the outer bolts:
   a. Step 1: Tighten each bolt to 22 ft. lbs. (30 Nm)
   b. Step 2: Tighten each bolt to 61 ft. lbs. (85 Nm)
22. Install or connect the following:
   • Dowel pin in the No. 3 cylinder head camshaft holder with a new O-ring
23. Thoroughly clean the intake and exhaust camshaft oil control orifices. Reinstall them with new O-rings.
24. Install or connect the following:
   • Camshafts
   • Intake manifold with a new gasket, and tighten the nuts in a crisscross pattern in 2 or 3 steps to 17 ft. lbs. (24 Nm) starting with the inner nuts
   • Bolts that secure the intake manifold to its bracket: 17 ft. lbs. (24 Nm)
   • Power steering pump bracket: 33 ft. lbs. (44 Nm)
   • Exhaust manifold with a new gasket. Apply anti-seize paste to the studs, and tighten the nuts to 23 ft. lbs. (31 Nm) in a crisscross sequence. Tighten the exhaust manifold bracket bolts to 17 ft. lbs. (24 Nm)
   • Exhaust manifold to the front exhaust pipe. Tighten the self-locking nuts to 40 ft. lbs. (55 Nm)
25. Verify that the engine is at TDC/compression for the No. 1 cylinder.
   • Timing belt. After the timing belt has been properly tensioned, tighten the adjusting bolt to 40 ft. lbs. (55 Nm). Refer to the timing belt unit repair section
   • Lower timing belt cover
   • Crankshaft pulley and tighten its bolt to 130 ft. lbs. (180 Nm)
26. Adjust the valves.
27. Inspect the VTEC rocker arms for free and smooth motion.
28. Remove the VTEC solenoid valve and its filter. Install a new filter, then reinstall the VTEC solenoid valve and tighten its bolts to 108 inch lbs. (12 Nm).
29. Install or connect the following:

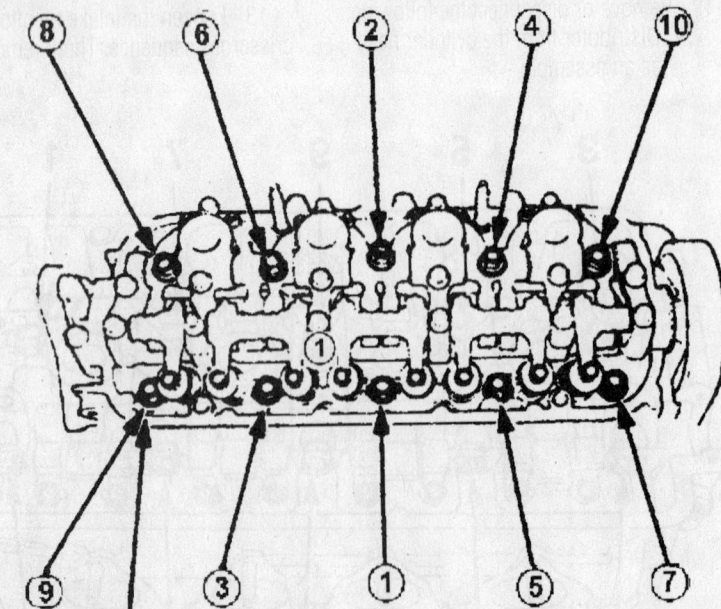

CYLINDER HEAD BOLTS
12 x 1.25 mm
100 N·m (10.0 kg-m, 72 lb-ft)
Apply clean engine oil to bolt threads and under bolt heads.

7923FG17

Cylinder head torque sequence—1.6L (B16A2) engine

• Distributor

30. Be sure all the spark plug sealing gaskets are fully seated.

• New gasket onto to the valve cover. Apply liquid gasket to the corners of the gasket that meet the camshaft holders. Don't let the sealant cure before installing the valve cover onto the cylinder head
• Valve cover. Gently wiggle the valve cover to be sure it is fully seated. Tighten the valve cover bolts in a crisscross pattern to 84 inch lbs. (10 Nm)
• New spark plugs
• Ignition wires
• Upper radiator hose, heater hoses, and intake manifold coolant bypass hoses
• Intake manifold vacuum lines, PCV, EVAP canister, and breather hoses
• Fuel lines to the fuel rail. Use new sealing washers on the banjo fitting. Carefully tighten the banjo fitting to 25 ft. lbs. (33 Nm). Tighten the service bolt to 11 ft. lbs. (15 Nm)
• Throttle cable. Adjust its tension so the cable has a deflection of 0.39–0.47 in. (10–12mm)

31. Installation of the remaining components is the reverse of removal.

32. After the installation procedure is complete, check that all tubes, hoses, and connectors are installed correctly.

2.0L Engine

1. Before servicing the vehicle, refer to the precautions in the beginning of this section.
2. Drain the cooling system.
3. Relieve the fuel system pressure.
4. Remove or disconnect the following:

• Negative battery cable
• Air cleaner housing
• Accessory drive belt
• Throttle cable
• Fuel lines
• Brake booster vacuum hose
• Evaporative Emissions (EVAP) canister hose
• Intake manifold bracket and air hose
• Water outlet housing
• Water bypass hose
• Fuel injector connectors
• Intake Air Temperature (IAT) sensor connector

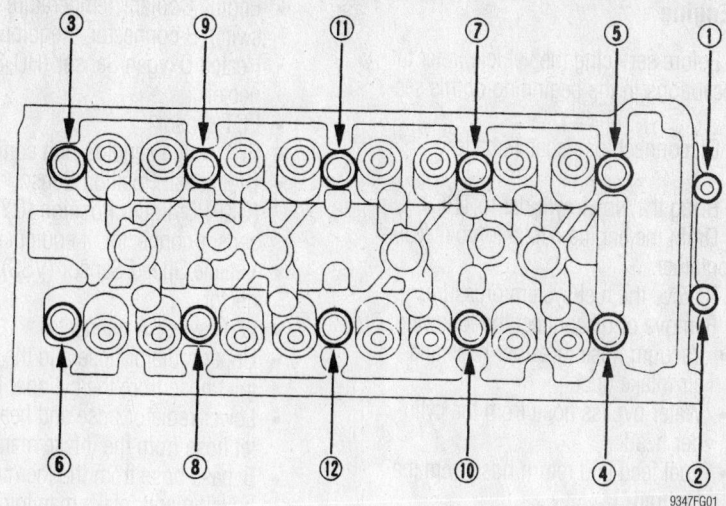

Cylinder head loosening sequence—2.0L engine

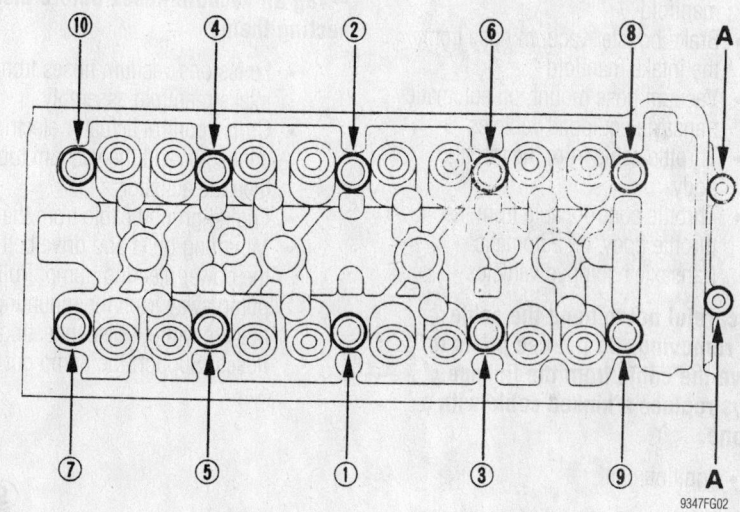

Cylinder head torque sequence—2.0L engine

• Idle Air Control (IAC) valve connector
• Throttle Position (TP) sensor connector
• Manifold Absolute Pressure (MAP) sensor connector
• Engine Coolant Temperature (ECT) sensor connector
• Heated Oxygen (HO$_2$S) sensor connector
• VTEC solenoid valve connector
• VTEC pressure switch connector
• Crankshaft Position (CKP) sensor connector
• Exhaust manifold cover
• Exhaust manifold heat shield
• Exhaust manifold
• Oil level dipstick
• Positive Crankcase Ventilation (PCV) valve and hose
• Ignition coil cover

• Ignition coils
• Intake manifold
• Valve cover
• Timing chain auto-tensioner
• Camshafts
• Timing chain idler gear
• Cylinder head. Loosen the bolts in sequence and in 1/3 turns.

To install:

5. Install the cylinder head with a new gasket. Tighten the bolts in sequence as follows:

a. Step 1: 22 ft. lbs. (29 Nm)
b. Step 2: Plus 90 degrees
c. Step 3: Plus 90 degrees
d. Step 4: If using new cylinder head bolts, add an additional 90 degrees

6. The remainder of the installation is the reverse of the removal procedure.

2.2L Engine

1. Before servicing the vehicle, refer to the precautions in the beginning of this section.

2. Disconnect the negative battery cable.

3. Bring the No. 1 cylinder to TDC.

4. Drain the engine coolant into a sealable container.

5. Relieve the fuel system pressure.

6. Remove or disconnect the following:

- Vacuum hose, breather hose and air intake duct
- Water bypass hose from the cylinder head
- Fuel feed and return hose from the fuel rail
- Evaporative emissions (EVAP) control canister hose from the intake manifold
- Brake booster vacuum hose from the intake manifold
- Vacuum hose mount, on automatic transaxle equipped vehicles
- Throttle cable from the throttle body
- Throttle control cable from the throttle body, on automatic transaxle equipped vehicles

➡**Be careful not to bend the cable when removing. Do not use pliers to remove the cable from the linkage. Always replace a kinked cable with a new one.**

- Ignition coil

➡**Label the connectors before disconnecting them.**

- Electrical connectors from the distributor and the spark plug wires from the spark plugs

➡**Matchmark the installed position of the distributor before removal.**

- Distributor
- Ignition coil wire from the distributor
- Connector and the terminal from the alternator, then the engine wiring harness from the valve cover
- Fuel injector connectors
- Intake Air Temperature (IAT) sensor connector, if equipped
- Idle Air Control (IAC) valve connector
- Throttle Position (TP) sensor connector
- Exhaust Gas Recirculation (EGR) valve lift sensor
- Ground cable terminals

- Engine Coolant Temperature (ECT) switch B connector, if equipped
- Heated Oxygen Sensor (HO2S) connector
- ECT sensor
- ECT gauge sending unit connector
- Crankshaft Position Sensor (CKP)/Cylinder Position (CYP) sensor connector, if equipped
- Vehicle Speed Sensor (VSS) connector
- ECT switch **A** connector
- Upper radiator hose and the heater inlet hose from the cylinder head
- Lower radiator hose and heater outlet hose from the intake manifold
- Bypass hose from the thermostat housing and intake manifold
- Thermostat. Discard the O-rings

➡**Tag all vacuum hoses before disconnecting them.**

- Emissions vacuum hoses from the intake manifold assembly
- Cruise control actuator electrical connector and the vacuum tube, then the actuator
- Engine ground cable from the body
- Mounting bolts and drive belt from the power steering pump. Pull the pump away from the mounting bracket without disconnecting the hoses. Support the pump out of the way

7. Raise and safely support the vehicle.

8. Remove or disconnect the following:

- Front wheel and tire assemblies
- Splash shield
- Intake manifold bracket bolts
- Intake manifold
- Exhaust pipe from the exhaust manifold
- Exhaust manifold and the exhaust manifold heat insulator
- Power steering pump mounting bracket
- Positive Crankcase Ventilation (PCV) hose, then remove the cylinder head cover. Replace the rubber seals if damages or deteriorated
- Timing belt. Refer to the timing belt unit repair section
- Cylinder head bolts in the reverse order of installation

➡**To prevent warpage, unscrew the bolts in sequence 1/3 turn at a time. Repeat the sequence until all bolts are loosened.**

9. Separate the cylinder head from the engine block with a suitable flat bladed prytool.

To install:

10. Be sure all cylinder head and block gasket surfaces are clean. Check the cylinder head for warpage. If warpage is less than 0.002 in. (0.05mm), cylinder head resurfacing is not required. Maximum

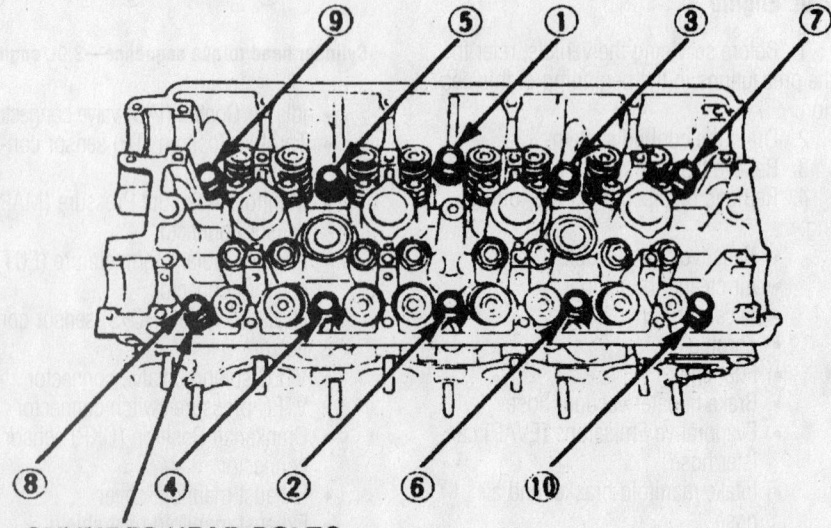

CYLINDER HEAD BOLTS
12 x 1.25 mm
100 N·m (10.0 kg-m, 72 lb-ft)
Apply clean engine oil bolt threads and under bolt heads.

Cylinder head torque sequence—2.2L engine

7923FG19

resurface limit is 0.008 in. (0.2mm) based on a cylinder head height of 3.94 in. (100mm).

11. Always use a new head gasket.

12. The **UP** mark on the camshaft pulley should be at the top.

13. Be sure the No. 1 cylinder is at TDC.

14. Clean the oil control orifice and install a new O-ring. Install and align the cylinder head dowel pins and oil control jet.

15. Install the bolts that secure the intake manifold to its bracket but do not tighten them.

16. Position the camshaft correctly.

17. Install the cylinder head, then tighten the cylinder head bolts sequentially in 3 steps:

    a. Step 1: 29 ft. lbs. (40 Nm).

    b. Step 2: 51 ft. lbs. (70 Nm).

    c. Step 3: 72 ft. lbs. (100 Nm).

18. Install or connect the following:

- Intake manifold and tighten the nuts in a crisscross pattern, in 2 or 3 steps, beginning with the inner nuts. Final torque should be 16 ft. lbs. (22 Nm). Always use a new intake manifold gasket.
- Intake manifold bracket to the intake manifold: 16 ft. lbs. (22 Nm)
- Heat insulator to the cylinder head and the block
- Power steering pump mounting bracket to the cylinder head. Tighten the 2 10mm bolts to 36 ft. lbs. (50 Nm). Torque the 8mm bolt to 16 ft. lbs. (22 Nm).
- Exhaust manifold and tighten the nuts in a crisscross pattern in 2 or 3 steps, beginning with the inner nut. Final torque should be 23 ft. lbs. (32 Nm). Always use a new exhaust manifold gasket.
- Exhaust manifold bracket, then the exhaust pipe, bracket and upper shroud.

➡**Be sure the camshaft sprocket and the crankshaft pulleys are aligned to TDC.**

- Timing belt. Refer to the timing belt unit repair section.
- Splash shield and the front wheels

19. Lower the vehicle.

20. Check and adjust the valves, as necessary.

21. Tighten the crankshaft pulley bolt to 181 ft. lbs. (250 Nm).

22. Installation of the remaining components is the reverse of removal.

23. Fill the cooling system.

24. Connect the negative battery cable and enter the radio security code.

25. Start the engine and check carefully for any leaks.

26. Check the ignition timing and tighten the distributor bolts to 13 ft. lbs. (18 Nm).

27. If equipped with engine and turn the steering wheel lock-to-lock to reset the 4WS control unit.

2.3L Engines

1. Before servicing the vehicle, refer to the precautions in the beginning of this section.

2. Drain the cooling system.

3. Relieve the fuel system pressure.

4. Remove or disconnect the following:

- Negative battery cable
- Air intake duct
- Throttle and cruise control cables
- Positive Crankcase Ventilation (PCV) valve and hose
- Fuel lines
- Brake booster vacuum hose
- Evaporative Emissions (EVAP) canister hoses
- Water bypass hoses
- Accessory drive belts
- Power steering pump
- Alternator wiring harness
- Radiator hoses
- Heater hoses
- Fuel injector connectors
- Intake Air Temperature (IAT) sensor connector
- Idle Air Control (IAC) valve connector
- Throttle Position (TP) sensor connector
- Manifold Absolute Pressure (MAP) sensor connector
- Heated Oxygen (HO2S) sensor connector (F23A1, F23A5 engines)
- Air/Fuel ratio sensor connector (F23A4 engine)
- Engine Coolant Temperature (ECT) sensor connector
- Radiator fan switch connector
- Coolant temperature gauge sender connector
- Exhaust Gas Recirculation (EGR) valve connector
- Crankshaft Position (CKP) sensor connector
- VTEC solenoid valve connector (F23A1, F23A4 engines)
- VTEC oil pressure switch connector (F23A1, F23A4 engines)
- Distributor
- Front motor mount bracket
- Valve cover
- Timing belt. Refer to the Timing Belt unit repair section.
- Camshaft pulley and back cover
- Intake manifold
- Exhaust manifold
- Cylinder head. Loosen the bolts in sequence and in 1/3 turns.

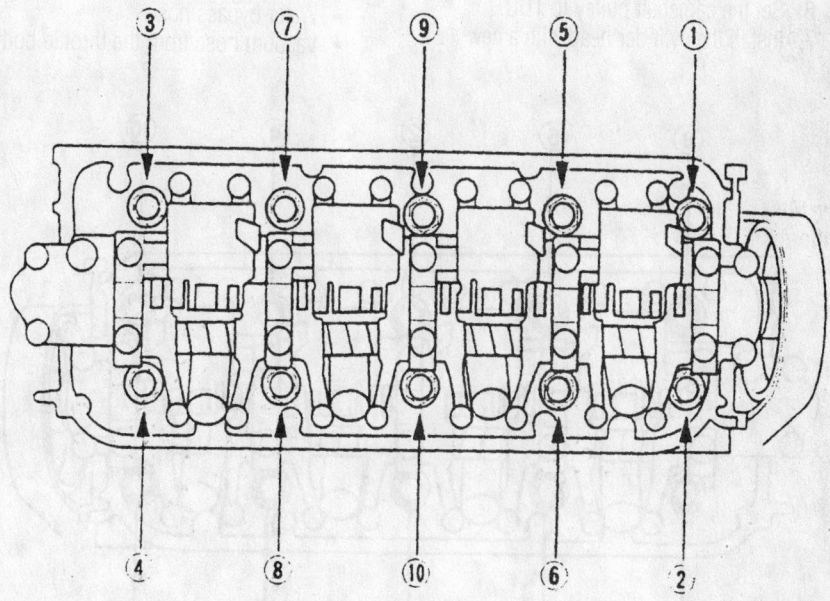

Cylinder head loosening sequence—2.3L engines

9347FG03

For Wheel Alignment specifications, see Section 1 of this manual

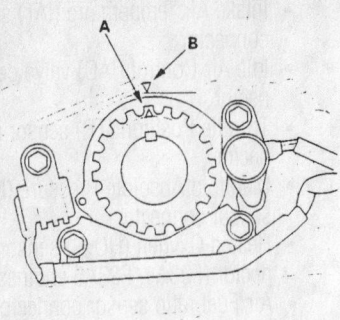

9347FG04

Set the crankshaft to TDC by aligning pointers A and B—2.3L engines

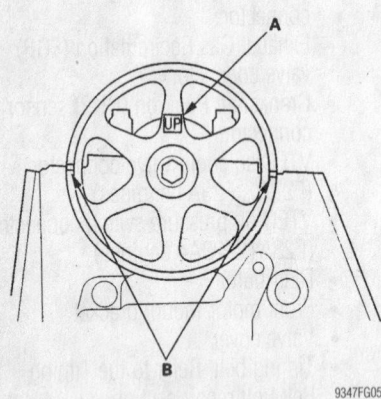

9347FG05

Align the camshaft pulley as shown prior to cylinder head installation—2.3L

To install:

5. Set the crankshaft pulley to Top Dead Center (TDC)

6. Set the camshaft pulley to TDC.

7. Install the cylinder head with a new gasket. Tighten the bolts in sequence as follows:

   a. Step 1: 22 ft. lbs. (29 Nm)

   b. Step 2: Plus 90 degrees

   c. Step 3: Plus 90 degrees

   d. Step 4: If using new cylinder head bolts, add an additional 90 degrees

8. The remainder of the installation is the reverse of the removal procedure.

3.0L Engine

1. Before servicing the vehicle, refer to the precautions in the beginning of this section.

2. Obtain the security code for the radio.

3. Disconnect the negative battery cable.

4. Drain the coolant.

5. Remove or disconnect the following:
- Evaporative emissions (EVAP) canister hose from the throttle body
- Air intake duct
- Upper engine covers
- Accelerator and cruise control cables from the throttle body
- Spark plug wire holder, cover, and intake manifold covers

6. Properly relieve the fuel system pressure.

7. Remove or disconnect the following:
- Fuel hoses from the supply rail
- Brake booster vacuum hose
- Positive Crankcase Ventilation (PCV) hose
- Breather hose
- Water bypass hose
- Vacuum hose from the throttle body
- Ground cable from the engine
- Alternator belt

8. Support the engine with a jack and a block of wood and remove the side engine mounting bracket.

9. Remove or disconnect the following:
- Power steering pump without disconnecting the hoses
- Alternator
- Wiring harness connectors from the components on the engine that may interfere with removing the cylinder head
- Distributor and spark plug wires
- Intake manifold
- Connectors from the fuel injectors
- Fuel supply rails
- Vacuum hoses from the fuel control valve

10. Set the engine to TDC by aligning the marks on the crankshaft and camshaft pulleys.

11. Remove or disconnect the following:
- Timing belt. Refer to the timing belt unit repair section.
- Upper and lower radiator hoses
- Heater hoses
- Both exhaust manifolds
- Water passage assembly
- Camshaft pulleys and rear timing belt covers
- Cylinder head bolts; loosen each cylinder head bolt ⅓ turn at a time in the correct sequence. This will take several passes.
- Cylinder heads

To install:

12. Clean the cylinder head and the surface of the cylinder block.

13. Install or connect the following:
- Oil control orifices and install them using new o-rings
- Dowel pins, if removed
- New cylinder head gaskets on the cylinder block

14. If moved, set the crankshaft and camshaft pulleys to TDC by aligning the marks on the pulley and oil pump.

15. Install or connect the following:
- Cylinder heads on the engine

16. Lubricate the cylinder head bolts with clean engine oil.

17. Tighten the cylinder head bolts in 3 separate steps, as follows:

   a. Step 1: Tighten each bolt in sequence to 29 ft. lbs. (39 Nm).

   b. Step 2: Tighten each bolt in sequence to 51 ft. lbs. (69 Nm).

   c. Step 3: Tighten each bolt a third time in sequence to a final torque of 72 ft. lbs. (98 Nm).

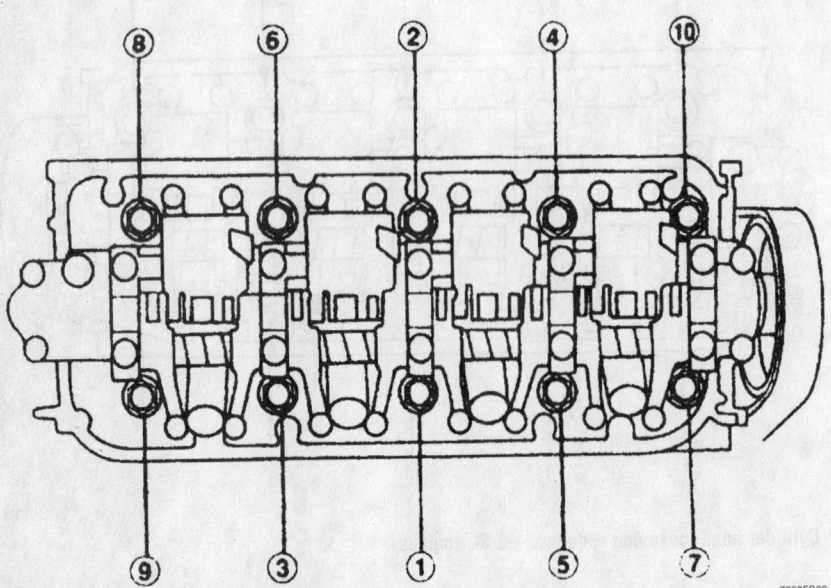

7923FG20

Cylinder head torque sequence—2.3L engines

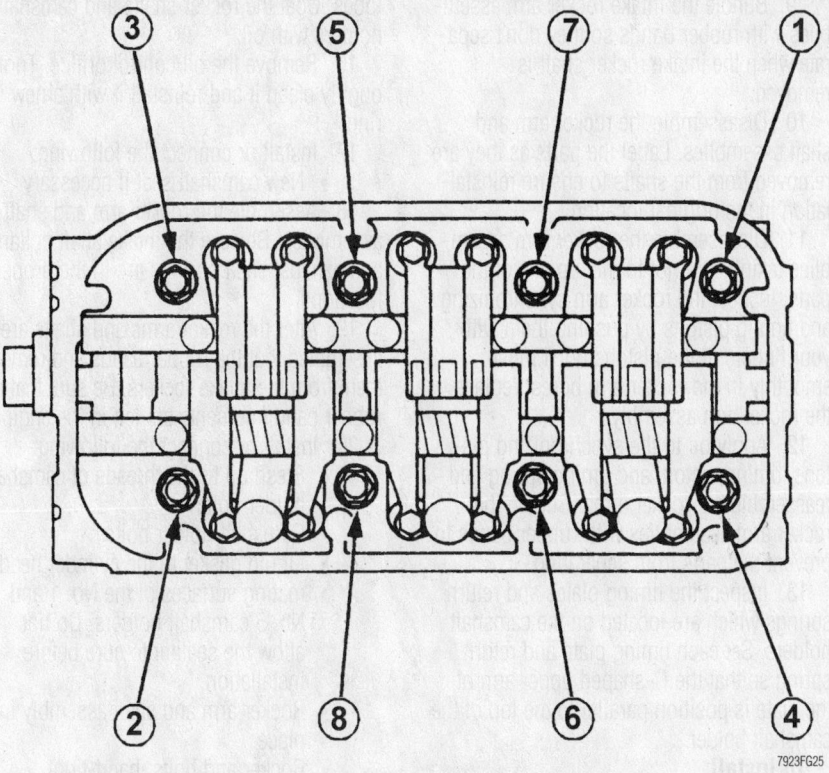

Loosen the cylinder head bolts in the sequence shown to prevent damage to the head—3.0L engine

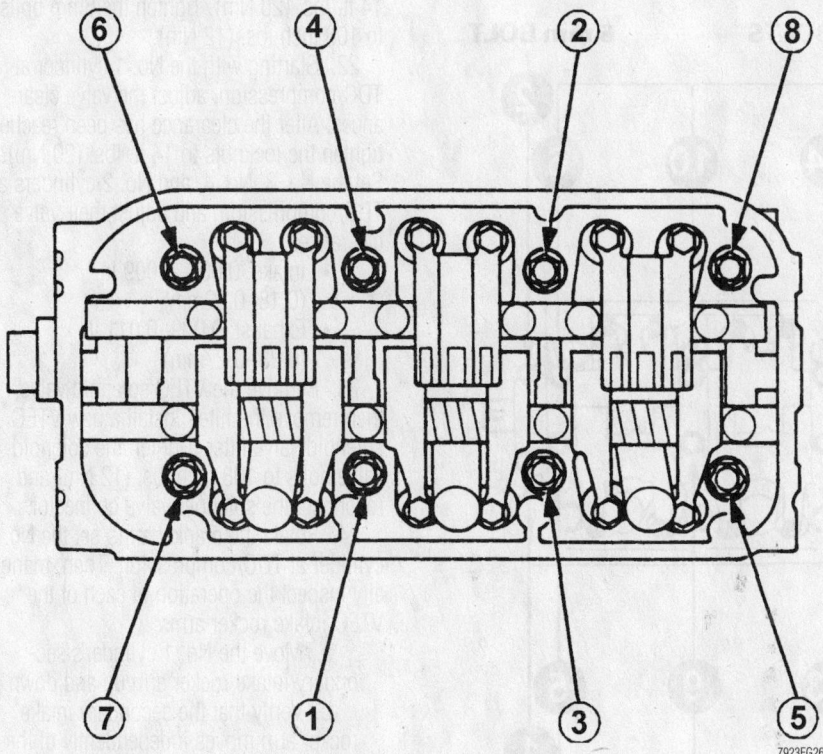

Cylinder head torque sequence—3.0L engine

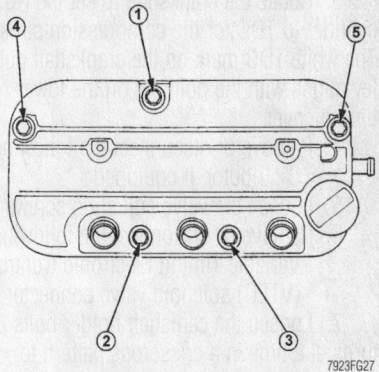

Tighten the cylinder head cover bolts in the sequence shown—3.0L engine

➡**If any cylinder head bolt makes noise while being tightened, loosen the bolts and begin the tightening sequence again.**

18. Install or connect the following:
   • Exhaust manifolds
   • Timing belt. Refer to the timing belt unit repair section.
19. Check and adjust the valve clearance if necessary.
20. Install or connect the following:
   • Cylinder head cover. Tighten the bolts in sequence to 108 inch lbs. (12 Nm).
   • Water passage. Be sure to use new gaskets and o-rings. Tighten the bolts to 16 ft. lbs. (22 Nm).
   • Intake manifold
   • All of the remaining hoses, tubes, and connectors are installed correctly.
   • Negative battery cable
21. Enter the security code for the radio.

Rocker Arms/Shafts

REMOVAL & INSTALLATION

1.6L (D16Y5) and 1.7L (D17A2) Engines

1. Before servicing the vehicle, refer to the precautions in the beginning of this section.
2. Remove or disconnect the following:
   • Negative battery cable

➡**Label the wires before disconnecting them.**

   • Ignition wires
   • Spark plugs and note their cylinder assignments.
   • Valve cover

3. Rotate the crankshaft to set the No. 1 cylinder to TDC for the compression stroke. The white TDC mark on the crankshaft pulley aligns with the pointers on the lower timing cover.

4. Remove or disconnect the following:
- Distributor, if equipped

5. Loosen the valve adjusting screws

6. Remove or disconnect the following:
- Variable Timing Electronic Control (VTEC) solenoid valve connector

7. Loosen the camshaft holder bolts 2 turns at a time in a crisscross pattern to prevent damaging the valves or rocker assembly.

8. Remove or disconnect the following:
- Rocker arm and shaft assemblies together with the camshaft holders. Do not remove the rocker shaft bolts yet. The bolts keep the bearing caps, springs, and rocker arms in place on the shafts.

➡ The rocker arms and shafts are an assembly. They must be removed from the engine as a unit. Always follow the torque sequence carefully when installing the rocker shaft assembly.

- Camshaft holder bolts from the rocker arm and shaft assembly

9. Bundle the intake rocker arm assemblies with rubber bands so they don't separate when the intake rocker shaft is removed.

10. Disassemble the rocker arm and shaft assemblies. Label the parts as they are removed from the shafts to ensure reinstallation in the original location.

11. Disassemble the rocker arm assemblies taking care not to mix up any of the parts. Inspect the rocker arm synchronizing and timing pistons by pushing them with your fingers. If the pistons don't move smoothly in the rocker arm bores, replace the rocker arm assembly.

12. Apply oil to the synchronizing pistons, timing piston, and timing spring and reassemble the rocker arms. Bundle the rocker arm assemblies with rubber bands to prevent the parts from separating.

13. Inspect the timing plates and return springs which are located on the camshaft holders. Set each timing plate and return spring so that the C-shaped upper arm of the plate is position parallel to the top of the camshaft holder.

To install:

14. Verify that the engine is set at TDC/compression for the No. 1 cylinder.

15. Lubricate the camshaft journals and lobes. Coat the rocker shafts and camshaft holders with oil.

16. Remove the oil control orifice. Thoroughly clean it and reinstall it with a new O-ring.

17. Install or connect the following:
- New camshaft seal if necessary

18. Assemble the rocker arm and shaft assemblies. Be sure the intake shaft collars and exhaust shaft springs are in the proper locations.

19. After the rocker arms and shafts are assembled, cut the rubber bands and remove them from the intake rockers. Be sure that no rubber band fragments are left in the engine.

20. Install or connect the following:
- Fresh oil to the threads of camshaft holder bolts
- Camshaft holder bolts
- Liquid gasket to the cylinder head mating surfaces of the No. 1 and No. 5 camshaft holders. Do not allow the sealant to cure before installation.
- Rocker arm and shaft assembly in place
- Rocker and bolts, hand-tight

21. Tighten each bolt 2 turns at a time in the crisscross sequence so that the rockers are evenly tightened and don't bind on the valves. Tighten the 8mm rocker arm bolts to 14 ft. lbs. (20 Nm). Tighten the 6mm bolts to 108 inch lbs. (12 Nm).

22. Starting with the No. 1 cylinder at TDC/compression, adjust the valve clearances. After the clearance has been reached, tighten the locknuts to 14 ft. lbs. (20 Nm). Set the No. 3, No. 4, and No. 2 cylinders at TDC/compression, and adjust their valve clearances.
- Intake: 0.007–0.009 in. (0.18–0.22mm)
- Exhaust: 0.009–0.011 in. (0.23–0.27mm)

23. Remove the VTEC solenoid valve, then remove the filter. Install a new VTEC solenoid valve filter. Tighten the solenoid valve bolts to 108 inch lbs. (12 Nm) and reconnect the solenoid valve connector.

24. Rotate the crankshaft to set the No. 1 cylinder at TDC/compression. Then, manually inspect the operation of each of the VTEC intake rocker arms:
- a. Move the No. 1 cylinder's secondary intake rocker arm up and down.
- b. Verify that the secondary intake rocker arm moves independently of the primary intake rocker arm.
- c. Repeat the rocker arm inspection for the other 3 cylinders with each cylinder set at TDC/compression.

25. Rotate the crankshaft back to

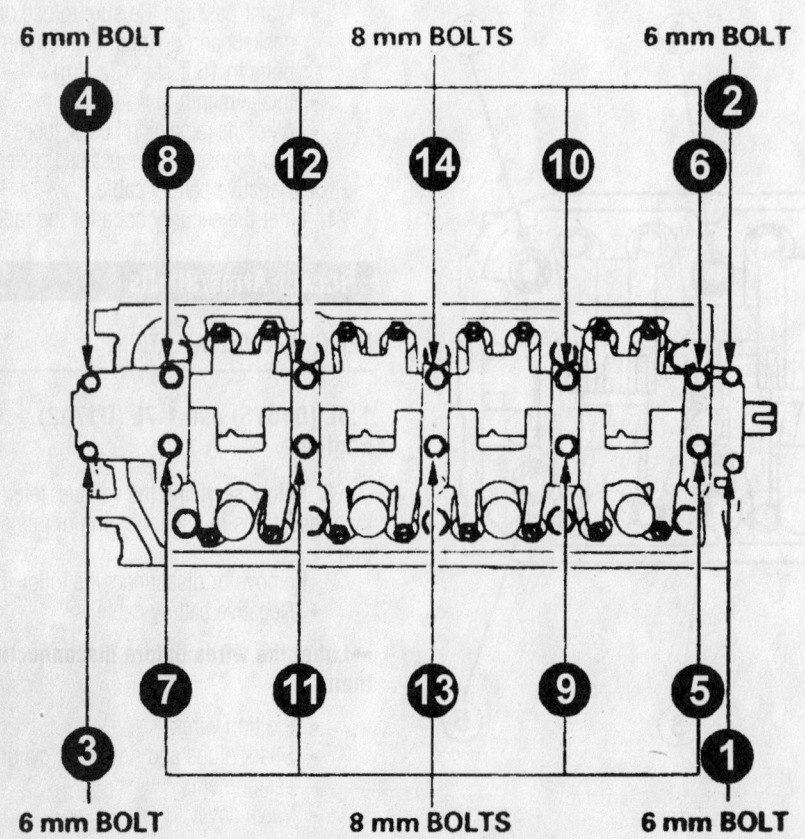

6 mm BOLT   8 mm BOLTS   6 mm BOLT

6 mm BOLT   8 mm BOLTS   6 mm BOLT

7923FG35

Rocker arm/shaft bolt loosening sequence—1.6L (D16Y5) and 1.7L (D17A2) engines

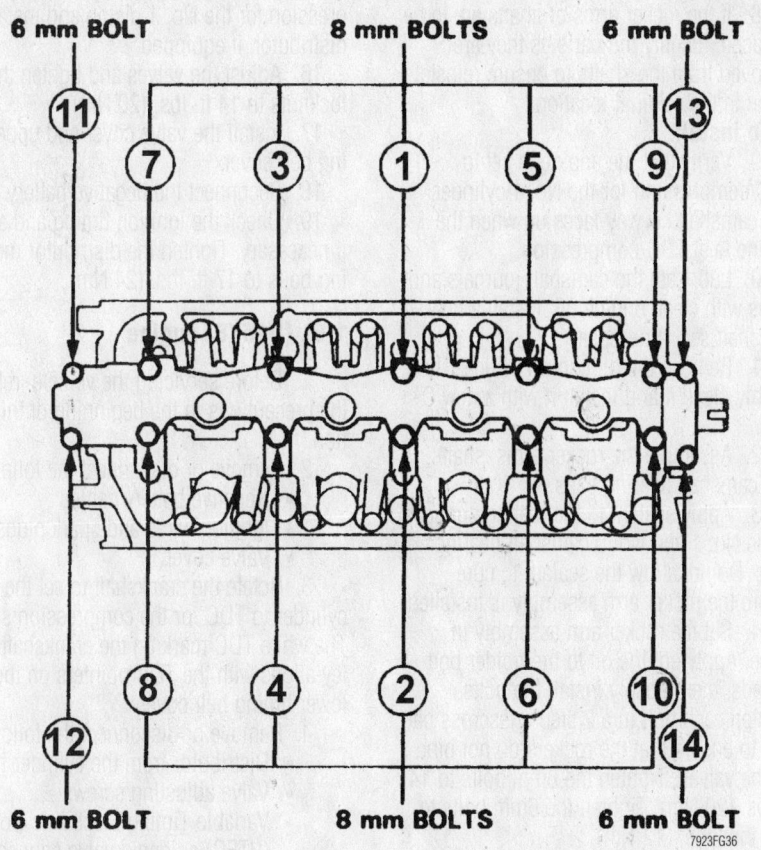

Rocker arm/shaft bolt torque sequence—1.6L (D16Y5) and 1.7L (D17A2) engine

7923FG36

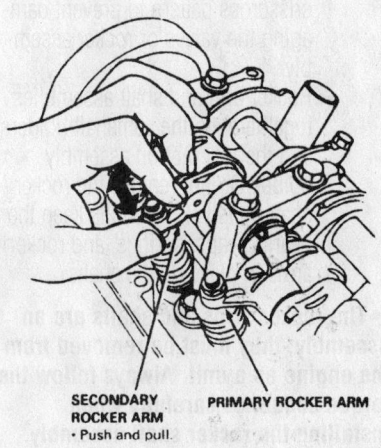

VTEC rocker arm inspection—1.6L (D16Y5) and 1.7L (D17A2) engines

7923FG37

TDC/compression for the No. 1 cylinder. Install the distributor, if equipped, but do not tighten the mounting bolts yet.

26. Tighten the crankshaft pulley bolt to 134 ft. lbs. (181 Nm).

27. Install or connect the following:
• Valve cover. Be sure the gasket is in good condition, and apply

sealant to the corners where the gasket meets the camshaft holders.
• Spark plugs and the ignition wires

28. Drain the engine oil and remove the oil filter. Install a new oil filter and refill the engine with fresh oil.

29. Install or connect the following:
• Negative battery cable

30. Warm the engine up to normal operating temperature.

31. Check the ignition timing and adjust it if necessary. Then, tighten the distributor mounting bolts to 17 ft. lbs. (24 Nm).

32. Check all fluid levels. Test drive the vehicle and observe the engine RPM changes at various speeds.

1.6L (D16Y7) and 1.7L (D17A1) Engine

1. Before servicing the vehicle, refer to the precautions in the beginning of this section.

2. Remove or disconnect the following:
• Negative battery cable

➡**Label the wires before disconnecting them.**

• Ignition wires and spark plugs
• Valve cover and the upper timing belt cover

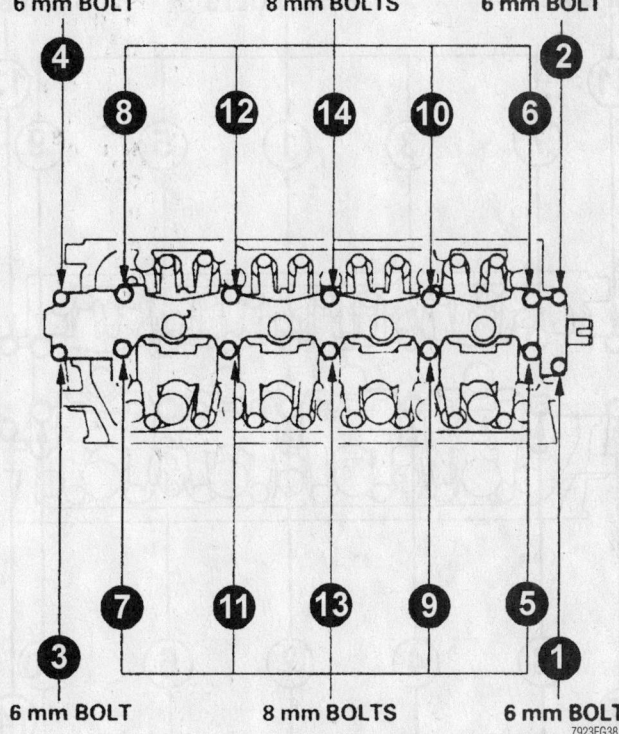

Rocker arm/shaft bolt loosening sequence—1.6L (D16Y7) and 1.7L (D17A1) engines

7923FG38

3. Set the No. 1 cylinder to TDC for the compression stroke. Verify that the TDC marks are correctly aligned. Once the engine is set in this position, it must not be disturbed.

4. Remove or disconnect the following:
- Distributor, if equipped

5. Loosen the valve adjusting screws.

6. Cover the timing belt with a clean shop towel to protect it from engine oil. If the belt is contaminated with oil, it must be replaced.

7. Remove or disconnect the following:
- Camshaft holder bolts. Unscrew the bolts 2 turns at a time in a criss-cross pattern to prevent damaging the valves, camshaft, or rocker arm assembly.

➡The rocker arms and shafts are an assembly; they must be removed from the engine as a unit. To prevent warpage, always follow the torque sequence carefully when removing or installing the rocker shaft assembly.

- Rocker arm and shaft assemblies. Do not remove the camshaft holder bolts. The bolts keep the camshaft bearing caps, springs, and rocker arms in place on the shafts.

8. If the rocker arms or shafts are to be replaced, identify the parts as they are removed from the shafts to ensure reinstallation in the original location.

To install:

9. Verify that the engine is set to TDC/compression for the No. 1 cylinder. The camshaft keyway faces up when the engine is at TDC/compression.

10. Lubricate the camshaft journals and lobes with clean engine oil. Install a new camshaft seal if necessary.

11. Remove the oil control orifice. Thoroughly clean it and install it with a new O-ring.

12. Assemble the rocker arms, shafts, and camshaft bearing caps.

13. Apply sealant to the mating surfaces of the No. 1 and No. 5 camshaft bearing caps. Do not allow the sealant to cure before the rocker arm assembly is installed.

14. Set the rocker arm assembly in place. Apply engine oil to the holder bolt threads, then loosely install the bolts. Tighten each bolt in a 2 step crisscross pattern to ensure that the rockers do not bind on the valves. Tighten the 8mm bolts to 14 ft. lbs. (20 Nm). Tighten the 6mm bolts to 104 inch lbs. (12 Nm).

15. Verify that the engine is at TDC/com-

pression for the No. 1 piston and install the distributor, if equipped.

16. Adjust the valves and tighten the locknuts to 14 ft. lbs. (20 Nm).

17. Install the valve cover and upper timing belt cover.

18. Reconnect the negative battery cable.

19. Check the ignition timing and adjust if necessary. Tighten the distributor mounting bolts to 17 ft. lbs. (24 Nm).

1.6L (D16Y8) Engine

1. Before servicing the vehicle, refer to the precautions in the beginning of this section.

2. Remove or disconnect the following:
- Negative battery cable
- Ignition wires and spark plugs
- Valve cover

3. Rotate the crankshaft to set the No. 1 cylinder to TDC for the compression stroke. The white TDC mark on the crankshaft pulley aligns with the TDC pointers on the lower timing belt cover.

4. Remove or disconnect the following:
- Distributor from the cylinder head
- Valve adjusting screws
- Variable Timing Electronic Control (VTEC) solenoid valve connector
- Camshaft holder bolts, by loosening them 2 turns at a time in a crisscross pattern to prevent damaging the valves or rocker assembly.
- Rocker arm and shaft assemblies together with the camshaft holders and the lost motion assembly holder. Do not remove the rocker shaft bolts yet. The bolts keep the bearing caps, springs, and rocker arms in place on the shafts.

➡The rocker arms and shafts are an assembly; they must be removed from the engine as a unit. Always follow the torque sequence carefully when installing the rocker shaft assembly.

- Camshaft holder bolts from the rocker arm and shaft assembly
- Lost motion assembly holder

5. Bundle the intake rocker arm assemblies with rubber bands so they don't separate when the intake rocker shaft is removed.

6. Disassemble the rocker arm and shaft assemblies. Label the parts as they are removed from the shafts to ensure reinstallation in the original location.

7. Disassemble the rocker arm assemblies taking care not to mix up any of the parts. Inspect the rocker arm synchronizing

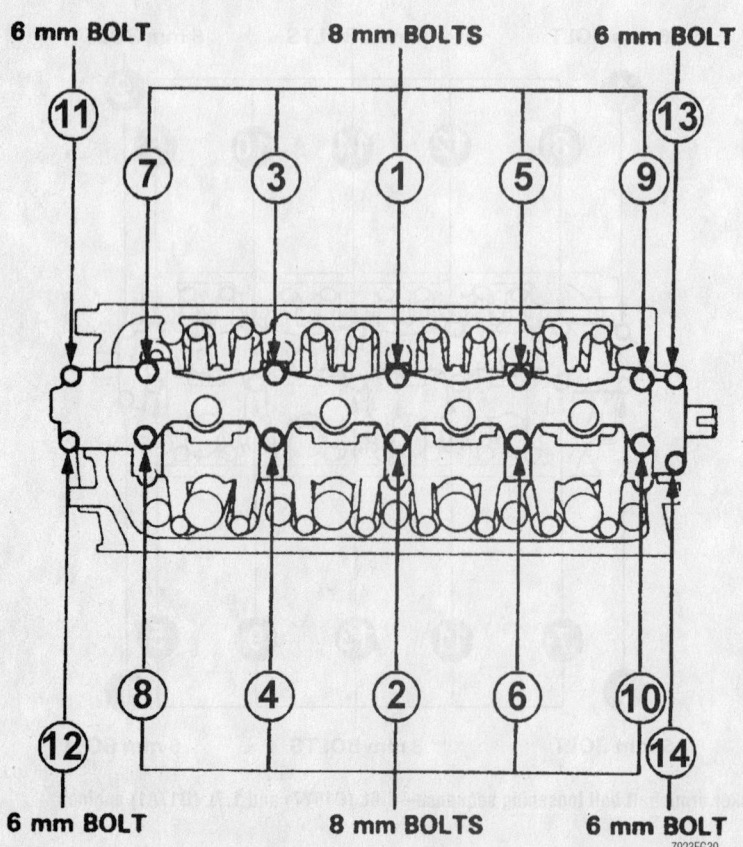

Rocker arm/shaft bolt tightening sequence—1.6L (D16Y7) and 1.7L (D17A1) engines

7923FG39

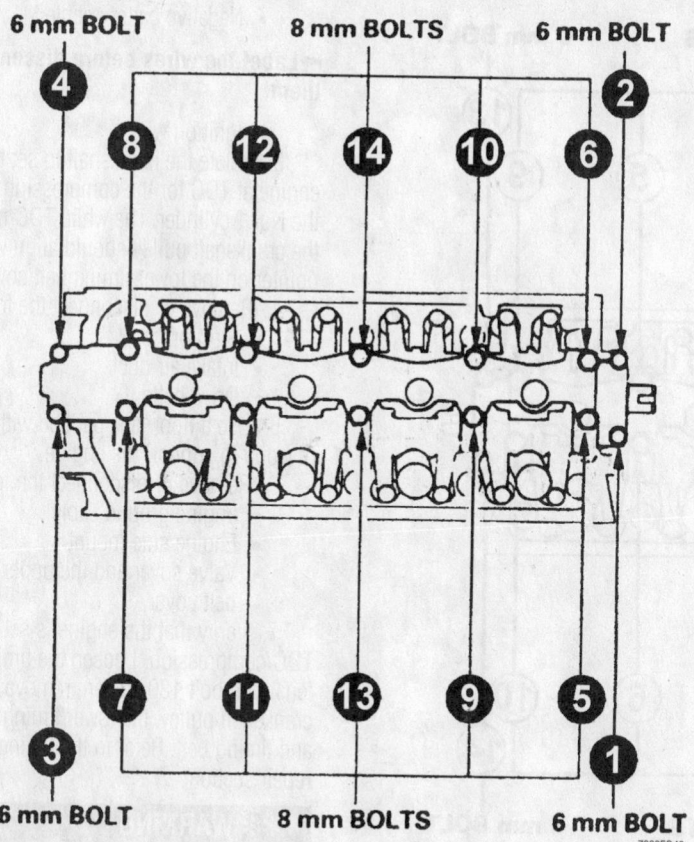

Rocker arm/shaft bolt loosening sequence—1.6L (D16Y8) engine

6 mm BOLT   8 mm BOLTS   6 mm BOLT

6 mm BOLT   8 mm BOLTS   6 mm BOLT

7923FG40

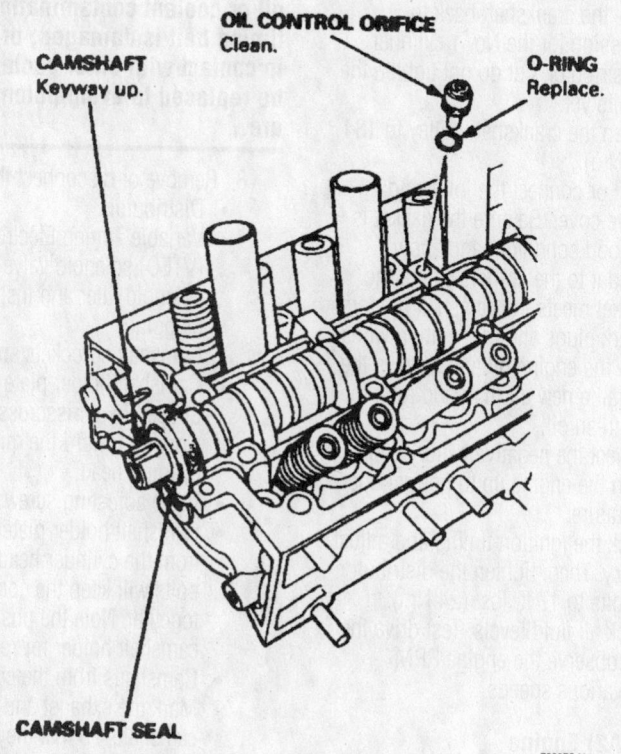

CAMSHAFT
Keyway up.

OIL CONTROL ORIFICE
Clean.

O-RING
Replace.

CAMSHAFT SEAL

Camshaft seal and oil control orifice—1.6L (D16Y8) engine

7923FG41

pistons by pushing them with your fingers. If the pistons don't move smoothly in their rocker arm bores, replace the rocker arm assembly.

8. Apply oil to the synchronizing pistons and reassemble the rocker arms. Bundle the rocker arm assemblies with rubber bands to prevent the parts from separating.

9. Remove or disconnect the following:
- Lost motion assembly from its port in the lost motion assembly holder. Inspect each lost motion assembly by pushing down on its piston. If the piston doesn't move smoothly, replace the lost motion assembly. Lost motion assemblies cannot be bled like hydraulic lash adjusters.

10. Install the lost motion assemblies back into the lost motion assembly holder.

To install:

11. Verify that the engine is set at TDC/compression for the No. 1 cylinder.

12. Lubricate the camshaft journals and lobes. Coat the rocker shafts and camshaft holders with fresh oil.

13. Remove the oil control orifice. Thoroughly clean it and reinstall it with a new O-ring.

14. Install or connect the following:
- New camshaft seal if necessary

15. Assemble the rocker arm and shaft assemblies. Be sure the intake shaft collars and exhaust shaft springs are in the proper locations.

16. After the rocker arms and shafts are assembled, cut the rubber bands and remove them from the intake rockers. Be sure that no rubber band fragments are left in the engine.

17. Install or connect the following:
- Lost motion assembly holder onto the camshaft holder

➡**Apply fresh oil to the threads of camshaft holder bolts before installation.**

- Camshaft holder bolts
- Liquid gasket to the cylinder head mating surfaces of the No. 1 and No. 5 camshaft holders. Don't allow the sealant to cure before installation.

18. Set the rocker arm and shaft assembly in place. Install and hand-tighten the bolts. Tighten each bolt 2 turns at a time in the crisscross sequence to ensure that the rockers do not bind on the valves. Tighten the 8mm rocker arm bolts to 14 ft. lbs. (20

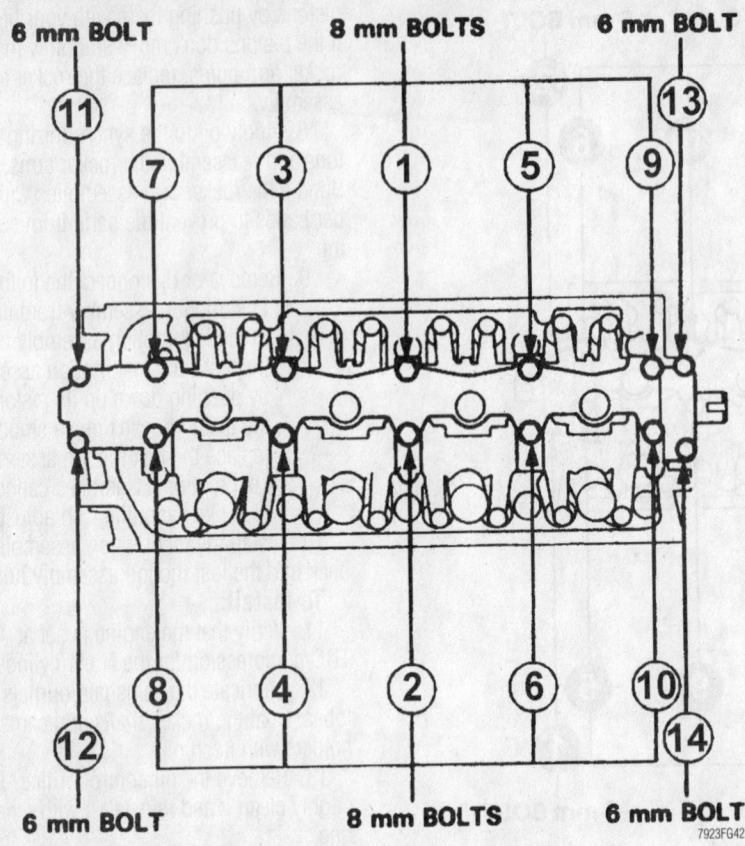

6 mm BOLT   **8 mm BOLTS**   **6 mm BOLT**

6 mm BOLT   **8 mm BOLTS**   **6 mm BOLT**

7923FG42

Rocker arm/shaft bolt tightening sequence—1.6L (D16Y8) engine

Nm). Tighten the 6mm bolts to 108 inch lbs. (12 Nm).

19. Starting with the No. 1 cylinder at TDC/compression, adjust the valve clearances. After the clearance has been reached, tighten the adjuster locknuts to 14 ft. lbs. (20 Nm). Set the No. 3, No. 4, and No. 2 cylinders at TDC/compression, then adjust their valve clearances.

- Intake: 0.007–0.009 in. (0.18–0.22mm)
- Exhaust: 0.009–0.011 in. (0.23–0.27mm)

20. Remove the VTEC solenoid valve, then remove the filter. Install a new VTEC solenoid valve filter. Tighten the solenoid valve bolts to 108 inch lbs. (12 Nm) and reconnect the solenoid valve connectors.

21. Rotate the crankshaft to set the No. 1 cylinder at TDC/compression. Then, manually inspect the operation of each of the VTEC intake rocker arms:

   a. Push the in on the No. 1 cylinder's mid-intake rocker arm.

   b. Verify that the mid-intake rocker arm moves independently of the primary and secondary intake rocker arms.

   c. Repeat the rocker arm inspection for the other 3 cylinders with each cylinder set at TDC/compression.

22. Rotate the crankshaft back to TDC/compression for the No. 1 cylinder. Install the distributor, but do not tighten the mounting bolts yet.

23. Tighten the crankshaft pulley to 134 ft. lbs. (181 Nm).

24. Install or connect the following:

- Valve cover. Be sure the gasket is in good condition, and apply sealant to the corners where the gasket meets the camshaft holders.
- Spark plugs and the ignition wires

25. Drain the engine oil and remove the oil filter. Install a new oil filter and refill the engine with fresh oil.

26. Connect the negative battery cable.

27. Warm the engine up to normal operating temperature.

28. Check the ignition timing and adjust it if necessary. Then, tighten the distributor mounting bolts to 17 ft. lbs. (24 Nm).

29. Check all fluid levels. Test drive the vehicle and observe the engine RPM changes at various speeds.

1.6L (B16A2) Engine

1. Before servicing the vehicle, refer to the precautions in the beginning of this section.

2. Remove or disconnect the following:

- Negative battery cable

➡ **Label the wires before disconnecting them.**

- Ignition wires

3. Rotate the crankshaft to set the engine at TDC for the compression stroke of the No. 1 cylinder. The white TDC mark on the crankshaft pulley should align with the pointer on the lower timing belt cover.

4. Remove or disconnect the following:

- Strut brace
- Intake air duct
- Drive belts

5. Use a floor jack padded with a block of wood to support the engine.

6. Remove or disconnect the following:

- Engine ground cable
- Engine side mount
- Valve cover and the upper timing belt cover

7. Verify that the engine is set at TDC/compression. Loosen the timing belt tensioner bolt 180°. Then, remove the crankshaft pulley, the lower timing cover, and timing belt. Refer to the timing belt unit repair section.

✳✳ WARNING

Inspect the timing belt for signs of cracked and broken teeth, as well as oil or coolant contamination. If the timing belt is damaged, or has been in contact with oil or coolant, it must be replaced to avoid potential failure.

8. Remove or disconnect the following:

- Distributor
- Variable Timing Electronic Control (VTEC) solenoid valve. Remove the solenoid filter and inspect it for clogging.
- Camshaft sprockets and back cover
- Camshaft holder plate bolts. Loosen in a crisscross sequence working toward the middle of the cylinder head.
- Valve adjusting screws
- Camshaft holder plates and holder from the cylinder head. The holder bolts will keep the components together. Note the positions of each camshaft holder for reassembly.
- Camshafts from the cylinder head. Mark the exhaust and intake camshafts so that they will not be confused.

9. Hold each rocker arm assembly together with a rubber band to prevent them from separating.

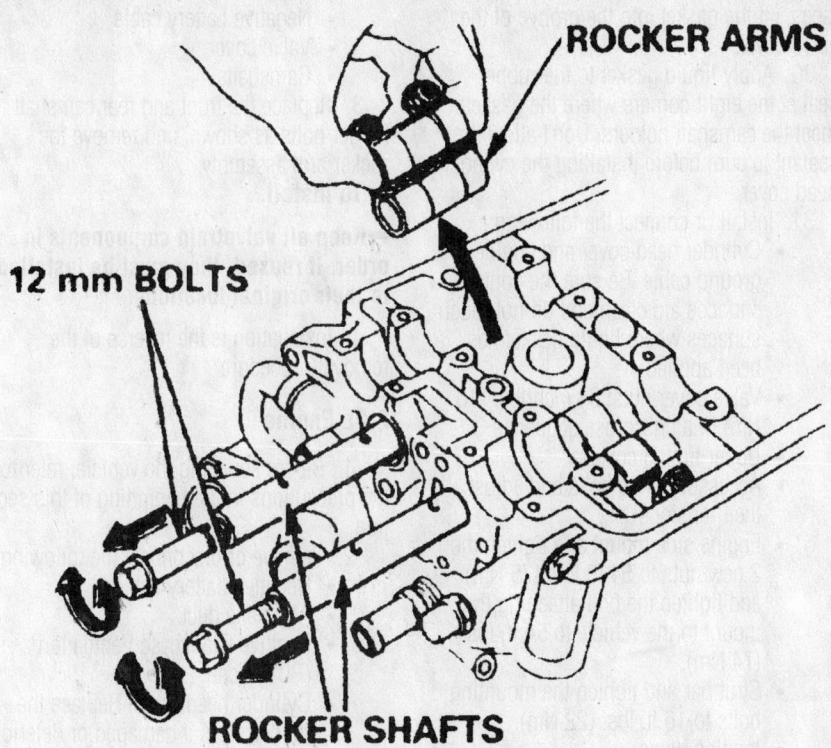

Removing the rocker arms—1.6L (B16A2) engines

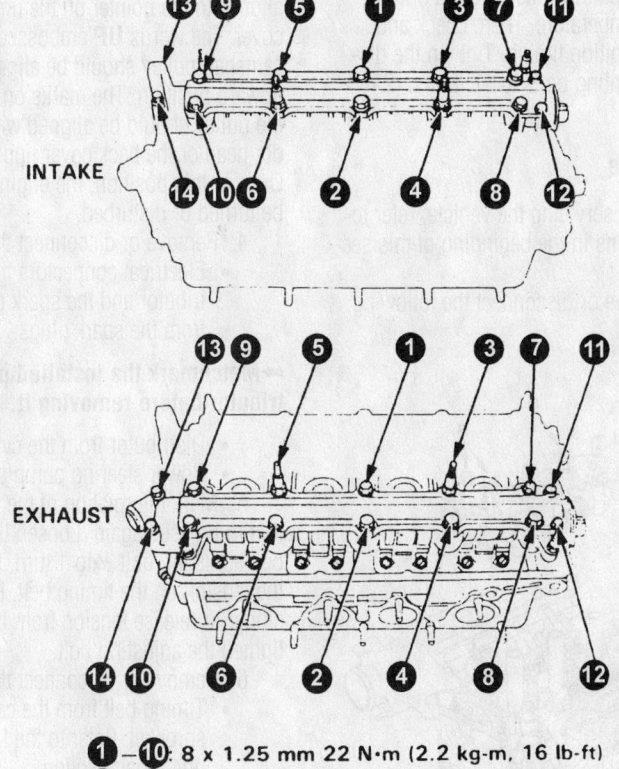

①–⑩: 8 x 1.25 mm 22 N·m (2.2 kg-m, 16 lb-ft)
⑪–⑭: 6 x 1.0 mm 11 N·m (1.1 kg-m, 8 lb-ft)

7923FG44

Camshaft holder bolt torque sequences—1.6L (B16A2) engines

10. Remove or disconnect the following:
- Intake and exhaust rocker shaft orifices from the cylinder head. The rocker shaft orifices are different and should be identified when removed. Thoroughly clean the orifices and reinstall them with new O-rings.
- Rocker arm shaft sealing bolts and discard the washers

11. Insert 12mm bolts into the rocker arm shafts. Remove each rocker arm set while slowly pulling out the rocker arm shaft.

➡ **Tag each rocker arm set to assure installation in their original locations.**

12. Inspect the rocker arm pistons. If they do not move smoothly, replace the rocker arm assembly.

13. Remove or disconnect the following:
- 2 lost motion assemblies from the cylinder head. Inspect each lost motion assembly by pushing the plunger with your finger. Replace the lost motion assembly if it does not move smoothly.

To install:

14. Install or connect the following:
- 2 lost motion assemblies to the cylinder head
- Engine oil to the rocker arm pistons, then bundle the rocker arms with a rubber band. Apply a light coat of clean engine oil to the rocker arms.
- Rocker arms in their original locations if they are being reused. If new assembles are being used, place them in the cylinder head.

15. Lightly coat the rocker arm shafts with clean engine oil, then install the rocker arm shafts into the cylinder head. A 12mm bolt can be installed into the end of the rocker arm shafts to aid in their installation. Be sure to install the shafts in the proper positions. Remove the 12mm bolts from the rocker arm shafts, if used.

16. Clean and install the rocker arm shaft orifices with new O-rings. If the holes in the rocker arm shafts are not aligned, screw a 12mm bolt into the end of the shaft to position the shaft.

17. Install or connect the following:
- Sealing bolts with new washers: 47 ft. lbs. (64 Nm)

18. Lubricate the camshaft lobes and journals with clean engine oil.

19. Install or connect the following:

- Camshafts into the cylinder head. Both the intake and exhaust camshafts should be installed with their keyways pointing straight up.
- New lubricated camshaft seals. Apply liquid gasket to a new camshaft end-plug and install it. If the end-plug is marked, the mark should be aligned with the cylinder head surface.

20. Apply liquid gasket to the cylinder head mating surfaces of the No. 1 and No. 5 camshaft holders, then install them, along with No. 2, 3, and 4 holders. Be sure to pay attention to the following points:

- Do not apply oil to the holder mating surface of camshaft seals.
- The arrows marked on the camshaft holders should point to the timing belt.

21. Install or connect the following:

- Camshaft holder plates

➡**Lubricate the threads of the 10mm holder bolts before installation.**

- Camshaft holder bolts, but don't tighten them yet

22. Evenly hand-tighten the camshaft holders. Be sure that the rocker arms are properly positioned on the valve stems.

23. Use a 2-step crisscross pattern to tighten the camshaft holder bolts. Begin tightening with the bolts in the middle of the cylinder head, and work toward the outer edges. Final torque specifications are as follows:

   a. 8mm bolts 20 ft. lbs. (28 Nm).
   b. 6mm bolts to 84–96 inch lbs. (10–11 Nm)

24. Verify that the camshaft keyways are pointing straight up and that the engine is at TDC/compression for the No. 1 cylinder. Fit the camshaft sprocket keys into their keyways.

25. Install or connect the following:

- Back cover and push the camshaft pulleys onto the camshafts. Then, tighten the sprocket retaining bolts to 41 ft. lbs. (57 Nm).
- Timing belt, then tension the belt. Refer to the timing belt unit repair section.
- Lower timing cover and the crankshaft pulley. Tighten the pulley bolt to 130 ft. lbs. (180 Nm).

26. Adjust the valve clearance.

27. Inspect the VTEC rocker arms for smooth and independent movement.

28. Install or connect the following:

- VTEC solenoid valve with a new filter. Tighten the valve mounting bolts to 108 inch lbs. (12 Nm).
- Distributor

29. Clean the valve cover gasket surfaces. Fit the gasket into the groove of the valve cover.

30. Apply liquid gasket to the rubber seal at the eight corners where the gaskets meet the camshaft holders. Don't allow the sealant to cure before installing the cylinder head cover.

31. Install or connect the following:

- Cylinder head cover and engine ground cable. Be sure the contact surfaces are clean and do not touch surfaces where liquid gasket has been applied.
- Valve cover nuts: 84 inch lbs. (10 Nm) in a crisscross sequence
- Upper timing cover.
- Accessory drive belts and adjust their tensions
- Engine side mount and tighten the 2 new nuts to 54 ft. lbs. (75 Nm) and tighten the bolt attaching the mount to the vehicle to 54 ft. lbs. (74 Nm).
- Strut bar and tighten the mounting bolts to 16 ft. lbs. (22 Nm).
- Ignition wires

32. Drain the engine oil. Install a new oil filter and refill the engine with fresh oil.

33. Reconnect the negative battery cable.

34. Warm the engine up to its normal operating temperature. Then, check and adjust the ignition timing. Tighten the distributor mounting bolts to 17 ft. lbs. (24 Nm).

2.0L Engine

1. Before servicing the vehicle, refer to the precautions in the beginning of this section.

2. Remove or disconnect the following:

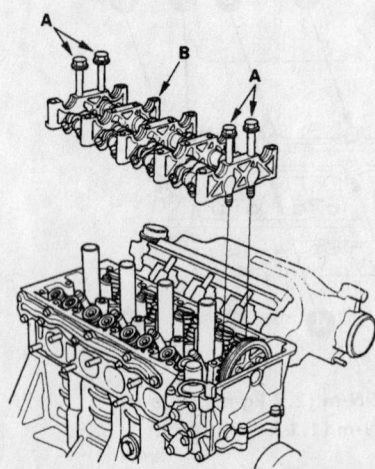

Use bolts (A) to hold the rocker arm assembly (B) together while removing or replacing the assembly—2.0L engine

- Negative battery cable
- Valve cover
- Camshafts

3. Replace the front and rear camshaft holder bolts as shown, and remove the rocker arm assembly.

   To install:

➡**Keep all valvetrain components in order. If reused, they must be installed in their original locations.**

4. Installation is the reverse of the removal procedure.

2.2L Engine

1. Before servicing the vehicle, refer to the precautions in the beginning of this section.

2. Remove or disconnect the following:

- Negative battery cable
- Air intake duct
- Positive Crankcase Ventilation (PCV) hose
- Cylinder head cover. Replace the rubber seals if damaged or deteriorated.
- Timing belt upper cover

3. Bring the No. 1 cylinder to TDC. The white mark on the crankshaft pulley should align with the pointer on the timing belt cover. The words **UP** embossed on the camshaft pulley should be aligned in the upward position. The marks on the edge of the pulley should be aligned with the cylinder head or the back cover upper edge. Once in this position, the engine must NOT be turned or disturbed.

4. Remove or disconnect the following:

- Electrical connectors from the distributor and the spark plug wires from the spark plugs.

➡**Matchmark the installed position distributor before removing it.**

- Distributor from the cylinder head
- Power steering pump drive belt

5. Mark the rotation of the timing belt if it is to be used again. Loosen the timing belt adjusting bolt ¾ to 1 turn, then release the tension on the timing belt. Push the tensioner to release tension from the belt, then tighten the adjusting bolt.

6. Remove or disconnect the following:

- Timing belt from the camshaft sprocket. Refer to the timing belt unit repair section.

⁂⁂ **WARNING**

Do not crimp or bend the timing belt more than 90°, or less than 1 inch (25mm) in diameter

NOTE:
- Identify parts as they are removed to ensure reinstallation in their original locations.
- Inspect rocker shafts and rocker arms
- Rocker arms must be installed in the same position if reused.
- Prior to reassembling, clean all the parts in solvent, dry them, and apply lubricant to any contact points.
- Bundle the rocker arms with rubber bands to keep them together as a set.

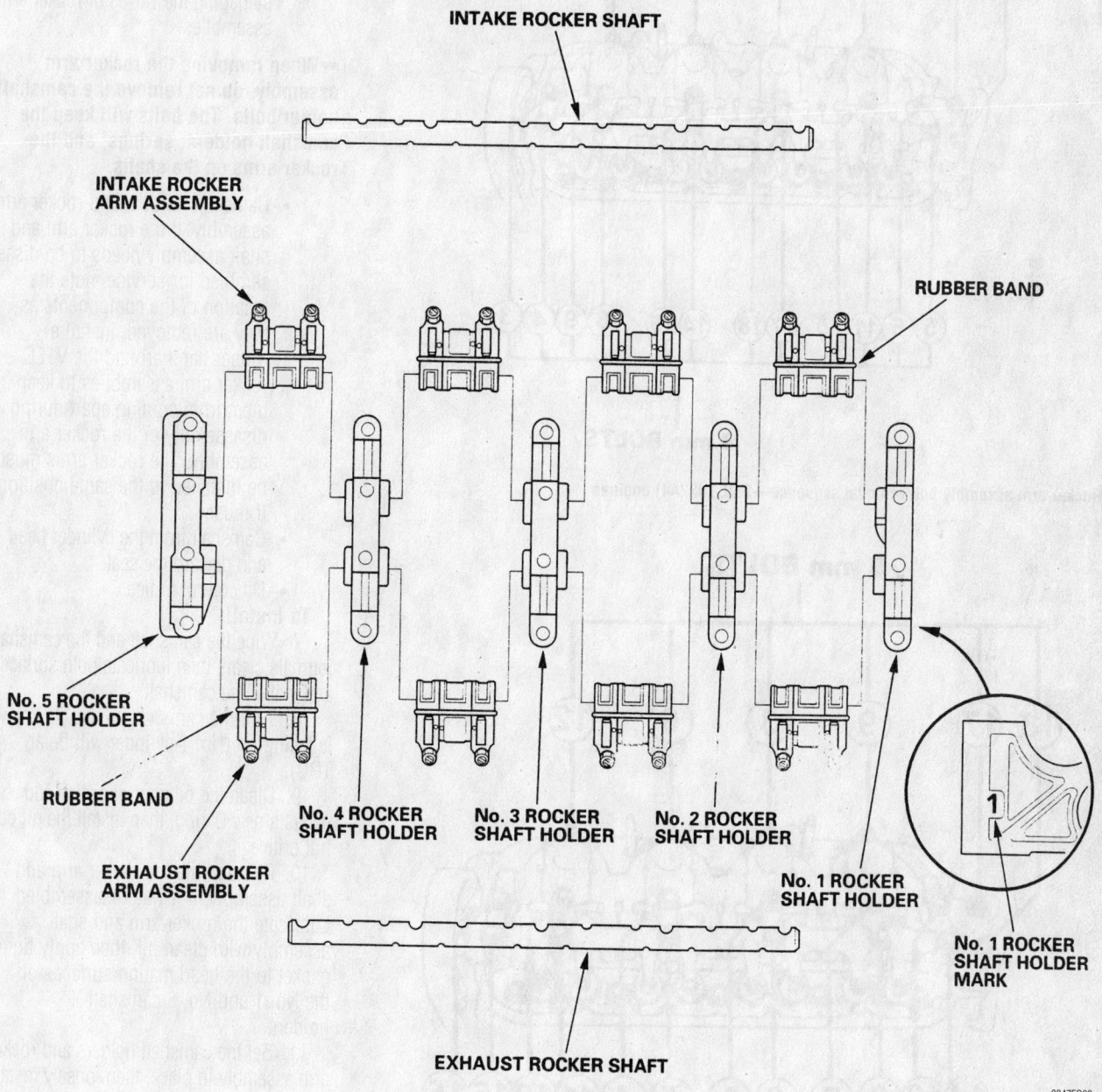

INTAKE ROCKER SHAFT

INTAKE ROCKER ARM ASSEMBLY

RUBBER BAND

No. 5 ROCKER SHAFT HOLDER

RUBBER BAND

EXHAUST ROCKER ARM ASSEMBLY

No. 4 ROCKER SHAFT HOLDER

No. 3 ROCKER SHAFT HOLDER

No. 2 ROCKER SHAFT HOLDER

No. 1 ROCKER SHAFT HOLDER

No. 1 ROCKER SHAFT HOLDER MARK

EXHAUST ROCKER SHAFT

9347FG08

Exploded view of the rocker arm assembly—2.0L engine

Timing belt service is covered in Section 3 of this manual

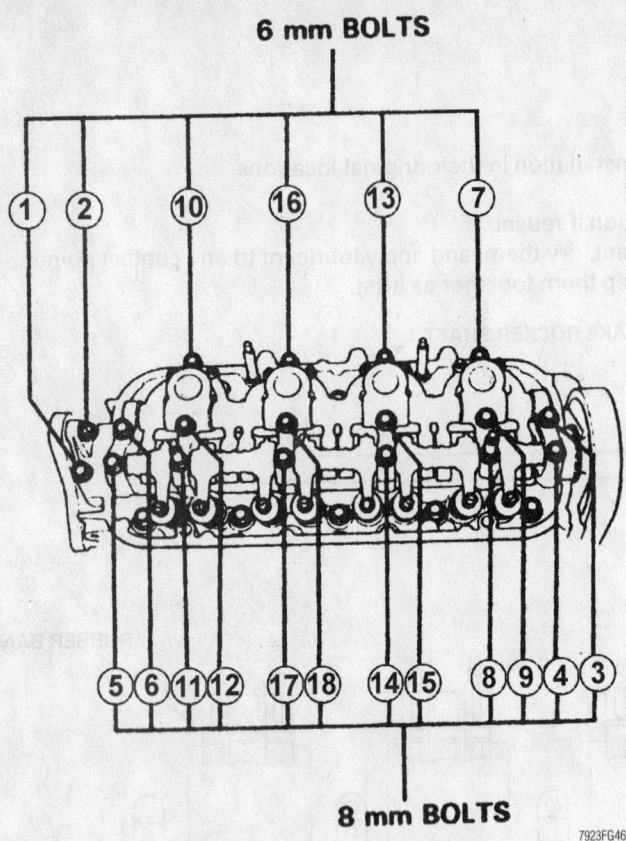

6 mm BOLTS

8 mm BOLTS

7923FG46

Rocker arm assembly bolt removal sequence—2.2L (H22A4) engines

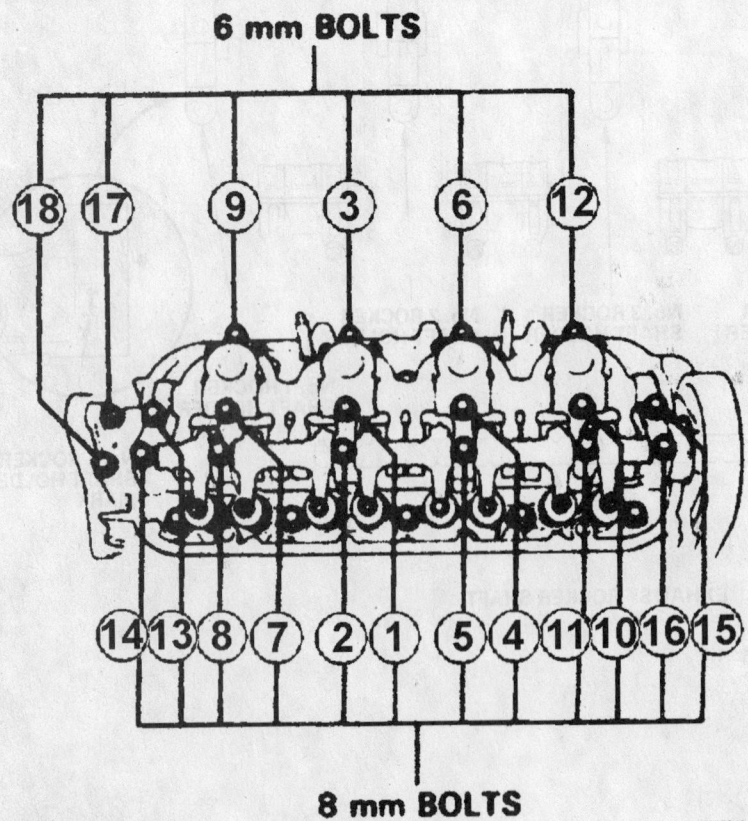

6 mm BOLTS

8 mm BOLTS

7923FG49

Rocker arm assembly torque sequence—2.2L (H22A4) engines

➡ Ensure the words UP embossed on the camshaft pulley is aligned in the upward position.

- Camshaft sprocket bolt, sprocket and sprocket key
- Timing belt back cover
- Valve adjusting screws
- Camshaft holder attaching bolts. Loosen the bolts 2 turns at a time in the proper sequence to prevent damaging the valves or rocker arm assemblies.

➡ When removing the rocker arm assembly, do not remove the camshaft holder bolts. The bolts will keep the camshaft holders, springs, and the rocker arms on the shafts.

- Camshaft holders and rocker arm assembly. If the rocker arm and shaft assembly needs to be disassembled for service, note the location of the components as they are removed. Install a rubber band around the VTEC rocker arm assemblies to keep them from coming apart during disassembly of the rocker arm assembly. The rocker arms must be installed in the same position if reused.
- Camshaft from the cylinder head and discard the seal.
- Oil control orifice

To install:

7. Wipe the camshaft and the camshaft journals clean, then lubricate both surfaces, and install the camshaft.

8. Turn the camshaft so that its keyway is facing up (No. 1 cylinder will be at TDC).

9. Clean the oil control orifice and install a new O-ring, then install the oil control orifice.

10. Reassemble the rocker arm and shaft assembly, if it was disassembled. Lubricate the rocker arm and shaft assembly with clean oil, then apply liquid gasket to the head mating surfaces of the No. 1 and No. 6 camshaft holders.

11. Set the camshaft holders and rocker arm assembly in place, then loosely install the attaching bolts.

12. Apply clean oil to the camshaft oil seal lip and the seal guide (part # 07NAG-PT0010A), then install the seal to the seal guide. Install the seal guide to the camshaft, then the installer cup (part # 07NAF-PT0010A), and the installer shaft (part # 07NAF-PT0020A). Tighten the nut on the

installer shaft to press the seal into the cylinder head.

13. Tighten the camshaft holder bolts 2 turns at a time in the proper sequence. The final torque for the 8mm bolts is 16 ft. lbs. (22 Nm) and the final torque for the 6mm bolts is 108 inch lbs. (12 Nm).

14. Install or connect the following:
- Timing belt back cover and a new gasket, if necessary. Tighten the bolt toward the exhaust manifold to 84 inch lbs. (10 Nm) and tighten the bolt toward the intake manifold to 108 inch lbs. (12 Nm).
- Camshaft sprocket key to the camshaft, then the camshaft sprocket and bolt. Tighten the bolt to 27 ft. lbs. (37 Nm).

15. Ensure the words **UP** embossed on the camshaft pulley is aligned in the upward position, then install the timing belt onto the camshaft sprocket. Loosen, then tighten the timing belt adjusting nut. Refer to the timing belt unit repair section.

16. Rotate the crankshaft pulley 5 or 6 turns to position the timing belt on the pulleys.

17. Set the No. 1 cylinder to TDC and loosen the timing belt adjusting nut 1 turn. Turn the crankshaft counterclockwise until the cam pulley has moved 3 teeth; this creates tension on the timing belt. Loosen, then tighten the adjusting nut, and tighten it to 33 ft. lbs. (45 Nm).

18. Adjust the valves.

19. Tighten the crankshaft pulley bolt to 181 ft. lbs. (245 Nm).

20. Install or connect the following:
- Upper timing belt cover. Tighten the bolt toward the exhaust manifold to 84 inch lbs. (10 Nm) and tighten the bolt toward the intake manifold to 108 inch lbs. (12 Nm).
- Cylinder head cover gasket cover to the groove of the cylinder head cover. Before installing the gasket, thoroughly clean the seal and the groove. Seat the recesses for the camshaft first, then work it into the groove around the outside edges. Be sure the gasket is seated securely in the corners of the recesses.

21. Apply liquid gasket to the 4 corners of the recesses of the cylinder head cover gasket. Do not install the parts if 5 minutes or more have elapsed since applying liquid gasket. After assembly, wait at least

20 minutes before filling the engine with oil.

22. Install or connect the following:
- Cylinder head (valve) cover. Tighten the bolts attaching the cylinder head cover in the proper sequence to 84 inch lbs. (10 Nm).
- PCV hose to the cylinder head cover
- Power steering belt, then adjust the belt.
- Distributor to the cylinder head. Snug the attaching bolts until the timing has been checked and adjusted.
- Spark plug wires and the distributor electrical connectors
- Air intake duct

23. Drain the oil from the engine into a sealable container. Install the drain plug and refill the engine with clean oil.

24. Connect the negative battery cable and enter the radio security code.

25. Start the engine and check carefully for any leaks.

26. Check and adjust the ignition timing as necessary, then tighten the distributor bolts to 13 ft. lbs. (18 Nm).

2.3L Engines

1. Before servicing the vehicle, refer to the precautions in the beginning of this section.

2. Disconnect the negative battery cable.

3. Turn the crankshaft so the No. 1 piston is at TDC.

➡️**The No. 1 piston is at top dead center when the pointer on the block aligns with the white painted mark on the flywheel (manual transaxle) or driveplate (automatic transaxle).**

4. Remove or disconnect the following:
- Air intake duct
- Engine ground cable from the cylinder head cover
- Connector and the terminal from the alternator
- Engine wiring harness from the valve cover
- Ignition coil

➡️**Label all electrical connectors before detaching them.**

- Electrical connectors from the distributor and the spark plug wires from the spark plugs.

➡️**Matchmark the installed position if the distributor before removal.**

- Distributor from the cylinder head
- Ignition coil wire from the distributor
- Positive Crankcase Ventilation (PCV) hose
- Cylinder head cover. Replace the rubber seals if damaged or deteriorated.
- Timing belt middle cover

5. Ensure the words **UP** embossed on the camshaft pulleys are aligned in the upward position.

6. Mark the rotation of the timing belt if it is to be used again. Loosen the timing belt adjusting nut ½ turn, then release the tension on the timing belt. Push the tensioner to release tension from the belt, then tighten the adjusting nut.

7. Remove the timing belt from the camshaft sprockets. Refer to the timing belt unit repair section.

✳✳ WARNING

Do not crimp or bend the timing belt more than 90°, or less than 1 inch (25mm) in diameter

8. Insert a 5.0mm pin punch in each of the camshaft caps, nearest to the sprockets, through the holes provided. Remove the camshaft sprocket attaching bolts, then remove the sprockets. Do not lose the sprocket keys.

9. Remove or disconnect the following:
- Side engine mount bracket B, then the timing belt back cover from behind the camshaft sprockets.
- Rocker arm adjusting screws, then the pin punches from the camshaft caps

➡️**Note the camshaft holders locations for ease of installation. Loosen the bolts in the reverse order of the holder bolts torque sequence.**

- Camshaft holders
- Camshafts from the cylinder head, then discard the camshaft seals.
- Rubber cap from the head, located at the end of the intake camshaft.
- Rocker arms from the cylinder head. Note the locations of the rocker arms.

Heater Core replacement is covered in Section 2 of this manual

Specified torque:
Except Intake ⑤, ⑦. Exhaust ⑥, ⑧:
10 N·m (1.0 kg-m, 7 lb-ft)

Intake ⑤, ⑦. Exhaust ⑥, ⑧:
12 N·m (1.2 kg-m, 9 lb-ft)

TIGHTENING SEQUENCE

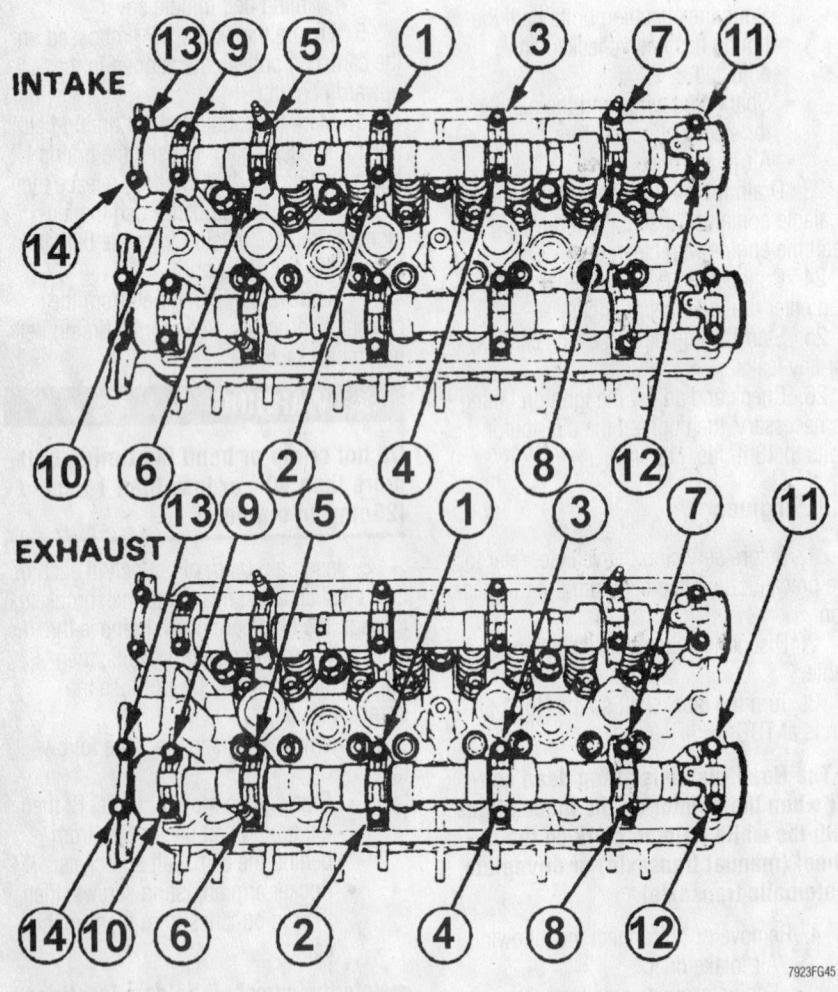

Camshaft holders torque sequence—2.3L engines

➡**The rocker arms have to be installed to their original locations if being reused.**

To install:

➡**Lubricate the rocker arms with clean oil before installation.**

10. Install or connect the following:
- Rocker arms on the pivot bolts and the valve stems. If the rocker arms are being reused, install them to their original locations. The locknuts and adjustment screws should

be loosened before installing the rocker arms.

11. Lubricate the camshafts with clean oil.

12. Install or connect the following:
- Camshaft seals to the end of the camshafts that the timing belt sprockets attach to. The open side (spring) should be facing into the cylinder head when installed.

➡**Be sure the keyways on the camshafts are facing up and install the camshafts to the cylinder head.**

- Rubber plug to the cylinder head at the end of the intake camshaft

13. Apply liquid gasket to the head mating surfaces of the No. 1 and No. 6 camshaft holders, then install them along with No. 2, 3, 4 and 5. I or E marks are stamped on the camshaft holders to identify them as Intake or Exhaust side holders. The arrows stamped on the holders should point toward the timing belt.

14. Snug the camshaft holders in place.

15. Press the camshaft seals securely into place.

16. Tighten the camshaft holder bolts in 2 steps, following the proper sequence, to ensure that the rockers do not bind on the valves. Tighten all the bolts, except the 4 studs, to 84 inch lbs. (10 Nm). Tighten the studs (number 5 and 7 bolts in the correct sequence) to 108 inch lbs. (12 Nm).

17. Install or connect the following:
- Timing belt back cover.
- Side engine mount bracket B. Tighten the bolt attaching the bracket to the cylinder head to 33 ft. lbs. (45 Nm). Tighten the bolts attaching the bracket to the side engine mount to 16 ft. lbs. (22 Nm).

18. Insert a 5.0mm pin punch in each of the camshaft caps, nearest to the pulleys, through the holes provided. Install the keys into the camshaft grooves.

19. Push the camshaft sprockets onto the camshafts, then tighten the retaining bolts to 27 ft. lbs. (38 Nm).

20. Ensure the words **UP** embossed on the camshaft pulleys are aligned in the upward position. Install the timing belt to the camshaft sprockets, then remove the 2, 5.0mm pin punches from the camshaft bearing caps. Refer to the timing belt unit repair section.

21. Loosen, then tighten the timing belt adjuster nut.

22. Turn the crankshaft counterclockwise until the cam pulley has moved 3 teeth; this creates tension on the timing belt. Loosen, then tighten the adjusting nut and tighten it to 33 ft. lbs. (45 Nm).

23. Adjust the valves.

24. Tighten the crankshaft pulley bolt to 181 ft. lbs. (250 Nm).

25. Install or connect the following:
- Middle timing belt cover and tighten the attaching bolts to 108 inch lbs. (12 Nm).
- Cylinder head cover and tighten

the cap nuts to 84 inch lbs. (10 Nm).

- PCV hose to the cylinder head cover
- Distributor. Snug the attaching bolts until the timing has been checked and adjusted.
- Spark plug wires, then the distributor electrical connectors
- Ignition coil wire to the distributor.
- Ignition coil
- Alternator wiring harness to the cylinder head cover.
- Terminal and connector to the alternator.
- Engine ground cable to the cylinder head cover.
- Air intake duct

26. Drain the oil from the engine into a sealable container. Install the drain plug and refill the engine with clean oil.

27. Connect the negative battery cable and enter the radio security code.

28. Start the engine and check carefully for any leaks.

29. Check and adjust the ignition timing. Tighten the distributor bolts to 13 ft. lbs. (18 Nm).

30. If equipped with 4WS, turn the steering wheel lock-to-lock to reset the 4WS control unit.

3.0L Engine

1. Before servicing the vehicle, refer to the precautions in the beginning of this section.

2. Remove or disconnect the following:

- Cylinder head cover
- Jam nuts and screws
- Rocker arm shaft bolts, 2 turns at a time in the sequence shown.

3. Lift the rocker arm assembly from the cylinder head. Leave the bolts in the shafts to retain the rocker arms and springs.

To install:

4. Clean all parts in solvent, dry with compressed air and lubricate with clean engine oil.

5. Place the rocker arm assemblies on the cylinder head and install the bolts loosely. Be sure that all rocker arms are in alignment with their valves.

6. Tighten each bolt 2 turns at a time in the correct sequence. Tighten the bolts to 17 ft. lbs. (24 Nm).

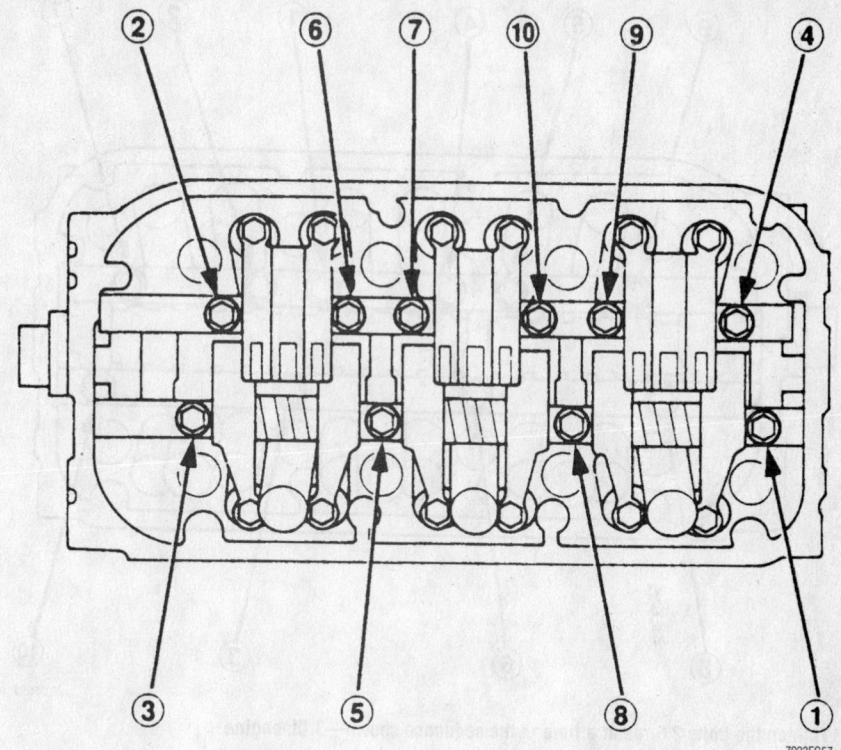

Be sure to loosen the rocker arm shaft bolts in the correct order as shown—3.0L engine

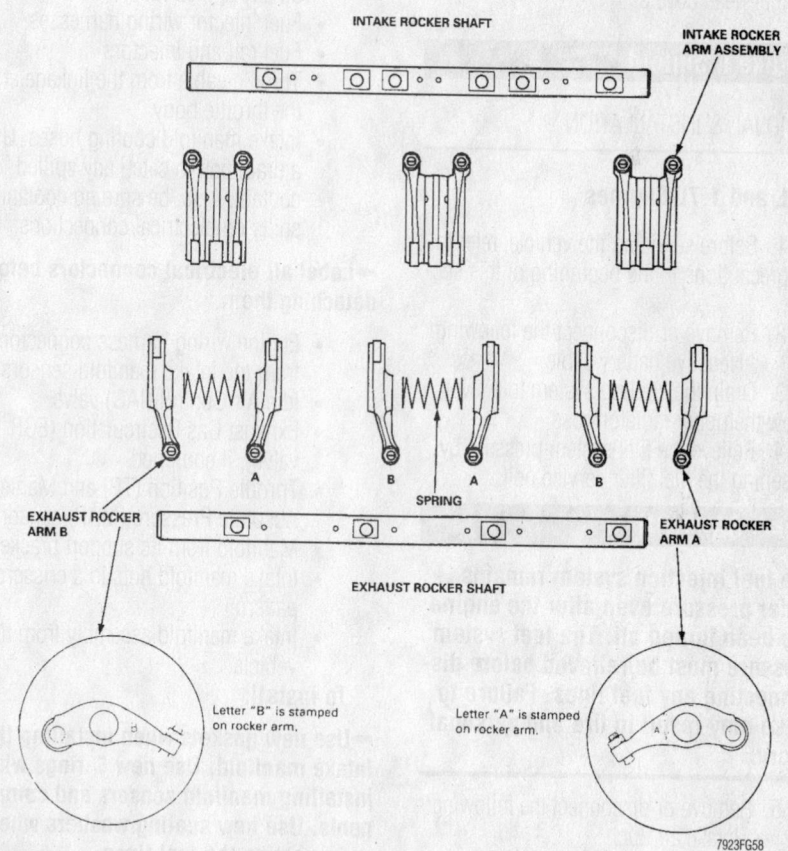

Exploded view of the rocker arms and related components—3.0L engine

Brake service is covered in Section 4 of this manual

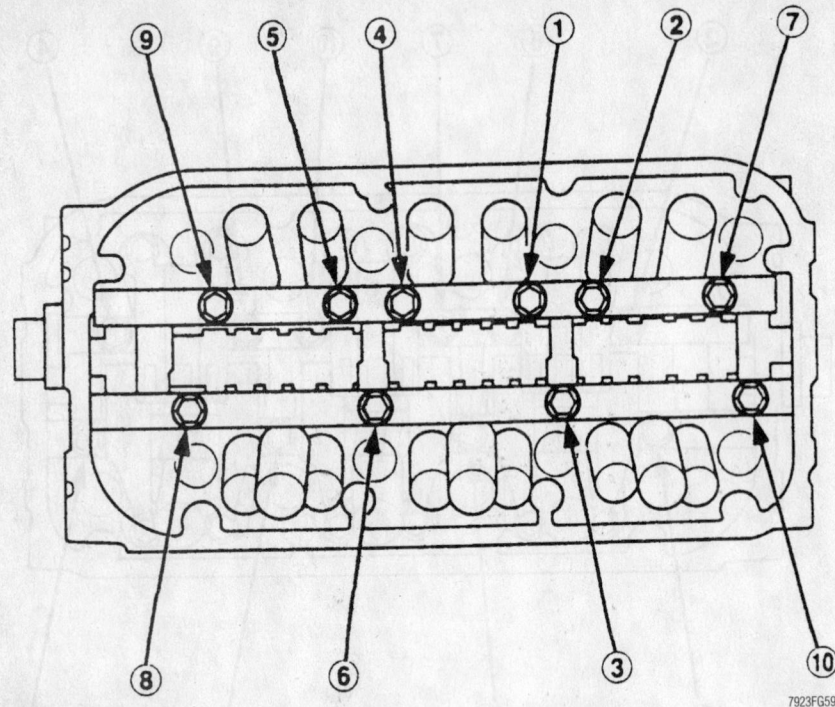

Tighten the bolts 2 turns at a time in the sequence shown—3.0L engine

7. Adjust the valves and install the cylinder head covers.

Intake Manifold

REMOVAL & INSTALLATION

1.6L and 1.7L Engines

1. Before servicing the vehicle, refer to the precautions in the beginning of this section.

2. Remove or disconnect the following:
  - Negative battery cable

3. Drain the cooling system to a level below the upper radiator hose.

4. Relieve the fuel system pressure by loosening the fuel filter service bolt.

✳✳ CAUTION

The fuel injection system remains under pressure even after the engine has been turned off. The fuel system pressure must be relieved before disconnecting any fuel lines. Failure to do so may result in fire and personal injury.

5. Remove or disconnect the following:
  - Intake air duct
  - Air cleaner assembly, if equipped with the D16Y7 engine

➡**Cover the throttle body opening to keep dirt out.**

- Fuel line from the fuel rail. Clean up any spilled fuel.
- Fuel injector wiring harnesses
- Fuel rail and injectors
- Throttle cable from the linkage at the throttle body
- Intake manifold cooling hoses. Use a drain pan to catch any spilled coolant. Also, be sure no coolant spills on electrical connections.

➡**Label all electrical connectors before detaching them.**

- Engine wiring harness connectors from the intake manifold sensors
- Idle Air Control (IAC) valve
- Exhaust Gas Recirculation (EGR valve), if equipped
- Throttle Position (TP) and Manifold Absolute Pressure (MAP) sensor
- Manifold from its support bracket
- Intake manifold nuts in a crisscross pattern.
- Intake manifold assembly from the vehicle.

To install:

➡**Use new gaskets when installing the intake manifold. Use new O-rings when installing manifold sensors and components. Use new sealing washers when reconnecting the fuel lines.**

6. Clean all gasket mating surfaces.
7. Install or connect the following:
  - New intake manifold gaskets

- Intake manifold
8. Tighten the intake manifold nuts in 2 or 3 steps in a crisscross pattern starting with the inside nuts. Tighten the nuts to 17 ft. lbs. (23 Nm).
9. Install or connect the following:
  - Support bracket bolts: 17 ft. lbs. (24 Nm)
  - Fuel rail and injectors
  - Fuel line using new washers
  - EGR valve and tighten its nuts to 15 ft. lbs. (21 Nm).
  - IAC valve. Tighten its mounting bolts to 16 ft. lbs. (22 Nm).
  - Fuel injector wiring harnesses
  - Intake manifold wiring harnesses
  - Intake manifold cooling hoses
  - Throttle cable
  - Intake air duct and air cleaner assembly

10. Refill and bleed the cooling system.
11. Connect the negative battery cable.
12. Verify that all sensors, valves, and vacuum lines are installed and connected properly. Be sure there are no loose electrical connections.
13. Turn the ignition on and off several times without starting the engine to pressurize the fuel system. Run the engine and check for proper operation. Check for vacuum leaks.
14. After the engine has warmed up, check the operation of the throttle cable and adjust it if necessary.

2.0L Engine

1. Before servicing the vehicle, refer to the precautions in the beginning of this section.
2. Drain the cooling system.
3. Relieve the fuel system pressure.
4. Remove or disconnect the following:
  - Negative battery cable
  - Cooling hoses from the intake manifold
  - Vacuum hoses and electrical connectors from the manifold and throttle body
  - Throttle cable from the throttle body
  - Fuel rail and fuel injectors
  - Intake manifold support brackets
  - Intake manifold

To install:

5. Installation is the reverse of the removal procedure, while using the following torque values:
  - Intake manifold fasteners: 16 ft. lbs. (22 Nm)
  - Throttle body fasteners: 16 ft. lbs. (22 Nm)

NOTE: Use new O-rings and gaskets when reassembling.

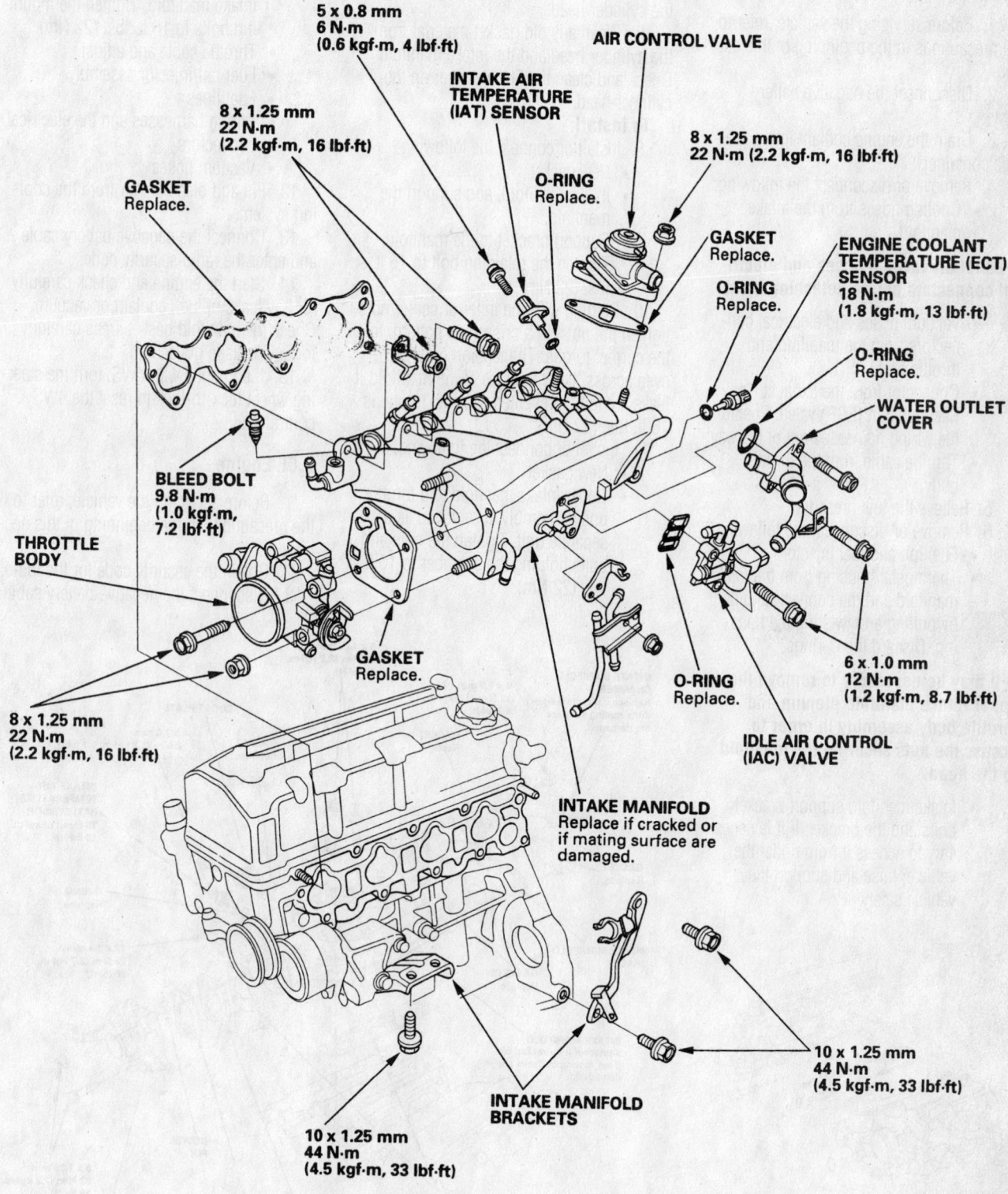

5 x 0.8 mm
6 N·m
(0.6 kgf·m, 4 lbf·ft)

INTAKE AIR
TEMPERATURE
(IAT) SENSOR

AIR CONTROL VALVE

8 x 1.25 mm
22 N·m
(2.2 kgf·m, 16 lbf·ft)

O-RING
Replace.

8 x 1.25 mm
22 N·m (2.2 kgf·m, 16 lbf·ft)

GASKET
Replace.

GASKET
Replace.

O-RING
Replace.

ENGINE COOLANT
TEMPERATURE (ECT)
SENSOR
18 N·m
(1.8 kgf·m, 13 lbf·ft)

O-RING
Replace.

WATER OUTLET
COVER

BLEED BOLT
9.8 N·m
(1.0 kgf·m,
7.2 lbf·ft)

THROTTLE
BODY

8 x 1.25 mm
22 N·m
(2.2 kgf·m, 16 lbf·ft)

GASKET
Replace.

O-RING
Replace.

6 x 1.0 mm
12 N·m
(1.2 kgf·m, 8.7 lbf·ft)

IDLE AIR CONTROL
(IAC) VALVE

INTAKE MANIFOLD
Replace if cracked or
if mating surface are
damaged.

INTAKE MANIFOLD
BRACKETS

10 x 1.25 mm
44 N·m
(4.5 kgf·m, 33 lbf·ft)

10 x 1.25 mm
44 N·m
(4.5 kgf·m, 33 lbf·ft)

9347FG09

Exploded view of the intake manifold—2.0L engine

For complete Engine Mechanical specifications, see Section 1 of this manual

- Intake manifold bracket bolts: 33 ft. lbs. (44 Nm)

2.2L and 2.3L Engines

1. Before servicing the vehicle, refer to the precautions in the beginning of this section.

2. Disconnect the negative battery cable.

3. Drain the engine coolant into a sealable container.

4. Remove or disconnect the following:
- Cooling hoses from the intake manifold

➡ **Label all vacuum hoses and electrical connectors before detaching them.**

- Vacuum hoses and electrical connectors from the manifold and throttle body
- Connector from the Exhaust Gas Recirculation (EGR) valve. Position the wiring harnesses out of the way.
- Throttle cable from the throttle body

5. Relieve the fuel pressure.

6. Remove or disconnect the following:
- Fuel rail and fuel injectors
- Thermostat housing from the intake manifold and the connecting pipe by pulling and twisting the housing. Discard the O-rings.

➡ **It may be necessary to remove the upper intake manifold plenum and throttle body assembly in order to access the nuts securing the manifold to the head.**

- Intake manifold support bracket bolts and the bracket. If it is necessary to access it from under the vehicle; raise and support the vehicle safely.

7. While supporting the intake manifold, remove the nuts attaching the intake manifold to the cylinder head, then remove the manifold. Remove the old gasket from the cylinder head.

8. Clean any old gasket material from the cylinder head and the intake manifold. Check and clean the FIA chamber on the cylinder head.

To install:

9. Install or connect the following:
- New gasket
- Intake manifold, and support the manifold.
- Support bracket to the manifold. Tighten the retaining bolt to 16 ft. lbs. (22 Nm).

10. Starting with the inner or center nuts, tighten the nuts, in a crisscross pattern, to the correct torque. The tension must be even across the entire face of the manifold if leaks are to be prevented. Correct torque is 16 ft. lbs. (22 Nm).

11. Install or connect the following:
- New gasket
- Upper intake manifold and throttle body assembly, if removed as a separate unit. Tighten the nuts and bolts holding the chamber to 16 ft. lbs. (22 Nm).

- New O-ring to the coolant connecting pipe, and to the thermostat housing.
- Housing to the coolant pipe and the intake manifold. Tighten the mounting bolts to 16 ft. lbs. (22 Nm).
- Throttle cable and adjust.
- Fuel rail/injector assembly
- Fuel lines
- Wiring harnesses and the electrical connectors
- Vacuum hoses

12. Fill and bleed the air from the cooling system.

13. Connect the negative battery cable and enter the radio security code.

14. Start the engine and check carefully for any leaks of fuel, coolant or vacuum. Check the manifold gasket areas carefully for any leakage of vacuum.

15. If equipped with 4WS, turn the steering wheel lock-to-lock to reset the 4WS control unit.

3.0L Engine

1. Before servicing the vehicle, refer to the precautions in the beginning of this section.

2. Obtain the security code for the radio.

3. Disconnect the negative battery cable.

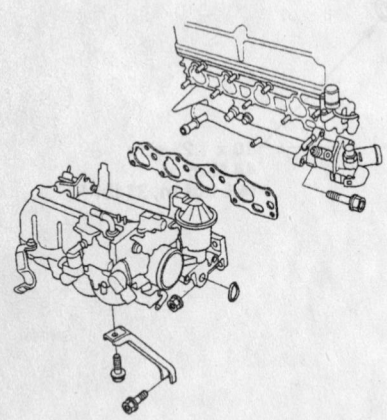

Intake manifold and related components—2.3L engine

7923FG60

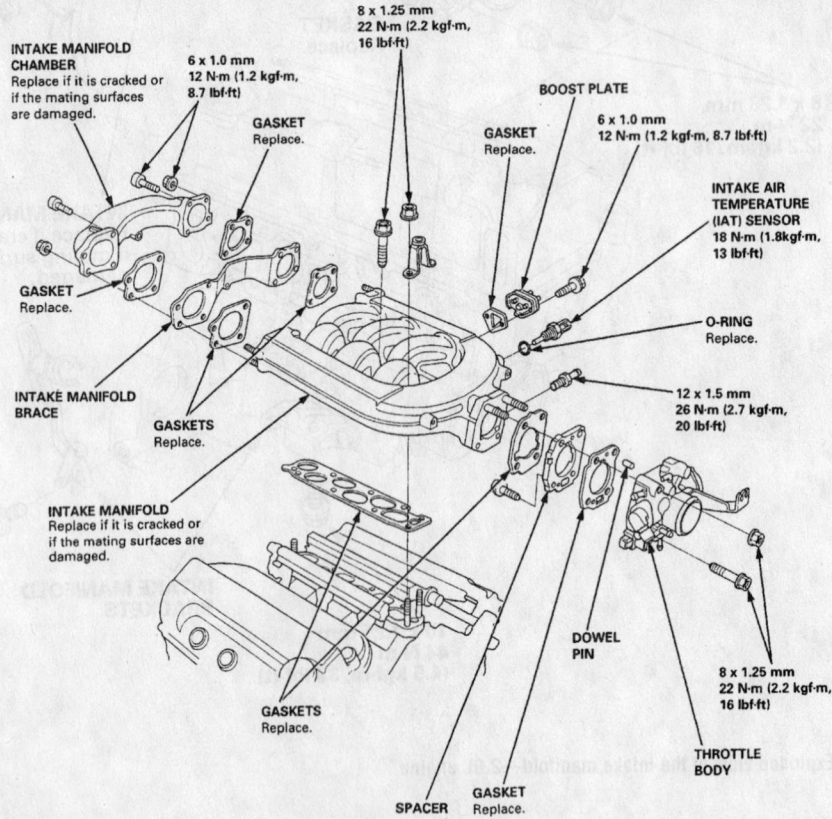

INTAKE MANIFOLD CHAMBER
Replace if it is cracked or if the mating surfaces are damaged.

6 x 1.0 mm
12 N·m (1.2 kgf·m, 8.7 lbf·ft)

8 x 1.25 mm
22 N·m (2.2 kgf·m, 16 lbf·ft)

GASKET
Replace.

BOOST PLATE

GASKET
Replace.

6 x 1.0 mm
12 N·m (1.2 kgf·m, 8.7 lbf·ft)

INTAKE AIR TEMPERATURE (IAT) SENSOR
18 N·m (1.8 kgf·m, 13 lbf·ft)

GASKET
Replace.

O-RING
Replace.

INTAKE MANIFOLD BRACE

GASKETS
Replace.

12 x 1.5 mm
26 N·m (2.7 kgf·m, 20 lbf·ft)

INTAKE MANIFOLD
Replace if it is cracked or if the mating surfaces are damaged.

GASKETS
Replace.

DOWEL PIN

SPACER

GASKET
Replace.

8 x 1.25 mm
22 N·m (2.2 kgf·m, 16 lbf·ft)

THROTTLE BODY

7923FGC4

Exploded view of the intake manifold and related components—3.0L engine

4. Drain the coolant.
5. Remove or disconnect the following:
   - Evaporative emissions (EVAP) canister hose from the throttle body.
   - Air intake duct
   - Upper engine covers
   - Accelerator and cruise control cables from the throttle body.

➡**Ensure that all components have been removed from the intake manifold.**

   - Intake manifold

To install:
6. Clean the mounting surfaces.
7. Install or connect the following:
   - New gasket
   - Intake manifold. Tighten the bolts to 16 ft. lbs. (22 Nm).
   - All removed hoses and wiring on the intake manifold and throttle body.
   - Engine covers
   - Intake air duct
8. Refill the cooling system.
9. Connect the negative battery cable, start the engine, and check for leaks.

Exhaust Manifold

REMOVAL & INSTALLATION

1.6L and 1.7L Engines

1. Before servicing the vehicle, refer to the precautions in the beginning of this section.
2. Disconnect the negative battery cable.
3. Raise and support the front of the vehicle and block the rear wheels.
4. Remove or disconnect the following:
   - Front exhaust pipe from the exhaust manifold/catalytic converter.
   - Exhaust manifold support brackets if their bolts are accessible from this angle. The splash shield may be removed for better access.
5. Lower the vehicle.

➡**Remove any rust or dirt from the exhaust manifold before removal. This will prevent dirt from entering the exhaust pipes.**

6. Remove or disconnect the following:
   - Manifold heat shield
   - Heated Oxygen Sensor (HO2S) harness

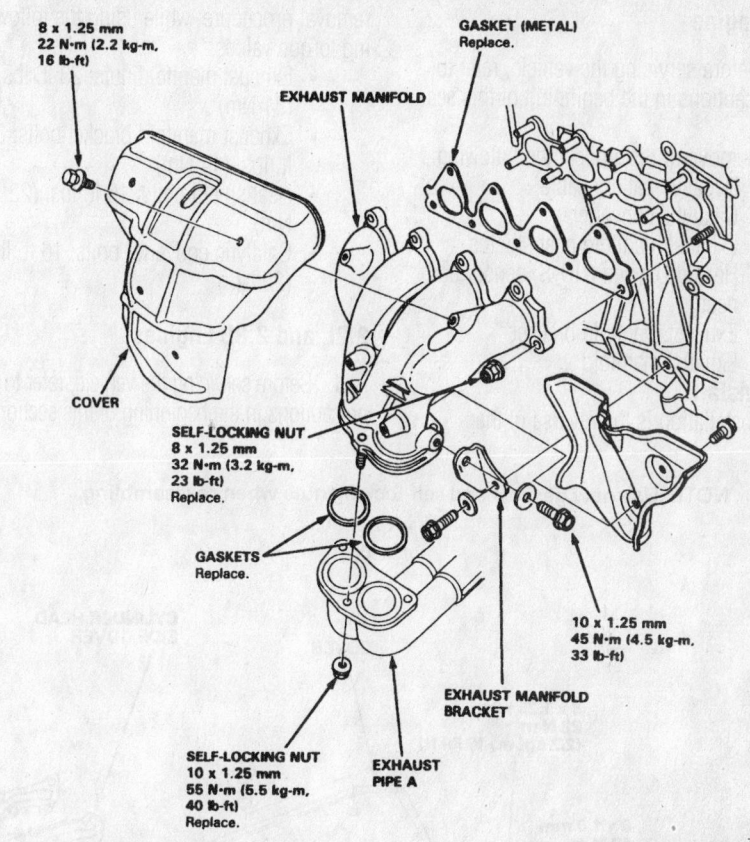

Exhaust manifold components—1.6L engine shown

   - HO2S, using an oxygen sensor socket or box end wrench to unscrew the sensor from the manifold. Handle the sensor carefully.
   - Exhaust manifold brackets
   - Exhaust manifold and separate it from the cylinder head.
   - Exhaust manifold and gasket.

To install:

➡**Use new gaskets and self-locking nuts when installing the exhaust manifold.**

7. Clean the gasket mating surfaces of the manifold and cylinder head ports.
8. Install or connect the following:
   - New gasket onto the cylinder head.
   - New gaskets onto the exhaust pipe flange.
   - Exhaust manifold. Apply anti-seize paste to the studs. Tighten the self-locking nuts to 23 ft. lbs. (32 Nm) in a crisscross pattern starting in the center of the manifold and working outward.
   - Manifold brackets and tighten their bolts to 17 ft. lbs. (24 Nm) for the

B16A2 engines and 33 ft. lbs. (45 Nm) for all other engines.
9. Carefully coat only the threads of the oxygen sensor body with anti-seize paste. Don't get any anti-seize on the sensor probe.
10. Install or connect the following:
   - HO2S and carefully tighten it to 33 ft. lbs. (45 Nm).
   - Heat shield and tighten the bolts to 16 ft. lbs. (22 Nm).
   - HO2S connector
11. Raise and support the front of the vehicle and block the rear wheels.
12. Install or connect the following:
   - Front exhaust pipe and the exhaust manifold/catalytic converter. Tighten the self-locking nuts to 40 ft. lbs. (55 Nm), if the converter is not attached to the manifold. If the converter is attached, tighten to 25 ft. lbs., (34 Nm). Install any manifold brackets and tighten them to 33 ft. lbs. (45 Nm).
   - Splash shield if it was removed.
13. Lower the vehicle and connect the negative battery cable.
14. Run the engine and check for exhaust leaks.

For Accessory Drive Belt illustrations, see Section 1 of this manual

2.0L Engine

1. Before servicing the vehicle, refer to the precautions in the beginning of this section.
2. Remove or disconnect the following:
   - Negative battery cable
   - Catalytic converter
   - Exhaust manifold heat shields
   - Heated Oxygen (HO2S) sensor connector
   - Exhaust manifold bracket
   - Exhaust manifold

To install:

3. Installation is the reverse of the removal procedure, while using the following torque values:
   - Exhaust manifold nuts: 23 ft. lbs. (31 Nm)
   - Exhaust manifold bracket bolts: 33 ft. lbs. (44 Nm)
   - Heat shield bolts: 16 ft. lbs. (22 Nm)
   - Catalytic converter bolts: 16 ft. lbs. (22 Nm)

2.2L and 2.3L Engines

1. Before servicing the vehicle, refer to the precautions in the beginning of this section.

2. Remove or disconnect the following:
   - Negative battery cable
3. Safely raise and support the vehicle.
4. Remove or disconnect the following:
   - Oxygen Sensor (O2S) connector, if it is located in the manifold.
   - Exhaust manifold upper cover
   - Heat insulator from the manifold, if equipped with air conditioning.
   - Nuts attaching the exhaust manifold to the front exhaust pipe.
   - Pipe from the manifold and discard the gasket. Support the pipe with

NOTE: Use new gaskets and self-locking nuts when reassembling.

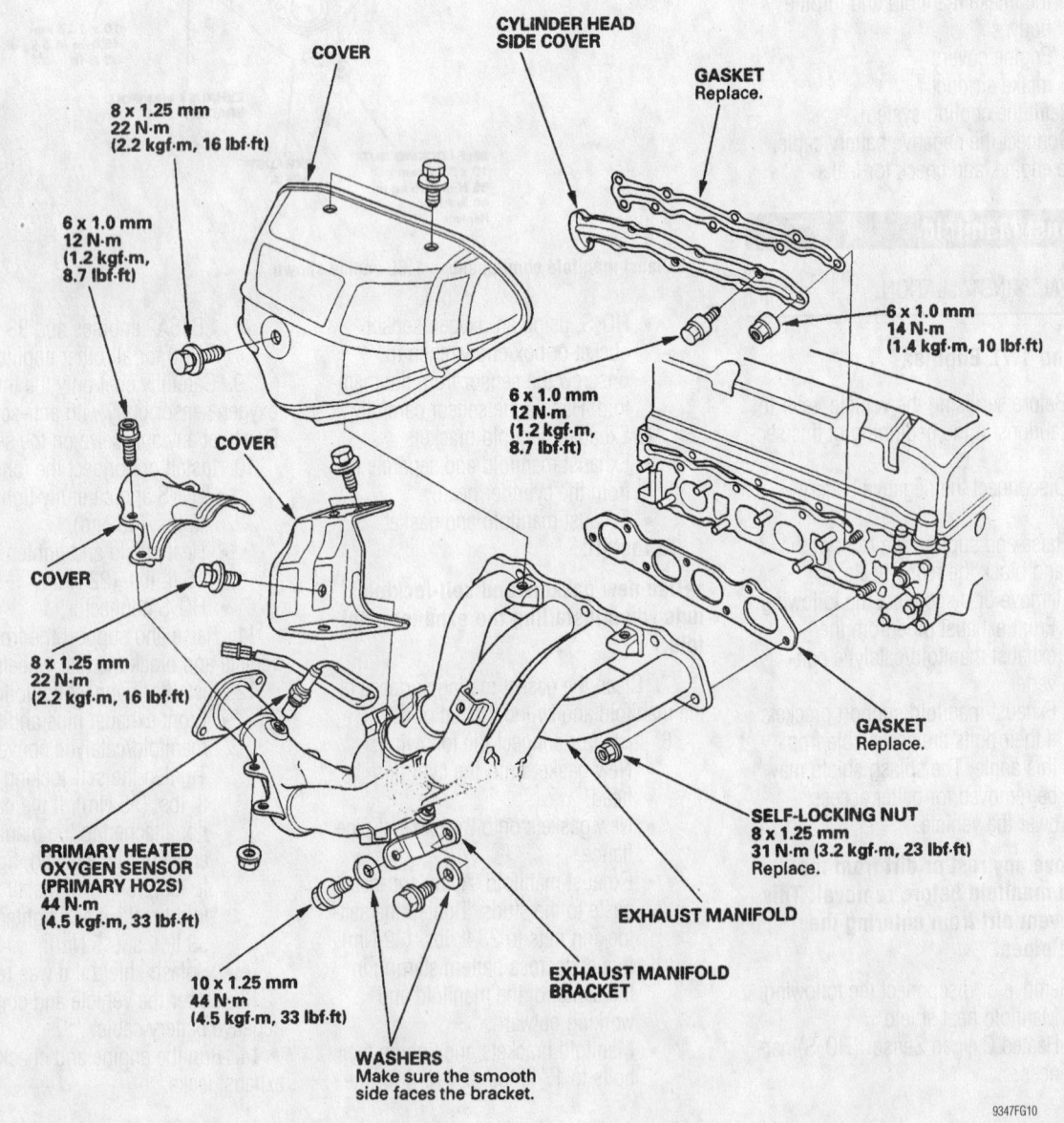

Exploded view of the exhaust manifold—2.0L engine

9347FG10

wire; do not allow it to hang by itself.

- Exhaust manifold bracket(s) bolts and bracket(s).
- Exhaust manifold attaching nuts, using a crisscross pattern (starting from the center).
- Manifold and discard the gasket. Clean the manifold and cylinder head mating surfaces.
- Lower manifold cover from the manifold, if equipped.

To install:

5. Install or connect the following:
- Lower manifold cover, if equipped, and tighten the attaching bolts to 16 ft. lbs. (22 Nm).
- New gasket
- Exhaust manifold into position and support it
- New nuts snug on the studs
- Support bracket(s) below the manifold. Tighten the bracket(s) mounting bolts to 33 ft. lbs. (44 Nm).

6. Starting with the manifold inner or center nuts, tighten the nuts in a crisscross pattern to the correct torque. The tension must be even across the entire face of the manifold if leaks are to be prevented. Tighten the nuts to 23 ft. lbs. (31 Nm).

7. Install or connect the following:
- Heat insulator to the manifold, if equipped with air conditioning. Tighten the attaching bolts to 84 inch lbs. (10 Nm) on Prelude models and 108 inch lbs. (12 Nm) on Accord models.
- Upper manifold cover and tighten the bolts to 16 ft. lbs. (22 Nm).
- Oxygen sensor connector, if detached
- Front exhaust pipe using new gaskets and nuts. Tighten the exhaust pipe attaching nuts to 40 ft. lbs. (55 Nm).
- Negative battery cable and enter the radio security code.

8. Start the engine and check for exhaust leaks.

9. If equipped with 4WS, turn the steering wheel lock-to-lock to reset the 4WS control unit.

3.0L Engine

1. Before servicing the vehicle, refer to the precautions in the beginning of this section.

2. Raise and safely support the vehicle.

3. Remove or disconnect the following:

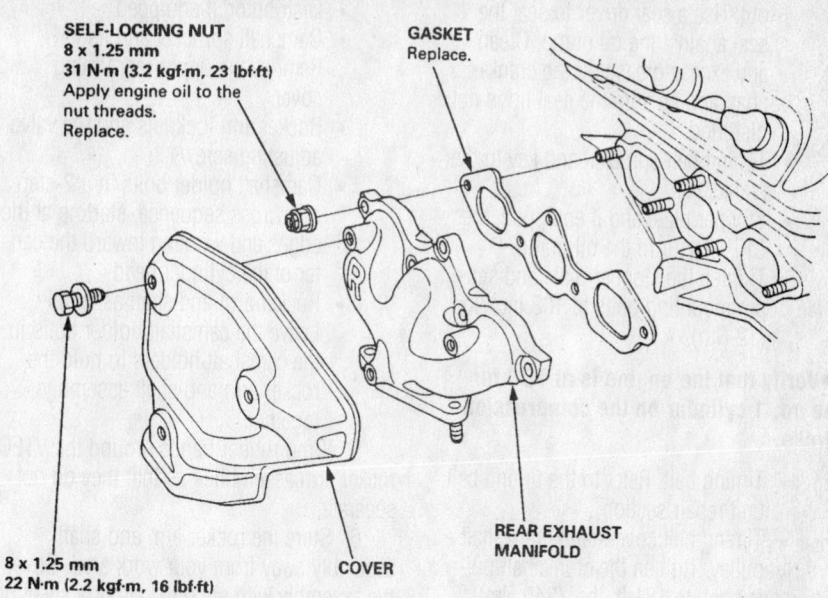

SELF-LOCKING NUT
8 x 1.25 mm
31 N·m (3.2 kgf·m, 23 lbf·ft)
Apply engine oil to the nut threads.
Replace.

GASKET
Replace.

REAR EXHAUST MANIFOLD

COVER

8 x 1.25 mm
22 N·m (2.2 kgf·m, 16 lbf·ft)

7923FG93

Exploded view of the rear exhaust manifold mounting—3.0L engine

- Engine undercover
- Exhaust pipe from the manifold to be removed

4. Lower the vehicle.

5. Remove or disconnect the following:
- Exhaust manifold heat shield
- Mounting nuts and the exhaust manifold.

To install:

6. Clean the mounting surfaces.

7. Install or connect the following:
- New gasket on the cylinder head
- Exhaust manifold. Tighten the nuts to 23 ft. lbs. (31 Nm).
- Heat shield. Tighten the bolts to 16 ft. lbs. (22 Nm).

8. Raise the vehicle and connect the exhaust pipe to the manifold using a new gasket. Tighten the nuts to 40 ft. lbs. (54 Nm).

Front Crankshaft Seal

REMOVAL & INSTALLATION

➡**The original radio may contain a coded anti-theft circuit. Obtain the security code number before disconnecting the battery cables.**

1. Before servicing the vehicle, refer to the precautions in the beginning of this section.

2. Disconnect the negative battery cable.

3. Safely raise and support the vehicle.

4. Remove or disconnect the following:
- Splash shield
- Engine accessory drive belts

5. Turn the engine to align the timing marks and set cylinder No.1 to TDC. The white mark on the crankshaft pulley should align with the pointer on the timing belt cover. Remove the inspection caps on the upper timing belt covers to check the alignment of the timing marks. The pointers for the camshafts should align with the green marks on the camshaft sprockets.

6. Remove or disconnect the following:
- Upper timing belt covers and crankshaft pulley.
- Lower timing belt cove

➡**Mark the direction of the timing belt rotation if it is to be reinstalled.**

- Timing belt. Refer to the timing belt unit repair section.
- Crankshaft Position (CKP) sensor from the oil pump, if equipped.
- Stopper plate
- Timing belt sprocket from the crankshaft. Do not lose the sprocket key.
- Seal from the front of the engine, using a suitable seal removal tool.

To install:

7. Clean the seal mounting surfaces on the engine block.

8. Apply a thin coat of grease on the crankshaft and seal lips.

9. Install or connect the following:
- Seal with the part number facing

out. Use a seal driver to seat the seal against the oil pump. Clean any excess grease off the crankshaft and be sure the seal lip is not distorted.

- Timing belt sprocket and key to the crankshaft.
- Stopper plate and if equipped, the CKP sensor to the oil pump. Tighten the stopper plate and sensor mounting bolts to 108 inch lbs. (12 Nm).

➡**Verify that the engine is at TDC for the no. 1 cylinder on the compression stroke.**

- Timing belt. Refer to the timing belt unit repair section.
- Timing belt covers and crankshaft pulley. Tighten the crankshaft pulley bolt to 181 ft. lbs. (245 Nm), with the aid of a crank pulley holder.
- Accessory drive belts, then adjust.

➡**Verify that all engine components that may have been removed have been reinstalled correctly.**

- Splash shield and lower the vehicle.
- Negative battery cable
10. Top up the engine oil if necessary.
11. Run the engine and check for leaks.

Camshaft

REMOVAL & INSTALLATION

1.6L (D16Y5, D16Y7, D16Y8) and 1.7L (D17A1, D17A2) Engines

1. Before servicing the vehicle, refer to the precautions in the beginning of this section.
2. Remove or disconnect the following components:
- Negative battery cable
- Ignition wires
- Valve cover and the upper timing belt cover.
3. Rotate the crankshaft to set the No. 1 cylinder at TDC for the compression stroke. Once the engine is in this position, it shouldn't be disturbed.
4. Remove or disconnect the following components:
- Timing belt. If the timing belt is contaminated with oil or coolant, it must be replaced. If the timing belt is to be reused, mark its direction of rotation. Refer to the timing belt unit repair section.

- Distributor, if equipped
- Camshaft sprocket and its key. Remove the upper rear timing cover.
- Rocker arm locknuts and the valve adjusting screws.
- Camshaft holder bolts in a 2-step crisscross sequence, starting at the edges and working toward the center of the cylinder head.
- Rocker arm and shaft assembly. Leave the camshaft holder bolts in the camshaft holders to hold the rocker arm and shaft assembly together.
5. Wrap rubber bands around the VTEC rocker arm assemblies so that they do not separate.
6. Store the rocker arm and shaft assembly away from your work area. Cover the assembly with shop towels or a sheet of plastic to protect it from dust.
7. Lift the camshaft from the cylinder head. Remove the camshaft seal.
8. Inspect the camshaft journals and lobes for signs of scoring or other damage.

To install:

9. Remove the oil control orifice. Thoroughly clean it and reinstall it with a new O-ring.

10. Clean and inspect the camshaft bearing caps in the cylinder head.
11. Lubricate the lobes and journals of the camshaft prior to installation.
12. Install or connect the following:
- Camshaft with the keyway facing up so that the camshaft will be at TDC/compression for the No. 1 cylinder.
- New, lightly lubricated, camshaft seal
13. Install the rocker arm and shaft assembly as follows:
   a. Remove the rubber bands from the VTEC rocker arms.
   b. Lubricate the rocker arm contact surfaces.
   c. Apply liquid gasket to the head mating surfaces of the No. 1 and No. 5 camshaft holders. Don't allow the sealant to cure before installing the rocker arm assembly.
   d. Set the rocker arm and shaft assembly in place. If equipped, install the lost motion assembly holder.
   e. Coat the threads of the camshaft holder bolts with clean oil and loosely install them.
   f. Tighten each bolt 2 turns at a time in the crisscross sequence to ensure that

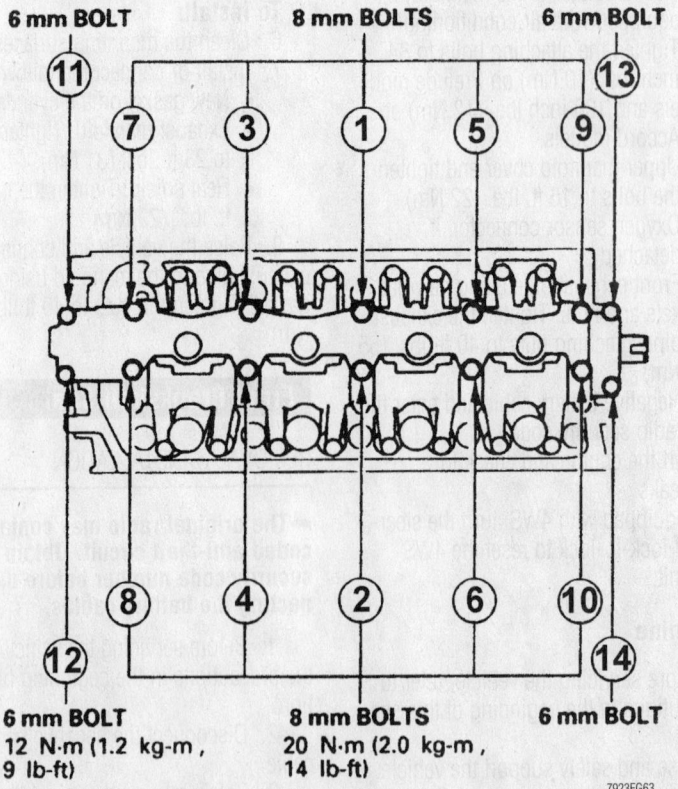

6 mm BOLT   **8 mm BOLTS**   **6 mm BOLT**

6 mm BOLT
12 N·m (1.2 kg-m, 9 lb-ft)

8 mm BOLTS
20 N·m (2.0 kg-m, 14 lb-ft)

6 mm BOLT

7923FG63

Camshaft holder bolt tightening sequence—1.6L (D16Y5, D16Y7, D16Y8) and 1.7L (D17A1, D17A2) engines

the rockers and camshaft holder do not bind on the camshaft journals.

    g. Tighten the 8mm camshaft holder bolts to 14 ft. lbs. (20 Nm), and the 6mm camshaft holder bolts to 108 inch lbs. (12 Nm).

14. Install or connect the following:
- Camshaft sprocket and key. Tighten the retaining bolt to 27 ft. lbs. (38 Nm).

➡**Verify that the engine remains at TDC/compression for the No. 1 cylinder.**

- Distributor, if equipped
- Timing belt. Tighten the tensioner bolt to 33 ft. lbs. (44 Nm) once the belt has been properly tensioned. Refer to the timing belt unit repair section.
- Lower timing cover. Tighten the crankshaft pulley bolt to 134 ft. lbs. (181 Nm).

15. Adjust the valves.

16. Manually inspect the VTEC rocker arms for smooth motion.

17. Be sure all the spark plug tube sealing gaskets are fully seated.

18. Apply liquid gasket to the corner recesses of a new valve cover gasket.

19. Install or connect the following:
- Gasket to the valve cover. Don't allow the sealant to cure before installation.
- Valve cover. Gently wiggle the valve cover to be sure it is fully seated. Tighten the valve cover bolts in a crisscross pattern to 84 inch lbs. (10 Nm).
- Ignition wires

20. Refill the engine with fresh oil and install a new filter.

21. Reconnect the battery cable.

22. Warm the engine up to normal operating temperature. Check for oil leaks.

23. Check the ignition timing and adjust it if necessary. Then, tighten the distributor mounting bolts to 17 ft. lbs. (24 Nm).

1.6L (B16A2) Engine

1. Before servicing the vehicle, refer to the precautions in the beginning of this section.

2. Remove or disconnect the following:
- Negative battery cable
- Ignition wires

3. Rotate the crankshaft to set the engine at TDC for the compression stroke of the No. 1 cylinder. The white TDC mark on the crankshaft pulley should align with the pointer on the lower timing belt cover.

4. Remove or disconnect the following:
- Strut brace
- Intake air duct
- Accessory drive belts

5. Use a floor jack padded with a block of wood to support the engine.

6. Remove or disconnect the following:
- Engine ground cable
- Engine side mount
- Valve cover and the upper timing belt cover

7. Verify that the engine is set at TDC/compression. Loosen the timing belt tensioner bolt 180°. Then, remove the crankshaft pulley, the lower timing cover, and timing belt. Refer to the timing belt unit repair section.

⁕⁕ WARNING

Inspect the timing belt for signs of cracked and broken teeth, as well as oil or coolant contamination. If the timing belt is damaged, or has been in contact with oil or coolant, it must be replaced to avoid potential failure.

8. Remove or disconnect the following:
- Distributor
- Variable Timing Electronic Control (VTEC) solenoid valve.
- Solenoid valve's filter and inspect it for clogging.
- Camshaft sprockets and back cover.
- Camshaft holder plate bolts in a crisscross sequence working toward the middle of the cylinder head.
- Valve adjusting screws
- Camshaft holder plates and holders from the cylinder head. The holder bolts will keep the components together. Note the positions of each camshaft holder for reassembly.
- Camshafts from the cylinder head. Mark the exhaust and intake camshafts so that they will not be confused.
- Intake and exhaust oil control orifices. Thoroughly clean each, and reinstall them with new O-rings.

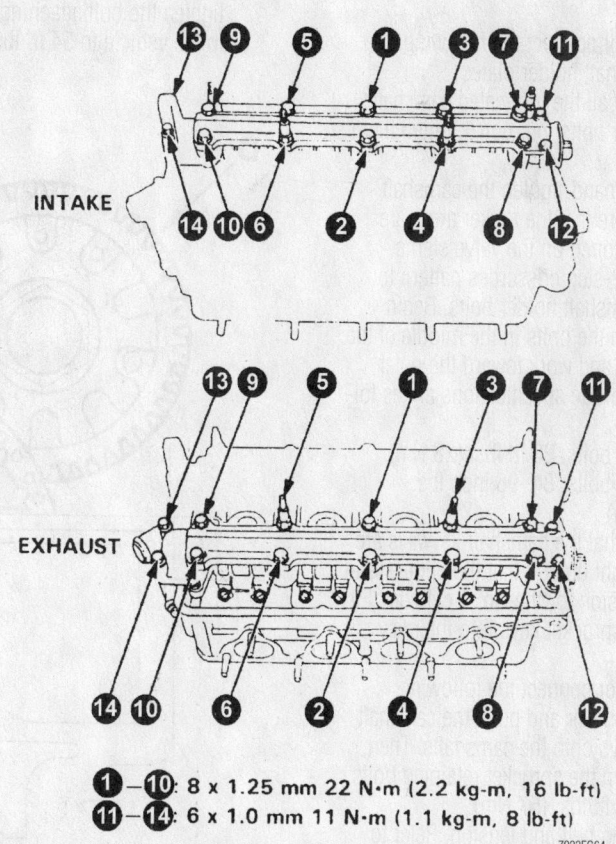

1—**10**: 8 x 1.25 mm 22 N·m (2.2 kg-m, 16 lb-ft)
11—**14**: 6 x 1.0 mm 11 N·m (1.1 kg-m, 8 lb-ft)

7923FG64

Camshaft holder bolt tightening sequences—1.6L (B16A2) engines

9. Inspect the camshaft lobes and journals for any signs of damage.

To install:

10. Install or connect the following:
- New O-ring and the dowel pin to the oil passage of the No. 3 camshaft holder.

➡️**Lubricate the camshaft lobes and journals with clean engine oil.**

- Camshafts into the cylinder head. Both the intake and exhaust camshafts should be installed with their keyways pointing straight up.
- New camshaft seals. Apply liquid gasket to a new camshaft end-plug and install it. If the end-plug has a mark, the mark should be aligned with the cylinder head surface.

11. Apply liquid gasket to the cylinder head mating surfaces of the No. 1 and No. 5 camshaft holders, then install them, along with No. 2, 3, and 4 holders. Be sure to pay attention to the following points:
- Do not apply oil to the holder mating surface of camshaft seals.
- The arrows marked on the camshaft holders should point to the timing belt.

12. Install or connect the following:
- Camshaft holder plates
- Install all the lubricated camshaft holder bolts, but don't tighten them yet.

13. Evenly hand-tighten the camshaft holders. Be sure that the rocker arms are properly positioned on the valve stems.

14. Use a 2-step crisscross pattern to tighten the camshaft holder bolts. Begin tightening with the bolts in the middle of the cylinder head, and work toward the outer edges. Final torque specifications are as follows:
 a. 8mm bolts: 20 ft. lbs. (28 Nm).
 b. 6mm bolts: 84–96 inch lbs. (10–11 Nm)

15. Verify that the camshaft keyways are pointing straight up and that the engine is at TDC/compression for the No. 1 cylinder. Fit the camshaft sprocket keys into their keyways.

16. Install or connect the following:
- Back cover and push the camshaft pulleys onto the camshafts. Then, tighten the sprocket retaining bolts to 41 ft. lbs. (57 Nm).
- Timing belt and tension. Refer to the timing belt unit repair section.
- Lower timing cover and the crankshaft pulley. Tighten the pulley bolt to 130 ft. lbs. (180 Nm).

17. Adjust the valve clearance.

18. Inspect the VTEC rocker arms for smooth and independent movement.

19. Install or connect the following:
- VTEC solenoid valve with a new filter. Tighten the valve mounting bolts to 108 inch lbs. (12 Nm).
- Distributor

20. Clean the valve cover gasket surfaces. Fit the gasket into the groove of the valve cover.

21. Apply liquid gasket to the rubber seal at the 8 corners where the gasket meets the camshaft holders. Don't allow the sealant to cure before installing the cylinder head cover.

22. Install or connect the following:
- Cylinder head cover and engine ground cable. Be sure the contact surfaces are clean and do not touch surfaces where liquid gasket has been applied.
- Valve cover nuts: 84 inch lbs. (10 Nm) in a crisscross sequence.
- Upper timing cover
- Accessory drive belts and adjust their tensions.
- Engine side mount, tighten the 2 new nuts to 54 ft. lbs. (75 Nm) and tighten the bolt attaching the mount to the vehicle to 54 ft. lbs. (74 Nm).

- Strut bar and tighten the mounting bolts to 16 ft. lbs. (22 Nm).
- Ignition wires

23. Drain the engine oil. Install a new oil filter and refill the engine with fresh oil.

24. Reconnect the negative battery cable.

25. Warm the engine up to its normal operating temperature. Then, check and adjust the ignition timing. Tighten the distributor mounting bolts to 17 ft. lbs. (24 Nm).

2.0L Engine

1. Before servicing the vehicle, refer to the precautions in the beginning of this section.

2. Loosen the valve adjustment screws so that all valves are closed and all rocker arms are loose.

3. Remove or disconnect the following:
- Negative battery cable
- Valve cover
- Camshaft bearing caps
- Camshafts

To install:

4. Set the engine to Top Dead Center (TDC) so that the timing chain sprocket timing marks are aligned with the cylinder head surface as shown.

5. Install or connect the following:

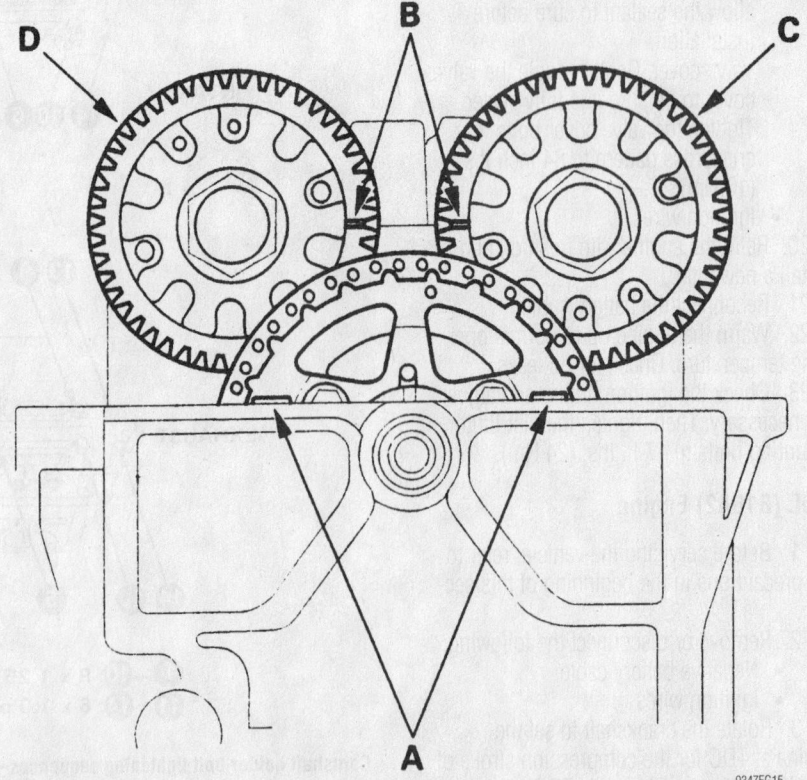

Timing chain sprocket alignment marks (A) and camshaft sprocket alignment marks (B)—2.0L engine

9347FG15

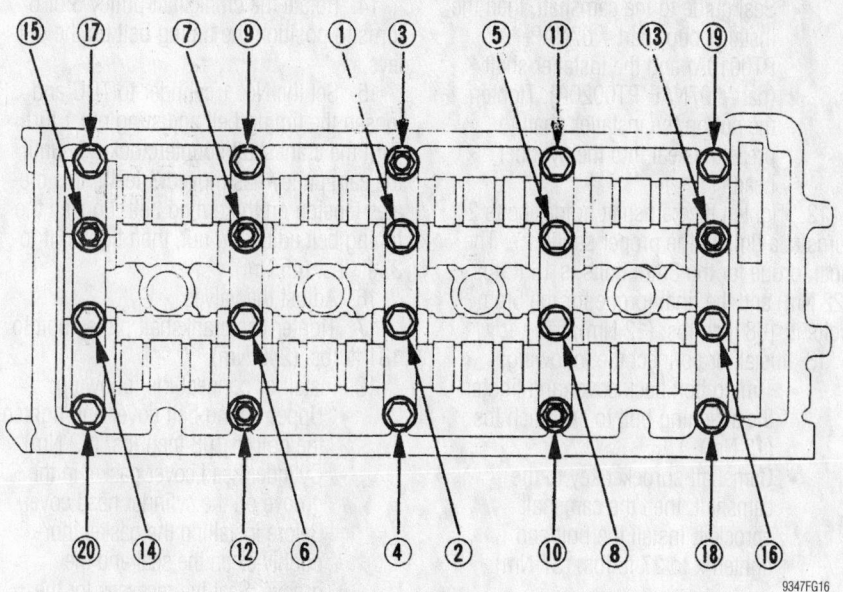

Camshaft bearing cap torque sequence—2.0L engine

- Camshafts with the sprocket timing marks aligned as shown
- Camshaft bearing caps and tighten the bolts in sequence to 16 ft. lbs. (22 Nm). Adjust the valve clearance.
- Valve cover
- Negative battery cable

2.2L Engine

1. Before servicing the vehicle, refer to the precautions in the beginning of this section.

2. Remove or disconnect the following:
- Negative battery cable
- Air intake duct
- Cylinder head cover and replace the rubber seals if damaged or deteriorated.
- Timing belt upper cover

3. Turn the engine to align the timing marks and set cylinder No.1 to TDC/compression. Once in this position, the engine must NOT be turned or disturbed.

4. Remove or disconnect the following:
- Electrical connectors from the distributor and the spark plug wires from the spark plugs.

➡ **Matchmark the installed position of the distributor before removing it.**

- Distributor from the cylinder head
- Power steering pump drive belt

5. Mark the rotation of the timing belt if it is to be used again. Loosen the timing belt adjusting bolt ¾ to 1 turn, then release the tension on the timing belt. Push the tensioner to release tension from the belt, then tighten the adjusting bolt.

6. Remove or disconnect the following:
- Timing belt from the camshaft

sprocket. Refer to the timing belt unit repair section.

✳✳ WARNING

Do not crimp or bend the timing belt more than 90˚, or less than 1 inch (25mm) in diameter

➡ Ensure the words UP embossed on the camshaft pulley is aligned in the upward position before removing the sprocket bolt.

- Camshaft sprocket bolt, sprocket and sprocket key.
- Timing belt back cover
- Valve adjusting screws
- Camshaft holder attaching bolts 2 turns at a time, in the proper sequence to prevent damaging the valves or rocker arm assemblies.

➡ When removing the rocker arm assembly, do not remove the camshaft holder bolts. The bolts will keep the camshaft holders, springs, and the rocker arms on the shafts.

- Camshaft holders and rocker arm

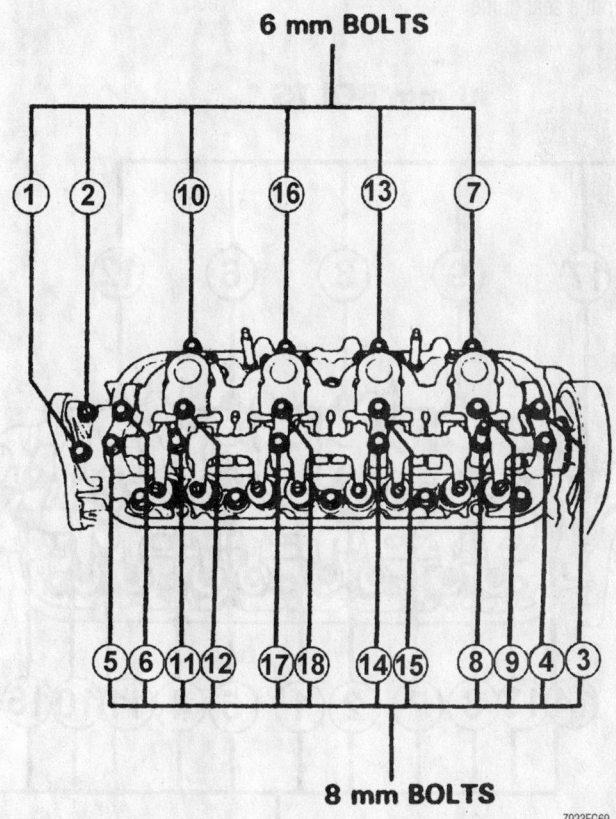

Rocker arm assembly bolt removal sequence—2.2L (H22A4) engines

assembly. If the rocker arm and shaft assembly needs to be disassembled for service, note the location of the components as they are removed. The rocker arms must be installed in the same position if reused.

- Camshaft from the cylinder head and discard the seal.

To install:

7. Wipe the camshaft and the camshaft journals clean, then lubricate both surfaces.

8. Install or connect the following:
- Camshaft

9. Turn the camshaft so that its keyway is facing up (No. 1 cylinder will be at TDC).

10. Reassemble the rocker arm and shaft assembly if it was disassembled. Lubricate the rocker arm and shaft assembly with clean oil, then apply liquid gasket to the head mating surfaces of the No. 1 and No. 6 camshaft holders.

11. Install or connect the following:
- Camshaft holders and rocker arm assembly in place, then loosely install the attaching bolts.
- Clean oil to the camshaft oil seal lip and the seal guide (part # 07NAG-PT0010A)
- Seal to the seal guide

- Seal guide to the camshaft, then the installer cup (part # 07NAF-PT0010A) and the installer shaft (part # 07NAF-PT0020A). Tighten the nut on the installer shaft to press the seal into the cylinder head.

12. Tighten the camshaft holder bolts 2 turns at a time in the proper sequence. The final torque for the 8mm bolts is 16 ft. lbs. (22 Nm) and the final torque for the 6mm bolts is 108 inch lbs. (12 Nm).

13. Install or connect the following:
- Timing belt back cover and tighten the attaching bolt to 108 inch lbs. (12 Nm).
- Camshaft sprocket key to the camshaft, then the camshaft sprocket. Install the bolt and tighten it to 27 ft. lbs. (37 Nm).

➡ **Ensure the words UP embossed on the camshaft pulley is aligned in the upward position, before installing the timing belt.**

- Timing belt onto the camshaft sprocket. Refer to the timing belt unit repair section. Loosen, then tighten the timing belt adjusting nut.

14. Rotate the crankshaft pulley 5 or 6 turns to position the timing belt on the pulleys.

15. Set the No. 1 cylinder to TDC and loosen the timing belt adjusting nut 1 turn. Turn the crankshaft counterclockwise until the cam pulley has moved 3 teeth; this creates tension on the timing belt. Loosen the timing belt adjusting nut, then tighten it to 33 ft. lbs. (45 Nm).

16. Adjust the valves.

17. Tighten the crankshaft pulley bolt to 181 ft. lbs. (245 Nm)

18. Install or connect the following:
- Upper timing belt cover and tighten the bolt to 108 inch lbs. (12 Nm).
- Cylinder head cover gasket in the groove on the cylinder head cover. Before installing the gasket thoroughly clean the seal and the groove. Seat the recesses for the camshaft first, then work it into the groove around the outside edges. Be sure the gasket is seated securely in the corners of the recesses.
- Cylinder head (valve) cover and tighten the cap nuts to 84 inch lbs. (10 Nm).
- Power steering belt, then adjust.
- Distributor to the cylinder head. Snug the attaching bolts until the timing has been checked and adjusted.
- Spark plug wires and the distributor electrical connectors.
- Ignition coil wire to the distributor.
- Air intake duct

19. Drain the oil from the engine into a sealable container. Install the drain plug and refill the engine with clean oil.

20. Connect the negative battery cable and enter the radio security code.

21. Start the engine and check carefully for any leaks.

22. Check and adjust the ignition timing as necessary, then tighten the distributor bolts to 16 ft. lbs. (22 Nm).

2.3L Engines

1. Before servicing the vehicle, refer to the precautions in the beginning of this section.

2. Disconnect the negative battery cable.

3. Turn the crankshaft so the No. 1 piston is at TDC.

➡ **The No. 1 piston is at TDC when the pointer on the block aligns with the white painted mark on the flywheel (manual transaxle) or driveplate (automatic transaxle).**

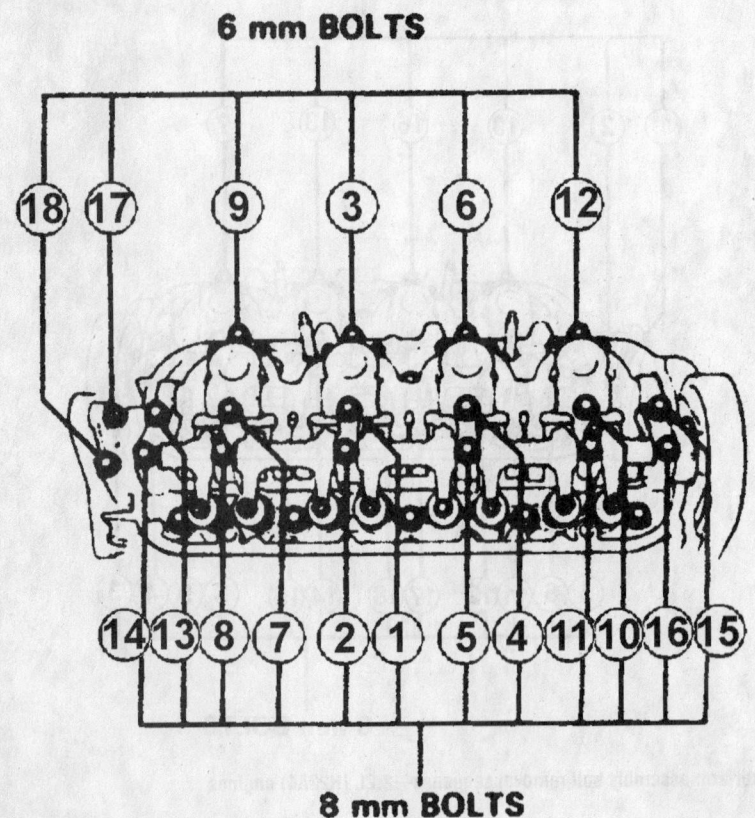

Rocker arm assembly torque sequence—2.2L (H22A4) engines

7923FG71

4. Remove or disconnect the following:
- Air intake duct
- Engine ground cable from the cylinder head cover
- Connector and the terminal from the alternator
- Engine wiring harness from the valve cover.
- Ignition coil

➡ **Tag all electrical connectors before disconnecting them.**

- Electrical connectors from the distributor
- Spark plug wires from the spark plugs
- Distributor from the cylinder head
- Ignition coil wire from the distributor
- Positive Crankcase Ventilation (PCV) hose
- Cylinder head cover. Replace the rubber seals if damaged or deteriorated.
- Timing belt middle cover

5. Ensure the words **UP** embossed on the camshaft pulleys are aligned in the upward position.

6. Mark the rotation of the timing belt if it is to be used again. Loosen the timing belt adjusting nut ½ turn, then release the tension on the timing belt. Push the tensioner to release tension from the belt, then tighten the adjusting nut.

7. Remove or disconnect the following:
- Timing belt from the camshaft sprockets. Refer to the timing belt unit repair section.

✳✳ WARNING

Do not crimp or bend the timing belt more than 90°, or less than 1 inch (25mm) in diameter

8. Insert a 5.0mm pin punch in each of the camshaft caps nearest to the sprockets through the holes provided.

9. Remove or disconnect the following:
- Camshaft sprocket attaching bolts, then the sprockets. Do not lose the sprocket keys.
- Side engine mount bracket B, then the timing belt back cover from behind the camshaft sprockets.
- Rocker arm adjusting screws, then the pin punches from the camshaft caps.
- Camshaft holders, note the holders locations for ease of installation.

Loosen the bolts in the reverse order of the installation.
- Camshafts from the cylinder head, then discard the camshaft seals.
- Rubber cap from the head, located at the end of the intake camshaft.

10. Remove the rocker arms from the cylinder head. Note the locations of the rocker arms.

➡ **The rocker arms have to be installed to their original locations if being reused.**

Specified torque:
Except Intake ⑤, ⑦. Exhaust ⑥, ⑧:
10 N·m (1.0 kg-m, 7 lb-ft)
Intake ⑤, ⑦. Exhaust ⑥, ⑧:
12 N·m (1.2 kg-m, 9 lb-ft)

To install:
11. Lubricate the rocker arms with clean oil.

12. Install or connect the following:
- Rocker arms on the pivot bolts and the valve stems. If the rocker arms are being reused, install them to their original locations. The locknuts and adjustment screws should be loosened before installing the rocker arms.

13. Lubricate the camshafts with clean oil.

TIGHTENING SEQUENCE

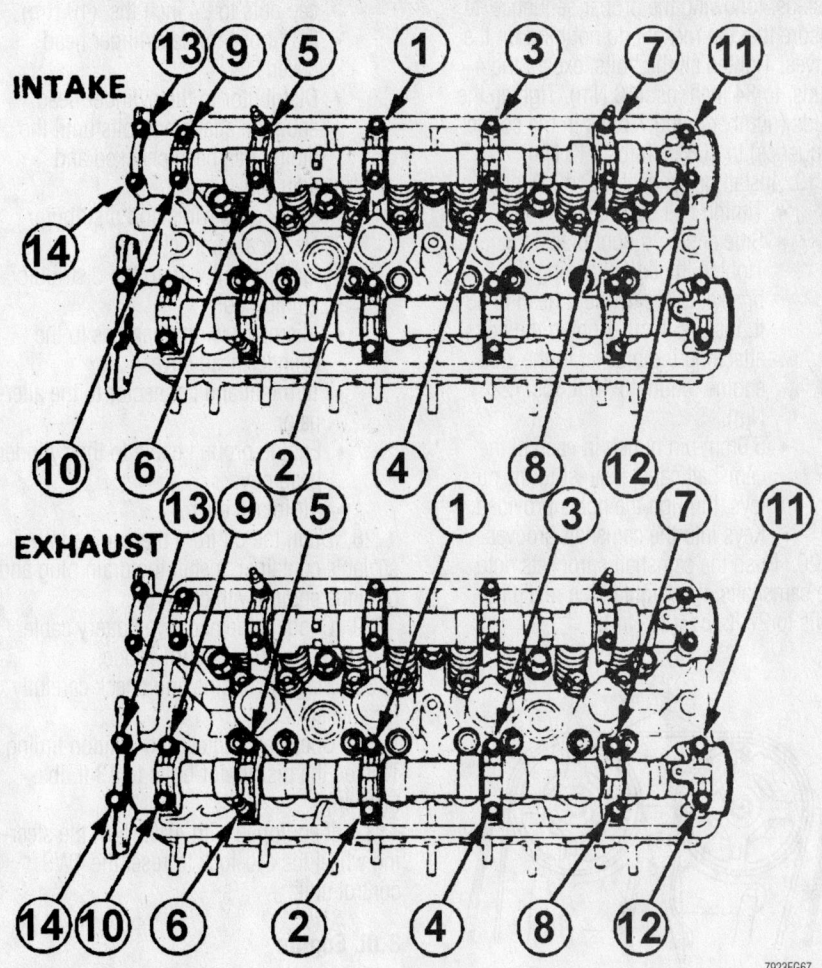

Camshaft holders torque sequence—2.3L engines

7923FG67

For Tune-up, Capacities and Firing orders, see Section 1 of this manual

14. Install or connect the following:
- Camshaft seals to the end of the camshafts that the timing belt sprockets attach to. The open side (spring) should be facing into the cylinder head when installed.

➡**Be sure the keyways on the camshafts are facing up and install the camshafts to the cylinder head.**

- Rubber plug to the cylinder head at the end of the intake camshaft.

15. Apply liquid gasket to the head mating surfaces of the No. 1 and No. 6 camshaft holders, then install them along with No. 2, 3, 4 and 5. **I** or **E** marks are stamped on the camshaft holders to identify them as Intake or Exhaust side holders. The arrows stamped on the holders should point toward the timing belt.

16. Snug the camshaft holders in place.

17. Press the camshaft seals securely into place.

18. Tighten the camshaft holder bolts in 2 steps, following the proper sequence, to ensure that the rockers do not bind on the valves. Tighten all the bolts, except the 4 studs, to 84 inch lbs. (10 Nm). Tighten the studs (number 5 and 7 bolts in the correct sequence) to 108 inch lbs. (12 Nm).

19. Install or connect the following:
- Timing belt back cover
- Side engine mount bracket B. Tighten the bolt attaching the bracket to the cylinder head to 33 ft. lbs. (45 Nm). Tighten the bolts attaching the bracket to the side engine mount to 16 ft. lbs. (22 Nm).
- 5.0mm pin punch in each of the camshaft caps, nearest to the pulleys, through the holes provided.
- Keys into the camshaft grooves.

20. Push the camshaft sprockets onto the camshafts, then tighten the retaining bolts to 27 ft. lbs. (38 Nm).

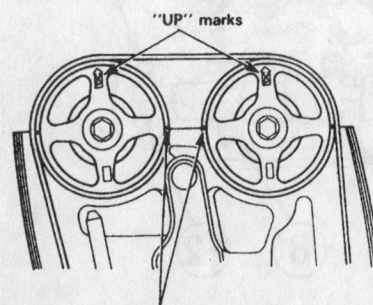

"UP" marks

Align the marks on the pulleys.

7923FG68

Camshaft sprockets alignment—2.3L engines

21. Ensure the words **UP** embossed on the camshaft pulleys are aligned in the upward position.

22. Install or connect the following:
- Timing belt to the camshaft sprockets, then remove the 2, 5.0mm pin punches from the camshaft bearing caps. Refer to the timing belt unit repair section.

23. Loosen, then tighten the timing belt adjuster nut.

24. Turn the crankshaft counterclockwise until the cam pulley has moved 3 teeth; this creates tension on the timing belt. Loosen the adjusting nut, then tighten it to 33 ft. lbs. (45 Nm).

25. Adjust the valves.

26. Tighten the crankshaft pulley bolt to 181 ft. lbs. (250 Nm).

27. Install or connect the following:
- Middle timing belt cover and tighten the attaching bolts to 108 inch lbs. (12 Nm).
- Cylinder head cover and tighten the cap nuts to 84 inch lbs. (10 Nm).
- PCV hose to the cylinder head cover.
- Distributor to the cylinder head, snug the attaching bolts until the timing has been checked and adjusted.
- Spark plug wires and distributor electrical connectors.
- Ignition coil wire to the distributor.
- Ignition coil
- Alternator wiring harness to the cylinder head cover
- Terminal and connector to the alternator.
- Engine ground cable to the cylinder head cover.
- Air intake duct

28. Drain the oil from the engine into a sealable container. Install the drain plug and refill the engine with clean oil.

29. Connect the negative battery cable and enter the radio security code.

30. Start the engine and check carefully for any leaks.

31. Check and adjust the ignition timing. Tighten the distributor bolts to 13 ft. lbs. (18 Nm).

32. If equipped with 4WS, turn the steering wheel lock-to-lock to reset the 4WS control unit.

3.0L Engine

1. Before servicing the vehicle, refer to the precautions in the beginning of this section.

2. Disconnect the negative battery cable.

3. Turn the engine to align the timing marks and set cylinder No.1 to TDC. The white mark on the crankshaft pulley should align with the pointer on the timing belt cover. Remove the inspection caps on the upper timing belt covers to check the alignment of the timing marks. The pointers for the camshafts should align with the green marks on the camshaft pulleys.

4. Remove or disconnect the following:
- Intake air duct
- Starter cable from the strut brace
- Strut brace
- Intake manifold covers
- Breather hose from the cylinder head cover
- Positive Crankcase Ventilation (PCV) hose from the cylinder head cover.
- Air conditioning compressor belt
- Alternator drive belt
- Power steering drive belt

➡**Label all electrical connectors before detaching them.**

- Electrical connectors from the distributor and spark plug wires from the spark plugs
- Distributor from the cylinder head
- Cylinder head covers and the side covers
- Timing belt covers and the timing belt. Refer to the timing belt unit repair section.
- Camshaft sprockets and the timing belt back covers.
- Bolts attaching the camshaft holder plates and camshaft holders in the opposite order of the installation sequence.
- Camshaft holder plates, camshaft holders, and the dowel pins. Discard the O-rings.
- Camshafts from the cylinder heads and the rubber cap from the rear cylinder head. Discard the camshaft seals.

To install:

5. Apply clean engine oil to the rocker arms and the camshafts.

6. Loosen the exhaust rocker arm adjusting screws and locknuts.

7. Be sure the rocker arms are properly positioned on the valve stems. Advance the crankshaft 30° from TDC to prevent interference between the pistons and valves, then install the camshafts. Position the rear camshaft on the cylinder head so the cam is not pushing on any valves.

8. Apply liquid gasket around the rubber cap, then install it to the cylinder head.

9. Install or connect the following:

- Camshaft seals to the camshafts with the open side (spring) facing in.
- Liquid gasket the cylinder head and camshaft holder mating surfaces
- Camshaft holders and the camshaft plates with the dowel pins
- New O-rings to the camshaft holder plates
- Clean oil to the camshaft holder bolts
- Bolts and tighten them in the proper sequence to 17 ft. lbs. (24 Nm)
- Timing belt back covers and tighten the attaching bolts to 108 inch lbs. (12 Nm).
- Camshaft sprockets and tighten the attaching bolts to 23 ft. lbs. (31 Nm).

10. Set the camshaft sprockets so that the No. 1 piston is at TDC. Align the TDC marks (green mark) on the camshaft pulleys to the pointers on the back covers.

11. Turn the crankshaft counterclockwise to set it at TDC. Align the TDC mark on the tooth of the timing belt drive pulley with the pointer on the oil pump.

12. Install or connect the following:
- Timing belt and timing belt covers. Refer to the timing belt unit repair section.

13. Set No. 1 cylinder to TDC.

14. Tighten the valve adjusting screws for No. 1, No. 2 and No. 4 cylinders. Tighten the screw until it contacts the valve, then tighten the screw 1⅛ turns. Hold the screw in place and tighten the locknut to 14 ft. lbs. (20 Nm).

15. Rotate the crankshaft pulley 1 turn clockwise, then tighten the adjusting screws for No. 3, No. 5, and No. 6 cylinders. Tighten the screw until it contacts the valve, then tighten the screw 1⅛ turns. Hold the screw in place and tighten the locknut to 14 ft. lbs. (20 Nm).

16. Install or connect the following:
- Cylinder head cover gasket into the groove of the cylinder head cover. Seat the recesses for the camshaft first, then work it into the groove around the outside edges.

➡Before installing the cylinder head cover gasket, thoroughly clean the seal groove.

- Liquid gasket to the cylinder head cover gasket at the 4 corners of the recesses. Use a shop towel and wipe the cylinder heads where the cylinder head covers will come in contact.
- Cylinder head covers, hold the gasket in the groove by placing your fingers on the camshaft contacting surfaces. With the cylinder head cover on the cylinder heads, slide the covers slightly back and forth to seat the cylinder head cover gaskets. Replace the washers if damaged or deteriorated

17. Tighten the cylinder head cover bolts in 2 or 3 steps. In the final step, tighten all the bolts, in sequence, to 11 ft. lbs. (15 Nm).

18. Install or connect the following:
- Cylinder head side covers with new O-rings and tighten the bolts to 108 inch lbs. (12 Nm).
- Distributor to the cylinder head and tighten the mounting bolt to 16 ft. lbs. (22 Nm).
- Spark plug wires to the correct spark plugs and the distributor electrical connectors.
- Power steering belt, and adjust the tension.
- Alternator belt and adjust the tension. Tighten the alternator mounting nut and bolt to 16 ft. lbs. (22 Nm).
- Air conditioning belt, and adjust the belt tension. Tighten the idler center nut to 33 ft. lbs. (44 Nm).
- PCV hose to the cylinder head cover
- Breather hose to the cylinder head cover
- Intake manifold cover and tighten the bolts to 108 inch lbs. (12 Nm).
- Intake air duct
- Strut brace and tighten the mounting bolts to 16 ft. lbs. (22 Nm).
- Starter cable to the strut brace

19. Drain the engine oil into a sealable container, then refill the engine with clean oil.

20. Connect the negative battery cable and enter the radio security code.

21. Start the engine, allowing it to idle and check for any signs of leakage.

Specified torque:
8 x 1.25 mm
24 N·m (2.4 kgf·m, 17 lbf·ft)
Apply engine oil to the bolt threads and flange

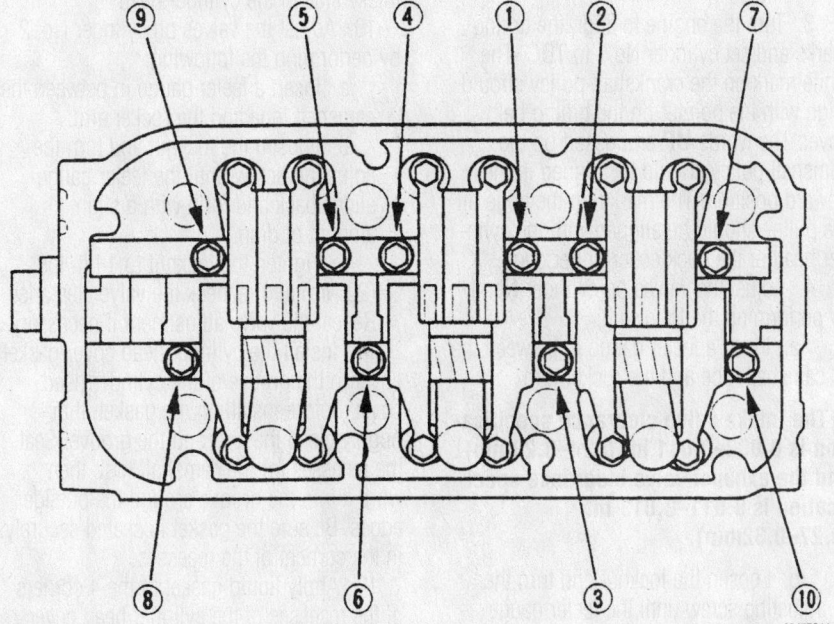

Camshaft holders torque sequence—3.0L engine

9347FG11

Valve Clearance

ADJUSTMENT

➡ **The radio may contain a coded theft protection circuit. Always obtain the code number before disconnecting the battery. If the vehicle is equipped with 4WS, the steering control unit is shut down when the battery is disconnected. After connecting the battery, turn the steering wheel lock-to-lock to reset the steering control unit.**

Civic

1. Before servicing the vehicle, refer to the precautions in the beginning of this section.

2. Disconnect the negative battery cable.

3. Remove the cylinder head cover and the upper timing belt cover.

4. Rotate the crankshaft to align the white TDC mark on the crankshaft pulley with the pointer on the cover for the No. 1 cylinder compression stroke. Be sure the **UP** mark on the camshaft sprocket is up and the TDC marks align with the edge of the cylinder head.

5. Hold a No. 1 cylinder rocker arm against the camshaft and use a feeler gauge to check the clearance at the valve stem. Except on B16A2 engines, intake valve clearance should be 0.007–0.009 in. (0.18–0.26mm), exhaust valve clearance should be 0.009–0.011 in. (0.23–0.27mm). On B16A2 engines, the intake valve clearance should be 0.006–0.007 in. (0.15–0.19mm), exhaust valve clearance should be 0.007–0.008 in. (0.17–0.21mm). Loosen the locknut and turn the adjusting screw to adjust the clearance. Tighten the locknut to 14 ft. lbs. (20 Nm) and recheck the clearance. Don't overtighten the locknut, the aluminum rockers will strip easily.

6. The adjustment order is 1–3–4–2. Rotate the crankshaft counterclockwise 180° (the camshaft sprocket will rotate 90°) to bring each cylinder to TDC/compression. Adjust each set of valves.

    a. At TDC for the No. 3 cylinder, the UP mark is pointed to the exhaust side of the cylinder head.

    b. At TDC for the No. 4 cylinder, the UP mark is pointed down, and the TDC marks align with the edge of the cylinder head.

    c. At TDC for the No. 2 cylinder, the UP mark is pointed to the intake side of the cylinder head.

7. After adjusting the valves of a VTEC engine, inspect the intake rocker arms for smooth and independent motion.

8. Apply sealant to the edges of the valve cover gasket where it meets the camshaft holders. Be sure the spark plug tube seals are properly seated.

9. Install the cylinder head and timing belt covers.

10. Tighten the crankshaft pulley bolt to 134 ft. lbs. (185 Nm).

11. Reconnect the negative battery cable. Enter the radio security code.

Prelude and Accord

➡ **The valve clearance should be adjusted when the engine is cold, the cylinder head temperature should be less than 100°F (38°C).**

➡ **The radio may contain a coded theft protection circuit. Always obtain the code number before disconnecting the battery.**

1. Before servicing the vehicle, refer to the precautions in the beginning of this section.

2. Remove or disconnect the following:
- Negative battery cable

➡ **Label the wires before disconnecting them.**

- Spark plug wires from the spark plugs
- Positive Crankcase Ventilation (PCV) hose
- Cylinder head cover. Replace the rubber seals if damaged or deteriorated.

3. Turn the engine to align the timing marks and set cylinder No.1 to TDC. The white mark on the crankshaft pulley should align with the pointer on the timing belt cover. The words **UP** embossed on the camshaft pulley should be aligned in the upward position. The marks on the edge of the pulley should be aligned with the cylinder head or the back cover upper edge.

4. Adjust the valves on cylinder No. 1 by performing the following:

    a. Insert a feeler gauge in between the camshaft lobe and the rocker arm.

➡ **The intake valve clearance specification is 0.009–0.011 in (0.24–0.28mm) and the exhaust valve clearance specification is 0.011–0.013 in. (0.27–0.32mm).**

    b. Loosen the locknut and turn the adjusting screw until the feeler gauge slides back and forth with a slight amount of drag.

    c. Tighten the locknut to 14 ft. lbs. (20 Nm) and recheck the valve clearance. Repeat the valve adjustment if necessary.

5. Rotate the crankshaft 180° counterclockwise (the camshaft pulleys will turn 90°) The **UP** arrow marks should be pointing to the exhaust side of the cylinder head.

6. Adjust the valves on cylinder No. 3 by performing the following:

    a. Insert a feeler gauge in between the camshaft lobe and the rocker arm.

    b. Loosen the locknut and turn the adjusting screw until the feeler gauge slides back and forth with a slight amount of drag.

    c. Tighten the locknut to 14 ft. lbs. (20 Nm) and recheck the valve clearance. Repeat the valve adjustment if necessary.

7. Rotate the crankshaft 180° counterclockwise (the camshaft pulleys will turn 90°) to bring No. 4 piston to TDC. The **UP** arrow marks should be pointing down, toward the crankshaft.

8. Adjust the valves on cylinder No. 4 by performing the following:

    a. Insert a feeler gauge in between the camshaft lobe and the rocker arm.

    b. Loosen the locknut and turn the adjusting screw until the feeler gauge slides back and forth with a slight amount of drag.

    c. Tighten the locknut to 14 ft. lbs. (20 Nm) and recheck the valve clearance. Repeat the valve adjustment if necessary.

9. Rotate the crankshaft 180° counterclockwise (the camshaft pulleys will turn 90°) to bring piston No. 2 to TDC. The **UP** arrow marks should be pointing to the intake side of the cylinder head.

10. Adjust the valves on cylinder No. 2 by performing the following:

    a. Insert a feeler gauge in between the camshaft lobe and the rocker arm.

    b. Loosen the locknut and turn the adjusting screw until the feeler gauge slides back and forth with a slight amount of drag.

    c. Tighten the locknut to 14 ft. lbs. (20 Nm) and recheck the valve clearance. Repeat the valve adjustment if necessary.

11. Install the cylinder head cover gasket cover to the groove of the cylinder head cover. Before installing the gasket, thoroughly clean the seal and the groove. Seat the recesses for the camshaft first, then work it into the groove around the outside edges. Be sure the gasket is seated securely in the corners of the recesses.

12. Apply liquid gasket to the 4 corners of the recesses of the cylinder head cover gasket. Do not install the parts if 5 minutes or more have elapsed since applying liquid gasket. After assembly, wait at least 20 minutes before filling the engine with oil.

13. Install or connect the following:

- Cylinder head (valve) cover. Tighten the bolts attaching to 84 inch lbs. (10 Nm).
- Spark plug wires to the correct spark plugs.
- Positive, then the negative battery cable and enter the radio security code.

14. If equipped with 4WS, start the engine and turn the steering wheel lock-to-lock to reset the 4WS control unit.

S2000

➡The valve clearance is checked with the engine COLD.

1. Before servicing the vehicle, refer to the precautions at the beginning of this section.
2. Disconnect the negative battery cable.
3. Remove the valve covers
4. Rotate the crankshaft so that the camshaft lobe is directed away from the valve tappet to be measured.
5. Insert a feeler gauge under the camshaft lobe at a 90 degree angle to the camshaft. Clearance for the intake valves should be 0.008–0.010 inch (0.21–0.25mm). Clearance for the exhaust valves should be 0.010–0.011 inch (0.25–0.29mm).
6. If adjustment is necessary, loosen the locknut and turn the adjusting screw until the clearance is correct.
7. Tighten the locknut to 14 ft. lbs. (20 Nm) and recheck the valve clearance.
8. Repeat for each valve requiring adjustment.

Starter Motor

1.6L and 1.7L Engines

1. Before servicing the vehicle, refer to the precautions in the beginning of this section.
2. Remove or disconnect the following:
   - Negative battery cable
   - air intake resonator
   - Starter motor wiring harness
   - Starter motor

To install:

3. Install or connect the following:
   - Starter motor and tighten the bolts to 33 ft. lbs. (44 Nm)
   - Starter motor wiring harness and tighten the battery cable terminal bolt to 84 inch lbs. (9 Nm)
   - Negative battery cable

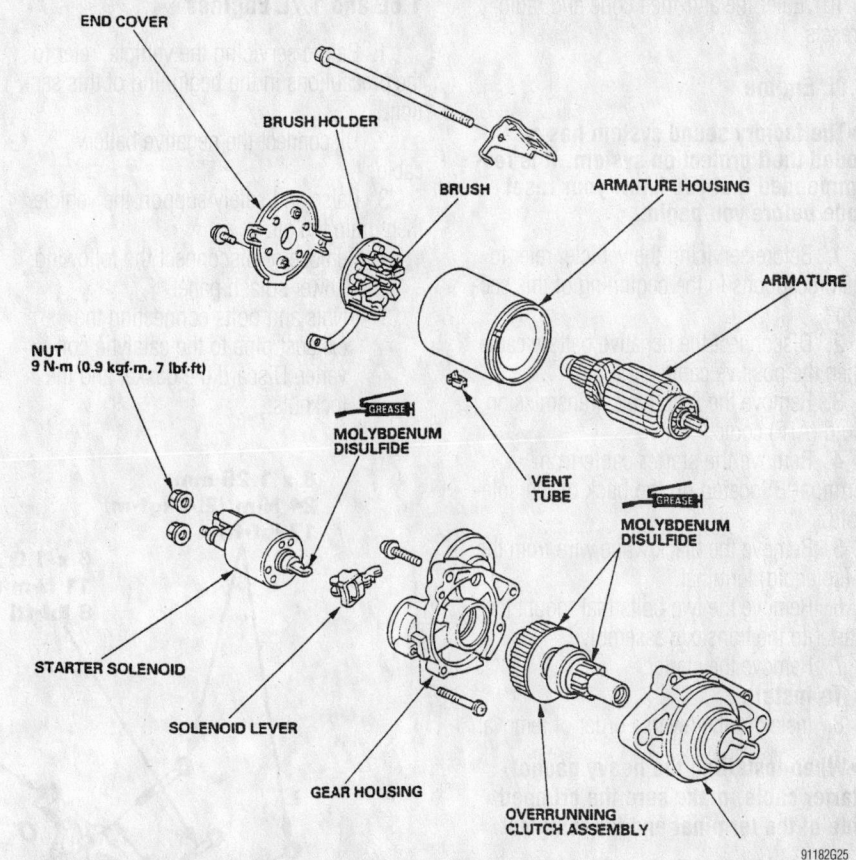

Exploded view of a typical Honda Starter

2.0L Engine

1. Before servicing the vehicle, refer to the precautions in the beginning of this section.
2. Remove or disconnect the following:
   - Negative battery cable
   - Accessory drive belt and tensioner
   - Alternator
   - Starter motor wiring harness
   - Starter motor

To install:

3. Install or connect the following:
   - Starter motor and tighten the bolts to 33 ft. lbs. (44 Nm)
   - Starter motor wiring harness and tighten the battery cable terminal bolt to 84 inch lbs. (9 Nm)
   - Alternator
   - Accessory drive belt tensioner and tighten the bolts to 16 ft. lbs. (22 Nm)
   - Accessory drive belt
   - Negative battery cable

2.2L and 2.3L Engines

➡The factory sound system has a coded theft protection system. It is recommended that you know your reset code before you begin.

1. Before servicing the vehicle, refer to the precautions in the beginning of this section.
2. Disconnect the negative battery cable.
3. Remove all wiring from harness.
4. Remove the lower radiator hose from the bracket on the starter motor.
5. Remove the starter cable from terminal B located on the back of the solenoid.
6. Remove the black/white wire from the S (solenoid) terminal.
7. Remove the two bolts that mount the starter to the transaxle assembly.
8. Remove the starter.

To install:

9. Install in the reverse order of removal.

➡When installing the heavy gauge starter cable, make sure the crimped side of the terminal end is facing out.

10. Enter the anti-theft code and radio presets.

3.0L Engine

➡ **The factory sound system has a coded theft protection system. It is recommended that you know your reset code before you begin.**

1. Before servicing the vehicle, refer to the precautions in the beginning of this section.
2. Disconnect the negative battery cable. Then the positive cable.
3. Remove the Automatic Transmission Fluid (ATF) cooler.
4. Remove the starter cable from terminal B located on the back of the solenoid.
5. Remove the black/white wire from the S (solenoid) terminal.
6. Remove the two bolts that mount the starter to the transaxle assembly.
7. Remove the starter.

To install:

8. Install in the reverse order of removal.

➡ **When installing the heavy gauge starter cable, make sure the crimped side of the terminal end is facing out.**

9. Enter the anti-theft code and radio presets.

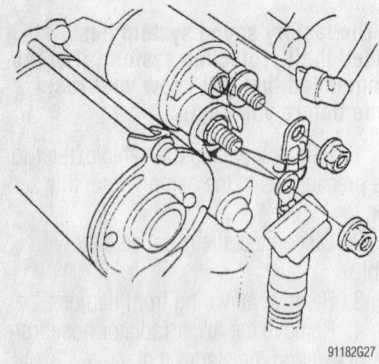

91182G27

Location if starter wiring

Oil Pan

REMOVAL & INSTALLATION

➡ **The radio may contain a coded theft protection circuit. Always obtain the code number before disconnecting the battery. If the vehicle is equipped with 4WS, the steering control unit is shut down when the battery is disconnected. After connecting the battery, turn the steering wheel lock-to-lock to reset the steering control unit.**

1.6L and 1.7L Engines

1. Before servicing the vehicle, refer to the precautions in the beginning of this section.
2. Disconnect the negative battery cable.
3. Raise and safely support the vehicle, then drain the oil.
4. Remove or disconnect the following:
   - Lower splash panel
   - Nuts and bolts connecting the exhaust pipe to the catalytic converter. Discard the gasket and the locknuts.

- Nuts attaching the exhaust pipe to the exhaust hanger.
- Locknuts attaching the exhaust pipe to the exhaust manifold, then discard the nuts
- Remove the exhaust pipe from the vehicle. Discard the exhaust gaskets.
- Oil pan bolts in a crisscross pattern.
- Oil pan. Lightly tap the corners of the oil pan with a rubber or plastic faced mallet. Clean off all the old gasket material.

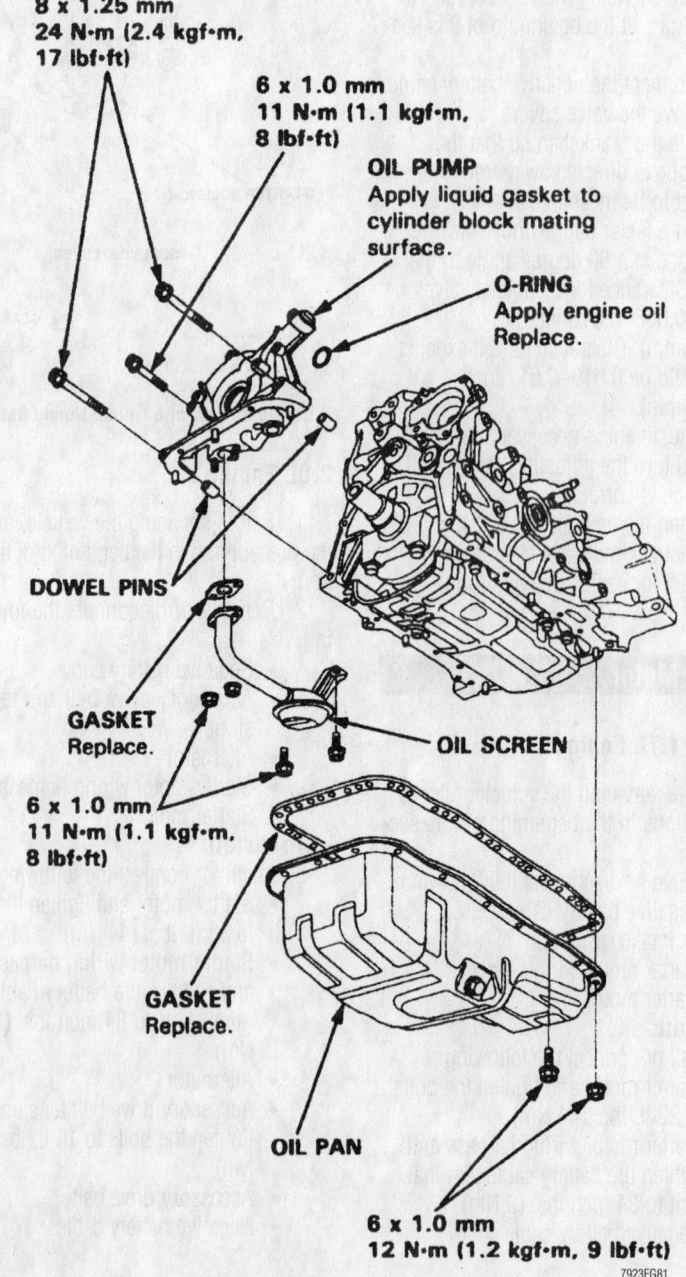

8 x 1.25 mm
24 N·m (2.4 kgf·m, 17 lbf·ft)

6 x 1.0 mm
11 N·m (1.1 kgf·m, 8 lbf·ft)

OIL PUMP
Apply liquid gasket to cylinder block mating surface.

O-RING
Apply engine oil
Replace.

DOWEL PINS

OIL SCREEN

GASKET
Replace.

6 x 1.0 mm
11 N·m (1.1 kgf·m, 8 lbf·ft)

GASKET
Replace.

OIL PAN

6 x 1.0 mm
12 N·m (1.2 kgf·m, 9 lbf·ft)

7923FG81

Oil pan and oil screen—1.6L (B16A2) engine shown

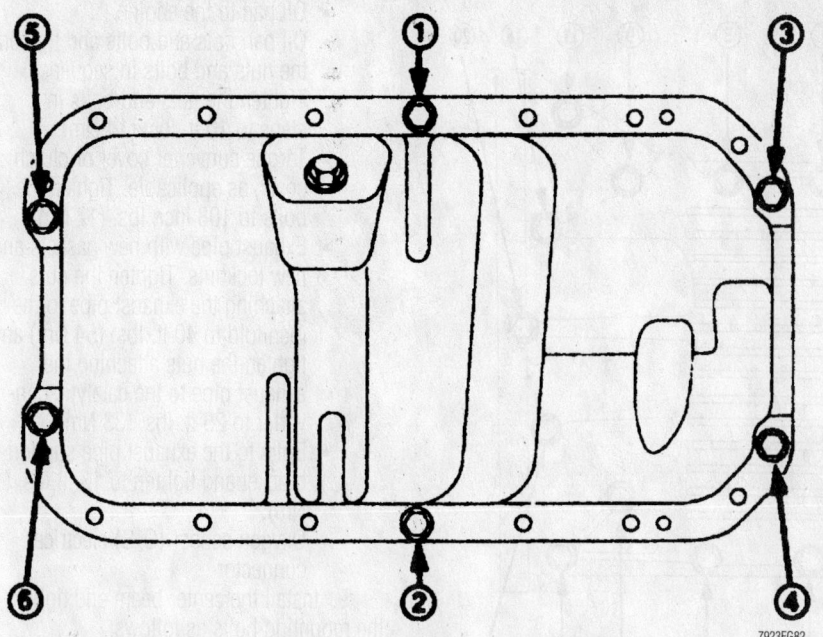

Oil pan bolt tightening sequence—1.6L (B16A2) engine shown

7923FG82

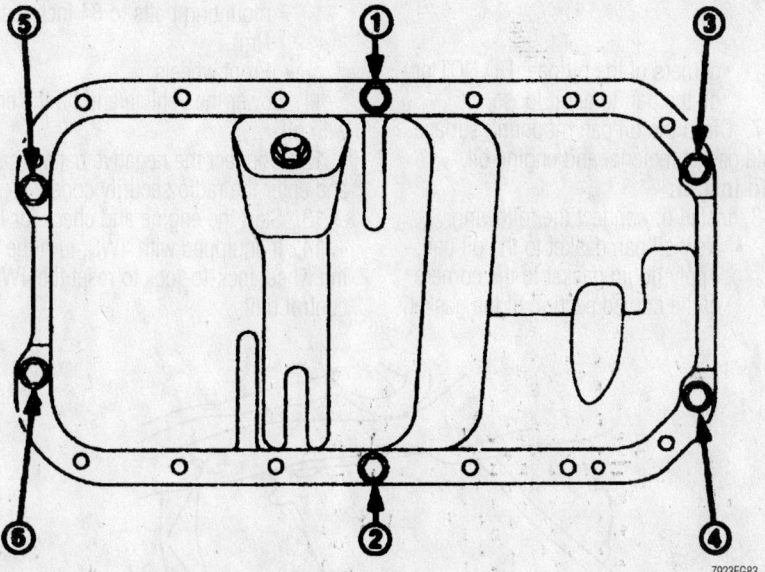

Oil pan bolt tightening sequence—1.6L (D16Y5, D16Y7, D16Y8) and 1.7L (D17A1, D17A2) engines

7923FG83

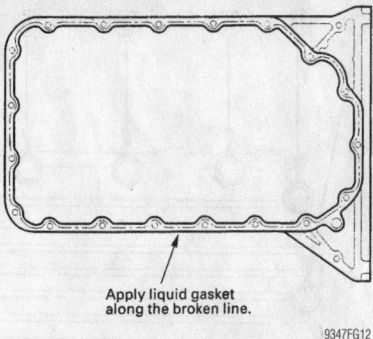

Apply liquid gasket along the broken line.

9347FG12

Apply liquid gasket along the broken line—2.0L engine

starting with the center bolt next to the oil drain plug. The final torque value for the nuts and bolts is 10 ft. lbs. (14 Nm).

➡**Excessive tightening can cause distortion of the oil pan gasket and oil leakage.**

8. Install or connect the following:
- Oil drain plug with a new crush washer. Tighten the plug to 33 ft. lbs. (44 Nm).
- Exhaust pipe using new gaskets and locknuts. Tighten the nuts attaching the exhaust pipe to the exhaust manifold to 40 ft. lbs. (54 Nm). Tighten the nuts attaching the exhaust pipe to the catalytic converter and the exhaust pipe hanger to 16 ft. lbs. (22 Nm).
- Lower splash panel.
9. Lower the vehicle.
10. Refill the engine with clean oil.
11. Connect the negative battery cable and enter the radio security code.
12. Run the engine and check for leaks.
13. Turn off the engine and check the oil level. Top off the oil level if necessary.

2.0L Engine

1. Before servicing the vehicle, refer to the precautions in the beginning of this section.
2. Drain the engine oil.
3. Remove the oil pan bolts and the oil pan.

To install:
4. Remove old liquid gasket material from the oil pan mating surfaces and the bolt holes.
5. Clean and dry the oil pan mating surfaces.
6. Apply liquid gasket (PN 08718-0009) as shown.
7. Install the oil pan. Tighten the bolts in

5. Inspect the oil screen and pick-up tube for damaged and clogging. If the screen and tube are clogged with oil residue, they should be thoroughly cleaned or replaced.

To install:
6. Install or connect the following:
- Oil screen and tube with a new gasket, if removed. Tighten the mounting nuts and bolts to 96 inch lbs. (11 Nm).

- Liquid gasket to the oil pan mating surface where the oil pump and the right side cover meet the engine block.
- Oil pan gasket to the oil pan.
- Oil pan.
- Center and end mounting nuts and bolts. Evenly hand-tighten the oil pan nuts and bolts.
7. Tighten the oil pan mounting nuts and bolts in a 3-step clockwise pattern

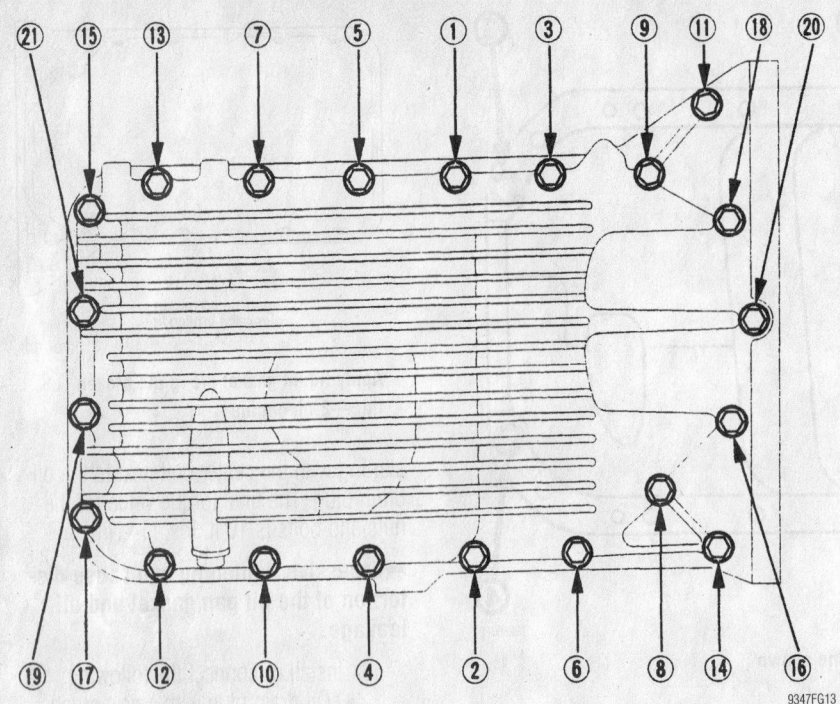

Oil pan torque sequence—2.0L engine

9347FG13

sequence and in two or three steps to 104 inch lbs. (12 Nm).

8. Fill the crankcase to the correct level.

9. Start the engine and check for leaks.

2.2L and 2.3L Engines

1. Before servicing the vehicle, refer to the precautions in the beginning of this section.

2. Disconnect the negative battery cable.

3. Raise and safely support the vehicle.

4. Drain the engine oil into a sealable container.

5. Install the drain bolt with a new gasket. Tighten the bolt to 33 ft. lbs. (44 Nm).

6. Remove or disconnect the following:
- Front wheels and the splash shield
- Center beam
- Oxygen sensor electrical connector
- Bolts from the support bracket on the exhaust pipe.
- Nuts attaching the exhaust pipe to the exhaust manifold and the catalytic converter.
- Exhaust pipe and discard the gaskets.
- Converter cover, if equipped with an automatic transaxle.
- Clutch cover, if equipped with a manual transaxle.
- Oil pan nuts and bolts (in a criss-cross pattern) and the oil pan; if necessary, use a mallet to tap the

corners of the oil pan. DO NOT pry on the pan to get it loose.

7. Clean the oil pan mounting surface of old gasket material and engine oil.

To install:

8. Install or connect the following:
- New oil pan gasket to the oil pan. Apply liquid gasket to the corners of the curved section of the gasket.
- Oil pan to the engine
- Oil pan nuts and bolts and tighten the nuts and bolts in sequence. Tighten the nuts and bolts in 2 steps to 10 ft. lbs. (14 Nm).
- Torque converter cover or clutch cover, as applicable. Tighten the bolts to 108 inch lbs. (12 Nm).
- Exhaust pipe with new gaskets and new locknuts. Tighten the nuts attaching the exhaust pipe to the manifold to 40 ft. lbs. (54 Nm) and tighten the nuts attaching the exhaust pipe to the catalytic converter to 25 ft. lbs. (33 Nm).
- Bolts to the exhaust pipe support bracket and tighten to 13 ft. lbs. (18 Nm).
- Oxygen sensor (O2S) electrical connector

9. Install the center beam and tighten the mounting bolts as follows:

  a. Prelude: 43 ft. lbs. (60 Nm)

10. Install or connect the following:
- Splash shield and tighten the mounting bolts to 84 inch lbs. (10 Nm).
- Front wheels

11. Lower the vehicle and fill the engine with oil.

12. Connect the negative battery cable and enter the radio security code.

13. Start the engine and check for leaks.

14. If equipped with 4WS, turn the steering wheel lock-to-lock to reset the 4WS control unit.

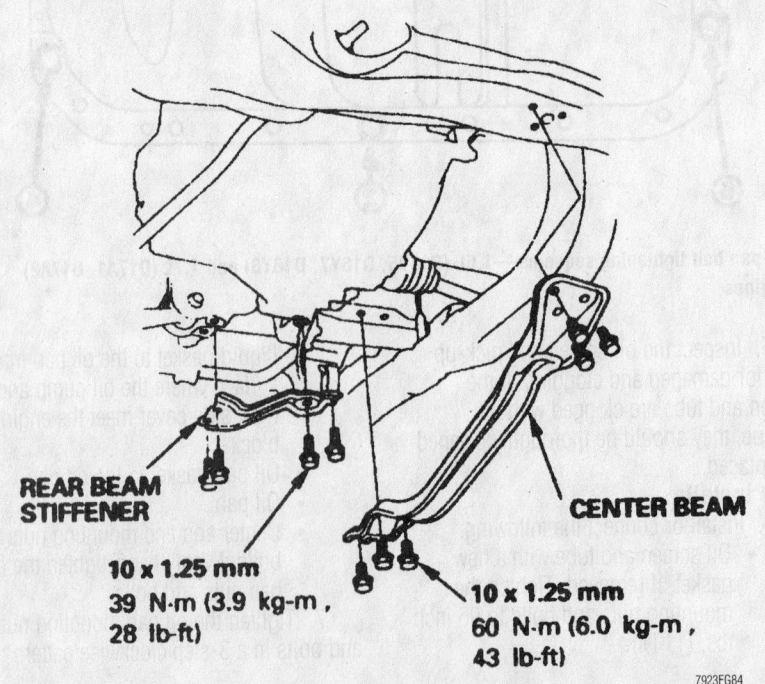

REAR BEAM STIFFENER

10 x 1.25 mm
39 N·m (3.9 kg-m, 28 lb-ft)

CENTER BEAM

10 x 1.25 mm
60 N·m (6.0 kg-m, 43 lb-ft)

7923FG84

To gain access to the oil pan, remove the center beam—2.2L and 2.3L engines

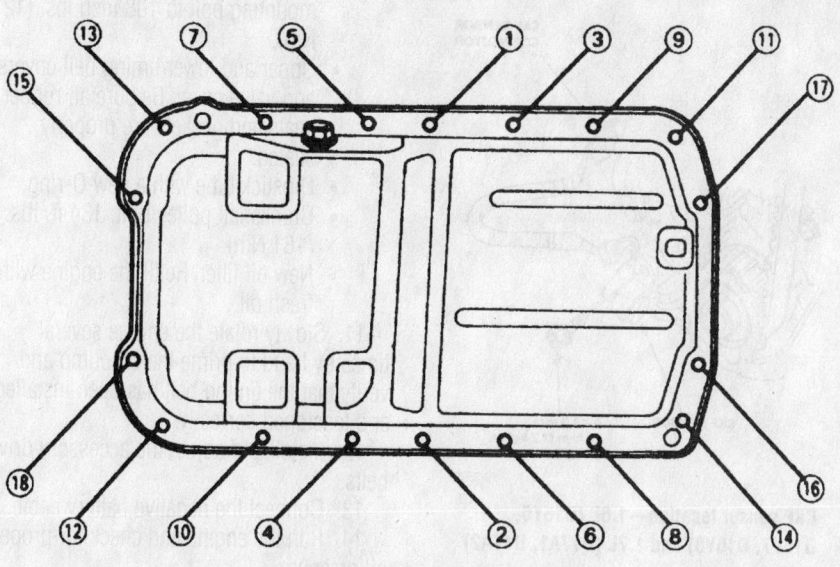

Oil pan mounting bolt tightening sequence—2.2L (H22A4) engines

3.0L Engine

1. Before servicing the vehicle, refer to the precautions in the beginning of this section.
2. Remove or disconnect the following:
   • Negative battery cable

3. Raise and safely support the vehicle.
4. Remove or disconnect the following:
   • Undercover
5. Drain the engine oil and replace the drain plug.
6. Remove or disconnect the following:

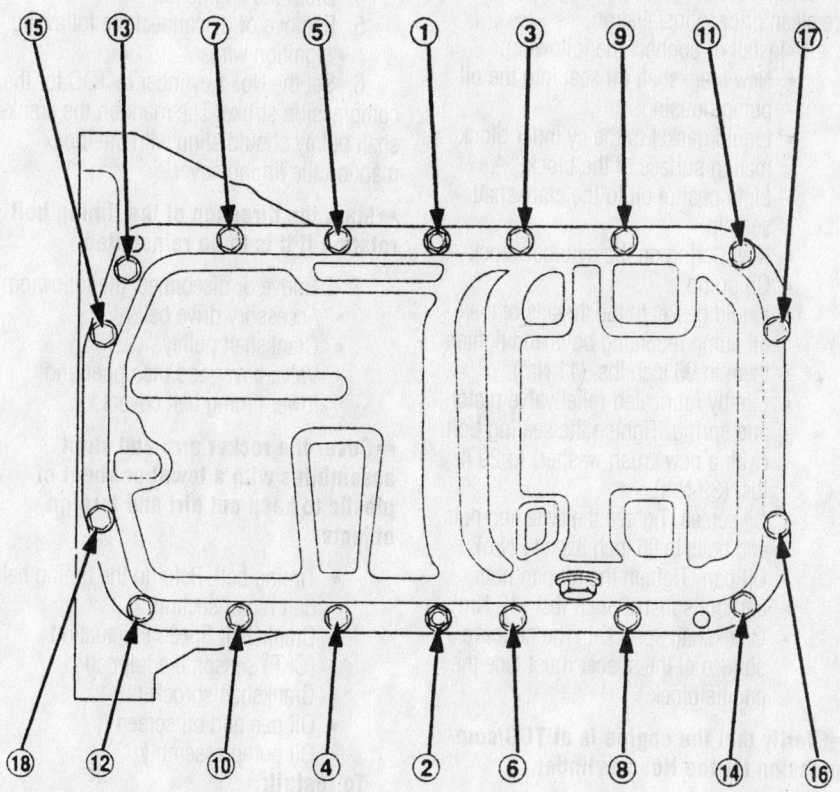

Oil pan mounting bolt tightening sequence—3.0L engine

• Front exhaust pipe
• Oil pan mounting bolts
7. Hammer a seal cutter between the engine block and oil pan to break the seal.
8. Remove the oil pan.

To install:

9. Clean the oil pan flange and engine block mounting surface.
10. Install or connect the following:
   • Sealant to the oil pan flange. Be sure to apply sealant toward the inside of the bolt holes.
   • Oil pan on the engine. Tighten the bolts in sequence to 10 ft. lbs. (14 Nm).
   • Exhaust pipe
   • Undercover
11. Lower the vehicle.

✱✱ WARNING

Operating the engine without the proper amount and type of engine oil will result in severe engine damage.

12. Refill the engine with the correct amount of oil.
13. Connect the negative battery cable.
14. Start the engine and check for leaks.

Oil Pump

REMOVAL & INSTALLATION

➡**The original radio may contain a coded anti-theft circuit. Always obtain the security code number before disconnecting the battery cables.**

1.6L (D16Y5, D16Y7, D16Y8) and 1.7l (D17A1, D17A2) Engines

1. Before servicing the vehicle, refer to the precautions in the beginning of this section.
2. Disconnect the negative battery cable.
3. Raise and safely support the vehicle.
4. Drain the engine oil.
5. Rotate the crankshaft to set the No. 1 cylinder to TDC for the compression stroke. The white TDC mark on the crankshaft pulley should align with the TDC pointers on the lower timing cover.

➡**Mark the direction of the timing belt rotation if it is to be reinstalled.**

6. Remove or disconnect the following:
   • Accessory drive belts and the crankshaft pulley
   • Valve cover and the upper and lower timing belt covers

Heater Core replacement is covered in Section 2 of this manual

CRANKSHAFT PULLEY:

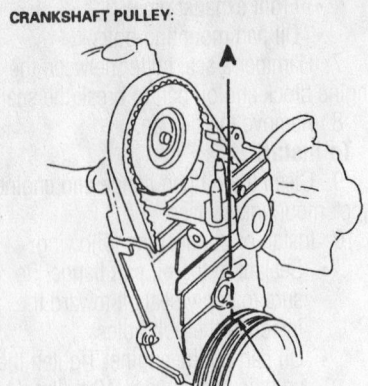

TDC MARK (WHITE)

CAMSHAFT PULLEY:

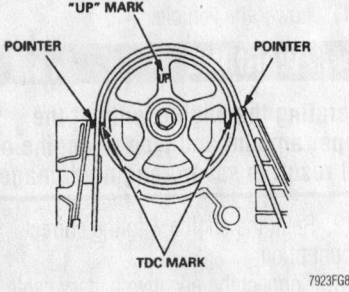

"UP" MARK

POINTER   POINTER

TDC MARK

7923FG88

**Crankshaft and camshaft TDC marks—
1.6L (D16Y5, D16Y7, D16Y8) and 1.7L
(D17A1, D17A2) engines**

➡**Cover the rocker arm and shaft
assemblies with a towel or sheet of
plastic to keep out dirt and foreign
objects.**

- Dipstick and its tube from the oil
  pump housing
- Timing belt. Refer to the timing belt
  unit repair section.
- Crankshaft Speed Fluctuation
  (CKF) sensor from the oil pump
  cover, and position it out of the
  way so that it will not come in contact with oil or become damaged.
- Crankshaft sprocket
- Oil pan
- Oil screen and pick-up tube from the
  oil pump housing and crankshaft
  buttress. If the screen and pick-up
  tube are blocked with oil residue,
  clean or replace them as necessary.
- Oil pump assembly

➡**If the rotors are to be reused, matchmark them with a felt-tipped marker for
assembly.**

To install:

➡**Replace the rotors if they are worn or
damaged. Use new O-rings when
assembling and installing the oil
pump.**

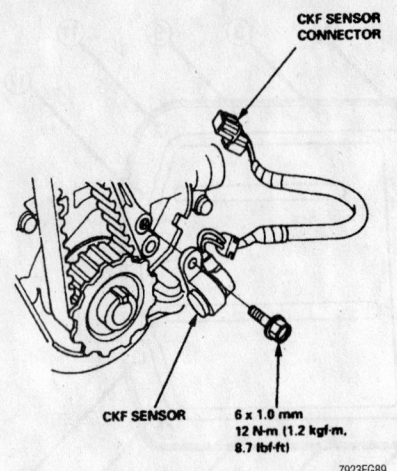

CKF SENSOR
CONNECTOR

CKF SENSOR   6 x 1.0 mm
                12 N·m (1.2 kgf·m,
                8.7 lbf·ft)

7923FG89

**CKF sensor location—1.6L (D16Y5,
D16Y7, D16Y8) and 1.7L (D17A1, D17A2)**

7. Install or connect the following:
- Rotors back into their original positions. Be sure they move without
  binding. Pack the rotor cavity with
  petroleum jelly to prevent oil starvation damage when the engine is
  initially started.

8. Assemble the oil pump and tighten
the rotor cover bolts to 60 inch lbs. (7 Nm).

9. Be sure all gasket mating surfaces
are clean prior to installation.

10. Install or connect the following:
- New crankshaft oil seal into the oil
  pump housing.
- Liquid gasket to the cylinder block
  mating surface of the block.
- Light coat of oil to the crankshaft
  seal lip.
- New O-ring on the cylinder block
- Oil pump
- Liquid gasket to the threads of the
  oil pump mounting bolts and tighten
  them to 96 inch lbs. (11 Nm).
- Lightly lubricated relief valve piston
  and spring. Tighten the sealing bolt
  (with a new crush washer) to 29 ft.
  lbs. (39 Nm).
- Oil screen. Tighten the fastening nuts
  and bolts to 96 inch lbs. (11 Nm).
- Oil pan. Tighten the oil pan nuts
  and bolts to 108 inch lbs. (12 Nm).
- Crankshaft sprocket. The concave
  surface of the spacer must face the
  engine block.

➡**Verify that the engine is at TDC/com-
pression for the No. 1 cylinder.**

- Timing belt. Refer to the timing belt
  unit repair section. Tighten the tensioner adjusting bolt to 33 ft. lbs.
  (44 Nm).
- CKF sensor. Tighten the sensor

mounting bolt to 108 inch lbs. (12
Nm).
- Upper and lower timing belt covers
  and valve cover. Be sure all rubber
  seals and gaskets are properly
  seated.
- Dipstick tube with a new O-ring.
- Crankshaft pulley bolt: 134 ft. lbs.
  (181 Nm).
- New oil filter. Refill the engine with
  fresh oil.

11. Slowly rotate the engine several
times by hand to prime the oil pump and
verify that the timing belt has been installed
and tensioned correctly.

12. Install and adjust the accessory drive
belts.

13. Connect the negative battery cable.

14. Run the engine and check for proper
oil pressure.

15. Check for leaks. Top up the engine
oil if necessary.

1.6L (B16A2) Engines

1. Before servicing the vehicle, refer to
the precautions in the beginning of this section.

2. Disconnect the negative battery cable.

3. Raise and safely support the vehicle.

4. Drain the engine oil.

5. Remove or disconnect the following:
- Ignition wires

6. Set the No. 1 cylinder to TDC for the
compression stroke. The mark on the crankshaft pulley should align with the index
mark on the timing cover.

➡**Mark the direction of the timing belt
rotation if it is to be reinstalled.**

7. Remove or disconnect the following:
- Accessory drive belts
- Crankshaft pulley
- Valve cover and the upper and
  lower timing belt covers.

➡**Cover the rocker arm and shaft
assemblies with a towel or sheet of
plastic to keep out dirt and foreign
objects.**

- Timing belt. Refer to the timing belt
  unit repair section.
- Crankshaft Speed Fluctuation
  (CKF) sensor, if equipped
- Crankshaft sprocket
- Oil pan and oil screen
- Oil pump assembly

To install:

➡**Replace the rotors if they are worn or
damaged. Use new O-rings when
assembling and installing the oil
pump.**

8. Install the rotors back into their original positions. Be sure they rotate without binding. Pack the rotor cavity with petroleum jelly to prevent oil starvation damage when the engine is initially started.

9. Assemble the oil pump and tighten the rotor cover bolts to 60 inch lbs. (7 Nm).

10. Be sure all gasket mating surfaces are clean prior to installation. Replace the crankshaft oil seal prior to installing the oil pump.

11. Install or connect the following:
- New crankshaft oil seal into the oil pump housing
- Liquid gasket to the oil pump mating surface of the cylinder block.
- Light coat of oil to the crankshaft seal lip.
- New oil passage O-ring on the cylinder block
- Oil pump. Apply liquid gasket to the threads of the oil pump mounting bolts and tighten them the 6mm bolts to 96 inch lbs. (11 Nm). Tighten the 8mm bolts to 17 ft. lbs. (24 Nm).
- Lubricated relief valve piston and spring
- Sealing bolt (with a new crush washer) and tighten it to 29 ft. lbs. (39 Nm).
- Oil screen
- Oil pan. Wait for the sealant to cure before refilling the engine with oil.
- Crankshaft sprocket. The concave surface of the spacer faces out.
- Timing belt. Refer to the timing belt unit repair section. Tighten the tensioner adjusting bolt to 40 ft. lbs. (55 Nm).
- CKF sensor. Tighten the sensor mounting bolt to 108 inch lbs. (12 Nm).
- Upper and lower timing belt covers and valve cover. Be sure all rubber seals and gaskets are properly seated.
- Crankshaft pulley bolt: 130 ft. lbs. (180 Nm).
- New oil filter. Refill the engine with fresh oil.
- Accessory drive belts, and adjust the belt tension.

12. Connect the negative battery cable.

13. Run the engine and check for proper oil pressure.

14. Check for leaks. Top up the engine oil if necessary.

2.0L Engine

1. Before servicing the vehicle, refer to the precautions in the beginning of this section.

2. Drain the engine oil.

3. Remove or disconnect the following:
- Negative battery cable
- Oil pan
- Timing chain
- Oil pump chain guide and tensioner
- Baffle plate
- Oil pump, chain, and crankshaft sprocket

To install:

➡Use a new oil pump chain tensioner for assembly.

4. Compress the oil pump chain tensioner and install the supplied retaining clip.

5. Install or connect the following:
- Oil pump, chain, and crankshaft sprocket. Tighten the 8mm bolts to 16 ft. lbs. (22 Nm) and the 6mm bolt to 104 inch lbs. (12 Nm).
- Baffle plate and tighten the bolts to 104 inch lbs. (12 Nm)
- Oil pump chain guide and tensioner. Tighten the 8mm bolt to 16 ft. lbs. (22 Nm) and the 6mm bolts to 104 inch lbs. (12 Nm).
- Timing chain
- Oil pan
- Negative battery cable

6. Fill the crankcase to the correct level.

7. Start the engine and check for leaks.

2.2L and 2.3L Engines

1. Before servicing the vehicle, refer to the precautions in the beginning of this section.

2. Disconnect the negative battery cable.

3. Drain the engine oil into a sealable container.

4. Turn the engine to align the timing marks and set cylinder No.1 to TDC. The white mark on the crankshaft pulley should align with the pointer on the timing belt cover.

5. Remove or disconnect the following:
- Valve cover and upper timing belt cover
- Power steering pump belt and the alternator belt, also the air conditioning belt if so equipped
- Crankshaft pulley and the lower timing belt cover.
- Balancer belt and the timing belt. Be sure to mark the rotation of the timing belt if it is going to be

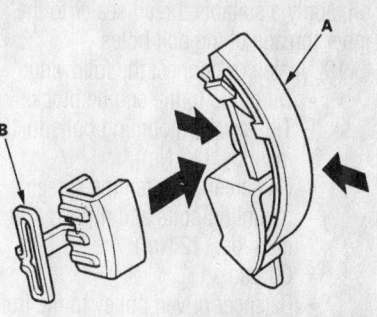

9347FG14

Compress the oil pump chain tensioner (A) and install the retaining clip (B)—2.0L engine

reused. Refer to the timing belt unit repair section.
- Timing belt and balancer belt tensioners.
- Crankshaft Position (CKP) sensor, if equipped
- Timing belt drive pulley and key from the crankshaft.
- Balancer driven pulley, by inserting a suitable tool into the maintenance hole in the front balancer shaft.

6. Align the rear timing balancer pulley using a 6 x 100mm bolt or rod. Mark the bolt or rod at a point 2.9 in. (74mm) from the end. Remove the bolt from the maintenance hole on the side of the block; insert the bolt/rod into the hole. Align the 74mm mark with the face of the hole. This pin will hold the shaft in place.

7. Remove or disconnect the following:
- Balancer gear case and the dowel pins. Discard the O-ring.
- Balancer driven gear attaching bolt and the balancer driven gear.
- Oil pan and the oil screen. Discard the screen gasket.
- Oil pump mounting bolts and oil pump assembly.
- Dowel pins from the engine and clean the oil pump mating surfaces of old gasket material and oil. Discard the O-rings.

To install:

8. Install the 2 dowel pins and new O-rings to the cylinder block.

9. Be sure that the mating surfaces are clean and dry. Apply a liquid gasket evenly in a narrow bead, centered on the mating surface. Once the sealant is applied, do not wait longer than 20 minutes to install the parts; the sealant will become ineffective. After final assembly, wait at least 30 minutes before adding oil to the engine, giving the sealant time to set. To prevent leakage of

Brake service is covered in Section 4 of this manual

oil, apply a suitable thread sealer to the inner threads of the bolt holes.

10. Install or connect the following:
   • Oil pump to the engine block. Tighten the mounting bolts to 108 inch lbs. (12 Nm).
   • Oil screen. Tighten the screen mounting bolts and nuts to 108 inch lbs. (12 Nm).
   • Oil pan
   • Balancer driven pulley to the front balancer belt, hold the balancer shaft in place with a suitable tool. Tighten the attaching bolt to 22 ft. lbs. (29 Nm).
   • Balancer driven gear to the rear balancer shaft. Tighten the bolt to 18 ft. lbs. (25 Nm).

➡ **Before installing the balancer driven gear and the gear case, apply molybdenum disulfide (lithium grease) to the thrust surfaces of the balancer gears.**

11. Align the groove on the pulley edge to the pointer on the balancer gear case.
12. Install or connect the following:
   • Balancer gear case to the engine and the mounting bolts and nut. The rear balancer shaft is being held in place with a 6 x 100mm bolt. Tighten the mounting bolts and nut to 18 ft. lbs. (25 Nm).
13. Check the alignment of the pointer on the balancer pulley to the pointer on the oil pump.
14. Install or connect the following:
   • Drive pulley to the crankshaft
   • CKP sensor. Tighten the mounting bolts to 108 inch lbs. (12 Nm).
   • Timing belt tensioners
   • Timing belt and the balancer belt. Refer to the timing belt unit repair section.
   • Crankshaft pulley and the lower timing belt cover.
   • Drive belts for the alternator, power steering, and air conditioning compressor. Tension the belts properly.
   • Valve cover and upper timing belt cover.
15. Refill the engine with clean, fresh oil.
16. Connect the negative battery cable and enter the radio security code.

3.0L Engine

1. Before servicing the vehicle, refer to the precautions in the beginning of this section.
2. Raise and safely support the vehicle.
3. Drain the engine oil.
4. Turn the crankshaft to position the

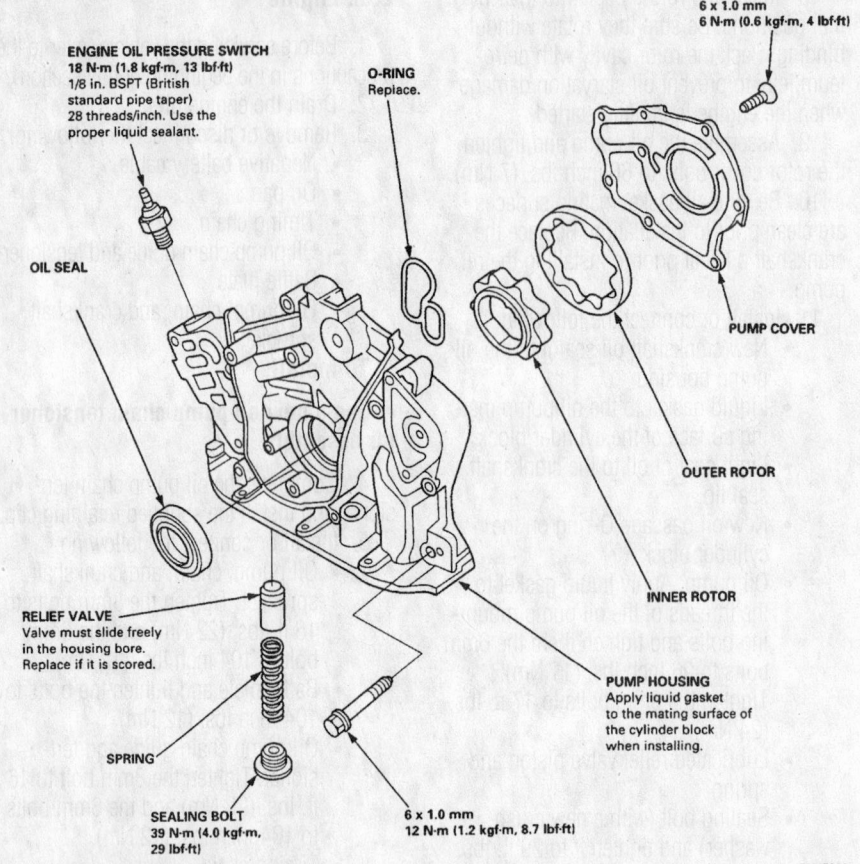

Exploded view of the oil pump—3.0L engine

No. 1 piston at TDC on the compression stroke.

5. Remove or disconnect the following:
   • Timing belt. Refer to the timing belt unit repair section.
   • Idler pulley.
   • Crankshaft Position (CKP) sensor.
   • Variable Timing Electronic Control (VTEC) solenoid valve
   • Oil filter
   • Oil pan and pick-up
   • Oil pump assembly

To install:
6. Install or connect the following:
   • New crankshaft seal in the oil pump
   • Sealant to the oil pump mounting surface and bolt holes on the engine block.
   • Grease to the lip of the new seal and engine oil to the o-ring.
   • Dowel pin and oil pump while aligning the inner rotor with the crankshaft. Tighten the bolts to 108 inch lbs. (12 Nm).
   • Oil pump pick-up. Tighten the mounting bolts to 108 inch lbs. (12 Nm).

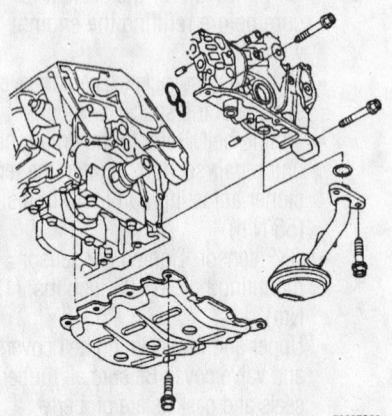

Oil pump mounting—3.0L engine

   • Oil pan, VTEC solenoid, oil filter, CKP, and idler pulley.
   • Timing belt. Refer to the timing belt unit repair section.

✱✱ WARNING

Operating the engine without the proper amount and type of engine oil will result in severe engine damage.

7. Fill the crankcase with the proper amount of new engine oil.

Rear Main Seal

REMOVAL & INSTALLATION

1. Before servicing the vehicle, refer to the precautions in the beginning of this section.

2. Remove the transmission.

3. Remove the driveplate from the crankshaft.

4. Carefully pry the crankshaft seal out of the retainer.

To install:

5. Apply clean engine oil to the lip of the new seal.

6. Install the seal onto the crankshaft and into the retainer using the appropriate seal driver.

7. Install the driveplate and the transmission.

Timing Chain, Sprockets, Front Cover and Seal

REMOVAL & INSTALLATION

2.0L Engine

1. Before servicing the vehicle, refer to the precautions in the beginning of this section.

2. Set the engine to Top Dead Center (TDC).

3. Drain the cooling system.

4. Relieve the fuel system pressure.

5. Remove or disconnect the following:
- Negative battery cable
- Air cleaner housing
- Vacuum tank

- Accessory drive belt
- Water bypass hose and tube
- Water pump pulley
- Accessory drive belt tensioner
- Alternator
- Idler pulley
- Throttle cable
- Intake manifold
- Exhaust manifold

6. Remove the timing chain auto tensioner as follows:

a. Remove the end cover and nozzle from the auto tensioner.

b. Use a 5 x 0.8mm bolt and locknut as shown to compress the auto-tensioner.

c. Remove the timing chain auto-tensioner.

- Valve cover
- Camshafts
- Timing chain idler gear
- Crankshaft sprocket
- Front crankshaft seal

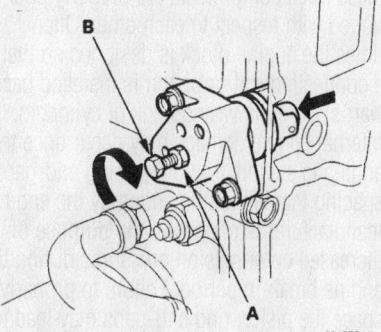

Use a 5 x 0.8mm bolt (B) and locknut (A) to compress the auto-tensioner—2.0L engine

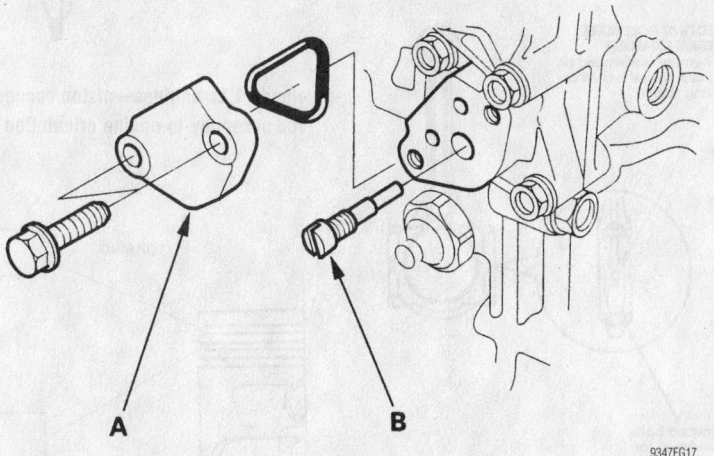

Timing chain auto-tensioner cover (A) and nozzle (B)—2.0L engine

- Front cover
- Oil pump chain guide
- Timing chain

To install:

7. Ensure that the crankshaft sprocket is set to TDC.

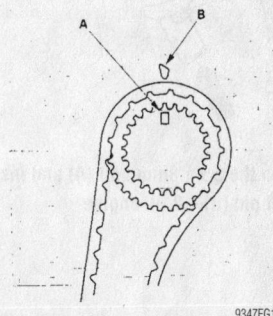

Align the sprocket key (A) with the cylinder block pointer (B) to set the engine to TDC—2.0L engine

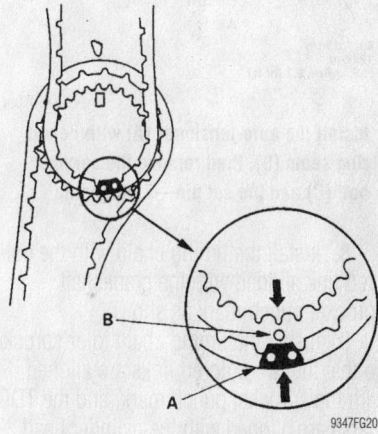

Install the timing chain with the colored link (A) aligned with the crankshaft sprocket punch mark (B)—2.0L engine

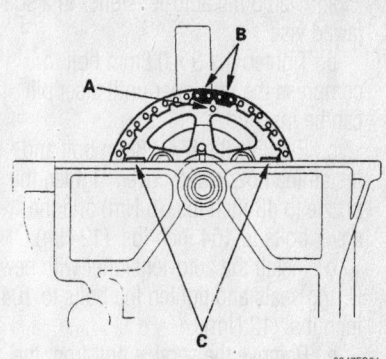

Timing chain idler sprocket punch mark (A), colored links (B) and TDC marks (C) in proper alignment—2.0L engine

For complete Engine Mechanical specifications, see Section 1 of this manual

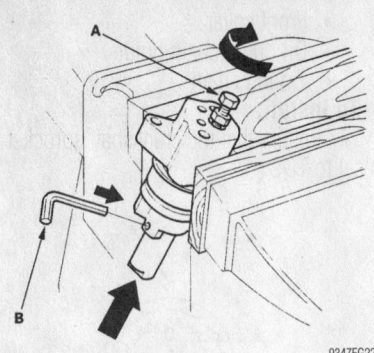

Tighten the 5 x).8mm bolt (A) and insert the set pin (B)—2.0L engine

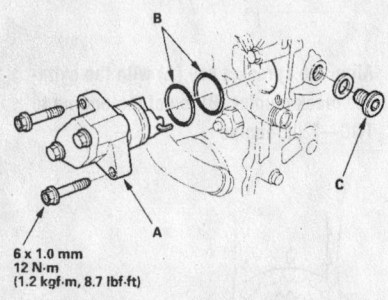

6 x 1.0 mm
12 N·m
(1.2 kgf·m, 8.7 lbf·ft)

Install the auto-tensioner (A) with new O ring seals (B), then remove the service bolt (C) and the set pin—2.0L engine

8. Install the timing chain with the colored link aligned with the crankshaft sprocket punch mark as shown.

9. Install the timing chain idler sprocket so that the two colored links are aligned with the sprocket punch mark, and the TDC marks are aligned with the cylinder head surface as shown. Tighten the idler sprocket bolt to 36 ft. lbs. (49 Nm).

10. Prepare the timing chain auto-tensioner for installation as follows:

a. Clamp the auto-tensioner in a soft-jawed vise.

b. Tighten the 5 x 0.8mm bolt to compress the tensioner until a set pin can be inserted.

c. Remove the 5 x 0.8mm bolt and install the nozzle and cover. Tighten the nozzle to 48 inch lbs. (5 Nm) and the cover bolts to 104 inch lbs. (12 Nm).

d. Install the auto-tensioner with new O ring seals and tighten the bolts to 104 inch lbs. (12 Nm).

e. Remove the service bolt from the cylinder head and remove the set pin.

f. Replace the service bolt and tighten it to 22 ft. lbs. (29 Nm).

11. The remainder of the installation is the reverse of the removal procedure, while using the following torque values:

- Oil pump chain guide bolts: 104 inch lbs. (12 Nm)
- Front cover: 10mm bolts to 33 ft. lbs. (44 Nm) and the 6mm bolts to 104 inch lbs. (12 Nm)
- Crankshaft sprocket bolt: 181 ft. lbs. (245 Nm)
- Idler pulley: 10mm bolts to 33. Ft. lbs. (44 Nm) and the 6mm bolt to 104 inch lbs. (12 Nm)
- Water pump pulley bolts: 10 ft. lbs. (14 Nm)
- Bypass tube bolts: 104 inch lbs. (12 Nm)

Piston and Ring

POSITIONING

When assembling the pistons, piston rings and connecting rods, and when installing these assemblies into the engine block, it is vitally important to ensure that these three components are properly positioned with respect to each other. Often times the engine block is designed so that if a connecting rod or piston is installed backwards, or in the wrong bank of cylinders, internal engine damage may occur once the engine is started. The piston ring end-gap spacing that is recommended by the engine manufacturer is often with the purpose of increased compression pressures during the engine break-in period. Failure to properly space the piston ring end-gaps may lead to increased oil consumption and extended break-in time. Therefore, always be sure to position the pistons, rings and connecting rods as shown in the accompanying illustrations.

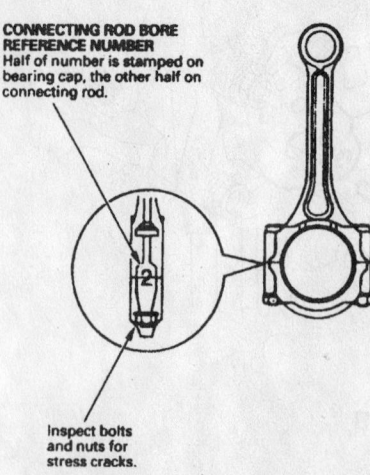

CONNECTING ROD BORE REFERENCE NUMBER Half of number is stamped on bearing cap, the other half on connecting rod.

Inspect bolts and nuts for stress cracks.

Honda engines—before removing the caps from the connecting rods, be sure to matchmark them as shown

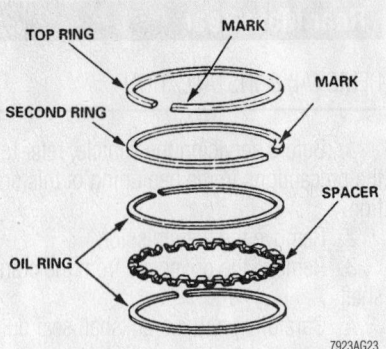

Honda engines—piston ring positioning

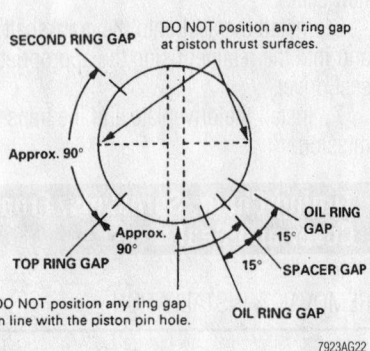

Honda engines—piston ring end-gap spacing

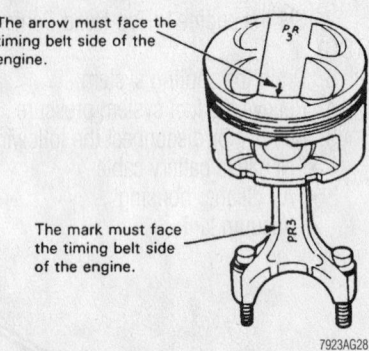

The arrow must face the timing belt side of the engine.

The mark must face the timing belt side of the engine.

Honda 1.6L engines—piston/connecting rod assembly-to-engine orientation

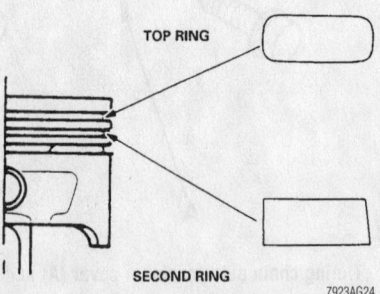

Honda 2.2L (H22A4) engines—compression ring locations

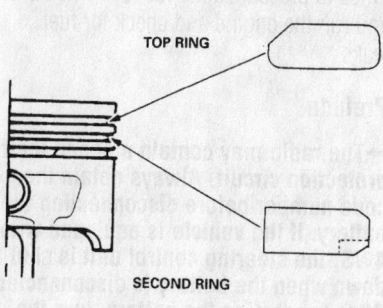

Honda 2.2L and 2.3L engines—compression ring locations

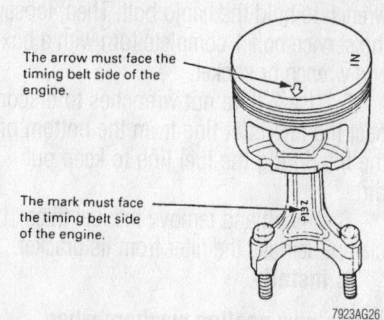

Honda 2.2L (H22A4) engines—piston/connecting rod assembly-to-engine orientation

✳✳ WARNING

Always be sure to matchmark the connecting rods and caps prior to disassembly so that they may be reassembled with their original counterparts. If the caps are not installed on their original connecting rods, the assemblies will most likely need machining to avoid bearing, connecting rod and/or crankshaft damage.

FUEL SYSTEM

Fuel System Service Precautions

Safety is the most important factor when performing not only fuel system maintenance but any type of maintenance. Failure to conduct maintenance and repairs in a safe manner may result in serious personal injury or death. Maintenance and testing of the vehicle's fuel system components can be accomplished safely and effectively by adhering to the following rules and guidelines.

• To avoid the possibility of fire and personal injury, always disconnect the negative battery cable unless the repair or test procedure requires that battery voltage be applied.

• Always relieve the fuel system pressure prior to disconnecting any fuel system component (injector, fuel rail, pressure regulator, etc.), fitting or fuel line connection. Exercise extreme caution whenever relieving fuel system pressure, to avoid exposing skin, face and eyes to fuel spray. Please be advised that fuel under pressure may penetrate the skin or any part of the body that it contacts.

• Always place a shop towel or cloth around the fitting or connection prior to loosening to absorb any excess fuel due to spillage. Ensure that all fuel spillage (should it occur) is quickly removed from engine surfaces. Ensure that all fuel soaked cloths or towels are deposited into a suitable waste container.

• Always keep a dry chemical (Class B) fire extinguisher near the work area.

• Do not allow fuel spray or fuel vapors to come into contact with a spark or open flame.

• Always use a back-up wrench when loosening and tightening fuel line connection fittings. This will prevent unnecessary stress and torsion to fuel line piping. Always follow the proper torque specifications.

• Always replace worn fuel fitting O-rings with new. Do not substitute fuel hose or equivalent, where fuel pipe is installed.

Fuel System Pressure

RELIEVING

✳✳ CAUTION

The fuel injection system remains under pressure after the engine has been turned OFF. Properly relieve fuel pressure before disconnecting any fuel lines. Failure to do so may result in fire or personal injury.

➡**The radio may contain a coded theft protection circuit. Always obtain the code number before disconnecting the battery. If the vehicle is equipped with 4WS, the steering control unit is shut down when the battery is disconnected. After connecting the battery, turn the steering wheel lock-to-lock to reset the steering control unit.**

1. Before servicing the vehicle, refer to the precautions in the beginning of this section.

2. Disconnect the negative battery cable.

3. Remove the fuel filler cap.

4. Use a box wrench to loosen the 6mm service bolt while holding the special banjo bolt with another wrench. On 1.6L engines, it is located on the fuel filter. On other engines, it is found on the fuel rail.

5. Place a rag or shop towel over the 6mm service bolt.

6. Slowly loosen the 6mm service bolt 1 complete turn.

✳✳ CAUTION

Do not allow fuel spray or fuel vapors to come in contact with a spark or open flame. Keep a dry chemical fire extinguisher nearby. Never store fuel in an open container due to risk of fire or explosion.

➡A fuel pressure gauge may be attached at the 6mm service bolt location. Always replace the washer between the service bolt and the banjo bolt whenever the service bolt is loosened.

7. Remove the service bolt and install a new washer. Tighten the 6mm service bolt to 108 inch lbs. (12 Nm). Don't overtighten the service bolts, their threads may strip and cause leaks.

8. Clean up any fuel spilled on the engine and intake manifold.

9. Install the fuel filler cap.

10. Reconnect the negative battery cable.

11. Turn the ignition **ON**, but don't start the engine. Repeat this 2 or 3 times to pressurize the fuel system. Check for fuel leaks.

12. Enter the radio security code.

13. If equipped with 4WS, start the engine and turn the steering wheel lock-to-lock to reset the 4WS control unit.

For Accessory Drive Belt illustrations, see Section 1 of this manual

Fuel Filter

REMOVAL & INSTALLATION

Civic

❋❋ CAUTION

The fuel injection system remains under pressure, even after the engine has been turned OFF. The fuel system pressure must be relieved before disconnecting any fuel lines. Failure to follow this procedure may result in fire or explosion.

➡The original radio contains a coded anti-theft circuit. Obtain the security code number before disconnecting the battery.

1. Before servicing the vehicle, refer to the precautions in the beginning of this section.
2. Disconnect the negative battery cable.
3. Place a rag under the fuel filter to catch fuel spray.
4. Relieve the fuel pressure by first loosening the fuel filler cap. Use a flare nut wrench to hold the banjo bolt. Then, loosen the service bolt 1 complete turn with a box-end wrench or socket.
5. Use 2 flare nut wrenches to disconnect the fuel inlet line from the bottom of the filter. Plug the fuel line to keep out dirt.
6. Unbolt and remove the fuel filter clamp. Remove the filter from its bracket.

To install:

➡Use new sealing washers when installing the fuel filter to prevent fuel leaks and the possibility of fire.

7. Clean the fuel line fittings before installing the filter.
8. Install the fuel filter and its clamp. Tighten the clamp bolt to 84 inch lbs. (10 Nm).
9. Connect the fuel inlet line and carefully tighten its fitting to 27 ft. lbs. (38 Nm).
10. Connect the fuel line with new washers and install the banjo bolt. Tighten the banjo bolt to 16 ft. lbs. (22 Nm). Install the service bolt and tighten it to 108 inch lbs. (12 Nm).
11. Connect the battery cable. Tighten the fuel filler cap.
12. Turn the ignition on and off several times to pressurize the fuel system. Start and run the engine and check for fuel leaks.

Prelude

➡The radio may contain a coded theft protection circuit. Always obtain the code number before disconnecting the battery. If the vehicle is equipped with 4WS, the steering control unit is shut down when the battery is disconnected. After connecting the battery, turn the steering wheel lock-to-lock to reset the steering control unit.

1. Before servicing the vehicle, refer to the precautions in the beginning of this section.
2. Disconnect the negative battery cable.
3. Place a shop towel under and around the fuel rail, then relieve the fuel pressure.

❋❋ CAUTION

Do not allow fuel spray or fuel vapors to come in contact with a spark or open flame. Keep a dry chemical fire extinguisher nearby. Never store fuel in an open container due to risk of fire or explosion.

4. Remove or disconnect the following:
   • 12mm banjo bolt and the fuel feed pipe from the fuel filter. Discard the washers.
   • Fuel filter clamp and the fuel filter.

To install:

5. Install or connect the following:
   • Fuel filter on the bracket and the filter clamp. Tighten the clamp bolts to 84 inch lbs. (10 Nm).

➡Clean the fuel fittings thoroughly before reconnecting them.

   • Fuel feed pipe to the filter and tighten the fitting to 28 ft. lbs. (38 Nm).
   • Fuel outlet pipe to the filter using new gaskets around the fitting. Tighten the banjo bolt to 20 ft. lbs. (28 Nm) on all Prelude models and 16 ft. lbs. (22 Nm) on all other models.
   • Negative battery cable and enter the radio security code.
6. Turn the ignition **ON** and check for fuel leaks.
7. If equipped with 4WS, start the engine and turn the steering wheel lock-to-lock to reset the 4WS control unit.

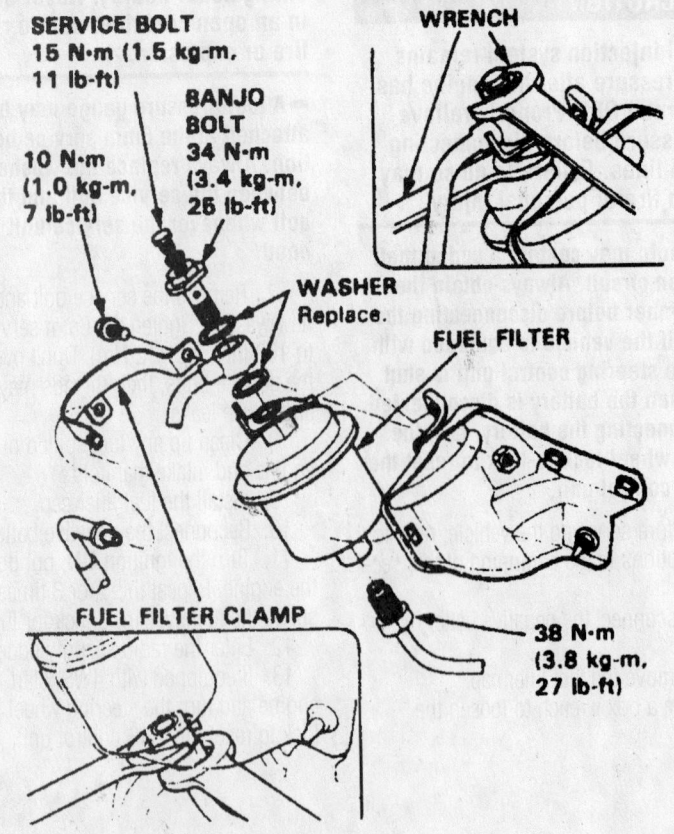

SERVICE BOLT
15 N·m (1.5 kg-m, 11 lb-ft)

BANJO BOLT
34 N·m (3.4 kg-m, 25 lb-ft)

10 N·m (1.0 kg-m, 7 lb-ft)

WRENCH

WASHER
Replace.

FUEL FILTER

FUEL FILTER CLAMP

38 N·m (3.8 kg-m, 27 lb-ft)

Fuel filter components—1.6L engines

7923FG94

28 N·m (2.8 kg-m, 20 lb-ft)

38 N·m (3.8 kg-m, 28 lb-ft)

WASHER Replace

10 N·m (1.0 kg-m, 7 lb-ft)

7923FG95

Fuel filter mounting—Prelude

Accord

1. Before servicing the vehicle, refer to the precautions in the beginning of this section.

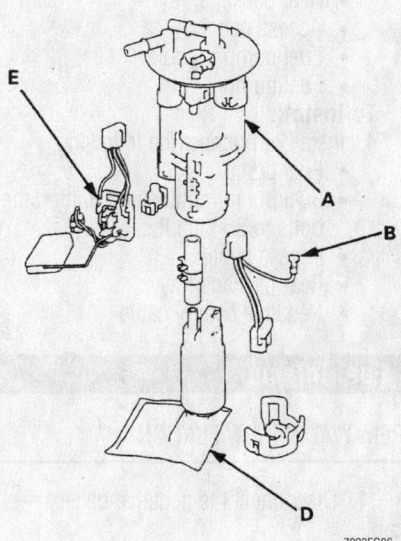

7923FG96

Fuel filter (A), wiring harness (B), suction filter (D) and sending unit (E)—Accord

2. Remove the fuel pump from the tank.
3. Remove the fuel filter from the pump module.

To install:

4. Install the new filter on the pump module.
5. Install the fuel pump in the tank.

S2000

1. Before servicing the vehicle, refer to the precautions in the beginning of this section.
2. Relieve the fuel system pressure.
3. Remove or disconnect the following:
   • Negative battery cable
   • Rear package tray
   • Access panel
   • Fuel pump module
   • Fuel filter

To install:

4. Install or connect the following:
   • Fuel filter
   • Fuel pump module and tighten the bolts to 36 inch lbs. (4 Nm)
   • Access panel
   • Rear package tray
   • Negative battery cable

Fuel Pump

REMOVAL & INSTALLATION

➡The radio may contain a coded theft protection circuit. Always obtain the code number before disconnecting the battery. If the vehicle is equipped with 4WS, the steering control unit is shut down when the battery is disconnected. After connecting the battery, turn the steering wheel lock-to-lock to reset the steering control unit.

Civic

❋❋ CAUTION

The fuel injection system remains under pressure, even after the engine has been turned OFF. The fuel system pressure must be relieved before disconnecting any fuel lines. Failure to follow this procedure may result in fire, explosion, or personal injury.

1. Before servicing the vehicle, refer to the precautions in the beginning of this section.
2. Disconnect the negative battery cable.
3. Loosen the fuel filler cap. Then, loosen the fuel filter service bolt to relieve the fuel pressure.
4. Remove or disconnect the following:
   • Rear seat cushions
   • Fuel pump access panel
   • 2-wire fuel pump harness

➡Clean the fuel line fittings before disconnecting them.

   • Fuel line and the hose from the fuel pump.
   • Fuel pump bolts and fuel pump from the fuel tank. Allow the fuel in the pump drain into the tank before removing the pump from the vehicle.
   • Fuel pump motor from its bracket.

To install:

➡Use new sealing washers when reconnecting the fuel line banjo bolt.

5. Install or connect the following:
   • Fuel pump into the fuel tank with a new O-ring. Then, tighten the mounting nuts to 48 inch lbs. (6 Nm).
   • Hose and the fuel line. Carefully tighten the banjo bolt to 20 ft. lbs. (28 Nm).

For Tire, Wheel and Ball Joint specifications, see Section 1 of this manual

- Fuel pump harness
- Fuel filler cap
- Fuel filter service bolt: 11 ft. lbs. (15 Nm)
- Battery cable and turn the ignition switch **ON** and **OFF** several times to pressurize the fuel system. Check the connections at the fuel pump for any leaks. Check the fuel filter service bolt for leaks.
- Fuel pump access cover
- Rear seat cushions or rear compartment trim. Be sure the clips are properly seated.

Prelude

✳✳ CAUTION

The fuel injection system remains under pressure after the engine has been turned OFF. Properly relieve fuel pressure before disconnecting any fuel lines. Failure to do so may result in fire or personal injury.

1. Before servicing the vehicle, refer to the precautions in the beginning of this section.
2. Disconnect the negative battery terminal.
3. Relieve the fuel pressure.
4. Remove or disconnect the following:

- Carpet in the luggage area.
- Fuel pump maintenance access cover in the floor.
- Electrical connector at the pump unit.

➡**Label the fuel lines before disconnecting them.**

- Fuel lines. Discard the washers from the fuel feed connection.
- Retaining nuts holding the pump
- Pump up and out of the tank.

➡**The pump sits on an angle and may require some manipulation to remove. If the pump still won't come out, loosen the fuel tank mounting nuts under the car, slide the tank downward a bit to give more clearance at the top.**

To install:
5. Install or connect the following:
- New sealing ring
- Fuel pump, making certain it is correctly seated and not wedged or jammed.
- Retaining nuts and tighten them evenly and alternately to 48 inch lbs. (6 Nm).

- Fuel lines. Make certain the clamp is secure; use new ones if necessary.
- New washers to the fuel feed connection before installing the attaching bolt. Tighten the fuel feed attaching bolt to 20 ft. lbs. (28 Nm).
- Fuel pump connector
- Negative battery cable and enter the radio security code.

6. Switch the ignition **ON** but do not engage the starter. The fuel pump should run for approximately 2 seconds, building pressure within the lines. Switch the ignition **OFF**, then **ON** 2 or 3 more times to build full system pressure. Check for fuel leaks.
7. If equipped with 4WS, start the engine and turn the steering wheel lock-to-lock to reset the 4WS control unit.
8. Install the maintenance access cover and seal or gasket, if used.
9. Reposition the carpeting in the luggage compartment.

Accord

1. Before servicing the vehicle, refer to the precautions in the beginning of this section.
2. Remove or disconnect the following:

- Negative battery cable.
- Spare tire cover

A
4 N·m
(0.4 kgf·m, 3 lbf·ft)

B

7923FG98

Exploded view of the fuel pump assembly—Accord

- Access panel from the floor
- 5-pin connector from the pump assembly.
- Fuel cap and relieve the fuel system pressure.
- Quick-connect connections from the pump assembly.
- Mounting bolts and the pump assembly from the tank.

To install:
3. Install or connect the following:
- Fuel pump, using a new gasket.
- Quick-connect fuel lines to the pump assembly.
- Fuel cap
- 5-pin connector to the pump.
- Negative battery cable and enter the radio security code.

4. Switch the ignition **ON** but do not engage the starter. The fuel pump should run for approximately 2 seconds, building pressure within the lines. Switch the ignition **OFF**, then **ON** 2 or 3 more times to build full system pressure. Check for fuel leaks.
5. If there are no leaks, install the access panel cover and the tire cover.

S2000

1. Before servicing the vehicle, refer to the precautions in the beginning of this section.
2. Relieve the fuel system pressure.
3. Remove or disconnect the following:

- Negative battery cable
- Rear package tray
- Access panel
- Fuel pump module
- Fuel pump

To install:
4. Install or connect the following:
- Fuel pump
- Fuel pump module and tighten the bolts to 36 inch lbs. (4 Nm)
- Access panel
- Rear package tray
- Negative battery cable

Fuel Injector

REMOVAL & INSTALLATION

1. Disconnect the negative battery cable.
2. Relieve the fuel system pressure.
3. Remove the fuel rail assembly
4. Carefully pull the injectors from the intake manifold.
5. Discard the seal rings, cushion rings and O-rings.

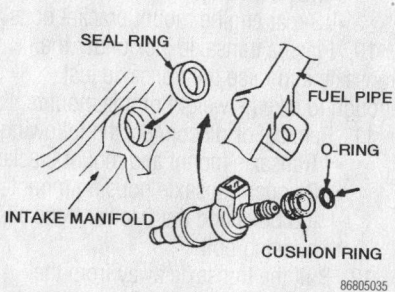

Always use new cushion rings, seal rings and O-rings

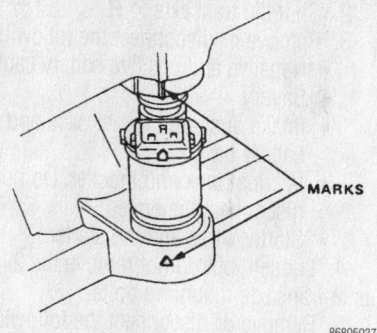

Be sure to align the center line on the injector with the mark on the fuel rail

To install:

6. Slide new cushion rings onto the injectors.

7. Coat new O-rings with clean engine oil and put them on the injectors.

8. Insert the injectors into the fuel rail. Be sure to align the center line on the injector with the mark on the fuel rail.

9. Coat new seal rings with clean engine oil and insert them into the intake manifold.

10. Install the fuel rail assembly.

DRIVE TRAIN

Transaxle Assembly

REMOVAL & INSTALLATION

Manual

CIVIC

✳✳ WARNING

Use only genuine Honda manual transaxle fluid (MTF)-it is specially formulated for use in Honda transaxles. If Honda MTF is not available, API SG/SJ 10W-30 or 10W-40 motor oil may be used as a temporary lubricant. However, motor oil will cause increased transaxle wear and shifting effort. Refill the transaxle with Honda MTF as soon as possible.

1. Before servicing the vehicle, refer to the precautions in the beginning of this section.

2. Remove or disconnect the following:
   - Negative and positive battery cables
   - Resonator, the air cleaner box, and the air intake duct.
   - Starter cables and the transaxle ground cable.
   - Back-up light switch connection
   - Upper radiator hose out of its bracket.
   - Vehicle Speed Sensor (VSS) connector.
   - Clutch fluid line bracket
   - Slave cylinder. It isn't necessary to disconnect the clutch fluid line.

3. Raise and safely support the vehicle.

4. Drain the transaxle fluid.

5. Remove or disconnect the following:
   - Front wheels
   - Strut pinch bolt and fork bolt
   - Lower ball joint from the steering knuckle using a ball joint remover.
   - Halfshaft inboard joints out of the transaxle case. Swing the steering knuckles out to free the halfshafts from the transaxle.

6. Tie the halfshafts up and out of the way with wire so that the joints will not be stressed. Tie plastic bags over the inboard joints to prevent damage to the CV-boots and splined shafts.

7. Remove or disconnect the following:
   - Shift rod and extension rod from the transaxle case. Drive the shift rod retaining pin out with a pin punch.
   - Front exhaust pipe
   - Engine-to-transaxle stiffener brackets and the clutch cover plate

8. Attach a lifting chain to the engine and lift slightly to ease the tension on the mounts.

9. Remove or disconnect the following:
   - Splash shield from underneath the vehicle.
   - Right-front mount/bracket assembly

10. Place a jack under the transaxle to support its weight.

11. Remove or disconnect the following:
   - Transaxle side mount and bracket
   - Starter's lower mounting bolt and the upper 3 transaxle case bolts.
   - 3 rear transaxle mount bracket bolts, then the lower 3 transaxle case bolts.
   - Pull the transaxle away from the engine until it clears the mainshaft.

Lower the transaxle out of the vehicle Be careful not to bend the clutch hydraulic line.

To install:

➡**Use new self-locking nuts and color-coded bolts when installing the transaxle and suspension components.**

12. Apply high temperature grease to the mainshaft splines, release fork contact points, and throw-out bearing. The manufacturer recommends part No. 08798-9002, Honda Super High temp Urea Grease.

13. Place the transaxle on a transaxle jack and raise it to the level of the engine.

14. Align the transaxle and engine. Be sure the transaxle case dowel pins are securely seated, and fit the transaxle onto the engine. Install the upper and lower transaxle case bolts and the 14mm rear mount bolts and washers. Only hand-tighten them at this time.

15. Raise the transaxle and install the side mount. Tighten the upper and lower transaxle case bolts to 47 ft. lbs. (64 Nm). Tighten the 14mm rear mount bracket bolts to 61 ft. lbs. (84 Nm).

16. First, tighten the transaxle side mount bracket nuts and bolt to 47 ft. lbs. (64 Nm) each. Next, tighten the mount bushing bolts to 47 ft. lbs. (64 Nm). Finally, tighten the through-bolt to 54 ft. lbs. (74 Nm).

17. Install or connect the following:
   - Right-front mount/bracket assembly. Use 3 new 12mm bolt and washers, and tighten them to 47 ft. lbs. (64 Nm). Tighten the 2 10mm bolts to 33 ft. lbs. (45 Nm).
   - Clutch cover
   - Engine-to-transaxle stiffener brackets and tighten the 8mm bolts to 17

For Wheel Alignment specifications, see Section 1 of this manual

ft. lbs. (24 Nm). Tighten the 10mm bolts to 33 ft. lbs. (44 Nm).

18. Once the transaxle is bolted to the engine, and the transaxle mounts are securely tightened, the engine lifting chain may be removed.

19. Install or connect the following:
- Shift rod with a new spring pin and clip. Then, fit the shift rod boot back into place. Connect the torque rod and tighten the bolt to 16 ft. lbs. (22 Nm).
- Front exhaust pipe. Use new self-locking nuts and gaskets. Tighten the rear flange nuts to 16 ft. lbs. (22 Nm). Tighten the front flange nuts to 40 ft. lbs. (54 Nm) for D16Y8 and D16Y7 engines, or 25 ft. lbs. (33Nm) for all others.
- New set rings on the inboard CV-joint splines
- Halfshafts into the transaxle case and intermediate shaft. The inboard joints must snap into place.
- Lower ball joint and damper fork
- Front wheels
- Slave cylinder and clutch pipe stay. Coat the slave cylinder's tip with high temperature grease. Be sure it snaps into the release fork. Tighten the slave cylinder mounting bolts to 16 ft. lbs. (22 Nm).
- VSS connector and back-up light switch connectors
- Wiring harness clamps and starter cables.
- Resonator, air cleaner box, and air intake duct. Fit the upper radiator hose back into its bracket.

20. Lower the vehicle and tighten the strut pinch bolts to 32 ft. lbs. (44 Nm). Tighten the fork bolts to 47 ft. lbs. (65 Nm). Tighten the ball joint castle nuts to 40 ft. lbs. (55 Nm), then tighten them only enough to install new cotter pins.

21. Turn the breather cap so that the arrow with the **F** mark points toward the front of the vehicle.

22. Refill the transaxle with the Honda MTF fluid.

23. Reconnect the positive and negative battery cables.

24. Bleed the clutch hydraulic system.

25. Check the clutch and transaxle for smooth operation.

26. Check and adjust the front wheel alignment.

PRELUDE

1. Before servicing the vehicle, refer to the precautions in the beginning of this section.

2. Shift the transaxle to **R**.
3. Remove or disconnect the following:
- Negative and positive battery cables
- Battery
- Intake duct, air cleaner case and battery base
- Vacuum tank and bracket. Do not disconnect the hoses.
- Starter wires and the starter.

4. Loosen, but do not remove the 2 upper transaxle mounting bolts.

5. Remove or disconnect the following:
- Transaxle ground cable and the back-up light switch wire.
- Engine harness clamp
- Shift cables from the transaxle case, leaving them attached to their bracket, and wire them safely out of the work area.
- Vehicle Speed Sensor (VSS) connector. Leave the sensor hoses connected and remove the sensor from the transaxle case.
- Slave cylinder mounting bolts.
- Slave cylinder from the release fork and move it out of the work area, leaving the hydraulic line connected to the slave cylinder

➡ **Do not depress the clutch pedal once the slave cylinder has been removed. Be careful not to kink the metal hydraulic line.**

6. Raise and safely support the vehicle. Drain the transaxle fluid.

7. Remove or disconnect the following:
- Clutch damper mounting bolts and raise the clutch damper.
- Remove the rear engine mount bracket stay, if equipped.
- Front wheels
- Cotter pins and lower arm ball joint nuts
- Ball joints and lower arms using a press-type ball joint tool.
- Damper fork bolt and the radius rod on the right side of the vehicle only.

8. Use a suitable tool to pry the right and left halfshafts out of the differential and the intermediate shaft. Pull on the inboard joint and remove the right and left halfshafts.

9. Remove or disconnect the following:
- Intermediate shaft from the differential. Tie plastic bags over the halfshaft inboard joints to prevent damage to the boots and splines. Wire the halfshafts to the underbody of the vehicle so that their weight doesn't hang on their outboard joints.
- Center beam and remove the clutch cover

- Rear beam stiffener and the intake manifold stay.
- 3 rear engine mount bracket bolts.

10. Place a transaxle jack under the transaxle and raise the transaxle just enough to take its weight off the mounts.

11. Remove or disconnect the following:
- Transaxle mount and mount bracket.
- 2 upper transaxle housing mounting bolts and the 3 lower transaxle housing bolts.

12. Pull the transaxle away from the engine to clear the mainshaft.

13. Lower the transaxle from the vehicle.

To install:

➡ **Use new self-locking nuts and set rings when assembling the front suspension components and halfshafts. Use new self-locking bolts when installing the center beam and rear engine mount bracket. These fasteners can be purchased from a Honda dealer.**

14. Be sure the dowel pins are installed into the transaxle case.

15. Apply heavy duty high temperature grease (use Honda part number 08798–9002) to the mainshaft splines, release fork bolt and paws, and the throw-out bearing. Install the bearing and release fork. Be sure the release fork snaps into place.

16. Raise the transaxle into position with a transaxle jack.

17. Install or connect the following:
- 3 lower and 2 upper transaxle mounting bolts and evenly tighten them to 47 ft. lbs. (65 Nm).
- Transaxle mount and mount bracket
- Through-bolt and tighten temporarily. Be sure the engine is level. First tighten the 3 bracket-to-mount nuts and 2 bolts to 28 ft. lbs. (39 Nm). Then, tighten the through-bolt to 47 ft. lbs. (65 Nm).
- 3 new rear engine mount bracket bolts on the engine side and tighten them to 40 ft. lbs. (55 Nm).
- Rear beam stiffener and tighten the bolts to 28 ft. lbs. (39 Nm).
- Intake manifold stay and tighten the bolts to 16 ft. lbs. (22 Nm).
- Clutch cover and tighten the bolts to 108 inch lbs. (12 Nm).
- Center beam and tighten the bolts to 43 ft. lbs. (60 Nm).
- Intermediate shaft. Tighten its mounting bolts to 28 ft. lbs. (39 Nm).
- New set rings onto the halfshaft inboard joint splines
- Halfshafts, making sure that they lock into place.

- Radius rod and damper fork. Only hand-tighten their fasteners at this time.
- Ball joint to the lower arm. Tighten the castle nut to 36–43 ft. lbs. (50–60 Nm). Then, only tighten the nut enough to install a new cotter pin.
- Rear engine mount bracket stay, if equipped. Tighten the nut to 15 ft. lbs. (21 Nm) and the bolt to 28 ft. lbs. (39 Nm).
- Clutch damper and tighten the bolts to 16 ft. lbs. (22 Nm).
- Front wheels

18. Lower the vehicle.

19. Use a floor jack placed under the right front control arm to raise the vehicle enough so that its weight is supported by the jack. Tighten the radius rod mounting bolts to 76 ft. lbs. (105 Nm) and the radius rod nut to 32 ft. lbs. (44 Nm). Tighten the damper pinch bolt to 32 ft. lbs. (44 Nm). Tighten the damper fork bolt to 47 ft. lbs. (65 Nm). After pre-loading the suspension, lower the vehicle and remove the floor jack.

20. Coat the tip of the slave cylinder with heavy duty high temperature grease.

21. Install or connect the following:
- Clutch hose pipe and clutch slave cylinder to the transaxle housing. Be sure the slave cylinder snaps into the release fork. Tighten the slave cylinder mounting bolts to 16 ft. lbs. (22 Nm).
- Speed sensor. Tighten the mounting bolt to 14 ft. lbs. (19 Nm).
- Shift cable and select cable to the shift arm lever.
- Shift cable assembly onto the transaxle case. Tighten the cable bracket mounting bolts to 16 ft. lbs. (22 Nm).
- New cotter pins
- Back-up light switch coupler and the transaxle ground cable.
- Harness clamp
- Starter. Tighten the 10mm bolt to 32 ft. lbs. (45 Nm) and the 12mm bolt to 54 ft. lbs. (75 Nm).
- Starter wires

22. Loosen the 3 front engine mount bracket bolts. Tighten them to 28 ft. lbs. (39 Nm).

23. Install or connect the following:
- Vacuum tank and its bracket
- Air cleaner case and intake duct

24. Fill the transaxle with the proper type and quantity of oil.

25. Install or connect the following:

- Battery base stay and the battery base. Tighten the battery base bolts to 16 ft. lbs. (22 Nm).

26. Install or connect the following:
- Battery and connect the battery cables.

27. Check the clutch pedal free-play.

28. Start the vehicle and check the transaxle and clutch for smooth operation.

29. On Preludes equipped with 4WS, and turn the steering wheel lock-to-lock to reset the steering control unit.

30. Check and adjust the front wheel alignment.

31. Enter the radio security code.

ACCORD

1. Before servicing the vehicle, refer to the precautions in the beginning of this section.

2. Shift the transaxle into **R**.

3. Remove or disconnect the following:
- Negative and positive battery cables and the battery.
- Idle Air Control (IAC) solenoid connector
- Intake duct, resonator, air cleaner case, and battery base.
- Starter wires and starter
- Transaxle ground cable and the back-up light switch wire.
- Cable stay, then the cables from the top housing of the transaxle
- Both cables and the stay together
- Vehicle Speed Sensor (VSS). Leave the speed sensor hoses connected.
- Shift cable bracket
- Shift and select cables from the top of the transaxle case. Leave the cables and bracket together, and wire them out of the work area.
- Mounting bolts and clutch slave cylinder with the clutch pipe and pushrod.
- Mounting bolt and clutch hose joint with the clutch pipe and clutch hose.

➡Do not depress the clutch pedal once the slave cylinder has been removed. Be careful not to kink the metal hydraulic lines.

- 2 upper transaxle case bolts

4. Raise and safely support the vehicle.

5. Remove or disconnect the following:
- Front wheels
- Engine splash shield

6. Drain the transaxle fluid.

7. Remove or disconnect the following:

- Clutch damper bracket and raise it out of the way.
- Subframe center beam
- Cotter pins and lower arm ball joint nuts
- Ball joints and lower arms using a press type ball joint tool.
- Right damper fork bolt
- Right damper pinch bolt, then separate the damper fork and damper.
- Radius rod bolts and nut, then the right radius rod.
- Right and left halfshafts from the differential and the intermediate shaft, using a suitable prytool.
- Left halfshaft
- Intermediate shaft from the differential by removing its 3 bearing shaft mounting bolts.

8. Swing the right halfshaft out and wire it up inside the right fender well. Tie plastic bags over the inboard CV-joints to protect the boots and splines from damage.

9. Remove or disconnect the following:
- Engine stiffener and the clutch cover
- Intake manifold bracket
- Rear engine mount bracket
- 3 rear engine mount bracket mounting bolts, then discard.

10. Place a transaxle jack under the transaxle. Raise the transaxle just enough to take the weight off the its mounts.

➡A chain hoist may be attached the transaxle lifting hooks to steady it and aid in lowering it from the vehicle.

11. Remove or disconnect the following:
- Transaxle housing mounting bolt on the engine side
- Transaxle mount bolt and loosen the mount bracket nuts.
- 3 transaxle housing mounting bolts, then the transaxle from the vehicle.

To install:

➡Use new self-locking nuts when assembling the front suspension. Install new set rings onto the inboard CV-joints. Use new self-locking bolts when installing transaxle rear mount bracket (the bolts are color coded by type). New fasteners are available from a Honda dealer.

12. Be sure the 2 dowel pins are installed into the transaxle case.

13. Apply heavy duty high temperature grease (use Honda part No. 08798–9002) to the release bearing, mainshaft splines, and

the release fork pawls. Install the release fork and release bearing.

14. Raise the transaxle into position.

15. Install or connect the following:
- 3 lower transaxle case bolts and tighten to 47 ft. lbs. (65 Nm).
- Transaxle mount and mount bracket
- Through-bolt and tighten temporarily. Be sure the engine is level and tighten the 3 mount bracket nuts to 40 ft. lbs. (55 Nm). Tighten the through-bolt to 47 ft. lbs. (65 Nm).
- Upper transaxle case bolts on the engine side and tighten to 47 ft. lbs. (65 Nm).
- 3 new rear engine bracket mounting bolts and tighten to 40 ft. lbs. (55 Nm).
- Intake manifold bracket and tighten the bolts to 16 ft. lbs. (22 Nm).
- Clutch cover and tighten the bolts to 108 inch lbs. (12 Nm).
- Subframe center beam with new self-locking bolts. Evenly tighten the bolts to 37 ft. lbs. (50 Nm).
- Engine stiffener plate, if equipped. Loosely install the mounting bolts. Tighten the stiffener-to-transaxle case mounting bolt to 28 ft. lbs. (39 Nm), then tighten the 2 stiffener-to-engine block mounting bolts to 28 ft. lbs. (39 Nm) beginning with the bolt closest to the transaxle.
- Radius rod and the damper fork. Hand-tighten all the fasteners.
- Intermediate shaft and tighten its mounting bolts to 28 ft. lbs. (39 Nm).
- New set ring on the end of each halfshaft.
- Right and left halfshafts. Turn the right and left steering knuckle fully outward and slide the axle into the differential, until the set ring is felt engaging the differential side gear.
- Lower control arm ball joints. Tighten the castle nuts to 40 ft. lbs. (50 Nm). Then, tighten them only enough to install a new cotter pin.
- Clutch damper and tighten its mounting bolts to 16 ft. lbs. (22 Nm).
- Front wheels. Lower the vehicle.

16. Place a floor jack under the right front knuckle, and raise the jack until it is supporting the vehicle's weight.

17. Tighten the radius rod mounting bolts to 76 ft. lbs. (105 Nm) and the radius rod nut to 32 ft. lbs. (44 Nm). Tighten the damper fork nut while holding the damper fork bolt to 40 ft. lbs. (55 Nm). Tighten the damper pinch bolt to 32 ft. lbs. (44 Nm).

18. Coat the tip of the slave cylinder with high temperature grease. Install the clutch hose joint and clutch slave cylinder to the transaxle housing. Be sure the slave cylinders tip snaps into the release fork. Tighten the slave cylinder mounting bolts to 16 ft. lbs. (22 Nm).

19. Install or connect the following:
- Speed sensor. Tighten the mounting bolt to 13 ft. lbs. (18 Nm).
- Shift cable and select cable to the shift arm lever. Tighten the cable bracket mounting bolts to 20 ft. lbs. (27 Nm).
- New cotter pins
- Back-up light switch connector
- Starter. Tighten the 10mm bolt to 32 ft. lbs. (45 Nm) and the 12mm bolt to 54 ft. lbs. (75 Nm).
- Starter wires
- Transaxle ground cable

20. Fill the transaxle with the proper type and quantity of oil.

21. Install or connect the following:
- Air cleaner case and the resonator, then the intake duct.
- Battery tray bracket and battery tray and tighten the bolts to 16 ft. lbs. (22 Nm).
- Battery and the battery cables.

22. Check the clutch pedal free-play.

23. Check and adjust the front wheel alignment.

24. Road test the vehicle and check the transaxle for smooth operation.

25. Loosen the 3 front engine mount bracket mounting bolts, then retighten them to 28 ft. lbs. (38 Nm).

26. Enter the radio security code.

S2000

1. Before servicing the vehicle, refer to the precautions in the beginning of this section.

2. Remove or disconnect the following:
- Battery cables and the battery
- Shift lever knob
- Center console
- Shift lever boot
- Shift lever
- Air cleaner housing
- Steering shaft from the steering gear box
- Alternator
- A/C compressor
- Exhaust manifold heat shields
- Upper starter mounting bolt
- Upper intake manifold bracket mounting bolt
- Suction valve hose
- Camshaft Position (CMP) sensor connectors

- Splash shield
- Steering gear box electrical connector
- Torque sensor connector
- Intake manifold bracket
- Heated Oxygen (HO$_2$S) sensor connectors
- Catalytic converter
- Exhaust manifold
- Driveshaft
- Shifter boot holder bolts
- Clutch slave cylinder
- Clutch release fork
- Lower transmission flange bolts

3. Support the front subframe with a floor jack and remove the two center mounting bolts.

4. Loosen the four outer mounting bolts 3 inches (75 mm).

5. Lower the front subframe until it is supported by the loosened bolts.

6. Support the transmission with the floor jack.

7. Remove or disconnect the following:
- Rear transmission mount
- Speed sensor connector and wiring harness
- Upper transmission flange bolts
- Transmission

To install:

➡Use new subframe bolts for assembly.

8. Installation is the reverse of the removal procedure, while using the following torque values:
- Transmission flange bolts: 47 ft. lbs. (64 Nm)
- Rear transmission mount bolts: 28 ft. lbs. (38 Nm)
- Subframe mounting bolts: 14mm bolts to 85 ft. lbs. (116 Nm) and 12mm bolts to 43 ft. lbs. (59 Nm)
- Clutch slave cylinder bolts: 16 ft. lbs. (22 Nm)
- A/C compressor bolts: 33 ft. lbs. (44 Nm)
- Steering shaft pinch bolt: 16 ft. lbs. (22 Nm)
- Shift lever bolts: 86 inch lbs. (10 Nm)

Automatic

CIVIC

1. Before servicing the vehicle, refer to the precautions in the beginning of this section.

2. Remove or disconnect the following:
- Negative and positive battery cables
- Resonator, the air cleaner box, and the air intake duct.

- Starter cables and the transaxle ground cable
- Engine wiring harness clip

➡**Label all electrical connectors before removing them.**

- Lock-up control solenoid connector
- Vehicle Speed Sensor and counter-shaft speed sensor connectors
- Upper transaxle case bolts and the rear engine mounting bolt

3. Raise and safely support the vehicle.

4. Remove or disconnect the following:
- Front wheels

5. Drain the automatic transaxle fluid. Then, install the drain plug with a new crush washer. Note the color, consistency, and odor of the drained fluid.

6. Remove or disconnect the following:
- Front splash shield
- Shift control and linear solenoid connectors
- Mainshaft speed sensor connector
- Strut pinch bolt and fork bolt
- Lower ball joint using a ball joint remover.

7. Pry the halfshaft inboard joints out of the transaxle case and intermediate shaft. Swing the steering knuckles out to free the halfshafts from the transaxle.

8. Tie the halfshafts up and out of the way with wire. Tie plastic bags over the inboard joints to prevent damage to the CV-boots and splined shafts.

9. Remove or disconnect the following:
- Front exhaust pipe
- Shift cable cover
- Shift cable from the transaxle control shaft. Move the shift cable out of the way, and tie it up with wire.
- Automatic Transaxle Fluid (ATF) cooler hoses from the cooler lines. Cap the lines to prevent fluid lose and contamination.
- Right-front mount and bracket assembly
- Engine stiffener and the torque converter cover plate.
- 8 torque converter-to-driveplate bolts 1 at a time by rotating the crankshaft pulley.

➡**There are no gear teeth on the drive-plate; the starter motor engages a ring gear on the inner edge of the torque converter.**

10. After unbolting the torque converter from the driveplate, rotate the crankshaft to set the engine at TDC/compression for the No. 1 cylinder.

11. Remove or disconnect the following:
- Ignition wires
- Distributor, if equipped

12. Attach a lifting chain to the engine and lift slightly to ease the tension on the mounts.

13. Place a transaxle jack under the transaxle and remove the transaxle side mount and bracket.

14. With the transaxle supported, remove the transaxle rear mount bracket bolts and transaxle case bolts.

15. Pull the transaxle away from the engine until it clears the locating dowel pins. Carefully lower the transaxle from the vehicle with the torque converter angled upward so it doesn't drop out of the transaxle.

16. Remove the torque converter from the transaxle. Inspect the ring gear teeth for breakage and inspect the converter's hub for burrs and scoring. Check the condition of the converter fluid. Replace the torque converter if necessary.

17. Inspect the transaxles front oil pump bearing and seal for signs of leakage and scoring. Inspect the mainshaft for burrs, scoring, and roughness.

18. With the transaxle removed, carefully inspect the driveplate for stress cracks, enlarged bolt holes, and other defects. Replace it if necessary.

To install:

➡**Use new self-locking nuts and color-coded bolts when installing the transaxle and suspension components.**

19. Flush the transaxle cooler lines to remove any contaminated fluid and residual clutch material:
- a. Use a pressurized flusher (Honda J38405-A or equivalent). Use only Honda flushing fluid (Honda J35944–20); other fluids may damage the system.
- b. Fill the flusher with 21 ounces of fluid. Pressurize the flusher to 80–120 psi, following the procedure on the fluid container and flusher.
- c. Clamp the discharge hose of the flusher to the cooler return line. Clamp the drain hose to the cooler inlet line and route it into a bucket or drain tank.
- d. Connect the flusher to air and water lines. The air line use a water trap to keep excess moisture out.
- e. Open the flusher water valve and flush the cooler for 10 seconds.

- f. Depress the flusher trigger to mix flushing fluid with the water. Flush for 2 minutes, turning the air valve on and off for 5 seconds every 15–20 seconds to create a surging action.
- g. After finishing 1 flushing cycle, reverse the hose and flush in the opposite direction.
- h. Dry the cooler lines with compressed air for 2 minutes or longer to remove all excess moisture from the system.

20. Install or connect the following:
- Starter motor onto the transaxle case and tighten its mounting bolts to 33 ft. lbs. (45 Nm).
- Torque converter with a new hub O-ring.

21. Place the transaxle on a transaxle jack and raise it to the level of the engine.

22. Align the transaxle and engine. Install the transaxle case bolts. Install new 14mm rear mount bolts and washers.

23. Raise the transaxle and install the side mount. Tighten the case bolts to 47 ft. lbs. (64 Nm). Tighten all of the 14mm rear mount bolts to 61 ft. lbs. (85 Nm).

24. Install or connect the following:
- Transaxle side mount and bracket. Tighten the bracket nuts to 47 ft. lbs. (64 Nm). Tighten the mount through-bolt to 54 ft. lbs. (74 Nm).

25. Remove the transaxle jack.

26. Rotate the crankshaft and install the torque converter-to-driveplate bolts. Tighten the bolts to 108 inch lbs. (12 Nm) in a crisscross pattern. Tighten the bolts to the specification in 2 steps.

27. Rotate the crankshaft to reset the engine at TDC/compression for the No. 1 cylinder. After the engine is set at TDC, it must not be disturbed until the distributor, if equipped, has been reinstalled.

28. Install or connect the following:
- Torque converter cover and tighten the bolts to 108 inch lbs. (12 Nm).
- Engine stiffener and tighten the 8mm bolts to 17 ft. lbs. (24 Nm). Tighten the 10mm bolts to 33 ft. lbs. (45 Nm).
- Right-front mount and bracket assembly. Tighten the 10mm bolt to 33 ft. lbs. (44 Nm), and the 12mm bolts to 47 ft. lbs. (64 Nm).

29. Remove the lifting chain and chain hooks.

30. Verify that the engine is at TDC/compression for the No. 1 cylinder. Align the tabs on the distributor drive with the

grooves on the end of the camshaft. Install the distributor and hand-tighten the mounting bolts. Reconnect the ignition wires.

31. Tighten the crankshaft pulley to 134 ft. lbs. (181 Nm).

32. Install or connect the following:
   - Transaxle cooler lines
   - New set rings on the inboard CV-joint splines
   - Halfshafts into the transaxle case and intermediate shaft. The inboard joints must snap into place.
   - Lower ball joint and damper fork
   - Front wheels
   - Shift cable linkage to the transaxle control shaft.
   - New lockwasher and tighten the linkage bolt to 10 ft. lbs. (14 Nm).
   - Shift cable cover and tighten its bolt to 16 ft. lbs. (22 Nm).
   - Front exhaust pipe. Use new self-locking nuts and gaskets. Tighten the rear flange nuts to 16 ft. lbs. (22 Nm), and the front flange nuts to 47 ft. lbs. (64 Nm).
   - VSS and countershaft speed sensor connectors
   - Lock-up control solenoid connector.
   - Shift control and linear solenoid connectors
   - Mainshaft speed sensor connector
   - Wiring harness clamps and starter cables
   - Resonator, air cleaner box, and air intake duct
   - Front splash shield

33. Lower the vehicle and tighten the strut pinch bolts to 32 ft. lbs. (44 Nm). Tighten the fork bolts to 47 ft. lbs. (65 Nm). Tighten the ball joint castle nuts to 40 ft. lbs. (55 Nm), then tighten them only enough to install new cotter pins.

34. Refill the transaxle with fresh ATF. Use only Honda Premium ATF or DEXRON®II or III ATF. Reconnect the positive and negative battery cables.

   a. Leave the flusher drain hose attached to the cooler return line.

   b. With the transaxle in park, run the engine for 30 seconds, or until approximately 1 quart of fluid is discharged. Immediately shut the engine off. This completes the cooler flushing process.

   c. Remove the drain hose and reconnect the cooler return line.

   d. Refill the transaxle to the proper level.

35. Check shift cable and throttle cable adjustments.

36. Check the ignition timing. Rotate the distributor counterclockwise to advance the timing, or clockwise to retard the timing.

When the timing has been set, tighten the distributor mounting bolts to 13 ft. lbs. (18 Nm).

37. Start the engine and shift through all the gears 3 times.

38. Let the engine warm up to operating temperature and check the fluid level with the transaxle in the **P** or **N** position.

39. Check and adjust the front wheel alignment.

40. Road test the vehicle. Recheck the transaxle fluid level.

PRELUDE

1. Before servicing the vehicle, refer to the precautions in the beginning of this section.

2. Disconnect both cables from the battery.

3. Shift the transaxle into **N**.

4. Remove or disconnect the following:
   - Battery hold-down and the battery

5. Drain the transaxle fluid and reinstall the drain plug with a new crush washer.

6. Remove or disconnect the following:
   - Air intake duct, air cleaner case, and resonator
   - Connector from the vacuum tank and the vacuum tank and tank bracket. Do not remove the vacuum tube from the vacuum tank.
   - Transaxle-to-body ground cable
   - Battery base with the ground cable and the battery base stay.
   - Lock-up control solenoid valve and shift control solenoid valve connectors.
   - Throttle control cable from the throttle control lever.
   - Countershaft speed sensor connector
   - Vehicle Speed Sensor (VSS) connector
   - Rear stiffener, then remove the VSS and Power Steering Sensor (PSS).

➡ **Do not disconnect the power steering pressure hoses from the VSS and PSS.**

   - Automatic Transaxle Fluid (ATF) cooler hoses at the joint pipes. Turn the ends of the cooler hoses upward to prevent fluid loss. Plug the joint pipes.
   - Starter motor
   - Upper transaxle housing mounting bolts
   - Front engine mount bracket bolts
   - Transaxle mount

7. Raise and support the vehicle safely.

8. Remove or disconnect the following:
   - Front wheels
   - Splash shield, subframe center beam and rear beam stiffener.

   - Cotter pins and castle nuts from the lower ball joints. Use a press-type ball joint tool to separate the ball joints from the lower arm.
   - Damper fork bolts, then separate the damper fork and the damper.

9. Use a suitable prytool to separate the right and left halfshafts from the differential.

10. Pull on the inboard joint and remove the right and left halfshafts. Tie plastic bags over the halfshaft ends to protect the boots and splined shafts from damage.

11. Remove or disconnect the following:
   - Right damper pinch bolt, then separate the right damper fork from the strut.
   - Right radius rod bolts and nut, then the radius rod
   - Torque converter cover and the shift cable cover.
   - Control lever lockbolt and the shift cable with the lever. Do not bend the shift control cable during removal. Wire the cable to the underbody of the vehicle our of the work area.
   - Driveplate bolts while rotating the crankshaft.

12. Place a transaxle jack below the transaxle and raise it enough to take the weight off the mounts.

13. Remove or disconnect the following:
   - Intake manifold bracket
   - Lower transaxle housing mounting bolts and lower rear engine mounting bolts.

14. Pull the transaxle away from the engine until it clears the dowel pins. Lower the transaxle out of the vehicle.

To install:

➡**Use new self-locking nuts when assembling the front suspension components. Use new set rings on the halfshaft inboard joints. Use new self-locking bolts for the subframe beams. These fasteners are available from a Honda dealer.**

15. Flush the transaxle cooling lines before installing the transaxle. Use a pressurized flushing canister, such as Honda tool No. J38405-A, or its equivalent. Use only biodegradable flushing fluid, Honda part No. J35944–20.

   a. Fill the flusher with 21 ounces of fluid. Pressurize the flusher to 80–120 psi, following the procedure on the fluid container and flusher.

   b. Clamp the discharge hose of the flusher to the cooler return line. Clamp the drain hose to the cooler inlet line and route it into a bucket or drain tank.

c. Connect the flusher to air and water lines. Open the flusher water valve and flush the cooler for 10 seconds.

d. Depress the flusher trigger to mix flushing fluid with the water. Flush for 2 minutes, turning the air valve on and off for 5 seconds every 15–20 seconds.

e. After finishing 1 flushing cycle, reverse the hose and flush in the opposite direction.

f. Dry the cooler lines with compressed air so that no moisture remains in the cooler lines.

16. Install the starter motor onto the transaxle case. Install the torque converter with a new hub O-ring. Tighten the starter bolts to 33 ft. lbs. (45 Nm).

17. Place the transaxle on a transaxle jack and raise it to the level of the engine.

18. Align the transaxle to the engine and install the transaxle housing mounting bolts and lower rear engine mounting bolts. Tighten the rear engine mounting bolts to 40 ft. lbs. (55 Nm) and the transaxle mounting bolts to 47 ft. lbs. (65 Nm). Install the intake manifold bracket and tighten the bolts to 16 ft. lbs. (22 Nm).

19. Tighten the front engine mount bracket bolts to 28 ft. lbs. (39 Nm).

20. Install or connect the following:
- Transaxle mount. Tighten the bolt to 47 ft. lbs. (65 Nm) and the nuts to 28 ft. lbs. (39 Nm).

21. Remove the transaxle jack.

22. Install or connect the following:
- Torque converter to the driveplate and the mounting bolts. Turn the crankshaft to rotate the driveplate. Tighten the bolts in 2 steps, first to 50 inch lbs. (6 Nm) in a crisscross pattern and finally to 108 inch lbs. (12 Nm). Check for free rotation after tightening the last bolt.
- Shift cable onto the control shaft and tighten the lockbolt to 10 ft. lbs. (14 Nm).
- Torque converter cover and the shift cable cover.
- New set ring onto the inboard joint of each halfshaft.
- Damper fork bolts and ball joint nuts to the lower arms. Tighten the ball joint nut to 47 ft. lbs. (65 Nm)
- New cotter pin
- Radius rod and the damper fork. Only hand-tighten the radius rod and damper fork fasteners at this point.

23. Turn the right steering knuckle fully

outward and slide the axle into the differential until the spring clip is felt engaging the differential side gear. Repeat the procedure on the left side.

24. Install or connect the following:
- Subframe rear beam stiffener and the center beam. Tighten the stiffener bolts to 28 ft. lbs. (39 Nm). Tighten the subframe center beam bolts to 43 ft. lbs. (60 Nm).
- Front wheels and lower the vehicle.

25. Use a floor jack to place the weight of the vehicle onto the right front knuckle. Tighten the radius rod bolts to 76 ft. lbs. (105 Nm) and the nut to 40 ft. lbs. (55 Nm). Tighten the damper pinch bolt to 32 ft. lbs. (44 Nm). Tighten the nut to 47 ft. lbs. (65 Nm) while holding the damper fork bolt. Remove the floor jack.

26. Install or connect the following:
- Speedometer sensor. Tighten the sensor bolt to 108 inch lbs. (12 Nm).
- ATF cooler hoses to the joint pipes
- Lock-up control solenoid and shift control solenoid valve connectors.
- VSS and PSS sensor connectors
- Starter motor cables and battery base and base stay.
- Ground cables on the body and on the transaxle.
- Vacuum tank, tank bracket, and tank connector.
- Resonator, air cleaner case, and air intake duct.

27. Refill the transaxle with ATF. Use only Honda Premium ATF or DEXRON®II ATF. Connect the negative and positive battery cables.

a. Leave the flusher drain hose attached to the cooler return line.

b. With the transaxle in park, run the engine for 30 seconds, or until approximately 1 quart of fluid is discharged. This completes the cooler flushing process.

c. Remove the drain hose and reconnect the cooler return line.

d. Refill the transaxle to the proper level with ATF.

28. Start the engine, set the parking brake, and shift the transaxle through all gears 3 times. Check for proper control cable adjustment.

29. On Preludes equipped with 4WS, and turn the steering wheel lock-to-lock to reset the steering control unit.

30. Check and adjust the front wheel alignment.

31. Let the engine reach operating tem-

perature with the transaxle in **N** or **P**, then turn the engine OFF and check the fluid level

32. After road testing the vehicle, loosen the front engine mount bolts, and tighten them to 28 ft. lbs. (39 Nm).

33. Enter the radio security code.

ACCORD

2.3L ENGINES

1. Before servicing the vehicle, refer to the precautions in the beginning of this section.

2. Remove or disconnect the following:
- Negative, then the positive battery cables
- Battery

3. Shift the transaxle into **N**.

4. Remove or disconnect the following:
- Air intake hose, air cleaner housing, and the resonator assembly.
- Battery base and the base stay
- Throttle cable from the throttle control lever.
- Transaxle ground cable and the speed sensor connectors.
- Solenoid valve connectors
- Lock-up control solenoid valve and shift control solenoid valve connectors.
- Transaxle cooler hoses from the joint pipes and plug the hoses.
- Starter cables and starter
- Countershaft Speed Sensor (CSS) and Vehicle Speed Sensor (VSS) connectors

5. Install a hoist to the engine.

6. Remove or disconnect the following:
- 4 upper bolts attaching the transaxle to the engine block.
- 3 bolts attaching the front engine mount bracket to the engine.
- Transaxle mount

7. Raise and safely support the vehicle. Remove the front wheels.

8. Drain the transaxle fluid and reinstall the drain plug with a new washer.

9. Remove or disconnect the following:
- Splash shield
- Subframe center beam
- Cotter pins and lower arm ball joint nuts, then separate the ball joints from the lower arms using a suitable tool.
- Right damper pinch bolt, then separate the damper fork and damper.
- Bolts and nut, then the right radius rod.

10. Using a small prying device, carefully pry the right and left halfshafts out of

the differential. Remove the right and left halfshafts. Tie plastic bags over the halfshaft ends to prevent damage to the CV boots and splines.

11. Remove or disconnect the following:
- Bolts mounting the intermediate shaft, then the intermediate shaft from the differential.
- Torque converter cover and shift cable cover.
- Shift control cable by removing the lockbolt.
- Shift cable lever from the control shaft. Don't disconnect the control lever from the shift cable. Wire the shift cable out of the work area and be careful not to kink it.
- 8 drive plate bolts, one at a time while rotating the crankshaft pulley.

12. Place a suitable jack under the transaxle and raise the jack just enough to take weight off of the mounts.

13. Remove or disconnect the following:
- Intake manifold bracket.
- Transaxle housing mounting bolts
- Mounting bolts from the rear engine mount bracket.
- 4 transaxle housing mounting bolts and 3 mount bracket nuts.

14. Pull the transaxle away from the engine until it clears the 14mm dowel pins, then lower it using the jack.

To install:

➡**Use new self-locking nuts when assembling the front suspension components. Install new set rings onto the halfshaft inboard joint splines. Replace any color-coded self-locking bolts.**

15. Flush the transaxle cooler lines before installing the transaxle. Use a pressurized flushing unit such as Honda J38405-A or equivalent. Use only Honda biodegradable flushing fluid, Honda J35944-20. Other fluids will damage the automatic transmission cooling system.

a. Fill the flusher with 21 ounces of fluid. Pressurize the flusher to 80–120 psi, following the procedure on the fluid container and flusher.

b. Clamp the discharge hose of the flusher to the cooler return line. Clamp the drain hose to the cooler inlet line and route it into a bucket or drain tank.

c. Connect the flusher to air and water lines. Open the flusher water valve and flush the cooler for 10 seconds. The air line should be equipped with a water trap to keep the system dry.

d. Depress the flusher trigger to mix flushing fluid with the water. Flush for 2 minutes, turning the air valve on and off

for 5 seconds every 15–20 seconds to create a surging action.

e. After finishing 1 flushing cycle, reverse the hose and flush in the opposite direction following the same steps.

f. Dry the cooler lines with compressed air so that no moisture is left in the cooler system.

16. Be sure the 2, 14mm dowel pins are installed into the torque converter housing.

17. Install or connect the following:
- Torque converter onto the transaxle mainshaft with a new hub O-ring.
- Starter motor onto the transaxle case and tighten the mounting bolts to 33 ft. lbs. (44 Nm).
- Transaxle and transaxle housing mounting bolts: 47 ft. lbs. (65 Nm)
- Rear engine mounting bolts: 40 ft. lbs. (54 Nm)
- Intake manifold bracket and tighten the bolts to 16 ft. lbs. (22 Nm).
- Upper bolts attaching the transaxle to the engine: 47 ft. lbs. (64 Nm)
- Front engine mount bracket bolts: 28 ft. lbs. (38 Nm)
- Transaxle mount and the nuts and bolt that attach the mount. Tighten the nuts first to 28 ft. lbs. (38 Nm), then tighten the bolt to 47 ft. lbs. (64 Nm).

18. Remove the jack from the transaxle.

19. Install or connect the following:
- Torque converter to the drive plate with the 8 bolts. Tighten the bolts in 2 steps in a crisscross pattern: first to 54 inch lbs. (6 Nm), and finally to 108 inch lbs. (12 Nm). Check for free rotation after tightening the last bolt.
- Shift control cable and control cable holder. Tighten the shift cable lockbolt to 10 ft. lbs. (14 Nm). Tighten the shift cable cover bolts to 13 ft. lbs. (18 Nm).
- Torque converter cover and tighten the bolts to 108 inch lbs. (12 Nm).

20. Remove the engine hoist.

21. Install or connect the following:
- Radius rod and damper fork
- Intermediate shaft into the differential and tighten the mounting bolts to 28 ft. lbs. (38 Nm).
- New set ring on the end of each halfshaft.

22. Turn the right steering knuckle fully outward and slide the axle into the differential until the set ring snaps into the differential side gear. Repeat the procedure on the left side.

23. Install or connect the following:
- Damper fork bolts and ball joint

nuts to the lower arms: 40 ft. lbs. (55 Nm) with a new cotter pin.
- Subframe center beam and tighten the center beam bolts to 28 ft. lbs. (39 Nm).
- Splash shield
- Front wheels and lower the vehicle.
- Speed sensor connector

24. Support the right front knuckle with a floor jack, until the weight of the vehicle is held by the jack. Tighten the damper fork pinch bolt to 32 ft. lbs. (44 Nm). Tighten the radius rod bolts to 76 ft. lbs. (105 Nm), and the radius rod nut to 32 ft. lbs. (44 Nm). Hold the damper fork bolt with a wrench, and tighten the nut to 40 ft. lbs. (55 Nm). Remove the floor jack.

25. Install or connect the following:
- Cables to the starter
- Throttle control cable
- Lock-up control solenoid valve and shift control solenoid valve connectors
- Speed sensor connectors and the transaxle ground cable
- Transaxle cooler inlet hose to the joint pipe. Attach a drain hose to the return line.
- Battery base stay and the battery base

26. Install the resonator assembly, the air cleaner assembly, and the air intake hose.
- Battery, positive then the negative battery cables

27. Refill the transaxle with ATF. Use only Honda Premium ATF or DEXRON®II ATF.

a. With the flusher drain hose attached to the cooler return line.

b. Place the transaxle in **P**, run the engine for 30 seconds, or until approximately 1 quart of fluid is discharged. Immediately shut off the engine. This completes the cooler flushing process.

c. Remove the drain hose and reconnect the cooler return line.

d. Refill the transaxle to the proper level with ATF.

28. Start the engine, set the parking brake, and shift the transaxle through all gears 3 times. Check for proper shift cable adjustment.

29. Let the engine reach operating temperature with the transaxle in **P** or **N**. Then, shut off the engine and check the fluid level.

30. Road test the vehicle.

31. After road testing the vehicle, loosen the front engine mount bracket bolts, then retighten them to 28 ft. lbs. (39 Nm).

32. Check and adjust the vehicle's front end alignment.

33. Enter the radio security code.

3.0L Engine

34. Before servicing the vehicle, refer to the precautions in the beginning of this section.

35. Remove or disconnect the following:
- Negative, then the positive battery cables
- Battery and tray
- Clamps securing the battery cables to the base.
- Intake air duct and the air cleaner assembly

36. Raise the vehicle and drain the transaxle fluid. Replace the drain plug with a new washer.

37. Remove or disconnect the following:
- Starter wiring and harness clamps, then the breather and radiator hoses from the retainer.
- Wiring connectors from the transaxle assembly
- Cooler lines; point them up to prevent fluid drainage.
- Bolt and nut, then the rear stiffener.
- Bolts securing the transaxle to the engine.
- Front mounting bracket bolts
- Engine under cover
- Lower shock absorber mounting and the lower ball joints from the control arms.
- Bolts securing the radius rods to the lower arms.
- Halfshafts. Keep the splined ends of the shafts clean.

➡️**Matchmark the installed position of the sub-frame on the main-frame before removing it.**

- Sub-frame from the main-frame
- Engine brace from the rear of the engine
- Shift cable cover, bracket and cable
- 8 bolts securing the drive plate to the torque converter

38. Attach a chain hoist to the engine and raise it slightly.

39. Place a jack under the transaxle.

40. Remove or disconnect the following:
- Transaxle mount bracket
- Intake manifold support bracket
- Rear mount bracket

41. Pull the transaxle back slightly until it comes off the dowels and lower it from the vehicle. Do not let the torque converter fall out of the transaxle.

To install:

42. Install or connect the following:

- Torque converter using a new O-ring, if removed.
- Dowel pins in the torque converter housing.
- Transaxle to the engine and the rear mount bracket. Tighten the 8mm bolt to 16 ft. lbs. (22 Nm) and the 12mm bolts to 40 ft. lbs. (54 Nm).
- Transaxle-to-engine bolts: 47 ft. lbs. (64 Nm)
- Breather tube with the dot facing up
- Transaxle mount bracket. Tighten the nuts to 28 ft. lbs. (38 Nm) and the through-bolt to 40 ft. lbs. (54 Nm).
- Driveplate-to-torque converter bolts: 108 inch lbs. (12 Nm) in a crisscross pattern.
- Shift cable, bracket and cover
- Engine brace on the rear of the engine.
- Halfshafts
- Sub-frame after aligning the matchmarks. Tighten the rear bolts to 47 ft. lbs. (64 Nm) and the front bolts to 76 ft. lbs. (103 Nm).
- Front mount. Tighten the bolts to 28 ft. lbs. (38 Nm).
- Shock absorbers and the radius rods to the lower control arms.
- Engine under cover
- All the wiring connectors
- Starter wiring and harness clamps
- Battery
- Air cleaner assembly and intake duct

43. Refill the transaxle with Genuine Honda® premium automatic transmission fluid.

Clutch

REMOVAL & INSTALLATION

1. Before servicing the vehicle, refer to the precautions in the beginning of this section.

2. Remove or disconnect the following:
- Negative battery cable

3. Raise and safely support the vehicle.

4. Remove or disconnect the following:
- Transmission from the vehicle. Matchmark the flywheel and clutch for reassembly.

5. Use a flywheel ring-gear holder to lock the flywheel in position.

6. Remove or disconnect the following:
- Pressure plate bolts, 2 turns at a time working in a crisscross pattern to prevent warping the pressure plate.
- Pressure plate and clutch disc

7. Inspect the flywheel, disc, and pressure plate for wear, cracks, and warpage. Light scoring of the flywheel may be polished out; gouges, warpage, burn marks, cracks, or chipped teeth require replacement of the flywheel.

➡️**If the flywheel is to be removed, but is going to be reused, matchmark it to the engine block prior to removal. Aligning the matchmarks upon reassembly will preserve driveline balance.**

8. Inspect the flywheel's ball bearing: turn the inner race of the bearing with your finger, and be sure it turns smoothly and quietly. If the bearing is loose or noisy, or exhibits rough motion, replace it.

9. Remove or disconnect the following:

- Release fork boot. Squeeze the release fork retaining spring to disengage the fork from its pivot.
- Release fork from the clutch housing.
- Release bearing. Spin the bearing by hand to check its degree of play. Replace the release bearing if it has excessive play or is leaking grease.

10. Inspect the rear main bearing oil seal for signs of leakage. If necessary, replace the seal to prevent oil leakage onto the clutch's friction surfaces.

To install:

11. If necessary, drive out the flywheel bearing, then use a suitably-sized bearing driver to install a new one. Use a crisscross pattern to tighten the flywheel mounting bolts in several steps to 87 ft. lbs. (118 Nm) for vehicles with SOHC engines. If equipped with the B16A2 engine, tighten the flywheel bolts to 76 ft. lbs. (105 Nm). For 2.0L engines, tighten the flywheel bolts to 94 ft. lbs. (127 Nm).

12. Install or connect the following:
- Clutch disc and pressure plate by aligning the dowels on the flywheel with the dowel holes in the pressure plate. If a new pressure plate is not being installed, align the matchmarks that were made during removal.
- Pressure plate bolts, hand-tight

13. Insert a suitable clutch disc alignment tool into the splined hole in the clutch disc. Align the clutch and pressure plate.

14. Tighten the pressure plate bolts in a crisscross pattern 2 turns at a time to pre-

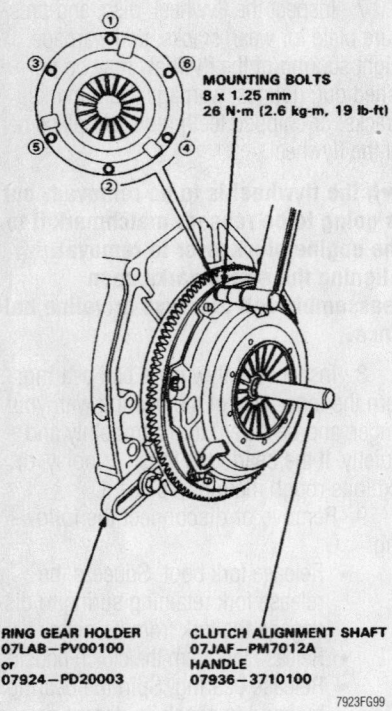

MOUNTING BOLTS
8 x 1.25 mm
26 N·m (2.6 kg-m, 19 lb-ft)

RING GEAR HOLDER
07LAB—PV00100
or
07924—PD20003

CLUTCH ALIGNMENT SHAFT
07JAF—PM7012A
HANDLE
07936—3710100

7923FG99

Clutch alignment tools and pressure plate torque sequence

vent warping the pressure plate. The final torque is 19 ft. lbs. (26 Nm).

15. Remove the alignment tool and ring gear holder.

16. Coat the mainshaft with heavy-duty high-temperature grease. The manufacturer recommends part No. 08798—9002, Honda super high-temp urea grease.

17. Coat the release fork pawls and the inner race of the release bearing with high temperature grease and install them into the clutch housing. Be sure the release fork retainer spring snaps into place on the pivot. The bearing and fork must fit together properly and slide back and forth smoothly.

18. Coat the tip of the slave cylinder with grease. Install the release fork boot.

19. Install or connect the following:
- Transmission, making sure the mainshaft is properly aligned with the clutch disc splines, and the transmission case dowels are properly aligned with the engine block.
- Transmission case bolts: 47 ft. lbs. (65 Nm), sequentially

20. Bleed the clutch hydraulic system.

21. Adjust the clutch pedal free-play.

22. Verify that all engine and transaxle components are installed and connected properly.

23. Reconnect the negative battery cable.

24. Road test the vehicle.

Hydraulic Clutch System

BLEEDING

1. Before servicing the vehicle, refer to the precautions in the beginning of this section.

2. Fill the clutch master cylinder reservoir with clean DOT 3 or 4 brake fluid.

3. Attach a rubber tube to the clutch slave cylinder bleed screw. Route the tube into a container of clean brake fluid.

4. Loosen the bleed screw.

5. Slowly pump the clutch pedal until the fluid draining from the slave cylinder is free of air bubbles.

6. Tighten the bleed screw to 72–84 inch lbs. (8–10 Nm).

7. Refill the clutch master cylinder reservoir with brake fluid.

Halfshaft

REMOVAL & INSTALLATION

Except S2000

1. Before servicing the vehicle, refer to the precautions in the beginning of this section.

2. Loosen the front spindle nut.

3. Raise and safely support the vehicle.

4. Remove or disconnect the following:
- Front wheels and the spindle nut

5. Drain the transaxle fluid and install the drain plug with a new washer. If the halfshaft to be removed is installed into the intermediate shaft, the transaxle fluid does not need to be drained.

6. Remove or disconnect the following:
- Damper fork nut and damper pinch bolt
- Damper fork
- Cotter pin and castle nut from the lower arm ball joint.

7. Install a hex nut flush onto the ball joint stud to prevent the ball joint tool from damaging the stud threads.

8. Using a ball joint tool, separate the lower arm from the knuckle.

9. Pull the knuckle outward.

10. Remove or disconnect the following:
- Halfshaft outboard joint from the hub by tapping it with a plastic hammer.
- Inner CV-joint away from the transaxle case to force the halfshaft set ring out of the groove.
- Halfshaft from the differential case or intermediate shaft by pulling on the inboard CV-joint

➡ Do not pull on the halfshaft as the CV-joint may come apart. Use care when prying out the assembly and pull it straight to avoid damaging the differential oil seal or intermediate shaft oil or dust seals.

To install:

11. Replace the differential oil seal or intermediate shaft seal if either were damaged during removal.

12. Install or connect the following:
- New set rings on the ends of the halfshafts
- Halfshafts and be sure the set ring locks in the differential gear groove and the halfshaft bottoms in the differential or intermediate shaft.
- Outboard joint into the hub. Be sure the splines mesh together and the joint is fully seated into the hub.
- Ball joint stud into the lower control arm.
- Damper fork into position. Tighten the upper damper pinch bolt to 32 ft. lbs. (44 Nm) and the fork nut to 47 ft. lbs. (65 Nm).

13. Tighten the ball joint castle nut to 40 ft. lbs. (55 Nm), then tighten the nut just enough to install a new cotter pin.

14. Install or connect the following:
- Front wheels
- New spindle nut; don't tighten it yet.

15. Lower the vehicle.

16. Tighten the spindle nut to 181 ft. lbs. (245 Nm) and stake its tab. Tighten the wheel nuts to 80 ft. lbs. (110 Nm).

17. Fill the transaxle with the proper type and quantity of fluid.

18. Warm the engine up, check the transaxle fluid level, and road test the vehicle.

S2000

1. Before servicing the vehicle, refer to precautions in the beginning of this section.

2. Remove or disconnect the following:
- Rear wheel
- Spindle nut
- Lower ball joint
- Wheel speed sensor harness
- Inboard joint mounting bolts

3. Pull the knuckle outward to separate the inboard joint from the differential.

4. Remove the outboard joint from the wheel hub by tapping it with a plastic-faced hammer.

To install:

5. Installation is the reverse of the removal procedure, while using the following torque values:
- Inboard joint mounting bolts: 61 ft. lbs. (83 Nm)

- Lower ball joint nut: 43–51 ft. lbs. (69–78 Nm)
- Wheel speed sensor harness bolts: 88 inch lbs. (10 Nm)
- Spindle nut: 181 ft. lbs. (245 Nm)
- Wheel lug nuts: 80 ft. lbs. (108 Nm)

CV-Joints

OVERHAUL

1. Remove the halfshaft.
2. Remove the large retaining band

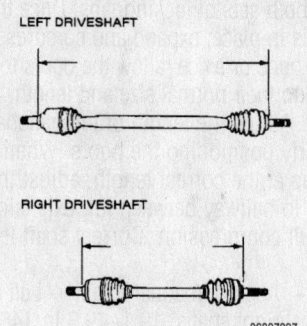

LEFT DRIVESHAFT

RIGHT DRIVESHAFT

86807027

Halfshafts must be set to the correct length before installing boot bands

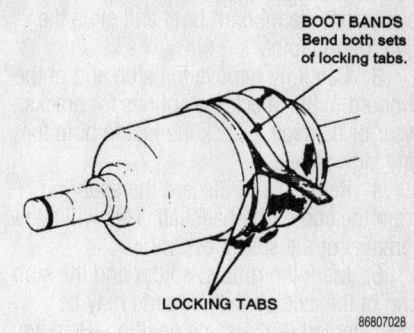

BOOT BANDS
Bend both sets of locking tabs.

LOCKING TABS

86807028

Always use new boot bands

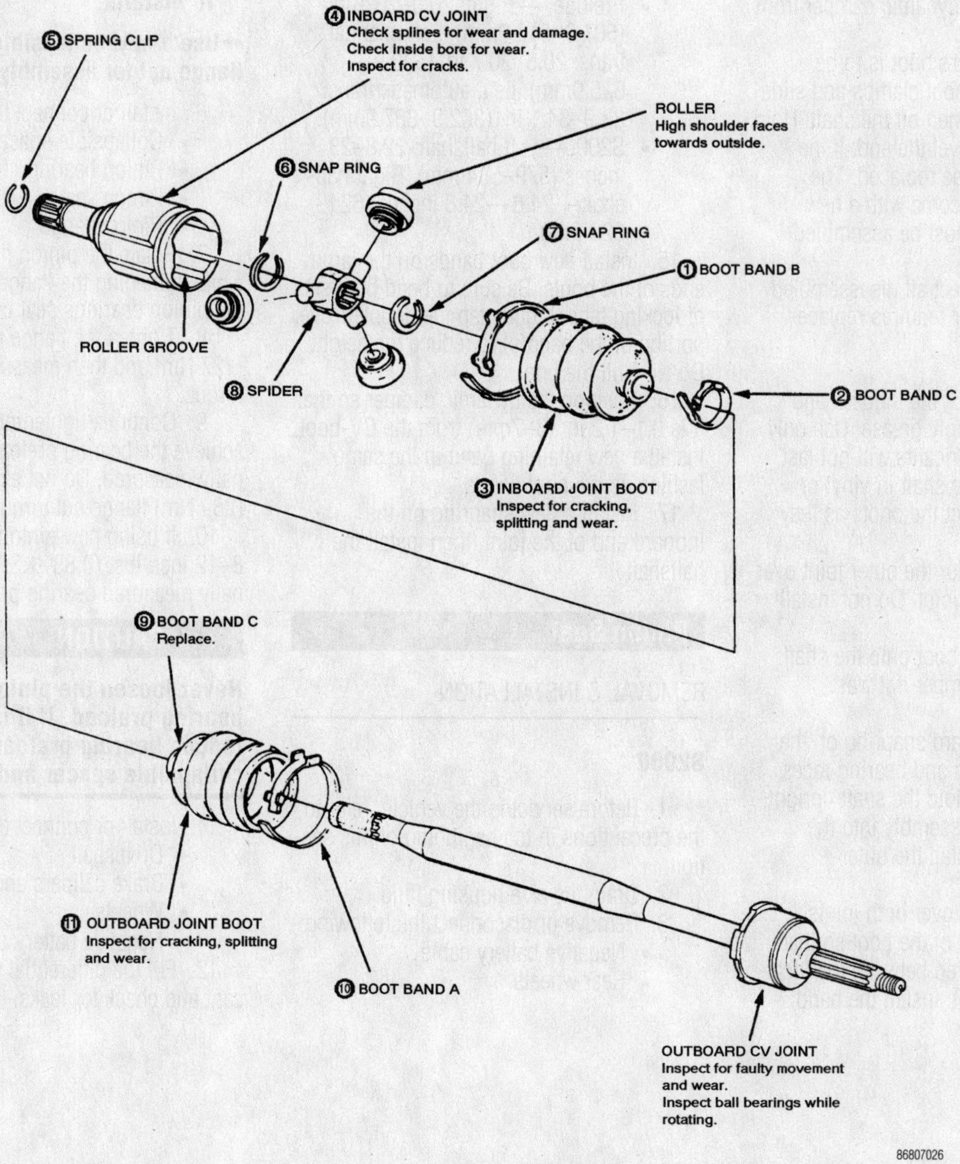

⑤ SPRING CLIP

④ INBOARD CV JOINT
Check splines for wear and damage.
Check inside bore for wear.
Inspect for cracks.

ROLLER
High shoulder faces towards outside.

⑥ SNAP RING

⑦ SNAP RING

① BOOT BAND B

ROLLER GROOVE

⑧ SPIDER

② BOOT BAND C

③ INBOARD JOINT BOOT
Inspect for cracking, splitting and wear.

⑨ BOOT BAND C
Replace.

⑪ OUTBOARD JOINT BOOT
Inspect for cracking, splitting and wear.

⑩ BOOT BAND A

OUTBOARD CV JOINT
Inspect for faulty movement and wear.
Inspect ball bearings while rotating.

86807026

Exploded view of the halfshaft

Timing belt service is covered in Section 3 of this manual

from the inboard boot. Remove the smaller band from the inboard boot and slide the boot off the joint.

3. Carefully remove the stub end of the inboard joint. Check the splines for cracks, wear or damage. Check the inside bore for any sign of wear.

4. Remove and discard the snapring from the end of the halfshaft. This will allow removal of the spider assembly.

5. Mark the rollers, spider and the stub end of the axle so that all parts may be reassembled in the same position. Remove the rollers from the spider.

6. Remove the second snapring from the shaft. Remove the joint boot. If equipped, remove the dynamic damper from the shaft.

7. If the outer joint's boot is to be replaced, remove the boot clamps and slide the boot off the joint, then off the shaft. Hold the outer joint and swivel the end. If the joint is noisy, it must be replaced. The replacement joint will come with a new shaft; the inner joint must be assembled onto the shaft.

8. Clean and inspect all disassembled parts. Any sign of wear requires replacement.

To install:

9. Thoroughly pack the inboard and outboard joints with moly grease. Use only moly grease; other lubricants will not last. Wrap the splines of the shaft in vinyl or electrical tape to protect the boots as they are installed.

10. Slide the boot for the outer joint over the shaft and onto the joint. Do not install the bands yet.

11. Slide the inner boot onto the shaft. Install the dynamic damper if it was removed.

12. Install the inboard snapring on the shaft. Install the rollers and bearing races on the spider shafts. Hold the shaft upright, then slide the spider assembly into the inboard shaft joint. Install the outer snapring.

13. Slide the boots over both joints. Position the small end of the boot so that the band will be centered between the locating humps on the shaft. Install the band;

bend both sets of locking tabs. Once the band is in place, expand and compress the boots once or twice; allow the boots to return to their normal size and length.

14. Adjust the length of the halfshaft by properly positioning the boots. When the shaft is at the correct length, adjust the boots to halfway between full extension and full compression. Correct shaft lengths are:

- Accord, manual trans — Left and right shafts: 19.1–19.3 in. (486–491mm).
- Accord, automatic trans — 33.3–33.5 in. (845–850mm). Right: 19.1–19.9 in. (486–491mm).
- Prelude — Right: 20.0–20.1 in. (507.9–512.9mm). Left, manual trans: 20.5–20.7 in. (520.9–525.9mm). Left, automatic trans: 33.9–34.1 in. (862.9–867.9mm).
- S2000—Left halfshaft: 22.8–23 inches (579–584 mm). Right halfshaft—24.6—24.8 inches (624–629 mm).

15. Install new boot bands on the large ends of the boots. Be sure to bend both sets of locking tabs. Lightly tap the doubled-over portion of the band(s) to reduce the height. Do NOT hit the boot.

16. Position the dynamic damper so that it is 0.1–1.2 in. (3–7mm) from the CV-boot. Install a new retaining band in the same fashion as the boot bands.

17. Install a new snapring on the inboard end of the joint, then install the halfshaft.

Pinion Seal

REMOVAL & INSTALLATION

S2000

1. Before servicing the vehicle, refer to the precautions in the beginning of this section.

2. Drain the axle housing fluid.

3. Remove or disconnect the following:
- Negative battery cable
- Rear wheels

- Driveshaft
- Brake calipers and pads

➡The brake calipers and pads must be removed so that there is no additional drag when measuring pinion bearing preload.

4. Use an inch lb. torque wrench and measure and record the amount of torque required to maintain pinion rotation through several revolutions.

5. Remove or disconnect the following:
- Pinion flange
- Pinion seal
- Pinion bearing
- Collapsible spacer

To install:

➡Use a new collapsible spacer and flange nut for assembly.

6. Install or connect the following:
- Collapsible spacer
- Pinion bearing
- Pinion seal
- Pinion flange

7. Rotate the pinion flange occasionally while tightening the flange nut to make sure the pinion bearings seat correctly.

8. Tighten the flange nut to 94 ft. lbs. (127 Nm) and then measure bearing preload torque.

9. Continue tightening the flange nut to achieve the bearing preload torque originally measured. Do not exceed 210 ft. lbs. (284 Nm) flange nut torque.

10. If using new pinion bearings, add 8–12 inch lbs. (0.88–1.37 Nm) to the originally measured bearing preload.

✷✷ CAUTION

Never loosen the pinion nut to reduce bearing preload. If it is necessary to reduce bearing preload, install a new collapsible spacer and pinion nut.

11. Install or connect the following:
- Driveshaft
- Brake calipers and pads
- Wheels
- Negative battery cable

12. Fill the differential with gear lubricant and check for leaks.

STEERING AND SUSPENSION

Air Bag

The air bag modules must be disabled if they, or any other part of the SRS, must be serviced or disconnected. Failing to disable the SRS before servicing its components may cause accidental air bag deployment and possible personal injury.

PRECAUTIONS

Several precautions must be observed when handling the inflator module to avoid accidental deployment and possible personal injury.

- Never carry the inflator module by the wires or connector on the underside of the module.
- When carrying a live inflator module, hold securely with both hands, and ensure that the bag and trim cover are pointed away.
- Place the inflator module on a bench or other surface with the bag and trim cover facing up.
- With the inflator module on the bench, never place anything on or close to the module which may be thrown in the event of an accidental deployment.

DISARMING

➡**The radio may contain a coded theft protection circuit. Always obtain the code number before disconnecting the battery.**

Driver's Side

1. Before servicing the vehicle, refer to the precautions in the beginning of this section.
2. Remove or disconnect the following:
   - Negative and positive battery cables

✳✳ CAUTION

Always wait at least 3 minutes after disconnecting the battery before working around the air bag.

- Steering wheel lower access cover
- Clip securing the air bag module/cable reel connection to the steering column.

➡**Spring-loaded air bag connectors contain a self-disabling contact. A shorting connector doesn't need to be installed on the driver's air bag connector.**

3. Uncouple the spring-loaded connectors:
   a. Hold the connector body, not the wiring.
   b. Pull the spring-loaded locking sleeve toward its stop while holding the opposite half of the connector.
   c. After releasing the locking sleeve, uncouple the connectors.

Passenger's Side

1. Before servicing the vehicle, refer to the precautions in the beginning of this section.
2. Remove or disconnect the following:
   - Negative and positive battery cables

✳✳ CAUTION

Always wait at least 3 minutes after disconnecting the battery before working around the air bag.

- Glove box door and frame
- Lower mounting brackets that may cover the air bag connection, if equipped.
- Passenger's air bag connector. Pull the spring-loaded sleeve toward the stop while holding the opposite half of the connector and pull the connector apart.

REARMING

Driver's Side

1. After servicing has been completed, couple the air bag and cable reel connectors. Press the sleeve side of the connector into the pawl side until the sleeve locks the connectors together.
2. Install or connect the following:
   - Clip securing the air bag/cable reel connection to the steering column.
   - Access cover
   - Positive and negative battery cables
3. Turn the ignition switch to the **ON** position, but don't start the engine. The air bag indicator light should turn on for 6 seconds, then turn off. If the air bag indicator light doesn't come on, or stays on longer than 6 seconds, the system fault must be diagnosed.
4. Enter the radio security code.

Passenger's Side

1. After servicing has been completed, immediately couple the air bag and cable reel connectors.
2. Install or connect the following:
   - Any lower mounting brackets that may have been removed.
   - Glove box frame and glove box door
   - Positive and negative battery cables
3. Turn the ignition switch to the **ON** position, but don't start the engine. The air bag indicator light should turn on for 6 seconds, then turn off. If the air bag indicator light doesn't come on, or stays on longer than 6 seconds, the system fault must be diagnosed.
4. Enter the radio security code.

Rack and Pinion Steering Gear

REMOVAL & INSTALLATION

Manual

CIVIC

✳✳ CAUTION

The air bag must be disabled before removing the steering wheel to center the cable reel. Failure to disarm the air bag system may cause accidental air bag deployment, resulting in unnecessary air bag system repairs and the risk of personal injury.

1. Before servicing the vehicle, refer to the precautions in the beginning of this section.
2. Position the front wheels straight ahead. Lock the steering column and remove the ignition key.
3. Remove or disconnect the following:
   - Negative, then the positive battery cables.
4. Disable the air bag.
5. Remove or disconnect the following:
   - Steering joint cover
   - Upper and lower steering joint bolts
6. Raise and support the vehicle safely.
7. Remove or disconnect the following:
   - Front wheels
   - Tie-rod end cotter pins and castle nuts

Heater Core replacement is covered in Section 2 of this manual

- Tie-rod ends from the steering knuckles, using a ball joint tool.
- Left tie-rod end and slide the rack all the way to the right.
- Self-locking nuts, then separate the catalytic converter or front exhaust pipe from the rear exhaust pipes.
- Catalytic converter or front exhaust pipe

8. If equipped with a manual transaxle, remove or disconnect the following:
- Shift lever extension rod from the clutch housing
- Pin retainer out of the way, then drive out the spring pin
- Shift rod

9. If equipped with an automatic transaxle, remove or disconnect the following:
- Shift cable bracket and holder
- Shift cable from the control shaft. Suspend the cable from the underbody with a piece of wire.

10. Remove or disconnect the following:
- Steering rack stiffener plate
- Steering rack mounting bracket
- Steering rack from the pinion shaft, by pulling the rack down.

11. Drop the steering rack far enough to permit the end of the pinion shaft and the grommet to come out of the hole in the bulkhead.

12. Slide the gearbox to the right until the left tie rod clears the subframe, then drop it down and out of the vehicle to the left.

To install:

→Use new self-locking nuts and gaskets when installing the catalytic converter.

13. Install or connect the following:
- Steering rack into position
- Pinion shaft grommet, insert the pinion through the hole in the bulkhead.
- Steering rack mounting cushion, bracket, and bolts. The arrow on the bracket faces the front of the vehicle. Tighten the bracket bolts to 28 ft. lbs. (39 Nm).
- Steering rack stiffener plate. Tighten the steering rack mounting bolts to 43 ft. lbs. (59 Nm). Tighten the stiffener plate bolts to 28 ft. lbs. (39 Nm).

14. Center the rack ends within their steering strokes.

15. Install or connect the following:
- Tie rod ends onto the rack ends
- Tie rod ends to the steering knuckles, then the castle nuts.
- Front wheels

- Catalytic converter using new gaskets and self-locking nuts. Tighten the front nuts to 16 ft. lbs. (22 Nm), and the rear nuts to 25 ft. lbs. (34 Nm).

16. If equipped with a manual transaxle, install or connect the following:
- Shift linkage with a new spring pin and clip.
- Extension rod and tighten its bolt to 16 ft. lbs. (22 Nm).

17. If equipped with an automatic transaxle, install or connect the following:
- Shift cable and brackets. Tighten the bracket bolts to 108 inch lbs. (12 Nm). Tighten the cable lockbolt to 10 ft. lbs. (14 Nm). Tighten the cable holder bolts to 16 ft. lbs. (22 Nm).

18. Verify that the rack is centered within its strokes. Lower the vehicle.

19. Center the air bag cable reel as follows:
   a. Remove the steering wheel.
   b. Turn the cable reel clockwise until it stops.
   c. Turn the steering wheel counterclockwise, approximately 2 turns, until the arrow on the label points straight up.
   d. Install the steering wheel.

20. During steering wheel installation, verify that the slot on the steering wheel shaft engages with the tabs on the turn signal canceling sleeve. The pins on the cable reel fit into the holes on the steering wheel body. Install a new steering wheel nut and tighten it to 36 ft. lbs. (50 Nm).

21. Line up the bolt hole in the steering joint with the groove in the pinion shaft. Slip the joint onto the pinion shaft. Pull the joint up and down to be sure the splines are fully seated. Tighten the joint bolts to 16 ft. lbs. (22 Nm).

→Connect the steering joint and pinion shaft with the cable reel and steering rack centered. Verify that the lower joint bolt is securely seated in the pinion shaft groove. If the steering wheel and rack are not centered, reposition the serrations at the lower end of the steering joint.

22. Install or connect the following:
- Steering joint cover

23. Tighten the ball joint castle nuts to 29–35 ft. lbs. (40–48 Nm). Then, tighten them only enough to install new cotter pins.

24. Enable the air bag.

25. Install or connect the following:
- Steering wheel's lower access cover
- Negative and positive battery cables

26. Turn the ignition switch to the **ON**

position. The air bag indicator light should come on for 6 seconds, then turn off. This light sequence indicates that the air bag system is enabled and functioning normally. If the air bag light stays on longer, or doesn't turn on, the system must be diagnosed.

27. Check the front wheel alignment and steering wheel spoke angle. Make adjustments by turning the left and right tie-rod ends equally.

28. Road test the vehicle.

Power

→The radio may contain a coded theft protection circuit. Always obtain the code number before disconnecting the battery. If the vehicle is equipped with 4WS, the steering control unit is shut down when the battery is disconnected. After connecting the battery, turn the steering wheel lock-to-lock to reset the steering control unit.

CIVIC

❊❊ CAUTION

The air bag must be disabled before removing the steering wheel to center the cable reel. Failure to disarm the air bag system may cause accidental air bag deployment, resulting in unnecessary air bag system repairs and the risk of personal injury.

1. Before servicing the vehicle, refer to the precautions in the beginning of this section.

2. Remove or disconnect the following:
- Power steering reservoir off of its mount
- Inlet hose

3. Insert a length of tubing into the inlet hose and route the tubing into a drain container.

4. With the engine running at idle, turn the steering wheel lock-to-lock several times until fluid stops running out of the hose.

5. Position the front wheels straight ahead. Shut off the engine and lock the steering column and remove the ignition key. Reconnect the reservoir inlet hose.

6. Remove or disconnect the following:
- Negative and positive battery cables. Wait 3 minutes before working around the air bags.
- Steering wheel's lower access cover

7. Uncouple the air bag connector from the cable reel connector as follows:
   a. Hold the cable reel connector. With your other hand, slide the spring-loaded

sleeve toward the stop tab on the air bag connector.

b. Separate the 2 connectors. There is no need to install a shorting connector, as the connectors are automatically grounded when they are uncoupled.

8. Remove or disconnect the following:
- Steering joint cover, then the upper and lower steering joint bolts.

9. Raise and support the vehicle safely.

10. Remove or disconnect the following:
- Front wheels
- Tie rod end cotter pins and castle nuts
- Tie rod ends from the steering knuckles, using a ball joint tool

11. If equipped with a manual transaxle, remove or disconnect the following:
- Shift lever extension rod from the clutch housing
- Pin retainer out of the way, then drive out the spring pin
- Shift rod

12. If equipped with an automatic transaxle, remove or disconnect the following:
- Shift cable bracket and holder
- Shift cable from the control shaft. Suspend the cable from the underbody with a piece of wire.

13. Remove or disconnect the following:
- Self-locking nuts and separate the catalytic converter from the exhaust pipes.
- Catalytic converter
- Hydraulic line and hose from the rack valve body using a flare nut wrench.
- Left tie rod end and slide the rack all the way to the right.
- Steering rack mounting bolts
- Steering rack from the pinion shaft by pulling the rack downward

14. Drop the gearbox far enough to permit the end of the pinion shaft to come out of the hole in the frame channel.

15. Slide the gearbox to the right until the left tie rod clears the subframe, then drop it down and out of the vehicle to the left.

To install:

➡**Use new self-locking nuts when installing the catalytic converter.**

✳✳ WARNING

Use only genuine Honda power steering fluid. Any other type or brand of fluid will damage the power steering pump.

16. Install or connect the following:
- Steering rack into position
- Pinion shaft grommet, insert the pinion through the hole in the bulkhead.
- Rack mounting bolts. Tighten the bracket bolts to 28 ft. lbs. (39 Nm). Tighten the mounting bolt under the valve body to 43 ft. lbs. (59 Nm).
- 2 hydraulic lines to the rack valve body. Carefully tighten the hydraulic line fitting to 28 ft. lbs. (38 Nm). Securely tighten the return hose clamp.

17. Center the rack ends within their steering strokes.

18. Install or connect the following:
- Tie rod ends onto the rack ends
- Tie rod ends to the steering knuckles, then the castle nuts.
- Front wheels
- Catalytic converter using new gaskets and self-locking nuts. Tighten the front nuts to 16 ft. lbs. (22 Nm), and the rear nuts to 25 ft. lbs. (34 Nm).

19. If equipped with a manual transaxles, install or connect the following:
- Shift linkage with a new spring pin and clip.
- Extension rod and tighten its bolt to 16 ft. lbs. (22 Nm).

20. If equipped with an automatic transaxles, install or connect the following:
- Shift cable and brackets. Tighten the bracket bolts to 108 inch lbs. (12 Nm). Tighten the cable lockbolt to 10 ft. lbs. (14 Nm). Tighten the cable holder bolts to 16 ft. lbs. (22 Nm).

21. Verify that the rack is centered within its strokes. Lower the vehicle.

22. Center the air bag cable reel as follows:

a. Remove the steering wheel.

b. Turn the cable reel clockwise until it stops.

c. Turn the steering wheel counterclockwise (approximately 2 turns) until the arrow on the label points straight up.

d. Install the steering wheel.

23. During steering wheel installation, verify that the slot on the steering wheel shaft engages with the tabs on the turn signal canceling sleeve. The pins on the cable reel fit into the holes on the steering wheel body. Install a new steering wheel nut and tighten it to 36 ft. lbs. (50 Nm).

24. Line up the bolt hole in the steering joint with the groove in the pinion shaft. Slip the joint onto the pinion shaft. Pull the joint up and down to be sure the splines are fully seated. Tighten the joint bolts to 16 ft. lbs. (22 Nm).

➡**Connect the steering joint and pinion shaft with the cable reel and steering rack centered. Verify that the lower joint bolt is securely seated in the pinion shaft groove. If the steering wheel and rack are not centered, reposition the serrations at the lower end of the steering joint.**

25. Install or connect the following:
- Steering joint cover

26. Tighten the ball joint castle nuts to 29–35 ft. lbs. (40–48 Nm). Then, tighten them only enough to install new cotter pins.

27. Install or connect the following:
- Air bag and cable reel connectors: Be sure the connectors fit squarely together. Then, press the connectors to couple them. The spring-loaded sleeve will lock into place as the 2 connectors are coupled.
- Steering wheel lower access cover
- Negative and positive battery cables

28. Turn the ignition switch to the **ON** position. The air bag indicator light should come on for 6 seconds, then turn off. This light sequence indicates that the air bag system is enabled and functioning normally. If the air bag light stays on longer, or doesn't turn on, the system must be diagnosed.

29. Be sure the reservoir inlet line has been reconnected. Fill the reservoir to the upper line with Honda power steering fluid. Run the engine at idle and turn the steering wheel lock-to-lock several times to bleed air from the system and fill the rack valve body. Recheck the fluid level and add more if necessary.

30. Check the power steering system for leaks.

31. Check the front wheel alignment and steering wheel spoke angle. Make adjustments by turning the left and right tie rod ends equally.

32. Road test the vehicle.

PRELUDE

➡**The electronic neutral check must be performed on 4WS equipped Preludes any time the steering rack, steering wheel, or steering column is removed, and before the wheels are aligned.**

1. Before servicing the vehicle, refer to the precautions in the beginning of this section.
2. Remove or disconnect the following:
   - Power steering reservoir off of its mount
   - Inlet hose
3. Insert a length of tubing into the inlet hose and route the tubing into a drain container.
4. With the engine running at idle, turn the steering wheel lock-to-lock several times until fluid stops running out of the hose. Shut off the engine.
5. Position the front wheels straight ahead. Lock the steering column with the ignition key. Reconnect the reservoir inlet hose.
6. Remove or disconnect the following:
   - Negative battery cable
   - Steering joint cover, then the upper and lower steering joint bolts.
7. Raise and support the vehicle safely.
8. Remove or disconnect the following:
   - Front wheels
   - Tie rod end cotter pins and castle nuts. Install a 12mm nut onto the end of the ball joint stud to protect the threads from damage.
   - Tie rod ends from the steering knuckles, using a ball joint tool
   - Heated Oxygen Sensor (HO2S) sensor connector
   - Self-locking nuts, then separate the catalytic converter from the exhaust pipe.
   - Exhaust pipe from the intake manifold
   - Exhaust pipe from the vehicle
9. If equipped with an automatic transaxle, remove or disconnect the following:
   - Remove the shift cable cover
   - Shift cable, and wire it up and out of the way.

➡**Clean any oil or dirt off of the valve body with solvent.**

   - Center beam from the subframe
   - Valve body shield
   - 4 hydraulic lines from the rack valve body, using a flare nut wrench. Plug the lines to keep dirt and moisture out.
10. On models with 4-Wheel Steering (4WS) remove or disconnect the following:
   - Carefully cut the wire tie securing the cover to the front sub-steering angle sensor
   - Cover
   - Sensor wiring harness from the 2 securing clamps

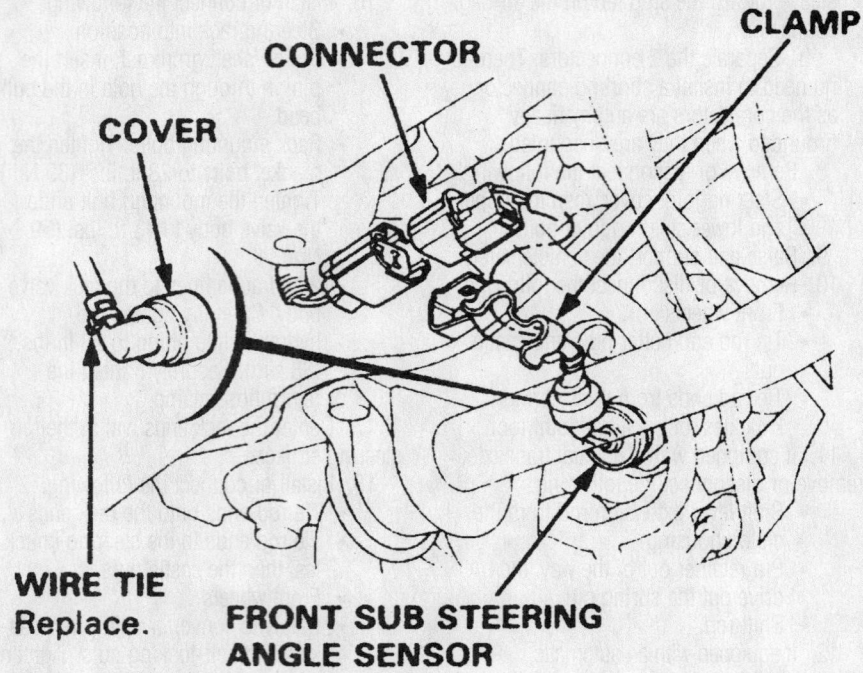

4WS front sub-steering angle sensor—Prelude

   - Sensor connector from the 4WS steering main wiring harness.
11. Remove or disconnect the following:
   - Steering joint bolt, then slide the pinion shaft out of the joint.
   - Left mounting bracket, then the right mounting brackets.
   - Left tie rod end and slide the rack all the way to the right.

12. Pull the steering rack down to release it from the pinion shaft.
13. Slide the steering rack to the right until the left tie rod clears the subframe, then drop it down and out of the vehicle to the left.

To install:

➡**Use new gaskets and self-locking nuts when installing the exhaust pipe.**

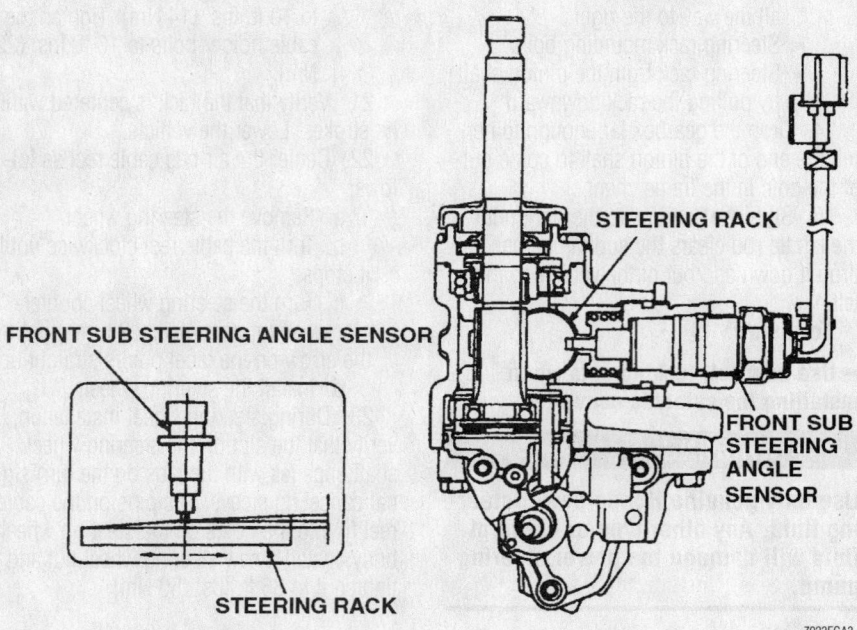

Front sub-steering angle sensor, harness, and steering rack—Prelude

** WARNING

Use only genuine Honda power steering fluid. Any other type or brand of fluid will damage the power steering pump.

14. Install or connect the following:
   - Steering rack into position
   - Pinion shaft grommet and insert the pinion through the hole in the firewall.
   - Right and left mounting brackets. Tighten the short bolts to 28 ft. lbs. (39 Nm), and the long bolts to 32 ft. lbs. (44 Nm).
15. Center the rack ends within their steering strokes.
16. Center the air bag cable reel as follows:
   a. Turn the steering wheel clockwise until it stops.
   b. Turn the steering wheel counterclockwise until the yellow gear tooth lines up with the alignment mark on the lower column cover.
17. Line up the bolt hole in the steering joint with the groove in the pinion shaft. Slip the joint onto the pinion shaft. Pull the joint up and down to be sure the splines are fully seated. Tighten the joint bolts to 16 ft. lbs. (22 Nm).

➡**Connect the steering joint and pinion shaft with the cable reel and steering rack centered. Verify that the lower joint bolt is securely seated in the pinion shaft groove. If the steering wheel and rack are not centered, reposition the serrations at the lower end of the steering joint.**

18. Install or connect the following:
   - 4 hydraulic lines to the rack valve body. Carefully tighten the 12mm fittings to 108 inch lbs. (13 Nm), the 14mm inlet fitting to 28 ft. lbs. (37 Nm), and the 17mm oil cooler fitting to 21 ft. lbs. (29 Nm).
   - Front sub-steering angle sensor to the 4WS harness.
   - Wire back into its clamps, making sure that it doesn't interfere with the stabilizer bar.
   - Sensor cover with a new wire tie
   - Valve body shield
   - Center beam. Use new self-locking bolts and tighten them to 43 ft. lbs. (60 Nm).
19. If equipped with an automatic transaxle, install or connect the following:

   - Shift cable and tighten the locknut to 10 ft. lbs. (14 Nm).
   - Cable holder and tighten its bolts to 13 ft. lbs. (18 Nm).
20. Install or connect the following:
   - Catalytic converter using new gaskets and self-locking nuts. Tighten the exhaust manifold nuts to 40 ft. lbs. (55 Nm), and the rear nuts to 25 ft. lbs. (34 Nm).
   - HO_2S connector
   - Tie rod ends onto the rack ends
   - Tie rod ends to the steering knuckles, then the castle nuts.
   - Front wheels
21. Verify that the rack is centered within its strokes. Lower the vehicle.
22. Install the steering joint cover.
23. Tighten the ball joint castle nuts to 36–43 ft. lbs. (50–60 Nm). Then, tighten them only enough to install new cotter pins.
24. Reconnect the negative battery cable.
25. Be sure the reservoir inlet line has been reconnected. Fill the reservoir to the upper line with Honda power steering fluid. Run the engine at idle and turn the steering wheel lock-to-lock several times to bleed air from the system and fill the rack valve body. Recheck the fluid level and add more if necessary.
26. Check the power steering system for leaks.
27. On Preludes without 4WS, check and adjust the front wheel alignment. On Preludes with 4WS, the electronic neutral check must be performed on the 4WS system.

ACCORD

1. Before servicing the vehicle, refer to the precautions in the beginning of this section.
2. Lift the power steering reservoir off of its mount and disconnect the inlet hose.
3. Insert a length of tubing into the inlet hose and route the tubing into a drain container.
4. With the engine running at idle, turn the steering wheel lock-to-lock several times until fluid stops running out of the hose. Immediately shut off the engine.
5. Position the front wheels straight ahead. Lock the steering column with the ignition key. Reconnect the reservoir inlet hose.
6. Remove or disconnect the following:
   - Negative battery cable
   - Steering joint cover and the upper and lower steering joint bolts.
7. Raise and support the vehicle safely.
8. Remove or disconnect the following:
   - Front wheels
   - Tie rod end cotter pins and castle nuts
   - Tie rod ends from the steering knuckles, using a ball joint tool.
   - Left tie rod end and slide the rack all the way to the right.
   - Heated Oxygen Sensor (HO_2S) connector
   - Self-locking nuts, then separate the catalytic converter from the exhaust pipe.

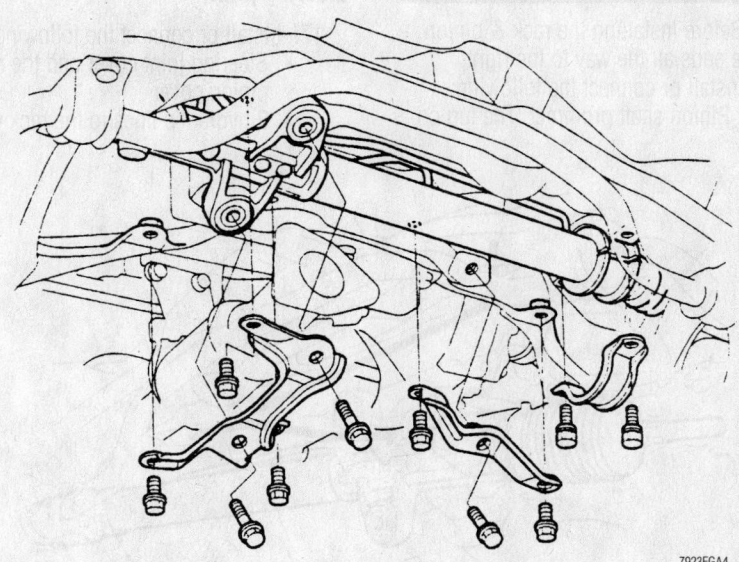

Power rack and pinion steering gear mounting—Accord

7923FGA4

For complete Engine Mechanical specifications, see Section 1 of this manual

- Catalytic converter
- Shift linkage from the transaxle case, if equipped with a manual transaxle.
- Shift cable cover and cable (wire it up and out of the way), if equipped with an automatic transaxle.
- 2 hydraulic lines from the rack valve body, using a flare nut wrench. Plug the lines to keep dirt and moisture out. Carefully move the disconnected lines to the rear of the rack assembly so that they are not damaged when the rack is removed.
- Rack stiffener plate, then the steering rack mounting bolts.

9. Pull the steering rack down to release it from the pinion shaft.

10. Drop the steering rack far enough to permit the end of the pinion shaft to come out of the hole in the frame channel.

11. Slide the steering rack to the right until the left tie rod clears the subframe, then drop it down and out of the vehicle to the left.

To install:

➡**Use new gaskets and self-locking nuts when installing the catalytic converter.**

❊❊ WARNING

Use only genuine Honda power steering fluid. Any other type or brand of fluid will damage the power steering pump.

12. Before installing the rack & pinion, slide the ends all the way to the right.

13. Install or connect the following:
- Pinion shaft grommet. The lug on

the pinion shaft grommet aligns with the slot on the valve body.
- Steering rack into position
- Pinion shaft grommet and insert the pinion through the hole in the bulkhead.
- Rack mounting bolts. Tighten the bracket bolts to 28 ft. lbs. (39 Nm). Tighten the stiffener plate mounting bolts to 32 ft. lbs. (43 Nm).

14. Center the rack ends within their steering strokes.

15. Center the air bag cable reel, as follows:

a. Turn the steering wheel left approximately 150°, to check the cable reel position with the indicator.

b. If the cable reel is centered, the yellow gear tooth lines up with the alignment mark on the cover.

c. Return the steering wheel right approximately 150° to position the steering wheel in the straight-ahead position.

16. Line up the bolt hole in the steering joint with the groove in the pinion shaft. Slip the joint onto the pinion shaft. Pull the joint up and down to be sure the splines are fully seated. Tighten the joint bolts to 16 ft. lbs. (22 Nm).

➡**Connect the steering joint and pinion shaft with the cable reel and steering rack centered. Verify that the lower joint bolt is securely seated in the pinion shaft groove. If the steering wheel and rack are not centered, reposition the serrations at the lower end of the steering joint.**

17. Install or connect the following:
- Steering joint cover and the rack & pinion cover
- 2 hydraulic lines to the rack valve

body. Carefully tighten the 14mm inlet fitting to 27 ft. lbs. (37 Nm) and the 16mm outlet fitting to 21 ft. lbs. (28 Nm).
- Shift cable and the select cable to the transaxle with new cotter pins, if equipped with a manual transaxle.
- Shift cable to the transaxle using a new lockwasher, if equipped with an automatic transaxle. Tighten the lockbolt to 10 ft. lbs. (14 Nm).
- Catalytic converter using new gaskets and self-locking nuts. Tighten the front nuts to 16 ft. lbs. (22 Nm), and the rear nuts to 25 ft. lbs. (34 Nm).
- HO2S sensor connector
- Tie rod ends onto the rack ends
- Tie rod ends to the steering knuckles, then the castle nuts.

18. Tighten the ball joint castle nuts to 29–35 ft. lbs. (40–48 Nm). Then, tighten them only enough to install new cotter pins.

19. Install the front wheels.

20. Lower the vehicle.

21. Reconnect the negative battery cable.

22. Be sure the reservoir inlet line has been reconnected. Fill the reservoir to the upper line with Honda power steering fluid. Run the engine at idle and turn the steering wheel lock-to-lock several times to bleed air from the system and fill the rack valve body. Recheck the fluid level and add more if necessary.

23. Check the power steering system for leaks.

24. Check the front wheel alignment and steering wheel spoke angle. Make adjustments by turning the left and right tie rod ends equally.

25. Road test the vehicle.

S2000

1. Before servicing the vehicle, refer to the precautions in the beginning of this section.

2. Remove or disconnect the following:
- Negative battery cable
- Front wheels
- Driver's air bag
- Steering wheel
- Steering coupler
- Outer tie rod ends
- Splash shield
- Stabilizer bar brackets
- Steering gear wiring connectors
- Steering gear mounting bolts

3. Move the steering gear forward and to the right to remove the steering gear.

To install:

4. Installation is the reverse of the

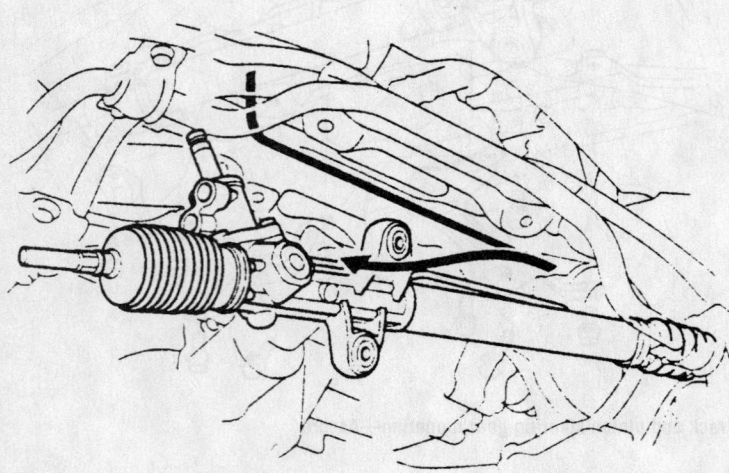

7923FGA5

Move the steering rack to the right, then down and out of the vehicle—Accord

removal procedure, while using the following torque values:

- Steering gear mounting bolts: 33 ft. lbs. (44 Nm)
- Steering gear ground cable bolt: 88 inch lbs. (10 Nm)
- Stabilizer bar bracket bolts: 61 ft. lbs. (83 Nm)
- Splash shield bolts: 88 inch lbs. (10 Nm)
- Outer tie rod end nuts: 40 ft. lbs. (54 Nm)
- Steering coupler pinch bolts: 16 ft. lbs. (22 Nm)

Strut

REMOVAL & INSTALLATION

Front

CIVIC

1. Before servicing the vehicle, refer to the precautions in the beginning of this section.
2. Raise and safely support the vehicle.

3. Remove or disconnect the following:
   - Front wheels
   - Brake hose brackets from the bottom of the strut tube. Do not disconnect the brake hoses.

➡**Some Civic models may not have brake hose brackets on their struts. In these cases, there is no need to unbolt the brackets.**

   - Damper pinch bolt
   - Damper fork nut and bolt
   - Damper fork
   - 2 strut mounting bolts from the shock tower
   - Strut from the vehicle

To install:

➡**Use new self-locking nuts when installing the strut.**

4. Install or connect the following:
   - Strut into the vehicle. Hand-tighten the strut mounting bolts. The alignment mark on the strut tube faces away from the wheel.
   - Damper fork onto the strut and lower control arm

   - Pinch and fork bolts
   - Brake hose brackets to the strut tube and tighten them to 16 ft. lbs. (22 Nm).
   - Front wheels and lower the vehicle.

5. Tighten the strut mount bolts to 36 ft. lbs. (50 Nm).
6. Tighten the pinch bolt to 32 ft. lbs. (44 Nm). Tighten the damper fork nut to 47 ft. lbs. (65 Nm).
7. Tighten the wheel nuts to 80 ft. lbs. (110 Nm).
8. Check the vehicle's front end alignment and adjust it if necessary.

PRELUDE AND ACCORD

1. Before servicing the vehicle, refer to the precautions in the beginning of this section.
2. Raise and safely support the vehicle.
3. Remove or disconnect the following:
   - Front wheels
   - Brake hose clamp bolts from the strut
   - Damper fork bolts, then the damper fork.
   - 3 strut mounting nuts
   - Strut from the vehicle

To install:

➡**Use new self-locking bolts when installing the struts and assembling the damper forks.**

4. Install or connect the following:
   - Strut into the vehicle. Hand-tighten the mounting nuts.

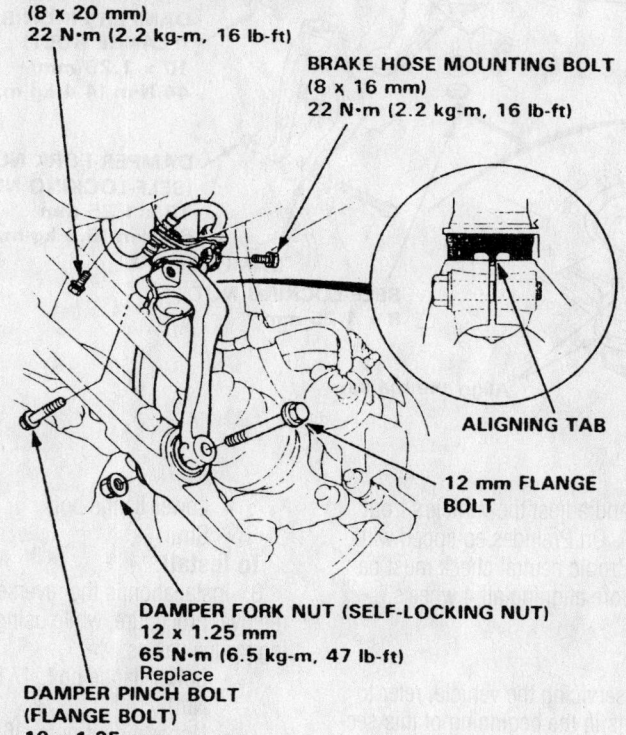

BRAKE HOSE MOUNTING BOLT
(8 x 20 mm)
22 N·m (2.2 kg-m, 16 lb-ft)

BRAKE HOSE MOUNTING BOLT
(8 x 16 mm)
22 N·m (2.2 kg-m, 16 lb-ft)

ALIGNING TAB

12 mm FLANGE BOLT

DAMPER FORK NUT (SELF-LOCKING NUT)
12 x 1.25 mm
65 N·m (6.5 kg-m, 47 lb-ft)
Replace

DAMPER PINCH BOLT
(FLANGE BOLT)
10 x 1.25 mm
44 N·m (4.4 kg-m, 32 lb-ft)

7923FGA6

Damper fork components—Civic

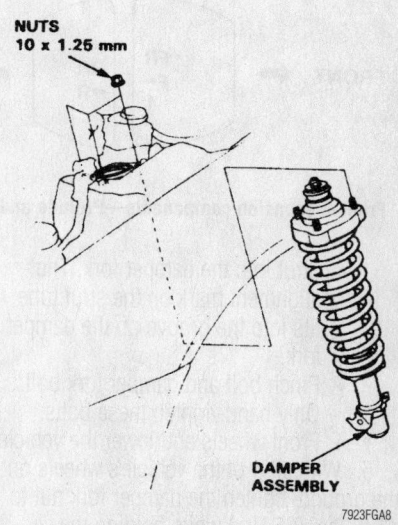

NUTS
10 x 1.25 mm

DAMPER ASSEMBLY

7923FGA8

Front strut and strut mount—Prelude and Accord

For Accessory Drive Belt illustrations, see Section 1 of this manual

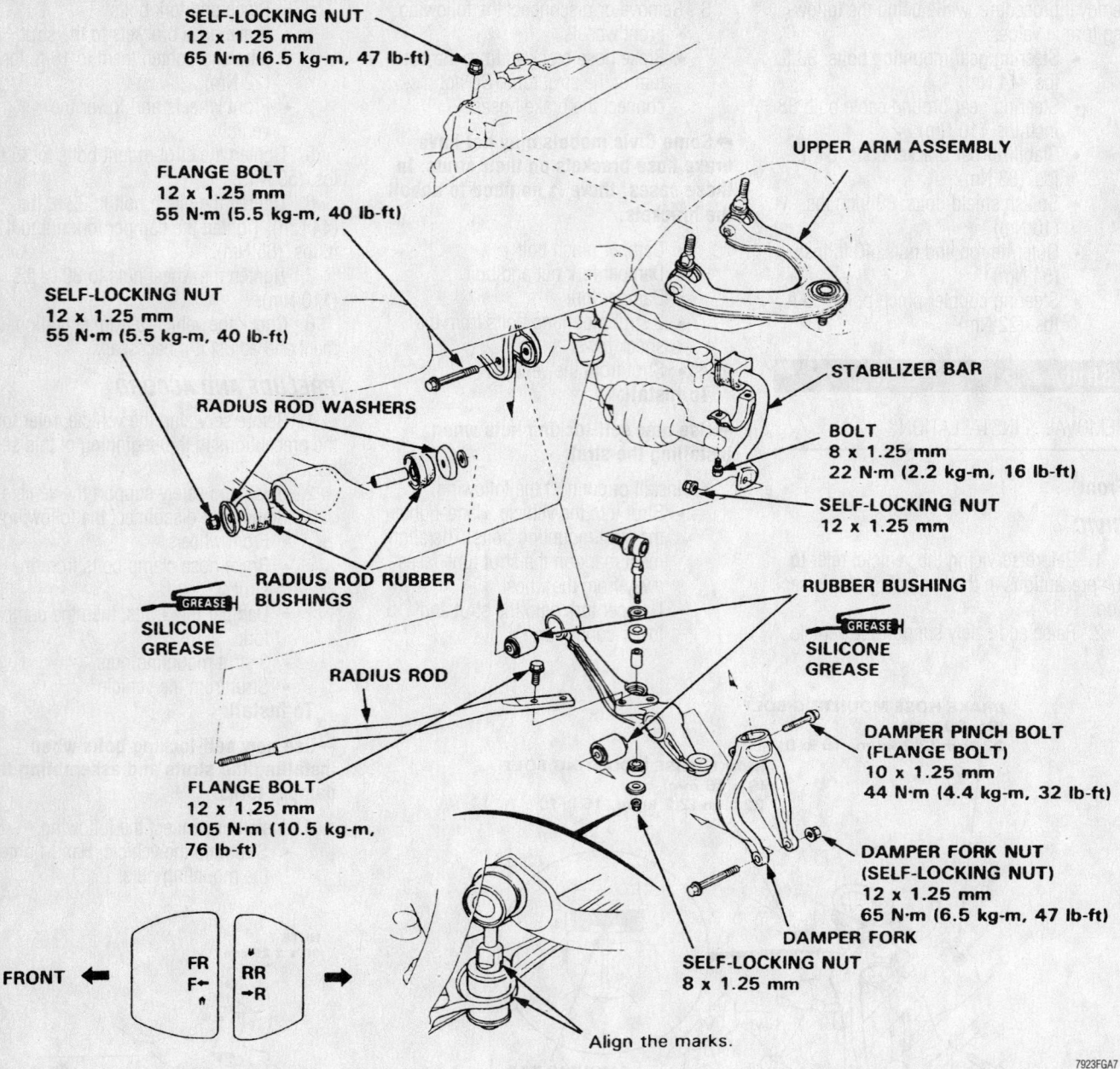

SELF-LOCKING NUT
12 x 1.25 mm
65 N·m (6.5 kg-m, 47 lb-ft)

FLANGE BOLT
12 x 1.25 mm
55 N·m (5.5 kg-m, 40 lb-ft)

SELF-LOCKING NUT
12 x 1.25 mm
55 N·m (5.5 kg-m, 40 lb-ft)

RADIUS ROD WASHERS

RADIUS ROD RUBBER BUSHINGS

GREASE SILICONE GREASE

RADIUS ROD

FLANGE BOLT
12 x 1.25 mm
105 N·m (10.5 kg-m, 76 lb-ft)

UPPER ARM ASSEMBLY

STABILIZER BAR

BOLT
8 x 1.25 mm
22 N·m (2.2 kg-m, 16 lb-ft)

SELF-LOCKING NUT
12 x 1.25 mm

RUBBER BUSHING

GREASE SILICONE GREASE

DAMPER PINCH BOLT (FLANGE BOLT)
10 x 1.25 mm
44 N·m (4.4 kg-m, 32 lb-ft)

DAMPER FORK NUT (SELF-LOCKING NUT)
12 x 1.25 mm
65 N·m (6.5 kg-m, 47 lb-ft)

DAMPER FORK

SELF-LOCKING NUT
8 x 1.25 mm

Align the marks.

FRONT ← FR F↑ RR →R →

7923FGA7

Front suspension components—Prelude and Accord

- Strut into the damper fork. The alignment mark on the strut tube fits into the groove on the damper fork.
- Pinch bolt and damper fork bolt. Only hand-tighten these bolts.
- Front wheels and lower the vehicle.

5. With all 4 of the vehicle's wheels on the ground, tighten the damper fork nut to 47 ft. lbs. (65 Nm) while holding the damper fork bolt. Tighten the damper fork pinch bolt to 32 ft. lbs. (44 Nm). Tighten the strut mounting nuts to 28 ft. lbs. (39 Nm).

6. Tighten the wheel nuts to 80 ft. lbs. (110 Nm).

7. Check and adjust the vehicle's front end alignment. On Preludes equipped with 4WS, the electronic neutral check must be performed before aligning all 4 wheels.

S2000

1. Before servicing the vehicle, refer to the precautions in the beginning of this section.

2. Remove or disconnect the following:
- Front wheel
- Lower ball joint
- Brake caliper bracket bolt
- Upper strut mount nuts

- Lower flange bolt
- Strut

To install:

3. Installation is the reverse of the removal procedure, while using the following torque values:
- Lower flange bolt: 47 ft. lbs. (64 Nm)
- Upper mount nuts: 36 ft. lbs. (49 Nm)
- Lower ball joint nut: 43–51 ft. lbs. (59–69 Nm)
- Brake caliper bracket bolt: 16 ft. lbs. (22 Nm)

Rear

CIVIC

✳✳ CAUTION

Removing rear suspension components may make the vehicle front-heavy and cause it to tip forward when raised on a hoist. Use under-lift support stands, or place additional weight in the trunk of the vehicle before hoisting it.

1. Before servicing the vehicle, refer to the precautions in the beginning of this section.

2. Remove the interior or trunk trim pieces that cover the strut mount, as follows:

   a. **Sedan and Coupe models:** Fold down the upper rear seat cushion. Carefully pry out the clips that secure the trunk and shock tower trim to the body.

Remove the trunk trim to expose the strut mounts.

   b. **Hatchback models:** Fold down the rear seat. Unbolt and remove the rear side shelf/speaker grille assemblies. Disconnect and remove the speaker. Carefully pry out the clips and remove the screws to remove shock tower trim panel.

3. Raise and support the vehicle.

4. Remove or disconnect the following:
   • Rear wheels
   • 2 upper mounting nuts
   • Wheel sensor bracket from the lower control arm.
   • Lower strut bolt and the knuckle flange bolt.
   • Strut from the vehicle

To install:

➡ **All suspension nuts and bolts should be tightened with the vehicle on the ground. Alternatively, raise the lower control arm with a floor jack until the**

jack is supporting the weight of the vehicle. This method pre-loads the suspension and allows room to work.

5. Install or connect the following:
   • Strut into the vehicle with the locknut facing the front of the vehicle. Hand-tighten the upper mounting nuts.
   • Wheel sensor bracket onto the lower control arm. Tighten the bolts to 84 inch lbs. (10 Nm).
   • Knuckle flange bolt and the lower strut bolt. Hand-tighten the bolts.
   • Wheels and lower the vehicle.
   • Upper mounting nuts to 36 ft. lbs. (50 Nm). Tighten the knuckle flange bolt and strut bolts to 40 ft. lbs. (55 Nm). Tighten the wheel nuts to 80 ft. lbs. (110 Nm).
   • Trunk side trim panels

6. Check and adjust the vehicle's rear wheel alignment.

PRELUDE

1. Before servicing the vehicle, refer to the precautions in the beginning of this section.

2. Raise and safely support the vehicle.

3. Remove or disconnect the following:
   • Trunk side trim and the 2 top strut nuts.
   • Upper ball joint cover
   • Cotter pin and upper ball joint nut

4. Fit a 10mm nut on the ball joint and separate the ball joint and the knuckle by using a ball joint removal tool.

5. Remove or disconnect the following:
   • Lower strut mounting bolt and lower the suspension.
   • Strut from the vehicle

To install:

➡ **Use new self-locking nuts when installing the rear struts.**

6. Install or connect the following:
   • Strut; loosely install the lower mounting bolt. Do not tighten.
   • Upper strut mounting bolts: 28 ft. lbs. (39 Nm).
   • Upper arm and knuckle and tighten the castle nut to 29–35 ft. lbs. (40–48 Nm).
   • Upper ball joint cover

7. Raise the rear suspension with a floor jack until the weight is on the strut.

8. Tighten the lower strut mounting bolt to 47 ft. lbs. (65 Nm).

9. Install the rear wheels and lower the vehicle.

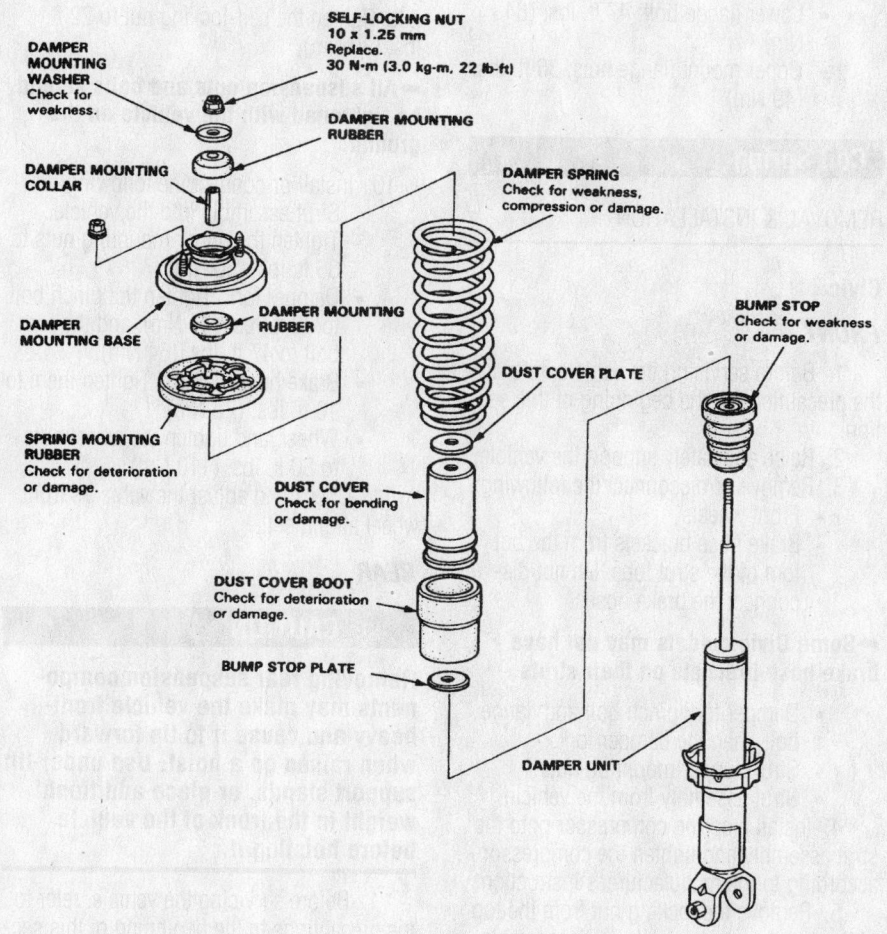

DAMPER MOUNTING WASHER
Check for weakness.

SELF-LOCKING NUT
10 x 1.25 mm
Replace.
30 N·m (3.0 kg-m, 22 lb-ft)

DAMPER MOUNTING RUBBER

DAMPER MOUNTING COLLAR

DAMPER MOUNTING BASE

DAMPER MOUNTING RUBBER

SPRING MOUNTING RUBBER
Check for deterioration or damage.

DAMPER SPRING
Check for weakness, compression or damage.

BUMP STOP
Check for weakness or damage.

DUST COVER PLATE

DUST COVER
Check for bending or damage.

DUST COVER BOOT
Check for deterioration or damage.

BUMP STOP PLATE

DAMPER UNIT

7923FGA9

Exploded view of the rear suspension strut—Civic

10. Tighten the rear wheel nuts to 80 ft. lbs. (110 Nm).

11. Check and adjust the vehicle's rear wheel alignment.

ACCORD

1. Before servicing the vehicle, refer to the precautions in the beginning of this section.

2. Fold the rear seat forward.

3. Remove or disconnect the following:
- Side bolster cushions. The side bolster cushions are secured by 1 screw at the bottom, and 2 clips at the top.
- Strut mount cap and upper strut mounting nuts.

4. Raise and safely support the vehicle.

5. Remove or disconnect the following:
- Rear wheels, then support the knuckle with a floor jack.
- Strut mounting bolt, then lower the jack.
- Strut

To install:

➡**Use new self-locking nuts when installing the strut.**

6. Install or connect the following:
- Strut into the upper mount. Only hand-tighten the upper mounting nuts.
- Strut into position on the knuckle, then the mounting bolt.

7. Place a jack under the lower strut mount. Raise the jack until the weight of the vehicle is on the jack.

8. With the suspension under load, tighten the lower mount bolt to 40 ft. lbs. (55 Nm). Tighten the upper nuts to 28 ft. lbs. (39 Nm).

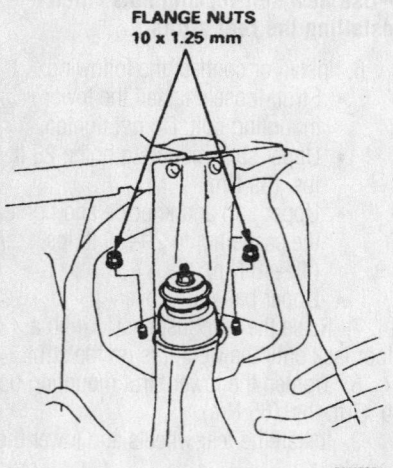

FLANGE NUTS
10 x 1.25 mm

7923FGB1

Rear strut upper mounting nut locations—Accord

9. Install or connect the following:
- Rear wheel. Lower the vehicle to the ground.

10. Tighten the wheel nuts to 80 ft. lbs. (110 Nm).

11. Install the rear seat side bolsters and fold the seat back into place.

12. Check and adjust the vehicle's rear wheel alignment.

S2000

1. Before servicing the vehicle, refer to the precautions in the beginning of this section.

2. Remove or disconnect the following:
- Rear wheel
- Spare tire
- Upper mount flange nuts
- Lower flange bolt
- Strut

To install:

3. Installation is the reverse of the removal procedure, while using the following torque values:
- Lower flange bolt: 47 ft. lbs. (64 Nm)
- Upper mount flange nuts: 36 ft. lbs. (49 Nm)

Coil Spring

REMOVAL & INSTALLATION

Civic

FRONT

1. Before servicing the vehicle, refer to the precautions in the beginning of this section.

2. Raise and safely support the vehicle.

3. Remove or disconnect the following:
- Front wheels
- Brake hose brackets from the bottom of the strut tube. Do not disconnect the brake hoses.

➡**Some Civic models may not have brake hose brackets on their struts.**

- Damper fork pinch bolt and flange bolt, then the damper fork.
- Strut's upper mounting nuts
- Strut assembly from the vehicle.

4. Install a spring compressor onto the strut assembly and tighten the compressor according to the manufacturer's instructions.

5. Remove the locking nut from the top of the shock absorber piston. Disassemble the strut and remove the coil spring.

To install:

➡**Use new self-locking nuts when assembling the strut.**

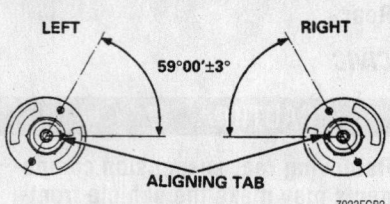

LEFT   **RIGHT**

59°00'±3°

ALIGNING TAB

7923FGB2

Strut bearing installation direction—Civic

6. Install or connect the following:
- Spring compressor onto the coil spring

7. Assemble the lower strut mounts, dust covers, coil spring, and upper strut mount onto the shock absorber. Position the strut bearing mounting studs so that they will line up with the mounting holes in the shock tower.

8. Install or connect the following:
- Mounting washer, and a new self-locking nut (loosely).

9. Hold the shock absorber piston with a hex wrench and tighten the self-locking nut. Tighten the self-locking nut to 22 ft. lbs. (30 Nm).

➡**All suspension nuts and bolts should be tightened with the vehicle on the ground.**

10. Install or connect the following:
- Strut assembly into the vehicle. Tighten the upper mounting nuts to 36 ft. lbs. (50 Nm).
- Damper fork. Tighten the pinch bolt to 32 ft. lbs. (44 Nm), and the fork bolt to 47 ft. lbs. (64 Nm).
- Brake hose clamps. Tighten them to 16 ft. lbs. (22 Nm).
- Wheel, and tighten the wheel nuts to 80 ft. lbs. (110 Nm).

11. Check and adjust the vehicle's front wheel alignment.

REAR

✳✳ CAUTION

Removing rear suspension components may make the vehicle front-heavy and cause it to tip forward when raised on a hoist. Use under-lift support stands, or place additional weight in the trunk of the vehicle before hoisting it.

1. Before servicing the vehicle, refer to the precautions in the beginning of this section.

2. Remove the interior or trunk trim pieces that cover the strut mount:

a. **Sedan and Coupe models:** Fold down the upper rear seat cushion.

Carefully pry out the clips that secure the trunk and shock tower trim to the body. Remove the trunk trim to expose the strut mounts.

b. **Hatchback models:** Fold down the rear seat. Unbolt and remove the rear side shelf/speaker grille assemblies. Disconnect and remove the speaker. Carefully pry out the clips and remove the screws to remove shock tower trim panel.

3. Raise and safely support the vehicle.

4. Remove or disconnect the following:
- 2 strut mounting bolts
- Wheel sensor brackets from the lower control arm. Do not disconnect the sensor.

5. Support the lower control arm with a floor jack.

6. Remove or disconnect the following:
- Strut mounting flange bolt and the knuckle flange bolt, then lower the floor jack
- Strut from the vehicle.

7. Install a spring compressor onto the strut assembly and tighten the compressor according to the manufacturer's instructions.

8. Remove the locking nut from the top of the shock absorber. Disassemble the strut and remove the coil spring.

To install:

➡**Use new self-locking nuts when assembling the strut.**

9. Install or connect the following:
- Spring compressor onto the coil spring.
- Upper and lower strut mounts, dust covers, and coil spring onto the shock absorber.
- Mounting washer, and a new self-locking nut (loosely)

10. Hold the shock absorber piston with a hex wrench and tighten the self-locking nut. Tighten the self-locking nut to 22 ft. lbs. (30 Nm).

➡**All suspension nuts and bolts should be tightened with the vehicle on the ground. Alternatively, raise the lower control arm with a floor jack until the jack is supporting the weight of the vehicle. This method pre-loads the suspension and allows room to work.**

11. Install or connect the following:
- Strut assembly into the vehicle. Tighten the upper mounting nuts to 36 ft. lbs. (50 Nm).
- Shock mounting bolt at the knuckle and tighten to 40 ft. lbs. (55 Nm).

- Knuckle flange bolt and tighten it to 40 ft. lbs. (55 Nm).
- Wheel sensor brackets
- Wheel, and tighten the wheel nuts to 80 ft. lbs. (110 Nm).
- Trunk side trim

12. Check and adjust the rear wheel alignment.

Accord and Prelude

FRONT

1. Before servicing the vehicle, refer to the precautions in the beginning of this section.

2. Raise and safely support the vehicle.

3. Remove or disconnect the following:
- Front wheels
- Brake hose clamp from the strut.
- Damper fork bolts and damper fork
- 3 strut mounting nuts
- Strut from the vehicle

4. Place the strut in vice and install a spring compressor onto the coil spring.

Follow the spring compressor manufacturer's instructions.

5. Compress the spring and remove the self-locking nut from the top of the strut. Disassemble the strut mounts and remove the coil spring.

6. Inspect the strut mounts for wear and damage. Replace any damaged or worn parts.

To install:

➡**Use new self-locking nuts when assembling and installing the struts.**

7. Install or connect the following:
- Spring compressor onto the coil spring. Set the spring onto the strut cartridge. The flat part of the coil spring is its top.

8. Assemble the strut mount and washer onto the strut. Tighten the self-locking nut to 22 ft. lbs. (29 Nm). Remove the spring compressor.

9. Install or connect the following:
- Strut into the vehicle. Hand-tighten the mounting nuts.

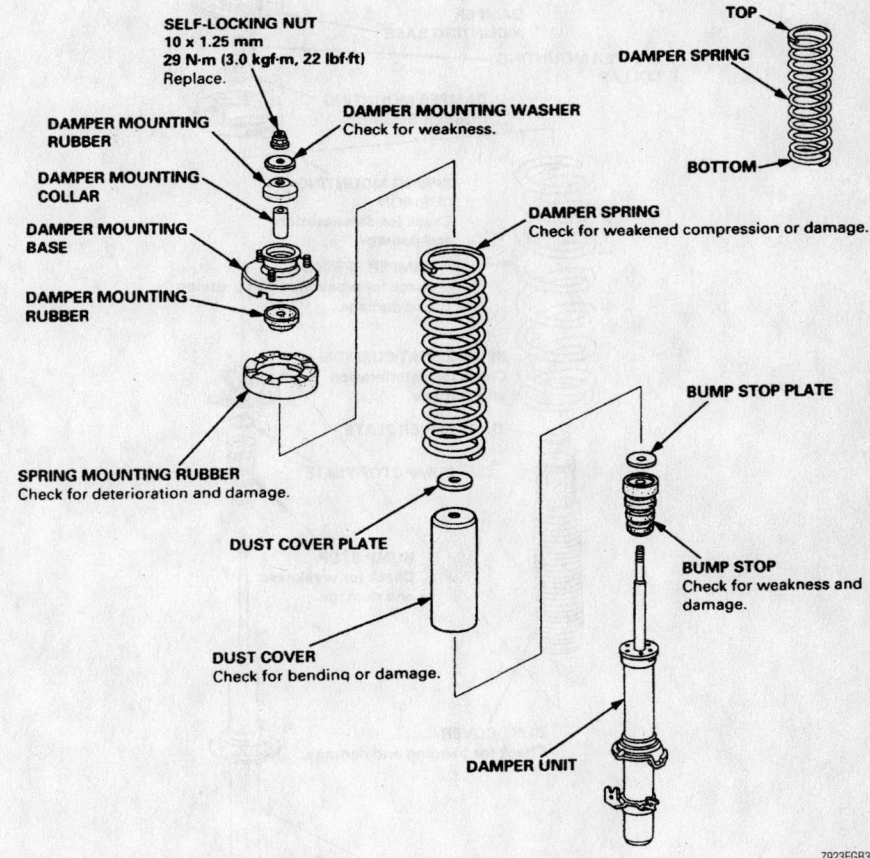

Coil spring, strut cartridge, and strut mount components—Accord and Prelude

7923FGB3

- Strut into the damper fork. The alignment mark on the strut tube fits into the groove on the damper fork.
- Pinch bolt and damper fork bolt. Only hand-tighten these bolts.
- Front wheels and lower the vehicle.

10. With all 4 of the vehicle's wheels on the ground, tighten the damper fork nut to 47 ft. lbs. (65 Nm) while holding the damper fork bolt. Tighten the damper fork pinch bolt to 32 ft. lbs. (44 Nm). Tighten the strut mounting nuts to 28 ft. lbs. (39 Nm).

11. Tighten the wheel nuts to 80 ft. lbs. (110 Nm).

12. Check and adjust the vehicle's front wheel alignment. On Preludes equipped with 4WS, the electronic neutral check must be performed before all 4 wheels are aligned.

REAR ACCORD

1. Before servicing the vehicle, refer to the precautions in the beginning of this section.

2. Remove or disconnect the following:
- Strut

3. Place the strut in a vice and install a spring compressor onto the coil spring. Follow the spring compressor manufacturer's instructions.

4. Compress the spring and remove the self-locking nut from the strut. Disassemble the strut mounts and remove the coil spring.

5. Inspect the strut mounts for wear and damage. Replace any damaged or worn parts.

To install:

➡**Use new self-locking nuts when assembling and installing the struts.**

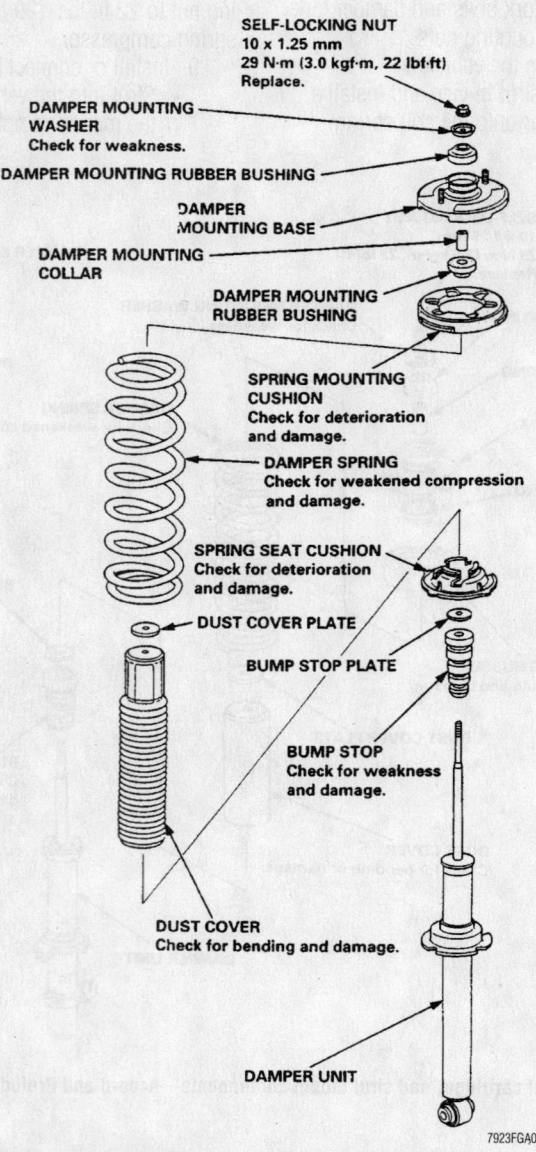

SELF-LOCKING NUT
10 x 1.25 mm
29 N·m (3.0 kgf·m, 22 lbf·ft)
Replace.

DAMPER MOUNTING WASHER
Check for weakness.

DAMPER MOUNTING RUBBER BUSHING

DAMPER MOUNTING BASE

DAMPER MOUNTING COLLAR

DAMPER MOUNTING RUBBER BUSHING

SPRING MOUNTING CUSHION
Check for deterioration and damage.

DAMPER SPRING
Check for weakened compression and damage.

SPRING SEAT CUSHION
Check for deterioration and damage.

DUST COVER PLATE

BUMP STOP PLATE

BUMP STOP
Check for weakness and damage.

DUST COVER
Check for bending and damage.

DAMPER UNIT

7923FGA0

Exploded view of the rear suspension strut assembly—Accord

6. Install or connect, the following:
- Spring compressor onto the coil spring. Set the spring onto the strut cartridge. The flat part of the coil spring is its top.
- Strut mount and washer onto the strut. Tighten the self-locking nut to 22 ft. lbs. (29 Nm). Remove the spring compressor.
- Strut into the vehicle. Hand-tighten the mounting nuts.
- Strut into position on the knuckle.
- Mounting bolt

7. Place a jack under the lower strut mount. Raise the jack until the weight of the vehicle is on the jack.

8. With the suspension under load, tighten the lower mount bolt to 40 ft. lbs. (55 Nm). Tighten the upper nuts to 28 ft. lbs. (39 Nm).

9. Install or connect, the following:
- Rear wheel. Lower the vehicle to the ground.
- Wheel nuts to 80 ft. lbs. (110 Nm).
- Rear seat side bolsters and fold the seat back into place.

10. Check and adjust the vehicle's rear wheel alignment.

REAR PRELUDE

1. Before servicing the vehicle, refer to the precautions in the beginning of this section.

2. Raise and safely support the vehicle.

3. Remove or disconnect the following:
- Trunk side trim and remove the 2 strut mounting nuts.
- Upper ball joint cover
- Cotter pin and upper ball joint nut

4. Fit a 10mm nut on the ball joint and separate the ball joint and the knuckle by using a ball joint removal tool.

5. Remove or disconnect the following:
- Lower strut mounting bolt and lower the suspension.
- Strut from the vehicle.

6. Place the strut in vice and install a spring compressor onto the coil spring. Follow the spring compressor manufacturer's instructions.

7. Compress the spring and remove the self-locking nut from the strut. Disassemble the strut mounts and remove the coil spring.

8. Inspect the strut mounts for wear and damage. Replace any damaged or worn parts.

To install:

➡**Use new self-locking nuts when installing the rear struts.**

9. Install or connect the following:
- Spring compressor onto the coil spring. Set the spring onto the strut

cartridge. The flat part of the coil spring is its top.

- Strut mount and washer onto the strut. Tighten the self-locking nut to 22 ft. lbs. (29 Nm). Remove the spring compressor.
- Strut to the vehicle and the lower mounting bolt (loosely). Do not tighten at this time.
- Upper strut mounting bolts. Tighten the bolts to 28 ft. lbs. (39 Nm).
- Upper arm and knuckle and tighten

the castle nut to 29–35 ft. lbs. (40–48 Nm).
- Upper ball joint cover

10. Raise the rear suspension with a floor jack until the weight is on the strut.

11. Tighten the lower strut mounting bolt to 47 ft. lbs. (65 Nm).

12. Install or connect the following:
- Rear wheels and lower the vehicle.
- Rear wheel nuts to 80 ft. lbs. (110 Nm)
- Trunk trim

13. Check and adjust the vehicle's rear wheel alignment. On Preludes equipped with 4WS, the electronic neutral check must be performed before all 4 wheels are aligned.

S2000

FRONT AND REAR

1. Before servicing the vehicle, refer to the precautions at the beginning of this section.

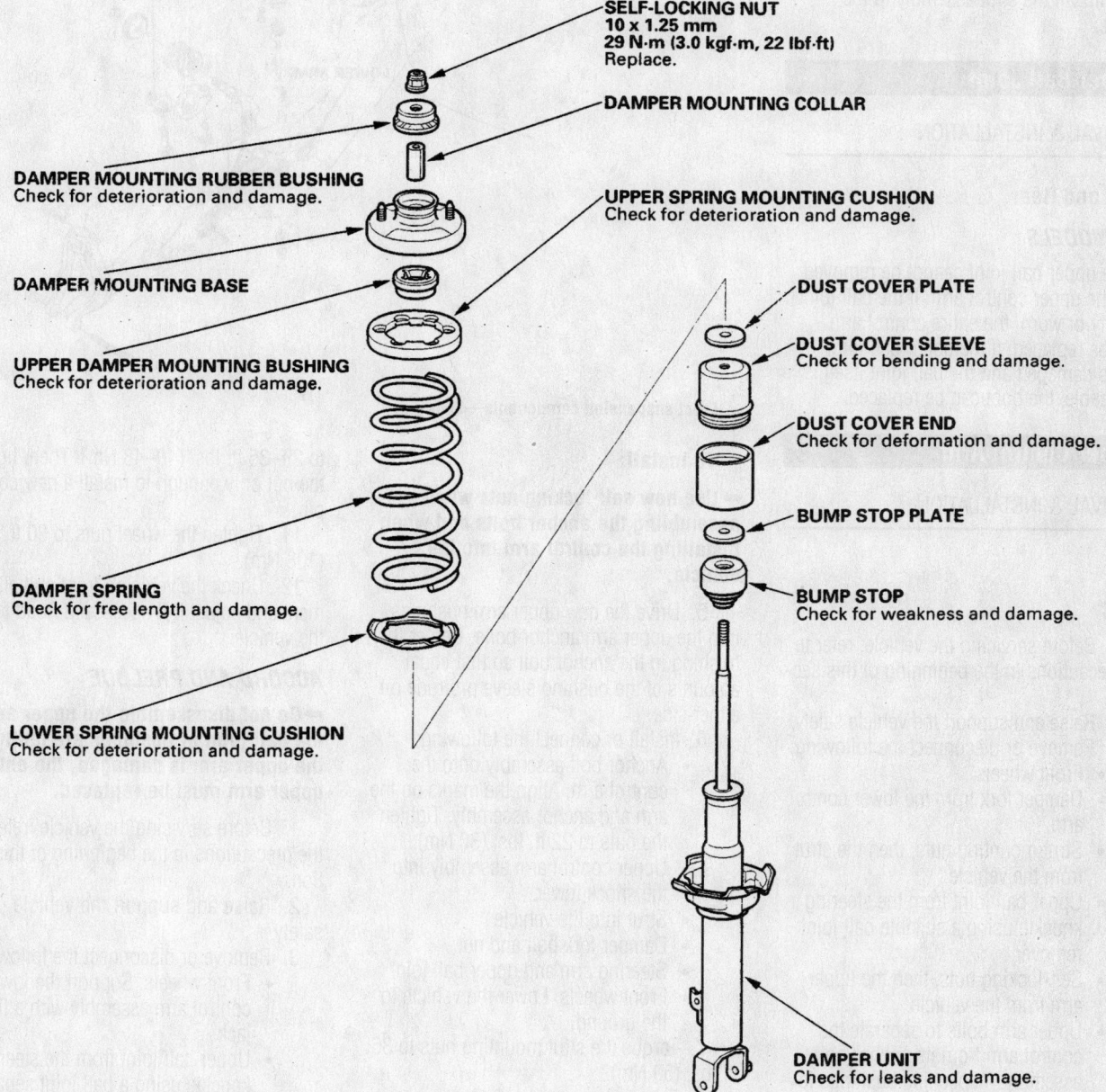

SELF-LOCKING NUT
10 x 1.25 mm
29 N·m (3.0 kgf·m, 22 lbf·ft)
Replace.

DAMPER MOUNTING COLLAR

DAMPER MOUNTING RUBBER BUSHING
Check for deterioration and damage.

UPPER SPRING MOUNTING CUSHION
Check for deterioration and damage.

DAMPER MOUNTING BASE

UPPER DAMPER MOUNTING BUSHING
Check for deterioration and damage.

DAMPER SPRING
Check for free length and damage.

LOWER SPRING MOUNTING CUSHION
Check for deterioration and damage.

DUST COVER PLATE

DUST COVER SLEEVE
Check for bending and damage.

DUST COVER END
Check for deformation and damage.

BUMP STOP PLATE

BUMP STOP
Check for weakness and damage.

DAMPER UNIT
Check for leaks and damage.

9347FG24

Exploded view of the strut and spring assembly—S2000—front shown

2. Remove the strut from the vehicle.

3. Compress the coil spring using a suitable spring compressor until the spring comes away from the seat.

4. Remove the center nut and slowly release the spring compressor.

To install:

5. Compress the spring and install it on the strut.

6. Install the lower washer and mounting bracket.

7. Install the upper washer and a new nut. Tighten the nut to 22 ft. lbs. (29 Nm).

8. Install the strut assembly in the vehicle.

Upper Ball Joint

REMOVAL & INSTALLATION

Front and Rear

ALL MODELS

The upper ball joint cannot be removed from the upper control arm. If the ball joint is faulty or worn, the entire control arm must be replaced. If the upper ball joint boot is damaged and the ball joint itself is still usable, the boot can be replaced.

Upper Control Arm

REMOVAL & INSTALLATION

Front

CIVIC

1. Before servicing the vehicle, refer to the precautions in the beginning of this section.

2. Raise and support the vehicle safely.

3. Remove or disconnect the following:
   - Front wheels
   - Damper fork from the lower control arm
   - Strut mounting nuts, then the strut from the vehicle
   - Upper ball joint from the steering knuckle using a suitable ball joint remover.
   - Self-locking nuts, then the upper arm from the vehicle.
   - Upper arm bolts to separate the control arm from its anchor bolt assembly. Inspect the bushings for signs of deterioration and replace them if they are damaged.

4. Place the upper control arm anchor bolt assembly into a vice and drive out the upper arm bushings.

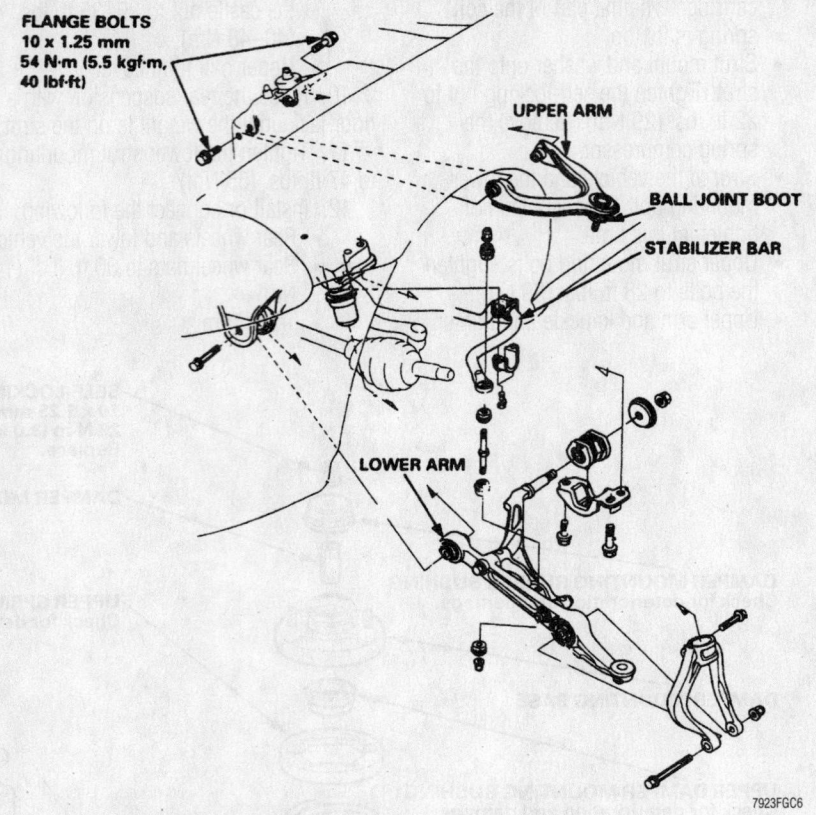

FLANGE BOLTS
10 x 1.25 mm
54 N·m (5.5 kgf·m, 40 lbf·ft)

UPPER ARM

BALL JOINT BOOT

STABILIZER BAR

LOWER ARM

7923FGC6

Front suspension components—Civic

To install:

➡ **Use new self-locking nuts when assembling the anchor bolts and when installing the control arm into the vehicle.**

5. Drive the new upper arm bushings into the upper arm anchor bolts. Center the bushing in the anchor bolt so that equal amounts of the bushing sleeve protrude on either side.

6. Install or connect the following:
   - Anchor bolt assembly onto the control arm. Align the marks on the arm and anchor assembly. Tighten the nuts to 22 ft. lbs. (30 Nm).
   - Upper control arm assembly into the shock tower.
   - Strut into the vehicle
   - Damper fork bolt and nut
   - Steering arm and upper ball joint
   - Front wheels. Lower the vehicle to the ground.

7. Torque the strut mounting nuts to 36 ft. lbs. (50 Nm).

8. Torque the upper control arm mounting nuts to 47 ft. lbs. (65 Nm).

9. Torque the damper fork nut to 47 ft. lbs. (65 Nm).

10. Torque the upper ball joint castle nut to 29–35 ft. lbs. (40–48 Nm). Then, tighten the nut only enough to install a new cotter pin.

11. Tighten the wheel nuts to 80 ft. lbs. (108 Nm).

12. Check the vehicle's front end alignment and adjust it if necessary. Road test the vehicle.

ACCORD AND PRELUDE

➡ **Do not disassemble the upper arm. If the ball joint or bushings are faulty, or the upper arm is damaged, the entire upper arm must be replaced.**

1. Before servicing the vehicle, refer to the precautions in the beginning of this section.

2. Raise and support the vehicle safely.

3. Remove or disconnect the following:
   - Front wheels. Support the lower control arm assembly with a floor jack.
   - Upper ball joint from the steering knuckle using a ball joint separator tool.
   - Self-locking nuts from the upper arm anchor bolts.
   - Upper arm from the vehicle

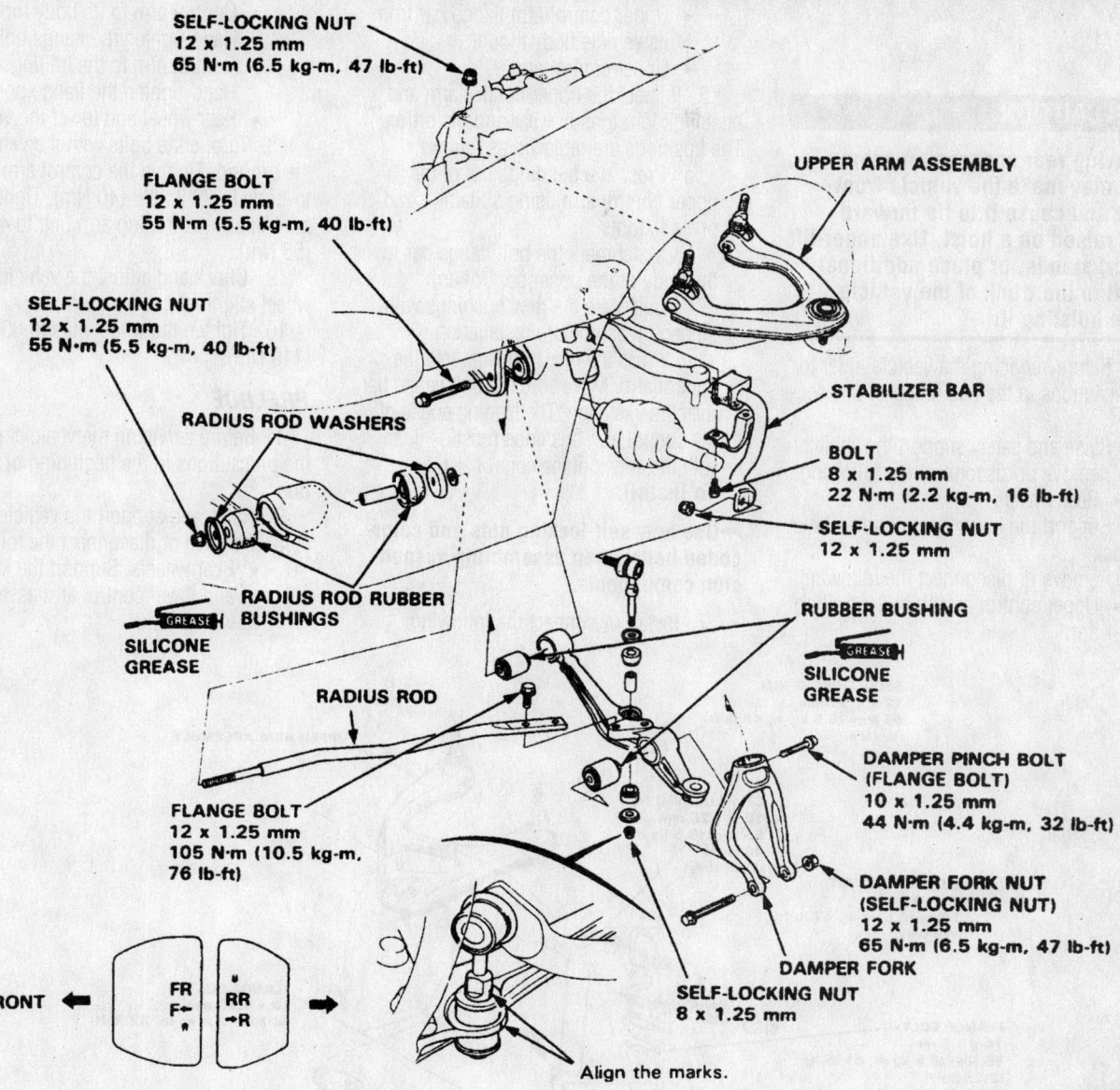

SELF-LOCKING NUT
12 x 1.25 mm
65 N·m (6.5 kg-m, 47 lb-ft)

FLANGE BOLT
12 x 1.25 mm
55 N·m (5.5 kg-m, 40 lb-ft)

SELF-LOCKING NUT
12 x 1.25 mm
55 N·m (5.5 kg-m, 40 lb-ft)

RADIUS ROD WASHERS

RADIUS ROD RUBBER BUSHINGS

SILICONE GREASE

RADIUS ROD

FLANGE BOLT
12 x 1.25 mm
105 N·m (10.5 kg-m, 76 lb-ft)

UPPER ARM ASSEMBLY

STABILIZER BAR

BOLT
8 x 1.25 mm
22 N·m (2.2 kg-m, 16 lb-ft)

SELF-LOCKING NUT
12 x 1.25 mm

RUBBER BUSHING

SILICONE GREASE

DAMPER PINCH BOLT (FLANGE BOLT)
10 x 1.25 mm
44 N·m (4.4 kg-m, 32 lb-ft)

DAMPER FORK NUT (SELF-LOCKING NUT)
12 x 1.25 mm
65 N·m (6.5 kg-m, 47 lb-ft)

DAMPER FORK

SELF-LOCKING NUT
8 x 1.25 mm

FRONT

FR F← ←RR →R

Align the marks.

7923FGC7

Front suspension components—Prelude and Accord

➡️**Do not disassemble the upper arm. If the ball joint or bushings are faulty, or the upper arm is damaged, the entire upper arm must be replaced.**

To install:

➡️**Use new self-locking nuts when installing the upper arm and strut.**

4. Install or connect the following:
  • Upper control arm assembly into the strut tower.
  • Upper ball joint
  • Front wheels and lower the vehicle.
5. With all 4 of the vehicle's wheels on the ground, torque the upper control arm

nuts to 47 ft. lbs. (65 Nm). Torque the castle nut to 32 ft. lbs. (44 Nm); then, only tighten it only enough to install a new cotter pin.

6. Tighten the wheel nuts to 80 ft. lbs. (110 Nm).

7. Check and adjust the vehicle's front end alignment. On Preludes equipped with 4WS, the electronic neutral check must be performed before all 4 wheels are aligned.

S2000

1. Before servicing the vehicle, refer to the precautions in the beginning of this section.

2. Remove or disconnect the following:

  • Front wheel
  • Wheel speed sensor harness
  • Upper ball joint
  • Inner flange bolts and the upper control arm

To install:

3. Installation is the reverse of the removal procedure, while using the following torque values:

  • Inner flange bolts: 76 ft. lbs. (103 Nm)
  • Upper ball joint nut: 36–43 ft. lbs. (49–59 Nm)
  • Wheel speed sensor harness bolts: 88 inch lbs. (10 Nm)

Rear

CIVIC

✳✳ CAUTION

Removing rear suspension components may make the vehicle front-heavy and cause it to tip forward when raised on a hoist. Use under-lift support stands, or place additional weight in the trunk of the vehicle before hoisting it.

1. Before servicing the vehicle, refer to the precautions in the beginning of this section.

2. Raise and safely support the vehicle.

3. Remove or disconnect the following:
   • Rear wheels

4. Support the lower control arm with a floor jack.

5. Remove or disconnect the following:
   • Upper control arm from the trailing arm

   • Upper control arm flange bar from its vehicle body mount
   • Upper control arm

6. Inspect the upper control arm and bushings for signs of wear and distortion. The bushings are replaced as follows:

   a. Press the bushings out of the upper control arm using suitably sized press fixtures.

   b. Matchmark the bolt flange bar to the body of the upper control arm.

   c. Lubricate the new bushings with silicon grease before installation.

   d. Press the new bushings into the control arm. Make sure the bolt flange bar matchmarks align. The leading edges of the control arm bushings must be flush with the edges of the control arm body.

To install:

➡ **Use new self-locking nuts and color-coded bolts when assembling suspension components.**

7. Install or connect the following:

   • Control arm to its body mount. Hand-tighten the flange bolts.
   • Control arm to the trailing arm. Hand-tighten the flange bolt.
   • Rear wheel and lower the vehicle.

8. Torque the bolts with the vehicle on the ground. Tighten the control arm bolts-to-body to 29 ft. lbs. (40 Nm). Tighten the control arm-to-trailing arm bolt to 40 ft. lbs. (55 Nm).

9. Check and adjust the vehicle's rear wheel alignment.

10. Tighten the wheel nuts to 80 ft. lbs. (110 Nm).

PRELUDE

1. Before servicing the vehicle, refer to the precautions in the beginning of this section.

2. Raise and support the vehicle safely.

3. Remove or disconnect the following:
   • Rear wheels. Support the knuckle and lower control arm assembly with a jack.

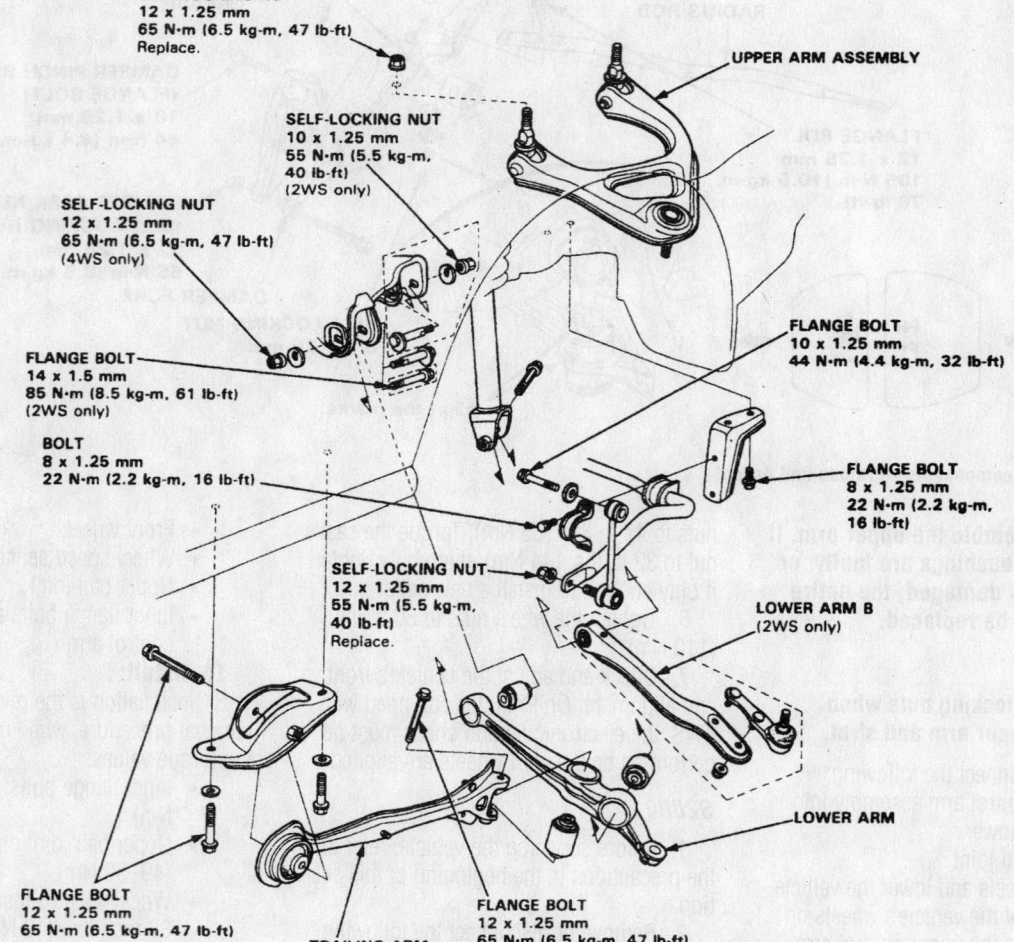

SELF-LOCKING NUT
12 x 1.25 mm
65 N·m (6.5 kg-m, 47 lb-ft)
Replace.

SELF-LOCKING NUT
10 x 1.25 mm
55 N·m (5.5 kg-m, 40 lb-ft)
(2WS only)

SELF-LOCKING NUT
12 x 1.25 mm
65 N·m (6.5 kg-m, 47 lb-ft)
(4WS only)

FLANGE BOLT
14 x 1.5 mm
85 N·m (8.5 kg-m, 61 lb-ft)
(2WS only)

BOLT
8 x 1.25 mm
22 N·m (2.2 kg-m, 16 lb-ft)

SELF-LOCKING NUT
12 x 1.25 mm
55 N·m (5.5 kg-m, 40 lb-ft)
Replace.

FLANGE BOLT
12 x 1.25 mm
65 N·m (6.5 kg-m, 47 lb-ft)

TRAILING ARM

FLANGE BOLT
12 x 1.25 mm
65 N·m (6.5 kg-m, 47 lb-ft)

UPPER ARM ASSEMBLY

FLANGE BOLT
10 x 1.25 mm
44 N·m (4.4 kg-m, 32 lb-ft)

FLANGE BOLT
8 x 1.25 mm
22 N·m (2.2 kg-m, 16 lb-ft)

LOWER ARM B
(2WS only)

LOWER ARM

Rear suspension components—Prelude

7923FGC8

- Upper ball joint from the knuckle using a ball joint separator tool.
- Trunk side trim
- 2 strut mounting nuts
- Self-locking nuts from the upper arm anchor bolts
- Upper arm from the vehicle

➡**Do not disassemble the upper arm. If the ball joint or bushings are faulty, or the upper arm is damaged, the entire upper arm must be replaced.**

To install:

➡**Use new self-locking nuts when installing the upper arm and strut.**

4. Install or connect the following:
- Upper control arm assembly into the strut tower.
- Upper ball joint
- Rear wheels and lower the vehicle.

5. With all 4 of the vehicle's wheels on the ground, torque the upper control arm nuts to 47 ft. lbs. (65 Nm). Torque the castle nut to 32 ft. lbs. (44 Nm); then, only tighten it only enough to install a new cotter pin.

6. Tighten the wheel nuts to 80 ft. lbs. (110 Nm).

7. Put the trunk side trim back into position.

8. Check and adjust the vehicle's rear end wheel alignment. On Preludes equipped with 4WS, the electronic neutral check must be performed before all 4 wheels are aligned.

ACCORD

1. Raise and safely support the vehicle.
2. Remove or disconnect the following:
- Rear wheels

3. Support the knuckle and lower control arm with a floor jack to compress the strut.

4. Remove or disconnect the following:
- Castle nut cap, cotter pin, and castle nut from the upper ball joint. Use a ball joint separator tool to separate the ball joint from the knuckle.
- Upper control arm

5. Check upper control arm and bushing for signs of wear and damage. Replace the upper control arm if the ball joint is faulty.

To install:

➡**Use new self-locking nuts when assembling suspension components.**

6. Install or connect the following:
- Upper arm into the vehicle
- Mounting bolts and only hand-tighten them.

- Upper arm to the knuckle.
- Castle nut at the ball joint to 32 ft. lbs. (44 Nm). Tighten the castle nut only enough to install a new cotter pin.
- Castle nut cap
- Rear wheels and lower the vehicle.

7. Tighten the upper mounting bolts to 28 ft. lbs. (39 Nm).

8. Tighten the wheel nuts to 80 ft. lbs. (110 Nm).

9. Check and adjust the vehicle's rear wheel alignment.

S2000

1. Before servicing the vehicle, refer to the precautions in the beginning of this section.

2. Remove or disconnect the following:
- Rear wheel
- Wheel speed sensor harness
- Upper ball joint
- Inner flange bolts and the upper control arm

To install:

3. Installation is the reverse of the removal procedure, while using the following torque values:
- Inner flange bolts: 98 ft. lbs. (132 Nm)
- Upper ball joint nut: 36–43 ft. lbs. (49–59 Nm)
- Wheel speed sensor harness bolts: 88 inch lbs. (10 Nm)

Lower Ball Joint

REMOVAL & INSTALLATION

Civic

➡**The steering knuckle must be removed from the vehicle for the ball joint to be replaced. The following special tools or their equivalents are needed to press the ball joint in and out of the knuckle: ball joint installer base tool 07965-SB00200, ball joint installer/remover tool 07965-SB00100, and ball joint remover base tool 07965-SH20200. A large vise will be required to hold the knuckle and the press tools. A ball joint clip guide tool 07974-SA50700 or 07GAG-SD40700 is used to install the retaining clip on the joint boot.**

1. Before servicing the vehicle, refer to the precautions in the beginning of this section.

2. Remove or disconnect the following:

- Steering knuckle assembly from the vehicle.
- Ball joint boot snapring and the boot.
- Snapring out of the groove in the ball joint body.

3. Install the ball joint removal tool onto the ball joint with the large end facing out. Install the ball joint nut to attach the tool to the joint.

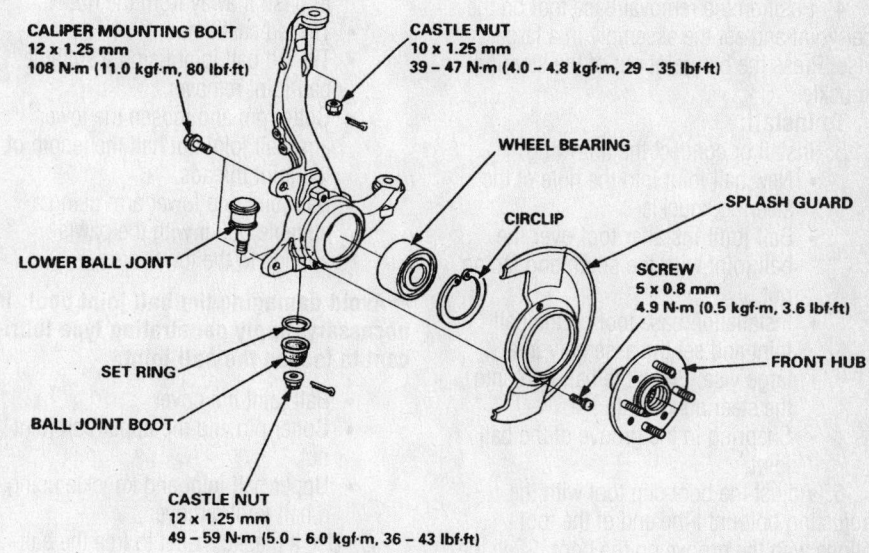

CALIPER MOUNTING BOLT
12 x 1.25 mm
108 N·m (11.0 kgf·m, 80 lbf·ft)

CASTLE NUT
10 x 1.25 mm
39 – 47 N·m (4.0 – 4.8 kgf·m, 29 – 35 lbf·ft)

WHEEL BEARING

CIRCLIP

SPLASH GUARD

LOWER BALL JOINT

SCREW
5 x 0.8 mm
4.9 N·m (0.5 kgf·m, 3.6 lbf·ft)

FRONT HUB

SET RING

BALL JOINT BOOT

CASTLE NUT
12 x 1.25 mm
49 – 59 N·m (5.0 – 6.0 kgf·m, 36 – 43 lbf·ft)

Knuckle components—Civic

7923FGB4

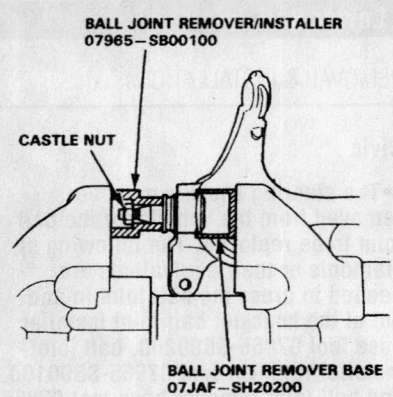

Ball joint removal tools—Civic

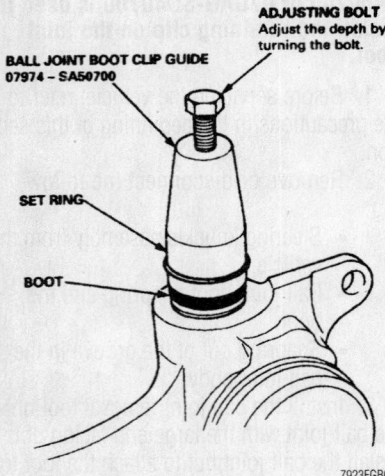

Ball joint boot clip guide—Civic

4. Position the removal base tool on the ball joint and set the assembly in a large vise. Press the ball joint out of the steering knuckle.

To install:

5. Install or connect the following:
  • New ball joint into the hole of the steering knuckle.
  • Ball joint installer tool over the ball joint with the small end facing out.
  • Installation base tool on the ball joint and set the assembly in a large vise. Press the ball joint into the steering knuckle.
  • Snapring in the groove of the ball joint.

6. Adjust the boot clip tool with the adjusting bolt until the end of the tool aligns with the groove on the boot. Slide the clip over the tool and into position.

7. Install the ball joint stud in the steering knuckle. Tighten the nut to 44 ft. lbs. (60 Nm).

Wheel Bearings

ADJUSTMENT

All Models

The wheel bearings are not adjustable or repairable and should be replaced if found defective.

REMOVAL & INSTALLATION

Front

CIVIC

➡ **A hydraulic press and several bearing drivers and attachments are needed to remove and install the hub and bearing.**

1. Before servicing the vehicle, refer to the precautions in the beginning of this section.

2. Pry the spindle nut stake away from the spindle, then loosen the nut.

3. Raise and safely support the vehicle.

4. Remove or disconnect the following:
  • Front wheel and the spindle nut
  • Wheel sensor wire bracket from the knuckle, but don't disconnect it.
  • Caliper mounting bolts and the caliper. Support the caliper out of the way with a length of wire. Do not let the caliper hang from the brake hose.
  • 6mm brake disc retaining screws. Screw 2, 12mm bolts into the disc to push it away from the hub.
  • Tie rod castle nut
  • Tie rod ball joint using a suitable ball joint remover.
  • Cotter pin and loosen the lower arm ball joint nut half the length of the joint threads.
  • Ball joint and lower arm using a suitable puller with the pawls applied to the lower arm.

➡ **Avoid damaging the ball joint boot. If necessary, apply penetrating type lubricant to loosen the ball joint.**

  • Ball joint nut cover
  • Cotter pin and the upper ball joint nut.
  • Upper ball joint and knuckle using a ball joint remover.

5. Use a plastic mallet to free the halfshaft from the knuckle. Pull the knuckle out to remove it.

➡ **A new wheel bearing must be used when the hub is removed.**

6. Place the knuckle in a press and use a base and pilot to press the hub assembly out of the wheel bearing.

7. Remove the knuckle ring seal and circlip. Remove the splash guard from the knuckle.

8. Press the wheel bearing out of the knuckle using a driving attachment.

To install:

9. Clean the knuckle and hub assembly and inspect them for damage.

10. Install or connect the following:
  • New wheel bearing into the hub using a driving tool.
  • Circlip in the outer groove of the knuckle.
  • Splash guard
  • Hub assembly into the steering knuckle using a base and a driving and guide tool.
  • Knuckle ring seal
  • Knuckle onto the spindle
  • Knuckle onto the upper and lower ball joints and tighten the castle nuts.
  • Tie rod ball joint onto the steering knuckle.

11. Tighten the upper ball joint nut and tie rod nut to 29–35 ft. lbs. (40–48 Nm) and the lower ball joint castle nut to 36–43 ft. lbs. (50–60 Nm).

12. Install or connect the following:
  • Anti-lock Brake System (ABS) wheel sensor wire brackets onto the knuckle. Tighten the mounting bolts to 84 inch lbs. (10 Nm).
  • Brake disc; use 2 lug nuts to evenly draw the disc onto the hub.
  • Retainer screws: 84 inch lbs. (10 Nm)
  • Spindle washer and nut. Don't tighten the nut until the vehicle is on the ground.
  • Brake caliper and tighten the bolts to 80 ft. lbs. (110 Nm).
  • Front wheels and lower the vehicle.

13. Tighten the spindle nut to 134 ft. lbs. (185 Nm), stake the nut, and install the grease cap.

14. Check and adjust the vehicle's front wheel alignment.

➡ **Avoid damaging the ball joint boot. If necessary, apply penetrating-type lubricant to loosen the ball joint.**

PRELUDE AND ACCORD

➡ **Once the hub has been removed, the wheel bearings must be replaced. A hydraulic press and bearing drivers must be used to remove and install the bearing.**

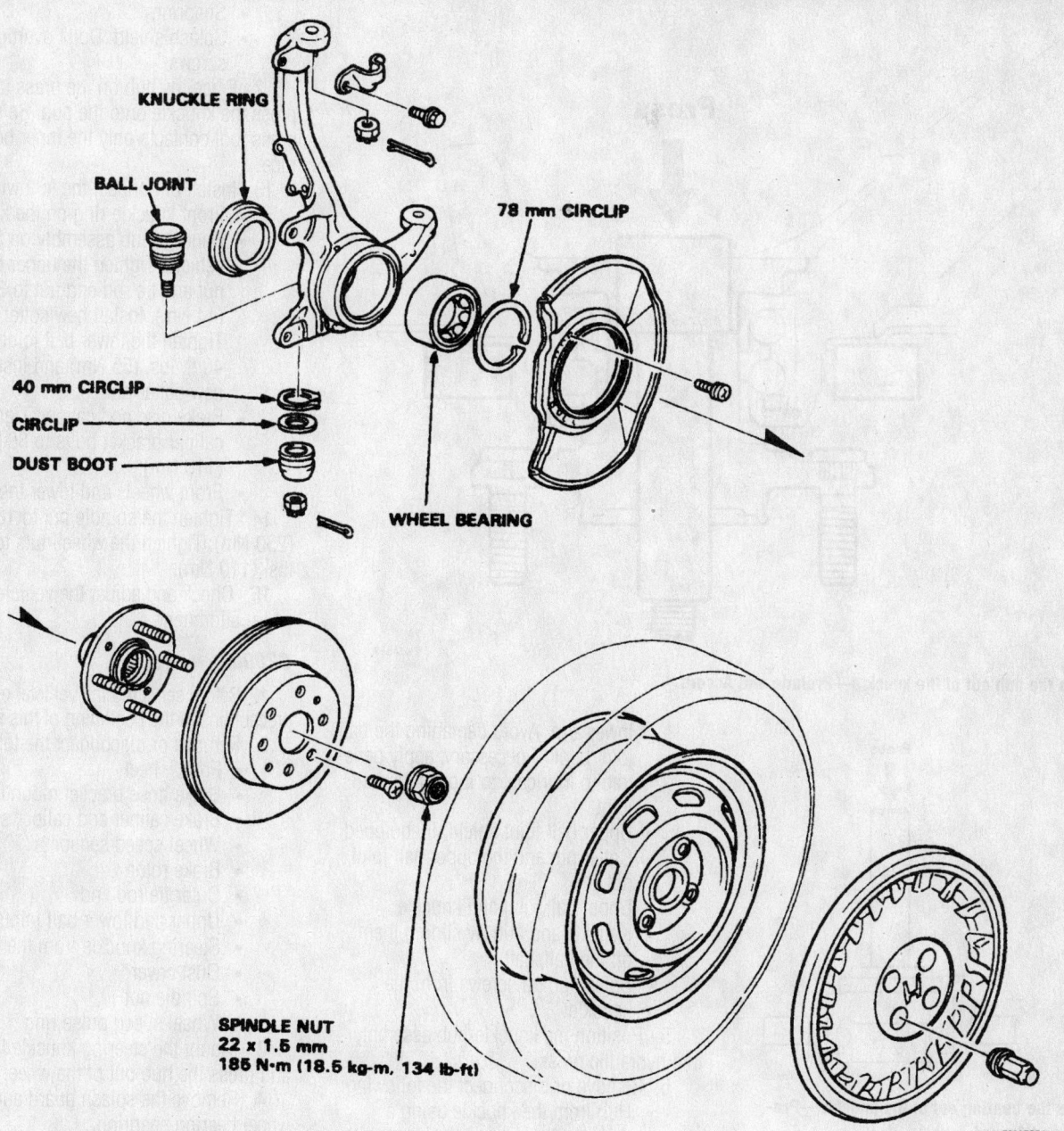

KNUCKLE RING

BALL JOINT

78 mm CIRCLIP

40 mm CIRCLIP

CIRCLIP

DUST BOOT

WHEEL BEARING

SPINDLE NUT
22 x 1.5 mm
185 N·m (18.5 kg-m, 134 lb-ft)

7923FGB8

Hub and steering knuckle components—Prelude and Accord

1. Before servicing the vehicle, refer to the precautions in the beginning of this section.

2. Pry the spindle nut stake away from the spindle and loosen the nut. Do not tighten or loosen a spindle nut unless the vehicle is sitting on all 4 wheels. The torque required is high enough to cause the vehicle to fall off the stands even when properly supported.

3. Raise and safely support the vehicle.

4. Remove or disconnect the following:

- Wheel and the spindle nut
- Caliper mounting bolts and the caliper. Support the caliper out of the way with a length of wire. Do not let the caliper hang from the brake hose.
- 6mm brake disc retaining screws. Screw 2, 8 x 1.25mm bolts into the disc to push it away from the hub.

➡**Turn each bolt 2 turns at a time to prevent cocking the brake disc.**

- Cotter pin from the tie rod castle nut, then the nut.
- Tie rod ball joint using a ball joint remover, then lift the tie rod out of the knuckle.
- Cotter pin, then loosen the lower arm ball joint nut half the length of the joint threads. The nut will retain the arm when the joint comes loose.
- Ball joint and lower arm using a puller with the pawls applied to the

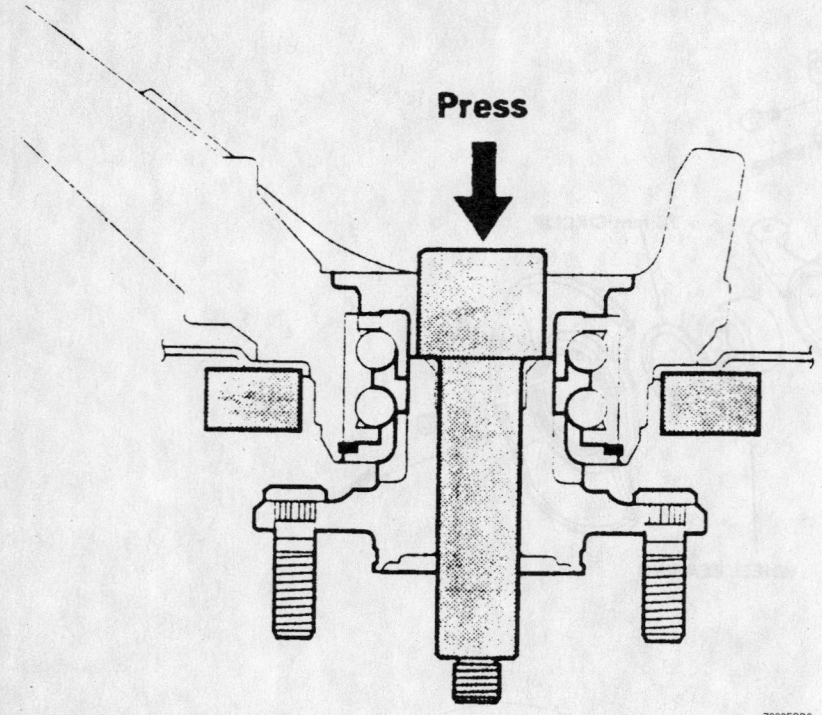

Press the hub out of the knuckle—Prelude and Accord

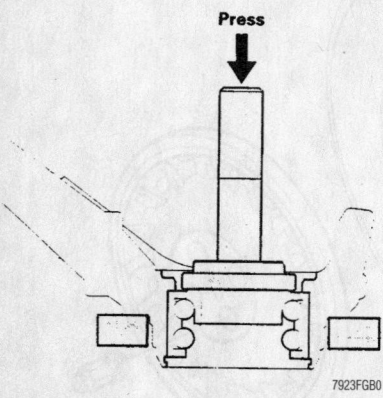

Press the bearing out of the knuckle—Prelude and Accord

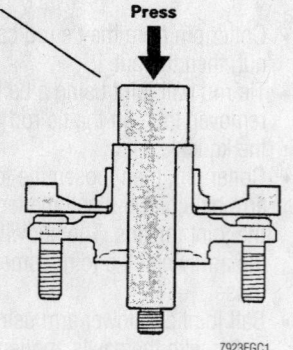

Use a press to remove the inner bearing race from the hub—Prelude and Accord

lower arm. Avoid damaging the ball joint boot. If necessary, apply penetrating lubricant to loosen the ball joint.
- Upper ball joint shield, if equipped.
- Cotter pin and the upper ball joint nut.
- Upper ball joint and knuckle
- Knuckle and hub by sliding them off the halfshaft.
- Splash guard screws from the knuckle.

5. Position the knuckle/hub assembly in a hydraulic press.

6. Remove or disconnect the following:
- Hub from the knuckle using a driver of the proper diameter while supporting the knuckle. The inner bearing race may stay on the hub.
- Splash guard and snapring from the knuckle.

7. Press the wheel bearing out of the knuckle while supporting the knuckle.

8. If necessary, remove the outboard bearing inner race from the hub using a bearing puller.

To install:

9. Clean the knuckle and hub thoroughly.

10. Press a new wheel bearing into the knuckle. Be sure the press tool contacts only the outer bearing race and properly support the knuckle so it is stable.

11. Install or connect the following:

- Snapring
- Splash shield. Don't overtighten the screws.

12. Place the hub on the press table and press the knuckle onto the hub. Be sure the press tool contacts only the inner bearing race.

13. Install or connect the following:
- Front knuckle ring on the knuckle
- Knuckle/hub assembly on the vehicle. Tighten the upper ball joint nut and tie rod end nut to 32 ft. lbs. (44 Nm). Install new cotter pins. Tighten the lower ball joint nut to 40 ft. lbs. (55 Nm) and install a new cotter pin.
- Brake disc and caliper. Tighten the caliper bracket bolts to 80 ft. lbs. (110 Nm).
- Front wheels and lower the vehicle.

14. Tighten the spindle nut to 180 ft. lbs. (250 Nm). Tighten the wheel nuts to 80 ft. lbs. (110 Nm).

15. Check and adjust the vehicle's front wheel alignment.

S2000

1. Before servicing the vehicle, refer to the precautions in the beginning of this section.

2. Remove or disconnect the following:
- Front wheel
- Brake hose bracket mounting bolts
- Brake caliper and caliper support
- Wheel speed sensor
- Brake rotor
- Outer tie rod end
- Upper and lower ball joints
- Steering knuckle from the vehicle
- Dust cover
- Spindle nut
- Wheel speed pulse ring

3. Mount the steering knuckle in a press and press the hub out of the wheel bearing.

4. Remove the splash guard and the wheel bearing snapring.

5. Press the wheel bearing out of the steering knuckle.

To install:

6. Installation is the reverse of the removal procedure, while using the following torque values:
- Splash guard screws: 48 inch lbs. (5 Nm)
- Spindle nut: 242 ft. lbs. (329 Nm)
- Upper ball joint nut: 36–43 ft. lbs. (49–59 Nm)
- Lower ball joint nut: 43–51 ft. lbs. (56–69 Nm)
- Outer tie rod end nut: 40 ft. lbs. (54 Nm)
- Brake caliper support bolts: 83 ft. lbs. (113 Nm)

Rear

CIVIC

1. Before servicing the vehicle, refer to the precautions in the beginning of this section.
2. Remove or disconnect the following:
   - Hub dust cap and loosen the spindle nut.
3. Raise and safely support the vehicle.
4. Remove or disconnect the following:
   - Rear wheels.
   - 2 brake rotor or drum retaining screws
   - Brake drum, if equipped with drum brakes.
5. If equipped with disc brakes, remove or disconnect the following:
   - Caliper shield and brake hose bracket
   - Caliper bracket and hang the caliper out of the way with a piece of wire.
   - Brake rotor
6. Remove or disconnect the following:
   - Hub assembly from the spindle
7. Clean the hub assembly in solvent.

8. Inspect the hub assembly for any signs of wear or damage. If the wheel bearings are damaged, the hub assembly must be replaced.

To install:

9. Clean the spindle and the brake rotor/drum mounting surfaces.
10. Install or connect the following:
    - Hub assembly onto the spindle
    - Spindle washer
    - Brake rotor or brake drum. Apply anti-seize paste to the retaining screws and tighten them to 84 inch lbs. (10 Nm). Don't overtighten the retaining screws.
11. If equipped with disc brakes, install or connect the following:
    - Brake caliper and tighten the mounting bolts to 28 ft. lbs. (39 Nm).
    - Brake hose bracket onto its mount.
    - Caliper dust shield and tighten the bolts to 84 inch lbs. (10 Nm).
    - New spindle nut and wheel assembly.
12. Lower the vehicle.
13. Tighten the spindle nut to 134 ft. lbs. (185 Nm). Tighten the wheel nuts to 80 ft.

lbs. (110 Nm). Stake the spindle nut with a punch. If the dust cap was bent during removal, install a new one.

PRELUDE AND ACCORD

➡**The rear wheel bearing and hub unit are replaced as a unit.**

1. Before servicing the vehicle, refer to the precautions in the beginning of this section.
2. Loosen the spindle nut.
3. Raise the vehicle and support it safely.
4. Remove or disconnect the following:
   - Rear wheels
   - Brake disc retaining screws
   - Brake hose brackets from the knuckle
   - Caliper bracket mounting bolts and hang the caliper out of the way with a piece of wire.
   - Brake disc. If the disc is frozen on the hub, screw 2, 8 x 1.25mm bolts evenly into the disc to push it away from the hub.
   - Spindle nut and pull the hub unit off of the spindle.

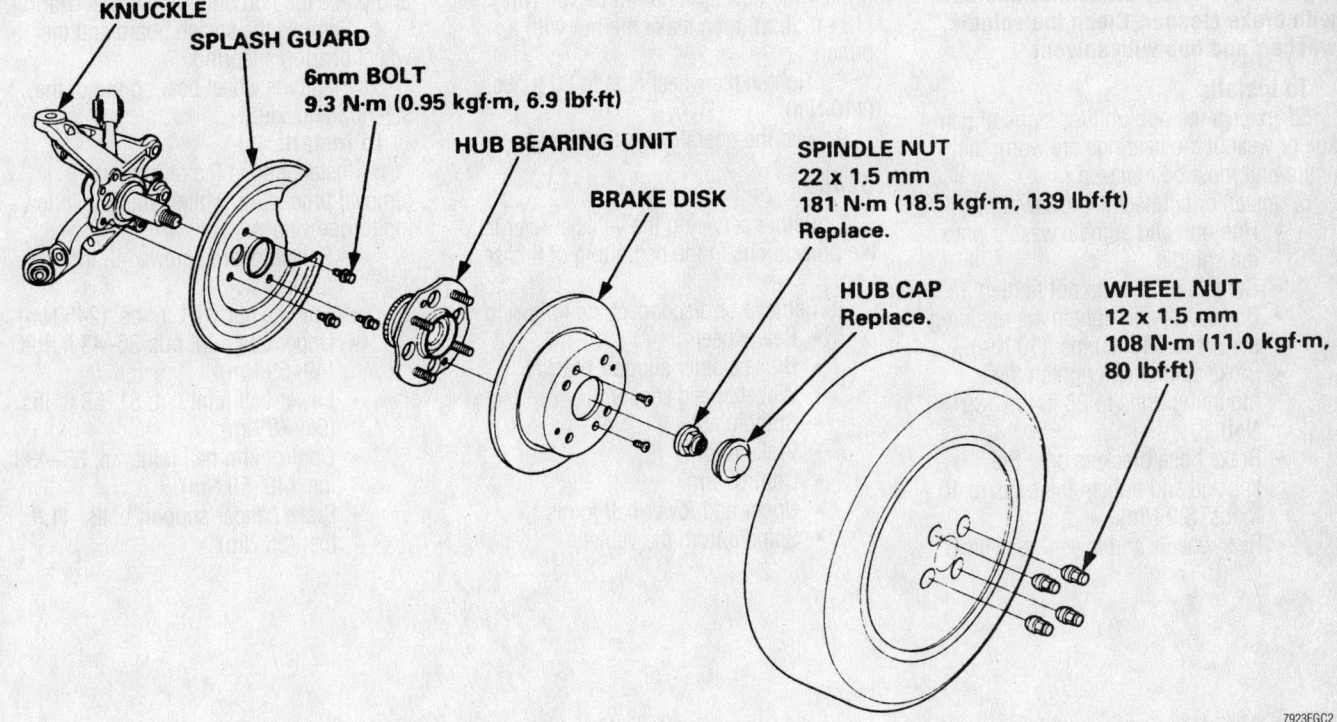

KNUCKLE
SPLASH GUARD
6mm BOLT
9.3 N·m (0.95 kgf·m, 6.9 lbf·ft)
HUB BEARING UNIT
BRAKE DISK
SPINDLE NUT
22 x 1.5 mm
181 N·m (18.5 kgf·m, 139 lbf·ft)
Replace.
HUB CAP
Replace.
WHEEL NUT
12 x 1.5 mm
108 N·m (11.0 kgf·m, 80 lbf·ft)

7923FGC2

Exploded view of the hub unit, drum brakes—Accord

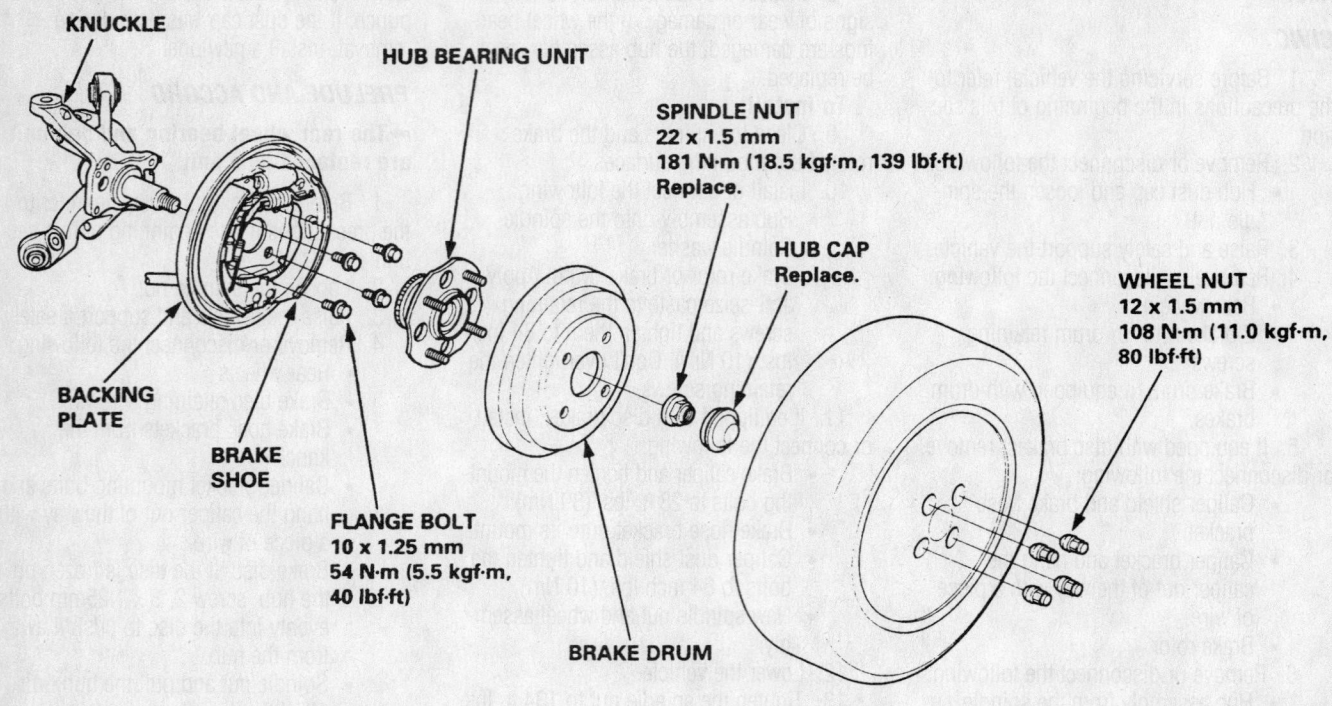

Hub unit, disc brakes—Accord and Prelude

➡ **Clean the backing plate and the mating surfaces of the brake disc and hub with brake cleaner. Clean the spindle, washer, and hub with solvent.**

To install:

5. Inspect the hub unit for signs of damage or wear. If the bearings are worn, the entire unit must be replaced.

6. Install or connect the following:
- Hub unit and spindle washer onto the spindle.
- Spindle nut but do not tighten it.
- Brake disc and tighten the retaining screws to 84 inch lbs. (10 Nm).
- Brake caliper and tighten the mounting bolts to 28 ft. lbs. (39 Nm).
- Brake hose brackets onto the knuckle and tighten the bolts to 16 ft. lbs. (22 Nm).
- Rear wheels and lower the vehicle.

7. With the vehicle on the ground, tighten the new spindle nut to 185 Nm (134 ft. lbs.), then stake the nut with a punch.

8. Tighten the wheel nuts to 80 ft. lbs. (110 Nm).

9. Test the operation of the brakes.

S2000

1. Before servicing the vehicle, refer to the precautions in the beginning of this section.

2. Remove or disconnect the following:
- Rear wheel
- Brake caliper support bracket
- Wheel speed sensor
- Spindle nut
- Brake rotor
- Control arm
- Upper and lower ball joints
- Spindle from the vehicle

3. Mount the steering knuckle in a press and press the hub out of the wheel bearing.

4. Remove the splash guard and the wheel bearing snapring.

5. Press the wheel bearing out of the steering knuckle.

To install:

6. Installation is the reverse of the removal procedure, while using the following torque values:
- Splash guard screws: 48 inch lbs. (5 Nm)
- Spindle nut: 181 ft. lbs. (245 Nm)
- Upper ball joint nut: 36–43 ft. lbs. (49–59 Nm)
- Lower ball joint nut: 51–58 ft. lbs. (68–78 Nm)
- Control arm ball joint nut: 36–43 ft. lbs. (49–59 Nm)
- Brake caliper support bolts: 41 ft. lbs. (55 Nm)

HONDA

2000–01
Insight

10

PRECAUTIONS

Before servicing any vehicle, please be sure to read all of the following precautions, which deal with personal safety, prevention of component damage, and important points to take into consideration when servicing a motor vehicle:

• Never open, service or drain the radiator or cooling system when the engine is hot; serious burns can occur from the steam and hot coolant.

• Observe all applicable safety precautions when working around fuel. Whenever servicing the fuel system, always work in a well-ventilated area. Do not allow fuel spray or vapors to come in contact with a spark, open flame, or excessive heat (a hot drop light, for example). Keep a dry chemical fire extinguisher near the work area. Always keep fuel in a container specifically designed for fuel storage; also, always properly seal fuel containers to avoid the possibility of fire or explosion. Refer to the additional fuel system precautions later in this section.

• Fuel injection systems often remain pressurized, even after the engine has been turned **OFF**. The fuel system pressure must be relieved before disconnecting any fuel lines. Failure to do so may result in fire and/or personal injury.

• Brake fluid often contains polyglycol ethers and polyglycols. Avoid contact with the eyes and wash your hands thoroughly after handling brake fluid. If you do get brake fluid in your eyes, flush your eyes with clean, running water for 15 minutes. If eye irritation persists, or if you have taken brake fluid internally, IMMEDIATELY seek medical assistance.

• The EPA warns that prolonged contact with used engine oil may cause a number of skin disorders, including cancer! You should make every effort to minimize your exposure to used engine oil. Protective gloves should be worn when changing oil. Wash your hands and any other exposed skin areas as soon as possible after exposure to used engine oil. Soap and water, or waterless hand cleaner should be used.

• All new vehicles are now equipped with an air bag system, often referred to as a Supplemental Restraint System (SRS) or Supplemental Inflatable Restraint (SIR) system. The system must be disabled before performing service on or around system components, steering column, instrument panel components, wiring and sensors. Failure to follow safety and disabling procedures could result in accidental air bag deployment, possible personal injury and unnecessary system repairs.

• Always wear safety goggles when working with, or around, the air bag system. When carrying a non-deployed air bag, be sure the bag and trim cover are pointed away from your body. When placing a non-deployed air bag on a work surface, always face the bag and trim cover upward, away from the surface. This will reduce the motion of the module if it is accidentally deployed. Refer to the additional air bag system precautions later in this section.

• Clean, high quality brake fluid from a sealed container is essential to the safe and proper operation of the brake system. You should always buy the correct type of brake fluid for your vehicle. If the brake fluid becomes contaminated, completely flush the system with new fluid. Never reuse any brake fluid. Any brake fluid that is removed from the system should be discarded. Also, do not allow any brake fluid to come in contact with a painted surface; it will damage the paint.

• Never operate the engine without the proper amount and type of engine oil; doing so WILL result in severe engine damage.

• Timing belt maintenance is extremely important! Many models utilize an interference-type, non-freewheeling engine. If the timing belt breaks, the valves in the cylinder head may strike the pistons, causing potentially serious (also time-consuming and expensive) engine damage. Refer to the maintenance interval charts in the front of this manual for the recommended replacement interval for the timing belt, and to the timing belt section for belt replacement and inspection.

• Disconnecting the negative battery cable on some vehicles may interfere with the functions of the on-board computer system(s) and may require the computer to undergo a relearning process once the negative battery cable is reconnected.

• When servicing drum brakes, only disassemble and assemble one side at a time, leaving the remaining side intact for reference.

• Only an MVAC-trained, EPA-certified automotive technician should service the air conditioning system or its components.

• The Insight utilizes an Integrated Motor Assisted (IMA) system that uses a high voltage circuit of 144 Volts. The IMA system must be properly turned off before servicing any components. Wear insulated gloves whenever the IMA system is inspected or serviced. Turn the battery switch to the OFF position with the locking cover in place and wait approximately 5 minutes before servicing the system. Before disconnecting the high voltage cable terminals check that the voltage between the terminals is 0 volts when measured with a voltmeter.

ENGINE REPAIR

Distributor

The Insight uses a Direct Ignition System (DIS).

Ignition Timing

ADJUSTMENT

Ignition timing is controlled by the Engine Control Module (ECM). If the ignition timing is not at 12 degrees Before Top Dead Center (BTDC) during idling in neutral inspect the ECM.

Engine Assembly

REMOVAL & INSTALLATION

1. Before servicing the vehicle, refer to the precautions in the beginning of this section.

2. Turn the battery switch **OFF** and measure the voltage before proceeding.

3. Disconnecting the negative battery cable will clear the radio's preset buttons. Write down the radio frequencies.

4. Properly relieve the fuel system pressure.

5. Drain the cooling system.

6. Drain the engine oil.

7. Drain the transmission fluid.

8. Remove or disconnect the following:
   • Battery and battery box
   • Engine cover
   • Breather pipe brake booster vacuum hose bracket from the air cleaner
   • Air cleaner/intake air duct assembly
   • Throttle cover and cable
   • Brake booster vacuum hose
   • Evaporative Emissions (EVAP) control canister hose

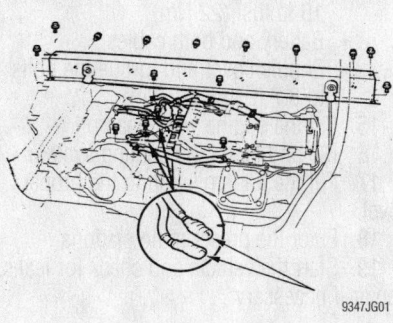

Remove the motor power cables

- Ground cable
- Starter cable and wire harness clamp
- Intelligent Power Unit (IPU) lid
- Motor power cables

✳✳ CAUTION

When servicing the Integrated Motor Assisted (IMA) system extreme caution must be observed. Severe personal injury may result.

- Engine Control Module (ECM) cover, electrical connectors and the main wire harness connectors
- Fuel feed and return hoses

9. Loosen the idler pulley center bolt and turn the adjusting bolt clockwise. Turn the adjusting bolt counterclockwise if not equipped with A/C

- A/C compressor/water pump belt
- Clutch slave cylinder/line assembly
- Shift cables
- Motor power cable clamps
- Both front wheels
- Splash shields and brackets
- Lower arm ball joints
- Driveshafts
- A/C compressor. Do not disconnect the A/C lines
- Nitris Oxide (NOX) adsorptive three way catalytic converter
- Left side under cover
- Grommet mounting bolts and pull the motor power cables out
- Motor power cable holder from the underbody
- Radiator and heater hoses and attach a chain hoist to the engine

- Rear engine mounting bolts
- Side mounting bolts
- Transmission mount bracket
- Lower the engine assembly away from the vehicle

To install:

10. Install the engine mount and torque the bolts to 25 ft. lbs. (33 Nm).

11. Install the idler pulley and torque the bolts to 16 ft. lbs. (22 Nm).

12. Position the engine to the vehicle.

13. Install or connect the following:

- Transmission mount bracket. Torque the bolts to 41 ft. lbs. (56 Nm). Loosen the upper transmission mount bolt.
- Side engine mount. Torque the bolts to 38 ft. lbs. (52 Nm). Loosen the side bolt
- Rear engine mount. Torque the bolts to 54 ft. lbs. (74 Nm). Loosen the upper bolt. Remove the engine hoist.

14. Torque the transmission side engine mount bolts to 40 ft. lbs. (54 Nm) and the rear mount bolt to 66 ft. lbs. (89 Nm).

- Motor power cable holder and torque the bolts to 7 ft. lbs. (9.3 Nm)
- Motor power cable through the opening and install the grommet. Torque the bolts to 7 ft. lbs. (9.3 Nm)
- Left side under cover and torque the bolts to 7 ft. lbs. (9.3 Nm)
- Catalytic converter with new gaskets. Torque the 10 x 1.25 mm bolt to 25 ft. lbs. (33 Nm) and the 8 x 1.25 mm bolt to 16 ft. lbs. (22 Nm).
- A/C compressor and torque the bolts to 16 ft. lbs. (22 Nm)
- Driveshafts with new spring clips
- Lower arm ball joints with new cotter pins
- Splash shield bracket and torque the bolts to 7 ft. lbs. (9.3 Nm)
- Splash shields and torque the bolts to 7 ft. lbs. (9.3 Nm)
- Motor power clamps
- Shift cables
- Clutch slave cylinder
- Water pump/A/C compressor belt and adjust as necessary
- Radiator and heater hoses
- ECM connectors through the bulkhead and install the grommets and harness clamps. Torque the bolts to 7 ft. lbs. (9.3 Nm)
- Main wire harness connectors,

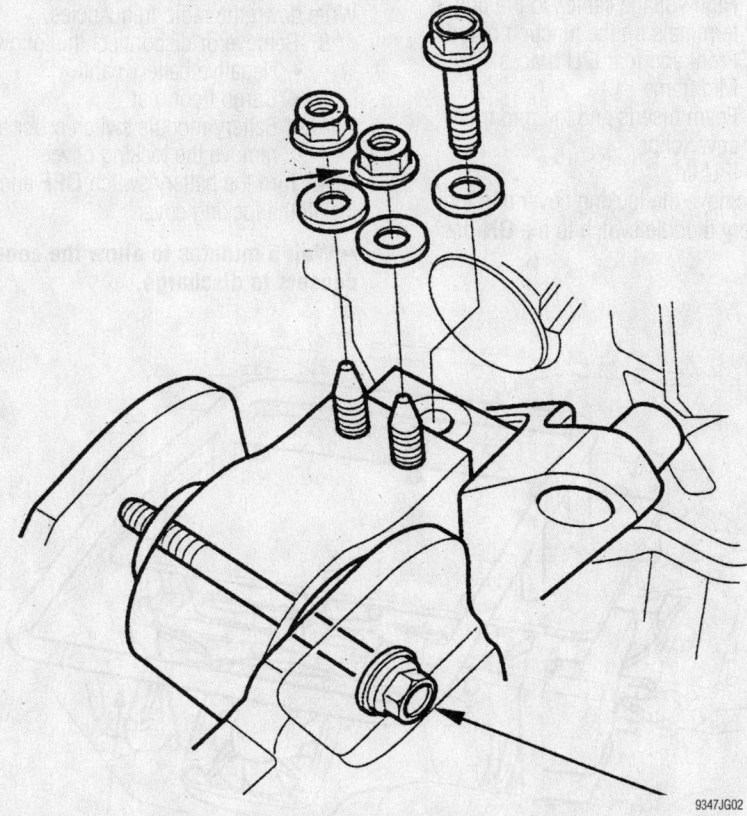

Install the transmission mounting bolts and loosen bolt C.

ECM connectors and the ECM cover. Torque the bolts to 8.7 ft. lbs. (12 Nm)
- Fuel feed and return hoses and torque the fasteners to 16 ft. lbs. (22 Nm)
- Motor power cables

✳✳ CAUTION

When servicing the IMA system extreme caution must be observed. Severe personal injury may result.

Battery Module

REMOVAL & INSTALLATION

1. Before servicing the vehicle, refer to the precautions in the beginning of this section.

2. Disconnecting the negative battery cable will clear the radio's preset buttons. Write down the radio frequencies.

3. Remove or disconnect the following:
- Negative battery cable
- Cargo floor mat
- Battery module switch cover and remove the locking cover

4. Turn the battery switch **OFF** and install the locking cover.

➡**Wait 5 minutes to allow the condensers to discharge.**

- Intelligent Power Unit (IPU) lid and measure the voltage at the junction board terminals. Voltage should be 0.1 v or less.
- Foam inserts and mid-frame
- Front and rear IPU braces from the junction board
- High voltage cables from the output terminals and wrap them in insulated tape

✳✳ CAUTION

Wear insulated gloves when working with the high voltage cables.

- Battery module air duct mounting bolt and move the module forward
- Connectors from the condenser terminal

5. Install the battery module lifting tool, 07YAK-001010A and install the 6 knurled bolts.

6. With the help of an assistant, lift the module out of the vehicle and place it on a flat surface.

- IPU lid
- Starter cable, ground cable and the harness clamp. Torque the 8 x 1.25 mm bolt to 7 ft. lbs. (9.3 Nm) and the 6 x 1.0 mm bolt to 9 ft. lbs. (12 Nm)
- EVAP control canister and the brake booster vacuum hose
- Throttle cable and adjust as necessary
- Air cleaner/intake air duct assembly and breather pipe and bracket and torque the bolts to 8.7 ft. lbs. (12 Nm)

ELECTRIC MOTOR REPAIR

To install:

7. Install or connect the following:
- Battery module to the vehicle and remove the special tools
- Condenser terminal connectors
- Battery module air duct and remove the tape from the high voltage cables

✳✳ CAUTION

Wear insulated gloves when working with the high voltage cables.

- High voltage cables to the output terminals on the junction board
- Front and rear IPU braces
- Mid frame
- Foam inserts and the mid frame cover clips
- IPU lid

8. Remove the locking cover and turn the battery module switch to the **ON** position.

- Battery box and torque the bolts to 16 ft. lbs. (22 Nm)
- Battery and both cables
- Engine cover and torque the bolts to 8.7 ft. lbs. (12 Nm)

15. Fill the engine with clean oil.

16. Fill and bleed the cooling system.

17. Fill the transmission to the proper level.

18. Enter the preset radio stations.

19. Start the vehicle and check for leaks, repair if necessary.

9. Install the cargo floor mat and the negative battery cable.

Power Control Unit

REMOVAL & INSTALLATION

1. Before servicing the vehicle, refer to the precautions in the beginning of this section.

2. Disconnecting the negative battery cable will clear the radio's preset buttons. Write down the radio frequencies.

3. Remove or disconnect the following:
- Negative battery cable
- Cargo floor mat
- Battery module switch cover and remove the locking cover

4. Turn the battery switch **OFF** and install the locking cover.

➡**Wait 5 minutes to allow the condensers to discharge.**

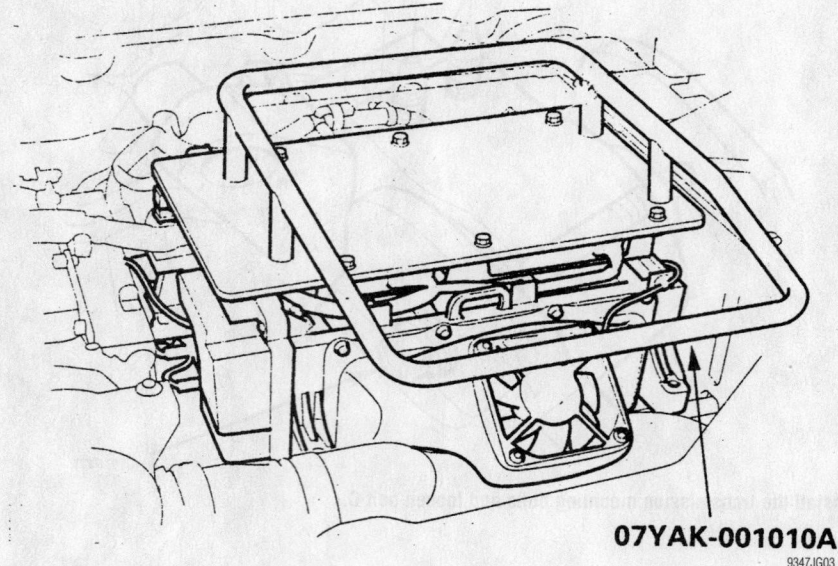

07YAK-001010A

9347JG03

Remove the battery module

- Intelligent Power Unit (IPU) lid and measure the voltage at the junction board terminals. Voltage should be 0.1 v or less
- Foam inserts and mid-frame
- Front and rear IPU braces from the junction board
- High voltage cables from the output terminals and wrap them in insulated tape

✲✲ CAUTION

Wear insulated gloves when working with the high voltage cables.

- Rear connector from the cooling fan assembly
- Harness clip from the fan shroud
- Fan bracket
- Fan duct and fan
- Raise the relay pack from the holder and disconnect the harness from the resistors
- Y condenser ground
- Power Control Unit (PCU) connector and the 12 v battery cables
- High voltage DC to DC converter 2P connector
- Move the intake duct away from the PCU
- PCU terminal cover
- PCU mounting bolts
- PCU

To install:

5. Install or connect the following:
- PCU to the vehicle and install the 4 mounting bolts
- PCU cables and terminal cover
- Intake duct to the PCU
- High voltage DC to DC converter 2P connectors
- 12 v battery cables and the PCU connecter
- Y condenser ground
- Relay pack to the holder and connect the harness to the resistors
- Fan and fan duct
- Fan bracket
- Cooling fan assembly connector

✲✲ CAUTION

Wear insulated gloves when working with the high voltage cables.

- High voltage cables to the output terminals on the junction board
- Front and rear IPU braces
- Mid frame

- Foam inserts and the mid frame cover clips
- IPU lid

6. Remove the locking cover and turn the battery module switch to the **ON** position.

7. Install the cargo floor mat and the negative battery cable.

Intelligent Motor Assist Motor

REMOVAL & INSTALLATION

1. Before servicing the vehicle, refer to the precautions in the beginning of this section.

2. Disconnecting the negative battery cable will clear the radio's preset buttons. Write down the radio frequencies.

✲✲ CAUTION

The motor rotor contains very powerful magnets and must be handled with care. Technicians with sensitive medical devices should not handle the motor rotor.

✲✲ WARNING

If the motor rotor is being installed by hand it is possible that the rotor will be pulled toward the stator with significant force causing personal injury. Follow the following precautions while working with the motor rotor:

3. Do not use the rotor motor if the fiberglass band is damaged. Magnets may come loose if the band breaks during installation.

4. Keep the motor rotor clear of any magnetically sensitive devices.

5. Store the rotor in a designated storage box and keep it away from magnetically sensitive devices.

6. Do not blow air near the rotor, causing metal particles to attach to the magnet.

7. Remove or disconnect the following:
- Negative battery cable
- Transmission assembly
- Clutch assembly
- 3 of the six attaching bolts and install guide pins

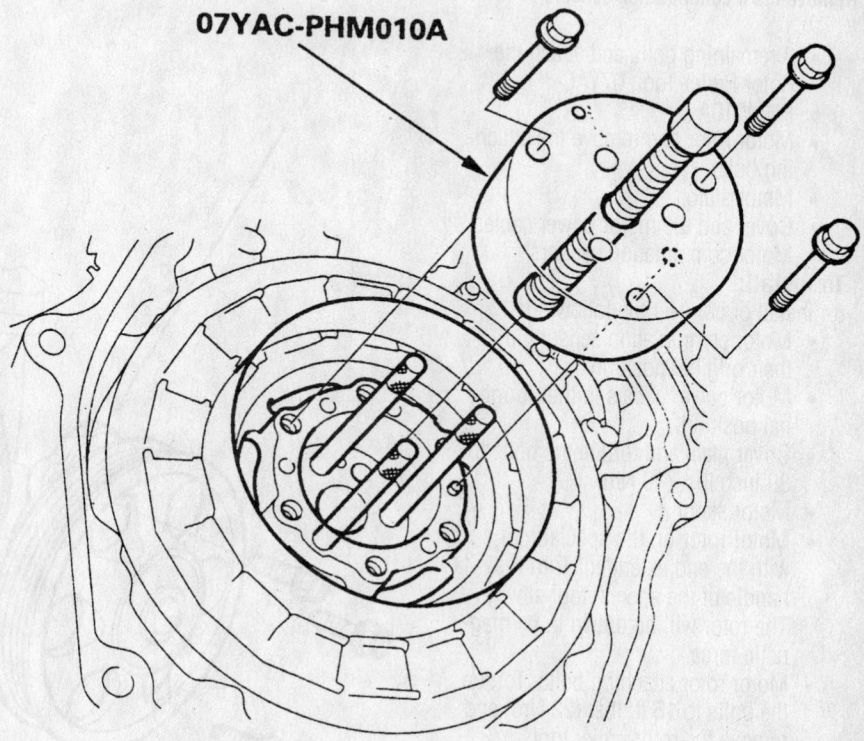

07YAC-PHM010A

9347JG04

Install the Rotor Puller Tool

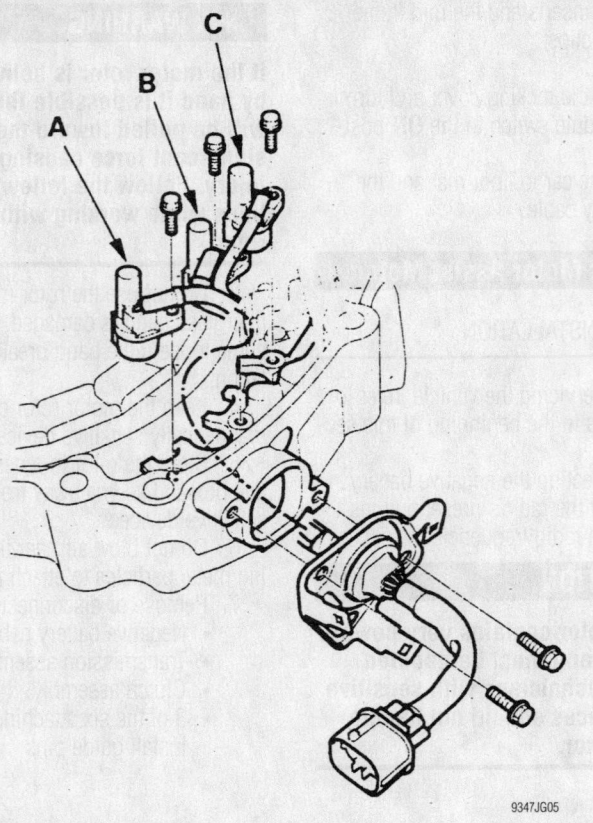

Remove the 3 commutation sensors

- 3 remaining bolts and attach the Rotor Puller Tool, 07YAC-PHM010A
- Motor rotor and remove the attaching bolts
- Motor stator
- Cover and the motor power cables
- Motor commutation sensors

To install:

8. Install or connect the following:
- Motor commutation sensors in their original position
- Motor power cables in their original position
- Cover plate and torque the bolts to 89 inch lbs. (10 Nm)
- Motor stator
- Motor rotor on the special tool with the end extended. Turn the handle of the special tool slowly. The rotor will be drawn in by magnetic force.
- Motor rotor attaching bolts. Torque the bolts to 16 ft. lbs. (22 Nm) and remove the rotor puller tool.
- Stator cover and torque the bolts to 9 ft. lbs. (12 Nm)
- Clutch assembly
- Transmission assembly
- Negative battery cable

Water Pump

REMOVAL & INSTALLATION

1. Before servicing the vehicle, refer to the precautions in the beginning of this section.
2. Drain the cooling system.
3. Remove or disconnect the following:
- Loosen the idler pulley center bolt and turn the adjusting bolt clockwise. Turn the adjusting bolt counterclockwise if not equipped with A/C.
- A/C compressor/water pump belt
- Water pump and discard the O-ring
4. Clean the mating surface where the water pump mounts to the thermostat housing.

To install:

5. Install or connect the following:
- Water pump with a new O-ring and torque the bolts to 8.7 ft. lbs. (12 Nm)
- A/C compressor/water pump belt and torque the idler pulley center bolt to 16 ft. lbs. (22 Nm)
6. Refill the cooling system.
7. Start the vehicle and check for leaks, repair if necessary.

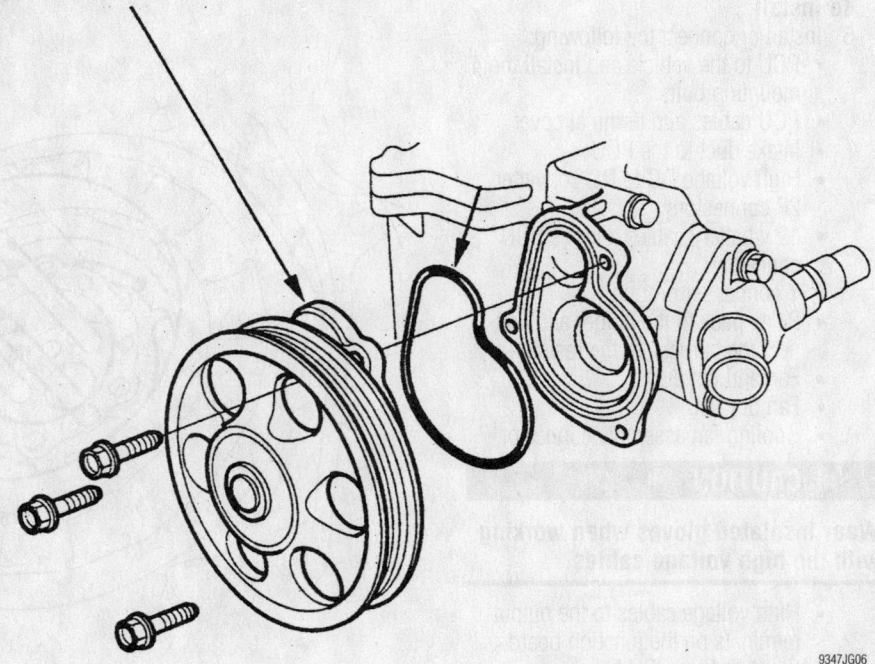

Exploded view of the water pump

Cylinder Head

REMOVAL & INSTALLATION

1. Before servicing the vehicle, refer to the precautions in the beginning of this section.

2. Turn the battery switch **OFF** and measure the voltage before proceeding.

3. Disconnecting the negative battery cable will clear the radio's preset buttons. Write down the radio frequencies.

4. Properly relieve the fuel system pressure.

5. Drain the cooling system.

6. Drain the engine oil.

7. Remove or disconnect the following:
- Both battery cables
- Breather pipe brake booster vacuum hose bracket from the air cleaner
- Air cleaner/intake air duct assembly
- Throttle cover and cable
- Brake booster vacuum hose
- Evaporative Emissions (EVAP) control canister hose
- Fuel feed and return hoses
- Positive Crankcase Ventilation (PCV) hose
- Heater hoses

8. Remove the engine wire harness connectors from the cylinder head and intake manifold for the following components:
- Fuel injectors
- Ignition coils
- Intake Air Temperature (IAT) sensor
- Idle Air Control (IAC) valve
- Throttle Position (TP) sensor
- Manifold Absolute Pressure (MAP) sensor
- Engine Coolant Temperature (ECT) sensor
- Primary Heated Oxygen Sensor (HO$_2$S)
- Secondary Heated Oxygen Sensor (HO$_2$S)
- VTEC solenoid valve
- VTEC pressure switch
- Top Dead Center (TDC1/TDC2) sensors

9. Remove or disconnect the following:
- Intake manifold
- Exhaust Gas Recirculation (EGR) plate
- Upper radiator hose and connecting pipes
- Catalytic converter
- Dipstick

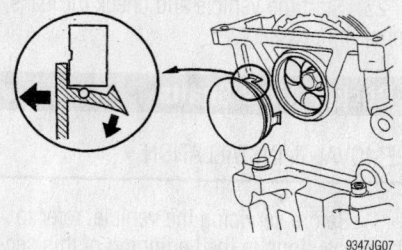

Remove the cylinder head plug

- Cylinder head cover
- Timing chain auto tensioner
- Cylinder head plug and hold the camshaft tight while loosening the sprocket mounting bolt
- Camshaft sprocket
- Engine mount bracket
- Cylinder head

To install:

10. Clean the mating surface where the head gasket attaches to the engine block.

11. Install or connect the following:
- Dowel pins and a new cylinder head gasket and apply liquid gasket to the cylinder head mating surface
- Cylinder head and apply clean engine oil to the threads and under the heads of the cylinder head bolts

12. Torque the cylinder head bolts is sequence as follows:
  a. Step 1: 29 ft. lbs. (39 Nm).
  b. Step 2: Turn the bolts an additional 90 degrees.
  c. Step 3: Torque the 6 mm bolts to 8.7 ft. lbs. (12 Nm).

➡**If any of the bolts make noise during installation, remove the bolt and retighten it.**

13. Install or connect the following:
- Camshaft sprocket into the cylinder head

- Timing chain on the camshaft sprocket and align the sprocket on the camshaft

14. Turn the camshaft sprocket counter-clockwise to relieve any timing chain free play. Check the alignment of the TDC mark (A) on the camshaft sprocket with the cylinder head.

15. Hold the camshaft in position and torque the mounting bolt to 41 ft. lbs. (56 Nm).

16. Install or connect the following:
- New cylinder head plug. Press the rod to pump any oil from the timing chain auto tensioner
- New O-ring (A) into the spacer (B). Align the spacer and new gasket (C) on the auto tensioner (D). Torque the bolts to 8.7 ft. lbs. (12 Nm) while pressing the timing chain auto tensioner against the cylinder head.

17. Adjust the valve clearance if necessary
- Cylinder head cover and torque the bolts to 8.7 ft. lbs. (12 Nm)
- Catalytic converter with a new gasket and nuts and torque the upper nut to 17 ft. lbs. (24 Nm), the upper bolts to 33 ft. lbs. (44 Nm) and the lower bolt to 16 ft. lbs. (22 Nm).
- Heater hoses and connecting pipes with new O-rings and torque the bolts to 8.7 ft. lbs. (12 Nm)
- Upper radiator hose
- EGR plate and intake manifold with a new gasket and torque the bolts to 16 ft. lbs. (22 Nm)
- PCV hose
- Fuel feed and return hoses. Torque the return hose bracket bolt to 8.7 ft. lbs. (12 Nm) and the feed hose fastener to 16 ft. lbs. (22 Nm)

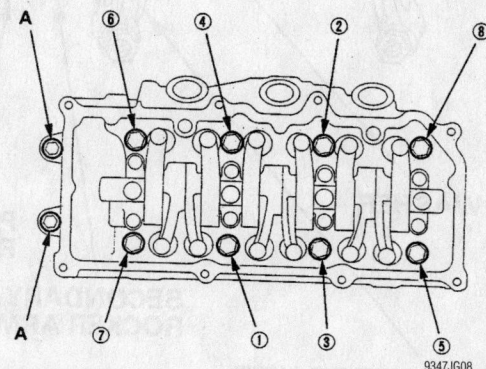

Cylinder head torque sequence

Timing belt service is covered in Section 3 of this manual

- Brake booster vacuum hose
- EVAP hoses and water bypass hoses

18. Install the engine wire harness connectors to the following components:
- TDC1 and TDC2 sensors
- VTEC solenoid and pressure switch
- Primary and secondary HO2S
- ECT sensor
- MAP sensor
- TP sensor
- IAC valve
- IAT sensor
- Ignition coils
- Fuel injectors

19. Install or connect the following:
- Air cleaner and intake air duct assembly then torque the bolts to 8.7 ft. lbs. (12 Nm)
- Throttle cable and adjust as necessary
- Engine cover and torque the bolts to 8.7 ft. lbs. (12 Nm)
- Both battery cables

20. Allow the engine to sit for approximately 30 minutes before refilling any fluids.

21. Fill the engine with clean oil.

22. Fill the cooling system.

23. Turn the ignition switch to the **ON position** and allow the fuel pump to run for a few seconds to pressurize the fuel system.

24. Enter the preset radio frequencies.

25. Start the vehicle and check for leaks, repair if necessary.

Rocker Arms/Shafts

REMOVAL & INSTALLATION

1. Before servicing the vehicle, refer to the precautions in the beginning of this section.

2. Turn the battery switch **OFF** and measure the voltage before proceeding.

3. Disconnecting the negative battery cable will clear the radio's preset buttons. Write down the radio frequencies.

4. Remove or disconnect the following:

- Negative battery cable
- Cylinder head
- Loosen the adjusting screws
- Rocker arm assembly in the proper sequence by turning the bolts two turns at a time

5. Do not remove the mounting bolts from the rocker arm.

To install:

6. Install or connect the following:
- Rocker arm assembly to the camshaft holders and torque the mounting bolts in the proper sequence to 16 ft. lbs. (22 Nm)
- Cylinder head
- Negative battery cable

7. Enter the preset radio frequencies.

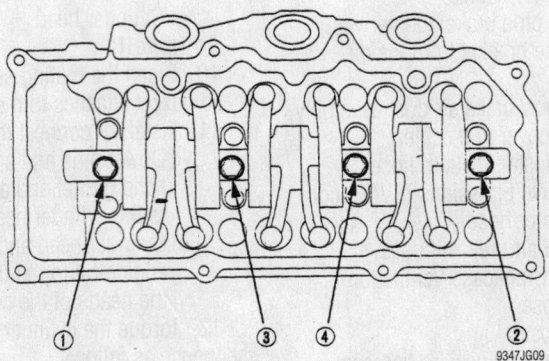

Loosen the rocker arm mounting bolts in the proper sequence

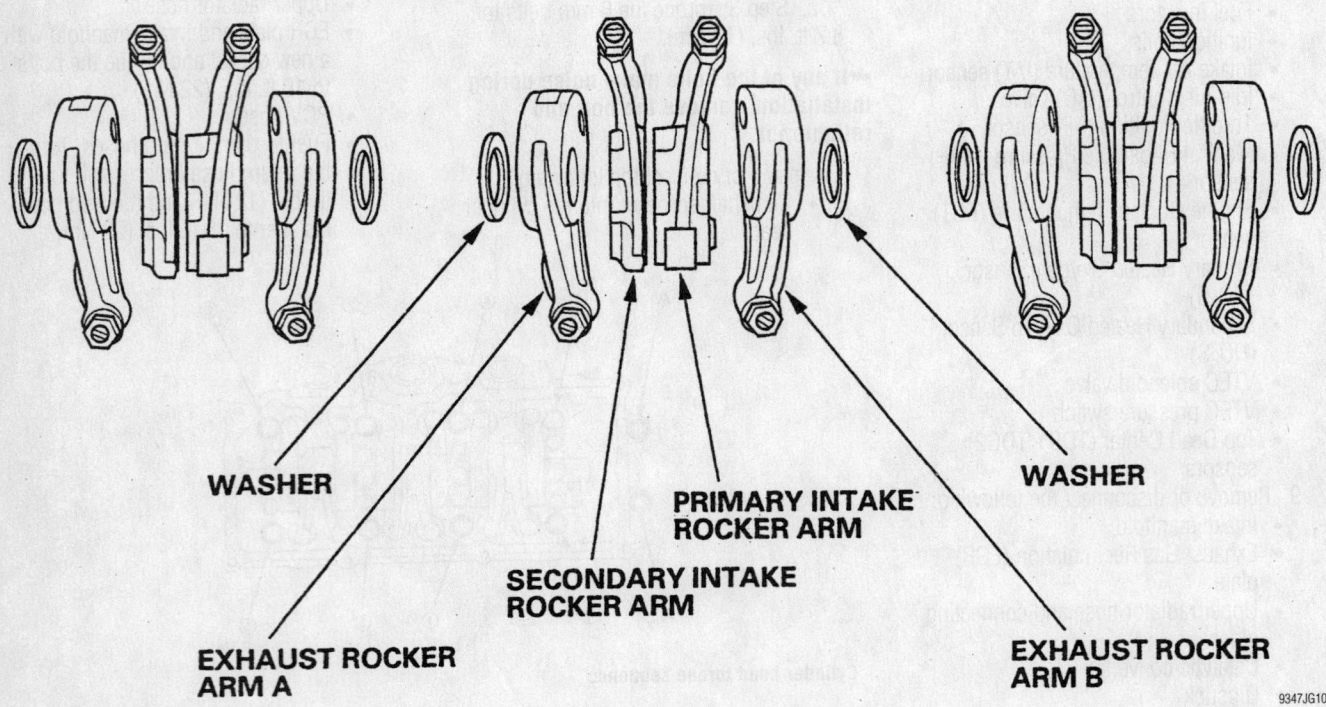

WASHER

PRIMARY INTAKE ROCKER ARM

WASHER

SECONDARY INTAKE ROCKER ARM

EXHAUST ROCKER ARM A

EXHAUST ROCKER ARM B

Exploded view of the rocker arm assembly—1.0L engine

8. Start the vehicle and check for leaks, repair if necessary.

Intake Manifold

REMOVAL & INSTALLATION

Upper

1. Before servicing the vehicle, refer to the precautions in the beginning of this section.

2. Turn the battery switch **OFF** and measure the voltage before proceeding.

3. Disconnecting the negative battery cable will clear the radio's preset buttons. Write down the radio frequencies.

4. Properly relieve the fuel system pressure.

5. Remove or disconnect the following:
   - Negative battery cable
   - Air cleaner/intake air duct assembly
   - Throttle body assembly
   - Exhaust Gas Recirculation (EGR) valve and port
   - Intake manifold
   - EGR plate
   - Intake manifold gasket

To install:

6. Install or connect the following:
   - New intake manifold gasket and EGR plate
   - New O-rings and install the intake manifold and torque the bolts to 16 ft. lbs. (22 Nm)
   - Throttle body assembly with a new O-ring and gasket and torque the bolts to 16 ft. lbs. (22 Nm)
   - EGR port with a new gasket and torque the bolts to 16 ft. lbs. (22 Nm)
   - EGR valve with new gaskets and

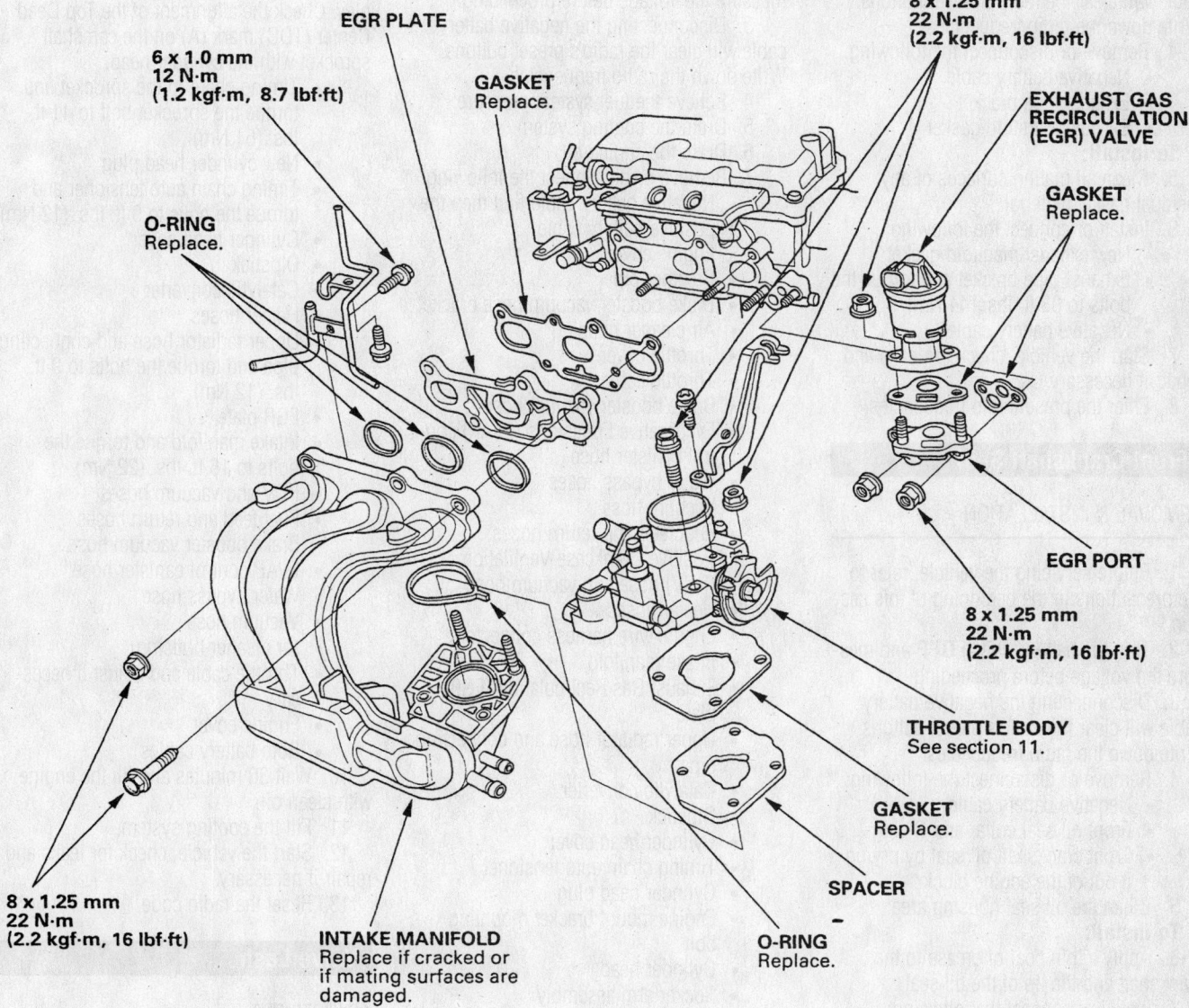

EGR PLATE

**6 x 1.0 mm
12 N·m
(1.2 kgf·m, 8.7 lbf·ft)**

GASKET
Replace.

O-RING
Replace.

**8 x 1.25 mm
22 N·m
(2.2 kgf·m, 16 lbf·ft)**

**EXHAUST GAS
RECIRCULATION
(EGR) VALVE**

GASKET
Replace.

EGR PORT

**8 x 1.25 mm
22 N·m
(2.2 kgf·m, 16 lbf·ft)**

THROTTLE BODY
See section 11.

GASKET
Replace.

SPACER

O-RING
Replace.

**8 x 1.25 mm
22 N·m
(2.2 kgf·m, 16 lbf·ft)**

INTAKE MANIFOLD
Replace if cracked or if mating surfaces are damaged.

9347JG11

Exploded view of the intake manifold assembly—1.0L engine

Heater Core replacement is covered in Section 2 of this manual

torque the bolts to 165 ft. lbs. (22 Nm)
  • Air cleaner/intake air duct assembly
  • Negative battery cable
7. Enter the preset radio frequencies.

Exhaust Manifold

REMOVAL & INSTALLATION

1. Before servicing the vehicle, refer to the precautions in the beginning of this section.
2. Turn the battery switch **OFF** and measure the voltage before proceeding.
3. Disconnecting the negative battery cable will clear the radio's preset buttons. Write down the radio frequencies.
4. Remove or disconnect the following:
  • Negative battery cable
  • Exhaust pipe bracket
  • Exhaust manifold gasket
To install:
5. Clean all mating surfaces of any residual gasket material.
6. Install or connect the following:
  • New exhaust manifold gasket
  • Exhaust pipe bracket and torque the bolts to 33 ft. lbs. (44 Nm)
  • Negative battery cable
7. Start the vehicle, check for leaks and repair if necessary.
8. Enter the preset radio frequencies.

Front Crankshaft Seal

REMOVAL & INSTALLATION

1. Before servicing the vehicle, refer to the precautions in the beginning of this section.
2. Turn the battery switch **OFF** and measure the voltage before proceeding.
3. Disconnecting the negative battery cable will clear the radio's preset buttons. Write down the radio frequencies.
4. Remove or disconnect the following:
  • Negative battery cable
  • Front oil seal collar and O-ring
  • Front crankshaft oil seal by prying it out of the engine block
5. Clean the oil seal housing area.
To install:
6. Apply a thin coat of grease to the crankshaft and the lip of the oil seal.
7. Install or connect the following:
  • New oil seal by using Oil Seal Driver Tool 07746-0030100. Drive the oil seal in until the driver bottoms out against the oil pump and

make certain that the seal is not distorted
  • New O-ring and oil seal collar
  • Negative battery cable
8. Enter the preset radio frequencies.
9. Start the vehicle and check for leaks, repair if necessary.

Camshaft and Lifters

REMOVAL & INSTALLATION

1. Before servicing the vehicle, refer to the precautions in the beginning of this section.
2. Turn the battery switch **OFF** and measure the voltage before proceeding.
3. Disconnecting the negative battery cable will clear the radio's preset buttons. Write down the radio frequencies.
4. Relieve the fuel system pressure.
5. Drain the cooling system.
6. Drain the engine oil.
7. Remove or disconnect the following:
  • Negative battery cable first then the positive battery cable
  • Engine cover
  • Breather pipe
  • Brake booster vacuum hose bracket
  • Air cleaner housing
  • Throttle cover
  • Throttle cable
  • Brake booster vacuum hose
  • Evaporative Emissions (EVAP) control canister hose
  • Water bypass hoses
  • Vacuum hose
  • Fuel feed and return hoses
  • Positive Crankcase Ventilation (PCV) hose and vacuum hose
  • Heater hoses
  • Engine wire harness connectors
  • Intake manifold
  • Exhaust Gas Recirculation (EGR) plate
  • Upper radiator hose and connecting pipe
  • Catalytic converter
  • Dipstick
  • Cylinder head cover
  • Timing chain auto tensioner
  • Cylinder head plug
  • Engine mount bracket mounting bolt
  • Cylinder head
  • Rocker arm assembly
  • Camshaft holders
  • Camshaft
To install:
8. Install or connect the following:
  • Camshaft and holders to the cylin-

der head and torque the bolts to 9 ft. lbs. (12 Nm)
  • Rocker arm assembly to the camshaft and torque the bolts to 16 ft. lbs. (22 Nm)
  • Cylinder head with a new gasket. Torque the bolts to 29 ft. lbs. (39 Nm) and an additional 90 degrees. Torque the 6mm bolts to 9 ft. lbs. (12 Nm).
  • Camshaft sprocket to the cylinder head
  • Timing chain to the camshaft sprocket
  • Camshaft sprocket to the camshaft
9. Turn the camshaft sprocket counterclockwise to relieve any timing chain free play. Check the alignment of the Top Dead Center (TDC) mark (A) on the camshaft sprocket with the cylinder head.
  • Timing chain to the sprocket and torque the sprocket bolt to 41 ft. lbs. (61 Nm)
  • New cylinder head plug
  • Timing chain auto tensioner and torque the bolts to 9 ft. lbs. (12 Nm)
  • Cylinder head cover
  • Dipstick
  • Catalytic converter
  • Heater hoses
  • Upper radiator hose and connecting pipe and torque the bolts to 9 ft. lbs. (12 Nm)
  • EGR plate
  • Intake manifold and torque the bolts to 16 ft. lbs. (22 Nm)
  • PCV and vacuum hoses
  • Fuel feed and return hoses
  • Brake booster vacuum hose
  • EVAP control canister hose
  • Water bypass hose
  • Vacuum hose
  • Air cleaner housing
  • Throttle cable and adjust if necessary
  • Engine cover
  • Both battery cables
10. Wait 30 minutes and fill the engine with clean oil.
11. Fill the cooling system.
12. Start the vehicle, check for leaks and repair if necessary.
13. Reset the radio code.

Valve Lash

ADJUSTMENT

1. Before servicing the vehicle, refer to the precautions in the beginning of this section.

2. Turn the battery switch **OFF** and measure the voltage before proceeding.

3. Disconnecting the negative battery cable will clear the radio's preset buttons. Write down the radio frequencies.

4. Disconnect the negative battery cable.

5. Remove the engine cover.

6. Remove the cylinder head cover.

7. Set cylinder No.1 at the Top Dead Center (TDC) position. The No. 1 piston TDC mark on the camshaft sprocket should be aligned with the cylinder head surface.

8. Insert the proper feeler gauge between the adjusting screw and the end of the valve stem.

9. Slide the feeler gauge back and forth. There should be a slight drag on the gauge.

10. If the valve is not adjusted properly, loosen the locknut and turn the adjusting screw until the drag on the feeler gauge is accurate.

11. Tighten the locknut and recheck the clearance.

12. Rotate the crankshaft 240 degrees clockwise (camshaft sprocket turn 120 degrees). The second TDC mark should be aligned with the cylinder head surface.

13. Insert the proper feeler gauge between the adjusting screw and the end of the valve stem.

14. Slide the feeler gauge back and forth. There should be a slight drag on the gauge.

15. If the valve is not adjusted properly, loosen the locknut and turn the adjusting screw until the drag on the feeler gauge is accurate.

16. Tighten the locknut and recheck the clearance.

17. Repeat this procedure for cylinder No. 2.

18. The proper adjustment order is 1–3–2.

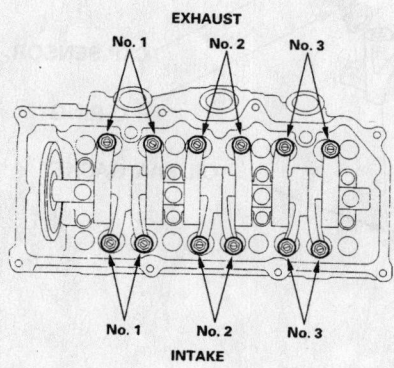

Exploded view of the adjusting screw locations–1.0L engine

19. Install the cylinder head cover.
20. Install the engine cover.
21. Connect the negative battery cable.
22. Enter the preset radio frequencies.

Starter Motor

REMOVAL & INSTALLATION

1. Before servicing the vehicle, refer to the precautions in the beginning of this section.

2. Turn the battery switch **OFF** and measure the voltage before proceeding.

3. Disconnecting the negative battery cable will clear the radio's preset buttons. Write down the radio frequencies.

4. Remove or disconnect the following:
- Negative battery cable first then the positive battery cable. Wait 3 minutes before starting any repairs
- Air cleaner/intake air duct assembly
- Starter cable and the black/white wire
- Wire harness clamps
- Starter

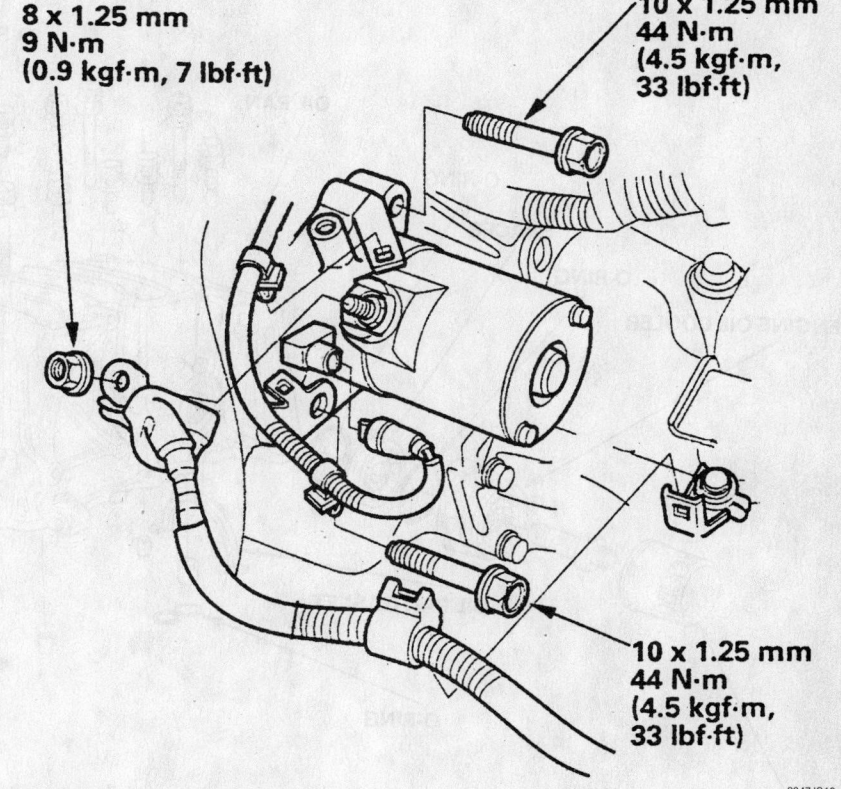

**8 x 1.25 mm
9 N·m
(0.9 kgf·m, 7 lbf·ft)**

**10 x 1.25 mm
44 N·m
(4.5 kgf·m,
33 lbf·ft)**

**10 x 1.25 mm
44 N·m
(4.5 kgf·m,
33 lbf·ft)**

Removal of the starter motor

To install:

5. Install or connect the following:
- Starter and torque the bolts to 33 ft. lbs. (44 Nm)
- Starter electrical connectors. Torque the nut to 7 ft. lbs. (9 Nm) and install the wire harness clamps
- Air cleaner/intake air duct assembly and torque the bolts to 8.7 ft. lbs. (9 Nm)
- Both battery cables

6. Enter the preset radio frequencies.

Oil Pan

REMOVAL & INSTALLATION

1. Before servicing the vehicle, refer to the precautions in the beginning of this section.

2. Drain the engine oil.

3. Support the powertrain assembly.

4. Remove or disconnect the following:
- Negative battery cable
- Right inner splash shield
- Engine structural collar
- Lower torque strut

Brake service is covered in Section 4 of this manual

- Oil filter adapter
- Oil pan and gasket

5. Thoroughly clean the gasket mating surfaces.

To install:

6. Apply silicone sealer to the oil pump-to-engine block parting line.

7. Install or connect the following:
- New gasket on the oil pan
- Oil pan and torque the retainers to 105 inch lbs. (12 Nm)
- Oil filter and adapter and torque the screws to 105 inch lbs. (12 Nm)

✷✷ WARNING

Follow the proper tightening sequence for the structural collar or damage to the collar or oil pan may occur!

8. Install the structural collar and torque the retainers as follows:
   a. Collar-to-oil pan bolts: 30 inch lbs. (3 Nm).
   b. Collar-to-transaxle bolts: 80 ft. lbs. (108 Nm).
   c. Collar-to-oil pan bolts: 40 ft. lbs. (54 Nm), final torque.

9. Install or connect the following:
- Lower torque strut
- Right inner splash shield
- Negative battery cable

10. Fill the engine with clean oil.

11. Start the vehicle and check for leaks, repair if necessary.

Oil Pump

REMOVAL & INSTALLATION

1. Before servicing the vehicle, refer to the precautions in the beginning of this section.

2. Drain the engine oil.

3. Remove or disconnect the following:
- Negative battery cable
- Oil pan
- Crankshaft sprocket, using Tool 6795 and Insert Tool C-4685-C2
- Crankshaft key
- Oil pickup tube
- Oil pump

To install:

4. Wash all parts in a solvent; then, inspect carefully for damage or wear, as follows:

   a. Inspect the mating surface of the oil pump should be smooth. Replace the pump cover, if scratched or grooved.
   b. Apply Mopar® gasket maker to the oil pump.
   c. Install the O-ring into the oil pump body discharge passage.

5. Prime the oil pump before installation by filling the rotor cavity with engine oil.

6. Install or connect the following:
- Oil pump, align the rotor flats with the crankshaft flats and torque the pump bolts to 21 ft. lbs. (28 Nm)

✷✷ WARNING

The front crankshaft seal MUST be out of the pump to align or damage may result.

- New front crankshaft seal, using Seal Driver Tool 6780
- Crankshaft key
- Crankshaft sprocket, using a Crankshaft Sprocket Installer Tool 6792
- Oil pump pickup tube
- Oil pan
- Negative battery cable

7. Refill the engine with clean oil.

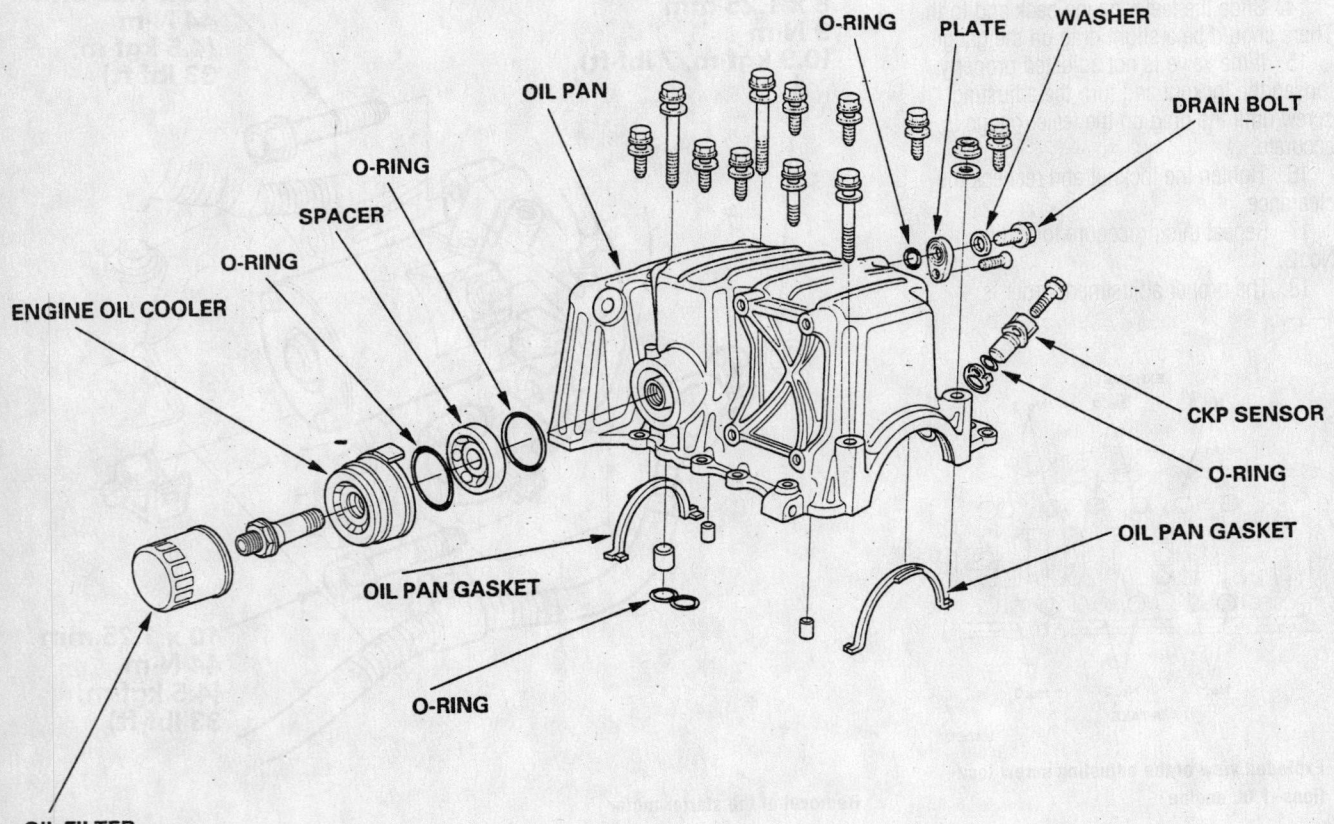

Exploded view of the oil pan assembly

9347JG14

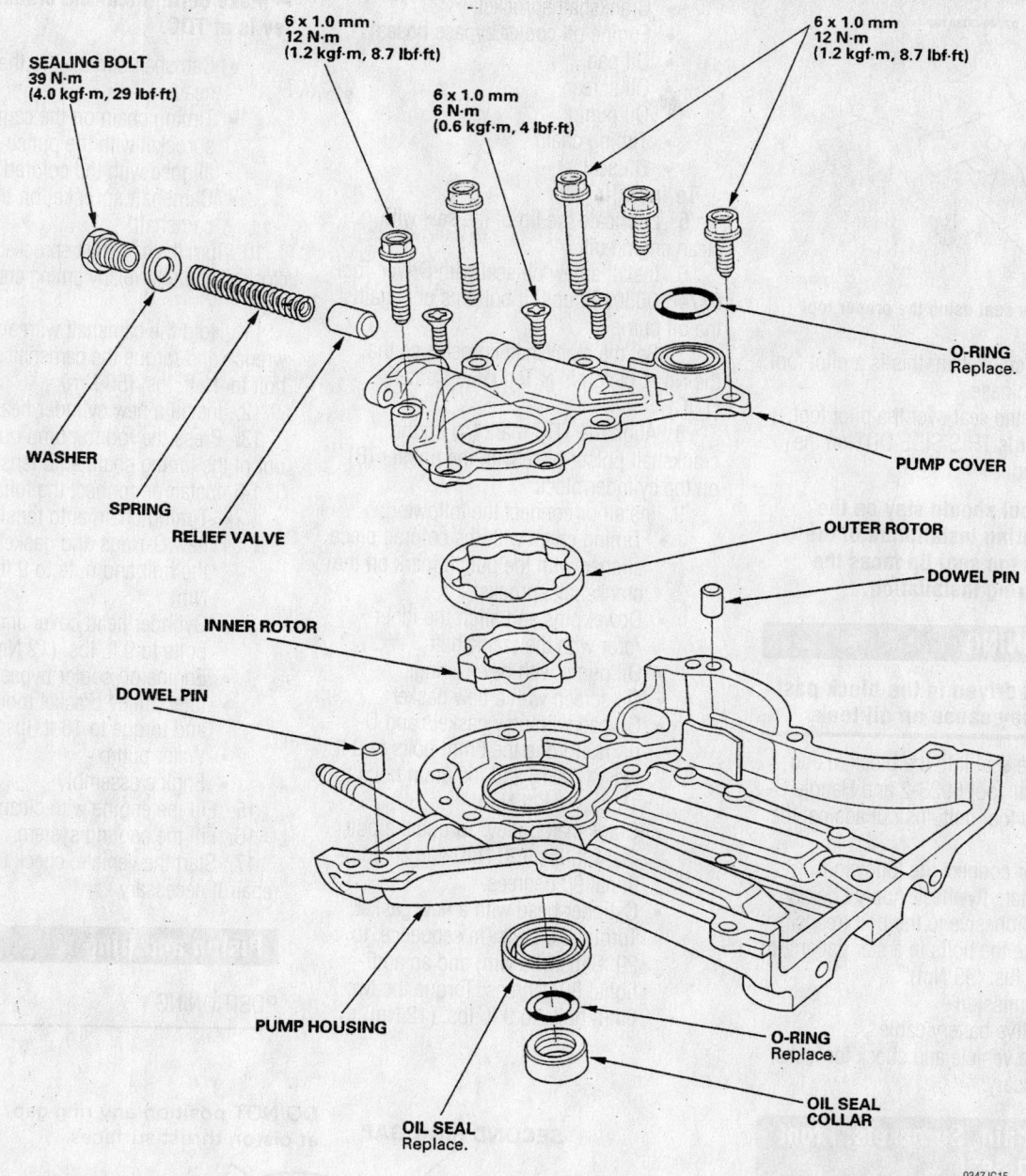

SEALING BOLT
39 N·m
(4.0 kgf·m, 29 lbf·ft)

6 x 1.0 mm
12 N·m
(1.2 kgf·m, 8.7 lbf·ft)

6 x 1.0 mm
6 N·m
(0.6 kgf·m, 4 lbf·ft)

6 x 1.0 mm
12 N·m
(1.2 kgf·m, 8.7 lbf·ft)

WASHER

SPRING

RELIEF VALVE

O-RING
Replace.

PUMP COVER

OUTER ROTOR

DOWEL PIN

INNER ROTOR

DOWEL PIN

PUMP HOUSING

OIL SEAL
Replace.

O-RING
Replace.

OIL SEAL COLLAR

9347JG15

Exploded view of the oil pump assembly

8. Start the engine and check for leaks; repair if necessary.

Rear Main Seal

REMOVAL & INSTALLATION

1. Before servicing the vehicle, refer to the precautions in the beginning of this section.
2. Remove or disconnect the following:
   • Transmission

• Flexplate/flywheel
• Rear main seal
3. Insert a seal remover between the dust lip and the metal case of the crankshaft seal. Angle the tool through the dust lip against the metal case of the seal. Pry out the seal.

✳✳ WARNING

DO NOT let the prytool contact the crankshaft seal surface. Contact of the tool blade against the crankshaft edge (chamfer) is permitted.

To install:

✳✳ WARNING

If the crankshaft edge (chamfer) has any burrs or scratches on the, clean it up with 400 grit sand paper to prevent seal damage during installation of the new seal.

➡No lubrication is necessary when installing the seal.

4. Place Crankcase Seal Pilot Tool

For complete Engine Mechanical specifications, see Section 1 of this manual

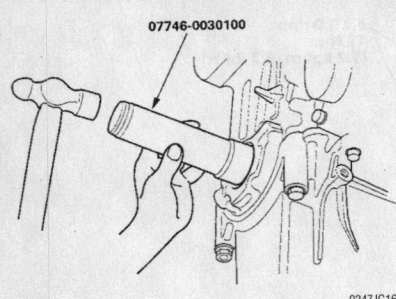

07746-0030100

9347JG16

Install the new seal using the proper tool

6926-1 on the crankshaft; this is a pilot tool with a magnetic base.

5. Position the seal over the pilot tool; be sure the words THIS SIDE OUT on the seal can be read.

➡ **The pilot tool should stay on the crankshaft during installation of the seal. Be sure the seal lip faces the crankcase during installation.**

❋❋ WARNING

If the seal is driven in the block past flush, this may cause an oil leak.

6. Drive the seal into the block, using Crankshaft Seal Tool 6926-2 and Handle C-4171, until the tool bottoms out against the block.

7. Install or connect the following:
- Flexplate/flywheel. Apply Lock & Seal Adhesive to the bolt treads. Torque the bolts in a star pattern, to 70 ft. lbs. (95 Nm).
- Transmission
- Negative battery cable

8. Start the vehicle and check for leaks, repair if necessary.

Timing Chain, Sprockets, Front Cover and Seal

REMOVAL & INSTALLATION

1. Before servicing the vehicle, refer to the precautions in the beginning of this section.

2. Drain the engine oil.

3. Drain the cooling system.

4. Remove or disconnect the following:
- Engine assembly
- Cylinder head
- Idler pulley bracket mounting bolt
- Water pump
- Idler pulley bolt with Special Tools 07JAB-001020A, 07MAB-PY301A and 07JAA-00101A

- Crankshaft sprocket
- Engine oil cooler bypass hoses
- Oil pan
- Oil screen
- Oil pump
- Timing chain
- Oil seal

To install:

5. Lubricate the lip of the seal with clean engine oil.

6. Install a new oil seal with Driver Tool 07746-0030100 until it bottoms out against the oil pump

7. Set the crankshaft sprocket so that the No. 1 piston is at Top dead Center (TDC).

8. Align the TDC mark (A) on the crankshaft pulser plate with the pointer (B) on the cylinder block.

9. Install or connect the following:
- Timing chain with the colored piece aligned with the punch mark on the crankshaft sprocket
- Dowel pins and align the inner rotor with the crankshaft
- Oil pump with new O-rings
- Oil screen with a new gasket
- Oil pan with new gaskets and O-rings. Torque the 6mm bolts to 9 ft. lbs. (12 Nm) and the 8mm bolts to 16 ft. lbs. (22 Nm).
- Crankshaft pulley. Torque the bolt to 14 ft. lbs. (20 Nm) and an additional 90 degrees.
- Cylinder head with a new gasket. Torque the bolts, in sequence, to 29 ft. lbs. (39 Nm) and an additional 90 degrees. Torque the two 6mm bolts to 9 ft. lbs. (12 Nm).

➡ **Make certain that the crankshaft pulley is at TDC.**

- Camshaft sprocket to the cylinder head
- Timing chain on the camshaft sprocket with the punch mark aligned with the colored link
- Camshaft sprocket on the camshaft

10. Turn the camshaft sprocket counterclockwise to relieve any timing chain free play.

11. Hold the camshaft with an open end wrench and torque the camshaft mounting bolt to 41 ft. lbs. (56 Nm).

12. Install a new cylinder head plug.

13. Press the rod to pump out any oil out of the timing chain auto tensioner.

14. Install or connect the following:
- Timing chain auto tensioner with new O-rings and gasket. Torque the bolt and nuts to 9 ft. lbs. (12 Nm).
- Cylinder head cover and torque the bolts to 9 ft. lbs. (12 Nm)
- Engine oil cooler bypass hoses
- Idler pulley bracket mounting bolt and torque to 16 ft. lbs. (22 Nm)
- Water pump
- Engine assembly

15. Fill the engine with clean oil.

16. Fill the cooling system.

17. Start the vehicle, check for leaks and repair if necessary.

Piston and Ring

POSITIONING

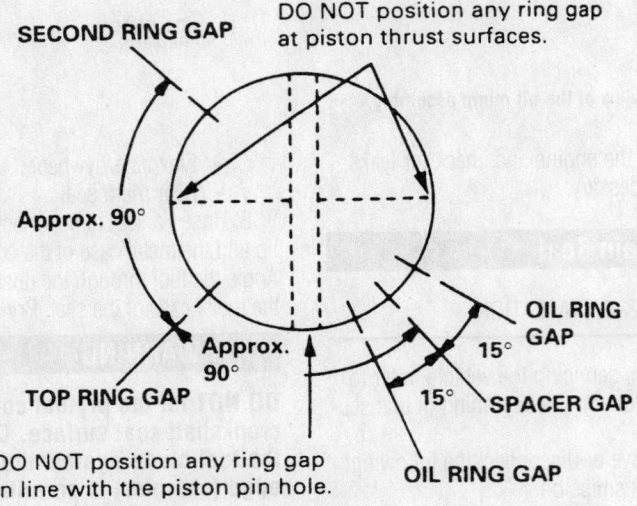

Piston ring end-gap spacing –1.0L engine

7923AG22

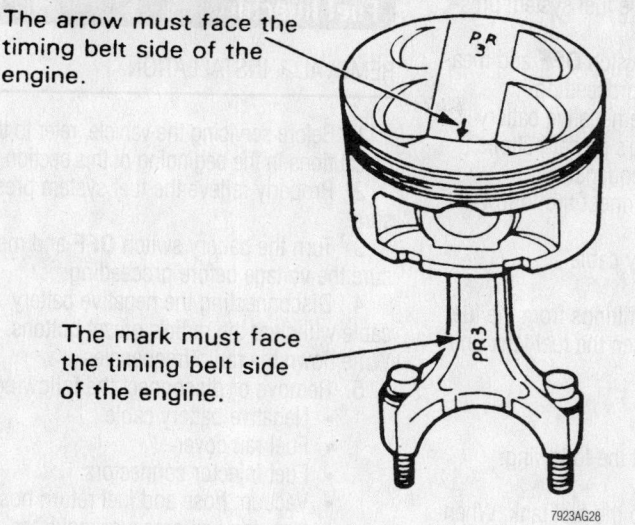

The arrow must face the timing belt side of the engine.

The mark must face the timing belt side of the engine.

Piston ring identification mark locations

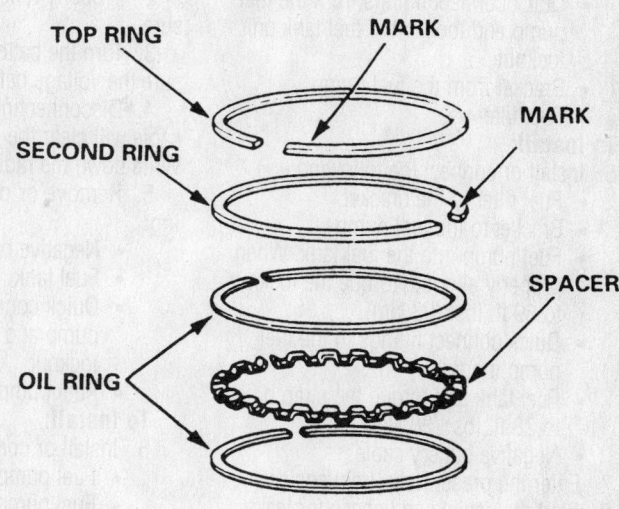

TOP RING

MARK

SECOND RING

MARK

OIL RING

SPACER

Piston ring positioning–1.0L engine

FUEL SYSTEM

Fuel System Service Precautions

Safety is an important factor when servicing the fuel system. Failure to conduct maintenance and repairs in a safe manner may result in serious personal injury. Maintenance and testing of the vehicle's fuel system components can be accomplished safely and effectively by adhering to the following rules and guidelines:

• To avoid the possibility of fire and personal injury, always disconnect the negative battery cable unless the repair or test procedure requires that battery voltage be applied.

• Always relieve the fuel system pressure prior to disconnecting any fuel system component (injector, fuel rail, pressure regulator, etc.), fitting or fuel line connection. Exercise extreme caution whenever relieving fuel system pressure, to avoid exposing skin, face and eyes to fuel spray. Please be advised that fuel under pressure may penetrate the skin or any part of the body that it contacts.

• Always place a shop towel or cloth around the fitting or connection prior to loosening to absorb any excess fuel due to spillage. Ensure that all fuel spillage is quickly removed from engine surfaces. Ensure that all fuel soaked cloths or towels are deposited into a suitable waste container.

• Always keep a dry chemical (Class B) fire extinguisher near the work area.

• Do not allow fuel spray or fuel vapors

to come into contact with a spark or open flame.

• Always use a back-up wrench when loosening and tightening fuel line connection fittings. This will prevent unnecessary stress and torsion to fuel line piping.

• Always replace worn fuel fitting O-rings. Do not substitute fuel hose where fuel pipe is installed.

Fuel System Pressure

RELIEVING

✳✳ CAUTION

Relieve the fuel system pressure before servicing any components of the fuel system. Service vehicles in well ventilated areas and avoid ignition sources. NEVER smoke while servicing the vehicle!

1. Before servicing the vehicle, refer to the precautions in the beginning of this section.

2. Turn the battery switch **OFF** and measure the voltage before proceeding.

3. Disconnecting the negative battery cable will clear the radio's preset buttons. Write down the radio frequencies.

4. Disconnect the negative battery cable.

5. Remove the fuel filler cap.

6. Slowly loosen the sealing nut on the fuel rail one complete turn.

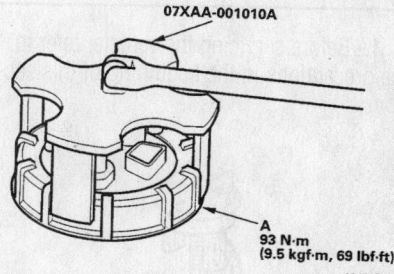

07XAA-001010A

A
93 N·m
(9.5 kgf·m, 69 lbf·ft)

Loosen the sealing nut on the fuel rail

7. Replace the washers whenever the sealing nut is loosened or removed.

Fuel Filter

The fuel filter is part of the fuel pump module located in the fuel tank.

REMOVAL & INSTALLATION

1. Before servicing the vehicle, refer to the precautions in the beginning of this section.

2. Properly relieve the fuel system pressure.

3. Turn the battery switch **OFF** and measure the voltage before proceeding.

4. Disconnecting the negative battery cable will clear the radio's preset buttons. Write down the radio frequencies.

5. Remove or disconnect the following:
• Negative battery cable
• Fuel tank

For Accessory Drive Belt illustrations, see Section 1 of this manual

- Quick connect fittings from the fuel pump and loosen the fuel tank unit locknut
- Bracket from the fuel pump
- Fuel filter

To install:

6. Install or connect the following:
- Fuel filter to the bracket
- Bracket to the fuel pump
- Fuel pump into the fuel tank. When properly aligned, torque the locknut to 69 ft. lbs. (93 Nm).
- Quick connect fittings to the fuel pump module
- Fuel tank and torque the strap bolts to 28 ft. lbs. (38 Nm)
- Negative battery cable

7. Enter the preset radio frequencies
8. Start the vehicle and check for leaks, repair if necessary.

Fuel Pump

REMOVAL & INSTALLATION

1. Before servicing the vehicle, refer to the precautions in the beginning of this section.

2. Properly relieve the fuel system pressure.

3. Turn the battery switch **OFF** and measure the voltage before proceeding.

4. Disconnecting the negative battery cable will clear the radio's preset buttons. Write down the radio frequencies.

5. Remove or disconnect the following:
- Negative battery cable
- Fuel tank
- Quick connect fittings from the fuel pump and loosen the fuel tank unit locknut
- Fuel pump

To install:

6. Install or connect the following:
- Fuel pump
- Fuel pump into the fuel tank. When properly aligned, torque the locknut to 69 ft. lbs. (93 Nm).
- Quick connect fittings to the fuel pump module
- Fuel tank and torque the strap bolts to 28 ft. lbs. (38 Nm)
- Negative battery cable

7. Enter the preset radio frequencies
8. Start the vehicle and check for leaks, repair if necessary.

Fuel Injector

REMOVAL & INSTALLATION

1. Before servicing the vehicle, refer to the precautions in the beginning of this section.

2. Properly relieve the fuel system pressure.

3. Turn the battery switch **OFF** and measure the voltage before proceeding.

4. Disconnecting the negative battery cable will clear the radio's preset buttons. Write down the radio frequencies.

5. Remove or disconnect the following:
- Negative battery cable
- Fuel rail cover
- Fuel injector connectors
- Vacuum hose and fuel return hose from the fuel pressure regulator
- Fuel rail retaining nuts
- Positive Crankcase Ventilation (PCV) valve
- Fuel rail
- Fuel injectors from the cylinder head

To install:

6. Install or connect the following:
- New cushion rings on the fuel injectors

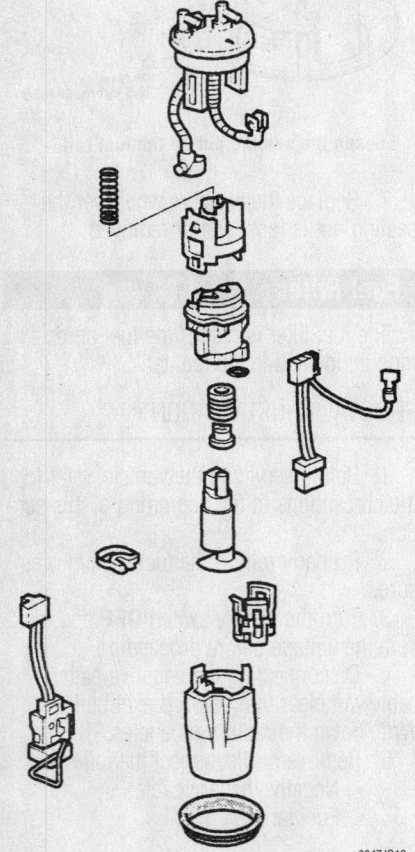

Exploded view of the fuel pump module

9347JG19

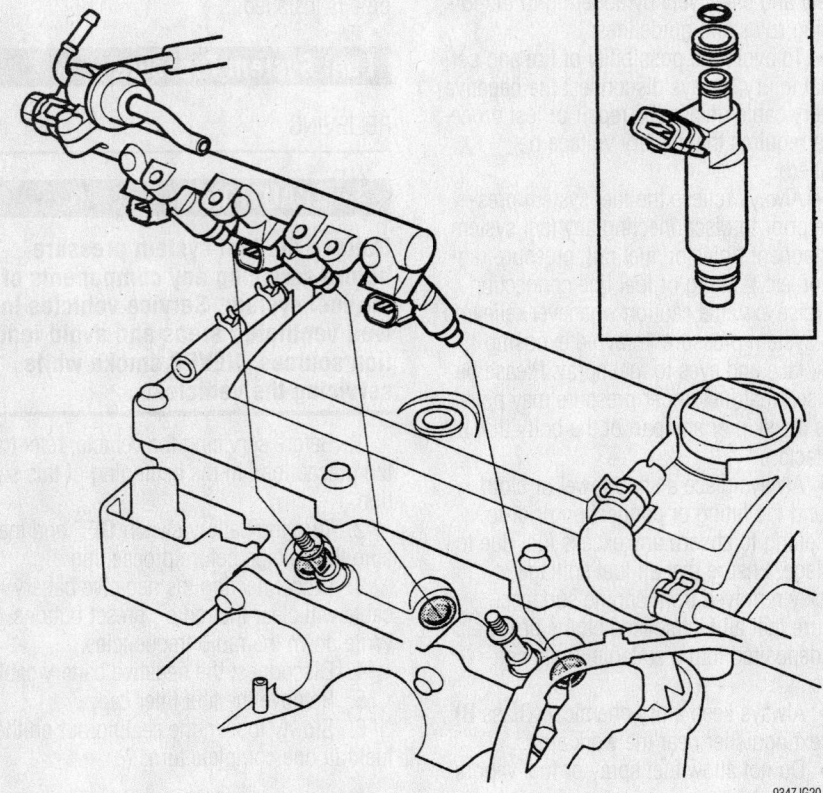

Exploded view of the fuel rail and fuel injectors–1.0L engine

9347JG20

- New O-rings on the fuel injectors after coating them with clean engine oil
- Fuel injectors into the fuel rail
- New seal rings coated with clean engine oil into the cylinder

- Fuel injectors into the cylinder head and torque the fuel rail retaining nuts to 8.7 ft. lbs. (12 Nm)
- Vacuum hose and fuel return hose to the pressure regulator

- Fuel injector electrical connector
- Fuel rail cover
- Negative battery cable

7. Enter the preset radio frequencies
8. Start the vehicle and check for leaks, repair if necessary.

DRIVE TRAIN

Transaxle Assembly

REMOVAL & INSTALLATION

Manual Transmission

1. Before servicing the vehicle, refer to the precautions in the beginning of this section.
2. Turn the battery switch **OFF** and measure the voltage before proceeding.
3. Disconnecting the negative battery cable will clear the radio's preset buttons. Write down the radio frequencies.
4. Drain the transmission fluid.
5. Remove or disconnect the following:

- Battery cables
- Battery
- Air cleaner/intake air duct assembly
- Engine cover
- Splash shields
- Starter motor cables
- Transmission and engine ground cables
- Back-up lamp switch wiring from the transmission
- Vehicle Speed Sensor (VSS) wire
- Neutral switch electrical connectors
- Air cleaner bracket
- Clutch line bracket
- Cable bracket and disconnect the cables from the top of the transmission
- Throttle drum cover
- Clutch slave cylinder
- Starter

6. Lift and support the powertrain assembly to take the weight off the transmission mounts.

- Transmission mount bracket
- Upper transmission mounting bolts
- Splash shield bracket
- Both driveshafts and place a stand under the transmission
- Rear engine mount bracket

- Four under transmission mounting bolts

7. Pull the transmission away from the engine until the mainshaft clears the clutch pressure plate. Remove the transmission from the vehicle.

To install:

8. Place the transmission under the vehicle and align it with the engine.
9. Install or connect the following:

- Four under transmission mounting bolts and torque the bolts to 47 ft. lbs. (64 Nm)
- Rear engine mount bracket. Torque the 10 x 1.25 mm bolt to 36 ft. lbs. (49 Nm) and the 12 x 1.25 mm bolt to 66 ft. lbs. (89 Nm).
- Both driveshafts
- Splash shield bracket and torque the bolts to 89 inch lbs. (10 Nm)
- Upper transmission mounting bolts and torque the bolts to 47 ft. lbs. (64 Nm)
- Transmission mount bracket and torque the bolts to 40 ft. lbs. (54 Nm). Remove the powertrain support.
- Starter and torque the bolts to 33 ft. lbs. (44 Nm)
- Clutch slave cylinder after lubricating the end of the rod with high temperature grease. Torque the bolts to 16 ft. lbs. (22 Nm).
- Throttle drum cover and torque the bolt to 89 inch lbs. (10 Nm)
- Cable bracket and cables and torque the bolts to 20 ft. lbs. (27 Nm)
- Clutch line bracket and wire harness clamp and torque the bolt to 89 inch lbs. (10 Nm)
- Air cleaner bracket and torque the bolt to 89 inch lbs. (10 Nm)
- Neutral switch electrical connector
- Back up light switch electrical connector
- VSS electrical connector
- Starter motor cables and torque the fasteners to 89 inch lbs. (10 Nm)

- Splash shields and torque the bolts to 89 inch lbs. (10 Nm)
- Engine cover
- Air cleaner/intake air duct assembly
- Battery and connect the terminals

10. Fill the transmission to the proper level.
11. Enter the preset radio frequencies.
12. Start the vehicle and check for leaks, repair if necessary.

Clutch

ADJUSTMENT

The clutch assembly is self-adjusting.

REMOVAL & INSTALLATION

1. Before servicing the vehicle, refer to the precautions in the beginning of this section.
2. Turn the battery switch **OFF** and measure the voltage before proceeding.
3. Disconnecting the negative battery cable will clear the radio's preset buttons. Write down the radio frequencies.
4. Drain the transmission fluid.
5. Remove or disconnect the following:

- Negative battery cable

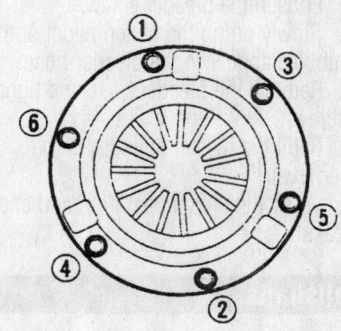

9347JG21

Pressure plate torque sequence–1.0L engine

- Transmission assembly and use a ring gear holder to lock the flywheel in position
- Pressure plate mounting bolts in a crisscross pattern to prevent warpage
- Pressure plate
- Clutch disc

6. Inspect the flywheel, pressure plate and disc for signs of wear, cracks and warpage. Replace if necessary.

To install:

7. Install or connect the following:
- Ring gear holding tool and lubricate the splines of the clutch disc with a high temperature grease
- Clutch disc
- Pressure plate and hand tighten the bolts

8. When properly aligned, torque the bolts in several passes in a crisscross pattern to 19 ft. lbs. (25 Nm).

9. Remove the special tools.
- Transmission assembly
- Negative battery cable

10. Fill the transmission to the proper level.

11. Enter the preset radio frequencies.

12. Start the vehicle and check for leaks, repair if necessary.

Hydraulic Clutch System

BLEEDING

1. Before servicing the vehicle, refer to the precautions in the beginning of this section.

2. Fill the clutch master cylinder reservoir with clean DOT 3 or 4 brake fluid.

3. Attach a hose to the bleeder screw and place the hose in a container of brake fluid.

4. Loosen the bleeder screw.

5. Slowly pump the clutch pedal until no air bubbles appear in the bleeder hose.

6. Remove the bleeder hose and tighten the screw.

7. Refill the master cylinder to the proper level.

8. Verify the clutch operation and check for leaks.

Halfshafts

REMOVAL & INSTALLATION

1. Before servicing the vehicle, refer to the precautions in the beginning of this section.

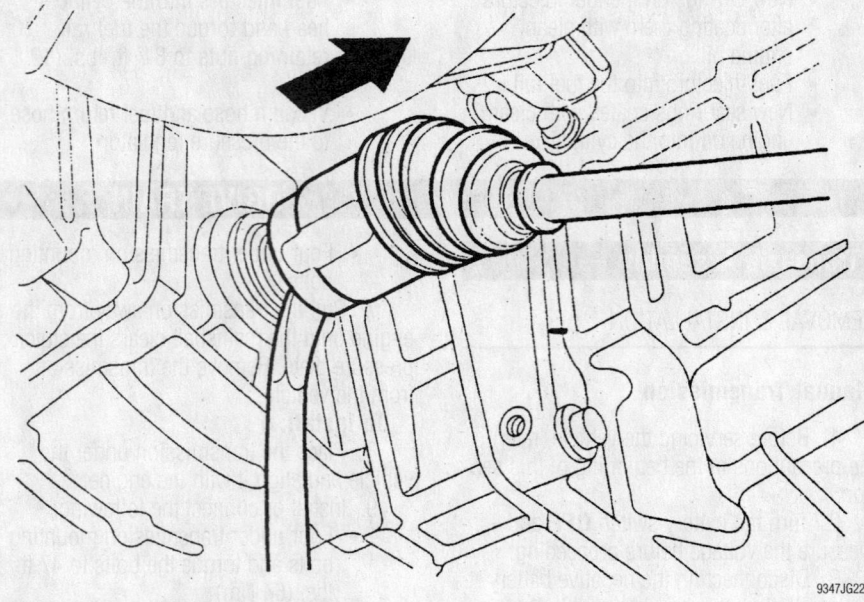

9347JG22

Remove the driveshaft from the differential

2. Turn the battery switch **OFF** and measure the voltage before proceeding.

3. Disconnecting the negative battery cable will clear the radio's preset buttons. Write down the radio frequencies.

4. Drain the transmission fluid.

5. Remove or disconnect the following:
- Negative battery cable
- Front wheels
- Spindle nut after lifting up on the locking tab
- Hold the ball joint pin and remove the flange nut. Separate the front stabilizer link and front stabilizer
- Brake hose clamp
- Antilock Brake System (ABS) wheel sensor harness clamp
- Driveshaft outboard joint from the front hub
- Cotter pin and the lower ball joint castle nut
- Separate the ball joint from the steering knuckle
- Inboard boot heat cover
- Driveshaft from the differential case
- Driveshaft straight out to avoid damaging the differential oil seal

To install:

6. Apply a thin coat of grease to the splined surface of the driveshaft. Remove the grease from the grooves and the set ring groove so that air can bleed from the differential.

7. Install or connect the following:
- New set ring on the driveshaft
- Inboard end of the driveshaft into the differential until the set ring locks into the groove

- Inboard boot heat cover and torque the bolts to 89 inch lbs. (10 Nm)
- Outboard joint to the front hub
- Steering knuckle to the lower arm and torque the castle nut to 40 ft. lbs. (54 Nm)
- Cotter pin
- ABS wheel sensor harness clamp and torque the bolt to 89 inch lbs. (10 Nm)
- Brake hose clamp and torque the bolt to 17 ft. lbs. (22 Nm)
- Front stabilizer link to the stabilizer and torque the new nut to 22 ft. lbs. (29 Nm)
- New spindle nut and torque the nut to 134 ft. lbs. (181 Nm)
- Front wheels
- Negative battery cable

8. Fill the transmission to the proper level.

9. Enter the preset radio frequencies.

10. Start the vehicle and check for leaks, repair if necessary.

CV-Joints

OVERHAUL

Inboard Joint

1. Before servicing the vehicle, refer to the precautions in the beginning of this section.

2. Turn the battery switch **OFF** and measure the voltage before proceeding.

Welded Type

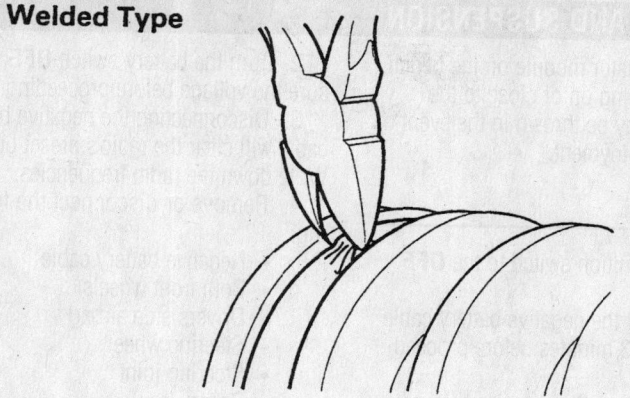

Double Loop Type

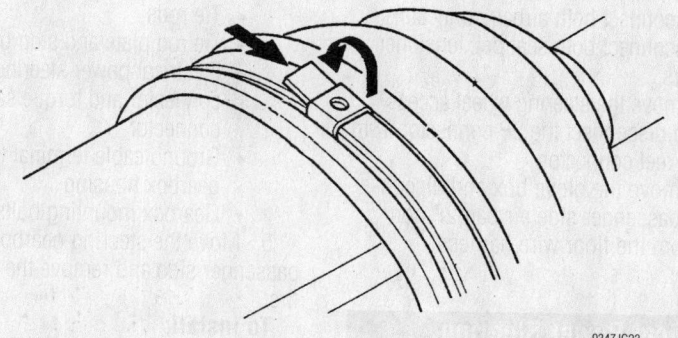

Remove the boot band from the inboard joint

9347JG23

Left driveshaft:	481 – 486 mm (18.9 – 19.1 in.)
Right driveshaft:	761 – 766 mm (30.0 – 30.2 in.)

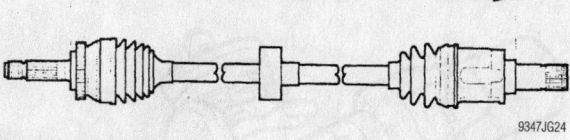

9347JG24

Adjust the length of the driveshaft

3. Disconnecting the negative battery cable will clear the radio's preset buttons. Write down the radio frequencies.

4. Remove or disassemble the following:
- Negative battery cable
- Driveshaft
- Set ring from the inboard joint
- Boot bands. If the band is welded, cut the band. If it is a double loop style, lift up the bend and push it into the clip.
- Inboard joint
- Inboard joint rollers
- Circlip
- Spider
- Inboard boot and dynamic damper

To install:
5. Install or connect the following:
- Dynamic damper and boot to the driveshaft
- Spider to the driveshaft by aligning the marks on the spider and the end of the driveshaft
- Circlip into the groove
- Rollers onto the spider with the high shoulder pointing outward and pack the inboard joint with grease
- Inboard joint to the driveshaft by aligning the marks

6. Adjust the length of the driveshaft for the proper side and adjust the boots halfway between full compression and extension. Properly position the dynamic damper between 16.3 and 16.5 inches.

7. Fit the boot ends over the driveshaft and inboard joint and install a band on the boot.

8. Tighten the boot band.

9. Install or connect the following:
- New set ring to the end of the driveshaft
- Driveshaft
- Negative battery cable

10. Enter the preset radio frequencies.

11. Start the vehicle and check for leaks, repair if necessary.

Outboard Joint

1. Before servicing the vehicle, refer to the precautions in the beginning of this section.

2. Turn the battery switch **OFF** and measure the voltage before proceeding.

3. Disconnecting the negative battery cable will clear the radio's preset buttons. Write down the radio frequencies.

4. Remove or disassemble the following:
- Halfshaft
- Boot bands from the outboard boot. Slide the outboard boot to the inboard joint side

5. Carefully place the driveshaft in a vise and using a slide hammer remove the outboard joint.

6. Remove the driveshaft from the vise.

7. Remove the stop ring from the driveshaft and remove the outboard boot.

To install:
8. Install or connect the following:
- New boot bands
- Outboard boot
- New stop ring into the driveshaft groove and pack the outboard joint with grease
- Outboard boot ends to the driveshaft and joint
- Tighten the boot bands
- Driveshaft
- Negative battery cable

9. Enter the preset radio frequencies.

10. Start the vehicle and check for leaks, repair if necessary.

STEERING AND SUSPENSION

Air Bag

✻✻ CAUTION

These vehicles are equipped with an air bag system. The system MUST BE disabled before performing service on or around system components, steering column, instrument panel components, wiring and sensors. Failure to follow safety and disabling procedures could result in accidental air bag deployment, possible personal injury and unnecessary system repairs.

PRECAUTIONS

Several precautions must be observed when handling the inflator module to avoid accidental deployment and possible personal injury:

- Never carry the inflator module by the wires or connector on the underside of the module.
- When carrying a live inflator module, hold securely with both hands, and ensure that the bag and trim cover are pointed away.
- Place the inflator module on a bench or other surface with the bag and trim cover facing up.

- With the inflator module on the bench, never place anything on or close to the module which may be thrown in the event of an accidental deployment.

DISARMING

1. Turn the ignition switch to the **OFF** position.
2. Disconnect the negative battery cable and wait at least 3 minutes before proceeding
3. Disconnecting the negative battery cable will clear the radio's preset buttons. Write down the radio frequencies.
4. Disconnect both airbag connectors.
5. Disconnect both seat belt tensioner connectors.
6. Remove the steering wheel access panel and disconnect the 2P connector from the cable reel connector
7. Remove the glove box and disconnect the passenger side air bag 2P connector from the floor wire harness connector.

Electrical Power Steering

REMOVAL & INSTALLATION

1. Before servicing the vehicle, refer to the precautions in the beginning of this section.

2. Turn the battery switch **OFF** and measure the voltage before proceeding.
3. Disconnecting the negative battery cable will clear the radio's preset buttons. Write down the radio frequencies.
4. Remove or disconnect the following:

- Negative battery cable
- Both front wheels
- Drivers side airbag
- Steering wheel
- Steering joint
- Battery box
- Front damper base beam and loosen the stop plate bolts
- Tie rods
- Tie rod plate and stop plate
- Electrical power steering motor 2P connector and torque sensor 7P connector
- Ground cable terminal from the gearbox housing
- Gearbox mounting bolts

5. Move the steering gearbox to the passenger side and remove the drivers side first.

To install:

6. Install or connect the following:
- Steering gearbox
- Steering gearbox mounting bolts and torque the bolts to 43 ft. lbs. (58 Nm)

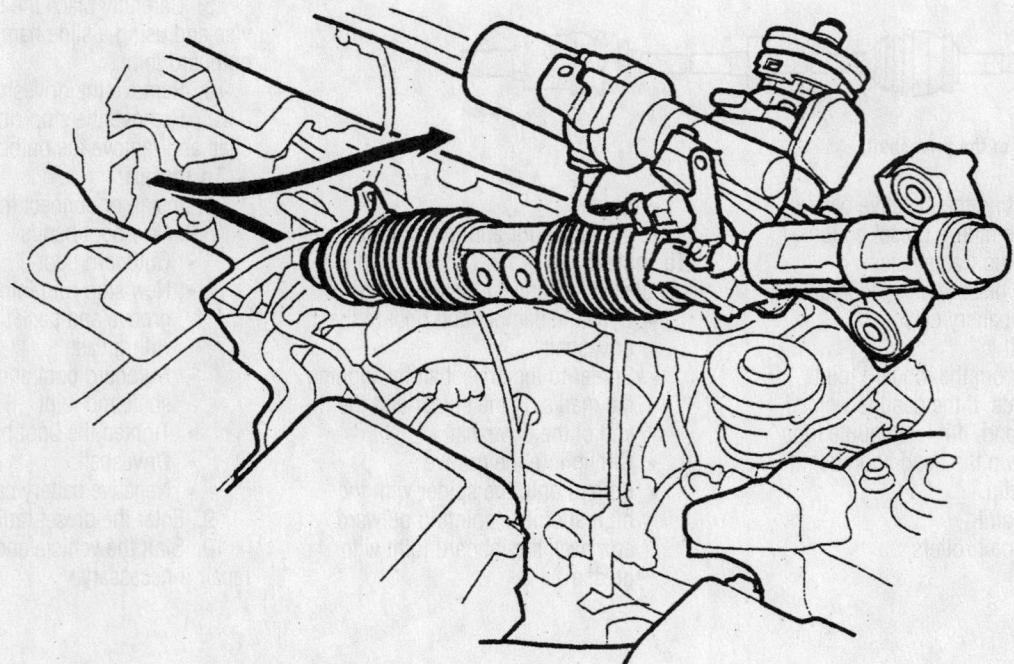

9347JG25

Remove the electrical power steering gearbox and motor as a unit

- Ground cable terminal to the gear-box housing
- Motor connector and torque sensor connector and torque the ground cable bolt to 89 inch lbs. (10 Nm)
- Tie rod ends
- Tie rod and stop plates and torque the bolts to 72 ft. lbs. (98 Nm) and bend the locking tabs into position
- Front damper base beam and battery box
- Steering joints
- Steering wheel and airbag module
- Front wheels
- Negative battery cable
7. Enter the preset radio frequencies.

Electric Steering Gear Motor

REMOVAL & INSTALLATION

1. Before servicing the vehicle, refer to the precautions in the beginning of this section.

2. Turn the battery switch **OFF** and measure the voltage before proceeding.
3. Disconnecting the negative battery cable will clear the radio's preset buttons. Write down the radio frequencies.
4. Remove or disconnect the following:
   - Negative battery cable
   - Steering gearbox
   - Electrical power steering motor 2P connector and torque sensor 7P connector
   - Ground cable terminal from the gearbox housing
   - Steering gear motor and O-ring

To install:
5. Install or connect the following:
   - New O-ring lubricated with silicone grease to the motor
   - Motor to the gearbox by engaging the motor shaft and pinion shaft. Torque the bolts to 89 inch lbs. (10 Nm)
   - Ground cable terminal to the gearbox housing
   - Motor connector and torque sensor connector and torque the

ground cable bolt to 89 inch lbs. (10 Nm)
   - Steering gearbox
   - Negative battery cable
6. Enter the preset radio frequencies.

Strut

REMOVAL & INSTALLATION

Front

1. Before servicing the vehicle, refer to the precautions in the beginning of this section.
2. Turn the battery switch **OFF** and measure the voltage before proceeding.
3. Disconnecting the negative battery cable will clear the radio's preset buttons. Write down the radio frequencies.
4. Remove or disconnect the following:
   - Negative battery cable
   - Front wheels
   - Mark each one right or left, as applicable, if both struts are being removed
   - Cotter pin from the tie rod end ball joint and remove the nut
   - Tie rod end ball joint from the strut using Special Tool 07MAC-SL00200
   - Stabilizer link from the strut
   - Wheel sensor harness bracket
   - Brake hose bracket
   - Strut pinch bolts while holding the nuts tight
   - Upper nuts from the strut
5. Lower the lower control arm and remove the strut assembly
To install:
6. Install or connect the following:
   - Strut to the body
   - New self-locking nuts to the top of the strut and hand tighten the nuts
   - Strut bottom to the steering knuckle and hand tighten the nuts
   - Stabilizer link to the strut and hand tighten the nut.
7. Using a jack stand and a block of wood, raise the suspension to load the front suspension.
8. Torque the upper mounting bolts to 40 ft. lbs. (54 Nm) and the lower pinch bolts to 72 ft. lbs. (98 Nm).
9. Torque the stabilizer link to strut flange nut to 22 ft. lbs. (29 Nm).

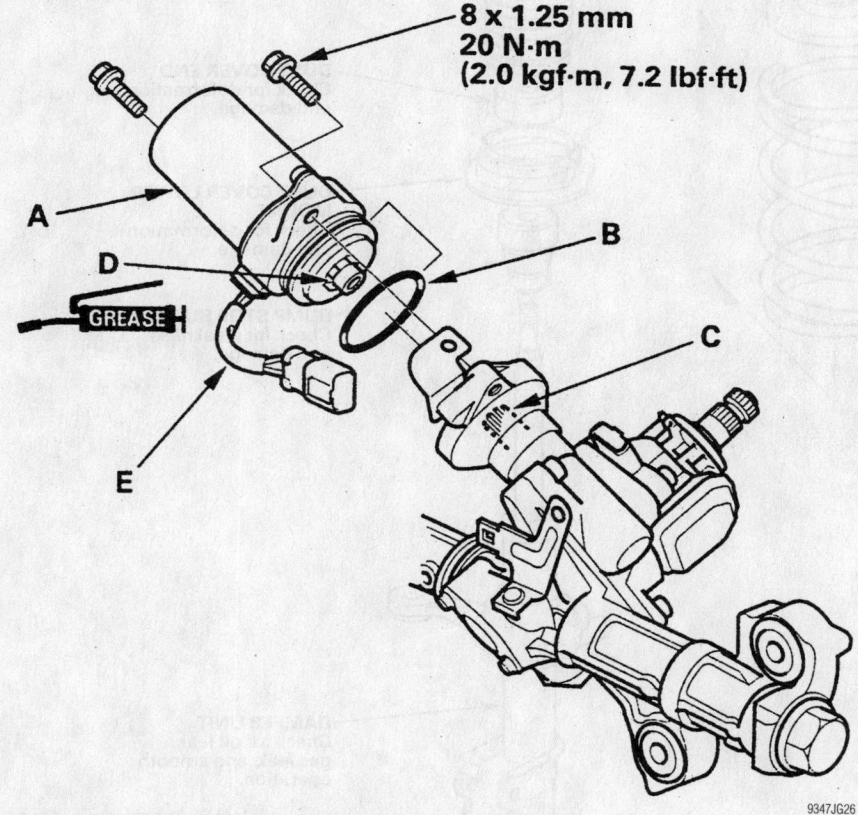

**8 x 1.25 mm
20 N·m
(2.0 kgf·m, 7.2 lbf·ft)**

A
D
GREASE
E
B
C

9347JG26

Exploded view of the electric power steering motor assembly

10. Install or connect the following:
   - Tie rod to the steering arm and torque the nut to 22 ft. lbs. (29 Nm)
   - Brake hose bracket to the strut and torque the bolt to 16 ft. lbs. (22 Nm)
   - Wheel sensor harness bracket and torque the bolts to 89 inch lbs. (10 Nm)
   - Front wheel
   - Negative battery cable
11. Enter the preset radio frequencies.

Rear

1. Before servicing the vehicle, refer to the precautions in the beginning of this section.

2. Turn the battery switch **OFF** and measure the voltage before proceeding.

3. Disconnecting the negative battery cable will clear the radio's preset buttons. Write down the radio frequencies.

4. Remove or disconnect the following:
   - Negative battery cable

   - Fender skirts and place a jack under each end of the rear axle beam
   - Rear wheels
   - Lower flange bolt
   - Upper mounting bolts
   - Strut

To install:

5. Install or connect the following:
   - Strut assembly and hand tighten the upper and lower mounting bolts.

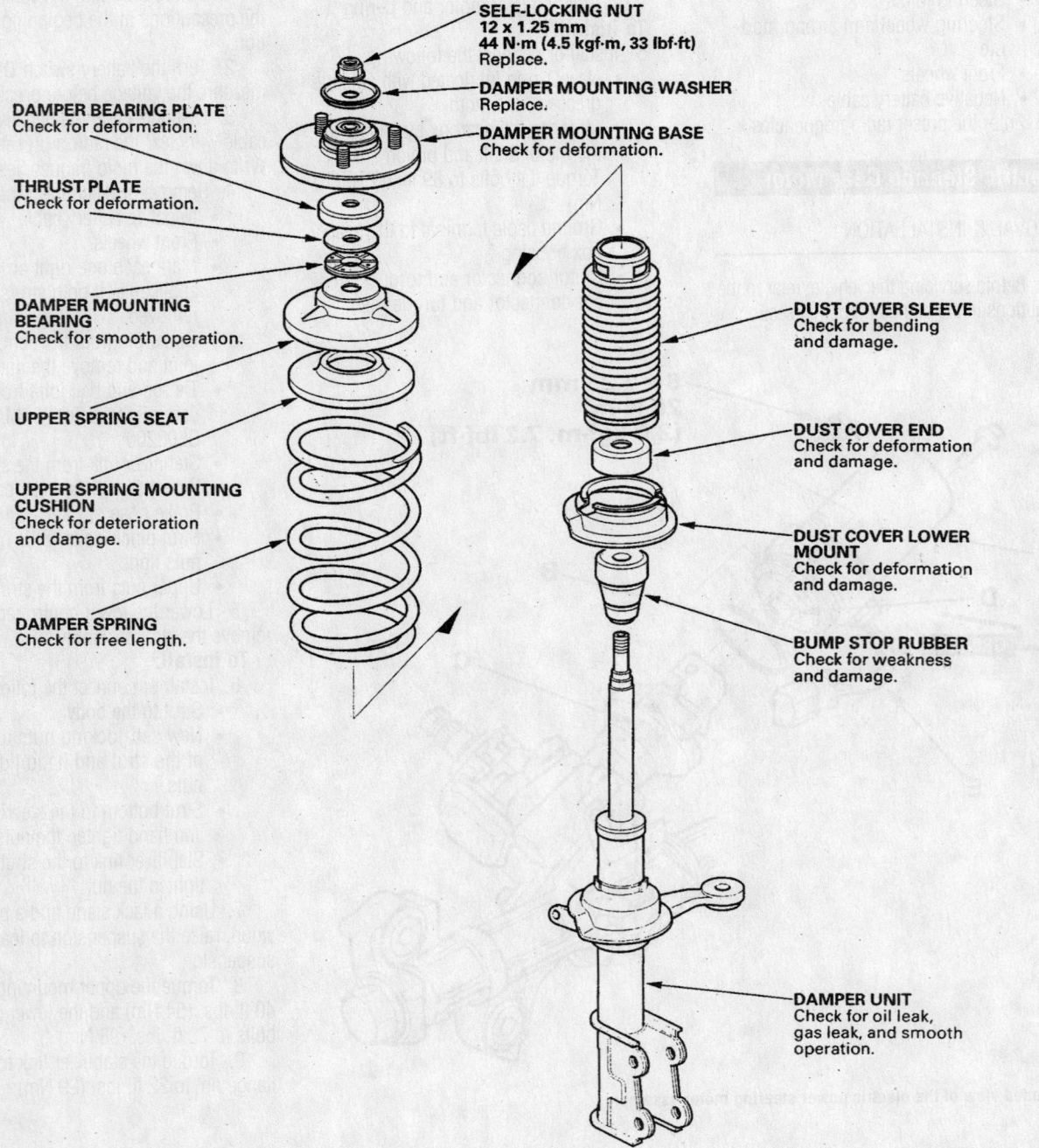

SELF-LOCKING NUT
12 x 1.25 mm
44 N·m (4.5 kgf·m, 33 lbf·ft)
Replace.

DAMPER MOUNTING WASHER
Replace.

DAMPER MOUNTING BASE
Check for deformation.

DAMPER BEARING PLATE
Check for deformation.

THRUST PLATE
Check for deformation.

DAMPER MOUNTING BEARING
Check for smooth operation.

UPPER SPRING SEAT

UPPER SPRING MOUNTING CUSHION
Check for deterioration and damage.

DAMPER SPRING
Check for free length.

DUST COVER SLEEVE
Check for bending and damage.

DUST COVER END
Check for deformation and damage.

DUST COVER LOWER MOUNT
Check for deformation and damage.

BUMP STOP RUBBER
Check for weakness and damage.

DAMPER UNIT
Check for oil leak, gas leak, and smooth operation.

9347JG27

Exploded view of the front suspension

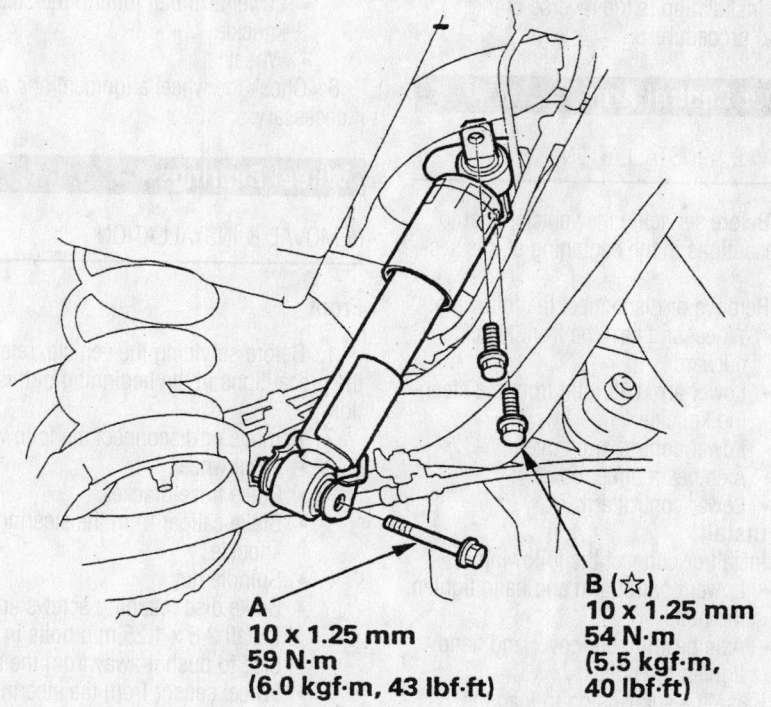

A
10 x 1.25 mm
59 N·m
(6.0 kgf·m, 43 lbf·ft)

B (☆)
10 x 1.25 mm
54 N·m
(5.5 kgf·m, 40 lbf·ft)

9347JG28

Remove the rear strut assembly

6. Raise the rear suspension to the load the vehicle weight.

7. Torque the upper mounting bolts to 40 ft. lbs. (54 Nm) and the lower mounting bolt to 43 ft. lbs. (59 Nm).
 • Rear wheels
 • Fender skirts and remove the support from the axle beam
 • Negative battery cable

8. Enter the preset radio frequencies.

Coil Spring

REMOVAL & INSTALLATION

Front

1. Before servicing the vehicle, refer to the precautions in the beginning of this section.

2. Turn the battery switch **OFF** and measure the voltage before proceeding.

3. Disconnecting the negative battery cable will clear the radio's preset buttons. Write down the radio frequencies.

4. Remove or disconnect the following:
 • Negative battery cable
 • Front wheel
 • Strut assembly and place it in a Strut Spring Compressor tool

5. Compress the spring as much as necessary to remove the nut.

6. Release the pressure from the strut compressor tool

7. Remove or disconnect the following:
 • Strut mounting washer and base
 • Strut bearing plate, thrust plate and mounting bearing
 • Upper spring seat and mounting cushion
 • Coil spring from the strut assembly

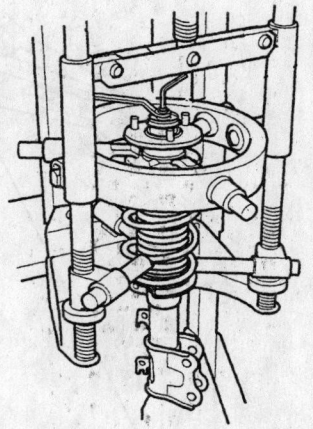

9347JG29

Coil spring mounted in the coil spring compressor tool

8. Inspect the coil spring for any signs of damage

To install:

9. Install or connect the following:
 • Coil spring to the strut
 • Mounting cushion and upper spring seat
 • Mount bearing, thrust plate and bearing plate
 • Mounting base and position the assembly is a strut compressor tool

10. Compress the strut spring and install a new washer and nut. Torque the nut to 33 ft. lbs. (44 Nm).

11. Remove the assembly from the compressor tool.

12. Install or connect the following:
 • Strut assembly to the vehicle
 • Front wheel
 • Negative battery cable

13. Enter the preset radio frequencies.

14. Check and adjust the front end alignment.

Rear

1. Before servicing the vehicle, refer to the precautions in the beginning of this section.

2. Turn the battery switch **OFF** and measure the voltage before proceeding.

3. Disconnecting the negative battery cable will clear the radio's preset buttons. Write down the radio frequencies.

4. Remove or disconnect the following:
 • Negative battery cable
 • Fender skirt
 • Rear wheel
 • Rear speed sensor connectors
 • Wheel sensor brackets and the brake hose bracket

5. Place a jack stand under each side of the axle beam.

6. Remove the lower flange bolt from the strut.

7. Lower the jacks evenly.

8. Remove the spring and upper cushion.

9. Remove the nut and bump stop.

To install:

10. Install or connect the following:
 • Nut and bump stop and torque the nut to 29 ft. lbs. (39 Nm)

11. Align the spring upper end with the stepped part of the spring cushion.

12. Align the spring lower end with the stepped part of the spring seat on the axle beam.
 • Upper spring cushion and spring

- Strut lower flange bolts and hand tighten them

13. Raise the suspension to load the vehicle weight.

14. Torque the flange bolt to 43 ft. lbs. (59 Nm).
- Wheel sensor bracket and torque the bolt to 89 inch lbs. (10 Nm)
- Brake hose bracket and torque the bolt to 16 ft. lbs. (22 Nm)
- Wheel sensor connectors
- Rear wheel and fender skirt
- Negative battery cable

15. Enter the preset radio frequencies.

Lower Ball Joint

REMOVAL & INSTALLATION

→**Always use a Ball Joint Removal Tool, 28 MM 07MAC-SL00200 to disconnect the ball joint.**

1. Before servicing the vehicle, refer to the precautions in the beginning of this section.

2. Remove the wheel.

3. Install a hex nut onto the threads of the ball joint. Make certain that the nut is flush with the ball joint pin end.

4. Lubricate the pressure bolt and the jaws on the ball joint removal tool.

5. Install the ball joint removal tool by installing the jaws to the ball joint. Adjust the jaw spacing by turning the pressure bolt.

6. Turn the adjusting bolt to make the jaws parallel and hand tighten the pressure bolt.

7. Tighten the pressure bolt until the ball joint pin pops loose from the steering arm.

8. Remove the ball joint removal tool and remove the nut from the ball joint.

9. Remove the ball joint from the steering arm.

To install:

10. Install or connect the following:
- Ball joint to the steering arm and torque the bolts to 51 ft. lbs. (69 Nm)

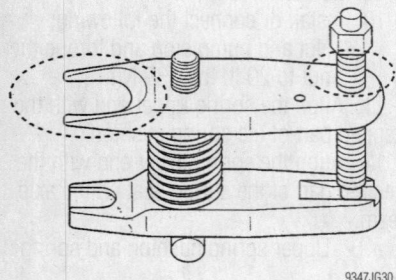

9347JG30

Install the ball joint removal tool to the ball joint

11. Installation is the reverse of the removal procedure.

Lower Control Arm

REMOVAL & INSTALLATION

1. Before servicing the vehicle, refer to the precautions in the beginning of this section.

2. Remove or disconnect the following:
- Wheel and turn the front knuckle outward
- Lower arm ball joint from the steering knuckle
- Lower control arm cover
- Axle beam under cover
- Lower control arm

To install:

3. Install or connect the following:
- Lower control arm and hand tighten the bolts
- Axle beam under cover and hand tighten the bolts

4. Raise the suspension to load the weight of the vehicle.

5. Torque the lower control arm and axle beam under cover bolts to 51 ft. lbs. (69 Nm).
- Lower control arm cover. Torque the lower bolts to 89 inch lbs. (10 Nm) and the upper bolt to 16 ft. lbs. (22 Nm).

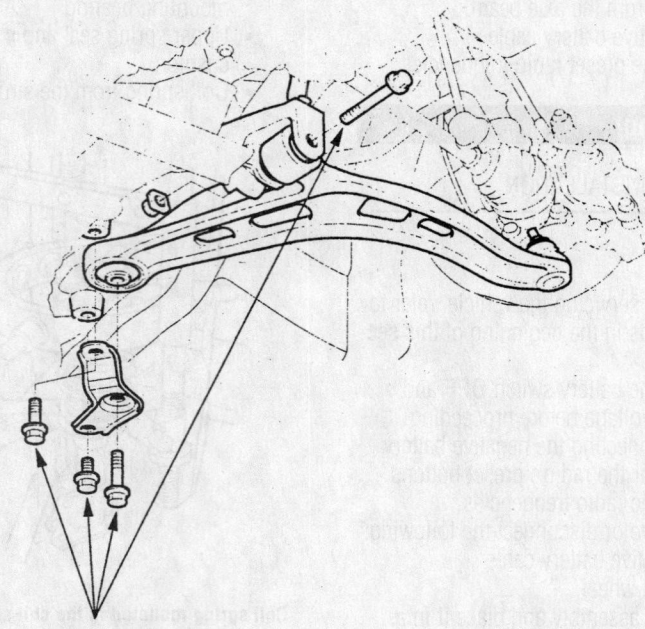

12 x 1.25 mm
69 N·m (7.0 kgf·m, 51 lbf·ft)

9347JG31

Remove the lower control arm

- Lower arm ball joint to the steering knuckle
- Wheel

6. Check the wheel alignment and adjust if necessary.

Wheel Bearings

REMOVAL & INSTALLATION

Front

1. Before servicing the vehicle, refer to the precautions in the beginning of this section.

2. Remove or disconnect the following:
- Front wheel
- Brake hose bracket
- Brake caliper from the steering knuckle
- Spindle nut
- Brake disc retaining screws and install 2 8 x 1.25 mm bolts in the disc to push it away from the hub
- Wheel sensor from the steering knuckle
- Lower arm ball joint
- Strut pinch bolts
- Steering knuckle off the driveshaft while supporting the outboard joint
- Steering knuckle from the strut

3. Separate the hub from the bearing unit using Special Tool 07965-SA70100 and a press.

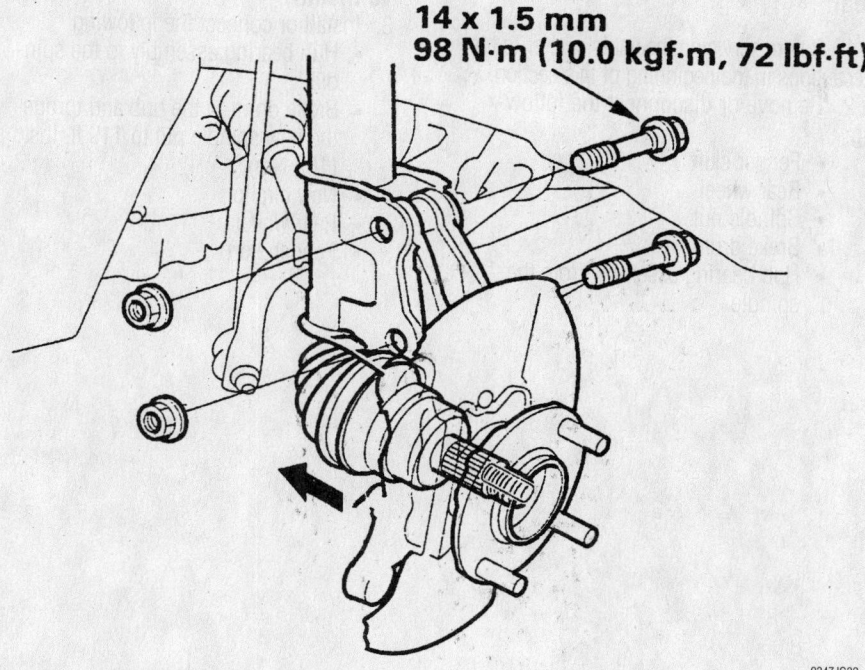

14 x 1.5 mm
98 N·m (10.0 kgf·m, 72 lbf·ft)

9347JG32

Remove the steering knuckle from the driveshaft

4. Press the inner race from the hub.
   - Splash guard from the steering knuckle
   - Hub bearing from the steering knuckle

To install:

5. Install or connect the following:
   - New hub bearing into the hub with Special Tools 07GAF-SD40200 and 07965-SD90100
   - Splash guard and torque the bolts to 89 inch lbs. (10 Nm)
   - Hub to the steering knuckle and torque the bolts to 43 ft. lbs. (59 Nm)
   - Steering knuckle to the driveshaft and torque the bolts to 72 ft. lbs. (98 Nm)
   - Strut pinch bolt and torque the bolt to 72 ft. lbs. (98 Nm)
   - Lower arm ball joint with a new castle nut and torque the nut to 40 ft. lbs. (54 Nm)
   - Wheel sensor and torque the bolts to 89 inch lbs. (10 Nm)

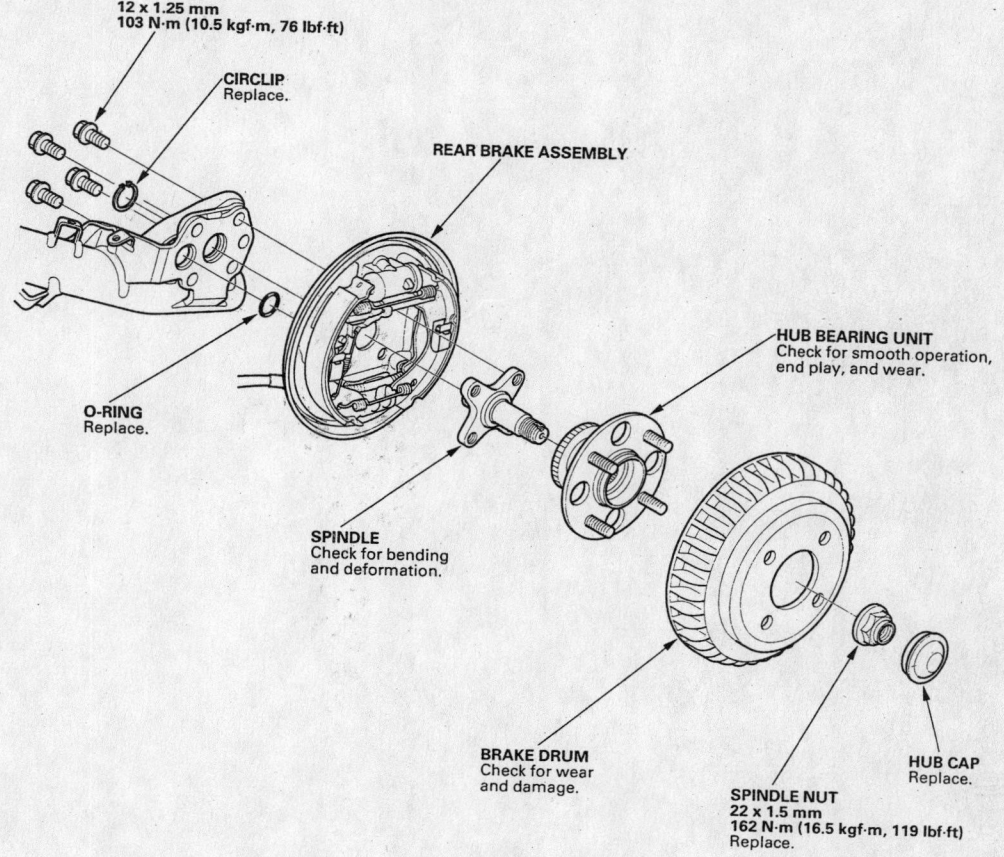

12 x 1.25 mm
103 N·m (10.5 kgf·m, 76 lbf·ft)

CIRCLIP
Replace.

REAR BRAKE ASSEMBLY

O-RING
Replace.

SPINDLE
Check for bending and deformation.

HUB BEARING UNIT
Check for smooth operation, end play, and wear.

BRAKE DRUM
Check for wear and damage.

SPINDLE NUT
22 x 1.5 mm
162 N·m (16.5 kgf·m, 119 lbf·ft)
Replace.

HUB CAP
Replace.

9347JG33

Exploded view of the hub bearing unit assembly

- Brake disc and torque the bolts to 89 inch lbs. (10 Nm)
- New spindle nut and torque the nut to 134 ft. lbs. (181 Nm)
- Caliper and bracket and torque the mounting larger bolts to 80 ft. lbs. (108 Nm)
- Brake hose bracket and torque the bolt to 22 ft. lbs. (26 Nm)
- Front wheel

6. Check the front end alignment and adjust if necessary.

Rear

1. Before servicing the vehicle, refer to the precautions in the beginning of this section.

2. Remove or disconnect the following:

- Fender skirt
- Rear wheel
- Spindle nut
- Brake drum
- Hub bearing assembly from the spindle

To install:

3. Install or connect the following:

- Hub bearing assembly to the spindle
- Brake drum to the hub and torque the new spindle nut to 119 ft. lbs. (162 Nm)
- New circlip
- Rear wheel
- Fender skirt

HYUNDAI

1998–01

Accent • Elantra • Sonata • Tiburon • XG 300

11

PRECAUTIONS

Before servicing any vehicle, please be sure to read all of the following precautions, which deal with personal safety, prevention of component damage, and important points to take into consideration when servicing a motor vehicle:

• Never open, service or drain the radiator or cooling system when the engine is hot; serious burns can occur from the steam and hot coolant.

• Observe all applicable safety precautions when working around fuel. Whenever servicing the fuel system, always work in a well-ventilated area. Do not allow fuel spray or vapors to come in contact with a spark, open flame, or excessive heat (a hot drop light, for example). Keep a dry chemical fire extinguisher near the work area. Always keep fuel in a container specifically designed for fuel storage; also, always properly seal fuel containers to avoid the possibility of fire or explosion. Refer to the additional fuel system precautions later in this section.

• Fuel injection systems often remain pressurized, even after the engine has been turned **OFF**. The fuel system pressure must be relieved before disconnecting any fuel lines. Failure to do so may result in fire and/or personal injury.

• Brake fluid often contains polyglycol ethers and polyglycols. Avoid contact with the eyes and wash your hands thoroughly after handling brake fluid. If you do get brake fluid in your eyes, flush your eyes with clean, running water for 15 minutes. If

eye irritation persists, or if you have taken brake fluid internally, IMMEDIATELY seek medical assistance.

• The EPA warns that prolonged contact with used engine oil may cause a number of skin disorders, including cancer. You should make every effort to minimize your exposure to used engine oil. Protective gloves should be worn when changing oil. Wash your hands and any other exposed skin areas as soon as possible after exposure to used engine oil. Soap and water, or waterless hand cleaner should be used.

• All new vehicles are now equipped with an air bag system. The system must be disabled before performing service on or around system components, steering column, instrument panel components, wiring and sensors. Failure to follow safety and disabling procedures could result in accidental air bag deployment, possible personal injury and unnecessary system repairs.

• Always wear safety goggles when working with, or around, the air bag system. When carrying a non-deployed air bag, be sure the bag and trim cover are pointed away from your body. When placing a non-deployed air bag on a work surface, always face the bag and trim cover upward, away from the surface. This will reduce the motion of the module if it is accidentally deployed. Refer to the additional air bag system precautions later in this section.

• Clean, high quality brake fluid from a sealed container is essential to the safe and

proper operation of the brake system. You should always buy the correct type of brake fluid for your vehicle. If the brake fluid becomes contaminated, completely flush the system with new fluid. Never reuse any brake fluid. Any brake fluid that is removed from the system should be discarded. Also, do not allow any brake fluid to come in contact with a painted surface; it will damage the paint.

• Never operate the engine without the proper amount and type of engine oil; doing so WILL result in severe engine damage.

• Timing belt maintenance is extremely important. Many models utilize an interference-type, non-freewheeling engine. If the timing belt breaks, the valves in the cylinder head may strike the pistons, causing potentially serious (also time-consuming and expensive) engine damage. Refer to the maintenance interval charts in the front of this manual for the recommended replacement interval for the timing belt, and to the timing belt section for belt replacement and inspection.

• Disconnecting the negative battery cable on some vehicles may interfere with the functions of the on-board computer system(s) and may require the computer to undergo a relearning process once the negative battery cable is reconnected.

• When servicing drum brakes, only disassemble and assemble one side at a time, leaving the remaining side intact for reference.

ENGINE REPAIR

➡**Disconnecting the negative battery cable on some vehicles may interfere with the functions of the on board computer system. The computer may undergo a relearning process once the negative battery cable is reconnected.**

Distributor

All engines except the 3.0L (VIN T) V6 are equipped with a Distributorless Ignition System (DIS).

REMOVAL

1. Before servicing the vehicle, refer to the precautions in the beginning of this section.
2. Remove the distributor cap.
3. Remove the distributor electrical connectors.

4. Matchmark the rotor to the distributor housing.
5. Matchmark the distributor housing to the engine.
6. Unbolt and remove the distributor.

INSTALLATION

Timing not disturbed

1. Before servicing the vehicle, refer to the precautions in the beginning of this section.
2. Install the distributor with the rotor-to-housing and the housing-to-engine matchmarks aligned.
3. Install the distributor electrical connectors.
4. Install the distributor cap.

5. Check the ignition timing and adjust as necessary.

Timing disturbed

1. Before servicing the vehicle, refer to the precautions in the beginning of this section.
2. Set the crankshaft to Top Dead Center (TDC) of the compression stroke for the No. 1 cylinder.
3. Align the timing marks on the distributor gear and the distributor housing.
4. Install the distributor by aligning the groove of the installation flange with the center of the installation stud.
5. Connect the distributor electrical connectors.
6. Install the distributor cap.
7. Check the ignition timing and adjust as necessary.

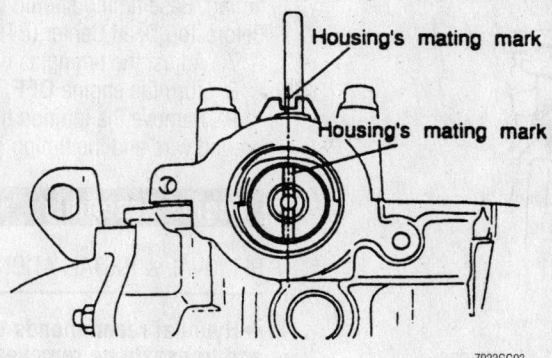

Aligning the distributor housing and gear mating marks—3.0L Sonata

Alternator

REMOVAL

1. Before servicing the vehicle, refer to the precautions in the beginning of this section.

2. Remove or disconnect the following:
   - Negative battery cable
   - Radiator mounting bolts, 1.6L only
   - Alternator drive belt
   - Alternator wiring harness connectors
   - Alternator. It may be necessary to raise the radiator

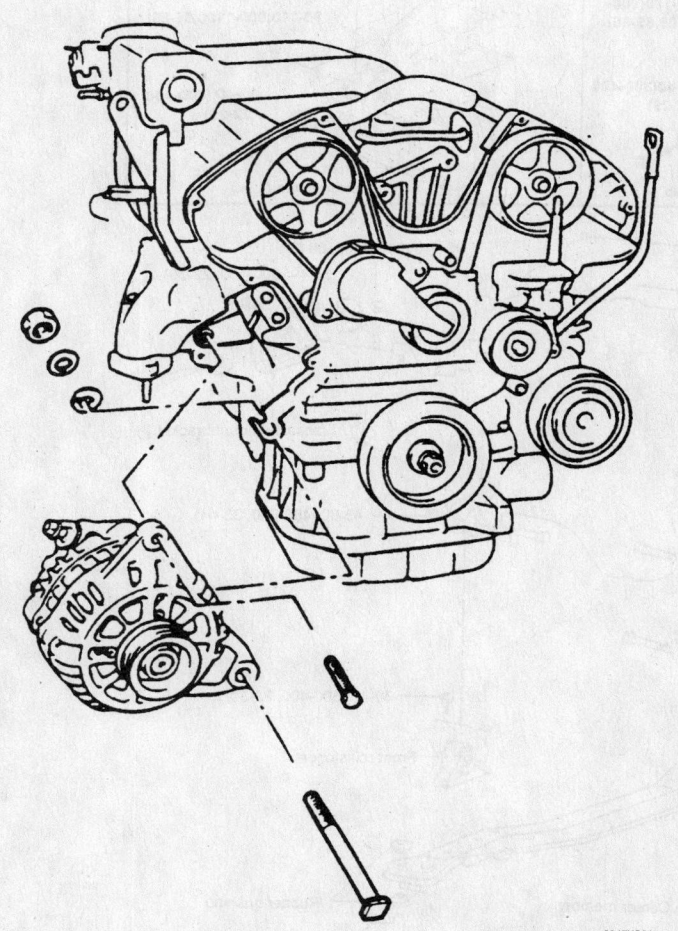

Exploded view of the alternator—XG 300

INSTALLATION

1. Before servicing the vehicle, refer to the precautions in the beginning of this section.

2. Install or connect the following:
   - Alternator and reposition the radiator, if raised
   - Alternator wiring harness connectors
   - Alternator drive belt

3. Adjust the alternator belt and torque the bolts to the following specifications:

   a. Accent: Pivot bolt to 14–18 ft. lbs. (19–25 Nm) and adjustment bolt to 14–20 ft. lbs. (19–28 Nm)

   b. 2.0L and 2.5L Sonata, Elantra and Tiburon: Pivot bolt to 14–18 ft. lbs. (19–25 Nm) and adjustment bolt to 105–132 inch lbs. (12–15 Nm)

   c. 2.4L Sonata: Pivot bolt to 26–41 ft. lbs. (34–54 Nm) and the adjustment bolt to 14–18 ft. lbs. (19–25 Nm)

   d. 3.0L Sonata and XG 300: Pivot bolt to 14–18 ft. lbs. (19–25 Nm) and adjustment bolt to 11–16 ft. lbs. (15–22 Nm)

4. Install the radiator mounting bolts, if removed.

5. Connect the negative battery cable.

Ignition Timing

ADJUSTMENT

Except 1.8L and 3.0L (VIN T) Engines

These engines are equipped with a Distributorless Ignition System (DIS). No adjustment is necessary.

1.8L and 3.0L (VIN T) Engines

➡**Timing is adjusted with the engine at normal operating temperature and all electrical accessories OFF.**

Connect the jumper wire from ground to the ignition timing terminal—3.0L engine

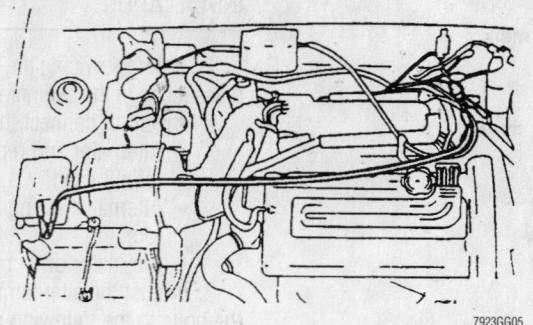

Connect the jumper wire from ground to the ignition timing terminal—1.8L engines

1. Before servicing the vehicle, refer to the precautions in the beginning of this section.

2. Start the engine and allow it to reach operating temperature.

3. Turn the engine **OFF**.

4. Ground the ignition timing terminal.

5. Connect a timing light to the No. 1 spark plug wire.

6. Start the engine and check the base timing. Base timing should be 3–7 degrees Before Top Dead Center (BTDC).

7. Adjust the timing as necessary.

8. Turn the engine **OFF**.

9. Remove the ignition timing connector ground wire and the timing light.

Engine Assembly

REMOVAL & INSTALLATION

➡**Hyundai recommends that the engine and transaxle be removed as a single unit on all models.**

1. Before servicing the vehicle, refer to the precautions in the beginning of this section.

2. Drain the cooling system.

3. Drain the transaxle.

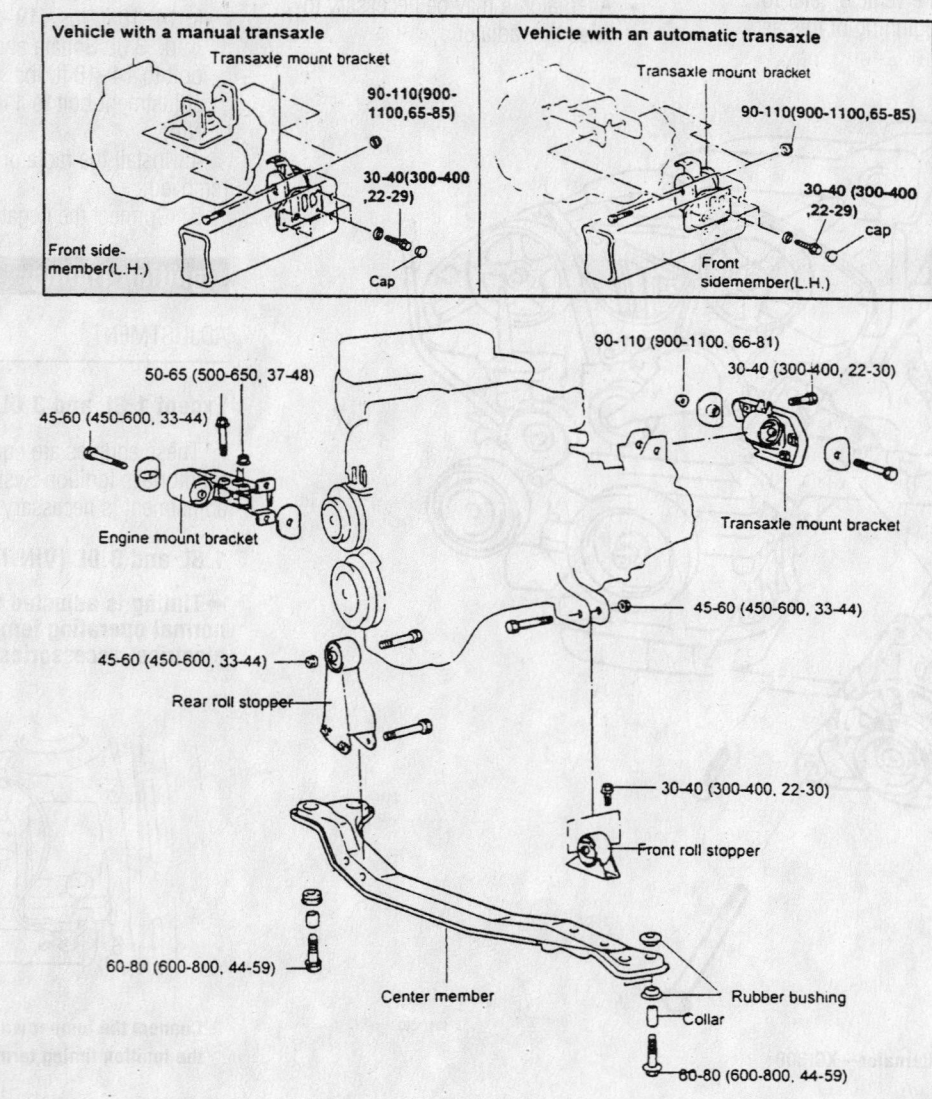

Exploded view of the engine mounts and torque specifications—Accent

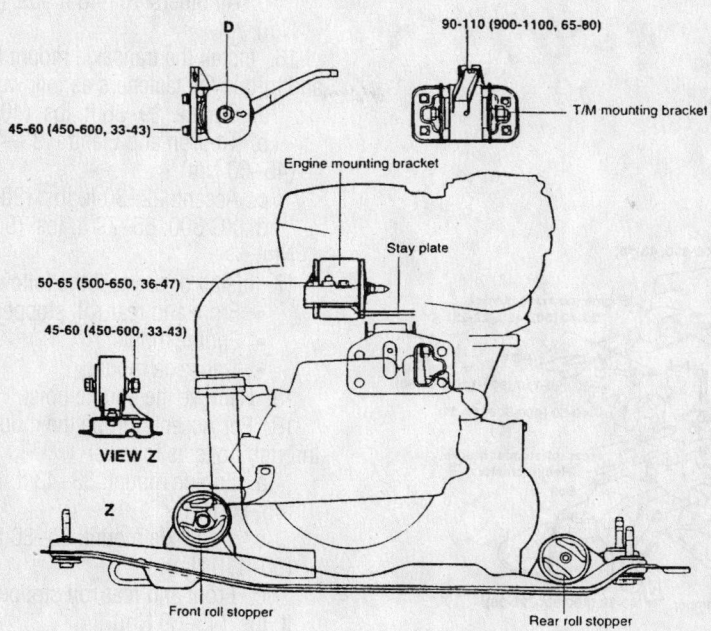

TORQUE: Nm (kg.cm, lb.ft)

7923GG13

Exploded view of the engine mounts and torque specifications—Tiburon and Elantra

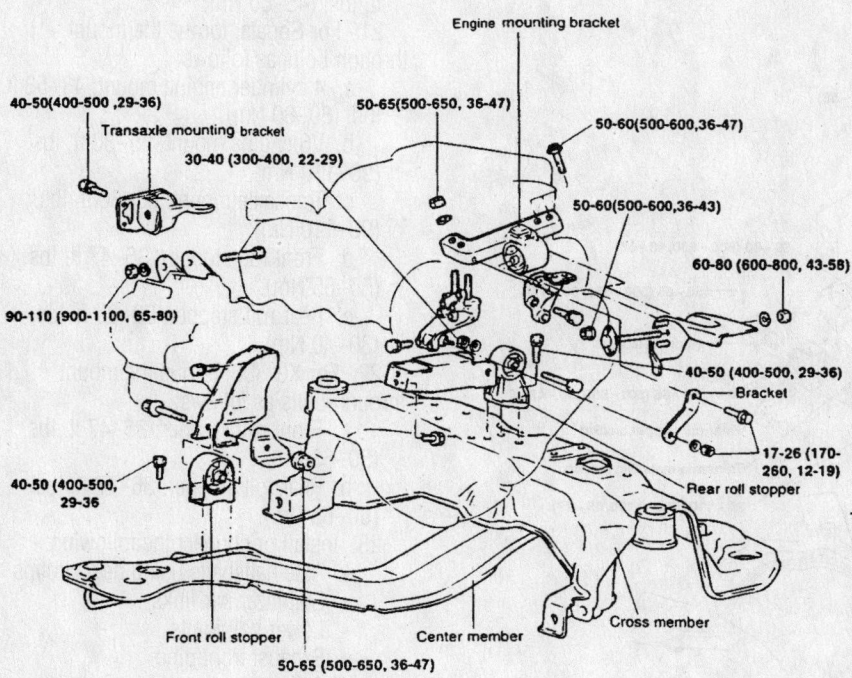

TORQUE : Nm (kg.cm, lb.ft)

7923GG15

Exploded view of the engine mounts and torque specifications—4 cylinder Sonata

4. Drain the engine oil.

5. Relieve fuel system pressure.

6. Remove or disconnect the following:
   • Battery
   • Hood
   • Air intake assembly
   • Accessory drive belts
   • Engine wiring harness connectors
   • Reverse lamp switch connector, if equipped
   • Speedometer cable
   • Alternator harness connectors
   • Oil pressure gauge sender connector
   • Radiator hoses
   • Cooling fan
   • Fuel lines
   • Control cable, if equipped
   • Brake booster vacuum line
   • Intake manifold vacuum lines
   • Heater hoses
   • Accelerator cable
   • Cruise control cable, if equipped
   • Engine ground cable

7. If equipped with a manual transaxle, disconnect or remove the following:
   • Clutch cable
   • Select control valve connector
   • Shift linkage rods

8. If equipped with an automatic transaxle, disconnect or remove the following:
   • Transaxle oil cooler lines
   • Shift cable
   • Transaxle wiring connectors

9. For all vehicles, remove or disconnect the following:
   • Radiator
   • Power steering pump
   • A/C compressor, if equipped
   • Exhaust front pipe
   • Lower ball joints
   • Stabilizer bar links

10. Separate the inner CV-joints from the transaxle and suspend the halfshafts out of the work area with safety wire.

11. Attach a hoist to the engine lifting eyes.

12. Remove or disconnect the following:
   • Front and rear roll stoppers
   • Engine mount and bracket
   • Transaxle mount and bracket

13. Lift the powertrain out of the vehicle.

To install:

14. Lower the powertrain into position.

15. Install the motor mount bracket and torque the fasteners as follows:

   a. V6 engines: 43–58 ft. lbs. (60–80 Nm).

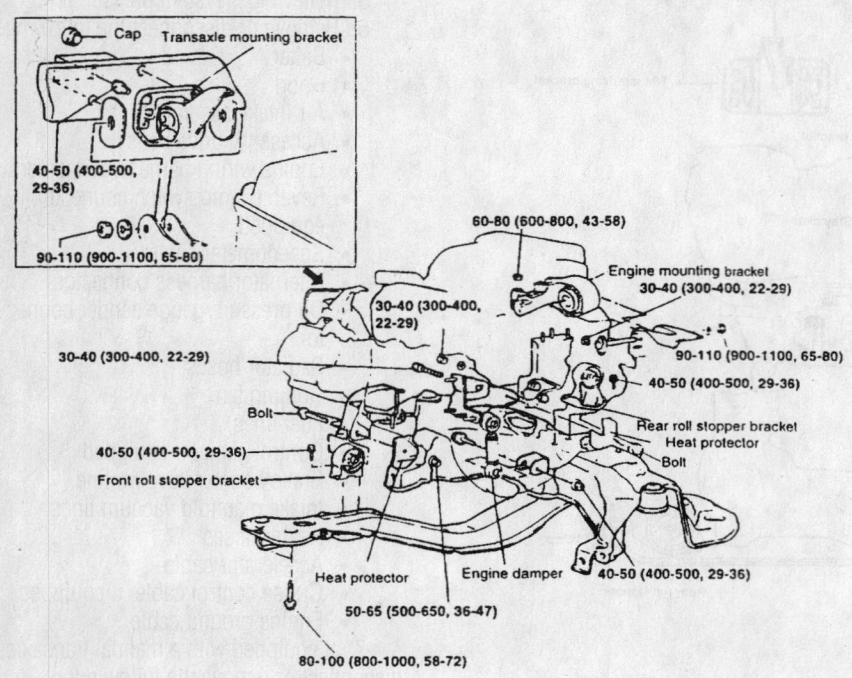

TORQUE : Nm (kg.cm, lb.ft)

7923GG16

Exploded view of the engine mounts and torque specifications—V6 Sonata

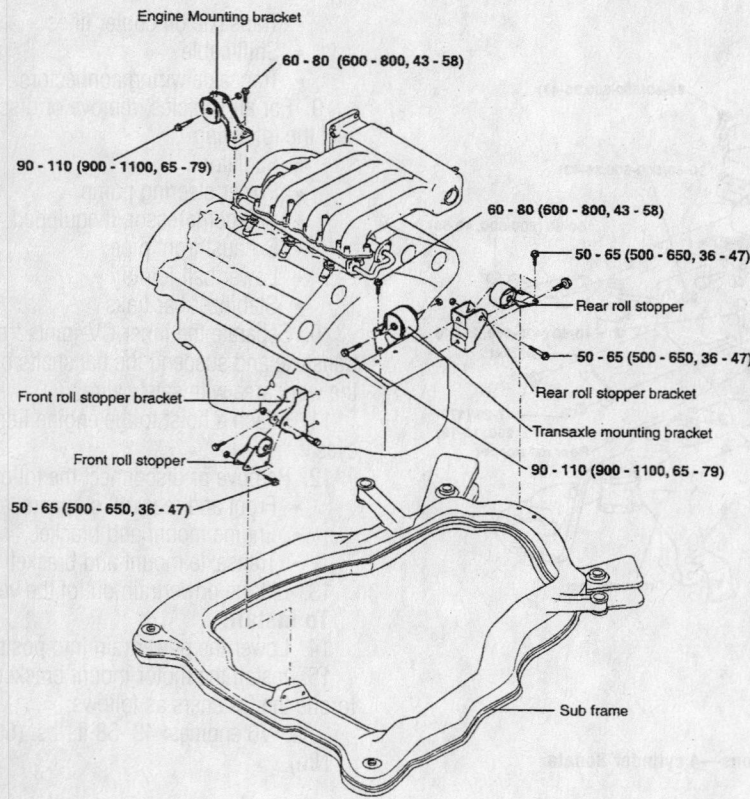

9347KG02

Exploded view of the engine mounts and torque specifications—XG 300

b. All others: 37–48 ft. lbs. (45–60 Nm).

16. Install the transaxle mount bracket and torque the fasteners as follows:

    a. Sonata: 29–36 ft. lbs. (40–50 Nm).

    b. Tiburon and Elantra: 33–43 ft. lbs. (45–60 Nm).

    c. Accent: 22–30 ft. lbs. (30–40 Nm).

    d. XG 300: 65–79 ft. lbs. (90–110 Nm).

17. Install or connect the following:
- Front and rear roll stoppers
- Engine mount
- Transaxle mount

18. Remove the engine hoist.

19. For Accent, torque the mount through bolts as follows:

    a. Engine mount: 33–43 ft. lbs. (45–60 Nm).

    b. Transaxle mount: 65–80 ft. lbs. (90–110 Nm).

    c. Front and rear roll stoppers: 33–43 ft. lbs. (45–60 Nm).

20. For Elantra and Tiburon, torque the mount through bolts as follows:

    a. Engine mount: 36–47 ft. lbs. (50–65 Nm).

    b. Transaxle mount: 65–80 ft. lbs. (90–110 Nm).

    c. Front and rear roll stoppers: 33–43 ft. lbs. (45–60 Nm).

21. For Sonata, torque the mount through bolts as follows:

    a. 4 cylinder engine mount: 43–58 ft. lbs. (60–80 Nm).

    b. V6 engine mount: 65–80 ft. lbs. (90–110 Nm).

    c. Transaxle mount: 65–80 ft. lbs. (90–110 Nm).

    d. Front roll stopper: 36–47 ft. lbs. (50–65 Nm).

    e. Rear roll stopper: 22–29 ft. lbs. (30–40 Nm).

22. For XG 300, torque the mount through bolts as follows:

    a. Front roll stopper: 36–47 ft. lbs. (50–65 Nm).

    b. Rear roll stopper: 36–47 ft. lbs. (50–65 Nm).

23. Install or connect the following:
- Axle halfshafts using new circlips
- Stabilizer bar links
- Lower ball joints
- Exhaust front pipe
- A/C compressor, if equipped
- Power steering pump
- Radiator

24. If equipped with a manual transaxle, install or connect the following:
- Clutch cable
- Select control valve connector
- Shift linkage rods

25. If equipped with an automatic transaxle, install or connect the following:
- Transaxle oil cooler lines
- Shift cable
- Transaxle wiring connectors

26. For all vehicles, install or connect the following:
- Engine ground cable
- Cruise control cable, if equipped
- Accelerator cable
- Heater hoses
- Intake manifold vacuum lines
- Brake booster vacuum line
- Fuel lines
- Cooling fan
- Radiator hoses
- Oil pressure gauge sender connector
- Alternator harness connectors
- Speedometer cable
- Reverse lamp switch connector
- Engine wiring harness connectors
- Accessory drive belts
- Air intake assembly
- Hood
- Battery

27. Fill the engine with clean oil.
28. Fill the transaxle to the correct level.
29. Fill the cooling system to the proper level.
30. Start the engine and check for leaks.

Water Pump

REMOVAL & INSTALLATION

1. Before servicing the vehicle, refer to the precautions in the beginning of this section.
2. Drain the cooling system.
3. Remove or disconnect the following:
- Negative battery cable
- Accessory drive belts
- Radiator hose
- Bypass hose, if equipped
- Water pump pulley
- Front cover
- Timing belt
- Alternator bracket
- Water pump

➡The water pump bolts are different lengths. Note the bolt location for assembly.

To install:

4. Install the water pump with new gaskets and O-rings. Tighten the bolts to the following specifications:
   a. 1.5L SOHC engine: 105–132 inch lbs. (12–15 Nm).

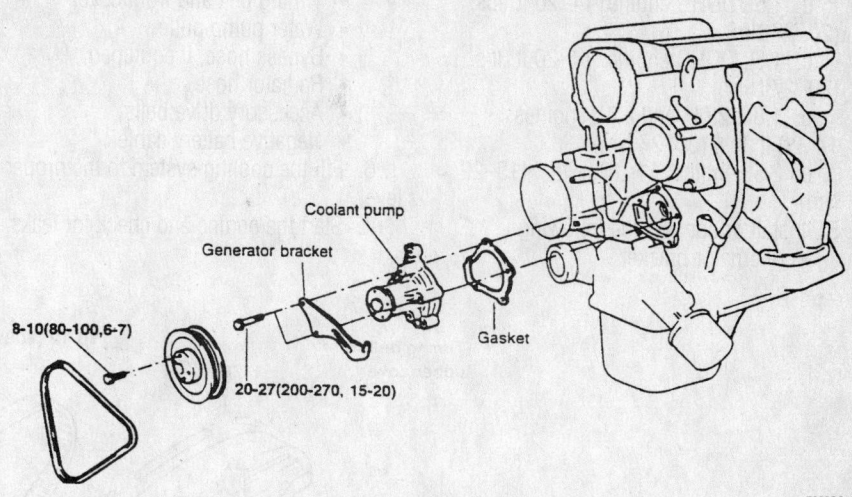

Exploded view of the water pump assembly—1.5L engines

8-10(80-100,6-7)
20-27(200-270, 15-20)
Coolant pump
Generator bracket
Gasket

7923GG17

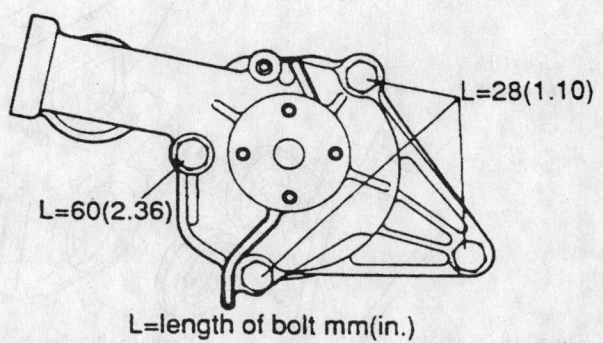

L=28(1.10)
L=60(2.36)
L=length of bolt mm(in.)

Water pump bolt lengths—1.5L—1.6L engines

7923GG19

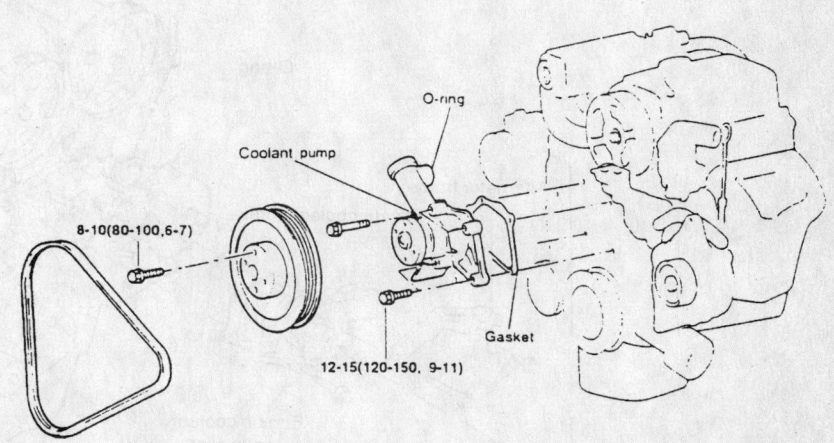

O-ring
Coolant pump
Gasket
8-10(80-100,6-7)
12-15(120-150, 9-11)

TORQUE : Nm (kg.cm, lb.ft)

Water pump assembly—1.8L and 2.0L engines

7923GG20

Timing belt service is covered in Section 3 of this manual

b. 1.5L DOHC engine: 14–20 ft. lbs. (20–27 Nm).

c. 1.6L DOHC engine: 14–20 ft. lbs. (20–27 Nm).

d. 1.8L, 2.0L, and 2.4L engines: 14–20 ft. lbs. (20–27 Nm).

e. 2.5L engine: 11–16 ft. lbs. (15–22 Nm).

5. Install or connect the following:
   • Alternator bracket

• Timing belt and front cover
• Water pump pulley
• Bypass hose, if equipped
• Radiator hose
• Accessory drive belts
• Negative battery cable

6. Fill the cooling system to the proper level.

7. Start the engine and check for leaks.

XG 300

1. Before servicing the vehicle, refer to the precautions in the beginning of this section.

2. Drain the cooling system.

3. Remove or disconnect the following:
   • Negative battery cable
   • Accessory drive belts
   • Water pump pulley

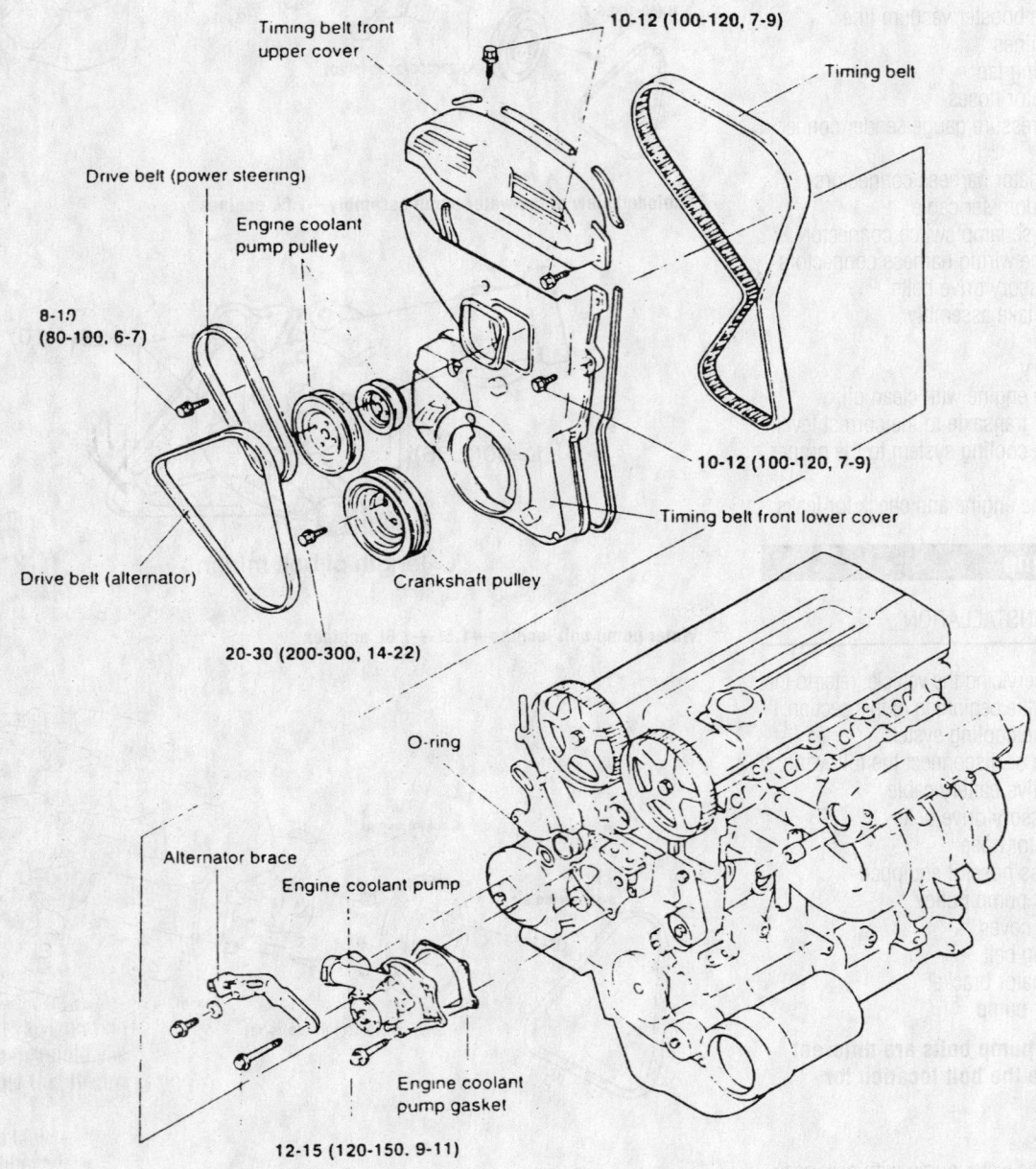

Exploded view of the water pump assembly and related components—2.0L engines

TORQUE : Nm (kg.cm, lb.ft)

7923GG21

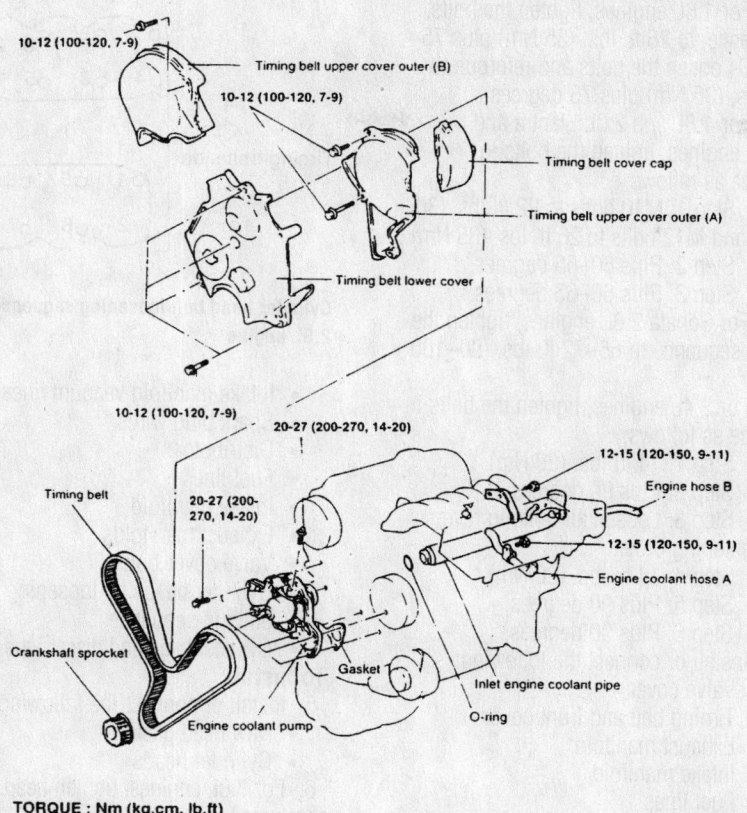

10-12 (100-120, 7-9)
Timing belt upper cover outer (B)
10-12 (100-120, 7-9)
Timing belt cover cap
Timing belt upper cover outer (A)
Timing belt lower cover
10-12 (100-120, 7-9)
20-27 (200-270, 14-20)
12-15 (120-150, 9-11)
Engine hose B
Timing belt
20-27 (200-270, 14-20)
12-15 (120-150, 9-11)
Engine coolant hose A
Crankshaft sprocket
Gasket
Inlet engine coolant pipe
Engine coolant pump
O-ring

TORQUE : Nm (kg.cm, lb.ft)

7923GG22

Water pump assembly—3.0L engine Sonata

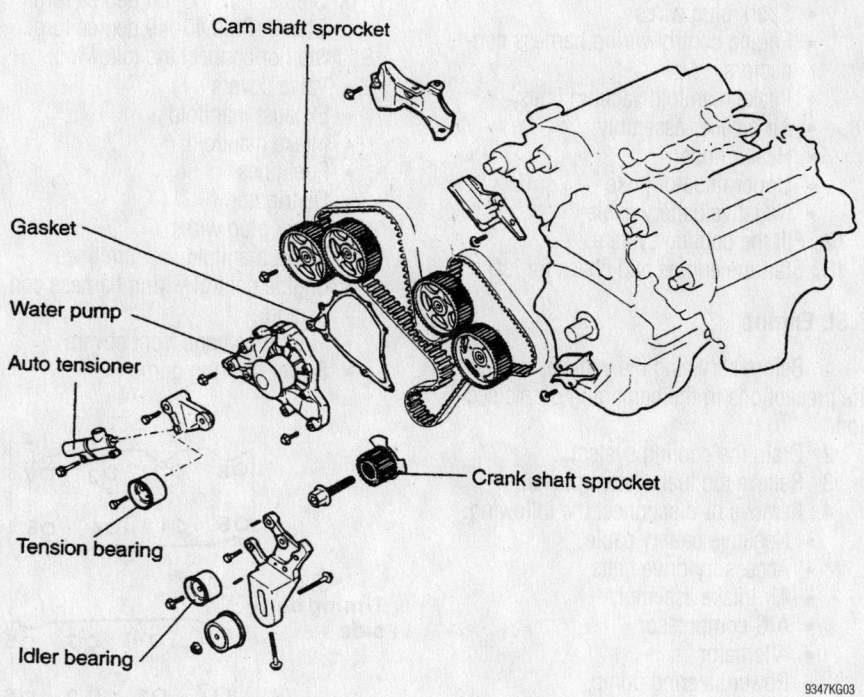

Cam shaft sprocket
Gasket
Water pump
Auto tensioner
Crank shaft sprocket
Tension bearing
Idler bearing

9347KG03

Water pump assembly—XG 300

Heater Core replacement is covered in Section 2 of this manual

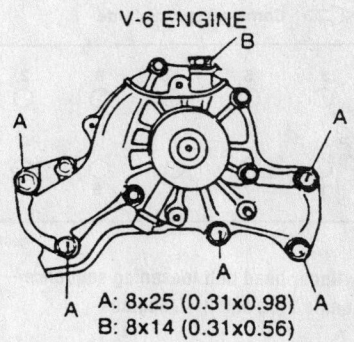

V-6 ENGINE
B
A   A
A   A
A: 8x25 (0.31x0.98)
B: 8x14 (0.31x0.56)

7923GG23

Water pump bolt lengths—3.0L Sonata

- Timing belt, tensioner and idler pulley
- Water pump

To install:

4. Clean all mating surfaces of any residual gasket material.

5. Install or connect the following:
- Water pump with a new gasket and torque the bolts to 16 ft. lbs. (22 Nm)
- Tensioner and timing belt and idler pulley
- Water pump pulley and drive belts
- Negative battery cable

6. Fill the cooling system to the proper level.

7. Start the vehicle, check for leaks and repair if necessary.

Cylinder Head

REMOVAL & INSTALLATION

4 Cylinder Engines

1. Before servicing the vehicle, refer to the precautions in the beginning of this section.

2. Drain the cooling system.

3. Relieve the fuel system pressure.

4. Remove or disconnect the following:

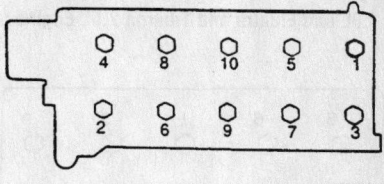

4   8   10   5   1
2   6   9   7   3

7923GG25

Cylinder head bolt loosening sequence— 1.5L, 1.6L, 1.8L, Elantra and Tiburon 2.0L engines

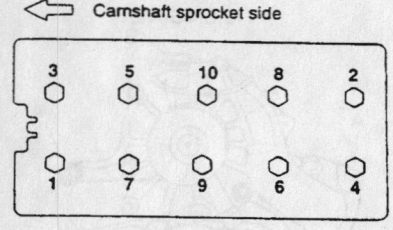

Cylinder head bolt loosening sequence—
Sonata 2.0L and 2.4L engines

- Negative battery cable
- Upper radiator hose
- Heater hose
- Air cleaner assembly
- Intake manifold vacuum lines
- Engine control wiring harness connectors
- Spark plug wires
- Distributor, if equipped
- Ignition coil
- Accessory drive belts
- Power steering pump and bracket
- Fuel lines
- Intake manifold
- Exhaust manifold
- Front cover
- Timing belt
- Valve cover bolts
- Cylinder head by loosening the bolts in sequence
- Cylinder head and discard the gasket

To install:

5. Install the cylinder head with a new gasket.

6. For 1.5L engines, tighten the bolts in sequence to 51–54 ft. lbs. (71–75 Nm).

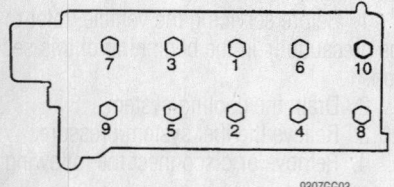

Cylinder head torque sequence—1.5L,
1.8L and Elantra and Tiburon 2.0L Engines

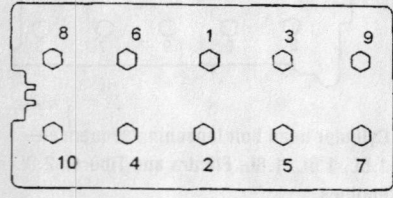

Cylinder head torque sequence—Sonata
2.0L and 2.4L Engines

7. For 1.6L engines, tighten the bolts, in sequence, to 26 ft. lbs. (35 Nm) plus 75 degrees. Loosen the bolts and retorque to 26 ft. lbs. (35 Nm) plus 75 degrees.

8. For 1.8L and 2.0L Elantra and Tiburon engines, tighten the bolts in sequence as follows:

    a. Step 1: M10 bolts to 22 ft. lbs. (30 Nm) and M12 bolts to 26 ft. lbs. (35 Nm).

    b. Step 2: Plus 60–65 degrees.

    c. Step 3: Plus 60–65 degrees.

9. For Sonata 2.0L engines, tighten the bolts in sequence to 65–72 ft. lbs. (90–100 Nm).

10. For 2.4L engines, tighten the bolts in sequence as follows:

    a. Step 1: 14 ft. lbs. (20 Nm).

    b. Step 2: Plus 90 degrees.

    c. Step 3: Loosen all bolts in reverse of tightening order.

    d. Step 4: 14 ft. lbs. (20 Nm).

    e. Step 5: Plus 90 degrees.

    f. Step 6: Plus 90 degrees.

11. Install or connect the following:
- Valve cover
- Timing belt and front cover
- Exhaust manifold
- Intake manifold
- Fuel lines
- Power steering pump and bracket
- Accessory drive belts
- Ignition coil
- Distributor, if equipped
- Spark plug wires
- Engine control wiring harness connectors
- Intake manifold vacuum lines
- Air cleaner assembly
- Heater hose
- Upper radiator hose
- Negative battery cable

12. Fill the cooling system.

13. Start the engine and check for leaks.

2.5L Engine

1. Before servicing the vehicle, refer to the precautions in the beginning of this section.

2. Drain the cooling system.

3. Relieve the fuel system pressure.

4. Remove or disconnect the following:
- Negative battery cable
- Accessory drive belts
- Air intake assembly
- A/C compressor
- Alternator
- Power steering pump
- Front covers
- Timing belt
- Engine control wiring harness connectors

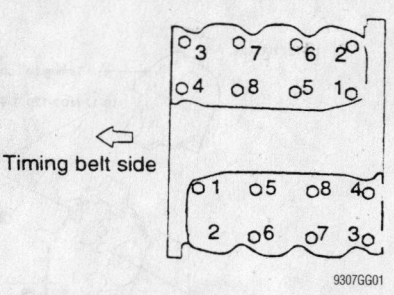

Cylinder head bolt loosening sequence—
2.5L engine

- Intake manifold vacuum lines
- Spark plug wires
- Distributor
- Fuel lines
- Intake manifold
- Exhaust manifolds
- Valve cover bolts
- Cylinder heads by loosening the bolts in sequence
- Cylinder head and discard the gasket

To install:

5. Install or connect the following:
- New head gaskets
- Cylinder heads

6. For 3.0L engines, tighten head bolts in sequence to 76–83 ft. lbs. (105–115 Nm).

7. For 2.5L engines, tighten the bolts in sequence as follows:

    a. Step 1: 18 ft. lbs. (25 Nm).

    b. Step 2: Plus 60–64 degree turn.

    c. Step 3: Plus 45–49 degree turn.

8. Install or connect the following:
- Valve covers
- Exhaust manifolds
- Intake manifold
- Fuel lines
- Distributor
- Spark plug wires
- Intake manifold vacuum lines
- Engine control wiring harness connectors
- Timing belt and front covers
- Power steering pump

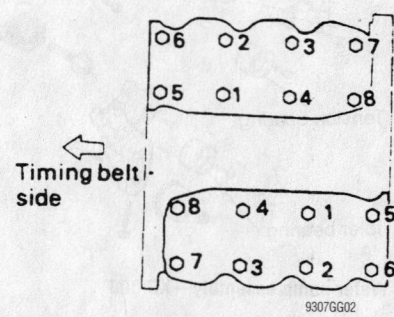

Cylinder head torque sequence—2.5L and
3.0L Sonata

- Alternator
- A/C compressor
- Air intake assembly
- Accessory drive belts
- Negative battery cable
9. Fill the cooling system.
10. Start the engine and check for leaks.

3.0L XG 300

1. Before servicing the vehicle, refer to the precautions in the beginning of this section.
2. Drain the cooling system.
3. Relieve the fuel system pressure.
4. Remove or disconnect the following:
- Negative battery cable
- Upper radiator hose
- Breather hose
- Air intake hose
- Vacuum hose
- Fuel hoses
- Intake manifold
- Spark plug wires
- Ignition coil
- Upper and lower timing belt covers
- Timing belt
- Camshaft sprockets
- Heat protector and exhaust manifold
- Water pump
- Rocker arm cover
- Camshafts
- Cylinder head and discard the gasket
5. Clean all mating surfaces of any residual gasket material.

To install:
6. Install or connect the following:
- New gasket with the identification mark facing the cylinder head
- Cylinder head and torque the bolts, in sequence, to 82 ft. lbs. (115 Nm)
- Camshafts
- Rocker arm cover and torque the bolts to 89 inch lbs. (10 Nm)
- Water pump and torque the bolts to 16 ft. lbs. (22 Nm)
- Heat protector and exhaust manifold and torque the bolts to 14 ft. lbs. (19 Nm)
- Camshaft sprockets
- Timing belt
- Upper and lower timing belt covers
- Ignition coil and spark plug wires
- Intake manifold and torque the bolts to 10 ft. lbs. (15 Nm)
- Fuel hoses
- Vacuum hose
- Air intake hose
- Breather hose
- Upper radiator hose
- Negative battery cable

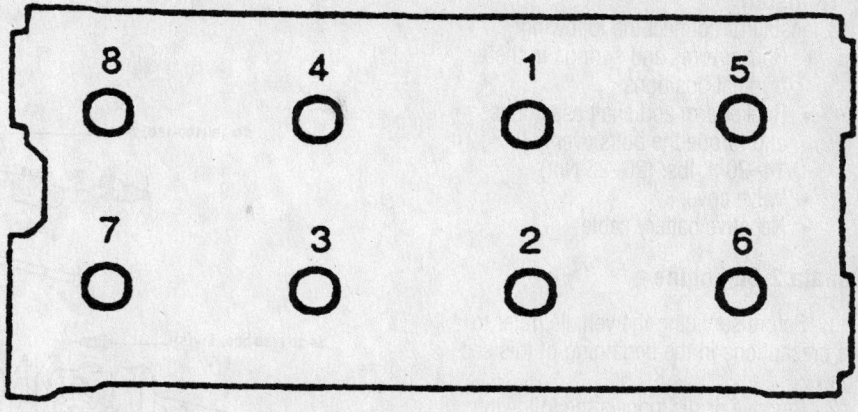

Cylinder head torque sequence—3.0L XG300 Engine

9347KG04

7. Fill the cooling system to the proper level.
8. Start the vehicle, check for leaks and repair if necessary.

Rocker Arms/Shafts

REMOVAL & INSTALLATION

Except 1.5L SOHC and Sonata 2.0L Engines

These engines are not equipped with rocker arms. The camshaft acts directly on the valves through hydraulic lash adjusters.

1.5L SOHC Engine

1. Before servicing the vehicle, refer to the precautions in the beginning of this section.
2. Remove or disconnect the following:
- Negative battery cable
- Valve cover
- Rocker arm shaft bolts by loosening them evenly in several steps
- Rocker arm and shaft assemblies

➡ **Keep all valvetrain components in order for assembly.**

- Rocker arms and springs from the shafts

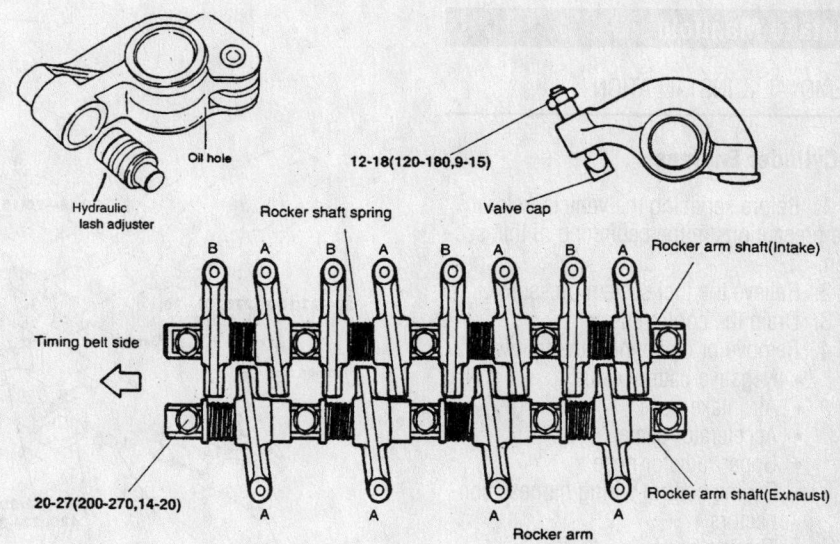

Rocker assembly components and arrangement. Rockers marked "A" and "B" must be returned to their original positions—1.5L SOHC engines

7923GG28

To install:

3. Install or connect the following:
- Rocker arms and springs in their original positions
- Rocker arm and shaft assemblies and torque the bolts evenly to 14–20 ft. lbs. (20–26 Nm)
- Valve cover
- Negative battery cable

Sonata 2.0L Engine

1. Before servicing the vehicle, refer to the precautions in the beginning of this section.

2. Remove or disconnect the following:
- Negative battery cable
- Valve cover
- Accessory drive belts
- Front cover
- Timing belt
- Camshaft
- Rocker arms

➡**Keep all valvetrain components in order for assembly.**

To install:

3. Install or connect the following:
- Rocker arms in their original positions
- Camshaft
- Timing belt
- Front cover
- Accessory drive belts
- Valve cover
- Negative battery cable

Intake Manifold

REMOVAL & INSTALLATION

4 Cylinder Engines

1. Before servicing the vehicle, refer to the precautions in the beginning of this section.

2. Relieve the fuel system pressure.
3. Drain the cooling system.
4. Remove or disconnect the following:
- Negative battery cable
- Air intake hose
- Accelerator cable
- Upper radiator hose
- Engine control wiring harness connectors
- Throttle body
- Positive Crankcase Ventilation (PCV) valve and hose
- Brake booster vacuum line
- Intake manifold vacuum hoses
- Fuel lines
- Surge tank

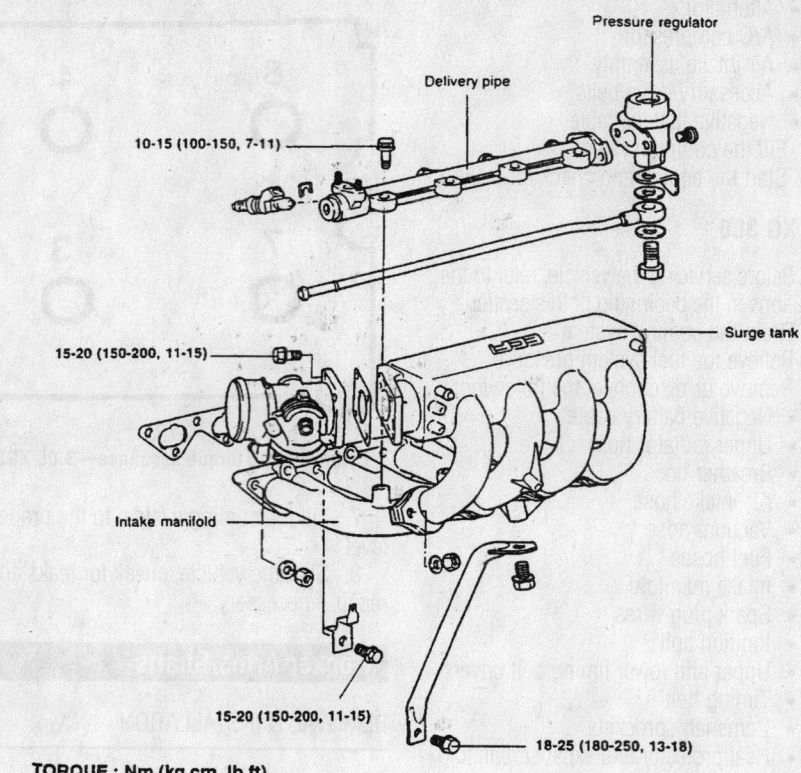

TORQUE : Nm (kg.cm, lb.ft)

Surge tank and intake manifold components—1.5L engine

7923GG34

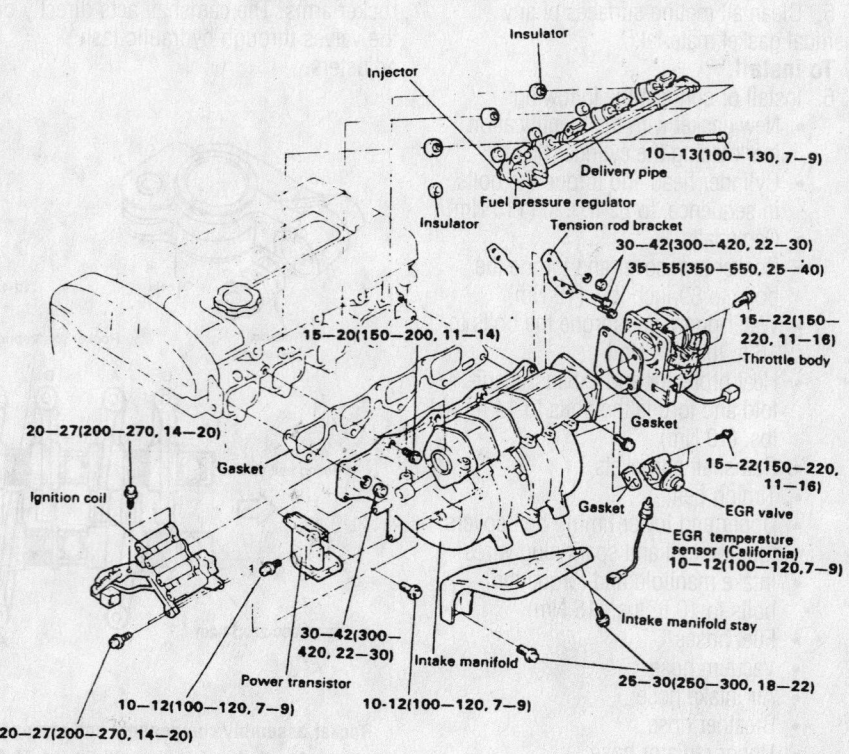

Surge tank and intake manifold components—1.8L engines

7923GG37

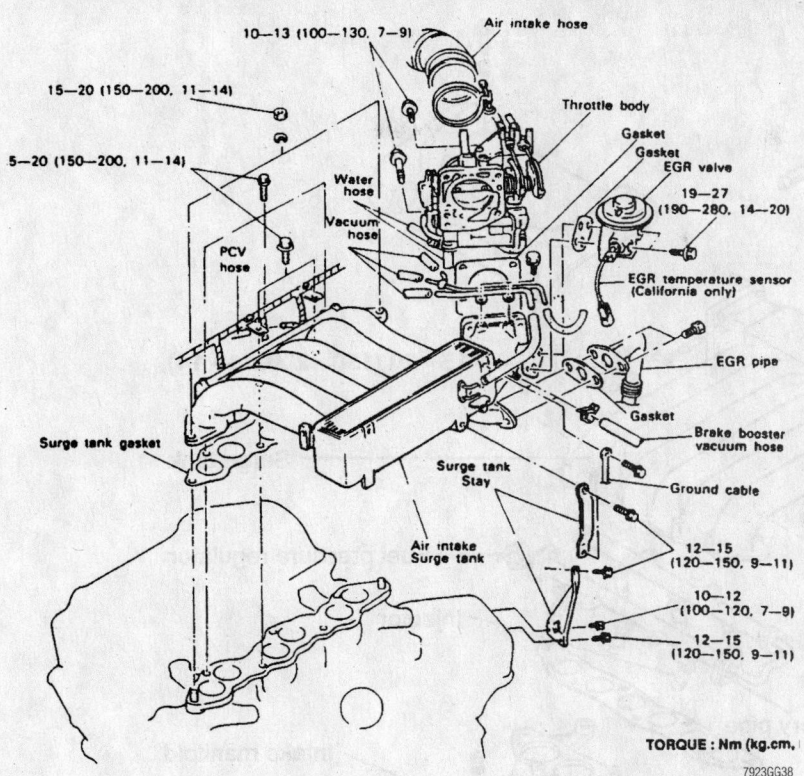

Surge tank and intake manifold components—Sonata 2.0L and 2.4L engines

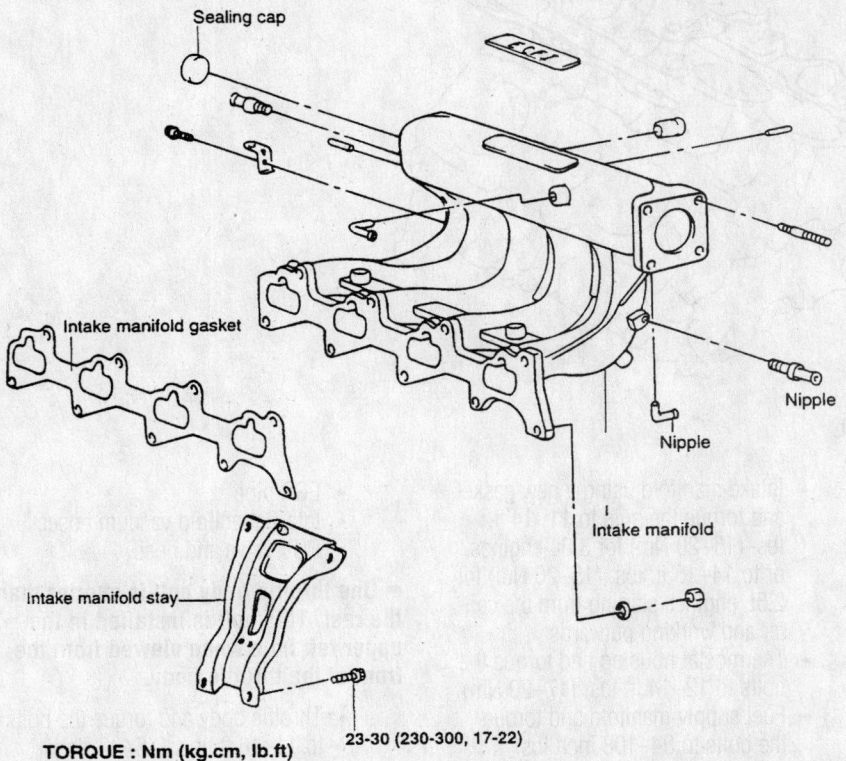

Surge tank and intake manifold components—Elantra and Tiburon 2.0L engine

- Fuel injector connectors
- Fuel supply manifold
- Heater hose
- Engine Coolant Temperature (ECT) sensor connector
- Spark plug wires
- Thermostat housing
- Distributor, if equipped
- Ignition coil
- Intake manifold bracket
- Intake manifold and discard the gasket

To install:

5. Install or connect the following:
   - Intake manifold using a new gasket and torque the nuts to 11–14 ft. lbs. (15–20 Nm), starting from the center and working outwards
   - Intake manifold bracket and torque the bolts to 13–18 ft. lbs. (18–25 Nm)
   - Ignition coil
   - Distributor, if equipped
   - Thermostat housing and torque the bolts to 12–14 ft. lbs. (17–20 Nm)
   - Spark plug wires
   - ECT sensor connector
   - Heater hose
   - Fuel supply manifold
   - Fuel injector connectors
   - Surge tank using a new gasket and torque the bolts to 11–16 ft. lbs. (15–22 Nm)
   - Fuel lines
   - Intake manifold vacuum hoses
   - Brake booster vacuum line
   - PCV valve and hose
   - Throttle body
   - Engine control wiring harness connectors
   - Upper radiator hose
   - Accelerator cable
   - Air intake hose
   - Negative battery cable
6. Fill the cooling system.
7. Start the engine and check for leaks.

V6 Engines

1. Before servicing the vehicle, refer to the precautions in the beginning of this section.
2. Relieve the fuel system pressure.
3. Drain the cooling system.
4. Remove or disconnect the following:
   - Negative battery cable
   - Air intake hose
   - Accelerator cable
   - Upper radiator hose
   - Engine control wiring harness

For complete Engine Mechanical specifications, see Section 1 of this manual

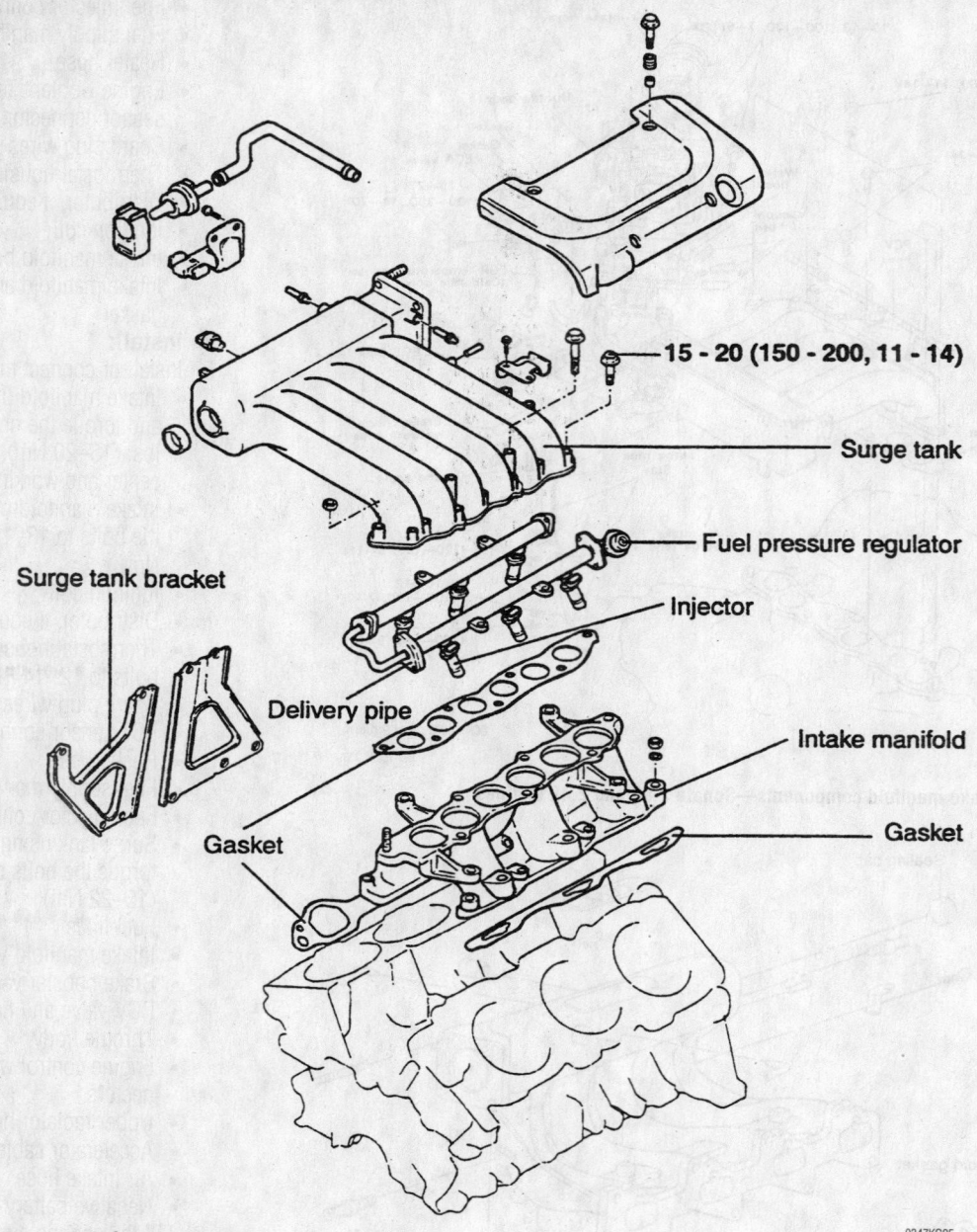

15 - 20 (150 - 200, 11 - 14)

Surge tank

Fuel pressure regulator

Injector

Intake manifold

Gasket

Surge tank bracket

Delivery pipe

Gasket

9347KG05

Surge tank and intake manifold components—XG 300

- Throttle body
- Positive Crankcase Ventilation (PCV) valve and hose
- Intake manifold vacuum hoses
- Exhaust Gas Recirculation (EGR) pipe
- Surge tank
- Fuel lines
- Fuel injector connectors
- Fuel supply manifold
- Thermostat housing
- Intake manifold and discard the gasket

To install:

5. Install or connect the following:

- Intake manifold using a new gasket and torque the nuts to 11–14 ft. lbs. (15–20 Nm) for 3.0L engines or to 14–15 ft. lbs. (15–20 Nm) for 2.5L engines starting from the center and working outwards
- Thermostat housing and torque the bolts to 12–14 ft. lbs. (17–20 Nm)
- Fuel supply manifold and torque the bolts to 84–108 inch lbs. (10–13 Nm)
- Fuel injector connectors
- Fuel lines
- Surge tank and torque the bolts to 11–14 ft. lbs. (15–20 Nm)

- EGR pipe
- Intake manifold vacuum hoses
- PCV valve and hose

➡ **One throttle body bolt is shorter than the rest. This bolt is installed in the upper left hole when viewed from the front of the throttle body.**

- Throttle body and torque the bolts to 11–16 ft. lbs. (15–22 Nm)
- Engine control wiring harness
- Upper radiator hose
- Accelerator cable
- Air intake hose
- Negative battery cable

6. Fill the cooling system.
7. Start the engine and check for leaks.

Exhaust Manifold

REMOVAL & INSTALLATION

4 Cylinder Engines

1. Before servicing the vehicle, refer to the precautions in the beginning of this section.
2. Remove or disconnect the following:
   - Negative battery cable
   - Heated Oxygen (HO2S) sensor
   - Exhaust manifold heat shield
   - Exhaust front pipe
   - Exhaust manifold and discard the gasket

To install:

3. Install the exhaust manifold with a new gasket and torque the nuts to the following specifications:
   a. 1.5L engine: 11–15 ft. lbs. (15–20 Nm).
   b. 1.8L and Sonata 2.0L engines: 18–22 ft. lbs. (25–30 Nm).
   c. Elantra and Tiburon 2.0L engine: 32–41 ft. lbs. (43–50 Nm).
   d. 2.4L engine: M8 fasteners to 18–22 ft. lbs. (25–30 Nm) and M10 fasteners to 25–40 ft. lbs. (34–55 Nm).
4. Install or connect the following:
   - Exhaust front pipe
   - Exhaust manifold heat shield
   - HO2S sensor
   - Negative battery cable
5. Start the engine and check for leaks.

1.6L Engines

1. Before servicing the vehicle, refer to the precautions in the beginning of this section.
2. Remove or disconnect the following:
   - Negative battery cable
   - Heated Oxygen (HO2S) sensor
   - Exhaust manifold heat protector
   - Exhaust manifold and discard the gasket

To install:

3. Install or connect the following:
   - Exhaust manifold with a new gasket and torque the bolts to 41 ft. lbs. (55 Nm)
   - Exhaust manifold heat protector and torque the bolts to 15 ft. lbs. (20 Nm)
   - HO2S sensor
   - Negative battery cable

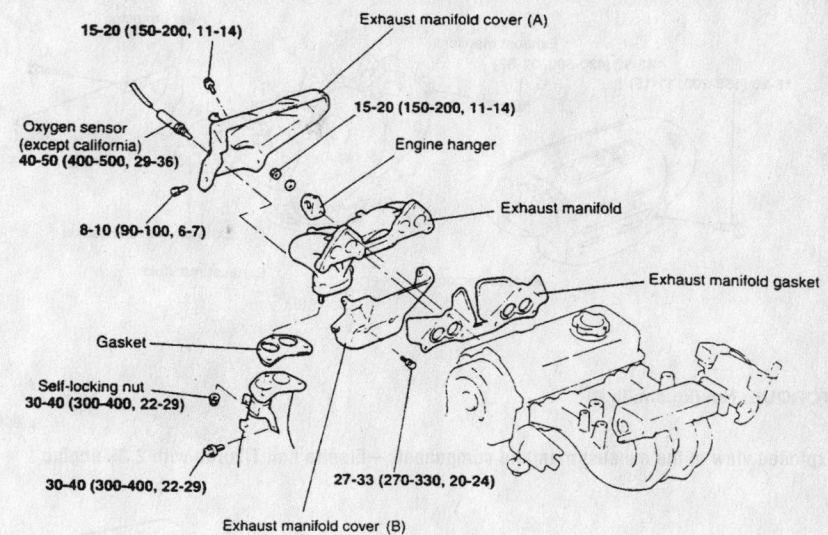

15-20 (150-200, 11-14)
Exhaust manifold cover (A)
15-20 (150-200, 11-14)
Engine hanger
Oxygen sensor (except california)
40-50 (400-500, 29-36)
Exhaust manifold
8-10 (90-100, 6-7)
Exhaust manifold gasket
Gasket
Self-locking nut
30-40 (300-400, 22-29)
30-40 (300-400, 22-29)
27-33 (270-330, 20-24)
Exhaust manifold cover (B)

TORQUE : Nm (kg.cm, lb.ft)

7923GG41

Exploded view of the exhaust manifold and related components—1.5L engines

6 Cylinder Sonata

1. Before servicing the vehicle, refer to the precautions in the beginning of this section.
2. Remove or disconnect the following:
   - Negative battery cable
   - Exhaust front pipe
   - Exhaust Gas Recirculation (EGR) tube

   - Oil dipstick tube
   - Exhaust manifold heat shields
   - Exhaust manifolds and discard the gasket

To install:

3. Install or connect the following:
   - Exhaust manifolds and torque the nuts to 11–16 ft. lbs. (15–22 Nm) for 3.0L engines or to 18–22 ft.

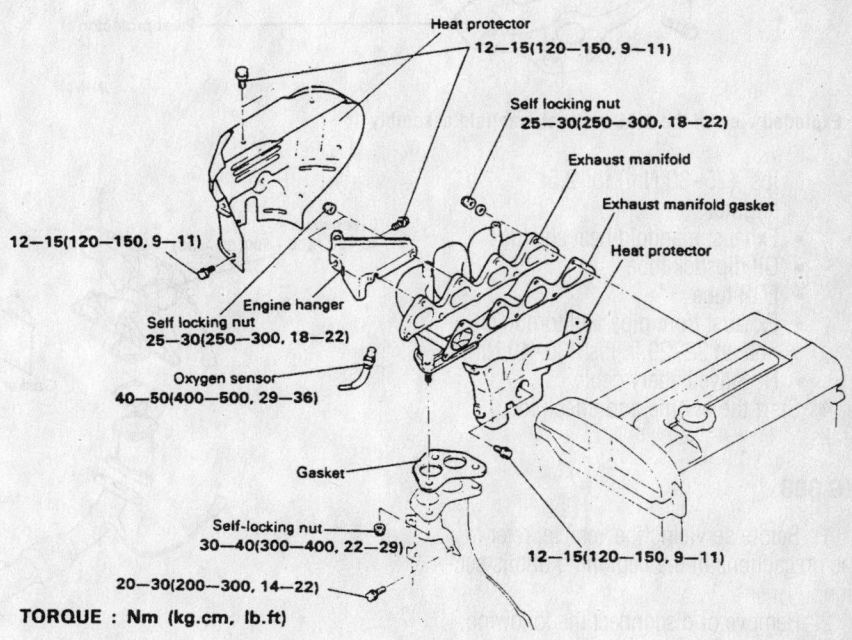

Heat protector
12-15(120-150, 9-11)
Self locking nut
25-30(250-300, 18-22)
Exhaust manifold
Exhaust manifold gasket
Heat protector
12-15(120-150, 9-11)
Engine hanger
Self locking nut
25-30(250-300, 18-22)
Oxygen sensor
40-50(400-500, 29-36)
Gasket
Self-locking nut
30-40(300-400, 22-29)
12-15(120-150, 9-11)
20-30(200-300, 14-22)

TORQUE : Nm (kg.cm, lb.ft)

7923GG42

Exploded view of the exhaust manifold components—1.8L and Sonata 2.0L engines

For Accessory Drive Belt illustrations, see Section 1 of this manual

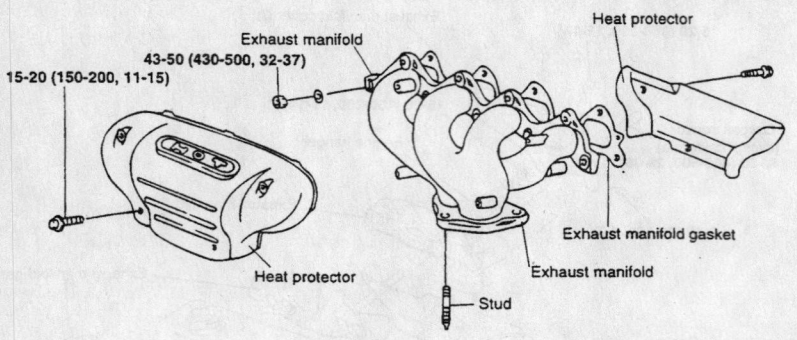

TORQUE: Nm (kg.cm, lb.ft)

Exploded view of the exhaust manifold components—Elantra and Tiburon with 2.0L engine

7923GG43

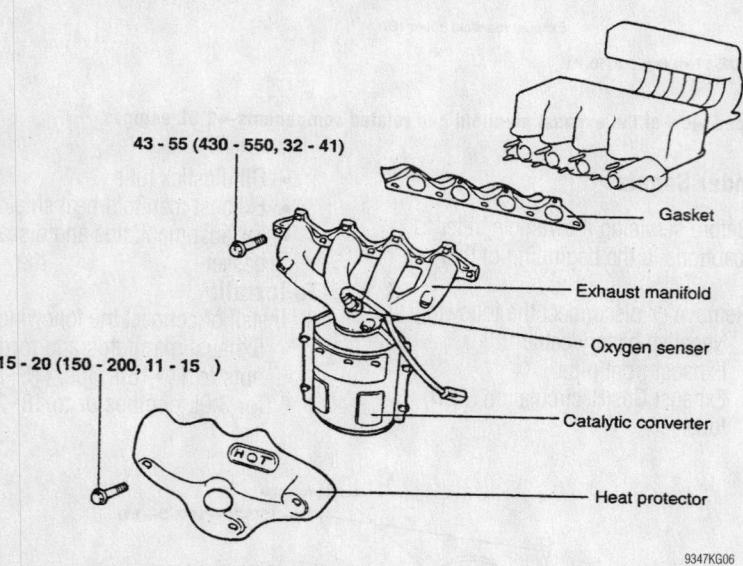

Exploded view of the 1.6L exhaust manifold assembly

9347KG06

lbs. (25–30 Nm) for 2.5L engines
- Exhaust manifold heat shields
- Oil dipstick tube
- EGR tube
- Exhaust front pipe and torque the nuts to 22–29 ft. lbs. (30–40 Nm)
- Negative battery cable

4. Start the engine and check for leaks.

XG 300

1. Before servicing the vehicle, refer to the precautions in the beginning of this section.

2. Remove or disconnect the following:
- Negative battery cable
- Heat protector
- Heated Oxygen (HO2S) sensor
- Exhaust manifold and discard the gaskets

3. Clean and residual gasket material from all mating surfaces.

To install:

4. Install or connect the following:
- Exhuast manifold with a new gasket and torque the bolts to 22 ft. lbs. (30 Nm)
- Heat protector and torque the bolts to 11 ft. lbs. (15 Nm)
- HO2S sensor
- Negative battery cable

Front Crankshaft Seal

REMOVAL & INSTALLATION

1. Before servicing the vehicle, refer to the precautions in the beginning of this section.

2. Remove or disconnect the following:
- Negative battery cable
- Accessory drive belts
- Front cover
- Timing belt
- Crankshaft timing sprocket
- Front crankshaft seal

To install:

3. Install the front crankshaft seal so that it is flush with the oil pump housing.

4. Install or connect the following:
- Crankshaft timing sprocket
- Timing belt
- Front cover
- Accessory drive belts
- Negative battery cable

5. Start the engine and check for leaks.

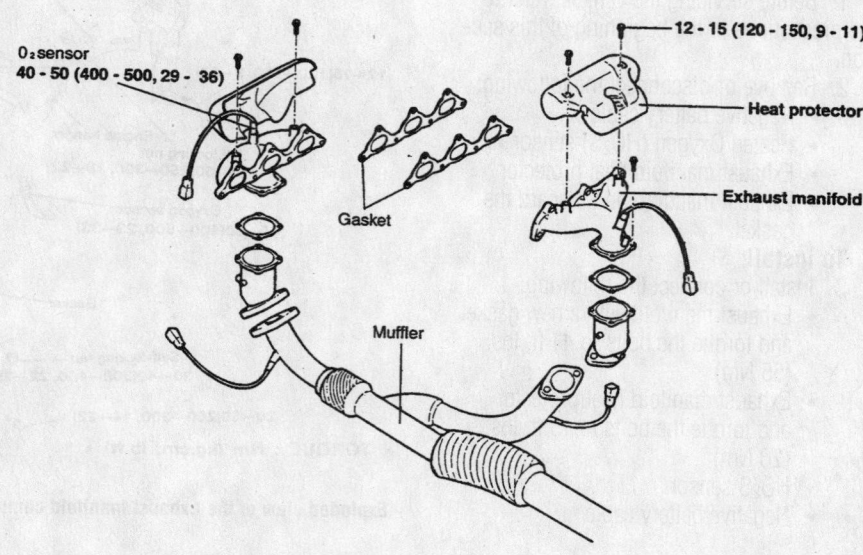

Exploded view of the exhaust manifold—XG 300

9347KG07

Camshaft and Valve Lifters

REMOVAL & INSTALLATION

1.5L SOHC Engines

➡**The hydraulic lash adjusters are housed in the rocker arms.**

1. Before servicing the vehicle, refer to the precautions in the beginning of this section.
2. Remove or disconnect the following:
   • Negative battery cable
   • Accessory drive belts
   • Ignition coil
   • Valve cover
   • Front cover
   • Timing belt
   • Camshaft timing belt sprocket
   • Rocker arm and shaft assembly
   • Camshaft bearing caps
   • Camshaft

To install:

3. Install or connect the following:
   • Camshaft
   • Camshaft bearing caps
   • Rocker arm and shaft assembly and torque the bolts evenly to 14–20 ft. lbs. (20–26 Nm)
   • Camshaft timing belt sprocket and torque the bolt to 58–72 ft. lbs. (80–100 Nm)
   • Timing belt
   • Front cover
   • Valve cover
   • Ignition coil
   • Accessory drive belts
   • Negative battery cable

1.5L DOHC, 1.6L DOHC, 1.8L, 2.0L and 2.4L Engines

1. Before servicing the vehicle, refer to the precautions in the beginning of this section.

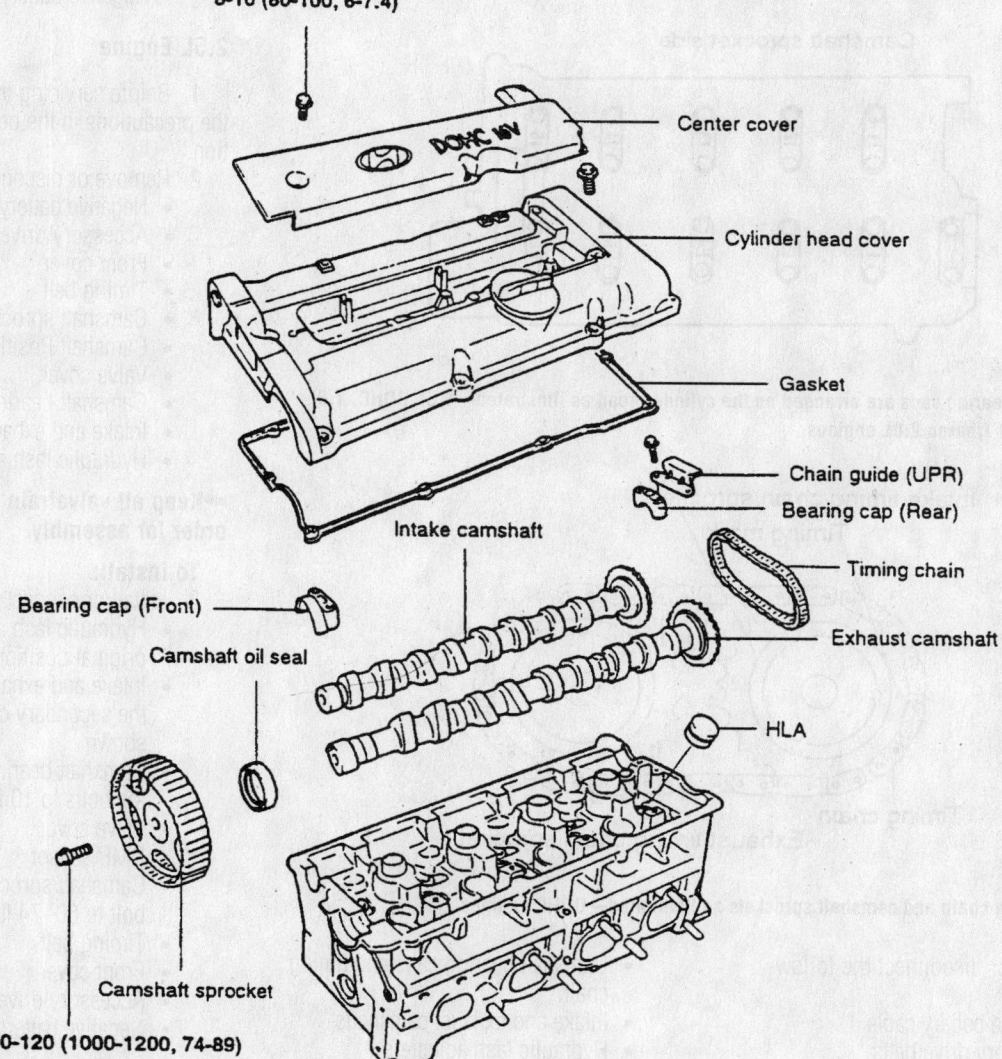

8-10 (80-100, 6-7.4)

Center cover

Cylinder head cover

Gasket

Chain guide (UPR)

Bearing cap (Rear)

Timing chain

Intake camshaft

Bearing cap (Front)

Camshaft oil seal

Exhaust camshaft

HLA

Camshaft sprocket

100-120 (1000-1200, 74-89)

Camshaft assembly components—1.5L DOHC, 1.8L and Elantra and Tiburon 2.0L engines

7923GG48

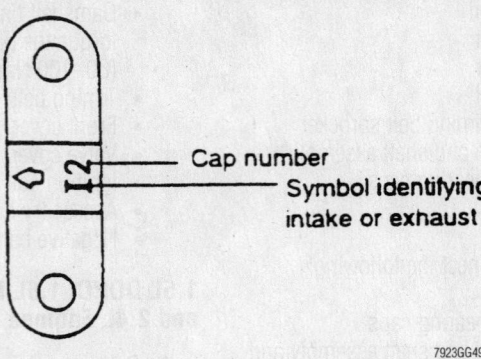

The camshaft bearing caps are identified with a letter and number stamp. The letter indicates either intake or exhaust and the number is sequential from the cylinder head end opposite the timing chain—1.5L DOHC, 1.8L and Elantra and Tiburon 2.0L engines

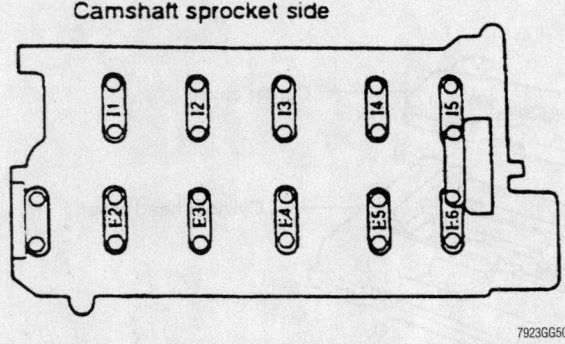

The camshaft bearing caps are arranged on the cylinder head as illustrated—1.5L DOHC, 1.8L and Elantra and Tiburon 2.0L engines

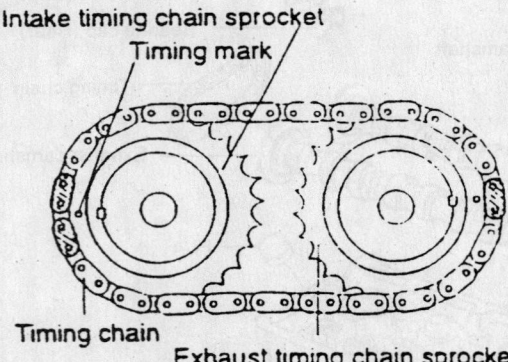

Align the timing chain and camshaft sprockets as illustrated—DOHC engines

2. Remove or disconnect the following:
- Negative battery cable
- Accessory drive belts
- Front cover
- Timing belt
- Camshaft sprocket
- Distributor, if equipped
- Camshaft position sensor, if equipped
- Valve cover
- Camshaft bearing caps and timing chain
- Intake and exhaust camshafts
- Hydraulic lash adjusters

➡**Keep all valvetrain components in order for assembly.**

To install:
3. Install or connect the following:
- Hydraulic lash adjusters in their original positions
- Intake and exhaust camshafts with the secondary chain aligned as shown
- Camshaft bearing caps and timing chain and torque the bolts to 15 ft. lbs. (21 Nm) for 2.0L (VIN P) engines or to 10 ft. lbs. (14 Nm) for all others
- Valve cover
- Distributor, if equipped
- Camshaft position sensor, if equipped
- Camshaft sprocket and torque the bolt to 60–74 ft. lbs. (80–100 Nm)
- Timing belt
- Front cover
- Accessory drive belts
- Negative battery cable

2.5L Engine

1. Before servicing the vehicle, refer to the precautions in the beginning of this section.
2. Remove or disconnect the following:
- Negative battery cable
- Accessory drive belts
- Front cover
- Timing belt
- Camshaft sprocket
- Camshaft Position Sensor (CMP)
- Valve cover
- Camshaft bearing caps
- Intake and exhaust camshafts
- Hydraulic lash adjusters

➡**Keep all valvetrain components in order for assembly.**

To install:
3. Install or connect the following:
- Hydraulic lash adjusters in their original positions
- Intake and exhaust camshafts with the secondary chain aligned as shown
- Camshaft bearing caps and torque the bolts to 10 ft. lbs. (14 Nm)
- Valve cover
- CMP sensor
- Camshaft sprocket and torque the bolt to 60–74 ft. lbs. (80–100 Nm)
- Timing belt
- Front cover
- Accessory drive belts
- Negative battery cable

3.0L Sonata

1. Before servicing the vehicle, refer to the precautions in the beginning of this section.
2. Remove or disconnect the following:
- Negative battery cable
- Accessory drive belts

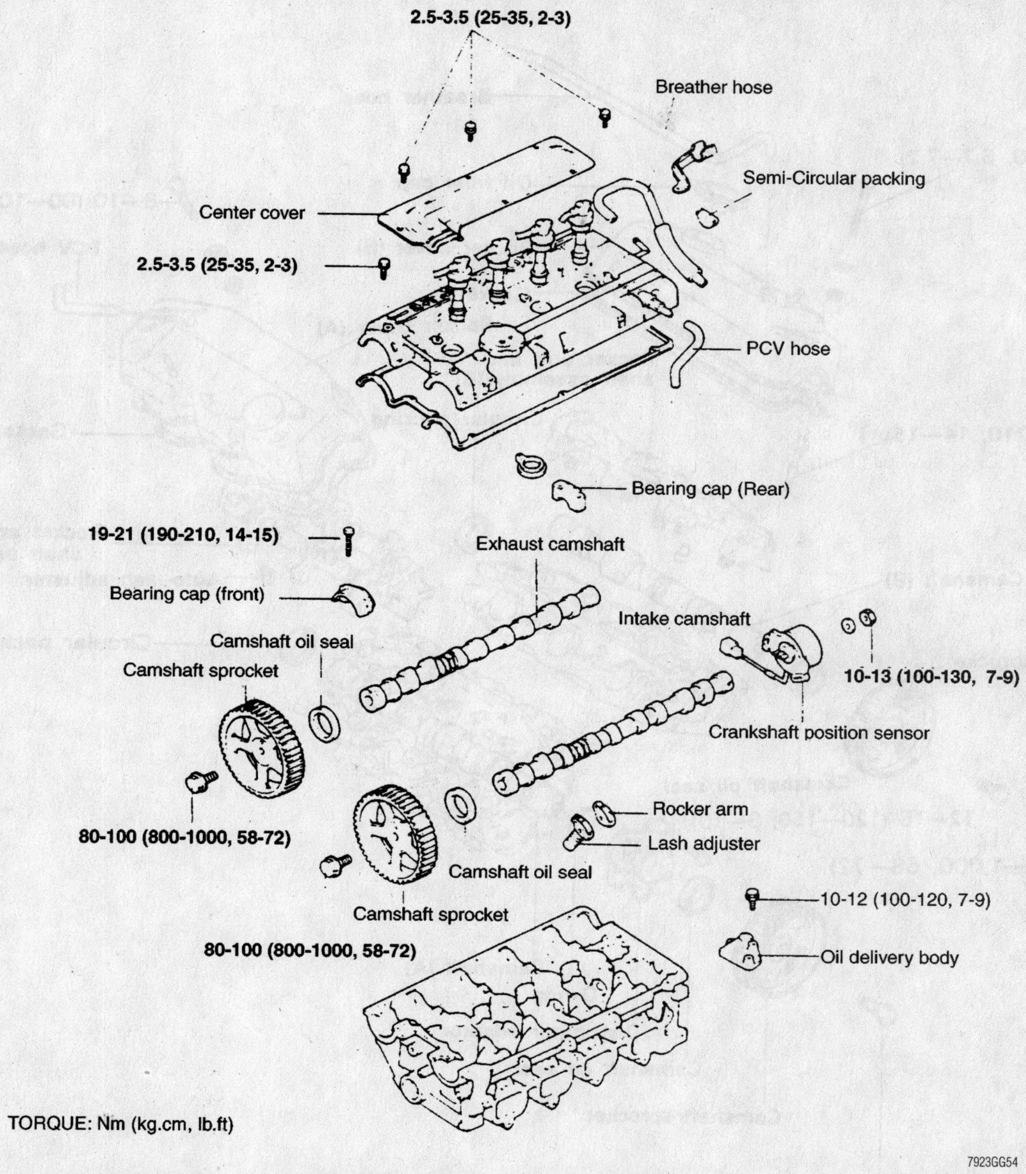

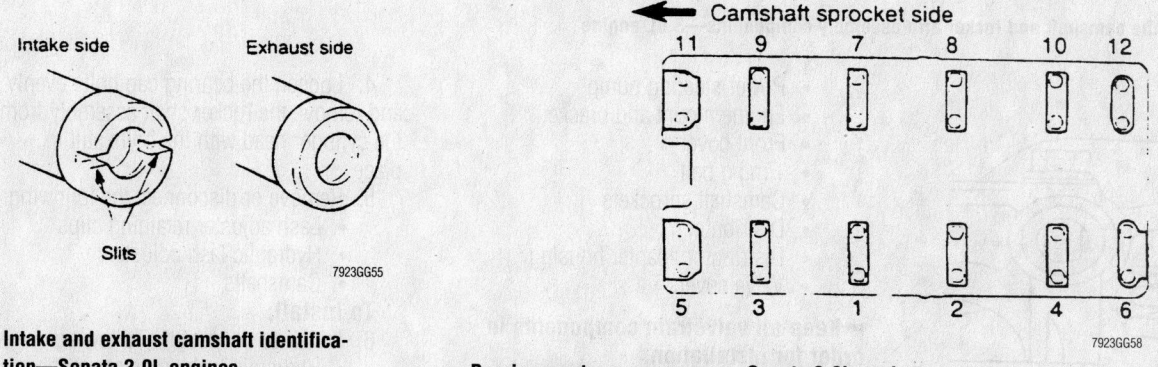

2.5-3.5 (25-35, 2-3)

Breather hose

Semi-Circular packing

Center cover

2.5-3.5 (25-35, 2-3)

PCV hose

Bearing cap (Rear)

19-21 (190-210, 14-15)

Exhaust camshaft

Bearing cap (front)

Intake camshaft

Camshaft oil seal

Camshaft sprocket

10-13 (100-130, 7-9)

Crankshaft position sensor

80-100 (800-1000, 58-72)

Rocker arm

Lash adjuster

Camshaft oil seal

Camshaft sprocket

10-12 (100-120, 7-9)

Oil delivery body

80-100 (800-1000, 58-72)

TORQUE: Nm (kg.cm, lb.ft)

7923GG54

Exploded view of the camshaft and rocker arm assembly components—Sonata 2.0L engines

Intake side

Exhaust side

Slits

7923GG55

Intake and exhaust camshaft identification—Sonata 2.0L engines

← Camshaft sprocket side

11	9	7	8	10	12
5	3	1	2	4	6

7923GG58

Bearing cap torque sequence—Sonata 2.0L engines

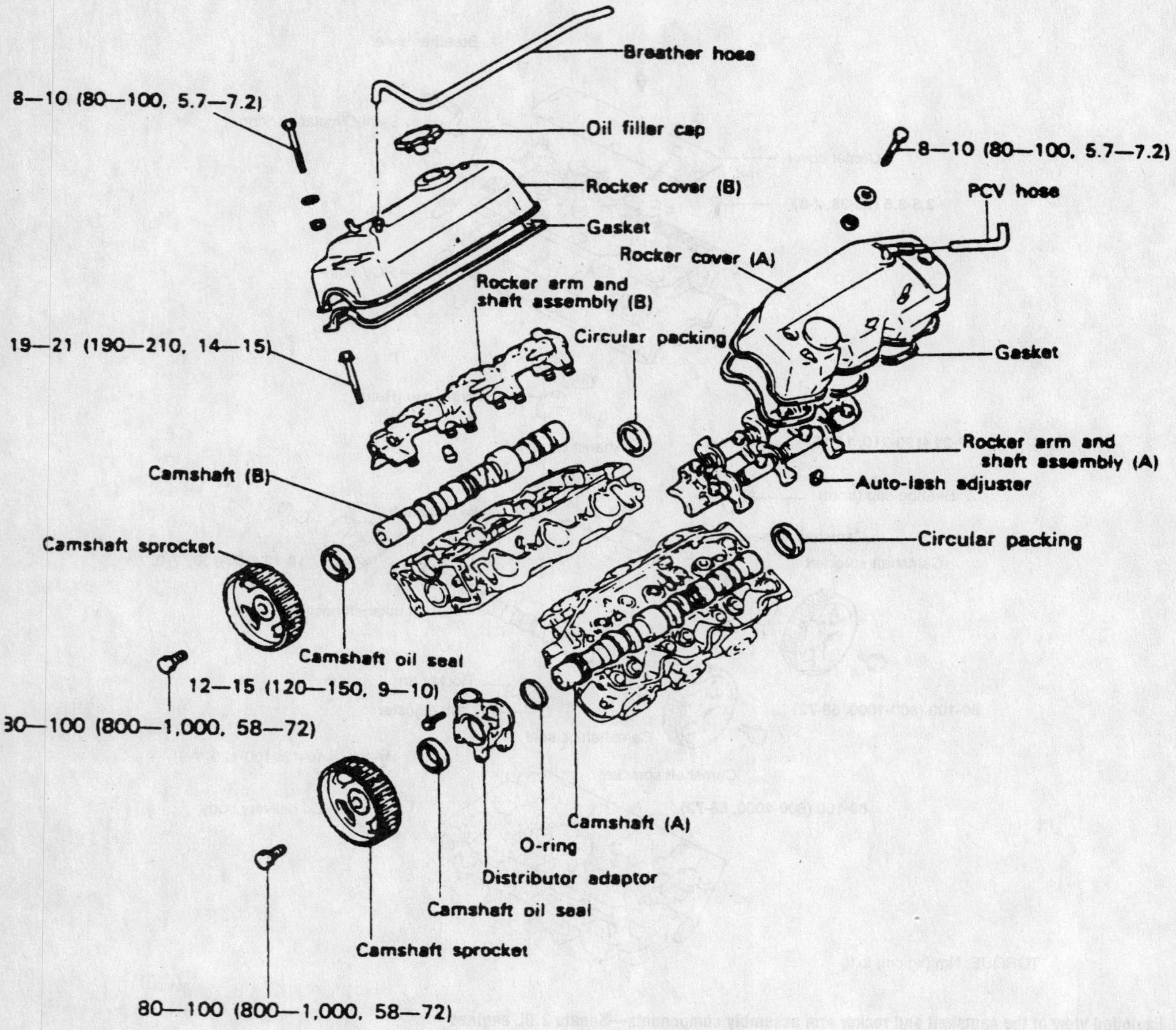

Breather hose

8—10 (80—100, 5.7—7.2)

Oil filler cap

8—10 (80—100, 5.7—7.2)

Rocker cover (B)

PCV hose

Gasket

Rocker cover (A)

Rocker arm and shaft assembly (B)

Circular packing

Gasket

19—21 (190—210, 14—15)

Rocker arm and shaft assembly (A)

Camshaft (B)

Auto-lash adjuster

Camshaft sprocket

Circular packing

Camshaft oil seal

12—15 (120—150, 9—10)

30—100 (800—1,000, 58—72)

Camshaft (A)

O-ring

Distributor adaptor

Camshaft oil seal

Camshaft sprocket

80—100 (800—1,000, 58—72)

7923GG61

Exploded view of the camshaft and rocker arm assembly components—3.0L engine

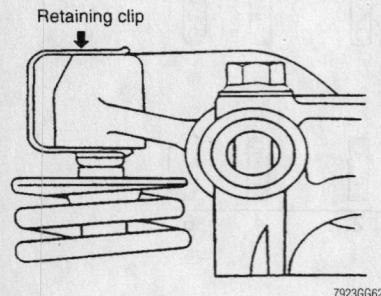

Retaining clip

7923GG62

Before removing the rocker arm assemblies, install lash adjuster retaining clips (PN 09246-32000)—3.0L engine

- Power steering pump
- Engine mount and bracket
- Front cover
- Timing belt
- Camshaft sprockets
- Distributor
- Distributor adapter housing
- Valve covers

➡**Keep all valvetrain components in order for installation.**

3. Install special holding clip (PN 09246 3200) on each rocker arm to hold the hydraulic lash adjusters in place.

4. Loosen the bearing cap bolts evenly and remove the rocker shaft assembly from the cylinder head with the bolts still in place.

5. Remove or disconnect the following:
- Lash adjuster retaining clips
- Hydraulic lash adjusters
- Camshafts

To install:

6. Install or connect the following:
- Hydraulic lash adjusters in their original locations
- Lash adjuster retaining clips
- Camshafts

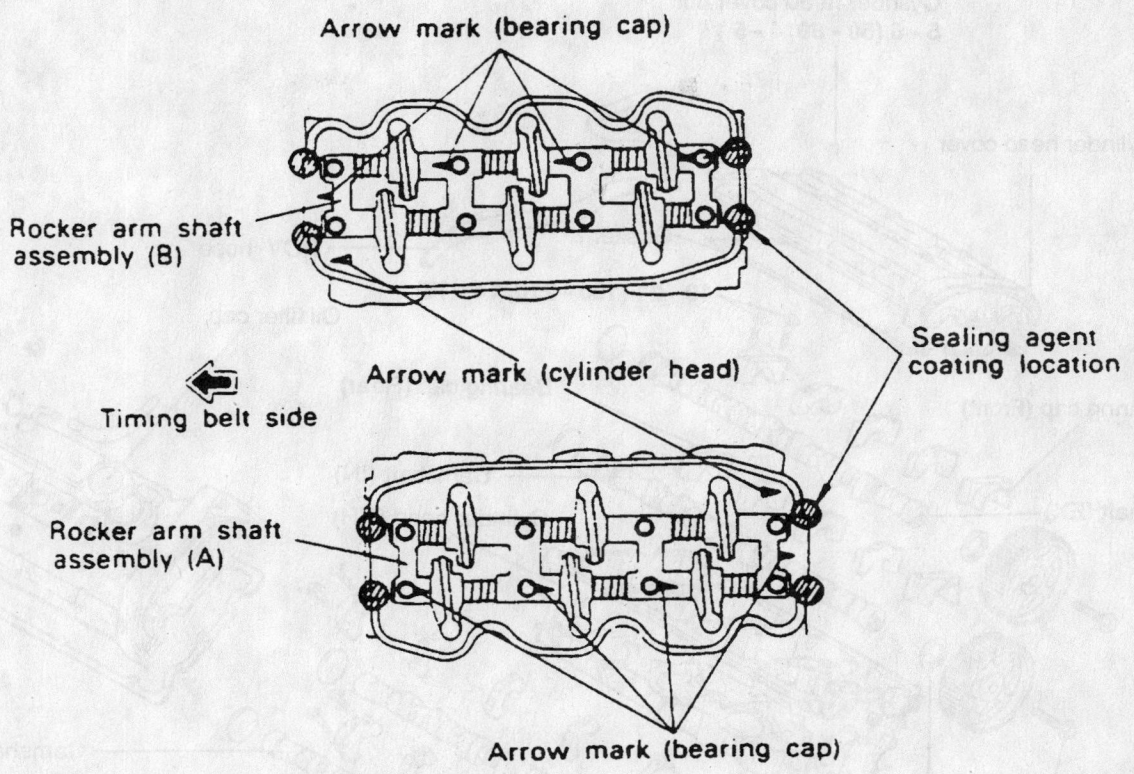

Install the rocker arm assemblies with the arrows as indicated. Place sealer at the corners—3.0L engine

- Rocker arm and shaft assemblies and torque the bearing cap bolts 14–15 ft. lbs. (19–21 Nm) starting from the center and working out
7. Remove the lash adjuster retaining clips.
8. Install or connect the following:
   - Valve covers
   - Distributor adapter housing
   - Distributor
   - Camshaft sprockets
   - Timing belt
   - Front cover
   - Engine mount and bracket
   - Power steering pump
   - Accessory drive belts
   - Negative battery cable

XG 300

1. Before servicing the vehicle, refer to the precautions in the beginning of this section.

2. Remove or disconnect the following:
   - Negative battery cable
   - Engine cover
   - Intake manifold
   - Breather hose and engine harness
   - Power steering pulley

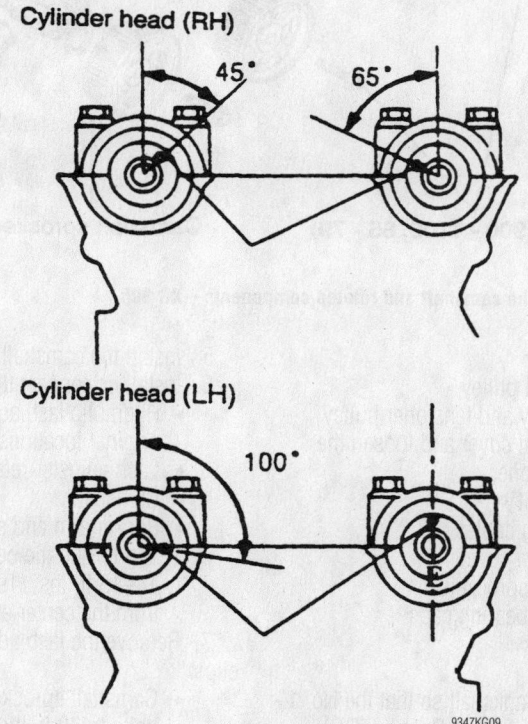

Install the camshaft dowel pin as shown—XG 300

For Maintenance Interval recommendations, see Section 1 of this manual

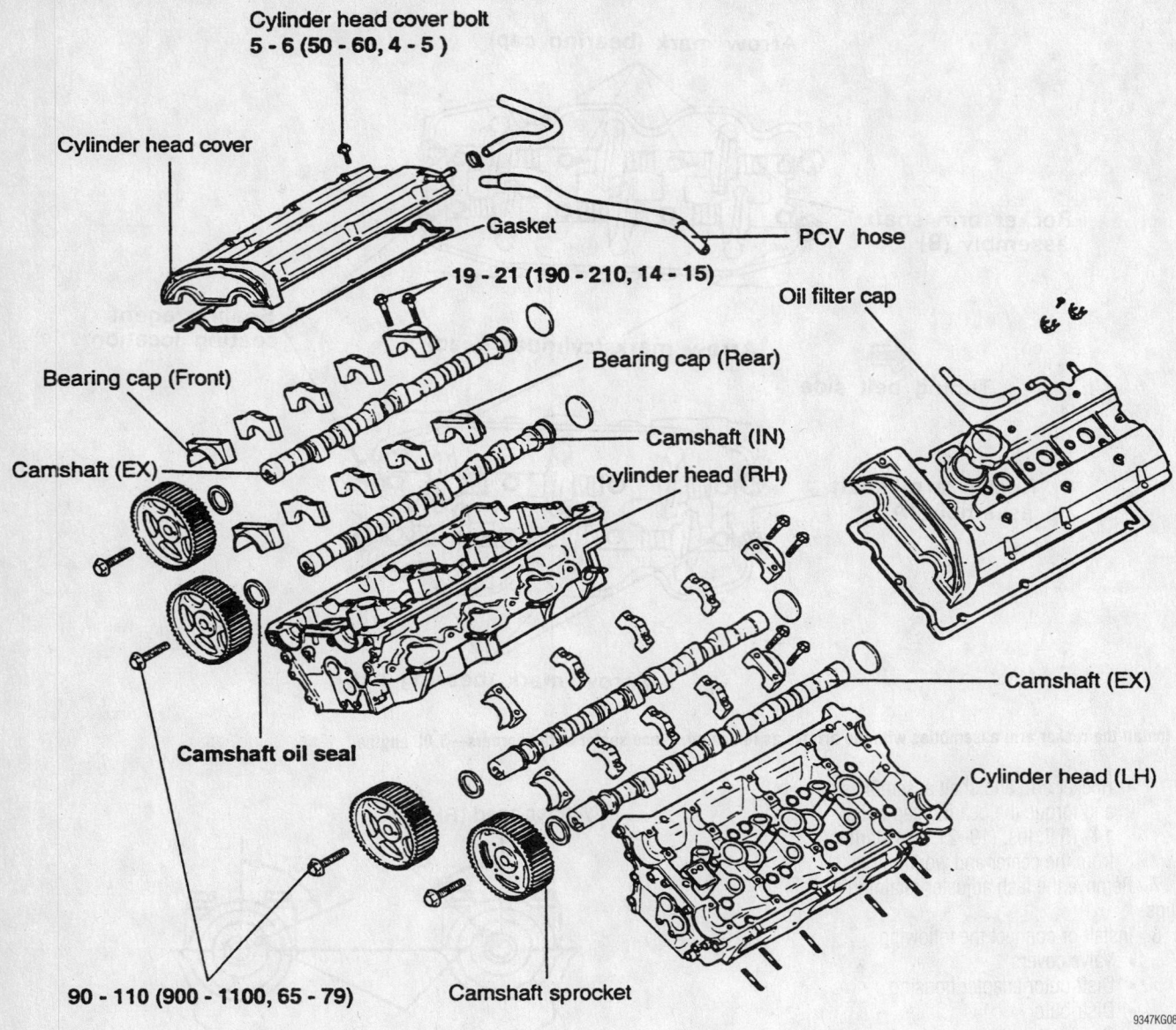

Cylinder head cover bolt
5 - 6 (50 - 60, 4 - 5)

Cylinder head cover

Gasket

PCV hose

19 - 21 (190 - 210, 14 - 15)

Oil filter cap

Bearing cap (Front)

Bearing cap (Rear)

Camshaft (EX)

Camshaft (IN)

Cylinder head (RH)

Camshaft oil seal

Camshaft (EX)

Cylinder head (LH)

90 - 110 (900 - 1100, 65 - 79)

Camshaft sprocket

9347KG08

Exploded view of the camshaft and related components—XG 300

- A/C pulley
- Crankshaft pulley
- Idler pulley and tensioner pulley
- Timing belt cover and loosen the auto tensioner
- Timing belt
- Spark plug cables
- Rocker arm cover
- Camshaft sprockets
- Camshaft bearing caps
- Camshafts

To install:

3. Rotate the crankshaft so that the No. 1 cylinder is at the Top Dead Center (TDC) position.

4. Make certain that the rocker arm is installed properly on the lash adjuster and valve.

5. Install the camshaft dowel pin.
6. Install or connect the following:
   - Hydraulic lash adjusters in their original locations
   - Lash adjuster retaining clips
   - Camshafts
   - Rocker arm and shaft assemblies and torque the bearing cap bolts 14–15 ft. lbs. (19–21 Nm) starting from the center and working out

7. Remove the lash adjuster retaining clips.

   - Camshaft sprockets and torque the bolts to 79 ft. lbs. (110 Nm)
   - Rocker arm cover with a new gasket
   - Spark plug cables
   - Timing belt, cover and tensioner
   - Idler pulley

- Tensioner pulley
- Crankshaft pulley
- A/C pulley
- Power steering pump pulley
- Breather hose and engine harness
- Intake manifold
- Engine cover
- Negative battery cable

Valve Lash

ADJUSTMENT

All engines use hydraulic valve lash adjusters. Valve lash adjustments are not necessary or possible on these engines.

Starter Motor

REMOVAL & INSTALLATION

1998–2000 Models

1. Before servicing the vehicle, refer to the precautions in the beginning of this section.
2. Remove or disconnect the following:
   - Negative battery cable
   - Air intake assembly
   - Speedometer cable
   - Shift cable
   - Starter wiring connectors
   - Starter motor

To install:

3. Install or connect the following:
   - Starter motor and torque the bolts to 20–25 ft. lbs. (26–33 Nm)
   - Starter wiring connectors and torque the battery cable nut to 88–140 inch lbs. (10–16 Nm)
   - Shift cable
   - Speedometer cable
   - Air intake assembly
   - Negative battery cable

2001–02 Models

1. Before servicing the vehicle, refer to the precautions in the beginning of this section.

2. Remove or disconnect the following:
   - Negative battery cable
   - Speedometer and shift control cables
   - Starter electrical connectors
   - Starter motor

To install:

3. Install or connect the following:
   - Starter motor and torque the bolts to 25 ft. lbs. (34 Nm)
   - Starter electrical connectors
   - Speedometer and shift control cables
   - Negative battery cable

Oil Pan

REMOVAL & INSTALLATION

All Except XG 300

1. Before servicing the vehicle, refer to the precautions in the beginning of this section.
2. Drain the engine oil.
3. Remove or disconnect the following:
   - Negative battery cable
   - Engine splash shield
   - Exhaust front pipe
   - Timing belt
   - Oil pan

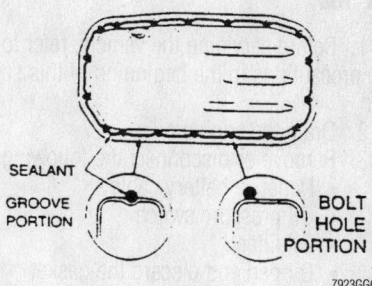

Oil pan sealant applications points—except 3.0L engine

To install:

4. Apply a ⅛ in. (3mm) bead of RTV sealer along the groove in the oil pan.
5. Install or connect the following:
   - Oil pan and torque the bolts to 48–72 inch lbs. (6–8 Nm)
   - Timing belt
   - Exhaust front pipe
   - Engine splash shield
   - Negative battery cable
6. Fill the engine with clean oil.
7. Start the vehicle, check for leaks and repair if necessary.

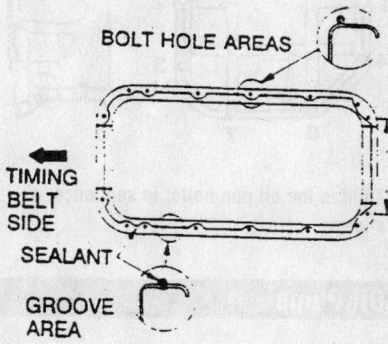

LIQUID-GASKET COATING AREA (TOP VIEW)

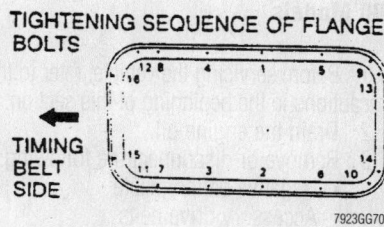

Oil pan sealant applications points and tightening sequence—3.0L Sonata

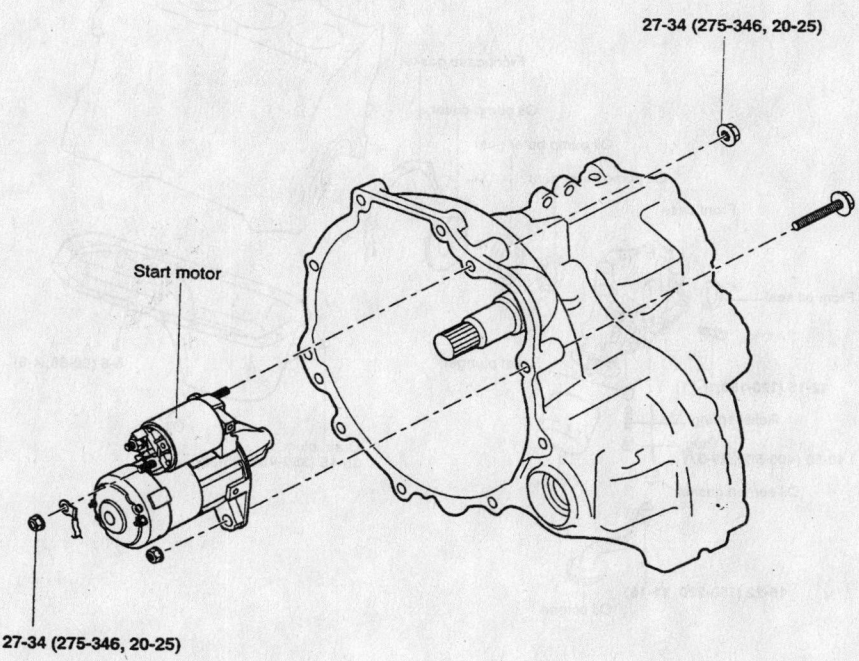

27-34 (275-346, 20-25)

Start motor

27-34 (275-346, 20-25)

Exploded view of the starter—XG 300

9347KG10

XG 300

1. Before servicing the vehicle, refer to the precautions in the beginning of this section.
2. Drain the engine oil.
3. Remove or disconnect the following:
   • Negative battery cable
   • Oil pressure switch
   • Oil filter
   • Oil pan and discard the gasket
4. Clean all mating surfaces of any residual gasket material.

To install:

5. Apply a 1/8 in. (3mm) bead of RTV sealer along the groove in the oil pan.
6. Install or connect the following:
   • Oil pan and torque the bolts to 89 inch lbs. (10 Nm)
   • Oil filter
   • Oil pressure switch and torque the bolt to 89 inch lbs. (10 Nm)
   • Negative battery cable
7. Fill the engine with clean oil.
8. Start the vehicle, check for leaks and repair if necessary.

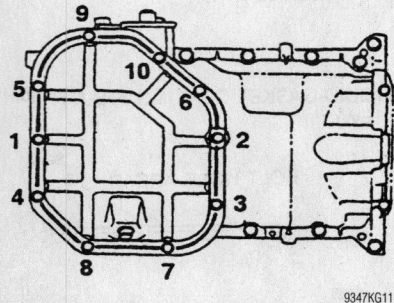

Tighten the oil pan bolts, in sequence, as shown

Oil Pump

REMOVAL & INSTALLATION

All Except Sonata 2.0L, 2.4L and XG 300 Models

1. Before servicing the vehicle, refer to the precautions in the beginning of this section.
2. Drain the engine oil.
3. Remove or disconnect the following:
   • Negative battery cable
   • Accessory drive belts
   • Front cover
   • Timing belt
   • Crankshaft timing sprocket
   • Oil pan
   • Oil pickup tube
   • Oil pump

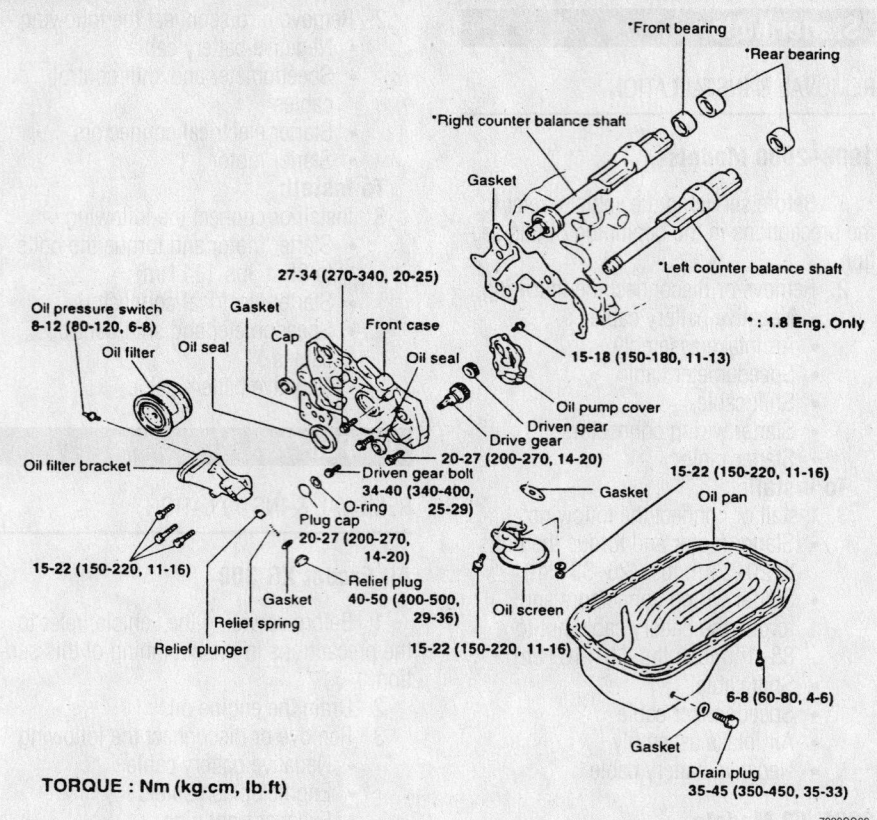

TORQUE : Nm (kg.cm, lb.ft)

Exploded view of the oil pump and pan—Sonata 2.0L engine

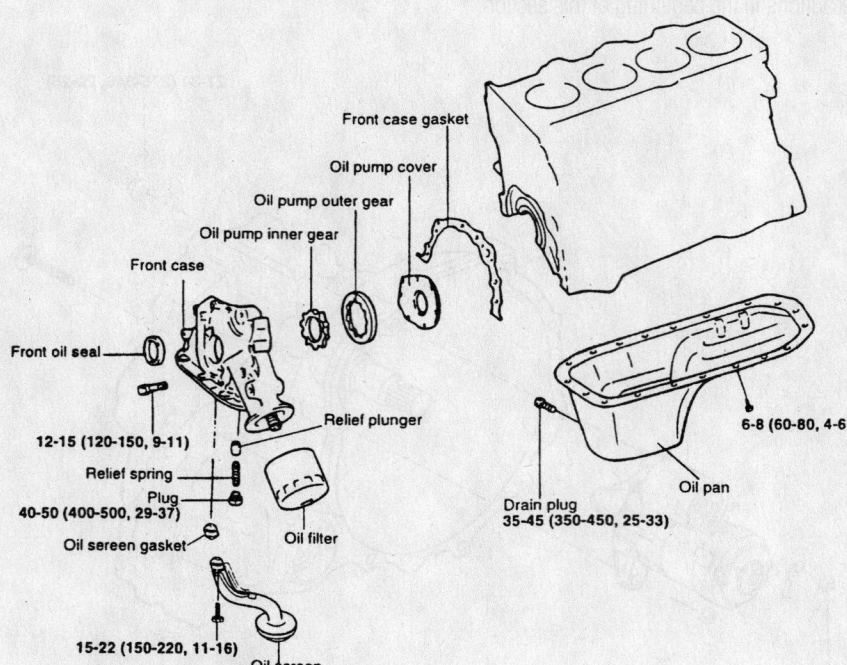

TORQUE : Nm (kg.cm, lb.ft)

Exploded view of the oil pump and pan—1.5L, 1.8L and Elantra and Tiburon with 2.0L engine

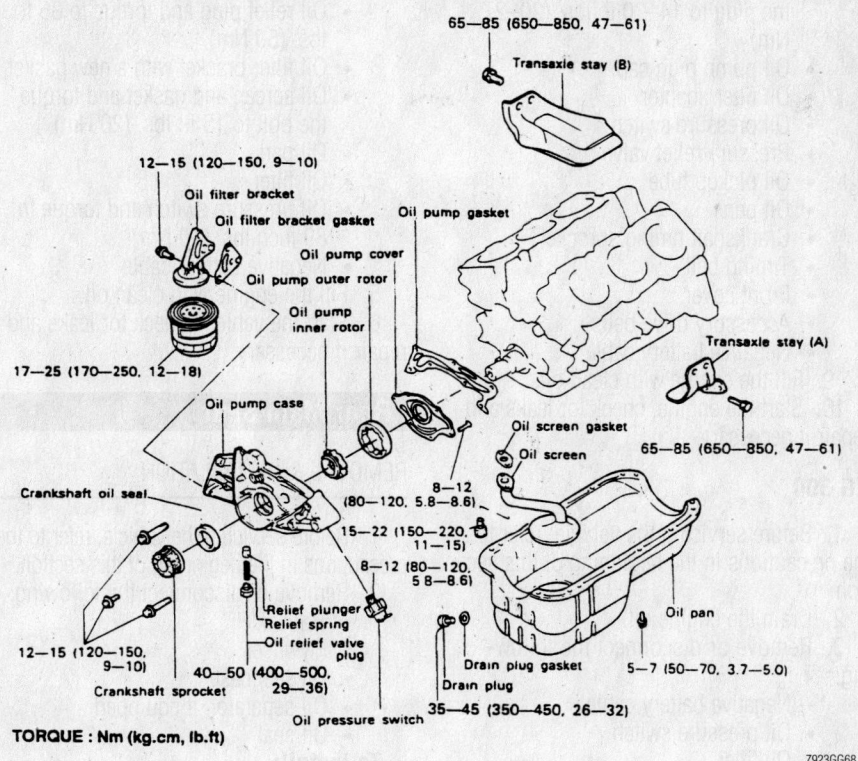

65—85 (650—850, 47—61)
Transaxle stay (B)

12—15 (120—150, 9—10)
Oil filter bracket
Oil filter bracket gasket
Oil pump gasket
Oil pump cover
Oil pump outer rotor
Oil pump inner rotor

17—25 (170—250, 12—18)
Oil pump case

Crankshaft oil seal

Transaxle stay (A)
65—85 (650—850, 47—61)
Oil screen gasket
Oil screen

8—12 (80—120, 5.8—8.6)
15—22 (80—220, 11—15)
8—12 (80—120, 5.8—8.6)

12—15 (120—150, 9—10)
Relief plunger
Relief spring
Oil relief valve plug
40—50 (400—500, 29—36)
Crankshaft sprocket
Drain plug
Oil pressure switch
35—45 (350—450, 26—32)

Oil pan
Drain plug gasket
5—7 (50—70, 3.7—5.0)

TORQUE : Nm (kg.cm, lb.ft)

7923GG68

Exploded view of the oil pump and pan—3.0L engine

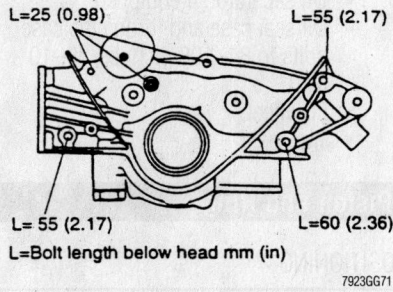

L=25 (0.98)   L=55 (2.17)

L=55 (2.17)   L=60 (2.36)
L=Bolt length below head mm (in)
7923GG71

Oil pump cover bolt lengths and locations—3.0L engine

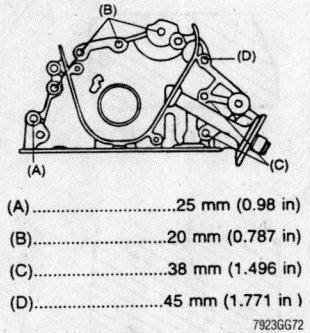

(A) ....25 mm (0.98 in)
(B).............................20 mm (0.787 in)
(C).............................38 mm (1.496 in)
(D) ....45 mm (1.771 in)
7923GG72

Oil pump cover bolt lengths and locations—1.8L and Elantra and Tiburon with 2.0L engines

(A) .... 25 mm (0.98 in.)
(B) .... 30 mm (1.18 in.)
(C) .... 45 mm (1.77 in.)
(D) .... 60 mm (2.36 in.)

7923GG73

Oil pump cover bolt lengths and locations—1.5L engines

To install:
4. Install or connect the following:
  • Oil pump using a new gasket and torque the bolts to 11 ft. lbs. (15 Nm)
  • Oil pickup tube and torque the bolts to 11 ft. lbs. (15 Nm)
  • Oil pan
  • Crankshaft timing sprocket
  • Timing belt
  • Front cover
  • Accessory drive belts
  • Negative battery cable
5. Fill the engine with clean oil.

6. Start the engine, check for leaks and repair if necessary.

Sonata 2.0L and 2.4L Engines

1. Before servicing the vehicle, refer to the precautions in the beginning of this section.
2. Drain the engine oil.
3. Remove or disconnect the following:
  • Negative battery cable
  • Accessory drive belts
  • Front cover

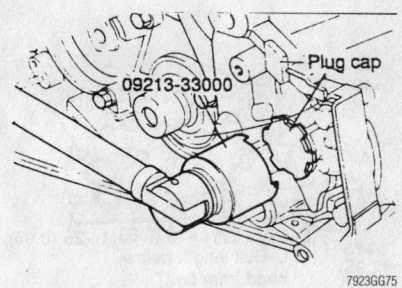

09213-33000
Plug cap
7923GG75

A special socket (PN 09213-33000) is available to remove the plug cap from the oil pump portion of the case—Sonata 2.0L and 2.4L engines

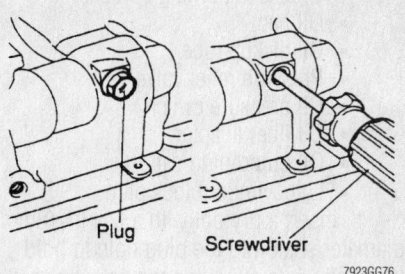

Plug   Screwdriver
7923GG76

Remove the left side cylinder block plug and insert a screwdriver into the hole to hold the balance shaft from turning—Sonata 2.0L and 2.4L engines

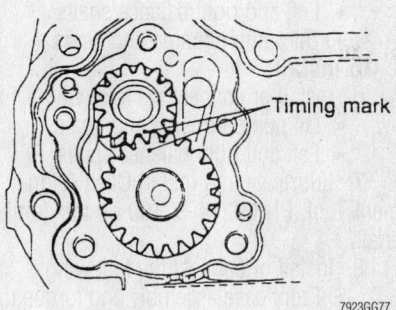

Timing mark
7923GG77

Align the timing marks on the gears during assembly—Sonata 2.0L and 2.4L engines

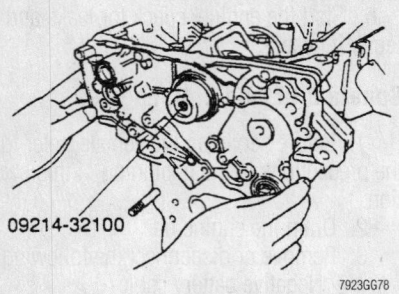

09214-32100

7923GG78

A special tool (PN 09214-32100) is used to center the front case hole on the crankshaft—Sonata 2.0L engines

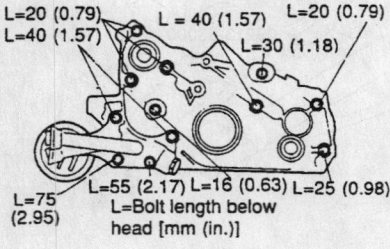

L=20 (0.79) L=40 (1.57) L=20 (0.79) L=40 (1.57) L=30 (1.18) L=75 (2.95) L=55 (2.17) L=16 (0.63) L=25 (0.98) L=Bolt length below head [mm (in.)]

7923GG79

Oil pump cover bolt lengths and locations—Sonata 2.0L engines

- Timing belt
- Crankshaft timing sprocket
- Oil pan
- Oil pickup tube
- Pressure relief valve
- Oil pressure switch
- Oil filter adapter
- Oil pump plug cap
- Left cylinder block plug

4. Insert a prytool with a ⁵⁄₁₆ in. (8mm) diameter shaft into the plug hole to hold the shaft while removing the balance shaft bolt.

5. Remove or disconnect the following:
- Left balance shaft retaining bolt
- Front case assembly
- Left and right balance shafts
- Oil pump housing and gears

To install:

6. Install or connect the following:
- Oil pump housing and gears
- Left and right balance shafts

7. Lubricate and install Case Alignment Tool, PN 09214-32100 on the crankshaft.

8. Install or connect the following:
- Front case assembly and torque the bolts to 20 ft. lbs. (27 Nm)
- Left balance shaft retaining bolt and torque the bolt to 25–29 ft. lbs. (34–40 Nm)
- Left cylinder block plug and torque

the plug to 14–20 ft. lbs. (20–27 Nm)
- Oil pump plug cap
- Oil filter adapter
- Oil pressure switch
- Pressure relief valve
- Oil pickup tube
- Oil pan
- Crankshaft timing sprocket
- Timing belt
- Front cover
- Accessory drive belts
- Negative battery cable

9. Fill the engine with clean oil.

10. Start the engine, check for leaks and repair if necessary.

XG 300

1. Before servicing the vehicle, refer to the precautions in the beginning of this section.

2. Drain the engine oil.

3. Remove or disconnect the following:
- Negative battery cable
- Oil pressure switch
- Oil filter
- Oil pan
- Oil screen and gasket
- Oil filter bracket and gasket
- Oil relief plug
- Oil pump case
- Oil pump rotor and both covers

To install:

4. Install or connect the following:
- Oil pump inner cover and torque the bolt to 11 ft. lbs. (15 Nm)
- Oil pump outer cover and rotor
- Oil pump case and torque the bolts to 11 ft. lbs. (15 Nm)

- Oil relief plug and torque to 36 ft. lbs. (50 Nm)
- Oil filter bracket with a new gasket
- Oil screen and gasket and torque the bolt to 15 ft. lbs. (20 Nm)
- Oil pan
- Oil filter
- Oil pressure switch and torque to 89 inch lbs. (10 Nm)
- Negative battery cable

5. Fill the engine with clean oil.

6. Start the vehicle, check for leaks and repair if necessary.

Rear Main Seal

REMOVAL & INSTALLATION

1. Before servicing the vehicle, refer to the precautions in the beginning of this section.

2. Remove or disconnect the following:
- Transaxle
- Flywheel
- Oil seal case
- Oil separator, if equipped
- Oil seal

To install:

3. Install or connect the following:
- Oil seal
- Oil separator, if equipped
- Oil seal case and torque the case bolts to 84–108 inch lbs. (8–10 Nm)
- Flywheel
- Transaxle

Piston and Ring

POSITIONING

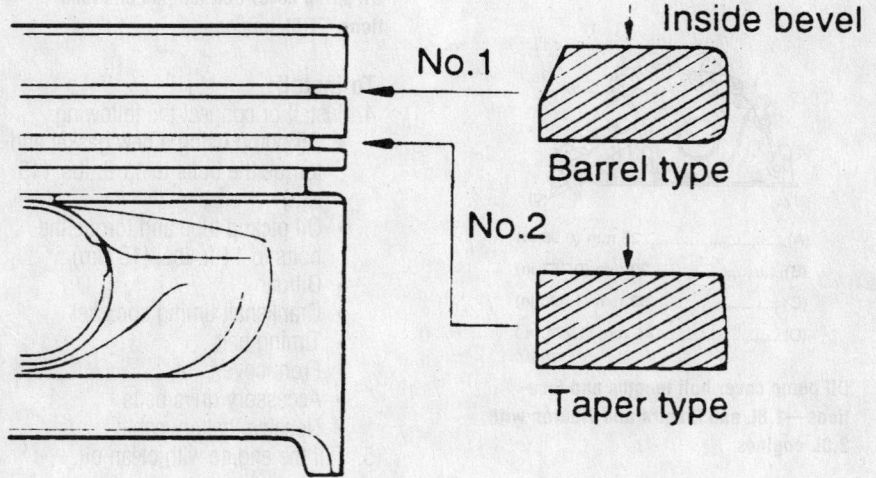

No.1 → Inside bevel Barrel type

No.2 → Inside bevel Taper type

7923AG35

Compression ring identification

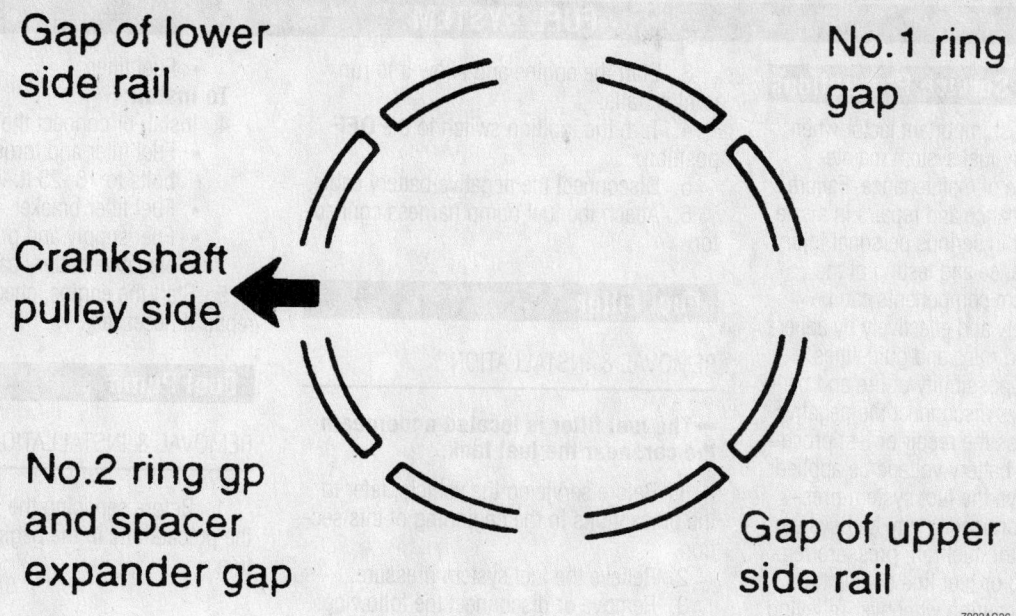

Gap of lower side rail

No.1 ring gap

Crankshaft pulley side

No.2 ring gp and spacer expander gap

Gap of upper side rail

7923AG36

Piston ring end-gap spacing

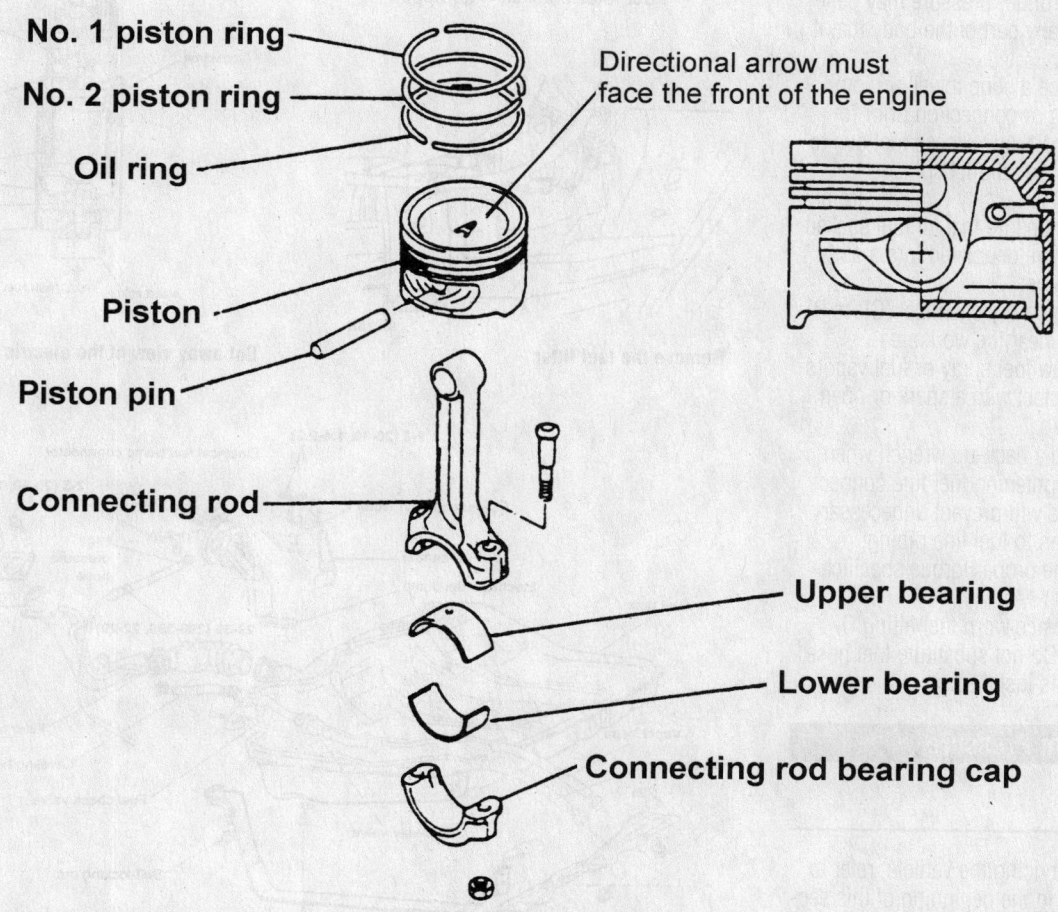

No. 1 piston ring

No. 2 piston ring

Oil ring

Directional arrow must face the front of the engine

Piston

Piston pin

Connecting rod

Upper bearing

Lower bearing

Connecting rod bearing cap

7923AG38

Piston and connecting rod assembly

FUEL SYSTEM

Fuel System Service Precautions

Safety is the most important factor when performing not only fuel system maintenance, but any type of maintenance. Failure to conduct maintenance and repairs in a safe manner may result in serious personal injury or death. Maintenance and testing of the vehicle's fuel system components can be accomplished safely and effectively by adhering to the following rules and guidelines.

• To avoid the possibility of fire and personal injury, always disconnect the negative battery cable unless the repair or test procedure requires that battery voltage be applied.

• Always relieve the fuel system pressure prior to disconnecting any fuel system component (injector, fuel rail, pressure regulator, etc.), fitting or fuel line connection. Exercise extreme caution whenever relieving the fuel system pressure, to avoid exposing skin, face and eyes to fuel spray. Please be advised that fuel under pressure may penetrate the skin or any part of the body that it contacts.

• Always place a shop towel or cloth around the fitting or connection prior to loosening to absorb any excess fuel due to spillage. Ensure that all fuel spillage (should it occur) is quickly removed from engine surfaces. Ensure that all fuel soaked cloths or towels are deposited into a suitable waste container.

• Always keep a dry chemical (Class B) fire extinguisher near the work area.

• Do not allow fuel spray or fuel vapors to come into contact with a spark or open flame.

• Always use a back-up wrench when loosening and tightening fuel line connection fittings. This will prevent unnecessary stress and torsion to fuel line piping. Always follow the proper torque specifications.

• Always replace worn fuel fitting O-rings with new. Do not substitute fuel hose where fuel pipe is installed.

Fuel System Pressure

RELIEVING

1. Before servicing the vehicle, refer to the precautions in the beginning of this section.
2. Remove or disconnect the following:
   • Rear seat cushion
   • Access panel
   • Fuel pump module connector

3. Start the engine and allow it to run until it stalls.
4. Turn the ignition switch to the **OFF** position.
5. Disconnect the negative battery cable.
6. Attach the fuel pump harness connector.

Fuel Filter

REMOVAL & INSTALLATION

➡**The fuel filter is located underneath the car, near the fuel tank.**

1. Before servicing the vehicle, refer to the precautions in the beginning of this section.
2. Relieve the fuel system pressure.
3. Remove or disconnect the following:
   • Negative battery cable
   • Fuel supply and pressure lines
   • Fuel filter bracket, if equipped

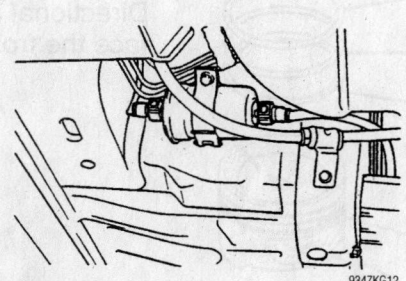

Remove the fuel filter   9347KG12

• Fuel filter

To install:
4. Install or connect the following:
   • Fuel filter and torque the mounting bolts to 18–25 ft. lbs. (25–35 Nm)
   • Fuel filter bracket, if equipped
   • Fuel supply and pressure lines
   • Negative battery cable
5. Start the engine, check leaks and repair if necessary.

Fuel Pump

REMOVAL & INSTALLATION

1. Before servicing the vehicle, refer to the precautions in the beginning of this section.

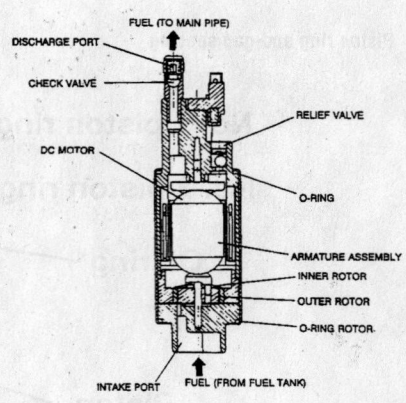

Cut away view of the electric fuel pump   7923GG81

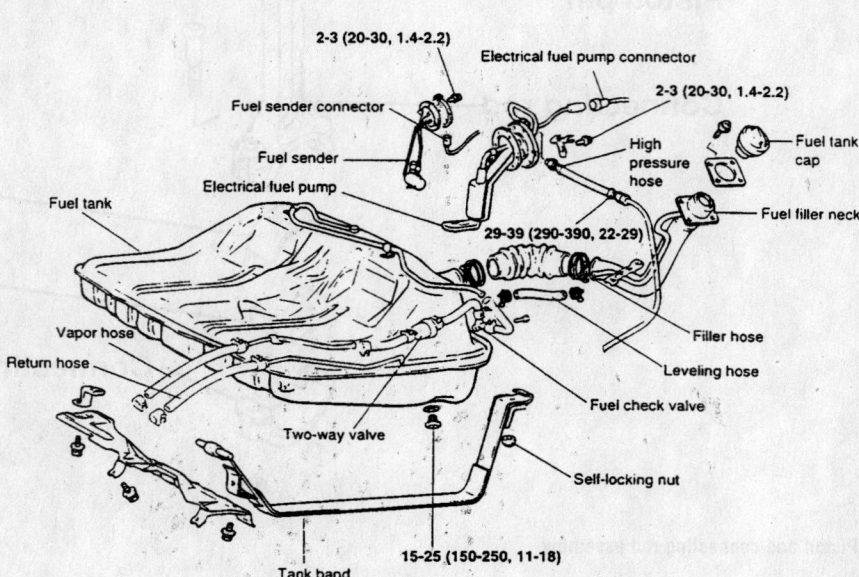

Exploded view of the fuel tank and fuel pump assembly—Sonata   7923GG84

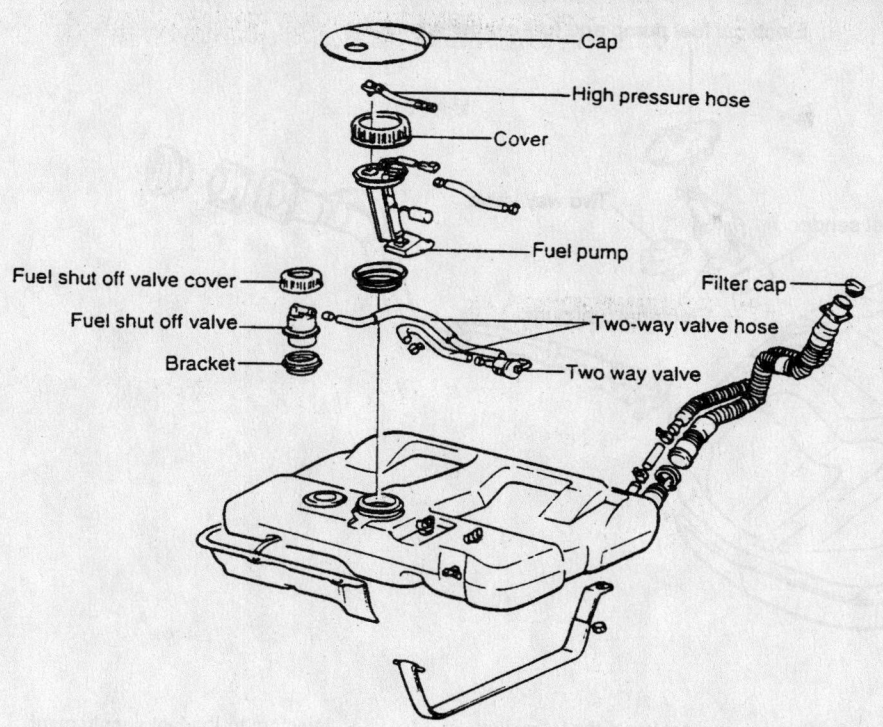

Cap
High pressure hose
Cover
Fuel pump
Fuel shut off valve cover
Fuel shut off valve
Bracket
Filter cap
Two-way valve hose
Two way valve

7923GG85

Exploded view of the fuel tank and fuel pump assembly—Accent

2. Relieve the fuel system pressure.
3. Drain the fuel tank.
4. Remove or disconnect the following:
   - Negative battery cable
   - Fuel supply, return and vapor lines
   - Fuel fill and vent hoses
   - Fuel pump module connector
   - Fuel level sender connector
   - Fuel tank straps
   - Fuel tank
   - Fuel pump module

To install:

5. Install or connect the following:
   - Fuel pump module and torque the mounting bolts to 12–24 inch lbs. (1–3 Nm)
   - Fuel tank
   - Fuel tank straps
   - Fuel level sender connector
   - Fuel pump module connector
   - Fuel fill and vent hoses
   - Fuel supply, return and vapor lines
   - Negative battery cable

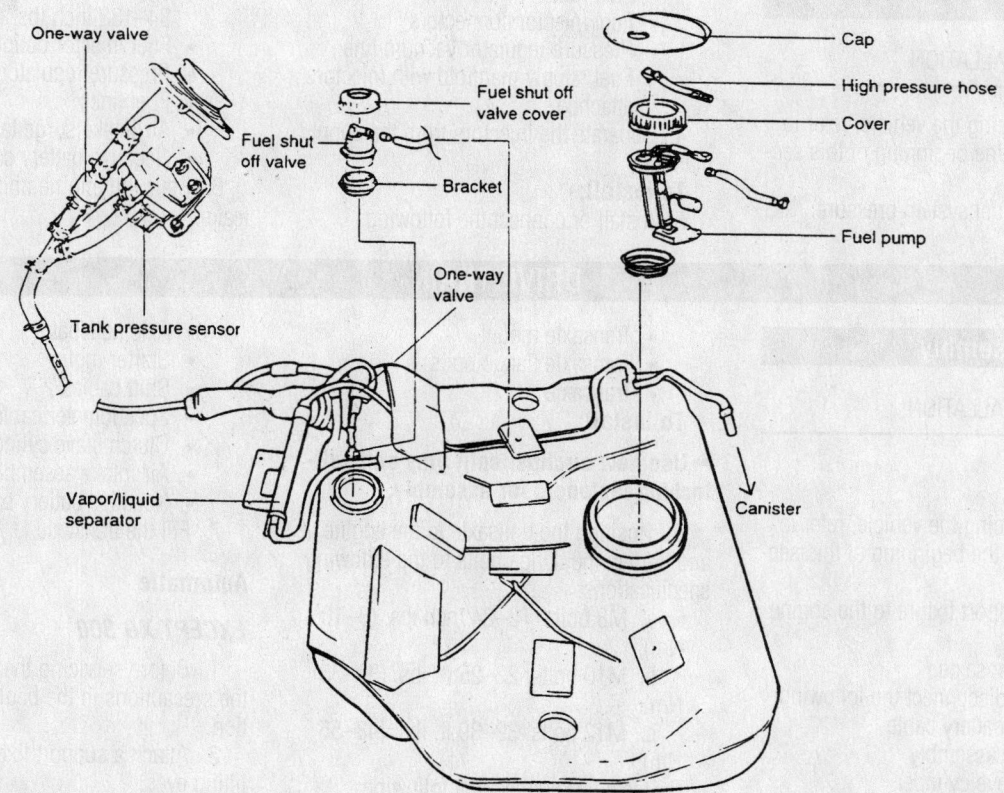

One-way valve
Fuel shut off valve cover
Fuel shut off valve
Bracket
One-way valve
Tank pressure sensor
Vapor/liquid separator
Canister
Cap
High pressure hose
Cover
Fuel pump

7923GG86

Fuel tank and fuel pump assembly—Tiburon and Elantra

Timing belt service is covered in Section 3 of this manual

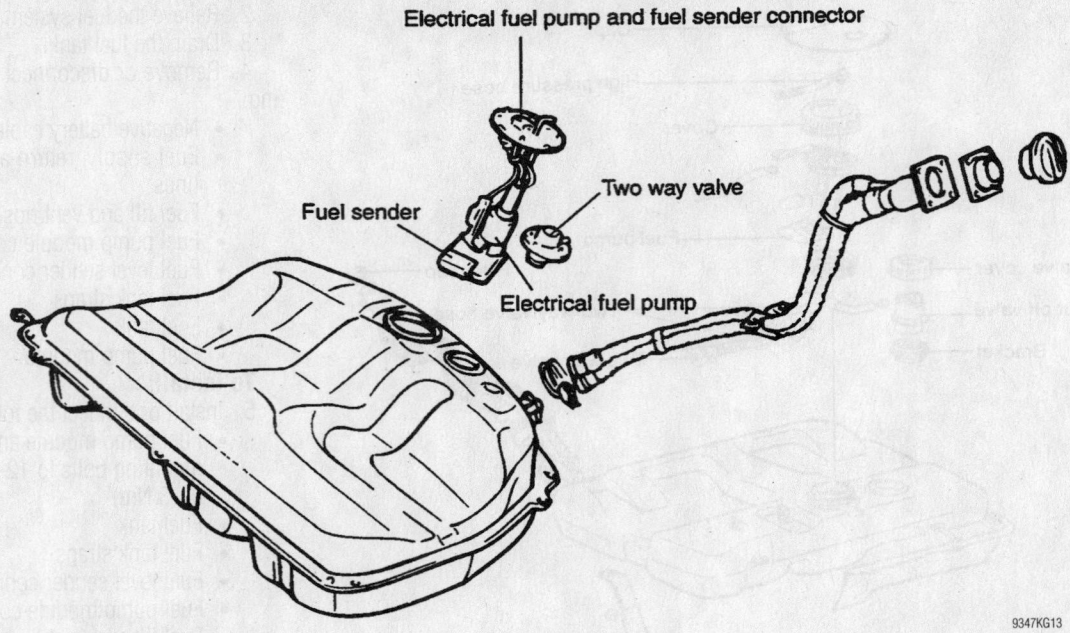

Electrical fuel pump and fuel sender connector

Fuel sender

Two way valve

Electrical fuel pump

9347KG13

Fuel tank and fuel pump assembly—XG 300

6. Fill the tank with fuel and check for proper fuel pump operation.

Fuel Injector

REMOVAL & INSTALLATION

1. Before servicing the vehicle, refer to the precautions in the beginning of this section.
2. Relieve the fuel system pressure.

3. Remove or disconnect the following:
   • Negative battery cable
   • Air intake surge tank, if necessary
   • Fuel lines
   • Fuel injector connectors
   • Pressure regulator vacuum line
   • Fuel supply manifold with injectors attached
4. Separate the injectors from the supply manifold.
   To install:
5. Install or connect the following:

   • Injectors to the fuel supply manifold using new O-rings
   • Fuel supply manifold with injectors attached and torque the bolts to 84–132 inch lbs. (10–15 Nm)
   • Fuel injector connectors
   • Pressure regulator vacuum line
   • Fuel lines
   • Air intake surge tank, if removed
   • Negative battery cable
6. Start the engine and check for leaks.

DRIVE TRAIN

Transaxle Assembly

REMOVAL & INSTALLATION

Manual

1. Before servicing the vehicle, refer to the precautions in the beginning of this section.
2. Attach a support fixture to the engine lifting eyes.
3. Drain the transaxle.
4. Remove or disconnect the following:
   • Negative battery cable
   • Air intake assembly
   • Clutch slave cylinder
   • Speedometer cable
   • Shift cables
   • Starter motor
   • Axle halfshafts
   • Flywheel cover

   • Transaxle mount
   • Transaxle flange bolts
   • Transaxle
   To install:

➡**Use new circlips, split pins and self-locking fasteners for assembly.**

5. Position the transaxle to the engine and tighten the flange bolts to the following specifications:
   a. M8 bolts: 72–84 inch lbs. (8–10 Nm).
   b. M10 bolts: 22–25 ft. lbs. (30–35 Nm).
   c. M12 bolts: 32–39 ft. lbs. (43–55 Nm).
6. Install or connect the following:
   • Transaxle mount and torque the bolts to 65–80 ft. lbs. (90–110 Nm)
   • Flywheel cover and torque the bolts to 72–84 inch lbs. (8–10 Nm)

   • Axle halfshafts
   • Starter motor
   • Shift cables
   • Speedometer cable
   • Clutch slave cylinder
   • Air intake assembly
   • Negative battery cable
7. Fill the transaxle.

Automatic

EXCEPT XG 300

1. Before servicing the vehicle, refer to the precautions in the beginning of this section.
2. Attach a support fixture to the engine lifting eyes.
3. Drain the transaxle.
4. Remove or disconnect the following:
   • Negative battery cable
   • Air intake assembly
   • Transaxle oil cooler lines

- Shift cable
- Speedometer cable
- Pulse generator connector
- Inhibitor connector
- Kickdown servo connector
- Solenoid valve connector
- Oil temperature sensor connector
- Starter motor
- Axle halfshafts
- Flywheel cover
- Torque converter
- Transaxle mount
- Transaxle flange bolts
- Transaxle

To install:

5. Position the transaxle to the engine and tighten the flange bolts to the following specifications:

    a. M8 bolts: 72–84 inch lbs. (8–10 Nm).

    b. M10 bolts: 22–25 ft. lbs. (30–35 Nm).

    c. M12 bolts: 32–39 ft. lbs. (43–55 Nm).

6. Install or connect the following:

- Transaxle mount and torque the bolts to 65–80 ft. lbs. (90–110 Nm)
- Torque converter and torque the bolts to 34–39 ft. lbs. (46–53 Nm)
- Flywheel cover and torque the bolts to 72–84 inch lbs. (8–10 Nm)
- Axle halfshafts
- Starter motor
- Oil temperature sensor connector
- Solenoid valve connector
- Kickdown servo connector
- Inhibitor connector
- Pulse generator connector
- Speedometer cable
- Shift cable
- Transaxle oil cooler lines
- Air intake assembly
- Negative battery cable

7. Fill the transaxle to the correct level.

XG 300

1. Before servicing the vehicle, refer to the precautions in the beginning of this section.

2. Attach a support fixture to the engine lifting eyes.

3. Drain the transaxle.

4. Remove or disconnect the following:

- Negative battery cable
- Air cleaner assembly
- Control cable
- Speedometer sensor connector
- Transaxle range switch connector

- Solenoid connector
- Oil temperature sensor connector
- Oil cooler lines
- Steering gear
- Sway bar
- Tie rod end
- Ball joints and drive shafts
- Steering U-joint and return tube mounting bolts
- Subframe
- Starter
- Engine-to-transaxle bolts
- Transaxle

To install:

5. Install or connect the following:

- Engine-to-transaxle and torque the bolts to 39 ft. lbs. (54 Nm)
- Starter
- Subframe and torque the subframe to transaxle bolts to 58 ft. lbs. (80 Nm) and the roll stopper bolts to 38 ft. lbs. (55 Nm)
- Steering U-joint and return tube mounting bolts
- Ball joints and driveshafts
- Tie rod end
- Sway bar
- Gear box
- Oil cooler lines
- Oil temperature sensor connector
- Solenoid connector

- Transaxle range switch connector
- Speedometer sensor connector and torque to19 ft. lbs. (26 Nm)
- Control cable and torque the bracket bolt to 18 ft. lbs. (25 Nm)
- Air cleaner assembly
- Negative battery cable

6. Fill the transaxle fluid to the proper level.

7. Start the vehicle, check for leaks and repair if necessary.

Clutch

ADJUSTMENTS

These vehicles are equipped with a hydraulic clutch system. No adjustment is necessary.

REMOVAL & INSTALLATION

1. Before servicing the vehicle, refer to the precautions in the beginning of this section.

2. Remove or disconnect the following:

- Transaxle
- Pressure plate bolts by loosening them evenly in a crossing pattern
- Pressure plate and clutch disc

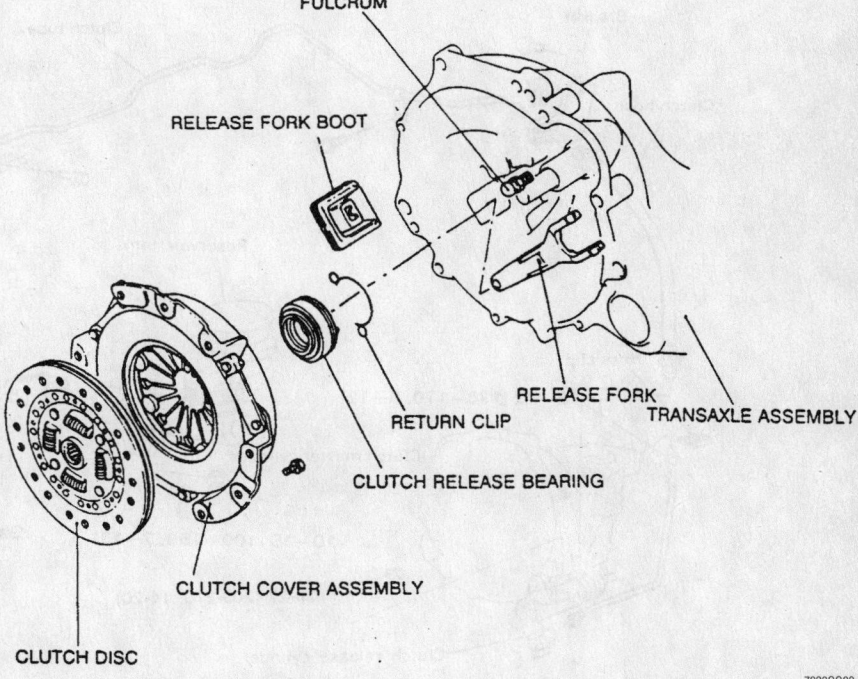

FULCRUM

RELEASE FORK BOOT

RETURN CLIP

RELEASE FORK

TRANSAXLE ASSEMBLY

CLUTCH RELEASE BEARING

CLUTCH COVER ASSEMBLY

CLUTCH DISC

7923GG93

Exploded view of the clutch disc and pressure plate components

Heater Core replacement is covered in Section 2 of this manual

To install:

3. Install or connect the following:
- Clutch disc on the flywheel
- Pressure plate and torque the bolts evenly in a crossing pattern to 11–15 ft. lbs. (15–21 Nm)
- Transaxle

Hydraulic Clutch System

BLEEDING

1. Connect a hose to the bleeder screw and place the other end of hose into a container of clean brake fluid. Open the bleeder screw.

2. Have an assistant pump the clutch pedal slowly until no air bubbles are present at the bleeder screw.

3. Close the bleeder screw.

4. Fill the clutch master cylinder.

5. Check the clutch operation.

Halfshaft

REMOVAL & INSTALLATION

All Models Except XG 300

1. Before servicing the vehicle, refer to the precautions in the beginning of this section.

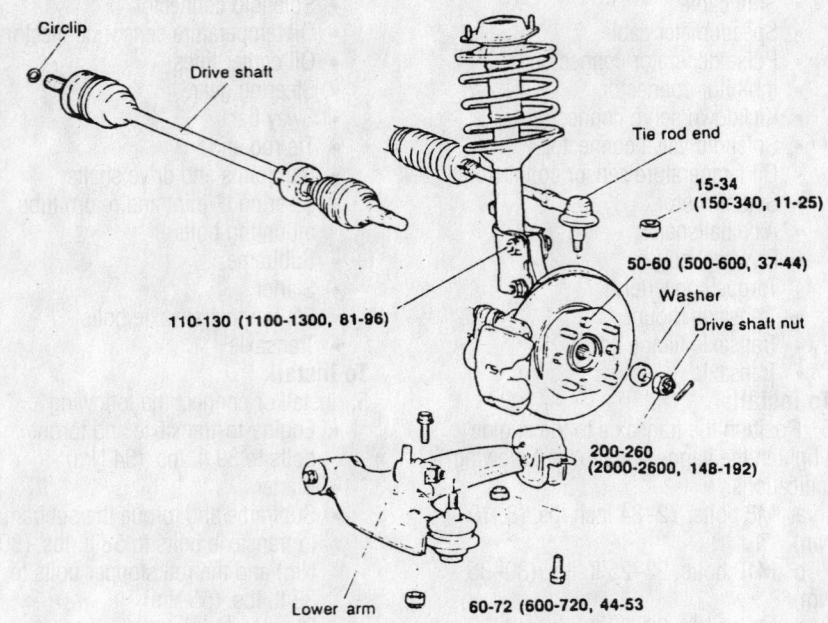

TORQUE : Nm (kg.cm, lb.ft)

Halfshaft components—except V6 Sonata

2. Remove or disconnect the following:
- Front wheel
- Spindle nut
- Wheel speed sensor, if equipped
- Outer tie rod end
- Stabilizer bar link
- Lower ball joint

3. Press the stub shaft out of the hub.

4. Pry the inner joint out of the transaxle or intermediate shaft.

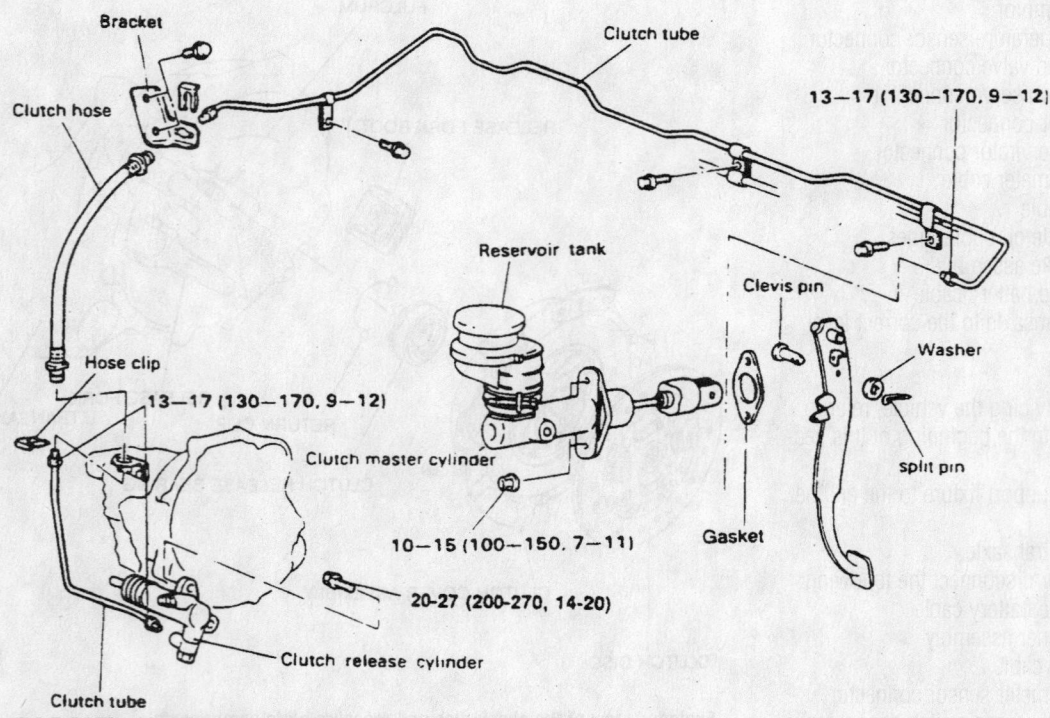

TORQUE : Nm (kg.cm, lb.ft)

Exploded view of the clutch hydraulic system

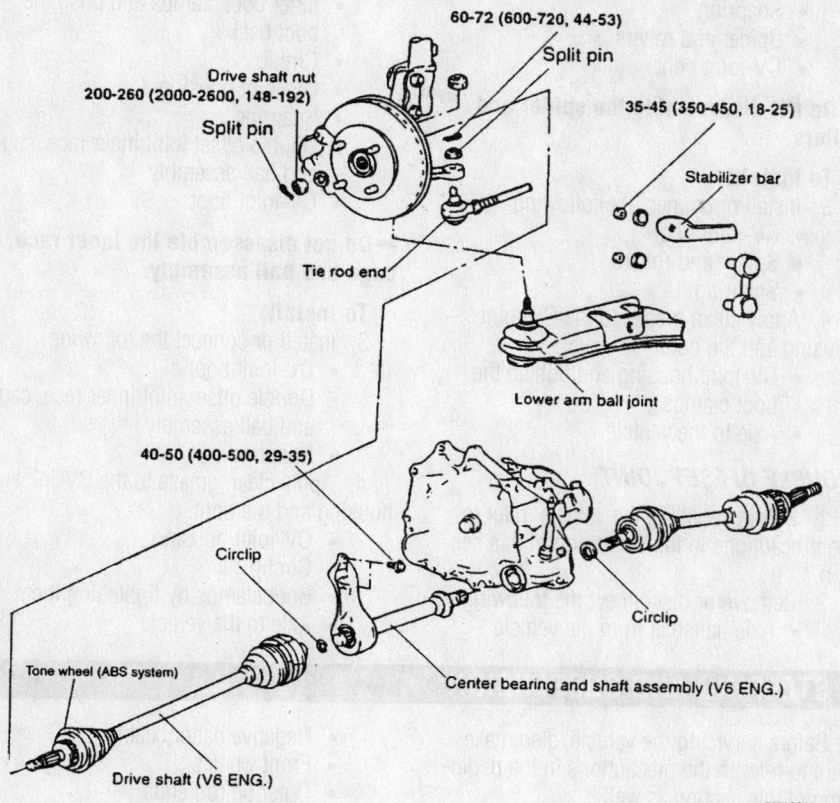

Halfshaft components—V6 Sonata

To install:

➡**Use new circlips, split pins and self-locking nuts for assembly.**

5. Install the inner joint so that the circlip is felt to seat in the retaining groove.

6. Guide the stub shaft into the hub.

7. Install or connect the following:

- Lower ball joint and torque the nut to 43–52 ft. lbs. (58–70 Nm)
- Stabilizer bar link
- Outer tie rod end and torque the nut to 17–25 ft. lbs. (23–34 Nm)
- Wheel speed sensor, if equipped
- Spindle nut and torque the nut to 144–187 ft. lbs. (195–253 Nm)
- Front wheel

8. Check and/or adjust the wheel alignment.

XG 300

1. Before servicing the vehicle, refer to the precautions in the beginning of this section.

2. Drain the transaxle fluid.

3. Remove or disconnect the following:

- Front wheel

- Split pin and halfshaft nut
- Ball joint from the steering knuckle
- Halfshaft from the axle hub

4. Install a pry bar between the center bearing bracket and the halfshaft and pry the halfshaft from the transaxle.

5. Remove the center bearing bracket bolts and install a pry bar between bracket and the engine and disconnect the bracket from the engine.

6. Remove the inner shaft from the transaxle.

To install:

7. Install or connect the following:

- Inner halfshaft to the transaxle
- Center bearing bracket and torque the bolt to 36 ft. lbs. (50 Nm)
- Halfshaft to the axle hub
- Ball joint to the steering knuckle and torque 88 ft. lbs. (110 Nm)
- Split pin and halfshaft nut and torque the nut to 206 ft. lbs. (280 Nm)
- Front wheel

8. Fill the transaxle fluid to the proper level.

9. Check and/or adjust the wheel alignment.

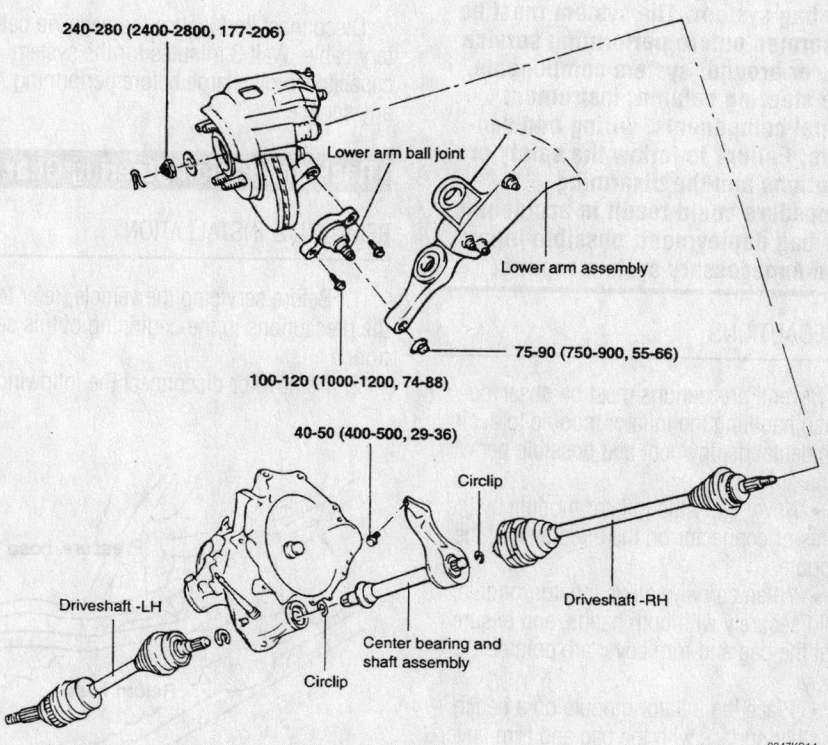

Exploded view of the halfshaft assembly—XG 300

Brake service is covered in Section 4 of this manual

CV-Joints

OVERHAUL

Outer CV-Joint

The outer CV-joint is serviced with the axle halfshaft as an assembly. The outer CV-joint boot may be replaced by first removing the inner joint.

Inner CV-Joint

TRIPOD JOINT

1. Before servicing the vehicle, refer to the precautions in the beginning of this section.
2. Remove or disconnect the following:
   - Axle halfshaft from the vehicle
   - Inner boot clamps and push the boot back
   - CV-joint housing

- Snapring
- Spider and rollers
- CV-joint boot

➡**Do not disassemble the spider and rollers.**

To install:

3. Install or connect the following:
   - CV-joint boot
   - Spider and rollers
   - Snapring
4. Apply clean grease to the CV-joint housing and the boot.
   - CV-joint housing and tighten the boot clamps
   - Axle to the vehicle

DOUBLE OFFSET JOINT

1. Before servicing the vehicle, refer to the precautions in the beginning of this section.
2. Remove or disconnect the following:
   - Axle halfshaft from the vehicle

- Inner boot clamps and push the boot back
- Circlip
- CV-joint housing
- Snapring
- Double offset joint inner race, cage and ball assembly
- CV-joint boot

➡**Do not disassemble the inner race, cage and ball assembly.**

To install:

3. Install or connect the following:
   - CV-joint boot
   - Double offset joint inner race, cage and ball assembly
   - Snapring
4. Apply clean grease to the CV-joint housing and the boot.
   - CV-joint housing
   - Circlip
   - Boot clamps by tightening them
   - Axle to the vehicle

STEERING AND SUSPENSION

Air Bag

✳✳ CAUTION

Some vehicles are equipped with an air bag system. The system must be disarmed before performing service on, or around, system components, the steering column, instrument panel components, wiring and sensors. Failure to follow the safety precautions and the disarming procedure could result in accidental air bag deployment, possible injury and unnecessary system repairs.

PRECAUTIONS

Several precautions must be observed when handling the inflator module to avoid accidental deployment and possible personal injury.

- Never carry the inflator module by the wires or connector on the underside of the module.
- When carrying a live inflator module, hold securely with both hands, and ensure that the bag and trim cover are pointed away.
- Place the inflator module on a bench or other surface with the bag and trim cover facing up.
- With the inflator module on the bench, never place anything on or close to the module which may be thrown in the event of an accidental deployment.

Before servicing the vehicle, also make sure to refer to the precautions in the beginning of this section as well.

DISARMING

Disconnect and isolate the negative battery cable. Wait 3 minutes for the system capacitor to discharge before performing any service.

Rack and Pinion Steering Gear

REMOVAL & INSTALLATION

1. Before servicing the vehicle, refer to the precautions in the beginning of this section.
2. Remove or disconnect the following:

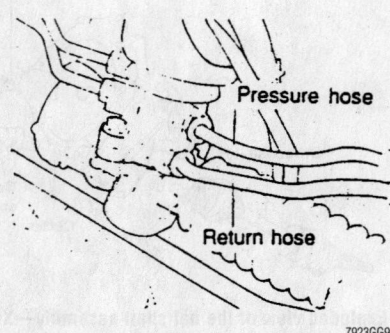

Pressure and return hose location on the rack

- Negative battery cable
- Front wheels
- Outer tie rod ends
- Steering column flexible coupler
- Power steering fluid hoses, if equipped with power steering
- Subframe center beam
- Exhaust front pipe
- Left lower control arm
- Stabilizer bar
- Steering gear

To install:

3. Install or connect the following:
   - Steering gear and torque the bolts to 44–59 ft. lbs. (60–80 Nm)
   - Stabilizer bar
   - Left lower control arm
   - Exhaust front pipe
   - Subframe center beam
   - Power steering fluid hoses, if equipped with power steering
   - Steering column flexible coupler

Power rack and pinion mounting bolt locations

COMPONENTS

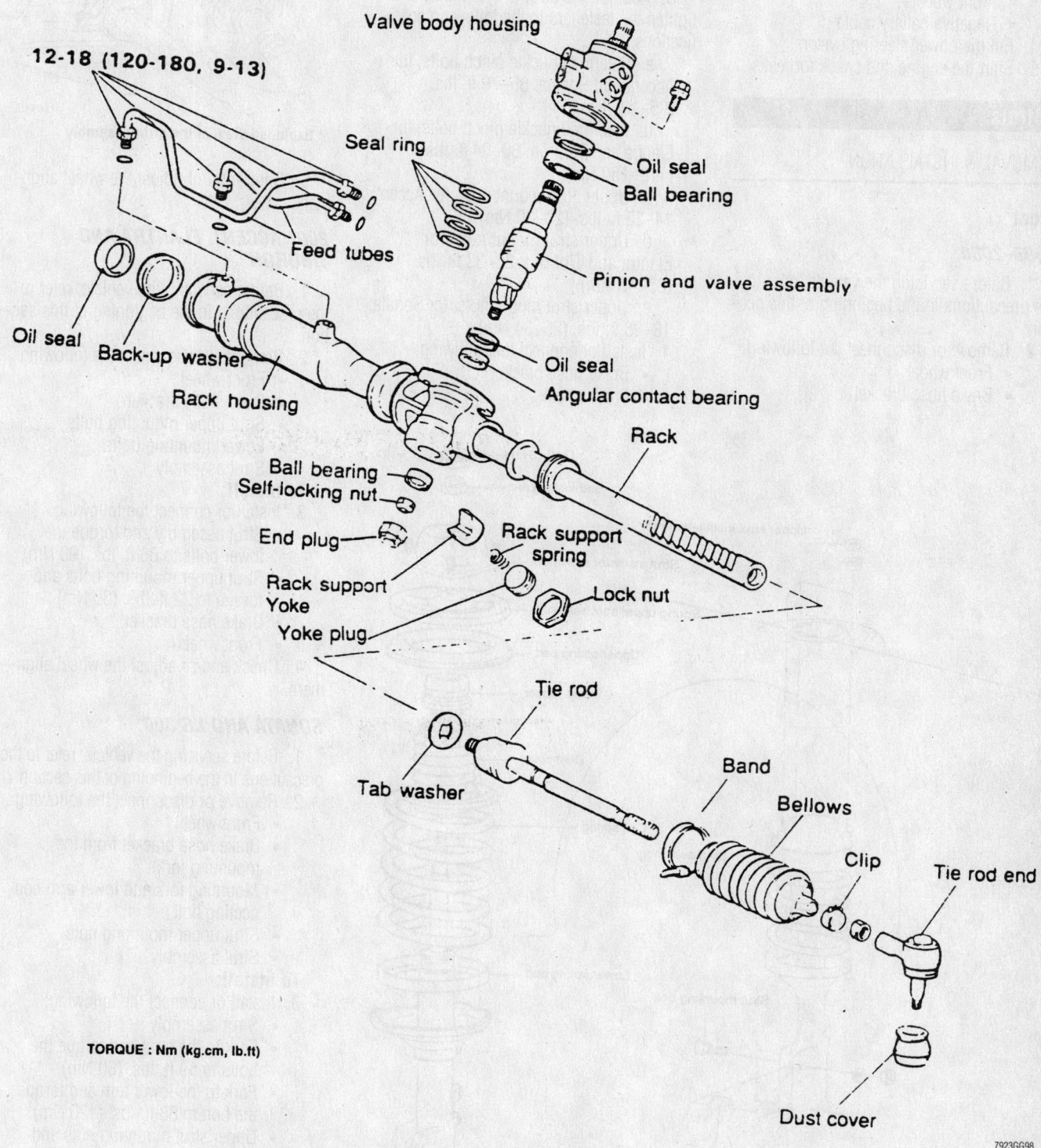

12-18 (120-180, 9-13)

Valve body housing

Seal ring

Oil seal

Ball bearing

Pinion and valve assembly

Feed tubes

Oil seal

Back-up washer

Oil seal

Angular contact bearing

Rack housing

Rack

Ball bearing

Self-locking nut

End plug

Rack support spring

Rack support Yoke

Lock nut

Yoke plug

Tie rod

Tab washer

Band

Bellows

Clip

Tie rod end

Dust cover

TORQUE : Nm (kg.cm, lb.ft)

Exploded view of the rack and pinion assembly

7923GG98

and torque the bolt to 11–14 ft. lbs. (15–19 Nm)
- Outer tie rod ends and torque the nuts to 17–25 ft. lbs. (23–34 Nm)
- Front wheels
- Negative battery cable

4. Fill the power steering system.
5. Start the engine and check for leaks.

Struts

REMOVAL & INSTALLATION

Front

1998–2000

1. Before servicing the vehicle, refer to the precautions in the beginning of this section.
2. Remove or disconnect the following:
   - Front wheel
   - Brake hose bracket

- Steering knuckle pinch bolts
- Upper strut mount
- Strut

To install:

3. Position the strut to the vehicle and tighten the fasteners to the following specifications:

   a. Steering knuckle pinch bolts, for Accent and Sonata: 65–76 ft. lbs. (95–105 Nm).

   b. Steering knuckle pinch bolts, for Elantra and Tiburon: 80–94 ft. lbs. (110–130 Nm).

   c. Upper strut mount nuts, for Accent: 14–22 ft. lbs. (20–30 Nm).

   d. Upper strut mount nuts, for Elantra and Tiburon: 25–33 ft. lbs. (35–45 Nm).

   e. Upper strut mount nuts, for Sonata: 18–25 ft. lbs. (25–34 Nm).

4. Install or connect the following:
   - Brake hose bracket
   - Front wheel

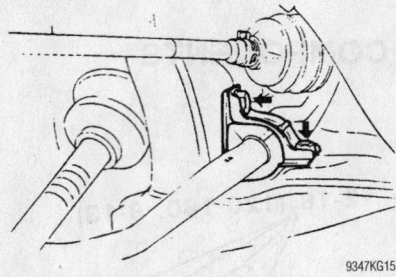

Exploded view of the strut assembly

5. Check and/or adjust the wheel alignment.

2001 ACCENT, ELANTRA AND TIBURON

1. Before servicing the vehicle, refer to the precautions in the beginning of this section.
2. Remove or disconnect the following:
   - Front wheel
   - Brake hose bracket
   - Strut upper mounting bolts
   - Lower mounting bolts
   - Strut assembly

To install:

3. Install or connect the following:
   - Strut assembly and torque the lower bolts to 66 ft. lbs. (90 Nm)
   - Strut upper mounting bolts and torque to 22 ft. lbs. (30 Nm)
   - Brake hose bracket
   - Front wheel

4. Check and/or adjust the wheel alignment.

SONATA AND XG 300

1. Before servicing the vehicle, refer to the precautions in the beginning of this section.
2. Remove or disconnect the following:
   - Front wheel
   - Brake hose bracket from the mounting fork
   - Mounting fork and lower arm connecting bolt
   - Strut upper mounting nuts
   - Strut assembly

To install:

3. Install or connect the following:
   - Strut assembly
   - Fork to the strut and torque the bolts to 59 ft. lbs. (80 Nm)
   - Fork to the lower arm and torque the bolt to 88 ft. lbs. (120 Nm)
   - Upper strut mounting nuts and torque to 36 ft. lbs. (50 Nm)
   - Brake hose bracket
   - Front wheel

4. Check and/or adjust the wheel alignment.

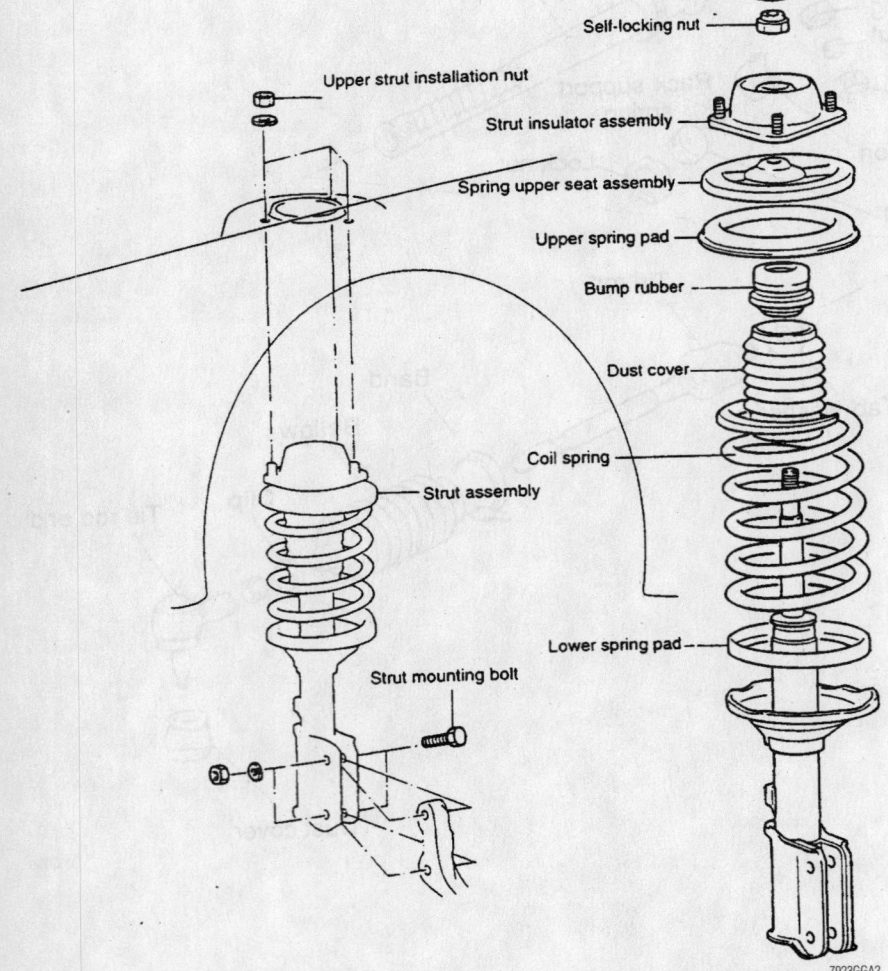

Dust cover
Self-locking nut
Upper strut installation nut
Strut insulator assembly
Spring upper seat assembly
Upper spring pad
Bump rubber
Dust cover
Coil spring
Strut assembly
Lower spring pad
Strut mounting bolt

Exploded view of the strut assembly components

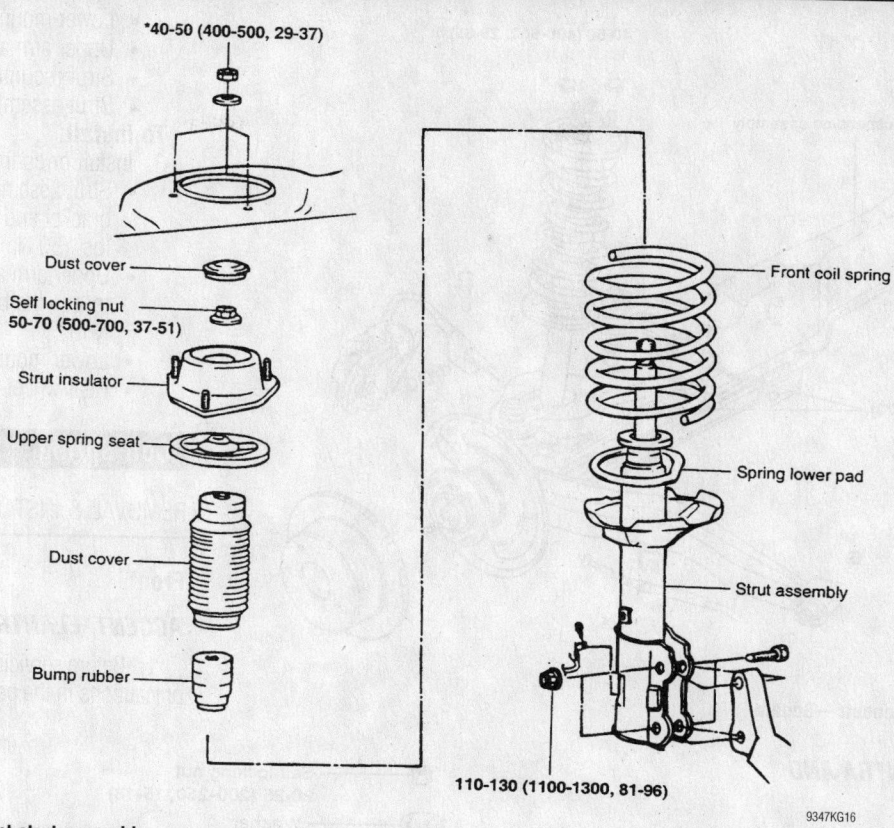

*40-50 (400-500, 29-37)

Dust cover

Self locking nut
50-70 (500-700, 37-51)

Strut insulator

Upper spring seat

Dust cover

Bump rubber

Front coil spring

Spring lower pad

Strut assembly

110-130 (1100-1300, 81-96)

9347KG16

Exploded view of the front strut assembly

Rear

1998–2000

1. Before servicing the vehicle, refer to the precautions in the beginning of this section.

2. Remove or disconnect the following:

- Rear wheel
- Upper strut mount access panel
- Upper strut mount nuts
- Wheel speed sensor connector
- Stabilizer bar link
- Hub knuckle pinch bolts
- Strut

To install:

3. Install or connect the following:

- Strut and torque the pinch bolts to 80–90 ft. lbs. (110–130 Nm) and the upper mount nuts to 14–22 ft. lbs. (20–30 Nm)
- Stabilizer bar link and torque the bolt to 25–33 ft. lbs. (35–45 Nm)
- Wheel speed sensor connector
- Upper strut mount access panel
- Rear wheel

4. Check and/or adjust the wheel alignment.

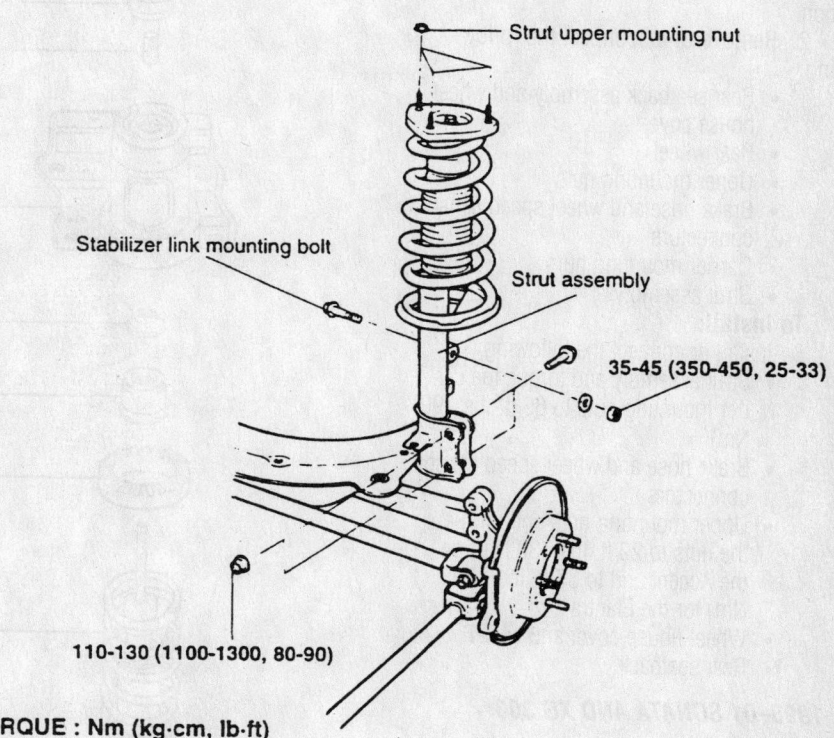

Strut upper mounting nut

Stabilizer link mounting bolt

Strut assembly

35-45 (350-450, 25-33)

110-130 (1100-1300, 80-90)

TORQUE : Nm (kg·cm, lb·ft)

Exploded view of the rear strut assembly

7923GGA3

For Accessory Drive Belt illustrations, see Section 1 of this manual

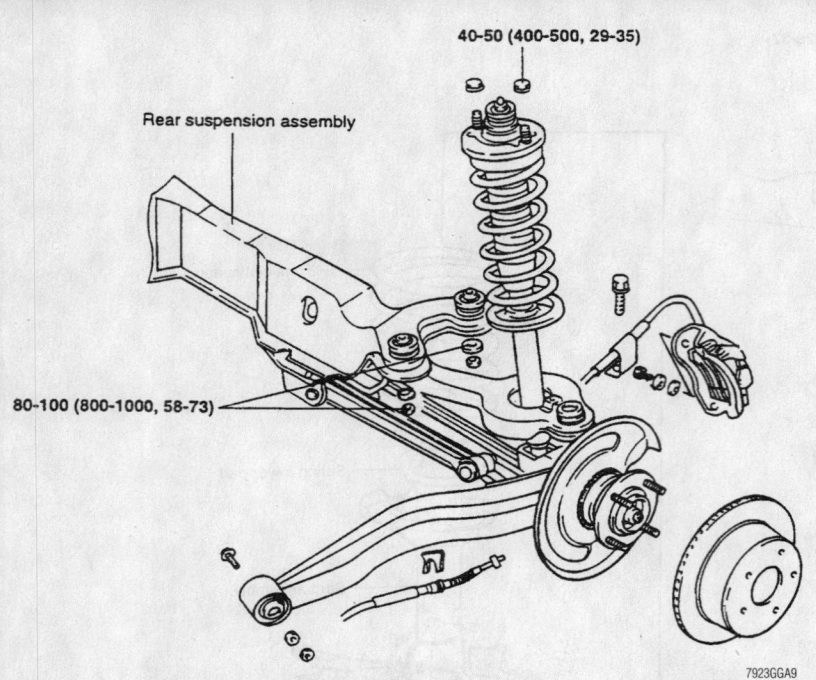

40-50 (400-500, 29-35)

Rear suspension assembly

80-100 (800-1000, 58-73)

7923GGA9

Rear suspension components—Sonata

2001 ACCENT, ELANTRA AND TIBURON

1. Before servicing the vehicle, refer to the precautions in the beginning of this section.

2. Remove or disconnect the following:

- Rear seatback assembly and wheel house cover
- Rear wheel
- Upper mounting nuts
- Brake hose and wheel speed sensor connectors
- Carrier mounting nuts
- Strut assembly

To install:

3. Install or connect the following:

- Strut assembly and torque the carrier mounting nuts to 66 ft. lbs. (90 Nm)
- Brake hose and wheel speed sensor connectors
- Upper mounting nuts and torque the nuts to 22 ft. lbs. (30 Nm) on the Accent and to 37 ft. lbs. (50 Nm) for the Elantra
- Wheel house cover and wheel
- Rear seatback

1998–01 SONATA AND XG 300

1. Before servicing the vehicle, refer to the precautions in the beginning of this section.

2. Remove or disconnect the following:

- Rear wheel

- Lower mounting bolt
- Upper arm and rear carrier bolt
- Strut mounting bracket
- Strut assembly

To install:

3. Install or connect the following:

- Strut assembly and mounting bracket and torque the bolt to 36 ft. lbs. (50 Nm)
- Upper arm and rear carrier bolt and torque the bolt to 88 ft. lbs. (120 Nm)
- Lower mounting bolt
- Rear wheel

Coil Spring

REMOVAL & INSTALLATION

Front

ACCENT, ELANTRA AND TIBURON

1. Before servicing the vehicle, refer to the precautions in the beginning of this section.

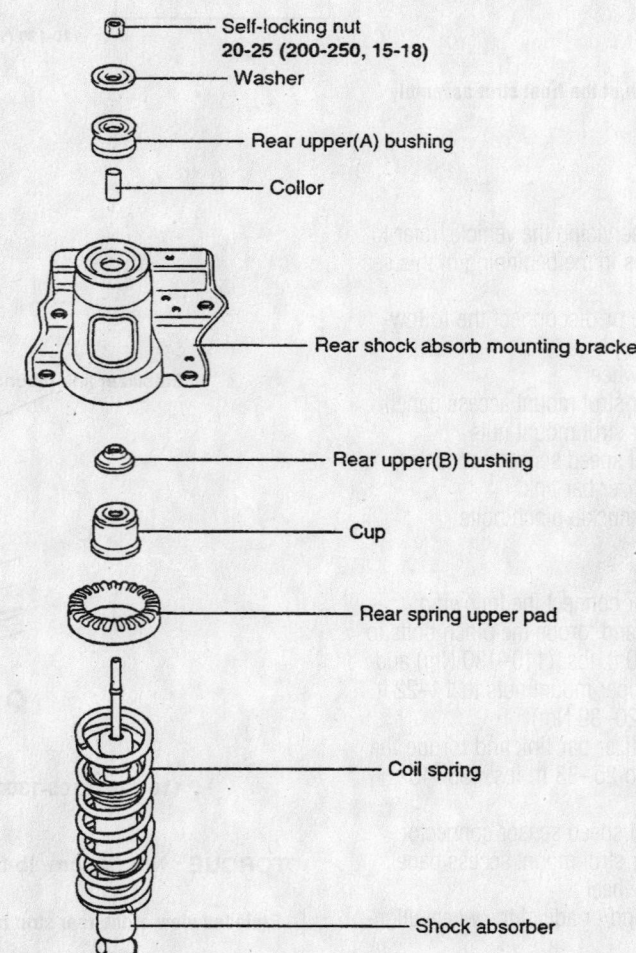

Self-locking nut
20-25 (200-250, 15-18)

Washer

Rear upper(A) bushing

Collor

Rear shock absorb mounting bracket

Rear upper(B) bushing

Cup

Rear spring upper pad

Coil spring

Shock absorber

9347KG17

Exploded view of the rear strut assembly

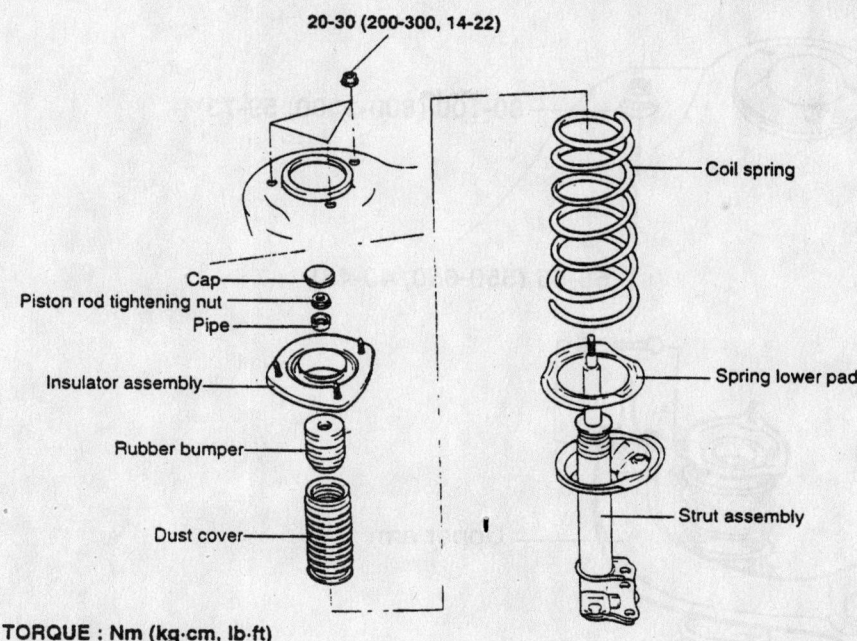

20-30 (200-300, 14-22)

Coil spring

Cap
Piston rod tightening nut
Pipe

Insulator assembly

Spring lower pad

Rubber bumper

Dust cover

Strut assembly

TORQUE : Nm (kg·cm, lb·ft)

7923GGA4

Rear strut components

2. Remove the strut from the vehicle and install a spring compressor.

3. Compress the coil spring so that the end of the spring comes away from the spring seat.

4. Remove or disconnect the following:
- Upper strut mount
- Upper spring seat
- Compressed spring from the strut
- Spring from the spring compressor

To install:

5. Compress the spring and install it on the strut.

6. Install or connect the following:
- Upper spring seat and the upper strut mount and torque the nut to 29–36 ft. lbs. (40–50 Nm)
- Strut to the vehicle

7. Check and/or adjust the wheel alignment.

SONATA AND XG 300

1. Before servicing the vehicle, refer to the precautions in the beginning of this section.

2. Remove the strut from the vehicle and install a Spring Compressor Tool, such as J38402.

3. Compress the coil spring so that the end of the spring comes away from the spring seat.

4. Remove or disconnect the following:
- Self locking nut

5. Install the Compressor Tool, J38402.

6. Remove the bracket, spring pad and coil spring.

To install:

7. Ccompress the coil spring with Compressor Tool J38402.

8. Install or connect the following:
- Coil spring to the strut
- Dust cover, upper spring pad, bushing and hand tighten the lock nut

9. Remove the compressor tool when the coil spring is properly aligned and torque the lock nut to 18 ft. lbs. (25 Nm).

10. Install the strut assembly.

Upper Ball Joint

The upper ball joints are replaced with the upper control arms as an assembly.

Lower Ball Joint

REMOVAL & INSTALLATION

Bolt-On Type

1. Before servicing the vehicle, refer to the precautions in the beginning of this section.

2. Remove or disconnect the following:
- Front wheel
- Ball joint stud from the knuckle

- Ball joint from the lower control arm

To install:

➡**Use a new split pin for assembly.**

3. Install or connect the following:
- Ball joint and torque the mounting bolts to 69–87 ft. lbs. (95–120 Nm)
- Stud nut and torque to 43–52 ft. lbs. (60–72 Nm)
- Front wheel

4. Check and/or adjust the wheel alignment.

Press-In Type

1. Before servicing the vehicle, refer to the precautions in the beginning of this section.

2. Remove or disconnect the following:
- Front wheel
- Lower control arm
- Ball joint dust cover

3. Press the ball joint out of the lower control arm.

To install:

4. Press the ball joint into the control arm.

5. Install or connect the following:
- Ball joint dust cover
- Lower control arm and torque the stud nut to 43–52 ft. lbs. (60–72 Nm)
- Front wheel

6. Check and/or adjust the wheel alignment.

Upper Control Arm

REMOVAL & INSTALLATION

Sonata and XG 300

1. Before servicing the vehicle, refer to the precautions in the beginning of this section.

2. Support the lower control arm assembly with a floor jack.

3. Remove or disconnect the following:
- Front wheel
- Ball joint nut, loosen only
- Upper arm ball joint from the steering knuckle with Special Tool 09568-34000
- Wheel house panel nuts
- Upper arm assembly
- Upper arm shaft

For Tire, Wheel and Ball Joint specifications, see Section 1 of this manual

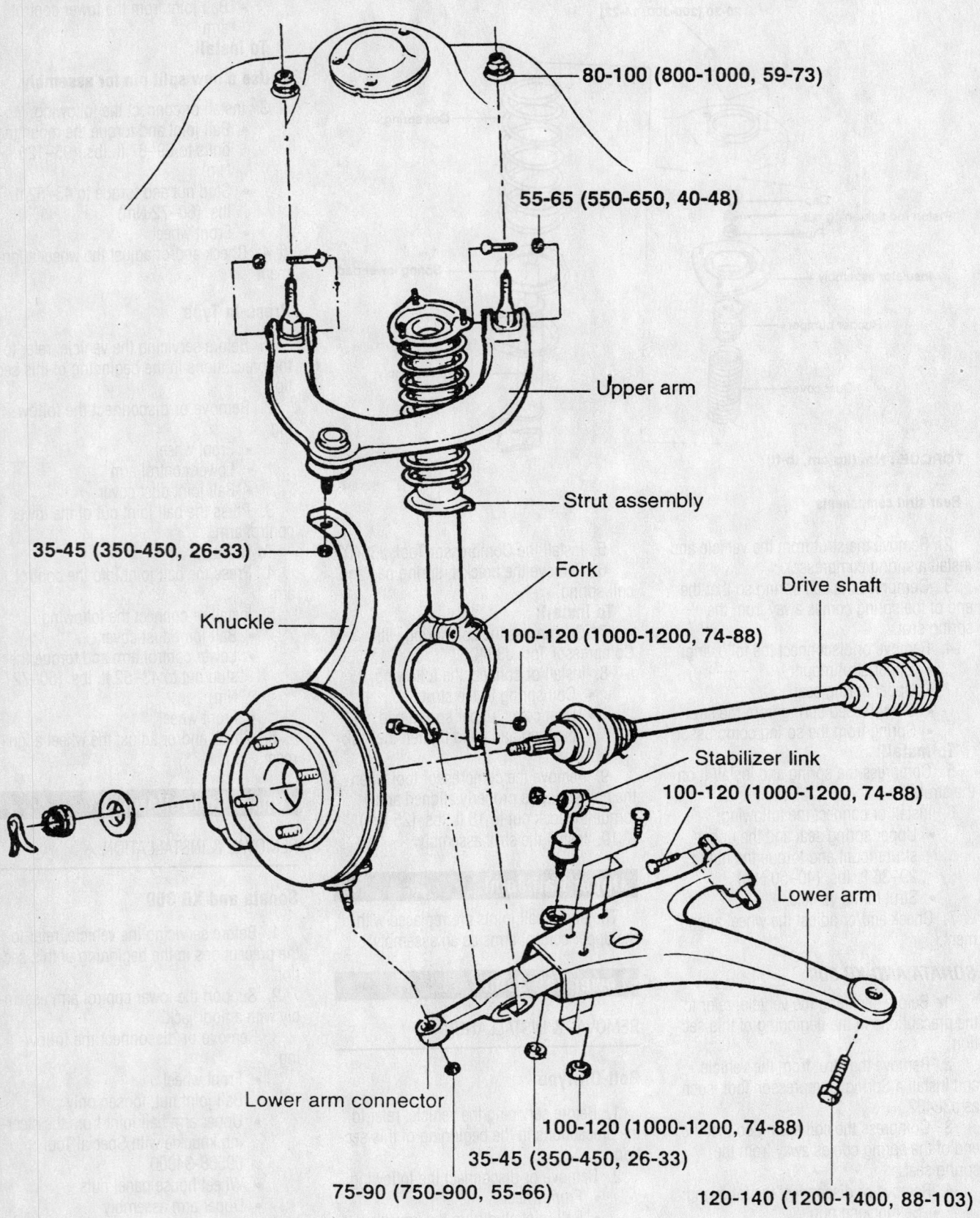

80-100 (800-1000, 59-73)

55-65 (550-650, 40-48)

Upper arm

Strut assembly

35-45 (350-450, 26-33)

Fork

Drive shaft

Knuckle

100-120 (1000-1200, 74-88)

Stabilizer link

100-120 (1000-1200, 74-88)

Lower arm

Lower arm connector

100-120 (1000-1200, 74-88)

35-45 (350-450, 26-33)

75-90 (750-900, 55-66)

120-140 (1200-1400, 88-103)

9347KG18

Exploded view of the upper control arm assembly—Sonata

To install:

4. Install or connect the following:
   - Upper control arm shaft
   - Upper control arm assembly and torque the bolts to 73 ft. lbs. (100 Nm)
   - Wheel house panel nuts and torque the nuts to 48 ft. lbs. (65 Nm)
   - Upper arm ball joint to the steering knuckle and torque the bolts to 33 ft. lbs. (45 Nm)
   - Front wheel

CONTROL ARM BUSHING REPLACEMENT

1. Before servicing the vehicle, refer to the precautions in the beginning of this section.
2. Remove or disconnect the following:
   - Control arm from the vehicle
   - Control arm bushings by unbolting them

To install:

3. Install or connect the following:
   - New bushings and torque the bolts to 40–48 ft. lbs. (55–65 Nm)
   - Control arm to the vehicle

Lower Control Arm

REMOVAL & INSTALLATION

Except Sonata and XG 300

1. Before servicing the vehicle, refer to the precautions in the beginning of this section.
2. Remove or disconnect the following:
   - Front wheel
   - Stabilizer bar link
   - Lower ball joint
   - Rear bushing bracket
   - Front bolt
   - Lower control arm

To install:

3. Install or connect the following:
   - Lower control arm and torque the front bolt to 72–87 ft. lbs. (100–120 Nm)
   - Rear bushing bracket. Tighten the bolts to 58–72 ft. lbs. (80–100 Nm)
   - Lower ball joint and torque the nut to 43–52 ft. lbs. (60–72 Nm)
   - Stabilizer bar link and torque the nut to 25–33 ft. lbs. (35–45 Nm)
   - Front wheel
4. Check and/or adjust the wheel alignment.

Sonata and XG 300

1. Before servicing the vehicle, refer to the precautions in the beginning of this section.
2. Remove or disconnect the following:
   - Front wheel
   - Lower ball joint nut, loosen only
   - Lower arm ball joint from the lower arm connector with Special Tool 09445-21000
   - Ball joint
   - Fork from the lower arm connector
   - Stabilizer bar link
   - Control arm inner bushing bolts
   - Lower control arm

To install:

3. Install or connect the following:
   - Lower control arm and torque the front bushing bolts to 74–88 ft. lbs. (100–120 Nm) and the rear bushing bolt to 88–103 ft. lbs. (120–140 Nm)
   - Stabilizer bar link and torque the nut to 26–33 ft. lbs. (35–45 Nm)
   - Damper fork lower bolt and torque the nut to 74–88 ft. lbs. (100–120 Nm)
   - Lower ball joint and torque the nut to 55–66 ft. lbs. (75–90 Nm)
   - Front wheel
4. Check and/or adjust the wheel alignment.

CONTROL ARM BUSHING REPLACEMENT

Except Sonata and XG 300

FRONT BUSHING

1. Before servicing the vehicle, refer to the precautions in the beginning of this section.
2. Remove the lower control arm from the vehicle.
3. Press the front bushing out of the control arm.

To install:

4. Lubricate the front bushing with soap and press into the control arm.
5. Install the control arm to the vehicle.
6. Check and/or adjust the wheel alignment.

REAR BUSHING

1. Before servicing the vehicle, refer to the precautions in the beginning of this section.
2. Remove or disconnect the following:
   - Front wheel
   - Rear bushing bracket
   - Rear bushing nut
   - Rear bushing

To install:

3. Install or connect the following:
   - Rear bushing and torque the nut to 25–33 ft. lbs. (35–45 Nm)
   - Rear bushing bracket and torque the bolts to 58–72 ft. lbs. (80–100 Nm)
   - Front wheel
4. Check and/or adjust the wheel alignment.

Sonata and XG 300

FRONT BUSHING

The front control arm bushing is serviced with the control arm as an assembly.

REAR BUSHING AND DAMPER FORK BUSHING

1. Before servicing the vehicle, refer to the precautions in the beginning of this section.
2. Remove the control arm from the vehicle.
3. Press the rear bushing and the damper fork bushing out of the control arm.

To install:

4. Press the rear bushing and the damper fork bushing into the control arm.
5. Install the control arm to the vehicle.
6. Check and/or adjust the wheel alignment.

Wheel Bearings

ADJUSTMENT

Front

The front wheel bearing is a sealed unit and is not adjustable.

Rear

WITH REAR DRUM BRAKES

1. Before servicing the vehicle, refer to the precautions in the beginning of this section.
2. Remove the rear wheels.
3. Loosen the spindle nut.
4. Torque the nut to 108–145 ft. lbs. (150–200 Nm). Check for correct bearing end-play by placing a dial indicator on the hub surface and moving the hub outward. Note the movement of the gauge and compare it to the desired reading of 0.008 in. (0.2mm) or less. If end-play exceeds the desired reading, retighten the rear hub bearing nut and recheck the end-play. If the reading is still excessive, replace the hub unit.
5. If end-play is correct, check the starting torque by attaching a spring balance to

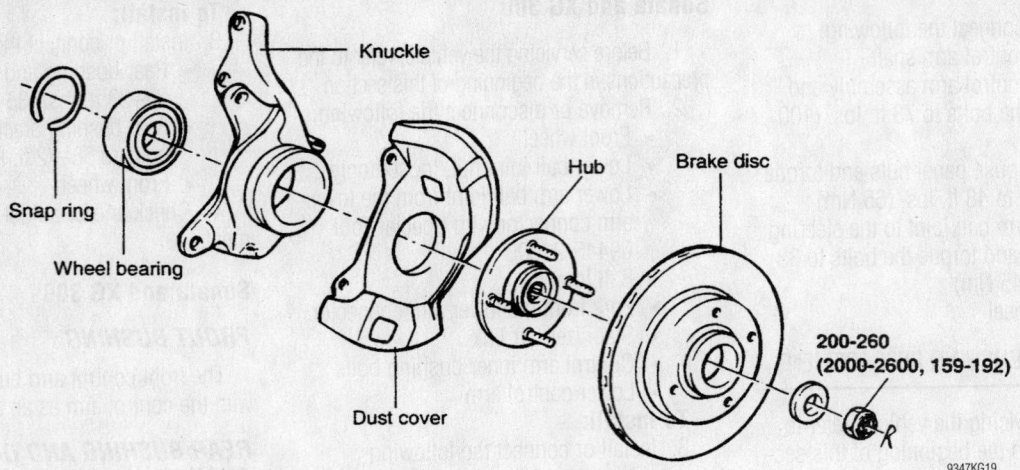

Exploded view of the front hub assembly

the hub lug bolts and pulling at a 90 degree angle while noting the required force to turn the hub. If the force required is above the desired reading of 5 lbs. (2.3 kg) or less, loosen the nut and again tighten to the desired torque. Recheck the starting torque. If the torque is still above the desired reading, replace the rear bearings.

6. Install the rear wheels.

WITH REAR DISC BRAKES

The rear wheel bearing is an integral part of the rear hub. No adjustment is possible.

REMOVAL & INSTALLATION

Front

1. Before servicing the vehicle, refer to the precautions in the beginning of this section.

2. Remove or disconnect the following:
   • Front wheel
   • Brake caliper
   • Lower ball joint
   • Spindle nut
   • Knuckle pinch bolts
   • Steering knuckle
3. Press the hub out of the wheel bearing.
4. Press the wheel bearings out of the steering knuckle.
5. If necessary, press the inner race off the hub.
   To install:
6. Press the wheel bearings into the steering knuckle.
7. Install the outer grease seal and press the hub into the wheel bearings.
8. Install or connect the following:
   • Inner grease seal
   • Steering knuckle and torque the

knuckle pinch bolts to 65–76 ft. lbs. (95–105 Nm)
   • Lower ball joint and torque the stud nut to 43–52 ft. lbs. (60–72 Nm)
   • Spindle nut and torque the nut to 144–187 ft. lbs. (195–253 Nm)
   • Brake caliper and torque the bracket bolts to 50 ft. lbs. (68 Nm)
   • Front wheel
9. Check and/or adjust the wheel alignment.

Rear

DRUM BRAKES

1. Before servicing the vehicle, refer to the precautions in the beginning of this section.
2. Remove or disconnect the following:
   • Rear wheel

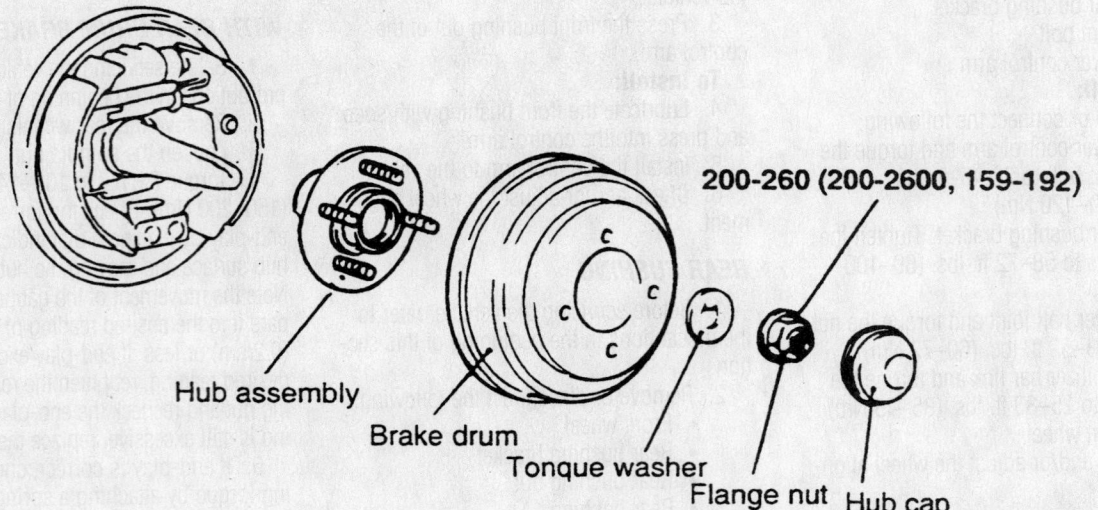

Exploded view of the rear hub assembly—with drum brakes

- Speed sensor, if equipped
- Grease cap
- Flange nut
- Outer bearing
- Brake drum
- Inner grease seal
- Inner bearing

3. Drive the bearing races out of the drum hub.

To install:

4. Install the inner and outer bearing races.

5. Apply grease to the bearings and to the cavity in the hub.

6. Install or connect the following:
- Inner bearing
- Inner grease seal
- Brake drum
- Outer bearing
- Flange nut and torque the nut to 159–192 ft. lbs. (200–260 Nm)

- Grease cap
- Wheel speed sensor, if equipped
- Rear wheel

DISC BRAKES

1. Before servicing the vehicle, refer to the precautions in the beginning of this section.

2. Release the parking brake.

3. Remove or disconnect the following:

- Rear wheel
- Wheel speed sensor, if equipped
- Brake caliper and rotor
- Rear axle hub bolts
- Tone wheel with Tool 09445-21000
- Carrier assembly
- Nut after unstaking it

4. Press out the rear axle hub.

5. Remove the bearing inner race with Tool 09445-21000.

6. Remove the bushings from the carrier with Tools 09453-33000B and 09545-21100.

To install:

7. Press in the bushings to the carrier with Tools 09453-33000B and 09545-21100.

8. Press in the bearing to the hub with Tool 09221-21000.

9. Tighten the flange nut to meet the concave portion of the spindle.

10. Press in the tone wheel with Tool 09221-21000. Torque the nut to 191 ft. lbs. (260 Nm).

11. Install the hub and bearing assembly to the backing plate and torque the bolts to and torque the bolts to 88 ft. lbs. (120 Nm).

12. Install the brake caliper and rotor.

13. Install the wheel speed sensor, if equipped.

14. Install the rear wheel.

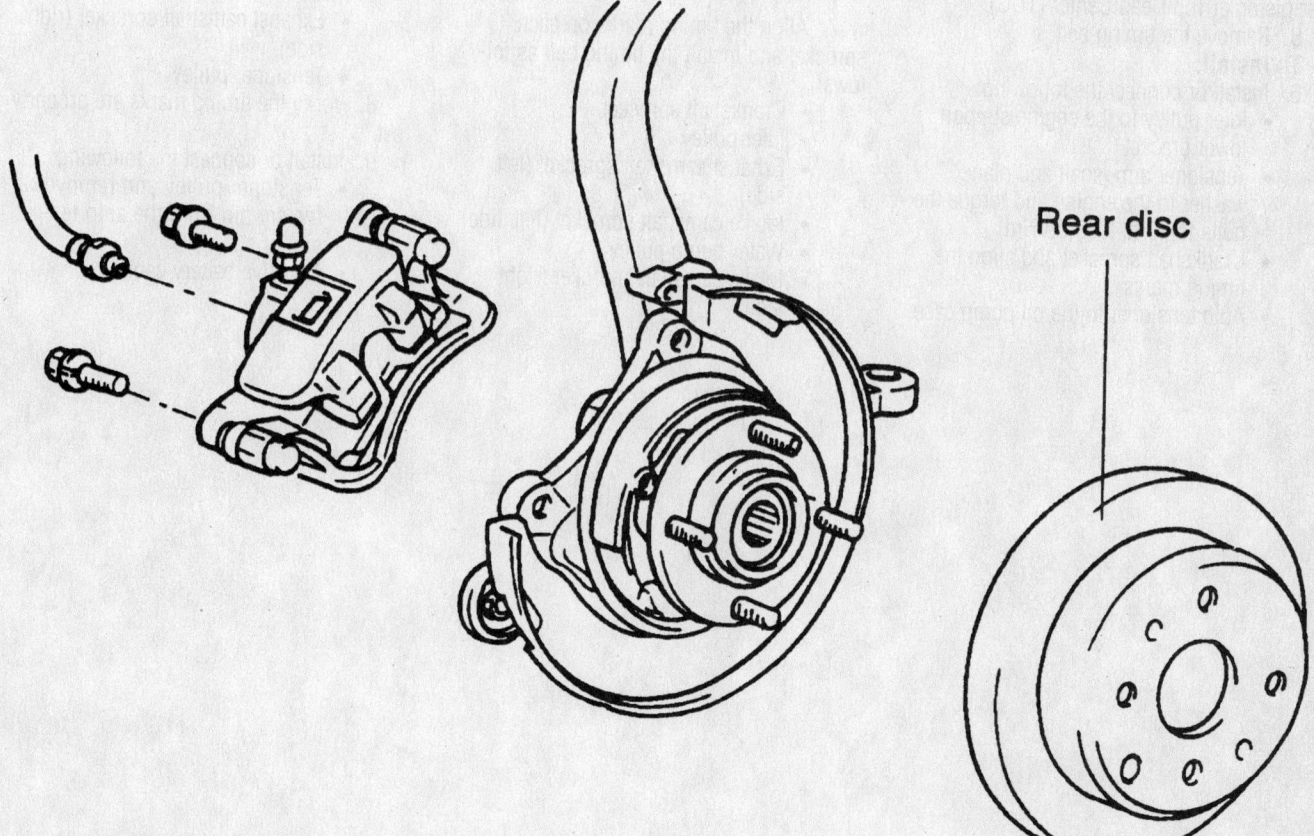

Rear disc

Exploded view of the rear wheel bearing assembly—with disc brakes

9734LG25

TIMING BELT

REMOVAL & INSTALLATION

3.0L (VIN G)

1. Before servicing the vehicle, refer to the precautions in the beginning of this section.
2. Remove or disconnect the following:
   - Negative battery cable
   - Engine cover
3. Rotate the drive belt tensioner clockwise approximately 14 degrees and remove the belt from the pulley.
   - Power steering pump pulley
   - Idler pulley
   - Tensioner pulley
   - Crankshaft pulley
   - Upper and lower timing belt covers
   - Auto tensioner
4. Rotate the crankshaft clockwise and align the timing mark to set the No. 1 cylinder piston at Top Dead Center (TDC).
5. Remove the timing belt.

To install:

6. Install or connect the following:
   - Idler pulley to the engine support lower bracket
   - Tensioner arm, shaft and plane washer to the engine and torque the bolts to 40 ft. lbs. (55 Nm)
   - Crankshaft sprocket and align the timing marks
   - Auto tensioner to the oil pump case

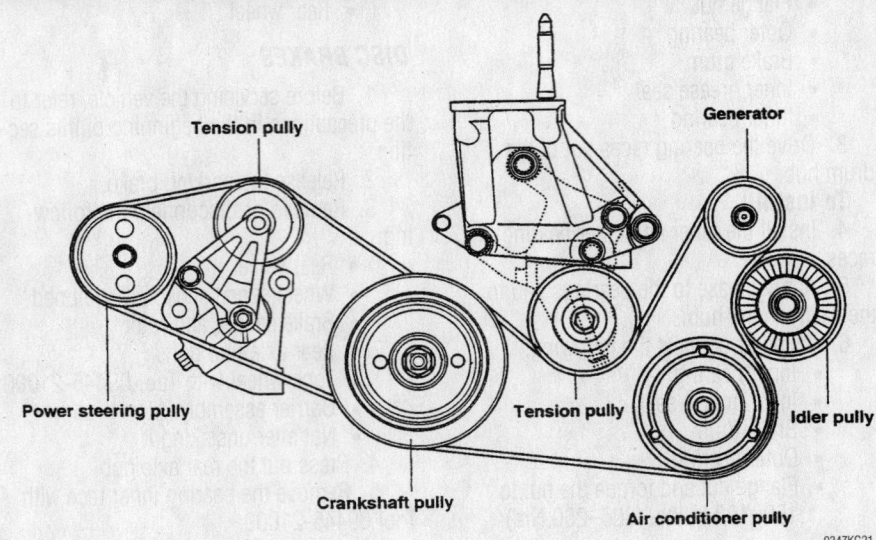

Installation of the timing belt—3.0L (VIN G) engine

7. Align the timing marks on each sprocket and install the timing belt as follows:
   - Crankshaft sprocket
   - Idler pulley
   - Exhaust camshaft sprocket (left side)
   - Intake camshaft sprocket (left side)
   - Water pump pulley
   - Intake camshaft sprocket (right side)
   - Exhaust camshaft sprocket (right side)
   - Tensioner pulley
8. Verify the timing marks are properly set.
9. Install or connect the following:
   - Tensioner pulley and remove the set pin from the auto tensioner
   - Negative battery cable

INFINITI

1998–00

G20 • I30 • Q45

12

PRECAUTIONS

Before servicing any vehicle, please be sure to read all of the following precautions, which deal with personal safety, prevention of component damage, and important points to take into consideration when servicing a motor vehicle:

• Never open, service or drain the radiator or cooling system when the engine is hot; serious burns can occur from the steam and hot coolant.

• Observe all applicable safety precautions when working around fuel. Whenever servicing the fuel system, always work in a well-ventilated area. Do not allow fuel spray or vapors to come in contact with a spark, open flame, or excessive heat (a hot drop light, for example). Keep a dry chemical fire extinguisher near the work area. Always keep fuel in a container specifically designed for fuel storage; also, always properly seal fuel containers to avoid the possibility of fire or explosion. Refer to the additional fuel system precautions later in this section.

• Fuel injection systems often remain pressurized, even after the engine has been turned **OFF**. The fuel system pressure must be relieved before disconnecting any fuel lines. Failure to do so may result in fire and/or personal injury.

• Brake fluid often contains polyglycol ethers and polyglycols. Avoid contact with the eyes and wash your hands thoroughly after handling brake fluid. If you do get brake fluid in your eyes, flush your eyes with clean, running water for 15 minutes. If

eye irritation persists, or if you have taken brake fluid internally, IMMEDIATELY seek medical assistance.

• The EPA warns that prolonged contact with used engine oil may cause a number of skin disorders, including cancer! You should make every effort to minimize your exposure to used engine oil. Protective gloves should be worn when changing oil. Wash your hands and any other exposed skin areas as soon as possible after exposure to used engine oil. Soap and water, or waterless hand cleaner should be used.

• All new vehicles are now equipped with an air bag system. The system must be disabled before performing service on or around system components, steering column, instrument panel components, wiring and sensors. Failure to follow safety and disabling procedures could result in accidental air bag deployment, possible personal injury and unnecessary system repairs.

• Always wear safety goggles when working with, or around, the air bag system. When carrying a non-deployed air bag, be sure the bag and trim cover are pointed away from your body. When placing a non-deployed air bag on a work surface, always face the bag and trim cover upward, away from the surface. This will reduce the motion of the module if it is accidentally deployed. Refer to the additional air bag system precautions later in this section.

• Clean, high quality brake fluid from a sealed container is essential to the safe and

proper operation of the brake system. You should always buy the correct type of brake fluid for your vehicle. If the brake fluid becomes contaminated, completely flush the system with new fluid. Never reuse any brake fluid. Any brake fluid that is removed from the system should be discarded. Also, do not allow any brake fluid to come in contact with a painted surface; it will damage the paint.

• Never operate the engine without the proper amount and type of engine oil; doing so WILL result in severe engine damage.

• Timing belt maintenance is extremely important! Many models utilize an interference-type, non-freewheeling engine. If the timing belt breaks, the valves in the cylinder head may strike the pistons, causing potentially serious (also time-consuming and expensive) engine damage. Refer to the maintenance interval charts in the front of this manual for the recommended replacement interval for the timing belt, and to the timing belt section for belt replacement and inspection.

• Disconnecting the negative battery cable on some vehicles may interfere with the functions of the on-board computer system(s) and may require the computer to undergo a relearning process once the negative battery cable is reconnected.

• When servicing drum brakes, only disassemble and assemble one side at a time, leaving the remaining side intact for reference.

ENGINE REPAIR

Distributor

REMOVAL

2.0L Engine

1. Before servicing the vehicle, refer to the precautions in the beginning of this section.
2. Remove or disconnect the following:
   • Negative battery cable
   • Splash shield, if equipped
   • Distributor connections but leave the ignition wires in place
   • Distributor cap hold-down screws and lift off the distributor cap with all ignition wires still connected
3. Matchmark the rotor to the distributor housing, and the distributor housing to the engine.

➡Do not crank the engine during this procedure. If the engine is cranked, the matchmark must be disregarded.

   • Hold-down bolt
   • Distributor from the engine

3.0L and 4.1L Engines

These engines are equipped with a distributorless ignition.

INSTALLATION

2.0L Engine

ENGINE NOT DISTURBED

1. If the engine was not disturbed, install or connect the following:
   • New distributor housing O-ring
   • Distributor in the engine so the

rotor is aligned with the matchmark on the housing and the housing is aligned with the matchmark on the engine. Be sure the distributor is fully seated and the distributor gear is fully engaged.
   • Snug the hold-down bolt
   • Distributor pick-up lead wires
   • Distributor cap and tighten the screws
   • Splash shield
   • Negative battery cable
2. Check and/or adjust the ignition timing and tighten the hold-down bolt.

ENGINE DISTURBED

1. If the engine was disturbed (cranked or turned over with the distributor removed), install or connect the following:
   • New distributor housing O-ring

2. Position the engine so the No. 1 piston is at TDC of its compression stroke and the mark on the vibration damper is aligned with **0** on the timing indicator.
- Distributor in the engine so the rotor is aligned with the position of the No. 1 ignition wire on the distributor cap. Be sure the distributor is fully seated and that the distributor shaft is fully engaged.

➡️**There are distributor cap runners inside the cap on 2.0L engine. Be sure the rotor is pointing to where the No. 1 runner originates inside the cap.**

- Snug the hold-down bolt
- Distributor pick-up lead wires
- Distributor cap and tighten the screws
- Splash shield, if equipped
- Negative battery cable

3. Check and/or adjust the ignition timing and tighten the hold-down bolt to 10–12 ft. lbs. (14–16 Nm).

Alternator

REMOVAL

2.0L Engine

1. Before servicing the vehicle, refer to the precautions in the beginning of this section.
2. Remove or disconnect the following:
- Negative battery cable
- Drive belt
- Alternator harness connector
- Alternator bracket, if necessary
- Alternator retainers
- Alternator

3.0L Engine

1. Before servicing the vehicle, refer to the precautions in the beginning of this section.
2. Remove or disconnect the following:
- Negative battery cable
- Engine right-hand undercover
- Right-hand side inspection cover
3. Loosen the belt idler pulley.
- Drive belt
- 4 A/C compressor mounting bolts
- Cooling fan and fan shroud
4. Slide the A/C compressor forward.
- Alternator harness connector
- Upper and lower alternator bolts
- Alternator

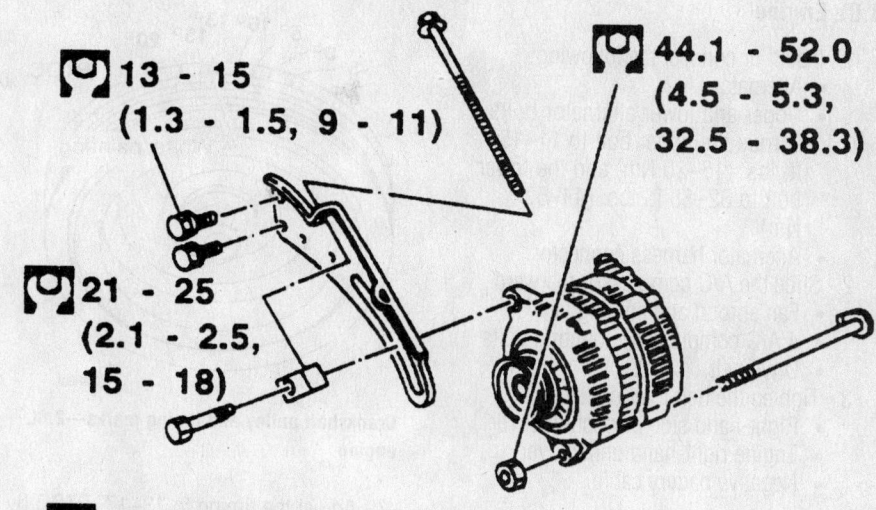

: N•m (kg-m, ft-lb)

9307HG20

Alternator and bracket retainer locations and torque specifications—G20 models

4.1L Engine

1. Before servicing the vehicle, refer to the precautions in the beginning of this section.
2. Remove or disconnect the following:
- Negative battery cable
- Engine upper cover
- Engine drive belt
- Alternator electrical connector
- Alternator

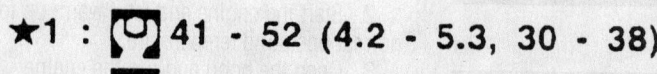

★1 : 41 - 52 (4.2 - 5.3, 30 - 38)
★2 : 21 - 26 (2.1 - 2.7, 15 - 20)

INSTALLATION

2.0L Engine

1. Install or connect the following:
- Alternator
- Alternator bracket, if removed
- Alternator retainers as shown in the accompanying illustration

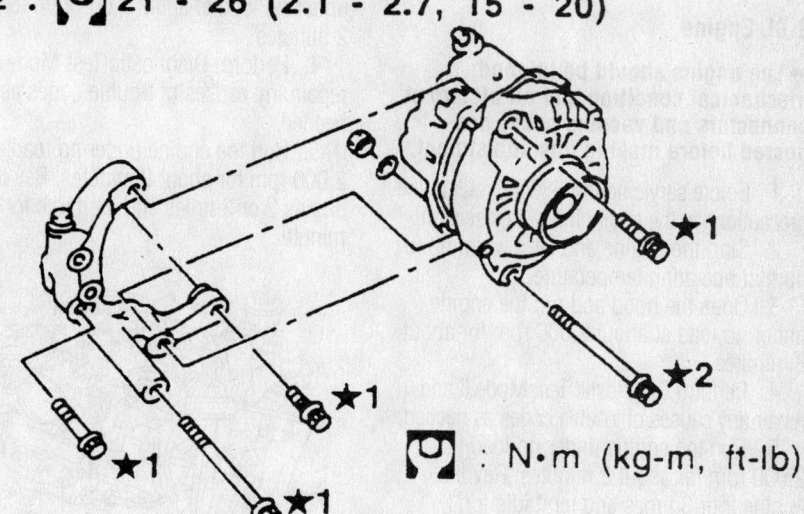

: N•m (kg-m, ft-lb)

9307HG01

Alternator bolt locations—Q45 models

3.0L Engine

1. Install or connect the following:
   - Alternator
   - Upper and lower alternator bolts. Torque the upper bolt to 11–15 ft. lbs. (15–20 Nm) and the lower bolt to 32–38 ft. lbs. (44–52 Nm)
   - Alternator harness connector
2. Slide the A/C compressor rearward.
   - Fan shroud and cooling fan
   - 4 A/C compressor mounting bolts
   - Drive belt
3. Tighten the belt idler pulley.
   - Right-hand side inspection cover
   - Engine right-hand undercover
   - Negative battery cable

4.1L Engine

1. Install or connect the following:
   - Alternator
   - Alternator bolts. Torque the alternator bolts marked **1**, (shown in the accompanying illustration) to 30–38 ft. lbs. (41–52 Nm) and the bolt marked **2**, (shown in the same illustration) to 15–20 ft. Lbs. (21–26 Nm)
   - Alternator electrical connector
   - Engine drive belt
   - Engine upper cover
   - Negative battery cable

Ignition Timing

ADJUSTMENT

2.0L Engine

➡The engine should be in good mechanical condition and all electrical connectors and vacuum hoses connected before making this adjustment.

1. Before servicing the vehicle, refer to the precautions in the beginning of this section.
2. Start the engine and let it warm up to normal operating temperature.
3. Open the hood and run the engine under no load at about 2,000 rpm for about 2 minutes.
4. Perform Diagnostic Test Mode II and repair any causes of trouble codes as needed.
5. Run the engine under no load at 2,000 rpm for about 2 minutes. Rev the engine 2 or 3 times and let it idle for 1 minute.
6. Turn **OFF** the engine and disconnect the Throttle Position (TP) sensor connector. Connect a timing light to the No. 1 spark plug wire. Start the engine.

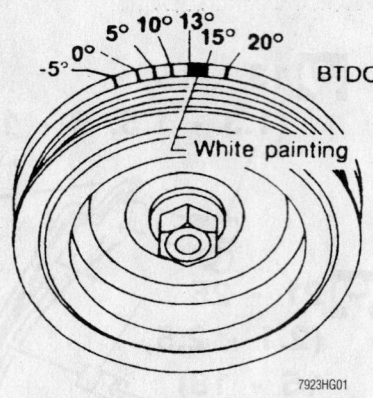

Crankshaft pulley and timing marks—2.0L engine

7. Adjust the timing to 13–17° BTDC by loosening the distributor mounting bolts and turning the distributor. When the timing is correct, tighten the mounting bolts and turn the engine **OFF**.
8. Reconnect the TP sensor connector. Start the engine and check the ignition timing again.

3.0L and 4.1 Engines

➡The engine should be in good mechanical condition and all electrical connectors and vacuum hoses attached before making this adjustment.

1. Before servicing the vehicle, refer to the precautions in the beginning of this section.
2. Start the engine and let it warm up to normal operating temperature.
3. Open the hood and run the engine under no load at about 2,000 rpm for about 2 minutes.
4. Perform Diagnostic Test Mode II and repair any causes of trouble codes as needed.
5. Run the engine under no load at 2,000 rpm for about 2 minutes. Rev the engine 2 or 3 times and let it idle for 1 minute.

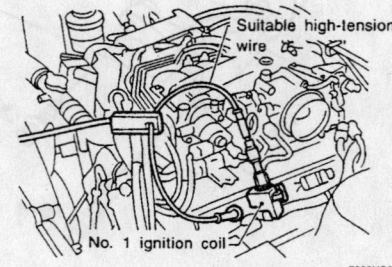

Connect the No. 1 ignition coil to the spark plug with the spare piece of high-tension wire—4.1L engine shown

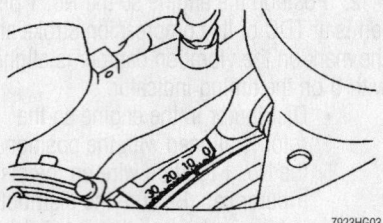

Location of timing marks—3.0L engine

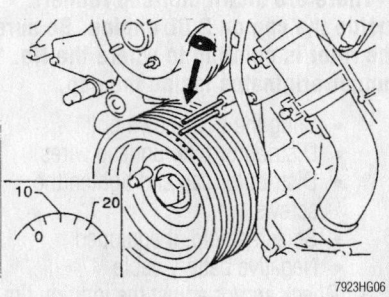

Location of timing marks—4.1L engine

6. Turn **OFF** the engine and disconnect the Throttle Position sensor connector. Remove the No. 1 ignition coil. Connect the coil to the spark plug using a spare piece of high-tension wire so you have a place to connect your timing light. Start the engine.
7. Run the engine under no load at 2,000 rpm for about 2 minutes. Rev the engine 2 or 3 times and let it idle.
8. Check the ignition timing and adjust if needed.
9. The correct ignition timing is as follows:
   a. Correct ignition timing for 3.0L engines is 8–12° BTDC.
   b. Correct ignition timing for the 4.1L engine is 13–17° BTDC.
   - Adjustment is made by loosening the screws and turning the Camshaft Position (CMP) sensor until the mark on the crankshaft pulley is pointing at 10° BTDC. Tighten the mounting screws and confirm ignition timing has not changed.
   - Turn the engine **OFF** and connect the TP sensor connector.

Engine Assembly

REMOVAL & INSTALLATION

2.0L Engine

1. Before servicing the vehicle, refer to the precautions in the beginning of this section.
2. Drain the coolant system.
3. Drain the engine oil.

4. Drain the transaxle fluid.

5. Release fuel system pressure and remove fuel line.

6. Remove or disconnect the following:
- Hood and hinges
- Negative battery cable
- Both front wheels
- Engine under cover
- Air cleaner assembly and duct
- Battery and battery tray
- All vacuum lines and wiring harness connectors
- Heater hoses
- Oil cooler lines, if equipped
- Power steering hoses
- Fuel lines
- Throttle cable
- Cruise control cable, if equipped
- A/T control cable, if equipped
- Cooling fans, radiator and recovery tank
- Drive shafts
- Front exhaust pipe
- Starter and intake manifold support
- Drive belts
- Alternator, A/C compressor, and the power steering pump from the engine and lay them aside. Do not disconnect the compressor or power steering pump lines.

7. Support the engine with a hoist and the transaxle with a suitable jack. Raise the engine and transaxle slightly and remove the center member.
- Engine mounting bolts from both sides and slowly lower the hoist and transaxle jack
- Engine and transaxle from beneath the vehicle

To install:

8. Install or connect the following:
- Center member bracket (manual transmission) on the engine, if removed. Ensure that all insulators are correctly positioned on the brackets. Torque the insulator through-bolts to 32–41 ft. lbs. (43–55 Nm)

9. If equipped with manual transaxle, ensure that the distance between the center of the insulator through-bolt and the center member is 2.28–2.36 in. (58–60mm). Torque the through-bolt to 46–58 ft. lbs. (62–78 Nm)
- Engine. Torque the center member-to-frame bolts to 57–72 ft. lbs. (77–98 Nm)
- Alternator, air conditioning compressor, and power steering pump
- Drive belts

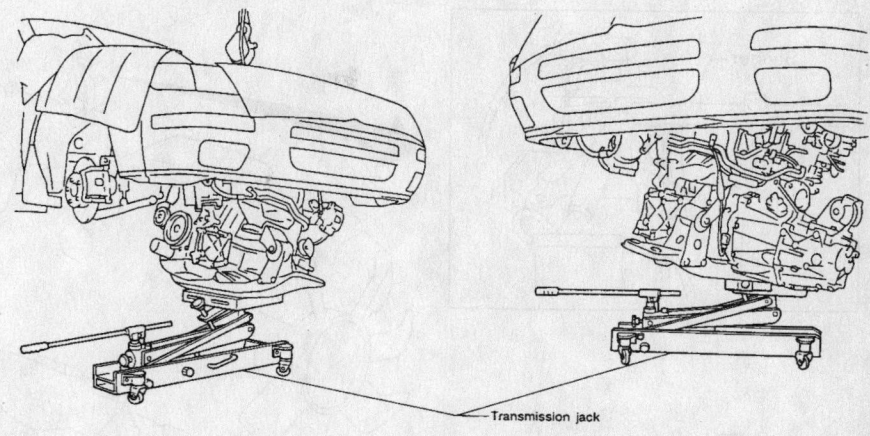

Remove the engine and the transaxle as an assembly from beneath the vehicle using a transaxle jack and a slinger

9307HG02

- Starter and intake manifold support
- Front exhaust pipe
- Drive shafts
- Cooling fans, radiator and recovery tank
- A/T control cable, if equipped
- Cruise control cable, if equipped
- Throttle cable
- Fuel lines
- Power steering hoses
- Oil cooler lines, if equipped
- Heater hoses
- All vacuum lines and wiring harness connectors
- Engine under cover
- Front wheels
- Battery and battery tray
- Air cleaner assembly and duct
- Negative battery cable
- Hood and hinges

10. Fill and bleed the cooling system.

11. Fill the engine with clean oil.

12. Fill the transaxle to the proper level.

13. Start the vehicle, check for leaks and repair if necessary.

3.0L Engine

It is recommended the engine and transaxle be removed as a single unit. If necessary, the units may be separated after removal.

➡**The engine and transaxle assembly must be removed from the under side of the vehicle.**

1. Drain the cooling system.
2. Drain the engine oil.
3. Drain the transaxle fluid.
4. Properly relieve the fuel system pressure.

5. Before servicing the vehicle, refer to the precautions in the beginning of this section.

6. Remove or disconnect the following:
- Negative battery cable
- Hood
- Both front wheels
- Engine undercover
- All necessary vacuum hoses, fuel hoses and electrical connections that would interfere with engine removal
- Front exhaust tubes
- Ball joints
- Driveshafts
- Radiator and fans
- Drive belts
- Alternator, compressor and power steering oil pump from the engine compartment

7. Set a suitable transmission jack under the transaxle. Hoist the engine with a suitable engine slinger.
- Rear engine mounting
- Control cable
- Front engine mounting
- Center member and slowly lower the transmission jack

8. Remove the engine and the transaxle assembly from the engine as shown in the accompanying illustration.

To install:

9. Install or connect the following:
- Center member bracket (manual transmission) on the engine, if removed. Ensure that all insulators are correctly positioned on the brackets. Torque the insulator through-bolts to 72 ft. lbs. (98 Nm)
- Engine. Torque the center member-

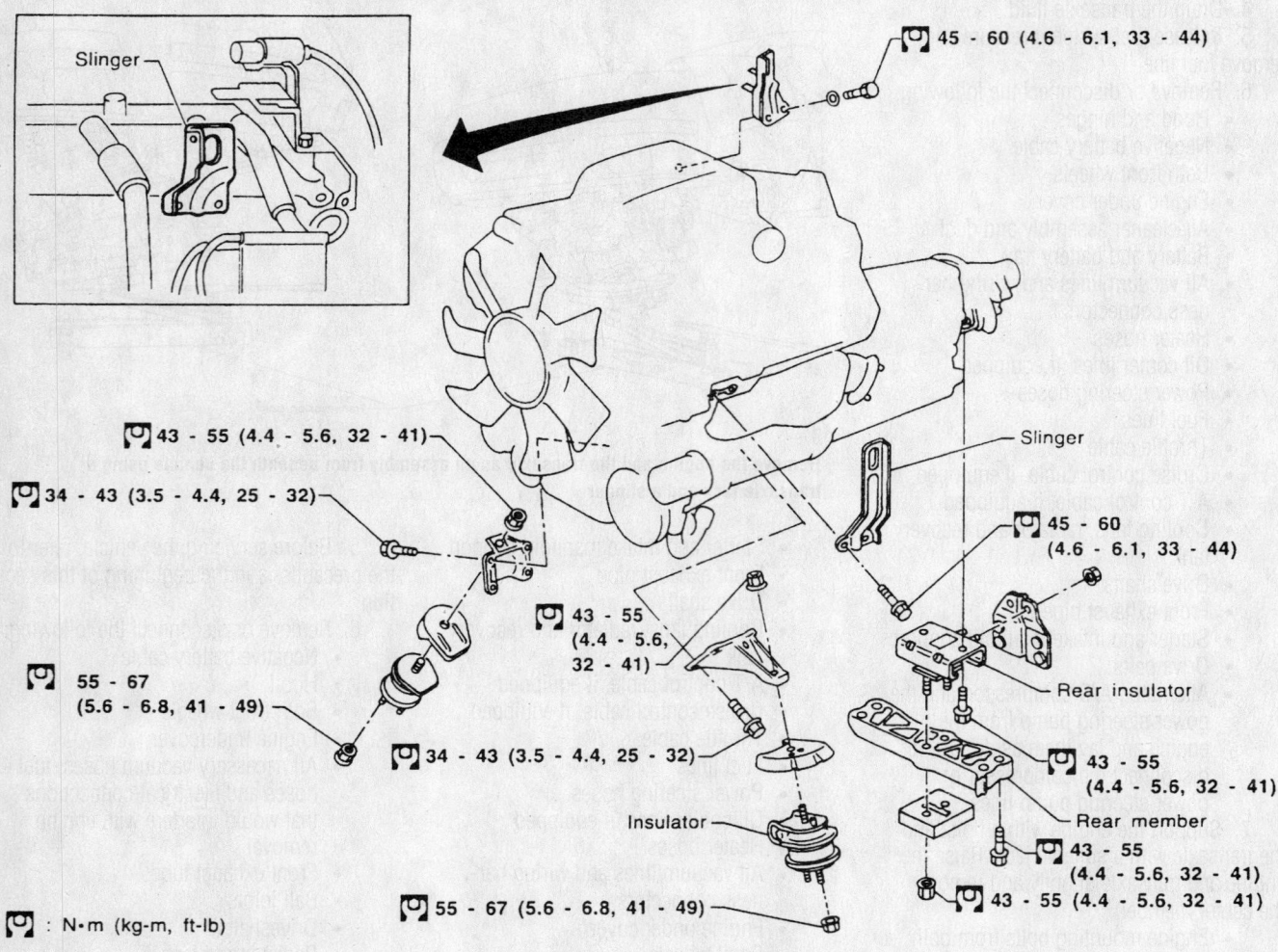

Slinger

45 - 60 (4.6 - 6.1, 33 - 44)

43 - 55 (4.4 - 5.6, 32 - 41)

34 - 43 (3.5 - 4.4, 25 - 32)

55 - 67
(5.6 - 6.8, 41 - 49)

43 - 55
(4.4 - 5.6,
32 - 41)

34 - 43 (3.5 - 4.4, 25 - 32)

Insulator

55 - 67 (5.6 - 6.8, 41 - 49)

Slinger

45 - 60
(4.6 - 6.1, 33 - 44)

Rear insulator

43 - 55
(4.4 - 5.6, 32 - 41)

Rear member

43 - 55
(4.4 - 5.6, 32 - 41)

43 - 55 (4.4 - 5.6, 32 - 41)

 : N•m (kg-m, ft-lb)

9307HG04

Engine mounting and torque specifications—4.1L engines

to-frame bolts to 57–72 ft. lbs. (77–98 Nm)
- Front engine mount. Torque the bolts to 72 ft. lbs. (98 Nm)
- Control cable
- Rear engine mount. Torque the bolts to 72 ft. lbs. (98 Nm) and remove the transaxle jack
- Alternator, air conditioning compressor, and power steering pump
- Radiator and fans
- Drive belts
- Driveshafts
- Ball joints
- Front exhaust tubes
- Vacuum hoses, electrical connectors, fuel hoses and wiring which was removed
- Engine under cover
- Front wheels
- Battery and battery tray
- Negative battery cable
- Hood and hinges

10. Fill and bleed the cooling system.
11. Fill the engine with clean oil.
12. Fill the transaxle to the proper level.
13. Start the vehicle, check for leaks and repair if necessary.

4.1L Engines

1. Before servicing the vehicle, refer to the precautions in the beginning of this section.
2. Evacuate the A/C system.
3. Release the fuel system pressure.
4. Drain the engine oil.
5. Drain the cooling system.
6. Drain the transaxle fluid.
7. Remove or disconnect the following:
- Negative battery cable
- Hood
- Engine under cover
- Transmission
- All necessary vacuum hoses, fuel hoses and electrical connections that would interfere with engine removal

- Front exhaust tubes
- Radiator and shroud
- Drive belts
- Power steering oil pump from the engine compartment

8. Attach engine slingers to the cylinder head and attach a suitable hoist to the slinger
- Engine mounting bolts from both sides and then slowly raise the engine
- Engine from the engine compartment

To install:

9. Install or connect the following:
- Engine. Torque the front mounting bolts to 41 ft. lbs. (55 Nm) and the rear mounting bolts to 21 ft. lbs. (28 Nm). Remove the engine supports
- Power steering pump
- Radiator and shroud
- Drive belts
- Front exhaust tubes
- Vacuum hoses, electrical connectors, fuel hoses and wiring

- Transmission
- Engine under cover
- Negative battery cable
- Hood

10. Fill the transaxle to the proper level.
11. Fill and bleed the cooling system.
12. Fill the engine with clean oil.
13. Recharge the A/C system.
14. Start the vehicle, check for leaks and repair if necessary.

Water Pump

REMOVAL & INSTALLATION

2.0L Engine

1. Before servicing the vehicle, refer to the precautions in the beginning of this section.
2. Drain the coolant from the radiator and engine block. The drain plug in the engine block is located at the left front of the cylinder block.
3. Remove or disconnect the following:
   - Negative battery cable
   - Right front wheel
   - Engine side cover
   - Drive belts
   - Right front engine mount
   - Water pump

To install:

4. Clean all mating surfaces and place a

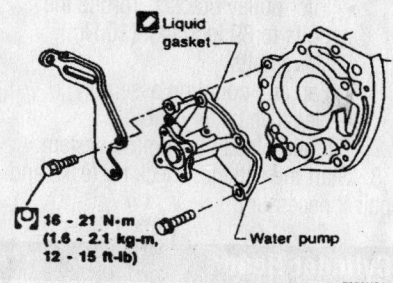

16 - 21 N·m
(1.6 - 2.1 kg-m,
12 - 15 ft-lb)

Liquid gasket

Water pump

7923HG08

Exploded view of the water pump mounting—2.0L engine

2–3mm bead of liquid gasket on the water pump mating surface.

5. Install or connect the following:
   - Water pump. Torque the bolts to 15 ft. lbs. (21 Nm)
   - Drive belts
   - Front engine mount
   - Engine side cover
   - Right front wheel
   - Negative battery cable
6. Fill and bleed the cooling system.
7. Start the vehicle, check for leaks and repair if necessary.

3.0L Engine

1. Before servicing the vehicle, refer to the precautions in the beginning of this section.

2. Drain the cooling system.
3. Remove or disconnect the following:
   - Negative battery cable
   - Right side engine mount and bracket
   - Drive belts
   - Idler pulley bracket
   - Timing chain tensioner cover
   - Water pump cover
4. Push the timing chain tensioner sleeve and apply a stopper pin so it does not return.
   - Timing chain tensioner assembly
   - 3 bolts that secure the water pump
5. Rotate the crankshaft 20° counter-clockwise to provide timing chain slack.
6. Put the 2 grade M8 bolts in the 2 M8 threaded holes of the water pump.
7. Tighten each bolt by turning alternately ½ turn until they reach the timing chain rear case. Be sure to turn each bolt ½ turn at a time to prevent damage.
8. Lift up the water pump and remove it.
9. When removing the water pump, do not allow the water pump gear to hit the timing chain.
10. Remove and discard the O-rings from the water pump.
11. Clean all traces of liquid gasket from the water pump and covers.

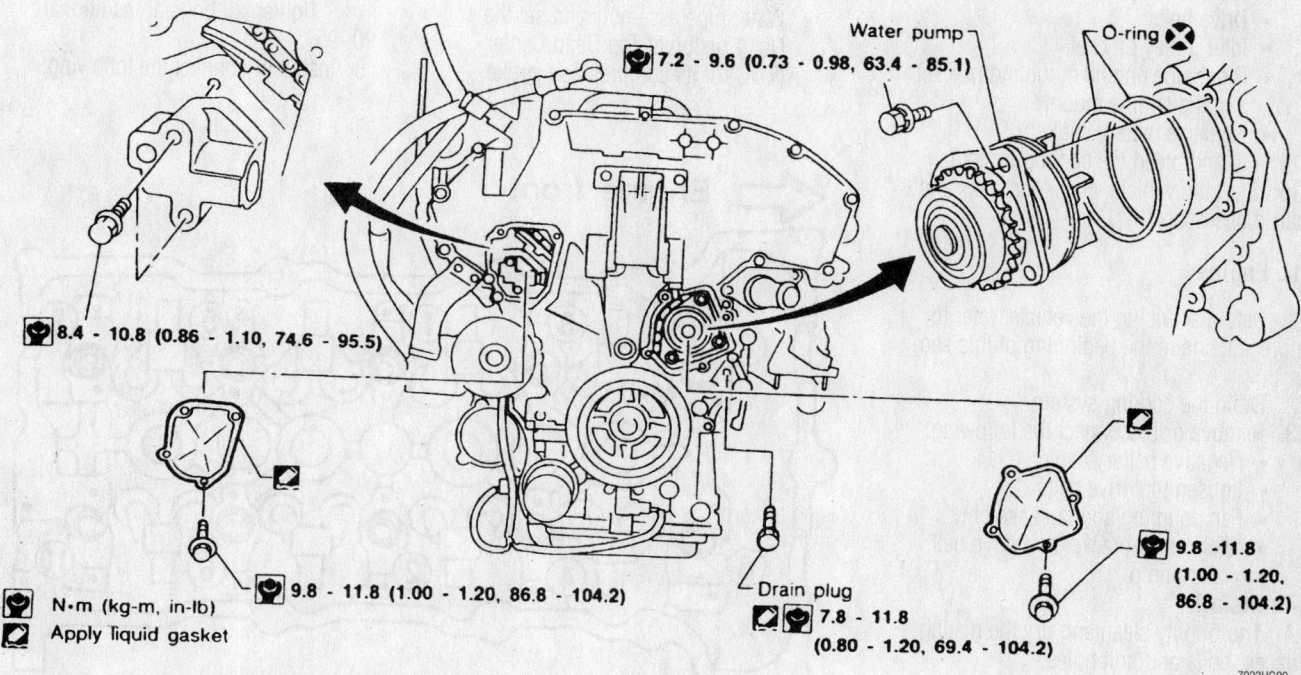

Water pump — 7.2 - 9.6 (0.73 - 0.98, 63.4 - 85.1) — O-ring

8.4 - 10.8 (0.86 - 1.10, 74.6 - 95.5)

9.8 - 11.8 (1.00 - 1.20, 86.8 - 104.2)

N·m (kg-m, in-lb)
Apply liquid gasket

9.8 - 11.8 (1.00 - 1.20, 86.8 - 104.2)

Drain plug
7.8 - 11.8
(0.80 - 1.20, 69.4 - 104.2)

9.8 -11.8
(1.00 - 1.20,
86.8 - 104.2)

7923HG09

Water pump and timing cover assembly—3.0L (VQ30DE) engine

Timing belt service is covered in Section 3 of this manual

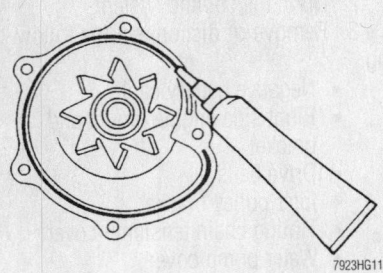

Apply a continuous bead of RTV sealant to the mounting surface of the water pump assembly

To install:

12. Install or connect the following:
- Water pump with new O-rings. Torque the bolts to 89 inch lbs. (10 Nm) and rotate the crankshaft pulley to its original position by turning it 20° clockwise
- Timing chain tensioner. Torque the bolts to 89 inch lbs. (10 Nm). Remove the stopper pin from the timing chain tensioner

13. Apply a continuous 0.091–0.130 in. (2–3mm) bead of liquid sealant to the mating surfaces of the timing chain tensioner and water pump covers.
- Timing chain tensioner and water pump covers to the engine block. Torque the cover bolts to 89 inch lbs. (10 Nm)
- Drive belts
- Idler pulley bracket
- Right side engine mounting bracket and the engine mount
- Negative battery cable

14. Fill and bleed the cooling system.

15. Start the vehicle, check for leaks and repair if necessary.

4.1L Engine

1. Before servicing the vehicle, refer to the precautions in the beginning of this section.

2. Drain the cooling system.

3. Remove or disconnect the following:
- Negative battery cable
- Loosen the drive belts
- Fan coupling and fan assembly
- Idler pulley bracket and drive belt
- Water pump

To install:

4. Thoroughly clean and dry the mating surfaces, bolts and bolt holes.

5. Apply liquid gasket to the water pump.

6. Install or connect the following:
- Water pump. Torque the bolts to 13 ft. lbs. (18 Nm)

- Idler pulley bracket. Torque the bolts to 89 inch lbs. (10 Nm)
- Drive belts
- Fan and coupling assembly
- Negative battery cable

7. Fill and bleed the cooling system.

8. Start the vehicle, check for leaks and repair if necessary.

Cylinder Head

REMOVAL & INSTALLATION

2.0L Engine

1. Before servicing the vehicle, refer to the precautions in the beginning of this section.

2. Relieve the fuel system pressure.

3. Drain the cooling system.

4. Remove or disconnect the following:
- Negative battery cable
- Right front wheel
- Engine side cover
- Radiator
- Air duct to intake manifold
- ASCD actuator
- Vacuum and fuel hoses
- Electrical connectors and wiring
- Spark plugs
- Rocker cover bolts in sequence
- Rocker cover
- Steering pump
- Intake manifold supports
- Water pipe assembly and set the No. 1 piston at Top Dead Center (TDC) of its compression stroke

➡Rotate the crankshaft until the mating mark on the camshaft sprocket is properly set.

- Timing chain tensioner
- Distributor. Do not turn the rotor with the distributor removed
- Camshaft sprockets and brackets
- Starter
- Heater hoses
- Cylinder head bolts in the proper sequence

To install:

5. Apply liquid gasket to the top of the chain cover where it meets the cylinder block before installing the head gasket.

6. Install the gasket and cylinder head on the block.

➡**Cylinder head bolts may be reused providing the dimension from the bottom of the head to the end of the bolt does not exceed 6.228 in. (158.2mm). If the dimension exceeds the specification, replace the cylinder head bolts.**

7. Tighten the cylinder head bolts as follows:
   a. Tighten all bolts to 29 ft. lbs. (39 Nm) using the proper sequence.
   b. Tighten all bolts to 58 ft. lbs. (78 Nm) using the proper sequence.
   c. Loosen all bolts completely.
   d. Tighten all bolts to 25–33 ft. lbs. (34–44 Nm) using the proper sequence.
   e. Tighten all bolts 90–95°.
   f. Tighten all bolts an additional 90–95°.

8. Install or connect the following:

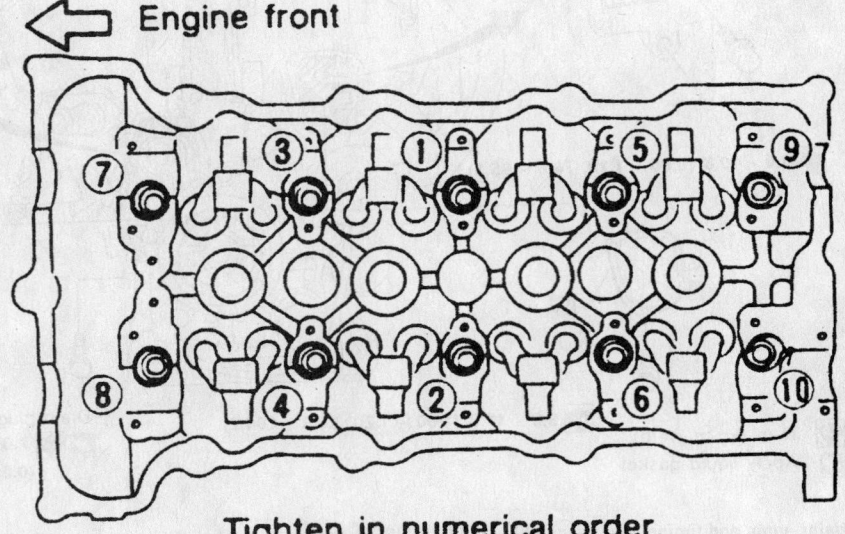

Engine front

Tighten in numerical order.

Cylinder head torque sequence—2.0L engine

- Heater hoses
- Camshafts and brackets. Ensure that the camshaft keys are at 12 o'clock

9. The procedure for tightening camshaft bolts must be followed exactly to prevent camshaft damage. Tighten the bolts as follows:

a. Tighten the right camshaft bolts Nos. 9 and 10 to 18 inch lbs. (2 Nm). Tighten bolts 1 through 8 to the same amount.

b. Tighten the left camshaft bolts No. 11 and No. 12 to 18 inch lbs. (2 Nm). Tighten bolts 1 through 10 to the same amount.

c. Tighten all bolts in sequence to 54 inch lbs. (6 Nm).

d. Tighten all bolts in sequence again. Tighten type A, B, and D bolts to 78–102 inch lbs. (9–12 Nm) and type C bolts to 13–19 ft. lbs. (18–25 Nm).

10. Line up the mating marks on the timing chain and camshaft sprockets and install the sprockets. Tighten the sprocket bolts to 101–116 ft. lbs. (137–157 Nm).

11. Install or connect the following:
- Timing chain guide, distributor, chain tensioner, oil filter bracket and power steering oil pump bracket
- Intake manifold supports
- Rocker cover. Torque the bolts, in sequence, to 89 inch lbs. (10 Nm)
- Spark plugs, power steering pump, alternator, water pump pulley and drive belts, air duct to the intake manifold and the radiator
- Vacuum and fuel hoses and reconnect all electrical connections
- Engine under cover
- Right front wheel
- Negative battery cable

12. Fill and bleed the cooling system.

13. Start the vehicle, check for leaks and repair if necessary.

3.0L Engine

1. Before servicing the vehicle, refer to the precautions in the beginning of this section.

2. Relieve the fuel system pressure.

3. Drain the engine oil.

4. Drain the cooling system.

➡ **Before disconnecting any hoses or connectors, note the locations for reassembly.**

5. Remove or disconnect the following:
- Negative battery cable

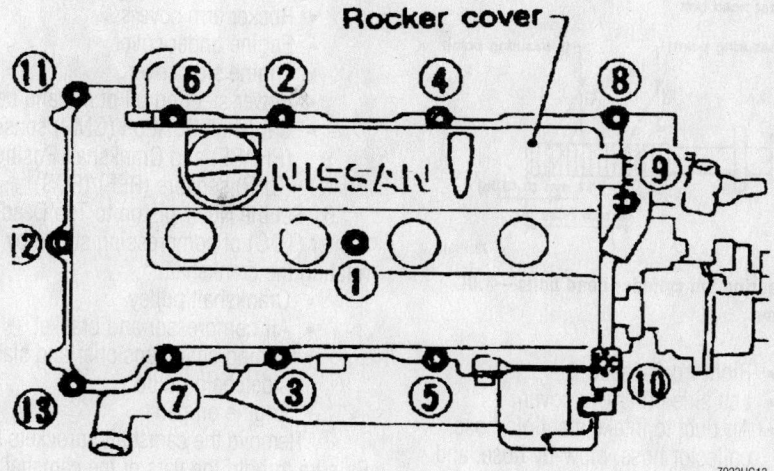

Tighten the rocker cover bolts according to the sequence shown—2.0L engine

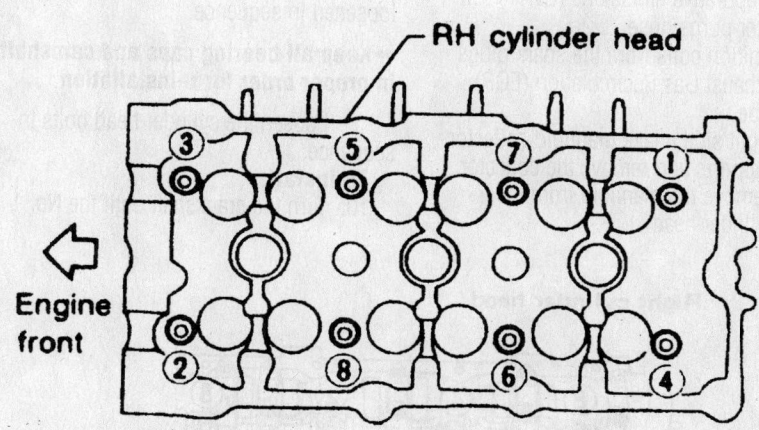

Loosen in numerical order.

Right cylinder head bolt loosening sequence—3.0L engine

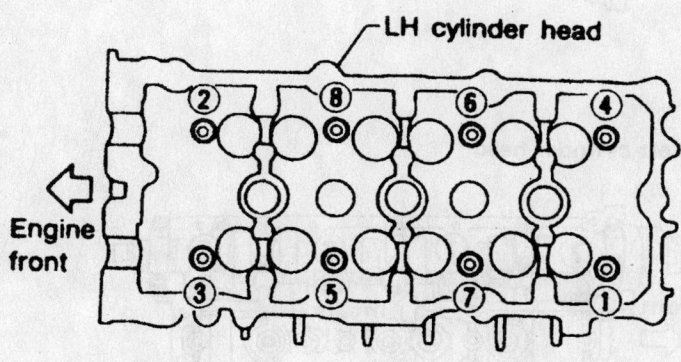

Loosen in numerical order.

Left cylinder head bolt loosening sequence—3.0L engine

Heater Core replacement is covered in Section 2 of this manual

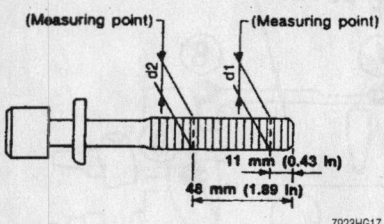

Measuring the cylinder head bolts—3.0L engine

- Right front wheel
- Left side ornament cover
- Air duct to intake manifold hose, collector hose, blow-by hose, and vacuum hoses
- Fuel hoses and the harness connections
- Evaporative emissions (EVAP) canister purge hose
- Ignition coils from the spark plugs
- Exhaust Gas Recirculation (EGR) tube
- Right side intake manifold collector supports and remove the collector. Remove the manifold from the cylinder head

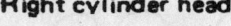

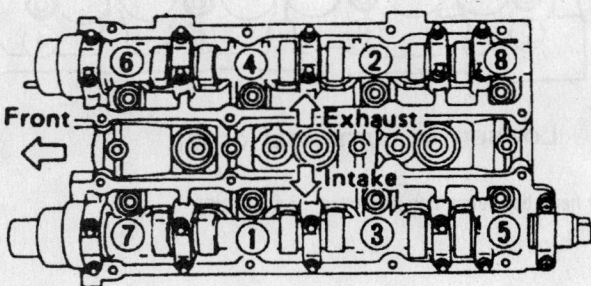

Right cylinder head

- Fuel tube assembly
- Rocker arm covers
- Engine under cover
- Engine side cover
- Power steering oil pump and belt
- Camshaft Position (CMP) sensor (PHASE) and Crankshaft Position (CKP) sensors (REF)/(POS).

6. Set the No. 1 piston to Top Dead Center (TDC) of compression stroke by rotating the crankshaft.
   - Crankshaft pulley
   - Air compressor and bracket
   - Timing chain tensioner and slack side chain guide
   - Engine oil pan

7. Remove the camshaft sprockets first. Be sure to hold the flats of the camshafts while removing the sprocket bolts.

8. Loosen the camshaft bearing caps in several steps. The bearing caps MUST be loosened in sequence.

➡ **Keep all bearing caps and camshafts in proper order for reinstallation.**

9. Loosen the cylinder head bolts in sequence.

To install:

10. Turn the crankshaft until the No. 1

piston is set 240° Before Top Dead Center (TDC) on compression stroke.

11. Using new head gaskets, install the cylinder heads.

➡ **If possible, replacement of the head bolts is suggested.**

12. If replacement of the head bolts is not possible, perform the following bolt measurement:

   a. Measure the diameter of the head bolt (11mm) from the bottom of the bolt.

   b. Measure the diameter of the head bolt (48mm) from the bottom of the bolt.

   c. Whenever the size difference between the 2 measurements exceeds 0.0043 in. (0.11mm) the head bolts must be replaced.

13. Lubricate the head bolt threads and the bolt seat surfaces with new engine oil.

14. Tighten the cylinder head bolts in sequence using the following steps:

   a. Step 1: All bolts in sequence to 72 ft. lbs. (98 Nm).

   b. Step 2: Completely loosen all bolts.

   c. Step 3: Tighten all bolts in sequence to 25–33 ft. lbs. (34–44 Nm).

   d. Step 4: Tighten all bolts in sequence an additional 90° clockwise.

   e. Step 5: Tighten all bolts in sequence an additional 90° clockwise.

15. Install or connect the following:
   - Camshaft tensioners. Tighten the tensioner mounting bolts to 75–96 inch lbs. (8.4–10.8 Nm)

➡ **The camshafts can be identified by the paint marks on the camshaft. The left cylinder head camshafts have a YELLOW paint mark and the right cylinder head camshafts have a WHITE paint mark.**

➡ **When installing the camshafts, position the camshaft keys at the 12 o'clock position in respect to the cylinder head angle.**

   - Camshafts and the bearing caps
   - New O-rings to the front of the engine block
   - Crankshaft sprocket with the mating mark facing out

16. Rotate the crankshaft clockwise and position the crankshaft to TDC of compression stroke and align the dowels of the camshaft sprockets to the 12 o'clock position.

   - Lower chain guide on the dowel pin with the front mark on the guide facing upward
   - Timing chains and sprockets to the intake camshafts. Be sure to align the timing chain and sprocket mating marks.

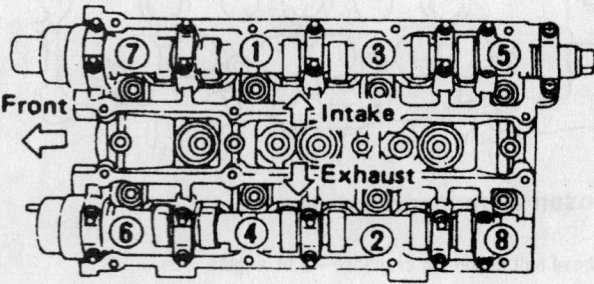

Left cylinder head

Cylinder head torque sequence—3.0L engine

17. Remove the left and right camshaft tensioner stopper pins.

18. Align the mating mark on the crankshaft with the matchmark (gold link) on the lower timing chain.

- Lower timing chain to the water pump sprocket

19. Working counterclockwise, install the lower timing chain camshaft sprockets. Be sure to align the sprocket marks with the blue links of the timing chain during installation.

20. Intake sprocket bolts and tighten to 88–95 ft. lbs. (119–128 Nm). Be sure to secure the camshafts while tightening the bolts

21. Timing chain guide, upper timing chain guide, lower timing chain tensioner and slack side timing chain guide

22. Timing cover evenly and gently. Be sure to align the dowel pin holes. Tighten the mounting bolts in sequence

- Front exhaust pipe and its support
- A/C compressor and bracket
- Crankshaft pulley to the crankshaft and install the mounting bolt

23. Torque the mounting bolt to 14–22 ft. lbs. (20–29 Nm). Torque the crankshaft bolt an additional 60–66° clockwise. This is approximately the angle from 1 hexagon bolt head corner to another.

- Ring gear cover plate
- CKP (PHASE) and CMP sensors (REF/POS)
- Power steering pump assembly
- Drive belts and the idler pulley
- Right front inner wheel cover and the right front wheel
- Engine undercovers
- Intake manifold, using new gaskets. Torque the nuts and bolts in sequence
- Intake manifold collector gasket with the arrow facing forward
- Intake manifold collector assembly and torque the mounting bolts to 16–18 ft. lbs. (22–25 Nm)
- Intake manifold collector support brackets
- EGR tube, using new gaskets and torque the mounting bolts to 15–20 ft. lbs. (21–26 Nm) in 2 progressive steps
- Spark plugs
- Ignition coils and torque the mounting bolts to 27–33 inch lbs. (3–4 Nm)
- Cylinder head cover ornament on the left side

- Water hoses to the cylinder head and intake manifold
- EVAP canister purge hoses
- Fuel hoses and wiring harness connections to the fuel rail
- Air duct-to-intake manifold hose, collector hose, blow-by hose, and vacuum hoses
- Negative battery cable

24. Fill the engine with clean oil.

25. Fill and bleed the cooling system.

26. Connect the negative battery cable.

27. Start the engine and run at 3000 rpm under no load to purge the air from the high pressure chamber. The engine may produce a rattling noise. This indicates that air still remains in the chamber and is not a matter of concern.

28. Inspect the vehicle for leaks and repair if necessary.

4.1L Engine

1. Before servicing the vehicle, refer to the precautions in the beginning of this section.

2. Remove or disconnect the following:

3. Properly relieve the fuel system pressure.

4. Drain the cooling system.

5. Drain the engine oil.

- Both battery cables
- Engine assembly from the vehicle and place it on a workstand
- Exhaust manifold
- Drain plugs on the sides of the engine and drain the coolant
- Intake manifold collector
- Ignition coil sub-harness
- Ignition coils
- Spark plugs
- Fuel rail with injectors. Do not disassemble the fuel hose
- Intake manifold

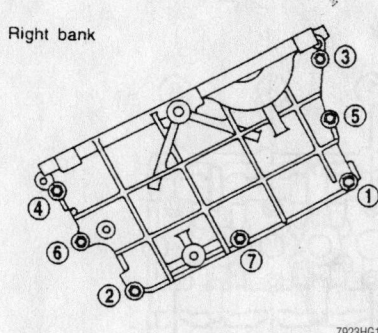

Right bank

7923HG19

Upper front cover bolt removal sequence—4.1L engine, right bank

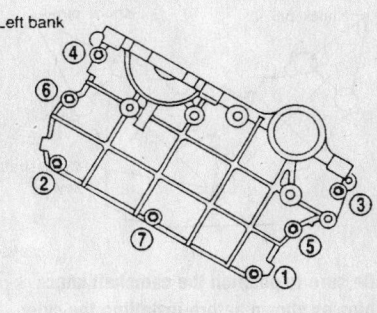

Left bank

7923HG20

Upper front cover bolt removal sequence—4.1L engine, left bank

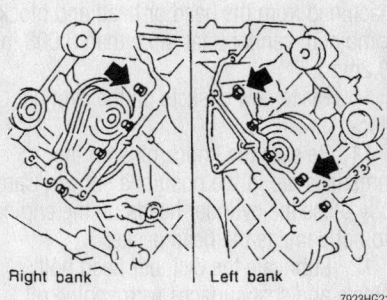

Right bank **Left bank**

7923HG21

Be sure to install the long bolts in the positions indicated by the arrows—4.1L engine

- Rocker arm covers and bring the No. 1 piston to Top Dead Center (TDC) on the compression stroke
- Camshaft Position (CMP) sensor
- Right and left upper front covers. Be sure to remove the bolts in the correct sequence to prevent damage to the cover
- Upper chain tensioners

6. Apply paint marks on the upper timing chains, camshaft and idler sprockets so they can be installed in their original positions.

- Camshaft sprocket
- Idler sprocket bolt
- Cylinder head sub-bolts. The bolts are different lengths, note their differences so they can be installed in their original positions

7. Loosen the cylinder head bolts gradually in the proper removal sequence, then remove the cylinder head.

To install:

8. Be sure all mating surfaces are clean before installation.

9. Check the cylinder head surface for warpage using a feeler gauge and a suitable straightedge. If the cylinder head is warped

Brake service is covered in Section 4 of this manual

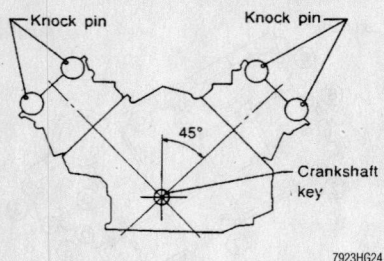

Be sure to position the camshaft knock pins as shown before installing the cylinder heads—4.1L engine

more than 0.004 in. (0.1mm), it must be resurfaced or replaced. The total amount machined from the head or head and block combined, cannot total more than 0.008 in. (0.2mm).

10. Place new gaskets on the cylinder block.

11. Be sure the knock pins on the camshafts are in the positions shown. Carefully place the cylinder heads on the engine. Do not damage the head gasket.

12. Lubricate the cylinder head bolt threats and seat surfaces with engine oil. Torque the bolts in sequence using the following sub-steps:
   a. Step 1: 22 ft. lbs. (29 Nm).

b. Step 2: 69 ft. lbs. (93 Nm).
c. Step 3: loosen the bolts completely.
d. Step 4: 18–25 ft. lbs. (25–35 Nm).
e. Step 5:Plus 90–95˚.

13. Install or connect the following:
   • Cylinder head sub-bolts. Torque the bolts to 56–74 inch lbs. (6–8 Nm). Be sure the long bolts are returned to the positions shown.
   • Idler and camshaft sprockets
   • Chain tensioners

14. Using new gaskets, install the cylinder head covers. Be sure to apply RTV silicone sealant to the gasket arch and rubber plugs.

15. Torque the cylinder head cover bolts in sequence using the following sub-steps:
   a. Step 1: Nos. 1 through 16: 35–52 inch lbs. (4–6 Nm).
   b. Step 2: Nos. 1 through 16: 61–78 inch lbs. (7–9 Nm).
   c. Step 3: Nos. 1 and 2: 61–78 inch lbs. (7–9 Nm).

16. Install or connect the following:
   • Intake valve timing control solenoid with a new O-ring. Torque the solenoid to 18–25 ft. lbs. (25–34 Nm)

17. Apply a bead of RTV silicone sealant to the upper front covers, then install them.

Torque the bolts to 56–74 inch lbs. (6–8 Nm).
   • CMP sensor
   • Rocker arm covers
   • Intake manifold with new gaskets
   • Fuel rail and injectors
   • Spark plugs and ignition coils
   • Intake manifold collector
   • Exhaust manifold with new gaskets
   • Engine assembly to the vehicle
   • Battery and cables

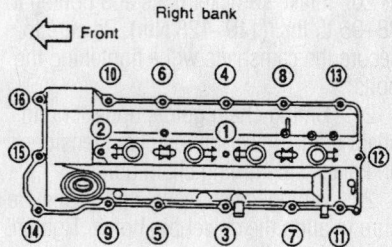

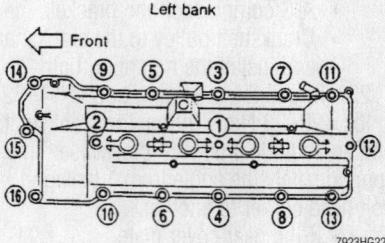

Cylinder head cover bolt tightening sequence—4.1L engine

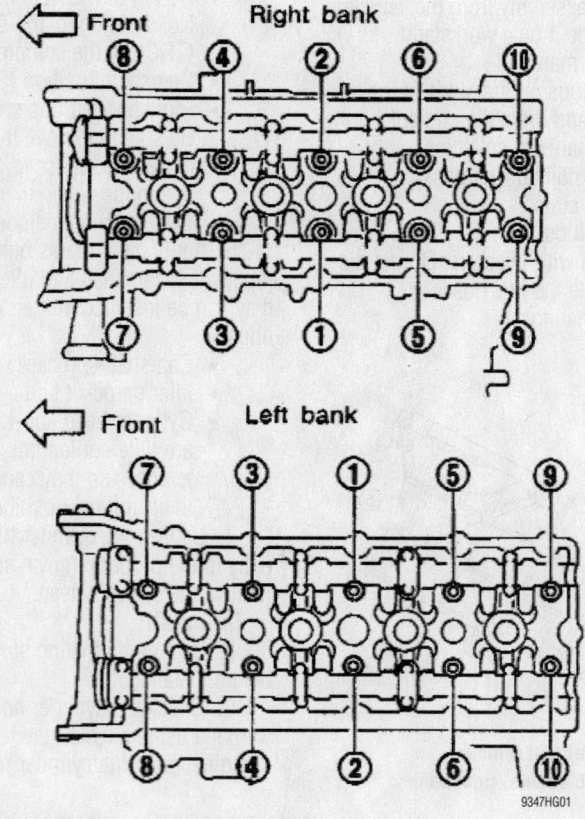

Cylinder head torque sequence—4.1L engine

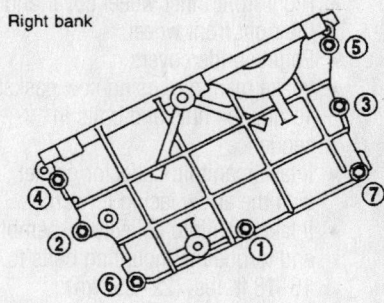

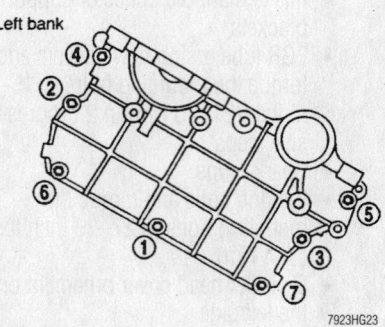

Torque the upper front cover bolts in the sequence shown—4.1L engine

18. Fill and bleed the cooling system.

19. Fill the engine with clean oil.

20. Start the vehicle, check for leaks and repair if necessary.

Rocker Arms/Shafts

REMOVAL & INSTALLATION

2.0L Engine

1. Before servicing the vehicle, refer to the precautions in the beginning of this section.

2. Relieve the fuel system pressure.

3. Drain the cooling system.

4. Remove or disconnect the following:

- Negative battery cable
- Right front wheel
- Engine side cover
- Radiator
- Air duct to the intake manifold
- Drive belts, water pump pulley, alternator and power steering pump
- Vacuum hoses, fuel hoses and wiring harness connectors
- Spark plugs, the AIV valve and resonator
- Rocker cover and oil separator. Loosen rocker cover bolts, using 2 to 3 steps, in the opposite sequence of tightening
- Intake manifold supports
- Oil filter bracket and power steering oil pump bracket

5. Set No. 1 piston at Top Dead Center (TDC) on the compression stroke by rotating the crankshaft.

- Chain tensioner
- Distributor. Do not turn the rotor with the distributor removed
- Timing chain guide, camshaft sprockets, camshafts, brackets, oil tubes and baffle plate. The camshaft bracket bolts must be loosened in sequence to prevent damage to the camshafts or the head.
- Rocker arm assembly

To install:

6. Check the hydraulic lash adjusters to ensure they did not bleed down during disassembly by trying to compress them. If the lash adjuster can be compressed 0.04 in. (1mm), air has entered and it must be bleed.

➡**Air cannot be bled from the lash adjusters by running the engine.**

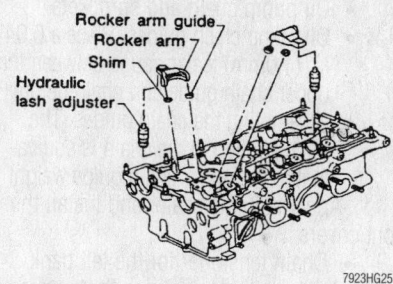

Exploded view of the rocker arms and related components—2.0L engine

7. Clean the camshaft end bracket and coat with liquid gasket. Install the camshafts, camshaft brackets, oil tubes and baffle plate. Ensure the left camshaft key is at 12 o'clock and the right camshaft key is at 10 o'clock.

➡**The procedure for tightening camshaft bracket bolts must be followed exactly to prevent camshaft damage.**

8. Line up the mating marks on the timing chain and camshaft sprockets and install the sprockets. Torque the sprocket bolts to 101–116 ft. lbs. (137–157 Nm).

9. Install or connect the following:

- Timing chain guide, distributor (ensure that rotor head is at 5 o'clock position) and chain tensioner
- Intake manifold supports. Clean the rocker cover and mating surfaces and apply a continuous bead of liquid gasket to the mating surface
- Rocker cover and oil separator. Tighten the rocker cover bolts in sequence

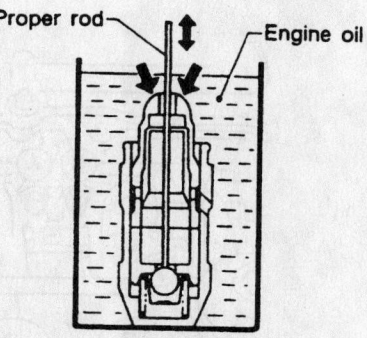

Submerge the lash adjuster in engine oil, lightly unseat the check ball with a thin rod and push on the plunger to release the air

- Oil filter bracket and the power steering pump bracket
- Spark plugs, AIV valve and the resonator
- Fuel lines, vacuum hoses and wiring connectors
- Water pump pulley, alternator and power steering pump
- Drive belts
- Intake manifold air duct, engine side cover and right wheel
- Radiator
- Negative battery cable

10. Fill and bleed the cooling system.

11. Start the vehicle, check for leaks and repair if necessary.

3.0L Engine

➡**The valves in the 3.0L engine are actuated directly by the camshaft. No rocker arms are used in this engine.**

4.1 Engines

1. Before servicing the vehicle, refer to the precautions in the beginning of this section.

2. Remove or disconnect the following:

- Negative battery cable
- Engine and transmission assembly from the vehicle
- Suspension member and engine mounts from the engine
- Air compressor bracket
- Cooling fan with coupling and the engine gusset

3. Separate the engine from the transmission and mount the engine on a suitable workstand.

- Oil pan
- Crank angle sensor and the Valve Timing Control (VTC) solenoid
- Chain tensioners and the upper front covers
- Front timing chain cover

➡**The timing chain will not be disengaged or dislocated from the crankshaft sprocket unless the front cover is removed. The cast portion of the front cover is located on the lower side of the crankshaft sprocket so the timing chain is not disengaged from the sprocket.**

- VTC assembly and the camshaft sprocket
- Oil pump chain and the timing chains

For complete Engine Mechanical specifications, see Section 1 of this manual

➡**Do not attempt to disassemble the VTC assembly since they are difficult to reassemble accurately in the field. If it should be disassembled, the VTC assembly must be replaced with a new one.**

- Camshaft brackets and the camshafts. Mark the parts so they can be reinstalled in their original positions
- Rocker arms. Be sure to identify each rocker arm so it can be reinstalled in it's original position.

To install:

4. Be sure all mating surfaces are clean before installation.

5. Install or connect the following:
- Rocker arms, camshafts and camshaft brackets on the right bank. Properly lubricate the rocker arms and camshafts prior to installation
- VTC assembly and the exhaust cam sprocket on the right bank

6. Be sure the camshafts are still correctly positioned and the piston in the No. 1 cylinder is still at Top Dead Center (TDC).
- Timing chain on the right bank, aligning the mating marks on the chain with those on the crankshaft and camshaft sprockets
- Chain tensioner on the right bank

7. Turn the crankshaft approximately 120° clockwise from the point where the No. 1 piston is at TDC on the compression stroke. At this point, the valves on the left bank still remain closed.

8. Correctly position the camshafts and rocker arms for the left cylinder head. Properly lubricate the rocker arms and camshafts prior to installation. Install the VTC assembly and the exhaust cam sprocket.

9. Install the timing chain on the left bank, aligning the mating marks on the chain with those on the crankshaft and camshaft sprockets.

10. Install or connect the following:

- Oil pump chain and sprockets
- Oil pump chain guides. Place a 0.04 in. (1.0mm) feeler gauge between the upper chain guide and chain before assembling the chain guides. The force applied to the chain is equivalent to the upper chain guide weight

11. Apply suitable sealer and install the front covers.
- Chain tensioner for the left bank

12. Apply suitable sealer to the rubber plugs and install them on the cylinder head.
- Crank angle sensor, VTC solenoid, rocker cover and crank pulley
- Transmission on the engine and install the engine assembly in the vehicle

Intake Manifold

REMOVAL & INSTALLATION

2.0L Engine

1. Before servicing the vehicle, refer to the precautions in the beginning of this section.

2. Properly relieve the fuel system pressure.

3. Drain the cooling system.

4. Remove or disconnect the following:
- Negative battery cable
- Fuel lines, vacuum hoses and electrical connectors
- Throttle linkage
- Intake manifold supports from the front and rear
- Intake manifold collector. Loosen the bolts in the sequence illustrated
- Injector tube assembly
- Power steering oil pump and the oil filter bracket

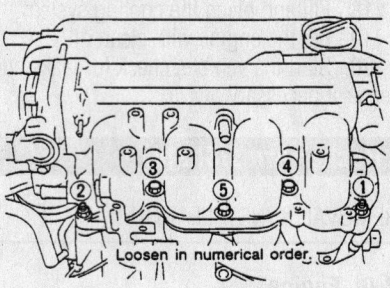

Intake manifold collector bolt loosening sequence—2.0L engine

Intake manifold bolt loosening sequence—2.0L engine

- Intake manifold-to-cylinder head bolts. Loosen the bolts in the sequence illustrated.
- Intake manifold and discard the gasket

To install:

5. Be sure all mating surfaces are clean prior to installation.

6. Fit a new gasket and the manifold into place. Start the support bolts to hold the manifold in place.

7. Install or connect the following:
- Intake manifold bolts. Torque the

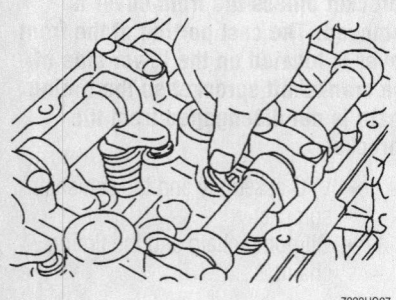

Press down on the lash adjuster to check for bleed-down—4.1L engines

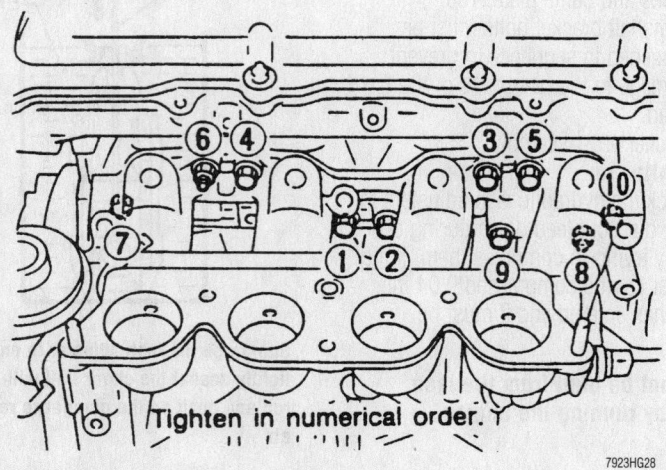

Lower intake manifold torque sequence—2.0L engine

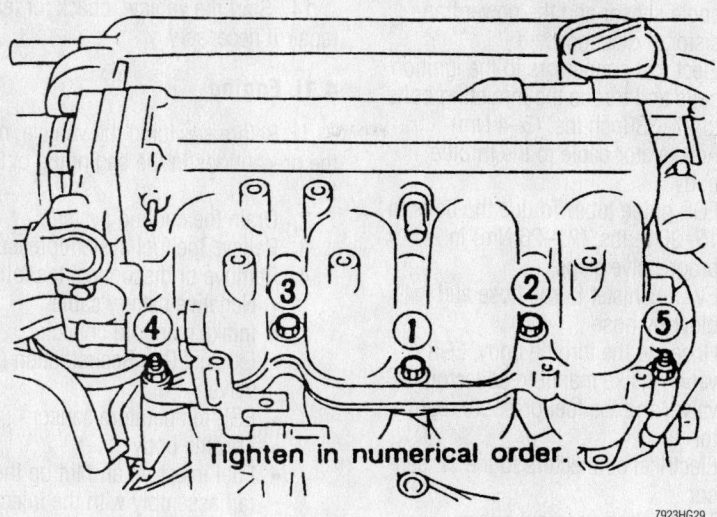

Upper intake manifold torque sequence—2.0L

7923HG29

bolts in sequence to 13–15 ft. lbs. (18–21 Nm). Torque the bolts in 2 steps, starting at the center and working towards the ends
- Injector tube assembly. Torque the bolts first to 84–96 inch lbs. (9–11 Nm), then to 15–20 ft. lbs. (21–26 Nm)
- Power steering oil pump and the oil filter bracket
- Intake manifold collector using a new gasket. Torque the bolts in sequence to 13–15 ft. lbs. (18–21 Nm)
- Fuel lines, vacuum hoses, electrical connectors and the throttle linkage
- Negative battery cable

8. Fill and bleed the cooling system.
9. Start the vehicle, check for leaks and repair if necessary.

3.0L Engine

1. Before servicing the vehicle, refer to the precautions in the beginning of this section.
2. Release the fuel system pressure.
3. Drain the cooling system.
4. Remove or disconnect the following:
- Negative battery cable
- Throttle body coolant hoses
- Electrical connectors from the Throttle Position (TP) sensor
- Hoses from the throttle body, the Exhaust Gas Recirculation (EGR) valve, intake manifold collector, Idle Air Control (IAC) valve, and the fuel pressure regulator
- Evaporative emissions (EVAP)canister purge hose and blow-by hose

- EGR guide tube
- Accelerator cable from the throttle body
- Intake manifold collector support brackets
- Right side electrical connectors from the ignition coils
- Electrical connector from the crank angle sensor and the power transistor, if necessary
- Intake manifold collector-to-intake manifold bolts/nuts and remove the intake manifold collector

5. Fuel injector assembly by removing or disconnecting the following:
- Electrical connectors from the fuel injectors
- Fuel lines from the fuel injector assembly

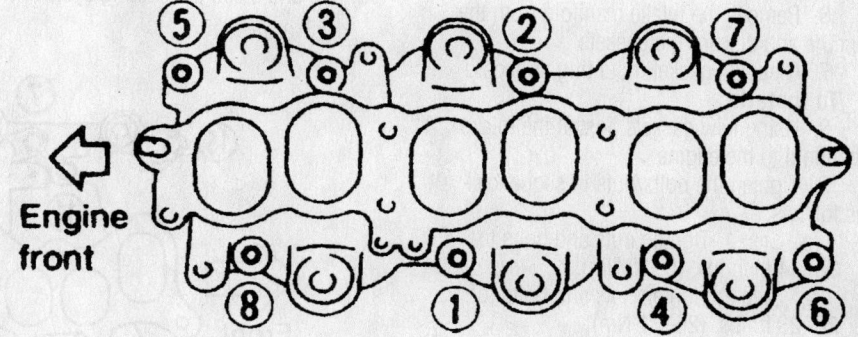

Tighten in numerical order.

7923HG31

Lower intake manifold torque sequence—3.0L engine

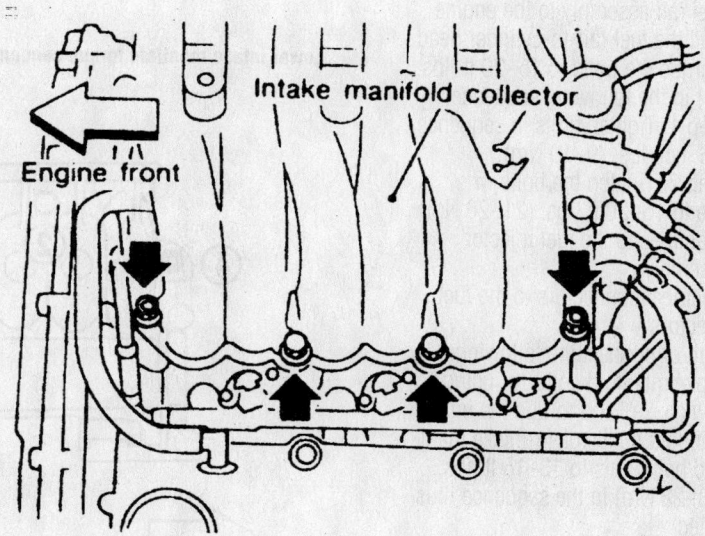

7923HG30

Upper intake manifold torque sequence—3.0L engine

For Accessory Drive Belt illustrations, see Section 1 of this manual

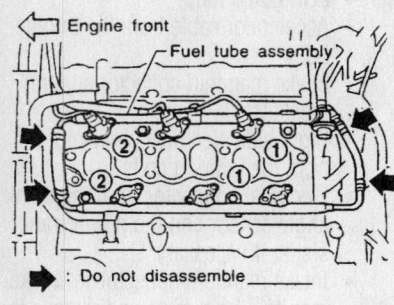

Tighten the fuel rail (tube) bolts in the sequence illustrated—3.0L engine

- Fuel rail-to-cylinder head bolts
- Fuel rail assembly from the engine
- Intake manifold bolts/nuts in the reverse sequence of the torque procedure

6. Remove the intake manifold from the engine and discard the gaskets

7. Clean all gasket mounting surfaces.

To install:

8. Using new gaskets, install the intake manifold to the engine.

9. Tighten the bolts/nuts in sequence as follows:

a. Step 1: Tighten nuts and bolts to 36–84 inch lbs. (5–10 Nm).

b. Step 2: Tighten nuts and bolts to 20–23 ft. lbs. (26–31 Nm).

c. Step 3: Repeat step 2 at least 5 times until all nuts and bolts have a final torque of 20–23 ft. lbs. (26–31 Nm).

10. Install the fuel injector assembly by installing or connecting the following:

- Fuel rail assembly to the engine

11. Install the fuel rail-to-cylinder head bolts and torque the bolts to 15–20 ft. lbs. (21–26 Nm) in the following sequence:

a. Step 1: Tighten bolts in sequence to 84–96 inch lbs. (9–10 Nm).

b. Step 2: Tighten the bolts in sequence to 15–20 ft. lbs. (21–26 Nm).

- Fuel lines to the fuel injector assembly
- Electrical connectors to the fuel injectors

12. Install or connect the following:

- Intake manifold collector using a new gasket, and torque the intake manifold collector-to-intake manifold bolts/nuts to 13–16 ft. lbs. (18–22 Nm) in the sequence illustrated
- Intake manifold collector supports. Torque the bolts to 14–18 ft. lbs. (20–25 Nm)
- Electrical connector to the crank

angle sensor and the power transistor, if disconnected

- Electrical connectors to the ignition coils and torque the mounting bolts to 27–33 inch lbs. (3–4 Nm)
- Accelerator cable to the throttle body
- EGR guide tube. Torque the bolts to 15–20 ft. lbs. (21–26 Nm) in 2 progressive steps
- EVAP canister purge hose and blow-by hose
- Hoses to the throttle body, EGR valve, intake manifold collector, IAC valve, and the fuel pressure regulator
- Electrical connectors to the TP sensor
- Throttle body coolant hoses
- Negative battery cable

13. Fill and bleed the cooling system.

14. Start the vehicle, check for leaks and repair if necessary.

4.1L Engine

1. Before servicing the vehicle, refer to the precautions in the beginning of this section.

2. Drain the cooling system.

3. Relieve the fuel system pressure.

4. Remove or disconnect the following:

- Negative battery cable
- Intake manifold collector
- Exhaust Gas Recirculation (EGR) valve
- EGR temperature sensor
- Throttle body
- Fuel injectors and lift up the fuel rail assembly with the injectors. Do not disconnect the fuel hose.
- Intake manifold and discard the gaskets

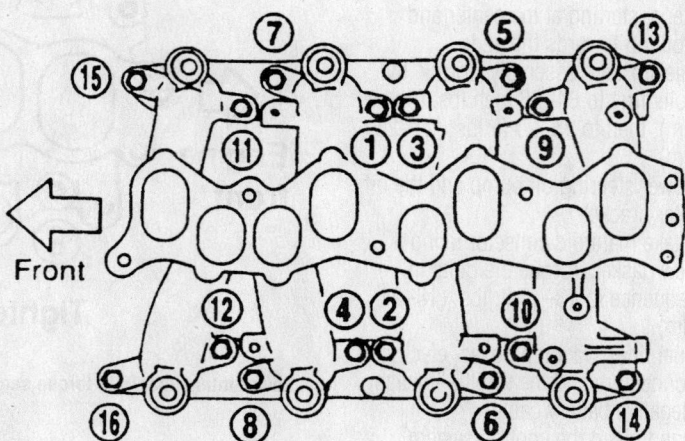

Lower intake manifold torque sequence—4.1L engine

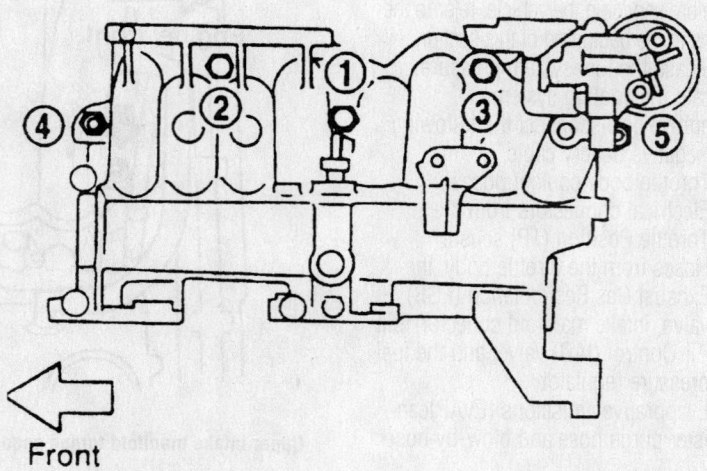

Upper intake manifold torque sequence—4.1L engine

To install:

5. Clean the intake manifold and intake manifold collector mounting surfaces

6. Install or connect the following:

- Intake manifold using new gaskets. Torque the bolts in sequence to 13–17 ft. lbs. (18–21 Nm).
- Fuel tube assembly. Torque the bolts in sequence first to 84–96 inch lbs. (9–11 Nm), then to 15–20 ft. lbs. (21–26 Nm).
- Throttle body. Torque the bolts in the following steps:

a. Step 1: bolts in sequence to 80–96 inch lbs. (9–11 Nm).

b. Step 2: Bolts in sequence to 13–16 ft. lbs. (18–22 Nm).

7. Install or connect the following:

- EGR temperature sensor
- EGR valve
- Intake manifold collector using a new gasket. Torque the bolts in sequence to 13–16 ft. lbs. (18–22 Nm).
- Negative battery cable

8. Fill and bleed the cooling system.

9. Start the vehicle, check for leaks and repair if necessary.

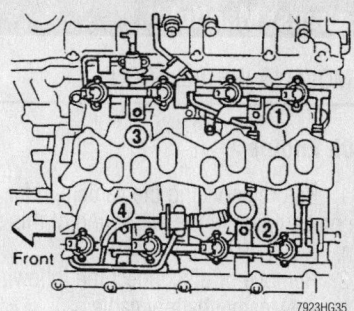

To prevent possible tube breakage, be sure to tighten the fuel tube mounting bolts according to the sequence shown

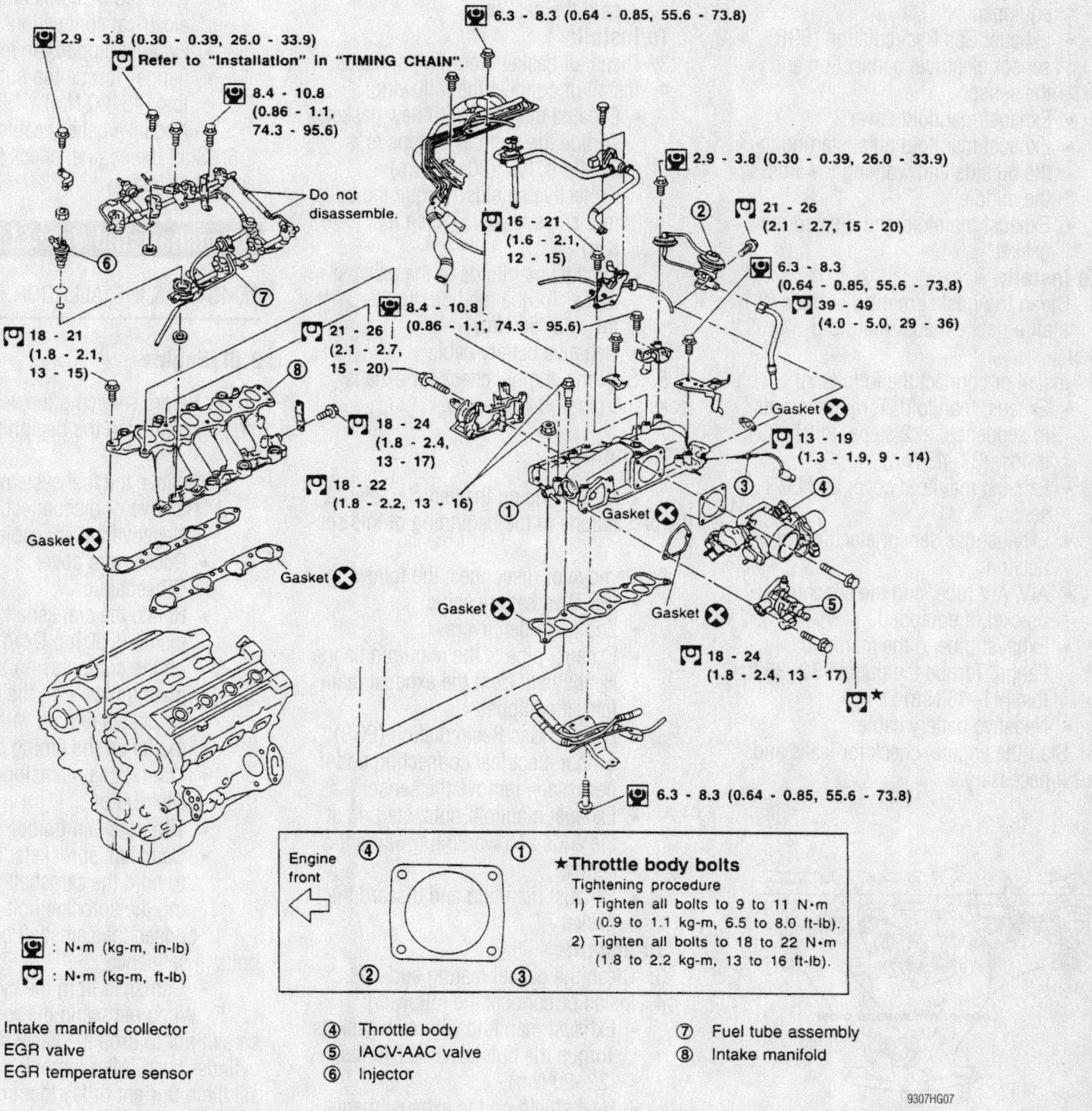

2.9 - 3.8 (0.30 - 0.39, 26.0 - 33.9)

Refer to "Installation" in "TIMING CHAIN".

8.4 - 10.8 (0.86 - 1.1, 74.3 - 95.6)

6.3 - 8.3 (0.64 - 0.85, 55.6 - 73.8)

Do not disassemble.

2.9 - 3.8 (0.30 - 0.39, 26.0 - 33.9)

16 - 21 (1.6 - 2.1, 12 - 15)

21 - 26 (2.1 - 2.7, 15 - 20)

6.3 - 8.3 (0.64 - 0.85, 55.6 - 73.8)

8.4 - 10.8 (0.86 - 1.1, 74.3 - 95.6)

39 - 49 (4.0 - 5.0, 29 - 36)

18 - 21 (1.8 - 2.1, 13 - 15)

21 - 26 (2.1 - 2.7, 15 - 20)

Gasket ⊗

13 - 19 (1.3 - 1.9, 9 - 14)

18 - 24 (1.8 - 2.4, 13 - 17)

18 - 22 (1.8 - 2.2, 13 - 16)

Gasket ⊗

Gasket ⊗

Gasket ⊗

Gasket ⊗

Gasket ⊗

18 - 24 (1.8 - 2.4, 13 - 17) ★

6.3 - 8.3 (0.64 - 0.85, 55.6 - 73.8)

Engine front

★**Throttle body bolts**
Tightening procedure
1) Tighten all bolts to 9 to 11 N·m (0.9 to 1.1 kg-m, 6.5 to 8.0 ft-lb).
2) Tighten all bolts to 18 to 22 N·m (1.8 to 2.2 kg-m, 13 to 16 ft-lb).

: N·m (kg-m, in-lb)

: N·m (kg-m, ft-lb)

① Intake manifold collector
② EGR valve
③ EGR temperature sensor
④ Throttle body
⑤ IACV-AAC valve
⑥ Injector
⑦ Fuel tube assembly
⑧ Intake manifold

Exploded view of the intake manifold assembly, related components and the throttle body torque sequence—4.1L engines

Exhaust Manifold

REMOVAL & INSTALLATION

2.0L Engine

1. Before servicing the vehicle, refer to the precautions in the beginning of this section.
2. Remove or disconnect the following:
   - Negative battery cable
   - Undercover and dust covers, if equipped
   - Exhaust pipe at the manifold flange
   - Air Injection Valve (AIV), AIV tube, and the attaching bracket, if equipped
   - Exhaust Gas Recirculation (EGR) sensor electrical connection and the sensor
   - Exhaust manifold cover
   - Exhaust manifold nuts, starting at the outside and working towards the middle
   - Exhaust manifold and discard the gasket

To install:

3. Clean the gasket mating surface and install a new exhaust manifold gasket.
4. Install or connect the following:
   - Exhaust manifold. Torque the nuts in sequence, in 2 steps, to 27–35 ft. lbs. (37–48 Nm)
   - Exhaust manifold cover and EGR sensor
   - Exhaust gas sensor electrical connection
   - AIV, AIV tube, and the attaching bracket, if equipped
   - Exhaust pipe to the manifold flange. Torque the nuts to 30–35 ft. lbs. (41–48 Nm)
   - Negative battery cable

5. Start the engine, check for leaks and repair if necessary.

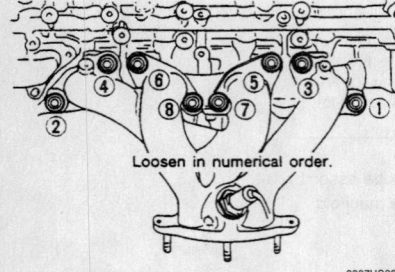

Loosen in numerical order.

9307HG23

Be sure to tighten the exhaust manifold nuts in the proper sequence—2.0L

3.0L Engine

1. Before servicing the vehicle, refer to the precautions in the beginning of this section.
2. Remove or disconnect the following:
   - Negative battery cable

➡**If necessary, soak the exhaust pipe retaining nuts with penetrating oil to loosen them.**

   - Exhaust manifolds from the exhaust pipes
   - Protective covers from the manifolds
   - Exhaust manifold-to-engine mounting nuts
   - Manifold from the engine and discard the gaskets

To install:

3. Clean all gasket mounting surfaces.
4. Install or connect the following:
   - Exhaust manifold with new gaskets. Torque the bolts in 2 steps to 22–24 ft. lbs. (30–32 Nm)
   - Protective shields. Torque the bolts in 2 steps to 46–57 inch lbs. (5–6 Nm)
   - Exhaust manifolds to the exhaust pipes. Torque the bolts to 32–37 ft. lbs. (43–50 Nm)
   - Negative battery cable

5. Start the engine, check for exhaust leaks and repair if necessary.

4.1L Engines

1. Before servicing the vehicle, refer to the precautions in the beginning of this section.
2. Remove or disconnect the following:
   - Negative battery cable
   - Engine undercovers
   - Exhaust pipe at the manifold flange
   - Heat shield from the exhaust manifold, if equipped
   - Exhaust Gas Recirculation (EGR) sensor electrical connection and if necessary, remove the sensor
   - Exhaust manifold nuts, starting at the ends and working towards the center
   - Exhaust manifold and discard the gasket

To install:

3. Clean the gasket mating surfaces.
4. Install or connect the following:
   - Exhaust manifold with new gaskets. Torque the nuts to 16–21 ft. lbs. (22–28 Nm)
   - Heat shield on the exhaust manifold, if equipped
   - EGR sensor. Torque the fastener to 30–37 ft. lbs. (40–50 Nm), if

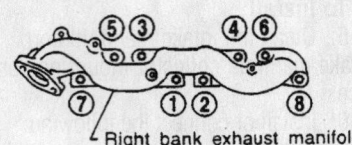

Right bank exhaust manifold

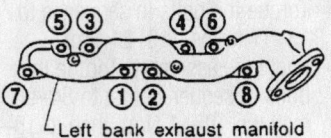

Left bank exhaust manifold

7923HG39

To avoid leaks, tighten the exhaust manifold nuts in the sequence shown—4.1L engine

removed. Reconnect the sensor electrical connection
   - Exhaust pipe to the manifold flange. Torque the nuts to 33–44 ft. lbs. (45–60 Nm)
   - Negative battery cable

5. Start the engine, check for leaks and repair if necessary.

Camshaft and Valve Lifters

REMOVAL & INSTALLATION

2.0L Engine

1. Before servicing the vehicle, refer to the precautions in the beginning of this section.
2. Relieve the fuel system pressure.
3. Remove or disconnect the following:
   - Negative battery cable
   - Rocker arm cover
   - Oil separator
   - Rotate the crankshaft until the No. 1 piston is at Top Dead Center (TDC) on the compression stroke and the mating marks on the camshaft sprockets line up with the mating marks on the timing chain
   - Timing chain tensioner
   - Distributor
   - Timing chain guide
   - Camshaft sprockets. Use a wrench to hold the camshaft while loosening the sprocket bolt

4. Loosen the camshaft bearing cap bolts in sequence.
   - Camshaft from the cylinder head

5. When removing the rocker arm, be careful not to drop the valve shims into the cylinder head. After removing the adjuster, set them upright or lay them down in a pan of clean engine oil. Do not lay them down on the bench or the oil will drain out and the adjuster will become air bound. Keep all

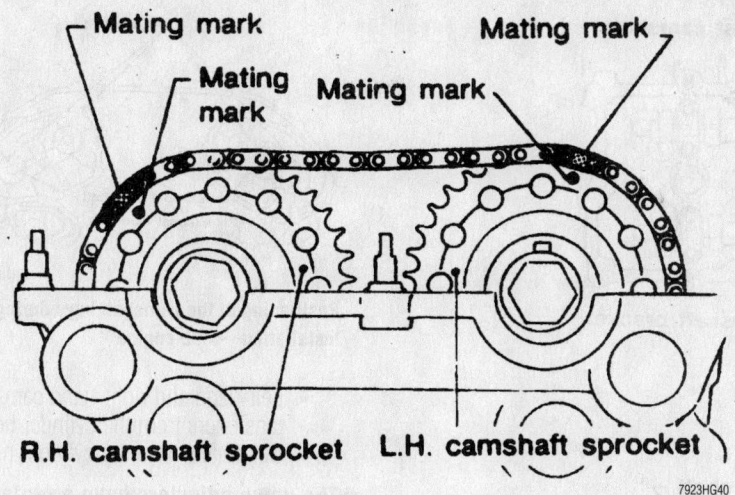

Be sure to align the marks on the sprockets with the marks on the chain—2.0L engine

7923HG40

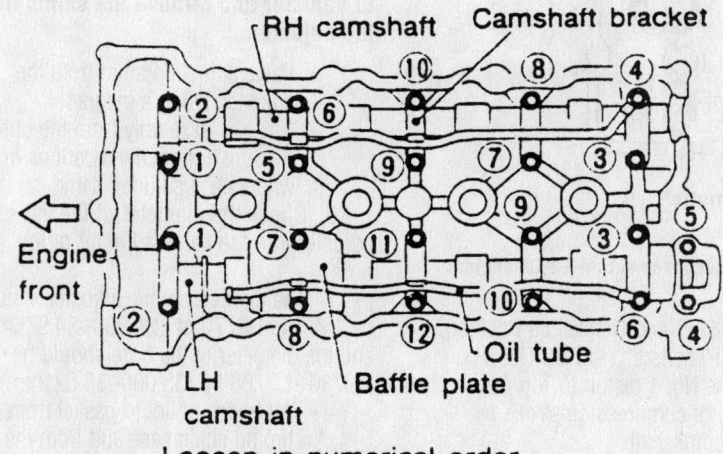

Loosen in numerical order.

9307HG24

Camshaft bearing cap bolt loosening sequence—2.0L engine

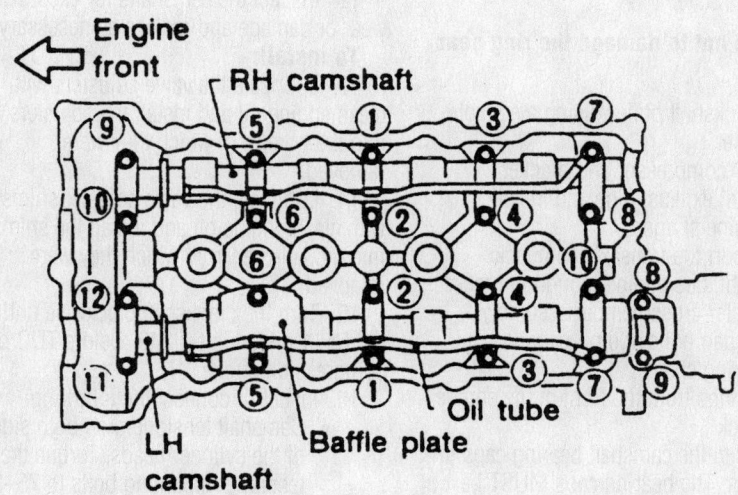

Camshaft bearing cap bolt torque sequence—2.0L engine

7923HG41

of these parts in order so they can be installed in the same locations.

To install:

6. Install the adjusters, shims and rockers into their original locations.

7. Clean the left-hand camshaft end bearing cap and coat the mating surface with liquid gasket. Install the camshafts, bearing caps, oil tubes and baffle plate. Ensure the left camshaft key is at 12 o'clock and the right camshaft key is at 10 o'clock.

8. The procedure for tightening bearing cap bolts must be followed exactly to prevent camshaft damage. Torque bolts as follows:

   a. Torque the right camshaft bolts 9 and 10 (in that order) to 17 inch lbs. (2 Nm), then bolts 1 through 8 (in that order) to the same specification.

   b. Torque the left camshaft bolts 11 and 12 (in that order) to 17 inch lbs. (2 Nm), then bolts 1 through 10 (in that order) to the same specification.

   c. Torque all bolts in sequence to 52 inch lbs. (6 Nm).

   d. Torque all bolts again in sequence to 78–102 inch lbs. (9–12 Nm), then bolts 8 and 9 on the left camshaft to 13–19 ft. lbs. (18–25 Nm).

9. Line up the mating marks on the timing chain and camshaft sprockets and install the sprockets. Torque sprocket bolts to 101–116 ft. lbs. (137–157 Nm).

10. Install or connect the following:
   • Timing chain guide and chain tensioner
   • Distributor making certain that the rotor head is at the 5 o'clock position

11. Clean the rocker cover and mating surfaces and apply a continuous bead of liquid gasket to the mating surface.

12. Install the rocker cover and oil separator. Tighten the rocker cover bolts as follows:

   a. Torque the nuts 1, 10, 11 and 8, in that order to 36 inch lbs. (4 Nm).

   b. Torque the nuts 1 through 13 as indicated in the figure to 72–84 inch lbs. (8–10 Nm).

13. Connect the negative battery cable.

3.0L Engine

1. Before servicing the vehicle, refer to the precautions in the beginning of this section.

2. Relieve the fuel system pressure.

3. Drain the engine oil.

4. Drain the cooling system.

5. Remove or disconnect the following:

For Wheel Alignment specifications, see Section 1 of this manual

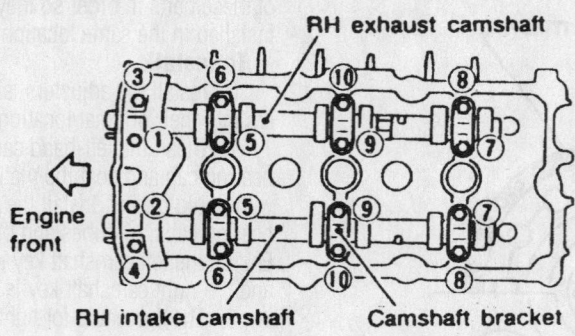

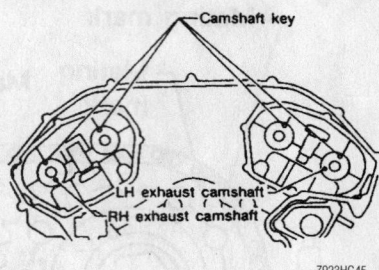

Positioning of the camshaft keys during installation—3.0L engine

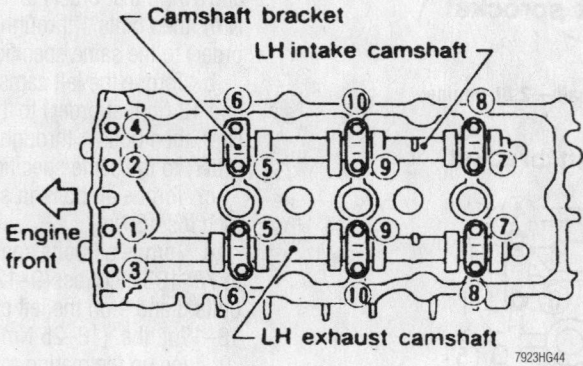

To avoid camshaft damage, loosen the bearing cap bolts in the sequence shown—3.0L engine

- Negative battery cable
- Left side rocker cover ornament

→**Before disconnecting any hoses or connectors, note the locations for reassembly.**

- Air duct to intake manifold hose, collector hose, blow-by hose, and vacuum hoses
- Fuel hoses and disconnect the harness connection
- Evaporative emissions (EVAP) canister purge hoses
- Water hoses from the cylinder head and intake manifold
- Ignition coils
- Spark plugs
- Exhaust Gas Recirculation (EGR) tube
- Intake manifold collector supports and the collector
- Fuel tube
- Intake manifold
- Rocker arm covers
- Engine undercovers
- Right front wheel
- Engine side covers
- Drive belts and idler pulley
- Power steering oil pump and belt
- Camshaft Position (CMP) sensor

(PHASE) and Crankshaft Position (CKP) sensors (REF)/(POS)

6. Set the No. 1 piston to Top Dead center (TDC) of compression stroke by rotating the crankshaft.
- Ring gear cover access plate

7. Loosen the crankshaft pulley bolt while securing the ring gear so the crankshaft cannot rotate.

→**Use care not to damage the ring gear teeth.**

- Crankshaft pulley, using a suitable puller
- A/C compressor and bracket
- Front exhaust pipe and install engine slingers

8. Support the transaxle with jack.
- Right side engine mounting bracket
- Center crossmember assembly
- Oil pan bolts and oil pans
- Timing chain
- O-rings from the front of the engine block

9. Loosen the camshaft bearing caps in several steps. The bearing caps MUST be loosened in sequence.

→**Keep all bearing caps and camshafts in proper order for reinstallation.**

- Left-hand and right-hand camshaft tensioners from the cylinder head
- Camshafts from the cylinder heads

→**The valve adjusters have a replaceable shim on the top of the adjuster. Note the proper locations of each shim to adjuster and remove the shims from the adjusters.**

- Valve adjusting shim from the adjuster, using a magnet
- Adjuster assembly from the bore. Be sure to note the locations from where each adjuster came

10. Check the diameter of the valve adjuster and the valve adjuster guide bore.

11. The diameter of the adjuster should be 1.3764–1.3770 in. (34.960–34.975mm) and the diameter of the bore should be 1.3780–1.3788 in. (35.000–35.021mm).

- All traces of liquid gasket from the timing chain case and from the water pump covers
- All traces of liquid gasket from the engine block

12. Inspect the camshafts for excessive wear or damage and replace as necessary.

To install:

13. Lubricate the valve adjusters with clean engine oil and install the adjusters into the bore from which they were removed.

14. Lubricate the valve adjuster shims with clean engine oil and install the shims into the adjuster from which they were removed.

15. Turn the crankshaft clockwise until the No. 1 piston is set 240° before TDC on compression stroke.

16. Install or connect the following:
- Camshaft tensioners on both sides of the cylinder heads. Torque the tensioner mounting bolts to 75–96 inch lbs. (8–11 Nm)

→**The camshafts can be identified by the paint marks on the camshaft. The**

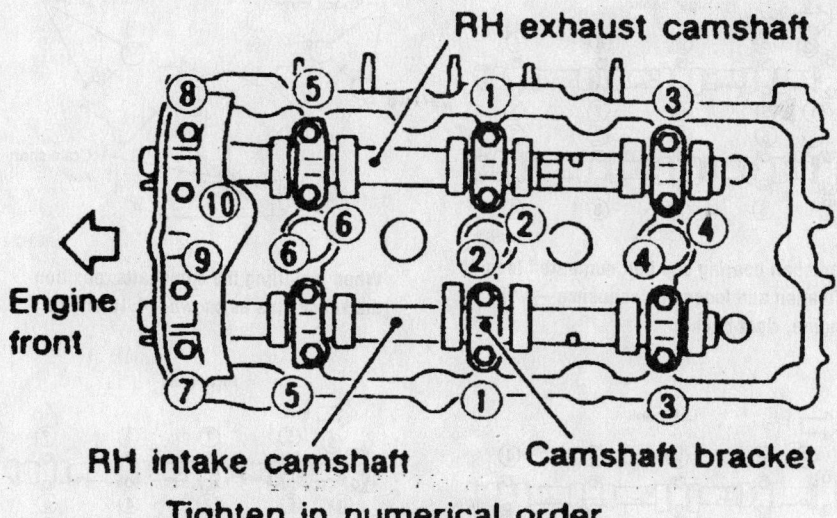

Be sure to tighten the camshaft bearing cap bolts in the correct sequence—3.0L engine, right cylinder head

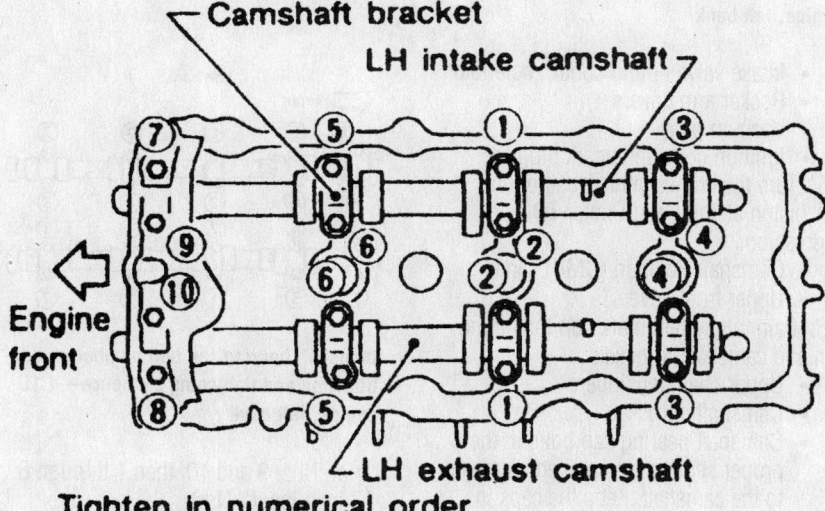

Tighten the camshaft bearing cap bolts in the correct sequence—3.0L engine, left cylinder head

left cylinder head camshafts have a YELLOW paint mark and the right cylinder head camshafts have a WHITE paint mark. When installing the camshafts, position the camshaft keys at the 12 o'clock position in respect to the cylinder head angle.

- Exhaust and intake camshafts and install the bearing caps. Before installing the No. 1 bearing cap, apply liquid gasket to the corners of the cap

17. Torque the camshaft bearing caps as follows:

a. Nos. 7 through 10 then, Nos. 1 through 6 to 17 inch lbs. (2 Nm).

b. All bolts in order to 52 inch lbs. (6 Nm).

c. All bolts in order to 80–104 inch lbs. (9–11 Nm).

- New O-rings to the front of the engine block

18. Apply sealant to the hatched portion of the of the rear timing chain case.

19. Align the rear timing chain case with the dowel pins and install onto the cylinder heads and engine block. Torque the rear timing chain case mounting bolts in sequence to 105–121 inch lbs. (11.8–13.7 Nm).

- Crankshaft sprocket with the mating mark facing out

20. Rotate the crankshaft clockwise and position the crankshaft to TDC of compression stroke and align the dowels of the camshaft sprockets to the 12 o'clock position in respect to the cylinder head

- Lower chain guide on the dowel pin with the front mark on the guide facing upward

21. On a workbench, align the marks on the intake and exhaust camshaft sprockets with the marks of the chain.

- Exhaust camshaft sprockets onto the dowel pin. Torque the bolts to 88–95 ft. lbs. (120–129 Nm). Be sure to secure the camshafts while tightening the bolts
- Align and install the timing chains and sprockets to the camshafts
- Timing cover evenly and gently. Be sure to align the dowel pin holes

➡**Leave the bolts unattended for 30 minutes or more after tightening.**

22. Apply a 0.091–0.130 in. (2.3–3.3mm) continuous bead of liquid gasket to the water pump cover and install the cover. Tighten the bolts to 84–108 inch lbs. (10–13 Nm).

- Rocker covers

23. Apply sealant to the front and rear seal of the oil pan and install the oil pan.

- Center crossmember assembly
- Right side engine mounting bracket and mount assembly
- Engine slinger assembly
- Front exhaust pipe and its support
- Air conditioning compressor and bracket
- Crankshaft pulley to the crankshaft and install the mounting bolt
- Torque the mounting bolt to 14–22 ft. lbs. (20–29 Nm). Torque the crankshaft bolt an additional 60–66° clockwise. This is approximately the angle from 1 hexagon bolt head corner to another
- Ring gear cover plate

For Maintenance Interval recommendations, see Section 1 of this manual

- CMP sensor (PHASE) and CKP sensors (REF)/(POS)
- Power steering pump assembly, drive belts and the idler pulley
- Engine side cover
- Right front wheel
- Engine undercovers

24. Using new gaskets, install the intake manifold. Torque the bolts in 2 stages as follows:

    a. 44–86 inch lbs. (5–10 Nm).
    b. 16–18 ft. lbs. (22–25 Nm).

25. Install or connect the following:

- Fuel tube assembly using new insulators. Torque the bolts in several steps to 15–20 ft. lbs. (21–26 Nm).
- Intake manifold collector gasket with the arrow facing forward
- Intake manifold collector assembly and support bracket. Torque the bolts to 16–18 ft. lbs. (22–25 Nm)
- EGR tube using new insulators. Torque the bolts in two steps to 15–20 ft. lbs. (21–26 Nm).
- Spark plugs and ignition coils. Torque the bolts to 27–33 inch lbs. (2.9–3.8 Nm)
- Water hoses to the cylinder head and intake manifold
- Fuel hoses and wiring harness connections to the fuel rail
- Air duct to intake manifold hose, collector hose, blow-by hose, and vacuum hoses
- Negative battery cable

26. Fill the engine with clean oil.
27. Fill and bleed the cooling system.
28. Start the engine and run at 3000 RPM under no load to purge the air from the high pressure chamber. The engine may produce a rattling noise. This indicates that air still remains in the chamber and is not a matter of concern.

4.1L Engine

1. Before servicing the vehicle, refer to the precautions in the beginning of this section.
2. Drain the cooling system.
3. Relieve the fuel system pressure.
4. Remove or disconnect the following:

- Negative battery cable
- Ornament cover
- Undercover
- Radiator and cooling fan
- Inlet and outlet hoses
- Alternator belt and idler bracket
- Air duct and intake manifold collector

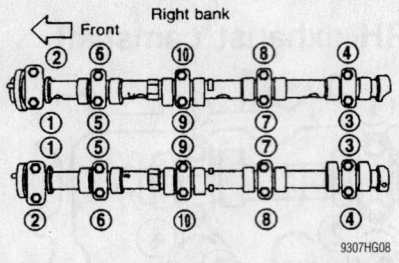

Camshaft bearing cap bolt numbered identification and loosening sequence—4.1L engine, right bank

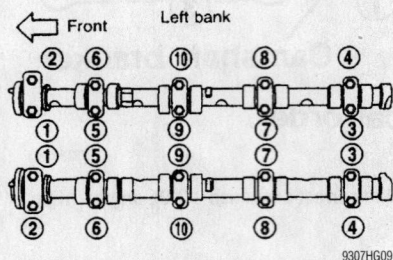

Camshaft bearing cap bolt numbered identification and loosening sequence—4.1L engine, left bank

- Intake valve timing control solenoid
- Rocker arm covers
- Vacuum pipe
- Ignition coils and spark plugs

5. Turn the crankshaft to position the No. 1 piston at Top Dead Center TDC on compression.

- Camshaft Position (CMP) sensor
- Upper front covers

6. Paint alignment marks on the timing chain and camshaft sprockets.

- Upper chain tensioners
- Camshaft sprockets
- Camshaft bearing cap bolts in the proper sequence to prevent damage to the camshaft. Keep the caps in order so they can be installed in the correct locations.
- Camshafts, rocker arms and lash adjusters. The lash adjusters and rocker arms must be installed in their original positions. Keep them in order

To install:

7. Install or connect the following:

- Lash adjusters and rocker arms

8. Lubricate the camshafts with engine oil and place them on the cylinder head with the knock pins facing away from the crankshaft.

9. Install the bearing caps in their original positions and torque the bolts as follows:

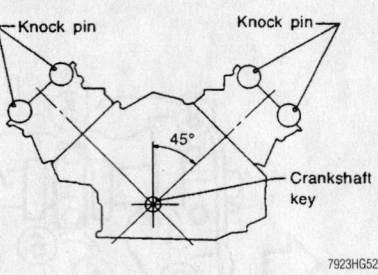

When installing the camshafts, position the knock pins as shown—4.1L engine

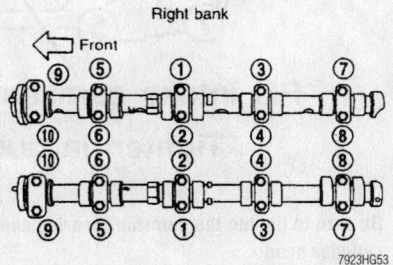

Camshaft bearing cap bolt numbered identification and tightening sequence—4.1L engine, right bank

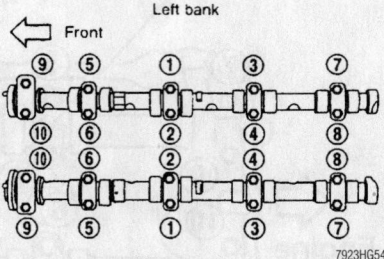

Camshaft bearing cap bolt numbered identification and tightening sequence—4.1L engine, left bank

    a. Nos. 9 and 10, then 1 through 8: 17 inch lbs. (2 Nm).
    b. All bolts in order: 52 inch lbs. (6 Nm).
    c. All bolts in order: 96–121 inch lbs. (11–14 Nm).

10. Install or connect the following:

- Camshaft sprockets. Torque the intake sprocket bolt to 76–83 ft. lbs. (103–113 Nm) and the exhaust sprocket bolt to 12–15 ft. lbs. (16–21 Nm)
- Chain tensioner. Torque the bolts to 82–95 ft. lbs. (9–11 Nm)
- Upper chain covers
- Rocker covers using new gaskets
- Valve timing control solenoid valve using a new O-ring. Torque the

solenoid to 18–25 ft. lbs. (25–34 Nm)
- Air duct and intake manifold collector
- Alternator belt and idler bracket
- Inlet and outlet hoses
- Spark plugs and ignition coils
- Rocker arm covers
- Vacuum pipe
- Radiator and cooling fan
- Engine under cover
- Ornament cover
- Negative battery cable

11. Fill and bleed the cooling system.
12. Start the vehicle, check for leaks and repair if necessary.

Valve Lash

ADJUSTMENT

2.0L Engine

➡A special gauge plate and collar will be needed to complete this procedure.

1. Before servicing the vehicle, refer to the precautions in the beginning of this section.
2. Remove the camshafts.
3. Install the J38957–1 gauge plate to the cylinder head. Use the bolts supplied in the kit to secure the plate to the cam bearing journals.
4. Install the collar J38957–2 on the dial indicator. Be sure the dished side of the collar is toward the gauge and tighten the setscrew.
5. Place the gauge on the No. 1 intake valve (shim side). Be sure the shim has been removed. Place the tip of the dial gauge on the top of the valve stem and the collar on the gauge plate. Zero the dial gauge.
6. Move the dial gauge to the other intake valve (rocker guide side). Place the tip of the dial gauge on the rocker guide and the collar of the gauge plate. Record the measurement.
7. Select the correct size shim using the chart. Shims are available in 17 different sizes ranging from 0.1102 in. (2.800mm) to 0.1260 in. (3.200mm) in increments of 0.001 in. (0.025mm).

3.0L and 4.1L Engines

➡Check and adjust the valve clearances while the engine is cold and not running.

Available shim

Thickness mm (in)	Identification mark
2.800 (0.1102)	28 00
2.825 (0.1112)	28 25
2.850 (0.1122)	28 50
2.875 (0.1132)	28 75
2.900 (0.1142)	29 00
2.925 (0.1152)	29 25
2.950 (0.1161)	29 50
2.975 (0.1171)	29 75
3.000 (0.1181)	30 00
3.025 (0.1191)	30 25
3.050 (0.1201)	30 50
3.075 (0.1211)	30 75
3.100 (0.1220)	31 00
3.125 (0.1230)	31 25
3.150 (0.1240)	31 50
3.175 (0.1250)	31 75
3.200 (0.1260)	32 00

7923HG56

Select the correct valve lash adjusting shim using the chart—2.0L engine

1. Before servicing the vehicle, refer to the precautions in the beginning of this section.
2. Remove the intake manifold collector.
3. Remove the left and right rocker covers.
4. Remove the spark plugs.
5. Set the No. 1 cylinder at Top Dead Center (TDC) on its compression stroke. Align the pointer with the TDC mark on the crankshaft pulley. Check that the valve adjusters on the No. 1 cylinder are loose and valve adjusters on the No. 4 cylinder are tight. If not, turn the crankshaft 1 revolution (360°) and align the pointer with the TDC mark on the crankshaft pulley.
6. Check the following valves:
- Both No. 1 intake valves
- Both No. 2 exhaust valves
- Both No. 3 exhaust valves
- Both No. 6 intake valves
7. Using a feeler gauge, measure the

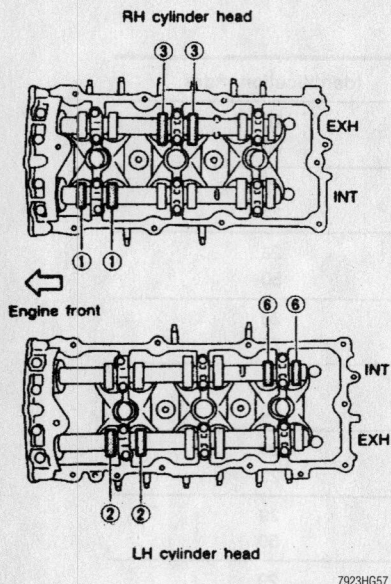

Valve lash checking sequence at TDC of cylinder No. 1—3.0L engine

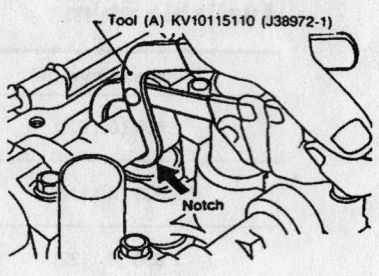

Install the depressor tool around the camshaft being careful not to damage the surfaces—3.0L engine

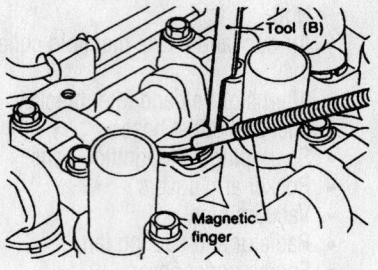

Use a magnet to remove the shim from the adjuster. Sometimes a shot of compressed air can help lift the shim up—3.0L engine

clearance between the valve adjuster and the camshaft. Record any valve clearance measurements that are out of specification. Intake valve clearance (cold) is 0.010–0.013 in. (0.26–0.34mm) and exhaust valve clearance (cold) is 0.011–0.015 in. 0.29–0.37mm).

8. Turn the crankshaft 240° and set the No. 3 cylinder to TDC of its compression stroke.

9. Check the following valves:
- Both No. 2 intake valves
- Both No. 3 intake valves
- Both No. 4 exhaust valves
- Both No. 5 exhaust valves

10. Using a feeler gauge, measure the clearance between the valve adjuster and the camshaft. Record any valve clearance measurements that are out of specification. Intake valve clearance (cold) is 0.010–0.013 in. (0.26–0.34mm) and exhaust valve clearance (cold) is 0.011–0.015 in. (0.29–0.37mm).

11. Turn the crankshaft 240° and set the No. 5 cylinder to TDC of its compression stroke.

12. Check the following valves:
- Both No. 1 exhaust valves
- Both No. 4 intake valves
- Both No. 5 intake valves
- Both No. 6 exhaust valves

13. Using a feeler gauge, measure the clearance between the valve adjuster and the camshaft. Record any valve clearance measurements that are out of specification. Intake valve clearance (cold) is 0.010–0.013 in. (0.26–0.34mm) and exhaust valve clear-

ance (cold) is 0.011–0.015 inches (0.29–0.37mm).

14. If all the valve clearances are within specification, install the cylinder head cover, spark plugs, and the intake manifold collector.

15. If an adjustment is necessary, adjust the valve clearance while engine is cold by removing the adjusting shim. The adjusting shim can be removed by using the following procedures:

a. Turn the crankshaft so the camshaft lobe of the valve to be adjusted is pointed straight up.

b. Turn the adjuster so the notch is pointed towards the center of the cylinder head; this will facilitate the shim removal process.

c. Using a depressor tool No. KV10115110 push down on the adjuster and insert a keeper tool on the edge of the adjuster to keep the adjuster in the depressed position.

d. Remove the depressor tool and remove the shim with a magnet.

➡**Compressed air can be blown into the hole of the adjuster to separate the adjusting shim from the adjuster.**

16. Determine the replacement adjusting shim size by using the following procedures and formula:

a. Using a micrometer determine thickness of the removed shim.

b. Calculate the thickness of a new adjusting shim so valve clearance is within the specified values.
- R= thickness of the removed shim
- N= thickness of the new shim
- M= measured valve clearance
- Calculate the Intake Shim as follows: N = R + M—0.0118 in. (0.30mm)
- Calculate the Exhaust Shim as follows: N = R + M—0.0130 in. (0.33mm)

17. Shims are available in 64 sizes from

0.0913–0.1161 in. (2.32–2.95mm) in steps of 0.004 in. (0.01mm). The thickness is stamped on the shim; this side is always installed facing down. Select new shims with thickness as close as possible to calculated valve and install it in the adjuster.

18. Install the new shim onto the adjuster.

19. Depress the adjuster and remove the keeper tool. Remove the depressor tool and recheck the valve clearance. Repeat this procedure for any other valves requiring adjustment.

20. When all valve adjustments are finished, install the cylinder head cover, spark plugs, and the intake manifold collector.

Starter Motor

REMOVAL & INSTALLATION

2.0L Engine

1. Before servicing the vehicle, refer to the precautions in the beginning of this section.

2. Remove or disconnect the following:
- Negative battery cable
- Starter insulator
- Starter harness connector and cable
- Starter mounting bolt and nut
- Starter

To install:

3. Install or connect the following:
- Starter. Torque the bolts to 27 ft. lbs. (36 Nm)
- Starter harness connector and cable
- Starter insulator
- Negative battery cable

3.0L Engine

1. Before servicing the vehicle, refer to the precautions in the beginning of this section.

2. Remove or disconnect the following:
- Negative battery cable
- Air duct assembly

- Harness protector
- Starter harness
- Both starter bolts
- Starter

To install:

3. Install or connect the following:
- Starter
- Both starter bolts. Tighten the long bolt to 57–72 ft. lbs. (77–98 Nm) and the short bolt to 22–30 ft. lbs. (30–41 Nm)
- Starter harness
- Harness protector
- Air duct assembly
- Negative battery cable

4.1L Engine

1. Before servicing the vehicle, refer to the precautions in the beginning of this section.
2. Remove or disconnect the following:
- Negative battery cable
- Steering gear and linkage assembly
- Harness connector
- Starter retainers
- Starter

To install:

3. Install or connect the following:
- Starter
- Starter retainers. Tighten the starter retainers to 30–37 ft. lbs. (40–50 Nm)
- Harness connector
- Steering gear and linkage assembly
- Negative battery cable

Oil Pan

REMOVAL & INSTALLATION

2.0L Engine

1. Before servicing the vehicle, refer to the precautions in the beginning of this section.
2. Raise and support the vehicle safely.
3. Drain the engine oil.
4. Remove or disconnect the following:
- Negative battery cable
- Engine undercover
- Steel oil pan bolts in the proper sequence
- Steel oil pan. Insert tool KV10111100 between steel oil pan and aluminum oil pan to break the seal.
- Front exhaust tube and support the transaxle with a suitable jack and

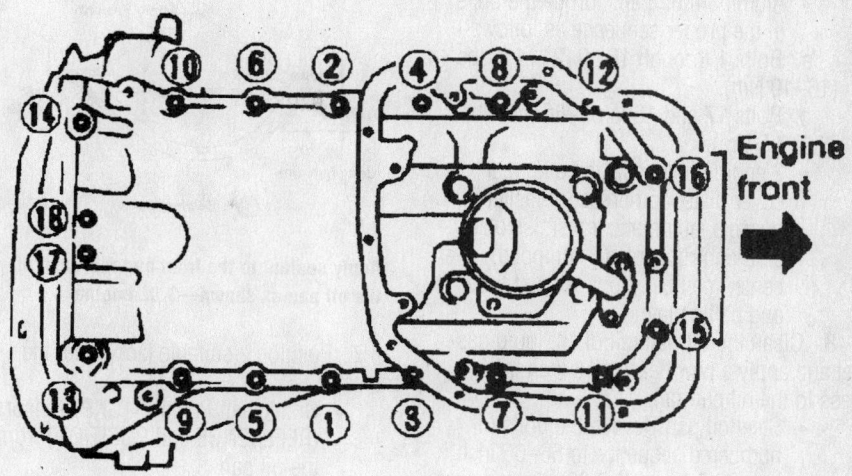

Tighten the aluminum oil pan mounting bolts in the sequence shown—2.0L engine

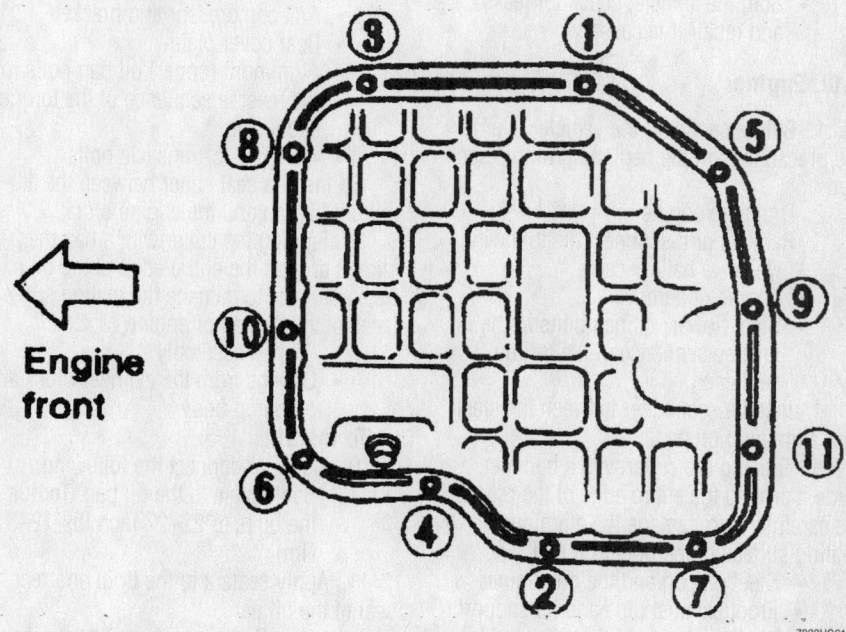

Be sure to tighten the steel oil pan mounting bolts in the proper order to prevent leakage—2.0L engine

raise the engine with an engine hoist
- Center crossmember
- Transaxle shift control cable, if equipped with an automatic transaxle
- A/C compressor bracket gussets and the rear cover plate
- Aluminum oil pan bolts in sequence
- Baffle plate
- 2 engine-to-transaxle bolts and

install them into vacant bolt holes on the oil pan. Tighten the bolts to release the oil pan from the cylinder block. Use tool KV10111100 to break the remaining seal

To install:

5. Remove the 2 bolts previously installed in the oil pan.

6. Clean the oil pan rail of all liquid gasket and apply a new bead of 1/8 inch thickness to the oil pan rail.

7. Install or connect the following:

- Aluminum oil pan. Torque the bolts in the proper sequence as follows:

a. Bolts 1 through 16 to 12–14 ft. lbs. (16–19 Nm).

b. Bolts 17 and 18 to 56–66 inch lbs. (6.5–7.5 Nm).

- 2 engine-to-transaxle bolts, rear cover plate, compressor bracket gussets, automatic transmission shift control cable (if equipped), center member, front exhaust tube and baffle plate

8. Clean the oil pan rail of all liquid gasket and apply a new bead of 1/8 inch thickness to the oil pan rail.

- Steel oil pan. Torque the bolts in numbered sequence to 56–66 inch lbs. (6–8 Nm). Wait 30 minutes before refilling engine case with oil
- Negative battery cable

9. Fill the engine with clean oil.

- Start the vehicle, check for leaks and repair if necessary.

3.0L Engine

1. Before servicing the vehicle, refer to the precautions in the beginning of this section.

2. Drain the engine oil.

3. Remove or disconnect the following:

- Negative battery cable
- Engine undercovers
- Steel (lower) oil pan bolts in the reverse sequence of the torque sequence

4. Insert a seal cutter between the steel and aluminum oil pan.

5. Tapping the cutter with a hammer, slide it around the entire edge of the oil pan. Be careful not to damage the aluminum mating surface of the upper oil pan.

- Steel oil pan and the oil strainer
- Front exhaust pipe and its support

6. Hang the engine at the right and left side engine slingers with a suitable hoist.

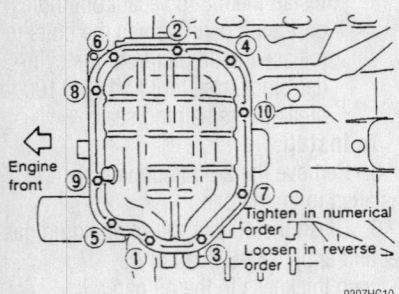

Steel (lower) oil pan loosening sequence—3.0L engine

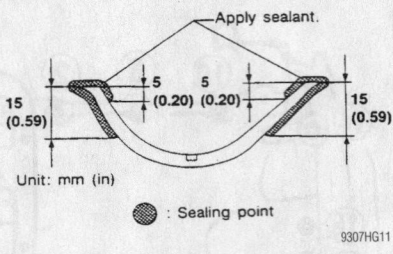

: Sealing point

Apply sealant to the front and rear seal of the oil pan as shown—3.0L engine

7. Position a suitable jack under the transaxle.

- Crankshaft Position (CKP) sensors (REFERENCE and POSITION) from the oil pan
- Front and rear engine mounting nuts and bolts
- Center crossmember assembly
- Engine drive belts
- A/C compressor and bracket
- Rear cover plate
- Aluminum (upper) oil pan bolts in the reverse sequence of the torque sequence
- 4 engine-to-transaxle bolts

8. Insert a seal cutter between the aluminum oil pan and the engine block.

9. Tapping the cutter with a hammer, slide it around the entire edge of the oil pan. Be careful not to damage the mating surfaces of the oil pan or engine block.

- Oil pan assembly
- O-rings from the cylinder block and oil pump body

To install:

10. Install or connect the following:

- Baffle plate to the oil pan. Torque the bolts to 22–27 inch lbs. (2–3 Nm)

11. Apply sealant to the front and rear seal of the oil pan.

- New O-rings to the cylinder block and the oil pump body

12. Apply a 0.177–0.217 in. (4.5–5.5mm) continuous bead of liquid gasket to the upper oil pan mating surface and install the oil pan. Torque the bolts in sequence to 12–14 ft. lbs. (16–19 Nm).

- Oil pan strainer. Torque the bolts to 12–14 ft. lbs. (16–19 Nm)
- Rear cover plate and the lower transaxle bolts
- A/C compressor and bracket
- Engine drive belts and adjust as necessary
- Center crossmember assembly
- Front and rear engine mounting nuts and bolts and remove the support jack and the engine hoist

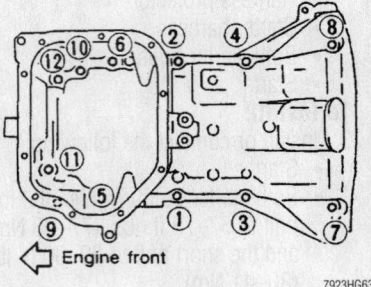

Aluminum oil pan torque sequence (loosen in reverse sequence)—3.0L engine

- CKP sensors (REFERENCE and POSITION) to the oil pan. Torque the bolts to 75–96 inch lbs. (9–10 Nm)
- Front exhaust pipe and its support
- Oil strainer

13. Apply a 0.177–0.217 in. (4.5–5.5mm) continuous bead of liquid gasket to the lower oil pan mating surface and install the oil pan. Tighten the mounting bolts in sequence to 57–66 inch lbs. (6–8 Nm).

➡**Wait at least 30 minutes before refilling the engine oil.**

- Engine undercovers
- Negative battery cable

14. Fill the engine with clean oil.

15. Start the engine, check for leaks and repair if necessary.

4.1L Engine

1. Before servicing the vehicle, refer to the precautions in the beginning of this section.

2. Drain the engine oil.

3. Attach an engine support fixture to the engine so the right and left engine mounts can be removed.

4. Remove or disconnect the following:

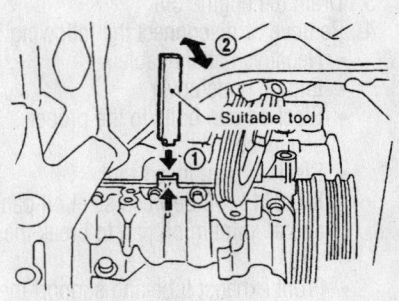

Use a suitable tool to break the seal between the oil pan and engine block—4.1L engine

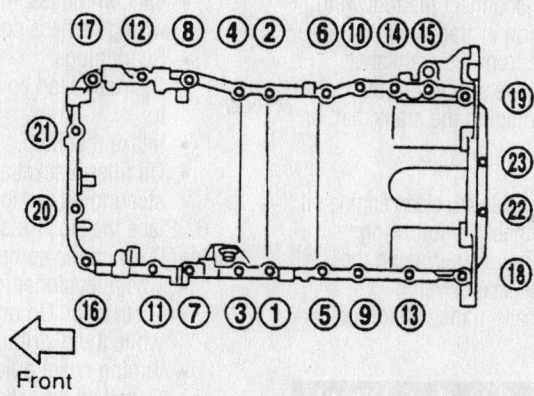

7923HG66

Tighten the oil pan bolts in the sequence shown—4.1L engine

- Negative battery cable
- Drive belts
- Cooling fan and coupling
- Power steering oil pump
- Front stabilizer bar brackets from the side members
- Right and left engine mounting bolts
- Steering shaft lower joint
- Power steering tube bracket and support the front suspension member
- Lower the front suspension member
- A/C compressor and bracket
- Oil pan mounting bolts, then insert a tool into the notch on the oil pan and break the seal between the pan and engine block. Be careful not to damage the sealing surface
- Pull the oil pan out from the front while lowering the suspension as needed

To install:

5. Clean all gasket mating surfaces thoroughly.

6. Apply a continuous bead of liquid gasket to the oil pan mating surface. Be sure the bead is ⅛ inch (3mm) wide.

7. Install the oil pan and tighten the retainers as follows:

    a. Bolts 1 through 21 in sequence: 12–14 ft. lbs. (16–19 Nm).

    b. Bolts 22 and 23: 56–65 in lbs. (6–7 Nm).

➡**Wait at least 30 minutes for the sealant to cure before filling the engine with oil.**

8. Install or connect the following:
- A/C compressor and bracket
- Front suspension member. Torque

the nuts to 87–101 ft. lbs. (147–167 Nm).
- Power steering tube on the suspension member
- Lower steering shaft joint
- Stabilizer bar to the suspension member. Torque the nuts to 35–46 ft. lbs. (47–62 Nm)
- Engine mounting bolts. Torque the nuts to 41–49 ft. lbs. (55–67 Nm)
- Cooling fan and coupling
- Drive belts and adjust as required
- Negative battery cable

9. Fill the engine with clean oil.

10. Start the vehicle, check for leaks and repair if necessary.

Oil Pump

REMOVAL & INSTALLATION

2.0L Engine

1. Before servicing the vehicle, refer to the precautions in the beginning of this section.

2. Relieve the fuel system pressure.

3. Drain the engine oil.

4. Remove or disconnect the following:
- Negative battery cable
- Drive belts
- Cylinder head with the intake and exhaust manifolds attached
- Oil pans
- Oil strainer and baffle plate
- Crankshaft pulley and the front cover assembly
- Oil pump from the inside of the front cover

To install:

5. Coat the oil pump gears with oil and fit the pump to the cover, using a new oil seal and O-ring.

6. Clean the mating surfaces of liquid gasket and apply a fresh bead of ⅛ inch (3mm) sealer to the sealing surface of the front cover.

7. Install or connect the following:
- Front cover assembly
- Crankshaft pulley
- Oil strainer, baffle plate, oil pans, cylinder head and drive belts
- Negative battery cable

8. Fill the engine with clean oil.

9. Start the vehicle, check for leaks and repair if necessary.

3.0L Engine

➡**The oil pump bolts to the front of the engine block and is driven by the crankshaft. Removal of the timing cover and chains are necessary for oil pump service.**

1. Before servicing the vehicle, refer to the precautions in the beginning of this section.

2. Drain the engine oil.

3. Rotate the engine and position it to Top Dead Center (TDC) compression stroke of cylinder No. 1.

4. Remove or disconnect the following:
- Negative battery cable
- Drive belts
- Camshaft Position (CMP) sensor (PHASE) and the Crankshaft Position (CKP) sensor (REF/POS)
- Right front wheel and inner fender cover
- Engine undercovers
- Crankshaft pulley
- Front exhaust pipe and its support and support the engine at the left and right side slingers with a suitable hoist
- Engine right side mounting insulator and bracket nuts and bolts
- Center crossmember assembly
- A/C compressor and mounting bracket
- Lower and upper oil pans
- Oil strainer from the oil pump
- Water pump cover and the front cover assembly
- Lower timing chain assembly
- Oil pump

To install:

➡**When installing the oil pump, be sure to apply engine oil to the gears.**

5. Install or connect the following:
- Oil pump. Torque the bolts to 57 inch lbs. (6.5 Nm) and the mounting screws to 33–44 inch lbs. (4–5 Nm)

- Lower timing chain assembly
- Front timing cover and water pump covers
- Oil strainer using a new gasket. Torque the bolts to 12–14 ft. lbs. (16–19 Nm)
- Upper and lower oil pans. Be sure to use new O-rings at the oil pump to upper oil pan mating surface
- A/C compressor and mounting bracket
- Center crossmember assembly
- Engine right side mounting insulator and bracket and remove the engine support hoist
- Front exhaust pipe and its support
- Crankshaft pulley
- Engine undercovers and the right side inner fender cover
- Right front wheel
- CMP sensor (PHASE) and the CKP sensor (REF/POS)
- Engine drive belts and adjust as necessary
- Negative battery cable

6. Fill the engine with clean oil.
7. Start the engine, check the oil pressure, and check for oil leaks.

4.1L Engine

➥The oil pump is mounted in the cylinder block below the left bank and behind the left timing chain.

1. Before servicing the vehicle, refer to the precautions in the beginning of this section.
2. Drain the engine oil.
3. Remove or disconnect the following:
   - Negative battery cable
   - Timing chains
   - Oil pump assembly from the front of the engine

To install:
4. Clean the oil pump mounting surface.
5. Install or connect the following:
   - Oil pump using a new gasket. Torque the short bolt to 56–66 ft. lbs. (6–8 Nm) and the long bolt to 12–14 ft. lbs. (16–19 Nm)
   - Timing chains
   - Negative battery cable

6. Fill the engine with clean oil.
7. Start the vehicle, check for leaks and repair if necessary.

Rear Main Seal

REMOVAL & INSTALLATION

1. Before servicing the vehicle, refer to the precautions in the beginning of this section.

2. Remove or disconnect the following:
   - Transmission or transaxle
   - Drive plate from the crankshaft

3. Carefully pry the seal out of the retainer without damaging the crankshaft or the seal retainer.

To install:
4. Lubricate the seal with clean engine oil.
5. Install or connect the following:
   - Seal into the retainer using the appropriate seal driver
   - Driveplate and transmission or transaxle

Timing Chain, Sprockets, Front Cover and Seal

REMOVAL & INSTALLATION

2.0L Engine

1. Before servicing the vehicle, refer to the precautions in the beginning of this section.
2. Relieve the fuel system pressure.
3. Raise and support the vehicle safely.
4. Drain the cooling system.
5. Remove or disconnect the following:
   - Negative battery cable
   - Engine under covers
   - Right front wheel
   - Engine side cover and lower the vehicle
   - Radiator
   - Intake manifold air duct
   - Drive belts, water pump pulley, alternator and power steering pump
   - Vacuum hoses, fuel hoses and wiring harness connectors
   - Spark plugs
   - Cylinder head cover and oil separator
   - Intake manifold supports
   - Oil filter bracket and the power steering oil pump bracket

6. Place the No. 1 piston at Top Dead Center (TDC) on the compression stroke.
   - Chain tensioner
   - Distributor. Do not turn the rotor while the distributor is removed
   - Timing chain guide
   - Camshaft sprockets
   - Camshafts, camshaft brackets, oil tubes and baffle plate
   - Starter
   - Heater hoses and the water hoses from the cylinder head
   - Knock sensor harness connector
   - Cylinder head outside bolts
   - Cylinder head with the intake and exhaust manifolds and raise and support the vehicle safely
   - Oil pans
   - Oil strainer and baffle plate
   - Crankshaft pulley and place a transmission jack under the main bearing beam and raise the engine slightly to take the weight off of the front engine mount
   - Front engine mount
   - Timing chain cover. Tap the seal out of the cover with a suitable seal driver
   - Timing chain sprocket bolts

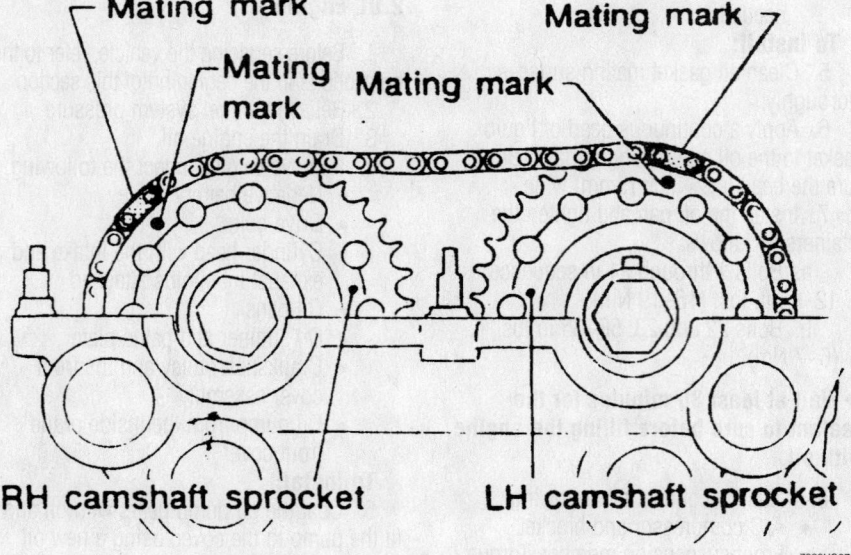

During disassembly, be sure to align the timing chain and camshaft sprocket mating marks—2.0L engine

7923HG67

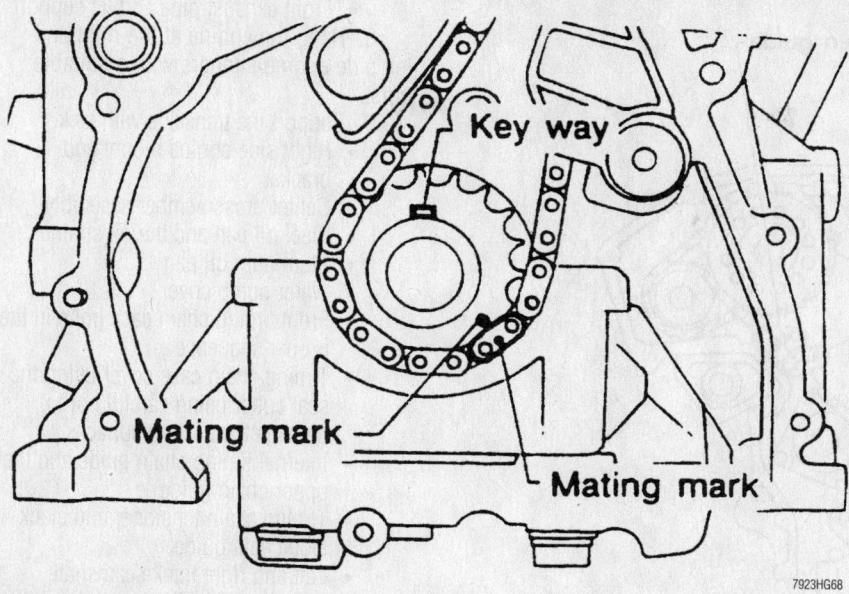

Crankshaft sprocket and timing chain alignment marks—2.0L engine

- Timing chain guides, timing chain and sprockets

To install:

7. Be sure all sealing surfaces are clean and prepared for assembly.

8. Install or connect the following:

- Crankshaft sprocket. Position the crankshaft so No. 1 piston is set at TDC (keyway at 12 o'clock, mating mark at 4 o'clock)

9. Fit the timing chain to crankshaft sprocket with the gold mating mark on the chain aligned with the mark on the sprocket. (The mating marks for the camshaft sprockets are silver).

- Timing chain guides and hang the chain off the left (front) guide. If necessary, secure the chain so it does not disengage from the crankshaft sprocket during assembly
- New seal in the front cover and apply engine oil to the lip of the seal and apply a bead of liquid gasket to the front cover
- Oil pump drive spacer and front cover. Torque the bolts evenly to 60 inch lbs. (6.7 Nm) and wipe away any excess liquid gasket
- Front engine mount
- Crankshaft pulley and temporarily tighten the bolt to hold the sprocket in place. The timing mark should align with the TDC mark
- Oil strainer, baffle plate and oil pan
- Cylinder head, camshafts, oil tubes and baffles. Position the left

camshaft key at 12 o'clock and the right camshaft key at 10 o'clock

- Camshaft sprockets by lining up the mating marks on the timing chain with the mating marks on the camshaft sprockets. Torque the camshaft bolts to 101–116 ft. lbs. (137–157 Nm) and the crankshaft pulley bolt to 105–112 ft. lbs. (142–152 Nm)
- Upper timing chain guide and distributor. Ensure that the rotor is at the 5 o'clock position

10. Before installing the chain tensioner, press the cam stopper down and the push in the sleeve until the hook can be engaged on the pin. When tensioner is bolted in position, the hook will release automatically. Ensure the arrow on the outside faces the front of the engine.

- Oil filter bracket and the power steering pump bracket
- Intake manifold supports
- Oil separator and the cylinder head cover
- Spark plugs
- Vacuum hoses, fuel hoses, and wiring harness connectors
- Alternator and power steering pump
- Water pump pulley
- Drive belts
- Radiator
- Engine under covers
- Right front wheel
- Negative battery cable

11. Fill and bleed the cooling system.

12. Start the vehicle, check for leaks and repair if necessary.

3.0L Engine

1. Before servicing the vehicle, refer to the precautions in the beginning of this section.

2. Drain the engine oil.

3. Drain the cooling system.

4. Relieve the fuel system pressure.

5. Remove or disconnect the following:

- Negative battery cable
- Left side ornament cover
- Air duct to intake manifold hose, collector hose, blow-by hose, and vacuum hoses
- Fuel hoses and the harness connections
- Evaporative emissions (EVAP) canister purge hoses
- Water hoses from the cylinder head and intake manifold
- Ignition coils from the spark plugs
- Exhaust Gas Recirculation (EGR) tube
- Intake manifold collector supports and the collector
- Bolts that secure the fuel tube and the fuel tube from the vehicle
- Bolts that secure the intake manifold to the engine block and the manifold. Loosen the bolts in the reverse sequence of the tightening procedure.
- Left-hand and right-hand rocker covers from the cylinder head
- Engine undercovers
- Right front wheel and the engine side covers
- Drive belts and the idler pulley
- Power steering oil pump belt and the power steering oil pump assembly
- Camshaft Position (CMP) sensor (PHASE) and Crankshaft Position (CKP) sensors (REF)/(POS)

6. Set the No. 1 piston to Top Dead Center (TDC) of compression stroke by rotating the crankshaft.

- Ring gear cover access plate

7. Loosen the crankshaft pulley bolt while securing the ring gear so the crankshaft cannot rotate.

➡**Use care not to damage the ring gear teeth.**

- Crankshaft pulley using a suitable puller
- Air conditioning compressor and bracket

Timing belt service is covered in Section 3 of this manual

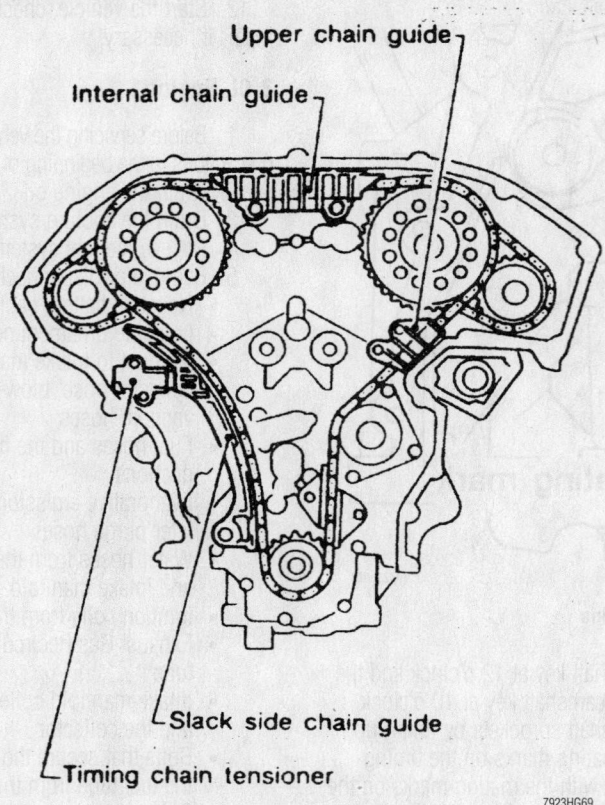

Timing chain tensioner and guide locations—3.0L engine

Upper chain guide

Internal chain guide

Slack side chain guide

Timing chain tensioner

7923HG69

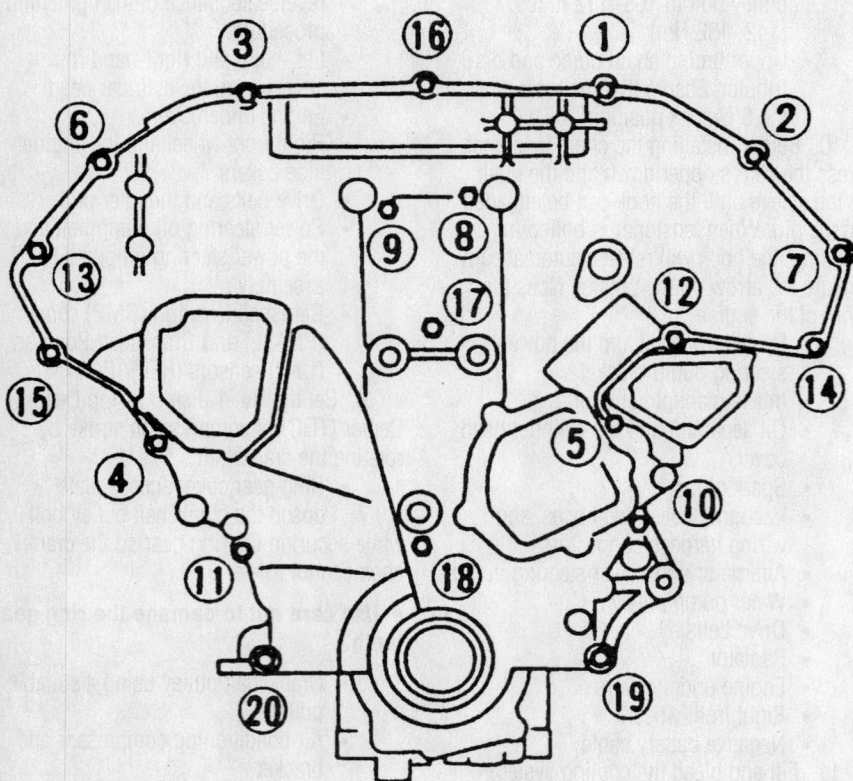

Front timing chain case bolt loosening sequence—3.0L engine

9307HG12

- Front exhaust pipe and its support

8. Hang the engine at the right and left side engine slingers with a suitable hoist.

9. Support the transaxle with jack.
- Right side engine mount and bracket
- Center crossmember assembly
- Steel oil pan and the oil strainer
- Aluminum oil pan
- Water pump cover
- Front timing chain case bolts in the proper sequence
- Timing chain case cover using the seal cutter being careful not to damage the sealing surfaces
- Internal timing chain guide and the upper chain guide
- Timing chain tensioner and slack side chain guide
- Left and right intake camshaft sprockets first. Be sure to hold the flats of the camshafts while removing the sprocket bolts
- Lower timing chain assembly. Be sure to note the aligning marks of the chain before removal

10. Insert a suitable stopper pin for the left and right camshaft tensioners.
- Left and right exhaust camshaft sprocket bolts. Be sure to hold the flats of the camshafts while removing the sprocket bolts
- Upper timing chain assembly. Be sure to note the aligning marks of the chain before removal
- Lower timing chain guide
- Crankshaft sprocket
- All traces of liquid gasket from the front timing chain case and from the water pump

11. Inspect the timing chain for excessive wear or damage and replace if necessary.

To install:

12. Install or connect the following:
- Crankshaft sprocket with the mating mark facing out

13. Position the crankshaft to TDC of compression stroke and align the dowels of the camshaft sprockets to the 12 o'clock position in respect to the cylinder head
- Lower timing chain guide. The front mark on the guide should face upwards

14. On a workbench, align the marks on the intake and exhaust camshaft sprockets with the marks of the chain.

15. Put the exhaust camshaft sprockets onto the dowel pin and torque the mounting bolts to 88–95 ft. lbs. (119–128 Nm). Be sure to secure the camshafts while tightening the bolts.

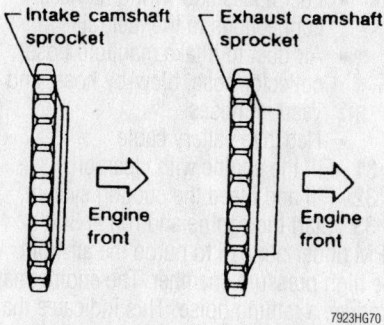

Identification of the intake and exhaust camshaft sprockets—3.0L engine

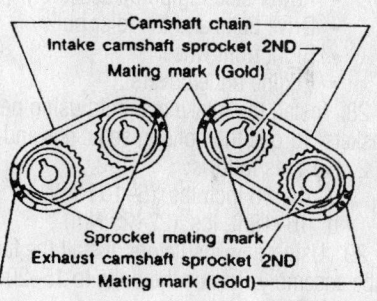

Upper timing chain alignment marks— 3.0L engine

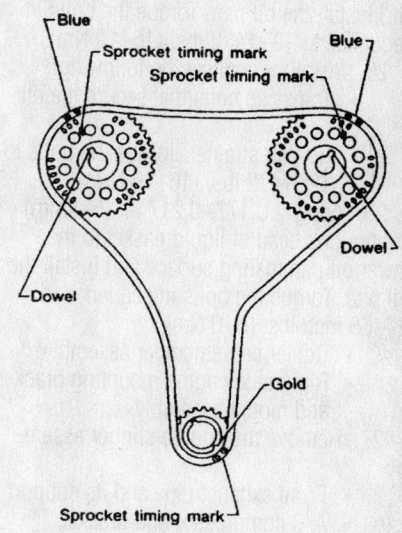

Lower timing chain alignment marks— 3.0L engine

- Timing chains and sprockets to the intake camshafts. Be sure to align the timing chain and sprocket mating marks

16. Remove the left and right camshaft tensioner stopper pins.

17. Align the mating mark on the crankshaft with the matchmark (gold link) on the lower timing chain.

- Lower timing chain to the water pump sprocket

18. Working counterclockwise, install the lower timing chain camshaft sprockets. Be sure to align the sprocket marks with the blue links of the timing chain during installation.

- Intake sprocket bolts. Torque the bolts to 88–95 ft. lbs. (119–128 Nm). Be sure to secure the camshafts while tightening the bolts
- Internal timing chain guide, upper timing chain guide, lower timing chain tensioner and slack side timing chain guide
- Torque the tensioner mounting bolt to 75–96 inch lbs. (8–11 Nm) and the guide bolts to 108–168 inch lbs. (13–19 Nm)

19. Apply a 0.102–0.142 in. (3–4mm) continuous bead of liquid gasket to all necessary areas on the front timing cover.

- Timing cover evenly and gently. Be sure to align the dowel pin holes

20. Torque the mounting bolts in sequence as follows:

   a. Bolts No. 1 and 2: 19–23 ft. lbs. (26–31 Nm).

   b. Bolts No. 3 to 20: 105–121 inch lbs. (12–14 Nm).

➡ **Leave the bolts unattended for 30 minutes or more after tightening. This will allow the liquid gasket to cure sufficiently.**

21. Apply a 0.091–0.130 in. (2–3mm) continuous bead of liquid gasket to the water pump cover and install the cover. Torque the bolts to 84–108 inch lbs. (10–13 Nm).

22. Apply a 0.12 in. (3mm) continuous bead of liquid gasket to the rocker covers and install the covers. Torque the mounting bolts in sequence as follows:

   a. Bolts No. 1 to 10: 9–26 inch lbs. (1–3 Nm).

   b. Bolts No. 1 to 10: 52–69 inch lbs. (6–8 Nm).

23. Apply sealant to the front and rear seal of the oil pan.

24. Apply a 0.177–0.217 in. (4.5–5.5mm) continuous bead of liquid gasket to the upper oil pan mating surface

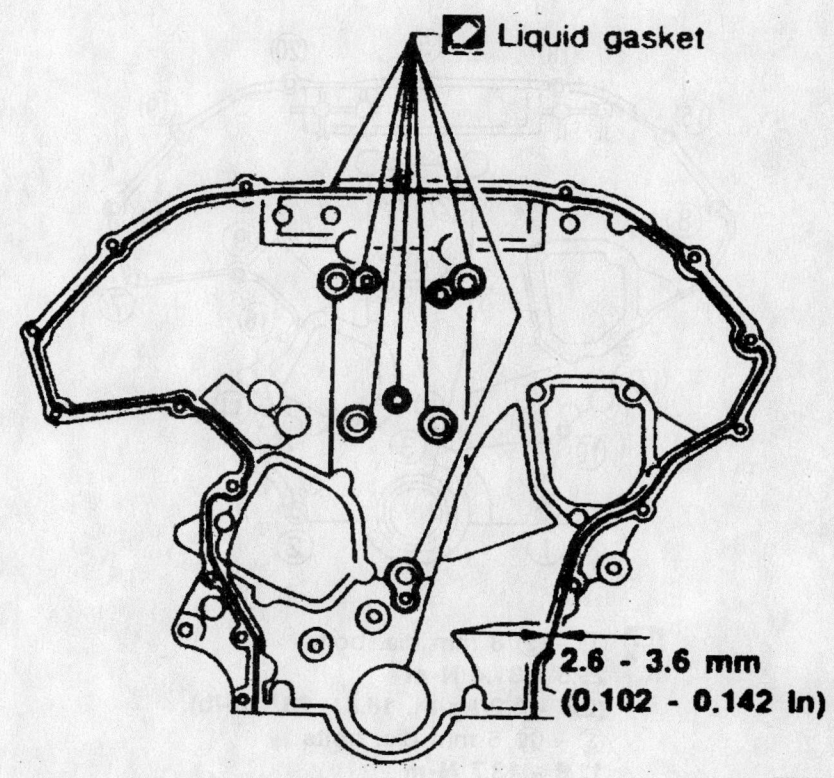

Application of liquid gasket to the front timing case—3.0L engine

Heater Core replacement is covered in Section 2 of this manual

and install the oil pan. Torque the bolts in sequence to 12–14 ft. lbs. (16–19 Nm).

25. Install or connect the following:
- Transaxle bolts that secure the oil pan
- Oil pan strainer. Torque the bolts to 12–14 ft. lbs. (16–19 Nm)

26. Apply a 0.177–0.217 in. (5–6mm) continuous bead of liquid gasket to the lower oil pan mating surface and install the oil pan. Torque the bolts in sequence to 57–66 inch lbs. (6–8 Nm)
- Center crossmember assembly
- Right side engine mounting bracket and mount assembly

27. Remove the engine slinger assembly.

- Front exhaust pipe and its support
- A/C compressor and bracket
- Crankshaft pulley to the crankshaft and the mounting bolt. Torque the mounting bolt to 14–22 ft. lbs. (20–29 Nm). Torque the crankshaft bolt an additional 60–66° clockwise. This is approximately the angle from 1 hexagon bolt head corner to another
- Ring gear cover plate
- CMP sensor (PHASE) and CKP sensors (REF)/(POS)

- Power steering pump assembly
- Drive belts and the idler pulley
- Right front wheel
- Engine undercovers

28. Install the intake manifold using new gaskets. Torque the bolts in sequence and in 2 stages as follows:
   a. 44–86 inch lbs. (5–10 Nm).
   b. 16–18 ft. lbs. (22–25 Nm).

29. Using new insulators, install the fuel tube assembly. Torque the bolts to 15–20 ft. lbs. (21–26 Nm).

30. Install or connect the following:
- Intake manifold collector gasket with the arrow facing forward
- Intake manifold collector. Torque the bolts to 16–18 ft. lbs. (22–25 Nm)
- Intake manifold collector support brackets
- EGR tube using new gaskets. Torque the bolts to 15–20 ft. lbs. (21–26 Nm) in 2 progressive steps
- Ignition coils. Torque the bolts to 27–33 inch lbs. (3–4 Nm)
- Rocker cover ornament on the left side
- Water hoses to the cylinder head and intake manifold
- EVAP canister purge hoses

- Fuel hoses and wiring harness connections to the fuel rail
- Air duct to intake manifold hose, collector hose, blow-by hose, and vacuum hoses
- Negative battery cable

31. Fill the engine with clean oil.
32. Fill and bleed the cooling system.
33. Start the engine and run at 3000 RPM under no load to purge the air from the high pressure chamber. The engine may produce a rattling noise. This indicates that air still remains in the chamber and is not a matter of concern.

4.1L Engine

1. Before servicing the vehicle, refer to the precautions in the beginning of this section.
2. Properly relieve the fuel system pressure.
3. Remove or disconnect the following:
- Negative battery cable
- Engine from the vehicle
- Alternator
- A/C compressor
- Exhaust manifold and place the engine on a suitable stand
- Intake manifold collector
- Injector harness and the fuel tube assembly with the injector

➡**Do not disassemble the fuel hose.**

- Intake manifold
- Valve cover

4. Set the No. 1 piston to Top Dead Center (TDC) on its compression stroke. Align the timing mark (orange paint) on the crankshaft pulley with the timing indicator on the front cover.

5. make sure the intake camshaft lobe for the No. 1 cylinder faces the intake port and the exhaust lobe faces the exhaust port.

➡**It is possible to check the camshaft positions by checking the notch positions on the camshaft when the No. 1 piston is at TDC of the compression stroke. At this position the cylinder head bolts can be removed.**

6. Remove or disconnect the following:
- Crankshaft pulley
- Crankshaft Position (CKP) sensor
- Upper front cover bolts and nuts in the sequence illustrated
- Front cover

7. Remove the upper chain tensioner by pressing the tensioner in and inserting a 0.04 inch (1mm) diameter pin in the pin

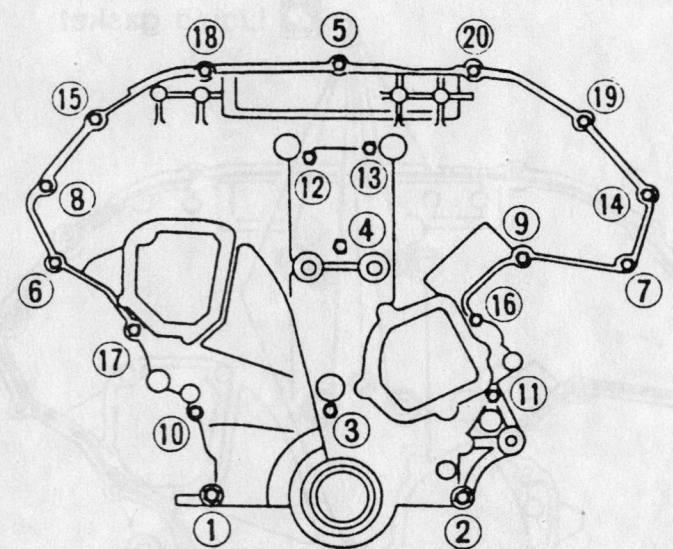

① - ② 8 mm dia. bolts
25.5 - 31.4 N·m
(2.6 - 3.2 kg-m, 18.8 - 23.1 ft-lb)
③ - ⑳ 6 mm dia. bolts
11.8 - 13.7 N·m
(1.2 - 1.4 kg-m, 8.7 - 10.1 ft-lb)

9307HG13

Front timing chain case bolt tightening sequence—3.0L engine

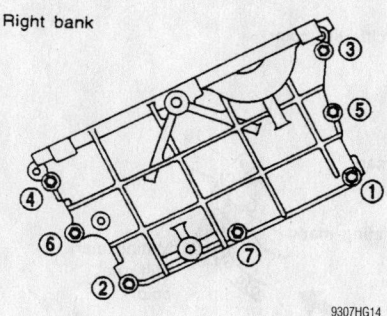

Right bank

Upper front cover (right bank) nut and bolt removal sequence—4.1L engine

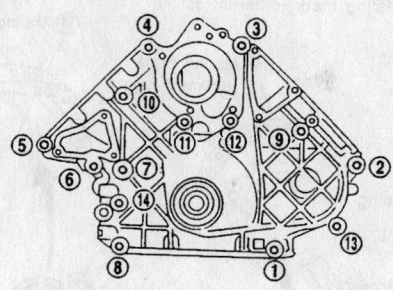

Remove the front cover bolts in the correct sequence—4.1L engine

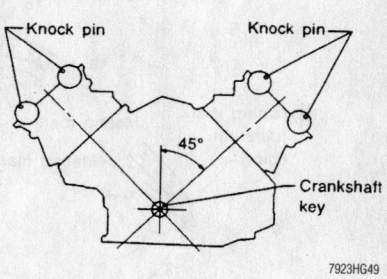

Knock pin   Knock pin   Crankshaft key

45°

Before assembly, be sure to turn the crankshaft key towards the left cylinder head—4.1L engine

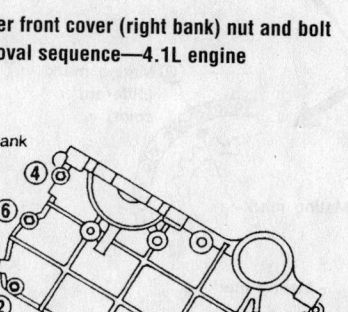

Left bank

Upper front cover (left bank) nut and bolt removal sequence—4.1L engine

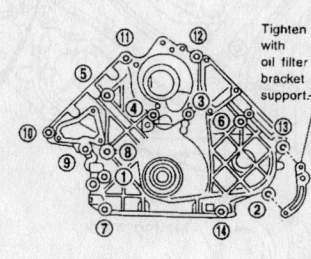

Tighten with oil filter bracket support

Position	Bolt dimensions	Tightening torque
①,③,⑤,⑦	M6 x 45	
②	M6 x 47	
⑥,⑩	M6 x 65	6.3 - 8.3 N·m (0.64 - 0.85 kg-m, 55.6 - 73.8 in-lb)
⑬	M6 x 67	
⑫	M6 x 84	
④,⑧	M8 x 50	16 - 21 N·m (1.6 - 2.1 kg-m, 12 - 15 ft-lb)
⑨	M10 x 52	
⑭	M10 x 60	30 - 40 N·m (3.1 - 4.1 kg-m, 22 - 30 ft-lb)
⑪	M10 x 62	

Front cover bolt location and torque specifications—4.1L engine

hole. Once secured remove the bolts and the tensioner

8. Matchmark the upper timing chain and camshaft sprockets to aid reassembly.

- Upper camshaft sprocket bolt while holding the hexagonal part of the camshaft with a wrench
- Camshaft sprockets

9. Remove the left and right tensioner covers from the front covers by pressing the tensioner in and inserting a 0.04 inch (1mm) diameter pin in the pin hole.

- Idler sprocket bolts
- Chain guide between the No. 1 camshaft bracket
- Cylinder head
- Upper timing chain

10. Matchmark the lower timing chain and idler sprocket to aid reassembly.

- Oil pan
- Front cover bolts in the proper sequence

11. Compress the lower chain tensioners and install a pin through the hole to secure it, then remove the tensioner.

- Oil pump drive chain

- Slack and chain guides. Be sure to note the locations of the bolts so they can be installed in their original positions
- Lower timing chains with the crankshaft sprockets

To install:

12. Be sure the crankshaft key is pointing toward the center of the left bank. This should be a 45° angle from the center.

13. Install or connect the following:
- Lower right bank timing chain by aligning the mark on the chain

with the mark on the sprocket and installing the sprocket with chain on the crankshaft. Be sure the thick side of the sprocket faces the cylinder block to provide clearance between the block and chain

- Slack and chain guides. Be sure to install the bolts in the correct locations. Torque the bolts to 10–14 ft. lbs. (13–19 Nm)
- Lower chain for the left bank in the same manner as the right one

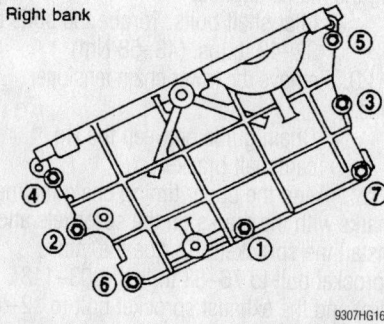

Right bank

Upper front cover (right bank) nut and bolt tightening sequence—4.1L engine

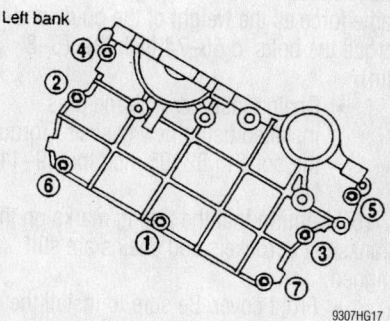

Left bank

Upper front cover (left bank) nut and bolt tightening sequence—4.1L engine

Brake service is covered in Section 4 of this manual

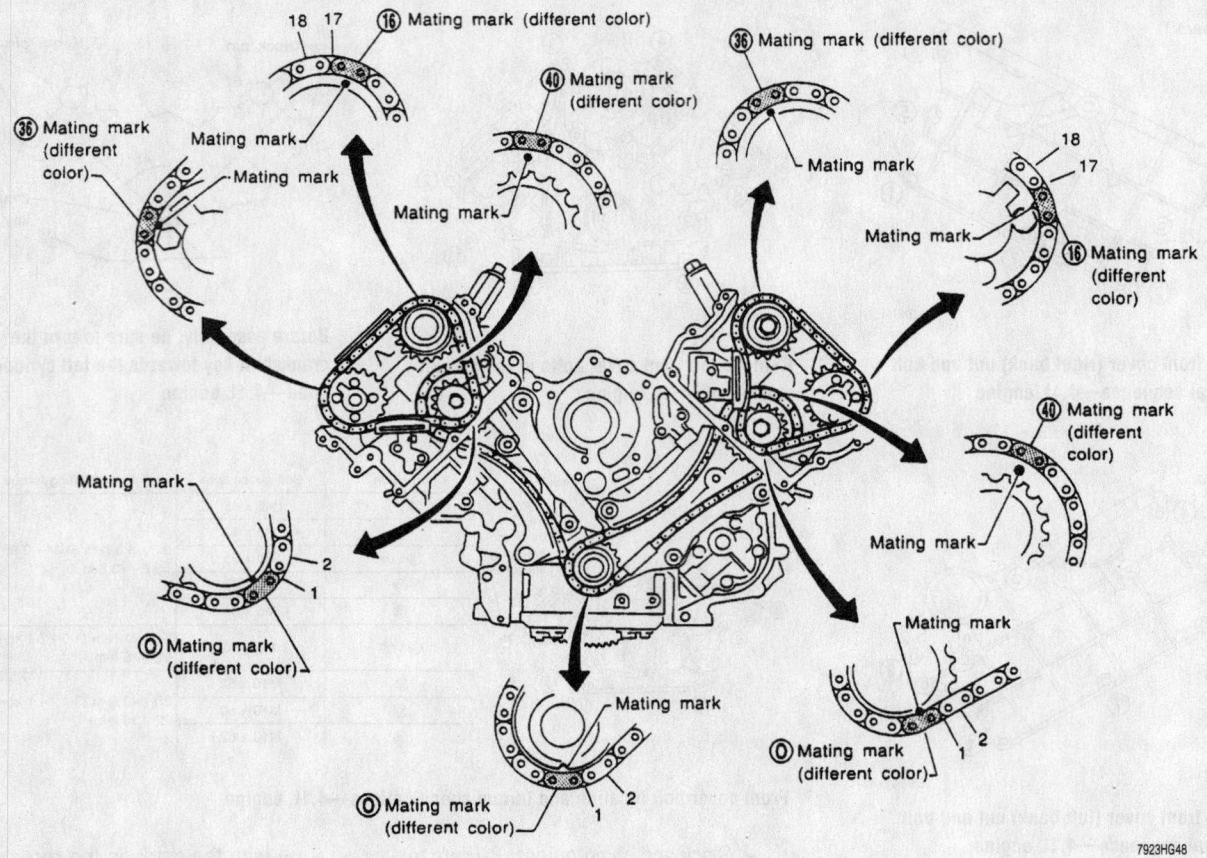

Timing chain and sprocket alignment marks—4.1L engine

- Left slack and chain guide. Torque the bolts to 10–14 ft. lbs. (13–19 Nm)
- Oil pump drive chain and sprockets. Torque the bolt on the driven gear to 22–30 ft. lbs. (30–40 Nm)
- Lower oil pump drive chain guide

14. Install the upper guide by installing the bolts loosely, then inserting a 0.04 in. (1mm) feeler gauge between the guide and chain. Press on the guide lightly with the same force as the weight of the guide and torque the bolts to 56–74 inch lbs. (6–8 Nm).

- Chain tensioner with the pins installed using new gaskets. Torque the bolts to 82–95 inch lbs. (9–11 Nm)

15. Confirm that the timing marks on the crankshaft sprockets and chains are still aligned.

- Front cover. Be sure to install the bolts in the correct locations

16. Apply engine oil to the idler shaft and install it on the idler sprocket.

17. Align the mark on the chain with the mark on the idler sprocket and install the sprocket.

- Place the upper chains on the idler sprockets. It is not necessary to align the mating marks at this time. The marks can be aligned after the cylinder head is installed

18. Install the cylinder heads

19. Align the marks on the upper chains with the marks on the sprockets, then install the sprockets on the camshafts while keeping the marks aligned.

- Idler shaft bolts. Torque the bolts to 32–43 ft. lbs. (43–58 Nm)

20. Remove the lower chain tensioner pins.

- Chain guide between the No. 1 camshaft bracket

21. Align the upper timing chain mating marks with the marks on the sprockets and install the sprockets. Torque the intake sprocket bolt to 76–83 ft. lbs. (103–113 Nm) and the exhaust sprocket bolt to 12–15 ft. lbs. (16–21 Nm)

22. Compress the upper chain tensioners and install a pin through it to secure it in position.

- Tensioners. Torque the bolts to 82–95 inch lbs. (9–11 Nm)

23. Lubricate the timing chains and related parts with clean engine oil and install the upper covers. Torque the cover bolts in sequence.

- All remaining parts in the reverse of removal

Piston and Ring

POSITIONING

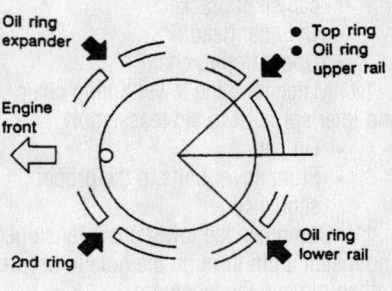

Infiniti engines—piston ring end-gap spacing

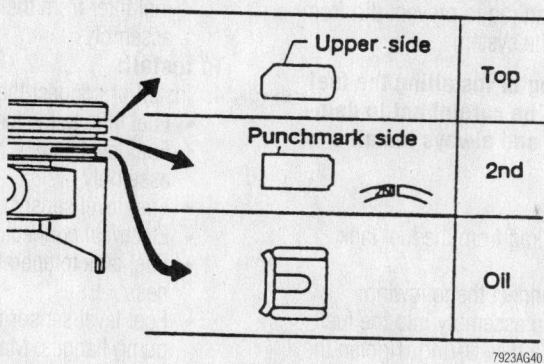

Infiniti 2.0L, 3.0L and 4.1L engines—piston ring positioning

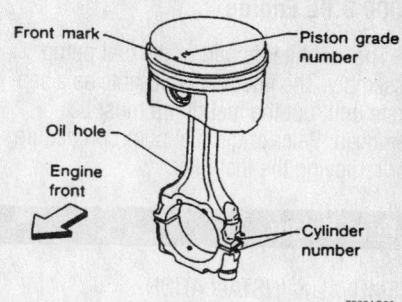

Infiniti 2.0L, 3.0L and 4.1L engines—piston/connecting rod assembly-to-engine orientation

FUEL SYSTEM

Fuel System Service Precautions

Safety is the most important factor when performing not only fuel system maintenance but any type of maintenance. Failure to conduct maintenance and repairs in a safe manner may result in serious personal injury or death. Maintenance and testing of the vehicle's fuel system components can be accomplished safely and effectively by adhering to the following rules and guidelines.

1. To avoid the possibility of fire and personal injury, always disconnect the negative battery cable unless the repair or test procedure requires that battery voltage be applied.

2. Always relieve the fuel system pressure prior to disconnecting any fuel system component (injector, fuel rail, pressure regulator, etc.), fitting or fuel line connection. Exercise extreme caution whenever relieving fuel system pressure, to avoid exposing skin, face and eyes to fuel spray. Please be advised that fuel under pressure may penetrate the skin or any part of the body that it contacts.

3. Always place a shop towel or cloth around the fitting or connection prior to loosening to absorb any excess fuel due to spillage. Ensure that all fuel spillage (should it occur) is quickly removed from engine surfaces. Ensure that all fuel soaked cloths or towels are deposited into a suitable waste container.

4. Always keep a dry chemical (Class B) fire extinguisher near the work area.

5. Do not allow fuel spray or fuel vapors to come into contact with a spark or open flame.

6. Always use a back-up wrench when loosening and tightening fuel line connection fittings. This will prevent unnecessary stress and torsion to fuel line piping. Always follow the proper torque specifications.

7. Always replace worn fuel fitting O-rings with new. Do not substitute fuel hose where fuel pipe is installed.

Fuel System Pressure

RELIEVING

1. Before servicing the vehicle, refer to the precautions in the beginning of this section.

2. Remove the fuel pump fuse.

3. Start the engine.

4. Allow the engine to run until it stalls.

5. After the engine stalls, crank the engine 2 or 3 times to release the remaining fuel pressure.

6. Turn the ignition switch **OFF**. Reinstall the fuel pump fuse into the fuse block.

➡**Do not crank the engine or turn the ignition switch ON after the fuel pump fuse has been reinstalled, or the fuel pressure will be reestablished.**

Fuel Filter

REMOVAL & INSTALLATION

All Except 2000 3.0L Engines

❄❄ CAUTION

Do not use conventional fuel filters, hoses or clamps when servicing this fuel system. They are not compatible with the high pressures of the injection system and could fail, causing personal injury. Use only components specifically designed for fuel injection.

1. Before servicing the vehicle, refer to the precautions in the beginning of this section.

2. Relieve the fuel system pressure.

3. Remove or disconnect the following:
   • Negative battery cable
   • Fuel hoses from the fuel filter, located at the right side of the engine compartment
   • Filter mounting screws
   • Filter from the vehicle

To install:

4. Inspect all hoses and clamps for damage of any type. Replace parts, as required.

➡**The fuel filters are directional and should be installed with the arrow facing the direction of fuel flow.**

5. Install or connect the following:
   • New filter in the bracket and install new hose clamp
   • Negative battery cable

6. Start the vehicle, check for fuel leaks and repair if necessary.

➡**On some vehicles, a code will be set and/or the check engine light will remain on after starting the vehicle. This is because a code was set for an open fuel pump circuit when the fuel pressure was released. If you did not disconnect the negative battery cable during this procedure, do it now so the code will be erased. The negative battery cable should be disconnected for at least 1 minute. Also, remember to reset the clock and radio stations when finished.**

For complete Engine Mechanical specifications, see Section 1 of this manual

2000 3.0L Engine

The fuel filter is part of the fuel pump assembly. The filter is serviceable as a separate unit, but the fuel pump must be removed. Refer to the fuel pump procedure for removing the fuel filter.

Fuel Pump

REMOVAL & INSTALLATION

2.0L Engine

1. Before servicing the vehicle, refer to the precautions in the beginning of this section.
2. Release the fuel system pressure.
3. Remove or disconnect the following:
   - Negative battery cable
   - Rear seat back and bottom
   - Inspection hole cover located beneath the rear seat
   - Connectors and fuel tubes
   - Six screws
   - Fuel pump/gauge assembly and disconnect the tubes and connector
   - Fuel pump by sliding it out on an angle

To install:
4. Install or connect the following:
   - Fuel pump/gauge assembly
   - All fuel lines and connectors
   - Six screws
   - Negative battery cable
5. Start the vehicle, check for leaks and repair if necessary.
6. Install the inspection cover.
7. Rear seat back and bottom.

3.0L Engine 1998–99

1. Before servicing the vehicle, refer to the precautions in the beginning of this section.
2. Relieve the fuel system pressure
3. Remove or disconnect the following:
   - Negative battery cable
   - Access panel under the rear seat

➡**If the vehicle has no fuel pump access cover, the fuel tank must be lowered or removed to gain access to the in-tank fuel pump.**

   - Fuel gauge electrical connector and pump electrical connector
   - Fuel outlet and the return hoses
   - Fuel pump assembly-to-fuel tank bolts and lift the fuel pump assembly from the fuel tank. Discard the O-ring. Plug the fuel tank opening

with a clean rag to prevent dirt from entering the system

➡**When removing or installing the fuel pump assembly, be careful not to damage or deform it and always install a new O-ring.**

To install:
4. Remove the rag from the fuel tank opening.
5. Install or connect the following:
   - Fuel pump assembly into the fuel tank with a new O-ring. Tighten the bolts to 17–23 inch lbs. (2.0–2.5 Nm)
   - Fuel lines and the electrical connectors. Always use new clamps when reconnecting fuel line hoses.
   - Fuel pump access cover
   - Negative battery cable
6. Start the engine, check for fuel leaks and repair if necessary.

➡**On some models, the Check Engine Light will stay ON after installation is completed. The memory code in the control unit must be erased. This code is stored for an open fuel pump circuit, this is caused when the fuel pressure is released. To erase the code, disconnect the battery cable for 10 seconds, then reconnect after installation of fuel pump.**

2000 3.0L Engine

1. Before servicing the vehicle, refer to the precautions in the beginning of this section.
2. Relieve the fuel system pressure
3. Remove or disconnect the following:
   - Negative battery cable
   - Rear seat bottom
   - Inspection hole cover from under the rear seat
   - Electrical and quick connectors
   - Six screws
   - Fuel level sensor/fuel pump assembly

➡**If replacement of the fuel filter is required, proceed with the following steps.**

   - Fuel level sensor/fuel pump assembly flange
   - Fuel tank temperature sensor harness
   - Fuel level sensor flange and raise the fuel level sensor
   - Fuel pump electrical connector
   - Quick connectors from the fuel level sensor
   - Fuel level sensor from the assembly
   - Snap fit connectors and remove the

fuel filter from the fuel pump assembly

To install:
4. Install or connect the following:
   - Fuel filter to the fuel pump assembly
   - Fuel level sensor to the fuel pump assembly
   - Fuel level sensor quick connectors
   - Electrical connectors
   - Fuel tank temperature sensor harness
   - Fuel level sensor unit and fuel pump flanges. Make certain that they snap together
   - Fuel pump assembly to the fuel tank
   - Six screws
   - Quick and electrical connectors
   - Negative battery cable
5. Start the vehicle, check for leaks and repair if necessary.
6. Install the inspection hole cover plate.
7. Install the rear seat bottom.

4.1L Engine

1. Before servicing the vehicle, refer to the precautions in the beginning of this section.
2. Relieve the fuel system pressure.
3. Remove or disconnect the following:
   - Negative battery cable
   - Trunk front finish panel
   - Wiring harness connector and fuel tubes
   - Fuel tank sender unit attaching bolts
   - Fuel tank sender and discard the O-ring
   - Fuel pump from the sender unit

To install:
4. Install or connect the following:
   - New fuel pump on the sender unit assembly
   - Sender unit in the fuel tank, using a new O-ring. Torque the bolts to 17–23 inch lbs. (2–2.5 Nm)
   - Wiring harness connectors and fuel tubes
   - Trunk room finish panel
   - Negative battery cable
5. Start the engine, check for leaks and repair if necessary.

Fuel Injector

REMOVAL & INSTALLATION

1. Before servicing the vehicle, refer to the precautions in the beginning of this section.

2. Relieve the fuel system pressure

3. Remove or disconnect the following:

- Negative battery cable
- Intake manifold collector
- Vacuum hose from the fuel pressure regulator
- Fuel hoses from fuel rail
- Fuel rail bolts
- Injector harness connectors
- Injectors and the fuel rail as an assembly

- Injector(s) from the fuel rail by pushing them out

4. Remove and discard the fuel injector O-rings

To install:

5. Lubricate the new O-rings with clean engine oil and install the O-rings on the injector(s).

6. Install or connect the following:
- Fuel injectors to the fuel rail
- Fuel rail and injectors as an assembly to the intake manifold

7. Tighten the fuel rail bolts in the following sequence;
   a. Step 1: 7–8 ft. lbs. (9–10 Nm).
   b. Step 2: 15–20 ft. lbs. (21–26 Nm).
- Fuel hoses to the fuel rail
- Vacuum hose to the fuel pressure regulator
- Intake manifold collector
- Negative battery cable

8. Start the vehicle, check for leaks and repair if necessary.

DRIVE TRAIN

Transmission Assembly

REMOVAL & INSTALLATION

Manual

G20 MODELS

1. Before servicing the vehicle, refer to the precautions in the beginning of this section.

2. Drain the transaxle fluid.

3. Remove or disconnect the following:
- Negative battery cable
- Air cleaner and air duct assembly
- Clutch operating cylinder from the transaxle
- Back-up light switch, neutral switch and ground harness connectors
- Speedometer sensor
- Starter
- Air bleeder hose
- Shift control and support rods
- Front exhaust tube
- Halfshafts and support the engine with a suitable jack under the oil pan
- Rear and left engine mount

4. Raise the jack and remove the lower transaxle housing bolts. Lower jack and remove the upper housing bolts. Keep the bolts in order as they are different lengths and must be returned to the same position.

5. Lower the transaxle.

To install:

6. Raise the transaxle into place and install the attaching bolts. Torque the shortest bolt to 22–30 ft. lbs. (30–40 Nm) and the remaining bolts to 51–59 ft. lbs. (70–79 Nm).

7. Install or connect the following:
- Rear and left engine mounts
- Driveshafts

- Shift control rods, support rod, bleeder air hose and starter
- Air bleeder hose
- Starter
- Speedometer sensor
- Back-up light switch, neutral switch and ground harness connectors
- Clutch operating cylinder. Torque the bolts to 22–27 ft. lbs. (29–37 Nm)
- Negative battery cable

8. Fill the transaxle to the proper level.

9. Start the vehicle, check for leaks and repair if necessary.

I30 MODELS

1. Before servicing the vehicle, refer to the precautions in the beginning of this section.

2. Drain the transaxle fluid.

3. Remove or disconnect the following:
- Battery, battery bracket and tray
- Air cleaner assembly with the Mass Air Flow (MAF) sensor
- Clutch operating cylinder; do not disconnect the hydraulic line from the cylinder
- Clutch hose clamp
- Speedometer pinion and the neutral position switch connectors and the ground harness connectors
- Starter
- Back-up lamp switch and the neutral position switch
- Crankshaft Position (CKP) sensor (POS) from the transaxle front side
- Shifter control rod and the support rod bracket from the transaxle
- Both driveshafts from the transaxle assembly and support the transaxle

4. Support the engine of the transaxle by placing a jack under the oil pan. Be sure to use a block of wood between the oil pan and jack.

- Bolts that secure the center crossmember
- Left-hand engine mounts

➡**The transaxle bolts are of different lengths, be sure to note the location of the bolts for reassembly.**

- Transaxle bolts
- Transaxle from the vehicle by sliding the transaxle input shaft out of the clutch, lowering the rear of the transaxle, then lowering the transaxle from of the vehicle

To install:

5. Install or connect the following:
- Transaxle assembly to the bell housing while aligning the output shaft of the transaxle with the clutch disc. Tighten the transaxle bolts to the specifications illustrated
- Left-hand engine mount. Torque the through-bolt to 32–41 ft. lbs. (43–55 Nm.)
- Center crossmember assembly. Torque the bolts to 57–72 ft. lbs. (77–98 Nm)
- Both driveshafts to the transaxle assembly
- Shifter control rod and the support rod bracket to the transaxle
- CKP sensor to the transaxle front side
- Back-up lamp switch and the neutral position switch
- Starter motor assembly to the transaxle
- Speedometer pinion and the ground harness connectors
- Clutch hose clamp
- Clutch operating cylinder. Torque the bolts to 22–30 ft. lbs. (30–40 Nm)
- Air cleaner assembly with the MAF sensor

For Accessory Drive Belt illustrations, see Section 1 of this manual

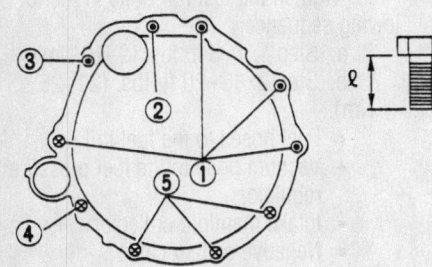

Bolt No.	Tightening torque N·m (kg-m, ft-lb)	"ℓ" mm (in)
①	70 - 79 (7.1 - 8.1, 51 - 59)	52 (2.05)
②	70 - 79 (7.1 - 8.1, 51 - 59)	65 (2.56)
③	70 - 79 (7.1 - 8.1, 51 - 59)	124 (4.88)
④	35.1 - 47.1 (3.58 - 4.80, 25.89 - 34.74)	40 (1.57)
⑤	35.1 - 47.1 (3.58 - 4.80, 25.89 - 34.74)	40 (1.57)

◉ M/T to engine
⊗ Engine to M/T

③ with starter
④ with support rod bracket

9307HG31

Manual transaxle-to-engine bolt torque sequence and specifications—I30 models

- Battery tray, bracket and battery
- Positive and the negative battery cables

6. Fill the transaxle with proper amount and type of fluid.

7. Start the vehicle, check for leaks and repair if necessary.

Automatic

G20 MODELS

1. Before servicing the vehicle, refer to the precautions in the beginning of this section.

2. Drain the transaxle fluid.

3. Remove or disconnect the following:

- Battery and bracket
- Air duct assembly
- Transaxle solenoid harness connector, Park Neutral Position (PNP) switch connector and revolution switch connector
- Crankshaft Position (CKP) sensor from the transaxle
- Control cable and transaxle coolant lines
- Halfshafts, the intake manifold support bracket and the starter
- Upper bolts attaching transaxle to the engine

4. Support the engine with a suitable stand and use a suitable jack to support the transaxle.

➡**Bolts are of different lengths, note the locations that the bolts are removed from.**

- Center member
- Rear cover plate and the bolts securing the torque converter to the driveplate. Rotate the crankshaft to gain access to the bolts
- Transaxle mounts
- Lower transaxle mounting bolts and lower the transaxle

To install:

5. Place a straightedge across the bell housing of the transaxle and measure the distance to the mounting bosses on the torque converter. The distance should be 0.626 in. (16mm). If not, the torque converter is not installed correctly.

6. Check the driveplate run-out with a dial indicator. Maximum allowable run-out is 0.008 in. (0.2mm).

7. Raise the transaxle into position and install the transaxle mounting bolts. Tighten the 50, 55, and 65mm long bolts to 51–59 ft. lbs. (70–79 Nm). Torque the 35 and 45 mm long bolts to 12–15 ft. lbs. (16–21 Nm).

8. Install or connect the following:

- Torque converter bolts. Torque the bolts to 33–43 ft. lbs. (44–59 Nm). Rotate the crankshaft to gain access to the bolts
- Rear cover and center member
- Transaxle mounts
- Halfshafts, intake manifold support bracket and the starter
- Control cable and transaxle coolant lines
- CKP sensor to the transaxle
- Transaxle solenoid harness connector, PNP switch connector and revolution switch connector
- Air duct assembly
- Battery and bracket

9. Fill the transaxle with fluid.

10. Start the vehicle, check for leaks and repair if necessary.

I30 MODELS

➡**The radio may contain a coded theft protection circuit. Always obtain the code number from the customer before disconnecting the battery.**

1. Before servicing the vehicle, refer to the precautions in the beginning of this section.

2. Drain the transaxle fluid.

3. Remove or disconnect the following:

- Battery and bracket
- Air cleaner and resonator
- Terminal cord assembly harness connector and Park Neutral Position (PNP) switch harness connector
- Revolution and Vehicle Speed Sensor (VSS) electrical connections
- Crankshaft Position (CKP) sensor from the transaxle
- Left-hand mounting bracket from transaxle and body
- Control cable from the transaxle
- Driveshafts
- Oil cooler pipes and cap pipes to avoid contamination
- Starter motor and place a jack under the oil pan to support the engine. Do not place the jack under the oil pan drain plug
- Crossmember
- Rear cover plate and bolts attaching the torque converter to the drive plate

4. Support the transaxle with a jack.

5. Remove the transaxle-to-engine bolts and lower the transaxle using the jack.

To install:

6. Install or connect the following:

- Torque converter in the transmission. Be sure the torque converter is fully seated in the front pump assembly. The distance from the front edge of the transmission to the bolt hole of the torque converter should be 0.75 in. (19mm)
- Position the transmission to the engine and install a few bolts to hold the transmission in place. Do not fully tighten the bolts at this time
- Torque converter-to-flexplate bolts. Torque the bolts to 33–43 ft. lbs. (44–59 Nm)

7. Secure the transmission to the

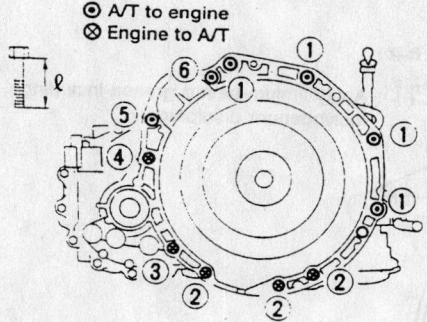

⊙ A/T to engine
⊗ Engine to A/T

Bolt No.	Tightening torque N·m (kg-m, ft-lb)	ℓ mm (in)
①	39 - 49 (4.0 - 5.0, 29 - 36)	45 (1.77)
②	30 - 36 (3.1 - 3.7, 22 - 27)	30 (1.18)
③	30 - 36 (3.1 - 3.7, 22 - 27)	40 (1.57)
④	74 - 83 (7.5 - 8.5, 54 - 61)	45 (1.77)
⑤	30 - 36 (3.1 - 3.7, 22 - 27)	80 (3.15)
⑥	30 - 36 (3.1 - 3.7, 22 - 27)	65 (2.56)

9307HG19

Transaxle bolts tightening sequence and torque specifications—1998–00 I30 models with automatic transaxle

engine. Torque the bolts in the sequence illustrated

8. install all remaining components in the reverse order of removal.

9. Fill the transmission with new fluid. Use the same amount of fluid that was drained before removal.

10. Connect negative battery cable and start the engine. Allow the engine to reach normal operating temperature and check the transmission fluid level. Add fluid as needed.

Q45 Models

1. Before servicing the vehicle, refer to the precautions in the beginning of this section.

2. Remove or disconnect the following:
- Negative battery cable
- Crankshaft Position (CKP) sensor
- Rear Heated O₂S sensor connector
- Exhaust tubes
- Fluid charging pipe
- Oil cooler lines. Plug fluid charging and oil cooler fittings after removing lines.
- Control linkage from the selector lever
- Neutral safety switch and solenoid harness connectors
- Speed sensor connection
- Driveshaft (make matchmarks to ease in installation). Insert plug into rear seal opening to prevent loss of fluid.

3. Support the transmission safely.
- Bolts securing the torque converter to the flexplate
- Gussets securing the transmission to the engine
- Bolts attaching the transmission to the engine

➡ **The bolts securing the transmission to the engine are of different lengths. Note the length of the bolts as they are removed.**

4. Support the engine safely. Avoid jacking directly under the oil pan drain plug.

5. Remove the transmission from the vehicle.

To install:

6. Install or connect the following:
- Transmission in the vehicle and install the torque converter-to-flex-plate bolts. Torque the bolts to 33–43 ft. lbs. (44–59 Nm)

7. Secure the transmission to the engine. Torque the bolts as follows:
   a. 70mm bolts: 80–87 ft. lbs. (108–118 Nm).
   b. 90mm bolts: 51–58 ft. lbs. (69–78 Nm).

8. Install the torque converter-to-drive plate bolts and torque in 2 steps to 33–43 ft. lbs. (44–59 Nm).
- Driveshaft, aligning the matchmarks made before removal

9. Install or connect the following:
- Speed sensor connection
- Neutral safety switch and solenoid harness connectors
- Control linkage to the selector lever
- Fluid charging and oil cooler lines
- Exhaust tubes

10. Lower the vehicle.

11. Connect negative battery cable.

12. Start the vehicle, check for leaks and repair if necessary.

Clutch

ADJUSTMENT

All models are equipped with a hydraulic clutch, which is self-adjusting.

REMOVAL & INSTALLATION

G20 Models

1. Before servicing the vehicle, refer to the precautions in the beginning of this section.

2. Raise and support the vehicle safely.

3. Remove or disconnect the following:
- Negative battery cable
- Transaxle

4. Insert alignment tool KV30101000 into the clutch disc hub and loosen the pressure plate bolts in small increments using a star-type pattern.
- Pressure plate and clutch disc as an assembly
- Release bearing by pulling the bearing retainers outward from the transaxle case

5. Inspect the clutch disc for surface wear. Measure from the friction surface to the top of the rivets. Wear limit is 0.012 in. (0.3mm). Replace clutch disc as necessary.

6. Inspect the contact surface of the flywheel for burns or discoloration. Check flywheel run-out. Maximum run-out is 0.0059 in. (0.15mm).

7. Using tools ST20050100 and ST20050010, check the pressure plate diaphragm springs. Measure from the pressure plate/flywheel mating surface to the top of the diaphragm spring. Height should be 1.201–1.280 in. (30.5–32.5mm). Replace pressure plate as necessary.

8. Inspect the release bearing for damage. Spin the bearing to see that it rolls freely.

To install:

9. Lightly lubricate the transaxle input shaft, input shaft collar, clutch lever assembly and the clutch release bearing with a lithium based grease.

For Wheel Alignment specifications, see Section 1 of this manual

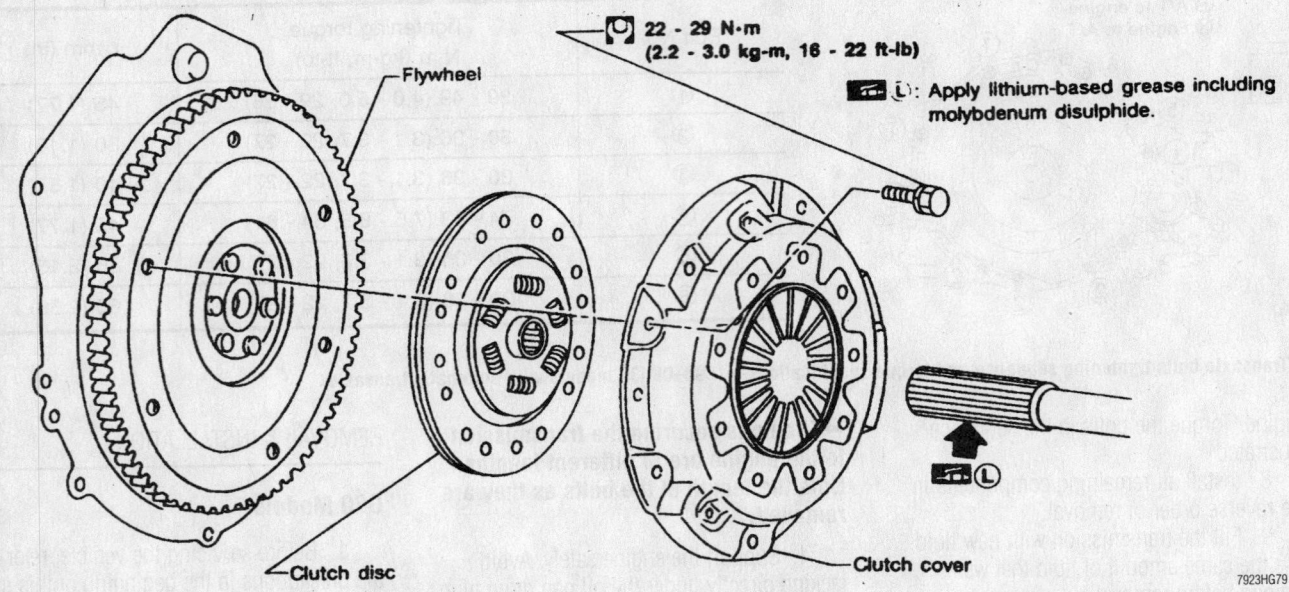

22 - 29 N·m
(2.2 - 3.0 kg-m, 16 - 22 ft-lb)

L: Apply lithium-based grease including molybdenum disulphide.

7923HG79

Clutch disc and pressure plate—G20 models

➡**Keep clutch disc and all clutch components clean during installation. Do not allow grease to contact the clutch disc.**

10. Insert alignment tool KV30101000 into the clutch disc hub. Install the clutch disc and pressure plate on the tool and torque the pressure plate bolts to 16–22 ft. lbs. (22–29 Nm) in 2–3 steps using a criss-cross pattern. Remove the tool.

11. Install or connect the following:
- Release bearing in the transaxle. Ensure that the bearing retainer clips are fully engaged
- Transaxle
- Negative battery cable

12. If necessary, adjust clutch pedal height and free-play.

I30 Models

1. Before servicing the vehicle, refer to the precautions in the beginning of this section.

2. Drain the transaxle fluid.

3. Remove or disconnect the following:
- Battery and battery bracket
- Air cleaner and the air flow meter

4. Raise and safely support the vehicle so there is clearance to remove the transaxle from underneath. Securely support the engine via the oil pan using a cushioning wooden block and jack.
- Transaxle from the engine and lower to the floor

5. Insert a clutch aligning bar or similar tool all the way into the clutch disc hub. This must be done so as to support the

weight of the clutch disc during removal. Mark the clutch assembly-to-flywheel relationship with paint or a center punch so the clutch assembly can be assembled in the same position from which it is removed.
- Bolts in reverse order of tightening sequence, a turn at a time
- Pressure plate and clutch disc
- Release mechanism from the transaxle housing

6. Inspect the pressure plate for wear, scoring, etc., and resurface or replace, as necessary.

7. Measure the thickness of the clutch plate lining to the rivet heads; if the it is worn to a minimum of 0.012 in. (0.3mm), replace the clutch plate.

8. Inspect the release bearing and replace as necessary.

9. Using a dial indicator, mount it to the engine and inspect the flywheel run-out; if the run-out exceeds 0.0059 in. (0.15mm), replace it.

To install:

10. Apply a small amount of grease to the transaxle input shaft splines.

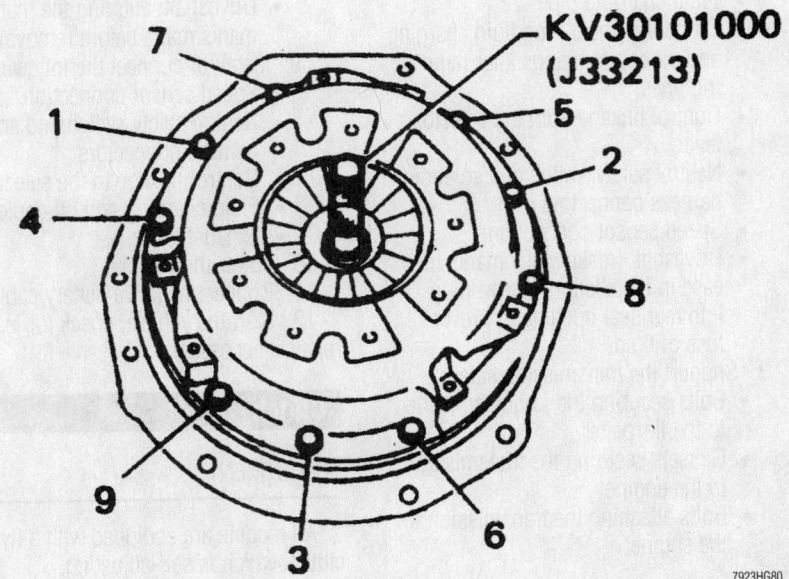

KV30101000 (J33213)

7923HG80

Tighten the pressure plate bolts according to the sequence shown—I30 models

11. Install the disc on the splines and slide back and forth a few times. Remove the disc and remove excess grease on hub. Be sure no grease contacts the disc or pressure plate.

12. Apply lithium based molybdenum disulfide grease to the bearing sleeve inside groove, the contact point of the withdrawal lever and bearing sleeve, the contact surface of the lever ball pin and lever.

13. Install or connect the following:
- Release mechanism and release bearing
- Pressure plate and clutch disc, aligning it with a splined dummy shaft tool KV301010000
- Torque the pressure plate bolts in sequence as follows:
a. Step 1: torque in sequence to: 7–14 ft. lbs. (10–20 Nm).
b. Step 2: torque in sequence to: 25–33 ft. lbs. (34–44 Nm).

14. Remove the dummy shaft.
- Transaxle in the correct position. Tighten the transaxle-to-engine bolts
- Rear and left-hand mounts
- Speedometer cable
- Electrical harness connector
- Clutch release cylinder
- Starter assembly

15. Securely support the transaxle and install the driveshafts.

16. Fill the transaxle with the required amount of approved fluid.
- Air flow meter and the air cleaner
- Battery and battery bracket

17. Road test the vehicle for proper shift operation.

Hydraulic Clutch System

BLEEDING

I30 and G20 Models

Bleeding is required to remove air trapped in the hydraulic system. The bleed screw is located on the clutch slave (release) cylinder.

1. Before servicing the vehicle, refer to the precautions in the beginning of this section.

2. Remove the bleed screw dust cap.

3. Attach a transparent vinyl tube to the bleed screw, immersing the free end in a clean container of clean brake fluid.

4. Fill the clutch master cylinder with the proper fluid.

5. Slowly depress the clutch pedal all the way several times and hold it down.

6. Have an assistant open the bleeder valve about ¾ turn to release the air. Then, close the bleeder valve while the pedal is still depressed.

7. Repeat the above procedure until no more air bubbles are seen in the fluid container.

8. Remove the bleed tube.

9. Replace the dust cap and refill the master cylinder.

10. Bleed the clutch damper, if equipped.

Halfshaft

REMOVAL & INSTALLATION

G20 Models

FRONT

1. Before servicing the vehicle, refer to the precautions in the beginning of this section.

2. Raise and support the vehicle safely.

3. Remove or disconnect the following:
- Front wheel
- Wheel bearing locknut
- Brake caliper assembly and rotor. Using a piece of wire, position the caliper so that it is not supported by the brake line
- Tie-rod from the ball joint
- Kingpin from the knuckle
- Halfshaft from the wheel hub/knuckle by lightly tapping it with a wood drift. Take care not to damage the CV-boots
- Halfshaft from the transaxle by prying outward with a suitable tool at the transaxle case

4. On automatic transaxle models, remove the left halfshaft by tapping it out with a drift from the right side of the transaxle case. Take care not to damage the pinion mate shaft and side gear.

To install:

5. Drive a new oil seal into the transaxle. For the right side use tool KV38106800 along the inner circumference of the oil seal. For the left side use tool KV38106700.

6. Install or connect the following:
- Halfshaft into the transaxle. Ensure that the serration's are aligned. Remove the tool

7. Push the halfshaft inward and install the circular clip in the groove of the side gear. After inserting the clip, pull outward on the flange of the slide joint to ensure the clip is properly meshed with the side gear. If it pulls out, the clip was not installed properly.
- Halfshaft into the wheel hub/knuckle. Torque the upper knuckle nut to 72–87 ft. lbs. (98–118 Nm) and wheel bearing locknut to 174–231 ft. lbs. (235–314 Nm)

8. Using a dial indicator, check wheel bearing axial end-play. Specification calls for 0.0020 in. (0.05mm) or less.
- Rotor and brake caliper
- Wheel

Q45 Models

REAR

> ※※ **CAUTION**
>
> **The amount of force need to loosen the rear wheel bearing nut is high enough to cause the vehicle to fall off the jack. Loosen and tighten this nut with the vehicle on the ground.**

1. Before servicing the vehicle, refer to the precautions in the beginning of this section.

2. Remove or disconnect the following:
- Rear wheel cotter pin, adjusting cap and insulator. Loosen the wheel bearing nut with the brakes applied and the vehicle sitting on the ground.

3. Raise the vehicle and support safely.
- Rear wheel
- Differential side flange bolts and nuts and separate shaft from the differential
- Wheel bearing locknut and washer from halfshaft
- Halfshaft by lightly tapping it with a copper hammer
- Halfshaft assembly from the vehicle

To install:

4. Install or connect the following:
- Halfshaft into wheel hub and install washer and wheel bearing locknut. Temporarily tighten the locknut
- Halfshaft with the differential side flange. Install the nuts and bolts and tighten to 61–69 ft. lbs. (83–93 Nm)
- Wheels and lower the vehicle to the ground
- Torque the wheel bearing locknut with the brakes applied to 152–203 ft. lbs. (206–275 Nm)
- Insulator, adjusting cap and a new cotter pin

I30 Models

RIGHT HALFSHAFT

1. Before servicing the vehicle, refer to the precautions in the beginning of this section.

2. Raise and support the front of the vehicle safely.

3. Remove or disconnect the following:
- Front wheel
- ABS wheel sensor and move it out of the way
- Brake hose from the strut
- Wheel bearing locknut

4. Matchmark and remove the bolts attaching the steering knuckle to the strut

➡ **Cover axle boots with waste cloth so as not to damage them when removing halfshaft.**

- Halfshaft from the knuckle by slightly tapping it
- Halfshaft from the transaxle using a suitable flat bladed tool
- Circlip on the end of the halfshaft and discard circlip
- Seal from the transaxle

To install:

5. Install or connect the following:
- New seal into the transaxle and install a halfshaft alignment tool

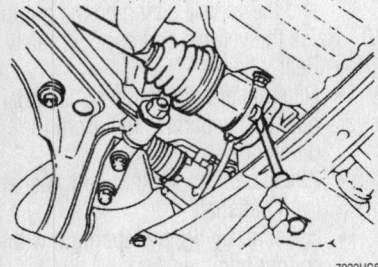

Separating the right halfshaft from the transaxle—I30 models

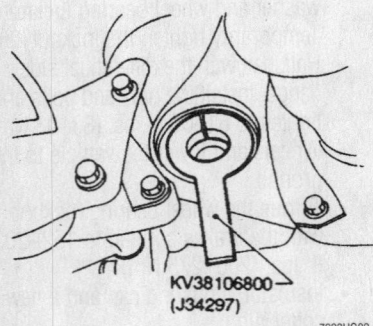

KV38106800 (J34297)

7923HG82

Right halfshaft alignment tool—I30 models

KV38106800 into the transaxle seal
- New circlip to the halfshaft, then insert the halfshaft into the transaxle

6. With the serration's aligned remove the alignment tool.

7. Push the halfshaft fully into the transaxle to seat the circlip. Try to pull the halfshaft from the transaxle by hand to verify that the circlip is properly seated.

- Halfshaft into the steering knuckle and install the hub locknut, do not tighten the hub nut at this time
- Steering knuckle to the strut
- Strut mounting bolts and align the matchmarks. Torque the bolts to 103–117 ft. lbs. (140–159 Nm)
- Brake hose to the strut
- ABS wheel sensor. Torque the attaching bolt to 13–17 ft. lbs. (18–24 Nm)
- Front wheels, lower the vehicle and torque hub locknut to 174–231 ft. lbs. (235–314 Nm)

8. Check and/or adjust the wheel alignment as necessary.

LEFT HALFSHAFT

1. Before servicing the vehicle, refer to the precautions in the beginning of this section.

2. Raise and support the front of the vehicle safely.

3. Remove or disconnect the following:
- Front wheel
- ABS wheel sensor and move it out of the way
- Brake hose from the strut
- Wheel bearing locknut

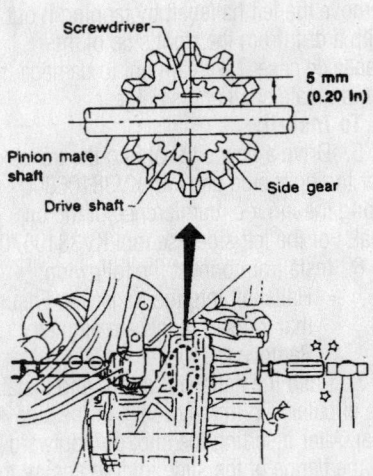

Separating the left halfshaft from an automatic transaxle—I30 models

- Bolts attaching the steering knuckle to the strut. Matchmark the bolts prior to removal

➡ **Cover axle boots with waste cloth so as not to damage them when removing halfshaft.**

- Halfshaft from the knuckle by slightly tapping it
- Bolts attaching the support bearing to the support bearing bracket
- Halfshaft from the transaxle using a suitable prytool, if equipped with a manual transaxle

4. If equipped with a automatic transaxle perform the following:

a. Remove the right halfshaft from the vehicle.

b. Insert a flat bladed tool into the transaxle where the right halfshaft was, place the end of the tool on the halfshaft and drive the left shaft from the pinion side gear.

- Support bearing bolts
- Halfshaft from the vehicle
- Circlip on the end of the halfshaft and discard circlip
- Seal from the transaxle

To install:

5. Install or connect the following:
- New seal into the transaxle and install a halfshaft alignment tool KV38106700 into the transaxle seal
- New circlip to the halfshaft, then insert the halfshaft into the transaxle

6. With the serration's aligned remove the alignment tool.

- Halfshaft fully into the transaxle to seat the circlip. Try to pull the halfshaft from the transaxle by hand to verify that the circlip is properly seated
- Support bearing bolts and torque the bolts to 10–14 ft. lbs. (13–19 Nm)
- Halfshaft into the steering knuckle and install the hub locknut, do not tighten the hub nut at this time
- Steering knuckle to the strut
- Strut mounting bolts and align the matchmarks. Torque the bolts to 103–117 ft. lbs. (140–159 Nm)
- Brake hose to the strut
- ABS wheel sensor. Torque the attaching bolt to 13–17 ft. lbs. (18–24 Nm)
- Front wheels, lower the vehicle and torque hub locknut to 174–231 ft. lbs. (235–314 Nm)

7. Check and/or adjust the wheel alignment as necessary.

CV-Joints

OVERHAUL

G20 Models

TRANSAXLE SIDE—DS83 TYPE

1. Before servicing the vehicle, refer to the precautions in the beginning of this section.
2. Disassemble the joint as follows:
   a. Remove the boot bands.
   b. Matchmark the slide joint housing and inner race before separating the assembly.
   c. Using a suitable prytool, remove the stopper ring and pull out the slide joint
   d. Matchmark the inner race and drive shaft.
   e. Remove the snap-ring.

89617G07

The inner CV joint uses a large C-clip to retain the ball and cage assembly in the outer housing

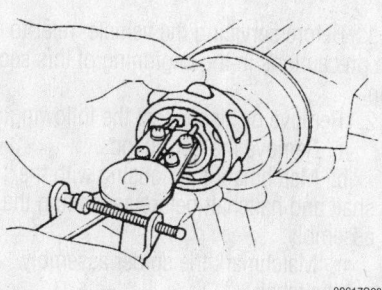

89617G08

After the outer housing is removed, the ball and cage assembly can slide from the shaft by removing the C-clip

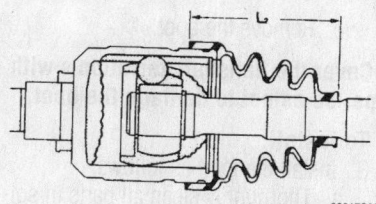

89617G02

Make sure to properly position the boot before tightening the boot clamps

f. Remove the ball cage, inner race and balls as a unit.
   g. Remove the boot.

➡ **Cover the halfshaft serration's with tape, so as not to damage the boot.**

To install:
3. Assemble the joint as follows:
   a. Thoroughly clean all parts in solvent and dry with compressed air. Check parts for evidence of damage, and replace as necessary.
   b. Install the boot and new boot band on the halfshaft.
   c. Install a new inner snap ring.
   d. Install the ball cage, inner race and balls as a unit. Confirm that the matchmarks are aligned.
   e. Install a new outer snap ring.
   f. Pack the CV joint with 5.0–6.0 ounces of grease.
   g. Ensure that the boot is properly installed on the halfshaft groove.
   h. Set the boot so that it does not swell or deform when its length is 3.82–3.90 in. (97–99mm).
   i. Lock the new boot bands securely.

TRANSAXLE SIDE—TS83 TYPE

1. Before servicing the vehicle, refer to the precautions in the beginning of this section.
2. Disassemble the joint as follows:
   a. Remove the boot bands.
   b. Matchmark the slide joint housing and inner race before separating the assembly.
   c. Matchmark the spider assembly and driveshaft
   d. Remove the snap-ring and the spider assembly.

➡ **Do not disassemble the spider assembly.**

   e. Remove the boot.

➡ **Cover the halfshaft serration's with tape, so as not to damage the boot.**

To install:
3. Assemble the joint as follows:
   a. Thoroughly clean all parts in solvent and dry with compressed air. Check parts for evidence of damage, and replace as necessary.
   b. Install the boot and new boot band on the halfshaft.
   c. Install the spider assembly making sure the matchmarks made during removal are properly aligned.
   d. Install a new snap ring.

e. Pack the joint with 4.5–5.11 ounces (124–145g) of grease.
   f. Install the slide joint housing.
   g. Ensure that the boot is properly installed on the halfshaft groove.
   h. Set the boot so that it does not swell or deform when its length is 3.90 in. (99mm).
4. Lock the new boot bands securely.

WHEEL SIDE

1. Before servicing the vehicle, refer to the precautions in the beginning of this section.
The joint on the wheel side cannot be disassembled.
2. Prior to separating the joint assembly, matchmark the halfshaft and joint assembly.
3. Separate the joint using a slide hammer.
4. Remove the boot bands.

To assemble:
5. Thoroughly clean all parts in solvent and dry with compressed air. Check parts for evidence of damage and replace as necessary.

➡ **Cover the halfshaft serration's with tape, so as not to damage the boot.**

6. Install the boot and small boot band on the halfshaft.
7. Set the joint assembly onto the halfshaft and align the matchmarks.

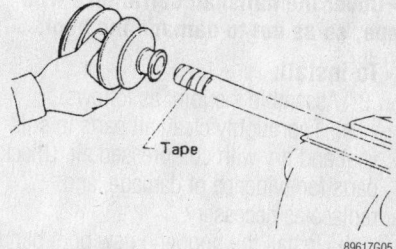

89617G05

Use vinyl tape and wrap the end of the shaft to protect the boot during installation

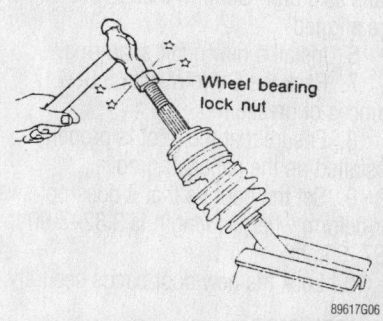

89617G06

Use an old nut to protect the threads when tapping the outer CV joint onto the shaft

8. Attach the joint assembly to the half-shaft by lightly tapping the serrated end with a plastic hammer.

➡**Using a metal hammer may damage the threads on the end of the joint.**

9. Pack the CV joint with 3.5–4.0 ounces of grease.

10. Ensure that the boot is properly installed on the halfshaft groove.

11. Set the boot so that it does not swell or deform when its length is 3.327–3.406 in. (84.5–86.5mm).

12. Lock the new boot bands securely.

I30 Models

TRANSAXLE SIDE

1. Before servicing the vehicle, refer to the precautions in the beginning of this section.

2. Disassemble the joint as follows:
   a. Remove the boot bands.
   b. Matchmark the slide joint housing and inner race before separating the assembly.
   c. Using a suitable prytool, remove the stopper ring and pull out the slide joint
   d. Matchmark the inner race and drive shaft.
   e. Remove the snap-ring.
   f. Remove the ball cage, inner race and balls as a unit.
   g. Remove the boot.

➡**Cover the halfshaft serration's with tape, so as not to damage the boot.**

To install:

3. Assemble the joint as follows:
   a. Thoroughly clean all parts in solvent and dry with compressed air. Check parts for evidence of damage, and replace as necessary.
   b. Install the boot and new boot band on the halfshaft.

4. Install a new inner snap ring.

5. Install the ball cage, inner race and balls as a unit. Confirm that the matchmarks are aligned.

6. Install a new outer snap ring.

7. Pack the CV joint with 5.8–6.17 ounces of grease.

8. Ensure that the boot is properly installed on the halfshaft groove.

9. Set the boot so that it does not swell or deform when its length is 3.82–3.90 in. (97–99mm).

10. Lock the new boot bands securely.

WHEEL SIDE

1. Before servicing the vehicle, refer to the precautions in the beginning of this section.

The joint on the wheel side cannot be disassembled.

2. Prior to separating the joint assembly, matchmark the halfshaft and joint assembly.

3. Separate the joint using a slide hammer.

4. Remove the boot bands.

To assemble:

5. Thoroughly clean all parts in solvent and dry with compressed air. Check parts for evidence of damage and replace as necessary.

➡**Cover the halfshaft serration's with tape, so as not to damage the boot.**

6. Install the boot and small boot band on the halfshaft.

7. Set the joint assembly onto the halfshaft and align the matchmarks.

8. Attach the joint assembly to the halfshaft by lightly tapping the serrated end with a plastic hammer.

➡**Using a metal hammer may damage the threads on the end of the joint.**

9. Pack the CV joint with 4.7–5.11 ounces of grease.

10. Ensure that the boot is properly installed on the halfshaft groove.

11. Set the boot so that it does not swell or deform when its length is 3.78–3.86 in. (96–98mm).

12. Lock the new boot bands securely.

Q45 Models

TRANSMISSION SIDE

1. Before servicing the vehicle, refer to the precautions in the beginning of this section.

2. Remove or disconnect the following:
   • Plug seal from the slide joint by gently tapping around the joint with a hammer
   • Boot bands

3. Put matchmarks on the slide joint housing and halfshaft prior to separating joint assembly.

4. Matchmark the spider assembly and driveshaft.
   • Snap-ring and the spider assembly

➡**Do not disassemble the spider assembly.**
   • Slide joint housing and the boot

➡**Cover the halfshaft serration's with tape, so as not to damage the boot.**

To assemble:

5. Thoroughly clean all parts in solvent and dry with compressed air. Check parts for evidence of damage and replace as necessary.

6. Install or connect the following:
   • Boot and small boot band on the halfshaft
   • Joint housing onto halfshaft
   • Spider assembly making sure the matchmarks are properly aligned

➡**The spider is press fit with the serration chamfer facing the shaft.**
   • Snap-ring
   • Coil spring, spring cap and new plug seal to the slide joint housing. Apply a suitable sealant to the plug seal prior to installation

➡**Hold the plug seal horizontally when pressing it into place. This will prevent the spring inside from falling down or tilting.**

7. Move the shaft in an axial direction to make sure that the spring is installed properly. If there is a drag or the spring is installed improperly, replace the plug seal with a new one.

8. Pack the halfshaft with 5.82–6.17 ounces (165–175g) of grease.

9. Ensure that the boot is properly installed on the halfshaft groove.

10. Set the boot so that it does not swell or deform when its length is 3.66–3.74 in. (93–95mm).

11. Lock the new boot bands securely.

Wheel Side

1. Before servicing the vehicle, refer to the precautions in the beginning of this section.

2. Remove or disconnect the following:
   a. Remove the boot bands.
   b. Matchmark the housing with the shaft and halfshaft before separating the assembly.
   c. Matchmark the spider assembly and halfshaft.
   d. Remove the snap-ring and the spider assembly.

➡**Do not disassemble the spider assembly.**

   e. Remove the boot.

➡**Cover the halfshaft serration's with tape, so as not to damage the boot.**

To install:

3. Install the joint as follows:
   a. Thoroughly clean all parts in solvent and dry with compressed air. Check parts for evidence of damage, and replace as necessary.
   b. Install the boot and new boot band on the halfshaft.

c. Install the spider assembly making sure the matchmarks made during removal are properly aligned.

➡ **The spider is press fit with the serration chamfer facing the shaft.**

d. Install a new snap ring.

4. Pack the joint with 4–4.34 ounces (113–123g) of grease.

a. Install the slide joint housing and the snap-ring.

b. Ensure that the boot is pro-

perly installed on the halfshaft groove.

5. Set the boot so that it does not swell or deform when its length is 3.78–3.86 in. (96–98mm).

6. Lock the new boot bands securely.

STEERING AND SUSPENSION

Air Bag

PRECAUTIONS

Several precautions must be observed when handling the inflator module to avoid accidental deployment and possible personal injury.

1. Never carry the inflator module by the wires or connector on the underside of the module.

2. When carrying a live inflator module, hold securely with both hands, and ensure that the bag and trim cover are pointed away.

3. Place the inflator module on a bench or other surface with the bag and trim cover facing up.

4. With the inflator module on the bench, never place anything on or close to the module that may be thrown in the event of an accidental deployment.

DISARMING

➡ **All Air Bag electrical wiring harnesses and connectors are covered with YELLOW outer insulation. Do not use electrical test equipment on any circuit related to the Air Bag sensors. When installing Air Bag components, always install with the arrow marks facing the front of the vehicle.**

1. Before servicing the vehicle, refer to the precautions in the beginning of this section.

2. Turn the ignition switch to the **OFF** position.

3. Disconnect both battery cables starting with the negative cable first and wait at least 10 minutes after the cables are disconnected. Be sure to insulate the battery terminal ends.

REARMING

1. Before servicing the vehicle, refer to the precautions in the beginning of this section.

2. Turn the ignition switch to the **OFF** position.

3. Connect both battery cables starting with the positive cable first.

➡ **The Air Bag or Air Bag system is equipped with a self-diagnostic operation. After turning the ignition key to the ON or START position, the AIR BAG warning lamp will illuminate for 7 seconds. After 7 seconds, the AIR BAG lamp will extinguish if no malfunction is detected. If the AIR BAG lamp does not extinguish after 7 seconds, check the Air Bag self-diagnostic system for a malfunction.**

Power Rack and Pinion Steering Gear

REMOVAL & INSTALLATION

G20 Models

1. Before servicing the vehicle, refer to the precautions in the beginning of this section.

✳✳ CAUTION

The air bag system must be disarmed before removing the steering wheel. Failure to do so may cause accidental deployment, property damage or personal injury.

2. Point the front tires straight ahead and lock the steering in this position.

✳✳ WARNING

Do not turn the steering wheel or column with the lower joint removed from the steering column or the spiral cable may be damaged.

3. Remove the steering wheel.

➡ **The steering wheel must be removed before disconnecting the steering column lower joint to avoid damaging the SRS spiral cable.**

4. Raise and support the vehicle safely and remove the front wheels.

5. Remove or disconnect the following:
   • Tie rod ends from the steering knuckles

   • Carbon canister and properly support the engine
   • Bolts attaching the engine mounts to the engine mounting center member
   • Engine mounting center member
   • Front stabilizer bar from the vehicle, if necessary
   • Nuts attaching the hole cover to the bulkhead

6. Move the hole cover aside and disconnect the lower joint from the rack and pinion. Matchmark the pinion shaft and the pinion housing to record the steering neutral position.
   • Power steering fluid pipes from the rack and pinion
   • Bolts attaching the mounting brackets and the rack and pinion from the vehicle

To install:

7. Install or connect the following:
   • Rack and pinion in the vehicle
   • Mounting brackets and tighten the mounting nuts and bolts in the proper sequence
   • New O-rings to the power steering fluid pipes and connect them to the rack and pinion. Torque the low pressure line 20–29 ft. lbs. (27–39 Nm) and the high pressure line to 11–18 ft. lbs. (15–25 Nm)

8. Align the lower steering joint to the pinion shaft and install the joint onto the pinion shaft. Torque the bolt to 17–22 ft. lbs. (24–29 Nm).

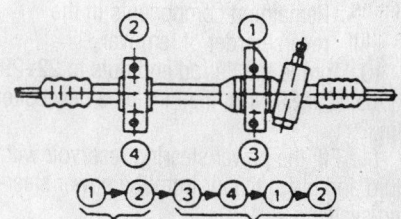

Temporary tightening   **Secure tightening**

9307HG26

Exploded view of the steering gear assembly—G20 Models

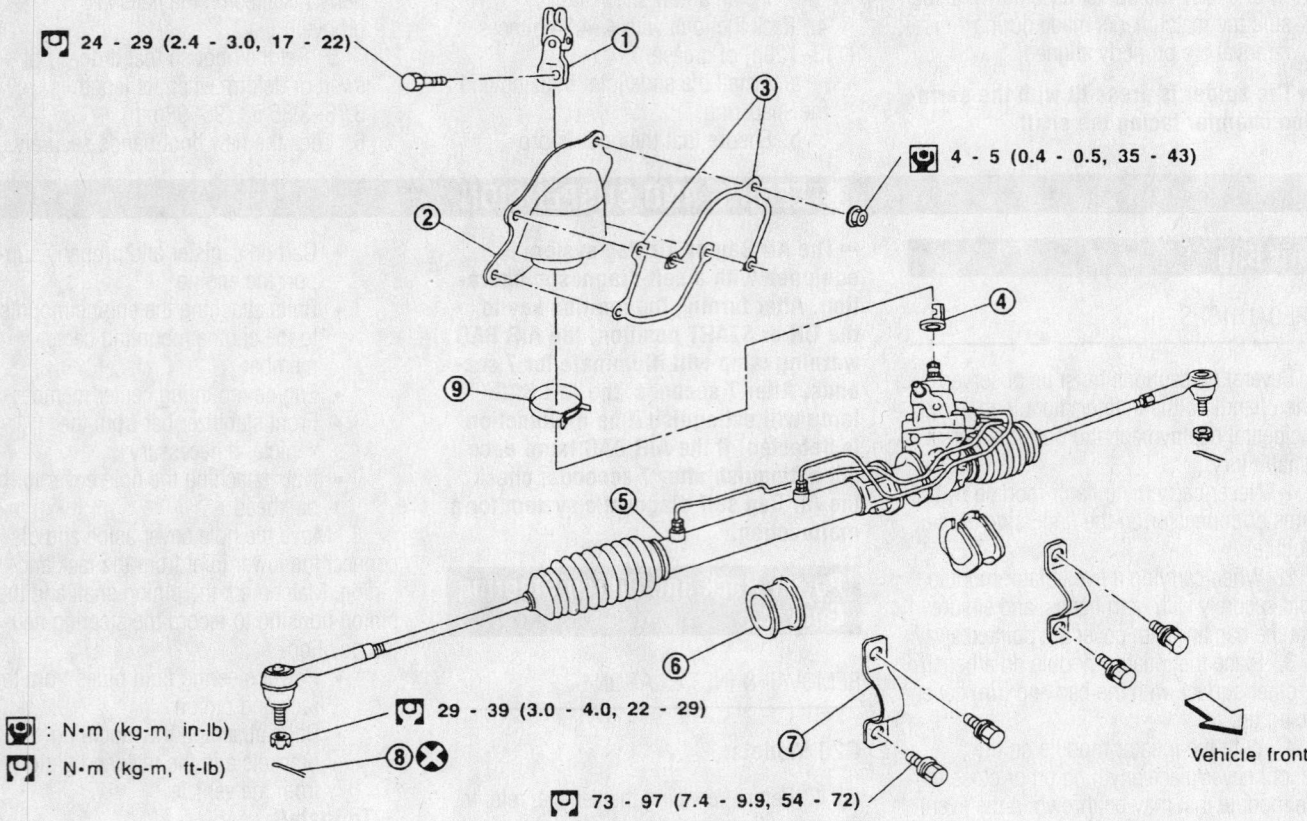

24 - 29 (2.4 - 3.0, 17 - 22)

4 - 5 (0.4 - 0.5, 35 - 43)

Vehicle front

: N•m (kg-m, in-lb)

: N•m (kg-m, ft-lb)

29 - 39 (3.0 - 4.0, 22 - 29)

73 - 97 (7.4 - 9.9, 54 - 72)

1. Lower joint
2. Hole cover
3. Insulator bracket
4. Rear cover cap
5. Gear and linkage assembly
6. Rack mounting insulator
7. Gear housing mounting bracket
8. Cotter pin
9. Clamp

9307HG25

Tighten the steering rack fasteners in this order—G20 Models

9. Properly position the hole cover. Torque the nuts to 2.9–3.6 ft. lbs. (4–5 Nm).
   - Front stabilizer
   - Engine mounting center member and tighten the attaching bolts. Attach the engine mounts to the center member and tighten the bolts. Remove the support from the engine
   - Remaining components in the reverse order of removal

10. Torque the tie rod end nuts to 22–29 ft. lbs. (29–39 Nm), then install a new cotter pin.

11. Fill the power steering reservoir with fluid and bleed the air from the power steering system.

12. Check the vehicle front end alignment and adjust as necessary.

I30 Models

1. Before servicing the vehicle, refer to the precautions in the beginning of this section.

✳✳ CAUTION

The air bag system must be disarmed before removing the steering wheel. Failure to do so may cause accidental deployment, property damage or personal injury.

2. Point the front tires straight ahead and lock the steering in this position.

✳✳ WARNING

Do not turn the steering wheel or column with the lower joint removed from the steering column or the spiral cable may be damaged.

3. Remove the steering wheel.

➡**The steering wheel must be removed before disconnecting the steering column lower joint to avoid damaging the SRS spiral cable.**

4. Remove or disconnect the following:
   - Both front wheels

   - Tie rod ends from the steering knuckles
   - Carbon canister from the vehicle and properly support the engine
   - Bolts attaching the engine mounts to the engine mounting center member
   - Engine mounting center member
   - Front stabilizer bar from the vehicle, if necessary
   - Nuts attaching the hole cover to the bulkhead

5. Move the hole cover aside and disconnect the lower joint from the rack and pinion. Matchmark the pinion shaft and the pinion housing to record the steering neutral position.
   - Power steering fluid pipes from the rack and pinion
   - Bolts attaching the mounting brackets and the rack and pinion from the vehicle

To install:

6. Install or connect the following:
   - Rack and pinion in the vehicle
   - Mounting brackets and tighten the

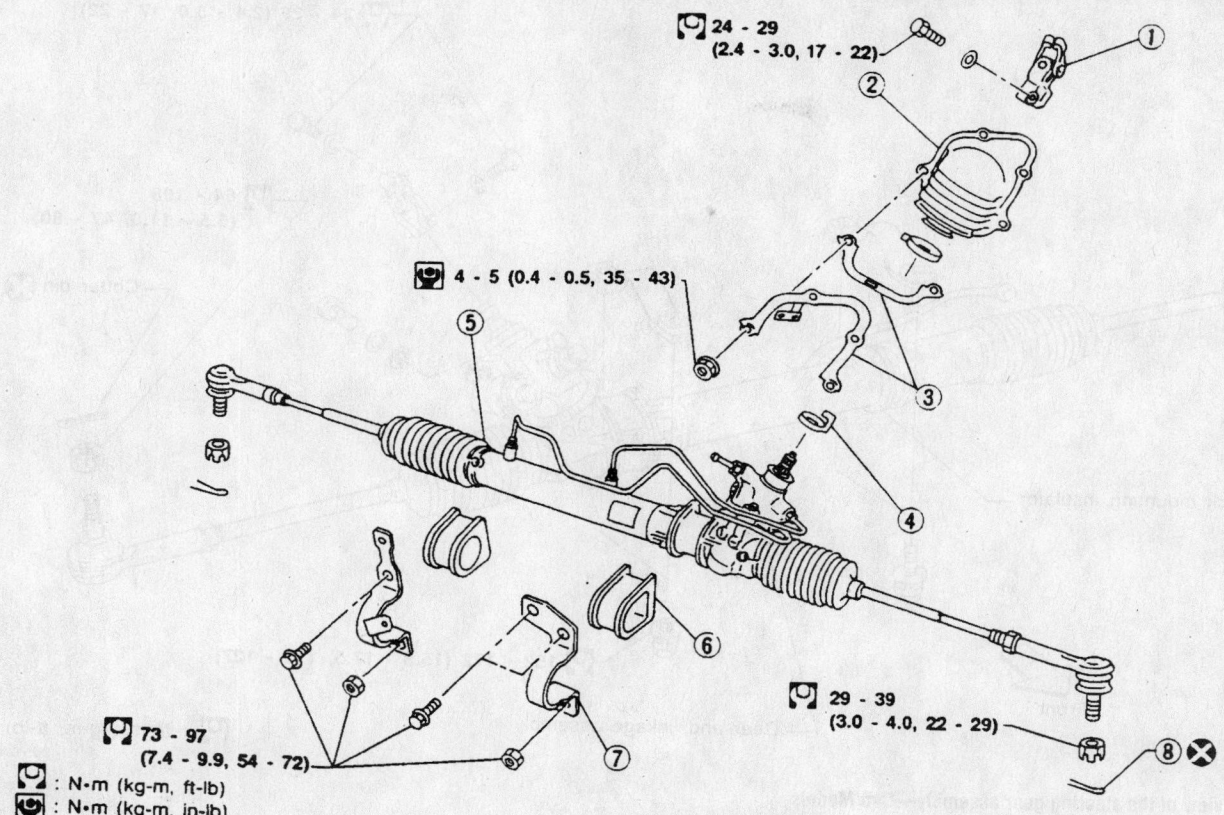

24 - 29
(2.4 - 3.0, 17 - 22)

4 - 5 (0.4 - 0.5, 35 - 43)

73 - 97
(7.4 - 9.9, 54 - 72)

29 - 39
(3.0 - 4.0, 22 - 29)

: N•m (kg-m, ft-lb)

: N•m (kg-m, in-lb)

①	Lower joint	④	Rear cover cap	⑦	Gear housing mounting bracket
②	Hole cover	⑤	Gear and linkage assembly	⑧	Cotter pin
③	Insulator bracket	⑥	Rack mounting insulator		

9307HG32

Exploded view of the steering gear assembly—I30 Models

mounting nuts and bolts in the proper sequence
- New O-rings to the power steering fluid pipes and connect them to the rack and pinion. Torque the low pressure line 20–29 ft. lbs. (27–39 Nm) and the high pressure line to 11–18 ft. lbs. (15–25 Nm)
7. Align the lower steering joint to the

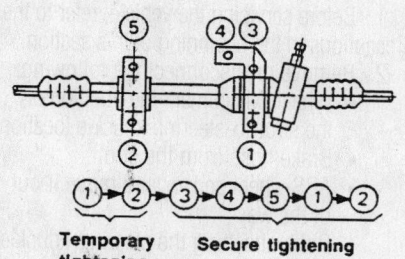

Temporary tightening   **Secure tightening**

9307HG33

Tighten the steering rack fasteners in this order—I30 Models

pinion shaft and install the joint onto the pinion shaft. Torque the bolt to 17–22 ft. lbs. (24–29 Nm).
8. Properly position the hole cover and install the attaching nuts. Torque the nuts to 2.9–3.6 ft. lbs. (4–5 Nm).
- Front stabilizer
- Engine mounting center member and tighten the attaching bolts. Attach the engine mounts to the center member and tighten the bolts. Remove the support from the engine
- Remaining components in the reverse order of removal
9. Torque the tie rod end nuts to 22–29 ft. lbs. (29–39 Nm), then install a new cotter pin.
10. Fill the power steering reservoir with fluid and bleed the air from the power steering system.
11. Check the vehicle front end alignment and adjust as necessary.

Q45 Models

✵✵ WARNING

Do not turn the steering wheel or column with the steering gear is removed.

1. Remove or disconnect the following:
- Both front wheels
- Tie rod ends from the steering knuckles
- Carbon canister from the vehicle and properly support the engine
- Bolts attaching the engine mounts to the engine mounting center member
- Front stabilizer bar from the vehicle
- Lower joint bolts
- Power steering fluid pipes from the rack and pinion
- Bolts attaching the mounting brackets and the rack and pinion from the vehicle

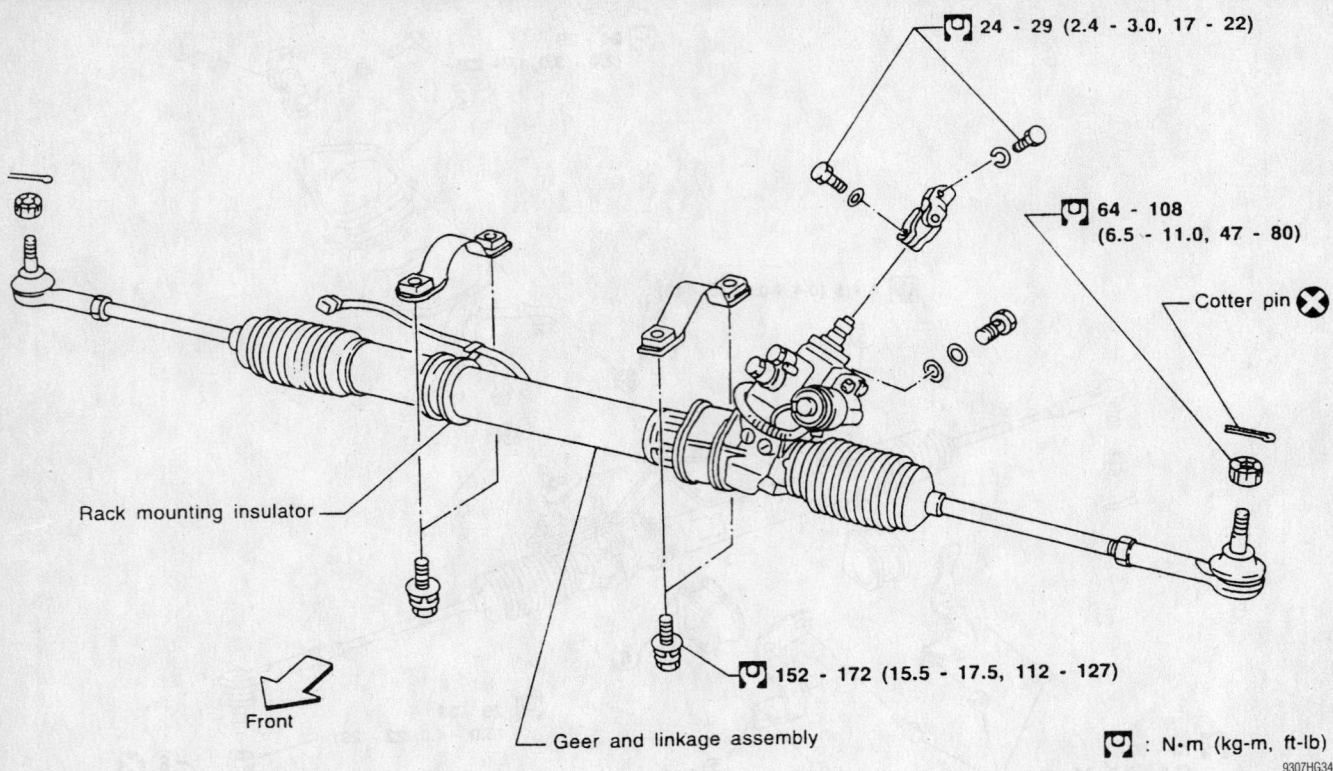

24 - 29 (2.4 - 3.0, 17 - 22)

64 - 108
(6.5 - 11.0, 47 - 80)

Cotter pin ⊗

Rack mounting insulator

152 - 172 (15.5 - 17.5, 112 - 127)

Front

Geer and linkage assembly

☐ : N•m (kg-m, ft-lb)

9307HG34

Exploded view of the steering gear assembly—Q45 Models

To install:

2. Install or connect the following:
- Rack and pinion in the vehicle.
- Mounting brackets. Torque the bolts to 112–127 ft. lbs. (152–172 Nm)
- New O-rings to the power steering fluid pipes and connect them to the rack and pinion. Torque the low pressure line 30–33 ft. lbs. (40–44 Nm) and the high pressure line to 11–18 ft. lbs. (15–25 Nm)

3. Align the lower steering joint to the pinion shaft and install the joint onto the pinion shaft. Install the bolt and torque to 17–22 ft. lbs. (24–29 Nm).
- Front stabilizer, if removed
- Engine mounting center member and tighten the attaching bolts. Attach the engine mounts to the center member and tighten the bolts if removed. Remove the support from the engine
- Remaining components in the reverse order of removal

4. Torque the tie rod end nuts to 47–80 ft. lbs. (64–108 Nm) and install a new cotter pin.

5. Fill the power steering reservoir with fluid and bleed the air from the power steering system.

6. Check the front end alignment and adjust as necessary.

Strut

REMOVAL & INSTALLATION

Front

G20 MODELS

1. Before servicing the vehicle, refer to the precautions in the beginning of this section.

2. Raise and support the vehicle safely.

3. Remove or disconnect the following:
- Strut mounting bolt at the lower suspension member and the 3 nuts inside the engine compartment. Do not remove the piston rod locknut
- Strut assembly and place in a suitable holding device

4. Using a prybar to hold the upper spring mount, loosen but do not remove the piston rod locknut.

5. Compress the spring with a spring compressor so the strut mounting insulator can be turned by hand.
- Piston rod locknut
- Coil spring from strut assembly

To install:

6. Inspect all components carefully for damage or wear. Replace as necessary.

7. Install or connect the following:
- Compressed coil spring on the

strut. Torque the locknut to 13–17 ft. lbs. (18–24 Nm)
- Strut. Ensure the bend in the lower strut bracket faces rearward on the left side and forward on the right side of the vehicle
- Upper spring seat with the cutout facing the inside of the vehicle. Torque the upper mounting bolts to 31–40 ft. lbs. (42–54 Nm) and the lower through-bolt to 82–93 ft. lbs. (112–126 Nm). Final tightening must take place with the suspension loaded (vehicle at normal ride height)

I30 MODELS

1. Before servicing the vehicle, refer to the precautions in the beginning of this section.

2. Remove or disconnect the following:
- Wheel. Matchmark the position of the strut-to-steering knuckle location
- Brake hose from the strut
- ABS wheel sensor and move it out of the way
- Bolts attaching the steering knuckle to the strut. Matchmark the assembly prior to removing the bolts
- Strut attaching nuts while holding the strut from inside the engine compartment

⁂ CAUTION

Do not remove the center locknut from the strut assembly until the strut is safely compressed.

- Strut from the vehicle
3. Place the strut assembly in a vise with the special holding tool ST35652000 or in a spring compressor.
4. Loosen the piston rod locknut.

⁂ CAUTION

Do not remove the piston rod locknut, the spring is under tension and can cause serious personal injury.

5. Compress the spring with the spring compressor, then remove the piston rod locknut.

➡**Before removing the strut from the coil spring, note the positioning of the strut in relationship to the coil spring for reassembly.**

- Strut mounting insulator bracket, strut mounting bearing, upper spring seat, and the upper spring rubber seat
- Strut, leaving the coil spring compressed
- Piston boot and rebound bumper from the strut

To install:
6. Install or connect the following:
- Rebound bumper and the boot to the strut piston
- Strut into the coil spring, be sure the strut and spring are properly positioned
- Upper spring rubber seat, upper spring seat, strut mounting bearing, and the strut mounting insulator bracket. Be sure that the cutout on the upper spring seat is facing the outside of the vehicle
- Piston rod locknut, then remove the spring compressor
- Torque the piston rod locknut to 43–65 ft. lbs. (59–88 Nm)
- Strut into the strut tower and install new attaching nuts. Torque the nuts to 29–40 ft. lbs. (39–54 Nm)
- Bolts attaching the steering knuckle to the strut and align the matchmarks. Torque the bolts to 103–117 ft. lbs. (140–159 Nm)
- ABS wheel sensor. Torque the bolt to 13–17 ft. lbs. (18–24 Nm)

- Brake hose to the strut
- Front wheels and lower the vehicle
7. Check and/or adjust the wheel alignment as necessary.

Q45 MODELS WITH A STANDARD SUSPENSION

1. Before servicing the vehicle, refer to the precautions in the beginning of this section.
2. Remove or disconnect the following:
- Front wheel
- Brake caliper and rotor
- Tie rod ball joint and lower ball joint with tool ST29020001
- Stabilizer connecting rod upper nut and separate the strut from the connecting rod
3. Remove the upper mounting insulator bolts.
4. Strut assembly
5. Secure the strut in a suitable holding fixture.
6. Loosen the piston rod locknut. Do not remove the locknut.
7. Compress the spring with the proper tool so the strut assembly mounting insulator can be turned by hand.
8. Remove the piston rod locknut. Remove the spring assembly, dust cover and rubber seat. Remove the strut insert.

To install:
9. Inspect the rubber parts for deterioration. If the rubber is pulling away from the metal, the mounting insulator should be replaced.
10. Fit the spring into the lower seat, install the dust cover/bumper and upper seat and mounting insulator.
11. Install the piston rod locknut and torque to 13–17 ft. lbs. (17–23 Nm).
12. Fit the strut into place and torque the upper mounting nuts to 30–35 ft. lbs. (40–47 Nm).
13. Torque the lower mounting bolt to 80–94 ft. lbs. (108–128 Nm).
14. Install the brake rotor and caliper.
15. Install the front wheel.

Q45 MODELS WITH A ACTIVE SUSPENSION

➡**The Nissan Consult or a scan tool that can issue commands to the control unit is required for bleeding the hydraulics in the Full Active Suspension system.**

1. Before servicing the vehicle, refer to the precautions in the beginning of this section.
2. Relieve the hydraulic pressure as follows:

a. Raise all 4 wheels off the ground and wait at least 3 minutes for the system to stabilize.
b. Remove both front inner fenders and the rear pressure control unit cover.
c. Loosen the locknut and slowly open the bypass valve on each pressure control unit. Open the valves all the way and leave them open until the job is finished.
3. Remove the flange joint from the top of the actuator.
4. Install 2, 15mm bolts into the actuator in the flange joint mounting bolt holes.
5. Insert a bar between the bolts and loosen the joint adapter. Do not remove it yet.
6. Remove the upper mount insulator nuts.
7. Disconnect the hydraulic lines. Cap the lines to keep the system clean.
8. Remove the lower actuator mounting nut and remove the assembly.
9. Secure the actuator/spring assembly in a suitable holding fixture. Scribe alignment marks on the spring, upper mount insulator and actuator unit.
10. Compress the spring with the proper tool so the joint adapter can be turned by hand. Remove the joint adapter and lift off the mount insulator, spring, and any other components necessary.

To install:
11. If the actuator is being replaced, the rubber bumper should also be replaced. Fit the bumper, dust cover and rubber seat onto the actuator.
12. Fit the spring into the lower seat with the matchmarks aligned. Install the upper seat/mounting insulator with the marks aligned and start the joint adapter. The joint adapter will be tightened after installing the actuator assembly.
13. Fit the strut into place and torque the upper mounting nuts to 30–41 ft. lbs. (40–55 Nm).
14. Torque the lower mounting bolt to 76–94 ft. lbs. (103–128 Nm).
15. Torque the joint adapter to 63–72 ft. lbs. (85–98 Nm).
16. Install the flange adapter and torque the bolts to 11–13 ft. lbs. (15–18 Nm).
17. Close the bypass valves on the pressure control units.
18. Bleed the system as follows:
a. With all 4 wheels about 2 in. (50mm) off the ground, run the engine for about 2 minutes.
b. Connect the Consult scan tool and

enter "WORK SUPPORT" mode. Select "4. AIR BLEEDING".

c. Check the fluid level in the reservoir. It should be slightly overfilled.

d. Touch "START" on the scan tool. The display will show a regular rise and fall in system pressure. When the pressure stabilizes, stop the engine.

e. Connect a clear tube to the air bleeder at the actuator and place the other end in a container.

➡**Do not allow the fluid to contact the body or the paint will be damaged.**

f. Open the bleeder and watch the fluid move through the tube. If there are still air bubbles in the fluid when the flow stops, check the fluid level, pressurize the system again and repeat the process.

Rear

G20 MODELS

1. Before servicing the vehicle, refer to the precautions in the beginning of this section.

2. Remove or disconnect the following:
- Wheels

✱✱ WARNING

Be sure to disconnect the ABS wheel sensor from the assembly. Failure to do so may result in damage to the sensor wire and the sensor becoming inoperative.

- Brake calipers and suspend them with a piece of wire. Do not let them hang by the hose

3. Using a transmission jack, raise the torsion beam slightly
- Strut lower mounting bolt

4. Open the trunk, remove the trim and remove the two nuts attaching the strut to the vehicle.
- Strut

✱✱ CAUTION

Do not remove the center locknut from the strut assembly until the strut is safely compressed.

5. Place the strut assembly in a vise with the special holding tool HT71780000 or in a spring compressor.

6. Loosen the piston rod locknut.

✱✱ CAUTION

Do not remove the piston rod locknut, the spring is under tension and can cause serious personal injury.

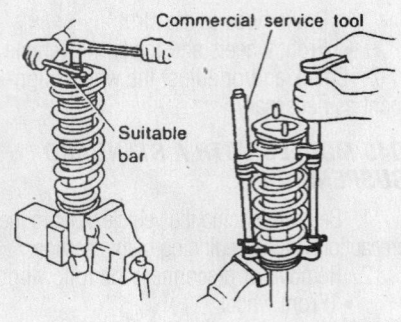

89618G30

Be sure the spring is compressed before removing the piston locknut—G20 models

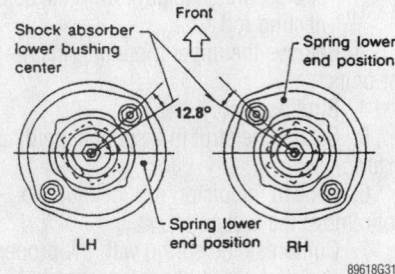

89618G31

Align the spring seats as shown—G20 models

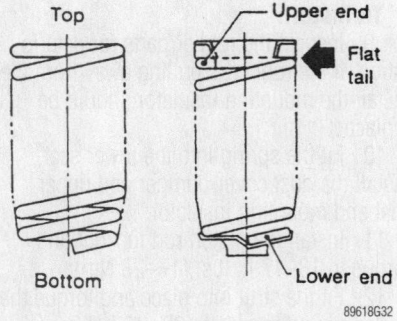

89618G32

Make sure the springs are installed as shown—G20 models

7. Compress the spring with the spring compressor then remove the piston rod locknut.

➡**Before removing the strut from the coil spring, note the positioning of the strut in relationship to the coil spring for reassembly.**

8. Remove or disconnect the following:

- Bushing, strut mounting bracket, and the upper spring seat rubber
- Strut, leaving the coil spring compressed
- Bushing, bound bumper cover and the bound bumper

To install:

9. Install or connect the following:
- Bound bumper, bound bumper cover and the bushing
- Strut into the coil spring, make sure the strut and spring are properly positioned
- Upper spring seat rubber, strut mounting bracket, and the bushing. Make sure that the mounting bracket is properly positioned
- Piston rod locknut then remove the spring compressor. Torque the locknut to 13–17 ft. lbs. (18–24 Nm)
- Strut with new attaching nuts. Torque the nuts to 14–16 ft. lbs. (19–22 Nm)
- Strut on the rear torsion beam. Torque the bolt to 80–94 ft. lbs. (108–127 Nm)

10. Remove the support from the rear torsion beam.

11. Install the rear wheels and lower the vehicle.

12. Check the vehicle's alignment and adjust as necessary.

I30 MODELS

1. Before servicing the vehicle, refer to the precautions in the beginning of this section.

2. Remove or disconnect the following:
- Rear wheels

3. Support the rear torsion beam assembly with a jack.

✱✱ CAUTION

Do not remove the center locknut from the strut assembly until the strut is safely compressed.

- 2 nuts attaching the strut to the vehicle located inside the trunk
- Bolt attaching the strut to the rear torsion beam assembly and remove the strut

4. Place the strut assembly in a vise with the special holding tool HT71780000 or in a spring compressor.
- Piston rod locknut

✱✱ CAUTION

Do not remove the piston rod locknut, the spring is under tension and can cause serious personal injury.

5. Compress the spring with the spring compressor, then remove the piston rod locknut.

➡**Before removing the strut from the coil spring, note the positioning of the** strut in relationship to the coil spring for reassembly.

6. Remove or disconnect the following:
   - Bushing, strut mounting bracket, and the upper spring seat rubber
   - Strut, leaving the coil spring compressed
   - Bushing, bound bumper cover, and the bound bumper

To install:

7. Install or connect the following:
   - Bound bumper, bound bumper cover, and the bushing
   - Strut into the coil spring, be sure the strut and spring are properly positioned
   - Upper spring seat rubber, strut

mounting bracket, and the bushing. Be sure that the mounting bracket is properly positioned
- Piston rod locknut, then remove the spring compressor. Torque the locknut to 13–17 ft. lbs. (18–24 Nm)
- Strut with new attaching nuts. Torque the nuts to 12–16 ft. lbs. (16–22 Nm)
- Strut on the rear torsion beam. Torque the bolt to 80–94 ft. lbs. (108–127Nm)
- Support from the rear torsion beam
- Rear wheels and lower the vehicle

8. Check the vehicle's alignment and adjust as necessary.

Q45 MODELS WITH A STANDARD SUSPENSION

✳✳ CAUTION

Do not remove piston rod locknut with the shock absorber on vehicle.

1. Before servicing the vehicle, refer to the precautions in the beginning of this section.
2. Remove the upper strut mounting nuts.
3. Raise and safely support the vehicle and remove the lower mounting bolt. Remove coil spring/strut absorber assembly.
4. Place the assembly into a suitable holding fixture and matchmark the spring,

Figure shows rear left actuator.

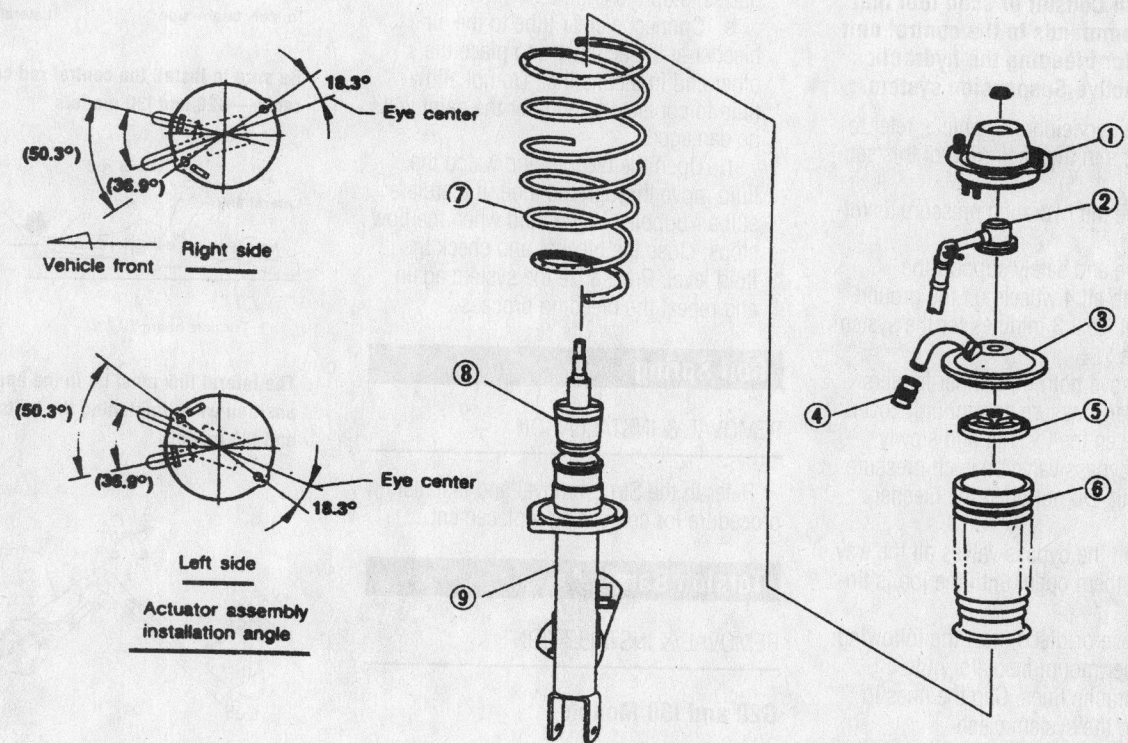

① Mount insulator	④ Air tube connector
② Rear joint hose	⑤ Bound bumper cover
③ Spring upper seat	⑥ Rear actuator dust cover

⑦ Coil spring
⑧ Bound bumper
⑨ Rear actuator

7923HG85

Rear actuator removal with Active Suspension—Q45 models

Heater Core replacement is covered in Section 2 of this manual

strut and upper seat. Loosen but do not remove the piston rod locknut.

5. Install a spring compressor and compress the spring until the upper spring seat can be turned by hand.

6. Remove the locknut, spring seat components, spring, bushings and bumper.

To install:

7. Fit the bumper, spring, upper seat and other components onto the strut with the matchmarks aligned. The top of the spring is flat.

8. Install the locknut and torque it to 13–17 ft. lbs. (18–24 Nm) and remove the spring compressor.

9. Install strut assembly. Torque the upper shock mounting nuts to 12–14 ft. lbs. (16–19 Nm) and the lower bolt to 57–72 ft. lbs. (77–98 Nm).

Q45 MODELS WITH A FULL ACTIVE SUSPENSION

➡ **The Nissan Consult or scan tool that can issue commands to the control unit is required for bleeding the hydraulics in the Full Active Suspension system.**

1. Before servicing the vehicle, refer to the precautions in the beginning of this section.

2. Relieve the hydraulic pressure as follows:

   a. Raise and safely support the vehicle with all 4 wheels off the ground and wait at least 3 minutes for the system to stabilize.

   b. Remove both front inner fenders and the rear pressure control unit cover.

   c. Loosen the locknut and slowly open the bypass valve on each pressure control unit. Do not open the bleeder valves.

   d. Open the bypass valves all the way, and leave them open, until the job is finished.

3. Remove or disconnect the following:
   - Upper mount insulator nuts
   - Hydraulic lines. Cap the lines to keep the system clean
   - Lower actuator mounting bolt and remove the actuator/spring assembly

4. Secure the actuator/spring assembly in a suitable holding fixture. Scribe alignment marks on the spring, upper mount insulator and actuator unit.

5. Compress the spring with the proper tool. Remove the piston rod locknut lift off the mount insulator, hose joint adapter, spring, and any other components necessary.

To install:

6. Fit the bumper and dust cover onto the actuator.

7. Fit the spring into the lower seat with the matchmarks aligned. Install the upper seat, mounting insulator and other components with the marks aligned.

8. Install the locknut and torque to 43–54 ft. lbs. (59–74 Nm).

9. Fit the assembly onto the vehicle. Torque the upper mounting nuts to 12–24 ft. lbs. (16–19 Nm) and the lower mounting bolt to 58–72 ft. lbs. (78–98 Nm).

10. Bleed the system as follows:

   a. With all 4 wheels off the ground, run the engine for about 2 minutes.

   b. Connect the scan tool and enter "WORK SUPPORT" mode. Select "4. AIR BLEEDING".

   c. Check the fluid level in the reservoir and make it slightly overfilled.

   d. Touch "START" on the scan tool. The display will show a regular rise and fall in system pressure that may last for several minutes. When the pressure stabilizes, stop the engine.

   e. Connect a clear tube to the air bleeder at the actuator and place the other end in a container. Do not allow fluid to contact the body or the paint will be damaged.

   f. Open the bleeder and watch the fluid move through the tube. If there are still air bubbles in the fluid when the flow stops, close the bleeder and check the fluid level. Pressurize the system again and repeat the bleeding process.

Coil Spring

REMOVAL & INSTALLATION

Refer to the Strut removal and installation procedure for coil spring replacement.

Torsion Bars

REMOVAL & INSTALLATION

G20 and I30 Models

1. Before servicing the vehicle, refer to the precautions in the beginning of this section.

2. Loosen the lug nuts.

3. Remove or disconnect the following:
   - Wheels

☼ WARNING

Be sure to disconnect the ABS wheel sensor from the assembly. Failure to do so may result in damage to the sensor wire and the sensor becoming inoperative.

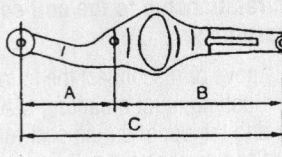

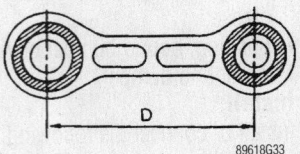

Measure the control rod and lateral links at these points—G20 and I30 models

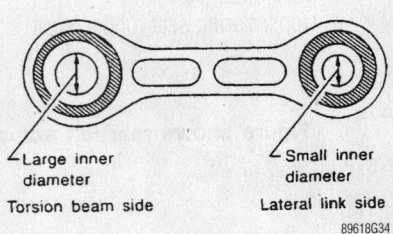

Be sure to install the control rod correctly—G20 and I30 models

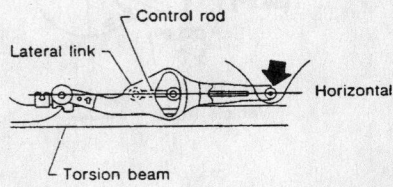

The lateral link must be in the horizontal position when tightening the bolts—G20 and I30 models

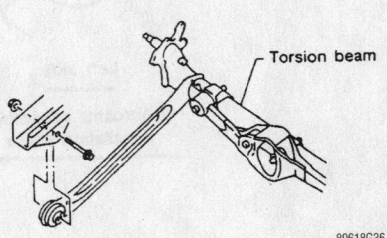

Tighten the torsion beam-to-chassis bolts with the suspension unloaded—G20 and I30 models

- Brake calipers and suspend them with a piece of wire. Do not let them hang by the hose

4. Using a transmission jack, raise the torsion beam a little, then remove the suspension mounting bolts.

5. Lower the jack and remove the suspension assembly.

6. The lateral link and control rod can now be removed.

7. Inspect the torsion beam and control rod for cracks, wear and deformation. The length of the lateral link and control rod is as follows:

    a. A—8.15–8.19 in. (207–208mm).
    b. B—15.51–15.55 in. (394–395mm).
    c. C—23.66–23.74 in. (601–603mm).
    d. D—4.17–4.25 in. (106–108mm).

To install:

8. When installing the control rod, connect the bushing with the smaller inner diameter to the lateral link. Install the lateral link and the control rod on the torsion beam. Place the lateral link with the arrow topside.

9. Place the lateral link and control rod horizontally against the beam, and tighten the bolts. Refer to the illustration.

10. Secure the torsion beam to the vehicle. Make sure the lateral link is horizontal , then tighten the link to the chassis.

11. Attach the struts to the torsion beam and tighten the fasteners.

12. Tighten the torsion beam-to-chassis bolts.

13. Install the calipers, ABS sensor and wheels. Lower the vehicle to the ground. Final tighten the lug nuts.

Lower Ball Joint

REMOVAL & INSTALLATION

The lower ball joint assembly is part of the lower control arm/transverse link. If replacement of the ball joint is required, the lower control arm needs to be replaced.

Lower Control Arms

REMOVAL & INSTALLATION

G20 Models

1. Before servicing the vehicle, refer to the precautions in the beginning of this section.

2. Remove or disconnect the following:
   - Stabilizer bar

➡**Take note of paint mark and clamp position when removing stabilizer bar for correct reinstallation.**

3. Support the steering knuckle with a suitable jack and remove the lower ball joint nut. Separate the ball joint from the knuckle.

- Bolts attaching the lower control arm to the chassis
- Lower control arm

To install:

4. If the lower ball joint is worn or damaged, the lower control arm must be replaced. The ball joint is not serviceable separately.

5. Install or connect the following:
   - Lower control arm to the chassis with the attaching bolts and nut
   - Ball joint stud in the knuckle. Torque the nut to 52–64 ft. lbs. (71–86 Nm)
   - Stabilizer bar and wheel
6. Lower the vehicle.

➡**Final tightening must be done with the vehicle at normal ride height, tires on the ground and the chassis loaded.**

7. Torque front control arm bolts to 87–108 ft. lbs. (118–147 Nm) and rear gusset nut to 69–87 ft. lbs. (93–118 Nm).

I30 Models

1. Before servicing the vehicle, refer to the precautions in the beginning of this section.

2. Remove or disconnect the following:
   - Front wheels
   - ABS wheel sensor and move it out of the way
   - Wheel bearing locknut
   - Tie rod from the steering knuckle
   - Bolts attaching the strut to the steering knuckle. Matchmark prior to removal
   - Halfshaft from the steering knuckle by lightly tapping the end of the shaft
   - Steering knuckle and the lower ball joint
   - Stabilizer bar from the lower control arm
   - Bolts attaching the link bushing pin to the chassis
   - Nut attaching the link to the control arm and remove the link
   - Bolts attaching the compression rod bushing clamp
   - Lower control arm/traverse link

To install:

3. Install or connect the following:
   - Lower control arm and the compression rod bushing clamp into the vehicle
   - Link bushing pin, if removed from the control arm

4. Tighten all bolts and nuts until they are snug enough to support the weight of the vehicle but not fully tight, the bolts should be torqued to specification with the vehicle on the floor.

   - Steering knuckle to the lower control arm and connect the ball joint. Torque the nut to 46–56 ft. lbs. (62–76 Nm)

➡**Always use a new nut when installing the ball joint to the control arm.**

   - Steering knuckle to the strut and to the halfshaft
   - Strut mounting bolts and align the matchmarks. Torque the bolts to 103–117 ft. lbs. (140–159 Nm)
   - Tie rod ball joint and tighten the nut to 46–54 ft. lbs. (63–73 Nm)
   - Wheel bearing locknut
   - ABS wheel sensor. Torque the attaching bolt to 13–17 ft. lbs. (18–24 Nm)
   - Front wheels

5. Lower the vehicle and torque the hub locknut to 174–231 ft. lbs. (235–314 Nm).

6. Torque the bolts attaching the compression rod bushing clamp and the link bushing pin, in the proper sequence to 87–108 ft. lbs. (118–147 Nm).

7. If the link bushing pin was removed from the control arm, torque the attaching nut to 87–108 ft. lbs. (118–147 Nm).

8. Tighten the sway bar attaching nut to 30–35 ft. lbs. (41–47 Nm).

9. Check the vehicle alignment.

Q45 Models

1. Before servicing the vehicle, refer to the precautions in the beginning of this section.

2. Remove or disconnect the following:
   - Negative battery cable
   - Front wheel
   - Nuts securing the tension rod to the transverse link (control arm)
   - Nut and separate the ball joint stud from the knuckle
   - Transverse link from the sub-frame

To install:

3. Install or connect the following:
   - Transverse link on the sub-frame. Temporarily install the bolt and nut
   - Tension rod on the transverse link. Torque the nuts to 87–94 ft. lbs. (118–127 Nm)
   - Nut on the ball joint stud. Torque the nut to 71–88 ft. lbs. (96–120 Nm)
   - Front wheel and lower the vehicle to the floor
   - Transverse link mounting bolt.

Brake service is covered in Section 4 of this manual

Torque the bolt to 72–87 ft. lbs. (98–118 Nm)
- Negative battery cable

CONTROL ARM BUSHING REPLACEMENT

The bushing are part of the transverse assembly, if they are defective the whole assembly must be replaced.

Wheel Bearings

ADJUSTMENT

The front and rear wheel bearing assemblies on all models are pressed in and are not adjustable. If the bearing assembly does not turn smoothly or has more than 0.002 in. (0.05mm) of axial play, replace the bearing assembly.

REMOVAL & INSTALLATION

Front

G20 MODELS

1. Before servicing the vehicle, refer to the precautions in the beginning of this section.
2. The axle nut torque is very high and should be loosened and tightened with the vehicle on the ground. Remove the cotter pin, adjusting cap and insulator and loosen the front axle nut.
3. Remove or disconnect the following:
- Brake caliper, carrier, and the rotor. Hang the caliper from the body with wire; do not let it hang by the brake hose.
- Cotter pin and nut and use a ball joint press to disconnect the tie rod end
- Cap and the upper king pin mounting nut and separate the kingpin from the third link
4. Hold a block of wood against the axle stub and strike it with a hammer to release it from the hub. Withdraw the axle from the hub and fold the steering knuckle down on the ball joint.
- Cotter pin and nut and use a ball joint press to disconnect the ball joint
- Steering knuckle

➡ **Wheel bearings must be replaced any time the hub is removed.**

5. Pry the grease seals out of the steering knuckle.
6. Support the steering knuckle and press the hub out of the bearing.

7. Remove the snap-rings and press the bearing out towards the inside of the knuckle.

To install:

8. Be sure all parts are clean and dry. The hub and steering knuckle should be inspected for cracks using dye or a magnetic crack detection process.
9. Install or connect the following:
- Inner snapring and carefully press the new bearing into the steering knuckle. Be sure the press tool contacts only the outer bearing race or the bearing will be damaged
- Outer snapring. Pack the new grease seals with clean grease and install them. If removed, install the splash guard
10. Support the inner race on the press table and carefully press the hub into the bearing. Be sure the hub turns smoothly in both directions.
- Steering knuckle onto the lower ball joint and start the nut. Fit the axle shaft through the hub and start the nut
11. Pack the king pin bearing housing with grease and fit the third link into place. Torque the kingpin nut to 72–87 ft. lbs. (98–118 Nm) and install the dust cap.
12. Torque the lower ball joint nut to 52–64 ft. lbs. (71–86 Nm). Install a new cotter pin.
- Tie rod end. Torque the nut to 22–29 ft. lbs. (29–39 Nm). Torque as needed to install a new cotter pin but do not exceed 36 ft. lbs. (49 Nm)
- Brake caliper, carrier, rotor and the wheel
13. Lower the vehicle to the ground.
14. Torque the front axle nut to 174–231 ft. lbs. (235–314 Nm). Install the insulator, adjusting cap and cotter pin.

I30 MODELS

➡ **Whenever the hub or bearing assembly is removed, the wheel bearing assembly must be replaced. Never reuse the old bearing assembly.**

1. Before servicing the vehicle, refer to the precautions in the beginning of this section.
2. Remove the knuckle assembly from the vehicle by separating the ball joint and tie rod end, then removing the retaining hardware securing the knuckle to the strut.
3. Using a shop press and a suitable tool, press the hub with the inner race from the steering knuckle.

4. Using a shop press and a suitable tool, press the bearing inner race from the hub and remove the outer grease seal.
5. Use snapring pliers to remove the snaprings from the steering knuckle.
6. Inspect the hub, steering knuckle and snaprings for cracks and/or wear; if necessary, replace the damaged part(s).

To install:

7. Install the inner snapring in the steering knuckle groove.
8. Using a shop press and a suitable tool, press the new wheel bearing assembly into the steering knuckle, until it seats, using a maximum pressure of 3 tons (2722kg).
9. Install the outer snapring.
10. Pack the new grease seal lips with multi-purpose grease.
11. Using a shop press and a suitable tool, press the new outer grease seal into the steering knuckle.
12. Using a shop press and a suitable tool, press the new inner grease seal into the steering knuckle.
13. Using a shop press and a suitable tool, press the hub into the steering knuckle, until it seats, using a maximum pressure of 5.5 tons (4990kg); be careful not to damage the grease seal.
14. To check the bearing operation, perform the following procedures:
   a. Increase the press pressure to 3.5–5.0 tons (3175–4536kg).
   b. Spin the steering knuckle, several turns, in both directions.
   c. Be sure the wheel bearings operate smoothly.
15. If the wheel bearings do not operate smoothly, replace the wheel bearing assembly.
16. Install the knuckle assembly.
17. Install the halfshaft into the hub. Torque the locknut to 174–231 ft. lbs. (235–314 Nm).
18. Install the wheel assembly and lower the vehicle.
19. Road test the vehicle and verify proper operation.

Q45 MODELS

1. Before servicing the vehicle, refer to the precautions in the beginning of this section.
2. Support the hub assembly with a suitable jack.
3. Remove or disconnect the following:
- Brake caliper, carrier and rotor. Hang the caliper from the body with wire, do not let it hang by the brake hose.
- Cotter pins and nuts and use a ball joint press to disconnect the lower ball joint and tie rod end

- Kingpin lower mounting nut to remove the steering knuckle assembly

➡**Wheel bearings must be replaced any time the hub is removed.**

4. Use a vise or a wheel to hold the hub and remove the hub cap and nut from the back of the hub. Remove the wheel speed sensor rotor.

5. Use a press or large drift pin to press the hub out of the steering knuckle.

6. Remove the snapring and press the bearings and grease seal out of the steering knuckle.

To install:

7. Be sure all parts are clean. Carefully press the new bearing into the steering knuckle. Be sure the press tool contacts only the outer bearing race or the bearing will be damaged.

8. Install a new grease seal and the snapring. If removed, install the splash guard.

9. Lightly lubricate the lips of the seal with clean grease. Be careful not to grease the bearing or hub mating surfaces.

10. Carefully press the hub into the bearing. Support the inner race on the press table or the bearing will be damaged. Do not exceed 3.9 tons (3538kg) pressure.

11. Install or connect the following:
- Speed sensor rotor and nut on the hub and torque to 152–210 ft. lbs. (206–284 Nm). Stake the nut into place

12. Lightly tap the cap into place and install the bolts.
- Steering knuckle to the king pin. Torque the nut to 108–137 ft. lbs. (88–108 Nm)
- Lower ball joint. Torque the nut to

65–80 ft. lbs. (88–108 Nm). Install a new cotter pin
- Tie rod end. Torque the nut to 22–29 ft. lbs. (29–39 Nm). Torque as needed to install a new cotter pin but do not exceed 36 ft. lbs. (49Nm)
- Rotor and the brake caliper
- Wheel and tire assembly

Rear

G20 MODELS

1. Before servicing the vehicle, refer to the precautions in the beginning of this section.

2. Raise and support the vehicle safely.

3. Remove or disconnect the following:
- Rear caliper and rotor. Hang the caliper from the body with wire, do not let hang by the brake hose

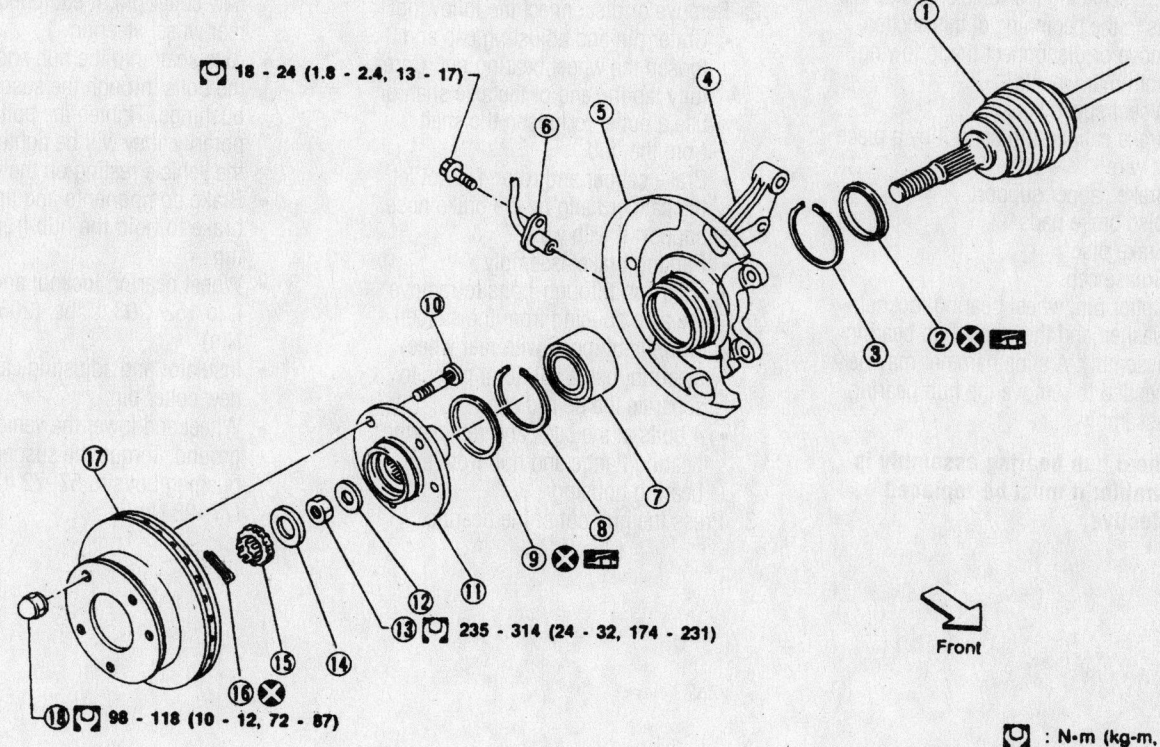

18 - 24 (1.8 - 2.4, 13 - 17)

235 - 314 (24 - 32, 174 - 231)

98 - 118 (10 - 12, 72 - 87)

Front

: N•m (kg-m, ft-lb)

①	Drive shaft	⑦	Wheel bearing assembly	⑬	Wheel bearing lock nut
②	Inner grease seal	⑧	Snap ring	⑭	Insulator
③	Snap ring	⑨	Outer grease seal	⑮	Adjusting cap
④	Knuckle	⑩	Hub bolt	⑯	Cotter pin
⑤	Baffle plate	⑪	Wheel hub	⑰	Disc rotor
⑥	ABS sensor	⑫	Plain washer	⑱	Wheel nut

7923HG87

Exploded view of the front knuckle assembly—I30 models

For complete Engine Mechanical specifications, see Section 1 of this manual

- Rear wheel hub cap, cotter pin and locknut
- Hub off the stub axle

➡The wheel bearing is integral with the hub and cannot be serviced separately.

To install:

4. Install or connect the following:
   - New hub assembly onto the axle stub
5. Replace the washer and wheel bearing locknut. Torque the locknut to 137–174 ft lbs. (186–235 Nm). Install a new cotter pin
   - Brake rotor, caliper and wheel
6. Lower the vehicle to the ground.

I30 MODELS

➡If the vehicle is equipped with ABS, the sensor must be removed to protect the sensor and its wiring.

1. Before servicing the vehicle, refer to the precautions in the beginning of this section.
2. Remove or disconnect the following:
   - Both rear wheels
   - Wheel speed sensor
   - Brake caliper and hang it by a piece of wire
   - Brake caliper support
   - Disc brake pads
   - Brake disc
   - Grease cap
   - Cotter pin, wheel bearing locknut, washer, and the wheel hub bearing assembly. A slide hammer may be needed to remove the hub bearing assembly.

➡The wheel hub bearing assembly is not repairable; it must be replaced when defective.

To install:

3. Install or connect the following:
   - Wheel hub bearing assembly, the washer and the wheel bearing locknut. Torque the wheel bearing locknut to 137–188 ft. lbs. (186–255 Nm)
4. Verify that the wheel bearings operate smoothly.
   - New cotter pin into the spindle to hold the wheel bearing locknut
5. Install a dial micrometer to the rear wheel hub bearing assembly and check the axial end-play; it should be less than 0.0020 in. (0.05mm).
   - Grease cap
   - ABS wheel sensor and its wiring
   - Brake assembly and the wheels

Q45 MODELS

1. Before servicing the vehicle, refer to the precautions in the beginning of this section.
2. Remove or disconnect the following:
   - Cotter pin and adjusting cap and loosen the wheel bearing nut. Carefully tap the end of the axle shaft or use a puller to loosen the shaft from the hub
   - Brake caliper and rotor. Do not let the caliper hang by the brake hose, support it with wire.
   - Parking brake assembly
   - Nuts and through-bolts to remove the axle housing from the suspension. If equipped with rear wheel steering, use a ball joint press to separate the tie rod end
   - 4 bolts at the back and remove the bearing flange and hub from the bearing housing
3. Press the hub out of the bearing

flange and use a puller to remove the bearing from the hub. If it is not damaged, the hub can be used again but the bearing and flange are supplied as a single unit.

To install:

➡The wheel bearing and flange are supplied as an assembly.

4. Place the hub on a press table and press the new bearing and flange onto the hub. Be sure the press tool contacts only the inner bearing race and take care not to damage the seal.
5. Install or connect the following:
   - Bearing flange onto the axle housing. Torque the bolts to 58–72 ft. lbs. (78–98 Nm)
   - Axle housing onto the lower ball joint, torque the nut to 58–69 ft. lbs. (78–93 Nm) and install a new cotter pin
   - Torque the tie rod end nut to 33–44 ft. lbs. (45–60 Nm) and install a new cotter pin, if equipped with rear wheel steering
   - Axle shaft into the hub and install the bolts through the suspension bushings. Tighten the bolts temporarily, they will be tightened with the vehicle resting on the wheels
   - Brake components and apply the brake to hold the hub from turning
   - Wheel bearing locknut and torque it to 152–203 ft. lbs. (206–275 Nm)
   - Insulator and adjusting cap and a new cotter pin
   - Wheel and lower the vehicle to the ground. Torque the suspension bushing bolts to 57–72 ft. lbs. (77–98 Nm)

KIA

1998–00
Sephia

13

PRECAUTIONS

Before servicing any vehicle, please be sure to read all of the following precautions, which deal with personal safety, prevention of component damage, and important points to take into consideration when servicing a motor vehicle:

- Never open, service or drain the radiator or cooling system when the engine is hot; serious burns can occur from the steam and hot coolant.
- Observe all applicable safety precautions when working around fuel. Whenever servicing the fuel system, always work in a well-ventilated area. Do not allow fuel spray or vapors to come in contact with a spark, open flame, or excessive heat (a hot drop light, for example). Keep a dry chemical fire extinguisher near the work area. Always keep fuel in a container specifically designed for fuel storage; also, always properly seal fuel containers to avoid the possibility of fire or explosion. Refer to the additional fuel system precautions later in this section.
- Fuel injection systems often remain pressurized, even after the engine has been turned **OFF**. The fuel system pressure must be relieved before disconnecting any fuel lines. Failure to do so may result in fire and/or personal injury.
- Brake fluid often contains polyglycol ethers and polyglycols. Avoid contact with the eyes and wash your hands thoroughly after handling brake fluid. If you do get brake fluid in your eyes, flush your eyes with clean, running water for 15 minutes. If eye irritation persists, or if you have taken

brake fluid internally, IMMEDIATELY seek medical assistance.

- The EPA warns that prolonged contact with used engine oil may cause a number of skin disorders, including cancer! You should make every effort to minimize your exposure to used engine oil. Protective gloves should be worn when changing oil. Wash your hands and any other exposed skin areas as soon as possible after exposure to used engine oil. Soap and water, or waterless hand cleaner should be used.
- All new vehicles are now equipped with an air bag system. The system must be disabled before performing service on or around system components, steering column, instrument panel components, wiring and sensors. Failure to follow safety and disabling procedures could result in accidental air bag deployment, possible personal injury and unnecessary system repairs.
- Always wear safety goggles when working with, or around, the air bag system. When carrying a non-deployed air bag, be sure the bag and trim cover are pointed away from your body. When placing a non-deployed air bag on a work surface, always face the bag and trim cover upward, away from the surface. This will reduce the motion of the module if it is accidentally deployed. Refer to the additional air bag system precautions later in this section.
- Clean, high quality brake fluid from a sealed container is essential to the safe and proper operation of the brake system. You should always buy the correct type of brake

fluid for your vehicle. If the brake fluid becomes contaminated, completely flush the system with new fluid. Never reuse any brake fluid. Any brake fluid that is removed from the system should be discarded. Also, do not allow any brake fluid to come in contact with a painted surface; it will damage the paint.

- Never operate the engine without the proper amount and type of engine oil; doing so WILL result in severe engine damage.
- Timing belt maintenance is extremely important! Many models utilize an interference-type, non-freewheeling engine. If the timing belt breaks, the valves in the cylinder head may strike the pistons, causing potentially serious (also time-consuming and expensive) engine damage. Refer to the maintenance interval charts in the front of this manual for the recommended replacement interval for the timing belt, and to the timing belt section for belt replacement and inspection.
- Disconnecting the negative battery cable on some vehicles may interfere with the functions of the on-board computer system(s) and may require the computer to undergo a relearning process once the negative battery cable is reconnected.
- When servicing drum brakes, only dissemble and assemble one side at a time, leaving the remaining side intact for reference.
- Only an MVAC-trained, EPA-certified automotive technician should service the air conditioning system or its components.

ENGINE REPAIR

Ignition Timing

This vehicle is equipped with a Distributorless Ignition System (DIS). No adjustment is necessary or possible.

Alternator

REMOVAL

1. Before servicing the vehicle, refer to the precautions in the beginning of this section.
2. Remove or disconnect the following:
   - Negative battery cable
   - Front air intake inlet pipe bolts
   - Top hose from the air intake inlet pipe

- Air intake inlet pipe clamp and the pipe
- Alternator **B** terminal cover cap
- Alternator **B** terminal nut and the **B** terminal lead
- Alternator **L** and **S** electrical connections
- Alternator pivot bolt and the tensioner mounting bolt, loosen but do not remove
- Drive belt(s); relieve tension on the belt by rotating the adjustment bolt
- Alternator tensioner mounting bolt and the belt tensioner
- Alternator pivot bolt

3. Loosen the bolt at the base of the adjusting bracket and rotate the bracket up.

4. Remove or disconnect the following:
   - Alternator

INSTALLATION

1. Before servicing the vehicle, refer to the precautions in the beginning of this section.

2. Install or connect the following:
   - Alternator
   - Alternator pivot bolt and hand-tighten at this time

3. Rotate the adjusting bracket into position, place the belt tensioner into position. And hand-tighten the mounting bolt.

4. Install or connect the following:
   - Drive belt

5. Adjust the belt tension by rotating the adjustment bolt. The belt deflection is as follows:
   a. New belt: 0.23–0.31 in. (6–8mm).
   b. Used belt: 0.28–0.35 in. (7–9mm).

6. Tighten the tensioner bolt to 14–19 ft. lbs. (19–26 Nm) and the pivot bolt to 28–38 ft. lbs. (38–51 Nm).

7. Install or connect the following:
- Alternator **L** and **S** electrical connections
- Alternator **B** terminal lead and nut
- Alternator **B** terminal cover cap
- Air intake inlet pipe and fasten the clamp
- Top hose to the air intake inlet pipe
- Front air intake inlet pipe bolts
- Negative battery cable

Engine Assembly

REMOVAL & INSTALLATION

1. Before servicing the vehicle, refer to the precautions in the beginning of this section.

2. Properly relieve the fuel system pressure.

3. Remove or disconnect the following:
- Battery cables
- Battery and tray

4. Drain and recycle the engine coolant.

5. Remove or disconnect the following:
- Data Link (DLC) from the Mass Air Flow (MAF) sensor bracket
- Intake Air Temperature (IAT) and Mass Air Flow (MAF) sensor connectors
- Air intake hose from the throttle body
- Ventilation hose and fresh air duct
- Accelerator and, if equipped, cruise control cables
- Air cleaner assembly
- Brake booster and purge control vacuum hoses from the intake manifold
- Upper and lower radiator hoses
- Fuel hose from the fuel injector rail
- Heater hoses
- Idle Air Control (IAC) and Throttle Position (TP) sensor connectors
- Fuel injector electrical connectors
- Starter and generator electrical connectors
- Engine ground strap
- Left and right splash shields
- Exhaust manifold heat shield and 3 power steering pump bracket bolts
- Air conditioning compressor mounting bolts and position it aside leaving the hoses attached
- Ground strap from the top of the transaxle

- No. 4 engine mount
- Input/turbine speed sensor connector, if equipped with an automatic transaxle
- Back-up light switch, if equipped with a manual transaxle
- Vehicle Speed Sensor (VSS)
- U-clip from the selector cable and the nut and washer from the transaxle linkage, if equipped with an automatic transaxle
- Linkage and extension bar, then the clutch release cylinder and hydraulic hose, if equipped with a manual transaxle
- Transaxle range switch and solenoid valve connectors, then the 2 transaxle oil cooler hoses, if equipped with an automatic transaxle
- Front wheels
- Oxygen (O_2) sensor electrical connectors
- Front exhaust pipe
- Halfshafts

6. Properly support the engine/transaxle assembly.

7. Remove or disconnect the following:
- 2 No. 2 engine mount-to-mounting member mounting bolts
- 1 bolt from the No. 3 engine mount
- 3 No. 1 engine mounting bolts, then carefully lift the engine/transaxle assembly from the vehicle

To install:

8. Installation is the reverse of the removal procedure. Note the following important steps.

9. When possible, leave the engine mounting nuts/bolts loose (hand tight) until all mounts are aligned and bolted. This may help in aligning the engine/transaxle assembly in the vehicle.

10. Tighten the engine mount bolts/nuts as follows:
- No. 1 mounting bolts: 50–70 ft. lbs. (67–93 Nm)
- No. 3 mounting bolts: 63–86 ft. lbs. (85–116 Nm)
- No. 2 mounting nuts: 28–38 ft. lbs. (38–51 Nm)
- No. 4 mounting bolts: 47–66 ft. lbs. (64–89 Nm)
- No. 4 mounting nuts: 49–68 ft. lbs. (68–93 Nm)

11. Install new circlips on the inner CV-joint stub shafts, if equipped, intermediate shaft. Grease the shaft splines before installing the halfshaft/intermediate shaft into the transaxle.

12. Always install new gaskets and/or O-rings. Use new self-locking nuts, especially on the exhaust.

13. Fill the engine and the transaxle with the proper types and quantities of oil. Fill the cooling system.

14. Connect the negative battery cable, start the engine and check for leaks. Check all fluid levels.

Water Pump

REMOVAL & INSTALLATION

1. Before servicing the vehicle, refer to the precautions in the beginning of this section.

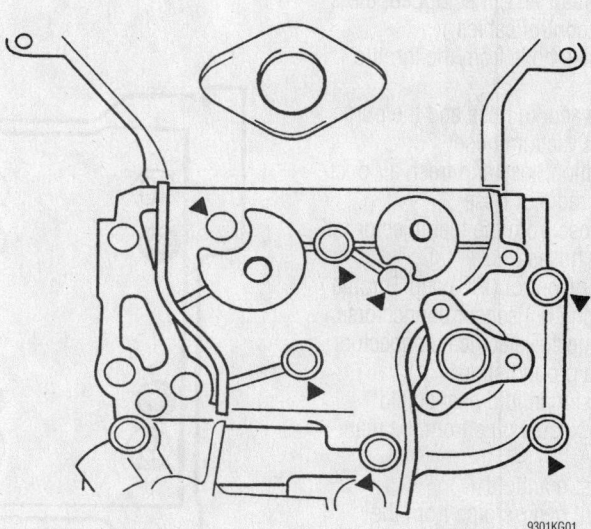

9301KG01

Water pump mounting bolt locations (arrows)—1.8L engine

2. Disconnect the negative battery cable.
3. Drain and recycle the engine coolant.
4. Remove or disconnect the following:
   • Timing belt, tensioner and idler pulleys
   • Water pump mounting bolts, then the pump
5. Clean all gasket mating surfaces.

To install:

6. Install or connect the following:
   • Water pump, using a new gasket and tighten the mounting bolts to 14–19 ft. lbs. (19–26 Nm)
   • Tensioner and idler pulleys
   • Timing belt
7. Fill the engine coolant.
8. Connect the negative battery cable.
9. Start the engine and check for leaks.

Cylinder Head

REMOVAL & INSTALLATION

1. Before servicing the vehicle, refer to the precautions in the beginning of this section.
2. Disconnect the negative battery cable.
3. Properly relieve the fuel system pressure.
4. Drain and recycle the engine coolant.
5. Drain and recycle the engine oil.
6. Remove or disconnect the following:
   • Positive Crankcase Ventilation (PCV) and crankcase ventilation hoses
   • Accelerator and, if equipped, the cruise control cables
   • Air intake hose from the throttle body
   • Brake vacuum hose and the purge control vacuum hose
   • Ventilation hose and fresh air duct
   • Upper radiator hose
   • Fuel hose from the fuel injector rail
   • Heater hoses
   • Idle Air Control (IAC) and Throttle Position (TP) sensor connectors
   • Fuel injector electrical connectors
   • Engine ground strap
   • Exhaust manifold heat shield
   • Front exhaust pipe from the manifold
   • Exhaust manifold
   • Coolant bypass pipe from the cylinder head
   • Intake manifold support bracket
   • Camshaft and Hydraulic Lash Adjusters (HLA's)

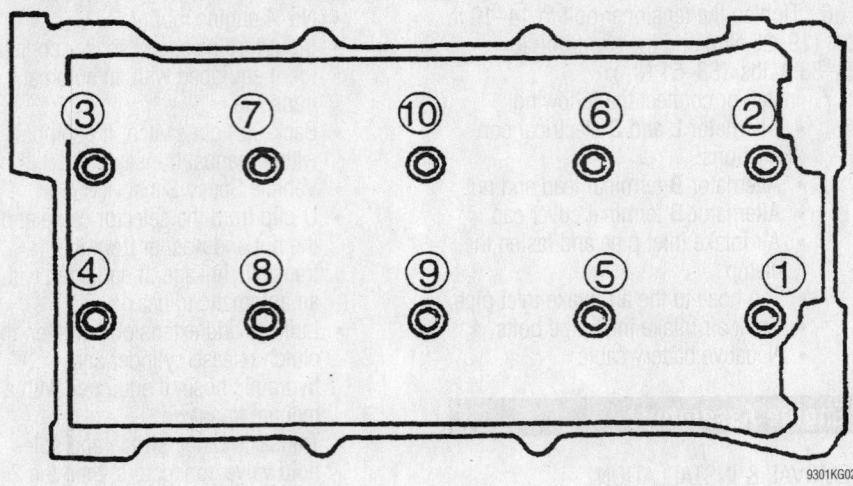

Cylinder head bolt removal sequence—1.8L engine

• Cylinder head bolts in several steps, in the order illustrated
• Cylinder head bolts, then the cylinder head

To install:

7. Thoroughly, clean the cylinder head and the block contact surfaces. Examine the head gasket and check the cylinder head for cracks. Check the cylinder head for warpage using a feeler gauge and straightedge. The maximum allowable distortion is 0.006 in. (0.15mm).
8. Clean the cylinder head bolts and the threads in the block. Be sure the bolts turn freely in the block.
9. Install or connect the following:
   • New head gasket on the engine block
   • Cylinder head
   • Cylinder head bolts

10. Tighten the head bolts in the following step using the proper sequence:
   a. Step 1: Tighten to 36 ft. lbs. (49 Nm).
   b. Step 2: Loosen the bolts in the reverse order shown.
   c. Step 3: Tighten to 29 ft. lbs. (39 Nm).
   d. Step 4: Tighten 90° (¼ turn).
   e. Step 5: Tighten 90° (¼ turn).
11. Install or connect the following:
   • Camshaft and HLA's
   • Intake manifold support bracket and tighten the mounting bolts to 28–38 ft. lbs. (37–52 Nm)
   • Coolant bypass pipe and tighten the mounting bolt to 66–86 ft. lbs. (89–117 Nm)
   • Exhaust manifold, tighten the manifold-to-cylinder head mounting

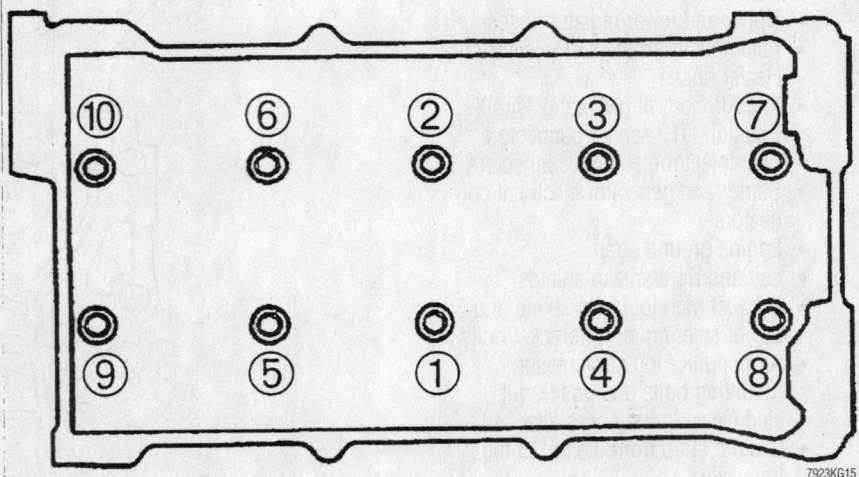

Cylinder head torque sequence—1.8L engine

nuts to 28–34 ft. lbs. (38–46 Nm) and the manifold-to-exhaust pipe mounting nuts to 16–21 ft. lbs. (22–28 Nm)

- Exhaust manifold heat shield and tighten the mounting bolts to 13–22 ft. lbs. (19–30 Nm)

12. The remaining steps of the installation procedure is the reverse of the removal, while keeping in mind the following:
- Attach all hoses and connectors
- Fill the engine oil and coolant

- Start the vehicle and check for leaks

Rocker Arms/Shafts

REMOVAL & INSTALLATION

The engine covered in this section are not equipped with rocker arms/shafts. The camshafts directly actuate the valve through a bucket type follower.

Intake Manifold

REMOVAL & INSTALLATION

1. Before servicing the vehicle, refer to the precautions in the beginning of this section.

2. Properly relieve the fuel system pressure. Disconnect the negative battery cable and drain the cooling system.

3. Remove or disconnect the following:

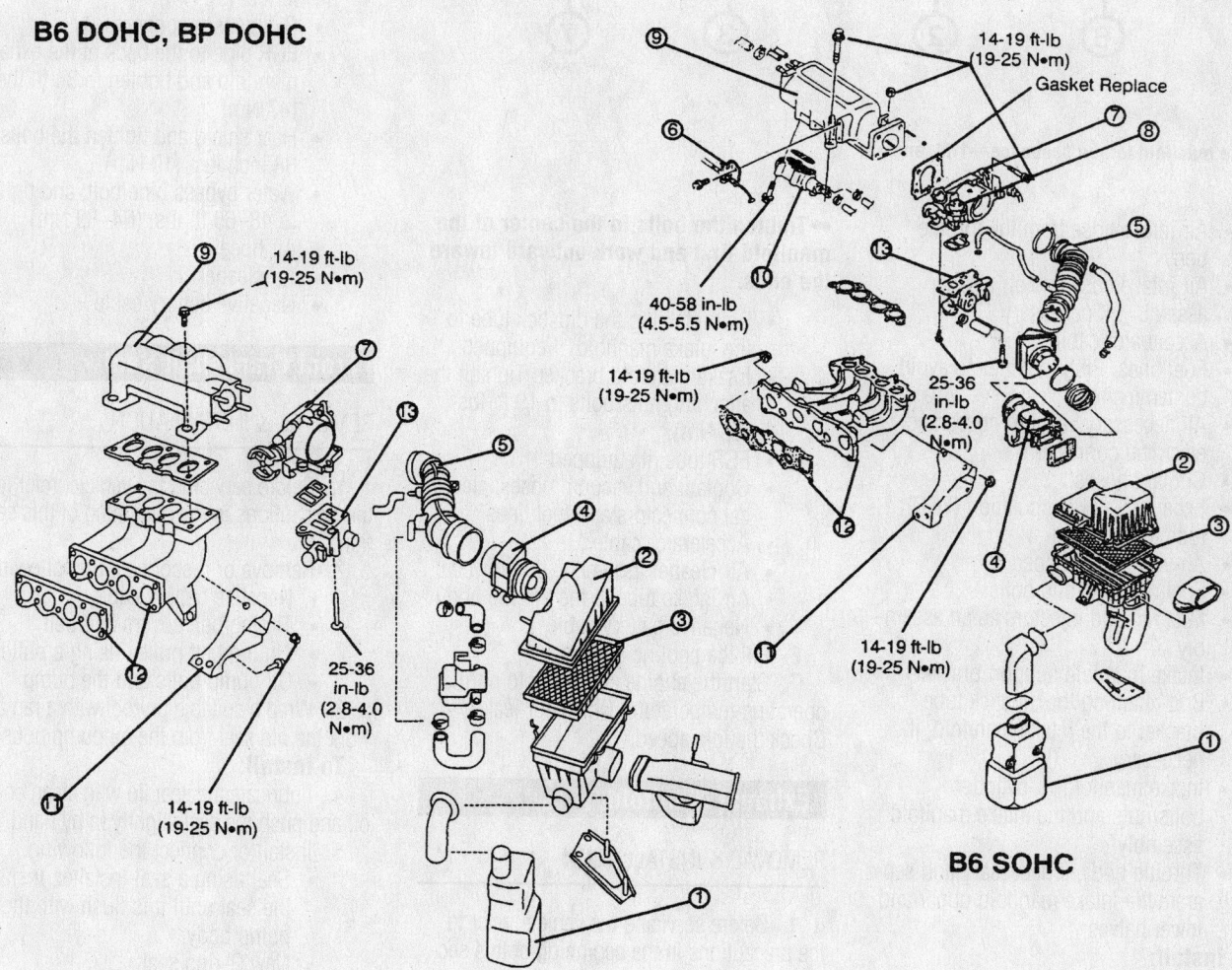

B6 DOHC, BP DOHC

B6 SOHC

1. Resonance chamber
2. Upper air filter housing
3. Air filter (B6 SOHC)
4. Mass air flow (MAF) sensor (B6 DOHC, BP DOHC)/ Volume air flow (VAF) sensor (B6 SOHC)
5. Intake air hose
6. Throttle cable
7. Throttle body
8. Dashpot (B6 SOHC)
9. Dynamic chamber
10. Air valve (B6 SOHC)
11. Intake manifold support bracket
12. Intake manifold and gasket (Replace)
13. Idle air control valve (B6 SOHC)/ Bypass air control (BAC) valve (B6 DOHC, BP DOHC)

Exploded view of the intake manifold assembly

7923KG17

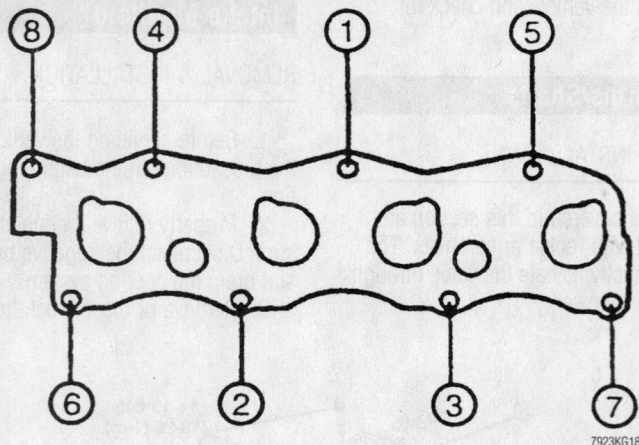

Intake manifold torque sequence—1.8L engine

- Air intake hose from the throttle body
- Air intake hose and air cleaner assembly, if necessary
- Accelerator cable
- Fuel lines. Plug the lines to avoid contamination.
- All necessary vacuum hoses and electrical connectors
- Coolant hoses
- Exhaust Gas Recirculation (EGR) tube, if equipped
- Air valve, if equipped
- Fuel rail attaching bolts
- Fuel rail and injectors as an assembly
- Intake manifold support bracket
- Bolt retaining the dipstick tube bracket to the intake manifold, if necessary
- Intake manifold-to-cylinder bolts/nuts and the intake manifold assembly
- Throttle body, if necessary and separate the intake manifold upper and lower halves

To install:

4. Clean all gasket mating surfaces.
5. Install or connect the following:
    - Upper and lower intake manifolds using a new gasket, if separated. Tighten the nuts/bolts to 19 ft. lbs. (25 Nm).
    - Throttle body using a new gasket, if removed. Tighten the retaining nuts/bolts to 19 ft. lbs. (25 Nm).
    - Intake manifold assembly to the cylinder head using a new gasket. Tighten the nuts/bolts to 19 ft. lbs. (25 Nm).

➡ **Tighten the bolts in the center of the manifold first and work outward toward the ends.**

- Bolt retaining the dipstick tube to the intake manifold, if equipped
- Intake manifold bracket. Tighten the attaching nuts/bolts to 19 ft. lbs. (25 Nm).
- EGR tube, if equipped
- Coolant and vacuum hoses, electrical connectors and fuel lines
- Accelerator cable
- Air cleaner assembly, if removed
- Air intake tube to the throttle body
- Negative battery cable

6. Fill the cooling system.
7. Start the engine and bring to normal operating temperature. Check for leaks. Check the idle speed.

Exhaust Manifold

REMOVAL & INSTALLATION

1. Before servicing the vehicle, refer to the precautions in the beginning of this section.
2. Remove or disconnect the following:
    - Negative battery cable
    - Air cleaner
    - Air hose
    - Water bypass pipe bolt
    - Exhaust manifold heat shield bolts and the heat shield
    - Oxygen (O2S) sensor electrical connector
    - Exhaust pipe-to-exhaust manifold nuts and discard the nuts. Suspend the exhaust system with wire.

- Exhaust Gas Recirculation (EGR) pipe from the exhaust manifold
- Nuts, bolts and the exhaust manifold. Discard the nuts.

To install:

3. Clean all gasket mating surfaces.
4. Install or connect the following:
    - New exhaust manifold gasket and the exhaust manifold. Tighten the mounting nuts and bolts to 29–34 ft. lbs. (39–47 Nm).
    - Exhaust pipe to the manifold. Install new nuts and tighten to 38 ft. lbs. (52 Nm).
    - O2S sensor connector
    - EGR pipe to the back of the exhaust manifold and tighten to 34 ft. lbs. (47 Nm)
    - Heat shield and tighten the bolts to 88 inch lbs. (10 Nm).
    - Water bypass pipe bolt, and tighten to 48–65 ft. lbs. (64–89 Nm)
    - Air hose
    - Air cleaner
    - Negative battery cable

Front Crankshaft Seal

REMOVAL & INSTALLATION

1. Before servicing the vehicle, refer to the precautions in the beginning of this section.
2. Remove or disconnect the following:
    - Negative battery cable
    - Timing belt covers and belt
    - Timing belt pulley using a puller
    - Oil pump bolts and the pump
3. Wrap a suitable prytool with a rag and work the old seal from the oil pump housing.

To install:

4. Lubricate the seal lip with clean engine oil and push the seal slightly in by hand.
5. Install or connect the following:
    - Seal using a seal installer. Install the seal until it is flush with the oil pump body.
    - New O-ring seal
6. Apply a 0.04–0.08 in. (1–2mm) bead of silicone to the oil pump body as shown in the accompanying illustration.

7. Install or connect the following:
    - Oil pump
    - Timing belt pulley. Align the keyway groove on the pulley to the keyway on the crankshaft.
    - Woodruff key with the tapered side towards the oil pump body
    - Remaining components in the reverse order of removal

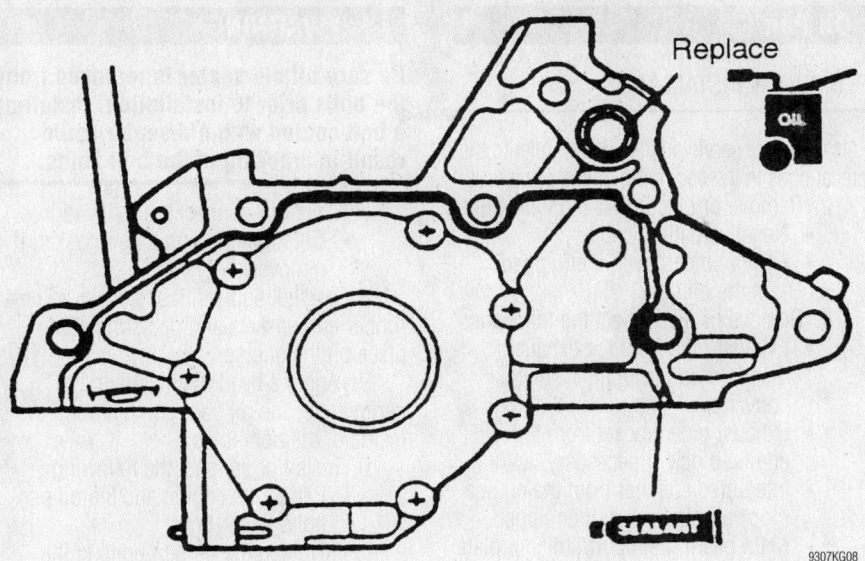

Apply a 0.04–0.08 in. (1–2mm) bead of silicone to the oil pump body

To install:

9. Apply a coat of the clean engine oil to the sides of the HLA's.

10. Install or connect the following:
- HLA's into the cylinder head bore

11. Check that the HLA moves freely in its bore.

12. Lubricate the camshaft journals and lobes with clean engine oil.

13. Install or connect the following:
- Camshafts in the cylinder head making sure that the camshaft dowel pins point straight up
- Camshaft caps in their original positions. Loosely install the cap bolts.
- Camshaft cap bolts in 5–6 steps to 13–20 ft. lbs. (18–27 Nm) in the proper sequence

14. Oil the lip of the new camshaft oil seal and, using a seal installer. Drive the

Camshaft

REMOVAL & INSTALLATION

1. Before servicing the vehicle, refer to the precautions in the beginning of this section.

2. Remove or disconnect the following:
- Negative battery cable
- Ignition coils and high-tension cords
- Positive Crankcase Ventilation (PCV) valve and ventilation hoses

3. Position the engine so that No. 1 cylinder is at Top Dead Center (TDC).

4. Remove or disconnect the following:
- Timing belt
- Cylinder head cover
- Camshaft Position (CMP) sensor
- Camshaft sprockets

5. Loosen the camshaft bearing cap retaining bolts in several steps, following the proper sequence.

6. Remove or disconnect the following:
- Camshafts
- Camshaft oil seal using a seal removal tool

➡The Hydraulic Lash Adjusters (HLA's) must be installed in the location from which they were removed.

7. Mark the HLA's to identify their original positions.

8. Remove the HLA's from the cylinder head using a magnet, and store upside-down in a oil-filled container

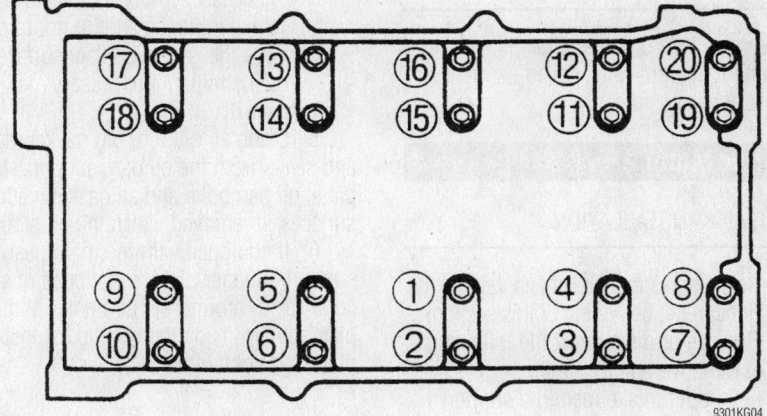

Camshaft bearing cap mounting bolt removal sequence—1.8L engine

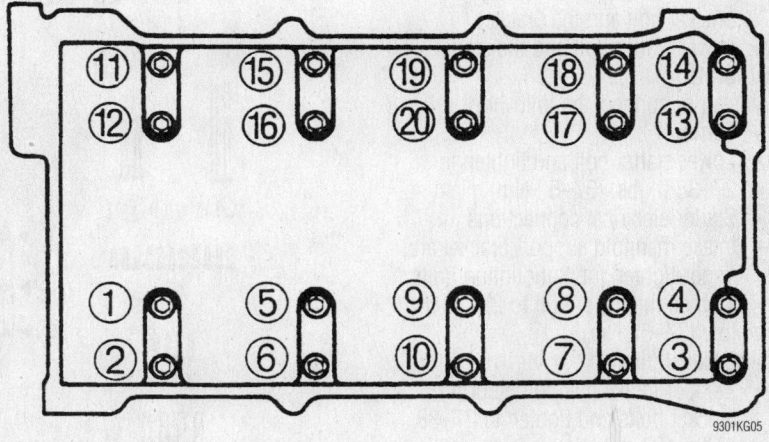

Camshaft bearing cap mounting bolt tightening sequence—1.8L engine

Timing belt service is covered in Section 3 of this manual

seal into the cylinder head until it is flush with the edge of the camshaft bearing cap.

15. Install or connect the following:
- Camshaft sprockets onto the camshaft. Be sure to align the **I** mark with the intake camshaft dowel pin and the **E** mark with the exhaust camshaft dowel pin, then tighten the retaining bolt to 36–44 ft. lbs. (49–61 Nm).
- CMP sensor
- Cylinder head cover
- Timing belt
- PCV and ventilation hoses
- Ignition coils and high-tension cords
- Negative battery cable

16. Check the engine fluid levels, start the vehicle and check for leaks.

Valve Lash

ADJUSTMENT

The valve lash on all engines is kept in adjustment hydraulically. No adjustment is necessary or possible.

Starter Motor

REMOVAL & INSTALLATION

1. Before servicing the vehicle, refer to the precautions in the beginning of this section.
2. Remove or disconnect the following:
- Negative battery cable
- 2 upper intake manifold support bracket bolts
- Starter electrical connections
- 2 upper starter bolts
- Exhaust pipe
- Lower intake manifold support bracket bolt and the bracket
- Lower starter bolt and the starter

To install:
3. Install or connect the following:
- Starter
- Lower starter bolt and tighten to 27–38 ft. lbs. (37–52 Nm)
- Starter electrical connections
- Intake manifold support bracket and finger-tighten the 3 mounting bolts. Tighten the lower bolt to 27–38 ft. lbs. (37–52 Nm).
- Exhaust pipe
- 2 upper intake manifold support bracket bolts and tighten to 27–38 ft. lbs. (37–52 Nm)
- 2 upper starter bolts and tighten to 27–38 ft. lbs. (37–52 Nm)
- Negative battery cable

Oil Pan

REMOVAL & INSTALLATION

1. Before servicing the vehicle, refer to the precautions in the beginning of this section.
2. Remove or disconnect the following:
- Negative battery cable
- Engine undercover, if equipped
3. Drain the oil.
4. Remove or disconnect the following:
- Exhaust pipe from the exhaust manifold and from the catalytic converter
- Exhaust pipe bracket from the engine block, if necessary
- Integrated stiffener from the engine block and transaxle, if equipped
- Main bearing support/stiffener plate that is installed between the oil pan and engine block, if equipped
- Bolts and the oil pan. It may be necessary to pry the pan away from the engine; be careful not to damage the gasket contact surfaces.
- Oil strainer, if necessary

To install:
5. Clean all oil, dirt, old gasket material and sealer from the oil pan, support/stiffener plate, oil pan bolts and all gasket mating surfaces. If removed, clean the oil strainer.

6. If equipped with the main bearing support/stiffener plate, run a bead of silicone sealer around the perimeter of the plate, going inside the bolt holes. Install the plate and tighten the bolts.

✳✳ WARNING

Be sure all old sealer is removed from the bolts prior to installation. Installing a bolt coated with old sealer could result in cracking of the bolt holes.

7. Install or connect the following:
- Oil strainer using a new gasket, if removed
8. If used, apply silicone sealer to new rubber end gaskets and press them into place on the engine.
9. Apply a bead of silicone to the perimeter of the oil pan, going around the inside of the bolt holes.
10. Install or connect the following:
- Pan to the engine and the oil pan bolts finger-tight
- Tighten the oil pan bolts to the specifications shown in the accompanying illustrations

✳✳ WARNING

Be sure all old sealer is removed from the bolts prior to installation. Installing a bolt coated with old sealer could result in cracking of the bolt holes.

- Integrated stiffener to the engine block and transaxle, if removed. Tighten the bolts to 38 ft. lbs. (52 Nm).
- Transverse member, if removed. Tighten the bolts to 93 ft. lbs. (126 Nm).
- Front exhaust pipe bracket, if equipped

Tightening torque:
Ⓐ Ⓑ Ⓒ Ⓓ : 5.8~8.0 lb-ft
(7.8~10.8 N·m, 0.8~1.1 kg-m)
Ⓔ : 28~38 lb-ft (38~51 N·m, 3.8~5.3 kg-m)

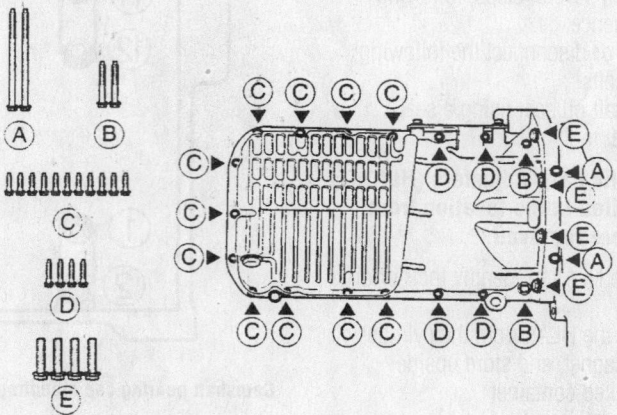

9307KG14

Oil pan mounting bolt locations and torque specifications—1.8L engine

- Front exhaust pipe, using new gaskets. Tighten the exhaust manifold flange nuts to 34 ft. lbs. (46 Nm).
- Oil pan drain plug using a new gasket. Tighten the drain plug to 30 ft. lbs. (41 Nm).
- Engine undercover

11. Fill the engine with the proper type and quantity of oil.

12. Connect the negative battery cable. Start the engine and bring to normal operating temperature. Check for leaks.

Oil Pump

REMOVAL & INSTALLATION

1. Before servicing the vehicle, refer to the precautions in the beginning of this section.

2. Remove or disconnect the following:
- Negative battery cable
- Crankshaft pulley, timing belt cover, belt and the crankshaft sprocket
- Oil pan
- Oil pick-up tube and discard the gasket
- Oil pump attaching bolts
- Oil pump
- Front crankshaft seal from the oil pump if the pump is being replaced

To install:

3. Clean the oil, dirt and old sealant from all contact surfaces.

4. Install or connect the following:
- New O-rings on the oil pump

5. If the oil seal was removed from the oil pump, apply clean engine oil to the lip of the seal. Push the seal in lightly by hand. Press the seal, with a protrusion of

0.02–0.04 in. (0.5–1.0mm), into the oil pump with a suitable tool (49 B014 401).

6. Apply a bead of silicone to the oil pump at the cylinder block contact surface, going inside the bolt holes.

7. Install or connect the following:
- Oil pump and tighten the bolts to 14–18 ft. lbs. (19–25 Nm)
- New gasket and the oil pump pick-up tube. Tighten the mounting bolts to 70–95 inch lbs. (8–11 Nm).
- Oil pan
- Crankshaft sprocket, timing belt and cover
- Crankshaft pulley
- Negative battery cable

8. Fill the engine with the proper type and quantity of oil. Run the engine and check for leaks.

SOHC shown (DOHC similar)

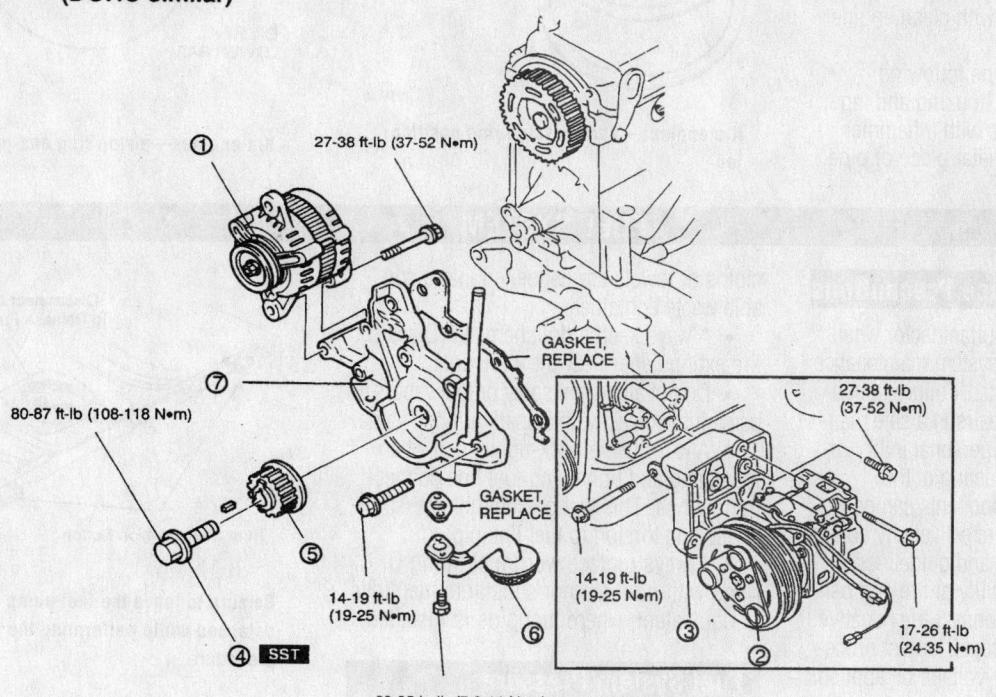

1. Generator
2. A/C compressor (if equipped)
3. A/C compressor bracket (if equipped)
4. Crankshaft pulley lock bolt
5. Timing belt pulley
6. Oil strainer
7. Oil pump

Exploded view of the oil pump

Heater Core replacement is covered in Section 2 of this manual

Rear Main Seal

REMOVAL & INSTALLATION

1. Before servicing the vehicle, refer to the precautions in the beginning of this section.
2. Remove or disconnect the following:
   - Negative battery cable
   - Transaxle assembly
   - Clutch and flywheel assembly, if equipped with a manual transaxle
   - Flexplate-to-crankshaft bolts, the flexplate and shim plates, if equipped with an automatic transaxle
3. Cut the oil seal lip with a knife. Install a rag to the housing and using a screwdriver, carefully pry the oil seal from the oil seal housing. Clean the gasket mounting surfaces.

To install:

4. Clean the oil seal housing. Coat the oil seal and the housing with clean engine oil.
5. Install or connect the following:
   - Oil seal into the housing and tap it evenly into place with a hammer and a large diameter piece of pipe.

The seal must be flush with the edge of the rear cover.
   - Flywheel assembly or the flexplate, as applicable, and tighten the mounting bolts to 71–76 ft. lbs. (97–102 Nm)
   - Clutch assembly, if applicable
   - Transaxle
   - Negative battery cable

Piston and Ring

POSITIONING

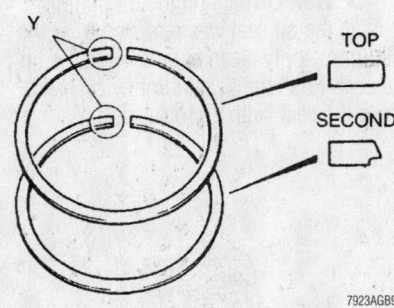

Kia engines—compression ring positioning

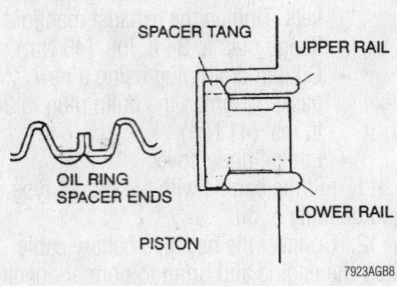

Kia engines—oil control ring positioning

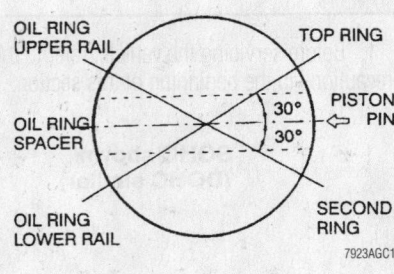

Kia engines—piston ring end-gap spacing

FUEL SYSTEM

Fuel System Service Precautions

Safety is the most important factor when performing not only fuel system maintenance but any type of maintenance. Failure to conduct maintenance and repairs in a safe manner may result in serious personal injury or death. Maintenance and testing of the vehicle's fuel system components can be accomplished safely and effectively by adhering to the following rules and guidelines.

• To avoid the possibility of fire and personal injury, always disconnect the negative battery cable unless the repair or test procedure requires that battery voltage be applied.

• Always relieve the fuel system pressure prior to disconnecting any fuel system component (injector, fuel rail, pressure regulator, etc.), fitting or fuel line connection. Exercise extreme caution whenever relieving fuel system pressure, to avoid exposing skin, face and eyes to fuel spray. Please be advised that fuel under pressure may penetrate the skin or any part of the body that it contacts.

• Always place a shop towel or cloth around the fitting or connection prior to loosening to absorb any excess fuel due to spillage. Ensure that all fuel spillage (should it occur) is quickly removed from engine surfaces. Ensure that all fuel soaked cloths or towels are deposited into a suitable waste container.

• Always keep a dry chemical (Class B) fire extinguisher near the work area.

• Do not allow fuel spray or fuel vapors to come into contact with a spark or open flame.

• Always use a back-up wrench when loosening and tightening fuel line connection fittings. This will prevent unnecessary stress and torsion to fuel line piping.

• Always replace worn fuel fitting O-rings with new. Do not substitute fuel hose or equivalent, where fuel pipe is installed.

Fuel System Pressure

RELIEVING

1. Before servicing the vehicle, refer to the precautions in the beginning of this section.
2. Release the rear seat retainers (clips or catches) and remove rear seat cushion.
3. Remove the fuel pump cover.
4. Detach the fuel pump electrical connector.
5. Start the engine, allowing it to idle until it runs out of fuel.

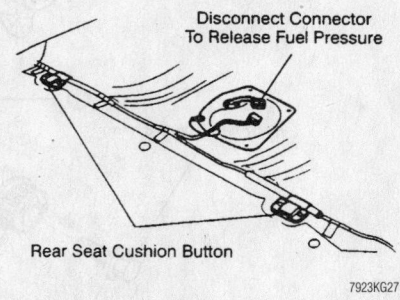

Be sure to leave the fuel pump connector detached while performing the service procedure

6. After the engine stalls, reattach the fuel pump connector and turn the ignition switch **OFF**.

Fuel Filter

REMOVAL & INSTALLATION

1. Before servicing the vehicle, refer to the precautions in the beginning of this section.
2. Relieve the fuel system pressure.
3. Remove or disconnect the following:
   - Negative battery cable

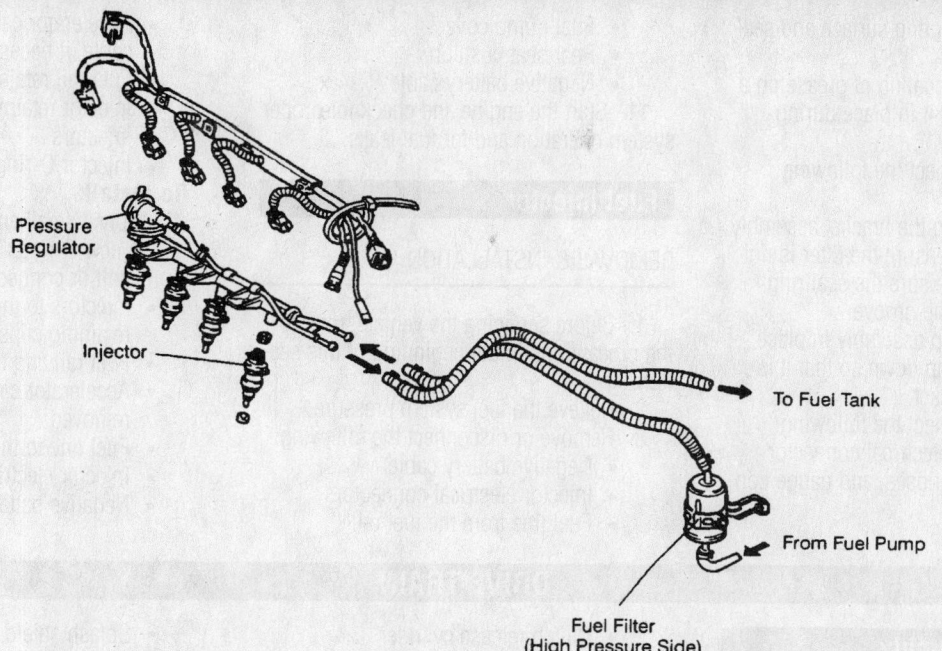

Pressure Regulator

Injector

To Fuel Tank

From Fuel Pump

Fuel Filter
(High Pressure Side)

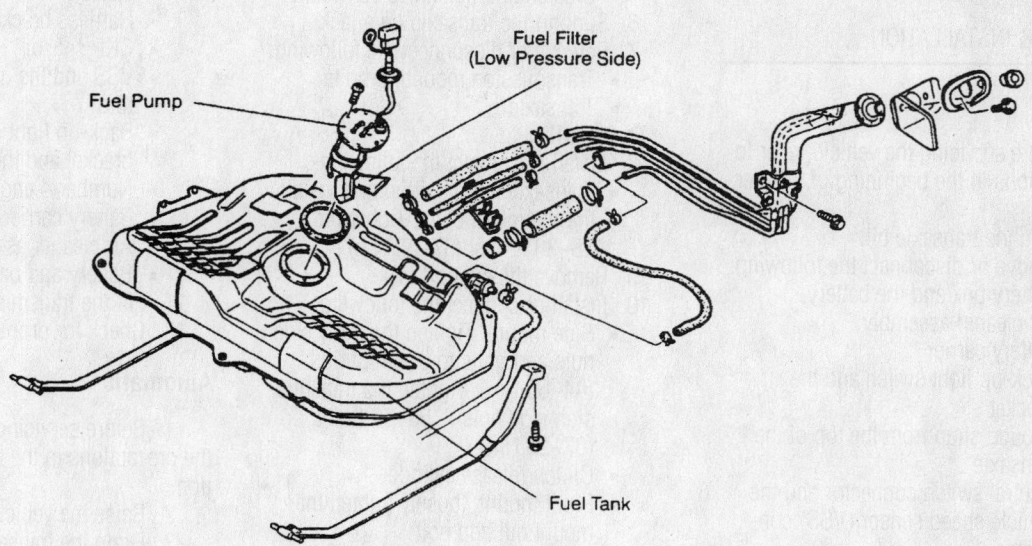

Fuel Filter
(Low Pressure Side)

Fuel Pump

Fuel Tank

7923KG28

Exploded view of the fuel system

- Fuel lines from both ends of the fuel filter. Plug the lines to prevent leakage.
- Filter and the mounting bracket

To install:

4. Install or connect the following:
- Filter in the mounting bracket
- Fuel lines to the filter
- Bracket nuts and tighten to 9.5 ft. lbs. (13 Nm)
- Negative battery cable

5. Run the engine and check for any fuel leaks.

Fuel Pump

REMOVAL & INSTALLATION

1. Before servicing the vehicle, refer to the precautions in the beginning of this section.

2. Relieve the fuel system pressure.

3. Remove or disconnect the following:
- Negative battery cable
- Rear seat cushion from the vehicle

4. Clean any dirt that has accumulated around the fuel pump cover so it will not enter the tank during pump removal and installation.

5. Remove or disconnect the following:
- Fuel pump cover
- Fuel gauge connector, hoses, and the gauge
- Fuel pump electrical connector
- Fuel pump from the bracket assembly
- Seal ring and discard

To install:

6. Clean the fuel pump mounting

Brake service is covered in Section 4 of this manual

flange, fuel tank mounting surface and seal ring groove.

7. Apply a light coating of grease on a new seal ring to hold it in place during assembly

8. Install or connect the following:
- Seal ring
- Fuel pump to the bracket assembly carefully to ensure the filter is not damaged. Be sure the seal ring remains in the groove.

9. Hold the pump assembly in place, and pull the fuel pump down so that it is tight against the bracket.

10. Install or connect the following:
- Fuel pump electrical connector
- Fuel gauge, hoses, and gauge connector

- Fuel pump cover
- Rear seat cushion
- Negative battery cable

11. Start the engine and check for proper system operation and for fuel leaks.

Fuel Injector

REMOVAL & INSTALLATION

1. Before servicing the vehicle, refer to the precautions in the beginning of this section.

2. Relieve the fuel system pressure.

3. Remove or disconnect the following:
- Negative battery cable
- Injector electrical connectors
- Fuel line from the fuel rail

- Accelerator cable bracket and the cable, if necessary
- Fuel rail retainers and the fuel rail
- Injector retaining clips and the injectors
- Injector O-rings and discard

To install:

4. Apply a small amount of clean engine oil to the new O-rings and install them.

5. Install or connect the following:
- Injectors to the fuel rail and the retaining clips
- Fuel rail and the fuel rail retainers
- Accelerator cable and bracket, if removed
- Fuel line to the fuel rail
- Injector electrical connectors
- Negative battery cable

DRIVE TRAIN

Transaxle Assembly

REMOVAL & INSTALLATION

Manual

1. Before servicing the vehicle, refer to the precautions in the beginning of this section.

2. Drain the transaxle oil.

3. Remove or disconnect the following:
- Battery box and the battery
- Air cleaner assembly
- Battery carrier
- Back-up light switch and the bracket
- Ground strap from the top of the transaxle
- Neutral switch connector and the vehicle speed sensor (VSS) connector
- Wire harness bracket and ground cable
- Crankshaft Position (CKP) sensor
- Wheels
- Splash shield
- Transverse member
- Extension bar and the change control rod
- Tie-rod ends
- Stabilizer control link
- Halfshaft and the joint shaft
- Intake manifold bracket
- Starter
- Front exhaust pipe

4. Support the engine and remove the engine mounting member.

5. Remove or disconnect the following:
- Rear engine/transmission mount
- Front engine/transmission mount

- Clutch release cylinder
- Side engine/transmission mount

6. Support the transaxle on a jack

7. Remove or disconnect the following:
- Transmission mounting bolts
- Transaxle

To install:

8. Install or connect the following:
- Transaxle into position and install the mounting bolts. Tighten to 28–38 ft. lbs. (37–52 Nm).

9. Remove the support jack.

10. Install or connect the following:
- Side mount. Tighten the body side nuts and bolts to 32–44 ft. lbs. (44–60 Nm). Tighten the transmission side nuts to 50–68 ft. lbs. (67–93 Nm).
- Clutch release cylinder
- Front mount, loosely tighten the mount nut and bolt
- Rear mount, align and set all bolts, then tighten to 50–68 ft. lbs. (67–93 Nm)
- Engine mounting member. Tighten the 4 outer nuts and bolts to 50–65 ft. lbs. (67–89 Nm) and the 2 remaining nuts to 28–38 ft. lbs. (38–51 Nm).
- Tighten the front mount nut and bolt to 50–68 ft. lbs. (67–93 Nm)
- Starter
- Manifold bracket and front exhaust pipe
- Joint shaft and the halfshaft
- Stabilizer control link
- Tie-rod ends
- Change control rod and the extension bar
- Transverse member

- Splash shield
- Wheels
- Harness bracket
- CKP sensor
- VSS and the neutral switch connectors
- Back-up light switch connector bracket and the switch
- Number 4 engine mount
- Battery carrier
- Air cleaner assembly
- Battery and battery box

11. Fill the transmission with fluid.

12. Check for proper clutch operation.

Automatic

1. Before servicing the vehicle, refer to the precautions in the beginning of this section.

2. Raise the vehicle on a hoist.

3. Drain the transaxle fluid.

4. Remove or disconnect the following:
- Battery and battery cover
- Air cleaner assembly
- Battery carrier
- Solenoid and the Transaxle Range Switch (TRS) connectors
- Selector cable
- Vehicle Speed Sensor (VSS) connector
- Harness bracket
- Throttle cable
- Front wheels
- Splash shields
- Front exhaust pipe
- Transverse member, if equipped
- Tie-rod ends
- Stabilizer control links
- Lower arm by removing the cinch bolt from the lower arm ball joints.

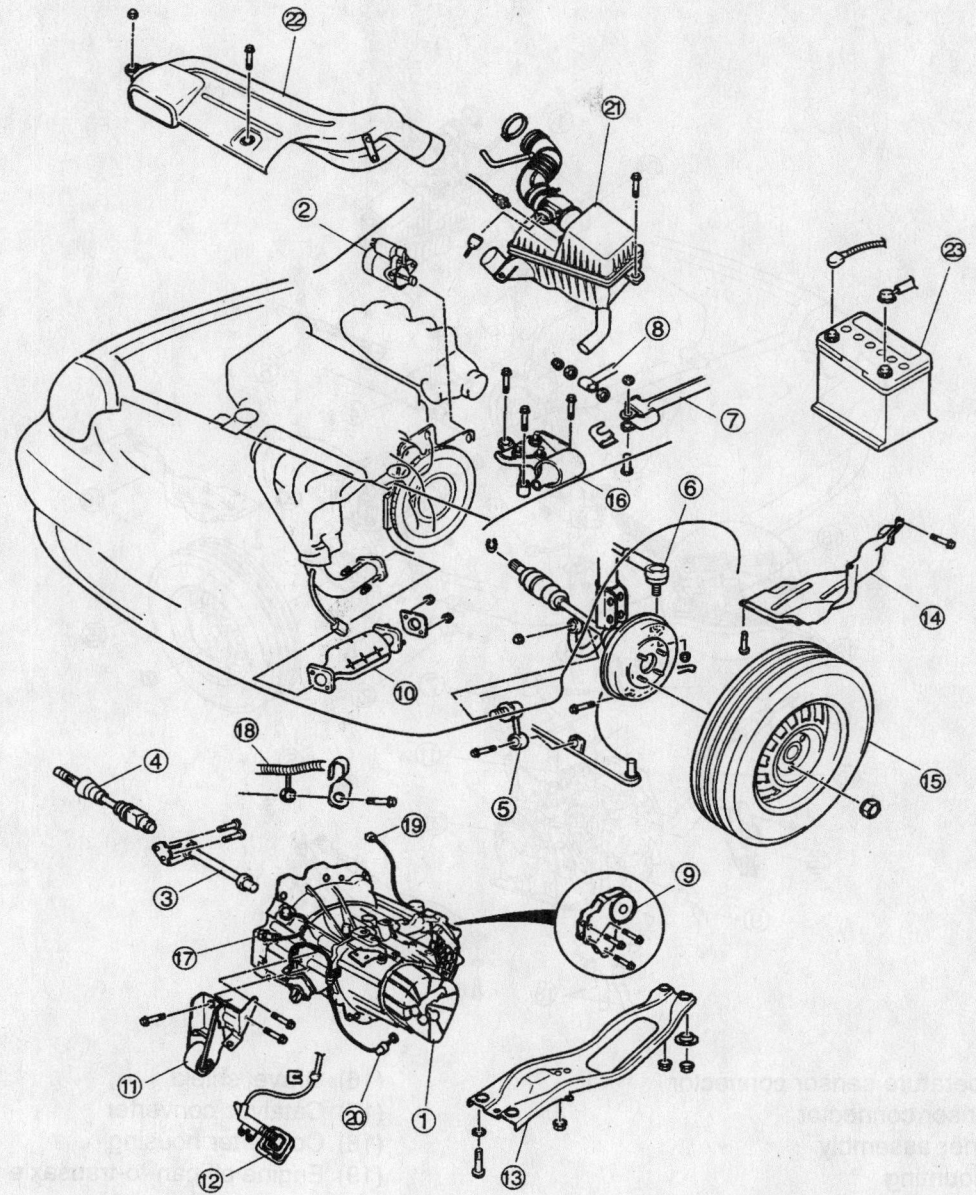

(1) Transaxle
(2) Starter
(3) Joint shaft
(4) Driveshaft
(5) Stabilizer control link
(6) Tie-rod end
(7) Change control rod
(8) Extension bar

(9) Engine mount No. 1
(10) Catalytic converter
(11) Engine mount No. 2
(12) Clutch release cylinder
(13) Engine mount member
(14) Splash shield
(15) Front wheel and tire
(16) Engine mount No. 4

(17) Crankshaft position sensor
(18) Ground
(19) Vehicle speed sensor connector
(20) Back-up switch connector
(21) Air cleaner assembly
(22) Fresh air duct
(23) Battery

9301KG07

Exploded view of the manual transaxle assembly mounting and related components—Sephia

For complete Engine Mechanical specifications, see Section 1 of this manual

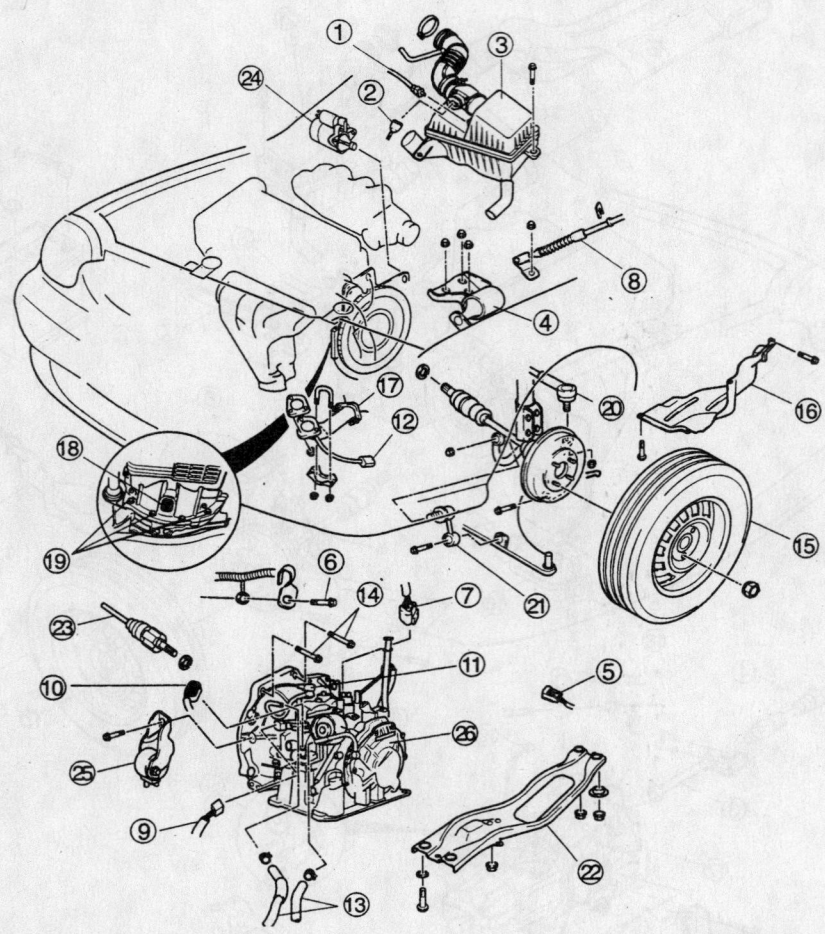

(1) Air temperature sensor connector
(2) MAF sensor connector
(3) Air cleaner assembly
(4) No. 4 mounting
(5) Input/turbine speed sensor connector
(6) Ground strap bolt
(7) Vehicle speed sensor connector
(8) Selector cable
(9) Transaxle range switch connector
(10) Solenoid valve connector
(11) Crankshaft position connector
(12) Oxygen sensor connector
(13) ATF cooler hose
(14) Upper converter housing bolts
(15) Wheel and tire

(16) Gravel shield
(17) Catalytic converter
(18) Converter housing
(19) Engine oil pan-to-transaxle mounting bolt
(20) Tie-rod end
(21) Stabilizer control link
(22) Engine mounting member
(23) Driveshaft
(24) Starter
(25) No. 2 engine mounting
(26) Auto transaxle

9301KG08

Exploded view of the automatic transaxle assembly mounting and related components—Sephia

Pry the lower arm out of the knuckle.

5. Support the engine

6. Remove or disconnect the following:
- Engine mounting member
- Left and right halfshaft. Install Differential Side Gear holder K49A-4208-AT to hold the side gears.
- Joint shaft
- Intake manifold bracket
- Starter
- Front engine/transmission mount
- Rear engine/transmission mount
- Inner and outer oil hoses
- Side engine/transmission mount

7. Hold the drive plate and remove the converter nuts.

8. Support the transaxle on a jack.

9. Remove or disconnect the following:
- Transaxle mounting bolts
- Transaxle

To install:

10. Support the transaxle on a jack and lift it into place. Align the transaxle with the engine and install the mounting bolts. Tighten to 65–86 ft. lbs. (89–116 Nm).

11. Hold the driveplate and install the torque converter mount nuts. 26–36 ft. lbs. (35–49 Nm).

12. Install or connect the following:
- Side engine/transmission mount. Loosely tighten the nuts of the transaxle side. Tighten the nuts and bolts of the body side to 32–44 ft. lbs. (44–60 Nm). Tighten the nuts of the transaxle side to 50–68 ft. lbs. (67–93 Nm).
- Inner and outer oil hoses
- Rear engine/transmission mount. Tighten the bolts to 50–68 ft. lbs. (67–93 Nm).
- Front engine/transmission mount. Tighten the mount bracket to the transaxle to 28–38 ft. lbs. (38–51 Nm). Loosely tighten the nuts and bolts of the engine mount rubber, then tighten to 50–68 ft. lbs. (67–98 Nm).
- Starter
- Manifold bracket
- Joint shaft into the transaxle
- Joint shaft to the cylinder block and tighten the bolts in sequence (counterclockwise). Tighten to 32–46 ft. lbs. (42–62 Nm).
- Halfshafts, be sure that the shafts are properly installed and do not pull out
- Engine mounting member, the mounting nuts/bolts and tighten the

nuts/bolts at the far corners to 48–65 ft. lbs. (64–89 Nm), tighten the remaining 2 nuts to 28–38 ft. lbs. (38–51 Nm).
- Front exhaust pipe
- Lower arm to the knuckle
- Stabilizer control link
- Tie-rod ends
- Transverse member, if removed
- Splash shields
- Wheels
- Throttle cable
- Harness bracket
- VSS connector
- Selector cable
- TRS connector
- Solenoid connector
- Battery carrier
- Air cleaner assembly
- Battery and battery cover

13. Test drive the vehicle. Check for proper operation in all gear ranges.

Clutch

ADJUSTMENTS

Pedal Height

1. Before servicing the vehicle, refer to the precautions in the beginning of this section.

2. Measure the distance from the upper surface of the pedal pad to the carpet.

3. The distance should be 7.83–8.15 in. (199–207mm).

4. If the distance is not as specified, loosen the locknut on the stopper bolt or switch.

5. Turn the switch or bolt until the distance is correct, then tighten the locknut.

Free-Play

1. Before servicing the vehicle, refer to the precautions in the beginning of this section.

2. Depress the clutch pedal by hand until resistance is felt. The free-play should be 0.12–0.20 in. (3.0–5.0mm).

3. If the free-play is not correct, loosen the clutch master cylinder pushrod locknut and turn the pushrod to adjust.

REMOVAL & INSTALLATION

1. Before servicing the vehicle, refer to the precautions in the beginning of this section.

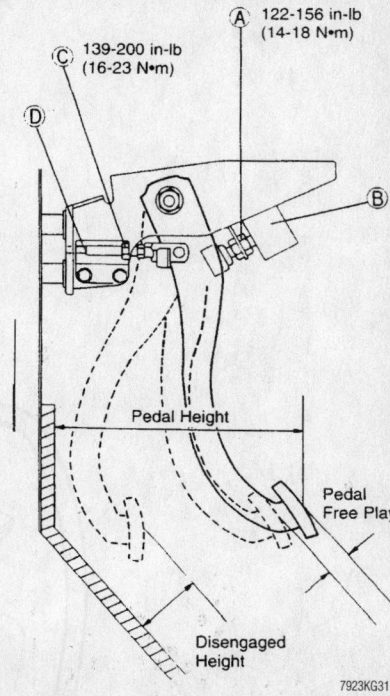

Ⓒ 139-200 in-lb (16-23 N•m)

Ⓐ 122-156 in-lb (14-18 N•m)

Ⓓ

Ⓑ

Pedal Height

Pedal Free Play

Disengaged Height

7923KG31

Clutch pedal measurement and adjustment points. (A) and (B) are for adjusting the pedal height, while (C) and (D) are for the free-play adjustment

2. Remove or disconnect the following:
- Negative battery cable
- Transaxle

3. Gradually loosen the clutch pressure plate bolts, in a crisscross pattern. Support the pressure plate and remove the bolts. Remove the pressure plate and clutch disc.

4. Inspect the pilot bearing. If it is worn or damaged and does not turn easily by hand, remove it using a puller/slide hammer.

5. Check the flywheel surface for scoring, cracks or burning and machine or replace, as necessary.

6. Install a flywheel holder to keep the flywheel from turning. Loosen the flywheel bolts evenly and gradually in a crisscross pattern. Remove the flywheel.

7. Inspect the clutch release bearing for wear. Replace it if it sticks or does not turn easily.

8. Inspect the release fork for wear or damage and replace as necessary.

To install:

9. Lubricate the release fork fingers and pivot with molybdenum grease and install in the release fork boot.

10. Install or connect the following:

For Accessory Drive Belt illustrations, see Section 1 of this manual

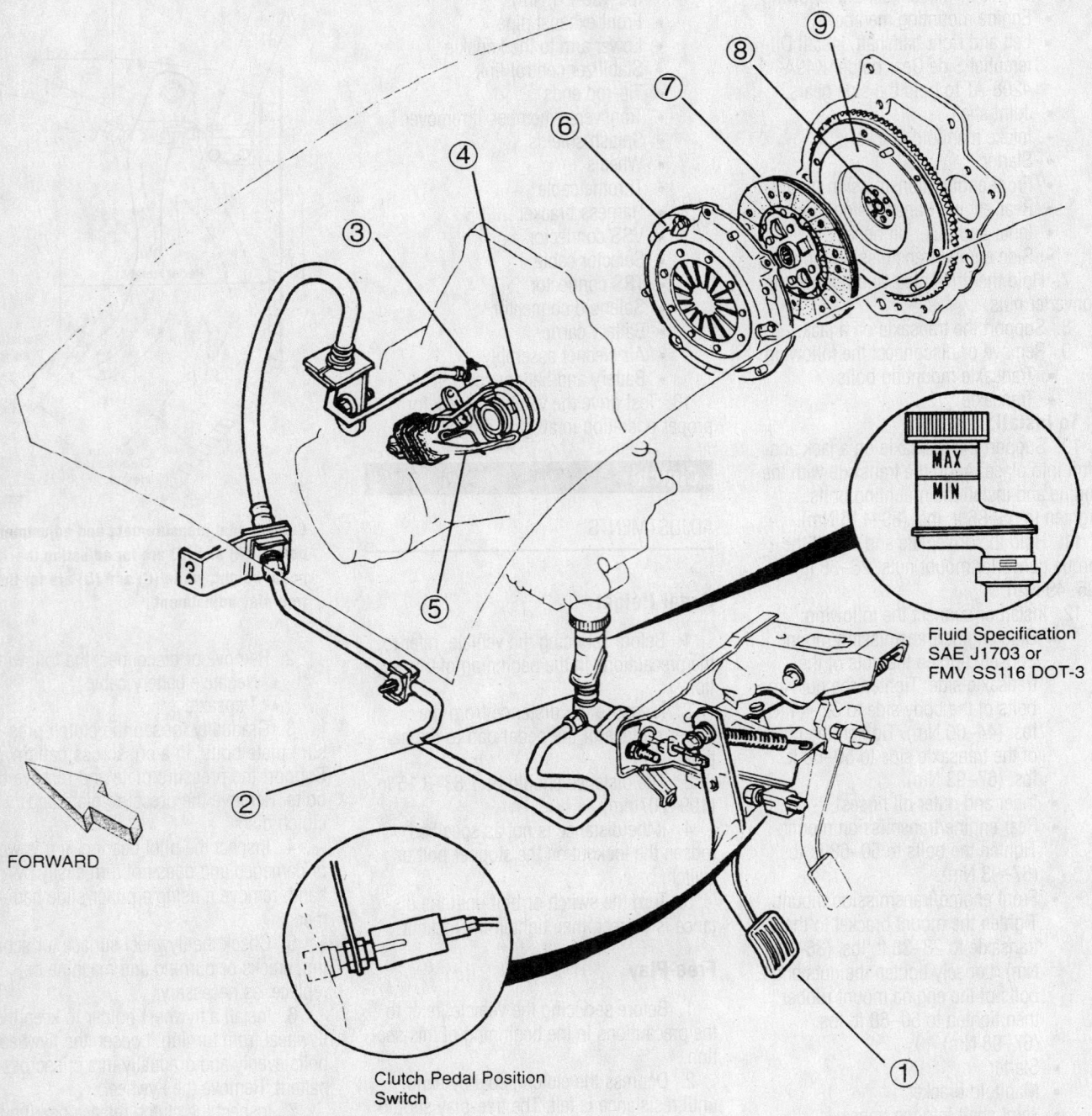

Fluid Specification
SAE J1703 or
FMV SS116 DOT-3

FORWARD

Clutch Pedal Position
Switch

1. Clutch pedal
2. Clutch master cylinder
3. Clutch release cylinder
4. Release bearing
5. Clutch release fork
6. Clutch cover
7. Clutch disc
8. Pilot bearing
9. Flywheel

7923KG32

Structural view of the hydraulic clutch system

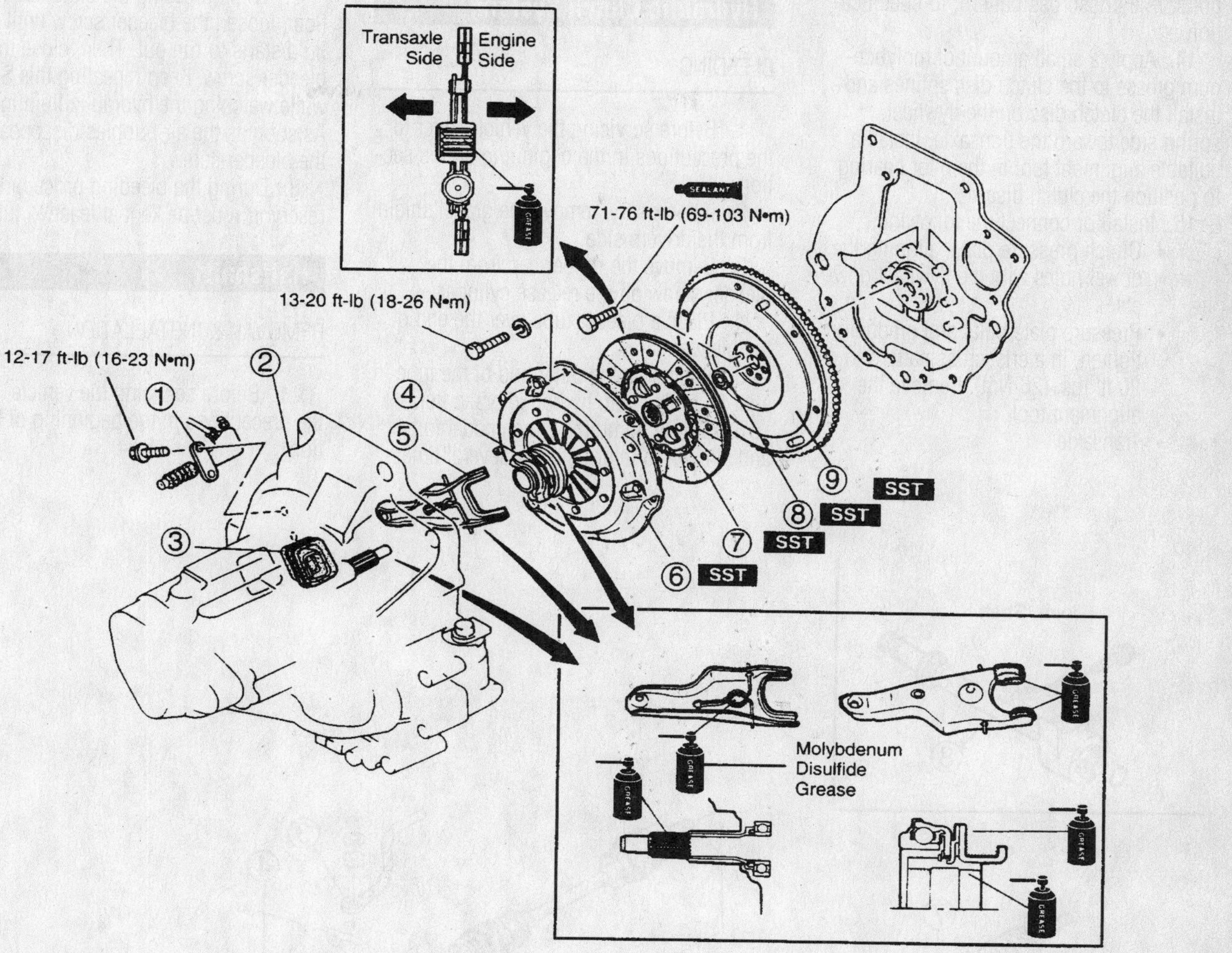

Transaxle
Side   Engine
Side

71-76 ft-lb (69-103 N•m)

13-20 ft-lb (18-26 N•m)

12-17 ft-lb (16-23 N•m)

Molybdenum
Disulfide
Grease

1. Clutch release cylinder
2. Transaxle housing
3. Boot
4. Release bearing
5. Clutch release fork

6. Clutch cover
7. Clutch disc
8. Pilot bearing
9. Flywheel

7923KG33

Exploded view of the clutch assembly

- Clutch release bearing on the
  release fork
- New pilot bearing in the flywheel, if
  removed

11. Be sure the flywheel mounting surface and the crankshaft or eccentric shaft
mounting surfaces are clean. Remove any
old sealant from the flywheel bolt hole
threads and the flywheel bolts.

12. Install or connect the following:
- Flywheel

13. Apply sealant to the flywheel bolt
threads and install them hand tight. Install
the flywheel holding tool. Tighten the

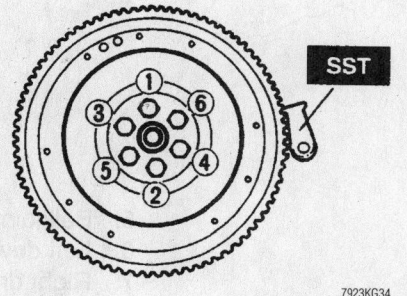

Flywheel tightening sequence

7923KG34

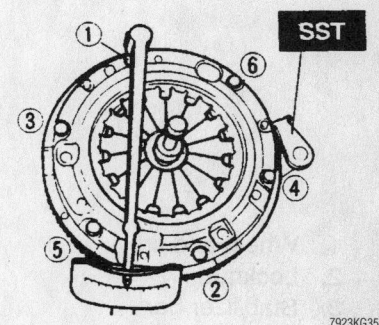

Pressure plate tightening sequence

7923KG35

For Tire, Wheel and Ball Joint specifications, see Section 1 of this manual

bolts, in a crisscross pattern, to specification.

14. Apply a small amount of molybdenum grease to the clutch disc splines and install the clutch disc on the flywheel, spring side toward the transaxle. Install a suitable alignment tool in the pilot bearing to position the clutch disc.

15. Install or connect the following:
- Clutch pressure plate, aligning the dowel holes with the flywheel dowels
- Pressure plate bolts and gradually tighten, in a crisscross pattern to 20 ft. lbs. (26 Nm). Remove the alignment tool.
- Transaxle

Hydraulic Clutch System

BLEEDING

1. Before servicing the vehicle, refer to the precautions in the beginning of this section.

2. If necessary, remove the gravel shield from the drivers side.

3. Remove the rubber cap from the bleeder screw on the release cylinder.

4. Place a bleeder tube over the end of the bleeder screw.

5. Submerge the other end of the tube in a jar half filled with hydraulic brake fluid.

6. Slowly pump the clutch pedal fully and allow it to return slowly, several times.

7. While pressing the clutch pedal to the floor, loosen the bleeder screw until the fluid starts to run out. Then, close the bleeder screw. Keep repeating this Step, while watching the hydraulic fluid in the jar. As soon as the air bubbles disappear, close the bleeder screw.

8. During the bleeding procedure the reservoir must be kept at least ¾ full.

Halfshaft

REMOVAL & INSTALLATION

1. Before servicing the vehicle, refer to the precautions in the beginning of this section.

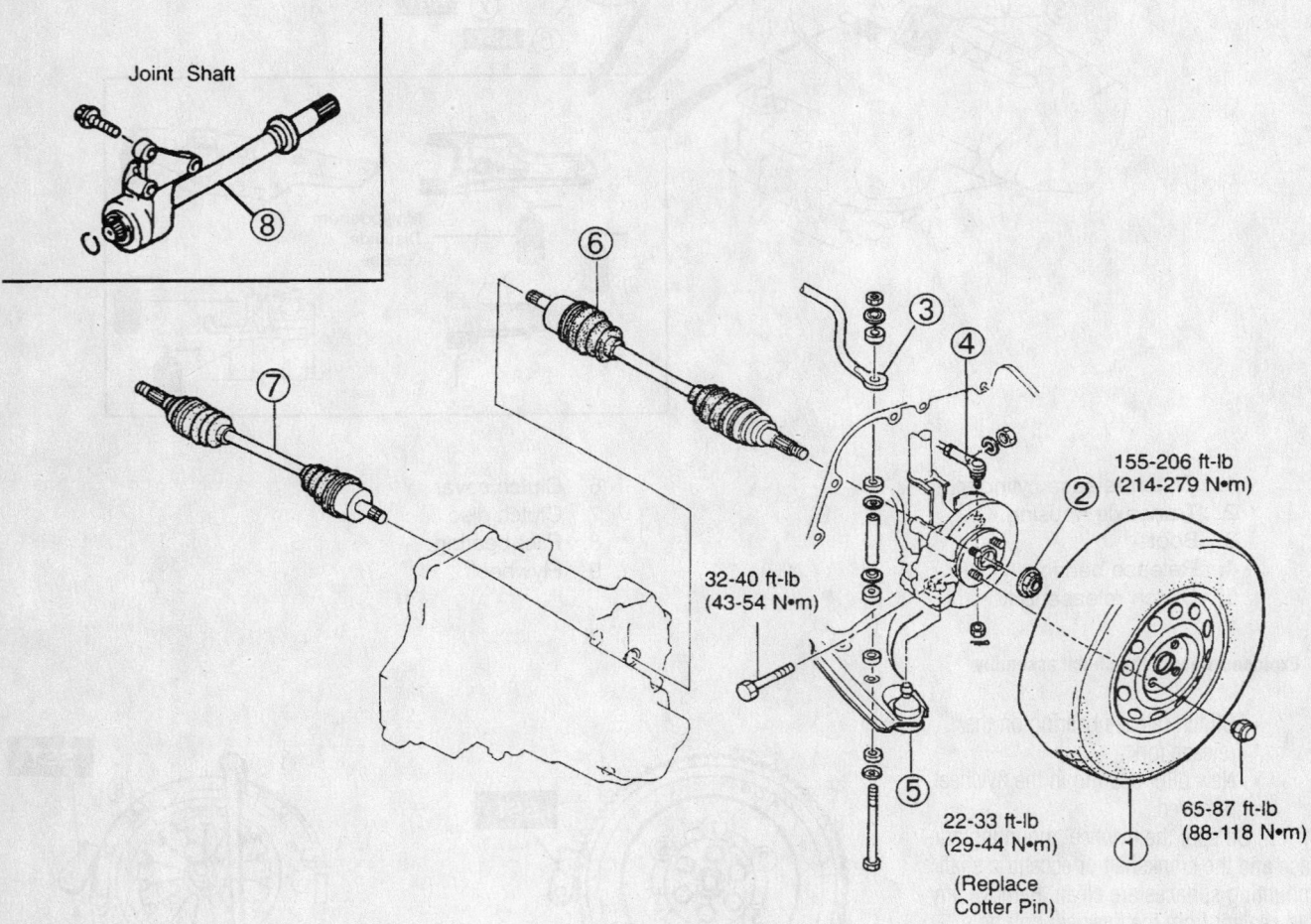

Joint Shaft

32-40 ft-lb (43-54 N•m)

155-206 ft-lb (214-279 N•m)

22-33 ft-lb (29-44 N•m)

(Replace Cotter Pin)

65-87 ft-lb (88-118 N•m)

1. Wheel and tire
2. Locknut
3. Stabilizer bar
4. Tie-rod end
5. Ball joint
6. Left driveshaft
7. Right driveshaft
8. Joint shaft

7923KG36

Exploded view of the halfshafts and related components

2. Remove or disconnect the following:
- Wheel and tire assemblies
- Splash shield, if equipped

3. Drain the transaxle.

4. Raise the staked portion of the hub locknut with a hammer and chisel. Lock the hub by applying the brakes and remove the nut.

5. Remove or disconnect the following:
- Stabilizer bar from the lower control arm
- Cotter pin and nut from the tie rod end ball stud.
- Tie rod end from the knuckle, using a suitable tool
- Lower ball joint pinch bolt and nut
- Ball joint from the knuckle, using a prybar on the control arm

6. Position a prybar between the inner CV-joint and transaxle case. Carefully pry the halfshaft from the transaxle being careful not damage the oil seal. If equipped with a right side intermediate shaft, insert the prybar between the halfshaft and intermediate shaft and tap on the bar to uncouple them.

7. Pull outward on the hub/knuckle assembly, push the outer CV-joint stub shaft through the hub, and remove the halfshaft. If the halfshaft is stuck in the hub, install the old hub nut to protect the stub shaft threads. Tap on the nut, using only a soft mallet, to remove the halfshaft.

➡**Install Differential Side Gear holder K49A-4208-AT, into the transaxle after removing the halfshaft, to keep the differential side gear in position. If the gear becomes out of position the differential may have to be removed to realign the gear.**

8. Remove the intermediate shaft, if necessary, by removing the support bearing bolts and pulling the shaft from the transaxle.

To install:

9. Install or connect the following:
- New circlip on the end of the intermediate shaft, with the end gap facing upward, if removed
- Intermediate shaft in the transaxle, being careful not to damage the oil seals
- Support bearing bolts and tighten, in sequence, to 45 ft. lbs. (61 Nm)
- New circlip on the end of the halfshaft, with the end gap facing upward
- Halfshaft into the transaxle, being careful not to damage the oil seal

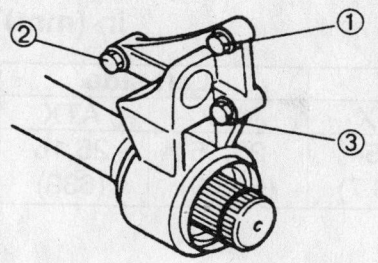

7923KG37

Support bearing bolt tightening sequence

- Halfshaft into the intermediate shaft, if equipped
- Other end of the halfshaft through the hub. Loosely install a new locknut.
- Lower ball joint into the knuckle
- Pinch bolt and nut and tighten to 40 ft. lbs. (54 Nm)
- Tie rod end to the steering knuckle and tighten the nut to 42 ft. lbs. (57 Nm). Install a new cotter pin. Tighten the nut, if necessary, to align the ball stud hole with the nut castellation.
- Stabilizer bar to the lower control arm
- Splash shield and the wheel and tire assemblies

10. Lock the hub with the brakes. Tighten the new hub nut to 155–206 ft. lbs. (214–279 Nm). After tightening, stake the locknut using a hammer and dull bladed chisel.

11. Fill the transaxle with the proper type and quantity of fluid.

CV-Joints

OVERHAUL

1. Before servicing the vehicle, refer to the precautions in the beginning of this section.

2. Remove the halfshaft.

3. Pry up the locking clip of the transaxle side boot retaining band with a suitable tool.

4. Using a pair of pliers, remove the retaining band.

5. Slide the boot away to access the CV-joint.

6. Matchmark the CV-joint housing and the shaft to ensure proper positioning during assembly.

7. Remove the outer ring.

8. Matchmark the shaft and tripod assembly to ensure proper positioning during assembly.

9. Remove the snapring.

➡**Be careful not to damage the needle bearings.**

10. Using a brass drift and a hammer, drive the tripod joint from the shaft.

➡**Cover the halfshaft serration's with tape, so as not to damage the boot.**

11. Slide the boot off the shaft.

➡**It is not necessary to remove the dynamic damper from the shaft unless replacement or repair is required.**

12. Pry up the locking clip of the dynamic damper retaining band with a suitable tool.

13. Using a pair of pliers, remove the retaining band.

14. Remove the dynamic damper.

➡**Do not remove the outer boot from the shaft unless it is necessary.**

15. Pry up the locking clip of the outer boots small and large retaining bands with a suitable tool.

16. Using a pair of pliers, remove the retaining bands.

17. Cover the halfshaft serration's with tape and remove the boot.

To install:

➡**The boot on the wheel side of the driveshaft is larger than the boot on the differential side.**

18. Cover the halfshaft serration's with tape and install the inner boot.

➡**The bands should be installed so that their pointed ends initially point in the forward direction of rotation.**

19. Install the dynamic damper and band. Fold the band back by pulling on the end of it with a pair of pliers, then lock the end of the band by bending the locking clip.

20. Install the outer boot.

21. Align the marks made during

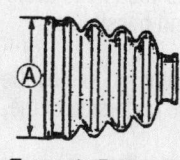
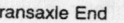
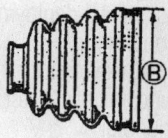

Transaxle End   **Wheel End**

9307KG12

Measure the boots as shown to ensure the larger boot is placed on the wheel side of the halfshaft.

For Wheel Alignment specifications, see Section 1 of this manual

Length of driveshaft

in (mm)

Item	Right side		Left side	
	MTX	ATX	MTX	ATX
TED	24.93 (633.3)	25.54 (648.7)	25.15 (638.9)	25.16 (639)

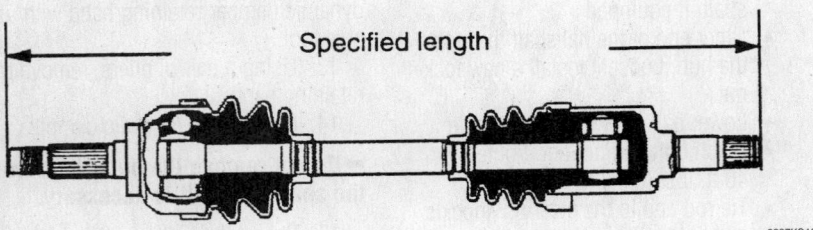

Make sure the driveshaft's specified length is correct

removal and install the tripod joint using a brass drift and a hammer.

22. Install the snapring.

23. Apply the grease supplied with the joint rebuild kit to the tripod joint, outer ring and the boot.

24. Install the outer ring.

25. If the outer boot was removed, fill it with the correct amount of grease as follows:
 a. Transaxle side: 4.94 oz. (140g).
 b. Wheel side: 4.58 oz. (130g).

26. Make sure the boots are not damaged, then carefully up the small end of the boots to release any trapped air.

27. Measure the length of the driveshaft to ensure it is the correct length. Refer to the accompanying illustration for the driveshaft measuring points and the correct specifications.

28. Install the boot retaining bands. Fold the bands back by pulling on the ends with a pair of pliers, then lock the end of the bands by bending the locking clips.

29. Install the halfshaft.

STEERING AND SUSPENSION

Air Bag

✳✳ CAUTION

Some vehicles are equipped with an air bag system. The system must be disabled before performing service on or around system components, steering column, instrument panel components, wiring and sensors. Failure to follow safety and disabling procedures could result in accidental air bag deployment, possible personal injury and unnecessary system repairs.

PRECAUTIONS

Several precautions must be observed when handling the inflator module to avoid accidental deployment and possible personal injury.

• Never carry the inflator module by the wires or connector on the underside of the module.

• When carrying a live inflator module, hold securely with both hands, and ensure that the bag and trim cover are pointed away.

• Place the inflator module on a bench or other surface with the bag and trim cover facing up.

• With the inflator module on the bench, never place anything on or close to the module which may be thrown in the event of an accidental deployment.

• An air bag is an explosive device. Handle with extreme caution.

• Always disconnect the battery and the air bag connector before removing the steering wheel or beginning work on the air bag system.

• Air bag components must not be repaired or opened. Always use new parts, including the wiring harness.

• Always place a removed air bag unit with the horn pad facing up. Put it in a safe place where it will not be disturbed.

• The air bag unit must not be exposed to grease, fluids, or cleaning agents.

• The air bag unit must not be exposed to temperatures above 194° F (90° C) at any time. Even the heat of a soldering iron can damage or ignite the charge.

• Storage and transport of air bags is subject to rules governing explosive devices and should be done only in the original package.

• Failure to follow proper safety precautions may result in personal injury through accidental firing of the air bag, or through failure of the air bag in an accident.

DISARMING

1. Before servicing the vehicle, refer to the precautions in the beginning of this section.

2. Turn the ignition switch to the **LOCK** position.

3. Disconnect the negative battery cable.

4. Wait 10 minutes for the battery backup power to discharge.

ARMING

1. Before servicing the vehicle, refer to the precautions in the beginning of this section.

2. Connect the negative battery cable.

3. Turn the ignition switch **ON**.

4. Verify that the air bag indicator illuminates for 4–8 seconds, then goes off.

Manual Rack and Pinion Steering Gear

REMOVAL & INSTALLATION

1. Before servicing the vehicle, refer to the precautions in the beginning of this section.

2. Remove or disconnect the following:
 • Negative battery cable
 • Front wheels
 • Cotter pins from both steering tie rod ends and the nuts
 • Tie rod out of the knuckle arm, using a suitable tool
 • Set plate from the firewall
 • Fixing bolt from the steering shaft to steering gear pinion shaft

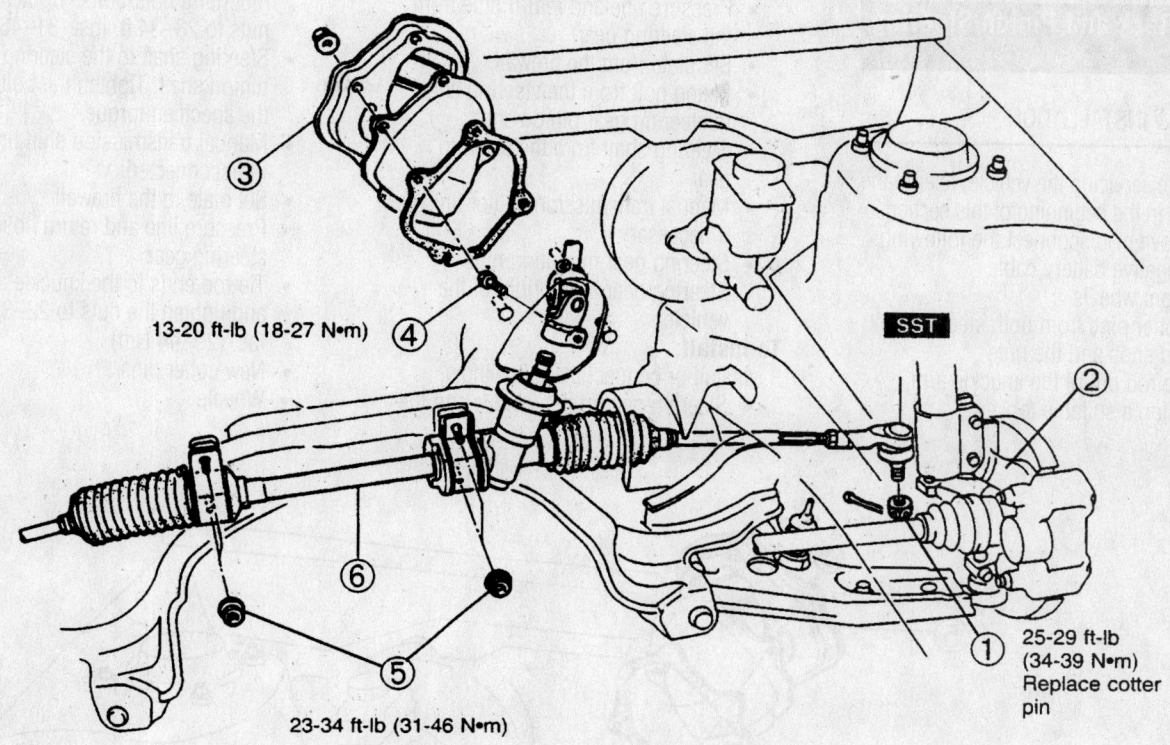

13-20 ft-lb (18-27 N•m)

23-34 ft-lb (31-46 N•m)

25-29 ft-lb
(34-39 N•m)
Replace cotter
pin

SST

1. Tie-rod end nut
2. Steering knuckle
3. Bulkhead sealing cover

4. Pinch bolts
5. Steering rack nuts
6. Steering rack

7923KG38

Exploded view of the manual steering gear assembly mounting

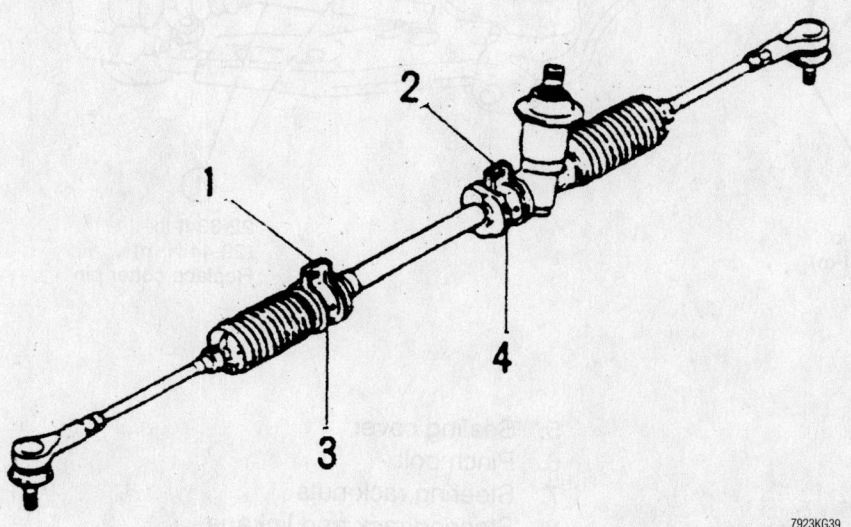

7923KG39

Rack mounting tightening sequence

- Steering shaft from the steering gear
- Steering gear mounting nuts

3. Move the steering gear to the right of the vehicle.

To install:

4. Install or connect the following:
- Steering gear to the vehicle
- Mounting nuts in the order shown. Tighten the nuts to 23–34 ft. lbs. (31–46 Nm).
- Steering shaft to the steering gear pinion shaft. Tighten the bolt/nut to 13–20 ft. lbs. (18–27Nm).
- Set plate to the firewall
- Tie rod ends to the knuckle arm and tighten the nuts to 25–29 ft. lbs. (34–39 Nm)
- New cotter pins
- Wheels
- Negative battery cable

5. Check the front end alignment.

For Maintenance Interval recommendations, see Section 1 of this manual

Power Rack and Pinion Steering Gear

REMOVAL & INSTALLATION

1. Before servicing the vehicle, refer to the precautions in the beginning of this section.
2. Remove or disconnect the following:
   - Negative battery cable
   - Front wheels
   - Cotter pins from both steering tie rod ends and the nuts
   - Tie rod out of the knuckle arm, using a suitable tool
   - Pressure line and return pipe from the steering gear
   - Set plate from the firewall
   - Fixing bolt from the steering shaft to steering gear pinion shaft
   - Steering shaft from the steering gear
   - Manual transmission shifter linkage if necessary
   - Steering gear mounting nuts
   - Steering gear to the right of the vehicle

To install:

3. Install or connect the following:
   - Steering gear to the vehicle and the mounting nuts/bolts. Tighten the nuts to 23–34 ft. lbs. (31–46 Nm).
   - Steering shaft to the steering gear pinion shaft. Tighten the bolt/nut to the specified torque.
   - Manual transmission shift linkage, if disconnected
   - Set plate to the firewall
   - Pressure line and return hose to the steering gear
   - Tie rod ends to the knuckle arm and tighten the nuts to 22–33 ft. lbs. (29–44 Nm).
   - New cotter pins
   - Wheels

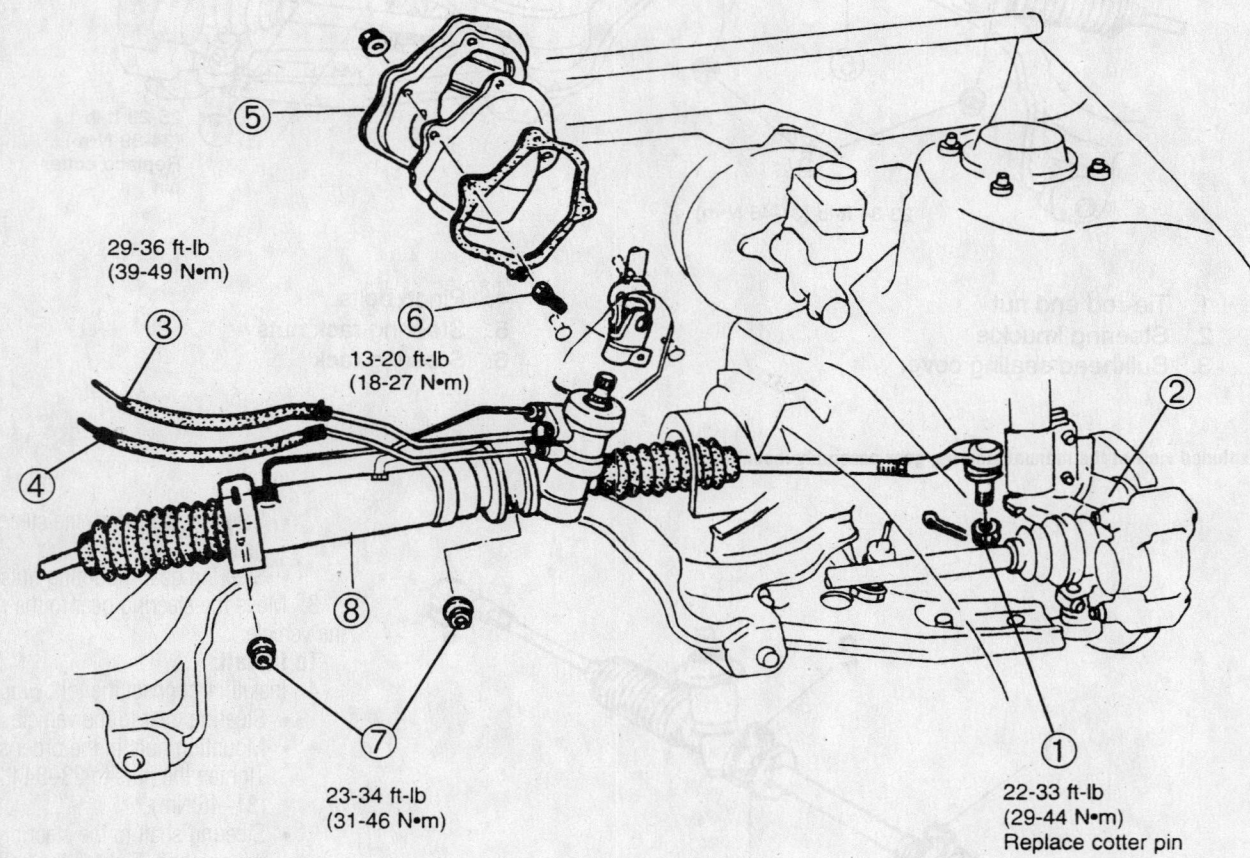

29-36 ft-lb (39-49 N•m)

13-20 ft-lb (18-27 N•m)

23-34 ft-lb (31-46 N•m)

22-33 ft-lb (29-44 N•m)
Replace cotter pin

1. Tie-rod end nut
2. Steering knuckle
3. High-pressure line
4. Return line
5. Sealing cover
6. Pinch bolt
7. Steering rack nuts
8. Steering rack and linkage

7923KG40

Exploded view of the power steering gear assembly mounting

- Negative battery cable
4. Check the power steering fluid and the front end alignment.

Strut

REMOVAL & INSTALLATION

Front

1. Before servicing the vehicle, refer to the precautions in the beginning of this section.

2. Remove or disconnect the following:
- Wheel and tire assembly
3. Support the lower control arm with a jack.
4. Remove or disconnect the following:
- Bolts or clips attaching the brake hose and/or Anti-lock Brake System (ABS) sensor harness to the strut
5. Paint alignment marks on the upper strut mounting block and strut tower, and on the lower strut mount-to-steering knuckle

so the strut can be reinstalled in the same position.
6. Remove or disconnect the folowing:
- Upper strut mounting block nuts and the strut-to-knuckle bolts
- Strut assembly

To install:
7. Install or connect the following:
- Strut into the strut tower, aligning the paint marks made during removal

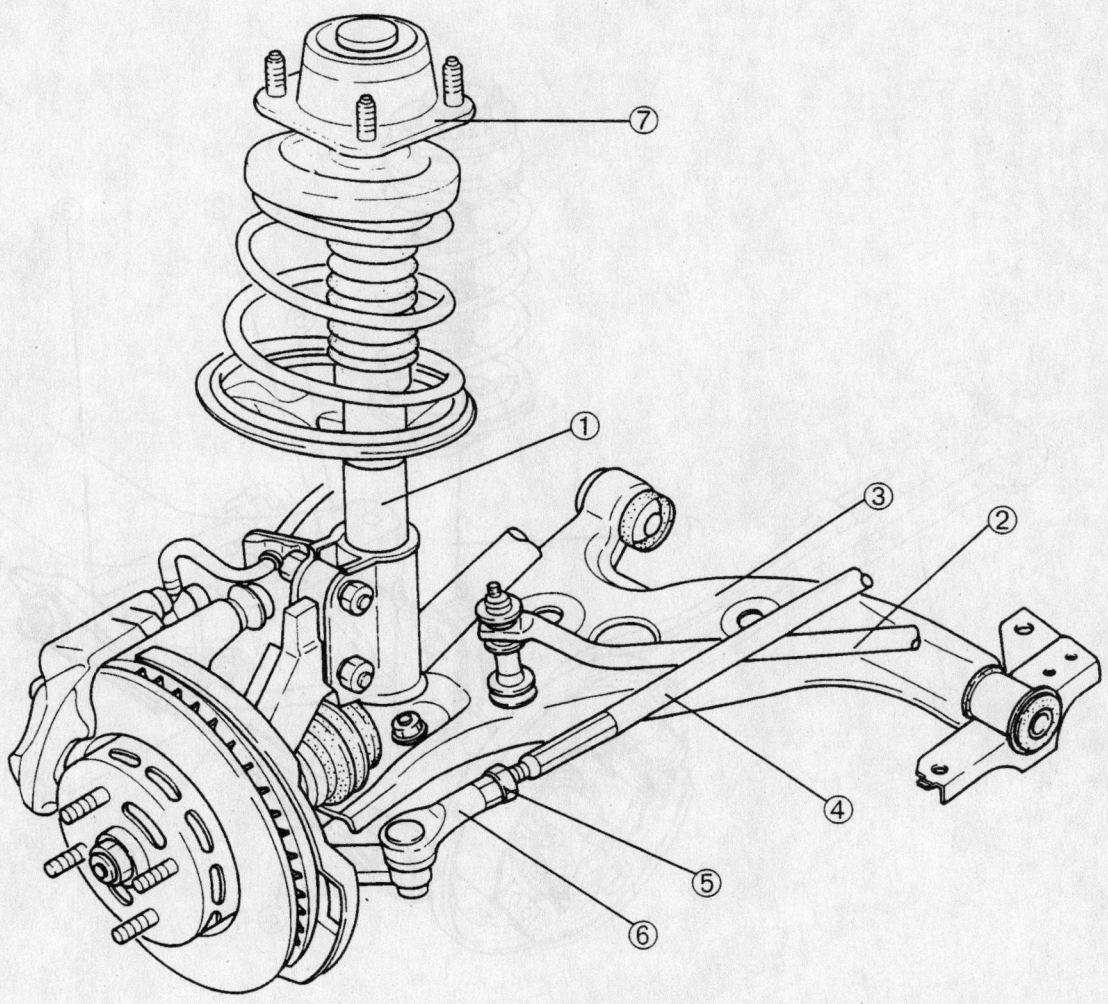

1. Front strut
2. Front stabilizer
3. Lower control arm
4. Tie-rod
5. Jam nut
6. Tie-rod end
7. Mounting block

7923KG41

Front suspension component identification

For Tune-up, Capacities and Firing orders, see Section 1 of this manual

- Strut mounting nuts and tighten to 17–22 ft. lbs. (23–29 Nm)
- Strut-to-knuckle bolts and tighten to 69–86 ft. lbs. (93–116 Nm)
- Clips or bolts attaching the brake hose and/or ABS sensor harness
- Wheel and tire assembly

8. Check the front end alignment.

Rear

1. Before servicing the vehicle, refer to the precautions in the beginning of this section.

2. Remove or disconnect the following:
- Side trim panels from the inside of the trunk or the rear seat and trim, as required
- Top mounting nuts from the strut mounting block assembly
- Rear wheels. The suspension will drop when the weight lifts off the wheels.
- Brake line or wiring retainers as required
- Bottom strut mount retainers
- Strut

To install:

3. Install or connect the following:
- Strut into the strut tower
- Strut mounting nuts and tighten to 17–22 ft. lbs. (23–29 Nm)
- Strut-to-knuckle bolts and tighten to 69–86 ft. lbs. (93–116 Nm)
- Clips or bolts attaching the brake hose and/or ABS sensor harness
- Wheel and tire assembly
- Side trim panels from the inside of the trunk or the rear seat and trim, if removed

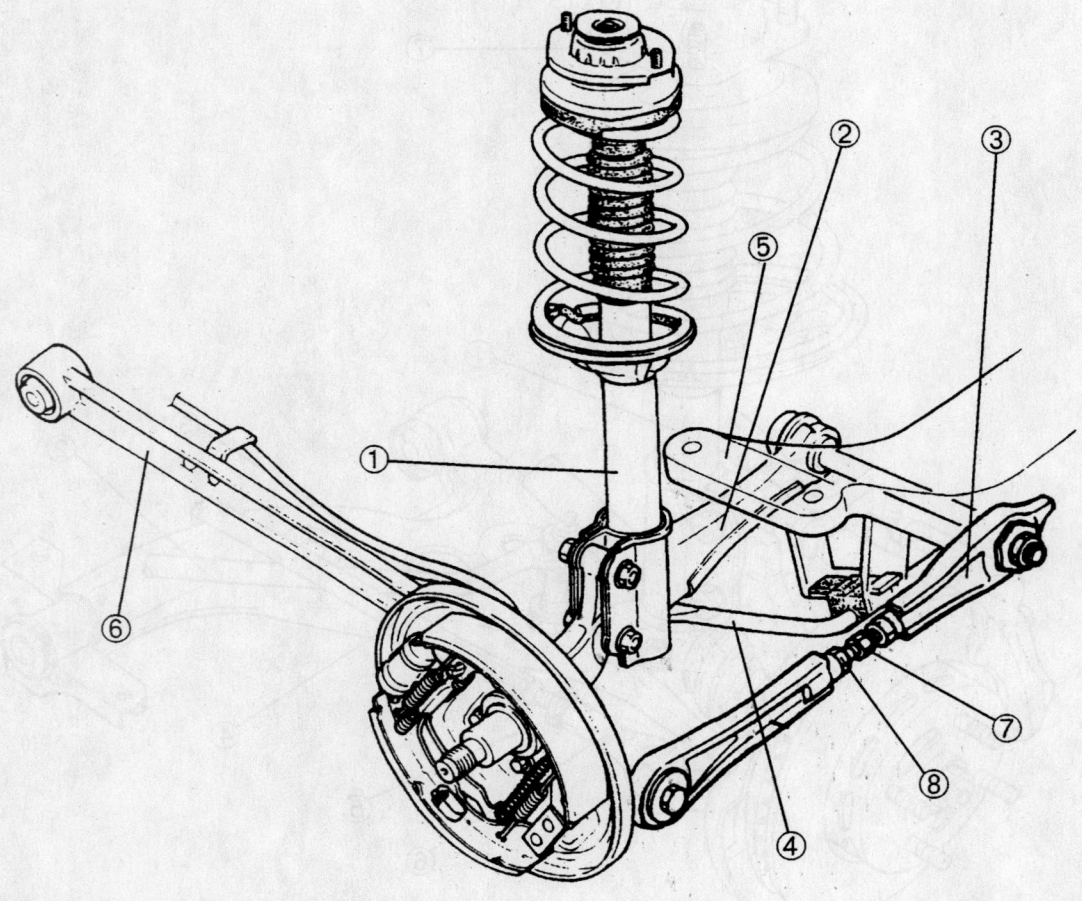

1. Rear strut
2. Front lateral link
3. Rear lateral link
. Rear stabilizer
5. Rear crossmember
6. Trailing link
7. Adjuster
8. Jam nuts

Rear suspension component identification

7923KG42

Coil Spring

REMOVAL & INSTALLATION

Front and Rear

1. Before servicing the vehicle, refer to the precautions in the beginning of this section.

2. Remove the strut from the vehicle.

3. Install the strut securely in a vise with either aluminum or copper plates to protect the strut.

4. Loosen the piston rod upper nut several turns but DO NOT REMOVE IT.

5. Install the lower end of the strut in the vise and install a coil spring compressor. Compress the coil spring and remove the upper nut.

✳✳ CAUTION

Failure to fully compress the spring and hold it securely can be extremely dangerous.

6. Slowly release the coil spring tension.

7. Remove the suspension support, dust seal, spring seat, spring insulators, coil spring and bumper.

8. While pushing on the piston rod, be sure that the pull stroke is even and that there is no unusual noise or resistance. Also inspect for any oil leakage around the piston rod.

9. Push the piston rod in, then release it. Be sure that the return rate is constant.

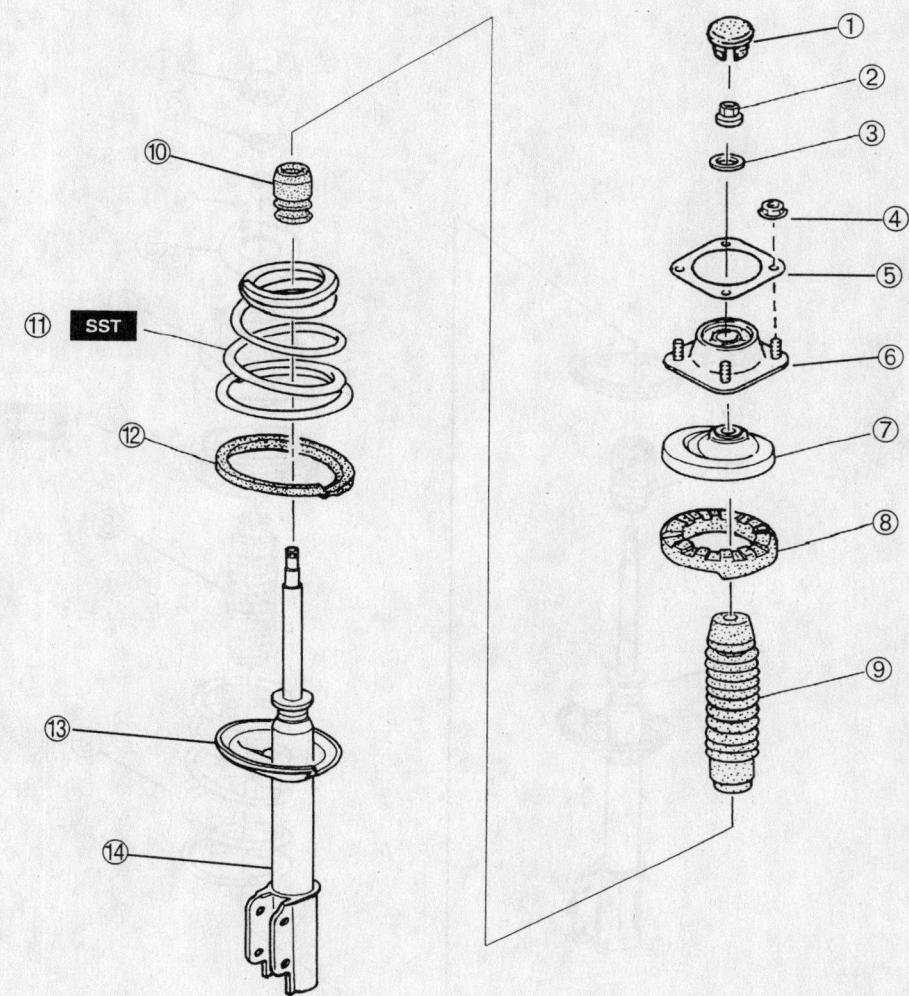

1. Dust cap
2. Piston retaining nut
3. Washer
4. Mounting nut
5. Gasket
6. Mounting block
7. Upper spring seat
8. Upper spring isolator
9. Dust boot
10. Rebound stopper
11. Coil spring
12. Lower spring isolator
13. Lower spring seat
14. Shock absorber

7932KG43

Exploded view of the front strut assembly

For complete service labor times, order the Chilton Labor Guide

10. If the shock absorber does not operate as described, replace it.

To assemble:

11. Install the strut assembly into a vise.

12. Install the bound stopper and dust boot onto the piston rod.

13. Install the coil spring and compress the coil spring with the spring compressor.

14. Install the rubber seat, the spring upper seat, the bearing and the mounting block. Be sure that the spring upper seat notched portion is facing inward and tighten the piston rod upper nut.

15. Remove the spring compressor from the strut. Secure the upper mounting block in the vise. Tighten the nut to 41–50 ft. lbs. (55–68 Nm) for the front strut and 47–59 ft. lbs. (64–80 Nm) for the rear strut.

16. Be sure that the spring is well seated in the upper seats.

17. Install the strut to the vehicle.

Lower Ball Joint

REMOVAL & INSTALLATION

1. Before servicing the vehicle, refer to the precautions in the beginning of this section.

2. Remove or disconnect the following:

- Wheel and tire assembly
- Ball joint stud pinch bolt and nut from the steering knuckle
- Ball joint from the knuckle, using a prytool
- Bolt, nut and the ball joint from the lower control arm

To install:

3. Installation is the reverse of the removal procedure. Tighten the ball joint-to-lower control arm bolt and nut to 86 ft. lbs. (117 Nm). Tighten the ball joint pinch bolt

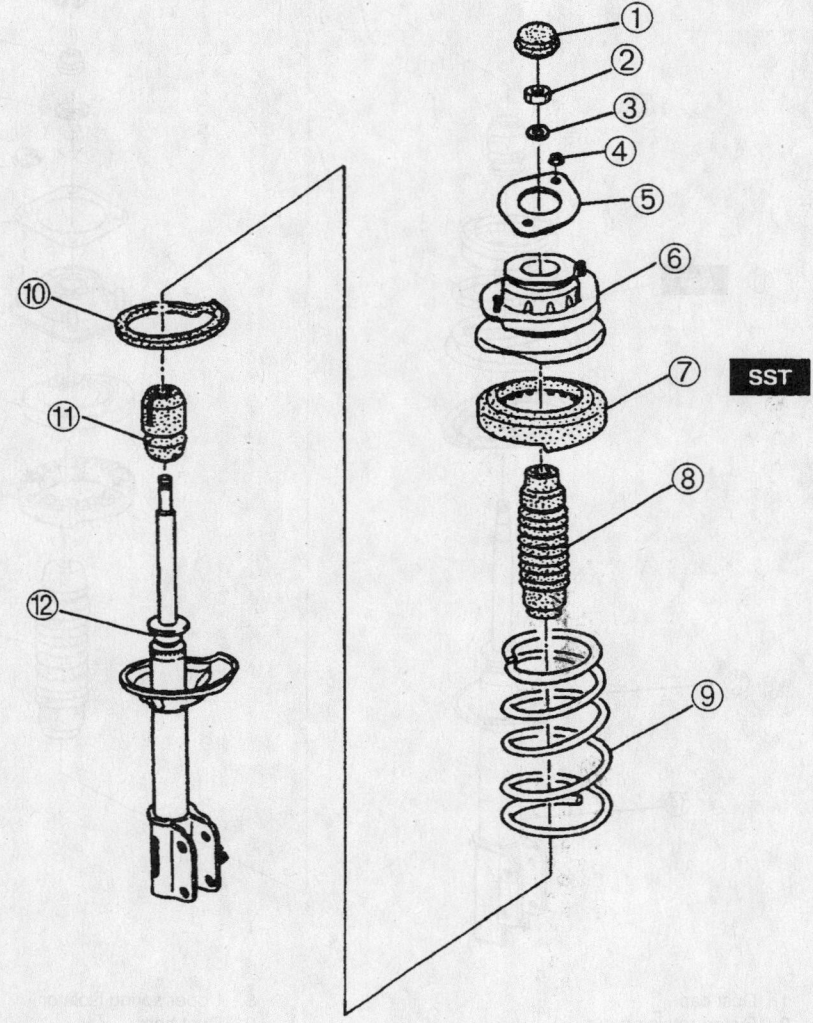

1. Dust cap
2. Piston retaining nut
3. Washer
4. Mounting nut
5. Gasket
6. Mounting block/upper spring seat
7. Upper spring isolator
8. Dust boot
9. Coil spring
10. Lower spring isolator
11. Rebound stopper
12. Strut

Exploded view of the rear strut assembly

7923KG44

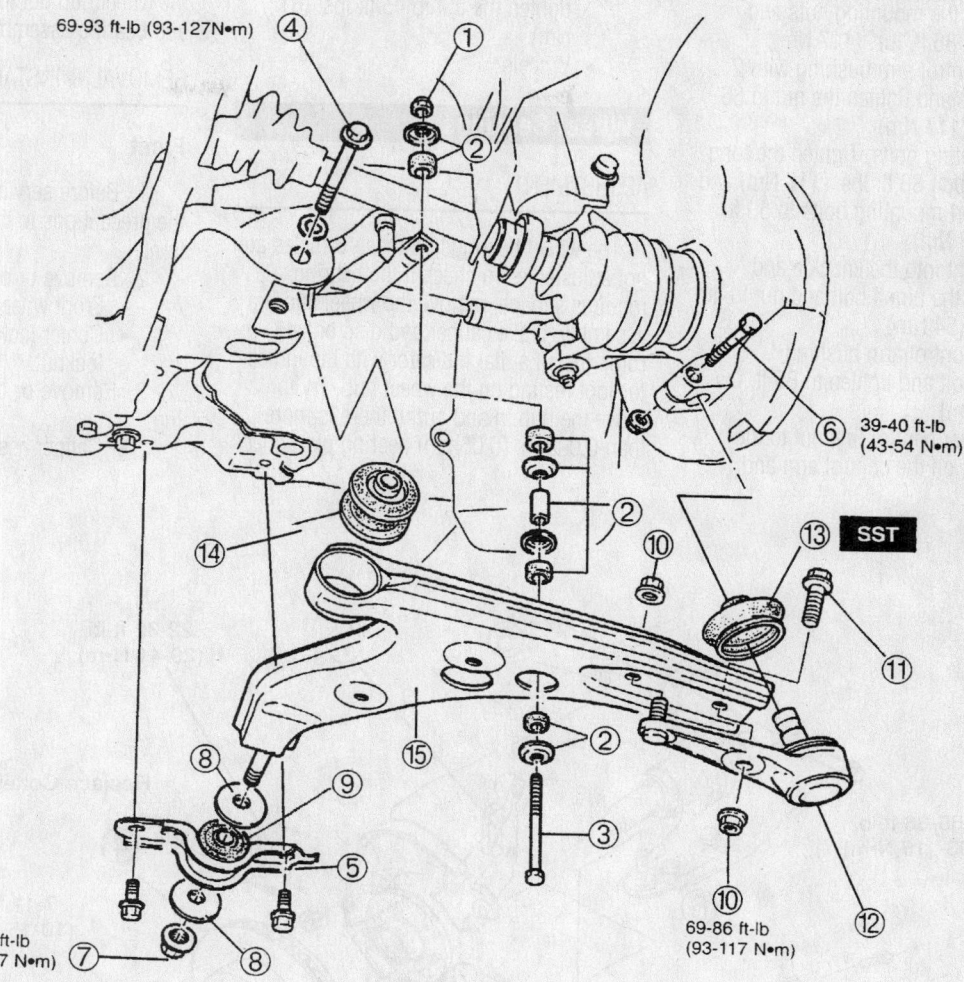

69-93 ft-lb (93-127N•m)

39-40 ft-lb (43-54 N•m)

SST

69-86 ft-lb (93-117 N•m)

69-86 ft-lb (93-117 N•m)

1. Stabilizer retaining nut
2. Stabilizer hardware - spacer, retainers, bushings
3. Stabilizer bolt
4. Pivot bolt
5. Mounting bolts
6. Pinch bolt,
7. Retaining nut
8. Washers
9. Control arm bushing - rear
10. Ball joint mounting nuts
11. Ball joint mounting bolt
12. Ball joint
13. Ball joint dust boot (Replace)
14. Control arm bushing - front
15. Control arm

7923KG45

Exploded view of the lower control arm with replaceable ball joint

and nut to 43 ft. lbs. (59 Nm). Check the front wheel alignment.

Lower Control Arm

REMOVAL & INSTALLATION

1. Before servicing the vehicle, refer to the precautions in the beginning of this section.
2. Remove or disconnect the following:
   - Front wheel
   - Stabilizer control link nut from the bracket on the lower control arm
   - Pivot bolt
   - Lower control arm ball joint bolt and nut from the steering knuckle
   - 3 mounting bolts
   - Retaining nut, washer and the rear control arm bushing
   - 2 ball joint mounting nuts and bolts from the lower control arm
3. Place the ball joint in a vise and use a chisel top carefully remove the dust boot from the ball joint.
4. Remove or disconnect the following:
   - Control arm front bushing
   - Control arm

   To install:
5. Apply a suitable general purpose grease to the new tire dust boot and press the boot onto the ball joint using a suitable tool.
6. Install or connect the following:
   - Ball joint onto the control arm.

Tighten the mounting nuts and bolts to 86 ft. lbs. (117 Nm).

- Rear control arm bushing with 2 washers and tighten the nut to 86 ft. lbs. (117 Nm)
- 3 mounting bolts. Tighten the long mount bolt 86 ft. lbs. (117 Nm) and the short mounting bolts to 50 ft. lbs. (68 Nm).
- Ball joint into the knuckle and tighten the pinch bolt and nut to 40 ft. lbs. (54 Nm)
- Front control arm bushing
- Pivot bolt and tighten to 86 ft. lbs. (117 Nm)
- Stabilizer control link nut to the bracket on the control arm and

tighten the nut to 45 ft. lbs. (61 Nm)
- Wheels

Wheel Bearings

ADJUSTMENT

The wheel bearings on these vehicles are not adjustable. To check if the bearing requires service, remove the wheel and tire assembly, brake caliper and disc brake rotor. Install a dial indicator with the indicator foot resting on the wheel hub. Try to move the hub in and out. If there is more than 0.002 in. (0.05mm) bearing play, check

the wheel hub nut torque or replace the hub and bearing assembly.

REMOVAL & INSTALLATION

Front

1. Before servicing the vehicle, refer to the precautions in the beginning of this section.
2. Remove or disconnect the following:
   - Front wheels
   - Center locknut. Discard the old locknut.
3. Remove or disconnect the following:
   - Caliper assembly from the knuckle.

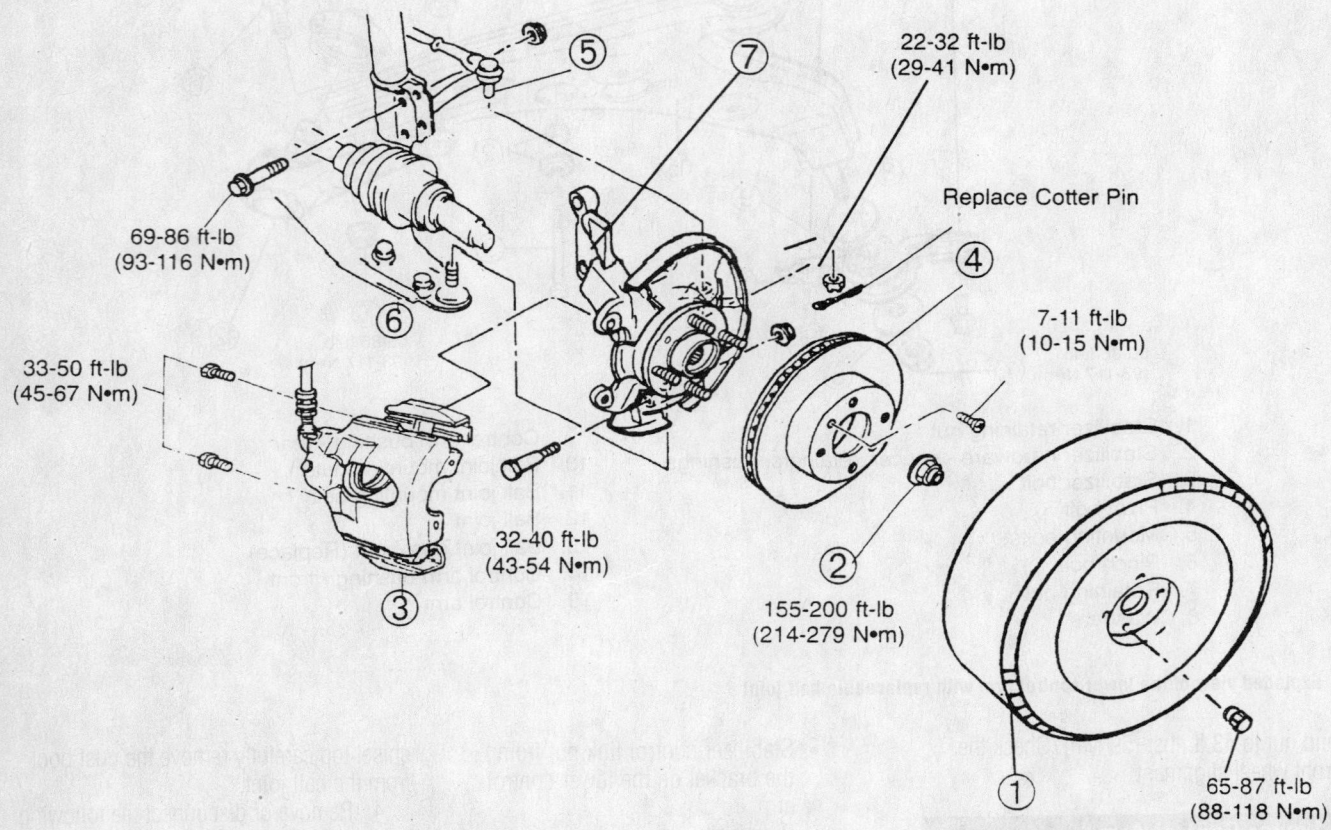

22-32 ft-lb (29-41 N•m)

Replace Cotter Pin

69-86 ft-lb (93-116 N•m)

7-11 ft-lb (10-15 N•m)

33-50 ft-lb (45-67 N•m)

32-40 ft-lb (43-54 N•m)

155-200 ft-lb (214-279 N•m)

65-87 ft-lb (88-118 N•m)

1. Wheel and tire
2. Locknut (Replace)
3. Brake caliper assembly
4. Brake rotor
5. Tie-rod end
6. Ball joint
7. Steering knuckle/wheel hub

7923KG46

Exploded view of the front steering knuckle and related components

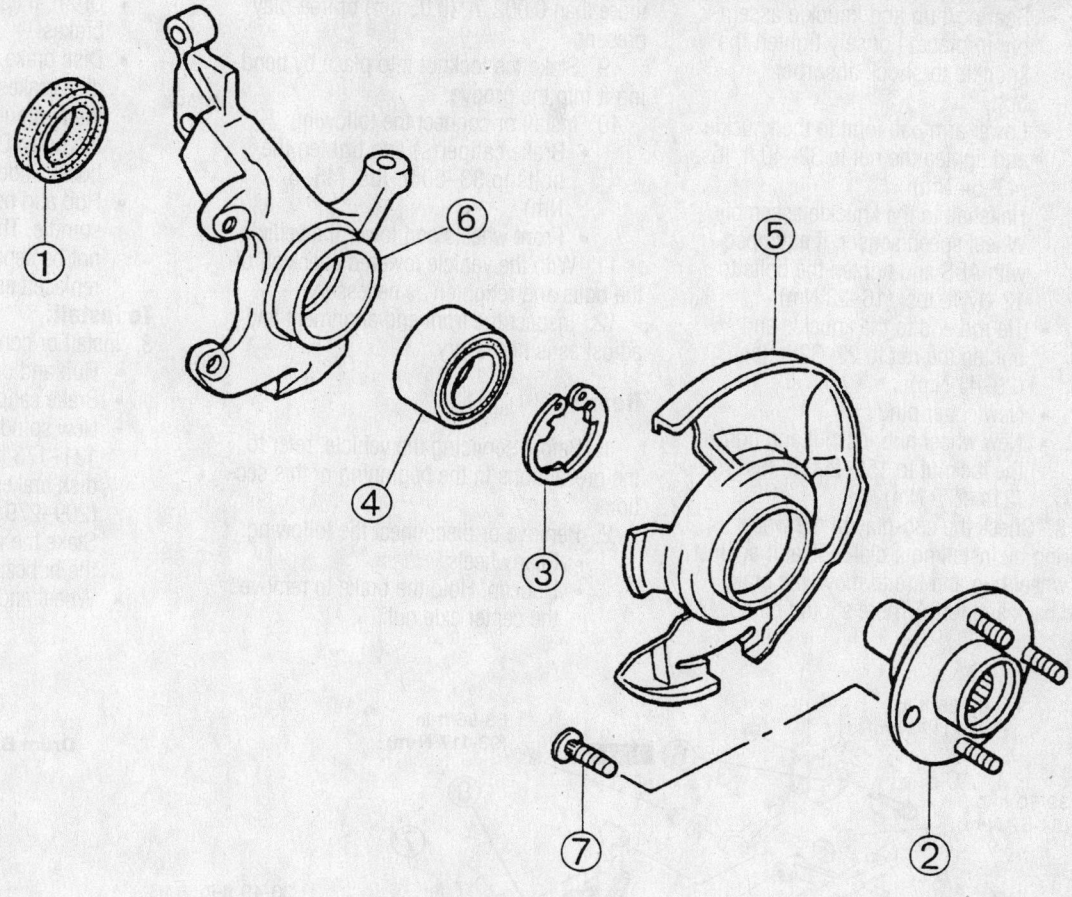

1. Oil seal (Replace)
2. Front wheel hub
3. Retaining ring (Replace)
4. Wheel bearing

5. Dust shield
6. Steering knuckle
7. Wheel stud

7923KG47

Exploded view of the front hub and bearing assembly

Do not disconnect the brake lines. Support the caliper with a piece of wire. Do not allow the caliper to hang by the hose at any time. Remove the brake disc.

- Anti-lock Brake System (ABS) speed sensor, if equipped
- Tie rod end cotter pin and nut
- Tie rod end out of the knuckle assembly
- Outer lower arm to ball joint mounting bolt and nut
- Lower arm from the knuckle assembly
- Knuckle assembly free of the half-shaft, using a plastic mallet
- Knuckle assembly

4. Clamp the knuckle in a vise with protected jaws.
5. Remove or disconnect the following:
- Inner oil seal from the knuckle
- Front wheel hub from the knuckle assembly, using an appropriate hub-puller
- Bearing inner race from the front wheel hub

➡**If the bearing inner race still remains on the hub assembly, grind a section of the bearing inner race until about 0.02 in. (0.50mm) remains. Remove with a chisel.**

- Retaining ring from within the knuckle

- Front wheel bearing from the knuckle, using a wheel bearing removal tool to press it out
6. Clean and inspect all parts but do not wash or clean the wheel bearing.
To install:
7. Install or connect the following:
- New wheel bearing into the knuckle assembly, using the press tools
- Wheel bearing retaining ring
- Front wheel hub, using a press and the correct bearing driver
- New oil seal using the appropriate seal driver and a hammer. Tap the oil seal in evenly until the special tool contacts the steering knuckle. Coat the lip of the oil seal with grease.

Timing belt service is covered in Section 3 of this manual

- Bearing/hub and knuckle assembly in place. Loosely tighten the knuckle to shock absorber bolt.
- Lower arm ball joint to the knuckle and tighten the nut to 32–40 ft. lbs. (43–54 Nm)
- Halfshaft to the knuckle assembly
- Wheel speed sensor, if equipped with ABS and tighten the bolts to 12–17 ft. lbs. (16–23 Nm)
- Tie rod end to the knuckle and tighten the nut to 22–32 ft. lbs. (29–41 Nm)
- New cotter pin
- New wheel hub locknut and tighten the locknut to 155–200 ft. lbs. (214–279 Nm)

8. Check the end-play of the wheel bearing by installing a dial indicator against the wheel hub and tire to move the brake disc back and forth. There should be no more than 0.002 in. (0.05mm) of free-play present.

9. Stake the locknut into place by bending it into the groove.

10. Install or connect the following:
- Brake caliper(s) and tighten the bolts to 33–50 ft. lbs. (45–67 Nm)
- Front wheels and lower the vehicle

11. With the vehicle lowered check all of the bolts and retighten as necessary.

12. Inspect the front end alignment and adjust as is necessary.

Rear

1. Before servicing the vehicle, refer to the precautions in the beginning of this section.

2. Remove or disconnect the following:
- Rear wheels
- Hubcap. Hold the brake to remove the center axle nut.
- Drum, if equipped with drum brakes
- Disc brake caliper, if equipped with disc brakes without disconnecting the hydraulic hose and hang it from the body. Do not let it hang by the hose. Slide the disc off the spindle.
- Hub and bearing assembly off the spindle. The hub and bearing cannot be separated and must be replaced as 1 piece.

To install:

3. Install or connect the following:
- Hub and drum or rotor
- Brake caliper, if equipped
- New spindle nut and tighten to 131–173 ft. lbs. (177–235 Nm) for disk brakes and 155–200 ft. lbs. (209–279 Nm) for drum brakes. Stake the nut into place. Replace the hubcap.
- Wheel and tire assembly

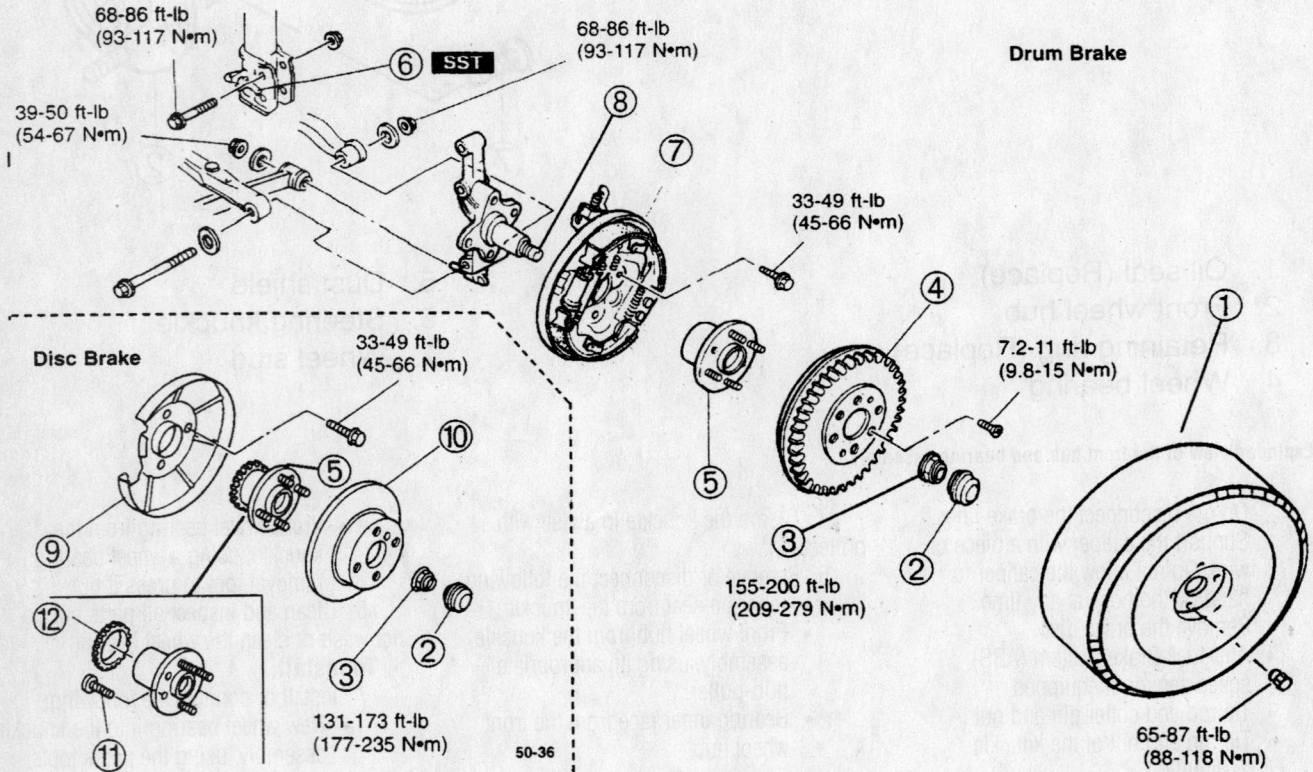

1. Wheel and tire
2. Dust cap
3. Locknut (Replace)
4. Brake drum (or disc)
5. Hub with bearing assembly
6. Brake line
7. Rear brake assembly (drum or disc)
8. Spindle
9. Dust cover
10. Brake rotor
11. Hub bolt
12. ABS sensor rotor

Exploded view of the rear axle assembly

7923KG48

LEXUS

1998–01
ES300 • IS300 • GS300 • GS400 • GS430 • LS400 • LS430 • SC300 • SC400

14

PRECAUTIONS

Before servicing any vehicle, please be sure to read all of the following precautions that deal with personal safety, prevention of component damage, and important points to take into consideration when servicing a motor vehicle:

• Never open, service or drain the radiator or cooling system when the engine is hot; serious burns can occur from the steam and hot coolant.

• Observe all applicable safety precautions when working around fuel. Whenever servicing the fuel system, always work in a well-ventilated area. Do not allow fuel spray or vapors to come in contact with a spark, open flame or excessive heat (a hot drop light, for example). Keep a dry chemical fire extinguisher near the work area. Always keep fuel in a container specifically designed for fuel storage; also, always properly seal fuel containers to avoid the possibility of fire or explosion. Refer to the additional fuel system precautions later in this section.

• Fuel injection systems often remain pressurized, even after the engine has been turned OFF. The fuel system pressure must be relieved before disconnecting any fuel lines. Failure to do so may result in fire and/or personal injury.

• Brake fluid often contains polyglycol ethers and polyglycols. Avoid contact with the eyes and wash your hands thoroughly after handling brake fluid. If you do get brake fluid in your eyes, flush your eyes with clean, running water for 15 minutes. If eye irritation persists, or if you have taken

brake fluid internally, IMMEDIATELY seek medical assistance.

• The EPA warns that prolonged contact with used engine oil may cause a number of skin disorders, including cancer. You should make every effort to minimize your exposure to used engine oil. Protective gloves should be worn when changing oil. Wash your hands and any other exposed skin areas as soon as possible after exposure to used engine oil. Soap and water, or waterless hand cleaner should be used.

• All new vehicles are now equipped with an air bag system. The system must be disabled before performing service on or around system components, steering column, instrument panel components, wiring and sensors. Failure to follow safety and disabling procedures could result in accidental air bag deployment, possible personal injury and unnecessary system repairs.

• Always wear safety goggles when working with, or around, the air bag system. When carrying a non-deployed air bag, be sure the bag and trim cover are pointed away from your body. When placing a non-deployed air bag on a work surface, always face the bag and trim cover upward, away from the surface. This will reduce the motion of the module if it is accidentally deployed. Refer to the additional air bag system precautions later in this section.

• Clean, high quality brake fluid from a sealed container is essential to the safe and proper operation of the brake system. You

should always buy the correct type of brake fluid for your vehicle. If the brake fluid becomes contaminated, completely flush the system with new fluid. Never reuse any brake fluid. Any brake fluid that is removed from the system should be discarded. Also, do not allow any brake fluid to come in contact with a painted surface; it will damage the paint.

• Never operate the engine without the proper amount and type of engine oil; doing so WILL result in severe engine damage.

• Timing belt maintenance is extremely important. Many models utilize an interference-type, non-freewheeling engine. If the timing belt breaks, the valves in the cylinder head may strike the pistons, causing potentially serious (also time-consuming and expensive) engine damage. Refer to the maintenance interval charts in the front of this manual for the recommended replacement interval for the timing belt, and to the timing belt section for belt replacement and inspection.

• Disconnecting the negative battery cable on some vehicles may interfere with the functions of the on-board computer system(s) and may require the computer to undergo a relearning process once the negative battery cable is reconnected.

• When servicing drum brakes, only disassemble and assemble one side at a time, leaving the remaining side intact for reference.

• Only an MVAC-trained, EPA-certified automotive technician should service the air conditioning system or its components.

ENGINE REPAIR

Alternator

REMOVAL

3.0L Engine (1MZ-FE)

1. Before servicing the vehicle, refer to the precautions in the beginning of this section.
2. Remove or disconnect the following:
   • Negative battery cable
   • Accessory drive belt
   • Alternator harness connectors
   • Alternator

3.0L Engine (2JZ-GE)

1. Before servicing the vehicle, refer to the precautions in the beginning of this section.

2. Remove or disconnect the following:
   • Negative battery cable
   • Engine under cover
   • Accessory drive belt
   • Alternator connector
   • Cap and nut
   • Alternator wire
   • Alternator wire clamp from the wire clip on the alternator
   • Bolt and pipe clamp
   • 2 automatic transmission oil cooler pipes from the alternator
   • Bolt, nut, pipe bracket and alternator

4.0L (1UZ-FE) and 4.3L (3UZ-FE) Engines

1. Before servicing the vehicle, refer to the precautions in the beginning of this section.

2. Remove or disconnect the following:
   • Negative battery cable
   • Air cleaner inlet
   • Accessory drive belt
   • Oil pan protector
   • Engine under cover
   • Power steering pump
   • Alternator harness connectors
   • Heated Oxygen (HO$_2$S) sensor wiring
   • Alternator

INSTALLATION

3.0L Engine (1MZ-FE)

1. Install or connect the following:
   • Alternator
   • Alternator harness connectors
   • Accessory drive belt. Tighten the

adjusting lock bolt to 13 ft. lbs. (18 Nm) and the pivot bolt to 41 ft. lbs. (56 Nm).
- Negative battery cable

3.0L Engine (2JZ-GE)

1. Install or connect the following:
- Bolt, nut, pipe bracket and alternator. Tighten the fasteners to 30 ft. lbs. (40 Nm).
- 2 automatic transmission oil cooler pipes to the alternator
- Bolt and pipe clamp
- Alternator wire clamp to the wire clip on the alternator
- Alternator wire
- Cap and nut
- Alternator connector
- Accessory drive belt
- Engine under cover
- Negative battery cable

4.0L (1UZ-FE) and 4.3L (3UZ-FE) Engines

1. Install or connect the following:
- Alternator. Tighten the fasteners to 29 ft. lbs. (39 Nm).
- HO_2S sensor wiring
- Alternator harness connectors
- Power steering pump
- Engine under cover
- Oil pan protector
- Accessory drive belt
- Air cleaner inlet
- Negative battery cable

Ignition Timing

ADJUSTMENT

The engines covered in this section are equipped with a Distributorless Ignition System (DIS). No timing adjustments are possible.

Engine Assembly

REMOVAL & INSTALLATION

ES300

1. Before servicing the vehicle, refer to the precautions in the beginning of this section.
2. Release the fuel pressure.
3. Drain the engine coolant and engine oil.
4. Remove or disconnect the following:

- Battery and tray
- Hood
- Accelerator cable and the throttle cable
- Air cleaner cover,
- Volume air flow meter and air cleaner duct as an assembly
- Cruise control actuator, if equipped
- Radiator
- Engine relay box and 2 bolts
- 5 connections from the relay box
- 2 igniter connectors
- Left fender apron connector
- Noise filter connector
- 2 ground straps
- Engine wiring harness from the engine.
- Vacuum hoses from the following connections: intake air control valve vacuum tank, charcoal canister, brake booster vacuum hose from the intake chamber.
- 2 heater hoses from the bulkhead
- Fuel feed and return lines
- Control cable from the transaxle
- Wiring harness from the PCM and route it through the bulkhead.
- Air conditioning compressor from the engine without disconnecting the lines and position it out of the way.
- Front exhaust pipe
- Halfshafts
- 2 power steering air hoses from the engine
- Hydraulic cooling fan pressure hose
- Power steering pump without disconnecting the lines and position it out of the way.
- Right and left lower engine mounts from the body
- Engine mounting shock absorber
- 3 front engine mounting bolts from the body
5. Attach a lifting device to the engine.
6. Remove or disconnect the following:
- Coolant reservoir tank.
- Right engine mounting bracket
- Engine moving control rod and right No. 2 engine mounting bracket
- Engine and transaxle as an assembly

✳✳ WARNING

Be careful not to hit the power steering or PNP switches.

- Engine mounting insulator below the oil filter
- Right rear engine-mounting insulator
- Front exhaust pipe stay
7. Label and detach the following connectors:
- Overdrive solenoid
- PNP switch
- Speedometer
- Starter terminal
- Speed sensor.
- 2 wire clamps from the transaxle
- Oil dipstick and guide
- Starter
- Flywheel housing cover
8. Turn the crankshaft pulley to gain access to the 8 torque converter bolts. Secure the crankshaft and remove them as they become accessible.
9. Install or connect the following:
- 2 exhaust manifold stays and plate
- 2 bolts attaching the transaxle to the oil pan
- 6 transaxle mounting bolts
- Transaxle

To install:
10. Position the transaxle to the engine. Transaxle mounting bolts: 47 ft. lbs. (64 Nm).
11. Install or connect the following:
- Bolts that attach the transaxle to the oil pan bolts and tighten them to 34 ft. lbs. (46 Nm).
- Exhaust manifold support. Tighten the bolts to 14 ft. lbs. (20 Nm).
- Bolts that attach the flywheel to the torque converter. Coat the threads with a locking compound. Rotate the engine and tighten the bolts alternately to 30 ft. lbs. (41 Nm).
- Starter to the engine
- Flywheel cover and tighten the bolts to 13 ft. lbs. (18 Nm).
- Dipstick and tube with a new O-ring
12. Attach the clamps and following connectors:
- Overdrive solenoid
- PNP switch
- Speedometer
- Starter terminal
- Speed sensor
13. Install or connect the following:
- Exhaust pipe's stay. Tighten the bolts to 15 ft. lbs. (21 Nm).
- Right rear insulator. Tighten the bolts to 47 ft. lbs. (64 Nm).
- Front engine mounting insulator. Tighten the bolts to 47 ft. lbs. (64 Nm).

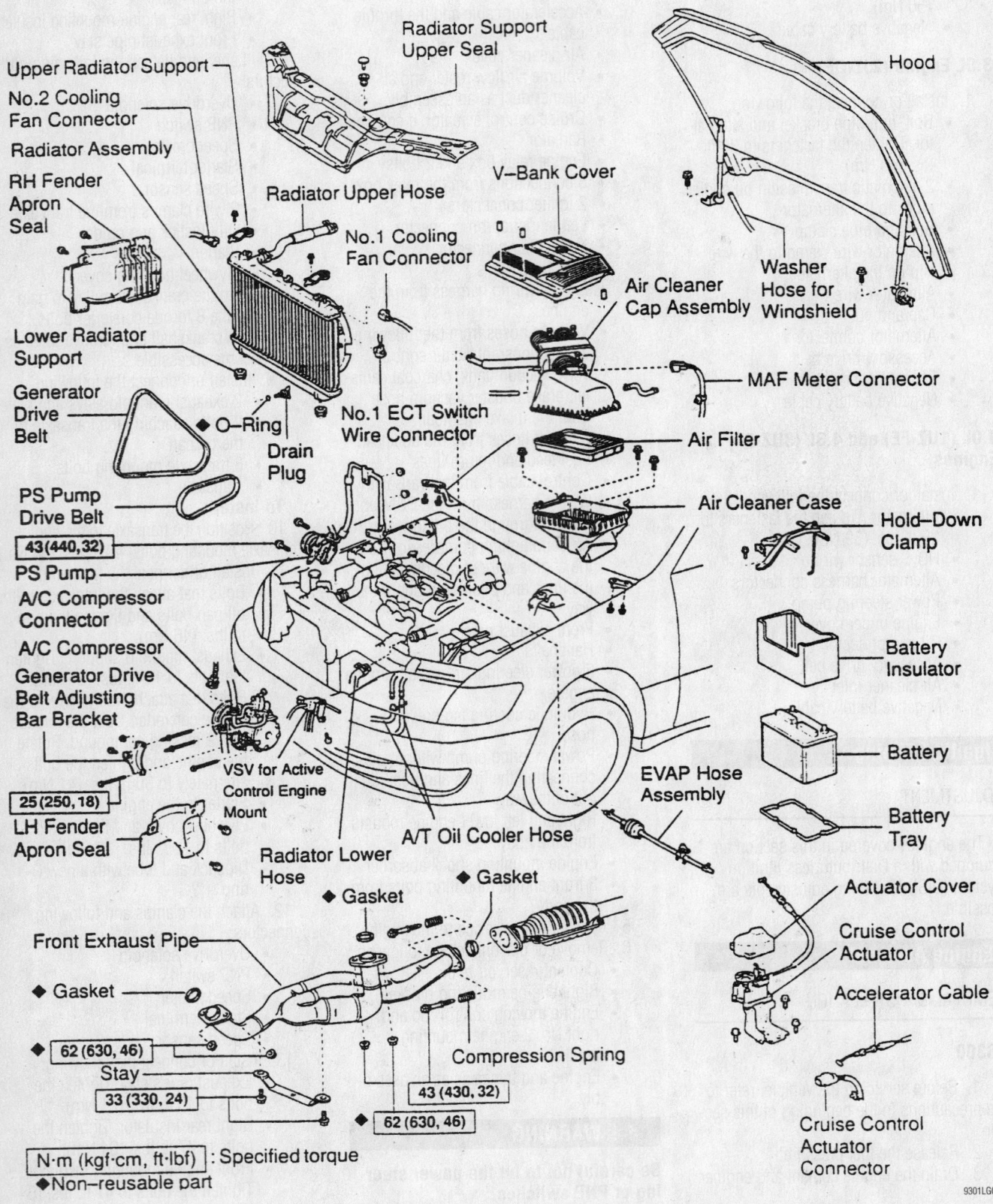

Upper Radiator Support

No.2 Cooling Fan Connector

Radiator Assembly

RH Fender Apron Seal

Radiator Support Upper Seal

Radiator Upper Hose

No.1 Cooling Fan Connector

V-Bank Cover

Hood

Air Cleaner Cap Assembly

Washer Hose for Windshield

MAF Meter Connector

Lower Radiator Support

Generator Drive Belt

◆ O-Ring

Drain Plug

No.1 ECT Switch Wire Connector

Air Filter

Air Cleaner Case

Hold-Down Clamp

PS Pump Drive Belt

43 (440, 32)

PS Pump

A/C Compressor Connector

A/C Compressor

Generator Drive Belt Adjusting Bar Bracket

25 (250, 18)

LH Fender Apron Seal

VSV for Active Control Engine Mount

Radiator Lower Hose

A/T Oil Cooler Hose

EVAP Hose Assembly

Battery Insulator

Battery

Battery Tray

Actuator Cover

Cruise Control Actuator

Accelerator Cable

Front Exhaust Pipe

◆ Gasket

◆ Gasket

◆ Gasket

Compression Spring

62 (630, 46)

Stay

33 (330, 24)

43 (430, 32)

62 (630, 46)

Cruise Control Actuator Connector

N·m (kgf·cm, ft·lbf) : Specified torque

◆Non-reusable part

9301LG01

Exploded view of the engine removal and related components—ES300

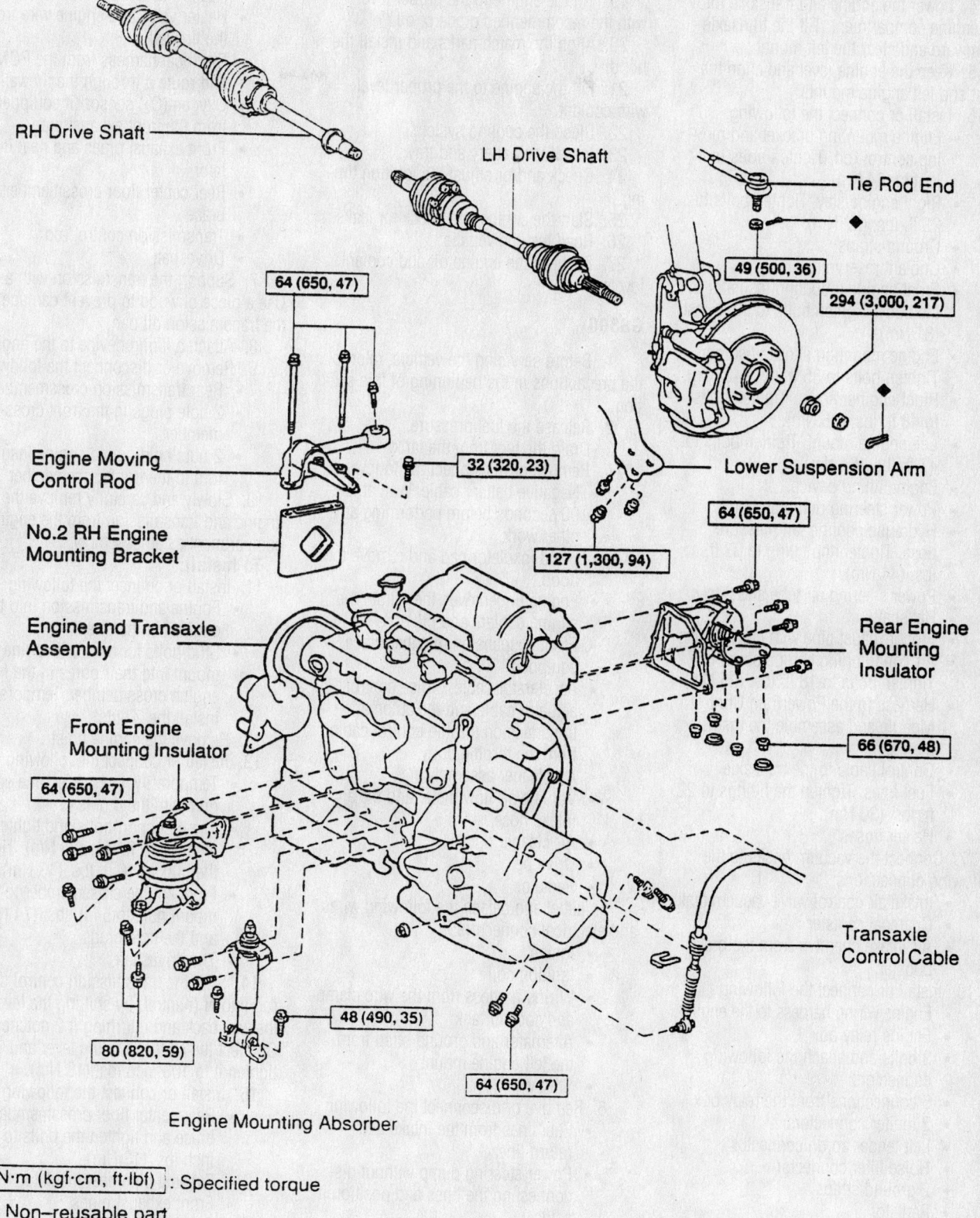

RH Drive Shaft

LH Drive Shaft

Tie Rod End

49 (500, 36)

294 (3,000, 217)

64 (650, 47)

Engine Moving
Control Rod

32 (320, 23)

Lower Suspension Arm

No.2 RH Engine
Mounting Bracket

64 (650, 47)

127 (1,300, 94)

Engine and Transaxle
Assembly

Rear Engine
Mounting
Insulator

66 (670, 48)

Front Engine
Mounting Insulator

64 (650, 47)

Transaxle
Control Cable

48 (490, 35)

80 (820, 59)

64 (650, 47)

Engine Mounting Absorber

N·m (kgf·cm, ft·lbf) : Specified torque

◆ Non–reusable part

9301LG02

Exploded view of the engine removal and related components (cont.)—ES300

14. Lower the engine and transaxle into the engine compartment. Tilt the transaxle downward and clear the left mount.

15. Keep the engine level and align the right and left engine mounts.

16. Install or connect the following:
- Engine mounting bracket and moving control rod. Tighten bolts to 47 ft. lbs. (64 Nm).
- Right engine stay. Tighten bolts to 23 ft. lbs. (32 Nm).
- Ground straps
- Coolant reservoir
- Front engine mounting insulator to the body. Tighten bolts to 59 ft. lbs. (81 Nm).
- Engine mounting shock absorber. Tighten bolts to 35 ft. lbs. (48 Nm).
- Right engine mount. Tighten bolts to 48 ft. lbs. (66 Nm).
- Left engine mount. Tighten bolts to 47 ft. lbs. (64 Nm).
- Engine lifting device
- Power steering pump and belt
- Hydraulic cooling fan pressure hose. Tighten the fitting to 33 ft. lbs. (44 Nm).
- Power steering air tube and hoses
- Halfshafts
- Front exhaust pipe with new gaskets
- Air conditioning compressor. Tighten bolts to 18 ft. lbs. (25 Nm).
- Harness to the Powertrain Control Module and assemble the instrument panel.
- Control cable to the transaxle
- Fuel lines. Tighten the fittings to 22 ft. lbs. (30 Nm).
- Heater hoses

17. Connect the vacuum hoses to the following connections:
- Intake air control valve vacuum tank
- Charcoal canister
- Air intake chamber from the brake booster

18. Install or connect the following:
- Engine wiring harness to the engine
- Engine relay box
- 2 bolts and attach the following connectors:
- 5 connections from the relay box
- 2 igniter connectors
- Left fender apron connector
- Noise filter connector
- 2 ground straps
- Radiator
- Cruise control actuator, if equipped
- Air cleaner cover
- Volume airflow meter and air cleaner duct assembly
- Throttle cable
- Accelerator cable

19. Fill the engine to the proper level with the recommended grade of oil.

20. Align the matchmarks and install the hood.

21. Fill the engine to the proper level with coolant.

22. Bleed the cooling system.

23. Install the battery and tray.

24. Check and/or adjust the ignition timing.

25. Start the engine and check for leaks.

26. Road test the vehicle.

27. Recheck the engine oil and coolant levels.

GS300

1. Before servicing the vehicle, refer to the precautions in the beginning of this section.

2. Release the fuel pressure.

3. Drain the fuel from the tank.

4. Remove or disconnect the following:
- Negative battery cable. Wait at least 90 seconds before performing any other work.
- Hood insulator pad and remove the hood.
- Engine undercover, then drain the engine coolant and oil.
- Front suspension member brace, if equipped
- Accelerator cable, cruise control actuator cable and the automatic transmission throttle control cable from the throttle body.
- Air cleaner assembly
- Volume air flow meter and the air intake hose
- Air cleaner duct
- Drive belt
- Radiator

5. Label and detach the following wires and electrical connectors:
- Igniter
- Ignition coil
- Wiring harness from the wire clamp and coolant tank
- Alternator and ground strap from the left engine mount
- Starter

6. Remove or disconnect the following:
- Fuel lines from the intake and return lines
- Power steering pump without disconnecting the lines and position it aside
- Air conditioning compressor without disconnecting the air conditioning lines and position it aside
- Brake booster vacuum hose
- Evaporative Emissions (EVAP) hose

- Heater hoses from the firewall
- Heater valve and engine wire from the firewall
- Electrical harness from the PCM and route it through the firewall
- Oxygen (O_2) sensor (if equipped) from the front exhaust pipe
- Front exhaust pipes and heat insulator
- Rear center floor crossmember brace
- Transmission control rod
- Driveshaft

7. Support the transmission with a jack. Use a piece of wood to prevent damage to the transmission oil pan.

8. Attach a lifting device to the engine.

9. Remove or disconnect the following:
- Rear transmission crossmember
- 2 hole plugs in the front crossmember
- 2 nuts holding the engine insulators to the front crossmember

10. Slowly and carefully remove the engine and transmission from the engine compartment as an assembly.

To install:

11. Install or connect the following:
- Engine and transmission into the engine compartment
- Stud bolts for the front engine mount into their bores in the front engine crossmember. Temporarily install the 2 nuts.

12. Remove the engine hoist

13. Install or connect the following:
- Temporarily, the rear engine support with the 4 nuts
- The 4 support bolts and tighten them to 19 ft. lbs. (25 Nm). Tighten the nuts to 10 ft. lbs. (13 Nm).
- Front engine crossmember to mount nuts to 54 ft. lbs. (74 Nm) and the hole plugs
- The driveshaft

14. Shift the transmission control shift rod into **N** (neutral) by shifting the lever all the way back and returning it 2 notches. Connect the shift rod to the lever and tighten it to 108 inch lbs. (13 Nm).

15. Install or connect the following:
- Rear center floor crossmember brace and tighten the bolts to 108 inch lbs. (13 Nm).
- Exhaust pipe heat insulator
- Front exhaust pipes
- Sub HO_2sensor, if equipped.
- Engine wiring harness to the PCM.
- PCM and its cover
- The lower portion of the passenger side instrument panel, the vent, the carpet and the scuff panel

- Heater water valve and engine wire to the cowl panel
- Heater hoses
- EVAP hose
- Brake booster hose

- The air conditioning compressor and tighten the Torx® bolt to 19 ft. lbs. (26 Nm). Tighten the nut and bolts to 38 ft. lbs. (52 Nm).
- Power steering pump

- Fuel lines with new gaskets and tighten the union bolts to 22 ft. lbs. (29 Nm).
- Igniter
- Ignition coil

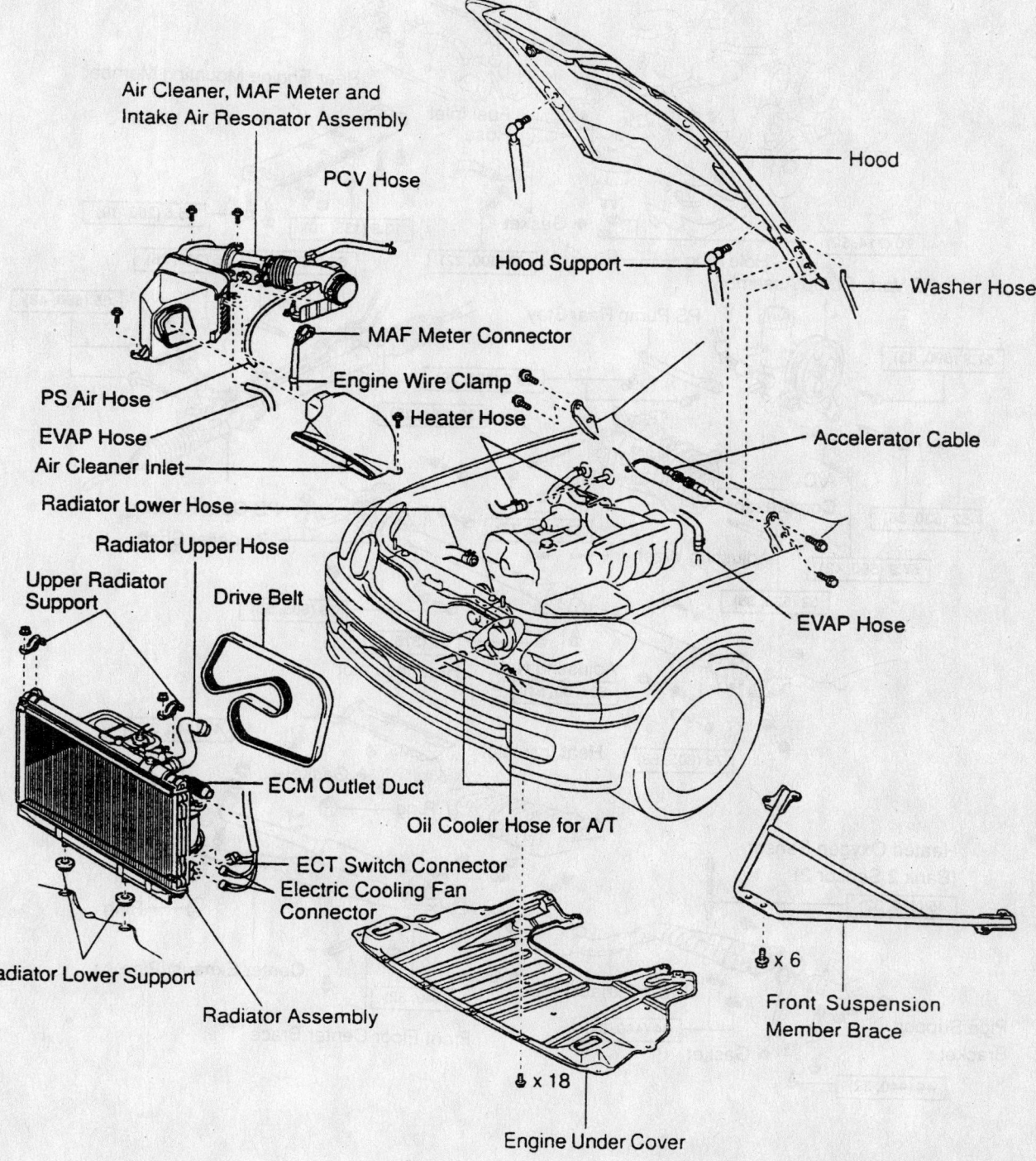

Air Cleaner, MAF Meter and Intake Air Resonator Assembly

PCV Hose

Hood

Hood Support

Washer Hose

MAF Meter Connector

Engine Wire Clamp

PS Air Hose

Heater Hose

EVAP Hose

Accelerator Cable

Air Cleaner Inlet

Radiator Lower Hose

Radiator Upper Hose

EVAP Hose

Upper Radiator Support

Drive Belt

ECM Outlet Duct

Oil Cooler Hose for A/T

ECT Switch Connector

Electric Cooling Fan Connector

☒ x 6

Radiator Lower Support

Front Suspension Member Brace

Radiator Assembly

☒ x 18

Engine Under Cover

9301LG03

Exploded view of the engine removal and related components—GS300

Timing belt service is covered in Section 3 of this manual

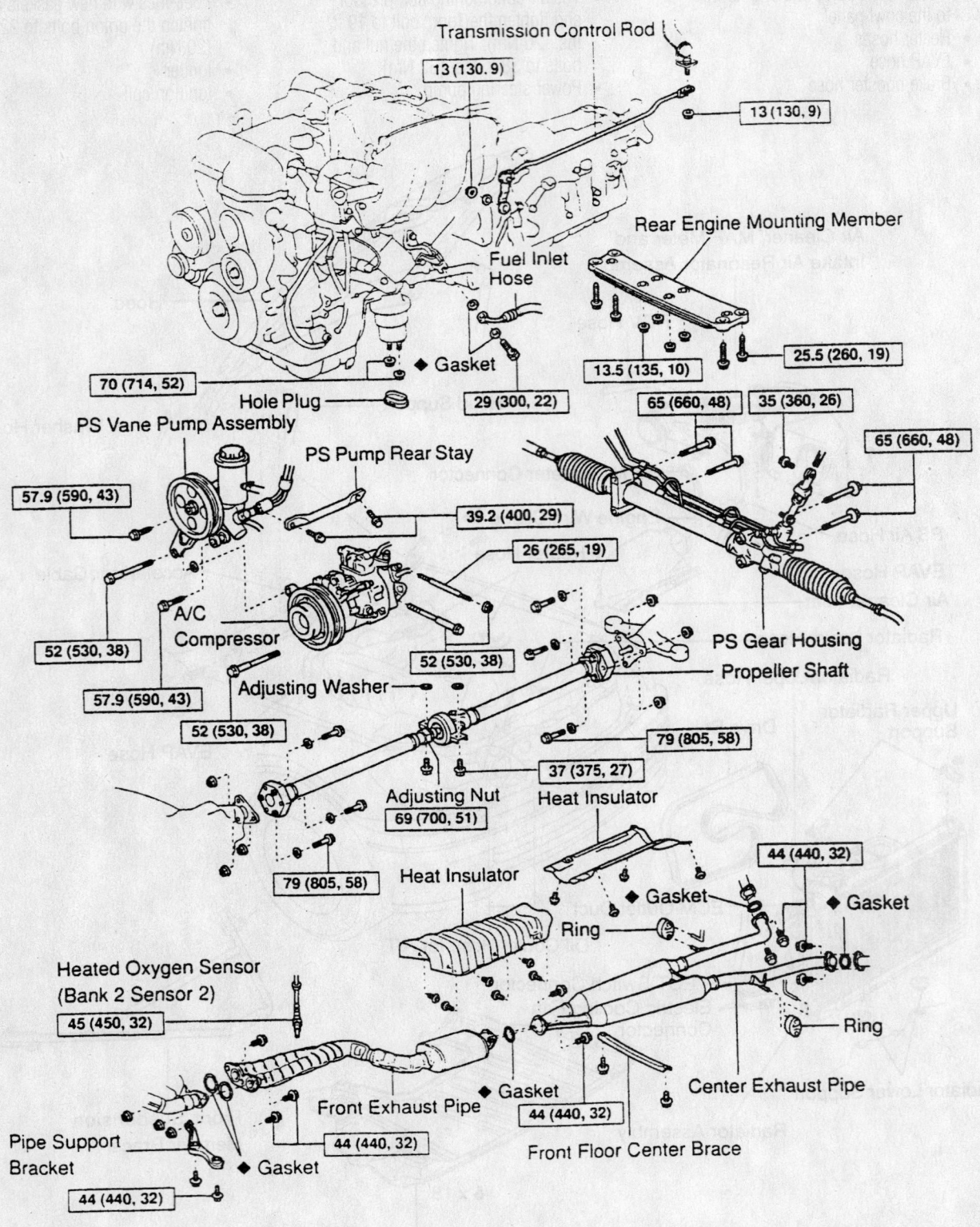

Transmission Control Rod

13 (130, 9)

13 (130, 9)

Rear Engine Mounting Member

Fuel Inlet Hose

25.5 (260, 19)

13.5 (135, 10)

65 (660, 48)

35 (360, 26)

70 (714, 52)

◆ Gasket

Hole Plug

29 (300, 22)

65 (660, 48)

PS Vane Pump Assembly

PS Pump Rear Stay

57.9 (590, 43)

39.2 (400, 29)

26 (265, 19)

A/C Compressor

52 (530, 38)

52 (530, 38)

PS Gear Housing

57.9 (590, 43)

Propeller Shaft

Adjusting Washer

52 (530, 38)

79 (805, 58)

Adjusting Nut

37 (375, 27)

69 (700, 51)

Heat Insulator

79 (805, 58)

Heat Insulator

44 (440, 32)

◆ Gasket

◆ Gasket

Ring

Heated Oxygen Sensor
(Bank 2 Sensor 2)

45 (450, 32)

Ring

Center Exhaust Pipe

Front Exhaust Pipe

◆ Gasket

Pipe Support Bracket

44 (440, 32)

44 (440, 32)

◆ Gasket

Front Floor Center Brace

44 (440, 32)

N·m (kgf·cm, ft·lbf) : Specified torque
◆ Non–reusable part

9301LG04

Exploded view of the engine removal and related components (cont.)—GS300

- Wiring harness from the wire clamp and coolant tank
- Alternator and ground strap from the left engine mount
- Starter
- Radiator
- Drive belt
- Air cleaner
- Volume air flow meter and intake air connector pipe as an assembly
- Air cleaner duct
- Accelerator, cruise control and the

automatic transmission throttle control cables
- Fuel
- Engine oil
- Coolant
- Negative battery cable

16. Start the engine and check for leaks.

17. Check the automatic transmission fluid level.

18. Check and/or adjust the ignition timing.

19. If equipped, install the front suspen-

sion brace and tighten the mounting bolts to 43 ft. lbs. (58 Nm).

20. Install the hood and the hood insulator pad.

21. Road test the vehicle.

22. Recheck the fluid levels.

IS300

1. Before servicing the vehicle, refer to the precautions in the beginning of this section.

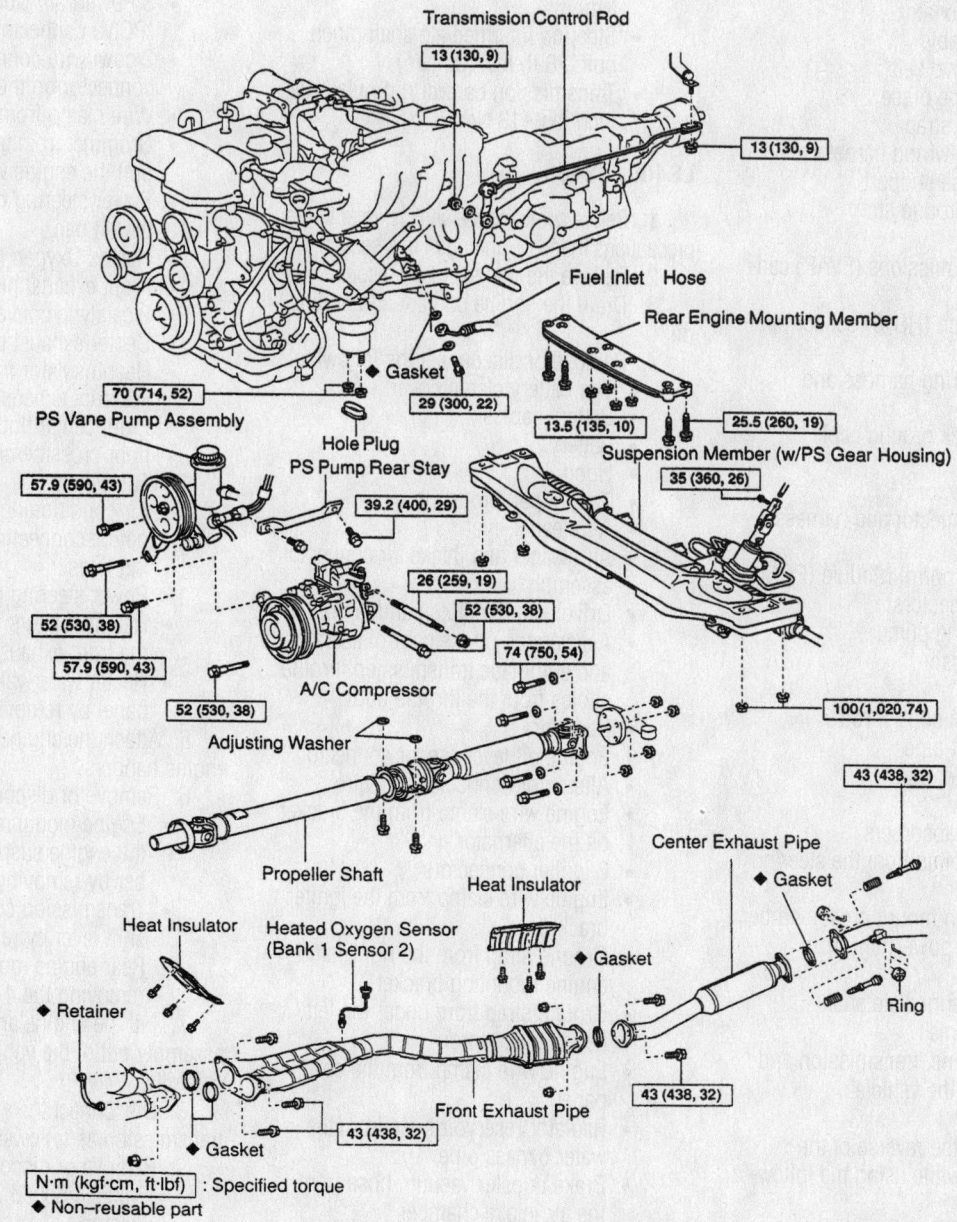

Exploded view of the engine mounting—IS300

N·m (kgf·cm, ft·lbf) : Specified torque
◆ Non-reusable part

9347LG03

Heater Core replacement is covered in Section 2 of this manual

2. Drain the cooling system.
3. Relieve the fuel system pressure.
4. Drain the engine oil.
5. Remove or disconnect the following:
- Negative battery cable
- Engine under cover
- Air cleaner inlet
- Brake booster vacuum hose
- Radiator hoses
- Mass Air Flow (MAF) sensor connector
- Positive Crankcase Ventilation (PCV) hose
- Air intake resonator
- Accelerator cable
- Accessory drive belt
- Front subframe brace
- Floor ground strap
- Starter motor wiring harness
- Fuel inlet hose support
- Dash panel ground strap
- Heater hoses
- Evaporative Emissions (EVAP) canister hose
- Heated Oxygen (HO2S) sensor connectors
- Alternator wiring harness and clamp
- Cylinder block ground cable bracket
- Igniter connector
- Data link connector and harness clamps
- Powertrain Control Module (PCM) harness connectors
- Power steering pump
- A/C compressor
- Drive shaft
- Transmission control rod
- Exhaust front pipe
- Exhaust center pipe
- Stabilizer bar
- Front shock absorbers
- Lower ball joints from the steering knuckles
- Transmission mount crossmember. Support the powertrain from below.
- Steering intermediate shaft
- Front subframe

6. Lower the engine, transmission and subframe away from the vehicle.

To install:

7. Installation is the reverse of the removal procedure, while using the following torque values:
- Transmission flange bolts: 53 ft. lbs. (72 Nm)
- Transmission flange-to-oil pan bolts: 27 ft. lbs. (37 Nm)
- Torque converter bolts: 74 ft. lbs. (100 Nm)

- Left and right motor mount nuts: 52 ft. lbs. (70 Nm)
- Front subframe bolts: 52 ft. lbs. (70 Nm)
- Transmission mount crossmember: 19 ft. lbs. (26 Nm)
- Lower ball joint bolts: 181 ft. lbs. (245 Nm)
- Lower shock absorber bolts: 47 ft. lbs. (64 Nm)
- Stabilizer bar bolts: 13 ft. lbs. (18 Nm)
- Stabilizer bar nuts: 36 ft. lbs. (49 Nm)
- Steering intermediate shaft pinch bolt: 26 ft. lbs. (35 Nm)
- Transmission control rod nuts: 108 inch lbs. (13 Nm)

LS400, LS430

1. Before servicing the vehicle, refer to the precautions in the beginning of this section.
2. Relieve the fuel system pressure
3. Drain the engine coolant and engine oil
4. Remove or disconnect the following:
- The battery clamp cover
- Battery cables
- Battery
- Hood
- The oil pan protector
- Air cleaner inlet
- Air cleaner and intake air connector assembly
- Drive belt, fan clutch and fan pulley
- Accelerator, cruise control actuator and automatic transmission throttle cables from the throttle body
- Radiator
- Engine oil level sensor connector
- Alternator connector and wire
- Engine wire clamp from the bracket on the alternator
- 2 igniter connectors
- Engine wire clamp from the igniter bracket
- Ground strap from the right-hand engine mounting bracket
- Ground strap from under the left-hand fender apron
- Engine wire clamp from the cowl panel
- Radiator reservoir hose from the water bypass pipe
- Brake booster vacuum hose from the air intake chamber
- Heater hose from the heater water valve and water bypass pipe
- Fuel inlet hose from the fuel inlet pipe
- Fuel return hose to the return pipe

- Power steering air hose from the air intake chamber
- 2 power steering hoses from the clamp on the right-hand No. 3 timing belt cover
- Evaporative Emission (EVAP) hose from the pipe (from the charcoal canister).
- Engine wire from the cabin as follows:
- Undercover from under the glove compartment
- Glove compartment
- 3 Powertrain Control Module (PCM) connectors
- 2 cowl wire connectors from the connector on the bracket
- Wire clamp from the bracket
- Grommet from the cowl panel, then pull the engine wire out.
- Power steering oil cooler pipe from the oil pan
- Heated Oxygen (HO2) sensors
- Front exhaust pipe
- 2 catalytic converters
- Center exhaust pipe
- Heat insulator from the rear side of the front exhaust pipe
- Front center floor and rear center floor crossmember braces
- Driveshaft
- Air conditioning compressor without disconnecting the air conditioning lines
- Power steering pump
- Heat insulators for the front side of the front exhaust pipe
- Heater water valve from the cowl panel by removing the 2 nuts.

5. Attach the engine chain hoist to the engine hangers.
6. Remove or disconnect the following:
- Engine mounting insulators from the engine suspension crossmember by removing the 2 nuts
- Transmission control rod from the shift lever by removing the nut
- Rear engine mounting member by removing the 4 nuts and 4 bolts

7. Lift the engine and transmission assembly out of the vehicle slowly and carefully.
8. Disconnect the engine from the transmission as follows:
9. Remove or disconnect the following:
- Vehicle Speed Sensor (VSS) connector
- Park/Neutral Position (PNP) switch connector
- Solenoid connector
- Direct clutch speed sensor connector

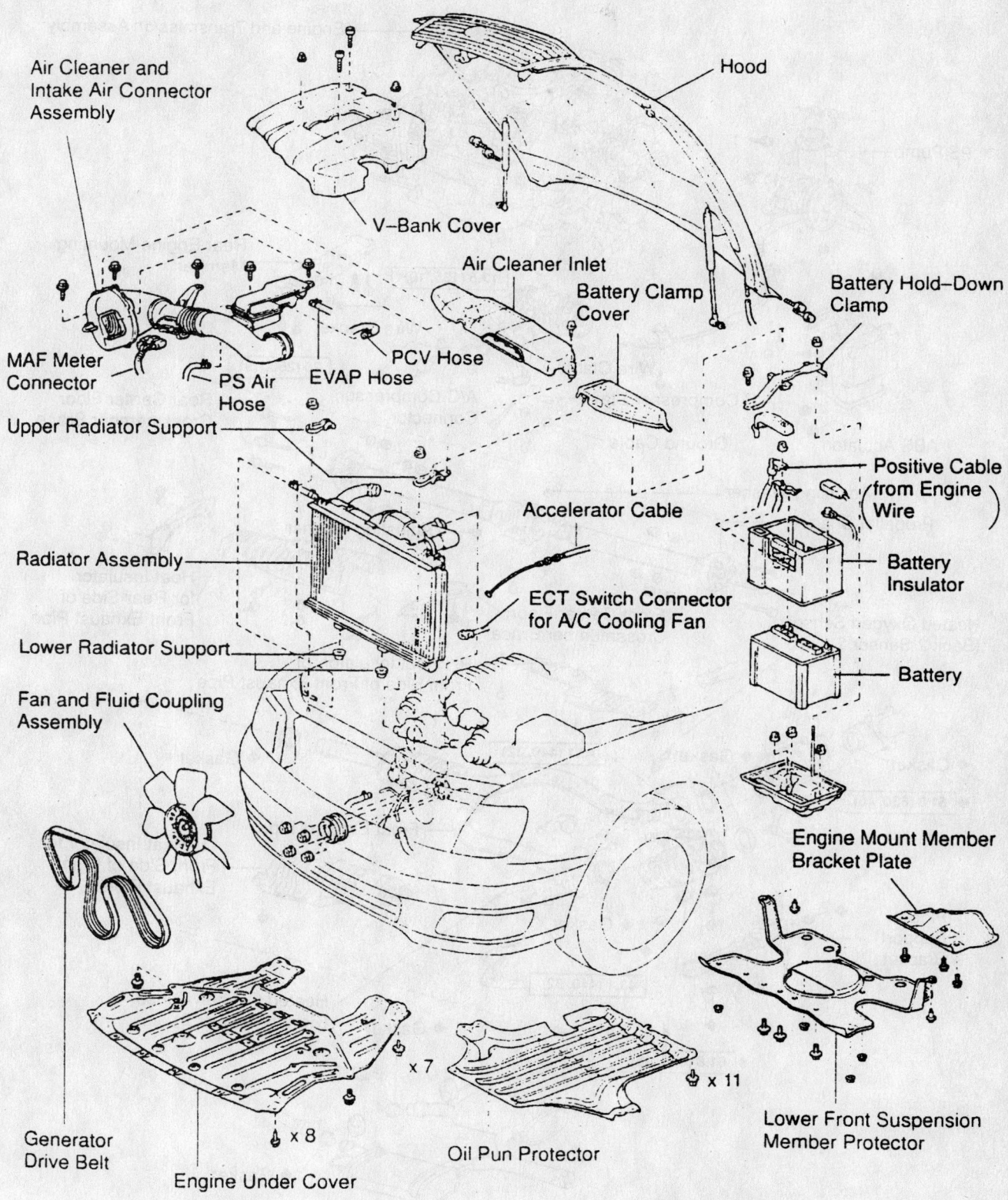

Air Cleaner and
Intake Air Connector
Assembly

Hood

V–Bank Cover

Air Cleaner Inlet

Battery Clamp
Cover

Battery Hold–Down
Clamp

MAF Meter
Connector

PS Air
Hose

EVAP Hose

PCV Hose

Upper Radiator Support

Accelerator Cable

Positive Cable
(from Engine
Wire)

Radiator Assembly

ECT Switch Connector
for A/C Cooling Fan

Battery
Insulator

Lower Radiator Support

Battery

Fan and Fluid Coupling
Assembly

Engine Mount Member
Bracket Plate

Generator
Drive Belt

x 7

Lower Front Suspension
Member Protector

x 8

Engine Under Cover

Oil Pun Protector

x 11

9301LG05

Exploded view of the engine removal and related components—LS400

Brake service is covered in Section 4 of this manual

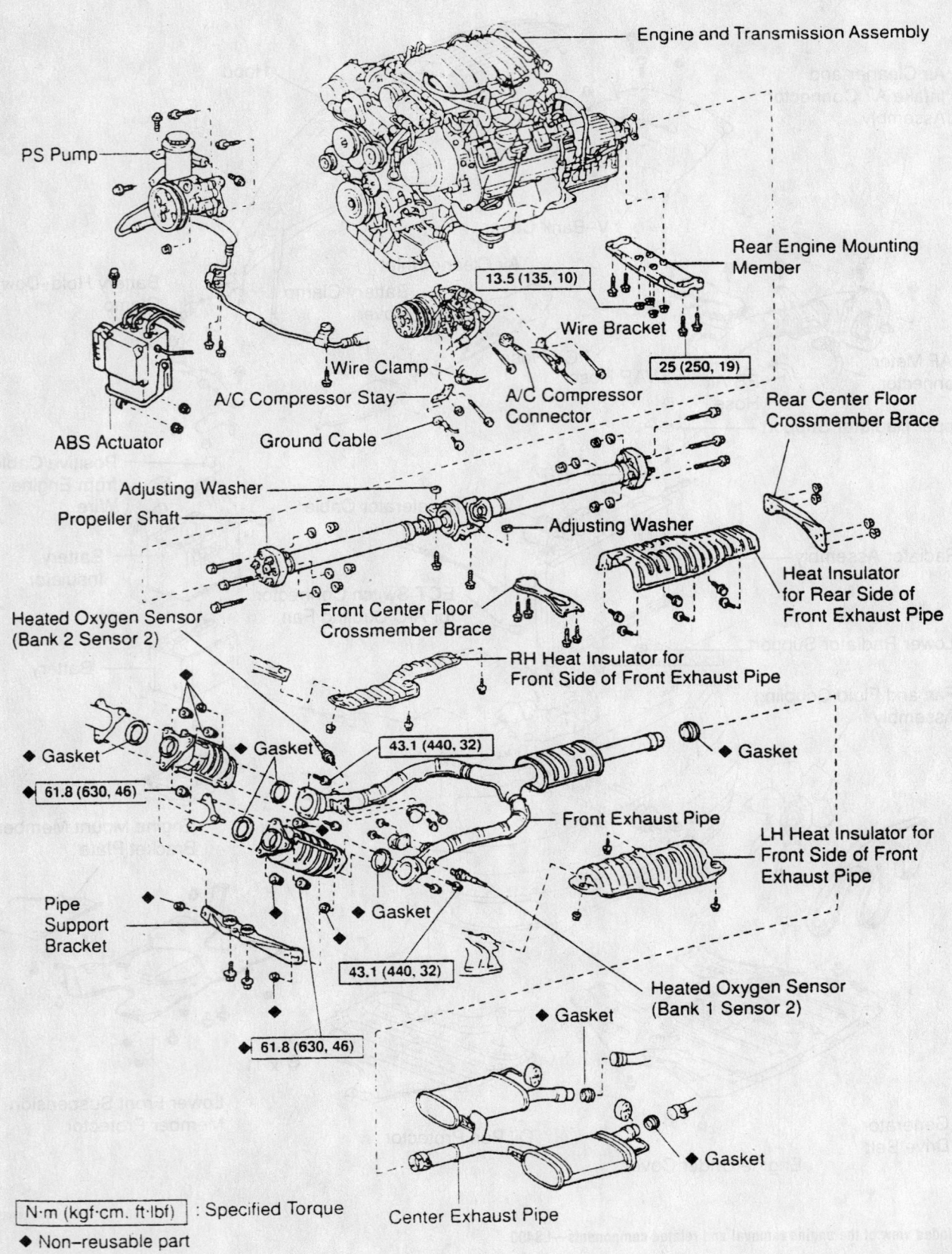

Engine and Transmission Assembly

PS Pump

Rear Engine Mounting Member

13.5 (135, 10)

Wire Bracket

25 (250, 19)

Wire Clamp

A/C Compressor Stay

A/C Compressor Connector

Ground Cable

ABS Actuator

Rear Center Floor Crossmember Brace

Adjusting Washer

Propeller Shaft

Adjusting Washer

Heat Insulator for Rear Side of Front Exhaust Pipe

Heated Oxygen Sensor (Bank 2 Sensor 2)

Front Center Floor Crossmember Brace

RH Heat Insulator for Front Side of Front Exhaust Pipe

◆ Gasket

◆ Gasket

43.1 (440, 32)

◆ Gasket

◆ 61.8 (630, 46)

Front Exhaust Pipe

LH Heat Insulator for Front Side of Front Exhaust Pipe

Pipe Support Bracket

◆ Gasket

43.1 (440, 32)

Heated Oxygen Sensor (Bank 1 Sensor 2)

◆ Gasket

◆ 61.8 (630, 46)

◆ Gasket

N·m (kgf·cm, ft·lbf) : Specified Torque

◆ Non-reusable part

Center Exhaust Pipe

9301LG06

Exploded view of the engine removal and related components (cont.)—LS400

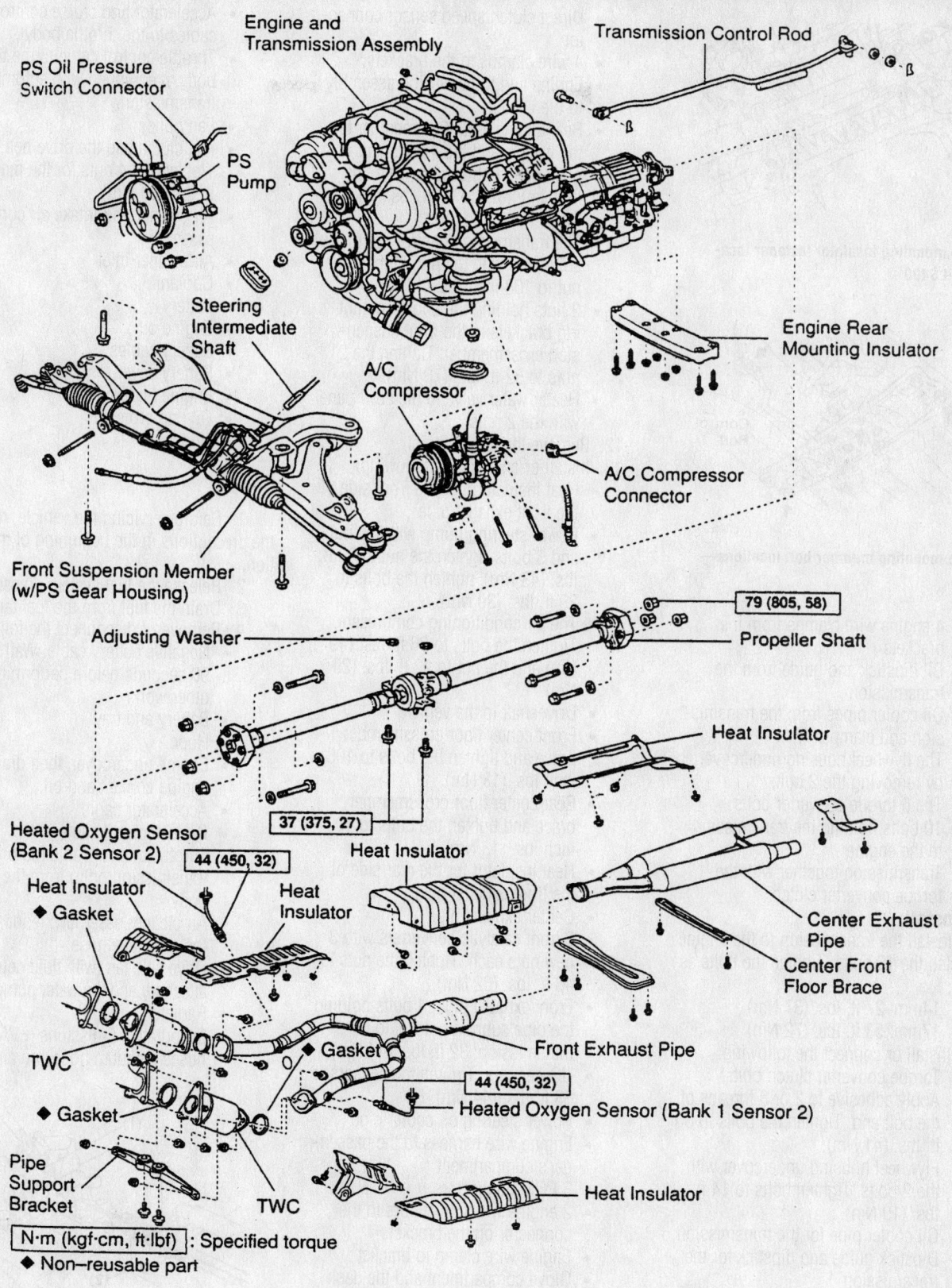

PS Oil Pressure Switch Connector

PS Pump

Engine and Transmission Assembly

Transmission Control Rod

Steering Intermediate Shaft

A/C Compressor

A/C Compressor Connector

Engine Rear Mounting Insulator

Front Suspension Member (w/PS Gear Housing)

Adjusting Washer

79 (805, 58)

Propeller Shaft

Heat Insulator

Heated Oxygen Sensor (Bank 2 Sensor 2)

44 (450, 32)

37 (375, 27)

Heat Insulator

Heat Insulator

Heat Insulator

Heat Insulator

◆ Gasket

Center Exhaust Pipe

Center Front Floor Brace

TWC

◆ Gasket

◆ Gasket

Front Exhaust Pipe

44 (450, 32)

Heated Oxygen Sensor (Bank 1 Sensor 2)

Pipe Support Bracket

TWC

Heat Insulator

N·m (kgf·cm, ft·lbf) : Specified torque

◆ Non–reusable part

Exploded view of the engine removal and related components—LS430

9347LG04

For complete Engine Mechanical specifications, see Section 1 of this manual

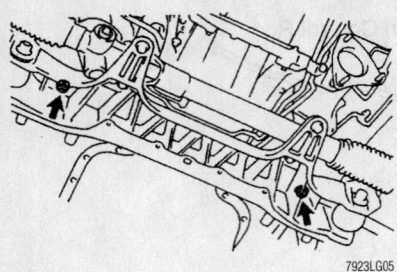

Engine mounting insulator fastener locations—LS400

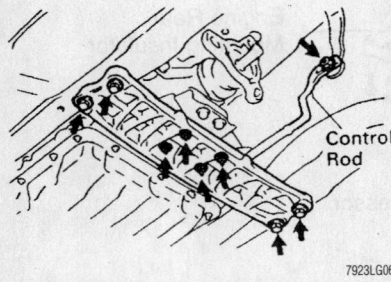

Engine mounting member bolt locations—LS400

- 4 engine wire clamps from the brackets
- Oil dipstick and guide from the transmission
- Oil cooler pipes from the transmission and clamps
- The flywheel housing undercover by removing the 2 bolts
- The 6 torque converter bolts
- 10 bolts holding the transmission to the engine
- Transmission together with the torque converter clutch

To install:

10. Install the transmission to the engine and install the 10 bolts. Tighten the bolts as follows:
- 14mm: 27 ft. lbs. (37 Nm)
- 17mm: 53 ft. lbs. (72 Nm)

11. Install or connect the following:
- Torque converter clutch bolts. Apply adhesive to 2 or 3 threads of the bolt end. Tighten the bolts to 30 ft. lbs. (41 Nm).
- Flywheel housing undercover with the 2 bolts. Tighten bolts to 14 ft. lbs. (19 Nm).
- Oil cooler pipe for the transmission
- Dipstick guide and dipstick for the transmission
- Engine wire to the transmission
- VSS connector
- PNP switch connector
- Solenoid connector

- Direct clutch speed sensor connector
- 4 wire clamps to the brackets
- Engine and transmission assembly to the vehicle
- Rear engine mounting member to the vehicle and the 4 bolts and 4 nuts; bolts tightened to 19 ft. lbs. (25 Nm), nuts to 10 ft. lbs. (14 Nm).
- The transmission control rod to the shift lever with the nut. Tighten the nut to 108 inch lbs. (13 Nm).
- 2 nuts holding the engine mounting brackets to the front suspension crossmember. Tighten the 2 nuts to 52 ft. lbs. (70 Nm).
- Heater water valve to the cowl panel with the 2 nuts

12. Remove the engine hoist.

13. Install or connect the following:
- Heat insulators for the front side of the front exhaust pipe
- Power steering pump with the nut and 3 bolts-tighten the nut to 32 ft. lbs. (43 Nm); tighten the bolts to 29 ft. lbs. (39 Nm).
- The air conditioning compressor. Tighten the bolts to 36 ft. lbs. (49 Nm) and the nut to 22 ft. lbs. (29 Nm).
- Driveshaft to the vehicle
- Front center floor crossmember brace and tighten the bolts to 108 inch lbs. (13 Nm).
- Rear center floor crossmember brace and tighten the bolts to 108 inch lbs. (13 Nm).
- Heat insulator for the rear side of the front exhaust pipe
- Center exhaust pipe
- 2 front catalytic converters with 3 new nuts each. Tighten the nuts to 46 ft. lbs. (62 Nm).
- Front exhaust pipe. 4 bolts holding the pipe support bracket to the transmission: 32 ft. lbs. (44 Nm).
- HO2sensors. Tighten the sensors to 33 ft. lbs. (44 Nm).
- Power steering oil cooler pipe
- Engine wire harness to the passenger's compartment
- 3 PCM connectors
- 2 engine wire connectors to the connector on the bracket
- Engine wire clamp to bracket
- Glove compartment and the dash undercover
- All of the engine assembly connectors, wires, straps, clamps and hoses
- Radiator

- Accelerator and cruise control cables to the throttle body
- Throttle control cable to the throttle body, if equipped with automatic transmission.
- Fan pulley
- Fan clutch and the drive belt. Tighten the 4 nuts for the fan to 16 ft. lbs. (21 Nm).
- Air cleaner and intake air connector assembly
- Air cleaner inlet
- Coolant
- Battery
- Engine oil
- Battery cables
- Battery cover
- Engine undercover
- Oil pan protector
- Hood

SC300

1. Before servicing the vehicle, refer to the precautions in the beginning of this section.

2. Release the fuel system pressure.

3. Drain the fuel from the fuel tank.

4. Remove or disconnect the following:
- Negative battery cable. Wait at least 90 seconds before performing any other work.
- Battery and tray
- Hood
- Engine undercover, then drain the engine coolant and oil.
- Accelerator cable
- Cruise control cable
- Throttle control cable (automatic transmission only) from the throttle body
- Air cleaner assembly, resonator and the air intake hose.
- Drive belt, fan (with fluid coupling attached) and the water pump pulley.
- Radiator
- Evaporative Emissions (EVAP) hoses (vacuum hose and air hose)

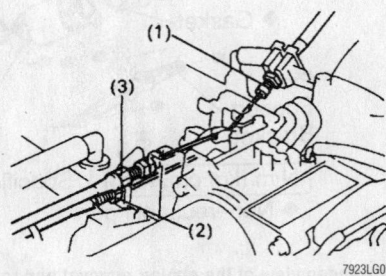

Accelerator cable (1), throttle control cable (2), cruise control cable (3) locations—SC300

from the charcoal canister
- Charcoal canister
- Power steering pump
- Air conditioning compressor
- All wires, electrical leads and vacuum hoses from the block
- Wiring clips or brackets
- Undercover beneath the glove box
- Lower instrument panel, the trim panel and the glove box door
- Right door sill trim (scuff plate). Lift the front edge of the carpet and remove the protective cover from the Powertrain Control Module (PCM)
- PCM connectors, the cowl wire connectors and the control unit connectors behind the glove box
- 2 nuts holding the harness to the firewall and carefully pull the engine harness into the engine compartment
- Power steering pipe and 2 clamp bolts from the engine block
- Union bolt and 2 gaskets
- Fuel inlet hose
- Starter wiring and unhook the starter wiring from the clip
- Front exhaust pipe and heat shield
- Transmission control rod at the shift lever
- Intermediate shaft

➡**Some vehicles are not equipped with adjusting washers.**

5. Attach an engine lift to the lift hooks
6. Remove or disconnect the following:
- The 2 nuts holding the engine to the front suspension crossmember
- The 4 bolts and 4 nuts holding the engine to the rear crossmember
- The rear engine mount
- Engine/transaxle assembly out of the engine compartment
7. Place the engine and transmission assembly onto a stand.
8. For vehicles with automatic transmission, remove or disconnect the following:
- Oil dipstick guide and dipstick for the transmission
- Oil cooler tubes
- Engine wire from the transmission
- Starter from the engine by removing the bolts
9. Separate the engine and transmission.

To install:
10. Install or connect the following:

- Engine and transmission
- Starter to the engine by installing the bolts
- Engine wire to the transmission
- With automatic transmission, the oil cooler tubes
- With automatic transmission, the oil dipstick guide and dipstick to the transmission
- Engine into the engine compartment
- The rear mount. Tighten the 4 nuts to 10 ft. lbs. (13 Nm). Tighten the 4 bolts to 19 ft. lbs. (25 Nm).
- Intermediate shaft to the rear differential and tighten the bolts and nuts to 54 ft. lbs. (74 Nm). Install the center support bearing set bolts with the adjusting washers. Tighten the bolts to 36 ft. lbs. (49 Nm).
- Transmission control rod to the shift lever (automatic transmission only) by installing the nut.
- Exhaust heat insulator by installing the 4 nuts
- No. 2 front exhaust pipe, tighten the nuts to 46 ft. lbs. (62 Nm).
- The pipe support bracket to the transmission with the 2 bolts. Tighten the bolts to 32 ft. lbs. (43 Nm).
- No. 2 front exhaust pipe to the front exhaust pipe with the 2 bolts and nuts. Tighten the bolts to 32 ft. lbs. (43 Nm).
- With transmissions, the clutch release cylinder and tighten the bolts to 108 inch lbs. (12 Nm).
- Starter wiring and secure the harness in the clips
- Fuel inlet hose with 2 new gaskets and tighten the union bolt to 22 ft. lbs. (29 Nm).
- Power steering pipe below the engine
- Engine harness
- Connectors to the proper PCM, controller, or relay. Make certain each connector is square and secure.
- PCM and cover; connect the wiring harnesses. Refit the carpet and install the scuff plate.
- Lower instrument panel trim, the glove box door and the undercover.
- Engine wiring harness in the clips and retainers
- All wires, electrical leads and vacuum hoses

- Air conditioning compressor. Through-bolt: 19 ft. lbs. (26 Nm). Other bolt and nut: 38 ft. lbs. (52 Nm).
- Power steering pump. Tighten the long bottom bolt to 43 ft. lbs. (58 Nm); tighten the others to 29 ft. lbs. (39 Nm). Connect the power steering air hoses.
- Charcoal canister and connect the hoses
- Radiator, coolant hoses and the transmission lines
- Water pump pulley, the fan and the drive belt. Tighten the 4 pulley nuts to 12 ft. lbs. (16 Nm).
- Air cleaner assembly, resonator and the air intake hose
- Accelerator, throttle control (automatic transmission only) and the cruise control cables to the throttle body.
- Battery tray and battery
- Battery cables
- All fluids, including fuel
11. Check the automatic transmission fluid level
12. Check the ignition timing
13. Shut the engine off and install the engine undercovers.
14. Install the hood

SC400, GS400, GS430

1. Before servicing the vehicle, refer to the precautions in the beginning of this section.
2. Relieve the fuel pressure from the fuel lines.
3. Drain the engine coolant from the cooling system.
4. Remove or disconnect the following:
- Battery cables and remove the battery. Wait at least 90 seconds before proceeding with any other work.
- Hood
- V-bank cover, if equipped
- Engine undercover and drain the engine oil. Lower the vehicle.
- Drive belt
- Throttle body
- Accelerator, transmission and cruise control cables from the throttle body.
- Air cleaner assembly
- Vacuum hose (from the power steering air control valve) from the air intake chamber.
- Intake air connector

For Accessory Drive Belt illustrations, see Section 1 of this manual

- Coolant reservoir tank
- Radiator
- Igniter connectors and wire clamp
- Engine wires located next to the relay box. The relay box is located next to the left strut tower.
- Engine ground cable
- Power steering solenoid valve connector
- Alternator
- Power steering tubes from the suspension crossmember
- Power steering reservoir tank and bracket from the body by removing the 3 bolts
- Power steering pump by removing the pump mounting bolts and nut. Do not disconnect the power steering lines and place the pump off to the side.
- Air conditioning compressor from the engine. Do not remove the compressor pressure lines.
- Heater water hose from the water bypass hose
- Heater water hose from the heater water valve
- Brake booster hose from the union on the air intake chamber
- Vacuum hose from the Vacuum Switching Valve (VSV) for the heater water valve from the air intake chamber
- Ground strap from the bracket on the body
- Fuel inlet hose from fuel tube
- Charcoal canister from the engine
- Engine wire from the cabin as follows:
- Passenger's side lower instrument panel undercover
- 4 screws to the lower instrument panel finish panel and glove compartment door assembly
- Glove compartment and finish panel.
- Right scuff plate
- Take out the front side of the floor carpet
- 2 nuts and the Powertrain Control Module (PCM) protector
- Mounting nut and disconnect the PCM from the floor panel.
- 2 connectors from the PCM
- Connector from the Anti-lock brake system (ABS) and Traction control electronic control unit (TRAC ECU)
- 2 connectors from the TRAC ECU
- 4 connectors from connector cassette
- Connector from air conditioning control assembly

- Bolt holding the engine wire clamp to the heater water valve bracket.
- 2 bolts holding the engine wire clamp to the body.
- Engine wiring harness (through the cowl panel) from the vehicle cabin
- Oxygen (O$_2$) sensors from the front exhaust pipe
- Front exhaust pipe
- Front catalytic converter by removing the 3 nuts and gasket
- Tailpipes
- Center exhaust pipe by disconnecting the 2 hooks
- Heat insulator by removing the 4 nuts
- Center floor crossmember brace by removing the 4 bolts
- Driveshaft from the vehicle using the proper tools (2 of tool SST 09922–10010), loosen the adjusting nut on the driveshaft. Place matchmarks on the transmission flange and the flexible coupling.
- The transmission control rod from the shift lever by removing the nut

5. Attach the engine chain hoist to the engine hangers.
6. Remove or disconnect the following:
   - 2 nuts holding the engine mounting insulators to the front suspension crossmember.
   - 4 bolts, 4 nuts and the rear engine mounting member.
   - Ground strap to the rear mounting member.
   - Engine out of the vehicle
   - Oil dipstick guide and dipstick for transmission
   - Oil cooler pipes for the transmission
   - All the engine wiring
   - Engine bolts holding the transmission to the engine
   - Engine from the transmission

To install:
7. Install or connect the following:
   - Transmission to the engine and install the bolts. Tighten the bolts to 42 ft. lbs. (57 Nm).
   - Engine wiring
   - Oil cooler pipe for the transmission. Tighten the unions on the pipes to 25 ft. lbs. (34 Nm).
   - Engine oil dipstick guide and the dipstick for the transmission
   - Engine and transmission to the vehicle
   - Rear engine mounting member with the 4 bolts and 4 nuts. Tighten the bolts to 19 ft. lbs. (25 Nm) and the nuts to 10 ft. lbs. (13 Nm).

- 2 nuts holding the engine mounting brackets to the front suspension crossmember. Tighten the nuts to 43 ft. lbs. (59 Nm).
- Engine chain hoist
- Transmission control rod to the shift lever by installing the nut
- Driveshaft
- Center floor crossmember brace by installing the 4 bolts. Tighten the bolts to 108 inch lbs. (13 Nm).
- Heat insulator for the front exhaust pipe by installing the 4 bolts
- Center exhaust pipe by installing the 2 hooks
- Tailpipe and tighten the 2 bolts to 14 ft. lbs. (19 Nm).
- Front catalytic converter and tighten the nuts to 46 ft. lbs. (62 Nm).
- Front exhaust. Tighten the 4 bolts and nuts holding the catalytic converter to the front exhaust pipe to 32 ft. lbs. (43 Nm). Tighten the 2 bolts and nuts holding the front exhaust pipe to the center exhaust pipe. Tighten the bolts to 32 ft. lbs. (43 Nm). Tighten the 4 bolts holding the pipe support bracket to the transmission. Tighten the bolt to 32 ft. lbs. (43 Nm).
- O$_2$ sensors to the front exhaust and tighten the sensors to 33 ft. lbs. (44 Nm).
- Engine wiring harness in through the cowl panel
- Engine wire retainer with the 3 bolts
- Reattach the connectors under the dash panel
- PCM with the nut
- PCM protector with the 2 nuts
- Floor carpet
- Scuff plate
- Lower instrument panel finish panel and glove compartment door assembly with the 4 screws.
- Instrument panel undercover with the 2 screws.
- Charcoal canister
- All hoses and grounds
- Air conditioning compressor with the nut and 3 bolts. Tighten the bolts to 36 ft. lbs. (49 Nm) and the nut to 22 ft. lbs. (29 Nm).
- Power steering pump with the nut and 3 bolts. Tighten the bolts to 29 ft. lbs. (39 Nm) and the nut to 32 ft. lbs. (43 Nm).
- Power steering reservoir tank and bracket with the 3 bolts
- Power steering tubes with the clamp and bolt

- Alternator, tighten the nut and bolt to 27 ft. lbs. (37 Nm).
- Power steering solenoid valve connector
- Engine wire connectors
- Theft deterrent horn connector.
- Ground cable to the body from the engine
- Igniter connectors
- Yellow taped connector to the igniter on the rear side.
- Radiator assembly
- The reservoir tank and the inlet pipe to the fan shroud and tighten the 4 bolts to 43 inch lbs. (5 Nm).
- The 2 hydraulic lines for the fan motor and tighten the bolts to 47 ft. lbs. (64 Nm).
- Upper and lower radiator hoses to the radiator
- 2 oil cooler hoses for the transmission to the radiator
- Coolant tank
- Intake air connector
- Vacuum hose (from the power steering air control valve) to the air intake chamber
- Air cleaner
- Accelerator, transmission throttle control and the cruise control actuator cables to the engine
- Throttle cover and hose clamp with the cap nut and 2 bolts
- Evaporative emission control (EVAP) hose to the hose clamp
- Drive belt to the engine
- Battery to the engine compartment and attach the electrical connectors
- Engine coolant
- V-bank cover if it was removed
- Engine undercover
- Hood.

8. Fill the engine oil and check the transmission oil.

9. Start the engine, bleed the cooling system, and check for leaks.

Water Pump

REMOVAL & INSTALLATION

3.0L (1MZ-FE) Engine

1. Before servicing the vehicle, refer to the precautions in the beginning of this section.

2. Remove or disconnect the following:
- Negative battery terminal
- Engine coolant

Water pump mounting bolts—3.0L (1MZ-FE) engine

- Timing belt
- Right and left camshaft pulleys
- No. 2 idler pulley by removing the bolt
- 3 clamps and engine wire from the rear timing belt cover
- 6 bolts holding the rear timing belt cover to the engine block
- 4 bolts and 2 nuts to the water pump
- Water pump
- All the old packing (sealant) and gasket material from the water pump and clean the mounting surfaces.
- All gasket material from the upper inner timing belt cover

To install:

3. Check that the water pump turns smoothly. Also, check the air hole for coolant leakage.

4. Using a new gasket, apply liquid sealer to the gasket, water pump and engine block.

5. Install or connect the following:
- Gasket and pump to the engine and install the 4 bolts and 2 nuts. Tighten the nuts and bolts to 53 inch lbs. (6 Nm).
- Rear timing belt cover and tighten the 6 bolts to 74 inch lbs. (9 Nm).
- Engine wire with the 3 clamps to the rear timing belt cover
- No. 2 idler pulley with the bolt. Tighten the bolt to 32 ft. lbs. (43 Nm). After tightening the bolt, be sure the idler pulley moves smoothly.
- Right-hand camshaft pulley, with the flange side **outward**. Be sure to align the knock pinhole on the camshaft pulley with the knock pin on the camshaft. Camshaft bolt to 65 ft. lbs. (88 Nm).
- Left-hand camshaft pulley with the flange side **inward**. Be sure to

align the knock pin hole on the camshaft pulley with the knock pin on the camshaft. Camshaft bolt to 94 ft. lbs. (125 Nm).
- Timing belt
- Engine coolant
- Negative battery cable to the battery and start the engine.

3.0L (2JZ-GE) Engine

1. Before servicing the vehicle, refer to the precautions in the beginning of this section.

2. Disconnect the negative battery cable. Wait at least 90 seconds before performing any work.

3. Drain the engine coolant.

4. Remove or disconnect the following:
- Radiator assembly
- Air cleaner
- Mass Air Flow (MAF) meter
- Intake air connector pipe assembly
- Timing belt
- Idler pulley
- Water inlet and the thermostat
- 2 bolts, the water bypass outlet and the No. 1 water bypass pipe
- 3 O-rings from the water bypass outlet and the No. 1 water bypass pipe.
- Generator
- Bolt and engine wire bracket
- Bolt and clamp bracket for the Crankshaft Sensor (CKP) connector
- Nuts and the No. 2 water bypass pipe from the water pump
- 6 bolts and the water pump and gasket
- Drain hose and the O-ring from the cylinder block

To install:

5. Install or connect the following:
- New O-ring to the cylinder block

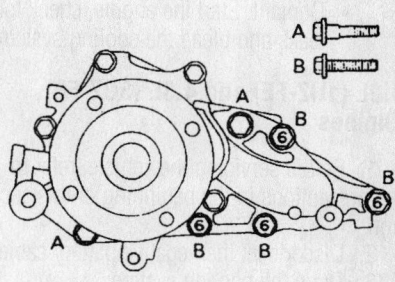

Water pump mounting bolt locations—3.0L (2JZ-GE) engine

7923LG10

Be sure to use new O-rings when installing the water bypass pipe—3.0L (2JZ-GE) engine

- Drain hose
- New gasket to the water pump
- Water pump to the water bypass pipe. Do not install the nut yet.
- Water pump with the 2 bolts (A) and the 4 bolts (B)

➡ Hand-tighten the bolts (A) first. Tighten all 6 bolts to 15 ft. lbs. (21 Nm).

- 2 nuts holding the No. 2 water bypass pipe to the water pump. Tighten the nuts to 15 ft. lbs. (21 Nm).
- Clamp bracket for the CKP sensor connector
- Engine wire bracket
- Generator
- O-rings to the No. 1 water bypass pipe.
- New O-ring and the water bypass outlet with the 2 bolts and tighten them to 78 inch lbs. (9 Nm).
- Thermostat and the water inlet
- Idler pulley
- Timing belt
- Air cleaner, the MAF meter and the intake air connector pipe assembly.
- Radiator assembly
- Negative battery cable
- Coolant. Start the engine, check for leaks and bleed the cooling system.

4.0L (1UZ-FE) and 4.3L (3UZ-FE) Engines

1. Before servicing the vehicle, refer to the precautions in the beginning of this section.
2. Disconnect the negative battery cable.
3. Drain the cooling system.
4. Remove or disconnect the following:
   - Timing belt
   - No. 2 idler pulley
   - Throttle body
   - Bypass hose(s) from the water inlet housing

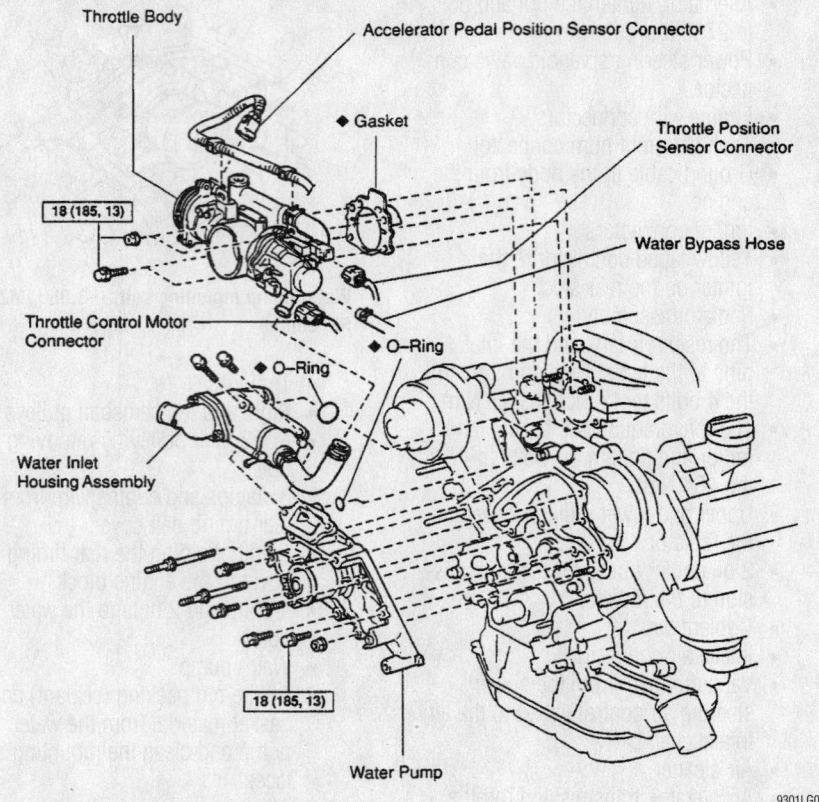

9301LG07

Exploded view of the water pump mounting and related components—4.0L engine shown

- 2 bolts holding the water inlet housing to water pump
- Water inlet housing and discard the gasket
- Mounting bolts, studs and the nut to the water pump
- Water pump by carefully prying between the pump and the cylinder block
- The old gasket and clean all mounting surfaces

To install:

5. Install new seal packing to the water pump groove and a new O-ring to the water bypass pipe end.
6. Install or connect the following:
   - Water pump to the water bypass pipe end

- Water pump and tighten the mounting bolts and nut to 13 ft. lbs. (18 Nm).
- Sealant to the groove of the water inlet housing
- A new O-ring to the water inlet housing
- Water inlet housing end into the water pump hole
- Water inlet and housing assembly with the 2 bolts. Alternately tighten the bolts to 13 ft. lbs. (18 Nm).
- Bypass hose(s) to the water inlet housing
- Throttle body. Tighten the mounting bolts to 13 ft. lbs. (18 Nm).
- No. 2 idler pulley

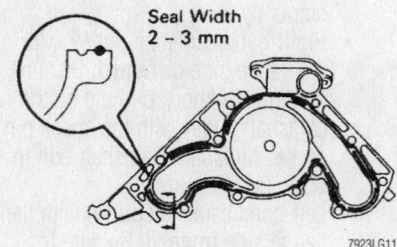

7923LG11

Apply silicone sealant to the water pump as shown—4.0L engine

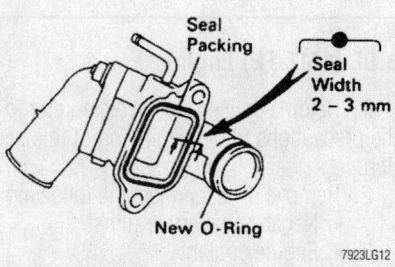

7923LG12

Apply sealant to the water inlet housing as shown—4.0L engine

- Timing belt
- Coolant
- Negative battery cable
7. Fill the cooling system.
8. Start the engine and check for leaks.

Cylinder Head

REMOVAL & INSTALLATION

3.0L (1MZ-FE) Engine

1. Before servicing the vehicle, refer to the precautions in the beginning of this section.
2. Drain the cooling system.
3. Relieve the fuel system pressure.
4. Remove or disconnect the following:

- Negative battery cable
- Accelerator and the throttle cables
- Air cleaner cover, air flow meter and the air duct
- Cruise control actuator and bracket, if equipped
- 2 engine ground straps
- Right engine mounting support
- Radiator hoses
- 2 heater hoses
- And plug the fuel feed and return lines from the fuel rail assembly.
- And plug the pressure hose from the hydraulic motor
- V-bank cover
- Fuel pressure control Vacuum Switching Valve (VSV)
- Fuel pressure regulator
- Cylinder head rear plate
- Intake air control valve VSV
- Exhaust Gas Recirculation (EGR) vacuum modulator
- EGR valve
- Intake Air Control (IAC) valve
- Fuel pressure regulator
- EGR VSV
- 2 nuts and the emission control valve set
- Brake booster vacuum hose
- Positive Crankcase Ventilation (PCV) hose
- Intake air control valve vacuum hose
- Data link connector from the mounting bracket
- 2 ground straps from the intake chamber
- Hydraulic motor pressure hose from the intake chamber
- Right Oxygen (O₂) sensor connec-

tor from the power steering pressure tube.
- 2 nuts and the power steering pressure tube from the intake chamber
- 2 power steering air hoses
- Engine hanger and the intake chamber support
- EGR pipe and gaskets
- Throttle Pressure (TP) sensor connector
- Idle Air Control (IAC) valve connector
- EGR gas temperature connector
- air conditioning idle up connector
- 2 vacuum hoses from the Thermal Vacuum Valve (TVV)
- Vacuum hose from the cylinder head rear plate
- Vacuum hose from the charcoal canister
- Air assist hose and the 2 water bypass hoses
- Air intake chamber
- Left engine wiring harness and position it out of the way

- Wiring harness from the rear of the engine
- Right engine wiring harness and position it out of the way
- Ignition coils and the spark plugs
- Timing belt
- Camshaft pulleys and the timing belt rear cover
- Cylinder head rear plate
- Water inlet pipe
- Air assist hose and vacuum hose
- Intake manifold and fuel rail assembly
- Water outlet
- EGR pipe from the right exhaust manifold
- Exhaust manifolds
- Dipstick assembly and the power steering pump bracket
- Valve covers and the Camshaft Position (CMP) sensor
- Camshafts

5. Be sure the engine is at or near ambient temperature and remove the 2 (one on each head) 8mm recessed hex bolts.

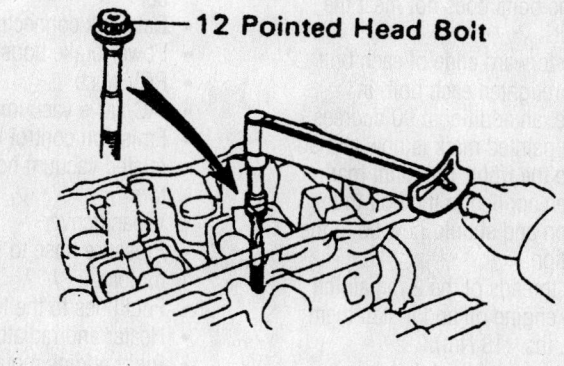

12 Pointed Head Bolt

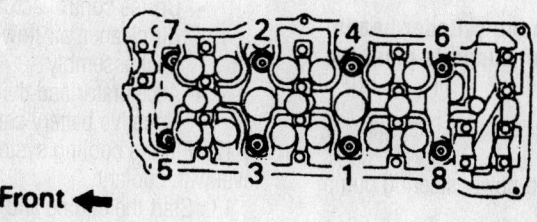

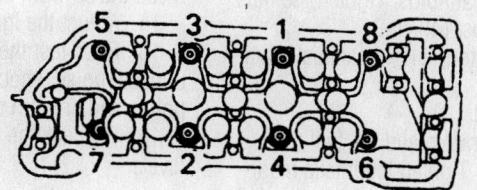

Front ←

Cylinder head torque sequence—3.0L (1MZ-FE) engine

7923LG13

For Wheel Alignment specifications, see Section 1 of this manual

Loosen and remove the 8 head bolts evenly, in 3 passes, in the reverse order of the tightening sequence. Carefully lift the head from the engine; if it is necessary to pry the head loose, take great care not to damage the mating surfaces. Place the head on wood blocks in a clean work area.

✳✳ WARNING

If the cylinder head bolts are loosened out of sequence, warpage or cracking could result.

6. Remove the cylinder head gasket. With a gasket scraper, carefully remove all the old gasket material from the cylinder head and engine block surfaces.

To install:

7. Place the new cylinder head gasket onto the cylinder block. Place the cylinder head onto the gasket.

8. Coat the threads of the 8 cylinder head bolts (12-sided) with clean engine oil and install the bolts into the cylinder head. Uniformly tighten the bolts in sequence in 3 steps to an ultimate tighten of 40 ft. lbs. (54 Nm). If any of the bolts does not meet the torque, replace it.

9. Mark the forward edge of each bolt with paint, then retighten each bolt, in proper sequence, an additional 90 degrees. Check that each painted mark is now at a 90 degrees angle to the front. The paint mark should have been applied to the bolt in the 9 o'clock position and should now be in the 12 o'clock position.

10. Coat the threads of the 2 remaining 8mm bolts with engine oil and install them. Tighten to 13 ft. lbs. (18 Nm).

11. Install or connect the following:
- Camshafts and adjust the valves

➡ **Apply sealant to the cylinder heads where the camshaft supports meet the cylinder heads.**

- Cylinder head covers. Use new gaskets
- Dipstick and power steering pump bracket
- Exhaust manifolds. Tighten the nuts to 36 ft. lbs. (49 Nm).
- EGR pipe to the right exhaust manifold
- Water outlet
- Intake manifold and the fuel rail assembly. Tighten the intake manifold nuts and bolts to 11 ft. lbs. (15 Nm).
- Air assist hose and the 2 water bypass hoses
- Water inlet pipe and the cylinder head rear plate

- Timing belt rear cover and the camshaft pulleys
- Timing belt
- Spark plugs and the ignition coils
- Right engine wiring harness
- Wiring harness to the rear of the engine
- Left engine wiring harness
- Air intake chamber
- EGR pipe. Use new gaskets
- 2 TVV vacuum hoses
- Vacuum hose to the rear cylinder head plate
- Charcoal canister vacuum hose
- TP sensor connector
- IAC valve connector
- EGR gas temperature connector
- Air conditioning idle up connector
- Engine hanger and the intake chamber support.
- 2 power steering air hoses
- Power steering pressure tube to the intake chamber
- O2sensor connector to the pressure tube
- 2 ground straps to the intake chamber
- Data link connector to the bracket
- Power brake booster vacuum hose
- PCV hose
- IAC valve vacuum hose
- Emission control valve set and related vacuum hoses and connectors
- V-bank cover
- Pressure hose to the hydraulic motor
- Fuel lines to the fuel rail assembly
- Heater and radiator hoses
- Right engine mounting support
- 2 engine ground straps
- Cruise control actuator and bracket
- Air cleaner, air flow meter and air duct assembly
- Accelerator and the throttle cables
- Negative battery cable.

12. Fill the cooling system to the proper level with coolant.

13. Start the engine and check for leaks. Bleed the air from the cooling system.

14. Adjust the ignition timing.

15. Road test the vehicle and check for unusual noise, shock, slippage, correct shift points and smooth operation.

16. Recheck the coolant and engine oil levels.

3.0L (2JZ-GE) Engine

1. Before servicing the vehicle, refer to the precautions in the beginning of this section.

2. Disconnect the negative battery cable. Wait at least 90 seconds before performing any other work.

3. Relieve the fuel pressure from the fuel lines.

4. Remove or disconnect the following:
- Coolant
- Undercovers
- Accelerator, throttle control (automatic transmission only) and cruise control cables from the throttle body
- Cleaner duct
- Air cleaner, airflow meter and the intake air pipe
- Drive belt, the fan and fluid coupling and the water pump pulley
- No. 2 front exhaust pipe
- Exhaust manifold cover
- 2 Heated Oxygen (HO2) sensor connector(s).
- Exhaust manifolds and gaskets by removing the 8 bolts
- Water bypass outlet and the No. 1 water bypass pipe
- Power steering air hose from the No. 4 timing belt cover
- Power steering hose from the air intake chamber
- 2 bolts and the vane pump from the pump bracket.
- 2 bolts, the pump rear stay. Put aside the vane pump and suspend it.
- Fuel return hose from the fuel return pipe. Plug the hose end.
- Fuel return hose from the oil dipstick guide.
- Bolt and bracket
- Engine wire from the intake manifold stay
- Throttle body and intake air connector assembly
- Bolt, pull out the oil dipstick guide with the dipstick and remove the O-ring from the dipstick guide.
- Transmission dipstick and guide, if equipped with automatic transmission
- Connector from the No. 2 vacuum pipe
- Exhaust Gas Recirculation (EGR) gas temperature sensor wiring harness
- 2 nuts and the vacuum pipe from the air intake chamber and intake manifold.
- No. 2 vacuum pipe and Vacuum Switching Valve (VSV) assembly.
- Nuts and the vacuum tank from the intake manifold.
- VSV connector and hoses

- Vacuum hose (from the air intake chamber) from port B of the vacuum tank
- Vacuum hose (from actuator) from the VSV
- Vacuum control valve set
- Data Link Connector (DLC1) bracket and VSV assembly
- Vacuum hose from the brake booster union and the Evaporative Emission (EVAP) hose from the No. 2 vacuum pipe.
- Bolt holding the engine wire protector to the air intake chamber
- 5 bolts, nut, air intake chamber and gasket
- No. 3 (top) timing belt cover by removing the oil filler cap and the 6 bolts using a 5mm hexagon wrench.
- 4 bolts, using a 5mm hexagon wrench, and the rear cylinder head cover
- Distributor with the spark plug wires attached
- Spark plugs.
- Drive belt tensioner by removing the 3 bolts

5. Set the engine to Top Dead Center (TDC)/compression for cylinder No. 1.

6. Remove or disconnect the following:
- Timing belt tensioner and dust boot. Remove the timing belt from the camshaft pulleys. Support the belt so that it remains in contact with the crankshaft pulley.
- Wire clamp from the bracket.
- HO2and the Crankshaft Position (CKP) sensors.
- 2 ground straps from the intake manifold.
- Engine Coolant Temperature (ECT) sender gauge
- Knock Sensor (KS)
- Oil pressure switch
- Oil level sensor
- air conditioning compressor
- 6 injector electrical connectors.
- 3 nuts and the engine wire protector from the intake manifold.
- Water bypass hose from the clamp on the oil filter bracket
- Water outlet, 2 nuts, and the bolt with the water bypass hose.
- 2 bolts and the intake manifold stay.
- Fuel pressure pulsation damper.
- Clamp bolt from the intake manifold
- Union bolt and gaskets

- Fuel inlet pipe
- 6 bolts, 2 nuts, the intake manifold
- Delivery pipe assembly and gasket
- Cylinder head covers (valve covers)
- Camshaft timing pulleys
- Rear (No. 4) timing belt cover
- Camshafts
- Cylinder head bolts in several passes and in the reverse order of the tightening sequence
- Head from the engine

7. Clean the head and block of all gasket material.

To install:

8. Install or connect the following:
- New gasket and cylinder head on the block
- Head bolts, lightly coated with engine oil and plate washers. Uniformly tighten the head bolts in several passes, in sequence to 26 ft. lbs. (35 Nm). Following the correct order, tighten each bolt an additional 90 degrees. Again following the correct order, tighten the bolts another 90 degrees of rotation.

⁂ WARNING

Correct bolt torque must be achieved in 3 steps; do not attempt to shorten the procedure by combining the two 90 degree steps.

- Camshafts. Coat the thrust portions of each with engine oil.

10 mm Bi–Hexagon Wrench

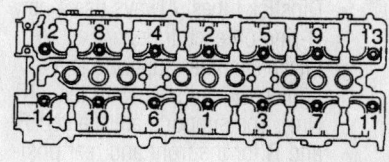

Cylinder head torque sequence—3.0L (2JZ-GE) engine

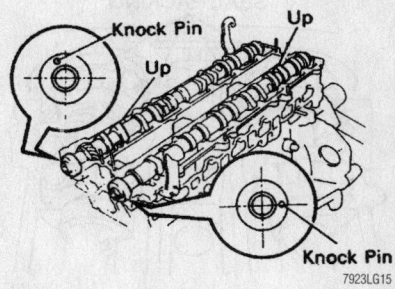

Position the knock pins as shown when installing the camshafts—3.0L (2JZ-GE) engine

- No. 3 and No. 7 bearing caps in place. Coat the bolt threads with oil, then uniformly and alternately tighten them temporarily.
- New oil seals, coated with multipurpose grease, over the camshafts
- Seal packing to the No. 1 bearing cap
- Remaining bearing caps in their proper locations. Coat the threads of each bolt with clean oil, then tighten them, in several passes, in the correct sequence, to 14 ft. lbs. (20 Nm).
- The 2 oil seals in as far as it will go

9. Rotate each camshaft until the forward straight (knock) pin is straight up. Loosen the exhaust Nos. 1, 2 and 6 bearing cap bolts until they can be turned by hand; retighten the bolts, in several passes, to 14 ft. lbs. (20 Nm). Loosen the intake Nos. 1, 2 and 5 bearing cap bolts and retighten the bolts, in several passes, to 14 ft. lbs. (20 Nm).

10. Turn each camshaft ⅓ of a revolution (120 degrees). Loosen the exhaust Nos. 4 and 7 bearing cap bolts; retighten the bolts, in several passes, to 14 ft. lbs. (20 Nm). Loosen the intake Nos. 4 and 6 bearing cap bolts; retighten the bolts, in several passes, to 14 ft. lbs. (20 Nm).

11. Turn each camshaft an additional ⅓ of a revolution, loosen the exhaust bearing cap bolts Nos. 3 and 5, then retighten the bolts, in several passes, to 14 ft. lbs. (20 Nm). Loosen the intake bearing cap bolts Nos. 3 and 7, then retighten the bolts, in several passes, to 14 ft. lbs. (20 Nm).

12. Check and adjust the valve clearance.

13. Install or connect the following:
- Rear (No. 4) timing belt cover. Tighten the bolts to 78 inch lbs. (9 Nm).

Apply sealant to the areas indicated on the cylinder head before installing the cover—3.0L (2JZ-GE) engine

- Camshaft timing pulleys. Align the shaft pin with the pulley groove and slide the pulley on. Install the bolt temporarily. Hold the hex portion of the camshaft with a wrench and tighten the pulley bolt to 59 ft. lbs. (79 Nm).
- Cylinder head covers
- Intake manifold and delivery pipe with a new gasket. Tighten the 6 bolts and 2 nuts to 20 ft. lbs. (27 Nm).
- Fuel inlet pipe to the fuel rail. Tighten the union bolt to 30 ft. lbs. (42 Nm).
- Clamp bolt to the intake manifold
- Fuel pressure pulsation damper
- Intake manifold stay and tighten the bolts to 29 ft. lbs. (39 Nm)
- Water outlet and the bypass hose. Tighten the bolts to 15 ft. lbs. (21 Nm).

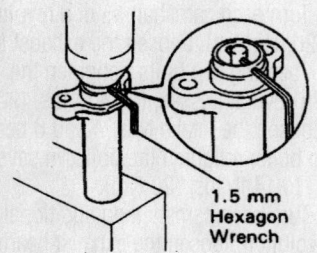

1.5 mm Hexagon Wrench

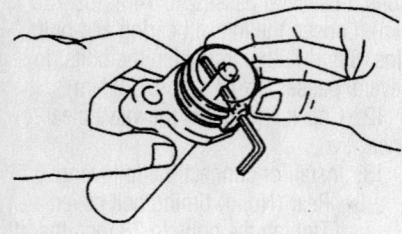

Compressing the timing belt tensioner—3.0L (2JZ-GE) engine

- Engine wiring harness. Secure the wiring in all clamps and retainers.
- Wiring leads to the proper sender, sensor or switch
- Injector leads.

14. Compress the timing belt tensioner in a vise and retain the pin with a 1.5mm hex wrench. Install the dust boot onto the tensioner.

15. Install the tensioner. Alternately tighten the bolts to 20 ft. lbs. (26 Nm). Remove the hex wrench with a pair of pliers, allowing the tensioner to be applied to the timing belt.

16. Turn the crankshaft 2 full revolutions clockwise. Check that all timing marks align as before. If the marks (cam and crankshaft) do not align, remove the timing belt and reinstall it.

17. Install the accessory drive belt tensioner. Take great care not to drop the bolts inside the lower timing cover. Tighten the bolts to 15 ft. lbs. (21 Nm).

18. Double check that the engine is still set to TDC/compression for cylinder No. 1. Check the alignment of both the crank and camshaft timing marks. Install the timing belt.

19. Install or connect the following:
- Spark plugs
- Wiring to the spark plugs
- No. 3 timing belt cover
- Cylinder head rear cover
- Air intake chamber with a new gasket. Tighten the bolts to 20 ft. lbs. (27 Nm). Install the bolt to hold the engine wire protector to the air intake chamber.
- Vacuum hose to the brake booster union and the EVAP hose to the No. 2 vacuum pipe.
- DLC connector and bracket and VSV connector
- Vacuum control set
- No. 2 vacuum pipe assembly and connect the hoses. Tighten the nuts to 20 ft. lbs. (27 Nm).
- EGR gas temperature sensor. Tighten it to 14 ft. lbs. (20 Nm).
- Vacuum hoses
- Dipstick tubes. Always use a new O-ring on each tube.
- Intake chamber supports and tighten the bolts to 13 ft. lbs. (18 Nm). The supports are marked **F** and **R** for the front and rear positions.
- Throttle body and intake air connector assembly
- Engine wire bracket
- Fuel return hose
- Vane pump to the pump bracket

- Power steering air hose to the No. 4 timing belt cover and intake chamber.
- Water bypass outlet and the bypass pipe. Always use new O-rings.
- Exhaust manifolds with new gaskets. Tighten the bolts to 29 ft. lbs. (39 Nm).
- O_2sensor leads.
- Front exhaust pipe. Tighten the bolts to 46 ft. lbs. (62 Nm).
- Manifold cover
- Water pump pulley
- Fan and coupling and the drive belt. Tighten the 4 nuts to 12 ft. lbs. (16 Nm).
- Air cleaner, airflow meter and the intake air connector pipe
- Air cleaner duct
- Control and accelerator cables to the throttle body
- Coolant
- Negative battery cable. Start the engine and check for leaks.
- Engine undercovers

4.0L (1UZ-FE) and 4.3L (3UZ-FE) Engines

1. Before servicing the vehicle, refer to the precautions in the beginning of this section.

2. Relieve the fuel system pressure.

3. Remove or disconnect the following:
- Negative battery cable. Wait at least 90 seconds before performing any other work.
- Oil pan protector
- Engine undercover
- Coolant
- Battery clamp cover
- Air cleaner inlet
- V bank cover by removing the bolt and 2 cap nuts
- Air cleaner and intake air connector assembly
- Drive belt, fluid coupling and the fan pulley. The drive belt tension may be slackened by turning the tensioner counterclockwise. The pulley bolt for the drive belt tensioner has a left-handed thread.
- Radiator
- Right-hand No. 3 timing belt cover
- Left-hand No. 3 timing belt cover
- Drive belt idler pulley by removing the pulley bolt and cover plate
- Right-hand No. 2 timing belt cover
- Left-hand No. 2 timing belt cover
- Distributor housings
- No. 1 ignition coil
- Air conditioning compressor from the engine

- Fan bracket by removing the 2 bolts and 2 nuts

4. Set the engine to Top Dead Center (TDC) on cylinder No. 1.

❄❄ WARNING

Since the thrust clearance of the camshaft is small, the camshaft must be kept level while it is being removed. If the camshaft is not kept level, the portion of the cylinder head receiving the shaft thrust may crack or be damaged, causing the camshaft to seize or break.

5. Turn the crankshaft pulley approximately 50 degrees clockwise and put the timing mark of the crankshaft pulley in line with the centers of the crankshaft pulley bolt and the idler pulley bolt.

❄❄ WARNING

If the timing belt is disengaged, having the crankshaft pulley at the wrong angle can cause the piston head and valve head to come into contact with each other when you remove the camshaft timing pulley. Always set the crankshaft pulley at the correct angle before removing the timing belt.

6. If the timing belt is to be reused, turn the crank pulley slowly; check that the 3 installation marks are present on the belt. If the marks are not present, make new installation marks before removing the belt. The marks should align with the timing marks on each camshaft pulley and the crank pulley.

7. Remove the timing belt tensioner. Alternately loosen the 2 bolts; remove the bolts, the tensioner and the dust protector.

8. Loosen the tension between the left side and the right side timing pulleys by slightly turning the left side camshaft clockwise.

9. Remove or disconnect the following:
- Timing belt from the camshaft timing pulleys. Using the proper tool, remove the bolt and the camshaft timing pulleys.
- Power steering pump from the engine. Do not disconnect the hoses or lines from the power steering pump. Support the power steering pump with a piece of wire. Do not allow the pump to hang.

- Front catalytic converter
- High tension spark plug wires, wire clamps and the wire cover assembly
- No. 2 ignition coil by removing the connector and the 2 bolts
- 2 bolts and the rear timing belt plate. Remove both plates
- Intake chamber assembly
- Throttle Position Sensor (TPS) connector
- With TRAC system, sub TP sensor connector
- With TRAC system, sub TP connector
- Idle Air Control (IAC) valve connector
- Exhaust Gas Recirculation (EGR) valve connector
- Vacuum Switching Valve (VSV) connector for fuel pressure control
- VSV connector for Evaporative Emissions (EVAP) system
- EGR gas temperature sensor connector
- Brake booster vacuum hose from the union on the air intake chamber
- Positive Crankcase Ventilation (PCV) hose from the PCV valve on the left-hand cylinder head
- Water bypass hose (from the EGR valve) from the rear water bypass joint
- Water bypass hose (from the throttle body) from the rear water bypass joint
- Vacuum hose (from the VSV for fuel pressure control) from the fuel pressure regulator
- EVAP hose (from charcoal canister) from the VSV for EVAP
- Heater hose from the water bypass pipe
- Fuel inlet hose from the delivery pipe
- Fuel return hose from the fuel return pipe.
- Engine wire from the delivery pipes and rear water bypass joint
- Fuel hose from the fuel pressure regulator
- 2 bolts and fuel return pipe from the intake manifold
- 8 injector connectors
- 6 bolts, 4 nuts, the intake manifold assembly and the 2 gaskets
- Water inlet and inlet housing
- Front water bypass joint
- Rear water bypass joint and No. 1

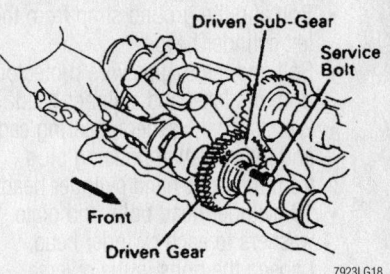

Securing the exhaust camshaft on the right cylinder head—4.0L (1UZ-FE) and 4.3L (3UZ-FE) Engines

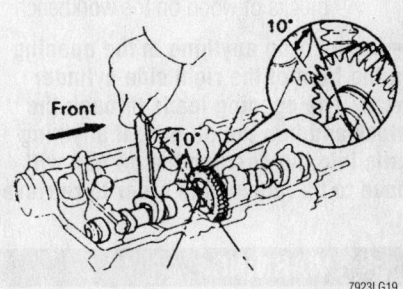

Turning the exhaust camshaft 10 degrees on the right cylinder head—4.0L (1UZ-FE) and 4.3L (3UZ-FE) Engines

EGR pipe assembly
- Oil dipstick and guide for the automatic transmission
- Oil dipstick and guide for the engine
- Engine hangers
- Right and left cylinder head covers by removing the 8 bolts, seal washers and gaskets
- If necessary, the semi-circular plug.
- Exhaust camshaft from the right side cylinder head. See the camshaft procedure for tightening sequence.
- Intake camshaft from the right side cylinder head. See the camshaft procedure for tightening sequence.
- Exhaust camshaft of the left side cylinder head. See the camshaft procedure for tightening sequence.

➡**When removing the camshaft, be sure the torsional spring force of the subgear has been eliminated.**

- Intake camshaft from the left side cylinder head. See the camshaft procedure for tightening sequence.
- Main Heated Oxygen (HO_2) sensor
- Bolt and HO_2 the ground cable from the right cylinder head

---

For Tune-up, Capacities and Firing orders, see Section 1 of this manual

- Bolt and the ground strap from the left cylinder head
- Bolt and the engine wire protector from the left-hand cylinder head.
- 2 bolts, seal washers, bearing cap and the camshaft housing plug from the right-hand cylinder head.
- 10 cylinder head bolts and plate washers to each cylinder head. Loosen the bolts in the reverse order of the tightening sequence. Lift the heads from the dowels on the block with the exhaust manifolds attached. Place the heads on blocks of wood on the workbench.

➡Do not drop anything in the opening in the front of the right side cylinder head. The opening leads through the block and into the oil pan. If anything falls into the opening the oil pan will have to be removed in order to retrieve it.

✳✳ WARNING

If necessary to pry the head loose, take great care not to damage the contact surfaces of the head or block.

- 2 bolts, seal washers, bearing cap

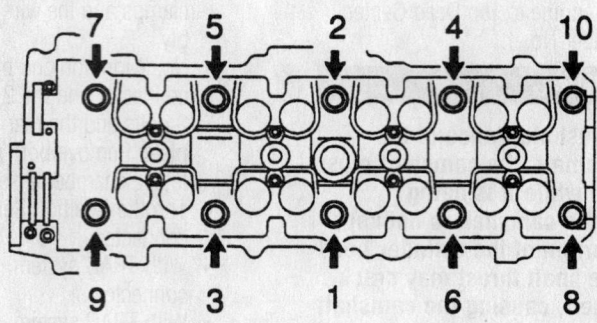

RH Bank

LH Bank

Cylinder head torque sequence—4.3L (3UZ-FE) engine

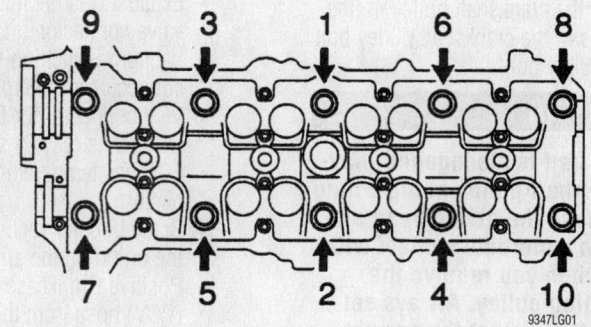

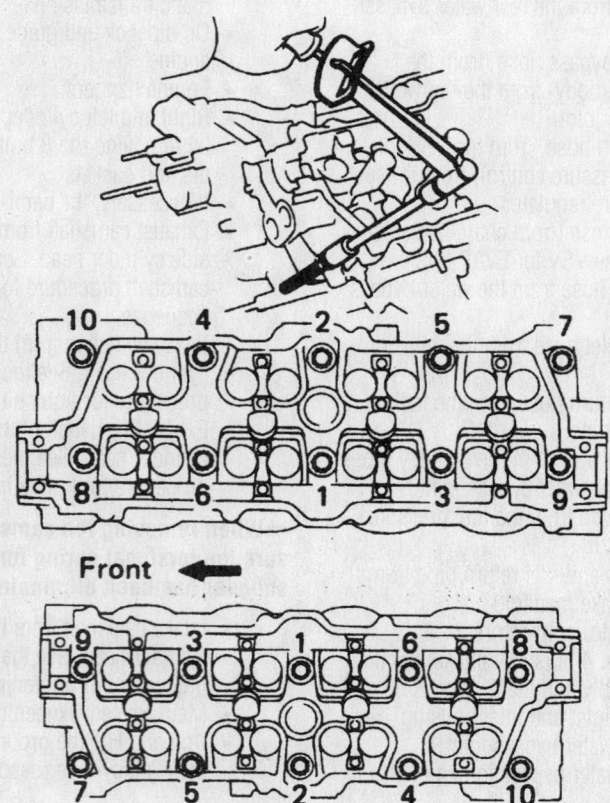

Front ⬅

Cylinder head torque sequence—4.0L (1UZ-FE) engine

and camshaft housing plug from the right-hand cylinder head.
- Right exhaust manifold from the cylinder head by removing the heat insulator, 8 nuts and the gasket.
- Left exhaust manifold from the cylinder head by removing the heat insulator, 8 nuts and the gasket.

To install:

10. Install or connect the following:
- Right exhaust manifold. The new gasket must be installed with the white marks facing the manifold side. Tighten the bolts to 33 ft. lbs. (44 Nm). Install the right O_2sensor.
- Left exhaust manifold. The new gasket must be installed with the white marks facing the manifold side. Tighten the bolts to 33 ft. lbs. (44 Nm). Install the left O_2sensor.
- The 2 new cylinder head gaskets in position on the engine block. Each gasket has a painted mark denoting the rear of the gasket. The gasket for the right bank has a white mark and the gasket for the left bank has a yellow mark. Double check the gasket position and placement.

11. For 4.0L engines, install the cylinder

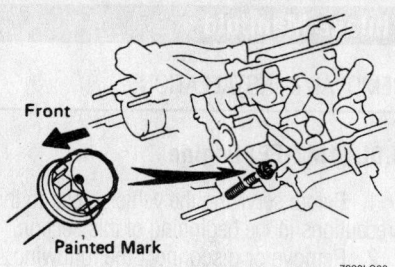

7923LG20

Apply a dot of paint at the front of each bolt—4.0L (1UZ-FE) and 4.3L (3UZ-FE) Engines

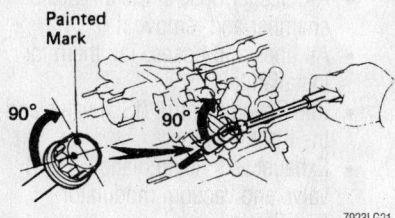

7923LG21

The paint mark must be 90 degrees from the starting point—4.0L (1UZ-FE) and 4.3L (3UZ-FE) Engines

heads and tighten the bolts in sequence as follows:
- Step 1: 29 ft. lbs. (39 Nm)
- Step 2: Plus 90 degrees

12. For 4.3L engines, install the cylinder heads and tighten the bolts in sequence as follows:
- Step 1: 44 ft. lbs. (60 Nm)
- Step 2: Plus 90 degrees
- HO$_2$.
- Engine wire to the right-hand cylinder head with the 2 bolt
- Ground cable to the right-hand cylinder head with the bolt
- The engine wire protector to the left-hand cylinder head with the bolt
- Ground cable to the left-hand cylinder head with the bolt
- Old packing and apply new seal packing to the bearing caps
- Bearing cap on the right side cylinder head, marked **I1**, in position with the arrow mark facing the rear. Install the bearing cap on the left side cylinder head, marked **I6**, in position with the arrow mark facing the front.
- Bearing cap bolts with new washers. Apply a light coat of oil on the threads of the cap bolts. Alternately tighten each bolt to 12 ft. lbs. (16 Nm).

➡ **Use silver colored bolts 1.50 in. (38mm) in length.**

- Camshaft housing plugs on the cylinder heads. Be sure to face the cupped side forward.

13. Turn the crankshaft pulley clockwise or counterclockwise and put the timing mark of the crankshaft pulley in line with the centers of the crankshaft pulley bolt and the idler pulley bolt.

✳✳ WARNING

Since the thrust clearance of the camshaft is small, the camshaft must be kept level while it is being installed. If the camshaft is not kept level, the portion of the cylinder head receiving the shaft thrust may crack or be damaged, causing the camshaft to seize or break.

14. Install or connect the following:
- Right side cylinder head intake camshaft. Tighten the bracket bolt in the reverse order of the loosening sequence.
- Right side cylinder head exhaust camshaft. Tighten the bracket bolt in the reverse order of the loosening sequence.
- Left side cylinder head intake camshaft. Tighten the bracket bolt in the reverse order of the loosening sequence.
- Left side cylinder head exhaust camshaft. Tighten the bracket bolt in the reverse order of the loosening sequence.

15. Check and adjust the valve clearance.
16. Install or connect the following:
- Camshaft oil seals with the proper tool (SST 09223-46011). Be sure to apply MP grease to the new oil seal lip.
- Semi-circular plugs, if removed

17. Clean the cylinder head covers. Apply new sealant in the correct locations and install the gaskets.
18. Install or connect the following:
- Right cylinder head cover and bolts. Tighten the bolts to 52 inch lbs. (6 Nm).
- Left cylinder head cover and bolts. Tighten the bolts to 52 inch lbs. (6 Nm).
- Engine hanger with the 2 bolts. Install both engine hangers. Tighten the bolts to 27 ft. lbs. (37 Nm).

- Oil dipstick guide for the engine
- Oil dipstick for the transmission
- Rear water bypass joint and No. 1 EGR pipe
- Front water bypass joints. Install 2 gaskets and alternately tighten the nuts to 13 ft. lbs. (18 Nm).
- Water inlet and inlet housing, alternately tighten the bolts to 13 ft. lbs. (18 Nm).
- Delivery pipe and intake manifold
- Return pipe with 2 new gaskets. Tighten the union bolt to 26 ft. lbs. (35 Nm).
- Engine wire to the delivery pipes and rear water bypass joint
- Fuel return hose to the fuel return pipe
- Fuel inlet hose to the left-hand delivery pipe
- Fuel hose to the fuel pressure regulator
- Air intake chamber assembly
- Brake booster vacuum hose to the union on the air intake chamber
- PCV hose to the PCV valve on the left-hand cylinder head
- Water bypass hose (from EGR valve) to the rear water bypass joint
- Water bypass hose (from throttle body) to the rear water bypass joint
- Vacuum hose (from VSV for fuel pressure control) to the fuel pressure regulator
- EVAP hose (from charcoal canister) from the VSV for EVAP
- TPS connector
- With TRAC system, sub TPS connector
- With TRAC system, sub throttle actuator connector
- IAC valve connector
- EGR valve connector
- EGR gas temperature sensor connector
- VSV connector for fuel pressure control
- VSV connector for EVAP
- Accelerator bracket with the 2 bolts.
- Accelerator, automatic transmission throttle control and the cruise control actuator cable.
- Spark plug wires and clamps to the right and left cylinder head cover
- Belt rear plates by installing the bolts. Tighten the bolts to 66 inch lbs. (8 Nm).
- No. 2 ignition coil
- A new gasket to the exhaust mani-

fold and install the catalytic converters. Tighten the 3 nuts to each converter to 46 ft. lbs. (62 Nm).
- Front exhaust pipe, tighten the bolts and nuts to 32 ft. lbs. (44 Nm). Tighten the 4 bolts holding the pipe support bracket to the transmission. Tighten the bolts to 32 ft. lbs. (44 Nm).
- Power steering pump with the nut and 3 bolts. Tighten the nut to 32 ft. lbs. (43 Nm) and the bolts to 29 ft. lbs. (39 Nm).

19. Align the knock pin on the right side camshaft with the knock pin of the timing pulley. Slide on the timing pulley with the right side mark facing forward. Tighten the bolt to 80 ft. lbs. (108 Nm).

20. Align the knock pin on the left side camshaft with the knock pin of the timing pulley. Slide on the timing pulley with the left side mark facing forward. Tighten the bolt to 80 ft. lbs. (108 Nm).

21. Install the timing belt to the left side camshaft timing pulley as follows:

a. Using the proper tool, slightly turn the left side timing pulley clockwise. Align the installation mark of the timing belt with the timing mark of the camshaft timing pulley and hang the timing belt on the left side camshaft pulley.

b. Align the timing marks of the left side camshaft pulley and the timing belt rear plate.

c. Check that the timing belt has tension between crankshaft timing pulley and the left side camshaft pulley.

22. Install the timing belt to the right side camshaft timing pulley as follows:

a. Using the proper tool, slightly turn the right side timing pulley clockwise. Align the installation mark of the timing belt with the timing mark of the camshaft timing pulley and hang the timing belt on the right side camshaft pulley.

b. Align the timing marks of the right side camshaft pulley and the timing belt rear plate.

c. Check that the timing belt has tension between crankshaft timing pulley and the right side camshaft pulley.

23. The timing belt tensioner must be set prior to installation. The tensioner can be set by:

a. Place a plate washer between the tensioner and a block. Using a press, press in the pushrod using 220–225 lbs. of pressure.

b. Align the holes of the pushrod and housing, pass a 0.05 inch (1.27mm) rod

through the holes to keep the setting position of the pushrod.

c. Release the press and install the dust boot to the tensioner.

24. Loosely install the tensioner. Evenly and alternately tighten the bolts to 20 ft. lbs. (26 Nm). Remove the tool from the tensioner.

25. Turn the crankshaft pulley 2 complete revolutions from TDC to TDC. Always turn the crankshaft clockwise. Check that all belt and pulley marks align with their reference marks. If any mark is out of perfect alignment, the timing belt must be removed and reinstalled.

26. Install or connect the following:
- Drive belt tensioner and tighten the bolt and nuts to 12 ft. lbs. (16 Nm).
- Both distributor housings and tighten the mounting bolts to 13 ft. lbs. (18 Nm). The distributors are marked L or R for correct installation.
- Distributor rotors and caps
- Fan bracket by installing the 2 bolts and 2 nuts. Tighten as follows:
- 12mm: 12 ft. lbs. (16 Nm)
- 14mm: 24 ft. lbs. (32 Nm)
- Air conditioning compressor. Tighten the bolts to 36 ft. lbs. (49 Nm) and the nut to 22 ft. lbs. (29 Nm).
- No. 1 ignition coil
- Right side No. 2 timing belt cover
- Left side No. 2 timing belt cover
- Drive belt idler pulley and cover plate. Tighten the bolt to 27 ft. lbs. (37 Nm).
- And secure the ignition wires. Make certain that all clips and retainers are securely engaged and that the wires are properly routed.
- Right side No. 3 timing belt
- Left-hand No. 3 timing belt cover
- Radiator assembly
- Fan pulley, fan, fluid coupling and the drive belt
- The air cleaner and intake air connector assembly
- V bank cover
- Coolant
- Negative battery cable to the battery
- Air cleaner inlet
- Battery clamp cover
- Engine undercover
- Oil pan protector

27. Start the engine and check for leaks

28. Bleed the cooling system and recheck the engine coolant level.

29. Make all the necessary engine adjustments.

Intake Manifold

REMOVAL & INSTALLATION

3.0L (1MZ-FE) Engine

1. Before servicing the vehicle, refer to the precautions in the beginning of this section.
2. Remove or disconnect the following:
- Negative battery cable
- Coolant
- Throttle/accelerator cable from the throttle body
- Air cleaner hose at the air intake chamber and remove it
- All lines and hoses. Tag them for installation.
- Idle Speed Control (ISC) valve and the throttle body.
- Exhaust Gas Recirculation (EGR) valve and vacuum modulator
- Distributor
- Cylinder head rear plate
- Intake chamber stays, any wires, then, the air intake chamber
- Fuel injection delivery pipe and the injectors.
- Water outlet and the bypass outlet.
- 2 bolts and the No. 2 idler pulley bracket stay.
- 8 bolts and 4 nuts, then lift out the intake manifold.

To install:

3. Thoroughly clean the intake manifold and cylinder head surfaces. Using a machinist's straight edge and a feeler gauge, check the surface of the intake manifold for warpage. If the warpage is greater than 0.0039 in. (0.10mm), replace the intake manifold.

4. Place new gaskets onto the intake manifold and position the intake manifold between the cylinder heads. Tighten the nuts and bolts to 13 ft. lbs. (18 Nm). Tighten the No. 2 pulley bracket bolts to 13 ft. lbs. (18 Nm).

5. Install or connect the following:
- Water bypass outlet and tighten the bolts to 74 inch lbs. (8.3 Nm). Tighten the water outlet to 74 inch lbs. (8 Nm).
- Injectors and delivery pipe.
- Air intake chamber and tighten the 2 bolts and 2 nuts to 32 ft. lbs. (43 Nm); use an 8mm hex wrench.
- Chamber stays and tighten the mounting bolts to 29 ft. lbs. (39 Nm).
- Remaining components. Tighten the emission control valve set to 73 inch lbs. (8 Nm).
- All hoses.
- Accelerator cable, if equipped with automatic transaxle

- Coolant
- Negative battery cable

3.0L (2JZ-GE) Engine

1. Before servicing the vehicle, refer to the precautions in the beginning of this section.
2. Remove or disconnect the following:
   - Negative battery cable
   - Coolant
   - Spark plug wires at the spark plugs

- Spark plugs
- Distributor with the spark plug leads attached.
- Radiator
- Water pump pulley
- Timing belt
- No. 2 front exhaust pipe
- 2 Oxygen (O$_2$) sensor leads
- 4 nuts, and the manifold heat shield
- Exhaust manifolds.

- Water bypass outlet and the No. 1 bypass pipe. Remove the 3 O-rings.
- Water outlet
- No. 1 bypass hose
- Vacuum Control Valve (VCV) set and the No. 2 vacuum pipe
- Fuel return hose from the oil dipstick guide, remove the mounting bolt and pull the guide and dipstick from the pan. Plug the hole.

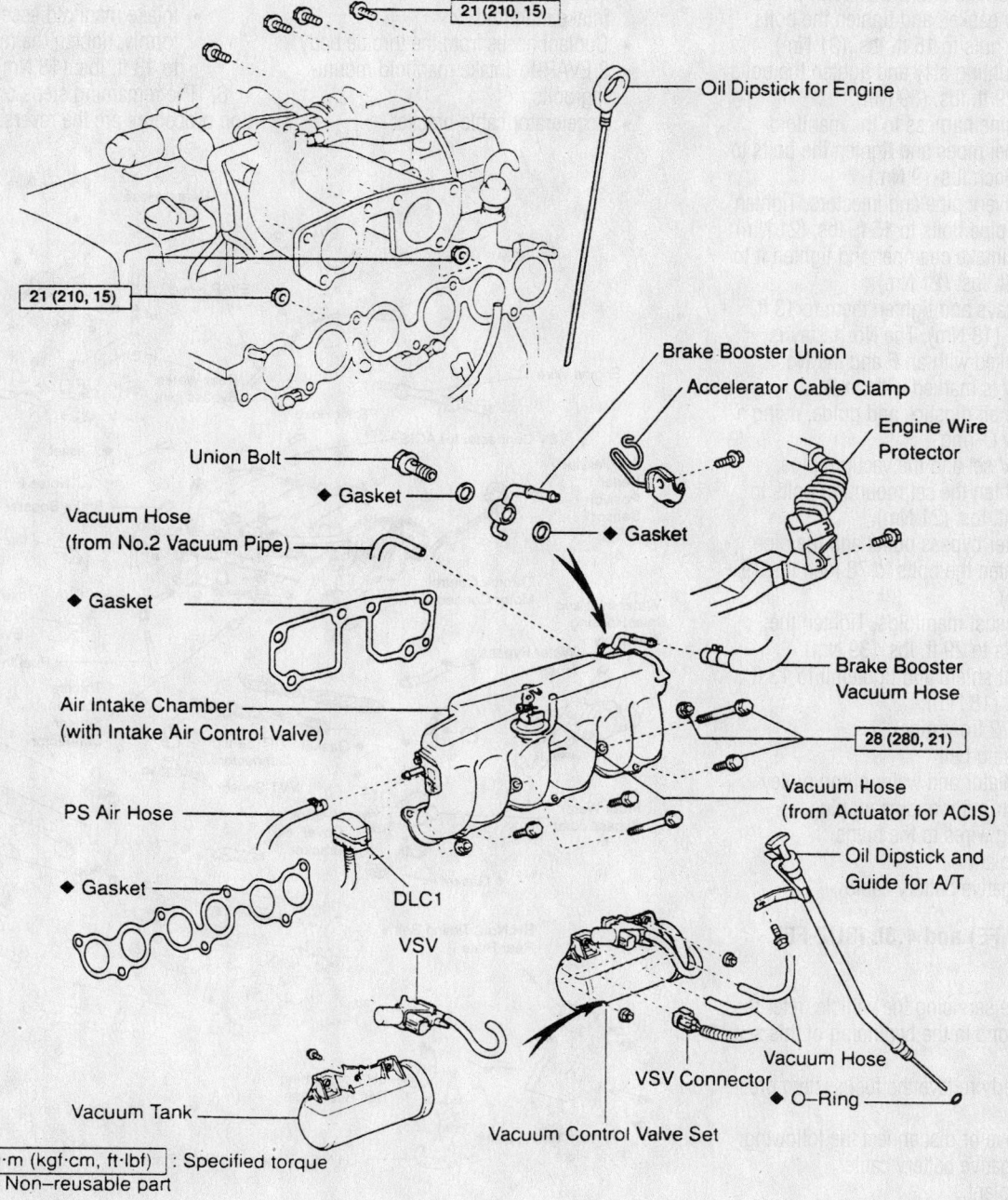

21 (210, 15)

Oil Dipstick for Engine

21 (210, 15)

Brake Booster Union

Accelerator Cable Clamp

Engine Wire Protector

Union Bolt

◆ Gasket

◆ Gasket

Vacuum Hose (from No.2 Vacuum Pipe)

◆ Gasket

Air Intake Chamber (with Intake Air Control Valve)

Brake Booster Vacuum Hose

28 (280, 21)

PS Air Hose

Vacuum Hose (from Actuator for ACIS)

◆ Gasket

DLC1

Oil Dipstick and Guide for A/T

VSV

Vacuum Tank

VSV Connector

Vacuum Control Valve Set

◆ O-Ring

N·m (kgf·cm, ft·lbf) : Specified torque
◆ Non−reusable part

9301LG09

Exploded view of the intake manifold mounting and related components—3.0L (2JZ-GE) engine

- Air intake chamber
- Fuel delivery pipe, then pull out the injectors.
- No. 1 and 2 fuel pipes
- Engine harness from the intake manifold
- Intake manifold stay
- Loosen the 6 bolts and 2 nuts, then lift out the intake manifold.

To install:

3. Install or connect the following:
- Install the intake manifold, with a new gasket, and tighten the bolts and nuts to 15 ft. lbs. (21 Nm).
- Mounting stay and tighten the bolts to 29 ft. lbs. (39 Nm).
- Engine harness to the manifold
- 2 fuel pipes and tighten the bolts to 78 inch lbs. (9 Nm)
- Delivery pipe and injectors. Tighten the pipe bolts to 15 ft. lbs. (21 Nm).
- Air intake chamber and tighten it to 15 ft. lbs. (21 Nm).
- 2 stays and tighten them to 13 ft. lbs. (18 Nm); The No. 1 stay is marked with an **F** and the No. 2 stay is marked with an **R**.
- The oil dipstick and guide, using a new O-ring
- VCV set and the vacuum pipe. Tighten the set mounting bolts to 15 ft. lbs. (21 Nm).
- Water bypass outlet and the pipe, tighten the bolts to 78 inch lbs. (9 Nm).
- Exhaust manifolds. Tighten the bolts to 29 ft. lbs. (39 Nm).
- Heat shield and tighten it to 13 ft. lbs. (18 Nm).
- No. 2 front pipe
- Timing belt
- Radiator and water pump pulley
- Distributor and spark plugs
- Plug wires to the plugs
- Coolant
- Negative battery cable

4.0L (1UZ-FE) and 4.3L (3UZ-FE) Engines

1. Before servicing the vehicle, refer to the precautions in the beginning of this section.

2. Properly relieve the fuel system pressure.

3. Remove or disconnect the following:
- Negative battery cable
- Coolant
- Accelerator cable
- Throttle and accelerator pedal position electrical connectors
- Throttle motor electrical connector

- Vacuum Switching Valve (VSV) electrical connectors
- Injector electrical connectors
- Noise filter electrical connector
- Brake booster vacuum hose from the intake manifold
- Positive Crankcase Ventilation (PCV) hose from the left-hand valve cover
- Evaporative Emission (EVAP) hoses and label for installation
- Power steering air hose from the intake manifold
- Coolant hoses from the throttle body
- 2 EVAP-to-intake manifold mounting bolts
- Accelerator cable bracket

- 3 V-bank cover brackets
- EVAP VSV
- Fuel supply and return lines
- 6 bolts and 4 nuts, then the intake manifold assembly

4. Clean the gasket mating surfaces of old gasket and sealant.

To install:

5. Install or connect the following:
- 2 intake manifold gaskets on the cylinder heads with the white painted mark facing upward.
- Intake manifold assembly and uniformly, tighten the mounting bolts to 13 ft. lbs. (18 Nm).

6. The remaining steps of the installation procedure are the reverse of the

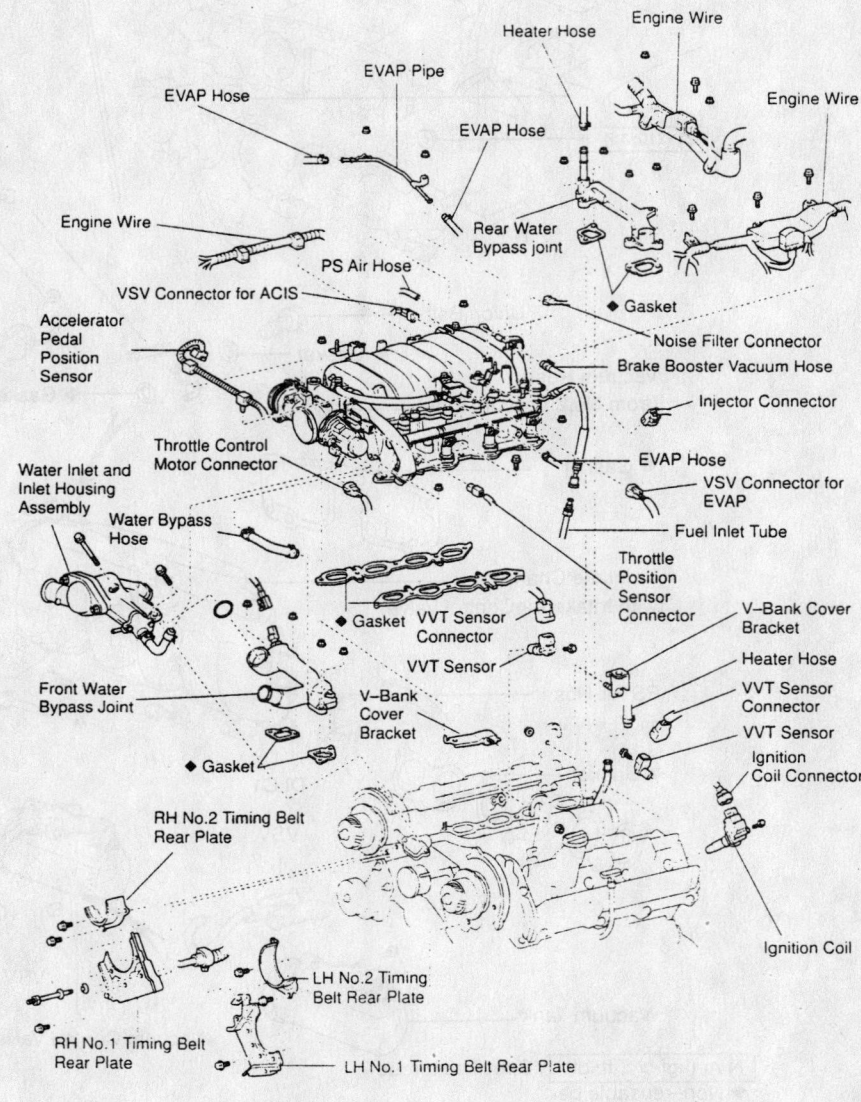

◆ Non-reusable part

Exploded view of the intake manifold mounting and related components—4.0L (1UZ-FE) and 4.3L (3UZ-FE) Engines

removal, while keeping in mind the following items:

- Tighten the V-bank cover brackets to 66 inch lbs. (8 Nm).
- Tighten the accelerator cable bracket mounting bolts to 13 ft. lbs. (18 Nm).
- Tighten the EVAP VSV to 13 ft. lbs. (18 Nm).
- All remaining components.
- Coolant
- Negative battery cable

Exhaust Manifold

REMOVAL & INSTALLATION

3.0L (1MZ-FE) Engine

1. Before servicing the vehicle, refer to the precautions in the beginning of this section.
2. Remove or disconnect the following:
   - Negative battery cable
   - Engine undercovers
   - 2 front exhaust pipe stay bolts
   - Front pipe from the center pipe
   - 3 nuts and the front pipe
   - Oxygen (O₂) sensor at the right side manifold
   - 3 mounting nuts and lift off the outside heat insulator
   - 6 nuts and lift off the right side manifold and gasket
   - Left side heat insulator
   - 6 nuts and lift off the left side manifold and gaskets

To install:

3. Install or connect the following:
   - Right manifold with a new gasket. Tighten the nuts to 29 ft. lbs. (39 Nm).
   - Outer insulator
   - Left manifold. Use a new gasket. Tighten the nuts to 29 ft. lbs. (39 Nm).
   - Outer insulator
   - Front exhaust pipe and tighten the manifold-to-pipe nuts to 46 ft. lbs. (62 Nm). Tighten the pipe-to-converter nuts to 32 ft. lbs. (43 Nm).
   - O₂ sensor
   - Undercovers
   - Battery cable

3.0L (2JZ-GE) Engine

1. Before servicing the vehicle, refer to the precautions in the beginning of this section.

2. Remove or disconnect the following:
   - Negative battery cable
   - Engine undercovers
   - No. 2 front exhaust pipe bolts and disconnect it from the front exhaust pipe. Loosen the 4 nuts, then remove the front pipe.
   - Both Oxygen (O₂) sensors at the manifold
   - 4 mounting nuts and lift off the outside heat insulator
   - 4 nuts and disconnect the manifolds from the pipe. Loosen the mounting bolts and remove the 2 manifolds and the gasket.

To install:

3. Install or connect the following:
   - Manifolds with a new gasket. Tighten the nuts to 30 ft. lbs. (40 Nm).
   - Outer insulator. Tighten the nuts to 13 ft. lbs. (18 Nm).
   - No. 2 front pipe. Tighten the nuts to 46 ft. lbs. (62 Nm).
   - Front exhaust pipe. Tighten the bolts/nuts to 32 ft. lbs. (43 Nm).
   - O₂ sensors
   - Undercovers
   - Battery cable

4.0L (1UZ-FE) and 4.3L (3UZ-FE) Engines

1. Before servicing the vehicle, refer to the precautions in the beginning of this section.

2. Remove or disconnect the following:
   - Negative battery cable
   - Coolant
   - Camshaft timing pulleys
   - Cooling fan hydraulic pump, on the SC400
   - Accelerator, throttle control and cruise control actuator cables
   - High tension cord cover and the right side ignition coil
   - Water inlet housing mounting bolts
   - Water bypass hose from the Idle Speed Control (ISC) valve
   - Water inlet and inlet housing assemblies
   - O-ring from the water inlet housing
   - Exhaust Gas Recirculation (EGR) pipe
   - Vacuum Switching Valve (VSV) connector
   - Vacuum pipe hose
   - EGR water bypass pipe
   - Fuel pressure VSV
   - EGR vacuum hoses and the EGR VSV

- Water bypass pipe hose from the ISC valve
- Water bypass joint hose
- Vacuum pipe hoses
- EGR gas temperature sensor
- EGR valve adapter
- Fuel pressure regulator vacuum hose
- Air intake chamber vacuum hose
- Vacuum hose from the Evaporative Emission (EVAP) BVSV
- Mounting bolts, hoses and the vacuum pipe
- ISC valve
- Throttle body sensor connectors and the water bypass pipe from the rear water bypass joint
- Positive Crankcase Ventilation (PCV) valve hose
- Throttle body and gasket
- Accelerator cable bracket and the brake booster vacuum union and hose
- Cold start injector connector and the cold start injector tube from the right side delivery pipe, if equipped
- Check connector from the intake chamber and remove the mounting nuts and bolts
- Air intake chamber and the cold start injector, tube and wire assembly, if equipped
- Engine wire from the intake manifold and from the right side cylinder head
- Heater hoses
- Delivery pipes and the fuel injectors
- Mounting bolts and nuts. Lift up the intake manifold
- Front and rear water bypass joint
- Front exhaust pipe and the main catalytic converters. Lower the vehicle.
- Right side Oxygen (O₂) sensor
- Mounting bolts and nuts and remove the right side exhaust manifold.
- Oil dipstick and guide
- Left side O₂ sensor
- Mounting bolts and nuts
- Left side exhaust manifold

To install:

3. Install or connect the following:
   - Right side exhaust manifold with a new gasket (the painted marks should face the manifold) and tighten the mounting bolts to 29 ft. lbs. (39 Nm).

Timing belt service is covered in Section 3 of this manual

- Right side O$_2$ sensor
- Left side exhaust manifold with a new gasket (the painted marks should face the manifold) and tighten the mounting bolts to 29 ft. lbs. (39 Nm).
- Left side O$_2$ sensor
- Oil dipstick and guide. Raise and safely support the vehicle.
- Catalytic converters and front exhaust pipe. Lower the vehicle.
- Front and rear water bypass joints. Tighten the mounting bolts to 13 ft. lbs. (18 Nm).
- Intake manifold, using new gaskets. Tighten the mounting nuts and bolts to 13 ft. lbs. (18 Nm).
- Delivery pipes and fuel injectors
- Fuel return pipe with new gaskets. Tighten the union bolt to 26 ft. lbs. (35 Nm).
- Fuel hoses and the injector connectors
- Engine wire to the delivery pipes
- Connectors on the left side delivery pipe
- ECT sensor
- Cold start injector time switch
- Water temperature sender gauge connectors
- Heater hoses and engine wire bracket
- Engine wire to the bracket.
- Cold start injector, tube and wire assembly. Tighten the mounting bolts to 69 inch lbs. (8 Nm), if equipped.
- Air intake chamber with new gaskets and tighten the mounting bolts to 13 ft. lbs. (18 Nm).
- Cold start injector tube to the right side delivery pipe and tighten the union bolt to 11 ft. lbs. (15 Nm), if equipped.
- Cold start injector connector, if necessary.
- Accelerator cable bracket
- Brake booster union and connect the vacuum hose. Tighten the union bolt to 22 ft. lbs. (29 Nm).
- Water bypass hose to the throttle body and the PCV hose to the cylinder head cover.
- Throttle body, using a new gasket. Tighten the mounting bolts to 13 ft. lbs. (18 Nm).
- Water bypass pipe
- Sensor connectors
- Idle Speed Control (ISC) valve and tighten the mounting bolts to 13 ft. lbs. (18 Nm).
- Water bypass hose.

- Vacuum pipe and the assorted hoses
- Remaining components.
- And adjust the accelerator cable, the automatic transmission throttle cable and the cruise control actuator cable.
- Cooling fan hydraulic pump on the SC400
- Camshaft timing pulleys.
- Coolant
- Negative battery cable

Camshaft and Valve Lifters

REMOVAL & INSTALLATION

The following procedures have the valve lash adjuster removal and installation incorporated.

3.0L (1MZ-FE) Engine

1. Before servicing the vehicle, refer to the precautions in the beginning of this section.

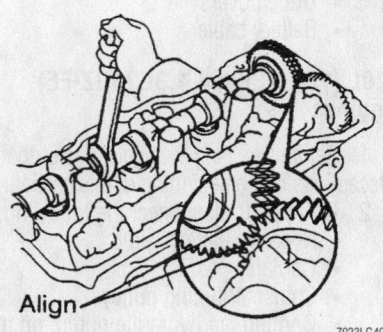

Aligning the right side camshaft timing marks—3.0L (1MZ-FE) engine

2. Remove or disconnect the following:
- Timing belt and idler pulley
- Camshaft timing pulleys
- Cylinder head covers

✳✳ WARNING

The thrust clearance on both the intake and exhaust camshafts is very small, the camshafts must be kept level during removal. If the camshafts are removed without being kept level, the camshaft may be caught in the cylinder head causing the head to break or the camshaft to seize.

3. To remove the exhaust and intake camshafts from the right side cylinder head:

4. Turn the camshaft with a wrench until the 2 pointed marks drive and driven gears are aligned. (The right camshaft gears have 2 marks apiece; the left side camshaft gears have 1 mark each.)

5. Secure the exhaust camshaft subgear to the main gear using a service bolt. A bolt 0.63–0.79 in. (16–20mm) long with a 6mm thread diameter and a 1mm pitch is recommended. When removing the exhaust

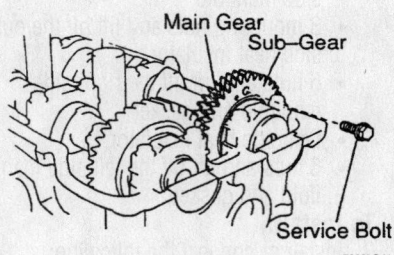

Securing the subgear and driven gear, right side—3.0L (1MZ-FE) engine

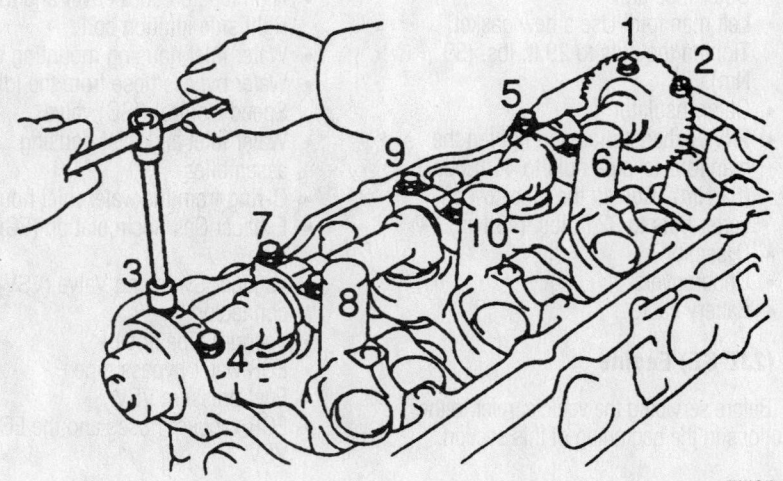

Right exhaust camshaft bearing loosening sequence—3.0L (1MZ-FE) engine

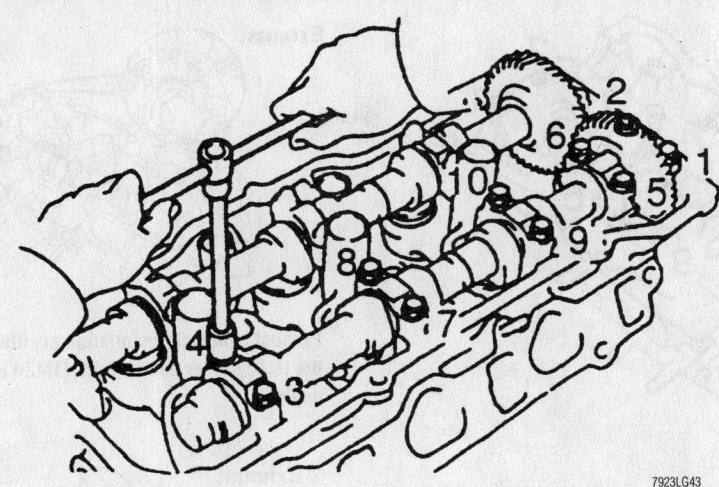

Right intake camshaft bearing loosening sequence—3.0L (1MZ-FE) engine

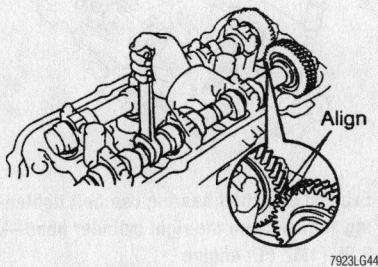

Aligning the left side camshaft timing marks—3.0L (1MZ-FE) engine

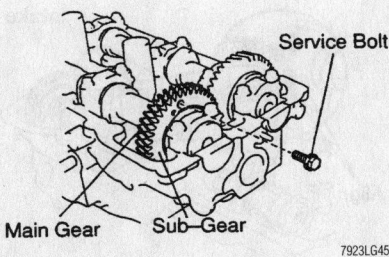

Securing the subgear and driven gear, left side—3.0L (1MZ-FE) engine

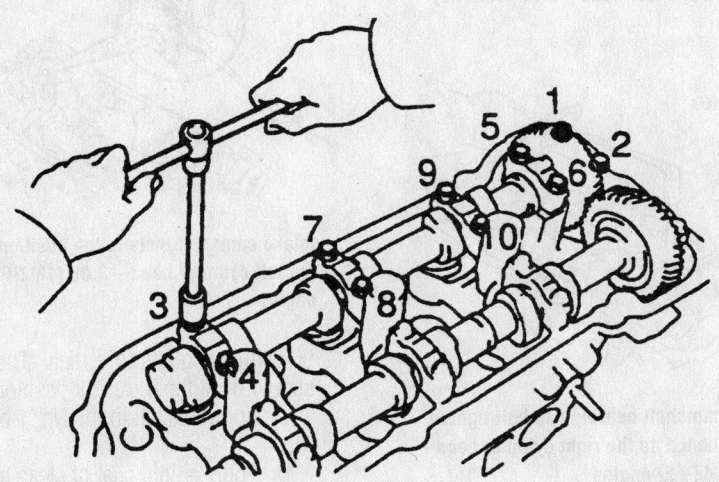

Left intake camshaft bearing cap bolt loosening sequence—3.0L (1MZ-FE) engine

keeping them in order, remove the oil seal, then lift out the intake camshaft.

9. To remove the exhaust and intake camshafts from the left side cylinder head:

a. Turn the camshaft with a wrench until the pointed marks on the drive and driven gears are aligned. (The right camshaft gears have 2 marks apiece; the left side camshaft gears have 1 mark each.)

b. Secure the exhaust camshaft sub-gear to the main gear using a service bolt. A bolt 0.63–0.79 in. (16–20mm) long with a 6mm thread diameter and a 1mm pitch is recommended. When removing the exhaust camshaft be sure the subgear is not loaded; all the force must be eliminated.

c. Uniformly loosen and remove the exhaust camshaft bearing cap bolts in several passes and in the proper sequence. Remove the 8 bearing cap bolts and remove the caps. Keep the caps in the correct order.

d. Remove the exhaust camshaft from the engine.

e. Uniformly loosen and remove the 10 bearing cap bolts in several passes, in the proper sequence. Remove the bearing caps, keeping them in order, remove the oil seal, then lift out the intake camshaft.

10. Remove the valve lash adjuster shims and hydraulic lash adjusters. Identify each lash adjuster and shim as it is removed so it can be reinstalled in the same position. If the lash adjusters are to be reused, store them upside down in a sealed container.

To install:

11. Install or connect the following:
- Valve lash adjusters and shims. Check valve clearance and replace the shims as necessary.

➡**Before installing the camshafts in either cylinder head, apply multi-purpose grease to the thrust portions of each camshaft.**

12. To install the right camshafts:

a. Position the intake camshaft on the head so that the alignment marks are at a 90° angle from vertical. The mark should be at the 3 o'clock position.

b. Apply sealant to the No. 1 bearing cap.

c. Apply a light coat of clean engine oil to the bolt threads and under the bolt head. Install the bearing caps to their

camshaft be sure the subgear is not loaded; all the force must be eliminated.

6. Uniformly loosen and remove the exhaust camshaft bearing cap bolts in several passes and in the proper sequence. Remove the 8 bearing cap bolts and remove

the caps, keeping them in the correct order.

7. Remove the exhaust camshaft from the engine.

8. Uniformly loosen and remove the 10 bearing cap bolts in several passes, in the proper sequence. Remove the bearing caps,

Heater Core replacement is covered in Section 2 of this manual

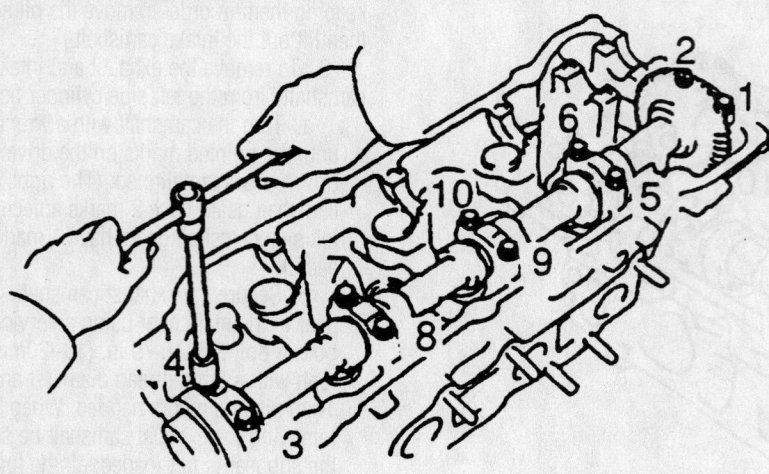

Left exhaust camshaft bearing cap bolt loosening sequence—3.0L (1MZ-FE) engine

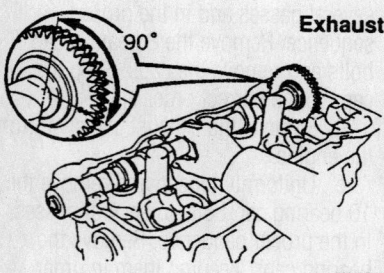

Exhaust camshaft installation position on the right cylinder head—3.0L (1MZ-FE) engine

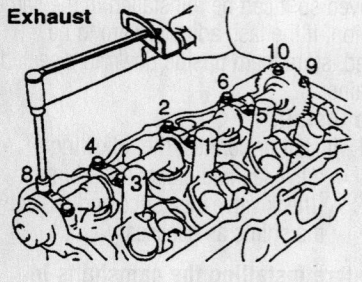

Exhaust camshaft bearing cap bolt tightening sequence on the right cylinder head—3.0L (1MZ-FE) engine

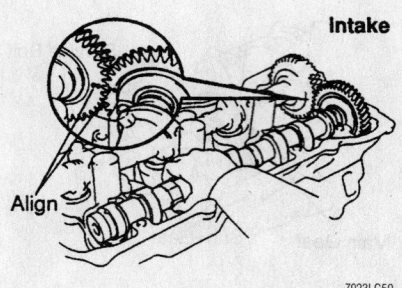

Intake camshaft installation position on the right cylinder head—3.0L (1MZ-FE) engine

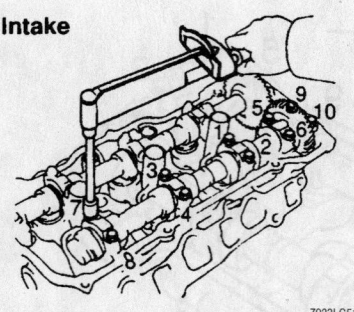

Intake camshaft bearing cap bolt tightening sequence on the right cylinder head—3.0L (1MZ-FE) engine

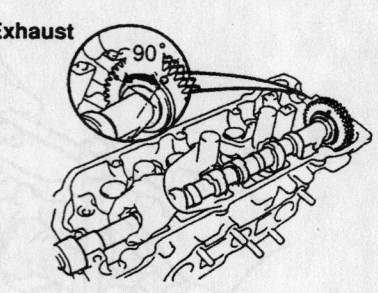

Exhaust camshaft installation position on the left cylinder head—3.0L (1MZ-FE) engine

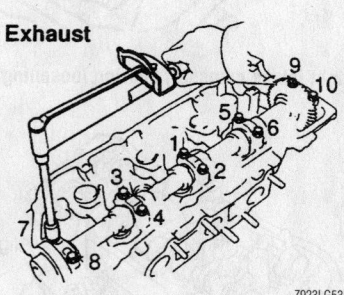

Exhaust camshaft bearing cap bolt tightening sequence on the right cylinder head—3.0L (1MZ-FE) engine

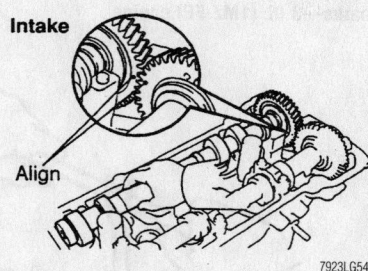

Intake camshaft installation position on the left cylinder head—3.0L (1MZ-FE) engine

proper position. Tighten the bolts evenly and in several passes in the reverse order of loosening to 12 ft. lbs. (16 Nm) in the proper sequence.

d. Position the exhaust camshaft on the head so that the alignment marks are at a 90° angle from vertical. The mark should be at the 9 o'clock position and must align with the marks on the other gear.

e. Apply a light coat of clean engine

oil to the bolt threads and under the bolt head. Install the bearing caps to their proper position. Tighten the bolts evenly and in several passes in the reverse order of loosening to 12 ft. lbs. (16 Nm) in the proper sequence.

f. Remove the service bolt.

13. To install the left camshafts:

a. Position the intake camshaft on the head so that the alignment mark is at a

90 degree angle from vertical. The mark should be at the 9 o'clock position.

b. Apply sealant to the No. 1 bearing cap.

c. Apply a light coat of clean engine oil to the bolt threads and under the bolt head. Install the bearing caps to their proper position. Tighten the bolts evenly and in several passes to 12 ft. lbs. (16 Nm) in the proper sequence.

d. Position the exhaust camshaft on the head so that the alignment marks are at a 90 degree angle from vertical. The mark should be at the 3 o'clock position and must align with the marks on the other gear.

Intake

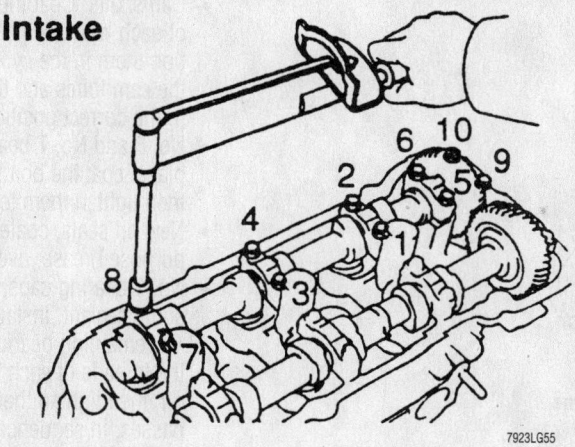

7923LG55

Intake camshaft bearing cap bolt tightening sequence on the right cylinder head—3.0L (1MZ-FE) engine

e. Apply a light coat of clean engine oil to the bolt threads and under the bolt head. Install the bearing caps to their proper position. Tighten the bolts evenly and in several passes to 12 ft. lbs. (16 Nm) in the proper sequence.

f. Remove the service bolt.

14. Apply multi-purpose grease to new camshaft oil seals. Install the seals.

15. Install or connect the following:
- No. 3 (rear) timing belt cover
- Camshaft timing gears
- Idler pulley, timing belt and covers

16. Check and adjust the valve clearance.

17. Install the cylinder head (valve) covers.

3.0L (2JZ-GE) Engine

1. Before servicing the vehicle, refer to the precautions in the beginning of this section.

2. Remove or disconnect the following:

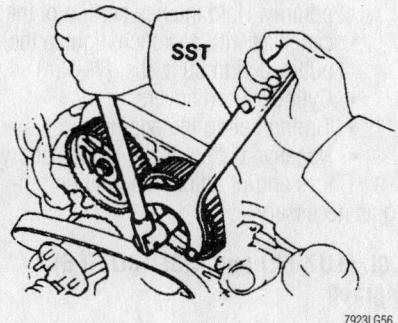

7923LG56

Removing the camshaft sprockets—3.0L (2JZ-GE) engine

- Negative battery cable from the battery
- Timing belt from the engine
- Cylinder head covers
- Camshaft sprocket
- 4 bolts and lift out the No. 4 (inner) timing belt cover.
- No. 1 camshaft bearing cap bolts. These are the bolts directly behind the sprockets.
- Bearing caps
- Remaining bearing cap bolts. Note that there are separate sequences for the exhaust and intake camshafts. Lift off all 12 bearing caps.
- Exhaust and intake camshafts
- Valve lash adjuster shims and hydraulic lash adjusters. Identify each lash adjuster and shim as it is

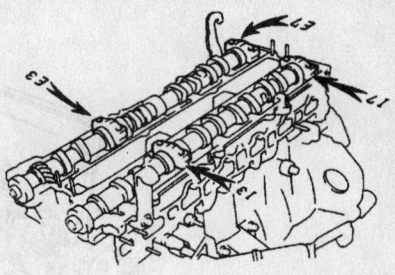

7923LG58

Installing No. 3 and 7 bearing caps—3.0L (2JZ-GE) engine

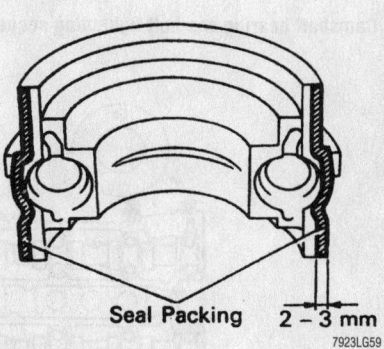

7923LG59

Applying sealant to the No. 1 bearing cap—3.0L (2JZ-GE) engine

removed so it can be reinstalled in the same position. If the lash adjusters are to be reused, store them upside down in a sealed container.

To install:

3. Install or connect the following:
- Valve lash adjusters and shims. Check valve clearance and replace the shims as necessary.

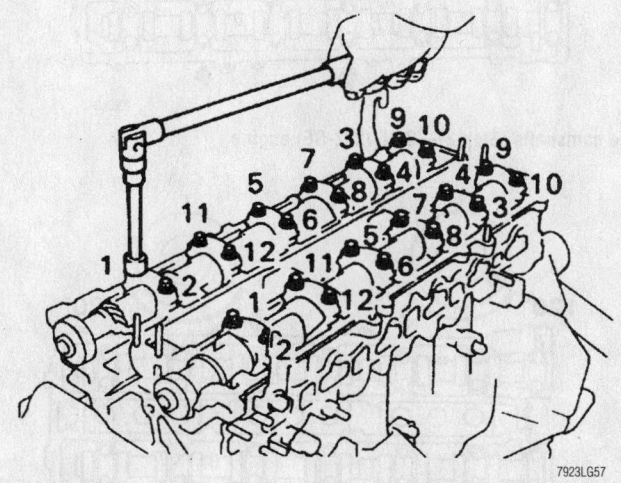

7923LG57

Camshaft bearing cap bolt removal sequence—3.0L (2JZ-GE) engine

Brake service is covered in Section 4 of this manual

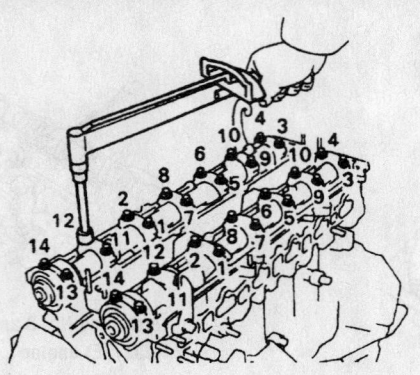

Camshaft bearing cap bolt tightening sequence—3.0L (2JZ-GE) engine

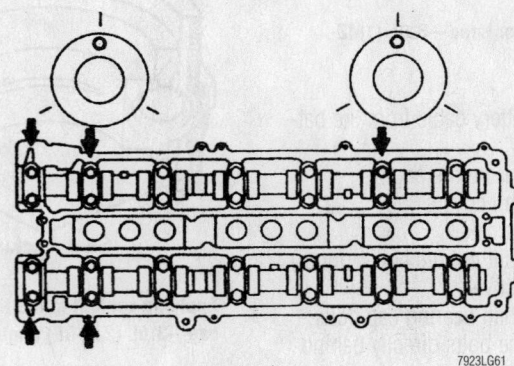

Tightening the camshafts (Step 1)—3.0L (2JZ-GE) engine

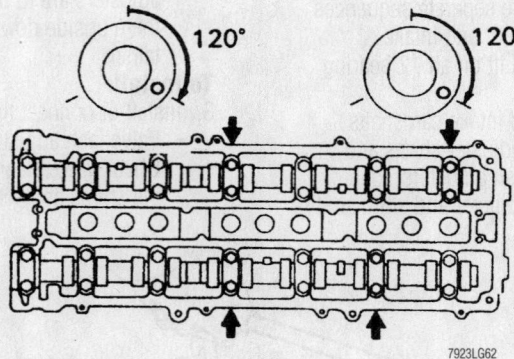

Tightening the camshafts (Step 2)—3.0L (2JZ-GE) engine

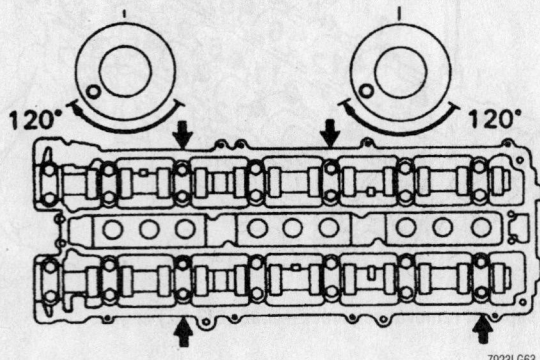

Tightening the camshafts (Step 3)—3.0L (2JZ-GE) engine

- Camshafts. Coat the thrust portions of each with engine oil, then position them in the cylinder head with the cam lobes and the knock pins in the correct position.
- No. 3 and No. 7 bearing caps in place, coat the bolt threads with oil, then tighten them temporarily.
- New oil seals, coated with multi-purpose grease, over the camshafts
- No. 1 bearing caps, then apply some sealant. Install the bolts.
- All remaining bearing caps. Coat the threads of each bolt with clean oil, then tighten them, in several passes, in sequence, to 14 ft. lbs. (20 Nm). Note that there are separate sequences for the intake and exhaust sides.
- Oil seal in as far as it will go

4. Rotate each camshaft until the forward straight (knock) pin is straight up. Loosen exhaust Nos. 1, 2 and 6 bearing cap bolts until they can be turned by hand; retighten them to 14 ft. lbs. (20 Nm). Loosen intake Nos. 1 and 2 and retighten to 14 ft. lbs. (20 Nm).

5. Turn each camshaft ⅓ of a revolution (120 degrees). Loosen exhaust Nos. 4 and 7 bearing cap bolts; retighten them to 14 ft. lbs. (20 Nm). Loosen intake Nos. 4 and 6 bearing cap bolts; retighten them to 14 ft. lbs. (20 Nm).

6. Turn each camshaft an additional ⅓ of a revolution, loosen exhaust bearing cap bolts Nos. 3 and 5, then retighten them to 14 ft. lbs. (20 Nm). Loosen intake bearing cap bolts Nos. 3 and 7, then retighten them to 14 ft. lbs. (20 Nm).

7. Check and adjust the valve clearance.
8. Install or connect the following:
- No. 4 inside timing belt cover and the camshaft pulleys. Align the shaft pin with the pulley groove and slide the pulley on. Install the bolt temporarily. Hold the hex portion of the camshaft with a wrench; tighten the pulley bolt to 59 ft. lbs. (79 Nm).
- Cylinder head covers
- Timing belt to the engine
- Negative battery cable to the battery

9. Check and/or adjust the ignition timing as necessary

4.0L (1UZ-FE) and 4.3L (3UZ-FE) Engines

1. Before servicing the vehicle, refer to the precautions in the beginning of this section.

2. Relieve the fuel pressure from the fuel lines.

3. Remove or disconnect the following:
- Negative battery cable. Wait at least 90 seconds before performing any other work
- Positive battery cable
- Battery.
- Air cleaner inlet
- V bank cover by removing the bolt and 2 cap nuts.
- Air cleaner and intake air connector assembly.
- Drive belt
- Fluid coupling and the fan pulley. The drive belt tension may be slackened by turning the tensioner counterclockwise. The pulley bolt for the drive belt tensioner has a left-handed thread.
- Radiator.
- The right-hand No. 3 timing belt cover
- Left-hand No. 3 timing belt cover
- Evaporative Emissions (EVAP) hose clamp from the timing belt cover. For all other vehicles, disconnect the EVAP hose from the hose clamp on the timing belt cover.
- 4 bolts to the left-hand timing belt cover.
- Cord grommet from the timing belt cover and remove the timing belt cover.
- Drive belt idler pulley by removing the pulley bolt and cover plate.
- Right-hand and left No. 2 timing belt covers.
- Distributor housings
- No. 1 ignition coil
- Air conditioning compressor from the engine. Do not disconnect the air conditioning pressure lines.
- Fan bracket by removing the 2 bolts and 2 nuts.
- Alternator from the engine
- Drive belt tensioner

4. Set the engine to Top Dead Center (TDC) on cylinder No. 1.

5. Turn the crankshaft pulley approximately 50° clockwise and put the timing mark of the crankshaft pulley in line with the centers of the crankshaft pulley bolt and the idler pulley bolt.

❊❊ WARNING

If the timing belt is disengaged, having the crankshaft pulley at the wrong angle can cause the piston head and valve head to come into contact with each other when you remove the
camshaft timing pulley. Always set the crankshaft pulley at the correct angle before removing the timing belt.

6. If the timing belt is to be reused, turn the crank pulley slowly; check that the 3 installation marks are present on the belt. If the marks are not present, make new installation marks before removing the belt. The marks should align with the timing marks on each camshaft pulley and the crank pulley.

7. Remove or disconnect the following:
- Timing belt tensioner
- Timing belt from the camshaft timing pulleys
- Camshaft timing pulleys
- No. 2 ignition coil
- Rear timing belt plates
- Throttle body
- Spark plug wires, wire clamps and the wire cover assembly from the right cylinder head.
- Right cylinder head cover by removing the 8 bolts and 8 washers.
- Transmission oil dipstick
- Evaporative Emission (EVAP) control hose from the Vacuum Switching Valve (VSV)
- Engine wire clamp from the wire bracket on the delivery pipe.
- Spark plug wires and clamps from the left-hand cylinder head cover.
- Left cylinder head cover by removing the 8 bolts and 8 seal washers.
- Semi-circular plugs, if necessary.

❊❊ WARNING

Since the thrust clearance of the camshaft is small, the camshaft must be kept level while it is being removed. If the camshaft is not kept level, the portion of the cylinder head receiving the shaft thrust may crack or be damaged, causing the camshaft to seize or break.

8. To remove the exhaust camshaft from the right side cylinder head:
   a. Position the service bolt hole of the drive subgear to the upright position. Secure the camshaft subgear to drive gear with a service bolt.
   b. Set the timing mark (single dot) on the camshaft drive gear at approximately 10 degrees. Turn the camshaft with a wrench on the hexagonal flats.

   c. Alternately loosen and remove the bearing cap bolts holding the intake camshaft side of the oil feed pipe to the cylinder head.
   d. Uniformly loosen (in several passes) and remove the bearing cap bolts, in sequence.
   e. Remove the oil feed pipe and the bearing caps. Remove the camshaft.
   f. To remove the intake camshaft from the right side cylinder head:
   g. Set the timing mark (single dot) on the camshaft drive gear at approximately 45 degrees. Turn the camshaft with a wrench on the hexagonal flats.
   h. Uniformly loosen (in several passes) and remove the bearing cap bolts in the proper sequence.
   i. Remove the bearing caps, oil seal and the intake camshaft.

9. To remove the exhaust camshaft of the left side cylinder head:
   a. Position the service bolt hole of the drive subgear to the upright position. Secure the camshaft subgear to drive gear with a service bolt.

➡ **When removing the camshaft, be sure the torsional spring force of the subgear has been eliminated.**

   b. Set the timing mark (2 dots) on the camshaft drive gear at approximately 15 degrees, by turning the camshaft with the proper tool.
   c. Alternately loosen and remove the bearing cap bolts holding the intake camshaft side of the oil feed pipe to the cylinder head.
   d. Uniformly loosen (in several passes) and remove the bearing cap bolts in the proper sequence.
   e. Remove the oil feed pipe and the bearing caps. Remove the camshaft.

10. To remove the intake camshaft from the left side cylinder head:
   a. Set the timing mark (single dot) of the camshaft drive gear at approximately 60 degrees, by turning the camshaft with the proper tool.
   b. Uniformly loosen (in several passes) and remove the bearing cap bolts, in sequence.
   c. Remove the bearing caps, oil seal and the intake camshaft.
   d. Remove the valve lash adjuster shims and hydraulic lash adjusters. Identify each lash adjuster and shim as it is removed so it can be reinstalled in the same position. If the lash adjusters are to

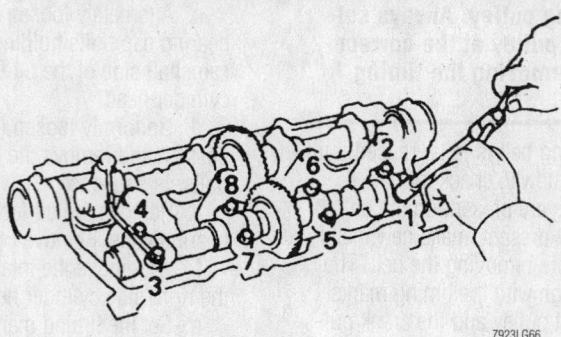

Bolt loosening sequence for the exhaust camshaft on the right cylinder head—4.0L (1UZ-FE) and 4.3L (3UZ-FE) Engines

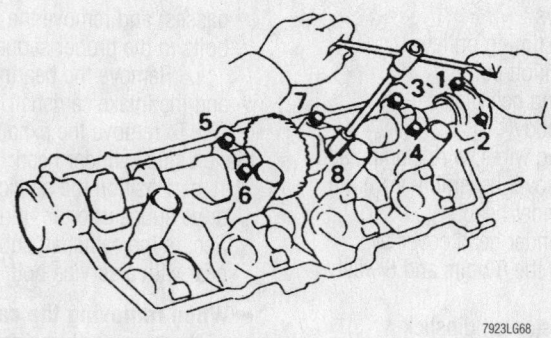

Loosen the bearing cap bolts for the intake camshaft on the right cylinder head in the sequence shown—4.0L (1UZ-FE) and 4.3L (3UZ-FE) Engines

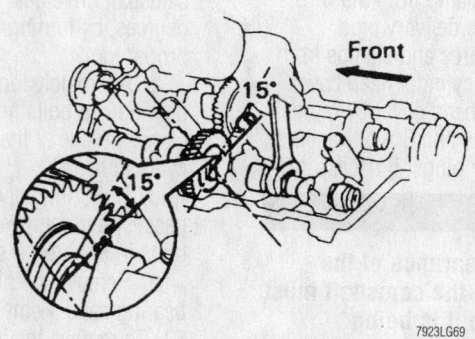

Turning the exhaust camshaft on the left cylinder head 15 degrees—4.0L (1UZ-FE) and 4.3L (3UZ-FE) Engines

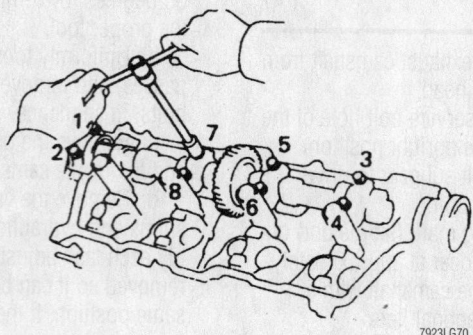

Bolt removal sequence for the intake camshaft on the left cylinder head—4.0L (1UZ-FE) and 4.3L (3UZ-FE) Engines

be reused, store them upside down in a sealed container.

To install:

11. Install or connect the following:
   • Valve lash adjusters and shims. Check the valve clearance and replace the shims as necessary.
   • New seal packing to the bearing caps
   • Bearing cap on the right side cylinder head, marked **I1**, in position with the arrow mark facing the rear. Install the bearing cap on the left side cylinder head, marked **I6**, in position with the arrow mark facing the front.
   • Oil on the threads of the cap bolts. Install the bearing cap bolts with new washers and tighten to 12 ft. lbs. (16 Nm).

12. To install the right side cylinder head intake camshaft:

   a. Apply grease to the thrust portion of the camshaft.

   b. Place the intake camshaft at a 45 degree angle of the timing mark (single dot) on the cylinder head.

   c. Remove any old packing and apply new seal packing to the bearing cap marked **I6** and install the front bearing cap, marked **I6** with the arrow facing rearward.

   d. Align the arrows at the front and rear of the cylinder head with the bearing cap.

   e. Install the remaining bearing caps in the proper sequence with the arrow mark facing rearward. Install the oil feed pipe and the mounting bolts.

   f. Uniformly tighten the bearing cap bolts in the proper sequence to 12 ft. lbs. (16 Nm).

13. To install the right side cylinder head exhaust camshaft:

   a. Set the timing mark (single dot) on the camshaft drive gear at a 10 degree angle by turning the intake camshaft with the proper tool.

   b. Apply grease to the thrust portion of the camshaft.

   c. Align the timing marks (single dots) on the camshaft drive and driven gears.

   d. Place the exhaust camshaft in the cylinder head. Install the rear bearing cap with the arrow mark facing rearward.

   e. Align the arrow marks at the front and rear of the cylinder head with the mark on the bearing cap. Apply a light coat of oil on the threads of the bearing cap bolts.

   f. Uniformly tighten the bearing cap

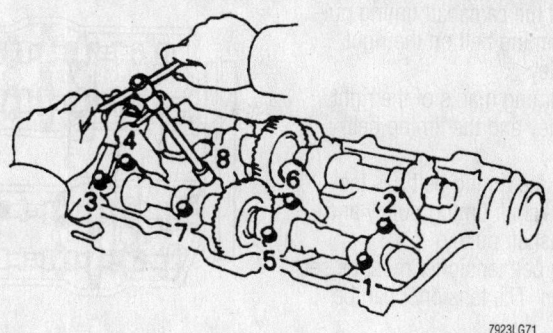

Bolt removal sequence for the exhaust camshaft on the left cylinder head—4.0L (1UZ-FE) and 4.3L (3UZ-FE) Engines

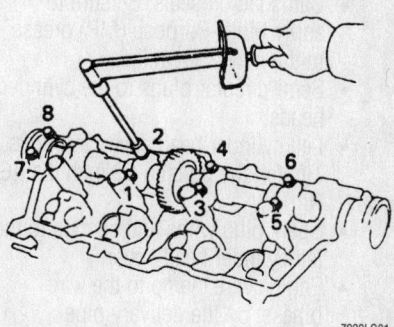

Left side intake camshaft bracket bolt tightening sequence—4.0L (1UZ-FE) and 4.3L (3UZ-FE) Engines

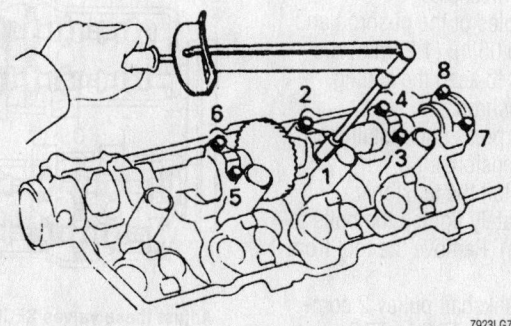

Right side intake camshaft bracket bolt tightening sequence—4.0L (1UZ-FE) and 4.3L (3UZ-FE) Engines

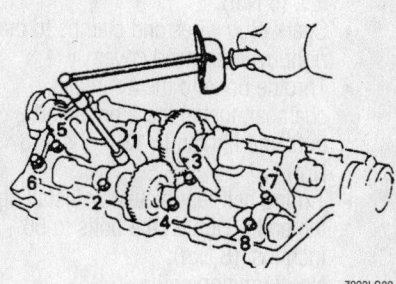

Left side exhaust camshaft bracket bolt tightening sequence—4.0L (1UZ-FE) and 4.3L (3UZ-FE) Engines

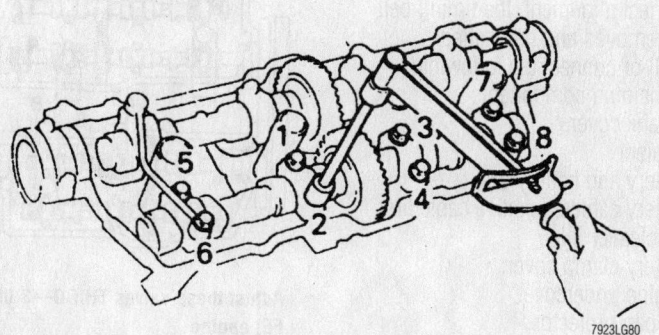

Right side exhaust camshaft bracket bolt tightening sequence—4.0L (1UZ-FE) and 4.3L (3UZ-FE) Engines

bolts in the proper sequence to 12 ft. lbs. (16 Nm).

g. Bring the service bolt installed upward by turning the camshaft with the proper tool. Remove the service bolt.

14. To install the left side cylinder head intake camshaft:

a. Apply grease to the thrust portion of the camshaft.

b. Place the intake camshaft with the timing mark (single dot) at a 60 degree angle on the cylinder head.

c. Remove any old packing and apply new seal packing to the bearing cap marked **I6** and install the front bearing cap, marked **I1** with the arrow facing rearward.

d. Align the arrows at the front and rear of the cylinder head with the bearing cap. Apply a light coat of oil on the threads of the bearing cap bolts.

e. Install the remaining bearing caps in the proper sequence with the arrow mark facing rearward. Install the oil feed pipe and the mounting bolts.

f. Uniformly tighten the bearing cap bolts in the proper sequence to 12 ft. lbs. (16 Nm).

15. Install the left side cylinder head exhaust camshaft by:

a. Set the timing mark (2 dots) on the camshaft drive gear at a 15 degree angle by turning the intake camshaft with the proper tool.

b. Apply grease to the thrust portion of the camshaft.

c. Align the timing marks (2 dots each) on the camshaft drive and driven gears.

d. Place the exhaust camshaft on the cylinder head. Install the rear bearing cap with the arrow mark facing rearward.

e. Align the arrow marks at the front and rear of the cylinder head with the mark on the bearing cap. Apply a light coat of oil on the threads of the bearing cap bolts.

f. Uniformly tighten the bearing cap bolts in the proper sequence to 12 ft. lbs. (16 Nm).

16. Bring the service bolt installed upward by turning the camshaft with the proper tool. Remove the service bolt.

17. Install or connect the following:

For Accessory Drive Belt illustrations, see Section 1 of this manual

- Camshaft oil seals. Be sure to apply Multi-Purpose (MP) grease to the new oil seal lip.
- Semi-circular plugs to the cylinder heads.
- Left cylinder head cover and bolts. Tighten the bolts to 52 inch lbs. (6 Nm).
- Spark plug wires and clamps to the left cylinder head cover
- Engine wire clamp to the wire bracket on the delivery pipe
- EVAP hose to the VSV
- Transmission oil dipstick
- Right cylinder head cover and bolts. Tighten the bolts to 52 inch lbs. (6 Nm).
- Spark plug wires and clamps to the right cylinder head cover
- Throttle body to the air intake chamber. Install the 2 bolts and 2 nuts and tighten to 13 ft. lbs. (18 Nm).
- Timing belt rear plates by installing the bolts. Tighten the bolts to 66 inch lbs. (8 Nm).
- No. 2 ignition coil

18. Align the knock pin on the right side camshaft with the knock pin of the timing pulley. Slide on the timing pulley with the right side mark facing forward. Tighten the bolt to 80 ft. lbs. (108 Nm).

19. Align the knock pin on the left side camshaft with the knock pin of the timing pulley. Slide on the timing pulley with the left side mark facing forward. Tighten the bolt to 80 ft. lbs. (108 Nm).

20. Turn the crankshaft pulley and align its groove with the **0** timing mark on the timing belt cover.

21. Turn each camshaft timing pulley and align the timing marks of the pulley with the timing belt rear plate.

22. Attach the timing belt to the left side camshaft timing pulley:

23. Using the proper tool, slightly turn the left side timing pulley clockwise. Align the installation mark of the timing belt with the timing mark of the camshaft timing pulley and hang the timing belt on the left side camshaft pulley.

24. Align the timing marks of the left side pulley and the timing belt rear plate.

25. Check that the timing belt has tension between crankshaft timing pulley and the left side camshaft pulley.

26. Install the timing belt to the right side camshaft timing pulley:

27. Using the proper tool, slightly turn the right side timing pulley clockwise. Align the installation mark of the timing belt with the timing mark of the camshaft timing pulley and hang the timing belt on the right side camshaft pulley.

28. Align the timing marks of the right side camshaft pulley and the timing belt rear plate.

29. Check that the timing belt has tension between crankshaft timing pulley and the right side camshaft pulley.

30. The timing belt tensioner must be set prior to installation. The tensioner can be set by:

    a. Place a plate washer between the tensioner and a block. Using a press, press in the pushrod using 220–225 lbs. (100–102 kg.) of pressure.

    b. Align the holes of the pushrod and housing, pass a 0.05 in. (1.27mm) rod through the holes to keep the setting position of the pushrod.

    c. Release the press and install the dust boot to the tensioner.

    d. Loosely install the tensioner. Evenly and alternately tighten the bolts to 20 ft. lbs. (26 Nm). Remove the tool from the tensioner.

    e. Turn the crankshaft pulley 2 complete revolutions from TDC to TDC. Always turn the crankshaft clockwise. Check that all belt and pulley marks align with their reference marks. If any mark is out of perfect alignment, the timing belt must be removed and reinstalled.

31. Install or connect the following:
- Remaining components
- V bank cover.
- Coolant
- Battery and battery tray
- Battery cables, positive cable first
- Air cleaner inlet
- Battery clamp cover
- Engine undercover
- Oil pan protector

Valve Lash

ADJUSTMENT

3.0L (1MZ-FE) Engine

1. Before servicing the vehicle, refer to the precautions in the beginning of this section.

➡ **Adjust the valve clearance when the engine is cold.**

2. Remove or disconnect the following:
- Negative battery cable
- Accelerator/throttle cable from the throttle linkage
- Air intake chamber

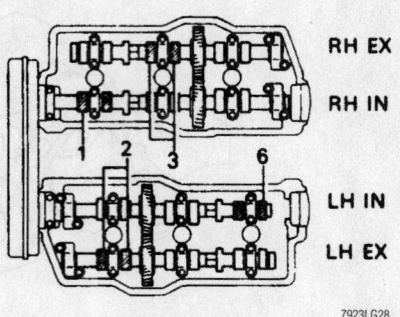

Adjust these valves FIRST—3.0L (1MZ-FE) engine

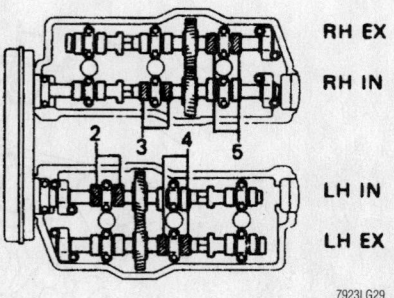

Adjust these valves SECOND—3.0L (1MZ-FE) engine

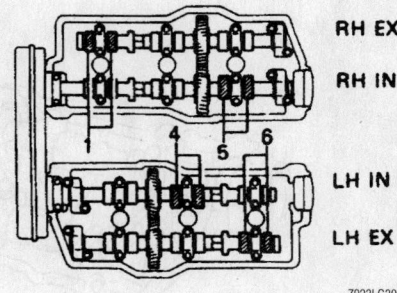

Adjust these valves THIRD—3.0L (1MZ-FE) engine

- Cylinder head covers

3. Turn the crankshaft pulley and align it's groove with the timing mark **0** of the No. 1 timing cover.

4. Check that the valve lash adjusters on the No. 1 intake are loose and the exhaust are tight. If not, turn the crankshaft on complete revolution (360 degrees).

5. Measure the clearance between the valve lash adjuster and the camshaft. Record the measurements on valves No. 1, 2, 3 and 6.

6. The intake valve clearance cold is 0.006–0.010 in. (0.15–0.25mm).

7. The exhaust valve clearance cold is 0.010–0.014 in. (0.25–0.35mm).

8. Turn the crankshaft ⅔ of a revolution

(240 degrees) and check the clearance on valves No. 2, 3, 4 and 5 and record.

9. Turn the crankshaft another ⅔ of a revolution and check valves; No. 1, 4, 5 and 6 and record.

10. Remove or disconnect the following:

- Adjusting shim and turn the crankshaft to position the cam lobe of the camshaft on the adjusting valve upward. Press down the valve lash adjuster with the proper tool and place the proper tool between the camshaft and the valve lash adjuster. Remove the tool.
- Adjusting shim with the proper tool.

11. Determine the thickness of the replacement shim as follows:

- T: Thickness of the used shim
- A: Measured valve clearance
- N: Thickness of new shim
- Intake: N = T + (A—0.006–0.010 in. (0.15–0.25mm))
- Exhaust: N = T + (A—0.010–0.014 in. (0.25–0.35mm))

12. Install the specified valve shim on the valve lash adjuster

13. Recheck the valve clearance.

14. Install the cylinder head covers and intake chamber.

15. Connect the negative battery cable.

3.0L (2JZ-GE) Engine

➡**Adjust the valve clearance when the engine is cold.**

1. Before servicing the vehicle, refer to the precautions in the beginning of this section.

2. Remove or disconnect the following:

- Negative battery cable
- Accelerator/throttle cable from the throttle linkage
- Cylinder head covers

3. Turn the crankshaft pulley and align it's groove with the timing mark **0** of the No. 1 timing cover.

4. Check that the timing marks on the camshaft sprockets are in alignment with the marks on the No. 4 timing cover. If not, turn the crankshaft 1 complete revolution (360 degrees).

5. Uniformly tighten the camshaft bearing cap bolts in several passes, in the sequence, to 14 ft. lbs. (20 Nm).

6. Measure the clearance between the valve lash adjuster and the camshaft. Record the measurements on valves No. 1, 4 and 5.

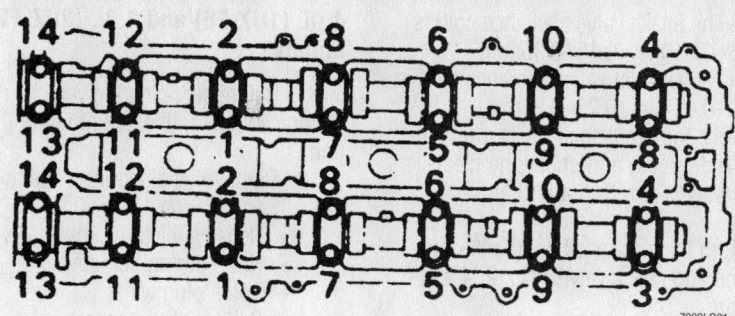

Camshaft bearing cap bolt tightening sequence—3.0L (2JZ-GE) engine

7923LG31

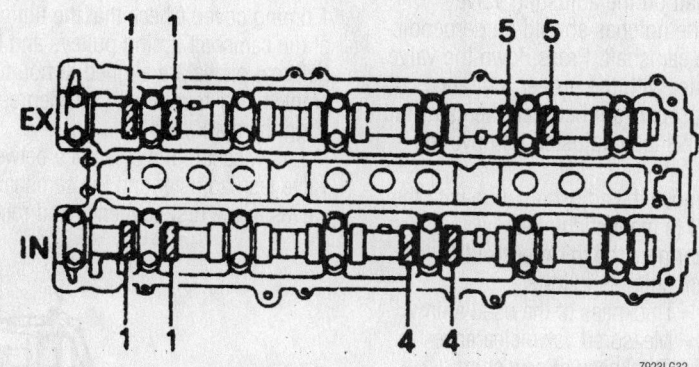

Adjust these valves FIRST—3.0L (2JZ-GE) engine

7923LG32

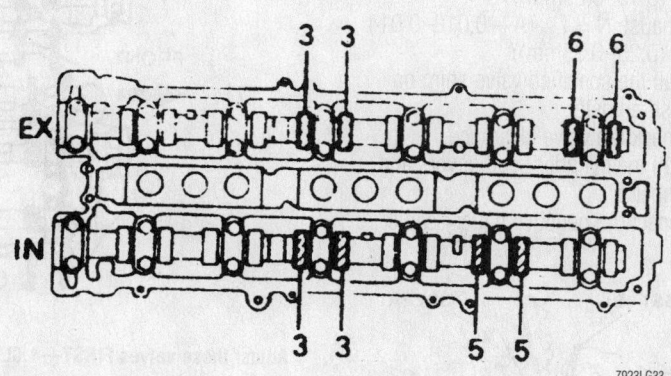

Adjust these valves THIRD—3.0L (2JZ-GE) engine

7923LG33

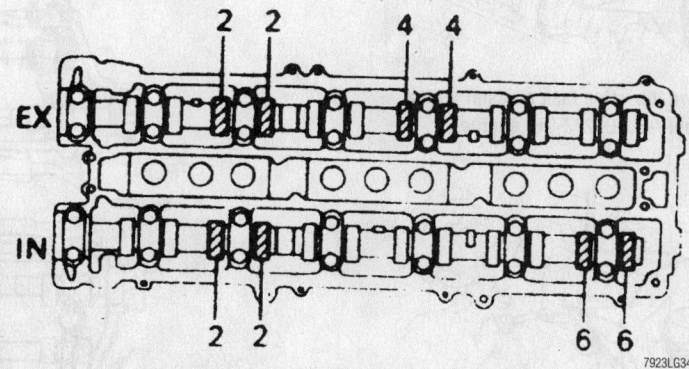

Adjust these valves THIRD—3.0L (2JZ-GE) engine

7923LG34

For Tire, Wheel and Ball Joint specifications, see Section 1 of this manual

a. The intake valve clearance cold is 0.006–0.010 in. (0.15–0.25mm).

b. The exhaust valve clearance cold is 0.010–0.014 in. (0.25–0.35mm).

7. Turn the crankshaft ⅔ of a revolution (240 degrees) and check the clearance on valves No. 3, 5 and 6 and record.

8. Turn the crankshaft another ⅔ of a revolution and check valves: No. 2, 4 and 6 and record.

9. Remove the adjusting shim and turn the crankshaft to position the cam lobe of the camshaft on the adjusting valve upward. The notches should be perpendicular to the camshaft. Press down the valve lash adjuster with the proper tool and place the proper tool between the camshaft and the valve lash adjuster. Remove the tool.

10. Remove the adjusting shim with the proper tool (a magnetic finger).

11. Determine the thickness of the replacement shim as follows:

- T = Thickness of the used shim
- A = Measured valve clearance
- N = Thickness of new shim
- Intake: $N = T + (A - 0.006-0.010$ in. $(0.15-0.25mm))$
- Exhaust: $N = T + (A - 0.010-0.014$ in. $(0.25-0.35mm))$

12. Install the specified valve shim on the valve lash adjuster.

13. Recheck the valve clearance.

14. Install the cylinder head covers and intake chamber.

15. Connect the negative battery cable.

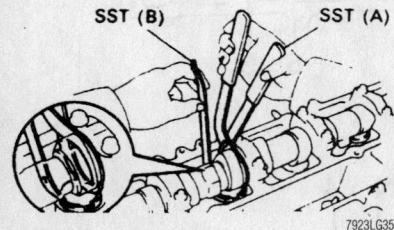

Press down the valve lash adjuster with a special tool—3.0L (2JZ-GE) engine

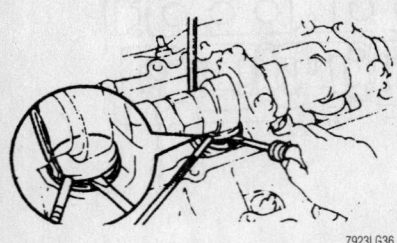

Removing the adjusting shim—3.0L (2JZ-GE) engine

4.0L (1UZ-FE) and 4.3L (3UZ-FE) Engines

1. Before servicing the vehicle, refer to the precautions in the beginning of this section.

2. Remove or disconnect the following:

- Negative battery cable
- No. 3 timing belt covers
- Spark plug wires
- Cylinder head covers.

3. Turn the crankshaft pulley and align its groove with the timing mark **0** of the No. 1 timing cover. Check that the timing marks of the camshaft timing pulleys and timing belt rear plates are aligned. If not, turn the crankshaft 1 revolution (360 degrees) and align the mark.

4. Measure the clearance between the valve lash adjuster and the camshaft on the valves in the first sequence and record.

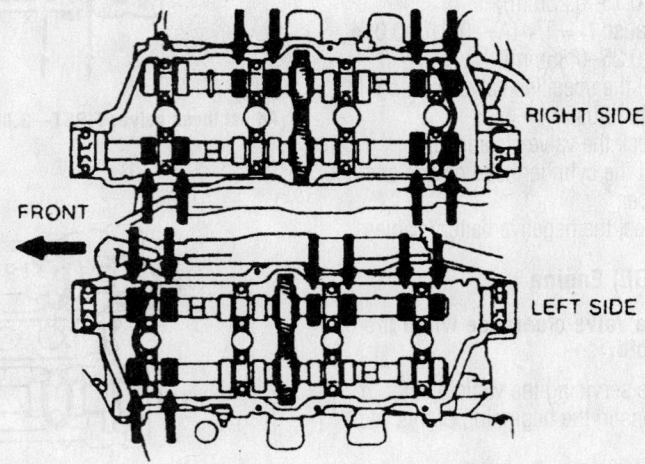

Adjust these valves FIRST—4.0L (1UZ-FE) and 4.3L (3UZ-FE) Engines

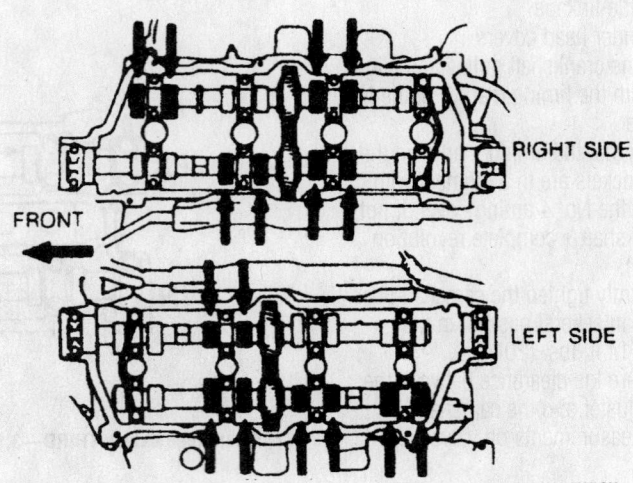

Adjust these valves SECOND—4.0L (1UZ-FE) and 4.3L (3UZ-FE) Engines

a. The intake valve clearance cold is 0.006–0.010 in. (0.15–0.25mm).

b. The exhaust valve clearance cold is 0.010–0.014 in. (0.25–0.35mm).

5. Turn the crankshaft 1 full revolution (360 degrees) and align the mark.

6. Measure the clearance between the valve lash adjuster and the camshaft on the valves in the second sequence and record.

7. Remove the adjusting shim and turn the crankshaft to position the cam lobe of the camshaft on the adjusting valve upward. Position the hole in the shim toward the outside of the cylinder head. Press down the valve lash adjuster with the proper tool and place the proper tool between the camshaft and the valve lash adjuster. Remove the tool.

8. Remove the adjusting shim with the proper tool.

9. Determine the thickness of the replacement shim as follows:

- T = Thickness of the used shim
- A = Measured valve clearance
- N = Thickness of new shim
- Intake: N = T + (A—0.006–0.010 in. (0.15–0.25mm))
- Exhaust: N = T + (A—0.010–0.014 in. (0.25–0.35mm))

10. Recheck the valve clearance. Install the cylinder head covers.

11. Connect the spark plug wires and install the No. 3 timing belt covers.

12. Connect the negative battery cable.

Starter Motor

REMOVAL & INSTALLATION

3.0L (1MZ-FE) Engine

1. Before servicing the vehicle, refer to the precautions in the beginning of this section.

2. Remove or disconnect the following:
   - Negative cable
   - Automatic transmission shift control cable
   - Engine wire
   - Starter connector
   - Nut, and disconnect the starter wire
   - 2 bolts, automatic transmission shift control cable clamp and starter

To install:

3. Installation is the reversal of the removal process.

4. Torque the starter bolts to 27 ft. lbs. (37 Nm).

3.0L (2JZ-GE) Engine

1. Before servicing the vehicle, refer to the precautions in the beginning of this section.

2. Remove or disconnect the following:
   - Negative cable
   - Starter connector
   - Nut, and disconnect the starter wire
   - 2 bolts and starter

To install:

3. Installation is the reversal of the removal procedure.

4. Tighten the starter bolts to 27 ft. lbs. (37 Nm).

4.0L (1UZ-FE) and 4.3L (3UZ-FE) Engines

1. Before servicing the vehicle, refer to the precautions in the beginning of this section.

2. Remove or disconnect the following:

   - V-bank cover
   - Accelerator cable
   - Intake air connector
   - Throttle Body
   - Intake manifold assembly
   - Rear water bypass joint
   - Water bypass pipe
   - 2 bolts
   - Water bypass pipe from the water pump
   - Wire clamp from the bracket on the water bypass pipe
   - O-ring from the water bypass pipe
   - Water bypass pipe bracket from the water bypass pipe
   - Starter
   - 2 bolts holding the starter to the cylinder block
   - Starter from the cylinder block
   - Starter connector
   - Nut, and disconnect the starter wire
   - Starter

To install:

3. Install or connect the following:
   - Wire clamp to the wire bracket with the bolt. Tighten to 87 inch lbs.
   - Starter wire with the nut. Tighten to 87 inch lbs.
   - Starter connector
   - Starter with the 2 bolts. Torque the bolts to 29 ft. lbs. (39 Nm).
   - Water bypass pipe bracket to the water bypass pipe
   - O-ring to the water bypass pipe
   - Water bypass pipe
   - Wire clamp to the bracket on the water bypass pipe
   - Water bypass pipe bolts. Torque the bolts to 13 ft. lbs. (18 Nm).
   - Rear water bypass joint

- Intake manifold assembly
- Throttle body
- Intake air connector
- Accelerator cable
- V-bank cover

Oil Pan

REMOVAL & INSTALLATION

3.0L (1MZ-FE) Engine

1. Before servicing the vehicle, refer to the precautions in the beginning of this section.

2. Drain the engine oil.

3. Remove or disconnect the following:
   - Negative battery cable from the battery
   - Right front wheel
   - Fender apron seal
   - Engine undercover
   - Front exhaust pipe
   - Front exhaust pipe bracket from the No. 1 oil pan
   - Flywheel housing undercover
   - 10 bolts and 2 nuts to the No. 2 oil pan

4. Insert a blade between the No. 1 and No. 2 oil pans. Tap the tool sideways to break the seal and remove the pan. Clean the surfaces of the oil pans.

   - Oil strainer and gasket from the engine by removing the 3 nuts.
   - No. 1 oil pan as follows. Make a note of the position of the each bolt. When replacing the bolts into the oil pan, place each bolt in the position from which it was removed.
   - Baffle plate from the No. 1 oil pan.

To install:

5. Clean all mating surfaces of the oil

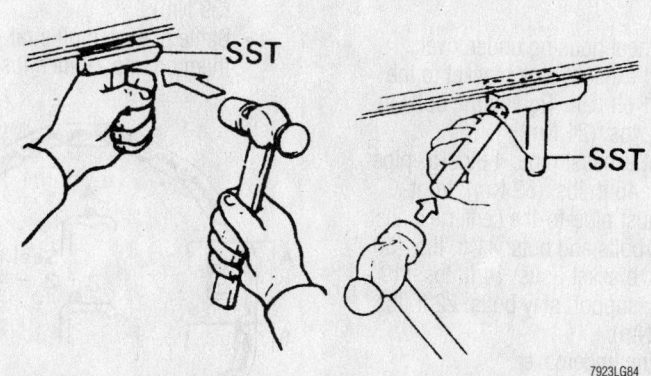

Use the special tool to break the seal and remove the oil pan—3.0L (1MZ-FE) engine

For Wheel Alignment specifications, see Section 1 of this manual

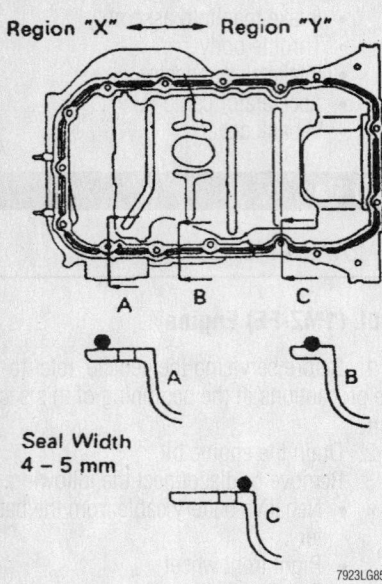

Apply sealant as shown to the No. 1 (upper) oil pan—3.0L (1MZ-FE) engine

pans. Using a non-residue solvent, clean both sealing surfaces to the oil pan.

6. Install or connect the following:
- Baffle plate to the No. 1 oil pan and tighten to 69 inch lbs. (8 Nm).
- No. 1 oil pan. Apply RTV sealant to the oil pan and engine block. Uniformly tighten the bolts and nuts in several passes to: 10mm: 69 inch lbs. (8 Nm); 12mm: 14 ft. lbs. (20 Nm); 14mm: 27 ft. lbs. (37 Nm)
- Flywheel housing undercover with the 2 bolts. Tighten the bolts to 69 inch lbs. (8 Nm).
- Oil strainer with the 3 nuts. Tighten the nuts to 69 inch lbs. (8 Nm).
- No. 2 oil pan. Apply RTV sealant to the oil pan and engine block. Uniformly tighten the bolts and nuts in several passes, to 69 inch lbs. (8 Nm).
- Flywheel housing undercover
- Front exhaust pipe bracket to the No. 1 oil pan. Tighten the bolts to 15 ft. lbs. (21 Nm).
- Front exhaust pipe. 4 pipe-to-pipe nuts: 46 ft. lbs. (62 Nm); front exhaust pipe-to-the center exhaust pipe bolts and nuts: 41 ft. lbs. (56 Nm); bracket bolts: 14 ft. lbs. (19 Nm); support stay bolts: 22 ft. lbs. (30 Nm).
- Engine undercover
- Right fender apron seal
- The right front wheel and lower the vehicle
- Engine with oil

3.0L (2JZ-GE) Engine

➡The No. 1 oil pan can not be removed with the engine in the vehicle. The engine/transmission assembly must be removed. The manufacturer does not provide any on vehicle information for the No. 2 oil pan removal and installation. If only the No. 2 oil pan is being serviced, the engine/transmission assembly can remain in the vehicle.

1. Before servicing the vehicle, refer to the precautions in the beginning of this section.
2. Remove or disconnect the following:
- Engine/transmission assembly
- Timing belt
- Idler pulley
- Crankshaft timing pulley
- Oil dipstick and guide
- Oil sensor lead
- 4 attaching bolts and lift off the oil level sensor. Be careful not to drop this sensor.
- 14 bolts (16 bolts for GS300) and 2 nuts and pry off the lower (No. 2) oil pan. Be careful not to damage the No. 1 pan while performing this procedure.
- Bolt and 2 nuts and drop down the oil strainer and gasket.
- 5 bolts and 2 nuts and drop down the baffle plate.
- 22 bolts and the carefully pry off the upper (No. 1) oil pan.
- O-ring from the cylinder block.

To install:
3. Install or connect the following:
- New O-ring in the block and scrape off any old sealant
- A ⅛ inch (3–4mm) bead of RTV sealant to the pan mating surface
- Upper pan. 12mm bolts: 15 ft. lbs. (21 Nm); 14mm bolts to 29 ft. lbs. (39 Nm)
- Baffle plate and oil strainer. Tighten them both to 78 inch lbs. (9 Nm).

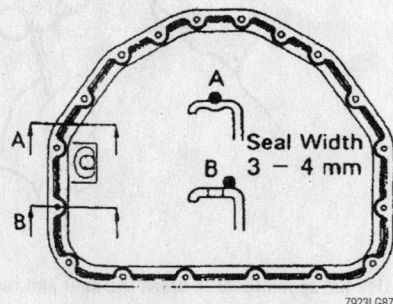

Lower oil pan sealant application—3.0L (2JZ-GE) engine

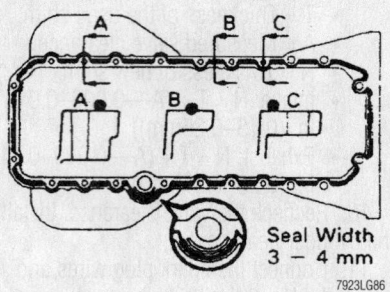

Upper oil pan sealant application—3.0L (2JZ-GE) engine

- Lower pan in the same manner as the upper pan and tighten the bolts to 78 inch lbs. (9 Nm).
- Oil level sensor and tighten it to 48 inch lbs. (5 Nm).
- Oil dipstick and guide
- Timing pulleys and belt
- Transmission to the engine
- Engine and transmission
- All fluids

4.0L (1UZ-FE) and 4.3L (3UZ-FE) Engines

LS400, LS430

1. Before servicing the vehicle, refer to the precautions in the beginning of this section.
2. Remove or disconnect the following:
- Engine/transmission assembly
- Remove the timing belt
- Idler pulleys
- Crankshaft timing pulley
- Oil dipstick and guide
- Oil level sensor lead
- 4 bolts and lift off the oil level sensor. Be careful not to drop this sensor.
- Oil filter and the bracket assembly by removing the stud bolt and 2 nuts
- Engine Crankshaft Position (CKP) sensor connector
- Sensor by removing the bolt
- 12 bolts and 2 nuts to the No. 2 oil pan. Use a gasket cutting tool to separate the No. 2 (lower) oil pan. Be careful not to damage the No. 1 pan while performing this procedure.
- 2 bolts and 3 nuts and drop down the baffle plate
- Oil strainer by removing the bolts and nuts
- Bolts, then carefully pry off the No. 1 oil pan. There are slots for inserting the prybar.

To install:

3. Install or connect the following:

- No. 1 pan. Apply a ⅛ inch (3–4mm) bead sealant to the pan mating surface. Bolts: 10mm: 66 inch lbs. (8 Nm); 12mm: 21 ft. lbs. (28 Nm)
- Oil strainer. Bolts and nuts: 66 inch lbs. (8 Nm).
- Baffle plate. Bolts and nuts: 66 inch lbs. (8 Nm).
- No. 2 pan in the same manner as the No. 1 oil pan and tighten the bolts to 66 inch lbs. (8 Nm). Be sure the bolts are 14mm in length.
- CKP sensor. Tighten the bolt to 56 inch lbs. (6 Nm).
- New O-ring in position on the oil filter bracket
- Bracket and tighten the bolt and nuts to 13 ft. lbs. (18 Nm).
- Wiring to the pressure switch.
- Oil level sensor and tighten the 4 bolts to 48 inch lbs. (5 Nm). Use a new gasket.
- Dipstick and guide
- Timing belt pulleys and the timing belt components
- Transaxle to the engine
- Engine and transaxle
- All fluids

SC400, GS400, GS430

➡The No. 1 oil pan cannot be removed with the engine in the vehicle. The engine and transmission must be removed as a unit, then separated. It may be possible to remove the No. 2 oil pan from the vehicle while the engine is still in the vehicle.

1. Before servicing the vehicle, refer to the precautions in the beginning of this section.

2. Remove or disconnect the following:

- Engine/transmission assembly
- Oil dipstick and guide
- 12 bolts and 2 nuts. Use a gasket-cutting tool to separate the No. 2 (lower) oil pan. Be careful not to damage the No. 1 pan while performing this procedure.
- 6 bolts and 2 nuts; remove the baffle plate
- 16 bolts, then carefully pry off the No. 1 oil pan

➡There are slots for inserting the pry-bar.

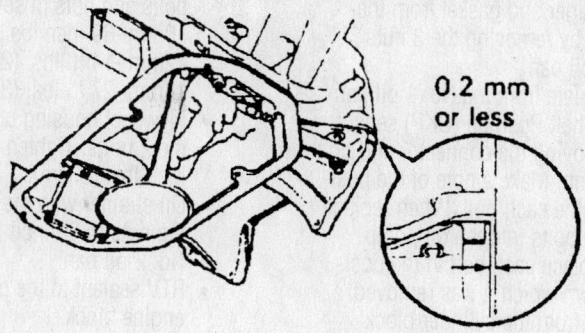

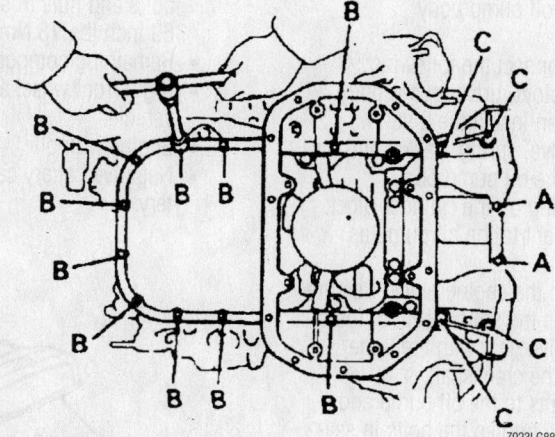

SC400 and GS400 oil pan bolt locations—(A) 0.78 in (20mm), (B) 1.38 in. (35mm), (C) 0.78 in (20mm) with 12mm bolt heads

To install:

3. Install or connect the following:

- No. 1 pan. Apply a ⅛ inch (3–4mm) bead on RTV sealant to the pan mating surface. Bolts: 12mm: 69 inch lbs. (8mm); 14mm: 20 ft. lbs. (28 Nm)
- Baffle plate. Bolts and nuts: 69 inch lbs. (8 Nm).
- RTV sealant to the pan mating surface
- No. 2 oil pan. Bolts: 69 inch lbs. (8 Nm)
- Dipstick and guide
- Engine/transaxle assembly
- All fluids

Oil Pump

REMOVAL & INSTALLATION

3.0L (1MZ-FE) Engine

1. Before servicing the vehicle, refer to the precautions in the beginning of this section.

2. Remove or disconnect the following:

- Negative battery cable from the battery
- Right front wheel
- Fender apron seal
- Engine undercover
- Engine oil
- Front exhaust pipe
- Front exhaust pipe bracket from the No. 1 oil pan
- Alternator drive belt from the engine
- Air conditioning compressor from the engine, without disconnecting the compressor lines
- Power steering pump drive belt and adjusting strut
- Timing belt from the engine
- Timing belt pulleys
- Rear timing belt cover from the engine by removing the wire clamps and 6 bolts
- Air conditioning compressor housing bracket by removing the 3 bolts.
- 10 bolts and 2 nuts to the No. 2 oil pan
- No. 2 oil pan from the engine

- Oil strainer and gasket from the engine by removing the 3 nuts
- No. 1 oil pan
- Baffle plate from the No. 1 oil pan
- Crankshaft Position (CKP) sensor by removing the connector and bolt
- Oil pump. Make a note of the position of the each bolt. When replacing the bolts into the oil pump body, place each bolt in the position from which it was removed.
- O-ring from the cylinder block
- Plug, gasket, spring and relief valve from the oil pump body

To install:

3. Install or connect the following:
- Driven rotors, drive, pump body cover, then install the 9 screws
- Relief valve, spring, gasket and the plug to the oil pump body
- New O-ring on the cylinder block
- RTV sealant to the oil pump as shown
- Pump on the engine block. Be sure to engage the spline teeth of the oil pump drive gear with the large teeth of the crankshaft.
- The 9 bolts to the oil pump and uniformly tighten the bolts in several passes. Tighten the bolts to: 10mm: 69 inch lbs. (8 Nm); 12mm: 14 ft. lbs. (20 Nm)
- CKP sensor and bolt. Tighten the bolt to 69 inch lbs. (8 Nm).
- Baffle plate to the No. 1 oil pan and tighten to 69 inch lbs. (8 Nm).
- No. 1 oil pan Uniformly tighten the

bolts and nuts in several passes: 10mm–69 inch lbs. (8 Nm); 12mm–14 ft. lbs. (20 Nm); 14mm–27 ft. lbs. (37 Nm)
- Flywheel housing undercover with the 2 bolts. Tighten the bolts to 69 inch lbs. (8 Nm).
- Oil strainer with the 3 nuts. Tighten the nuts to 69 inch lbs. (8 Nm).
- No. 2 oil pan
- RTV sealant to the oil pan and engine block
- No. 2 oil pan. Uniformly tighten the bolts and nuts in several passes to 69 inch lbs. (8 Nm).
- Remaining components
- Right front wheel and lower the vehicle.
- Engine with oil
- Negative battery cable to the battery.

3.0L (2JZ-GE) Engine

1. Before servicing the vehicle, refer to the precautions in the beginning of this section.

2. Remove or disconnect the following:
- Engine and transmission
- Timing belt
- Idler pulley
- Crankshaft timing pulley
- Oil dipstick and tube
- Oil level sensor
- No. 2 (lower) oil pan
- Oil strainer by removing the bolt and 2 nuts
- Oil baffle plate by removing the 6 bolts
- No. 1 (upper) oil pan by removing the 22 bolts. Take note of bolt size and placement for correct re-installation.

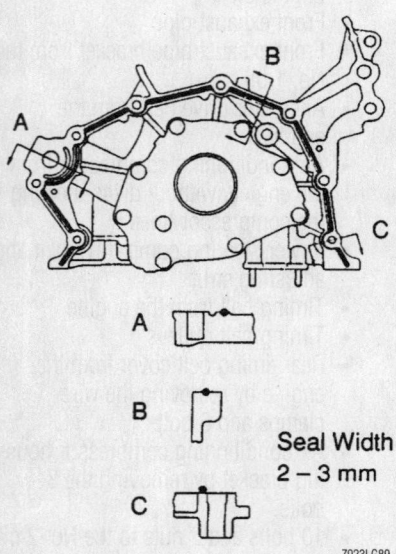

Apply sealant to the mounting surface of the oil pump in the areas shown—3.0L (1MZ-FE) engine

Seal Width 2 – 3 mm

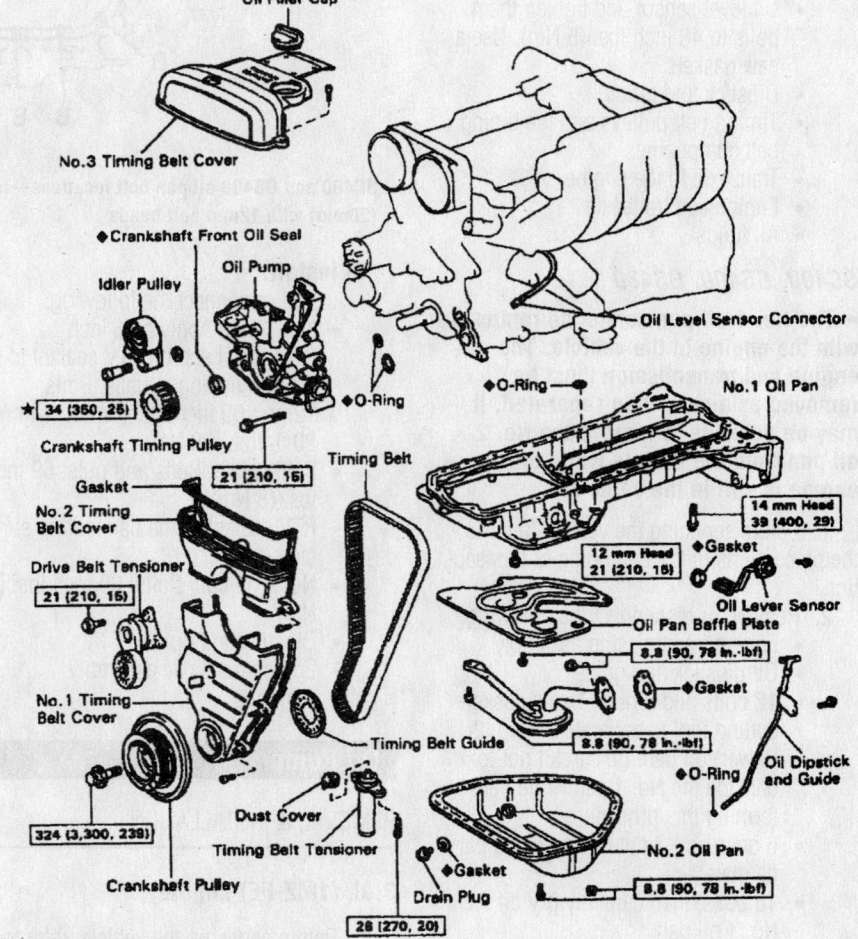

N·m (kgf·cm, ft·lbf) : Specified torque
♦ Non-reusable part
★ Precoated part

Exploded view of the oil pump and related component mountings—3.0L (2JZ-GE) engine

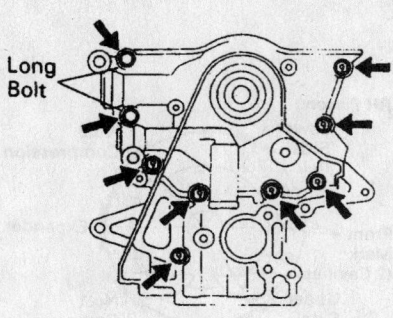

Oil pump mounting bolt installation locations—3.0L (2JZ-GE) engine

- 9 mounting bolts to the oil pump body. Carefully drive the pump off the cylinder block using a brass drift.
- 2 O-rings

To install:

3. Install or connect the following:
- 2 new O-rings in the cylinder block
- A ⅛ inch (3–4mm) bead of RTV sealant around the pump mating surface, taking great care around the oil passages.
- Pump and tighten the bolts to 15 ft. lbs. (21 Nm).
- A new O-ring on the block
- RTV sealant around the No. 1 oil pan
- No. 1 oil pan. Bolts: 12mm–15 ft. lbs. (21 Nm); 14mm–29 ft. lbs. (39 Nm)
- Oil baffle plate and tighten the nuts and bolts to 78 inch lbs. (9 Nm).
- Oil strainer and tighten the nuts and bolts to 78 inch lbs. (9 Nm).
- RTV sealant around the No. 2 oil pan
- No. 2 oil pan and tighten the bolts to 78 inch lbs. (9 Nm).
- Oil lever sensor with a new gasket and tighten the bolts to 48 inch lbs. (6 Nm).
- Oil dipstick with a new O-ring.
- Remaining components
- All fluids
- Negative battery cable.

4.0L (1UZ-FE) and 4.3L (3UZ-FE) Engines

➡The oil pump cannot be removed with the engine in the vehicle. The engine and transmission must be removed as a unit, then separated.

1. Before servicing the vehicle, refer to the precautions in the beginning of this section.
2. Remove or disconnect the following:
- Engine/transmission assembly
- Timing belt
- Idler pulleys
- Crankshaft timing pulley
- Oil dipstick and guide
- Oil level sensor lead
- 4 bolts and lift off the oil level sensor. Be careful not to drop this sensor.
- Main Oxygen (O_2) sensor bracket, if necessary
- Oil filter and filter bracket assembly by removing the stud bolt and 2 nuts.
- Engine Crankshaft Position (CKP) sensor. Remove the sensor by removing the bolt.
- 12 bolts and 2 nuts from the No. 2 oil pan
- No. 2 (lower) oil pan. Use a gasket-cutting tool
- 2 bolts and 3 nuts and drop down the baffle plate
- Oil strainer
- No. 1 oil pan. There are slots for inserting the prybar.
- 8 bolts holding the oil pump to the engine.

➡**Make certain to observe bolt position during removal. The bolts are different lengths and sizes. Record their position for proper reassembly.**

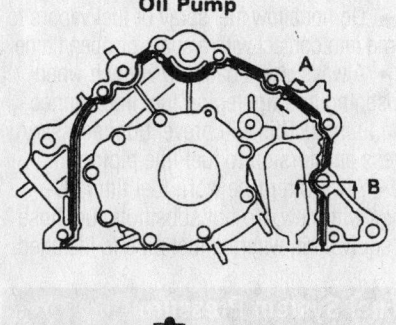

Oil Pump

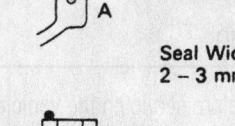
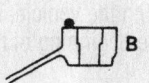

Seal Width
2 – 3 mm

Apply sealant to the oil pump and the No. 1 oil pan, as shown, before installing the oil pump—4.0L (1UZ-FE) and 4.3L (3UZ-FE) Engines

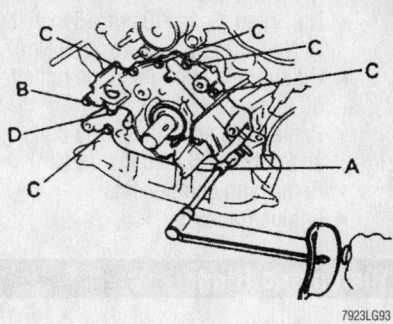

4.0L engine oil pump mounting bolt locations, according to bolt lengths—(A) 1.97 in. (50mm), (B) 4.17 in. (106mm), (C) 1.18 in. (30mm) and (D) 1.57 in. (40mm)

- Oil pump from the engine block
- O-ring from the block

To install:

➡**Prior to installing the oil pump, lubricate the gears with clean engine oil.**

3. Install or connect the following:
- A 2–3mm wide (0.08–0.12 in.) bead of RTV sealant to the oil pump
- New O-ring in position on the block
- Oil pump on the engine
- The 8 bolts in their correct locations. Tighten the bolts with 12mm heads to 12 ft. lbs. (16 Nm) and the bolts with 14mm heads to 22 ft. lbs. (30 Nm).
- A ⅛ inch (3–4mm) bead of RTV sealant to the pan mating surface.

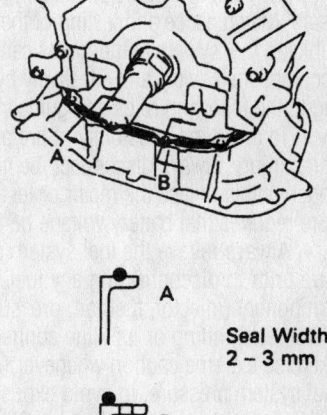

No.1 Oil Pan

Seal Width
2 – 3 mm

- No. 1 pan. Bolts–10mm: 66 inch lbs. (8 Nm); 12mm: 21 ft. lbs. (28 Nm)
- Oil strainer and tighten the bolts to 66 inch lbs. (8 Nm)
- Baffle plate and tighten the bolts and nuts to 66 inch lbs. (8 Nm)
- Remaining components
- Engine/transaxle

Piston and Ring

POSITIONING

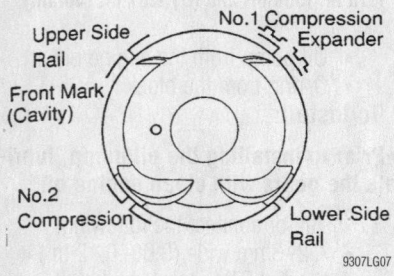

Piston ring positioning—3.0L (2JZ-GE) engine

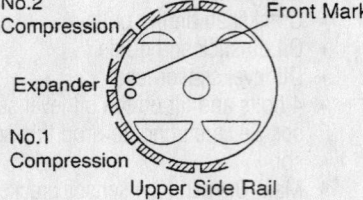

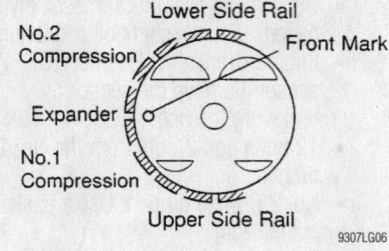

Piston ring positioning—3.0L (1MZE-FE) engine

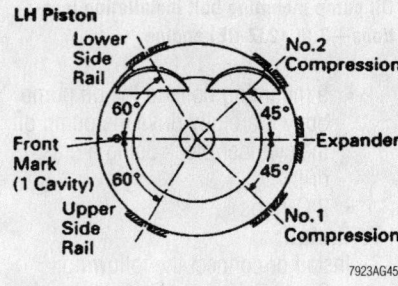

Piston ring positioning—4.0L (1UZ-FE) and 4.3L (3UZ-FE) engines

FUEL SYSTEM

Fuel System Service Precautions

Safety is the most important factor when performing not only fuel system maintenance but any type of maintenance. Failure to conduct maintenance and repairs in a safe manner may result in serious personal injury or death. Maintenance and testing of the vehicle's fuel system components can be accomplished safely and effectively by adhering to the following rules and guidelines.

- To avoid the possibility of fire and personal injury, always disconnect the negative battery cable unless the repair or test procedure requires that battery voltage be applied.
- Always relieve the fuel system pressure prior to disconnecting any fuel system component (injector, fuel rail, pressure regulator, etc.), fitting or fuel line connection. Exercise extreme caution whenever relieving fuel system pressure, to avoid exposing skin, face and eyes to fuel spray. Please be advised that fuel under pressure may penetrate the skin or any part of the body that it contacts.
- Always place a shop towel or cloth around the fitting or connection prior to loosening to absorb any excess fuel due to spillage. Ensure that all fuel spillage (should it occur) is quickly removed from engine surfaces. Ensure that all fuel soaked cloths or towels are deposited into a suitable waste container.
- Always keep a dry chemical (Class B) fire extinguisher near the work area.
- Do not allow fuel spray or fuel vapors to come into contact with a spark or open flame.
- Always use a back-up wrench when loosening and tightening fuel line connection fittings. This will prevent unnecessary stress and torsion to fuel line piping.
- Always replace worn fuel fitting O-rings with new. Do not substitute fuel hose or equivalent, where a fuel pipe is installed.

Fuel System Pressure

RELIEVING

1. Before servicing the vehicle, refer to the precautions in the beginning of this section.
2. Remove the fuse for the electronic fuel pump.
3. Start the engine until the engine stalls.
4. Disconnect the negative battery terminal.
5. Place a catch-pan under the joint to be disconnected. A large quantity of fuel may be released when the joint is opened.

6. Wear eye or full-face protection.
7. Place a shop towel over the area and slowly release the joint using a wrench of the correct size.
8. Allow the any fuel left in the line to bleed off slowly before fully disconnecting the joint.
9. Plug the opened lines immediately to prevent fuel spillage or the entry of dirt.
10. Dispose of the released fuel properly.
11. After connecting fuel lines, install the fuse for the fuel pump and start the engine.
12. Check for leaks and repair as needed.

Fuel Filter

REMOVAL & INSTALLATION

The fuel filter on the ES300 is located under the hood, on the driver's side, by the fenderwell. The SC300 fuel filter is located under the vehicle, on the driver's side, in front of the rear axle.

The fuel filter for the LS400 and SC400 is located under the vehicle on the left side before the rear axle. The fuel filter on the GS300 and GS400 is located under the vehicle, next to the left rear exhaust resonator.

1. Before servicing the vehicle, refer to

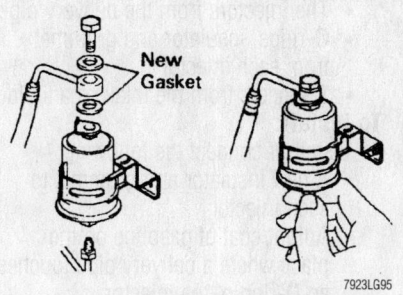

Exploded view of a typical fuel line connection at the filter

the precautions in the beginning of this section.

2. Remove or disconnect the following:
   • Negative battery cable. Wait at least 90 seconds before performing any other work.
   • On the GS300, he rear body protector

3. Slowly loosen the lower flare nut fitting until all the pressure is relieved and all the fuel is collected.

4. Loosen the union bolt on the upper portion of the filter and remove the banjo fitting and 2 metal gaskets. Discard the gaskets.

5. Loosen the fuel filter bracket bolt, remove the fuel line with the flared nut from the filter and pull the filter from the mounting bracket.

To install:

6. Install or connect the following:
   • A new fuel filter to the vehicle and tighten the bracket bolt.
   • Banjo fitting with a new metal gasket on each side
   • Union bolt. Tighten the union bolt to 22 ft. lbs. (30 Nm).
   • Flare nut to the lower connection. Tighten the flare nut to 22 ft. lbs. (30 Nm).

7. On the GS300, install the body protector.

8. Lower the vehicle if raised.

9. Remove the drain pan and/or rags and connect the negative battery cable.

10. Start the engine and visually inspect the upper and lower connections for leaks.

Fuel Pump

REMOVAL & INSTALLATION

ES300 and IS300

1. Before servicing the vehicle, refer to the precautions in the beginning of this section.

2. Relieve the fuel system pressure.

3. With the ignition switch in the **LOCK** position, disconnect the negative battery terminal.

4. Remove or disconnect the following:
   • Rear seat cushion
   • Fuel pump connector
   • Floor service hole cover

➡ **Do not lift the fuel pump assembly up using the wiring harness.**

   • Fuel filler cap
   • Fuel outlet pipe and the return hose from the pump bracket
   • 8 screws and lift out the pump/bracket assembly with gasket
   • Fuel pump lead wire
   • Lower end of the pump off the bracket
   • Fuel hose from the pump and remove the pump
   • Rubber cushion from the pump

To install:

5. Install or connect the following:
   • Filter and rubber cushion on the new pump
   • Pump on the bracket
   • Fuel hose and the wire connector on the pump
   • Pump, using a new gasket and tighten the 8 screws to 35 inch lbs. (4 Nm).
   • Fuel pipe and return hose to the

pump and tighten the bolts to 22 ft. lbs. (29 Nm).
   • Wire
   • Service cover, and replace the rear seat.
   • Negative battery cable

6. Start the engine and check for leaks.

Except ES300

1. Before servicing the vehicle, refer to the precautions in the beginning of this section.

2. Remove or disconnect the following:
   • Negative battery cable. Wait at least 90 seconds before performing any other work.
   • Trunk floor mat
   • Trunk trim cover
   • Fuel pump electrical connector
   • Rear seat bottom and seat back
   • Partition cover
   • Mounting bolts and remove the fuel pump set plate.
   • 3 nuts and disconnect the fuel pump bracket from the tank.
   • Fuel hose from the bracket
   • Pump, bracket and set plate as an assembly

To install:

3. Install or connect the following:
   • A new gasket on the set plate
   • Fuel hose to the pump and bracket

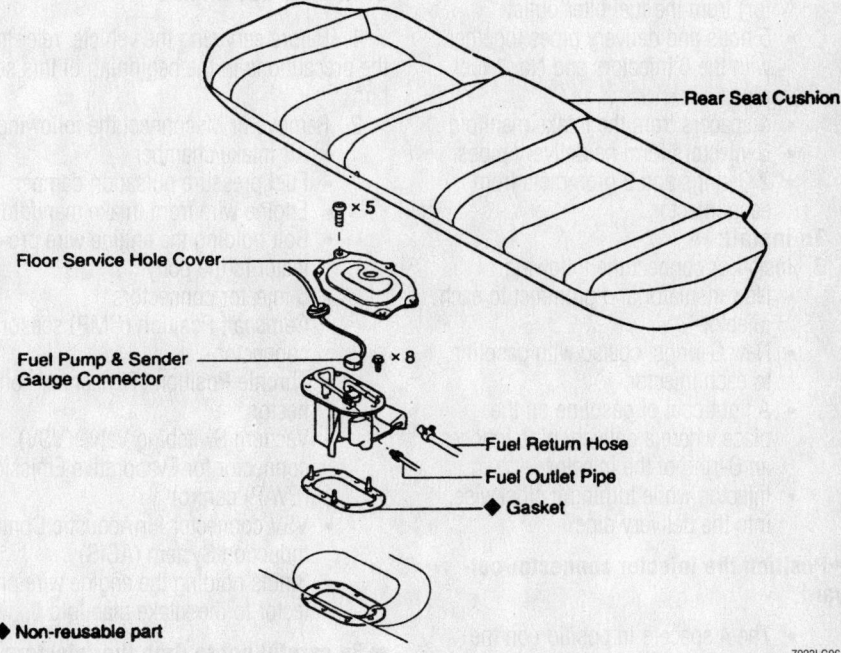

Rear Seat Cushion

Floor Service Hole Cover

× 5

Fuel Pump & Sender Gauge Connector

× 8

Fuel Return Hose

Fuel Outlet Pipe

◆ **Gasket**

◆ **Non-reusable part**

7923LG96

Exploded view of the fuel pump assembly—ES300

- Pump and bracket assembly with the 3 nuts; tighten the nuts to 48 inch lbs. (5 Nm). Install the set plate and tighten the bolts to 26 inch lbs. (3 Nm).
- Panel partition
- Rear seat cushion and back
- Fuel pump electrical connector
- Trim panel
- Spare tire and the trunk floor mat
- Negative battery cable

4. Start the engine; check the fuel system for leaks

Fuel Injector

REMOVAL & INSTALLATION

3.0L (1MZ-FE) Engine

1. Before servicing the vehicle, refer to the precautions in the beginning of this section.
2. Remove or disconnect the following:
- 3 cap nuts, using a 5mm hexagon wrench. Loosen the V-bank cover fastener counterclockwise.
- V-Bank cover
- Air cleaner hose with resonator
- air intake chamber assembly
- Injector connectors
- Air assist hoses and pipe
- No. 1 fuel pipe and remove the fuel hose clamp
- No. 1 fuel pipe (fuel tube connector) from the fuel filter outlet
- 5 bolts and delivery pipes together with the 6 injectors and No. 1 fuel pipe.
- 4 spacers from the intake manifold
- 6 injectors form he delivery pipes.
- 2 O-rings and 2 grommets from each injector

To install:
3. Install or connect the following:
- New insulator and grommet to each injector
- New O-rings, coated with gasoline, to each injector
- A light coat of gasoline on the place where a delivery pipe touches an O-ring of the injector
- Injector, while turning it clockwise, into the delivery pipe

➡️**Position the injector connector outward.**

- The 4 spacers in position on the intake manifold
- A light coat of gasoline on the place where the intake manifold touches an O-ring

- The delivery pipe and fuel pipe together with the 6 injectors in position on the intake manifold. Position the injector connector outward.
- Temporarily, the 4 bolts holding the delivery pipes to the intake manifold
- Temporarily install the bolt holding the No. 1 fuel pipe to the intake manifold.

➡️**Check that the injectors rotate smoothly. If the injectors do not rotate smoothly, the probable cause is incorrect installation of the O-rings. Replace the O-rings.**

- 4 bolts holding the delivery pipes to the intake manifold. Torque to 14 ft. lbs. (19.5 Nm)
- No. 1 fuel pipe (fuel tube connector) to the fuel filter
- Fuel hose clamp to the fuel filter with a "click" sound. After installing the clamp, check that the clamp is fixed by pulling up the clamp.
- Air assist hoses
- Injector connectors
- Air intake chamber assembly
- Air cleaner hose with resonator
- V-bank cover, using a 5mm hexagon wrench with the 3 cap nuts.
- Press down the V-bank cover.

3.0L (2JZ-GE) Engine

1. Before servicing the vehicle, refer to the precautions in the beginning of this section.
2. Remove or disconnect the following:
- Air intake chamber
- Fuel pressure pulsation damper
- Engine wire from intake manifold
- Bolt holding the engine wire protector to the body
- 6 injector connectors
- Camshaft Position (CMP) sensor connector
- Throttle Position (TP) sensor connector
- Vacuum Switching Valve (VSV) connector for Evaporative Emission (EVAP) control
- VSV connector for Acoustic Control Induction System (ACIS)
- 3 nuts holding the engine wire protector to the intake manifold

➡️**Be careful not to drop the injectors when removing the delivery pipe.**

- 3 bolts and delivery pipe together with the 6 injectors

- The injectors from the delivery pipe
- O-rings, insulator and grommet from each injector
- 3 spacers from the intake manifold

To install:
3. Install or connect the following:
- A new insulator and grommet to each injector
- A light coat of gasoline on the place where a delivery pipe touches an O-ring of the injector
- Injector, while turning clockwise and counterclockwise, into the delivery pipe

➡️**Position the injector connector outward.**

- The 3 spacers in position on the intake manifold
- A light coat of gasoline on the place where an intake manifold touches an O-ring
- Injectors together with the delivery pipe and 3 bolts in position on the intake manifold. Check that the injectors rotate smoothly. Position the injector connector upward.

➡️**If the injectors do not rotate smoothly, the probable cause is incorrect installation of the O-rings. Replace the O-rings.**

4. Tighten the 3 bolts holding the delivery pipe to the intake manifold. Tighten the bolts to 15 ft. lbs. (21 Nm).
5. Install or connect the following:
- Engine wire protector with the 3 nuts.
- 6 injector connectors.

➡️**The No. 1, No. 3 and No. 5 injector connectors are dark gray, and the No. 2, No. 4, and the No. 6 injectors connectors are brown.**

6. Install or connect the following:
- Camshaft position sensor connector
- Throttle position sensor connector
- VSV connector for EVAP
- Bolt holding the engine wire protector to the body.

4.0L (1UZ-FE) and 4.3L (3UZ-FE) Engines

1. Before servicing the vehicle, refer to the precautions in the beginning of this section.
2. Remove or disconnect the following:
- V-bank cover
- Intake air connector
- Accelerator cable.

- Fuel pressure pulsation dampers.
- VVT sensor connectors
- Vacuum Switching Valve (VSV) for Evaporative Emissions (EVAP)
- 2 nuts and accelerator cable bracket
- 2 nuts and accelerator cable bracket
- 3 V-bank cover brackets
- VSV connector for Acoustic Control Induction System (ACIS) from the No. 1 V-bank cover bracket.
- 4 bolts and 3 V-bank cover brackets
- Engine wire from the delivery pipe
- 2 wire clamps from the wire clamp bracket on the right-hand deliver pipe
- 8 injector connectors
- 4 nuts holding the delivery pipe to the intake manifold

- 2 delivery pipes and 8 injectors assembly and 4 spacers
- 2 O-rings, grommet and insulator from each injector

To install:

3. Install or connect the following:
   - A new insulator and grommet to each injector
   - A light coat of gasoline to new O-rings and install them to each injector
   - A light coat of gasoline on the place where a delivery pipe touches an O-ring of the injector
   - Injector, while turning the clockwise and counterclockwise, into the delivery pipe

➡**Position the injector connector outward.**

   - The 4 spacers in position on the intake manifold

- A light coat of gasoline on the place where an intake manifold touches an O-ring
- The delivery pipes in position on the intake manifold
- Temporarily, the 3 bolts holding the delivery pipe to the intake manifold

➡**Check that the injectors rotate smoothly. If the injectors do not rotate smoothly, the probable cause is incorrect installation of the O-rings. Replace the O-rings.**

4. Tighten the 3 bolts holding the delivery pipe to the intake manifold. Tighten the bolts to 15 ft. lbs. (21 Nm).
5. Install or connect the following:
   - Engine wire protector with the 3 nuts.
   - Injector connectors.
   - Remaining components

DRIVE TRAIN

Transmission Assembly

REMOVAL & INSTALLATION

Automatic

GS300

1. Before servicing the vehicle, refer to the precautions in the beginning of this section.
2. Turn the ignition switch to the **LOCK** position and disconnect the negative battery cable. Wait at least 90 seconds or longer before doing any work on the vehicle.
3. Remove or disconnect the following:
   - Transmission level gauge
   - Transmission dipstick and tube
   - Throttle cable from the throttle body
   - Oxygen (O_2) sensor from the exhaust system
   - Left and right tail pipes
   - Front and center exhaust pipe
   - Exhaust heat insulator
   - Rear center floor crossmember brace
   - Shift control rod from the shift lever
   - Driveshaft
   - Overdrive and direct clutch speed sensor
   - No. 1 Vehicle Speed Sensor (VSS)
   - No. 2 VSS
   - Solenoid wire
   - Park/Neutral Position (PNP) switch
   - Wiring from the starter
   - 2 oil cooler union nuts

- Oil cooler hoses from the oil cooler pipes
- Front oil cooler pipe bracket
- Center and rear oil cooler pipe brackets
- 2 oil cooler pipes
- Torque converter inspection plate
- Torque converter bolts

4. Support the transmission with a suitable jack.
5. Support the engine with a jack and a block of wood.
6. Remove or disconnect the following:
   - Rear transmission mount
   - Wiring harness clamps
   - Starter
   - 9 transmission mounting bolts and transmission

To install:

7. Install or connect the following:
   - Transmission and tighten the bolts to 53 ft. lbs. (52 Nm)
   - Starter and tighten the bolts to 27 ft. lbs. (37 Nm).
   - Rear transmission mount and tighten the bolts to 19 ft. lbs. (25 Nm).
   - And tighten the torque converter bolts to 30 ft. lbs. (41 Nm) while rotating the crankshaft.
   - Converter inspection plate
   - 2 oil cooler pipes
   - Center and rear oil cooler pipe brackets
   - Front oil cooler pipe bracket and tighten to 49 inch lbs. (5.5 Nm).

- Oil cooler hoses to the oil cooler pipes
- 2 oil cooler union nuts and tighten to 32 ft. lbs. (44 Nm).
- Wiring to the starter
- Transmission electrical connectors
- Driveshaft
- Remaining components
- Negative battery cable
- Transmission level gauge

8. Fill the transmission to the proper level with Dexron®II or equivalent.

IS300

1. Before servicing the vehicle, refer to the precautions in the beginning of this section.
2. Drain the cooling system.
   - Negative battery cable
   - Transmission oil dipstick and tube
   - Air cleaner
   - Mass Air Flow (MAF) sensor
   - Exhaust manifold
   - Engine under covers
   - Upper radiator hose
   - Exhaust front pipe
   - Exhaust center pipe
   - Shift control rod
   - Drive shaft
   - Oil cooler lines
   - Torque converter
   - Rear transmission mount crossmember. Support the transmission with a jack.

- Transmission wiring harness connectors
- Starter motor
- Transmission flange bolts
- Transmission

To install:

3. Installation is the reverse of the removal procedure, while using the following torque values:

- 17mm transmission flange bolts: 53 ft. lbs. (72 Nm)

- 14mm transmission flange bolts: 27 ft. lbs. (37 Nm)
- Starter motor bolts: 27 ft. lbs. (37 Nm)
- Rear transmission mount crossmember bolts: 19 ft. lbs. (25 Nm)
- Torque converter bolts: 35 ft. lbs. (48 Nm)
- Oil cooler lines: 33 ft. lbs. (44 Nm)
- Drive shaft center support bolts: 36 ft. lbs. (49 Nm)

- Drive shaft U-joint flange bolts: 54 ft. lbs. (74 Nm)
- Shift control rod nuts: 108 inch lbs. (13 Nm)
- Exhaust manifold nuts: 29 ft. lbs. (39 Nm)

LS400, LS430

1. Before servicing the vehicle, refer to the precautions in the beginning of this section.

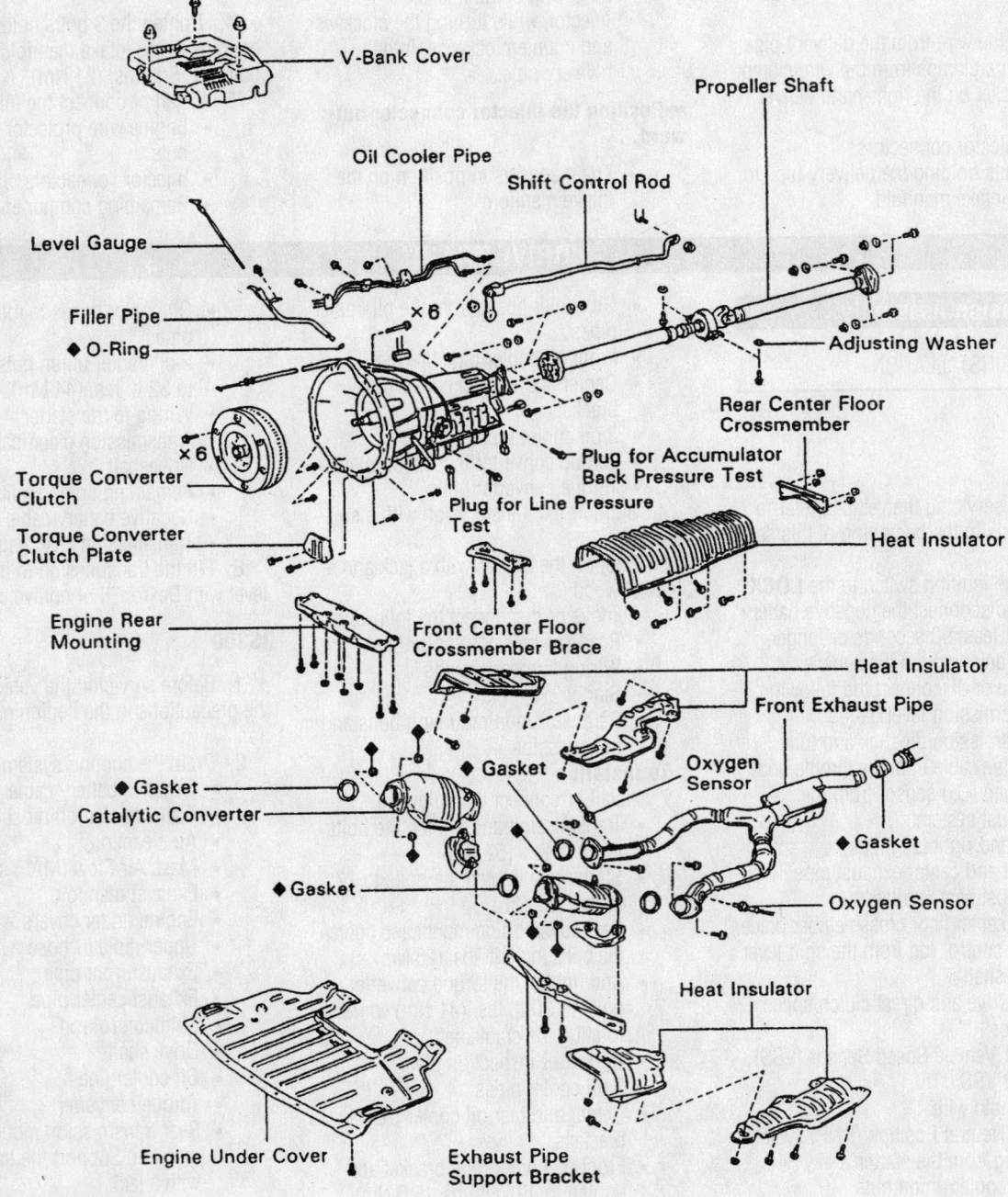

Exploded view of the transmission mounting—LS400

- V-Bank Cover
- Propeller Shaft
- Oil Cooler Pipe
- Shift Control Rod
- Level Gauge
- Filler Pipe
- ◆ O-Ring
- ×6
- Adjusting Washer
- Rear Center Floor Crossmember
- Plug for Accumulator Back Pressure Test
- ×6
- Torque Converter Clutch
- Torque Converter Clutch Plate
- Plug for Line Pressure Test
- Heat Insulator
- Engine Rear Mounting
- Front Center Floor Crossmember Brace
- Heat Insulator
- Front Exhaust Pipe
- Oxygen Sensor
- ◆ Gasket
- Catalytic Converter
- ◆ Gasket
- ◆ Gasket
- Oxygen Sensor
- ◆ Gasket
- Heat Insulator
- Engine Under Cover
- Exhaust Pipe Support Bracket

◆ **Non-reusable part**

7923LG97

2. Remove or disconnect the following:

- Negative battery cable. Wait at least 90 seconds before performing any other work.
- Transmission dipstick and tube
- Throttle cable
- Driveshaft
- Engine undercover
- Shift control rod
- Exhaust pipe support bracket by removing the 2 bolts
- Catalytic converters by removing the 6 nuts
- Both side heat insulators
- Oil cooler tube clamps and disconnect the tubes
- Torque converter inspection plate by removing the 2 bolts.
- Torque converter bolts

3. Support the transmission with a suitable jack.

4. Remove or disconnect the following:

- Rear transmission mount
- Overdrive direct clutch speed sensor connector
- Vehicle Speed Sensor (VSS) connector
- Park/Neutral Position (PNP) switch connector
- Solenoid connector
- 3 wiring harness clamps from the bracket on the transmission.
- 10 transmission mounting bolts and the transmission.

To install:

5. Install or connect the following:

- Transmission and tighten the bolts as follows: 14mm–27 ft. lbs. (37 Nm); 17mm–53 ft. lbs. (72 Nm)
- 3 wiring harness clamp to the bracket on the transmission.
- Solenoid connector
- PNP switch connector
- VSS connector
- Overdrive direct clutch speed sensor connector
- Rear transmission mount and tighten the bolts to 19 ft. lbs. (20 Nm) and the nuts to 10 ft. lbs. (13 Nm).
- And tighten the torque converter bolts to 30 ft. lbs. (41 Nm).

- Converter inspection plate
- Support from the transmission
- Oil cooler pipes and tighten the union nuts to 32 ft. lbs. (44 Nm).
- Side heat insulators
- Catalytic converters with new gaskets and new nuts. Tighten the nuts to 46 ft. lbs. (62 Nm).
- Exhaust pipe support bracket with the 2 bolts and tighten the bolts to 32 ft. lbs. (44 Nm).
- Shift control rod
- Engine undercover
- Driveshaft
- Throttle control cable
- Transmission tube and dipstick
- Negative battery cable

6. Fill the transmission to the proper level with Dexron®II, or equivalent.

SC300, SC400, GS400, GS430

1. Before servicing the vehicle, refer to the precautions in the beginning of this section.

2. Remove or disconnect the following:

- Negative battery cable.
- V-bank cover, if equipped
- Automatic transmission oil level gauge if equipped.
- Transmission dipstick and tube
- Throttle cable and clamps
- Exhaust pipe and converters
- Exhaust heat insulator
- Rear center floor crossmember brace
- Shift control rod
- Driveshaft

➡ **The bolts inserted from the driveshaft side should not be removed.**

- The electrical harness from the transmission.
- Oil cooler tube clamp and disconnect the tubes.
- Lower engine cover
- Torque converter inspection plate
- Torque converter bolts

3. Support the transmission with a suitable jack.

4. Remove or disconnect the following:

- Starter, if necessary
- Rear transmission mount

- Transmission mounting bolts and the transmission.

To install:

5. Before installing the transmission, use calipers and a straightedge to check the distance between the installed surface of the torque converter and the front edge of the transmission case. Correct distance is 0.673 in. (17.1mm). If this distance is not correct, check the torque converter installation.

6. Install or connect the following:

- Transmission and tighten the bolts to 14mm: 29 ft. lbs. (39 Nm); 17mm: 42 ft. lbs. (57 Nm)
- If removed, the starter. Tighten the bolts to 27 ft. lbs. (37 Nm).
- Rear transmission mount and tighten the bolts to 19 ft. lbs. (20 Nm).
- And tighten the torque converter bolts to 25 ft. lbs. (33 Nm) while rotating the crankshaft.
- Converter inspection plate
- Lower engine cover
- Oil cooler lines and tighten the lines to 25 ft. lbs. (34 Nm)
- Oil cooler pipe bracket and tighten the bolt
- Transmission electrical connectors
- Shift control rod and adjust the shift linkage. Tighten the nut to 12 ft. lbs. (16 Nm).
- Rear center floor crossmember brace. Tighten the bolts to 108 inch lbs. (13 Nm).
- Heat insulator
- Transmission filler tube and dipstick
- Front exhaust pipe and converters with new gaskets
- Throttle control cable
- Driveshaft. Flange bolts: 58 ft. lbs. (79 Nm). Center bearing support bolts: 36 ft. lbs. (49 Nm). Adjusting nut: 35 ft. lbs. (48 Nm).
- Crossmember brace and tighten to 96 inch lbs. (13 Nm).
- Automatic transmission oil level gauge
- V-bank cover
- Negative battery cable

Timing belt service is covered in Section 3 of this manual

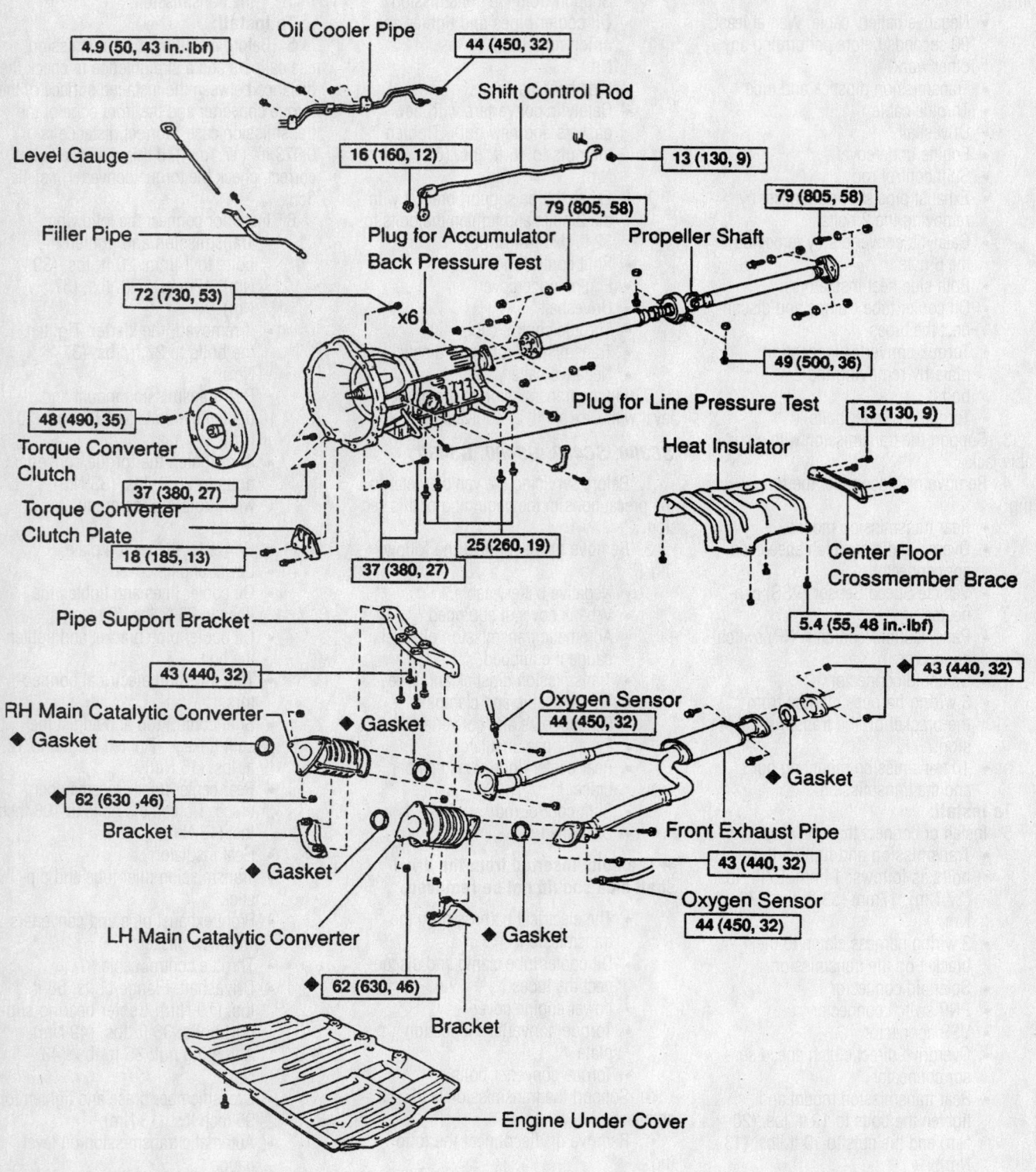

Oil Cooler Pipe

4.9 (50, 43 in.·lbf)

44 (450, 32)

Shift Control Rod

Level Gauge

16 (160, 12)

13 (130, 9)

Filler Pipe

79 (805, 58)

Plug for Accumulator Back Pressure Test

79 (805, 58)

Propeller Shaft

72 (730, 53)

x6

49 (500, 36)

48 (490, 35)

Plug for Line Pressure Test

13 (130, 9)

Torque Converter Clutch

Heat Insulator

37 (380, 27)

Torque Converter Clutch Plate

18 (185, 13)

37 (380, 27)

25 (260, 19)

Center Floor Crossmember Brace

5.4 (55, 48 in.·lbf)

Pipe Support Bracket

43 (440, 32)

43 (440, 32)

RH Main Catalytic Converter

◆ Gasket

Oxygen Sensor

44 (450, 32)

◆ Gasket

62 (630 ,46)

Bracket

◆ Gasket

◆ Gasket

Front Exhaust Pipe

43 (440, 32)

LH Main Catalytic Converter

Oxygen Sensor

44 (450, 32)

◆ Gasket

62 (630, 46)

Bracket

Engine Under Cover

N·m (kgf·cm, ft·lbf) : Specified torque
◆ Non–reusable part

7923LG98

Exploded view of the transmission mounting—SC300, SC400 and GS400

7. Adjust the PNP switch
8. Fill the transmission with Dexron® II.

Transaxle Assembly

REMOVAL & INSTALLATION

Automatic

ES300

1. Before servicing the vehicle, refer to the precautions in the beginning of this section.

2. Turn the ignition switch to the **LOCK** position and disconnect the negative battery cable. Wait at least 90 seconds or longer before doing any work on the vehicle.

3. Remove or disconnect the following:
   • Battery
   • Air cleaner assembly
   • Throttle cable from the throttle body
   • Cruise control actuator cover and detach the connector
   • Ground wire
   • Starter
   • Vehicle Speed Sensor (VSS) connectors
   • Direct clutch speed sensor
   • Park/Neutral Position (PNP) switch connector on the transaxle
   • Solenoid connector on the transaxle
   • Shift control cable
   • Oil cooler hoses
   • 2 front side transaxle mounting bolts
   • 2 front engine mounting bolts
   • Oil cooler line mounting bolts from the front frame
   • 3 upper transaxle to engine mounting bolts

4. Install an engine support fixture. Tie steering gear housing to engine support fixture.

5. Raise and safely support the vehicle.
6. Drain the transaxle/differential fluid.
7. Remove or disconnect the following:
   • Front wheels
   • Front exhaust pipe
   • Engine side covers and undercovers
   • Both halfshafts
   • Front side engine mounting nut
   • Rear side engine mounting bolts (remove hole plugs)
   • 4 left side transaxle mounting bolts
   • Steering gear housing

   • Front frame assembly
8. Properly support the transaxle assembly.
9. Remove or disconnect the following:
   • Rear end plate mounting bolts
   • Torque converter cover
   • Torque converter retaining bolts
   • Remaining transaxle mounting bolts
10. Carefully remove the transaxle assembly from the vehicle.

To install:
11. Install or connect the following:
   • Transaxle aligning the 2 dowel pins on the block with the converter housing. Tighten the bolts as follows: 10mm–34 ft. lbs. (46 Nm); 12mm–47 ft. lbs. (64 Nm).
   • Torque converter bolts. Coat the threads of the with sealer. Install the bolts starting with the green bolt followed by the rest and tighten the bolts evenly to 20 ft. lbs. (27 Nm).
   • End plate and tighten the bolts to 27 ft. lbs. (37 Nm).
   • Front frame assembly and tighten the fasteners as follows: 12mm–24 ft. lbs. (32 Nm); 19mm–134 ft. lbs. (181 Nm); nut–27 ft. lbs. (36 Nm).
   • 2 fender liner set screws.
   • Steering gear to the frame and tighten the bolts and nuts to 134 ft. lbs. (181 Nm).
   • Sway bar brackets and toque the bolts to 14 ft. lbs. (19 Nm).
   • Left transaxle mounting bolts and tighten them to 38 ft. lbs. (52 Nm).
   • Rear side mounting bolts and nuts and tighten them to 48 ft. lbs. (66 Nm). Install the plugs.
   • Front engine mounting nut and tighten it to 59 ft. lbs. (80 Nm).
   • Halfshafts
   • Right and left engine side covers
   • Lower engine cover
   • Exhaust pipe to the engine with new gaskets and tighten the nuts to 46 ft. lbs. (62 Nm). Connect the exhaust pipe to the converter with a new gasket and tighten the nuts and bolts to 32 ft. lbs. (43 Nm).
   • Wheel
   • Engine support
   • 4 upper transaxle mounting bolts and tighten them to 47 ft. lbs. (64 Nm).
   • Oil cooler clamping bolts to the front frame

   • 2 front side engine mounting bolts and tighten them to 59 ft. lbs. (80 Nm).
   • 2 front side transaxle mounting bolts and tighten them to 59 ft. lbs. (80 Nm).
   • Remaining components
   • Battery and connect the battery cables.
12. Fill the transaxle/differential to the proper level with Dexron®II, or equivalent.
13. Check the transaxle/differential fluid level.
14. Check the front wheel alignment.

Halfshaft

REMOVAL & INSTALLATION

ES300

1. Before servicing the vehicle, refer to the precautions in the beginning of this section.

2. Remove or disconnect the following:
   • Negative battery cable
   • Front wheel(s)
   • Front fender apron seal
   • Transaxle fluid
   • Tie rod end from the steering knuckle by removing the cotter pin and nut. Separate the tie rod from the steering knuckle.
   • Stabilizer bar link from the lower control arm. Make note of the washers and cushions positions.
   • Lower ball joint from the steering knuckle by removing the bolt and 2 nuts. Push down on the lower control arm and separate the steering knuckle from the ball joint.
   • Cotter pin, lock cap and locknut holding the halfshaft to the steering knuckle
   • Left halfshaft from the steering knuckle
   • Halfshaft from the transaxle
   • Snapring from the halfshaft
   • Right halfshaft bearing lockbolt. The lockbolt is located in the center of the halfshaft, near the dampener.
   • Snapring and pull the halfshaft from the transaxle.

To install:
3. Install or connect the following:
   • Right halfshaft to the transaxle. Coat the side gear shaft and differential case sliding surface with gear oil.

Heater Core replacement is covered in Section 2 of this manual

- Snapring to the halfshaft
- Bearing lockbolt. Tighten the lock-bolt to 24 ft. lbs. (32 Nm).
- New snapring to the inner spline of the left halfshaft. Coat the side gear shaft and differential case sliding surface with gear oil. Install the halfshaft to the transaxle with the snapring opening facing down. The halfshaft should click into place when installing.
- Halfshaft to the steering knuckle, then install the locknut. Tighten the locknut to 217 ft. lbs. (294 Nm).

- Lock cap and a new cotter pin to the halfshaft.
- Steering knuckle to the lower ball joint. Install the 2 nuts and bolt. Tighten the nuts and bolt to 94 ft. lbs. (127 Nm).
- Stabilizer bar link to the lower control arm. Tighten the nut to 29 ft. lbs. (39 Nm).
- Tie rod to the steering knuckle and tighten the nut to 36 ft. lbs. (49 Nm). Install a new cotter pin to the tie rod end.
- Front fender apron seal

- Wheel(s) and lower the vehicle. Tighten the lug nuts to 76 ft. lbs. (103 Nm).
- Transaxle fluid
- Negative battery cable

Except ES300

1. Before servicing the vehicle, refer to the precautions in the beginning of this section.
2. Remove or disconnect the following:
   - Negative battery cable
   - Rear tire and wheel assembly

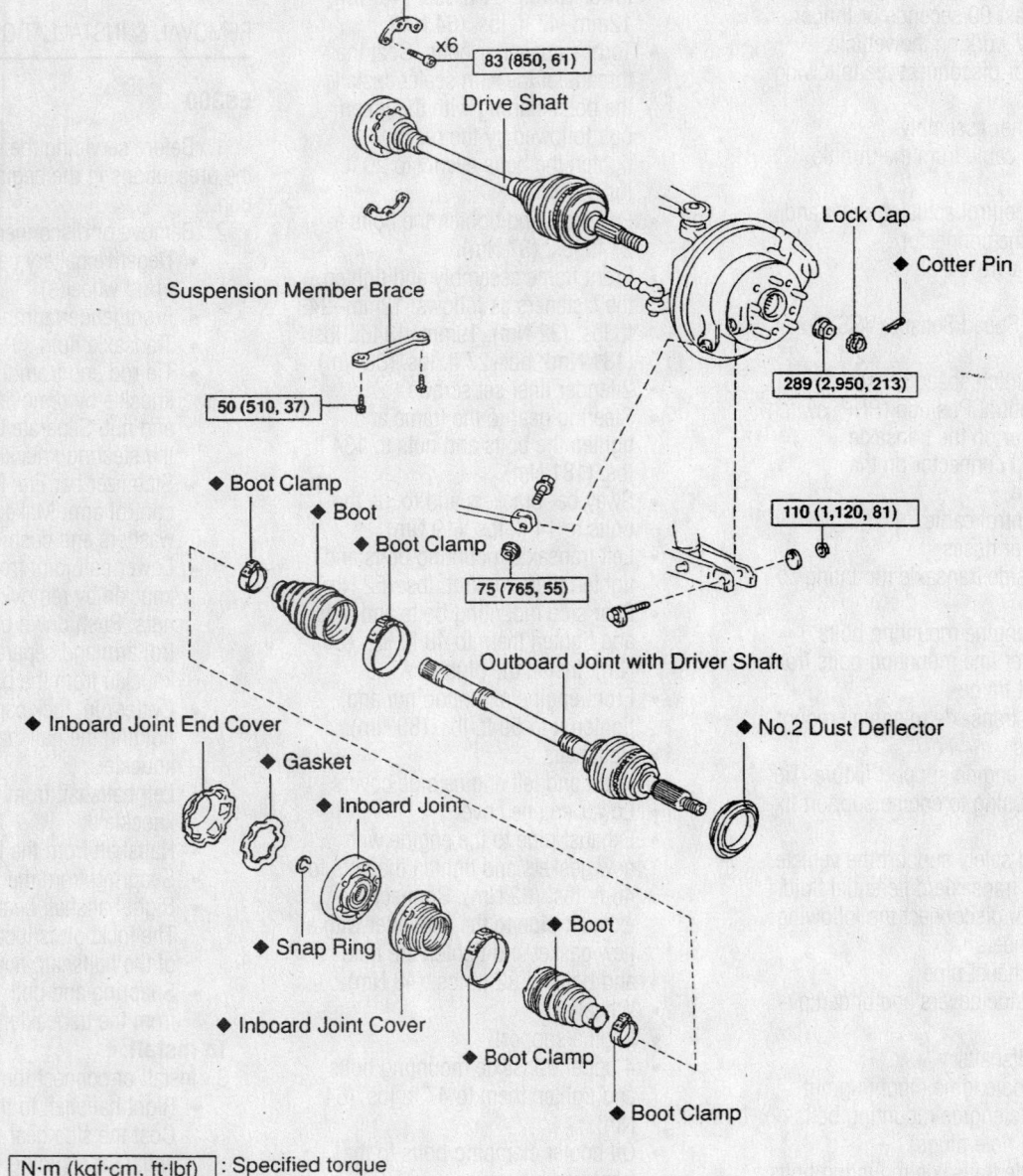

Exploded view of the rear halfshaft and related components—except ES300

9301LG12

- Cotter pin, locknut cap, and lock-nut
- Height control sensor, if equipped
- 2 exhaust pipe support brackets, if necessary

3. Place matchmarks on the halfshaft and the side gear shaft. Remove the 6 hex bolts and 2 washers.

4. Hold the inboard joint side of the halfshaft so the outboard joint side does not bend too much. Tap the end of the halfshaft with a rubber mallet to loosen it from the axle hub and remove the halfshaft.

To install:

5. Insert the outboard joint side of the halfshaft through the axle hub. Align the matchmarks on the side gear shaft and the halfshaft.

6. Coat the threads with clean oil and install the hex bolts. Tighten the bolts to 61 ft. lbs. (83 Nm).

7. Install or connect the following:
- Exhaust pipe support brackets, if removed, and tighten to 14 ft. lbs. (19 Nm).
- Bearing locknut, if removed, and have a helper apply the brakes. Tighten the locknut to 213 ft. lbs. (289 Nm).
- Lockcap and a new cotter pin.
- Height control sensor, if removed.
- Rear tire and wheel assembly.
- Negative battery cable

CV-Joints

OVERHAUL

ES300

1. Before servicing the vehicle, refer to the precautions in the beginning of this section.

2. Once the driveshaft is removed from the vehicle, place matchmarks on the outboard and inboard joints and the shaft. Do not use a punch to make the marks.

3. Remove or disconnect the following:
- Boot clamps. Use a side cutter or pliers
- Outboard joint shaft expanding the snapring
- 2 boots
- Dust cover on the left-hand driveshaft, using a hammer and suitable chisel.
- Dust cover from the inboard joint shaft, using a press
- Dust cover

- Snapring. Use a snapring expander.
- Bearing, using the press
- Snapring
- No. 2 dust deflector, using a hammer and suitable chisel

➡ **Be careful not to damage the Anti-lock Brake System (ABS) speed sensor rotor.**

To install:

4. Install or connect the following:
- No. 2 dust deflector
- New snapring to the inboard joint shaft
- New bearing
- New dust cover
- Dust cover on the left-hand driveshaft
- A new dust cover, using a press
- Outboard and inboard joint boots and new boot clamps as a temporary measure. Before installing the boot, place 3 new clamps to the small boot ends and large end (wheel side) and install it to the driveshaft
- Inboard joint shaft to outboard joint shaft.
- Using a snapring expander, put in the inboard joint shaft expanding the snapring.
- Boot to outboard joint, before assembling the boot, pack the outboard joint and boot with grease in the boot kit.
- Boot to inboard joint shaft. Pack the inboard joint, and boot with grease in the boot kit, and install the boot to the inboard joint shaft.
- Boot clamps to both boots, make sure that the 2 boots are on the shaft groove. Hold the clamp near the clamp's free end over the closing hooks.

5. Secure the clamp by drawing the closing hooks together. Secure the clamp onto the boot.

GS300, GS400, GS430, IS300

1. Before servicing the vehicle, refer to the precautions in the beginning of this section.

2. Remove or disconnect the following:
- End cover. Use nuts and bolts to keep the inboard joint together. Hand tighten only.
- 4 Boot clamps, using a side cutter or pliers.
- Inboard joint, place matchmarks on

the inboard joint and driveshaft; do not use punch marks.
- Snapring, using a snapring expander
- Using a press, the inboard joint from the driveshaft
- Inboard and outboard boot
- Inboard joint cover from the inboard joint

To install:

3. Install or connect the following:
- Inboard and outboard joint boots
- New No. 2 Dust deflector, using a press

➡ **Be careful not to damage the Anti-lock Brake System (ABS) speed sensor rotor.**

- Inboard joint
- Inner race to the cage so that the indented beveled part of the inner race is on the opposite side to the beveled top of the cage.
- Outer race so that the indented side of the outer race is facing the same side as the beveled surface of the cage.

4. Match the narrow projections of the inner race with the wide projections of the outer race.

5. Tilt the cage and inner race to the side and insert the balls one by one.

6. Install or connect the following:
- New boots and new boot clamps, temporarily
- 4 new boot clamps to the boots
- 2 boots to the driveshaft.
- Inboard joint cover and apply RTV to the inboard joint cover.

7. Remove grease from the surface of the inboard joint facing the cover

8. Align the bolt holes of the cover with those of the inboard joint, then insert the hexagon bolts.

9. Use a plastic hammer to tap the rim of the inboard joint cover into place

10. To install the inboard joint, align the matchmarks placed before removal.

11. Using a brass bar and hammer, tap the inboard joint onto the driveshaft.

12. Install or connect the following:
- New snapring
- Boots to joints, pack with the proper grease. 3.5–3.7 oz. (100–105g).
- New boot clamps to both boots
- 6 hexagon bolts and washers from the end cover side, install the 6 nuts to the boot side.

Brake service is covered in Section 4 of this manual

SC300, SC400

1. Before servicing the vehicle, refer to the precautions in the beginning of this section.

2. Using a suitable prytool, remove the end cover.

3. Use nuts and bolts to keep the inboard joint together. Hand tighten only.

4. Remove or disconnect the following:
- 4 Boot clamps, using a side cutter or pliers.
- Inboard joint, place matchmarks on the inboard joint and driveshaft; do not use punch marks.
- Snapring, using a snapring expander
- Using a press, the inboard joint from the driveshaft
- Inboard joint cover from the inboard joint
- Inboard and outboard boot
- No. 3 dust deflector

To install:

5. Install or connect the following:
- New No. 3 dust deflector, using a press.

6. If the joint has come apart, reassemble it in the following order:

a. Align the matchmarks placed before removal

b. Inner race to the cage so that the indented beveled part of the inner race is on the opposite side to the beveled top of the cage.

c. Outer race so that the indented side of the outer race is facing the same side as the beveled surface of the cage.

7. Match the narrow projections of the inner race with the wide projections of the outer race.

8. Tilt the cage and inner race to the side and insert the balls one by one.

9. Install or connect the following:
- New boots and new boot clamps, temporarily
- 4 new boot clamps to the boots
- 2 boots to the driveshaft
- Inboard joint cover and apply Formed In Place Gasket (FIPG) to the inboard joint cover. Avoid applying an excessive amount to the surface.

10. Remove grease from the surface of the inboard joint facing the cover

11. Align the bolt holes of the cover with those of the inboard joint, then insert the hexagon bolts.

12. Use a plastic hammer to tap the rim of the inboard joint cover into place

13. To install the inboard joint, align the matchmarks placed before removal.

14. Using a brass bar and hammer, tap the inboard joint onto the driveshaft.

15. Install or connect the following:
- New snapring
- Boots to joints, pack with 3.5–3.7 oz. (100–105g)
- Boot clamps onto the boot. Clearance is 0.031 inch (0.8mm)
- End cover, and pack with grease. SC 400 uses 1.8–1.9 oz. (50–55g); SC300 uses 1.5–1.7 oz. (42–47g)

16. Remove grease from the surface of the inboard joint facing the cover

17. Glue on a new gasket, with the side with adhesive on it facing toward the outer race side of the inboard joint

18. Align the bolt holes of the cove with those of the inboard joint

19. Install or connect the following:
- 6 hexagon bolts an washer from the end cover side.
- 6 nuts to the boot side.

20. Check that the claw of the end cover touches the inboard joint

LS400, LS430

1. Before servicing the vehicle, refer to the precautions in the beginning of this section.

2. Using a suitable prytool, remove the end cover.

3. Use nuts and bolts to keep the inboard joint together. Hand tighten only.

4. Remove or disconnect the following:

- 4 Boot clamps, using a side cutter or pliers.
- Inboard joint, place matchmarks on the inboard joint and driveshaft; do not use punch marks.
- Snapring, using a snapring expander
- Using a press, the inboard joint from the driveshaft
- Inboard joint cover from the inboard joint
- Inboard and outboard boot
- No. 2 dust deflector, using a suitable chisel and hammer
- New No. 3 dust deflector, using a press

To install:

5. If the joint has come apart, reassemble it in the following order.

a. Align the matchmarks placed before removal

b. Inner race to the cage so that the indented beveled part of the inner race is on the opposite side to the beveled top of the cage.

c. Outer race so that the indented side

of the outer race is facing the same side as the beveled surface of the cage.

6. Match the narrow projections of the inner race with the wide projections of outer race.

7. Tilt the cage and inner race to the side and insert the balls one by one.

8. Install or connect the following:
- New boots and new boot clamps, temporarily
- 4 new boot clamps to the boots
- 2 boots to the driveshaft.
- Inboard joint cover and apply Formed In Place Gasket (FIPG) to the inboard joint cover. Avoid applying an excessive amount to the surface.

9. Remove grease from the surface of the inboard joint facing the cover

10. Align the bolt holes of the cover with those of the inboard joint, then insert the hexagon bolts.

11. Use a plastic hammer to tap the rim of the inboard joint cover into place

12. To install the inboard joint, align the matchmarks placed before removal.

13. Using a brass bar and hammer, tap the inboard joint onto the driveshaft.

14. Install or connect the following:
- New snapring
- Boots to joints, pack with 3.5–3.7 oz. (100–105g) of grease
- A new gasket, with the side with adhesive on it facing toward the outer race side of the inboard joint
- 6 hexagon bolts an washer from the end cover side.
- 6 nuts to the boot side.

15. Check that the claw of the end cover touches the inboard joint

Axle Shaft, Bearing and Seal

REMOVAL & INSTALLATION

GS300, GS400, GS430, IS300

1. Drain the gear oil.
2. Remove or disconnect the following:
- Rear driveshaft
- Side gear shaft
- Snapring from the side gear, using a suitable tool
- Side gear shaft oil seal

To install:
- Side gear shaft oil seal
- New oil seal

3. Check installation of side gear shaft. There should be 0.08–0.12 inch (2–3mm) of play in the axial direction. Check that the side gear shaft will not come out by pulling on it.

SC300, SC400

1. Remove or disconnect the following:
   - Gear oil
   - Rear driveshaft
   - Side gear shaft
   - Snapring from the side gear, using a suitable tool
   - Side gear shaft oil seal

To install:

2. Install or connect the following:
   - New oil seal until it is flush with the carrier end surface
   - Multipurpose grease to the oil seal lip
   - Side gear shaft, and install a new snapring
   - Side gear shaft to the differential

➡**Check that the side gear shaft does not come out by trying to pull it out by hand.**

   - Reconnect the rear driveshaft
   - Gear oil

LS400, LS430

1. Remove or disconnect the following:
   - Gear oil
   - Rear driveshaft
   - Side gear shaft
   - Snapring from the side gear
   - Side gear shaft oil seal

To install:

2. Install or connect the following:
   - Side gear shaft oil seal
   - New oil seal
   - Multi-Purpose (MP) grease to the oil seal lip
   - Side gear shaft
   - Snapring from the side gear, using a suitable tool
   - Side gear shaft oil seal
   - Gear oil

Pinion Seal

REMOVAL & INSTALLATION

GS300, GS400, GS430

1. Drain the gear oil.
2. Remove the driveshaft and the companion flange
3. Remove the oil seal
4. Check oil slinger

To install:

5. Installation is the reversal of the removal procedure.

SC300, SC400, LS400, LS430

1. Remove or disconnect:
   - Gear oil
   - Driveshaft
   - Companion flange
   - Oil seal and slinger

To install:

2. Install or connect:
   - Install the oil slinger.
   - New oil seal

➡**Oil seal drive-in depth: 0.079 inch (2.0mm).**

   - Multi-Purpose (MP) grease to the oil seal lip
   - Companion flange on the shaft
   - Gear oil on the threads of a new nut. Torque to 80 ft. lbs. (108 Nm).

3. Adjust the drive pinion preload as necessary, stake drive the pinion nut, and install the driveshaft.
4. Fill the differential with hypoid gear oil

IS300

1. Before servicing the vehicle, refer to the precautions in the beginning of this section.

2. Remove or disconnect the following:
   - Driveshaft
   - Rear wheels
   - Rear brake calipers
   - Pinion flange
   - Pinion seal

➡**The rear brake calipers must be removed so that there is no additional drag when measuring pinion bearing preload.**

To install:

3. Install or connect the following:
   - Pinion seal and flange
   - New pinion flange nut

4. Rotate the pinion flange occasionally while tightening the flange nut to make sure the pinion bearings seat correctly. Do not exceed 249 ft. lbs. (338 Nm).
5. Take frequent bearing preload torque readings.
6. The pinion bearing preload specifications are as follows:
   a. Used bearings: 4.3–6.9 inch lbs. (0.49–0.78 Nm).
   b. New bearings: 8.7–13.9 inch lbs. (0.98–1.57 Nm).

❊❊ CAUTION

Never loosen the pinion nut to reduce bearing preload. If it is necessary to reduce bearing preload, install a new collapsible spacer and pinion nut.

7. Install or connect the following:
   - Driveshaft
   - Brake calipers
   - Rear wheels
8. Fill the differential with gear lubricant and check for leaks.

STEERING AND SUSPENSION

Air Bag

❊❊ CAUTION

These vehicles are equipped with an air bag system. The system must be disabled before performing service on or around system components, steering column, instrument panel components, wiring and sensors. Failure to follow safety and disabling procedures could result in accidental air bag deployment, possible personal injury and unnecessary system repairs.

PRECAUTIONS

Several precautions must be observed when handling the inflator module to avoid accidental deployment and possible personal injury.

   - Never carry the inflator module by the wires or connector on the underside of the module.
   - When carrying a live inflator module, hold securely with both hands and ensure that the bag and trim cover are pointed away.
   - Place the inflator module on a bench or other surface with the bag and trim cover facing up.
   - With the inflator module on the bench, never place anything on or close to the

module which may be thrown in the event of an accidental deployment.

DISARMING

To avoid personal injury when working on vehicles equipped with an air bag, the negative battery cable must be disconnected and at least 90 seconds must elapse before working on the system. Failure to do so may result in deployment of the air bag.

ARMING

To rearm the air bag system, simply reconnect the battery cable(s).

Power Rack and Pinion Steering Gear

REMOVAL & INSTALLATION

ES300

1. Before servicing the vehicle, refer to the precautions in the beginning of this section.
2. Remove or disconnect the following:
   - Negative battery cable and wait at least 90 seconds before working on the vehicle to disarm the air bag.
   - Front wheels
   - Left and right front fender apron seals
   - Cotter pin and nut holding the steering knuckle to the tie rod end. Using a tie rod puller, disconnect the tie rod end from the steering knuckle.
3. Place matchmarks on the intermediate shaft and the control valve shaft.
4. Loosen the upper bolt and remove the lower bolt holding the control valve shaft to the intermediate shaft. Disconnect the intermediate shaft from steering rack housing.
5. Remove or disconnect the following:
   - Tube clamp
   - Return line and the pressure line from the control valve housing
   - 4 stabilizer bar bolts and 2 nuts. Position the stabilizer bar out of the way. Do not remove the sway bar from the vehicle.
   - Heated Oxygen (HO2) sensor (bank 1 sensor 1).
   - 2 steering gear mounting bolts and nuts. Remove the steering gear through the left side of the vehicle.

To install:

6. Install or connect the following:
   - Steering gear on the vehicle and

install the 2 mounting bolts and nuts. Tighten the nuts and bolts to 134 ft. lbs. (181 Nm).
   - HO2 sensor. Tighten the sensor to 33 ft. lbs. (44 Nm).
   - Stabilizer bar bolts and nuts and tighten as follows: Bolts: 14 ft. lbs. (19 Nm); Nuts: 29 ft. lbs. (39 Nm).
   - Pressure and return lines and tighten the connectors to 18 ft. lbs. (25 Nm).
   - Tube clamp and tighten the nut to 84 inch lbs. (10 Nm).
   - Intermediate shaft to the steering rack and tighten the retaining bolts to 26 ft. lbs. (35 Nm).
   - Tie rods to the steering knuckles with the castellated nuts. Tighten the nut to 36 ft. lbs. and install a new cotter pin. The prongs of the cotter pin should be firmly wrapped around the flats of the nut.
   - Front fender apron seals by installing the 2 bolts
   - Front wheels and lower the vehicle
   - Power steering fluid
   - Negative battery cable
7. Release the steering wheel
8. Bleed the system
9. Check for leaks, adjust the toe-in and check the steering wheel center point

GS300, GS400, GS430

1. Before servicing the vehicle, refer to the precautions in the beginning of this section.
2. Remove or disconnect the following:
   - Negative battery cable
   - Front wheels
3. Matchmark the steering column universal joint to the control valve shaft.
4. Loosen the upper bolt and remove the lower bolt to the intermediate shaft universal joint.
5. Remove or disconnect the following:

   - Intermediate shaft from the control valve shaft.
   - Tie rod ends from the steering knuckle.
   - Fluid lines from the rack and pinion and cap the lines.
   - 2 tube clamps by removing the bolt.
   - Mounting bolts and nuts. Remove the rack and pinion.

To install:

6. Center the rack and pinion to the following dimensions:
   - Dimension A: 1.14 in. (28.9mm)
   - Dimension B: 23.54 in. (589mm)
7. Install or connect the following:
   - Rack and tighten the bolts to 72 ft. lbs. (98 Nm).
   - 2 tube clamps and tighten the bolt to 12 ft. lbs. (17 Nm).
8. Align the matchmarks on the intermediate shaft and control valve shaft. Tighten the intermediate shaft bolts to 26 ft. lbs. (35 Nm).
9. Install or connect the following:
   - Fluid lines to the rack and pinion with new washers. Tighten the union bolts to 36 ft. lbs. (49 Nm).
   - Tie rod ends
   - Wheels

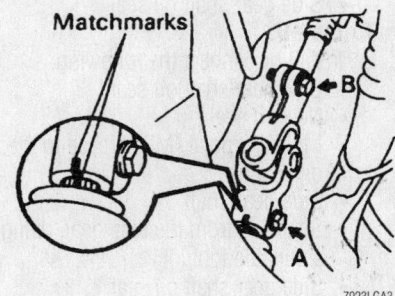

Matchmarking the intermediate shaft to the control valve shaft—GS300, GS400, GS430

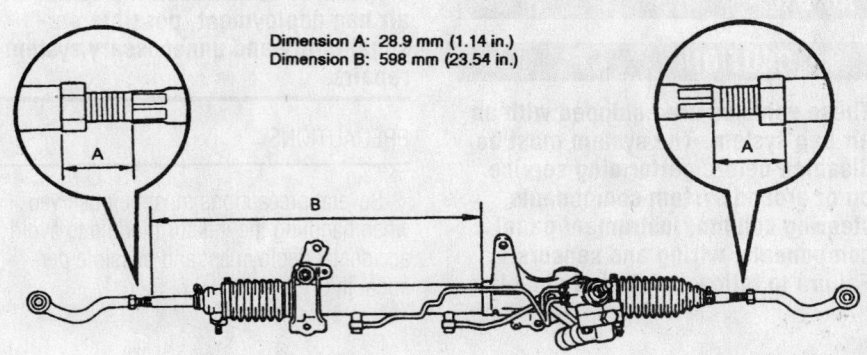

Dimension A: 28.9 mm (1.14 in.)
Dimension B: 598 mm (23.54 in.)

Centering the rack and pinion—GS300, GS400, GS430

- Negative battery cable.
10. Lower the vehicle.
11. Check the steering wheel center point.
12. Check the front wheel alignment.

IS300

1. Before servicing the vehicle, refer to the precautions in the beginning of this section.

2. Remove or disconnect the following:
- Negative battery cable
- Steering wheel
- Front wheels
- Brake calipers
- Outer tie rod ends
- Engine under cover
- Intermediate shaft
- Front subframe brace
- Pressure and return lines
- Steering gear

To install:

3. Installation is the reverse of the removal procedure, while using the following torque values:
- Steering gear mounting bracket bolts: 54 ft. lbs. (74 Nm)
- Fluid return line: 30 ft. lbs. (40 Nm)
- Fluid pressure line: 31 ft. lbs. (42 Nm)
- Front subframe brace large bolts: 88 ft. lbs. (119 Nm)

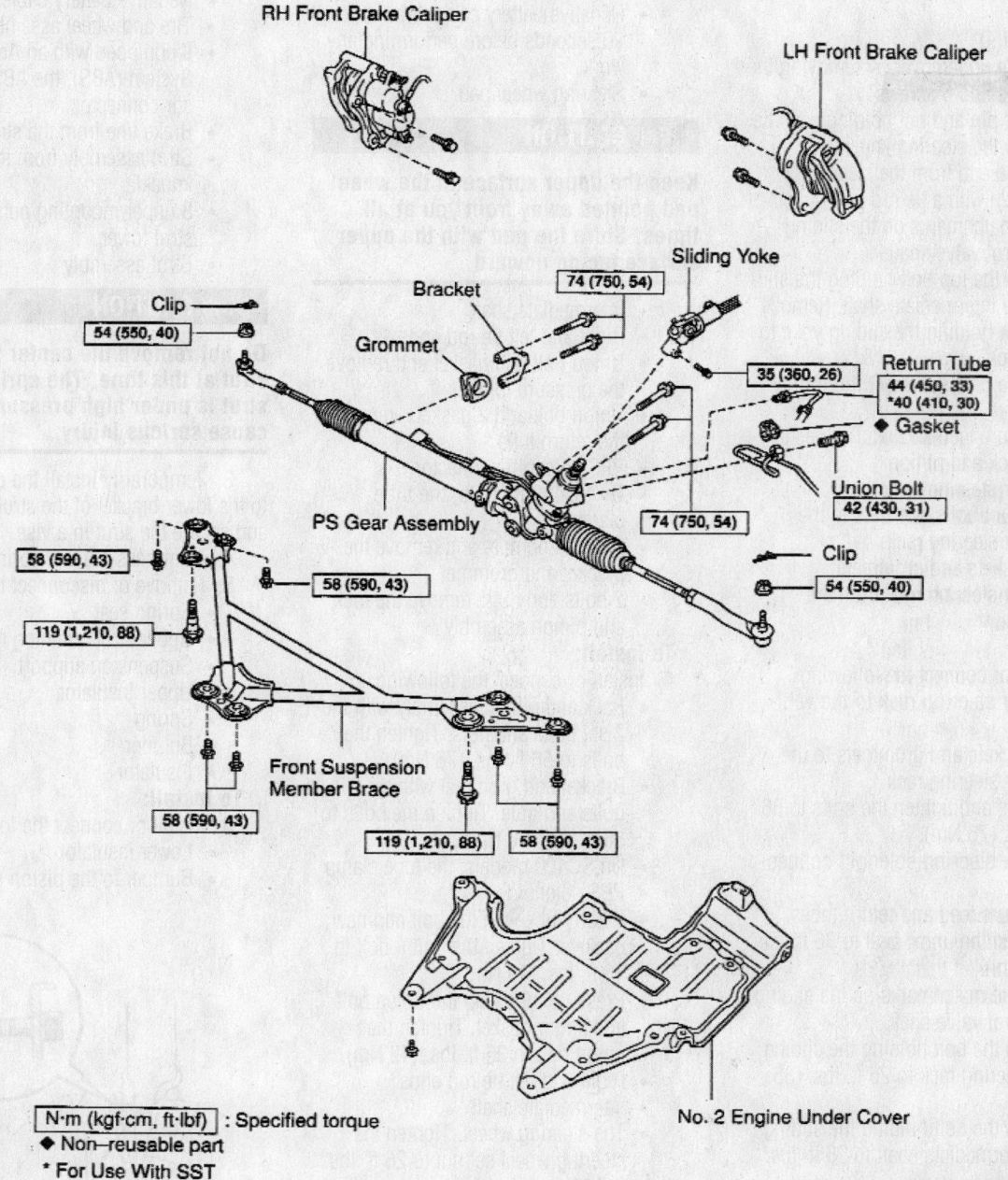

N·m (kgf·cm, ft·lbf) : Specified torque
◆ Non–reusable part
* For Use With SST

Exploded view of the steering gear mounting—IS300

9347LG02

For Accessory Drive Belt illustrations, see Section 1 of this manual

- Front subframe brace small bolts: 43 ft. lbs. (58 Nm)
- Tie rod end nuts: 40 ft. lbs. (54 Nm)
- Steering wheel nut: 26 ft. lbs. (35 Nm)

LS400, LS430

1. Before servicing the vehicle, refer to the precautions in the beginning of this section.
2. Remove or disconnect the following:
- Wheel(s)
- Engine undercover by removing the 8 bolts and 5 screws.
- Cotter pin and nut holding each tie rod to the steering knuckle.
- Tie rod end from the steering knuckle with a tie rod end puller.
3. Place matchmarks on the sliding yoke and control valve shaft.
4. Loosen the top bolt holding the sliding yoke to the intermediate shaft. Remove the bottom bolt holding the sliding yoke to the steering rack.
5. Remove or disconnect the following:
- Pressure feed and return lines to the rack and pinion
- Power steering connector
- 4 mount bolts and nuts to the power steering rack
- 2 brackets and grommets
- Power steering rack from the vehicle

To install:
6. Install or connect the following:
- Power steering rack to the vehicle.
- 2 brackets and grommets to the power steering rack.
- 4 bolts and tighten the bolts to 56 ft. lbs. (76 Nm).
- Power steering solenoid connector
- Pressure feed and return tubes. Tighten the union bolt to 36 ft. lbs. (49 Nm).
7. Align the matchmarks on the sliding yoke and control valve shaft.
8. Tighten the bolt holding the sliding yoke to the steering rack to 26 ft. lbs. (35 Nm).
9. Tighten the bolt holding the sliding yoke to the intermediate shaft to 26 ft. lbs. (35 Nm).
10. Install or connect the following:
- Tie rod end to the steering knuckle. Tighten the nut to 48 ft. lbs. (65 Nm). Install a new cotter pin.

- Engine undercover
- Wheel(s)
11. Bleed the power steering system and check the front end alignment.

SC300 and SC400

1. Before servicing the vehicle, refer to the precautions in the beginning of this section.
2. Place the front wheels facing straight ahead.
3. Remove or disconnect the following:
- Negative battery cable. Wait at least 90 seconds before performing any work.
- Steering wheel pad

> **✳✳ CAUTION**
>
> **Keep the upper surface of the wheel pad pointed away from you at all times. Store the pad with the upper surface facing upward.**

- Intermediate shaft
- Right and left tie rod ends
- Union bolt and gasket and remove the pressure tube
- Union bolt and 2 gaskets; remove the return tube
- PPS solenoid connector
- On SC400 models, the tube clamp
- 2 bolts and nuts and remove the bracket and grommet
- 2 bolts and nuts; remove the rack and pinion assembly

To install:
4. Install or connect the following:
- Rack and pinion assembly with the 2 set bolts and nuts. Tighten the bolts to 56 ft. lbs. (76 Nm).
- Bracket and grommet with the 2 bolts and nuts. Tighten the bolts to 56 ft. lbs. (76 Nm).
- On SC400 models, the tube clamp
- PPS solenoid
- Return tube with the bolt and new gaskets. Tighten the union bolt to 36 ft. lbs. (49 Nm).
- Pressure tube with the union bolt and a new gasket. Tighten the union bolt to 36 ft. lbs. (49 Nm).
- Right and left tie rod ends
- Intermediate shaft
- The steering wheel. Tighten the steering wheel set nut to 26 ft. lbs. (35 Nm).
- Steering wheel pad
- Negative battery cable
- Steering fluid and bleed the steering system.

Strut and Coil Spring

REMOVAL & INSTALLATION

Front

ES300

1. Before servicing the vehicle, refer to the precautions in the beginning of this section.
2. Remove or disconnect the following:
- Negative battery cable
- Tire and wheel assembly
- If equipped with an Anti-lock Brake System (ABS), the ABS speed sensor connector
- Brake line from the strut housing
- Strut assembly from the steering knuckle
- 3 upper mounting nuts from the strut tower
- Strut assembly

> **✳✳ CAUTION**
>
> **Do not remove the center nut to the strut at this time. The spring on the strut is under high pressure and can cause serious injury.**

3. Temporarily install the bolt and nuts to the lower bracket of the strut to support it and secure the strut in a vise.
4. Compress the coil spring.
5. Remove or disconnect the following:
- Spring seat
- Upper strut retaining nut
- Suspension support
- Upper insulator
- Spring
- Bumper
- Insulator

To install:
6. Install or connect the following:
- Lower insulator
- Bumper to the piston rod

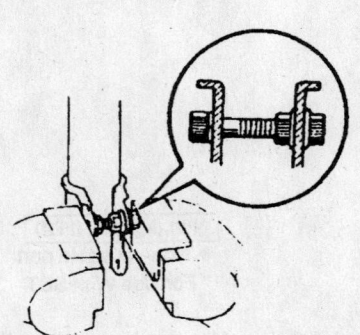

Temporarily install the support nuts and bolt to the strut—ES300

7923LGA5

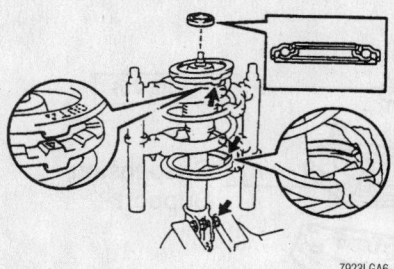

Align the out mark of the upper spring seat with the mark on the upper insulator— ES300 and LS400

- Coil spring end into the gap of the lower seat
- Upper insulator
- Upper support to the piston rod, aligning it with the groove in the strut rod.

7. Install or connect the following:
- Spring seat. Tighten the new upper strut retaining nut to 36 ft. lbs. (49 Nm).

8. Remove the strut from the vise and disassemble the securing nuts and bolt.

9. Rotate the upper support so the lowest bolt on the support aligns with the projection part of the lower spring.

10. Install or connect the following:
- Strut and tighten the strut to body bolts to 59 ft. lbs. (80 Nm).
- Strut to the steering knuckle and tighten the bolts to 156 ft. lbs. (211 Nm).

11. Run the brake hose through the brake hose bracket and install the clip.

12. Install or connect the following:
- ABS speed sensor and tighten the mounting bolt to 48 inch lbs. (5 Nm).
- Brake line to the strut housing and tighten the bolt to 22 ft. lbs. (29 Nm).
- Wheel
- Negative battery cable.

13. Check the front alignment.

GS300, GS400, GS430

1. Before servicing the vehicle, refer to the precautions in the beginning of this section.

2. Remove or disconnect the following:
- Negative battery cable.
- Front wheel
- Brake caliper, leaving the line attached

3. Loosen the 3 upper strut mounting nuts.

4. Loosen, but do not remove, the upper strut rod nut.

> **✳✳ CAUTION**
>
> **Do not remove the upper strut nut at this time.**

5. Remove or disconnect the following:
- Anti-lock Brake System (ABS) speed sensor and harness
- Upper suspension arm from the steering knuckle
- Stabilizer bar from the link and remove the bracket
- Strut from the lower suspension arm.
- 3 upper strut mounting nuts and remove the strut.

6. Compress the coil spring.

7. Remove or disconnect the following:
- Piston rod locknut
- Suspension support, coil spring and bumper.

8. If disposing the strut, perform the following procedure:
   a. Fully extend the strut rod.
   b. Drill a hole near the bottom of the shock to remove the gas inside.

> **✳✳ CAUTION**
>
> **The gas is harmless, but be careful of chips that may fly up when the gas is released.**

To install:

9. Install or connect the following:
- Spring bumper
- Coil spring
- Suspension support to the rod and temporarily install a new nut

10. Turn the suspension support so one of the bolts on the support faces the same direction as shown in the illustration.

➡️**Align the bolt so a line drawn between the rod and bolt would be at**

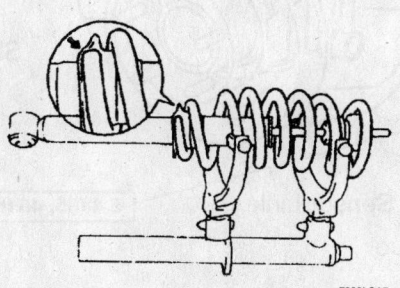

Matching the spring to the seat

90° to the direction of the lower bushing.

11. Install or connect the following:
- Spring compressor
- Strut and tighten the upper retaining nuts to 41 ft. lbs. (56 Nm).
- New upper strut rod nut to 20 ft. lbs. (27 Nm).
- Strut to the lower arm and temporarily tighten the nut and bolt.
- Stabilizer bar bracket and tighten the bolts to 21 ft. lbs. (28 Nm).
- The stabilizer bar to the link and tighten the bolts to 29 ft. lbs. (39 Nm).
- Upper suspension arm to the steering knuckle. Tighten the nut to 64 ft. lbs. (87 Nm) and install a new cotter pin.
- ABS speed sensor and tighten the bolt to 69 inch lbs. (8 Nm).
- Caliper
- Wheel

12. Bounce the vehicle several times to stabilize the suspension.

13. Tighten the lower strut bolt and nut to 116 ft. lbs. (157 Nm).

14. Check the front wheel alignment.

IS300

1. Before servicing the vehicle, refer to the precautions in the beginning of this section.

2. Remove or disconnect the following:

- Front wheel
- Wheel speed sensor and harness clamp
- Upper ball joint
- Level control sensor link
- Stabilizer bar link
- Lower strut bolt
- Upper strut mount cap
- Upper strut mount nuts
- Strut assembly

3. Install a suitable spring compressor and remove the center nut.

4. Remove the upper strut mount and the coil spring.

To install:

5. Installation is the reverse of the removal procedure, while using the following torque values:

- Upper strut mount center nut: 25 ft. lbs. (34 Nm)
- Upper strut mounting nuts: 26 ft. lbs. (35 Nm)
- Lower strut mount bolt: 47 ft. lbs. (64 Nm)

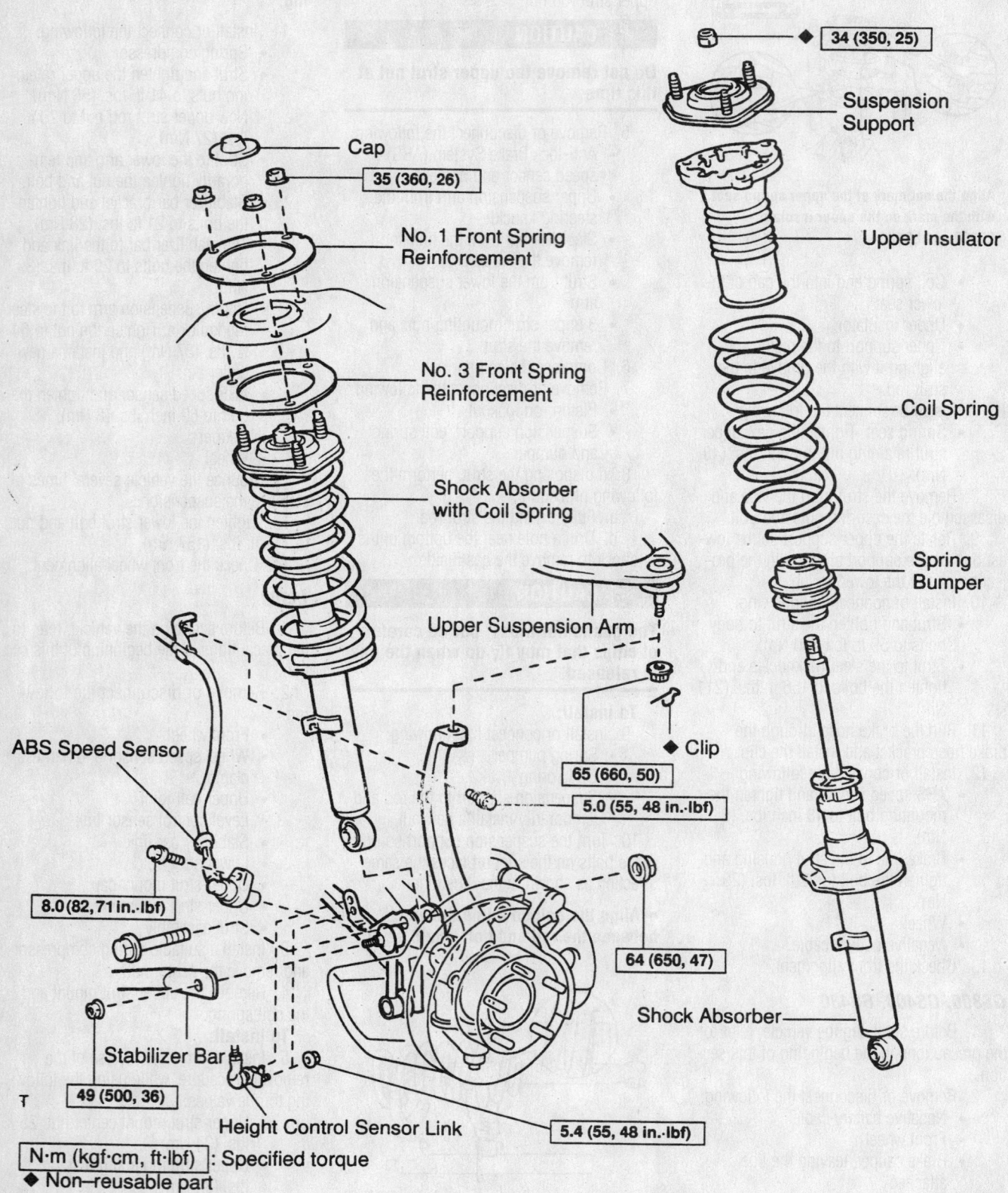

◆ 34 (350, 25)

Suspension Support

Cap

35 (360, 26)

No. 1 Front Spring Reinforcement

No. 3 Front Spring Reinforcement

Upper Insulator

Shock Absorber with Coil Spring

Coil Spring

Upper Suspension Arm

Spring Bumper

◆ Clip

65 (660, 50)

ABS Speed Sensor

5.0 (55, 48 in.-lbf)

8.0 (82, 71 in.-lbf)

64 (650, 47)

Shock Absorber

Stabilizer Bar

49 (500, 36)

Height Control Sensor Link

5.4 (55, 48 in.-lbf)

N·m (kgf·cm, ft·lbf) : Specified torque
◆ Non–reusable part

9347LG05

Exploded view of the front strut assembly mounting—IS300

- Upper ball joint nut: 50 ft. lbs. (65 Nm)
- Stabilizer bar link nut: 36 ft. lbs. (49 Nm)

LS400, LS430—WITHOUT AIR SUS-PENSION

1. Before servicing the vehicle, refer to the precautions in the beginning of this section.
2. Remove or disconnect the following:
   - Tire and wheel assembly
   - Steering knuckle from the upper ball joint
   - Strut assembly from the lower strut bracket

- Strut cover from the upper strut mount
- 3 mounting nuts and remove the strut assembly with the coil spring from the vehicle.

❋❋ CAUTION

Do not remove the center nut to the strut at this time.

3. Compress the coil spring.
4. Remove or disconnect the following:
   - Piston rod locknut
   - Suspension support, coil spring and the bumper

5. If disposing the strut, perform the following procedure:
   a. Fully extend the strut rod.
   b. Drill a hole within the shaded area shown in the illustration to remove the gas inside.

❋❋ CAUTION

The gas is harmless, but be careful of chips that may fly up when drilling.

   c. Properly dispose of the strut assembly.

To install:

6. Install or connect the following:
   - Spring bumper
   - Coil spring. Match the end of the coil into the recess of the strut spring seat.
   - Suspension support to the rod and temporarily install a new nut.
7. Turn the suspension support so one of the bolts on the support faces the same direction as shown in the illustration.

➥**Align the bolt so a line drawn between the rod and bolt would be at 90° to the direction of the lower bushing.**

8. Tighten the strut rod nut to 20 ft. lbs. (27 Nm) and install the cap.
9. Remove the spring compressor
10. Install or connect the following:
    - Strut and tighten the upper retaining nuts to 43 ft. lbs. (58 Nm).
    - Strut to the lower bracket and temporarily install the nut and bolt.
    - Upper control arm to the steering knuckle. Tighten the nut to 48 ft.

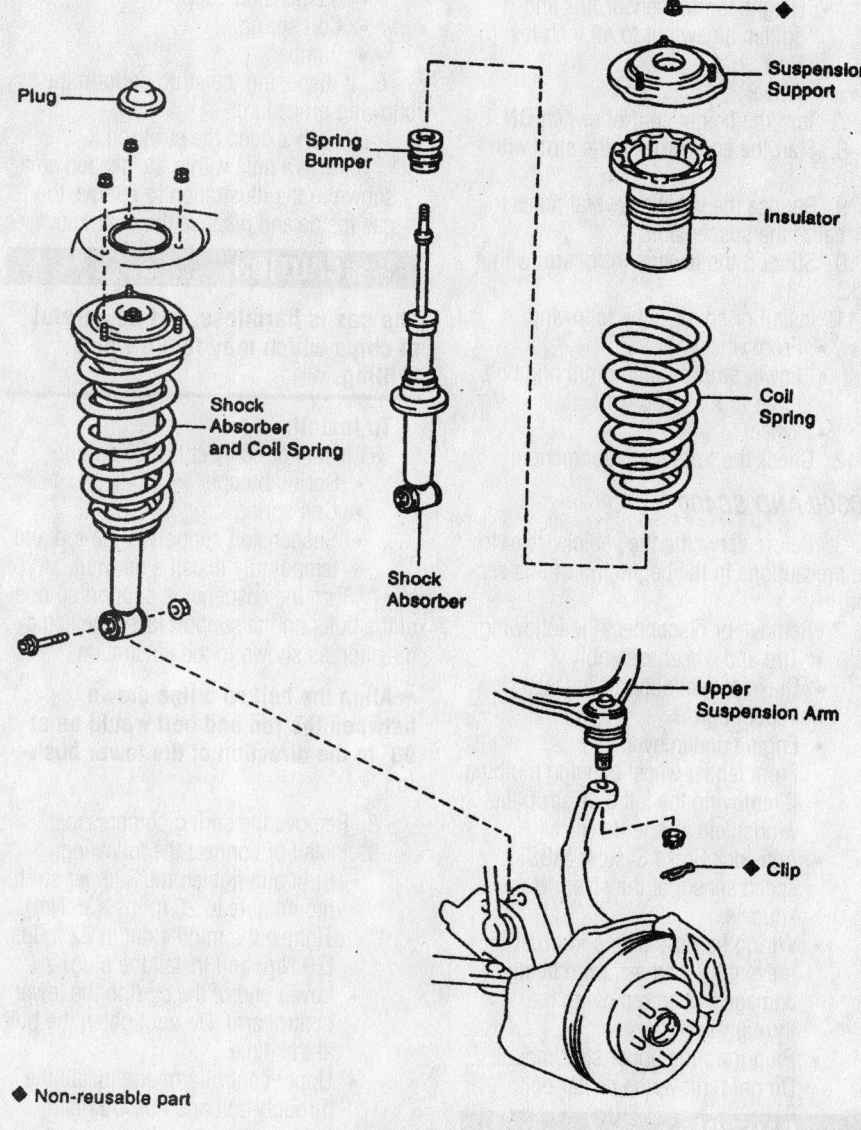

Plug

Spring Bumper

Shock Absorber and Coil Spring

Shock Absorber

Suspension Support

Insulator

Coil Spring

Upper Suspension Arm

◆ Clip

◆ Non-reusable part

7923LGA8

Exploded view of the strut and spring mounting—LS400 without air suspension shown

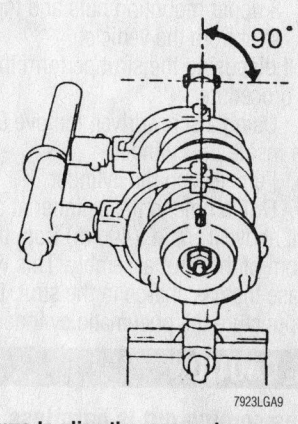

90°

7923LGA9

Be sure to align the suspension support with one of the upper mounting bolts as shown—LS400 without air suspension

For Wheel Alignment specifications, see Section 1 of this manual

lbs. (65 Nm) and install a new cotter pin.
- Wheel

11. Lower the vehicle.

12. Bounce the vehicle several times to stabilize the suspension.

13. Tighten the lower strut bolt and nut to 116 ft. lbs. (157 Nm).

14. Check the front wheel alignment.

LS400, LS430—WITH AIR SUSPENSION

1. Before servicing the vehicle, refer to the precautions in the beginning of this section.

2. Move the height control switch to **OFF**.

3. Bleed the air from the suspension.

4. Remove or disconnect the following:
- Wheel
- Height control sensor link from the lower strut bracket.
- Cotter pin and nut holding the upper control arm to the steering knuckle.
- Upper ball joint from the steering knuckle.
- Pneumatic cylinder from the lower bracket by removing the through-bolt.
- Air tube from the strut
- The actuator cover.

❊❊ CAUTION

Do not remove the center nut from the pneumatic cylinder.

- Actuator electrical connector
- 2 bolts to the suspension control actuator and position the actuator aside.
- 3 upper mounting nuts and the strut from the vehicle.

5. If disposing the strut perform the following procedure:

a. Using a screwdriver, remove the air from inside the cylinder.

b. Fully extend the cylinder.

c. Drill a hole in the cylinder at a point above 1.57 in. (40mm) from the bottom of the strut assembly. This will release the gas charge in the strut. Do not puncture the pneumatic cylinder.

❊❊ CAUTION

The gas coming out is harmless, but be careful of chips that may fly up while drilling.

To install:

6. Install or connect the following:

- Strut and tighten the upper mounting nuts to 43 ft. lbs. (58 Nm).
- Suspension control actuator and tighten the bolts. Tighten the 2 nuts to 13 ft. lbs. (17 Nm).
- Suspension control actuator cover and tighten the nuts to 43 ft. lbs. (58 Nm).
- 2 new O-rings to the air tube. Install the tube and tighten it to 13 ft. lbs. (17 Nm). Install the grommet.
- The strut to the lower strut bracket and temporarily install the nut and bolt.
- Steering knuckle to the upper ball joint. Tighten the nut to 48 ft. lbs. (65 Nm) and install a new cotter pin.
- Height control sensor link and tighten a new nut to 48 inch lbs. (5 Nm).
- Wheel

7. Turn the height control switch **ON**.

8. Start the engine to fill the strut with air.

9. Bounce the vehicle several times to normalize the suspension.

10. Support the lower control arm with a jack.

11. Install or connect the following:
- Front wheel
- Lower strut mounting nut and bolt to 76 ft. lbs. (106 Nm).
- Wheel

12. Check the front end alignment.

SC300 AND SC400

1. Before servicing the vehicle, refer to the precautions in the beginning of this section.

2. Remove or disconnect the following:
- Tire and wheel assembly
- Brake caliper support bracket
- Fender apron
- Engine undercover
- Front fender wheel opening molding
- If removing the left side strut, the windshield washer tank
- Anti-lock Brake System (ABS) speed sensor at the steering knuckle.
- Wiring harness clamp in order to prevent the harness from being damaged when removing the through-bolt.
- Plug from the upper strut mount. Do not remove the center bolt.

❊❊ CAUTION

Do not remove the center bolt to the strut at this time. The spring on the strut is under high pressure and can cause serious injury or vehicle damage.

- Upper control arm through-bolt from the subframe
- Upper control arm and turn the control arm completely around. It is not necessary to remove the upper ball joint.
- Strut at the lower control arm by removing the nut and bolt.
- 3 upper mounting nuts and remove the strut assembly with the coil spring from the vehicle.

3. Compress the coil spring.

4. Remove or disconnect the following:
- Piston rod locknut
- Suspension support
- Coil spring
- Bumper

5. If disposing the strut, perform the following procedure:

a. Fully extend the strut rod.

b. Drill a hole within the shaded area shown in the illustration to remove the gas inside and dispose the old strut.

❊❊ CAUTION

The gas is harmless, but be careful of chips which may fly up when drilling.

To install:

6. Install or connect the following:
- Spring bumper
- Coil spring
- Suspension support to the rod and temporarily install a new nut.

7. Turn the suspension support so one of the bolts on the support faces the same direction as shown in the illustration.

➡**Align the bolt so a line drawn between the rod and bolt would be at 90° to the direction of the lower bushing.**

8. Remove the spring compressor.

9. Install or connect the following:
- Strut and tighten the 3 upper strut mount nuts to 26 ft. lbs. (35 Nm). Tighten the middle nut to 22 ft. lbs. (29 Nm) and install the plug.
- Lower end of the strut to the lower control arm. Do not tighten the bolt at this time.
- Upper control arm and install the through-bolt and nut. Do not tighten the bolt at this time.
- VSS, wiring harness and the washer tank
- Fender apron and the engine undercover

- Caliper support bracket and tighten the bolts to 87 ft. lbs. (118 Nm).
- Tire and wheel assembly

10. Lower the vehicle.

11. Bounce the vehicle a few times to stabilize the suspension, then tighten the strut to lower arm bolt to 106 ft. lbs. (143 Nm). Tighten the upper arm to 121 ft. lbs. (164 Nm).

12. Check the front end alignment.

Rear

ES300

1. Before servicing the vehicle, refer to the precautions in the beginning of this section.

2. Remove or disconnect the following:
- Tire and wheel assembly
- Load sensing proportioning valve spring assembly from the lower arm.

- Anti-lock Brake System (ABS) speed sensor harness and brake line from the strut assembly.
- Stabilizer bar link from the strut.

3. Loosen the 2 nuts attaching the strut to the axle carrier.

4. Support the axle carrier.

5. Remove or disconnect the following:
- Rear seat back and package tray trim.

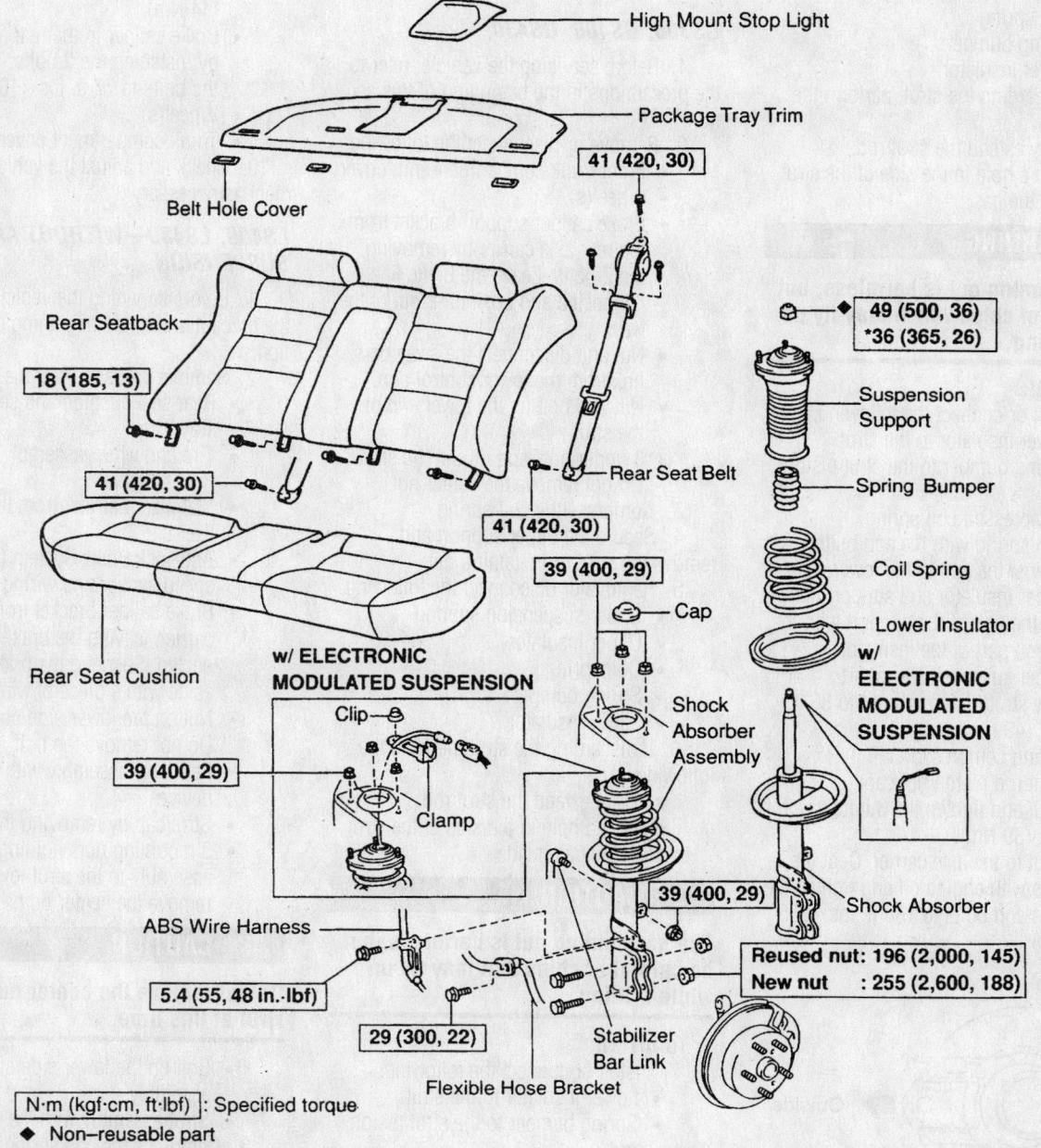

High Mount Stop Light

Package Tray Trim

Belt Hole Cover

41 (420, 30)

Rear Seatback

18 (185, 13)

41 (420, 30)

Rear Seat Cushion

w/ ELECTRONIC MODULATED SUSPENSION

Clip

39 (400, 29)

Clamp

ABS Wire Harness

5.4 (55, 48 in.·lbf)

29 (300, 22)

Flexible Hose Bracket

Rear Seat Belt

41 (420, 30)

39 (400, 29)

Cap

Shock Absorber Assembly

39 (400, 29)

Stabilizer Bar Link

◆ 49 (500, 36)
*36 (365, 26)

Suspension Support

Spring Bumper

Coil Spring

Lower Insulator

w/ ELECTRONIC MODULATED SUSPENSION

Shock Absorber

Reused nut: 196 (2,000, 145)
New nut : 255 (2,600, 188)

N·m (kgf·cm, ft·lbf) : Specified torque
◆ Non–reusable part
* For use with SST

Exploded view of the rear strut and coil spring mounting—ES300

7923LGA0

- Upper mounting nuts.
- 2 lower mounting bolts and remove the strut assembly.

6. Compress the coil spring.

7. Temporarily install a bolt and 2 nuts on the bracket at the lower end of the strut and secure it in a vise.

8. Secure the upper support and remove the strut rod retaining nut.

9. Remove or disconnect the following:
- Upper suspension support
- Upper insulator
- Coil spring
- Spring bumper
- Lower insulator

10. If discarding the strut, perform the following:
   a. Fully extend the strut rod.
   b. Drill a hole in the side of the strut to release the gas.

✳✳ WARNING

The gas coming out is harmless, but be careful of chips which may fly up while drilling.

To install:

11. Install or connect the following:
- Lower insulator to the strut.
- Spring bumper to the strut piston rod.
- Compressed coil spring.
- Coil spring with the end butted against the gap in the lower seat
- Upper insulator and support matching the bolt of the support with the cut-off part of the insulator.
- Upper suspension support
- New strut piston rod nut to 36 ft. lbs. (49 Nm)
- Spring compressor
- Strut rod piston nut cap
- Strut and tighten the 3 nuts to 29 ft. lbs. (39 Nm).
- Strut to the axle carrier. Coat the nuts with engine oil and tighten the nuts and bolts to 188 ft. lbs. (255 Nm).

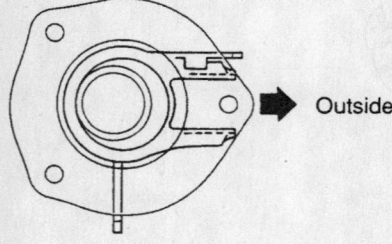

7923LGB1

Position the upper suspension support as shown when assembling the strut—ES300

- ABS harness to the strut and tighten the bolt to 48 inch lbs. (6 Nm).
- Brake line to the strut and tighten the retaining nut to 22 ft. lbs. (29 Nm).
- Spring to the lower arm and tighten the nut to 10 ft. lbs. (13 Nm).
- LSPV to the lower arm. Tighten the nut to 108 inch lbs. (12 Nm).
- Rear wheel
- Rear seat and package tray

GS300, GS400, GS430

1. Before servicing the vehicle, refer to the precautions in the beginning of this section.

2. Remove or disconnect the following:
- Front trunk compartment trim cover
- Wheel(s).
- Brake caliper support bracket from the rear axle carrier by removing the 2 bolts. Leave the brake line connected and position it out of the way.
- Nut and disconnect the sway bar link from the lower control arm.
- Nut and bolt on the lower end of the strut.
- 3 upper nuts and lift out the strut. Do not remove the center nut.

3. Compress the coil spring.

4. Secure the upper support and remove the strut rod retaining nut.

5. Remove or disconnect the following:
- Upper suspension support
- Upper insulator
- Coil spring
- Spring bumper
- Lower insulator

6. If discarding the strut, perform the following:
   a. Fully extend the strut rod.
   b. Drill a hole in the side of the strut to drain the gas inside

✳✳ CAUTION

The gas coming out is harmless, but be careful of chips that may fly up while drilling.

To install:

7. Install or connect the following:
- Lower insulator to the strut.
- Spring bumper to the strut piston rod.
- Compressed coil spring. Position the coil spring with the end butted against the gap in the lower seat.
- Upper insulator and suspension support.

- Upper suspension support and tighten a new strut piston rod nut to 20 ft. lbs. (27 Nm).

8. Remove the spring compressor.

9. Install or connect the following:
- Strut to the vehicle and tighten the 3 upper mounting nuts to 14 ft. lbs. (20 Nm). Install the cap.
- And tighten the lower strut bolt and nut to 101 ft. lbs. (137 Nm).
- Sway bar link to the lower control arm. Tighten the nut to 33 ft. lbs. (44 Nm).
- Brake caliper to the rear axle carrier by installing the 2 bolts. Tighten the bolts to 77 ft. lbs. (104 Nm).
- Wheel(s)
- Trunk compartment cover trim

10. Check and adjust the vehicle alignment as necessary.

LS400, LS430—WITHOUT AIR SUSPENSION

1. Before servicing the vehicle, refer to the precautions in the beginning of this section.

2. Remove or disconnect the following:
- Rear seat cushion and seat back.
- Tray trim
- Tire and wheel assembly
- Rear halfshaft
- Stabilizer bar link from the stabilizer bar
- Anti-lock Brake System (ABS) speed sensor and wiring harness
- Brake caliper bracket from the axle carrier, leaving the brake line connected. Suspend the brake caliper aside with a piece of wire.
- Nut on the lower side of the strut. Do not remove the bolt.
- Rear axle assembly with a lifting device
- Strut cap by removing the 3 nuts
- 3 mounting nuts holding the strut assembly to the strut tower. Do not remove the center bolt.

✳✳ CAUTION

Do not remove the center nut to the strut at this time.

- Bolt on the lower side of the strut assembly
- Strut assembly with the coil spring

3. Compress the coil spring.

4. Secure the strut housing in a vise.

5. Remove or disconnect the following:
- Strut rod retaining nut
- Upper suspension support
- Upper insulator

- Coil spring
- Spring bumper
- Lower insulator

6. If discarding the strut, perform the following:

   a. Fully extend the strut rod.

   b. Drill a hole in the strut (about 1 in. above the strut lower mount) and drain the gas inside

✳✳ CAUTION

The gas coming out is harmless, but be careful of chips which may fly up while drilling.

To install:

7. Install or connect the following:
   - Lower insulator to the strut.
   - Spring bumper to the strut piston rod.
   - Coil spring
   - Upper insulator and support
   - Upper suspension support

8. Temporarily install the upper strut rod retaining nut.

9. Rotate the suspension support so that the rod and one of the bolts on the suspension support are aligned with the lower bushing.

10. Remove the spring compressor.

11. Install or connect the following:
    - Strut assembly to the vehicle and tighten the 3 nuts to 47 ft. lbs. (64 Nm). Tighten the strut rod retaining nut to 20 ft. lbs. (27 Nm).
    - Strut assembly cap and install the 3 nuts.
    - Strut to the rear axle carrier. Install the bolt from the rear of the vehicle and temporarily tighten the nut.
    - Brake caliper and tighten the mounting bolts to 77 ft. lbs. (104 Nm)
    - ABS speed sensor and wiring harness
    - Stabilizer link to the stabilizer bar and tighten the nut to 48 ft. lbs. (65 Nm).
    - Rear halfshaft
    - Tire and wheel assembly

12. Bounce the vehicle up and down to stabilize the suspension.

13. Support the rear axle assembly with a lifting device. Tighten the lower strut bolt to 101 ft. lbs. (137 Nm).

14. Install or connect the following:
    - Rear seat cushion and rear seat back.
    - Package tray trim

15. Check the wheel alignment.

LS400, LS430—WITH AIR SUSPENSION

1. Before servicing the vehicle, refer to the precautions in the beginning of this section.

2. Bleed the air system from the suspension.

3. Remove or disconnect the following:
   - Rear seat cushion and seat back
   - Package tray trim
   - Trunk trim panel. Move the height control switch, located in the trunk area, to the **OFF** position.
   - Tire and wheel assembly
   - Rear halfshaft
   - Stabilizer links from the stabilizer bar
   - Anti-lock Brake System (ABS) speed sensor and wiring harness
   - Brake caliper bracket from the rear axle carrier. Do not disconnect the brake line.

4. Place matchmarks on the height control sensor link and bracket. Disconnect the height control sensor link from the No. 1 lower control arm.

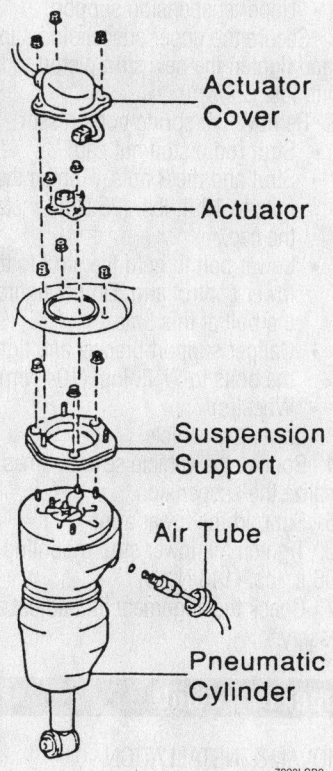

Pneumatic cylinder (strut) component overview (air suspension)

Actuator Cover

Actuator

Suspension Support

Air Tube

Pneumatic Cylinder

7923LGB2

5. Support the rear axle assembly with a lifting device.

6. Remove or disconnect the following:
   - Nut on the lower side of the shock absorber. Do not remove the bolt.
   - Grommet and disconnect the air tube from the shock absorber.
   - Actuator cover from the strut tower by removing the 3 nuts.
   - Actuator electrical connector from the top of the strut.
   - Actuator by removing the 2 nuts
   - 3 upper mounting nuts holding the strut to the strut tower.

7. Lower the rear axle assembly

8. Remove or disconnect the following:
   - Bolt on the lower side of the shock absorber.
   - Pneumatic cylinder strut assembly from the vehicle.
   - Suspension support from the strut assembly by removing the 3 nuts.

9. If discarding the pneumatic cylinder, perform the following:

   a. Using a screwdriver, depressurize the air from inside the cylinder.

   b. Drill a hole in the shaded area shown in the illustration and remove the gas inside.

✳✳ CAUTION

The gas coming out is harmless, but be careful of chips which may fly up when drilling.

To install:

10. Install or connect the following:
    - Suspension support to the pneumatic cylinder (strut) and tighten the nuts to 27 ft. lbs. (36 Nm).
    - Strut assembly to the vehicle and tighten the upper mounting nuts to 47 ft. lbs. (64 Nm).

11. Match the holes in the pneumatic cylinder with the holes in the suspension control actuator.

12. Install or connect the following:
    - Actuator and tighten the mounting nuts to 69 inch lbs. (8 Nm).
    - Actuator cover and tighten the 3 nuts to 18 ft. lbs. (25 Nm).
    - New O-rings and connect the air line to the shock absorber. Tighten the fitting to 13 ft. lbs. (18 Nm).
    - Strut to the rear axle carrier. Insert the bolt from the vehicle's rear and temporarily tighten the nut.
    - Height control sensor link to the

For Tune-up, Capacities and Firing orders, see Section 1 of this manual

No. 1 lower control arm. Mounting nut: 48 inch lbs. (5 Nm).
- Rear brake caliper to the rear axle carrier and tighten the mounting bolts to 77 ft. lbs. (104 Nm).
- ABS speed sensor and wiring harness.
- Stabilizer bar link and tighten the nut to 48 ft. lbs. (65 Nm).
- The halfshaft
- Actuator electrical connector to the top of the strut.
- Tire

13. Move the height control switch to the **ON** position. Start the engine and fill the pneumatic cylinder with air.

14. Bounce the vehicle up and down several times to stabilize the suspension.

15. Turn the suspension height control to the **OFF** position.

16. Remove the tire and wheel assembly.

17. Support the rear axle carrier with a lifting device. Tighten the lower strut bolt to 101 ft. lbs. (137 Nm).

18. Install or connect the following:
- Package tray trim
- Rear seat cushion and seat back

19. Turn the suspension control switch to the **ON** position.

20. Check the wheel alignment.

SC300 AND SC400

1. Before servicing the vehicle, refer to the precautions in the beginning of this section.

2. Raise the rear of the vehicle and support it with safety stands.

3. Remove or disconnect the following:
- Wheel(s).
- Brake caliper support bracket by removing the 2 bolts. Leave the brake line connected and position it aside.
- Nut and bolt on the lower end of the strut.
- Cap nut on the upper end of the strut. Remove the 3 upper nuts and lift out the strut. Do not remove the center nut from the strut.

✻✻ CAUTION

Do not remove the center nut on the strut at this time. The spring on the strut is under high pressure and can cause serious injury.

4. Compress the coil spring.

5. Secure the strut housing with 2 nuts and a bolt as shown in the illustration and secure it in a vise.

6. Secure the upper support and remove the strut rod retaining nut.

7. Remove the upper suspension support, upper insulator, coil spring, spring bumper and the lower insulator.

8. If discarding the strut, perform the following:
   a. Fully extend the strut rod.
   b. Drill a hole in the strut in the shaded area shown in the illustration and drain the gas inside

✻✻ CAUTION

The gas coming out is harmless, but be careful of chips which may fly up while drilling.

To install:

9. Install or connect the following:
- Lower insulator to the strut
- Spring bumper to the strut piston rod
- The (compressed) coil spring

10. Position the coil spring with the end butted against the gap in the lower seat.
- Upper insulator and support matching the bolt of the support with the cut off part of the insulator.
- Upper suspension support.

11. Secure the upper suspension support and tighten the new strut piston rod nut to 20 ft. lbs. (27 Nm).

12. Remove the spring compressor.
- Strut rod piston nut cap
- Strut and the 3 nuts. Tighten the nuts to 19 ft. lbs. (25 Nm). Install the cap.
- Lower bolt to hold the strut to the lower control arm. Do not tighten the bolt at this time.
- Caliper support bracket and tighten the bolts to 77 ft. lbs. (104 Nm).
- Wheel(s)

13. Lower the vehicle.

14. Bounce the vehicle several times to normalize the suspension.

15. Support the lower arm.

16. Tighten the lower strut mounting bolt to 106 ft. lbs. (143 Nm).

17. Check the alignment and adjust as necessary.

Upper Ball Joint

REMOVAL & INSTALLATION

The upper ball joint is an integral part of the upper arm and is not replaced separately. The upper ball joint replacement is accomplished by replacing the upper arm.

Upper Control Arm

REMOVAL & INSTALLATION

GS300

1. Before servicing the vehicle, refer to the precautions in the beginning of this section.

2. Remove or disconnect the following:
- Negative battery cable
- Wheel

3. Loosen the 3 upper strut mounting nuts.

4. Loosen, but do not remove, the upper strut rod nut.

✻✻ CAUTION

DO NOT completely remove the upper strut nut at this time.

5. Remove or disconnect the following:
- Brake caliper, leaving the line attached and secure it out of the way
- Anti-lock Brake System (ABS) speed sensor and harness
- Cotter pin and nut from the upper control arm
- Upper control arm from the steering knuckle
- Stabilizer bar from the link and remove the bracket
- Cotter pin and nut from the lower control arm
- Strut from the lower suspension arm
- 3 upper strut mounting nuts and remove the strut
- Mounting bolts holding the upper control arm to the frame
- Upper control arm from the vehicle.

To install:

6. Install or connect the following:
- Upper suspension arm and tighten the mounting bolts to 39 ft. lbs. (53 Nm).
- Strut and tighten the upper retaining nuts to 41 ft. lbs. (56 Nm). Tighten the new upper strut rod nut to 20 ft. lbs. (27 Nm).
- Strut to the lower arm and temporarily tighten the nut and bolt.
- Stabilizer bar bracket and tighten the bolts to 21 ft. lbs. (28 Nm).
- Stabilizer bar to the link and tighten the bolts to 29 ft. lbs. (39 Nm).
- Upper suspension arm to the steering knuckle. Tighten the nut to 64 ft. lbs. (87 Nm) and install a new cotter pin.

- ABS speed sensor and tighten the bolt to 69 inch lbs. (8 Nm).
- Caliper
- Front wheel

7. Lower the vehicle.

8. Bounce the vehicle several times to stabilize the suspension.

9. Tighten the lower strut bolt and nut to 116 ft. lbs. (157 Nm).

10. Check the front wheel alignment.

IS300

1. Before servicing the vehicle, refer to the precautions in the beginning of this section.

2. Remove or disconnect the following:
- Front wheel
- Strut and spring assembly
- Inner bolts and the control arm

To install:

3. Install or connect the following:
- Control arm and tighten the inner bolts to 44 ft. lbs. (59 Nm)
- Strut and spring assembly
- Front wheel

LS400, LS430

1. Before servicing the vehicle, refer to the precautions in the beginning of this section.

2. Raise and safely support the vehicle.

3. Remove or disconnect the following:
- Wheel
- Strut or if equipped with air suspension, remove the pneumatic cylinder.
- Anti-lock Brake System (ABS) speed sensor wire harness from the upper control arm by removing the bolt.
- Mounting bolts holding the upper control arm to the vehicle.
- Upper control arm

To install:

4. Install or connect the following:
- Upper control arm and tighten the 2 mounting bolts to 83 ft. lbs. (113 Nm).
- ABS speed sensor wire harness to the upper control arm with the attaching bolt.
- Strut, or if equipped with air suspension, install the pneumatic cylinder.
- Wheel

5. Lower the vehicle.

6. Check and adjust the wheel alignment as necessary.

SC300 and SC400

1. Raise the front of the vehicle and support it on safety stands.

2. Remove or disconnect the following:

- Wheel
- Caliper support bracket by removing the 2 bolts. Leave the brake line connected and suspend it aside.
- Rotor
- Front fender splash shield, fender liner and wheel opening molding.
- On the left side, the washer tank
- Bolt and disconnect the Anti-lock Brake System (ABS) speed sensor from the steering knuckle. Remove the 3 bolts and disconnect the wire harness clamp.
- Cotter pin and the nut from the upper ball joint; press the upper ball joint from the knuckle.
- Through-bolt, nut and the upper control arm.

To install:

3. Install or connect the following:
- Upper control arm. Connect the upper control arm to the subframe and install the through-bolt. Do not tighten the bolt at this time.

➡ The upper control arm mounting bolts are not tightened until the sus-

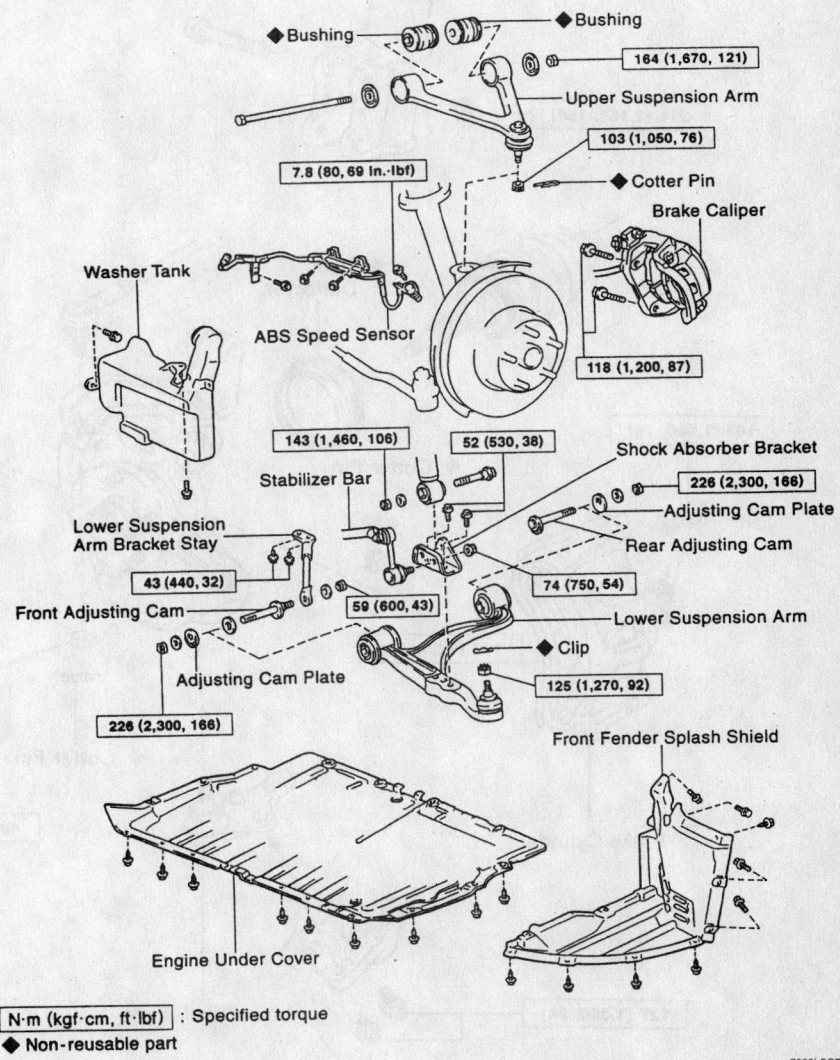

N·m (kgf·cm, ft·lbf) : Specified torque
◆ Non-reusable part

7923LGC7

Exploded view of the front suspension control arms and related components—SC300 and SC400 models

pension has been assembled and vehicle is on the ground.

- Ball joint to the knuckle and tighten the nut to 76 ft. lbs. (103 Nm). Install a new cotter pin.
- Wire harness and ABS speed sensor. Tighten the speed sensor to knuckle bolt to 69 inch lbs. (8 Nm).
- Washer tank, the fender liner, splash shield and molding.
- Rotor
- Caliper support bracket and tighten the bolts to 87 ft. lbs. (118 Nm).

- Wheel
4. Lower the vehicle.
5. Bounce the suspension several times to set the suspension.
6. Support the lower arm and tighten the upper control arm through-bolt and nut to 121 ft. lbs. (164 Nm).
7. Check the front wheel alignment and adjust as necessary.

CONTROL ARM BUSHING REPLACEMENT

The control arm bushings are serviced with the control arm as an assembly.

Lower Control Arm

REMOVAL & INSTALLATION

ES300

1. Before servicing the vehicle, refer to the precautions in the beginning of this section.
2. Remove or disconnect the following:

- Negative battery cable
- Front wheel(s)
- Side fender apron seal

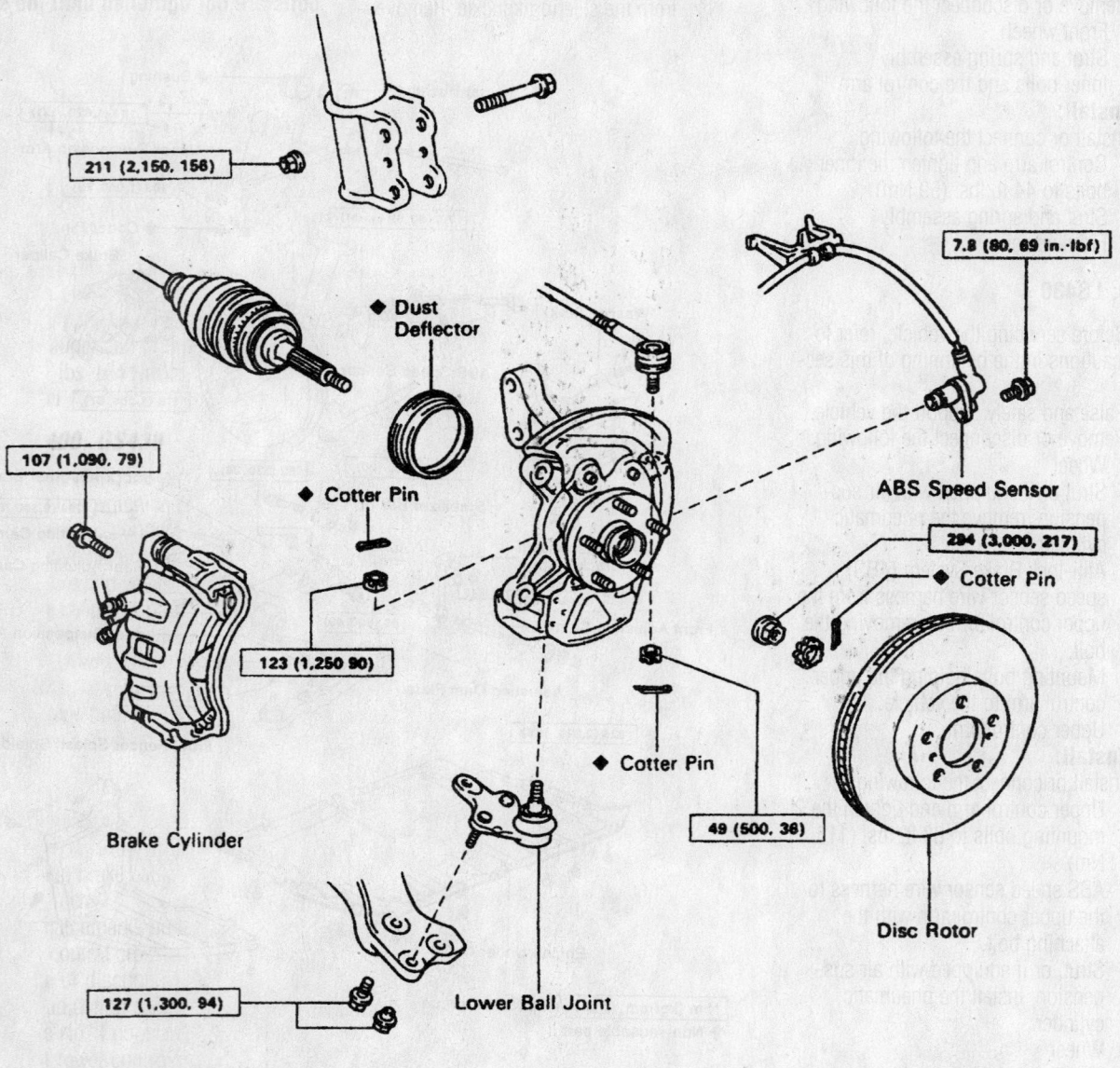

211 (2,150, 156)

7.8 (80, 69 in.·lbf)

◆ Dust Deflector

107 (1,090, 79)

◆ Cotter Pin

ABS Speed Sensor

294 (3,000, 217)

◆ Cotter Pin

123 (1,250 90)

◆ Cotter Pin

49 (500, 36)

Brake Cylinder

Disc Rotor

127 (1,300, 94)

Lower Ball Joint

N·m (kgf·cm, ft·lbf) : Specified torque
◆ Non-reusable part

Exploded view of the lower suspension—ES300

7923LGB3

- Steering knuckle with the axle hub, from the vehicle.
- Dust deflector from the knuckle
- Cotter pin and the nut from the ball joint stud.

3. Remove the lower ball joint from the steering knuckle.

To install:

4. Install the lower ball joint onto the steering knuckle and tighten nut to 90 ft. lbs. (123 Nm). Install new cotter pin.

5. Align the hole in the dust deflector with the ABS speed sensor. Using the appropriate driver, install a new dust deflector.

6. Install or connect the following:
- Steering knuckle and hub onto the vehicle.
- Fender apron seal
- Front wheel(s)
- The negative battery cable

GS300, GS400, GS430

1. Before servicing the vehicle, refer to the precautions in the beginning of this section.

2. Remove or disconnect the following:
- Negative battery cable
- Wheel(s)
- Caliper, leaving the brake line connected and suspend it out of the way.

✶✶ WARNING

Never allow the brake caliper to hang freely from the brake hose.

- Rotor
- Anti-lock Brake System (ABS) speed sensor and harness
- Tie rod end from the arm on the lower ball joint
- Cotter pin and nut. Disconnect the upper control arm from the steering knuckle.
- Cotter pin and nut. Disconnect the steering knuckle from the lower control arm.
- Steering knuckle and ball joint assembly from the vehicle.
- 2 ball joint mounting bolts, then remove the ball joint from the steering knuckle.

To install:

3. Install or connect the following:
- Ball joint and tighten the bolts to 83 ft. lbs. (113 Nm).
- Steering knuckle to the lower and

upper suspension arms. Tighten the lower control arm nut to 95 ft. lbs. (127 Nm) and install a new cotter pin. Tighten the upper control arm to 64 ft. lbs. (87 Nm) and install a new cotter pin.
- Tie rod end to the ball joint arm. Tighten the nut to 64 ft. lbs. (87 Nm) and install a new cotter pin.
- Rotor
- Caliper
- ABS speed sensor and harness. Tighten the sensor retaining bolt to 69 inch lbs. (8 Nm).
- Wheel(s)
- Negative battery cable

4. Check the front wheel alignment.

IS300

1. Before servicing the vehicle, refer to the precautions in the beginning of this section.

2. Remove or disconnect the following:
- Front wheel
- Engine under covers
- Level control sensor link
- Front subframe brace
- No. 2 lower control arm
- Brake caliper and rotor
- Outer tie rod end
- Stabilizer bar link
- Lower strut bolt
- Lower ball joint
- Steering gear
- No. 1 lower control arm

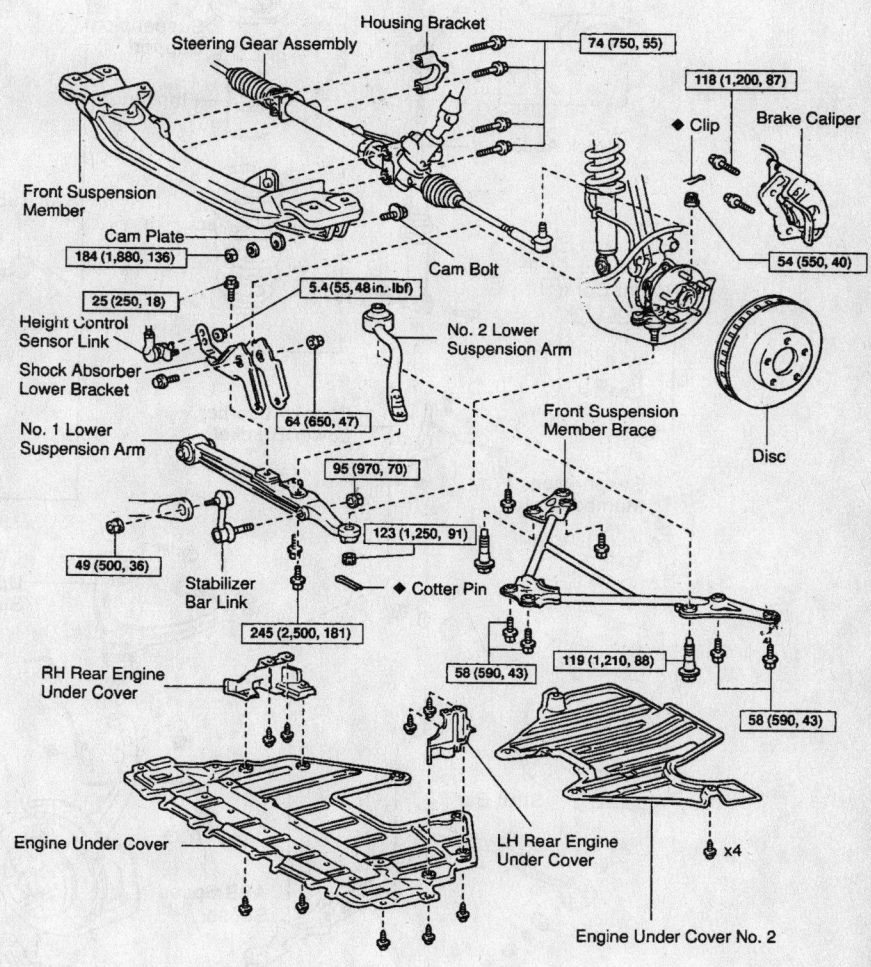

Exploded view of the front suspension—IS300

N·m (kgf·cm, ft·lbf) : Specified torque
◆ Non–reusable part

9347LG06

To install:

3. Installation is the reverse of the removal procedure, while using the following torque values:

- No. 1 lower control arm bolt: 136 ft. lbs. (184 Nm)
- Lower ball joint nut: 91 ft. lbs. (123 Nm)
- Outer tie rod end nut: 40 ft. lbs. (54 Nm)
- Brake caliper bolts: 87 ft. lbs. (118 Nm)
- No. 2 control arm-to-No. 1 control arm bolts: 181 ft. lbs. (245 Nm)

- Front subframe brace small bolts: 43 ft. lbs. (58 Nm)
- Front subframe large bolts: 88 ft. lbs. (119 Nm)

LS400, LS430

1. Before servicing the vehicle, refer to the precautions in the beginning of this section.

2. If equipped with air suspension, move the height control switch (located in the trunk) to the **OFF** position.

3. Remove or disconnect the following:
- Tire and wheel assembly

- Anti-lock Brake System (ABS) speed sensor and wiring harness from the steering knuckle.
- Brake caliper support bracket by removing the 2 bolts. Leave the brake line connected. Support the caliper aside by using a piece of wire.

4. Loosen the 2 lower ball joint mounting bolts.

➡ **Do not remove the bolts.**

5. Remove or disconnect the following:
- Clip and nut from the tie rod end

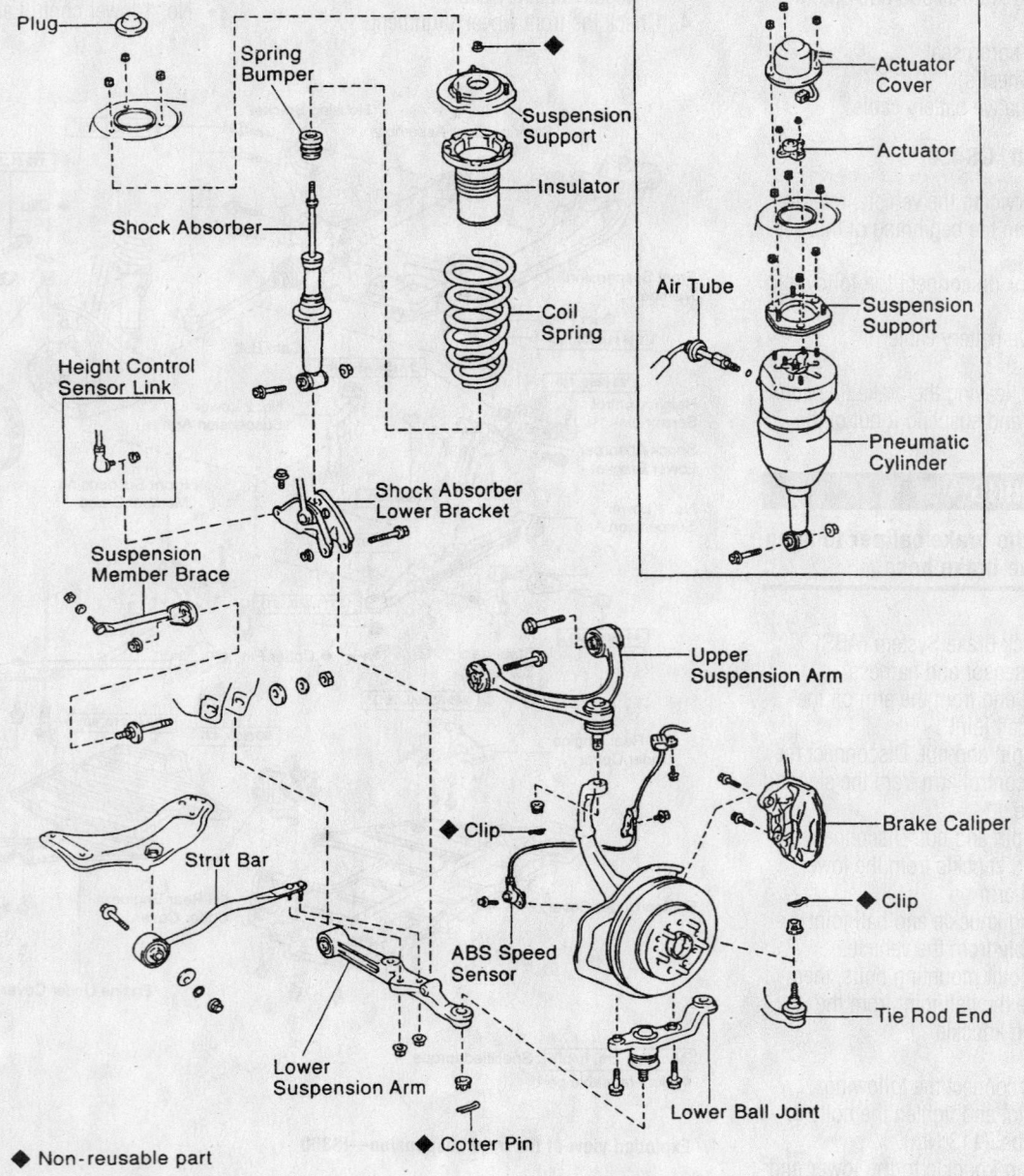

Plug — Spring Bumper — Suspension Support — Insulator — Shock Absorber — Coil Spring — Height Control Sensor Link — Shock Absorber Lower Bracket — Suspension Member Brace — Actuator Cover — Actuator — Air Tube — Suspension Support — Pneumatic Cylinder — Upper Suspension Arm — Clip — Brake Caliper — Clip — ABS Speed Sensor — Strut Bar — Lower Suspension Arm — Cotter Pin — Tie Rod End — Lower Ball Joint

◆ Non-reusable part

Exploded view of the lower ball joint mounting—LS400, LS430

7923LGB4

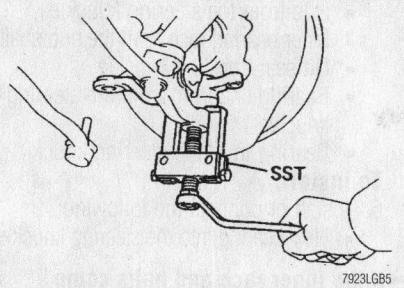

Disconnecting the ball joint from the lower suspension arm—LS400, LS430

- Tie rod end from the steering arm with the proper tool.
- Lower ball joint mounting bolts from the steering knuckle.
- Cotter pin and nut from the lower ball joint.
- Lower ball joint from the lower control arm

To install:

6. Install or connect the following:
   - Ball joint to the lower control arm. Tighten the nut to 112 ft. lbs. (152 Nm) and install a new cotter pin.
   - Mounting bolts, temporarily, holding the ball joint to the steering knuckle
   - Tie rod end to the steering knuckle. Tighten the nut to 48 ft. lbs. (65 Nm) and install a new cotter pin.
   - Lower ball joint bolts to 83 ft. lbs. (113 Nm).
   - Brake caliper support bracket and tighten the 2 bolts to 87 ft. lbs. (118 Nm).
   - ABS speed sensor and wiring harness to the steering knuckle.
   - Wheel
7. Lower the vehicle.
8. Turn the height control switch **ON**.

SC300 and SC400

The lower ball joint is not replaceable. If the lower ball joint is defective, replace the lower arm and ball joint as an assembly, as follows:

1. Before servicing the vehicle, refer to the precautions in the beginning of this section.
2. Raise the front of the vehicle and support it on safety stands.
3. Remove or disconnect the following:
   - Wheel and the engine undercover
   - Caliper support bracket from the vehicle by removing the 2 bolts. Support the caliper and bracket

with a wire. Do not let the assembly hang from the brake line.
   - Nut and disconnect the stabilizer bar from the lower control arm.
   - Cotter pin and nut from the lower ball joint. Press the lower ball joint out of the steering knuckle.
   - Lower end of the strut by removing the nut and bolt.
   - Nut, 2 bolts and the front lower arm bracket stay
4. Matchmark the front and rear adjustment cams to the body and then remove the nuts and adjusting cams.
5. Lift out the lower control arm.

To install:

6. Install or connect the following:
   - Bracket to the lower control arm by installing the 2 bolts. Tighten the bolts to 38 ft. lbs. (52 Nm).
   - Lower control arm to the body and temporarily install the adjusting cams and nuts. Do not tighten the nuts at this time.
   - Lower control arm to the knuckle and tighten the ball joint nut to 92 ft. lbs. (125 Nm). Install a new cotter pin.
   - Strut to the arm and tighten the bolt and nut to 106 ft. lbs. (143 Nm).
   - Stabilizer bar link and tighten the nut to 54 ft. lbs. (74 Nm).
   - Brake caliper support bracket to the vehicle and tighten the bolts to 87 ft. lbs. (118 Nm.
   - Wheel
7. Lower the vehicle.
8. Bounce it several times to set the suspension.
9. Support the lower arm, align the matchmarks on the adjusting cams and tighten the nuts to 166 ft. lbs. (226 Nm).
10. Check the wheel alignment.

CONTROL ARM BUSHING REPLACEMENT

The control arm bushings are serviced with the control arm as an assembly.

Wheel Bearings

ADJUSTMENT

Check the backlash in bearing shaft direction and the axle hub deviation. Maximum for backlash should be 0.0020 in. (0.05mm) and for axle hub deviation 0.020 in. (0.05mm).

➡**The front and rear wheel bearings are non-adjustable. If the wheel bearing is out of specifications, replace the wheel bearing.**

REMOVAL & INSTALLATION

Front

ES300

1. Before servicing the vehicle, refer to the precautions in the beginning of this section.
2. Remove or disconnect the following:
   - Negative battery cable
   - Front wheels
   - Fender apron seal
   - Cotter pin and lock cap from the end of the halfshaft
   - Halfshaft locknut.
   - Brake caliper and use a wire to support it out of the way.

✳✳ WARNING

Never allow the caliper to hang freely from the brake hose.

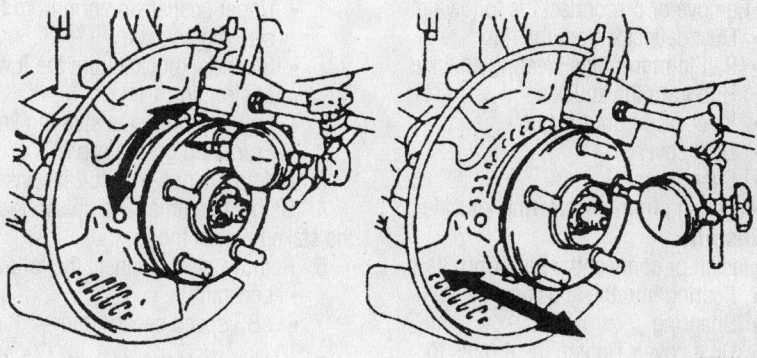

Checking wheel bearings for excessive play

Timing belt service is covered in Section 3 of this manual

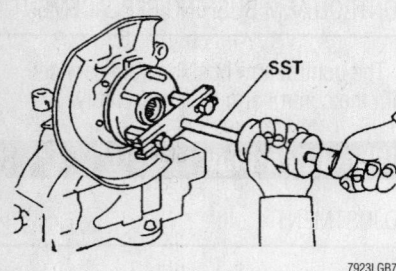

Removing the axle hub from the steering knuckle—ES300

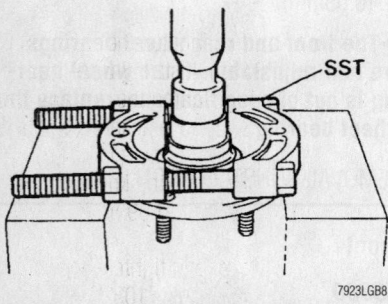

Remove the inner race from the hub—ES300

- Rotor
- Anti-lock Brake System (ABS) speed sensor from the steering knuckle

3. Loosen the nuts on the lower end of the strut.

4. Disconnect and separate the tie rod end from the steering knuckle.

5. Remove or disconnect the following:
- Lower control arm from the ball joint
- Driveshaft from the axle hub
- 2 nuts on the lower end of the strut
- Steering knuckle

6. Clamp the steering knuckle in a vise with soft jaws to protect the knuckle.

7. Remove or disconnect the following:
- Dust deflector from the hub
- Ball joint from the steering knuckle
- Hub from the knuckle
- Inner race from the hub
- Dust cover
- Snapring
- Bearing from the steering knuckle

To install:

8. Install or connect the following:
- Bearing into the knuckle
- Snapring
- Dust cover. Tighten the 4 bolts to 74 inch lbs. (8.3 Nm).
- Hub into the steering knuckle
- Lower ball joint to the steering knuckle. Tighten the nut to 90 ft.

lbs. (123 Nm) and install a new cotter pin.
- Dust deflector
- Knuckle on the lower strut
- Lower ball joint to the lower arm. Tighten the bolts to 94 ft. lbs. (127 Nm).
- Tie rod end to the steering knuckle. Tighten the nut to 36 ft. lbs. (49 Nm).
- Nuts on the lower strut to 156 ft. lbs. (211 Nm).
- ABS speed sensor. Tighten the mounting bolt to 69 inch lbs. (8 Nm).
- Rotor
- Caliper. Tighten the mounting bolts to 79 ft. lbs. (107 Nm).
- Axle locknut. Tighten the nut to 217 ft. lbs. (294 Nm). Install the lock cap and a new cotter pin.
- Front fender apron seal
- Wheel

9. Turn the wheel by hand, verify that the wheel turns without noise and without binding.

10. Lower the vehicle.

GS300, GS400, GS430

1. Before servicing the vehicle, refer to the precautions in the beginning of this section.

2. Remove or disconnect the following:
- Negative battery cable.
- Front wheel
- Caliper, leaving the brake line connected and suspend it out of the way.

❋❋ WARNING

Never allow the brake caliper to hang freely from the brake hose.

- Rotor
- Anti-lock Brake System (ABS) speed sensor and harness
- Tie rod from the arm on the lower ball joint
- Upper suspension arm from the steering knuckle
- Steering knuckle from the lower control arm
- Ball joint from the steering knuckle
- Front hub grease cap

3. Clamp the hub in a soft jaw vise.

4. Using a hammer and chisel, loosen the staked part of the locknut.

5. Remove or disconnect the following:
- Locknut
- ABS speed sensor rotor

➡ Do not scratch the serrations of the sensor rotor.

- Brake dust cover bolts and shift the cover toward the outside.

- Hub from the steering knuckle
- Inner bearing race from the hub shaft
- Oil seal from the knuckle
- Bearing snapring from the steering knuckle
- Bearing from the steering knuckle

To install:

6. Install or connect the following:
- New bearing into the steering knuckle

➡ If the inner race and balls come loose from the bearing outer race, be sure to install them on the same side as before.

- Snapring
- New outside inner race and tap in the new seal. Tap the seal until it is flush with the end surface of the steering knuckle.
- Brake dust cover to the knuckle and tighten the bolts to 74 inch lbs. (8 Nm).
- Hub into the steering knuckle
- ABS speed sensor rotor.
- Axle hub locknut. Tighten the nut to 147 ft. lbs. (199 Nm) and stake it.
- Grease cap to the steering knuckle by tapping lightly around the circumference of the cap with a hammer.
- Ball joint to the steering knuckle. Tighten the 2 bolts to 83 ft. lbs. (113 Nm).
- Steering knuckle to the upper and lower suspension arms. Tighten the upper nut to 64 ft. lbs. (87 Nm) and the lower nut to 95 ft. lbs. (127 Nm). Install a new cotter pin on the lower nut. Install the clip on the upper suspension arm nut.
- Tie rod end to the steering knuckle. Tighten the nut to 64 ft. lbs. (87 Nm) and install a new cotter pin.
- Rotor, disc brake pads and the brake caliper.
- ABS speed sensor and harness. Tighten the sensor retaining bolt to 69 inch lbs. (8 Nm).
- Wheel

7. Lower the vehicle and connect the negative battery cable.

8. Check the front wheel alignment.

LS400, LS430

1. Before servicing the vehicle, refer to the precautions in the beginning of this section.

2. If equipped with air suspension, move the height control switch in the trunk area to the **OFF** position.

3. Remove or disconnect the following:
- Front tire and wheel assembly
- Brake caliper bracket from the steering knuckle, leaving the brake

line connected. Support the caliper with a piece of wire.

- Brake rotor
- Anti-lock Brake System (ABS) speed sensor from the steering knuckle
- Steering knuckle from the lower ball joint by removing the 2 bolts.
- Steering knuckle from the upper ball joint
- Steering knuckle with the axle hub from the vehicle
- Grease cap from the hub
- Nut and the speed sensor rotor
- 4 bolts and shift the brake dust cover towards the hub side.
- Axle hub from the steering knuckle
- Outside inner race from the axle
- Oil seal from the steering knuckle
- Snapring and bearing from the steering knuckle.

To install:

4. Install or connect the following:
- Bearing in the steering knuckle
- Snapring
- Inner race (outside)
- New oil seal until it is flush with the end surface of the steering knuckle
- Brake dust cover to the steering knuckle and tighten the bolts to 74 inch lbs. (8.4 Nm).
- Axle hub to the steering knuckle
- ABS speed sensor
- New nut on the axle shaft. Tighten the nut to 147 ft. lbs. (199 Nm). Stake the nut and install the grease cap.
- Steering knuckle to the lower ball joint and tighten the bolts to 83 ft. lbs. (113 Nm).
- Steering knuckle to the upper ball joint and tighten the nut to 48 ft. lbs. (65 Nm).
- Brake rotor
- Brake caliper and tighten the 2 bolts to 87 ft. lbs. (118 Nm).

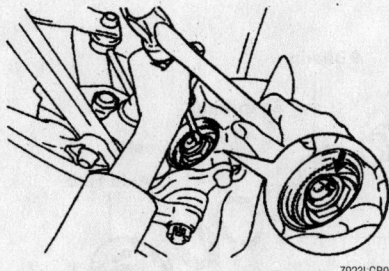

Axle hub nut is located on the inboard side of the knuckle assembly—LS400, LS430, SC300 and SC400

- Speed sensor to the steering knuckle
- Front tire and wheel assembly

5. If equipped with air suspension, turn the height control switch to the **ON** position.

SC300 AND SC400

1. Before servicing the vehicle, refer to the precautions in the beginning of this section.

2. Remove or disconnect the following:
- Front tire and wheel assembly
- Brake caliper support bracket, leaving the brake line connected
- Rotor by removing the 2 screws.
- Anti-lock Brake System (ABS) speed sensor
- Cotter pin and nut and disconnect the tie rod from the steering knuckle.
- Cotter pin and nut and disconnect the steering knuckle from the upper control arm.
- Clip and nut and press the knuckle off the lower control arm.
- Steering knuckle from the vehicle
- Hub bearing cap from the steering knuckle
- Hub nut
- ABS sensor rotor
- 4 bolts and shift the brake dust shield toward the hub (outside).
- Axle hub from the knuckle
- Inner bearing race from the axle hub
- Oil seal
- Snapring and bearing

To install:

3. Press the bearing into the knuckle. If the inner race and balls come loose from the outer race, be sure to install them on the same side as before.

4. Install or connect the following:
- Snapring and inner race, then tap in a new oil seal until it is flush with the end surface of the knuckle.

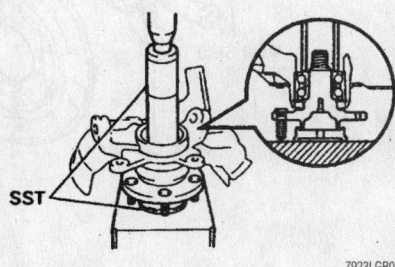

Pressing the hub into the knuckle—SC300 and SC400

- Brake dust cover and tighten the bolts to 74 inch lbs. (8.3 Nm).
- Hub into the knuckle
- Speed sensor
- New locknut and tighten it to 147 ft. lbs. (199 Nm). Stake the nut with a chisel. Tap the bearing cap into place.
- Knuckle to the upper control arm and tighten the nut to 76 ft. lbs. (103 Nm). Install a new cotter pin.
- Knuckle to the lower control arm and tighten the nut to 92 ft. lbs. (125 Nm). Install a new clip.
- Tie rod end to the steering knuckle with the nut. Tighten the nut to 36 ft. lbs. (49 Nm). Install a new cotter pin.
- Rotor by installing the 2 screws.
- Caliper support bracket and tighten the bolt to 87 ft. lbs. (118 Nm).
- Speed sensor to the knuckle and tighten the bolt to 69 inch lbs. (8 Nm).
- Front wheel and tighten the lug nuts to 76 ft. lbs. (103 Nm).

5. Lower the vehicle.
6. Check the front end alignment and ABS speed sensor signal.

IS300

1. Before servicing the vehicle, refer to the precautions in the beginning of this section.

2. Remove or disconnect the following:
- Front wheel
- Brake caliper and rotor
- Wheel speed sensor
- Upper and lower ball joints
- Steering knuckle from the vehicle
- Grease cap
- Hub locknut
- Brake dust cover
- Wheel speed sensor pulse ring

3. Press the hub out of the wheel bearing.

4. Remove the grease seal and the snapring, then press the wheel bearing out of the steering knuckle.

To install:

➡**Use a new hub locknut for assembly.**

5. Installation is the reverse of the removal procedure, while using the following torque values:
- Hub locknut: 108 ft. lbs. (147 Nm)
- Brake dust cover bolts: 74 inch lbs. (8.3 Nm)
- Lower ball joint bolts: 83 ft. lbs. (113 Nm)
- Upper ball joint nut: 50 ft. lbs. (65 Nm)

Heater Core replacement is covered in Section 2 of this manual

- Brake caliper support bolts: 87 ft. lbs. (118 Nm)
- Wheel lug nuts: 76 ft. lbs. (103 Nm)

Rear

ES300

1. Before servicing the vehicle, refer to the precautions in the beginning of this section.
2. Raise and safely support the vehicle.
3. Remove or disconnect the following:
   - Rear tire and wheel assembly
   - If equipped with rear disc brakes, the caliper mounting bolts. Leave the brake line connected and suspend the assembly out of the way.
   - Brake rotor or drum
   - 4 bolts and pull off the rear axle hub
   - O-ring

➡ **If it is necessary to replace the hub or bearing, replace the components as an assembly.**

To install:

4. Install or connect the following:
   - Hub on the carrier and tighten the bolts to 59 ft. lbs. (80 Nm).
   - Rotor or drum
   - If equipped with rear disc brakes, the caliper and tighten the bolts to 34 ft. lbs. (64 Nm).
   - Wheel

GS300, GS400, GS430

1. Before servicing the vehicle, refer to the precautions in the beginning of this section.
2. Remove or disconnect the following:
   - Negative battery cable.
   - Rear tire and wheel assembly
   - Brake caliper support from the rear axle carrier and support it with a piece of wire.
3. Place matchmarks on the disc brake rotor and the axle hub.
4. Remove or disconnect the following:
   - Brake rotor
   - Speed sensor
   - Rear halfshaft
   - Parking brake shoes
   - Parking brake cable
   - Strut rod
5. Place matchmarks on the adjusting cam and rear control crossmember.
6. Remove or disconnect the following:
   - Nut, adjusting cam and the washer to the No. 1 control arm.
   - No. 1 lower control arm from the crossmember.
   - Loosen the nut holding the lower control arm to the axle carrier.
   - No. 2 lower control arm from the axle carrier.

- Nut, then remove the No. 2 lower control arm from the axle carrier.
- Nut holding the upper control arm to the axle carrier.
- Axle carrier
- Nut holding the No. 1 control arm to the axle carrier
- No. 1 lower control arm from the axle carrier
- Dust deflector.
- Axle hub from the carrier
- Backing plate
- Inner race (outside)
- Oil seal
- Snapring
- Bearing

To install:

7. Install or connect the following:
   - Bearing to the axle carrier

➡ **If the inner races come loose from the bearing outer race, be sure to install them on the same side as before.**

- Snapring. Install the inner race (outside) and a new oil seal.
- Backing plate. Install the inner race (inside) and press in the axle hub with the proper tools.
- Inner oil seal. Align the holes for the speed sensor in the dust deflector and axle carrier. Install the dust deflector.
- No. 1 lower arm to the axle carrier and install a new nut. Tighten the nut to 43 ft. lbs. (59 Nm).
- Upper control arm to the axle carrier. Tighten the new nut and bolt to 80 ft. lbs. (109 Nm).

- No. 2 lower control arm to the axle carrier and tighten a new nut to 110 ft. lbs. (150 Nm).
- No. 1 lower control arm to the rear crossmember. Tighten the nut to 136 ft. lbs. (184 Nm).
- Strut rod to the axle carrier. Tighten the nuts and bolts to 134 ft. lbs. (184 Nm).
- Parking brake cable and slide the backing plate to the inside. Install the hex bolt and tighten it to 132 ft. lbs. (180 Nm).
- Shoe guide plate set bolt. Tighten the bolt to 13 ft. lbs. (18 Nm).
- 4 hub bolts and tighten them to 19 ft. lbs. (26 Nm).
- Bolts at the speed sensor and tighten them to 69 inch lbs. (8 Nm).
- Parking brake shoes
- Halfshafts. Apply the brakes and tighten the locknut to 213 ft. lbs. (289 Nm).
- Brake rotor
- Brake caliper support to the rear axle carrier. Tighten the bolts to 77 ft. lbs. (104 Nm).
- Rear tire and wheel assembly
- Negative battery cable

8. Lower the vehicle and bounce it a few times to stabilize the suspension.

LS400, LS430

1. Before servicing the vehicle, refer to the precautions in the beginning of this section.
2. If equipped with air suspension,

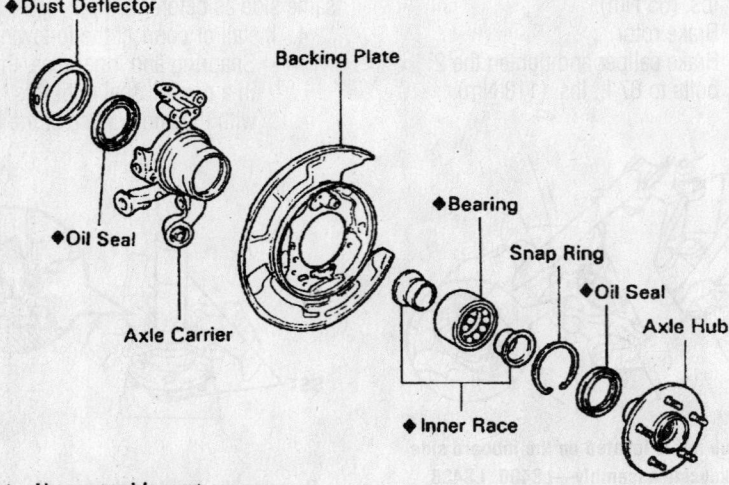

◆ Non-reusable part

Exploded view of the axle carrier — GS300

Exploded view of the axle carrier—GS300

7923LGC1

move the height control switch in the trunk area to the **OFF** position.

3. Remove or disconnect the following:
- Negative battery cable
- Rear wheel(s)
- Height control sensor link from the lower control arm
- Anti-lock Brake System (ABS) speed sensor and wiring harness
- Brake caliper bracket from the rear axle carrier by removing the 2 bolts. Support the caliper with a piece of wire.
- Brake rotor
- Parking brake shoes and cable
- Cotter pin, lock cap and the nut holding the halfshaft to the rear axle.
- Suspension member brace by removing the 2 bolts.
- Halfshaft bolts and washers
- Halfshaft from the vehicle
- Strut rod

4. Place matchmarks on the adjusting cam and body for the No. 1 control arm.

5. Remove or disconnect the following:
- Nut and adjusting cam
- Nut on the axle carrier side of the No. 1 lower control arm
- Separate the control arm from the axle carrier
- No. 1 lower control arm
- Stabilizer bar link from the No. 2 lower control arm.

6. Place matchmarks on the adjusting cam and body.

7. Remove or disconnect the following:
- Nut and adjusting cam from the No. 2 lower control arm.
- Nut and bolt holding the No. 2 lower control arm to the axle carrier.
- No. 2 control arm from the vehicle.
- Nut and bolt on the lower side of the strut assembly.
- 2 upper control arm set nuts and bolts.
- Axle carrier with the upper control arm.

8. Secure the axle carrier in a vise.

9. Remove or disconnect the following:
- Nut holding the upper control arm to the axle carrier and remove the control arm.
- Dust deflector. Use a suitable pry-tool.
- Oil seal

10. Remove the 2 bolts and nuts and shift the backing the plate towards the hub side (outside).

11. Remove or disconnect the following:

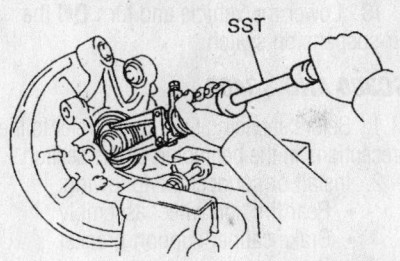

Removing the oil seal (inner)—LS400, LS430

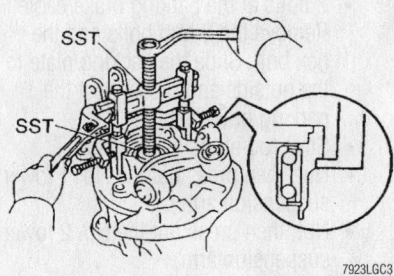

Removing the axle hub from the axle carrier—LS400, LS430

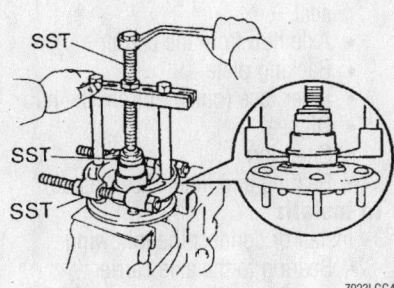

Removing the inner race (outside) from the axle hub—LS400, LS430

- Axle hub
- Backing plate.
- Inner race (outside) from the axle hub
- Oil seal (outer) from the axle
- Snapring from inside the axle housing
- Bearing from the axle housing

To install:

12. Install or connect the following:
- New bearing to the axle housing
- Snapring to the axle carrier, using snapring pliers.
- New outer oil seal. Coat the oil seal lip with multipurpose grease.
- Backing plate to the axle housing. Do not install the bolts or nuts at this time.

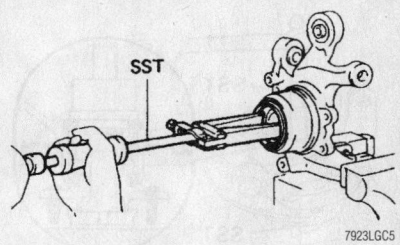

Removing the oil seal (outer)—LS400, LS430

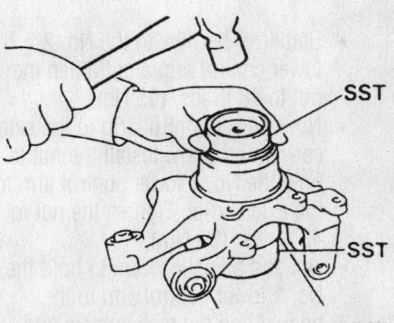

Installing the oil seal (outer)—LS400, LS430

- Inner race (inside) to the axle housing
- Axle hub to the axle housing
- Backing plate in position. Tighten the bolts and nuts to 43 ft. lbs. (59 Nm).
- New oil seal (inner) to the axle housing. Coat the oil seal lip with multipurpose grease.
- New dust deflector. Be sure to align the hose for the ABS speed sensor in the dust deflector and axle carrier.
- Upper control arm to the axle carrier by installing the nut. Tighten the nut to 80 ft. lbs. (108 Nm).
- Axle carrier and upper control arm to the vehicle as an assembly.
- 2 upper control arm set bolts and tighten the bolts to 121 ft. lbs. (164 Nm).
- Bolt and nut holding the strut to the axle carrier. Tighten to 101 ft. lbs. (137 Nm).
- Bolt and nut connecting the No. 2 lower control arm to the axle carrier. Tighten the bolt to 60 ft. lbs. (81 Nm).
- Nut and adjusting cam to hold the No. 2 lower control arm to the body. Align the adjusting cam marks and tighten the nut to 57 ft. lbs. (78 Nm).

Brake service is covered in Section 4 of this manual

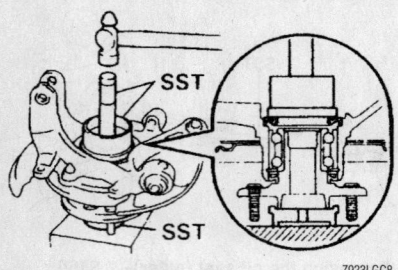

Installing the oil seal (inner)—LS400, LS430

7923LGC8

- Stabilizer bar link to the No. 2 lower control arm and tighten the nut to 48 ft. lbs. (65 Nm).
- No. 1 lower control arm to the axle carrier and body. Install the nut to hold the No. 1 lower control arm to the axle carrier. Tighten the nut to 43 ft. lbs. (59 Nm).
- Nut and adjusting cam to hold the No. 1 lower control arm to the body. Align the matchmarks and tighten the nut to 57 ft. lbs. (78 Nm).
- Strut rod to the axle carrier and body. Install the bolt and nut to hold the strut rod to the body. Tighten to 57 ft. lbs. (78 Nm).
- Bolt and nut to hold the strut rod to the axle carrier. Tighten to 136 ft. lbs. (184 Nm)
- Parking brake shoes and cable.
- Outboard joint side of the halfshaft and align the matchmarks on the side gear shaft and the halfshaft. Coat the threads with clean oil and install the hexagon bolts. Tighten bolts to 61 ft. lbs. (83 Nm).
- Suspension member brace with the 2 bolts. Tighten the 2 bolts to 37 ft. lbs. (50 Nm).
- Nut to hold the halfshaft to the rear axle. Tighten the nut to 213 ft. lbs. (289 Nm).
- Lock cap and cotter pin
- Brake disc to the axle hub with the matchmarks aligned. Install the 2 screws and tighten the screws to 48 inch lbs. (5 Nm).
- Brake caliper to the vehicle and install the 2 bolts. Tighten the bolts to 77 ft. lbs. (104 Nm).
- ABS speed sensor and wiring harness
- Height control sensor link with the matchmarks aligned. Tighten the nut to 48 inch lbs. (5 Nm).
- Rear wheel(s)
- Negative battery cable

13. Lower the vehicle and turn **ON** the air suspension switch.

SC300 AND SC400

1. Before servicing the vehicle, refer to the precautions in the beginning of this section.
2. Install or connect the following:
   - Rear tire and wheel assembly
   - Brake caliper support bracket
   - Brake rotor
   - Speed sensor
   - Rear halfshaft
   - Parking brake shoes
   - 2 bolts at the parking brake cable. Remove the 2 hub bolts and the hex bolt. Slide the backing plate to the outside and disconnect the parking brake cable.
   - Strut rod at the axle carrier
   - Nut, then press out the No. 1 lower suspension arm
   - Nut, then press out the No. 2 lower suspension arm
   - Nut, then press out the upper suspension arm.
   - Axle carrier
   - Dust deflector and pull out the oil seal.
   - Axle hub from the carrier
   - Backing plate
   - Inner race (outside) from the hub
   - Oil seal
   - Snapring
   - Bearing and inner race (inside)

To install:

3. Install or connect the following:
   - Bearing to the axle carrier

➡**If the inner races come loose from the bearing outer race, be sure to install them on the same side as before.**

- Snapring, the inner race (outside) and a new oil seal
- Backing plate. Install the inner race (inside) and press in the axle hub with the proper tools.
- Inner oil seal. Align the holes for the speed sensor in the dust deflector and axle carrier. Install the dust deflector.
- Upper arm to the axle carrier. Tighten the nut and bolt to 80 ft. lbs. (109 Nm).
- No. 2 lower arm to the carrier and tighten a new nut to 110 ft. lbs. (150 Nm).
- No. 1 lower arm to the carrier and tighten a new nut to 43 ft. lbs. (59 Nm).
- Strut rod to the carrier. Do not tighten the bolt at this time.

- Parking brake cable and slide the backing plate to the inside. Install the hex bolt and tighten it to 132 ft. lbs. (180 Nm). Install the 2 hub bolts and tighten them to 19 ft. lbs. (26 Nm).
- 2 bolts at the parking brake cable and tighten them to 69 inch lbs. (8 Nm). Install the parking brake shoes and the ABS sensor.
- Halfshaft. Tighten the locknut to 213 ft. lbs. (289 Nm).
- Brake rotor
- Brake caliper to the rear axle carrier by installing the 2 bolts. Tighten the bolts to 77 ft. lbs. (104 Nm).
- Rear tire and wheel assembly. Lower the vehicle and bounce it a few times to stabilize the suspension. Raise the vehicle again, support the axle carrier and tighten the strut rod to 136 ft. lbs. (184 Nm).

IS300

1. Before servicing the vehicle, refer to the precautions in the beginning of this section.
2. Remove or disconnect the following:
   - Rear wheel
   - Wheel speed sensor
   - Axle halfshaft
   - Brake caliper and rotor
   - Parking brake shoes
   - Parking brake cable
   - No. 1 lower suspension arm bolt
   - No. 2 lower suspension arm bolt
   - Toe control link
   - Upper ball joint
   - Axle carrier from the vehicle

3. Press the hub out of the wheel bearing, then remove the backing plate.

4. Remove the snapring, then press the wheel bearing out of the axle carrier.

To install:

➡**Use a new toe control link nut for assembly.**

5. Installation is the reverse of the removal procedure, while using the following torque values:
   - Backing plate bolts: 43 ft. lbs. (59 Nm)
   - No. 1 lower suspension arm bolt: 55 ft. lbs. (75 Nm)
   - No. 2 lower suspension arm bolt: 81 ft. lbs. (110 Nm)
   - Toe control link nut: 36 ft. lbs. (49 Nm)
   - Upper ball joint nut: 80 ft. lbs. (108 Nm)
   - Brake caliper support bolts: 77 ft. lbs. (104 Nm)
   - Rear wheel lug nuts: 76 ft. lbs. (103 Nm)

MAZDA

PRECAUTIONS

Before servicing any vehicle, please be sure to read all of the following precautions, which deal with personal safety, prevention of component damage, and important points to take into consideration when servicing a motor vehicle:

• Never open, service or drain the radiator or cooling system when the engine is hot; serious burns can occur from the steam and hot coolant.

• Observe all applicable safety precautions when working around fuel. Whenever servicing the fuel system, always work in a well-ventilated area. Do not allow fuel spray or vapors to come in contact with a spark, open flame, or excessive heat (a hot drop light, for example). Keep a dry chemical fire extinguisher near the work area. Always keep fuel in a container specifically designed for fuel storage; also, always properly seal fuel containers to avoid the possibility of fire or explosion. Refer to the additional fuel system precautions later in this section.

• Fuel injection systems often remain pressurized, even after the engine has been turned **OFF**. The fuel system pressure must be relieved before disconnecting any fuel lines. Failure to do so may result in fire and/or personal injury.

• Brake fluid often contains polyglycol ethers and polyglycols. Avoid contact with the eyes and wash your hands thoroughly after handling brake fluid. If you do get brake fluid in your eyes, flush your eyes with clean, running water for 15 minutes. If

eye irritation persists, or if you have taken brake fluid internally, IMMEDIATELY seek medical assistance.

• The EPA warns that prolonged contact with used engine oil may cause a number of skin disorders, including cancer! You should make every effort to minimize your exposure to used engine oil. Protective gloves should be worn when changing oil. Wash your hands and any other exposed skin areas as soon as possible after exposure to used engine oil. Soap and water, or waterless hand cleaner should be used.

• All new vehicles are now equipped with an air bag system. The system must be disabled before performing service on or around system components, steering column, instrument panel components, wiring and sensors. Failure to follow safety and disabling procedures could result in accidental air bag deployment, possible personal injury and unnecessary system repairs.

• Always wear safety goggles when working with, or around, the air bag system. When carrying a non-deployed air bag, be sure the bag and trim cover are pointed away from your body. When placing a non-deployed air bag on a work surface, always face the bag and trim cover upward, away from the surface. This will reduce the motion of the module if it is accidentally deployed. Refer to the additional air bag system precautions later in this section.

1. Clean, high quality brake fluid from a

sealed container is essential to the safe and proper operation of the brake system. You should always buy the correct type of brake fluid for your vehicle. If the brake fluid becomes contaminated, completely flush the system with new fluid. Never reuse any brake fluid. Any brake fluid that is removed from the system should be discarded. Also, do not allow any brake fluid to come in contact with a painted surface; it will damage the paint.

• Never operate the engine without the proper amount and type of engine oil; doing so WILL result in severe engine damage.

• Timing belt maintenance is extremely important! Many models utilize an interference-type, non-freewheeling engine. If the timing belt breaks, the valves in the cylinder head may strike the pistons, causing potentially serious (also time-consuming and expensive) engine damage. Refer to the maintenance interval charts in the front of this manual for the recommended replacement interval for the timing belt, and to the timing belt section for belt replacement and inspection.

• Disconnecting the negative battery cable on some vehicles may interfere with the functions of the on-board computer system(s) and may require the computer to undergo a relearning process once the negative battery cable is reconnected.

• When servicing drum brakes, only disassemble and assemble one side at a time, leaving the remaining side intact for reference.

ENGINE REPAIR

Distributor

REMOVAL

1998–00

1. Before servicing the vehicle, refer to the precautions in the beginning of this section.
2. Remove or disconnect the following:
   • Negative battery cable
   • Distributor cap and position it aside, leaving the ignition wires connected
   • Distributor electrical connector(s) from the side of the distributor
3. Using a wrench on the crankshaft pulley, rotate the crankshaft to position the No. 1 piston on Top Dead Center (TDC) of the compression stroke; the crankshaft pulley mark should align with the timing indi-

cator and the distributor rotor should point towards the No. 1 spark plug wire tower position of the cap.

4. Using chalk or paint, mark the position of the distributor housing on the cylinder head. Also mark the position of the distributor rotor in relation to the distributor housing.
5. Remove distributor hold-down bolt(s).
6. On distributors attached to the end of the cylinder head (or inline with the camshaft), remove it by pulling it straight outward.
7. On distributors attached to the side of the cylinder head (or perpendicular with the camshaft), slowly pull it outward while watching the rotor. These distributors are gear driven and as you remove it, the gears will disengage inside the engine, causing the rotor to rotate. when the rotor stops

moving, stop pulling outward. Re-align the distributor body-to-cylinder head match-mark (do not push it back in to do this, simply rotate the body to align the marks). Place a third mark indicating the new rotor position-to-distributor body relation. When installing the distributor, align this mark and the body-to-head mark to properly position the distributor.

8. Inspect the O-ring on the distributor housing and replace it, if it is damaged or worn.

INSTALLATION

Engine Not Disturbed

1998–00

1. Using engine oil, lubricate the O-ring.
2. Install or connect the following:

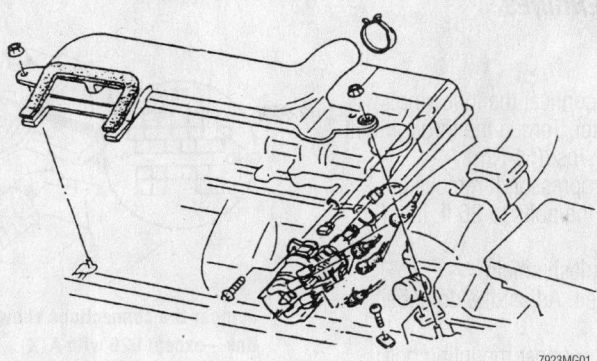

Exploded view of a typical side mounted distributor

- Distributor

➡**Be sure to engage the distributor drive gear or tangs with the camshaft gear or slot. Align the mark that was made on the distributor housing with the mark that was made on the cylinder head.**

- Distributor hold-down bolt(s)
- Electrical connector(s)
- Distributor cap
- Negative battery cable

3. Start the engine and check or adjust the ignition timing.

Engine Disturbed

1998–00

1. Remove or disconnect the following:
   - Spark plug wire from the No. 1 cylinder spark plug
   - Spark plug from the No. 1 cylinder
2. Press a thumb over the spark plug hole.
3. Using a wrench on the crankshaft pulley, rotate the crankshaft until pressure is felt at the spark plug hole, indicating the piston is approaching TDC on the compression stroke. Continue rotating the crankshaft until the crankshaft pulley mark aligns with the timing cover indicator.

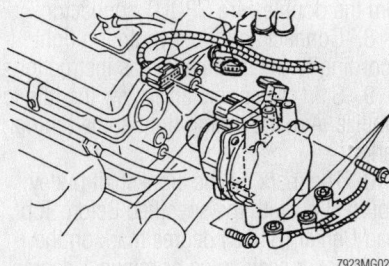

Exploded view of a typical end or inline mounted distributor

4. Place the distributor rotor in position so that it aligns with the No. 1 spark plug wire tower on the distributor cap.
5. Using engine oil, lubricate the O-ring.
6. Install or connect the following:
   - Distributor

➡**Be sure to engage the distributor drive gear or tangs with the camshaft gear or slot. Align the mark that was made on the distributor housing with the mark that was made on the cylinder head.**

- Distributor hold-down bolt(s)
- Electrical connector(s)
- Distributor cap
- Spark plug in the No. 1 cylinder and connect the spark plug wire
- Negative battery cable

7. Start the engine and check or adjust the ignition timing.

Alternator

REMOVAL

Miata and Protégé

1998–00

1. Before servicing the vehicle, refer to the precautions in the beginning of this section.
2. Remove or disconnect the following:
   - Negative battery cable
   - Vacuum hose
   - Solenoid bracket, if equipped
   - Pressure pipe bracket and the Exhaust Gas Recirculation (EGR) solenoid valve bracket
   - Electrical connectors from the alternator
   - Alternator drive belt
   - Alternator bracket

- Alternator pivot and adjusting bar bolts
- Alternator

626

4 CYLINDER ENGINES

1998–00

1. Before servicing the vehicle, refer to the precautions in the beginning of this section.
2. Remove or disconnect the following:
   - Negative battery cable
   - Alternator upper mounting bolt
   - Alternator adjusting bolt
   - Drive belt from the alternator pulley
   - Transverse member
   - Electrical connectors from the alternator
   - Front exhaust pipe at the catalytic converter and suspend it on a piece of wire
   - Oxygen (O_2S) sensor
   - 3 exhaust manifold flange nuts and the hold-down bracket clamp
   - Exhaust pipe
   - Alternator lower through-bolt
   - Alternator

6 CYLINDER ENGINES

1998–00

1. Before servicing the vehicle, refer to the precautions in the beginning of this section.
2. Remove or disconnect the following:
   - Negative battery cable
   - Fresh air duct
   - Radiator upper bracket
   - Condenser fan, if equipped
   - Electrical connectors from the alternator
   - Loosen the belt tensioner locknut and tension adjusting bolt
   - Alternator upper mounting bolt
   - Right splash shield
   - Drive belt from the alternator pulley
   - A/C compressor and support it aside, leaving the refrigerant lines connected, if necessary
   - Alternator through-bolt
   - Alternator

Millenia

1998–00

1. Before servicing the vehicle, refer to the precautions in the beginning of this section.

2. Disconnect the negative battery cable.

3. If equipped with the 2.3L engine, remove the front charge air cooler, radiator upper seal board and the condenser fan assembly.

4. Remove or disconnect the following:
- Electrical connectors from the alternator
- Right splash shield
- Drive belt from the alternator pulley
- A/C compressor and support it aside, leaving the refrigerant lines connected
- Upper and lower alternator mounting bolts
- Alternator

INSTALLATION

Miata and Protégé

1998–00

1. Install or connect the following:
- Alternator with the pivot bolt
- Alternator bracket
- Alternator electrical connectors
- Drive belt
- Upper mounting bolt

2. Adjust the belt tension. Torque the lower through bolt to 27–38 ft. lbs. (37–52 Nm) and the upper mounting bolt to 12–19 ft. lbs. (16–26 Nm)

3. Install or connect the following:
- Pressure pipe bracket and the EGR solenoid valve bracket
- Negative battery cable

626

4 CYLINDER ENGINES

1998–00

Install or connect the following:
- Alternator
- Alternator through-bolt
- Exhaust pipe using new gaskets. Torque the exhaust pipe-to-converter nuts to 28–38 ft. lbs. (38–51 Nm), the exhaust manifold flange nuts to 28–38 ft. lbs. (38–51 Nm) and the exhaust clamp nuts to 14–18 ft. lbs. (19–25 Nm)
- O_2S. Torque it to 36 ft. lbs. (49 Nm)
- O_2S electrical connector
- Alternator electrical connectors
- Transverse member. Torque the bolts to 68–96 ft. lbs. (94–131 Nm)
- Alternator. Adjust the belt tension. Torque the lower through bolt to 24–33 ft. lbs. (32–46 Nm) and the upper mounting bolt to 12–16 ft. lbs. (16–22 Nm)
- Negative battery cable

6 CYLINDER ENGINES

1998–00

1. Install or connect the following:
- Alternator. Torque the through bolt to 38 ft. lbs. (51 Nm)
- A/C compressor, if removed. Torque the bolts to 26 ft. lbs. (35 Nm)
- Right splash shield
- Drive belt. Adjust the drive belt tension
- Alternator upper mounting bolt. Torque the bolt to 18 ft. lbs. (25 Nm)

2. Tighten the belt tensioner locknut and tension adjusting bolt.

3. Install or connect the following:
- Electrical connectors to the alternator
- Condenser fan, if removed
- Radiator upper bracket
- Fresh air duct
- Negative battery cable

Millenia

1998–00

1. Install or connect the following:
- Alternator. Torque the through bolt to 24–33 ft. lbs. (32–46 Nm) and the upper bolt to 12–16 ft. lbs. (16–22 Nm)
- A/C compressor. Torque the bolts to 12–16 ft. lbs. (16–22 Nm)
- Right splash shield
- Drive belt
- Electrical connectors to the alternator

2. If equipped with the 2.3L engine, install the condenser fan assembly, radiator upper seal board and the front charge air cooler using new O-rings. Torque the mounting bolts to 12–16 ft. lbs. (16–22 Nm)

3. Connect the negative battery cable.

Ignition Timing

ADJUSTMENT

➡If the information given in the following procedures differs from that on the emission information label located in the engine compartment, follow the directions given on the label. The label often reflects production changes made during the model year.

Except 2.3L Engines

1998–00

1. Apply the parking brake. If equipped with a manual transaxle, place the shifter in

Jumper the connections shown on the data link—except 626 with ATX

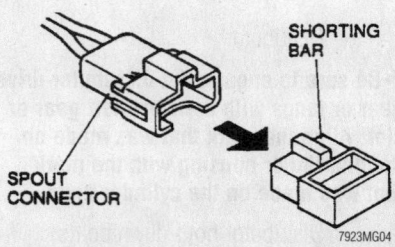

Remove the shorting bar from the spout connector—626 with ATX

the neutral position. If equipped with an automatic transaxle, place the shift lever in **P**.

2. Locate the timing marks on the crankshaft pulley and timing belt lower cover. The engine may have to be cranked slightly to see the mark on the crankshaft pulley.

3. Start the engine and allow it to come to normal operating temperature. Be sure all accessories are **OFF**.

4. Check the idle speed and adjust, if necessary.

5. Turn the engine OFF.

6. On all engines except 2.0L engine with an automatic transaxle, connect a jumper wire between the TEN terminal and the GND terminal at the underhood diagnosis connector.

7. On the 2.0L engine with an automatic transaxle, remove the shorting bar from the double wire SPOUT connector.

8. Connect an inductive timing light according to the manufacturer's instructions.

9. Start the engine and allow the idle to stabilize. Aim the timing light at the timing marks.

10. The mark on the crankshaft pulley should align with the specified Before Top Dead Center (BTDC) degree mark on the timing cover scale, plus or minus 1 degree. If the marks are within alignment proceed with step 13. If the marks are not aligned, proceed to Step 12.

11. Loosen the distributor lockbolts just

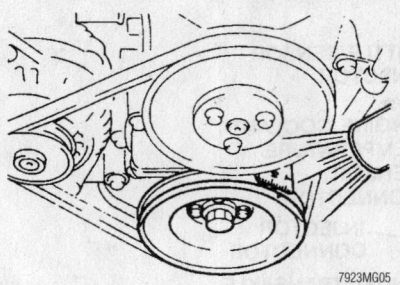

Connect an inductive timing light and aim it at the crankshaft pulley. Read the pulley mark against the scale

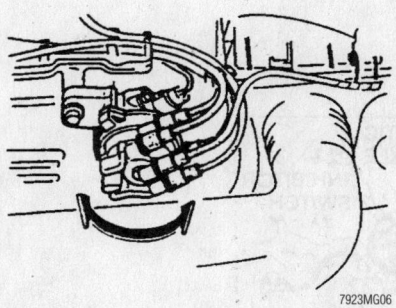

If adjustment is necessary, loosen the distributor lockbolts and rotate it until the mark is aligned

enough to turn the distributor. While aiming the timing light at the timing marks, turn the distributor until the marks are aligned. Tighten the distributor lockbolts to 14–19 ft. lbs. (19–25 Nm) and recheck the timing.

12. The ignition timing is now set. Disconnect the jumper wire from the underhood diagnosis connector or install the shorting bar from the double wire SPOUT connector.

13. Remove all test equipment.

2.3L Engines

The 2.3L engine utilizes individual ignition coils for each cylinder. The timing is controlled by the computer. Ignition timing adjustment is not possible or necessary.

Engine Assembly

REMOVAL & INSTALLATION

Except Miata

1998–00

➥**The procedure for pulling the engine requires removing the transaxle along**

with it. **As a result, when the halfshafts are pulled from the transaxle, a special plug/side gear holding tool is recommended.**

1. Before servicing the vehicle, refer to the precautions in the beginning of this section.

2. Properly relieve the fuel system pressure.

3. Drain the engine oil.

4. Drain the transaxle fluid.

5. Drain the cooling system.

6. Remove or disconnect the following:

- battery cables and battery
- Battery tray
- hood
- front wheels
- splash shield(s) from under the vehicle
- air cleaner assembly and resonance chamber, including the air flow meter and all of the ducting
- Oil dipstick

7. On the Millenia with the 2.3L engine, remove the charge air cooler, front grille, upper seal board (panels that the grille mounts to) and coolant overflow tank.

8. Remove or disconnect the following:

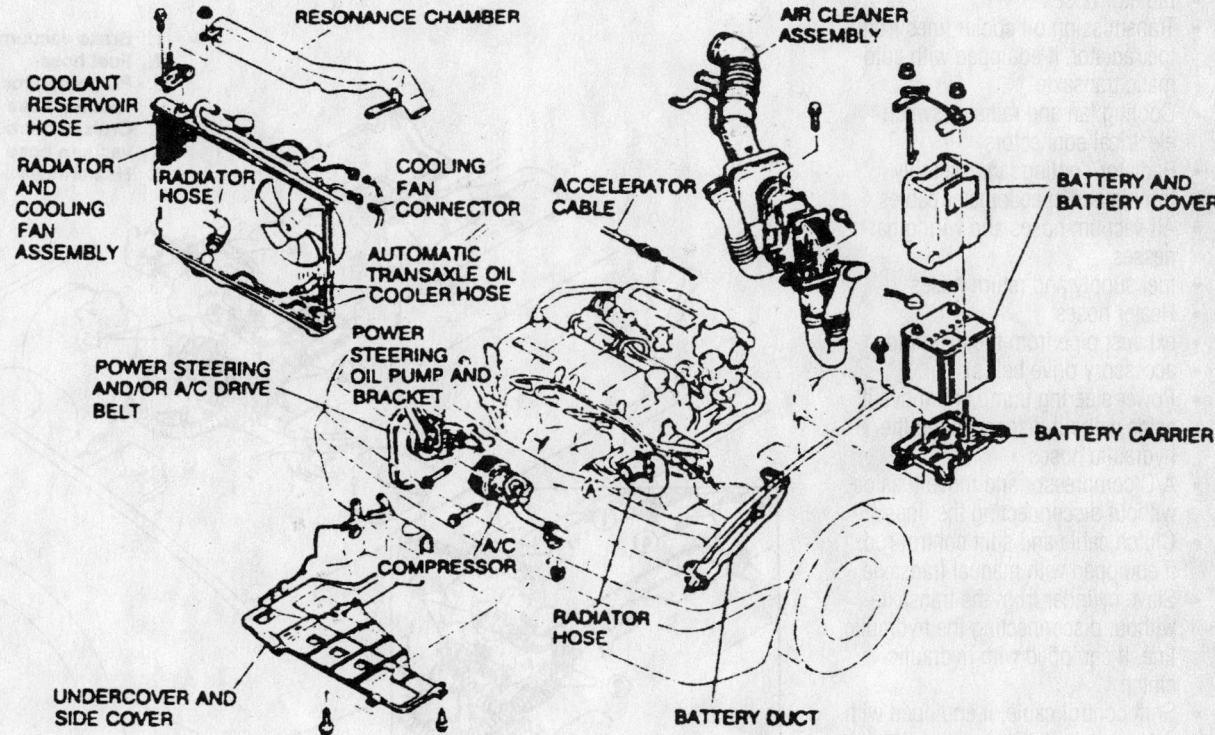

View of typically removed external components for engine removal—front wheel drive models

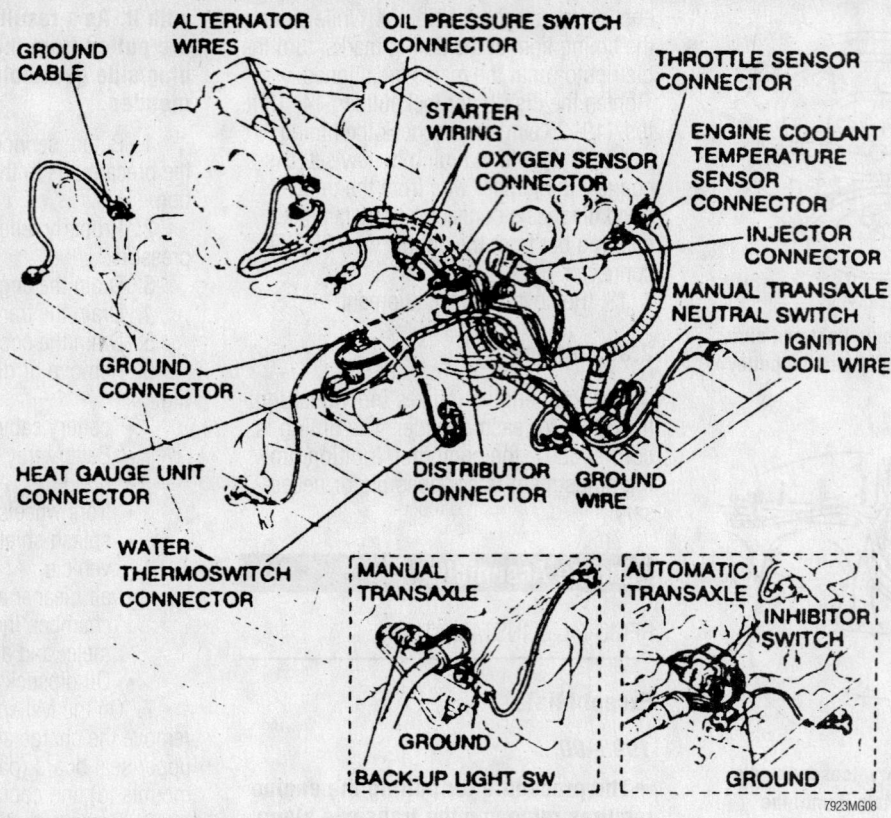

View of the common electrical harness plug connector points—front wheel drive models

- radiator hoses
- Transmission oil cooler lines from the radiator, if equipped with automatic transaxle
- Cooling fan and radiator switch electrical connectors
- Radiator/cooling fan assembly
- throttle and speedometer cables
- All vacuum hoses and wiring harnesses
- fuel supply and return hoses
- Heater hoses
- exhaust pipe from the manifold
- accessory drive belt(s)
- Power steering pump and move it aside without disconnecting the hydraulic hoses
- A/C compressor and move it aside without disconnecting the lines
- Clutch cable and shift control rod, if equipped with manual transaxle
- Slave cylinder from the transaxle without disconnecting the hydraulic line, if equipped with hydraulic clutch
- Shift control cable, if equipped with automatic transaxle
- Tie rod ends from the steering knuckles
- Stabilizer bar from the lower control arms

1. Brake vacuum hose
2. Fuel hose
3. Purge control vacuum hose
4. Cruise control vacuum hose
5. Heater hose

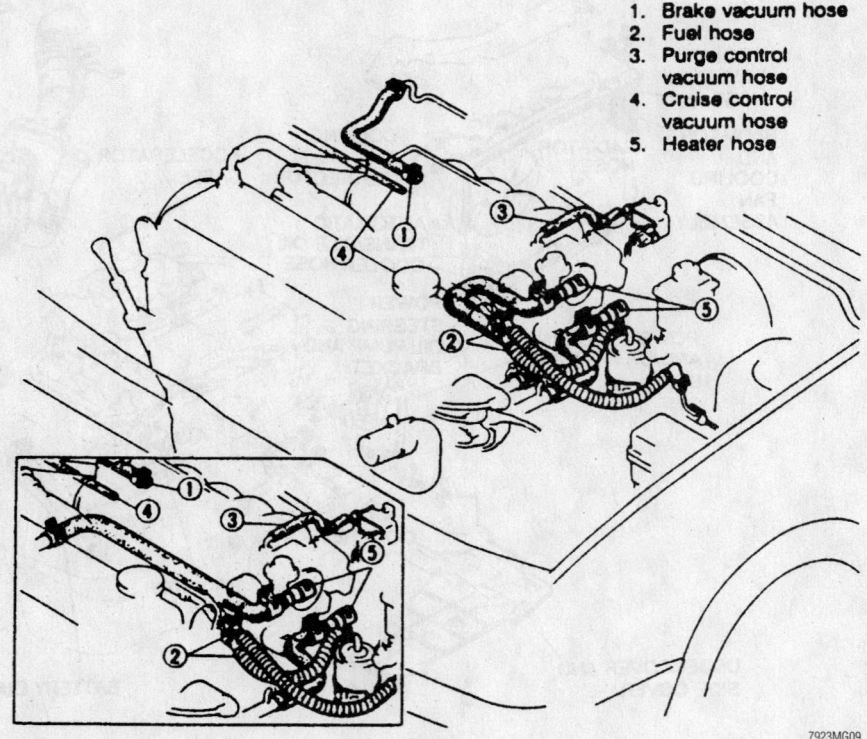

Typical vacuum, fuel and water hose disconnect points for engine removal—front wheel drive models

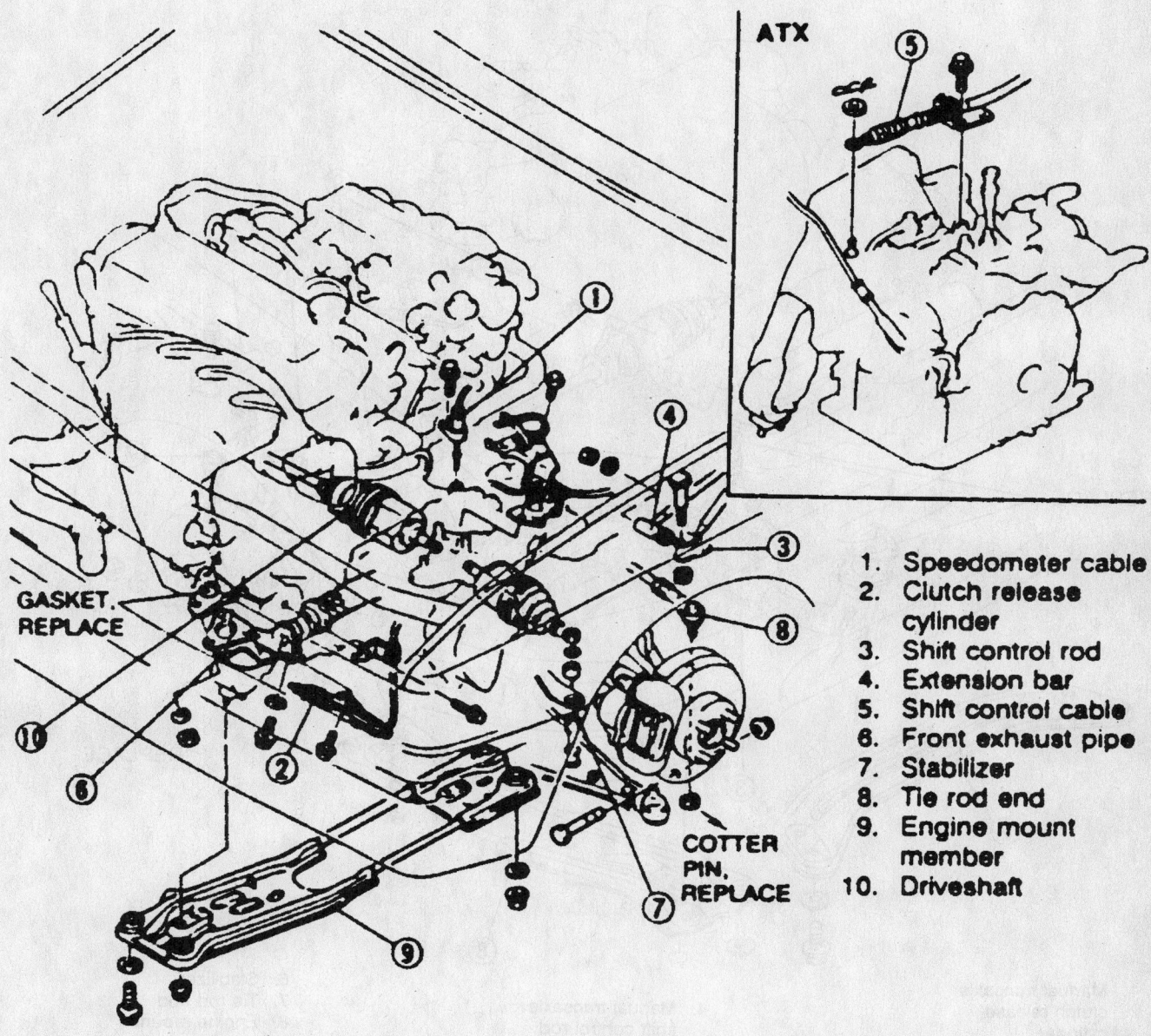

ATX

1. Speedometer cable
2. Clutch release cylinder
3. Shift control rod
4. Extension bar
5. Shift control cable
6. Front exhaust pipe
7. Stabilizer
8. Tie rod end
9. Engine mount member
10. Driveshaft

GASKET, REPLACE

COTTER PIN, REPLACE

7923MG10

Typically removed 4-cylinder engine under-vehicle components for engine removal—front wheel drive models

Timing belt service is covered in Section 3 of this manual

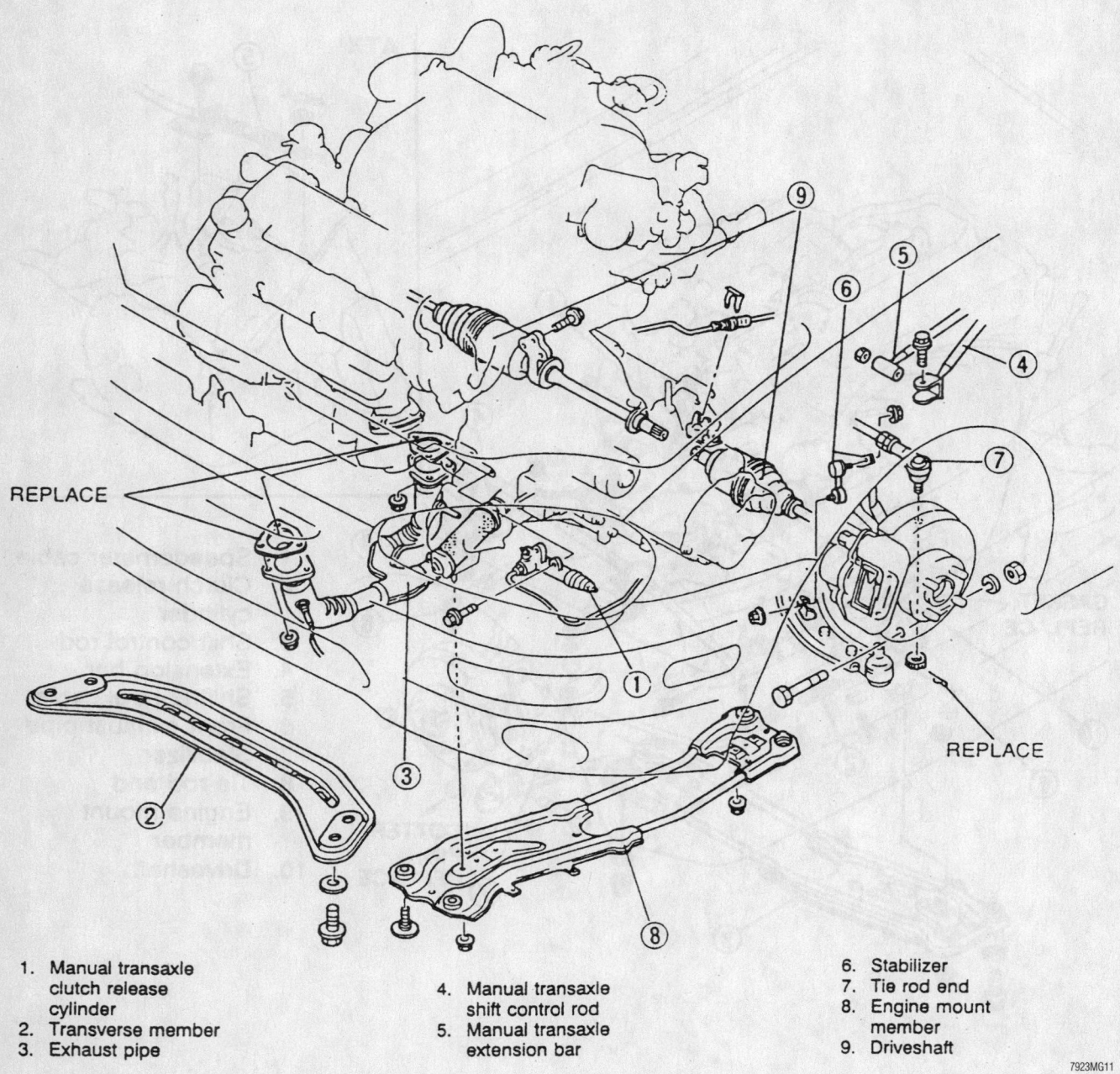

1. Manual transaxle
   clutch release
   cylinder
2. Transverse member
3. Exhaust pipe
4. Manual transaxle
   shift control rod
5. Manual transaxle
   extension bar
6. Stabilizer
7. Tie rod end
8. Engine mount
   member
9. Driveshaft

7923MG11

Typically removed 6-cylinder engine under-vehicle components for engine removal—front wheel drive models

9. Attach an engine lifting chain to the engine lifting eyes. Attach the chain to a suitable engine hoist and raise the hoist until there is tension on the chain.

10. Remove or disconnect the following:
- engine mount member
- Lower ball joint from the steering knuckle
- Bolts from the right side intermediate shaft support, if equipped
- Intermediate shaft by prying it from the transaxle

11. Suspend the halfshafts with wire.

12. Remove or disconnect the following:
- dynamic damper from the right side engine mount, if equipped.
- Engine/transaxle mount nuts/bolts and right engine (if equipped) and left transaxle mounts

13. Lift the engine/transaxle assembly from the vehicle.

14. Remove or disconnect the following:
- intake manifold bracket
- Starter
- Torque converter nuts

- Engine-to-transaxle stiffener, if equipped
- No. 2 engine mount
- Throttle valve cable
- transaxle mounting bolts
- Transaxle from the engine

To install:

15. Installation is the reverse of the removal procedure. Note the following important steps.

16. When possible, leave the engine mounting nuts/bolts loose (hand tight) until all mounts are aligned and bolted. This may help in aligning the engine and transmission assembly in the vehicle.

17. Install new circlips on the inner CV-joint stub shafts, if equipped, and the intermediate shaft. Grease the shaft splines before installing the halfshaft/intermediate shaft into the transaxle.

18. Always install new gaskets and/or O-rings. Use new self-locking nuts, especially on the exhaust.

19. Fill the engine and the transaxle with oil. Fill the cooling system.

20. Connect the negative battery cable, start the engine and check for leaks.

21. Check the ignition timing and the idle speed.

22. Check all fluid levels.

Miata

1998–00

1. Before servicing the vehicle, refer to the precautions in the beginning of this section.

2. Properly relieve the fuel system pressure.

3. Drain the engine oil.

4. Drain the cooling system.

5. Remove or disconnect the following:
- Negative battery cable
- Fresh air duct and the air cleaner/air flow meter assembly
- transmission
- accelerator cable from the throttle body
- Undercover
- radiator hoses and the coolant reservoir hose
- Cooling fan electrical connector
- Transmission oil cooler hose, if equipped with automatic transmission
- radiator/cooling fan assembly
- accessory drive belts
- Power steering pump and move it aside without disconnecting the hydraulic hoses
- Air conditioner compressor and move it aside without disconnecting the refrigerant lines

6. Disconnect the following electrical connectors:
- Steering pressure sensor electrical connector
- Throttle position sensor electrical connector
- Idle air control valve electrical connector
- Heated Oxygen (HO2S) sensor electrical connector
- Ignition coil electrical connectors
- Crankshaft position sensor electrical connector
- Ground electrical connectors
- Fuel injector electrical connectors
- Alternator electrical connectors
- Oil pressure sensor electrical connector
- Starter electrical connectors

7. Disconnect the following hoses:
- Brake vacuum hose
- Fuel hose

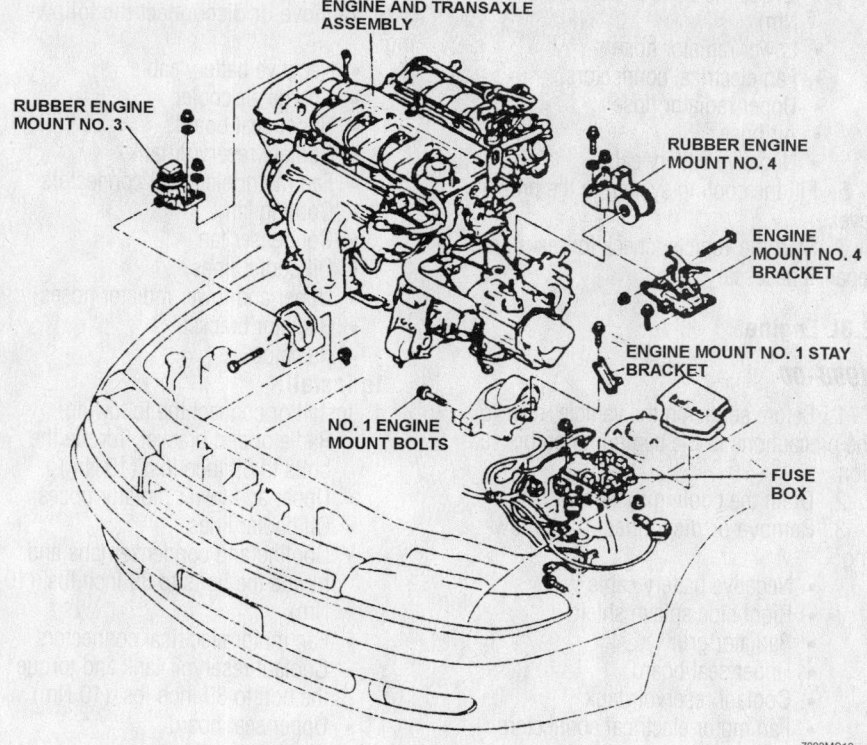

ENGINE AND TRANSAXLE ASSEMBLY

RUBBER ENGINE MOUNT NO. 3

RUBBER ENGINE MOUNT NO. 4

ENGINE MOUNT NO. 4 BRACKET

ENGINE MOUNT NO. 1 STAY BRACKET

NO. 1 ENGINE MOUNT BOLTS

FUSE BOX

7923MG12

Once all components and mounts are unfastened, the engine is removed with the transaxle attached—front wheel drive

Heater Core replacement is covered in Section 2 of this manual

- Purge control vacuum hose
- Cruise control vacuum hose
- Water inlet hose
- Heater hose
- exhaust pipe from the exhaust manifold and install suitable lifting equipment onto the engine.
- Engine from the vehicle

To install:

8. Install or connect the following:
- Engine assembly by tilting the engine downward and aligning the engine mounts with the crossmember holes. Torque the nuts to 42–57 ft. lbs. (57–78 Nm)
- Exhaust pipe to the manifold using a new gasket. Torque the nuts to 34 ft. lbs. (46 Nm)

9. Connect the following hoses:
- Brake vacuum hose
- Fuel hose
- Purge control vacuum hose
- Cruise control vacuum hose
- Water inlet hose
- Heater hose

10. Connect the following electrical connectors:
- Steering pressure sensor electrical connector
- Throttle position sensor electrical connector
- Idle air control valve electrical connector
- HO2S electrical connector
- Ignition coil electrical connectors
- Crankshaft position sensor electrical connector
- Ground electrical connectors
- Fuel injector electrical connectors
- Alternator electrical connectors
- Oil pressure sensor electrical connector
- Starter electrical connectors

11. Install or connect the following:
- air conditioner compressor and power steering pump
- Drive belt(s)
- radiator and fans and all cooling system hoses
- Accelerator cable
- air cleaner and air flow meter assembly
- transmission
- negative battery cable

12. Fill and bleed the cooling system.

13. Fill the engine and transmission.

14. Start the engine and check for leaks.

15. Check the ignition timing and idle speed.

Radiator

REMOVAL & INSTALLATION

1.5L, 1.6L 1.8L and 2.0L Engines

1998–00

1. Before servicing the vehicle, refer to the precautions in the beginning of this section.

2. Drain the cooling system.

3. Remove or disconnect the following:
- Negative battery cable
- Air hose
- Upper radiator hose
- Cooling and condenser fan electrical connectors
- Lower radiator hose
- Cooling fan
- Condenser fan
- Oil cooler lines, if equipped
- Radiator

To install:

4. Install or connect the following:
- Radiator and torque the mounting bolts to 18 ft. lbs. (25 Nm)
- Oil cooler lines, if equipped
- Cooling and condenser fans and torque the bolts to 89 inch lbs. (10 Nm)
- Lower radiator hose
- Fan electrical connectors
- Upper radiator hose
- Air hose
- Negative battery cable

5. Fill the cooling system to the proper level.

6. Start the vehicle, check for leaks and repair if necessary.

2.3L Engine

1998–00

1. Before servicing the vehicle, refer to the precautions in the beginning of this section.

2. Drain the cooling system.

3. Remove or disconnect the following:
- Negative battery cable
- Right side splash shield
- Radiator grill
- Upper seal board
- Coolant reservoir tank
- Fan motor electrical connectors
- Cooling fan
- Condenser fan
- Upper and lower radiator hoses
- Oil cooler lines
- Radiator bracket
- Radiator

To install:

4. Install or connect the following:
- Radiator and bracket. Torque the bracket bolts to 89 inch lbs. (10 Nm)
- Oil cooler lines
- Upper and lower radiator hoses
- Condenser fan and torque the bolts to 89 inch lbs. (10 Nm)
- Cooling fan and torque the bolts to 89 inch lbs. (10 Nm)
- Fan motor electrical connectors
- Coolant reservoir tank and torque the bolt to 89 inch lbs. (10 Nm)
- Upper seal board
- Radiator grill
- Right side splash shield
- Negative battery cable

5. Fill the cooling system to the proper level.

6. Start the vehicle, check for leaks and repair if necessary.

2.5L Engine

1998–00

1. Before servicing the vehicle, refer to the precautions in the beginning of this section.

2. Drain the cooling system.

3. Remove or disconnect the following:
- Negative battery cable
- Charge air cooler
- Upper seal board
- Coolant reservoir tank
- Fan motor electrical connectors
- Cooling fan
- Condenser fan
- Oil cooler lines
- Upper and lower radiator hoses
- Radiator bracket
- Radiator

To install:

4. Install or connect the following:
- Radiator and bracket. Torque the bolts to 89 inch lbs. (10 Nm)
- Upper and lower radiator hoses
- Oil cooler lines
- Cooling and condenser fans and torque the bolts to 89 inch lbs. (10 Nm)
- Fan motor electrical connectors
- Coolant reservoir tank and torque the bolt to 89 inch lbs. (10 Nm)
- Upper seal board
- Charge air cooler
- Negative battery cable

5. Fill the cooling system to the proper level.

6. Start the vehicle, check for leaks and repair if necessary.

Water Pump

REMOVAL & INSTALLATION

1.5L, 1.6L and 1.8L (BP) Engines

1998–00

1. Before servicing the vehicle, refer to the precautions in the beginning of this section.
2. Disconnect the negative battery cable.
3. Drain the engine coolant.
4. Remove or disconnect the following:
   - Timing belt covers and timing belt
   - Coolant inlet pipe and gasket
   - timing belt with idler pulleys still attached to the water pump
   - water pump

To install:

5. Clean all gasket mating surfaces.
6. Using a new gasket, install the water pump on the engine. Torque the mounting bolts to 14–18 ft. lbs. (19–25 Nm). Torque the bolt from the water pump to the alternator bracket to 28–38 ft. lbs. (38–51 Nm).
7. Install or connect the following:
   - timing belt idler pulleys
   - coolant inlet pipe with a new gasket. Torque the coolant inlet pipe bolts to 14–18 ft. lbs. (19–25 Nm)
   - timing belt and the timing belt covers
   - Negative battery cable
8. Fill and bleed the cooling system.
9. Start the engine and bring to normal operating temperature. Check for leaks.

1.8L (FP) and 2.0L Engines

1998–00

1. Before servicing the vehicle, refer to the precautions in the beginning of this section.
2. Drain the cooling system.
3. Remove or disconnect the following:
   - negative battery cable
   - timing belt
   - power steering oil pump adjuster
   - 5 water pump mounting bolts
   - Water pump

To install:

4. Clean all gasket mating surfaces.
5. Install or connect the following:
   - Water pump using a new gasket. Torque the bolts to 14–18 ft. lbs. (19–25 Nm)
   - power steering oil pump adjuster. Torque the bolts to 12–16 ft. lbs. (16–22 Nm)

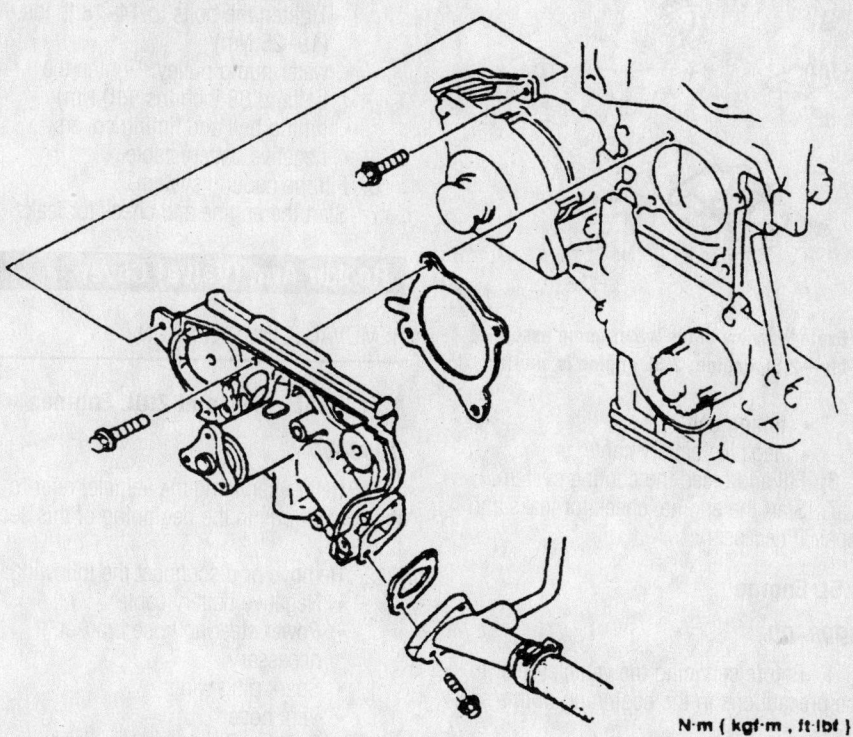

N·m { kgf·m , ft·lbf }
7923MG13

Exploded view of the water pump assembly—1.5L, 1.6L and 1.8L (BP) engines

7923MG14

Exploded view of the water pump assembly—1.8L (FP) and 2.0L engines

Brake service is covered in Section 4 of this manual

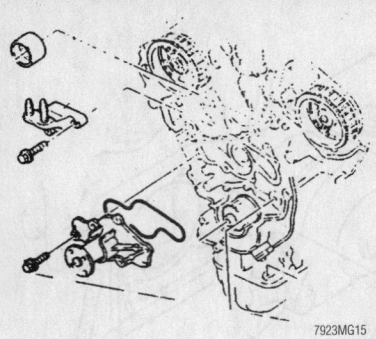

Exploded view of the water pump assembly—2.5L engine, 2.3L engine is similar

- timing belt
- negative battery cable
6. Fill and bleed the cooling system.
7. Start the engine, check for leaks and repair if necessary.

2.5L Engine

1998–00

1. Before servicing the vehicle, refer to the precautions in the beginning of this section.
2. Drain the cooling system.
3. Remove or disconnect the following:
   - Negative battery cable
   - timing belt
   - No. 3 engine mount bracket
   - 5 water pump mounting bolts
   - Water pump

To install:
4. Clean the mating surfaces.
5. Install or connect the following:
   - Water pump using a new gasket. Torque the bolts to 14–18 ft. lbs. (19–25 Nm)
   - engine mount bracket. Torque the bolts to 32–44 ft. lbs. (44–60 Nm)
   - timing belt
   - negative battery cable
6. Fill the cooling system.
7. Start the engine, check for leaks and repair if necessary.

2.3L Engine

1998–00

1. Before servicing the vehicle, refer to the precautions in the beginning of this section.
2. Disconnect the negative battery cable.
3. Drain the cooling system.
   - timing belt covers and timing belt
   - Water pump pulley bolts the pulley
   - water pump

To install:
4. Clean the mating surfaces.
5. Install or connect the following:

- Water pump using a new gasket. Tighten the bolts to 14–18 ft. lbs. (19–25 Nm)
- water pump pulley. Tighten the bolts to 88 inch lbs. (10 Nm)
- timing belt and timing covers
- negative battery cable
6. Fill the cooling system.
7. Start the engine and check for leaks.

Rocker Arm (Valve) Cover

REMOVAL & INSTALLATION

1.5L, 1.6L, 1.8L and 2.0L Engines

1998–00

1. Before servicing the vehicle, refer to the precautions in the beginning of this section.
2. Remove or disconnect the following:
   - Negative battery cable
   - Power steering hose bracket, if necessary
   - Spark plug wires
   - Vent hose
   - Positive Crankcase Ventilation (PCV) hose
   - Rocker arm cover and discard the gasket
3. Clean all mating surfaces of any residual gasket material.

To install:
4. Install or connect the following:
   - Rocker arm cover with a new gasket. Torque the bolts, in sequence, to 80 inch lbs. (9 Nm)
   - PCV hose
   - Vent hose
   - Spark plug wires
   - Power steering hose bracket, if removed
   - Negative battery cable

2.3L Engine

1998–00

1. Before servicing the vehicle, refer to the precautions in the beginning of this section.
2. Remove or disconnect the following:
   - Negative battery cable
   - Ignition coils
   - Vent hose
   - Positive Crankcase Ventilation (PCV) hose
   - Rocker arm cover and discard the gasket
3. Clean all mating surfaces of any residual gasket material.

To install:
4. Install or connect the following:
   - Rocker arm cover with a new gasket. Torque the bolts, in sequence, to 80 inch lbs. (9 Nm)
   - PCV hose
   - Vent hose
   - Ignition coils
   - Negative battery cable

2.5L Engine

1998–00

1. Before servicing the vehicle, refer to the precautions in the beginning of this section.
2. Remove or disconnect the following:
   - Negative battery cable
   - Intake manifold stay bracket
   - Vent hoses
   - Spark plug wires
   - Intake manifold
   - Positive Crankcase Ventilation (PCV) hose
   - Rocker arm cover and discard the gasket
3. Clean all mating surfaces of any residual gasket material.

To install:
4. Install or connect the following:
   - Rocker arm cover with a new gasket. Torque the bolts, in sequence, to 80 inch lbs. (9 Nm)
   - PCV hose
   - Intake manifold
   - Spark plug wires
   - Vent hose
   - Intake manifold stay bracket and torque the bolts to 19 ft. lbs. (25 Nm)
   - Negative battery cable

Cylinder Head

REMOVAL & INSTALLATION

1.5L and 1.8L (BP) Engines

1998–00

1. Before servicing the vehicle, refer to the precautions in the beginning of this section.
2. Relieve the fuel system pressure.
3. Drain the cooling system.
4. Remove or disconnect the following:
   - Negative battery cable
   - right front wheel and splash shield
   - Water pump pulley bolts, loosen them
   - power steering belt shield
   - Power steering belt

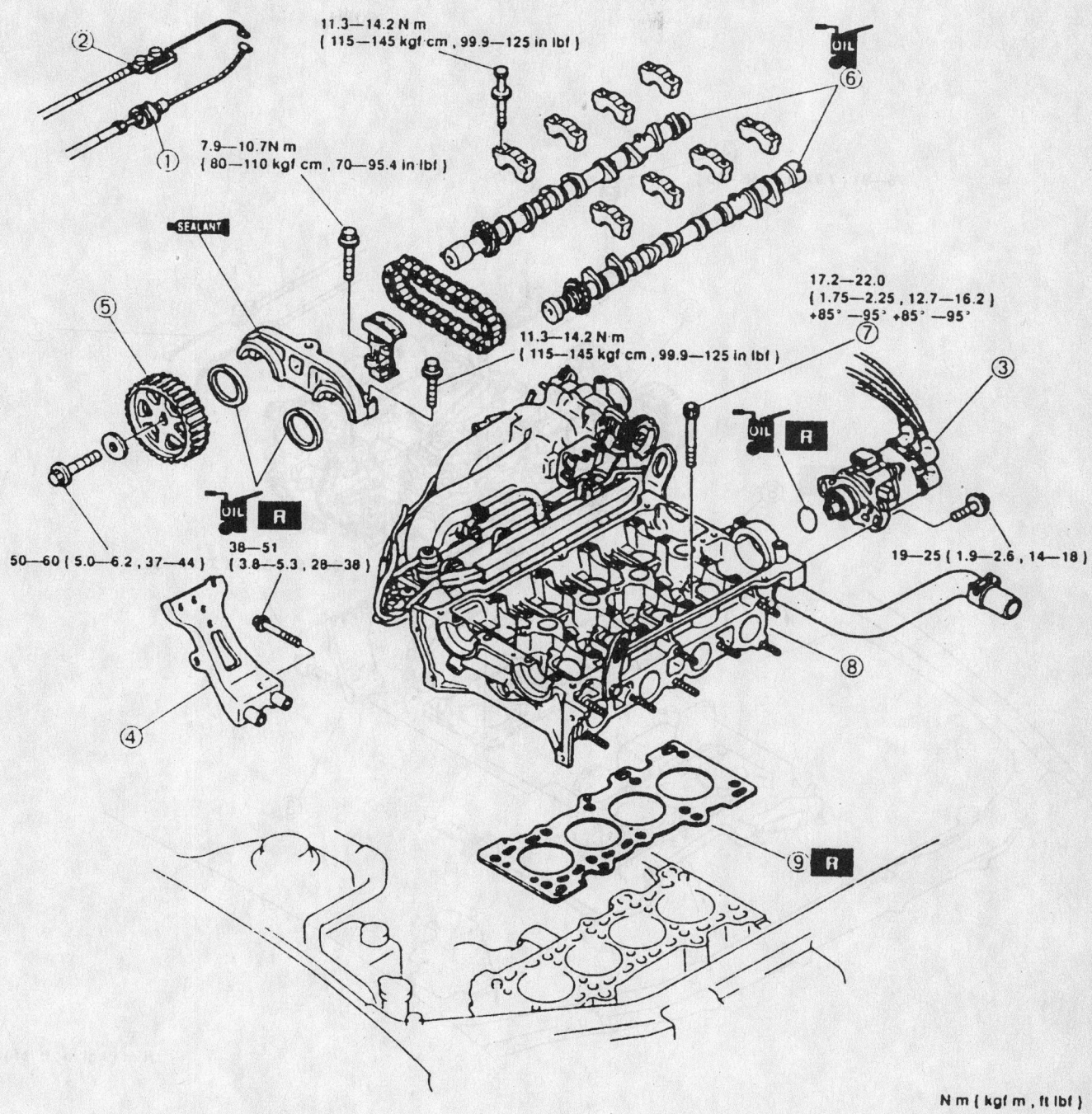

11.3—14.2 N·m
(115—145 kgf·cm , 99.9—125 in lbf)

7.9—10.7N·m
{ 80—110 kgf·cm , 70—95.4 in lbf }

SEALANT

17.2—22.0
(1.75—2.25 , 12.7—16.2)
+85° —95° +85° —95°

11.3—14.2 N·m
{ 115—145 kgf·cm , 99.9—125 in lbf }

50—60 (5.0—6.2 , 37—44)

OIL R
38—51
(3.8—5.3 , 28—38)

OIL R
19—25 (1.9—2.6 , 14—18)

N·m (kgf·m , ft lbf)

1. Accelerator cable
2. Throttle cable (ATX)
3. Distributor
4. Intake manifold stay
5. Camshaft pulley

6. Camshaft
7. Cylinder head bolt
8. Cylinder head assembly
9. Cylinder head gasket

7923MG19

Exploded view of the cylinder head assembly—1.5L engine

For complete Engine Mechanical specifications, see Section 1 of this manual

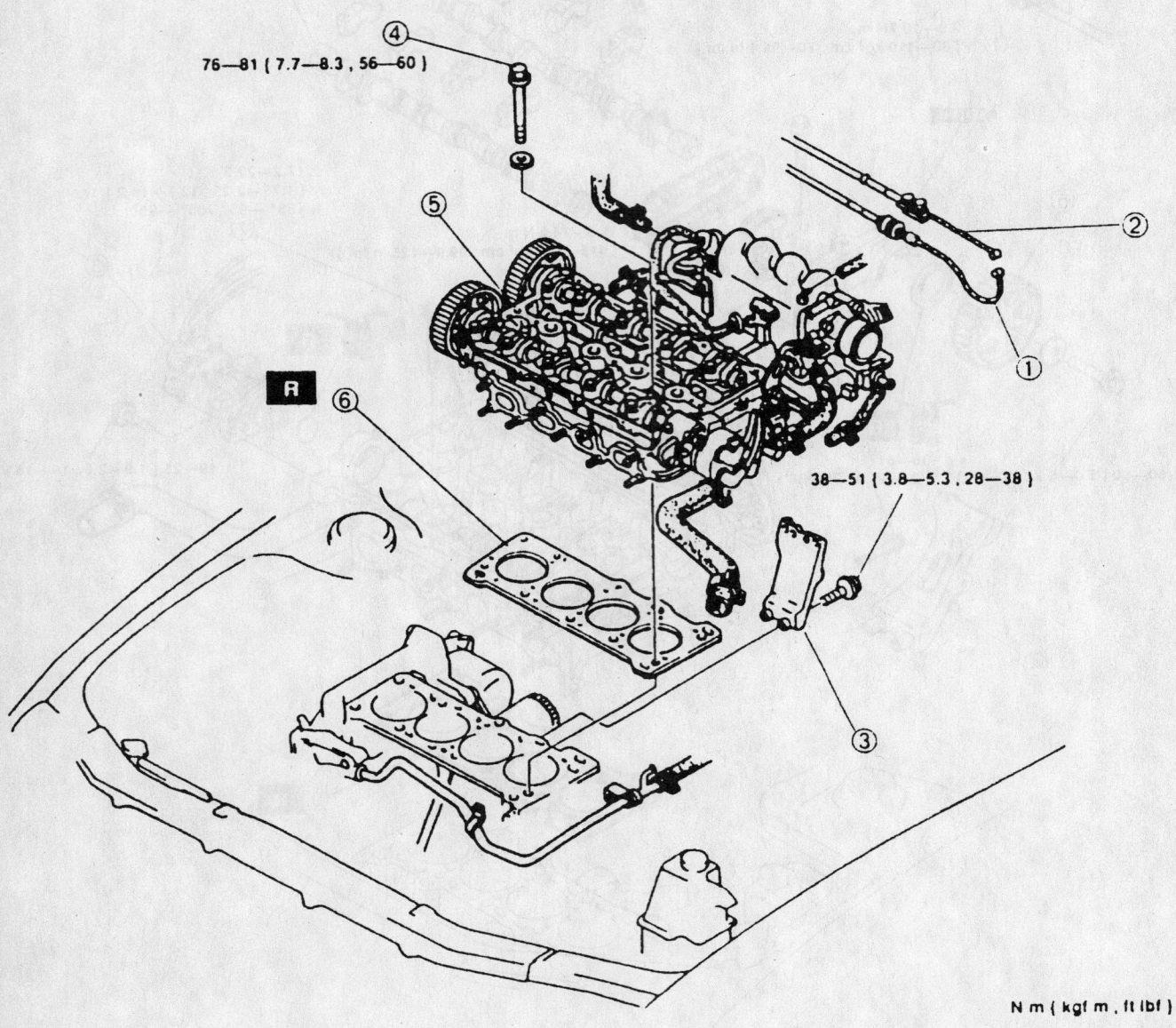

76—81 (7.7—8.3 , 56—60)

38—51 (3.8—5.3 , 28—38)

R

N m (kgf m , ft lbf)

1. Accelerator cable
2. Throttle cable (ATX)
3. Intake manifold stay
4. Cylinder head bolt
5. Cylinder head assembly
6. Cylinder head gasket

7923MG20

Exploded view of the cylinder head assembly for the 1.8L (BP) engine

- Alternator belt
- water pump pulley
- Crankshaft pulley bolt
- Crankshaft pulley using a puller
- Guide plate
- power steering hose brackets from the cylinder head cover
- Spark plug wires and wire clips
- breather tube and PCV valve from the cylinder head cover
- Cylinder head cover bolts, in 2 steps, by reversing the tightening sequence
- Cylinder head cover
- oil dipstick and bracket
- timing belt upper cover and install a suitable engine support tool
- No. 3 engine mount bracket
- timing belt middle and lower covers

5. Rotate the crankshaft, in the normal direction of rotation, until the No. 1 cylinder piston is at Top Dead Center (TDC) on the compression stroke. Be sure the timing marks on the crankshaft and camshaft sprocket(s) are properly aligned and mark the direction of rotation of the belt.

6. Remove or disconnect the following:
- Timing belt tensioner, loosen it
- Timing belt.

➡**Do not rotate the crankshaft until the timing belt is reinstalled.**

- Air cleaner assembly and front pipe
- exhaust manifold
- accelerator and throttle cables
- Distributor/coil
- Engine coolant temperature sensor
- Cooling fan coolant temperature sensor
- Temperature gauge sensor
- Spark plug wires from the spark plugs
- Spark plugs
- Distributor cap and wires assembly
- Distributor
- Heater hoses
- Brake vacuum hose
- Purge hose
- Fuel hoses
- Water hose
- Upper radiator hose
- camshaft sprockets and the camshaft

7. On the 1.5L engine, remove the tappets. Identify each tappet as it is removed so it can be reinstalled in the same position.

8. On the 1.8L engine, remove the hydraulic lifters. Identify each lifter as it is removed so it can be reinstalled in the same position. If the lifters are to be reused, store

them upside down in an oil-filled sealed container.

9. Remove or disconnect the following:
- intake manifold bracket
- Intake manifold
- Cylinder head bolts, in 2–3 steps, in sequence
- Cylinder head

To install:

10. Thoroughly, clean the cylinder head and the block contact surfaces. Examine the head gasket and check the cylinder head for cracks. Check the cylinder head for warpage using a feeler gauge and straightedge. The maximum allowable distortion is 0.004 in. (0.10mm).

11. Clean the cylinder head bolts and the threads in the block. Be sure the bolts turn freely in the block.

12. Install new head gasket on the engine block.

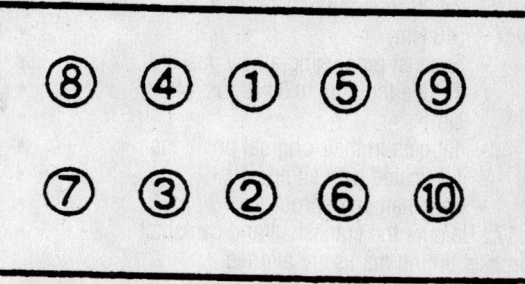

CRANKSHAFT PULLEY SIDE

7923MG16

Cylinder head bolt tightening sequence—4-cylinder engines

13. Be sure the camshaft sprocket timing marks are still aligned, as set during the removal procedure.

14. Install the cylinder head.

15. Lubricate the bolt threads and seat surfaces with clean engine oil and install them as follows:

   a. On the 1.5L engine, torque the bolts in 2–3 steps to 13–16 ft. lbs. (17–22 Nm) in the proper sequence. Paint a reference mark on each bolt head and turn the bolts, in sequence, 90 degrees, plus an additional 90 degrees.

   b. On the 1.8L (BP) engine, torque the bolts in 2–3 steps to 56–60 ft. lbs. (75–81 Nm) in the proper sequence.

16. Install or connect the following:
- Intake manifold. Torque the bolts to 19 ft. lbs. (25 Nm)
- Intake manifold bracket
- Exhaust manifold using a new gas-

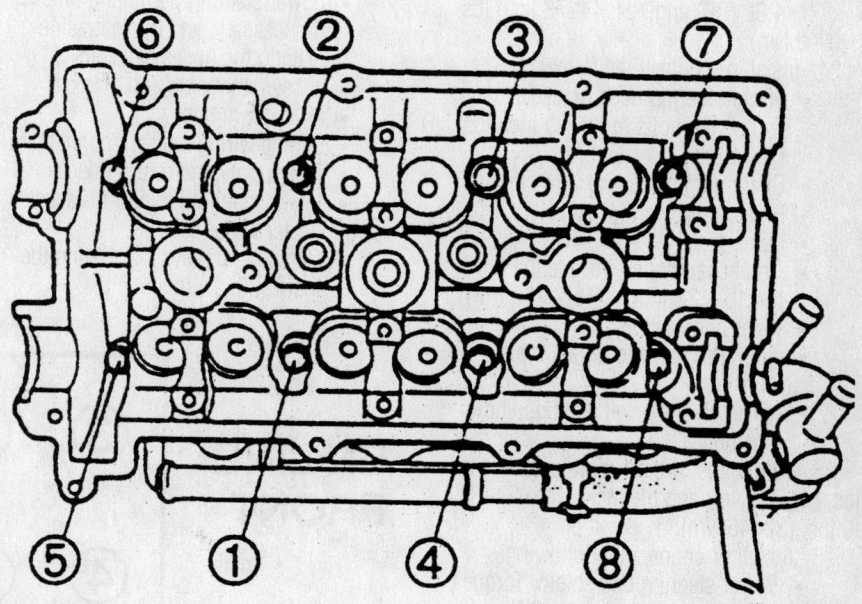

Cylinder head bolt tightening sequence—6-cylinder engines

7923MG18

For Accessory Drive Belt illustrations, see Section 1 of this manual

ket. Torque the nuts to 34 ft. lbs. (46 Nm)
- Exhaust pipe using a new gasket. Torque the nuts to 34 ft. lbs. (46 Nm)
- Tappets in their original positions lubricated with engine oil
- camshaft and sprockets

17. Be sure the crankshaft and camshaft sprocket timing marks are aligned.

18. Install or connect the following:
- Timing belt and set the tension. Carefully rotate the crankshaft 2 turns to be sure the timing marks still aligned
- timing belt middle and lower covers. Torque the bolts to 70–95 inch lbs. (8–11 Nm)
- no. 3 engine mount bracket. Torque the nut to 70–95 inch lbs. (8–11 Nm) and the bolt to 14–16 ft. lbs. (19–22 Nm)

19. Remove the engine support tool.

20. Install or connect the following:
- upper timing belt cover. Torque the bolts to 70–95 inch lbs. (8–11 Nm)
- oil dipstick and bracket

21. Apply silicone sealant to the cylinder surface in the area adjacent to the front camshaft bearing caps.

22. Install the cylinder head cover using a new gasket and sealant.

23. Torque the cylinder head cover bolts, in 5–6 steps, in sequence, to:
- a. 1.5L (Z5D) engines: 61–95 inch lbs. (7–11 Nm).
- b. 1.8L (BP) engines: 44–78 inch lbs. (5–9 Nm).

24. Install or connect the following:
- power steering hose brackets. Torque the bolts to 70–95 inch lbs. (7–11 Nm)
- Spark plug wires and wire clips
- Breather tube and PCV valve
- guide plate
- Crankshaft pulley. Torque the bolt to 116–122 ft. lbs. (157–166 Nm)
- water pump pulley
- alternator belt and adjust the tension
- power steering belt and adjust the tension

25. Torque the through-bolt to 32–44 ft. lbs. (44–60 Nm) and the lockbolt to 24–33 ft. lbs. (32–46 Nm).

26. Install or connect the following:
- Power steering belt shield. Torque the bolts to 86 inch lbs. (9 Nm)
- splash shield and wheel
- electrical engine harness connectors
- heater hose

- Brake vacuum hose
- Purge hose
- Fuel hoses
- Water hose
- Upper radiator hoses
- distributor
- Spark plugs
- Distributor cap and wires
- Accelerator and throttle cables, adjust if necessary
- Air cleaner assembly
- negative battery cable

27. Fill the cooling system.

28. Adjust the ignition timing.

29. Start the engine, check for leaks and repair if necessary.

1.6L and 1.8L (FP) Engines

1998–00

1. Before servicing the vehicle, refer to the precautions in the beginning of this section.

2. Properly relieve the fuel system pressure.

3. Drain the engine coolant.

4. Remove or disconnect the following:
- negative battery cable
- timing belt
- front exhaust pipe
- exhaust manifold insulator and the Exhaust Gas Recirculation (EGR) pipe
- fresh air duct and air cleaner assembly
- power steering pump and bracket, if necessary, and move it aside leaving the hoses attached
- accelerator cable and bracket
- all vacuum hoses
- engine wiring harness connectors
- fuel supply and return hoses
- intake manifold support bracket
- heater hoses
- Camshaft pulleys by holding the them with a wrench

- camshafts
- Cylinder head bolts in the order
- Cylinder head

To install:

5. Thoroughly, clean the cylinder head and the block contact surfaces.

6. Clean the cylinder head bolts and the threads in the block. Be sure the bolts turn freely in the block.

7. Measure the length of the cylinder head bolts, as shown, maximum bolt length is 3.957 in. (100.5mm) for 1.6L engine or 4.154 in. (105.5mm) for 1.8L (FP) engine.

8. Install or connect the following:
- new head gasket
- cylinder head

9. Torque the cylinder head bolts, in sequence, as follows.
- a. Step 1: 13–16 ft. lbs. (17–22 Nm).
- b. Step 2: Turn 85–95 degrees.
- c. Step 3: Turn an additional 85–95 degrees.

10. Install or connect the following:
- camshafts
- Camshaft pulleys. Torque the bolts to 37–44 ft. lbs. (50–60 Nm)
- heater hoses
- intake manifold support bracket
- fuel hoses
- engine wiring harness connector
- any vacuum hoses that were removed
- accelerator cable and bracket
- Power steering pump and bracket, if removed
- air cleaner assembly and fresh air duct
- EGR system
- exhaust manifold insulator and front exhaust pipe
- timing belt
- negative battery cable

11. Fill the cooling system.

12. Start the vehicle, check for leaks and repair if necessary.

ENGINE FRONT

Cylinder head bolt removal sequence—1.6L and 1.8L (FP) engines

9301MG01

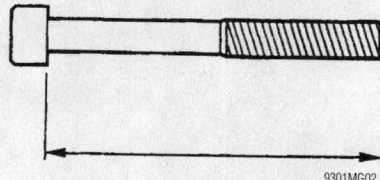

Replace any bolts that exceed the maximum length—1.6L and 1.8L engines

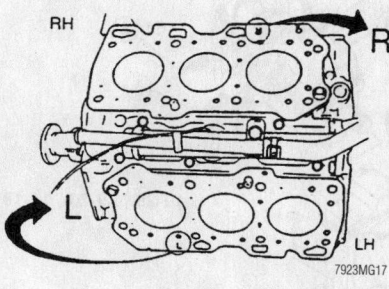

Cylinder head gasket positioning—6-cylinder engines

2.0L Engine

1998–00

1. Before servicing the vehicle, refer to the precautions in the beginning of this section.
2. Relieve the fuel system pressure.
3. Drain the cooling system.
4. Remove or disconnect the following:
   - Negative battery cable
   - splash shield, fresh air duct and air cleaner assembly
   - accelerator cable
   - The following hoses: heater, brake vacuum, purge, fuel, water and upper radiator
   - All accessory drive belts
   - Power steering pump and move it aside, leaving the hoses attached
   - alternator bracket and move aside
   - Exhaust pipe from the manifold
   - Spark plug wires
   - Power steering hose brackets from the cylinder head cover
   - The following electrical connectors: distributor/coil, engine coolant temperature sensor, cooling fan coolant temperature sensor and temperature gauge sensor
   - All hoses from the cylinder head cover
5. Loosen the cover bolts in 5–6 steps,

in the reverse order of the tightening sequence.

6. Remove or disconnect the following:
   - Cylinder head cover
   - timing belt cover and the timing belt
   - coolant temperature sensor housing from the cylinder head
   - Distributor
   - camshaft sprockets and the camshaft
   - hydraulic lifters

7. Identify each lifter as it is removed so it can be reinstalled in the same position. If the lifters are to be reused, store them upside down in an oil-filled sealed container.

8. Remove or disconnect the following:
   - Cylinder head bolts by loosening them, in 2–3 steps, in sequence
   - Cylinder head

9. Clean all gasket mating surfaces. Inspect the cylinder head for damage, cracks and water and oil leakage. Check the head gasket surface for distortion using a straightedge and feeler gauge. Maximum allowable distortion is 0.004 in. (0.10mm).

To install:

10. Install or connect the following:
    - New cylinder head gasket
    - Cylinder head

11. Apply clean engine oil to the bolt threads and seating faces.

12. Install new cylinder head bolts and torque in 2–3 steps, in sequence, to 13–16 ft. lbs. (17–22 Nm).

13. Paint a mark on the edge of each cylinder head bolt to use as a reference. Turn each bolt, in sequence, 90 degrees. Again, turn each bolt, in sequence, an additional 90 degrees.

14. Apply clean engine oil to the hydraulic lifters and install them in their original positions. Be sure they move freely in the bores.

15. Install or connect the following:
    - camshafts and sprockets
    - Distributor
    - Distributor/coil electrical connectors
    - timing belt and cover
    - new cylinder head cover gasket with sealant on the cylinder head cover; apply sealant to the cylinder head surface in the area adjacent to the front camshaft caps
    - Cylinder head cover. Torque the bolts in 5–6 steps, in sequence, to 61–95 inch lbs. (7–11 Nm)

- hoses to the cylinder head cover
- Spark plug wires
- exhaust manifold
- Alternator bracket. Torque the nut/bolt to 19 ft. lbs. (25 Nm)
- alternator belt and adjust the tension
- Power steering pump through and lockbolts
- Pump pressure switch connector
- power steering pump belt and adjust the tension. Torque the through bolt to 45 ft. lbs. (61 Nm) and the lockbolt to 34 ft. lbs. (46 Nm)
- power steering pump belt shield. Torque the bolts to 86 inch lbs. (9 Nm)
- Power steering hose brackets to the cylinder head cover. Torque the bolts to 88 inch lbs. (10 Nm)
- coolant temperature sensor housing using a new gasket. Torque the bolts to 19 ft. lbs. (25 Nm)
- Electrical connectors at the housing
- Accelerator cable, adjust as necessary
- heater, brake vacuum, purge, fuel, water and upper radiator hoses
- air cleaner assembly, fresh air duct and splash shield
- negative battery cable

16. Fill and bleed the cooling system.
17. Run the engine and check for proper operation.

2.3L Engines

1998–00

1. Before servicing the vehicle, refer to the precautions in the beginning of this section.
2. Relieve the fuel system pressure.
3. Drain the engine coolant.
4. Remove or disconnect the following:
   - Negative battery cable
   - oxygen (O2S) sensor connectors
   - Exhaust pipe-to-manifold nuts and lower the exhaust pipes
   - right-hand 3-way catalytic converter
   - compressor (supercharger)
   - intake manifold
   - timing belt covers and timing belt
   - spacer and O-ring from the front of the camshaft
   - ignition coils
   - cylinder head cover mounting bolts, in 5–6 steps, using the reverse of the tightening sequence

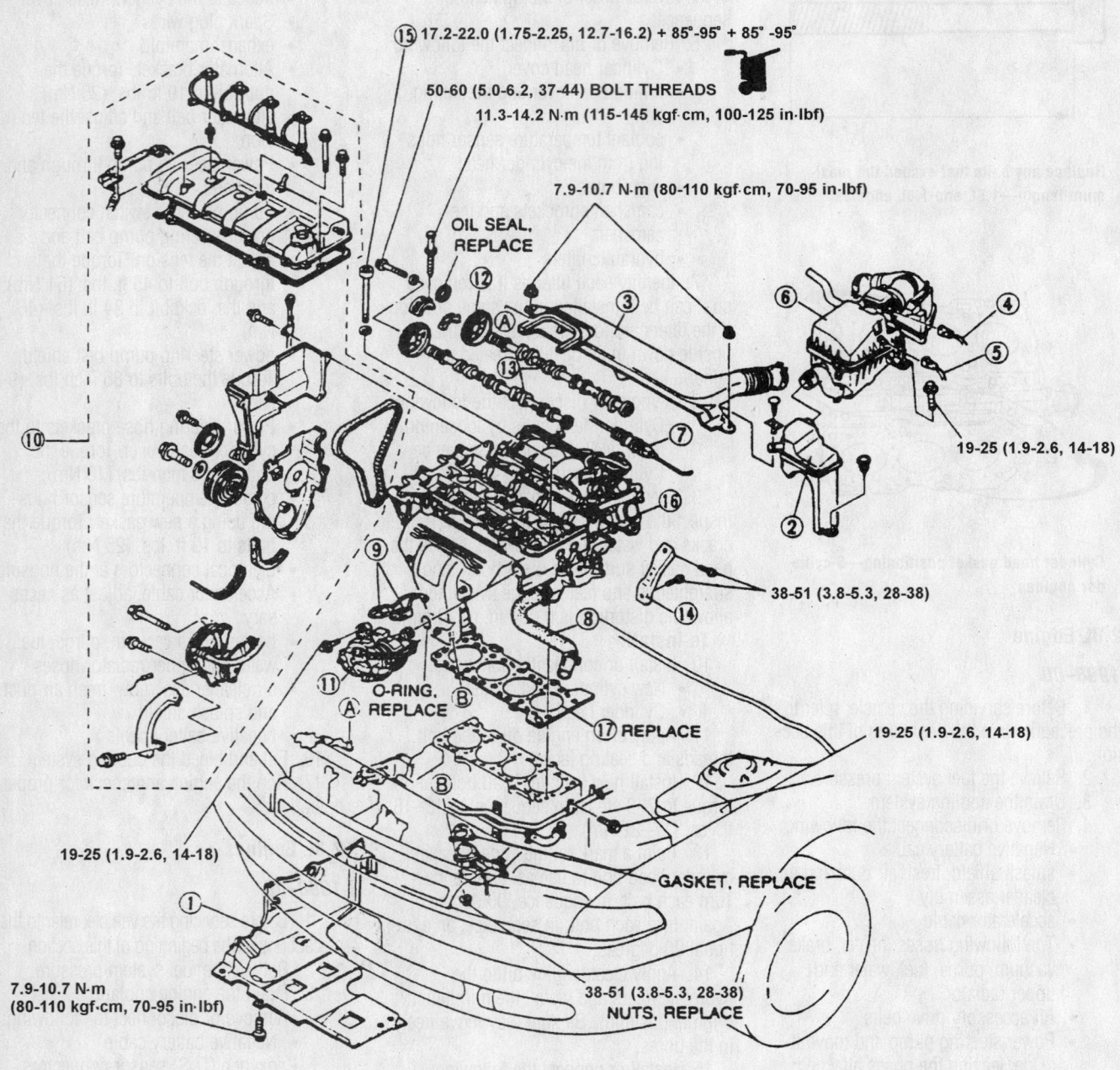

15 17.2-22.0 (1.75-2.25, 12.7-16.2) + 85°-95° + 85° -95°

50-60 (5.0-6.2, 37-44) BOLT THREADS

11.3-14.2 N·m (115-145 kgf·cm, 100-125 in·lbf)

7.9-10.7 N·m (80-110 kgf·cm, 70-95 in·lbf)

OIL SEAL,
REPLACE

19-25 (1.9-2.6, 14-18)

38-51 (3.8-5.3, 28-38)

O-RING,
REPLACE

19-25 (1.9-2.6, 14-18)

REPLACE

19-25 (1.9-2.6, 14-18)

7.9-10.7 N·m
(80-110 kgf·cm, 70-95 in·lbf)

38-51 (3.8-5.3, 28-38)
NUTS, REPLACE

GASKET, REPLACE

N·m (kgf·m, ft·lbf)

1. Undercover
2. Resonance chamber No.1
3. Fresh-air duct
4. Mass airflow sensor connector
5. Intake air temperature sensor connector
6. Air cleaner housing and resonance chamber No.2
7. Accelerator cable
8. Hose
9. Harness connectors
10. Timing belt
11. Distributor
12. Camshaft pulley
13. Camshaft
14. Intake manifold bracket
15. Cylinder head bolt
16. Cylinder head
17. Cylinder head gasket

Exploded view of the cylinder head assembly—2.0L engine

7923MG21

- Cylinder head cover
- camshaft sprockets

5. Turn the camshafts so the knock pins are aligned with the marks on the camshaft caps. This will reduce the pressure on the adjustment shims.

6. Note the markings on the camshaft caps prior to removal, so they can be reinstalled in the same positions. The right-hand (rear) caps are marked with numbers and the left-hand (front) caps are marked with letters.

7. Loosen the front camshaft cap bolts in sequence, in 5–6 steps. Remove the front camshaft caps. remove the remaining camshaft cap bolts in the proper sequence. Remove the caps, being sure to remove the thrust caps last. Do not damage the cylinder head thrust bearing support.

8. Remove or disconnect the following:
- camshafts and oil seals
- lifters and adjustment shims

9. Identify and mark each lifter as it is removed so it can be reinstalled in the same position.

10. Remove or disconnect the following:
- lower radiator hose
- Water inlet pipe
- compressor bracket
- alternator bracket bolt to gain additional clearance
- rubber insulator from the left-hand cylinder head

11. Temporarily install the No. 3 engine mount, which was removed with the timing belt, to support the engine.

12. Remove or disconnect the following:

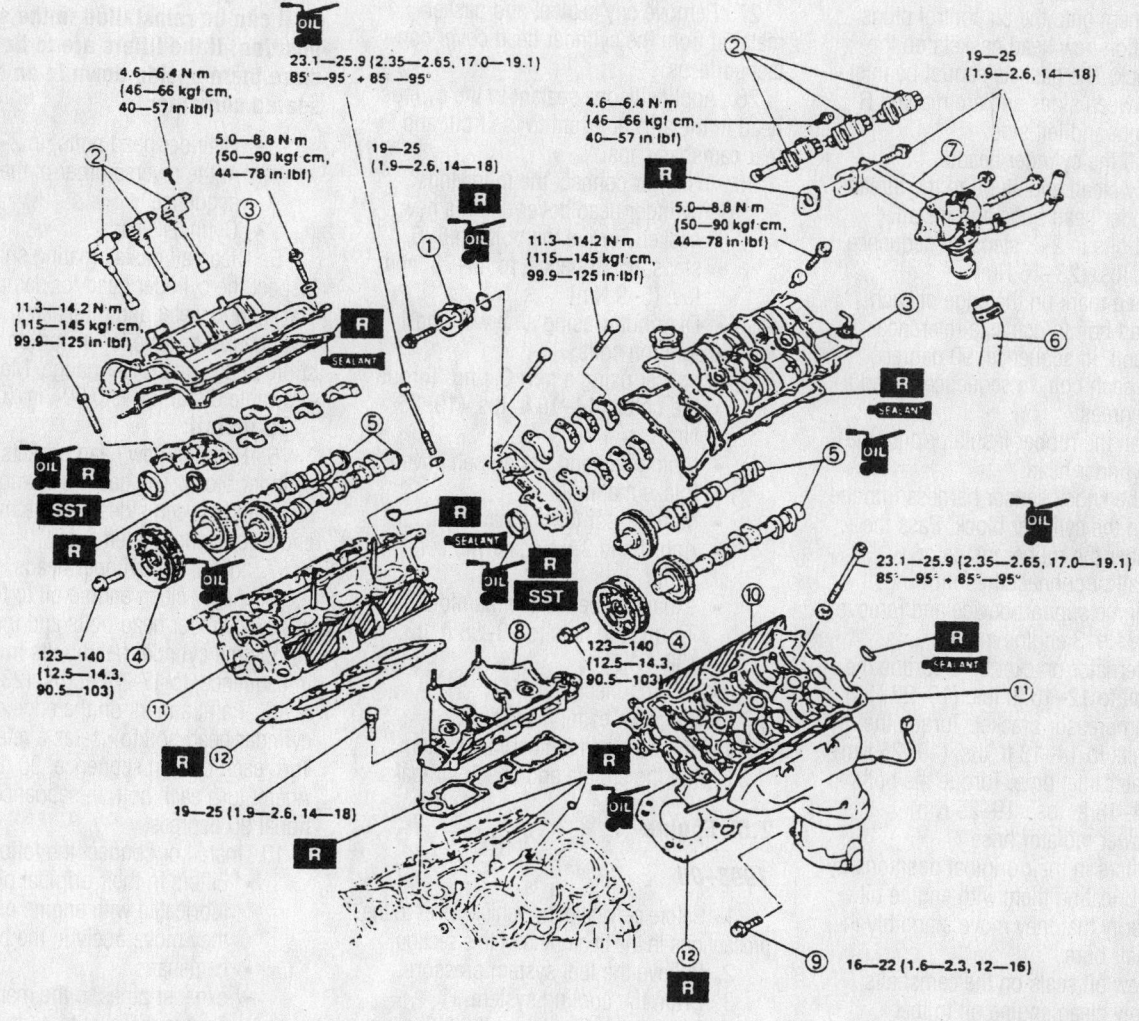

1. Spacer
2. Ignition coil
3. Cylinder head cover
4. Camshaft pulley
5. Camshaft
6. Lower radiator hose

7. Water inlet pipe
8. Lysholm compressor bracket
9. Generator bolt
10. Rubber insulator (LH)
11. Cylinder head
12. Cylinder head gasket

Exploded view of the cylinder head and related components—2.3L engine

7923MG24

- Engine support device
- Cylinder head bolts, in 2–3 steps, by the reversing order of the torque sequence
- Cylinder heads
- oil control plug O-rings

13. Clean all gasket mating surfaces. Inspect the cylinder head for damage, cracks, and water and oil leakage. Check the head gasket surface for distortion using a straightedge and feeler gauge. Maximum allowable distortion is 0.004 in. (0.10mm).

To install:

14. Apply clean engine oil to the O-rings, and install them onto the oil control plugs.

15. Position new head gaskets on the cylinder block. The gaskets cannot be interchanged between sides and are marked **R** and **L** for right and left side.

16. Install the cylinder heads.

17. Apply clean engine oil to the threads of new cylinder head bolts and install. Torque the bolts in 2–3 steps, in sequence, to 17–19 ft. lbs. (23–26 Nm).

18. Paint a mark on the edge of each cylinder head bolt to use as a reference. Turn each bolt, in sequence, 90 degrees. Again, turn each bolt, in sequence, an additional 90 degrees.

19. Install the rubber insulator onto the left-hand cylinder head.

20. Fit the knock sensor harness into the drill hole on the cylinder block. Pass the harness under the rubber insulator.

21. Install or connect the following:
- Engine support device and remove the No. 3 engine mount
- alternator bracket bolt. Torque the bolt to 12–16 ft. lbs. (16–22 Nm)
- compressor bracket. Torque the bolts to 14–18 ft. lbs. (19–25 Nm)
- water inlet pipe. Torque the bolts to 14–18 ft. lbs. (19–25 Nm)
- Lower radiator hose
- Lifters in their original positions by lubricating them with engine oil. Verify that they move smoothly in their bore
- New oil seals on the camshafts

22. Apply clean engine oil to the camshaft lobes, journals and supports.

23. Install or connect the following:
- camshafts so the gear marks align
- thrust caps. Torque the bolts, in 5–6 steps, until they are fully seated on the cylinder head

24. Apply silicone sealant, at a thickness of 0.06–0.09 in. (1.5–2.5mm), to the cylinder head surface in the area forward of the camshaft gear cavity.

25. Install or connect the following:
- remaining camshaft caps in their

original positions. Torque the caps, in sequence, in 5 equal steps, with the final step being 100–125 inch lbs. (11–14 Nm)
- New camshaft oil seal lubricated with engine oil. Tap the seal in evenly with a Seal Installer tool 49 F401 337A with a final protrusion of 0–0.02 in. (0–0.5mm). Tap in a new blind cap
- camshaft sprockets. Torque the bolts to 91–103 ft. lbs. (123–140 Nm)

26. Measure and adjust the valve clearances.

27. Remove any sealant and gasket material from the cylinder head cover contact surfaces.

28. Apply silicone sealant to the cylinder head in the area adjacent to the front and rear camshaft caps.

29. Install or connect the following:
- cylinder head cover using a new gasket. Torque the bolts in 5–6 steps, in sequence, to 44–78 inch lbs. (5–9 Nm)
- Distributor using a new O-ring
- ignition coils
- spacer using a new O-ring. Torque the bolt to 14–18 ft. lbs. (19–25 Nm)
- timing belt and timing belt cover
- intake manifold
- compressor (supercharger)
- right-hand 3-way catalytic converter
- exhaust pipes to the manifolds. Torque the nuts to 28–38 ft. lbs. (38–51 Nm)
- O2S connectors
- negative battery cable

30. Fill and bleed the coolant system.

31. Run the engine and check for leaks.

2.5L Engine

1998–00

1. Before servicing the vehicle, refer to the precautions in the beginning of this section.

2. Relieve the fuel system pressure.

3. Drain the cooling system.

4. Remove or disconnect the following:
- negative battery cable
- fresh air duct and the air cleaner assembly
- Battery, if additional clearance space is needed.
- accelerator cable
- wiring harness from the cylinder heads
- fuel, heater and vacuum hoses
- intake manifold
- distributor

- ventilation pipe from the left cylinder head cover
- Cylinder head covers
- timing belt covers and the timing belt
- camshafts
- 3 bolts and the seal plate from the front of the engine
- upper radiator hose
- oxygen (O2S) sensor connectors
- Exhaust pipe-to-manifold nuts and lower the pipes
- hydraulic lifters

➡**Identify each lifter as it is removed so it can be reinstalled in the same position. If the lifters are to be reused, store them upside down in an oil-filled, sealed container.**

- Cylinder head bolts, in 2–3 steps, in the reverse order of the torque sequence
- Cylinder heads

5. Clean all gasket mating surfaces. Inspect the cylinder head for damage, cracks, and water and oil leakage. Check the head gasket surface for distortion using a straightedge and feeler gauge. Maximum allowable distortion is 0.004 in. (0.10mm).

To install:

6. Position new head gaskets on the cylinder block. The gaskets cannot be interchanged between sides and are marked **R** and **L** for right and left side.

7. Install the cylinder heads.

8. Apply clean engine oil to the threads of new cylinder head bolts and install. Torque the cylinder head bolts in 2–3 steps, in sequence, to 17–19 ft. lbs. (23–26 Nm).

9. Paint a mark on the edge of each cylinder head bolt to use as a reference. Turn each bolt, in sequence, 90 degrees. Again, turn each bolt, in sequence, an additional 90 degrees.

10. Install or connect the following:
- Lifters in their original positions lubricated with engine oil. Be sure they move freely in the bores
- camshafts
- exhaust pipes to the manifolds. Torque the nuts to 41 ft. lbs. (55 Nm)
- O2S connectors
- timing belt and timing belt covers

11. Apply sealant to the cylinder head surface in the area of the front and rear camshaft caps.

12. Install or connect the following:
- Cylinder head covers using new gaskets. Torque the bolts in 5–6 steps, in sequence, to 44–78 inch lbs. (5–9 Nm)

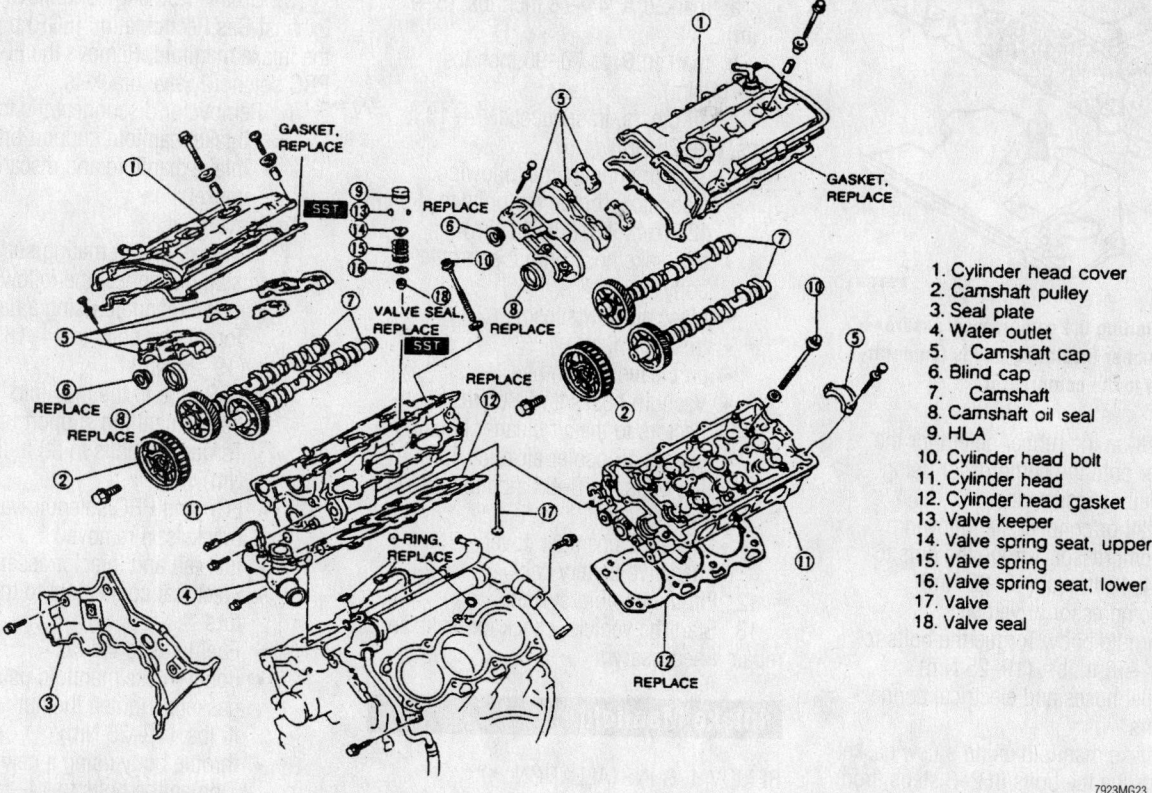

1. Cylinder head cover
2. Camshaft pulley
3. Seal plate
4. Water outlet
5. Camshaft cap
6. Blind cap
7. Camshaft
8. Camshaft oil seal
9. HLA
10. Cylinder head bolt
11. Cylinder head
12. Cylinder head gasket
13. Valve keeper
14. Valve spring seat, upper
15. Valve spring
16. Valve spring seat, lower
17. Valve
18. Valve seal

7923MG23

Exploded view of the cylinder head assemblies—2.5L engine

- intake manifold using new gaskets. Torque the bolts to 14–18 ft. lbs. (19–25 Nm)
- distributor using a new O-ring lubricated with engine oil. Hand-tighten the mounting bolts
- vacuum, heater and fuel hoses
- wiring harness to the cylinder heads
- Accelerator cable, adjust as necessary
- Battery, if removed
- air cleaner assembly and the fresh air duct
- negative battery cable

13. Fill and bleed the cooling system.
14. Adjust the ignition timing and idle speed.
15. Run the engine and check for proper operation.

Rocker Arms/Shafts

REMOVAL & INSTALLATION

All Mazda engines covered in this manual are not equipped with rocker arms/shafts, the camshafts directly actuate

the valves through a bucket type cam follower.

Supercharger

REMOVAL & INSTALLATION

2.3L Engine

1998–00

1. Before servicing the vehicle, refer to the precautions in the beginning of this section.
2. Relieve the fuel system pressure.
3. Drain the cooling system.
4. Remove or disconnect the following:
   - Negative battery cable
   - dynamic chamber cover
   - charge air cooler air duct
   - Vacuum hoses and electrical connectors from the air cleaner housing
   - Air cleaner assembly
   - Fresh air ducts
   - mass air flow sensor and the air intake hose from the throttle body
   - resonator
   - right-hand charge air cooler

- left-hand charge air cooler
- accelerator cable
- Vacuum hoses from the rear of the intake manifold and Exhaust Gas Recirculation (EGR) valve
- EGR valve
- air intake pipe assembly
- charge air cooler pipe
- Fuel supply line at the fuel rails and discard the copper washers
- fuel and vacuum lines from the fuel pressure regulator
- Coolant hoses
- Wiring harness from the intake manifold
- intake manifold mounting nuts and bolts in 2–3 steps
- Intake manifold
- Fuel hoses and electrical connectors from the throttle body
- Throttle body
- drive belt from the compressor (supercharger)
- mounting bolts from the compressor
- Compressor

To install:
5. Clean all gasket mating surfaces.

For Maintenance Interval recommendations, see Section 1 of this manual

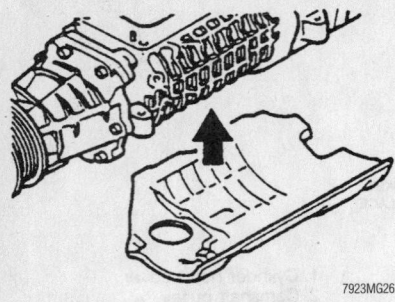

7923MG26

When installing the compressor, ensure that the rubber insulating pad is temporarily affixed to the compressor

6. Position the rubber shield for the compressor onto the compressor using double sided adhesive tape.

7. Install or connect the following:
- Compressor. Torque the nuts to 14–18 ft. lbs. (19–25 Nm)
- Compressor drive belt
- throttle body. Torque the bolts to 14–18 ft. lbs. (19–25 Nm)
- Fuel hoses and electrical connectors
- Intake manifold using a new gasket. Torque the bolts in 2–3 steps, from the center to the ends, to 14–18 ft. lbs. (19–25 Nm)
- Wiring harness onto the intake manifold
- Coolant hoses
- fuel and vacuum lines to the fuel pressure regulator
- Fuel supply line to the fuel rail using new copper crush washers
- charge air cooler pipe
- Position the air intake pipe assembly using new gaskets

8. Hand-tighten the nuts/bolts in the order shown until the air intake pipe contacts the intake manifold. Verify that the rubber gaskets are not twisted or distorted. Torque the bolts marked **A** to 70–95 inch lbs. (8–11 Nm) and all others, in sequence, to 14–18 ft. lbs. (19–25 Nm).

9. Install or connect the following:
- EGR valve using a new gasket
- vacuum hoses to the intake manifold and EGR valve
- Accelerator cable and adjust it
- Left and right-hand charge air coolers using new gaskets. Hand-tighten the nuts/bolts in the order shown until the air intake pipes and charge air coolers contact the intake manifold. Verify that the rubber gaskets are not twisted or distorted.

10. Torque the charge air cooler bolts to:

a. marked **A**: 44–78 inch lbs. (5–9 Nm).

b. marked **B**: to 70–95 inch lbs. (8–11 Nm).

c. all others, in sequence: 14–18 ft. lbs. (19–25 Nm).

11. Install or connect the following:
- resonator. Torque the bolts to 12–16 ft. lbs. (16–22 Nm)
- air intake hose onto the throttle body
- Mass air flow sensor
- fresh air ducts
- air cleaner assembly
- Vacuum hoses and electrical connectors to the air cleaner housing
- charge air cooler air duct. Torque the bolts to 70–95 inch lbs. (8–11 Nm)
- dynamic chamber cover
- negative battery cable

12. Fill the cooling system.

13. Start the vehicle, check for leaks and repair if necessary.

Intake Manifold

REMOVAL & INSTALLATION

1.5L and 1.8L (BP) Engines

1998–00

1. Before servicing the vehicle, refer to the precautions in the beginning of this section.

2. Relieve the fuel system pressure.

3. Drain the cooling system.

4. Remove or disconnect the following:
- Negative battery cable
- mass air flow sensor electrical connector
- Air ducts, air cleaner assembly, mass air flow sensor and resonance chamber
- throttle and accelerator cables
- Fuel lines
- Coolant hoses from the throttle body
- Idle Air Control (IAC) valve and the Throttle Position Sensor (TPS) electrical connectors
- Throttle body
- Vacuum lines at the intake manifold
- dynamic chamber (upper intake manifold)
- Fuel hoses from the fuel rail
- Electrical connectors for the fuel injectors

➡**On some models, it may be necessary to remove the fuel rail with the injectors connected.**

5. On the 1.5L engine, remove the Exhaust Gas Recirculation (EGR) pipe from the intake manifold. Remove the EGR and PRC solenoid valve brackets.

6. Remove or disconnect the following:
- intake manifold support bracket
- Intake manifold and discard the gasket

To install:

7. Clean all gasket mating surfaces.

8. Install or connect the following:
- intake manifold using a new gasket. Torque the bolts to 14–18 ft. lbs. (19–25 Nm)
- EGR pipe to the manifold
- Intake manifold support bracket. Torque the bolts to 38 ft. lbs. (51 Nm)
- EGR and PRC solenoid valve brackets, if removed
- fuel rail and injector assembly
- Electrical connectors to the injectors
- Fuel lines to the rail
- upper intake manifold using new gaskets. Tighten the nuts to 14–18 ft. lbs. (19–25 Nm)
- throttle body using a new gasket. Tighten the bolts to 14–18 ft. lbs. (19–25 Nm)
- electrical connectors for the IAC valve and the TPS
- vacuum and coolant lines
- Connect and adjust the throttle and accelerator cables
- fuel lines
- resonance chamber
- Air cleaner assembly, mass air flow sensor and ducts
- Mass air flow sensor connector
- negative battery cable

9. Fill the cooling system.

10. Run the engine and check for leaks.

1.6L and 1.8L (FP) Engines

1998–00

1. Before servicing the vehicle, refer to the precautions in the beginning of this section.

2. Relieve the fuel system pressure. Drain the cooling system.

3. Remove or disconnect the following:
- Negative battery cable
- Fuel hoses from the fuel rail
- Fuel injector electrical connectors
- fuel rail with the injectors connected
- components in the order illustrated

To install:

4. Clean all gasket mating surfaces.

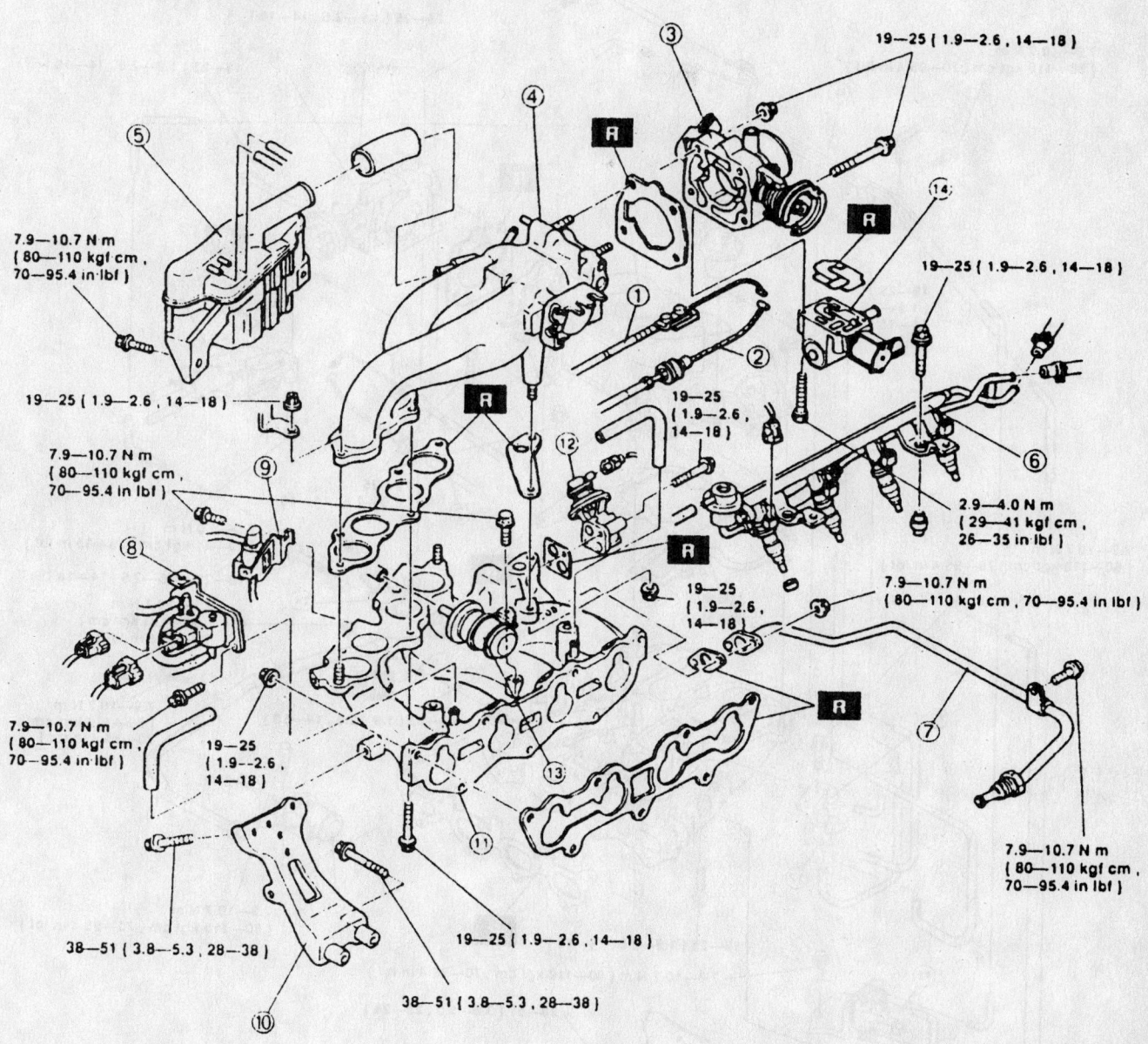

7.9—10.7 N m
(80—110 kgf cm,
70—95.4 in lbf)

19—25 (1.9—2.6 , 14—18)

19—25 (1.9—2.6 , 14—18)

7.9—10.7 N m
(80—110 kgf cm,
70—95.4 in lbf)

19—25 (1.9—2.6 , 14—18)

19—25
(1.9—2.6 ,
14—18)

2.9—4.0 N m
(29—41 kgf cm ,
26—35 in lbf)

7.9—10.7 N m
(80—110 kgf cm , 70—95.4 in lbf)

7.9—10.7 N m
(80—110 kgf cm,
70—95.4 in lbf)

19—25
(1.9—2.6 ,
14—18)

7.9—10.7 N m
(80—110 kgf cm ,
70—95.4 in lbf)

38—51 (3.8—5.3 , 28—38)

19—25 (1.9—2.6 , 14—18)

38—51 (3.8—5.3 , 28—38)

N m (kgf m , ft lbf)

1. Throttle cable (ATX)
2. Accelerator cable
3. Throttle body
4. Dynamic chamber
5. Resonance chamber
6. Fuel distributor
7. EGR pipe
8. EGR solenoid valve stay
9. PRC solenoid valve stay
10. Intake manifold stay
11. Intake manifold
12. EGR valve
13. Air filter
14. BAC valve

7923MG27

Exploded view of the intake manifold and related components—1.5L engine

For Tune-up, Capacities and Firing orders, see Section 1 of this manual

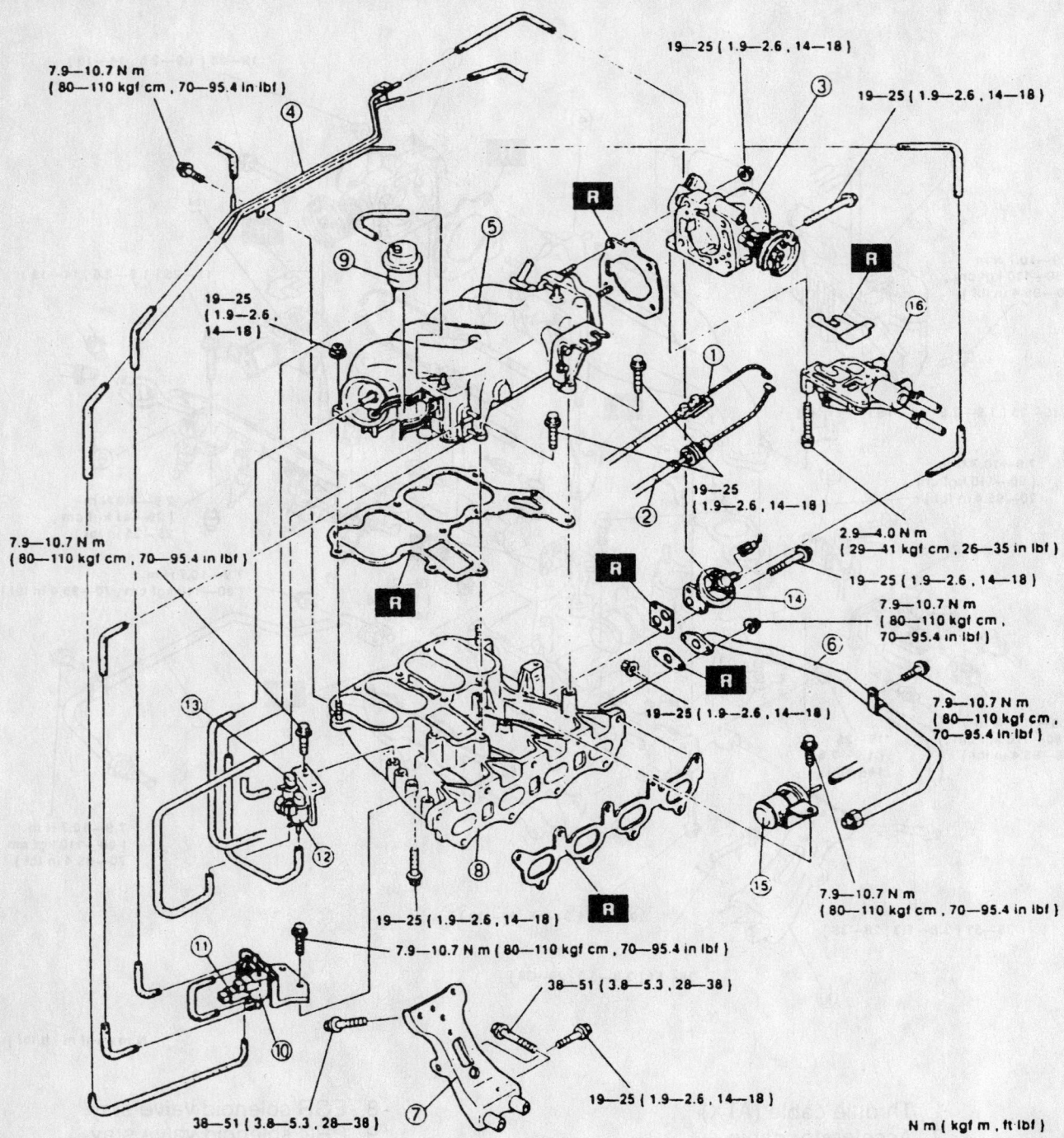

7.9—10.7 N m
(80—110 kgf cm , 70—95.4 in lbf)

19—25 (1.9—2.6 , 14—18)

19—25 (1.9—2.6 , 14—18)

19—25
(1.9—2.6 ,
14—18)

7.9—10.7 N m
(80—110 kgf cm , 70—95.4 in lbf)

19—25 (1.9—2.6 , 14—18)

2.9—4.0 N m
(29—41 kgf cm , 26—35 in lbf)

19—25 (1.9—2.6 , 14—18)

7.9—10.7 N m
(80—110 kgf cm ,
70—95.4 in lbf)

7.9—10.7 N m
(80—110 kgf cm ,
70—95.4 in lbf)

19—25 (1.9—2.6 , 14—18)

7.9—10.7 N m
(80—110 kgf cm , 70—95.4 in lbf)

19—25 (1.9—2.6 , 14—18)

7.9—10.7 N m (80—110 kgf cm , 70—95.4 in lbf)

38—51 (3.8—5.3 , 28—38)

38—51 (3.8—5.3 , 28—38)

19—25 (1.9—2.6 , 14—18)

N m (kgf m , ft lbf)

1. Throttle cable (ATX)
2. Accelerator cable
3. Throttle body
4. Vacuum pipe
5. Dynamic chamber
6. EGR pipe
7. Intake manifold stay
8. Intake manifold
9. Vacuum chamber
10. EGR solenoid valve (vacuum)
11. EGR solenoid valve (vent)
12. VICS solenoid valve
13. PRC solenoid valve
14. EGR valve
15. Air filter
16. BAC valve

Exploded view of the intake manifold and related components—1.8L (BP) engine

7923MG28

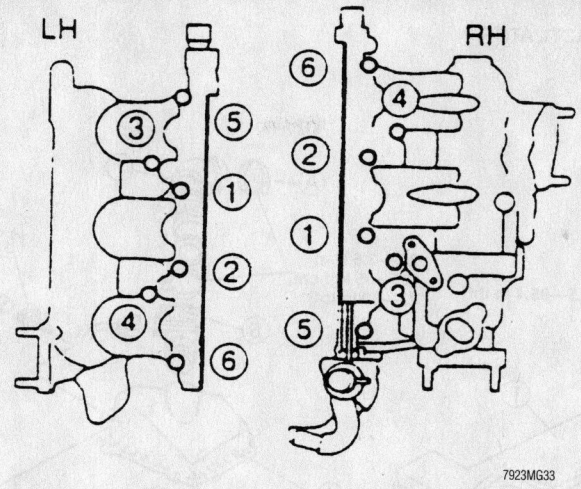

LH **RH**

7923MG33

Be sure to tighten the intake manifold bolts in the sequence shown—2.3L engine

➡ **Be sure that the convex side of the intake manifold gasket is facing the manifold side, as shown.**

5. Install or connect the following:
    - intake manifold using a new gasket. Torque the bolts to 14–18 ft. lbs. (19–25 Nm)
    - components in the reverse order of the removal sequence
    - fuel rail. Tighten the bolts to 14–18 ft. lbs. (19–25 Nm)
    - fuel lines and electrical connectors to the fuel rail
    - negative battery cable
6. Fill the cooling system.
7. Run the engine and check for leaks.

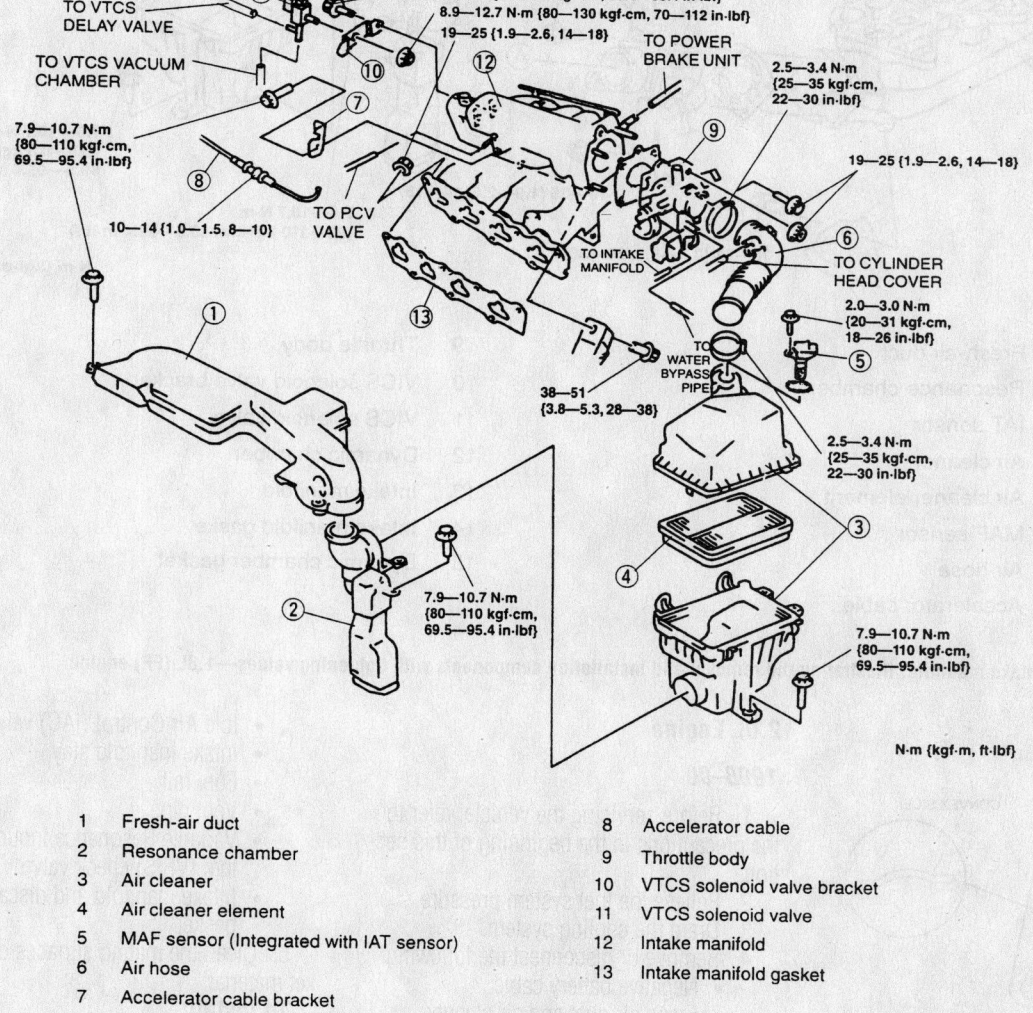

TO VTCS DELAY VALVE

TO VTCS VACUUM CHAMBER

7.9—10.7 N·m {80—110 kgf-cm, 69.5—95.4 in·lbf}

7.9—10.8 N·m {80—110 kgf-cm, 69.4—95.4 in·lbf}
8.9—12.7 N·m {80—130 kgf-cm, 70—112 in·lbf}
19—25 {1.9—2.6, 14—18}

TO POWER BRAKE UNIT

2.5—3.4 N·m {25—35 kgf-cm, 22—30 in·lbf}

19—25 {1.9—2.6, 14—18}

TO PCV VALVE

10—14 {1.0—1.5, 8—10}

TO INTAKE MANIFOLD

TO CYLINDER HEAD COVER

TO WATER BYPASS PIPE

38—51 {3.8—5.3, 28—38}

2.0—3.0 N·m {20—31 kgf-cm, 18—26 in·lbf}

2.5—3.4 N·m {25—35 kgf-cm, 22—30 in·lbf}

7.9—10.7 N·m {80—110 kgf-cm, 69.5—95.4 in·lbf}

7.9—10.7 N·m {80—110 kgf-cm, 69.5—95.4 in·lbf}

N·m {kgf·m, ft·lbf}

1	Fresh-air duct	8	Accelerator cable
2	Resonance chamber	9	Throttle body
3	Air cleaner	10	VTCS solenoid valve bracket
4	Air cleaner element	11	VTCS solenoid valve
5	MAF sensor (Integrated with IAT sensor)	12	Intake manifold
6	Air hose	13	Intake manifold gasket
7	Accelerator cable bracket		

9301MG03

Exploded view of the intake manifold, illustrating the removal and installation components with tightening values—1.6L engine

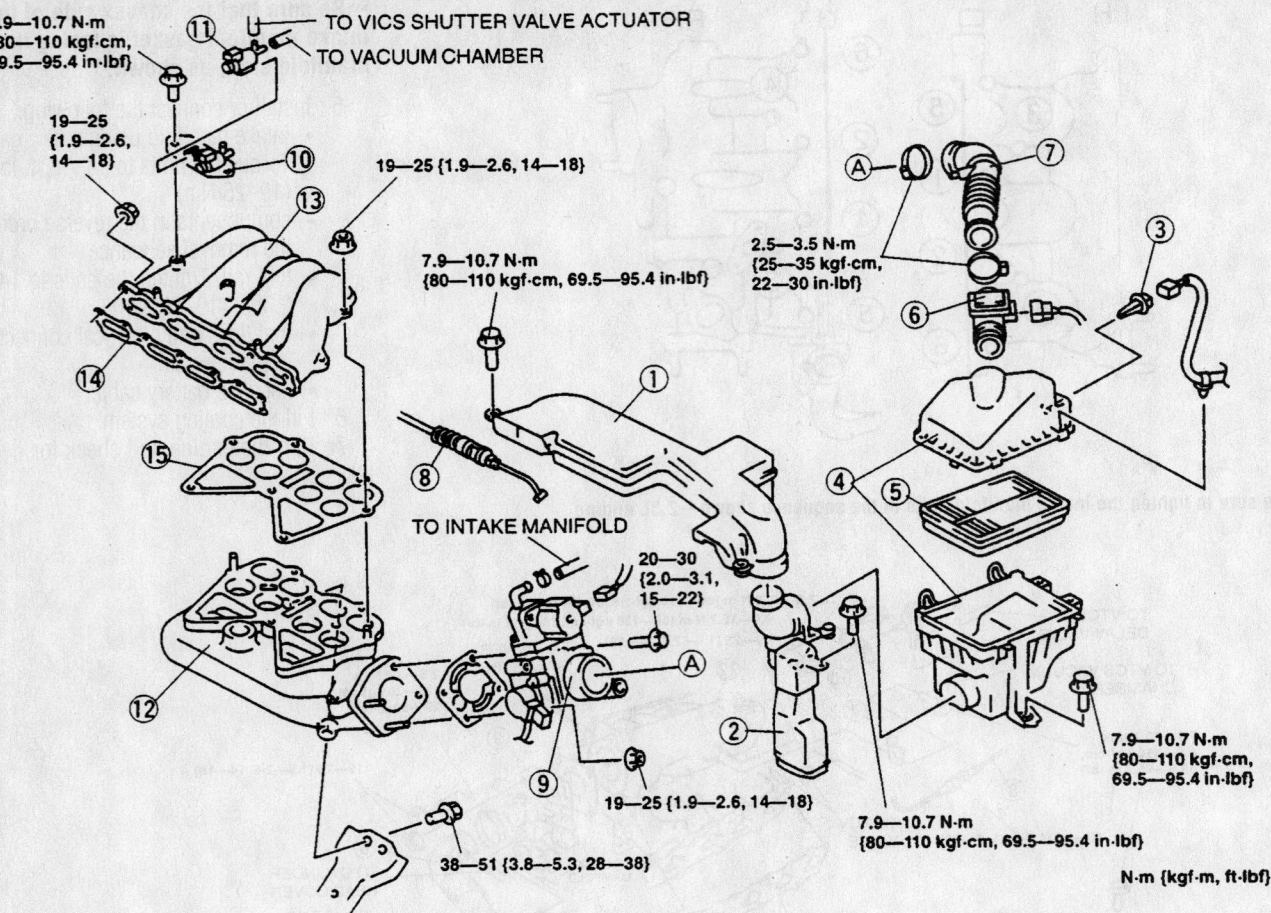

7.9—10.7 N·m
{80—110 kgf·cm,
69.5—95.4 in·lbf}

TO VICS SHUTTER VALVE ACTUATOR

TO VACUUM CHAMBER

19—25
{1.9—2.6,
14—18}

19—25 {1.9—2.6, 14—18}

7.9—10.7 N·m
{80—110 kgf·cm, 69.5—95.4 in·lbf}

2.5—3.5 N·m
{25—35 kgf·cm,
22—30 in·lbf}

TO INTAKE MANIFOLD

20—30
{2.0—3.1,
15—22}

19—25 {1.9—2.6, 14—18}

38—51 {3.8—5.3, 28—38}

7.9—10.7 N·m
{80—110 kgf·cm,
69.5—95.4 in·lbf}

7.9—10.7 N·m
{80—110 kgf·cm, 69.5—95.4 in·lbf}

N·m {kgf·m, ft·lbf}

1	Fresh-air duct	9	Throttle body
2	Resonance chamber	10	VICS solenoid valve bracket
3	IAT sensor	11	VICS solenoid valve
4	Air cleaner	12	Dynamic chamber
5	Air cleaner element	13	Intake manifold
6	MAF sensor	14	Intake manifold gasket
7	Air hose	15	Dynamic chamber gasket
8	Accelerator cable		

9301MG05

Exploded view of the intake manifold, illustrating the removal and installation components with tightening values—1.8L (FP) engine

CONVEX SIDE

9301MG04

Cross-sectional view of the intake manifold gasket—1.6L and 1.8L engines

2.0L Engine

1998–00

1. Before servicing the vehicle, refer to the precautions in the beginning of this section.
2. Relieve the fuel system pressure.
3. Drain the cooling system.
4. Remove or disconnect the following:
   • Negative battery cable
   • Fresh air duct and air cleaner assembly
   • Mass Air Flow (MAF) sensor
   • Accelerator cable
   • Throttle body assembly

• Idle Air Control (IAC) valve
• Intake manifold stay
• Fuel rail
• Vent pipe
• Variable Resonance Induction System (VRIS) check valve
• Intake manifold and discard the gasket

5. Clean the mating surfaces of any gasket material

To install:

6. Install or connect the following:
   • Intake manifold with a new gasket. Make certain that the convex side of the new gasket faces the intake

manifold. Torque the bolts to 18 ft. lbs. (25 Nm)

- VRIS check valve. Torque the bolt to 89 inch lbs. (10 Nm)
- Vent pipe. Torque the bolts to 18 ft. lbs. (25 Nm)
- Intake manifold stay
- IAC valve. Torque the bolt to 18 ft. lbs. (25 Nm)
- Throttle body. Torque the nuts to 22 ft. lbs. (30 Nm)
- Accelerator cable
- MAF sensor. Torque the bolt to 89 inch lbs. (10 Nm)
- Air cleaner/air duct assembly. Torque the bolts to 30 inch lbs. (3.5 Nm)
- Negative battery cable

7. Fill the cooling system.
8. Start the vehicle, check for leaks and repair if necessary.

2.3L Engine

1998–00

1. Before servicing the vehicle, refer to the precautions in the beginning of this section.

2. Relieve the fuel system pressure.
3. Drain the cooling system.
4. Remove or disconnect the following:
   - Negative battery cable
   - dynamic chamber cover
   - charge air cooler air duct
   - Vacuum hoses and electrical connectors from the air cleaner housing
   - Air cleaner assembly
   - Fresh air ducts
   - mass air flow sensor and the air intake hose from the throttle body
   - resonator
   - right-hand charge air cooler
   - left-hand charge air cooler
   - accelerator cable
   - Vacuum hoses from the rear of the intake manifold and Exhaust Gas Recirculation (EGR) valve
   - EGR valve
   - air intake pipe assembly
   - charge air cooler pipe
   - Fuel supply line at the fuel rails and discard the copper washers
   - Fuel and vacuum lines from the fuel pressure regulator
   - Coolant hoses

- Wiring harness from the intake manifold
- intake manifold mounting nuts and bolts in 2–3 steps
- Intake manifold

5. If necessary, label and disconnect the fuel hoses and electrical connectors from the throttle body. Remove the throttle body.

To install:

6. Clean all gasket mating surfaces.
7. Install or connect the following:
   - Throttle body, if removed. Tighten the nuts/bolts to 14–18 ft. lbs. (19–25 Nm)
   - Fuel hoses and electrical connectors.
   - Intake manifold using new gaskets. Tighten the nuts/bolts in 2–3 steps, from the center to the ends, to 14–18 ft. lbs. (19–25 Nm)
   - Wiring harness onto the intake manifold
   - Coolant hoses
   - fuel and vacuum lines to the fuel pressure regulator
   - Fuel supply line to the fuel rail using new copper washers
   - charge air cooler pipe

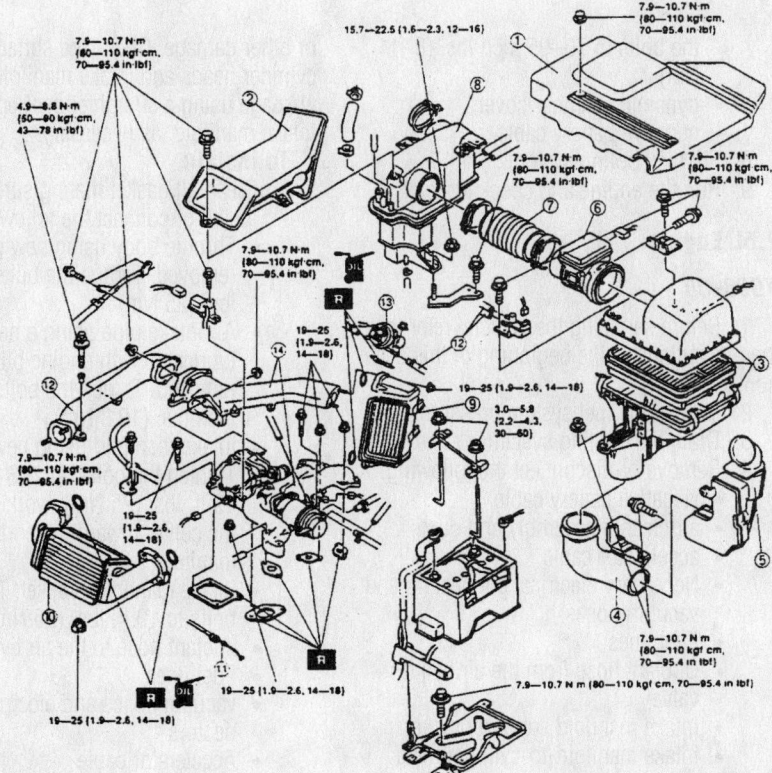

1. Dynamic chamber cover
2. Charge air cooler air duct
3. Air cleaner assembly
4. Air duct
5. Fresh air duct
6. Mass air flow sensor
7. Air intake hose
8. Resonator
9. Charge air cooler (RH)
10. Charge air cooler (LH)
11. Accelerator cable
12. Vacuum hose assembly
13. EGR control valve
14. Air intake pipe assembly

Exploded view of the intake manifold assembly (1 of 2)—2.3L engine

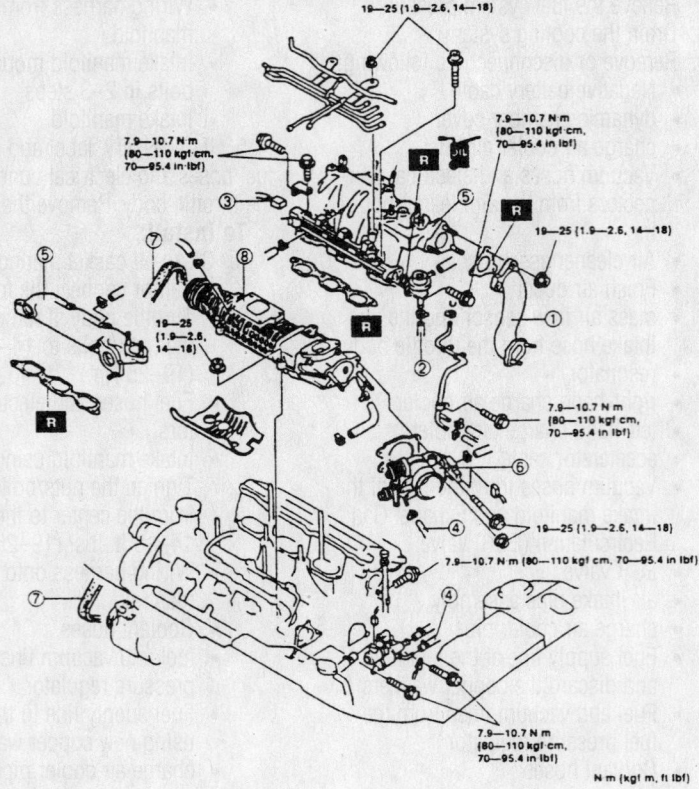

Exploded view of the intake manifold assembly (2 of 2)—2.3L engine

1. Charge air cooler pipe
2. Fuel hose
3. Fuel distributor connector
4. Coolant hose
5. Intake manifold assembly
6. Throttle body assembly
7. Drive belt
8. Lysholm compressor

N·m (kgf·m, ft·lbf)

7923MG32

- Air intake pipe assembly using new gaskets. Verify that the rubber gaskets are not twisted or distorted. Tighten the bolts marked **A** to 70–95 inch lbs. (8–11 Nm) and all other bolts, in sequence, to 14–18 ft. lbs. (19–25 Nm)
- EGR valve using a new gasket
- vacuum hoses to the intake manifold and EGR valve
- Accelerator cable, adjust as necessary
- Left and right-hand charge air coolers using new gaskets. Verify that the rubber gaskets are not twisted or distorted. Tighten the bolts marked **A** to 44–78 inch lbs. (5–9 Nm). Tighten the bolts marked **B** to 70–95 inch lbs. (8–11 Nm) and all other bolts, in sequence, to 14–18 ft. lbs. (19–25 Nm)
- resonator. Tighten the nuts/bolts to 12–16 ft. lbs. (16–22 Nm)
- air intake hose onto the throttle body
- Mass air flow sensor
- fresh air and air ducts
- air cleaner assembly
- Vacuum hoses and electrical connectors to the air cleaner housing
- charge air cooler air duct. Tighten

the bolts to 70–95 inch lbs. (8–11 Nm)
- dynamic chamber cover
- negative battery cable
8. Fill the cooling system.
9. Run the engine and check for leaks.

2.5L Engine

1998–00

1. Before servicing the vehicle, refer to the precautions in the beginning of this section.
2. Relieve the fuel system pressure.
3. Drain the cooling system.
4. Remove or disconnect the following:
- Negative battery cable
- air cleaner assembly and ducts
- accelerator cable
- Necessary electrical connectors and vacuum hoses
- Fuel lines
- Coolant hose from the air bypass valve
- intake manifold support bracket
- Intake manifold-to-cylinder head bolts
- Intake manifold
- Throttle body and air intake pipe from the manifold, if necessary
5. Check the intake manifold for cracks

or other damage. Check the surface of the cylinder heads and intake manifold for warpage using a straightedge. Replace the intake manifold, as necessary.
To install:
6. Clean all gasket mating surfaces.
7. Install or connect the following:
- Throttle body using new gaskets, if removed. Torque the bolts to 19 ft. lbs. (25 Nm)
- Air intake pipe using a new O-rings lubricated with engine oil, if removed. Torque the bolts to 95 inch lbs. (10.8 Nm)
- Intake manifold using new gaskets. Tighten the bolts, in 2–3 steps, to 19 ft. lbs. (25 Nm), working from the center toward the ends of the manifold.
- intake manifold bracket. Tighten the bolts to 19 ft. lbs. (25 Nm)
- coolant hose to the air bypass valve
- Fuel lines
- vacuum hoses and electrical connectors
- Accelerator cable
- air cleaner assembly and ducts
- negative battery cable
8. Fill the cooling system.
9. Start the engine and check for leaks. Check the idle speed.

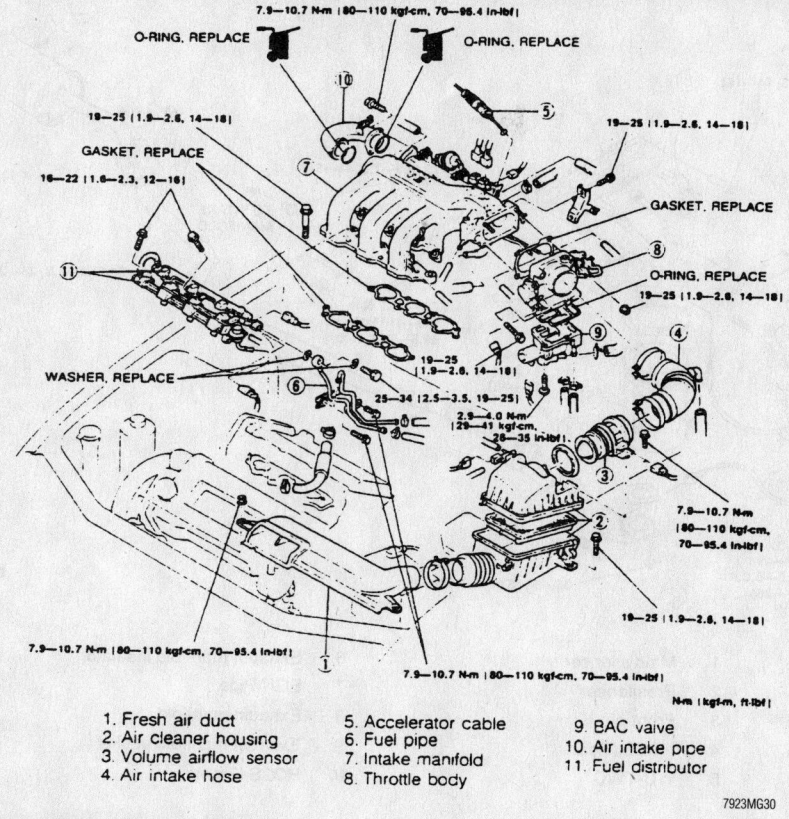

1. Fresh air duct
2. Air cleaner housing
3. Volume airflow sensor
4. Air intake hose
5. Accelerator cable
6. Fuel pipe
7. Intake manifold
8. Throttle body
9. BAC valve
10. Air intake pipe
11. Fuel distributor

Exploded view of the intake manifold assembly—2.5L engine

Exhaust Manifold

REMOVAL & INSTALLATION

1.5L and 1.8L (BP) Engines

1998–00

1. Before servicing the vehicle, refer to the precautions in the beginning of this section.
2. Remove or disconnect the following:
   - negative battery cable
   - air cleaner and air hose
   - water bypass pipe bolt
   - exhaust manifold heat shield bolts and the heat shield
   - oxygen (O$_2$S) sensor electrical connector
   - Exhaust pipe-to-exhaust manifold nuts and discard them
   - Exhaust Gas Recirculation (EGR) pipe from the exhaust manifold
   - Exhaust manifold nuts and bolts and discard the nuts
   - Exhaust manifold

To install:
3. Clean all gasket mating surfaces.

4. Install or connect the following:
   - Exhaust manifold. Torque the bolts to 14–16 ft. lbs. (19–22 Nm) on the 1.5L engine or to 29–34 ft. lbs. (39–47 Nm) on the 1.8L engine
   - exhaust pipe. Torque the new nuts to 38 ft. lbs. (52 Nm)
   - (O$_2$S) connector
   - EGR pipe. Torque the pipe to 34 ft. lbs. (47 Nm)
   - heat shield. Torque the bolts to 88 inch lbs. (10 Nm)
   - water bypass pipe. Torque the bolt to 48–65 ft. lbs. (64–89 Nm)
   - air hose and air cleaner
   - negative battery cable

1.6L and 1.8L (FP) Engines

1998–00

1. Before servicing the vehicle, refer to the precautions in the beginning of this section.
2. Remove or disconnect the following:
   - negative battery cable
   - Air cleaner and hose assembly
   - Water bypass pipe-to-engine block bolt

   - Exhaust Gas Recirculation (EGR) pipe
   - Oxygen (O$_2$S) sensor from the exhaust system
   - front exhaust pipe from the Warm-Up Three Way Catalytic (WU-TWC) converter
   - exhaust manifold insulator
   - Warm-Up Three Way Catalytic (WU-TWC) converter from the manifold
   - Exhaust manifold

To install:
3. Be sure all gasket mating surfaces are clean prior to assembly.
4. Tighten the components following the illustration.
5. Install or connect the following:
   - Exhaust manifold
   - Warm-Up Three Way Catalytic (WU-TWC) converter to the manifold
   - exhaust manifold insulator
   - front exhaust pipe from the Warm-Up Three Way Catalytic (WU-TWC) converter
   - Oxygen (O$_2$S) sensor to the exhaust system

Timing belt service is covered in Section 3 of this manual

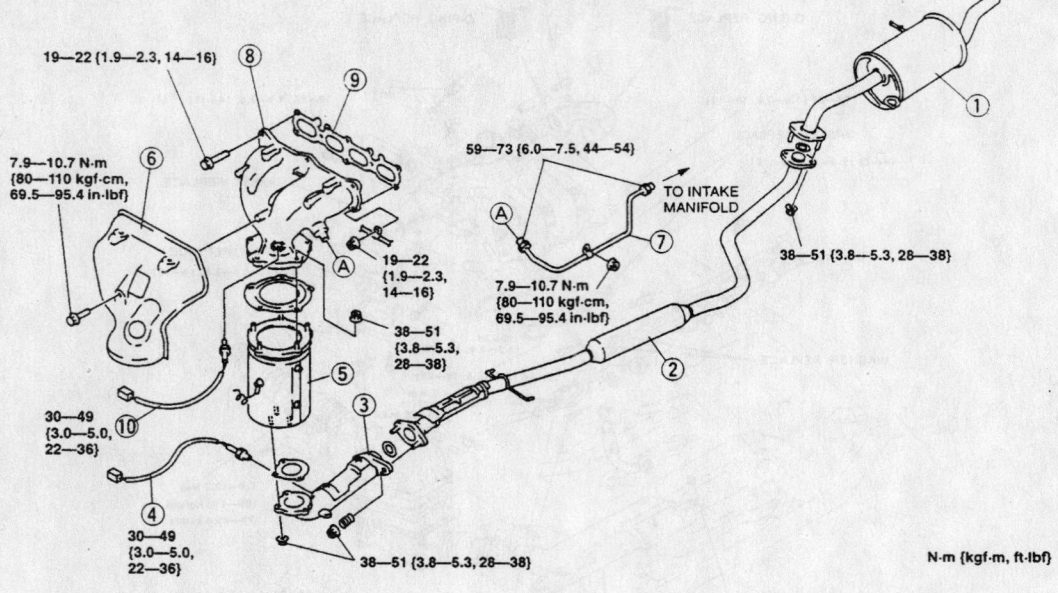

19—22 {1.9—2.3, 14—16}

7.9—10.7 N·m
{80—110 kgf-cm,
69.5—95.4 in·lbf}

59—73 {6.0—7.5, 44—54}

TO INTAKE
MANIFOLD

7.9—10.7 N·m
{80—110 kgf-cm,
69.5—95.4 in·lbf}

38—51 {3.8—5.3, 28—38}

19—22
{1.9—2.3,
14—16}

38—51
{3.8—5.3,
28—38}

30—49
{3.0—5.0,
22—36}

30—49
{3.0—5.0,
22—36}

38—51 {3.8—5.3, 28—38}

N·m {kgf·m, ft·lbf}

1	Main silencer	6	Exhaust manifold insulator
2	Presilencer	7	EGR Pipe
3	Front pipe	8	Exhaust manifold
4	HO2S (Rear)	9	Exhaust manifold gasket
5	WU-TWC	10	HO2S (Front)

9301MG06

Exploded view of the exhaust system—1.6L and 1.8L (FP) engines

- EGR pipe
- Water bypass pipe-to-engine block bolt
- Air cleaner and hose assembly
- negative battery cable

2.0L Engine

1998–00

1. Before servicing the vehicle, refer to the precautions in the beginning of this section.
2. Remove or disconnect the following:

- negative battery cable
- Exhaust manifold insulator
- Oxygen (O$_2$S) sensor electrical connector
- Oxygen (O$_2$S) sensor, if necessary
- Exhaust Gas Recirculation (EGR) pipe, if equipped
- Exhaust pipe flange nuts
- Exhaust pipe from the manifold
- Exhaust manifold

To install:

3. Be sure all gasket mating surfaces are clean prior to assembly.
4. Install or connect the following:

- Exhaust manifold using a new gasket. Torque the bolts to 17 ft. lbs. (23 Nm)

- Exhaust pipe flange using a new gasket. Torque the nuts to 34 ft. lbs. (46 Nm)
- Exhaust Gas Recirculation (EGR) pipe, if equipped
- Oxygen (O$_2$S) sensor, if necessary
- Oxygen (O$_2$S) sensor electrical connector
- Exhaust manifold insulator
- negative battery cable

2.3L Engine

1998–00

1. Before servicing the vehicle, refer to the precautions in the beginning of this section.
2. Remove or disconnect the following:

- negative battery cable
- Front and rear exhaust pipe nuts and lower the exhaust system

➡ **Both pipes must be disconnected, even if only one manifold is to be removed.**

- Exhaust Gas Recirculation (EGR) pipe, if removing the rear (right side) manifold
- Charge air cooler and coolant/condenser fans, If removing the front (left side) manifold

- front and rear Oxygen (O$_2$S) sensor connectors
- 3 heat shield bolts and the heat shield
- Exhaust manifold

To install:

3. Clean all gasket mating surfaces.
4. Install or connect the following:

- exhaust manifold using a new gasket. Torque the bolts to 12–16 ft. lbs. (16–22 Nm)
- heat shield. Torque the bolts to 70–95 inch lbs. (8–11 Nm)
- O$_2$S connectors
- Coolant/condenser fans and the charge air cooler, if installing the front (left side) manifold. Torque the bolts to 14–18 ft. lbs. (19–25 Nm)
- EGR pipe, if installing the rear (right side) manifold
- exhaust pipes using new gaskets. Torque the nuts to 28–38 ft. lbs. (38–51 Nm).
- negative battery cable

2.5L Engine

1998–00

1. Before servicing the vehicle, refer to the precautions in the beginning of this section.

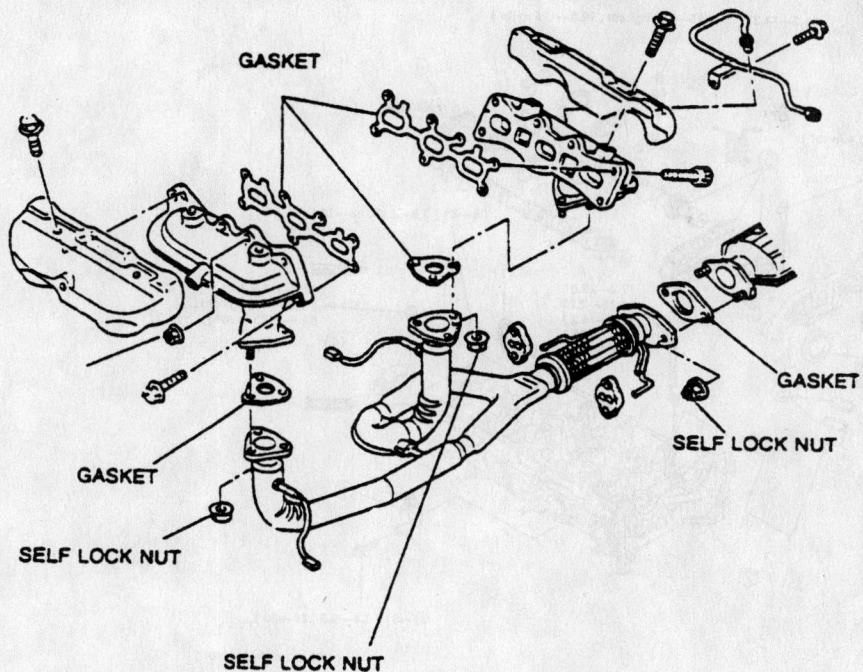

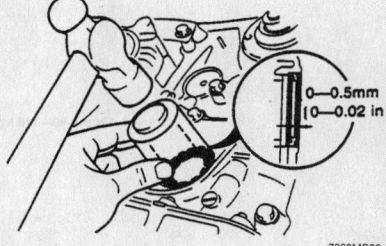

Exploded view of the exhaust manifolds—2.5L engine, 2.3L engine is similar

2. Remove or disconnect the following:
- negative battery cable
- oxygen (O2S) sensor connectors
- Front and rear exhaust pipe nuts and lower the exhaust system

➡**Both pipes must be disconnected, even if only one manifold is to be removed.**

- Exhaust Gas Recirculation (EGR) pipe, if removing the rear (right side) manifold
- 3 heat shield bolts and the heat shield
- 2 nuts and 5 bolts and the exhaust manifold

To install:
3. Clean all gasket mating surfaces.
4. Install or connect the following:
- exhaust manifold using a new gasket. Torque the nuts to 15–20 ft. lbs. (20–28 Nm) and the bolts to 12–16 ft. lbs. (16–22 Nm)
- heat shield. Torque the bolts to 89 inch lbs. (10 Nm)
- EGR pipe, If installing the rear (right side) manifold
- exhaust pipes using new gaskets. Torque the nuts to 38 ft. lbs. (51 Nm)
- O2S connectors
- negative battery cable

Front Crankshaft Seal

REMOVAL & INSTALLATION

1998–00

1. Before servicing the vehicle, refer to the precautions in the beginning of this section.
2. Remove or disconnect the following:
- negative battery cable
- timing belt
- Crankshaft damper bolt and damper
- Timing belt sprocket
- Sprocket key from the crankshaft

Install the seal using an appropriate driver, which fits over the crankshaft snout and presses on the outside edge of the

- Oil seal from the engine block using a prybar

✳✳ WARNING

Be careful not to score the crankshaft or the seal seat.

3. Clean the seal bore.
To install:
4. Install or connect the following:
- New oil seal lubricated with clean engine oil, drive it into the engine using an installation tool until it seats
- Sprocket key onto the crankshaft
- Timing belt sprocket
- Crankshaft damper
- timing belt
- negative battery cable

Camshaft and Valve Tappet

REMOVAL & INSTALLATION

1.5L Engine

1998–00

1. Before servicing the vehicle, refer to the precautions in the beginning of this section.

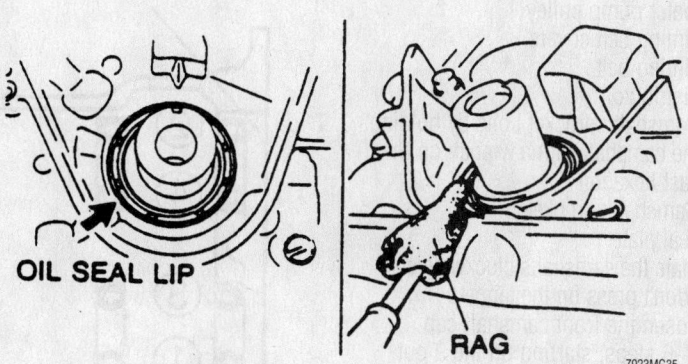

Remove the front engine seal by cutting the seal lip, then, so as not to damage the crankshaft, carefully pry the seal out with a prybar

Heater Core replacement is covered in Section 2 of this manual

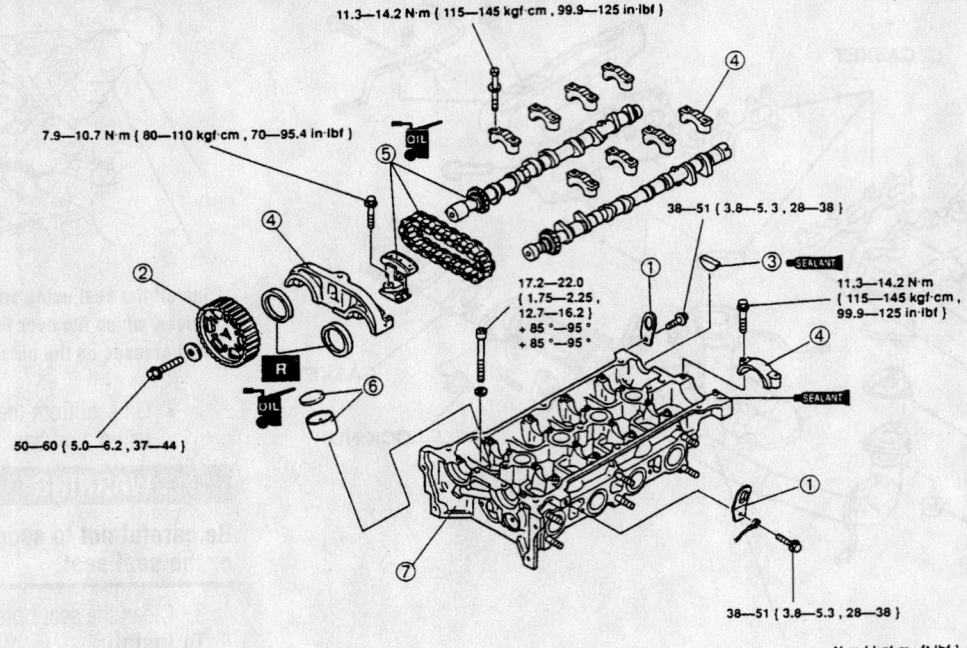

11.3—14.2 N·m (115—145 kgf·cm , 99.9—125 in·lbf)

7.9—10.7 N·m (80—110 kgf·cm , 70—95.4 in·lbf)

38—51 (3.8—5.3 , 28—38)

17.2—22.0
(1.75—2.25 ,
12.7—16.2)
+ 85°—95°
+ 85°—95°

11.3—14.2 N·m
(115—145 kgf·cm ,
99.9—125 in·lbf)

50—60 (5.0—6.2 , 37—44)

38—51 (3.8—5.3 , 28—38)

N·m (kgf·m , ft·lbf)

1. Engine hanger
2. Camshaft pulley
3. Seal cap
4. Camshaft cap

5. Camshaft, timing chain, and chain adjuster
6. Tappet and adjustment shim
7. Cylinder head

7923MG37

Exploded view of the camshaft assemblies—1.5L engine

2. Remove or disconnect the following:
- negative battery cable.
- power steering hose brackets from the cylinder head cover
- Spark plug wires and plug wire clips
- breather tube and Positive Crankcase Ventilation (PCV) valve from the cylinder head cover
- Cylinder head cover by loosening in 2–3 steps
- accessory drive belts
- Water pump pulley
- Timing belt covers
- Timing belt
- distributor
- Camshaft sprocket bolts by holding the camshaft with a wrench on the cast hexagon
- Camshaft sprockets
- seal plate

3. Rotate the camshafts clockwise so the cams don't press on the tappets

4. Loosen the front camshaft cap bolts in 5–6 steps, starting on the 2 outside bolts and finishing on the 2 inside bolts.

5. Remove or disconnect the following:
- Front camshaft bolts and caps

➡**Note the location of the numbers on top of the camshaft caps, so the caps can be reinstalled in their original positions.**

- Camshaft cap bolts in 5–6 steps
- Camshaft caps
- camshafts
- Chain and oil seals from the camshafts

- Lifters and adjustment shims, if necessary

➡**Identify and mark each lifter as it is removed so it can be reinstalled in the same position.**

To install:

6. Install or connect the following:
- Lifters in their original positions, lubricated with engine oil

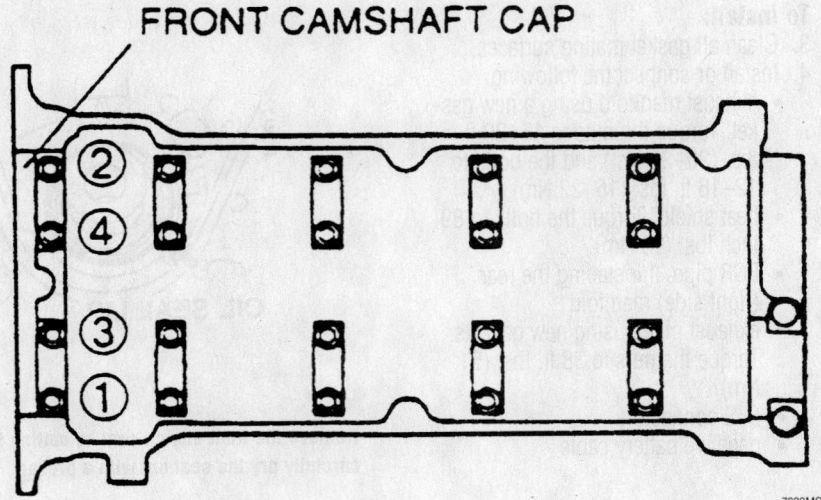

FRONT CAMSHAFT CAP

7923MG38

Front camshaft cap bolt loosening sequence—1.5L engine

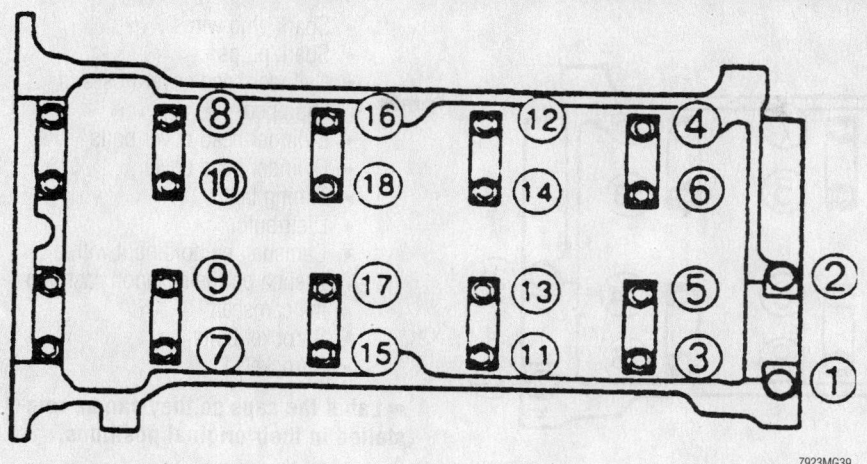

Camshaft cap bolt loosening sequence—1.5L engine

7923MG39

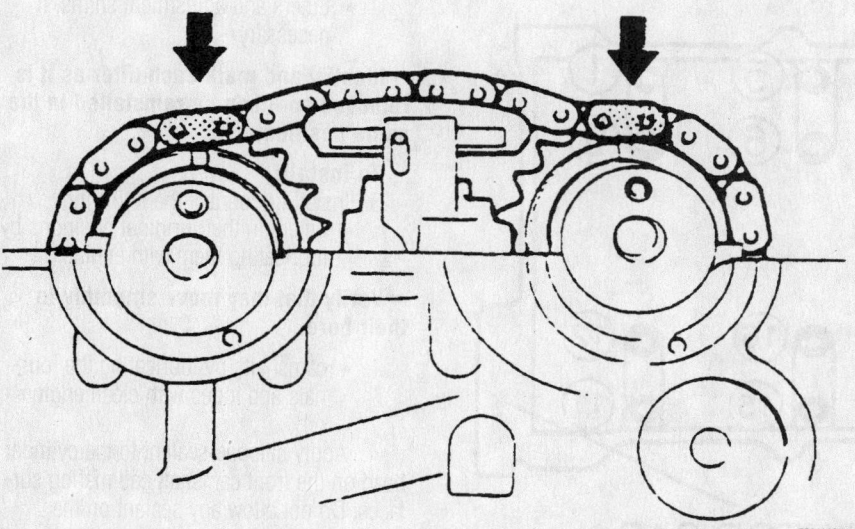

When installing the camshafts, align the marks on the camshaft gears with the colored/marked links of the chain—1.5L engine

7923MG40

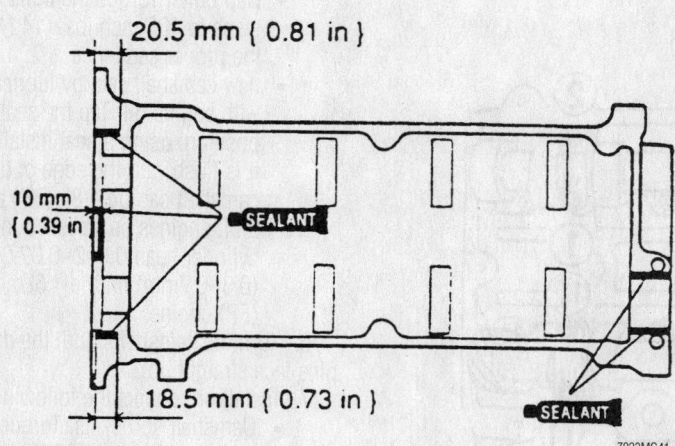

Apply sealant in the positions shown before installing the camshaft bearing caps—1.5L engine

7923MG41

➡**Verify that they move smoothly in their bore.**

- Chain adjuster between the camshafts
- Camshafts by lubricating the camshaft lobes and journals with engine oil

➡**Be sure none of the lobes are located directly on the tappets.**

7. Align the camshaft gear and timing chain marks.

8. Apply silicone sealant to the cylinder head on the front camshaft caps mating surface. Do not get sealant on the camshaft journals.

9. Install the camshaft bearing caps in their original locations.

10. Hand tighten the camshaft cap bolts numbered: 5, 7, 2 and 4. Torque, in sequence, in 5–6 steps with a final torque of 100–125 inch lbs. (11–14 Nm).

11. Install or connect the following:

- New camshafts seals lubricated with engine oil, using a seal installer until the seals are flush with the edge of the camshaft cap
- seal plate
- camshaft sprockets
- Timing belt and timing belt covers
- Water pump pulley
- Accessory drive belts

12. Adjust the drive belt tension.

13. Apply silicone sealant to a new cylinder head cover gasket, and install the gasket on the cylinder head cover.

14. Apply silicone sealant to the cylinder head in the area adjacent to the front camshaft caps.

15. Install or connect the following:

- distributor
- cylinder head cover. Torque the bolts in 2 steps, by reversing of the loosening sequence, to 61–95 inch lbs. (7–11 Nm)
- power steering hose brackets. Torque the bolts to 89 inch lbs. (10 Nm)
- Spark plug wires and clips
- breather hose and PCV valve
- Adjust the valve clearance

1.6L, 1.8L and 2.0L Engines

1998–00

1. Before servicing the vehicle, refer to the precautions in the beginning of this section.

2. Remove or disconnect the following:

- Negative battery cable

Brake service is covered in Section 4 of this manual

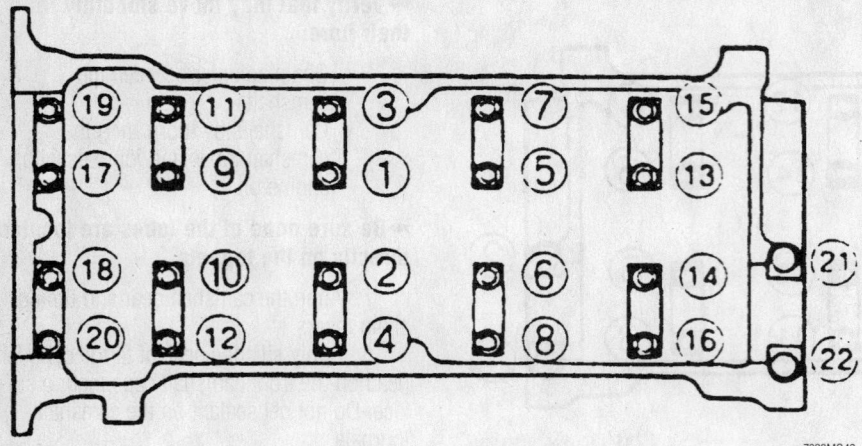

Tighten the camshaft cap bolts according to the sequence shown—1.5L engine

7923MG42

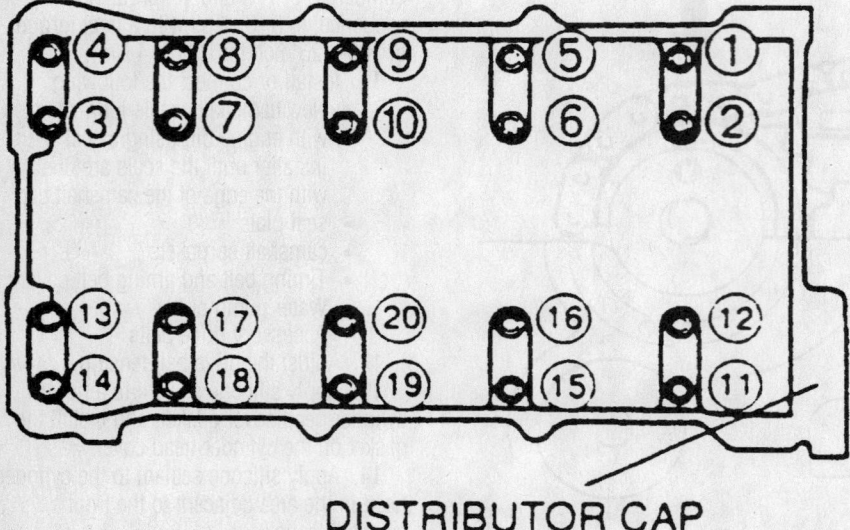

DISTRIBUTOR CAP

7923MG45

Camshaft cap bolt loosening sequence—1.8L (BP) and 2.0L engines

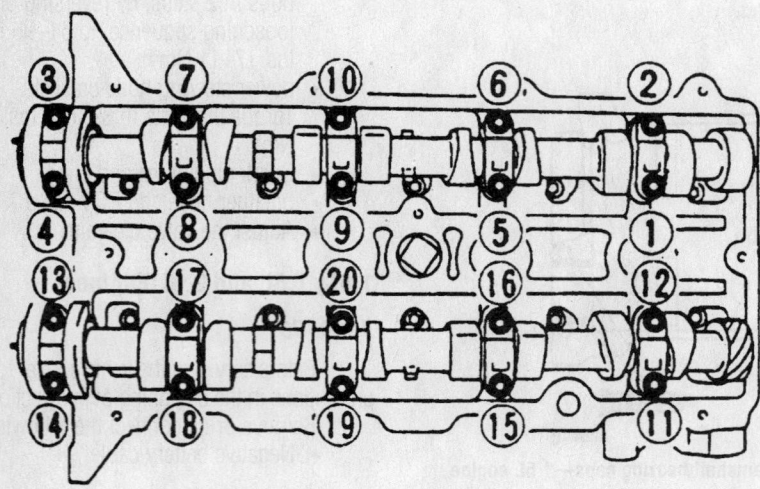

9301MG07

Camshaft cap bolt loosening sequence—1.6L and 1.8L (FP) engines

- Spark plug wires
- Spark plugs
- Cylinder head cover hoses, if equipped
- Cylinder head cover bolts
- Cylinder head cover
- Timing belt
- Distributor
- Camshaft by holding it with a wrench on the hexagon cast into the camshaft
- Sprocket bolts
- Sprockets

➡ **Label the caps so they can be reinstalled in their original positions.**

- Camshaft cap bolts by loosening in 2–3 steps in the sequence shown
- Camshaft caps
- Lifters and adjustment shims, if necessary

➡ **Identify and mark each lifter as it is removed so it can be reinstalled in the same position.**

To install:

3. Install or connect the following:
- Lifters in their original positions by lubricating them with engine oil

➡ **Verify that they move smoothly in their bore.**

- camshafts by lubricating the journals and lobes with clean engine oil

4. Apply silicone sealant to the cylinder head on the front camshaft cap mating surfaces. Do not allow any sealant on the camshaft journals.

5. Install or connect the following:
- camshaft caps in their original positions
- Cap bolts. Torque the bolts in 2–3 steps to 125 inch lbs. (14 Nm) in the proper sequence
- new camshaft seal by lubricating it with engine oil. Tap the seal into position, using a seal installer, until it is flush with the edge of the camshaft cap on 1.8L (BP) and 2.0L engines or recessed into the cylinder head 0.012–0.027 in. (0.3–0.7mm) on the 1.6L and 1.8L (FP) engine

6. Turn the camshafts until the dowel pins face straight up.

7. Install or connect the following:
- Camshaft sprockets. Torque the bolts to 44 ft. lbs. (60 Nm) by holding the camshaft with the wrench on the cast hexagon.
- remaining components

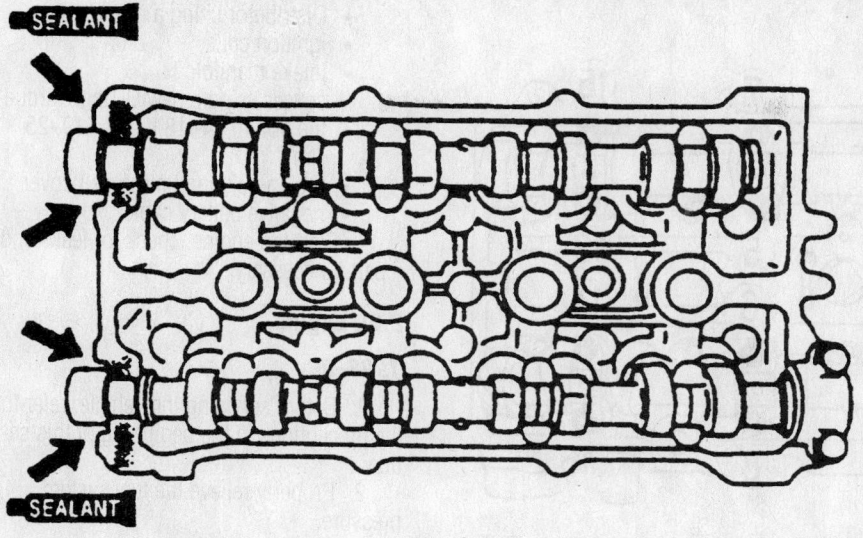

Apply silicone sealant to the cylinder head in the positions shown—1.8L (BP) and 2.0L engines

7923MG46

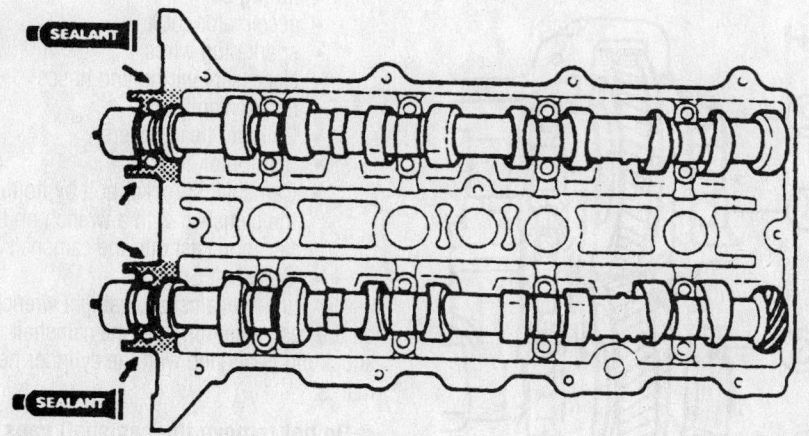

Apply silicone sealant to the cylinder head in the positions shown—1.6L and 1.8L (FP) engines

9301MG08

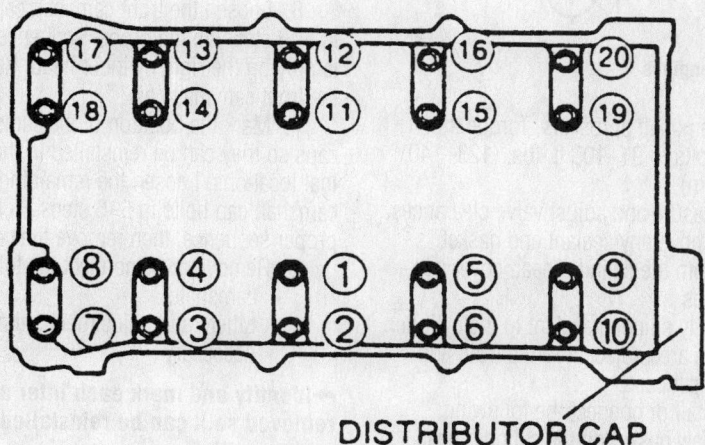

Camshaft cap bolt tightening sequence—1.8L (BP) and 2.0L engines

7923MG47

2.3l Engine

1998–00

1. Before servicing the vehicle, refer to the precautions in the beginning of this section.

2. Relieve the fuel system pressure.

3. Remove or disconnect the following:
   - Negative battery cable
   - timing belt covers and timing belt
   - spacer and O-ring from the front of the camshaft
   - ignition coils
   - intake manifold
   - Cylinder head cover bolts, in 5–6 steps, by reversing of the tightening sequence.
   - Cylinder head cover
   - camshaft sprockets

4. Turn the camshafts so the knock pins are aligned with the marks on the camshaft caps. This will reduce the pressure on the adjustment shims.

5. Note the markings on the camshaft caps prior to removal, so they can be reinstalled in the same positions. The right-hand (rear) caps are marked with numbers and the left-hand (front) caps are marked with letters.

6. Remove or disconnect the following:
   - Front camshaft caps by reversing of the torque sequence, in 5–6 steps
   - remaining camshaft caps using the proper sequence

➡**Remove the thrust caps last. Do not damage the cylinder head thrust bearing support.**

   - camshafts and oil seals
   - Lifters and adjustment shims, if necessary.

➡**Identify and mark each lifter as it is removed so it can be reinstalled in the same position.**

To install:

7. Install or connect the following:
   - Lifters in their original positions, lubricated with engine oil. Verify that they move smoothly in their bore.
   - New oil seals on the camshafts
   - camshafts with the camshaft lobes, journals and supports lubricated with engine oil and the gear marks aligned
   - thrust caps. Torque the bolts, in 5–6 steps, until the caps are fully seated on the cylinder head

For complete Engine Mechanical specifications, see Section 1 of this manual

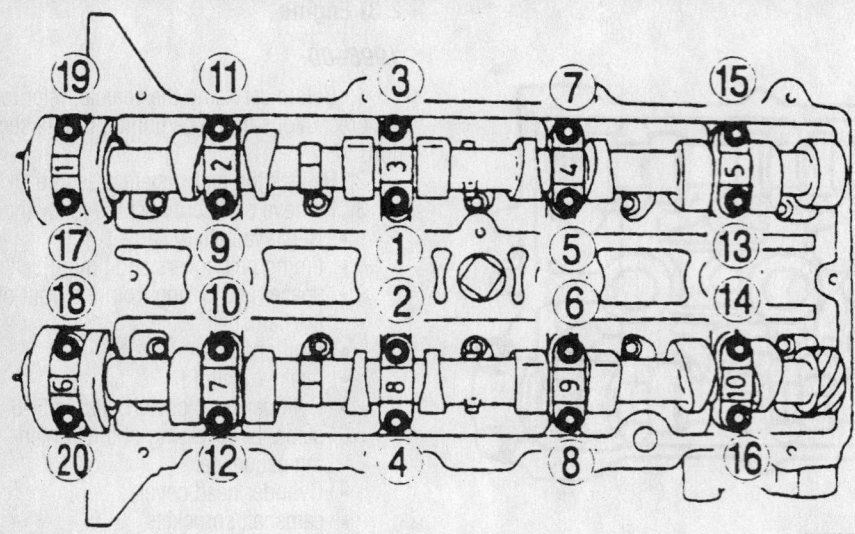

Camshaft cap bolt tightening sequence—1.6L and 1.8L (FP) engines

9301MG09

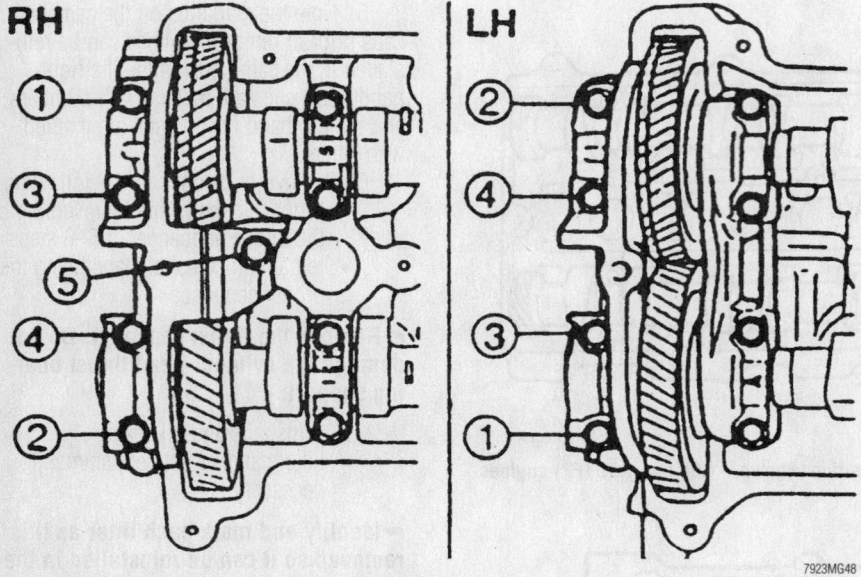

Front camshaft cap bolt loosening sequence—2.3L and 2.5L engines

7923MG48

8. Apply silicone sealant, at a thickness of 0.06–0.09 in. (1.5–2.5mm), to the cylinder head surface in the area forward of the camshaft gear cavity.

9. Install or connect the following:
- remaining camshaft caps in their original positions. Torque the bolts, in sequence, in 5 equal steps, with the final step being 100–125 inch lbs. (11–14 Nm)
- New camshaft oil seal lubricated with engine oil. Tap the seal in evenly with a Seal Installer tool 49 F401 337A with a final protrusion of 0–0.02 in. (0–0.5mm)
- New blind cap by tapping it in.

- camshaft sprockets. Torque the bolts to 91–103 ft. lbs. (123–140 Nm)

10. Measure and adjust valve clearances.

11. Remove any sealant and gasket material from the cylinder head cover contact surfaces.

12. Apply silicone sealant to the cylinder head in the area adjacent to the front and rear camshaft caps.

13. Install or connect the following:
- New gasket on the cylinder head
- cylinder head cover. Torque the cover bolts in 5–6 steps, in sequence, to 44–78 inch lbs. (5–9 Nm)

- Distributor using a new O-ring
- ignition coils
- intake manifold
- spacer using a new O-ring. Torque the bolt to 14–18 ft. lbs. (19–25 Nm)
- timing belt and timing belt cover
- negative battery cable

14. Start the engine, check for leaks and repair if necessary.

2.5L Engine

1998–00

1. Before servicing the vehicle, refer to the precautions in the beginning of this section.

2. Properly relieve the fuel system pressure.

3. Drain the cooling system

4. Remove or disconnect the following:
- negative battery cable
- timing belt
- accelerator cable
- spark plug wires
- Necessary wiring and hoses
- intake manifold
- Cylinder head covers
- distributor
- Camshaft sprocket bolt by holding the camshaft with a wrench on the hexagon cast into the camshaft
- Camshaft Sprocket

5. Turn the camshaft, using a wrench on the cast hexagon, until the camshaft knock pin is aligned with the cylinder head marks.

➡Do not remove the camshaft caps when the camshaft lobe is pressing on a lifter, as the thrust journal support may become damaged.

6. Loosen the front camshaft cap bolts in 5–6 steps, in the proper sequence. Bolt **A** is only on the right cylinder head. Remove the front camshaft cap.

7. Mark the position of the camshaft caps so they can be reinstalled in their original locations. Loosen the remaining camshaft cap bolts in 5–6 steps, in the proper sequence, then remove the caps.

8. Remove or disconnect the following:
- camshafts
- Lifters and adjustment shims, if necessary

➡Identify and mark each lifter as it is removed so it can be reinstalled in the same position.

To install:

9. Install or connect the following:
- Lifters in their original positions

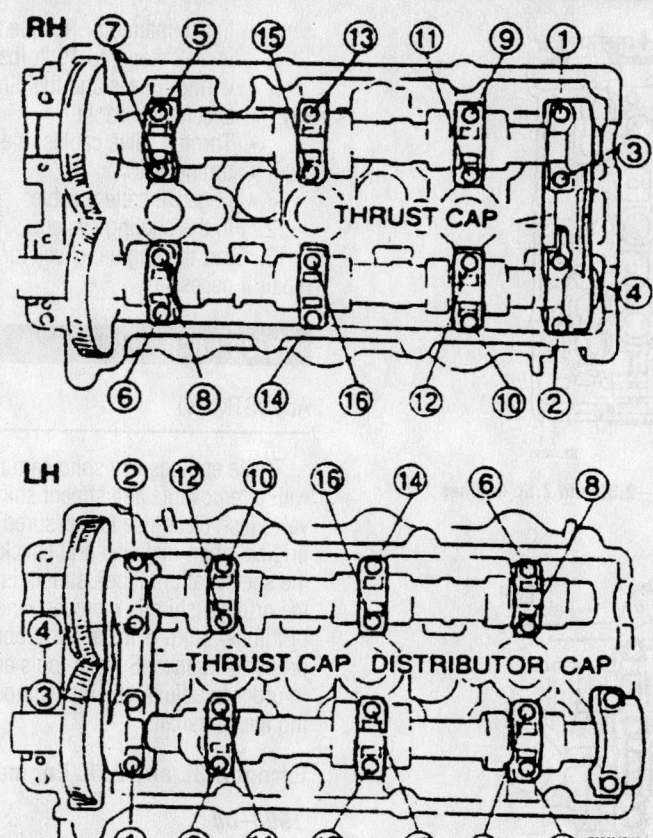

Camshaft cap bolt loosening sequence—2.3L and 2.5L engines

ALIGN THE MARKS

When installing the camshafts, ensure that the marks on the cam gears are aligned—2.3L and 2.5L engines

lubricated with engine oil. Verify that they move smoothly in their bore

• Camshafts by lubricating the camshaft journals, lobes and gears with clean engine oil and aligning the timing marks

➡**The thrust plate positions for the right and left cylinder head camshafts are different.**

10. Be sure the camshaft cap and cylinder head surfaces are clean. Apply a small amount of sealant to the mating surface of the front camshaft cap on both cylinder heads and the rear exhaust camshaft cap on the left cylinder head. Do not get any sealant on the camshaft rotating surfaces.

11. Install or connect the following:

• front camshaft caps and thrust plate caps. Torque the bolts until the cap seats fully to the cylinder head
• Remaining caps in their original locations. Torque the bolts in 5–6 steps to 126 inch lbs. (14 Nm), in the proper sequence
• New oil seals in the cylinder head using an installer
• New blind cap coated with sealant, tap it in place using a plastic hammer
• camshaft sprockets

➡**On the right cylinder head, install the sprocket so the R mark can be seen and the timing mark aligns with the camshaft knock pin. On the left cylinder head, install the sprocket so the L mark can be seen and the timing mark aligns with the camshaft knock pin.**

• Camshaft sprocket bolts lubricated with engine oil, by holding the camshaft with a wrench on the cast hexagon. Torque the bolt to 103 ft. lbs. (140 Nm)
• Cylinder head cover using a new gasket coated with sealant. Torque the bolts, in sequence, in 2–3 steps, to 78 inch lbs. (8.8 Nm)
• Ventilation pipe to the left cover
• Distributor using a new O-ring lubricated with engine oil
• Distributor with the blade fitting into the camshaft groove and loosely tighten the retaining bolt
• intake manifold, using a new gasket, loosely tighten the bolts
• Instake manifold stay. Torque the bolts to 19 ft. lbs. (25 Nm)

For Accessory Drive Belt illustrations, see Section 1 of this manual

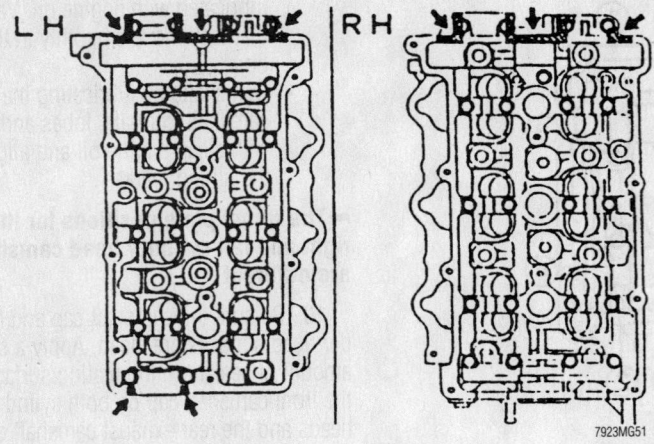

Put silicone sealant on the cylinder head at the positions shown—2.3L and 2.5L engines

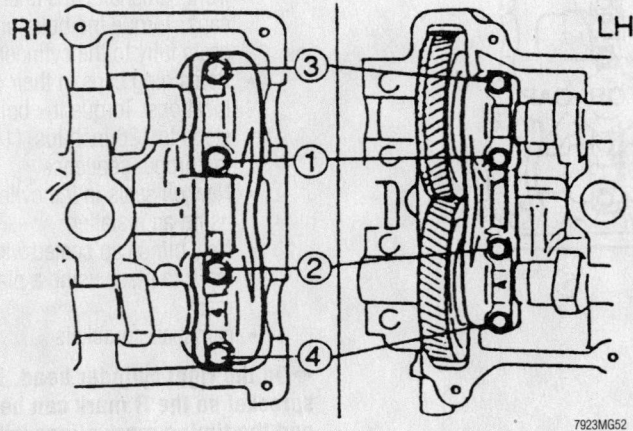

Front camshaft cap bolt tightening sequence—2.3L and 2.5L engines

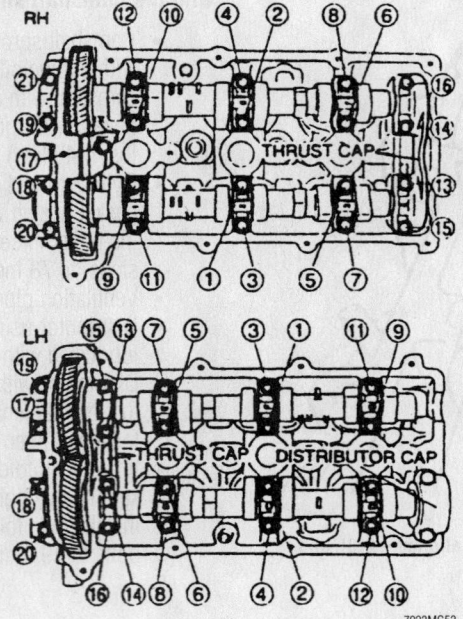

Camshaft cap bolt tightening sequence—2.3L and 2.5L engines

- Intake manifold . Torque the bolts in 2–3 steps, to 19 ft. lbs. (25 Nm)
- wiring, hoses and fuel lines
- accelerator cable
- Throttle valve cables, if equipped
- timing belt
- negative battery cable

12. Fill the cooling system.

13. Start the engine, check for leaks and repair if necessary.

Valve Lash

ADJUSTMENT

These engines use solid cam followers with a removable adjustment shim. The valve lash clearance is measured with the original shim installed and checked against the specification. If adjustment is necessary, the original shim is removed, and a thicker or thinner shim is installed to obtain the proper clearance. Special tools are required in order to adjust the shim without removing the camshaft.

Except 2.3L and 2.5L Engines

1998–00

➡With the engine cold, standard valve clearance is 0.010–0.012 in. (0.25–0.31mm) on intake and exhaust sides.

1. Before servicing the vehicle, refer to the precautions in the beginning of this section.

2. Remove the cylinder head cover.

3. Measure the valve clearance by turning the crankshaft clockwise until the No. 1 piston is at Top Dead Center (TDC).

4. Measure the valve clearance at **A**. If the clearance exceeds specifications, replace the adjustment shim.

5. Turn the crankshaft clockwise 360 degrees until the No. 4 piston is at TDC. Measure the valve clearance at **B**. If the

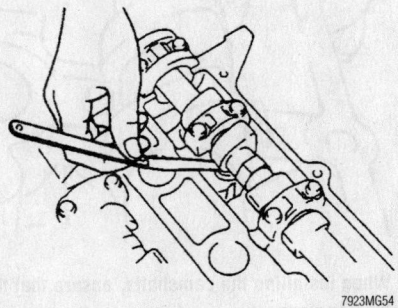

Ensure that the cam lobe faces away from the follower when checking the valve clearance

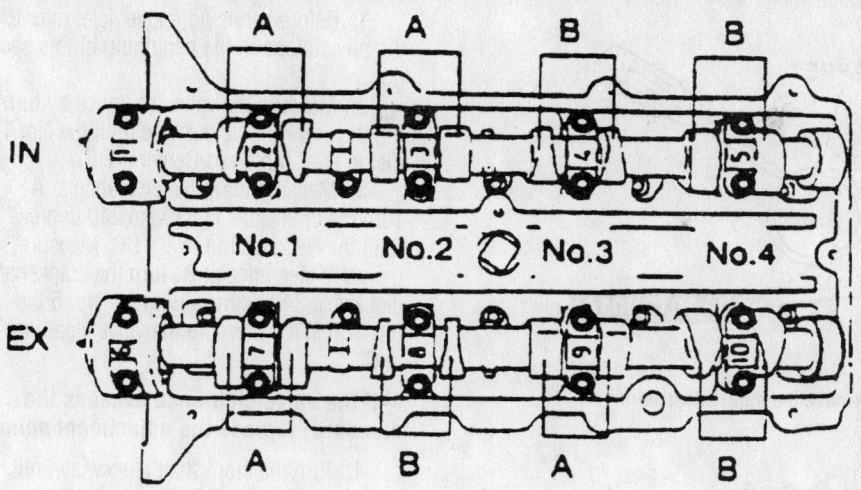

Valve clearance checking positions—4-cylinder engines

7923MG55

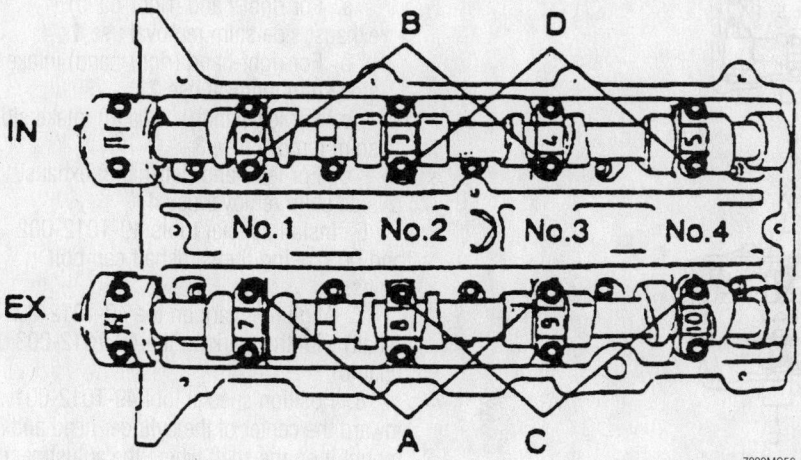

Cam bearing cap bolt removal positions—4-cylinder engine

7923MG56

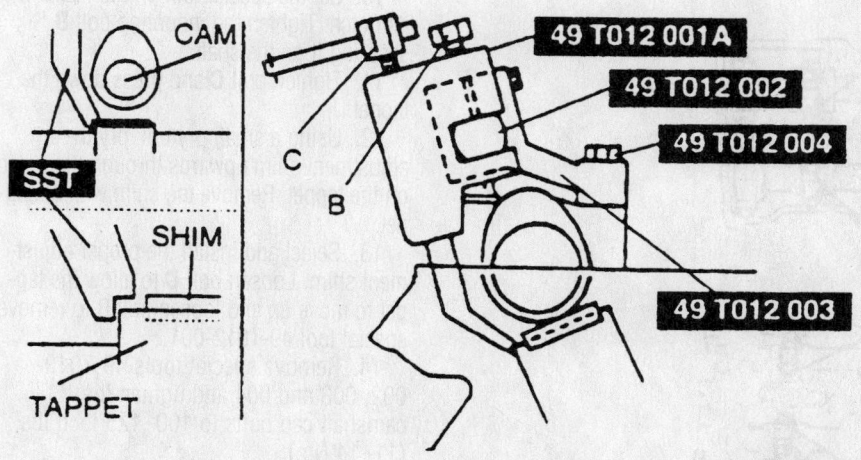

7923MG57

Mount the tappet depressor tool onto the shaft above the tappet which needs adjustment

clearance exceeds specifications, replace the adjustment shim.

6. Repeat this procedure for all the camshafts.

7. Turn the crankshaft clockwise until the cam on the camshaft requiring the adjustment is positioned straight up.

8. Remove the camshaft cap bolts as follows:

a. For exhaust side No. 1, 2 and 3 cylinder adjustment shim removal use **A**.

b. For intake side No. 1, 2 and 3 cylinder adjustment shim removal use **B**.

c. For exhaust side No. 2, 3 and 4 cylinder adjustment shim removal use **C**.

d. For intake side No. 2, 3 and 4 cylinder adjustment shim removal use **D**.

9. Install special tools 49-T012-002 and 003, using the camshaft cap bolt holes. Tighten the bolts to 100–125 inch lbs. (11–14 Nm).

10. Align the mark on the 49-T012-002 (shaft) with the mark on the 49-T012-003 (clamp). Tighten special tool 49-T012-004 (bolt) to secure the shaft.

11. Position special tool 49-T012-001A toward the center of the cylinder head and mount it on the shaft where the adjustment shim needs replacement.

12. Position the notch of the tappet to allow a small prytool to be inserted.

13. Set the special tool on the tappet by its notch. Tighten the mounting bolt **B** securing it on the shaft.

14. Tighten bolt **C**, and press down the tappet.

15. Using a small prytool, pry the adjustment shim upwards through the notch on the tappet. Remove the shim with a magnet.

16. Select and install the proper adjustment shim. Loosen bolt **C** to allow the tappet to move up, and loosen bolt **B** to remove special tool 49-T012-001A.

17. Remove special tools 49-T012-002, 003 and 004, and torque the camshaft cap bolts to 100–125 inch lbs. (11–14 Nm).

18. Repeat the procedure for all necessary adjustment shims. Check the valve clearance.

2.3L and 2.5L Engines

1998–00

➡️With the engine cold, standard valve clearance is 0.011–0.012 in. (0.27–0.31mm) on intake and exhaust sides.

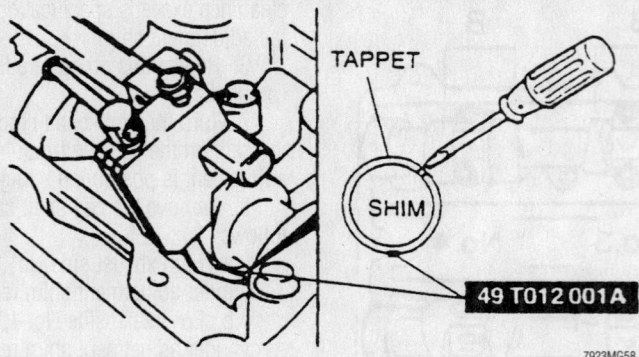

With the tappet depressed, use a small prytool to remove the adjustment shim

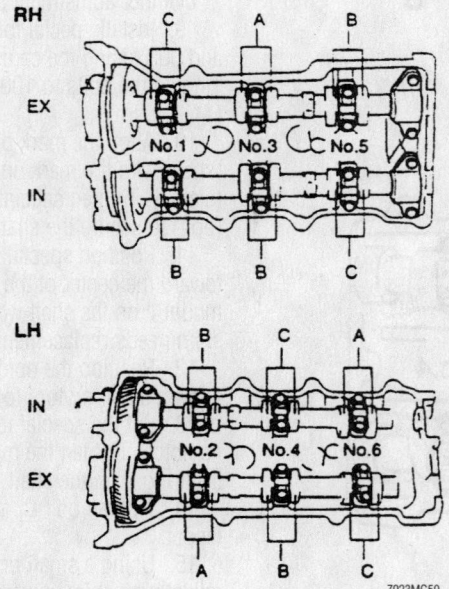

Valve clearance checking positions—6-cylinder engine

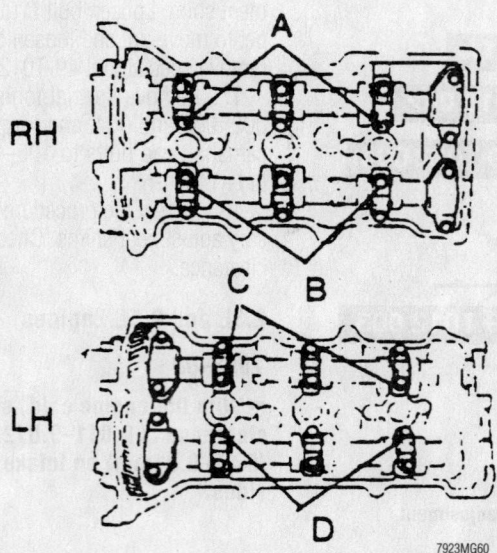

Camshaft cap bolt removal positions—6-cylinder engine—refer to text

1. Before servicing the vehicle, refer to the precautions in the beginning of this section.

2. Measure the valve clearance by turning the crankshaft clockwise until the No. 1 piston is at Top Dead Center (TDC).

3. Measure the valve clearance at **A**. Turn the crankshaft clockwise 240 degrees until the No. 3 piston is at TDC. Measure the valve clearance at **B**. Turn the crankshaft clockwise 240 degrees until the No. 5 piston is at TDC. Measure the valve clearance at **C**.

➡**If the valve clearance exceeds the standard, replace the adjustment shim.**

4. Turn the crankshaft clockwise until the cam, on the camshaft requiring the adjustment shim replacement, is positioned straight up.

5. camshaft cap bolts as follows:

   a. For right-hand (right-hand) exhaust side shim removal use **1**.

   b. For right-hand (right-hand) intake side shim removal use **2**.

   c. For left-hand (left-hand) intake side shim removal use **3**.

   d. For left-hand (left-hand) exhaust side shim removal use **4**.

6. Install special tools 49-T012-002 and 003, using the camshaft cap bolt holes.

7. Align the mark on the 49-T012-002 (shaft) with the mark on the 49-T012-003 (clamp).

8. Position special tool 49-T012-001 toward the center of the cylinder head and mount it on the shaft where the adjustment shim needs replacement.

9. Position the notch of the tappet to allow a small prytool to be inserted.

10. Set the special tool on the tappet by its notch. Tighten the mounting bolt **B** securing it on the shaft.

11. Tighten bolt **C** and press down the tappet.

12. Using a small prytool, pry the adjustment shim upwards through the notch on the tappet. Remove the shim with a magnet.

13. Select and install the proper adjustment shim. Loosen bolt **C** to allow the tappet to move up and loosen bolt B to remove special tool 49-T012-001.

14. Remove special tools 49-T012-002, 003 and 004 and tighten the camshaft cap bolts to 100–125 inch lbs. (11–14 Nm).

15. Repeat the procedure for all necessary adjustment shims. Check the valve clearance.

Starter Motor

REMOVAL & INSTALLATION

Miata and Protégé

1998–00

1. Remove or disconnect the following:
   - Negative battery cable
   - Air cleaner
   - Catalytic converter pipe from the front pipe
   - Intake manifold support bracket bolts
   - Starter electrical connectors
   - Starter

To install:

2. Install or connect the following:
   - Starter and loosely tighten the lower starter mounting bolt
   - Starter electrical connectors
   - Intake manifold support bracket. Torque the bolts to 28–38 ft. lbs. (38–51 Nm)
   - Upper starter bolts. Torque the bolts to 24–33 ft. lbs. (32–46 Nm). The upper mounting bolts must be tightened first
   - Catalytic converter pipe to the front pipe. Torque the bolts to 28–38 ft. lbs. (38–51 Nm)
   - Air cleaner
   - Negative battery cable

626

4 CYLINDER

1998–00

1. Remove or disconnect the following:
   - Negative battery cable
   - Fresh air duct and resonance chamber
   - Air cleaner assembly electrical connectors
   - Air cleaner assembly
   - Intake manifold bracket
   - Wiring at the starter
   - Starter

To install:

2. Install or connect the following:
   - Starter. Torque the bolts to 33 ft. lbs. (46 Nm)
   - Wiring at the starter
   - Intake manifold bracket. Torque the bolts to 38 ft. lbs. (51 Nm)
   - Air cleaner assembly
   - Air cleaner assembly electrical connectors

- Fresh air duct and resonance chamber
- Negative battery cable

6 CYLINDER

1998–00

1. Remove or disconnect the following:
   - Negative battery cable
   - Fresh air duct
   - Air cleaner assembly electrical connector
   - Air cleaner assembly

2. If equipped with automatic transaxle, proceed as follows:
   a. Relieve the fuel system pressure. Drain the cooling system.
   b. Disconnect the accelerator cable from the throttle body. Label and disconnect the electrical connectors, vacuum hoses and coolant hoses from the throttle body.
   c. Remove the throttle body.
   d. Disconnect and plug the fuel supply and return lines.
   e. Disconnect the transaxle selector cable from the transaxle and remove the cable bracket.
   f. Remove the starter bracket.

3. Remove or disconnect the following:
   - Wiring at the starter
   - Starter mounting bolts
   - Starter

To install:

4. Install or connect the following:
   - Starter
   - Starter mounting bolts. Torque the starter mounting bolts to 38 ft. lbs. (51 Nm)
   - Wiring at the starter.

5. If equipped with automatic transaxle, proceed as follows:
   a. Install the starter bracket.
   b. Install the cable bracket and connect the transaxle selector cable to the transaxle.
   c. Connect and fuel supply and return lines.
   d. Install the throttle body.
   e. Connect the accelerator cable to the throttle body. Attach the electrical connectors, vacuum hoses and coolant hoses to the throttle body.

6. Install or connect the following:
   - Air cleaner assembly
   - Air cleaner assembly electrical connector
   - Fresh air duct
   - Negative battery cable

7. Fill the cooling system.
8. Start the vehicle, check for leaks and repair if necessary.

Millenia S Models

1998–00

1. Remove or disconnect the following:
   - Negative battery cable
   - Charge air cooler duct
   - Battery clamp, box and battery
   - Battery tray
   - Rear charge air cooler
   - Pipe bracket
   - Starter electrical connectors
   - Starter

To install:

2. Install or connect the following:
   - Starter. Torque the bolts to 24–33 ft. lbs. (32–46 Nm)
   - B-terminal wire. Torque the nut to 12–16 ft. lbs. (16–22 Nm)
   - S-terminal wire
   - Pipe bracket. Torque the bolt to 14–18 ft. lbs. (19–25 Nm)
   - Rear charge air cooler using new O-rings. Torque the nuts to 14–18 ft. lbs. (19–25 Nm)
   - Battery tray
   - Battery, box and clamp
   - Charge air cooler duct
   - Negative battery cable

Millenia

1998–00

1. Remove or disconnect the following:
   - Negative battery cable
   - Battery clamp, box and battery
   - Battery tray
   - Shift cable from the selector lever
   - Cable from the bracket by squeeze the lock tabs
   - Electrical connectors from the starter solenoid
   - 2 selector cable bracket bolts and the bracket
   - 2 nuts and the bolt from the starter bracket and the bracket
   - Starter electrical connectors
   - Starter

To install:

2. Install or connect the following:
   - Starter. Torque the bolts to 24–33 ft. lbs. (32–46 Nm)
   - B-terminal wire. Torque the nut to 12–16 ft. lbs. (16–22 Nm)
   - S-terminal wire to the solenoid
   - Starter bracket

For Wheel Alignment specifications, see Section 1 of this manual

- Selector cable bracket. Torque the bolts to 5–7 ft. lbs. (7–9 Nm)
- Starter solenoid electrical connectors
- Shift cable into the cable bracket and into the selector lever
- Battery tray
- Battery, box and clamp
- Negative battery cable

Oil Pan

REMOVAL & INSTALLATION

1.5L and 1.8L (BP) Engines

1998–00

1. Before servicing the vehicle, refer to the precautions in the beginning of this section.
2. Drain the engine oil.
3. Remove or disconnect the following:
   - negative battery cable
   - right-hand splash shield

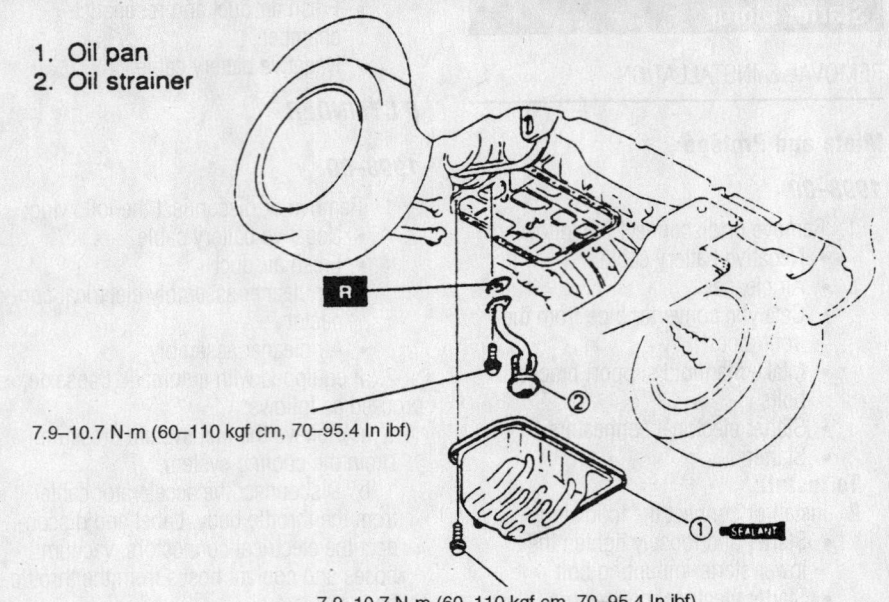

1. Oil pan
2. Oil strainer

7.9–10.7 N-m (60–110 kgf cm, 70–95.4 In ibf)

7.9–10.7 N-m (60–110 kgf cm, 70–95.4 In ibf)

7923MG61

Exploded view of the oil pan and related components—1.5L engine, 2.0L engine is similar

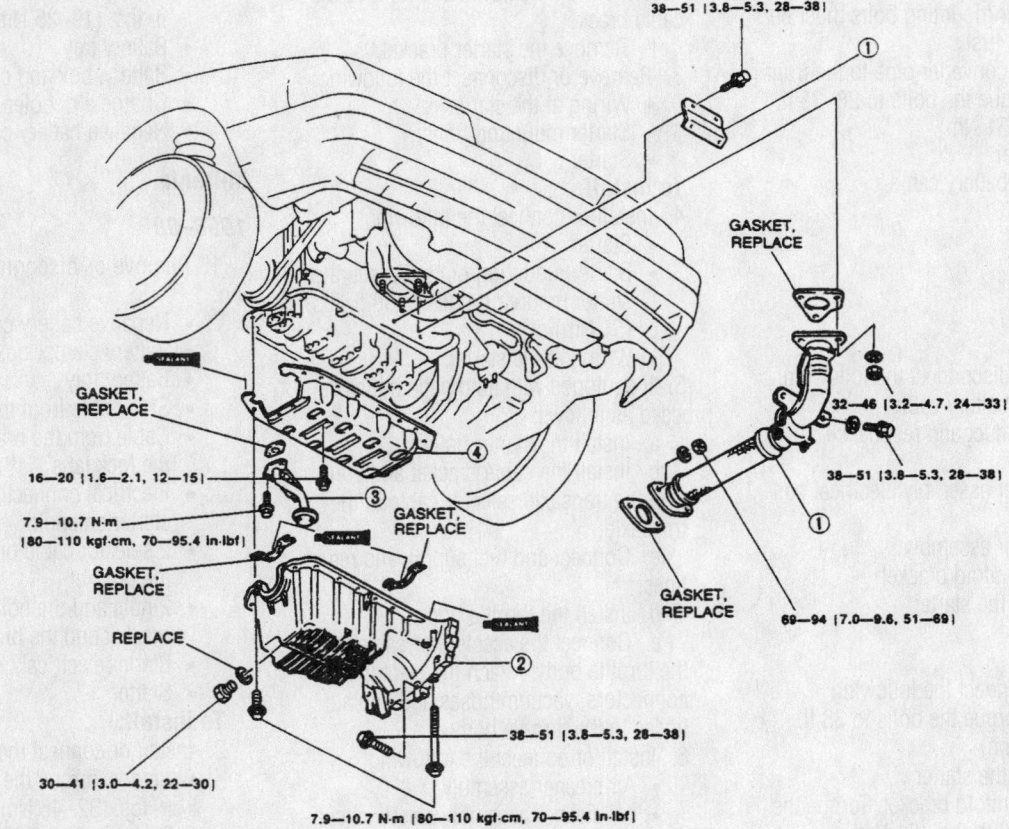

1. Front exhaust pipe and bracket
2. Oil pan
3. Oil strainer
4. Main bearing support plate (MBSP)

N-m [kgf-m, ft-lbf]

7923MG62

Exploded view of the oil pan and related components—1.8L (BP) engine

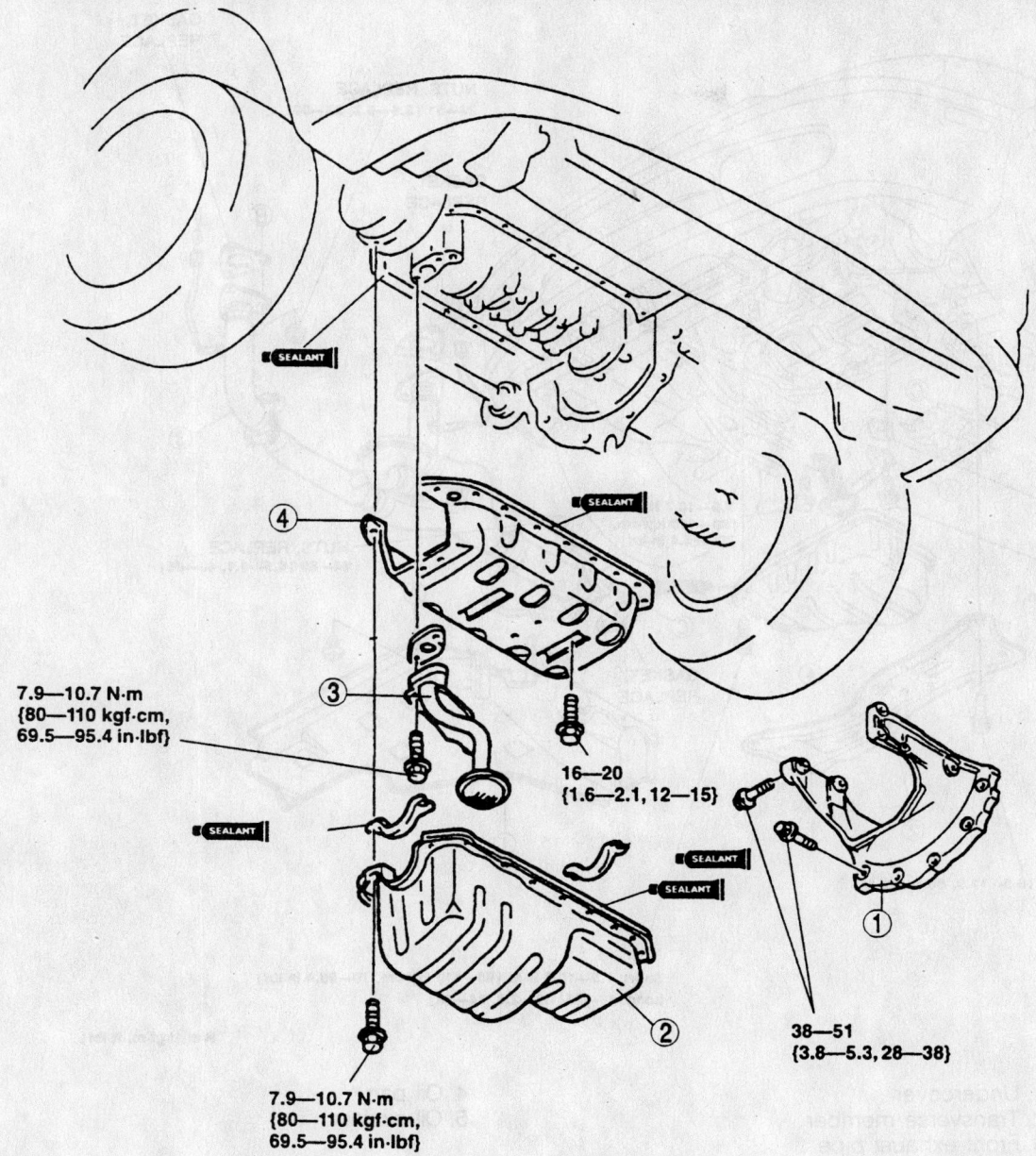

7.9—10.7 N·m
{80—110 kgf·cm,
69.5—95.4 in·lbf}

16—20
{1.6—2.1, 12—15}

38—51
{3.8—5.3, 28—38}

7.9—10.7 N·m
{80—110 kgf·cm,
69.5—95.4 in·lbf}

N·m {kgf·m, ft·lbf}

9301MG10

Exploded view of the oil pan and related components—1.6L engine

- transverse member
- oxygen (O₂S) sensor connector
- Exhaust pipe-to-manifold nuts and discard them
- oil pan bolts and the oil pan

To install:

4. Clean the oil pan. Clean all dirt, oil, gasket and old sealant from the oil pan and cylinder block contact surfaces.

5. Apply a continuous bead of silicone sealant on the gaskets and around the oil pan, going on the inside of the bolt holes.

6. Install or connect the following:

- New oil pan gaskets
- Oil pan. Torque the vertical bolts to 70–95 inch lbs. (8–11 Nm) and the horizontal bolts to 28–38 ft. lbs. (38–51 Nm)
- exhaust pipe. Torque the new nuts to 28–38 ft. lbs. (38–51 Nm)
- O₂S connector
- transverse member. Torque the bolts to 69–97 ft. lbs. (94–131 Nm)
- right-hand splash shield. Torque the bolts to 70–95 inch lbs. (8–11 Nm)

- Negative battery cable

7. Fill the engine with clean oil.

8. Start the vehicle, check for leaks and repair if necessary.

1.6L and 1.8L (FP) Engines

1998–00

1. Before servicing the vehicle, refer to the precautions in the beginning of this section.

2. Drain the engine oil.

3. Remove or disconnect the following:

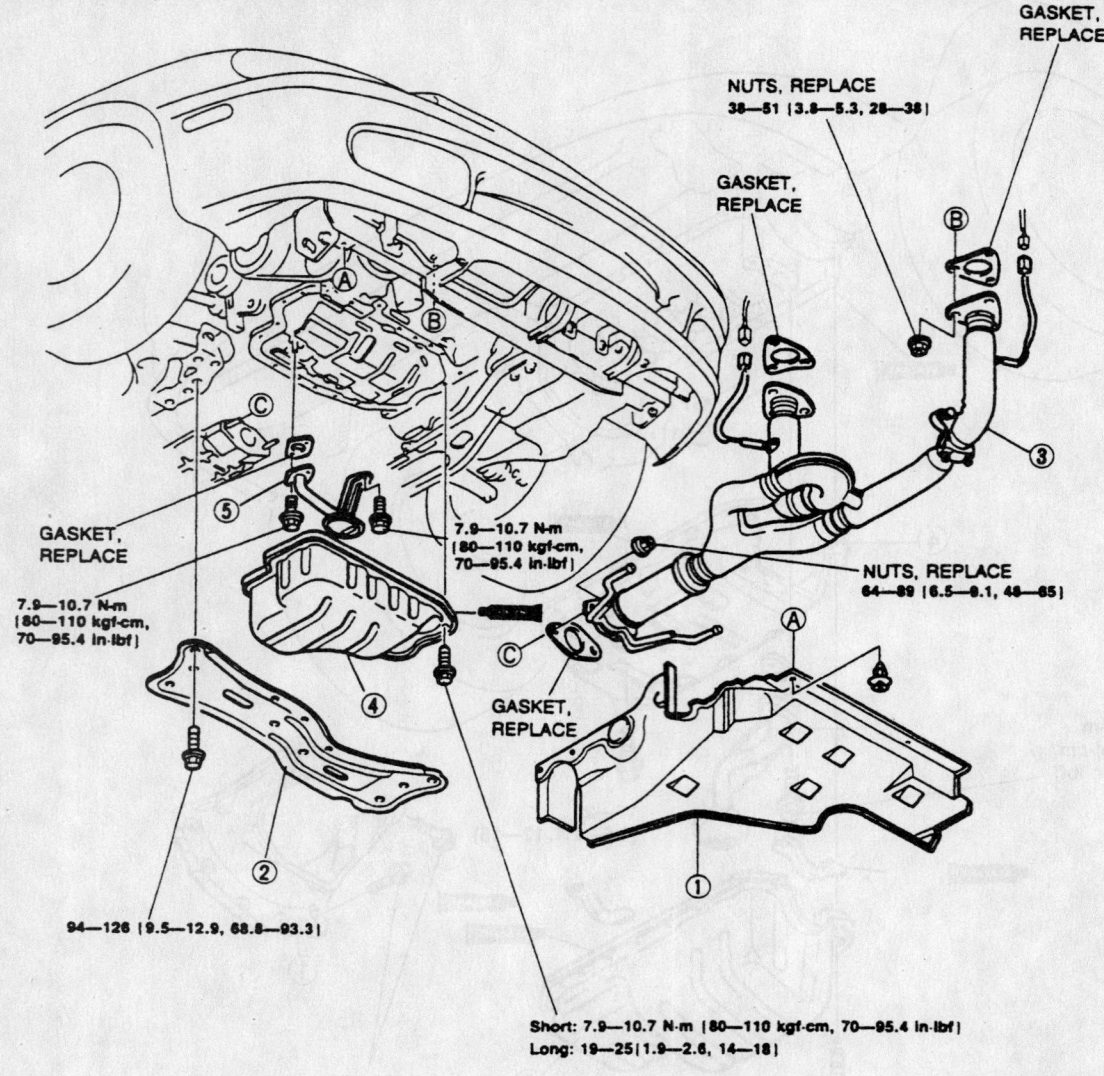

GASKET, REPLACE

NUTS, REPLACE
38—51 (3.8—5.3, 28—38)

GASKET, REPLACE

GASKET, REPLACE

GASKET, REPLACE

7.9—10.7 N·m (80—110 kgf-cm, 70—95.4 in·lbf)

7.9—10.7 N·m (80—110 kgf-cm, 70—95.4 in·lbf)

NUTS, REPLACE
64—89 (6.5—9.1, 48—65)

GASKET, REPLACE

94—126 (9.5—12.9, 68.8—93.3)

Short: 7.9—10.7 N·m (80—110 kgf-cm, 70—95.4 in·lbf)
Long: 19—25 (1.9—2.6, 14—18)

N·m (kgf·m, ft·lbf)

7923MG63

1. Undercover
2. Transverse member
3. Front exhaust pipe
4. Oil pan
5. Oil strainer

Exploded view of the oil pan and related components for the 2.3L and 2.5L engines

- Negative battery cable
- oxygen (O_2S) sensors
- Front exhaust pipe
- Integrated stiffener (1), for the 1.6L engine
- Oil pan, for the 1.8L (FP) engine
- Oil pan (2) and oil strainer (3) and the Main Bearing Support Plate (MBSP) (4), for the 1.6L engine

4. Clean the oil pan. Clean all dirt, oil, gasket and old sealant from the oil pan and cylinder block contact surfaces.

To install:

5. On the 1.8L (FP) engine, apply a continuous bead of silicone sealant on the gas-

kets and around the oil pan, going on the inside of the bolt holes.

6. On the 1.6L engine, apply silicone sealant as shown.

7. Install or connect the following:
- New oil pan gasket
- Oil pan. Torque the bolts to 14–18 ft. lbs. (19–25 Nm) for the 1.8L engine or to 69.5–95.4 inch lbs. (7.9–10.7 Nm) for the 1.6L engine
- front exhaust pipe
- O_2S
- negative battery cable

8. Fill the engine with clean oil.

9. Start the vehicle, check for leaks and repair if necessary.

2.0L, 2.3L and 2.5L Engines

1998–00

1. Before servicing the vehicle, refer to the precautions in the beginning of this section.

2. Drain the engine oil.

3. Remove or disconnect the following:
- negative battery cable
- passenger side splash shield
- oxygen (O_2S) sensor

- Front exhaust pipe
- oil pan bolts and the oil pan

To install:

4. Clean the oil pan. Clean all dirt, oil and old sealant from the oil pan and cylinder block contact surfaces.

5. Apply a continuous bead of silicone sealant around the oil pan, going on the inside of the bolt holes.

6. Install or connect the following:
   - oil pan. Torque the bolts to 14–18 ft. lbs. (19–25 Nm)
   - front pipe. Torque the nuts to 28–38 ft. lbs. (38–51 Nm)
   - O$_2$S connector
   - splash shield. Torque the bolts to 70–95 inch lbs. (8–11 Nm)
   - Negative battery cable

7. Fill the engine with clean oil.

8. Start the engine, check for leaks and repair if necessary.

Oil Pump

REMOVAL & INSTALLATION

Protégé

1998–00

1. Before servicing the vehicle, refer to the precautions in the beginning of this section.

2. Remove or disconnect the following:
   - negative battery cable
   - crankshaft pulley
   - Timing belt cover and belt
   - Crankshaft sprocket
   - oil pan
   - oil pickup tube and discard the gasket
   - oil pump attaching bolts
   - front crankshaft seal from the oil pump, if the pump is being replaced

To install:

3. Clean the oil, dirt and old sealant from all contact surfaces.

4. If the oil seal was removed from the oil pump, apply clean engine oil to the lip of the seal. Push the seal in lightly be hand. Press the seal, with a protrusion of 0.02–0.04 inch (0.5–1.0mm), into the oil pump with a Seal Installer tool 49 B014 401.

5. Apply a bead of silicone to the oil pump at the cylinder block contact surface, going inside the bolt holes.

6. Install or connect the following:
   - New O-rings on the oil pump

- oil pump. Torque the bolts to 14–18 ft. lbs. (19–25 Nm)
- Oil pump pickup tube, using a new gasket. Torque the bolts to 70–95 inch lbs. (8–11 Nm)
- oil pan
- crankshaft sprocket
- Timing belt and cover
- crankshaft pulley
- negative battery cable

7. Fill the engine with clean oil.

8. Start the engine, check for leaks and repair if necessary.

626

6 CYLINDER

1998–00

1. Before servicing the vehicle, refer to the precautions in the beginning of this section.

2. Discharge and recover the air conditioning refrigerant.

3. Remove or disconnect the following:
   - negative battery cable
   - oil pan
   - air conditioning compressor and bracket
   - power steering pump and tensioner, move the pump aside without disconnecting the lines
   - crankshaft pulley
   - Timing belt cover
   - Timing belt
   - Crankshaft sprocket
   - oil pump bolts (4 long and 5 short)
   - 2 oil strainer-to-pump bolts
   - Oil pump body
   - Oil seal from the housing
   - Oil pump cover and rotors

To install:

4. Clean the oil, dirt and old sealant from all contact surfaces.

5. Install or connect the following:
   - New oil seal by pressing it into the pump housing with a protrusion of 0–0.03 inch (0–0.7mm)
   - rotors into the oil pump body with the alignment marks aligned
   - Pump cover. Torque the bolts to 53–78 inch lbs. (6–9 Nm)

6. Apply a continuous bead of silicone sealant to the oil pump mating surface.

7. Install or connect the following:
   - oil pump body. Torque the bolts to 14–18 ft. lbs. (19–25 Nm)
   - New oil strainer-to-pump gasket. Torque the bolts to 70–95 inch lbs. (8–11 Nm)

- crankshaft sprocket and key
- Timing belt
- Timing belt cover
- Crankshaft pulley
- power steering pump and tensioner. Torque the 2 tensioner upper bolts and the power steering pump rear bracket bolt to 33 ft. lbs. (46 Nm) and the tensioner lower bolt to 18 ft. lbs. (25 Nm)
- air conditioning compressor bracket. Torque the bolts to 38 ft. lbs. (51 Nm)
- Air conditioning compressor. Torque the bolts to 38 ft. lbs. (51 Nm)
- oil pan
- negative battery cable

8. Fill the engine with clean oil.

9. Evacuate and charge the air conditioning system.

10. Start the engine, check for leaks and repair if necessary.

Miata 1.8L (BP) and 4-Cylinder 626 Engines

1998–00

1. Before servicing the vehicle, refer to the precautions in the beginning of this section.

2. Remove or disconnect the following:
   - negative battery cable
   - crankshaft pulley
   - Timing belt cover
   - Timing belt
   - Crankshaft sprocket
   - a/C compressor and move it aside, leaving the refrigerant lines attached
   - A/C compressor mounting bracket
   - oil pan
   - oil pickup tube and discard the gasket
   - oil pump body bolts and the oil pump body
   - front crankshaft seal from the oil pump, if the pump is being replaced
   - Oil pump cover bolts, cover and rotors, on the 2.0L engine

To install:

3. Clean the oil, dirt and old sealant from all contact surfaces.

4. If the oil seal was removed from the oil pump, apply clean engine oil to the lip of the seal. Push the seal in lightly be hand. Press the seal, with a protrusion of 0–0.02 inch (0–0.5mm), into the oil pump.

5. On the 2.0L engine, install or connect the following:

- oil pump rotors into the pump body with the rotor marks aligned with each other
- Oil pump cover. Torque the bolts to 53–78 inch lbs. (6–9 Nm)

6. Apply a bead of silicone to the oil pump body-to-cylinder block contact surface, going inside the bolt holes.

7. Install or connect the following:
- New O-rings on the oil pump
- oil pump body. Torque the bolts to 14–18 ft. lbs. (19–25 Nm)
- Oil pump pickup tube using a new gasket. Torque the bolts to 88 inch lbs. (10 Nm)
- oil pan
- air conditioning compressor bracket. Torque the bolts to 38 ft. lbs. (52 Nm)
- Air conditioning compressor. Torque the bolts to 26 ft. lbs. (35 Nm)
- crankshaft sprocket
- Timing belt
- Timing belt cover
- crankshaft pulley
- negative battery cable

8. Fill the engine with clean oil.

9. Start the vehicle, check for leaks and repair if necessary.

Millenia

1998–00

Due to space requirements, the engine assembly must be removed in order to replace the oil pump.

1. Before servicing the vehicle, refer to the precautions in the beginning of this section.

2. Disconnect the negative battery cable.

> ✳✳ **CAUTION**

Wait at least 90 seconds before performing any work. The backup power supply system for the Supplemental Restraint System (SRS) must deplete its stored energy.

3. Drain the engine oil.
4. Drain the cooling system.
5. Remove or disconnect the following:

- All electrical connections, hoses and cables necessary to remove the engine assembly
- Front exhaust pipe
- engine assembly
- oil pan
- Oil strainer
- Oil pan baffle
- timing belt
- Front timing belt pulley and key
- Vacuum pump, if equipped
- front seal
- Oil pump and discard the O-rings

To install:

6. Make sure that the oil pump mating surfaces are clean.

7. Install or connect the following:
- New O-rings coated with engine oil in the oil pump cavity

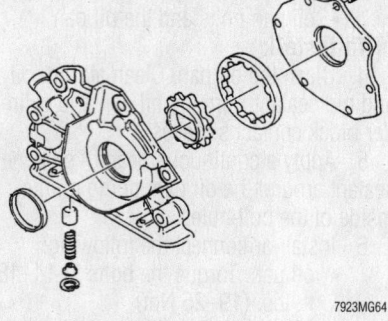

7923MG64

Exploded view of a typical oil pump assembly

- New oil seal coated with engine oil to the oil pump

8. Apply silicone sealant to the oil pump contact surface.

9. Install or connect the following:
- Oil pump. Torque the bolts, in sequence, to 16–22 ft. lbs. (22–30 Nm) for the "A" bolts and to 14–18 ft. lbs. (19–25 Nm) for all other bolts
- Vacuum pump, if equipped, using a new gasket and O-ring
- Key and the timing belt pulley
- Timing belt
- Oil pan baffle
- Oil pump strainer and oil pan
- Engine assembly into the vehicle
- Front exhaust pipe
- All electrical connectors, cables, hoses and any components necessary to complete the engine installation
- Negative battery cable

10. Fill the engine with clean oil.
11. Fill the cooling system.
12. Start the engine, bleed the cooling system, make any necessary adjustments.
13. Check for leaks and road test for proper operation.

Rear Main Seal

REMOVAL & INSTALLATION

1998–00

1. Before servicing the vehicle, refer to the precautions in the beginning of this section.

2. Remove or disconnect the following:

- negative battery cable
- transaxle/transmission assembly
- Clutch/flywheel assembly, if equipped with a manual transaxle/transmission
- Flexplate/shim plates, if equipped

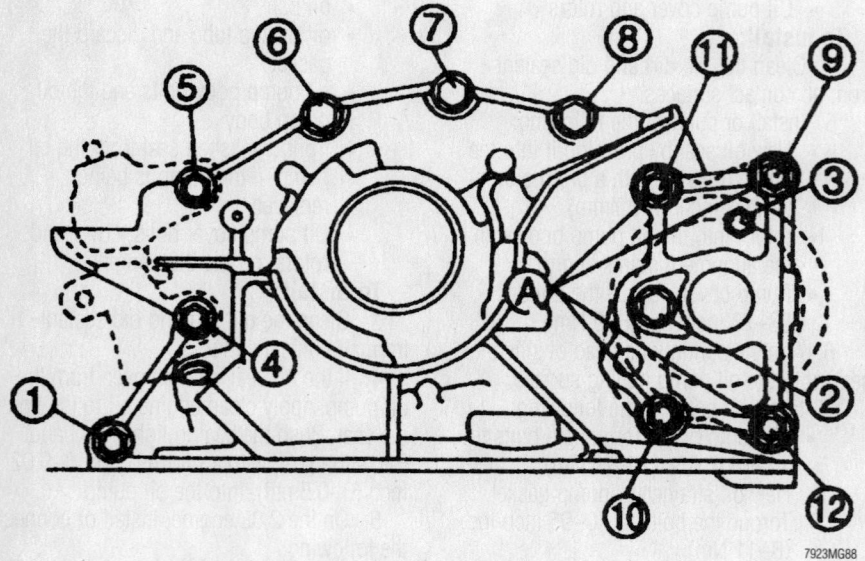

7923MG88

Tightening the oil pump mounting bolts in sequence—2.3L and 2.5L engines

with an automatic transaxle/transmission

3. Cut the oil seal lip with a knife. Install a rag to the housing and using a prytool, carefully pry the oil seal from the oil seal housing.

4. Clean the gasket mounting surfaces.

To install:

5. Clean the oil seal housing. Coat the oil seal and the housing with clean engine oil.

6. Install or connect the following:
- New oil seal into the housing by tapping it evenly into place with a hammer and a seal installer until it is flush with the edge of the rear cover
- clutch/flywheel assembly or the flexplate, as applicable
- transaxle/transmission
- Negative battery cable

Piston and Ring

POSITIONING

Before removing the caps from the connecting rods, be sure to matchmark them—Mazda engines

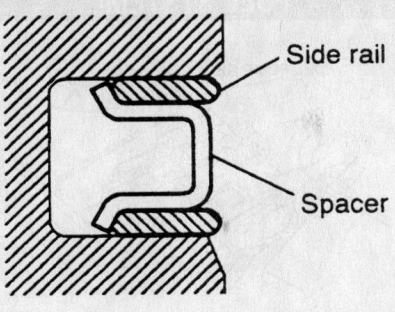

Compression ring identification and positioning—Mazda engines

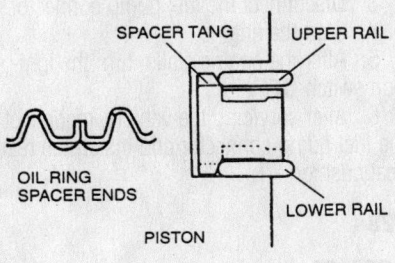

Upper, spacer and lower oil ring identification and positioning—Mazda engines

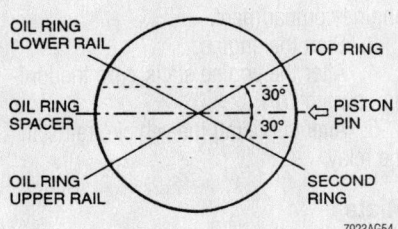

Piston ring end-gap spacing—Mazda engines

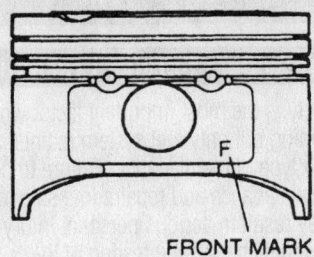

Piston-to-engine block mark location on the piston—Mazda 1.5L (Z5) and 1.8L (BP) engines

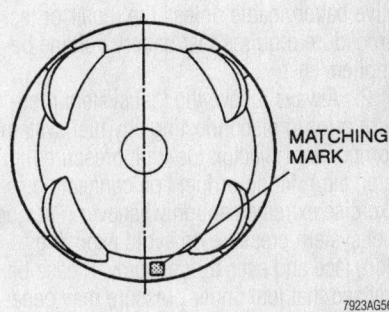

Piston-to-engine block mark location on the piston face—Mazda 2.0L (FS) and 2.5L (KL) engines

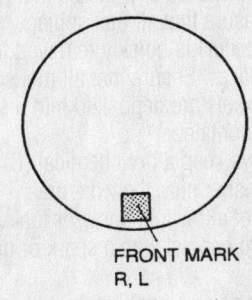

Piston-to-engine positioning mark location—Mazda 2.3L (KJ) engine

FUEL SYSTEM

Fuel System Service Precautions

Safety is the most important factor when performing not only fuel system maintenance but any type of maintenance. Failure to conduct maintenance and repairs in a safe manner may result in serious personal injury or death. Maintenance and testing of the vehicle's fuel system components can be accomplished safely and effectively by adhering to the following rules and guidelines.

1. To avoid the possibility of fire and personal injury, always disconnect the negative battery cable unless the repair or test procedure requires that battery voltage be applied.

2. Always relieve the fuel system pressure prior to disconnecting any fuel system component (injector, fuel rail, pressure regulator, etc.), fitting or fuel line connection. Exercise extreme caution whenever relieving fuel system pressure, to avoid exposing skin, face and eyes to fuel spray. Please be advised that fuel under pressure may penetrate the skin or any part of the body that it contacts.

3. Always place a shop towel or cloth around the fitting or connection prior to loosening to absorb any excess fuel due to spillage. Ensure that all fuel spillage (should it occur) is quickly removed from engine surfaces. Ensure that all fuel soaked cloths or towels are deposited into a suitable waste container.

4. Always keep a dry chemical (Class B) fire extinguisher near the work area.

5. Do not allow fuel spray or fuel vapors to come into contact with a spark or open flame.

6. Always use a back-up wrench when loosening and tightening fuel line connection fittings. This will prevent unnecessary stress and torsion to fuel line piping. Always follow the proper torque specifications.

7. lways replace worn fuel fitting O-rings with new. Do not substitute fuel hose where fuel pipe is installed.

Fuel System Pressure

RELIEVING

Protégé

1998–00

1. Before servicing the vehicle, refer to the precautions in the beginning of this section.

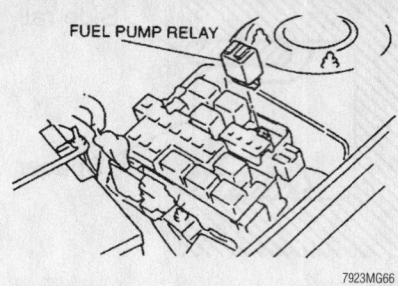

Fuel pump relay location—626 and with 2.0L and 2.5L engines

2. Remove the rear seat cushion and locate the fuel pump connector.
3. Disconnect the fuel pump connector.
4. Start the engine.
5. After the engine stalls, turn the ignition switch **OFF**.
6. After servicing the vehicle, reconnect the fuel pump connector and install the rear seat cushion.

626

1998–00

1. Before servicing the vehicle, refer to the precautions in the beginning of this section.
2. Remove the fuel pump relay from the relay box, located in the left side of the engine compartment.
3. Start the engine.
4. After the engine stalls, turn the ignition switch **OFF**.
5. After servicing the vehicle, reinstall the relay.

Miata

1998–00

1. Before servicing the vehicle, refer to the precautions in the beginning of this section.

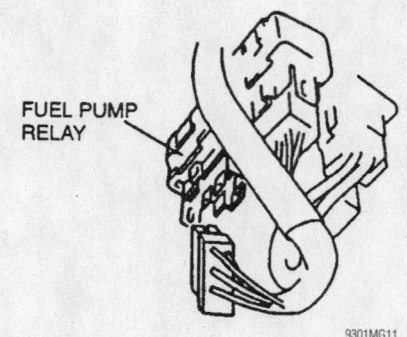

Fuel pump relay connector location—Miata

2. Loosen the fuel filler cap to release the pressure in the tank..
3. Remove the fuel pump relay connector, located above the accelerator pedal.
4. Start the engine.
5. After the engine stalls, turn the ignition switch **OFF**.
6. After servicing the vehicle, reinstall the relay and tighten the fuel filler cap.

Millenia

2.3L ENGINES

1. Before servicing the vehicle, refer to the precautions in the beginning of this section.
2. If necessary for clearance, remove the cruise control actuator and position aside.
3. Remove the fuel pump relay from the relay box, located in the right side of the engine compartment.
4. Start the engine.
5. After the engine stalls, turn the ignition switch **OFF**.
6. After servicing the vehicle, reinstall the relay and the cruise actuator, if necessary.

2.5L ENGINES

1998–00

To relieve the fuel system pressure on the 2.5L engine, refer to the 626 procedure.

Fuel Filter

The fuel filter on all Mazda cars can be located on a bracket in the left rear of the engine compartment, next to or beneath the brake master cylinder fluid reservoir or as part of the fuel pump assembly.

On the Millenia, the fuel filter is located beneath an access cover in the trunk. Access to the cover is achieved by removing the trunk mat to expose the cover.

REMOVAL & INSTALLATION

In-Line Fuel Filter

1998–00

1. Before servicing the vehicle, refer to the precautions in the beginning of this section.
2. Properly relieve the fuel system pressure.
3. Remove or disconnect the following:
   - Negative battery cable
   - Air intake hose and/or filter housing, if necessary
   - Fuel line clamps, if equipped

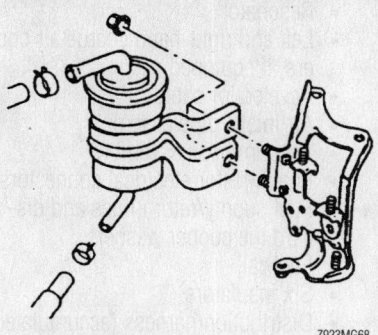

7923MG68

Exploded view of an in line fuel filter and related components

- Fuel lines from the filter and plug the openings
- Filter clamp bolt/nut by loosening them.
- Fuel filter from its mounting bracket

➡**Note the direction of the flow arrow on the filter so the replacement filter can be installed in the correct position.**

To install:

4. Install or connect the following:
   - fuel filter in its mounting bracket, making sure the flow arrow is pointing in the proper direction
   - Fuel lines to the fuel filter
   - Fuel line clamps, if equipped
   - Air intake hose and/or filter housing, if removed
   - negative battery cable
5. Pressurize the fuel system and check all connections for leaks.

Fuel Pump Filter

1998–00

1. Before servicing the vehicle, refer to the precautions in the beginning of this section.
2. Properly relieve the fuel system pressure.
3. Remove or disconnect the following:
   - negative battery cable
   - Fuel pump cover plate
   - Fuel lines from the fuel pump
4. Raise the fuel pump from the tank.
5. Remove the fuel filter assembly.

To install:

6. Install or connect the following:
   - Fuel filter to the fuel pump. Lower the pump assembly into position
   - Fuel lines
   - Negative battery cable

7. Start the vehicle, check for leaks and repair if necessary.
8. Install the fuel pump cover plate

Millenia

1998–00

1. Before servicing the vehicle, refer to the precautions in the beginning of this section.
2. Insure the ignition switch is **OFF**.
3. Relieve the fuel system pressure.
4. Remove or disconnect the following:
   - negative battery cable
   - Trunk mat
   - service hole cover
   - fuel lines from both ends of the fuel filter
   - Fuel filter and bracket
   - Fuel filter from the mounting bracket

To install:

5. Install or connect the following:
   - Filter in the bracket. Torque the nut to 70–95 inch lbs. (8–11 Nm)
   - Fuel lines to the filter
   - service hole cover
   - Trunk mat
   - negative battery cable
6. Run the engine and check for any fuel leaks.

Fuel Pump

REMOVAL & INSTALLATION

Protégé

1998–00

1. Before servicing the vehicle, refer to the precautions in the beginning of this section.
2. Relieve the fuel pressure.
3. Remove or disconnect the following:
   - Negative battery cable
   - Rear seat cushion
   - Fuel pump/sending unit electrical connector
   - Fuel pump/sending unit access cover
   - Fuel supply and return hoses from the fuel pump/sending unit
   - Fuel pump/sending unit from the fuel tank
   - Sending unit electrical connector
   - Sending unit from the fuel pump assembly

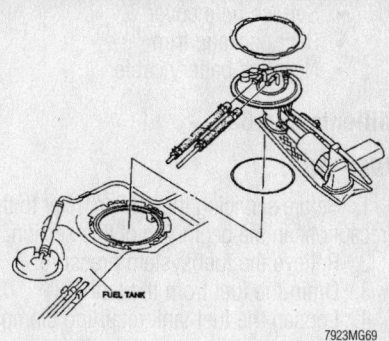

7923MG69

Exploded view of the fuel pump assembly—626

To install:

4. Install or connect the following:
   - Sending unit to the fuel pump assembly
   - Sending unit electrical connector
   - Fuel pump/sending unit into the fuel tank with a new gasket
   - Fuel supply and return lines
   - Access cover
   - Sending unit electrical connector
   - Rear seat cushion
   - Negative battery cable
5. Start the engine and check fuel leaks.

Miata

1998–00

1. Before servicing the vehicle, refer to the precautions in the beginning of this section.
2. Properly relieve the fuel pressure.
3. Remove or disconnect the following:
   - negative battery cable
   - rear package trim
   - Service hole cover
   - fuel pump cover
   - Fuel pump connector
   - Fuel hoses
   - fuel pump and gauge sender unit as an assembly
   - Fuel pump from the sender bracket

To install:

- New O-ring set
- Fuel pump to the sender bracket

➡**Pull the fuel pump down so that it is tight against the bracket.**

- Fuel pump and gauge sender unit as an assembly
- Fuel hoses
- Fuel pump connector
- Fuel pump cover

- Service hole cover
- Rear package trim
- Negative battery cable

Millenia and 626

1998–00

1. Before servicing the vehicle, refer to the precautions in the beginning of this section.
2. Relieve the fuel system pressure.
3. Drain the fuel from the tank.
4. Loosen the fuel tank retaining clamps.
5. Remove or disconnect the following:
   - Negative battery cable
   - Hoses from the fuel tank
   - Fuel pump electrical connector
   - Fuel tank heat shield
   - Fuel tank support brackets by supporting the fuel tank
   - Fuel tank
   - All fuel hoses from the fuel pump unit
   - Fuel pump ring by turning counterclockwise
   - Fuel pump and gaskets from the fuel tank

To install:
6. Install or connect the following:
   - Fuel pump using a new gasket
   - Fuel pump ring by turning it clockwise until the flange hits the stopper
   - Fuel hoses to the fuel pump
   - Fuel tank
   - Fuel tank support brackets
   - Fuel tank heat shield
   - Fuel pump electrical connector
   - Hoses to the fuel tank
   - Fuel tank retaining clamps
   - Negative battery cable
7. Add a minimum of 10 gallons of fuel to the tank and check for leaks.

Fuel Injector

REMOVAL & INSTALLATION

✳✳ CAUTION

Fuel injection systems remain under pressure after the engine has been turned OFF. Properly relieve fuel pressure before disconnecting any fuel lines. Failure to do so may result in fire or personal injury. Do not allow fuel spray or fuel vapors to come in contact with a spark or open flame. Keep a dry chemical fire extinguisher nearby. Never store fuel in an open container due to risk of fire or explosion.

Miata, Protégé and 4-Cylinder 626

1998–00

1. Before servicing the vehicle, refer to the precautions in the beginning of this section.
2. Relieve the fuel system pressure.
3. Remove or disconnect the following:
   - Negative battery cable
   - Throttle, accelerator cables and the cable bracket on Miata and Protégé
   - Fuel injector wiring harness
   - Fuel lines at the fuel rail
   - Vacuum hose from the fuel pressure regulator
   - Fuel line mounting bracket bolt
   - Fuel rail mounting bolts, spacers and insulators
   - Fuel rail, with the injectors attached
   - Fuel injectors, grommets and O-rings from the fuel rail
   - O-rings from the fuel injectors

To install:
4. Install or connect the following:
   - New O-rings and grommets lubricated with engine oil on the fuel injectors
   - Insulators and injectors on the intake manifold
   - Grommets and the fuel rail onto the injectors. Torque the bolts to 14–18 ft. lbs. (19–25 Nm)
   - Vacuum hose to the fuel pressure regulator
   - Fuel lines to the fuel rail
   - Fuel line bracket. Torque the bolts to 70–95 inch lbs. (8–11 Nm)
   - Fuel injector wiring harness
   - Cable bracket, if removed. Torque the bolt to 70–95 inch lbs. (8–11 Nm)
   - Throttle and accelerator cables, if removed, adjust as necessary
   - Negative battery cable
5. Turn the ignition switch **ON** to pressurize the fuel system.
6. Check for leaks and correct as necessary, before starting the engine.

Millenia and 6-Cylinder 626

1998–00

1. Before servicing the vehicle, refer to the precautions in the beginning of this section.
2. Relieve the fuel system pressure.
3. Remove or disconnect the following:
   - Negative battery cable
   - Charge air cooler air duct, if equipped
   - Air cleaner assembly, as necessary, for clearance

- Resonator
- Left and right-hand charge air coolers, if equipped
- Accelerator cable
- Air intake pipe assembly
- Vacuum hose assembly
- Fuel injector electrical connectors
- Fuel supply/return lines and discard the copper washers
- Fuel rail
- Six insulators
- Distribution harness (accumulated connector) from the fuel rails
- Spacer from the top of each fuel injector and discard it
- Fuel injectors from the fuel rails by rotating back and forth
- Fuel pressure regulator, if necessary

To install:
4. Install or connect the following:
   - Fuel pressure regulator, if removed. Torque the bolts to 61–86 inch lbs. (7–10 Nm)
   - New O-rings lubricated with engine oil on the injectors
   - Fuel injectors into the fuel rails
   - New spacers on the injectors
   - Distribution harness. Torque the screws to 22–31 inch lbs. (2.5–3.5 Nm)
   - 6 insulators and the fuel rails. Torque the bolts to 14–18 ft. lbs. (19–25 Nm)
   - Fuel supply and return lines using new copper washers
   - Fuel injector electrical connectors
   - Vacuum hose assembly. Torque the nuts to 70–95 inch lbs. (8–11 Nm)
   - Air intake pipe assembly. Torque the nuts to 70–95 inch lbs. (8–11 Nm) and the bolts to 44–78 inch lbs. (5–9 Nm)
   - Accelerator cable and adjust it. Torque the bolt to 70–95 inch lbs. (8–11 Nm)
   - Both charge air coolers with new O-rings. Torque the bolts to 14–18 ft. lbs. (19–25 Nm)
   - Resonator. Torque the nuts to 70–95 inch lbs. (8–11 Nm)
   - Air cleaner assembly. Torque the nuts to 70–95 inch lbs. (8–11 Nm)
   - Charge air cooler air duct. Torque the nuts to 70–95 inch lbs. (8–11 Nm)
   - Negative battery cable
5. Turn the ignition switch **ON** to pressurize the fuel system.
6. Check for fuel leaks and correct as necessary before starting the engine.

DRIVE TRAIN

Transaxle Assembly

REMOVAL & INSTALLATION

Manual Transaxle

1998–00

1. Before servicing the vehicle, refer to the precautions in the beginning of this section.

2. Drain the transaxle oil.
3. Remove or disconnect the following:
   - Negative battery cable
   - Air cleaner assembly
   - Speedometer cable and/or sensor wires from the transaxle
   - Clutch release (slave) cylinder from the transaxle
   - Water, secondary air and Exhaust Gas Recirculation (EGR) pipe brackets
   - Wiring harness clip
   - Coupler for the neutral switch and backup lamp switch
   - Body ground connector
   - 2 upper transaxle mounting bolts and mount an Engine Support tool 49-eR301-025A to the engine hanger
   - Front wheels
   - Engine under and side covers

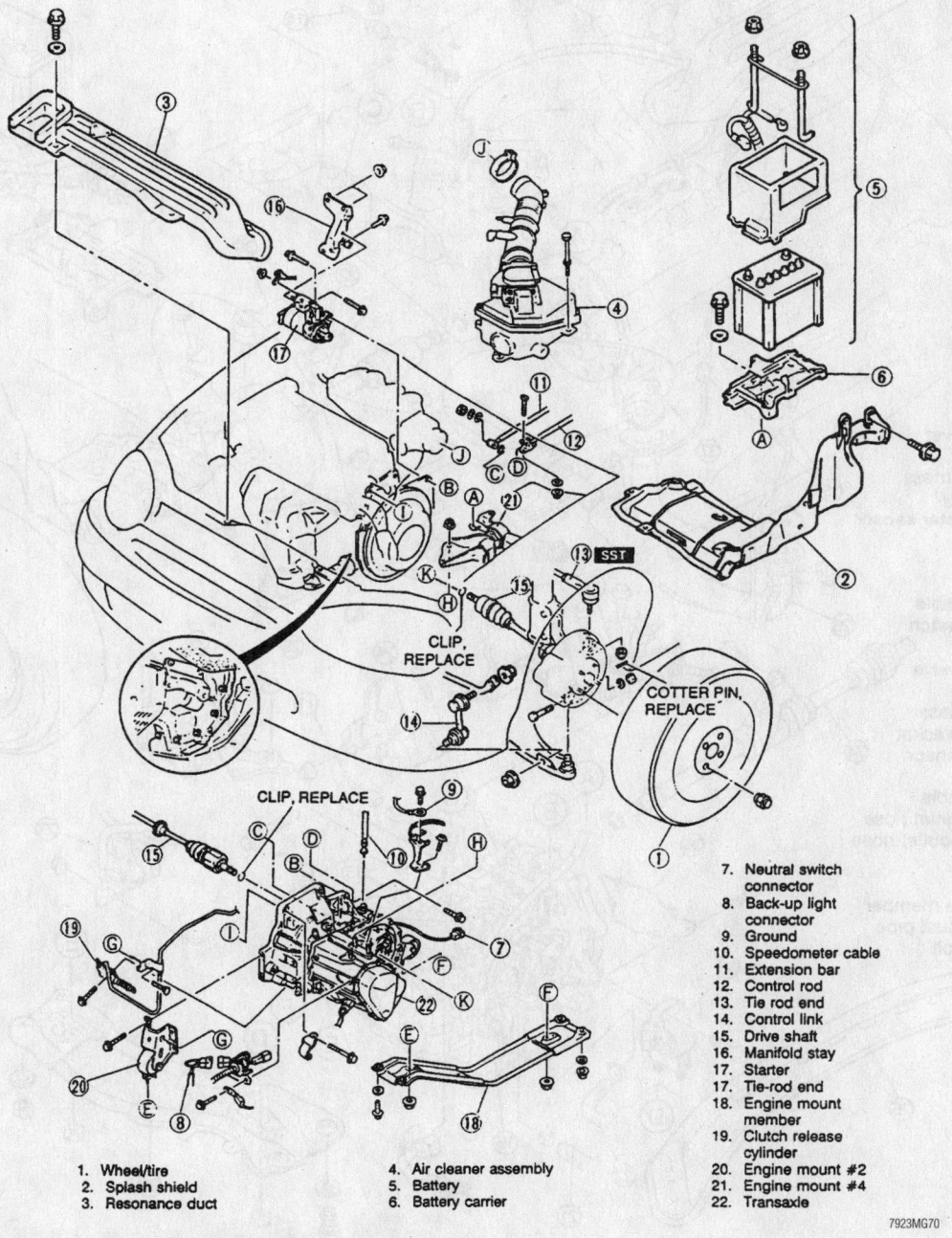

1. Wheel/tire
2. Splash shield
3. Resonance duct
4. Air cleaner assembly
5. Battery
6. Battery carrier
7. Neutral switch connector
8. Back-up light connector
9. Ground
10. Speedometer cable
11. Extension bar
12. Control rod
13. Tie rod end
14. Control link
15. Drive shaft
16. Manifold stay
17. Starter
18. Engine mount member
19. Clutch release cylinder
20. Engine mount #2
21. Engine mount #4
22. Transaxle

7923MG70

Exploded view of a typical transaxle assembly mounting and related components

Timing belt service is covered in Section 3 of this manual

1. Wheel/tire
2. Splash shield
3. Resonance duct
4. Airflow meter connector
5. Air cleaner assembly

28. Lower arm
29. Nut
30. Tie rod end
31. Stabilizer

32. Clip
33. Engine mounting member
34. Drive shaft

35. Engine mount no.4
36. Undercover
37. Nut
38. Engine mount no.2
39. Transaxle

6. Strut bar
7. Battery cover
8. Battery
9. Engine harness
10. Battery tray
11. Speedometer sensor connector
12. Clip
13. Snap pin
14. Selector cable
15. Inhibitor switch connector
16. Solenoid valve connector
17. Front harness
18. Harness bracket
19. Oxygen sensor connector
20. Throttle cable
21. Oil cooler inlet hose
22. Oil cooler outlet hose
23. Fuel filter
24. Starter
25. Transverse member
26. Front exhaust pipe
27. Nut and bolt

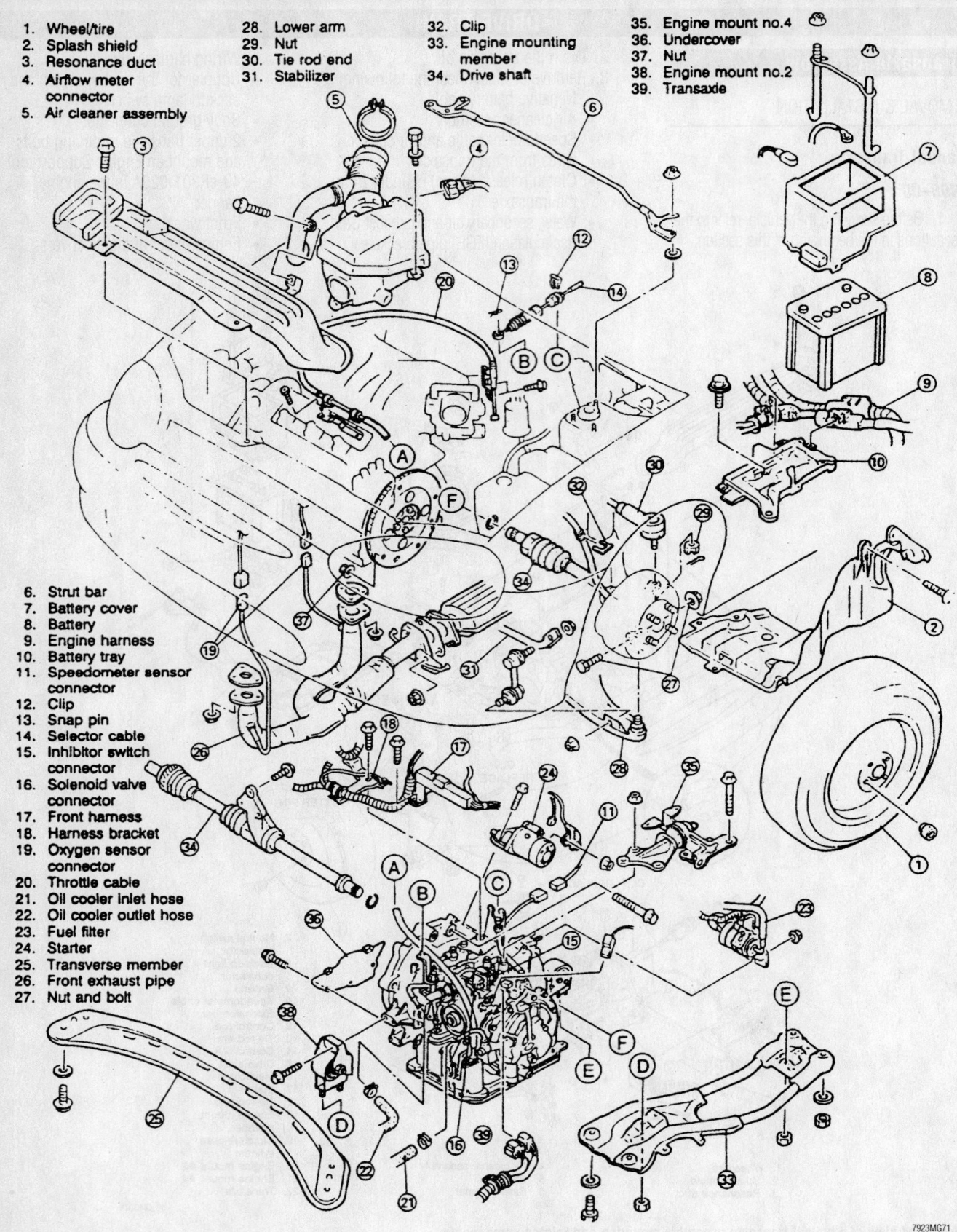

Exploded view of a typical automatic transaxle assembly mounting and related components

7923MG71

- Intake manifold support bracket, if necessary
- Halfshafts

4. Insert a Differential Side Gear Holder 49-B027-001 to hold the side gears in place and prevent misalignment.

5. Remove or disconnect the following:
- Transaxle crossmember
- Gearshift control rod from the transaxle
- Extension bar from the transaxle
- Wiring connectors
- Starter motor
- End plates

6. Lean the engine toward the transaxle side to lower the transaxle by loosening the engine support hook bolt. Support the transaxle with a transaxle jack.

7. Remove or disconnect the following:
- Engine brackets, as necessary
- Remaining transaxle bolt
- Transaxle

To install:

8. Before installing the transaxle, lightly coat the splines of the primary shaft gear with molybdenum disulfide grease.

9. Attach a thick rope to 2 places on the transaxle. Place a board on the jack and lower the transaxle onto the board. Using the jack, lift the transaxle into position and throw the end of the rope over the support fixture bar. Tension the rope to guide the transaxle onto its mounts while lifting the transaxle with the jack.

10. Once the transaxle is in place, have an assistant install and tighten all the transaxle-to-engine mounting bolts.

11. The remainder of the installation is the reverse of the removal procedure. Tighten all fasteners to specifications.

12. Fill the transaxle with fluid.

Automatic Transaxle

1998–00

1. Before servicing the vehicle, refer to the precautions in the beginning of this section.

2. Properly relieve the fuel system pressure.

3. Drain the transaxle oil.

4. Remove or disconnect the following:
- Front wheels
- Battery and battery box
- Air cleaner and ducting
- Splash shield
- Speedometer, throttle and shift cables
- Wiring connectors from the transaxle

- Fuel lines, if necessary

5. If the fuel filter is mounted to the transaxle, unbolt its mounting bracket and position it aside.

6. Remove or disconnect the following:
- Exhaust crossover, coolant, vacuum or EGR pipe mounting brackets from the transaxle
- Intake manifold support bracket, on the 2.0L engines
- Wiring and remove the starter
- Exhaust pipe
- Tie rod ends
- Lower ball joints
- Halfshafts and install tool 49 G030 455 to hold the differential side gears in place when the halfshafts are removed

7. Remove or disconnect the following:
- Torque converter-to-flywheel nuts and/or bolts
- Oil cooler hoses and plug them to prevent leakage

8. Attach lifting equipment and support the engine from above.

9. Remove the lower mounting frame

10. Support the transaxle with a jack.

11. Remove the front and left rear mounts.

12. Allow the engine/transaxle to tilt towards the left.

13. Install the remaining bolts and slide the transaxle away from the engine to lower it out of the vehicle. Do not let the torque converter fall out.

To install:

14. Be sure the torque converter is properly placed and carefully guide the transaxle into place. Start all the transaxle-to-engine bolts, then tighten them to specifications.

15. The remainder of the installation is the reverse of the removal procedure.

Transmission Assembly

REMOVAL & INSTALLATION

Manual Transmission

1998–00

1. Before servicing the vehicle, refer to the precautions in the beginning of this section.

2. Drain the transmission oil.

3. Remove or disconnect the following:
- Negative battery cable
- Shifter knob
- Center console
- Gearshift lever

- Engine undercover and the performance rod
- Exhaust pipe from the manifold
- Entire exhaust system as an assembly and matchmark the driveshaft flange at the rear
- Driveshaft
- Clutch release cylinder and set it aside, without disconnecting the hydraulic hose
- Starter
- Speedometer cable and wiring from the frame member

4. Support the transmission and differential with jacks.

5. Remove the transmission-to-differential frame, also called the Power Plant Frame (PPF), as follows:

a. Remove the frame-to-transmission bracket.

b. Remove the bolts from the underside of the frame at the differential end, noting their location. Pry out the spacer from the frame.

c. Remove the differential mounting spacer from the underside of the differential.

d. Insert a 14 x 1.5mm bolt through the frame hole and turn it into the sleeve. Twist and pull the bolt downward.

e. Install a 6 x 1mm bolt in the side hole to hold the sleeve and remove the long bolt. Remove the short bolt.

➡**Do not remove the spacers from the end of the PPF. Doing so will reduce the performance of the frame. If the spacers are removed, the PPF must be replaced as an assembly.**

f. Remove the transmission side bolts and the frame member.

6. Remove or disconnect the following:
- Bolts from the clutch housing
- Transmission from the vehicle

To install:

7. Lightly lubricate the main shaft spline and the release bearing fork contact points with molybdenum grease and install the fork. Place a wood block on a floor jack and use it to tilt the engine up in front.

8. Install or connect the following:
- Transmission. Torque the bolts alternately, in several passes, to 48–65 ft. lbs. (64–89 Nm)
- Transmission to the differential frame. Torque the bolts to 76–91 ft. lbs. (104–124 Nm)
- Bracket. Torque the bracket-to-transmission bolts to 27–40 ft. lbs.

Heater Core replacement is covered in Section 2 of this manual

(37–54 Nm) and the bracket-to-frame bolts to 77–91 ft. lbs. (104–124 Nm).

➡ After the frame installation, position a straightedge between the body frame members on each side of the vehicle. Measure the distance between the bottom of the frame to the straightedge; it should be 2.403–2.797 in. (61–71mm). If the distance is not as specified, reposition the frame member at the transmission.

- Speedometer cable and wiring
- Starter
- Clutch release cylinder
- Driveshaft. Torque the nuts to 22 ft. lbs. (30 Nm)
- Exhaust system using new gaskets. Torque the nuts to 34 ft. lbs. (46 Nm)
- Performance rod and engine undercover
- Shift lever
- Rear console
- Shift lever knob
- Negative battery cable

9. Fill the transmission with clean oil.

10. Verify proper operation of the transmission.

Automatic Transmission

1998–00

1. Before servicing the vehicle, refer to the precautions in the beginning of this section.

2. Shift the selector lever to the **N** position.

3. Drain the transmission fluid.

4. Remove or disconnect the following:
- Negative battery cable
- Engine undercover
- Shift rod
- Performance rod
- Complete exhaust system from the exhaust manifold and mark the position of the driveshaft on the rear axle flange

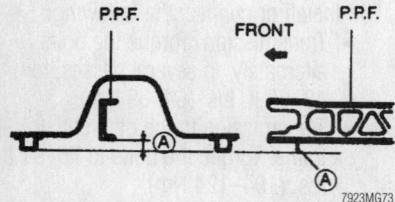

Measure distance "A" after installing the transmission-to-differential frame member—Miata with N4A-HL and NC4A-EL transmissions

- Driveshaft
- Speedometer cable
- Vacuum hose from the vacuum diaphragm
- Electrical connectors from inhibitor switch, kickdown solenoid, overdrive cancel solenoid, oil pressure switch and lockup solenoid
- Dipstick and dipstick tube
- Oil cooler lines

5. Support the transmission and differential with jacks.

6. Remove the transmission-to-differential frame as follows:

   a. Disconnect the wiring harness from the frame.

   b. Remove the bolts from the underside of the frame at the differential end, noting their location. Pry out the spacer from the frame.

   c. Remove the differential mounting spacer from the underside of the differential.

   d. Insert a 14 x 1.5mm bolt through the frame hole and turn it into the sleeve. Twist and pull the bolt downward.

   e. Install a 6 x 1mm bolt in the side hole to hold the sleeve and remove the long bolt. Remove the short bolt.

   f. Remove the transmission side bolts and remove the frame member.

7. Remove or disconnect the following:
- Torque converter bolts
- Starter
- Transmission mounting bolts
- Transmission, being careful not to drop the torque converter

To install:

8. Be sure the torque converter is fully installed in the transmission. The distance between 1 of the bolt hole lugs and a straightedge laid across the bell housing should be 0.89 in. (22.5mm).

9. Install or connect the following:
- Transmission. Torque the bolts to 48–66 ft. lbs. (64–89 Nm)
- Starter
- Torque converter. Torque the bolts to 27–40 ft. lbs. (36–54 Nm)

10. Install the transmission-to-frame member as follows:

   a. Install the differential mounting spacer on the underside of the differential. Torque the bolts to 38 ft. lbs. (52 Nm).

   b. Position the jack under the transmission so the transmission is level.

   c. Position the frame and install the transmission side bolts. Snug the bolts.

   d. Be sure the sleeve is installed in the block. Install the spacer and bolts with the reamer bolt in the front hole. Snug the bolts.

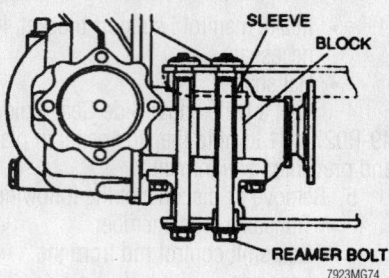

Sleeve and reamer bolt positioning—Miata with N4A-HL and NC4A-EL transmissions

   e. Torque the transmission side bolts to 77–91 ft. lbs. (104–123 Nm), and the differential bolts to the same specification.

   f. Connect the wiring harness to the frame member and remove the jacks.

➡ After the frame installation, position a straightedge between the body frame members on each side of the vehicle. Measure the distance between the bottom of the frame to the straightedge; it should be 2.023–2.417 in. (51.5–61.5mm). If the distance is not as specified, reposition the frame member at the transmission.

11. Install or connect the following:
- Oil cooler lines using new gaskets
- Dipstick tube and dipstick
- Electrical connectors and the vacuum hose
- Speedometer cable
- Driveshaft by aligning the matchmarks. Torque the bolts to 22 ft. lbs. (30 Nm)
- Exhaust system
- Front pipe to the exhaust manifold using a new gasket. Torque the nuts to 34 ft. lbs. (46 Nm)
- Performance rod
- Engine undercover
- Shift rod
- Negative battery cable

12. Fill the transmission with fluid.

13. Start the engine and check for leaks and proper operation.

Clutch

REMOVAL & INSTALLATION

1998–00

1. Before servicing the vehicle, refer to the precautions in the beginning of this section.

2. Remove or disconnect the following:

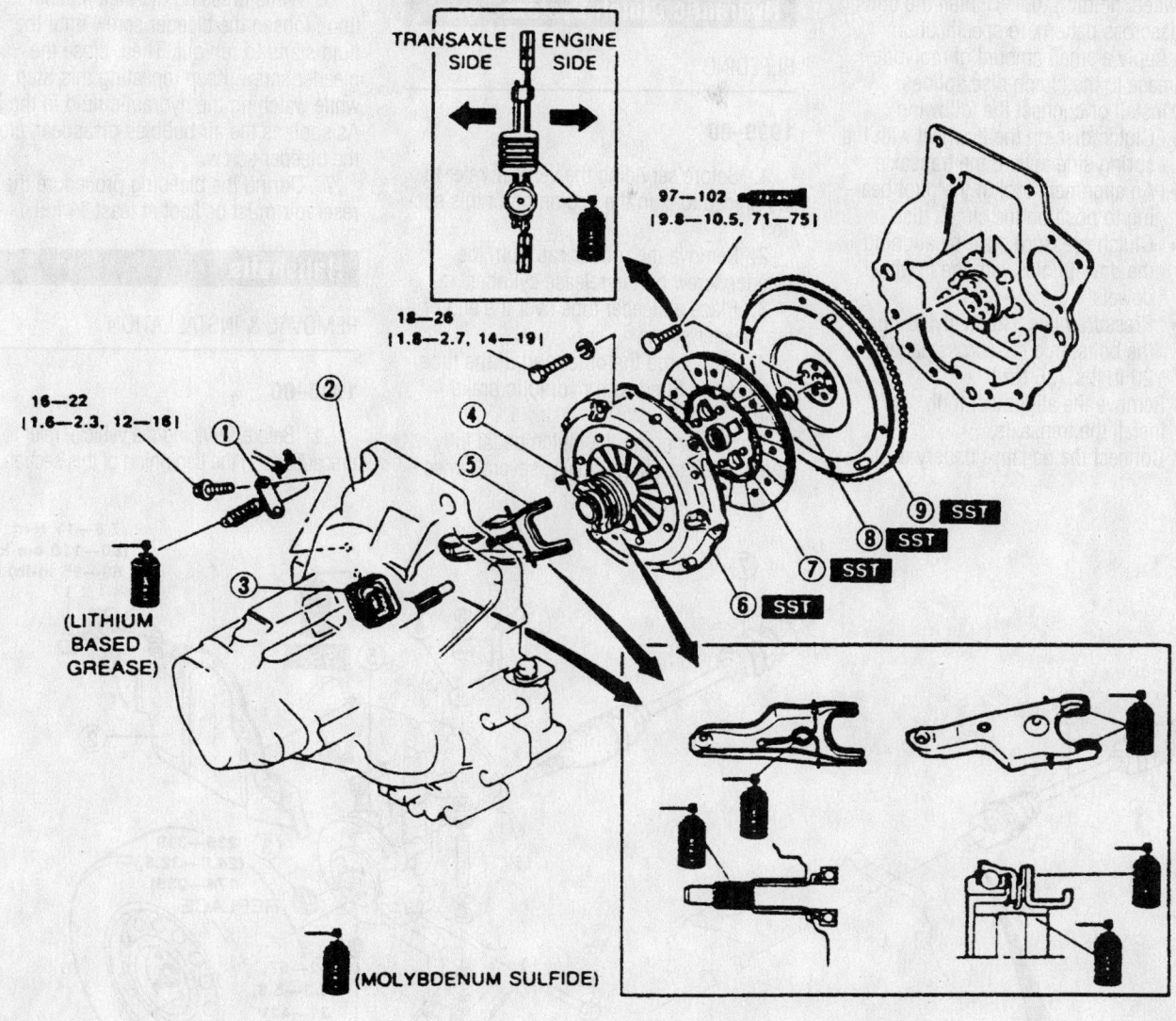

TRANSAXLE ‖ **ENGINE**
SIDE ‖ **SIDE**

97—102
(9.8—10.5, 71—75)

18—26
(1.8—2.7, 14—19)

16—22
(1.6—2.3, 12—16)

(LITHIUM BASED GREASE)

(MOLYBDENUM SULFIDE)

1. Clutch release cylinder
2. Transaxle
3. Boot
4. Clutch release collar
5. Clutch release fork
6. Clutch cover
7. Clutch disc
8. Pilot bearing
9. Flywheel

7923MG75

Exploded view of a typical clutch disc and pressure plate assembly with related components

- Negative battery cable
- Transaxle

3. Gradually loosen the clutch pressure plate bolts, in a crisscross pattern. Support the pressure plate and remove the bolts.

4. Remove the pressure plate and clutch disc.

5. Inspect the pilot bearing. If it is worn or damaged and does not turn easily by hand, remove it using a puller/slide hammer.

6. Check the flywheel surface for scoring, cracks or burning and machine or replace, as necessary.

7. Install Holder tool 49 E011 1A0 to keep the flywheel from turning. Loosen the flywheel bolts evenly and gradually in a crisscross pattern. Remove the flywheel.

8. Inspect the clutch release bearing for wear. Replace it if it sticks or does not turn easily.

9. Inspect the release fork for wear or damage and replace as necessary.

To install:

10. Lubricate the release fork fingers and pivot with molybdenum grease and install in the release fork boot.

11. Install or connect the following:
- Clutch release bearing on the release fork
- New pilot bearing in the flywheel, if removed, using a installation tool

12. Be sure the flywheel mounting surface and the crankshaft or eccentric shaft mounting surfaces are clean. Remove any old sealant from the flywheel bolt hole threads and the flywheel bolts.

13. Install the flywheel.

14. Apply sealant to the flywheel bolt threads and install them hand tight. Install

Brake service is covered in Section 4 of this manual

the flywheel holding tool. Tighten the bolts, in a crisscross pattern, to specification.

15. Apply a small amount of molybdenum grease to the clutch disc splines.

16. Install or connect the following:
- Clutch disc on the flywheel with the spring side toward the transaxle
- An alignment tool in the pilot bearing to position the clutch disc
- Clutch pressure plate by aligning the dowel holes with the flywheel dowels
- Pressure plate. Gradually, torque the bolts, in a crisscross pattern, to 20 ft. lbs. (26 Nm)

17. Remove the alignment tool.
18. Install the transaxle.
19. Connect the negative battery cable.

Hydraulic Clutch System

BLEEDING

1998–00

1. Before servicing the vehicle, refer to the precautions in the beginning of this section.

2. Remove the rubber cap from the bleeder screw on the release cylinder.

3. Place a bleeder tube over the end of the bleeder screw.

4. Submerge the other end of the tube in a jar half filled with hydraulic brake fluid.

5. Slowly pump the clutch pedal fully and allow it to return slowly, several times.

6. While pressing the clutch pedal to the floor, loosen the bleeder screw until the fluid starts to run out. Then, close the bleeder screw. Keep repeating this Step, while watching the hydraulic fluid in the jar. As soon as the air bubbles disappear, close the bleeder screw.

7. During the bleeding procedure the reservoir must be kept at least ¾ full.

Halfshafts

REMOVAL & INSTALLATION

1998–00

1. Before servicing the vehicle, refer to the precautions in the beginning of this section.

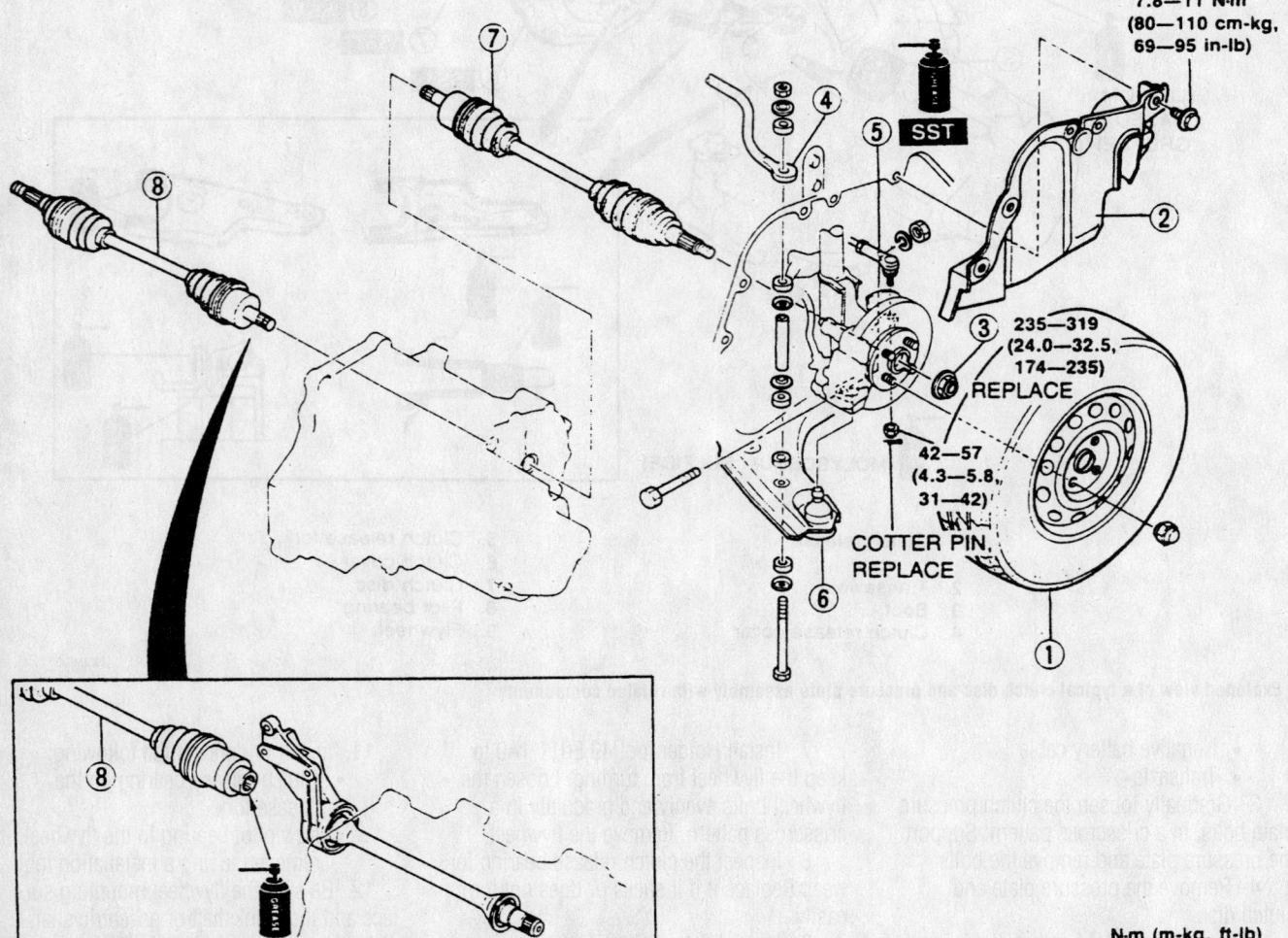

7.8—11 N·m
(80—110 cm-kg,
69—95 in-lb)

235—319
(24.0—32.5,
174—235)
REPLACE

42—57
(4.3—5.8,
31—42)

COTTER PIN,
REPLACE

N·m (m-kg, ft-lb)

1. Wheel and tire
2. Splash shield
3. Locknut
4. Stabilizer
5. Tie-rod end
6. Lower ball joint
7. Left driveshaft
8. Right driveshaft

Exploded view of a typical halfshaft mounting

7923MG76

2. Drain the transaxle oil.

3. Remove or disconnect the following:
- Wheels
- Splash shield, if equipped

4. Raise the staked portion of the hub locknut with a hammer and chisel.

5. Lock the hub by applying the brakes and remove the nut.

6. Remove or disconnect the following:
- Stabilizer bar from the lower control arm
- Cotter pin and nut from the tie rod end ball stud
- Tie rod end from the knuckle
- Transverse member, on the 626 and Millenia
- Lower ball joint pinch bolt and nut
- Lower ball joint from the knuckle

7. If removing the left side shaft on 626 or Millenia with automatic transaxle, proceed as follows:

a. Suspend the engine using engine support tool 49 G017 5A0.

b. Remove the engine mount member.

8. Position a prybar between the inner CV-joint and transaxle case. Carefully pry the halfshaft from the transaxle being careful not to damage the oil seal. If equipped with a right side intermediate shaft, insert the prybar between the halfshaft and intermediate shaft and tap on the bar to uncouple them.

9. Pull outward on the hub/knuckle assembly, push the outer CV-joint stub shaft through the hub and remove the halfshaft. If the halfshaft is stuck in the hub, install the old hub nut to protect the stub shaft threads. Tap on the nut, using only a soft mallet, to remove the halfshaft.

➡ **Install plug tool 49 G030 455 into the transaxle after removing the halfshaft, to keep the differential side gear in position. If the gear becomes positioned incorrectly, the differential may have to be removed to realign the gear.**

10. Remove the intermediate shaft, if necessary, by removing the support bearing bolts and pulling the shaft from the transaxle.

To install:

11. Install or connect the following:
- New circlip on the end of the intermediate shaft, if removed, with the end gap facing upward.
- Intermediate shaft in the transaxle, being careful not to damage the oil seals
- Intermediate shaft support bearing

bolts. Torque in sequence, to 45 ft. lbs. (61 Nm)
- New circlip on the end of the halfshaft, with the end gap facing upward
- Halfshaft into the transaxle, being careful not to damage the oil seal

➡ **If equipped, push the halfshaft into the intermediate shaft.**

- Other end of the halfshaft through the hub. Loosely install a new locknut

12. If installing the left side shaft on 626 and Millenia with automatic transaxle, proceed as follows:

a. Install the engine mount member. Torque the mount member-to-body bolts to 66 ft. lbs. (89 Nm).

b. Torque the front mount-to-mount member nuts to 77 ft. lbs. (104 Nm) and the side mount bolts to 44 ft. lbs. (60 Nm).

c. Remove the engine support tool.

13. Install or connect the following:
- Lower ball joint into the knuckle. Torque the pinch bolt to 40 ft. lbs. (54 Nm)
- Transverse member, on 626 and Millenia. Torque the bolts to 96 ft. lbs. (132 Nm)
- Tie rod end to the steering knuckle. Torque the nut to 42 ft. lbs. (57 Nm) on all except 626 and Millenia or to 32 ft. lbs. (44 Nm) for 626 and Millenia
- New cotter pin. Tighten the nut, if necessary, to align the ball stud hole with the nut castellation
- Stabilizer bar to the lower control arm
- Splash shield
- Wheels
- New hub nut. Torque it to 174–235 ft. lbs. (235–318 Nm). After tightening, stake the locknut using a hammer and dull bladed chisel

14. Fill the transaxle.

CV-Joints

OVERHAUL

1998–00

Two types of CV-joints are used. The inboard CV-joints are the tri-Pot type. All outboard CV-joints are Birfield type. The Birfield CV-joint cannot be disassembled; if

an outboard CV-joint boot needs replacement, the inboard CV-joint must be removed. If the outboard CV-joint needs to be replaced, replace the entire halfshaft as an assembly.

1. Remove the halfshaft from the vehicle and clamp it in a vise equipped with jaw caps, to prevent damage to the machined surfaces. Do not allow the vise to contact the boot or its clamps.

2. Remove the large boot clamp from the inboard CV-joint, using side cutters. After removing the clamp, roll the boot back over the shaft.

➡ **Check the grease for contamination by rubbing it between 2 fingers. Any gritty feeling indicates a contaminated CV-joint, in which case the entire CV-joint must be disassembled, cleaned and inspected. If the grease is not contaminated and the CV-joint has been operating satisfactorily, continue with the boot replacement procedure and add the required lubricant.**

3. Paint alignment marks on the outer race and shaft for assembly reference. Remove the wire ring bearing retainer and remove the outer race.

4. Paint alignment marks on the tri-pot bearing and shaft for assembly reference. Remove the tri-pot bearing snapring and, using a brass drift and hammer, remove the tri-pot bearing from the shaft.

5. Remove the small clamp and remove the inner boot from the halfshaft. If the boot is to be reused, wrap the shaft splines with tape before removing.

6. If the outer CV-joint boot is to be replaced, remove the clamps and slide the boot off the shaft from the inboard side.

To install:

7. If the outboard boot was removed, slide the boot onto the shaft from the inboard side. Wrap tape on the splines before installing to protect the boot.

8. Install the inboard boot and remove the tape from the shaft.

9. Install the tri-pot assembly on the halfshaft. Tap the assembly onto the shaft using a hammer and brass drift. Install the tri-pot assembly retaining ring.

10. Fill the CV-joint outer race with high temperature CV-joint grease. Install the outer race over the tri-pot joint and install the wire ring bearing retainer.

11. Position the CV-joint boot(s). Make sure the boot is fully seated in the grooves in the shaft and outer race.

For complete Engine Mechanical specifications, see Section 1 of this manual

12. Insert a small prybar with rounded edges between the boot and the outer bearing race to allow trapped air to escape from the boot. Install new boot clamps.

13. Wrap the clamps around the boots in a clockwise direction, pull tight with pliers and bend the locking tabs to secure in position.

14. Work the CV-joint through its full range of travel at various angles. The joint should flex, extend and compress smoothly.

15. Install the halfshaft into the vehicle.

STEERING AND SUSPENSION

Air Bag

PRECAUTIONS

1998–00

Several precautions must be observed when handling the inflator module to avoid accidental deployment and possible personal injury.

1. Never carry the inflator module by the wires or connector on the underside of the module.

2. When carrying a live inflator module, hold securely with both hands, and ensure that the bag and trim cover are pointed away.

3. Place the inflator module on a bench or other surface with the bag and trim cover facing up.

4. With the inflator module on the bench, never place anything on or close to the module which may be thrown in the event of an accidental deployment.

5. An air bag is an explosive device. Handle with extreme caution.

6. Always disconnect the battery and the air bag connector before removing the steering wheel or beginning work on the air bag system.

7. Air bag components must not be repaired or opened. Always use new parts, including the wiring harness.

8. Always place a removed air bag unit with the horn pad facing up. Put it in a safe place where it will not be disturbed.

9. The air bag unit must not be exposed to grease, fluids, or cleaning agents.

10. The air bag unit must not be exposed to temperatures above 194°F (90°C) at any time. Even the heat of a soldering iron can damage or ignite the charge.

11. Storage and transport of air bags is subject to rules governing explosive devices and should be done only in the original package.

12. Failure to follow proper safety precautions may result in personal injury through accidental firing of the air bag, or through failure of the air bag in an accident.

DISARMING

1998–00

1. Before servicing the vehicle, refer to the precautions in the beginning of this section.

2. If equipped, deactivate the audio anti-theft system.

3. Turn the ignition switch to LOCK.

4. Disconnect and isolate the negative battery cable and wait for more than 1 minute to allow the backup power supply to deplete its stored power.

Sway Bar

REMOVAL & INSTALLATION

1.5L, 1.6L and 1.8L Engines

1998–00

1. Before servicing the vehicle, refer to the precautions in the beginning of this section.

2. Remove or disconnect the following:
   - Negative battery cable
   - Cross member
   - Sway bar control link
   - Sway bar bracket
   - Sway bar bushing
   - Sway bar

To install:

3. Install or connect the following:
   - Sway bar and bushings
   - Sway bar brackets and torque the bolts to 44 ft. lbs. (60 Nm)
   - Sway bar control links and torque the nuts to 44 ft. lbs. (60 Nm)
   - Cross member
   - Negative battery cable

2.0L and 2.5L Engines

1998–00

1. Before servicing the vehicle, refer to the precautions in the beginning of this section.

2. Remove or disconnect the following:
   - Negative battery cable
   - Cross member
   - Sway bar control links
   - Sway bar brackets and bushings
   - Sway bar

To install:

3. Install or connect the following:
   - Sway bar
   - Sway bar bushings and brackets. Torque the bolts to 39 ft. lbs. (53 Nm)
   - Sway bar control links and torque the nuts to 39 ft. lbs. (53 Nm)
   - Cross member
   - Negative battery cable

2.3L Engine

1998–00

1. Before servicing the vehicle, refer to the precautions in the beginning of this section.

2. Remove or disconnect the following:
   - Negative battery cable
   - Splash shields
   - Transverse member
   - Power steering pressure and return hoses
   - Intermediate shaft mounting bolt
   - Engine mount member
   - Engine mount through bolt
   - Shock absorber and spring connecting bolt
   - Sway bar links
   - Sway bar brackets and bushings
   - Sway bar

To install:

3. Install or connect the following:
   - Sway bar
   - Sway bar bushings and brackets. Torque the bolts to 44 ft. lbs. (60 Nm)
   - Sway bar control links and torque the bolts to 44 ft. lbs. (60 Nm)
   - Shock absorber connecting bolt and torque to 101 ft. lbs. (137 Nm)
   - Engine mount through bolt and torque to 44 ft. lbs. (60 Nm)
   - Engine mount member and torque the bolts to 68 ft. lbs. (93 Nm)
   - Intermediate shaft bolt and torque to 19 ft. lbs. (26 Nm)
   - Power steering pressure and return hoses. Torque the fasteners to 43 ft. lbs. (58 Nm)
   - Transverse member and torque the bolts to 56 ft. lbs. (76 Nm)
   - Splash shields
   - Negative battery cable

Tie Rods

REMOVAL & INSTALLATION

1998–00

1. Before servicing the vehicle, refer to the precautions in the beginning of this section.

2. Remove or disconnect the following:
- Negative battery cable
- Tie rod nut

- Tie rod end from the steering knuckle
- Engine mount member
- Tie rod end

To install:

3. Install or connect the following:
- Tie rod end
- Engine mount member and torque the bolts to 68 ft. lbs. (93 Nm)
- Tie rod end to the steering knuckle
- Tie rod nut and torque the new castle nut to 32 ft. lbs. (44 Nm)
- Negative battery cable

Rack and Pinion Steering Gear

REMOVAL & INSTALLATION

1998–00

1. Before servicing the vehicle, refer to the precautions in the beginning of this section.

2. Remove or disconnect the following:
- Negative battery cable
- Front wheels

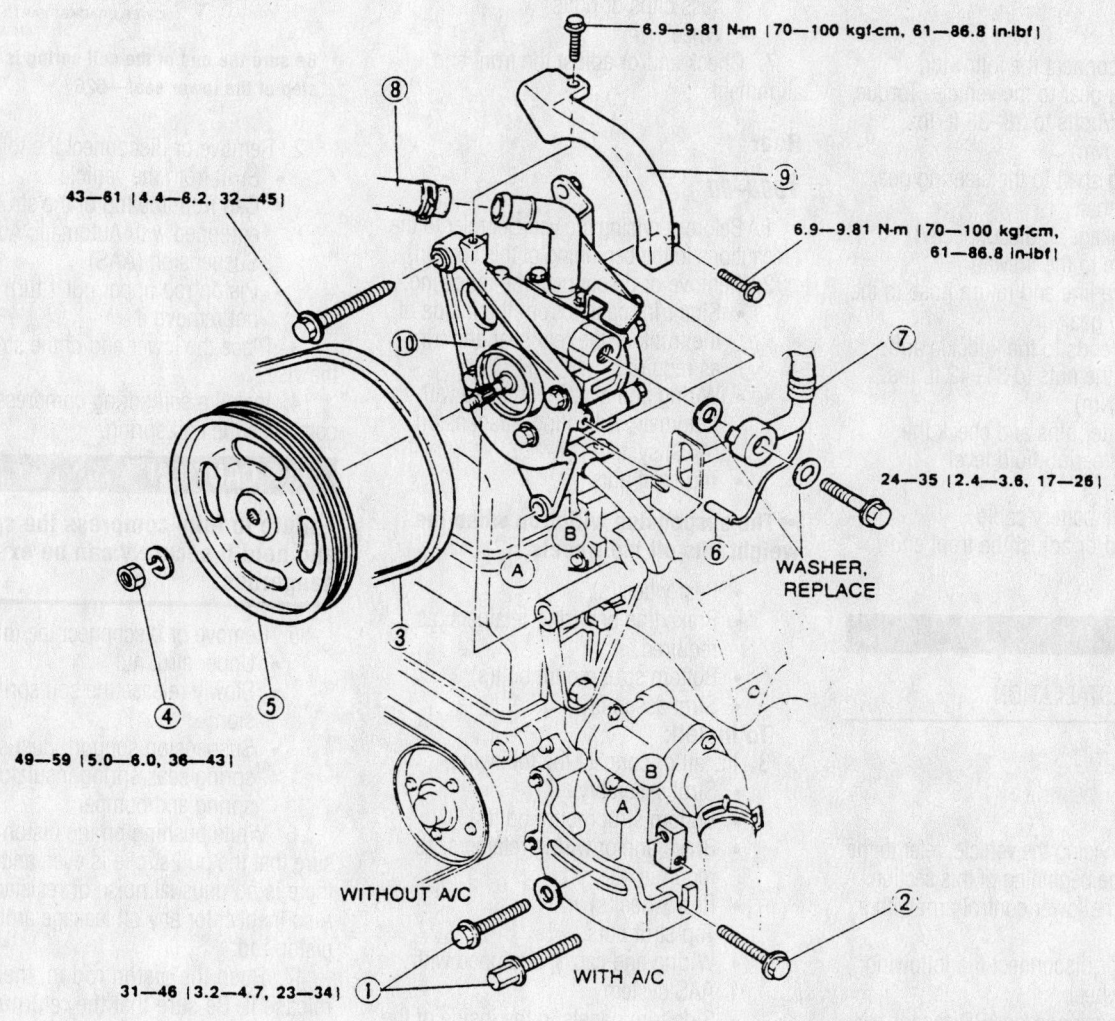

6.9—9.81 N-m (70—100 kgf-cm, 61—86.8 in-lbf)

43—61 (4.4—6.2, 32—45)

6.9—9.81 N-m (70—100 kgf-cm, 61—86.8 in-lbf)

24—35 (2.4—3.6, 17—26)

WASHER, REPLACE

49—59 (5.0—6.0, 36—43)

31—46 (3.2—4.7, 23—34)

WITHOUT A/C

WITH A/C

N-m (kgf-m, ft-lbf)

1. Lock bolt
2. Adjusting bolt
3. Drive belt
4. Nut
5. Pulley
6. SPS connector
7. Pressure pipe
8. Return hose
9. Pulley cover
10. Power steering oil pump and bracket

7923MG80

Exploded view of the power steering gear assembly

For Accessory Drive Belt illustrations, see Section 1 of this manual

- Cotter pins and nuts from both steering tie rod ends

3. Press the tie rod out from the knuckle arm.

4. Remove or disconnect the following:
- Pressure line and return pipe from the steering gear
- Set plate from the firewall
- Steering shaft to steering gear pinion shaft bolt
- Shaft from the steering gear
- Shifter linkage, if necessary
- Steering gear mounting nuts
- Steering gear from the right of the vehicle

To install:

5. Install or connect the following:
- Steering gear to the vehicle. Torque the nuts/bolts to 28–38 ft. lbs. (37–52 Nm)
- Steering shaft to the steering gear pinion shaft
- Shift linkage, if disconnected
- Set plate to the firewall
- Pressure line and return hose to the steering gear
- Tie rod ends to the knuckle arm. Torque the nuts to 31–42 ft. lbs. (42–57Nm)
- New cotter pins and check the power steering fluid level
- Wheels
- Negative battery cable

6. Check and/or adjust the front end alignment.

Strut

REMOVAL & INSTALLATION

Front

1998–00

1. Before servicing the vehicle, refer to the precautions in the beginning of this section.

2. Support the lower control arm with a jack.

3. Remove or disconnect the following:
- Front wheel
- Brake hose and/or ABS sensor harness to the strut bolts or clips
- Actuator from the top of the strut, if equipped with Automatic Adjusting Suspension (AAS)

4. Paint alignment marks on the upper strut mounting block and strut tower, and on the lower strut mount-to-steering knuckle so the strut can be reinstalled in the same position.

5. Remove or disconnect the following:
- Upper strut mounting nuts

- Strut-to-knuckle bolts
- Strut assembly

To install:

6. Install or connect the following:
- Strut into the strut tower, aligning the paint marks made during removal
- Upper mounting nuts and tighten to specifications
- Strut-to-knuckle bolts and tighten to specifications
- Actuator and engage the electrical connector, if equipped with AAS
- Brake hose and/or ABS sensor harness clips or bolts
- Wheel

7. Check and/or adjust the front end alignment.

Rear

1998–00

1. Before servicing the vehicle, refer to the precautions in the beginning of this section.

2. Remove or disconnect the following:
- Side trim panels from the inside of the trunk or the rear seat and trim, as required
- Wiring and cap, if equipped with Automatic Adjusting Suspension (AAS) system
- Top strut nuts

➡**The suspension will drop when the weight lifts off the wheels.**

- Rear wheel(s)
- Brake line or wiring retainers, as required
- Bottom strut mount bolt(s)
- Strut assembly

To install:

3. Install or connect the following:
- Strut assembly
- Bottom strut mount bolt(s)
- Brake line or wiring retainers, as required
- Rear wheel(s)
- Top strut nuts
- Wiring and cap, if equipped with AAS system
- Side trim panels to the inside of the trunk or the rear seat and trim, as required

Coil Spring

REMOVAL & INSTALLATION

1998–00

1. Before servicing the vehicle, refer to the precautions in the beginning of this section.

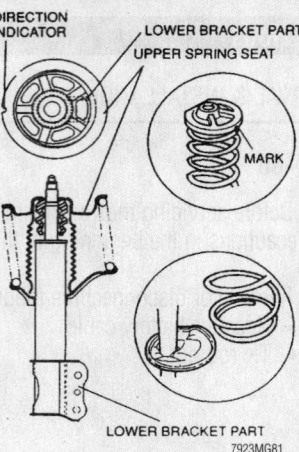

Be sure the end of the coil spring is in the step of the lower seat—626

2. Remove or disconnect the following:
- Strut from the vehicle
- Cap from the top of the strut, if not equipped with Automatic Adjusting Suspension (AAS)
- Piston rod upper nut 1 turn but do not remove it

3. Place the lower end of the strut in the vise.

4. Install a coil spring compressor and compress the coil spring.

✳✳ CAUTION

Failure to fully compress the spring and hold it securely can be extremely dangerous.

5. Remove or disconnect the following:
- Upper strut nut
- Slowly release the coil spring tension
- Suspension support, dust seal, spring seat, spring insulators, coil spring and bumper

6. While pushing on the piston rod, be sure that the pull stroke is even and that there is no unusual noise or resistance. Also inspect for any oil leakage around the piston rod.

7. Push the piston rod in, then release it. Be sure that the return rate is constant.

8. If the shock absorber does not operate as described, replace it.

To assemble:

9. Install or connect the following:
- Strut assembly into a vise
- Bump stopper and dust boot onto the piston rod
- Coil spring

10. Compress the coil spring with the spring compressor

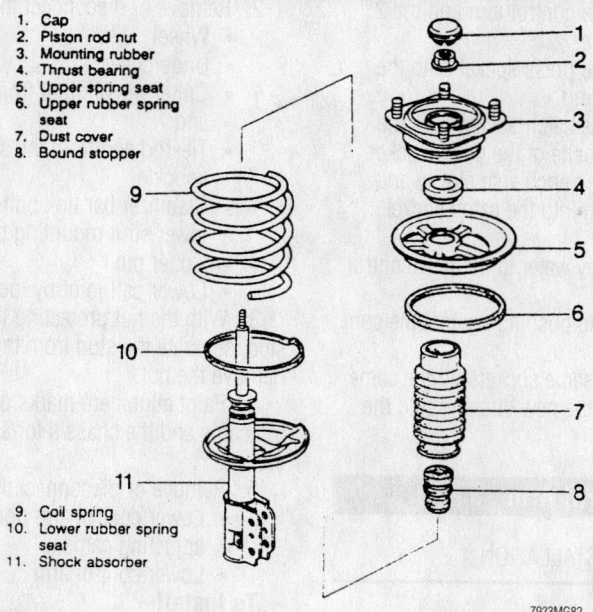

1. Cap
2. Piston rod nut
3. Mounting rubber
4. Thrust bearing
5. Upper spring seat
6. Upper rubber spring seat
7. Dust cover
8. Bound stopper

9. Coil spring
10. Lower rubber spring seat
11. Shock absorber

7923MG82

Exploded view of the front strut assembly—rear strut is similar

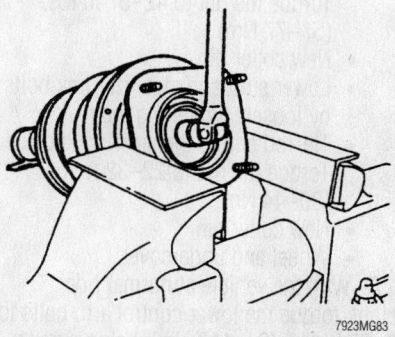

7923MG83

Secure the upper strut mount in a vise and loosen the piston rod nut one turn but do not remove it

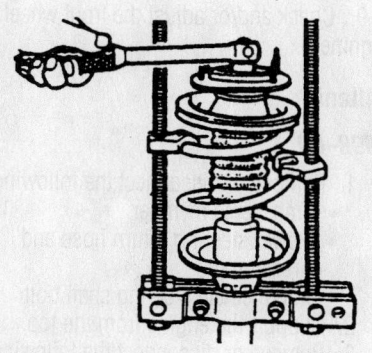

7923MG84

Use a coil spring compressor and relieve the spring tension from the upper mount, then remove the piston rod nut

11. Install or connect the following:
   - Rubber seat, the spring upper seat, the bearing and the mounting block
   - Piston rod upper nut
12. Be sure that the spring upper seat notched portion is facing inward and tighten the piston rod upper nut.
13. Remove the spring compressor from

1. **Stabilizer nut**
2. **Retainer, bushing and spacer**
3. **Stabilizer bolt**
4. **Bolt, washer**
5. **Bolt**
6. **Bolt, nut**
7. **Nut**
8. **Washer**
9. **Lower control arm bushing (rear)**
10. **Nut**
11. **Bolt**
12. **Lower arm ball joint**
13. **Ball joint dust boot**
14. **Lower arm bushing (front)**
15. **Lower arm**

the strut. Secure the upper mounting block in the vise and tighten the nut to specification.

14. Be sure that the spring is well seated in the upper seats.
15. Install the strut to the vehicle.

Lower Ball Joint

REMOVAL & INSTALLATION

Except Protégé, Miata and Millenia

1998–00

The lower ball joint is an integral part of the lower control and cannot be replaced separately. If the lower ball joint is defective, the entire lower control arm must be replaced. Refer to the lower control arm procedure.

Protege, Miata and Millenia

1998–00

1. Before servicing the vehicle, refer to the precautions in the beginning of this section.
2. Remove or disconnect the following:
   - Wheel
   - Ball joint stud pinch bolt and nut from the steering knuckle
   - Ball joint by prying it from the knuckle
   - Ball joint to lower control arm bolt and nut

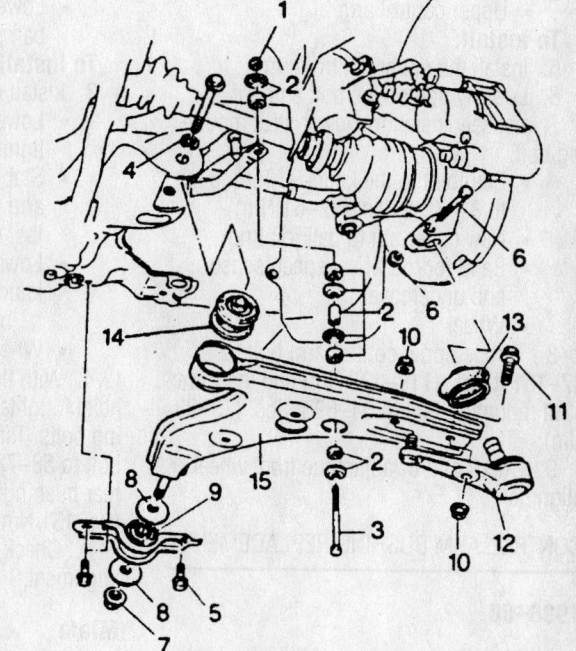

7923MG85

Exploded view of a common lower control arm with replaceable ball joint

For Tire, Wheel and Ball Joint specifications, see Section 1 of this manual

To install:

3. Install or connect the following:
- Ball joint to lower control arm. Torque the bolt to 86 ft. lbs. (117 Nm)
- Ball joint to the knuckle
- Ball joint to the steering knuckle. Torque the bolt to 43 ft. lbs. (59 Nm)
- Wheel

4. Check and/or adjust the front wheel alignment.

Upper Control Arm

REMOVAL & INSTALLATION

Miata

1998–00

1. Before servicing the vehicle, refer to the precautions in the beginning of this section.
2. Remove or disconnect the following:
- Wheel
- Undercover and band for the wheel speed sensor harness

3. Support the lower control arm with a jack.
4. Remove or disconnect the following:
- Cotter pin
- Upper ball joint nut by loosening it
- Ball joint by pressing it from the knuckle
- Upper ball joint nut
- Lower strut mounting bolt
- Upper control arm bolt and nut
- Upper control arm

To install:

5. Install the upper control arm.
6. Loosely tighten the bolt and nut.
7. Loosely install the lower strut mounting bolt.
- Ball joint to the knuckle. Torque nut to 31–45 ft. lbs. (42–61 Nm)
- New cotter pin to ball joint nut
- Band for the wheel speed sensor and undercover
- Wheel

8. Torque upper control arm bolt to 87–101 ft. lbs. (118–137 Nm) and the lower strut mounting bolt to 54–69 ft. lbs. (73–93 Nm).
9. Check and/or adjust the front wheel alignment.

CONTROL ARM BUSHING REPLACEMENT

1998–00

All Mazda's use a pressed in control arm bushing, and the pressing can be done using two appropriately sized sockets (a press socket and a catch socket) and a large bench vise.

1. Position the control arm and the 2 sockets into a vise.
2. Position the press socket onto the control arm bushing.
3. Position the catch socket onto the control arm, opposite of the press socket.
4. Tighten the bench vise slowly and press the bushing into the catch socket.

To install:

5. Apply soapy water to the new control arm bushing.
6. Position the bushing against the control arm.
7. Using the same sockets, in the same positions, press the new bushing into the control arm.

Lower Control Arm

REMOVAL & INSTALLATION

Except Miata and Millenia

1998–00

1. Before servicing the vehicle, refer to the precautions in the beginning of this section.
2. Remove or disconnect the following:
- Wheel
- Lower ball joint pinch bolt from the steering knuckle
- Stabilizer bar link from the lower control arm
- Lower control arm bolts and nuts
- Lower control arm with the lower ball joint

To install:

3. Install or connect the following:
- Lower control arm and loosely tighten the mounting nuts and bolts
- Stabilizer link to the lower control arm. Torque the nut to 27–39 ft. lbs. (37–53 Nm)
- Lower ball joint to the steering knuckle. Torque the bolt to 26–41 ft. lbs. (35–56 Nm)
- Wheel

4. With the vehicle at normal ride height, tighten the lower control arm mounting bolts. Torque the front bushing through-bolt to 58–78 ft. lbs. (79–106 Nm) and the rear bushing strap bolts to 69–96 ft. lbs. (94–131 Nm).
5. Check and/or adjust the front wheel alignment.

Miata

1998–00

1. Before servicing the vehicle, refer to the precautions in the beginning of this section.

2. Remove or disconnect the following:
- Wheel
- Undercover
- Cotter pin and nut from the tie-rod end
- Tie-rod end from the steering knuckle
- Stabilizer bar link bolt and the lower strut mounting bolt
- Cotter pin
- Lower ball joint by loosening it

3. With the nut protecting the ball joint stud, separate the stud from the knuckle. Remove the nut.
4. Paint alignment marks on the adjusting cams and the chassis for assembly reference.
5. Remove or disconnect the following:
- Lower control arm bolts, nuts and adjusting cams
- Lower control arm

To install:

6. Install or connect the following:
- Lower control arm by loosely tightening the bolts and nuts
- Lower ball joint to the knuckle. Torque the nut to 42–57 ft. lbs. (57–77 Nm)
- New cotter pin
- Lower strut and stabilizer link bolts by loosely tightening them
- Tie-rod end to the steering knuckle. Torque the nut to 22–32 ft. lbs. (30–44 Nm)
- New cotter pin
- Wheel and undercover

7. With the vehicle at normal ride height, torque the lower control arm bolts to 69–83 ft. lbs. (94–113 Nm), being sure to align the marks on the cam plates and chassis made during removal.
8. Torque the lower strut mounting bolt to 69 ft. lbs. (93 Nm) and the stabilizer link bolt to 40 ft. lbs. (54 Nm).
9. Check and/or adjust the front wheel alignment.

Millenia

1998–00

1. Remove or disconnect the following:
- Transverse member
- Power steering return hose and pressure pipe
- Intermediate steering shaft bolt
2. Support the engine from the top.
3. Remove or disconnect the following:
- Engine mount member
- Bolts for engine mount No. 1
- Lower strut mounting bolt
- Tie-rod end from the steering knuckle

- Upper lateral link ball joint
- Lower ball joint
- Stabilizer control link nut
- Gusset

4. Support the crossmember using a jack.

5. Remove or disconnect the following:
- Crossmember mounting nuts and lower the crossmember to gain clearance
- Lower arm assembly

To install:

6. Install or connect the following:
- Lower arm assembly to the vehicle
- Crossmember mounting bolts
- Gusset. Torque the bolts to 58–86 ft. lbs. (79–116 Nm)
- Stabilizer control link. Torque the nut to 32–44 ft. lbs. (44–60 Nm)
- Lower ball joint. Torque the bolts to 58–86 ft. lbs. (79–116 Nm) and the nut to 86–115 ft. lbs. (116–156 Nm)
- Upper lateral link nut and bolt. Torque to 58–86 ft. lbs. (79–116 Nm)
- Tie-rod end. Torque the nut to 41–59 ft. lbs. (55–80 Nm)
- Strut lower mounting bolt. Torque to 73–101 ft. lbs. (98–137 Nm)
- No. 1 engine mount. Torque the bolts to 32–44 ft. lbs. (44–60 Nm)
- Engine mount member. Torque the bolts to 50–68 ft. lbs. (67–93 Nm) and the nuts to 55–77.3 ft. lbs. (75–104 Nm)

7. Remove the engine support tool.
8. Install or connect the following:
- Intermediate steering shaft. Torque the bolt to 14–19 ft. lbs. (18–26 Nm)
- Power steering pressure pipe and return hose
- Transverse member

9. Check the power steering fluid and fill to proper level, bleed if necessary.
10. Check and/or adjust the front end alignment.

CONTROL ARM BUSHING REPLACEMENT

1998–00

All Mazda's use a pressed in control arm bushing, and the pressing can be done using 2 appropriately sized sockets (a press socket and a catch socket) and a large vise.

1. Position the control arm and the 2 sockets into a vise.

2. Position the press socket onto the control arm bushing.

3. Position the catch socket onto the control arm, opposite of the press socket.

4. Tighten the vise slowly and press the bushing into the catch socket.

To install:

5. Apply soapy water to the new control arm bushing.

6. Position the bushing against the control arm.

7. Using the same sockets, in the same positions, press the new bushing into the control arm.

Wheel Bearings

ADJUSTMENT

The front and rear wheel bearings are not adjustable. If the bearings become loose or make noise, they must be replaced.

REMOVAL & INSTALLATION

Except Miata

FRONT

1998–00

1. Before servicing the vehicle, refer to the precautions in the beginning of this section.

2. Remove or disconnect the following:
- Steering knuckle from the vehicle
- Inner oil seal from the knuckle

3. Using Mazda Hub Puller tools 49 G033 102, 49 G033 104 and 49 G033 105, pull the front wheel hub from the knuckle assembly.

4. Remove or disconnect the following:
- Bearing inner race from the front wheel hub
- Retaining ring from within the knuckle
- Front wheel bearing from the knuckle, using the hub puller tools
- Brake dust shield

5. Clean and inspect all parts but do not wash or clean the wheel bearing. The bearing must be replaced.

To install:

6. Using Mazda Press tools 49 G033 107 and 49 H026 103, install a new dust shield cover assembly to the knuckle.

7. Using the press tools, press a new wheel bearing into the knuckle assembly.

8. Install or connect the following:
- Wheel bearing retaining ring

- New oil seal using Installation tool 49 V001 795
- Front wheel hub using the Mazda press tools

REAR

1998–00

1. Before servicing the vehicle, refer to the precautions in the beginning of this section.

➡The wheel bearings are not serviceable. If the bearings are bad, a new hub/bearing assembly must be installed.

2. Remove or disconnect the following:
- Rear wheels
- Brake drum, if equipped
- Rear caliper and rotor assembly from the hub, if equipped
- Hub dust cover

3. Raise the staked portion of the hub retaining nut with a hammer and chisel.

4. Remove or disconnect the following:
- Hub retaining nut and discard it
- Hub and bearing assembly from the spindle

To install:

5. Install or connect the following:
- Bearing assembly on the spindle. Torque the new nut to 131–173 ft. lbs. (177–235 Nm)
- Stake the nut into the groove in the spindle
- Dust cover
- Assemble the brakes
- Rear wheel

Miata

FRONT

1998–00

1. Before servicing the vehicle, refer to the precautions in the beginning of this section.

➡The wheel bearings are not serviceable. If the bearings are bad, a new hub/bearing assembly must be installed.

2. Remove or disconnect the following:
- Front wheels
- Hub center dust cap

3. Raise the staked portion of the hub locknut with a hammer and chisel. Lock the hub by applying the brakes and loosen the nut.

For Wheel Alignment specifications, see Section 1 of this manual

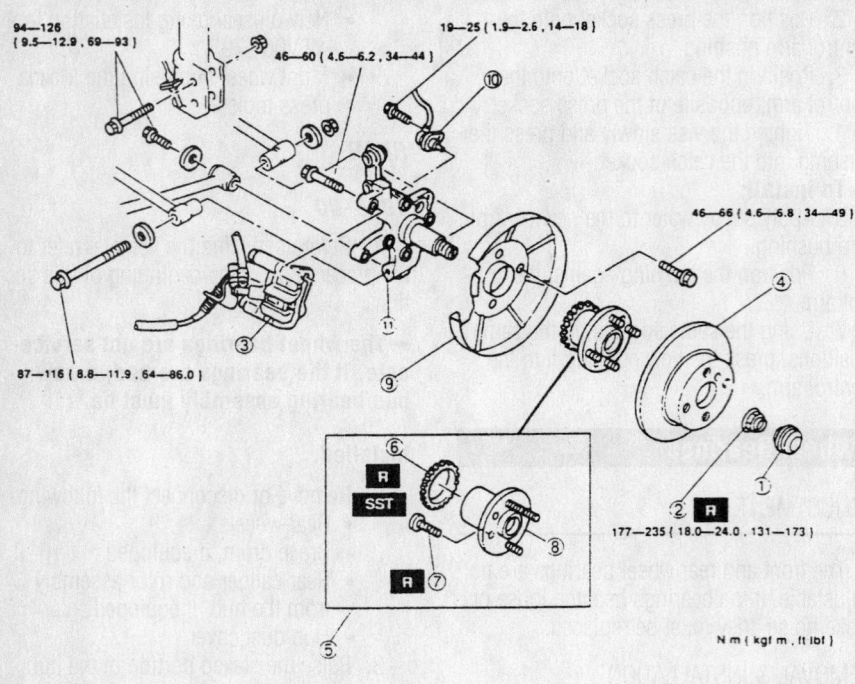

94—126
{ 9.5—12.9 , 69—93 }

46—60 { 4.6—6.2 , 34—44 }

19—25 { 1.9—2.6 , 14—18 }

46—66 { 4.6—6.8 , 34—49 }

87—116 { 8.8—11.9 , 64—86.0 }

177—235 { 18.0—24.0 , 131—173 }

N m (kgf m , ft lbf)

1. Hub cap
2. Locknut
3. Brake caliper assembly
4. Disc plate
5. Wheel hub assembly
   Inspect for damage
   Inspect bearing for damage and rough rotation
6. ABS sensor rotor
7. Hub bolt
8. Wheel hub
9. Dust cover
   Inspect for damage and cracks
10. ABS wheel-speed sensor
11. Hub spindle
    Inspect for damage and cracks

7923MG87

Exploded view of the rear wheel hub and bearing assembly (disc brake model shown, drum is similar)

4. Remove or disconnect the following:
   - Brake caliper
   - Brake rotor
   - Hub locknut and discard it
   - Hub from the spindle

To install:

5. Install or connect the following:
   - Hub to the spindle
   - Hub locknut. Torque the nut to 123–159 ft. lbs. (167–215 Nm)
   - Stake the new nut into the spindle groove using a dull chisel
   - Brake rotor
   - Brake caliper
   - Hub center dust cap
   - Front wheels

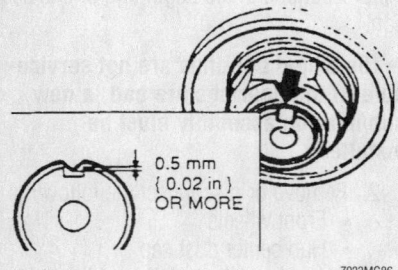

0.5 mm
{ 0.02 in }
OR MORE

7923MG86

When installing the new hub nut, be sure to stake it into the notch on the spindle

REAR

1998–00

1. Before servicing the vehicle, refer to the precautions in the beginning of this section.
2. Remove the rear wheels.
3. Raise the staked portion of the hub locknut with a hammer and chisel. Lock the hub by applying the brakes and loosen the nut.
4. Remove or disconnect the following:
   - Brake caliper and position it aside
   - Brake rotor
   - Hub locknut and discard it
   - Anti-lock Brake System (ABS) wheel speed sensor, if equipped
   - Speed sensor bracket from the rear knuckle
   - Upper and lower knuckle mounting through bolts
   - Knuckle assembly from vehicle
   - Rear bearing oil seal by prying it from the knuckle
   - Wheel hub by pressing it from the knuckle
   - Retaining snapring from within the knuckle
   - Bearing assembly from the knuckle, once the wheel hub is removed
5. The inner race of the bearing may remain on the hub. Use a chisel and move the bearing race away from the rear hub

flange. Once there is enough clearance, press the race from the hub.

To install:

6. Press the new bearing into the rear knuckle assembly.
7. Apply some grease to the wheel bearing inner race and press the rear hub into the bearing. Make sure to position a suitable support on the backside of the bearing.
8. Install or connect the following:
   - Retaining snapring to the knuckle
   - New rear oil seal by lubricating it with grease and pressing it into the rear knuckle
   - Knuckle assembly to the vehicle
   - Upper and lower knuckle mounting through bolts. Torque the upper bolt to 34–49 ft. lbs. (47–66 Nm) and the lower bolt to 47–54 ft. lbs. (63–74 Nm)
   - Speed sensor bracket to the rear knuckle
   - ABS wheel speed sensor, if equipped
   - New hub locknut. Torque the nut to 160–216 ft. lbs. (216–294 Nm)
   - Stake the new nut into the spindle groove using a dull chisel
   - Brake rotor
   - Brake caliper
   - Rear wheels

MITSUBISHI

1998–01
3000GT • Eclipse • Galant • Mirage

16

PRECAUTIONS

Before servicing any vehicle, please be sure to read all of the following precautions, which deal with personal safety, prevention of component damage, and important points to take into consideration when servicing a motor vehicle:

• Never open, service or drain the radiator or cooling system when the engine is hot; serious burns can occur from the steam and hot coolant.

• Observe all applicable safety precautions when working around fuel. Whenever servicing the fuel system, always work in a well-ventilated area. Do not allow fuel spray or vapors to come in contact with a spark, open flame, or excessive heat (a hot drop light, for example). Keep a dry chemical fire extinguisher near the work area. Always keep fuel in a container specifically designed for fuel storage; also, always properly seal fuel containers to avoid the possibility of fire or explosion. Refer to the additional fuel system precautions in this section.

• Fuel injection systems often remain pressurized, even after the engine has been turned **OFF**. The fuel system pressure must be relieved before disconnecting any fuel lines. Failure to do so may result in fire and/or personal injury.

• Brake fluid often contains polyglycol ethers and polyglycols. Avoid contact with the eyes and wash your hands thoroughly after handling brake fluid. If you do get brake fluid in your eyes, flush your eyes with clean, running water for 15 minutes. If

eye irritation persists, or if you have taken brake fluid internally, IMMEDIATELY seek medical assistance.

• The EPA warns that prolonged contact with used engine oil may cause a number of skin disorders, including cancer. You should make every effort to minimize your exposure to used engine oil. Protective gloves should be worn when changing oil. Wash your hands and any other exposed skin areas as soon as possible after exposure to used engine oil. Soap and water, or waterless hand cleaner should be used.

• All new vehicles are now equipped with an air bag system. The system must be disabled before performing service on or around system components, steering column, instrument panel components, wiring and sensors. Failure to follow safety and disabling procedures could result in accidental air bag deployment, possible personal injury, and unnecessary system repairs.

• Always wear safety goggles when working with, or around, the air bag system. When carrying a non-deployed air bag, be sure the bag and trim cover are pointed away from your body. When placing a non-deployed air bag on a work surface, always face the bag and trim cover upward, away from the surface. This will reduce the motion of the module if it is accidentally deployed. Refer to the additional air bag system precautions later in this section.

• Clean, high quality brake fluid from a sealed container is essential to the safe and

proper operation of the brake system. You should always buy the correct type of brake fluid for your vehicle. If the brake fluid becomes contaminated, completely flush the system with new fluid. Never reuse any brake fluid. Any brake fluid that is removed from the system should be discarded. Also, do not allow any brake fluid to come in contact with a painted surface; it will damage the paint.

• Never operate the engine without the proper amount and type of engine oil; doing so will result in severe engine damage.

• Timing belt maintenance is extremely important. Many models utilize an interference-type, non-freewheeling engine. If the timing belt breaks, the valves in the cylinder head may strike the pistons, causing potentially serious (also time-consuming and expensive) engine damage. Refer to the maintenance interval charts in the front of this manual for the recommended replacement interval for the timing belt, and to the timing belt section for belt replacement and inspection.

• Disconnecting the negative battery cable on some vehicles may interfere with the functions of the on-board computer system(s) and may require the computer to undergo a relearning process once the negative battery cable is reconnected.

• When servicing drum brakes, only disassemble and assemble one side at a time, leaving the remaining side intact for reference.

ENGINE REPAIR

➡ **Disconnecting the negative battery cable on some vehicles may interfere with the functions of the on board computer systems and may require the computer to undergo a relearning process, once the negative battery cable is reconnected.**

Distributor

REMOVAL

Before removing the distributor, position No. 1 cylinder at TDC on the compression stroke and align the timing marks.

1. Before servicing the vehicle, refer to the precautions in the beginning of this section.

2. Remove or disconnect the following:
• Negative battery cable

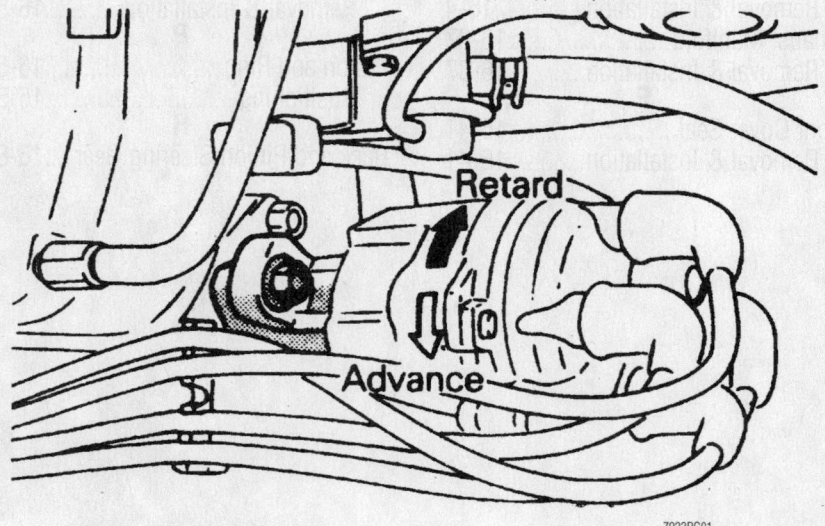

Adjusting the distributor—1.5L Mirage shown

7923PG01

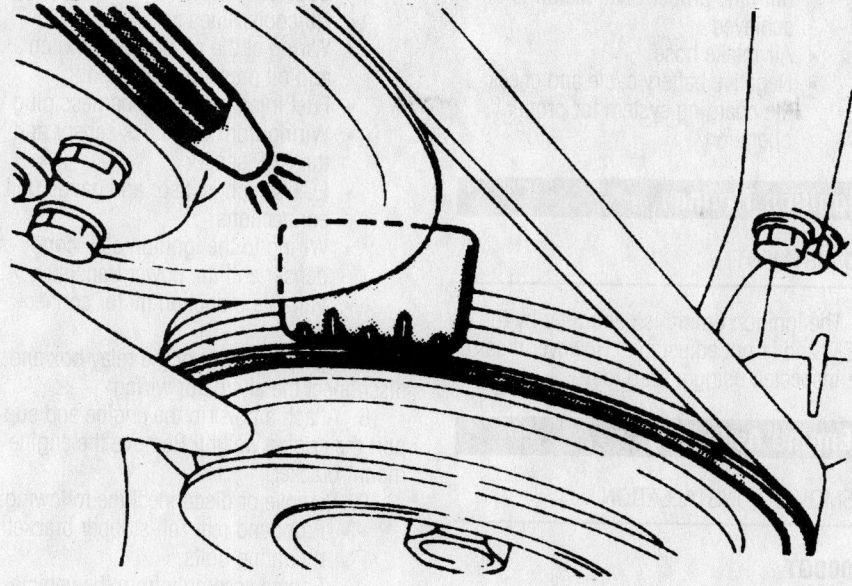

7923PG02

Checking the ignition timing—1.5L Mirage shown

- Ignition wire cover, if equipped
- Distributor harness connector
- Distributor cap with all ignition wires still connected
- Coil wire, if necessary

3. Matchmark the rotor to the distributor housing, and the distributor housing to the engine.

4. Remove or disconnect the following:
- Hold-down nut
- Distributor from the engine

INSTALLATION

Timing Not Disturbed

1. Install or connect the following:
- New distributor housing O-ring and lubricate with clean oil
- Distributor in the engine, match-marks aligned
- Hold-down nut
- Distributor harness connectors
- Distributor cap
- Coil wire, if removed
- Negative battery cable

2. Adjust the ignition timing and tighten the hold-down nut to 96 inch lbs. (11 Nm).

Timing Disturbed

1. Install a new distributor housing O-ring and lubricate with clean oil.

2. Position the engine so the No. 1 piston is at TDC of its compression stroke and the mark on the vibration damper is aligned with **0** on the timing indicator.

3. Align the distributor housing and gear mating marks. Install the distributor in the engine so the slot or groove of the distributor's installation flange aligns with the distributor installation stud in the engine block. Be sure the distributor is fully seated. Inspect alignment of the distributor rotor making sure the rotor is aligned with the position of the No. 1 ignition wire in the distributor cap.

4. Install or connect the following:
- Hold-down nut
- Distributor harness connectors
- Distributor cap
- Negative battery cable

5. Adjust the ignition timing and tighten the hold-down nut to 96 inch lbs. (11 Nm).

Alternator

REMOVAL & INSTALLATION

1.5L, 1.6L, 1.8L, 2.0L and 2.4L Engines

1. Remove or disconnect the following:
- Negative battery cable
- Left side cover panel under the vehicle
- Air intake hose, on turbocharged Galant models
- Drive belts
- Water pump pulleys

- Alternator upper bracket/brace

2. On the 1.6L engine remove the battery, windshield washer reservoir and battery tray.

3. On the 1.6L engine, remove the attaching bolts at the top of the radiator and lift up the radiator. Do not disconnect the radiator hoses.

4. Remove or disconnect the following:
- Alternator wiring connectors
- Alternator mounting bolts and remove the alternator

INSTALLATION

1. Position the alternator on the lower mounting fixture and install the lower mounting bolt and nut. Tighten nut just enough to allow for movement of the alternator.

2. On the 1.6L engine, lower the radiator and reinstall the upper attaching bolts.

3. On the 1.6L engine, install the battery, windshield washer reservoir and battery tray.

4. Install or connect the following:
- Alternator upper bracket/brace and connect the alternator electrical harness
- Water pump pulleys
- Drive belts and adjust to the proper tension

5. On turbocharged Galant models, install the air intake hose.

6. Install or connect the following:
- Left side cover panel under the vehicle as required
- Negative battery cable and check for proper operation

3.0L DOHC Engine

1. Remove or disconnect the following:
- Negative battery cable
- Headlamp washer reservoir tank
- Condenser fan and upper radiator insulator
- Alternator drive belt
- Alternator upper and lower mounting bolts
- Alternator support bracket mounting bolts
- Alternator support bracket from the vehicle
- Alternator wiring harness
- Alternator from the vehicle

To install:

2. Install or connect the following:
- Alternator to the vehicle and connect the wiring harness

- Alternator support bracket to the vehicle and tighten the bracket mounting bolts to specifications

3. Position the alternator on the mounting bracket. Install and tighten the mounting bolt and nut to 17 ft. lbs. (24 Nm).

4. Install or connect the following:
- Drive belt and adjust the tensioner until the proper belt tension is achieved
- Upper radiator insulator and condenser fan
- Headlamp washer reservoir tank
- Negative battery cable and check the charging system for proper operation

3.0L SOHC and 3.5L Engines

1. Remove or disconnect the following:
- Negative battery cable
- Air intake hose
- Alternator drive belt

2. On California models, remove the rear bank converter assembly.

3. Remove the engine roll stopper stay bracket assembly.

4. On the 3.0L SOHC engine, disconnect the EGR temperature sensor wire and remove the EGR pipe assembly.

5. On the 3.0L SOHC engine, remove the intake plenum stay bracket assembly.

6. Remove or disconnect the following:
- Alternator wiring harness connectors
- Alternator upper and lower mounting bolts

7. From beneath the vehicle, remove the alternator.

To install:

8. Position the alternator on the lower mounting fixture. Install and tighten the mounting bolt and nut to 14–18 ft. lbs. (20–25 Nm).

9. Install or connect the following:
- Alternator wiring harness

10. On the 3.0L SOHC engine, install the intake plenum stay bracket and tighten the mounting bolt to 13 ft. lbs. (18 Nm).

11. On the 3.0L SOHC engine, install the EGR pipe and tighten the fitting connections to 43 ft. lbs. (60 Nm).

12. On the 3.0L SOHC engine, connect the EGR temperature sensor wire.

13. Install or connect the following:
- Engine roll stopper stay and tighten the mounting bolt to 35 ft. lbs. (45 Nm) and the nut to 36–43 ft. lbs. (50–60 Nm).
- Rear converter assembly, if removed
- Drive belt and adjust the tensioner

until the proper belt tension is achieved
- Air intake hose
- Negative battery cable and check the charging system for proper operation

Ignition Timing

ADJUSTMENT

The ignition timing is controlled by the ECM and is not adjustable. However it can be inspected using a scan tool.

Engine Assembly

REMOVAL & INSTALLATION

3000GT

1. Before servicing the vehicle, refer to the precautions in the beginning of this section.

2. Relieve fuel system pressure.

3. Remove or disconnect the following:
- Negative battery cable
- Hood assembly
- Air cleaner assembly and all adjoining air intake duct work
- Cruise control linkage and actuator assemblies

4. Drain the engine coolant.

5. Remove or disconnect the following:
- Radiator assembly, coolant reservoir and intercooler
- HO2S sensor connection at the front exhaust pipe
- Front exhaust pipe assembly
- Transaxle assembly
- Accelerator cable, breather hose and heater hose connections from the engine
- Vacuum hoses, note locations
- Fuel feed and return hoses
- Solenoid valve assembly and disconnect the ground cable
- Purge hose and EGR temperature sensor, if equipped
- Air conditioning and power steering drive belts
- Air conditioning compressor and the power steering pump assemblies
- Harness connections for the ISC motor position sensor and P sensor
- EGR temperature sensor (California)

6. For the turbocharger, disconnect the following:

- Booster vacuum hose
- Oil cooler lines
- Wiring at the oil pressure switch and oil pressure gauge unit
- Fuel injection wiring harness plug
- Wiring from the knock sensor and the CKP sensor
- ECT switch, sensor and gauge unit connections
- Wiring to the ignition coil, condenser and the power transistor
- Variable induction motor connection

7. Open the cover of the relay box and disconnect the alternator wiring.

8. Attach a hoist to the engine and support the engine weight. Remove the engine mount bracket.

9. Remove or disconnect the following:
- Front and rear roll stopper bracket mounting bolts
- Engine assembly from the vehicle

To install:

10. Install the engine and secure into position. Secure the engine mount bracket to block and tighten bolts to 72–87 ft. lbs. (100–120 Nm). Install through-bolt and tighten bolt to 51 ft. lbs. (70 Nm).

11. Install the front and rear roll stopper through-bolt and tighten to 36–43 ft. lbs. (50–60 Nm).

12. Open the cover of the relay box and connect alternator wiring.

13. Install or connect the following:
- Variable induction motor connection
- Fuel feed and return hoses, using a new sealing ring. Tighten pressure hose connection to 48 inch lbs. (5 Nm).
- Wiring to the ignition coil, condenser and power transistor
- ECT switch, sensor and gauge unit connections
- Wiring from the knock sensor and the CKP sensor
- Fuel injection wiring harness plug
- Wiring at the oil pressure switch and oil pressure gauge unit

14. Connect the following for the turbocharger:
- Booster vacuum hose
- Oil cooler lines
- EGR temperature sensor (California)
- Harness connections for the ISC motor position sensor and TP sensor

15. Install or connect the following:
- Air conditioning compressor and the power steering pump assemblies

- Engine drive belts
- Purge hose and the EGR temperature sensor, if equipped
- Solenoid valve assembly and connect ground cable to engine block
- Vacuum hoses to the engine
- Accelerator cable, breather hose and heater hose connections to the engine
- Transaxle assembly
- Front exhaust pipe assembly, using new gaskets. Tighten manifold mounting bolts to 36 ft. lbs. (50 Nm).
- HO2S sensor connection at the front exhaust pipe

16. Replace the radiator assembly, coolant reservoir and intercooler. Refill the cooling system.

17. Install or connect the following:
- Cruise control linkage and the actuator assemblies
- Hood assembly, air cleaner assembly and all adjoining air intake duct work
- Negative battery cable

18. Start the engine and check for leaks.

Diamante

1. Before servicing the vehicle, refer to the precautions in the beginning of this section.

2. Remove the hood assembly.

3. Relieve fuel system pressure.

4. Remove or disconnect the following:
- Negative, then the positive battery cable
- Battery
- Air cleaner assembly and all adjoining air intake duct work

5. Drain the engine coolant and remove the radiator assembly and coolant reservoir (and bracket).

6. Remove or disconnect the following:
- Engine undercover, if equipped
- Front exhaust pipe
- Transaxle assembly
- Accelerator cable from the throttle body
- Vacuum hoses from the intake manifold, label for installation
- High pressure fuel line and the fuel return line
- Vacuum hoses from the solenoid valves
- Vacuum hoses from the purge canister
- Heater hose connections from the engine

- Harness for the EGR temperature sensor connection, if equipped
- Engine drive belts
- Power steering pump oil pressure switch connection from the pump
- Power steering pump and secure away from the work area
- Air conditioning compressor. Wire the compressor aside. Do not discharge or disconnect the air conditioning lines.
- Wiring to the alternator
- Harness plugs for the BARO sensor, ISC motor, TP sensor, fuel injectors and knock sensor
- Harness plugs for the ECT switch, sensor and gauge
- Harness plugs for the ignition coil, condenser and ignition power transistor
- Harness plugs for the variable induction control motor and the MAP sensor
- Harness plugs for the CKP and CMP sensors
- Radiator overflow tank and remove the mounting bracket
- Ground cable connections

7. Attach a hoist to the engine and take up the engine weight. Remove the engine mount bracket. Remove any torque control brackets (roll stoppers).

8. Lift the engine slowly and remove from the engine compartment.

To install:

9. Install or connect the following:
- Engine and secure all control brackets
- Transaxle assembly
- Engine ground cable connections
- Harness plugs for the CKP and CMP sensors
- Harness plugs for the variable induction control motor and the MAP sensor
- Harness plugs for the ignition coil, condenser and ignition power transistor
- Harness plugs for the ECT switch, sensor and gauge
- Harness plugs for the BARO sensor, ISC motor, TP sensor, fuel injectors and knock sensor
- Wiring to the alternator
- Air conditioning compressor assembly
- Power steering pump assembly
- Power steering pump oil pressure switch harness plug to the pump
- Engine drive belts, adjust
- Harness for the EGR temperature sensor
- Heater hose connections to the engine, using new hose clamps
- Vacuum hoses to the purge canister
- Vacuum hoses to the solenoid valves
- High pressure fuel line and the fuel

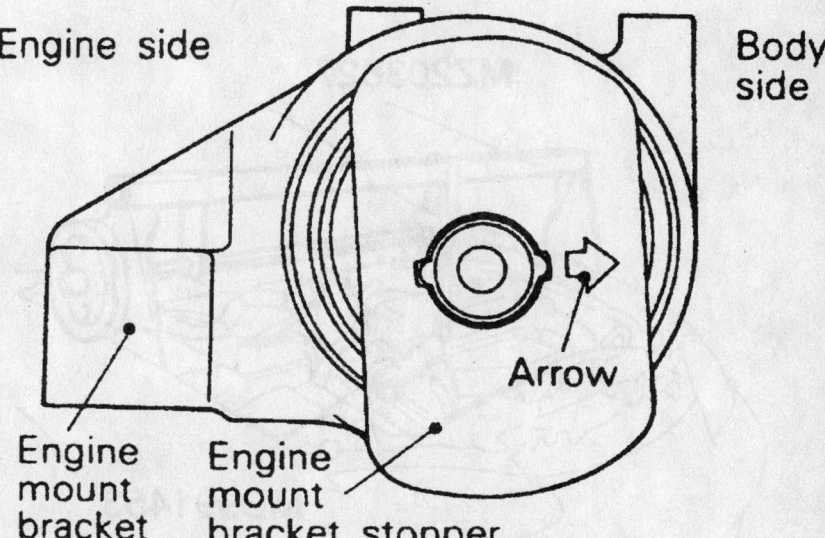

Alignment of the engine mount stopper bracket—Diamante shown

7923PG06

return line, using new clamps or O-rings
- Vacuum hoses
- Accelerator cable to the throttle body
- Air cleaner assembly and all adjoining air intake duct work
- Radiator and coolant reservoir assembly
- Transaxle assembly
- Exhaust system to the engine, using new gaskets
- Battery to the vehicle
- Positive, then the negative battery cables
- Engine undercover, if equipped

10. Fill the engine with the proper amount of engine oil and coolant.

11. Install the hood.

12. Start the engine and check for leaks.

Eclipse

2.0L (4G63) ENGINE

1. Before servicing the vehicle, refer to the precautions in the beginning of this section.

2. Relieve the fuel system pressure.

3. Remove or disconnect the following:
- Negative battery cable
- Hood
- Intake air duct

4. Drain the engine coolant.

5. Remove or disconnect the following:
- Radiator
- Engine undercover

6. Attach an engine lifting fixture to the engine and remove the transaxle assembly.
- Power steering pressure switch, oil pressure switch, oil pressure gauge sender and the alternator wiring connectors
- Alternator
- Power steering pump from the bracket and position the pump out of the way
- Air conditioning compressor from the bracket and position it out of the way. Do not disconnect the hoses.
- Accelerator cable from the throttle body and mounting bracket

7. Disconnect the following connectors:
- IAC motor
- Knock sensor
- HO$_2$S sensor
- ECT gauge sender
- ECT sensor
- Ignition module (power transistor)
- TP sensor
- Condenser
- MAP sensor
- Injectors
- Ignition coil
- CMP sensor
- CKP sensor
- air conditioning compressor connector
- Engine control wiring harness

8. Remove or disconnect the following:
- Brake booster vacuum hose
- Fuel lines from the fuel supply rail

- The A and B water hose connections
- Vacuum hoses, label
- Front exhaust pipe from the turbocharger

9. Support the engine under the oil pan with a floor jack and a piece of wood.

10. Remove the engine support fixture and replace it with a hoist. Lift up the engine to take the weight off of the engine mount bracket.

11. Remove the engine mount bracket.

12. Lift the engine up slowly out of the engine compartment.

To install:

13. Slowly lower the engine assembly into the vehicle.

14. Position the floor jack under the oil pan with a piece of wood in between. Use the floor jack to adjust the height of the engine while installing the engine mount bracket.

15. Remove the chain hoist and install the engine support fixture.

16. Install or connect the following:
- Front exhaust pipe to the turbocharger
- Vacuum hoses
- The **A** and **B** water hoses
- New O-ring on the high pressure fuel line. Apply a small amount of clean engine oil to the O-ring and connect the fuel lines to the fuel supply rail.
- Brake booster vacuum hose

17. Connect the following connectors:
- IAC motor
- Knock sensor
- HO$_2$S sensor
- ECT gauge sender
- ECT sensor
- Ignition module (power transistor)
- TP sensor
- Condenser
- MAP sensor
- Injectors
- Ignition coil
- CMP sensor
- CKP sensor
- air conditioning compressor connector
- Engine control wiring harness

18. Install or connect the following:
- Air conditioning compressor and the power steering pump in their brackets
- Alternator
- Oil pressure gauge sender, oil pressure switch and the power steering pressure switch connectors
- Transaxle assembly and remove the engine support fixture

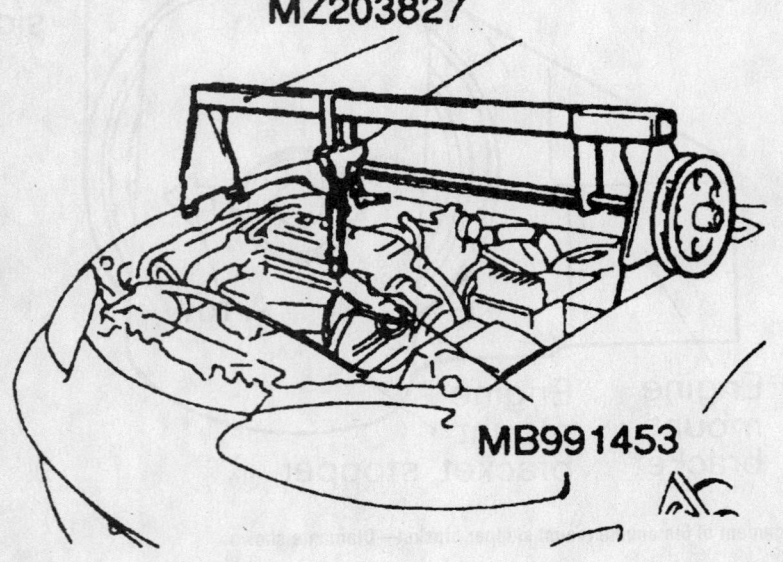

MZ203827

MB991453

7923PG05

Common method of supporting the engine using a fixture specifically designed for that purpose

- Engine undercover
- Radiator and connect the hoses
- Intake air duct
19. Refill the engine with coolant.
20. Install the hood.
21. Connect the negative battery cable.

2.0L (420A) AND 2.4L (4G64) ENGINES

1. Before servicing the vehicle, refer to the precautions in the beginning of this section.
2. Relieve the fuel system pressure.
3. Remove or disconnect the following:
   - Negative battery cable
   - Hood
   - Intake air duct
4. Drain the engine coolant.
5. Remove or disconnect the following:
   - Hoses and remove the radiator
   - Engine undercover
6. Attach an engine lifting fixture to the engine and remove the transaxle assembly.
7. Disconnect the following connectors:
   - air conditioning compressor
   - Power steering pressure switch
   - HO2S sensor
   - ECT gauge sender
   - ECT sensor
   - MAP sensor
   - IAT sensor
8. Remove or disconnect the following:
   - Power steering pump from the bracket and position the pump out of the way
   - Air conditioning compressor from the bracket and position it out of the way. Do not disconnect the hoses.
   - Accelerator cable from the throttle body and mounting bracket
9. Disconnect the following connectors:
   - IAC motor
   - Knock sensor
   - Ignition module (power transistor)
   - EGR solenoid
   - Oil pressure switch
   - TP sensor
   - Condenser
   - Injectors
   - Ignition coil
   - CMP sensor
   - CKP sensor
   - Engine control wiring harness
10. Remove or disconnect the following:
   - Heater hoses from the engine
   - Fuel lines from the fuel supply rail
   - Purge air hose and the brake booster vacuum hose

- Front exhaust pipe from the manifold
11. Place a floor jack against the oil pan with a piece of wood in between to protect the oil pan.
12. Raise the engine with the jack and remove the engine support fixture.
13. Install a chain hoist to the top of the engine.
14. Remove the engine mount bracket.
15. Lift the engine up slowly out of the engine compartment.

To install:

16. Slowly lower the engine assembly into the vehicle
17. Position the floor jack under the oil pan with a piece of wood in between. Use the floor jack to adjust the height of the engine while installing the engine mount bracket.
18. Remove the chain hoist and install the engine support fixture.
19. Install or connect the following:
   - Front exhaust pipe to the manifold
   - Brake booster vacuum hose
   - New O-ring on the high pressure fuel line. Apply a small amount of clean engine oil to the O-ring and connect the fuel lines to the fuel supply rail.
20. Connect the following connectors:
   - IAC motor
   - Knock sensor
   - Ignition module (power transistor)
   - EGR solenoid
   - Oil pressure switch
   - TP sensor
   - Condenser
   - Injectors
   - Ignition coil
   - CMP sensor
   - CKP sensor
   - Engine control wiring harness
21. Install or connect the following:
   - Accelerator cable, adjust
   - Air conditioning compressor and the power steering pump in their brackets
   - IAT sensor, MAP sensor, ECT sensor and gauge sender, HO2S sensor, power steering pressure switch and the air conditioning compressor harness connectors
   - Radiator and hoses
   - Transaxle and remove the engine support fixture
   - Engine undercovers
   - Intake air duct
   - Negative battery cable

- Hood
22. Refill the engine with the proper amount of coolant.

3.0L ENGINE

1. Before servicing the vehicle, refer to the precautions in the beginning of this section.
2. Disconnect the negative battery cable.
3. Drain the engine coolant.
4. Drain the engine oil and the transmission oil.
5. Relieve the fuel system pressure.
6. Remove or disconnect the following:
   - All wires, cables and hoses connected to the engine
   - Hood
   - Air intake and breather hoses
   - Radiator hoses and remove the radiator
   - Front exhaust pipe
   - Power steering pump and position it aside
   - Air conditioning compressor drive belt
   - Compressor from its mount and hang it out of the way. Do not disconnect the hoses and do not allow the compressor to hang by the hoses.
   - Engine hoist equipment and make certain the attaching points on the engine are secure
7. Raise the hoist enough to support the engine.
8. Remove or disconnect the following:
   - Front and rear engine roll stoppers
   - Left engine mount and support bracket
9. Slowly lift the engine and remove it from the vehicle.

Galant

1. Before servicing the vehicle, refer to the precautions in the beginning of this section.
2. Disconnect the negative battery cable.
3. Drain the engine coolant.
4. Drain the engine oil and the transmission oil.
5. Relieve the fuel system pressure.
6. Remove or disconnect the following:
   - Hood
   - Transaxle assembly
   - Radiator hoses and remove the radiator
   - Accelerator cable and remove the bracket

Timing belt service is covered in Section 3 of this manual

- Air intake and breather hoses
- Heater hoses
- Brake booster vacuum hose at the engine
- Vacuum hoses at the throttle body, label
- Fuel feed and return hoses

7. Disconnect the following:
- Power steering pressure switch
- Alternator
- Oil pressure switch
- air conditioning compressor
- Each injector
- Power transistor
- Ignition coil
- TP sensor
- IAC motor
- ECT switch
- ECT sensor
- EGR temperature sensor
- Engine control wiring harness
- HO$_2$S sensor
- CKP sensor
- CMP sensor
- Refrigerant temperature switch
- Condenser connection

8. Remove or disconnect the following:
- Power steering pump and position it aside
- Air conditioning compressor drive belt
- Compressor from its mount and hang it out of the way. Do not disconnect the hoses and do not allow the compressor to hang by the hoses.
- Front exhaust pipe
- Engine hoist equipment and make certain the attaching points on the engine are secure

9. Raise the hoist enough to support the engine.

10. Remove or disconnect the following:
- Front and rear engine roll stoppers
- Left engine mount and support bracket

11. Slowly lift the engine and remove it from the vehicle.

To install:

12. Lower the engine into the vehicle.

13. Install the front and rear roll stoppers and the left engine mount. Do not torque the through-bolts at this time.

14. Remove the lifting apparatus from the engine.

15. Connect the exhaust system to the manifold, using a new gasket and new locking nuts. Tighten the nuts and the small bolt to 33 ft. lbs. (44 Nm)

16. Tighten the engine mount nuts and bolts. Correct torque values are:

- Upper mount to engine nuts: 42 ft. lbs. (57 Nm)
- Upper mount to engine bolt: 108 inch lbs. (12 Nm)
- Upper mount through-bolt: 72–87 ft. lbs. (98–118 Nm)
- Rear roll stopper through-bolt: 32 ft. lbs. (44 Nm)
- Front roll stopper through-bolt: 41 ft. lbs. (57 Nm)

17. Install or connect the following:
- Air conditioning compressor, tightening the mounting bolts to 18 ft. lbs. (25 Nm)
- Power steering pump, tightening the front bolts to 21 ft. lbs. (28 Nm) and the rear bolt to 16 ft. lbs. (22 Nm).
- Accessory drive belts

18. Connect the following:
- Power steering pressure switch
- Alternator
- Oil pressure switch
- air conditioning compressor
- Each injector
- Power transistor
- Ignition coil
- TP sensor
- IAC motor
- ECT switch
- ECT sensor
- EGR temperature sensor
- Engine control wiring harness
- HO$_2$S sensor
- CKP sensor
- CMP sensor
- Refrigerant temperature switch
- Condenser connection

19. Install or connect the following:
- Fuel return hose and secure with the retaining clamp
- New O-ring, connect the high pressure fuel line and tighten the bolts to 48 inch lbs. (6 Nm)
- Vacuum lines running to the throttle body
- Heater hoses
- Accelerator cable bracket, tightening the bolts to 48 inch lbs. (6 Nm), and connect the accelerator cable
- Radiator and connect the hoses
- Transaxle

20. Fill the coolant system.

21. Connect the negative battery cable.

22. Start the engine and check for leaks.

23. Install the hood.

Mirage

1. Before servicing the vehicle, refer to the precautions in the beginning of this section.

2. Relieve fuel system pressure.

3. Remove or disconnect the following:
- Negative battery cable
- Undercover, if equipped
- Hood assembly
- Air cleaner assembly and all adjoining air intake duct work

4. Drain the engine coolant.

5. Remove or disconnect the following:
- Radiator assembly and coolant reservoir
- Transaxle assembly
- Ground cable, accelerator cable, breather hose and heater hose connections from the engine

6. Note locations and remove vacuum hoses from engine.

7. Remove or disconnect the following:
- Fuel feed and return hoses
- CKP and CMP sensor wiring
- HO$_2$S sensor, ECT gauge and ECT sensor connections
- Oil pressure switch
- Thermo switch, with automatic transmissions
- Harness connections for the ISC motor and TP sensor
- IAT sensor
- EGR temperature sensor (California)
- Injector harness plugs
- Power transistor and the ignition coil connections
- Alternator and power steering switch wiring
- Air conditioning compressor and hang it out of the way–Do not allow the compressor to hang by the hoses.
- Power steering pump and hang it out of the way–Do not allow the pump to hang by the hoses.
- Starter and alternator harness clamp, for 1.8L engines
- Exhaust manifold to head pipe nuts

8. Attach a hoist to the engine and support the engine weight. Remove the engine mount bracket. Remove any torque control brackets (roll stoppers):

9. Remove the engine assembly from the vehicle.

To install:

10. Install the engine and secure in position. The front lower mount through-bolt nut should not be tightened until the full weight of the engine is on the mount. Tighten through-bolt to 72 ft. lbs. (100 Nm) and bracket mounting bolts to 42 ft. lbs. (58 Nm). Tighten bracket mounting nut to 38 ft. lbs. (53 Nm).

11. Using a new gasket, position exhaust pipe onto the manifold and tighten the flange nuts to 36 ft. lbs. (50 Nm).

12. Install or connect the following:
- Power steering pump, alternator and air conditioning compressor
- Accessory drive belts
- Alternator and power steering wiring
- Alternator and starter harness clamp for 1.8L engines
- Ignition coil and power transistor connections
- Fuel injector harness connections
- EGR temperature sensor plug—California models
- IAT sensor
- IAC and TPS connectors
- Thermo switch, automatic transmission
- Oil pressure switch wiring
- HO2S sensor, ECT gauge and ECT sensor
- CKP and CMP sensors
- Fuel feed hose and tighten bolts to 44 inch lbs. (5 Nm), using new O-rings
- Fuel return hose, using a new hose clamp
- Vacuum hoses and the brake booster vacuum supply
- Breather hose, heater hoses, accelerator cable and ground cables. Inspect accelerator cable for proper adjustment.
- Transaxle assembly
- Radiator assembly and refill the cooling system
- Air cleaner and hood assembly
- Negative battery cable

Water Pump

REMOVAL & INSTALLATION

3000GT and Diamante

1. Before servicing the vehicle, refer to the precautions in the beginning of this section.
2. Drain the cooling system.
3. Remove or disconnect the following:
- Negative battery cable
- Engine undercover
- Clamp bolt from the power steering hose
4. Support the engine with the appropriate equipment and remove the engine mount bracket.
5. Remove or disconnect the following:

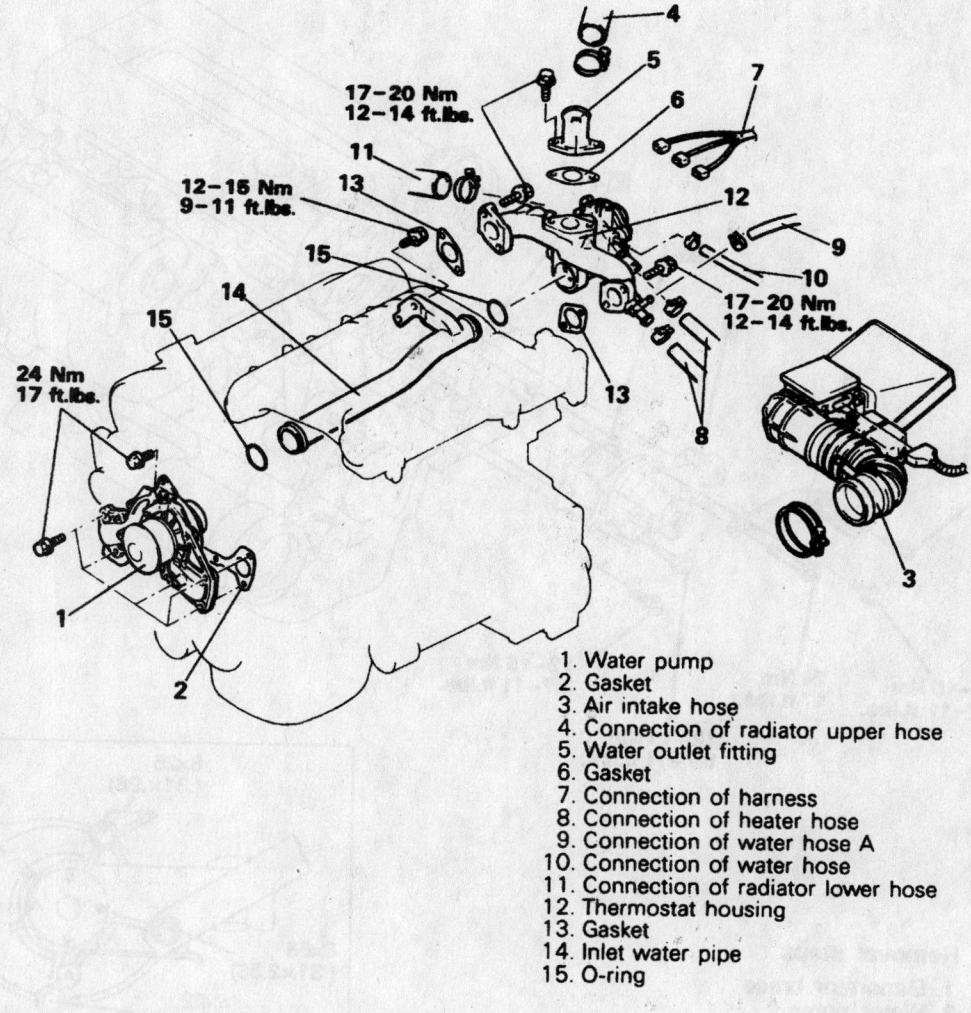

17–20 Nm
12–14 ft.lbs.

12–15 Nm
9–11 ft.lbs.

24 Nm
17 ft.lbs.

17–20 Nm
12–14 ft.lbs.

1. Water pump
2. Gasket
3. Air intake hose
4. Connection of radiator upper hose
5. Water outlet fitting
6. Gasket
7. Connection of harness
8. Connection of heater hose
9. Connection of water hose A
10. Connection of water hose
11. Connection of radiator lower hose
12. Thermostat housing
13. Gasket
14. Inlet water pipe
15. O-ring

7923PG12

Water pump and related components—DOHC Diamante shown, 3000GT similar

Heater Core replacement is covered in Section 2 of this manual

- Timing belt
- Coolant hoses from the pump, if equipped
- Alternator brace

➡ **The water pump bolts are different in size. Note their locations for installation.**

6. Remove the water pump, gasket and O-ring where the water inlet pipe joins the pump.

To install:

7. Thoroughly clean both gasket surfaces of the water pump and block.

8. Install or connect the following:
- New O-ring into the groove on the front end of the water inlet pipe. Do not apply oils or grease to the O-ring. Wet with water only.

- Water pump assembly to the engine block, with new gasket. Torque the mounting bolts to 17 ft. lbs. (24 Nm)
- Hoses to the pump
- Timing belt
- Engine drive belts

9. Fill the system with coolant.

10. Connect the negative battery cable, run the vehicle until the thermostat opens and fill the radiator completely.

11. Once the vehicle has cooled, recheck the coolant level.

Eclipse

1. Before servicing the vehicle, refer to the precautions in the beginning of this section.

2. Disconnect the negative battery cable.
3. Drain the engine coolant.
4. Remove or disconnect the following:
- Timing belt
- Alternator brace from the water pump
- Timing belt rear cover
- Water pump mounting bolts
- Water pump, gasket and O-ring

To install:

5. Install or connect the following:
- New O-ring on the water inlet pipe. Coat the O-ring with water or coolant. Do not allow oil or other grease to contact the O-ring.
- Water pump to the engine block, with new gasket. Torque the mounting bolts to 10 ft. lbs. (13 Nm)
- Alternator brace on the water pump.

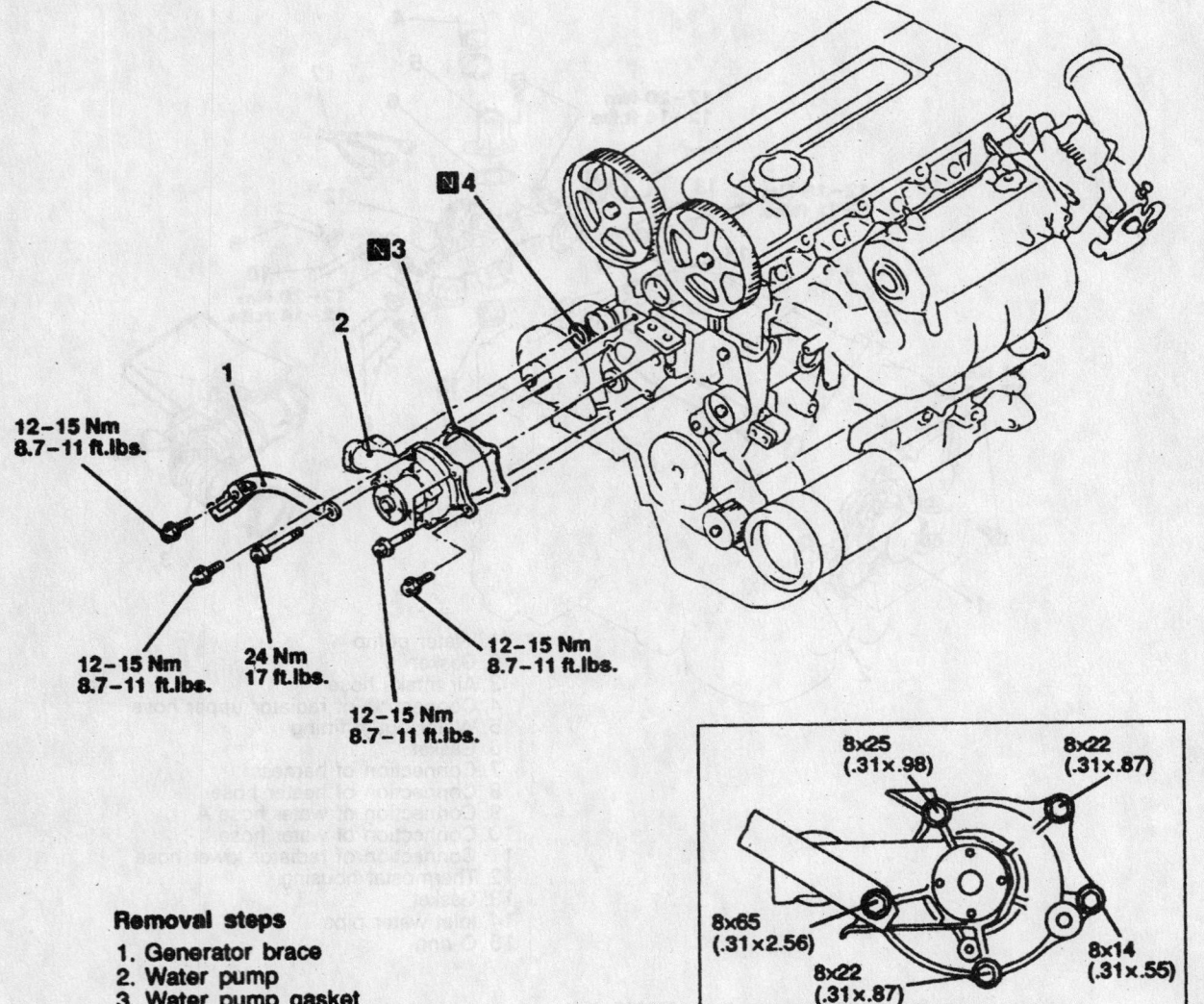

12–15 Nm
8.7–11 ft.lbs.

12–15 Nm
8.7–11 ft.lbs.

24 Nm
17 ft.lbs.

12–15 Nm
8.7–11 ft.lbs.

12–15 Nm
8.7–11 ft.lbs.

Removal steps

1. Generator brace
2. Water pump
3. Water pump gasket
4. O-ring

8x25 (.31x.98)
8x22 (.31x.87)
8x65 (.31x2.56)
8x22 (.31x.87)
8x14 (.31x.55)

Bolt diameter x length: mm (in.)

7923PG09

Water pump mounting—2.0L (4G63) engine

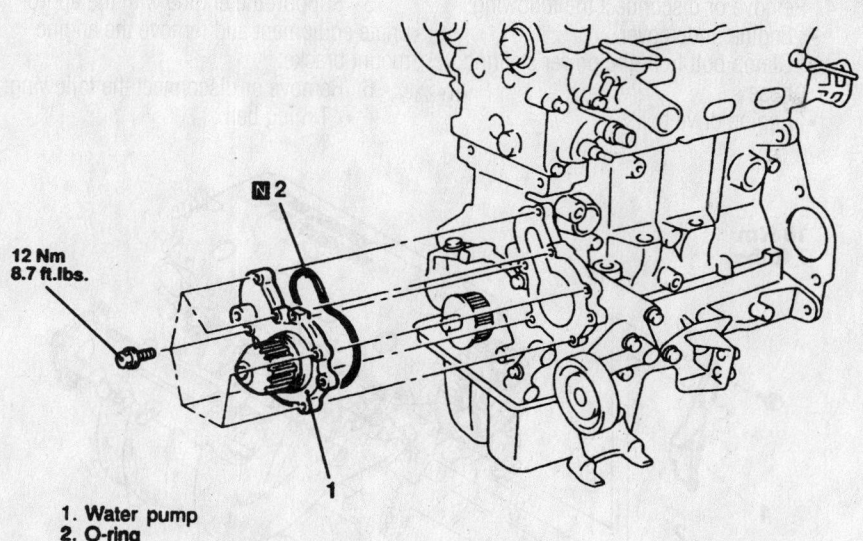

12 Nm
8.7 ft.lbs.

1. Water pump
2. O-ring

7923PG10

Water pump mounting—2.0L (420A) engines

Torque the brace pivot bolt to 17 ft. lbs. (24 Nm).
- Timing belt rear cover
- Timing belt

- Remaining components
6. Refill the engine with coolant.
7. Connect the negative battery cable, start the engine and check for leaks.

Galant

1. Before servicing the vehicle, refer to the precautions in the beginning of this section.
2. Disconnect the negative battery cable.
3. Drain the cooling system.
4. Remove or disconnect the following:
   - Engine undercover
   - Clamp bolt from the power steering hose
5. Support the engine with the appropriate equipment and remove the engine mount bracket.
6. Remove or disconnect the following:
   - Engine drive belts and the air conditioning tensioner bracket
   - Timing belt covers from the front of the engine
   - Camshaft and silent shaft timing belts
   - Alternator brace
   - Water pump, gasket and O-ring where the water inlet pipe(s) joins the pump

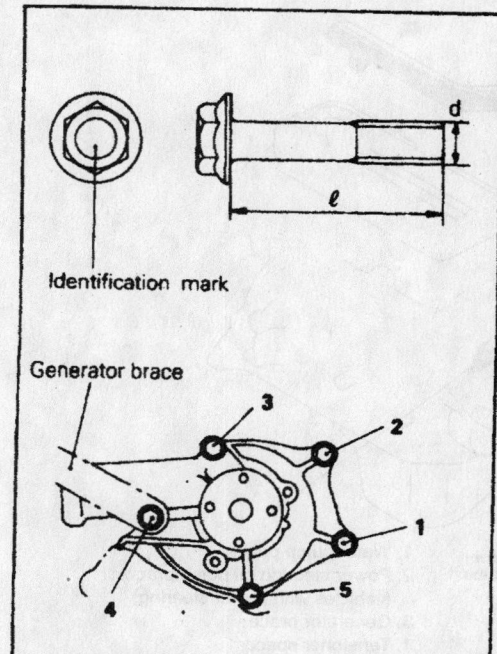

Identification mark

Generator brace

Water pump bolt identification—Galant

No.	Identification mark	Bolt diameter (d) x length (ℓ) mm (in.)	Torque Nm (ft.lbs.)
1	4	8 x 14 (.31 x .55)	
2	4	8 x 22 (.31 x .87)	12–15 (9–10)
3	4	8 x 30 (.31 x 1.18)	
4	7	8 x 65 (.31 x 2.56)	20–27 (15–19)
5	4	8 x 28 (.31 x 1.10)	12–15 (9–10)

7923PG11

Brake service is covered in Section 4 of this manual

To install:

7. Thoroughly clean both gasket surfaces of the water pump and block.

8. Install a new O-ring into the groove on the front end of the water inlet pipe and wet with clean antifreeze only. Do not apply oils or grease to the O-ring.

9. Using a new gasket, install the water pump assembly. Tighten bolts with the head mark **4** to 10 ft. lbs. (14 Nm) and bolts with the head mark **7** to 18 ft. lbs. (24 Nm).

10. Install or connect the following:
- Timing belts
- Engine drive belts
- Engine mount bracket
- Engine undercover

11. Fill the system with coolant.

12. Connect the negative battery cable, run the vehicle until the thermostat opens and fill the radiator completely.

13. Once the vehicle has cooled, recheck the coolant level.

Mirage

1. Before servicing the vehicle, refer to the precautions in the beginning of this section.

2. Disconnect the negative battery cable.

3. Drain the cooling system.

4. Remove or disconnect the following:
- Engine undercover
- Clamp bolt from the power steering hose
- Engine drive belts

5. Support the engine with the appropriate equipment and remove the engine mount bracket.

6. Remove or disconnect the following:
- Timing belt

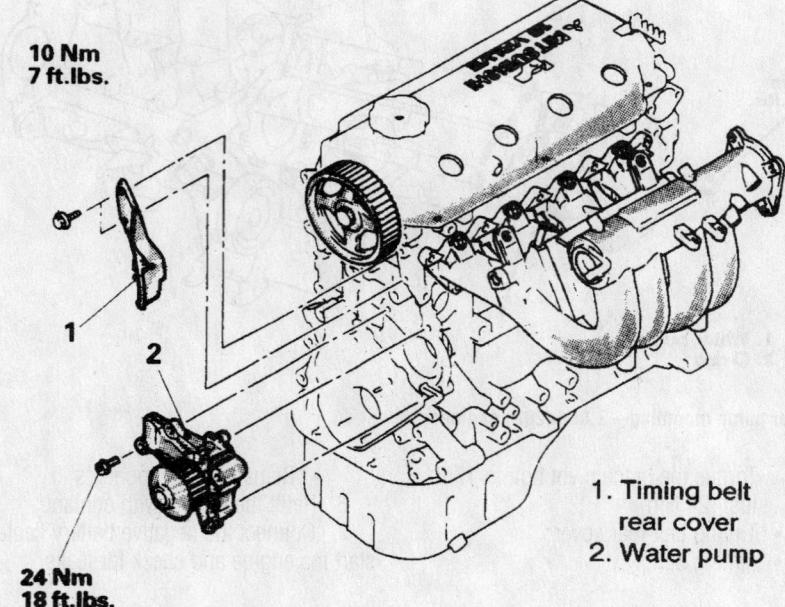

**10 Nm
7 ft.lbs.**

**24 Nm
18 ft.lbs.**

1. Timing belt rear cover
2. Water pump

7923PG08

Water pump and related components—Mirage with 1.8L (4G93) engines

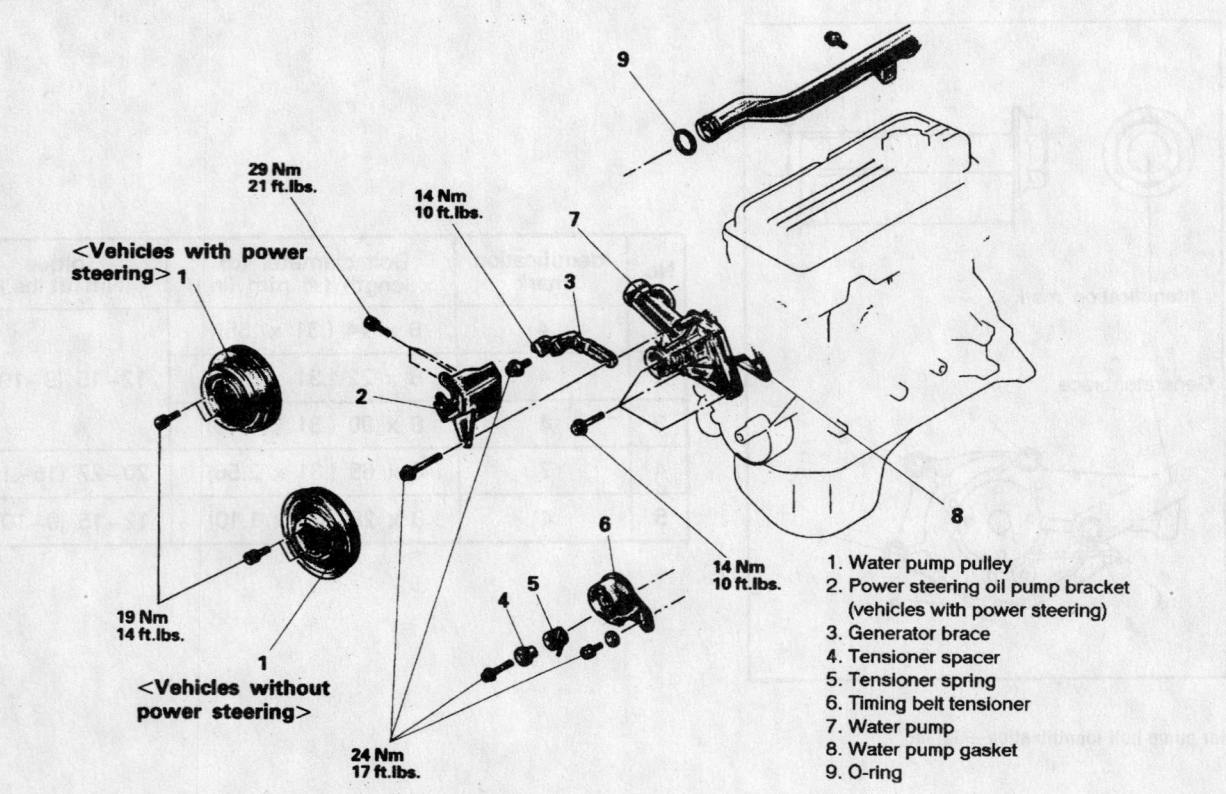

**29 Nm
21 ft.lbs.**

**14 Nm
10 ft.lbs.**

<Vehicles with power steering>

**19 Nm
14 ft.lbs.**

<Vehicles without power steering>

**14 Nm
10 ft.lbs.**

**24 Nm
17 ft.lbs.**

1. Water pump pulley
2. Power steering oil pump bracket (vehicles with power steering)
3. Generator brace
4. Tensioner spacer
5. Tensioner spring
6. Timing belt tensioner
7. Water pump
8. Water pump gasket
9. O-ring

7923PG07

Water pump and related components—Mirage with 1.5L (4G15) engine

- Power steering pump bracket
- Alternator brace

➡**The water pump mounting bolts are different in length, note their positioning for reassembly.**

7. Remove the water pump, gasket and O-ring where the water inlet pipe(s) joins the pump.

To install:

8. Thoroughly clean both gasket surfaces of the water pump and block.

9. For 1.5L engines, install a new O-ring into the groove on the front end of the water inlet pipe. Do not apply oils or grease to the O-ring. Wet the O-ring with water only.

10. For 1.8L engines, apply a 0.09–0.12 in. (2.5–3.0mm) continuous bead of sealant to water pump and install the pump assembly. Install the water pump within 15 minutes of the application of the sealant. Wait 1 hour after installation of the water pump to refill the cooling system or starting the engine.

11. Install or connect the following:
- Gasket and pump assembly and tighten the bolts to 17 ft. lbs. (24 Nm)
- Remaining components in the reverse order of removal

12. Fill the system with coolant.

13. Connect the negative battery cable, run the vehicle until the thermostat opens and fill the radiator completely.

14. Once the vehicle has cooled, recheck the coolant level.

Cylinder Head

REMOVAL & INSTALLATION

3000GT

1. Before servicing the vehicle, refer to the precautions in the beginning of this section.

2. Relieve fuel system pressure.

3. Disconnect the negative battery cable.

4. Drain the cooling system.

5. Remove or disconnect the following:
- Air intake hoses
- Air intake plenum and intake manifold
- Turbocharger, if equipped
- Exhaust manifold
- Timing belt
- Triple pipe assembly across the top of the engine

- Breather hose
- Spark plug wire center cover and remove the spark plug wires

➡**When removing the valve cover, note that bolts for the front head cover are black and bolts for the rear head cover are green. All bolts are 10mm long except for the one closest to the sprockets on the rear head, which is 20mm long.**

6. For SOHC engine, remove the camshaft sprockets. For DOHC engines, remove the intake camshaft sprockets.

7. Remove or disconnect the following:
- Rear timing belt cover
- Ignition coil
- All water hoses from the thermostat housing
- Thermostat housing
- Water inlet from the front head and discard O-ring

8. Loosen the cylinder head mounting bolts in 3 steps, starting from the outside and working inward. Lift off the cylinder head assembly and remove the head gasket.

To install:

9. Thoroughly clean the sealing surfaces of the head and block.

10. Place a new head gasket on the cylinder block with the identification marks in the front top (upward) position. Do not use sealer on the gasket.

11. Carefully install the cylinder head on

the block. Be sure the head bolt washers are installed with the chamfered edge upward.

12. Using 3 even steps, torque the head bolts in sequence, to 76–83 ft. lbs. (105–115 Nm) for non-turbocharged engine or 87–94 ft. lbs. (120–130 Nm) for turbocharged engine.

13. On turbocharged models, loosen all cylinder head bolts and retighten in sequence to 87–94 ft. lbs. (120–130 Nm).

14. Install or connect the following:
- New O-ring and connect the water inlet to the front head
- Gaskets and install the thermostat housing and connect the hoses
- Ignition coil and the rear timing belt cover

15. For SOHC engine, install the camshaft sprockets. For DOHC engines, install the intake camshaft sprockets. Tighten the retaining bolt to 65 ft. lbs. (90 Nm).

16. Apply sealer to the lower edges of the half-round portions of the belt-side of the new gasket and install the valve cover. Tighten the bolts in the proper sequence to 26 inch lbs. (3 Nm). Then, retighten bolts Nos. 1–6 to 35 inch lbs. (4 Nm).

17. Install or connect the following:
- Spark plug wires and install the center cover
- Breather hose
- Triple pipe assembly across the top of the engine and tighten the retaining bolts to 84 inch lbs. (10 Nm).

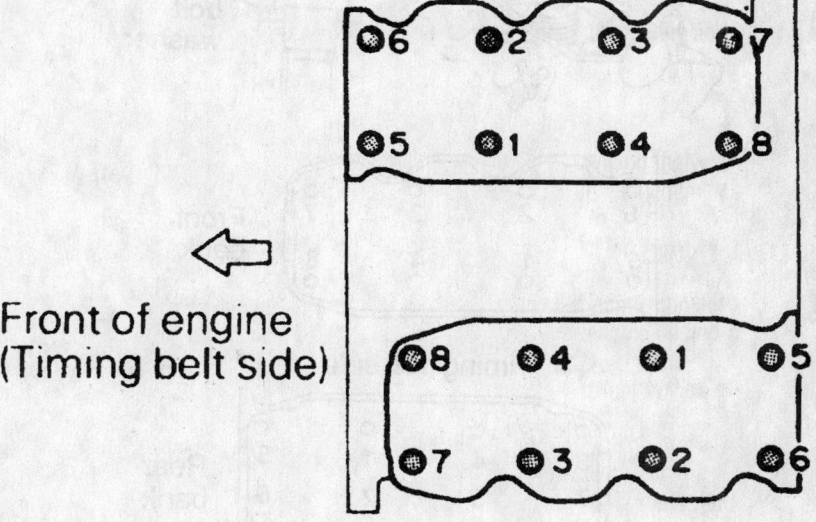

Front of engine (Timing belt side)

Cylinder head bolt torque sequence—3000GT

7923PG24

- Timing belt
- Intake manifold, air intake plenum, turbocharger and exhaust manifold, using new gaskets
- Air intake hoses

18. Fill the system with coolant.

19. Adjust the accelerator cable. Check and adjust the idle speed and ignition timing.

20. Once the vehicle has cooled, recheck the coolant level.

Diamante

3.0L DOHC ENGINE

1. Before servicing the vehicle, refer to the precautions in the beginning of this section.

2. Relieve fuel system pressure. Disconnect the negative battery cable.

3. Drain the cooling system.

4. Remove or disconnect the following:
- Air intake hoses
- Air intake plenum and intake manifold
- Exhaust manifold
- Timing belt

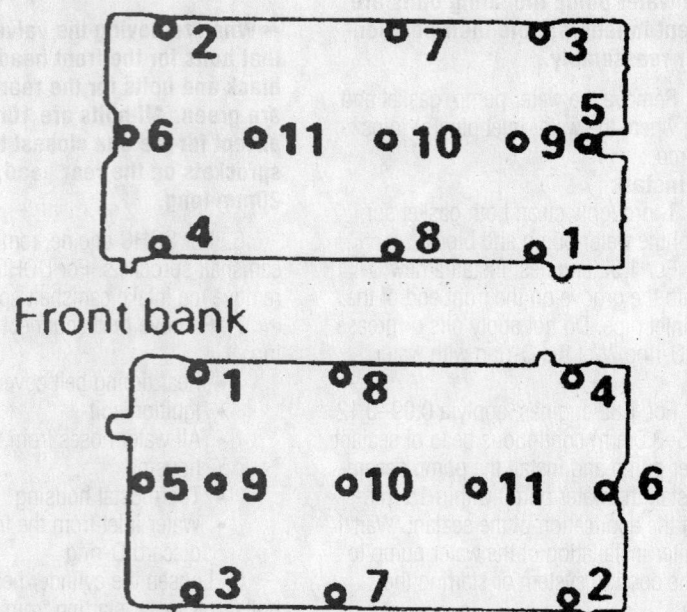

Rocker cover bolt torque sequence—Diamante 3.0L DOHC engine

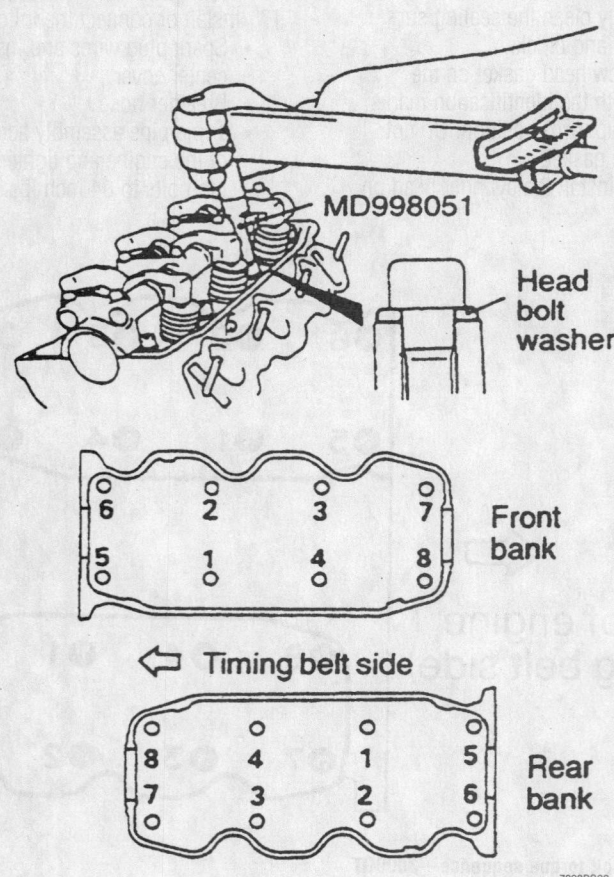

Tighten the cylinder head bolts according to the sequence shown—3.0L (SOHC and DOHC) engines

- Breather hose
- Spark plug wire center cover and remove the spark plug wires
- Rocker covers
- Intake camshaft sprockets
- Rear timing belt cover
- Ignition coil assembly
- Water hoses from the thermostat housing
- Thermostat housing
- Water inlet from the front head

5. Loosen the cylinder head mounting bolts in the reverse of the torque sequence and loosen the bolts in 3 steps. Lift off the cylinder head assembly and remove the head gasket.

To install:

6. Thoroughly clean the sealing surfaces of the head and block.

7. Place a new head gasket on the cylinder block with the identification marks in the front top (upward) position. Do not use sealer on the gasket.

8. Carefully install the cylinder head on the block. Be sure the head bolt washers are installed with the chamfered edge upward. Using 3 even steps, torque the head bolts in sequence, to 76–83 ft. lbs. (105–115 Nm).

9. Install new O-ring and connect the water inlet to the head. Tighten the mounting bolt to 10 ft. lbs. (13 Nm).

10. Replace the gaskets and install the thermostat housing. Tighten the mounting bolts to 12–14 ft. lbs. (17–20 Nm).

11. Install or connect the following:
- Hoses to the thermostat housing, using new clamps
- Ignition coil and torque the mounting bolts to 84 inch lbs. (10 Nm)
- Rear timing belt cover and torque the mounting bolts to 17 ft. lbs. (24 Nm)
- Intake camshaft sprockets. Tighten the retaining bolt to 65 ft. lbs. (90 Nm).

12. Apply sealer to the lower edges of the valve cover. Tighten the bolts in the proper sequence to 44–51 inch lbs. (5–6 Nm).

13. Install or connect the following:
- Spark plug wires and install the center cover. Tighten the bolts that secure the center cover to 27 inch lbs. (3 Nm)
- Breather hose
- Timing belt
- Exhaust manifold assembly
- Intake manifold and air intake plenum, using all new gaskets
- Air intake hoses

14. Change the engine oil and oil filter.
15. Fill the system with coolant.
16. Connect the negative battery cable.
17. Adjust the accelerator cable. Start the engine. Check and adjust the idle speed and ignition timing.
18. Once the vehicle has cooled, recheck the coolant level.

3.0L SOHC ENGINE

1. Before servicing the vehicle, refer to the precautions in the beginning of this section.
2. Relieve the fuel system pressure. Disconnect the negative battery cable.
3. Drain the cooling system.
4. Remove or disconnect the following:
- Air intake hose
- Exhaust manifold
- Air intake plenum and intake manifold
- Timing belt
- Camshaft sprockets and the rear timing belt cover
- Power steering pump bracket. If removing the rear head, remove the alternator brace.
- Water inlet pipe
- Purge pipe assembly
- Valve cover

5. Using the reverse sequence of the installation sequence, loosen the cylinder

head mounting bolts in 3 steps. Lift off the cylinder head assembly and remove the head gasket.

To install:

6. Thoroughly clean the sealing surfaces of the head and block.
7. Place a new head gasket on the cylinder block making sure the identification mark on the cylinder head gasket is in the front top (upward) location. Do not use sealer on the gasket.
8. Carefully install the cylinder head on the block. Be sure the head bolt washers are installed with the chamfered edge upward. Using 3 even steps, torque the head bolts in sequence, to 76–83 ft. lbs. (105–115 Nm).
9. Apply sealer to the lower edges of the half-round portions and install the valve cover. Tighten valve cover bolts to 84 inch lbs. (9 Nm).
10. Install or connect the following:
- Purge pipe assembly
- Water inlet pipe
- Power steering pump bracket and alternator brace
- Rear timing belt cover and camshaft sprockets. Torque the retaining bolt to 65 ft. lbs. (90 Nm).
- Timing belt
- Intake manifold, air intake plenum and exhaust manifold, using new gaskets
- Air intake hose

11. Fill the system with coolant.
12. Connect the negative battery cable.
13. Start the engine.

14. Check and adjust the idle speed and ignition timing.
15. Once the vehicle has cooled, recheck the coolant level.

3.5L ENGINE

1. Before servicing the vehicle, refer to the precautions in the beginning of this section.
2. Disconnect the negative battery cable.
3. Drain the engine coolant
4. Remove or disconnect the following:
- Timing belt
- Intake and exhaust manifolds
- Spark plug wires
- Cylinder head covers
- Timing belt rear center cover

5. Loosen the cylinder head bolts gradually in 3 stages, in the opposite of the installation sequence.
6. Remove the cylinder head.

To install:

7. Clean the cylinder head and mounting surface on the engine block.
8. Install the cylinder head using a new gasket.
9. Tighten the bolts in sequence using 3 stages to 76–83 ft. lbs. (103–113 Nm).
10. Install or connect the following:
- Timing belt rear center cover
- Cylinder head covers using new gaskets. Tighten the bolts to 24–36 inch lbs. (3–4 Nm).
- Spark plug wires
- Intake and exhaust manifolds

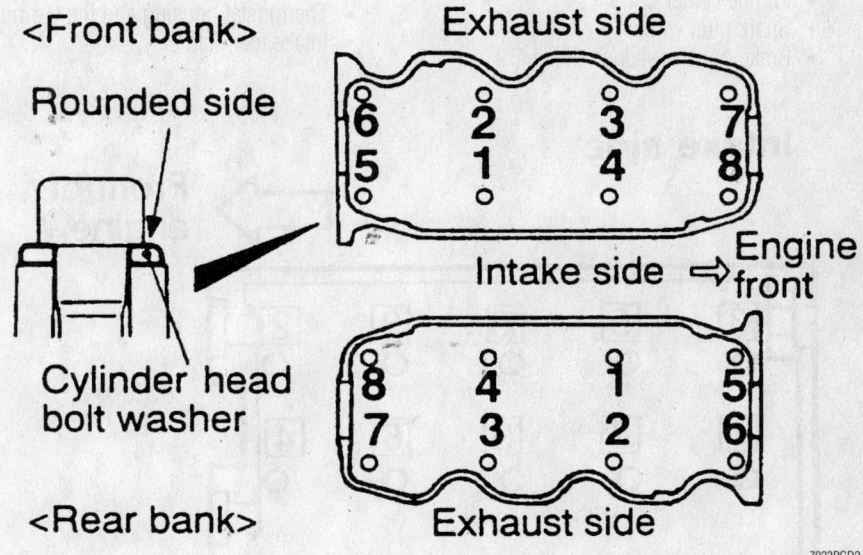

Cylinder head bolt tightening sequence—3.5L engine

For Accessory Drive Belt illustrations, see Section 1 of this manual

- Timing belt
- Remaining components
11. Refill the cooling system.
12. Connect the negative battery cable.

Eclipse

2.0L (4G63) ENGINE

1. Before servicing the vehicle, refer to the precautions in the beginning of this section.
2. Relieve the fuel system pressure.
3. Disconnect the negative battery cable.
4. Drain the engine coolant.
5. Remove or disconnect the following:
- Accelerator cable and remove the mounting bracket
- Intake air duct (hose) from the throttle body
6. Disconnect the following connectors:
- IAC motor
- Knock sensor
- HO$_2$S sensor
- ECT gauge sender
- ECT sensor
- Ignition module (power transistor)
- TP sensor
- Condenser
- MAP sensor
- Injectors
- Ignition coil
- CMP sensor
- CKP sensor
- air conditioning compressor
- Engine control wiring harness
7. Remove or disconnect the following:
- Engine center cover
- Spark plug wires
- Brake booster vacuum hose

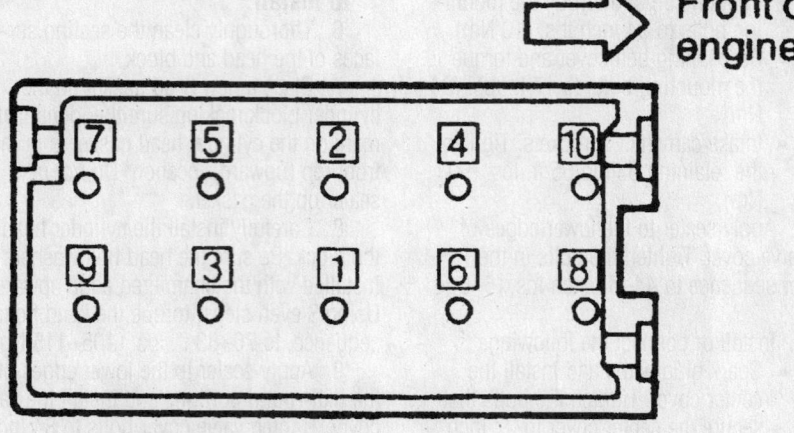

Cylinder head bolt torque sequence—2.0L (4G63) engine

- Fuel lines from the fuel supply rail
- Bypass hose and the water hose connections
- Vacuum hoses, breather hose and the PCV hose
- Timing belt
- Power steering pump
- Cylinder head cover and the semi-circular packing
- Heat protector

8. Mark the position of the hose clamps on the hoses and disconnect the water hoses and the radiator hoses.
9. Remove or disconnect the following:
- Thermostat housing and the O-ring
- Intake manifold stay

- Turbocharger assembly from the exhaust manifold
10. Gradually loosen the cylinder head bolts in 2 or 3 steps using the specified sequence and remove the bolts.
11. Remove the cylinder head and the gasket.

To install:
12. Thoroughly clean the deck surface of the engine block and the sealing surface of the cylinder head.
13. Measure the length of the cylinder head bolts from below the head to the end, if the bolt measures more than 3.913 in. (99.4mm), replace the bolt.
14. Install a new gasket on the engine block with the identification mark facing upwards.
15. Carefully place the cylinder head on the engine. Apply clean engine oil to the bolts and install the bolts finger-tight.
16. Torque the bolts using the following procedure:
   a. Torque the bolts in sequence to 58 ft. lbs. (78 Nm).
   b. Loosen the bolts completely in the reverse order.
   c. Torque the bolts in sequence to 15 ft. lbs. (20 Nm).
   d. Mark the head bolts and cylinder head as shown.
   e. Tighten the bolts in sequence 90 degrees.
   f. Tighten the bolts in sequence an additional 90 degrees.
17. Install or connect the following:
- Turbocharger to the exhaust manifold, using a new gasket

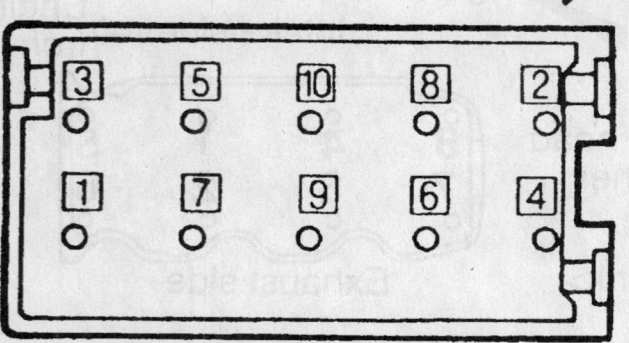

Cylinder head bolt removal sequence—2.0L (4G63) engine

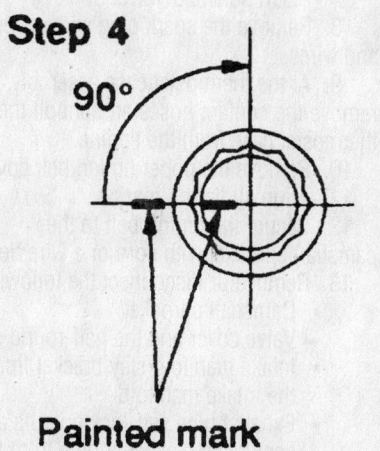

Step 4
90°
Painted mark

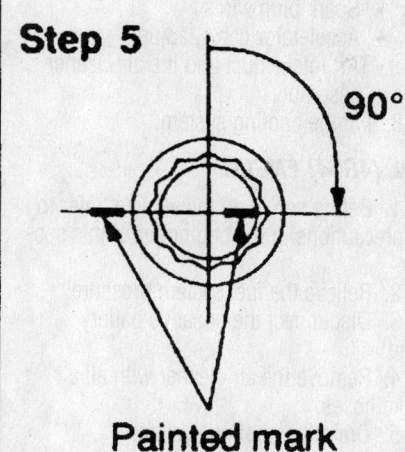

Step 5
90°
Painted mark

7923PG19

To ensure that the bolts are tightened exactly 180 total degrees, mark the head bolt and cylinder head as shown—2.0L (4G63) engine

- Intake manifold stay
- Thermostat housing and connect the hoses
- Heat protector

18. Apply sealant to the semi-circular packing and install it on the cylinder head.

19. Apply sealant at the front of the cylinder head where the camshaft oil seal retainer and the cylinder head come together and install the cylinder head cover using a new gasket.

20. Install or connect the following:
- Power steering pump
- Timing belt
- PCV, breather and vacuum hoses
- Water hose and the bypass hose

21. Use a new O-ring and connect the lines to the fuel supply rail. Apply a small amount of engine oil to the new O-ring.

22. Install or connect the following:
- Brake booster vacuum hose
- Spark plug wires and install the center cover

23. Connect the following connectors:
- IAC motor
- Knock sensor
- HO$_2$S sensor
- ECT gauge sender
- ECT sensor
- Ignition module (power transistor)
- TP sensor
- Condenser
- MAP sensor
- Injectors
- Ignition coil

- CMP sensor
- CKP sensor
- air conditioning compressor
- Engine control wiring harness

24. Install or connect the following:
- Intake air hose to the throttle body
- Accelerator cable, adjust

25. Refill the engine with coolant.
26. Connect the negative battery cable.
27. Start the engine and check for leaks.

2.0L (420A) ENGINE

1. Before servicing the vehicle, refer to the precautions in the beginning of this section.

2. Relieve the fuel system pressure.

3. Disconnect the negative battery cable.

4. Drain the engine coolant.

5. Remove the air cleaner and air intake duct.

6. Disconnect the following connectors:
- air conditioning compressor
- Power steering pressure switch
- HO$_2$S sensor
- ECT gauge sender
- ECT sensor
- MAP sensor
- IAT sensor
- TP sensor
- IAC motor
- Injector harness
- Ignition coil
- CMP sensor
- EGR solenoid valve

7. Install or connect the following:
- Spark plug wires
- Accelerator cable from the throttle body
- Heater hoses from the rear of the engine
- Fuel lines from the fuel supply rail
- Purge air hose and the brake booster vacuum hose connections

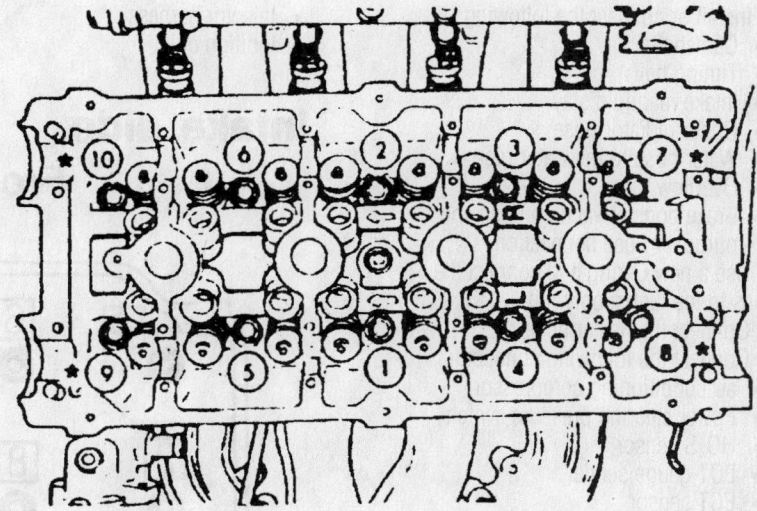

*Location of 110 mm (4.330 in.) short bolts.

7923PG20

Cylinder head bolt torque sequence—2.0L (420A) engine

- Overflow tube connection
- Upper radiator hose and the water hose connections
- Timing belt
- Intake manifold stay
- Intake and exhaust camshafts
- Exhaust pipe connection from the exhaust manifold
- Cylinder head mounting bolts and the cylinder head

To install:

8. Thoroughly clean the cylinder head and engine block sealing surfaces.

9. Place a new head gasket on the engine block and carefully place the cylinder head on the engine.

10. Coat the threads of the bolts with clean engine oil and install the bolts finger-tight in the engine block.

➡**The short bolts go in the corners.**

11. Torque the cylinder head bolts in the following sequence:
- Torque bolts 1–6 to 25 ft. lbs. (33 Nm), then torque bolts 7–10 to 20 ft. lbs. (27 Nm)
- Torque bolts 1–6 to 50 ft. lbs. (67 Nm), then torque bolts 7–10 to 20 ft. lbs. (27 Nm)
- Again, torque bolts 1–6 to 50 ft. lbs. (67 Nm), then torque bolts 7–10 to 20 ft. lbs. (27 Nm)
- Turn all fasteners 1–10 ¼ turn (90 degrees) more in sequence.

12. Use a new gasket and connect the front exhaust pipe to the exhaust manifold.

13. Install or connect the following:
- Camshafts
- Timing belt
- Intake manifold stay
- Upper radiator hose
- Water hose to the water pipe
- Overflow tube
- Brake booster vacuum hose and purge air hose connection

14. Use a new O-ring and connect the fuel lines to the fuel supply rail.

15. Connect the heater hose.

16. Connect the following connectors:
- air conditioning compressor
- Power steering pressure switch
- HO$_2$S sensor
- ECT gauge sender
- ECT sensor
- MAP sensor
- IAT sensor
- TP sensor
- IAC motor
- Injector harness
- Ignition coil
- CMP sensor
- EGR solenoid valve

17. Install or connect the following:
- Spark plug wires
- Accelerator cable, adjust
- Air intake duct and the air cleaner assembly

18. Fill the cooling system.

2.4L (4G64) ENGINE

1. Before servicing the vehicle, refer to the precautions in the beginning of this section.

2. Relieve the fuel system pressure.

3. Disconnect the negative battery cable.

4. Remove the air cleaner with all air intake hoses.

5. Drain the cooling system.

6. Remove or disconnect the following:
- Accelerator cable
- Cable mounting brackets and position the cable aside
- Breather hose
- Vacuum lines at the throttle body, label for identification
- High pressure fuel line, plug
- Fuel return hose, plug

7. Disconnect the following connectors:
- air conditioning compressor
- Power steering pressure switch
- HO$_2$S sensor
- ECT gauge sender
- ECT sensor
- MAP sensor
- IAT sensor
- TP sensor
- IAC motor
- Injector harness
- Ignition coil

- CMP sensor
- EGR solenoid valve

8. Remove the spark plug wire cover and wires.

9. At the thermostat case assembly, remove the coolant hoses and unbolt the thermostat case from the engine.

10. Remove the upper timing belt cover.

11. Align all timing marks.

12. Secure the timing belt to the camshaft sprocket with cord or a wire tie.

13. Remove or disconnect the following:
- Camshaft sprocket
- Valve cover and the half-round seal
- Intake manifold stay bracket from the intake manifold
- Exhaust pipe self-locking nuts and separate the exhaust pipe from the exhaust manifold. Discard the gasket.

14. Loosen the cylinder head mounting bolts in 3 steps, starting from the outside and working inward. Lift off the cylinder head assembly and remove the head gasket.

To install:

15. Thoroughly clean the mating surfaces of the head and block.

16. Place a new head gasket on the cylinder block with the identification marks at the front top (upward) position. Do not use sealer on the gasket.

17. Inspect the cylinder head bolt length prior to installation. If the length exceeds 3.91 in. (99.4mm), the bolt must be replaced. Install the washer onto the bolt so the chamfer on the washer faces towards the head of the bolt.

18. Carefully install the cylinder head on

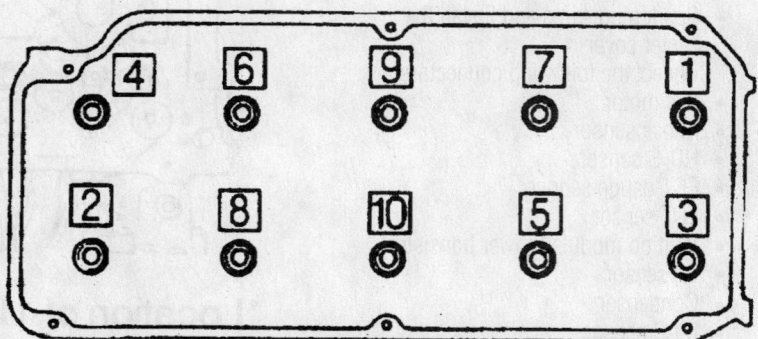

Intake side

Front of engine ⇨

Exhaust side

7923PG21

Cylinder head bolt removal sequence—2.4L (4G64) engine

Intake side

Front of engine ⇨

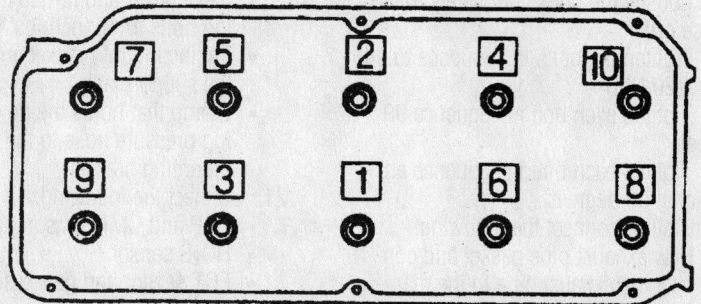

Exhaust side

7923PG22

Cylinder head bolt installation sequence—2.4L (4G64) engine

- MAP sensor
- IAT sensor
- TP sensor
- IAC motor
- Injector harness
- Ignition coil
- CMP sensor
- EGR solenoid valve

23. Install the spark plug wires and cover.

24. Replace the O-rings and connect the fuel lines.

25. Install the air cleaner and intake hose. Connect the breather hose.

26. Fill the cooling system.

27. Connect the negative battery cable

1.5L ENGINE

1. Before servicing the vehicle, refer to the precautions in the beginning of this section.

the block and tighten the cylinder head bolts as follows:

   a. Following the proper tightening sequence, tighten the cylinder head bolts to 58 ft. lbs. (78 Nm).

   b. Loosen all bolts completely.

   c. Torque bolts to 15 ft. lbs. (20 Nm).

   d. Tighten bolts an additional ¼ turn.

   e. Tighten bolts an additional ¼ turn.

19. Install or connect the following:

- New exhaust pipe gasket and connect the exhaust pipe to the manifold. Tighten the bolts to 33 ft. lbs. (44 Nm).
- Thermostat case and tighten the mounting bolts to 18 ft. lbs. (24 Nm)
- Coolant hoses to the thermostat case

20. Apply sealer to the perimeter of the half-round seal and to the lower edges of the half-round portions of the belt-side of the new gasket. Install the valve cover.

21. Install or connect the following:

- Camshaft sprocket with the timing belt attached. Remove the cord or wire tie.
- Upper timing belt cover
- Intake manifold stay and tighten the mounting bolts to 22 ft. lbs. (30 Nm)

22. Connect the following connectors:

- air conditioning compressor
- Power steering pressure switch
- HO₂S sensor
- ECT gauge sender
- ECT sensor

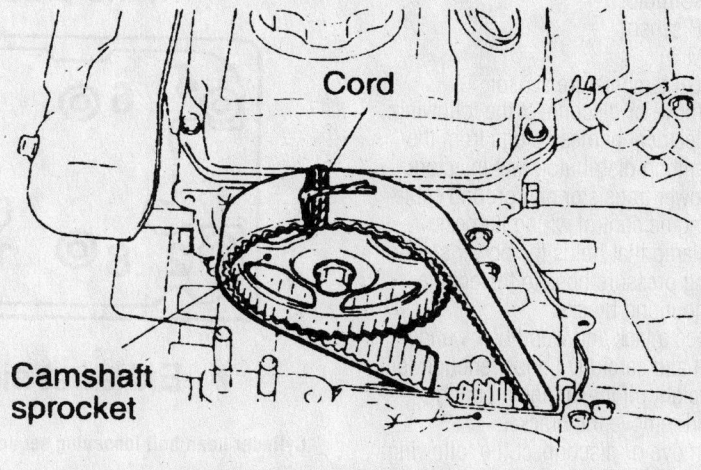

Cord

Camshaft
sprocket

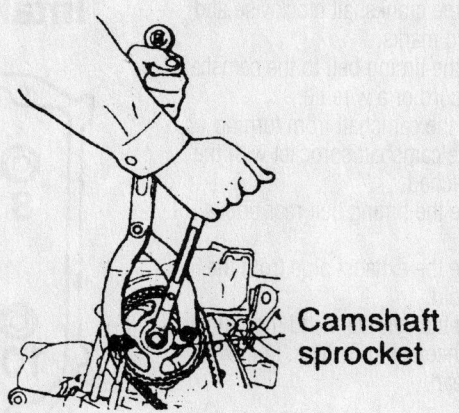

Camshaft
sprocket

7923PG23

Secure the timing belt to the camshaft sprocket and remove the sprocket—2.4L (4G64) engine

For Wheel Alignment specifications, see Section 1 of this manual

2. Relieve the fuel system pressure. Disconnect the negative battery cable.

3. Drain the cooling system.

4. Remove or disconnect the following:
- Air intake hose and the air cleaner assembly
- Ground cable connection and the accelerator cable
- PCV and the breather hose connection
- Vacuum hoses from the intake and throttle body, label for reference
- Vacuum line for the brake booster
- Upper radiator hose, throttle body hoses, bypass hose and heater hose connections
- Fuel feed and return lines
- Spark plug wires

5. Disconnect the electrical harness plugs from the following:
- CKP and CMP sensors
- HO2S sensor
- ECT sensor and gauge sender
- ISC motor
- TP sensor
- IAT
- EGR temperature sensor

6. Remove or disconnect the following:
- Electrical harness plugs from the ignition distributor, fuel injectors, power transistor and ground cable
- Engine control wiring harness
- Clamp that holds the power steering pressure hose to the engine mounting bracket

7. Place a jack and wood block under the oil pan and carefully lift just enough to take the weight off the engine mounting bracket and remove the bracket.

8. Remove or disconnect the following:
- Valve cover
- Timing belt upper cover

9. Rotate the crankshaft clockwise and align the timing marks.

10. Attach the timing belt to the camshaft sprocket with cord or a wire tie.

11. Secure the camshaft from turning and remove the camshaft sprocket with the timing belt attached.

12. Remove the timing belt rear upper cover.

13. Remove the exhaust pipe from the exhaust manifold.

14. Loosen the cylinder head mounting bolts in sequence using 3 steps. Remove the cylinder head.

To install:

15. Thoroughly clean the mating surfaces of the head and block.

16. Place a new head gasket on the cylinder block with the identification marks facing upward. Do not use sealer on the gasket.

17. Carefully install the cylinder head on the block. Tighten the cylinder head bolts as follows:

a. 36 ft. lbs. (49 Nm) in the correct sequence

b. Loosen the bolts completely in the reverse order.

c. Tighten the bolts in sequence to 14 ft. lbs. (20 Nm)

d. Tighten each bolt in sequence 90 degrees

e. Tighten each bolt in sequence an additional 90 degrees

18. Install or connect the following:
- New exhaust pipe gasket and connect the exhaust pipe to the manifold
- Upper rear timing cover

19. Align the timing marks and install the cam sprocket. Torque the retaining bolt to 51 ft. lbs. (70 Nm). Check the belt tension and adjust, if necessary. Install the outer timing cover.

20. Install or connect the following:
- Valve cover and torque the retaining bolts to 16 inch lbs. (1.8 Nm)
- Engine mount bracket and remove the support jack
- Clamp that holds the power steering pressure hose to the engine mounting bracket

21. Connect the following:
- CKP and CMP sensors
- HO2S sensor
- ECT sensor and gauge sender
- ISC motor
- TP sensor
- IAT

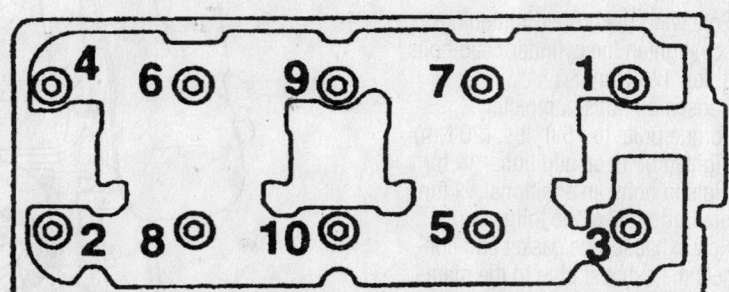

Cylinder head bolt loosening sequence—Mirage with 1.5L (4G15) engine

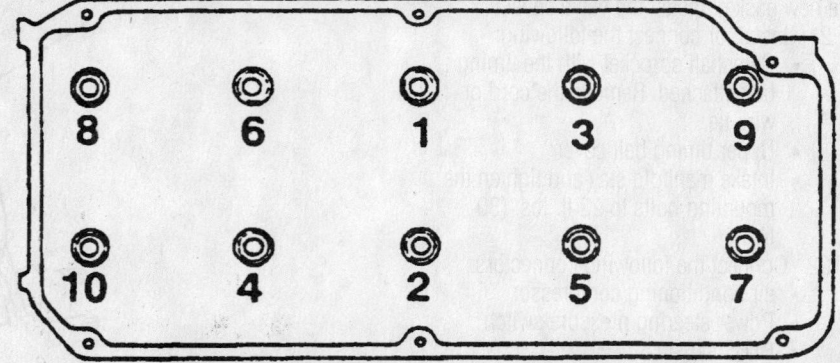

Cylinder head bolt tightening sequence—Mirage with 1.5L (4G15) engine

- EGR temperature sensor
22. Install or connect the following:
- Ignition distributor, fuel injectors, power transistor and ground cable
- Engine control wiring harness
23. Replace the O-rings and connect the fuel lines.
24. Install the air cleaner assembly. Connect the breather hose.
25. Fill the system with coolant.
26. Connect the negative battery cable.

1.8L ENGINE

1. Before servicing the vehicle, refer to the precautions in the beginning of this section.
2. Relieve fuel system pressure. Disconnect the negative battery cable.
3. Remove the air cleaner assembly.
4. Drain the cooling system.
5. Disconnect the brake booster vacuum hose and PVC valve connection.
6. Note the locations and disconnect the vacuum hoses from the intake and throttle body.
7. Remove or disconnect the following:

- Upper radiator hose, overflow tube and the water hose from the thermostat to the throttle body
- Fuel feed and return lines
- Accelerator cable connection from the throttle body
- Oil pressure switch
8. Disconnect the following:
- HO_2S sensor
- ECT sensor and gauge sender
- IAC motor
- EGR temperature sensor
- TP sensor
- Knock sensor
- Fuel injectors
- Spark plug wires
- Control harness assembly and position aside
- Thermostat housing, thermostat and the thermostat case with O-ring from the engine
- Rocker cover
- Timing belt upper cover
9. Rotate the crankshaft clockwise and align the timing marks.
10. Attach the timing belt to the camshaft sprocket with cord or a wire tie.
11. Secure the camshaft from turning and remove the camshaft sprocket with the timing belt attached.
12. Remove the timing belt rear upper cover.

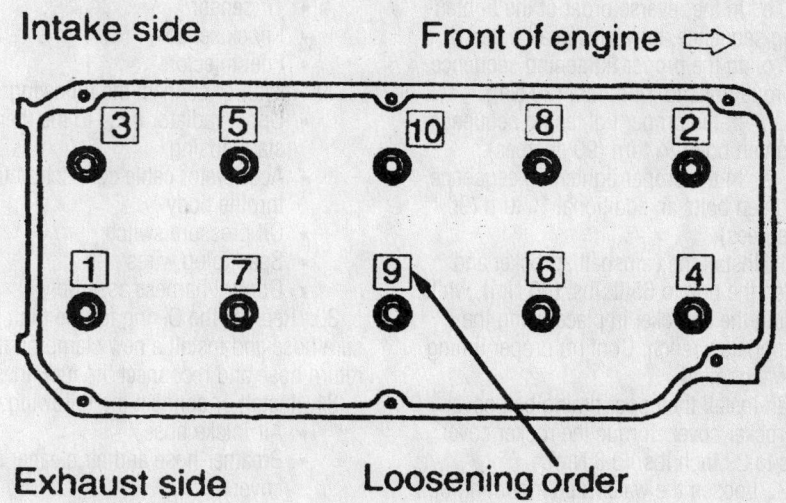

Cylinder head bolt loosening sequence—1.8L engine

7923PG15

Cylinder head bolt torque sequence—1.8L engine

7923PG16

13. Loosen the cylinder head bolts in 2 or 3 steps in the proper sequence.
14. Remove the cylinder head from the engine.

✳✳ CAUTION

When removing the cylinder head, take care not to bend or damage the plug guide. The plug guide can not be replaced.

To install:
15. Thoroughly clean the mating surfaces of the head and block.
16. Place a new head gasket on the cylinder block with the identification marks facing upward. Do not use sealer on the gasket.
17. Carefully install the cylinder head on the block.
18. Measure the cylinder head bolts prior to installation. Replace any that exceed 3.795 in. (96.4mm).
19. Apply a small amount of engine oil to the thread section of the bolt and install so the chamfer of the washer faces upward.
20. Tighten the cylinder head bolts as follows:

   a. In the proper tightening sequence, torque bolts to 54 ft. lbs. (75 Nm).

b. In the reverse order of the tightening sequence, fully loosen all bolts.

c. In the proper tightening sequence, torque bolts to 14 ft. lbs. (20 Nm).

d. In the proper tightening sequence, tighten bolts ¼ turn (90 degrees).

e. In the proper tightening sequence, tighten bolts an additional ¼ turn (90 degrees).

21. Install the camshaft sprocket and tighten the bolt to 65 ft. lbs. (90 Nm), while holding the sprocket in place using the appropriate wrench. Confirm proper timing mark alignment.

22. Install the upper timing belt cover and rocker cover. Torque the rocker cover bolts to 29 inch lbs. (3.3 Nm).

23. Loosen the water pipe mounting bolt for ease of thermostat housing installation.

24. Apply a thin bead of RTV sealant to the water tube connection on the thermostat case.

25. Apply a small amount of water to the O-ring of the water inlet pipe and press the thermostat case assembly onto the water inlet pipe. Install the thermostat case assembly mounting bolt tightening to 16 ft. lbs. (22 Nm).

26. Tighten the water pipe mounting bolt.

27. Install the thermostat into the housing so the jiggle valve is located at the top. Tighten the housing bolts to 10 ft. lbs. (14 Nm).

28. Connect the following:
- HO2S sensor
- ECT sensor and gauge sender
- IAC motor
- EGR temperature sensor

- TP sensor
- Knock sensor
- Fuel injectors

29. Install or connect the following:
- Upper radiator hose to the thermostat housing
- Accelerator cable connection to the throttle body
- Oil pressure switch
- Spark plug wires
- Control harness assembly

30. Replace the O-ring for the high pressure hose and install a new clamp on the return hose and reconnect the fuel lines.

31. Install or connect the following:
- Air intake hose
- Breather hose and air cleaner case cover
- Brake booster and the PCV vacuum hoses

32. Fill the system with coolant.

33. Connect the negative battery cable

Rocker Arms/Shafts

REMOVAL & INSTALLATION

3000GT and Diamante

3.0L SOHC ENGINE

On this engine, the hydraulic lash adjusters are built into the rocker arms.

1. Before servicing the vehicle, refer to the precautions in the beginning of this section.

2. Disconnect the negative battery cable.

3. Remove the valve cover. Install lash adjuster retainer tools to prevent the auto-lash adjuster from falling out of the rocker arm.

4. Loosen rocker arm and shaft assembly evenly in several steps. Remove the rocker arm and shaft assembly as a complete unit.

5. Remove the rear camshaft bearing cap and slide the rocker arms, springs and washers from the shaft. If they are to be reused, note the location and positioning of all rocker shaft components. It is recommended that all lash adjusters and rockers be replaced as a complete set.

To install:

6. Immerse the lash adjusters in clean diesel fuel. Using a small wire, move the plunger of the lash adjuster up and down 4 or 5 times while pushing down lightly on the check ball in order to bleed out the air. Install the lash adjusters in the rocker arms.

7. Using a light coat of engine oil, assemble the rocker arms to the shaft. Install the rear camshaft bearing cap.

8. Lubricate the camshaft and rocker shaft with clean engine oil and position on the cylinder head.

9. Apply a drop of sealant to the rear edges of the end caps.

10. Install or connect the following:
- Assembly making sure the notches in the rocker shafts are facing up
- Cap bolts and tighten evenly and gradually to 14 ft. lbs. (20 Nm). Remove the lash adjuster retainers.
- Valve cover
- Negative battery cable

13. Bearing cap
14. Rocker arm
15. Spring
16. Rocker arm
17. Spring
18. Bearing cap no. 3
19. Rocker arm
20. Spring
21. Rocker arm
22. Spring
23. Bearing cap no. 2
24. Rocker arm
25. Spring
26. Rocker arm
27. Spring
28. Rocker arm shaft
29. Rocker arm shaft
30. Bearing cap no. 1

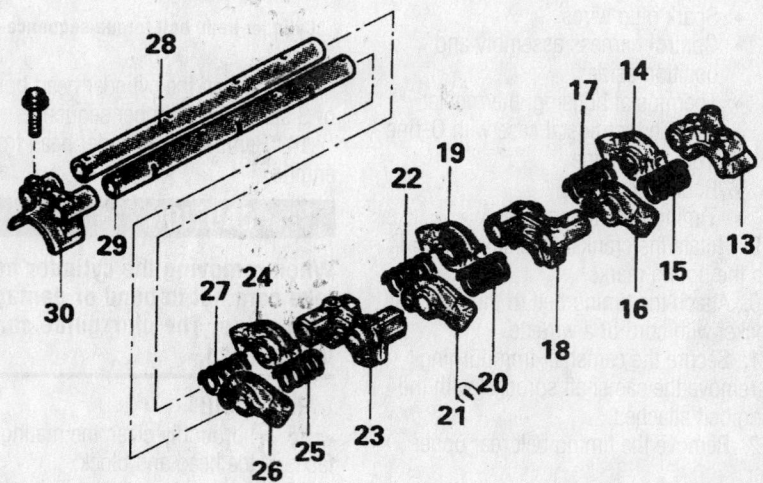

Rocker arm assembly—Diamante 3.0L SOHC engine

7923PG34

3.0L DOHC ENGINE

1. Before servicing the vehicle, refer to the precautions in the beginning of this section.

2. Relieve the fuel system pressure.

3. Remove or disconnect the following:
- Battery negative cable
- Timing belt cover and timing belt
- Center cover, breather and PCV hoses, and spark plug cables
- Rocker cover
- Throttle body stay, both camshaft sprockets, and oil seals
- CMP sensor and adapter from the rear of the camshaft
- Intake and exhaust camshafts
- Rocker arms and lash adjusters from the head

➡ **It is recommended that all lash adjusters and rockers be replaced as a complete set.**

To install:

4. Immerse the lash adjusters in clean diesel fuel. Using a small wire, move the plunger of the lash adjuster up and down 4 or 5 times while pushing down lightly on the check ball in order to bleed out the air. Lubricate and install the lash adjusters in the cylinder head.

5. Lubricate the camshafts with clean engine oil and position the camshafts on the cylinder head.

6. Install the bearing caps. Tighten the caps in sequence and in 2 or 3 steps. Caps 2, 3 and 4 have a front mark. Install with the mark aligned with the front mark on the cylinder head. Intake caps have **I** stamped on the cap and exhaust caps have **E**. Also, be sure the rocker arm is correctly mounted on the lash adjuster and the valve stem end. Torque the front and rear retaining cap bolts to 14 ft. lbs. (20 Nm) and tighten the center 3 retaining cap bolts to 96 inch lbs. (11 Nm).

7. Apply a coating of engine oil to the oil seals and install.

8. Install the timing belt, valve cover and all related parts. Refer to the timing belt unit repair section.

9. Connect the negative battery cable and check for leaks.

3.5L ENGINE

1. Before servicing the vehicle, refer to the precautions in the beginning of this section.

2. Disconnect the negative battery cable.

3. Remove the rocker arm cover.

4. Install the lash adjuster clips on the rocker arms, then loosen the bearing cap bolts. Do not remove the bolts from the bearing caps.

5. Remove the rocker arms, shafts and bearing caps as an assembly.

To install:

6. Install the bearing caps/rocker arm assemblies. Tighten the bolts to 23 ft. lbs. (31 Nm).

7. Remove the lash adjuster clips.

8. Install the rocker arm cover using a new gasket.

9. Connect the negative battery cable.

Eclipse and Galant

2.4L ENGINE

1. Before servicing the vehicle, refer to the precautions in the beginning of this section.

2. Remove or disconnect the following:
- Negative battery cable
- Accelerator cable, remove the cable clamp mounting screws and position the accelerator cable out of the way.

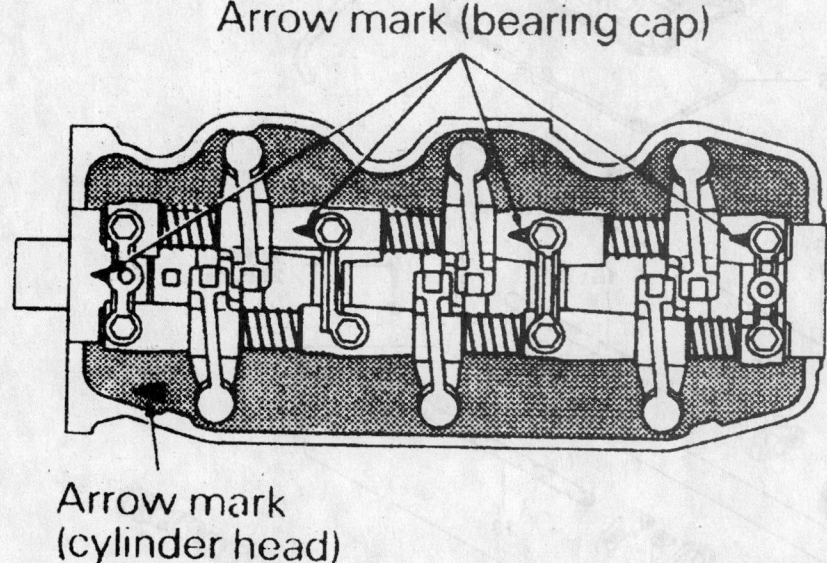

Arrow mark (bearing cap)

Arrow mark (cylinder head)

7923PG35

When installing the rocker arm/shaft assemblies, ensure that the arrow marks point in the same direction as the arrow stamped into the cylinder head—Diamante 3.0L SOHC engine

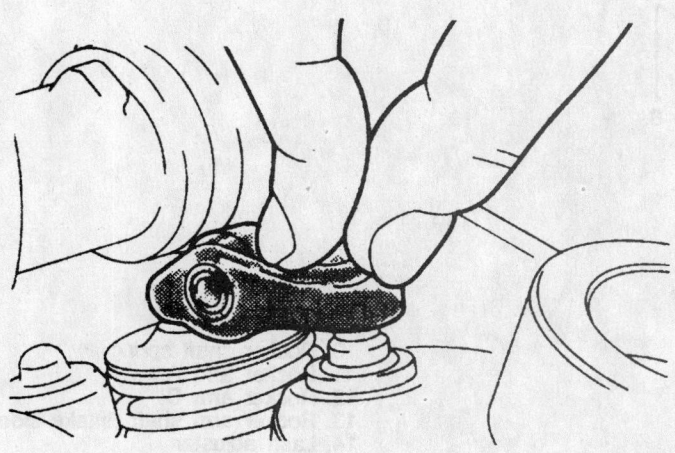

7923PG36

The rocker arms sit beneath the camshaft and are supported on one end by the valve stem and on the other end by the hydraulic lash adjuster—Diamante 3.0L DOHC engine

For Tune-up, Capacities and Firing orders, see Section 1 of this manual

- Air intake hose
- Breather hose and the PCV hose
- Spark plug cables from the spark plugs
- Rocker cover and gasket

3. Install lash adjuster retainer tools to the rocker arm.

4. Remove the rocker shaft hold-down bolts gradually and evenly and remove the rocker shaft/arm assemblies.

5. Disassemble the rockers and the rocker shaft springs from the rocker shafts. If they are to be reused, note the location and positioning of all rocker shaft components. It is recommended that all lash

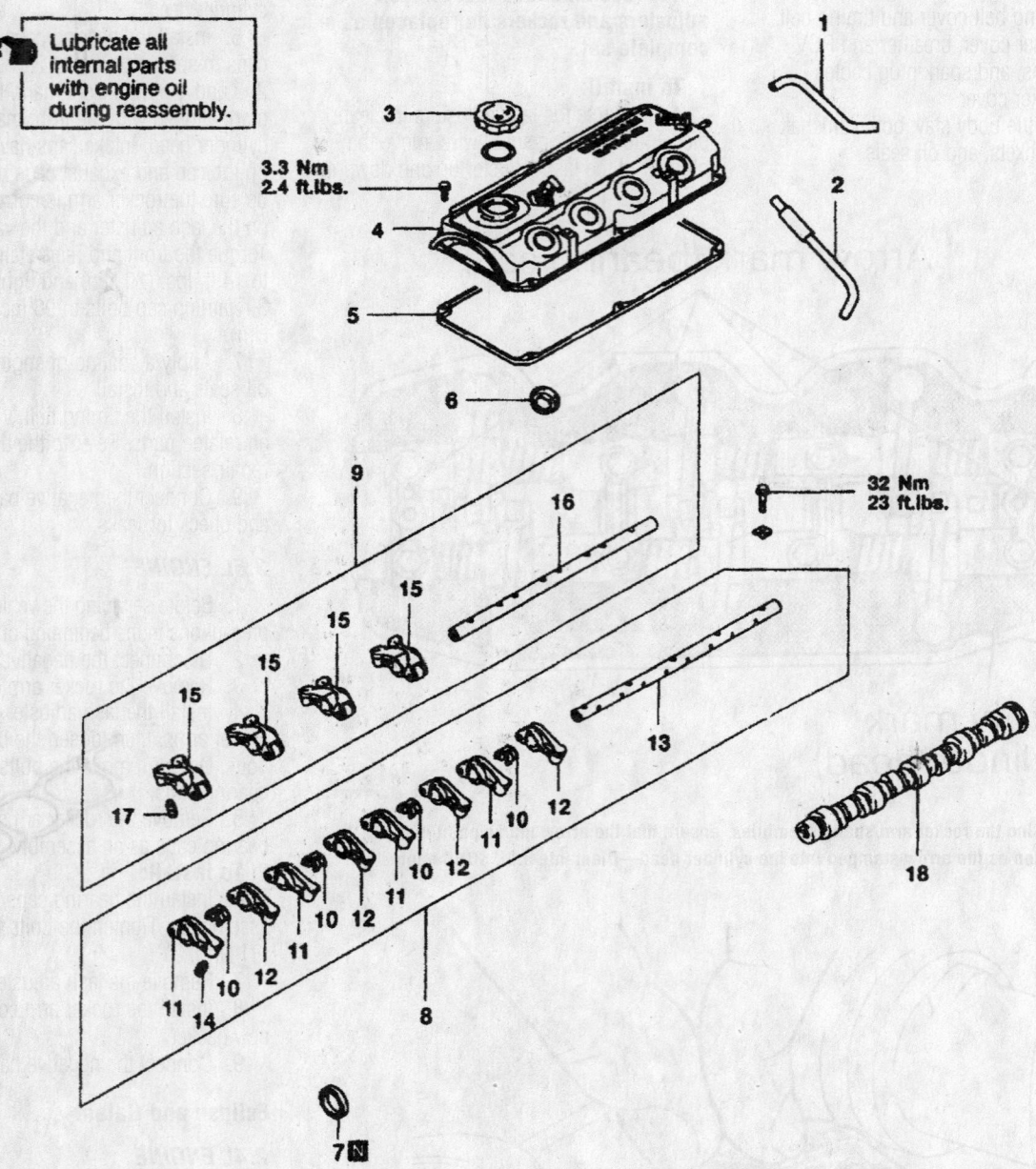

Lubricate all internal parts with engine oil during reassembly.

3.3 Nm 2.4 ft.lbs.

32 Nm 23 ft.lbs.

1. Breather hose
2. P.C.V. hose
3. Oil filler cap
4. Rocker cover
5. Rocker cover gasket
6. Oil seal
7. Oil seal
8. Rocker arms and rocker arm shaft
9. Rocker arms and rocker arm shaft
10. Rocker shaft spring
11. Rocker arm A
12. Rocker arm B
13. Rocker arm shaft (Intake side)
14. Lash adjuster
15. Rocker arm C
16. Rocker arm shaft (Exhaust side)
17. Lash adjuster
18. Camshaft

Rocker arm shafts and components—Eclipse and Galant 2.4L (4G64) engines

7923PG32

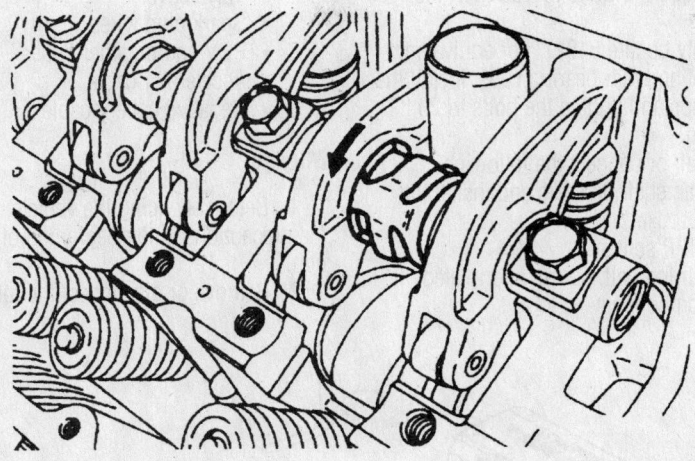

Installing the rocker shaft springs—Eclipse and Galant 2.4L (4G64) engines

adjusters and rockers be replaced as a complete set.

To install:

6. Immerse the lash adjusters in clean diesel fuel, and using a small wire, move the plunger up and down 4 or 5 times. While pushing down lightly on the check ball in order to bleed the air from the adjuster.

7. Install the lash adjusters to the rocker arms and attach the special holding tool.

8. Lubricate the rocker shaft with clean engine oil and install the rocker arms.

9. Temporarily tighten the rocker shaft assembly with the mounting bolts so that all rocker arms on the inlet valve side do not push on the valves.

10. Fit the rocker shaft springs from above and position them so that they are at right angles to the plug side. Install the rocker springs before installing the exhaust side rocker shaft and rocker arm assembly.

11. Install the exhaust side rocker shaft assembly in the engine. Tighten the rocker shaft mounting bolts gradually and evenly to 23 ft. lbs. (32 Nm).

12. Remove the lash adjuster retaining tools.

13. Install or connect the following:
- Rocker cover and tighten the mounting bolts to 29 inch lbs. (3.3 Nm)
- Spark plug wires to the spark plugs
- PCV and breather hoses
- Air intake hose
- Accelerator cable brackets and reconnect the accelerator cable
- Negative battery cable

2.0L (4G63) ENGINE

1. Before servicing the vehicle, refer to the precautions in the beginning of this section.

2. Relieve the fuel system pressure.

3. Remove or disconnect the following:
- Negative battery cable
- Accelerator cable
- PCV hose and breather hose
- Spark plug wires
- Valve cover
- Timing belt upper and lower covers
- Timing belt
- CMP sensor

➡ **Valvetrain components that are to be reused must be installed in the same locations from which they were removed.**

4. Remove the camshafts, rocker arms and lash adjusters.

To install:

5. Install the lash adjusters and rocker arms into the cylinder head. Lubricate lightly with clean oil prior to installation.

6. Apply engine oil to the lobes and journals of each camshaft. Install the camshafts into the cylinder head.

7. Install or connect the following:
- Camshaft bearing caps and tighten the bolts in the proper sequence to specifications in 3 even steps
- Camshaft oil seals and install the sprockets
- CMP sensor
- Timing belt, covers and related components
- Valve cover and all related components
- Negative battery cable

2.0L (420A) ENGINE

1. Before servicing the vehicle, refer to the precautions in the beginning of this section.

2. Remove or disconnect the following:
- Negative battery cable
- Accelerator cable
- PCV hose and breather hose
- Spark plug wires
- Cylinder head cover
- Timing belt upper and lower covers
- Timing belt
- CMP sensor
- Camshaft sprockets

➡ **Valvetrain components that are to be reused must be installed in the same locations from which they were removed.**

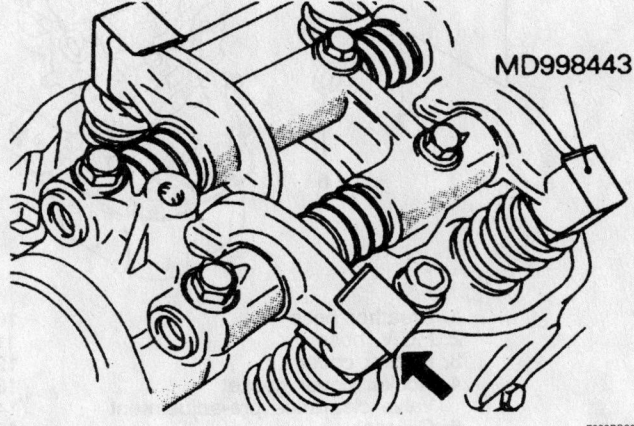

MD998443

Install the auto lash adjuster holder to prevent them from falling out—2.0L (4G63) and 2.4L (4G64) engines

3. Remove the camshafts, rocker arms and lash adjusters.

To install:

4. Install the lash adjusters and rocker arms into the cylinder head. Lubricate lightly with clean oil prior to installation.

5. Apply engine oil to the lobes and journals of each camshaft. Install the camshafts into the cylinder head.

6. Install and tighten the 4 center camshaft bearing caps in the proper sequence to specifications in 3 even steps. Torque the bolts to 108 inch lbs. (12 Nm).

7. Apply Loctite 51817®, or equivalent, to the front and rear bearing caps. Install the bearing caps and tighten the bolts to 20 ft. lbs. 28 Nm.

8. Install or connect the following:
- Camshaft oil seals and install the sprockets
- CMP sensor
- Timing belt, covers and related components

- Valve cover
- Spark plug wires
- PCV and breather hoses
- Accelerator cable
- Negative battery cable

Mirage

1. Before servicing the vehicle, refer to the precautions in the beginning of this section.

2. Remove or disconnect the following:

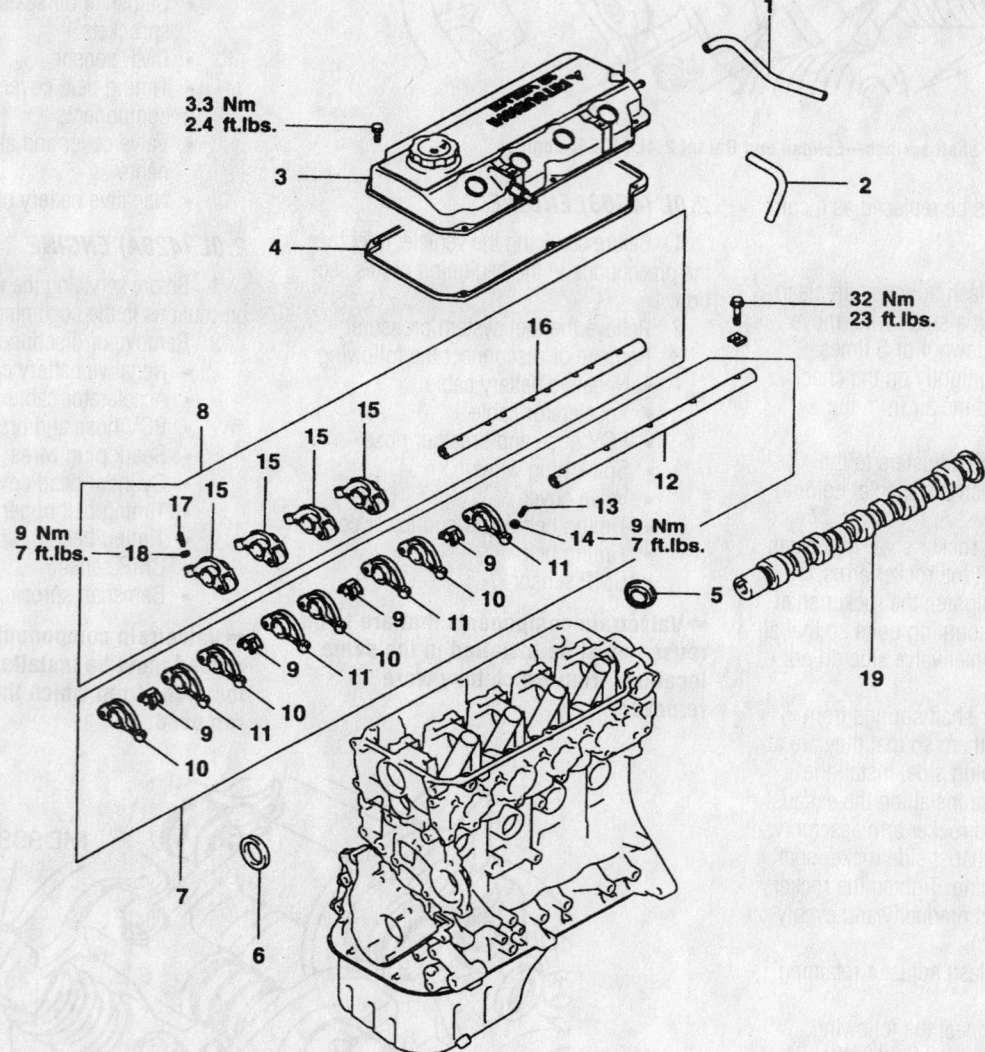

1. Breather hose
2. P.C.V. hose
3. Rocker cover
4. Rocker cover gasket
   Valve clearance pre-adjustment
5. Oil seal
6. Oil seal
7. Rocker arms and rocker arm shaft
8. Rocker arms and rocker arm shaft
9. Rocker shaft spring
10. Rocker arm A
11. Rocker arm B
12. Rocker arm shaft (Intake side)
13. Adjusting screw
14. Nut
15. Rocker arm C
16. Rocker arm shaft (Exhaust side)
17. Adjusting screw
18. Nut
19. Camshaft

Camshaft, rocker arm and shaft assemblies—Mirage 1.8L (4G93) engine

7923PG27

- Negative battery cable
- Spark plug cables—for 1.8L engine
- Accelerator cable, breather hose and PCV hose connections
- Rocker cover

3. Loosen both rocker arm shaft assemblies gradually and evenly and remove the rocket shafts from the vehicle.

4. If disassembly is required, keep all parts in the exact order of removal.

To install:

5. Lubricate the rocker shaft with clean engine oil and install the rockers and springs.

6. Install the rocker arm and shaft assemblies. Tighten the rocker arm shaft retainer bolts to 23 ft. lbs. (32 Nm).

7. Check valve adjustment and install the valve cover. Tighten the valve cover bolts to 16 inch lbs. (1.8 Nm) for the 1.5L engine or to 29 inch lbs. (3.3 Nm) for the 1.8L engine.

8. Install or connect the following:
   - Spark plug cables, if detached
   - Accelerator cable, breather hose and PCV hose
   - Negative battery cable

Turbocharger

REMOVAL & INSTALLATION

3000GT

RIGHT SIDE (FRONT)

1. Before servicing the vehicle, refer to the precautions in the beginning of this section.

2. Remove or disconnect the following:
   - Negative battery cable
   - Radiator
   - Right side transaxle bracket
   - Front exhaust pipe
   - All air intake hoses and pipes along the front of the engine
   - Alternator
   - Oil dipstick tube
   - Turbocharger heat shield
   - Water feed pipes
   - Heated Oxygen (HO2S) sensor
   - Oil return line
   - Exhaust extension fitting and bracket
   - All air conditioning components preventing removal of the turbocharger
   - Oil feed tube

- Turbocharger to exhaust manifold bolts and remove the turbocharger assembly

To install:

3. Clean all mating surfaces. Pour clean engine oil through the oil pipe feed hole in the turbocharger.

4. Install or connect the following:
   - Turbocharger to the manifold, with new gaskets. Torque the bolts to 40–47 ft. lbs. (55–65 Nm).
   - Oil feed pipe
   - Air conditioning components
   - Exhaust extension fitting and bracket. Torque the nuts to 40–47 ft. lbs. (55–65 Nm).
   - Oil return line
   - HO2S sensor
   - Water feed pipes
   - Turbocharger heat shield
   - Dipstick tube
   - Alternator
   - Air intake hoses and pipes along the front of the engine
   - New gasket and connect the front exhaust pipe
   - Right side transaxle bracket
   - Radiator

5. Fill the system with coolant.

6. Connect the negative battery cable.

LEFT SIDE (REAR)

1. Before servicing the vehicle, refer to the precautions in the beginning of this section.

2. Remove the battery.

3. Drain the coolant.

4. Remove or disconnect the following:
   - Front exhaust pipe
   - Accelerator cable from the throttle body
   - Intake air hose, the air pipe across the top of the engine and its heat shield
   - Clutch booster vacuum hose and disconnect the accelerator cable from the pedal
   - Air intake hoses coming from the air cleaner box
   - HO2S sensor and the turbocharger heat shield
   - EGR pipe, if equipped
   - Oil feed pipe
   - EGR valve, if equipped
   - Water feed pipes
   - Exhaust extension fitting and bracket
   - Inner heat shield
   - Oil return tube

- Turbocharger to exhaust manifold nuts and remove the turbocharger assembly

To install:

5. Clean all mating surfaces. Pour clean engine oil through the oil pipe feed hole in the turbocharger.

6. Install or connect the following:
   - New gasket and ring and install the turbocharger to the manifold. Torque the nuts to 40–47 ft. lbs. (55–65 Nm).
   - Oil return line
   - Inner heat shield
   - Exhaust extension fitting and bracket. Torque the nuts to 40–47 ft. lbs. (55–65 Nm).
   - Water feed pipes
   - EGR valve, if equipped
   - Oil feed pipe
   - EGR pipe, if equipped
   - Turbocharger heat shield and HO2S sensor
   - Air intake hoses coming from the air cleaner box. Be sure the triangular aligning marks are engaged.
   - Clutch booster vacuum hose
   - Heat shield, the air pipe across the top of the engine and the air intake hose
   - Accelerator cable to the throttle body
   - Front exhaust pipe

7. Fill the system with coolant.

8. Install the battery.

Eclipse

❋❋ CAUTION

The air bag system must be disarmed before removing the turbocharger.

1. Before servicing the vehicle, refer to the precautions in the beginning of this section.

2. Disconnect the negative battery cable.

3. Drain the engine coolant.

4. Remove or disconnect the following:
   - Condenser fan motor assembly, if equipped with air conditioning
   - HO2S sensor
   - Dipstick and tube assembly
   - Air cleaner and air intake hose assembly
   - Air intake hose from the turbocharger
   - Engine coolant hoses from the turbocharger
   - Oil supply pipe connection

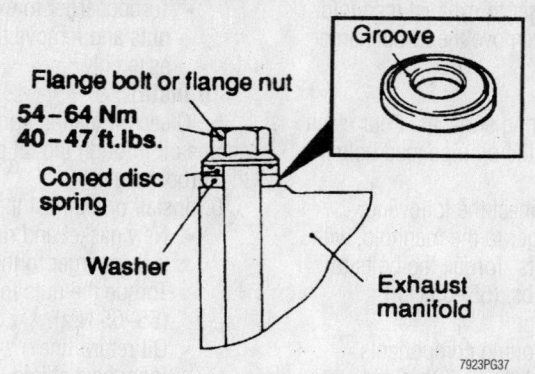

Install the groove of the cone-shaped disc spring toward the flange bolt or nut—Eclipse

- Heat shields
- Engine hanger
- Front exhaust pipe from the turbocharger
- Oil return pipe and gaskets
- Flange bolts and nut that attach the turbo to the exhaust manifold. Take note of the positions of the coned disc springs and the washers.
- Turbocharger, gasket and ring

To install:

5. Use a new gasket and install the turbo to the exhaust manifold. Be sure the coned disc spring and the washers are installed in

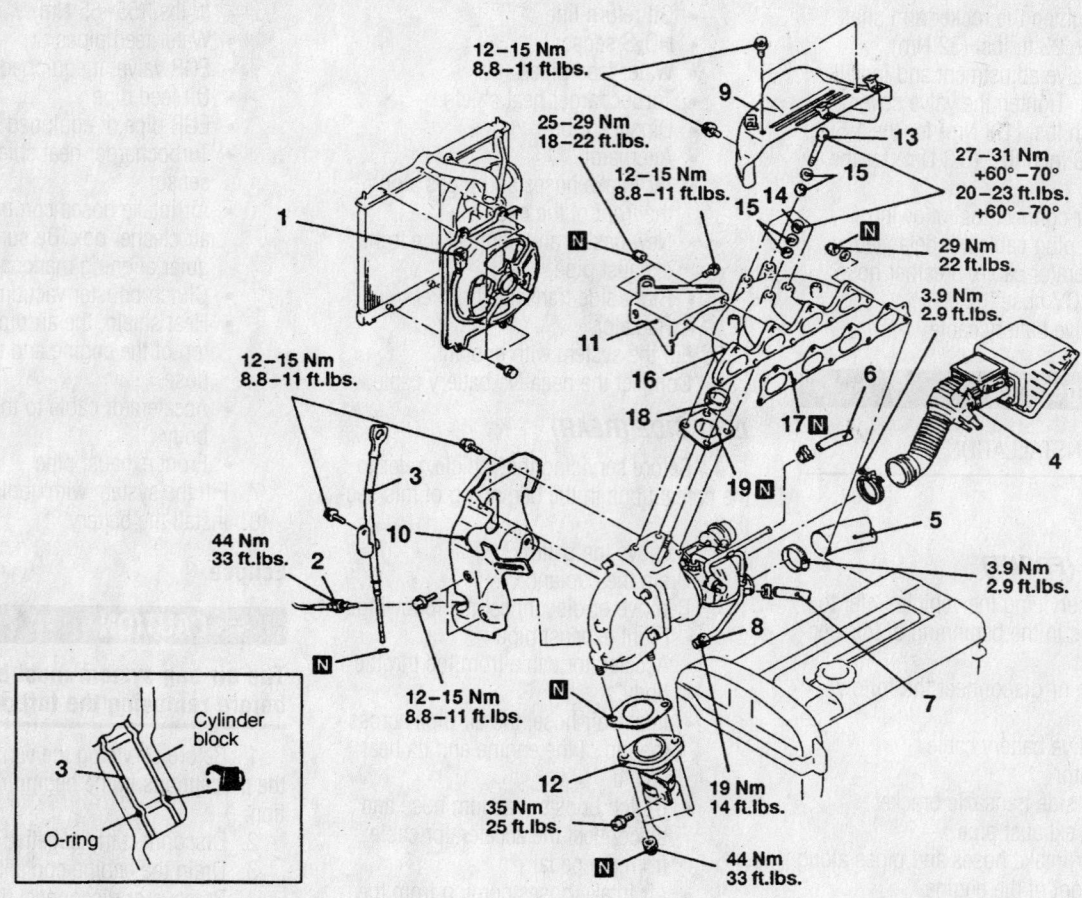

Removal steps

1. Condenser fan motor assembly <Vehicles with air conditioning>
2. Heated oxygen sensor <front>
3. Engine oil level gauge guide
4. Air cleaner and air intake hose assembly
5. Air hose (A) connection
6. Water hose connection
7. Water hose connection
8. Oil pipe (A) connection
9. Heat protector (A)
10. Heat protector (B)
11. Engine hanger
12. Front exhaust pipe connection
13. Flange bolts
14. Flange nut
15. Coned disc spring
16. Exhaust manifold
17. Exhaust manifold gasket
18. Ring
19. Gasket (A)

Exploded view of the turbocharger mounting—Eclipse 2.0L engine

7923PG47

their original positions. Torque the bolts and nut to 20–23 ft. lbs. (27–31 Nm). Further tighten the bolts and nuts 60–70 degrees.

6. Install or connect the following:
   - Exhaust pipe to the turbo—Torque the mounting bolts to 40–47 ft. lbs. (54–64 Nm).
   - Oil return pipe
   - Engine hanger
   - Heat shields
   - Oil supply pipe. Torque the flare nut fittings to 14 ft. lbs. (19 Nm).
   - Engine coolant hoses to the turbo
   - Air hose
   - Air cleaner and duct assembly
   - Dipstick tube and dipstick, with new O-ring
   - HO$_2$S sensor
   - Condenser fan assembly if removed
   - Negative battery cable and refill the engine with coolant

Intake Manifold

REMOVAL & INSTALLATION

3000GT

1. Before servicing the vehicle, refer to the precautions in the beginning of this section.
2. Relieve the fuel system pressure.
3. Disconnect battery negative cable.
4. Drain the cooling system.
5. Remove or disconnect the following:
   - Air intake hoses
   - Accelerator cables from the throttle body
   - Vacuum hoses, label
   - Harness connectors from the upper intake plenum
   - Clutch booster vacuum hose connection, if equipped
   - EGR components on California vehicles
   - Plenum retaining bracket
   - Plenum retaining nuts and bolts and remove the air intake plenum
   - High pressure and return fuel hoses
   - Fuel injector connectors and the pressure regulator vacuum line
   - Fuel rail with the injectors attached
   - Timing belt upper cover
   - Intake manifold mounting nuts
   - Intake manifold

To install:

6. Thoroughly clean the mating sur-

faces of the heads, intake manifold and air intake plenum.

7. Install new intake manifold gaskets to the heads with the adhesive side facing up.

8. Place the manifold on the heads and install the cone disc springs and/or lockwashers.

9. Lubricate the studs lightly with oil, then install the nuts following this procedure:
   - Front bank: 48–72 inch lbs. (3–5 Nm).
   - Rear bank: 14–17 ft. lbs. (20–23 Nm).
   - Front bank: 14–17 ft. lbs. (20–23 Nm).
   - Repeat both banks: 14–17 ft. lbs. (20–23 Nm).

10. Install or connect the following:
   - Timing belt upper cover
   - Fuel rail assembly
   - Fuel injector connectors and the pressure regulator vacuum line
   - Fuel hoses, with new O-rings
   - Plenum. Tighten the retaining nuts and bolts evenly and gradually to 13 ft. lbs. (18 Nm).
   - Retaining bracket
   - EGR components on California vehicles
   - Upper intake harness connectors and vacuum hoses
   - Adjust the accelerator cables
   - Air intake hoses

11. Fill the system with coolant.
12. Connect the negative battery cable.

Diamante

1. Before servicing the vehicle, refer to the precautions in the beginning of this section.
2. Relieve the fuel system pressure.
3. Remove or disconnect the following:
   - Negative cable and drain the cooling system
   - Air intake hose(s)
   - Accelerator control cables from the throttle body
   - Vacuum hoses including the brake booster hose
   - Wiring harness connectors
   - High pressure and return fuel hoses
   - EGR pipe and remove the EGR valve and EGR temperature sensor from the intake plenum assembly
   - MAP sensor, if equipped
   - Plenum retaining bracket

- Plenum retaining nuts and bolts and remove the air intake plenum from the intake manifold
- Upper timing belt covers
- Water pump stay bracket

➡ **It is not necessary to remove the fuel injectors from the intake unless the manifold assembly is being replaced.**

4. Remove or disconnect the following:
   - Fuel rail with the injectors attached
   - Coolant hoses from the intake manifold
   - Intake manifold mounting nuts and remove the intake manifold
5. Clean the gasket mounting surfaces.

To install:

6. Thoroughly clean the mating surfaces of the heads, intake manifold and air intake plenum.

7. Install new intake manifold gaskets to the cylinder heads with the adhesive side facing up.

8. Place the manifold on the cylinder heads.

9. Lubricate the studs lightly with oil and install the nuts.

10. Tighten the mounting nuts as follows:
   - Front bank nuts: 48–72 inch lbs. (5–8 Nm).
   - Rear bank nuts: 14–17 ft. lbs. (20–23 Nm).
   - Front bank nuts: 14–17 ft. lbs. (20–23 Nm).

11. Connect the coolant hoses to the intake manifold.

12. Using new O-rings, install the fuel rail assembly, if removed. Tighten the mounting bolts to 84–108 inch lbs. (10–13 Nm).

13. Install or connect the following:
   - Plenum, with new gasket. Tighten the retaining nuts and bolts evenly and gradually to 13 ft. lbs. (18 Nm).
   - Retaining bracket and tighten the retaining bolts to 13 ft. lbs. (18 Nm)
   - MAP sensor, if removed
   - EGR valve, using a new gasket. Tighten the bolts to 16 ft. lbs. (22 Nm)
   - EGR temperature sensor and tighten the fitting to 84–108 inch lbs. (10–12 Nm)
   - EGR pipe and tighten the fittings to 43 ft. lbs. (60 Nm)
   - High pressure fuel hose, use a new O-ring. Tighten the retaining bolts

Timing belt service is covered in Section 3 of this manual

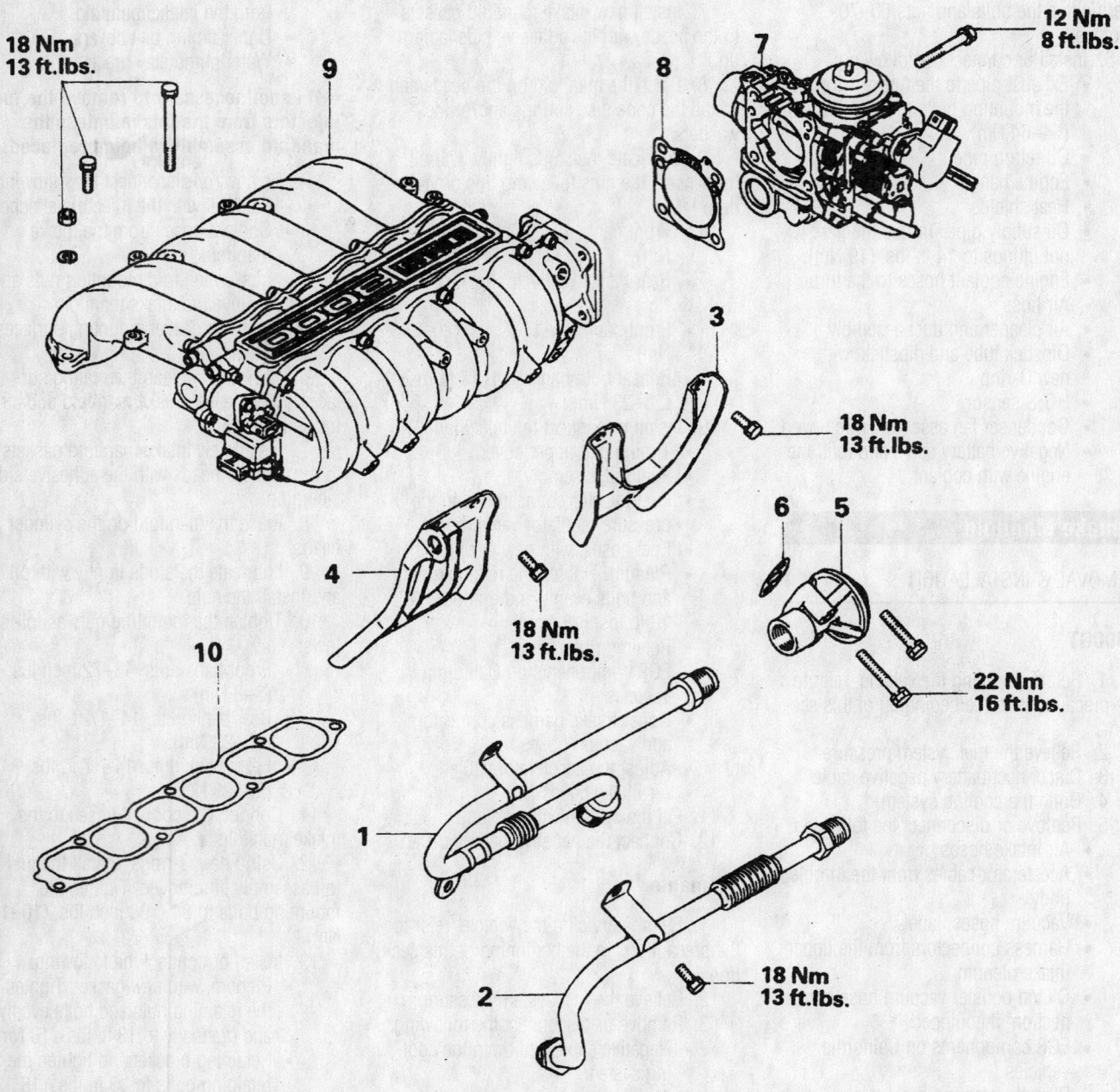

18 Nm
13 ft.lbs.

12 Nm
8 ft.lbs.

18 Nm
13 ft.lbs.

18 Nm
13 ft.lbs.

22 Nm
16 ft.lbs.

18 Nm
13 ft.lbs.

1. EGR pipe -- Up to 1993 <California> model
2. EGR pipe -- From 1994 <California> model
3. Intake manifold plenum stay, rear
4. Intake manifold plenum stay, front
5. EGR valve
6. EGR valve gasket } <For California>
7. Throttle body
8. Throttle body gasket
9. Intake manifold plenum
10. Intake manifold plenum gasket

Exploded view of air intake plenum assembly—Diamante shown, 3000GT similar

7923PG42

1. Connection for high-pressure fuel hose
2. O-ring
3. Connection for fuel return hose
4. Connection for vacuum hoses
5. Wiring harness connector
6. Oxygen sensor <For California from 1994 models>
7. Fuel rail (with injectors)
8. Insulators
9. Timing belt upper cover
10. Water pump stay mounting bolt
11. Intake manifold mounting nut
12. Intake manifold mounting nut
13. Cone disc spring
14. Intake manifold
15. Intake manifold gasket

Intake manifold and related components—Diamante shown, 3000GT similar

to 48 inch lbs. (5 Nm).
- Fuel return hose, using a new clamp
- Water pump stay bracket
- Upper timing belt covers
- Harness connectors and vacuum hoses
- Accelerator cables, adjust
- Air intake hose(s)
14. Fill the system with coolant.
15. Connect the negative battery cable.

Eclipse

2.0L (4G63) ENGINE

1. Before servicing the vehicle, refer to the precautions in the beginning of this section.
2. Relieve the fuel system pressure.
3. Remove the battery.
4. Drain the engine coolant.
5. Remove or disconnect the following:
   - Accelerator cable

- Air intake hose
- Ignition coil and the module wiring connectors
- MAP sensor
- Condenser
- TP sensor and the IAC motor connectors
- Knock sensor and the ECT sensor connectors
- CMP sensor and the CKP sensor connectors
- Air conditioning compressor connector
- Engine control wiring harness bracket and position the harness out of the way
- Vacuum hoses, label for reference
- Spark plug wires from the ignition coil
- Fuel lines from the fuel rail
- Heater hoses
- Fuel rail assembly and insulators
- Ignition coil and module

- EGR valve assembly
- Intake manifold stay and the engine hanger
- Intake manifold and gasket

To install:
6. Install or connect the following:
   - Intake manifold
   - Intake manifold stay and the engine hanger
   - EGR assembly
   - Ignition coil and module
   - Fuel rail and insulators
   - Heater hoses and fuel lines
   - Spark plug wires to the coil towers
   - Vacuum hoses
   - MAP sensor
   - TP sensor and the IAC motor connectors
   - Knock sensor and the ECT sensor connectors
   - CMP sensor and the CKP sensor connectors.
   - Ignition condenser

Heater Core replacement is covered in Section 2 of this manual

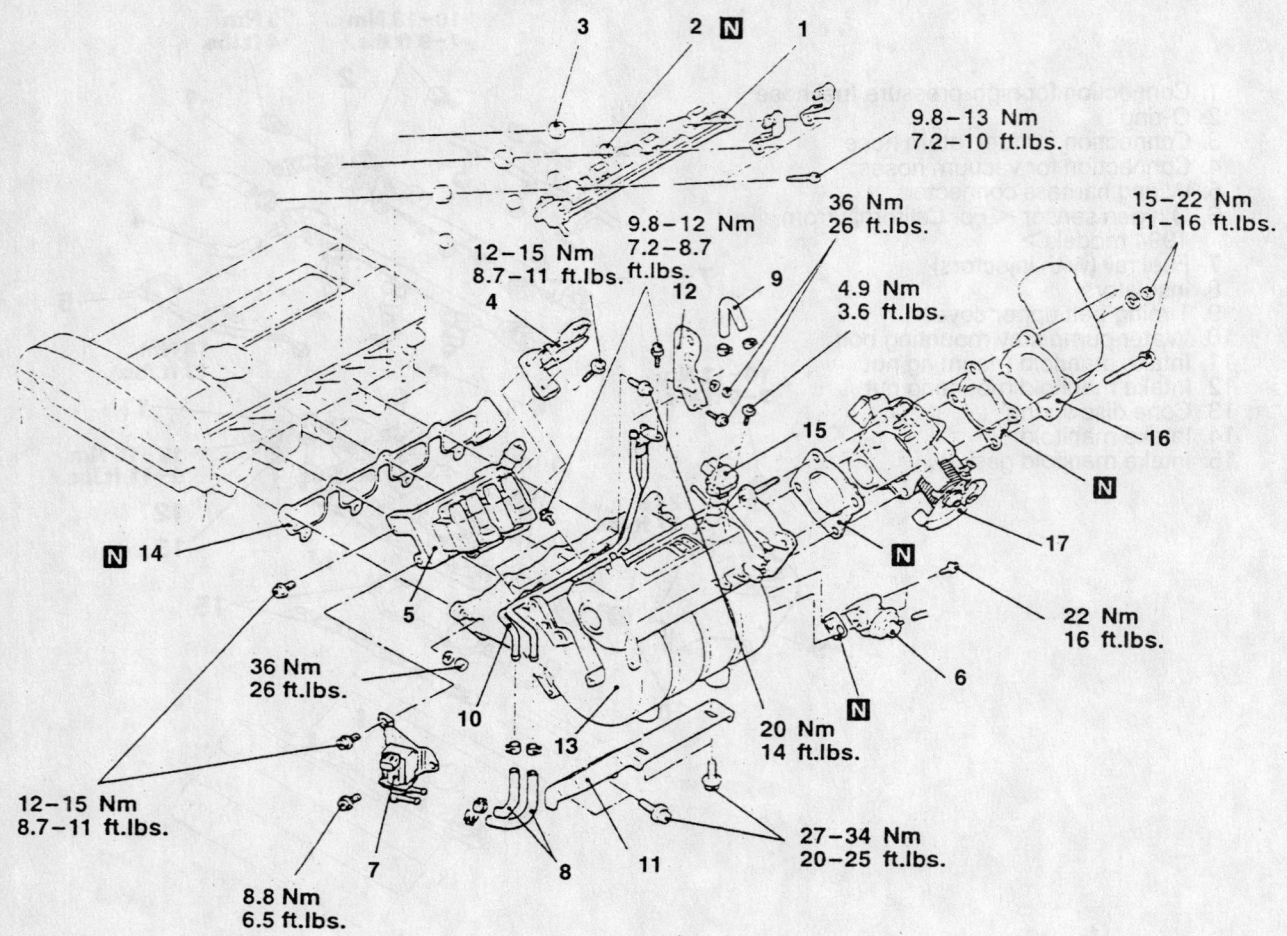

1. Fuel rail, fuel injector and pressure regulator assembly
2. Insulator
3. Insulator
4. Ignition power transistor
5. Ignition coil
6. EGR valve assembly
7. Evaporative emission purge solenoid valve assembly
8. Purge hoses
9. Hose
10. Vacuum pipe
11. Intake manifold stay
12. Engine hanger
13. Intake manifold
14. Intake manifold gasket
15. Manifold differential pressure sensor
16. Charge air cooler fitting
17. Throttle body

7923PG40

Exploded view of the intake manifold and related components—2.0L (4G63) engine

- Accelerator cable, adjust
- Battery
7. Refill the engine with coolant.
8. Start the engine and check for leaks.

2.0L (420A) ENGINE

1. Before servicing the vehicle, refer to the precautions in the beginning of this section.

2. Disconnect the negative battery cable.

3. Drain the engine coolant.

4. Remove or disconnect the following:
- Vacuum reservoir, if equipped with cruise control
- Air intake hose and breather hose
- Accelerator cable from the bracket
- Engine harness retaining clips

- MAP sensor
- IAT sensor
- Vacuum hose connection
- TP sensor and the IAC motor connectors

5. Position the engine control wiring harness out of the way.

6. Remove or disconnect the following:
- Alternator wiring harness connector

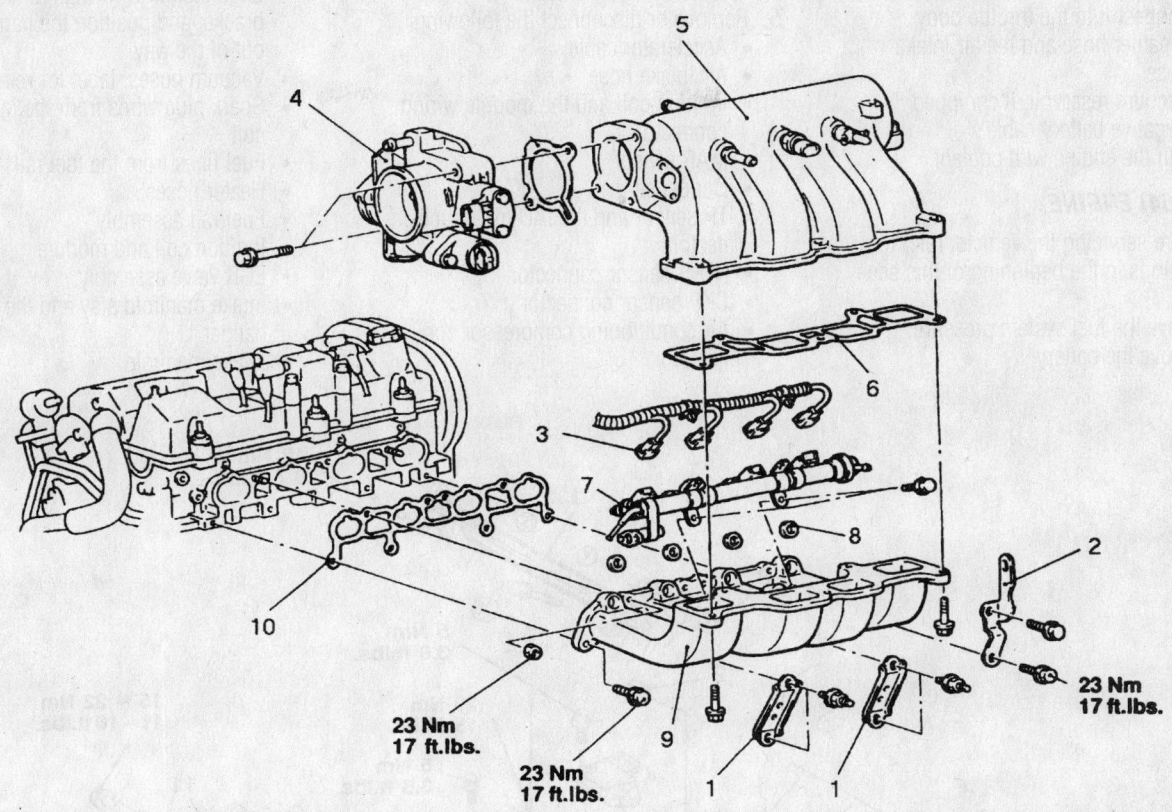

1. Intake manifold stay
2. Engine hanger
3. Injector connector
4. Throttle body
5. Intake manifold plenum
6. Intake manifold plenum gasket
7. Fuel rail, injector and pressure regulator assembly
8. O-ring
9. Intake manifold
10. Intake manifold gasket

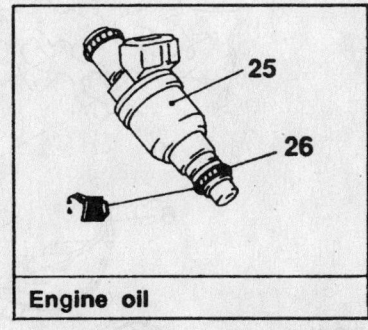

Engine oil

7923PG39

Intake manifold and related components—2.0L (420A) engine

- PCV hose assembly
- Vacuum hoses, label for reference
- EGR pipe connection
- Fuel lines from the fuel rail
- Intake manifold stay and the engine hanger
- Throttle body
- Intake manifold plenum and gasket
- Injector connectors
- Fuel rail with the injectors
- Intake manifold

To install:

7. Install or connect the following:
- Intake manifold
- Fuel rail assembly and connect the injectors
- Intake plenum
- Throttle body
- Intake manifold stay and the engine hanger
- Fuel lines to the fuel rail, use new O-ring(s)
- EGR pipe

- Vacuum hoses and the PCV hose assembly
- Alternator wiring harness connector

8. Reposition the engine control wiring harness and install the brackets and clips.

9. Install or connect the following:
- IAC motor and the TP sensor connectors
- Vacuum hose to the throttle body
- MAP and the IAT sensor connectors

Brake service is covered in Section 4 of this manual

- Accelerator cable in the bracket and connect it to the throttle body
- Breather hose and the air intake hose
- Vacuum reservoir, if equipped
- Negative battery cable

10. Refill the engine with coolant.

2.4L (4G64) ENGINE

1. Before servicing the vehicle, refer to the precautions in the beginning of this section.
2. Relieve the fuel system pressure.
3. Remove the battery.

4. Drain the engine coolant.
5. Remove or disconnect the following:
   - Accelerator cable
   - Air intake hose
   - Ignition coil and the module wiring connectors
   - MAP sensor
   - Condenser
   - TP sensor and the IAC motor connectors
   - HO2S sensor connector
   - CKP sensor connector
   - Air conditioning compressor connector

- Engine control wiring harness bracket and position the harness out of the way
- Vacuum hoses, label for reference
- Spark plug wires from the ignition coil
- Fuel lines from the fuel rail
- Heater hoses
- Fuel rail assembly
- Ignition coil and module
- EGR valve assembly
- Intake manifold stay and the engine hanger
- Intake manifold

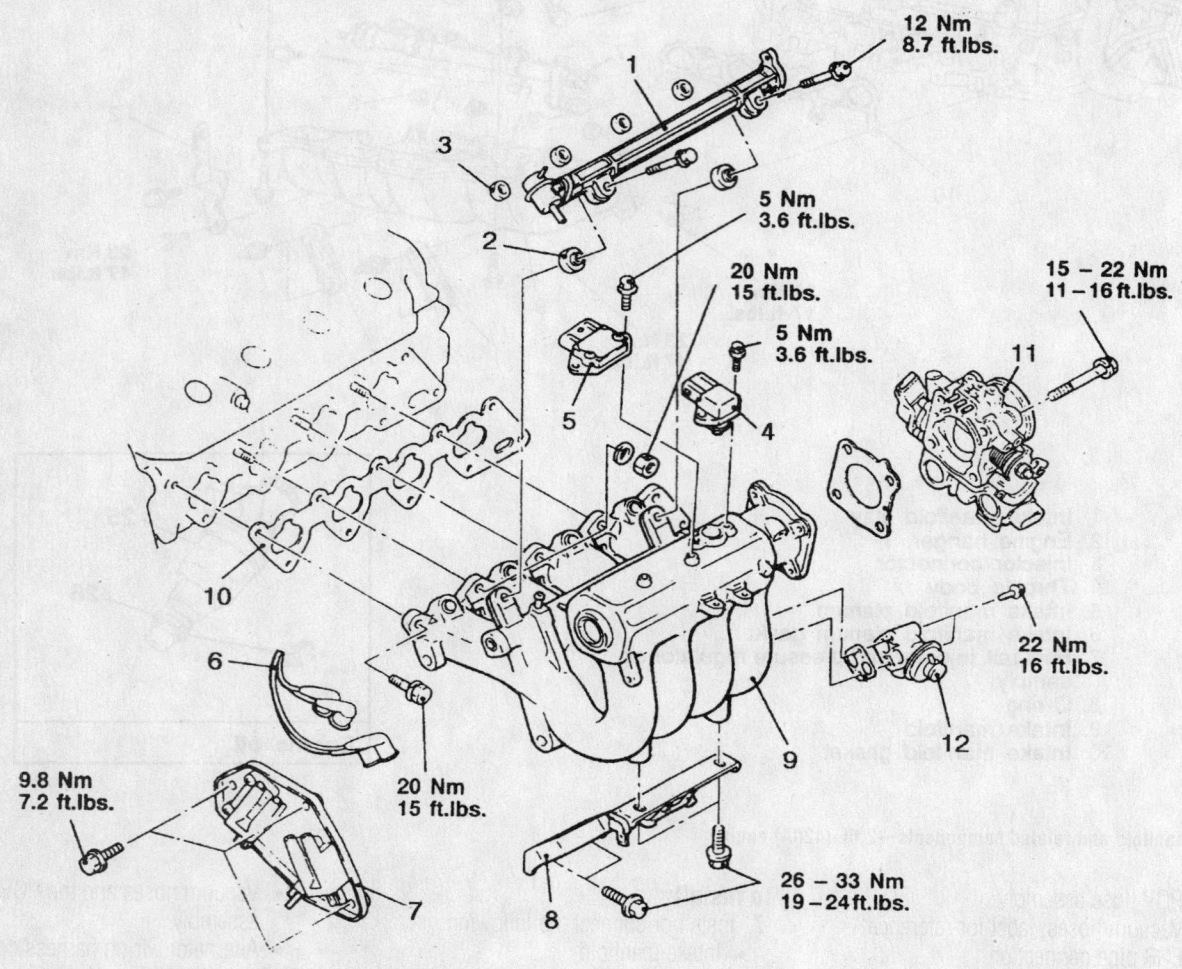

1. Fuel rail, fuel injector and pressure regulator assembly
2. Insulator
3. Insulator
4. Manifold differential pressure sensor
5. Ignition power transistor
6. Spark plug cable connection

7. Ignition coil
8. Intake manifold stay
9. Intake manifold
10. Intake manifold gasket
11. Throttle body
12. EGR valve assembly

Intake manifold and related components—Eclipse and Galant 2.4L (4G64) engines

7923PG41

To install:

6. Install or connect the following:
   - Intake manifold. Torque the intake manifold bolts to 15 ft. lbs. (20 Nm).
   - Intake manifold stay and the engine hanger. Torque the mounting bolts to 19–24 ft. lbs. (26–33 Nm).
   - EGR assembly
   - Ignition coil and module
   - Fuel rail and insulators and reconnect the high-pressure fuel hose
   - Heater hoses and fuel lines
   - Spark plug wires to the coil towers
   - Vacuum hoses
   - Engine harness in the proper position
   - MAP sensor
   - TP sensor and the IAC motor connectors
   - HO$_2$S
   - Ignition condenser
   - Accelerator cable, adjust
   - Battery
7. Refill the engine with coolant.

Galant

1. Before servicing the vehicle, refer to the precautions in the beginning of this section.
2. Relieve the fuel system pressure.
3. Disconnect battery negative cable.
4. Drain the cooling system.
5. Remove or disconnect the following:
   - Accelerator cable
   - Air intake hose
   - Coolant hose from the throttle housing
   - Vacuum lines, label for reference
   - High pressure fuel line and fuel return hose
   - Throttle control cable brackets
6. Disconnect the following:
   - ECT sensor and gauge sender
   - IAC motor
   - EGR temperature sensor
   - Ignition coil
   - Knock sensor
   - HO$_2$S sensor
   - TP sensor
   - Distributor (if equipped)
   - air conditioning temperature sensor
   - Ignition power transistor
   - Fuel injectors
7. Remove or disconnect the following:
   - Spark plug wires
   - Intake manifold stay bracket
   - Intake manifold mounting bolts and remove the intake manifold assembly

To install:

8. Clean all gasket material from the cylinder head intake mounting surface and intake manifold assembly.
9. Install or connect the following:
   - Intake manifold, using a new gasket. Torque the manifold in a crisscross pattern, starting from the inside and working outwards to 15 ft. lbs. (20 Nm).
   - Fuel injectors, fuel rail and pressure regulator to the engine. Torque the retaining bolts to 48 inch lbs. (6 Nm).
   - Intake manifold brace bracket and tighten bolts to 21 ft. lbs. (29 Nm)
   - Spark plug wires
10. Connect the following:
    - ECT sensor and gauge sender
    - IAC motor
    - EGR temperature sensor
    - Ignition coil
    - Knock sensor
    - HO$_2$S sensor
    - TP sensor
    - Distributor (if equipped)
    - air conditioning temperature sensor
    - Ignition power transistor
    - Fuel injectors
11. Install or connect the following:
    - Throttle control cable brackets
    - High pressure fuel line and fuel return hose
    - Vacuum lines
    - Coolant hoses
    - Accelerator cable
    - Air intake hose
12. Fill the system with coolant.
13. Connect the negative battery cable.

Mirage

1.5L (4G15) ENGINE

1. Before servicing the vehicle, refer to the precautions in the beginning of this section.
2. Relieve the fuel system pressure.
3. Remove or disconnect the following:
   - Battery negative cable and drain the cooling system
   - Upper radiator hose, heater hose and water bypass hose
   - Thermostat housing from intake manifold
   - Accelerator cable, breather hose and air intake hose
   - Vacuum hoses, label for reference
   - Throttle body assembly

- High pressure fuel line and the fuel return hose
4. Disconnect the following:
   - HO$_2$S sensor
   - ECT sensor
   - IAC motor
   - IAT sensor
   - Distributor (if equipped)
   - EGR temperature sensor
5. Remove or disconnect the following:
   - Spark plug wires
   - Fuel rail, fuel injectors, pressure regulator and insulators
   - EGR valve from the intake manifold
   - Intake manifold support bracket and remove the engine mount support bracket
   - Intake manifold mounting bolts and remove the intake manifold assembly

To install:

6. Clean all gasket material from the cylinder head intake mounting surface and intake manifold assembly.
7. Install or connect the following:
   - Intake manifold gasket, using a new gasket. Torque the manifold in a crisscross pattern, starting from the inside and working outwards to 13 ft. lbs. (18 Nm).
   - Intake manifold support bracket and tighten the mounting bolts to 16 ft. lbs. (22 Nm)
   - Engine mount support bracket and tighten the mounting bolts to 26 ft. lbs. (36 Nm)
   - EGR valve and tighten the mounting bolts to 15 ft. lbs. (21 Nm)
   - Install the fuel rail, fuel injectors and pressure regulator to the engine, using new insulators and O-rings. Torque the retaining bolts to 84–108 inch lbs. (10–13 Nm).
   - Spark plug wires
8. Connect the following:
   - HO$_2$S sensor
   - ECT sensor
   - IAC motor
   - IAT sensor
   - Distributor (if equipped)
   - EGR temperature sensor
9. Install or connect the following:
   - Fuel feed and return lines
   - Throttle body assembly
   - Vacuum hoses and pipes as necessary, including the brake booster vacuum line
   - Accelerator cable
   - Breather and air intake hose

For complete Engine Mechanical specifications, see Section 1 of this manual

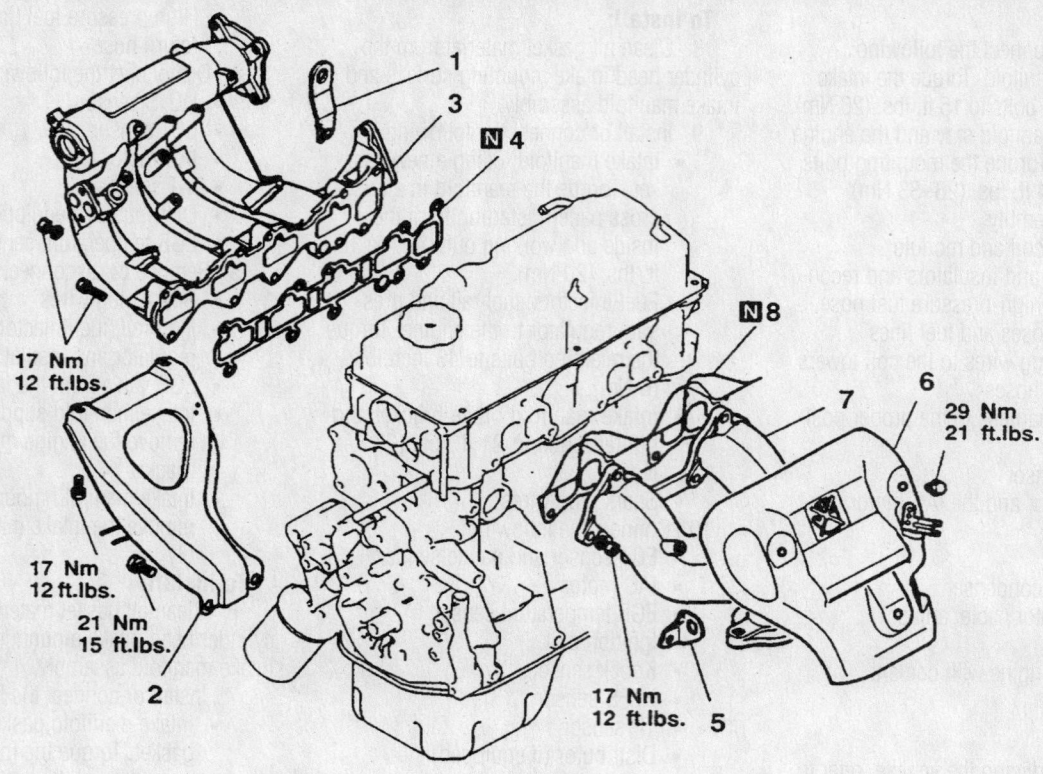

**17 Nm
12 ft.lbs.**

**17 Nm
12 ft.lbs.**

**21 Nm
15 ft.lbs.**

**17 Nm
12 ft.lbs.**

**29 Nm
21 ft.lbs.**

1. Engine hanger
2. Intake manifold stay
3. Intake manifold
4. Intake manifold gasket
5. Engine hanger
6. Exhaust manifold cover
7. Exhaust manifold
8. Exhaust manifold gasket

7923PG38

Exploded view of the intake and exhaust manifold mounting—Mirage 1.5L (4G15) engine

- Thermostat housing to the intake manifold and tighten the mounting bolts to 13 ft. lbs. (18 Nm)
- Upper radiator hose, heater hose and water bypass hose
10. Fill the system with coolant.
11. Connect the negative battery cable.

1.8L (4G93) ENGINE

1. Before servicing the vehicle, refer to the precautions in the beginning of this section.
2. Relieve the fuel system pressure.
3. Remove or disconnect the following:
- Battery negative cable and drain the cooling system
- Accelerator cable and the air intake hose
4. Disconnect the following:
- HO2S sensor
- ECT sensor
- IAC motor
- EGR temperature sensor
- TP sensor
- Oil pressure switch

- Distributor (if equipped)
- Fuel injectors
5. Label and remove all vacuum hoses.
6. Remove or disconnect the following:
- Upper radiator hose, heater hose and water bypass hose
- High pressure fuel line and the fuel return hose
- Fuel rail, fuel injectors, pressure regulator and insulators
- Intake manifold support bracket
- Thermostat housing, if necessary for clearance
- Intake manifold mounting bolts/nuts and remove the intake manifold assembly

To install:

7. Clean all gasket material from the cylinder head intake mounting surface and intake manifold assembly.
8. Install or connect the following:
- Intake manifold, using a new gasket. Torque the manifold in a criss-cross pattern, starting from the

inside and working outwards to 14 ft. lbs. (20 Nm).
- Thermostat housing
- Intake manifold brace bracket
- Fuel rail, fuel injectors and pressure regulator to the engine. Torque the retaining bolts to 108 inch lbs. (12 Nm).
- Fuel feed and return lines
- Upper radiator hose, heater hose and water bypass hoses
- Vacuum hoses
9. Connect the following:
- HO2S sensor
- ECT sensor
- IAC motor
- EGR temperature sensor
- TP sensor
- Oil pressure switch
- Distributor (if equipped)
- Fuel injectors
10. Connect and adjust the accelerator cable and install the air intake hose.
11. Fill the system with coolant.
12. Connect the negative battery cable.

Exhaust Manifold

REMOVAL & INSTALLATION

3000GT

NON-TURBOCHARGED ENGINES

1. Before servicing the vehicle, refer to the precautions in the beginning of this section.
2. Remove or disconnect the following:
   • Battery negative cable
   • Exhaust pipe from the exhaust manifold
   • Electric cooling fan assembly, if necessary
   • Dipstick tube, if removing the front manifold
   • Alternator, if removing the front manifold from 3.0L DOHC engine
   • EGR pipe
   • Electrical connector and remove the HO$_2$S sensor
   • Exhaust manifold mounting bolts, the inner heat shield and the exhaust manifold

To install:

3. Clean all gasket material from the mating surfaces.
4. Install or connect the following:
   • New gasket and install the manifold. Tighten the nuts in a crisscross pattern to 22 ft. lbs. (30 Nm).
   • Heat shields
   • EGR pipe
   • HO$_2$S sensor
   • Electric cooling fan assembly, dipstick tube and alternator, as required
   • New flange gasket and connect the exhaust pipe
   • Negative battery cable and check for exhaust leaks

TURBOCHARGED ENGINES

1. Before servicing the vehicle, refer to the precautions in the beginning of this section.
2. Disconnect the negative battery cable.
3. Drain the engine coolant.
4. Remove or disconnect the following:
   • Turbocharger assembly
   • Heat shield
   • Mounting nuts
   • Exhaust manifold

➡**The cone disc springs are installed at all lower mounting points.**

To install:

5. Clean all gasket material from the mating surfaces.
6. Install new gaskets and install the manifold. Be sure all cone disc springs are in their original locations with the grooved side facing the nut. Tighten the manifold nuts using the following procedure:
   a. Tighten all but the outer 2 nuts to 22 ft. lbs. (30 Nm).
   b. Tighten the outer 2 nuts to 34–38 ft. lbs. (47–53 Nm).
   c. Loosen the outer 2 nuts, then tighten them to 22 ft. lbs. (30 Nm).
7. Install or connect the following:
   • Heat shield
   • Turbocharger assembly
8. Fill the cooling system.
9. Connect the negative battery cable and check for exhaust leaks.

Diamante

1. Before servicing the vehicle, refer to the precautions in the beginning of this section.
2. Remove or disconnect the following:
   • Battery negative cable
   • Exhaust pipe from the exhaust manifold
   • Condenser electric cooling fan assembly
3. For the DOHC engine, remove the following:
   • Alternator and mounting bracket
   • Separate the air conditioning compressor from the mounting bracket. Leaving the hoses connected, position the compressor aside.
   • Heat protector
4. If removing the front manifold, remove the oil dipstick and tube from the engine.
5. If removing the rear manifold, disconnect the EGR tube.
6. For the SOHC engine, if removing the rear manifold, remove the intake plenum stay and the roll stopper bracket.
7. Remove or disconnect the following:
   • Electrical connector and remove the HO$_2$S sensor
   • Exhaust manifold mounting bolts the manifold

To install:

8. Clean all gasket material from the mating surfaces.
9. Install or connect the following:
   • New gasket and install the manifold. Tighten the nuts in a crisscross pattern to 21 ft. lbs. (30 Nm)

for the J- engine or to 14 ft. lbs. (19 Nm) for the H- engine.
   • Heat shields
   • EGR tube and intake plenum stay and roll stopper bracket, if removed
   • HO$_2$S sensor
   • Electric cooling fan assembly, air conditioning compressor, dipstick tube and alternator, as required
   • New flange gasket and connect the exhaust pipe or converter assembly
   • Negative battery cable and check for exhaust leaks

Eclipse

2.0L (4G63) ENGINE

✳✳ **CAUTION**

The air bag system must be disarmed before removing the exhaust manifold or turbocharger.

1. Before servicing the vehicle, refer to the precautions in the beginning of this section.
2. Disconnect the negative battery cable.
3. Drain the engine coolant.
4. Remove or disconnect the following:
   • Condenser fan motor assembly, if equipped with air conditioning
   • HO$_2$S sensor
   • Dipstick and tube assembly
   • Air cleaner and air intake hose assembly
   • Air intake hose from the turbocharger
   • Engine coolant hoses from the turbocharger
   • Oil supply pipe connection.
   • Heat shields
   • Engine hanger
   • Front exhaust pipe from the turbocharger
   • Oil return pipe and gaskets
   • Flange bolts and nuts that attach the turbo to the exhaust manifold. Take note of the positions of the coned disc springs and the washers.
   • Turbocharger, gasket and ring
   • Exhaust manifold and gasket

To install:

5. Use a new gasket and install the exhaust manifold.
6. Use a new gasket and install the turbo to the exhaust manifold. Be sure the coned disc spring and the washers are installed in their original positions. Torque the bolts

For Accessory Drive Belt illustrations, see Section 1 of this manual

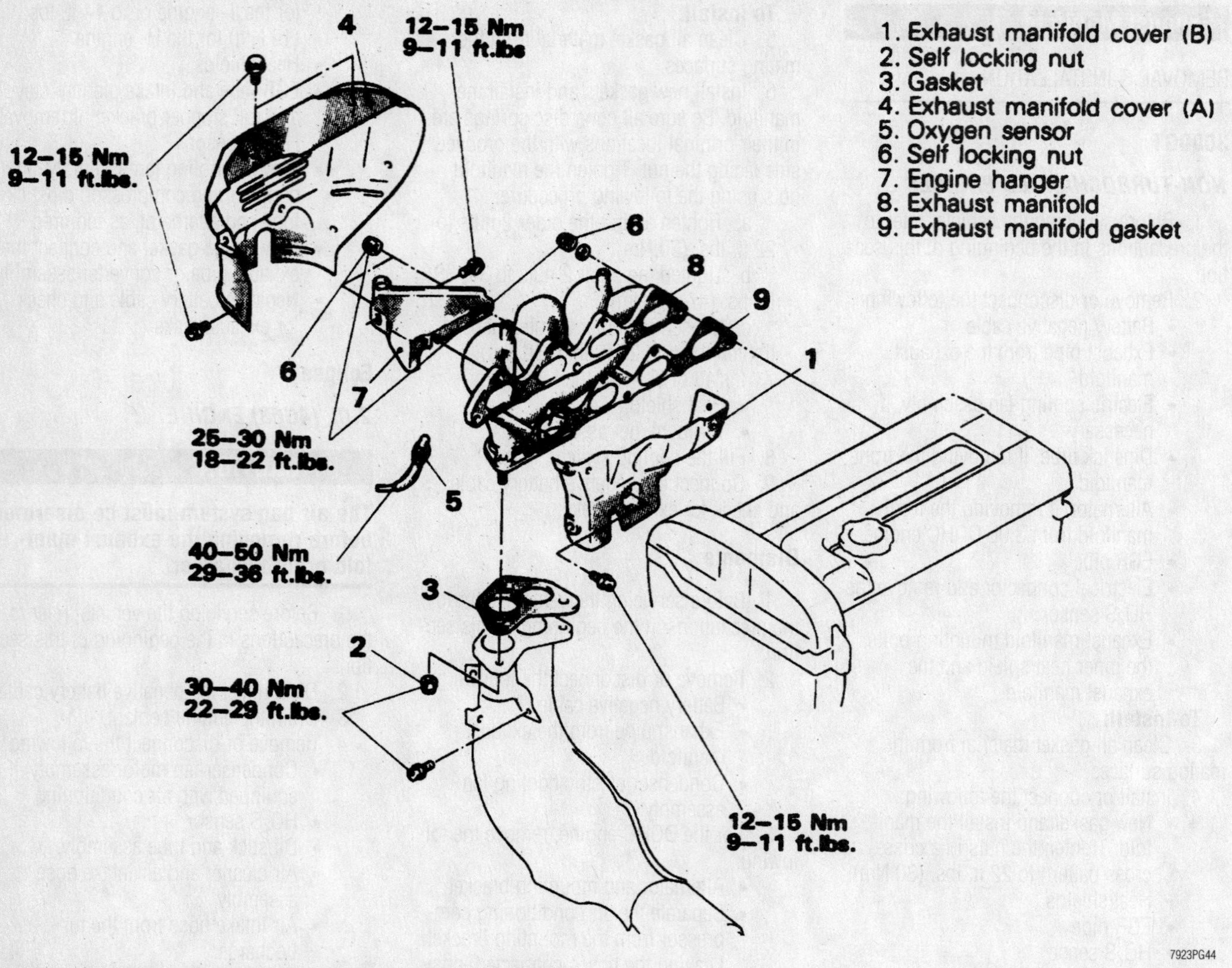

12–15 Nm
9–11 ft.lbs

12–15 Nm
9–11 ft.lbs.

25–30 Nm
18–22 ft.lbs.

40–50 Nm
29–36 ft.lbs.

30–40 Nm
22–29 ft.lbs.

12–15 Nm
9–11 ft.lbs.

1. Exhaust manifold cover (B)
2. Self locking nut
3. Gasket
4. Exhaust manifold cover (A)
5. Oxygen sensor
6. Self locking nut
7. Engine hanger
8. Exhaust manifold
9. Exhaust manifold gasket

7923PG44

Exhaust manifold and related parts—Eclipse 2.0L (4G63) engine

and nuts to specification, loosen them and tighten them again.

7. Use a new gasket and connect the exhaust pipe to the turbo.

8. Using new gaskets, install the oil return pipe.

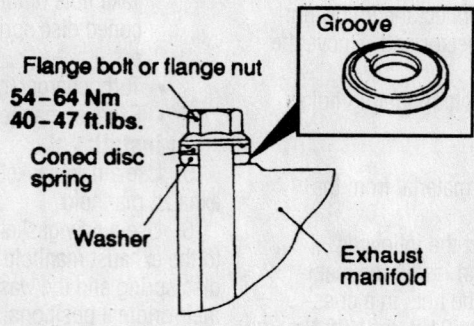

Groove

Flange bolt or flange nut
54–64 Nm
40–47 ft.lbs.

Coned disc spring

Washer

Exhaust manifold

7923PG37

Install the groove of the cone-shaped disc spring toward the flange bolt or nut—Eclipse 2.0L (4G63) engine

9. Install or connect the following:
• Engine hanger
• Heat shields
• Oil supply pipe
• Engine coolant hoses to the turbo
• Air hose

• Air cleaner and duct assembly
• Dipstick tube, using a new O-ring
• HO_2S sensor
• Condenser fan assembly if removed
• Negative battery cable and refill the engine with coolant

2.0L (420A) ENGINE

❈❈ CAUTION

The air bag system must be disarmed before removing the exhaust manifold.

1. Before servicing the vehicle, refer to the precautions in the beginning of this section.

2. Disconnect the negative battery cable.

3. Drain the engine coolant.

4. Remove or disconnect the following:
• Air intake hose

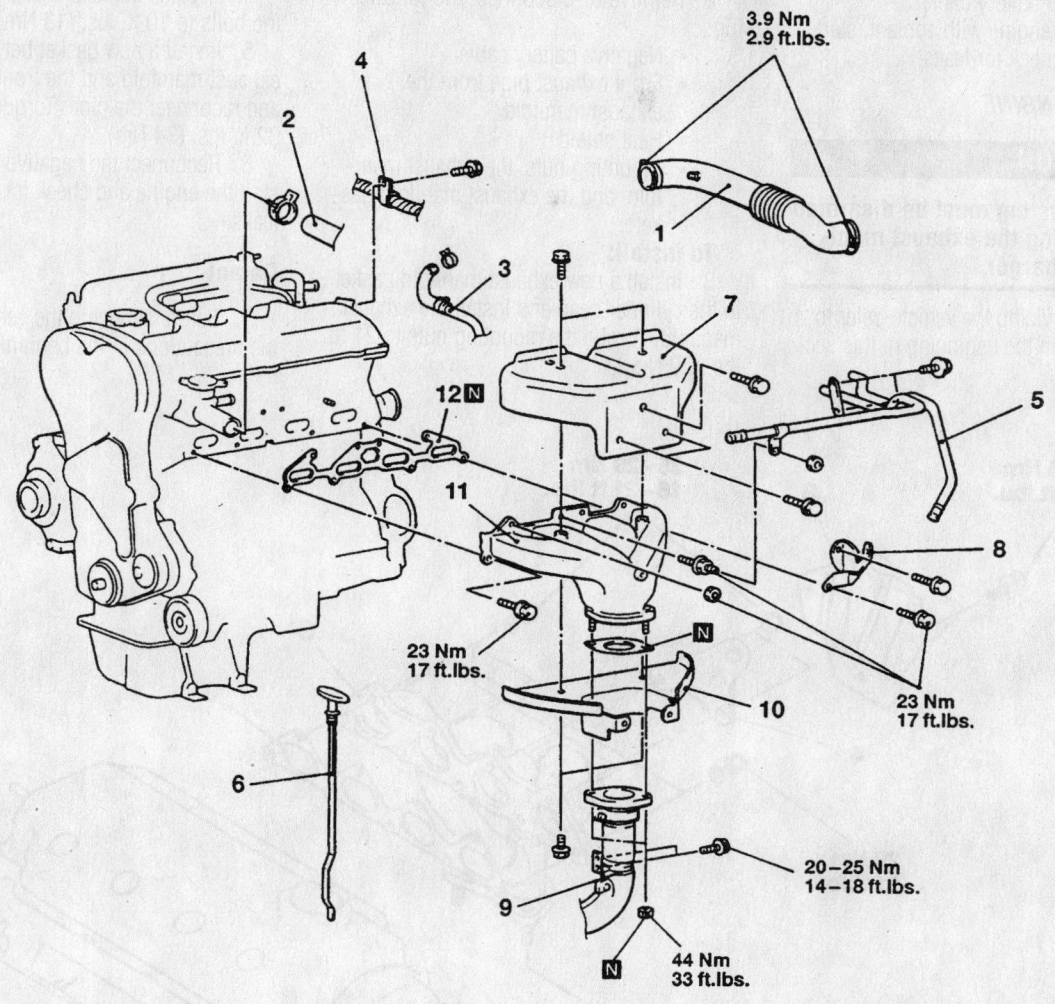

**3.9 Nm
2.9 ft.lbs.**

**23 Nm
17 ft.lbs.**

**23 Nm
17 ft.lbs.**

**20—25 Nm
14—18 ft.lbs.**

**44 Nm
33 ft.lbs.**

Removal steps

1. Air intake hose
2. Radiator upper hose connection
3. Air hose connection
4. Control wiring harness connection
5. Water pipe assembly
6. Engine oil level gauge
7. Heat protector
8. Engine hanger
9. Front exhaust pipe connection
10. Heat protector
11. Exhaust manifold
12. Exhaust manifold gasket

7923PG45

Exploded view of the exhaust manifold and related components—2.0L (420A) engine

- Upper radiator hose from the water outlet
- Air hose connection
- Engine control wiring harness from the rear of the engine
- Water pipe assembly
- Oil dipstick
- Upper heat shield
- Engine hanger
- Pulsed secondary air injection (check valve) valve from the exhaust pipe (manual transaxle only)
- Front exhaust pipe from the manifold
- Lower heat shield
- Exhaust manifold and gasket

To install:

5. Use a new gasket and install the exhaust manifold. Torque the nuts and bolts to 17 ft. lbs. (23 Nm).
6. Install the lower heat shield.
7. Use a new gasket and connect the front exhaust pipe to the manifold.
8. On vehicles with manual transaxles, connect the pulsed secondary air injection valve to the exhaust pipe.
9. Install or connect the following:
   - Engine hanger
   - Upper heat shield
   - Dipstick and the water pipe
10. Attach the engine wiring harness to the rear of the engine.
11. Install or connect the following:
    - Air hose and the upper radiator hose
    - Air intake hose

For Tire, Wheel and Ball Joint specifications, see Section 1 of this manual

- Negative battery cable

12. Refill the engine with coolant, start the engine and check for leaks.

2.4L (4G64) ENGINE

❊❊ CAUTION

The air bag system must be disarmed before removing the exhaust manifold or turbocharger.

1. Before servicing the vehicle, refer to the precautions in the beginning of this section.

2. Remove or disconnect the following:

- Negative battery cable
- Front exhaust pipe from the exhaust manifold
- Heat shield
- Mounting nuts, the exhaust manifold, and the exhaust manifold gasket

To install:

3. Install a new exhaust manifold gasket to the cylinder head and install the exhaust manifold. Torque the mounting nuts to 21 ft. lbs. (29 Nm).

4. Replace the heat shield and tighten the bolts to 10 ft. lbs. (13 Nm).

5. Install a new gasket between the exhaust manifold and the front exhaust pipe and reconnect the pipe. Torque the nuts to 32 ft. lbs. (34 Nm).

6. Reconnect the negative battery cable, start the engine and check for any exhaust leaks.

Galant

1. Before servicing the vehicle, refer to the precautions in the beginning of this section.

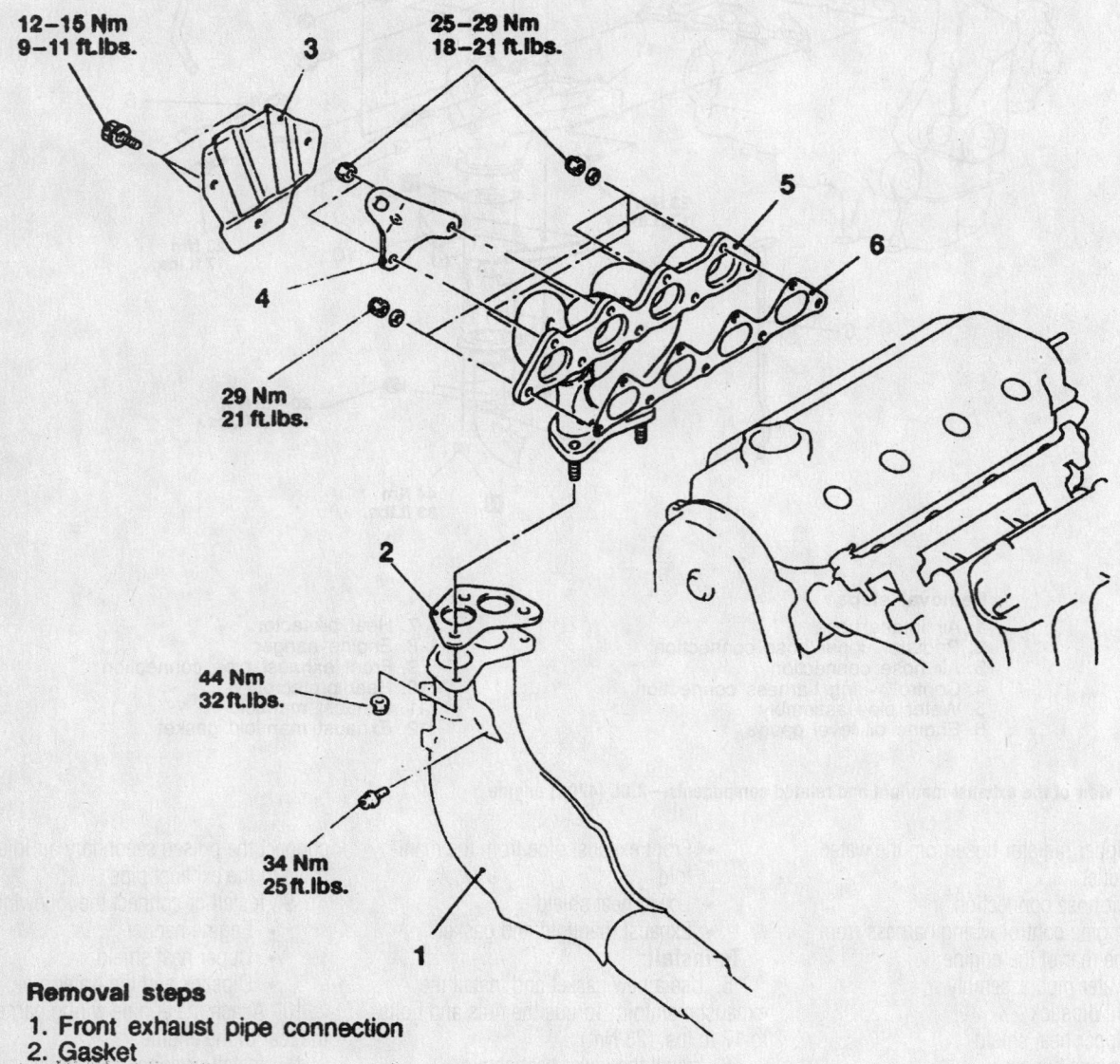

12–15 Nm
9–11 ft.lbs.

25–29 Nm
18–21 ft.lbs.

29 Nm
21 ft.lbs.

44 Nm
32 ft.lbs.

34 Nm
25 ft.lbs.

Removal steps

1. Front exhaust pipe connection
2. Gasket
3. Heat protector
4. Engine hanger
5. Exhaust manifold
6. Exhaust manifold gasket

Exploded view of the exhaust manifold mounting—Eclipse 2.4L (4G64) engine shown, Galant similar

2. Remove or disconnect the following:
- Battery negative cable
- Exhaust pipe from the exhaust manifold
- Outer exhaust manifold heat shield and engine hanger
- Exhaust manifold mounting nuts and the exhaust manifold from the engine

To install:

3. Clean all gasket material from the mating surfaces.

4. Install a new gasket and install the manifold. Tighten the nuts to in a crisscross pattern to 18–21 ft. lbs. (25–29 Nm).

5. Install the heat shields and tighten the mounting bolts to 10 ft. lbs. (14 Nm).

6. Install a new flange gasket and connect the exhaust pipe. Tighten the mounting nuts to 32 ft. lbs. (44 Nm).

7. Connect the negative battery cable and check for exhaust leaks.

Mirage

1. Before servicing the vehicle, refer to the precautions in the beginning of this section.

2. Remove or disconnect the following:
- Battery negative cable
- Exhaust pipe from the exhaust manifold
- Electric cooling fan assembly
- HO$_2$S sensor
- EGR pipe
- Outer exhaust manifold heat shield and engine hanger
- Exhaust manifold mounting bolts, the inner heat shield and the exhaust manifold

To install:

3. Clean all gasket material from the mating surfaces.

4. Using a new gasket and install the manifold. For 1.5L engines, tighten the nuts on a crisscross patter to 13 ft. lbs. (18 Nm). For 1.8L engines, tighten the inner nuts to in a crisscross pattern to 13 ft. lbs. (18 Nm) and tighten the 2 outer (larger) nuts to 22 ft. lbs. (30 Nm).

5. Install or connect the following:
- Heat shields
- EGR pipe
- HO$_2$S sensor
- Electric cooling fan assembly
- New flange gasket and connect the exhaust pipe
- Negative battery cable and check for exhaust leaks

Front Cover Seal

REMOVAL & INSTALLATION

3000GT

1. Before servicing the vehicle, refer to the precautions in the beginning of this section.

2. Remove or disconnect the following:
- Negative battery cable
- Undercover
- Cruise control pump and link assembly
- Alternator assembly

3. Raise and support the engine to take the weight off of the engine mount.

4. Remove or disconnect the following:
- Air hose and the air pipe
- Power steering tensioner and drive belt
- Timing belt covers and the timing belt
- Crankshaft sprocket

5. Pry out the crankshaft seal using a suitable tool.

To install:

6. Using a driver tool, install the new crankshaft seal.

7. Install or connect the following:
- Timing belt and the timing belt covers
- Crankshaft sprocket
- Air hose and the air pipe
- Alternator and the cruise control link and pump assembly
- Undercover
- Negative battery cable

Diamante

1. Before servicing the vehicle, refer to the precautions in the beginning of this section.

2. Remove or disconnect the following:
- Negative battery cable
- Drive belts
- Crankshaft pulley
- Timing belt covers and the timing belt
- CKP sensor
- Crankshaft sprocket, the sensing blade, spacer and Woodruff® key

3. Pry the seal from the bore, using a suitable tool.

To install:

4. Using a seal driver, install the new crankshaft seal. Lubricate the lips of the seal with clean engine oil.

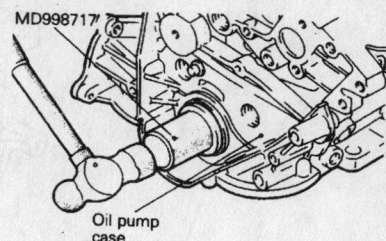

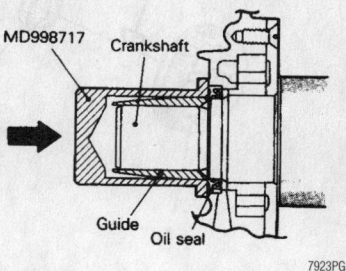

Crankshaft seal installation—Diamante and 3000GT

5. Install or connect the following:
- Woodruff® key, spacer, sensing blade and the crankshaft sprocket
- CKP sensor and tighten the retaining bolts to 84 inch lbs. (9 Nm)
- Timing belt and the timing belt covers
- Crankshaft pulley and retaining bolt. Torque the retaining bolt to 130–137 ft. lbs. (180–190 Nm) for the DOHC engine or to 108–116 ft. lbs. (150–160 Nm) for the SOHC engine.
- Drive belts, adjust
- Negative battery cable

Eclipse

2.0L AND 2.4L ENGINES

1. Before servicing the vehicle, refer to the precautions in the beginning of this section.

2. Remove or disconnect the following:
- Negative battery cable
- Timing belt
- Crankshaft sprocket

3. Carefully pry the oil seal out of the front case. Be careful not to damage the oil seal bore or the crankshaft sealing surface.

To install:

4. Apply clean engine oil to the oil seal lip. Using a seal driver, install the oil seal.

5. Install or connect the following:
- Crankshaft sprocket. If equipped, tighten the crankshaft bolt to 87 ft. lbs. (118 Nm).

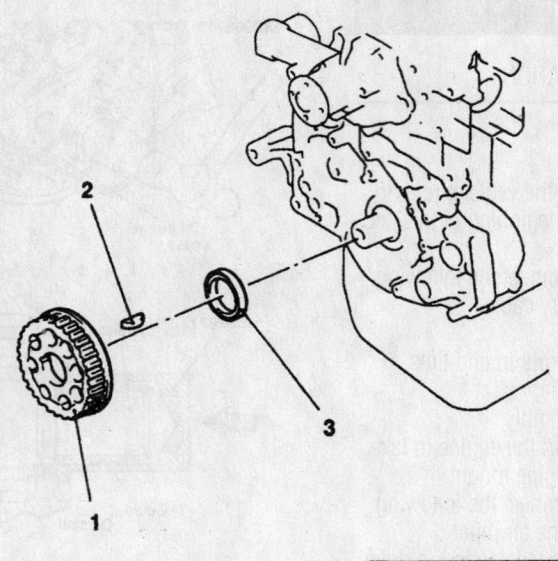

1. Crankshaft sprocket B
2. Key
3. Crankshaft front oil seal

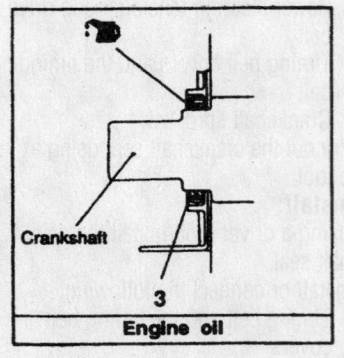

Front crankshaft oil seal—Eclipse 2.0L (4G63) engine shown

7923PG48

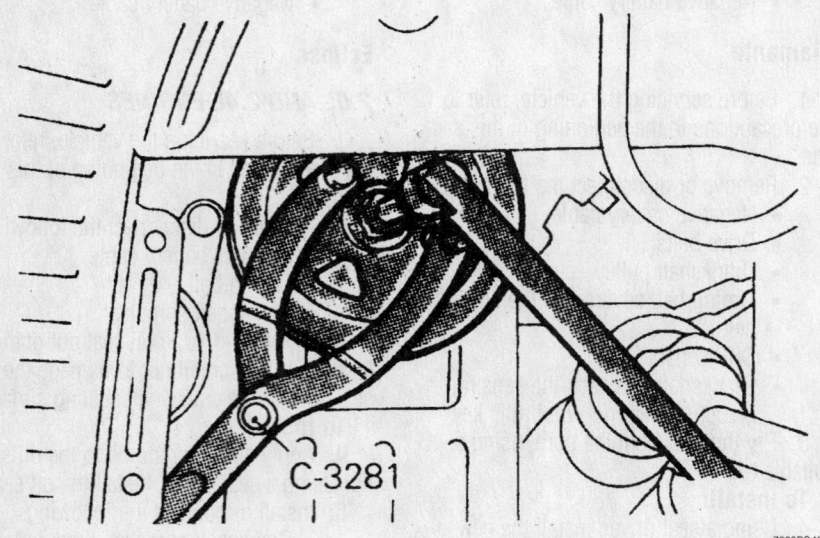

C-3281

7923PG49

The crankshaft pulley pin spanner tool should be used to hold pulley while bolt is removed—Diamante and 3000GT

- Timing belt
- Negative battery cable

Mirage

1. Before servicing the vehicle, refer to the precautions in the beginning of this section.
2. Remove or disconnect the following:
   - Negative battery cable.
   - Crankshaft pulley retainer bolts and remove the pulley
   - Vibration damper retainer bolt and washer and remove damper
   - Timing belt
   - Crankshaft sprocket
3. Pry out the oil seal from front of engine.

To install:

4. Using proper size driver, install a new front seal.
5. Lubricate the lips of the new seal with clean engine oil.
6. Install or connect the following:
   - Timing belt, timing covers, valve cover and remaining components
   - Crankshaft sprocket and vibration damper
   - Engine undercover and connect the negative battery cable

Camshaft and Valve Lifters

REMOVAL & INSTALLATION

3000GT Engine

1. Before servicing the vehicle, refer to the precautions in the beginning of this section.
2. Relieve the fuel system pressure.
3. Remove or disconnect the following:
   - Battery negative cable
   - Timing belt cover and timing belt
   - Center cover, breather and PCV hoses, and spark plug cables
   - Rocker cover, semi-circular packing, throttle body stay, both camshaft sprockets, and oil seals
   - CMP sensor and adapter
   - Intake and exhaust camshafts

To install:

4. Lubricate the camshafts with clean engine oil and position the camshafts on the cylinder head.

➡**The intake camshaft has a J stamped on the hexagon and the exhaust camshaft has a K or N.**

5. Be sure the dowel pins on both camshaft sprocket ends are positioned properly.
6. Install the bearing caps. Tighten the

caps gradually and in 2 or 3 steps. Caps 2, 3 and 4 have a front mark. Install with the mark aligned with the front mark on the cylinder head. Intake caps have **I** stamped on the cap and exhaust caps have **E**. Torque the front and rear retaining cap bolts to 15 ft. lbs. (20 Nm) and tighten the center retaining cap bolts to 96 inch lbs. (11 Nm).

7. Apply a coating of engine oil to the oil seals and install.

8. Install the timing belt, valve cover and all related parts. Refer to the timing belt unit repair section.

9. Connect the negative battery cable and check for leaks.

Diamante

3.0L DOHC ENGINE

1. Before servicing the vehicle, refer to the precautions in the beginning of this section.

2. Relieve the fuel system pressure.

3. Remove or disconnect the following:
- Negative battery cable
- Intake manifold plenum
- Timing belt cover and the timing belt
- Center cover, breather, PCV hoses, and the spark plug cables
- Rocker cover and the semi-circular packing
- CMP sensor
- Camshaft sprockets

➡**Be sure to keep the valvetrain components labeled and in proper order for reassembly.**

4. Loosen the bearing cap bolts in 2–3 steps. Label and remove all camshaft bearing caps.

5. Mark the components and remove the intake and the exhaust camshafts.

6. Remove the rocker arms and the lash adjusters. Be sure to note the location of the valvetrain components for reinstallation purposes.

To install:

➡**Lubricate the valvetrain components with clean engine oil.**

7. Bleed and install the lash adjusters in their original bores in the cylinder head.

8. Install the rocker arms to the cylinder head.

9. Lubricate the camshafts with clean engine oil and position the camshafts on the cylinder head.

✳✳ WARNING

Be sure to properly position the knock pins of the camshaft to prevent valve to piston interference.

➡**The intake camshaft on the Diamante has a B or J stamped on the hexagon. The exhaust camshaft on the Diamante has a D or K stamped on the hexagon.**

10. Install the bearing caps. Tighten the caps in sequence and in 2 or 3 steps. Caps 2, 3 and 4 have a front mark. Install with the mark aligned with the front mark on the cylinder head. Torque the retaining bolts for caps No. 2, 3 and 4 to 96 inch lbs. (11 Nm) and tighten the retaining bolts for the front and rear caps to 14 ft. lbs. (20 Nm).

11. Apply a coating of engine oil to the oil seals and install the oil seals to the front and rear of the camshafts.

12. Install or connect the following:
- Camshaft sprockets and tighten the sprocket bolts to 65 ft. lbs. (90 Nm)
- CMP sensor
- Timing belt

- Rocker cover and the semi-circular packing
- Intake manifold plenum
- Spark plug cables, center cover, breather and PCV hoses
- Negative battery cable and check for leaks

3.0L SOHC ENGINE

1. Before servicing the vehicle, refer to the precautions in the beginning of this section.

2. Remove or disconnect the following:
- Negative battery cable
- Intake manifold plenum stay bracket
- CMP sensor
- Valve covers and the timing belt

3. Using a camshaft sprocket holding tool, hold the sprocket and loosen the bolt.

4. Remove the bolt and note the positioning of the knock pin at the end of the camshaft and remove the sprocket.

5. Install auto lash adjuster retainer tools on the rocker arms.

➡**Be sure to note the position of the rocker arms, rocker shafts and bearing caps for reinstallation purposes.**

6. Remove or disconnect the following:
- Camshaft bearing caps but do not remove the bolts from the caps
- Rocker arms, rocker shafts and bearing caps, as an assembly
- Camshaft from the cylinder head

7. Inspect the bearing journals on the camshaft, cylinder head, and bearing caps.

To install:

➡**The right bank camshaft is identified by a 4mm slit at the rear end of the camshaft.**

8. Lubricate the camshaft journals and

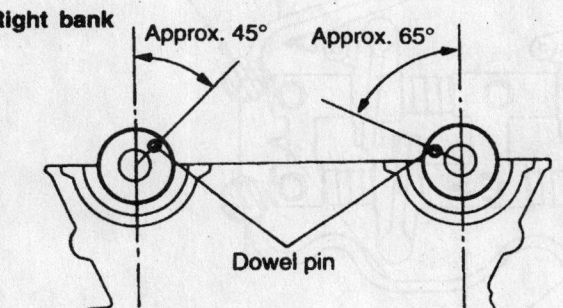

Proper positioning of the camshaft knock pins—Diamante 3.0L DOHC engine

7923PG58

Slit
7923PG59

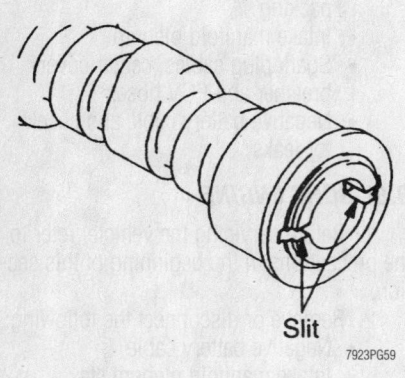

Right bank camshaft identification—Diamante 3.0L SOHC engine

Right bank
Left bank
Approx. 60°
Approx 71°
7923PG60

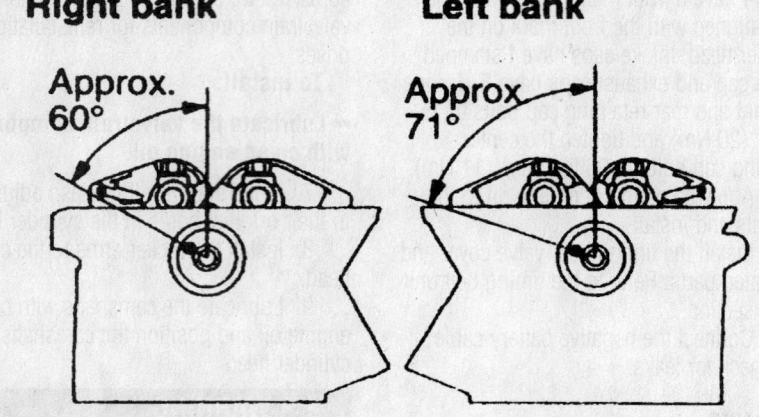

Proper positioning of the camshafts—Diamante 3.0L SOHC engine

Arrow mark (bearing cap)
Timing belt side
Arrow mark (cylinder head)
Arrow mark (bearing cap)
7923PG61

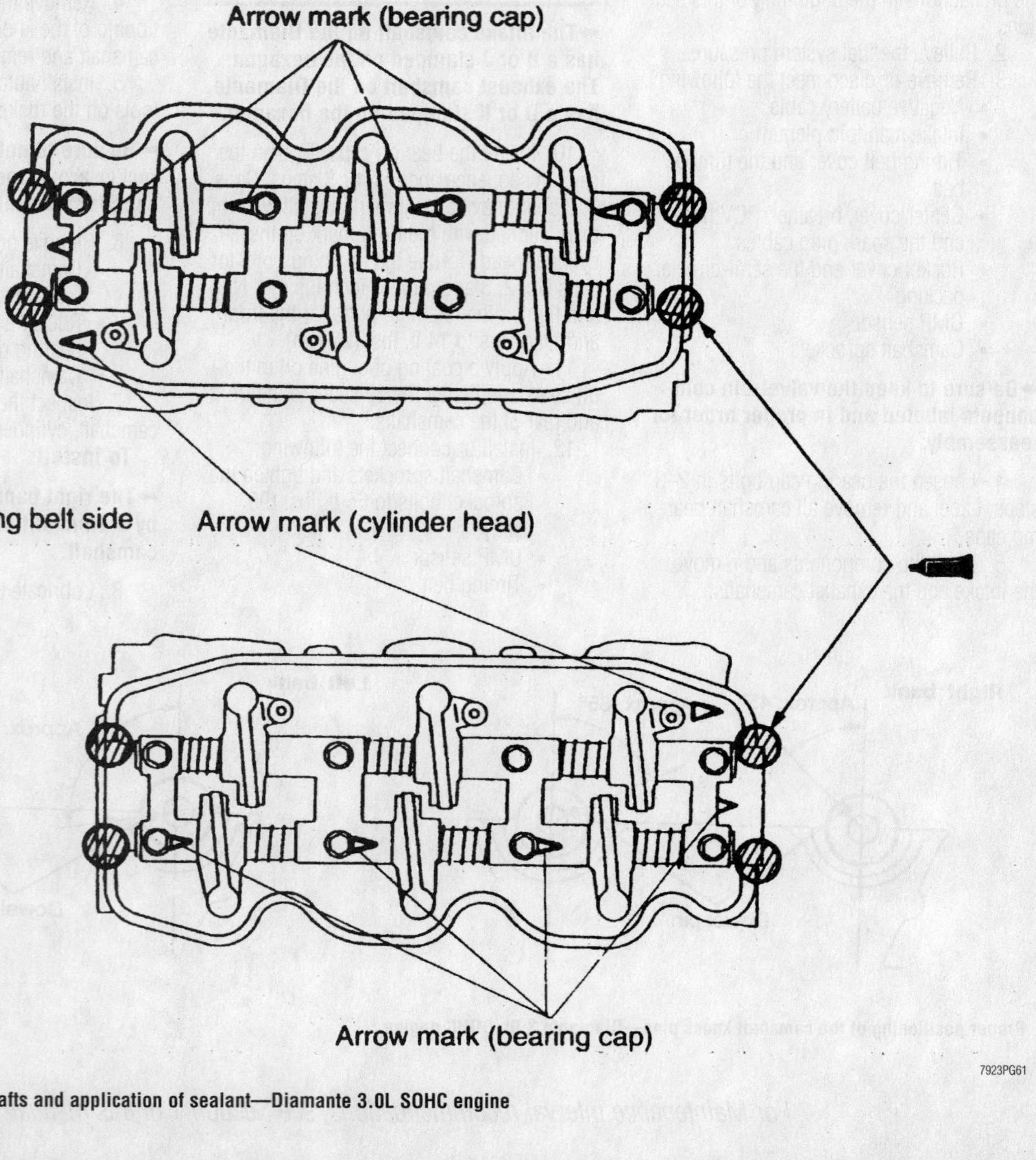

Alignment of the rocker shafts and application of sealant—Diamante 3.0L SOHC engine

camshaft with clean engine oil and install the camshaft in the cylinder head. Be sure to properly position the knock pin of the camshaft as noted during removal.

9. Apply sealer at the ends of the bearing caps and install the rocker arms, rocker shafts and bearing caps as an assembly. Properly position the arrows on the bearing caps.

10. Torque the bearing cap bolts in the following sequence: No. 3, No. 2, No. 1 and No. 4 to 85 inch lbs. (10 Nm).

11. Repeat the sequence increasing the torque to 14 ft. lbs. (20 Nm).

12. Remove the auto lash adjuster retainer tools from the rocker arms.

13. Install the camshaft sprocket and bolt.

14. Using a camshaft sprocket holding tool, hold the sprocket and tighten the bolt to 65 ft. lbs. (90 Nm).

15. Install or connect the following:
  • Timing belt and valve covers

  • CMP sensor
  • Intake manifold plenum stay bracket
  • Negative battery cable and check for leaks

3.5L ENGINE

1. Before servicing the vehicle, refer to the precautions in the beginning of this section.

2. Remove or disconnect the following:

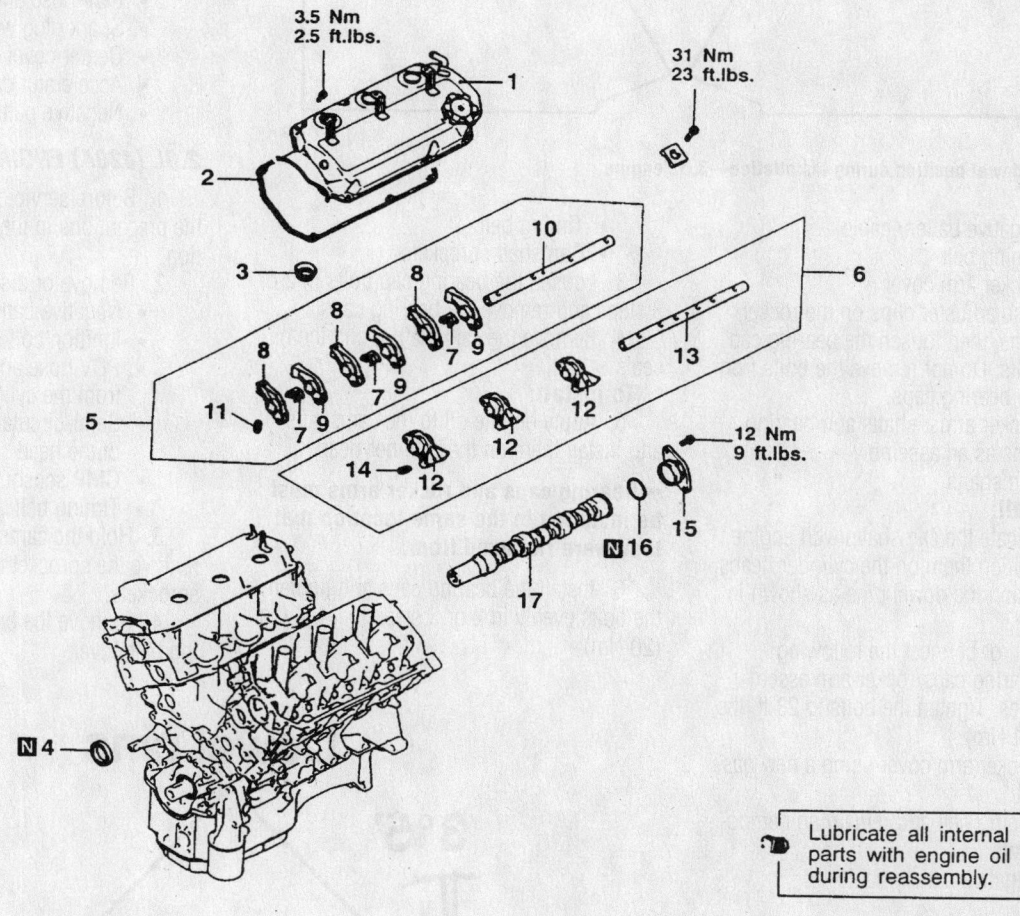

Lubricate all internal parts with engine oil during reassembly.

Removal steps

1. Rocker cover
2. Rocker cover gasket
3. Oil seal
4. Camshaft oil seal
5. Rocker arm, rocker arm shaft
6. Rocker arm, rocker arm shaft
7. Rocker shaft spring
8. Rocker arm A
9. Rocker arm B
10. Rocker arm shaft
11. Lash adjuster
12. Rocker arm C
13. Rocker arm shaft
14. Lash adjuster
15. Thrust case
16. O-ring
17. Camshaft

7923PGD3

Exploded view of the camshaft mounting—3.5L engine

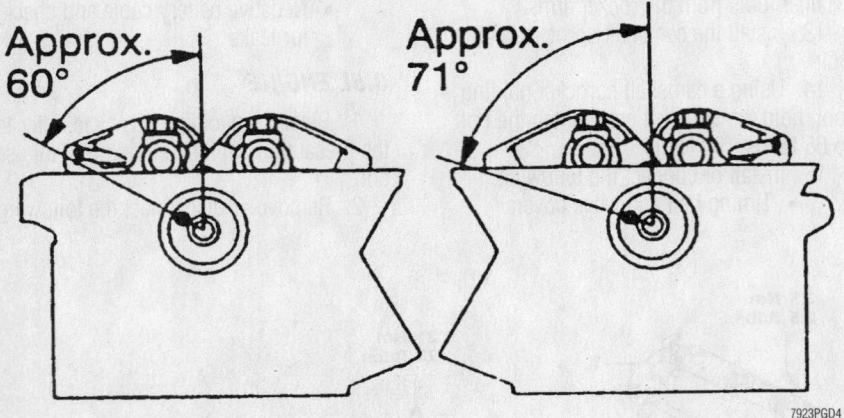

Rear bank

Approx. 60°

Front bank

Approx. 71°

7923PGD4

Camshaft dowel position during installation—3.5L engine

- Negative battery cable
- Timing belt
- Rocker arm cover
- Lash adjuster clips on the rocker arms, then loosen the bearing cap bolts. Do not remove the bolts from the bearing caps.
- Rocker arms, shafts and bearing caps as an assembly
- Camshafts

To install:

3. Lubricate the camshafts with engine oil and position them on the cylinder heads.

4. Position the dowel pins as shown in the drawing.

5. Install or connect the following:
- Bearing caps/rocker arm assemblies. Tighten the bolts to 23 ft. lbs. (31 Nm).
- Rocker arm cover using a new gasket
- Timing belt and remaining components
- Negative battery cable

Eclipse

2.0L (4G63) ENGINE

1. Before servicing the vehicle, refer to the precautions in the beginning of this section.

2. Remove or disconnect the following:
- Negative battery cable
- Accelerator cable from the throttle body and remove the cable bracket from the intake plenum
- Engine center cover
- Spark plug cables
- Breather hose and the PCV hose from the rocker cover
- Rocker cover

- Timing belt
- Camshaft sprockets

3. Loosen the bearing cap bolts in 2 or 3 steps and remove the bearing caps.

4. Remove the camshaft(s) and the oil seals.

To install:

5. Apply engine oil to the camshafts and install them on the cylinder head.

➡ **Bearing caps and rocker arms must be installed in the same location that they were removed from.**

6. Install the bearing caps and tighten the bolts evenly in 2 or 3 steps to 14 ft. lbs. (20 Nm).

7. Apply engine oil to the lip of the seal. Using MB998713, install the front oil seal.

8. Install or connect the following:
- Camshaft sprockets
- Timing belt

9. Apply sealant to the semi-circular packing and install it in the cylinder head.

10. Apply sealant to the lower part of the front and rear bearing caps where they meet the cylinder head.

11. Install or connect the following:
- Rocker cover
- PCV hose and the breather hose
- Spark plug wires
- Center cover
- Accelerator cable
- Negative battery cable

2.0L (420A) ENGINE

1. Before servicing the vehicle, refer to the precautions in the beginning of this section.

2. Remove or disconnect the following:
- Negative battery cable
- Ignition coil pack
- PCV hose and the breather hose from the cylinder head cover
- Semi-circular packing from the rear of the head
- CMP sensor
- Timing belt

3. Hold the camshaft sprockets and remove the sprocket mounting bolts and the sprockets.

4. Remove the bracket and the rear timing belt cover.

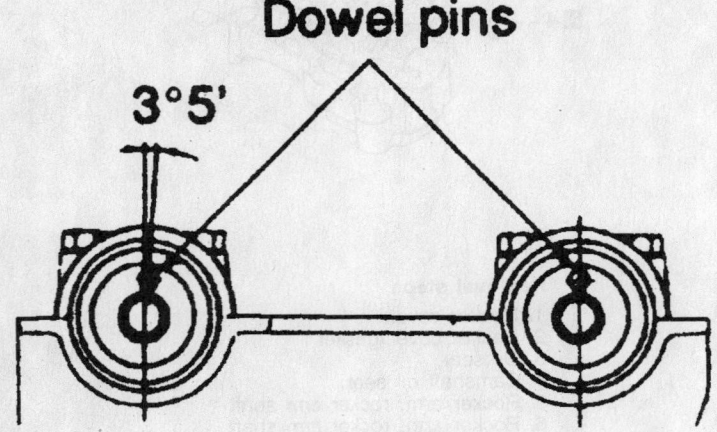

Dowel pins

3°5'

Exhaust side   **Intake side**

7923PG53

Position the camshafts with the dowels facing up—Eclipse 2.0L (4G63) engine

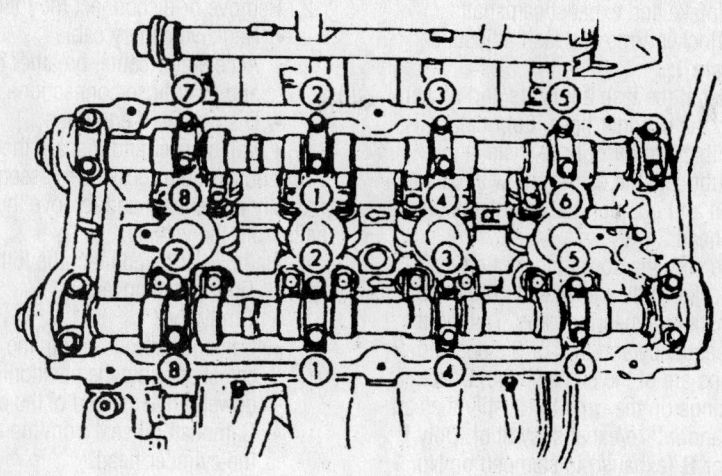

Camshaft bearing cap bolt removal sequence—Eclipse 2.0L (420A) engine

7923PG54

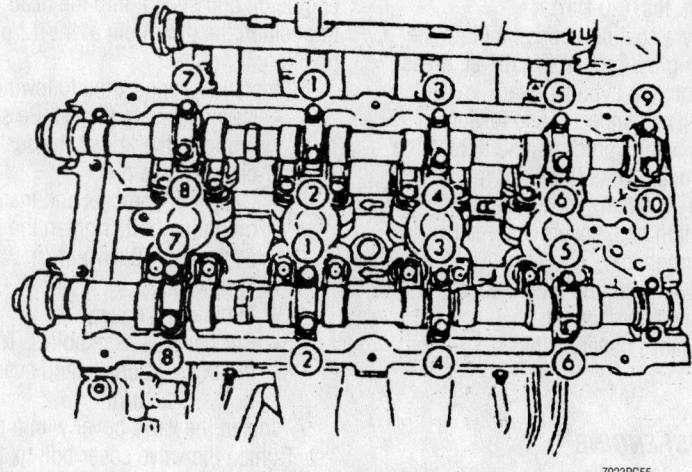

Camshaft bearing cap bolt installation sequence—Eclipse 2.0L (420A) engine

7923PG55

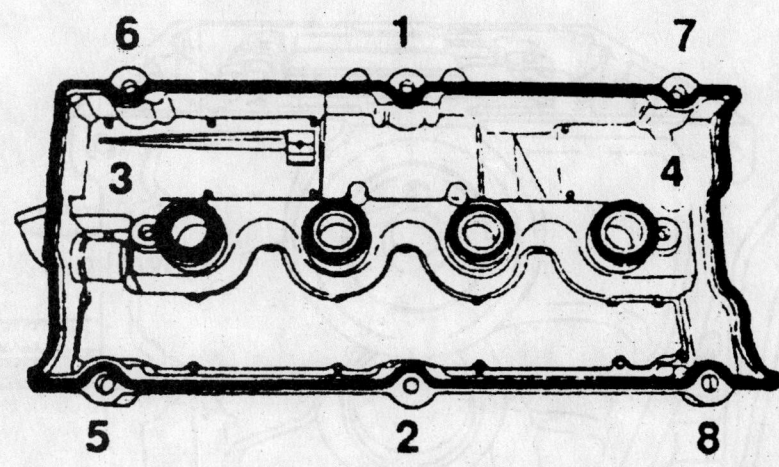

Tighten the cylinder head cover bolts in the sequence shown—Eclipse 2.0L (420A) engine

7923PG56

5. Gradually loosen the camshaft bearing caps in sequence, one camshaft at a time and remove the bearing caps.

➡**Keep the bearing caps in order. They must be installed in the location that they were removed from.**

6. Mark the camshafts for later identification and remove the camshafts. The camshafts are not interchangeable.

To install:

7. Apply engine oil to the camshaft and install the camshafts.

8. Install the bearing caps. Torque the bolts evenly and in sequence.

9. Apply Loctite 518® to the outside camshaft bearing caps and install them.

10. Install or connect the following:
- Camshaft oil seal
- Rear timing belt cover and the bracket
- Camshaft sprockets, using special tool(s)
- Timing belt

11. Apply Loctite 5699®, or equivalent, to the semi-circular packing and install it in the rear of the cylinder head.

12. Install or connect the following:
- CMP sensor
- Cylinder head cover. Torque the bolts evenly in 3 steps in the proper sequence.
- Air, breather and PCV hoses
- Coil pack
- Negative battery cable

2.4L (4G64/G) ENGINE

1. Before servicing the vehicle, refer to the precautions in the beginning of this section.

2. Remove or disconnect the following:
- Remove the battery
- Accelerator cable bracket and position the cable aside
- Air intake hose
- Breather hose and disconnect the PCV hose
- Spark plug cables
- Rocker cover

3. Install lash adjuster retainer tools to the rocker arm.

4. Remove or disconnect the following:
- Timing belt covers and the timing belt
- Camshaft sprocket retainer bolt and remove the sprocket from the shaft
- Camshaft oil seal
- Both rocker arm shaft assemblies from the head

• Camshaft from the cylinder head

To install:

5. Lubricate the camshaft journals and camshaft with clean engine oil and install the camshaft in the cylinder head.

6. Install the rocker arm and shaft assemblies. Tighten the rocker arm shaft retainer bolts to 21–25 ft. lbs. (29–35 Nm).

7. Apply a coating of engine oil to the oil seal. Using the proper size driver, press-fit the seal into the cylinder head.

8. Install or connect the following:
• Camshaft sprocket and retainer bolt to 65 ft. lbs. (90 Nm)
• Timing belt and belt covers

9. Remove the lash adjuster retaining tools.

10. Install or connect the following:
• Rocker cover using new gasket material on mating surfaces
• Spark plug cables
• Air intake hose
• Breather hose and connect the PCV hose
• Battery

11. Run the engine at idle until normal operating temperature is reached. Check idle speed and ignition timing; adjust as required.

Galant

2.4L (4G64/L) ENGINE

1. Before servicing the vehicle, refer to the precautions in the beginning of this section.

2. Relieve the fuel system pressure.

3. Remove or disconnect the following:
• Negative battery cable
• Accelerator cable, PCV hoses, breather hoses, spark plug cables
• Valve cover
• Timing belt upper and lower covers
• Timing belt
• Camshaft sprockets

4. Loosen the bearing cap bolts in 2–3 steps. Label and remove all camshaft bearing caps.

5. Remove or disconnect the following:

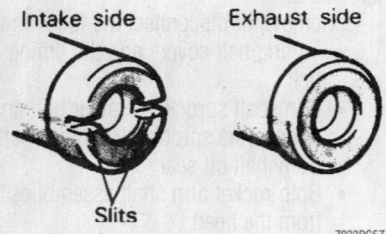

Camshaft identification—2.4L (4G64) engine

• Intake and exhaust camshafts
• Rocker arms and lash adjusters

To install:

6. Install the lash adjusters and rocker arms into the cylinder head. Lubricate lightly with clean oil prior to installation.

7. Lubricate the camshafts with clean engine oil and position the camshafts on the cylinder head.

8. Be sure the dowel pin on both camshaft sprocket ends are located on the top.

9. Install the bearing caps. Tighten the caps in sequence and in 2 or 3 steps. No. 2 and 5 caps are of the same shape. Check the markings on the caps to identify the cap number and intake/exhaust symbol. Only **L** (intake) or **R** (exhaust) is stamped on No. 1 bearing cap. Also, be sure the rocker arm is correctly mounted on the lash adjuster and the valve stem end. Torque the retaining bolts to 15 ft. lbs. (20 Nm).

10. Apply a coating of engine oil to the oil seal. Using the proper size driver, press-fit the seal into the cylinder head.

11. Install or connect the following:
• Camshaft sprockets and tighten the sprocket bolts to 58–72 ft. lbs. (80–100 Nm)
• Timing belt, covers and related components
• Valve cover and reconnect all related components
• Negative battery cable

Mirage

1.5L (4G15) ENGINE

1. Before servicing the vehicle, refer to the precautions in the beginning of this section.

2. Remove or disconnect the following:
• Negative battery cable
• Accelerator cable, breather hose and PCV hose connections
• Distributor, if equipped
• Valve cover and discard the gasket

3. Loosen both rocker arm assemblies gradually and evenly and remove the rocker shafts from the vehicle.

4. Remove or disconnect the following:
• Timing belt covers
• Timing belt
• Camshaft sprocket from the camshaft. Note the positioning of the dowel pin at the end of the camshaft.
• Camshaft oil seal from the front of the cylinder head
• Camshaft from the head

To install:

5. Lubricate the camshaft with clean engine oil and slide it into the head. Be sure to position the dowel pin at the 12 o'clock position.

6. Install or connect the following:
• New camshaft oil seal. Be sure to lubricate the lips of the seal with clean engine oil.
• Camshaft sprocket and install the mounting bolt. Tighten the bolt to 51 ft. lbs. (70 Nm)
• Timing belt
• Timing belt covers
• Rocker shaft assemblies. Torque the bolts gradually and evenly to 23 ft. lbs. (32 Nm).

7. Install the valve cover with a new gasket. Tighten the valve cover bolt to 16 inch lbs. (1.8 Nm).

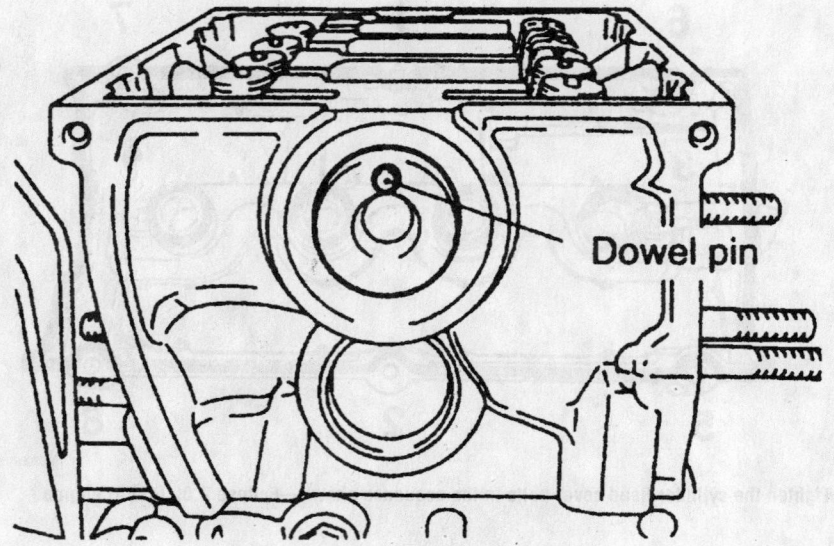

Positioning of the camshaft dowel pin—Mirage 1.5L (4G15) engine

8. Install or connect the following:
- Distributor, if equipped
- Accelerator cable, breather hose and PCV hose
- Negative battery cable and check the ignition timing

1.8L (4G93) ENGINE

1. Before servicing the vehicle, refer to the precautions in the beginning of this section.

2. Remove or disconnect the following:
- Negative battery cable
- Spark plug cables
- MAF sensor connector and remove the air cleaner case cover
- Accelerator cable, breather hose and PCV hose connections
- Rocker cover and discard the gasket

3. Loosen both rocker arm shaft assemblies gradually and evenly and remove the rocket shafts from the vehicle.

4. Remove or disconnect the following:
- Timing belt covers
- Timing belt
- Camshaft sprocket from the camshaft. Note the positioning of the dowel pin at the end of the camshaft.
- Camshaft oil seal from the front of the cylinder head
- Camshaft from the head

To install:

5. Lubricate the camshaft journals and camshaft with clean engine oil and install the camshaft in the cylinder head. Be sure to position the dowel pin at the end of the camshaft as noted during the removal procedure.

6. Install or connect the following:
- New camshaft oil seal.
- Camshaft sprocket and tighten the retainer bolt to 65 ft. lbs. (90 Nm)
- Timing belt
- Timing belt covers
- Rocker arm and shaft assemblies. Tighten the rocker arm shaft retainer bolts to 23 ft. lbs. (32 Nm).
- Valve cover with a new gasket. Tighten the valve cover bolts to 29 inch lbs. (3.3 Nm).
- Spark plug cables
- Accelerator cable, breather hose and PCV hose
- MAF sensor connector and install the air cleaner case cover
- Negative battery cable

Valve Lash

ADJUSTMENT

Valve clearance is not adjustable on these vehicles.

Starter Motor

REMOVAL & INSTALLATION

1. Remove or disconnect the following:
- Negative battery cable
- Air-flow sensor assembly connector and remove the breather hose
- Resonator retaining nuts and remove the air intake hose and resonator assembly as required

➡**Use care when removing the air cleaner cover because the air-flow sensor is attached and is a sensitive component.**

2. If equipped with Active-ECS suspension, remove the air compressor as follows:
   a. Disconnect the 2 electrical connectors, from the compressor.
   b. Disconnect the air line at the compressor.
   c. Remove the 3 mounting bolts, securing the compressor to the chassis.

3. Raise the vehicle and support safely.

4. Remove or disconnect the following:
- Engine undercover
- Hat shield from beneath the intake manifold on the 1.5L engine
- Speedometer cable connector at the transaxle end, if necessary
- Starter motor electrical connections
- Starter motor mounting bolts and remove the starter

5. Install or connect the following:
- Starter motor mounting bolts and remove the starter. Tighten the starter mounting bolts to 22 ft. lbs. (31 Nm).
- Starter motor electrical connections
- Speedometer cable connector at the transaxle end, if necessary
- Hat shield from beneath the intake manifold on the 1.5L engine
- Engine undercover

6. Lower the vehicle.

7. Install or connect the following:
- Air compressor, if equipped with Active-ECS suspension
- Resonator retaining nuts and

remove the air intake hose and resonator assembly as required
- Air-flow sensor assembly connector and remove the breather hose
- Negative battery cable and check the starter for proper operation

Oil Pan

REMOVAL & INSTALLATION

3000GT

1. Before servicing the vehicle, refer to the precautions in the beginning of this section.

2. Remove or disconnect the following:
- Negative battery cable
- Oil pan drain plug and drain the engine oil
- Transfer case assembly, on vehicles equipped with AWD
- Exhaust pipe, and on turbocharged engines, the return pipe for the turbocharger from the side of the oil pan
- Oil pan mounting bolts

3. Separate and remove the engine oil pan.

To install:

4. Thoroughly clean the oil pan, cylinder block bolts and bolt holes.

5. Apply a thin bead of sealer around the surface of the oil pan.

6. Assemble the oil pan to the cylinder block within 15 minutes after applying the sealant.

7. Install the oil pan mounting bolts and tighten to 48–72 inch lbs. (6–8 Nm).

8. Fill the engine with the proper amount of oil.

9. Connect the negative battery cable and check for leaks.

Diamante

3.0L ENGINES

1. Before servicing the vehicle, refer to the precautions in the beginning of this section.

2. Remove or disconnect the following:
- Negative battery cable
- Oil pan drain plug and drain the engine oil
- Left side crossmember. If equipped with 4WS, it will also be necessary to remove the right side crossmember.
- Starter motor

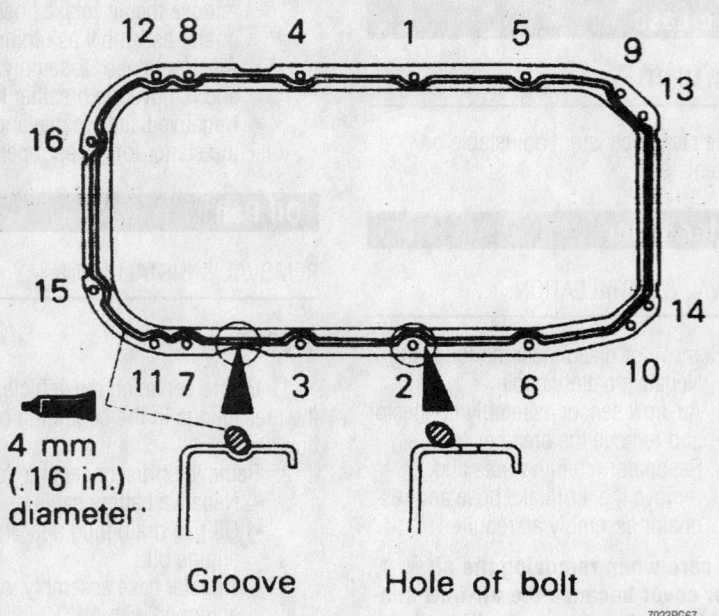

Oil pan bolt tightening sequence and application of sealant to the pan—Diamante 3.0L (J- and H-) engines

- Roll stopper stay bracket, from the rear transaxle stay bracket
- Transaxle stay brackets
- Bell housing lower cover
- Oil pan mounting bolts
- The engine oil pan

To install:

3. Apply a 0.16 in. (4mm) continuous bead of sealer around the surface of the oil pan.

➡️**Assemble the oil pan to the cylinder block within 15 minutes after applying the sealant.**

4. Install the oil pan mounting bolts. Following proper sequence, tighten mounting bolts to 48 inch lbs. (6 Nm).

5. Install or connect the following:
- Lower bell housing cover and the starter motor
- Transaxle stay brackets and connect the roll stopper bracket
- Crossmember(s) and tighten the mounting bolts to 43–51 ft. lbs. (60–70 Nm)

6. Fill the engine with the proper amount of oil.

7. Connect the negative battery cable and check for leaks.

3.5L ENGINE

1. Before servicing the vehicle, refer to the precautions in the beginning of this section.

2. Disconnect the negative battery cable.

3. Drain the engine oil.

4. Remove the mounting bolts from the lower oil pan.

5. Place a block of wood against the side of the pan and tap the block with a hammer to break the seal and remove the lower pan.

6. Remove or disconnect the following:
- Starter
- Dipstick tube
- Upper oil pan

☀️☀️ WARNING

Do not pry or use seal breaker tool to remove the oil pan. Damage to the aluminum surface can result.

7. Screw a bolt into the threaded hole to force the oil pan from the engine block and remove the pan.

8. Remove the bolt used to remove the pan.

To install:

9. Clean and degrease the sealing surfaces of the upper oil pan and engine block.

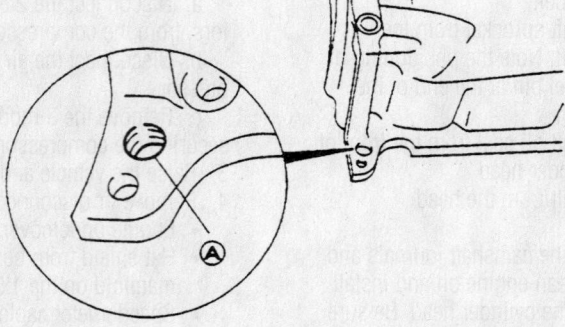

Install a bolt in the threaded hole to force the oil pan from the engine block—3.5L engine

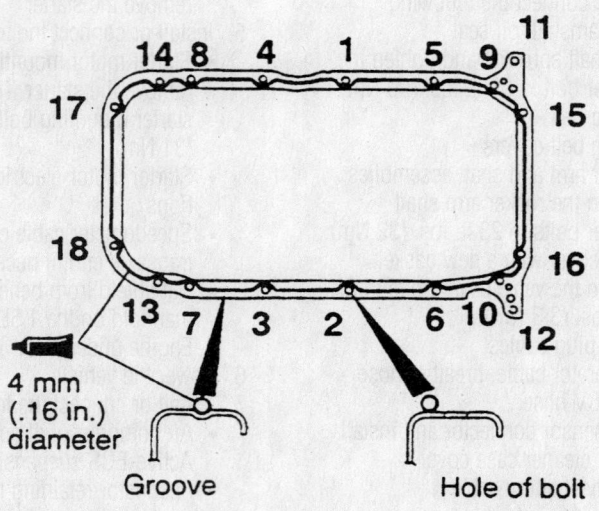

Apply sealant and tighten the bolts in the order shown—3.5L engine, upper oil pan

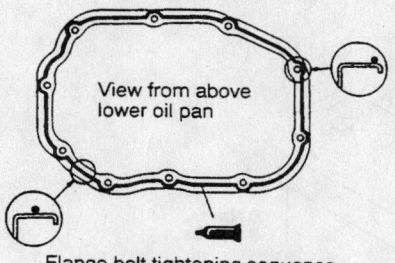

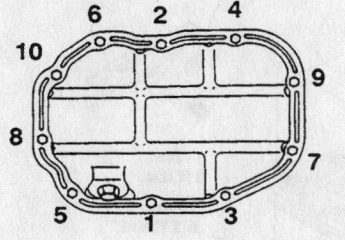

Apply sealant and tighten the bolts in the order shown—3.5L engine, lower oil pan

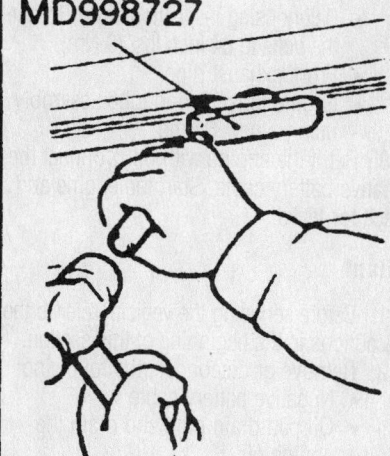

Using the special tool MB998727 to remove the oil pan—Eclipse 2.0L (4G63) engine

10. Apply a bead of silicone sealant along the mounting surface of the upper oil pan.

11. Install or connect the following:
- Upper oil pan. Tighten the bolts in sequence to 48 inch lbs. (6 Nm).
- Dipstick tube using a new O-ring
- Starter assembly

12. Clean and degrease the sealing surface of the lower oil pan.

13. Place a bead of sealant on the mounting surface of the lower oil pan. Install the lower pan. Tighten the bolts in sequence to 84–108 inch lbs. (10–12 Nm).

14. Install the drain plug using a new washer. Tighten the drain plug to 29 ft. lbs. (39 Nm).

15. Lower the vehicle and fill the crankcase to the correct level.

16. Connect the negative battery cable.

17. Start the engine and check for leaks.

Eclipse

2.0L (4G63) ENGINE

1. Before servicing the vehicle, refer to the precautions in the beginning of this section.

2. Disconnect the negative battery cable.

3. Safely raise and support the vehicle.

4. Remove or disconnect the following:
- Front exhaust pipe
- Exhaust pipe and muffler assembly

5. Drain the engine oil.

6. Remove or disconnect the following:

- Dipstick and tube
- Transfer case assembly (AWD)
- Bell housing cover
- Oil return pipe from the oil pan
- Oil pan mounting bolts

7. Tap between the oil pan and the engine block to break the seal and remove the oil pan.

To install:

8. Clean the sealing surface on the oil pan and engine block. Apply a continuous bead of sealant to the oil pan.

9. Clean the oil pan mounting bolt holes in the oil seal case.

10. Install the oil pan. Torque the mounting bolts to 60 inch lbs. (6.9 Nm).

11. Use a new gasket and connect the oil return pipe to the oil pan.

12. Install or connect the following:
- Bell housing cover
- Transfer case assembly if equipped
- Dipstick and tube assembly
- Front exhaust pipe
- Exhaust pipe and muffler

13. Lower the vehicle and fill the crankcase to the correct level.

14. Connect the negative battery cable.

15. Start the engine and check for leaks.

2.0L (420A) ENGINE

1. Before servicing the vehicle, refer to the precautions in the beginning of this section.

2. Disconnect the negative battery cable.

3. Drain the engine oil.

4. Remove or disconnect the following:
- Front exhaust pipe

- Dipstick and tube assembly
- Front plate
- Oil pan mounting bolts
- Oil pan and gasket

To install:

5. Apply sealant at the point where the engine block meets the oil pump.

6. Use a new gasket and install the oil pan. Torque the mounting bolts to 105 inch lbs. (12 Nm).

7. Install or connect the following:
- Front plate
- Front exhaust pipe
- Dipstick and tube assembly

8. Refill the crankcase with oil to the proper level.

9. Connect the negative battery cable.

10. Start the engine and check for leaks.

2.4L (4G64) ENGINE

1. Before servicing the vehicle, refer to the precautions in the beginning of this section.

2. Remove the negative battery cable.

3. Drain the engine oil.

4. Remove or disconnect the following:
- Engine dipstick and tube assembly
- Front exhaust pipe
- Bell housing inspection cover
- Bolts attaching the oil pan to the cylinder block

5. Remove the oil pan assembly.

To install:

6. Clean the sealing surface on the oil pan and engine block. Apply a continuous bead of sealant to the oil pan.

7. Install or connect the following:
- Oil pan to the cylinder block and

tighten the bolts to 60 inch lbs. (7 Nm)
- Bell housing inspection cover. Torque the bolts to 84 inch lbs. (9 Nm).
- Front exhaust pipe
- Engine dipstick and tube assembly using a new O-ring

8. Refill the engine with oil. Connect the negative battery cable. Start the engine and check for leaks.

Galant

1. Before servicing the vehicle, refer to the precautions in the beginning of this section.
2. Remove or disconnect the following:
   - Negative battery cable
   - Oil pan drain plug and drain the engine oil
   - Oil dipstick and tube assembly
   - HO2S sensor connector
   - Front exhaust pipe from the vehicle
   - Bell housing cover
   - Oil pan retainer bolts
3. Tap in between the engine block and the oil pan.

➡ Do not use a pry tool when removing the oil pan. Damage to engine components may occur.

To install:

4. Apply sealant around the gasket surfaces of the oil pan.
5. Install or connect the following:
   - Oil pan onto the cylinder block within 15 minutes after applying sealant. Tighten to 72 inch lbs. (8 Nm).
   - Oil drain plug and tighten to 29 ft. lbs. (39 Nm)
   - Bell housing cover. Tighten the mounting bolts to 84 inch lbs. (9 Nm).
   - Front exhaust pipe and tighten the bolts at the catalytic converter to 36 ft. lbs. (49 Nm). Tighten the nuts at the exhaust manifold to 32 ft. lbs. (44 Nm).
   - HO2S sensor connector
6. Fill the crankcase to the proper level.
7. Connect the negative battery cable. Start the engine and check for leaks.

Mirage

1.5L (4G15) ENGINE

1. Before servicing the vehicle, refer to the precautions in the beginning of this section.
2. Disconnect the negative battery cable.
3. Drain the engine oil.
4. Remove or disconnect the following:
   - Bell housing lower cover
   - Oil pan retainer bolts

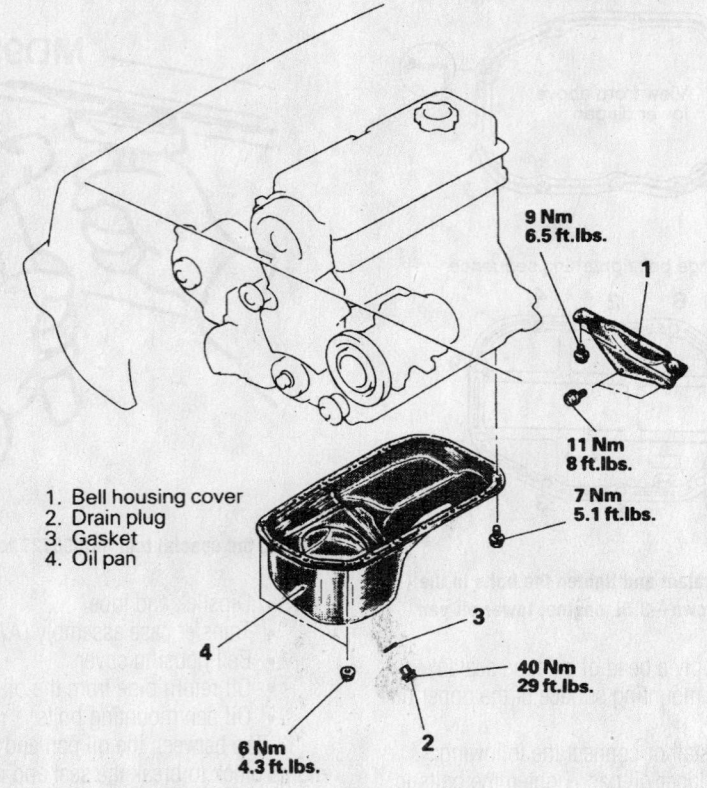

1. Bell housing cover
2. Drain plug
3. Gasket
4. Oil pan

9 Nm 6.5 ft.lbs.
11 Nm 8 ft.lbs.
7 Nm 5.1 ft.lbs.
40 Nm 29 ft.lbs.
6 Nm 4.3 ft.lbs.

7923PG65

Oil pan and related components—Mirage 1.5L (4G15) engine

➡ Do not use a pry tool when removing the oil pan.

To install:

5. Clean all gasket surfaces of the cylinder block and the oil pan.
6. Apply sealant to the gasket surfaces of the oil pan.
7. Install or connect the following:
   - Oil pan onto the cylinder block within 15 minutes after applying sealant. Tighten to 60 inch lbs. (7 Nm).
   - Bell housing cover
   - Oil drain plug with a new seal and tighten to 29 ft. lbs. (40 Nm)
8. Lower the vehicle and fill the crankcase to the proper level with clean engine oil.
9. Connect the negative battery cable. Start the engine and check for leaks.

1.8L (4G93) ENGINE

1. Before servicing the vehicle, refer to the precautions in the beginning of this section.
2. Disconnect the negative battery cable.
3. Raise the vehicle and support safely.
4. Remove or disconnect the following:
   - Oil pan drain plug and drain the engine oil

- Exhaust pipe from the engine manifold
- Bell housing lower cover
- Oil pan retainer bolts and remove the oil pan

➡ Do not use a pry tool when removing the oil pan.

To install:

5. Clean all gasket surfaces of the cylinder block and the oil pan.
6. Apply sealant around the gasket surfaces of the oil pan.
7. Install or connect the following:
   - Oil pan onto the cylinder block within 15 minutes after applying sealant. Tighten to 60 inch lbs. (5 Nm).
   - Bell housing cover
   - Exhaust pipe to the engine manifold with new gasket in place. Tighten the exhaust pipe to manifold flange nuts to 33 ft. lbs. (45 Nm). Install and tighten the support bolt to 18 ft. lbs. (25 Nm).
   - Oil drain plug and tighten to 29 ft. lbs. (40 Nm)
8. Fill the crankcase to the proper level.
9. Connect the negative battery cable. Start the engine and check for leaks.

Oil Pump

REMOVAL & INSTALLATION

3.0L Engines

➡Whenever the oil pump is disassembled or the cover removed, the gear cavity must be filled with petroleum jelly to seal the pump and act as a prime. Do not use grease.

1. Before servicing the vehicle, refer to the precautions in the beginning of this section.
2. Disconnect the negative battery cable.
3. Drain the engine oil.
4. Remove or disconnect the following:
- Front engine mount bracket and accessory drive belts
- Timing belt upper and lower covers
- Timing belt and crankshaft sprocket
- Oil pan
- Oil screen and gasket
- Front cover mounting bolts. Note the lengths of the mounting bolts as they are removed for proper installation.
- Front case cover and oil pump assembly

To install:

5. Thoroughly clean all gasket material from all mounting surfaces.
6. Apply engine oil to the entire surface of the gears or rotors.
7. Assemble the front case cover and oil pump assembly to the engine block.
8. Install or connect the following:
- Oil screen with new gasket
- Oil pan
- Crankshaft sprocket and timing belt
- Timing belt covers
- Drive belts and the front engine mount bracket
- Negative battery cable, refill the crankcase and check for adequate oil pressure

3.5L Engine

1. Before servicing the vehicle, refer to the precautions in the beginning of this section.
2. Remove or disconnect the following:
- Negative battery cable
- Timing belt
3. Drain the engine oil.
4. Remove or disconnect the following:
- Splash shield from the wheel well
- Oil filter adapter

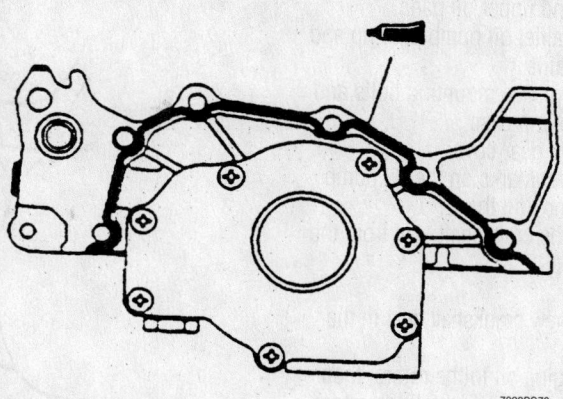

Apply sealant to the rear of the oil pump case—3.5L engine

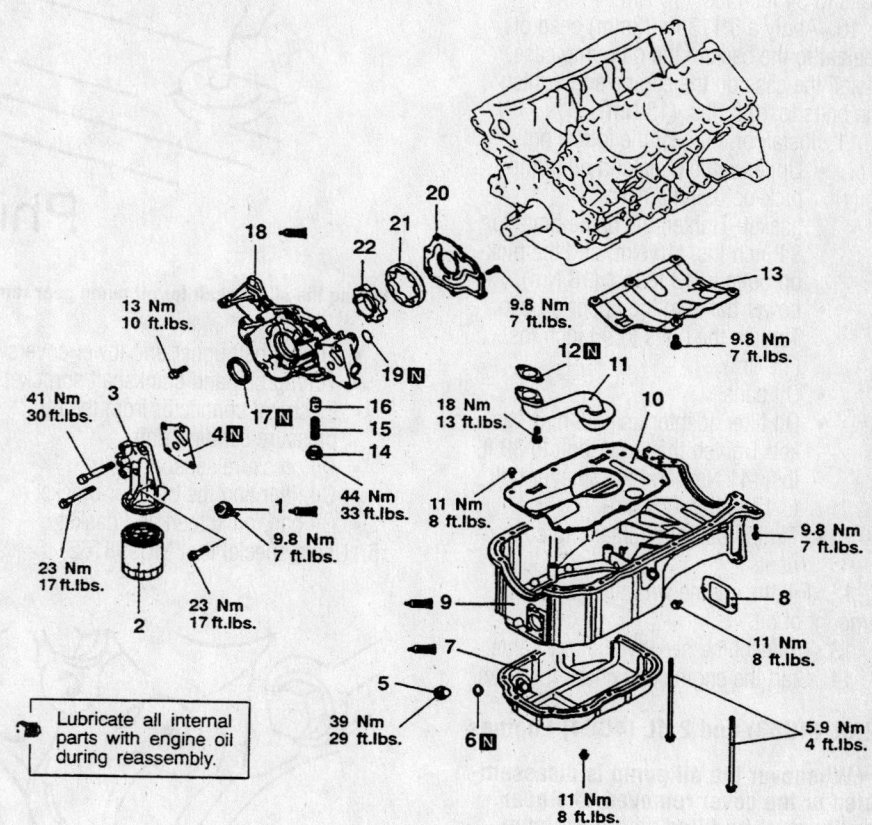

Lubricate all internal parts with engine oil during reassembly.

Removal steps

1. Oil pressure gauge unit
2. Oil filter
3. Oil filter bracket
4. Oil filter bracket gasket
5. Drain plug
6. Drain plug gasket
7. Oil pan, lower
8. Cover
9. Oil pan, upper
10. Baffle plate
11. Oil screen
12. Oil screen gasket
13. Baffle plate
14. Plug
15. Relief spring
16. Relief plunger
17. Crankshaft oil seal
18. Oil pump case
19. O-ring
20. Oil pump cover
21. Oil pump outer rotor
22. Oil pump inner rotor

Exploded view of the oil pump mounting—3.5L engine

Heater Core replacement is covered in Section 2 of this manual

- Lower and upper oil pans
- Lower baffle, oil pump pick-up and upper baffle
- Oil pump case mounting bolts and the oil pump case
- Oil pump gear cover

5. Make matchmarks on the oil pump rotors before removing them.

6. Remove the crankshaft seal from the oil pump case.

To install:

7. Install a new crankshaft seal in the oil pump cover.

8. Apply engine oil to the rotors, then align the matchmarks and install the rotors in the oil pump case.

9. Install the rotor cover. Tighten the bolts to 84 inch lbs. (10 Nm).

10. Apply a 0.113 in. (3mm) bead of sealant to the back of the oil pump case. Install the case on the engine and tighten the bolts to 10 ft. lbs. (13 Nm).

11. Install or connect the following:
- Upper baffle plate and oil pump pick-up using a new gasket—Tighten the baffle bolts to 84 inch lbs. (10 Nm) and the pick-up bolts to 13 ft. lbs. (18 Nm).
- Lower baffle in the upper oil pan. Tighten the bolts to 96 inch lbs. (11 Nm).
- Oil pans
- Oil filter adapter using a new gasket. Tighten the larger bolt to 30 ft. lbs. (41 Nm) and the smaller bolt to 17 ft. lbs. (23 Nm).
- Timing belt and remaining components

12. Fill the engine with the correct amount of oil.

13. Connect the negative battery cable.

14. Start the engine and check for leaks.

2.0L (4G63) and 2.4L (4G64) Engines

➡Whenever the oil pump is disassembled or the cover removed, the gear cavity must be filled with petroleum jelly to seal the pump and act as a prime. Do not use grease.

1. Before servicing the vehicle, refer to the precautions in the beginning of this section.

2. Disconnect the negative battery cable. Rotate the engine so No. 1 cylinder is on TDC of its compression stroke.

3. Drain the engine oil.

4. Using the proper equipment, support the weight of the engine. Remove the front engine mount bracket and accessory drive belts.

5. Remove or disconnect the following:

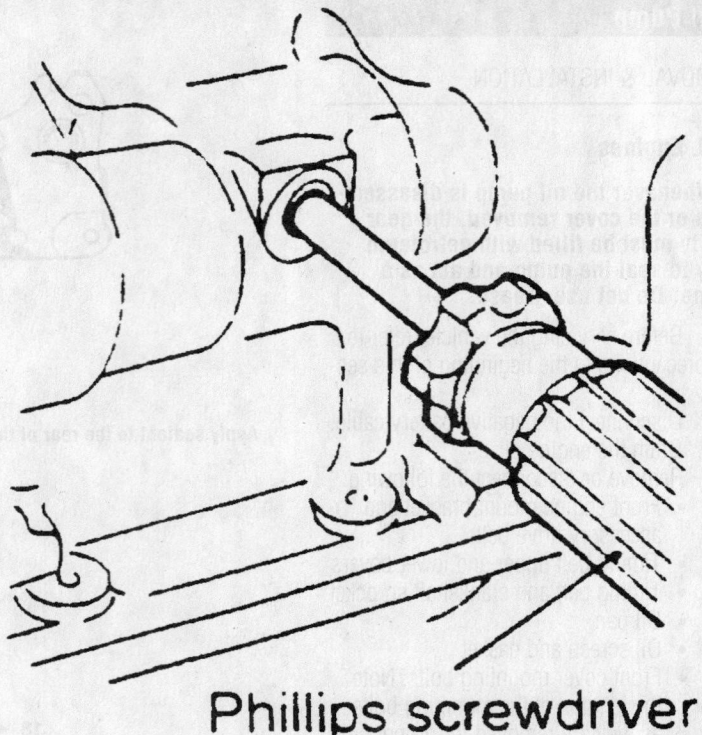

Phillips screwdriver

7923PG71

Holding the silent shaft for oil pump gear removal—Eclipse 2.0L (4G63) engine

- Timing belt upper and lower covers
- Timing belt and crankshaft sprocket
- Electrical connector from the oil pressure sending unit
- Oil pressure sensor
- Oil filter and the oil filter bracket
- Oil pan, oil screen and gasket

6. Using special tool MD998162, remove the plug cap in the engine front cover.

7. Remove or disconnect the following:
- Plug on the side of the engine block. Insert a steel rod with a shank diameter of 0.32 in. (8mm) into the plug hole. This will hold the silent shaft.

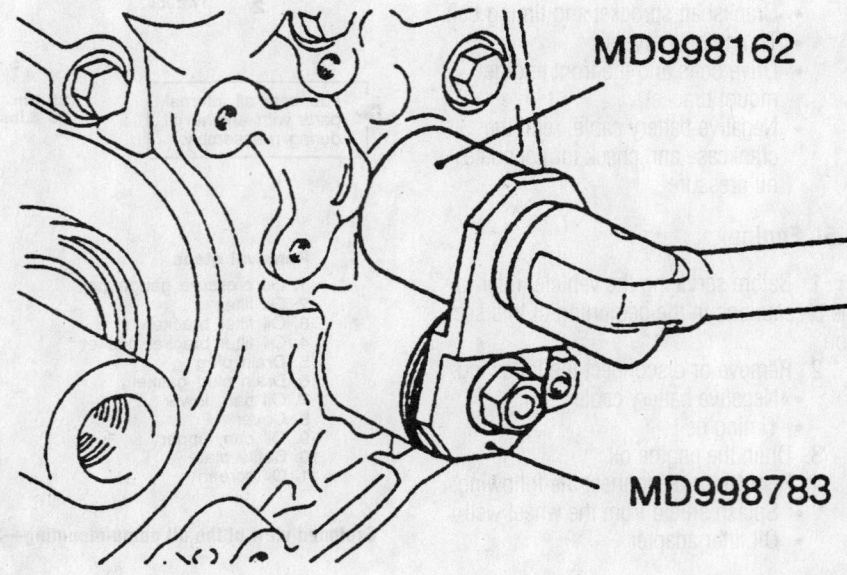

MD998162

MD998783

7923PG72

Use the special socket and holder to remove the balance shaft plug—Eclipse 2.0 (4G63) engine

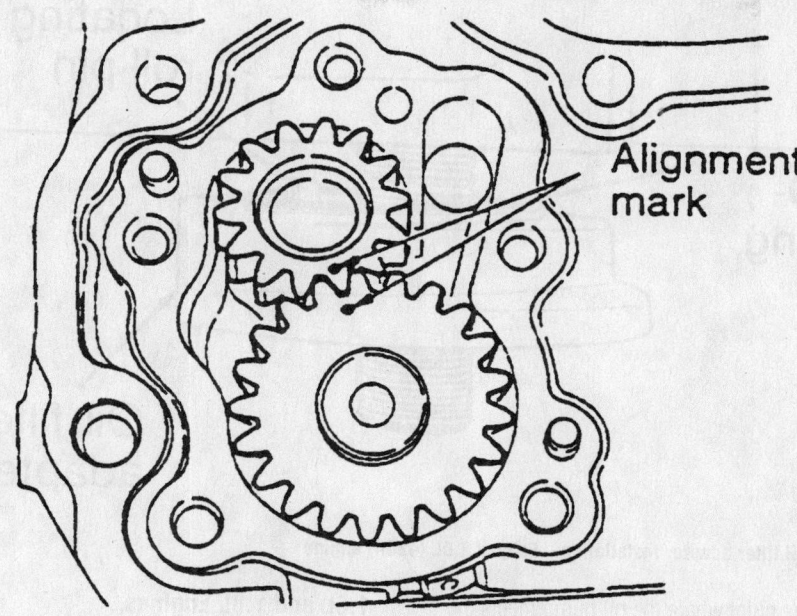

Alignment mark

7923PG73

Aligning oil pump timing marks—Eclipse 2.0L (4G63) engine

- Driven gear bolt that secures the oil pump driven gear to the silent shaft
- Front cover mounting bolts. Note the lengths of the mounting bolts as they are removed for proper installation.
- Front case cover and oil pump assembly. If necessary, the silent shaft can come out with the cover assembly.

- Oil pump cover, located on the back of the engine front cover. Remove the oil pump drive and driven gears.

8. After disassembling the oil pump, clean all components and remove gasket material from mating surfaces.

9. Assemble the oil pump gears into the front case and rotate it to ensure smooth rotation and no looseness. Be sure there is

no ridge wear on the contact surface between the front case and the gear surface of the oil pump front cover.

To install

10. Align the timing mark on the oil pump drive gear with that on the driven gear and install them into the engine front case. Apply engine oil to the gears.

11. Install the oil pump cover and tighten the retainer bolts to 13 ft. lbs. (18 Nm) on Eclipse models and 17 ft. lbs. (24 Nm) on Galant models.

12. Using the appropriate driver, install a new crankshaft seal into the front case.

13. Position new front case gasket in place. Set seal guide tool MD998285 on the front end of the crankshaft to protect the seal from damage. Apply a thin coat of oil to the outer circumference of the seal pilot tool.

14. Install the front case assembly through a new front case gasket and temporarily tighten the flange bolts.

15. Mount the oil filter on the bracket with new oil filter bracket gasket in place. Install the bolts with washers and tighten to 14 ft. lbs. (19 Nm).

16. Insert a Phillips screwdriver into the hole in the left side of the engine block to lock the silent shaft in place.

17. Install or connect the following:
- Oil pump drive gear onto the left silent shaft. Tighten the driven gear bolt to 27 ft. lbs. (37 Nm).
- New O-ring to the groove in the front case and install the plug cap. Tighten the cap to 17 ft. lbs. (24 Nm).
- Oil screen in position with new gasket in place

18. Clean both mating surfaces of the oil pan and the cylinder block. Apply sealant in the groove in the oil pan flange.

➡ **After applying sealant to the oil pan, do not exceed 15 minutes before installing the oil pan.**

19. Install or connect the following:
- Oil pan to the engine and secure with the retainers. Tighten bolts to 60 inch lbs. (7 Nm).
- Oil pressure gauge unit and the oil pressure switch
- Electrical harness connector
- Oil cooler. Oil cooler bolt to 31 ft. lbs. (43 Nm).

20. Refill the crankcase. Install new oil filter.

21. Connect the negative battery cable

L = 20 (.79)

L = 40 (1.57)

L = 40 (1.57)

*L = 30 (1.18)

L = 20 (.79)

L = 25 (.98)

L = 75 (2.95)

L = 55 (2.17)

Tighten together with belt tensioner

L = Bolt length below head [mm (in.)]

7923PG74

Front case bolt identification—Eclipse 2.0L (4G63) and 2.4L (4G64) engines

Brake service is covered in Section 4 of this manual

and start the engine. Verify correct oil pressure. Inspect for leaks.

2.0L (420A) Engine

1. Before servicing the vehicle, refer to the precautions in the beginning of this section.
2. Disconnect the negative battery cable.
3. Drain the engine oil.
4. Remove or disconnect the following:
   - Rear plate
   - Oil filter and adapter
   - Oil pan
   - Oil pick-up tube
   - Timing belt
5. Remove the crankshaft sprocket.
6. Remove the crankshaft oil seal.
7. Remove or disconnect the following:
   - Oil pump mounting bolts
   - Oil pump

To install:

8. Apply a bead of sealant to the sealing surface of the oil pump and install a new O-ring into the counterbore on the oil pump discharge passage.
9. Carefully install the oil pump on the crankshaft until seated to the engine block. Torque the bolts to 17 ft. lbs. (23 Nm).
10. Install or connect the following:
    - New crankshaft oil seal in the oil pump
    - Crankshaft sprocket using the proper installation tools
    - Timing belt and related components
    - Oil pick-up tube
11. Apply Loctite® 18718, or equivalent,

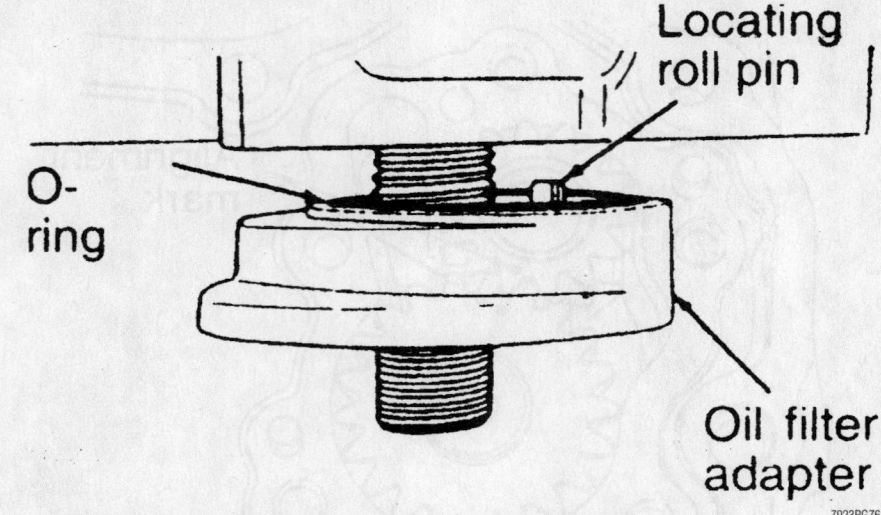

Oil filter adapter installation—Eclipse 2.0L (420A) engine

at the point where the oil pump meets the engine block.

12. Install the oil pan using a new gasket. Torque the mounting bolts to 108 inch lbs. (12 Nm).
13. Use a new O-ring and install the oil filter adapter to the engine. Made sure the roll pin aligns with the hole. Torque the assembly to 40 ft. lbs. (55 Nm).
14. Install or connect the following:
    - New oil filter
    - Rear plate
15. Refill the engine with the proper amount of oil.
16. Start the engine and check for leaks.

1.5L and 1.8L Engines

➡**Whenever the oil pump is disassembled or the cover removed, the gear cavity must be filled with petroleum jelly to seal the pump and act as a prime. Do not use grease.**

1. Before servicing the vehicle, refer to the precautions in the beginning of this section.
2. Remove or disconnect the following:
   - Negative battery cable
   - Front engine mount bracket and accessory drive belts
   - Timing belt upper and lower covers
   - Timing belt and crankshaft sprocket
   - Oil pan and remove the oil screen
   - Front cover mounting bolts. Note the lengths of the mounting bolts as they are removed for proper installation.
   - Front case assembly and oil pump assembly
   - Oil pump cover
   - Inner and outer gears from the front case

To install

3. Remove all gasket material from the mating surfaces and clean all parts.
4. Thoroughly coat both oil pump gears with clean engine oil and install them in the correct direction of rotation.
5. Install the pump cover and tighten the bolts to 84 inch lbs. (10 Nm).
6. Coat the relief valve and spring with clean engine oil. Install them and tighten the plug to 33 ft. lbs. (45 Nm).
7. Install or connect the following:

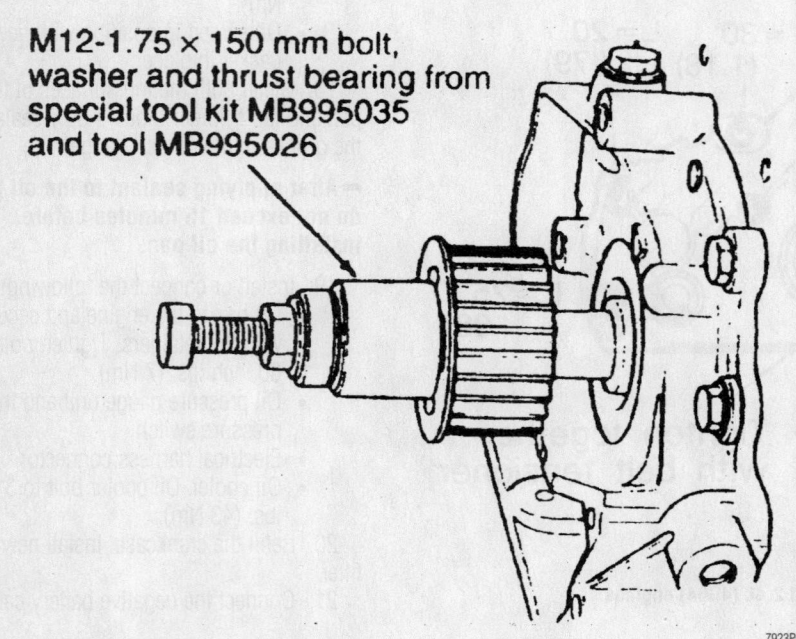

M12-1.75 × 150 mm bolt, washer and thrust bearing from special tool kit MB995035 and tool MB995026

Crankshaft sprocket installation—Eclipse 2.0L (420A) engine

- New front crankshaft seal and coat the lips of the seal with clean engine oil
- Front case and oil pump assembly to the engine block using a new gasket. Tighten the bolts to 10 ft. lbs. (14 Nm).
- Oil screen with new gasket. Torque the screen bolts to 14 ft. lbs. (19 Nm).
- Oil pan
- Crankshaft sprocket and timing belt

8. Fill the crankcase to the proper level.

9. Connect the negative battery cable.

Rear Main Seal

REMOVAL & INSTALLATION

1. Before servicing the vehicle, refer to the precautions in the beginning of this section.

2. Remove or disconnect the following:
- Transaxle
- Flywheel or flexplate from the crankshaft

3. Carefully pry the seal out of the oil seal case without damaging the sealing surface of the crankshaft.

To install:

4. Apply engine oil to the lip of the new seal and install the seal in the case using the proper size seal driver.

5. Install or connect the following:
- Flywheel or flexplate
- Transaxle

Piston and Ring

POSITIONING

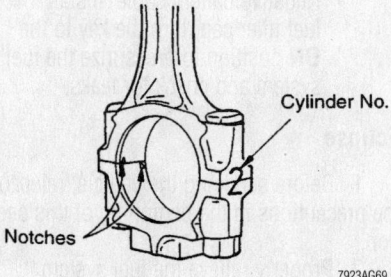

Before removing the caps from the connecting rods, be sure to matchmark them as shown

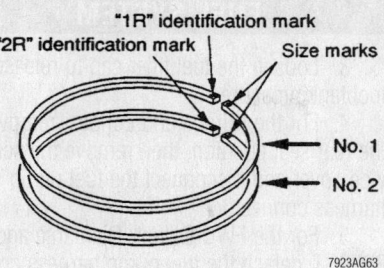

1.5L and 2.4L engines—compression ring identification mark locations

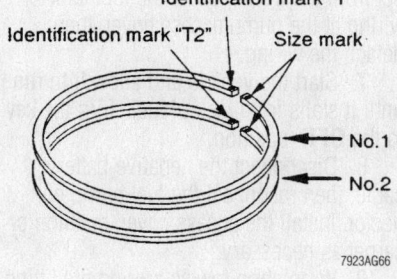

1.8L engine—compression ring identification mark locations

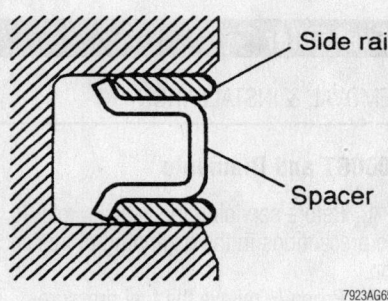

1.8L engine—oil side and spacer ring positioning

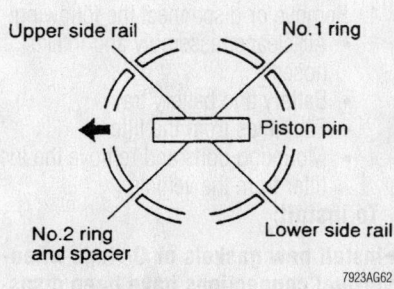

Except 2.0L (420A) engines—piston ring end-gap spacing

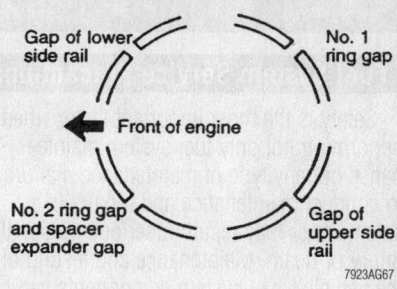

2.0L (420A) engines—piston ring end-gap spacing

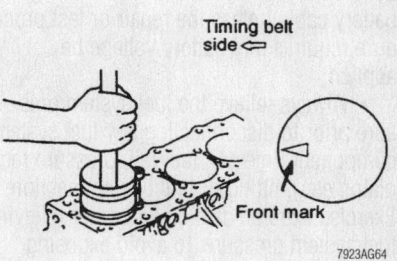

1.5L, 1.8L. 2.0L and 2.4L engines—piston-to-engine block mark location on the piston face

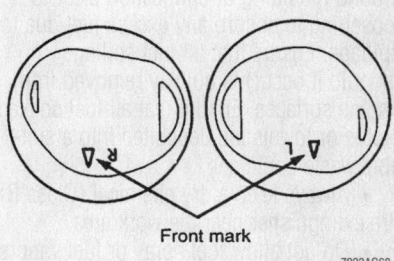

3.0L engine—piston-to-engine block mark locations

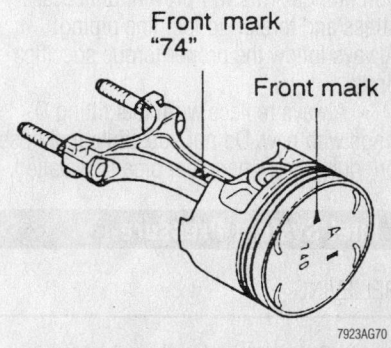

3.5L engine—piston and connecting rod assembly positioning

For complete Engine Mechanical specifications, see Section 1 of this manual

FUEL SYSTEM

Fuel System Service Precautions

Safety is the most important factor when performing not only fuel system maintenance but any type of maintenance. Failure to conduct maintenance and repairs in a safe manner may result in serious personal injury or death. Maintenance and testing of the vehicle's fuel system components can be accomplished safely and effectively by adhering to the following rules and guidelines.

• To avoid the possibility of fire and personal injury, always disconnect the negative battery cable unless the repair or test procedure requires that battery voltage be applied.

• Always relieve the fuel system pressure prior to disconnecting any fuel system component (injector, fuel rail, pressure regulator, etc.), fitting or fuel line connection. Exercise extreme caution whenever relieving fuel system pressure, to avoid exposing skin, face and eyes to fuel spray. Please be advised that fuel under pressure may penetrate the skin or any part of the body that it contacts.

• Always place a shop towel or cloth around the fitting or connection prior to loosening to absorb any excess fuel due to spillage. Ensure that all fuel spillage (should it occur) is quickly removed from engine surfaces. Ensure that all fuel soaked cloths or towels are deposited into a suitable waste container.

• Always keep a dry chemical (Class B) fire extinguisher near the work area.

• Do not allow fuel spray or fuel vapors to come into contact with a spark or open flame.

• Always use a back-up wrench when loosening and tightening fuel line connection fittings. This will prevent unnecessary stress and torsion to fuel line piping. Always follow the proper torque specifications.

• Always replace worn fuel fitting O-rings with new. Do not substitute fuel hose or equivalent, where fuel pipe is installed.

Fuel System Pressure

RELIEVING

1. Before servicing the vehicle, refer to the precautions in the beginning of this section.
2. Turn the ignition to the **OFF** position.

3. Loosen the fuel filler cap to release fuel tank pressure.
4. For the Mirage and Eclipse, remove the rear seat cushion, then remove the service cover and disconnect the fuel pump harness connector.
5. For the FWD Galant, Diamante and 3000GT, detach the fuel pump harness connector located near the fuel tank. It may be necessary to raise the vehicle to access the connector.
6. For the AWD Galant, remove the carpet from the trunk, locate the fuel tank wiring at the pump access cover, then detach the wiring.
7. Start the vehicle and allow it to run until it stalls from lack of fuel. Turn the key to the **OFF** position.
8. Disconnect the negative battery cable, then reconnect the fuel pump connector. Install the access cover, cushion or carpet as necessary.
9. Wrap shop towels around the fitting that is being disconnected to absorb residual fuel in the lines.
10. Place shop towels into proper safety container.

Fuel Filter

REMOVAL & INSTALLATION

3000GT and Diamante

1. Before servicing the vehicle, refer to the precautions in the beginning of this section.
2. Properly relieve the fuel pressure.
3. Disconnect the negative battery cable.

➡**The filter is located in the engine compartment, mounted on the inner fender panel.**

4. Remove or disconnect the following:
   • Air cleaner assembly and intake hoses
   • Battery and battery tray
   • Fuel lines from the filter
   • Mounting bolts and remove the fuel filter from the vehicle

To install:

➡**Install new gaskets or O-rings whenever fuel connections have been disassembled.**

5. Install or connect the following:
   • Filter to its bracket finger-tight
   • New gaskets and connect the high pressure hose and eye bolt, then

the main pipe and eye bolt. Tighten the eye bolts to 22 ft. lbs. (30 Nm). Tighten the flare nut to 25 ft. lbs. (35 Nm).
6. Tighten the mounting bolts fully.
7. Install or connect the following:
   • Air cleaner assembly
   • Battery and battery tray
   • Negative battery cable, install the fuel filler cap, turn the key to the **ON** position to pressurize the fuel system and check for leaks.

Mirage and Galant

➡**The fuel filter is located in the engine compartment.**

1. Before servicing the vehicle, refer to the precautions in the beginning of this section.
2. Properly relieve the fuel system pressure.
3. Remove or disconnect the following:
   • Negative battery cable
   • Air intake hose and the battery
   • Fuel lines from the filter
   • Mounting bolts and the fuel filter from the vehicle

To install:

4. If equipped with flare fitting, tighten the fitting by hand before installing the filter to the vehicle.
5. Install or connect the following:
   • Filter to its bracket finger-tight
   • New gaskets and connect the high pressure hose and eye bolt, then the main pipe. Tighten the eye bolts to 22 ft. lbs. (30 Nm). Tighten the flare nut to 27 ft. lbs. (37 Nm).
6. Tighten the filter mounting bolts fully.
7. Install or connect the following:
   • Air intake hose
   • Battery
   • Negative battery cable, install the fuel filler cap, turn the key to the **ON** position to pressurize the fuel system and check for leaks.

Eclipse

1. Before servicing the vehicle, refer to the precautions in the beginning of this section.
2. Properly relieve the fuel system pressure.
3. Disconnect the negative battery cable.
4. On turbo models and models equipped with the 2.4L engine, remove the battery and the air intake hose.
5. Remove the fuel lines from the filter.

6. On the 2.0L non-turbo engine models, remove the clamp and the hose from the fuel pressure regulator.

7. Remove or disconnect the following:
- Fuel filter mounting bracket bolts and remove the fuel filter
- Bracket screw and remove the fuel filter from the mounting bracket

8. On 2.0L non-turbo models, remove the following from the filter:
  a. The eye bolt and washer.
  b. The fuel connector and washer with the fuel pressure regulator.

To install:

9. On 2.0L non-turbo models, install the fuel connector with the fuel pressure regulator to the filter with 2 new washers and tighten the eye bolt to 22 ft. lbs. (36 Nm).

10. Install or connect the following:
- Fuel filter to the mounting bracket with the screw
- Fuel filter to the vehicle with the bracket mounting bolts
- Main fuel pipe to the fuel filter connector or the filter itself. Torque the flare nut to 27 ft. lbs. (36 Nm).

11. On the 2.0L non-turbo engine models reconnect the hose and clamp to the fuel pressure regulator.

12. Reconnect the high pressure fuel hose to the fuel filter. Torque the eye bolt to 22 ft. lbs. (29 Nm).

13. On 2.0L turbo and 2.4L engine models, install the battery and the air intake hose.

14. Reconnect the negative battery cable, start the engine and check for fuel leaks.

Fuel Pump

REMOVAL & INSTALLATION

3000GT

1. Before servicing the vehicle, refer to the precautions in the beginning of this section.

2. Relieve fuel system pressure. Remove the fuel filler cap.

3. Remove or disconnect the following:
- Negative battery cable
- Fuel gauge cover located in the rear floor pan
- Fuel pump and gauge electrical connector
- Overfill limiter (2-way valve)
- Both sides of the high pressure fuel hose
- Fuel pump and gauge assembly from the tank

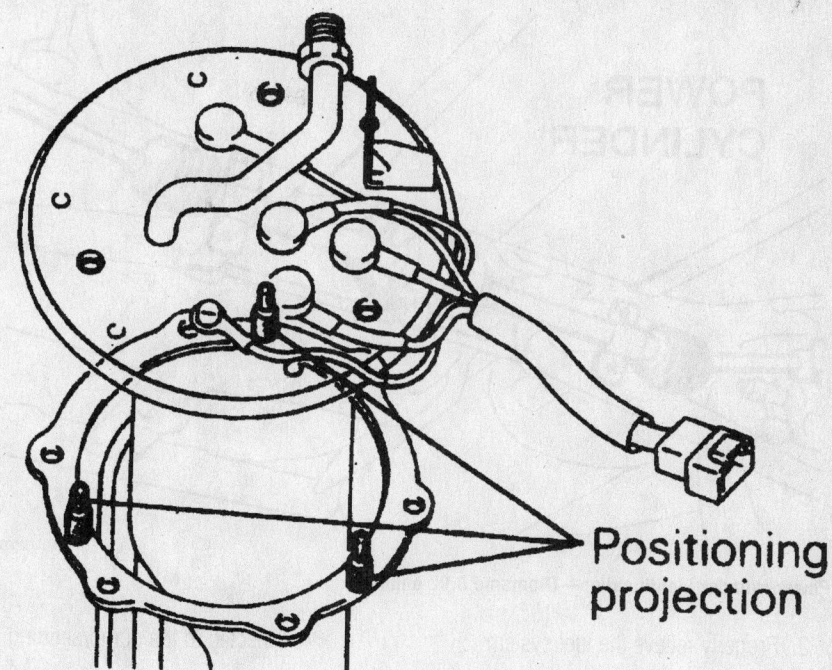

Align the positioning projections—3000GT 3.0L engine

7923PG81

To install:

4. Align the 3 projections on the packing with the holes on the fuel pump and the nipples on the pump facing the same direction as before removal.

5. Temporarily tighten the flare nut on the high pressure hose by hand. Making sure the hose does not twist, tighten body side nut to 22 ft. lbs. (30 Nm) and the fuel pump side nut to 25 ft. lbs. (35 Nm).

6. Install or connect the following:
- Overfill limiter (2-way valve) with the long shouldered side of the valve facing the canister
- Electrical connector to the pump assembly
- Negative battery cable and check the entire system for leaks
- Sealer to the rear floor pan and install the cover into place

Diamante

1. Before servicing the vehicle, refer to the precautions in the beginning of this section.

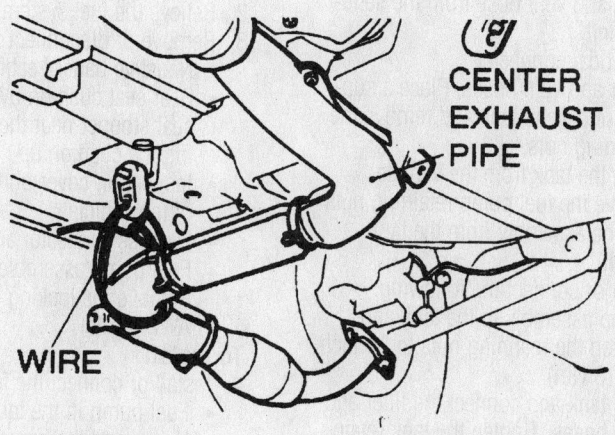

Proper method of supporting rear exhaust system—Diamante 3.0L engine

7923PG79

For Accessory Drive Belt illustrations, see Section 1 of this manual

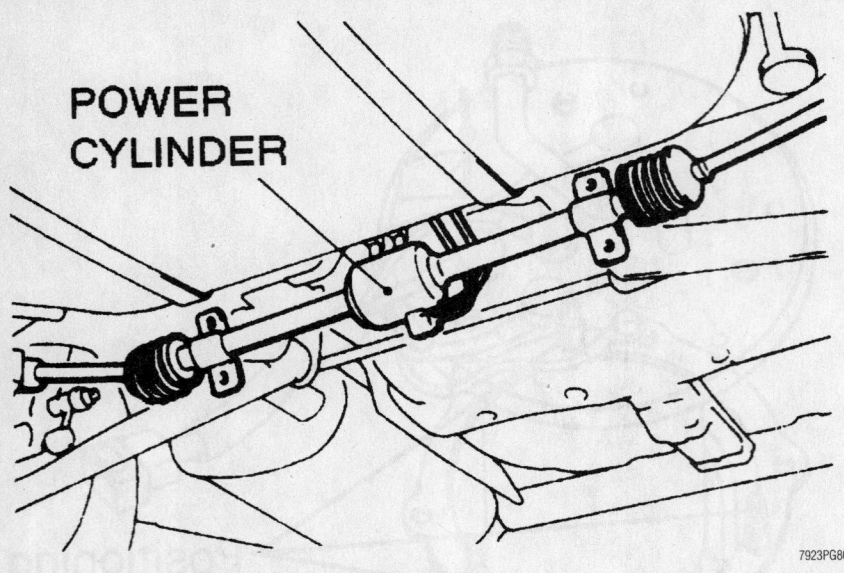

POWER CYLINDER

Power cylinder identification—Diamante 3.0L engine

7923PG80

2. Properly relieve the fuel system pressure.

3. Disconnect the negative battery cable.

4. Raise the vehicle and support it safely.

5. Remove the left rear wheel well liner, if equipped.

6. Disconnect the center exhaust system from the main muffler. Disconnect the rear exhaust hangers, lower the exhaust and secure aside.

7. Remove the tank drain plug and drain the fuel into an approved container.

8. On models equipped with 4WS, remove the mounting bolts and lower the rear steering gear.

9. Remove or disconnect the following:
- Fuel return hose, high pressure hose and vent hose from the sending unit
- Electrical connector
- Filler and vent hoses. Place a support under the tank and remove the retaining nuts.

10. Lower the tank from the vehicle.

11. Remove the fuel pump retaining nuts and remove the assembly from the tank.

To install:

12. Install or connect the following:
- Pump assembly to the tank and tighten the retaining nuts to 24 inch lbs. (3 Nm)
- Fuel tank and connect the filler and vent hoses. Tighten the tank retaining nuts and bolts to 19 ft. lbs. (26 Nm).
- Return hose, high pressure hose and all other hoses and connectors

connected to the pump/sending unit
- Power cylinder unit and tighten the mounting bolts to 31 ft. lbs. (43 Nm), if equipped with 4WS
- Exhaust pipe and secure the rear hangers
- Left rear wheel well liner, if removed

13. Lower the vehicle and return fuel to the gas tank.

14. Connect the negative battery cable and check the entire system for proper operation and leaks.

Eclipse

1. Before servicing the vehicle, refer to the precautions in the beginning of this section.

2. Relieve the fuel system pressure.

3. Remove or disconnect the following:
- Negative battery cable
- Rear seat cushion by pulling the seat stopper near the floor and lifting the cushion up
- Inspection cover on the right side of the vehicle
- Harness connector and the fuel lines
- Fuel pump assemble from the tank. Remove the locking ring on the AWD model.

To install:

4. Install or connect the following:
- Fuel pump in the tank
- Hoses and the harness connector
- Inspection cover
- Rear seat
- Negative battery cable

Galant

1. Before servicing the vehicle, refer to the precautions in the beginning of this section.

2. Properly relieve the fuel system pressure.

3. Remove or disconnect the following:
- Negative battery cable
- Rear seat cushion, by pulling the seat stopper outward and lifting the lower cushion upward
- Access cover
- Fuel pump wiring
- Return hose and the high pressure fuel hose
- Pump mounting nuts and remove the pump assembly

To install:

4. Install the fuel pump assembly to the tank and tighten the retaining nuts to 22 inch lbs. (2.5 Nm).

➡ **Tilt the float to the left of the vehicle, when installing the pump assembly.**

5. Install or connect the following:
- High pressure hose, return hose and the fuel tank wiring
- Negative battery cable

6. Check the fuel pump for proper pressure and inspect the entire system for leaks.

7. Apply sealant to the access cover and install the cover.

8. Install the rear seat cushion.

Mirage

1. Before servicing the vehicle, refer to the precautions in the beginning of this section.

2. Properly relieve the fuel system pressure.

3. Disconnect the negative battery cable.

4. Raise and safely support the vehicle.

5. Drain the fuel from the fuel tank into an approved container.

6. Disconnect the filler and vent hoses.

7. Support the tank with a transmission jack. Disconnect the retainer straps and lower the tank to gain access to the fitting on top of the tank.

8. Remove or disconnect the following:
- Return hose, high pressure hose and vapor hoses from the pump/sending unit
- Electrical connectors at the pump/sending unit
- Fuel tank from the vehicle
- Access plate to the fuel tank and remove the pump assembly

To install:

9. Install fuel pump into fuel tank, with

new packing gasket, and tighten mounting nuts.

10. Raise the tank in position under the vehicle.

11. Attach all connections to the top of the tank.

12. Raise the tank completely and position the retainer straps around the fuel tank. Install new fuel tank self-locking nuts and tighten to 22 ft. lbs. (31 Nm).

13. Connect the return hose and high pressure hoses.

14. Install the vapor hose and the filler hose. Install the filler hose retainer screws to the fender, if removed.

15. Lower the vehicle and pour the drained fuel into the gas tank.

16. Connect the negative battery cable. Check the fuel pump for proper pressure and inspect the entire system for leaks.

Fuel Injector

REMOVAL & INSTALLATION

1.5L, 1.8L, And 2.0L SOHC Engines

1. Relieve the fuel system pressure as described in this section.
2. Remove or disconnect the following:
   • PCV hose from the valve cover
   • Breather hose at the opposite end of the valve cover
   • High pressure fuel line

❊❊ CAUTION

Observe all applicable safety precautions when working around fuel. Whenever servicing the fuel system, always work in a well ventilated area. Do not allow fuel spray or vapors to come in contact with a spark or open flame. Keep a dry chemical fire extinguisher near the work area. Always keep fuel in a container specifically designed for fuel storage; also, always properly seal fuel containers to avoid the possibility of fire or explosion.

3. Remove or disconnect the following:
   • Vacuum hose from the fuel pressure regulator
   • Fuel return hose from the pressure regulator
   • Electrical connector from each injector. Label for reference
   • Bolt(s) holding the fuel rail to the manifold. Carefully lift the rail up

and remove it with the injectors attached. Take great care not to drop an injector. Place the rail and injectors in a safe location on the workbench; protect the tips of the injectors from dirt and/or impact.
   • Injector insulators from the intake manifold, discard. The insulators are not reusable.
   • Injectors from the fuel rail by pulling gently in a straight outward motion. Make certain the grommet and O-ring come off with the injector.

To install:

4. Install a new insulator in each injector port in the manifold.

5. Remove the old grommet and O-ring from each injector. Install a new grommet and O-ring; coat the O-ring lightly with clean, thin oil.

6. If the fuel pressure regulator was removed, replace the O-ring with a new one and coat it lightly with clean, thin oil. Insert the regulator straight into the rail, then check that it can be rotated freely. If it does not rotate smoothly, remove it and inspect the O-ring for deformation or jamming. When properly installed, align the mounting holes and tighten the retaining bolts to 84 inch lbs. (9 Nm). This procedure must be followed even if the fuel rail was not removed.

7. Install or connect the following:
   • Injector into the fuel rail, constantly turning the injector left and right during installation. When fully installed, the injector should still turn freely in the rail. If it does not, remove the injector and inspect the O-ring for deformation or damage.
   • Delivery pipe and injectors to the engine. Make certain that each injector fits correctly into its port and that the rubber insulators for the fuel rail mounts are in position.
   • Fuel rail retaining bolts and tighten them to 108 inch lbs. (12 Nm)
   • Wiring harnesses to the appropriate injector
   • Fuel return hose to the pressure regulator, then connect the vacuum hose

8. Replace the O-ring on the high pressure fuel line, coat the O-ring lightly with clean, thin oil and install the line to the fuel rail. Tighten the mounting bolts.
   • PCV hose and the breather hose
   • Negative battery cable

9. Pressurize the fuel system and inspect all connections for leaks.

1.6L and 2.0L DOHC Engines

1. Relieve the fuel system pressure as described in this section.
2. Disconnect the negative battery cable.
3. Wrap the connection with a shop towel and disconnect the high pressure fuel line at the fuel rail.

❊❊ CAUTION

Observe all applicable safety precautions when working around fuel. Whenever servicing the fuel system, always work in a well ventilated area. Do not allow fuel spray or vapors to come in contact with a spark or open flame. Keep a dry chemical fire extinguisher near the work area. Always keep fuel in a container specifically designed for fuel storage; also, always properly seal fuel containers to avoid the possibility of fire or explosion.

4. Remove or disconnect the following:
   • Fuel return hose and remove the O-ring
   • Vacuum hose from the fuel pressure regulator
   • PCV hose. On 2.0L engine, remove the center cover
   • Electrical connectors from each injector, label for reference
   • Injector rail retaining bolts. Make sure the rubber mounting bushings do not get lost.

5. Lift the rail assembly up and away from the engine.

6. Remove the injectors from the rail by pulling gently. Discard the lower insulator. Check the resistance through the injector. The specification for 2.0L turbocharged engine is 2–3 ohms at 70°F (20°C). The specification for the others is 13–15 ohms at 70°F (20°C).

To install:

7. Install or connect the following:
   • New grommet and O-ring to the injector. Coat the O-ring with light oil.
   • Injector to the fuel rail

8. Replace the seats in the intake manifold. Install the fuel rail and injectors to the manifold. Make sure the rubber bushings are in place before tightening the mounting bolts.

For Tire, Wheel and Ball Joint specifications, see Section 1 of this manual

9. Tighten the retaining bolts to 72 inch lbs. (11 Nm).

10. Install or connect the following:
- Connectors to the injectors and install the center cover
- PCV hose
- Fuel pressure regulator vacuum hose
- Fuel return hose

11. Replace the O-ring, lightly lubricate it and connect the high pressure fuel line.

12. Connect the negative battery cable and check the entire system for proper operation and leaks.

2.4L Engine

1. Relieve the fuel system pressure as described in this section.

2. Label and disconnect the spark plug wires. Position the wires aside.

3. Remove or disconnect the following:
- PCV hose from the valve cover
- High pressure fuel line to the fuel rail and disconnect the line. Be prepared to contain fuel spillage; plug the line to keep out dirt and debris.

❊❊ CAUTION

Observe all applicable safety precautions when working around fuel. Whenever servicing the fuel system, always work in a well ventilated area. Do not allow fuel spray or vapors to come in contact with a spark or open flame. Keep a dry chemical fire extinguisher near the work area. Always keep fuel in a container specifically designed for fuel storage; also, always properly seal fuel containers to avoid the possibility of fire or explosion.

4. Remove or disconnect the following:
- Vacuum hose from the fuel pressure regulator
- Fuel return hose from the pressure regulator
- Electrical connector from each injector, label for reference
- Bolt(s) holding the fuel rail to the manifold. Carefully lift the rail up and remove it with the injectors attached. Take great care not to drop an injector. Place the rail and injectors in a safe location on the workbench; protect the tips of the injectors from dirt and/or impact.
- Injector insulators from the intake manifold, discard. The insulators are not reusable.
- Injectors from the fuel rail by

pulling gently in a straight outward motion. Make certain the grommet and O-ring come off with the injector.

To install:

5. Install a new insulator in each injector port in the manifold.

6. Remove the old grommet and O-ring from each injector. Install a new grommet and O-ring; coat the O-ring lightly with clean, thin oil.

7. If the fuel pressure regulator was removed, replace the O-ring with a new one and coat it lightly with clean, thin oil. Insert the regulator straight into the rail, then check that it can be rotated freely. If it does not rotate smoothly, remove it and inspect the O-ring for deformation or jamming. When properly installed, align the mounting holes and tighten the retaining bolts to 84 inch lbs. (9 Nm). This procedure must be followed even if the fuel rail was not removed.

8. Install or connect the following:
- Injector into the fuel rail, constantly turning the injector left and right during installation. When fully installed, the injector should still turn freely in the rail. If it does not, remove the injector and inspect the O-ring for deformation or damage.
- Delivery pipe and injectors to the engine. Make certain that each injector fits correctly into its port and that the rubber insulators for the fuel rail mounts are in position.
- Fuel rail retaining bolts and tighten them to 108 inch lbs. (12 Nm)
- Wiring harnesses to the appropriate injector
- Fuel return hose to the pressure regulator, then connect the vacuum hose
- O-ring on the high pressure fuel line, coat the O-ring lightly with clean, thin oil and install the line to the fuel rail. Tighten the mounting bolts to 48 inch lbs. (6 Nm).
- PCV hose and spark plug wires
- Negative battery cable

9. Pressurize the fuel system and inspect all connections for leaks.

3.0L and 3.5L Engines

1. Relieve the fuel system pressure.

2. Disconnect the negative battery cable.

❊❊ CAUTION

Work MUST NOT be started until at least 90 seconds after the ignition

switch is turned to the LOCK position and the negative battery cable is disconnected from the battery. This will allow time for the air bag system backup power supply to deplete its stored energy preventing accidental air bag deployment which could result in unnecessary air bag system repairs and/or personal injury.

3. Drain the cooling system.

4. Disconnect all components from the air intake plenum and remove the plenum from the intake manifold.

5. Wrap the connection with a shop towel and disconnect the high pressure fuel line at the fuel rail.

❊❊ CAUTION

Observe all applicable safety precautions when working around fuel. Whenever servicing the fuel system, always work in a well ventilated area. Do not allow fuel spray or vapors to come in contact with a spark or open flame. Keep a dry chemical fire extinguisher near the work area. Always keep fuel in a container specifically designed for fuel storage; also, always properly seal fuel containers to avoid the possibility of fire or explosion.

6. Remove or disconnect the following:
- Fuel return hose and remove the O-ring
- Vacuum hose from the fuel pressure regulator
- Electrical connectors from each injector
- Fuel pipe connecting the fuel rails
- Injector rail retaining bolts. Make sure the rubber mounting bushings do not get lost.

7. Lift the rail assemblies up and away from the engine.

8. Remove the injectors from the rail by pulling gently. Discard the lower insulator.

To install:

➡**Some of the vehicles may have a clip that secures the injector to the fuel rail. Be sure to remove or install the injector clip where necessary.**

9. Install or connect the following:
- New grommet and O-ring to the injector. Coat the O-ring with light oil.
- Injector to the fuel rail

10. Replace the seats in the intake manifold. Install the fuel rails and injectors to the

manifold. Make sure the rubber bushings are in place before tightening the mounting bolts.

11. Tighten the retaining bolts to 84–108 inch lbs. (10–13 Nm). Install the fuel pipe with new gasket.

12. Install or connect the following:
- Electrical connectors to the injectors
- Fuel return hose
- O-ring, lightly lubricate it and connect the high pressure fuel line

- Intake plenum and all related items, using new gaskets
13. Fill the cooling system.
14. Connect the negative battery cable and check the entire system for proper operation and leaks.

DRIVE TRAIN

Transaxle assembly

REMOVAL & INSTALLATION

Manual

3000GT

1. Before servicing the vehicle, refer to the precautions in the beginning of this section.
2. Remove the battery and battery tray.
3. Raise the vehicle and support safely.
4. If equipped with AWD, disconnect the exhaust pipe. Remove the mounting bolts and lower the transfer case from the vehicle.
5. Remove or disconnect the following:
- Left side splash shield and engine undercover
- Air cleaner assembly and all adjoining duct work
- Shifter control cables and speedometer connector
- Clutch release cylinder
- Reverse light switch
6. Support the weight of the transaxle and remove the transaxle mount through-bolt. Remove the access plug, remove the bolts for the bracket and remove the brackets.
7. Remove or disconnect the following:
- Transaxle ground cable
- Tie rod end and ball joint from the steering knuckle
- Right frame member
- Starter motor
- Halfshafts by inserting a prybar between the transaxle case and the driveshaft and prying the shaft from the transaxle. Do not pull on the driveshaft. Doing so damages the inboard joint. Tie the shafts aside.
- Transaxle brackets
- Transaxle assembly.

To install:
8. Install or connect the following:
- Transaxle to the engine and install the mounting bolts
- Halfshafts. Use new circlips on the axle ends.
- Starter motor and cover

- Right side frame member
- Ball joint and tie rod to the steering knuckle
- Transaxle ground cable
- Side mount brackets and install the access plug
- Reverse light switch
- Clutch release cylinder
- Shifter control cables and speedometer connector
- Transfer case and related items on AWD vehicles
- Air cleaner assembly and all adjoining ductwork
- Left side splash shield
- Battery tray and battery
- Negative battery cable and check the transaxle and transfer case for proper operation

ECLIPSE

1. Before servicing the vehicle, refer to the precautions in the beginning of this section.

2. Remove or disconnect the following:
- Battery and the air intake hoses
- Battery tray and support
- Auto-cruise actuator and bracket, if equipped with cruise control
- Charcoal canister and bracket
- Shift and select cables from the transaxle
- Back-up light switch and the Vehicle Speed Sensor (VSS) connectors
- Starter assembly
- Engine support fixture to the engine and remove the transaxle mounting bolts
3. Remove or disconnect the following:
- Rear roll stopper bracket mounting bolts
- Transaxle mounting bracket mounting nuts
4. Raise the vehicle and remove the engine undercovers.
5. If equipped with all wheel drive, remove the transfer case assembly.

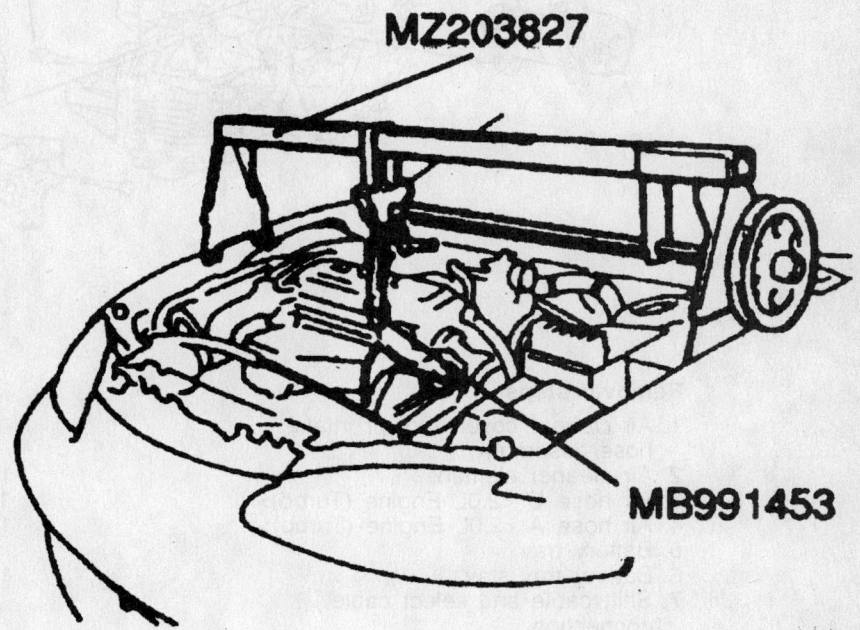

MZ203827

MB991453

Proper method of supporting the engine assembly for transaxle removal

7923PG82

For Wheel Alignment specifications, see Section 1 of this manual

6. Remove or disconnect the following:
- Axle shafts
- Slave cylinder from the bell housing without disconnecting the fluid line. Position it out of the way.
- Bell housing cover and the right-hand center member stay (support)
- Center member

7. Place a transmission jack under the transaxle and remove the transaxle mounting bolt.

8. Remove the transaxle mounting and lower the transaxle.

To install:

9. Raise the transaxle into position and install the transaxle mounting. Torque the through-bolt to 50 ft. lbs. (69 Nm).

10. Install or connect the following:

- Transaxle assembly mounting bolt. Torque the bolt to 22–25 ft. lbs. (30–34 Nm).
- Center member assembly and the right-hand stay
- Bell housing cover and the slave cylinder
- Axle shafts. Be sure to install the washer in the proper direction.

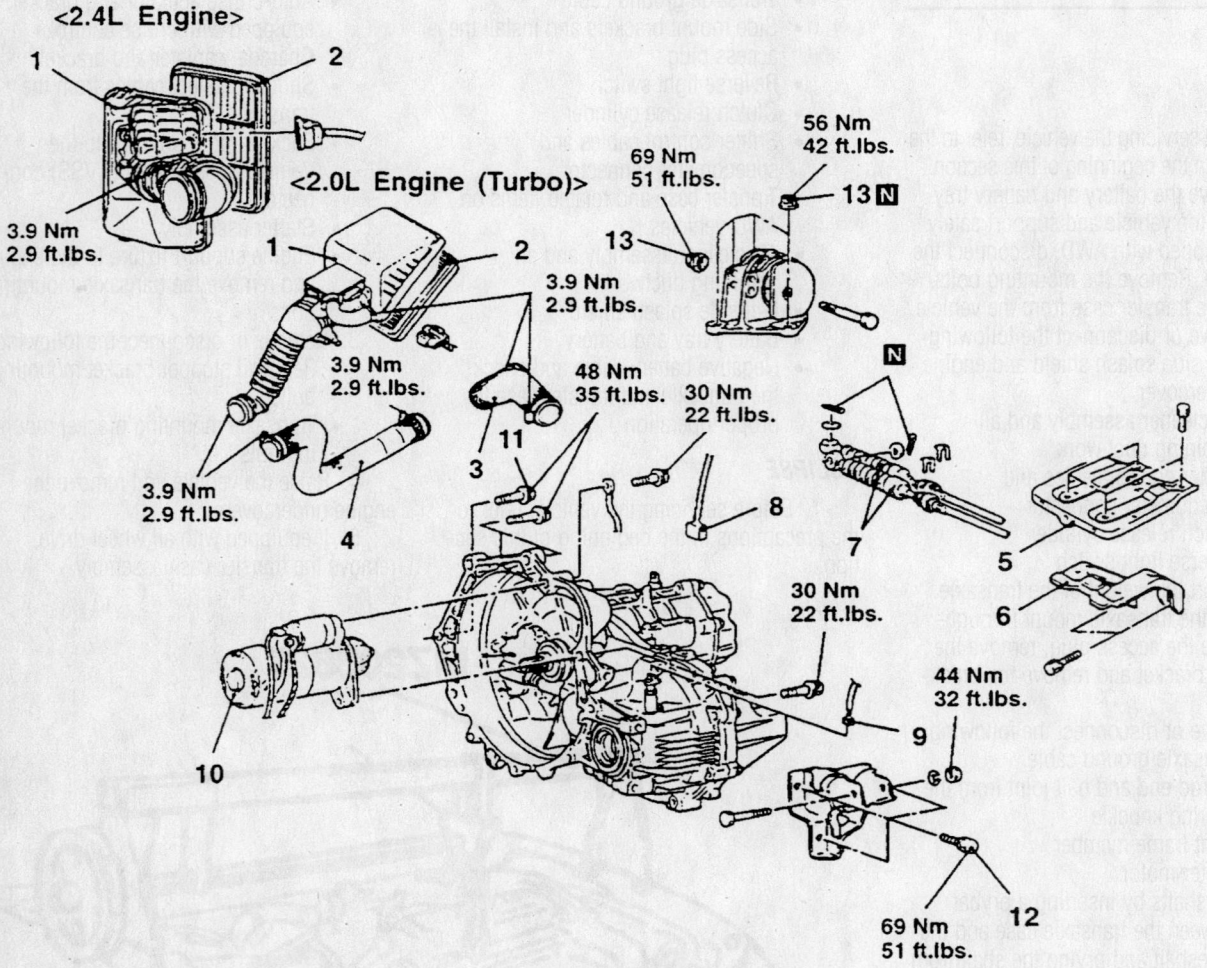

Removal steps

1. Air cleaner cover and air intake hose assembly
2. Air cleaner element
3. Air hose C <2.0L Engine (Turbo)>
4. Air hose A <2.0L Engine (Turbo)>
5. Battery tray
6. Battery tray stay
7. Shift cable and select cable connection
8. Backup light switch connector
9. Vehicle speed sensor connector
10. Starter motor
11. Transaxle assembly mounting bolts
12. Rear roll stopper bracket mounting bolts
13. Transaxle mounting bracket mounting nuts
- Supporting engine assembly

7923PG86

Exploded view of the manual transaxle mounting (1 of 2)—FWD Eclipse with 2.4L engine

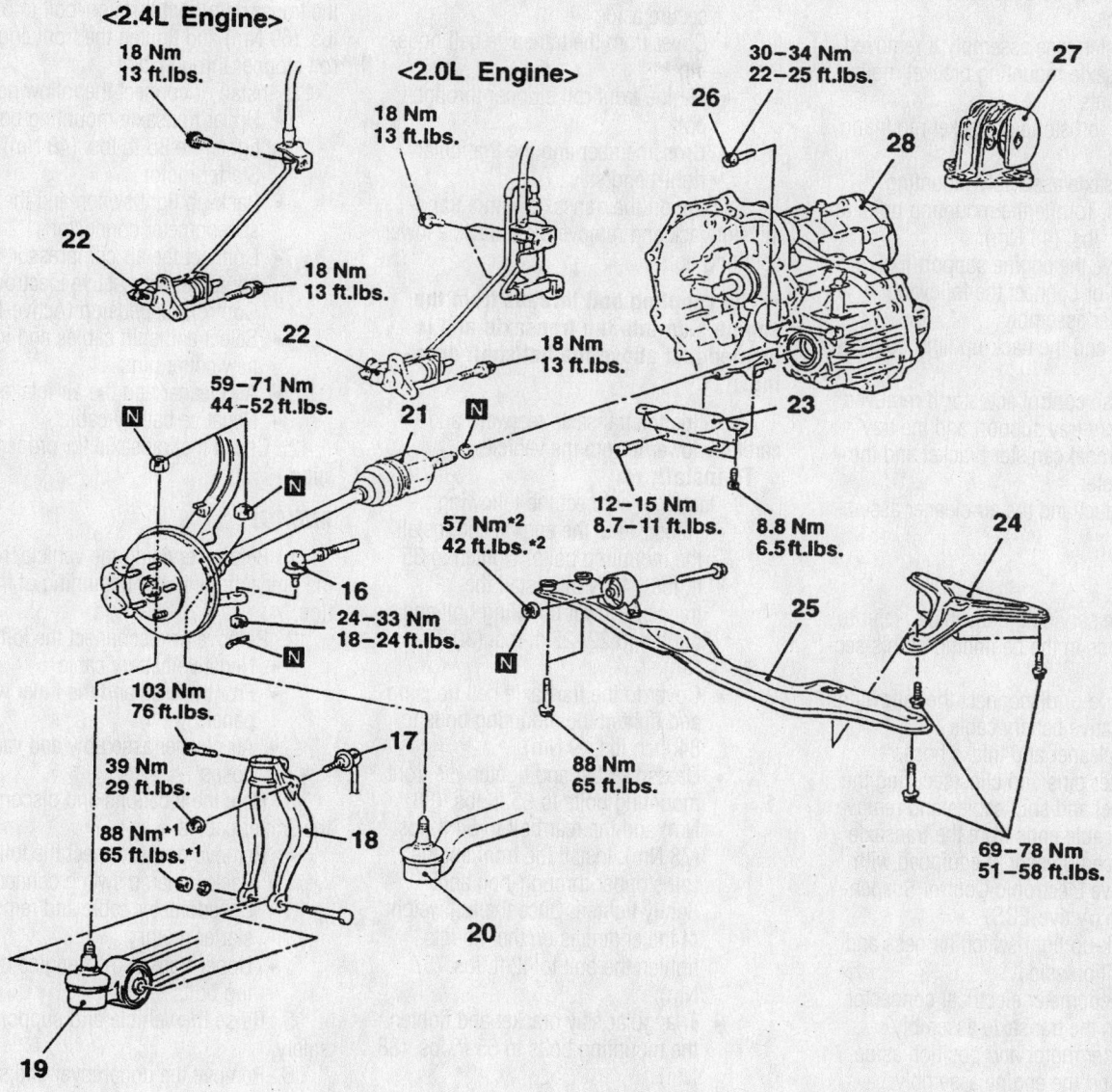

<2.4L Engine>
18 Nm
13 ft.lbs.

<2.0L Engine>
18 Nm
13 ft.lbs.

30–34 Nm
22–25 ft.lbs.

27

26

28

22

18 Nm
13 ft.lbs.

22

59–71 Nm
44–52 ft.lbs.

21

18 Nm
13 ft.lbs.

23

12–15 Nm
8.7–11 ft.lbs.

8.8 Nm
6.5 ft.lbs.

24

57 Nm*2
42 ft.lbs.*2

16

24–33 Nm
18–24 ft.lbs.

25

103 Nm
76 ft.lbs.

17

39 Nm
29 ft.lbs.

88 Nm*1
65 ft.lbs.*1

18

88 Nm
65 ft.lbs.

20

69–78 Nm
51–58 ft.lbs.

19

Lifiting up of the vehicle
16. Tie rod end ball joint and kunckle connection
17. Stabilizer link connection
18. Damper fork
19. Lateral lower arm ball joint and kunckle connection
20. Compression lower arm ball joint and kunckle connection
21. Drive shaft connection
22. Clutch release cylinder connection
23. Bell housing cover
24. Stay (R.H.)
25. Center member assembly
26. Transaxle assembly mounting bolt
27. Transaxle mounting
28. Transaxle assembly

Caution
*1: Indicates parts which should be temporarily tightened, and then fully tightened with the vehicle on the ground in the unladen condition.
*2: For tightening locations indicated by the symbol, first tighten temporarily, and then make the final tightening with the entire weight of the engine applied to the vehicle body.

7923PG87

Exploded view of the manual transaxle mounting (2 of 2)—FWD Eclipse with 2.4L engine

- Engine undercovers and lower the vehicle
- Transfer case assembly if removed
- Transaxle mounting bracket mounting nuts
- Rear roll stopper bracket mounting bolts
- Transaxle assembly mounting bolts. Torque the mounting bolts to 35 ft. lbs. (48 Nm).
11. Remove the engine support fixture.
12. Install or connect the following:
- Starter assembly
- VSS and the back-up light connectors
- Cruise control actuator if removed
- Battery tray support and the tray
- Charcoal canister bracket and the canister
- Air duct and the air cleaner assembly

GALANT

1. Before servicing the vehicle, refer to the precautions in the beginning of this section.
2. Remove or disconnect the following:
- Negative battery cable
- Air cleaner and intake hoses
- Cotter pins and clips securing the select and shift cables and remove the cable ends from the transaxle
- Air compressor, if equipped with Active Electronic Control Suspension (Active-ECS)
- Back-up light switch harness and position aside
- Speedometer electrical connector from the transaxle assembly
- Starter motor and position aside
3. Support the engine assembly.
4. Remove or disconnect the following:
- Rear roll stopper mounting bracket
- Transaxle mount bracket
- Upper transaxle mounting bolts
5. Raise and safely support the vehicle.
6. Remove or disconnect the following:
- Front wheel assemblies
- Right-hand undercover
- Cotter pin and disconnect the tie rod end, from the steering knuckle
- Stabilizer bar link from the damper fork
- Damper fork from the lateral lower control arm
- Lateral lower arm and the compression arm lower ball joints from the steering knuckle
- Halfshafts from the transaxle, and secure aside.
- Clutch release cylinder without dis-

connecting the hydraulic line and secure aside

- Cover from the transaxle bell housing
- Engine front roll stopper through-bolt
- Crossmember and the triangular right-hand stay
7. Support the transaxle with a transmission jack and remove the transaxle lower coupling bolt.

➡ The coupling bolt threads from the engine side into the transaxle and is located just above the halfshaft opening.

8. Slide the transaxle rearward and carefully lower it from the vehicle.

To install:
9. Install or connect the following:
- Transaxle to the engine and install the mounting bolts. Tighten to 35 ft. lbs. (48 Nm). Install the transaxle lower coupling bolt and tighten to 22–25 ft. lbs. (30–34 Nm).
- Cover to the transaxle bell housing and tighten the mounting bolts to 84 inch lbs. (9 Nm)
- Crossmember and tighten the front mounting bolts to 65 ft. lbs. (88 Nm) and the rear bolt to 54 ft. lbs. (73 Nm). Install the front engine roll stopper through-bolt and lightly tighten. Once the full weight of the engine is on the mounts, tighten the bolt to 42 ft. lbs. (57 Nm).
- Triangular stay bracket and tighten the mounting bolts to 65 ft. lbs. (88 Nm)
- Clutch release cylinder
- Halfshafts, using new circlips on the axle ends
- Tie rod and ball joints to the steering knuckle. Tighten the ball joint self-locking nuts to 48 ft. lbs. (65 Nm). Tighten the tie rod end nut to 21 ft. lbs. (28 Nm) and secure with a new cotter pin.
- Damper fork to the lower control arm and tighten the through-bolt to 65 ft. lbs. (88 Nm)
- Stabilizer link to the damper fork and tighten the self-locking nut to 29 ft. lbs. (39 Nm)
- Underpan
- Wheels and lower vehicle
- Transaxle mount bracket to the transaxle and tighten the mounting nuts to 32 ft. lbs. (43 Nm)
- Rear roll stopper mounting bracket

10. Remove the engine support. Tighten the transaxle mount through-bolt to 51 ft. lbs. (69 Nm) and tighten the front engine roll stopper through-bolt.
11. Install or connect the following:
- Upper transaxle mounting bolts and tighten to 35 ft. lbs. (48 Nm)
- Starter motor
- Back-up light switch and the speedometer connector
- Connect the air compressor, if equipped with Active Electronic Control Suspension (Active-ECS)
- Select and shift cables and install new cotter pins
- Air cleaner and the air intake hose
- Negative battery cable
12. Check the transaxle for proper operation.

MIRAGE

1. Before servicing the vehicle, refer to the precautions in the beginning of this section.
2. Remove or disconnect the following:
- Negative battery cable
- Front wheels and the inner wheel panels
- Air cleaner assembly and vacuum hoses
3. Note the locations and disconnect the shifter cables.
4. Remove or disconnect the following:
- Back-up lamp switch connector
- Speedometer cable and remove the starter motor
- Upper transaxle-to-engine mounting bolts
5. Raise the vehicle and support safely.
6. Remove the undercover and splash pan.
7. Drain the transaxle oil.
8. Support the engine and remove the crossmember.
9. Remove or disconnect the following:
- Upper transaxle mounting bolt and bracket
- Stabilizer bar, tie rod ends and the lower ball joint connections
- Clutch release cylinder and clutch oil line bracket. Disconnect the clutch cable, if equipped with cable controlled clutch system.
- Halfshafts by inserting a prybar between the transaxle case and the driveshaft and prying the shaft from the transaxle. Do not pull on the driveshaft.
- Bell housing lower cover
- Transaxle to engine bolts and lower the transaxle from the vehicle

To install:

10. Install or connect the following:
- Transaxle to the engine and install the mounting bolts
- Bell housing cover

➡**When installing the halfshafts, use new circlips on the axle ends.**

11. Install or connect the following:
- Halfshafts into the transaxle
- Slave cylinder or connect the clutch cable
- Ball joints, tie rod ends and stabilizer bar connections
- Upper transaxle mounting bracket and bolt
- Crossmember
- Undercover
- Upper transaxle-to-engine mounting bolts
- Starter motor
- Back-up light switch connector and speedometer cable
- Shifter cables, adjust
- Air cleaner assembly
- Front wheels
- Negative battery cable and check the transaxle for proper operation

Automatic

3000GT

1. Before servicing the vehicle, refer to the precautions in the beginning of this section.
2. Disarm the air bag, if equipped. Remove the battery, battery tray and washer tank.
3. Remove or disconnect the following:
- Air cleaner assembly and adjoining duct work
- Shifter control cable
- Oil cooler hoses, plug
4. Disconnect the following
- Inhibitor switch
- Kickdown servo switch
- Pulse generator
- Oil temperature sensor
- Shift control solenoid valve
- Ground cable.
- Speedometer cable.
5. Raise and safely support the vehicle. Remove the undercovers.
6. Support the weight of the transaxle and remove the mount bracket. Remove the upper bell housing bolts.
7. Remove or disconnect the following:
- Tie rod end and ball joint from the steering knuckle
- Right frame member

- Starter
- Halfshafts by inserting a prybar between the transaxle case and the driveshaft and prying the shaft from the transaxle
- Remaining mounting brackets
- Bell housing cover plate
- Bolts holding the flexplate to the torque converter
- Lower transaxle to engine bolts and remove the transaxle assembly

To install:

8. After the torque converter has been mounted on the transaxle, install the transaxle assembly on the engine. Tighten the driveplate bolts to 34–38 ft. lbs. (46–53 Nm). Install the bell housing cover.
9. Install the mounting brackets.
10. Replace the circlips and install the halfshafts to the transaxle.
11. Install or connect the following:
- Starter and frame member
- Tie rods and ball joint to the steering arm
- Upper bell housing bolts
- Transaxle mounting bracket
- Undercovers
12. Connect the following
- Inhibitor switch
- Kickdown servo switch
- Pulse generator
- Oil temperature sensor
- Shift control solenoid valve
- Ground cable.
- Speedometer cable.

13. Install or connect the following:
- Oil cooler hoses
- Shifter control cable
- Air cleaner assembly and adjoining ductwork
- Washer tank, battery tray and battery
14. Start the engine and check for leaks.

DIAMANTE

1. Before servicing the vehicle, refer to the precautions in the beginning of this section.
2. Properly disarm the SRS system.
3. Raise and safely support the vehicle.
4. Remove or disconnect the following:
- Front wheels
- Engine side cover and undercovers
5. Drain the transaxle assembly.
6. If equipped, remove the front catalytic converter.
7. Remove or disconnect the following:
- Exhaust pipe, main muffler and catalytic converter
- Tie rod end and ball joint from the steering knuckle
- Support bearing for the left side halfshaft
- Halfshafts by inserting a prybar between the transaxle case and the driveshaft and prying the shaft from the transaxle
- Air cleaner assembly and adjoining ductwork
- Engine harness connection

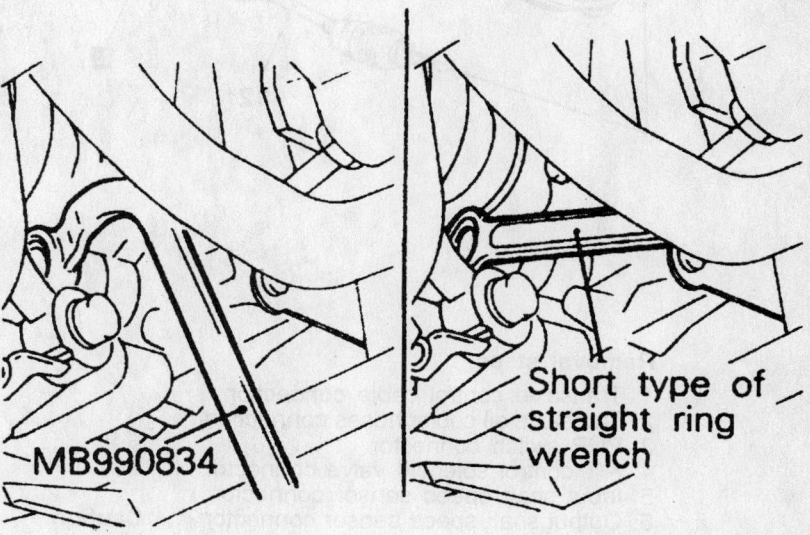

Location of 4-wheel steering oil pump mounting bolts—Diamante with a F4A33 automatic transaxle

For Tune-up, Capacities and Firing orders, see Section 1 of this manual

- Compressor assembly, if the vehicle is equipped with Active Electronic Controlled Suspension (Active-ECS)—suspend with wire—Do not allow the compressor to hang from the air hose.
- Roll stopper stay bracket, if equipped
- Speedometer cable from the transaxle
- The clip that secures the shifter
- Shifter control cable from the transaxle
- Plug the oil cooler hoses from the transaxle

8. Disconnect the following:
- Park/neutral switch electrical harness

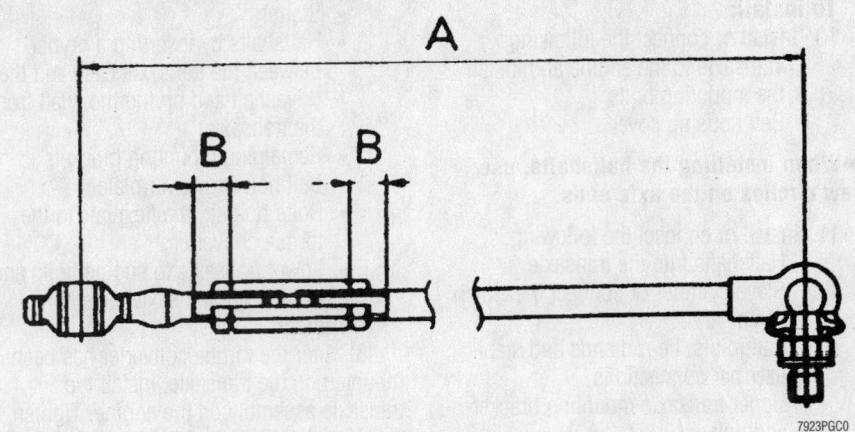

Height sensor rod adjustment—Diamante with a F4A33 automatic transaxle

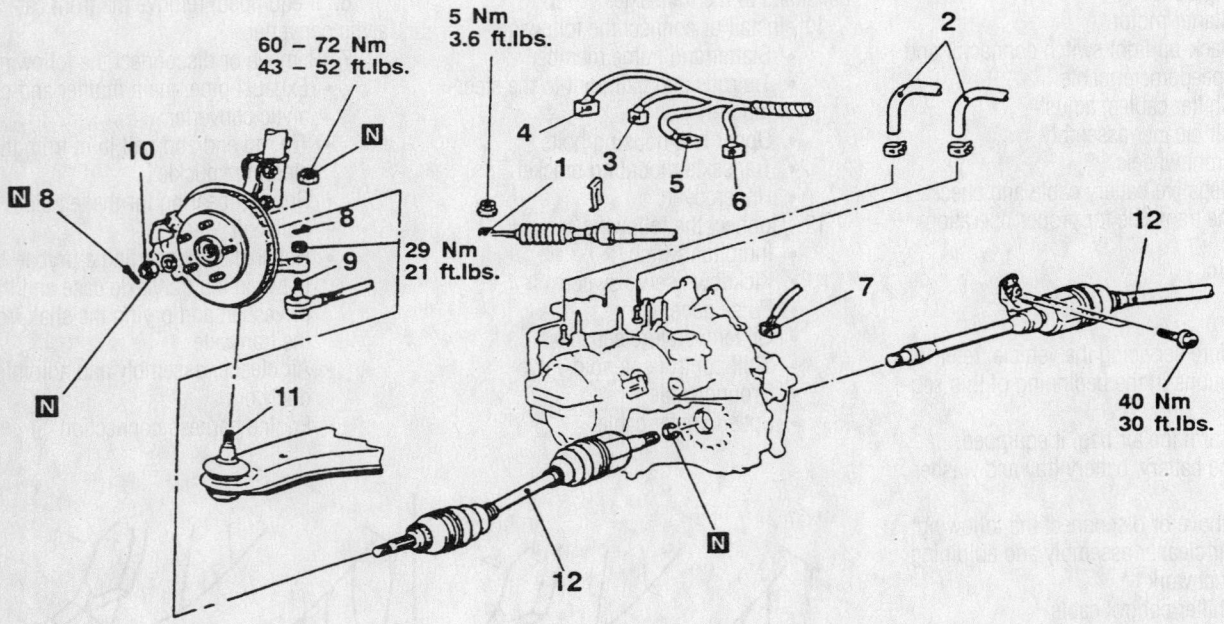

Removal steps

1. Transaxle control cable connection
2. Transaxle oil cooler hoses connection
3. PNP switch connector
4. A/T control solenoid valve connector
5. Input shaft speed sensor connector
6. Output shaft speed sensor connector
7. Vehicle speed sensor connector
8. Split pin
9. Connection of the tie rod end
10. Drive shaft nut
11. Connection for the lower arm ball joint
12. Drive shaft and inner shaft assembly (RH) and the drive shaft (LH)

Caution
Mounting locations marked by * should be provisionally tightened, and then fully tightened when the body is supporting the full weight of the engine.

Transaxle removal—Diamante with F4A51 transaxle 1 of 2

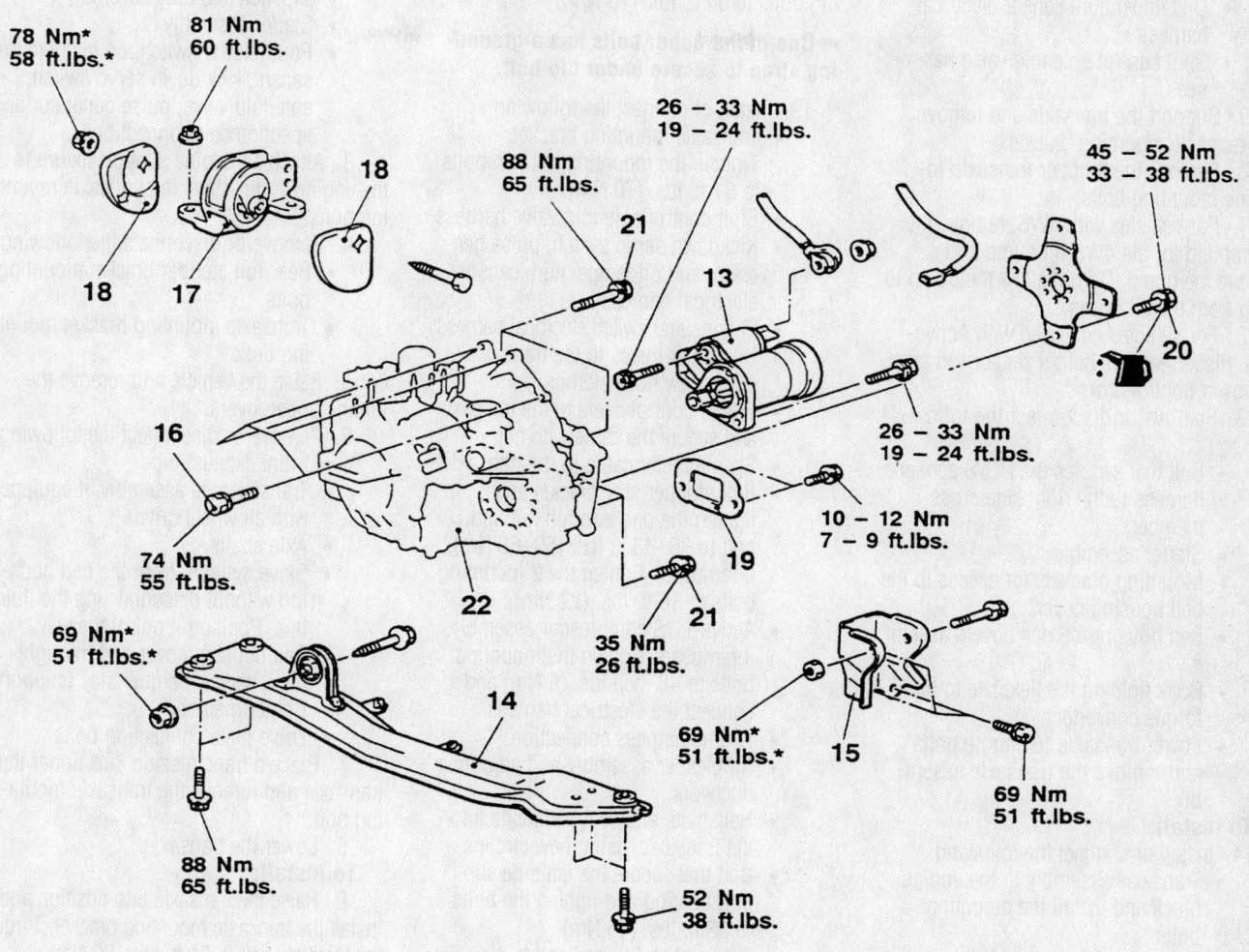

78 Nm*
58 ft.lbs.*

81 Nm
60 ft.lbs.

18

18   17

26 – 33 Nm
19 – 24 ft.lbs.

88 Nm
65 ft.lbs.

21

13

45 – 52 Nm
33 – 38 ft.lbs.

20

16

26 – 33 Nm
19 – 24 ft.lbs.

74 Nm
55 ft.lbs.

22

19

10 – 12 Nm
7 – 9 ft.lbs.

21

69 Nm*
51 ft.lbs.*

14

35 Nm
26 ft.lbs.

69 Nm*
51 ft.lbs.*

15

69 Nm
51 ft.lbs.

88 Nm
65 ft.lbs.

52 Nm
38 ft.lbs.

Lifting up of the vehicle

13. Starter motor
14. Center member assembly
15. Rear roll stopper bracket
16. Transaxle upper portion fixing bolt
17. Transaxle mounting bracket
18. Transaxle mount stopper
● Support the engine and transaxle assembly
19. Bell housing cover

20. Drive plate attaching bolt
21. Transaxle lower portion fixing bolt
22. Transaxle assembly

Caution
Mounting locations marked by * should be provisionally tightened, and then fully tightened when the body is supporting the full weight of the engine.

7923PG85

Transaxle removal—Diamante with F4A51 transaxle 2 of 2

- Kickdown servo switch
- Pulse generator
- Oil temperature sensor electrical harness
- Shift control solenoid valve harness.

9. Support the transaxle and remove the transaxle mounting bracket.

10. Remove the 3 upper transaxle-to-engine mounting bolts.

11. For vehicles with 4WS, remove the heat shield for the 4WS oil pump and remove the pump. Do not allow the pump to hang from the oil hoses.

12. For vehicles equipped with Active-ECS, disconnect the height sensor rod from the lower control arm.

13. Remove or disconnect the following:

- Bolt that secures the HO$_2$S sensor harness to the right side cross-member
- Starter assembly
- Mounting brackets for access to the bell housing cover
- Bell housing/oil pan covers assembly
- Bolts holding the flexplate to the torque converter
- Lower transaxle to engine bolts and remove the transaxle assembly

To install:

14. Install or connect the following:

- Transaxle assembly to the engine block and install the mounting bolts
- Bolts that secure the torque converter to the driveplate. Tighten the bolts to 34–38 ft. lbs. (46–53 Nm).
- Bell housing/oil pan covers
- Transaxle stay brackets that were removed for access to the bell housing cover
- Starter assembly and connect the wiring
- Bolt that secures the HO$_2$S sensor harness to the right side cross-member and tighten the bolt to 84–108 inch lbs. (10–12 Nm)

15. For vehicles equipped with Active-ECS, connect the height sensor rod from the lower control arm. Check the height sensor rod for a length (A) of 10.59–10.63 in. (269–270mm).

16. If removed, install the 4WS oil pump and tighten the mounting bolts to 17 ft. lbs. 24 Nm).

17. If removed, install the 4WS oil pump heat shield and tighten the mounting bolts to 17 ft. lbs. 24 Nm).

18. Install the 3 upper transaxle-to-engine mounting bolts. Tighten the mounting bolts to 54 ft. lbs. (75 Nm).

➡**One of the upper bolts has a grounding strap to secure under the bolt.**

19. Install or connect the following:

- Transaxle mounting bracket. Tighten the mounting nut and bolts to 51 ft. lbs. (70 Nm).
- Shift control solenoid valve harness
- Kickdown servo switch, pulse generator and oil temperature sensor electrical harness
- Park/neutral switch electrical harness
- Oil cooler hoses to the transaxle, using new hose clamps
- Shifter control cable to the transaxle and secure the cable with clip
- Speedometer cable to the transaxle
- Roll stopper stay bracket and tighten the one through nut and bolt to 36–43 ft. lbs. (50–60 Nm), if removed. Tighten the 2 mounting bolts to 16 ft. lbs. (22 Nm).
- Active-ECS compressor assembly, if removed. Tighten the mounting bolts to 48 inch lbs. (5 Nm) and connect the electrical harness.
- Engine harness connection
- Air cleaner assembly and adjoining ductwork
- Halfshafts and seat halfshafts into the transaxle, using new circlips
- Bolt that secure the left side support bearing and tighten the bolts to 33 ft. lbs. (45 Nm)
- Ball joint and tie rod end to the steering knuckle. Using new nuts, tighten the ball joint castle nut to 43–52 ft. lbs. (60–72 Nm) and tighten the tie rod castle nut to 22 ft. lbs. (30 Nm). Install new cotter pins.
- Exhaust system, using new gaskets
- Front catalytic converter, if removed
- Engine undercovers
- Negative battery cable

20. Fill the transaxle to the correct level.
21. Start the engine and check for leaks.

ECLIPSE

1. Before servicing the vehicle, refer to the precautions in the beginning of this section.

2. Remove or disconnect the following:

- Battery and the air intake hoses
- Battery tray and support
- Auto-cruise actuator and bracket, if equipped with cruise control
- Charcoal canister and bracket
- Shift and select cables from the transaxle

- Back-up light switch and the vehicle speed sensor connectors
- Dipstick and tube assembly
- Starter assembly
- Park/neutral switch, oil temperature sensor, kick down servo switch, solenoid valve, pulse generator and speedometer connections

3. Attach an engine support fixture to the engine and remove the transaxle mounting bolts.

4. Remove or disconnect the following:

- Rear roll stopper bracket mounting bolts
- Transaxle mounting bracket mounting nuts

5. Raise the vehicle and remove the engine undercovers.

6. Remove or disconnect the following:

- Front exhaust pipe
- Transfer case assembly, if equipped with all wheel drive
- Axle shafts
- Slave cylinder from the bell housing without disconnecting the fluid line. Position it out of the way.
- Bell housing cover and the right-hand center member stay (support)
- Center member
- Drive plate connecting bolts

7. Place a transmission jack under the transaxle and remove the transaxle mounting bolt.

8. Lower the transaxle.

To install:

9. Raise the transaxle into position and install the transaxle mounting bracket. Torque the through-bolt to 51 ft. lbs. (69 Nm).

10. Install or connect the following:

- Transaxle assembly mounting bolt. Torque the bolt to 22–25 ft. lbs. (29–34 Nm).
- Drive plate connecting bolts. Torque the bolts to 33–38 ft. lbs. (45–52 Nm).
- Center member assembly and the right-hand stay
- Bell housing cover and the slave cylinder
- Axle shafts.
- Front exhaust pipe
- Transfer case assembly if removed
- Engine undercovers and lower the vehicle
- Transaxle mounting bracket mounting nuts
- Rear roll stopper bracket mounting bolts
- Transaxle assembly mounting bolts. Torque the bolts to 35 ft. lbs. (48 Nm).

11. Remove the engine support fixture.

12. Install or connect the following:
- Park/neutral switch, oil temperature sensor, kick down servo switch, solenoid valve, pulse generator and speedometer connections
- Starter assembly
- Dipstick and tube assembly
- Vehicle speed sensor and the back-up light connectors
- Cruise control actuator if removed
- Battery tray support and the tray
- Charcoal canister bracket and the canister
- Air duct and the air cleaner assembly

13. Refill the transaxle and the transfer case, if equipped, with the proper fluid.

GALANT

1. Before servicing the vehicle, refer to the precautions in the beginning of this section.

2. Remove or disconnect the following:
- Negative battery cable
- Air cleaner and intake hoses

3. Drain the transaxle into a suitable waste container.

4. Remove or disconnect the following:
- Nut securing the shifter lever to the transaxle
- Cable retaining clip and remove the cable from the transaxle
- Shifter cable mounting bracket
- Electrical connectors for the speedometer, neutral safety switch (inhibitor switch), the pulse generator, kickdown servo switch, and the oil temperature sensor
- Oil cooler lines, at the transaxle
- Dipstick and tube from the transaxle
- Starter motor and position it aside

5. Support the engine assembly.

6. Remove or disconnect the following:
- Rear roll stopper mounting bracket
- Transaxle mount bracket
- Upper transaxle mounting bolts

7. Raise and safely support the vehicle.

8. Remove or disconnect the following:
- Front wheel assemblies
- Right-hand undercover
- Tie rod end from the steering knuckle
- Stabilizer bar link from the damper fork
- Damper fork from the lateral lower control arm
- Later lower arm and the compres-

sion arm lower ball joints, from the steering knuckle
- Halfshafts from the transaxle and secure aside
- Cover from the transaxle bell housing
- Engine front roll stopper through-bolt
- Crossmember and the triangular right-hand stay
- Bolts holding the flexplate to the torque converter

9. Support the transaxle using a transmission jack, and remove the transaxle lower coupling bolt.

➡**The coupling bolt threads from the engine side into the transaxle and is located just above the halfshaft opening.**

10. Slide the transaxle rearward and carefully lower it from the vehicle.

To install:

11. After the torque converter has been mounted on the transaxle, install the transaxle assembly to the engine. Install the mounting bolts and tighten to 35 ft. lbs. (48 Nm). Install the transaxle lower coupling bolt and tighten to 21–25 ft. lbs. (29–34 Nm).

12. Install or connect the following:
- Torque converter to the flexplate and tighten the bolts to 33–38 ft. lbs. (45–52 Nm)
- Cover to the transaxle bell housing and tighten the mounting bolts to 84 inch lbs. (9 Nm)
- Crossmember and tighten the front mounting bolts to 65 ft. lbs. (88 Nm) and the rear bolt to 54 ft. lbs. (73 Nm)
- Front engine roll stopper through-bolt and lightly tighten. Once the full weight of the engine is on the mounts, tighten the bolt to 42 ft. lbs. (57 Nm).
- Triangular stay bracket and tighten the mounting bolts to 65 ft. lbs. (88 Nm)
- Halfshafts, using new circlips on the axle ends
- Tie rod and ball joints to the steering knuckle. Tighten the ball joint self-locking nuts to 48 ft. lbs. (65 Nm). Tighten the tie rod end nut to 21 ft. lbs. (28 Nm) and secure with a new cotter pin.
- Damper fork to the lower control arm and tighten the through-bolt to 65 ft. lbs. (88 Nm)

- Stabilizer link to the damper fork, and tighten the self-locking nut to 29 ft. lbs. (39 Nm)
- Underpan
- Wheels and lower the vehicle
- Transaxle mount bracket to the transaxle, and tighten the mounting nuts to 32 ft. lbs. (43 Nm)
- Rear roll stopper mounting bracket
- Engine support. Tighten the transaxle mount through-bolt to 51 ft. lbs. (69 Nm) and tighten the front engine roll stopper through-bolt.
- Upper transaxle mounting bolts and tighten to 35 ft. lbs. (48 Nm)
- Starter motor
- Dipstick tube and the dipstick
- Shifter cable mounting bracket
- Shifter lever and tighten the retaining nut to 14 ft. lbs. (19 Nm)
- Oil cooler lines and secure with clamps
- Electrical connectors for the speedometer, neutral safety switch (inhibitor switch), pulse generator, kickdown servo switch and oil temperature sensor
- Air cleaner and the air intake hose
- Negative battery cable

13. Fill the transaxle to the correct level.

MIRAGE

1. Before servicing the vehicle, refer to the precautions in the beginning of this section.

2. Remove or disconnect the following:
- Negative battery cable
- Battery and battery tray
- Air hose and air cleaner assembly

3. Raise the vehicle and support safely.

4. Remove the under guard pan.

5. Drain the transaxle oil.

6. Remove or disconnect the following:
- Control cable and cooler lines
- Shift control solenoid valve connector
- Inhibitor switch, kickdown servo switch, the pulse generator and oil temperature sensor, if equipped
- Speedometer cable and remove the starter
- Transaxle mounting bolts and bracket
- Stabilizer bar from the lower control arm
- Steering tie rod end and the ball joint from the steering arm
- Halfshafts at the inboard side from

the transaxle. Tie the joint assembly aside.

7. Support the engine and remove the center member.

8. Remove or disconnect the following:
- Bell housing cover and remove the driveplate bolts
- Transaxle assembly lower connecting bolt, located just over the half-shaft opening
- Transaxle

To install:

9. Install the transaxle assembly on the engine.

10. Tighten the driveplate bolts to 33–38 ft. lbs. (46–53 Nm). Install the bell housing cover.

11. Install or connect the following:
- Center member
- Halfshafts to the transaxle, using new circlips
- Tie rods, ball joints and stabilizer links to the steering arm
- Transaxle mounting bracket and bolts
- Starter
- Speedometer cable
- Inhibitor switch, kickdown servo switch, the pulse generator and oil temperature sensor, if disconnected
- Shift control solenoid valve connector
- Control cables and oil cooler lines
- Air cleaner assembly
- Battery tray and battery. Connect the positive, then the negative terminal.

12. Fill the transaxle to the correct level.

13. Start the engine and check for leaks.

Clutch

ADJUSTMENT

➡ The following adjustment is for cable actuated clutch systems. Hydraulic systems are self-adjusting.

Mirage

1. Before servicing the vehicle, refer to the precautions in the beginning of this section.

2. Measure the clutch pedal height (measurement A). The specification is 6.38–6.50 in. (162–165mm).

➡ The clutch pedal height is not adjustable. If not within specifications, part replacement is required.

3. Depress clutch pedal several times and check the pedal free-play (measurement B).

4. If measurement is not 0.67–0.87 in. (17–22mm), adjustment is required.

Clutch pedal height

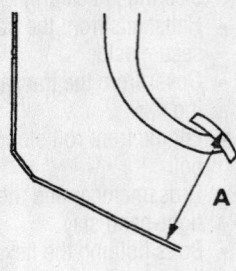

7923PGD1

Clutch pedal height (A) measurement— Mirage

5. To adjust turn the outer cable adjusting nut, located at the firewall, until free-play is within range.

6. Depress clutch pedal several times and recheck measurement.

REMOVAL & INSTALLATION

3000GT

1. Before servicing the vehicle, refer to the precautions in the beginning of this section.

2. Disconnect the negative battery cable.

3. Raise and safely support the vehicle.

4. Remove or disconnect the following:
- Transaxle assembly from the vehicle
- Pressure plate attaching bolts. If the pressure plate is to be reused, loosen the bolts in succession, 1 or 2 turns at a time to prevent warping the cover flange.
- Pressure plate release bearing assembly and the clutch disc. Do not use solvent to clean the bearing.

5. Inspect the condition of the clutch components and replace any worn parts.

To install:

6. Inspect the flywheel for heat damage or cracks. Resurface or replace the flywheel as required.

7. Using the proper alignment tool, install the clutch disc to the flywheel. Install the pressure plate assembly and tighten the pressure plate bolts evenly to 11–15 ft. lbs. (15–21 Nm). Remove the alignment tool.

8. Apply a light coat of high temperature grease to the clutch fork at the ball pivot and where the fork contacts the bearing. Also a little bit of grease can be applied to end of the release cylinder pushrod and to the pushrod hole on the fork. Apply a

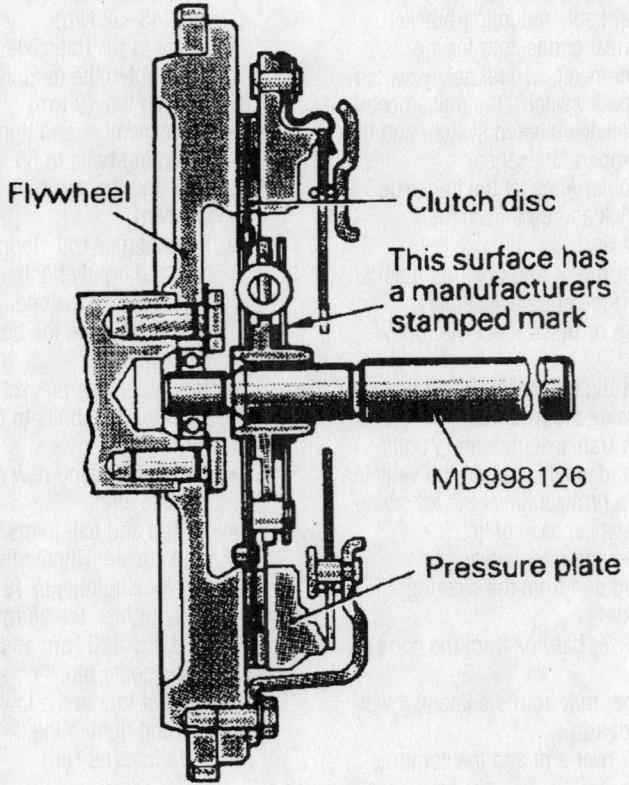

Flywheel
Clutch disc
This surface has a manufacturers stamped mark
MD998126
Pressure plate

7923PG88

Use the alignment dowel to center the disc on the flywheel—Mirage light coat of grease on the transaxle input shaft splines.

9. Install or connect the following:
 • New clutch release bearing. Pack its inner surface with high temperature grease
 • Transaxle assembly

10. Lower the vehicle and connect the negative battery cable.

11. Check the clutch for proper operation.

Eclipse, Galant and Mirage

1. Before servicing the vehicle, refer to the precautions in the beginning of this section.

2. Remove or disconnect the following:
 • Negative battery cable
 • Transaxle assembly from the vehicle
 • Pressure plate attaching bolts, pressure plate and clutch disc. If the pressure plate is to be reused,

loosen the bolts in a diagonal pattern, 1 or 2 turns at a time. This will prevent warping the clutch cover assembly.
 • Return clip and the pressure plate release bearing. Do not use solvent to clean the bearing.

3. Inspect the clutch release fork and fulcrum for damage or wear. If necessary, remove the release fork and unthread the fulcrum from the transaxle.

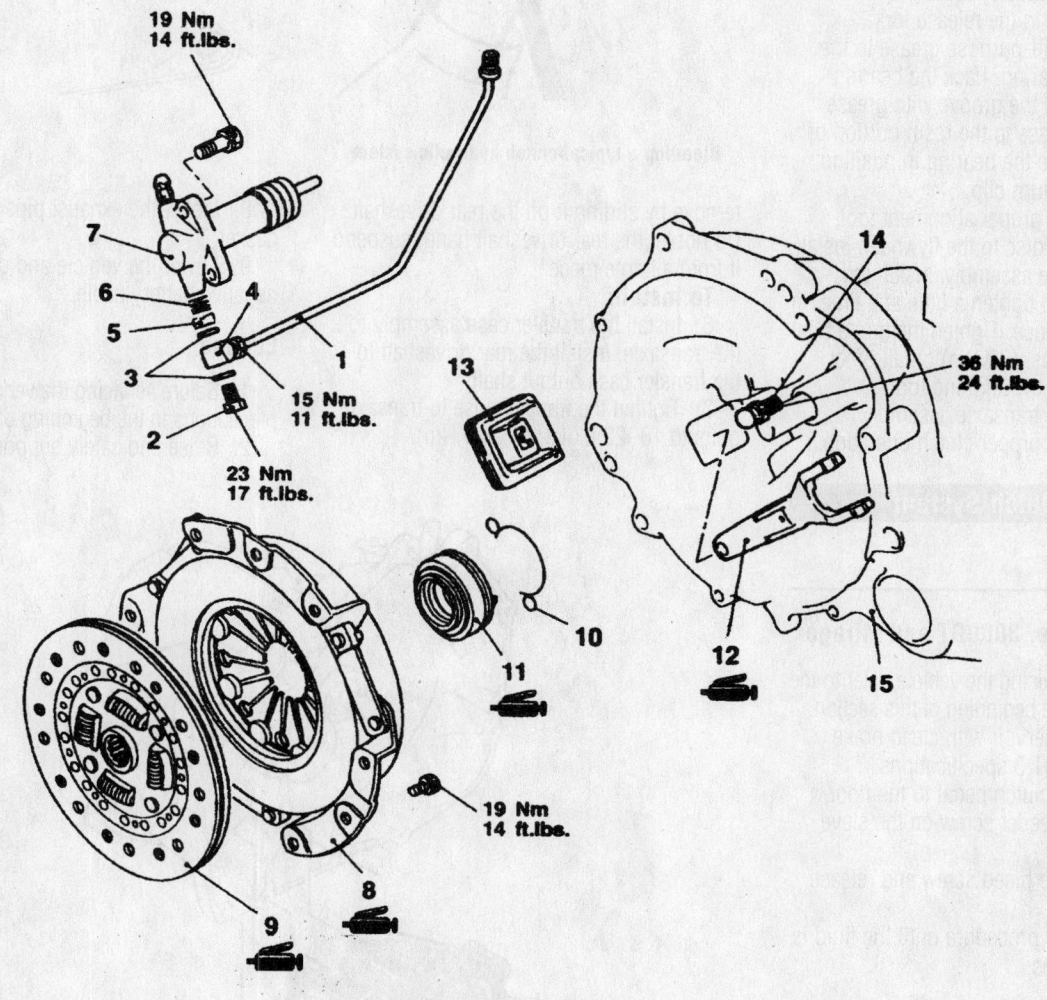

19 Nm
14 ft.lbs.

15 Nm
11 ft.lbs.

23 Nm
17 ft.lbs.

36 Nm
24 ft.lbs.

19 Nm
14 ft.lbs.

1. Clutch oil tube
2. Union bolt
3. Gasket
4. Union
5. Valve plate
6. Valve plate spring
7. Clutch release cylinder
8. Clutch cover
9. Clutch disc
10. Return clip
11. Clutch release bearing
12. Release fork
13. Release fork boot
14. Fulcrum
15. Transaxle

7923PG89

Exploded view of non-turbo clutch assembly—Eclipse with F5M31, F5M33, F5Mc1 and W5M33 manual transaxles

Timing belt service is covered in Section 3 of this manual

4. Carefully inspect the condition of the clutch components and replace any worn or damaged parts.

To install:

5. Inspect the flywheel for heat damage or cracks. Resurface or replace the flywheel as required.

6. Install the fulcrum and tighten to 25 ft. lbs. (35 Nm). Install the release fork. Apply a coating of multi-purpose grease to the point of contact with the fulcrum and the point of contact with the release bearing. Apply a coating of multi-purpose grease to the end of the release cylinder pushrod and the pushrod hole in the release fork.

7. Apply multi-purpose grease to the clutch release bearing. Pack the bearing inner surface and the groove with grease. Do not apply grease to the resin portion of the bearing. Place the bearing in position and install the return clip.

8. Using the proper alignment tool, install the clutch disc to the flywheel. Install the pressure plate assembly. Install the retainer bolts and tighten a little at a time, in a diagonal sequence. Tighten them to a final torque of 16 ft. lbs. (22 Nm) on all other models. Remove the aligning tool.

9. Install the transaxle assembly.

10. Check for proper clutch operation.

Hydraulic Clutch System

BLEEDING

Galant, Eclipse, 3000GT and Mirage

1. Before servicing the vehicle, refer to the precautions in the beginning of this section.

2. Fill the reservoir with clean brake fluid meeting DOT 3 specifications.

3. Press the clutch pedal to the floor, then open the bleeder screw on the slave cylinder.

4. Tighten the bleed screw and release the clutch pedal.

5. Repeat the procedure until the fluid is free of air bubbles.

Transfer Case Assembly

REMOVAL & INSTALLATION

3000GT

1. Before servicing the vehicle, refer to the precautions in the beginning of this section.

2. Disconnect the negative battery cable.

3. Raise the vehicle and support safely.

4. Disconnect the front exhaust pipe.

5. Unbolt the transfer case assembly and

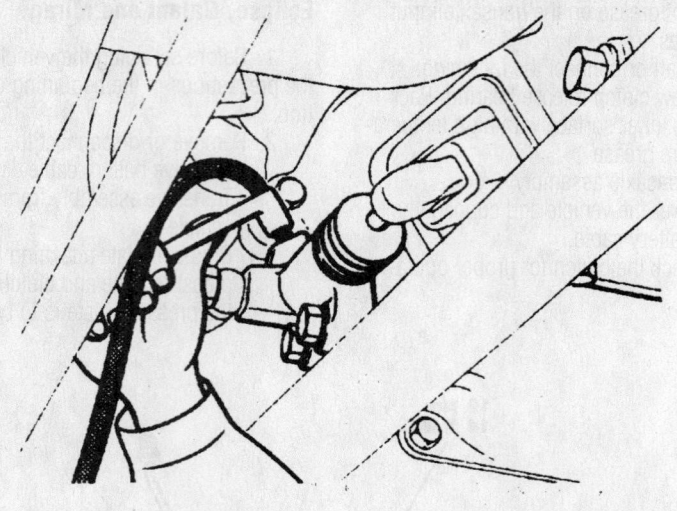

Bleeding a typical clutch hydraulic system

7923PG91

remove by sliding it off the rear driveshaft. Do not let the rear driveshaft hang; suspend it from a frame piece.

To install:

6. Install the transfer case assembly to the transaxle. Install the rear driveshaft to the transfer case output shaft.

7. Tighten the transfer case to transaxle bolts to 18–22 ft. lbs. (25–29 Nm).

8. Install the exhaust pipe using a new gasket.

9. Lower the vehicle and connect the negative battery cable.

Eclipse

1. Before servicing the vehicle, refer to the precautions in the beginning of this section.

2. Raise and safely support the vehicle.

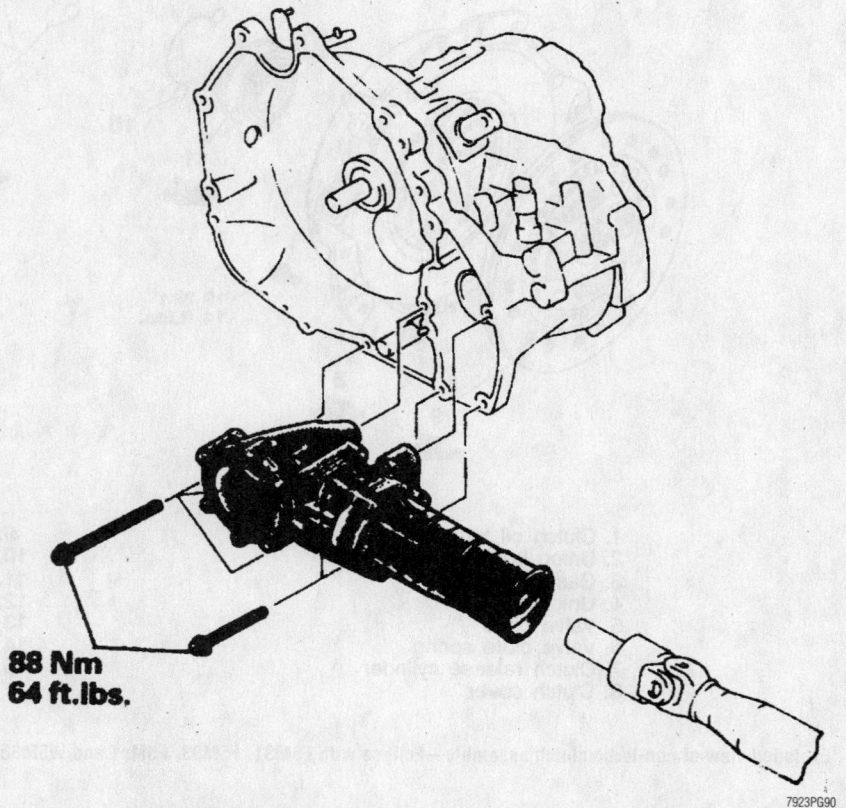

88 Nm
64 ft.lbs.

7923PG90

Transfer case assembly—3000GT with F5M33, W5MG1 and W6MG1 transfer cases

3. Remove or disconnect the following:
   - Engine undercovers
   - Front exhaust pipe
   - Transfer case mounting bolts

   4. Support the driveshaft with wire or string and remove the transfer case from the transaxle.

   To install:

   5. Slide the driveshaft into the transfer case and install the transfer case to the transaxle. Torque the bolts to 40–44 ft. lbs. (54–59 Nm).
   6. Install the front exhaust pipe.
   7. Install the engine undercover.
   8. Lower the vehicle.

Halfshaft

REMOVAL & INSTALLATION

3000GT

FRONT

1. Before servicing the vehicle, refer to the precautions in the beginning of this section.
2. Disconnect the negative battery cable.
3. Raise the vehicle and support safely.

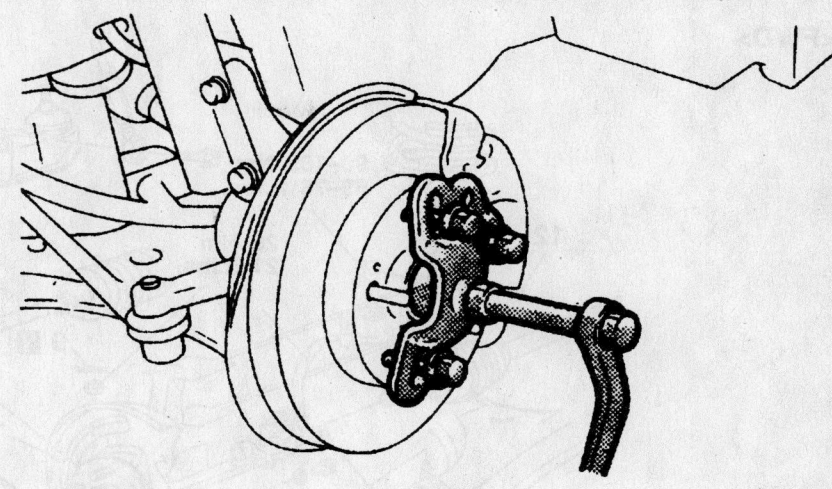

Front halfshaft removal—3000GT

Remove the lower ball joint and the tie rod end from the steering knuckle.

4. Remove the cotter pin, halfshaft nut and the washer.
5. On vehicles with an inner shaft, remove the center support bearing bracket bolts and washers.
6. On vehicles with an inner shaft, remove the halfshaft by setting up a puller on the outside wheel hub and pushing the halfshaft from the front hub. Then, tap the shaft union at the joint case with a plastic hammer to remove the halfshaft shaft and inner shaft from the transaxle.
7. On vehicles without an inner shaft, remove the halfshaft by setting up a puller

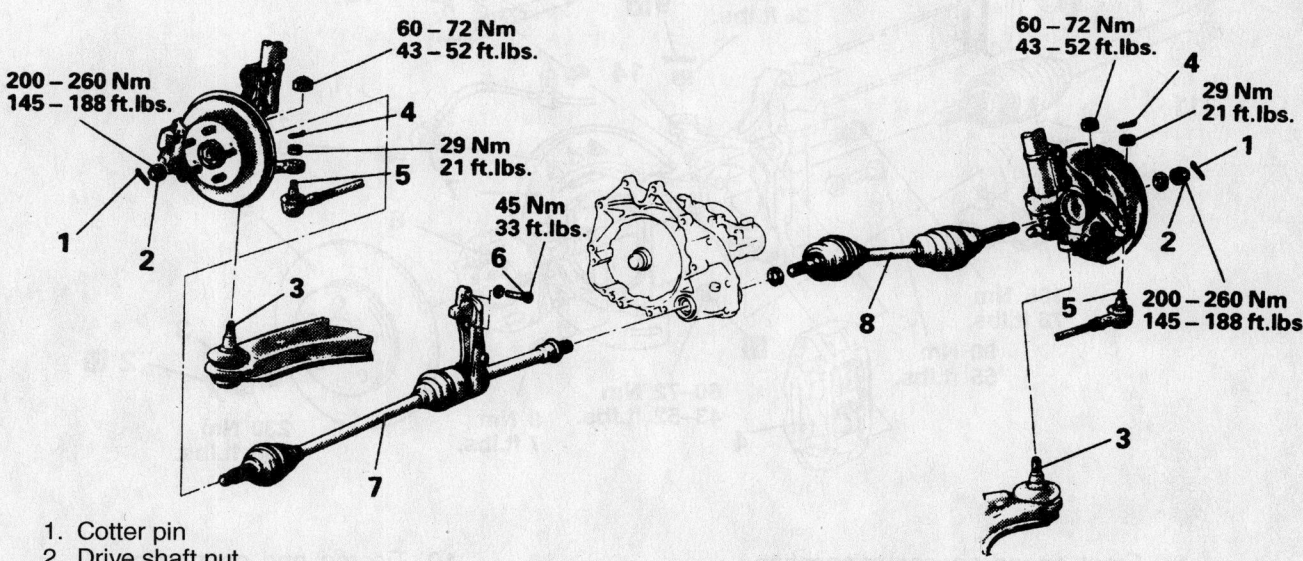

1. Cotter pin
2. Drive shaft nut
3. Lower arm ball joint connection
4. Cotter pin
5. Tie rod end connection
6. Center bearing bracket installation bolt
7. Drive and inner shaft assembly (L.H.)
8. Drive shaft (R.H.)
9. Circlip

Front halfshafts and related components—3000GT

Caution
In the case of AWD-vehicles with A.B.S., take care not to damage the rotor for A.B.S. installed to the B.J. outer race.

7923PG95

Heater Core replacement is covered in Section 2 of this manual

<FWD>

<AWD>

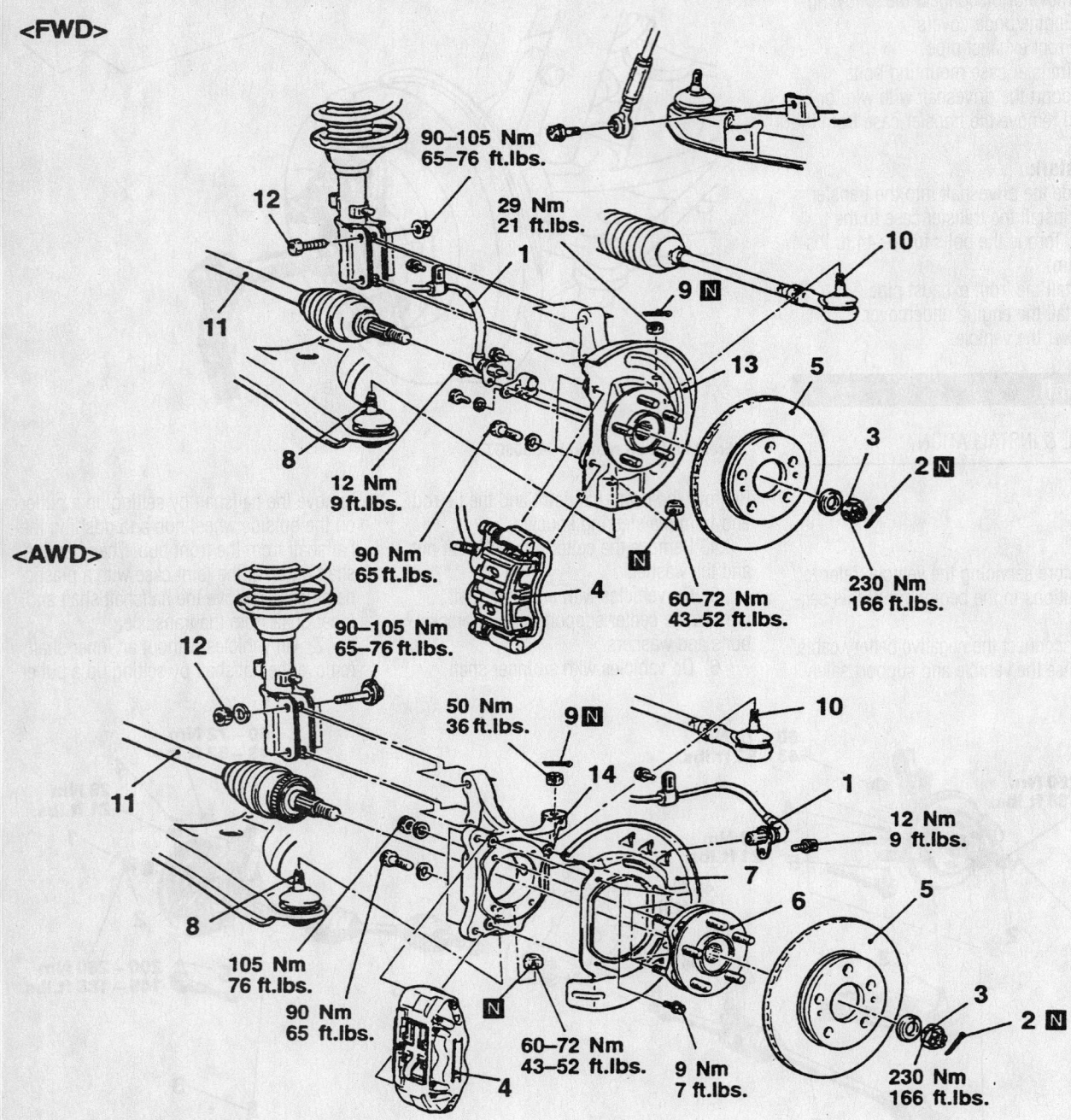

1. Front speed sensor connection
   <Vehicles with ABS*>
2. Cotter pin
3. Drive shaft nut
4. Caliper assembly
5. Brake disc
6. Front hub unit bearing
7. Dust shield
8. Lower arm ball joint connection
9. Cotter pin

10. Tie rod end connection
11. Drive shaft
12. Front strut mounting bolt or nut
13. Hub and knuckle
14. Hub

NOTE
*: Anti-lock braking system

7923PG97

Exploded view of the front suspension components—3000GT

on the outside wheel hub and pushing the halfshaft from the front hub. After pressing the outer shaft, insert a prybar between the transaxle case and the halfshaft and pry the shaft from the transaxle. Do not pull on the shaft.

To install:

8. Replace the circlips on the ends of the halfshafts.

9. Insert the halfshaft into the transaxle.

10. Pull the strut assembly out and install the other end to the hub.

11. Install or connect the following:

- Center bearing bracket bolts and tighten to 33 ft. lbs. (45 Nm)
- Washer so the chamfered edge faces outward. Tighten the nut to 188 ft. lbs. (260 Nm)
- Tie rod end and ball joint
- Wheel and lower the vehicle

REAR

➡ **On vehicles with a limited slip differential, the right and left halfshafts are not the same.**

If both halfshafts are to be removed, mark them for proper installation.

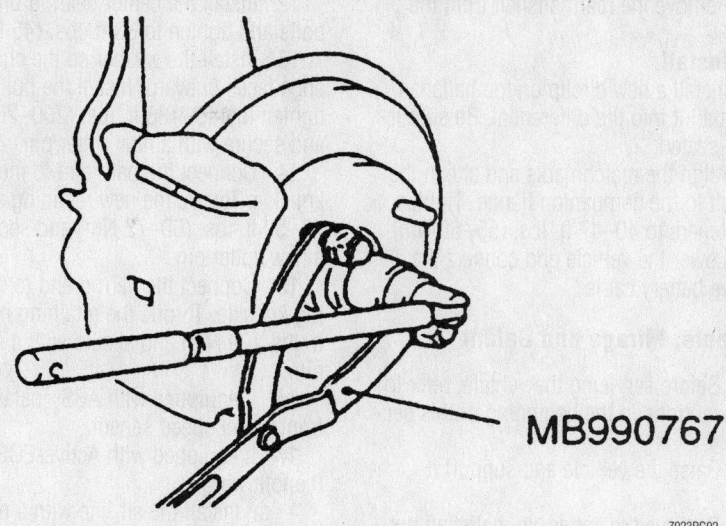

MB990767

7923PG92

Proper method for disconnecting ball joint studs and tie rod ends

1. Before servicing the vehicle, refer to the precautions in the beginning of this section.

2. Disconnect the negative battery cable. Raise the vehicle and support it safely.

3. Matchmark the halfshaft and the companion flange.

4. Remove the bolts that attach the rear halfshaft to the companion flange.

5. Use a prybar to pry the inner shaft out of the differential case.

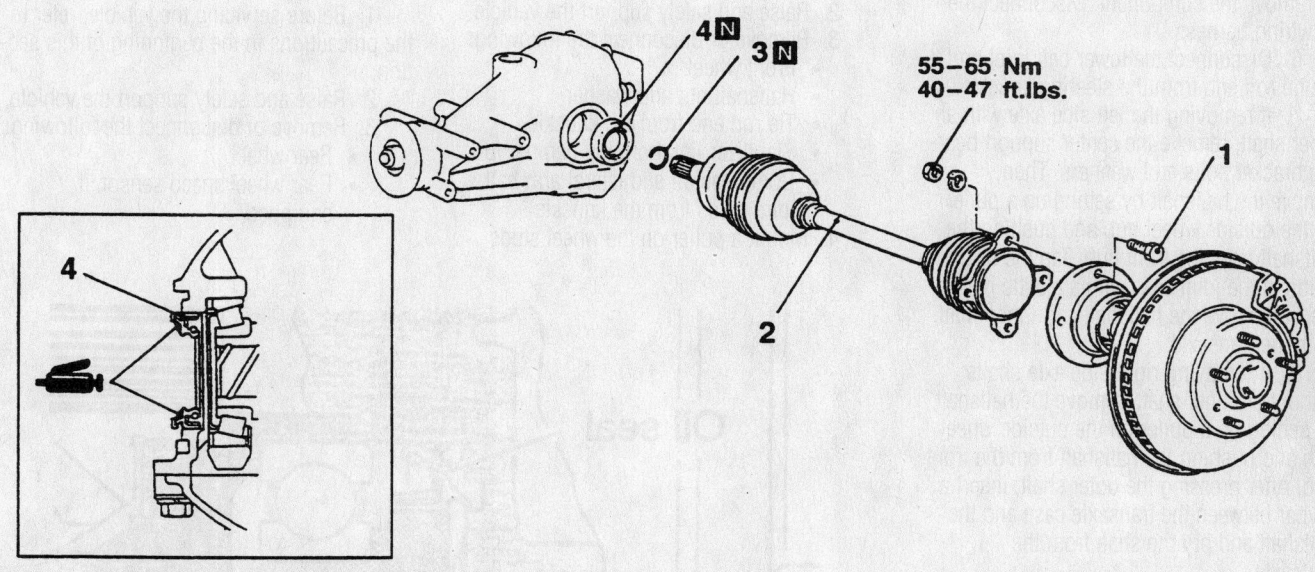

55–65 Nm
40–47 ft.lbs.

Removal steps

1. Bolt
2. Drive shaft
3. Circlip
4. Oil seal

7923PG98

Rear halfshaft removal—AWD 3000GT

Brake service is covered in Section 4 of this manual

6. Remove the rear halfshaft from the vehicle.

To install:

7. Install a new circlip on the halfshaft and install it into the differential. Be sure it is fully seated.

8. Align the matchmarks and attach the halfshaft to the companion flange. Tighten the fasteners to 40–47 ft. lbs. (55–65 Nm).

9. Lower the vehicle and connect the negative battery cable.

Diamante, Mirage and Galant

1. Before servicing the vehicle, refer to the precautions in the beginning of this section.

2. Raise the vehicle and support it safely.

3. Remove the cotter pin, halfshaft nut and washer.

4. If equipped with ABS, remove the front wheel speed sensor.

5. If equipped with Active Electronic Control Suspension (Active-ECS) perform the following:

   a. Loosen the nut that secures the air line to the to the top of the strut and discard the O-ring.

   b. Remove the bolts that secure the actuator to the top of the strut and remove the component. Disconnect the wiring harness.

6. Disconnect the lower ball joint and the tie rod end from the steering knuckle.

7. If removing the left side axle with an inner shaft, remove the center support bearing bracket bolts and washers. Then, remove the halfshaft by setting up a puller on the outside wheel hub and pushing the halfshaft from the front hub. Tap the shaft union at the joint case with a plastic hammer to remove the halfshaft and inner shaft from the transaxle.

8. If removing right side axle shafts without an inner shaft, remove the halfshaft by setting up a puller on the outside wheel hub and pushing the halfshaft from the front hub. After pressing the outer shaft, insert a prybar between the transaxle case and the halfshaft and pry the shaft from the transaxle.

➡ **Do not pull on the shaft; doing so damages the inboard joint.**

To install:

9. Replace the circlips on the ends of the halfshafts.

10. Insert the halfshaft into the transaxle. Be sure it is fully seated.

11. Pull the strut assembly out and install the other end to the hub.

12. Install the center bearing bracket bolts and tighten to 33 ft. lbs. (45 Nm).

13. Install the washer so the chamfered edge faces outward. Install the nut and tighten to 145–188 ft. lbs. (200–260 Nm) and secure with a new cotter pin.

14. Connect the ball joint to the steering knuckle. Torque the new retaining nut to 43–52 ft. lbs. (60–72 Nm) and secure with a new cotter pin.

15. Connect the tie rod end to the steering knuckle. Torque the retaining nut to 21 ft. lbs. (29 Nm) and secure with a new cotter pin.

16. If equipped with ABS, install the front wheel speed sensor.

17. If equipped with Active-ECS, perform the following:

   a. Install the air line with a new O-ring.

   b. Install the actuator to the top of the strut. Connect the wiring harness.

18. Install the wheel and lower the vehicle to the floor.

Eclipse

FRONT

1. Before servicing the vehicle, refer to the precautions in the beginning of this section.

2. Raise and safely support the vehicle.

3. Remove or disconnect the following:
- Front wheel
- Halfshaft nut and washer
- Tie rod end from the knuckle
- Stabilizer link from the damper fork
- Compression and lateral arm ball joint studs from the knuckle

4. Mount a puller on the wheel studs

and push the halfshaft through the hub assembly.

5. Detach the inner halfshaft from the transaxle by carefully prying the CV-joint housing out.

6. Pull the knuckle assembly outward and remove the halfshaft.

To install:

7. Place a new circlip on the inner halfshaft and install the halfshaft in the transaxle.

8. Push out on the knuckle assembly and install the halfshaft through the hub.

9. Using new nuts, install the lateral and compression arm ball joint studs in the knuckle. Tighten the nuts to 43–52 ft. lbs. (59–71 Nm). Install new cotter pins.

10. Install the damper fork on the knuckle. Do nut tighten the nut at this time.

11. Attach the stabilizer link to the damper fork. Tighten the nut to 29 ft. lbs. (39 Nm).

12. Install the washer and nut on the halfshaft. Prevent the hub from turning and tighten the nut to 145–188 ft. lbs. (196–255 Nm).

13. Install the wheel and lower the vehicle to the floor. Tighten the damper fork nut to 65 ft. lbs. (88 Nm).

REAR

1. Before servicing the vehicle, refer to the precautions in the beginning of this section

2. Raise and safely support the vehicle.

3. Remove or disconnect the following:
- Rear wheel
- Rear wheel speed sensor, if equipped

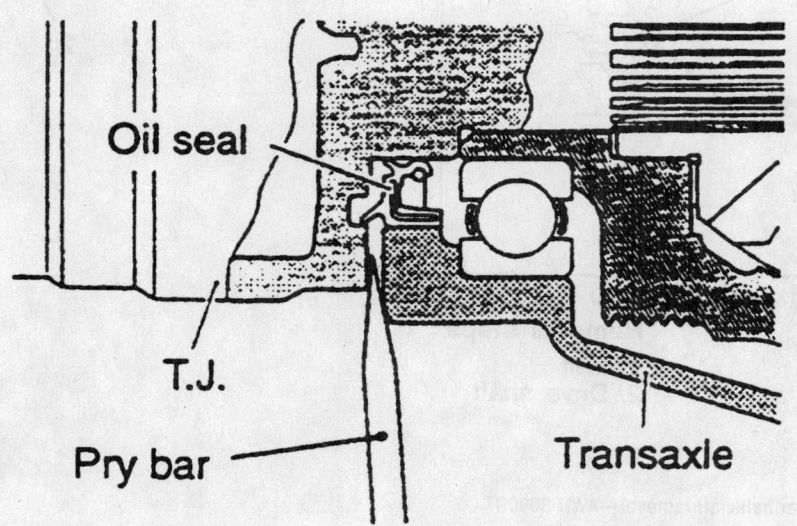

Proper method for removing the inner halfshaft from the transaxle or differential

7923PG94

- Caliper and rotor, if equipped with disc brakes
- Brake drum and shoes, If equipped with drum brakes
- Brake hydraulic line from the wheel cylinder
- Parking brake cable from the rear brakes
- Lower end of the shock absorber from the knuckle
- Trailing and lower arms from the knuckle
- Toe control arm ball joint from the knuckle

4. Prevent the hub assembly from turning by using a tool such as MB990767 and remove the halfshaft nut and washer.

5. Remove the differential mount support.

※ WARNING

Do not pull on the halfshaft to remove it from the differential. Damage to the inner CV-joint will occur.

6. Push the lower part of the knuckle outward and pry the inner halfshaft out of the differential.

7. Push the outer end of the halfshaft through the hub/knuckle and remove it.

To install:

8. Install the outer end of the halfshaft through the hub/knuckle.

9. Place a new circlip on the inner halfshaft and install the halfshaft in the differential.

10. Install or connect the following:
- Differential mount support

- Washer and a new nut on the end of the halfshaft. Tighten the nut to 145–188 ft. lbs. (196–255 Nm).
- Toe control arm to the knuckle. Tighten the new nut to 20 ft. lbs. (28 Nm).
- Lower and trailing arms to the knuckle. Do not tighten the fasteners at this time.
- Shock absorber. Tighten the bolt to 71 ft. lbs. (98 Nm).

11. Assemble the brake components.
12. Install or connect the following:
- Rear wheel speed sensor
- Rear wheel and lower the vehicle to the floor. Tighten the lower arm nut to 71 ft. lbs. (98 Nm) and the trailing arm nut to 85–99 ft. lbs. (118–137 Nm).

STEERING AND SUSPENSION

Air Bag

※ CAUTION

All vehicles are equipped with an air bag system. The system must be disabled before performing service on or around system components, steering column, instrument panel components, wiring and sensors. Failure to follow safety and disabling procedures could result in accidental air bag deployment, possible personal injury and unnecessary system repairs.

PRECAUTIONS

Several precautions must be observed when handling the inflator module to avoid accidental deployment and possible personal injury.

- Never carry the inflator module by the wires or connector on the underside of the module.
- When carrying a live inflator module, hold securely with both hands, and ensure that the bag and trim cover are pointed away.
- Place the inflator module on a bench or other surface with the bag and trim cover facing up.
- With the inflator module on the bench, never place anything on or close to the module which may be thrown in the event of an accidental deployment.

DISARMING

1. Before servicing the vehicle, refer to the precautions in the beginning of this section.

2. Position the front wheels in the straight-ahead position and place the key in the **LOCK** position. Remove the key from the ignition lock cylinder.

3. Disconnect the negative battery cable and insulate the cable end with high-quality electrical tape or similar non-conductive wrapping.

4. Wait at least 1 minute before working on the vehicle. The air bag system is designed to retain enough voltage to deploy the air bag for a short period of time after the battery has been disconnected.

REARMING

1. Connect the negative battery cable, turn the ignition switch to the **ON** position and check the SRS warning light for proper operation.

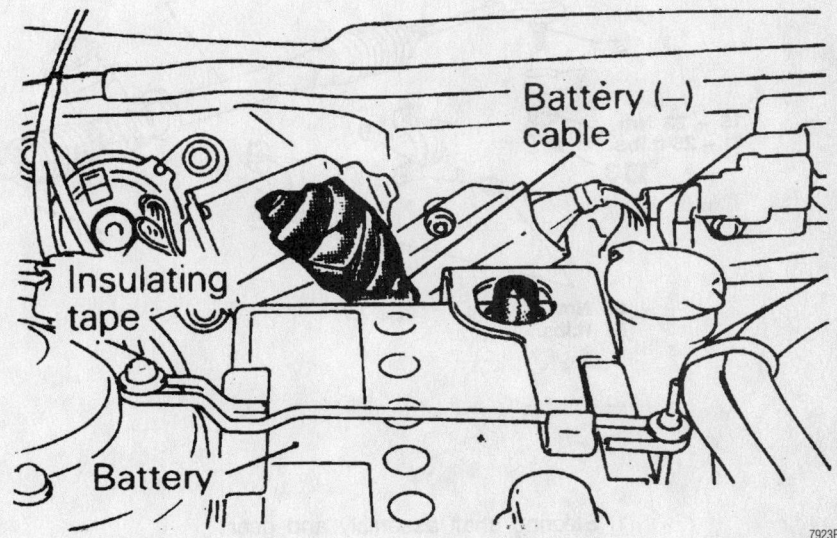

7923PG99

Insulate the negative battery cable to prevent accidental deployment of the air bag

For complete Engine Mechanical specifications, see Section 1 of this manual

Rack and Pinion Steering Gear

REMOVAL & INSTALLATION

Manual

MIRAGE

➡Prior to removal of the steering rack, center the front wheels and remove the ignition key. Failure to do so may damage the SRS clockspring and render SRS system inoperative. Be sure to properly disarm the air bag system.

1. Before servicing the vehicle, refer to the precautions in the beginning of this section.

2. Remove or disconnect the following:
   • Battery negative cable
   • Wheels
   • HO2S sensor and remove the front exhaust pipe

3. Properly support the engine. Remove both roll stopper mounting bolts and the 4 center member installation bolts.

4. Remove the center member.

➡Matchmark the pinion input shaft of the rack to the lower steering column joint for installation purposes.

5. Remove or disconnect the following:
   • Pinch bolt holding the lower steering column joint to the rack and pinion input shaft
   • Cotter pins and disconnect the tie rod ends from the steering knuckle
   • Rack and pinion steering assembly and its rubber mounts from the right side of the vehicle

To install:

6. Align the matchmarks of the input shaft and install the rack to the vehicle.

7. Secure the rack using the retainer clamps and bolts. Torque the bolts to 51 ft. lbs. (70 Nm).

8. Torque the steering column pinch bolt to 13 ft. lbs. (18 Nm).

9. Install or connect the following:
   • Center member
   • Front exhaust pipe
   • HO2S sensor
   • Tie rod ends to the steering knuckles and tighten the castle nuts to 25

ft. lbs. (34 Nm). Install new cotter pins.
   • Wheels and connect the negative battery cable

10. Perform a front end alignment.

Power

3000GT AND DIAMANTE

➡Prior to removal of the steering gear box, center the front wheels and remove the ignition key. Failure to do so may damage the SRS clock spring and render SRS system inoperative.

1. Before servicing the vehicle, refer to the precautions in the beginning of this section.

2. Remove or disconnect the following:
   • Negative battery cable.
   • Front exhaust pipe
   • Transfer case assembly, if equipped with AWD
   • Bolt holding the lower steering column joint to the rack and pinion input shaft
   • Tie rod ends
   • Left and right frame members

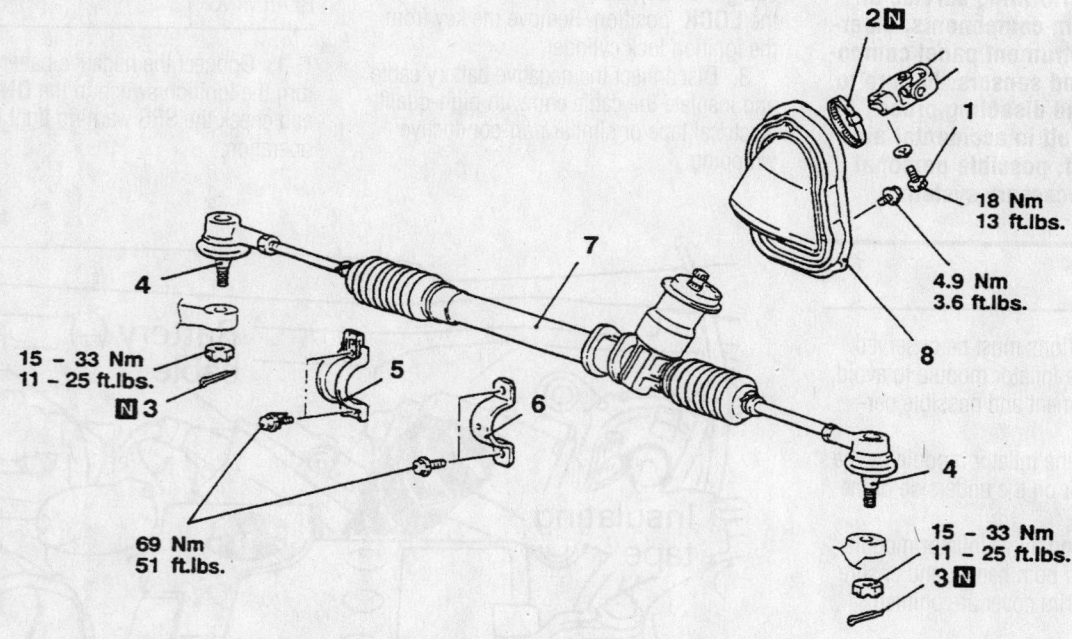

4 — Tie-rod end
15 – 33 Nm
11 – 25 ft.lbs.
3 N
69 Nm
51 ft.lbs.
7
2 N
1
18 Nm
13 ft.lbs.
4.9 Nm
3.6 ft.lbs.
8
5
6
4
15 – 33 Nm
11 – 25 ft.lbs.
3 N

1. Steering shaft assembly and gear box connecting bolt
2. Band
3. Cotter pin
4. Tie-rod end and knuckle connection
5. Cylinder clamp
6. Gear housing clamp
7. Gear box assembly
8. Steering cover assembly

Exploded view of the manual steering gear mounting—Mirage

7923PG00

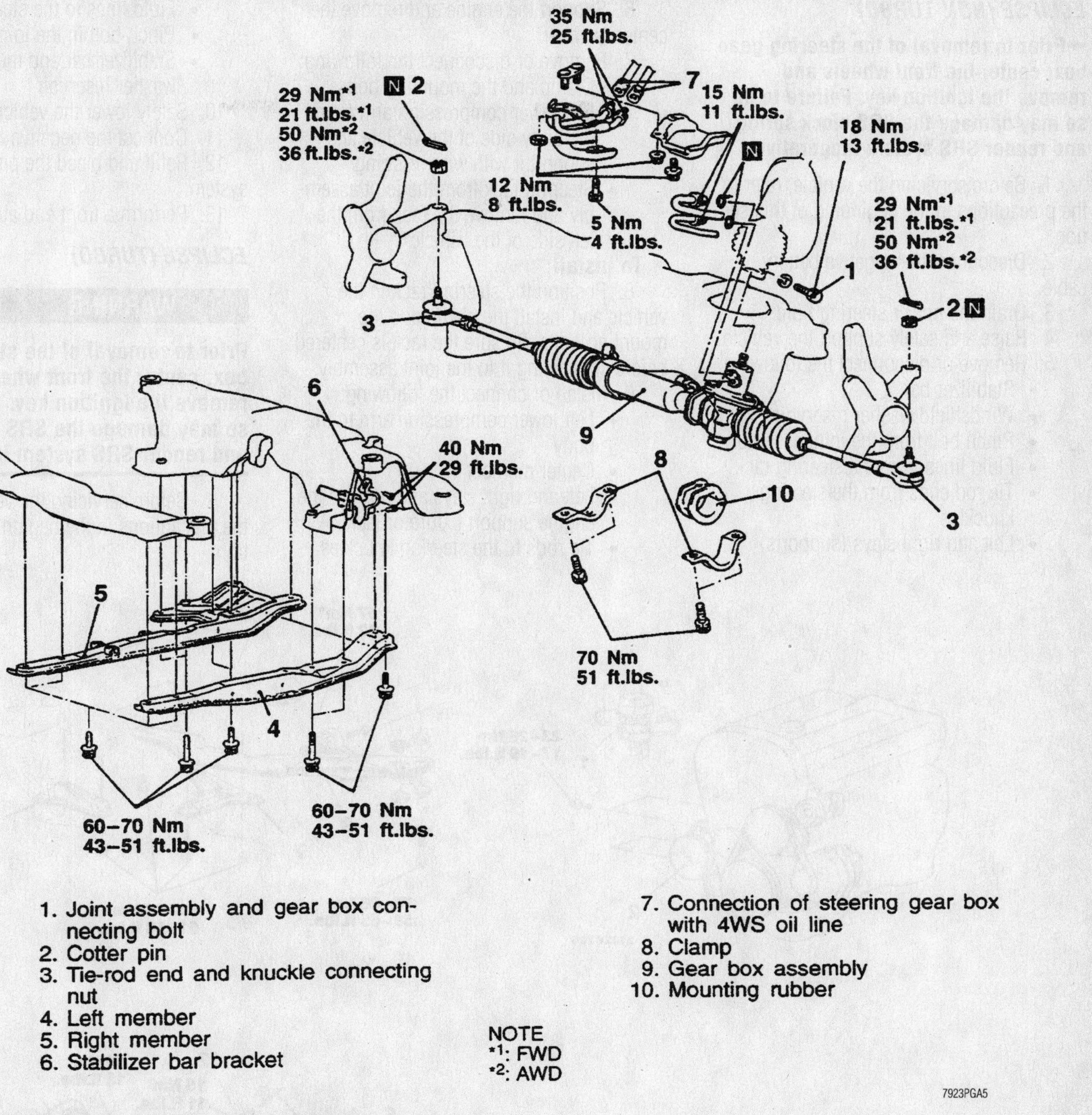

35 Nm
25 ft.lbs.

29 Nm*1
21 ft.lbs.*1
50 Nm*2
36 ft.lbs.*2

N 2

15 Nm
11 ft.lbs.

18 Nm
13 ft.lbs.

12 Nm
8 ft.lbs.

5 Nm
4 ft.lbs.

29 Nm*1
21 ft.lbs.*1
50 Nm*2
36 ft.lbs.*2

2 N

3

9

8

10

3

6

40 Nm
29 ft.lbs.

70 Nm
51 ft.lbs.

5

4

60–70 Nm
43–51 ft.lbs.

60–70 Nm
43–51 ft.lbs.

1. Joint assembly and gear box connecting bolt
2. Cotter pin
3. Tie-rod end and knuckle connecting
   nut
4. Left member
5. Right member
6. Stabilizer bar bracket

7. Connection of steering gear box
   with 4WS oil line
8. Clamp
9. Gear box assembly
10. Mounting rubber

NOTE
*1: FWD
*2: AWD

7923PGA5

Exploded view of the power steering gear removal—3000GT

- Stabilizer bar bracket
- Lines going to the rear pump, if
  equipped with 4-wheel steering
- Rack and pinion steering assembly
  and its rubber mounts. Move the
  rack to the right to remove it from
  the crossmember.

To install:
3. Install or connect the following:
   - Rack and install the mounting
     bolts, tightening bolts to 51 ft. lbs.
     (70 Nm). When installing the rub-
     ber rack mounts, align the projec-

tion of the mounting rubber with
the indentation in the crossmember.
Install the pinch bolt.
- Pressure and return lines to the rack
  and to the rear pump, if equipped
- Frame members and tighten the
  bolts to 43–51 ft. lbs. (60–70 Nm)
- Tie rods and install new cotter pins
- Transfer case and front exhaust
  pipe
4. Refill the reservoir and bleed the system.
5. Perform a front end alignment.

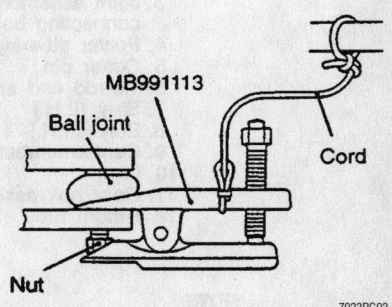

MB991113

Ball joint

Cord

Nut

7923PG93

Proper tie-rod end removal method

For Accessory Drive Belt illustrations, see Section 1 of this manual

ECLIPSE (NON-TURBO)

➡ Prior to removal of the steering gear box, center the front wheels and remove the ignition key. Failure to do so may damage the SRS clock spring and render SRS system inoperative.

1. Before servicing the vehicle, refer to the precautions in the beginning of this section.

2. Disconnect the negative battery cable.

3. Drain the power steering fluid.

4. Raise and safely support the vehicle.

5. Remove or disconnect the following:
- Stabilizer bar
- Windshield washer reservoir
- Pinch bolt from the joint assembly
- Fluid lines from the steering rack
- Tie rod ends from the steering knuckles
- Left and right stays (supports)

6. Support the engine and remove the center member.

7. Remove or disconnect the following:
- Clamp and the mounting bolts
- Left lower compression arm from the body side of the vehicle and support it with wire or string
- Steering rack from the joint assembly and remove the rack from the left side of the vehicle

To install:

8. Position the steering rack in the vehicle and install the clamp and the mounting bolts. Be sure the rack is centered before connecting it to the joint assembly.

9. Install or connect the following:
- Left lower compression arm to the body
- Center member
- Left and right stays and remove the engine support fixture or jack
- Tie rods to the steering knuckles
- Fluid lines to the steering rack
- Pinch bolt in the joint assembly
- Stabilizer bar and the windshield washer reservoir

10. Safely lower the vehicle.

11. Connect the negative battery cable.

12. Refill and bleed the power steering system.

13. Perform a front end alignment.

ECLIPSE (TURBO)

✳✳ WARNING

Prior to removal of the steering gear box, center the front wheels and remove the ignition key. Failure to do so may damage the SRS clock spring and render SRS system inoperative.

1. Before servicing the vehicle, refer to the precautions in the beginning of this section.

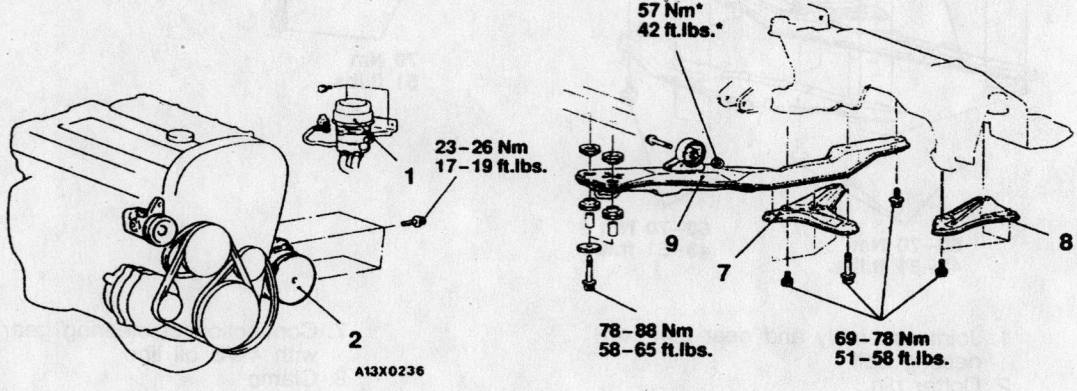

57 Nm*
42 ft.lbs.*

23–26 Nm
17–19 ft.lbs.

78–88 Nm
58–65 ft.lbs.

69–78 Nm
51–58 ft.lbs.

A13X0236

1. Brake fluid reservoir assembly
2. A/C compressor
3. Joint assembly and gear box connecting bolt
4. Power steering pipe connection
5. Cotter pin
6. Tie-rod end and knuckle connection
7. Stay (L.H.)
8. Stay (R.H.)
9. Centermember assembly
10. Clamp
11. Gear box assembly
12. Return tube

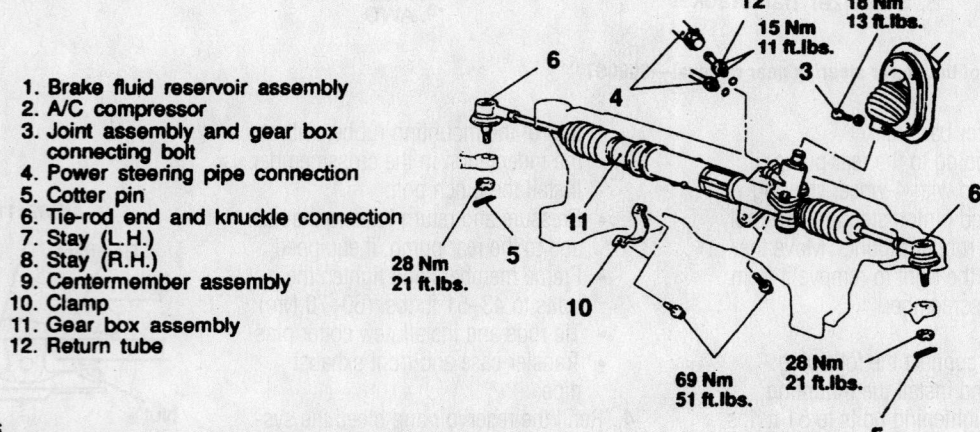

15 Nm
11 ft.lbs.

18 Nm
13 ft.lbs.

28 Nm
21 ft.lbs.

69 Nm
51 ft.lbs.

28 Nm
21 ft.lbs.

NOTE
The fasteners marked * should be temporarily tightened before they are finally tightened once the total weight of the engine has been placed on the vehicle body.

Power steering rack assembly and related components—Eclipse

7923PGA2

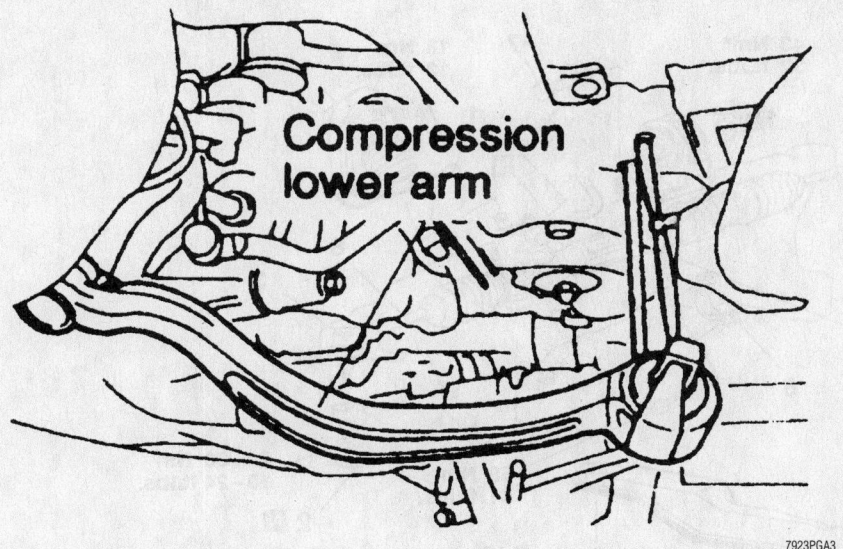

Disconnect the lower compression arm from the body—Eclipse

7923PGA3

2. Disconnect the negative battery cable.

3. Drain the power steering fluid.

4. Raise and safely support the vehicle.

5. Remove or disconnect the following:
- Stabilizer bar
- Fluid level sensor and remove the brake fluid reservoir and position it out of the way. Do not disconnect the brake hose.
- Electrical connector from the air conditioning compressor
- Air conditioning compressor from the bracket and position it out of the way. Do not disconnect the hoses.
- Pinch bolt from the joint assembly
- Fluid lines from the steering rack
- Tie rod ends from the steering knuckles
- Left and right stays (supports)

6. Support the engine and remove the center member assembly.

7. Remove or disconnect the following:
- Clamp and the mounting bolts
- Left lower compression arm from the body side of the vehicle and support it with wire or string

8. Disconnect the steering rack from the joint assembly and remove the rack from the left side of the vehicle.

To install:

9. Position the steering rack in the vehicle and install the clamp and the mounting bolts. Be sure the rack is centered before connecting it to the joint assembly.

10. Install or connect the following:

- Left lower compression arm to the body
- Center member assembly
- Left and right stays
- Tie rod ends to the steering knuckles
- Fluid lines to the steering rack
- Pinch bolt in the joint assembly
- Stabilizer bar

11. Safely lower the vehicle.

12. Install or connect the following:
- Air conditioning compressor and connect the harness connector
- Brake fluid reservoir and connect the fluid level sensor
- Negative battery cable

13. Refill and bleed the power steering system.

14. Perform a front end alignment.

GALANT

> **✳✳ WARNING**
>
> **Prior to removal of the steering gear box, center the front wheels and remove the ignition key. Failure to do so may damage the SRS clock spring and render SRS system inoperative.**

1. Before servicing the vehicle, refer to the precautions in the beginning of this section.

2. Disconnect the negative battery cable.

3. Raise and properly support the vehicle.

4. Remove or disconnect the following:

- Both front wheel assemblies
- Bolt holding lower steering column joint to the rack and pinion input shaft
- Stabilizer bar
- Cotter pins and the tie rod ends from the steering knuckle

5. On vehicles equipped with Electronic Control Power steering (EPS), disconnect the wiring harness from the solenoid connector.

6. Locate the 2 triangular braces near the crossmember and remove both.

7. Support the center crossmember. Remove the through-bolt from the front round roll stopper and remove the bolts securing the center crossmember.

8. Remove the center crossmember.

9. Properly support the engine and remove the rear roll stopper through-bolt.

10. Remove or disconnect the following:
- Power steering fluid pressure pipe and return hose from the rack fittings. Plug the fittings to prevent excessive fluid leakage.
- Clamp bolts and the 2 bolts securing the rack assembly to the chassis
- Rack and pinion steering assembly and its rubber mounts

➡**When removing the rack and pinion assembly, tilt the assembly to the vehicle side of the compression lower arm and remove from the left side of the vehicle.**

To install:

11. Center the rack assembly and insert the pinion into the steering column shaft.

12. Install or connect the following:
- Rack and with the mounting bolts. Torque the mounting bolts to 51 ft. lbs. (69 Nm).
- Pinch bolt and tighten the bolt to 13 ft. lbs. (18 Nm)
- Power steering fluid lines to the rack and tighten the pressure hose fitting to 11 ft. lbs. (15 Nm). Secure the return hose with the clamp.

13. Raise the engine into position. Install the rear roll stopper through-bolt and tighten to 32 ft. lbs. (43 Nm).

14. Raise the crossmember into position. Install the center member mounting bolts and tighten the front bolts to 58–65 ft. lbs. (78–88 Nm) and the rear bolt to 51–58 ft. lbs. (69–78 Nm).

15. Install or connect the following:
- Front roll stopper bolt and tighten the nut to 32 ft. lbs. (43 Nm)

For Tire, Wheel and Ball Joint specifications, see Section 1 of this manual

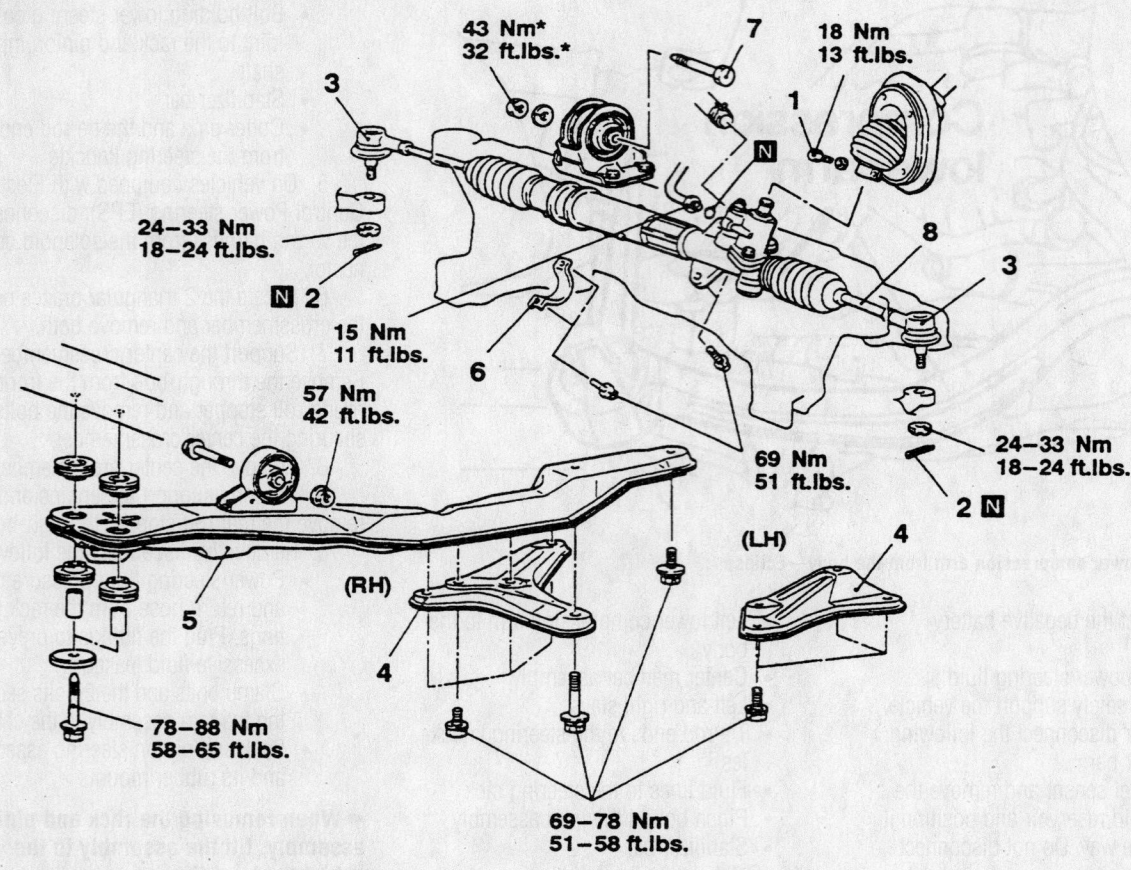

43 Nm*
32 ft.lbs.*

18 Nm
13 ft.lbs.

24–33 Nm
18–24 ft.lbs.

15 Nm
11 ft.lbs.

57 Nm
42 ft.lbs.

69 Nm
51 ft.lbs.

24–33 Nm
18–24 ft.lbs.

78–88 Nm
58–65 ft.lbs.

69–78 Nm
51–58 ft.lbs.

(RH)

(LH)

7923PGA4

1. Joint assembly and gear box connecting bolt
2. Cotter pin
3. Connection for tie rod end and knuckle
4. Stay
5. Center member assembly
6. Clamp
7. Bolt
8. Gear box assembly

Caution
The fasteners marked * should be temporarily tightened before they are finally tightened once the total weight of the engine has been placed on the vehicle body.

Exploded view of the power steering gear removal procedure—Galant

- 2 triangular braces and tighten the mounting bolts to 50–56 ft. lbs. (69–78 Nm)
- Stabilizer bar
- Tie rod ends and tighten nuts to 20 ft. lbs. (27 Nm)

16. On vehicles equipped with EPS, connect the wiring harness to the solenoid connector.

17. Install the wheel assemblies and lower the vehicle.

18. Refill the reservoir with power steering fluid and bleed the system.

19. Perform a front end alignment.

MIRAGE

➡ Prior to removal of the steering gear box, center the front wheels and remove the ignition key. Failure to do so may damage the SRS clockspring and render SRS system inoperative.

1. Before servicing the vehicle, refer to the precautions in the beginning of this section.

2. Drain the power steering system.

3. Remove or disconnect the following:

- Battery negative cable. Raise the vehicle and support safely

- HO2S sensor and remove the front exhaust pipe

4. Properly support the engine.

- Both roll stopper mounting bolts and the 4 center member installation bolts. Remove the center member.
- Center member

➡ Matchmark the pinion input shaft of the rack to the lower steering column joint for installation purposes.

5. Remove or disconnect the following:

- Pinch bolt holding the lower steering column joint to the rack and pinion input shaft

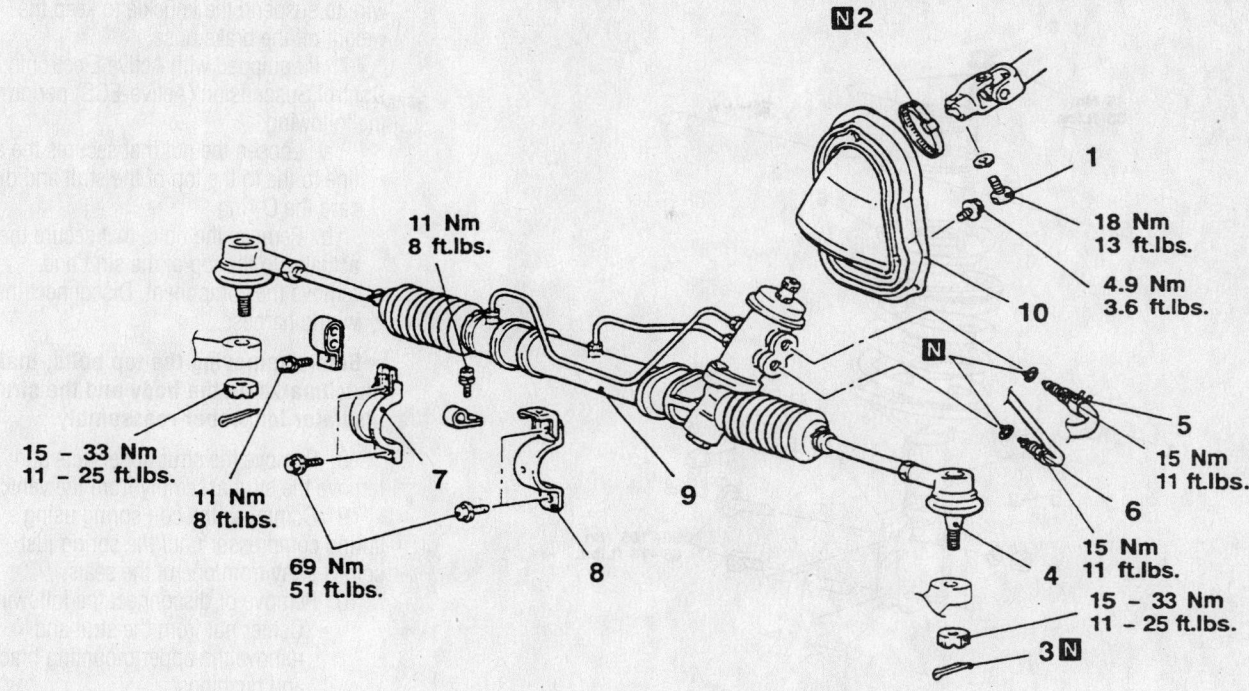

1. Steering shaft assembly and gear box connecting bolt
2. Band
3. Cotter pin
4. Tie-rod end and knuckle connection
5. Return tube connection
6. Pressure tube connection
7. Cylinder clamp
8. Gear housing clamp
9. Gear box assembly
10. Steering cover assembly

7923PGA1

Exploded view of the power steering gear assembly—Mirage

- Cotter pins and disconnect the tie rod ends from the steering knuckle
- Power steering fluid pressure pipe and return hose from the rack fittings
- Rack and pinion steering assembly and its rubber mounts from the right side of the vehicle

To install:

6. Align the matchmarks of the input shaft and install the rack to the vehicle.

7. Secure the rack using the retainer clamps and bolts. Torque the bolts to 51 ft. lbs. (70 Nm).

8. Torque the steering column pinch bolt to 13 ft. lbs. (18 Nm).

9. Using new O-rings, connect the power steering fluid lines to the rack fittings.

10. Install or connect the following:
- Center member
- Front exhaust pipe

- HO$_2$S sensor
- Tie rod ends to the steering knuckles and tighten the castle nuts to 25 ft. lbs. (34 Nm). Install new cotter pins.
- Wheels and connect the negative battery cable

11. Refill the reservoir and bleed the system.

12. Perform a front end alignment.

Strut and Coil Spring

REMOVAL & INSTALLATION

Front

3000GT

1. Before servicing the vehicle, refer to the precautions in the beginning of this section.

2. Disconnect the negative battery cable.

3. Raise and safely support vehicle.

4. Remove the brake hose and tube bracket. Do not pry the brake hose and tube clamp away when removing it.

5. If equipped with ABS, disconnect the front speed sensor mounting clamp from the strut.

6. Support the lower arm and remove the strut to knuckle bolts. Use a piece of wire to suspend the knuckle to keep the weight off the brake hose.

7. If equipped with Active Electronic Control Suspension (Active-ECS) perform the following:

a. Loosen the nut that secures the air line to the to the top of the strut and discard the O-ring.

b. Remove the bolts that secure the actuator to the top of the strut and remove the component. Disconnect the wiring harness.

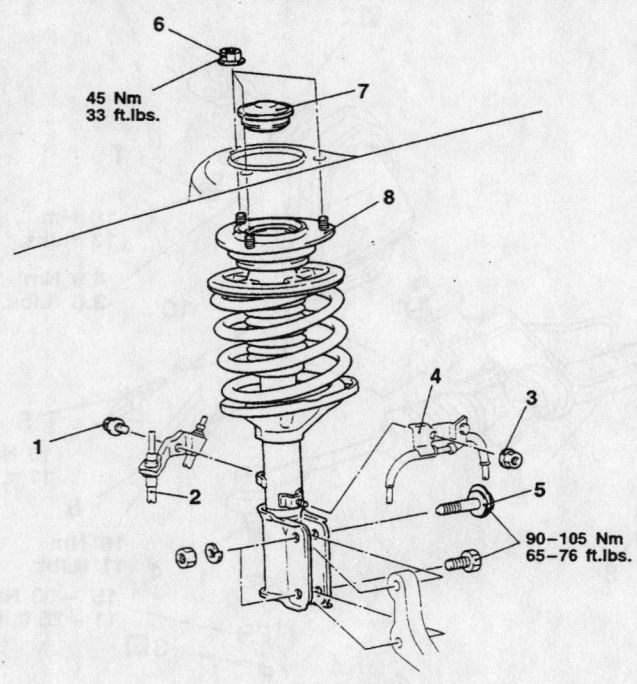

45 Nm
33 ft.lbs.

90–105 Nm
65–76 ft.lbs.

1. Brake hose tube clamp mounting bolt
2. Brake hose tube clamp
3. Front speed sensor clamp mounting nut <ABS>
4. Front speed sensor clamp <ABS>
5. Strut lower mounting bolt
6. Strut upper mounting bolt
7. Dust cover
8. Strut assembly

7923PGA7

Exploded view of the front strut assembly removal—3000GT

8. Before removing the top bolts, make matchmarks on the body and the strut insulator for proper reassembly.

9. Remove the strut upper bolts and remove the strut assembly from the vehicle.

10. Compress the coil spring using a spring compressor until the spring just comes away from one of the seats.

11. Remove or disconnect the following:
- Center nut from the strut and remove the upper mounting bracket and bushings
- Coil spring

To install:

12. Install or connect the following:
- Compressed spring on the strut assembly
- Upper bushings and the mounting bracket
- Nut and tighten it to 16 ft. lbs. (22 Nm)
- Strut to the vehicle and install the top bolts
- Strut to the knuckle and install the bolts

13. If equipped with Active-ECS, perform the following:

a. Install the air line with a new O-ring.

b. Install the actuator to the top of the strut. Connect the wiring harness.

14. Install or connect the following:
- Brake hose bracket and the ABS clamp
- Wheel and tire assembly
- Negative battery cable

15. Perform a front end alignment.

DIAMANTE

1. Before servicing the vehicle, refer to the precautions in the beginning of this section.

2. Disconnect the negative battery cable.

3. Raise and safely support the vehicle.

4. Remove the brake hose and the tube bracket.

➡ **Do not pry the brake hose and tube clamp away when removing it.**

5. If equipped with ABS, disconnect the front speed sensor mounting clamp from the strut.

6. Support the lower arm and remove the strut to knuckle bolts. Use a piece of

wire to suspend the knuckle to keep the weight off the brake hose.

7. If equipped with Active Electronic Control Suspension (Active-ECS) perform the following:

a. Loosen the nut that secures the air line to the to the top of the strut and discard the O-ring.

b. Remove the bolts that secure the actuator to the top of the strut and remove the component. Disconnect the wiring harness.

➡ **Before removing the top bolts, make matchmarks on the body and the strut insulator for proper reassembly.**

8. Remove the strut upper nuts and remove the strut assembly from the vehicle.

9. Compress the coil spring using a spring compressor until the spring just comes away from one of the seats.

10. Remove or disconnect the following:
- Center nut from the strut and remove the upper mounting bracket and bushings
- Coil spring

To install:

11. Install or connect the following:
- Compressed spring on the strut assembly
- Upper bushings and the mounting bracket
- Nut and tighten it to 43 ft. lbs. (59 Nm)
- Strut to the vehicle and tighten the upper mounting nuts to 33 ft. lbs. (45 Nm)

12. Align the strut to the knuckle and connect with the mounting bolts. Torque the mounting bolts to 70–76 ft. lbs. (90–105 Nm).

13. If equipped with Active-ECS, perform the following:

a. Install the air line with a new O-ring.

b. Install the actuator to the top of the strut. Connect the wiring harness.

14. Install or connect the following:
- Brake hose bracket and the ABS clamp, if equipped
- Wheel and tire assembly

15. Perform a front end alignment.

MIRAGE

1. Before servicing the vehicle, refer to the precautions in the beginning of this section.

2. Disconnect the negative battery cable.

3. Raise and safely support vehicle.

4. Remove the brake hose and tube bracket retainer bolt and bracket from the front strut. Do not pry the brake hose and tube clamp away when removing.

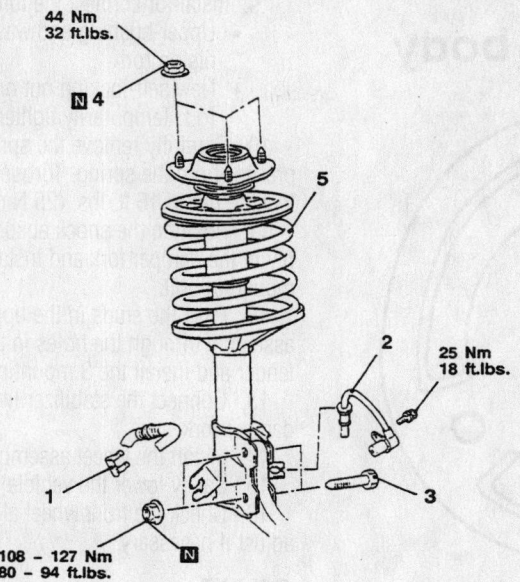

Removal steps
1. Brake hose clamp
2. Front speed sensor <Vehicles with ABS>
3. Bolts
4. Self-locking nut
5. Strut assembly

Caution
For vehicles with ABS, be careful when handling the pole piece at the tip of the speed sensor so as not to damage it by striking against other parts.

7923PGA6

Front strut assembly and related parts—Mirage

5. If equipped with ABS, disconnect the front speed sensor mounting clamp from the strut.

6. Support the lower arm using floor jack. Remove the lower strut to knuckle bolts.

➡ **Before removing the top bolts, make matchmarks on the body and the strut insulator for proper reassembly.**

7. Remove the strut upper mounting bolts. Remove the strut assembly from the vehicle.

8. Compress the coil spring using a spring compressor until the spring just comes away from one of the seats.

9. Remove or disconnect the following:
- Center nut from the strut and remove the upper mounting bracket and bushings
- Coil spring

To install:

10. Install or connect the following:
- Compressed spring on the strut assembly
- Upper bushings and the mounting bracket
- Nut and tighten it to 43 ft. lbs. (59 Nm)

- Strut to the vehicle and install the top mounting bolts. Tighten the mounting bolts to 29 ft. lbs. (40 Nm).

11. Position the strut on the knuckle and install the mounting bolts. While holding

the head of the lower mounting bolt, tighten the nuts to 80–94 ft. lbs. (110–130 Nm).

12. Install or connect the following:
- Brake hose bracket and the ABS clamp, if equipped
- Wheel and tire assembly

13. Perform a front end alignment.

Shock Absorber and Coil Spring

REMOVAL & INSTALLATION

Front

ECLIPSE

1. Before servicing the vehicle, refer to the precautions in the beginning of this section.

2. Raise and safely support the vehicle.

3. Remove or disconnect the following:
- Front wheel
- 3 upper shock absorber mounting nuts. Do not remove the larger nut in the center of the strut at this time.
- Stabilizer link from the damper fork
- Damper fork mounting bolt
- Shock absorber assembly from the vehicle

4. Use a coil spring compressor and compress the coil spring.

5. While holding the piston rod, remove the self-locking nut.

6. Remove or disconnect the following:
- Upper bracket assembly and spring pad

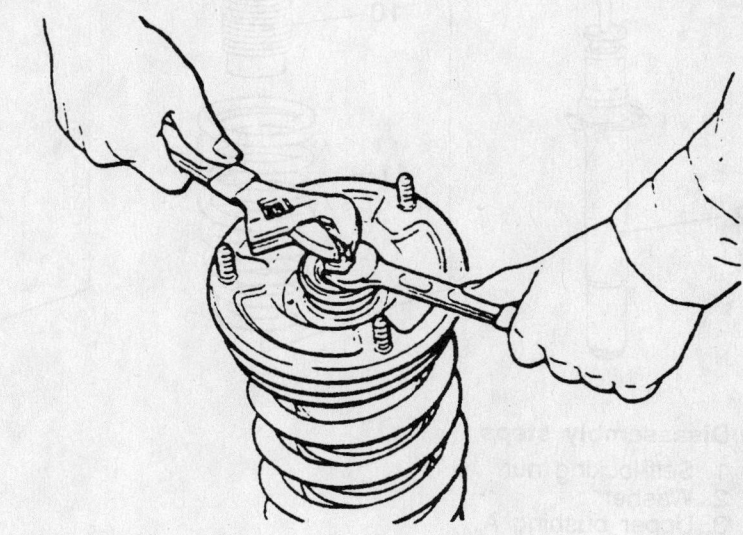

Removing the self-locking nut—Eclipse

7923PGB1

Inside of the body

Damper fork installation bolt

7923PGB2

Upper bracket assembly alignment—Eclipse

- Collar, upper bushing, cup assembly, bump rubber and dust cover
- Coil spring from the shock absorber

To install:

7. Align the end of the coil spring with the stepped part of the spring seat and install the compressed coil spring on the shock.

8. Install the dust cover, bump rubber, cup assembly, upper bushing, collar, upper spring pad and bracket assembly on the strut.

9. Install or connect the following:
- Upper bushing and washer on the piston rod
- New self-locking nut on the piston rod. Temporarily tighten the nut.

10. Carefully remove the spring compressor from the spring. Torque the self-locking nut to 16 ft. lbs. (25 Nm).

11. Position the shock absorber assembly in the damper fork and install the mounting bolt.

12. Pass the studs in the upper bracket assembly through the holes in the inner fender and install the 3 mounting nuts.

13. Connect the stabilizer link to the damper fork.

14. Install the wheel assembly.

15. Safely lower the vehicle to the floor.

16. Check the front wheel alignment and adjust if necessary.

GALANT

1. Before servicing the vehicle, refer to the precautions in the beginning of this section.

2. Disconnect the negative battery cable.

3. Raise and safely support vehicle.

4. Remove or disconnect the following:
- Appropriate wheel assembly

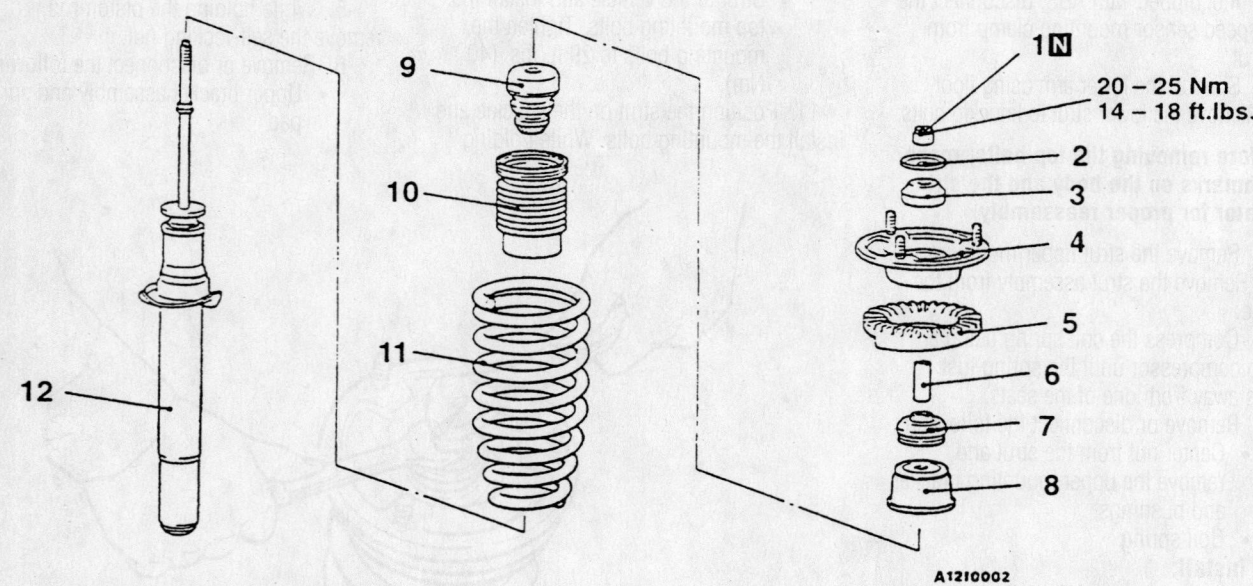

A1210002

Disassembly steps

1. Self-locking nut
2. Washer
3. Upper bushing A
4. Upper bracket assembly
5. Upper spring pad
6. Collar
7. Upper bushing B
8. Cup assembly
9. Bump rubber
10. Dust cover
11. Coil spring
12. Shock absorber assembly

20 – 25 Nm
14 – 18 ft.lbs.

Exploded view of the coil spring removal procedure—Eclipse

7923PGB3

- Sway bar link from the damper fork
- Damper fork lower through-bolt and upper pinch bolt
- Damper fork assembly
- Shock absorber upper nuts and remove the strut assembly from the vehicle

5. Compress the coil spring with a special compression tool.

6. Remove or disconnect the following:
- Self-locking nut and washer
- Upper bushing, upper bracket assembly, the upper spring pad, and the collar
- Other upper bushing, cup assembly, bump rubber, dust cover, and the coil spring. Carefully remove the coil spring compression tool

To install:

7. Install or connect the following:
- Compressed coil spring to the shock absorber assembly. Be sure to align the edge of coil spring to

the stepped part of the spring seat. Install the dust cover, bump rubber, cup assembly, upper bushing, collar, and upper spring pad.
- Upper bracket assembly and position it so that the 3 bolts are in the correct position
- Upper bushing, washer, and locknut. Torque the locknut to 18 ft. lbs. (24 Nm).
- Shock absorber and tighten the upper mounting nuts to 32 ft. lbs. (44 Nm)

8. Align the shock to the damper fork and install the damper fork. Tighten the lower through-bolt/nut to 65 ft. lbs. (88 Nm) and the upper pinch bolt to 76 ft. lbs. (103 Nm).

9. Connect the sway bar link to the damper fork and tighten the link nut to 29 ft. lbs. (39 Nm).

10. Install the wheel and tire assembly.

11. Perform a front end alignment.

Rear

_3000GT_

1. Before servicing the vehicle, refer to the precautions in the beginning of this section.

2. Disconnect the negative battery cable.

3. Raise and safely support the vehicle.

4. Remove the rear side trim in the luggage compartment and remove the Active Electronic Control Suspension (Active-ECS) connector and cap.

5. Support the suspension and remove the upper shock absorber mounting bolts.

6. Remove or disconnect the following:
- Wheel and tire assembly and the lower shock mounting bolt
- Shock absorber from the vehicle

To install:

7. Position the shock absorber in the trailing arm and temporarily install the lower mounting bolt. After the vehicle is on the

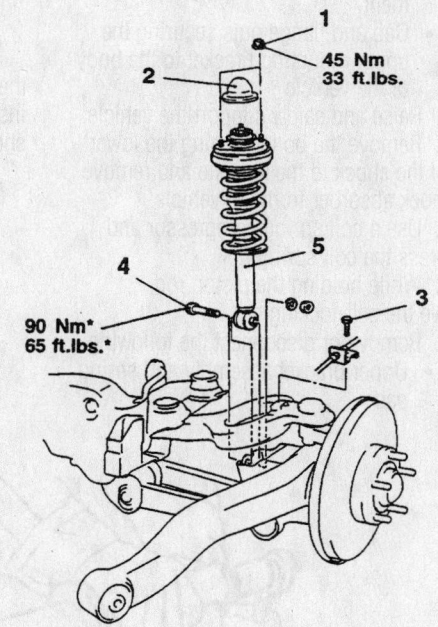

1. Shock absorber upper mounting nut
2. Cap
3. Brake tube clamp bolt
4. Shock absorber lower mounting bolt
5. Shock absorber

Caution
*: Indicates parts which should be temporarily tightened, and then fully tightened with the vehicle on the ground in an unladen condition.

7923PGA8

Rear shock absorber removal—3000GT

For Tune-up, Capacities and Firing orders, see Section 1 of this manual

ground at normal ride height, tighten the bolt to 65 ft. lbs. (90 Nm).

8. Guide the upper mounting studs through the body and tighten the upper mounting nuts to 33 ft. lbs. (45 Nm).

9. Install or connect the following:
- Cap and the Active-ECS connector
- Wheel and tire assembly and connect the negative battery cable

10. Lower the vehicle.

DIAMANTE

1. Before servicing the vehicle, refer to the precautions in the beginning of this section.

2. Disconnect the negative battery cable.

3. Raise and properly support vehicle. Remove both rear wheels.

4. Support the lower control arm with a jack.

5. Matchmark the positioning of the upper spring plate to the vehicle for reinstallation purposes.

6. If equipped with Active Electronic Control Suspension (Active-ECS), perform the following:

  a. Loosen the nut that secures the air line to the to the top of the strut and discard the O-ring.

  b. Remove the bolts that secure the actuator to the top of the strut and remove the component. Disconnect the wiring harness.

7. Remove the shock absorber lower mounting bolt and remove the 2 nuts that secure the shock upper plate to the vehicle.

8. Lower the support jack and remove the shock from the vehicle.

To install

9. Position the upper spring plate and install the strut. Use the support jack to assist with installation.

10. Tighten the upper strut mounting nuts to 33 ft. lbs. (45 Nm).

11. Tighten the lower strut mounting bolt to 71 ft. lbs. (98 Nm).

12. If equipped with Active-ECS perform the following:

  a. Using a new O-ring, tighten the nut that secures the air line to the to the top of the strut to 84 inch lbs. (9 Nm).

  b. Install the actuator to the top of the shock absorber and secure with mounting bolts. Connect the wiring harness.

13. Remove the support jack, install wheels and lower vehicle.

14. Connect the negative battery cable.

ECLIPSE

1. Before servicing the vehicle, refer to the precautions in the beginning of this section.

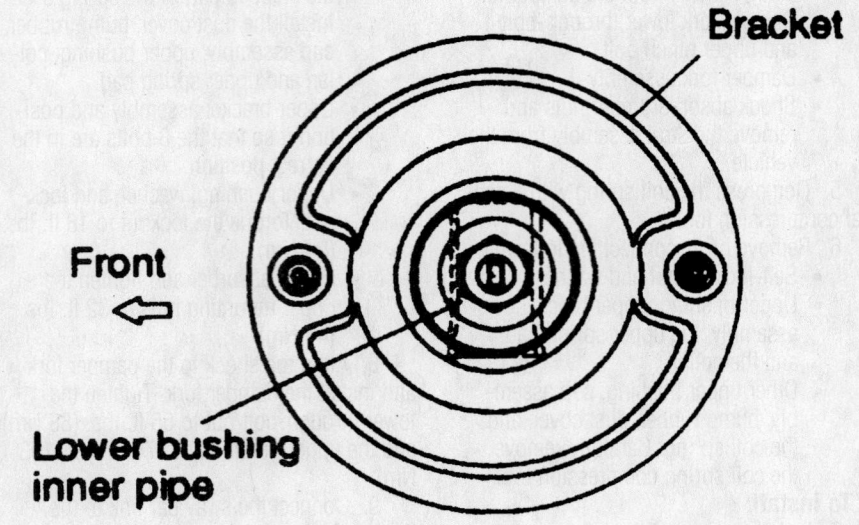

Correct upper bracket installed position—Eclipse

2. Remove or disconnect the following:
- Service lid in the luggage compartment
- Cap and flange nuts securing the upper mounting bracket to the body of the vehicle

3. Raise and safely support the vehicle.

4. Remove the bolt attaching the lower end of the shock to the knuckle and remove the shock absorber from the vehicle.

5. Use a coil spring compressor and compress the coil spring.

6. While holding the piston rod, remove the self-locking nut.

7. Remove or disconnect the following:
- Upper bracket assembly and spring pad

- Collar, upper bushing, cup assembly, bump rubber and dust cover
- Coil spring from the shock absorber

To install:

8. Align the end of the coil spring with the stepped part of the spring seat and install the compressed coil spring on the shock absorber.

9. Install or connect the following:
- Dust cover, bump rubber, cup assembly, upper bushing, collar, upper spring pad and bracket assembly on the shock absorber
- Upper bushing and washer on the piston rod
- New self-locking nut on the piston rod. Temporarily tighten the nut.

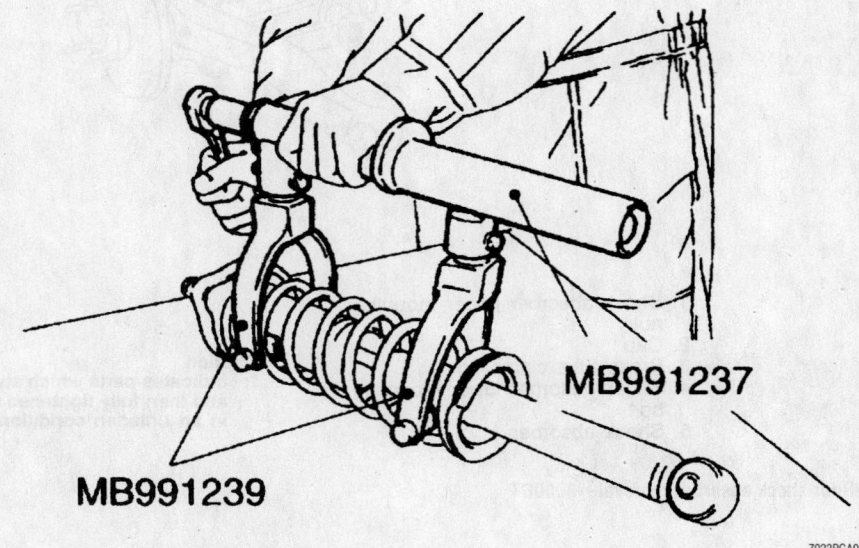

Correct method for compressing the coil spring

10. Remove the spring compressor from the spring. Torque the self-locking nut to 16 ft. lbs. (25 Nm).

11. Install the upper bracket of the shock to the vehicle. Torque the mounting nuts to 32 ft. lbs. (44 Nm).

12. Raise the suspension up with a jack or adjustable stand to align the shock absorber lower mounting holes.

13. Install the lower mounting bolt. Torque the bolt to 71 ft. lbs. (96 Nm).

14. Remove the jack or stand and safely lower the vehicle to the floor.

15. Install the cap and service lid.

GALANT

1. Before servicing the vehicle, refer to the precautions in the beginning of this section.

2. Raise and support the vehicle chassis.

3. Raise and support the lower control arm assembly slightly.

4. In order to gain access to the top mounting nuts, remove the rear seat as follows:

   a. While pulling the rear seat stopper outward, lift the lower cushion upward. Remove the lower cushion.

   b. Remove the seat back mounting bolts.

   c. Lift the seat back upward and remove the seat.

5. Remove or disconnect the following:
- Shock upper mounting nuts
- Shock lower mounting bolt and remove the assembly from the vehicle

6. Use a coil spring compressor and compress the coil spring.

7. Remove the shock cap.

8. While holding the piston rod, remove the self-locking nut.

9. Remove or disconnect the following:
- Upper bracket assembly and spring pad
- Collar, upper bushing, cup assembly, bump rubber and dust cover
- Coil spring from the shock

To install

10. Align the end of the coil spring with the stepped part of the spring seat and install the compressed coil spring on the shock.

11. Install or connect the following:
- Dust cover, bump rubber, cup assembly, upper bushing, collar, upper spring pad and bracket assembly on the shock

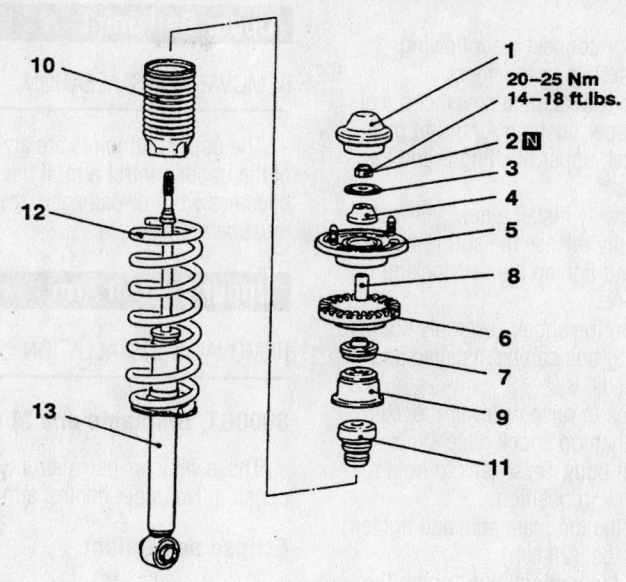

1. Cap
2. Self-locking nut
3. Washer
4. Upper bushing A
5. Bracket
6. Spring pad
7. Upper bushing B
8. Collar
9. Cup
10. Dust cover
11. Bump rubber
12. Coil spring
13. Shock absorber assembly

20–25 Nm
14–18 ft.lbs.

7923PGB4

Exploded view of the rear shock absorber assembly—Galant and Mirage, 3000GT is similar

- Upper bushing and washer on the piston rod
- New self-locking nut on the piston rod. Temporarily tighten the nut.

12. Remove the spring compressor from the spring. Torque the self-locking nut to 16 ft. lbs. (25 Nm).

13. Install the shock cap.

14. Position the shock assembly so that the lower mounting bolt can be installed and lightly tightened.

15. Use a jack to raise or lower the lower control arm, so that the top shock plate studs align through the body. Raise the jack to hold the shock assembly in position.

16. Install the top plate nuts on the studs and tighten the mounting nuts to 32 ft. lbs. (44 Nm).

17. With the vehicle on the ground, tighten the lower mounting bolt to 71 ft. lbs. (98 Nm).

18. Install the rear seat back and cushion.

MIRAGE

1. Before servicing the vehicle, refer to

the precautions in the beginning of this section.

2. Remove or disconnect the following:
- Trunk interior trim to gain access to the top mounting nuts
- Top cap and upper shock mounting nuts

3. Raise and support vehicle chassis.

4. Support the trailing arm assembly with a jack.

5. Matchmark the upper spring plate to the vehicle chassis for reassembly and remove the upper spring plate mounting nuts.

6. Remove the shock lower mounting bolt and remove the assembly from the vehicle.

7. Compress the coil spring using the proper spring compressor.

8. Hold the piston rod with a wrench and remove the self-locking nut.

9. Remove or disconnect the following:
- Washer, upper bushing A, bracket, spring pad, upper bushing B, collar, cup, dust cover and bump rubber
- Coil spring

To install

10. Install or connect the following:
- Coil spring on the shock
- Bump rubber, dust cover, cup, collar, upper bushing A, spring pad, bracket, upper bushing B and the washer

11. Temporarily install a new self-locking nut, carefully release the spring from the compressor and tighten the self-locking nut to specifications.

12. Position the shock assembly so that lower mounting bolt can be installed and lightly tightened.

13. Use jack to raise or lower the axle assembly so that top shock plate studs aligns through body. Raise jack to hold the shock assembly in position.

14. Install the top plate nuts and tighten them to 20 ft. lbs. (28 Nm).

15. Lower the vehicle and tighten the lower mounting bolt to 65 ft. lbs. (90 Nm).

16. Install top cap and interior trim.

Upper Ball Joint

REMOVAL & INSTALLATION

The upper ball joints are an integral part of the upper control arm. If the ball joint becomes worn or damaged, the control arm must be replaced.

Upper Control Arm

REMOVAL & INSTALLATION

3000GT, Diamante and Mirage

These vehicles use a strut type front suspension. No upper control arm is used.

Eclipse and Galant

1. Before servicing the vehicle, refer to the precautions in the beginning of this section.

2. Raise and safely support the vehicle.

3. Remove or disconnect the following:
- Front wheel
- Upper arm ball joint from the steering knuckle
- Upper arm shaft mounting nuts from the body
- Upper arm
- Through-bolts that attach the upper arm to the shafts

To install:

4. Assembly the upper arm to the shafts at the proper angle. Torque the through-bolts and nuts to 41 ft. lbs. (57 Nm). The proper angle is 84–86°. After the arm and the shafts are connected at the right angle, measure dimensions A and B to insure correct assembly.
- A O-ring: 11.8 in. (299.9mm)
- B O-ring: 9.2 in. (234.0mm)

5. Install or connect the following:
- Control arm assembly to the body with new self-locking nuts. Torque the self-locking nuts to 62 ft. lbs. (86 Nm).

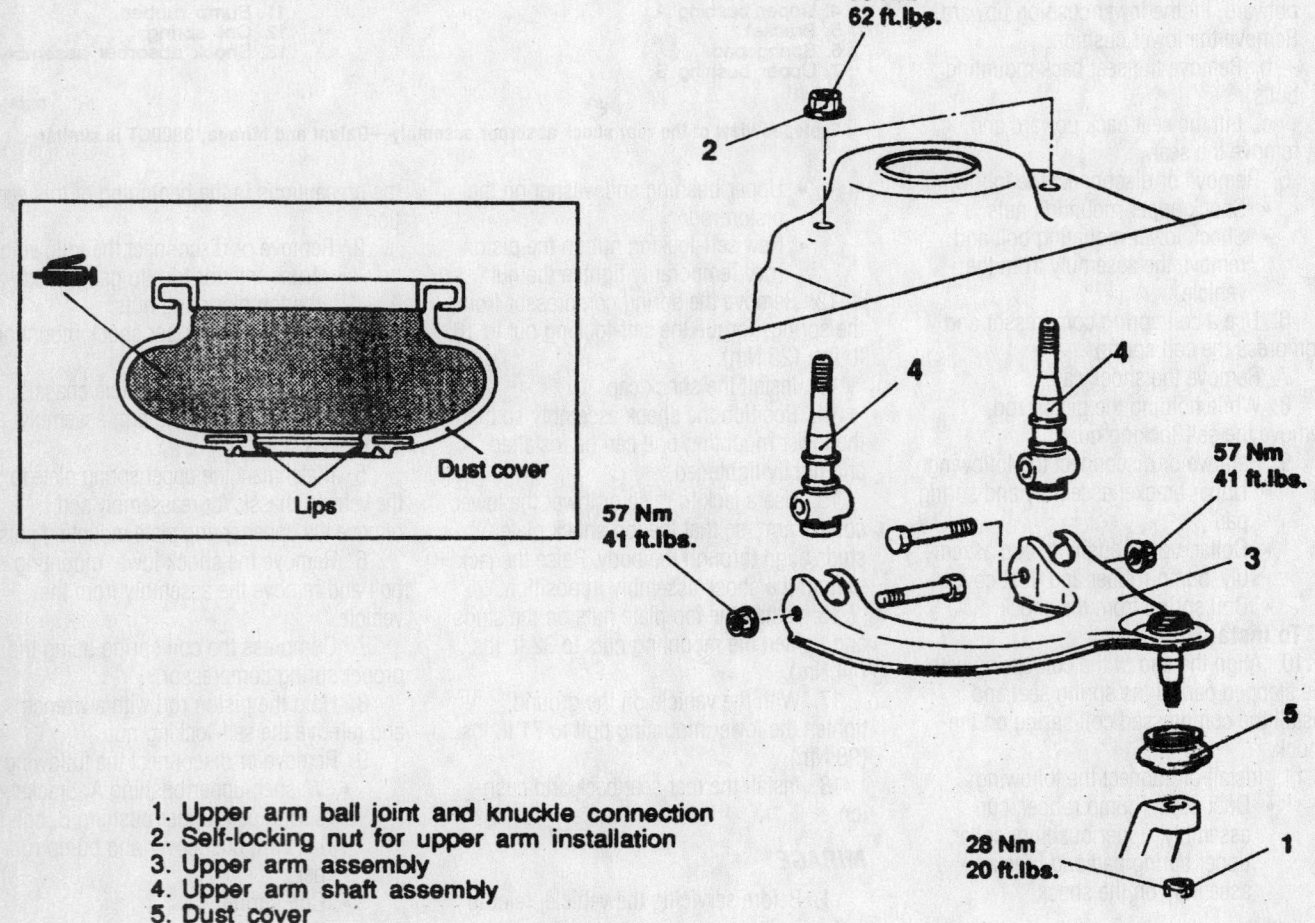

1. Upper arm ball joint and knuckle connection
2. Self-locking nut for upper arm installation
3. Upper arm assembly
4. Upper arm shaft assembly
5. Dust cover

Upper control arm assembly—Eclipse and Galant

7923PGB7

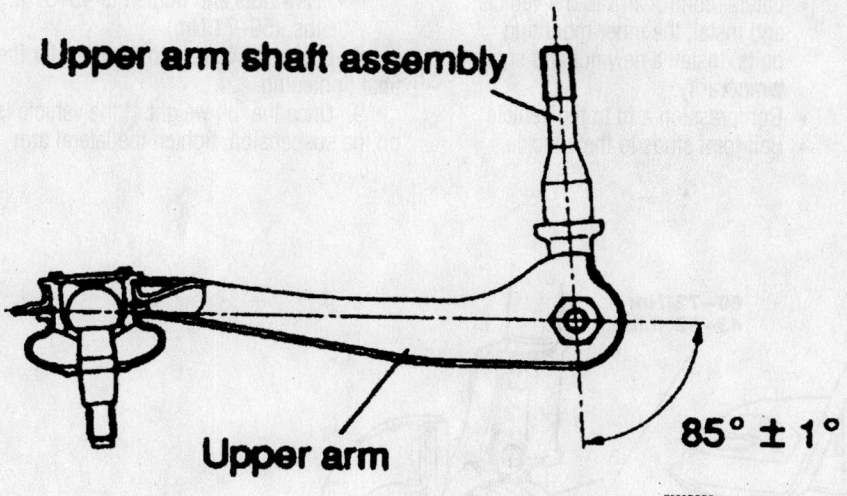

Correct angle of control arm and shafts—Eclipse and Galant

A : 299.9 mm (11.8 in.)
B : 234.0 mm (9.2 in.)

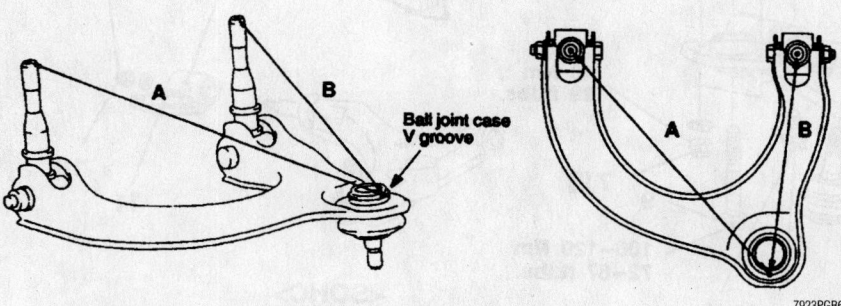

Measure the dimensions A and B as shown—Eclipse and Galant

- Upper arm ball joint to the steering knuckle with a new self-locking nut. Torque the locking nut to 20 ft. lbs. (28 Nm).
- Front wheel
6. Check the front wheel alignment and adjust if necessary.
7. Lower the vehicle.

Lower Ball Joint

REMOVAL & INSTALLATION

The lower ball joint is an integral part of the lower control arm assembly, and can not be serviced separately. A worn or damaged ball joint requires replacement of lower control arm assembly.

Lower Control Arm

REMOVAL & INSTALLATION

Front

3000GT AND DIAMANTE

1. Before servicing the vehicle, refer to the precautions in the beginning of this section.
2. Disconnect the negative battery cable.
3. Raise the vehicle and support safely allowing wheels and suspension to hang freely.
4. Remove or disconnect the following:
- Sway bar links from the lower control arm

- Ball joint stud from the steering knuckle
5. Remove or disconnect the following:
- Inner mounting frame through-bolt and nut
- Rear mount bolts. Remove the clamp if equipped
- Rear rod bushing if servicing

To install:
6. Assemble the control arm and bushing.
7. Install or connect the following:
- Control arm to the vehicle and install the through-bolt. Replace the nut and snug temporarily.
- Rear mount clamp, bolts and replacement nuts. Torque the bolts to 72–87 ft. lbs. (100–120 Nm). Torque the nuts to 29 ft. lbs. (40 Nm).
- Ball joint stud to the knuckle
- New nut and tighten to 43–52 ft. lbs. (60–72 Nm)
- Sway bar and links
8. Lower the vehicle to the floor for the final tightening of the frame mount through-bolt.
9. Once the full weight of the vehicle is on the floor, tighten the frame mount through-bolt nuts to 75–90 ft. lbs. (102–122 Nm).
10. Connect the negative battery cable.
11. Check the wheel alignment and adjust if necessary.

ECLIPSE AND GALANT

The lower lateral arm ball joint and the compression arm ball joint are integral components of the lateral arm and the compression arm respectively. If the ball joints are to be serviced, the arms must be replaced.
1. Before servicing the vehicle, refer to the precautions in the beginning of this section.
2. Raise and support the vehicle safely.
3. Disconnect both ball joint studs from the steering knuckle.
4. To remove the lower lateral arm, remove the crossmember brackets.
5. Remove or disconnect the following:
- Inner lateral arm mounting bolts and nut
- Arm from the vehicle

For Accessory Drive Belt illustrations, see Section 1 of this manual

- 2 bolts holding the compression arm
- Compression arm

To install:

6. Assemble the control arms and bushings.

7. Install or connect the following:

- Lateral control arm to the vehicle and install the inner mounting bolts. Install a new nut and snug temporarily.
- Compression arm to the vehicle
- Ball joint studs to the knuckle

- New nuts and tighten to 43–51 ft. lbs. (59–71 Nm)

8. Lower the vehicle to the floor for the final tightening.

9. Once the full weight of the vehicle is on the suspension, tighten the lateral arm

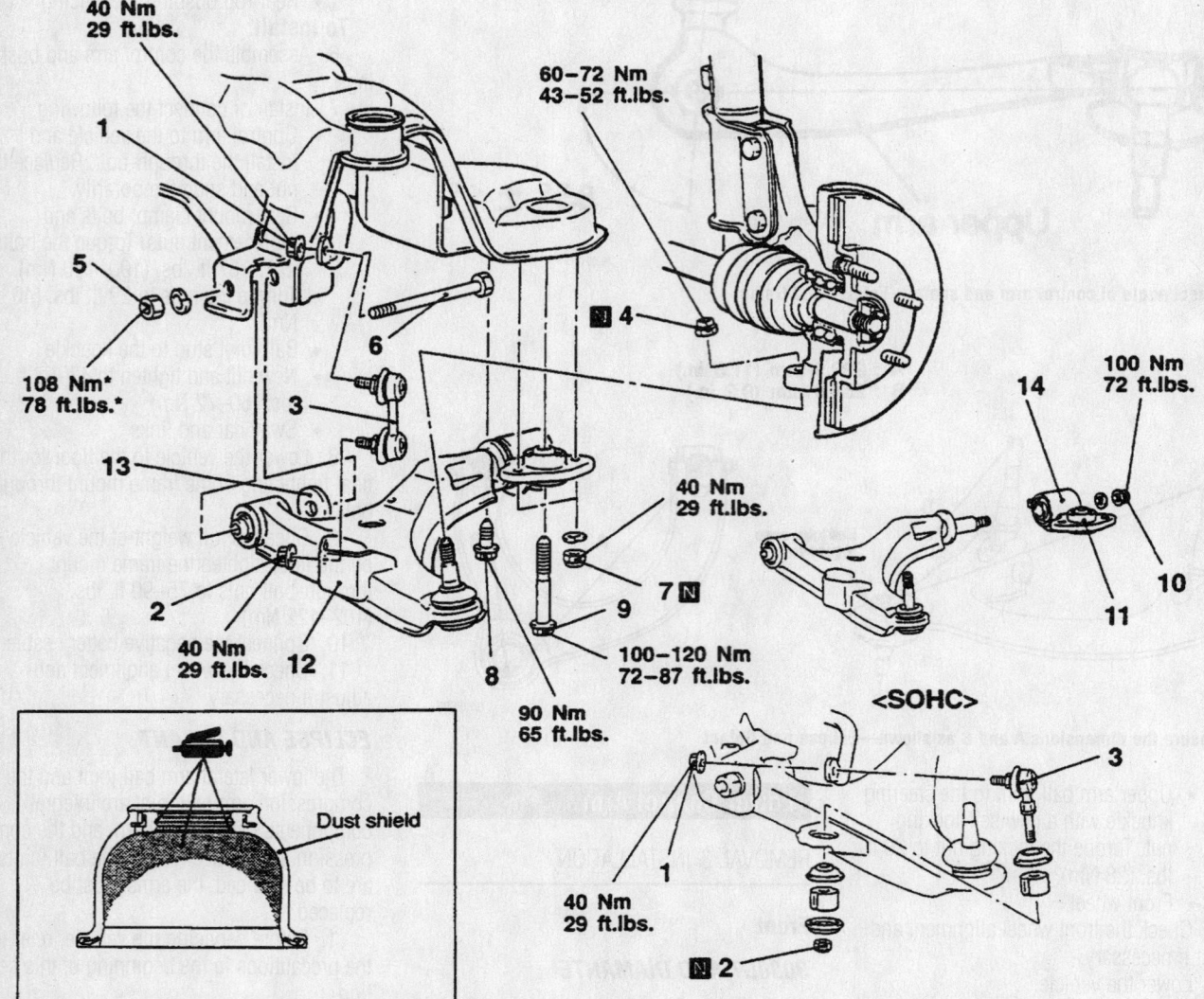

Removal steps

1. Stabilizer link mounting nut (stabilizer bar side)
2. Stabilizer link mounting nut (lower arm side)
3. Stabilizer link
4. Self-locking nut connecting lower arm ball joint to knuckle
5. Lower arm mounting nut
6. Lower arm mounting bolt
7. Clamp mounting self-locking nut
8. Clamp mounting bolt (small)
9. Clamp mounting bolt (large)

10. Lower arm clamp mounting self-locking nut
11. Lower arm mounting clamp
12. Lower arm
13. Stopper
14. Rod bushing

Caution

*: Indicates parts which should be temporarily tightened, and then fully tightened with the vehicle on the ground in an unladen condition.

Exploded view of the lower control arm removal procedure—3000GT shown, Diamante is similar

7923PGB9

rear bolt to 71–85 ft. lbs. (98–118 Nm) and the front bolt to the damper fork to 64 ft. lbs. (88 Nm).

10. Torque the bolts for the compression arm to 60 ft. lbs. (83 Nm).

11. Reinstall the crossmember brackets with their mounting bolts. Torque the mounting bolts to 51–58 ft. lbs. (69–78 Nm).

12. Perform an alignment on the vehicle.

MIRAGE

➡ **The suspension components should not be tightened until the vehicle's weight is resting on its wheels.**

1. Before servicing the vehicle, refer to the precautions in the beginning of this section.

2. Raise the vehicle and support safely.

3. Remove or disconnect the following:

- Wheel and tire assembly
- Sway bar links or mounting nuts and bolts from lower control arm. Remove the joint cups and bushings.
- Ball joint stud from the steering knuckle
- Inner lower arm mounting bolt and nut
- Rear mount bolts from the retaining

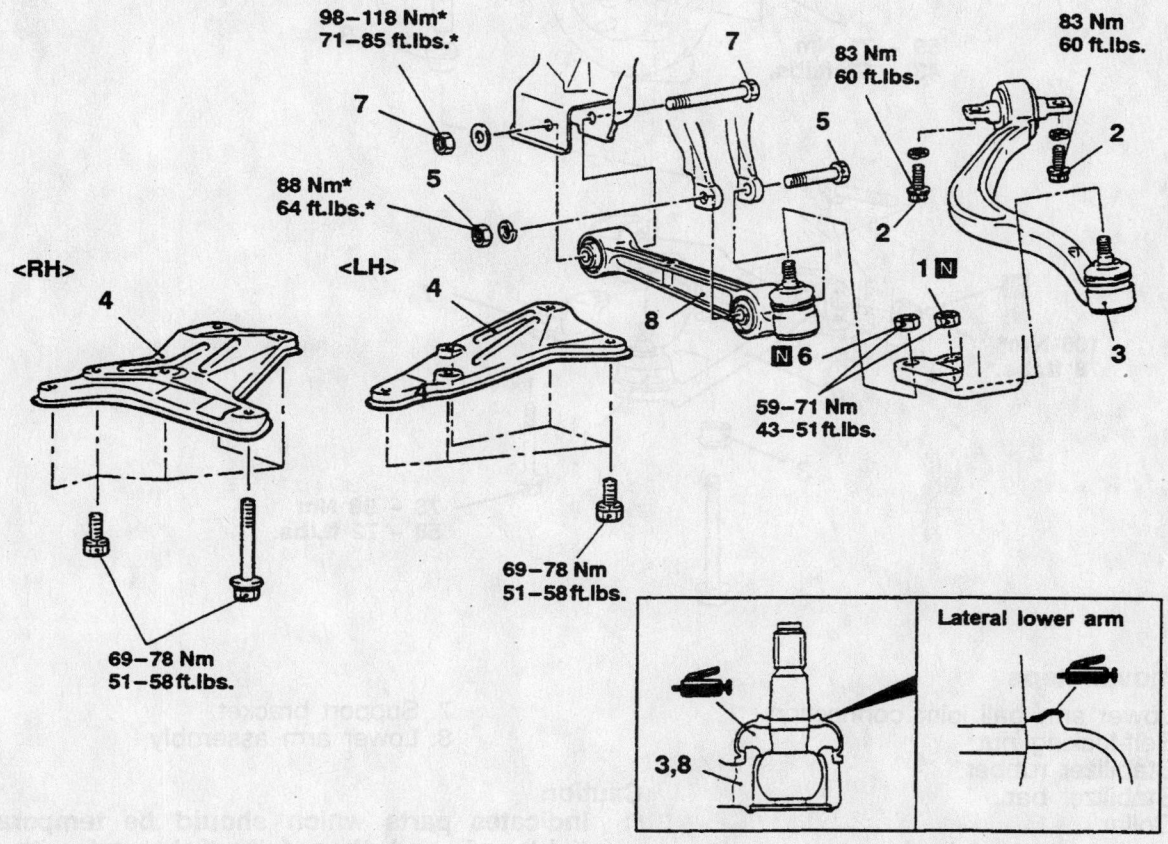

Compression lower arm assembly removal steps

1. Connection for compression lower arm ball joint and knuckle
2. Compression lower arm mounting bolt
3. Compression lower arm assembly

Lateral lower arm assembly removal steps

4. Stay
5. Shock absorber lower mounting bolt and nut
6. Connection for lateral lower arm ball joint and knuckle
7. Lateral lower arm mounting bolt and nut
8. Lateral lower arm assembly

Caution
*: Indicates parts which should be temporarily tightened, and then fully tightened with the vehicle on the ground in the unladen condition.

Exploded view of the lower control arms—Eclipse and Galant

7923PGB8

For Tire, Wheel and Ball Joint specifications, see Section 1 of this manual

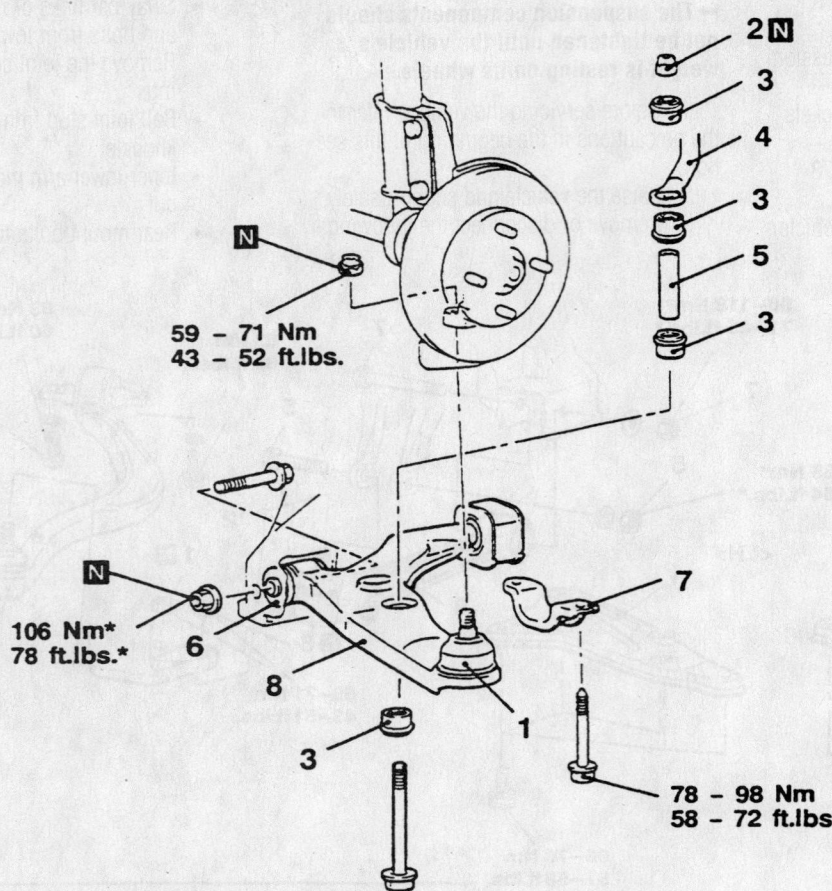

2
3
4
3
5
3

59 – 71 Nm
43 – 52 ft.lbs.

7

106 Nm*
78 ft.lbs.*

6

8

3

1

78 – 98 Nm
58 – 72 ft.lbs.

Removal steps

1. Lower arm ball joint connection
2. Self-locking nut
3. Stabilizer rubber
4. Stabilizer bar
5. Collar
6. Lower arm front bushing
   connection
7. Support bracket
8. Lower arm assembly

Caution
*: Indicates parts which should be temporarily tightened, and then fully tightened with the vehicle on the ground in the unladen condition.

7923PGB0

Lower control arm assembly and related components—Mirage

clamp. Remove the rear retainer clamp if equipped.
- Arm from the vehicle

To install:

4. Install or connect the following:
- Control arm to the vehicle and install the inner mounting bolt. Install new nut and tighten to 78 ft. lbs. (108 Nm).
- Rear mount clamp and bolts. Torque the clamp mounting bolts to 65 ft. lbs. (90 Nm).
- Ball joint stud to the knuckle. Install a new nut and tighten to 43–52 ft. lbs. (60–72 Nm).

- Sway bar and links

5. Lower the vehicle to the floor for the final tightening of the inner frame mount bolt.

6. Install the wheel and tire assembly.

Wheel Bearings

ADJUSTMENT

The front and rear wheel bearings on these vehicles are not adjustable. If the bearings are noisy or become loose, they must be replaced.

REMOVAL & INSTALLATION

Front

3000GT

1. Before servicing the vehicle, refer to the precautions in the beginning of this section.

2. Remove or disconnect the following:
- Negative battery cable
- Cotter pin, halfshaft nut and washer

3. Raise the vehicle and support safely. If equipped with ABS, remove the front

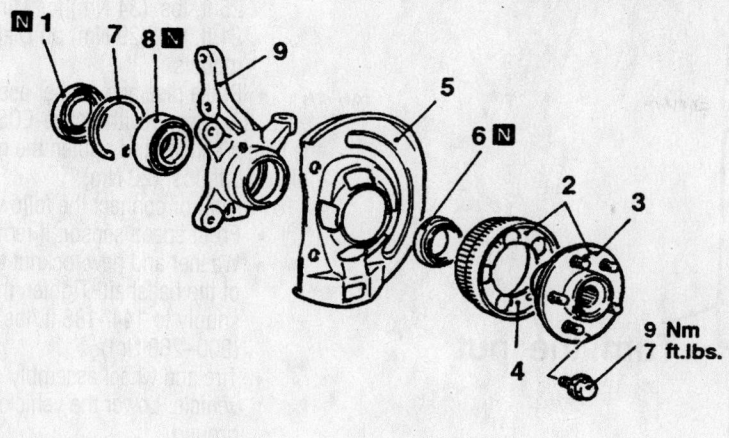

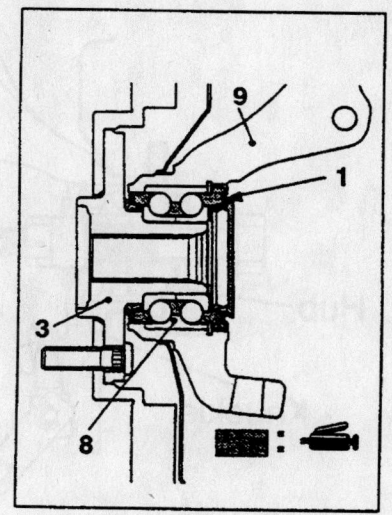

1. Oil seal (drive shaft side)
2. Hub and rotor

3. Hub
4. Rotor <Vehicles with ABS>

5. Dust shield
6. Oil seal (hub side)
7. Snap ring
8. Wheel bearing
9. Knuckle

9 Nm
7 ft.lbs.

7923PGC6

Exploded view of the front hub and bearing—3000GT

wheel speed sensor. Remove the ball joint and tie rod end from the steering knuckle.

4. Remove or disconnect the following:
  • Caliper and brake pads and suspend with a wire
  • Halfshaft from the hub

5. Unbolt the lower end of the strut and remove the hub and steering knuckle assembly.

6. Press the hub from the bearing and remove the bearing races from the knuckle.

To install:

7. Install or connect the following:
  • Bearing and hub into the knuckle assembly with pressing tools, using new parts as required. Check the bearing turning torque. It should be 16 inch lbs. (1.8 Nm) or less.
  • Knuckle assembly to the vehicle and install the strut bolts

8. Apply a thin coat of grease to the lip of the halfshaft side axle seal and drive into place until it contacts the inner bearing outer race.

9. Pull the strut assembly out and install the halfshaft to the hub.

10. Install or connect the following:
  • Washer so the chamfered edge

faces outward. Install the nut and tighten to 166 ft. lbs. (230 Nm).
  • Tie rod end and ball joint
  • Wheel and lower the vehicle

DIAMANTE AND MIRAGE

1. Before servicing the vehicle, refer to the precautions in the beginning of this section.

2. Disconnect the negative battery cable.

3. Raise the vehicle and support safely. Remove the halfshaft nut.

4. If equipped with ABS, remove the front wheel speed sensor.

5. If equipped with Active Electronic Control Suspension (Active-ECS), disconnect the height sensor from the lower control arm.

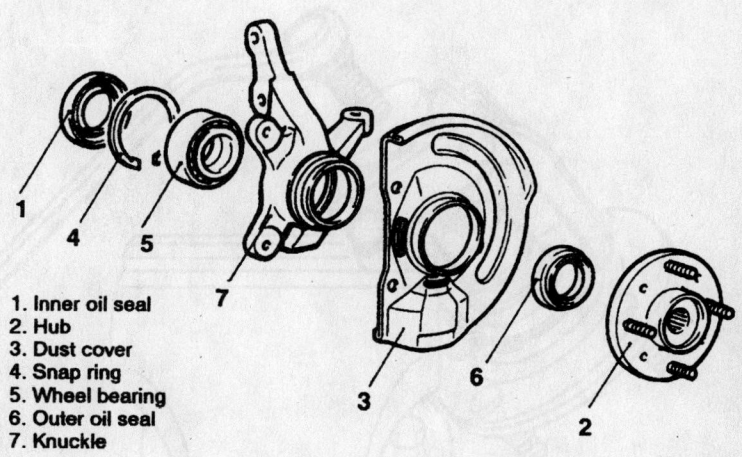

1. Inner oil seal
2. Hub
3. Dust cover
4. Snap ring
5. Wheel bearing
6. Outer oil seal
7. Knuckle

7923PGC1

Front wheel bearing assembly exploded view—Mirage and Diamante

For Wheel Alignment specifications, see Section 1 of this manual

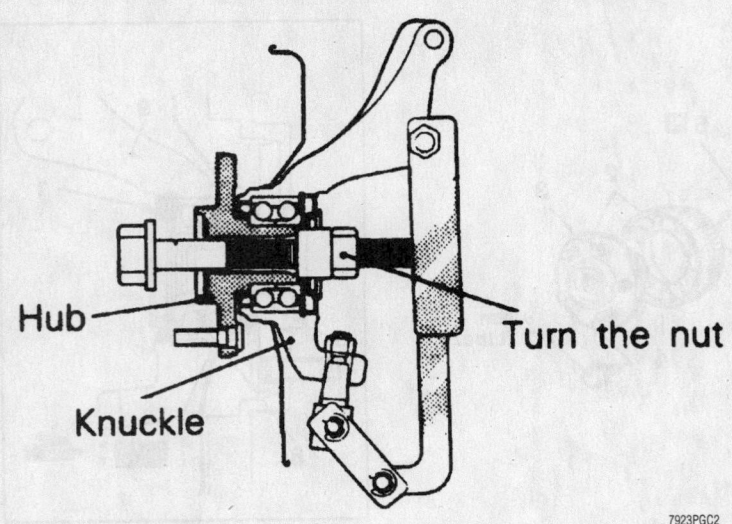

Use of press tool for hub removal—Mirage and Diamante

6. Remove the caliper assembly and brake pads. Suspend the caliper with a wire.

7. Ball joint and tie rod end from the steering knuckle.

8. Remove the halfshaft from the hub.

9. Unbolt the lower end of the strut and remove the hub and steering knuckle assembly from the vehicle.

10. Press the hub from the bearing and remove the bearing races from the knuckle.

To install:

11. Press the wheel bearing into the knuckle. Once the bearing is installed, install the inner race.

12. Install the grease seal.

13. Using a pressing, mount the front hub assembly into the knuckle. Tighten the nut of the pressing tool to 144–188 ft. lbs. (200–260 Nm). Rotate the hub to seat the bearing.

14. Install or connect the following:
- Hub and knuckle assembly onto the vehicle. Install the lower ball joint stud into the steering knuckle and install a new nut. Tighten to 52 ft. lbs. (72 Nm).
- Halfshaft into the hub/knuckle assembly
- 2 front strut lower mounting bolts and tighten to 80–94 ft. lbs. (110–130 Nm) on Mirage or 65–76 ft. lbs. (90–105 Nm) on Diamante models.
- Tie rod end and tighten the nut to 25 ft. lbs. (34 Nm) for Mirage and 21 ft. lbs. (29 Nm) on Diamante models
- Brake disc and caliper assembly

15. If equipped with Active-ECS, connect the height sensor and tighten the mounting bolt to 15 ft. lbs. (20 Nm).

16. Install or connect the following:
- Front speed sensor, if removed
- Washer and new locknut to the end of the halfshaft. Tighten the locknut snugly to 144–188 ft. lbs. (200–260 Nm).
- Tire and wheel assembly onto the vehicle. Lower the vehicle to the ground.

ECLIPSE

1. Before servicing the vehicle, refer to the precautions in the beginning of this section.

2. Raise and safely support the vehicle.

3. Remove or disconnect the following:
- Front wheel
- Axle nut
- Wheel speed sensor, vehicles with ABS
- Caliper and suspend it out of the way with wire or string
- Brake rotor
- Steering knuckle from the upper arm

4. Pull the knuckle away from the vehicle to access the hub mounting bolts on the inboard side of the hub. Be careful not to damage the ball joint boot or the ABS rotor if equipped.

5. Remove the mounting bolts and the front hub assembly.

➡**Do not disassemble the hub assembly. If binding or damaged, it must be replaced as a unit.**

To install:

6. Install or connect the following:
- Hub to the knuckle. Torque the mounting bolts to 65 ft. lbs. (88 Nm).
- Knuckle to the upper arm
- Brake rotor and the caliper
- Wheel speed sensor if removed
- Axle nut and tighten to 145–188 ft. lbs. (196–255 Nm)
- Wheel and lower the vehicle to the floor

GALANT

1. Before servicing the vehicle, refer to the precautions in the beginning of this section.

2. Raise the vehicle and support safely.

3. Remove or disconnect the following:

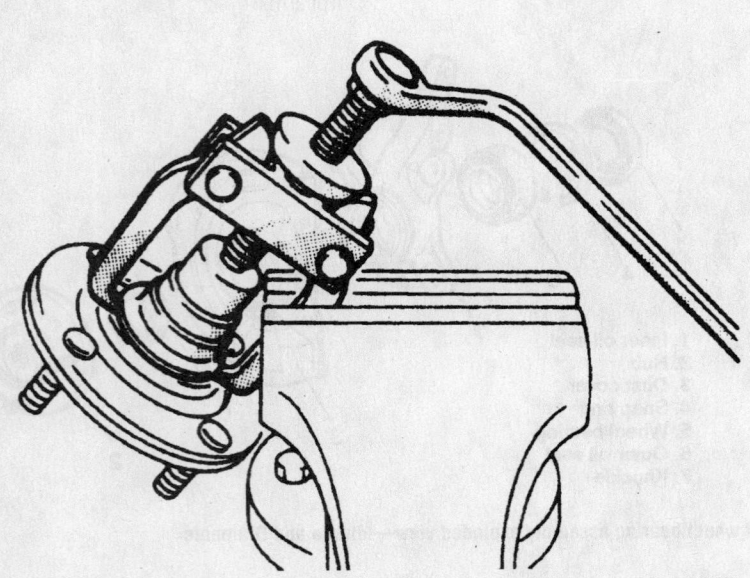

Removing inner race from hub—Mirage and Diamante

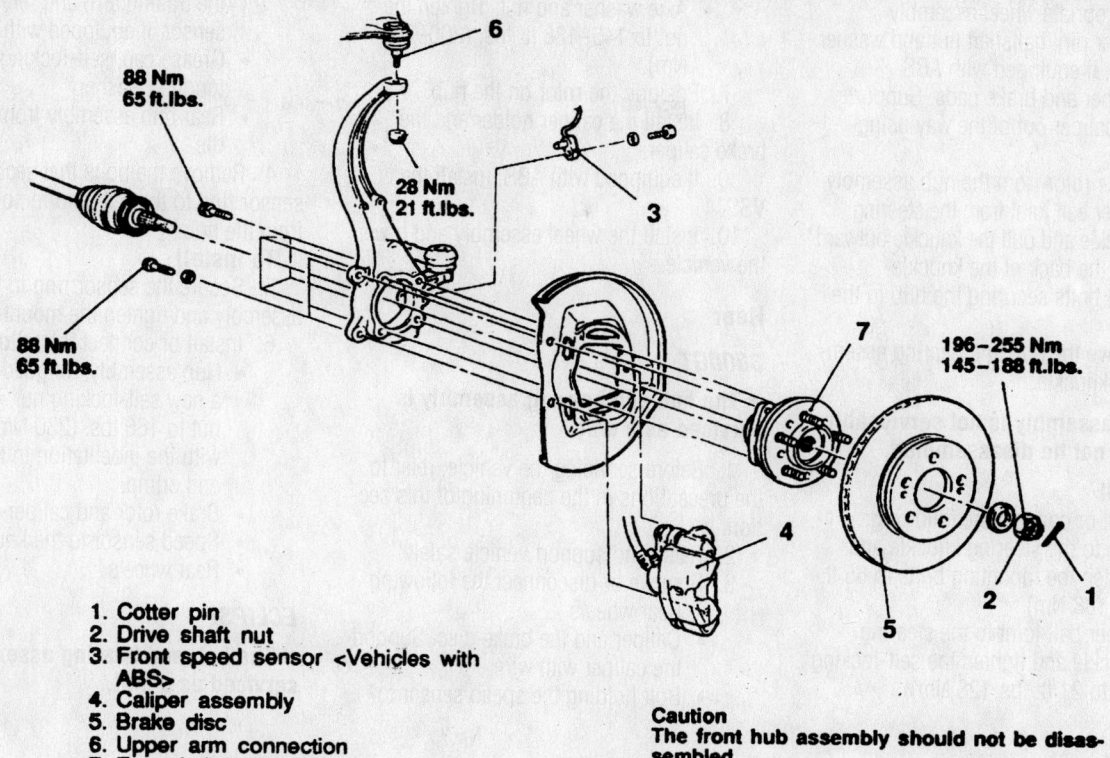

88 Nm
65 ft.lbs.

28 Nm
21 ft.lbs.

6

3

88 Nm
65 ft.lbs.

7

196–255 Nm
145–188 ft.lbs.

4

2   1

5

1. Cotter pin
2. Drive shaft nut
3. Front speed sensor <Vehicles with ABS>
4. Caliper assembly
5. Brake disc
6. Upper arm connection
7. Front hub assembly

Caution
The front hub assembly should not be disassembled.

7923PGC4

Front hub and related components—Eclipse

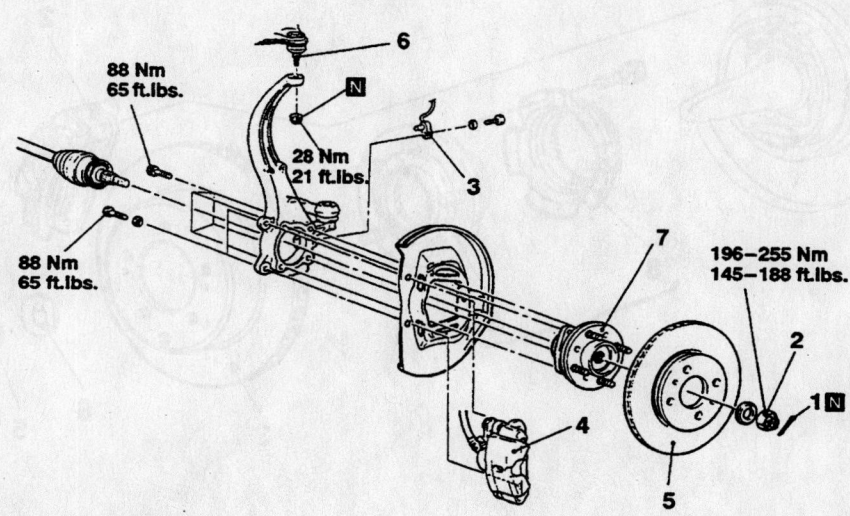

88 Nm
65 ft.lbs.

6

28 Nm
21 ft.lbs.

3

88 Nm
65 ft.lbs.

7

196–255 Nm
145–188 ft.lbs.

4

2

1

5

Removal steps

1. Cotter pin
2. Drive shaft nut
3. Front speed sensor <Vehicles with ABS>
4. Caliper assembly

5. Brake disc
6. Connection for upper arm
7. Front hub assembly

Caution
The front hub assembly should not be disassembled.

7923PGC5

Exploded view of the front hub removal—Galant

- Appropriate wheel assembly
- Cotter pin, halfshaft nut and washer
- VSS, if equipped with ABS
- Caliper and brake pads. Support the caliper out of the way using wire.
- Brake rotor from the hub assembly
- Upper ball joint from the steering knuckle and pull the knuckle outward

4. From the back of the knuckle, remove the 4 bolts securing the hub to the knuckle.

5. Remove the hub and bearing assembly from the knuckle.

➡**The hub assembly is not serviceable and should not be disassembled.**

To install

6. Install or connect the following:
- Hub to the steering knuckle and tighten the mounting bolts to 65 ft. lbs. (88 Nm)
- Upper ball joint to the steering knuckle and tighten the self-locking nut to 21 ft. lbs. (28 Nm)

- Axle washer and nut. Tighten the nut to 145–188 ft. lbs. (200–260 Nm).

7. Position the rotor on the hub.

8. Install the caliper holder and the brake caliper.

9. If equipped with ABS, install the VSS.

10. Install the wheel assembly and lower the vehicle.

Rear

3000GT

➡**The hub and bearing assembly is serviced as a unit.**

1. Before servicing the vehicle, refer to the precautions in the beginning of this section.

2. Raise and support vehicle safely.

3. Remove or disconnect the following:
- Rear wheels
- Caliper and the brake disc. Support the caliper with wire.
- Bolt holding the speed sensor to

the trailing arm and remove the sensor, if equipped with ABS
- Grease cap, self-locking nut and tongued washer
- Rear hub assembly from the spindle

4. Remove the bolts that secure the ABS sensor ring to the hub and remove the ring from the hub.

To install

5. Secure the sensor ring to the hub assembly and tighten the mounting bolts.

6. Install or connect the following:
- Hub assembly, tongued washer and a new self-locking nut. Torque the nut to 166 lbs. (230 Nm), align with the indentation in the spindle, and crimp.
- Brake rotor and caliper assembly
- Speed sensor to the knuckle
- Rear wheels

ECLIPSE

➡**The hub and bearing assembly is serviced as a unit.**

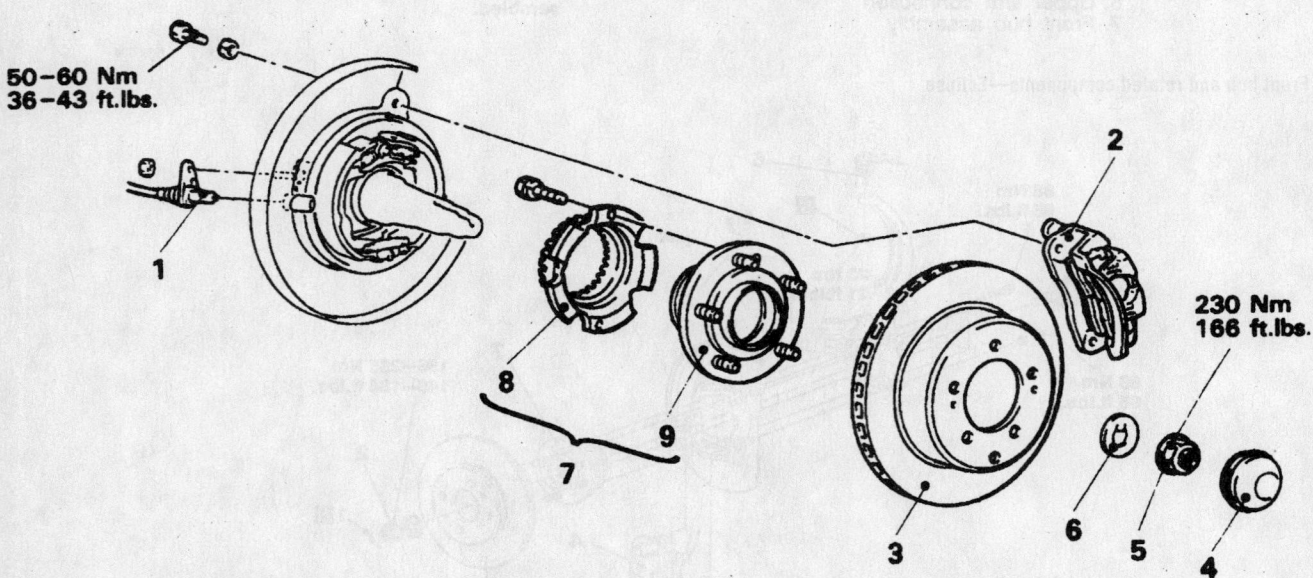

50–60 Nm
36–43 ft.lbs.

230 Nm
166 ft.lbs.

1. Rear speed sensor <Vehicles with ABS>
2. Caliper assembly
3. Brake disc
4. Hub cap
5. Flange nut
6. Tongued washer
7. Rear hub assembly
8. Rear rotor <Vehicles with ABS>
9. Rear hub unit bearing

Exploded view of the rear hub/bearing assembly removal—Eclipse

7923PGC9

1. Before servicing the vehicle, refer to the precautions in the beginning of this section.

2. Disconnect the negative battery cable.

3. Raise and safely support the vehicle.

4. Remove or disconnect the following:
- Wheel and tire assembly
- Rear wheel speed sensor if equipped with ABS
- Brake drum. Or, if equipped with disc brakes, remove the caliper assembly and rotor. Suspend the caliper out of the way with wire.

5. On vehicles with rear disc brakes, remove the parking brake shoes.

6. On vehicles equipped with AWD, remove the axle shaft locking nut, and using a suitable tool, separate the hub from the axle shaft.

7. Remove the hub mounting bolts from behind the backing plate and remove the hub.

➡**The rotor for the ABS must be removed and installed using a press.**

To install:

8. Press the rotor (ABS) to the hub.

9. On vehicles with AWD, engage the splines of the axle shaft with the hub assembly and tighten the axle shaft locking nut to 145–188 ft. lbs. (196–255 Nm).

10. Install or connect the following:
- Hub and tighten the mounting bolts to 54–65 ft. lbs. (74–88 Nm)
- Parking brake shoes if equipped

- Rotor and caliper or drum
- Speed sensor if equipped
- Wheel and tire assembly

11. Lower the vehicle to the floor.

12. Connect the negative battery cable.

GALANT AND DIAMANTE

1. Before servicing the vehicle, refer to the precautions in the beginning of this section.

2. Raise the vehicle and support safely.

3. Remove the appropriate wheel assembly.

4. If equipped with ABS, remove the vehicle speed sensor.

5. Remove the brake drum from the hub assembly.

6. From the back of the knuckle, remove the 4 bolts securing the hub to the knuckle.

7. Remove the hub and bearing assembly from the knuckle.

➡**The hub assembly is not serviceable and should not be disassembled.**

8. If replacing the hub, use special socket MB991248 and a press, to remove the wheel sensor rotor from the hub.

To install

9. Press the wheel sensor rotor onto the hub.

10. Install or connect the following:
- Hub to the knuckle and tighten the mounting bolts to 54–65 ft. lbs. (74–88 Nm)

- Brake drum on the hub
- Vehicle speed sensor, if equipped with ABS
- Wheel assembly and lower the vehicle

MIRAGE

➡**The wheel bearing is serviced by replacement of the hub.**

1. Before servicing the vehicle, refer to the precautions in the beginning of this section.

2. If equipped with ABS, remove the wheel speed sensor.

3. Raise and safely support the vehicle.

4. Remove or disconnect the following:
- Rear wheel
- Caliper and brake disc or brake drum
- Dust cap and flange nut
- Rear hub assembly

To install:

5. Install or connect the following:
- Rear hub assembly using a new flange nut. Torque the flange nut to 130 ft. lbs. (180 Nm).
- Dust cap
- Wheel speed sensor if removed. The air gap should be 0.012–0.035 in. (0.3–0.9mm).
- Brake disc and caliper, or brake drum
- Rear wheel assembly and lower the vehicle to the floor

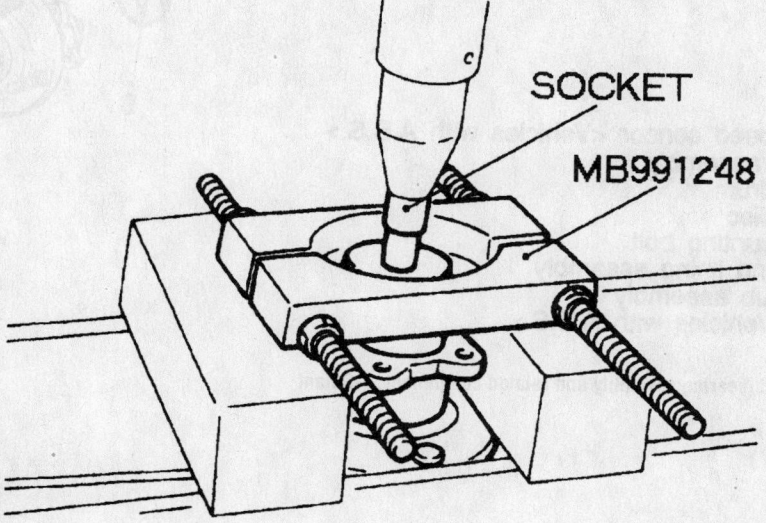

SOCKET

MB991248

7923PGC7

Use a press to remove the speed sensor rotor from the hub—Galant

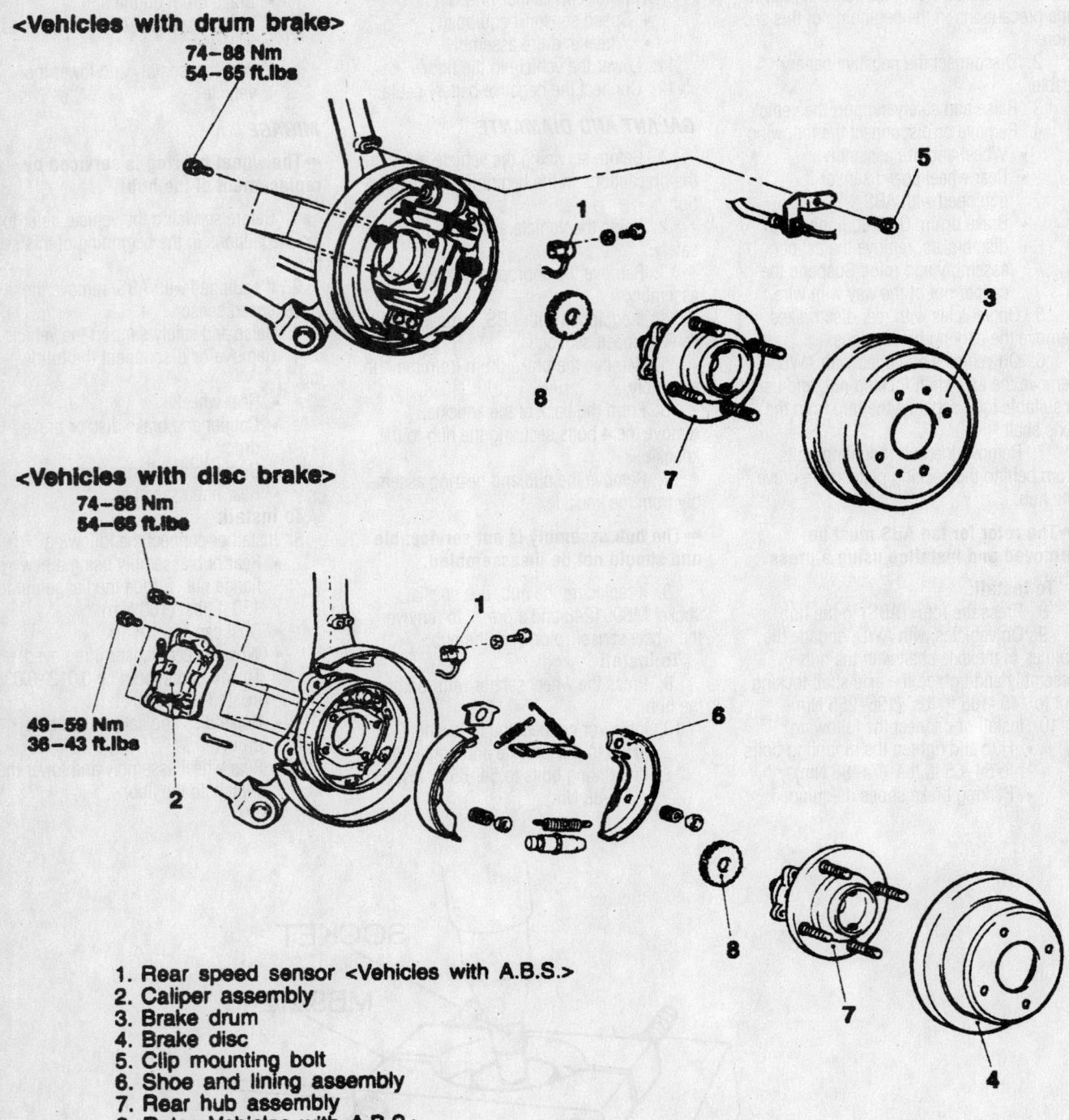

<Vehicles with drum brake>

74—88 Nm
54—65 ft.lbs

<Vehicles with disc brake>

74—88 Nm
54—65 ft.lbs

49—59 Nm
36—43 ft.lbs

1. Rear speed sensor <Vehicles with A.B.S.>
2. Caliper assembly
3. Brake drum
4. Brake disc
5. Clip mounting bolt
6. Shoe and lining assembly
7. Rear hub assembly
8. Rotor<Vehicles with A.B.S.>

7923PGC8

Exploded view of the rear hub/bearing assembly and related components—Galant

PRECAUTIONS

Before servicing any vehicle, please be sure to read all of the following precautions, which deal with personal safety, prevention of component damage and important points to take into consideration when servicing a motor vehicle:

• Never open, service or drain the radiator or cooling system when the engine is hot; serious burns can occur from the steam and hot coolant.

• Observe all applicable safety precautions when working around fuel. Whenever servicing the fuel system, always work in a well-ventilated area. Do not allow fuel spray or vapors to come in contact with a spark, open flame, or excessive heat (a hot drop light, for example). Keep a dry chemical fire extinguisher near the work area. Always keep fuel in a container specifically designed for fuel storage; also, always properly seal fuel containers to avoid the possibility of fire or explosion. Refer to the additional fuel system precautions later in this section.

• Fuel injection systems often remain pressurized, even after the engine has been turned OFF. The fuel system pressure must be relieved before disconnecting any fuel lines. Failure to do so may result in fire and/or personal injury.

• Brake fluid often contains polyglycol ethers and polyglycols. Avoid contact with the eyes and wash your hands thoroughly after handling brake fluid. If you do get brake fluid in your eyes, flush your eyes with clean, running water for 15 minutes. If

eye irritation persists, or if you have taken brake fluid internally, IMMEDIATELY seek medical assistance.

• The EPA warns that prolonged contact with used engine oil may cause a number of skin disorders, including cancer! You should make every effort to minimize your exposure to used engine oil. Protective gloves should be worn when changing oil. Wash your hands and any other exposed skin areas as soon as possible after exposure to used engine oil. Soap and water, or waterless hand cleaner should be used.

• All new vehicles are now equipped with an air bag system. The system must be disabled before performing service on or around system components, steering column, instrument panel components, wiring and sensors. Failure to follow safety and disabling procedures could result in accidental air bag deployment, possible personal injury and unnecessary system repairs.

• Always wear safety goggles when working with, or around, the air bag system. When carrying a non-deployed air bag, be sure the bag and trim cover are pointed away from your body. When placing a non-deployed air bag on a work surface, always face the bag and trim cover upward, away from the surface. This will reduce the motion of the module if it is accidentally deployed. Refer to the additional air bag system precautions later in this section.

• Clean, high quality brake fluid from a sealed container is essential to the safe and

proper operation of the brake system. You should always buy the correct type of brake fluid for your vehicle. If the brake fluid becomes contaminated, completely flush the system with new fluid. Never reuse any brake fluid. Any brake fluid that is removed from the system should be discarded. Also, do not allow any brake fluid to come in contact with a painted surface; it will damage the paint.

• Never operate the engine without the proper amount and type of engine oil; doing so WILL result in severe engine damage.

• Timing belt maintenance is extremely important! Many models utilize an interference-type, non-freewheeling engine. If the timing belt breaks, the valves in the cylinder head may strike the pistons, causing potentially serious (also time-consuming and expensive) engine damage. Refer to the maintenance interval charts in the front of this manual for the recommended replacement interval for the timing belt and to the timing belt section for belt replacement and inspection.

• Disconnecting the negative battery cable on some vehicles may interfere with the functions of the on-board computer system(s) and may require the computer to undergo a relearning process once the negative battery cable is reconnected.

• When servicing drum brakes, only disassemble and assemble one side at a time, leaving the remaining side intact for reference.

ENGINE REPAIR

Distributor

REMOVAL

The Nissan 3.0L engines are equipped with a Distributorless Ignition System (DIS).

1.6L, 1.8L, 2.0L and 2.4L Engines

1. Before servicing the vehicle, refer to the precautions in the beginning of this section.

2. Set the engine to Top Dead Center (TDC) with the No. 1 piston on compression stroke.

3. Remove or disconnect the following:
   • Negative battery cable
   • Distributor spark plug wires from the distributor cap
   • Distributor cap. Scribe a mark on the engine block to show the rotor

and distributor position prior to removal.
   • Wiring connections to the distributor
   • Bolt(s) holding distributor to engine
   • Distributor by pulling it upward from the cylinder block

➡Do not disturb the camshaft or crankshaft position after the distributor is removed from the engine. If any of these components are moved, TDC on cylinder No. 1 will have to be found again before reinstalling the distributor.

INSTALLATION

1.6L, 1.8L, 2.0L and 2.4L Engines

ENGINE NOT DISTURBED

1. Install or connect the following:
   • New distributor housing O-ring

   • Distributor so the rotor is aligned with the matchmark on the housing and the housing is aligned with the matchmark on the engine.

➡Be sure the distributor is fully seated and the distributor gear is fully engaged.

   • Snug the hold-down bolt
   • Distributor pick-up lead wires
   • Distributor cap and tighten the screws
   • Splash shield
   • Spark plug wires
   • Negative battery cable

2. After the ignition timing has been adjusted, tighten the hold-down bolt(s) as follows:
   • GA16DE and VG30E engines: 80–104 inch lbs. (9–11 Nm)

- SR20DE and KA24DE (Altima) engines: 108–144 inch lbs. (13–16 Nm)
- KA24DE (240SX) engine: 34–39 inch lbs. (4–5 Nm)

ENGINE DISTURBED

1. Install the a new distributor housing O-ring

2. Position the engine so the No. 1 piston is at Top Dead Center (TDC) of its compression stroke and the mark on the vibration damper is aligned with 0 on the timing indicator.

- Distributor in the engine so the rotor is aligned with the position of the No. 1 ignition wire on the distributor cap. Be sure the distributor is fully seated and that the distributor shaft is fully engaged.
- Snug the hold-down bolt
- Distributor pick-up lead wires
- Distributor cap and tighten the screws. Install the splash shield, if equipped.
- Spark plug wires
- Negative battery cable.

3. After the ignition timing has been adjusted, tighten the hold-down bolt(s) as follows:

- GA16DE and VG30E engines: 80–104 inch lbs. (9–11 Nm)
- SR20DE and KA24DE (Altima) engines: 108–144 inch lbs. (13–16 Nm)
- KA24DE (240SX) engine: 34–39 inch lbs. (4–5 Nm)

Ignition Timing

ADJUSTMENT

1.6L, 1.8L, 2.0L and 2.4L Engines

Visually check the air cleaner, intake hoses, ducts, Exhaust Gar Recirculation (EGR) valve operation and electrical connections prior to the adjustment of the ignition timing. Correct or repair any problem as required. Be sure to inspect the throttle valve and the Throttle Position (TP) sensor for proper operation.

1. Before servicing the vehicle, refer to the precautions in the beginning of this section.

2. Locate the timing marks on the crankshaft pulley and the front of the engine.

3. Clean the timing marks.

4. Using chalk or white paint, color the mark on the crankshaft pulley and the mark on the scale which will indicate the correct timing when aligned with the notch on the crankshaft pulley.

5. Attach a tachometer to the engine.

6. Attach a timing light to the engine, to No. 1 cylinder ignition wire.

7. Check to be sure all of the wires clear the fan; start the engine and allow it to reach normal operating temperatures.

8. Block the front wheels and set the parking brake. Shift the transmission into NEUTRAL for automatic and manual transaxles; do not stand in front of the vehicle when making adjustments.

9. Perform the following procedures:
   a. Race the engine at 2000 rpm for about 2 minutes under a no-load condition; be sure all of the accessories are turned off.
   b. Perform on board engine diagnostics and repair any fault code.
   c. Race the engine 2–3 times under no-load, then run the engine it for 1 minute at idle.
   d. Stop the engine and disconnect the Throttle Position (TP) sensor.
   e. Race the engine at 2000 rpm for about 2 minutes under a no-load condition; be sure all of the accessories are turned OFF.
   f. Run the engine at idle speed.

10. Aim the timing light at the timing marks. If the marks on the pulley and the engine are aligned when the light flashes, the timing is correct. The correct ignition timing is as follows:

- 1.6L (GA16DE): 6–10 degrees Before Top Dead Center (BTDC)
- 1.8L (QG18DE): 4–14 degrees Before Top Dead Center (BTDC)
- 2.0L (SR20DE): 13–17 degrees BTDC
- 2.4L (KA24DE): 18–22 degrees BTDC

11. Turn the engine OFF and remove the tachometer and the timing light. If the marks are not in alignment, proceed with the following steps.

12. Turn the engine OFF.

13. Loosen the bolts that secure the distributor just enough so it can be turned.

14. Start the engine. Keep the wires of the timing light clear of the cooling fan.

15. With the timing light aimed at the pulley and the marks on the engine, turn the distributor for the proper adjustment.

16. Race the engine 2–3 times under no-load, then run the engine it for 1 minute at idle.

17. Aim the timing light at the timing marks. If the marks on the pulley and the engine are aligned when the light flashes, the timing is correct.

18. Tighten the bolt that secures the distributor and recheck the timing.

19. Turn the engine OFF and remove the tachometer and the timing light.

20. Connect the TP sensor.

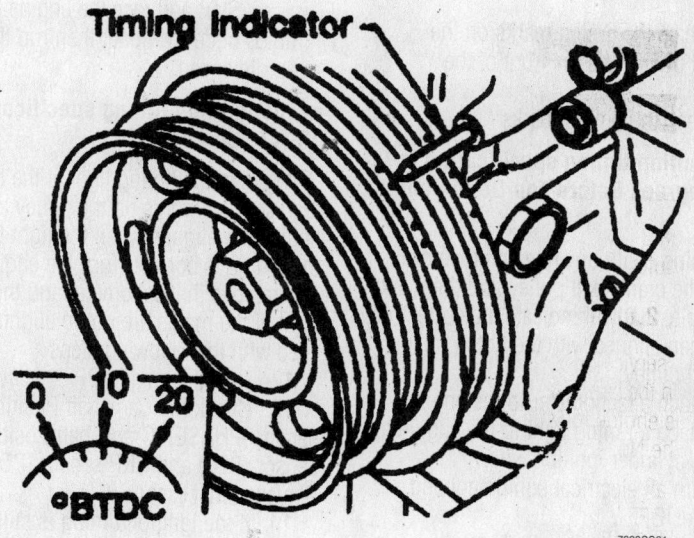

Point the timing light at the crankshaft pulley to see the timing marks—1.6L, 1.8L and 2.0L engines

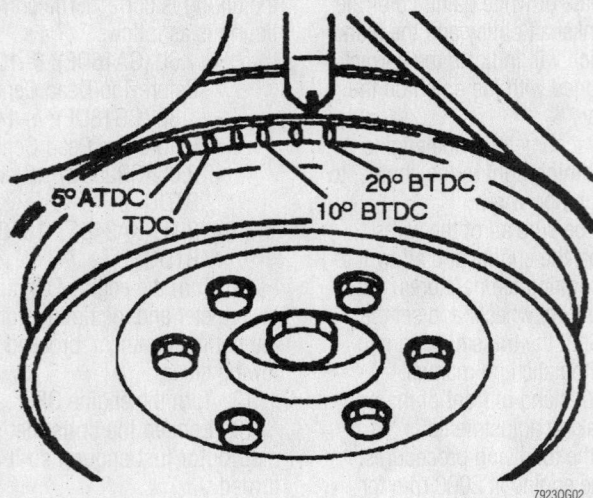

The timing marks are located on the crankshaft pulley—2.4L engine

3.0L Engine

➡**The ignition timing is not adjustable. If not within specifications, further diagnostic inspection is required. The following procedure is for viewing the ignition timing setting.**

Visually check the air cleaner, intake hoses, ducts, Exhaust Gas Recirculation (EGR) valve operation and electrical connections prior to the adjustment of the ignition timing. Correct or repair any problem as required. Be sure to inspect the throttle valve and Throttle Position (TP) sensor for proper operation.

1. Before servicing the vehicle, refer to the precautions in the beginning of this section.

2. Locate the timing marks on the crankshaft pulley and the front of the engine.

3. Clean the timing marks.

➡**The ignition timing specification is 13–17 degrees Before Top Dead Center (BTDC).**

4. Using chalk or white paint, color the mark on the crankshaft pulley and the mark on the scale, that will indicate the correct timing when aligned with the notch on the crankshaft pulley.

5. Attach a tachometer to the engine.

6. Attach a timing light to the engine to number 1 cylinder ignition wire.

7. Turn all electrical equipment and accessories OFF.

8. Check to be sure all of the wires clear the fan, then, start the engine and allow it to reach normal operating temperatures.

9. Block the front wheels and set the parking brake. Shift the transmission into NEUTRAL for manual transmission and automatic transmissions. Do not stand in front of the vehicle when making adjustments.

10. Perform the following procedures:

a. Race the engine at 2000 rpm for about 2 minutes under a no-load condition; be sure all of the accessories are turned OFF.

b. Perform on board engine diagnostics and repair any fault code.

c. Race the engine at 2000 rpm for about 2 minutes under a no-load condition.

d. Turn the engine OFF and disconnect the TP sensor.

e. Start and race the engine 2–3 times under no-load, then run the engine at idle speed.

➡**The ignition timing specification is 13–17 degrees BTDC.**

11. Aim the timing light at the timing marks. If the marks on the pulley and the engine are aligned when the light flashes, the timing is correct. Turn the engine OFF and remove the tachometer and the timing light. If the marks are not in alignment, proceed with the following steps.

12. Turn the engine OFF.

13. Check the Camshaft Position (CMP) sensor (PHASE), Crankshaft Position (CKP) sensor (REF) and CKP sensor (POS). Replace if necessary.

14. If the ignition timing is still not correct, substitute a known good Electronic Control Module (ECM).

➡**The ECM may be the cause of the problem but this is rarely the case.**

15. Turn the engine OFF and remove the tachometer and the timing light.

Alternator

REMOVAL

240SX

1. Before servicing the vehicle, refer to the precautions in the beginning of this section.

2. Remove or disconnect the following:
- Negative battery cable
- Engine undercover
- Drive belt
- Harness connector from alternator
- Cooling fan lower shroud
- Alternator

1.6L, 1.8L and 2.0L Engines

2.0L Engines
1. Before servicing the vehicle, refer to the precautions in the beginning of this section.

2. Remove or disconnect the following:
- Negative battery cable
- 2 lead wires and connector from the alternator
- Drive belt adjusting bolt, loosen only
- Drive belt
- Alternator

3.0L Engines

1. Before servicing the vehicle, refer to the precautions in the beginning of this section.

2. Remove or disconnect the following:
- Negative battery cable
- Splash guard on the right side of the vehicle
- Dive belt
- Four A/C mounting bolts
- Radiator fan and shroud
- A/C compressor forward
- Alternator harness connector
- Alternator mounting bolts and lower the alternator from the vehicle

Altima

1. Before servicing the vehicle, refer to the precautions in the beginning of this section.

2. Drain the cooling system below the upper radiator hose level.

3. Remove or disconnect the following:
- Negative battery cable
- Upper radiator hose

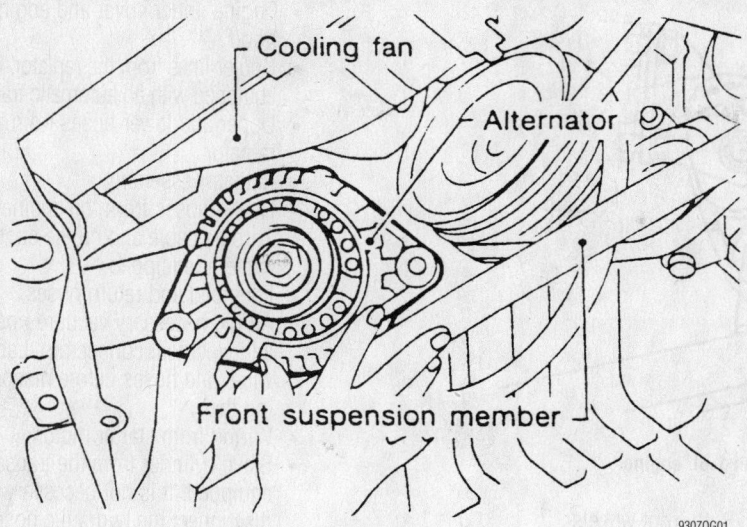

Removing the alternator—240SX engines

- Alternator electrical harness, harness stay and the harness-to-A/C compressor
- Throttle cable
- Loosen the adjusting bolt
- Accessory drive belt
- Alternator mounting bolts
- Alternator from the engine

INSTALLATION

240SX

1. Install or connect the following:
   - Alternator and torque the mounting bolts to 15 inch lbs. (20 Nm)
   - Cooling fan lower shroud
   - Harness connector from alternator
   - Drive belt
   - Engine undercover
   - Negative and battery cable

1.6L, 1.8L and 2.0L Engines

1. Install or connect the following:
   - Alternator and retaining bolts loosely
   - Belt and connect the wiring
2. Adjust the drive belt.
3. Torque the retaining bolts to 25 ft. lbs. (34 Nm).
4. Connect the negative battery cable.

3.0L Engines

Install or connect the following:
- Alternator and torque the bolts to 38 ft. lbs. (52 Nm)
- Alternator harness connector

- A/C compressor back into location
- Radiator cooling fan and shroud
- A/C compressor mounting bolts
- Drive belt and tension the belt
- Splash guard
- Negative battery cable

➡**Proper belt tension is important. A belt that is too tight may cause alternator bearing failure; one that is too loose will cause a gradual battery discharge and/or belt slippage, resulting in belt breakage from overheating.**

Altima

1. Install or connect the following:
   - Alternator and torque the bolts to 11–15 ft. lbs. (16–20 Nm)
   - Harness-to-A/C compressor, harness stay and the alternator electrical harness
   - Throttle cable
   - Drive belt. Properly tension the belt
   - Upper radiator hose
   - Negative battery cable
2. Top off the cooling system.
3. Start the vehicle, check for leaks and repair if necessary.

Engine Assembly

REMOVAL & INSTALLATION

Sentra and 200SX

➡**The engine and transaxle are removed as one unit from the underside of the vehicle.**

1. Before servicing the vehicle, refer to the precautions in the beginning of this section.
2. Relieve the fuel system pressure.
3. Drain the coolant from the radiator and the engine block.
4. Drain the engine oil.
5. Remove or disconnect the following:
   - Negative and positive battery cables
   - Battery and tray from the vehicle
   - Both front wheels
   - Engine undercovers and the engine side covers
   - Air cleaner assembly and air duct
   - Vacuum hoses. Make sure to note the locations prior to disconnection them
   - Heater hoses from the engine
   - Automatic transmission cooler hoses from the transaxle, if equipped
   - Power steering hoses
   - Fuel hoses from the engine
   - Harness and wiring connections. Make sure to note the locations prior to disconnection them.
   - Throttle cable and the cruise control cable
   - Control cable, if equipped with an automatic transmission
   - Cooling fans, radiator and the recovery tank
   - Front halfshafts from the vehicle
   - Front exhaust pipe
   - Control rod and support rod from the transaxle, 1.6L engine only
   - Starter motor and intake manifold support brackets
   - Engine drive belts
   - Alternator and adjusting brackets
   - Power steering pump and A/C compressor. It is not necessary to disconnect the lines.
   - Cylinder head front mounting bracket, 1.6L engine only
6. Position a transmission jack under the transaxle and support the engine with engine slinger.
7. Remove or disconnect the following:
   - Center crossmember
   - Front stabilizer bar, if necessary
   - Engine mounting bolts from both sides of the engine
8. Slowly lower the jacking devices and remove the engine and transaxle from the vehicle.

To install:
9. Install or connect the following:

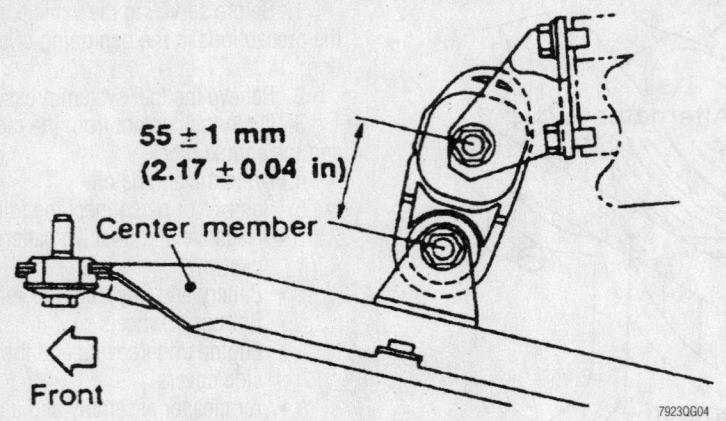

55 ± 1 mm
(2.17 ± 0.04 in)

Center member

Front

7923QG04

Height adjustment of the buffer rod—Sentra and 200SX with 1.6L engine

- Engine and transaxle assembly
- Mounting bolts to both sides of the engine and torque the bolts to 44 ft. lbs. (60 Nm)

10. For vehicles with manual transaxles, adjust the height of the mounting bracket (buffer rod). The distance between the 2 through-bolts should be 2.13–2.20 in. (54–56mm).

11. Install or connect the following:
- Center crossmember and torque the bolts to 40 ft. lbs. (54 Nm)

12. Remove the engine support jacks and engine slinger.

13. Install or connect the following:
- Cylinder head front mounting bracket, 1.6L engine only
- A/C compressor and power steering pump
- Alternator and brackets
- Starter motor and intake manifold support bracket
- Control and support rod, 1.6L engine only. Torque the control rod bolt to 10–13 ft. lbs. (14–18 Nm) and the support rod bolt to 26–35 ft. lbs. (35–47 Nm)
- Front exhaust pipe
- Drive belts
- Both front halfshafts
- Radiator, cooling fans and recovery tank
- Control cable, automatic transmissions only
- Throttle and cruise control cables, if equipped
- Wiring harness and electrical connections
- Power steering hoses and fuel line
- Transmission cooler lines, if equipped
- Vacuum hoses
- Air cleaner assembly
- Engine side and under covers

- Both front wheels
- Battery tray and battery
- Both battery cables

14. Fill the engine with clean oil.
15. Fill the cooling system.
16. Start the engine and check for leaks. Make all the necessary adjustments.

Altima

➡**The engine and transaxle must be removed as a single unit. The engine and transaxle are removed from under the vehicle.**

1. Before servicing the vehicle, refer to the precautions in the beginning of this section.
2. Release fuel system pressure.
3. Drain the cooling system.
4. Drain the engine oil.
5. Remove or disconnect the following:
- Battery cables and the battery tray
- Air cleaner assembly
- Both front wheels

- Engine under cover and engine hood
- Cooler lines from the radiator, if equipped with an automatic transaxle
- Upper and lower hoses from the radiator
- Radiator assembly
- Heater hoses from the engine
- Throttle cable and cruise control cable, if equipped
- Fuel feed and return hoses
- All the necessary vacuum hoses and electrical connectors. Label all wires and hoses before disconnecting them
- Wiring from starter motor
- Slave cylinder from the transaxle, if equipped. It is not necessary to disconnect the hydraulic hose.
- Engine drive belts. Be sure to mark belts for reinstallation
- Alternator, A/C compressor and the power steering pump
- Both halfshafts from the transaxle and support the engine with slinger and support the transaxle with proper jack
- Left and right engine mounting through-bolts
- Crossmember
- Front and rear engine mounts

6. Lower the transaxle and engine assembly from the vehicle.

➡**The engine and transaxle assembly should be removed through the bottom of the vehicle. Do not attempt to remove the assembly from above.**

To install:

7. Raise the transaxle and engine assembly to the vehicle.

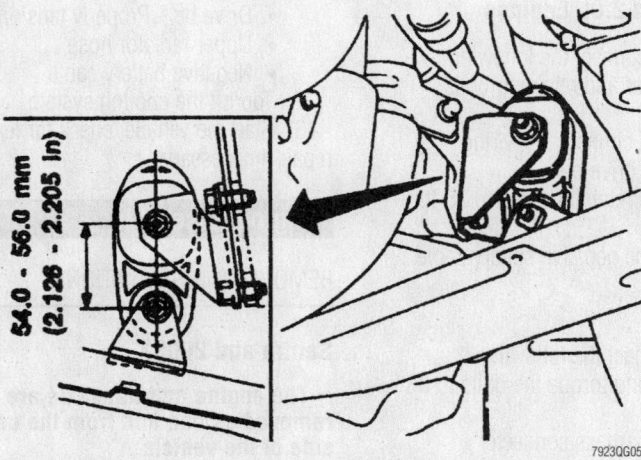

54.0 - 56.0 mm
(2.126 - 2.205 in)

7923QG05

Be sure to adjust the height of the engine mount for manual transmission vehicles—2.0L engine

8. Install or connect the following:
- Front and rear engine mounts. Torque the mounting bolts to 55 ft. lbs. (75 Nm)
- Crossmember and torque the bolts to 57–72 ft. lbs. (77–98 Nm)
- Left and right engine mounting through bolts. Torque the bolts to 72 ft. lbs. (98 Nm)

9. Remove the engine and transaxle support jacks.
- Both halfshafts
- Power steering pump, A/C compressor and alternator
- Slave cylinder, if equipped
- Starter motor
- Drive belts
- Vacuum hoses and electrical connectors
- Fuel feed and return lines
- Throttle and cruise control cables, if equipped
- Radiator
- Heater and radiator hoses
- Cooler lines, if equipped
- Engine side and under covers
- Both front wheels
- Air cleaner assembly
- Battery tray and battery
- Both battery cables
- Hood

10. Fill the cooling system.
11. Fill the engine with clean oil.
12. Start the vehicle, check for leaks and repair if necessary.

Maxima

It is recommended the engine and transaxle be removed as a single unit. If need be, the units may be separated after removal.

➡**The engine and transaxle assembly must be removed from the underside of the vehicle.**

1. Before servicing the vehicle, refer to the precautions in the beginning of this section.
2. Release the fuel system pressure.
3. Drain the cooling system.
4. Drain the engine oil.
5. Drain the automatic transaxle, if equipped.
6. Remove or disconnect the following:
- Negative battery cable
- Hood
- Engine under cover
- Air cleaner, the air intake tube, the air flow meter and the throttle linkage

- Drive belts
- Engine ground cable
- Electrical connector from the crank angle sensor
- Engine electrical harness connectors
- Fuel feed and fuel return hoses
- Upper and lower radiator hoses
- Heater inlet and outlet hoses
- Engine vacuum hoses
- Power steering pump, A/C compressor and alternator
- Carbon canister
- Auxiliary fan, washer tank and the radiator (with the fan assembly)
- Clutch release cylinder from the clutch housing, if equipped with a manual transaxle
- Shift control rod and the shift support rod, on some models with a manual transaxle
- Control cable from the transaxle, on models with an automatic transaxle

7. Install engine slingers to the block and connect a suitable lifting device to the slingers. Do not tension the lifting device at this point.

8. Remove or disconnect the following:
- Exhaust pipe at both the manifold connections
- Front exhaust pipe from the vehicle and support the engine and transaxle assembly with proper jack
- Right and left side halfshafts from their side flanges
- Bolt holding the radius link support

9. Lower the shifter and selector rods and remove the bolts from the motor mount brackets. Remove the nuts holding the front and rear motor mounts to the frame.
10. On some models it will be necessary to remove the center crossmember assembly from the vehicle.
11. Lower the engine/transaxle assembly onto an engine stand.

To install:
12. Raise the engine/transaxle assembly into the vehicle. When raising the engine onto the mounts, be sure to keep it as level as possible.
13. After installing the motor mounts, adjust and install the buffer rods; the front should be 3.50–5.58 in. (89–91mm) and the rear should be 3.90–3.98 in. (99–101mm).
14. Check the clearance between the frame and clutch housing and be sure the engine mount bolts are seated in the groove of the mounting bracket.

15. Remove the transaxle and engine jack assembly.
16. Install or connect the following:
- Center crossmember, if removed. Torque the bolts to 72 ft. lbs. (98 Nm)
- Halfshafts
- Radius link support
- Front exhaust pipe and remove the engine slingers and supports
- Control cable, if equipped
- Shift control and support rods, if equipped
- Clutch release cylinder, if equipped
- Radiator, auxiliary fan and washer tank
- Carbon canister
- Power steering pump, A/C compressor and alternator
- All engine vacuum hoses
- Heater and radiator hoses
- Fuel feed and return lines
- Engine electrical connectors and ground cables
- Drive belts
- Air cleaner, air intake tube and air flow meter
- Throttle linkage
- Engine under cover
- Negative battery cable
- Hood

17. Fill the transmission fluid.
18. Fill the cooling system.
19. Fill the engine with clean oil.
20. Start the vehicle, check for leaks and repair if necessary.

240SX

➡**The engine assembly is removed from the top of the vehicle.**

1. Before servicing the vehicle, refer to the precautions in the beginning of this section.
2. Release fuel pressure from the fuel system before attempting to disconnect any fuel lines.
3. Drain the cooling system.
4. Drain the engine oil from the vehicle.
5. Remove or disconnect the following:
- Battery cables
- Battery
- Hood
- Engine undercover
- Transmission from the vehicle
- Oil cooler lines, if equipped
- Radiator and shroud
- Air cleaner
- Engine drive belts
- Fan and pulley

- Electrical harness connectors at the water temperature sensor, oil pressure sending unit and the starter motor
- Primary ignition wires
- Fuel hoses
- Electrical connections at the alternator
- Heater hoses and throttle connections
- Engine ground cable, thermal transmitter wire, wire to the fuel cut-off solenoid and the vacuum cut solenoid wire
- Front exhaust pipe
- A/C compressor
- Power steering pump
- Power brake booster

6. Attach a hoist to the lifting hooks on the engine (at either end of the cylinder head). Support the engine.

7. Remove or disconnect the following:
- Engine mounting nuts from both lower sides of the engine mounts
- Engine from the vehicle

➡ **When lifting the engine out, guide it carefully to avoid hitting parts such as the master cylinder.**

To install:

8. With the engine assembly safely secured to the hoist, lower the assembly into the vehicle.

9. Torque the engine mounting nuts to 51–58 ft. lbs. (69–78 Nm). It may be necessary to lower or raise the engine hoist to correctly position the engine assembly to line up with the mount holes. Remove the engine hoist.

10. Install or connect the following:
- Power brake booster
- Power steering pump
- A/C compressor
- Front exhaust pipe
- Engine ground cable
- Thermal transmitter wire, fuel cut-off solenoid and vacuum cut off solenoid wiring
- Radiator and shroud
- Radiator and heater hoses
- Fuel feed and return hoses
- Ignition wires
- Water temperature sensor, oil pressure sending unit and starter electrical connectors
- Fan and pulley
- Drive belts
- Oil cooler lines, if equipped
- Air cleaner
- Transmission assembly
- Engine under cover
- Battery and both cables
- Hood

11. Fill the cooling system.
12. Fill the engine with clean oil.
13. Start the vehicle, check for leaks and repair if necessary.

Water Pump

REMOVAL & INSTALLATION

1.6L and 1.8L Engines

1. Before servicing the vehicle, refer to the precautions in the beginning of this section.

2. Drain the cooling system.

3. Remove or disconnect the following:
- Negative battery cable
- Cylinder head front mounting bracket and loosen the water pump pulley bolts
- Engine drive belts
- Water pump pulley
- Coolant hoses from the water inlet and thermostat housing
- Water pump and thermostat housing

4. Remove all traces of gasket material from sealing surfaces.

To install:

5. Apply a continuous bead of liquid sealer to the sealing surface of the thermostat housing. The sealant should be 0.079–0.118 in. (2–3mm) diameter.

6. Install or connect the following:
- Water pump. Torque the bolts to 56–73 inch lbs. (7–8 Nm)
- Pulley to the water pump and tighten the mounting bolts to 56–73 inch lbs. (7–8 Nm)
- Coolant hoses to the thermostat housing
- Drive belts and adjust as needed
- Cylinder head front mounting bracket
- Negative battery cable

7. Fill the cooling system

8. Start the engine, check for leaks and repair if necessary.

2.0L Engine

1. Before servicing the vehicle, refer to the precautions in the beginning of this section.

2. Drain the cooling system.

3. Remove or disconnect the following:
- Negative battery cable
- Right front wheel
- Engine side and front covers
- Loosen the water pump pulley bolts
- Drive belts
- 3 lower water pump bolts and position a jack stand under the engine
- Front engine mount
- Water pump

4. Remove all traces of liquid gasket material from sealing surfaces.

To install:

5. Apply a continuous bead of liquid sealer to the mating surface of the water pump. Sealer should be 0.079–0.118 in. (2–3mm) wide.

6. Install or connect the following:
- Water pump and torque the bolts to 12–15 ft. lbs. (16–21 Nm)
- Front engine mount and remove the engine support
- Water pump pulley. Torque the mounting bolts to 55–73 inch lbs. (6–8 Nm)
- Drive belts and adjust as needed
- Engine front and side cover
- Right front wheel
- Negative battery cable

7. Fill the cooling system.

8. Start the vehicle, check for leaks and repair if necessary.

2.4L Engine

ALTIMA

1. Before servicing the vehicle, refer to the precautions in the beginning of this section.

2. Drain the cooling system.

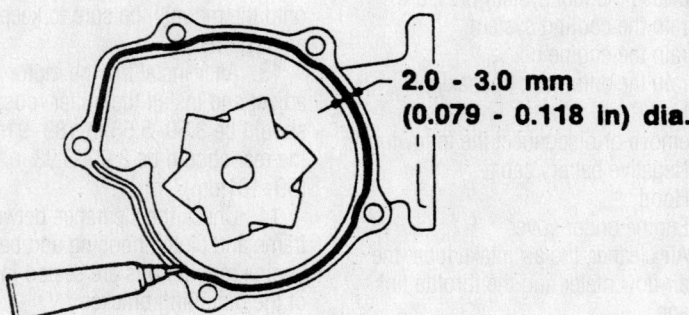

2.0 - 3.0 mm (0.079 - 0.118 in) dia.

7923QG06

Apply RTV sealant to the water pump sealing surface as shown—1.6L and 1.8L engines

3. Remove or disconnect the following:
- Negative battery cable
- Right lower splash cover
- Alternator and A/C compressor
- Coolant tube
- Water pump

➡**Do not disconnect the air conditioning compressor lines. Unbolt the compressor and lay it off to the side.**

➡**The mounting bolts are different sizes and must be reinstalled in the correct location; therefore it is a good idea to arrange the bolts so that they can be easily identified during installation.**

To install:

4. Be sure all gasket surfaces are clean and properly apply a continuous bead of silicone sealer to the pump.
5. Install or connect the following:
- Water pump and torque the 6mm bolts to 57–66 inch lbs. (6–8 Nm) and the 8mm bolts to 12–14 ft. lbs. (16–19 Nm)
- Coolant tube
- Alternator and A/C compressor
- Right side lower splash shield
- Negative battery cable
6. Fill the cooling system.
7. Start the engine, check for leaks and repair if necessary.

240SX

1. Before servicing the vehicle, refer to the precautions in the beginning of this section.

2. Drain the cooling system.
3. Remove or disconnect the following:
- Negative battery cable
- Fan coupling with the fan
- Drive belts
- Water pump

To install:

4. Be sure all gasket surfaces are clean and properly apply liquid sealer to the pump.
5. Install or connect the following:
- Water pump and torque the bolts to 12–14 ft. lbs. (16–19 Nm)
- Fan with the coupling and torque the bolts to 66 inch lbs. (8 Nm)
- Drive belts
- Negative battery cable
6. Fill the cooling system.
7. Start the engine, check for leaks and repair if necessary.

3.0L Engine

1. Before servicing the vehicle, refer to the precautions in the beginning of this section.
2. Drain the cooling system.
3. Position a jack under the oil pan for support. Be sure to place a block of wood on the jack for protection to the engine parts.
4. Remove or disconnect the following:

- Negative battery cable
- Right side engine mount and bracket

- Drive belts and the idler pulley bracket
- Chain tensioner cover and the water pump cover

5. Push the timing chain tensioner sleeve and apply a stopper pin so it does not return.
6. Remove or disconnect the following:
- Timing chain tensioner assembly
- 3 bolts that secure the water pump

7. Rotate the crankshaft 20 degrees counterclockwise to provide timing chain slack.
8. Put M8 bolts in 2 M8 threaded holes of the water pump.
9. Tighten each bolt by turning alternately ½ turn until they reach the timing chain rear case. Be sure to turn each bolt ½ turn at a time to prevent damage.
10. Lift up the water pump and remove it.
11. When removing the water pump, do not allow the water pump gear to hit the timing chain.
12. Remove and discard the O-rings from the water pump.
13. Clean all traces of liquid gasket from the water pump and covers.

To install:

14. Install or connect the following:
- Water pump using new O-rings to the engine block. Torque the 3 water pump mounting bolts evenly to 62–89 inch lbs. (7–10 Nm)
15. Rotate the crankshaft pulley to its original position by turning it 20 degrees clockwise.

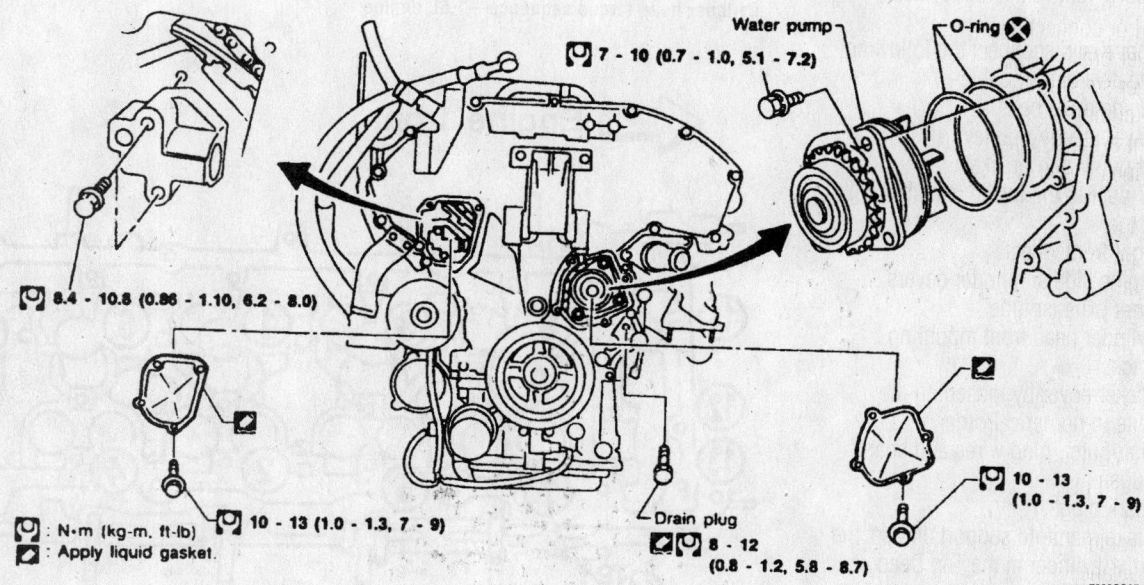

Water pump and timing cover assembly—3.0L engine

Heater Core replacement is covered in Section 2 of this manual

16. Install or connect the following:
- Timing chain tensioner and torque the bolts to 75–89 inch lbs. (9–10 Nm)

17. Remove the stopper pin from the timing chain tensioner.

18. Apply a continuous 0.091–0.130 in. (2.3–3.3mm) bead of liquid sealant to the mating surfaces of the timing chain tensioner and water pump covers.
- Timing chain tensioner and water pump covers to the engine block. Torque the bolts to 89–108 inch lbs. (10–13 Nm)
- Drive belts and the idler pulley bracket
- Right side engine mounting bracket and the engine mount
- Negative battery cable

19. Remove the jack from under the engine and install the drain plugs to the cylinder block.

20. Fill the cooling system.

21. Start the engine, check for leaks and repair if necessary.

Cylinder Head

REMOVAL & INSTALLATION

1.6L and 1.8L Engines

1. Before servicing the vehicle, refer to the precautions in the beginning of this section.

2. Drain the cooling system.

3. Properly relieve the fuel system pressure.

4. Remove or disconnect the following:
- Negative battery cable
- Engine drive belts
- Power steering pulley
- Oil pump and bracket
- Air duct to the intake manifold collector
- Right front wheel
- Engine side and under covers
- Front exhaust tube
- Cylinder head front mounting bracket
- Rocker cover by loosening the bolts in numerical order
- Distributor, plug wires and spark plugs
- Spark plugs
- Intake manifold support and set the No. 1 cylinder at the Top Dead Center (TDC) position
- Idler pulley, camshaft sprockets and timing chains
- Camshafts

5. Loosen the cylinder head bolts in 2–3 steps in the reverse order of the tightening sequence to prevent warpage or cracking of the cylinder head assembly.

6. Remove or disconnect the following:
- Cylinder head (carefully), from the block, pulling the head up evenly from both ends. If the head seems stuck, do not pry it off. Tap lightly around the lower perimeter of the head with a rubber mallet to help break the seal. The cylinder head and the intake and exhaust manifolds are removed together.
- Cylinder head gasket

To install:

7. Thoroughly clean both the cylinder block and head mating surfaces. Avoid scratching either surface.

8. Coat the threads and the seating sur-face of the head bolts with clean engine oil. Install the cylinder head assembly (always replace the head gasket). Install head bolts (with washers) in their proper locations.

9. For 1.6L engines, tighten the bolts in sequence as follows:
- Step 1: Bolts 1–10 to 22 ft. lbs. (29 Nm)
- Step 2: Bolts 1–10 to 43 ft. lbs. (59 Nm)
- Step 3: Loosen bolts 1–10 completely
- Step 4: Bolts 1–10 to 22 ft. lbs. (29 Nm)
- Step 5: Bolts 1–10 plus 50–55 degrees
- Step 6: Bolts 11–15 to 74 inch lbs. (8 Nm)

10. For 1.8L engines, tighten the bolts in sequence as follows:

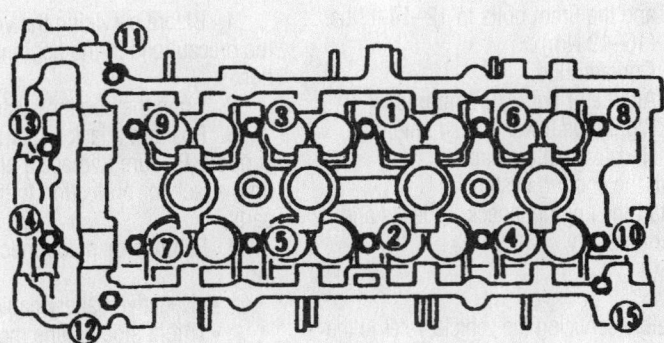

Tighten in numerical order.

7923QG08

Cylinder head torque sequence—1.6L engine

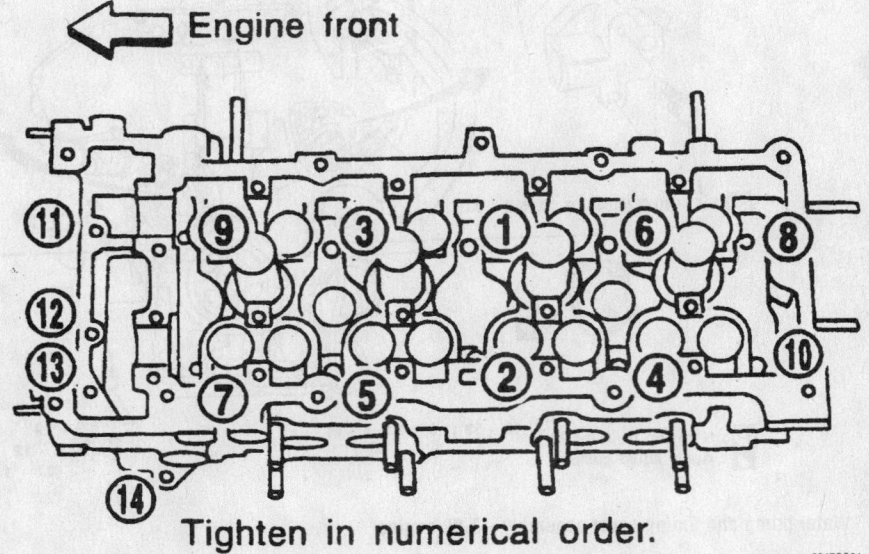

Tighten in numerical order.

9347QG01

Cylinder head torque sequence—1.8L engine

- Step 1: Bolts 1–10 to 22 ft. lbs. (29 Nm)
- Step 2: Bolts 1–10 to 43 ft. lbs. (59 Nm)
- Step 3: Loosen bolts 1–10 completely
- Step 4: Bolts 1–10 to 22 ft. lbs. (29 Nm)
- Step 5: Bolts 1–10 plus 50–55 degrees
- Step 6: Bolts 11–14 to 74 inch lbs. (8 Nm)

11. Install or connect the following:
- Camshafts
- Idler pulley, camshaft sprockets and timing chains
- Intake manifold support
- Distributor
- Spark plugs and wires
- Distributor cap
- Rocker arm cover and torque the bolts to 34 inch lbs. (4 Nm)
- Cylinder head front mounting bracket
- Front exhaust tube
- Engine side and under covers
- Right front wheel
- Air duct to the intake manifold collector
- Oil pump and bracket
- Power steering pulley
- Drive belts
- Negative battery cable

12. Fill the cooling system.
13. Start the vehicle, check for leaks and repair if necessary.

2.0L Engine

1. Before servicing the vehicle, refer to the precautions in the beginning of this section.
2. Release the fuel system pressure.
3. Drain the cooling system.
4. Remove or disconnect the following:
- Negative battery cable
- Engine under covers
- Right front wheel and engine side cover
- Radiator assembly
- Air duct and intake manifold
- Drive belts and water pump pulley
- Alternator
- Power steering pump
- Cylinder head cover and oil separator
- Oil filter and power steering pump brackets
- Front exhaust pipe from the exhaust manifold

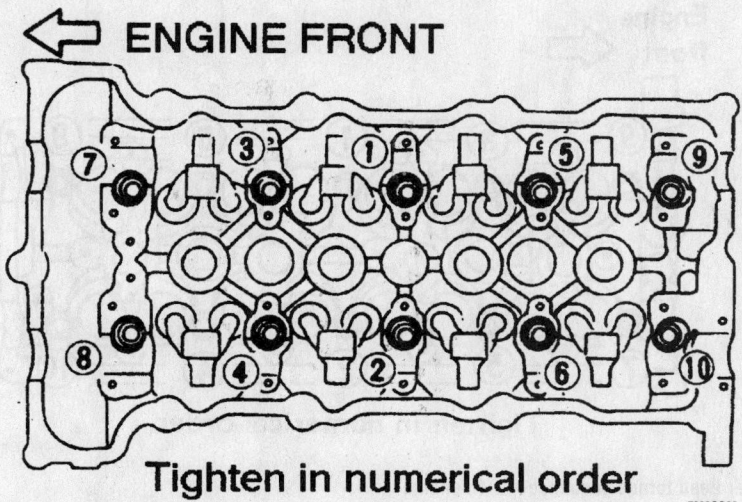

Tighten in numerical order

79230G09

Cylinder head torque sequence—2.0L engine

- Distributor assembly
- Timing chain, tensioner, chain guide and camshaft sprockets
- Camshafts
- Water hose from the cylinder block and water hose from the heater
- Starter motor
- Water pipe bolt
- Knock Sensor (KS) harness connector and the Exhaust Gas Recirculation (EGR) tube
- Cylinder outside bolts. Remove the cylinder head bolts in 2 or 3 steps.
- Cylinder head completely with manifolds attached

To install:

5. Check all components for wear. Replace as necessary. Clean all mating surfaces and replace the cylinder head gasket.

➡**If the length of any cylinder head bolt exceeds 158.2mm, replace the bolt.**

6. Install cylinder head. Torque the cylinder head bolts in the following sequence:
- a. Step 1: 29 ft. lbs. (39 Nm).
- b. Step 2: 58 ft. lbs. (78 Nm).
- c. Step 3: loosen all bolts in sequence completely.
- d. Step 4: 25–33 ft. lbs. (34–44 Nm).
- e. Step 5: plus 90–95 degrees clockwise in sequence.
- f. Step 6: plus additional 90–95 degrees.

➡**Do not turn any bolt 180–200 degrees clockwise all at once.**

7. Install or connect the following:
- KS connector and EGR tube

- Starter motor and the wiring
- Water hoses to the engine block and heater
- Camshafts
- Camshaft sprockets timing chain guide, tensioner and timing chain
- Distributor assembly
- Front exhaust pipe to the exhaust manifold
- Oil filter and power steering pump brackets
- Cylinder head cover and oil separator
- Power steering pump
- Alternator
- Drive belts
- Air duct and intake manifold
- Radiator
- Engine side cover and right front wheel
- Engine under covers
- Negative battery cable

8. Fill the cooling system.
9. Start the vehicle, check for leaks and repair if necessary.

2.4L Engine

1. Before servicing the vehicle, refer to the precautions in the beginning of this section.
2. Drain cooling system.
3. Relieve the fuel system pressure.
4. Remove or disconnect the following:
- Negative battery cable
- Intake manifold collector, exhaust manifold and all related components
- Distributor assembly. Using a block

Brake service is covered in Section 4 of this manual

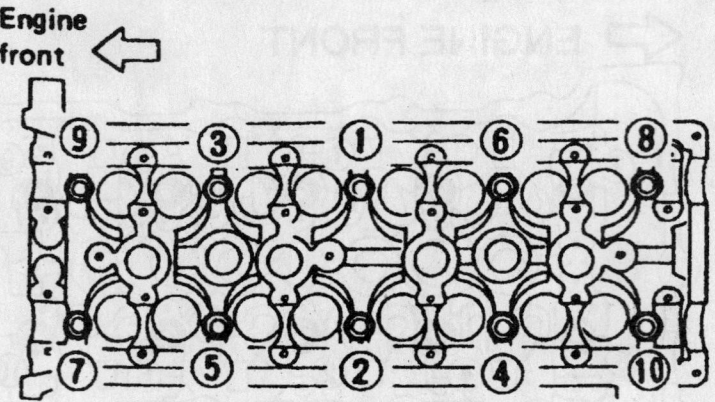

Tighten in numerical order.

7923QG10

Cylinder head torque sequence—2.4L engine

of wood, set a jack under the aluminum oil pan and remove the front engine mount.
- Cylinder head cover
- Timing chain and camshaft sprockets
- Camshafts

➡**The valvetrain components must be reassembled in their original positions.**

5. Loosen the cylinder head bolts in reverse order of tightening.

➡**A warped or cracked cylinder head could result from loosening in incorrect order. The cylinder head bolts should be loosened in 2 or 3 steps.**

6. Remove the cylinder head and the intake manifold. Remove the cylinder head gasket. The lower timing chain will not be disengaged from crankshaft sprocket.

To install:
7. Clean the gasket surfaces.
8. Install or connect the following:
- New cylinder head gasket
- Cylinder head and temporarily tighten the cylinder head bolts. This is necessary to avoid damaging the cylinder head gasket. Be sure to install washers between the bolts and cylinder head
- Idler shaft assembly
- Upper timing chain and cover

9. Tighten the cylinder head bolts in sequence as follows:
   a. Step 1: 22 ft. lbs. (29 Nm).
   b. Step 2: 59 ft. lbs. (79 Nm).
   c. Step 3: loosen all the bolts completely.
   d. Step 4: 18–25 ft. lbs. (25–34 Nm).
   e. Step 5: plus 86–91 degrees clockwise.
10. Install or connect the following:

- Camshafts
- Timing chains, chain tensioner and camshaft sprockets
- Cylinder head cover
- Distributor assembly
- Intake manifold collector
- Negative battery cable
11. Fill the cooling system.
12. Start the vehicle, check for leaks and repair if necessary.

3.0L Engine

1. Before servicing the vehicle, refer to the precautions in the beginning of this section.
2. Relieve the fuel system pressure.
3. Drain the engine oil.
4. Drain the cooling system.

➡**Before detaching any hoses or connectors, note the locations for reassembly.**

5. Remove or disconnect the following:
- Negative battery cable
- Intake manifold collector
- Fuel tube
- Intake manifold
- Cylinder head covers
- Ignition coils
- Exhaust Gas Recirculation (EGR) guide tube
- Engine under cover
- Right front wheel and engine side cover
- Drive belts and idler pulley
- Steel (lower) and aluminum (upper) oil pans
- Water pump cover
- Timing chain case cover
- Timing chains, camshaft sprockets and related components
- Crankshaft sprocket
6. Loosen the bolts that secure the rear timing chain case. The bolts must be loosened in the reverse order of installation sequence.
7. Remove or disconnect the following:
- Rear timing case cover using seal cutter tool

➡**Remove the O-rings from the front of the engine block.**

- Camshafts
- Cylinder head bolts in the reverse order of the tightening sequence. The bolts should be loosened in 2–3 steps.

➡**A warped or cracked cylinder head could result from removing the bolts in incorrect order.**

- Cylinder heads from the vehicle
- Discard the head gaskets
8. Remove all traces of liquid gasket

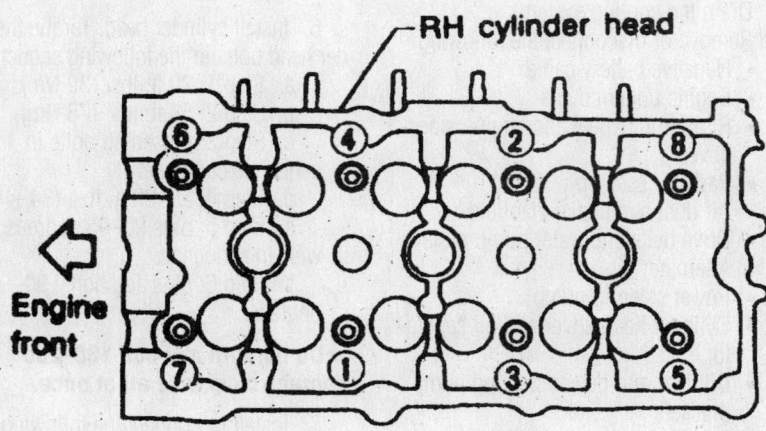

Tighten in numerical order.

7923QG11

Right cylinder head torque sequence—3.0L engine

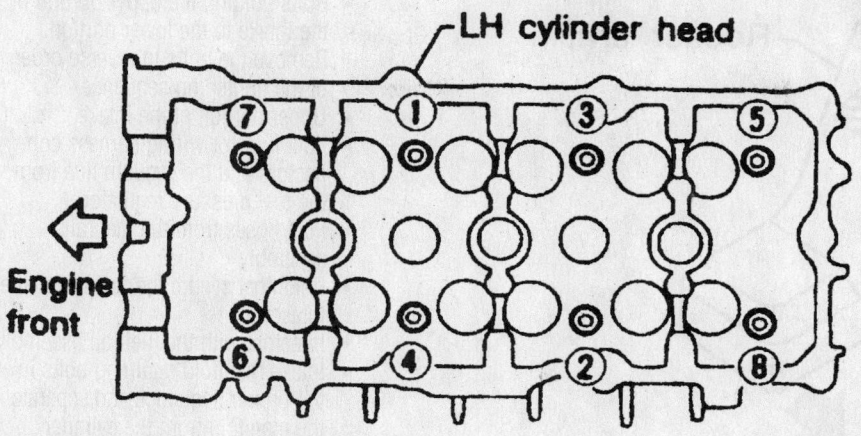

LH cylinder head

Engine front

Tighten in numerical order.

79230G12

Left cylinder head torque sequence—3.0L engine

from the timing chain case and from the water pump covers.

9. Remove all traces of liquid gasket from the engine block.

10. Inspect the timing chain for excessive wear or damage and replace as necessary.

To install:

11. Turn the crankshaft until the No. 1 piston is set 240 degrees before Top Dead Center (TDC) on compression stroke.

12. Using new head gaskets, install the cylinder heads.

➡**If possible, replacement of the head bolts is suggested.**

13. If replacement of the head bolts is not possible, perform the following bolt measurement:

   a. Measure the diameter of the head bolt 11mm from the bottom of the bolt.

   b. Measure the diameter of the head bolt 48mm from the bottom of the bolt.

   c. Whenever the size difference between the 2 measurements exceeds 0.11mm the head bolts must be replaced.

14. Install the cylinder head bolts and torque in sequence as follows:

   a. Step 1: 72 ft. lbs. (98 Nm).

   b. Step 2: Completely loosen all bolts.

   c. Step 3: 25–33 ft. lbs. (24–44 Nm).

   d. Step 4: plus 90–95 degrees clockwise.

   e. Step 5: plus 90–95 degrees clockwise.

15. Install or connect the following:
   • Camshafts and related components

   • New O-rings to the front of the engine block

16. Apply sealant to the hatched portion of the of the rear timing chain case.

17. Align the rear timing chain case with the dowel pins and install onto the cylinder heads and engine block.

18. Torque the rear timing chain case mounting bolts in sequence to 105–121 inch lbs. (11.8–13.7 Nm).
   • Water pump cover
   • Upper and lower oil pans
   • Idler pulley and drive belts
   • Cylinder head covers
   • Intake manifold
   • Fuel tube
   • Intake manifold collector
   • Negative battery cable

19. Fill the cooling system.

20. Fill the engine with clean oil.

21. Start the vehicle, check for leaks and repair if necessary.

Rocker Arms

REMOVAL & INSTALLATION

Except 2.0L Engine

Nissan engines, with the exception of the 2.0L engine, do not utilize rocker arms. The valves are actuated directly by the camshafts.

2.0L Engine

1. Before servicing the vehicle, refer to the precautions in the beginning of this section.

2. Release the fuel pressure following the proper procedure.

3. Remove or disconnect the following:
   • Negative battery cable
   • Rocker arm cover, gasket and the oil separator
   • Intake manifold supports, oil filter bracket and the power steering pump

4. Set the No. 1 cylinder at Top Dead Center (TDC) on the compression stroke.

5. Remove or disconnect the following:
   • Timing chain tensioner from the side of the head

6. Matchmark the position of the rotor and housing and remove the distributor.

7. Remove or disconnect the following:
   • Timing chain guide
   • Camshaft sprockets while holding the camshaft stationary with a large wrench. Secure the timing chain with wire so the timing is not lost. The front cover will have to be removed if the chain timing is lost.

➡**When removing the camshafts, loosen the journal caps in the opposite sequence of tightening. Camshaft damage may result if this step is not followed.**

   • Camshafts, brackets, oil tubes and the baffle plate. Label all components for proper installation.

➡**It is essential that all parts be kept in the same order and orientation for reinstallation. Be sure to mark and separate parts to keep them from getting mixed. This will aid assembly.**

   • Rocker arms, shims, rocker arm guides and the hydraulic lash adjusters. Label all components for proper installation.

➡**The valve lifters must be stored in the vertical position or submersed in clean oil to prevent air from entering the lifters.**

8. Inspect the surfaces of the rockers and replace if there are any signs of damage.

To install:

9. Lubricate the rocker arms, shims, rocker arm guides and the hydraulic lash adjusters.

10. Install or connect the following:
   • Rocker arms, shims, rocker arm guides and the hydraulic lash adjusters in their original locations
   • Camshafts, brackets, oil tubes and the baffle plate in the proper location

For complete Engine Mechanical specifications, see Section 1 of this manual

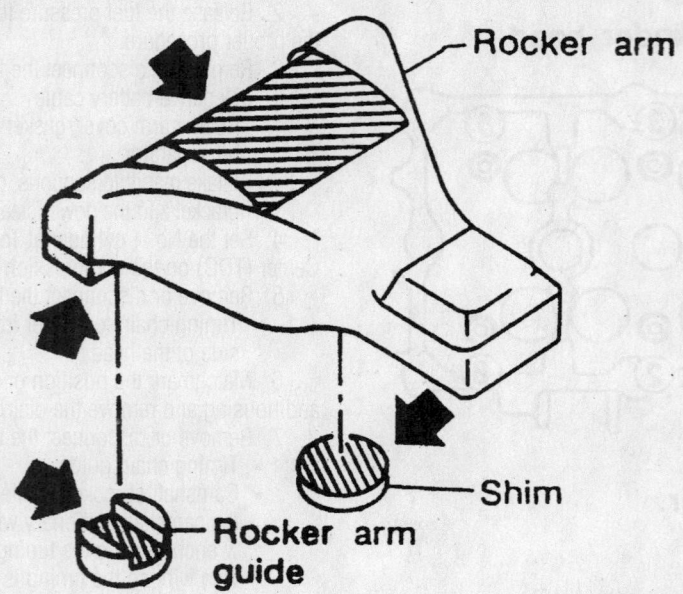

Rocker arm, guide and shim—2.0L engine

7923QG15

11. Tighten the bolts in sequence as follows:

    a. Right camshaft bolts Nos. 9 and No. 10: 17 inch lbs. (2 Nm).

    b. Right camshaft bolts No. 1 through 8: 17 inch lbs. (2 Nm).

    c. Left camshaft bolts Nos. 11 and 12: 17 inch lbs. (2 Nm).

    d. Left camshaft bolts Nos. 1 through 10: 17 inch lbs. (2 Nm).

    e. All camshaft bolts in numerical sequence: 52 inch lbs. (6 Nm).

    f. All camshaft bolts in numerical sequence: 87–104 inch lbs. (10–11 Nm).

    g. Rear 2 bolts of the left-hand camshaft: 13–19 ft. lbs. (18–25 Nm).

12. Install or connect the following:
- Camshaft sprockets while holding the camshaft stationary with a large wrench
- Remaining components in the reverse order from which they were removed
- Negative battery cable

13. Check and adjust the ignition and valve timing. If there is air in the lifters, bleed the air by running the engine at 1000 rpm for 10 minutes.

Intake Manifold

REMOVAL & INSTALLATION

1.6L and 1.8L Engines

1. Before servicing the vehicle, refer to the precautions in the beginning of this section.

2. Relieve the fuel system pressure.

3. Drain the cooling system.

4. Remove or disconnect the following:
- Negative battery cable
- Air cleaner assembly
- Throttle linkage, electrical connections and vacuum lines from the throttle body
- Intake manifold collector support brackets

5. The throttle body can be removed from the manifold at this point or can be removed as an assembly with the intake manifold.

6. Remove or disconnect the following:

- Bolts holding the upper portion of the intake to the lower portion. Remove the bolts in reverse order of the tightening sequence
- Upper portion of the intake
- Fuel injector wiring harness connectors and the vacuum line from the fuel pressure regulator
- Fuel hoses from the fuel rail assembly
- Bolts that secure the fuel rail to the intake
- Injectors with the fuel rail assembly
- Intake manifold retaining bolts in the proper sequence and separate the manifold from the cylinder head. Remove the bolts in reverse order of the tightening sequence
- Intake manifold gasket and clean all the gasket contact surfaces thoroughly with a gasket scraper and suitable solvent. All traces of old gasket material must be removed to ensure proper sealing.
- Inspect the intake manifold for cracks. Using a metal straightedge, check the surface of the intake manifold for warpage.

To install:

7. Install or connect the following:
- New intake manifold gasket onto the cylinder head and position the lower intake manifold over the mounting studs and onto the gasket.
- Intake manifold and torque the bolts to 13–15 ft. lbs. (18–21 Nm) in sequence
- Injectors with the fuel rail assembly. Be sure to install the fuel rail

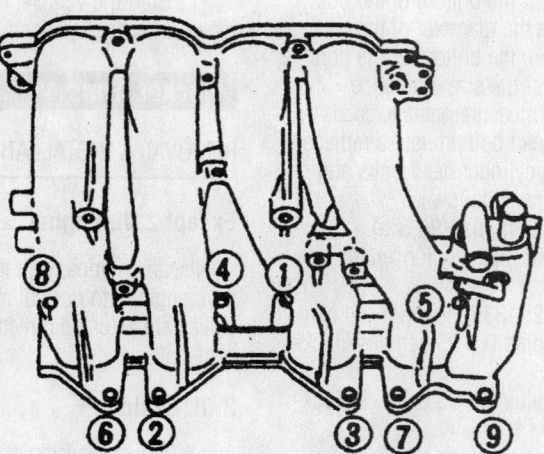

Tighten in numerical order.

Lower intake manifold torque sequence—1.6L engine

7923QG16

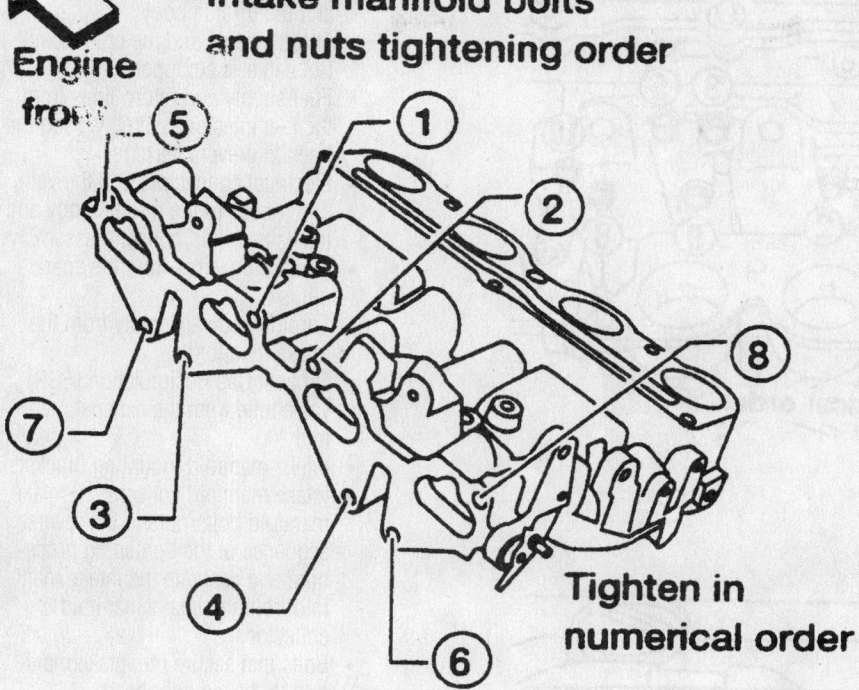

★ **Intake manifold bolts and nuts tightening order**

Engine front

Tighten in numerical order

9347QG02

Lower intake manifold torque sequence—1.8L engine

insulators. Torque the bolts in 2 steps to 13–15 ft. lbs. (18–21 Nm)
- Fuel injector wiring harness connectors and the vacuum line to the fuel pressure regulator
- Fuel hoses to the fuel rail assembly using new hose clamps

- Intake manifold collector using a new gasket. Torque the bolts to 13–15 ft. lbs. (18–21 Nm) in sequence
- Throttle body or throttle chamber, if removed. Torque the bolts in a crisscross pattern to 13–16 ft. lbs. (18–22 Nm)

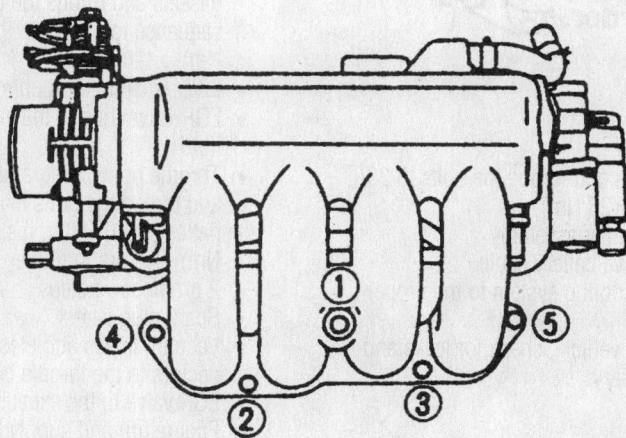

Tighten in numerical order.

79230G17

Upper intake manifold torque sequence—1.6L and 1.8L engines

➡ Be sure to properly position the throttle body gasket with the cut out facing down.
- Intake manifold collector support brackets
- Throttle linkage, electrical connections and vacuum lines
- Air cleaner
- Negative battery cable

8. Fill the cooling system to the proper level.
9. Start the engine, check for leaks and repair if necessary.

2.0L Engine

1. Before servicing the vehicle, refer to the precautions in the beginning of this section.
2. Relieve the fuel system pressure.
3. Drain the cooling system.
4. Remove or disconnect the following:
- Negative battery cable
- Air cleaner assembly
- Manifold support brackets
- Throttle linkage, electrical connections and vacuum lines from the throttle body. Be sure to note the locations of all connections.
- Exhaust Gas Recirculation (EGR) tube from the manifold
- Fuel rail assembly
- Dive belts and water pump pulley
- Alternator and power steering pump
- Oil filter bracket and power steering bracket
- Intake manifold collector retaining bolts in the reverse of installation sequence and separate the collector from the manifold
- Intake manifold assembly retaining bolts in the reverse order of the tightening sequence and separate the manifold from the cylinder head
- All gasket material and clean all the gasket contact surfaces thoroughly with a gasket scraper and a suitable solvent. All traces of old gasket material must be removed to ensure proper sealing.
- Inspect the intake manifold for cracks. Using a metal straightedge, check the surface of the intake manifold for warpage.

To install:

5. Install or connect the following:
- Intake manifold with new gaskets and torque the bolts, in sequence, to 13–15 ft. lbs. (18–21 Nm)

For Accessory Drive Belt illustrations, see Section 1 of this manual

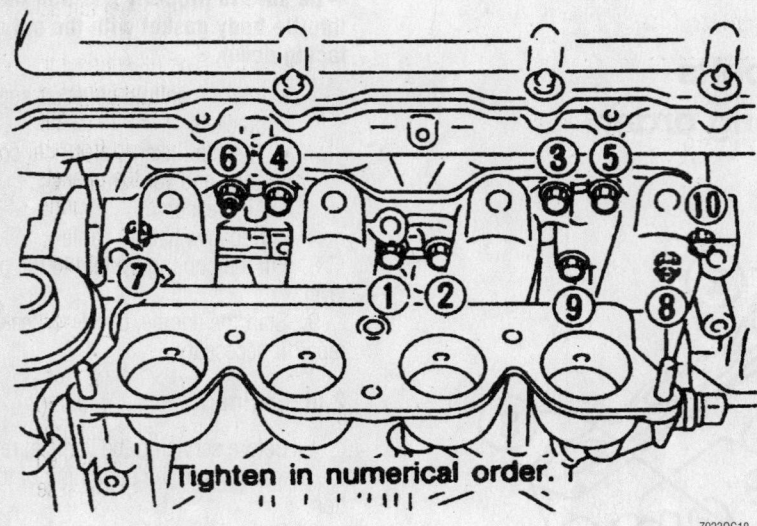

Tighten in numerical order.

7923QG18

Lower intake manifold torque sequence—2.0L engine

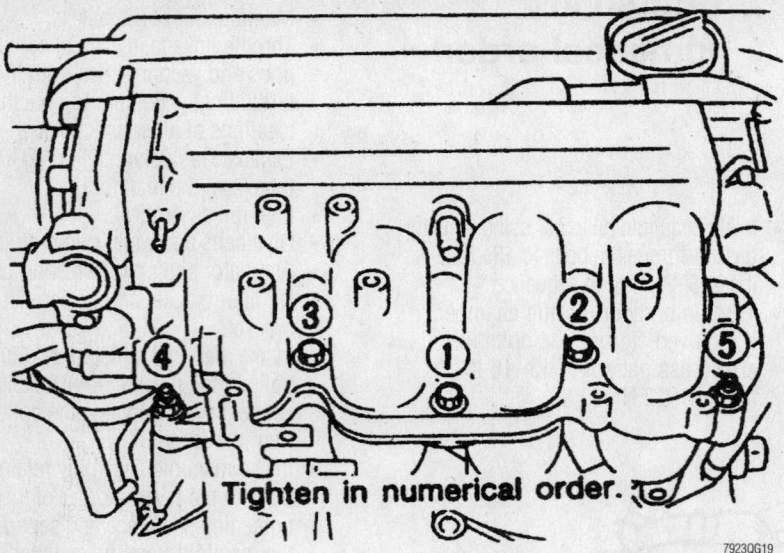

Tighten in numerical order.

7923QG19

Upper intake manifold torque sequence—2.0L engine

- Intake manifold collector with new gaskets and torque the bolts, in sequence, to 13–15 ft. lbs. (18–21 Nm)
- Oil filter bracket
- Power steering bracket
- Alternator and power steering pump
- Water pump pulley and drive belts
- Fuel rail assembly with new insulators and torque the bolts to 35 inch lbs. (4 Nm)
- EGR tube
- Throttle linkage, vacuum lines and electrical connections to the throttle body
- Manifold support brackets with new

gaskets and torque the bolts to 20 ft. lbs. (26 Nm)
- Air cleaner assembly
- Negative battery cable

6. Fill the cooling system to the proper level.

7. Start the vehicle, check for leaks and repair if necessary.

2.4L Engine

1. Before servicing the vehicle, refer to the precautions in the beginning of this section.

2. Relieve the fuel system pressure.

3. Drain the cooling system.

4. Remove or disconnect the following:
- Negative battery cable
- Air duct between the air flow meter and the throttle body
- Throttle cable and the cruise control cable, if equipped
- Fuel supply and return lines from the fuel injector assembly. Plug the lines to prevent leakage.
- Electrical connectors and the vacuum hoses to the throttle body and intake manifold/collector assembly
- Spark plug wires from the spark plugs
- Throttle body assembly from the intake manifold
- Exhaust Gas Recirculation (EGR) valve tube from the exhaust manifold
- Intake manifold mounting brackets
- Intake manifold collector-to-intake manifold bolts/nuts in the reverse sequence of the tightening procedure and separate the intake manifold from the intake manifold collector
- Bolts that secure the intake manifold to the cylinder head.
- Manifold. Be sure to loosen the bolts in the reverse sequence of the tightening procedure

5. Using a putty knife, clean the gasket mounting surfaces. Check the intake manifold/collector for cracks and warpage.

To install:

6. Install or connect the following:
- Intake manifold with new gaskets and torque the bolts, in sequence, to 12–14 ft. lbs. (16–19 Nm)
- Intake manifold collector using new gaskets and torque the bolts in sequence to 12–14 ft. lbs. (16–19 Nm)
- Intake manifold mounting brackets
- EGR valve tube to the exhaust manifold
- Throttle body using a new gasket and torque the bolts in a crisscross pattern to 13–16 ft. lbs. (18–22 Nm). Be sure to tighten the bolts in 2 progressive steps
- Spark plug wires
- Vacuum hoses and electrical connectors to the throttle body
- EGR valve to the exhaust manifold
- Fuel return and supply lines
- Throttle and cruise control cables, if equipped
- Air duct
- Negative battery cable

7. Fill the cooling system to the proper level.

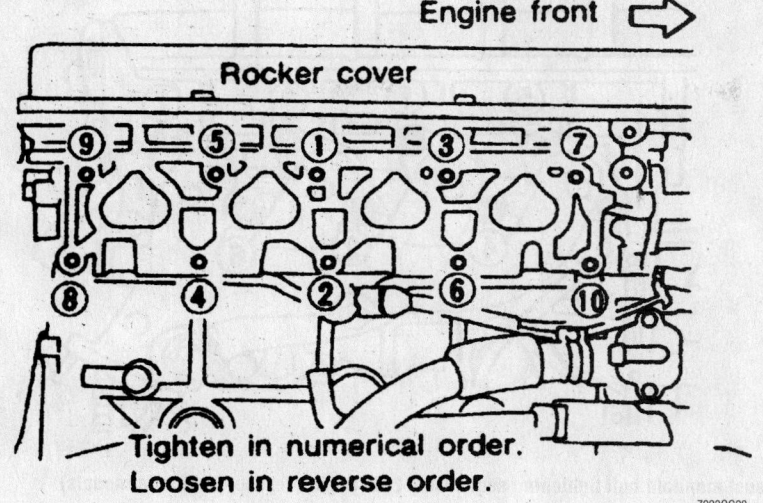

Engine front ⇨

Rocker cover

Tighten in numerical order.
Loosen in reverse order.

7923QG20

Intake manifold torque sequence—2.4L engine

8. Start the vehicle, check for leaks and repair if necessary.

3.0L Engine

1. Before servicing the vehicle, refer to the precautions in the beginning of this section.

2. Drain the cooling system.

3. Release the fuel system pressure.

4. Remove or disconnect the following:
- Negative battery cable
- Throttle body coolant hoses
- Electrical connectors from the Throttle Position (TP) sensor
- Hoses from the throttle body, the Exhaust Gas Recirculation (EGR) valve, intake manifold collector, Idle Air Control (IAC) valve and the fuel pressure regulator
- Canister purge hose and blow-by hose
- EGR guide tube
- Accelerator cable from the throttle body
- Intake manifold collector support brackets
- Right side electrical connectors from the ignition coils
- Electrical connector from the crank angle sensor and the power transistor, if necessary
- Intake manifold collector-to-intake manifold bolts/nuts and the intake manifold collector

5. Remove the fuel injector assembly by performing the following procedures:

   a. Detach the electrical connectors from the fuel injectors.

   b. Disconnect the fuel lines from the fuel injector assembly.

   c. Remove the fuel rail-to-cylinder head bolts.

   d. Remove the fuel rail assembly from the engine.

6. Remove or disconnect the following:
- Intake manifold bolts/nuts in the reverse of the installation sequence
- Intake manifold from the engine and discard the gaskets

7. Clean all gasket mounting surfaces.

To install:

8. Using new gaskets, install the intake manifold to the engine.

9. Torque the bolts in sequence as follows:

   a. Step 1: 44–89 inch lbs. (5–10 Nm)

   b. Step 2: 20–23 ft. lbs. (26–31 Nm).

10. Install the fuel injector assembly by performing the following procedures:

   a. Install the fuel rail assembly to the engine.

   b. Install the fuel rail-to-cylinder head bolts and tighten the bolts to 15–20 ft. lbs. (21–26 Nm) in 2 progressive steps.

   c. Connect the fuel lines to the fuel injector assembly.

   d. Connect the electrical connectors to the fuel injectors.

11. Install the intake manifold collector. Torque the bolts to 8–11 ft. lbs. (11–15 Nm).

12. Install or connect the following:
- Crank angle sensor and transmitter electrical connectors
- Right side ignition coil electrical connectors
- Intake manifold collector support brackets
- Accelerator cable to the throttle body
- EGR guide tube
- Canister purge and blow by hoses
- Throttle body, EGR valve and intake manifold collector hoses
- IAC valve and fuel pressure regulator hoses
- TP sensor electrical connector
- Throttle body coolant hose
- Negative battery cable

13. Fill the cooling system.

14. Start the vehicle, check for leaks and repair if necessary.

Exhaust Manifold

REMOVAL & INSTALLATION

1.6L, 1.8L and 2.0L Engine

1. Before servicing the vehicle, refer to the precautions in the beginning of this section.

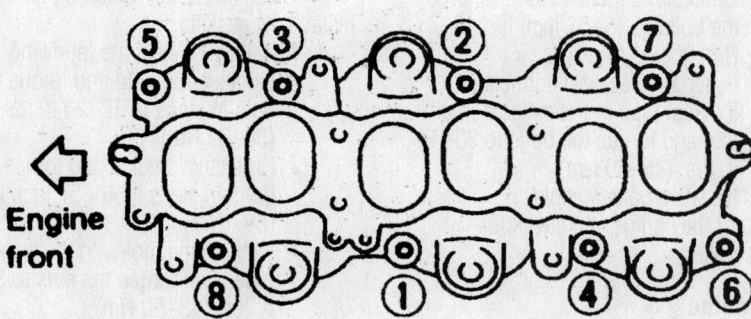

Engine front

Tighten in numerical order.

7923QG22

Intake manifold torque sequence—3.0L engine

2. Remove or disconnect the following:
- Negative battery cable
- Engine undercovers
- Air cleaner or collector assembly
- Heat shields from the manifold and front exhaust pipe
- Front exhaust pipe from the exhaust manifold
- Temperature sensors, Oxygen (O_2) sensors and air induction pipes from the manifold
- Manifold support brackets
- Exhaust manifold attaching nuts and the manifold from the block. Discard the exhaust manifold gaskets.

3. Clean the gasket surfaces and check the manifold for cracks and warpage.

To install:

4. Install the exhaust manifold with a new gasket.

5. On 2.0L engines, torque the fasteners from the center outward in several stages to 27–35 ft. lbs. (37–48 Nm).

6. On 1.6L and 1.8L engines, tighten the mounting nuts with washers in sequence to 19–21 ft. lbs. (26–29 Nm).

7. Install or connect the following:
- Temperature sensors, O_2 sensors and air induction pipes
- Manifold support brackets
- Exhaust pipe to the manifold using a new gasket. Torque the nuts to 21–25 ft. lbs. (28–33 Nm) for 1.6L and 1.8L engine models or 32–37 ft. lbs. (43–50 Nm) for 2.0L engines
- Heat shields
- Air cleaner or collector assembly
- Engine undercovers
- Negative battery cable

8. Start the engine and check for exhaust leaks.

2.4L Engine

1. Before servicing the vehicle, refer to the precautions in the beginning of this section.

2. Remove or disconnect the following:
- Negative battery cable
- Exhaust pipe from the exhaust manifold
- Oxygen (O_2) sensor electrical connector
- Exhaust manifold cover
- Exhaust Gas Recirculation (EGR) tube from the exhaust manifold
- Exhaust manifold-to-engine bolts and nuts and discard the gaskets
- Retaining bolts and nuts in reverse of the tightening sequence

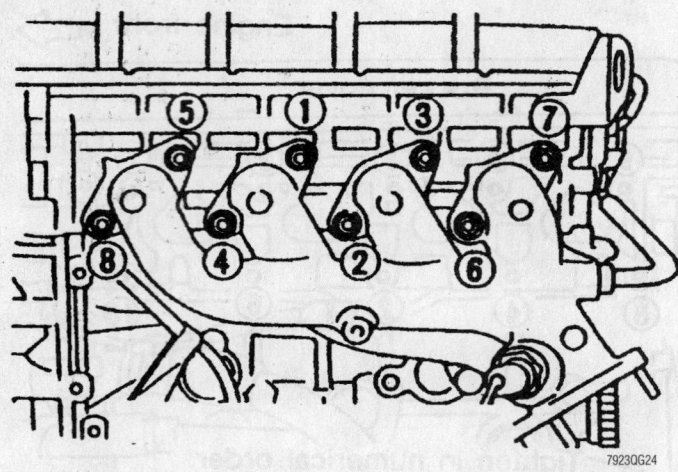

Exhaust manifold bolt tightening sequence—2.4L engine (except California models)

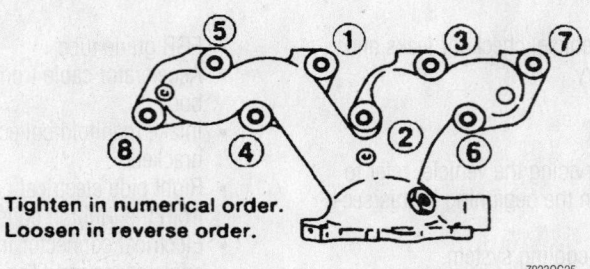

**Tighten in numerical order.
Loosen in reverse order.**

Exhaust manifold bolt tightening sequence—2.4L engine (California models)

- Exhaust manifold from the vehicle

To install:

3. Clean all gasket mounting surfaces and install new gaskets.

4. Install or connect the following:
- Exhaust manifold to the engine and torque the nuts to 27–35 ft. lbs. (37–48 Nm)
- EGR tube to the exhaust manifold and torque the nuts to 29–36 ft. lbs. (39–49 Nm)
- Exhaust manifold cover and torque the bolts to 45–57 inch lbs. (5–7 Nm)
- O_2 sensor electrical connector
- Exhaust pipe to the exhaust manifold and torque the bolts to 33–44 ft. lbs. (45–60 Nm)
- Negative battery cable

5. Start the engine and check for exhaust leaks.

3.0L Engine

1. Before servicing the vehicle, refer to the precautions in the beginning of this section.

2. Remove or disconnect the following:
- Negative battery cable

- Exhaust manifolds from the exhaust pipes
- Protective covers from the manifolds
- Heated Oxygen (HO_2S) sensor from the manifold, if equipped
- Exhaust manifold-to-engine mounting nuts
- Manifolds from the engine and discard the gaskets

To install:

3. Clean all gasket mounting surfaces. Install new gaskets.

4. Install or connect the following:
- Exhaust manifold and torque the nuts in steps to 22–24 ft. lbs. (30–32 Nm)
- Protective shields and torque the bolts in steps to 46–57 inch lbs. (5–7 Nm)
- Exhaust manifolds to the exhaust pipes and torque the nuts to 32–37 ft. lbs. (43–50 Nm)
- HO_2S sensor to the manifold and torque the fastener to 30–44 ft. lbs. (40–60 Nm), if equipped
- Negative battery cable

5. Start the engine and check for exhaust leaks.

Front Crankshaft Seal

REMOVAL & INSTALLATION

➡ The front crankshaft seal procedure is applicable to timing belt-equipped engines only. For the front seal on engines equipped with timing chains, refer to the timing chain, sprockets, front cover and seal procedure later in this section.

Camshaft and Valve Lifters

REMOVAL & INSTALLATION

1.6L and 1.8L Engines

1. Before servicing the vehicle, refer to the precautions in the beginning of this section.
2. Drain the cooling system.
3. Relieve the fuel system pressure.
4. Remove or disconnect the following:
   - Negative battery cable
   - All engine drive belts
   - Exhaust pipe from the exhaust manifold
   - Power steering pulley and pump with the mounting bracket
   - Cylinder head cover
   - Distributor assembly
   - Timing chain tensioners and camshaft sprocket

➡ Before the camshafts are removed from the cylinder head, note the positioning of the pins at the end of the camshafts for reassembly purposes.

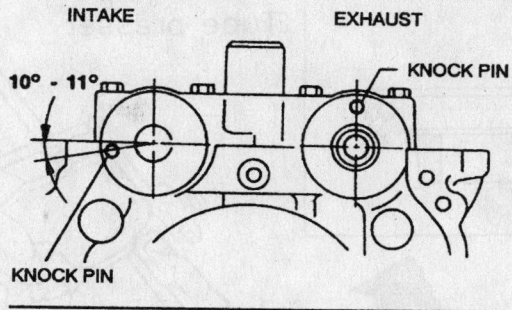

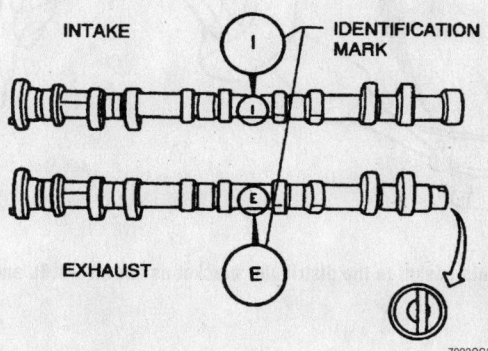

Positioning and identification of the camshafts—1.6L and 1.8L engines

- Camshaft bearing caps in sequence
- Camshafts from the cylinder head
- Idler sprocket bolt. These parts should be reassembled in their original position.
- Shims from the tops of the lifters. Be sure to note the position of each shim.
- Valve lifters from the bores in the cylinder head. Note the positioning of the lifters for reassembly.

5. Measure the diameter of the lifters. The diameter should be 1.1795–1.1801 in. (29.960–29.975mm).
6. Measure the diameter of the lifter bores. The diameter should be 1.1811–1.1819 in. (30.000–30.021mm).
7. Clearance between the lifter and bore should be 0.0010–0.0024 in. (0.025–0.061mm).

To install:

8. Install or connect the following:
   - Lifters and shims to the cylinder head in the proper locations as noted during removal

➡ The exhaust and intake camshafts are marked with identification stamps. (E for exhaust and I for intake).

   - Camshafts to the cylinder head and position the intake camshaft knock pin at the 9 o'clock position and the exhaust camshaft at the 12 o'clock position

9. Install the camshaft bearing caps and tighten the mounting bolts as follows:
   a. Bolts 11 through 15, then bolts 1 through 10: 18 inch lbs. (2 Nm).
   b. Bolts 1 through 15: 53 inch lbs. (6 Nm).
   c. Bolts 1 through 14: 87–105 inch lbs. (10–12 Nm).
   d. Bolt 15: 56–73 inch lbs. (7–8 Nm).

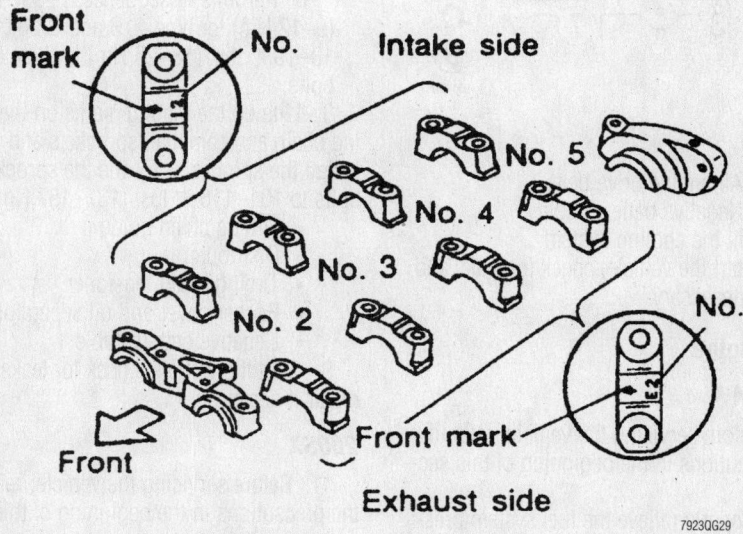

Be sure to install the camshaft bearing caps in their original positions—1.6L and 1.8L engines

For Wheel Alignment specifications, see Section 1 of this manual

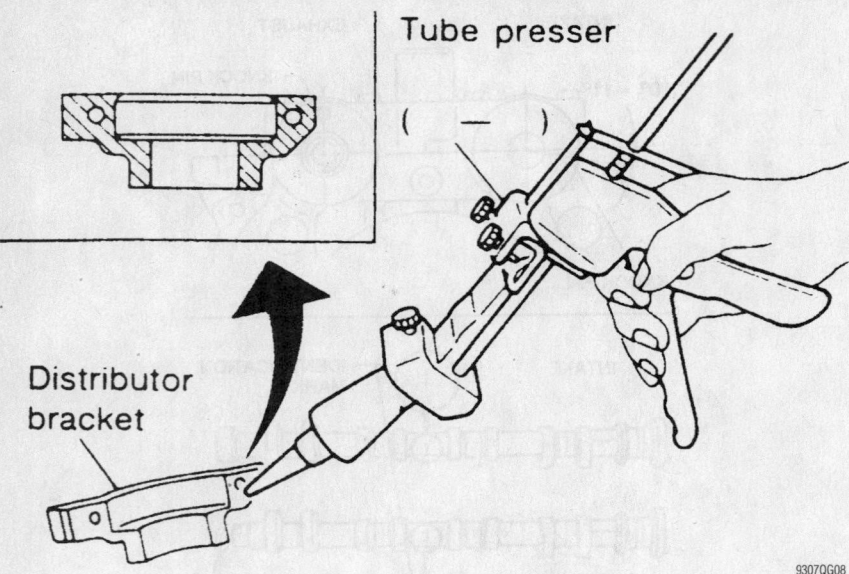

If removed, apply liquid gasket to the distributor bracket as shown—1.6L and 1.8L engines

Camshaft bolt torque sequence—1.6L and 1.8L engines

➡If any part of the valvetrain has been has been replaced, the valve adjustment must be checked. DO NOT adjust the valves or rotate the camshafts at this point. Internal engine damage will result.

10. Install or connect the following:
   • Camshaft sprockets with timing chains
   • Distributor assembly
11. Check and adjust the valve clearance.
12. Install or connect the following:
   • Cylinder head cover
   • Power steering pulley pump
   • Exhaust pipe from the exhaust manifold
   • All engine drive belts
   • Negative battery cable
13. Fill the cooling system.
14. Start the vehicle, check for leaks and repair if necessary.

2.0L Engine

SENTRA

1. Before servicing the vehicle, refer to the precautions in the beginning of this section.
2. Properly relieve the fuel system pressure.
3. Remove or disconnect the following:
   • Negative battery cable
   • Rocker cover and oil separator
4. Rotate the crankshaft until the No.1 piston is at Top Dead Center (TDC) on the compression stroke. Then, rotate the crankshaft until the mating marks on the camshaft sprockets line up with the mating marks on the timing chain.
5. Remove or disconnect the following:
   • Timing chain tensioner
   • Distributor
   • Timing chain guide
   • Camshaft sprockets. Use a wrench to hold the camshaft while loosening the sprocket bolt.
   • Camshaft bracket bolts in the opposite order of the tightening sequence
   • Camshaft

To install:
6. Clean the left-hand camshaft end bracket and coat the mating surface with liquid gasket. Install the camshafts, camshaft brackets, oil tubes and baffle plate. Ensure the left camshaft key is at 12 o'clock and the right camshaft key is at 10 o'clock.
7. The procedure for tightening camshaft bolts must be followed exactly to prevent camshaft damage. Torque the bolts as follows:
   a. Right bolts 9 and 10 (in that order): 18 inch lbs. (2 Nm).
   b. Right bolts 1–8 (in that order): 18 inch lbs. (2 Nm).
   c. Left bolts 11 and 12 (in that order): 18 inch lbs. (2 Nm).
   d. Left bolts 1–10 (in that order): 18 inch lbs. (2 Nm).
   e. All bolts in sequence: 52 inch lbs. (6 Nm).
   f. All bolts in sequence: 7–9 ft. lbs. (9–12 Nm) for type A, B and C bolts and 13–19 ft. lbs. (18–25 Nm) for type D bolts.
8. Line up the mating marks on the timing chain and camshaft sprockets and install the sprockets. Torque the sprocket bolts to 101–116 ft. lbs. (137–157 Nm).
   • Timing chain guide
   • Distributor
   • Timing chain tensioner
   • Rocker cover and oil separator
   • Negative battery cable
9. Start the vehicle, check for leaks and repair if necessary.

200SX

1. Before servicing the vehicle, refer to the precautions in the beginning of this section.
2. Release the fuel system pressure.
3. Drain the cooling system.
4. Remove or disconnect the following:

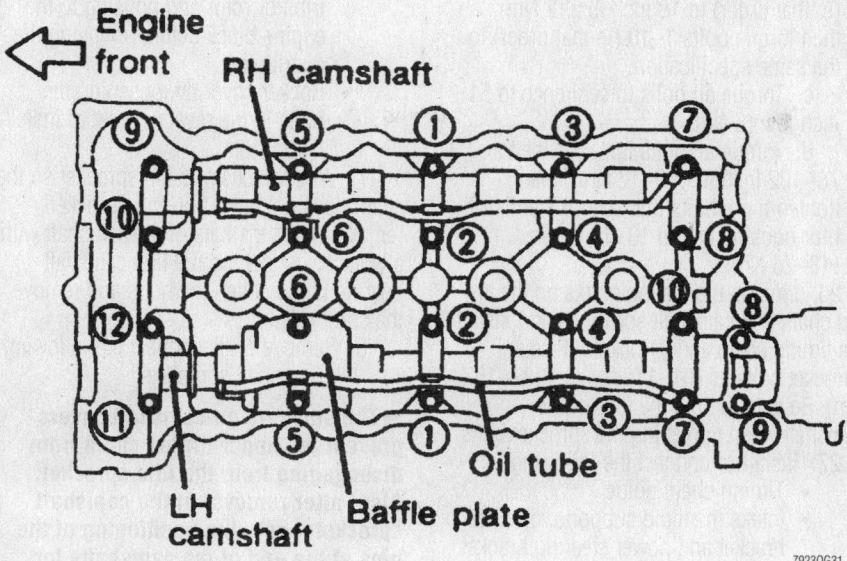

To prevent damage to the camshafts, torque the bearing caps in the numbered.......

7923QG31

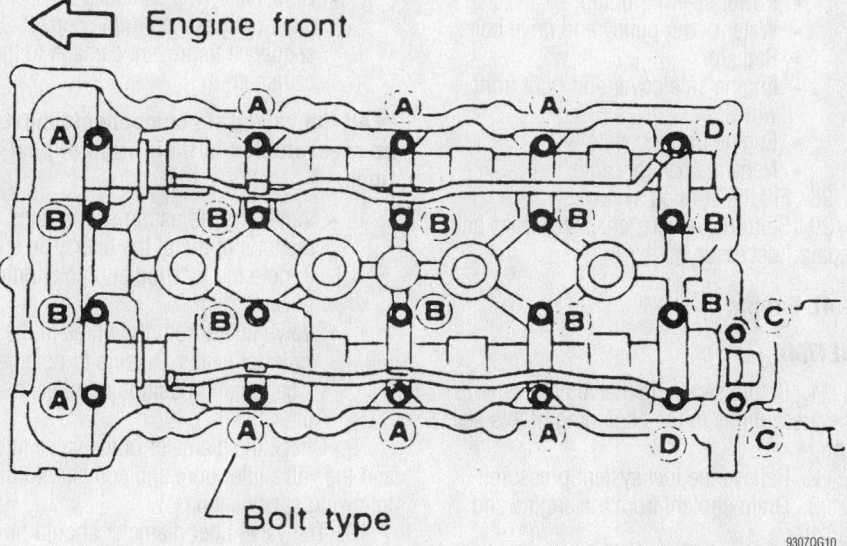

9307QG10

.........and lettered sequence shown. Refer to the text for the proper torque—2.0L engines

- Negative battery cable
- Engine undercovers
- Right front wheel and the engine side cover
- Radiator assembly
- Air duct from the intake manifold
- Drive belts and water pump pulley
- Alternator and the power steering pump from the engine
- Vacuum hoses, fuel hoses, wires and the electrical connections
- Distributor cap and ignition wires from the engine

- All spark plugs
- Rocker cover nuts in sequence
- Rocker cover and oil separator
- Intake manifold supports, oil filter bracket and the power steering bracket

5. Rotate the engine and set the No. 1 piston at Top Dead Center (TDC) on the compression stroke. Rotate crankshaft until mating marks on camshaft sprockets are in the correct position.

6. Remove or disconnect the following:

- Timing chain tensioner
- Distributor assembly. Matchmark the position of the distributor rotor and housing to the engine block for reinstallation purposes.
- Timing chain guide

➡ **Wire the camshaft sprockets to the timing chain to maintain proper timing chain position.**

7. Holding the flats of the camshaft sprockets, remove the mounting bolts. Remove the sprockets from the cam shafts.

➡ **Note the positioning of the pins on the end of the camshafts for installation purposes.**

8. Remove the oil tubes, baffle plate, camshaft brackets and the camshafts. It is important that all parts are kept in order for correct installation.

➡ **It is essential that the valvetrain components are kept in specific order for reassembly.**

9. Remove the rocker arms, shims, rocker arm guides and the hydraulic lash adjusters.

➡ **When the lifters are removed, keep them straight up or soak them in clean engine oil to prevent air from entering them.**

10. Measure the diameter of the lifters. The diameter should be 0.6685–0.6690 in. (16.980–16.993mm).

11. Measure the diameter of the lifter bores. The diameter should be 0.6693–0.6701 in. (17.000–17.020mm).

12. Standard clearance between the lash adjuster and the guide hole should be 0.0003–0.0016 in. (0.007–0.040mm).

To install:

➡ **The hydraulic lifters must be bled to remove the air from them. Air can not be bled from this type of lifter by running the engine.**

13. Bleed the lifters as follows:
   a. Submerse the lifter into a container of clean engine oil.
   b. While pushing the plunger, insert a thin rod into the check ball and lightly push the check ball.
   c. Air is completely bled when the plunger no longer moves.

14. Check all components for wear, replace as necessary and clean all mating surfaces.

For Maintenance Interval recommendations, see Section 1 of this manual

➡**Apply clean engine oil to all components prior to installation. Always replace the rocker arm guide with a new one.**

15. At this point it is necessary to perform valve adjustment. Adjust the valves as follows:

➡**It will be necessary to determine the proper shim size when replacing the valve, cylinder head, shim, rocker arm guide, or the valve seat.**

16. Insert a dial gauge into the lifter bore.
17. Before measuring, be sure the following parts are installed in the cylinder head.

- Valve
- Valve spring
- Collet
- Retainer
- Rocker arm guide (except shim)

18. On the shim side, measure the difference between contact surfaces of the rocker arm guide and the valve stem end.

➡**When measuring, lightly pull dial indicator rod toward you to eliminate play in tool.**

19. Using this reading, select the proper shim size.
20. Shims are available in thicknesses from 0.1102–0.1260 in. (2.800–3.200mm) in steps of 0.0010 in. (0.025mm).
21. Measure all the valves and select the proper shim sizes.
22. Remove the tool from cylinder head.
23. Install or connect the following:

- Valve lifters to the bores in the cylinder head
- Rocker arm guides, shims and the rocker arms to the cylinder head

24. Clean the left-hand camshaft end bracket and coat the mating surface with liquid gasket. Install the camshafts, camshaft brackets, oil tubes and the baffle plate.

❊❊ WARNING

Ensure that the left camshaft key is at the 12 o'clock position and the right camshaft key is also at the 12 o'clock position.

25. The procedure for tightening camshaft bolts must be followed exactly to prevent camshaft damage. Torque the bolts as follows:

   a. Right camshaft bolts No. 9 and 10 (in that order) to 18 inch lbs. (2 Nm), then torque bolts 1–8 (in that order) to the same specification.

   b. Left camshaft bolts No. 11 and 12

(in that order) to 18 inch lbs. (2 Nm), then torque bolts 1–10 (in that order) to the same specification.

   c. Torque all bolts in sequence to 51 inch lbs. (6 Nm).

   d. Torque all bolts in sequence to 78–102 inch lbs. (9–12 Nm), then tighten the 2 bolts that secure the distributor housing cap to 13–19 ft. lbs. (18–25 Nm).

26. Line up the mating marks on the timing chain and camshaft sprockets and install the timing chain and sprockets. Torque sprocket bolts to 101–116 ft. lbs. (137–157 Nm). Be sure to hold the flats of the camshaft when tightening the sprocket bolts.

27. Install or connect the following:

- Timing chain guide
- Intake manifold supports, oil filter bracket and power steering bracket
- Rocker cover and oil separator
- Spark plugs, distributor and plug wires
- Alternator
- Power steering pump
- Water pump pulley and drive belts
- Radiator
- Engine side cover and right front wheel
- Engine under covers
- Negative battery cable

28. Fill the cooling system.
29. Start the vehicle, check for leaks and repair if necessary.

2.4L Engine

ALTIMA

1. Before servicing the vehicle, refer to the precautions in the beginning of this section.
2. Relieve the fuel system pressure.
3. Drain coolant from the engine and radiator.
4. Remove or disconnect the following:

- Negative battery cable
- Air intake ducts and the air cleaner assembly
- Vacuum hoses, fuel hoses, wires, harness and connectors that are necessary for removal of the rocker cover
- Alternator and mounting bracket
- Upper radiator hose and cooling fan

5. Set the No. 1 piston at Top Dead Center (TDC) on its compression stroke.
6. Remove or disconnect the following:

- Spark plug wires from the spark plugs
- Distributor assembly. Matchmark and note the positioning of the dis-

tributor rotor and housing to the engine block before removing the distributor

- Rocker cover by loosening the bolts in the reverse order of installation

7. Wire the chain to the sprocket so the chain does not fall off during sprocket removal. Hold the flats of the camshaft with a wrench just behind the first camshaft bearing cap. Loosen the bolts and remove the sprockets.

8. Remove or disconnect the following:

- Camshaft sprockets

➡**The stoppers on camshaft covers prevent the upper timing chain from disengaging from the idle sprocket. Also, after removal of the camshaft sprockets, note the positioning of the pins at the end of the camshafts for reinstallation purposes.**

- Camshaft bearing caps in reverse order of installation
- Camshafts. The camshaft brackets must be loosened in the correct sequence to prevent damage to the camshaft

➡**All the valvetrain components must be reassembled in their original positions.**

- Valve lifter adjusting shims from the tops of the of the lifters. Be sure to note the location and positioning of each shim.
- Valve lifters from the bores in the cylinder heads. Be sure to note the location and positioning of each lifter

9. Check the diameter of the valve lifter and the valve lifter bore and compare to the following specifications.
10. The valve lifter diameter should be 1.3370–1.3376 in. (33.960–33.975mm).
11. The lifter guide bore diameter should be 1.3386–1.3394 in. (33.960–33.975mm).
12. The valve lifter to lifter guide bore clearance should be 0.0010–0.0024 in. (0.025–0.061mm).

To install:

➡**When installing the valve components, apply a coat of clean engine oil to the component.**

13. Install or connect the following:

- Lifters into the lifter bores from which they were removed
- Valve shims to the lifters from which they came

14. Install the camshafts in the same position as noted during removal and

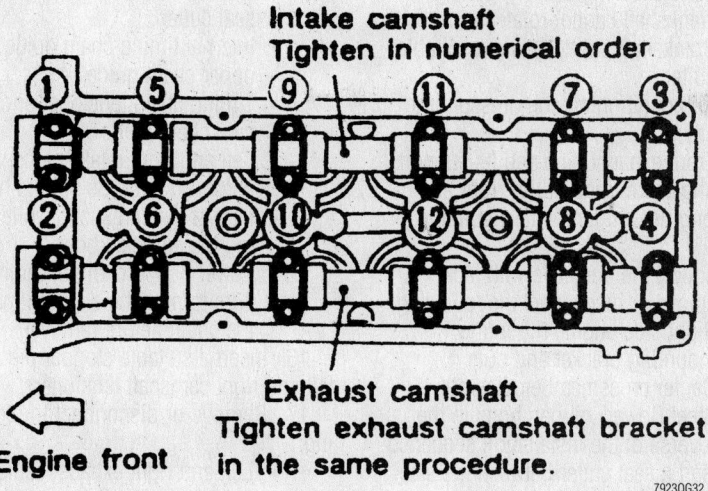

Intake camshaft
Tighten in numerical order.

Exhaust camshaft
Tighten exhaust camshaft bracket
in the same procedure.

Engine front

7923QG32

Tighten the camshaft bearing caps in sequence to prevent damage to the camshaft and cylinder head—2.4L engines

camshaft bearing caps; torque cap bolts in the proper sequence as follows:
   a. Step 1: 17 inch lbs. (2 Nm).
   b. Step 2: 81–104 inch lbs. (9–12 Nm).

➥**When installing the timing chain and sprockets, align the marks on the sprockets with the colored links of the chain.**

   15. Install or connect the following:
   • Camshaft sprockets with the timing chain and torque the bolts to 123–130 ft. lbs. (167–177 Nm). Install the chain guide between both camshaft sprockets. The alignment marks on the upper portion of the timing chain should now be aligned with the marks on the sprockets
   • Rocker cover
   • Distributor
   • Spark plugs and wires
   • Upper radiator hose and cooling fan
   • Alternator
   • Vacuum hoses, fuel lines and electrical connectors
   • Air cleaner assembly
   • Negative battery cable
   16. Fill the cooling system.
   17. Start the vehicle, check for leaks and repair if necessary.

240SX

   1. Before servicing the vehicle, refer to the precautions in the beginning of this section.

   2. Release fuel system pressure.
   3. Remove or disconnect the following:
   • Negative battery cable
   • Air intake ducts and the air cleaner assembly
   • Vacuum hoses from valve cover. The fuel rail may need to be removed to access valve cover nuts.
   • Cooling fan and the radiator fan shroud
   • Spark plug wires
   • Valve cover and turn the crankshaft to align the timing marks at Top Dead Center (TDC) on No. 1 cylinder
   • Distributor and matchmark the position of the distributor rotor and the distributor before removing the distributor
   • Upper timing chain cover
   • Upper timing chain guide and tensioner
   4. Wire the chain to the sprocket so the chain does not fall off during sprocket removal. Hold the flats of the camshaft with a wrench just behind the first camshaft bearing cap. Loosen the bolts and remove the sprockets.

➥**The stoppers on camshaft covers prevent the upper timing chain from disengaging from the idle sprocket.**

➥**After removal of the camshaft sprockets, note the positioning of the pins at the end of the camshafts for reinstallation purposes.**

   5. Remove the cam bearing caps in reverse order of the tightening sequence and

remove the camshafts. The camshaft brackets must be loosened in the correct sequence to prevent damage to the camshaft.

➥**These parts should be reassembled in their original position. Bolts should be loosened in 2–3 steps (loosen all bolts in the reverse of the tightening order).**

   6. Remove or disconnect the following:
   • Valve lifter adjusting shims from the tops of the lifters
   • Valve lifters from the bores in the cylinder heads
   7. Check the diameter of the valve lifter and the valve lifter bore and compare to the following specifications.
   8. The valve lifter diameter should be 1.3370–1.3376 in. (33.960–33.975mm).
   9. The lifter guide bore diameter should be 1.3386–1.3394 in. (33.960–33.975mm).
   10. The valve lifter to lifter guide bore clearance should be 0.0010–0.0024 in. (0.025–0.061mm).
   To install:

➥**Apply a clean coat of new engine oil to the valvetrain components.**

   11. Install or connect the following:
   • Lifters into the lifter bores from which they were removed
   • Valve shims to the lifters from which they came
   • Camshafts in the same position as noted during removal and install the camshaft bearing caps
   12. Torque bearing cap bolts in the proper sequence as follows:
   a. Step 1: 17 inch lbs. (2 Nm).
   b. Step 2: 80–104 inch lbs. (9–12 Nm).

➥**When installing the timing chain and sprockets, align the marks on the sprockets with the colored links of the chain.**

   13. Install or connect the following:
   • Camshaft sprockets and torque the bolts to 123–130 ft. lbs. (167–177 Nm)
   • Timing chain
   • Upper timing chain guide and tensioner
   • Timing chain cover
   • Distributor
   • Valve cover
   • Spark plug wires
   • Cooling fan and shroud
   • Vacuum hoses and fuel lines, if removed

- Air cleaner assembly
- Negative battery cable

14. Start the engine and make the necessary adjustments. Check for proper operation and leaks.

3.0L Engine

1. Before servicing the vehicle, refer to the precautions in the beginning of this section.
2. Relieve the fuel system pressure.
3. Drain the engine oil.
4. Drain the cooling system.
5. Remove or disconnect the following:
   - Negative battery cable
   - Left side rocker cover ornament

➡ **Before detaching any hoses or connectors, note the locations for reassembly.**

- Air duct to intake manifold hose, collector hose, blow-by hose and vacuum hoses
- Fuel hoses and detach the harness connections
- Canister purge hoses
- Water hoses from the cylinder head and intake manifold
- All 6 ignition coils from the spark plugs
- Spark plugs
- Bolts that secure the Exhaust Gas Recirculation (EGR) tube
- EGR tube
- Intake manifold collector supports and the collector
- Bolts that secure the fuel tube and the fuel tube
- Bolts that secure the intake manifold to the engine block and the manifold. Loosen the bolts in the reverse sequence of the tightening procedure
- Left-hand and right-hand rocker covers from the cylinder head
- Engine undercovers
- Right front wheel and engine side covers
- Drive belts and idler pulley
- Power steering oil pump belt and the power steering oil pump assembly
- Camshaft Position (CMP) sensor (PHASE) and Crankshaft Position (CKP) sensors (REF)/(POS)

6. Set the No. 1 piston to Top Dead Center (TDC) of compression stroke by rotating the crankshaft.
7. Remove or disconnect the following:
   - Ring gear cover access plate. Loosen the crankshaft pulley bolt

while securing the ring gear so the crankshaft cannot rotate
- Crankshaft pulley, using a suitable puller
- Air conditioning compressor and bracket
- Front exhaust pipe and its support

8. Hang the engine at the right and left side engine slingers with a suitable hoist.
9. Support the transaxle with jack.
10. Remove or disconnect the following:
   - Right side engine mounting, mounting bracket and nuts
   - Center crossmember assembly
   - Steel (lower) oil pan bolts in the reverse of the installation sequence

11. Insert a seal cutter between the steel and aluminum oil pan
12. Tapping the cutter with a hammer, slide it around the entire edge of the oil pan. Be careful not to damage the aluminum mating surface of the upper oil pan.
13. Remove or disconnect the following:
   - Steel oil pan and the oil strainer
   - Aluminum (upper) oil pan bolts in the reverse of the installation sequence
   - Transaxle bolts that secure the oil pan

14. Insert a seal cutter between the aluminum oil pan and the engine block.
15. Tapping the cutter with a hammer, slide it around the entire edge of the oil pan. Be careful not to damage the mating surfaces of the oil pan or engine block.
   - Oil pan from the vehicle
   - Water pump cover and the bolts that secure the front timing chain case cover

- Timing chain case cover, using the seal cutter
- Internal timing chain guide and the upper chain guide
- Timing chain tensioner and slack side chain guide
- Left and right intake camshaft sprockets first. Be sure to hold the flats of the camshafts while removing the sprocket bolts
- Lower timing chain assembly. Be sure to note the aligning marks of the chain before removal

16. Insert a suitable stopper pin for the left and right camshaft tensioners.
17. Remove or disconnect the following:

- Left and right exhaust camshaft sprocket bolts. Be sure to hold the flats of the camshafts while removing the sprocket bolts
- Upper timing chain assembly. Be sure to note the aligning marks of the chain before removal
- Lower timing chain guide
- Crankshaft sprocket
- Bolts that secure the rear timing chain case. The bolts must be loosened in sequence
- Rear timing case cover, using the seal cutter

➡ **Remove the O-rings from the front of the engine block.**

- Camshaft bearing caps in several steps. The bearing caps MUST be loosened in sequence

➡ **Keep all bearing caps and camshafts in proper order for installation.**

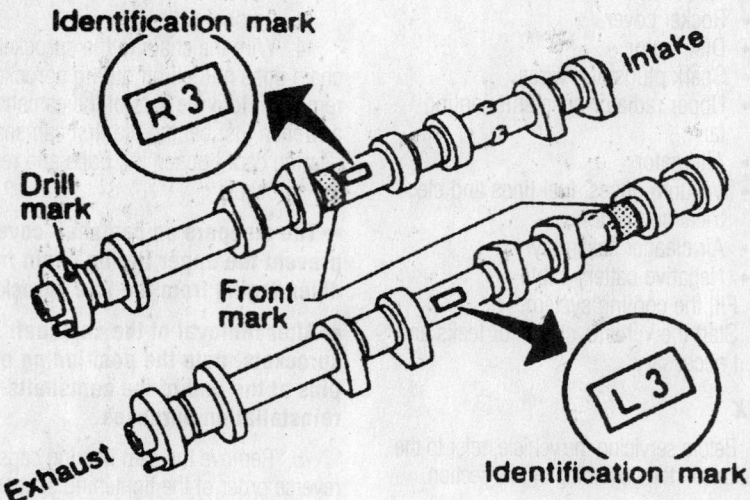

Camshaft identification marks—3.0L engines

- Left-hand and right-hand camshaft tensioners from the cylinder head
- Camshafts from the cylinder heads

➡**The valve lifters have a replaceable shim on the top of the lifter. Note the proper locations of each shim to lifter and remove the shims from the lifters.**

- Valve adjusting shim from the lifter, using a magnet
- Lifter assembly from the bore. Be sure to note the locations from where each lifter came

18. Check the diameter of the valve lifter and the valve lifter guide bore.

19. The diameter of the lifter should be 1.3764–1.3770 in. (34.960–34.975mm) and the diameter of the bore should be 1.3780–1.3788 in. (35.000–35.021mm).

20. Remove all traces of liquid gasket from the timing chain case and from the water pump covers.

21. Remove all traces of liquid gasket from the engine block.

22. Inspect the camshafts for excessive wear or damage and replace as necessary.

To install:

➡**Before installing the camshaft brackets, apply RTV sealant to the mating surface of the No. 1 journal head.**

23. Lubricate the valve lifters with clean engine oil and install the lifters into the bore from which they were removed.

24. Lubricate the valve lifter shims with clean engine oil and install the shims into the lifter from which they were removed.

25. Turn the crankshaft clockwise until the No. 1 piston is set 240 degrees before TDC on compression stroke.

26. Install or connect the following:

- Camshaft tensioners on both sides of the cylinder heads and torque the bolts to 75–96 inch lbs. (8.4–10.8 Nm)

➡**The camshafts can be identified by the paint marks on the camshaft. The left cylinder head camshafts have a YELLOW paint mark and the right cylinder head camshafts have a WHITE paint mark.**

- Exhaust and intake camshafts and install the bearing caps. Before installing the No. 1 bearing cap, apply liquid gasket to the corners of the cap

➡**When installing the camshafts, position the camshaft keys at the 12 o'clock**

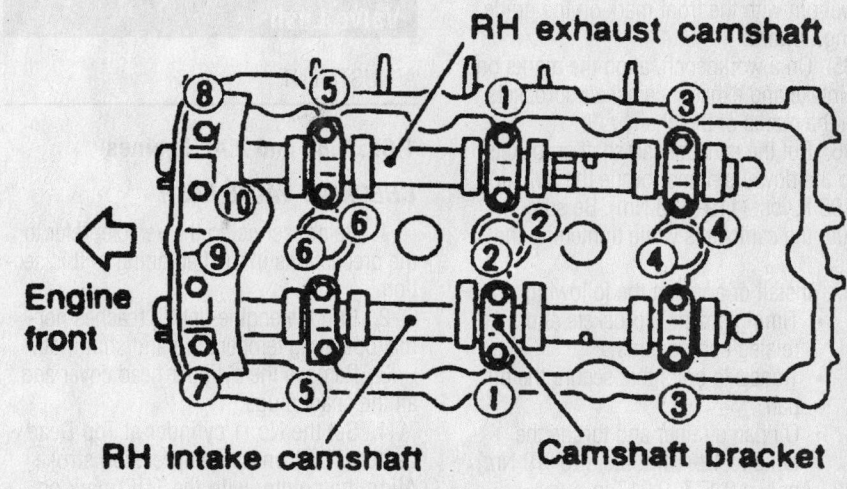

Right cylinder head camshaft bearing cap tightening sequence—3.0L engines

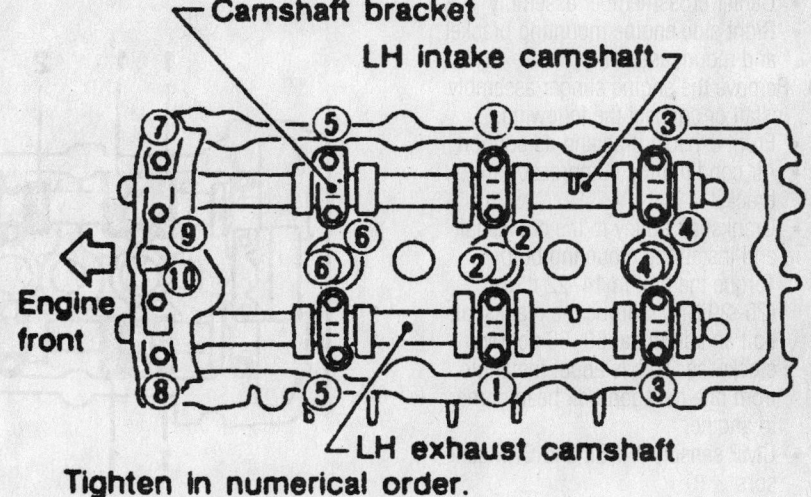

Left cylinder head camshaft bearing cap tightening sequence—3.0L engines

position in respect to the cylinder head angle.

27. Torque the camshaft bearing caps as follows:

  a. Bolts No. 7–10: 17 inch lbs. (2 Nm).

  b. Bolts No. 1–6: 17 inch lbs. (2 Nm).

  c. Bolts No. 1–10: 52 inch lbs. (6 Nm).

  d. Bolts No. 1–10: 81–104 inch lbs. (9–11 Nm).

28. Install new O-rings to the front of the engine block.

29. Apply sealant to the hatched portion of the of the rear timing chain case.

30. Align the rear timing chain case with the dowel pins and install onto the cylinder heads and engine block.

31. Torque the rear timing chain case mounting bolts in sequence to 105–121 inch lbs. (11.8–13.7 Nm).

32. Install the crankshaft sprocket with the mating mark facing out.

33. Rotate the crankshaft clockwise and position the crankshaft to TDC of compression stroke and align the dowels of the camshaft sprockets to the 12 o'clock position in respect to the cylinder head.

34. Install the lower chain guide on the

dowel pin with the front mark on the guide facing upward.

35. On a workbench, align the marks on the intake and exhaust camshaft sprockets with the marks of the chain.

36. Put the exhaust camshaft sprockets onto the dowel pin and torque the bolts to 88–95 ft. lbs. (119–128 Nm). Be sure to secure the camshafts while tightening the bolts.

37. Install or connect the following:
- Timing chains, sprockets and related components
- Transaxle bolts that secure the oil pan
- Oil pan strainer and torque the bolts to 12–14 ft. lbs. (16–19 Nm)

38. Apply a 0.177–0.217 in. (4.5–5.5mm) continuous bead of liquid gasket to the lower oil pan mating surface and install the oil pan. Torque the bolts in sequence to 57–66 inch lbs. (6.4–7.5 Nm).

39. Install or connect the following:
- Center crossmember assembly
- Right side engine mounting bracket and mount assembly

40. Remove the engine slinger assembly.

41. Install or connect the following:
- Front exhaust pipe and its support
- Air conditioning compressor and bracket
- Crankshaft pulley to the crankshaft and install the mounting bolt. Torque the bolt to 14–22 ft. lbs. (20–29 Nm). Torque the crankshaft bolt an additional 60–66 degrees clockwise. This is about the angle from one hexagon bolt head corner to another
- CMP sensor, PHASE and CKP sensors
- Power steering pump
- Idler pulley and all belts
- Engine side and under covers
- Right front wheel
- Intake manifold
- Rocker covers
- Fuel tube
- Intake manifold support and collector
- EGR tube
- Spark plugs and ignition coils
- Coolant hoses
- Canister purge hoses
- Fuel feed and return lines
- Vacuum hoses
- Negative battery cable

42. Fill the cooling system.

43. Fill the engine with clean oil.

44. Start the vehicle, check for leaks and repair if necessary.

Valve Lash

ADJUSTMENT

1.6L, 1.8L and 2.4L Engines

CHECKING VALVE LASH

1. Before servicing the vehicle, refer to the precautions in the beginning of this section.

2. Run the engine until it reaches normal operating temperature and shut if off.

3. Remove the cylinder head cover and all the spark plugs.

4. Set the No. 1 cylinder at Top Dead center (TDC) on its compression stroke. Align the pointer with the TDC mark on the crankshaft pulley. Check that the valve lifters on the No. 1 cylinder are loose and valve lifters on the No. 4 cylinder are tight. If not, turn the crankshaft 1 revolution (360 degrees) and align the pointer with the TDC mark on the crankshaft pulley.

5. Check the following valves:
- Both No. 1 intake valves
- Both No. 1 exhaust valves
- Both No. 2 intake valves
- Both No. 3 exhaust valves

6. Using a feeler gauge, measure the clearance between the valve lifter and the camshaft. Record any valve clearance measurements which are out of specification.

7. Turn the crankshaft 1 revolution (360 degrees) and align the mark on the crankshaft pulley with the pointer. Check the following valves:
- Both No. 2 exhaust valves
- Both No. 3 intake valves
- Both No. 4 intake valves
- Both No. 4 exhaust valves

8. Using a feeler gauge, measure the clearance between the valve lifter and the camshaft. Record any valve clearance measurements which are out of specification.

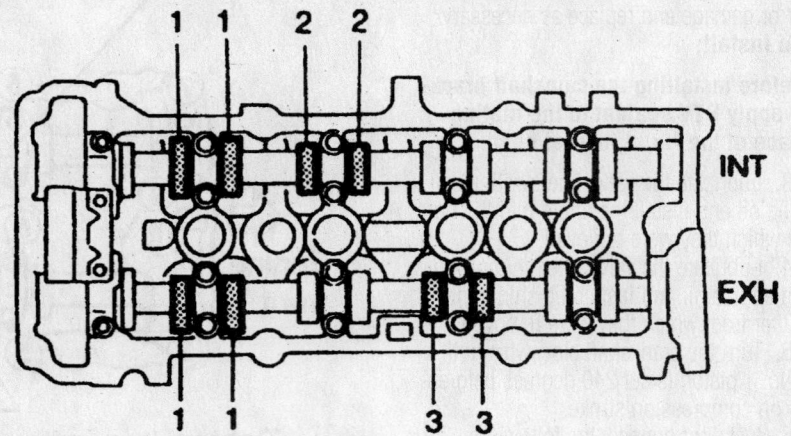

Measure the clearance of the valves indicated when the No. 1 piston is at TDC on compression—1.6L, 1.8L and 2.4L engines

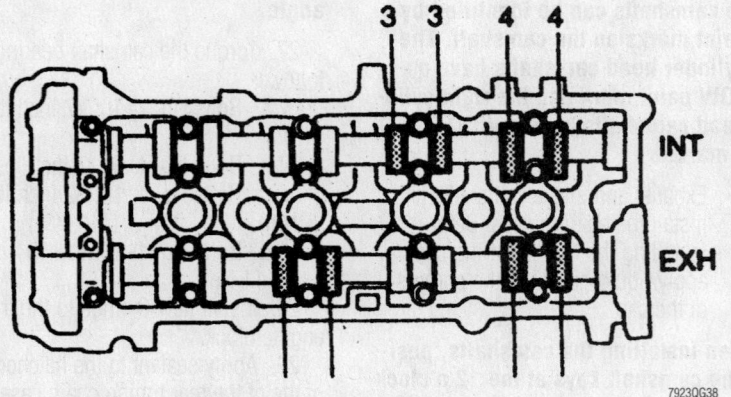

Measure the clearance of the valves indicated when the No. 4 piston is at TDC on compression—1.6L and 2.4L engines

9. If all the valve clearances are within specification, install the cylinder head cover and the spark plugs.

ADJUSTING VALVE LASH

1. Before servicing the vehicle, refer to the precautions in the beginning of this section.

2. If an adjustment is necessary, adjust the valve clearance while engine is cold by removing the adjusting shim. The adjusting shim can be removed by using the following procedures:

    a. Turn the crankshaft so the camshaft lobe of the valve to be adjusted is pointed straight up.

    b. Turn the lifter so the notch is pointed towards the center of the cylinder head; this will facilitate the shim removal process.

    c. Using a depressor tool, push down on the lifter and insert a keeper tool on the edge of the lifter to keep the lifter in the depressed position.

    d. Remove the depressor tool and remove the shim with a magnet.

3. Determine the replacement adjusting shim size by using the following procedures and formula:

    a. Using a micrometer determine thickness of the removed shim.

    b. Calculate the thickness of a new adjusting shim so valve clearance is within the specified values.

    c. R = thickness of the removed shim.

    d. N = thickness of the new shim.

    e. M = measured valve clearance.

- 1.6L and 1.8L engines: Intake shim determination formula: N = R + (M—0.0146 in. or 0.37mm)
- 1.6L and 1.8L engines: Exhaust shim determination formula: N = R + (M—0.0157 in. or 0.40mm)
- 2.4L engine (240SX models): Intake shim determination formula: N = R + (M—0.0146 in. or 0.37mm)
- 2.4L engine (240SX models): Exhaust shim determination formula: N = R + (M—0.0146 in. or 0.37mm)
- 2.4L engine (Altima models): Intake shim determination formula: N = R + (M—0.0138 in. or 0.35mm)
- 2.4L engine (Altima models): Exhaust shim determination formula: N = R + (M—0.0146 in. or 0.37mm)

Shims are available in different sizes from 0.0772–0.1055 in. (1.96–2.68mm) in increments of 0.0008 in. (0.02mm). The thickness is stamped on the shim; this side is always installed facing down. Select new shims with thickness as close as possible to calculated valve and install it in the lifter.

4. Install the new shim onto the lifter.

5. Depress the lifter and remove the keeper tool. Remove the depressor tool and recheck the valve clearance. Repeat this procedure for any other valves requiring adjustment.

6. Install the cylinder head cover and spark plugs when all valve adjustments are finished.

2.0L Engines

The engine is equipped with hydraulic lash adjusters. The valve lash is not adjustable.

3.0L Engine

➡**Check and adjust the valve clearances while the engine is cold and not running.**

CHECKING VALVE LASH

1. Before servicing the vehicle, refer to the precautions in the beginning of this section.

2. Remove or disconnect the following:
- Intake manifold collector
- Left and right rocker covers
- Spark plugs

3. Set the No. 1 cylinder at Top Dead center (TDC) on its compression stroke. Align the pointer with the TDC mark on the crankshaft pulley. Check that the valve lifters on the No. 1 cylinder are loose and valve lifters on the No. 4 cylinder are tight. If not, turn the crankshaft 1 revolution (360 degrees) and align the pointer with the TDC mark on the crankshaft pulley.

4. Check the following valves:
- Both No. 1 intake valves
- Both No. 2 exhaust valves
- Both No. 3 exhaust valves
- Both No. 6 intake valves

5. Using a feeler gauge, measure the clearance between the valve lifter and the

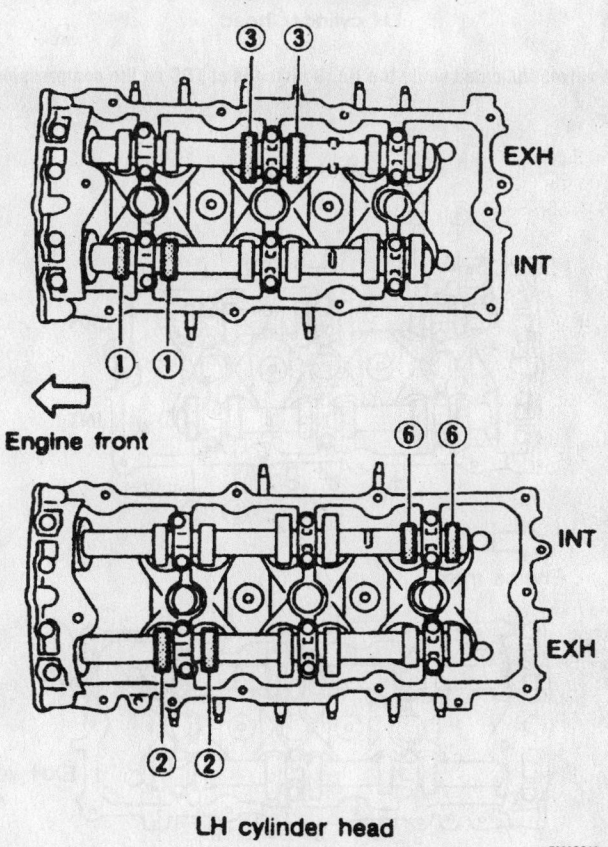

RH cylinder head

Engine front

LH cylinder head

7923QG39

Measure the valves indicated while the No. 1 piston is at TDC on the compression stroke—3.0L engine

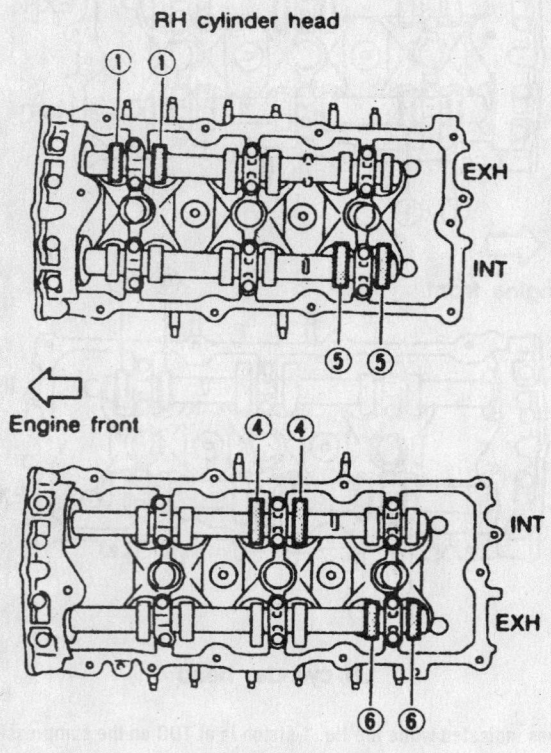

RH cylinder head

Engine front

LH cylinder head

79230G40

Measure the valves indicated while the No. 3 piston is at TDC on the compression stroke—3.0L engine

RH cylinder head

Engine front

LH cylinder head

79230G41

Measure the valves indicated while the No. 5 piston is at TDC on compression—3.0L engine

camshaft. Record any valve clearance measurements that are out of specification. Intake valve clearance (cold) is 0.010–0.013 in. (0.26–0.34mm) and exhaust valve clearance (cold) is 0.011–0.015 in. (0.29–0.37mm).

6. Turn the crankshaft 240 degrees and set the No. 3 cylinder to TDC of its compression stroke.

7. Check the following valves:
   • Both No. 2 intake valves
   • Both No. 3 intake valves
   • Both No. 4 exhaust valves
   • Both No. 5 exhaust valves

8. Using a feeler gauge, measure the clearance between the valve lifter and the camshaft. Record any valve clearance measurements that are out of specification. Intake valve clearance (cold) is 0.010–0.013 in. (0.26–0.34mm) and exhaust valve clearance (cold) is 0.011–0.015 in. (0.29–0.37mm).

9. Turn the crankshaft 240 degrees and set the No. 5 cylinder to TDC of its compression stroke.

10. Check the following valves:
   • Both No. 1 exhaust valves
   • Both No. 4 intake valves
   • Both No. 5 intake valves
   • Both No. 6 exhaust valves

11. Using a feeler gauge, measure the clearance between the valve lifter and the camshaft. Record any valve clearance measurements that are out of specification. Intake valve clearance (cold) is 0.010–0.013 in. (0.26–0.34mm) and exhaust valve clearance (cold) is 0.011–0.015 in. (0.29–0.37mm)

12. If all the valve clearances are within specification, install the cylinder head cover, spark plugs and the intake manifold collector.

ADJUSTING VALVE LASH

1. Before servicing the vehicle, refer to the precautions in the beginning of this section.

2. If an adjustment is necessary, adjust the valve clearance while engine is cold by removing the adjusting shim. The adjusting shim can be removed by using the following procedures:

   a. Turn the crankshaft so the camshaft lobe of the valve to be adjusted is pointed straight up.

   b. Turn the lifter so the notch is pointed towards the center of the cylinder head; this will facilitate the shim removal process.

   c. Using a depressor tool, push down on the lifter and insert a keeper tool on the edge of the lifter to keep the lifter in the depressed position.

d. Remove the depressor tool and remove the shim with a magnet.

➡**Compressed air can be blown into the hole of the lifter to separate the adjusting shim from the lifter.**

3. Determine the replacement adjusting shim size by using the following procedures and formula:

a. Using a micrometer determine thickness of the removed shim.

b. Calculate the thickness of a new adjusting shim so valve clearance is within the specified values.

c. R = thickness of the removed shim.

d. N = thickness of the new shim.

e. M = measured valve clearance.

- Intake shim determination formula: N = R + (M—0.0118 in. or 0.30mm)

- Exhaust shim determination formula: N = R + (M—0.0130 in. or 0.33mm)

4. Shims are available in 64 sizes from 0.0913–0.1161 in. (2.32–2.95mm) in steps of 0.004 in. (0.01mm). The thickness is stamped on the shim; this side is always installed facing down. Select new shims with thickness as close as possible to calculated valve and install it in the lifter.

5. Install the new shim onto the lifter.

6. Depress the lifter and remove the keeper tool. Remove the depressor tool and recheck the valve clearance. Repeat this procedure for any other valves requiring adjustment.

7. When all valve adjustments are finished, install the cylinder head cover, spark plugs and the intake manifold collector.

Starter Motor

REMOVAL & INSTALLATION

240SX

WITH AUTOMATIC TRANSMISSION

1. Before servicing the vehicle, refer to the precautions in the beginning of this section.

2. Support the transmission with a jack and remove the 4 rear mounting bracket bolts.

3. Lightly lower the transmission to make room and pull out the dipstick tube.

4. Remove or disconnect the following:

- Negative battery cable
- Connector bracket from the front mount bracket

- Harness connector
- Starter bolts
- Starter

To install:

5. Install or connect the following:

- Starter and torque the bolts to 30–37 ft. lbs. (40–50 Nm)
- Harness connector
- Connector bracket to the front mount bracket
- Dipstick tube
- Transmission into position
- 4 rear mounting bracket bolts
- Negative battery cable

6. Remove the transmission jack

WITH MANUAL TRANSMISSION

1. Before servicing the vehicle, refer to the precautions in the beginning of this section.

2. Remove or disconnect the following:

- Negative battery cable
- Connector bracket from the front mount bracket
- Harness connector
- Starter bolts
- Starter

To install:

3. Install or connect the following:

- Starter and torque the bolts to 30–37 ft. lbs. (40–50 Nm)
- Harness connector
- Connector bracket to the front mount bracket
- Negative battery cable

1.6L, 1.8L and 2.0L Engines

1. Before servicing the vehicle, refer to the precautions in the beginning of this section.

2. Remove or disconnect the following:

- Negative battery cable
- Wiring at the starter
- Bolts attaching the starter to the engine
- Starter

To install:

3. Install or connect the following:

- Starter and torque the bolts to 70 inch lbs. (8 Nm)
- Starter electrical connections
- Negative battery cable

Maxima

1. Remove or disconnect the following:

- Negative battery cable
- Air duct
- Harness protector from the harness
- Starter wiring at the starter
- Starter-to-engine bolts
- Starter from the vehicle

To install:

2. Install or connect the following:

- Starter and torque the long bolt to 57–72 ft. lbs. (77–98 Nm) and the short bolt to 22–30 ft. lbs. (30–41 Nm)
- Starter wiring
- Harness protector

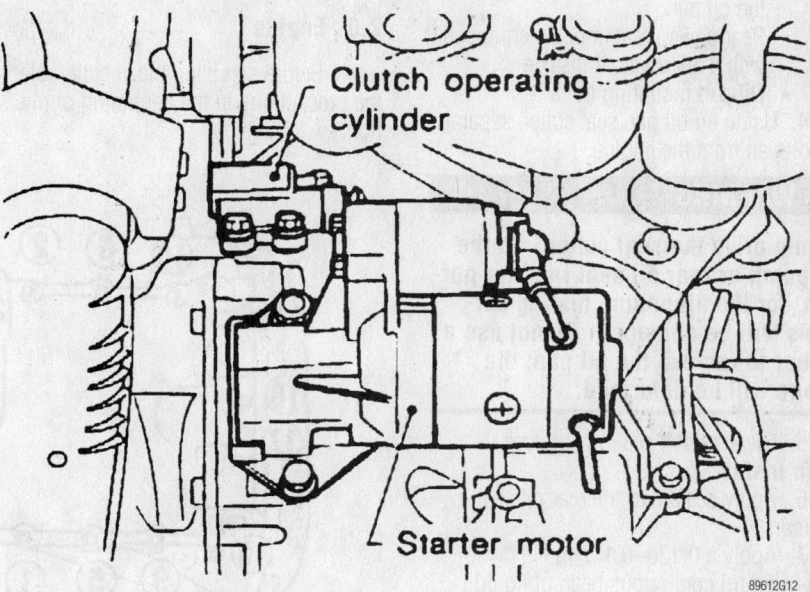

Starter location and mounting detail—Maxima with a manual transaxle

Timing belt service is covered in Section 3 of this manual

- Air duct
- Negative battery cable

Altima

1. Remove or disconnect the following:
   - Negative battery cable
   - Air inlet tube
   - Harness bracket
   - Wiring at the starter
   - Starter

To install:

2. Install or connect the following:
   - Starter and torque the bolts to 60 inch lbs. (7 Nm)
   - Starter electrical connectors
   - Harness bracket
   - Air inlet tube
   - Negative battery cable

Oil Pan

REMOVAL & INSTALLATION

1.6L and 1.8L Engines

1. Before servicing the vehicle, refer to the precautions in the beginning of this section.
2. Drain the engine oil.
3. Remove or disconnect the following:
   - Negative battery cable
   - Engine undercovers
   - Front exhaust tube and properly support the transaxle assembly
   - Center crossmember
   - Support brackets from the sides of the oil pan
   - Rear cover plate, models equipped with a automatic transaxle
   - Oil pan mounting bolts
4. Using an oil pan seal cutter, separate the oil pan from the engine.

❊❊ WARNING

Do not drive the seal cutter into the oil pump or rear oil seal retainer portion, for the aluminum mating surfaces will be damaged. Do not use a prybar to remove the oil pan; the flange will be deformed.

5. Clean all the sealing surfaces.

To install:

6. Apply sealant to the rear oil seal retainer.
7. Apply a 0.128–0.177 in. (3.5–4.5mm) continuous bead of liquid gasket to the oil pan mating surface.
8. Install or connect the following:
   - Oil pan and torque bolts, in

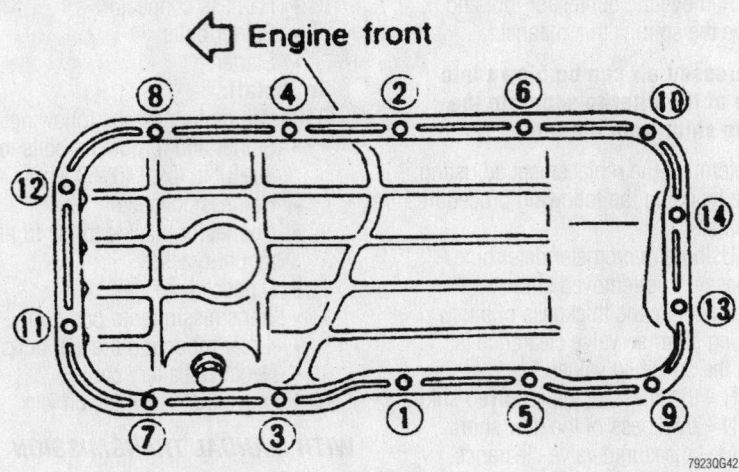

Tighten the oil pan bolts in the correct sequence to prevent oil leakage—1.6L and 1.8L engines

79230G42

sequence, to 56–73 inch lbs. (6.3–8.3 Nm)
   - Rear cover plate, models equipped with a automatic transaxle
   - Oil pan support brackets
   - Center crossmember
   - Front exhaust tube
   - Engine undercovers
   - Oil pan plug, using a new gasket and tighten the plug to 21–28 ft. lbs. (7–8 Nm)
   - Negative battery cable
9. After 30 minutes of gasket curing time, refill the oil pan with the specified quantity of clean oil.
10. Start the vehicle, check for leaks and repair if necessary.

2.0L Engine

1. Before servicing the vehicle, refer to the precautions in the beginning of this section.

2. Drain the engine oil.
3. Remove or disconnect the following:
   - Negative battery cable
   - Engine undercover
   - Lower steel oil pan bolts in the reverse of installation sequence
   - Steel oil pan. Insert a cutting tool between steel oil pan and aluminum oil pan. Tap the tool around the perimeter of the pan to cut the gasket material.
   - Oil baffle bolts and oil baffle
   - Front exhaust tube and set a suitable jack under the transaxle and raise the engine
   - Center crossmember from the vehicle
   - Transaxle shift control cable, if equipped with an automatic transaxle
   - Compressor gussets and the rear cover plate
   - Aluminum oil pan bolts. Loosen

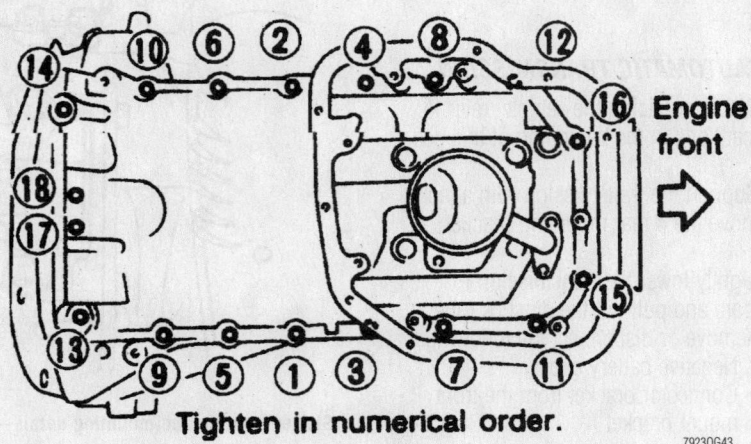

Tighten in numerical order.

79230G43

Aluminum oil pan bolt tightening sequence—2.0L engine

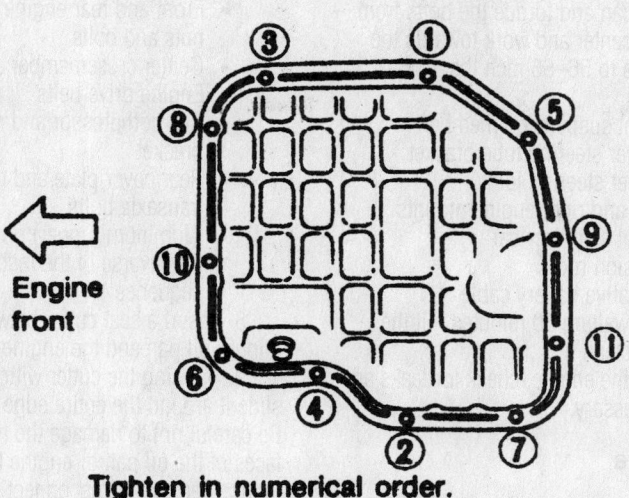

Tighten in numerical order.

7923QG44

Steel oil pan bolt tightening sequence—2.0L engine

aluminum oil pan bolts in reverse order of the tightening sequence.
- 2 transaxle mounting bolts and refit the them into vacant holes at the bottom of the oil pan. Use a cutting tool to cut the gasket material.
- 2 transaxle mounting bolts that were relocated and the pan from the vehicle

To install:

4. Clean the oil pan rail of all liquid gasket and apply a new bead of 5/32 in. (4.5mm) thickness to the aluminum oil pan rail.

5. Install the aluminum oil pan. Torque the bolts in the opposite order of removal as follows:

   a. Bolts No. 1–16: 12–14 ft. lbs. (16–19 Nm).

   b. Bolts No. 17–18: 60–72 inch lbs. (6–8 Nm).

6. Install or connect the following:
- 2 transaxle mounting bolts, rear cover plate and the compressor gussets
- Automatic transmission shift control cable (if equipped)
- Center crossmember member, front exhaust tube and the baffle plate and torque the bolts to 56–66 inch lbs. (6.4–7.5 Nm)

7. Clean the steel oil pan rail of all liquid gasket and apply a new bead of 5/32 in. (4.5mm) thickness to the steel oil pan rail.

8. Install or connect the following:
- Steel oil pan and torque the bolts in the proper sequence to 56–66 inch lbs. (6.4–7.5 Nm)
- Negative battery cable

9. After 30 minutes, refill the engine with clean oil

2.4L Engine

ALTIMA

1. Before servicing the vehicle, refer to the precautions in the beginning of this section.

2. Drain the engine oil.

3. Remove or disconnect the following:
- Negative battery cable
- Engine undercover
- Bolts securing the steel oil pan to

the aluminum oil pan in reverse order of the tightening sequence

4. Install a seal cutter between the steel oil pan and the aluminum oil pan

5. Tapping the cutter with a hammer, slide it around the entire edge of the oil pan. Take care not to damage the aluminum oil pan.

6. Remove or disconnect the following:
- Steel oil pan
- Baffle plate and oil strainer
- Front exhaust tube
- Front suspension member
- A/C compressor gussets
- Rear cover plate
- Aluminum oil pan retaining bolts in reverse order of the tightening sequence

7. Insert a seal cutter between the oil pan and the cylinder block.

8. Tapping the cutter with a hammer, slide it around the entire edge of the oil pan. Take care not to damage the aluminum oil pan.

9. Lower the oil pan from the cylinder block and remove it from the engine.

To install:

10. Carefully scrape the old gasket material away from the pan and cylinder block mounting surfaces, then apply a continuous 3.5–4.5mm wide bead of liquid gasket around the oil pan. Install the pan within 5 minutes or else this step will have to be repeated.

11. Install or connect the following:

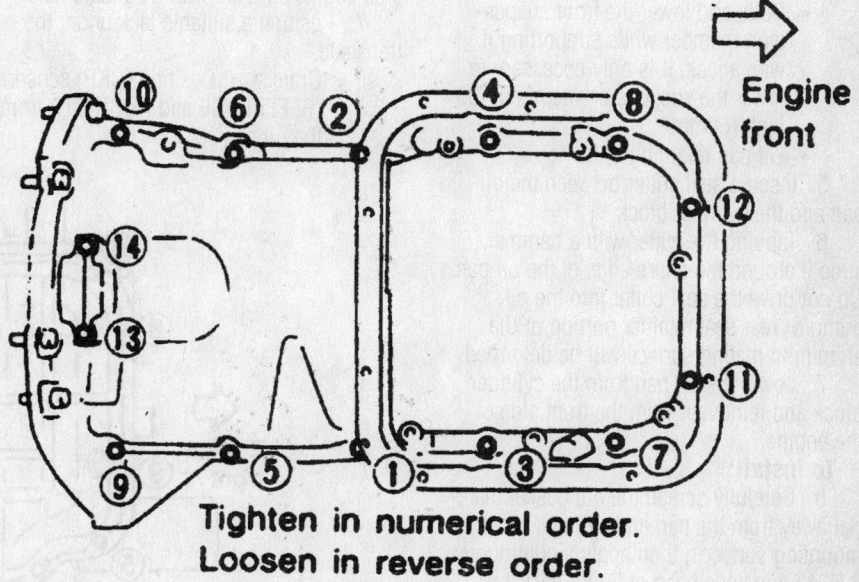

Tighten in numerical order.
Loosen in reverse order.

7923QG45

Oil pan bolt loosening and tightening sequence—2.4L engine

Heater Core replacement is covered in Section 2 of this manual

- Aluminum oil pan and torque the bolts, in sequence, to 13 ft. lbs. (17.5 Nm)
- Baffle plate
- Steel oil pan. Torque the bolts in sequence to 61 inch lbs. (7 Nm)
- Rear cover plate
- Front suspension member
- Front exhaust tube
- A/C compressor gussets
- Front suspension member
- Engine undercovers
- Negative battery cable

12. Wait 30 minutes before refilling the crankcase with clean oil.

13. Fill the engine with clean oil.

14. Start the vehicle, checks for leaks and repair if necessary.

240SX

1. Before servicing the vehicle, refer to the precautions in the beginning of this section.

2. Position a hoist on the engine and support the engine.

3. Drain the engine oil.

4. Remove or disconnect the following:
- Negative battery cable
- Tension rod bolts at the transverse link
- Front stabilizer bar from the side member
- Left and right engine mounting bolts
- Lower steering joint
- Power steering tube bracket at the left tension rod
- Bolts and lower the front suspension member while supporting it with a jack. It is only necessary to lower the suspension member 2.36 inch (60mm).
- Oil pan retaining bolts

5. Insert a seal cutter between the oil pan and the cylinder block.

6. Tapping the cutter with a hammer, slide it around the entire edge of the oil pan. Do not drive the seal cutter into the oil pump or rear seal retainer portion or the aluminum mating surface will be deformed.

7. Lower the oil pan from the cylinder block and remove it from the front side of the engine.

To install:

8. Carefully scrape the old gasket material away from the pan and cylinder block mounting surfaces, then apply a continuous 3.5–4.5mm wide bead of liquid gasket around the oil pan. Install the pan within 5 minutes or this step will have to be repeated.

9. Install or connect the following:

- Oil pan and torque the bolts from the center and work towards the ends to 56–66 inch lbs. (6.5–7.5 Nm)
- Front suspension member
- Power steering tube bracket
- Lower steering joint
- Left and right engine mounts
- Front stabilizer bar
- Tension rod
- Negative battery cable

10. After waiting 30 minutes, fill the engine with clean oil.

11. Start the engine, check for leaks and repair if necessary.

3.0L Engine

1. Before servicing the vehicle, refer to the precautions in the beginning of this section.

2. Drain the engine oil

3. Remove or disconnect the following:

- Negative battery cable
- Engine undercovers
- Steel (lower) oil pan bolts in the reverse of the installation sequence

4. Insert a seal cutter between the steel and aluminum oil pan.

5. Tapping the cutter with a hammer, slide it around the entire edge of the oil pan. Be careful not to damage the aluminum mating surface of the upper oil pan.

- Steel oil pan and the oil strainer
- Front exhaust pipe and its support

6. Hang the engine at the right and left side engine slingers with a suitable hoist.

7. Position a suitable jack under the transaxle.

- Crankshaft Position (CKP) sensors (REFERENCE and POSITION) from the oil pan

- Front and rear engine mounting nuts and bolts
- Center crossmember assembly
- Engine drive belts
- A/C compressor and mounting bracket
- Rear cover plate and the lower transaxle bolts
- Aluminum (upper) oil pan bolts in the reverse of the installation sequence

8. Insert a seal cutter between the aluminum oil pan and the engine block.

9. Tapping the cutter with a hammer, slide it around the entire edge of the oil pan. Be careful not to damage the mating surfaces of the oil pan or engine block.

10. Remove or disconnect the following:

- Oil pan assembly
- Bolts that secure the baffle plate and the baffle plate
- O-rings from the cylinder block and oil pump body

To install:

11. Install or connect the following:

- Baffle plate to the oil pan and torque the bolts to 22–27 inch lbs. (2.5–3.1 Nm) and apply sealant to the front and rear seal of the oil pan
- New O-rings to the cylinder block and the oil pump body

12. Apply a 4.5–5.5mm wide continuous bead of liquid gasket to the upper oil pan mating surface and install the oil pan. Torque the bolts in sequence to 12–14 ft. lbs. (16–19 Nm).

13. Install or connect the following:

- Oil pan strainer and torque the bolts to 12–14 ft. lbs. (16–19 Nm)
- Rear cover plate and lower transaxle bolts
- A/C compressor and bracket

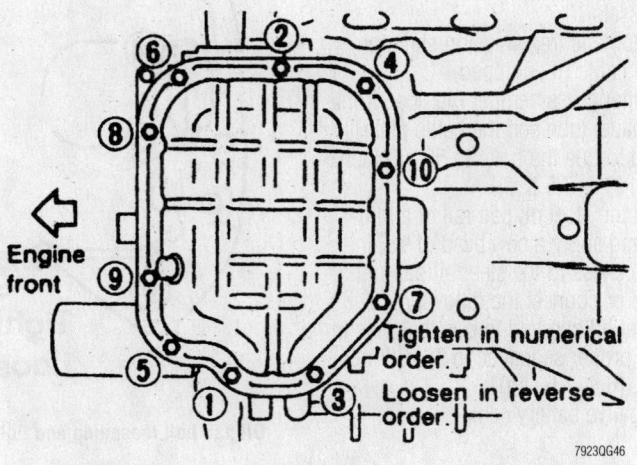

Engine front

Tighten in numerical order.

Loosen in reverse order.

7923QG46

Bolt tightening sequence for the steel oil pan—3.0L engines

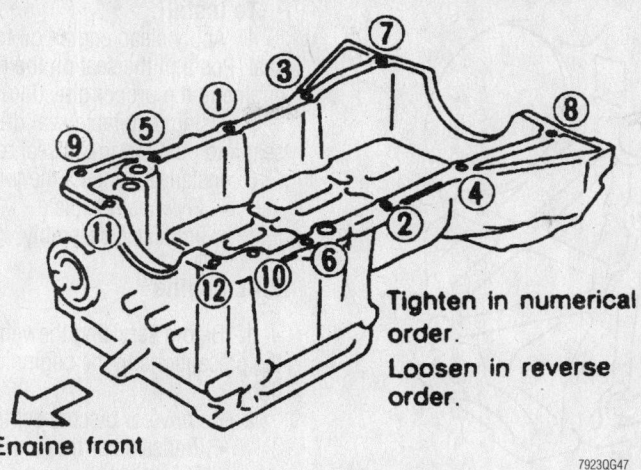

Tighten in numerical order.

Loosen in reverse order.

Engine front

79230G47

To prevent pan warpage, tighten the bolts in the sequence shown—3.0L engines

- Drive belts
- Center crossmember
- Front and rear engine mount hardware
- CKP sensors
- Front exhaust tube and support
- Oil strainer
- Steel oil pan and torque the bolts, in sequence, to 66 inch lbs. (7.5 Nm)
- Engine under covers
- Negative battery cable

14. After waiting approximately 30 minutes, fill the engine with clean oil.

15. Start the vehicle, check for leaks and repair if necessary.

Oil Pump

REMOVAL & INSTALLATION

1.6L, 1.8L and 2.0L Engines

1. Before servicing the vehicle, refer to the precautions in the beginning of this section.
2. Drain the engine oil
3. Remove or disconnect the following:
   - Negative battery cable
   - Drive belts
   - Cylinder head
   - Oil pan and strainer
   - Engine front cover
   - Oil pump from the front cover

To install:
4. Install or connect the following:
   - Oil pump cover and torque the long bolt to 70 inch lbs. (8 Nm) and the short bolt to 44 inch lbs. (5 Nm)
   - Front cover and torque the bolts to 43 ft. lbs. (58 Nm)

- Oil strainer and oil pan
- Cylinder head

➡Refer to Section 1 of this manual for the cylinder head torque sequence illustration. The illustration is located after the Torque Specification Chart.

- Drive belts
- Negative battery cable
5. Fill the engine with clean oil.
6. Start the vehicle, check for leaks and repair if necessary.

2.4L Engine

1. Before servicing the vehicle, refer to the precautions in the beginning of this section.
2. Drain the engine oil
3. Remove or disconnect the following:
   - Negative battery cable
   - Engine front cover
   - Oil pump cover

To install:
4. Install or connect the following:
   - Oil pump cover and torque the long bolt to 15 ft. lbs. (20 Nm) and shorter bolt to 69 inch lbs. (8 Nm)
   - Front cover
   - Negative battery cable
5. Fill the engine with clean oil.
6. Start the vehicle, check for leaks and repair if necessary

3.0L Engine

1. Before servicing the vehicle, refer to the precautions in the beginning of this section.
2. Drain the engine oil

3. Remove or disconnect the following:
   - Negative battery cable
   - Drive belts
   - Camshaft Position (CMP) sensor (PHASE) and the Crankshaft Position (CKP) sensor (REF)/(POS)
   - Engine lower covers
   - Crankshaft pulley
   - Front exhaust tube and support
   - Right side mounting insulator and bracket
   - Center member
   - A/C compressor and move it aside
   - Oil pans
   - Water pump cover
   - Front cover
   - Timing chain
   - Oil pump assembly
4. Clean all mating surfaces.

To install:
5. Install or connect the following:
   - Oil pump
   - Timing chain
   - Front cover and torque the long bolt to 95 inch lbs. (11 Nm) and the short bolt to 71 inch lbs. (8 Nm)
   - Water pump cover
   - Oil pans
   - A/C compressor
   - Center member
   - Right side mounting insulator and bracket
   - Front exhaust tube and support
   - Crankshaft pulley
   - CMP and CKP sensors
   - Engine lower covers
   - Drive belts
   - Negative battery cable
6. Fill the engine with clean oil.
7. Start the vehicle, check for leaks and repair if necessary.

Rear Main Seal

REMOVAL & INSTALLATION

1.6L, 1.8L and 2.0L Engines

1. Before servicing the vehicle, refer to the precautions in the beginning of this section.

2. Remove or disconnect the following:
   - Transaxle
   - Driveplate/ flywheel
   - Oil seal retainer
3. Carefully pry the seal from the retainer. Be sure not to scratch the sealing surface of the crankshaft or oil seal bore.

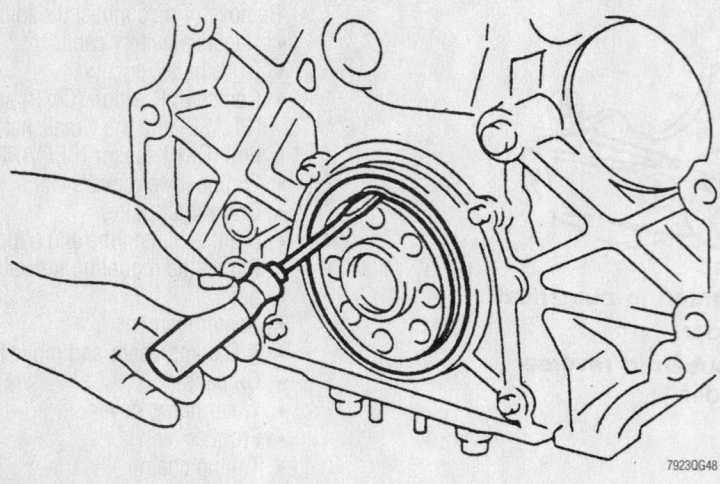

Install the rear main seal using a suitable driver—1.6L, 1.8L and 2.0L engines

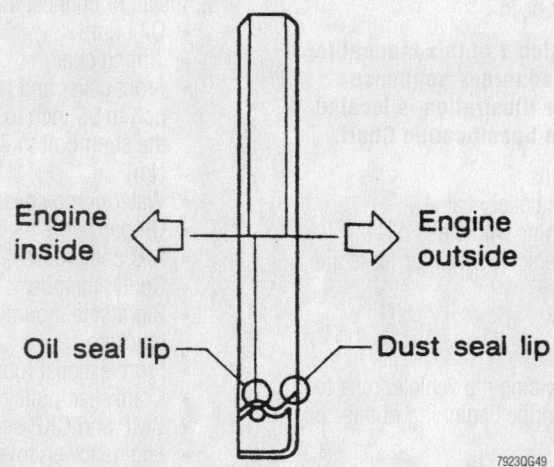

Engine inside ← → Engine outside

Oil seal lip — Dust seal lip

Be sure to install the seal in the correct orientation—1.6L, 1.8L and 2.0L engines

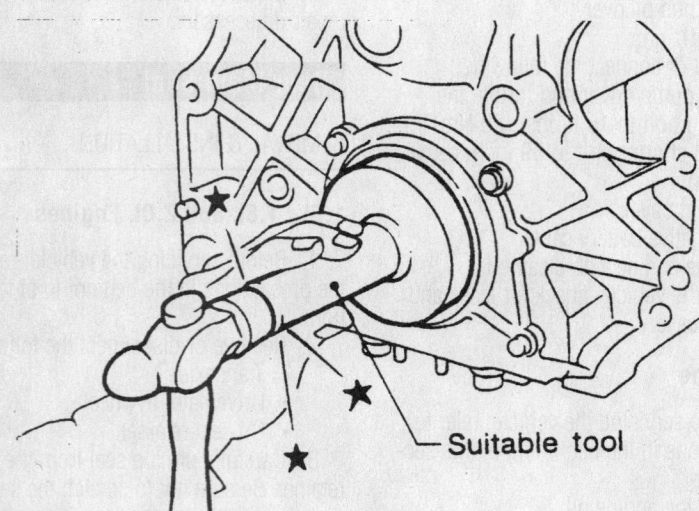

— Suitable tool

Carefully pry the rear main seal out of the retainer on the rear of the engine—1.6L, 1.8L and 2.0L engines

To install:

4. Apply clean engine oil to the new seal. Position the seal on the rear of the engine in the proper direction.

5. Using a suitable seal driver, tap the seal into position in the seal retainer.

6. Install or connect the following:
  • Flywheel/flexplate
  • Transaxle assembly

2.4L Engine

1. Before servicing the vehicle, refer to the precautions in the beginning of this section.

2. Remove or disconnect the following:
  • Transaxle or transmission
  • Driveplate/flywheel
  • Rear oil seal retainer with the oil seal

3. Tap the oil seal out of the retainer with a hammer and drift.

To install:

4. Apply clean engine oil to the new seal.

5. Install the new seal in the retainer with a suitable seal driver.

6. Apply a continuos bead of RTV silicone sealant, 2–3mm wide, to the seal retainer. Be sure to apply around the inner side of the bolt holes.

7. Using a suitable seal driver, tap the seal into position in the seal retainer.

8. Install the oil seal retainer and torque the bolts to 56–66 inch lbs. (6.5–7.5 Nm)

  • Driveplate/flywheel
  • Transmission or transaxle

3.0L Engine

1. Before servicing the vehicle, refer to the precautions in the beginning of this section.

2. Drain the engine oil.

3. Remove or disconnect the following:
  • Transaxle or transmission
  • Driveplate/flywheel
  • Oil pan
  • Oil seal retainer

4. Tap the oil seal out of the retainer with a hammer and drift.

5. Clean all mating surfaces of any residual liquid gasket.

To install:

6. Install or connect the following:
  • New seal into the retainer
  • Oil seal retainer
  • Oil pan
  • Driveplate/flywheel
  • Transaxle/transmission

7. Fill the engine with clean oil.

8. Start the vehicle, check for leaks and repair if necessary.

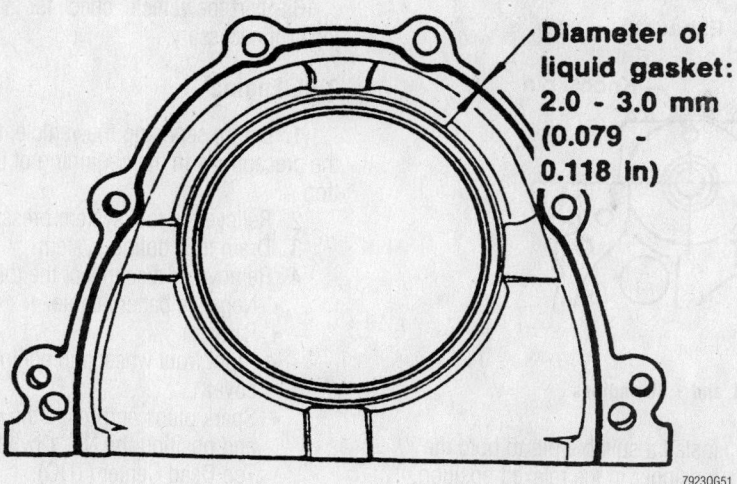

Diameter of liquid gasket: 2.0 - 3.0 mm (0.079 - 0.118 in)

Apply sealant to the seal retainer as shown—2.4L engine

Timing Chain, Sprockets, Front Cover and Seal

REMOVAL & INSTALLATION

1.6L and 1.8L Engine

1. Before servicing the vehicle, refer to the precautions in the beginning of this section.
2. Relieve the fuel system pressure.
3. Drain the cooling system.
4. Remove or disconnect the following:
   - Negative battery cable
   - Upper radiator hose
   - Engine drive belts
   - Power steering pulley and the pump with bracket
   - Air duct from the intake manifold collector
   - Right front wheel and engine side covers
   - Engine undercovers
   - Front exhaust pipe
   - Cylinder head front mounting bracket
   - Cylinder head cover from the engine
   - Rocker cover
   - Distributor cap
   - Spark plugs
   - Intake manifold support and set the No. 1 piston at the Top Dead Center (TDC) compression stroke
   - Distributor
   - Cylinder head front cover
   - Water pump pulley
   - Thermostat housing
   - Lower timing chain tensioner
   - Upper timing chain tensioner and slack side timing chain guide
   - Idler sprocket bolt
   - Camshaft sprocket bolts and the sprockets from the camshafts. Be sure to mark the sprockets for proper reinstallation
   - Camshaft mounting caps by loosening the bolts in 2 or 3 steps
   - Camshafts from the engine
   - Idler sprocket bolt
   - Cylinder head with the manifolds
   - Idler sprocket shaft from the rear side
   - Upper timing chain and support the engine assembly
   - Center crossmember
   - Oil pan and strainer assembly
   - Crankshaft pulley
   - Engine front mount and bracket
   - Bolts that secure the front timing cover and the cover from the

engine. Once the timing chain cover is removed, drive out the old oil seal.
   - Idler sprocket and the lower timing chain
   - Oil pump drive spacer and the crankshaft sprocket
5. Remove the timing chain guide.

To install:

6. Drive a new oil seal into the front cover. Lubricate the oil seal lip with clean engine oil.
7. Confirm that No. 1 piston is set at Top Dead Center (TDC) on compression stroke.
8. Install or connect the following:
   - Crankshaft sprocket with the marks of the sprocket facing the front of the engine
   - Oil pump drive spacer and the chain guide
   - Lower timing chain. Set the chain by aligning its mating mark with the one on the crankshaft sprocket. Be sure the sprocket's mating mark faces the front of the engine.

➡**The number of links between the alignment marks are the same for the left and the right side.**

   - Crankshaft sprocket and the lower timing chain. Set the timing chain by aligning its mating mark with the one on the crankshaft sprocket. Be sure sprocket's mating mark faces engine front.
   - Front cover assembly, using liquid gasket
   - Engine front mounting bracket and the engine mount
   - Oil strainer, oil pan assembly and the crankshaft pulley

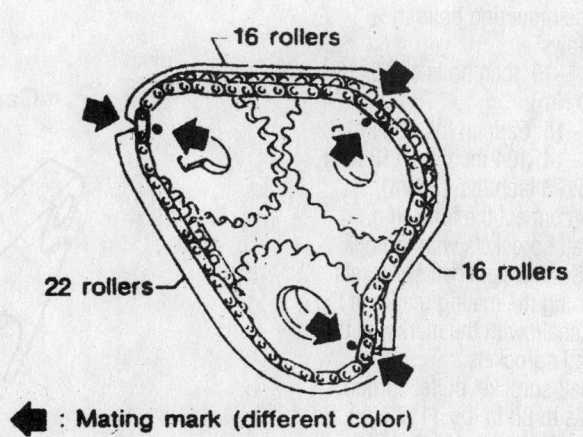

: Mating mark (different color)

Be sure to align the camshaft sprockets with the timing chain—1.6L and 1.8L engines

For complete Engine Mechanical specifications, see Section 1 of this manual

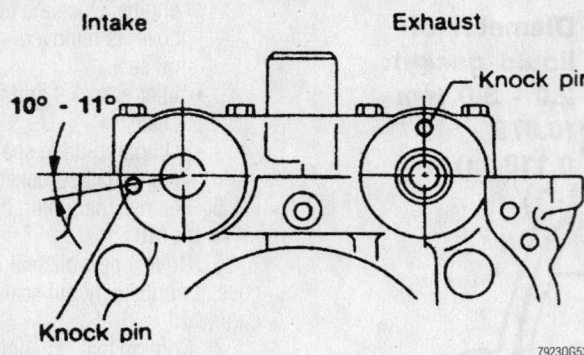

Positioning of camshaft knock pins during assembly—1.6L and 1.8L engines

- Center crossmember

9. Set the idler sprocket by aligning the mating mark on the larger sprocket with the silver mating mark on the lower timing chain.

10. Install or connect the following:
- Upper timing chain and set it by aligning the mating mark on the smaller sprocket with the silver mating marks on the upper timing chain. Be sure sprocket marks face engine front.
- Idler sprocket shaft to the rear side
- Cylinder head assembly
- Idler sprocket bolt. Be sure to lubricate the bolt with clean engine oil
- Exhaust and intake camshafts. The camshafts and marked I for intake and E for exhaust

11. Position the intake camshaft knock pin at the 9 o'clock position and the exhaust camshaft knock pin at the 12 o'clock position.

12. Install or connect the following:
- Camshaft bearing caps and distributor bracket. Apply liquid sealant to the distributor bracket.

13. Torque the mounting bolts in sequence as follows:
   - a. Bolts 11–15, then bolts 1–10: 18 inch lbs. (2.0 Nm).
   - b. Bolts 1–15: 52 inch lbs. (6 Nm).
   - c. Bolts 1–14: 104 inch lbs. (12 Nm).
   - d. Bolt 15: 73 inch lbs. (8 Nm).

14. Install or connect the following:
- Camshaft sprockets with timing chain. Set the camshaft sprockets by aligning the mating marks of the timing chain with the marks on the camshaft sprockets.
- Camshaft sprocket bolts. Torque the bolts to 86 ft. lbs. (117 Nm). Be sure to lubricate the bolts with clean engine oil
- Upper timing chain tensioner. Before installation of the tensioner,

install a suitable pin to hold the tensioner in the relaxed position. After installing the chain tensioner, remove the pin
- Lower timing chain tensioner. Be sure the notch of the gasket is positioned down
- Thermostat housing
- Water pump pulley
- Cylinder head front cover
- Distributor
- Intake manifold support
- Spark plugs and leads
- Distributor cap
- Rocker cover
- Cylinder head cover
- Front exhaust pipe
- Engine under covers
- Engine side covers and the right front wheel
- Air duct to the intake manifold collector
- Power steering pulley and oil pump
- Drive belts
- Upper radiator hose
- Negative battery cable

15. Fill the cooling system.

16. Start the vehicle, check for leaks and repair if necessary.

2.0L Engine

1. Before servicing the vehicle, refer to the precautions in the beginning of this section.

2. Relieve the fuel system pressure.

3. Drain the cooling system.

4. Remove or disconnect the following:
- Negative battery cable
- Radiator
- Right front wheel and engine side cover
- Spark plugs and rotate the engine and position the No. 1 cylinder to Top Dead Center (TDC).
- Air duct to the intake manifold
- Drive belts and the water pump pulley
- Alternator and the power steering pump from the engine
- Vacuum hoses, fuel hoses and the wiring harness connectors
- Cylinder head cover
- Intake manifold supports
- Oil filter bracket and the power steering pump bracket
- Timing chain tensioner
- Distributor
- Timing chain guide and holding the flats of the camshaft sprockets, remove the bolts that secure the sprockets
- Timing chain sprockets from the camshafts
- Oil tubes, baffle plate and camshaft brackets
- Camshafts from the cylinder head
- Starter motor
- Coolant hoses from the engine block

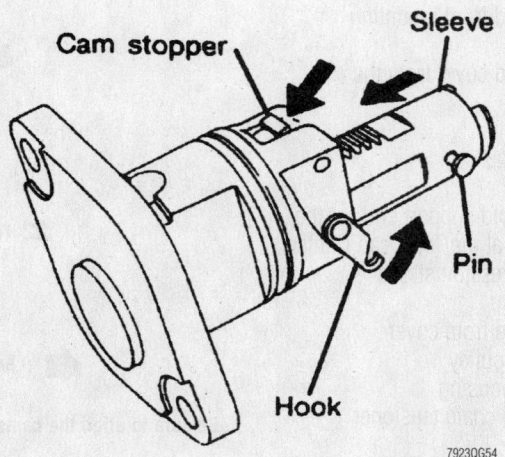

Timing chain tensioner—2.0L engines

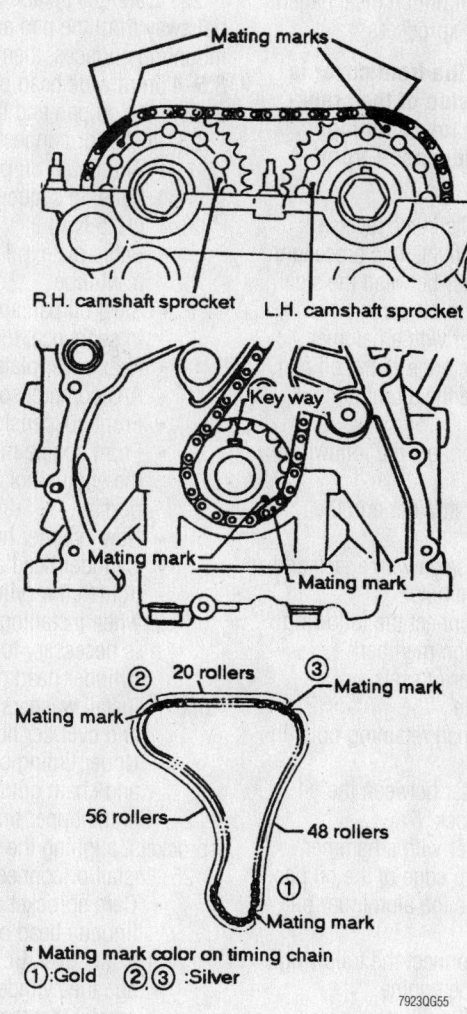

Timing chain sprocket alignment marks—2.0L (SR20DE) engines

- Knock sensor (KS) harness connector
- Exhaust Gas Recirculation (EGR) tube
- Cylinder head
- Oil pan, oil strainer and the baffle plate
- Crankshaft pulley using a suitable puller
- Engine front mount
- Front cover and oil pump drive spacer
- Timing chain guides and timing chain. Check the timing chain for excessive wear at the roller links

To install:

5. Clean all gasket mating surfaces.
6. Install or connect the following:
   - Crankshaft sprocket. Position the crankshaft so that No. 1 piston is set at TDC (keyway at 12 o'clock, mating mark at 4 o'clock) fit timing chain to crankshaft sprocket so the mating mark is in line with mating mark on crankshaft sprocket. The mating marks on timing chain for the camshaft sprockets should be silver. The mating mark on the timing chain for the crankshaft sprocket should be gold
   - Timing chain to the crankshaft sprocket and install the timing chain guides. Tighten the timing chain guides to 10–14 ft. lbs. (13–19 Nm). Drape the timing chain over the left chain guide
   - Oil pump drive spacer to the crankshaft
7. Apply a continuous bead of liquid sealant to the front timing cover and install the cover. Tighten the front cover mounting bolts to 57–66 inch lbs. (6.4–7.5 Nm).
8. Install or connect the following:
   - Right front engine mount
   - Crankshaft pulley and torque the bolt to 105–112 ft. lbs. (142–152 Nm). Be sure the No. 1 piston is at TDC
   - Oil strainer, baffle plate and the oil pan assembly
   - Cylinder head assembly. Be sure to apply a bead of sealant to the joint of the block and front timing cover
   - EGR tube
   - KS harness connector
   - Coolant hoses to the engine block, using new hose clamps
   - Starter motor
   - Camshafts, camshaft bearing caps, oil tubes and the baffle plate

➡ **When installing the camshafts, be sure to position the left-hand and right-hand camshaft keys at 12 o'clock. Also be sure the camshaft brackets are facing in the correct direction.**

   - Camshaft sprockets by lining up the mating marks on the timing chain with the mating marks on the camshaft sprockets and torque the bolts to 101–116 ft. lbs. (137–157 Nm)
   - Timing chain guide and distributor
   - Chain tensioner. Press the cam stopper down and the press-in sleeve until the hook can be engaged on the pin. When tensioner is bolted in position the hook will release automatically

➡ **Ensure the arrow on the outside of the tensioner faces the front of the engine.**

   - Oil filter bracket and power steering pump bracket
   - Intake manifold supports
   - Cylinder head cover
   - Vacuum hoses and fuel lines
   - Alternator and power steering pump
   - Water pump pulley and drive belts
   - Air duct
   - Spark plugs after making certain that the No. 1 piston is at the TDC position
   - Engine side cover and right front wheel
   - Radiator
   - Negative battery cable
9. Fill the cooling system.
10. Start the engine, check for leaks and repair if necessary.

2.4L Engine

ALTIMA

1. Before servicing the vehicle, refer to the precautions in the beginning of this section.
2. Drain the cooling system.
3. Drain the engine oil.
4. Remove or disconnect the following:
   - Negative battery cable
   - Spark plug wires and set the NO. 1 piston at the Top Dead Center (TDC) of the compression stroke
   - Engine undercover
   - Vacuum hoses, fuel hoses, wires, harness and connectors
   - Drive belts
   - Power steering reservoir
   - Alternator and bracket, the upper radiator hose, the air duct and the front exhaust tube
   - Intake manifold collector supports, intake manifold collector and the exhaust manifold
5. Set the No. 1 piston at Top Dead Center (TDC) on its compression stroke.
6. Remove or disconnect the following:
   - Distributor
7. Using a block of wood, set a transmission jack under the aluminum oil pan and remove the front engine mounting.
8. Remove or disconnect the following:
   - Rocker cover. Remove the rocker cover bolts in the proper sequence
   - Camshaft sprockets

➡**The stoppers on camshaft covers prevent the upper timing chain from disengaging from the idle sprocket.**

   - Cam bearing caps in sequence
   - Camshafts. The camshaft brackets must be loosened in reverse order of tightening to prevent damage to the camshaft

➡**These parts must be reassembled in their original positions.**

   - Cylinder head bolts in the reverse order of installation

➡**A warped or cracked cylinder head could result from loosening in incorrect order. The cylinder head bolts should be loosened in 2 or 3 steps.**

   - Cam sprocket cover
   - Upper chain tensioner and upper chain guides
   - Upper timing chain
   - Idler sprocket bolt
   - Cylinder head and the intake manifold
   - Cylinder head gasket. The lower

timing chain will not be disengaged from crankshaft sprocket

➡**The cast portion of the front cover is located on the lower side of the crankshaft sprocket, so the lower timing chain need not be disengaged from idler sprocket.**

   - Steel oil pan bolts in the reverse sequence of the tightening procedure
9. Install a seal cutter between the steel oil pan and the aluminum oil pan.
10. Tapping the cutter with a hammer, slide it around the entire edge of the oil pan. Take care not to damage the aluminum oil pan.
11. Remove or disconnect the following:
    - Steel oil pan
    - Baffle plate, oil strainer and the front tube
12. Support the transaxle with a jack and the engine with a engine hoist.
13. Remove or disconnect the following:
    - Front suspension member
    - A/C compressor gussets
    - Rear cover plate
    - Aluminum oil pan retaining bolts in sequence
14. Insert a seal cutter between the oil pan and the cylinder block.
15. Tapping the cutter with a hammer, slide it around the entire edge of the oil pan. Take care not to damage the aluminum oil pan.
16. Remove or disconnect the following:
    - Oil pan from the engine
    - Crankshaft pulley
    - Front timing chain cover
    - Oil pump drive spacer
    - Lower timing chain tensioner, tensioner arm and lower timing chain guide
    - Lower timing chain and idler sprocket

To install:

17. Install or connect the following:
    - Crankshaft sprocket and oil pump drive spacer
    - Idler sprocket and lower timing chain
18. Set the lower timing chain on the sprockets, aligning the mating marks. The mating marks on the timing chain assembly will be silver.
19. Install or connect the following:
    - Chain tension arm and chain guide
    - Lower timing chain tensioner
20. Apply a continuous bead of liquid gasket to the front cover and install the front cover. Install a new oil seal.
21. Install or connect the following:
    - Crankshaft pulley and torque tighten bolt to 105–112 ft. lbs. (142–152 Nm)

22. Carefully scrape the old gasket material away from the pan and cylinder block mounting surfaces, then apply a continuous 3.5–4.5mm wide bead of liquid gasket around the oil pan and the cylinder block.
23. Install or connect the following:
    - Aluminum oil pan and torque the bolts, in sequence, to 13 ft. lbs. (17.5 Nm)
    - Baffle plate, oil strainer and the front tube
    - Steel oil pan and torque the bolts, in sequence, to 61 inch lbs. (7 Nm)
    - Rear cover plate
    - A/C compressor gussets
    - Front suspension member
    - Front engine mounting and remove the engine hoist and transaxle support
    - New cylinder head gasket
    - Cylinder head and temporarily tighten the cylinder head bolts when installing the front cover. This is necessary to avoid damaging the cylinder head gasket. Be sure to install washers between the bolts and cylinder head
    - Upper timing chain, chain tensioner and chain guide
24. Set the upper timing chain on idler sprockets, aligning the mating marks.
25. Install or connect the following:
    - Cam sprocket cover. Apply a continuous bead of liquid gasket to front cover. Be careful not to damage the cylinder head gasket. Be careful that the upper timing chain does not slip or jump when installing cam sprocket cover
26. Tighten cylinder head bolts.
    - Camshafts and camshaft bearing caps
    - Camshaft sprockets, then torque the sprocket bolts to 123–130 ft. lbs. (167–176 Nm). Install the chain guide between both camshaft sprockets. The alignment marks on the upper portion of the timing chain should now be aligned
    - Rocker cover
    - Distributor
    - Intake manifold collector supports and collector
    - Exhaust manifold
    - Alternator and bracket
    - Upper radiator hose
    - Air duct assembly
    - Front exhaust tube
    - Power steering reservoir
    - Drive belts
    - Vacuum hoses and fuel lines
    - Engine under cover

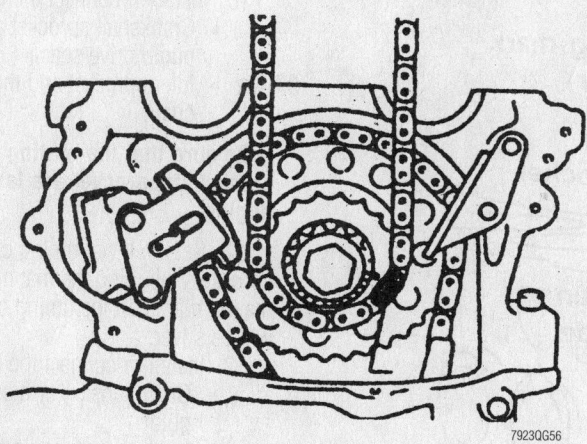

Be sure to align the mark on the idler sprocket with the mark on the upper chain—2.4L engines

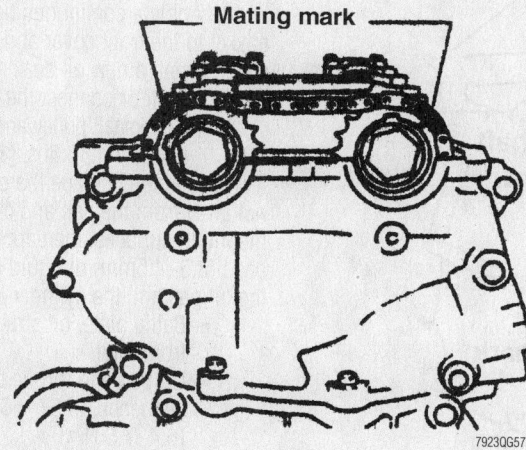

Align the marks on the camshaft sprockets with the upper portion of the upper timing chain and mating marks—2.4L engine

- Spark plug wires
- Negative battery cable
27. Fill the cooling system.
28. Fill the engine with clean oil.
29. Start the vehicle, check for leaks and repair if necessary.

240SX

1. Before servicing the vehicle, refer to the precautions in the beginning of this section.
2. Relieve the fuel system pressure.
3. Drain the coolant from engine and radiator.
4. Drain the engine oil.
5. Remove or disconnect the following:
- Negative battery cable
- Engine undercover
- Air duct assembly
- Fan shroud and the cooling fan
- Exhaust manifold cover
- Front exhaust tube and if equipped, the A.I.V. pipe
- Vacuum hoses, fuel lines and electrical connections
- Spark plugs
- Distributor cap with the spark plug wires attached
- Injector tube assembly with the injectors
- Rocker cover bolts, in reverse order of installation and set the No. 1 piston at Top Dead Center (TDC) on its compression stroke
- Distributor. Be sure to note the positioning of the distributor rotor before removing the distributor.
- Camshaft sprockets. Be sure to hold the flats of the camshafts when removing the sprocket bolts.

➡The stoppers on the inside of the camshaft covers prevent the upper timing chain from disengaging from the idle sprocket.

- Camshaft bearing caps, in reverse order of installation, then both of the camshafts

➡The camshaft bearing caps and camshafts should be kept in their original position for reassembly.

- Cylinder head bolts in the reverse order of installation

➡Head warpage or cracking could result from removing the head bolts in the incorrect order. The cylinder head bolts should be loosened in 2 or 3 steps.

- Camshaft sprocket cover
- Upper chain tensioner and upper chain guides

➡Compress the piston of the tensioner and insert a suitable pin into the pin hole.

- Upper timing chain
- Idler sprocket bolt
- Cylinder head with the intake manifold and exhaust manifold assembly
- Cylinder head gasket
- Oil pan bolts

6. Install a seal cutter between the oil pan and the engine block.
7. Tapping the cutter with a hammer, slide it around the entire edge of the oil pan. Take care not to damage the oil pan.
8. Remove or disconnect the following:
- Oil pan assembly
- Baffle plate, oil strainer and the front tube
- Engine drive belts
- A/C compressor idler pulley
- Crankshaft pulley
- Front timing chain cover
- Oil pump drive spacer
- Lower timing chain tensioner, tensioner arm and the lower timing chain guide

➡Compress the piston of the tensioner and insert a suitable pin into the pin hole.

- Lower timing chain and crankshaft sprocket

To install:
9. Check all components for wear. Replace as necessary. Clean all mating surfaces and replace the cylinder head gasket.

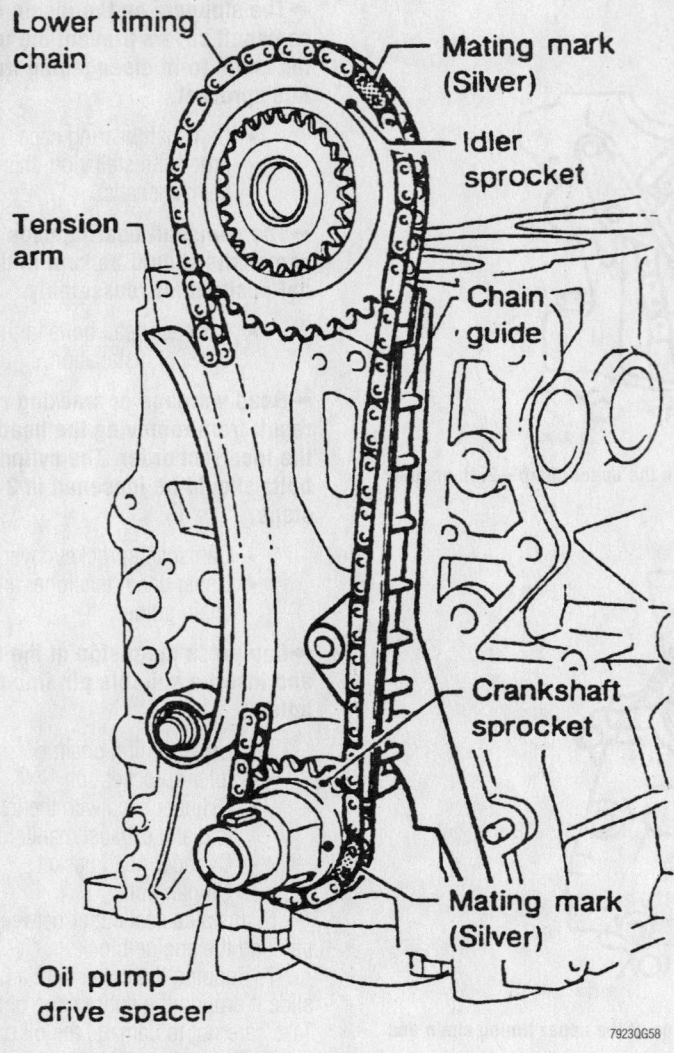

Lower timing chain alignment marks—2.4L engine

10. Install or connect the following:
- Crankshaft sprocket and the oil pump drive spacer
- Idler sprocket and the lower timing chain

➡ **Be sure that the mating marks of the crankshaft sprocket are facing front of engine.**

11. Set the lower timing chain on the sprockets, aligning the mating marks. The mating marks on the timing chain assembly will be silver.

12. Install or connect the following:
- Chain tension arm and the chain guide
- Lower timing chain tensioner

➡ **After installation of the tensioner, remove the pin to release the piston.**

13. Apply a continuous bead of liquid gasket to the front cover and install the front cover. Install a new oil seal.

14. Install or connect the following:
- Crankshaft pulley and torque the bolt to 105–112 ft. lbs. (142–152 Nm)

15. Carefully scrape the old gasket material away from the pan and cylinder block mounting surfaces, then apply a continuous bead (3.5–4.5mm) of liquid gasket around the oil pan and the cylinder block.

- Baffle plate, oil strainer and the front tube
- Oil pan and torque the bolts, in sequence, to 57–66 inch lbs. (6.4–7.5 Nm)
- A/C compressor idler pulley
- New cylinder head gasket and install the idler shaft

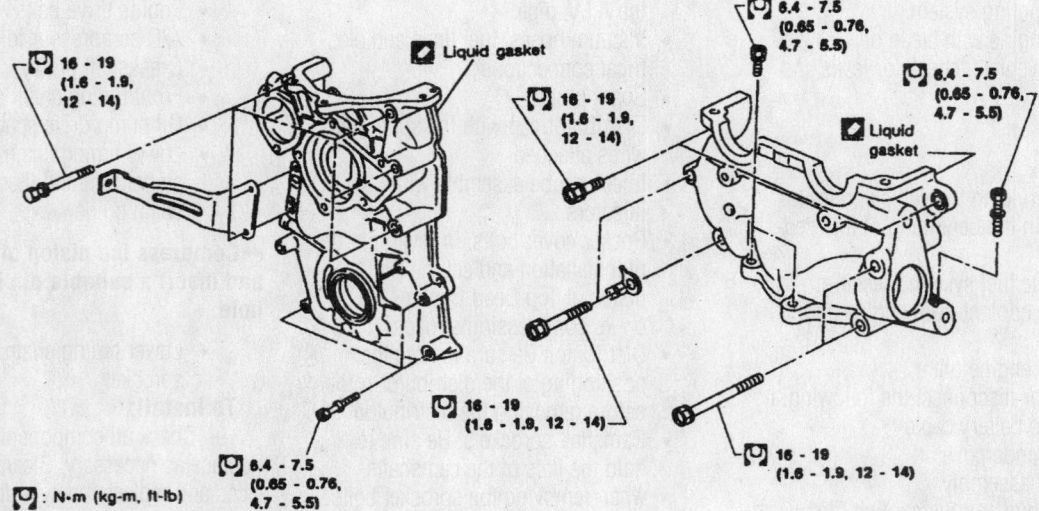

Exploded view of the timing chain front covers—2.4L engine

- Cylinder head and temporarily tighten the cylinder head bolts when installing the front cover. This is necessary to avoid damaging the cylinder head gasket. Be sure to install washers between the bolts and cylinder head

❄❄ CAUTION

Do not fully tighten any of the cylinder head bolts at this time. Tighten the bolts finger-tight only.

- Upper timing chain, chain tensioner and the chain guide. Be sure to align the mark on the timing chain with the idler

➡**After installation of the tensioner, remove the pin to release the piston.**

16. Apply a continuous bead of liquid sealant and install the camshaft sprocket cover. Tighten the mounting bolts to specifications.

17. Tighten the cylinder head bolts.
- Camshafts and the camshaft bearing caps
- Camshaft sprockets and torque the bolts to 123–130 ft. lbs. (167–176 Nm). Install the chain guide between both of the camshaft sprockets. The alignment marks on the upper portion of the timing chain should now be aligned
- Distributor
- Rocker cover
- Fuel injector tube assembly
- Spark plugs
- Distributor cap and plug wires
- Vacuum hoses and fuel lines
- Front exhaust tube and AIV, if equipped
- Exhaust manifold cover
- Cooling fan and shroud
- Air duct assembly
- Engine under cover
- Negative battery cable

18. Fill the engine with clean oil.
19. Fill the cooling system.
20. Start the vehicle, check for leaks and repair if necessary.
21. Check the ignition timing; adjust the timing as necessary.

3.0L Engine

1. Before servicing the vehicle, refer to the precautions in the beginning of this section.
2. Drain the engine oil.

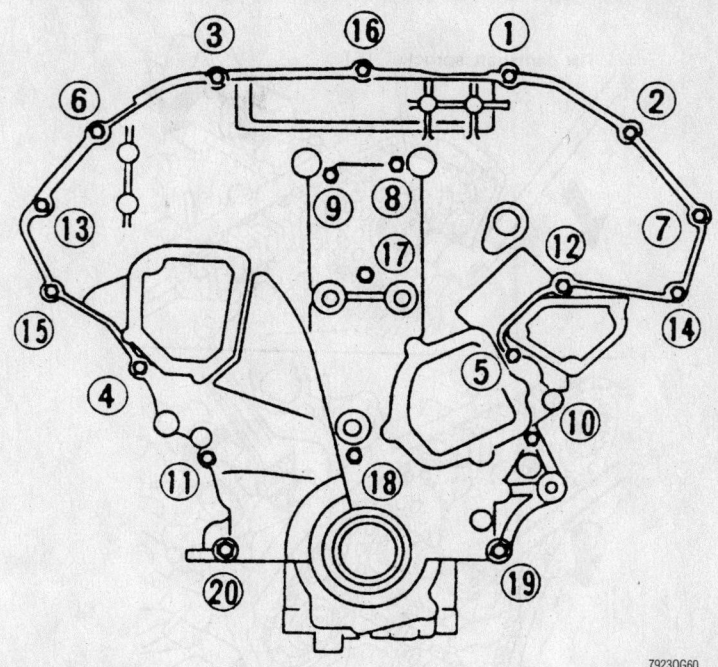

Remove the front timing chain case mounting bolts in the sequence shown—3.0L engines

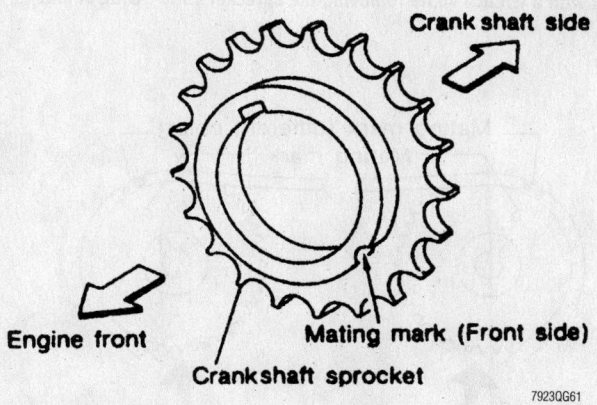

Crankshaft sprocket with mating marks—3.0L engine

3. Drain the cooling system.
4. Relieve the fuel system pressure.
5. Remove or disconnect the following:

- Negative battery cable
- Left side rocker cover ornament

➡**Before detaching any hoses or connectors, note the locations for reassembly.**

- Air duct to intake manifold hose, collector hose, blow-by hose and vacuum hoses
- Fuel hoses and detach the harness connections
- Canister purge hoses

- Water hoses from the cylinder head and intake manifold
- All 6 ignition coils from the spark plugs
- Spark plugs
- Bolts that secure the Exhaust Gas Recirculation (EGR) tube and remove the tube
- Intake manifold collector supports and the collector
- Fuel tube assembly
- Intake manifold. Loosen the bolts in the reverse sequence of the tightening procedure
- Left-hand and right-hand rocker covers from the cylinder head

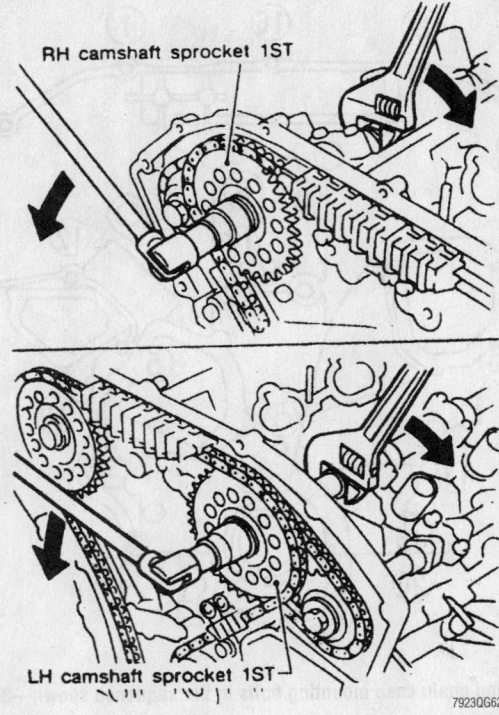

RH camshaft sprocket 1ST

LH camshaft sprocket 1ST

7923QG62

Hold the camshaft with a wrench while removing the sprocket bolts—3.0L engine

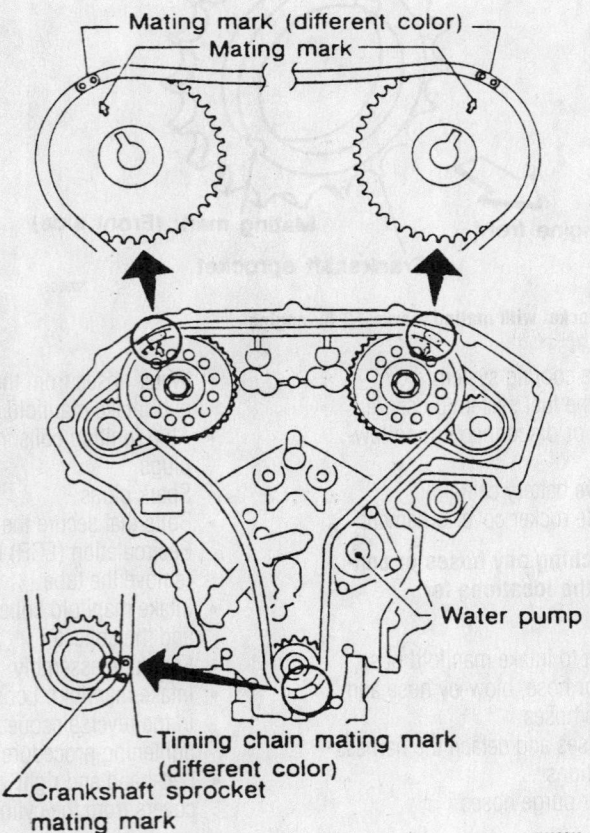

Mating mark (different color)
Mating mark

Water pump

Timing chain mating mark
(different color)

Crankshaft sprocket
mating mark

7923QG63

Timing chain alignment marks—3.0L engine

- Engine undercovers
- Right front wheel and the engine side covers
- Drive belts and the idler pulley
- Power steering oil pump belt and the power steering oil pump assembly
- Camshaft Position (CMP) sensor (PHASE) and Crankshaft Position (CKP) sensors (REF)/(POS)

6. Set the No. 1 piston to Top Dead Center (TDC) of compression stroke by rotating the crankshaft.

7. Loosen the crankshaft pulley bolt while securing the ring gear so the crankshaft cannot rotate.

- Ring gear cover access plate

➡**Use care not to damage the ring gear teeth.**

- Crankshaft pulley using a suitable puller
- A/C compressor and bracket
- Front exhaust pipe and its support

8. Hang the engine at the right and left side engine slingers with a suitable hoist.

9. Support the transaxle with jack.

- Right side engine mounting and bracket
- Center crossmember assembly
- Steel (lower) oil pan bolts in the reverse of the installation sequence

10. Insert a seal cutter between the steel and aluminum oil pan.

11. Tapping the cutter with a hammer, slide it around the entire edge of the oil pan. Be careful not to damage the aluminum mating surface of the upper oil pan.

- Steel oil pan and the oil strainer
- Aluminum (upper) oil pan bolts in the reverse of the installation sequence
- Transaxle bolts that secure the oil pan

12. Insert a seal cutter between the aluminum oil pan and the engine block.

13. Tapping the cutter with a hammer, slide it around the entire edge of the oil pan. Be careful not to damage the mating surfaces of the oil pan or engine block.

- Oil pan from the vehicle
- Water pump cover and the bolts that secure the front timing chain case
- Timing chain case cover using the seal cutter
- Internal timing chain guide and the upper chain guide
- Timing chain tensioner and slack side chain guide
- Left and right intake camshaft sprockets first. Be sure to hold the flats of the camshafts while removing the sprocket bolts

- Lower timing chain assembly. Be sure to note the aligning marks of the chain before removal

14. Insert a suitable stopper pin for the left and right camshaft tensioners.
- Left and right exhaust camshaft sprocket bolts. Be sure to hold the flats of the camshafts while removing the sprocket bolts
- Upper timing chain assembly. Be sure to note the aligning marks of the chain before removal
- Lower timing chain guide
- Crankshaft sprocket
- All traces of liquid gasket from the front timing chain case and from the water pump

15. Inspect the timing chain for excessive wear or damage and replace as necessary.

To install:

16. Install or connect the following:
- Crankshaft sprocket with the mating mark facing out

17. Position the crankshaft to TDC of compression stroke and align the dowels of the camshaft sprockets to the 12 o'clock position in respect to the cylinder head.
- Lower timing chain guide. The front mark on the guide should face upwards

18. On a work bench, align the marks on the intake and exhaust camshaft sprockets with the marks of the chain
- Exhaust camshaft sprockets onto the dowel pin and torque the mounting bolts to 88–95 ft. lbs. (119–128 Nm). Be sure to secure the camshafts while tightening the bolts
- Timing chains and sprockets to the intake camshafts. Be sure to align the timing chain and sprocket mating marks
- Left and right camshaft tensioner stopper pins

19. Align the mating mark on the crankshaft with the matchmark (gold link) on the lower timing chain.

20. Install the lower timing chain to the water pump sprocket.

21. Working counterclockwise, install the lower timing chain camshaft sprockets. Be sure to align the sprocket marks with the blue links of the timing chain during installation.
- Intake sprocket and torque the bolts to 88–95 ft. lbs. (119–128 Nm). Be sure to secure the camshafts while tightening the bolts

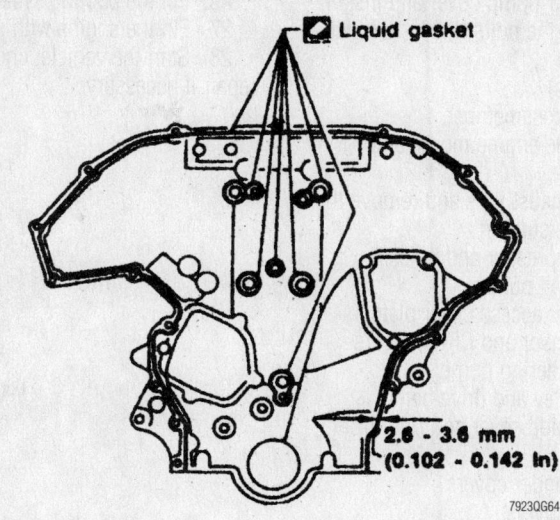

Application of liquid gasket to the front timing case—3.0L engines

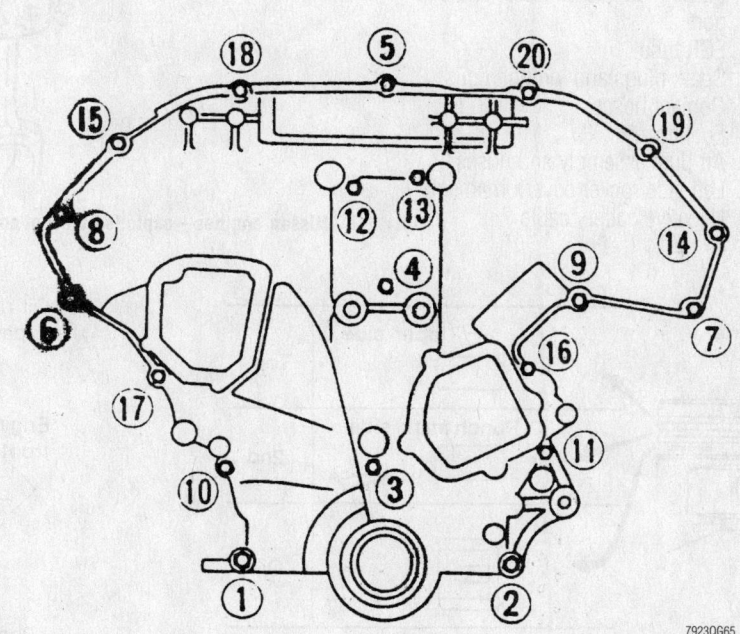

Tighten the front timing chain case bolts according to the sequence shown—3.0L engines

- Internal timing chain guide, upper timing chain guide, lower timing chain tensioner and slack side timing chain guide

22. Torque the tensioner mounting bolt to 75–96 inch lbs. (8.4–10.8 Nm) and the guide bolts to 108–168 inch lbs. (13–19 Nm).

23. Apply a 0.102–0.142 in. (2.6–3.6mm) continuous bead of liquid gasket to all necessary areas as shown on the front timing cover.
- Timing cover evenly and gently. Be sure to align the dowel pin holes

24. Torque the mounting bolts in sequence as follows:
   a. Bolts No. 1 and 2: 19–23 ft. lbs. (26–31 Nm).
   b. Bolts No. 3–20: 105–121 inch lbs. (11.8–13.7 Nm).

➡**Leave the bolts unattended for 30 minutes or more after tightening. This will allow the liquid gasket to cure sufficiently.**

25. Apply a 0.091–0.130 in. (2.3–3.3mm) continuous bead of liquid

gasket to the water pump cover and install the cover. Torque the bolts to 84–108 inch lbs. (10–13 Nm).

- Oil pan(s)
- Center crossmember
- Right side engine mount and bracket
- Front exhaust pipe and remove the transaxle support
- A/C compressor and bracket
- Crankshaft pulley
- Ring gear access cover plate
- CMP sensor and CKP sensors
- Power steering pump
- Idler pulley and drive belts
- Engine side cover and right front wheel
- Engine under covers
- Rocker covers
- Intake manifold
- Fuel tube assembly
- Intake manifold collector and support
- EGR tube
- Spark plugs and ignition coils
- Coolant hoses
- Fuel hoses
- Air duct assembly and hoses
- Left side rocker cover ornament
- Negative battery cable

26. Fill the cooling system.
27. Fill the engine with clean oil.
28. Start the vehicle, check for leaks and repair if necessary.

Piston and Ring

POSITIONING

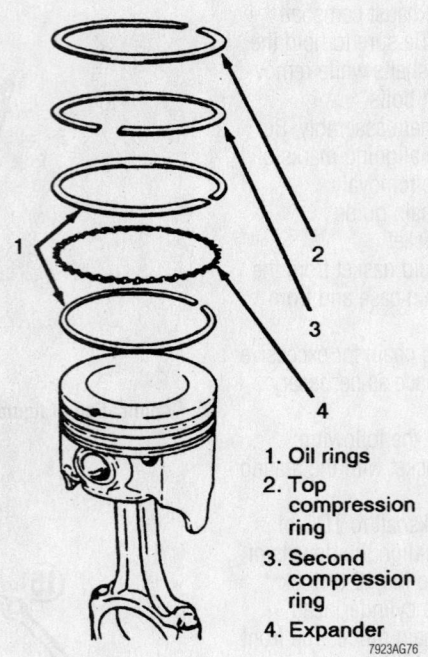

1. Oil rings
2. Top compression ring
3. Second compression ring
4. Expander

Nissan engines—exploded view of common piston ring mounting

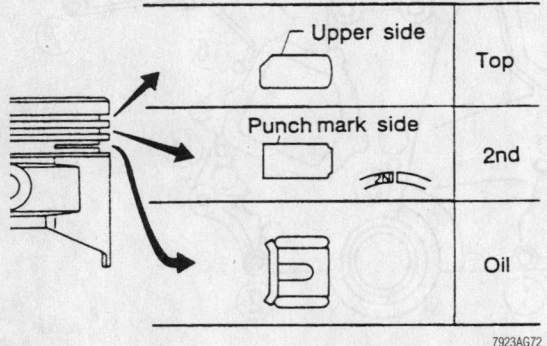

Nissan 1.6L, 1.8L, 2.0L and 2.4L engines—piston ring positioning

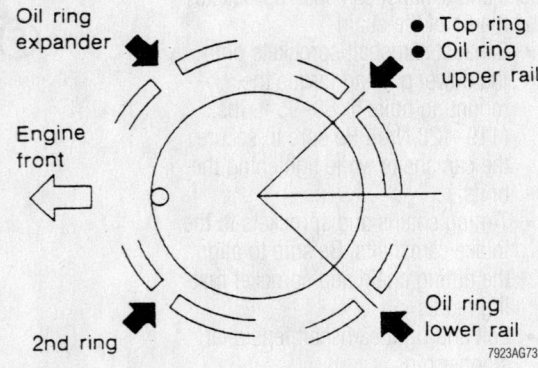

Nissan engines—piston ring end-gap spacing

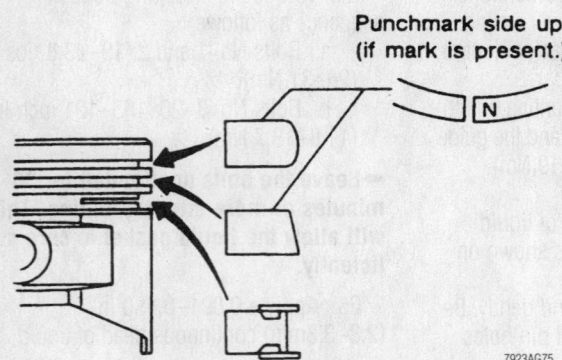

Nissan 3.0L engine—piston ring positioning

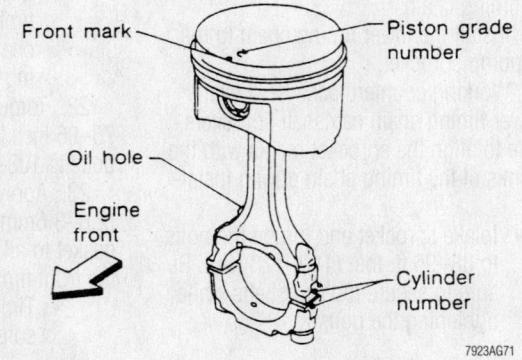

Nissan engines—piston and connecting rod assembly positioning

FUEL SYSTEM

Fuel System Service Precautions

Safety is the most important factor when performing not only fuel system maintenance but any type of maintenance. Failure to conduct maintenance and repairs in a safe manner may result in serious personal injury or death. Maintenance and testing of the vehicle's fuel system components can be accomplished safely and effectively by adhering to the following rules and guidelines.

• To avoid the possibility of fire and personal injury, always disconnect the negative battery cable unless the repair or test procedure requires that battery voltage be applied.

• Always relieve the fuel system pressure prior to disconnecting any fuel system component (injector, fuel rail, pressure regulator, etc.), fitting or fuel line connection. Exercise extreme caution whenever relieving fuel system pressure, to avoid exposing skin, face and eyes to fuel spray. Please be advised that fuel under pressure may penetrate the skin or any part of the body that it contacts.

• Always place a shop towel or cloth around the fitting or connection prior to loosening to absorb any excess fuel due to spillage. Ensure that all fuel spillage (should it occur) is quickly removed from engine surfaces. Ensure that all fuel soaked cloths or towels are deposited into a suitable waste container.

• Always keep a dry chemical (Class B) fire extinguisher near the work area.

• Do not allow fuel spray or fuel vapors to come into contact with a spark or open flame.

• Always use a back-up wrench when loosening and tightening fuel line connection fittings. This will prevent unnecessary stress and torsion to fuel line piping. Always follow the proper torque specifications.

• Always replace worn fuel fitting O-rings with new. Do not substitute fuel hose where fuel pipe is installed.

Fuel System Pressure

RELIEVING

The fuel pump fuse is located in the dash fuse box or in the engine compartment fuse box. Check the lid of the fuse box for exact location.

1. Before servicing the vehicle, refer to the precautions in the beginning of this section.
2. Remove the fuel pump fuse.
3. Start the engine.
4. Start the engine and run until the engine stalls.
5. After the engine stalls, try to restart the engine; if the engine will not start, the fuel pressure has been released.
6. Turn the ignition switch OFF. Reinstall the fuel pump fuse into the fuse block.

➡**Do not crank the engine or turn the ignition switch ON after the fuel pump fuse has been reinstalled, or the fuel pressure will be re-established.**

Fuel Filter

REMOVAL & INSTALLATION

All Models

EXCEPT MAXIMA 2000–01

1. Before servicing the vehicle, refer to the precautions in the beginning of this section.
2. Properly relieve fuel system pressure.
3. Remove or disconnect the following:
   • Negative battery cable
   • Fuel hose clamps
   • Hoses from the fuel filter
   • Bolt securing the filter to the bracket or the filter from the bracket clips
   • Filter
   To install:
4. Install or connect the following:
   • New filter and secure the filter in the bracket
5. If necessary, replace the fuel line hoses and hose clamps
   • Fuel hoses and tighten the clamps
   • Negative battery cable
   • Fuel pump fuse
6. Start the engine and check for leaks.

2000–01 Maxima

1. Before servicing the vehicle, refer to the precautions in the beginning of this section.
2. Properly relieve fuel system pressure.
3. Remove or disconnect the following:
   • Negative battery cable
   • Rear seat bottom

• Inspection hole cover
• Electrical and quick connectors
• Six screws
• Fuel level sensor unit and fuel pump assembly
• Flange and snap fit portion of the fuel pump
• Fuel tank temperature sensor harness
• Fuel level sensor flange
• Fuel pump connector
• Quick connectors from the fuel level sensor
• Fuel level sensor from the chamber
• Fuel filter from the chamber
To install:
4. Install or connect the following:
   • Fuel filter to the chamber
   • Fuel level sensor to the chamber
   • Quick connectors to the fuel level sensor
   • Fuel pump connector
   • Fuel level sensor flange
   • Fuel tank temperature sensor harness
   • Fuel pump assembly to the fuel tank
   • Screws and electrical connectors
   • Quick connectors
   • Negative battery cable
5. Start the vehicle, check for leaks and repair if necessary.
   • Inspection hole cover
   • Rear seat bottom

Fuel Pump

REMOVAL & INSTALLATION

Sentra and 200SX

The fuel pump is located in the fuel tank on all vehicles. In-tank fuel pumps are accessible by lifting up the rear seat to gain access to the inspection cover.

1. Before servicing the vehicle, refer to the precautions in the beginning of this section.
2. Relieve the fuel system pressure.
3. Remove or disconnect the following:
   • Negative battery cable
   • Rear seat from the vehicle
   • Inspection cover that is located under the rear seat
   • Inlet and outlet fuel lines from the fuel pump assembly
   • Fuel pump and gauge wiring connections

- 6 mounting bolts that secure the fuel pump assembly to the top of the fuel tank

4. Raise up the fuel pump assembly and detach the fuel tubes and connector.
   - Fuel gauge assembly
   - Fuel pump with the fuel chamber

5. Pull up the front of the fuel pump chamber and slide the chamber forward.
   - Fuel pump from the chamber
   - O-ring seal or gasket

To install:

6. Install or connect the following:
   - Fuel pump to the fuel pump chamber and slide chamber rearward
   - Fuel pump with the fuel pump chamber
   - Fuel gauge assembly using a new O-ring
   - Fuel tubes and connector. Use new hoses and clamps.
   - 6 mounting bolts to the top of the fuel gauge unit and torque the bolts to 17–22 inch lbs. (23–3 Nm)
   - Fuel pump and gauge wiring connections
   - New inlet and outlet fuel lines to the fuel pump assembly
   - Negative battery cable

7. Start the vehicle, check for leaks and repair if necessary.
   - Inspection cover and the rear seat

Altima

1. Before servicing the vehicle, refer to the precautions in the beginning of this section.

2. Relieve the pressure from the fuel system.

3. Remove or disconnect the following:
   - Negative battery cable
   - Rear seat and the access cover
   - Fuel pump electrical connector
   - Fuel lines from the fuel pump assembly
   - Locking ring
   - Fuel gauge assembly
   - Fuel tube and connector from the fuel gauge

➡**When the fuel sending unit needs to be removed, pull the tab upwards. The tab is located on the sending unit, opposite the end of the float. After the tab is pulled, the sending unit will lift straight out of the tank bracket.**

   - Fuel pump by pinching the 2 locking tabs together. Lift the fuel pump assembly straight upward and out of fuel tank.
   - O-ring and discard

4. Place a clean rag in the hole to keep out dirt.

To install:

5. Remove the rag

6. Install or connect the following:
   - New O-ring and fuel pump
   - Electrical connection and fuel tube to the fuel gauge sending unit
   - Fuel sending unit into the tank

➡**Verify that the mark on the fuel tank and the components are aligned when installing the pump and fuel gauge sending unit.**

   - Locking ring and torque the ring to 22–26 ft. lbs. (30–35 Nm)
   - Fuel lines and fuel pump electrical connector. Always install new clamps on the fuel lines.
   - Negative battery cable

7. Start the engine, check for leaks and repair if necessary.
   - Fuel pump access cover
   - Rear seat

Maxima and 240SX

1. Before servicing the vehicle, refer to the precautions in the beginning of this section.

2. Relieve the fuel system pressure

3. Remove or disconnect the following:
   - Negative battery cable
   - Rear seat or open the access panel in the trunk
   - Fuel gauge electrical connector and pump electrical connector
   - Fuel outlet and the return hoses
   - Fuel tank, if necessary

4. On some 240SX models you need to remove the fuel pump assembly-to-fuel tank bolts and lift the fuel pump assembly from the fuel tank.

5. On other models you need to remove the locking ring and raise the fuel pump from the tank. Disconnect the feed tube while raising the pump.

6. Discard the O-ring. Plug the fuel tank opening with a clean rag to prevent dirt from entering the system.

➡**When removing or installing the fuel pump assembly, be careful not to damage or deform it and always install a new O-ring.**

To install:

7. Remove the rag

8. Install or connect the following:
   - Fuel pump assembly into the fuel tank using a new O-ring
   - Fuel pump assembly-to-fuel tank bolts and torque the bolts to 17–22 inch lbs. (2.0–2.5 Nm)

   - Locking ring assembly and tighten
   - Fuel tank assembly, if removed
   - Fuel lines and the electrical connectors. Always use new clamps when reconnecting fuel line hoses

➡**When installing the upper plate, be sure to align the mark with the center marks on the fuel tank.**

   - Negative battery cable

9. Start the engine, check for fuel leaks and repair if necessary.

10. Install the fuel pump access cover.

➡**On some models, the Check Engine Light will stay ON after installation is completed. The memory code in the control unit must be erased. This code is stored for an open fuel pump circuit, this is caused when the fuel pressure is released. To erase the code, disconnect the battery cable for 10 seconds, then reconnect after installation of fuel pump.**

Fuel Injector

REMOVAL & INSTALLATION

1. Before servicing the vehicle, refer to the precautions in the beginning of this section.

2. Relieve the fuel system pressure

3. Remove or disconnect the following:
   - Negative battery cable
   - Intake manifold collector
   - Vacuum hose from the pressure regulator
   - Fuel hoses from the rail
   - Injector electrical connectors
   - Fuel rail bolts
   - Injector rail assembly with injectors from the intake manifold
   - Injector from the rail by pushing on the injector tail piece
   - Discard injector O-rings

To install:

4. Clean the injector tail piece and lubricate new O-rings with a smear of clean engine oil.

5. Install or connect the following:
   - New O-rings
   - Injector to the fuel rail
   - Fuel rail and the injectors as an assembly to the intake manifold

6. Install the fuel rail bolts and tighten in 2 steps as follows:
   a. Step 1: Bolts to 84–96 inch lbs. (9–10 Nm).
   b. Step 2: Bolts to 15–20 ft. lbs. (21–26 Nm).
   - Injector electrical connectors

- Vacuum hose to the pressure regulator
- Fuel hoses to the rail
- All remaining components in the reverse order of removal

Maxima

1. Remove or disconnect the following:

- Negative battery cable
- Intake manifold collector
- Vacuum hose from the pressure regulator
- Fuel hoses from the rail
- Injector electrical connectors
- Fuel rail bolts

2. To remove the fuel injector from the fuel rail, expand and remove the clips securing the injectors and press the fuel injector out from the fuel rail. Discard the O-rings.

To install:

3. Install new O-rings onto the fuel injector.

4. Wet the new O-rings with a thin coating of oil and press the injector into the fuel rail.

5. Install new injector retaining clips.

6. Install the fuel injector assembly by performing the following procedures:

a. Place new injector gaskets onto the manifold.

b. Install the fuel rail assembly to the engine.

c. Install the fuel rail-to-cylinder head bolts and torque the bolts to 84–96 inch lbs. (9.3–10.8 Nm). Then tighten them again to 16–19 ft. lbs. (21–26 Nm).

d. Connect the fuel rail assembly to the fuel lines.

e. Connect the vacuum hose to the fuel pressure regulator.

f. Connect the electrical connectors to the fuel injectors.

7. Install the intake manifold collector.

8. Connect the negative battery cable.

9. Start the engine and check for leaks.

DRIVE TRAIN

Transmission Assembly

REMOVAL & INSTALLATION

Automatic

240SX

1. Before servicing the vehicle, refer to the precautions in the beginning of this section.

2. Raise and support the vehicle safely.

3. Drain the transmission fluid.

4. Remove or disconnect the following:

- Negative battery cable
- Crankshaft Position sensor (CKP) from the transmission
- Transmission harness connector and clamps
- Right side oil cooler pipe
- Speed sensor harness connector
- Control linkage from the selector lever
- Propeller shaft

- Heat shield from the catalytic converter
- Exhaust tube bracket and separate the rear exhaust tube from the catalytic converter
- Starter
- Gussets and end plate
- Bolts securing the torque converter to the drive plate. Properly support the transmission assembly
- Rear mount and slightly lower the transmission
- Left side oil cooler pipe
- Engine to transmission bolts
- Transmission assembly

➡ **Tagging the different length transmission bolts upon removal is necessary to ensure that they are installed in their original position.**

To install:

5. Clean the engine and transmission mating surfaces.

6. Install the transmission to the engine.

Torque the transmission mounting bolts as follows:

a. Bolts No. 1 and No. 2: 29–36 ft. lbs. (39–49 Nm).

b. Bolt No. 3: 22–29 ft. lbs. (29–39 Nm).

c. Gusset mounting bolts: 22–29 ft. lbs. (29–39 Nm).

7. Install or connect the following:

- Left side oil cooler pipe
- Rear mount
- Torque converter to drive plate
- Engine gussets and end plate
- Starter
- Exhaust tube mounting bracket to the transmission
- Catalytic converter heat shield
- Propeller shaft
- Control linkage to the selector lever
- Speed sensor harness connector
- Right side oil cooler pipe
- Transmission harness connectors
- CKP sensor
- Negative battery cable

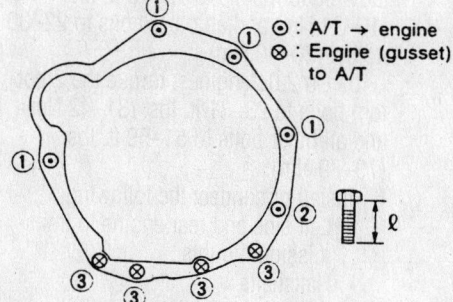

⊙ : A/T → engine
⊗ : Engine (gusset) to A/T

Bolt No.	Tightening torque N·m (kg-m, ft-lb)	Bolt length "ℓ" mm (in)
①	39 - 49 (4.0 - 5.0, 29 - 36)	40 (1.57)
②	39 - 49 (4.0 - 5.0, 29 - 36)	50 (1.97)
③	29 - 39 (3.0 - 4.0, 22 - 29)	25 (0.98)
Gusset to engine (4 bolts)	29 - 39 (3.0 - 4.0, 22 - 29)	20 (0.79)

7923QG67

Transmission bolt tightening specifications—240SX with automatic transmission

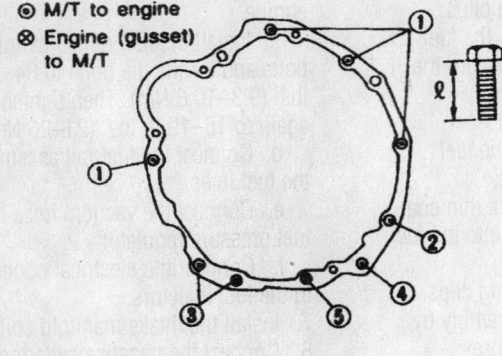

◉ M/T to engine
⊗ Engine (gusset) to M/T

Bolt No.	Tightening torque N·m (kg-m, ft-lb)
①	39 - 49 (4.0 - 5.0, 29 - 36)
②	39 - 49 (4.0 - 5.0, 29 - 36)
③*	29 - 39 (3.0 - 4.0, 22 - 29)
④*	29 - 39 (3.0 - 4.0, 22 - 29)
⑤	29 - 39 (3.0 - 4.0, 22 - 29)
Gusset to engine	29 - 39 (3.0 - 4.0, 22 - 29)

*: With nut.

79230G66

Transmission bolt tightening specifications—240SX with manual transmission

8. Fill the transmission with clean oil to the proper level.

9. Start the vehicle, check for leaks and repair if necessary.

Manual

1. Before servicing the vehicle, refer to the precautions in the beginning of this section.

2. Drain the transmission fluid.

3. Remove or disconnect the following:
- Negative battery cable
- Crankshaft Position sensor (CKP) from the transmission
- Shift lever and control housing
- Clutch operating cylinder
- Speed sensor, OD position switch and Back-up lamp switch
- Neutral position switch
- Rear Heated Oxygen Sensor (HO2S)
- Starter
- Gussets
- Exhaust tube mounting bracket and properly support the transmission
- Rear mount
- Transmission from the engine
- Transmission

To install:

4. Lightly lubricate the clutch disc splines and main drive gear splines. Also lubricate the control lever sliding surfaces with grease.

5. Properly support the transmission and install the transmission to the rear of the engine.

6. Use the following torque specifications to bolt the transmission to the engine:

a. Bolts No. 1 and 2: 29–36 ft. lbs. (39–49 Nm).

b. Bolts No. 3, 4 and 5: 22–29 ft. lbs. (29–39 Nm).

7. Install or connect the following:
- Rear mount and remove the transmission support
- Exhaust tube bracket
- Gussets
- Starter
- Rear Heated Oxygen Sensor (HO2S)
- Neutral position switch
- Speed sensor, OD position switch and Back-up lamp switch
- Clutch operating cylinder
- Shift lever and control housing
- CKP sensor
- Negative battery cable

8. Fill the transmission with clean oil to the proper level.

9. Start the vehicle, check for leaks and repair if necessary.

Transaxle Assembly

REMOVAL & INSTALLATION

Manual

SENTRA AND 200SX

1. Before servicing the vehicle, refer to the precautions in the beginning of this section.

2. Drain the fluid from the transaxle.

3. Remove or disconnect the following:
- Both battery cables
- Battery and bracket from the vehicle

- Crankshaft Position (CKP) sensor from the transaxle
- Air cleaner assembly
- All electrical connectors from the transaxles
- Control cable from the transaxle
- Speed sensor, OD position switch and Back-up lamp switch
- Neutral position switch
- Starter
- Shift control rod
- Halfshafts and properly support the transmission
- Left side and rear engine to transmission mounts

4. Slide the transmission away from the engine and lower the transmission assembly.

To install:

5. Install the transaxle mounting bolts in the proper location as noted during removal.

a. On 1.6L and 1.8L engines: torque the 2 bottom bolts to 12–15 ft. lbs. (16–21 Nm) and all other bolts to 22–30 ft. lbs. (30–40 Nm).

b. On 2.0L engines: torque the 2 bottom bolts to 23–31 ft. lbs. (31–42 Nm) and all other bolts to 51–59 ft. lbs. (70–79 Nm).

6. Install or connect the following:
- Left side and rear engine to transmission mounts
- Halfshafts
- Shift control rod
- Starter
- Neutral position switch
- Speed sensor, OD position switch and back-up lamp switch

GA engine models
⊙ M/T to engine

⊗ Engine (gusset) to M/T

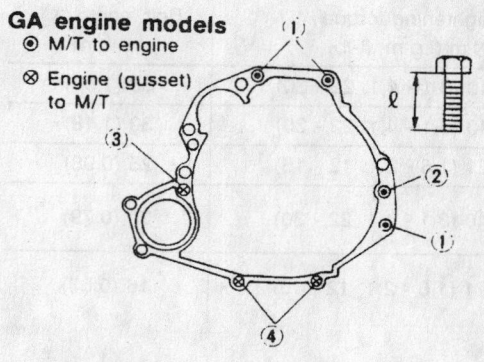

Bolt No.	Tightening torque N·m (kg-m, ft-lb)	"ℓ" mm (in)
①	30 - 40 (3.1 - 4.1, 22 - 30)	70 (2.76)
②	30 - 40 (3.1 - 4.1, 22 - 30)	85 (3.35)
③	30 - 40 (3.1 - 4.1, 22 - 30)	30 (1.18)
④	16 - 21 (1.6 - 2.1, 12 - 15)	25 (0.98)
Front gusset to engine	30 - 40 (3.1 - 4.1, 22 - 30)	20 (0.79)
Rear gusset to engine	16 - 21 (1.6 - 2.1, 12 - 15)	16 (0.63)

9307QG13

Bolt locations and torque specifications—1.6L and 1.8L engines with manual transaxle

SR engine models
⊙ M/T to engine

⊗ Engine to M/T

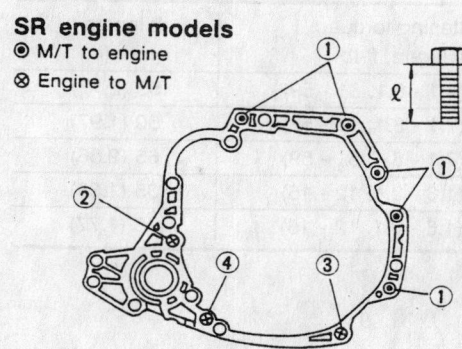

Bolt No.	Tightening torque N·m (kg-m, ft-lb)	"ℓ" mm (in)
①	70 - 79 (7.1 - 8.1, 51 - 59)	55 (2.17)
②	70 - 79 (7.1 - 8.1, 51 - 59)	65 (2.56)
③	31 - 42 (3.2 - 4.3, 23 - 31)	35 (1.38)
④	31 - 42 (3.2 - 4.3, 23 - 31)	45 (1.77)

9307QG14

Bolt locations and torque specifications—2.0L engine with manual transaxle

- Clutch control cable
- CKP sensor
- Air cleaner assembly
- Battery and both cables

7. Fill the transmission with clean oil to the proper level.

8. Start the vehicle, check for leaks and repair if necessary.

Automatic

1. Before servicing the vehicle, refer to the precautions in the beginning of this section.

2. Drain the fluid from the transaxle.

3. Remove or disconnect the following:
- Both battery cables
- Battery and bracket from the vehicle
- Crankshaft Position (CKP) sensor from the transaxle
- Air cleaner assembly
- Torque converter clutch solenoid valve electrical connector
- Inhibitor switch and Vehicle Speed Sensor (VSS) electrical connectors
- Throttle wire from the engine side
- Control cable

- Oil cooler hoses
- Halfshafts
- Intake manifold support bracket
- Starter
- Upper engine to transmission bolts and properly support the transmission
- Center member
- Front and rear gussets and engine rear plate
- Rear transmission to engine bracket
- Rear transmission mount
- Transmission assembly

To install:

When connecting the torque converter to the transaxle, be sure to measure the distance between the mounting lug of the converter and the front edge of the transaxle.

4. The measured distance between the converter and the front of the transaxle should be:

a. 0.831 in. (21.1mm) or more for GA16DE engine vehicles

b. 0.626 in. (15.9mm) or more for SR20DE engine vehicles

5. Raise the transaxle and install to engine drive plate.

6. Install the transaxle mounting bolts in the proper location as noted during removal.

a. On 1.6L and 1.8L engines torque the 2 bottom bolts to 12–15 ft. lbs. (16–21 Nm) and all other bolts to 22–30 ft. lbs. (30–40 Nm).

b. On 2.0L engines torque the 2 bottom bolts to 12–15 ft. lbs. (16–21 Nm) and all other bolts to 51–59 ft. lbs. (70–79 Nm).

7. Install or connect the following:
- Torque converter to the drive plate and torque the bolts to 33–43 ft. lbs. (44–59 Nm)
- Rear transmission mount
- Rear transmission bracket
- Front and rear gussets and the rear engine plate
- Center member
- Starter and torque the bolts to 31 ft. lbs. (42 Nm)
- Intake manifold support bracket
- Both half shafts

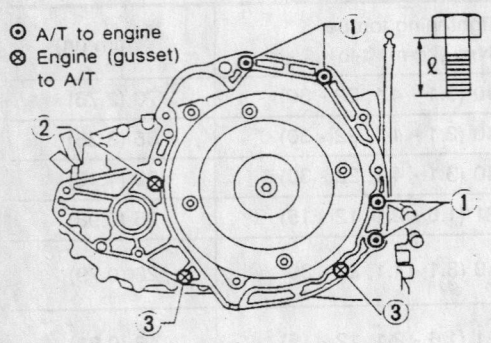

- ⊙ A/T to engine
- ⊗ Engine (gusset) to A/T

Bolt No.	Tightening torque N·m (kg-m, ft-lb)	Bolt length "ℓ" mm (in)
①	30 - 40 (3.1 - 4.1, 22 - 30)	50 (1.97)
②	30 - 40 (3.1 - 4.1, 22 - 30)	30 (1.18)
③	16 - 21 (1.6 - 2.1, 12 - 15)	25 (0.98)
Front gusset to engine	30 - 40 (3.1 - 4.1, 22 - 30)	20 (0.79)
Rear gusset to engine	16 - 21 (1.6 - 2.1, 12 - 15)	16 (0.63)

9307QG11

Bolt locations and torque specifications—1.6L and 1.8L engines with automatic transaxle

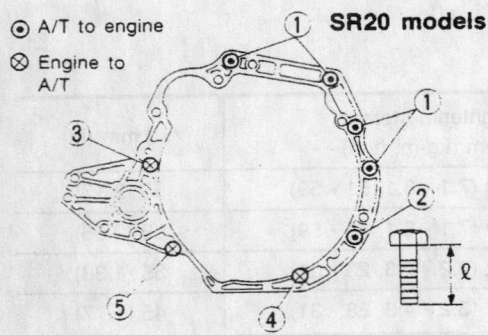

SR20 models

- ⊙ A/T to engine
- ⊗ Engine to A/T

Bolt No.	Tightening torque N·m (kg-m, ft-lb)	Bolt length "ℓ" mm (in)
①	70 - 79 (7.1 - 8.1, 51 - 59)	55 (2.17)
②	70 - 79 (7.1 - 8.1, 51 - 59)	50 (1.97)
③	70 - 79 (7.1 - 8.1, 51 - 59)	65 (2.56)
④	16 - 21 (1.6 - 2.1, 12 - 15)	35 (1.38)
⑤	16 - 21 (1.6 - 2.1, 12 - 15)	45 (1.77)

9307QG12

Bolt locations and torque specifications—2.0L engine with automatic transaxle

- Control cable
- Throttle wire
- CKP sensor
- Torque converter clutch solenoid valve, VSS and inhibitor switch electrical connectors
- Air duct
- Battery and both cables

8. Fill the transmission to the proper level.

9. Start the vehicle, check for leaks and repair if necessary.

Altima

MANUAL

1. Before servicing the vehicle, refer to the precautions in the beginning of this section.

2. Drain the transmission fluid.

3. Remove or disconnect the following:
- Battery cables
- Battery and tray
- Air cleaner box with the Mass Air Flow (MAF) sensor
- Air duct
- Clutch operating cylinder
- Speedometer pinion electrical connectors
- Park/Neutral Position (PNP) switch electrical connectors

- Starter
- Crankshaft Position (CKP) sensor
- Shift control rod
- Front wheels
- Halfshafts and properly support the engine
- Rear and left side engine mounts
- Transmission assembly

To install:

4. Install the transaxle assembly into the vehicle.

5. Torque the 4 lower mounting bolts to 22–30 ft. lbs. (30–40 Nm) and all remaining bolts to 29–36 ft. lbs. (39–49 Nm).

6. Install or connect the following:
- Rear and left side engine mounts
- Halfshafts and remove the engine support
- Both front wheels
- Shift control rod
- CKP sensor
- Starter and torque the bolts to 30 ft. lbs. (41 Nm)
- PNP switch electrical connectors
- Speedometer pinion connectors
- Clutch operating cylinder
- Air duct and air cleaner assembly
- Battery tray and battery
- Both battery cables

7. Fill the transmission with clean fluid.

8. Start the vehicle, check for leaks and repair if necessary.

AUTOMATIC

1. Before servicing the vehicle, refer to the precautions in the beginning of this section.

2. Drain the transmission fluid.

3. Remove or disconnect the following:
- Battery cables
- Battery and tray
- Air cleaner and resonator
- Park/Neutral Position (PNP) switch
- Revolution sensor and Vehicle Speed Sensor (VSS) electrical connectors
- Crankshaft Position (CKP) sensor
- Left hand mounting bracket from the transaxle and body
- Control cable
- Both front wheels
- Halfshafts
- Oil cooler pipes
- Starter and properly support the engine
- Center member
- Rear cover plate
- Torque converter
- Transaxle assembly

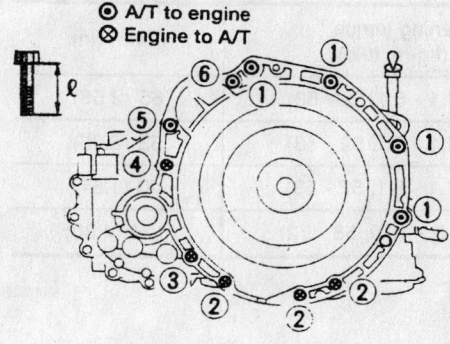

⊙ A/T to engine
⊗ Engine to A/T

Bolt No.	Tightening torque N·m (kg-m, ft-lb)	ℓ mm (in)
①	39 – 49 (4.0 – 5.0, 29 – 36)	45 (1.77)
②	30 – 36 (3.1 – 3.7, 22 – 27)	30 (1.18)
③	30 – 36 (3.1 – 3.7, 22 – 27)	40 (1.57)
④	74 – 83 (7.5 – 8.5, 54 – 61)	45 (1.77)
⑤	30 – 36 (3.1 – 3.7, 22 – 27)	80 (3.15)
⑥	30 – 36 (3.1 – 3.7, 22 – 27)	65 (2.56)

7923QG72

Be sure to install the bolts in the correct location and tighten them to specification—Altima with automatic transaxle

➡When removing the torque converter, turn the crankshaft for access to the bolts. Place alignment marks on the converter and drive plate, so the converter can be installed in its original position.

➡The transaxle mounting bolts are different lengths. Tagging the bolts upon removal will facilitate proper tightening during installation.

To install:

➡When installing the torque converter to the transaxle, measure the depth of the converter to ensure proper installation.

4. Using a straight edge across the mounting flange, measure the depth of the converter. The measurement is to the bolt mounting flange of the converter.

5. The depth measurement of the converter should be 0.75 in. (19mm) or more.

➡The transaxle mounting bolts are different lengths and require special torque specifications. Use care when installing and tightening these bolts.

6. Install the transaxle assembly into the vehicle.

7. Refer to the diagram for the automatic transaxle mounting bolt torque specifications.

8. Torque the bolts holding the converter to the flexplate to 33–43 ft. lbs. (44–59 Nm).

9. Install or connect the following:
- Rear cover plate
- Center member
- Starter and remove the engine support
- Oil cooler pipes
- Halfshafts and both front wheels
- Control cable
- Left hand mounting bracket
- CKP sensor
- Revolution and VSS sensor electrical connectors
- PNP switch
- Air cleaner and resonator
- Battery, tray and both cables

10. Fill the transaxle with the proper type and amount of fluid.

11. Start the vehicle, check for leaks and repair if necessary.

Maxima

MANUAL

1. Before servicing the vehicle, refer to the precautions in the beginning of this section.

2. Drain the fluid from the transaxle.

3. Remove or disconnect the following:
- Battery cables
- Battery and the battery tray
- Air cleaner and Mass Air Flow (MAF) sensor
- Clutch operating cylinder and hose clamps
- Speedometer pinion and Park/Neutral Position (PNP) switch connectors
- Crankshaft Position (CKP) sensor (POS) from the transaxle
- Starter
- Shift control rod and support rod bracket
- Front wheels
- Halfshafts and properly support the engine
- Center member

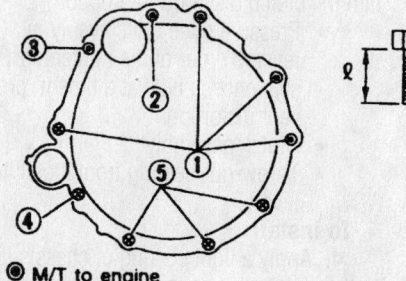

⊙ M/T to engine
⊗ Engine to M/T

Bolt No.	Tightening torque N·m (kg-m, ft-lb)	"ℓ" mm (in)
①	70 – 79 (7.1 – 8.1, 51 – 59)	52 (2.05)
②	70 – 79 (7.1 – 8.1, 51 – 59)	65 (2.56)
③	70 – 79 (7.1 – 8.1, 51 – 59)	124 (4.88)
④	35.1 – 47.1 (3.58 – 4.80, 25.89 – 34.74)	40 (1.57)
⑤	35.1 – 47.1 (3.58 – 4.80, 25.89 – 34.74)	40 (1.57)

③ with starter
④ with support rod bracket

7923QG73

Manual transaxle bolt torque specifications and locations—Maxima

Timing belt service is covered in Section 3 of this manual

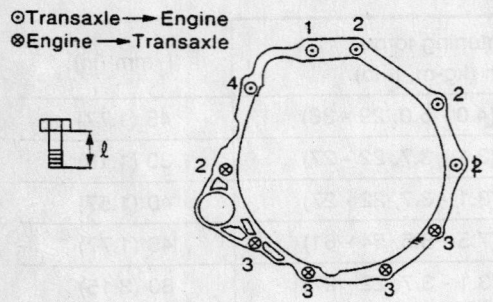

Bolt No.	Tightening torque N·m (kg-m, ft-lb)	ℓ mm (in)
1	70 - 79 (7.1 - 8.1, 52 - 58)	65 (2.56)
2	70 - 79 (7.1 - 8.1, 52 - 58)	52 (2.05)
3	70 - 79 (7.1 - 8.1, 52 - 58)	40 (1.57)
4	78 - 98 (7.9 - 10.0, 58 - 72)	124 (4.88)

9307QG20

Automatic transaxle bolt torque specifications and locations—1998–01 Maxima

- Left-hand mounting bracket from the transaxle and body
- Transaxle

To install:

➡The transaxle mounting bolts are different lengths and require special torque specifications. Use care when installing and tightening these bolts.

4. Install the transaxle assembly into the vehicle.

5. Install the transaxle mounting bolts in the proper location as noted during removal.

6. Install or connect the following:
- Left hand mounting bracket
- Center member
- Halfshafts and front wheels
- Shift control rod and support rod bracket
- Starter
- CKP sensor
- Speedometer pinion and PNP switch connectors
- Clutch operating cylinder and hose clamp
- Air cleaner and MAF sensor
- Battery and tray
- Battery cables

7. Fill the transaxle with clean fluid.

8. Start the vehicle, check for leaks and repair if necessary.

AUTOMATIC

1. Before servicing the vehicle, refer to the precautions in the beginning of this section.

2. Drain the transaxle fluid.

3. Remove or disconnect the following:
- Battery cables
- Battery and tray
- Air cleaner and resonator
- Park/Neutral Position (PNP) switch
- Revolution sensor and Vehicle Speed Sensor (VSS) electrical connectors
- Crankshaft Position (CKP) sensor
- Left hand mounting bracket from the transaxle and body

- Control cable
- Both front wheels
- Halfshafts
- Oil cooler pipes
- Starter and properly support the engine
- Center member
- Rear cover plate
- Torque converter
- Transaxle assembly

➡When removing the torque converter, turn the crankshaft for access to the bolts. Place alignment marks on the converter and drive plate, so the converter can be installed in its original position.

➡The transaxle mounting bolts are different lengths. Tagging the bolts upon removal will facilitate proper tightening during installation.

To install:

➡When installing the torque converter to the transaxle, measure the depth of the converter to ensure proper installation.

4. Using a straight edge across the mounting flange, measure the depth of the converter. The measurement is to the bolt mounting flange of the converter.

5. The depth measurement of the converter should be 0.75 in. (19mm) or more.

➡The transaxle mounting bolts are different lengths and require special torque specifications. Use care when installing and tightening these bolts.

6. Install the transaxle assembly into the vehicle.

7. Refer to the diagram for the automatic transaxle mounting bolt torque specifications.

8. Torque the bolts holding the converter to the flexplate to 33–43 ft. lbs. (44–59 Nm).

9. Install or connect the following:

- Rear cover plate
- Center member
- Starter and remove the engine support
- Oil cooler pipes
- Halfshafts and both front wheels
- Control cable
- Left hand mounting bracket
- CKP sensor
- Revolution and VSS sensor electrical connectors
- PNP switch
- Air cleaner and resonator
- Battery, tray and both cables

10. Fill the transaxle with the proper type and amount of fluid.

11. Start the vehicle, check for leaks and repair if necessary.

Clutch

REMOVAL & INSTALLATION

1. Before servicing the vehicle, refer to the precautions in the beginning of this section.

2. Remove or disconnect the following:

- Transmission/transaxle assembly

3. Insert a clutch disc centering tool into the clutch disc hub for support.

- Pressure plate bolts evenly in reverse order of the tightening sequence, a little at a time to prevent distortion
- Clutch assembly
- Throw-out bearing from the clutch lever

To install:

4. Apply a light coating of chassis lube to the clutch disc splines, input shaft and pilot bearing. Use a disc centering tool to aid installation.

5. Install or connect the following:
- Disc and pressure plate

6. On all except Maxima, torque the pressure plate bolts in a crisscross pattern

and in several steps to 16–22 ft. lbs. (20–26 Nm).

7. On Maxima, torque the pressure plate bolts in a crisscross pattern in the following 2 steps:

    a. Step 1: 7–14 ft. lbs. (10–20 Nm).

    b. Step 2: 25–33 ft. lbs. (34–44 Nm).

8. Install or connect the following:

- New throw-out bearing in the clutch release lever. Remove the clutch disc centering tool
- Transaxle into the vehicle. If the mating surfaces will not come together, do not force the units together. Remove the transaxle and recheck that the disc is centered

➡**DO NOT draw the transaxle to the engine with the bolts. This may damage the clutch and/or transaxle. Also, be careful not to move the throw-out bearing when installing the transaxle.**

9. After the transaxle is installed, connect the clutch cable and check operation before complete reassembly.

10. Adjust the clutch pedal as necessary.

Hydraulic Clutch System

BLEEDING

Bleeding is required to remove air trapped in the hydraulic system. The bleed screw is located on the clutch slave (operating) cylinder.

Some models are also equipped with a clutch damper mechanism. The clutch damper mechanism is bled in exactly the same manner as the operating cylinder. It should be bled along with the operating cylinder.

1. Before servicing the vehicle, refer to the precautions in the beginning of this section.

2. Remove the bleed screw dust cap.

3. Attach a transparent vinyl tube to the bleed screw, immersing the free end in a clean container of clean brake fluid.

4. Fill the master cylinder with the proper fluid.

5. Open the bleed screw about ¾ turn.

6. Depress the clutch pedal quickly. Hold it down. Have an assistant tighten the bleed screw. Allow the pedal to return slowly.

7. Repeat the above procedure until no more air bubbles are seen in the fluid container.

8. Remove the bleed tube.

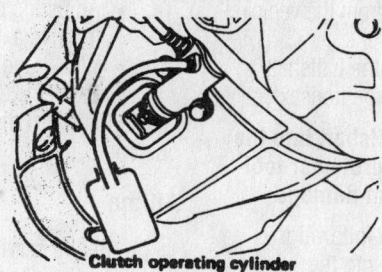

Clutch operating cylinder

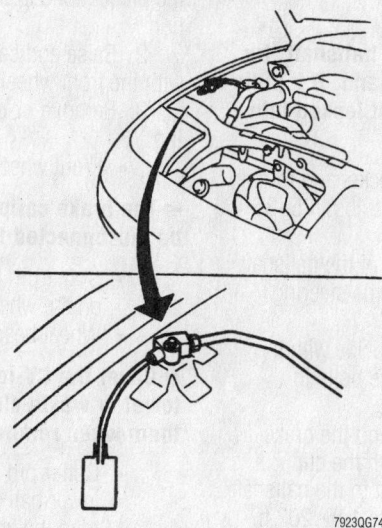

79230G74

Clutch system bleeding points—Altima, Maxima and 240SX

9. Replace the dust cap and refill the master cylinder.

10. Bleed the clutch damper, if equipped.

Halfshaft

REMOVAL & INSTALLATION

Sentra and 200SX

➡**The halfshafts will require a special tool for the spline alignment of the halfshaft end into the transaxle case. Do not perform this procedure without access to this tool. The Kent Moore tool Number is J-34296 and J-34297**

1. Before servicing the vehicle, refer to the precautions in the beginning of this section.

2. Raise the front of the vehicle and support it on jackstands, then remove the wheel and the tire assembly.

3. Remove or disconnect the following:

- Wheel
- Hub nut using a bar to hold the wheel from turning
- Clip and separate the brake hose from the strut
- Caliper assembly and support it with a wire. Do not allow the caliper to hang from the brake hose
- Bolts that secure the strut to the steering knuckle

➡**Cover the halfshaft boots with shop towels to protect them during removal of the shaft.**

- Halfshaft from the knuckle by lightly tapping it with a hammer. If it is hard to remove, use a puller

4. Remove the halfshaft from the transaxle as follows:

    a. Models without support bearing: Pry the halfshaft from the transaxle.

    b. Models with support bearing: Remove the support bearing bolts and pull the halfshaft from transaxle.

➡**When removing the halfshaft from the transaxle, do not pull on the halfshaft. The halfshaft will separate at the sliding joint (damaging the boot). Use a small prybar to remove it from the transaxle. Be sure to replace the oil seal in the transaxle.**

Heater Core replacement is covered in Section 2 of this manual

5. Remove the halfshaft from the vehicle.

To install:

6. Use a new circlip on the halfshaft and install a new oil seal to the transaxle.

➡**When installing the halfshaft into the transaxle, use a oil seal protector tool to protect the oil seal from damage.**

7. Install or connect the following:
- Halfshaft assembly into the transaxle

➡**After installation of the halfshaft, try to pull the flange out by hand. If it pulls out, the circular clip is not locked into the transaxle.**

- Support bearing bracket and torque the bolts to 19–26 ft. lbs. (25–35 Nm)

8. Lubricate the splines of the halfshaft and insert the shaft through the steering knuckle.

9. Align the steering knuckle with the lower strut mount. Torque the bolts to 68–82 ft. lbs. (92–111 Nm).
- Disc brake caliper and the brake hose to the strut with the clip
- Washer and hub nut to the halfshaft and torque the nut to 145–202 ft. lbs. (197–274 Nm)

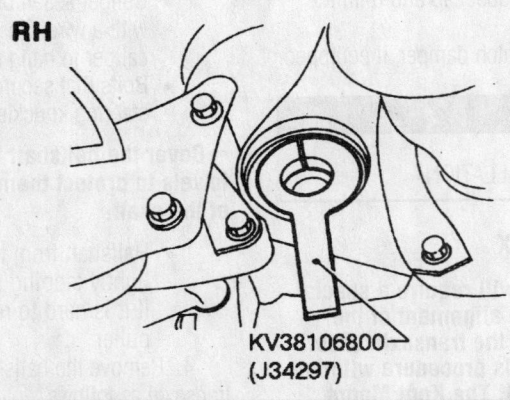

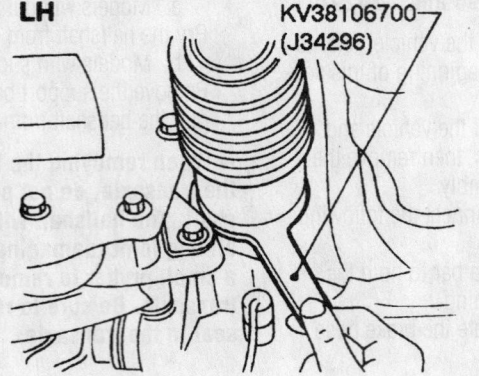

KV38106800 (J34297)

KV38106700 (J34296)

Halfshaft installation tools—Sentra and 200SX

- Adjusting cap and a new cotter pin in drive axle
- Wheel and tire assembly and lower the vehicle

10. Road test the vehicle for proper operation.

Altima

1. Before servicing the vehicle, refer to the precautions in the beginning of this section.

2. Raise and safely support the vehicle with the front wheels hanging freely.

3. Remove or disconnect the following:
- Front wheels from the vehicle

➡**The brake caliper does not need to be disconnected from the knuckle.**

- Cotter pin from the castellated nut on the wheel hub
- Wheel bearing locknut

➡**Cover the CV-joint boots with a shop towel or waste cloth so not to damage them when removing the halfshaft.**

- Cotter pin and castle nut from the lower ball joint

4. Strike the knuckle with a hammer and pull down on the transverse link to separate the lower ball joint from the knuckle.

- Tie rod end from the steering knuckle
- Halfshaft from the steering knuckle by tapping it with a block of wood and a mallet

5. Using a prybar, reach through the engine crossmember and carefully pry the right inner CV-joint from the transaxle.

6. If equipped with manual transaxle, carefully pry the left inner CV-joint from the transaxle.

7. If equipped with automatic transaxle, insert a long tool into the opening for the right halfshaft and strike the tool to with a hammer.

8. Remove the left halfshaft from the transaxle.

To install:

➡**Whenever the halfshafts are removed, the axle seals should be replaced.**

9. When installing the shafts into the transaxle, use a new oil seal and install an alignment tool along the inner circumference of the oil seal.

10. Install or connect the following:
- Halfshaft into the transaxle, align the serration's and remove the alignment tool

11. Push the halfshaft, then press-fit the circular clip on the shaft into the clip groove on the side gear.

➡**After insertion, attempt to pull the flange out of the side joint to be sure the circular clip is properly seated in the side gear and will not come out.**

- Halfshaft into the steering knuckle
- Lower ball joint and tie rod end and torque the lower ball joint-to-control arm nuts to 52–64 ft. lbs. (71–86 Nm) and the tie rod end-to-steering knuckle nut to 22–29 ft. lbs. (29–39 Nm)
- New cotter pins to the castle nuts
- Axle nut and torque the locknut to 174–231 ft. lbs. (235–314 Nm)
- New cotter pin on the wheel hub and install the wheel
- Front wheels to the vehicle

12. Road test the vehicle for proper operation.

13. Check the transaxle fluid level and top off as necessary.

Maxima

1. Before servicing the vehicle, refer to the precautions in the beginning of this section.

2. Raise and support the front of the vehicle safely and remove the wheels.

3. Remove or disconnect the following:
- Anti-Lock Brake (ABS) wheel sensor and move it out of the way
- Brake hose from the strut
- Wheel bearing locknut
- Bolts attaching the steering knuckle to the strut. Matchmark the bolts before removal.

➡ **Cover axle boots with waste cloth so as not to damage them when removing halfshaft.**

- Halfshaft from the knuckle by slightly tapping it
- Bolts attaching the support bearing to the support bearing bracket
- Halfshaft from the transaxle with a flat bladed tool, if equipped with a manual transaxle

4. If equipped with a automatic transaxle perform the following:

   a. Remove the right halfshaft from the vehicle.

   b. Insert a flat bladed tool into the transaxle where the right halfshaft was, place the end of the tool on the halfshaft, then, drive the left shaft from the pinion side gear.

5. Remove or disconnect the following:
- Support bearing bolts and the halfshaft from the vehicle
- Discard the circlip on the end of the halfshaft
- Seal from the transaxle

To install:

6. Install or connect the following:
- New seal into the transaxle and
- install a halfshaft alignment tool into the transaxle seal
- New circlip to the halfshaft, then insert the halfshaft into the transaxle

7. With the serration's aligned remove the alignment tool.

8. Push the halfshaft fully into the transaxle to seat the circlip. Try to pull the halfshaft from the transaxle by hand to verify that the circlip is properly seated.

- Support bearing and torque the bolts to 10–14 ft. lbs. (13–19 Nm)
- Halfshaft into the steering knuckle and install the hub locknut, do not tighten the hub nut
- Steering knuckle to the strut
- Strut mounting bolts to the matchmarks and torque the bolts to 103–117 ft. lbs. (140–159 Nm)
- Brake hose to the strut
- ABS wheel sensor and torque the bolt to 13–17 ft. lbs. (18–24 Nm)
- Front wheels and torque hub locknut to 174–231 ft. lbs. (235–314 Nm)

9. Check and/or adjust the wheel alignment as necessary.

240SX

1. Before servicing the vehicle, refer to the precautions in the beginning of this section.

2. Raise and safely support the vehicle.

3. Remove or disconnect the following:
- Rear wheel(s)

4. Loosen the wheel bearing locknut and

lightly tap on the axle shaft to loosen it from the steering knuckle. Remove the nut from the axle shaft.
- Brake caliper assembly

➡ **Support the brake caliper, do not allow the caliper to hang freely from the brake hose.**

- Anti-Lock Brake (ABS) sensor from the steering knuckle
- Separate the lower ball joint
- Pull the hub assembly outward
- Shaft from hub assembly
- Axle shaft from the differential

To install:

5. Install or connect the following:
- Shaft to the differential and torque the bolts to 25–33 ft. lbs. (34–44 Nm)
- Shaft into the hub assembly
- Lower control to the ball joint and torque the mounting nut to 52–64 ft. lbs. (71–86 Nm) and install a new cotter pin
- Wheel bearing nut and torque it to 152–202 ft. lbs. (206–274 Nm)
- ABS sensor to the steering knuckle and torque the bolt to 13–17 ft. lbs. (18–24 Nm)
- Brake caliper assembly. If the brake line was disconnected, bleed the brake system
- Wheel assembly

6. Lower the vehicle and perform a road test.

CV-Joints

OVERHAUL

240SX

TRANSMISSION SIDE

1. Before servicing the vehicle, refer to the precautions in the beginning of this section.

2. Remove or disconnect the following:
- Plug seal from the slide joint by gently tapping around the joint with a hammer
- Boot bands

3. Put matchmarks on the slide joint housing and halfshaft prior to separating the joint assembly.

4. Matchmark the spider assembly and halfshaft.
- Snapring and the spider assembly

➡ **Do not disassemble the spider assembly.**

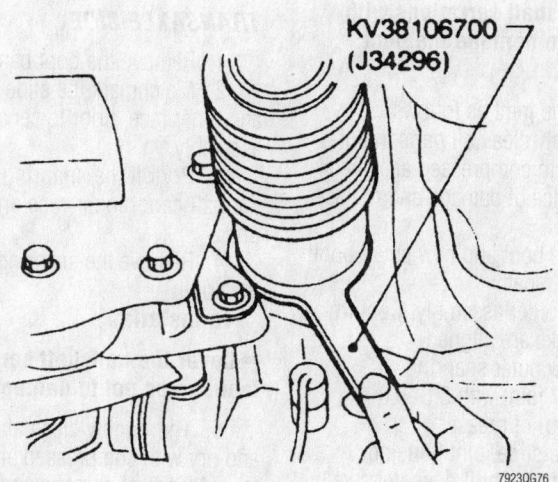

KV38106700
(J34296)

79230G76

Left halfshaft alignment tool—Maxima

For complete Engine Mechanical specifications, see Section 1 of this manual

- Slide joint housing and the boot

➡ **Cover the halfshaft serrations with tape, so as not to damage the boot.**

To install:

5. Thoroughly clean all parts in solvent and dry with compressed air. Check parts for evidence of damage and replace as necessary.

6. Install or connect the following:
- Boot and small boot band on the halfshaft
- Joint housing onto halfshaft
- Spider assembly making sure the matchmarks are properly aligned

➡ **The spider is press fit with the serration chamfer facing the shaft.**

- Snapring
- Coil spring, spring cap and new plug seal to the slide joint housing. Apply a suitable sealant to the plug seal prior to installation

➡ **Hold the plug seal horizontally when pressing it into place. This will prevent the spring inside from falling down or tilting.**

7. Move the shaft in an axial direction to make sure that the spring is installed properly. If there is a drag or the spring is installed improperly, replace the plug seal with a new one.

8. Pack the halfshaft with 3.60–3.77 ounces (102–107g) of grease.

9. Ensure that the boot is properly installed on the halfshaft groove.

10. Set the boot so that it does not swell or deform when its length is 3.74–3.82 in. (95–97mm).

11. Lock the new boot bands securely.

WHEEL SIDE

1. Before servicing the vehicle, refer to the precautions in the beginning of this section.

2. Remove or disconnect the following:
a. Remove the boot bands.
b. Matchmark the housing with the shaft and halfshaft before separating the assembly.
c. Matchmark the spider assembly and halfshaft.
d. Remove the snapring and the spider assembly.

➡ **Do not disassemble the spider assembly.**

e. Remove the boot.

➡ **Cover the halfshaft serrations with tape, so as not to damage the boot.**

To install:

3. Install the joint as follows:
a. Thoroughly clean all parts in solvent and dry with compressed air. Check parts for evidence of damage and replace as necessary.
b. Install the boot and new boot band on the halfshaft.
c. Install the spider assembly making sure the matchmarks made during removal are properly aligned.

➡ **The spider is press fit with the serration chamfer facing the shaft.**

d. Install a new snapring.
4. Pack the joint with 4–4.41 ounces (115–125g) of grease.
a. Install the slide joint housing and the snapring.
b. Ensure that the boot is properly installed on the halfshaft groove.
5. Set the boot so that it does not swell or deform when its length is 3.74–3.82 in. (95–97mm).
6. Lock the new boot bands securely.

Sentra and 200SX

TRANSAXLE SIDE

1. Before servicing the vehicle, refer to the precautions in the beginning of this section.
2. Disassemble the joint as follows:
a. Remove the boot bands.
b. Matchmark the slide joint housing and inner race before separating the assembly.
c. Matchmark the spider assembly and drive shaft.
d. Remove the snapring.
e. Remove the spider assembly.
f. Remove the boot.

➡ **Cover the halfshaft serrations with tape, so as not to damage the boot.**

To install:

3. Assemble the joint as follows:
a. Thoroughly clean all parts in solvent and dry with compressed air. Check parts for evidence of damage and replace as necessary.
b. Install the boot and new small boot band on the halfshaft.
4. Install the spider assembly. Confirm that the matchmarks are aligned.
5. Install a new outer snapring.
6. Pack the CV joint with 5.5–5.8 ounces (155–165g) of grease.
a. Install the slide joint housing.
7. Set the boot so that it does not swell or deform when its length is 4–4.07 in. (101.5–103.5mm).
8. Lock the new boot bands securely.

WHEEL SIDE

1. Before servicing the vehicle, refer to the precautions in the beginning of this section.

The joint on the wheel side cannot be disassembled.

2. Prior to separating the joint assembly, matchmark the halfshaft and joint assembly.

3. Separate the joint using a slide hammer.

4. Remove the boot bands.

To assemble:

5. Thoroughly clean all parts in solvent and dry with compressed air. Check parts for evidence of damage and replace as necessary.

➡ **Cover the halfshaft serrations with tape, so as not to damage the boot.**

6. Install the boot and small boot band on the halfshaft.

7. Set the joint assembly onto the halfshaft and align the matchmarks.

8. Attach the joint assembly to the halfshaft by lightly tapping the serrated end with a plastic hammer.

➡ **Using a metal hammer may damage the threads on the end of the joint.**

9. Pack the CV joint with 4.6–4.41 ounces (115–125g) of grease.

10. Ensure that the boot is properly installed on the halfshaft groove.

11. Set the boot so that it does not swell or deform when its length is 3.78–3.86 in. (96–98mm).

12. Lock the new boot bands securely.

Maxima

TRANSAXLE SIDE

1. Remove the boot bands.

2. Matchmark the slide joint housing and inner race, prior to separating the joint assembly.

3. Pry off the snapring and remove the ball cage, inner race and balls as a unit.

4. Remove the snapring and withdraw the boot.

To install:

➡ **Cover the halfshaft serrations with tape, so as not to damage the boot.**

5. Thoroughly clean all parts in solvent and dry with compressed air. Check parts for evidence of damage and replace as necessary.

6. Install the boot and new boot band on the halfshaft.

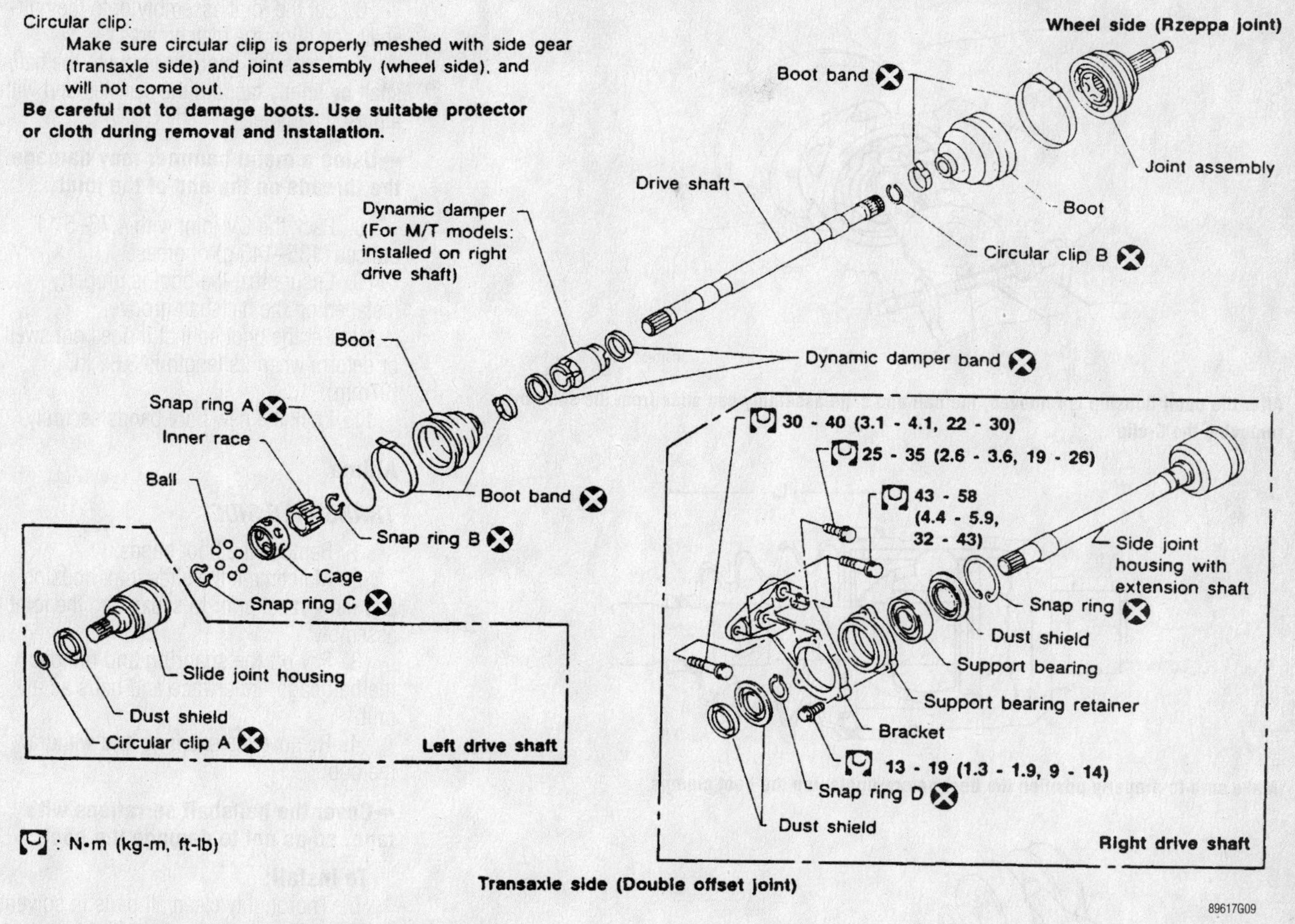

Circular clip:
   Make sure circular clip is properly meshed with side gear (transaxle side) and joint assembly (wheel side), and will not come out.
Be careful not to damage boots. Use suitable protector or cloth during removal and installation.

Wheel side (Rzeppa joint)

Boot band ⊗

Joint assembly

Boot

Circular clip B ⊗

Drive shaft

Dynamic damper
(For M/T models:
installed on right
drive shaft)

Boot

Dynamic damper band ⊗

Snap ring A ⊗

Inner race

Ball

Boot band ⊗

Snap ring B ⊗

Cage

Snap ring C ⊗

Slide joint housing

Dust shield

Circular clip A ⊗

Left drive shaft

⊡ : N•m (kg-m, ft-lb)

Transaxle side (Double offset joint)

⊡ 30 - 40 (3.1 - 4.1, 22 - 30)

⊡ 25 - 35 (2.6 - 3.6, 19 - 26)

⊡ 43 - 58
(4.4 - 5.9,
32 - 43)

Side joint
housing with
extension shaft

Snap ring ⊗

Dust shield

Support bearing

Support bearing retainer

Bracket

⊡ 13 - 19 (1.3 - 1.9, 9 - 14)

Snap ring D ⊗

Dust shield

Right drive shaft

89617G09

Exploded view of the halfshafts and related components

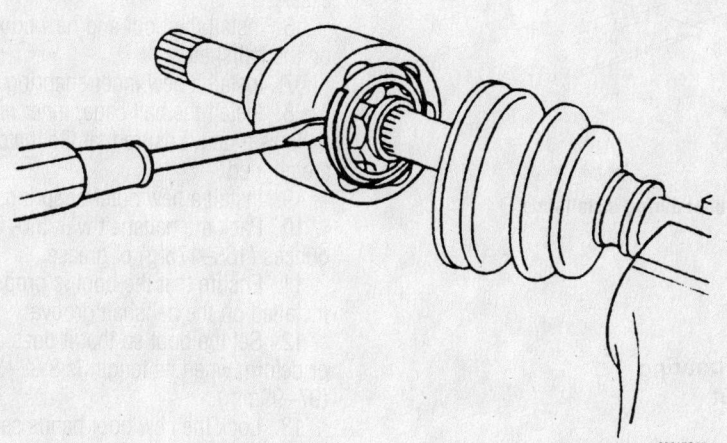

89617G07

The inner CV joint uses a large C-clip to retain the ball and cage assembly in the outer housing

7. Install a new inner snapring.
8. Install the ball cage, inner race and balls as a unit. Confirm that the matchmarks are aligned.
9. Install a new outer snapring.

10. Pack the CV joint with 5.0–6.0 ounces (165–175 g) of grease.
11. Ensure that the boot is properly installed on the halfshaft groove.
12. Set the boot so that it does not swell

or deform when its length is 3.86 in. (98mm).
   13. Lock the new boot bands securely.

WHEEL SIDE

The joint on the wheel side cannot be disassembled.
   1. Prior to separating the joint assembly, matchmark the halfshaft and joint assembly.
   2. Separate the joint using a slide hammer.
   3. Remove the boot bands and the boot.
To install:
   4. Thoroughly clean all parts in solvent and dry with compressed air. Check parts for evidence of damage and replace as necessary.

➡**Cover the halfshaft serrations with tape, so as not to damage the boot.**

   5. Install the boot and small boot band on the halfshaft.

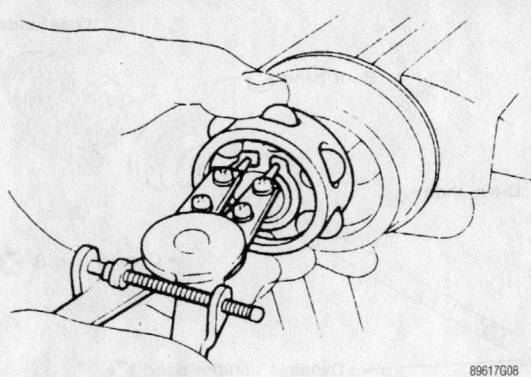

After the outer housing is removed, the ball and cage assembly can slide from the shaft by removing the C-clip

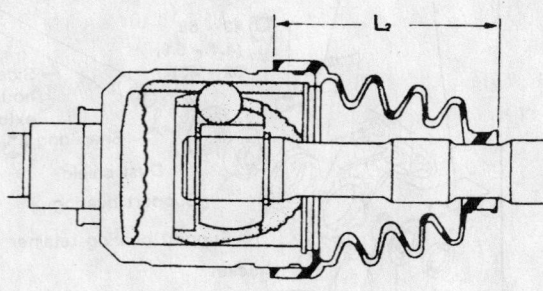

Make sure to properly position the boot before tightening the boot clamps

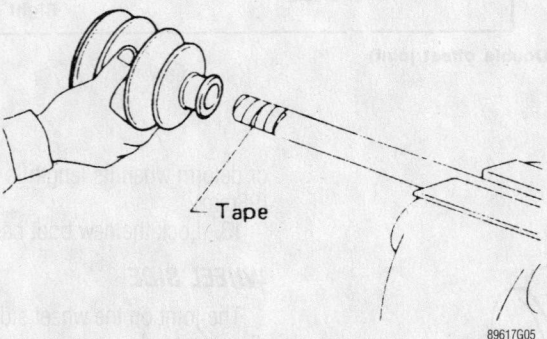

Use vinyl tape and wrap the end of the shaft to protect the boot during installation

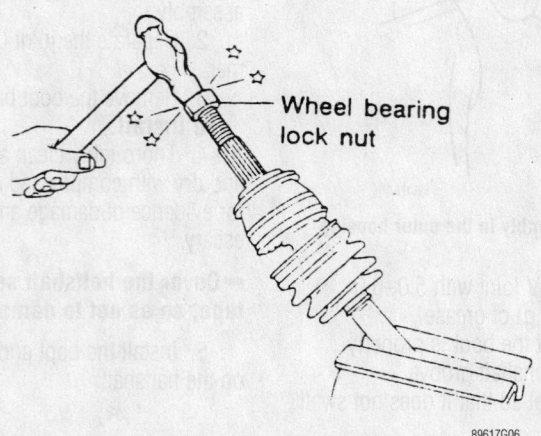

Use an old nut to protect the threads when tapping the outer CV joint onto the shaft

6. Set the joint assembly onto the half-shaft and align the matchmarks.

7. Attach the joint assembly to the half-shaft by lightly tapping the serrated end with a plastic hammer.

➡**Using a metal hammer may damage the threads on the end of the joint.**

8. Pack the CV joint with 4.76–5.11 ounces (135–145 g) of grease.

9. Ensure that the boot is properly installed on the halfshaft groove.

10. Set the boot so that it does not swell or deform when its length is 3.82 in. (97mm).

11. Lock the new boot bands securely.

Altima

TRANSAXLE SIDE

1. Remove the boot bands.

2. Matchmark the slide joint housing and inner race, prior to separating the joint assembly.

3. Pry off the snapring and remove the ball cage, inner race and balls as a unit.

4. Remove the snapring and withdraw the boot.

➡**Cover the halfshaft serrations with tape, so as not to damage the boot.**

To install:

5. Thoroughly clean all parts in solvent and dry with compressed air. Check parts for evidence of damage and replace as necessary.

6. Install the boot and new boot band on the halfshaft.

7. Install a new inner snapring.

8. Install the ball cage, inner race and balls as a unit. Ensure that the matchmarks are aligned.

9. Install a new outer snapring.

10. Pack the halfshaft with 5.0–6.0 ounces (165–175 g) of grease.

11. Ensure that the boot is properly installed on the halfshaft groove.

12. Set the boot so that it does not swell or deform when its length is 3.82–3.90 in. (97–99mm).

13. Lock the new boot bands securely.

WHEEL SIDE

The joint on the wheel side cannot be disassembled.

1. Prior to separating the joint assembly, matchmark the halfshaft and joint assembly.

2. Separate the joint using a slide hammer.

3. Remove the boot bands.

To assemble:

4. Thoroughly clean all parts in solvent and dry with compressed air. Check parts for evidence of damage and replace as necessary.

➡**Cover the halfshaft serrations with tape, so as not to damage the boot.**

5. Install the boot and small boot band on the halfshaft.

6. Set the joint assembly onto the halfshaft and align the matchmarks.

7. Attach the joint assembly to the halfshaft by lightly tapping the serrated end with a plastic hammer.

➡**Using a metal hammer may damage the threads on the end of the joint.**

8. Pack the halfshaft with 3.5–4.0 ounces (100–115 g) of grease.

9. Ensure that the boot is properly installed on the halfshaft groove.

10. Set the boot so that it does not swell

or deform when its length is 3.327–3.406 in. (84.5–86.5mm).

11. Lock the new boot bands securely.

Pinion Seal

REMOVAL & INSTALLATION

240SX

1. Before servicing the vehicle, refer to the precautions in the beginning of this section.

2. Remove or disconnect the following:
   - Driveshaft
   - Pinion nut with a suitable socket
   - Companion flange using a suitable puller
   - Seal using a suitable puller

To install:

3. Apply multi-purpose grease to the sealing lips of the seal.

4. Install or connect the following:

   - Seal into the carrier using a suitable driver
   - Companion flange, drive pinion nut and torque the nut to 137–217 ft. lbs. (186–294 Nm)
   - Driveshaft

Axle Housing Assembly

REMOVAL & INSTALLATION

240SX

1. Before servicing the vehicle, refer to the precautions in the beginning of this section.

2. Remove or disconnect the following:
   - Anti-Lock Brake (ABS) sensor from the axle assembly and move the wiring and sensor aside.
   - Driveshaft
   - Halfshafts
   - Nuts attaching the axle to the sus-

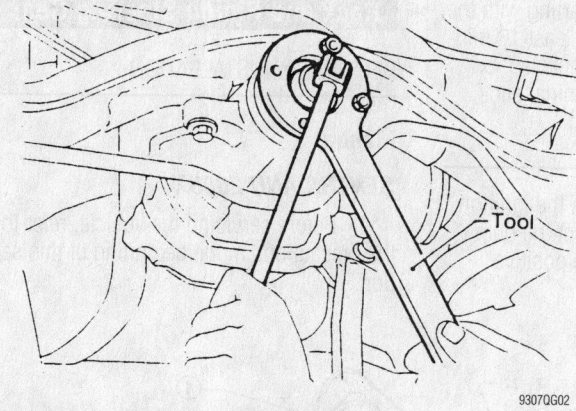

Loosening the pinion nut—Altima and 240SX

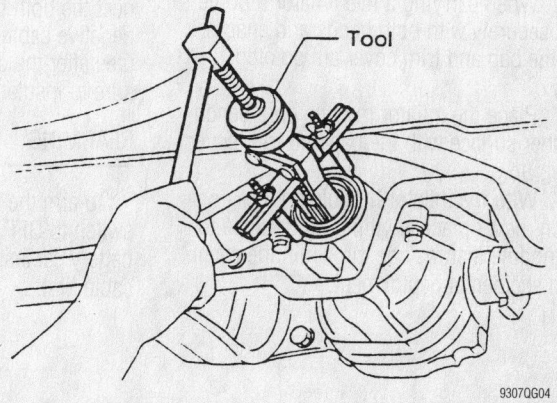

Remove the seal using a suitable puller—Altima and 240SX

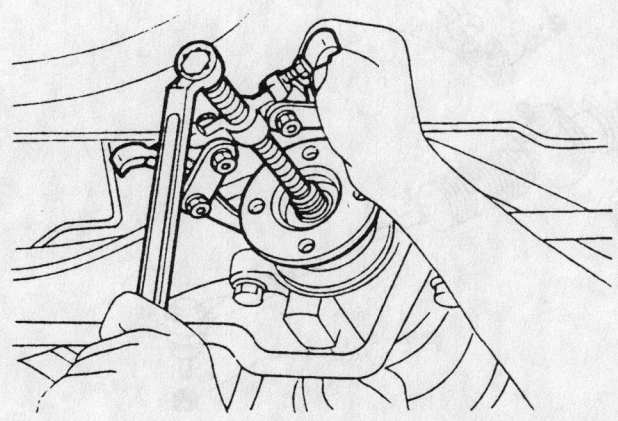

Use a puller to remove the companion flange—Altima and 240SX

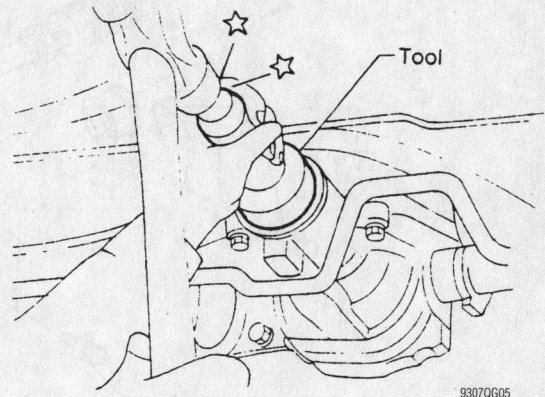

Use a suitable driver and a hammer to install the seal—Altima and 240SX

For Tire, Wheel and Ball Joint specifications, see Section 1 of this manual

pension member and support the weight of the axle housing with a jack

• Mounting member from the front of the axle and move the axle forward using the jack

• Rear cover stud bolts from the suspension member

• Lower the axle using the jack to remove it from the vehicle

3. Support the suspension member with a jackstand to prevent the insulators from being twisted or damaged.

To install:

• Axle into position using the jack

• Rear cover stud bolts to the suspension member and torque to 72–87 ft. lbs. (98–118 Nm)

• Mounting member to the front of the axle and torque to 72–87 ft. lbs. (98–118 Nm)

• Nuts attaching the axle to the suspension and torque to 72–87 ft. lbs. (98–118 Nm)

• Halfshafts

• Driveshaft

• ABS sensor to the axle assembly and reroute the wiring

STEERING AND SUSPENSION

Air Bag

PRECAUTIONS

Several precautions must be observed when handling the inflator module to avoid accidental deployment and possible personal injury.

1. Never carry the inflator module by the wires or connector on the underside of the module.

2. When carrying a live inflator module, hold securely with both hands and ensure that the bag and trim cover are pointed away.

3. Place the inflator module on a bench or other surface with the bag and trim cover facing up.

4. With the inflator module on the bench, never place anything on or close to the module that may be thrown in the event of an accidental deployment.

DISARMING

➡**All SRS electrical wiring harnesses and connectors are covered with YELLOW outer insulation. Do not use electrical test equipment on any circuit related to the SRS (air bag) sensors. When installing SRS components, always install with the arrow marks facing the front of the vehicle.**

To disarm the SRS system turn the ignition switch to OFF position. Then, disconnect the both battery cables starting with the negative cable first and wait at least 10 minutes after the cables are disconnected. Be sure to insulate the battery terminal ends.

REARMING

To arm the SRS system turn the ignition switch to OFF position. Connect the both battery cables starting with the positive cable first.

➡**The SRS or air bag system is equipped with a self-diagnostic operation. After turning the ignition key to the ON or START position, the AIR BAG warning lamp will illuminate for 7 seconds. After 7 seconds, the AIR BAG lamp will extinguish if no malfunction is detected. If the AIR BAG lamp does not extinguish after 7 seconds, check the SRS self-diagnostic system for a malfunction.**

Rack and Pinion Steering Gear

REMOVAL & INSTALLATION

Manual

SENTRA AND 200SX

1. Before servicing the vehicle, refer to the precautions in the beginning of this section.

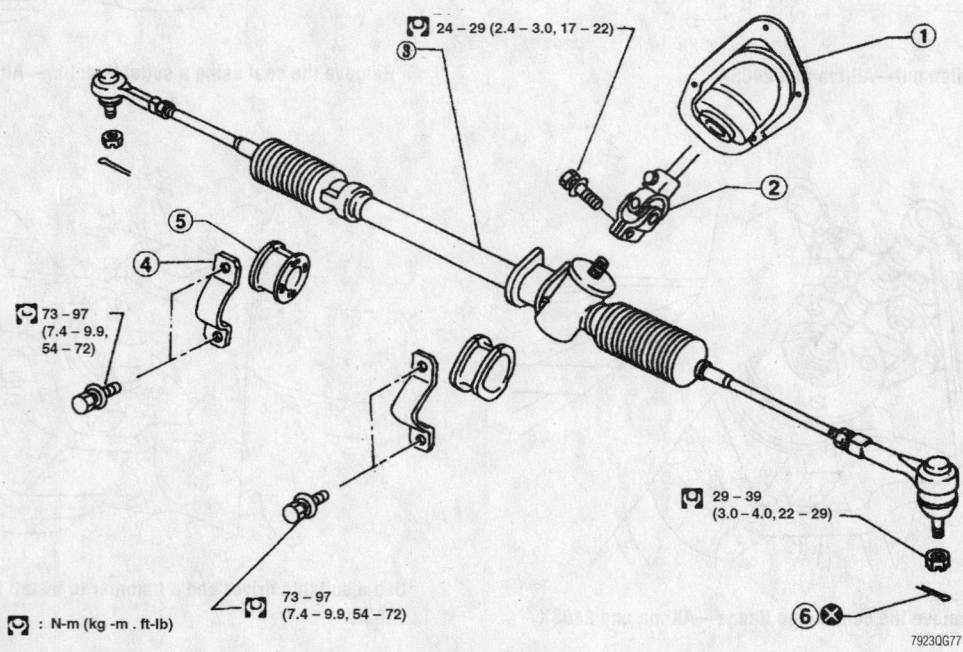

Exploded view of the manual rack and pinion steering gear mounting—Sentra and 200SX

2. Remove or disconnect the following:
- Front wheels
- Both tie rod ends from the steering knuckles

3. Matchmark the steering column shaft to the lower joint and remove the pinch bolt from the joint.
- Steering gear mounting bolts
- Mounting clamps from the steering gear
- Steering gear by sliding it off the steering shaft
- Steering gear from the vehicle

To install:

4. Install or connect the following:
- Steering gear assembly to the vehicle. Be sure to align the matchmarks of the rack with the marks on the steering shaft
- Steering gear mounting clamps and torque the bolts to 58 ft. lbs. (78 Nm)
- Lower joint-to-steering column pinch bolt and torque the bolt to 22 ft. lbs. (29 Nm)
- Tie rod end to the steering knuckle and torque the castle nut to 29 ft. lbs. (39 Nm) and install a new cotter pin

➡**If installing a new rack and pinion assembly, transfer the lower steering joint to the new rack and pinion prior to installation. When installing the lower steering joint to the steering gear, be sure that the wheels are aligned with the vehicle (straight-ahead position).**

5. To center the steering gear, turn it all the way to the lock position on one side. Now, count the number of turns it takes to get to the opposite side lock position. Turn the steering gear ½ the number of turns towards the original starting position. The steering rack should now be centered. When connecting the steering joint to the steering column shaft, be sure to align the matchmarks made during disassembly.

6. Install the front wheels, remove the jackstands and lower the vehicle.

7. Check the vehicle's alignment.

Power

SENTRA AND 200SX

1. Before servicing the vehicle, refer to the precautions in the beginning of this section.

2. Raise and support the vehicle safely.

3. Remove or disconnect the following:
- Low pressure hose clamp
- Low pressure hose at the steering

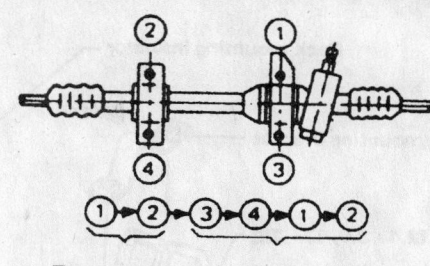

Temporary tightening   Secure tightening

79230G78

Tighten the power steering gear mounting bolts according to the sequence shown—Sentra and 200SX

gear. Be sure to use a pan to catch the fluid
- Flare nut and the high pressure tube at the steering gear, then drain the fluid from the gear
- Tie rod ends from the steering knuckle

4. Place a floor jack under the transaxle and support it.
- Front exhaust pipe and the rear engine mount

5. Position the front wheels so they are pointing straight ahead.

6. Matchmark the steering column lower joint to the steering gear.

➡**The steering gear splines have a flat spot or keyway. Be sure to note this during removal.**

7. Remove or disconnect the following:
- Bolt that secures steering column lower joint
- Bolts, steering gear unit and the linkage

To install:

8. Install or connect the following:
- Power steering gear assembly to the vehicle. Align the steering column to the steering gear

➡**Be sure to align the flat spot or keyway during installation.**

- Steering gear mounts and torque the bolts in sequence to 54–72 ft. lbs. (73–97 Nm)
- Pinch bolt for the steering column-to-gear connection and torque the bolt to 17–22 ft. lbs. (24–29 Nm)
- Tie rod ends to the steering knuckle and torque the nut to 22–29 ft. lbs. (29–39 Nm). Tighten the tie rod mounting nut further so the groves in the nut align with first cotter pin hole. Install a new cotter pin

- Power steering low pressure hose and torque the fitting to 20–29 ft. lbs. (27–39 Nm)
- Power steering high pressure and torque the fitting to 11–18 ft. lbs. (15–25 Nm)
- Rear engine mount and remove the floor jack
- Front exhaust pipe assembly using new gaskets

9. Fill the power steering system and start the engine.

10. Check the wheel alignment.

ALTIMA AND 240SX

1. Before servicing the vehicle, refer to the precautions in the beginning of this section.

2. Disconnect the negative battery cable and disarm the air bag.

3. Raise and safely support the vehicle as necessary.

4. Remove or disconnect the following:
- Bolt securing the lower steering column shaft to the power steering gear assembly. Be sure to matchmark the shaft from the steering gear to the steering column joint for correct installation
- Hoses from the power steering gear and plug the hoses to prevent leakage
- Cotter pins and castle nuts from the tie rod ends
- Tie rod ends from the steering knuckle, using a ball joint separating tool
- Front exhaust pipe mounting nuts and bolts
- Front exhaust pipe from the vehicle
- Control cable or linkage from the transmission and position it out of the way, if necessary
- Power steering gear mounting bolts or nuts

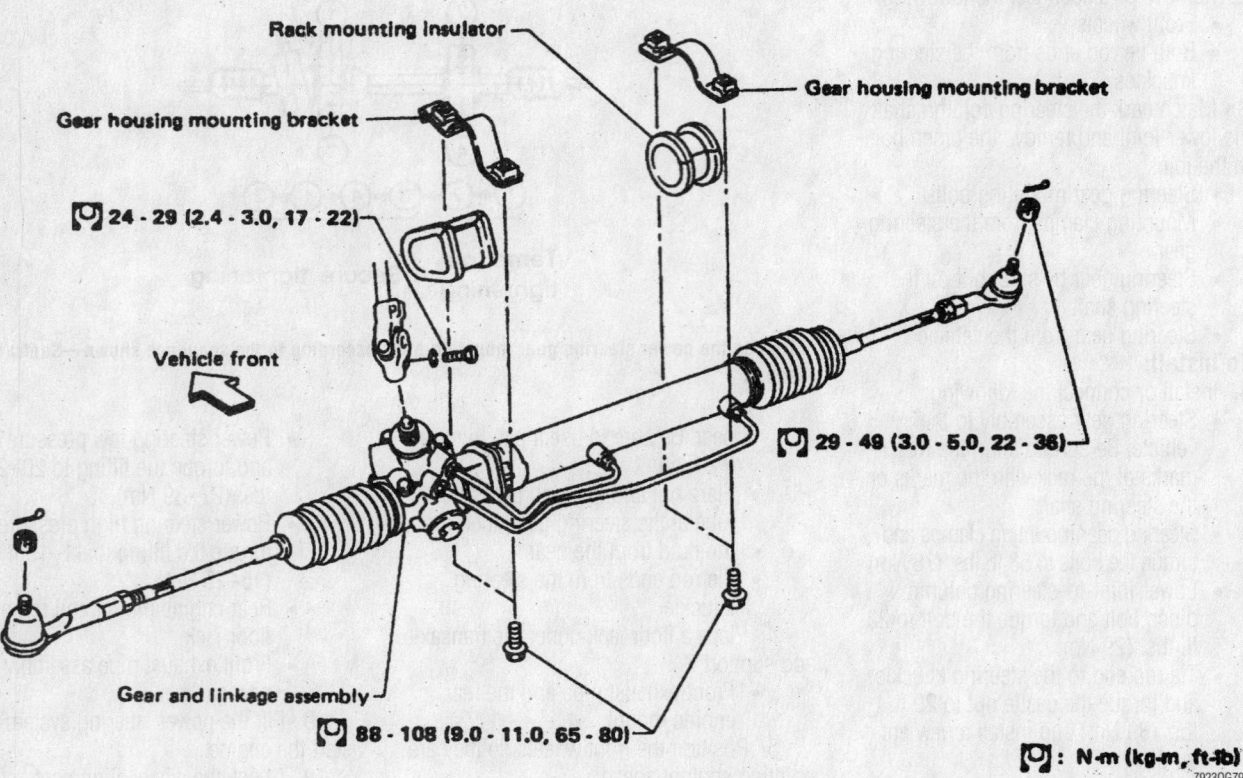

Rack mounting insulator

Gear housing mounting bracket

Gear housing mounting bracket

⊡ 24 - 29 (2.4 - 3.0, 17 - 22)

Vehicle front

⊡ 29 - 49 (3.0 - 5.0, 22 - 36)

Gear and linkage assembly

⊡ 88 - 108 (9.0 - 11.0, 65 - 80)

⊡ : N·m (kg-m, ft-lb)

7923QG79

Exploded view of the power steering gear mounting—240SX

- Steering gear from the vehicle. Use care when separating the steering column joint

5. Inspect the steering gear mount bushings and replace as necessary.

To install:

6. Align the steering column-to-steering gear matchmark and install the steering gear to the vehicle. Be sure to properly install the mounting bushings and hand-tighten the mounting nuts or bolts.

➡ **When installing the lower steering joint to the steering gear, be sure that the wheels are aligned straight and the steering joint slot is aligned.**

7. Torque the steering gear mounts as follows:

   a. 240SX: 65–80 ft. lbs. (88–108 Nm) in the sequence illustrated.

   b. Altima: 54–72 ft. lbs. (73–97 Nm) in the sequence illustrated.

8. Install or connect the following:

- Pinch bolt securing the lower steering column shaft to the power steering gear assembly and torque the pinch bolt to 17–22 ft. lbs. (24–29 Nm)

- Tie rod end to steering knuckle and torque the castle nut to 22–29 ft.

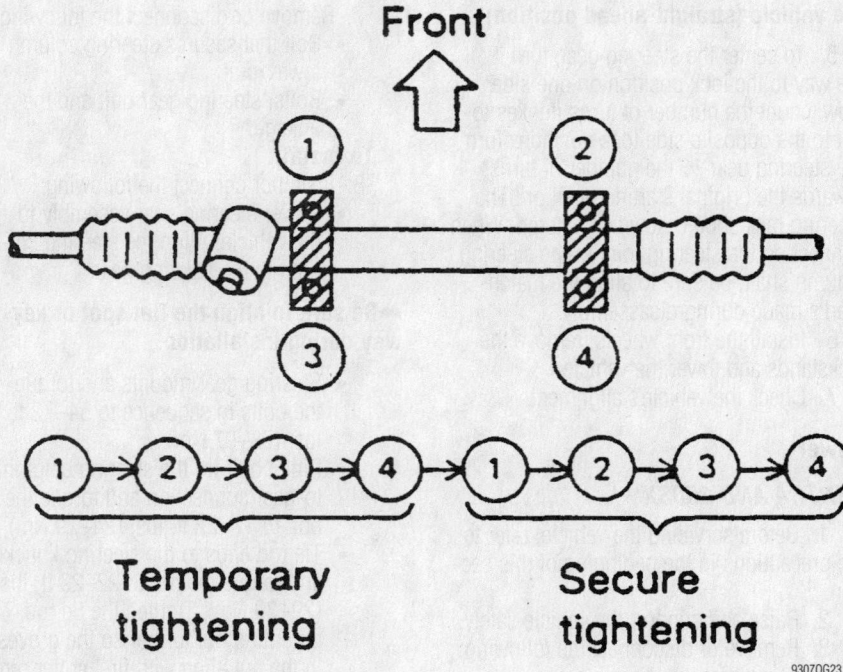

Front

1   2

3   4

1 → 2 → 3 → 4   1 → 2 → 3 → 4

Temporary tightening   **Secure tightening**

9307QG23

Tighten the mounting bolts using the illustrated procedure—240SX

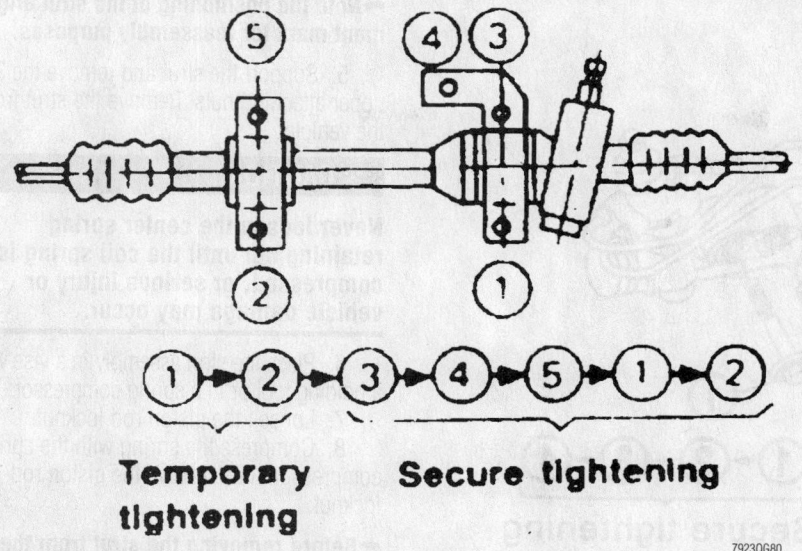

Temporary tightening

Secure tightening

7923QG80

Tighten the mounting bolts using the illustrated procedure—Altima and 1998 Maxima

lbs. (29–39 Nm). Tighten the castle nut further to align the slot in the castle nut with the cotter pin hole and install a new cotter pin

• Control cable or linkage to the transmission, if removed
• Front exhaust pipe assembly, using new gaskets
• Power steering hoses to the steering gear

9. Start the engine and fill the power steering reservoir.

10. Perform a front end alignment.

11. Connect the negative battery cable.

12. If equipped, enable the air bag system.

1998 MAXIMA

1. Before servicing the vehicle, refer to the precautions in the beginning of this section.

2. Disconnect both battery cables and wait at least 10 minutes after the battery cables are disconnected. This will disarm the air bag system so the steering wheel can be removed.

3. Point the front tires straight ahead and lock the steering in this position.

❊❊ WARNING

Do not turn the steering wheel or column with the lower joint removed from the steering column or the spiral cable may be damaged.

4. Remove the steering wheel.

➡ **The steering wheel must be removed before disconnecting the steering column lower joint to avoid damaging the SRS spiral cable.**

5. Raise and support the vehicle safely and remove the front wheels.

6. Remove or disconnect the following:
• Tie rod ends from the steering knuckles
• Carbon canister from the vehicle and support the engine
• Bolts attaching the engine mounts to the engine mounting center member
• Engine mounting center member
• Front stabilizer bar from the vehicle
• Nuts attaching the hole cover to the bulkhead

7. Move the hole cover aside and disconnect the lower joint from the rack and pinion. Matchmark the pinion shaft and the pinion housing to record the steering neutral position.
• Power steering fluid pipes from the rack and pinion
• Bolts attaching the mounting brackets
• Rack and pinion from the vehicle

To install:

8. Position the rack and pinion in the vehicle and install the mounting brackets. Torque the mounting nuts and bolts in the proper sequence to 54–72 ft. lbs. (73–97 Nm).

9. Install or connect the following:
• New O-rings to the power steering

fluid pipes and connect them to the rack and pinion. Torque the low pressure line 20–29 ft. lbs. (27–39 Nm) and the high pressure line to 11–18 ft. lbs. (15–25 Nm)

10. Align the lower steering joint to the pinion shaft and install the joint onto the pinion shaft. Torque the bolt to 17–22 ft. lbs. (24–29 Nm).
• Hole cover and torque the nuts to 36–43 inch lbs. (4–5 Nm)
• Front stabilizer
• Engine mounting center member and torque the bolts to 57–72 ft. lbs. (77–98 Nm)
• Engine mounts to the center member and torque the bolts to 57–72 ft. lbs. (77–98 Nm). Remove the support from the engine
• Remaining components in the reverse order of removal

11. Torque the tie rod end nuts to 22–29 ft. lbs. (29–39 Nm) and install a new cotter pin.

12. Fill the power steering reservoir with fluid and bleed the air from the power steering system.

13. Check the vehicle front end alignment and adjust as necessary.

1999–01 MAXIMA

1. Before servicing the vehicle, refer to the precautions in the beginning of this section.

2. Disconnect both battery cables and wait at least 10 minutes after the battery cables are disconnected. This will disarm the air bag system so the steering wheel can be removed.

3. Point the front tires straight ahead and lock the steering in this position.

❊❊ WARNING

Do not turn the steering wheel or column with the lower joint removed from the steering column or the spiral cable may be damaged.

4. Remove the steering wheel.

➡ **The steering wheel must be removed before disconnecting the steering column lower joint to avoid damaging the SRS spiral cable.**

5. Raise and support the vehicle safely and remove the front wheels.

6. Remove or disconnect the following:
• Tie rod ends from the steering knuckles

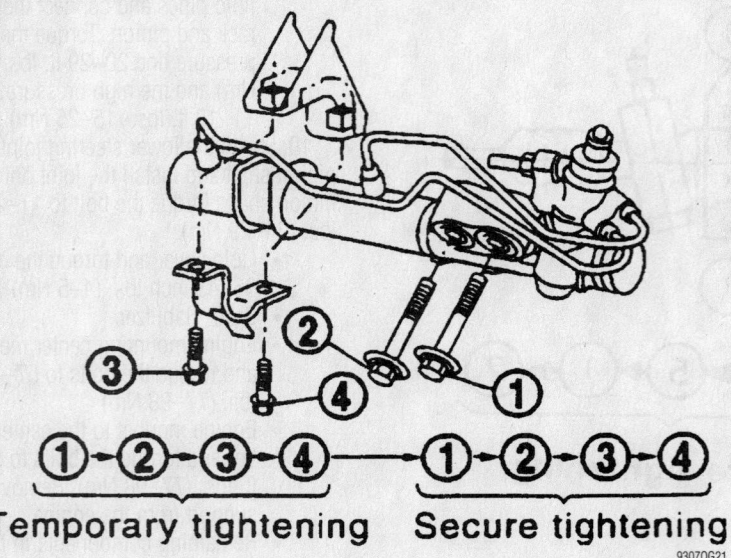

Temporary tightening   **Secure tightening**

9307QG21

Tighten the mounting bolts using the illustrated procedure—1999–01 Maxima

- Front exhaust tube and properly support the engine
- Bolts attaching the engine mounts to the engine mounting center member
- Engine mounting center member
- Front stabilizer bar from the vehicle
- Nuts attaching the hole cover to the bulkhead

7. Move the hole cover aside and disconnect the lower joint from the rack and pinion. Matchmark the pinion shaft and the pinion housing to record the steering neutral position.

- Power steering fluid pipes from the rack and pinion
- Bolts attaching the mounting brackets
- Rack and pinion from the vehicle

To install:

8. Position the rack and pinion in the vehicle and install the mounting brackets. Torque the mounting nuts and bolts in the proper sequence to 54–72 ft. lbs. (73–97 Nm).

9. Install or connect the following:

- New O-rings to the power steering fluid pipes and connect them to the rack and pinion. Torque the low pressure line 20–29 ft. lbs. (27–39 Nm) and the high pressure line to 11–18 ft. lbs. (15–25 Nm)

10. Align the lower steering joint to the pinion shaft and install the joint onto the pinion shaft. Torque the bolt to 17–22 ft. lbs. (24–29 Nm).

- Hole cover and torque the nuts to 36–43 inch lbs. (4–5 Nm)
- Front stabilizer

- Engine mounting center member and torque the bolts to 57–72 ft. lbs. (77–98 Nm).
- Engine mounts to the center member. Torque the bolts to 57–72 ft. lbs. (77–98 Nm). Remove the support from the engine
- Remaining components in the reverse order of removal

11. Torque the tie rod end nuts to 22–29 ft. lbs. (29–39 Nm), then install a new cotter pin.

12. Fill the power steering reservoir with fluid and bleed the air from the power steering system.

13. Check the vehicle front end alignment and adjust as necessary.

Strut and Coil Spring

REMOVAL & INSTALLATION

Front

SENTRA AND 200SX

1. Before servicing the vehicle, refer to the precautions in the beginning of this section.

2. Raise and support the vehicle on jackstands.

3. Remove or disconnect the following:

- Wheel
- Brake tube from the strut
- Anti-Lock Brake (ABS) wiring from the strut, if equipped with ABS

4. Support the transverse link with a jackstand.

- Steering knuckle from the strut

➡ Note the positioning of the strut alignment mark for reassembly purposes.

5. Support the strut and remove the 3 upper attaching nuts. Remove the strut from the vehicle.

✳✳ CAUTION

Never loosen the center spring retaining nut until the coil spring is compressed, or serious injury or vehicle damage may occur.

6. Place the strut assembly in a vise with a holding tool or in a spring compressor.

7. Loosen the piston rod locknut.

8. Compress the spring with the spring compressor, then remove the piston rod locknut.

➡ Before removing the strut from the coil spring, note the positioning of the strut in relationship to the coil spring for reassembly.

9. Remove or disconnect the following:

- Strut mounting insulator bracket, strut mounting bearing, upper spring seat and the upper spring rubber seat
- Strut, leaving the coil spring compressed
- Piston boot and rebound bumper from the strut

To install:

10. Install or connect the following:

- Rebound bumper and the boot to the strut piston
- Strut into the coil spring, be sure the strut and spring are properly positioned
- Upper spring rubber seat, upper spring seat, strut mounting bearing and the strut mounting insulator bracket. Be sure that the cutout on the upper spring seat is facing the outside of the vehicle
- Piston rod locknut. Remove the tool and torque the piston rod locknut to 43–54 ft. lbs. (59–74 Nm)

➡ When installing the strut, be sure to position the alignment mark toward the outside of the vehicle.

- Strut to the vehicle
- 3 upper attaching nuts and torque the nuts to 18–22 ft. lbs. (25–29 Nm)
- Steering knuckle to the strut and torque the bolts to 68–82 ft. lbs. (92–111 Nm)
- Brake tube to the strut and the ABS wiring to the strut, if it was removed

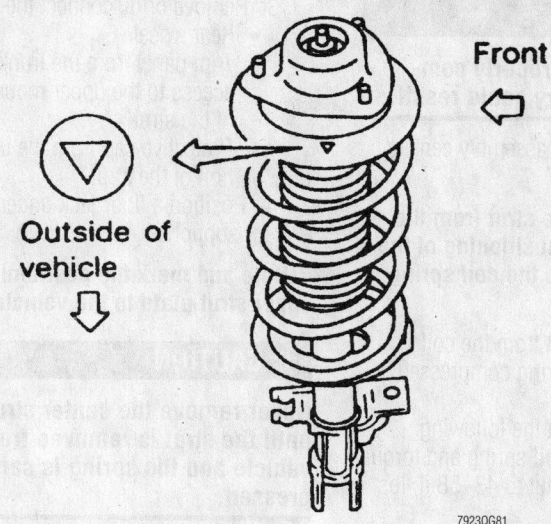

Front

Outside of vehicle

7923QG81

During assembly, be sure to point the alignment mark toward the outside of the vehicle—Sentra and 200SX

11. Bleed the brake system and install the wheel.

12. Perform a front end alignment.

ALTIMA

1. Before servicing the vehicle, refer to the precautions in the beginning of this section.

2. Raise and support the vehicle on jackstands.

3. Remove or disconnect the following:
- Wheel
- Brake tube from the strut
- Anti-Lock Brake (ABS) wiring from the strut, if equipped with ABS

4. Support the transverse link with a jackstand.
- Steering knuckle from the strut and properly support the strut
- 3 upper attaching nuts
- Strut from the vehicle

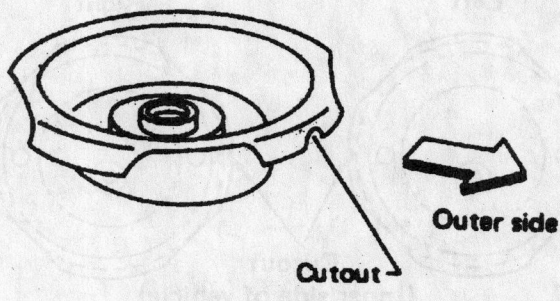

Outer side

Cutout

7923QG82

Position the alignment mark toward the outside of the vehicle—Altima

---

✳✳ WARNING

Never loosen the center spring retaining nut until the coil spring is compressed, or serious injury or vehicle damage may occur.

To install:

➡**When installing the strut, be sure to position the alignment mark toward the outside of the vehicle.**

5. Install or connect the following:
- Strut to the vehicle
- 3 upper attaching nuts and torque the nuts to 29–40 ft. lbs. (39–54 Nm)
- Steering knuckle to the strut and torque the nuts to 87–108 ft. lbs. (118–147 Nm) on 1998 models and 123–137 ft. lbs. (167–186 Nm) on 1999–2001 models
- Brake tube to the strut and the ABS wiring to the strut, if removed

---

6. Bleed the brake system and install the wheel.

7. Lower the vehicle and perform a front end alignment.

MAXIMA

1. Before servicing the vehicle, refer to the precautions in the beginning of this section.

2. Raise and safely support the vehicle.

3. Remove or disconnect the following:
- Wheel

4. Matchmark the position of the strut-to-steering knuckle location.
- Brake hose from the strut
- Anti-Lock Brake (ABS) wheel sensor and move it out of the way
- Bolts attaching the steering knuckle to the strut. Matchmark the bolts before removal

5. Open the hood and remove the strut attaching nuts while holding the strut.

✳✳ CAUTION

Do not remove the center locknut from the strut assembly until the strut is safely compressed.

- Strut from the vehicle

6. Place the strut assembly in a vise with a holding tool or in a spring compressor.

7. Loosen the piston rod locknut.

✳✳ CAUTION

Do not remove the piston rod locknut, the spring is under tension and can cause serious personal injury.

8. Compress the spring with the spring compressor, then remove the piston rod locknut.

➡**Before removing the strut from the coil spring, note the positioning of the strut in relationship to the coil spring for reassembly.**

9. Remove or disconnect the following:
- Strut mounting insulator bracket, strut mounting bearing, upper spring seat and the upper spring rubber seat
- Strut, leaving the coil spring compressed
- Piston boot and rebound bumper from the strut

To install:

10. Install or connect the following:

- Rebound bumper and the boot to the strut piston
- Strut into the coil spring, be sure the strut and spring are properly positioned
- Upper spring rubber seat, upper spring seat, strut mounting bearing and the strut mounting insulator bracket. Be sure that the cutout on the upper spring seat is facing the outside of the vehicle
- Piston rod locknut. Remove the tool and torque the piston rod locknut to 43–58 ft. lbs. (59–76 Nm) on 1998 models and 44–65 ft. lbs. (59–88 Nm) on 1999–01 models
- Strut into the strut tower
- New attaching nuts and torque to 29–40 ft. lbs. (39–54 Nm) on 1998 models and 32–38 ft. lbs. (43–51 Nm) on 1999–01 models
- Bolts attaching the steering knuckle to the strut and align the match-marks and torque to 103–117 ft. lbs. (140–159 Nm)
- ABS wheel sensor and torque to 13–17 ft. lbs. (18–24 Nm)
- Brake hose to the strut
- Front wheels

11. Lower the vehicle.

12. Check and/or adjust the wheel alignment as necessary.

240SX

1. Before servicing the vehicle, refer to the precautions in the beginning of this section.

2. Raise and support the vehicle on jackstands.

3. Remove or disconnect the following:
- Wheel
- Brake tube from the strut
- Anti-Lock Brake (ABS) wiring from the strut, if equipped with ABS

4. Support the transverse link with a jackstand.
- Steering knuckle from the strut by unfastening the 2 through-bolts and support the strut
- 3 upper attaching nuts
- Strut from the vehicle

❊❊ WARNING

Never loosen the center spring retaining nut until the coil spring is compressed, or serious injury or vehicle damage may occur.

5. Compress the strut coil spring with a spring compressor.

❊❊ CAUTION

If coil spring is not properly compressed serious injury could result.

6. Remove the strut assembly center locknut.

➡**Before removing the strut from the coil spring, note the positioning of the strut in relationship to the coil spring for reassembly.**

7. Separate the strut from the coil spring. Keep the coil spring compressed.

To install:

8. Install or connect the following:
- Strut into the coil spring and torque the center locknut to 43–58 ft. lbs. (59–78 Nm)

➡**When installing the strut, be sure to position the alignment mark toward the inside of the vehicle.**

- Strut to the vehicle with 3 new upper attaching nuts and torque the nuts to 29–40 ft. lbs. (39–54 Nm)
- Steering knuckle to the strut and torque the bolts to 90–112 ft. lbs. (123–152 Nm)
- Brake tube to the strut and the ABS wiring to the strut, if it was removed

9. Bleed the brake system and install the wheel.

10. Perform a front end alignment.

Rear

SENTRA AND 200SX

1. Before servicing the vehicle, refer to the precautions in the beginning of this section.

2. Raise the vehicle and support it safely.

3. Remove or disconnect the following:
- Rear wheel
- Trim panel from the trunk to gain access to the upper mounting nuts of the strut
- Protective cap from the upper portion of the strut

4. Position a floor jack under the rear axle for support.

➡**Note and mark the positioning of the upper strut plate to the vehicle body.**

❊❊ CAUTION

Never remove the center strut nut until the strut is removed from the vehicle and the spring is safely compressed.

5. Remove or disconnect the following:
- Lower strut mounting through-bolt
- 2 upper strut mounting nuts and the strut from the vehicle

6. Place the strut assembly in a vise with a holding tool or in a spring compressor.

7. Loosen the piston rod locknut.

8. Compress the spring with the spring compressor, then remove the piston rod locknut.

➡**Before removing the strut from the coil spring, note the positioning of the strut in relationship to the coil spring for reassembly.**

9. Remove or disconnect the following:
- Strut mounting insulator bracket, strut mounting bearing, upper spring seat and the upper spring rubber seat
- Strut, leaving the coil spring compressed
- Piston boot and rebound bumper from the strut

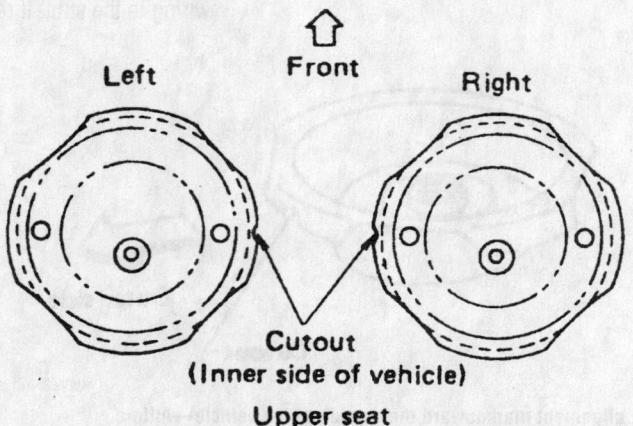

During assembly, position the alignment marks toward the inside of the vehicle—240SX

7923QG83

To install:

10. Install or connect the following:
- Rebound bumper and the boot to the strut piston
- Strut into the coil spring, be sure the strut and spring are properly positioned
- Upper spring rubber seat, upper spring seat, strut mounting bearing and the strut mounting insulator bracket. Be sure that the cutout on the upper spring seat is facing the outside of the vehicle
- Piston rod locknut and torque the locknut to 13–17 ft. lbs. (18–24 Nm)

11. Remove the spring compressor from the coil spring.
- Strut and torque the 2 upper mounting nuts to 12–14 ft. lbs. (16–19 Nm)
- Upper mount protective cap
- Through-bolt to the lower mount of the strut and torque the bolt to 72–87 ft. lbs. (98–118 Nm)
- Trunk trim panel
- Rear wheel

12. Lower the vehicle and perform an alignment.

ALTIMA

1. Before servicing the vehicle, refer to the precautions in the beginning of this section.
2. Raise and safely support the vehicle.
3. Remove or disconnect the following:
- Rear wheels from the vehicle and support the rear axle with a jack
- Strut lower mounting through-bolts

➡**Be sure to note the position the strut upper plate to the vehicle for reinstallation purposes.**

- 2 nuts from the top of the strut
- strut as an assembly

❄❄ CAUTION

Do not remove the center locknut from the strut assembly until the strut is safely compressed.

4. Compress the strut coil spring with a spring compressor.
- Strut assembly center locknut

➡**Before removing the strut from the coil spring, note the positioning of the strut in relationship to the coil spring for reassembly.**

- Strut leaving the coil spring compressed

➡**Mark the coil spring position to the strut assembly for reinstallation purposes.**

5. To remove the spring from the strut assembly, perform the following steps:
  a. Compress the coil spring with the proper compressor tool.
  b. Remove the center retaining nut holding strut mounting insulator.
  c. Slowly decompress the coil spring.
  d. Remove the strut mounting insulator.
  e. Remove coil spring.

To install:
6. Install or connect the following:
- Coil spring onto the strut assembly. Be sure to align the matchmarks made during the removal procedure
- Strut mounting insulator and compress the coil spring assembly

➡**It will be necessary to use a new locknut for the center retaining nut of the coil spring.**

- Center retaining nut and torque to 43–58 ft. lbs. (59–78 Nm). Be sure the spring is seated properly on the strut and in the mounting insulator

7. Slowly remove the spring compressor tool.
- Strut assembly and torque the upper nuts to 31–40 ft. lbs. (42–54 Nm)

8. Torque the lower strut through-bolt to 87–108 ft. lbs. (118–147 Nm) on 1998 models and 123–137 ft. lbs. (167–186 Nm) on 1999–01 models.

➡**Be sure to hold the through-bolt and tighten the nuts.**

9. Install the wheels, lower the vehicle and perform a front end alignment.

MAXIMA

1. Before servicing the vehicle, refer to the precautions in the beginning of this section.
2. Raise and safely support the vehicle.
3. Remove the rear wheels.
4. Support the rear torsion beam assembly with a jack.
5. Open the trunk and remove the 2 nuts attaching the strut to the vehicle.

❄❄ CAUTION

Do not remove the center locknut from the strut assembly until the strut is safely compressed.

6. Remove the bolt attaching the strut to the rear torsion beam assembly and remove the strut.
7. Place the strut assembly in a vise with a holding tool or in a spring compressor.
8. Loosen the piston rod locknut.

❄❄ CAUTION

Do not remove the piston rod locknut, the spring is under tension and can cause serious personal injury.

9. Compress the spring with the spring compressor, then remove the piston rod locknut.

➡**Before removing the strut from the coil spring, note the positioning of the strut in relationship to the coil spring for reassembly.**

10. Remove or disconnect the following:
- Bushing, strut mounting bracket and the upper spring seat rubber
- Strut, leaving the coil spring compressed
- Bushing, bound bumper cover and the bound bumper

To install:
11. Install or connect the following:
- Bound bumper, bound bumper cover and the bushing
- Strut into the coil spring, be sure the strut and spring are properly positioned
- Upper spring seat rubber, strut mounting bracket and the bushing. Be sure that the mounting bracket is properly positioned
- Piston rod locknut. Remove the tool and torque the piston rod locknut to 13–17 ft. lbs. (18–24 Nm) on 1998 models and 15–18 ft. lbs. (20–24 Nm) on 1999–01 models
- Strut and torque the new nuts to 12–14 ft. lbs. (16–19 Nm) on 1998 models and 18–25 ft. lbs. (25–34 Nm) on 1999–01 models

12. Position the strut on the rear torsion beam and install the bolt. Torque the bolt attaching the strut to the torsion beam assembly to 72–87 ft. lbs. (98–118 Nm) on 1998 models and 80–94 ft. lbs. (108–127 Nm) on 1999–01 models.
13. Remove the support from the rear torsion beam.
14. Install the rear wheels and lower the vehicle.
15. Check the vehicle's alignment and adjust as necessary.

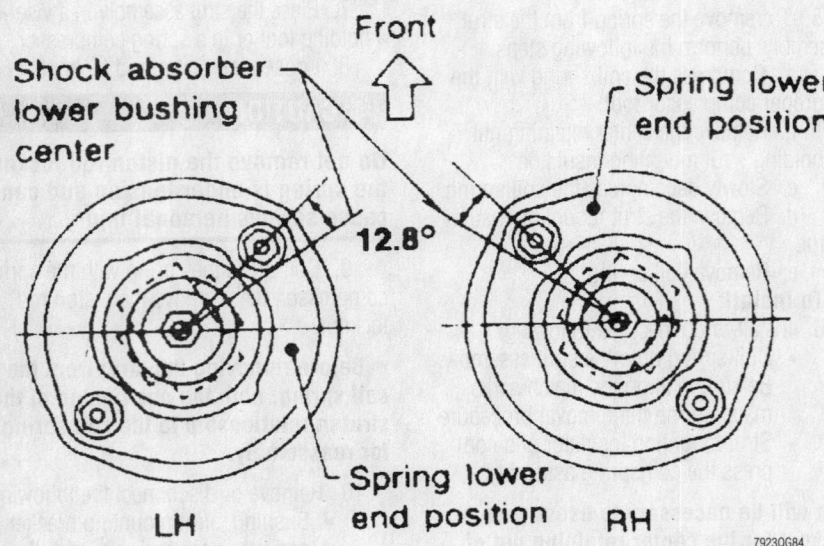

Positioning of the strut mounting brackets—Maxima

240SX

1. Before servicing the vehicle, refer to the precautions in the beginning of this section.
2. Raise and safely support the vehicle.
3. Remove or disconnect the following:
   - Rear wheels from the vehicle and support the rear control arm with a jack
   - Strut lower mounting bolt
4. Mark the upper spring plate to the vehicle for reinstallation purposes.
   - 2 strut upper mounting nuts

❈❈ WARNING

Do not remove the center piston locknut until the spring has been compressed.

   - Strut from the vehicle
5. Mark the positioning of the coil spring to the strut assembly.
6. Using a coil spring compressor, compress the spring.

❈❈ CAUTION

If coil spring is not properly compressed serious injury could result.

7. Remove the strut assembly center locknut.

➡ **Before removing the strut from the coil spring, note the positioning of the strut in relationship to the coil spring for reassembly.**

8. Remove or disconnect the following:
   - Upper strut plate
   - Strut leaving the coil spring compressed
9. Note the positioning of the spring and slowly release the spring compressor. Remove the coil spring assembly.

To install:

10. Properly position and compress the coil spring.
11. Install or connect the following:
   - Strut to the coil spring
   - Upper strut plate and torque the center locknut to 13–17 ft. lbs. (18–24 Nm). Verify that the coil spring is seated properly on the strut
12. Remove the coil spring compressor.
   - Strut assembly to the vehicle
   - Top mounting nuts and torque to 12–14 ft. lbs. (16–19 Nm). Make sure to use new locking nuts

➡ **Be sure that all final tightening is done with the full weight of the vehicle on the ground.**

   - Lower mounting bolt and torque to 72–87 ft. lbs. (98–118 Nm)
   - Rear wheels
13. Check the vehicle alignment and adjust it as necessary.

Shock Absorber and Coil Spring

REMOVAL & INSTALLATION

240SX

The 240SX uses a coil-over-shock absorber assembly that looks much like a MacPherson strut. It is removed, installed and overhauled in much the same manner.

1. Before servicing the vehicle, refer to the precautions in the beginning of this section.
2. Remove or disconnect the following:
   - Rear seat and package shelf to gain access to the upper shock attaching nuts
   - Upper attaching nuts
   - Rear wheel and support the lower control arm with a jack
   - Lower shock absorber mounting bolt
   - Shock absorber assembly from the vehicle

❈❈ CAUTION

Coil springs are under extreme loads when compressed. Be sure to properly align the spring with the compressing tool to prevent personal injury from the spring releasing unexpectedly.

3. Compress the coil spring with the proper spring compressor tool.
4. Remove or disconnect the following:
   - Center retaining nut holding the upper mounting insulator
5. Slowly decompress the coil spring.
   - Mounting insulator and the coil spring
   - Coil spring

To install:

6. Install or connect the following:
   - Compressed coil spring onto the strut assembly. Be sure to the end of the spring is in the notch on the lower seat
   - Upper mounting insulator

➡ **It will be necessary to use a new locknut for the center retaining nut of the coil spring.**

   - Center retaining nut and torque the nut to 13–17 ft. lbs. (18–24 Nm)
7. Slowly remove the spring compressor tool.
   - Shock absorber assembly in the vehicle and install the mounting fasteners and torque the lower mounting bolt to 72–87 ft. lbs.

CAUTION:
Do not jack up at lower link.
When installing rubber parts, final tightening must be carried out under unladen condition* with tires on ground.
* Fuel, radiator coolant and engine oil full.
   Spare tire, jack, hand tools and mats in designated positions.

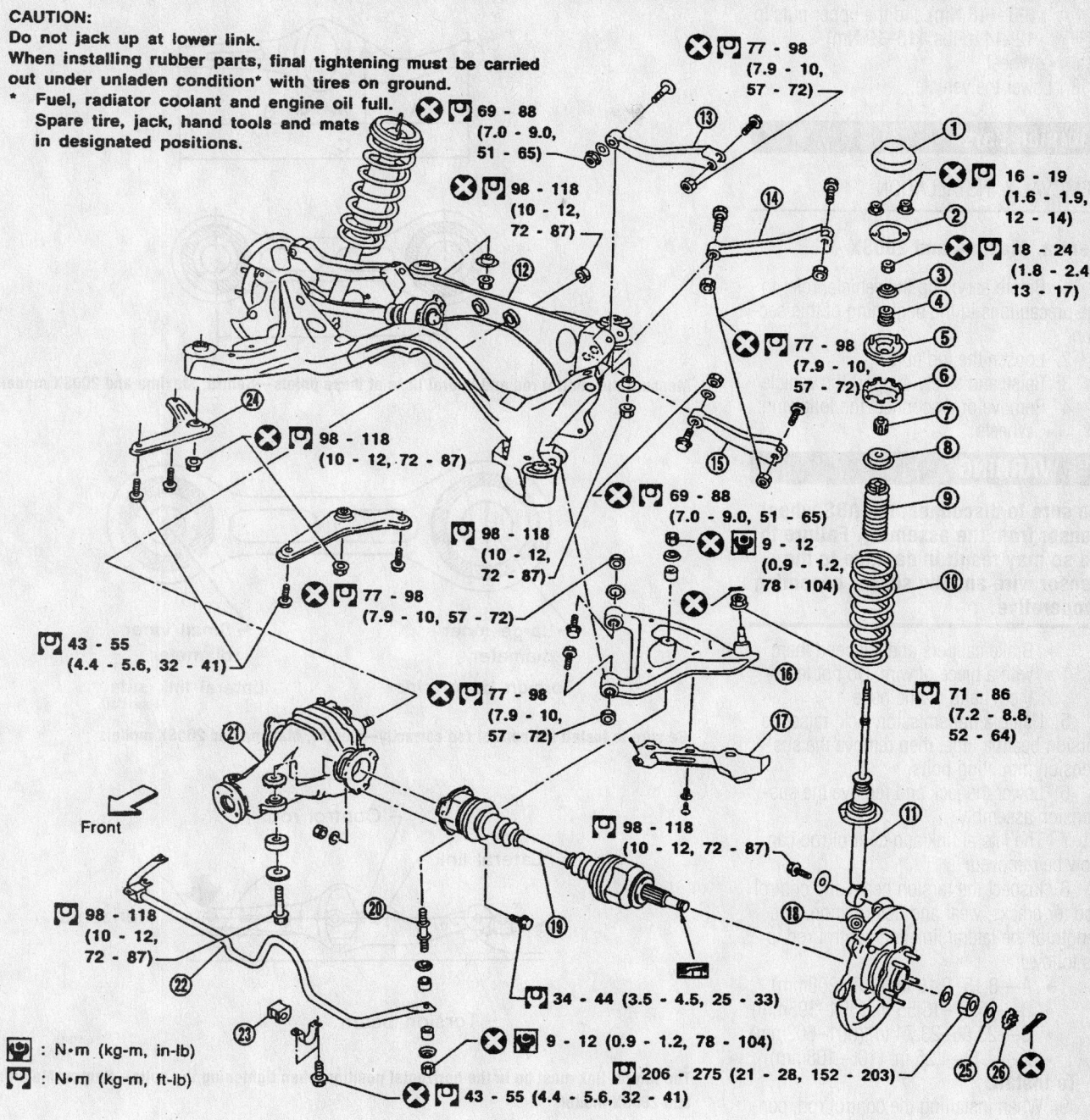

69 - 88 (7.0 - 9.0, 51 - 65)

98 - 118 (10 - 12, 72 - 87)

77 - 98 (7.9 - 10, 57 - 72)

77 - 98 (7.9 - 10, 57 - 72)

69 - 88 (7.0 - 9.0, 51 - 65)

9 - 12 (0.9 - 1.2, 78 - 104)

98 - 118 (10 - 12, 72 - 87)

98 - 118 (10 - 12, 72 - 87)

77 - 98 (7.9 - 10, 57 - 72)

43 - 55 (4.4 - 5.6, 32 - 41)

77 - 98 (7.9 - 10, 57 - 72)

16 - 19 (1.6 - 1.9, 12 - 14)

18 - 24 (1.8 - 2.4, 13 - 17)

77 - 98 (7.9 - 10, 57 - 72)

71 - 86 (7.2 - 8.8, 52 - 64)

98 - 118 (10 - 12, 72 - 87)

98 - 118 (10 - 12, 72 - 87)

34 - 44 (3.5 - 4.5, 25 - 33)

9 - 12 (0.9 - 1.2, 78 - 104)

206 - 275 (21 - 28, 152 - 203)

43 - 55 (4.4 - 5.6, 32 - 41)

Front

: N•m (kg-m, in-lb)

: N•m (kg-m, ft-lb)

①	Cap	⑩	Coil spring	⑲	Drive shaft
②	Gasket	⑪	Shock absorber	⑳	Connecting rod
③	Upper plate	⑫	Suspension member	㉑	Final drive
④	Bushing	⑬	Rear upper link	㉒	Stabilizer bar
⑤	Shock absorber mounting bracket	⑭	Front upper link	㉓	Bushing
⑥	Upper rubber seat	⑮	Lateral link	㉔	Member stay
⑦	Bushing	⑯	Lower arm	㉕	Insulator
⑧	Plate	⑰	Protector	㉖	Adjusting cap
⑨	Bumper rubber with dust cover	⑱	Axle housing		

7923QG88

Exploded view of the rear suspension—240SX

(98–118 Nm) and the upper nuts to 12–14 ft. lbs. (16–19 Nm)
- Wheel
8. Lower the vehicle.

Torsion Bars

REMOVAL & INSTALLATION

Sentra, Maxima and 200SX

1. Before servicing the vehicle, refer to the precautions in the beginning of this section.
2. Loosen the lug nuts.
3. Raise and safely support the vehicle
4. Remove or disconnect the following:
   - Wheels

❊❊ WARNING

Be sure to disconnect the ABS wheel sensor from the assembly. Failure to do so may result in damage to the sensor wire and the sensor becoming inoperative.

- Brake calipers and suspend them with a piece of wire. Do not let them hang by the hose
5. Using a transmission jack, raise the torsion beam a little, then remove the suspension mounting bolts.
6. Lower the jack and remove the suspension assembly.
7. The lateral link and control rod can now be removed.
8. Inspect the torsion beam and control rod for cracks, wear and deformation. The length of the lateral link and control rod is as follows:
   - A—8.15–8.19 in. (207–208mm)
   - B—15.51–15.55 in. (394–395mm)
   - C—23.66–23.74 in. (601–603mm)
   - D—4.17–4.25 in. (106–108mm)

To install:
9. When installing the control rod, connect the bushing with the smaller inner diameter to the lateral link. Install the lateral link and the control rod on the torsion beam. Place the lateral link with the arrow topside.
10. Place the lateral link and control rod horizontally against the beam and tighten the bolts. Refer to the illustration.
11. Secure the torsion beam to the vehicle. Make sure the lateral link is horizontal, then tighten the link to the chassis.
12. Attach the struts to the torsion beam and tighten the fasteners.
13. Tighten the torsion beam-to-chassis bolts.

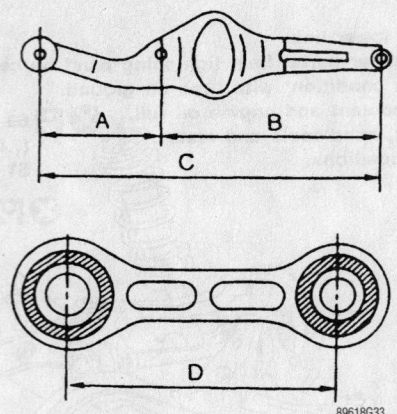

Measure the control rod and lateral links at these points—Sentra, Maxima and 200SX models

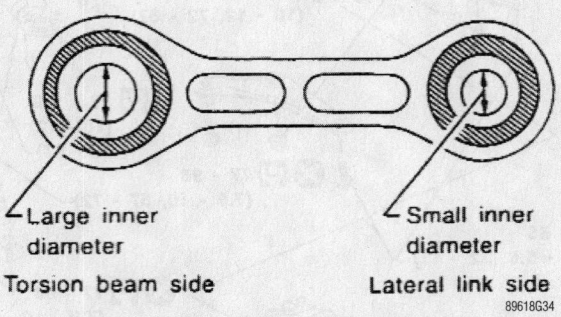

Be sure to install the control rod correctly—Sentra, Maxima and 200SX models

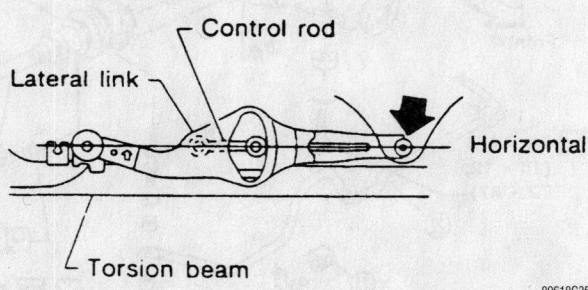

The lateral link must be in the horizontal position when tightening the bolts—Sentra, Maxima and 200SX models

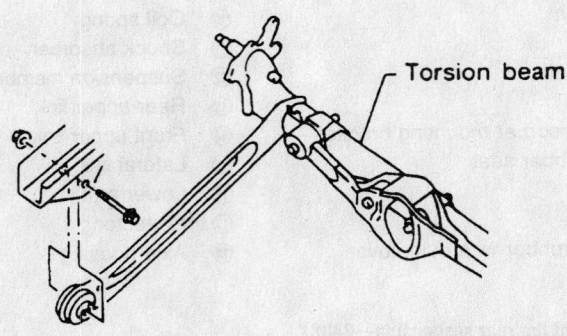

Tighten the torsion beam-to-chassis bolts with the suspension unloaded—Sentra, Maxima and 200SX models

14. Install the calipers, ABS sensor and wheels. Lower the vehicle to the ground.

Lower Ball Joint

REMOVAL & INSTALLATION

The ball joint is an integral part of the lower control arm. If the ball joint is defective the control arm must be replaced.

Lower Control Arm (Transverse Link)

REMOVAL & INSTALLATION

240SX

1. Before servicing the vehicle, refer to the precautions in the beginning of this section.
2. Raise and support the vehicle safely. Allow the lower control arm to hang free.
3. Remove or disconnect the following:
   - Front wheels
   - Cotter pin and castle nut from the tie rod end
   - Tie rod end from the steering knuckle
   - Cotter pin and castle nut from the ball joint stud
   - Knuckle from the ball joint, using the proper tools
   - Stabilizer bar link from the lower arm
   - Nuts and bolts that connect the lower control arm to the tension rod
   - Bolt that connects the lower control arm to the crossmember
   - Lower control arm from the vehicle

To install:

➡**Torquing of the suspension components should be performed with the full weight of the vehicle resting on the suspension.**

4. Install or connect the following:
   - Control arm to the crossmember and install the through bolt, lockwasher and torque the nut 80–94 ft. lbs. (108–127 Nm)
   - Nuts that connect the lower control arm to the tension rod. Secure the bolts and torque the mounting nuts to 69–80 ft. lbs. (93–108 Nm)
   - Stabilizer bar transverse link to the

lower arm and torque the link to 14–22 ft. lbs. (20–29 Nm)
   - Knuckle to the lower ball joint and torque the castle nut to 71–88 ft. lbs. (96–120 Nm) and install a new cotter pin
   - Tie rod end to the knuckle and torque the castle nut to 22–36 ft. lbs. (29–49 Nm). Install a new cotter pin
   - Front wheels and lower the vehicle
5. Tighten all nuts and bolts to specification and perform a wheel alignment.

Sentra and 200SX

1. Before servicing the vehicle, refer to the precautions in the beginning of this section.
2. Raise the vehicle and support it safely.
3. Remove or disconnect the following:
   - Front wheels
   - Disc brake caliper from the steering knuckle

✳✳ WARNING

DO NOT allow the disc brake caliper to hang from the brake hose. Support the disc caliper with safety wire.

   - Cotter pin and loosen the wheel bearing locknut
   - Cotter pin and the castle nut from the tie rod ball joint. Separate the tie rod with a suitable puller.
   - 2 bolts that secure the lower portion of the strut to the steering knuckle

4. Using a plastic or rubber mallet, tap on the loosened wheel bearing locknut to loosen the halfshaft in the knuckle. Remove the locknut and remove the halfshaft from the steering knuckle. Be sure to cover the CV-joints with a shop rag.

➡**Support the halfshaft assembly with wire. Do not allow the halfshaft to hang by the inner joint.**

5. Remove or disconnect the following:
   - Nut that secures the stabilizer link to the lower control arm
   - Link from the control arm. Note the positioning of the washers and spacers for reassembly
   - Cotter pin and castle nut from the lower ball joint
   - Lower ball joint from the knuckle
   - Knuckle from the vehicle
   - Mounting nuts/bolts that secure the lower control arm to the frame
   - Control arm from the vehicle

To install:

➡**Final tightening of all suspension components should take place with the weight of the vehicle on the wheels.**

6. Install the lower control arm assembly and torque mounting bolts/nuts as follows:
   a. Through bolt and nut: 76–90 ft. lbs. (103–123 Nm).
   b. 2 saddle bracket mounting bolts: 58–72 ft. lbs. (78–98 Nm).
7. Install or connect the following:
   - Steering knuckle to the lower ball joint and torque the castle nut to 43–54 ft. lbs. (59–74 Nm). Install a new cotter pin
   - Stabilizer link to the lower control arm and torque the nut to 12–16 ft. lbs. (16–22 Nm)
   - Halfshaft through the wheel bearing
   - Wheel bearing locknut. Do not torque the locknut at this time
   - Steering knuckle to the strut and torque the bolts to 68–82 ft. lbs. (92–111 Nm)
   - Tie rod end and torque the castle nut to 22–29 ft. lbs. (29–39 Nm). Install a new cotter pin
   - Disc brake caliper to the steering knuckle
8. Tighten the halfshaft mounting nut (hub nut) and torque the nut to 145–202 ft. lbs. (196–274 Nm). It may be necessary to have an assistant hold the brake pedal while tightening the locknut. Install the adjusting cap and a new cotter pin.
9. Install the front wheels, lower the vehicle and perform a front end alignment.

Maxima

1. Before servicing the vehicle, refer to the precautions in the beginning of this section.
2. Raise and safely support the vehicle.
3. Remove or disconnect the following:
   - Front wheels
   - Anti-Lock Brake (ABS) wheel sensor and move it out of the way
   - Wheel bearing locknut
   - Tie rod from the steering knuckle
   - Bolts attaching the strut to the steering knuckle. Matchmark the bolts before removal
   - Halfshaft from the steering knuckle by lightly tapping the end of the shaft
   - Steering knuckle and the lower ball joint

- Stabilizer bar from the lower control arm
- Bolts attaching the link bushing pin to the chassis
- Nut attaching the link to the control arm and the link, if necessary
- Bolts attaching the compression rod bushing clamp
- Lower control arm/traverse link

To install:

4. Install or connect the following:
- Lower control arm and the compression rod bushing clamp into the vehicle
- Link bushing pin, if removed from the control arm

5. Tighten all bolts and nuts until they are snug enough to support the weight of the vehicle but not fully tight, the bolts should be torqued to specification with the vehicle on the floor.

➡ **Always use a new nut when installing the ball joint to the control arm.**

- Steering knuckle to the strut and to the halfshaft
- Strut mounting bolts and torque the bolts to 103–117 ft. lbs. (140–159 Nm)
- Tie rod ball joint and torque the nut to 46–54 ft. lbs. (63–73 Nm) on 1998 models and 22–29 ft. lbs. (29–39 Nm) on 1999–01 models
- Wheel bearing locknut
- ABS wheel sensor and torque the bolt to 13–17 ft. lbs. (18–24 Nm)
- Front wheels, lower the vehicle and torque hub locknut to 174–231 ft. lbs. (235–314 Nm)

6. Torque the bolts attaching the compression rod bushing clamp and the link bushing pin, in the proper sequence to 87–108 ft. lbs. (118–147 Nm).

7. If the link bushing pin was removed from the control arm torque the attaching nut to 87–108 ft. lbs. (118–147 Nm).

8. Torque the sway bar attaching nut to 30–35 ft. lbs. (41–47 Nm).

9. Check the vehicle alignment.

Altima

➡ **The lower ball joint is integral with the lower control arm (transverse link). They are removed and replaced as an assembly.**

1. Before servicing the vehicle, refer to the precautions in the beginning of this section.

2. Raise and safely support the vehicle.

3. Remove or disconnect the following:

- Front wheels
- Stabilizer bar. The bar is removed by unfastening the nut that hold the bar to the transverse link gusset plate.

➡ **Take note of position of marks on clamp face and stabilizer bar for reassembling.**

- Lower ball joint to knuckle cotter pin and nut
- Ball joint stud from knuckle using the proper tool
- Transverse link mounting bolts and nuts
- Link

To install:

4. Install or connect the following:
- Transverse link with mounting bolts and torque nuts and bolts to 87–108 ft. lbs. (118–147 Nm)

➡ **The final tightening of suspension components must be done with wheels on the ground and vehicle at curb weight.**

- Lower ball joint to the knuckle and torque the nut to 52–64 ft. lbs. (71–86 Nm) and install a new cotter pin
- Stabilizer bar link to the transverse link and torque the nuts to 30–35 ft. lbs. (41–47 Nm)

5. Install wheels and safely lower vehicle to ground.

6. Check the front end alignment.

CONTROL ARM BUSHING REPLACEMENT

The bushing is an integral part of the lower control arm. If the bushing is defective the control arm must be replaced.

Wheel Bearings

ADJUSTMENT

Front

SENTRA, 200SX, ALTIMA AND MAXIMA

➡ **Whenever the hub or bearing assemblies are removed, the wheel bearing must be replaced. Never reuse the old bearing assembly.**

The wheel bearings are sealed and are not adjustable. If defective, replacement is the only option.

240SX

1. Before servicing the vehicle, refer to the precautions in the beginning of this section.

2. Raise and safely support the front wheels.

3. Remove the front wheels.

4. Remove the hub cap and wheel bearing nut cotter pin.

5. Torque the wheel bearing locknut to 151–210 ft. lbs. (206–284 Nm).

6. Check that wheel bearings operate smoothly.

7. Check axial end-play. Axial end-play must be 0.0020 in. (0.05mm) or less. If axial end-play is not within specification or wheel bearing does not turn smoothly, replace the wheel bearing assembly.

9301QG01

Bolt tightening sequence for the lower control arms—Maxima

8. Replace the wheel bearing locknut cotter pin and reinstall the hub cap.

9. Install the wheels and safely lower the vehicle.

Rear

The rear wheel bearings on the 240SX are not adjustable. Replace the bearing assembly if a growling noise is emitted during operation or if the bearing drags or turns roughly.

SENTRA, 200SX, ALTIMA AND MAXIMA

If the wheel hub bearing assembly is removed, it must be replaced.

➡**The wheel hub bearing assembly is not repairable; it must be replaced when defective.**

1. Before servicing the vehicle, refer to the precautions in the beginning of this section.

2. Torque the wheel bearing locknut to 138–188 ft. lbs. (187–255 Nm).

3. Verify that the wheel bearings operate smoothly.

4. Install a new cotter pin into the spindle to hold the wheel bearing locknut.

5. Install a dial indicator to the rear wheel hub bearing assembly and check the axial end-play; it should be less than 0.0020 in. (0.05mm).

6. Install the grease cap.

7. If the axial end-play exceeds specifications, the wheel bearing must be replaced.

REMOVAL & INSTALLATION

Front

SENTRA AND 200SX

➡**Whenever the hub or bearing assembly is removed, the wheel bearing assembly must be replaced. Never reuse the old bearing assembly.**

1. Before servicing the vehicle, refer to the precautions in the beginning of this section.

2. Raise the vehicle and support it safely.

3. Remove or disconnect the following:
- Front wheel
- Wheel bearing/axle shaft locknut while depressing the brake pedal
- Brake caliper and support it with a piece of wire. It is not necessary to disconnect the brake line from the caliper

- Anti-Lock Brake System (ABS) sensor from the steering knuckle

➡**Do not depress the brake pedal or twist the brake line.**

- Tie rod end
- Halfshaft from the knuckle by slightly tapping with a soft hammer. Position the axle shaft nut on the threads of the shaft to protect them when lightly tapping
- Lower ball joint nut and separate
- 2 strut-to-knuckle retaining bolts and separate
- Steering knuckle from the vehicle

4. Place the assembly in a vise. Drive the hub with the inner race from the knuckle with a suitable tool. Remove the inner and outer grease seals.
- Bearing inner race and outer grease seal from the hub
- Snapring and press the bearing outer race to remove the bearing from the steering knuckle

To install:

5. Press a new wheel bearing into the knuckle assembly not exceeding 3.3 tons (2994 kg) pressure.

6. Install or connect the following:
- Snapring and pack the grease seal lips with chassis grease
- Inner and outer grease seals

7. Press the wheel hub into the knuckle not exceeding 3.3 tons (2994 kg) pressure.

8. Check bearing operation and by applying 3.9–5.5 tons (3538–4990 kg) pressure to the hub assembly. Spin the hub several times in both directions.

9. Be sure the bearings rotate freely. If the bearings do not rotate freely, replace the bearings.

10. Install or connect the following:
- Knuckle and wheel hub assembly
- Lower ball joint and torque the nut to 43–54 ft. lbs. (59–74 Nm). Install a new cotter pin
- Strut and torque the bolts to 68–82 ft. lbs. (92–118 Nm)
- Tie rod end and tighten the nut to 22–29 ft. lbs. (29–39 Nm). Install a new cotter pin
- Disc brake caliper
- Torque the wheel bearing locknut to 145–203 ft. lbs. (196–275 Nm). Install a new cotter pin
- Front wheels and lower the vehicle

11. Check the vehicle's alignment.

12. Road test the vehicle and verify proper operation.

MAXIMA AND ALTIMA

➡**Whenever the hub or bearing assembly is removed, the wheel bearing assembly must be replaced. Never reuse the old bearing assembly.**

1. Before servicing the vehicle, refer to the precautions in the beginning of this section.

2. Remove or disconnect the following:
- Knuckle assembly from the vehicle
- Hub with the inner race from the steering knuckle, using a shop press and a suitable tool
- Bearing inner race from the hub, using a shop press and a suitable tool
- Outer grease seal
- Inner grease seal from the steering knuckle, using a prybar
- Inner and outer snaprings from the steering knuckle, using snapring pliers
- Sealed bearing assembly from the steering knuckle, using a shop press and a suitable tool

3. Inspect the hub, steering knuckle and snaprings for cracks and/or wear; if necessary, replace the damaged part(s).

To install:

4. Install or connect the following:
- Inner snapring in the steering knuckle groove
- New wheel bearing assembly into the steering knuckle, using a shop press and a suitable tool, until it seats, using a maximum pressure of 3 tons (2722 kg)
- Outer snapring

5. Pack the new grease seal lips with multi-purpose grease.
- New outer grease seal into the steering knuckle, using a shop press and a suitable tool
- Hub into the steering knuckle, using a shop press and a suitable tool, until it seats, using a maximum pressure of 5.5 tons (4990 kg); be careful not to damage the grease seal

6. To check the bearing operation, perform the following procedures:
   a. Increase the press pressure to 3.5–5.0 tons (3175–4536 kg).

Heater Core replacement is covered in Section 2 of this manual

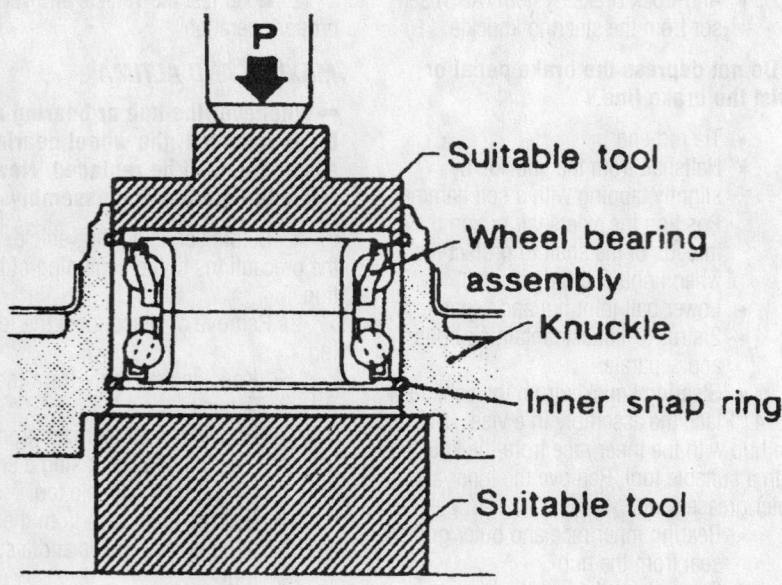

Typical method of installing the wheel bearing

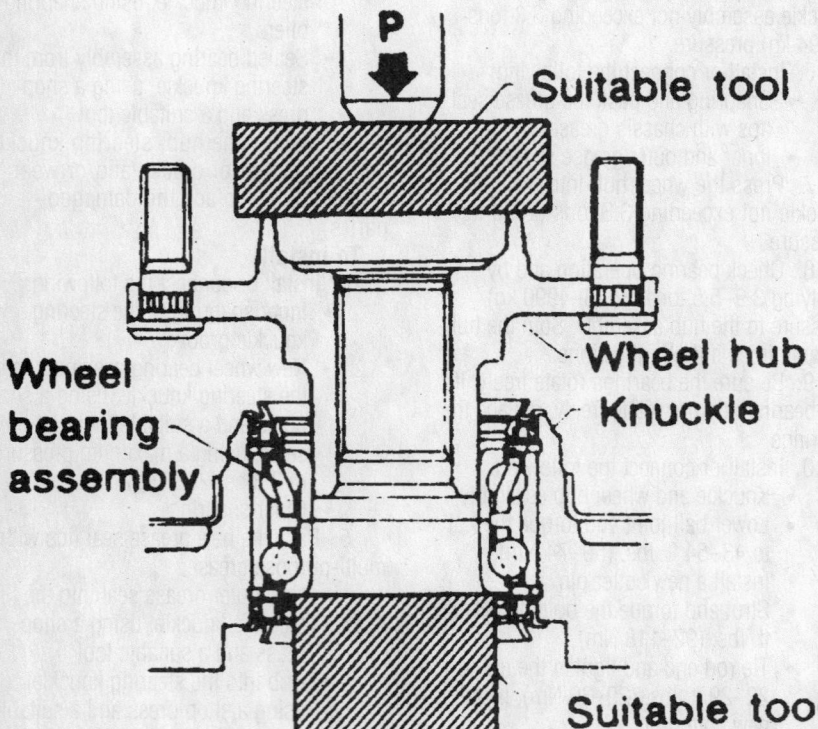

Use a press to install the hub into the knuckle assembly

b. Spin the steering knuckle, several turns, in both directions.

c. Be sure the wheel bearings operate smoothly.

7. If the wheel bearings do not operate smoothly, replace the wheel bearing assembly.

8. Install the knuckle assembly.

9. Install the halfshaft into the hub. Torque the locknut to 174–231 ft. lbs. (235–314 Nm).

10. Install the wheel assembly and lower the vehicle.

11. Road test the vehicle and verify proper operation.

240SX

➡ **If defective, the wheel bearing assembly can only be serviced by replacement of the hub.**

1. Before servicing the vehicle, refer to the precautions in the beginning of this section.

2. Raise and support the front of the vehicle safely.

3. Remove or disconnect the following:
- Wheels
- Brake caliper assembly. Be sure not to twist the brake hose

➡ **When removing the brake caliper it is not necessary to disconnect the brake hose. Be sure to support the brake caliper after removal.**

- Anti-Lock Brake (ABS) sensor from the steering knuckle, if applicable
- Brake rotor
- Wheel bearing dust cap
- Locking nut
- Wheel bearing locknut and washer

➡ **The hub bearing is a sealed type and will be removed with the hub assembly.**

4. Pull hub assembly toward you to remove the hub assembly from the spindle.
- Tie rod joint and lower ball joint
- Knuckle from the strut

To install:

5. Install or connect the following:
- Knuckle to the strut
- Tie rod joint and lower ball joint

6. Lubricate the spindle and slide hub assembly onto spindle.
- Wheel bearing locknut and torque the locknut to 152–210 ft. lbs. (206–284 Nm)

7. Turn the hub assembly several times in both directions to seat the hub.

8. Measure the wheel bearing axial end-play. Axial end-play should be 0.0020 in. (0.05mm) or less.

9. Clinch the locking nut with a hammer and chisel.

10. Install or connect the following:
- New dust cap
- ABS sensor and torque the bolt to 96–144 inch lbs. (11–16 Nm)
- Brake rotor and caliper assembly
- Front wheels

11. If the brake line was disconnected, bleed the brake system.

Rear

SENTRA, 200SX, ALTIMA AND MAXIMA

If the wheel hub bearing assembly is removed, it must be replaced.

→**If the vehicle is equipped with Anti-Lock Brake (ABS), the sensor must be removed to protect the sensor and its wiring.**

1. Before servicing the vehicle, refer to the precautions in the beginning of this section.

2. Raise and safely support the vehicle. Remove the rear wheel(s).

3. If equipped with disc brakes, remove or disconnect the following:
   - Brake caliper and hang it by a piece of wire
   - Brake caliper support
   - Disc brake pads
   - Brake rotor

4. If equipped with drum brakes, remove or disconnect the following:
   - Brake drum
   - Brake shoe assembly, if necessary
   - Grease cap

5. Remove the cotter pin, wheel bearing locknut, washer and the wheel hub bearing assembly. A slide hammer may be needed to remove the hub bearing assembly.

→**The wheel hub bearing assembly is not repairable; it must be replaced when defective.**

To install:

→**If the vehicle is equipped with ABS, the sensor ring must be removed and installed on the new hub.**

6. Apply oil to the threaded portion of the spindle and both sides of the plain washer.

7. Install the wheel hub bearing assembly, the washer and the wheel bearing locknut. Torque the wheel bearing locknut to 138–188 ft. lbs. (187–255 Nm).

8. Verify that the wheel bearings operate smoothly.

9. Install or connect the following:
   - New cotter pin into the spindle to hold the wheel bearing locknut

10. Install a dial micrometer to the rear wheel hub bearing assembly and check the axial end-play. It should be less than 0.0020 in. (0.05mm).
   - Grease cap
   - ABS sensor and its wiring, if removed
   - Brake assembly and the wheels

240SX

1. Before servicing the vehicle, refer to the precautions in the beginning of this section.

2. Raise and support the rear of the vehicle and remove the rear wheels. Remove the cotter pin, adjusting cap and insulator.

3. Apply the parking brake firmly to hold the rear halfshaft while removing the wheel bearing locknut. Remove the wheel bearing locknut.

4. Remove or disconnect the following:
   - Caliper and move it aside. Do not disconnect the hose from the caliper. Do not allow the caliper to hang by the hose; support the caliper with a length of wire or rest it on a suspension member. Remove the brake disc
   - Halfshaft from the axle housing by lightly tapping it. Cover the driveshaft boots with a shop towel to prevent damage
   - 4 bolts on the rear of the axle housing that hold the wheel bearing, flange and hub to the axle housing
   - Wheel bearing, flange and hub assembly
   - Wheel bearing from the axle hub, using a press

5. Mount the hub in a vise and remove the inner race, using bearing puller. Discard the inner race and grease seals.

6. Clean all parts in a suitable solvent. Check the wheel hub and axle housing for cracks, preferably using the dye penetrant

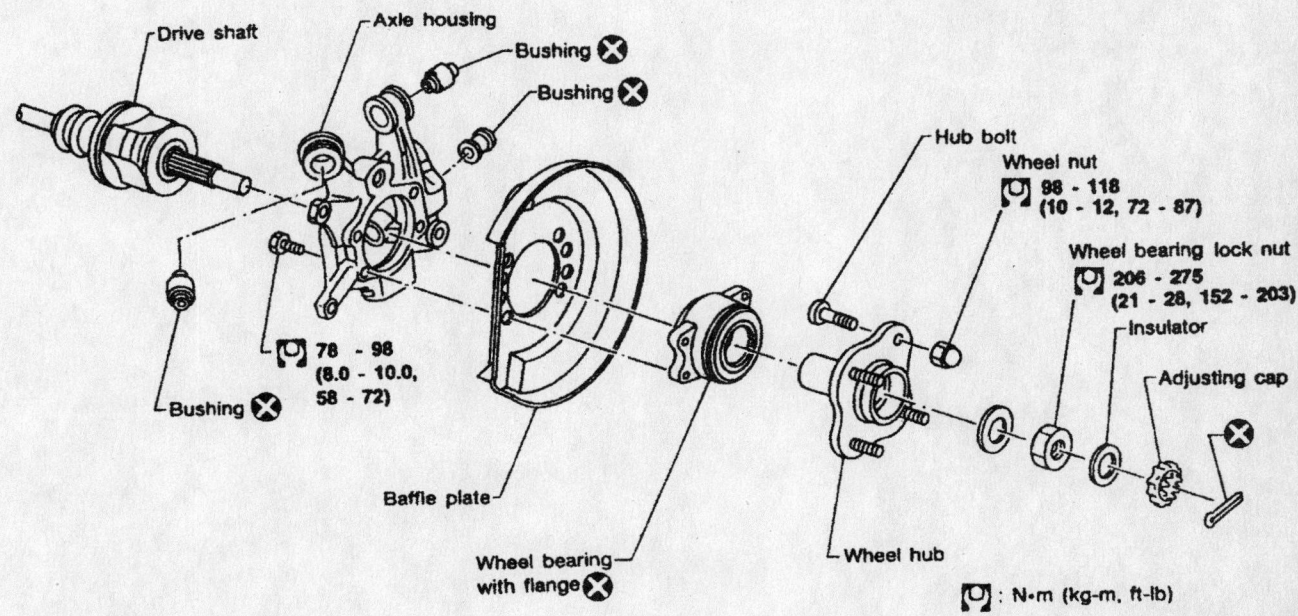

exploded view of rear wheel bearing assembly—240SX

Brake service is covered in Section 4 of this manual

method. Check the wheel bearing seating surface for roughness, seizure or other damage that may interfere with proper bearing function. Check the rubber bushing for wear.

To install:

7. Place the hub on a block of wood and seat the inner race using a suitable drift. Be careful not to damage the grease seals during installation of the inner race.

8. Install or connect the following:

- Bearing into the hub using a suitable drift and a press
- Bearing/hub assembly to the axle housing and torque the bolts to 58–72 ft. lbs. (78–98 Nm)

9. Lubricate the halfshaft splines. Properly align the splines and insert the halfshaft into the wheel hub.

- Wheel bearing locknut and torque the nut to 152–203 ft. lbs. (206–275 Nm). Install the insulator

and fit adjusting cap. Install a new cotter pin

10. Check the axial end-play as follows before mounting the rear wheels. Mount a dial indicator so the stylus of the dial rests on the face of the hub and check the wheel bearing axial end-play by attempting to rock the wheel hub in and out. The end-play should be 0.0020 in. (0.05mm) or less.

11. Install the rotor and caliper assembly.

12. Mount the rear wheels and lower the vehicle.

PRECAUTIONS

Before servicing any vehicle, please be sure to read all of the following precautions, which deal with personal safety, prevention of component damage, and important points to take into consideration when servicing a motor vehicle:

• Never open, service or drain the radiator or cooling system when the engine is hot; serious burns can occur from the steam and hot coolant.

• Observe all applicable safety precautions when working around fuel. Whenever servicing the fuel system, always work in a well-ventilated area. Do not allow fuel spray or vapors to come in contact with a spark, open flame, or excessive heat (a hot drop light, for example). Keep a dry chemical fire extinguisher near the work area. Always keep fuel in a container specifically designed for fuel storage; also, always properly seal fuel containers to avoid the possibility of fire or explosion. Refer to the additional fuel system precautions later in this section.

• Fuel injection systems often remain pressurized, even after the engine has been turned **OFF**. The fuel system pressure must be relieved before disconnecting any fuel lines. Failure to do so may result in fire and/or personal injury.

• Brake fluid often contains polyglycol ethers and polyglycols. Avoid contact with the eyes and wash your hands thoroughly after handling brake fluid. If you do get brake fluid in your eyes, flush your eyes with clean, running water for 15 minutes. If

eye irritation persists, or if you have taken brake fluid internally, IMMEDIATELY seek medical assistance.

• The EPA warns that prolonged contact with used engine oil may cause a number of skin disorders, including cancer! You should make every effort to minimize your exposure to used engine oil. Protective gloves should be worn when changing oil. Wash your hands and any other exposed skin areas as soon as possible after exposure to used engine oil. Soap and water, or waterless hand cleaner should be used.

• All new vehicles are now equipped with an air bag system. The system must be disabled before performing service on or around system components, steering column, instrument panel components, wiring and sensors. Failure to follow safety and disabling procedures could result in accidental air bag deployment, possible personal injury, and unnecessary system repairs.

• Always wear safety goggles when working with, or around, the air bag system. When carrying a non-deployed air bag, be sure the bag and trim cover are pointed away from your body. When placing a non-deployed air bag on a work surface, always face the bag and trim cover upward, away from the surface. This will reduce the motion of the module if it is accidentally deployed. Refer to the additional air bag system precautions later in this section.

• Clean, high quality brake fluid from a

sealed container is essential to the safe and proper operation of the brake system. You should always buy the correct type of brake fluid for your vehicle. If the brake fluid becomes contaminated, completely flush the system with new fluid. Never reuse any brake fluid. Any brake fluid that is removed from the system should be discarded. Also, do not allow any brake fluid to come in contact with a painted surface; it will damage the paint.

• Never operate the engine without the proper amount and type of engine oil; doing so WILL result in severe engine damage.

• Timing belt maintenance is extremely important! Many models utilize an interference-type, non-freewheeling engine. If the timing belt breaks, the valves in the cylinder head may strike the pistons, causing potentially serious (also time-consuming and expensive) engine damage. Refer to the maintenance interval charts in the front of this manual for the recommended replacement interval for the timing belt, and to the timing belt section for belt replacement and inspection.

• Disconnecting the negative battery cable on some vehicles may interfere with the functions of the on-board computer system(s) and may require the computer to undergo a relearning process once the negative battery cable is reconnected.

• When servicing drum brakes, only disassemble and assemble one side at a time, leaving the remaining side intact for reference.

ENGINE REPAIR

➡ **Disconnecting the negative battery cable on some vehicles may interfere with the functions of the on board computer systems and may require the computer to undergo a relearning process, once the negative battery cable is reconnected.**

Alternator

REMOVAL

900 and 9-3 MODELS

1. Before servicing the vehicle, refer to the precautions in the beginning of this section.

2. Remove or disconnect the following:
   • Negative battery cable

• Air cleaner
• Air intake hose to throttle body
• Drive belt
• Drive belt tensioner
• Right front wheel
• Belt drive cover
• Alternator electrical connections
• Alternator retaining bolts
• Exhaust manifold-to-exhaust pipe mounting hardware
• Catalytic converter mounting bolt. Push catalytic converter aside.
• Alternator

9000 MODELS

2.3L ENGINE

1. Before servicing the vehicle, refer to the precautions in the beginning of this section.

2. Remove or disconnect the following:
   • Negative battery cable
   • Right front wheel
   • Inner wheel well
   • Drive belt
   • Wheel arch to subframe support bar
   • Power steering pump
   • Alternator mounting bolts
   • Pump bracket
   • Alternator electrical connections
   • Alternator

9-5 MODELS

2.3L ENGINE

1. Before servicing the vehicle, refer to the precautions in the beginning of this section.

2. Remove or disconnect the following:

- Negative battery cable
- Intake manifold shield
- Crankcase ventilation solenoid and constant pressure valves
- Drive belt
- Belt tensioner with Tool 83–95–254
- Alternator upper mounting bolt
- Right side engine mount nut
- Rear engine mount nut
- Front exhaust pipe
- Hose between the oil trap and the oil sump
- Alternator electrical connectors
- Alternator lower bolt and insert a pry bar between the gearbox and subframe
- Alternator

3.0L ENGINE

1. Before servicing the vehicle, refer to the precautions in the beginning of this section.
   2. Remove or disconnect the following:
   - Negative battery cable
   - V belt and tensioner
   - Right front wheel and housing cover

Remove the alternator through the right front wheel well opening–3.0L Engine

9347UG01

- 6 outside hex screws from the pulley
- Engine oil pressure sensor
- Right front halfshaft
- Alternator electrical connectors
- Alternator air intake
- Alternator through the right side wheel well opening

INSTALLATION

900 and 9-3 MODELS

1. Before servicing the vehicle, refer to the precautions in the beginning of this section.
   2. Install or connect the following:
   - Alternator
   - Catalytic converter mounting bolt
   - Exhaust manifold to exhaust pipe mounting hardware
   - Alternator retaining bolts and torque them to 15 ft. lbs. (20 Nm)
   - Alternator electrical connections
   - Belt drive cover
   - Belt tensioner
   - Drive belt

- Air intake hose to throttle body
- Air cleaner
- Right front wheel
- Negative battery cable

9000 MODELS

2.3L ENGINE

1. Before servicing the vehicle, refer to the precautions in the beginning of this section.
   2. Install or connect the following:
   - Alternator electrical connections
   - Power steering pump bracket
   - Top alternator bolt. Do not tighten.
   3. Align alternator, then tighten bolts to 19 ft. lbs. (25 Nm)
   - Power steering pump
   - Wheel arch to subframe support bar
   - Belt
   - Inner wheel well
   - Wheel
   - Negative battery cable

9-5 MODELS

2.3L ENGINE

1. Before servicing the vehicle, refer to the precautions in the beginning of this section.
   - Alternator and loosely install the lower retaining bolt
   - Alternator electrical connectors
   - Hose between the oil trap and oil sump
   - Front exhaust pipe
   - Upper retaining bolts and torque the bolts to 33 ft. lbs. (45 Nm)
   - Belt tensioner and belt
   - Crankcase ventilation solenoid and constant pressure valves
   - Right side engine mount and torque the nut to 37 ft. lbs. (50 Nm)
   - Rear engine mount and torque the nut to 19 ft. lbs. (25 Nm)
   - Intake manifold shield
   - Negative battery cable

3.0L ENGINE

1. Before servicing the vehicle, refer to the precautions in the beginning of this section.
   - Alternator and torque the bolts to 33 ft. lbs. (45 Nm)
   - Alternator electrical connectors
   - Alternator air intake
   - Engine oil pressure sensor
   - Right front halfshaft
   - Belt pulley and torque the six bolts to 15 ft. lbs. (20 Nm)

- Wheel housing cover
- Right front wheel
- Belt tensioner and V belt. Torque the bolt to 30 ft. lbs. (40 Nm).
- Negative battery cable

Ignition Timing

ADJUSTMENT

Saab vehicles are equipped with a distributorless Ignition System (DIS). Ignition timing is controlled by the Electronic Control Module (ECM). No adjustment is necessary or possible.

Engine Assembly

REMOVAL & INSTALLATION

900 and 9-3 MODELS

➡ **The engine and transaxle are removed together as an assembly.**

1. Before servicing the vehicle, refer to the precautions in the beginning of this section.

2. With all 4 wheels on the ground, loosen the axle hub nuts.

3. Relieve the fuel system pressure.
4. Drain the engine coolant.
5. Remove or disconnect the following:
   - Battery
   - Air cleaner assembly and attached hoses
   - Resonator
   - Throttle cable and move it to one side
   - Cruise control wiring harness and cables from the throttle body
   - Cruise control unit retaining nuts
   - Cruise control unit
   - Fuel lines at their connections at the front of the fuel injection manifold and on the fuel pressure regulator
   - Turbocharger pressure line from the turbocharger compressor, if equipped
   - Intercooler/throttle housing, if equipped
   - Vacuum hose from the secondary injection
   - Brake booster vacuum hose from the intake manifold
   - Boost pressure control unit, if equipped
   - Drive belt using a ⅜ ratchet extension

- Pressure and return pipe from the steering servo pump, and plug the pipes to prevent oil from escaping.
- Cooling system hoses at the following connections:
- Heater control valve
- Expansion tank
- Bottom of the radiator
- Thermostat housing

6. Remove or disconnect the following:
   - A/C compressor and its mounting bracket without disconnecting any hoses. Place them on the filter housing for the heater system. Secure the alternator so it will not drop or become damaged.
   - Positive battery cable from the clips holding it to the body
   - Ground cable from the transaxle
   - Ignition cable
   - Electrical connections from the ignition coil
   - Wiring connectors from the transaxle
   - Oxygen (O₂S) sensor connector
   - Catalytic converter temperature sensor connector, if equipped

7. Inside the vehicle, pull back carpet under glove box to gain access to the central locking system and disconnect the con-

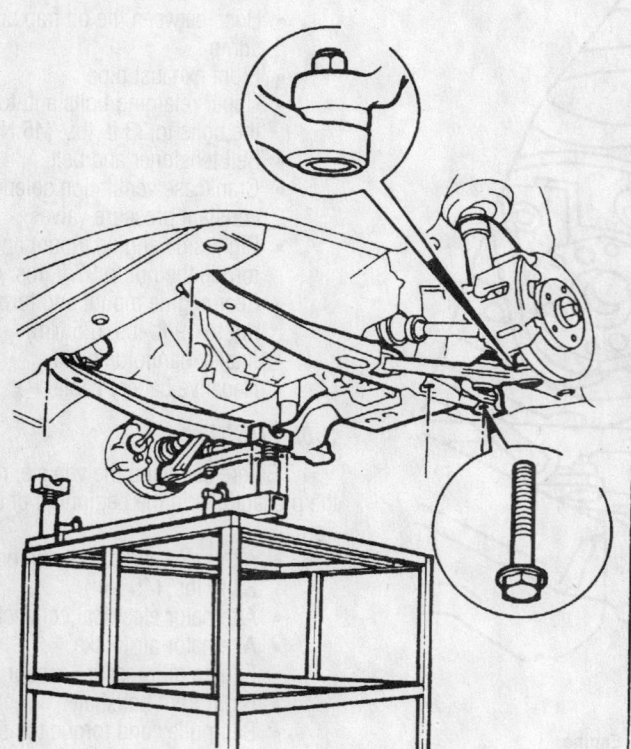
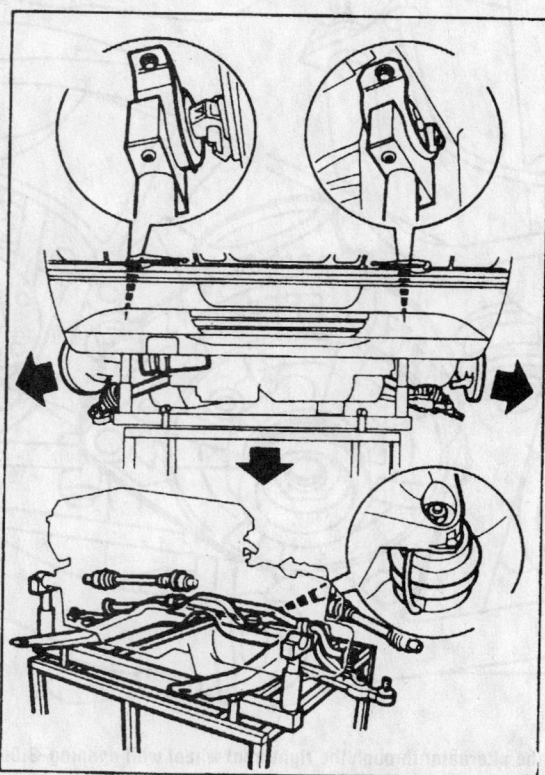

Lowering the entire powertrain assembly—900 and 9-3 models

7923SG02

trol module. Feed the wires to the engine compartment through the grommet.
- Hub nuts and both wheels
- Front splash shield
- Tapered pin from the gearshift rod joint located under the vehicle
- Clutch cable and clutch pipe if equipped
- Both ball joint nuts and the ball joints from the struts

➡**Be sure the halfshafts can slide out of the hubs. If necessary, use a wheel puller to push them out now.**

- Front pipe from the exhaust manifold
- Catalytic converter
- Oil cooler lines
- All the lines connected to the transaxle housing

8. Position the lifting table under the vehicle so it is directly under the front engine mount and gearbox. Take off the subframe bolts and the front engine mounts.

9. Lower the lifting table slightly and separate the halfshafts from the hubs.

10. Lower the lifting table fully and remove the subframe.

11. Place the engine on an engine stand.

To install:

12. Place the engine/transaxle on the lifting table.

13. Install the subframe on the rear engine mount and torque the bolts to 35 ft. lbs. (47 Nm).

14. Install the halfshafts into the hubs as the engine is raised into the engine compartment.

15. Install the engine mount and subframe bolts and torque them in sequence as follows:

    a. Front bolts: 85 ft. lbs. (115 Nm).

    b. Middle subframe bolts: 141 ft. lbs. (192 Nm).

    c. Rear subframe bolts: 90 ft. lbs. (122 Nm).

    d. Rear engine mount: 54 ft. lbs. (73 Nm).

16. Install or connect the following:
- All oil cooler and hydraulic lines
- All charge air cooler lines between the turbocharger and the cooler
- Front exhaust pipe to the exhaust manifold
- Catalytic converter to the rear exhaust section
- Engine shield to the underside of the vehicle
- Hub nuts and wheels and torque

the wheel nuts to 89 ft. lbs. (121 Nm) but do not tighten the hub nuts yet

17. Feed the wiring through the grommet and reconnect the central locking system control module.
- Clutch cable or bleed the clutch hydraulic cylinder, if equipped
- Tapered pin to the gearshift selector
- O2S sensor and catalytic converter temperature sensor
- All transaxle wiring
- A/C compressor, servo pump and drive belt
- Power brake booster vacuum hose and pressure sensor connections
- Cruise control including cables and the electrical harness connections
- Battery and cables

18. With all 4 wheels on the ground torque the hub nuts to 215 ft. lbs. (290 Nm).

19. Fill the coolant system.

20. Fill the engine with clean oil.

21. Start the vehicle, check for leaks and repair if necessary.

9000 MODELS

➡**The engine and transaxle are removed together as an assembly.**

1. Before servicing the vehicle, refer to the precautions in the beginning of this section.

2. Properly relieve the fuel system pressure.

3. Drain the engine coolant.

4. Drain the engine oil.

5. Remove or disconnect the following:
- Battery and tray
- Connecting bolt from the expansion tank
- Tank from the suction tube
- Overflow hoses from the radiator

6. Loosen the drive belt from the compressor by loosening the locknut, and loosening the adjusting nut under the locknut.
- Upper connection on the oil cooler
- Pipe clip on the radiator and slide the pipe down behind the radiator
- Connector to the electromagnetic clutch on the compressor

7. Loosen the compressor mounting complete with the belt tensioner.

8. Place a protective cloth over the radiator member and rest the compressor on the radiator member. Secure the compressor to the radiator member.

9. Remove or disconnect the following:
- Turbocharger pressure pipe situ-

ated between the turbocharger unit and the intercooler
- Oxygen (O2S) sensor connector
- Flange joint between the exhaust pipe and the exhaust manifold
- Exhaust pipe coming from the turbocharger, if equipped

10. Push the exhaust pipe to one side and unhook the rubber hangers from the exhaust system.
- Bottom coolant hose from the water pump
- Bottom retaining bolt from the radiator fan
- Speedometer drive from the gearbox

11. If equipped with a manual transmission select 4th gear and separate the rubber joint in the gear selector linkage.
- Clips on the rubber covers over the inboard universal joints and slide the covers off the drive axles
- Electrical leads from the alternator and the starter motor
- Oil pressure switch connector
- Clips securing the top radiator hose
- Top radiator hose at the cylinder head
- Solenoid valve from the bracket on the radiator after unplugging the electrical connections
- Bolts from the top of the radiator fan
- Wiring loom
- Radiator fan assembly
- Air mass meter from the air intake duct and the air cleaner. Leave the rubber socket connector attached to the turbocharger unit, if equipped.
- Air intake duct
- Air cleaner assembly
- Relief valve hose from the turbocharger pressure pipe and the pipe, if equipped
- Hall Effect transducer
- Ground lead from gearbox
- Back-up lights from gearbox
- Throttle cable and the throttle linkage.

12. Place a clamp on the hydraulic line to the slave cylinder and pinch the line tightly.

13. With proper wrenches, open the line to the clutch slave cylinder.

14. Remove the front wheels.

15. From both sides of the vehicle, slacken the lower bolts retaining the steering swivel member to the strut assembly

16. Remove or disconnect the following:
- 2 upper bolts holding the steering swivel member to the strut assembly, then pivot the steering swivel member outwards
- Inboard universal joint pulls out of the halfshaft. Position dust covers over the exposed halfshaft cups.
- Gear selector cable to the selector lever, on models equipped with an automatic transmission only
- Engine mount bolt
- Power steering fluid reservoir from the servo and position it within the engine compartment. Drain the fluid from the container
- Large bore and delivery hoses from the steering servo pump and plug the open ends
- Fuel return line from the pressure regulator
- Rear engine mount nut

17. Loosen, but do not remove front mount bolts.

18. Attach the lifting sling, 83-92-409 to the rear lifting lug.

19. Lift the engine sufficiently to provide access for the removal of the components located between the engine and the firewall.

20. Remove or disconnect the following:
- Vacuum hoses from the intake manifold. Tag them for identification prior to removal.
- Coolant hoses running between the heater core and the water pump pipe
- Connection between the fuel pipe and the fuel injection manifold. Do not allow the fuel to spill or collect.
- Any clips securing the wiring looms to the oil pipe or water pipe
- Vacuum hoses from the inlet manifold. Tag them for identification prior to removal.
- Wiring loom to the fuel injection manifold
- Grounding connections and electrical connectors from the wiring harness
- Air-cooled oil cooler and place it on top of the engine. The 2 lower bolts need only be loosened.
- Hood lifts

21. Install hood extenders for extra clearance

22. Carefully lift the engine from the vehicle, taking care not to damage the radiator.

To install:

23. Lower the engine into the vehicle.

24. Install or connect the following:
- Engine mounts and torque the bolts to 37 ft. lbs. (50 Nm)

- Air-cooled oil cooler then tighten the 2 lower bolts
- Grounds and electrical connectors to the wiring harness
- Wiring loom to the fuel injection manifold
- Wiring looms to the oil pipe, water pipe, intake manifold steady bar and the oil supply pipe
- Connection between the fuel pipe and the fuel injection manifold
- Coolant hoses between the heater core and water pump pipe
- Vacuum hoses to the inlet manifold
- Components located between the engine and the firewall, then remove the lifting sling
- Nut from the rear engine mounting, then tighten the front mount bolts
- Fuel return line to the pressure regulator
- Large bore hose and delivery hose to the steering servo pump
- Steering reservoir for the servo, then refill the fluid container
- Engine stay bolt
- Inboard universal joint into the halfshaft by pivoting the steering swivel member outwards
- Dust covers over exposed halfshaft cups and corresponding clamps
- Rubber joint to the gear selector linkage on models equipped with a manual transmission only. First select 4th gear.
- Gear selector cable to the selector lever, on models equipped with an automatic transmission only
- Upper bolts retaining the steering swivel member to the strut assembly and tighten the 2 lower bolts
- Front wheels
- Line to the clutch slave cylinder
- Throttle cable and throttle linkage
- Hall Effect transducer
- Negative battery cable to the gearbox
- Back-up light switch
- Relief valve hose to the turbocharger pressure pipe, if equipped
- Air cleaner
- Air intake duct
- Air mass meter connector
- Air intake duct socket connector to air mass meter and the air cleaner
- Top radiator hose to the cylinder head
- Electrical leads to the alternator
- Electrical leads to the starter motor
- Oil pressure switch connector
- Speedometer drive to the gearbox

- Flange joint between the exhaust pipe and the exhaust manifold
- Exhaust pipe coming from the turbocharger, if equipped
- Rubber hangers to the exhaust system
- Bottom coolant hose to the water pump
- O_2S sensor connector
- Electromagnetic clutch connector on the air conditioning compressor.

25. Tighten the compressor mounting complete with the belt tensioner.

26. Install the upper connection on the oil cooler if equipped

27. Tighten the pipe clip on the radiator.

28. Tighten the drive belt for the compressor.

29. Install the expansion tank and suction and overflow hoses to the expansion tank.

30. Install the battery and battery tray.

31. Verify that lift supports are no longer needed, then carefully remove them.

32. Fill the radiator with coolant, the engine with oil and the transmission with fluid.

33. Start the engine and allow it to reach operating temperature.

34. Check all fluid levels and top off if necessary.

9-5 MODELS

2.3L ENGINE

1. Before servicing the vehicle, refer to the precautions in the beginning of this section.

2. Properly relieve the fuel system pressure.

3. Drain the engine coolant.

4. Drain the engine oil.

5. Remove or disconnect the following:
- Battery and tray
- Ground cable
- Breather hose from under the fuse box and cables to the gearbox, automatic transaxle only
- Reverse light switch and remove the lower engine cover, manual transaxle only
- Steering column locking bolt and move the steering column upward. Make certain that the wheels are in the straight ahead position.
- Throttle cable
- Evaporative emissions (EVAP) purge valve vacuum hose from the intake manifold
- Brake servo vacuum hose by pressing the red sleeve downward
- Fuel connections and plug the lines

- Positive battery cable at the distribution terminal
- Ground cable at the gearbox
- Clutch slave cylinder hose, manual transaxle only and select 4th gear
- Selector rod after placing the vehicle in 3rd gear, manual transaxle only
- Selector lever cable, automatic transaxle only
- Upper radiator hose
- Bypass valve vacuum hose
- Hose between the throttle body and charge air cooler
- Pressure/temperature sensor electrical connector
- Bypass valve and intake manifold
- Radiator fan electrical connector
- Upper radiator hose to the oil cooler, if equipped
- Fan cowling
- Grille
- Hose between the turbocharger and charge air cooler
- Mass Air Flow (MAF) sensor
- Lower radiator from the water pump
- Heat exchanger hoses
- Vacuum hose from the bypass valve, if equipped
- Radiator
- Engine wire harness and bracket
- Wiper arms and covers
- Engine Control Module (ECM) electrical connectors
- Hub nuts and raise the vehicle
- Both front wheels. Drive a wedge between the gearbox and subframe, and between the oil sump and subframe.
- A/C compressor retaining bolts
- Air filter housing retaining nuts and air hose
- Condenser cooler, charge air cooler and A/C compressor and suspend them to the radiator member
- A/C compressor connector
- Quick release coupling to the wastegate
- Right side engine mount and yoke
- Belt and tensioner
- Left side engine mount
- Power steering reservoir and hoses
- Quick release couplings from the gearbox to the oil cooler, if equipped
- Steering swivel joint
- Hardware securing the anti-roll bar supports to the strut
- Outer steering links

- Halfshafts from the hub
- Oil cooler
- Exhaust system from behind the catalytic converter
- Temperature sensor and place a lifting trolley under the subframe
- Subframe supporting plate bolts

6. Lower the trolley slightly and unhook the oil cooler hose from the bracket

7. Move the trolley to the rear to allow access for the A/C compressor and remove the powertrain assembly.

To install:

8. Position the trolley under the vehicle with the engine in position.

9. Install or connect the following:

- Driveshafts to the steering swivel joint and torque the bolts to 22 ft. lbs. (30 Nm) plus 90 degrees
- Subframe and supporting plate. Torque the subframe bolts to 84 ft. lbs. (115 Nm) and the supporting plate to 48 ft. lbs. (65 Nm). Move the trolley away from the vehicle.
- Exhaust pipe and torque the bolts to 18 ft. lbs. (25 Nm)
- Temperature sensor
- Air cleaner
- Outer steering swivel members to the anti-roll bar support
- Hub nuts but do not tighten them
- A/C compressor and air filter housing and torque the bolt to 35 ft. lbs. (47 Nm)
- Steering gear and torque the locking nut to 19 ft. lbs. (25 Nm)
- Left side engine mount
- Right side engine mount and yoke and remove the wedges. Torque the mounting bolts to 36 ft. lbs. (50 Nm)
- Selector lever cable, if equipped

10. Lock 4th gear with Tool 87 92 335 in the gearbox and gear shift housing.

11. The installation is the reverse of removal.

- Selector rod on manual tranaxles and remove the locking pins
- Bleeder hose under the fuse box
- Hoses to the heat exchanger
- Coolant hose to the expansion tank
- Engine wire harness and bracket to the bulkhead partition
- ECM and wiring
- Wiper arms and cover plate
- Ground cable to the gearbox
- Positive cable to the positive terminal
- Connector with electrical leads to

the gear box, automatic transaxle only
- Reverse light connector, manual transaxle only
- Quick release couplings to the automatic transaxle fluid cooler
- Fuel connections and rubber protectors
- Vacuum hose for the brake servo and intake manifold
- EVAP canister purge valve vacuum hose
- Throttle cable
- Hose between the turbocharger and charge air cooler
- Engine oil cooler and cover
- Radiator fans
- Front grille
- Upper and lower radiator hoses
- Power steering reservoir
- Power steering pump pipes
- Quick release coupling for the turbocharger wastegate vacuum hose
- A/C compressor electrical connectors
- MAF sensor connector
- Bypass valve and intake manifold
- Pressure pipe and charge air cooler hose
- Pressure/temperature sensor connector
- Front wheels and torque the hub nuts to 215 ft. lbs. (290 Nm)
- Battery tray and battery
- Intake manifold cover and torque the bolts to 16 ft. lbs. (22 Nm)
- Shut off valve vacuum hose

12. Fill the engine with coolant.
13. Fill the engine with clean oil.
14. Fill the transaxle to the proper level.
15. Start the vehicle, check for leaks and repair if necessary.

3.0L ENGINE

1. Before servicing the vehicle, refer to the precautions in the beginning of this section.

2. Properly relieve the fuel system pressure.

3. Drain the engine coolant.

4. Drain the engine oil.

5. Remove or disconnect the following:
- Unbolt the steering column inside the vehicle
- Engine and battery cover
- Battery and tray
- Automatic transaxle electrical connectors
- Ground cable from the gearbox

Timing belt service is covered in Section 3 of this manual

- Gearbox vent hose
- Selector lever cable
- Positive cable from the junction box
- Throttle cable from the throttle body
- Turbocharger pressure hose from the throttle body
- Turbocharger pressure sensor electrical connectors
- Intake Air Temperature (IAT) sensor electrical connector
- Ground cable from the pressure hose pipe
- Turbocharger pressure hose from the charge air cooler
- Lower spoiler sections
- Air cleaner rubber mounts and install wedges between the oil sump to subframe and the gearbox to the subframe
- Rear Heated Oxygen (HO $_2$S) sensor, if equipped
- Front exhaust system after the catalytic converter
- Anti-roll bar link rods at the arms
- Outer tie rod ends from the strut
- Lower steering swivel joint bolts from the link arms
- Vacuum hose from the charge air bypass valve
- Turbocharger pressure hose with the pipe and valve
- Crankcase breather pipe from the turbocharger intake pipe
- Mass Air Flow (MAF) sensor and intake hose
- Turbocharger intake pipe
- Radiator fan
- Wiper arms and cover
- Engine Control Module (ECM) rubber seal and connector
- Engine wire harness cover
- Engie wire harness and place it on top of the engine
- Vacuum hose from the shut off valve
- Brake servo vacuum hose
- Vent hose from the throttle body
- Fuel rail lines
- Power steering fluid reservoir
- Upper, lower and expansion tank radiator hoses
- Automatic transaxle hydraulic hoses
- Radiator
- Heat exchanger hose connections
- Hose between the turbocharger outlet and charge air cooler
- Front grille
- Charge air cooler and condenser
- A/C compressor electrical connectors

- A/C compressor pipes from the subframe
- V-belt and tensioner
- Right side engine mount and bracket
- Left side engine mount and place a lifting trolley under the vehicle and raise it into position
- Subframe and triangular stiffener bolts and lower the powertrain slightly
- Halfshafts from the hub
- A/C compressor retaining bolts
- Engine assembly

To install:

6. Position the lifting trolley under the vehicle.

7. Install or connect the following:
- Engine assembly
- A/C compressor
- Halfshafts into the hubs
- Alternator cooling air duct
- Subframe and triangular stiffener and torque the bolts to 84 ft. lbs. (115 Nm)
- Left side engine mount and torque the bolts to 36 ft. lbs. (50 Nm)
- Right side engine mount and bracket and torque the bolts to 36 ft. lbs. (50 Nm)
- V-belt
- A/C compressor pipes to the subframe
- Hose between the turbocharger outlet and charge air cooler
- Front grille
- A/C compressor electrical connectors
- Radiator
- Automatic transaxle hydraulic hoses
- Upper and lower radiator hoses and the expansion tank hose
- Cabin heat exchanger hoses
- Power steering reservoir
- Fuel lines
- Vent hose to the throttle body
- Vacuum hose for the brake servo to the intake manifold
- Shut off valve vacuum hose
- Engine wire harness connector and cover
- Engine control module and rubber seal
- Wiper arms and cover plate
- Radiator fan
- Turbocharger intake pipe and hose
- MAF sensor and vent hose
- Crankcase breather pipe to the turbocharger intake pipe and torque the fasteners to 18 ft. lbs. (24 Nm)
- Exhaust manifold heat shield and

torque the bolts to 18 ft. lbs. (24 Nm)
- Turbocharger pressure hose with the charge air bypass pipe and valve
- Lower steering swivel joint and torque the bolts to 22 ft. lbs. (30 Nm)
- Link rods to the anti-roll bar
- Outer tie rod ends to the strut
- Front exhaust system and torque the bolts to 16 ft. lbs. (22 Nm)
- Air cleaner rubber mounts to the subframe
- HO $_2$Sensor electrical connector
- Lower spoiler sections
- Turbocharger pressure hose to the charge air cooler
- Turbocharger pressure sensor and IAT sensor connectors
- Ground cable for the pressure hose connecting pipe
- Turbocharger pressure hose to the throttle body
- Throttle cable
- Positive cable to the junction terminal
- Selector lever cable to the gear actuator
- Ground cable to the gearbox
- Gearbox vent hose
- Ground cable to the left side wheel housing
- Front wheels and hub nuts and torque the nuts to 148 ft. lbs. (200 Nm) plus an additional 30 degrees
- Battery tray and battery
- Battery cables and engine cover
- Automatic transaxle wiring harness
- Steering column locking bolt and torque the bolt to 18 ft. lbs. (25 Nm)

8. Fill the engine with coolant
9. Fill the engine with clean oil.
10. Fill the transaxle to the proper level.
11. Start the vehicle, check for leaks and repair if necessary.

Water Pump

REMOVAL & INSTALLATION

2.0L and 2.3L Engines

1. Before servicing the vehicle, refer to the precautions in the beginning of this section.

2. Loosen the expansion tank pressure cap.

3. Drain the coolant.

4. Remove or disconnect the following:

- Negative battery cable
- Center air deflector
- Right front wheel
- Front section of the inner fender panel
- Expansion tank, with its hoses and electrical connector

5. Pull hard on the drive belt and lock the tensioner.

- Water pump and A/C compressor belt

6. Place protective covers on the engine oil cooler and upper radiator crossmember.

- A/C compressor electrical connector
- A/C compressor with the lines attached and position it to the side
- A/C compressor bracket
- Coolant hoses to water pump
- Oxygen O_2S sensor wire from the clips
- Coolant pipe from the turbocharger
- Coolant pipe from water pump
- Three bolts attaching the water pump to the timing cover
- Water pump by prying first at the sleeve in the cylinder block, then around the rest of the pump

✳✳ WARNING

Use care not to damage the O_2S sensor.

- Water pump pulley then take the pump out of its housing

To install:

7. Clean the sealing surfaces and use a new gasket.

8. Install or connect the following:

- Water pump in the housing
- Water pump pulley and torque the bolts to 72 inch lbs. (8 Nm)
- Water pump to cylinder head sleeve and torque the bolts to 15 ft. lbs. (20 Nm)
- Coolant pipe to the water pump
- Coolant pipe to the turbocharger
- O_2S sensor wire in the clips
- Coolant hoses to water pump
- A/C compressor bracket
- A/C compressor
- Compressor electrical connector

9. Protective covers from the engine oil cooler and the upper radiator crossmember may now be removed.

- Water pump belt. Pull on it firmly and remove the locking pin from the tensioner. Check for proper belt

alignment and tension.

- Coolant hoses and electrical connector to expansion tank
- The expansion tank
- Center air deflector
- Front section of the inner fender panel
- Right front wheel
- Negative battery cable

10. Properly fill the cooling system.

11. Start the engine, check for leaks and repair if necessary.

3.0L Engine

1. Before servicing the vehicle, refer to the precautions in the beginning of this section.

2. Drain the cooling system.

3. Remove or disconnect the following:

- Negative battery cable
- Lower center air deflector
- Top engine covers
- Power steering reservoir
- Connection on the torque arm engine mount
- The torque arm
- Power steering line clamp from the torque arm engine mount
- The engine mount
- Upper coolant hose
- Coolant expansion tank hose
- Upper alternator air intake
- Power steering pump pulley and water pump pulley bolts, loosen only
- Drive belt and tensioner
- Power steering pump pulley
- Water pump pulley
- Timing cover
- Water pump bolts and pump

4. Check the timing belt for damage and replace if necessary.

To install:

5. Clean water pump mounting surface. Coat the water pump O-ring and sealing surface with acid-free petroleum jelly.

6. Install or connect the following:

- Water pump with new O-ring and torque the bolts to 18 ft. lbs. (25 Nm)
- Timing cover
- Water pump pulley and torque the water pump pulley bolts to 72 inch lbs. (8 Nm)
- Power steering pump pulley
- Drive belt tensioner and belt
- Upper alternator air intake
- Torque arm engine mount
- Power steering line clamp

- Torque arm
- The connection to the torque arm engine mount
- Upper coolant hose
- Coolant expansion tank to upper coolant hose
- Power steering reservoir
- Upper engine covers
- Lower center air deflector
- Negative battery cable

7. Fill the radiator with coolant and check for leaks.

9-5 MODELS

2.3L ENGINE

1. Before servicing the vehicle, refer to the precautions in the beginning of this section.

2. Drain the cooling system.

3. Remove or disconnect the following:

- Negative battery cable
- Front lower cover
- Mass Air Flow (MAF) sensor electrical connector
- MAF sensor and air hose
- V-belt tension with Tool 83–95–254
- V-belt from the power steering pump and water pump
- Crankcase breather pipe
- Camshaft cover
- Boost pressure control valve connector
- Engine lifting eye
- Turbocharger wastegate valve hoses, loosen only
- Bypass valve and pipe
- Exhaust manifold heat shield
- Quick release coupling on the vent hose with Tool 83–95–261
- Turbocharger intake pipe
- Power steering pump and move it aside
- Water pump inlet hoses
- 2 longitudinal coolant pipes from the engine block and water pump
- Water pump and connecting piece

To install:

4. Clean all mating surfaces and connecting piece seals. Replace the seals if necessary.

5. Install or connect the following:

- Connecting piece to the engine block
- Water pump and torque the bolts to 16 ft. lbs. (22 Nm)
- Coolant pipe to the water pump and turbocharger. Torque the water

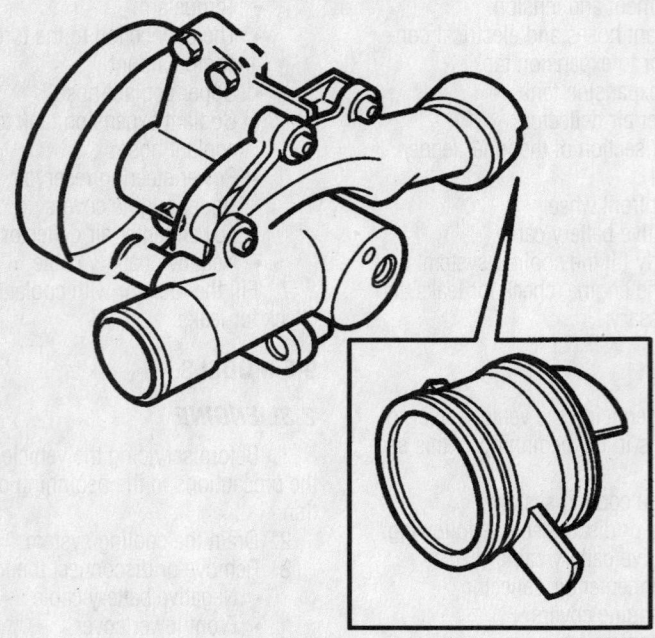

Remove the water pump and engine block connecting piece–2.3L Engine

9347UG02

pump bolt to 15 ft. lbs. (20 Nm) and the turbocharger bolt to 18 ft. lbs. (24 Nm).
- Longitudinal pipes to the water pump and torque the bolts to 89 inch lbs. (10 Nm)
- Water pump inlet hoses
- Power steering pump and torque the bolts to 18 ft. lbs. (24 Nm)
- Turbocharger intake pipe
- Turbocharger vent hose
- Exhaust manifold heat shield and torque the bolts to 15 ft. lbs. (20 Nm)
- Bypass pipe and valve and torque the bolts to 6 ft. lbs. (8 Nm)
- Turbocharger wastegate valve hoses
- Engine lifting eye
- Boost pressure control valve connector
- Crankcase breather pipe and camshaft cover and torque the pipe to 18 ft. lbs. (24 Nm)
- V-belt
- MAF sensor and air hose
- Front lower cover
- Negative battery cable

6. Fill the cooling system.

7. Start the vehicle, check for leaks and repair if necessary.

3.0L ENGINE

1. Before servicing the vehicle, refer to the precautions in the beginning of this section.

2. Drain the cooling system.
3. Remove or disconnect the following:
- Negative battery cable
- Lower front cover and raise the engine until the right front mount is unloaded
- Mass Air Flow (MAF) sensor and hose
- Power steering hose from the holder
- Right side engine mount and bracket
- Water pump and power steering pump pulley bolts and loosen the V-belt
- Water pump and power steering pump pulleys
- Belt tensioner
- Timing cover
- Water pump bolts and pump

To install:

4. Clean all gasket mating surfaces.
5. Install or connect the following:
- Water pump with a new O-ring and torque the bolts to 18 ft. lbs. (24 Nm)
- Timing cover
- Belt tensioner and torque the bolts to 30 ft. lbs. (40 Nm)
- Water pump and power steering pump pulleys. Do not tighten the bolts
- V-belt and torque the water pump pulley bolts to 6 ft. lbs. (8 Nm) and the power steering pump pulley bolts to 15 ft. lbs. (20 Nm)

- Right side engine mount and bracket
- Power steering hose to the holder
- MAF sensor
- Lower front cover
- Negative battery cable

6. Fill the cooling system.

7. Start the vehicle, check for leaks and repair if necessary.

Cylinder Head

REMOVAL & INSTALLATION

✴✴ CAUTION

The fuel system pressure must be relieved before disconnecting any fuel lines. Failure to do so may result in personal injury.

900 and 9-3 MODELS

2.0L AND 2.3L ENGINES

1. Before servicing the vehicle, refer to the precautions in the beginning of this section.

2. Properly relieve the fuel system pressure.
3. Drain the cooling system.
4. Drain the engine oil.
5. Remove or disconnect the following:
- Battery
- Exhaust manifold
- Turbocharger unit, if equipped
- Oil hoses from the oil cooler if equipped, and plug them
- Tensioning pulley and drive belt from the A/C compressor and slacken the securing bolts for the steering pump bracket
- Power steering pump drive belt, push the pump aside
- Wiring harness clips on the cylinder head
- Two bolts in the timing cover, which are screwed into the cylinder head from underneath
- Bolts in the right-hand engine mounting, which are screwed into the cylinder head, together with the spacer sleeves
- Hose between the thermostat housing and the radiator at the thermostat housing
- Fuel pressure regulator and ground leads for the fuel injection system
- A/C compressor. Leave lines still attached and carefully put it on the air intake for the heating system.
- Bracket for the air conditioning compressor

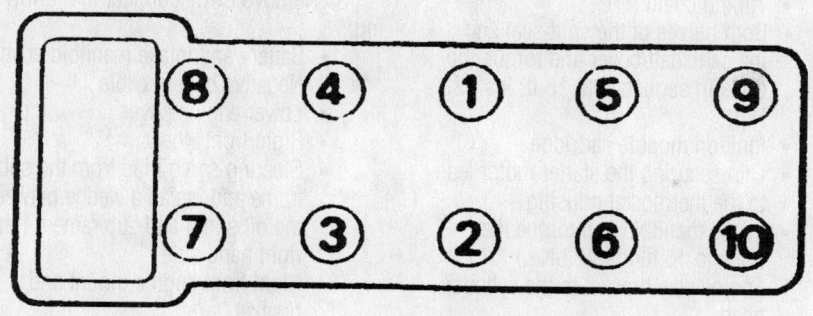

7923SG03

Cylinder head bolt torque sequence—2.0L and 2.3L engines

- Intake manifold complete with injectors
- Temperature sensor lead
- Ignition module cartridge
- Valve cover
- Crankcase ventilation hose
- Semicircular rubber plug halves from the cylinder head

6. Line up the **0** mark on the flywheel with the timing mark. Check that the camshaft timing marks are also in line.
- Timing chain tensioner
- Camshaft sprockets

➡**Do not rotate the camshaft or crankshaft after the sprockets have been removed, as rotation of one of the shafts can result in damage to the valves.**

- Cylinder head bolts

7. Siphon off the oil from the cylinder head.

8. Install Guide Pin 83-92-128 in one of the bolt holes and lift off the cylinder head, making sure the pivoting guide for the timing chain is not damaged.

To install:

9. Align the **0** mark on the flywheel with the timing mark on the housing. Align the marks on the camshafts with their respective timing marks.

10. Install the cylinder head gasket, making sure it is held in position by the guide sleeves in the cylinder head flange.

11. Position the timing chain and pivoting guide for cylinder head installation.

12. Install the cylinder head. Use the guide pin as a pivot for the head, which must be turned slightly to enable it to pass the pivoting guide. Thereafter, alignment will be determined by the guide sleeves.

13. Install the 2 M8 bolts in the underside of the cylinder head.

14. Install the remaining cylinder head bolts and torque them in sequence as follows:
   a. 30 ft. lbs. (44 Nm).
   b. 44 ft. lbs. (60 Nm).
   c. Plus 90 degrees (¼ turn).

15. Install or connect the following:
- Camshaft sprockets
- Timing chain
- Timing chain tensioner
- Semi-circular rubber plug halves in the cylinder head
- Valve cover and torque the bolts to 11 ft. lbs. (15 Nm)
- Crankcase ventilation hose
- Ignition module module and torque to 8 ft. lbs. (11 Nm)
- Temperature sensor electrical lead
- Intake manifold complete with injectors and torque the bolts to 16 ft. lbs. (22 Nm)
- A/C compressor bracket
- A/C compressor
- Fuel pressure regulator
- Ground leads for the fuel injection system
- Upper radiator hose to thermostat housing
- Bolts in the right-hand engine mounting. Screw them into the cylinder head, together with the spacer sleeves.
- Wiring harness clips on the cylinder head
- Power steering belt
- Tensioning pulley and drive belt for the air conditioner compressor
- Exhaust manifold and torque the bolts to 19 ft. lbs. (25 Nm)
- Negative battery cable

16. Fill the engine with clean oil.

17. Fill the cooling system.

18. Start the vehicle, check for leaks and repair if necessary.

9000 MODELS

1. Before servicing the vehicle, refer to the precautions in the beginning of this section.

2. Drain the cooling system.

3. Remove or disconnect the following:
- Negative battery cable
- Right front wheel
- Right front inner fender panel
- Radiator expansion tank
- Power steering reservoir, set aside and leave the hoses attached
- A/C compressor drive belt
- A/C compressor electrical leads
- Compressor from its mounting bracket. Disconnect the top pipe connected to the air-cooled oil cooler and push the pipe to one side. Carefully rest the compressor on the radiator crossmember.
- Compressor mounting bracket
- Front exhaust pipe flange, unhook the rubber hangers.
- Turbocharger unit stay bar and the oil return pipe, if equipped
- Hose from the intercooler at the turbocharger unit, if equipped
- Oil supply pipe from the turbocharger, if equipped
- Hose between the air mass meter and the turbocharger unit, if equipped
- Coolant hose from the thermostat housing and the hose from the cylinder head
- Oil supply hose or pipe so as not to obstruct the removal of the exhaust manifold. If necessary, remove the clip holding the pipe to the cylinder head and slave cylinder.
- Exhaust manifold complete with the turbocharger unit (if equipped), pushing the oil supply pipe aside at the same time
- Temperature transducer electrical connector
- Engine stay bracket
- Bolt securing the engine stay bracket to the cylinder head
- Intake manifold
- Breather hose for crankcase ventilation from the camshaft cover
- Ignition module cartridge
- Camshaft cover and align the crankshaft with the **0** timing mark and check that the camshaft timing marks also coincide
- Bolts securing exhaust then intake

camshaft sprockets. Using a wrench installed over the flats on the camshaft, hold the camshaft and remove the bolt securing the camshaft sprocket

✳✳ WARNING

To remove or refit the timing chain, the camshafts and crankshaft must be lined up with their respective timing marks, No. 1 cylinder at Top Dead Center, (TDC). Never rotate the crankshaft or the camshaft once the timing chain has been detached. A fully opened valve could come in contact with a piston at TDC.

- Timing chain tensioner
- Two cylinder head bolts adjacent to the timing cover, which are accessible from below
- Starter motor lead from the clip on the thermostat housing
- Cylinder head bolts

4. Install a guide pin in the drilled hole in the right top corner of the cylinder head. Be sure the timing chain is positioned such that the pivoting chain guide will not obstruct the cylinder head and carefully lift the cylinder head from the engine block.

To install:

5. Before installation, clean both the cylinder head and the engine block surfaces. Install a new gasket. Be sure the crankshaft is aligned in the **0** position and that the camshafts are aligned with their respective timing marks.

➡ **When the pistons of the No. 1 and No. 4 cylinders are at TDC, the crankshaft 0 mark on the flywheel must be aligned with the mark on the clutch cover or the end plate, if the clutch cover has been removed. The marks on the camshafts must be aligned with those on the cam bearing caps. This indicates the exhaust valves for No. 1 and No. 4 cylinders are closed.**

6. Install a guide pin in the drilled hole in the top of the right corner of the cylinder head and lower the cylinder head carefully into position on the engine block. Locate the cylinder head on the guide sleeves.

7. Install the cylinder head bolts and torque them in sequence, as follows:
   a. 44 ft. lbs. (60 Nm).
   b. 59 ft. lbs. (80 Nm).
   c. An additional 90 degrees (¼ turn).
8. Install or connect the following:

- Timing chain
- Both halves of the split seal and the camshaft cover and torque the bolts in sequence to 16 ft. lbs. (22 Nm)
- Ignition module cartridge
- Clip securing the starter motor lead to the thermostat housing
- Intake manifold and torque the bolts to 16 ft. lbs. (22 Nm)
- Engine stay bracket to the cylinder head
- Exhaust manifold and gasket to the cylinder head
- Oil supply pipe and clip
- Slave cylinder bolt
- Oil return line
- Steady bar for the turbocharger, if equipped
- Hose between turbocharger unit and intercooler, if equipped
- Cooler hose to the thermostat housing and cylinder head
- Air mass meter connector to the turbocharger, if equipped
- Hose between the intercooler and the turbocharger unit, if equipped
- Nuts securing the front section of the exhaust pipe to the turbocharger compressor, if equipped
- A/C compressor mounting bracket to both cylinder head and block
- A/C compressor. Leave the coolant hose in the bracket when installing the compressor
- A/C compressor electrical leads and be sure the lead is clear of the compressor pulley
- Steering servo reservoir
- Coolant expansion tank and clamp
- Top pipe to the air-cooled oil cooler, secure the cooler to the radiator
- Overflow line between the expansion tank and the radiator
- Compressor belt, adjust the tension and tighten the belt tensioner bolt
- Inner right wheel arch and wheel
- Negative battery cable
9. Fill the cooling system.
10. Start the engine, check for leaks and repair if necessary.

9-5 MODELS

2.3L ENGINE

1. Before servicing the vehicle, refer to the precautions in the beginning of this section.
2. Drain the cooling system.

3. Remove or disconnect the following:

- Battery and intake manifold cover
- Negative battery cable
- Lower engine cover
- Right front wheel
- Steering servo pipe from the subframe and install a wedge between the oil sump and subframe on the right hand side
- Right front engine mount and bracket
- Mass Air Flow (MAF) sensor
- Crankcase breather pipe
- Front lifting eye
- Charge air pipe and bypass valve
- Pressure/temperature sensor connector
- Bypass valve vacuum hose
- V-belt
- Belt tensioner and alternator. Move the bracket aside.
- Temperature sensor and ignition discharge module connectors
- Coolant hoses from the cylinder head
- Throttle body lever cover
- Throttle cable and dipstick tube
- Fuel hoses
- Crankcase breather nipple from the timing cover
- Intake manifold steady bar
- Turbocharger steady bar
- Turbocharger and exhaust manifold heat shield
- Servo pump and suspend it on the radiator crossmember
- Lower screw from the steering servo pump bracket
- Bracket for the ignition discharge module's connector from the rear lifting eye
- Heat exchanger pipe bracket
- Intake manifold and partition and move it rearward
- Ignition discharge module and spark plugs
- Camshaft cover and align the pulley marks with the timing cover and make certain that the camshafts are in line with their timing marks
- Camshaft sprocket bolts. Make certain that the camshaft does not turn.
- Idler sprocket bolt and remove the chain tensioner
- Camshaft sprockets and place a rubber band between the chain guides
- Inner gasket from the camshaft cover partition

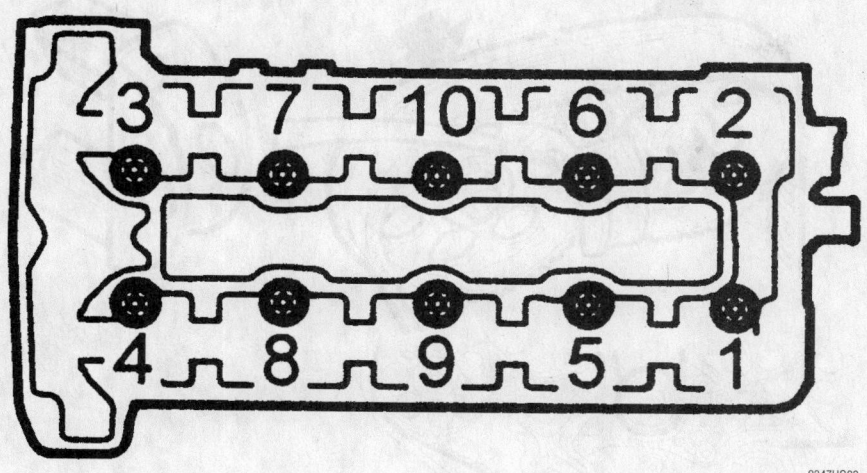

9347UG03

Remove the cylinder head bolts in sequence—9-5

- Cylinder head by first removing the bolts from the timing cover

To install:

4. Clean all mating surfaces.

5. Install or connect the following:
   - Inner gasket for the camshaft cover partition
   - Cylinder head gasket and turn the crankshaft 45 degrees in the rotational direction of the engine

6. Install the cylinder head. Torque the bolts in sequence as follows:

a. 30 ft. lbs. (40 Nm).
b. 44 ft. lbs. (60 Nm).
c. Plus an additional 90 degrees.

7. Install or connect the following:
   - Bolts between the timing cover and cylinder head and torque the bolts to 16 ft. lbs. (22 Nm). Make certain that the camshafts are aligned with the timing marks and reset the crankshaft to the **O**mark.
   - Camshaft sprocket and chain start-

ing with the exhaust camshaft. Do not tighten the bolts.
   - Timing chain tensioner and torque the bolts to 47 ft. lbs. (63 Nm)
   - Timing chain tensioner plug, push rod and spring and torque the bolt to 16 ft. lbs. (22 Nm)

8. Torque the idler pulley and camshaft sprocket bolts to 47 ft. lbs. (63 Nm). Rotate the crankshaft 2 revolutions and make check the settings of the crankshaft pulley and camshafts.

- Camshaft cover after lightly coating the opening with clean oil and torque the bolts, in sequence to 11 ft. lbs. (15 Nm)
- Spark plugs and torque the plugs to 21 ft. lbs. (28 Nm)
- Ignition discharge module and torque the bolt to 8 ft. lbs. (11 Nm)
- Intake manifold and intermediate partition. Remove the straps and torque the bolts to 16 ft. lbs. ()22 Nm).
- Alternator bracket and make certain that the adjuster sleeve is tapped out slightly
- Belt tensioner and tighten the ignition discharge module connector bracket
- Coolant hoses to the cylinder head and install the bolt securing the pipe to the thermostat housing cover
- Ignition discharge module electrical connector
- Temperature sensor electrical connector
- Turbocharger retaining nuts between the exhaust manifold and turbocharger and torque the bolts to 19 ft. lbs. (25 Nm)
- Turbocharger heat shield
- Dipstick tube
- Turbocharger and intake manifold steady bars
- Fuel hoses and rubber protectors
- Crankcase ventilation nipple on the valve cover
- Throttle cable and adjust as necessary
- Ground connections to the cylinder head
- Throttle body cover
- Servo pump
- V-belt and torque the pulley bolts to 15 ft. lbs. (20 Nm)
- Lifting eye
- Solenoid valve connector
- Charge air pipe and bypass valve

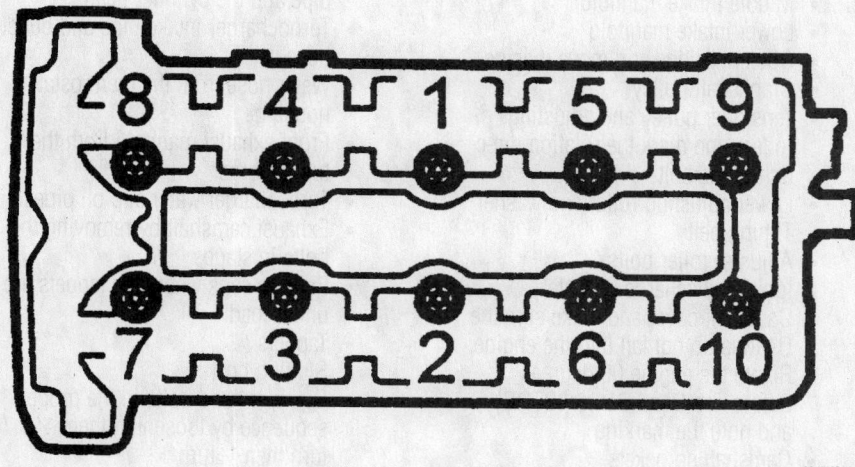

9347UG04

Cylinder head bolt torque sequence—9-5

For complete Engine Mechanical specifications, see Section 1 of this manual

- Bypass valve vacuum hose and pressure/temperature sensor connector
- MAF sensor
- Right side engine mount and bracket. Torque the bolts to 39 ft. lbs. (47 Nm) and the nuts to 78 ft. lbs. (105 Nm). Remove the wedge and secure the servo pump pipe to the subframe.
- Right front wheel
- Crankcase breather pipe and torque the banjo bolt to 18 ft. lbs. (24 Nm)
- Lower engine cover
- Negative battery cable
- Intake manifold and battery cover

9. Fill the cooling system and check all fluid levels.

10. Start the vehicle, check for leaks and repair if necessary.

3.0L ENGINE FRONT

1. Before servicing the vehicle, refer to the precautions in the beginning of this section.

2. Properly relieve the fuel system pressure.

3. Drain the cooling system.

4. Remove or disconnect the following:

- Engine cover
- Battery and vent hose
- Lower engine cover
- Turbocharger bracket and install wedges between the oil sump and subframe and gearbox and subframe
- Turbocharger delivery pipe from the charge air cooler
- Mass Air Flow (MAF) sensor and hose
- Power steering hose from the holder
- Right side engine mount and bracket
- V-belt tension
- Water pump and power steering pump pulleys
- Belt tensioner and alternator air intake
- Timing cover
- Throttle cable
- Upper intake manifold vacuum hoses and pressure sensor connector
- Vacuum hose and water hoses from the throttle body
- Turbocharger delivery pipe with the bypass pipe and valve from the throttle body
- Pressure/temperature sensor connector
- Throttle body connector and purge valve hose from the throttle body
- Ignition discharge module connector

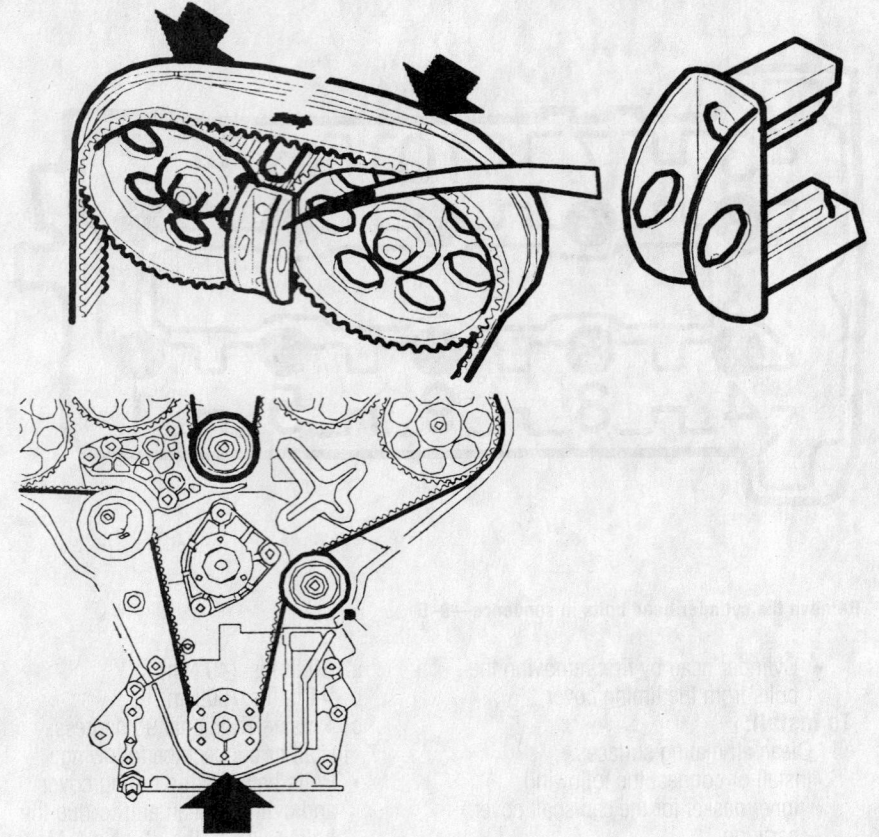

9347UG05

Make certain that the markings on the camshaft sprockets and timing cover are aligned along with the crankshaft.

- Fuel injector electrical connector
- Fuel lines
- Upper intake manifold and move it aside
- Middle intake manifold
- Lower intake manifold
- Coolant bridge and move it aside
- Crankshaft pulley
- Tensioner pulley and adjusting rollers and mark the rotation direction of the belt
- Lower adjusting roller and washer
- Timing belt
- Adjuster roller bolts
- Ignition discharge module
- Camshaft cover and make sure the O-rings do not fall into the engine. Rotate the engine 60 degrees Before Top Dead Center (BTDC) and note the marking
- Camshaft sprockets
- Water pump
- Inner timing cover
- Front exhaust manifold heat shield
- Hose connections for the front Secondary Air Injection (SAI) valve

- SAI pipe and valve from the exhaust manifold
- Crankcase ventilation pipe from between the turbocharger intake pipe and the cylinder head
- Turbocharger intake pipe and outlet hose
- Water hose from the thermostat housing
- Front exhaust manifold from the turbocharger
- Turbocharger water and oil pipes
- Exhaust camshaft by removing the bolts in stages
- Bearing caps where the tappets are under load
- Tappets
- Stuffing box
- Cylinder head bolts in the proper sequence by loosening them ¼ turn then ½turn
- Cylinder head gasket

To install:

5. Clean all mating surfaces of any residual gasket material.

6. Make certain that the crankshaft is at 60 degrees BTDC.

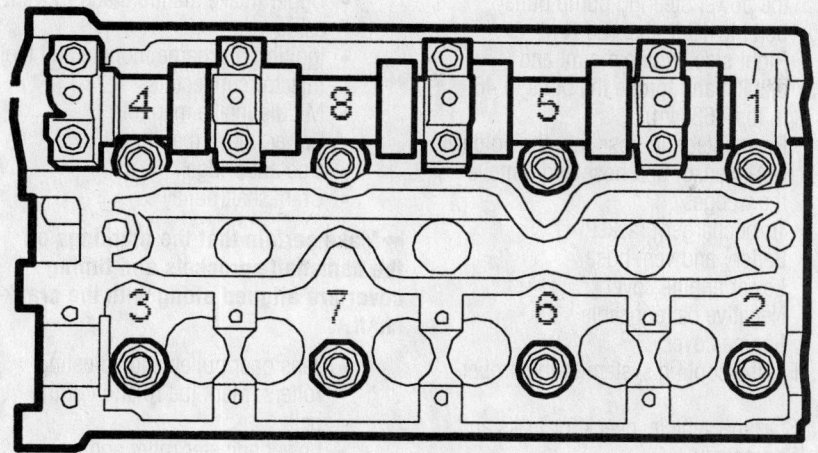

Remove the cylinder head bolts as shown—3.0L

9347UG06

Cylinder head bolt torque sequence—3.0L

9347UG07

7. Be sure that the proper gasket is used and is marked **OBEN/TOP**

8. Install or connect the following:
- Tappets

9. Cylinder head with a new gasket. Torque the new bolts in sequence as follows:
   a. 18 ft. lbs. (24 Nm).
   b. Plus an additional 90 degrees.
   c. Plus an additional 90 degrees.

10. Install or connect the following:
- Exhaust camshaft into position after lubricating it with clean engine oil. The locating pins for the intake camshaft should be in line with the inside bolts for the bearing caps. The guide pin for the exhaust camshaft should point straight up from the plane of the cylinder head.
- Stuffing box and bearing caps.

Torque the bolts in ½ to 1 turn to 6 ft. lbs. (8 Nm).
- Camshafts sprockets in relation to the locating pins. The number in the sprocket hub should match the number on the camshaft
- Rear timing cover
- Camshaft cover and make certain that the O-rings are in place.

11. Lubricate with a soap solution and place tool 10–81–52–381 in the corners at the camshaft bearing caps. Torque the bolts to 6 ft. lbs. (8 Nm).
- Ignition discharge module
- Turbocharger oil pipes and torque the oil pipe to 15 ft. lbs. (20 Nm); the oil pipe to filter adapter to 19 ft. lbs. (25 Nm) and the oil return pipe to 10 ft. lbs. (15 Nm).
- Turbocharger water pipes. Torque

the water pipes to 19 ft. lbs. (25 Nm) and the cylinder head retaining bolt to 10 ft. lbs. (15 Nm)
- Water hose to the thermostat housing
- Front exhaust pipe to the turbocharger and torque the bolts to 18 ft. lbs. (24 Nm)
- Hose between the turbocharger and charge air cooler
- Crankcase breather pipe to the turbocharger intake pipe and torque the bolt to 18 ft. lbs. (24 Nm)
- Turbocharger intake pipe and torque the V clamp to 2 ft. lbs. (3 Nm) and the U clamp to 10 ft. lbs. (15 Nm)
- SAI pipe with new gaskets to the front exhaust manifold
- SAI hose to the valve
- Crankcase ventilation pipe between the cylinder head turbocharger intake pipe and torque to 18 ft. lbs. (24 Nm)
- Water pump and torque the bolts to 19 ft. lbs. (25 Nm)
- Bracket with the tensioner roller and the upper adjusting roller

12. Install Locking Tool KM—800–1 on No. 1 and 2 camshaft sprockets and Tool KM—800–2 on No. 3 and 4 camshaft sprockets

13. Rotate the crankshaft to just before the 0 degree mark and install Tool KM—800–10. Carefully rotate the crankshaft in the direction of the engine until the arm is lying against the water pump flange.

14. Install the camshaft belt to the markings on the belt and the marked direction of rotation. Loosely install the lower tensioner pulley and make certain that the washer is in place.

15. Adjust the tensioner pulley counterclockwise by hand and make certain that the markings are aligned properly.

16. Tighten the center adjuster roller bolts lightly. Adjust the lower adjuster roller counterclockwise until a tension of 275–300 Nm is reached. When properly tightened, remove the belt tension meter.

17. Make certain that the adjusting nut on the upper adjuster roller rotates freely. Adjust the tensioning pulley until the lines are aligned. Torque the fastener to 15 ft. lbs. (20 Nm).

18. Remove the locking tool from camshafts No. 1–2 and install tool KM—

800–20. Adjust the upper roller counter-clockwise with tool 83–94–983 until the markings on the camshaft sprocket are aligned with the markings on tool KM—800–20. Torque the fastener to 30 ft. lbs. (40 Nm).

19. Remove all locking tools.

20. Rotate the engine in the proper direction 2 revolutions until just before the **0** mark. Install Tool KM—800–10 on the crankshaft.

21. Rotate the crankshaft in the proper direction until the arm lies against water pump and tighten it.

22. Install Tool KM—800–20 and make certain that the markings on the camshaft sprocket are opposite the markings on the tool and that the edge of the belt is aligned with the edge of the sprocket.

23. Install or connect the following:
- Coolant bridge after lubricate the bolts and torque them to 22 ft. lbs. (30 Nm)
- Lower intake manifold with a new gasket and torque the bolts to 15 ft. lbs. (20 Nm)
- Middle intake manifold and torque the bolts to 15 ft. lbs. (20 Nm)
- Upper intake manifold with a new gasket and torque the bolts to 15 ft. lbs. (20 Nm)
- Fuel lines
- Ignition discharge module and fuel injector connectors
- Turbocharger delivery pipe to the throttle body
- Turbocharger delivery pipe to the charge air cooler and tighten all clamps
- Pressure/temperature sensor electrical connector
- Bypass pipe and valve
- Water hoses and vacuum hose to the throttle body
- Throttle body connector and attach the hoses to the purge valve. Torque the bypass pipe to 6 ft. lbs. (8 Nm).
- Vacuum hoses and pressure/temperature sensor connectors to the intake manifold
- Throttle cable
- Crankshaft pulley and torque the bolts to 15 ft. lbs. (20 Nm)
- Timing cover and torque the bolts to 6 ft. lbs. (8 Nm)
- Belt tensioner and alternator air intake
- Water pump and power steering pump pulleys
- V-belt and torque the water pump pulley bolts to 6 ft. lbs. (8 Nm) and

the power steering pump pulley bolts to 15 ft. lbs. (20 Nm)
- Right side engine mount and bracket and torque the bolts to 46 ft. lbs. (63 Nm)
- Power steering hose into the holder
- MAF sensor and hose and remove the wedges
- Turbocharger bracket
- Battery and vent hose
- Lower engine cover
- Negative battery cable
- Engine cover

24. Fill the cooling system to the proper level.

25. Start the vehicle, check for leaks and repair if necessary.

3.0L ENGINE—REAR

1. Before servicing the vehicle, refer to the precautions in the beginning of this section.

2. Properly relieve the fuel system pressure.

3. Drain the cooling system.

4. Remove or disconnect the following:
- Engine and battery cover
- Lower engine cover and install wedges between the oil sump and subframe and the gearbox and subframe
- Mass Air Flow (MAF) sensor and hose
- Power steering hose from the holder
- Right side engine mount and bracket
- Water pump and power steering pump pulley bolts
- Tension from the V belt
- Water pump and power steering pump pulleys
- Belt tensioner and alternator air intake
- Timing cover
- Throttle cable
- Vacuum hoses and pressure sensor connector from the upper intake manifold
- Oxygen (O2S) sensor connector from the holder on the rear of the Secondary Air Injection (SAI) valve
- SAI pipe from the cylinder head and exhaust manifold connections
- Water hoses and vacuum hose from the throttle body
- Turbocharger delivery hose with the bypass pipe
- Throttle body connector and hose to the purge valve
- Fuel lines

- Upper intake manifold and move it aside
- Ignition discharge module and fuel injector connectors
- Middle intake manifold
- Lower intake manifold
- Coolant bridge
- Crankshaft pulley

➡ **Make certain that the markings on the camshaft sprockets and timing cover are aligned along with the crankshaft.**

- Tensioner pulley and adjusting rollers. Mark the rotation of the belt.
- Lower adjuster roller and washer
- Timing belt
- Bracket with the tensioning roller and the upper adjuster roller
- Ignition discharge module
- Camshaft cover with the O-rings. Rotate the crankshaft 60 degrees Before Top Dead Center (BTDC)
- Camshaft sprockets
- Inner timing cover
- Separate the exhaust pipes
- Exhaust camshaft
- Stuffing box
- Tappets and place them in the proper order for installation
- Cylinder head bolts in the proper sequence by loosening them ¼ turn then ½ turn
- Cylinder head gaskets

To install:

5. Clean all mating surfaces of any residual gasket material.

6. Make certain that the crankshaft is at 60 degrees BTDC.

7. Be sure that the proper gasket is used and is marked **OBEN/TOP**

8. Install the tappets.

9. Install the cylinder head with a new gasket. Torque the new bolts in sequence as follows:
   a. 18 ft. lbs. (24 Nm).
   b. Plus an additional 90 degrees.
   c. Plus an additional 90 degrees.

10. Install or connect the following:
- Exhaust camshaft into position after lubricating it with clean engine oil. The locating pins for the intake camshaft should be in line with the inside bolts for the bearing caps. The guide pin for the exhaust camshaft should point straight up from the plane of the cylinder head.
- Stuffing box and bearing caps. Torque the bolts in ½ to 1 turn to 6 ft. lbs. (8 Nm).

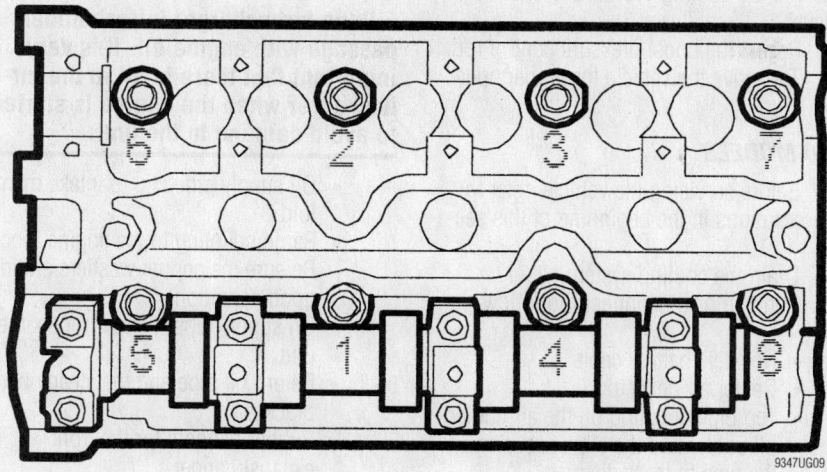

Cylinder head bolt torque sequence–3.0L engine

- Camshafts sprockets in relation to the locating pins. The number in the sprocket hub should match the number on the camshaft. Torque the bolt to 37 ft. lbs. (50 Nm) plus 60 degrees.
- Rear timing cover
- Camshaft cover and make certain that the O-rings are in place

11. Lubricate with a soap solution and place Tool 10–81–52–381 in the corners at the camshaft bearing caps. Torque the bolts to 6 ft. lbs. (8 Nm).

- Ignition discharge module
- Hose to the SAI pipe and install the pipe with new gaskets to the exhaust manifold and cylinder head
- O₂S connector in the holder
- Front and rear exhaust pipes and torque the bolts to 30 ft. lbs. (40 Nm)
- Water pump and torque the bolts to 19 ft. lbs. (25 Nm)
- Bracket with the tensioning roller and the upper adjuster roller

12. Install Locking Tool KM—800–1 on No. 1 and 2 camshaft sprockets and Tool KM—800–2 on No. 3 and 4 camshaft sprockets

13. Rotate the crankshaft to just before the 0 degree mark and install Tool KM—800–10. Carefully rotate the crankshaft in the direction of the engine until the arm is lying against the water pump flange.

14. Install the camshaft belt to the markings on the belt and the marked direction of rotation. Loosely install the lower tensioner pulley and make certain that the washer is in place.

15. Adjust the tensioner pulley counterclockwise by hand and make certain that the markings are aligned properly.

16. Tighten the center adjuster roller bolts lightly. Adjust the lower adjuster roller counterclockwise until a tension of 275–300 Nm is reached. When properly tightened, remove the belt tension meter.

17. Make certain that the adjusting nut o0n the upper adjuster roller rotates freely. Adjust the tensioning pulley until the lines are aligned. Torque the fastener to 15 ft. lbs. (20 Nm).

18. Remove the locking tool from camshafts No. 1–2 and install Tool KM—800–20. Adjust the upper roller counterclockwise with tool 83–94–983 until the markings on the camshaft sprocket are aligned with the markings on Tool KM—800–20. Torque the fastener to 30 ft. lbs. (40 Nm).

19. Remove all locking tools.

20. Rotate the engine in the proper direction 2 revolutions until just before the **0** mark. Install Tool KM—800–10 on the crankshaft.

21. Rotate the crankshaft in the proper direction until the arm lies against water pump and tighten it.

22. Install Tool KM—800–20 and make certain that the markings on the camshaft sprocket are opposite the markings on the tool and that the edge of the belt is aligned with the edge of the sprocket.

23. Install or connect the following:
- Coolant bridge after lubricating the bolts and torque them to 22 ft. lbs. (30 Nm)
- Lower intake manifold with a new gasket and torque the bolts to 15 ft. lbs. (20 Nm)
- Middle intake manifold and torque the bolts to 15 ft. lbs. (20 Nm)
- Upper intake manifold with a new gasket and torque the bolts to 15 ft. lbs. (20 Nm)
- Fuel lines
- Ignition discharge module and fuel injector connectors
- Turbocharger delivery pipe to the throttle body
- Bypass pipe and valve
- Water hoses and vacuum hose to the throttle body
- Throttle body connector and attach the hoses to the purge valve and torque the bypass pipe to 6 ft. lbs. (8 Nm)
- Vacuum hoses and pressure/temperature sensor connectors to the intake manifold
- Throttle cable
- Turbocharger delivery pipe to the charge air cooler
- Vacuum hose and pressure/temperature sensor connector to the upper intake manifold
- Turbocharger bypass pipe and valve
- Crankcase pulley and torque the bolts to 15 ft. lbs. (20 Nm)
- Timing cover and torque the bolts to 6 ft. lbs. (8 Nm)
- Belt tensioner and alternator air intake
- Water pump and power steering pump pulleys
- V-belt. Torque the water pump pulley bolts to 6 ft. lbs. (8 Nm) and the power steering pump pulley bolts to 15 ft. lbs. (20 Nm).
- Right side engine mount and bracket and torque the bolts to 46 ft. lbs. (63 Nm)
- Power steering hose into the holder
- MAF sensor and hose and remove the wedges
- Turbocharger bracket
- Battery and vent hose
- Lower engine cover
- Negative battery cable
- Engine cover

For Tire, Wheel and Ball Joint specifications, see Section 1 of this manual

24. Fill the cooling system to the proper level.

25. Start the vehicle, check for leaks and repair if necessary.

Turbocharger

REMOVAL & INSTALLATION

2.0L and 2.3L Engines

900 AND 9-3 MODELS

1. Before servicing the vehicle, refer to the precautions in the beginning of this section.

2. Drain the cooling system.

3. Remove or disconnect the following:
   - Negative battery cable
   - Hose from the charge air cooler
   - Front exhaust pipe from the engine. Take care not to damage the Oxygen (O$_2$S) Sensor electrical wiring harness.
   - Locking ring on the wastegate after breaking the seal
   - Oil return pipe from the turbocharger
   - Oil pipe from the oil filter housing
   - Hose to the boost control valve
   - Nuts from the wastegate diaphragm
   - Intake pipe and wastegate as a unit
   - Intake pipe and the bypass hose
   - Turbocharger coolant pipe with the oil pipe
   - Turbocharger and discard the gasket

To install:

4. Install or connect the following:
   - Turbocharger with a new gasket and torque the nuts to 16 ft. lbs. (22 Nm) and fill the turbocharger interchamber passage with engine oil.

❊❊ WARNING

It is very important that there is oil in the turbocharger when the engine is started to avoid damage to the unit.

- Turbocharger coolant and oil pipes
- Intake pipe for the turbocharger
- Wastegate with the boost pressure control valve and pipes
- Intake hose with the bypass hose
- Locking ring and seal for the wastegate
- Front pipe to the turbocharger
- Hose connecting the charge air cooler with the turbocharger

- Negative battery cable

5. Fill the cooling system and check the oil level.

6. Reseal the boost pressure control rod.

7. Test drive the vehicle for proper operation.

9000 MODELS

1. Before servicing the vehicle, refer to the precautions in the beginning of this section.

2. Drain the cooling system.

3. Remove or disconnect the following:
   - Negative battery cable
   - Center air deflector
   - Top pipe coupling on the air-cooled oil cooler and the clips securing the pipe to the radiator
   - Solenoid valve from its mounting on the radiator and the electrical leads
   - Radiator fan
   - Oxygen (O$_2$S) sensor cable clamps
   - Exhaust pipe from the turbocharger. Bend it down slightly
   - Electrical connectors for the air mass meter
   - Toggle fasteners securing the air mass meter to the air cleaner cover
   - Rubber socket connector off the turbocharger unit
   - Turbocharger pressure pipe from the turbocharger compressor
   - Oil pipe to the turbocharger unit
   - Clip securing the oil pipe to the cylinder head
   - Oil pipe banjo coupling from the block
   - Oil pipe clip on the intake manifold
   - Coolant pipes from the turbocharger unit
   - Steady bar bracket between the engine block and the turbocharger compressor
   - Securing bolts to the oil return lines. Cap the opening to prevent washers or nuts from the exhaust manifold dropping inside during the removal.
   - Turbocharger unit from the exhaust manifold and discard the gasket

To install:

4. Install or connect the following:
   - Turbocharger unit on the exhaust manifold with a new gasket. Torque the new nuts to 16 ft. lbs. (22 Nm).

❊❊ WARNING

Fill the turbocharger interchamber passage with engine oil. It is very important that there is oil in the turbocharger when the engine is started to avoid damage to the unit.

- Oil supply pipe clip to intake manifold
- Banjo coupling to the engine block. Be sure the copper washers are in good condition.
- Oil supply pipe to the turbocharger unit
- Return oil pipe and the steady bar bracket
- Rubber hangers for the front exhaust hanger
- Exhaust pipe to the turbocharger compressor with new locknuts. Lubricate the 3 studs on the turbocharger with Molykote 1000®. Torque the nuts to 19 ft. lbs. (25 Nm).
- Coolant pipes to the turbocharger
- Turbocharger pressure pipe to the turbocharger compressor
- Air mass meter and rubber socket connector between the air cleaner body and the inlet side of the turbocharger compressor
- Radiator solenoid valve and secure the electrical leads into their clips
- Return hose to the solenoid valve
- Oil pipe to the oil cooler and secure the pipe clip to the radiator
- Radiator drain plug
- Center air deflector
- Negative battery cable

5. Fill the cooling system.

6. Check the oil level and top off if necessary.

7. Start the vehicle, check for leaks and repair if necessary.

9-5 MODELS—2.3L ENGINE

1. Before servicing the vehicle, refer to the precautions in the beginning of this section.

2. Drain the cooling system.

3. Remove or disconnect the following:
   - Negative battery cable
   - Lower front cover
   - Loosen the oil return pipe from the turbocharger to the engine
   - Oil pipe between the oil filter adapter and the turbocharger
   - Bypass pipe and valve
   - Control valve connector

- Mass Air Flow (MAF) sensor electrical connector
- Loosen the hoses to the turbocharger and diaphragm housing
- Crankcase breather pipe
- Engine lifting eye
- MAF sensor and air hose
- Exhaust manifold heat shield
- Quick release coupling on the vent hose
- Intake manifold V-clamp
- Intake manifold
- Hose between the turbocharger and the charge air cooler
- Exhaust system from the turbocharger
- Oil pipe from the oil filter adapter and copper washers
- Coolant pipe to the turbocharger
- Coolant return pipe from the cylinder head and the pressure sensor bracket
- Turbocharger and discard the gasket

To install:

4. Install or connect the following:
- Turbocharger with a new gasket and torque the 2 upper bolts to 18 ft. lbs. (24 Nm)
- Coolant return pipe with new copper washers and torque the nipple to 26 ft. lbs. (35 Nm), the banjo bolt to 19 ft. lbs. (25 Nm), the coolant return pipe to 19 ft. lbs. (25 Nm) and the coolant return pipe clamp to 7 ft. lbs. (9 Nm).
- Turbocharger oil pipe with new copper washers and torque the banjo bolt to 15 ft. lbs. (20 Nm)
- Turbocharger intake pipe with a new O-ring and loosen the adjusting screw on the intake manifold
- Torque the V clamp between the turbocharger to intake pipe to 2.5 ft. lbs. (4 Nm) and the intake manifold and adjusting screw to 18 ft. lbs. (24 Nm).
- Hoses to the turbocharger and diaphragm housing
- Charge air cooler inlet hose and torque the bolts to 6 ft. lbs. (8 Nm)
- Heat shield and torque the bolts to 15 ft. lbs. (20 Nm)
- MAF sensor and air hose
- Engine lifting eye
- Crankcase breather pipe and torque the bolts to 18 ft. lbs. (24 Nm)
- Charge air bypass valve and pipe

with a new O-ring and torque the bolts to 6 ft. lbs. (8 Nm)
- Lower nut to the turbo charger and torque the stud to 16 ft. lbs. (22 Nm) and the lock nut to 18 ft. lbs. (24 Nm)
- Turbocharger oil pipe with new copper washers and torque the bolt to 19 ft. lbs. (25 Nm)
- Turbocharger stays and torque to 18 ft. lbs. (24 Nm)
- Oil return pipe between the turbocharger and engine and torque the bolts to 10 ft. lbs. (15 Nm)
- Lower engine cover
- Negative battery cable

5. Fill the engine with coolant.
6. Start the vehicle, check for leaks and repair if necessary.

9-5 MODELS—3.0L ENGINE

1. Before servicing the vehicle, refer to the precautions in the beginning of this section.

2. Drain the cooling system.
3. Remove or disconnect the following:

- Negative battery cable
- Upper engine cover
- Turbocharger pressure hose with the bypass pipe and valve
- Exhaust manifold heat shield
- Crankcase breather pipe
- Turbocharger intake pipe
- Turbocharger outlet hose to the charge air cooler
- Radiator fan and clamp off the coolant hoses to the turbocharger
- Coolant pipes
- Front pipe from the turbocharger
- Oxygen (O_2S) sensor electrical connector
- Front and rear exhaust systems
- Front pipe from the rear exhaust manifold
- Turbocharger oil return pipe
- Turbocharger mounting bracket

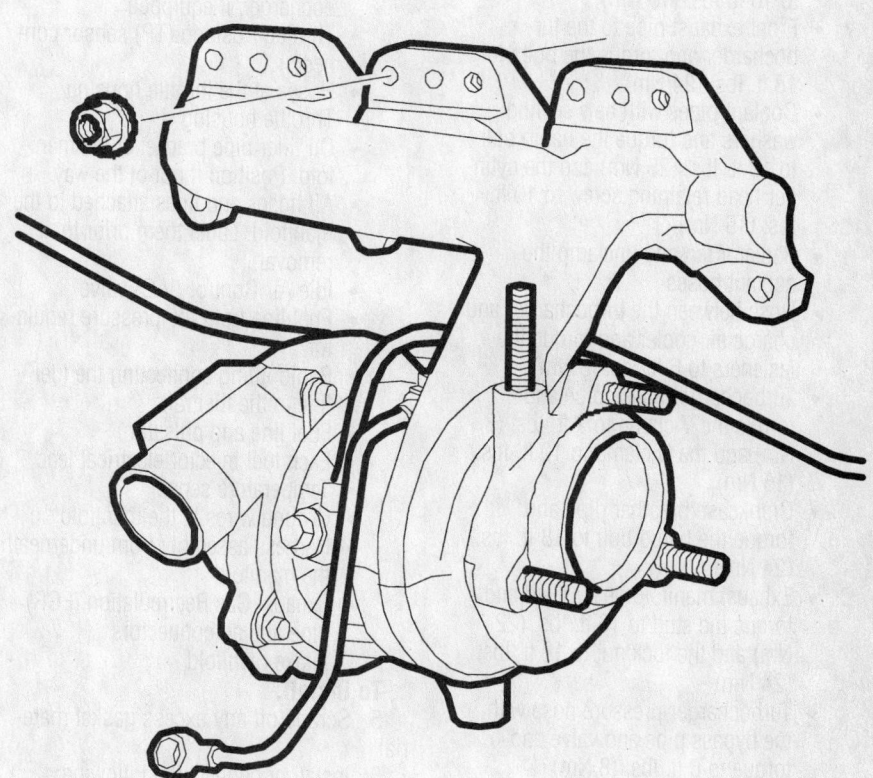

9347UG10

Remove the turbocharger and exhaust manifold as an assembly

- Front exhaust manifold and tur-bocharger as an assembly

To install:

4. Fill the turbo inlet with oil. Rotate the compressor wheel by hand several times to assure the bearings are well lubricated.

5. Install or connect the following:
- Turbocharger and exhaust manifold with a new gasket and torque the bolts to 15 ft. lbs. (20 Nm)
- Front pipe to the exhaust manifold with a new gasket and torque the bolts to 30 ft. lbs. (40 Nm)
- Pipe joint between the front and rear exhaust systems and torque the bolts to 16 ft. lbs. (22 Nm)
- O2S electrical connector
- Oil pipe to the oil filter adapter with new gaskets and sealing washers and torque the bolts to 15 ft. lbs. (20 Nm)
- Turbocharger mounting bracket and torque the bolts to 18 ft. lbs. (24 Nm)
- Oil return pipe and torque the bolts to 10 ft. lbs. (15 Nm)
- Front exhaust pipe to the tur-bocharger and torque the bolts to 18 ft. lbs. (24 Nm)
- Coolant pipes with new sealing washers and torque the banjo bolt to 19 ft. lbs. (25 Nm) and the cylinder head retaining screw to 10 ft. lbs. (15 Nm)
- Radiator fan and unclamp the coolant hoses
- Hose between the turbocharger and charge air cooler and torque the fasteners to 6 ft. lbs. (8 Nm)
- Turbocharger intake pipe and torque the V clamp to 2 ft. lbs. (3 Nm) and the U clamp to 10 ft. lbs. (15 Nm)
- Crankcase breather pipe and torque the banjo bolt to 18 ft. lbs. (24 Nm)
- Exhaust manifold heat shield and torque the stud to 16 ft. lbs. (22 Nm) and the lock nut to 18 ft. lbs. (24 Nm)
- Turbocharger pressure hose with the bypass pipe and valve and torque to 6 ft. lbs. (8 Nm)
- Negative battery cable
- Engine upper cover

6. Fill the cooling system to the proper level.

7. Start the vehicle, check for leaks and repair if necessary.

Intake Manifold

REMOVAL & INSTALLATION

2.0L and 2.3L Engines

❊❊ CAUTION

The fuel injection system remains under pressure, even after the engine has been turned OFF. The fuel system pressure must be relieved before disconnecting any fuel lines. Failure to do so may result in fire or personal injury.

1. Before servicing the vehicle, refer to the precautions in the beginning of this section.
2. Drain the cooling system.
3. Properly relieve the fuel system pressure.
4. Remove or disconnect the following:

- Negative battery cable
- Rubber elbow running between the throttle housing and the tur-bocharger, if equipped
- Throttle Position (TP) sensor connector
- Hoses at the throttle housing
- Throttle housing
- Oil filler pipe bracket at the manifold. Position it out of the way.
- All hoses and lines attached to the manifold. Label them prior to removal.
- Idle Air Control (IAC) valve
- Fuel line from the pressure regulator
- Banjo fitting connecting the fuel line to the fuel rail
- Fuel line and pulsator
- Each fuel injector electrical lead
- Temperature sensor
- Ground wires at the manifold
- Harness assembly from underneath the manifold
- Exhaust Gas Recirculation (EGR) pipe and all connectors
- Intake manifold

To install:

5. Scrape off any excess gasket material

6. Install or connect the following;
- Intake manifold with a new gasket and torque the bolts in a crisscross pattern to 16 ft. lbs. (22 Nm)
- Wire harness
- EGR pipe
- Ground wires

- Temperature sensor
- Injector leads
- Fuel line to the pressure regulator
- Fuel line/pulsator to the fuel rail
- Oil filler pipe bracket to the manifold
- IAC valve
- Throttle housing
- TP sensor connector
- Rubber elbow between the tur-bocharger and the intake manifold, if equipped
- Negative battery cable

7. Fill the cooling system.

8. Start the vehicle, check for leaks and repair if necessary.

3.0L Engine

1. Before servicing the vehicle, refer to the precautions in the beginning of this section.

2. Remove or disconnect the following:

- Negative battery cable
- Engine covers
- Cruise control cable
- Throttle cable
- Control rod
- Intake tube from the Mass Air Flow (MAF) sensor assembly
- Vacuum hoses and electrical connectors from MAF assembly
- MAF assembly
- Hoses and wiring from the upper intake manifold. Label prior to removal.
- Upper intake manifold
- Idle Air Control (IAC) valve
- Fuel pressure regulator hose
- Harness from under the throttle plate housing
- Injector electrical harness
- Camshaft Position (CMP) sensor connector
- Fuel line
- Lower intake manifold

To install:

3. Install or connect the following:
- Lower intake manifold with new gaskets and torque the bolts to 15 ft. lbs. (20 Nm)
- Fuel lines
- Injector wiring
- CMP sensor
- Upper intake manifold with new gaskets
- Throttle plate housing and torque the bolts to 15 ft. lbs. (20 Nm)
- Wiring and hoses
- IAC valve
- Hose to the fuel pressure regulator

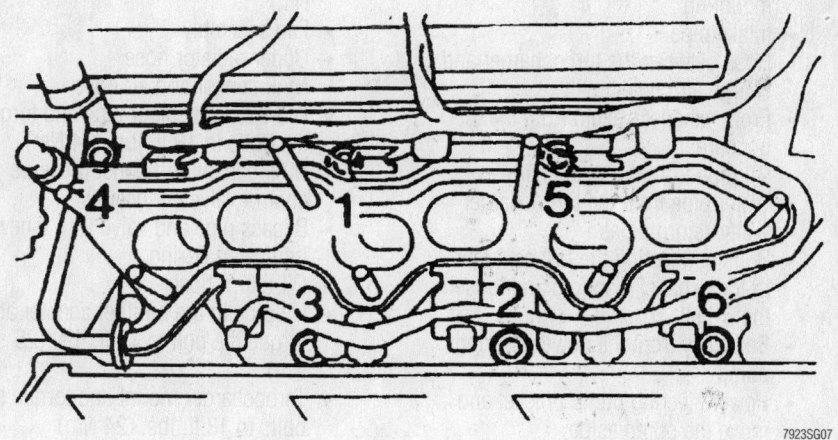

Intake manifold bolt tightening sequence—3.0L engine

- MAF assembly and hoses
- Throttle control rod
- Cruise control cable
- Engine covers
- Negative battery cable

Exhaust Manifold

REMOVAL & INSTALLATION

2.0L Engine

1. Before servicing the vehicle, refer to the precautions in the beginning of this section.
2. Drain the cooling system.
3. Remove or disconnect the following:

- Negative battery cable
- Turbocharger
- Drive belt
- Steering servo pump and bracket
- Exhaust manifold and discard the gasket

4. Clean all mating surfaces of any residual gasket material.

To install:

5. Install or connect the following:

- Exhaust manifold with a new gasket and torque the bolts to 19 ft. lbs. (25 Nm)
- Steering servo pump and bracket
- Drive belt
- Turbocharger
- Negative battery cable

6. Fill the cooling system.
7. Start the vehicle, check for leaks and repair if necessary.

2.3L Engine

900 MODELS

1. Before servicing the vehicle, refer to the precautions in the beginning of this section.
2. Remove or disconnect the following:

- Negative battery cable
- Oxygen (O₂S) sensor connector
- Front pipe from exhaust manifold

3. Using a ⅜ inch ratchet extension, relieve the drive belt tensioner and remove the belt.

- Power steering pump pulley
- Power steering pump bracket

4. Move the pump to one side without disconnecting the hoses.
5. Remove the middle part of the exhaust manifold first, then remove the outer tubes.

To install:

6. Fit the outer tubes first, then mount middle part of exhaust manifold. Tighten the nuts to 18 ft. lbs. (25 Nm).
7. Install or connect the following:

- Power steering pump bracket
- Power steering pump and pulley
- Drive belt
- Front pipe to the exhaust manifold
- O₂S electrical connector
- Negative battery cable

8. Start the vehicle, check for leaks and repair if necessary.

9000 MODELS

1. Before servicing the vehicle, refer to the precautions in the beginning of this section.
2. Drain the cooling system.
3. Remove or disconnect the following:

- Negative battery cable
- Center air deflector
- Top pipe coupling on the air-cooled oil cooler
- Clips securing the top pipe to the radiator
- Solenoid valve from its mounting on the radiator
- Radiator fan
- Oxygen (O₂S) sensor cable clamps
- Exhaust pipe from the turbocharger. Bend it down slightly.
- Air mass meter
- Air cleaner cover
- Turbocharger pressure pipe from the compressor
- Oil pipe to the turbocharger unit
- Clip securing the oil pipe to the cylinder head
- Oil pipe banjo coupling from the block
- Clip on the intake manifold
- Coolant pipes from the turbocharger unit
- Steady bar bracket between the engine block and the turbocharger compressor
- Oil return lines. Cap the opening to prevent washers or nuts from the exhaust manifold dropping inside during removal.
- Turbocharger unit from the exhaust manifold
- A/C drive belt. Insert a sheet of metal to protect the oil cooler.
- A/C compressor mounting bolts. Lift the air conditioning compressor towards the expansion tank.
- Exhaust manifold complete with gasket

To install:

4. Clean all mating surfaces
5. Install or connect the following:

- Exhaust manifold gasket and manifold and torque the nuts to 19 ft. lbs. (25 Nm)
- A/C compressor and belt, remove the locking tool and check for proper belt tension
- Turbocharger to the exhaust manifold and torque the new nuts to 16 ft. lbs. (22 Nm)

❋❋ WARNING

Fill turbocharger interchamber passage with engine oil. It is very important that there is oil in the turbocharger when the engine is started to avoid damage to the unit.

- Clip holding the turbocharger oil supply pipe to the intake manifold
- Banjo coupling to the engine block. Be sure the copper washers are in good condition.
- Pipe to the turbocharger unit
- Return oil pipe and the steady bar bracket between the turbocharger unit and the crankcase
- Rubber hangers for the front exhaust hanger
- Exhaust pipe to the turbocharger compressor with new locknuts. Lubricate the 3 studs on the turbocharger with Molykote 1000®. Torque the nuts to 19 ft. lbs. (25 Nm).
- Coolant pipes to the turbocharger unit
- Turbocharger pressure pipe to the turbocharger compressor
- Air mass meter
- Air cleaner assembly
- Radiator fan and solenoid valve
- Return hose to the solenoid valve
- A/C compressor
- Oil pipe to the oil cooler
- Center air deflector
- Negative battery cable

6. Properly fill the cooling system and check the oil level and quality.

7. Start the engine, check for leaks and repair if necessary.

9-5 MODELS—2.3L ENGINE

1. Before servicing the vehicle, refer to the precautions in the beginning of this section.

2. Drain the cooling system.

3. Remove or disconnect the following:
- Negative battery cable
- Lower front cover
- Bolt from the turbocharger mount on the engine
- Turbocharger return hose
- Oil pipe from the turbocharger to the oil filter adapter
- Bypass pipe and valve
- Exhaust manifold heat shield
- Radiator fan electrical connectors
- Upper radiator hose
- Radiator fan
- Boost pressure control valve connector
- Mass Air Flow (MAF) sensor connector
- Crankcase breather pipe
- MAF sensor and air hose
- Turbocharger to diaphragm hoses
- Quick release coupling on the vent hose
- Intake manifold V-clamp at the turbocharger

- Intake pipe from the steering pump mounting
- Intake pipe
- Hose between the turbocharger and charge air cooler
- Front exhaust system from the turbocharger
- Oil pipe from the turbocharger
- Water pipe from the turbocharger
- Water return pipes
- Bolts connecting the turbocharger to the exhaust manifold and loosen the V-belt
- Belt away from the power steering pump
- Power steering pump bracket and move the pump aside
- Exhaust manifold while saving the spacer sleeves and washers. Discard the gasket.

To install:
4. Install or connect the following:
- Exhaust manifold with a new gasket and torque the bolts to 19 ft. lbs. (25 Nm)
- Power steering pump bracket. Torque the upper left mounting bolt to 18 ft. lbs. (24 Nm) first then torque the remaining bolts to the same specification.
- V-belt and tighten it with the tensioner

➡ **Fill the turbo inlet with oil. Rotate the compressor wheel by hand several times to assure the bearings are well lubricated.**

- Turbocharger to the exhaust manifold and torque the lock nuts to 18 ft. lbs. (24 Nm) and the studs to 16 ft. lbs. (22 Nm)
- Front exhaust pipe to the turbocharger and torque the bolts to 18 ft. lbs. (24 Nm)

Water return pipe with new copper washers and torque the bolts to 19 ft. lbs. (25 Nm)

- Water pipe. Torque the turbo bolt to 19 ft. lbs. (25 Nm) and the water pump bolt to 15 ft. lbs. (20 Nm).
- Turbocharger intake pipe with a new lubricated O-ring to the power steering pump bracket. Torque the bolt and adjusting screw to 18 ft. lbs. (24 Nm).
- Hoses to the diaphragm and turbocharger
- Charge air cooler intake pipe to the turbocharger and torque the bolts to 6 ft. lbs. (8 Nm)
- Exhaust manifold heat shield and

torque the bolts to 15 ft. lbs. (20 Nm)
- Fan assembly
- Upper radiator hose
- MAF sensor and air hose
- Crankcase breather pipe and torque the bolts to 18 ft. lbs. (24 Nm)
- MAF sensor and boost pressure control valve connectors
- Bypass pipe and valve with a new lubricated O-ring
- Oil return hose
- Oil pipe to the oil filter adapter and torque the bolt to 19 ft. lbs. (25 Nm)
- Turbocharger mount and torque the bolts to 18 ft. lbs. (24 Nm)
- Lower front cover
- Negative battery cable

5. Fill the cooling system to the proper level.

6. Start the vehicle, check for leaks and repair if necessary.

9-5 MODELS—3.0L ENGINE—FRONT

1. Before servicing the vehicle, refer to the precautions in the beginning of this section.

2. Remove or disconnect the following:
- Negative battery cable
- Upper engine cover
- Turbocharger bypass pipe and valve
- Temperature/pressure sensor connector
- Turbocharger delivery pipe from the throttle body
- Turbocharger delivery pipe from the charge air cooler
- Exhaust manifold heat shield
- Secondary Air Injection (SAI) pipe and valve
- Crankcase ventilation pipe
- Turbocharger intake pipe
- Turbocharger outlet hose
- Fan motor wiring
- Radiator fan
- Turbocharger water pipes
- Dipstick tube
- Front pipe from the turbocharger
- Front exhaust pipe
- Turbocharger oil return pipe
- Turbocharger mounting bracket
- Oil pipe from the oil filter adapter
- Front exhaust manifold with the turbocharger and discard the gasket

3. Clean all mating surfaces of any residual gasket material.

To install:
4. Install or connect the following:
- Front exhaust manifold assembly with a new gasket and torque the bolts to 15 ft. lbs. (20 Nm)

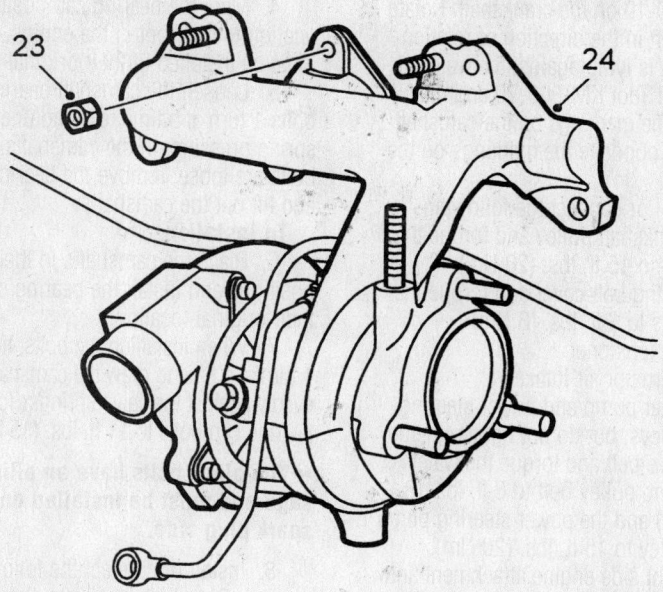

Remove the exhaust manifold and turbocharger as an assembly—3.0L engine

- Front exhaust pipe with a new gasket and torque the bolts to 30 ft. lbs. (40 Nm)
- Oil pipe to the oil filter adapter and torque the fastener to 15 ft. lbs. (20 Nm)
- Turbocharger mounting bracket and torque the bolts to 18 ft. lbs. (25 Nm)
- Turbocharger oil return pipe and torque the bolt to 10 ft. lbs. (15 Nm)
- Front exhaust pipe to the turbocharger and torque the bolts to 18 ft. lbs. (25 Nm)
- Dipstick tube
- Turbocharger water return pipe and torque the banjo bolt to 18 ft. lbs. (25 Nm) and the retaining screw to 10 ft. lbs. (15 Nm)
- Radiator and torque the bolts to 6 ft. lbs. (8 Nm)
- Radiator fan wiring
- Turbocharger-to-charge air cooler hose and torque to 6 ft. lbs. (8 Nm)
- Turbocharger intake pipe and torque to 10 ft. lbs. (15 Nm)
- Crankcase ventilation pipe
- SAI pipe and valve
- Turbocharger delivery pipe to the throttle body and charge air cooler
- Temperature/pressure sensor connector
- Exhaust manifold heat shield and torque the bolts to 18 ft. lbs. (25 Nm)

- Turbocharger by pass pipe and valve and torque to 6 ft. lbs. (8 Nm)
- Upper engine cover
- Negative battery cable

5. Start the engine, check for leaks and repair if necessary.

9-5 MODELS—REAR

1. Before servicing the vehicle, refer to the precautions in the beginning of this section.

2. Remove or disconnect the following:
- Negative battery cable
- Upper engine cover
- Air inlet to the alternator
- Secondary Air Injection (SAI) pipe and valve
- Outer nuts from the rear exhaust manifold
- Front exhaust pipe
- Lower nuts from the rear exhaust manifold
- Upper nuts from the rear exhaust manifold
- Rear exhaust manifold and discard the gasket

3. Clean all mating surfaces of any residual gasket material.

To install:

4. Install or connect the following:
- Rear exhaust manifold with a new gasket and torque all the bolts to 15 ft. lbs. (20 Nm)
- Front exhaust pipe and torque the turbocharger locknuts to 18 ft. lbs.

(25 Nm) and rear exhaust manifold bolts to 30 ft. lbs. (40 Nm)
- SAI pipe and valve
- Air intake to the alternator
- Upper engine cover
- Negative battery cable

5. Start the vehicle, check for leaks and repair if necessary.

Front Crankshaft Seal

REMOVAL & INSTALLATION

2.0L Engine

1. Before servicing the vehicle, refer to the precautions in the beginning of this section.

2. Remove or disconnect the following:
- Negative battery cable
- Right front wheel and inner cover
- Drive belt tension
- Protective plate and install Flywheel Locking Tool, 83-94-868, on the ring gear
- Crankshaft pulley
- Front crankshaft seal

To install:

3. Install or connect the following:
- New crankshaft seal using special Tool 83-94-876
- Crankshaft pulley and torque to 130 ft. lbs. (175 Nm)
- Drive belt tension
- Inner cover and right front wheel
- Negative battery cable

2.3L and 3.0L Engines

1. Before servicing the vehicle, refer to the precautions in the beginning of this section.

2. Remove or disconnect the following:
- Negative battery cable
- Lower spoiler sections and insert a wedge between the oil pump and sub frame
- Mass Air Flow (MAF) sensor and hose
- Power steering hose from the holder
- Right front engine attachment and mount
- Water pump and power steering pump pulley bolts and release the drive belt tension
- Water pump and power steering pump pulleys
- Belt tensioner

For Tune-up, Capacities and Firing orders, see Section 1 of this manual

- Alternator air intake
- Timing cover
- Crankshaft pulley

3. Zero the engine by installing special Tool KM-800-20. Fit Locking Tool KM-800-2 for the camshaft sprockets and Tool KM-800-10 for the crankshaft.

4. Undo the tensioning pulley and adjuster rollers by using special Tool 83-94-983 as a counterstay on the rollers.

5. Mark the direction of the belt rotation. Remove special Tool KM-800-10 from the crankshaft. Remove the lower adjuster roller and washer.

6. Remove the timing belt.

7. Rotate the crankshaft to 60 degree Before Top Dead Center (BTDC).

8. Install special Tool 83-95-063 and remove the crankshaft timing belt sprocket and spacer ring.

9. Remove the crankshaft seal.

To install:

10. Install the crankshaft seal.

11. Install the spacer ring and place special Tool 93-21-795 on the back of the crankshaft timing belt sprocket and install the sprocket. Torque the sprocket to 185 ft. lbs. (250 Nm) plus 45 degrees.

12. Install Locking Tool KM-800-1 on No. 1 and No. 2 camshaft sprockets Tool KM-800-2 on No. 3 and No. 4 camshaft sprockets.

13. Rotate the crankshaft to just before the "0" degree mark and install tool KM-800-10. Slowly rotate the crankshaft in the proper direction until the arm is lying against the water pump flange. Remove the special tool.

14. Install the camshaft belt and install Tool KM-800-30 to hold the belt in place. Loosely install the lower tensioning pulley. Adjust the pulley counterclockwise by hand to prevent the belt from jumping. Make certain that the markings on the belt align with the markings on the camshaft sprocket and crankshaft. Remove Tool KM-800-30.

15. Tighten the center bolts of the adjuster rollers slightly. Adjust the lower adjuster roller counterclockwise until a tension of 275–300 (NM) is reached. Tighten the adjuster roller bolt to 30 ft. lbs. (40 Nm).

16. Make certain that the nut on the upper adjuster roller rotates freely.

17. Remove the locking tools from the camshaft sprockets and adjust the upper adjuster roller counterclockwise with Tool KM-83-94-983 until the markings on the camshaft sprocket align with the markings on Tool KM-800-20.

18. Remove all locking tools.

19. Rotate the engine 2 full revolutions until just before the "0" mark and install

Tool KM-800-10 on the crankshaft. Rotate the crankshaft in the direction of rotation until the arm is lying against the water pump. Install Tool KM-800-20 and make certain that the markings on the camshaft sprocket are opposite the markings on the tool.

20. Install or connect the following:
- Crankshaft pulley and torque the bolt to 15 ft. lbs. (20 Nm)
- Timing belt cover and torque the bolts to 6 ft. lbs. (8 Nm)
- Belt tensioner
- Alternator air intake
- Water pump and power steering pulleys, but do not tighten them
- Drive belt and torque the water pump pulley bolt to 6 ft. lbs. (8 Nm) and the power steering pump pulley to 15 ft. lbs. (20 Nm).
- Right side engine attachment and mount and torque to 47 ft. lbs. (63 Nm)
- MAF sensor and hose
- Lower spoiler sections
- Negative battery cable

Camshaft

REMOVAL & INSTALLATION

2.0L and 2.3L Engines

1. Before servicing the vehicle, refer to the precautions in the beginning of this section.

2. Remove or disconnect the following:
- Negative battery cable
- Ignition module cartridge
- Valve cover
- Crankcase ventilation hose

3. Position the crankshaft at Top Dead Center (TDC). The **0** mark on the flywheel should align with the timing mark on the bell housing end plate. The camshafts should be lined up with their respective timing marks.
- Oil pipes
- Center bolts securing the camshaft sprockets

✶✶ WARNING

Use a proper holding tool to hold the camshafts from rotating. Always keep the camshafts in their correct basic setting. If the setting of the crankshaft or camshafts is altered at this stage the valves can be damaged.

- Timing chain tensioner
- Camshaft sprockets

4. Mark the bearing cap positions and relation to the front of the engine. The caps must be installed in their original location.

5. Loosen the camshaft bearing cap bolts 1 turn at a time to avoid uneven valve spring pressure on the camshafts. When all bolts are loose, remove the bearing caps and lift out the camshafts.

To install:

6. Place the camshafts in their proper positions and install the bearing caps in their original location.

7. When installing the bolts, tighten them 1 turn at a time to draw the camshaft down evenly against the valve springs. Torque the bearing cap bolts to 11 ft. lbs. (15 Nm).

➡**The black bolts have an oiling passage and must be installed on the spark plug side.**

8. Install or connect the following:
- Camshaft sprockets. Exhaust cam is installed first. Hand-tighten the center bolts securing the camshaft sprockets.
- Timing chain tensioner with the piston under tension. Be sure the copper gasket is in good condition and that the sealing surface is clean and free from burrs. Torque the tensioner to 47 ft. lbs. (63 Nm).

9. Trigger the chain tensioner by pressing the pivoting chain guide against it. Thereafter, press the pivoting guide against the chain to give the chain its basic tension. Check that the chain tensioner maintains tension on the chain when the pressure on the chain guide is released and that the basic setting stop for the tensioner holds the chain guide tight against the chain. A limited amount of play will be present until the hydraulic pressure takes over once the engine is running.

10. Depress the pivoting guide to check that the tensioner is working. Rotate the crankshaft 2 complete turns clockwise, viewed from the transmission end. Be sure the crankshaft and camshaft timing marks still align properly.

11. Hold the camshafts in their proper position and torque the cam sprocket bolts to 49 ft. lbs. (67 Nm).

12. Install or connect the following:
- Oil pipes
- Valve cover and torque the bolts to 11 ft. lbs. (14 Nm)
- Crankcase ventilation hose
- Ignition module cartridge
- Negative battery cable

13. Start the engine, check for leaks and repair if necessary.

9000 MODELS

✳✳ WARNING

To avoid damage to the valves, DO NOT rotate the camshafts. The crankshaft may only be turned between 0 Top Dead Center (TDC) and 60° Before Top Dead Center (BTDC) when the camshafts are locked in position with the appropriate locking tool.

1. Properly relieve the fuel system pressure.

2. Drain the cooling system.

3. Before servicing the vehicle, refer to the precautions in the beginning of this section.

4. Remove or disconnect the following:
- Negative battery cable
- Front exhaust pipe from the exhaust manifolds
- Bracket for the check valves
- Lower center air deflector
- Top engine covers
- Cruise control cable and throttle cable
- Throttle control rod from the bracket
- Bracket mounting bolt and set the bracket with cables attached to the side
- Clamps on the air intake pipes at the intake manifold and the Mass Air Flow (MAF) meter
- Air intake pipes and raise them slightly
- Vacuum hoses and electrical connections and label them for identification
- Pipes with resonator attached
- Intake plenum bolts
- Idle Air Control (IAC) valve connector
- Fuel pressure regulator hose
- Wiring harness from under the throttle body
- Throttle Position Sensor (TPS)
- Ignition coil
- Traction Control System (TCS) connectors
- Intake plenum and plug the intake runners
- Fuel injector
- Crankshaft Position (CKP) sensor connectors
- Fuel line connections
- Center intake manifold and fuel rails and set them aside
- Spark plug wires

- Ignition coil. Bend the ignition coil aside and disconnect the ignition coil bracket.
- Lifting eye
- Heat shield over the exhaust manifold
- Resonator bracket and secondary air injection pipe from the exhaust manifold
- Power steering reservoir
- Torque arm engine mount connection
- Torque arm
- Power steering line clamp from the torque arm engine mount
- Engine mount
- Hose from the coolant expansion tank
- Upper alternator air intake
- Power steering pump pulley
- Water pump pulley
- Six outer crankshaft pulley bolts

➡ **When removing the crankshaft pulley, remove the 6 outer bolts only, DO NOT remove the center bolt.**

- Drive belt and tensioner
- Power steering pump
- Timing cover
- Crankshaft pulley
- Timing chain, tensioner and camshaft sprockets

5. Rotate the crankshaft back to 60° BTDC to prevent damage to the valves.

6. Remove or disconnect the following:

- Valve cover. Be sure the O-rings stay in position and do not fall into the engine.
- Bearing caps. Note the position markings on the bearing caps and loosen the bearing cap bolts in stages of ½ to 1 turn at a time.

➡ **The bearing caps located where valve tappets are compressed should be removed last.**

- Camshafts

To install:

7. Be sure the crankshaft is still positioned at 60° BTDC.

8. Be sure all gasket contact surfaces are clean.

➡ **The camshaft bearing caps are marked L1-8 for the Front or Left bank and R1-8 for the Rear or Right bank and the camshaft bearing seats in each head are numbered 1–8.**

9. Thoroughly lubricate the camshafts

and install the camshafts with new front camshaft gaskets. Be sure the locating pins are properly positioned. Install the bearing caps in their proper location and position. Tighten the bearing cap bolts in sequence ½ to 1 turn at a time, to a torque of 72 inch lbs. (8 Nm).

10. Install or connect the following:
- Timing covers

11. Check to be sure that the camshaft locating pins are in the proper position. Check the locating pins, if they are hollow, replace them with solid pins.

➡ **The locating pins of both camshafts 1 & 2 should be point towards the inboard bolts of the camshaft bearing caps. The locating pin for the intake camshaft No, 3 should be pointing downwards, in line with the inboard bolt of the bearing caps. The locating pin for the exhaust No. 4 camshaft should be pointing upwards, in line with the edge of the camshaft sensor.**

- Camshaft sprockets
- Tensioner
- Timing chain
- Crankcase ventilation housing

12. Be sure the valve cover O-rings are still in position and clean. Lubricate the O-rings with soapy water, apply (81–52–381) sealer at the corners of the large end bearing caps.

- Valve cover
- Timing cover
- Crankshaft pulley and torque the bolts to 15 ft. lbs. (20 Nm)
- Water pump pulley
- Power steering pump pulley
- Drive belt tensioner, and drive belt
- Upper alternator air intake
- Torque arm engine mount
- Power steering line clamp
- Torque arm and bolt the connection to the torque arm engine mount
- Upper hose to the coolant expansion tank
- Power steering reservoir
- Lifting eye
- Secondary air injection pipe to the exhaust manifold
- Exhaust manifold heat shield
- Resonator bracket
- Ignition coil bracket
- Ignition coil
- Spark plugs and torque them to 19 ft. lbs. (25 Nm)
- Spark plug wires
- Center intake manifold with fuel rail

- Center intake manifold and torque the bolts to 15 ft. lbs. (20 Nm)
- Fuel lines
- CKP sensor
- Fuel injector connectors
- Intake plenum into position and connect the TCS throttle body and torque to 15 ft. lbs. (20 Nm)
- TPS connector
- Ignition coil connector
- TCS connector
- Wiring harness
- Fuel pressure regulator hose
- IAC valve connector
- Labeled electrical and vacuum connections on the intake manifold
- Air intake pipes, with resonator attached
- Vacuum hoses and electrical connection to air intake
- Air intake pipes to the mass airflow meter and intake manifold
- Throttle control bracket with cables
- Throttle control rod
- Throttle cable and cruise control cable. Adjust the kick-down cable and the throttle cable.
- Upper engine covers
- Lower center air deflector
- Bracket for the check valves
- Front exhaust pipe to the exhaust manifold
- Negative battery cable

13. Fill the cooling system.

14. Start the vehicle, check for leaks and repair if necessary.

9-5 MODELS

3.0L ENGINE

❋❋ WARNING

To avoid damage to the valves, DO NOT rotate the camshafts. The crankshaft may only be turned between 0 Top Dead Center (TDC) and 60° Before Top Dead Center (BTDC) when the camshafts are locked in position with the appropriate locking tool.

1. Properly relieve the fuel system pressure.

2. Drain the cooling system.

3. Before servicing the vehicle, refer to the precautions in the beginning of this section.

4. Remove or disconnect the following:
- Negative battery cable
- Lower spoiler sections and install wedges between the oil sump and the subframe and the gearbox and subframe

- Mass Air Flow (MAF) sensor and hose
- Power steering hose from the holder
- Right side engine mount and bracket
- Water pump and power steering pump pulley bolts
- Tension from the V belt
- Water pump and power steering pump pulleys
- Belt tensioner
- Alternator air intake
- Timing cover
- Crankshaft pulley
- Throttle cable
- Upper intake manifold vacuum hoses
- Pressure sensor connector
- Water hoses from the throttle body
- Turbocharger pressure hose
- Charge air bypass pipe
- Vent valve hose from the throttle body
- Ignition discharge module
- Fuel injector connectors
- Fuel lines
- Upper and middle intake manifolds
- Lower intake manifold and make certain that the markings on the camshaft sprockets and the timing cover are aligned
- Tensioning pulley and adjuster rollers
- Lower adjuster roller after marking the rotation of the belt
- Bracket and tensioning roller and upper adjuster roller. Rotate the crankshaft to 60 degrees BTDC.
- Camshaft sprockets
- Water pump
- Inner timing cover
- Camshaft cover
- Camshafts
- Tappets with special Tool 83–91–401

To install:

5. Install or connect the following:
- Valve tappets and rotate the crankshaft to 60 degrees BTDC
- Camshafts into position and fit the stuffing boxes and bearing caps

6. The locating pin on the camshafts should point toward the inside bolts for the camshaft bearing caps.

7. The locating pin for the intake camshaft should be in line (downward) with the inside bolts for the bearing caps.

8. The guide pin for the exhaust camshaft should point straight up from the plane of the cylinder head. Torque the bolts, in stages, to 6 ft. lbs. (8 Nm).

➡**The bearings caps marked L1—L8 are for the cylinder head for 1–3–5 cylinders and the caps marked R1—R8 are for cylinders 2–4–6. The camshaft bearing seats in each cylinder head are numbered 1–8.**

- Camshaft cover and torque the bolts to 6 ft. lbs. (8 Nm)
- Ignition discharge module
- Inner timing cover
- Camshaft sprockets properly in relation to the locating pins and torque the bolts to 37 ft. lbs. (50 Nm) plus 60 degrees.
- Water pump and torque the bolts to 19 ft. lbs. (25 Nm)
- Bracket with the tensioning roller and the upper adjusting roller
- Tool KM—800–1 on camshafts no. 1–2 sprockets and Tool KM—800–2 on no. 3–4 camshaft sprockets

9. Rotate the crankshaft just before the 0 degree make and install Tool KM—800–10. Rotate the crankshaft in the direction of the engine until the arm is lying against the water pump flane. Remove the tool.

- Camshaft belt according to the markings on the belt
- Loosely install the lower tensioning pulley and make certain that the washer is in place. Adjust the tensioning pulley by hand so that the belt does not jump.

➡**Make sure that the markings on the camshaft belt and correspond with the markings on the camshaft sprocket and the crankshaft.**

10. Install a Crankshaft locking tool. Install Tool 83-93-985 with a cut piece from an old timing belt, to measure belt tension.

11. Snug the center bolts of the adjusting rollers. Turn the lower adjusting roller counterclockwise, until a belt tension of 275–300 Nm is reached. Tighten adjusting roller center bolts to 30 ft. lbs. (40 Nm).

➡**This adjustment of the timing belt is just preparation and should not be used as a final check.**

12. Adjust the tensioner pulley until the marks are aligned. Tighten the tensioner pulley to 15 ft. lbs. (20 Nm).

13. Remove camshaft locking tool from camshaft sprockets 1 & 2. Adjust the upper adjusting roller until sprocket No. 2 moves 1–2mm clockwise. Tighten the upper adjusting roller to 30 ft. lbs. (40 Nm) and remove the upper locking tool.

14. Rotate the engine 2 complete revolutions to just before **0** TDC and install the locking tool on the crankshaft. Carefully turn the crankshaft until the arm of the locking tool is against the water pump flange and tighten the locking tool. Set a locking tool into position on the front of the camshaft sprockets. Be sure that the timing marks on the camshaft sprockets are aligned with the marks on the tool and that the edge of the timing belt is flush with the edge of the camshaft sprockets.

➡**Also, check that the alignment marks on the tensioner pulley are still aligned.**

15. Install or connect the following:
   - Lower intake manifold with new gaskets and torque the bolts to 15 ft. lbs. (20 Nm)
   - Middle intake manifold and torque the bolts to 15 ft. lbs. (20 Nm)
   - Upper intake manifold with a new gasket and torque the bolts to 15 ft. lbs. (20 Nm)
   - Fuel lines
   - Ignition discharge module connectors
   - Fuel injector connectors
   - Water hoses and vacuum hose to the throttle body
   - Turbocharger pressure hose and charge air bypass pipe
   - Throttle body connector and vent valve
   - Upper intake manifold vacuum hoses
   - Pressure sensor electrical connector
   - Throttle cable
   - Crankshaft pulley and torque the bolt to 15 ft. lbs. (20 Nm)
   - Timing cover and torque the bolts to 6 ft. lbs. (8 Nm)
   - Belt tensioner and alternator air intake
   - Water pump and power steering pump pulleys but do not tighten them
   - V belt and torque the water pump bolts to 6 ft. lbs. (8 Nm) and the power steering pump bolts to 15 ft. lbs. (20 Nm)
   - Right side engine mount and bracket and torque the bolts to 47 ft. lbs. (63 Nm)
   - Power steering hose in the holder
   - MAF sensor and hose and remove the wedges

   - Negative battery cable
   - Lower spoiler section
16. Fill the cooling system to the proper level.
17. Start the vehicle, check for leaks and repair if necessary.

Valve Lash

ADJUSTMENT

1. Before servicing the vehicle, refer to the precautions in the beginning of this section.

The hydraulic cam followers used in Saab engines do not require adjusting. The cam followers keep the valve clearance within 18.75–20.8mm. However, if the cam followers are making excessive noise or are diagnosed to be defective, perform the following procedure:

2. Before servicing the vehicle, refer to the precautions in the beginning of this section.
3. Disconnect the negative battery cable.
4. If a cam follower is noisy, it can be found by removing the valve cover and, using a screwdriver, gently pushing down on each cam follower until the defective follower(s) is found by exhibiting a spongy feeling.
5. Replace the defective cam follower(s); first removing the camshaft(s).
6. Reinstall the camshaft(s) and the valve cover.

Starter Motor

REMOVAL & INSTALLATION

900 and 9-3 Models

1. Before servicing the vehicle, refer to the precautions in the beginning of this section.
2. Remove or disconnect the following:
   - Negative battery cable
   - Upper mounting bolt
   - Starter motor electrical connections
   - Lower mounting bolt
   - Starter motor

To install:
3. Install or connect the following:
   - Starter motor and torque the bolts to 3 ft. lbs. (4 Nm)
   - Starter electrical connections
   - Upper mounting bolt
   - Negative battery cable

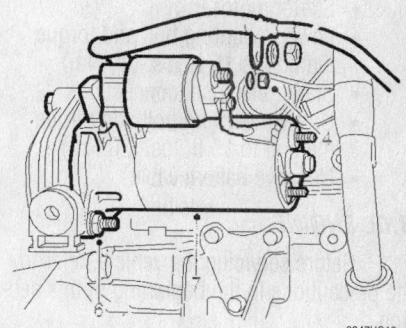

9347UG12

Remove the starter motor

9000 Models

1. Before servicing the vehicle, refer to the precautions in the beginning of this section.
2. Remove or disconnect the following:
   - Negative battery cable
   - Air mass meter-to-throttle housing elbow
   - Top bolt on the intake manifold support bar
   - Top starter bolt. Use an obstruction wrench and ratchet with a flexible extension
   - Starter electrical connections
   - Lower mounting bolt and support bar

➡**The starter comes out through the rear and between the intake manifold and master cylinder.**

To install:
3. Secure starter with top bolt first and torque the bolt to 10 ft. lbs. (15 Nm).
4. Install or connect the following:
   - Bottom bolt and support bar
   - Intake manifold support bar bolt
   - Air mass meter to throttle housing elbow
   - Negative battery cable

9-5

2.3L ENGINE

1. Before servicing the vehicle, refer to the precautions in the beginning of this section.
2. Remove or disconnect the following:
   - Negative battery cable
   - Upper mounting bolt
   - Starter motor electrical connections
   - Lower mounting bolt
   - Starter motor

To install:
3. Install or connect the following:

- Starter motor
- Lower mounting bolt and torque the bolt to 15 ft. lbs. (20 Nm)
- Starter electrical connections
- Upper mounting bolt and torque the bolt to 15 ft. lbs. (20 Nm)
- Negative battery cable

3.0L ENGINE

1. Before servicing the vehicle, refer to the precautions in the beginning of this section.
2. Remove or disconnect the following:
   - Negative battery cable
   - Alternator
   - Electrical connections at the starter
   - Upper retaining bolt. Access is through the right front wheel housing using a ratchet and extension.
   - Lower retaining bolt
   - Starter

To install:

3. Install or connect the following:
   - Starter motor
   - Lower retaining bolt and torque the bolt to 15 ft. lbs. (20 Nm)
   - Upper retaining bolt 13 ft. lbs. (18 Nm)
   - Electrical connections at starter
   - Alternator
   - Negative battery cable

Oil Pan

REMOVAL & INSTALLATION

900 and 9-3 Models

2.0L AND 2.3L ENGINES

1. Before servicing the vehicle, refer to the precautions in the beginning of this section.
2. Drain the engine oil.
3. Remove or disconnect the following:
   - Negative battery cable
   - Dipstick
4. Install Engine Lifting Beam 83–94–850 and slightly raise the engine.
   - Oxygen (O₂S) sensor connector. Unbolt the bracket and set a lifting table under the engine.
   - Front wheels
   - Front spoiler
   - Front exhaust pipe at the exhaust manifold and intermediate pipe
   - Front exhaust pipe from the turbocharger, if equipped. Disconnect the front pipe catalytic converter bracket and remove the front pipe.
   - Lower ball joints from the steering knuckles

- Rear engine mount and raise a lifting table into position
- Subframe
- Oil level sensor
- Sensor harness
- Flywheel inspection cover
- Oil pan

➡**Do not remove the guide sleeve from the block**

To install:

5. Thoroughly clean the flanges on the sump and block using a suitable solvent. Apply an even bead of Loctite® 518 sealant, along the oil pan flange.
6. Install or connect the following:
   - Oil pan and torque the bolts to 16 ft. lbs. (22 Nm)
   - Flywheel inspection cover
   - Oil level sensor harness
   - Oil level sensor
7. Install the sub frame and tighten the bolts as follows:
   a. Front bolts: 85 ft. lbs. (115 Nm).
   b. Center bolts: 140 ft. lbs. (190 Nm).
   c. Rear bolts: 81 ft. lbs. (110 Nm) plus75° additional torque.
   - Lower ball joints to the steering knuckles and torque the nuts to 55 ft. lbs. (75 Nm)
   - Front exhaust pipe catalytic converter bracket

➡**Apply Molykote 1000® to the studs of the exhaust manifold or turbocharger, if equipped).**

   - Front exhaust pipe
   - Catalytic converter bracket
   - Front pipe to the mid-pipe
   - Front spoiler
   - Front wheels
   - Dipstick
   - O₂S sensor
   - Negative battery cable
8. Fill the engine with oil and remove the engine lifting beam.
9. Start the vehicle, check for leaks and repair if necessary.

9000 Models

1. Before servicing the vehicle, refer to the precautions in the beginning of this section.
2. Drain the engine oil.
3. Remove or disconnect the following:
   - Negative battery cable
   - Oil dipstick
   - Right front wheel
   - Inner wheel well
   - Front and rear engine mount hardware

- Oxygen (O₂S) sensor
- Front section of the exhaust pipe
- Tie rod between the wheel well and the subframe and install a shop crane and slightly raise the engine
- Bottom bolt holding the transmission to the oil pan
- Oil level sensor
- Two rubber plugs in the back of the transmission case. Access these by bending down the splash plate.
- Two bolts securing the oil pan to the block under the plugs. Tap the guide sleeve into the block.
- Oil pan by working from back to front
- Guide sleeve from the cylinder block

To install:

4. Thoroughly clean the flanges on the sump and block using a suitable solvent. Apply an even bead of Permatex® Ultra Blue sealant along the oil pan flange.
5. Install or connect the following:
   - Rubber seal for the oil strainer in the groove on the oil pan
   - Oil pan by working from the front edge first, to the back

➡**The longer bolt with the washer should be installed in the middle on the right-hand side.**

   - Oil pan bolts and torque the bolts to 15 ft. lbs. (20 Nm)
   - Two rubber plugs in the back of the transmission
   - Bolt securing the oil pan to the transmission case at the bottom
   - Oil level sensor and align the engine over the mounts and lower it into position
   - Tie rod between the wheel well and the subframe
   - Oil dipstick
   - Front and rear engine mount hardware
   - O₂S sensor
   - Exhaust pipe
   - Inner wheel well
   - Front wheel
   - Negative battery cable
6. Fill the engine with clean oil.
7. Start the vehicle, check for leaks and repair if necessary.

9-5 Models

2.3L ENGINE

1. Before servicing the vehicle, refer to the precautions in the beginning of this section.
2. Drain the engine oil.

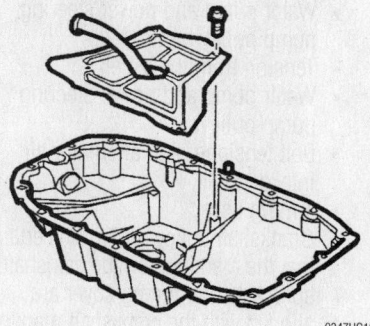

9347UG13

Exploded view of the oil pan– 2.3L engine 9–5 models

3. Remove or disconnect the following:
- Negative battery cable
- Upper engine cover
- Dipstick
- Oxygen (O_2S) sensor cables
- Turbocharger bypass pipe
- Exhaust manifold heat shield
- Lower engine cover
- Front exhaust pipe
- Gearbox cover plate
- Crankcase breather hose from the oil pan
- Oil pan

To install:

4. Transfer the splash guard and pipe to a new oil pan, if replacing the pan.

5. Clean all mating surfaces of gasket material.

6. Apply an even bead of flange sealant to the mating surface on the oil pan.

7. Install or connect the following:
- Oil pan and make certain that the pipe to the oil filter adapter is properly positioned in the oil pan and torque the oil pan bolts evenly to 16 ft. lbs. (22 Nm)
- Crankcase ventilation hose
- Gearbox cover plate
- Front exhaust pipe
- O_2S sensor cables
- Exhaust manifold heat shield
- Turbocharger bypass pipe
- Dipstick
- Upper and lower engine covers
- Negative battery cable

8. Fill the engine with clean oil.

9. Start the vehicle, check for leaks and repair if necessary.

3.0L ENGINE

1. Before servicing the vehicle, refer to the precautions in the beginning of this section.

2. Drain the engine oil.

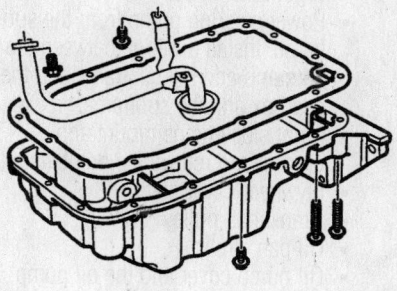

9347UG14

Exploded view of the oil pan, suction pipe and the antislosh baffle—3.0L engine

3. Remove or disconnect the following:
- Negative battery cable
- Oxygen (O_2S) sensor cables
- Lower rear spoiler section
- Front exhaust pipe
- Oil pan

4. If replacing the oil pan, remove the following components:
   a. Oil suction pipe and O-ring.
   b. Anti-slosh baffle.

To install:

5. Clean all mating surfaces of any gasket material.

6. Install or connect the following:
- Anti-slosh baffle to the oil pan and torque the bolts to 6 ft. lbs. (8 Nm)
- Oil suction pipe with new O-rings and torque the bolts to 6 ft. lbs. (8 Nm)
- Oil pan with a new gasket and

torque the bolts to 10 ft. lbs. (15 Nm)
- Front exhaust pipe and torque the front bolts to 18 ft. lbs. (24 Nm) and the exhaust manifold bolts to 30 ft. lbs. (40 Nm)
- O_2S sensor cables
- Lower rear spoiler section
- Negative battery cable

7. Fill the engine with clean oil.

8. Start the vehicle, check for leaks and repair if necessary.

Oil Pump

REMOVAL & INSTALLATION

900 and 9-3 Models

2.0L AND 2.3L ENGINES

1. Before servicing the vehicle, refer to the precautions in the beginning of this section.

2. Remove or disconnect the following:
- Negative battery cable
- Air cleaner and relieve the belt tension
- Right front wheel and inner wheel well cover
- Crankshaft pulley and oil pump circlip
- Oil pump cover and gears

To install:

3. Make certain that the markings on the oil pump ring gear faces outward and check

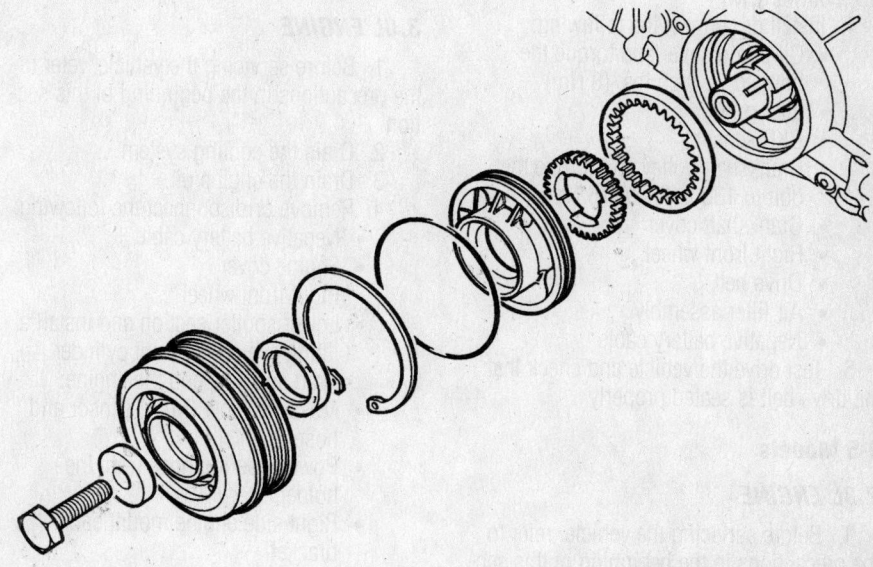

9347UG15

Exploded view of the oil pump assembly 2.0L—2.3L Engines

Timing belt service is covered in Section 3 of this manual

the condition of the O-ring, replace if necessary.

4. Install or connect the following:
- Oil pump gears and cover
- Oil pump circlip with the chamfer facing outward and the opening facing downward
- Crankshaft pulley and torque the bolt to 130 ft. lbs. (175 Nm)
- Right front inner wheel well and front tire
- Air cleaner and tension the drive belt
- Negative battery cable

9000 Models

1. Before servicing the vehicle, refer to the precautions in the beginning of this section.

2. Remove or disconnect the following:
- Negative battery cable
- Air filter assembly
- Drive belt. Using a ⅜ in. ratchet extension release the pressure. Place a short drill bit in tensioner housing to hold in the non-tension position.
- Right front wheel
- Crankshaft access cover
- Crankshaft pulley
- Crankshaft Position (CKP) sensor
- Idler pulley
- Oil pump

To install:

3. Insert oil pump gears making sure that the marking on the oil pump ring gear faces outward.

4. Install or connect the following:
- Oil pump cover and torque the bolts to 72 inch lbs. (8 Nm)
- Idler pulley
- CKP sensor
- Crankshaft pulley and torque the bolt to 130 ft. lbs. (175 Nm)
- Crankshaft cover
- Right front wheel
- Drive belt
- Air filter assembly
- Negative battery cable

5. Test drive the vehicle and check that the drive belt is seated properly.

9-5 Models

2.3L ENGINE

1. Before servicing the vehicle, refer to the precautions in the beginning of this section.

2. Remove or disconnect the following:
- Negative battery cable
- Intake manifold cover

- Right front wheel
- Power steering pump from the subframe. Install a wedge between the oil sump and the subframe and the gearbox and the subframe.
- Right side engine mount and bracket and relieve the belt tension
- Flywheel cover plate
- Crankshaft pulley
- Oil pan circlip
- Oil pump cover and the oil pump

To install:

3. Clean the oil pan circlip.

4. Make certain that the marking on the oil pump ring gear face out and the pump gear with flange are facing in.

5. Lubricate the O-ring with vaseline.

6. Install or connect the following:
- Oil pump gears
- Oil pump cover using the locating arrows
- Oil pan circlip with the chamfer facing outward and the opening downward
- Crankshaft pulley and torque the bolts to 130 ft. lbs. (175 Nm)
- Flywheel cover plate
- Drive belt
- Right side engine mount and bracket and torque the bolts to 37 ft. lbs. (50 Nm)
- Power steering pump pipe to the holder and remove the wedge
- Right front wheel
- Negative battery cable

7. Start the vehicle, check for leaks and repair if necessary.

3.0L ENGINE

1. Before servicing the vehicle, refer to the precautions in the beginning of this section.

2. Drain the cooling system.

3. Drain the engine oil.

4. Remove or disconnect the following:
- Negative battery cable
- Engine cover
- Right front wheel
- Lower spoiler section and install a lifting eye on the front cylinder head and suspend the engine.
- Mass Air Flow (MAF) sensor and hose
- Power steering hose from the holder
- Right side engine mount and bracket
- Front exhaust manifold heat shield
- Crankcase breather pipe from the turbocharger intake pipe
- Turbocharger intake pipe and vent valve

- Water pump and power steering pump pulley bolts
- Tension from the V belt
- Water pump and power steering pump pulleys
- Belt tensioner and alternator air intake
- Timing cover
- Crankshaft pulley and make certain that the markings on the camshaft sprockets and timing cover are aligned with the crankshaft marking.

5. Install locking Tool KM—800–1 and KM—800–2 for the camshaft sprockets and KM—800–10 for the crankshaft.
- Tensioning pulley and adjusting rollers. Mark the direction of rotation for the belt and remove the crankshaft tool.
- Lower adjusting roller and washer. Rotate the crankshaft to 60 degrees Before Top Dead Center (BTDC).
- Camshaft sprockets
- Water pump
- Inner timing cover
- Crankshaft timing belt pulley
- Alternator and oil pressure switch connector
- Front exhaust pipe
- Oil pan and strainer
- A/C compressor and bracket and move them aside
- Oil pump housing
- Cover and remove the 2 pump impellers

To install:

6. Install or connect the following:
- Oil pump impellers and cover. Torque the bolts to 53 inch lbs. (6 Nm). Make certain that the marks on the impellers are properly aligned.
- Oil pump housing with a new gasket and torque the bolts to 53 inch lbs. (6 Nm)

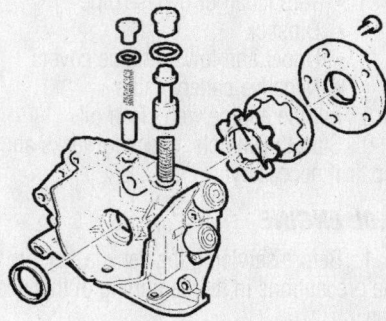

9347UG16

Exploded view of the 9-5— 3.0L oil pump assembly

- A/C compressor and bracket
- Right front wheel
- Oil strainer and oil pan and torque the oil strainer bolts to 6 ft. lbs. (8 Nm) and the oil pan bolts to 10 ft. lbs. (15 Nm)
- Front exhaust pipe and torque the front bolt to 16 ft. lbs. (22 Nm) and the front pipe to manifold bolt to 30 ft. lbs. (40 Nm)
- Alternator and oil pressure switch connector
- Crankshaft timing belt pulley and torque the bolts to 185 ft. lbs. (250 Nm) plus 45 degrees
- Water pump and torque the bolts to 19 ft. lbs. (25 Nm)
- Upper and lower adjusting rollers
- Camshaft sprockets and torque the bolts to 37 ft. lbs. (50 Nm) plus 60 degrees
- Locking tools between the camshaft sprockets to lock the camshafts of both heads in position.

7. Rotate the crankshaft forward to just before **0** Top Dead Center (TDC) and install a crankshaft locking tool on the crankshaft. Carefully rotate the engine until the arm of the tool is against the water pump flange. Be sure the crankshaft is at **0** TDC and all timing marks are aligned. Remove the locking tool.

8. If reusing the belt, install the timing belt according to its marked direction of rotation and timing marks. Adjust the tensioning roller loosely by hand to prevent the belt from slipping out of the cogs. Always adjust counterclockwise.

9. Measure the belt tension.

10. Tighten the center bolts of the adjusting roller lightly. Adjust the adjusting rollers counterclockwise. Begin with the lower roller and adjust it to a belt tension of 275–300 Newtons.

➡**Adjustment of the belt tension is only a preparatory measure and must not be used as a check when the belt is finally adjusted.**

11. Continue to carry out the adjustment by means of the tensioning roller, mark against mark. Remove the locking tool for camshaft sprockets 1 and 2. Carry out the final adjustment with the upper center adjusting roller until camshaft sprocket No. 2 moves 0.04–0.08 in. (1–2mm) forward.

12. Remove the locking tool for camshaft sprockets 3 and 4 and also remove the crankshaft locking tool.

13. Torque the tensioning roller to 15 ft. lbs. (20 Nm); the upper adjusting roller to 30 ft. lbs. (40 Nm) and the lower adjusting roller to 15 ft. lbs. (20 Nm).

14. Turn the engine over 2 revolutions to the **0** mark and refit the locking tool on the crankshaft. Check that the markings on the camshaft sprockets are in alignment with the markings on the timing cover. Check the positioning by installing the 2 camshaft locking tools, which should fit, and also by fitting tool 83–94–926 on camshaft sprockets 1 and 2 and 3 and 4. Also check the tensioning roller to ensure that the marks are still in alignment.

15. Install or connect the following:
- Crankshaft pulley and torque the bolts to 15 ft. lbs. (20 Nm)
- Timing belt cover and torque the bolts to 72 inch lbs. (8 Nm)
- Belt tensioner and alternator air intake
- Water pump and power steering pump pulleys
- V-belt and torque the water pump pulley bolts to 72 inch lbs. (8 Nm) and the power steering pulley bolts to 15 ft. lbs. (20 Nm)
- Right side engine mount and bracket and torque the bolts to 47 ft. lbs. (63 Nm)
- Power steering hose in the holder
- MAF sensor and hose and remove the engine lifting support
- Turbocharger intake pipe
- Crankcase breather pipe and torque the bolts to 18 ft. lbs. (24 Nm)
- Front exhaust manifold heat shield and torque the bolts to 18 ft. lbs. (24 Nm)
- Negative battery cable
- Lower spoiler section
- Engine cover

16. Fill the cooling system to the proper level.

17. Fill the engine with clean oil.

18. Start the vehicle, check for leaks and repair if necessary.

Rear Main Seal

REMOVAL & INSTALLATION

2.3L Engines

1. Before servicing the vehicle, refer to the precautions in the beginning of this section.

2. Remove or disconnect the following:

- Negative battery cable
- Transmission
- Flywheel or driven plate

To install:

3. Lubricate the sealing lips and install fitting Tool 83-94-884 and tap the seal in place.

4. Install or connect the following:
- Driven plate and torque the bolts to 70 ft. lbs. (95 Nm) and the flywheel to 59 ft. lbs. (80 Nm)
- Transmission
- Negative battery cable

3.0L Engines

1. Before servicing the vehicle, refer to the precautions in the beginning of this section.

2. Remove or disconnect the following:
- Negative battery cable
- Transmission
- Drive plate and pry out the rear main seal

To install:

3. Lubricate the sealing lips and install fitting Tools 83-94-967 and 83-94-975 and tap the seal in place.

4. Install or connect the following:
- Drive plate and torque the bolts to 48 ft. lbs. (65 Nm) plus an additional 30 degrees
- Transmission
- Negative battery cable

Timing Chain, Sprockets, Front Cover and Seal

REMOVAL & INSTALLATION

900 AND 9-3 MODELS

1. Before servicing the vehicle, refer to the precautions in the beginning of this section.

2. Drain the engine oil.

3. Drain the cooling system.

4. Remove or disconnect the following:
- Negative battery cable
- Bracket for the steering servo pump, complete with the pump and alternator
- Belt tensioner and the water pump pipe
- Crankshaft pulley
- Oil pipes
- Water pump pulley
- Timing chain cover
- Oil seal

Heater Core replacement is covered in Section 2 of this manual

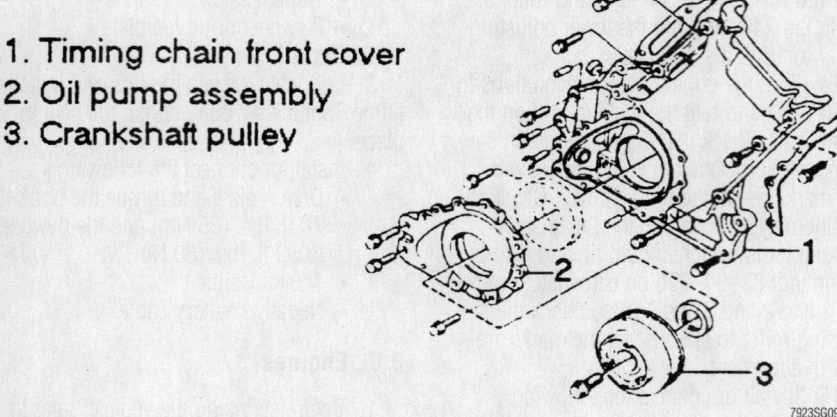

1. Timing chain front cover
2. Oil pump assembly
3. Crankshaft pulley

Exploded view of the timing chain front cover assembly—2.0L engine

5. The cam chain and crankshaft timing sprocket should now both be visible. From above, release the timing chain tensioner by pressing the pivoting guide firmly against it. Remove the chain tensioner.

6. Using a wrench to hold the cast hex bolt on the camshafts, remove the center bolts securing the camshaft sprockets. Throughout this procedure, keep the camshafts in their basic correct setting. If they are rotated out of position at any stage, especially without their sprockets and chain, the valves can be damaged.

7. Disconnect the timing chain from the sprockets and remove the chain, clearing it from the crankshaft sprockets.

To install:

8. Using a suitable seal installer, tap the new seal into the front cover until it is flush with the casting.

9. To install the timing chain, place the chain around the crankshaft sprocket. Run the chain up through the opening in the cylinder head. Install the chain and sprocket on the exhaust cam first. Be sure the chain is taut between the crankshaft and camshaft sprockets. Install and snug the bolt but do not tighten.

10. Install the chain and sprocket to the intake cam. Keep the chain taut between the cam sprockets while it is being installed. Install and snug the bolt but do not tighten. Be sure the chain is seated in the guide tensioner grooves.

11. Tension the chain tensioner by fully depressing the piston, then rotating it to the locked position.

12. Install the chain tensioner with the piston under tension. Be sure the copper gasket is in good condition and that the

sealing surface is clean and free from burrs. Torque the tensioner to 47 ft. lbs. (63 Nm).

13. Trigger the chain tensioner by pressing the pivoting chain guide against it, thereafter, press the pivoting guide against the chain to give the chain its basic tension. Check that the chain tensioner maintains tension on the chain when the pressure on the chain guide is released and that the basic setting stop for the tensioner holds the chain guide tight against the chain. A limited amount of play will be present until the hydraulic pressure takes over once the engine is running.

14. Check the setting by rotating the crankshaft 2 complete turns in its normal direction of rotation around to the timing mark. The basic setting of the cams should remain unaltered.

15. Lock the exhaust cam by using a wrench on the cast hex bolt and torque the sprocket bolt to 48 ft. lbs. (65 Nm). Repeat this on the intake cam.

16. Complete the procedure on the intake cam sprocket. When loosening or tightening the sprocket center bolts, hold the cam still using a wrench installed over the flats on the camshaft.

17. Install or connect the following:
- Timing cover and torque the bolts to 15 ft. lbs. (20 Nm)
- Oil pump
- Oil supply pipes
- Water pump pulley
- Crankshaft pulley and torque the nut to 140 ft. lbs. (190 Nm)
- Belt tensioner
- Water pump pipe
- Power steering pump
- Alternator
- Drive belt
- Negative battery cable

18. Fill the cooling system to the proper level.

19. Fill the engine with clean oil.

20. Start the engine, check for leaks and repair if necessary.

9000 MODELS

1. Before servicing the vehicle, refer to the precautions in the beginning of this section.

2. Drain the cooling system.

3. Drain the engine oil.

4. Remove or disconnect the following:
- Negative battery cable
- Right front wheel
- Inner wheel well
- Serpentine drive belt and belt tensioner
- Tie bar between the wheel arch and the subframe
- Power steering pump

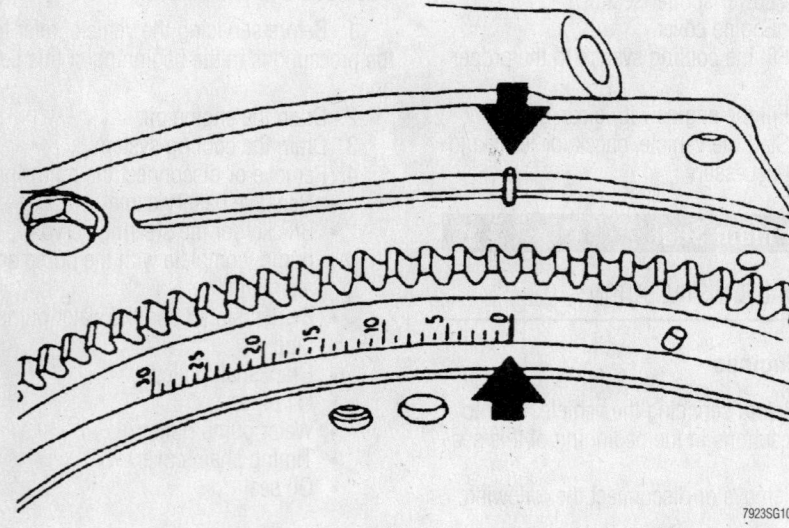

Be sure to align the crankshaft timing mark, as shown—2.0L engine

- Power steering pump bracket
- Alternator
- Top engine mounting bracket. Note the location of the bolts as the bolts are of different lengths.
- Torque arm
- Top belt tensioner bracket
- A/C compressor and bracket. Do not disconnect any hoses.
- Coolant hoses
- Water pump
- Crankshaft pulley
- Crankshaft Position (CKP) sensor
- Oil pan. It may be necessary to move the coolant pipe aside.
- Timing cover bolts, including the bolts holding the cover to the cylinder head. Note the locations of all bolts as they are of different lengths
- Timing cover by tapping it off the guide pin
- Lid on the valve cover
- Ignition wires
- Valve cover

➡**To remove or refit the timing chain, the camshafts and crankshaft must be lined up with their respective timing marks, No. 1 cylinder at Top Dead Center (TDC).**

- Top chain guide and chain tensioner for the balance shaft chain
- Idler sprocket and balance shaft chain

✳✳ WARNING

Throughout this procedure, keep the camshafts in their basic correct setting. If they are rotated out of position at any stage, especially without their sprockets and chain, the valves can be damaged.

5. The camshaft chain and crankshaft timing sprocket should both be visible. From above, release the timing chain tensioner by pressing the pivoting guide firmly against it. Remove the chain tensioner.

6. Remove or disconnect the following:
- Exhaust camshaft, using a wrench installed over the flats hold the camshaft and remove the center bolt securing the camshaft sprocket. Repeat this step for the intake camshaft.
- Timing chain from the sprockets

To install:

7. To install the timing chain, place the

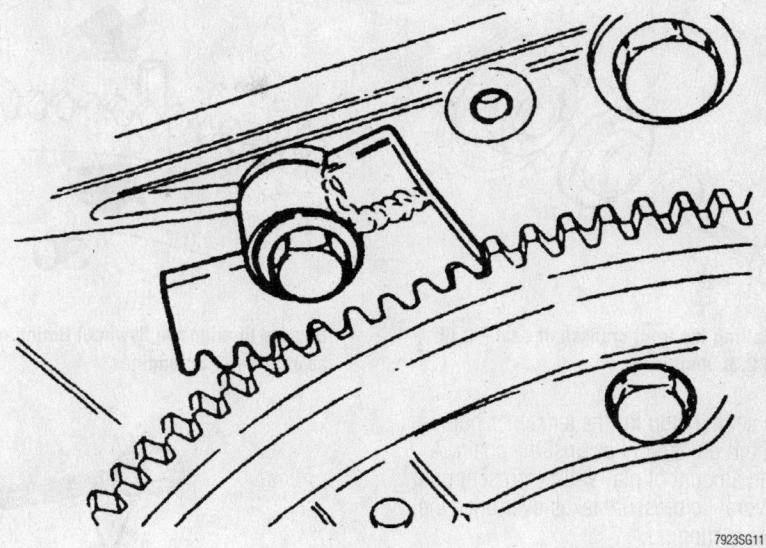

Secure the engine in position using a flywheel locking tool—2.0L and 2.3L engines

chain around the crankshaft sprocket. Run the chain up through the opening in the cylinder head. Install the chain and sprocket on the exhaust cam first. Be sure the chain is taut between the crankshaft and camshaft sprockets. Snug the bolt but do not tighten.

8. Install the chain and sprocket to the intake cam. Keep the chain taut between the cam sprockets while it is being installed. Snug the bolt but do not tighten. Be sure the chain is seated in the guide tensioner grooves.

9. Tension the chain tensioner by fully depressing the piston, then rotating it to the locked position.

10. Install the chain tensioner with the piston under tension. Be sure the copper gasket is in good condition and that the sealing surface is clean and free from burrs.

11. Trigger the chain tensioner by pressing the pivoting chain guide against it, thereafter, press the pivoting guide against the chain to give the chain its basic tension. Check that the chain tensioner maintains tension on the chain when the pressure on the chain guide is released and that the

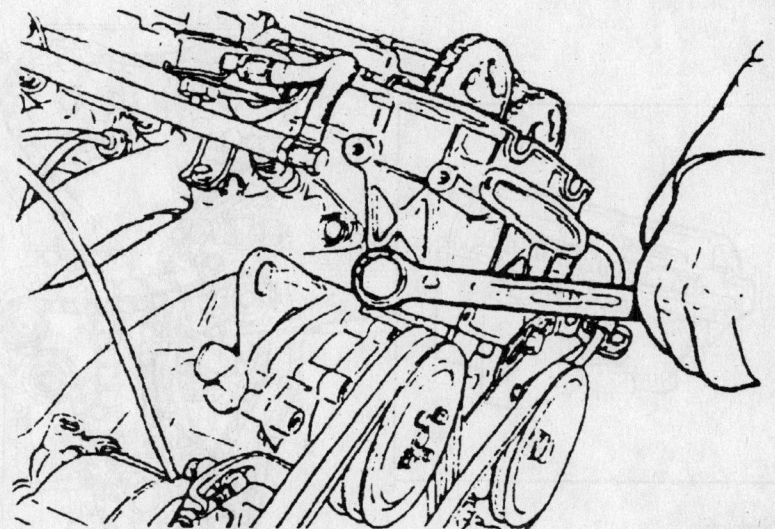

Unthread the timing chain tensioner to remove it from the front cover—2.0L and 2.3L engines

Brake service is covered in Section 4 of this manual

Installing the front crankshaft seal—2.0L and 2.3L engines

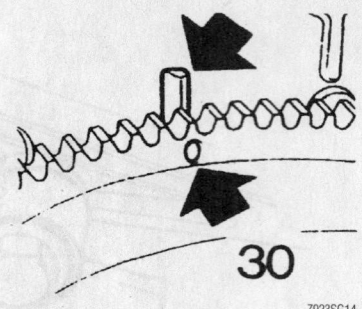

Be sure to align the flywheel timing mark, as shown—2.3L engines

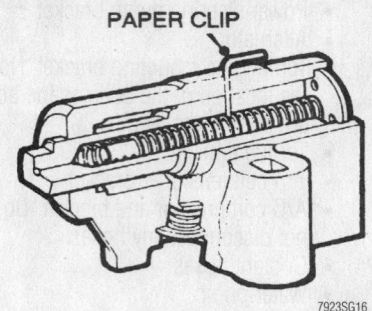

Preparing the balance shaft's tensioner for installation—2.3L engine

basic setting stop for the tensioner holds the chain guide tight against the chain. A limited amount of play will be present until the hydraulic pressure takes over once the engine is running.

12. Check the setting by rotating the crankshaft 2 complete turns in its normal direction of rotation around to the timing mark. The basic setting of the cams should remain unaltered.

13. Lock the exhaust cam by using a wrench installed over the flats on the camshaft and torque the sprocket bolt to 48 ft. lbs. (65 Nm). Repeat this step on the intake cam.

➡ **The accuracy of the timing chain adjustment will depend on the condition of the chain.**

14. Install the balance shaft chain and sprocket on the crankshaft.

15. Install or connect the following:
  • Oil pump drive dog

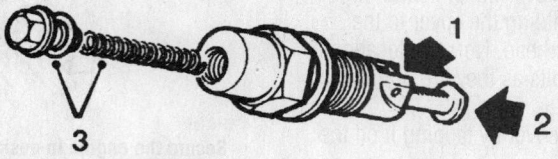

1. Push down on the catch
2. Push in on tensioner arm
3. Tensioner plug, push rod and spring

Exploded view of the timing chain tensioner—2.3L engines

• Balance shaft chain and idler wheel sprocket. Ensure that the aligning marks on the bearing housing and sprocket are in line. When installing, leave some slack in the chain in line with the tensioner, and keep the chain reasonably taut by means of the top chain guide.

➡ **There is an alternate way of installing the balance shaft chain. Install the top chain guide first, then adjust the run of the chain around the sprockets. Adjusting the chain is easier this way, although it will be more awkward to install the idler wheel sprocket.**

16. Cock the balance shaft chain tensioner and insert a paper clip through the hole in the cylinder to prevent the tensioner being triggered. Before installing the tensioner, be sure that the plunger is turned to the position in which the spring acts fully on it.

17. Install the balance shafts pivoting chain guide and tensioner. Torque the tensioner to 89 inch lbs. (10 Nm).

✳✳ WARNING

It is extremely important for the correct torque to be applied when installing the tensioner.

18. Install the top chain guide and trigger the tensioner by removing the paper clip.

19. Rotate the crankshaft a few times to ensure the balance shafts chain is installed correctly.

20. Ensure that the timing cover flange is absolutely clean.

21. Remove all traces of old sealant from

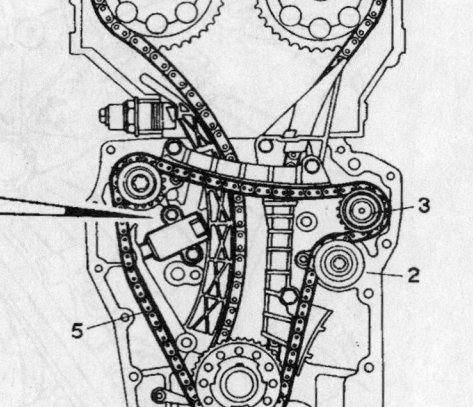

1. Timing chain
2. Balance shaft
3. Idler pulley
4. Balance shaft chain tensioner
5. Balance shaft chain

Timing and balance shaft chain assemblies' component identification—2.3L engines

the cover. Apply a 0.04 in. (1mm) bead of 45–3028972 sealant to the flanges of the cover. Use sealant sparingly as excess sealant can get into the oil ways and damage the engine.

22. Install or connect the following:
- Timing cover and torque the bolts in their correct positions to 15 ft. lbs. (20 Nm)

23. Apply an even bead of Permatex® Ultra Blue sealant and fit the rubber seal for the oil strainer in the groove on the sump.
- Oil pan and torque the bolts to 15 ft. lbs. (20 Nm)
- Crankshaft pulley and torque the retaining bolt to 140 ft. lbs. (190 Nm)
- Coolant pipe
- CKP sensor
- Engine mounts
- Water pump
- Coolant hoses
- A/C compressor
- Power steering pump and bracket
- Alternator
- Top engine mounting bracket
- Torque arm
- Serpentine drive belt
- Belt tensioner
- Tie bar between the wheel arch and the subframe
- Inner wheel well and wheel
- Negative battery cable

24. Fill the cooling system to the proper level.

25. Fill the engine with clean oil.

26. Start the engine and allow it to reach normal operating temperature and check for leaks.

9-5 Models

2.3L ENGINE

1. Before servicing the vehicle, refer to the precautions in the beginning of this section.

2. Drain the cooling system.

3. Drain the engine oil.

4. Remove or disconnect the following:
- Negative battery cable
- Belt circuit cover
- Upper engine cover
- Lower spoiler section
- Ignition discharge module and spark plugs
- Crankcase ventilation pipe
- Camshaft cover and align the

crankshaft and camshafts with the marks by turning the crankshaft clockwise
- Tension from the V-belt
- Idler sprocket so that the timing chain tensioner sleeve is free

✳✳ WARNING

Throughout this procedure, keep the camshafts in their basic correct setting. If they are rotated out of position at any stage, especially without their sprockets and chain, the valves can be damaged.

5. The camshaft chain and crankshaft timing sprocket should both be visible. From above, release the timing chain tensioner by pressing the pivoting guide firmly against it. Remove the chain tensioner.

6. Remove or disconnect the following:
- Exhaust camshaft, using a wrench installed over the flats hold the camshaft and remove the center bolt securing the camshaft sprocket. Repeat this step for the intake camshaft.
- Timing chain from the sprockets

To install:

7. Install or connect the following:
- Timing chain and make certain

that the chain lies over the camshaft sprockets on the intake side
- Connect the ends of the new chain together with a chain link and install a new chain lock
- Chain tensioner with a new washer. Torque the tensioner to 46 ft. lbs. (63 Nm). Turn the engine over 2 turns and check the **O** marks on the camshaft and crankshaft.
- Camshaft cover and torque the bolts to 10 ft. lbs. (15 Nm)
- Spark plugs and torque the plugs to 21 ft. lbs. (28 Nm)
- Ignition discharge module and torque the fastener to 8 ft. lbs. (11 Nm)
- Crankcase ventilation pipe
- Lower spoiler section
- Upper engine cover
- Negative battery cable

8. Fill the cooling system to the proper level.

9. Fill the engine with clean oil.

10. Start the vehicle, check for leaks and repair if necessary.

Piston and Ring

POSITIONING

Ensure that the notch in the piston crown points towards the timing cover and that the numbers on the connecting rod face the exhaust side.

7923AGB4

2.0L and 2.3L engines—piston and connecting rod assembly positioning

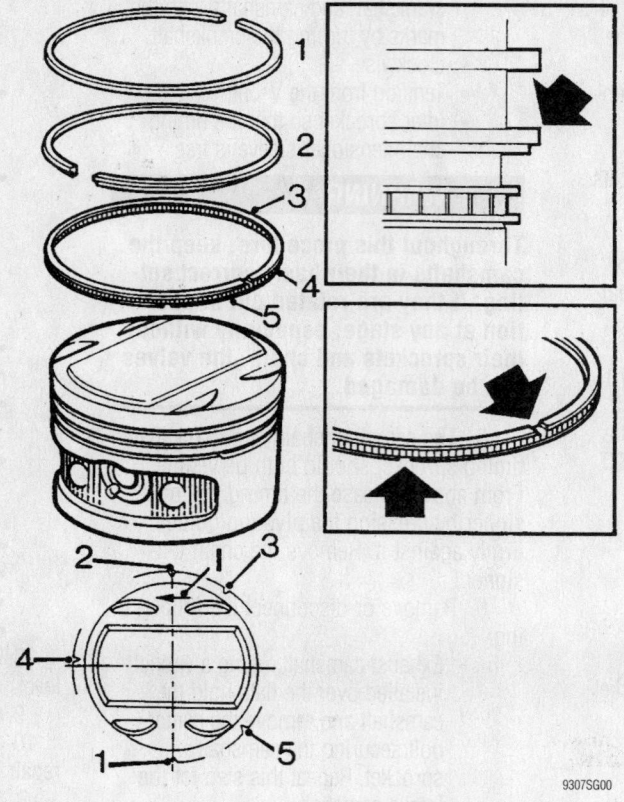

3.0L engine—piston and connecting rod assembly positioning

FUEL SYSTEM

Fuel System Service Precautions

Safety is the most important factor when performing not only fuel system maintenance but any type of maintenance. Failure to conduct maintenance and repairs in a safe manner may result in serious personal injury or death. Maintenance and testing of the vehicle's fuel system components can be accomplished safely and effectively by adhering to the following rules and guidelines.

1. To avoid the possibility of fire and personal injury, always disconnect the negative battery cable unless the repair or test procedure requires that battery voltage be applied.

2. Always relieve the fuel system pressure prior to disconnecting any fuel system component (injector, fuel rail, pressure regulator, etc.), fitting or fuel line connection. Exercise extreme caution whenever relieving fuel system pressure to avoid exposing skin, face, and eyes to fuel spray. Please be advised that fuel under pressure may penetrate the skin or any part of the body that it contacts.

3. Always place a shop towel or cloth around the fitting or connection prior to loosening to absorb any excess fuel due to spillage. Ensure that all fuel spillage

(should it occur) is quickly removed from engine surfaces. Ensure that all fuel soaked cloths or towels are deposited into a suitable waste container.

4. Always keep a dry chemical (Class B) fire extinguisher near the work area.

5. Do not allow fuel spray or fuel vapors to come into contact with a spark or open flame.

6. Always use a back-up wrench when loosening and tightening fuel line connection fittings. This will prevent unnecessary stress and torsion to fuel line piping. Always follow the proper torque specifications.

7. Always replace worn fuel fitting O-rings with new. Do not substitute fuel hose where fuel pipe is installed.

Fuel System Pressure

RELIEVING

900, 9000 and 9-3 Models

✷✷ CAUTION

The fuel injection system remains under pressure, even after the engine has been turned OFF. The fuel system

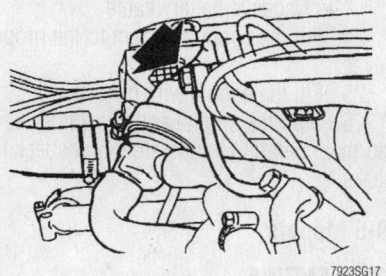

Location of fitting on fuel rail to relieve fuel pressure—900 and 9-3

pressure must be relieved before disconnecting any fuel lines. Failure to do so may result in fire and/or personal injury.

1. Before servicing the vehicle, refer to the precautions in the beginning of this section.

2. Disconnect the negative battery cable.

3. Place clean shop rags under and around the fuel line banjo coupling to absorb any released fuel or fuel spray.

4. With the ignition switch in the **OFF** position, carefully relieve the fuel system pressure by firmly cracking open the banjo coupling at the inlet to the fuel injection manifold.

5. Be sure to always replace the banjo coupling washers whenever you loosen or remove the banjo couplings.

9-5 Models

✳✳ CAUTION

The fuel injection system remains under pressure, even after the engine has been turned OFF. The fuel system pressure must be relieved before disconnecting any fuel lines. Failure to do so may result in fire and/or personal injury.

1. Remove fuse no. 19 from the fuse and relay panel with the engine running.
2. Turn the ignition to the "OFF" position when the engine stops.
3. Install the fuse.

Fuel Filter

REMOVAL & INSTALLATION

✳✳ CAUTION

The fuel injection system remains under pressure, even after the engine has been turned OFF. The fuel system pressure must be relieved before disconnecting any fuel lines. Failure to do so may result in fire and/or personal injury.

1. Before servicing the vehicle, refer to the precautions in the beginning of this section.
2. Properly relieve the fuel system pressure.
3. Be sure that the ignition switch is in the **OFF** position.
4. Locate the fuel filter, which is mounted under the vehicle and forward of the fuel tank.
5. Thoroughly clean the area around the banjo fittings before continuing.
6. Remove or disconnect the following:
   - Fuel filter splash guard, if equipped
   - Banjo fittings on both sides of the fuel filter. Contain the fuel with a shop towel. Properly dispose of the towel once the job is complete.
   - Band-type clamp on the filter
   - Fuel filter

To install:
7. Install or connect the following:
   - New fuel filter. Make sure that the

arrow is pointing in the correct direction of fuel flow.
   - Band type clamp
   - Banjo fittings to the filter with new sealing washers and torque to 16 ft. lbs. (21 Nm)
   - Fuel filter splash guard, if equipped
8. Start the vehicle, check for leaks and repair if necessary.

Fuel Pump

REMOVAL & INSTALLATION

900 and 9-3 Models

✳✳ CAUTION

The fuel injection system remains under pressure, even after the engine has been tuned OFF. The fuel system pressure must be relieved before disconnecting any fuel lines. Failure to do so may result in fire and/or personal injury.

1. Before servicing the vehicle, refer to the precautions in the beginning of this section.
2. Remove or disconnect the following:
   - Negative battery cable
   - Fuel into a suitable container
   - Rubber hoses from the tank. Plug the openings.
   - Fuel filter clamp
   - Metal straps holding fuel tank. Support the tank from below with a pole jack or other suitable device.
3. Carefully lower the tank, right side first, until the top is visible.
   - Two wiring connectors to pump
   - Fuel pressure and return lines
   - The fuel tank
   - Retaining ring
4. Lift the pump about 2 inches, then rotate clockwise about 80° and remove
To install:
5. Install or connect the following:
   - Fuel pump and new O-ring
   - Retaining ring and torque the ring to 55 ft. lbs. (75 Nm)
6. Raise fuel tank and support with appropriate stands.
   - Fuel lines
   - Electrical connectors
   - Metal straps
   - Rubber hoses to tank
   - Fuel filter clamp
7. Lower the car and refill the tank.

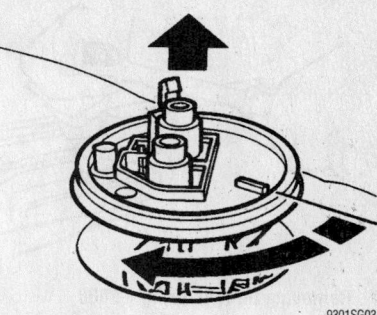

9301SG03

Removing the fuel pump

8. Reconnect the negative battery cable and check that the system is operating properly and that there are no fuel leaks.

9000 Models

BOSCH FUEL PUMP

✳✳ CAUTION

The fuel injection system remains under pressure, even after the engine has been tuned OFF. The fuel system pressure must be relieved before disconnecting any fuel lines. Failure to do so may result in fire and/or personal injury.

1. Properly relieve the fuel system pressure.
2. Before servicing the vehicle, refer to the precautions in the beginning of this section.
3. Remove or disconnect the following:
   - Negative battery cable. Tape or tie off the cable so that it cannot be reconnected during this procedure.
   - Carpet in the trunk
   - Floor panel
   - Oblong cover plate over fuel pump
   - Electrical harness connector and move the cover aside
   - Fuel lines at the fuel pump. Place the fuel lines off to the side. Soak up any fuel released with a shop towel.
   - Large threaded ring securing the pump to the tank
   - Fuel pump. Tilt it out of the tank and to the right, being careful not to damage the arm for the fuel level sender.
To install:
4. Install or connect the following:
   - New O-ring in the groove on top of tank
   - Fuel pump and align the mark on

For Accessory Drive Belt illustrations, see Section 1 of this manual

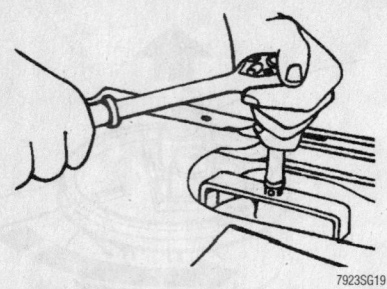

Removing threaded ring—9000

the pump with the mark on the tank. This must be done for the fuel sender to operate properly.

- Large threaded ring holding the pump to the tank and torque to 40 ft. lb. (55 Nm). Check that the alignment of the marks are still in line. A tolerance of 5° is permissible.
- New O-rings and fuel lines to the pump
- Electrical connector and the clip
- Negative battery cable
- Access plate in the trunk floor. Pay careful attention that the plate seals properly so that no exhaust fumes can enter the cabin.
- Floor panel
- Carpet in trunk

WALBRO FUEL PUMP

1. Before servicing the vehicle, refer to the precautions in the beginning of this section.
2. Properly relieve the fuel system pressure.
3. Remove or disconnect the following:

- Negative battery cable. Tape or tie off the cable so it cannot contact the battery terminals.
- Trunk carpet
- Floor panel in the trunk
- Cover plate to gain access to the fuel pump
- Electrical wiring at the fuel pump

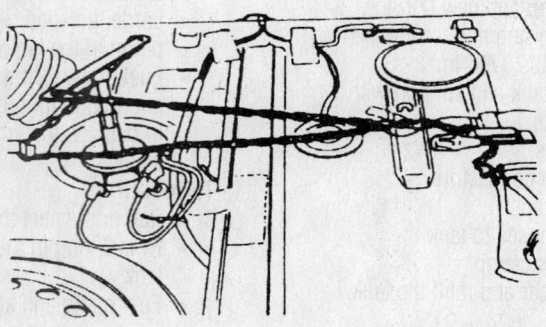

Positioning of tool and chains for tightening the retaining ring—Walbro pump

- Protective plate over the fuel line fittings
- Fuel lines at the fuel pump. Use angled needle-nose pliers or similar tool. Use a shop towel to absorb any fuel released.

4. Move the fuel lines and wiring harness to the side.
5. Place Tool 83–94–397 over the fuel pump. Place the chain supplied with the tool through the 2 load securing brackets on the trunk floor and secure each end with a screwdriver or similar tool.

- Large threaded ring that holds the pump in the tank. Remove the and tools and finish removing the ring by hand.
- Rubber seal
- Fuel pump. Tilt the top part of it forward and to the left. Allow the fuel to run out of the pump before removing it from the trunk.

To install:

6. Wipe clean the sealing area of the opening in the fuel tank.
7. Install or connect the following:

- Fuel pump. Be careful not to damage to the fuel level float arm. Once in the tank, line up the marks on the tank and pump.
- Large threaded ring. Coat it with petroleum jelly and place on top of the fuel pump with the mark on the ring lined up with the mark on the fuel tank.

8. Press down hard and rotate the ring 90°.
9. Reinstall special Tool 83–94–397, with the chain situated for tightening.
10. Tighten the ring until it will no longer turn.
11. Check that the pump did not rotate more than 30° from the alignment marks.
12. Install or connect the following:

- Fuel lines using new O-rings. The return line is connected to the nipple closest to the front of the

vehicle while the feed line is connected closest to the rear of the vehicle.

- Electrical connections
- Protective plate over the fuel line fittings
- Negative battery cable
- Access plate in the trunk floor. Pay careful attention that the plate seals properly so that no exhaust fumes can enter the cabin.
- Floor panel
- Trunk carpet

9-5 Models

1. Before servicing the vehicle, refer to the precautions in the beginning of this section.
2. Properly relieve the fuel system pressure.
3. Raise the rear seat cushions and fold the carpeting out of the way.
4. Remove or disconnect the following:

- Negative battery cable
- Fuel pump cover
- Upper connector from the fuel pump

5. Carefully loosen the check valves and fuel lines from the pump. Move the yellow hooks to one side. The check valves are connected to the pump with quick release couplings.
6. The white check valve is the delivery side of the valve and marked "Pressure". The black check valve is the return side of the valve and is marked "Return" on the fuel pump.

- Screw ring with Tool 83–94–462 and lift the fuel pump until the upper section is slightly above the fuel tank

7. Rotate the fuel pump approximately 80 degrees and remove the pump from the fuel tank.

To install:

8. Clean the sealing surfaces on the fuel pump and fuel tank.
9. Install or connect the following:

- New O-ring in the groove on the fuel tank and carefully lower the fuel pump into position. Make certain that the marks on the fuel pump and tank are opposite one another and press the pump into position.
- Screw ring and torque the ring to 55 ft. lbs. (75 Nm)
- Check valves to the fuel pump with new O-rings. Make certain that the valves are properly positioned
- Fuel lines and upper connector
- Negative battery cable

10. Start the vehicle, check for leaks and

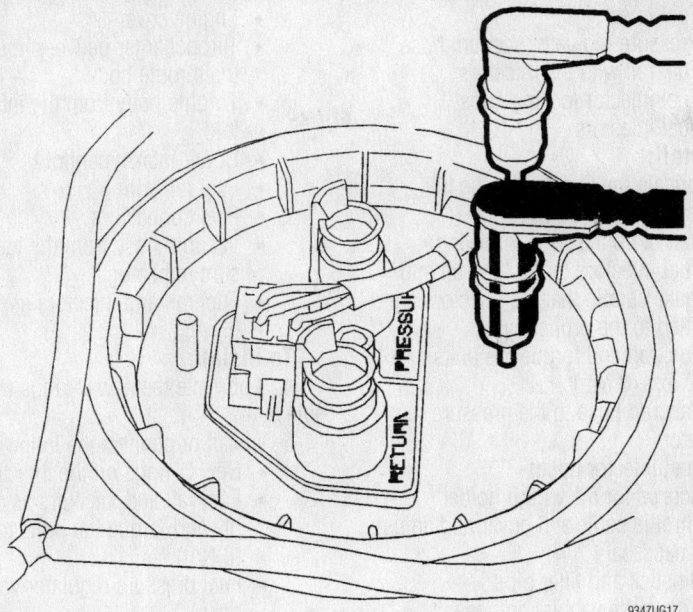

Remove the Pressure and Return valves from the top of the fuel pump

proper operation of the fuel pump and repair if necessary.

11. Install the fuel pump cover.

12. Reposition the carpeting and the rear seat cushions.

Fuel Injector

REMOVAL & INSTALLATION

900 and 9-3 Models

1. Before servicing the vehicle, refer to the precautions in the beginning of this section.

2. Properly relieve the fuel system pressure.

3. Remove or disconnect the following:
   - Negative battery cable
   - Resonator
   - Cowl on turbo engines only
   - Crankcase breather hose
   - Idle adjusting valve
   - Dipstick holder bracket
   - Throttle and cruise control cable bracket
   - Fuel pressure and return lines
   - Four retaining bolts and 2 covers on the fuel rail
   - Electrical connectors to fuel rail
   - Fuel rail
   - Injector locking clips
   - Injectors

To install:

4. Lubricate injector O-rings with petroleum jelly

5. Install or connect the following:
   - Injectors and clips to fuel rail and place the fuel rail into position
   - Electrical connections to fuel rail. Press the fuel rail into place.
   - Two covers and 4 bolts
   - Fuel pressure and return lines
   - Throttle and cruise control cable bracket
   - Dipstick holder bracket
   - Idle adjusting valve
   - Crankcase breather hose
   - Resonator
   - Cowl on turbo engines only
   - Negative battery cable

6. Start the vehicle, check for leaks and repair if necessary.

9000 Models

1. Before servicing the vehicle, refer to the precautions in the beginning of this section.

2. Properly relieve the fuel system pressure.

3. Remove or disconnect the following:
   - Negative battery cable
   - Plastic pipe between throttle body and resonator
   - Idle Air Control (IAC) valve from throttle body
   - Cruise control cable
   - Throttle body
   - Vacuum hoses to intake manifold. Label prior to removal
   - Upper half of intake manifold bracket retaining screws. There are 4 on each side. Remove the top screws and loosen the bottom.
   - Oil fill pipe bracket
   - Remaining screws on upper half of the intake manifold
   - Upper half of the intake manifold
   - IAC connector
   - Two injector support rails
   - Fuel injector connectors
   - Fuel rail with injectors
   - Injector retaining clips
   - Injectors

To install:

4. Lubricate injector O-rings with petroleum jelly

5. Install or connect the following:
   - Injectors and clips to fuel rail
   - Fuel rail and torque the bolts to 13 ft. lbs. (18 Nm)
   - Electrical connections to fuel rail
   - Two injector support rails
   - IAC connector
   - Upper half of intake manifold
   - Upper half of intake manifold retaining screws
   - Oil fill pipe bracket
   - Remaining screws to upper intake manifold bracket
   - Vacuum hoses to intake manifold

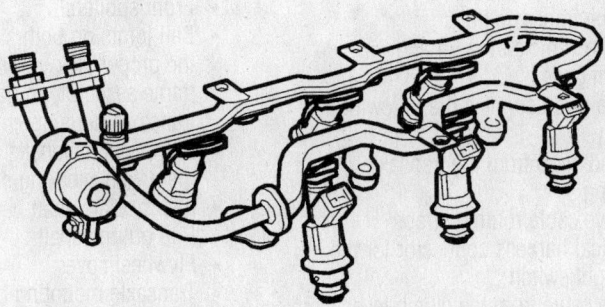

Remove the fuel rail and injectors as an assembly

For Tire, Wheel and Ball Joint specifications, see Section 1 of this manual

- Throttle body
- Cruise control cable
- IAC to throttle body
- Plastic pipe between throttle body and resonator
- Negative battery cable

6. Start the vehicle, check for leaks and repair if necessary.

9-5 Models

2.3L ENGINE

1. Before servicing the vehicle, refer to the precautions in the beginning of this section.

2. Properly relieve the fuel system pressure.

3. Remove or disconnect the following:
- Negative battery cable
- Upper engine cover
- Crankcase ventilation hoses
- Dipstick with the oil filler pipe
- Cover and detach the throttle cable from the spindle
- Turbocharger delivery pipe
- Fuel injector connectors
- Ignition discharge module
- Manifold Absolute Pressure (MAP) sensor
- Boost pressure control valve
- Turbocharger pressure sensor and slacken the upper bolt on the wiring holder
- Fuel rail retaining bolts and cable ties

- Fuel lines
- Pressure regulator vacuum hose
- Fuel rail with the injectors
- Fuel injector locking clips
- Fuel injectors

To install:

4. Lubricate the O-rings on the fuel injectors

5. Install or connect the following:
- Fuel injectors to the fuel rail and make certain that the injectors are fitted to the proper cables
- Fuel rail and torque the bolts to 6 ft. lbs. (8 Nm)
- Vacuum hose to the pressure regulator
- Fuel injector hoses
- Screws for the wiring holder
- Throttle cable and cover and adjust if necessary
- Dipstick and filler pipe
- Crankcase ventilation hoses
- Upper engine cover
- Negative battery cable

6. Start the vehicle, check for leaks and repair if necessary.

3.0L ENGINE

1. Before servicing the vehicle, refer to the precautions in the beginning of this section.

2. Properly relieve the fuel system pressure.

3. Remove or disconnect the following:
- Negative battery cable

- Engine cover
- Turbocharger delivery hose from the throttle body
- Throttle body from the intake manifold
- Upper intake manifold
- Fuel rail wiring
- Fuel connectors
- Vacuum hose from the fuel pressure regulator
- Fuel rail and injectors as an assembly

To install:

4. Lubricate the new O-rings with vaseline.

5. Install or connect the following:
- New O-rings on the injectors
- Fuel rail and injectors as an assembly and torque the bolts to 6 ft. lbs. (8 Nm)
- Fuel pressure regulator vacuum hose
- Fuel lines and wiring
- Upper intake manifold with a new gasket and torque the bolts to 15 ft. lbs. (20 Nm)
- Throttle body
- Throttle cable and adjust if needed
- Turbocharger delivery hose
- Engine cover
- Negative battery cable

6. Start the vehicle, check for leaks and repair if necessary.

DRIVE TRAIN

Manual Transaxle

REMOVAL & INSTALLATION

900 and 9-3 Models

1. Before servicing the vehicle, refer to the precautions in the beginning of this section.

2. Drain the transaxle fluid.

3. Place the vehicle on a lift and engage it in 4th gear.

4. Remove or disconnect the following:
- Battery
- Ground strap from the transaxle housing
- Positive cable routing straps
- Electrical harness connector for the rear light switch
- Clutch cable from the clutch lever. Release the cable's rubber damper from the fastener on the transaxle.
- Oxygen (O2S) sensor connectors and remove the securing straps

5. Install the engine holder. Insert the

lifting bar in the eyelets and connect to the engine holder.

6. Install the transaxle locking pin to ensure the vehicle does not move out of gear.

- Shift rod and linkage from the transaxle
- Front wheels
- Front exhaust pipes
- Front spoilers
- Ball joints on both sides and using the proper jack, remove the subframe assembly to gain access to transaxle housing
- Both front halfshafts and suspend them with securing straps
- Intermediate shaft bearing bracket. Pull out the shaft.
- Flywheel cover
- Transaxle mounting bracket
- Three transaxle bracket bolts on the transaxle
- Engine/transaxle surface mounting bolts
- Two outer bolts holding the gear-

box to the engine mounting surface

7. Install the lifting eye on the transaxle and connect the hoist to the transaxle.

8. Remove the remaining bolt. Pull out the transaxle assembly.

9. Lock the flywheel in position.

To install:

10. Install the clutch disc with pressure plate and torque the pressure plate retaining nuts to 16 ft. lbs. (22 Nm)

11. Using a hoist lift transaxle into position. To help with installation use a guide pin to line up the transaxle.

12. Install or connect the following:
- Three upper bolts into the engine housing
- Transaxle/engine surface mounting bolts and torque the bolts to 65 ft. lbs. (90 Nm)
- Transaxle mounting bracket
- Flywheel cover
- Intermediate driveshaft with bracket bearing assembly
- Halfshafts

13. Install the subframe as follows:
   a. Torque the ball joints to 55 ft. lbs. (75 Nm).
   b. Torque the front subframe bolts to 85 ft. lbs. (11 Nm).
   c. Torque the 10 middle bolts to 141 ft. lbs. (190 Nm).
   d. Torque rear bolts to 81 ft. lbs. (110 Nm) plus an additional 75 degrees.
14. Install or connect the following:
- Front exhaust pipes
- Front wheels
- Shift rod. Engage 4th gear.
- Shift rod mounting bolt and remove the engine holder and lifting rods
- O$_2$S sensor and harness connectors
- Clutch cable
- Rear light switch harness
- Negative battery cable
15. Fill the transaxle with fluid.
16. Start the vehicle, check for leaks and repair if necessary.

9000 Models

1. Before servicing the vehicle, refer to the precautions in the beginning of this section.
2. Place the vehicle in 4th gear.
3. Drain the transaxle.
4. Remove or disconnect the following:
- Battery and tray
- Locking clips holding the accelerator cable in the lead-though and turn the cable to one side
- Front distribution box
- Positive junction (without removing the cables)
- Positive cable's 2 clamps on battery shelf
- Connection from the Antilock Braking System (ABS) control module
- Bypass valve from the turbocharger pressure pipe
- Temperature sensor from the turbocharger pressure pipe
- Hose clips from the turbocharger pressure pipe
- Connector on the cable for the speedometer sensor
- Reverse light switch connector
- Clutch slave cylinder hydraulic line
- All the engine-to-transaxle bolts accessible from the top side, except the top bolt
5. Place the lifting beam on the edges of the engine compartment. Insert the hook into the engine lifting eye and tighten the lifting beam wing nut slightly to support the engine and transaxle from above.

6. Raise and support the vehicle.
- Left front wheel and the edging of the wheel housing
- Front part of the inner fender and both left and middle spoiler units
- Ground cable from the gearbox
- Nuts on the gear selector universal joint and disconnect the shift linkage
- 3 bolts connecting the ball joint to the lower control arm
- Anti-roll bar from the lower control arm
- Front engine mounting nut from the bolt
- Screw from the wheel housing stay and put the 2 washers in a safe place
- The transaxle side of the subframe must be folded down out of the way
- Bolt in the front link of the subframe assembly and the 2 bolts that hold the front link
- Bolt in the rear link of the subframe assembly
- 2 bolts that hold the rear link, one of which also holds the steering rack
- 2 bolts in the front corners of the subframe, then the 4 bolts in the rear corners
- Carefully fold down the subframe. Put the plate between the subframe and the chassis in a safe place.
- Clamp around the CV-joint boot and disconnect the CV-joint. With the ball joint disconnected, the strut will move with the halfshaft but be careful not to damage the boot
- Protective plate from the transaxle
7. Support the transaxle and remove the remaining engine-to-transaxle bolts. Carefully lower the transaxle out of the vehicle.

To install:
8. Fit 2 locating pins into the engine block.
9. Lift the transaxle into position and onto the locating pins. Push the transaxle against the engine, making sure to align the clutch with the crankshaft, then start some of the bolts. Use the bolts to draw the transaxle up against the engine block.
10. Remove the locating pins and safely raise and support the vehicle.
11. Install or connect the following:
- Protection plate for the transaxle and install the remaining lower

bolts. Torque the bolts to 50 ft. lbs. (70 Nm). Hang the subframe and secure each link with 1 bolt. Be sure the plate in the rear corner is positioned correctly.
- Bolt in the front engine mount, making sure the washer is in the correct position
- All the remaining subframe bolts but do not tighten any of them yet. When all bolts are installed, torque all the bolts in the subframe to 41 ft. lbs. (55 Nm).
- Bolt to the stay in the wheel housing and torque to 37 ft. lbs. (50 Nm)
- Bolts holding the anti-roll bar bearing
- Nut holding the anti-roll bar mount in the lower control arm
- Halfshaft
- Lower ball joint to the control arm and torque the bolts to 25 ft. lbs. (34 Nm)
- CV-joint boot clamp
12. Be sure the transaxle is in 4th gear, fit the 2 screws in the selector universal joint.
- Battery ground cable to the gearbox. Lower the vehicle and remove the lifting beam.
13. Torque all the remaining engine-to-transaxle bolts to 50 ft. lbs. (70 Nm). Check the position of the washer in the engine mount and torque the nut to 50 ft. lbs. (70 Nm).
- Slave cylinder hydraulic line
- Reverse light connector
- Speedometer cable
- Turbocharger pressure pipe
- Hose on the bypass valve
- Wiring on the temperature sensor
- Battery shelf
- Wiring on the ABS control module
- Positive cable and the 2 clamps on the battery shelf
- Front distribution box
- Accelerator cable and connect the locking clips in the lead-through and the clip on the cable
- Battery cables
- Left and middle spoiler units
- Front part of the inner fender
- Edging of the wheel housing
- Wheel
14. Fill the transaxle with clean fluid.
15. Bleed the clutch system.
16. Check for any leakage.
17. Test drive the vehicle.

For Wheel Alignment specifications, see Section 1 of this manual

9-5 Models

1. Before servicing the vehicle, refer to the precautions in the beginning of this section.
2. Place the vehicle in 4th gear and lock it in position with Tool 87–92–335.
3. Drain the transaxle.
4. Remove or disconnect the following:
   - Front grille
   - Intake manifold cover
   - Battery cover
   - Battery and tray
   - Clip from the selector rod
5. Place the vehicle in 3rd gear so the selector rod disengages from the linkage. Install Locking Pin 87–92–335 in the lever housing.
   - Clip from the slave cylinder and disconnect the delivery line
   - Reverse light connector
   - 3 upper gearbox bolts and install locating studs in the engines upper outer guide holes
   - Oxygen (O$_2$S) sensor
   - Selector linkage from the gearbox
   - Rear engine mount nut
6. Loosen the 3 bolts from the rear engine pad. Do not remove the bolts.
7. Install a lifting beam to the engine assembly.
   - Both front wheels and wheel well covers
   - Front exhaust pipe
   - Stay between the engine mount and engine
   - Rear engine mount and pad
   - Screws from the steering gear
   - Clips securing the steering servo line to the subframe
   - A/C lines from the subframe
   - Engine oil cooler from the charge air cooler
   - Screws securing the outer ball joints to the steering swivel
   - Loosen the upper ball joint from the anti-roll bar
   - Flywheel cover plate
   - Bolts between the gearbox and oil pan
   - Screws securing the rear support plate to the subframe and place a lifting trolley under the subframe
   - Subframe and front mounts for the steering servo delivery line
   - Left halfshaft
   - Ground cables from the gearbox
   - Left engine pad
8. Lower the transmission so that it clears the structural member.
9. Remove the transmission mounting.
10. Connect a lifting tool to the transmission.

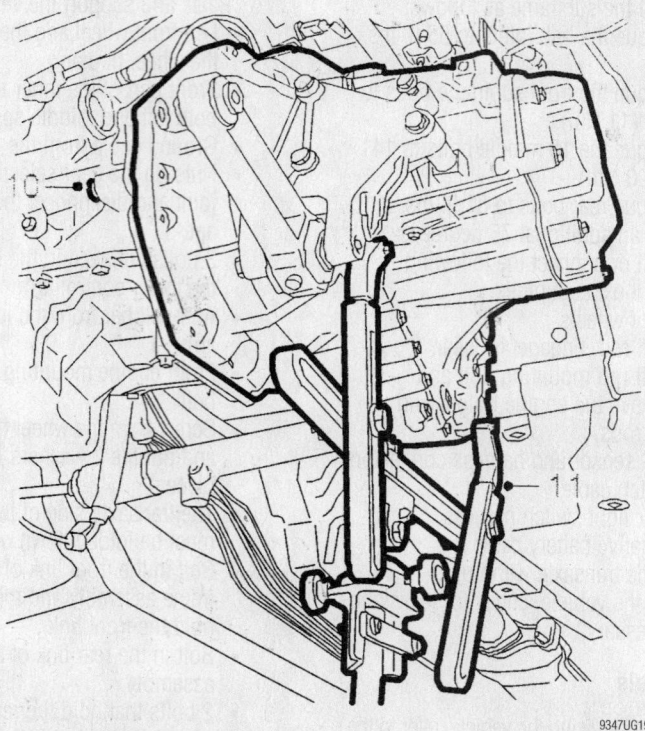

Install a lifting tool to the transmission—9-5 Models

9347UG19

11. Remove the transmission from the vehicle.

To install:

12. Lubricate the primary shaft splines.
13. Install a protective collar in the right shaft seal in the transmission.
14. Position the lifting trolley under the vehicle.
15. Install or connect the following:
   - Transmission and turn the engine shaft if needed to install the transmission
   - Flywheel cover plate
   - Bolts between the transmission and the oil pan and torque the bolts to 30 ft. lbs. (40 Nm)
   - Screws between the engine and transmission and torque to 50 ft. lbs. (70 Nm). Remove the lifting trolley.
   - Transmission bolts and torque the bolts to 18 ft. lbs. (24 Nm)
   - Ground cables to the transmission
   - Shaft until the circlip snaps into position
   - Transmission mountings and torque the bolts to 30 ft. lbs. (40 Nm)
   - Clutch delivery line and clips
   - Left engine pad. Torque the engine pad-to-transmission bolts to 62 ft. lbs. (85 Nm) and the engine pad to body bolts to 45 ft. lbs. (60 Nm).

   - Reverse light electrical connector
16. Position the lifting trolley under the vehicle with the subframe properly aligned.
   - Front ball joints and tighten the steering servo delivery lines
   - Radiator journals to the subframe
   - A/C lines into the brackets on the subframe
   - Screws and rear support plate for the subframe. Move the lifting trolley aside. Torque the subframe bolts to 74 ft. lbs. (100 Nm) plus an additional 45 degrees.
17. Torque the support plate bolts to 44 ft. lbs. (60 Nm).
18. Torque rod-to-subframe bolts and torque the bolts to 22 ft. lbs. (30 Nm).
   - Engine oil cooler
   - Rear attaching clamps for the steering servo line to the subframe
   - Steering gear and torque the bolts to 66 ft. lbs. (90 Nm)
   - Outer ball joints to the steering swivel member and torque the bolts to 64 ft. lbs. 85 Nm)
   - Anti-roll bar stay and torque the fastener to 64 ft. lbs. (85 Nm)
   - Cover plate between the engine and transmission
   - Rear engine pad and engine mount and torque the bolts to 50 ft. lbs. (70 Nm)

- Engine mount stay and torque the fastener to 16 ft. lbs. (22 Nm)
- Front exhaust pipe and torque the flange to the turbocharger to 19 ft. lbs. (25 Nm)
- Engine stay and torque the fastener to 16 ft. lbs. (22 Nm)
- Inner wheel well covers and remove the lifting devise from the top of the engine
- Upper engine bolts and torque the bolts to 50 ft. lbs. (70 Nm)
- Front torque arm to transmission and torque the arm to 34 ft. lbs. (47 Nm)
- Selector linkage to the transmission and torque the bolt to 18 ft. lbs. (24 Nm)
- Gear linkage to the selector rod. Place the vehicle in 4th gear and secure with a locking pin. Torque the clamp on the linakge to 16 ft. lbs. (22 Nm).
- O$_2$sensor connectors
- Battery and tray
- Battery cables
- Battery cover
- Front wheels
- Intake manifold cover
- Grille

19. Fill and bleed the clutch system.
20. Fill the transmission fluid to the proper level.
21. Start the vehicle, check for leaks and proper operation. Repair if necessary.

Automatic Transaxle

REMOVAL & INSTALLATION

900 and 9-3 Models

1. Before servicing the vehicle, refer to the precautions in the beginning of this section.
2. Drain the transaxle.
3. Remove or disconnect the following:
- Battery
- Ground strap from the transaxle housing
- Dipstick and sleeve
- Vent hose from the transaxle housing
- Positive cable routing straps
- Transaxle selector lever from housing
- Electrical harness on transaxle housing
- Oxygen (O$_2$S) sensor and securing straps

4. Install the engine lifting beam. Insert the lifting bar in the eyelets and connect to the engine holder.
5. Remove or disconnect the following:
- Front wheels
- Front exhaust pipes
- Front spoilers
- Ball joints on both sides
- Subframe assembly
- Cooler lines from the transaxle
- Left and right halfshafts and suspend the shafts with securing straps
- Intermediate shaft bearing bracket and pull out the shaft
- Actuator cover and release the torque converter's securing point on the actuator
- Transaxle mounting bracket
- Transaxle mounting bolts that secure the transaxle to the engine
- Engine/transaxle surface mounting bolts
- 2 outer bolts holding the gearbox to the engine mounting surface

6. Install the lifting eye to the transaxle and connect the hoist to the transaxle.
7. Remove the last bolt and pull out the transaxle assembly.

To install:

8. Using a hoist lift the transaxle into position.
9. To help with installation use a guide pin to line up the transaxle.
10. When the transaxle is in position, fit the 3 upper bolts into the engine housing.
11. Torque the upper transaxle-to-engine mounting bolts to 55 ft. lbs. (75 Nm). Install the transaxle/engine lifting beam to hold the unit in place.
12. Install or connect the following:
- Transaxle/engine surface mounting bolts and torque the bolts to 55 ft. lbs. (75 Nm)
- Transaxle mounting bracket to the transaxle housing and the frame assembly
- Torque converter to the actuator with the retaining bolts and torque the bolts to 37 ft. lbs. (55 Nm)
- Actuator cover
- Intermediate halfshaft with the bracket bearing assembly
- Outer halfshafts
- Transaxle cooler lines to the transaxle assembly
- Sub-frame assembly. Fit the sub-frame bolts and torque the front bolts to 85 ft. lbs. (116 Nm), middle bolts to 141 ft. lbs. (190 Nm),

and the rear bolts to 81 ft. lbs. (110 Nm) plus 75° bolt rotation. Torque the ball joints to 55 ft. lbs. (75 Nm).
- Front exhaust pipes
- Both front wheels

13. Fill the transaxle with fluid.
- Selector lever cable and selector lever to the transaxle housing
- O$_2$S sensor electrical harness
- Dipstick and sleeve to the transaxle housing
- Positive battery cable
- Battery and connect the ground cable

14. Test drive the vehicle.

9000 Models

➡ **The engine and transaxle are removed together.**

1. Before servicing the vehicle, refer to the precautions in the beginning of this section.
2. Drain the transaxle.
3. Drain the cooling system.
4. Remove or disconnect the following:
- Battery and tray
- Expansion tank
- Drive belt for the compressor
- Upper connection on the oil cooler, loosen the pipe clip on the radiator and slide the pipe down behind the radiator
- Connector to the electromagnetic clutch on the compressor and loosen the compressor mounting complete with the belt tensioner

5. Place a protective cloth over the radiator member and rest the compressor on the radiator member. Secure the compressor to the radiator member.
6. If turbocharged, remove the turbocharger pressure pipe, situated between the turbocharger unit and the intercooler.
7. Remove or disconnect the following:
- Oxygen (O$_2$S) sensor
- Flange joint between the exhaust pipe and the exhaust manifold. If turbocharged, disconnect the exhaust pipe coming from the turbocharger. Push the exhaust pipe to one side and unhook the rubber hangers from the exhaust system. Disconnect the bottom coolant hose from the water pump.
- Bottom retaining bolt for the radiator fan
- Speedometer drive from the gearbox

- Gear selector cable from the selector lever. Do not separate at the ball joint.
- Clips on the rubber covers over the inboard universal joints and slide the covers off the drive axles
- Electrical leads from the alternator and the starter motor
- Connector for the oil pressure switch
- Top radiator hose
- Solenoid valve from the bracket on the radiator and unplug the electrical connections
- Fan
- Air mass meter. Leave the rubber socket connector attached to the turbocharger unit
- Air intake duct by pulling it out of the aperture in the wing and twisting the ends inwards
- Air cleaner top section first, then the remaining section
- Relief valve hose from the turbocharger pressure pipe and remove the pipe, if equipped
- Hall effect transducer, the ground lead from the gearbox and the electrical connector for the back-up lights
- Throttle cable linkage and install a clamp to the hydraulic line to the slave cylinder and pinch the line tightly. With proper wrenches, open the line to the clutch slave cylinder.
- Both front wheels

8. From both sides of the vehicle, slacken the lower bolts retaining the steering swivel member to the strut assembly. Remove the 2 upper bolts.

9. Pivot the steering swivel member outwards to pull the inboard universal joint out of the halfshaft. Position dust covers over the exposed halfshaft cups.

- Engine mount bolt
- Steering reservoir from the servo and position it within the engine compartment. Drain the fluid from the container.
- Large bore hose and the delivery hose from the steering servo pump and plug the open ends
- Fuel return line from the pressure regulator
- Nut from the rear engine mounting and back off the front mount bolts a few turns

10. Attach a shop crane to the rear lifting lug.

11. Lift the engine sufficiently to provide access for the removal of the components located between the engine and the firewall.

- Vacuum hoses from the inlet manifold
- Coolant hoses running between the heater core and the water pump pipe
- Coupling between the fuel pipe and the fuel injection manifold

12. Cut the clips securing the wiring looms to the oil pipe and water pipe. Disconnect and tag the vacuum hoses from the inlet manifold.

13. Remove or disconnect the following:
- Wiring loom to the fuel injection manifold
- Grounding connections
- Air-cooled oil cooler and place it on top of the engine. The 2 lower bolts need only be loosened.
- Hood lifts and install hood extenders for extra clearance
- Engine from the vehicle, taking care not to damage the radiator.

To install:

14. Install or connect the following:
- Engine in the vehicle and connect all the engine mounts
- Air-cooled oil cooler
- Ground connections
- Wiring loom to the fuel injection manifold
- Wiring looms to the oil pipe, water pipe, inlet manifold steady bar and the oil supply pipe
- Coupling between the fuel pipe and the fuel injection manifold
- Coolant hoses running between the heater core and the water pump pipe
- Vacuum hoses to the correct connections on the inlet manifold

15. Before removing the lifting sling, connect the components located between the engine and the firewall.

- Nut at the rear engine mounting and tighten the front mount bolts
- Fuel return line to the pressure regulator
- Large bore hose and the delivery hose to the steering servo pump
- Steering reservoir for the servo. Fill the fluid container.
- Engine stay bolt and pivot the steering swivel member outwards to install the inboard universal joint into the halfshaft. Position dust covers over the exposed halfshaft cups and install clamps.
- Gear selector cable to the selector lever
- Lower bolts retaining the steering swivel member to the strut assembly. Tighten the 2 upper bolts

- Front wheels
- Line to the clutch slave cylinder
- Throttle linkage
- Hall Effect transducer
- Negative battery cable to the gearbox and the electrical connector for the back-up lights
- Relief valve hose to the turbocharger pressure pipe, if equipped
- Install the air cleaner
- Install the air intake duct
- Air mass meter
- Top radiator hose at the cylinder head
- Electrical leads to the alternator and the starter motor
- Connector for the oil pressure switch
- Connect the speedometer drive to the transmission
- Flange joint between the exhaust pipe and the exhaust manifold. If turbocharged, connect the exhaust pipe coming from the turbocharger. Hook the rubber hangers on the exhaust system
- Bottom coolant hose to the water pump
- O_2S sensor electrical harness
- Connector into the electromagnetic clutch on the compressor and tighten the compressor mounting complete with the belt tensioner
- Upper connection on the oil cooler and tighten the pipe clip on the radiator
- Drive belt for the compressor
- Expansion tank
- Battery and tray
- Both battery cables

16. Remove the lift supports and lower the vehicle.

17. Fill the cooling system to the proper level.

18. Fill the transmission with fluid.

19. Start the engine and allow it to reach operating temperature.

20. Check the ignition timing and all fluid levels.

21. Test drive the vehicle.

9-5 Models

1. Before servicing the vehicle, refer to the precautions in the beginning of this section.

2. Drain the transmission fluid.

3. Drain the cooling system.

4. Remove or disconnect the following:
- Front grille
- Intake manifold cover

- Battery and tray
- 16 pin and 10 pin wire connectors
- Transmission breather hose
- Gear selector arm from the transmission
- Shifting cable
- Cable channel from the engine and gear case, 3.0L engine only
- 3 upper bolts from the transmission
- Dipstick tube
- Oxygen (O $_2$S) sensor connectors
- Rear engine mount nut
- 3 bolts from the rear engine cushion, loosen only

5. Install a lifting beam to the engine and relieve the weight on the engine and transmission.

- Both front wheels
- Front exhaust system
- Rear engine mount and pad, 2.3L engine only
- Steering gear screws
- Rear clamps securing the power steering delivery pipe to the subframe
- A/C pipes from the subframe holder
- Air cleaner casing from the subframe
- Engine oil cooler from the charge air cooler, 2.3L engine only
- Outer ball joint to steering knuckle bolts
- Upper anti-roll bar ball joints
- Torque arm from the subframe
- Rear support plates

6. Position a lifting trolley with a holder under the vehicle. Align the trolley to the subframe.

7. Remove or disconnect the following:

- Remaining bolts from the subframe and slightly lower the subframe
- Power steering delivery pipe clamps. Lower the lifting trolley and move the subframe aside.
- Splash plate, 2.3L engine only
- Plug covering the torque converter bolts, 3.0L engine only
- Bolts securing the torque converter to the drive plate. Rotate the plate and pulley together.

➡**The crankshaft must be rotated to gain access for all the bolts**

8. Press the torque converter against the transmission to keep the converter in place during transmission removal.

- Torque arm and bracket

- Oil cooler inlet and outlet hoses
- Left drive shaft and suspend it
- Left side engine pad and lower the transmission slightly
- Transmission bracket
- Transmission from the engine assembly
- Transmission from the vehicle

To install:

9. Install or connect the following:

10. Rotate the torque converter so that the bolt holes align with the drive plate.

11. Make certain that the 2 guide sleeves are on the engine.

12. Lubricate and install new drive shaft seals.

13. Raise the transmission into position under the vehicle.

14. Install the transmission and torque the bottom bolts between the engine and transmission to 55 ft. lbs. (74 Nm) and the bolts between the transmission and oil pan to 34 ft. lbs. (47 Nm).

15. Remove the lifting beam from the transmission .

16. Install or connect the following:

- Transmission bolts and torque the bolts to 18 ft. lbs. (24 Nm)
- Engine pad mount to the transmission and torque the bolts to 62 ft. lbs. (84 Nm)
- Bolts securing the engine pad to the body and torque the bolts to 46 ft. lbs. (63 Nm)
- Drive shaft and make certain that the circlip snaps into position
- Torque converter to drive plate and torque the bolts to 22 ft. lbs. (30 Nm)
- Splash plate, 2.3L engine only
- Plug covering the torque converter bolts, 3.0L engine only
- Cooler hoses and torque the fasteners to 20 ft. lbs. (27 Nm)
- Torque arm bracket
- Ground cable to the bracket and raise the subframe into position
- Power steering delivery pipe clamps
- Outer ball joints and A/C pipes
- Subframe and rear support plates and torque the subframe bolts to 74 ft. lbs. (100 Nm) plus 45 degrees and the support plate bolts to 44 ft. lbs. (60 Nm)
- Outer ball joints to the steering knuckles and torque the bolts to 63 ft. lbs. (85 Nm)
- Anti-roll bar link and torque the

fastener to 68 ft. lbs. (92 Nm)
- Steering gear and torque the bolts to 66 ft. lbs. (90 Nm)
- Engine oil cooler and air filter housing, 2.3L engine only
- Rear engine cushion
- Rear engine mount and torque the bolts to 52 ft. lbs. (70 Nm)
- Torque arm to the subframe and torque the bolts to 19 ft. lbs. (25 Nm)
- Front exhaust system and torque the bolts to 18 ft. lbs. (24 Nm) on the 2.3L engine. On the 3.0L engine torque the turbo bolts to 18 ft. lbs. (24 Nm) and the exhaust manifold bolts to 30 ft. lbs. (40 Nm).
- O$_2$S sensor wiring connectors and remove the lifting beam

17. Torque the rear engine cushion bolts to 18 ft. lbs. (24 Nm) and the nut to 33 ft. lbs. (45 Nm).

- Engine to transmission and torque the bolts to 52 ft. lbs. (70 Nm)
- Cable channel to the transmission, 3.0L engine only
- Dipstick tube
- Transmission breather hose
- Shifter cable to the bracket
- Shifter cable to the selector lever. Adjust if necessary.
- Torque rod and torque the bolt to 34 ft. lbs. (47 Nm)
- Battery tray and battery
- Battery cables and cover
- Intake manifold cover
- Front grille
- Front wheels

18. Fill the cooling system to the proper level.

19. Fill the transmission to the proper level.

20. Start the vehicle, check for leaks and repair if necessary.

Clutch

ADJUSTMENT

The clutch cable utilized in the 900 Series vehicles is self-adjusting. Although it does not require periodic adjustments, when the clutch cable is first reinstalled in the vehicle a small procedure will insure that the self-adjuster is properly functioning.

1. Check the functioning of the clutch cable by moving the clutch lever forward in the car's direction of travel. The balancing

spring should, then be compressed and the length of the clutch cable cover reduced. When the clutch lever is released, the cover should regain its original length. Repeat 3 or 4 times.

2. Hold the balancing spring tightly and give it a small jerk to remove any free-play.

REMOVAL & INSTALLATION

900 and 9-3 Models

1. Before servicing the vehicle, refer to the precautions in the beginning of this section.

2. Place the vehicle on a lift and engage it in 4th gear.

3. Remove or disconnect the following:
- Transaxle assembly and lock the flywheel in position
- Pressure plate retaining nuts
- Pressure plate
- Clutch disc

To install:

4. Install or connect the following:
- Clutch disc with pressure plate and torque the nuts to 16 ft. lbs. (22 Nm)
- Transaxle

5. To help with installation use a guide pin to line up the transaxle.

6. Fill the transaxle with fluid.

7. Install the clutch cable.

8. Install the transmission assembly.

9. Test drive the vehicle.

9000 Models

1. Before servicing the vehicle, refer to the precautions in the beginning of this section.

2. Remove or disconnect the following:
- Battery and tray
- Washer fluid container and connectors
- Stay for the Anitlock Braking System (ABS) hydraulic unit, if equipped

- Bulkhead cover and separate the speedometer cable connector by removing the washer hose, then the speedometer cable through the rubber grommet
- Air mass meter and intake air temperature sensor connectors
- Bypass valve delivery pipe hose
- Delivery pipe between the throttle housing and the intercooler, if equipped
- Starter motor
- Selector rod universal joint and selector rod
- Slave cylinder pressure line and pinch off the hose with clamps
- Left-hand engine mount and attach a sling to the engine lifting beam. Raise and support the vehicle.
- Left wheel and inner wheel housing
- Reverse light connector
- Ball joint
- Sway bar
- Three bottom bolts from the joint between the engine and the transaxle
- Speedometer cable at the transaxle
- Center and left skirts under the spoiler
- Subframe
- Universal joint
- Top nut and bolt from the clutch housing face
- Transaxle and install a flywheel locking tool
- Clutch assembly

3. Inspect the flywheel for wear, scoring, cracking or other damage. Replace or resurface, as necessary.

To install:

4. Use a centering arbor type tool or an appropriate input shaft to center the clutch disc to the flywheel.

5. Install or connect the following:
- Clutch pressure plate. Torque the bolts, alternately and evenly, in

several steps, to 10–20 ft. lbs. (13–25 Nm). Remove the flywheel lock, if used.
- Transaxle assembly and engage the transaxle input shaft into the clutch plate splines

➡**Prior to installation, ensure that the halfshaft is in position and the aluminum tube is pressed into the seal.**

- Three bottom bolts into the joint face and torque the bolts to 40–74 ft. lbs. (54–100 m)
- Universal joint
- Subframe
- Sway bar
- Ball joint
- Starter
- Wheel housing bolts and torque the bolts to 32–43 ft. lbs. (43–57 Nm)
- Left engine mount and torque the bolt to37–67 ft. lbs. (41–91 Nm)
- Negative battery cable
- Reverse light switch
- Wheel housing liners
- Left wheel
- Selector rod universal joint
- Slave cylinder pressure pipe (remove the clamping tongs)
- Speedometer cable
- Bulkhead cover

6. Bleed the slave cylinder, check the transaxle fluid level and road test the vehicle.

9-5 Models

1. Before servicing the vehicle, refer to the precautions in the beginning of this section.

2. Remove or disconnect the following:
- Transmission and install Flywheel Tool 83–94–868
- Pressure plate
- Clutch assembly

To install:

3. Install or connect the following:
- Driven plate and drive plate on the flywheel. Hand tighten the bolts.
- Center the drive plate and torque the bolts to 16 ft. lbs. (22 Nm). Remove the flywheel locking tool
- Transmission assembly

4. Fill the transmission with clean fluid to the proper level.

Hydraulic Clutch System

BLEEDING

1. Before servicing the vehicle, refer to the precautions in the beginning of this section.

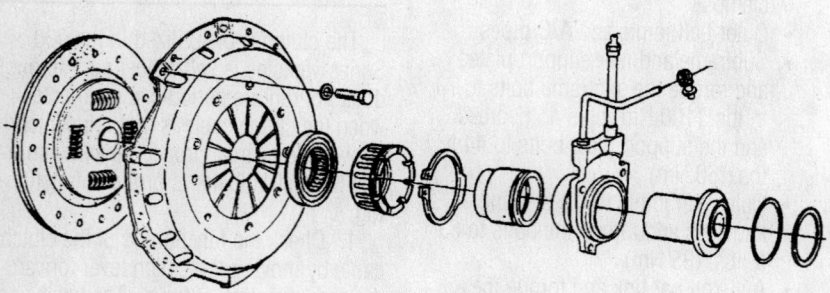

7923SG21

Exploded view of the clutch assembly—9000

2. Connect a hose to the slave cylinder bleeder valve. Place the other end of the hose in a suitable jar partially filled with brake fluid.

3. Fill the master cylinder with brake fluid.

4. Open the bleeder valve on the slave cylinder ½ turn.

5. Place a cooling system pressure tester gauge over the opening of the master cylinder.

6. Pump the tester until all air is expelled from the hydraulic clutch system.

7. Close the slave cylinder bleeder valve.

8. Check that all air was removed from the system and the clutch is functioning properly. Adjust the fluid level, as required.

Halfshaft

REMOVAL & INSTALLATION

900 and 9-3 Models

1. Before servicing the vehicle, refer to the precautions in the beginning of this section.

2. Remove or disconnect the following:
   - Negative battery cable
   - Wheel
   - Hub center nut
   - Ball joint
   - Sway bar hardware and rubber bushing

3. Push down on the lower control arm. Using a rubber mallet, tap the halfshaft out of the hub.

4. Move the strut to one side. Be extremely careful not to stretch or break the Antilock Braking System (ABS) sensor cables or brake hoses. Place a drain pan under the transaxle to catch any fluid spillage.

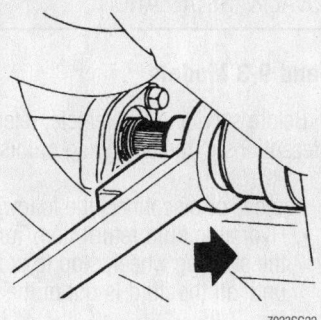

7923SG22

Removing the halfshaft joint from the intermediate shaft—900 and 9-3

- Halfshaft
- Inner halfshaft joint from the transaxle on the left-hand side
- Halfshaft joint from the intermediate shaft on the right-hand side

To install:

5. Install or connect the following:
   - Halfshaft
   - Hub and new center nut onto halfshaft. Do not tighten yet.
   - Ball joint and torque the ball joint nut to 55 ft. lbs. (75 Nm)
   - Sway bar to the lower control arm. Install the rubber bushing, a new washer and new retaining nut and torque the nut to 89 inch lbs. (10 Nm).
   - Wheel
   - Negative battery cable

6. Tighten the hub center nut to 214 ft. lbs. (290 Nm).

7. Check the transaxle fluid level and top off if necessary.

9000 Models

1. Before servicing the vehicle, refer to the precautions in the beginning of this section.

2. Remove or disconnect the following:
   - Negative battery cable
   - Front wheel
   - Inner fender panel

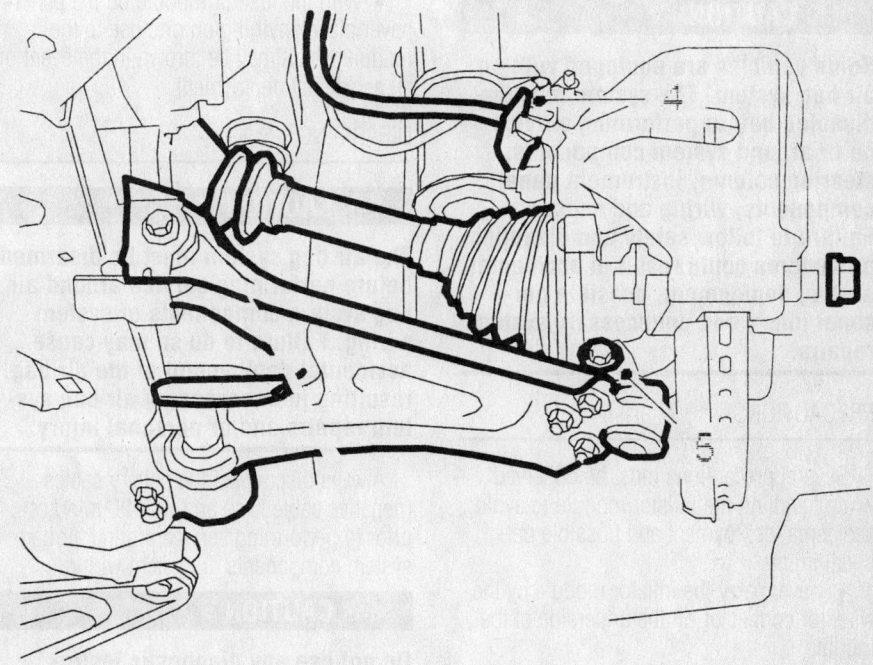

9347UG20

Remove the right side halfshaft from the transmission

- Strut from the steering swivel member
- Flexible brake hose from the clip on the strut
- Antilock Braking System (ABS) sensor and lead, if equipped
- Rubber boot clamp on the inboard CV-joint and separate the 2 halves of the joint
- Hub center nut
- Halfshaft

To install:

3. Install or connect the following:
   - Halfshaft and new hub center nut. Join the 2 halves of the joint.
   - New rubber boot and clamp and torque the strut-to-steering swivel bolts to 58–77 ft. lbs. (78–105 Nm)
   - ABS sensor, if equipped
   - Brake hose to the clip on the strut
   - Inner fender panel
   - Wheel
   - Negative battery cable

4. With all 4 wheels on the ground torque the hub center nut to 207–221 ft. lbs. (280–300 Nm). Check the alignment and test drive the vehicle.

9-5 Models

1. Before servicing the vehicle, refer to the precautions in the beginning of this section.

2. Remove or disconnect the following:

- Negative battery cable
- Front wheel
- Hub nut
- Halfshaft from the transmission, left side
- Cover and knock out the halfhsaft from the intermediate shaft, right side only
- Antilock Brake System (ABS) cable fasterners and clips
- Ball joint from the steering swivel and lower the suspension arm
- Strut
- Halfshaft from the hub by tapping lightly on it

To install:

3. Install or connect the following:

- Halfshaft to the hub and hand tighten the hub nut
- Halfshaft to the transmission and make certain that the circlip snaps into position
- Halfshaft to the intermediate shaft and make certain that the circlip snaps into position
- Ball joint to the steering swivel and

torque the fastener to 74 ft. lbs. (100 Nm)

- ABS cable to the clips
- Front wheel and torque the hub nut to 123 ft. lbs. (170 Nm) plus an additional 45 degrees
- Negative battery cable

4. Check and fill the transmission fluid to the proper level.

CV-Joints

REMOVAL & INSTALLATION

900 and 9-3 Models

1. Secure the driveshaft assembly in a vice

2. Remove or disconnect the following:

- Clamps and rubber boot
- Circlip

3. Tap the CV joint off the axle and take care not to damage splines

4. Rotate inner bearing to facilitate removal of steel balls and cage from outer race

To install:

5. Install or connect the following:

- New rubber boot onto shaft
- Inner bearing into outer race. Rotate so that steel balls can be inserted and pack with 80 grams Molycote Rapid G grease.
- Press joint assembly onto shaft
- Circlip
- Rubber boot retaining clamps

9000 Models

1. Secure the driveshaft assembly in a vice.

2. Remove or disconnect the following:

- Clamps and rubber boot
- Circlip

3. Tap the CV joint off the axle and take care not to damage splines

To install:

4. Install or connect the following:

- New rubber boot onto shaft
- CV joint and pack with 80 grams Esso HF Nebula EP2. 120 grams for Turbo model.
- Press joint assembly onto shaft
- Circlip
- Rubber boot retaining clamps

STEERING AND SUSPENSION

Air Bag

✻✻ CAUTION

Some vehicles are equipped with an air bag system. The system must be disabled before performing service on or around system components, steering column, instrument panel components, wiring and sensors. Failure to follow safety and disabling procedures could result in accidental air bag deployment, possible personal injury and unnecessary system repairs.

PRECAUTIONS

Several precautions must be observed when handling the inflator module to avoid accidental deployment and possible personal injury.

- Never carry the inflator module by the wires or connector on the underside of the module.

- When carrying a live inflator module, hold securely with both hands, and ensure that the bag and trim cover are pointed away.

- Place the inflator module on a bench

or other surface with the bag and trim cover facing up.

- With the inflator module on the bench, never place anything on or close to the module which may be thrown in the event of an accidental deployment.

DISARMING

✻✻ CAUTION

The air bag system must be disarmed before performing service around air bag system components or system wiring. Failure to do so may cause accidental deployment of the air bag, resulting in unnecessary air bag system repairs and/or personal injury.

Always disconnect the battery cables (negative cable first) and wait 20 minutes prior to performing service around air bag system components or system wiring.

✻✻ CAUTION

Do not use any diagnostic instruments that are battery powered, such as buzzers, ohmmeters or diode testers, to diagnose faults in the steering wheel or electronic control

unit. Using such devices may trigger the air bag. Also, ensure that the battery cables cannot accidentally come into contact with the battery terminals.

REARMING

To rearm the air bag system, reconnect the battery cables.

Rack and Pinion Steering Gear

REMOVAL & INSTALLATION

900 and 9-3 Models

1. Before servicing the vehicle, refer to the precautions in the beginning of this section.

2. Remove or disconnect the following:

- Hydraulic fluid return line. Turn the steering wheel, stop to stop, until all the fluid is out of the system.
- Negative battery cable
- Lower left dash finish panel
- Fuse box
- Steering column and pinion shaft

locknut. Make certain to straighten the steering wheel and mark it with chalk to center.
- Column shaft from pinion shaft
- Tracking rods
- Front wheels
- Tie rods
- Rack assembly retaining clamps
- Pressure and return lines from the rack assembly
- Rack assembly through the left wheel housing

To install:

➡**If a replacement rack assembly is being used, transfer all lines and bellows to the replacement unit as required.**

3. Install or connect the following:
- Steering rack assembly through the left wheel housing and torque the bolts to 18 ft. lbs. (24 Nm)
- Pressure and return lines to the steering rack and torque to 21 ft. lbs. (28 Nm)
- Rack assembly retaining clamps
- Tie rods and torque to 44 ft. lbs. (60 Nm)
- Tracking rods to the center of the rack assembly and torque the rods to 63–74 ft. lbs. (85–100 Nm)
- Column shaft and pinion shaft
- Lower left dash finish panel
- Fuse box
- Front wheels
- Negative battery cable
- Hydraulic fluid return line

4. Fill the reservoir and turn the steering wheel, stop to stop, until all the air is out of the system.

5. Check the toe-in and steering wheel alignment.

9000 Models

1. Before servicing the vehicle, refer to the precautions in the beginning of this section.

2. Remove or disconnect the following:
- Negative battery cable
- Padding from under the instrument panel
- Trim on the left side of the center tunnel, as required
- Carpet where the steering column passes through the bulkhead
- Rubber boot from the intermediate shaft
- Pinch bolt in the lower clamp of the intermediate shaft

- Intermediate shaft
- Cover panel from the bulkhead. Take care not to damage the gasket, seal and plastic bushing.
- Front wheels
- Inner fender panel rear section on the left side
- Tie rod ends
- Power steering fluid from the pump reservoir
- Hoses from the pump and reservoir. Plug the openings to prevent fluid from leaking out and dirt from entering.
- Retaining bolts from the rack and pinion assembly
- Vertical brace between the engine subframe and the body
- Rack and pinion unit. Pull through the left fender inner panel opening. Do not damage the rubber boots or brake hose.

To install:

3. Install or connect the following:
- Rack and pinion unit through the left fender opening and torque bolts to 46–56 ft. lbs. (60–80 Nm)
- Vertical brace between the engine subframe and the body
- Hoses from the pump and reservoir
- Tie rod ends
- Inner fender panel rear section on the left side
- Front wheels
- Cover panel to the bulkhead
- Pinch bolt in the lower clamp and torque the bolt to 27–32 ft. lbs. (35–42 Nm)
- Negative battery cable

4. Fill the reservoir and bleed the system by allowing the engine to run at idle and turning the steering wheel, stop to stop, 2 or more times.

9-5 Models

1. Before servicing the vehicle, refer to the precautions in the beginning of this section.

2. Drain the power steering reservoir.

3. Remove or disconnect the following:
- Negative battery cable
- Power steering reservoir and move it aside
- Return hose from the power steering reservoir
- Steering column shaft from the steering gear
- Intake manifold cover

- Rear engine cushion to the subframe
- Engine cushion to the engine mounting
- Both front wheels
- Track rod lock nut
- Track rod end from the steering swivel
- Outer track rod end from the steering swivel with Puller Tool 89–96–696
- Track rod end from the track rod and count the number of revolutions to ease installation
- Reinforcement from the rear attaching point on the subframe
- Exhaust system between the catalytic converter and the silencer
- Subframe center attaching point, lower the subframe
- Engine cushion
- Delivery and return pipes from the valve body
- Steering gear retaining bolts
- Steering gear through the passenger side wheel well housing

To install:

4. Install or connect the following:
- Steering gear into position. Hand tighten the bolts at this time.
- Valve body dust cover
- Delivery and return pipes to the valve body and torque the bolts to 25 ft. lbs. (30 Nm)
- Steering gear bolts and torque the bolts to 70 ft. lbs. (95 Nm)
- Engine cushion and raise the subframe and torque the center attaching bolts to 75 ft. lbs (100 Nm) plus 45 degrees
- Exhaust system between the catalytic converter and the silencer
- Subframe rear attaching points to the reinforcement and torque the bolts to 75 ft. lbs. (100 Nm) plus 45 degrees
- Outer track rod ends to the track rods. Hand tighten the lock nuts.
- Outer track rod ends on the steering swivel and torque the lock nuts to 45 ft. lbs. (60 Nm)
- Both front wheels
- Rear engine cushion to the subframe and torque the bolts to 20 ft. lbs. (25 Nm)
- Rear engine cushion to engine mount and torque the bolts to 35 ft. lbs. (50 Nm)
- Intake manifold cover

- Steering column shaft with the steering gear and torque the bolts to 25 ft. lbs. (30 Nm)
- Return hose to the power steering reservoir
- Power steering fluid reservoir
- Negative battery cable

5. Fill and bleed the power steering system.

6. Start the vehicle, check for leaks and repair if necessary.

7. Check the toe-in and adjust if necessary.

Strut

REMOVAL & INSTALLATION

Front

900 AND 9-3 MODELS

1. Before servicing the vehicle, refer to the precautions in the beginning of this section.

2. Remove or disconnect the following:
- Negative battery cable
- Front wheel
- Hub nut
- Wheel Speed Sensor (WSS)
- Caliper. Be sure the caliper is prop-

erly supported and not hanging from the brake hose.
- Rotor and backing plate
- Tie rod end screw with Puller Tool 89–96–696
- Sway bar and bushing
- Ball joint
- Three upper strut mounting bolts
- Strut housing
- Spring and shock absorber cartridge

3. Place the strut in a vise and compress the spring.
- Self-locking nut from the bearing plate and discard the nut
- Coil spring, bellows and rubber bumper
- Upper flange nut
- Strut cartridge from the housing

To install:

4. Install or connect the following:
- New strut cartridge and torque the spanner nut to 159 ft. lbs. (215 Nm)
- Coil spring. Be sure the end of the spring is up against the spring stop.
- Upper spring seat. Make sure the notches are properly aligned. Place the upper bearing assembly on the

upper seat and secure it using a new self-locking nut.

5. Remove the spring compressor and be sure the coil spring stays properly seated.
- Strut assembly and torque the upper mounting bolts 13 ft. lbs. (18 Nm)
- Ball joint and new self locking nut and torque the nut to 55 ft. lbs. (75 Nm)
- Tie rod end
- Sway bar and bushing and torque to 96 inch lbs. (12 Nm)
- Backing plate and rotor
- Caliper
- WSS
- Hub nut. Do not tighten yet
- Front wheel
- Negative battery cable

6. Torque the hub nut to 125 ft. lbs. (170 Nm) plus an additional 45 degrees.

7. Pump the brake pedal to position the brake caliper piston.

✳✳ CAUTION

Do not attempt to move the vehicle until a firm pedal is obtained.

8. Check front wheel alignment.

9000 MODELS

1. Before servicing the vehicle, refer to the precautions in the beginning of this section.

2. With all 4 wheels on the ground, loosen the front hub nut.

3. Remove or disconnect the following:
- Negative battery cable
- Front wheel
- Hub nut
- Wheel Speed Sensor (WSS)
- Caliper. Be sure the caliper is properly supported and not hanging from the brake hose.
- Rotor and backing plate
- Tie rod end
- Sway bar and rubber bushing
- Strut from lower steering swivel and pull strut from swivel assembly
- Three upper strut mounting bolts
- Strut housing
- Spring and shock absorber cartridge

4. Place the strut in a vise.

5. Compress the spring.
- Self locking nut
- Coil spring, bellows and rubber snubber
- Upper flange nut
- Strut cartridge from the housing

To install:

6. Install a new strut cartridge into the

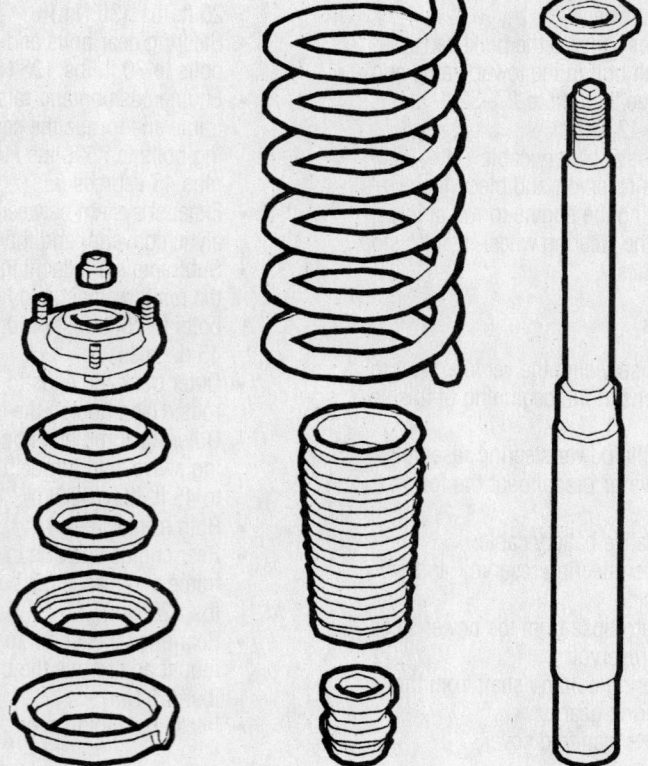

Exploded view of the strut assembly—9-5 Models

9347UG21

housing and torque the spanner nut to 159 ft. lbs. (215 Nm).

7. Position the compressed coil spring in place. Be sure the end of the spring is up against the spring stop.

8. Position the upper spring seat with the notches properly aligned. Place the upper bearing assembly on the upper seat and secure it using a new self-locking nut.

9. Remove the spring compressor and be sure the coil spring stays properly seated.

10. Position the strut assembly on the vehicle. Torque the upper mounting bolts 13 ft. lbs. (18 Nm).

11. Install or connect the following:
- Strut to steering swivel assembly bolts
- Tie rod end
- Sway bar at the lower control arm link and torque to 96 inch lbs. (12 Nm)
- Backing plate
- Brakes
- WSS
- Hub nut. Do not tighten.
- Wheel
- Negative battery cable

12. With all 4 wheels on the ground torque the hub nut to 215 ft. lbs. (290 Nm).

13. Pump the brake pedal to position the brake caliper piston.

❊❊ CAUTION

Do not attempt to move the vehicle until a firm pedal is obtained.

14. Check front wheel alignment.

9-5 MODELS

1. Before servicing the vehicle, refer to the precautions in the beginning of this section.

2. Remove or disconnect the following:
- Negative battery cable
- Front wheel
- Nut from the anti roll bar link
- Steering swivel member from the strut and move the Antilock Braking System (ABS) sensor cable and brake hose aside
- Upper retaining bolts from the strut
- Strut assembly

3. Place the strut in a vise.

4. Compress the spring with Tool 88–18–791 and Holder 88–18–817.
- Self locking nut
- Coil spring, bellows and rubber snubber

- Upper flange nut
- Strut cartridge from the housing

To install:

5. Install or connect the following:
- Compression stop, spring seat, rubber gaiter and spring. Make certain that the lower end of the spring is properly seated.
- Upper spring seat. Make sure the notches are properly aligned. Place the upper bearing assembly on the upper seat and secure it using a new self-locking nut and torque the nut to 55 ft. lbs. (75 Nm).

6. Remove the spring compressor and be sure the coil spring stays properly seated.

- Strut assembly and torque the bolts to 13 ft. lbs. (18 Nm)
- Press the steering swivel member inward and torque the bolts to 75 ft. lbs. (100 Nm) plus 45 degrees
- Anti roll bar link and torque the nut to 70 ft. lbs. (95 Nm)
- ABS sensor cable and brake hose
- Front wheel
- Negative battery cable

Shock Absorber

REMOVAL & INSTALLATION

Rear

900 AND 9-3 MODELS

1. Before servicing the vehicle, refer to the precautions in the beginning of this section.

2. Remove or disconnect the following:
- Upper mounting bolt in trunk. Cut out a flap in the carpeting to access the mounting bolt and bushings.
- Nut, washer and bushing from the mounting point
- Rear wheel
- Lower shock mounting bolt
- Shock

To install:

3. Install or connect the following:
- Shock to the upper mounting. Be sure to install the lower part of the bushing on the shock.
- Lower shock mounting bolt and torque the bolt to 46 ft. lbs. (62 Nm)

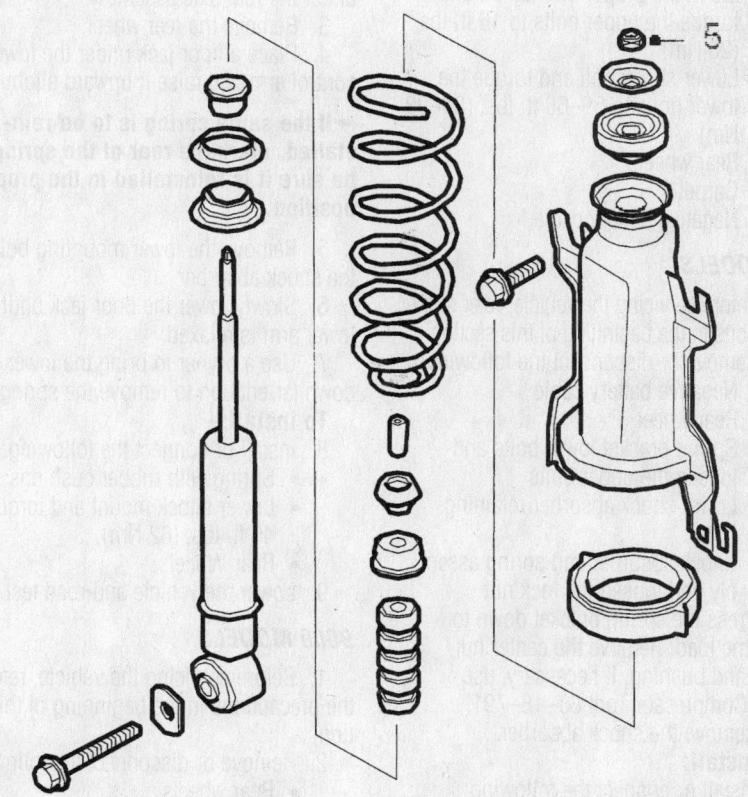

Exploded view of the shock absorber and spring assembly—9-5 Models

9347UG22

For Tire, Wheel and Ball Joint specifications, see Section 1 of this manual

- Rear wheel
- Upper bushing, washer and nut and torque to 15 ft. lbs. (20 Nm)
- Carpeting

9000 MODELS

1. Before servicing the vehicle, refer to the precautions in the beginning of this section.

2. Disconnect the negative battery cable.

3. Remove the rear wheel.

4. Place jackstands under the rear end of the trailing arms where they mount to the rear axle.

5. Position a jack at the rear jacking point. Raise the rear of the vehicle enough to relieve the load on the shock absorbers and anti-roll bar.

6. Remove the lower shock absorber bolts.

7. Remove the upper shock bolts, nut, washer and bushing. Carpet will have to be pulled back.

8. Remove the shock from the vehicle.

To install:

9. Install or connect the following:
- Shock and make sure the bushings are in the proper orientation and torque the upper bolts to 16 ft. lbs. (20 Nm)
- Lower shock bolt and torque the lower bolts to 59–66 ft. lbs. (80–90 Nm)
- Rear wheels
- Carpet
- Negative battery cable

9-5 MODELS

1. Before servicing the vehicle, refer to the precautions in the beginning of this section.

2. Remove or disconnect the following:
- Negative battery cable
- Rear wheel
- Spring bracket lower bolts and loosen the upper bolts
- Lower shock absorber retaining bolt
- Shock absorber and spring assembly and loosen the lock nut

3. Press the spring bracket down to relieve the load. Remove the center nut, washer and bushing. If necessary, use Spring Compressor Tool 88–18–791.

4. Remove the shock absorber.

To install:

5. Install or connect the following:
- Spring, spacer ring and bracket on the shock absorber

6. Press the bracket down to relieve the shock absorber load and install the bushing and washer.

- New locknut and torque the nut to 15 ft. lbs. (20 Nm)
- Shock absorber and torque the bolts to 40 ft. lbs. (55 Nm)
- Lower mounting bolt on the rear axle and torque the bolt to 140 ft. lbs. (190 Nm)
- Rear wheel
- Negative battery cable

Coil Spring

REMOVAL & INSTALLATION

Front

The front coil spring removal and installation procedure is covered in the strut removal and installation procedure.

Rear

900 AND 9-3 MODELS

1. Before servicing the vehicle, refer to the precautions in the beginning of this section.

2. Raise the vehicle and safely support it on jackstands. Do not place the stands under the rear axle assembly.

3. Remove the rear wheel.

4. Place a floor jack under the lower control arm and raise it upward slightly.

➡**If the same spring is to be reinstalled, mark the rear of the spring to be sure it is reinstalled in the proper position.**

5. Remove the lower mounting bolt from the shock absorber.

6. Slowly lower the floor jack until the lower arm is relaxed.

7. Use a prybar to bring the lower arm down far enough to remove the spring.

To install:

8. Install or connect the following:
- Spring with rubber cushions
- Lower shock mount and torque to 46 ft. lbs. (62 Nm)
- Rear Wheels

9. Lower the vehicle and road test.

9000 MODELS

1. Before servicing the vehicle, refer to the precautions in the beginning of this section.

2. Remove or disconnect the following:
- Rear wheels
- Hand brake cable from the retaining bracket connected to the lower control arm
- Antilock Braking System (ABS)

sensor wiring by removing the clip and releasing the cable, if equipped

3. Place a jack under the lower arm.

4. Lower the jack slowly and remove the coil spring.

To install:

5. Install or connect the following:
- Coil spring
- ABS wiring, if equipped
- Rear wheels

6. Road test the vehicle.

9-5 MODELS

Refer to the shock absorber removal and installation procedure for the rear coil spring.

Upper Ball Joint

REMOVAL & INSTALLATION

900 and 9-3 Models

1. Before servicing the vehicle, refer to the precautions in the beginning of this section.

2. Remove or disconnect the following:
- Wheel
- Anti-roll bar nut

3. Loosen the ball joint nut with Puller Tool 89 96 696 to press out the bolt from the swivel member.
- Ball joint nut and pull down the stanchion/ suspension arm.
- Lock ring securing the bellows
- Ball joint

To install:
- Bellows lock ring
- Stanchion/suspension arm into position
- Ball joint and torque the nut to 55 ft. lbs. (75 Nm)
- Anti-roll bar and torque the nut to 7 ft. lbs. (10 Nm)
- Wheel

9-5 Models

1. Before servicing the vehicle, refer to the precautions in the beginning of this section.

2. Remove or disconnect the following:
- Wheel
- Transverse link from the longitudinal link and install a spacer between the upper and lower links
- Press out the ball joint with Tool 89–96–761

To install:

3. Press in the ball joint with Tool 89–96–761.

4. Install or connect the following:

- Transverse link to the longitudinal link and torque the bolt to 70 ft. lbs. (90 Nm) plus 60 degrees
- Wheel

Lower Ball Joint

REMOVAL & INSTALLATION

900 and 9-3 Models

➡**The ball joint cannot be removed from the control arm. To replace the ball joint, the control arm must be replaced.**

1. Before servicing the vehicle, refer to the precautions in the beginning of this section.
2. Raise the vehicle.
3. Remove or disconnect the following:
   - Wheel
   - Sway bar link bolt
   - Ball joint out of the steering knuckle
   - Retaining nut at the support arm
   - Retaining bolt at the subframe

To install:
4. Install or connect the following:

- Arm and bolt at the subframe and torque the retaining bolt at the subframe to 85 ft. lbs. (115 Nm)
- Bolt at the support arm and torque the bolt to 68 ft. lbs. (92 Nm)
- Sway bar link and torque the link to 89 inch lbs. (10 Nm)
- Ball joint and torque to 55 ft. lbs. (75 Nm)
- Wheel
5. Check the front wheel alignment.

9000 Models

1. Before servicing the vehicle, refer to the precautions in the beginning of this section.
2. Remove or disconnect the following:
   - Wheel
   - Three mounting bolts securing the ball joint to the lower control arm
   - Lockwasher from atop the steering swivel member
   - Ball joint

To install:
3. Install or connect the following:
   - Ball joint and torque it to 20 ft. lbs. (28 Nm)

- Locking ring atop the ball joint
- Ball joint to the lower control arm and torque the 3 mounting bolts to 22 ft. lbs. (30 Nm)
- Wheel
4. Lower the vehicle and test drive.

9-5 Models

1. Before servicing the vehicle, refer to the precautions in the beginning of this section.
2. Remove or disconnect the following:

- Wheel
- Transverse link from the longitudinal link and place a spacer between the upper and lower links
- Press the brake pistons back
- Brake line bracket from the longitudinal link
- Brake caliper bolts and support the caliper with wire ties
- Antilock Braking System (ABS) sensor connector, if equipped
- Hub retaining nuts
- Handbrake return spring and cable
- Hub and brake disc

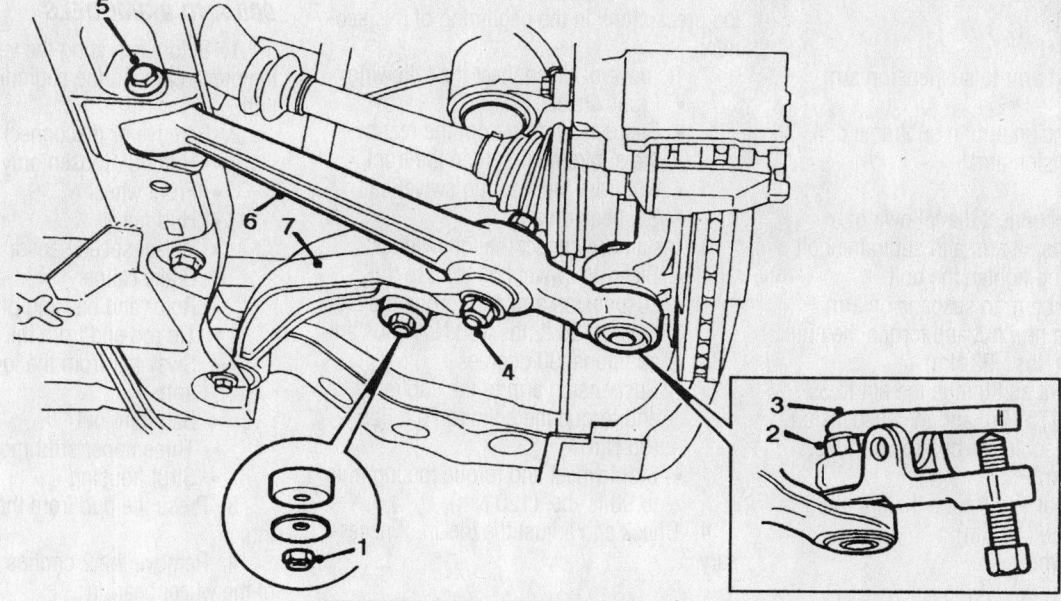

1. Sway bar nut
2. Ball joint nut
3. Ball joint press tool
4. Support arm connection
5. Subframe connection
6. Lower control arm
7. Support arm

7923SG23

Lower control arm connection points—900 and 9-3

For Wheel Alignment specifications, see Section 1 of this manual

- Press out the ball joint with Tool 89–96–761

To install:

3. Press in the ball joint with Tool 89–96–761.

4. Install or connect the following:
  - Hook the handbrake return spring and cable to the hub and brake disc
  - Hub on the back plate and secure it to the longitudinal link with new nuts and torque to 40 ft. lbs. (50 Nm) plus 30 degrees
  - ABS sensor connector
  - Brake caliper
  - Brake line bracket to the longitudinal link and torque the bolt to 70 ft. lbs. (90 Nm) plus 60 degrees
  - Wheel and torque the nuts to 80 ft. lbs. (110 Nm)

5. Depress the brake pedal and verify proper operation of the braking system.

Suspension Arm

REMOVAL & INSTALLATION

900 and 9-3 Models

1. Remove or disconnect the following:
  - Front wheel
  - Sway bar
  - Ball joint
  - Support arm to suspension arm nut
  - Suspension arm to subframe bolt
  - Suspension arm

To install:

2. Install or connect the following:
  - Suspension arm and subframe bolt and hand tighten the bolt
  - Support arm to suspension arm using a new nut and torque the nut to 68 ft. lbs. (92 Nm)
  - Ball joint and torque the nut to 55 ft. lbs. (75 Nm) and the suspension arm to subframe bolt to 85 ft. lbs. (115 Nm)
  - Sway bar and torque the nut to 89 inch lbs. (10 Nm)
  - Front wheel

9000 Models

1. Remove or disconnect the following:
  - Wheel
  - Bolts holding suspension arm to ball joint
  - Nut holding suspension arm to sway bar
  - Sway bar mounting nut, loosen only

2. Pull the steering knuckle out of the way

3. Pry down on the suspension arm and pull out the sway bar link
  - Two nuts holding the suspension arm to the subframe. Push the bolts towards the engine.
  - Suspension arm support bracket
  - Suspension arm

To install:

➡**Do not tighten the rear nut for the suspension arm until the car has been lowered to the floor. Damage to the rubber bushing may result.**

4. Place the suspension arm in position and rest its weight on the ball joint.

5. Install or connect the following:
  - Hardware holding the front of the suspension arm to the subframe
  - Support bracket
  - Sway bar link and torque the fastener to 18 ft. lbs. (24 Nm)
  - Bolts holding the suspension arm to the ball joint and torque the bolts to 22 ft. lbs. (30 Nm)
  - Wheel

9-5 Models

1. Before servicing the vehicle, refer to the precautions in the beginning of this section.

2. Remove or disconnect the following:
  - Front wheel
  - Suspension arm from the rear of the subframe first then the front
  - Nut from the steering swivel ball
  - Suspension arm

3. Install or connect the following:
  - Steering swivel ball joint to the suspension arm and torque the fastener to 23 ft. lbs. (30 Nm) plus an additional 90 degrees
  - Suspension arm to the subframe and torque the bolts to 70 ft. lbs. (95 Nm)
  - Front wheel and torque the lug nuts to 90 ft. lbs. (120 Nm)

4. Check and adjust the toe-in, if necessary.

Support Arm

REMOVAL & INSTALLATION

900 and 9-3 Models

1. Remove or disconnect the following:
  - Front wheel
  - Sway bar
  - Ball joint

  - Support arm to suspension arm bolt
  - Support arm to subframe bolt

2. Pry out support arm

To install:

3. Install or connect the following:
  - Support arm to subframe bolt and torque the bolt to 74 ft. lbs. (100 Nm) plus 75 to 90°
  - Support arm to suspension arm bolt and torque the bolt to 68 ft. lbs. (92 Nm)
  - Ball joint and torque the nut to 55 ft. lbs. (75 Nm)
  - Sway bar and torque the nut to 89 inch lbs. (10 Nm)
  - Wheel

Wheel Bearings

ADJUSTMENT

The wheel bearings found in all Saab vehicles are sealed units requiring no adjustment.

REMOVAL & INSTALLATION

Front

900 AND 9-3MODELS

1. Before servicing the vehicle, refer to the precautions in the beginning of this section.

2. Remove or disconnect the following:
  - Hub nut, loosen only
  - Front wheel
  - Hub nut
  - Wheel speed sensor
  - Brake caliper
  - Rotor and backing plate
  - Tie rod end from the knuckle
  - Sway bar from the lower control arm
  - Ball joint out
  - Three upper strut mounting bolts
  - Strut housing

3. Press the hub from the wheel bearing.

4. Remove the 2 circlips on each side of the wheel bearing.

5. Press out the wheel bearing.

To install:

6. Install one circlip and press in wheel bearing until it contacts the circlip. Be sure the press tool contacts only the outer bearing race.

7. Install the other circlip.

8. Clean the bearing surface of the hub on a wire wheel. If the hub is pitted or damaged, it should be replaced.

9. Support the inner bearing race and press hub into wheel bearing. If the inner race is not properly supported, the bearing will fail quickly.

10. Install or connect the following:
- Strut assembly and torque the upper mounting bolts 13 ft. lbs. (18 Nm)
- Ball joint and a new self locking nut and torque the nut to 55 ft. lbs. (75 Nm)
- Tie rod end
- Sway bar and torque at the lower control arm link to 89 inch lbs. (10 Nm)
- Backing plate and rotor
- Caliper
- Wheel sensor
- Hub nut. Do not tighten.
- Wheel

11. With all 4 wheels on the ground torque the hub nut to 215 ft. lbs. (290 Nm).

12. Pump the brake pedal to position the brake caliper piston.

✳✳ CAUTION

Do not attempt to move the vehicle until a firm pedal is obtained.

13. Check front wheel alignment.

9000 MODELS

1. Before servicing the vehicle, refer to the precautions in the beginning of this section.

2. Remove or disconnect the following:
- Hub nut. Loosen only while still on ground.
- Wheel assembly
- Hub center nut and thrust washer
- Flexible brake hose
- Caliper
- Locating studs for the rotors and rotors

3. Push in on the halfshaft. If the CV-joint shaft does not push in, use a puller to break it loose.

➡**Do not allow the puller to push the halfshaft in more than 2 in. or damage to the inboard joint may occur.**

- Four bolts securing the hub to the knuckle. All vehicles use socket head screws to retain the hub assembly. For easier removal, cut an Allen wrench to fit the head and turn it with an 8mm wrench.
- Hub and disc backing plate
- Two bolts connecting the knuckle to the strut

To install:

4. Clean the bearing seat and lightly coat with molybdenum grease.

5. Install or connect the following:
- Hub assembly. Draw the hub in by tightening the 4 retaining screws and torque the screws to 41–44 ft. lbs. (55–60 Nm).
- Knuckle. Connect the 2 strut bolts with the nuts facing to the front of the vehicle and torque the bolts to 58–78 ft. lbs. (78–105 Nm).
- Rotor and brake assembly and torque the caliper bolts to 52–80 ft. lbs. (70–110 Nm)
- Wheel

6. With all 4 wheels on the ground torque the center hub nut to 207–221 ft. lbs. (280–300 Nm).

7. Pump the brake pedal to seat the brake pads before road testing.

9-5 MODELS

1. Before servicing the vehicle, refer to the precautions in the beginning of this section.

2. Remove or disconnect the following:
- Hub center. Loosen only while still on ground
- Wheel assembly
- Hub center nut and thrust washer
- Brake caliper from the spindle and support it to the side
- Brake disc
- Antilock Braking System (ABS) sensor
- Steering link with Tool 89–96–696
- Steering swivel from the strut and move the bracket aside
- Ball joint from the steering swivel
- Backing plates
- Halfshaft from the hub
- Steering swivel and press off the hub with Tool 89–96–704
- Circlip and press out the wheel bearing with Tool 89–96–704

To install:

3. Install or connect the following:
- Press in a new bearing
- Circlip with the opening facing downward
- Press on the hub with Tool 89–96–704
- Halfshaft to the hub
- Holder for the ABS sensor and backing plates
- Ball joint to the steering swivel
- Press the steering swivel to the strut and ABS sensor bracket. Torque the bolts to 75 ft. lbs. (100 Nm) plus 45 degrees. Torque the ball joint nut to 65 ft. lbs. (85 Nm).
- Outer steering link to the steering swivel and torque the nut to 45 ft. lbs. (60 Nm)
- ABS sensor
- Brake disc and torque the bolts to 80 ft. lbs. (110 Nm)
- Front wheel
- New hub nut and torque the nut to 126 ft. lbs. (170 Nm) plus 45 degrees

4. Depress the brake pedal several times and verify the proper operation of the braking system.

Rear

900 AND 9-3 MODELS

1. Before servicing the vehicle, refer to the precautions in the beginning of this section.

2. Remove or disconnect the following:
- Rear wheels
- Caliper and back off the adjuster on the parking brake shoes.
- Parking brake return spring and lever
- Rotor retaining screws and the rotor
- Four hub retaining nuts
- Wheel speed sensor
- Hub and bearing

➡**There is a spacer behind the brake backing plate.**

To install:

3. Install or connect the following:
- Hub assembly, backing plate and spacer and torque the nuts to 37 ft. lbs. (50 Nm)
- Brake rotor. Apply low strength Loctite® to the rotor retaining screw.
- Parking brake lever and return spring
- Speed sensor
- Caliper and retaining bolts. Use Loctite® applied.

4. Screw in the brake shoe adjusting screw until the rotor cannot turn. Back off the screw until the rotor can rotate freely.

5. Install the wheels.

6. Pump the brakes to position the caliper piston.

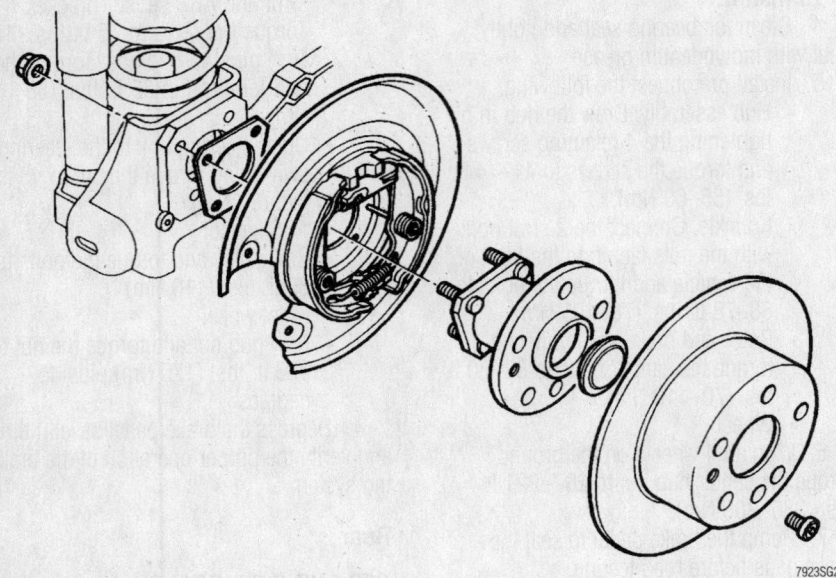

Exploded view of the rear hub and bearing components—900 and 9-3

⁜⁜ CAUTION

Do not attempt to move the vehicle until a firm pedal is obtained.

9000 MODELS

1. Before servicing the vehicle, refer to the precautions in the beginning of this section.
2. Remove or disconnect the following:
    *   Wheel
    *   Hand brake cable from the caliper. Remove the adjuster screw plug and loosen the adjusting screw enough to allow the brake piston to slide back.
    *   Caliper and backing plate
    *   Antilock Braking System (ABS) sensor, if equipped

*   Rotor
*   Hub nut dust cap
*   Hub nut and thrust washer.
*   Hub assembly

➡**Whenever the hub nut is removed, a new one must always be used because the locking device on the nut becomes ineffective.**

To install:

3. Check the spindle for damage and repair or replace as required.
4. Install or connect the following:
    *   Hub assembly
    *   Thrust washer and new nut and torque the nut to 214 ft. lbs. (290 Nm)
    *   Dust cover

*   Rotor
*   Caliper and torque the bolts to 51–65 ft. lbs. (70–90 Nm)
*   Wheel
5. Pump the brake pedal several times before driving the vehicle.

9-5 MODELS

1. Before servicing the vehicle, refer to the precautions in the beginning of this section.
2. Remove or disconnect the following:
    *   Wheel
    *   Brake line bracket from the longitudinal link
    *   Brake caliper and support it to the side
    *   Handbrake adjustment screw
    *   Brake disc
    *   Antilock Braking System (ABS) sensor electrical connector
3. Separate the wheel hub from the backing plate
To install:
4. Install or connect the following:
    *   Wheel hub to the backing plate and secure the assembly to the longitudinal link and torque the nuts to 40 ft. lbs. (50 Nm) plus 30 degrees
    *   ABS sensor electrical connector
    *   Brake disc to the hub
    *   Brake caliper
    *   Brake line bracket to the longitudinal link
    *   Handbrake adjustment screw. Screw in the adjustment 7 notches.
    *   Wheel
5. Depress the brake pedal several times to verify proper braking operation.
6. Adjust the handbrake, if necessary.

SUBARU

1998–01
Legacy • Legacy Outback • Legacy SUS • Impreza • Impreza Outback • Impreza Outback Sport

19

PRECAUTIONS

Before servicing any vehicle, please be sure to read all of the following precautions that deal with personal safety, prevention of component damage, and important points to take into consideration when servicing a motor vehicle:

• Never open, service or drain the radiator or cooling system when the engine is hot; serious burns can occur from the steam and hot coolant.

• Observe all applicable safety precautions when working around fuel. Whenever servicing the fuel system, always work in a well-ventilated area. Do not allow fuel spray or vapors to come in contact with a spark, open flame, or excessive heat (a hot drop light, for example). Keep a dry chemical fire extinguisher near the work area. Always keep fuel in a container specifically designed for fuel storage; also, always properly seal fuel containers to avoid the possibility of fire or explosion. Refer to the additional fuel system precautions later in this section.

• Fuel injection systems often remain pressurized, even after the engine has been turned **OFF**. The fuel system pressure must be relieved before disconnecting any fuel lines. Failure to do so may result in fire and/or personal injury.

• Brake fluid often contains polyglycol ethers and polyglycols. Avoid contact with the eyes and wash your hands thoroughly after handling brake fluid. If you do get brake fluid in your eyes, flush your eyes with clean, running water for 15 minutes. If

eye irritation persists, or if you have taken brake fluid internally, IMMEDIATELY seek medical assistance.

• The EPA warns that prolonged contact with used engine oil may cause a number of skin disorders, including cancer. You should make every effort to minimize your exposure to used engine oil. Protective gloves should be worn when changing oil. Wash your hands and any other exposed skin areas as soon as possible after exposure to used engine oil. Soap and water, or waterless hand cleaner should be used.

• All new vehicles are now equipped with an air bag system. The system must be disabled before performing service on or around system components, steering column, instrument panel components, wiring and sensors. Failure to follow safety and disabling procedures could result in accidental air bag deployment, possible personal injury, and unnecessary system repairs.

• Always wear safety goggles when working with, or around, the air bag system. When carrying a non-deployed air bag, be sure the bag and trim cover are pointed away from your body. When placing a non-deployed air bag on a work surface, always face the bag and trim cover upward, away from the surface. This will reduce the motion of the module if it is accidentally deployed. Refer to the additional air bag system precautions later in this section.

• Clean, high quality brake fluid from a

sealed container is essential to the safe and proper operation of the brake system. You should always buy the correct type of brake fluid for your vehicle. If the brake fluid becomes contaminated, completely flush the system with new fluid. Never reuse any brake fluid. Any brake fluid that is removed from the system should be discarded. Also, do not allow any brake fluid to come in contact with a painted surface; it will damage the paint.

• Never operate the engine without the proper amount and type of engine oil; doing so will result in severe engine damage.

• Timing belt maintenance is extremely important. Many models utilize an interference-type, non-freewheeling engine. If the timing belt breaks, the valves in the cylinder head may strike the pistons, causing potentially serious (also time-consuming and expensive) engine damage. Refer to the maintenance interval charts in the front of this manual for the recommended replacement interval for the timing belt, and to the timing belt section for belt replacement and inspection.

• Disconnecting the negative battery cable on some vehicles may interfere with the functions of the on-board computer system(s) and may require the computer to undergo a relearning process once the negative battery cable is reconnected.

• When servicing drum brakes, only disassemble and assemble one side at a time, leaving the remaining side intact for reference.

ENGINE REPAIR

➡**Disconnecting the negative battery cable on some vehicles may interfere with the functions of the on board computer systems and may require the computer to undergo a relearning process, once the negative battery cable is reconnected.**

Alternator

REMOVAL

1. Remove or disconnect the following:
   • Negative battery cable
   • Connector and terminal from the alternator
   • V-belt cover
   • Front side V-belt
   • Alternator to bracket bolts
   • Alternator from the vehicle

INSTALLATION

1. Install or connect the following:
   • Alternator into the vehicle
   • Alternator to bracket bolts
   • Front side V-belt and cover
   • Connector and terminal to the alternator
   • Negative battery cable
2. Check and adjust the belt tension.

Ignition Timing

ADJUSTMENT

All Subaru models are equipped with Distributorless Ignition System (DIS). The ignition timing is controlled by the Powertrain Control Module (PCM) and is not adjustable.

Engine Assembly

✳✳ CAUTION

Some models covered by this manual may be equipped with an air bag. Whenever working near any of the Supplemental Restraint System (SRS) components, such as the impact sensors, the air bag module, steering column and instrument panel, properly disable the SRS.

REMOVAL & INSTALLATION

2.2L and 2.5L Engines

1. Before servicing the vehicle, refer to the precautions in the beginning of this section.

2. Relieve the fuel system pressure.

3. Drain the engine oil and coolant into suitable containers.

4. Remove or disconnect the following:
- Battery cables
- Battery from the vehicle
- Radiator hoses
- Fan motor harness
- Radiator

5. If equipped with air conditioning, discharge the system using an approved recovery/recycling machine. Disconnect and cap the lines from the compressor.
- Air intake duct
- Air cleaner element and upper cover
- Evaporator canister and bracket
- Oxygen (O$_2$S) sensor connector
- Engine ground terminal
- Crankshaft Position (CKP) sensor connector
- Camshaft Position (CMP) sensor connector
- Knock Sensor (KS) connector
- Alternator connector and terminal
- Air conditioning compressor connectors, if equipped
- Accelerator cable
- Cruise control cable, if equipped
- Clutch release spring, clutch cable and hill holder cable, if equipped with a manual transaxle
- Brake booster hose(s)
- Heater inlet and outlet hoses
- Alternator drive belt
- Spark plug wires from left side of engine
- Power steering pump line bracket
- Power steering pump, leave the lines connected and position aside
- Exhaust Y-pipe
- Lower starter nuts
- Lower engine-to-transaxle nuts
- Front engine mount-to-crossmember nuts
- Starter

6. If equipped with an automatic transaxle, perform the following:

a. Remove the torque converter service hole plug.

b. Rotate the engine. Remove the torque converter-to-drive plate bolts as they become accessible.

7. Remove or disconnect the following:
- Pitching stopper
- Fuel delivery, return and evaporation hoses

8. Support the engine with a suitable lifting device attached to the engine lifting eyes.

9. Slightly raise the engine.

10. Raise the transaxle with a floor jack.

11. Slowly remove the engine from the vehicle.

To install:

12. Apply a small amount of grease to the splines of the mainshaft.

13. Position the engine in the engine compartment and align it with the transaxle.

14. Install or connect the following:
- Engine upper bolts: 34–40 ft. lbs. (44–54 Nm)

15. Remove the lifting device and floor jack.

16. Install or connect the following:
- Pitching stopper and tighten the bolts to the following specifications:
a. Body side: 49 ft. lbs. (67 Nm).
b. Bracket side: 40 ft. lbs. (54 Nm).

17. If equipped with an automatic transaxle, perform the following:

a. Install the torque converter-to-drive plate bolts while rotating the engine, and tighten to 20 ft. lbs. (26 Nm).

b. Install the service hole cover.

18. Install or connect the following:
- Evaporator canister and bracket
- Power steering pump. Torque retainers to 22–36 ft. lbs. (29–47 Nm).
- Drive belt, adjust tension
- Starter. Tighten bolts to 34–40 ft. lbs. (44–52 Nm).
- Lower engine-to-transaxle nuts. Tighten to 34–40 ft. lbs. (44–52 Nm).
- Lower engine mounting nuts. Tighten to 61 ft. lbs. (83 Nm) in the inner most elliptical hole in the front crossmember so the clearance is 0.16–0.24 in. (4–6mm).
- Exhaust Y-pipe with new gaskets and nuts
- Brake booster hose
- Heater inlet and outlet hoses
- Accelerator and the cruise control cables, if equipped

19. If equipped with a manual transaxle, install the following:
- Clutch release spring
- Clutch cable
- Hill holder cable

20. Install or connect the following:
- Engine harness connectors
- O$_2$sensor connector
- Engine ground terminal
- CKP sensor connector

- CMP sensor connector
- Knock sensor connector
- Alternator connector and terminal
- Air conditioning compressor connectors, if equipped
- Air cleaner element and cover
- Air conditioning lines, if equipped, with new O-rings. Tighten the bolts to 23 ft. lbs. (31 Nm).
- Radiator
- Engine cover
- Battery

21. Fill the engine with the recommended oil.

22. Fill and bleed the cooling system.

23. Charge the air conditioning system using an approved recovery/recycling machine.

24. Adjust the clutch cable.

25. If equipped, check the automatic transaxle fluid level and add Dexron®II if necessary.

26. Start the engine and allow it to reach normal operating temperature. Check for leaks.

Water Pump

REMOVAL & INSTALLATION

1. Before servicing the vehicle, refer to the precautions in the beginning of this section.

2. Remove or disconnect the following:
- Negative battery cable
- Engine undercover, if equipped

3. Drain the coolant into a suitable container.

4. Remove or disconnect the following:
- Radiator fan connector(s)
- Radiator outlet and heater hoses
- Heater bypass hose or overflow hose, if equipped
- Reservoir tank, on Legacy models
- Radiator fan motor assembly(ies)
- Accessory drive belts
- Timing belt
- Belt tension adjuster
- Belt idler No. 2
- Camshaft Position (CMP) sensor
- Left side camshaft pulley(s)
- Left side rear timing belt cover
- Tensioner bracket
- Radiator and heater hoses from water pump
- Water pump retainer bolts
- Water pump

5. Inspect the radiator hoses for deterioration and replace as necessary.

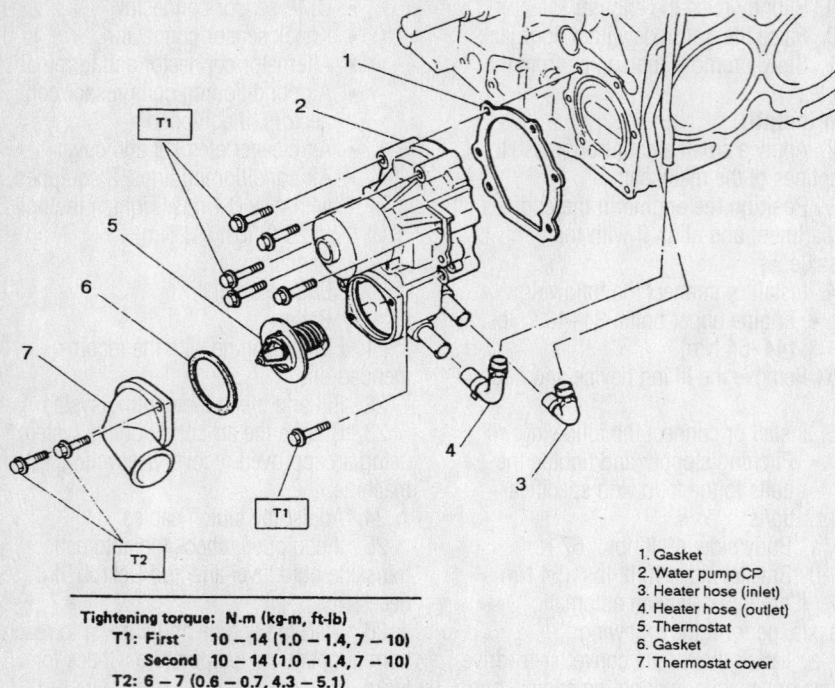

1. Gasket
2. Water pump CP
3. Heater hose (inlet)
4. Heater hose (outlet)
5. Thermostat
6. Gasket
7. Thermostat cover

Tightening torque: N.m (kg-m, ft-lb)
T1: First   10 – 14 (1.0 – 1.4, 7 – 10)
    Second 10 – 14 (1.0 – 1.4, 7 – 10)
T2: 6 – 7 (0.6 – 0.7, 4.3 – 5.1)

7923TG01

Exploded view of the water pump assembly—all engines

To install:

6. Clean the gasket mating surfaces thoroughly. Always use new gaskets during installation.

7. Install or connect the following:
- Water pump, tighten the bolts in sequence to 10 ft. lbs. (13 Nm). After tightening the bolts once, retighten to the same specification again.
- Radiator heater hoses to water pump
- Tensioner bracket
- Left side rear timing belt cover
- Left side camshaft pulley(s)
- CMP sensor
- Belt idler No. 2
- Belt tension adjuster
- Timing belt
- Accessory drive belts
- Radiator fan assembly(ies)
- Reservoir tank, if removed
- Heater bypass hose or overflow hose, if equipped
- Air intake duct
- Radiator outlet and heater hoses
- Radiator fan connector(s)
- Engine undercover, if removed

8. Fill the system with coolant and connect the negative battery cable.

9. Start the engine and allow it to reach operating temperature.

10. Check for leaks.

Cylinder Head

➡**On some models, engine compartment room is limited, so it may be necessary to remove the engine to service the cylinder heads.**

REMOVAL & INSTALLATION

2.2L Engines

1. Before servicing the vehicle, refer to the precautions in the beginning of this section.

2. Remove or disconnect the following:
- Negative battery cable
- Drive belt(s)
- Power steering pump
- Alternator and bracket
- Valve rocker cover
- Positive Crankcase Ventilation (PCV) hose and spark plug wires
- Connector bracket attaching bolt
- Crankshaft Position (CKP) and Camshaft Position (CMP) sensors
- Oil pressure switch
- Knock sensor
- Blow-by hose

3. Relieve the fuel system pressure and disconnect the fuel pipes.
- Intake manifold and gasket
- Water pipe
- Timing belt, camshaft sprocket and related components

- Oil level gauge guide attaching bolt on the left cylinder head
- Cylinder head bolts in the proper sequence. Leave bolts 1 and 3 installed loosely to prevent the cylinder head from falling.
- Cylinder head from the block, use a plastic-faced hammer, if needed, to separate the head from the cylinder block
- Bolts 1 and 3
- Cylinder head and gasket

4. Clean all gasket material from both mating surfaces.

To install:

5. Inspect the cylinder head for warpage. Warpage should not exceed 0.0020 in. (0.05mm).

6. Install a new head gasket and the cylinder head.

7. Secure the head in place with the mounting bolts. Coat each bolts with clean engine oil, and hand-tighten. Tighten the cylinder head bolts, in sequence, to the following specifications:
- a. Step 1: 22 ft. lbs. (29 Nm).
- b. Step 2: 51 ft. lbs. (69 Nm).
- c. Step 3: loosen all bolts by 180 degrees, then loosen an additional 180 degrees.
- d. Step 4: bolts 1 and 2: 25 ft. lbs. (34 Nm).
- e. Step 5: bolts 3, 4, 5 and 6: 11 ft. lbs. (15 Nm).
- f. Step 6: all bolts plus 80–90 degrees.
- g. Step 7: all bolts plus 80–90 degrees.

➡**Do not exceed 180 degrees total tightening.**

8. Install or connect the following:
- Oil level gauge guide attaching bolt on the left cylinder head
- Timing belt, camshaft sprocket and related components
- Water pipe
- Intake manifold and tighten the bolts to 21–25 ft. lbs. (28–34 Nm)
- Fuel delivery pipes
- Blow-by hose
- Knock sensor
- Oil pressure switch connector
- CKP and CMP sensors
- Connector bracket attaching bolt
- Spark plug wires
- PCV hose
- Valve rocker cover and tighten the bolts to 44 inch lbs. (5 Nm)
- Alternator
- Power steering pump
- Accessory drive belt
- Negative battery cable

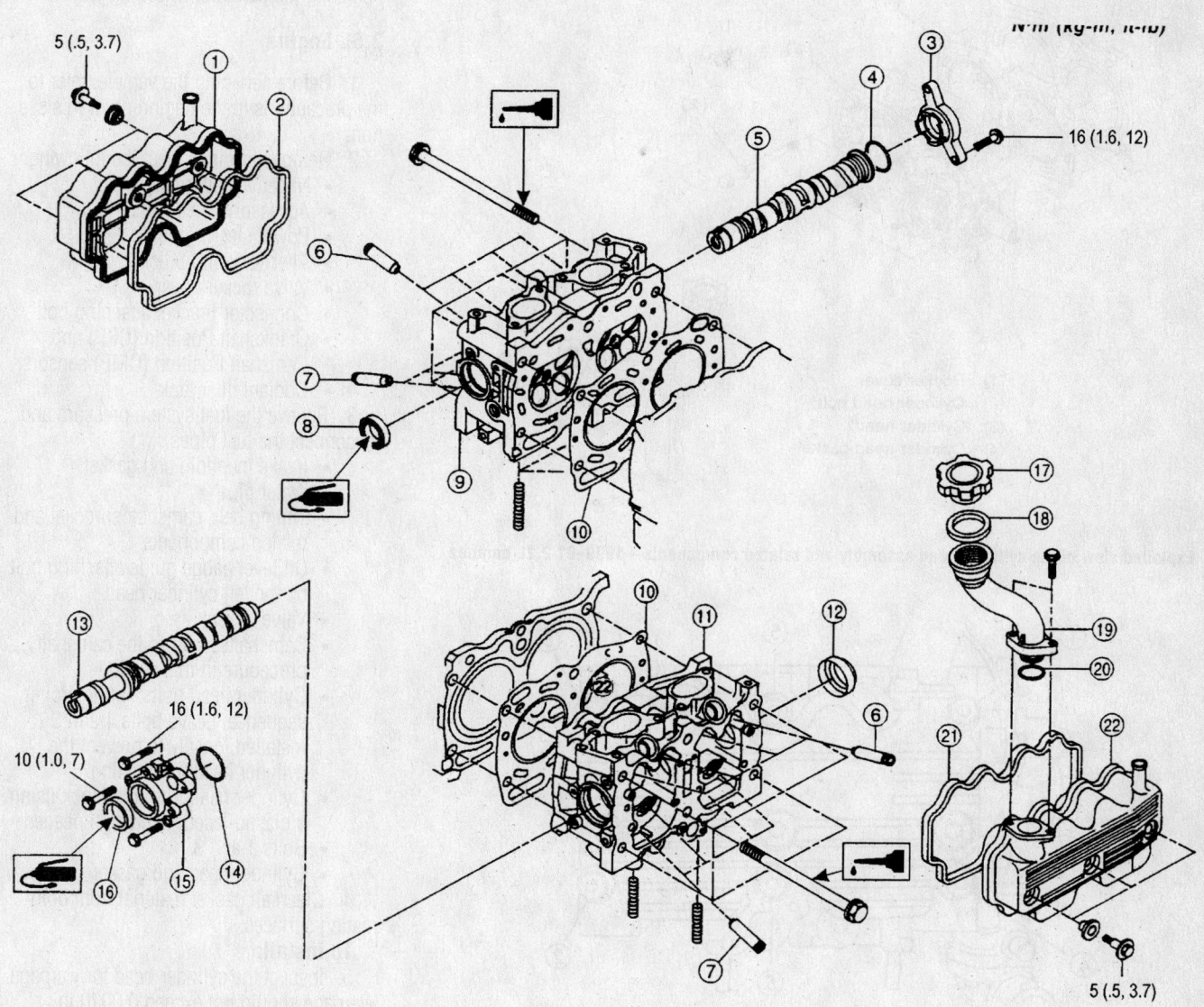

N·m (kg-m, ft-lb)

5 (.5, 3.7)

16 (1.6, 12)

16 (1.6, 12)

10 (1.0, 7)

5 (.5, 3.7)

① Rocker cover (RH)
② Rocker cover gasket
③ Camshaft support (RH)
④ O-ring
⑤ Camshaft (RH)
⑥ Intake valve guide
⑦ Exhaust valve guide
⑧ Oil seal
⑨ Cylinder head (RH)
⑩ Cylinder head gasket
⑪ Cylinder head (LH)

⑫ Plug
⑬ Camshaft (LH)
⑭ O-ring
⑮ Camshaft support (LH)
⑯ Oil seal
⑰ Oil filler cap
⑱ Gasket
⑲ Oil filler pipe
⑳ O-ring
㉑ Rocker gasket
㉒ Rocker cover (LH)

7923TG04

Exploded view of the cylinder head assembly and related components—1998 2.2L engines

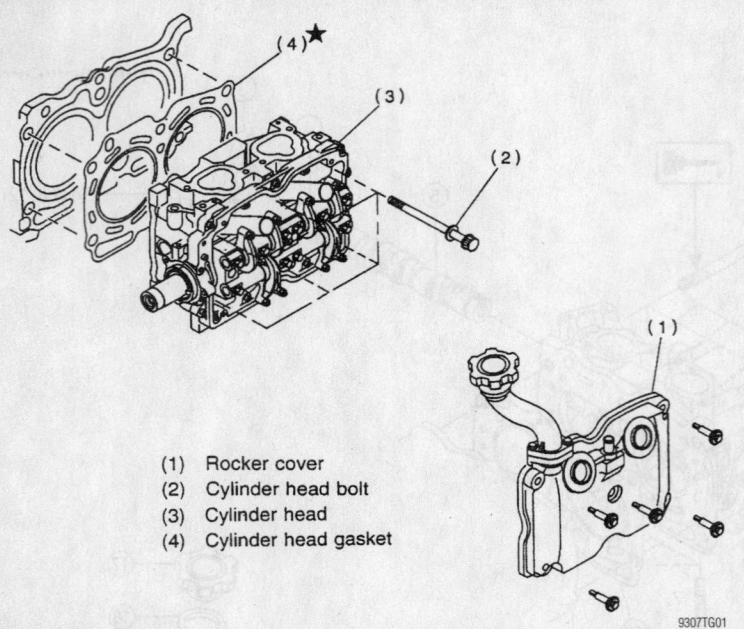

(1) Rocker cover
(2) Cylinder head bolt
(3) Cylinder head
(4) Cylinder head gasket

Exploded view of the cylinder head assembly and related components—1999–01 2.2L engines

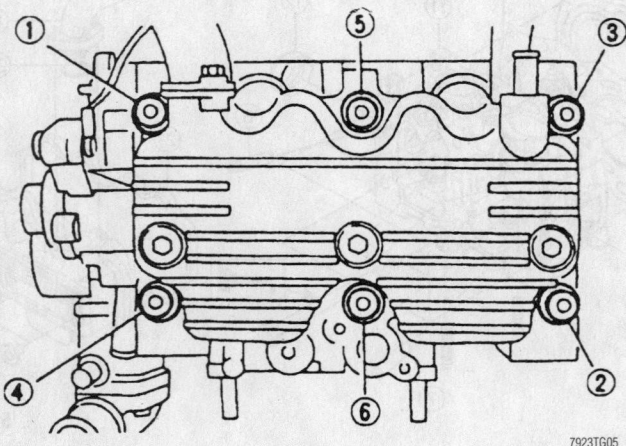

Cylinder head bolt loosening sequence—2.2L engines

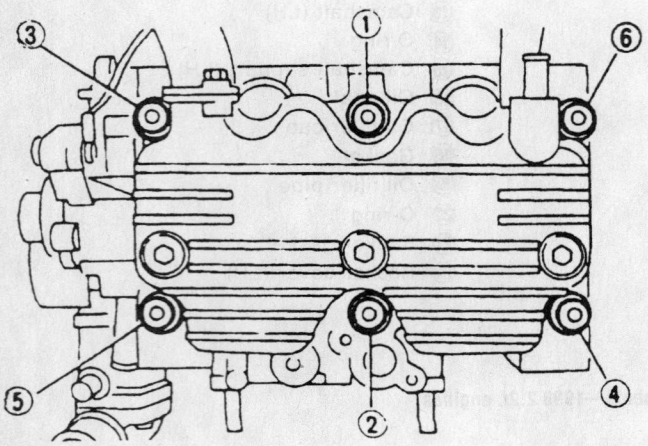

Cylinder head bolt tightening sequence—2.2L engines

9. Start the engine and allow it to reach operating temperature. Check for leaks.

2.5L Engine

1. Before servicing the vehicle, refer to the precautions in the beginning of this section.

2. Remove or disconnect the following:
- Negative battery cable
- Accessory drive belts
- Power steering pump
- Alternator and bracket
- Valve rocker cover
- Connector bracket attaching bolt
- Crankshaft Position (CKP) and Camshaft Position (CMP) sensors
- Coolant filler tank

3. Relieve the fuel system pressure and disconnect the fuel pipes.
- Intake manifold and gasket
- Water pipe
- Timing belt, camshaft sprocket and related components
- Oil level gauge guide attaching bolt on the left cylinder head
- Valve covers
- Camshafts, refer to the camshaft procedure in this section
- Cylinder head bolts, in the proper sequence. Leave bolts 1 and 3 installed loosely to prevent the cylinder head from falling.
- Cylinder head from the block using a plastic-faced hammer, if needed
- Bolts 1 and 3
- Cylinder head and gasket

4. Clean all gasket material from both mating surfaces.

To install:

5. Inspect the cylinder head for warpage. Warpage should not exceed 0.0020 in. (0.05mm).

6. Install a new head gasket and the cylinder head.

7. Secure the head in place with the mounting bolts. Coat each bolts with clean engine oil, and hand-tighten. Tighten the cylinder head bolts to the following specifications:

   a. Step 1: all bolts to 22 ft. lbs. (29 Nm).

   b. Step 2: all bolts to 51 ft. lbs. (69 Nm).

   c. Step 3: loosen all bolts by 180 degrees, then loosen an additional 180 degrees.

   d. Step 4: bolts 1 and 2 to 25 ft. lbs. (24 Nm).

   e. Step 5: bolts 3, 4, 5 and 6 to 11 ft. lbs. (15 Nm).

   f. Step 6: all bolts plus 80–90 degrees.

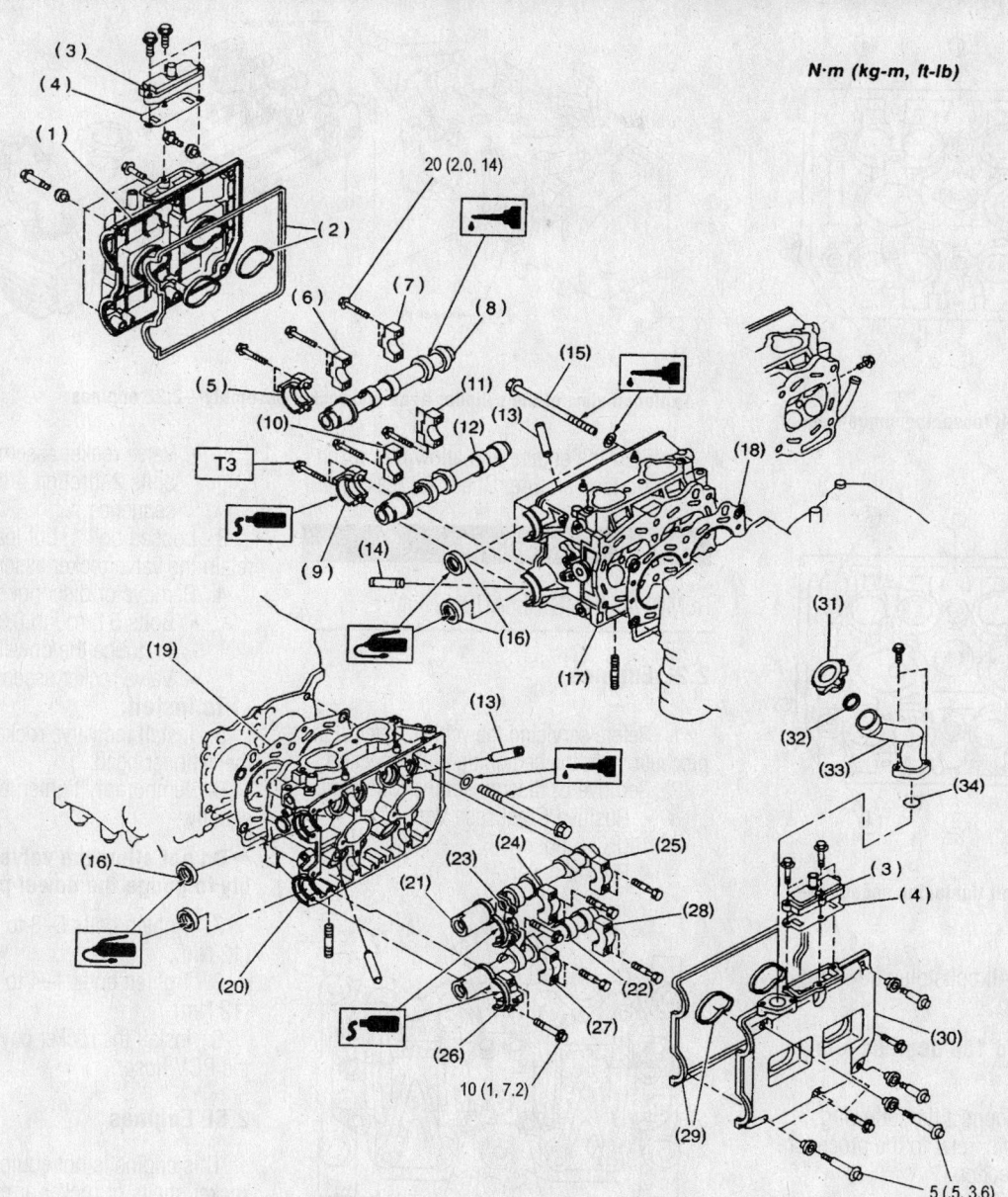

N·m (kg-m, ft-lb)

20 (2.0, 14)

T3

10 (1, 7.2)

5 (.5, 3.6)

(1) Rocker cover (RH)
(2) Rocker cover gasket (RH)
(3) Oil separator cover
(4) Gasket
(5) Intake camshaft cap (Front RH)
(6) Intake camshaft cap (Center RH)
(7) Intake camshaft cap (Rear RH)
(8) Intake camshaft (RH)
(9) Exhaust camshaft cap (Front RH)
(10) Exhaust camshaft cap (Center RH)
(11) Exhaust camshaft cap (Rear RH)
(12) Exhaust camshaft (RH)
(13) Intake valve guide
(14) Exhaust valve guide

(15) Cylinder head bolt
(16) Oil seal
(17) Cylinder head (RH)
(18) Cylinder head gasket (RH)
(19) Cylinder head gasket (LH)
(20) Cylinder head (LH)
(21) Intake camshaft (LH)
(22) Exhaust camshaft (LH)
(23) Intake camshaft cap (Front LH)
(24) Intake camshaft cap (Center LH)
(25) Intake camshaft cap (Rear LH)
(26) Exhaust camshaft (Front LH)
(27) Exhaust camshaft cap (Center LH)
(28) Exhaust camshaft cap (Rear LH)

(29) Rocker cover gasket (LH)
(30) Rocker cover (LH)
(31) Oil filler cap
(32) Gasket
(33) Oil filler duct
(34) O-ring

Exploded view of the cylinder head and related components—2.5L engine

7923TG07

Timing belt service is covered in Section 3 of this manual

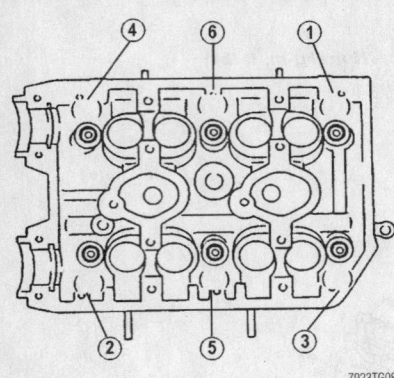

Cylinder head bolt loosening sequence—2.5L engine

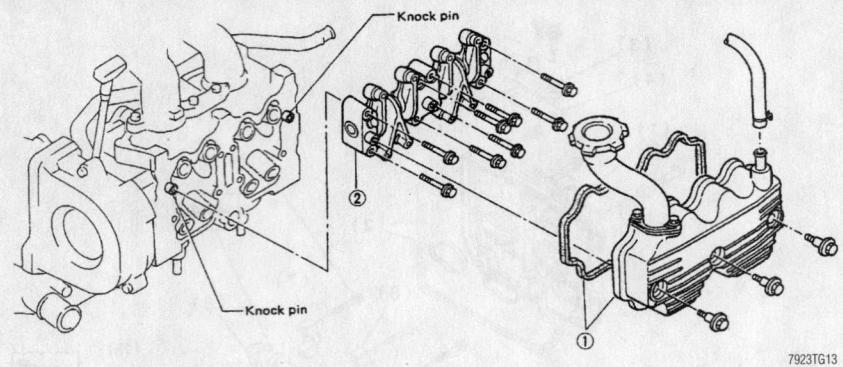

Exploded view of the cylinder head and rocker assembly—2.2L engines

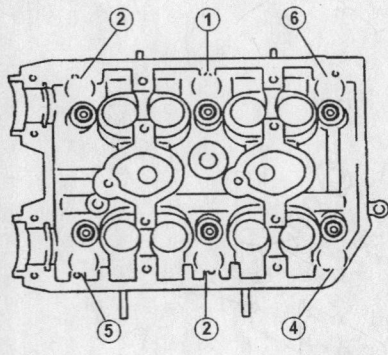

Cylinder head bolt tightening sequence—2.5L engines

g. Step 7: All bolts plus 80–90 degrees.

➡**Do not exceed 180 degrees total tightening.**

8. Install or connect the following:
- Camshafts, refer to the procedure in this section
- Valve covers
- Oil level gauge guide attaching bolt on the left cylinder head
- Timing belt, camshaft sprockets and related components
- Water pipe
- Intake manifold and tighten the bolts to 21–25 ft. lbs. (28–34 Nm)
- Fuel delivery pipes
- Blow-by hose
- Knock sensor
- CKP and CMP sensors
- Connector bracket attaching bolt
- Spark plug wires
- Valve rocker cover and tighten the bolts to 48 inch lbs. (9 Nm)
- Alternator
- Power steering pump
- Accessory drive belt
- Negative battery cable

9. Start the engine and allow it to reach operating temperature. Check for leaks.

Rocker Arms/Shafts

REMOVAL & INSTALLATION

2.2L Engines

1. Before servicing the vehicle, refer to the precautions in the beginning of this section.
2. Remove or disconnect the following:
- Positive Crankcase Ventilation (PCV) hose
- Rocker cover

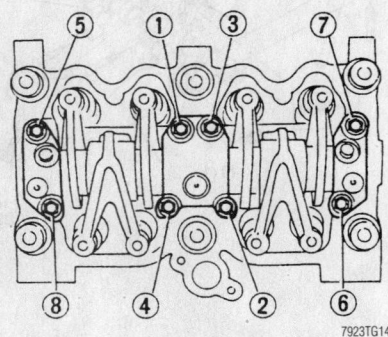

Rocker shaft bolt loosening/tightening sequence—1998 2.2L engines

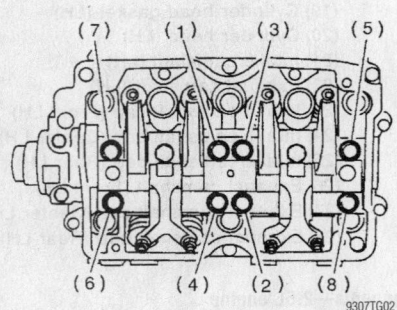

Rocker shaft bolt loosening/tightening sequence—1999–01 2.2L engines

- Valve rocker assembly by removing bolts 2 through 4 in numerical sequence
3. Loosen bolt 1, but leave it engaged to retain the valve rocker assembly.
4. Remove or disconnect the following:
- Bolts 5 through 8, taking care not to gouge the dowel pin
- Valve rocker assembly

To install:
5. Install the valve rocker assembly on the cylinder head.
6. Temporarily tighten bolts 1 through 4 equally.

➡**Do not allow the valve rocker assembly to gouge the dowel pins.**

7. Tighten bolts 5–8 to 108 inch lbs. (12 Nm).
8. Tighten bolts 1–4 to 108 inch lbs. (12 Nm).
9. Install the rocker cover and connect the PCV hose.

2.5L Engines

This engine is not equipped with either rocker shafts or rocker arms. Instead, the camshafts act directly on the individual valves.

Intake Manifold

REMOVAL & INSTALLATION

2.2L Engines

1. Before servicing the vehicle, refer to the precautions in the beginning of this section.
2. Release the fuel system pressure.
3. Remove or disconnect the following:
- Negative battery cable
- Engine cover, if necessary
4. Drain the cooling system into a suitable container.

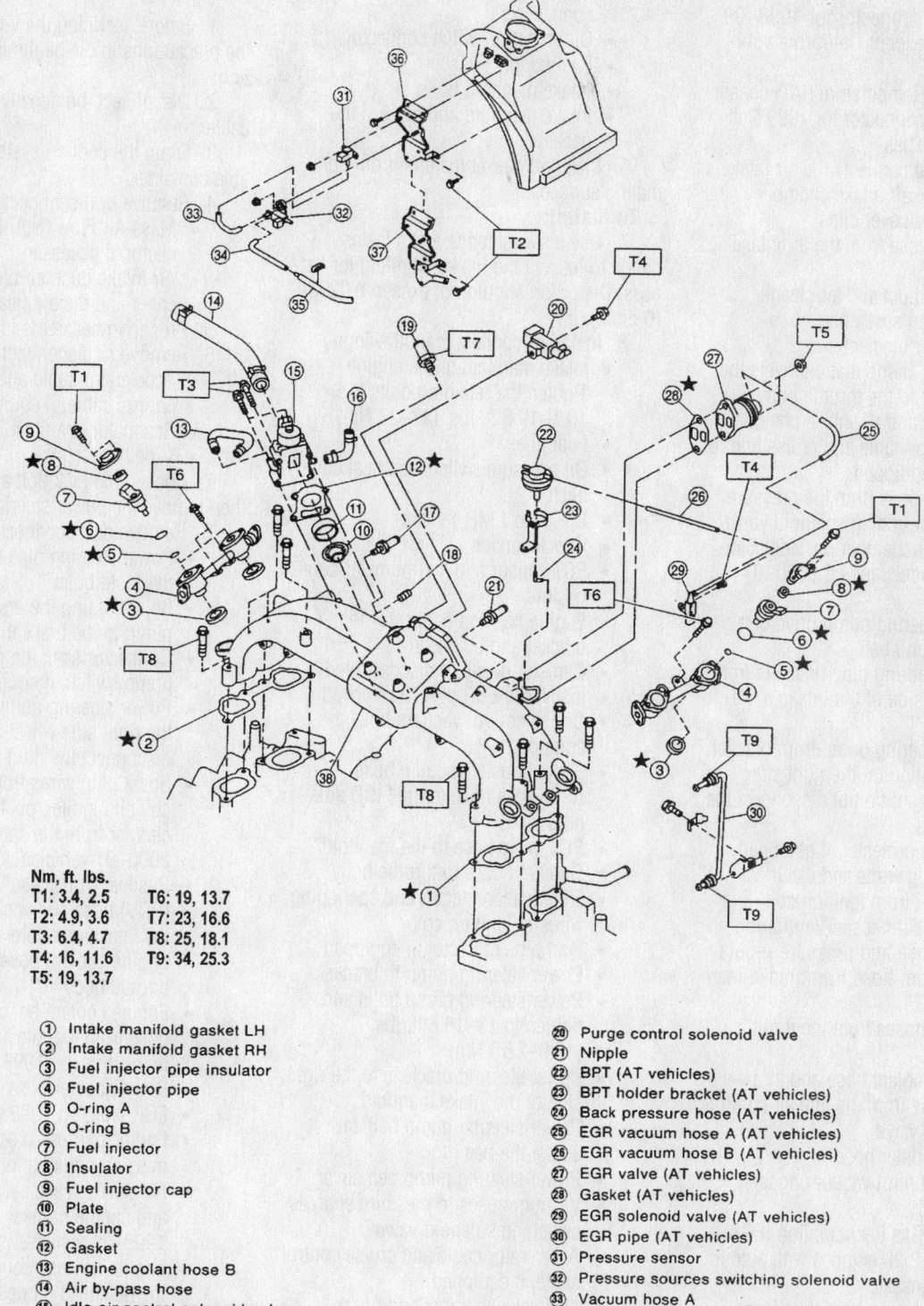

Nm, ft. lbs.

T1: 3.4, 2.5	T6: 19, 13.7
T2: 4.9, 3.6	T7: 23, 16.6
T3: 6.4, 4.7	T8: 25, 18.1
T4: 16, 11.6	T9: 34, 25.3
T5: 19, 13.7	

① Intake manifold gasket LH
② Intake manifold gasket RH
③ Fuel injector pipe insulator
④ Fuel injector pipe
⑤ O-ring A
⑥ O-ring B
⑦ Fuel injector
⑧ Insulator
⑨ Fuel injector cap
⑩ Plate
⑪ Sealing
⑫ Gasket
⑬ Engine coolant hose B
⑭ Air by-pass hose
⑮ Idle air control solenoid valve
⑯ Engine coolant hose A
⑰ Nipple (AT vehicles)
⑱ Plug
⑲ PCV valve

⑳ Purge control solenoid valve
㉑ Nipple
㉒ BPT (AT vehicles)
㉓ BPT holder bracket (AT vehicles)
㉔ Back pressure hose (AT vehicles)
㉕ EGR vacuum hose A (AT vehicles)
㉖ EGR vacuum hose B (AT vehicles)
㉗ EGR valve (AT vehicles)
㉘ Gasket (AT vehicles)
㉙ EGR solenoid valve (AT vehicles)
㉚ EGR pipe (AT vehicles)
㉛ Pressure sensor
㉜ Pressure sources switching solenoid valve
㉝ Vacuum hose A
㉞ Vacuum hose B
㉟ Vacuum hose C
㊱ Bracket (Except Canada spec. vehicles)
㊲ Bracket (For Canada spec. vehicles)
㊳ Intake manifold

7923TG16

Exploded view of the intake manifold assembly—2.2L engine

Heater Core replacement is covered in Section 2 of this manual

- Mass Air Flow (MAF) sensor electrical connector for 1998–99 vehicles, except California vehicles
- Intake Air Temperature (IAT) sensor electrical connector for 1999 California vehicles
- Clamp that connects the air intake duct to the air intake chamber
- Air cleaner cover clips
- Blow-by hose from the air intake duct
- Air intake duct and air cleaner cover as an assembly
- Air cleaner element

5. Loosen the clamp that connects the air intake chamber to the throttle body
- Air hoses and air intake chamber
- Accelerator cable and cruise control cable, if equipped
- Vacuum hoses from the pressure sources switching solenoid valve
- Resonator chamber on 1999 California models and all 2000–01 models
- Power steering pump drive belt cover(s) and belt
- Power steering pipe brackets from the right side of the intake manifold
- Power steering pump from bracket, then position on the right side wheel apron; do not disconnect the fluid lines
- Fuel pipe protector, if equipped
- Spark plug wires and electrical connector from ignition coil
- Positive Crankcase Ventilation (PCV) hose and pressure regulator vacuum hose from intake manifold
- Coolant hoses from the throttle body
- Engine coolant hose and air bypass hose from the idle air control solenoid valve
- Brake booster hose
- Cruise control vacuum hose, if equipped
- Exhaust Gas Recirculation (EGR) pipe, on 2.2L engines with automatic transaxle
- Canister hoses from the pipes
- Engine harness connectors from bulkhead harness connectors, then remove from bracket
- Engine Coolant Temperature (ECT) sensor connector and thermometer connector, if equipped
- Knock sensor connector
- Crankshaft Position (CKP) and

Camshaft Position (CMP) sensor connectors
- Oil pressure switch connector
- Fuel supply lines
- Intake manifold bolts
- Intake manifold and discard the gaskets

6. Clean all gasket material from both mating surfaces.

To install:

7. Use a straightedge and a feeler gauge to inspect the intake manifold for flatness. Distortion should not exceed 0.020 in. (0.5mm).

8. Install or connect the following:
- Intake manifold to the engine. Tighten the retaining bolts to 16.7–19.5 ft. lbs. (23–27 Nm).
- Fuel lines
- Oil pressure switch electrical connector
- CKP and CMP sensors
- Knock sensor
- ECT sensor and thermometer connectors
- Engine harness connector to the bracket and bulkhead
- Canister hoses, if disconnected
- Install the EGR pipe, if removed
- Cruise control vacuum hose, if equipped
- Brake booster vacuum hose
- Air bypass hose to the FICD solenoid valve
- PCV valve hose to the manifold
- Coolant hoses to throttle body
- Electrical connector and spark plug wires to ignition coil
- Fuel pipe protector, if equipped
- Power steering pump to bracket
- Power steering pump bolts and tighten to 13–16.6 ft. lbs. (17.6–22.6 Nm)
- Power steering brackets to the right side of the intake manifold
- Power steering pump belt and adjust the belt
- Power steering pump belt cover
- Vacuum hoses to pressure sources switching solenoid valve
- Accelerator cable and cruise control cable, if equipped
- Air intake chamber and hoses
- Air cleaner element, and the upper cover and duct as a unit
- MAF sensor electrical connector
- Engine cover, if equipped
- Negative battery cable

9. Fill the cooling system.
10. Start the engine and allow it to reach operating temperature. Check for leaks and test drive the vehicle.

2.5L Engine

1. Before servicing the vehicle, refer to the precautions in the beginning of this section.
2. Disconnect the negative battery cable.
3. Drain the cooling system into a suitable container.
4. Remove or disconnect the following:
- Mass Air Flow (MAF) sensor connector, if necessary
- Air intake duct, air cleaner upper cover and the air cleaner element

5. Properly release the fuel pressure.
6. Remove or disconnect the following:
- Accelerator cable and the cruise control cable, if equipped
- Resonator chamber, if equipped
- V-belt cover(s)

7. Loosen the lock bolt and slider bolt, then remove the power steering belt
8. Remove or disconnect the following:
- Power steering pipe bracket-to-manifold bolts
- Bolts holding the power steering pump to the bracket
- Connector from the power steering pump switch, if equipped
- Power steering pump, and place on the right side wheel apron. Do NOT disconnect the fluid lines.
- Spark plug wires from the ignition coil and igniter, on 1998–99 vehicles, or from the spark plugs, on 2000–01 vehicles
- Positive Crankcase Ventilation (PCV) hose and vacuum hose from the intake manifold
- Engine coolant hoses from the throttle body
- Engine coolant hose and air bypass hose from the idle air control solenoid valve, on 1998 vehicles
- Brake booster hose
- EGR pipe, on 1998 vehicles
- Engine harness bracket from transmission housing, on 1998 vehicles
- Air cleaner case stay (right side) and engine harness bracket, on 1999–01 vehicles
- Engine harness connectors form the bulkhead harness connectors
- Engine Coolant Temperature (ECT) sensor connector
- Knock sensor, Camshaft Position (CMP) sensor and Crankshaft Position (CKP) sensor electrical connectors
- Oil pressure switch
- Fuel hoses from the fuel pipes
- Intake manifold mounting bolts

- Intake manifold and discard the gaskets

➡️**The intake manifold sits on pins that protrude from the cylinder heads. Be sure the pins remain in the cylinder heads.**

To install:
9. Install or connect the following:
- New gaskets
- Intake manifold to the engine. Tighten the mounting bolts to 16.7–19.5 ft. lbs. (23–27 Nm).

- Fuel hoses to the fuel pipes, be sure to secure the hoses with new clamps
- Knock sensor, CMP sensor, CKP sensor, oil pressure switch and ECT sensor wiring

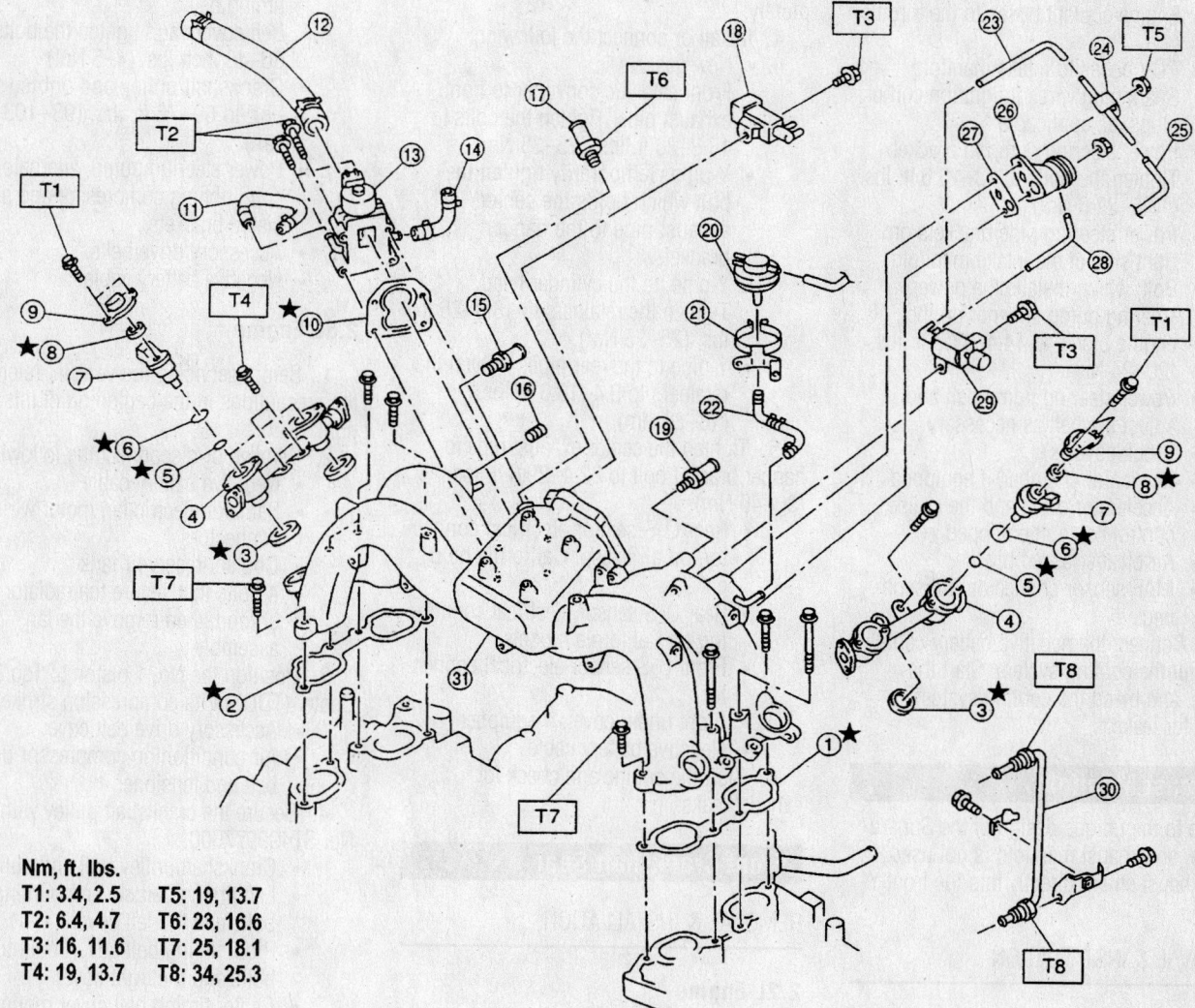

Nm, ft. lbs.

T1: 3.4, 2.5	T5: 19, 13.7
T2: 6.4, 4.7	T6: 23, 16.6
T3: 16, 11.6	T7: 25, 18.1
T4: 19, 13.7	T8: 34, 25.3

① Intake manifold gasket RH
② Intake manifold gasket LH
③ Fuel injector pipe insulator
④ Fuel injector pipe
⑤ O-ring A
⑥ O-ring B
⑦ Fuel injector
⑧ Insulator
⑨ Fuel injector cap
⑩ Gasket
⑪ Engine coolant hose B

⑫ Air by-pass hose
⑬ Idle air control solenoid valve
⑭ Engine coolant hose A
⑮ Nipple (AT model)
⑯ Plug
⑰ PCV valve
⑱ Purge control solenoid valve
⑲ Nipple
⑳ BPT
㉑ BPT holder bracket

㉒ Back pressure hose
㉓ EGR vacuum hose A
㉔ EGR vacuum pipe
㉕ EGR vacuum hose C
㉖ EGR valve
㉗ Gasket
㉘ EGR vacuum hose B
㉙ EGR solenoid valve
㉚ EGR pipe
㉛ Intake manifold

Exploded view of the intake manifold and related components—2.5L engine

7923TG17

Brake service is covered in Section 4 of this manual

- Engine harness bracket and engine harness connectors to bulkhead connectors
- EGR pipe, if removed
- Brake booster hose
- Engine coolant hose and air bypass hose to idle air control solenoid valve, if removed
- Engine coolant hoses to the throttle body
- PCV hoses to intake manifold
- Spark plug wires to ignition coil or plugs, as applicable
- Power steering pump to bracket. Tighten the bolts to 13–16.6 ft. lbs. (17.6–22.6 Nm).
- Power steering pipe brackets on right side of the intake manifold
- Bolt, which installs the power steering pump stiffener on the engine block, to 14.4–17.3 ft. lbs. (20–24 Nm)
- Power steering pump belt and adjust the belt as necessary
- V-belt cover(s)
- Resonator chamber, if equipped
- Accelerator cable and the cruise control cable, if equipped
- Air cleaner assembly
- MAF sensor connector, if disconnected

10. Connect the negative battery cable and refill the cooling system. Start the engine, and bleed the cooling system. Check for leaks.

Exhaust Manifold

Due to the unique design of the Subaru engine, an exhaust manifold is not used. The exhaust enters directly into the front Y-pipe.

REMOVAL & INSTALLATION

1. Before servicing the vehicle, refer to the precautions in the beginning of this section.
2. Remove or disconnect the following:

- Negative battery cable
- Front Oxygen (O$_2$S) sensor electrical connector
- Front under cover, if equipped
- Rear O$_2$S electrical connector, on California vehicles
- Rear O$_2$S sensor electrical connector on all vehicles, except California
- Y-pipe-to-rear pipe mounting nuts and separate the Y-pipe from the rear pipe

- Bolts that secure the front Y-pipe to the cylinder head
- Y-pipe from the hanger bracket
- Front exhaust pipe from the catalytic converter and discard the gaskets

To install:

3. Clean all gasket surfaces completely.
4. Install or connect the following:

- New gaskets
- Front catalytic converter to front exhaust pipe. Tighten the bolts to 18.8–26 ft. lbs. (25–35 Nm).
- Y-pipe. Temporarily tighten the bolt which holds the center exhaust pipe to the hanger bracket.
- Y-pipe, to the cylinder head. Tighten the retainers to 18.8–26 ft. lbs. (25–35 Nm).
- Y-pipe to the rear pipe. Tighten the retainers to 9.4–16.6 ft. lbs. (13–23 Nm).

5. Tighten the center exhaust pipe to hanger bracket bolt to 22.4–29.6 ft. lbs. (30–40 Nm).

- Rear O$_2$S sensor electrical connector, on all except California models
- Rear O$_2$S sensor electrical connector, on California models
- Front O$_2$S sensor electrical connector
- Front under cover, if equipped
- Negative battery cable

6. Start the engine and check for exhaust leaks.

Front Crankshaft Seal

REMOVAL & INSTALLATION

2.2L Engine

1. Before servicing the vehicle, refer to the precautions in the beginning of this section.
2. Remove or disconnect the following:

- Negative battery cable
- Accessory drive belts
- Power steering pump and alternator
- Air conditioner compressor brackets

3. Secure the crankshaft pulley with tool No. 499977000.
4. Remove or disconnect the following:

- Crankshaft pulley bolt and pulley
- Timing belt cover mounting bolts
- Belt covers
- Timing belt

- Timing belt crankshaft sprocket
- Crankshaft seal from the oil pump housing

To install:

5. Using a suitable seal driver, install a new crankshaft seal.
6. Install or connect the following:

- Timing belt crankshaft sprocket and timing belt
- Belt covers and tighten the bolts to 36–48 inch lbs. (4–5 Nm)
- Crankshaft pulley and tighten the bolt to 69–76 ft. lbs. (93–103 Nm)
- Power steering pump, alternator, air conditioning compressor and associated brackets
- Accessory drive belts
- Negative battery cable

2.5L Engine

1. Before servicing the vehicle, refer to the precautions in the beginning of this section.
2. Remove or disconnect the following:

- Negative battery cable
- Radiator electric fan motor wiring connectors
- Coolant reservoir tank
- 4 bolts that secure the radiator shroud, then remove the fan assembly

3. Position the No. 1 piston to Top Dead Center (TDC) of its compression stroke.

- Accessory drive belt cover
- Air conditioning compressor drive belt and tensioner

4. Secure the crankshaft pulley with tool No. ST499977000.

- Crankshaft pulley bolt and pulley
- Left timing belt cover mounting bolts and the left cover
- Right timing belt cover mounting bolts and the right cover
- Center timing belt cover mounting bolts and the center cover
- Timing belt
- Timing belt crankshaft sprocket
- Crankshaft seal from the oil pump housing

To install:

5. Install or connect the following:

- New crankshaft seal, using a suitable seal driver
- Timing belt crankshaft sprocket and the timing belt
- Center, right, then the left timing belt covers. Tighten the bolts to 44 inch lbs. (5 Nm).
- Crankshaft pulley and tighten the bolt to 94 ft. lbs. (127 Nm)

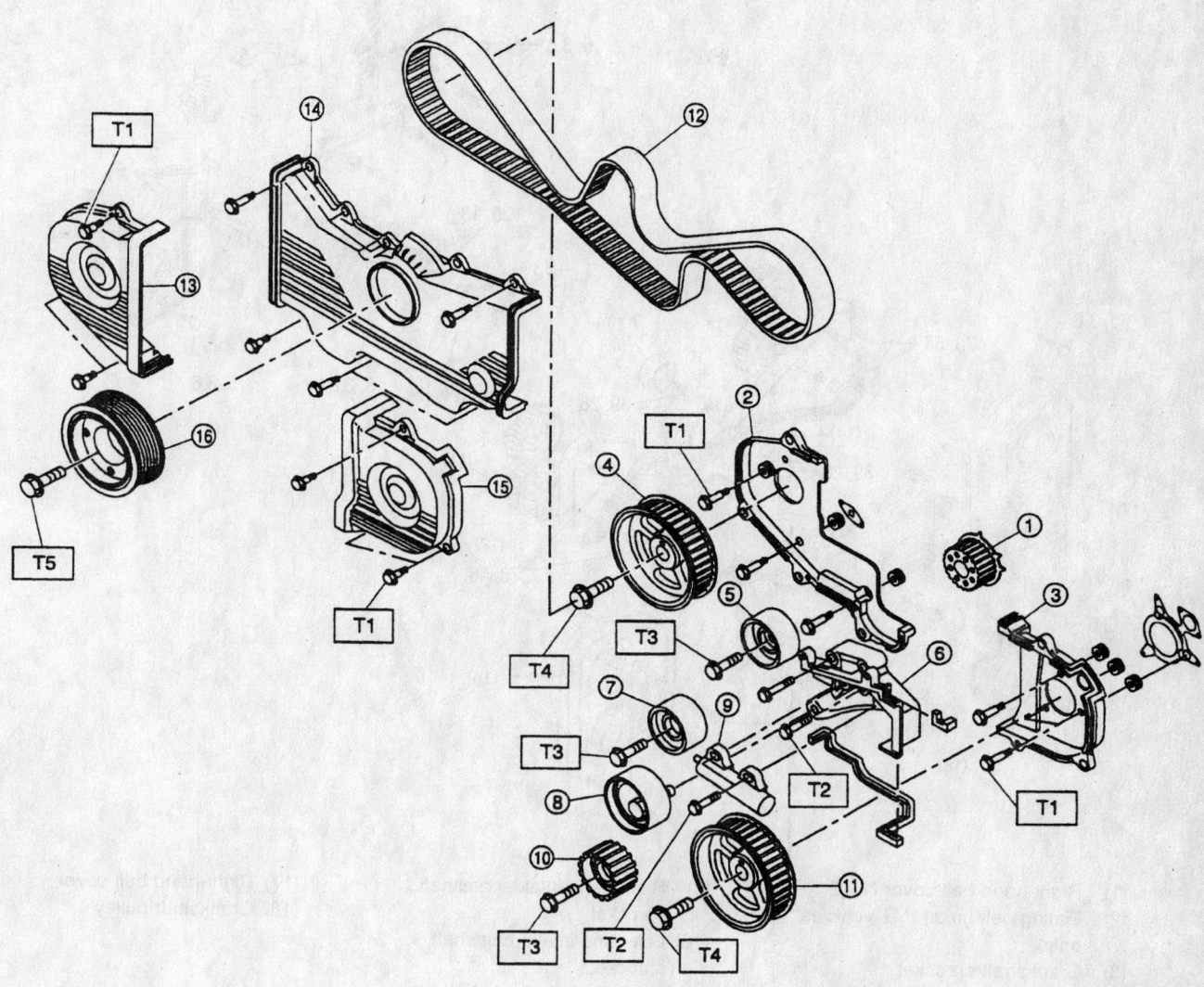

1. Crankshaft sprocket
2. Belt cover No. 2 (RH)
3. Belt cover No. 2 (LH)
4. Camshaft sprocket (RH)
5. Belt idler
6. Tensioner bracket
7. Belt idler
8. Belt tensioner
9. Tensioner adjuster
10. Belt idler No. 2
11. Camshaft sprocket (LH)
12. Timing belt

13. Belt cover (RH)
14. Front belt cover
15. Belt cover (LH)
16. Crankshaft pulley

Tightening torque: N·m (kg-m, ft-lb)
T1: 5 ± 1 (0.5 ± 0.1, 3.6 ± 0.7)
T2: 25 ± 2 (2.5 ± 0.2, 18.1 ± 1.4)
T3: 39 ± 4 (4.0 ± 0.4, 28.9 ± 2.9)
T4: 78 ± 5 (8.0 ± 0.5, 57.9 ± 3.6)
T5: $108 ^{+10}_{-5}$ ($11 ^{+1.0}_{-0.5}$, $79.6 ^{+7.2}_{-3.6}$)

7923TG20

Exploded view of the timing belt covers and components—2.2L engine

For complete Engine Mechanical specifications, see Section 1 of this manual

Nm, ft. lbs.

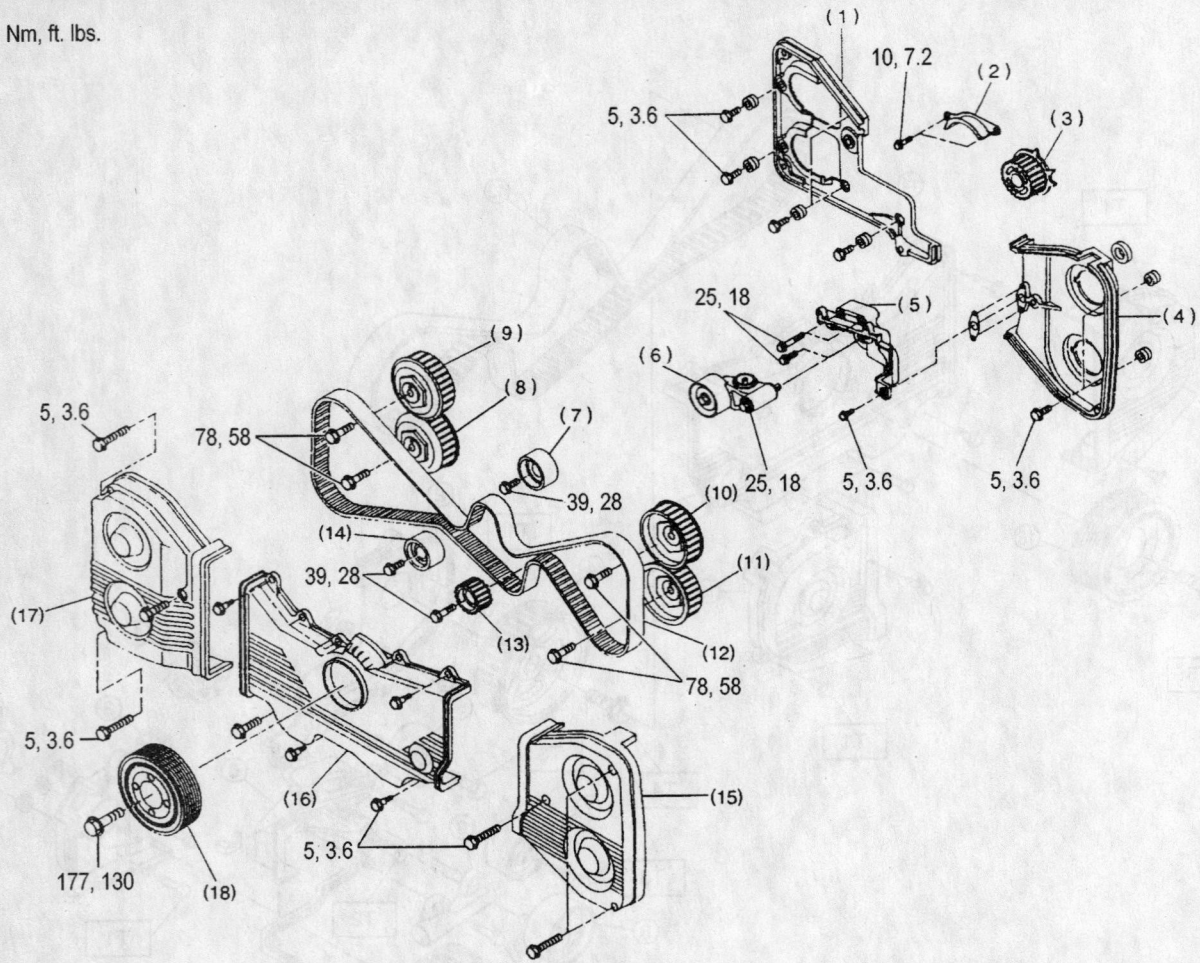

(1) Right-hand belt cover No. 2
(2) Timing belt guide (MT vehicles only)
(3) Crankshaft sprocket
(4) Left-hand belt cover No. 2
(5) Tensioner bracket
(6) Automatic belt tension adjuster ASSY
(7) Belt idler
(8) Right-hand exhaust camshaft sprocket

(9) Right-hand intake camshaft sprocket
(10) Left-hand intake camshaft sprocket
(11) Left-hand exhaust camshaft sprocket
(12) Timing belt
(13) Belt idler No. 2
(14) Belt idler
(15) Left-hand belt cover
(16) Front belt cover

(17) Right-hand belt cover
(18) Crankshaft pulley

7923TG21

Exploded view of the timing belt covers and components—2.5L engine

- Air conditioning compressor drive belt tensioner and the drive belts
- Fan shroud and fan motor assembly
- Accessory drive belt cover
- Negative battery cable

Camshaft

On some models, it may be necessary to remove the engine from the vehicle to perform this service.

REMOVAL & INSTALLATION

2.2L Engines

1. Before servicing the vehicle, refer to the precautions in the beginning of this section.
2. Remove or disconnect the following:
   - Negative battery cable
   - Timing belt covers, timing belt and camshaft sprockets

- Valve rocker covers
- Rocker arm assemblies. Refer to the rocker arms/shafts procedure in this section.
- Camshaft cap bolts in the proper sequence
- Camshaft cap

3. To remove the left camshaft, remove or disconnect the following:
   - Camshaft Position (CMP) sensor

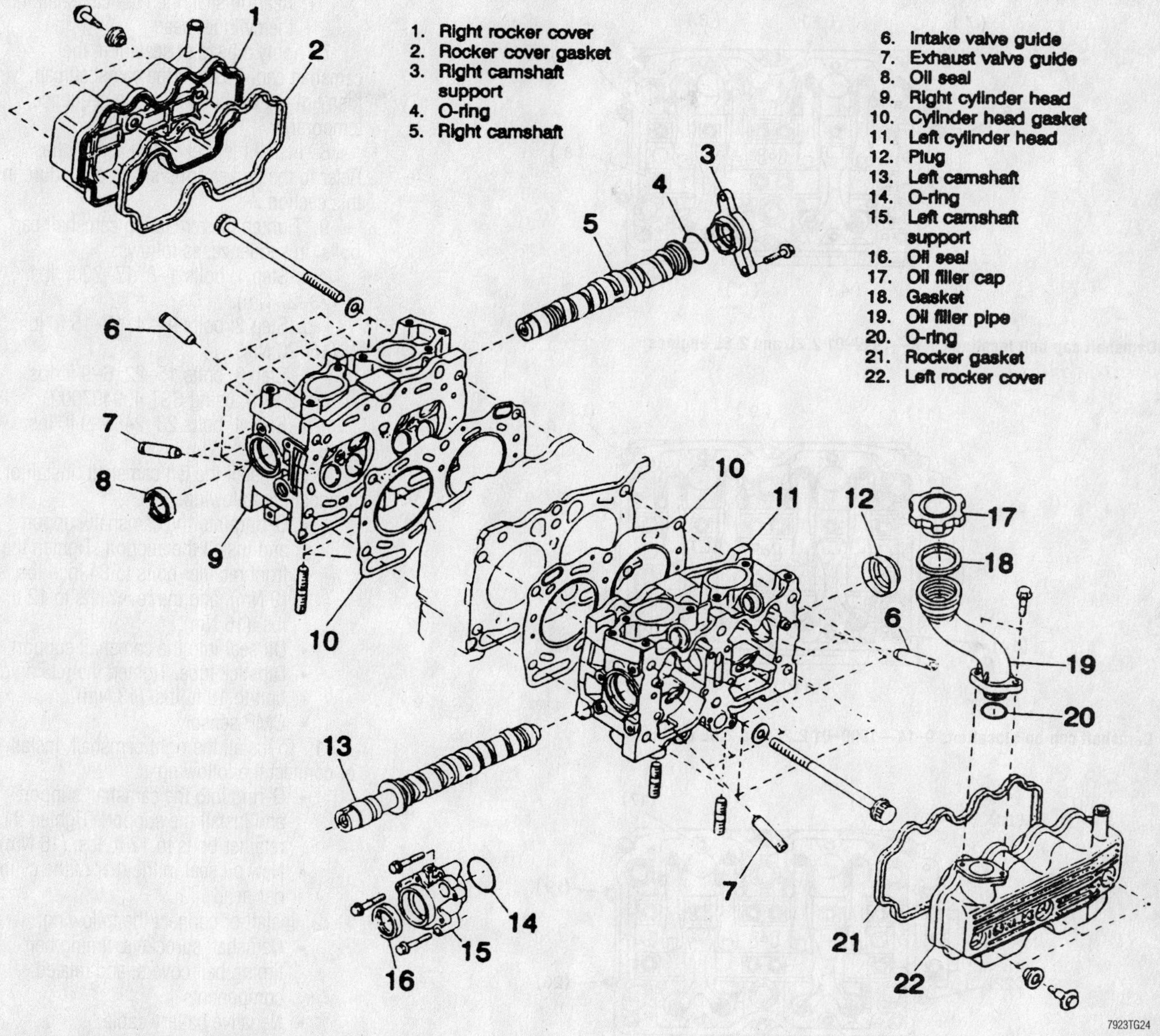

1. Right rocker cover
2. Rocker cover gasket
3. Right camshaft support
4. O-ring
5. Right camshaft

6. Intake valve guide
7. Exhaust valve guide
8. Oil seal
9. Right cylinder head
10. Cylinder head gasket
11. Left cylinder head
12. Plug
13. Left camshaft
14. O-ring
15. Left camshaft support
16. Oil seal
17. Oil filler cap
18. Gasket
19. Oil filler pipe
20. O-ring
21. Rocker gasket
22. Left rocker cover

7923TG24

Exploded view of the camshaft assembly—2.2L engine

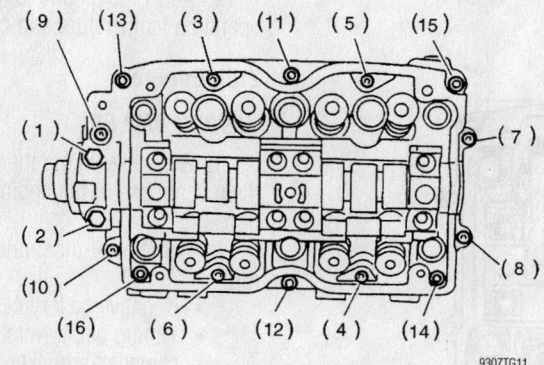

9307TG11

Camshaft cap bolt loosening sequence—1999–01 2.2L and 2.5L engines

- Oil dipstick tube attaching bolt and the dipstick tube
- CMP sensor support

4. To remove the right camshaft, remove the camshaft support on the right side.

5. Remove or disconnect the following:
- Camshaft O-ring
- Camshaft and rear seal
- Oil seal from the camshaft support

To install:

➡**Lubricate the camshaft journals with clean engine oil prior to installation.**

6. Install or connect the following:

For Accessory Drive Belt illustrations, see Section 1 of this manual

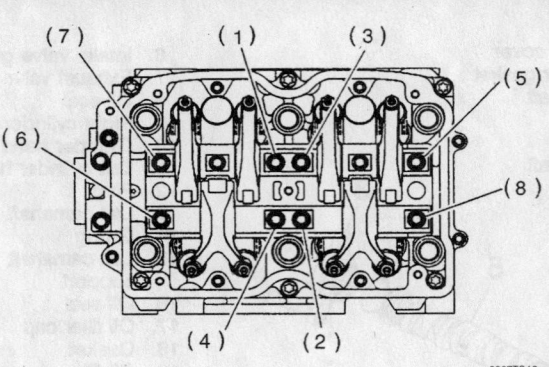

Camshaft cap bolt locations 1–8—1999–01 2.2L and 2.5L engines

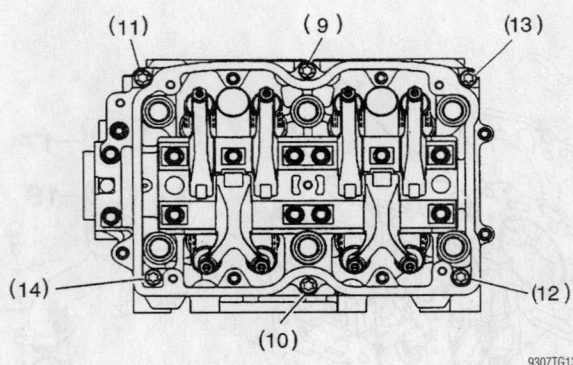

Camshaft cap bolt locations 9–14—1999–01 2.2L and 2.5L engines

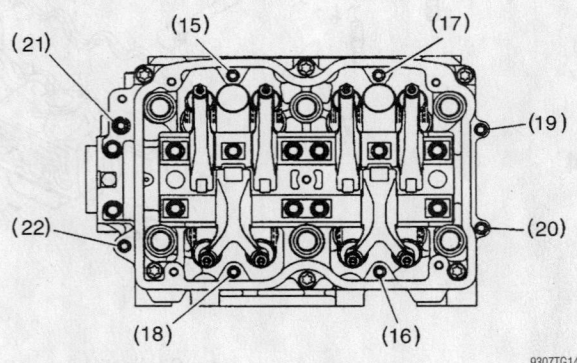

Camshaft cap bolt locations 15–22—1999–01 2.2L and 2.5L engines

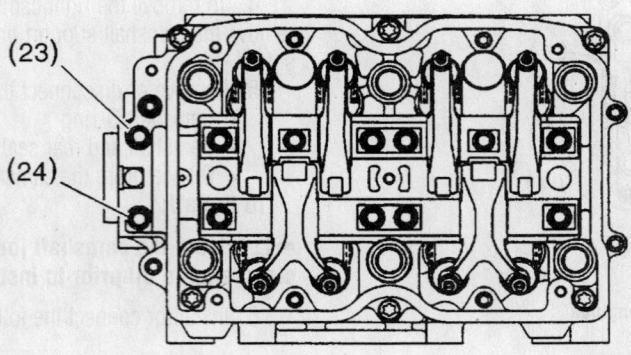

Camshaft cap bolt locations 23–24—1999–01 2.2L and 2.5L engines

- Rear oil seal, then the camshaft into the cylinder head

7. Apply a bead of sealant in the camshaft cap. Position the camshaft cap, then tighten the bolts 7–10, in sequence, temporarily

8. Install the rocker arm assemblies. Refer to the rocker arms/shafts procedure in this section.

9. Tighten the remaining camshaft cap bolts, in sequence, as follows:

    a. Step 1: bolts 1–8: 17–20 ft. lbs. (23–27 Nm).

    b. Step 2: bolts 9–14: 12–15 ft. lbs. (16–20 Nm).

    c. Step 3: bolts 15–22: 6–9 ft. lbs. (8–12 Nm), using SST 499497000.

    d. Step 4: bolts 23–24: 6–9 ft. lbs. (8–12 Nm).

10. To install the left camshaft, install or connect the following:

- O-ring into the camshaft support and install the support. Tighten the front retainer bolts to 84 inch lbs. (9 Nm), and the rear bolts to 12 ft. lbs. (16 Nm).
- Oil seal into the camshaft support
- Dipstick tube. Tighten the retaining bolt to 10 ft. lbs. (13 Nm).
- CMP sensor

11. To install the right camshaft, install or connect the following:

- O-ring into the camshaft support and install the support. Tighten the retainer bolts to 12 ft. lbs. (16 Nm).
- New oil seal in the rear of the cylinder head

12. Install or connect the following:

- Camshaft sprockets, timing belt, timing belt covers, and related components
- Negative battery cable

13. Check the fluid levels and start the engine.

14. Allow the engine to reach normal operating temperature and check for leaks.

2.5L Engine

1998 VEHICLES

1. Before servicing the vehicle, refer to the precautions in the beginning of this section.

2. Remove or disconnect the following:

- Negative battery cable
- Timing belt covers, timing belt and camshaft sprockets
- Camshaft Position (CMP) sensor
- Ignition coils
- Valve rocker covers and gaskets
- Loosen the intake camshaft cap

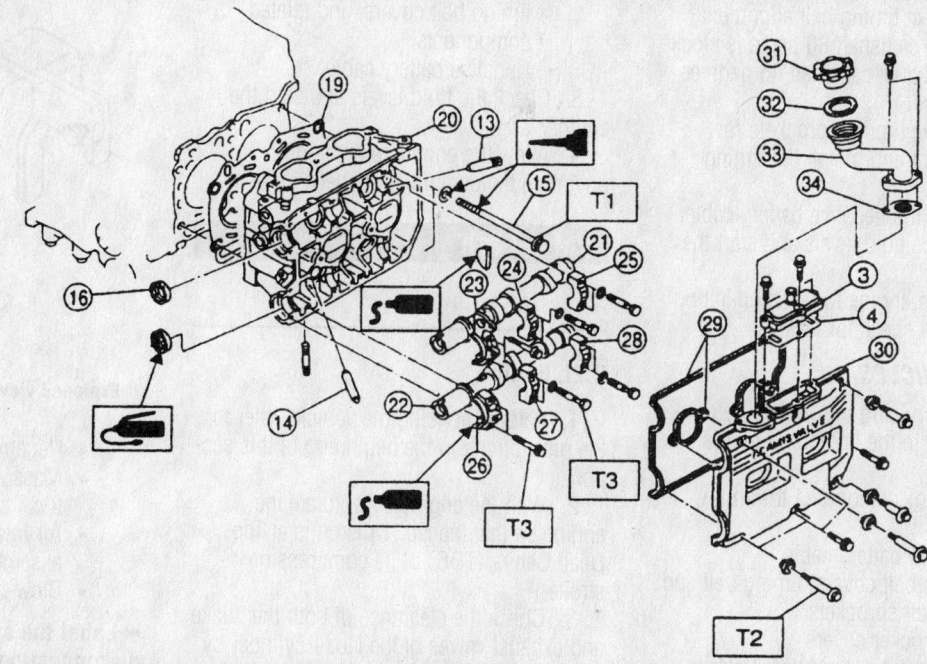

① Rocker cover (RH)
② Rocker cover gasket (RH)
⑨ Exhaust camshaft cap (Front RH)
⑩ Exhaust camshaft cap (Center RH)
⑫ Exhaust camshaft (RH)
⑮ Cylinder head bolt
⑯ Oil seal
⑲ Cylinder head gasket (LH)
⑳ Cylinder head (LH)
㉑ Intake camshaft (LH)
㉒ Exhaust camshaft (LH)
㉓ Intake camshaft cap (Front LH)
㉔ Intake camshaft cap (Center LH)

㉕ Intake camshaft cap (Rear LH)
㉖ Exhaust camshaft (Front LH)
㉗ Exhaust camshaft cap (Center LH)
㉘ Exhaust camshaft cap (Rear LH)
㉙ Rocker cover gasket (LH)
㉚ Rocker cover (LH)
㉛ Oil filler cap
㉜ Gasket
㉝ Oil filler duct
㉞ Gasket

Tightening torque: N·m (kg-m, ft-lb)
T1: Refer to [W4E1]☆2.
T2: 5 (0.5, 3.6)
T3: 10 (1.0, 7)

7923TG25

Exploded view of the camshaft and cylinder head assembly—2.5L engine

bolts in sequence, in small increments

➡**Be sure to keep the intake and exhaust bearing caps and camshafts in proper order for reassembly.**

- Intake camshaft bearing caps and the camshaft
- Loosen the exhaust camshaft cap bolts in sequence, in small increments
- Exhaust camshaft bearing caps and the camshaft

To install:

➡**Lubricate the camshaft bearings prior to camshaft installation.**

3. Position the camshafts so the base circle (non-lobe portion) of the camshafts are in contact with the lash adjusters. This will position the lobes of the camshafts away from the valves.

➡**The left camshaft will need to be rotated for timing belt alignment.**

4. Apply liquid sealant to the front bearing cap mating surfaces, then install the bearing caps. Tighten the caps in sequence in 2 progressive steps to 14 ft. lbs. (20 Nm).

5. Install or connect the following:
- New oil seals to the camshafts using a suitable seal installation tool
- Rocker covers using new gaskets.

Be sure to apply liquid sealant to the front edges of the gasket at the camshaft opening.
- Ignition coils
- CMP sensor
- Camshaft sprockets, tighten the retaining bolts to 58 ft. lbs. (78 Nm). Be sure to secure the sprockets when tightening the bolts.

✳✳ WARNING

Only rotate camshafts the specified amount. If the camshafts are rotated beyond the specified amount, the valves will contact each other and cause severe internal damage.

For Tire, Wheel and Ball Joint specifications, see Section 1 of this manual

6. For correct timing belt alignment, rotate the intake camshaft 80 degrees clockwise and the exhaust camshaft 45 degrees counterclockwise.

7. Check the timing sprockets for proper alignment and install the timing belt.

8. Connect the negative battery cable.

9. Check the fluid levels and start the engine.

10. Allow the engine to reach operating temperature and check for leaks.

1999–01 VEHICLES

1. Before servicing the vehicle, refer to the precautions in the beginning of this section.

2. Remove or disconnect the following:

- Negative battery cable
- Timing belt covers, timing belt and camshaft sprockets
- Valve rocker covers
- Rocker arm assemblies. Refer to the rocker arms/shafts procedure in this section.
- Camshaft cap bolts in the proper sequence
- Camshaft cap
- Camshaft and rear seal
- Oil seal from the rear side of the camshaft

To install:

➡ **Lubricate the camshaft journals with clean engine oil prior to installation.**

3. Install the camshaft into the cylinder head

4. Apply a bead of sealant in the camshaft cap. Position the camshaft cap, then tighten the bolts 7–10, in sequence, temporarily

5. Install the rocker arm assemblies. Refer to the rocker arms/shafts procedure in this section.

6. Tighten the remaining camshaft cap bolts, in sequence, as follows:

 a. Step 1: bolts 1–8: 17–20 ft. lbs. (23–27 Nm).

 b. Step 2: bolts 9–14: 12–15 ft. lbs. (16–20 Nm).

 c. Step 3: bolts 15–22: 6–9 ft. lbs. (8–12 Nm), using SST 499497000.

 d. Step 4: bolts 23–24: 6–9 ft. lbs. (8–12 Nm).

7. Install or connect the following:

➡ **Lubricate the seals lips with clean engine oil prior to installation**

- Oil seal and plug with suitable tools
- Camshaft sprockets, timing belt,

timing belt covers, and related components

- Negative battery cable

8. Check the fluid levels and start the engine.

9. Allow the engine to reach normal operating temperature and check for leaks.

Valve Lash

ADJUSTMENT

2.2L Engine

1. Before servicing the vehicle, refer to the precautions in the beginning of this section.

2. With the engine cold, rotate the engine so that the No. 1 piston is at Top Dead Center (TDC) of its compression stroke.

3. Check the clearance of both the intake and exhaust valves of the No. 1 cylinder by inserting a feeler gauge between each valve stem and rocker arm.

4. If the clearance is not within specifications, loosen the locknut with the proper size wrench and turn the adjusting stud either in or out until the valve clearance is correct.

➡ **Proper valve clearance is obtained when the feeler gauge slides between the valve stem and the rocker arm with a minimum amount of resistance.**

5. Tighten the locknut and recheck the valve stem-to-rocker clearance.

6. The rest of the valves are adjusted in the same way. Bring each piston to TDC of its compression stroke, then check and adjust the valves for that cylinder. The proper valve adjustment sequence is 1–3–2–4.

7. Rotate the crankshaft at least 2 revolutions, then recheck the valve clearance.

8. Tighten the rocker arm locknuts to 10–13 ft. lbs. (14–18 Nm).

9. Install the valve covers using new gaskets. Tighten the retaining nuts to 24–36 inch lbs. (3–4 Nm).

2.5L Engine

➡ **The valve adjustment should be performed while the engine is cold. A Shim Replacer Kit 498187100 will be needed to perform the valve adjustment.**

1. Before servicing the vehicle, refer to the precautions in the beginning of this section.

2. Remove or disconnect the following:

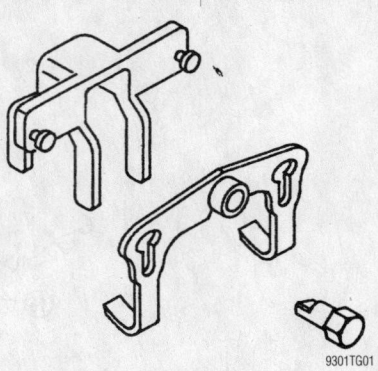

Exploded view of the shim replacer kit

- Negative battery cable
- Mass Air Flow (MAF) sensor electrical connector
- Air intake duct with the air cleaner assembly
- Blow-by hose

➡ **Label the spark plug wires before disconnecting them.**

- Spark plug wires
- Battery and tray
- Washer tank

3. Place a drip tray under the vehicle, and remove the valve covers.

4. Turn the crankshaft pulley clockwise until the arrow mark on the camshaft is set to the position as shown.

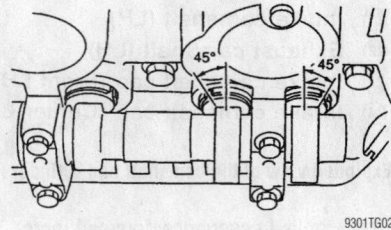

Position the lifter notch as shown, to remove the shim

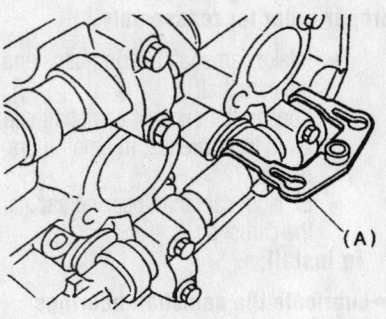

First install part "A" of the tool to the camshaft . . .

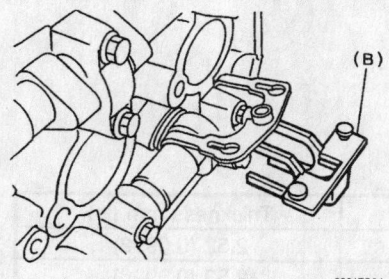

9301TG04

. . . then part "B" under part "A" as shown

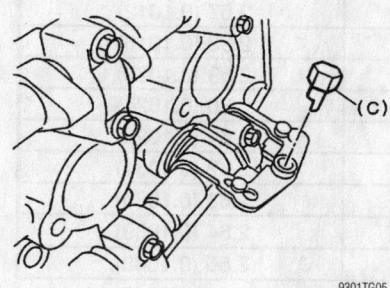

9301TG05

Turn part "C" until the adjusting shim can be removed

➡ **the checking or adjusting the exhaust valve is performed from under the vehicle.**

5. Check the valve clearance using the appropriate sized feeler gauge as follows:

   a. Intake valve clearance specification is 0.0071–0.0087 in. (0.18–0.22mm).

   b. Exhaust valve clearance specifica-

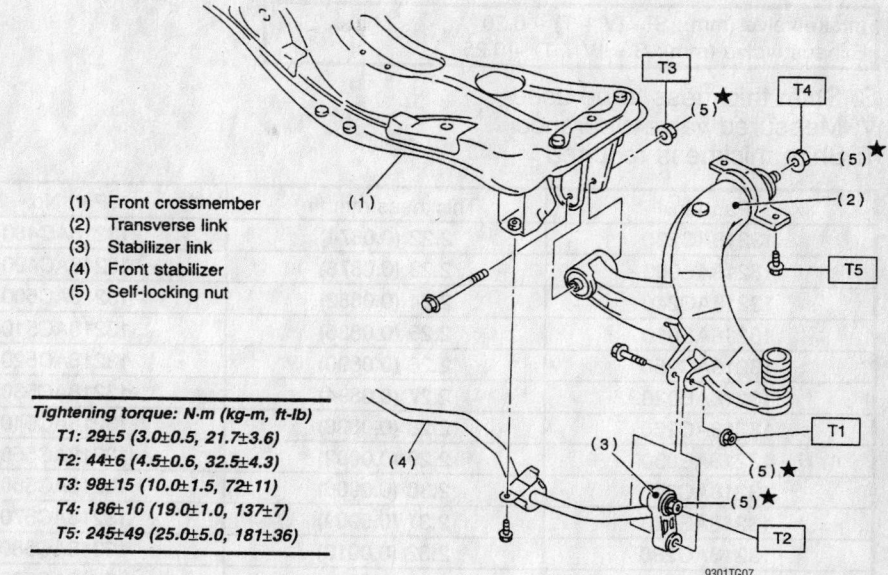

(1) Front crossmember
(2) Transverse link
(3) Stabilizer link
(4) Front stabilizer
(5) Self-locking nut

Tightening torque: N·m (kg-m, ft-lb)
T1: 29±5 (3.0±0.5, 21.7±3.6)
T2: 44±6 (4.5±0.6, 32.5±4.3)
T3: 98±15 (10.0±1.5, 72±11)
T4: 186±10 (19.0±1.0, 137±7)
T5: 245±49 (25.0±5.0, 181±36)

9301TG07

Position the camshaft as shown to adjust No. 3 intake valve and No. 2 exhaust valve

tion is 0.0090–0.0106 in. (0.23–0.27mm).

6. If any valve needs adjustment, perform the following, while referring to the accompanying illustrations:

   a. Rotate the notch of the lifter outward by 45 degrees.

   b. Install part "A" of the Replacer on to the camshaft.

   c. Install part "B" of the Replacer as shown.

   d. Install part "C" and turn until part "B" pushes the lifter away.

   e. Insert tweezers into the notch of the valve lifter, and take out the shim.

   f. Measure the thickness of the shim, then using the chart select and install a new shim and recheck the clearance.

7. Remove the adjusting tools.
8. Install or connect the following:
- Valve covers
- Spark plug wires
- All components removed to access the valve covers

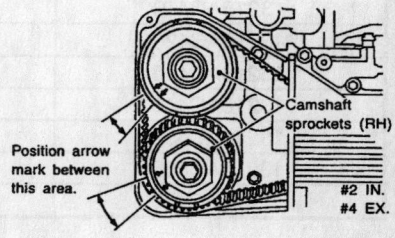

9301TG08

Position the camshaft as shown to adjust No. 2 intake valve and No. 4 exhaust valve

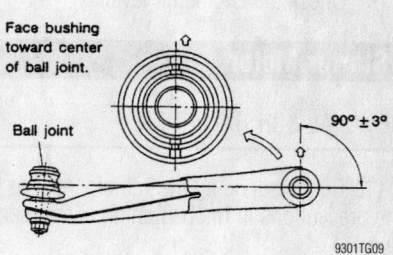

9301TG09

Position the camshaft as shown to adjust No. 4 intake valve and No. 1 exhaust valve

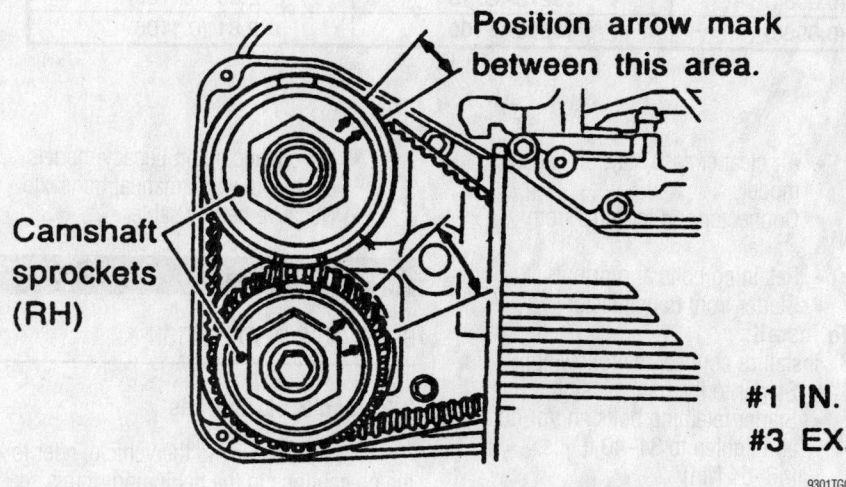

Camshaft sprockets (RH)

Position arrow mark between this area.

#1 IN.
#3 EX.

9301TG06

Position the camshaft as shown to adjust No. 1 intake valve and No. 3 exhaust valve

For Wheel Alignment specifications, see Section 1 of this manual

Intake valve (mm): $S = (V + T) - 0.20$
Exhaust valve (mm): $S = (V + T) - 0.25$

S: Shim thickness to be used
V: Measured valve clearance
T: Shim thickness required

Part No.	Thickness mm (in)	Part No.	Thickness mm (in)
13218AC230	2.22 (0.0874)	13218AC480	2.52 (0.0992)
13218AE000	2.23 (0.0878)	13218AC490	2.53 (0.0996)
13218AC240	2.24 (0.0882)	13218AC500	2.54 (0.1000)
13218AE010	2.25 (0.0886)	13218AC510	2.55 (0.1004)
13218AC250	2.26 (0.0890)	13218AC520	2.56 (0.1008)
13218AE020	2.27 (0.0894)	13218AC530	2.57 (0.1012)
13218AC260	2.28 (0.0898)	13218AC540	2.58 (0.1016)
13218AE030	2.29 (0.0902)	13218AC550	2.59 (0.1020)
13218AC270	2.30 (0.0906)	13218AC560	2.60 (0.1024)
13218AE040	2.31 (0.0909)	13218AC570	2.61 (0.1028)
13218AC280	2.32 (0.0913)	13218AC580	2.62 (0.1031)
13218AC290	2.33 (0.0917)	13218AC590	2.63 (0.1035)
13218AC300	2.34 (0.0921)	13218AC600	2.64 (0.1039)
13218AC310	2.35 (0.0925)	13218AC610	2.65 (0.1043)
13218AC320	2.36 (0.0929)	13218AC620	2.66 (0.1047)
13218AC330	2.37 (0.0933)	13218AC630	2.67 (0.1051)
13218AC340	2.38 (0.0937)	13218AC640	2.68 (0.1055)
13218AC350	2.39 (0.0941)	13218AC650	2.69 (0.1059)
13218AC360	2.40 (0.0945)	13218AC660	2.70 (0.1063)
13218AC370	2.41 (0.0949)	13218AE050	2.71 (0.1067)
13218AC380	2.42 (0.0953)	13218AC670	2.72 (0.1071)
13218AC390	2.43 (0.0957)	13218AE060	2.73 (0.1075)
13218AC400	2.44 (0.0961)	13218AC680	2.74 (0.1079)
13218AC410	2.45 (0.0965)	13218AE070	2.75 (0.1083)
13218AC420	2.46 (0.0969)	13218AC690	2.76 (0.1087)
13218AC430	2.47 (0.0972)	13218AE080	2.77 (0.1091)
13218AC440	2.48 (0.0976)	13218AC700	2.78 (0.1094)
13218AC450	2.49 (0.0980)	13218AE090	2.79 (0.1098)
13218AC460	2.50 (0.0984)	13218AC710	2.80 (0.1102)
13218AC470	2.51 (0.0988)	13218AE100	2.81 (0.1106)

9301TG10

Valve adjusting shim chart

- Battery tray and battery
9. Check the engine oil level.

Starter Motor

REMOVAL & INSTALLATION

1. Before servicing the vehicle, refer to the precautions in the beginning of this section.
2. Remove or disconnect the following:
   - Negative battery cable
   - Intake Air Temperature (IAT) connector, on Legacy models equipped with a manual transaxle
   - Air cleaner case and duct
   - Air cleaner case stay, on Legacy models
   - Connector and terminal from starter
   - Retaining bolts and/or nuts
   - Starter from transmission

To install:
3. Install or connect the following:
   - Starter to the transmission
   - Starter retaining bolts and/or nuts and tighten to 34–40 ft. lbs. (46–54 Nm)
   - Connector and terminal to starter
   - Air cleaner case stay, on Legacy models
   - Air cleaner case and duct
   - IAT connector, on Legacy models equipped with a manual transaxle
   - Negative battery cable

Oil Pan

REMOVAL & INSTALLATION

2.2L and 2.5L Engines

1. Before servicing the vehicle, refer to the precautions in the beginning of this section.
2. Remove or disconnect the following:
   - Negative battery cable
   - Air intake duct

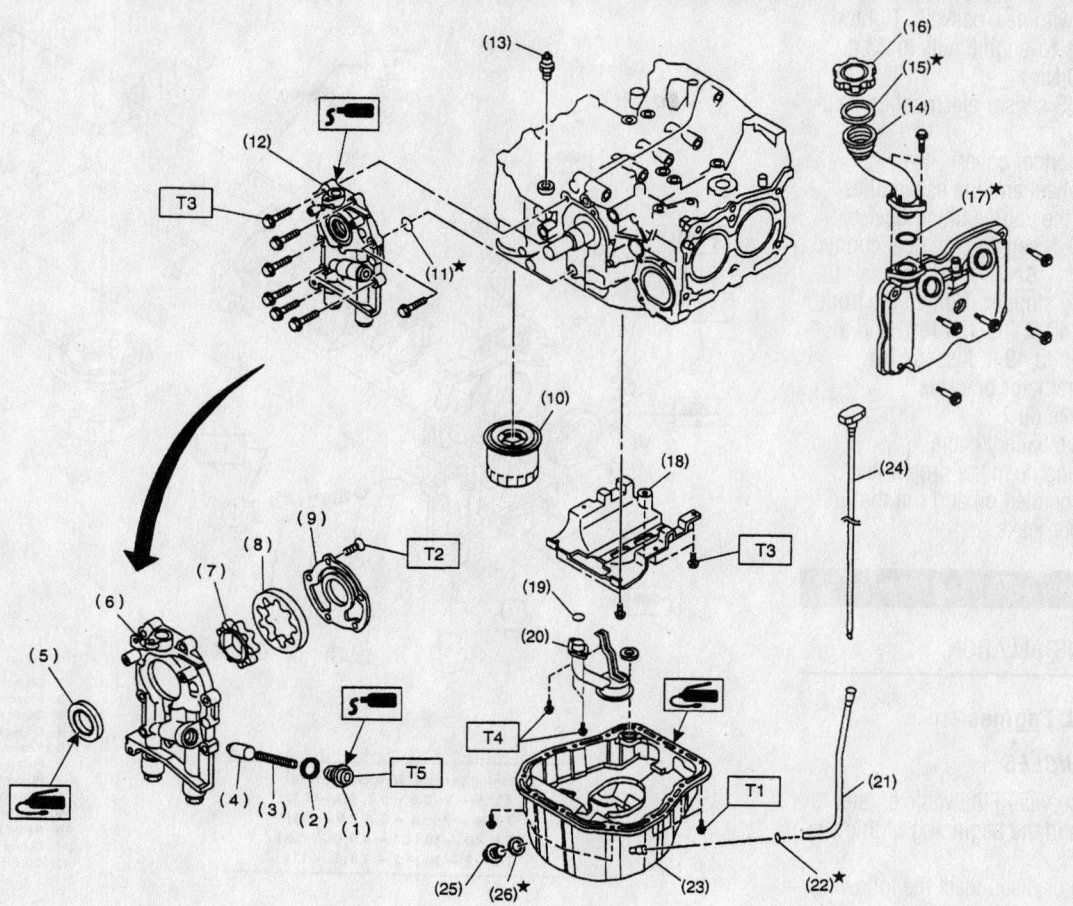

1) Plug
2) Washer
3) Relief valve spring
4) Relief valve
5) Oil seal
6) Oil pump case
7) Inner rotor
8) Outer rotor
9) Oil pump cover
10) Oil filter
11) O-ring
12) Oil pump ASSY

13) Oil pressure switch
14) Oil filler duct
15) O-ring
16) Oil filler cap
17) O-ring
18) Baffle plate
19) O-ring
20) Oil strainer
21) Oil level gauge guide
22) O-ring
23) Oil pan
24) Oil level gauge

25) Drain plug
26) Metal gasket

Tightening torque: N·m (kg-m, ft-lb)
T1: 5 (0.5, 3.6)
T2: $5^{+1}/_{-0}$ ($0.5^{+0.1}/_{-0}$, $3.6^{+0.7}/_{-0}$)
T3: 6.4 (0.65, 4.7)
T4: 10 (1.0, 7.2)
T5: 44.1±3.4 (4.5±0.35, 32.5±2.5)

9307TG03

Exploded view of the oil pan and lubrication components—2.2L and 2.5L engines

- Front Oxygen (O2S) sensor electrical connector
- Pitching stopper
- Upper radiator brackets

3. Support the engine with a suitable lifting device.

- Front wheel and tire assemblies

4. Lift up the engine slightly.

- Engine under cover

5. Drain the oil from the engine into a suitable container.

6. Install the drain plug with a new gasket and tighten it to 33–36 ft. lbs. (43–47 Nm).

7. Remove or disconnect the following:

- Rear O2S sensor electrical connector
- Exhaust Y-pipe
- Nuts that secure the front engine mounts to the front crossmember
- Oil pan mounting bolts

8. While supporting the oil pan, use a rubber mallet and tap the oil pan to free it from the engine.

9. Clean all gasket material from both mating surfaces.

To install:

10. Apply a continuous bead of sealer to a new oil pan gasket.

11. Install the oil pan assembly. Tighten the bolts to 36–48 inch lbs. (4–5 Nm).

12. Lower the engine onto the front crossmember.

13. Install or connect the following:

- Front engine mount nuts and tighten to 61 ft. lbs. (83 Nm)

For Maintenance Interval recommendations, see Section 1 of this manual

- Y-pipe with new gaskets. Tighten the pipe-to-engine nuts to 23 ft. lbs. (30 Nm)
- Rear O₂S sensor electrical connector
- Engine under cover
- Front wheel and tire assemblies
14. Remove the engine lifting device.
- Front O₂S sensor electrical connector
- Pitching stopper. Tighten the front bolt to 40 ft. lbs. (54 Nm) and the rear bolt to 49 ft. lbs. (67 Nm).
- Upper radiator brackets
- Air intake duct
- Negative battery cable
15. Fill the engine to the proper level with the recommended oil and run the engine. Check for leaks.

Oil Pump

REMOVAL & INSTALLATION

2.2L and 2.5L Engines

1998–99 VEHICLES

1. Before servicing the vehicle, refer to the precautions in the beginning of this section.
2. Remove or disconnect the following:
- Negative battery cable
3. Drain the coolant into a suitable separate container.
4. Remove or disconnect the following:
- Timing belt
- Belt tensioner bracket
- Engine coolant pipe, if necessary
- Left side camshaft sprocket and left side belt cover, if necessary
- Water pump
- Oil pump mounting bolts
- Oil pump by carefully prying it from the engine block

❋❋ WARNING

Use extreme care not to damage the engine block or the oil pump during removal of the pump.

To install:

5. Measure the tip clearance of the rotors. If clearance is greater than 0.0071 in. (0.18mm), replace the rotors.
6. Measure the clearance between the outer rotor and the cylinder block rotor housing. If clearance exceeds 0.0079 in. (0.20mm), replace the rotor.
7. Measure the side clearance between the oil pump inner rotor and the pump cover. If clearance exceeds 0.0047

in. (0.12mm), replace the rotor or pump body.
8. Install a new front oil seal on the pump cover using a driver.
9. Assemble the oil pump.
10. Apply sealant and a new O-ring to the oil pump.
11. Install or connect the following:
- Oil pump and tighten the bolts to 60 inch lbs. (7 Nm)
- Water pump
- Left side camshaft sprocket and left side belt cover
- Engine coolant pipe, if necessary
- Belt tensioner bracket
- Timing belt
- Negative battery cable
12. Fill and bleed the cooling system.
13. Start the engine and check for leaks.

2000–01 VEHICLES

1. Before servicing the vehicle, refer to the precautions in the beginning of this section.
2. Remove or disconnect the following:

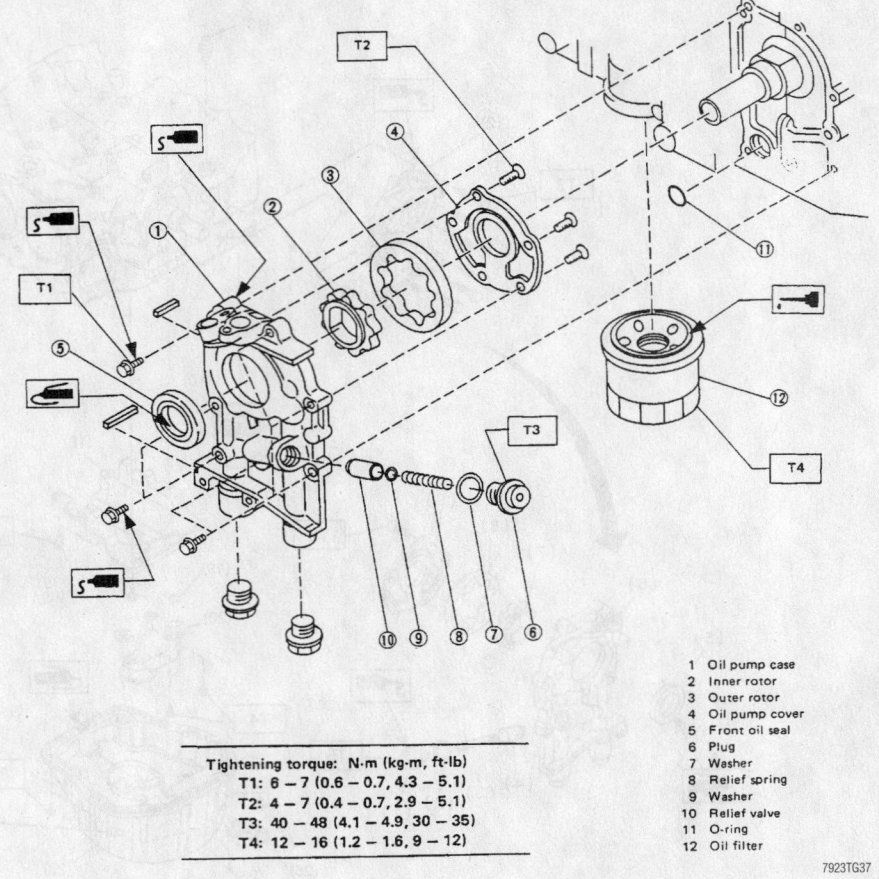

Tightening torque: N·m (kg-m, ft-lb)
T1: 6 – 7 (0.6 – 0.7, 4.3 – 5.1)
T2: 4 – 7 (0.4 – 0.7, 2.9 – 5.1)
T3: 40 – 48 (4.1 – 4.9, 30 – 35)
T4: 12 – 16 (1.2 – 1.6, 9 – 12)

1 Oil pump case
2 Inner rotor
3 Outer rotor
4 Oil pump cover
5 Front oil seal
6 Plug
7 Washer
8 Relief spring
9 Washer
10 Relief valve
11 O-ring
12 Oil filter

7923TG37

Oil pump and components

- Negative battery cable
- Engine under cover
3. Drain the coolant into a suitable separate container.
- Radiator main fan and sub fan assemblies, on Impreza models
- Radiator, on Legacy models
- Crankshaft Position (CKP) and Camshaft Position (CMP) sensors
- Drive belts
- Rear side V-belt tensioner
- Crankshaft pulley using a suitable tool
- Water pump
- Timing belt guide, if equipped with a manual transaxle
- Crankshaft sprocket
- Oil pump mounting bolts
- Oil pump by carefully prying it from the engine block

❋❋ WARNING

Use extreme care not to damage the engine block or the oil pump during removal of the pump.

To install:

4. Measure the tip clearance of the rotors. If clearance is greater than 0.0071 in. (0.18mm), replace the rotors.

5. Measure the clearance between the outer rotor and the cylinder block rotor housing. If clearance exceeds 0.0079 in. (0.20mm), replace the rotor.

6. Measure the side clearance between the oil pump inner rotor and the pump cover. If clearance exceeds 0.0059 in. (0.15mm), replace the rotor or pump body.

7. Assemble the oil pump.

8. Apply sealant and a new O-ring to the oil pump.

9. Install or connect the following:
   - Oil pump and tighten the bolts to 60 inch lbs. (7 Nm)
   - Crankshaft sprocket
   - Timing belt guide, if equipped with a manual transaxle
   - Water pump
   - Crankshaft pulley using a suitable tool
   - Rear side V-belt tensioner
   - Drive belts
   - CKP and CMP sensors
   - Radiator, on Legacy models
   - Radiator main fan and sub fan assemblies, on Impreza models
   - Engine under cover
   - Negative battery cable

10. Fill and bleed the cooling system.

11. Start the engine and check for leaks.

Rear Main Seal

REMOVAL & INSTALLATION

1. Before servicing the vehicle, refer to the precautions in the beginning of this section.

2. Remove or disconnect the following:
   - Engine from the vehicle.
   - Clutch assembly/flywheel using the Clutch Disc Guide tool 499747000, if equipped with a manual transmission
   - Torque converter flexplate from the crankshaft, if equipped with an automatic transmission
   - Oil seal from the cylinder block using a small prybar

To install:

3. Install or connect the following:
   - New oil seal by pressing it into the

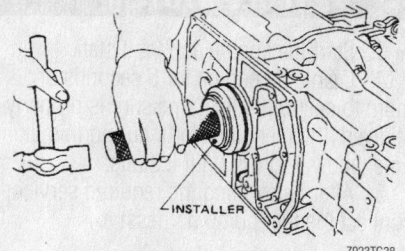

Installing the rear main seal

cylinder block using the appropriate driver
   - Flywheel housing using new gaskets and sealant where necessary.
   - Flywheel and tighten the bolts to 50–54 ft. lbs. (69–75 Nm).
   - Engine

Piston and Ring

POSITIONING

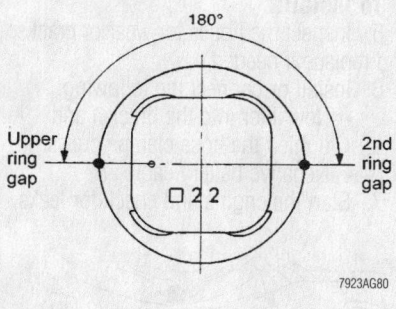

Subaru 2.2L engine—compression ring end-gap spacing

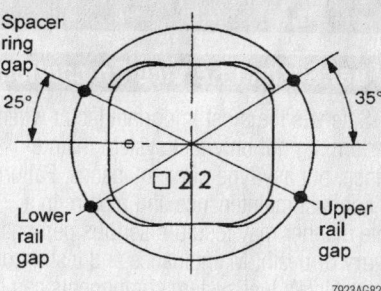

Subaru 2.2L engine—upper, spacer and lower oil ring end-gap spacing

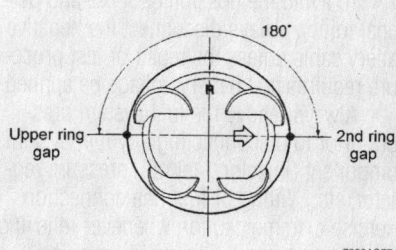

Subaru 2.5L engine—compression ring end-gap spacing

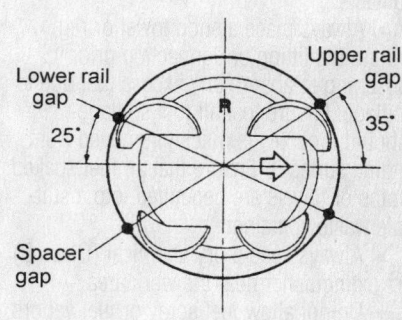

Subaru 2.5L engine—upper, spacer and lower oil ring end-gap spacing

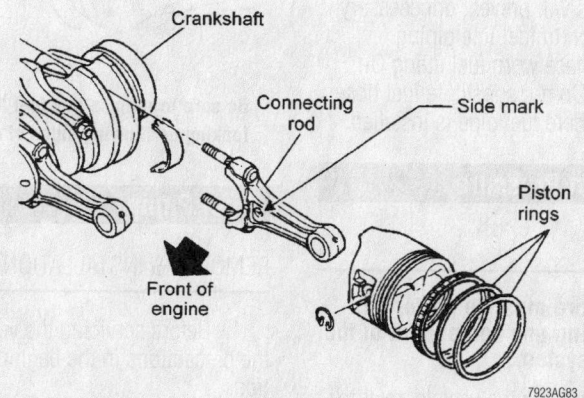

Subaru engines—piston and connecting rod assembly positioning

FUEL SYSTEM

Fuel System Service Precautions

Safety is the most important factor when performing not only fuel system maintenance, but any type of maintenance. Failure to conduct maintenance and repairs in a safe manner may result in serious personal injury or death. Maintenance and testing of the vehicle's fuel system components can be accomplished safely and effectively by adhering to the following rules and guidelines.

• To avoid the possibility of fire and personal injury, always disconnect the negative battery cable unless the repair or test procedure requires that battery voltage be applied.

• Always relieve the fuel system pressure prior to disconnecting any fuel system component (injector, fuel rail, pressure regulator, etc.), fitting or fuel line connection. Exercise extreme caution whenever relieving fuel system pressure, to avoid exposing skin, face and eyes to fuel spray. Please be advised that fuel under pressure may penetrate the skin or any part of the body that it contacts.

• Always place a shop towel or rag around the fitting or connection prior to loosening to absorb any excess fuel due to spillage. Ensure that all fuel spillage (should it occur) is quickly removed from engine surfaces. Ensure that all fuel soaked cloths or towels are deposited into a suitable waste container.

• Always keep a dry chemical (Class B) fire extinguisher near the work area.

• Do not allow fuel spray or fuel vapors to come into contact with a spark or open flame.

• Always use a back-up wrench when loosening and tightening fuel line connection fittings. This will prevent unnecessary stress and torsion to fuel line piping.

• Always replace worn fuel fitting O-rings with new. Do not substitute fuel hose or equivalent, where fuel pipe is installed.

Fuel System Pressure

RELIEVING

➡This procedure must be performed prior to servicing any component of the fuel injection system.

1. Before servicing the vehicle, refer to the precautions in the beginning of this section.
2. Disconnect the fuel pump connector from the fuel pump relay.

3. Start the engine and let it stall.
4. Crank the engine for 5 seconds or more to ensure the fuel pressure is properly relieved. If the engine starts during this time, allow it to run until it stalls.
5. After performing the required service, connect the fuel pump harness.

Fuel Filter

REMOVAL & INSTALLATION

1. Before servicing the vehicle, refer to the precautions in the beginning of this section.
2. Locate the fuel filter in the engine compartment on the left inside fender.
3. Properly relieve the fuel system pressure.
4. Remove or disconnect the following:
   • Negative battery cable
   • Hose clamp screws and slide the hoses off the filter
   • Filter from the bracket

To install:
5. Inspect the hoses for wear or cracks, and replace if needed.
6. Install or connect the following:
   • New filter into the bracket and tighten the hose clamp screws
   • Negative battery cable
7. Start the engine and check for leaks.

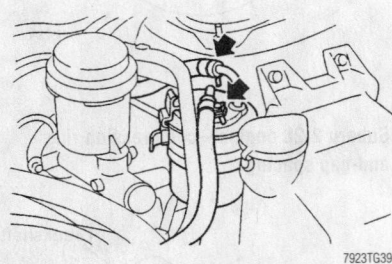

Be sure to replace any fuel lines that are leaking or showing signs of deterioration

Fuel Pump

REMOVAL & INSTALLATION

1. Before servicing the vehicle, refer to the precautions in the beginning of this section.
2. Relieve the fuel system pressure.
3. Disconnect the negative battery cable.
4. On the 2000–01 Legacy, perform the following:

a. Raise and safely support the vehicle.
   b. Remove the front side fuel tank cover.
   c. Drain the fuel tank into a suitable container.
   d. Tighten the drain plug to 14–24 ft. lbs. (19–33 Nm).
   e. Install the front side fuel tank cover. Tighten the retainers to 9.4–16.6 ft. lbs. (13–23 Nm).
5. Remove or disconnect the following:
   • Rear seat bottom, to reach the fuel pump access cover, if not already done
6. On Legacy models, fold the seat back, then roll the floor mat back.
   • Fuel pump cover mounting bolts and the cover
   • Electrical harness from the pump assembly

➡**Label the fuel lines before disconnecting them from the pump.**

   • Fuel lines from the fuel pump
   • 8 fuel pump mounting nuts
   • Fuel pump assembly from the tank
To install:
7. Install or connect the following:
   • New gasket
   • Fuel pump assembly into the fuel tank and secure with the mounting

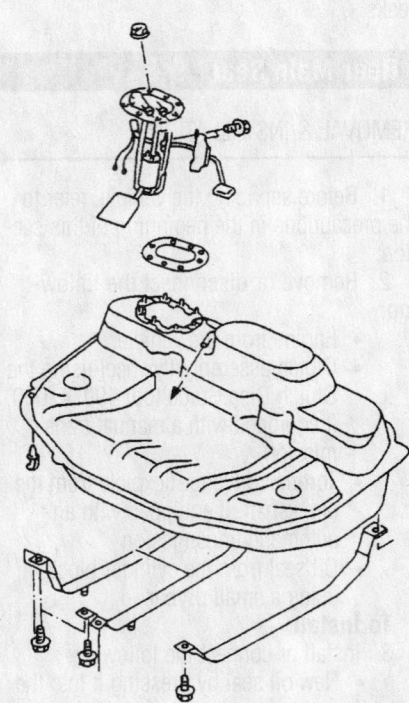

Exploded view of the fuel pump assembly

nuts. Tighten the nuts to 24–48 inch lbs. (3–6 Nm).
- Electrical harness to the fuel pump assembly
- Fuel lines to the pump assembly, then tighten the clamps and fittings
- Fuel pump service cover and cover mounting bolts
- Rear seat bottom
- Negative battery cable

8. Start the engine and check for leaks.

Fuel Injector

REMOVAL & INSTALLATION

1998 Vehicles

1. Before servicing the vehicle, refer to the precautions in the beginning of this section.
2. Relieve the fuel system pressure.
3. Remove or disconnect the following:
- Negative battery cable
- Fuel injector connector
- Fuel injector cup retaining screws
- Fuel injector from the fuel pipe assembly
- Injector O-rings and insulator. Discard the O-rings and the insulator.

To install:
4. Install or connect the following:

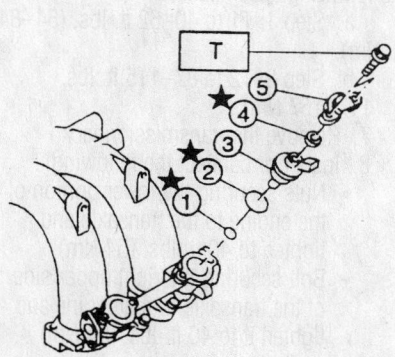

① O-ring B
② O-ring A
③ Fuel injector
④ Insulator
⑤ Fuel injector cup
★: Replacement part

9307TG04

Exploded view of the fuel injector—1998 engines

- New O-rings and insulators on the injector(s)
- Injector into the fuel pipe
- Fuel injector cup retaining screws and tighten to 2.1–2.9 ft. lbs. (2.9–3.9 Nm)
- Fuel injector electrical connector
- Negative battery cable

5. Start the engine and check for leaks.

1999–01 Vehicles

1. Before servicing the vehicle, refer to the precautions in the beginning of this section.
2. Relieve the fuel system pressure.
3. To remove the right side injectors, remove or disconnect the following:
- Air cleaner ducts and resonator chamber, on California vehicles
- Mass Air Flow (MAF) sensor connector and air intake duct and air cleaner upper cover as a unit, on non-California vehicles
- Air cleaner element on non-California vehicles
- Spark plug wires from the right side spark plugs
- V-belt covers and power steering pump belt
- Power steering pump brackets-to-intake manifold bolts

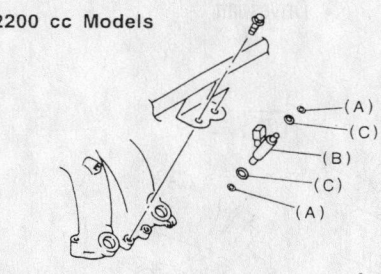

2200 cc Models

(A) O-ring
(B) Fuel injector
(C) Insulator

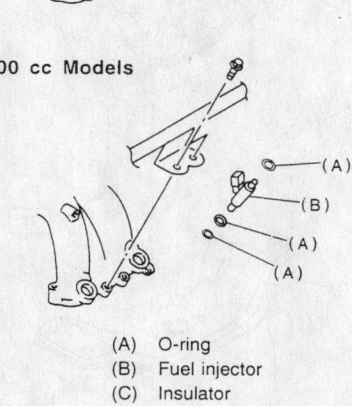

2500 cc Models

(A) O-ring
(B) Fuel injector
(C) Insulator

9307TG05

Exploded view of the fuel injector—1999–01 engines

- Power steering pump-to-bracket bolts, then position the pump on the right side wheel apron

4. To remove the injectors on the left side, remove or disconnect the following:
- Windshield washer motor electrical connector
- Electrical connector from the rear window washer, on station wagon only
- Rear window washer hose from the washer motor and plug or cap the line
- Two bolts that secure the washer tank to the body
- Washer tank and secure it out of the way
- Spark plug wires from the left side spark plugs
- Fuel pipe protector

5. Remove or disconnect the following (for either side):
- Band that secures the engine harness to the fuel injector pipe, if equipped
- Intake manifold protector, if equipped
- Fuel injector electrical connector(s)
- Bolts that hold the fuel injector pipe (fuel rail) to the intake manifold, if applicable

➡**Automatic transaxle equipped Legacy's may have a retaining clip that must be removed before the injector can be removed.**

6. Pull up on the injector pipe (fuel rail), then remove the fuel injector(s) from the intake manifold. Remove and discard the injector O-rings.

To install:
- Install or connect the following:
- New injector O-rings
- Fuel injector(s) into the intake manifold
- Retaining clips, if applicable
- Injector pipe (fuel rail) and secure with the retaining bolts. Tighten the bolts to 2.1–2.9 ft. lbs. (2.9—3.9 Nm).
- Fuel injector electrical connector(s)
- Intake manifold protector, if equipped
- Band that secures the engine harness to the fuel injector pipe, if equipped

7. To install the injectors on the left side, install or connect the following:

- Fuel pipe protector
- Spark plug wires to the left side spark plugs
- Washer tank and secure with the two mounting bolts
- Rear window washer hose to the washer motor
- Electrical connector to the rear window washer, on station wagon only

- Windshield washer motor electrical connector

8. To install the right side injectors, install or connect the following:
- Power steering pump into position
- Pump-to-bracket bolts
- Power steering pump brackets-to-intake manifold bolts
- Power steering pump belt and V-belt covers

- Spark plug wires to the right side spark plugs
- Air cleaner element, on non-California vehicles
- Air intake duct and upper cover and the MAF sensor connector
- Air cleaner ducts and resonator chamber, on California vehicles

DRIVE TRAIN

Transaxle Assembly

REMOVAL & INSTALLATION

Manual

1. Before servicing the vehicle, refer to the precautions in the beginning of this section.

2. Remove or disconnect the following:
- Negative battery cable
- Air intake duct and cleaner case
- Air cleaner stay, if equipped
- Front Oxygen (O2S) sensor connector
- Neutral position switch connector
- Back-up light switch connector
- Vehicle Speed Sensor (VSS) connector, if equipped

- Clutch cable, if equipped
- Clutch release spring
- Starter
- Pitching stopper
- Drive belt cover
- Slave (operating) cylinder, on 2.5L engines

3. Install engine support assembly 927670000.
- Bolt securing the right upper side of the transaxle to the engine
- Engine under cover, if equipped
- Rear O2S sensor connector
- Front Y-pipe
- Rear exhaust pipe and muffler, on Legacy models
- Heat shield cover
- Hanger bracket from the right side of the transaxle
- Driveshaft

- Spring, and disconnect the shifter stay and rod from the transaxle
- Bolts securing the sway bar clamps to the crossmember
- Ball joints from the steering knuckle
- Halfshafts from the transaxle
- 2 nuts securing the lower side of the transaxle to the engine

4. Support the transaxle with a jack.
- Rear transaxle crossmember
- Transaxle from the vehicle. Move the jack rearward until the mainshaft is withdrawn from the clutch cover.

To install:

5. Install or connect the following:
- Transaxle assembly and secure it to the engine block
- Crossmember

6. Tighten the crossmember retainers to the following specifications:
   a. Step 1: T1 to 40–62 ft. lbs. (54–84 Nm).
   b. Step 2: T2 to 87–115 ft. lbs. (117–157 Nm).

7. Remove the transmission jack.

8. Install or connect the following:
- Nuts securing the lower portion of the engine to the transaxle and tighten to 40 ft. lbs. (54 Nm)
- Bolt securing the right upper side of the transaxle to the engine and tighten it to 40 ft. lbs. (54 Nm)

9. Remove the engine support.
- Drive belt cover
- Slave (operating) cylinder, on 2.5L engines

10. Install the pitching stopper and tighten the bolts to the following specifications:
   a. Step 1: T1 to 33–40 ft. lbs. (44–54 Nm).
   b. Step 2: T2 to 35–49 ft. lbs. (47–67 Nm).

11. Install or connect the following:
- Halfshafts into the transaxle with new roll pins

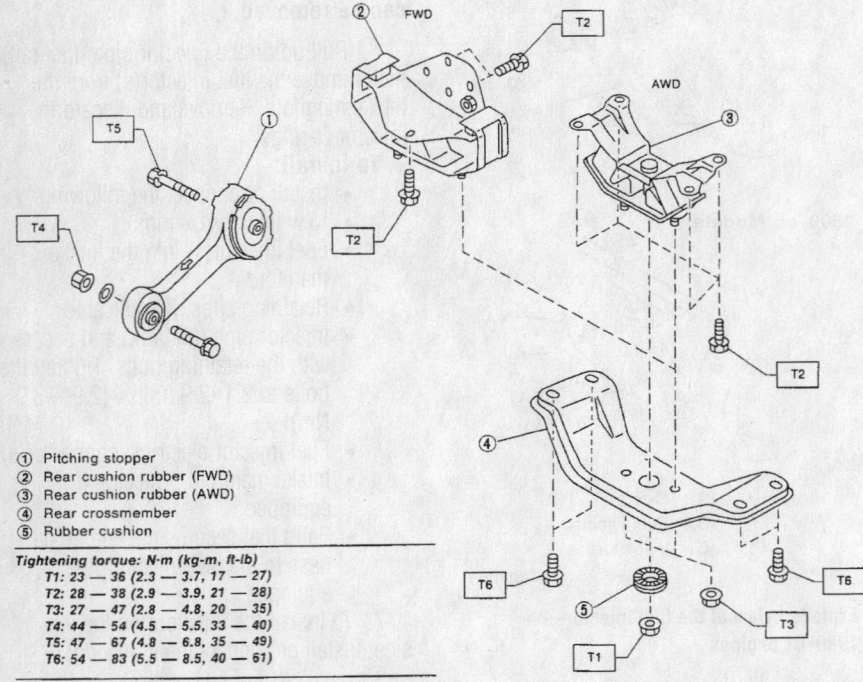

① Pitching stopper
② Rear cushion rubber (FWD)
③ Rear cushion rubber (AWD)
④ Rear crossmember
⑤ Rubber cushion

Tightening torque: N·m (kg-m, ft-lb)
T1: 23 — 36 (2.3 — 3.7, 17 — 27)
T2: 28 — 38 (2.9 — 3.9, 21 — 28)
T3: 27 — 47 (2.8 — 4.8, 20 — 35)
T4: 44 — 54 (4.5 — 5.5, 33 — 40)
T5: 47 — 67 (4.8 — 6.8, 35 — 49)
T6: 54 — 83 (5.5 — 8.5, 40 — 61)

7923TG44

Exploded view of the transaxle mounting—Impreza models

- Ball joint to the steering knuckle and tighten the bolt to 29–43 ft. lbs. (39–59 Nm)
- Sway bar to the crossmember and tighten the clamp bolts to 15–21 ft. lbs. (21–29 Nm)
- Shift control rod and stay to the transaxle and install the spring
- Driveshaft
- Heat shield cover, if removed
- Rear exhaust pipe and muffler, if removed
- Y-pipe with new gaskets and nuts
- Hanger bracket on the right side of the transaxle, if removed
- Rear O2S sensor connector
- Engine under cover, if removed
- Transaxle connectors bracket
- Drive belt cover
- Pitching stopper
- Starter
- Front O2S sensor connector
- VSS connector, if equipped
- Neutral position switch connector
- Back-up light switch connector
- Clutch cable (if equipped)
- Clutch release spring
- Air cleaner case stay and case

Automatic

1. Before servicing the vehicle, refer to the precautions in the beginning of this section.
2. Drain the transaxle fluid.
3. Remove or disconnect the following:
   - Negative battery cable
   - Air intake duct with air cleaner case
   - Air cleaner case stay
   - Speedometer cable or electronic wiring connector from the speed sensor
   - Front Oxygen (O2S) sensor connector
   - Transaxle harness connector
   - Inhibitor switch connector
   - Revolution sensor connector
   - Transaxle ground terminal
   - Clip band that secures the air breather hose to the pitching stopper
   - Starter and air intake boot
   - Timing hole inspection plug
   - 4 bolts that hold the torque converter to the driveplate

- Air intake duct and attach the air-flow sensor connector
- Negative battery cable

Automatic

1. Before servicing the vehicle, refer to the precautions in the beginning of this section.
2. Drain the transaxle fluid.
3. Remove or disconnect the following:

- Automatic Transaxle Fluid (ATF) level gauge
- Pitching stopper rod from the bracket
- Engine-to-transaxle mounting nut and bolt on the right side
- Buffer rod

4. Support the engine assembly with special engine support tool.
   - Exhaust system
   - Exhaust brackets or hangers that attach to the transaxle, as necessary
5. Drain the transmission fluid.
   - ATF cooler hoses from the pipes of the transmission side
   - ATF level gauge guide

➡ **Matchmark the installed position of the driveshaft before removal.**

- Driveshaft. Plug the opening at the rear of extension housing to prevent oil from flowing out.
- Gearshift cable from the transaxle select lever
- Stabilizer from the transverse link
- Parking brake cable bracket from the transverse link
- Transverse link bolts and lower the link
- Spring pins
- Halfshafts from the transaxle

➡ **Discard the old spring pin and always install a new pin.**

- Oil cooler hoses
6. Place a transaxle jack under the transaxle.
   - Engine to transaxle mounting nuts

➡ **Do not place the jack under the oil pan otherwise the oil pan may be damaged.**

- Rear cushion rubber mounting nuts and the rear crossmember
7. Move the torque converter and transaxle as a unit away from the engine and lower it from the vehicle.
   To install:
8. Install or connect the following:
   - Transaxle to the engine and temporarily tighten the engine-to-transaxle mounting nuts
   - Rear crossmember to the rear cushion rubber mounts. Align the rear cushion guide with the rear crossmember guide hole and tighten nuts.
   - Rear crossmember to the chassis. Tighten the rear crossmember bolts to 39–49 ft. lbs. (53–66 Nm).

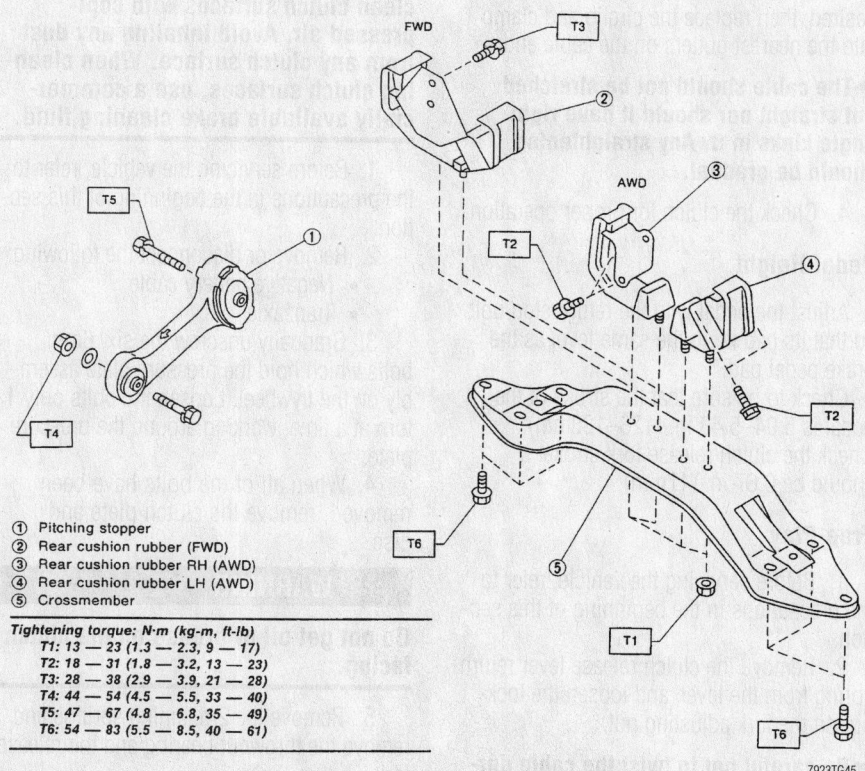

① Pitching stopper
② Rear cushion rubber (FWD)
③ Rear cushion rubber RH (AWD)
④ Rear cushion rubber LH (AWD)
⑤ Crossmember

Tightening torque: N·m (kg-m, ft-lb)
T1: 13 — 23 (1.3 — 2.3, 9 — 17)
T2: 18 — 31 (1.8 — 3.2, 13 — 23)
T3: 28 — 38 (2.9 — 3.9, 21 — 28)
T4: 44 — 54 (4.5 — 5.5, 33 — 40)
T5: 47 — 67 (4.8 — 6.8, 35 — 49)
T6: 54 — 83 (5.5 — 8.5, 40 — 61)

7923TG45

Exploded view of the engine and transaxle mounts—Impreza and Legacy models

- Engine to transaxle retaining nuts to 34–40 ft. lbs. (46–54 Nm)

9. Remove the transaxle jack from the vehicle.

10. Remove the engine support tool and install the buffer rod.

11. Install or connect the following:
   - Axle shafts to the transaxle using new spring pins
   - Transverse link temporarily to the front crossmember. Do not complete final torque at this point.
   - Stabilizer temporarily to the transverse link
   - Parking brake cable bracket to the transverse link
   - Transverse link-to-front crossmember mounting bolts and transverse link-to-stabilizer mounting bolts, with the tires placed on the ground
   - Transverse link to front crossmember (self-locking nuts) to 40–62 ft. lbs. (54–84 Nm) and the transverse link to stabilizer to 18–32 ft. lbs. (24–44 Nm).
   - Gearshift cable to the select lever. Be sure the lever operates smoothly all across the operating range.
   - Driveshaft. Tighten the driveshaft-to-rear differential retaining bolts to 17–24 ft. lbs. (23–33 Nm) and center bearing location retaining bolts to 25–33 ft. lbs. (34–45 Nm).
   - Oil cooler hoses
   - Engine to transaxle bolts to 34–40 ft. lbs. (46–54 Nm)
   - Starter
   - Pitching stopper. Be sure to tighten the bolt for the body side first. Tightening torque for the body side bolt is 35–49 ft. lbs. (47–67 Nm). The engine or transaxle side bolt is torque to 33–40 ft. lbs. (44–54 Nm).
   - Torque converter-to-driveplate mounting bolts to 17–20 ft. lbs. (23–27 Nm)
   - ATF level gauge guide
   - ATF cooler hoses to the pipes of the transmission side
   - Timing hole inspection plug, air intake boot and air breather hose to the pitching stopper
   - O$_2$S sensor connector.
   - Transaxle harness connector
   - Inhibitor switch connector
   - Revolution sensor connector, if equipped
   - Transaxle ground terminal, on Legacy models
   - Speedometer cable. Tighten the cable nut by hand, then turn it

approximately 30 degrees more with a tool.
   - Exhaust system and exhaust brackets or hangers that attach to the transaxle, as necessary
   - Air cleaner case stay
   - Air intake duct with air cleaner case
   - Battery ground cable

12. Refill and check transaxle oil level.

13. Road test the vehicle for proper operation across all operating ranges.

Clutch

ADJUSTMENT

Some models are equipped with a mechanical clutch system that is adjustable. Other models are equipped with a hydraulic system that is not adjustable.

Cable

The clutch cable can be adjusted at the cable bracket where the cable is attached to the side of the transaxle housing.

1. Before servicing the vehicle, refer to the precautions in the beginning of this section.

2. Remove the circlip and clamp.

3. Slide the cable end in the direction desired, then replace the circlip and clamp into the nearest gutters on the cable end.

➡**The cable should not be stretched out straight nor should it have right angle kinks in it. Any straightening should be gradual.**

4. Check the clutch for proper operation.

Pedal Height

Adjust the pedal with the return stop bolt, so that its pad is on the same level as the brake pedal pad.

Check to be sure that the stroke of the pedal is 5.04–5.43 in. (128–138mm). Check the clutch release fork stroke. It should be 0.67 in. (17mm).

Free-Play

1. Before servicing the vehicle, refer to the precautions in the beginning of this section.

2. Remove the clutch release lever return spring from the lever, and loosen the locknut on the fork adjusting nut.

➡**Be careful not to twist the cable during adjustment**

3. Turn the adjusting nut (spherical nut) until a release fork free-play of 0.14–0.18 in. (3.5–4.5mm) is obtained.

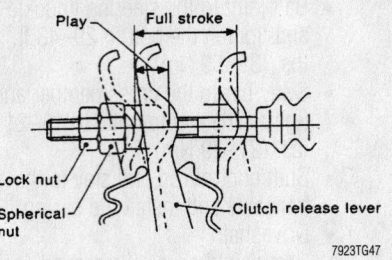

Be sure to tighten the locknut after making the necessary adjustments—mechanical clutch

4. Tighten the locknut.

5. Install the return spring on the lever. Hook the long hook side of the return spring with the lever.

6. Check the pedal free-play. It should be 0.12–0.16 in. (3.0–4.0mm).

7. Adjust the pedal free-play, as necessary, with the pedal adjusting bolt.

REMOVAL & INSTALLATION

✳✳ CAUTION

The clutch driven disc may contain asbestos that has been determined to be a cancer causing agent. Never clean clutch surfaces with compressed air. Avoid inhaling any dust from any clutch surface. When cleaning clutch surfaces, use a commercially available brake cleaning fluid.

1. Before servicing the vehicle, refer to the precautions in the beginning of this section.

2. Remove or disconnect the following:
   - Negative battery cable
   - Transaxle

3. Gradually unscrew the six, 6mm bolts which hold the pressure plate assembly on the flywheel. Loosen the bolts only 1 turn at a time, working around the pressure plate.

4. When all of the bolts have been removed, remove the clutch plate and disc.

✳✳ WARNING

Do not get oil or grease on the clutch facing.

5. Remove the 2 retaining springs and remove the throwout bearing and the release fork.

➡**Do not disassemble either the clutch cover or disc. Inspect the parts for wear or damage and replace any parts as necessary. Replace the clutch disc if**

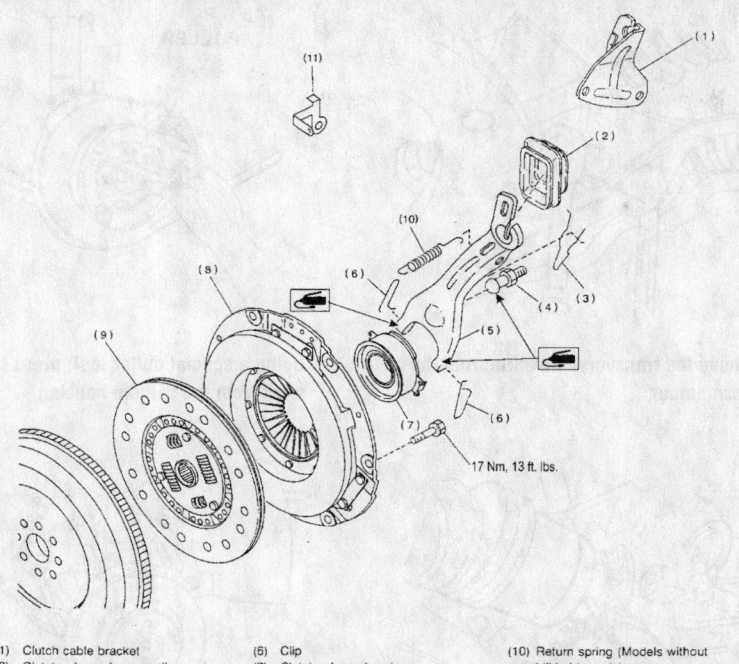

(1) Clutch cable bracket
(2) Clutch release lever sealing
(3) Retainer spring
(4) Pivot
(5) Clutch release lever

(6) Clip
(7) Clutch release bearing
(8) Clutch cover
(9) Clutch disc

(10) Return spring (Models without hill holder only)
(11) Clutch return spring bracket

17 Nm, 13 ft. lbs.

7923TG48

Exploded view of the clutch system components—mechanical clutch

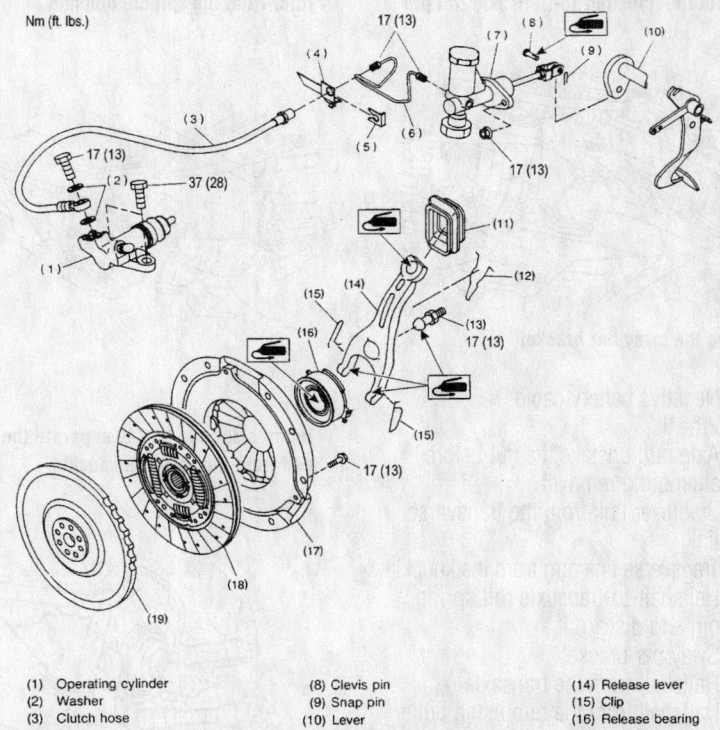

Nm (ft. lbs.)

17 (13)

17 (13)

37 (28)

17 (13)

17 (13)

17 (13)

17 (13)

(1) Operating cylinder
(2) Washer
(3) Clutch hose
(4) Bracket
(5) Clamp
(6) Pipe
(7) Master cylinder ASSY

(8) Clevis pin
(9) Snap pin
(10) Lever
(11) Clutch release lever sealing
(12) Retainer spring
(13) Pivot

(14) Release lever
(15) Clip
(16) Release bearing
(17) Clutch cover
(18) Clutch disc
(19) Flywheel

7923TG49

Exploded view of the clutch system components—hydraulic clutch

there is any oil or grease on the facing. Do not wash or attempt to lubricate the throwout bearing. If it requires replacement, the bearing may be removed and a new one installed in the holder by means of a press.

To install:

6. Fit the release fork boot on the front of the transaxle housing.

7. Install or connect the following:
- Release fork
- Throwout bearing assembly and secure it with the 2 springs. Coat the inside diameter of the bearing holder and the fork-to-holder contact points with grease.

8. Insert a pilot shaft through the clutch cover and disc, then insert the end of the pilot into the needle bearing.

9. Tighten the pressure plate bolts gradually, 1 turn at a time, until the proper torque is reached. Tighten to 13 ft. lbs. (17 Nm).

✷✷ WARNING

When installing the clutch pressure plate assembly, be sure that the O marks on the flywheel and the clutch pressure plate assembly are at least 120 degrees apart. These marks indicate the direction of residual unbalance. Also, be sure that the clutch disc is installed properly, noting the FRONT and REAR markings.

10. After installation of the transaxle in the car, perform the adjustments outlined above.

Hydraulic Clutch System

BLEEDING

➡ **To properly bleed the system, it must be bled at the slave cylinder and at the damper.**

1. Before servicing the vehicle, refer to the precautions in the beginning of this section.

2. Connect a vinyl tube to the air bleeder on the clutch damper (on 1998–99 vehicles) or the clutch operating (slave) cylinder (on 2000–01 vehicles). Put the other end in a jar with clean clutch fluid.

3. With the help of an assistant depressing the clutch pedal, slowly open the bleeder valve. Close the bleeder valve and release the pedal. Repeat this process until no air bubbles appear in the jar.

Timing belt service is covered in Section 3 of this manual

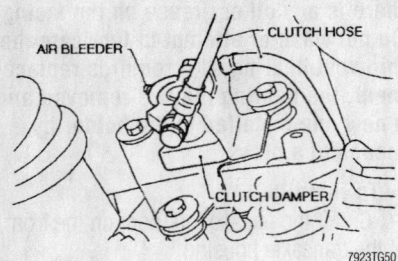

Bleeding the hydraulic clutch at the clutch damper

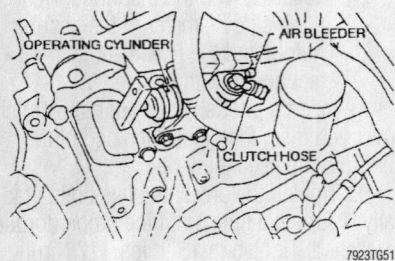

Bleeding the hydraulic clutch at the slave cylinder

4. Move the tube to the bleeder on the slave cylinder and repeat the process. Check the operation of the clutch after the bleed procedure is complete.

Transfer Case Assembly

REMOVAL & INSTALLATION

The transfer case must be removed as an assembly with the transaxle.

Halfshaft

REMOVAL & INSTALLATION

1. Before servicing the vehicle, refer to the precautions in the beginning of this section.
2. Remove or disconnect the following:

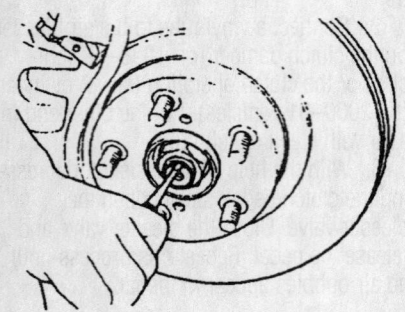

Unstaking the axle nut

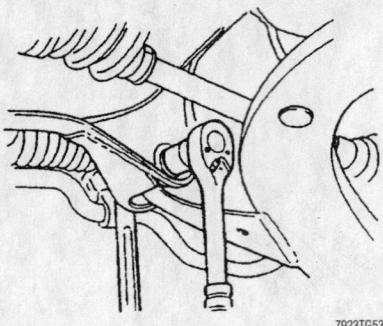

Remove the transverse link arm from the crossmember

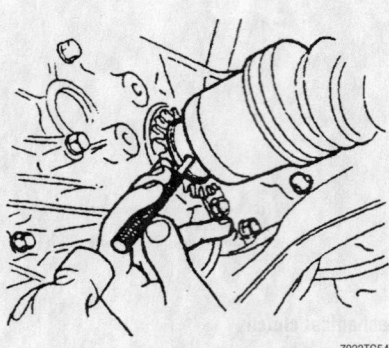

Drive out the halfshaft-to-transaxle roll pin

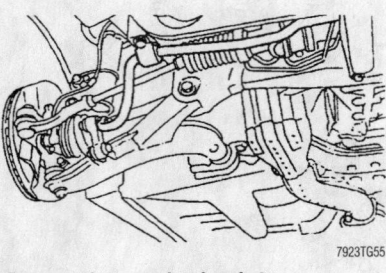

Remove the sway bar bracket

- Negative battery cable
- Wheel
- Axle nut, unstake the nut before attempting removal
- Stabilizer link from the transverse link
- Transverse link arm from the knuckle
- Halfshaft-to-transaxle roll/spring pin and discard it
- Sway bar bracket
- Halfshaft from the transaxle
- Halfshaft from the hub using puller 92707000

To install:
3. Install or connect the following:
- Halfshaft into the hub
4. Using installer 922431000 and adapter 927390000, pull the halfshaft through the hub.

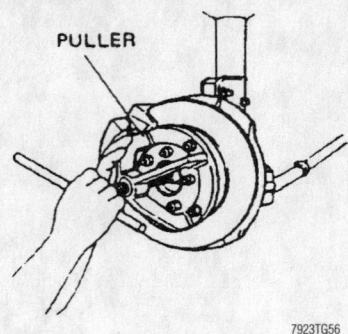

Using a special puller tool, press the axle shaft from the spindle housing

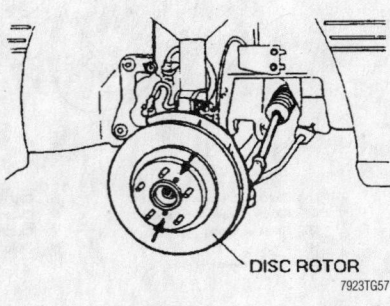

Use 2 8mm bolts (arrows) to loosen the rotor from the spindle housing

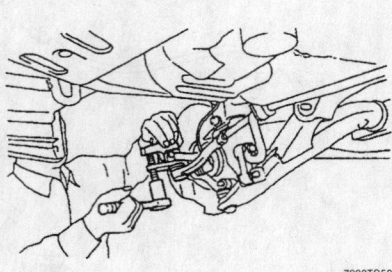

Using a special tool to separate the tie rod end from the steering knuckle

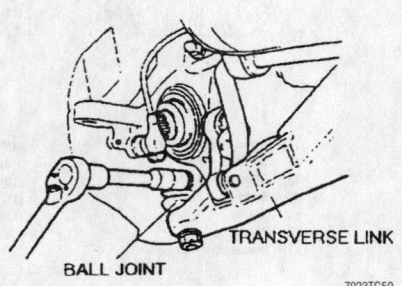

Remove the transverse link arm from the spindle housing

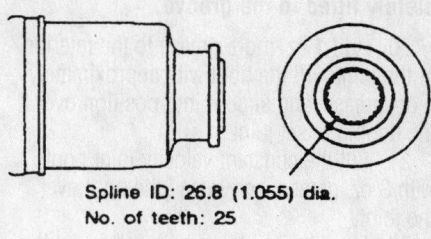

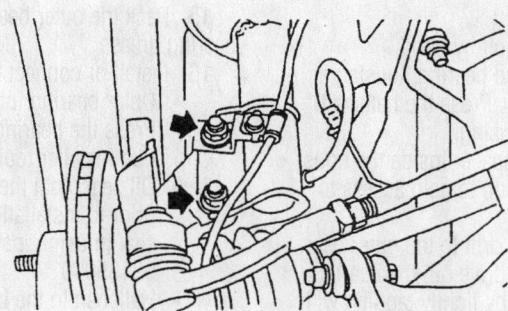

Be sure to identify the correct halfshaft

Spline ID: 26.8 (1.055) dia.
No. of teeth: 25

Unit: mm (in)

7923TG60

Before loosening the strut-to-housing bolts (arrows), matchmark the camber adjustment bolt and strut

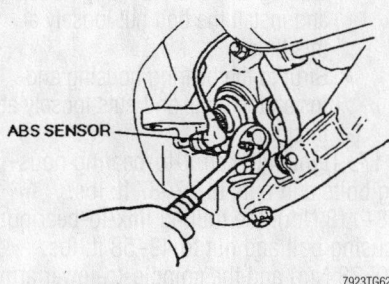

ABS SENSOR

7923TG62

Removing the ABS sensor

5. Install or connect the following:
- Temporarily tighten a new axle nut
- Align the halfshaft roll/spring pin hole
- Halfshaft onto the transaxle
- New pin.
- Transverse link to the knuckle and tighten a new self-locking nut to 29–43 ft. lbs. (39–59 Nm)
- Sway bar bracket
- New axle nut to 152 ft. lbs. (206 Nm) and stake the nut
- Wheel
- Negative battery cable

CV-Joints

OVERHAUL

1. Before servicing the vehicle, refer to the precautions in the beginning of this section.
2. Remove the bands from the boots at both the constant velocity and double offset joints, and slide the boots away from the joints.
3. Pry the circlip out of the double off-set joint, and slide the outer race of the joint off the shaft.
4. Remove the balls from the cage, rotate the cage slightly, and slide the cage inward on the axle shaft.
5. Using snapring pliers, remove the outer snapring which retains the inner race to the shaft.
6. Slide the inner race, cage, and boot off the axle shaft.

✳✳ WARNING

Exercise care to avoid damaging the boot on the inner snapring.

7. Pull back the constant velocity joint boot and pivot the stub axle around the joint far enough to expose a ball.
8. Remove the exposed ball, and continue this procedure until all balls are removed, at which time the outer race (stub axle) may be removed from the axle shaft.
9. Remove the retaining snapring, and slide the inner race off the shaft.
10. Inspect the parts of both joints for wear, damage, or corrosion, and replace if necessary. Examine the axle shaft for bending or distortion, and replace if evident. Should the boots be dried out, cracked, or distorted, they must be replaced.
11. Install the constant velocity joint inner race on the axle shaft, and retain with a snapring.
12. Assemble the joint in the opposite order of disassembly.
13. Slide the double offset joint cage onto the shaft, with the counterbore toward the end of the shaft.
14. Install the inner race on the shaft, and install the retaining snapring.
15. Position the cage over the inner race, and fill the cage pockets with grease.
16. Insert the balls into the cage.
17. Fill the well in the outer race with approximately 1 oz. grease, and slide the outer race onto the axle shaft.
18. Align the outer race track and ball positions and place it into the part where the shaft, inner cage and balls were previously installed, then install it into the outer race.
19. Install the retaining circlip in the grove on the outer race.

➡**Assure that the balls, cage and inner race are completely fitted in the outer race of the joint. Exercise care not to place the matched position of the circlip in the ball groove of the outer race. Finally, pull lightly on the shaft**

Heater Core replacement is covered in Section 2 of this manual

and assure that the circlip is completely fitted in the groove.

20. Add 1 oz. more grease to the interior of the joint. Fill the boot with approximately 1 oz. grease, and slide it into position over the double offset joint.

21. Fill the constant velocity joint boot with 3 oz. grease, and install the boot over the joint.

22. Band the boots on both joints tightly enough that they cannot be turned by hand.

✲✲ WARNING

Use only grease specified for use in constant velocity joints.

Axle Shaft

REMOVAL & INSTALLATION

1. Before servicing the vehicle, refer to the precautions in the beginning of this section.
2. Remove or disconnect the following:
   - Negative battery cable
   - Axle nut
   - Sway bar bracket
   - Lower control arm-to-rear housing bolt and nut
   - Trailing link assembly-to-rear housing bolt and nut
   - Halfshaft-to-differential roll pin
   - Halfshaft from the differential
   - Axleshaft from the hub using puller 92707000

To install:
3. Install or connect the following:
   - Axleshaft into the hub
   - Axleshaft into place using installer 922431000 and adapter 927390000
   - New axle nut and temporarily tighten the nut
4. Align the axleshaft-to-differential roll pin holes and slide the axleshaft onto the splines. Install a new roll pin.
   - Trailing link assembly to the rear housing
   - Trailing link bolt and new nut and tighten to 94 ft. lbs. (127 Nm)
   - Lower control arm to the rear housing
   - Bolt and new nut and torque them to 116 ft. lbs. (157 Nm)
   - Sway bar bracket
5. Torque the new axle nut to 152 ft. lbs. (206 Nm).
   - Wheel
   - Negative battery cable

Axle Bearings

REMOVAL & INSTALLATION

1. Before servicing the vehicle, refer to the precautions in the beginning of this section.
2. Remove or disconnect the following:
   - Tire and wheel assembly
   - Dust cap, cotter pin, castle nut, conical spring and center piece

➡ **Remove the center piece by wedging flat tool between the separation while using a hammer to tap the center piece free.**

   - Brake drum off by hand
   - Rear axle and bearing housing as an assembly. Press the halfshaft from the housing.
3. Position the spacer inside the housing in a radial direction to gain access to the outer bearing.
4. Place a brass drift to the outer bearing inner race. Then drive out the bearing, together with the oil by lightly tapping with a hammer.

➡ **Do not reuse the old bearing, the old bearing may develop abnormal noise if reinstalled.**

5. Remove or disconnect the following:
   - Spacer from the housing
6. Reverse the bearing housing and drive out the inner bearing and seal.

➡ **Always tap around the periphery of the bearing outer race.**

7. Clean and inspect the bearing inner and outer race for cracks or damage. Check for noise and binding when the outer race is turned slowly and while the inner race is held in position.
8. Clean and inspect the spacer. Check for damage and deformation. Replace as necessary.
9. For the bearing housing, wipe the inner surface with a clean cloth and check for cracks or damage. Replace as necessary.

To install:
10. Pack the inner bearing with wheel bearing grease.
11. Install or connect the following:
    - Bearing into the housing. Press the bearing using bearing hub installer tool 92135000.
12. Coat the oil seal with grease.
    - Oil seal using installer tool 92135000 to press it into place

➡ **A gap between the seal lip and bearing changes oil seal lip interference, resulting in excessive wear on the lip.**

13. Pack the housing with approximately 0.6 oz. (17 g) of wheel bearing grease.

✲✲ WARNING

Do not pack excessively, otherwise it may leak from the outer bearing in the inside of the brake drum which may cause brake failure.

14. Install or connect the following:
    - Spacer in the housing
15. Pack the outer bearing with wheel bearing grease
16. Install or connect the following:
    - Outer bearing into the housing. Press the bearing using bearing hub installer tool 92135000.
    - Oil seal (coat the seal with grease prior to installation), and press the seal in using installer tool 92135000
    - Halfshaft to the bearing housing and press in place using installer tool 92232000
    - Halfshaft and housing assembly into the vehicle
    - Trailing arm to the bearing housing and install the bolt nut loosely at this time
    - Strut to the bearing housing and install the bolts and nuts loosely at this time
17. Tighten the strut-to-bearing housing bolts and nuts to 72–87 ft. lbs. (98–118 Nm), the trailing link-to-bearing housing bolt and nut to 43–58 ft. lbs. (59–78 Nm) and the spindle-to-lower arm bolt and nut to 54–69 ft. lbs. (74–93 Nm).
18. Install or connect the following:
    - Brake backing plate assembly to bearing housing and tighten the retaining bolts to 13–23 ft. lbs. (18–31 Nm)
    - Brake line bracket and brake drum
    - Center piece, conical spring and castle nut
    - Wheels and lug nuts
19. Lower the vehicle and apply the brake.
20. Tighten the axle nut to 108 ft. lbs. (147 Nm). After tightening to specifications, tighten the axle shaft nut 30 degrees further. Install a new cotter pin.

STEERING AND SUSPENSION

Air Bag

✳✳ CAUTION

Some vehicles are equipped with an air bag system. The system must be disabled before performing service on or around system components, steering column, instrument panel components, wiring and sensors. Failure to follow safety and disabling procedures could result in accidental air bag deployment, possible personal injury and unnecessary system repairs.

PRECAUTIONS

Several precautions must be observed when handling the inflator module to avoid accidental deployment and possible personal injury.

• Never carry the inflator module by the wires or connector on the underside of the module.

• When carrying a live inflator module, hold it securely with both hands, and ensure that the bag and trim cover are pointed away.

• Place the inflator module on a bench or other surface with the bag and trim cover facing up.

• With the inflator module on the bench, never place anything on or close to the module that may be thrown in the event of an accidental deployment.

DISARMING

1. Before servicing the vehicle, refer to the precautions in the beginning of this section.
2. Disconnect the negative battery cable.

3. Disconnect the positive battery cable.
4. Wait more than 20 seconds before starting work.

Power Rack and Pinion Steering Gear

REMOVAL & INSTALLATION

Legacy and Legacy Outback Models

1. Before servicing the vehicle, refer to the precautions in the beginning of this section.
2. Remove or disconnect the following:
   • Negative battery cable
   • Air intake duct
   • Front axle nut, loosen only at this time
   • Front tire and wheel assemblies

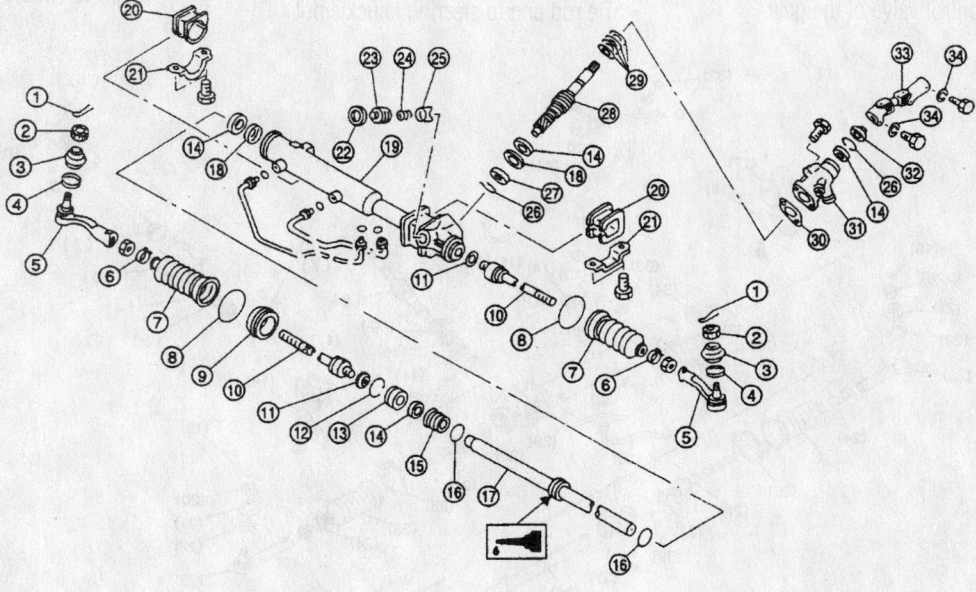

① Cotter pin	⑬ Rack stopper	㉔ Spring
② Castle nut	⑭ Oil seal	㉕ Sleeve
③ Dust cover	⑮ Rack bushing	㉖ C-ring
④ Clip	⑯ O-ring	㉗ Ball bearing
⑤ Tie-rod end	⑰ Rack	㉘ Valve
⑥ Clip	⑱ Back-up washer	㉙ Seal ring
⑦ Boot	⑲ Rack housing	㉚ Packing
⑧ Clip	⑳ Adapter	㉛ Valve housing
⑨ Spacer	㉑ Clamp	㉜ Dust seal
⑩ Tie-rod	㉒ Lock nut	㉝ Universal joint
⑪ Lock washer	㉓ Adjusting screw	㉞ Spring washer
⑫ Circlip		

Exploded view of the steering rack assembly—Legacy models

7923TG63

Brake service is covered in Section 4 of this manual

- Electrical connector from the Oxygen sensor (O2S)
- Front exhaust pipe assembly
- Tie rod end cotter pin and loosen the castle nut
- Tie rod ends from the steering knuckle arm using a ball joint puller
- Jack up plate and the front stabilizer bar

3. From the power steering rack, remove the center pressure pipe, connect a vinyl hose to the pipe and joint, then turn the steering wheel to discharge the fluid into a container.

➡️**When discharging the power steering fluid (line A and B), turn the steering wheel fully, left and right. Be sure to disconnect the other pipe and drain the fluid in the same manner.**

4. From the control valve of the gearbox assembly, remove the power steering **C** and **D** pressure pipes. Remove pipe **D** first and pipe **C** second.

5. If not disconnected when draining the fluid from the control valve of the gearbox assembly, remove the power steering **A** and **B** pressure pipes. Remove pipe **A** first and pipe **B** second.

6. Remove or disconnect the following:
- Universal joint assembly. Match-mark the assembly before removal.
- Power steering gearbox-to-crossmember assembly bolts
- Gearbox assembly from the vehicle

To install:
7. Install or connect the following:
- Power steering rack and tighten the rack to crossmember bolts to 35–52 ft. lbs. (47–70 Nm)
- Universal joint assembly making sure to align the matchmarks
- Power steering pressure pipes and tighten to 7–12 ft. lbs. (10–16 Nm)
- Universal joint assembly-to-power steering gearbox bolts and tighten to 16–19 ft. lbs. (22–24 Nm) and the universal joint assembly-to-steering shaft bolts 16–19 ft. lbs. (22–24 Nm)
- Tie rod end to steering knuckle nut

and tighten to 18–22 ft. lbs. (25–29 Nm). After tightening this nut, turn it no more than 60 degrees further to align the cotter pin hole.
- New cotter pin
- Front stabilizer into the vehicle
- Exhaust Y-pipe and O2S sensor connector
- Tires and tighten the wheel lug nuts to specification

8. Partially lower the vehicle, then refill and bleed the power steering system.

9. Check for fluid leaks and the fluid level, then install the jack up plate.

10. Check and adjust the toe-in and the steering angle.

Impreza, Impreza Outback and Impreza Outback Sport Models

1. Before servicing the vehicle, refer to the precautions in the beginning of this section.

2. Remove or disconnect the following:
- Negative battery cable

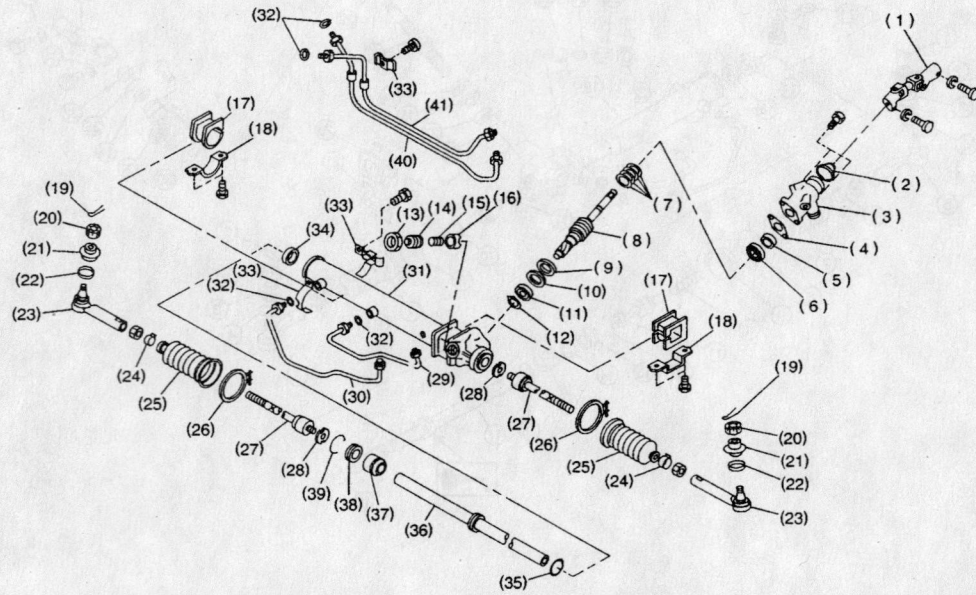

(1) Universal joint	(15) Spring	(29) Pipe B
(2) Dust cover	(16) Sleeve	(30) Pipe A
(3) Valve housing	(17) Adapter	(31) Steering body
(4) Gasket	(18) Clamp	(32) O-ring
(5) Oil seal	(19) Cotter pin	(33) Clamp
(6) Special bearing	(20) Castle nut	(34) Oil seal
(7) Seal ring	(21) Dust cover	(35) Piston ring
(8) Pinion and valve ASSY	(22) Clip	(36) Rack
(9) Oil seal	(23) Tie-rod end	(37) Rack bushing
(10) Back-up washer	(24) Clip	(38) Rack stopper
(11) Ball bearing	(25) Boot	(39) Circlip
(12) Snap ring	(26) Band	(40) Pipe E
(13) Lock nut	(27) Tie-rod	(41) Pipe F
(14) Adjusting screw	(28) Lock washer	

7923TG64

Exploded view of the steering rack assembly—Impreza models

- Front wheels
- Front Y-pipe
- Tie rod end cotter pin and nut
- Tie rod ends from the steering knuckle using a puller
- Jack-up plate and front sway bar
- Fluid lines from the rack and pinion

3. Matchmark the universal joint to the serration in the steering rack for installation reference.

- Lower and upper universal joint bolts and lift the joint upward, disconnecting it from the rack and pinion shaft
- Clamp bolts securing the rack and pinion to the crossmember
- Rack and pinion

To install:

4. Install the rack and pinion. Tighten the clamp bolts to 43 ft. lbs. (59 Nm).

5. Align the steering rack to the universal joint. Push the long yoke of the joint all the way into the serrated position of the steering shaft, setting the bolt hole in the cut-out. Pull the short yoke all the way out of the serrated portion of the rack and pinion, setting the bolt hole in the cut-out. Insert the bolt through the short yoke. Pull the yoke and ensure the bolt is properly engaged in the cut-out. Fasten the short yoke side with the spring washer and bolt, then fasten the yoke side. Tighten the bolts to 17 ft. lbs. (24 Nm).

6. Install or connect the following:

- Tie rod ends to the steering knuckle
- Sway bar and jack-up plate
- Y-pipe with new gaskets and nuts
- Wheels

7. Fill and bleed the steering system.

Strut

REMOVAL & INSTALLATION

Front

WITH STANDARD STRUTS

1. Before servicing the vehicle, refer to the precautions in the beginning of this section.

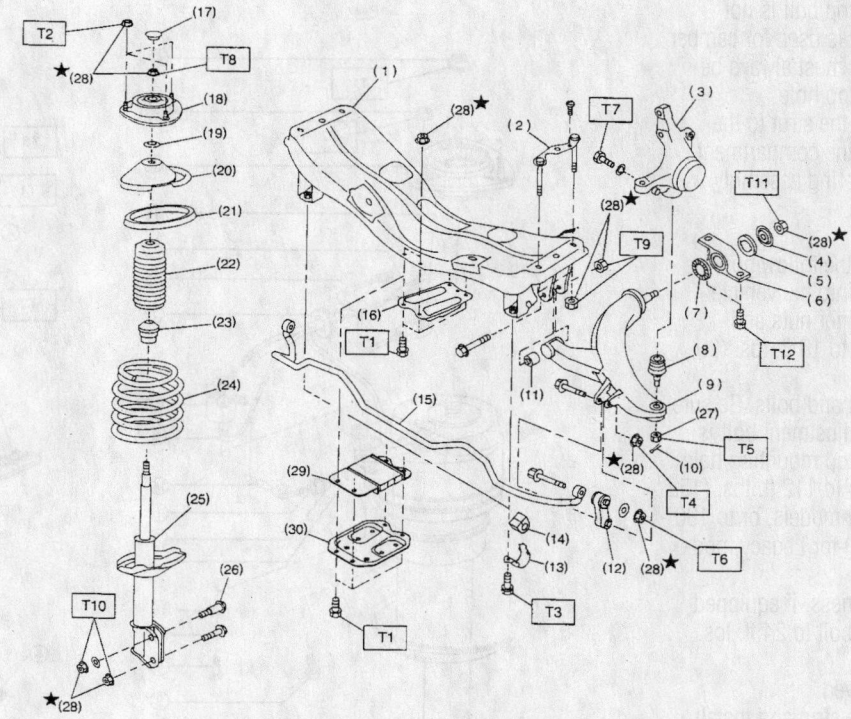

(1)	Crossmember	(16)	Jack-up plate (Except 2500 cc MT model)	(30)	Jack-up plate (2500 cc MT model)
(2)	Bolt ASSY	(17)	Dust seal		
(3)	Housing	(18)	Strut mount		
(4)	Washer	(19)	Spacer		
(5)	Stop rubber (Rear)	(20)	Upper spring seat		
(6)	Rear bushing	(21)	Rubber seat		
(7)	Stop rubber (Front)	(22)	Dust cover		
(8)	Ball joint	(23)	Helper		
(9)	Transverse link	(24)	Coil spring		
(10)	Cotter pin	(25)	Damper strut		
(11)	Front bushing	(26)	Adjusting bolt		
(12)	Stabilizer link	(27)	Castle nut		
(13)	Clamp	(28)	Self-locking nut		
(14)	Bushing	(29)	Dynamic damper (2500 cc MT model)		
(15)	Stabilizer				

Tightening torque: N·m (kg-m, ft-lb)
T1: *18±5 (1.8±0.5, 13.0±3.6)*
T2: *20±6 (2.0±0.6, 14.5±4.3)*
T3: *25±4 (2.5±0.4, 18.1±2.9)*
T4: *29±5 (3.0±0.5, 21.7±3.6)*
T5: *39 (4, 29)*
T6: *44±6 (4.5±0.6, 32.5±4.3)*
T7: *49±10 (5.0±1.0, 36±7)*
T8: *54±5 (5.5±0.5, 39.8±3.6)*
T9: *98±15 (10.0±1.5, 72±11)*
T10: *152±20 (15.5±2.0, 112±14)*
T11: *186±10 (19.0±1.0, 137±7)*
T12: *245±49 (25.0±5.0, 181±36)*

9307TG06

Exploded view of the front suspension assembly—Legacy and Impreza models

For complete Engine Mechanical specifications, see Section 1 of this manual

2. Remove or disconnect the following:
- Negative battery cable
- Front wheel assembly
- Caliper, if necessary, leaving the line connected, and suspend it out of the way with a piece of wire or string
- Clip or bolt attaching the brake line to the strut housing
- Bolt securing the Anti-lock Brake System (ABS) sensor harness, if equipped

➡ **Scribe a matchmark on the camber adjusting bolt which secures the strut to the housing.**

- 2 bolts and nuts securing the strut to the steering knuckle. Notice that the shaft of the top bolt is not round. This bolt is used for camber adjustment, and must always be installed in the top hole.
- 3 nuts securing the strut to the body in the engine compartment
- Strut and coil spring assembly from the vehicle

To install:

3. Install or connect the following:
- Strut assembly into the vehicle
- Upper strut retainer nuts and tighten the nuts to 15 ft. lbs. (20 Nm)
- Lower strut nuts and bolts. Be sure the alignment adjustment bolt is installed in the top mounting hole. Tighten the nuts to 112 ft. lbs. (152 Nm) for Impreza models, or to 130 ft. lbs. (177 Nm) for Legacy models.
- ABS sensor harness, if equipped and tighten the bolt to 24 ft. lbs. (32 Nm)
- Caliper, if removed
- Brake line to the strut and install the clip or bolt
- Front wheel
- Negative battery cable

4. Check the front end alignment and adjust as necessary.

WITH PNEUMATIC STRUTS

1. Before servicing the vehicle, refer to the precautions in the beginning of this section.

2. Remove or disconnect the following:
- Negative battery cable
- Air line and height sensor harness (from inside the engine compartment) from the strut assembly
- Front wheel assembly
- Anti-lock Brake System (ABS) sensor, if equipped

- Caliper, leaving the line connected, and suspend it out of the way with a piece of wire or string
- Clip attaching the brake line to the strut housing

3. Matchmark the camber adjustment bolt to the strut housing as reference for installation.

- Bolt securing the ABS sensor harness, if equipped
- 2 bolts and nuts securing the strut to the steering knuckle. Notice that the shaft of the top bolt is not round. This bolt is used for camber adjustment, and must always be installed in the top hole.

- 3 nuts securing the strut to the body in the engine compartment
- Strut and coil spring assembly

To install:

4. Install or connect the following:
- Strut assembly
- Upper strut retainer nuts and tighten to 15 ft. lbs. (20 Nm)
- ABS sensor harness (if equipped), and tighten the bolt to 14 ft. lbs. (20 Nm)
- Lower strut nuts and bolts. Be sure the alignment adjustment bolt is installed in the top mounting hole. Tighten the nuts to 112 ft. lbs. (152 Nm).

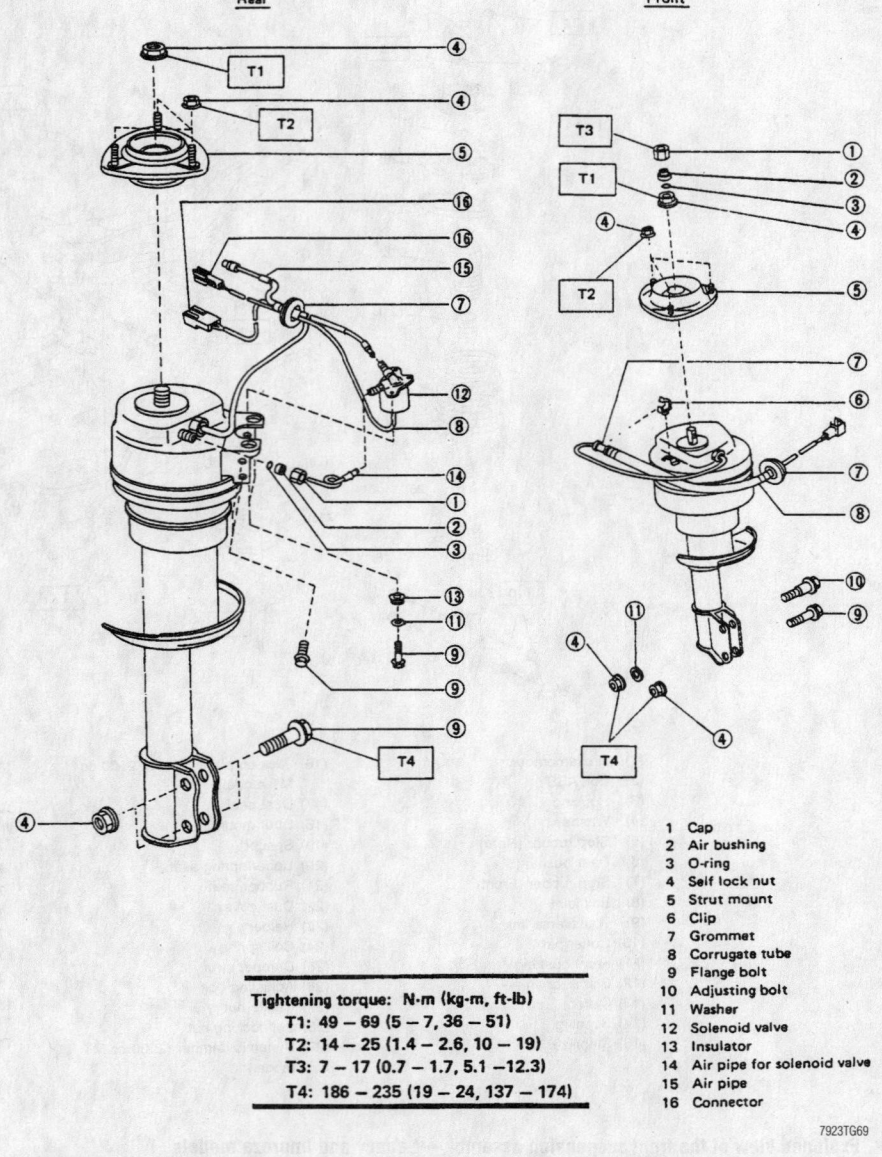

Tightening torque: N·m (kg-m, ft-lb)
T1: 49 – 69 (5 – 7, 36 – 51)
T2: 14 – 25 (1.4 – 2.6, 10 – 19)
T3: 7 – 17 (0.7 – 1.7, 5.1 – 12.3)
T4: 186 – 235 (19 – 24, 137 – 174)

1 Cap
2 Air bushing
3 O-ring
4 Self lock nut
5 Strut mount
6 Clip
7 Grommet
8 Corrugate tube
9 Flange bolt
10 Adjusting bolt
11 Washer
12 Solenoid valve
13 Insulator
14 Air pipe for solenoid valve
15 Air pipe
16 Connector

7923TG69

Exploded view of the front and rear pneumatic suspension assembly—Legacy and Impreza models

- Caliper
- Brake line to the strut and install the clip
- Height sensor harness and air line
- Front wheel
- Negative battery cable

5. Start the vehicle and allow enough time for the struts to pressurize before driving.

6. Check the front end alignment and adjust as necessary.

Rear

WITH STANDARD STRUTS

1. Before servicing the vehicle, refer to the precautions in the beginning of this section.

2. Remove or disconnect the following:
- Rear seat assembly, on Sedan only
- Rear speaker grille and service hole cap, on Wagon only
- Strut mount cap
- Wheel and tire assembly
- Brake hose clip
- Union bolt from the brake caliper. Move the brake hose out of the way.

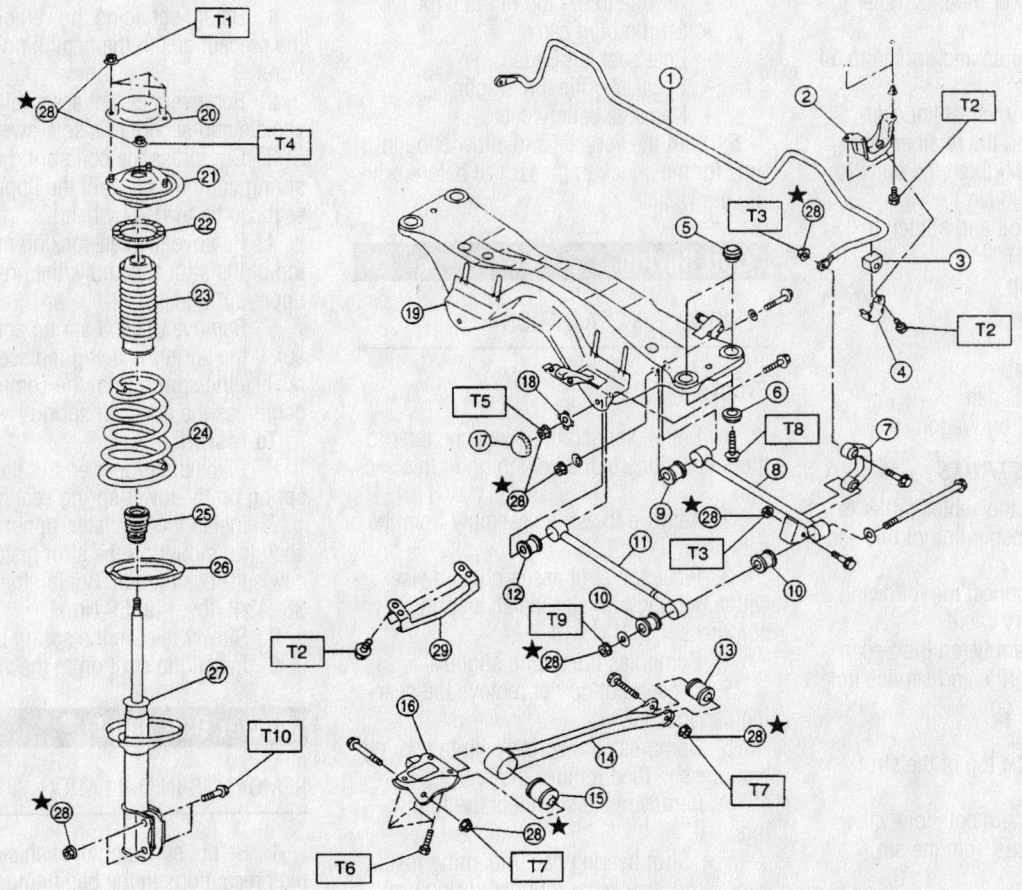

① Stabilizer	⑯ Trailing link bracket
② Stabilizer bracket	⑰ Cap (Protection)
③ Stabilizer bushing	⑱ Washer
④ Clamp	⑲ Crossmember
⑤ Floating bushing	⑳ Strut mount cap
⑥ Stopper	㉑ Strut mount
⑦ Stabilizer link	㉒ Rubber seat upper
⑧ Rear lateral link	㉓ Dust cover
⑨ Bushing (C)	㉔ Coil spring
⑩ Bushing (A)	㉕ Helper
⑪ Front lateral link	㉖ Rubber seat lower
⑫ Bushing (B)	㉗ Damper strut
⑬ Trailing link rear bushing	㉘ Self-locking nut
⑭ Trailing link	㉙ Crossmember reinforcement
⑮ Trailing link front bushing	lower (Sedan model only)

Tightening torque: N·m (kg-m, ft-lb)
T1: 20 ± 6 (2.0 ± 0.6, 14.5 ± 4.3)
T2: 25 ± 7 (2.5 ± 0.7, 18.1 ± 5.1)
T3: 44 ± 6 (4.5 ± 0.6, 32.5 ± 4.3)
T4: 59 ± 10 (6.0 ± 1.0, 43 ± 7)
T5: 98 ± 15 (10.0 ± 1.5, 72 ± 11)
T6: 98 ± 20 (10.0 ± 2.0, 72 ± 14)
T7: 113 ± 15 (11.5 ± 1.5, 83 ± 11)
T8: 127 ± 20 (13.0 ± 2.0, 94 ± 14)
T9: 137 ± 20 (14.0 ± 2.0, 101 ± 14)
T10: 196^{+39}_{-10} ($20.0^{+4.0}_{-1.0}$, 145^{+29}_{-7})

7923TG71

Exploded view of the rear standard strut assembly—Legacy and Impreza models

- Lower nuts and bolts securing the strut to the rear wheel housing
- Retainer nuts securing the strut bearing cap to the strut tower, from inside the vehicle
- Strut from the vehicle

To install:

3. Install or connect the following:
- Strut onto the vehicle, making sure to position the strut properly in the upper strut tower mounts. Refer to the illustration.
- Strut retainer nuts and tighten to 14 ft. lbs. (20 Nm)
- Strut to the rear wheel knuckle assembly using the retainer nuts and bolts, and tighten the bolts to 145 ft. lbs. (196 Nm)
- Brake union bolt and tighten to 13 ft. lbs. (18 Nm)
- Brake hose clip

4. Bleed the brakes.
- Wheel
- Strut mount cap
- Rear seat, on Sedan
- Speaker grille, on Wagon

WITH PNEUMATIC STRUTS

1. Before servicing the vehicle, refer to the precautions in the beginning of this section.

2. Remove or disconnect the following:
- Negative battery cable
- Rear seat assembly, on the Sedan
- Rear speaker grille and service hole cap, on the Wagon
- Strut mount cap
- Air line from the top of the strut assembly
- Height sensor and solenoid valve wiring harnesses from the strut assembly
- Wheel and tire assembly
- Brake hose clip
- Union bolt from the brake caliper. Move the brake hose out of the way.
- Lower nuts and bolts securing the strut to the rear wheel housing
- Retainer nuts securing the strut bearing cap to the strut tower (from inside the vehicle)
- Strut from the vehicle

To install:

3. Install or connect the following:
- Strut on to the vehicle, making sure to position the strut properly in the upper strut tower mounts. Refer to the illustration if needed. Install the retainer nuts, and tighten to 11 ft. lbs. (15 Nm).
- Strut to the rear wheel knuckle

assembly, using the retainer nuts and bolts, and tighten the bolts to 145 ft. lbs. (196 Nm).
- Brake union bolt and tighten to 13 ft. lbs. (18 Nm)
- Brake hose clip

4. Bleed the brakes.
- Wheel
- Height sensor and solenoid valve wiring harnesses to the strut
- Air line to the top of the strut
- Strut mount cap
- Rear seat, on Sedan
- Speaker grille, on Wagon
- Negative battery cable

5. Start the vehicle, and allow enough time for the shock to pressurize before driving the vehicle.

Coil Spring

REMOVAL & INSTALLATION

Front

1. Before servicing the vehicle, refer to the precautions in the beginning of this section.

2. Remove the strut assembly from the vehicle.

3. Place the strut assembly in a vise with a holding tool and install a spring compressor.

4. Compress the spring slightly.

5. Loosen but do not remove the bearing cap locknut.

6. Compress the spring with the spring compressor, then remove the locknut.

7. Remove or disconnect the following:

- Strut bearing cap, mounting insulator bracket and upper spring seat
- Coil assembly, leaving the spring compressed
- Strut boot and rebound bumper from the strut. Inspect and replace if worn.
- Strut retainer nut using a suitable wrench
- Strut insert from the assembly

To install:

8. Install or connect the following:
- Strut into the chamber and install the retainer nut. Tighten the nut snugly.
- Rebound bumper and the boot to the strut piston rod
- Coil spring on the strut assembly. Be sure the spring is properly positioned on the lower bracket.
- Upper spring seat, mounting insu-

lator and bearing cap. Be sure the upper spring seat is facing the proper direction.
- Locknut and tighten to 36–43 ft. lbs. (49–59 Nm)

9. Loosen and remove the spring compressor from the coil spring.

10. Install the strut to the vehicle.

Rear

1. Before servicing the vehicle, refer to the precautions in the beginning of this section.

2. Remove the strut assembly from the vehicle and secure in a soft jawed vise.

3. Compress the coil spring with a spring compressor until the upper spring seat can be turned by hand.

4. Remove the self-locking nut on the top of the strut assembly, then remove the upper spring seat.

5. Remove the coil spring and compressor. If the spring is being replaced, slowly release the spring from the compressor and compress the new coil spring.

To install:

6. Place the proper end of the coil spring on the lower spring seat on the strut.

7. Install the insulator, upper spring seat and strut mount on the strut piston. Install a new self-locking nut. Tighten the nut to 36–43 ft. lbs. (49–59 Nm).

8. Slowly release the spring compressor.

9. Install the strut on to the vehicle.

Lower Ball Joint

REMOVAL & INSTALLATION

1. Before servicing the vehicle, refer to the precautions in the beginning of this section.

2. Remove or disconnect the following:
- Negative battery cable
- Front wheel and tire assembly
- Ball joint castle nut cotter pin and discard the cotter pin
- Castle nut
- Ball joint from the lower control arm using a suitable puller or prytool
- Bolt securing the ball joint to the steering knuckle
- Ball joint using a suitable wedge to expand the steering knuckle connection point

To install:

3. Install or connect the following:
- Ball joint to the steering knuckle
- Retaining bolt and tighten to 36 ft. lbs. (49 Nm)

- Ball joint to the lower control arm and tighten the castle nut to 29 ft. lbs. (39 Nm). Then, tighten the castle nut an additional 60 degrees until the slot in the castle nut is aligned with the cotter pin hole in the ball joint.
- New cotter pin
- Wheel
- Negative battery cable

Front Lower Control Arm

REMOVAL & INSTALLATION

1. Before servicing the vehicle, refer to the precautions in the beginning of this section.
2. Remove or disconnect the following:

- Tire and wheel assembly
- Sway link from the lower control arm
- Bolt securing the ball joint to the steering knuckle
- Nuts (NOT the bolts) securing the lower control arm to the crossmember
- 2 bolts holding the bushing bracket of the control arm to the body
- Ball joint from the steering knuckle
- Bolts securing the lower control arm to the crossmember, then the lower control

To install:

3. Install or connect the following:
- Lower control arm; temporarily tighten the 2 bolts used to secure the rear bushing of the lower control arm to the body

➡**These bolts should be tightened so they can still move back and forth in the oblong shaped hole in the bracket which holds the bushing.**

- Bolts used to secure the lower control arm to the crossmember and temporarily tighten the nuts
- Ball joint into the steering knuckle and secure with the retaining bolt
- Sway link to the control arm and temporarily tighten the bolts

❊❊ WARNING

Discard loosened self-locking nut and replace with a new one.

4. Lower the vehicle, then tighten the bolts to the following specifications:
  a. Lower control arm-to-sway bar: 21 ft. lbs. (27 Nm).

➡**Move the rear bushing back and forth until the control arm-to-rear bushing clearance if established. Refer to the illustration for specifications.**

  b. Lower control arm-to-crossmember: 72 ft. lbs. (98 Nm).

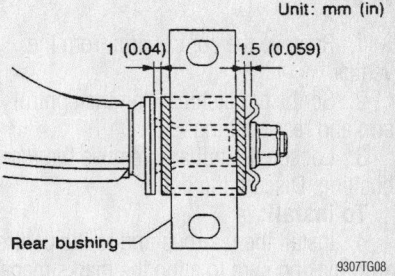

Unit: mm (in)

1 (0.04)   1.5 (0.059)

Rear bushing

9307TG08

Proper control arm-to-rear bushing clearance specifications—Impreza and Legacy

  c. Lower control arm-to-rear link bushing-to-body: 181 ft. lbs. (245 Nm).
5. Check the wheel alignment and adjust if necessary.

CONTROL ARM BUSHING REPLACEMENT

Front

1. Remove the control arm from the vehicle.
2. Mount the control arm in a soft jawed vise.
3. Use either a press or a control arm bushing fixture (C-clamp like tool) along with a slotted washer and a piece of pipe (slightly larger than the bushing) and press out the old bushing.
4. Clean the inside bushing contact surfaces of rust and old rubber.
   To install:
5. Apply a light coating of grease to both the replacement busing and bushing contact surfaces on the control arm.
6. Align the bushing according to the illustration.
7. Install the bushing using the press tool. A bushing install clamp can also be used to compress the bushing into the control arm.
8. Install the control arm on the vehicle.

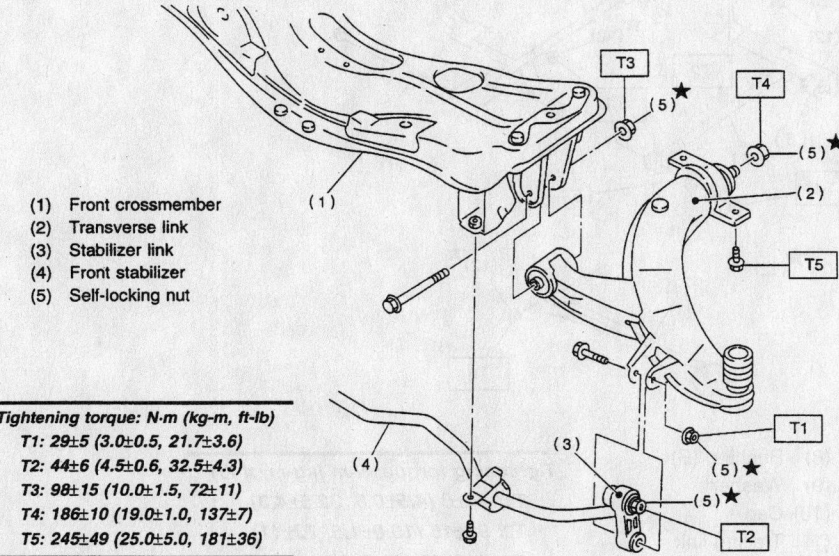

(1) Front crossmember
(2) Transverse link
(3) Stabilizer link
(4) Front stabilizer
(5) Self-locking nut

T3
T4
T5
(5)★
(5)★
(2)
(1)
(3)
(5)★
(5)★
T1
T2
(4)

Tightening torque: N·m (kg-m, ft-lb)
T1: 29±5 (3.0±0.5, 21.7±3.6)
T2: 44±6 (4.5±0.6, 32.5±4.3)
T3: 98±15 (10.0±1.5, 72±11)
T4: 186±10 (19.0±1.0, 137±7)
T5: 245±49 (25.0±5.0, 181±36)

9307TG07

Exploded view of the lower control arm (transverse link)—Impreza and Legacy

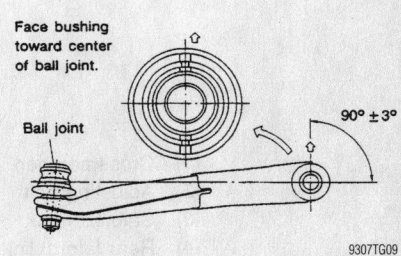

Face bushing toward center of ball joint.

Ball joint

90° ±3°

9307TG09

The front control arm bushing must be installed in the proper direction

For Tire, Wheel and Ball Joint specifications, see Section 1 of this manual

Rear

1. Remove the control arm from the vehicle.

2. Scribe a matchmark on the control arm and rear bushing.

3. Loosen the nut and remove the rear bushing. Discard the nut.

To install:

4. Install the rear bushing to the control arm, making sure to align the marks made during removal.

5. Install a new nut and tighten to 137 ft. lbs. (168 Nm).

Rear Lower Control Arm

REMOVAL & INSTALLATION

Impreza

TRAILING LINK

1. Before servicing the vehicle, refer to the precautions in the beginning of this section.

2. Remove or disconnect the following:
- Tire and wheel assembly
- Rear parking bracket clamp and Anti-lock Brake System (ABS) sensor harness, if equipped
- Bolts that secure the trailing link to the bracket
- Bolt that secures the trailing link to the rear housing
- Trailing link from the vehicle

To install:

3. Install or connect the following:
- Trailing link and the through-bolts and nuts. DO NOT tighten the nuts and bolts at this time.
- ABS sensor bracket, if equipped, and parking brake cable to the trailing link.
- Tire and wheel assembly

4. Tighten the trailing link-to-bracket bolt to 72 ft. lbs. (98 Nm) and the nut to 83 ft. lbs. (113 Nm).

5. Tighten the trailing link-to-rear housing to 83 ft. lbs. (113 Nm).

6. Check the wheel alignment and adjust if necessary.

LATERAL LINK

1. Before servicing the vehicle, refer to the precautions in the beginning of this section.

2. Remove or disconnect the following:
- Tire and wheel assembly
- Anti-lock Brake System (ABS) sensor harness from the trailing link, if equipped.
- Bolts that secure the lateral link to the rear housing.

➡**Discard the old self-locking nuts and replace with new ones during installation.**

- Bolts which secure the trailing link to the rear housing
- Halfshaft from the rear differential using a suitable tool.

➡**On all except 2.2L engines, do not remove the circlip attached to the**

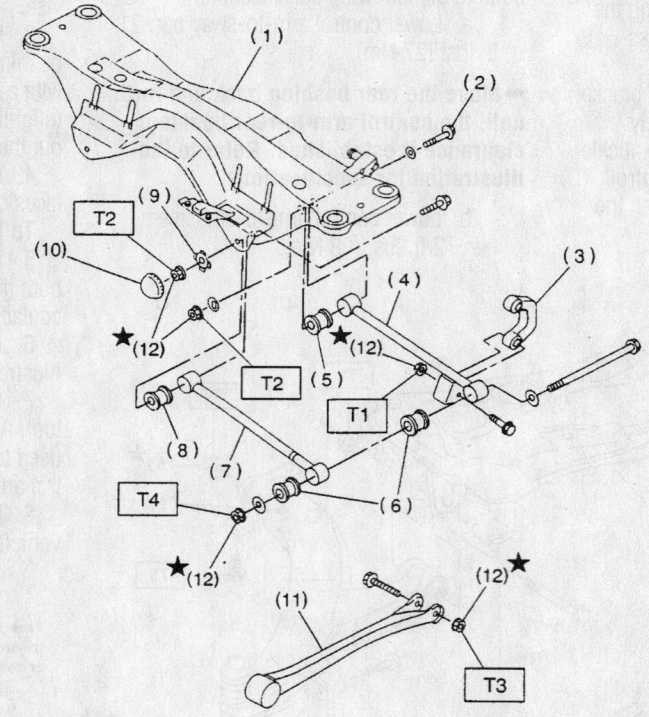

(1)	Crossmember	(8)	Bushing (B)
(2)	Adjusting bolt	(9)	Washer
(3)	Stabilizer link	(10)	Cap
(4)	Rear lateral link	(11)	Trailing link
(5)	Bushing (C)	(12)	Self-locking nut
(6)	Bushing (A)		
(7)	Front lateral link		

Tightening torque: N·m (kg-m, ft-lb)
T1: 44±6 (4.5±0.6, 32.5±4.3)
T2: 98±15 (10.0±1.5, 72±11)
T3: 113±15 (11.5±1.5, 83±11)
T4: 137±20 (14.0±2.0, 101±14)

9307TG10

Lateral link mounting and tightening specifications—Impreza

inside of the differential. On 2.2L engines, the side spline circlip comes out together with the shaft. Be careful not to damage the side bearing retainer.

3. Scribe an alignment mark on the rear lateral link adjusting bolt and crossmember.

4. Remove or disconnect the following:
- Outer lateral link bolt securing the lateral link to the housing
- Bolts securing the front and rear lateral links to the crossmember
- Lateral links from the vehicle

To install:

5. Install or connect the following:
- Bolts securing the front and rear lateral links to the crossmember and handtighten
- Outer lateral link bolt securing the

lateral link to the housing and handtighten
- Halfshaft to the rear differential
- Bolts that secure the lateral link to the rear housing
- Bolts which secure the trailing link to the rear housing
- ABS sensor harness to the trailing link, if equipped
- Tire and wheel assembly

6. Tighten the lateral link bolts as shown in the illustration.

7. Check the wheel alignment and adjust if necessary.

Legacy

1. Before servicing the vehicle, refer to the precautions in the beginning of this section.

2. Remove or disconnect the following:

- Tire and wheel assembly
- Wheel bearing assembly, refer to the procedure in this section
- Bolt holding the parking brake cable to the control arm
- Bolt securing the brake hose to the rear arm
- Bolt securing the Anti-lock Brake Sensor (ABS) sensor to the rear arm
- Brake line from the wheel cylinder with a flare nut wrench, if equipped with drum brakes. Plug the line to avoid contaminating the system. Suspend the brake backing plate from the subframe.
- Nut securing the stabilizer link to the rear arm
- Bolt holding the shock absorber to the rear arm

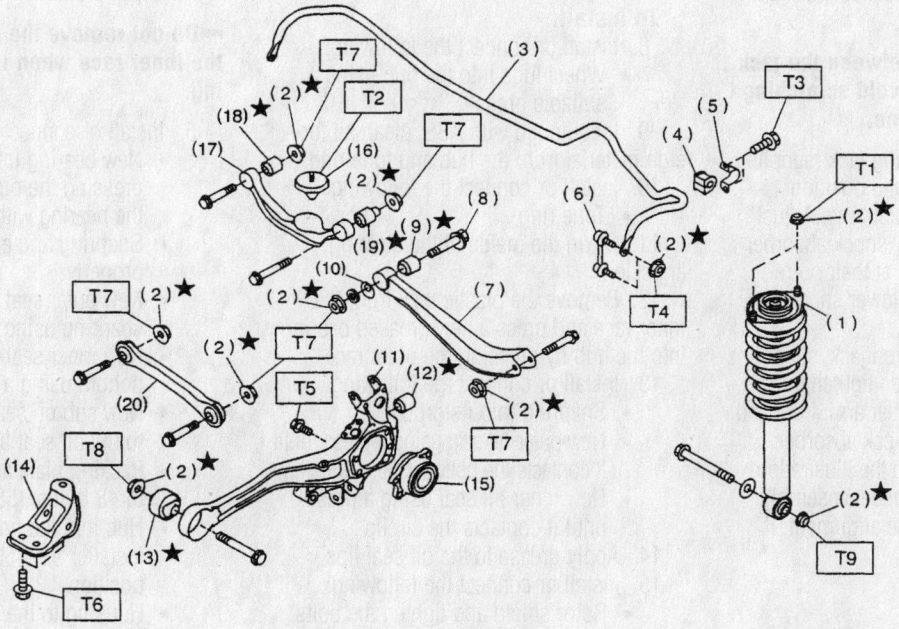

(1) Shock absorber	(12) Rear arm rear bushing	
(2) Self-locking nut	(13) Rear arm front bushing	
(3) Stabilizer	(14) Rear arm bracket	
(4) Stabilizer bushing	(15) Hub bearing unit	
(5) Clamp	(16) Helper	
(6) Stabilizer link	(17) Link upper	
(7) Link rear	(18) Link upper bushing (Inside)	
(8) Adjusting bolt	(19) Link upper bushing (Outside)	
(9) Link rear bushing	(20) Link front	
(10) Adjusting washer		
(11) Rear arm		

Tightening torque: N·m (kg-m, ft-lb)

T1: 30±7 (3.1±0.7, 22.4±5.1)
T2: 32±10 (3.3±1.0, 23.9±7.2)
T3: 39±7 (4.0±0.7, 28.9±5.1)
T4: 44±6 (4.5±0.6, 32.5±4.3)
T5: 66±10 (6.7±1.0, 48.5±7.2)
T6: 108±15 (11±1.5, 80±11)
T7: 123±15 (12.5±1.5, 90±11)
T8: 147±20 (15±2, 108±14)
T9: 157±20 (16±2, 116±14)

9307TG16

Rear control arm mounting and tightening specifications—Legacy

3. Use a suitable transaxle jack to support the rear arm horizontally.

4. Remove or disconnect the following:
- Bolt securing the rear arm to the body
- Nut securing the front link to the rear arm, loosen
- Nut securing the rear link to the rear arm, loosen
- Bolts holding the rear arm to the links
- Rear arm

To install:

5. Use a transaxle jack to support the rear arm.

6. Install or connect the following:
- Rear arm and temporarily tighten the bolts securing the rear arm to the link
- Wheel bearing unit
- Bolt securing the ABS sensor to the rear arm
- Brake hose-to-rear arm bolt
- Parking brake cable clamp-to-rear arm bolt

➡**Place a rag or cloth between the jack and its mating area to avoid scratching the rear link and subframe.**

7. Place the tire changing jack (supplied with the car) upside down and position between the rear link and subframe. Adjust the jack position so the rear shock absorber is aligned with the rear arm at their corresponding holes. Install the lower shock absorber bolts.

8. Using the transmission jack, support the rear arm horizontally, then tighten the nuts and bolts holding the rear arm, front and rear links, upper link and shock absorber. Refer to the specifications in the illustration.

9. Install the tire and wheel assembly.

10. Check and adjust the alignment, if necessary.

BUSHING REPLACEMENT

1. Use a suitable press to remove and install the bushings.

Wheel Bearings

ADJUSTMENT

The wheel bearings are not adjustable.

REMOVAL & INSTALLATION

Front

1. Before servicing the vehicle, refer to the precautions in the beginning of this section.

2. Remove or disconnect the following:
- Steering knuckle assembly from the vehicle

3. Position the steering knuckle in a soft-jawed vise.

4. Press the hub from the steering knuckle. If the inner bearing race remains in the hub, press it out.

5. Remove or disconnect the following:
- Rotor shield
- Inner and outer seals
- Snapring from the steering knuckle

6. Press the inner bearing race to remove the outer bearing.

7. Remove or disconnect the following:
- Tone ring, if equipped with Anti-lock Brake System (ABS)
- Wheel lugs from the hub using a suitable press

➡**To prevent deforming the hub, do not hammer the lugs out.**

To install:

8. Install or connect the following:
- Wheel lugs into the hub using a suitable press

9. If equipped with ABS, clean all foreign material from the hub and tone ring.

10. Install or connect the following:
- Tone ring

11. Clean the inside of the steering knuckle.

12. Remove the plastic lock from the inner race and press a new greased bearing into the hub by pressing the outer race.

13. Install or connect the following:
- Snapring into its groove
- New outer oil seal using a press, until it contacts the bottom of the housing
- New inner oil seal using a press, until it contacts the circlip

14. Apply grease to the oil seal lips.

15. Install or connect the following:
- Rotor shield and tighten the bolts to 10 ft. lbs. (14 Nm)
- Hub to the steering knuckle

16. Press a new bearing into the hub by driving the inner race.

17. Install the steering knuckle on the vehicle.

Rear

1. Before servicing the vehicle, refer to the precautions in the beginning of this section.

2. Loosen the parking brake adjustment.

3. Remove or disconnect the following:
- Wheel assembly
- Unstake and remove the axle nut
- Caliper, leaving the line connected, and suspend it aside
- Rotor
- Parking brake cable
- Sway bar clamp
- Bolt securing the lateral link to the housing
- Bolts securing the trailing link to the housing
- Halfshaft
- Bolts securing the strut to the housing
- Speed sensor from the backing plate, if equipped with Anti-lock Brake System (ABS)
- Housing assembly
- Hub from the rear housing using hub stand 92708000 and puller 927420000
- Backing plate from the housing
- Outer, inner and sub oil seals
- Snapring
- Bearing by pressing the inner race

To install:

4. Clean the housing thoroughly.

➡**Do not remove the plastic lock from the inner race when installing the bearing.**

5. Install or connect the following:
- New bearing into the housing by pressing the outer race and pack the bearing with grease
- Snapring and ensure that it fits properly
- New outer seal until it contacts the snapring using a press
- New inner seal until it contacts the bottom using a press
- New sub oil seal and apply grease to the oil seal lip
- Backing plate and tighten the bolts to 43 ft. lbs. (58 Nm)
- Hub into the housing using installer 927450000 to press it into position
- Housing to the strut and tighten the bolts to 119 ft. lbs. (162 Nm)
- Speed sensor, if equipped with ABS
- Halfshaft
- Trailing link to the housing and tighten the bolt and new nut to 94 ft. lbs. (127 Nm)
- Lateral link to the housing and tighten the bolt and new nut to 116 ft. lbs. (157 Nm)
- Sway bar clamp
- Parking brake cable
- Rear brake assembly
- New axle nut and tighten it to 152 ft. lbs. (206 Nm). Stake the nut.
- Wheel

6. Adjust the parking brake cable.

SUZUKI

1998–01
Esteem • Swift

20

PRECAUTIONS

Before servicing any vehicle, please be sure to read all of the following precautions, which deal with personal safety, prevention of component damage, and important points to take into consideration when servicing a motor vehicle:

• Never open, service or drain the radiator or cooling system when the engine is hot; serious burns can occur from the steam and hot coolant.

• Observe all applicable safety precautions when working around fuel. Whenever servicing the fuel system, always work in a well-ventilated area. Do not allow fuel spray or vapors to come in contact with a spark, open flame, or excessive heat (a hot drop light, for example). Keep a dry chemical fire extinguisher near the work area. Always keep fuel in a container specifically designed for fuel storage; also, always properly seal fuel containers to avoid the possibility of fire or explosion. Refer to the additional fuel system precautions later in this section.

• Fuel injection systems often remain pressurized, even after the engine has been turned **OFF**. The fuel system pressure must be relieved before disconnecting any fuel lines. Failure to do so may result in fire and/or personal injury.

• Brake fluid often contains polyglycol ethers and polyglycols. Avoid contact with the eyes and wash your hands thoroughly after handling brake fluid. If you do get brake fluid in your eyes, flush your eyes with clean, running water for 15 minutes. If eye irritation persists, or if you have taken

brake fluid internally, IMMEDIATELY seek medical assistance.

• The EPA warns that prolonged contact with used engine oil may cause a number of skin disorders, including cancer! You should make every effort to minimize your exposure to used engine oil. Protective gloves should be worn when changing oil. Wash your hands and any other exposed skin areas as soon as possible after exposure to used engine oil. Soap and water, or waterless hand cleaner should be used.

• All new vehicles are now equipped with an air bag system. The system must be disabled before performing service on or around system components, steering column, instrument panel components, wiring and sensors. Failure to follow safety and disabling procedures could result in accidental air bag deployment, possible personal injury and unnecessary system repairs.

• Always wear safety goggles when working with, or around, the air bag system. When carrying a non-deployed air bag, be sure the bag and trim cover are pointed away from your body. When placing a non-deployed air bag on a work surface, always face the bag and trim cover upward, away from the surface. This will reduce the motion of the module if it is accidentally deployed. Refer to the additional air bag system precautions later in this section.

• Clean, high quality brake fluid from a sealed container is essential to the safe and proper operation of the brake system. You should always buy the correct type of brake

fluid for your vehicle. If the brake fluid becomes contaminated, completely flush the system with new fluid. Never reuse any brake fluid. Any brake fluid that is removed from the system should be discarded. Also, do not allow any brake fluid to come in contact with a painted surface; it will damage the paint.

• Never operate the engine without the proper amount and type of engine oil; doing so WILL result in severe engine damage.

• Timing belt maintenance is extremely important! Many models utilize an interference-type, non-freewheeling engine. If the timing belt breaks, the valves in the cylinder head may strike the pistons, causing potentially serious (also time-consuming and expensive) engine damage. Refer to the maintenance interval charts in the front of this manual for the recommended replacement interval for the timing belt, and to the timing belt section for belt replacement and inspection.

• Disconnecting the negative battery cable on some vehicles may interfere with the functions of the on-board computer system(s) and may require the computer to undergo a relearning process once the negative battery cable is reconnected.

• When servicing drum brakes, only dissemble and assemble one side at a time, leaving the remaining side intact for reference.

• Only an MVAC-trained, EPA-certified automotive technician should service the air conditioning system or its components.

ENGINE REPAIR

➡**Disconnecting the negative battery cable on some vehicles may interfere with the functions of the on board computer systems and may require the computer to undergo a relearning process, once the negative battery cable is reconnected.**

Distributor

REMOVAL & INSTALLATION

These models utilize a Distributorless Ignition System (DIS). With this system, the Electronic Control Module (ECM) determines proper ignition timing and time for

the primary ignition coil circuit to turn **ON** and **OFF**.

Alternator

REMOVAL

Swift

1. Before servicing the vehicle, refer to the precautions in the beginning of this section.
2. Remove or disconnect the following:
   • Negative battery cable
   • **B** terminal wire and coupler from the alternator

• Alternator drive belt adjusting bolt and loosen the adjuster arm bolt
• Alternator cover
• Alternator mounting bolts and nut
• Alternator

Esteem

1. Before servicing the vehicle, refer to the precautions in the beginning of this section.
2. Remove or disconnect the following:
   • Negative battery cable
   • **B** terminal wire and coupler from the alternator
   • Alternator mounting bolts and loosen the drive belt adjusting bolt
   • Alternator cover

- Alternator bracket bolts and the bracket
- Alternator

INSTALLATION

Swift

1. Install or connect the following:
   - Alternator
   - Alternator mounting bolts and nut. Tighten to 13–20 ft. lbs. (18–28 Nm).
   - Alternator cover. Tighten the cover retainers to 36–60 inch lbs. (4–7 Nm).
   - Alternator drive belt adjusting bolt and tighten the adjuster arm bolt. Tighten to 13–20 ft. lbs. (18–28 Nm).
   - **B** terminal wire and coupler to the alternator
   - Negative battery cable

Esteem

1. Install or connect the following:
   - Alternator
   - Alternator bracket and bolts. Tighten the bolts to 16 ft. lbs. (23 Nm).
   - Alternator cover

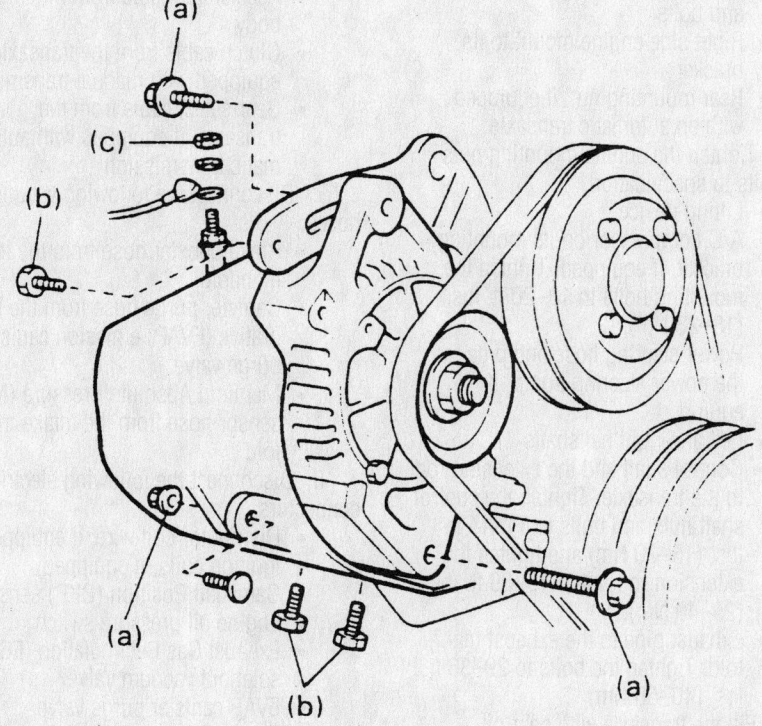

Alternator mounting—Swift models

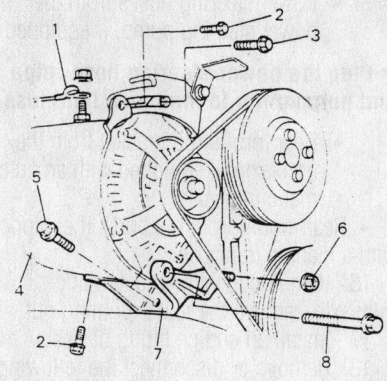

1. "B" terminal wire
2. Cover bolts
3. Drive belt adjusting bolt
4. Cover
5. Bracket bolts
6. Bracket nut
7. Bracket
8. Mounting bolt

9307UG12

Alternator mounting—Esteem models

- Alternator drive belt adjusting bolt and mounting bolts. Tighten to 16 ft. lbs. (23 Nm).
- **B** terminal wire and coupler to the alternator
- Negative battery cable

Ignition Timing

ADJUSTMENT

Ignition timing is controlled by the Electronic Control Module (ECM). The ECM receives signals from various sensors mounted on the engine. No ignition timing adjustment is possible.

Engine Assembly

REMOVAL & INSTALLATION

1.3L Engine

✳✳ CAUTION

The fuel injection system remains under pressure after the engine has been turned OFF. Properly relieve fuel pressure before disconnecting any fuel lines. Failure to do so may result in fire or personal injury.

1. Before servicing the vehicle, refer to the precautions in the beginning of this section.
2. Properly relieve the fuel system pressure.
3. Remove or disconnect the following:
   - Battery and tray
   - Windshield washer hose from the hood
4. Using a grease pencil or marker, mark the hood hinge to hood outline. With the aid of an assistant, remove the hood.
5. Drain the cooling system.
6. Remove or disconnect the following:
   - Air cleaner assembly with the Mass Air Flow (MAF) sensor outlet hose
   - Radiator and cooling fan
7. Disconnect the following electrical wires and release the wiring harness from the clamps:
   - Ignition coil wire from the distributor cap, if equipped
   - Distributor electrical wires, if equipped
   - Ignition coil assembly, if equipped
   - Exhaust Gas Recirculation (EGR) solenoid vacuum valve
   - Radiator fan temperature switch
   - Engine coolant temperature gauge sensor
   - Engine Coolant Temperature (ECT) sensor
   - Idle Air Control (IAC) actuator

- Ground wires from the intake manifold
- Throttle Position (TP) sensor
- Fuel injector
- Camshaft Position (CMP) sensor, if equipped
- Oxygen (O_2S) sensor
- Oil pressure gauge sensor
- Alternator
- Starter
- Back-up light switch
- Negative battery cable from the transaxle
- Vehicle Speed Sensor (VSS)
- Noise filter ground wire
- Evaporative (EVAP) emission canister purge valve, if equipped

8. Disconnect the following vacuum hoses:

- Brake booster hose from the intake manifold
- Canister purge hose
- Air conditioning valve hose

9. Remove or disconnect the following:

- Fuel return hose and the fuel feed hose from the throttle body
- Heater inlet and outlet hoses

10. Disconnect the following cables:

- Accelerator cable from the throttle body
- Clutch cable from the transaxle, if equipped
- Speedometer cable from the transaxle, if equipped
- Shift switch, if equipped with an automatic transaxle
- VSS, if equipped

11. Remove or disconnect the following:

- EVAP canister from the vehicle
- Fender apron extensions
- Exhaust pipe from the exhaust manifold
- Control shaft and extension rod from the transaxle

12. Drain the engine and transaxle oil.
13. Remove or disconnect the following:

- Left and right halfshafts

➡**For engine and transaxle removal, it is not necessary to remove the half-shafts from the steering knuckles.**

14. Remove or disconnect the following:

- Air conditioning compressor from its mounting bracket with the hoses still attached, if equipped with air conditioning

➡**Suspend the compressor where no damage will occur during engine removal and installation.**

15. Remove or disconnect the following:

- Power steering hoses from the power steering pump, if equipped

➡**Plug the power steering hose, pipe and pump ports to minimize fluid loss.**

- Rear torque rod bracket from the transaxle, if equipped with an automatic transaxle
- Rear mount from the body, if equipped with a manual transaxle

16. If equipped with an automatic transaxle, remove the rear mounting nut.
17. Install an engine lifting device.
18. Remove or disconnect the following:

- Rear mount from the body
- Left side engine mounting bracket bolts and bracket
- Right side engine mount from its bracket

19. Before lifting the engine and assembly check to be sure that all the hoses, electric wires and cables are disconnected.
20. Remove the engine with the transaxle from the vehicle.

To install:

21. Lower the engine and transaxle into the engine compartment but do not remove the lifting device.
22. Install or connect the following:

- Rear mount to the body
- Left side engine mounting bracket and bolts
- Right side engine mount to its bracket
- Rear mounting nut, if equipped with an automatic transaxle

23. Tighten the engine mounting nuts and bolts to specification.

- Lifting device
- A/C compressor on its mounting bracket, if equipped. Tighten the mounting bolts to 13–20 ft. lbs. (18–28 Nm).
- Power steering hose and pipe to the power steering pump, if equipped
- Left and right halfshafts
- Control shaft and the extension rod to the transaxle. Tighten the control shaft nuts and bolts to 11–14 ft. lbs. (15–20 Nm) and tighten the extension rod nut to 19–29 ft. lbs. (25–40 Nm).
- Exhaust pipe to the exhaust manifold. Tighten the bolts to 29–36 ft. lbs. (40–50 Nm)

24. Fill the transaxle with gear oil.
25. Install the remaining components in the reverse order of removal.
26. Adjust the clutch pedal free-play.
27. Adjust the accelerator cable free-play.

28. Fill the engine with engine oil and the cooling system with coolant.
29. Fill the power steering reservoir and bleed the power steering system.
30. Run the engine and verify that there are no fuel, coolant, transmission or exhaust leaks.

1.6L Engine

✳✳ CAUTION

The fuel system pressure must be relieved before disconnecting any fuel lines. Failure to do so may result in personal injury.

1. Before servicing the vehicle, refer to the precautions in the beginning of this section.
2. Relieve the fuel system pressure.
3. Mark the position of the hood on the hinges for installation reference, then remove the hood with the aid of an assistant.
4. Drain the cooling system.
5. Remove or disconnect the following:

- Radiator and cooling fan
- Air cleaner outlet hose
- Air cleaner case bolts and case

6. Disconnect the following cables:

- Accelerator cable from the throttle body
- Clutch cable from the transaxle, if equipped with manual transmission
- Gear select cable from the transaxle, if equipped with automatic transmission

7. Disconnect the following vacuum hoses:

- Brake booster hose from the intake manifold
- Canister purge hose from the Evaporative (EVAP) emission canister purge valve
- Manifold Absolute Pressure (MAP) sensor hose from the intake manifold

8. Disconnect the following electrical connectors:

- Distributor coil wire, if equipped
- Ignition coils, if equipped
- Camshaft Position (CKP) sensor
- Engine oil pressure switch
- Exhaust Gas Recirculation (EGR) solenoid vacuum valve
- EVAP canister purge valve
- Engine Coolant Temperature (ECT) sensor
- Fuel injectors .
- Power steering pressure switch
- Heated Oxygen Sensor (HO$_2$S)

- Back-up light switch, manual transmission
- Shift switch, automatic transmission
- Forward clutch revolution sensor, automatic transmission
- Automatic transmission Vehicle Speed Sensor (VSS)
- Alternator
- Starter
- Battery negative cable from the transaxle
- Throttle Position (TP) sensor
- Idle Air Control (IAC) valve
- Manifold Absolute Pressure (MAP) sensor
- Crankshaft Position (CKP) sensor, if equipped
- All engine wires from the engine

9. Remove or disconnect the following:
- Fuel feed hose from the feed pipe
- Return hose from the fuel pressure regulator
- Heater inlet and outlet hoses
- Right and left engine undercovers
- Front exhaust pipe from the exhaust manifold and center exhaust pipe
- Gear shift control shaft from the transaxle and the extension rod, if equipped with a manual transaxle

10. Drain the engine and transaxle oil.

11. Remove or disconnect the following:
- Left and right halfshafts
- A/C compressor from the compressor bracket with the hoses still attached, if equipped. Position the air conditioning out of the way from the engine.

12. If equipped with power steering, drain the power steering pump of fluid.

13. Remove or disconnect the following:
- Power steering hose from the power steering pump, if equipped

14. Install a lifting device on the engine.

15. Remove or disconnect the following:
- Center member from the vehicle by unfastening the 7 nuts and 4 bolts
- Left engine mount
- Right engine mount and bracket

16. Check to be sure all cooling hoses, vacuum hoses and electrical wires are disconnected from the engine.

17. Lower the engine with the transaxle from the vehicle.

To install:

18. Raise the engine and transaxle into the engine compartment.

19. Install or connect the following:

- Right engine mount with the bolts and nuts. Tighten the bolts and nuts to 40 ft. lbs. (55 Nm).
- Left engine mount and install the bolts and nuts. Tighten the bolts and nuts to 40 ft. lbs. (55 Nm).

20. Install the center member using the 7 nuts and 4 bolts. Tighten the bolts and nuts as follows:

a. Center member to the radiator support: 33 ft. lbs. (45 Nm).

b. Center member to crossmember: 33 ft. lbs. (45 Nm).

c. Engine mounts to center member nuts: 40 ft. lbs. (55 Nm).

21. Remove the lifting device.

22. Install or connect the following:
- Power steering hose to the power steering pump, if equipped
- A/C compressor on the air conditioning bracket on the engine, if equipped
- Left and right halfshafts
- Gear shift control shaft on the transaxle and the extension rod, if equipped with a manual transaxle
- Front exhaust pipe on the center exhaust pipe and exhaust manifold
- Remaining components in the reverse order of removal

23. Fill the cooling system, engine, transaxle and power steering pump.

24. Adjust all cables and check all connections.

25. Connect the negative battery cable to the battery.

26. Start the vehicle and check for leaks.

Water Pump

REMOVAL & INSTALLATION

1.3L Engines

1. Before servicing the vehicle, refer to the precautions in the beginning of this section.

2. Disconnect the negative battery cable.

3. Drain the cooling system into a suitable container and tighten the drain plug.

4. Remove or disconnect the following:
- Air cleaner assembly and the Mass Air Flow (MAF) sensor and outlet hose
- Air cleaner bracket
- Right side fender apron clips by pushing the center pin

➡**Do not push the center pin too far in, or it will fall off into the fender.**

- Power steering and air conditioning belt, if equipped
- Loosen the water pump pulley bolts
- Alternator drive belt
- Water pump pulley

5. To remove the crankshaft pulley perform the following:

a. If equipped with a manual transaxle, insert a suitable flat bladed tool into the hole in the bell housing next to the exhaust pipe. This will lock the crankshaft in place.

b. If equipped with an automatic

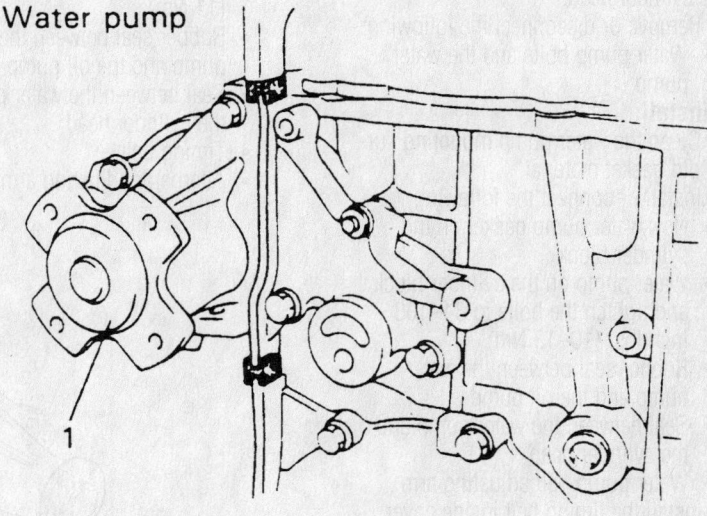

1. Water pump

1

Water pump location—1.3L engine

7923UG05

transaxle, hold a suitable flat bladed tool in line with the oil pan and insert into the teeth of the drive plate. This will lock the crankshaft in place.

   c. Loosen the crankshaft pulley bolts.

   d. Remove the crankshaft timing belt pulley bolt with a 17mm socket.

   e. Remove the pulley from the crankshaft.

   f. Install the crankshaft bolt.

   g. Remove the flat bladed tool that was used to lock the crankshaft in place.

➡ **To remove the crankshaft pulley with the engine assembly mounted on the body, it is necessary to remove the crankshaft timing belt pulley bolt. If the engine assembly is dismounted, the bolt does not need to be removed.**

  6. Remove or disconnect the following:
- Resonator and the timing belt outside cover
- Loosen the right engine mounting bolt
- Timing belt

✳✳ CAUTION

After the timing belt is removed never turn the camshafts or the crankshaft. Interference may occur between the pistons and the valves causing component damage.

- Timing belt inside cover
- Water pump belt adjusting arm

  7. Carefully remove the rubber seal between the water and oil pumps, and remove the seal between the water pump and the cylinder head.

  8. Remove or disconnect the following:
- Water pump bolts and the water pump

To install:

  9. Clean the water pump mounting surface of old gasket material.

  10. Install or connect the following:
- New water pump gasket on the cylinder block
- Water pump on the cylinder block and tighten the bolts to 84–108 inch lbs. (10–13 Nm)
- Rubber seal between the water pump and the oil pump
- Seal between the water pump and the cylinder head
- Water pump belt adjusting arm

  11. Install the timing belt inside cover.

  12. With the crankshaft locked in position, remove the crankshaft bolt and install the crankshaft pulley. Tighten the crankshaft pulley bolts to 10–13 ft. lbs. (14–18 Nm). Using a 17mm socket, tighten the crank-

shaft timing belt pulley bolt to 76–83 ft. lbs. (105–115 Nm).

  13. Install or connect the following:
- Timing belt
- Water pump pulley and drive belt. Tighten the water pump pulley bolts to 84–96 inch lbs. (9–12 Nm).
- Remaining components in the reverse order of removal

  14. Fill the cooling system.

  15. Connect the negative battery cable.

  16. Start the engine and top off the coolant as necessary.

  17. Check the cooling system for leaks.

  18. Check the ignition timing.

1.6L Engine

  1. Before servicing the vehicle, refer to the precautions in the beginning of this section.

  2. Disconnect the negative battery cable.

  3. Drain the cooling system and tighten the drain plug.

  4. Remove or disconnect the following:
- Timing belt
- Alternator adjusting arm
- Oil dipstick guide and dipstick
- Water pump bolts, gasket, water pump and rubber seal

To install:

  5. Clean the water pump mounting surface of old gasket material.

  6. Install or connect the following:
- New water pump gasket on the cylinder block
- Water pump on the cylinder block and tighten the bolts to 96 inch lbs. (11 Nm)
- Rubber seal between the water pump and the oil pump
- Seal between the water pump and the cylinder head
- Timing belt
- Alternator adjusting arm

- Oil dipstick guide and dipstick using a new O-ring

  7. Adjust drive belt tension.

  8. Fill the cooling system with engine coolant.

  9. Connect the negative battery cable.

  10. Start the engine and top off the coolant as necessary.

  11. Check the ignition timing.

Cylinder Head

REMOVAL & INSTALLATION

1.3L Engines

✳✳ CAUTION

The fuel injection system remains under pressure after the engine has been turned OFF. Properly relieve fuel pressure before disconnecting any fuel lines. Failure to do so may result in fire or personal injury.

  1. Before servicing the vehicle, refer to the precautions in the beginning of this section.

  2. Relieve the fuel system pressure.

  3. Drain the cooling system.

  4. Remove or disconnect the following:
- Air cleaner assembly
- Ignition coil wire from the distributor cap, if equipped
- Distributor electrical wires, if equipped
- Ignition coil assembly, if equipped
- Exhaust Gas Recirculation (EGR) solenoid vacuum valve
- Radiator fan thermo switch
- Engine Coolant Temperature (ECT) sensor
- ECT gauge sensor
- Idle Air Control (IAC) valve
- Throttle Position (TP) sensor
- Fuel injector

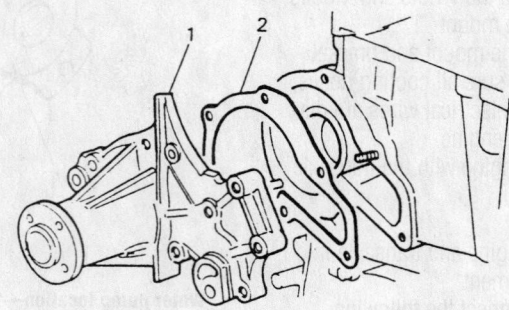

1. Water pump   2. Gasket

7923UG06

Exploded view of the water pump mounting—1.6L engine

- Ground wires from the intake manifold.
- Oxygen (O2S) sensor
- Radiator hose from the thermostat housing
- Heater hose from the intake manifold
- Throttle body coolant outlet hose from the throttle body
- Manifold Absolute Pressure (MAP) sensor hose from the intake manifold, if equipped
- Canister hose from its pipe
- Canister purge hose from the intake manifold
- Brake booster from the intake manifold
- Fuel return hose and the fuel feed hose from the throttle body
- Throttle cable from the throttle body
- Water pump and crankshaft pulleys
- Timing belt
- Rubber seal between the cylinder head and the water pump
- Exhaust pipe from the exhaust manifold
- Spark plug wire clamps from the cylinder head cover
- Positive Crankcase Ventilation (PCV) hose
- Cylinder head cover

5. Loosen all valve adjusting screws and allow the valves to close.
6. Remove or disconnect the following:
- Cylinder head bolts in the reverse order of the tightening sequence

- Cylinder head with the distributor, intake manifold and exhaust manifold
- Distributor, intake manifold and exhaust manifold from the cylinder head

7. Clean the cylinder block mating surface of any old gasket material and clean any engine coolant from the cylinders.

To install:

8. Install or connect the following:
- Intake and exhaust manifolds
- Distributor on the cylinder head
- New cylinder head gasket with the top mark facing up and toward the crankshaft pulley
- Cylinder head on the engine block

9. Coat the cylinder head mounting bolts threads with clean engine oil.
10. Tighten the bolts as follows:
   a. Step 1: 26 ft. lbs. (35 Nm).
   b. Step 2: 41 ft. lbs. (55 Nm).
   c. Step 3: 49 ft. lbs. (68 Nm).
11. Install or connect the following:
- Rubber seal between the cylinder head and the water pump
- Timing belt
- Exhaust pipe to the exhaust manifold, tighten the attaching bolts to 26–36 ft. lbs. (35–50 Nm)
- Crankshaft and water pump pulleys
- Throttle cable to the throttle body
- Fuel return hose and the fuel feed hose to the throttle body
- Remaining components in the reverse order of removal

12. Refill the cooling system with coolant.
13. Connect the negative battery cable.
14. Adjust the ignition timing.
15. With the engine running, be sure that there are no fuel, coolant or exhaust leaks.

1.6L Engine

❈ CAUTION

The fuel system pressure must be relieved before disconnecting any fuel lines. Failure to do so may result in personal injury.

1. Before servicing the vehicle, refer to the precautions in the beginning of this section.
2. Drain the cooling system.
3. Relieve the fuel system pressure.
4. Remove or disconnect the following:
- Air cleaner outlet hose
- Intake manifold rear stiffener bolt, alternator adjustment arm reinforcement bolt and right mounting bracket stiffener from the intake manifold
- Heated Oxygen (HO2S) sensor coupler and release its clamps
- Exhaust from the manifold

5. Disconnect the electrical connectors from the following components:
- Distributor, if equipped
- Ignition coils, if equipped
- Engine Coolant Temperature (ECT) sensor and gauge
- Engine ground wire from intake manifold
- Exhaust Gas Recirculation (EGR) solenoid vacuum valve
- Fuel injectors
- Throttle Position (TP) sensor
- Idle Air Control (IAC) valve
- Heated Oxygen (HO2S) sensor
- Evaporative Emissions (EVAP) canister purge valve, if equipped
- Manifold Absolute Pressure (MAP) sensor, if equipped
- Tank pressure control solenoid vacuum valve, if equipped
- Camshaft Position (CMP) sensor, if equipped
- Evaporative emissions solenoid purge valve

6. Label and disconnect the vacuum hoses from the following:
- Evaporative Emissions (EVAP) canister purge hose
- Brake booster supply hose

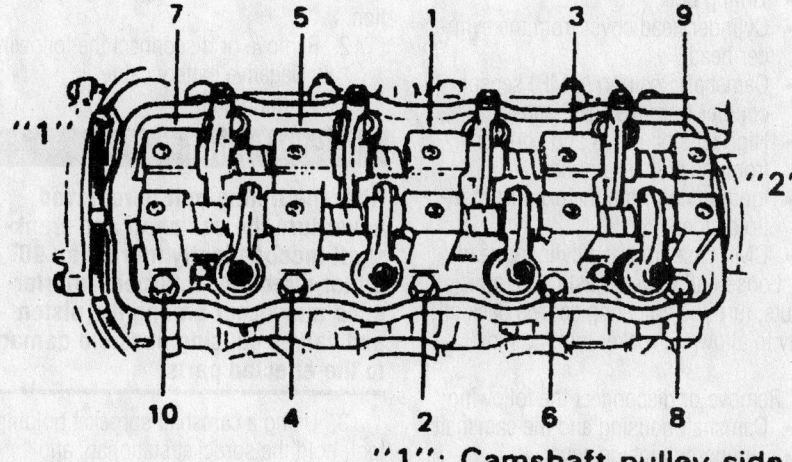

"1": Camshaft pulley side
"2": Distributor side

7923UG07

Cylinder head bolt torque sequence—1.3L engine

Timing belt service is covered in Section 3 of this manual

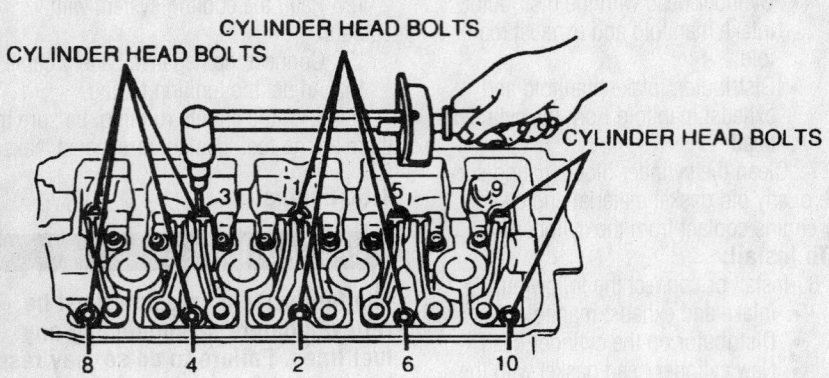

Cylinder head bolt torque sequence—1.6L engine

7. Remove or disconnect the following:
- Fuel feed and return hoses from each pipe
- Cylinder head cover
- Valve lash adjusting screws

8. Disconnect the following hoses:
- Radiator inlet hose
- Heater inlet hose
- IAC valve outlet

9. Remove or disconnect the following:
- Timing belt
- Cylinder head bolts in reverse order of tightening. Once each bolt is loose, remove the bolts from the cylinder head.

10. Check to be sure all components are removed or disconnected before removing the cylinder head.

11. Remove the cylinder head with the intake manifold, exhaust manifold and distributor as an assembly.

To install:

12. Install a new cylinder head gasket and the cylinder head, with the distributor case, onto the cylinder block.

13. Tighten the cylinder head bolts, in sequence, using the following 3 Steps:
   a. Step 1: 26 ft. lbs. (35 Nm).
   b. Step 2: 41 ft. lbs. (55 Nm).
   c. Step 3: 49 ft. lbs. (68 Nm).

14. Install or connect the following:
- Timing belt
- Engine cooling water hoses

15. Adjust the valve lash.

16. Install or connect the following:
- Cylinder head cover
- Fuel feed and return hoses to each pipe
- Vacuum hoses
- All remaining electrical components
- All remaining components in the reverse order of removal

17. Fill the engine coolant and check all fluids.

18. Connect the negative battery cable to the battery.

19. Start the engine and check for leaks.

Rocker Arms/Shafts

REMOVAL & INSTALLATION

Only the 1.3L engine uses rocker arms/shafts. On the 1.6L engine, the camshaft directly actuates the valves.

1.3L Engine

1. Before servicing the vehicle, refer to the precautions in the beginning of this section.

2. Drain the cooling system into a resealable container.

3. Remove or disconnect the following:
- Negative battery cable
- Timing belt
- Cylinder head cover from the cylinder head
- Camshaft Position (CMP) sensor coupler from the CMP sensor
- High tension cords and couplers from the coil assemblies
- Ignition coil from the exhaust manifold side
- CMP case from the cylinder head

4. Loosen all valve adjusting screw lock nuts, turn the adjusting screws back all the way to allow the rocker arms to move freely.

5. Remove or disconnect the following:
- Camshaft housing and the camshaft
- Timing belt inside cover
- Intake rocker arm with the clip from the rocker arm shaft. Make sure not to bend the clip when removing the intake rocker arm
- Rocker arm shaft bolts
- Exhaust rocker arms and spring by pulling out the shaft on the battery

side. If necessary, remove the battery for clearance.

6. Inspect the rocker arms and shafts for wear and/or damage and replace parts as necessary.

   a. If the tip of the valve lash adjuster is badly worn, replace the adjuster.

To install:

7. Apply engine oil to the rocker arms and the rocker arm shafts.

8. Install or connect the following:
- O-ring on the shaft
- Rocker arm shaft with its holes facing up
- Rocker arm (exhaust side) and the rocker arm spring
- Rocker arm shaft bolts and tighten the bolts to 96 inch lbs. (11 Nm)

9. Pour a small amount of engine oil into the arm pivot holding part of the shaft.

10. Install or connect the following:
- Rocker arm (intake side), with the clips to the rocker arm shaft
- Camshaft and housing
- Camshaft oil seal
- Timing belt inside cover
- Camshaft timing belt pulley, while fitting the pin on the camshaft into the slot. Tighten the pulley bolt to 43 ft. lbs. (60 Nm).
- Timing belt
- Remaining components in the reverse order of removal.

11. Check the ignition timing.

1.6L Engine

1. Before servicing the vehicle, refer to the precautions in the beginning of this section.

2. Remove or disconnect the following:
- Negative battery cable
- Timing belt

✳✳ WARNING

After the timing belt is removed, never turn the camshaft and crankshaft independently more than 90° in either direction. If turned, interference may occur among the piston and valves causing possible damage to the effected parts.

3. Using a camshaft sprocket holding tool, hold the sprocket stationary and remove the camshaft sprocket bolt.

4. Remove or disconnect the following:
- Cylinder head cover
- Distributor cap and distributor assembly from the engine, if equipped
- Ignition coils and the Camshaft

Position (CMP) sensor case from the cylinder head with the CMP sensor coupler disconnected

- Loosen all of the valve adjusting screw locknuts until the rocker arms move freely
- Camshaft housing and camshaft
- Rocker arm shaft plug from the cylinder head
- Timing belt inner cover-to-cylinder head bolts and the cover
- All intake rocker arms and clips from the rocker shaft. Keep all parts in order so they can be reinstalled in their original locations.
- 6 rocker arm shaft-to-cylinder head bolts. Push the rocker arm shaft through the rear of the cylinder head until the end of the rocker shaft appears.
- O-ring from the rear of the rocker arm shaft
- Exhaust rocker arms, rocker arm springs and rocker shaft by pulling the rocker arm shaft through the front of the cylinder head. Be sure to keep the parts in order for installation purposes.

5. Clean and inspect all parts for wear and/or damage. Replace parts as necessary.

To install:

6. Lubricate the rocker arms and shafts with clean engine oil before installation.

7. Install or connect the following:

- Push the rocker shaft into the front of the cylinder head
- Exhaust rocker arms and springs as the rocker arm shaft is being installed into the cylinder head
- Push the rocker arm shaft through the rear of the cylinder head
- New O-ring onto the rocker shaft

8. Rotate the rocker arm shaft so the flat machined surface is horizontal and facing downward, parallel with the cylinder head mating surface and slide the shaft back into the cylinder head.

9. Install or connect the following:

- 6 rocker arm shaft bolts and tighten the rocker arm-to-cylinder head bolts to 96 inch lbs. (11 Nm). Fill the rocker arm shaft bolt holes with clean engine oil.
- Intake rocker arms and clips onto the rocker arm shaft

➡ The camshaft carrier caps are embossed with numbers and arrows to ensure correct assembly. The No. 1 camshaft carrier cap must be installed

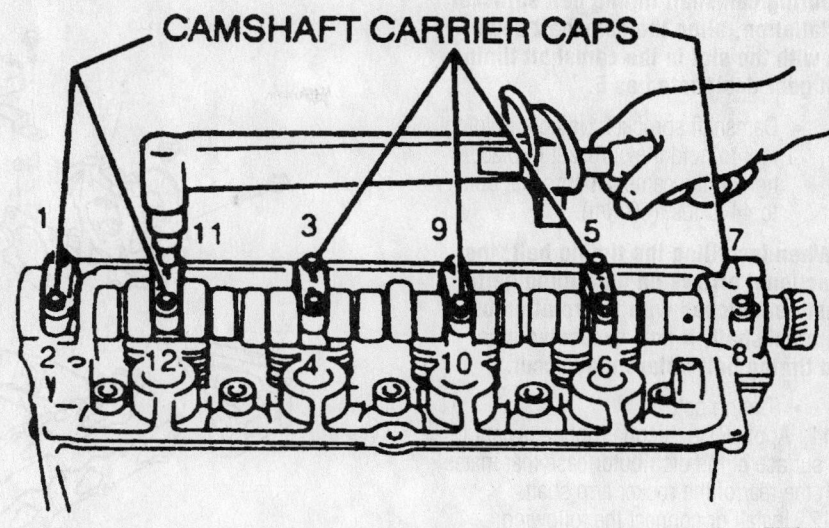

Camshaft carrier cap bolt torque sequence—1.6L engine

at the front of the cylinder head with the remaining carrier caps following in numerical order. The directional arrows must always point toward the front of the cylinder head.

- Camshaft

✶✶ WARNING

If the camshaft carrier cap bolts are tightened at random, damage to the camshaft may occur.

10. Lubricate the new camshaft seal lip with clean engine oil

- New Camshaft seal into the cylinder head until it is flush with the camshaft carrier surface
- Timing belt inner cover and tighten the cover-to-cylinder head bolts to 89 inch lbs. (10 Nm)
- Rocker arm shaft plug into the cylinder head and tighten to 24 ft. lbs. (33 Nm)

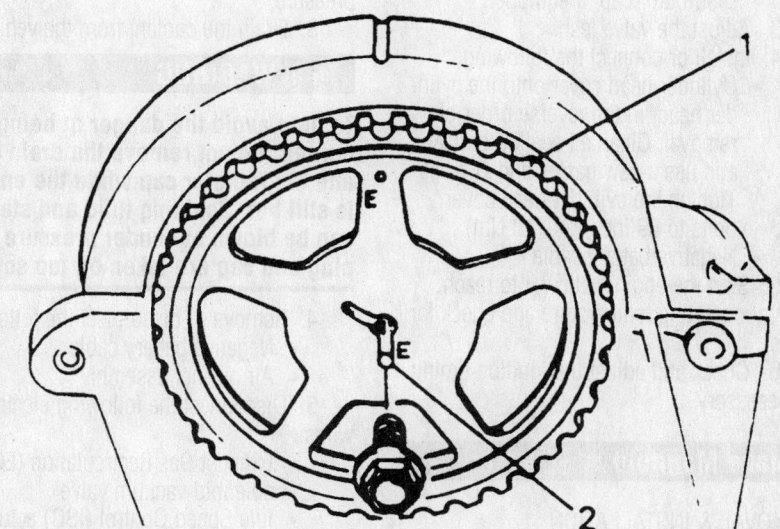

1. Camshaft timing belt pulley   2. Dowel pin

Align the dowel pin with the E-slot in the sprocket—1.6L engine

Heater Core replacement is covered in Section 2 of this manual

→During camshaft timing belt sprocket installation, align the camshaft dowel pin with the slot in the camshaft timing belt gear designated as E.

- Camshaft sprocket. Using a holding tool to hold the sprocket in place, tighten the camshaft sprocket bolt to 44 ft. lbs. (60 Nm).

→When installing the timing belt, the directional arrows on the timing belt must be matched with the rotation of the crankshaft. If not, excessive wear and timing belt failure may occur.

- Timing belt

11. Apply RTV silicone rubber sealant to the surface of the distributor case that mates with the rear of the rocker arm shaft.

12. Install or connect the following:
- Distributor case and tighten the 3 case-to-cylinder head bolts to 89 inch lbs. (10 Nm), if equipped
- CMP sensor case and tighten the bolts to 96 inch lbs. (11 Nm), if equipped
- Ignition coil, if equipped

→With the timing marks aligned on the sprockets and the timing belt installed, the number 4 piston is at Top Dead Center (TDC) of the compression stroke.

- Distributor into the distributor case, if equipped. Be sure the rotor is aligned with the No. 4 tower on the distributor cap.
- Distributor cap, if equipped

13. Adjust the valve lash.

14. Install or connect the following:
- Cylinder head cover onto the cylinder head, in the reverse order of removal. Clean all sealing surfaces and use a new gasket and O-rings. Tighten the cylinder head cover bolts to 89 inch lbs. (10 Nm).
- Negative battery cable

15. Start the engine, allow it to reach normal operating temperature and check for leaks.

16. Check and adjust the ignition timing as necessary.

Intake Manifold

REMOVAL & INSTALLATION

✳✳ CAUTION

The fuel system pressure must be relieved before disconnecting any fuel lines. Failure to do so may result in personal injury.

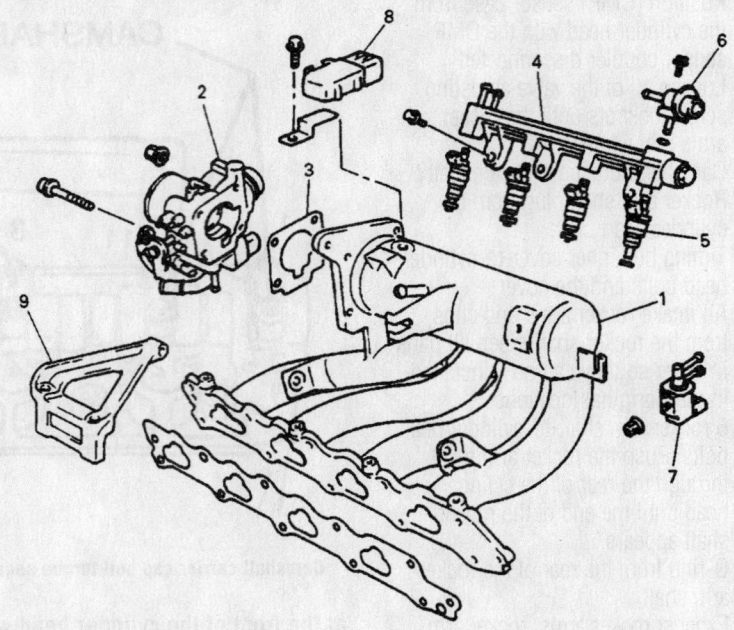

1. Intake manifold
2. Throttle body
3. Gasket
4. Fuel delivery pipe
5. Fuel injector
6. Fuel pressure regulator
7. EVAP canister purge valve
8. MAP sensor
9. Intake manifold upper stiffener

7923UG14

Intake manifold and related components—1.3L engine

1. Before servicing the vehicle, refer to the precautions in the beginning of this section.

2. Properly relieve the fuel system pressure.

3. Drain the coolant from the vehicle.

✳✳ WARNING

To help avoid the danger of being burned, do not remove the drain plug and the radiator cap while the engine is still hot. Scalding fluid and steam can be blown out under pressure if the plug and cap are taken off too soon.

4. Remove or disconnect the following:
- Negative battery cable
- Air cleaner assembly

5. Disconnect the following electrical wires:
- Exhaust Gas Recirculation (EGR) solenoid vacuum valve
- Idle Speed Control (ISC) actuator
- Ground wires from the intake manifold
- Fuel injectors
- Throttle Position (TP) sensor
- Idle Air Control (IAC) valve
- Tank pressure control solenoid vacuum valve, if equipped

- Manifold Absolute Pressure (MAP) sensor

6. Remove or disconnect the following:
- Fuel return and feed hoses from the fuel pipes
- Coolant hoses from the throttle body and the intake manifold

7. Disconnect the following vacuum hoses:
- Canister purge hose from the intake manifold
- Manifold Absolute Pressure (MAP) sensor hose from the intake manifold
- Brake booster hose from the intake manifold
- Positive Crankcase Ventilation (PCV) hose from the PCV valve
- Accelerator cable from the throttle body

8. Remove or disconnect the following:
- Intake manifold attaching nuts and bolts
- Intake manifold and throttle body

To install:

9. Before installing the gasket, make sure that the mating surfaces of the intake manifold and the cylinder head are clean and undamaged.

10. Install or connect the following:

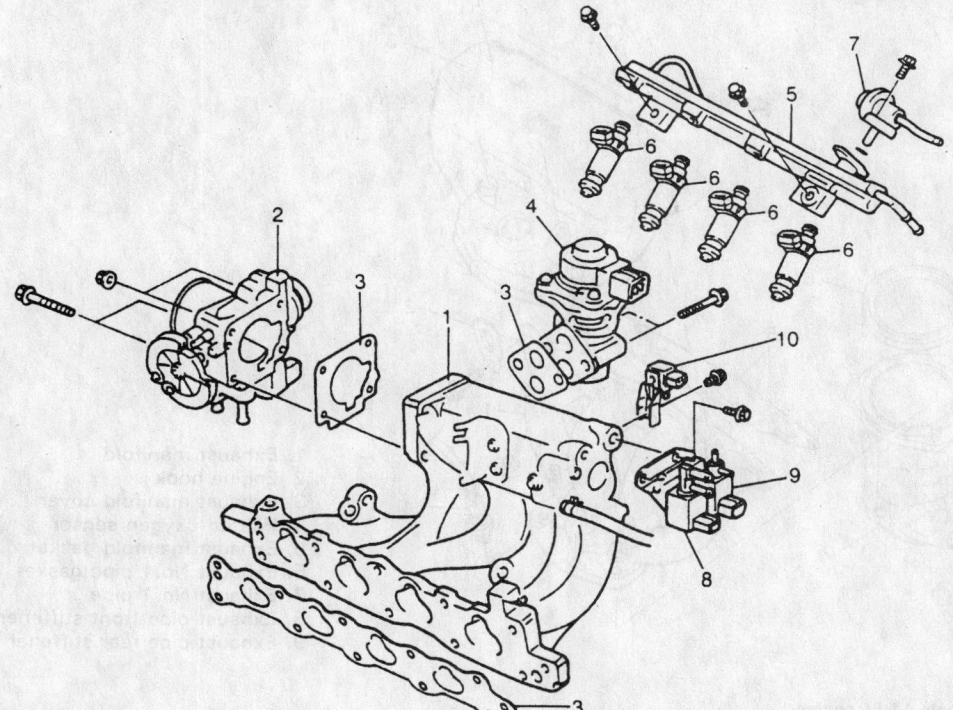

1. Intake manifold
2. Throttle body
3. Gasket
4. EGR valve
5. Fuel delivery pipe
6. Fuel injector
7. Fuel pressure regulator
8. EVAP canister purge valve
9. Tank pressure control solenoid valve
10. MAP sensor

7923UG15

Intake manifold and related components—1.6L engine

- New intake manifold gasket on the cylinder head
- Intake manifold and throttle body on the cylinder head
- Intake manifold mounting nuts and bolts. Tighten the nuts and bolts to 13–20 ft. lbs. (18–28 Nm). Be sure that the clamps are properly installed on the lower intake manifold bolts.
- Remaining components in the reverse order of removal
11. Refill the cooling system.
12. Connect the negative battery cable.
13. Start the engine and check for fuel and cooling system leaks.

Exhaust Manifold

REMOVAL & INSTALLATION

1.3L Engine

✴✴ CAUTION

To avoid the danger of being burned, do not service the exhaust system while it is hot.

Service should be performed only after the system cools down.

1. Before servicing the vehicle, refer to the precautions in the beginning of this section.
2. Remove or disconnect the following:
   - Negative battery cable
   - Heated oxygen (HO$_2$S) sensor electrical coupler and release the wire from its clamps
   - Exhaust manifold cover
   - Exhaust manifold stiffener bolt
   - 2 bolts attaching the exhaust pipe to the exhaust manifold
   - Exhaust manifold mounting nuts and bolts
   - Exhaust manifold and the gasket

To install:

3. Before installing any components check the exhaust manifold and the engine for deterioration or damage and replace as necessary.
4. Install or connect the following:
   - Manifold gasket and the exhaust manifold on the engine. Tighten the nuts and bolts to 13–20 ft. lbs. (18–28 Nm).
   - Exhaust pipe gasket on the exhaust pipe and position the exhaust pipe
   - 2 bolts that attach the exhaust pipe on the exhaust manifold and tighten the bolts to 25–36 ft. lbs. (35–50Nm)

- Exhaust manifold stiffener and tighten the bolt to 29–43 ft. lbs. (40–60 Nm)
- Remaining components in the reverse order of removal
- Negative battery cable
5. Run the engine and check for exhaust leaks.

1.6L Engine

1. Before servicing the vehicle, refer to the precautions in the beginning of this section.
2. Remove or disconnect the following:
   - Negative cable
   - 2 exhaust pipe bolts connecting the exhaust pipe to the exhaust manifold
   - Oxygen (O$_2$S) sensor lead wire at the coupler
   - Exhaust manifold heat shield from the manifold by unfastening the nut and bolt
   - Exhaust manifold mounting bolts and nuts
   - Exhaust manifold

To install:

3. Clean and inspect the sealing surfaces of the exhaust manifold and the cylinder head.

Brake service is covered in Section 4 of this manual

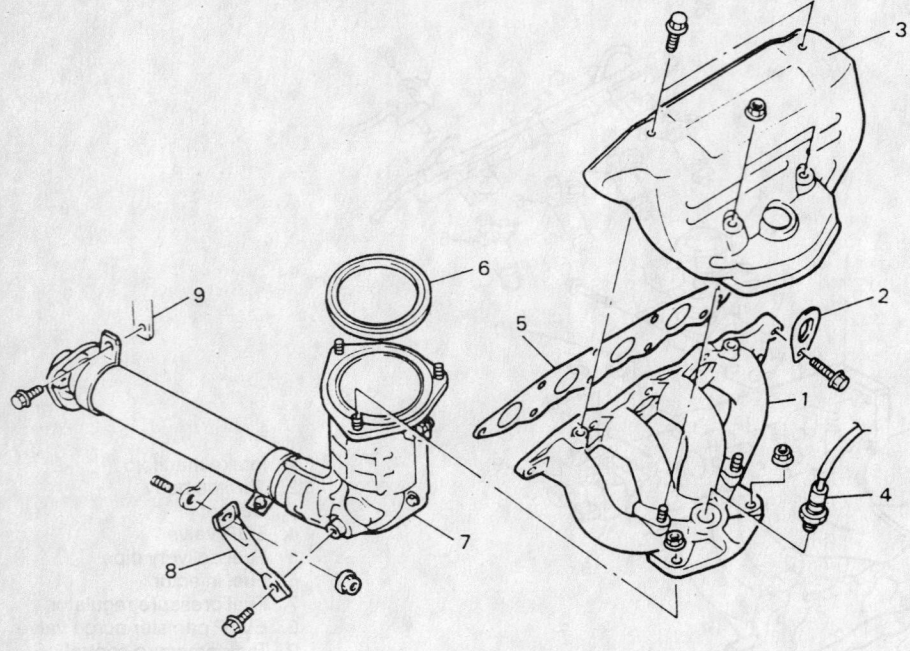

1. Exhaust manifold
2. Engine hook
3. Exhaust manifold cover
4. Heated oxygen sensor
5. Exhaust manifold gasket
6. Exhaust No. 1 pipe gasket
7. Exhaust No. 1 pipe
8. Exhaust pipe front stiffener
9. Exhaust pipe rear stiffener

7923UG16

Exhaust manifold and related components—1.6L engine

4. Install or connect the following:
- New gaskets
- Exhaust manifold and tighten the mounting bolts and nuts to 17 ft. lbs. (23 Nm)
- O_2S lead wire at the coupler
- 3 exhaust pipe bolts connecting the exhaust pipe to the exhaust manifold and tighten to 36 ft. lbs. (50 Nm)
- Manifold heat shield to the exhaust manifold using the nuts and bolts
- Negative battery cable
5. Check for exhaust leaks when finished.

Front Crankshaft Seal

REMOVAL & INSTALLATION

1.3L Engines

1. Remove or disconnect the following:
- Negative battery cable
- Timing belt
- Crankshaft sprocket bolt using a suitable gear stopper to hold the flywheel
- Sprocket bolt, sprocket and key
- Seal from the oil pump housing, using a suitable tool

➡ Be careful not to damage the crankshaft or the oil pump sealing surfaces when removing or installing the seal.

2. Clean and inspect the surfaces of the crankshaft and the oil pump assembly.
To install:
3. Lubricate the new seal with clean engine oil.

4. Install or connect the following:
- New seal over the crankshaft and into the oil pump, making sure the oil seal lip is not turned up. Use an oil seal guide tool.

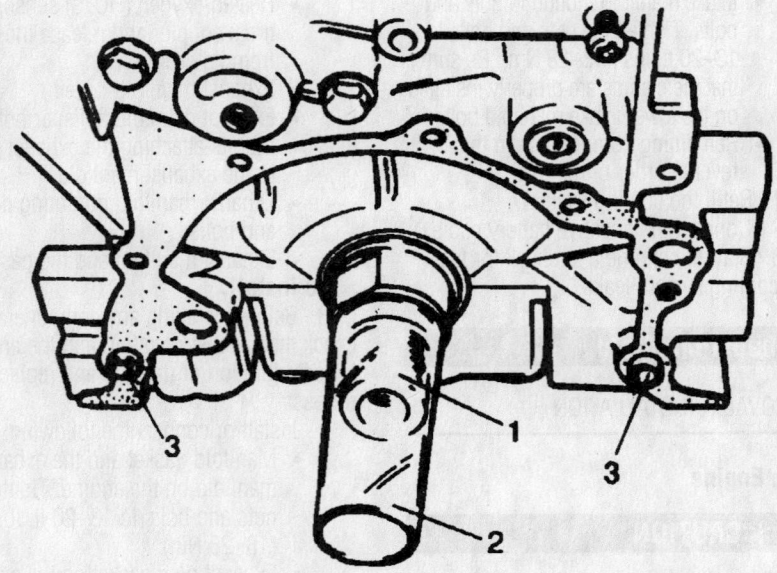

1 Crankshaft
2 Oil seal guide (Vinyl resin) (special tool 09926-18210)
3 Oil pump pin

Oil seal guide tool—1.3L engine

7923UG17

- Crankshaft sprocket and timing belt
- Negative battery cable

1.6L Engine

➡ **The front oil seal can be removed from the engine without removing the oil pump.**

1. Before servicing the vehicle, refer to the precautions in the beginning of this section.
2. Remove or disconnect the following:
   - Negative battery cable
   - Timing belt
   - Crankshaft timing belt sprocket

✳✳ WARNING

When removing the front seal, be extremely careful not to damage the crankshaft.

3. Using a knife, cut off the oil seal lip.
4. Tape the end of a flat bladed tool to avoid damaging the crankshaft. Pry out the oil seal using the taped end of the tool.
5. Inspect the oil seal contact surface on the crankshaft for signs of wear or damage.

To install:

6. Wipe the seal bore with a clean rag.
7. Apply multipurpose grease to the lip of a new oil seal.
8. Install or connect the following:
   - Oil seal into place using a seal installer tool. Be sure the seal surface is flush with the edge of the oil pump case. Work from the front of the cover. Be extremely careful not to damage the seal.
   - Crankshaft sprocket
   - Timing belt
   - Negative battery cable to the battery
9. Start the engine and check for leaks.

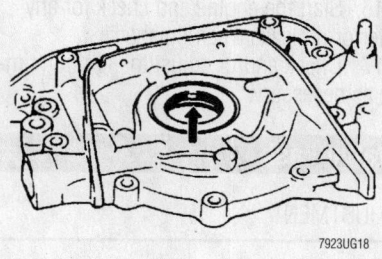

Front crankshaft oil seal location—1.6L engine

Camshaft and Valve Lifters

REMOVAL & INSTALLATION

1.3L Engine

1. Before servicing the vehicle, refer to the precautions in the beginning of this section.
2. Drain the cooling system into a resealable container.
3. Relieve the fuel system pressure.
4. Remove or disconnect the following:
   - Timing belt
   - Cylinder head cover from the cylinder head
   - Camshaft Position (CMP) sensor coupler from the CMP sensor
   - High tension cords and couplers from the coil assemblies
   - Ignition coil from the exhaust manifold side
   - CMP case from the cylinder head
5. Loosen all valve adjusting screw lock nuts, turn the adjusting screws back all the way to allow the rocker arms to move freely.

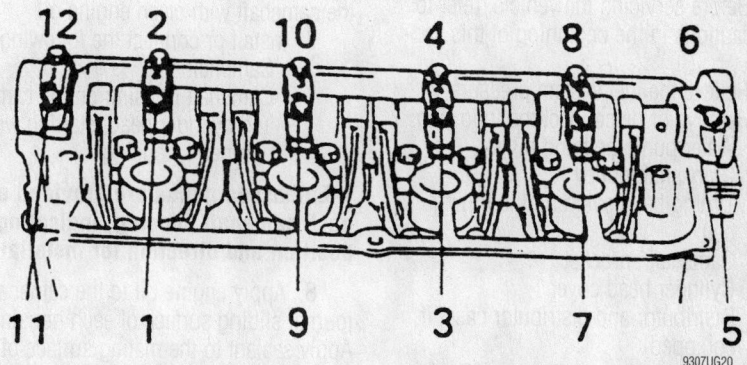

Loosen the camshaft bearing caps using several steps in the sequence shown—1.3L engines

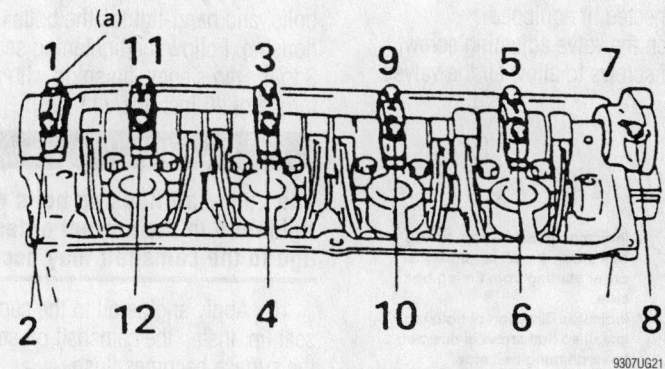

Tighten the camshaft bearing caps using several steps in the sequence shown—1.3L engines

6. Remove or disconnect the following:
   - Camshaft bearing caps in the sequence illustrated in several steps
   - Camshaft
   - Timing belt inside cover
   - Intake rocker arm with the clip from the rocker arm shaft. Make sure not to bend the clip when removing the intake rocker arm
   - Rocker arm shaft bolts
   - Exhaust rocker arms and spring by pulling out the shaft on the battery side. If necessary, remove the battery for clearance.
7. Inspect the rocker arms and shafts for wear and/or damage and replace parts as necessary.
   a. If the tip of the valve lash adjuster is badly worn, replace the adjuster.

To install:

8. Apply engine oil to the rocker arms and the rocker arm shafts.
9. Install or connect the following:
   - O-ring on the shaft
   - Rocker arm shaft with its holes facing up

For complete Engine Mechanical specifications, see Section 1 of this manual

- Rocker arm (exhaust side) and the rocker arm spring
- Rocker arm shaft bolts and tighten the bolts to 96 inch lbs. (11 Nm)

10. Pour a small amount of engine oil into the arm pivot holding part of the shaft.

11. Install or connect the following:
- Rocker arm (intake side), with the clips to the rocker arm shaft
- Camshaft and the bearing caps. Tighten the bearing caps in several passes to reach the final torque of 96 inch lbs. (11 Nm). Refer to the accompanying illustration for the proper camshaft bearing cap torque sequence.
- Camshaft oil seal
- Timing belt inside cover
- Camshaft timing belt pulley, while fitting the pin on the camshaft into the slot. Tighten the pulley bolt to 43 ft. lbs. (60 Nm).
- Timing belt
- Remaining components in the reverse order of removal.

12. Check the ignition timing.

1.6L Engine

1. Before servicing the vehicle, refer to the precautions in the beginning of this section.

2. Relieve the fuel system pressure.

3. Remove or disconnect the following:
- Water pump belt and pulley
- Crankshaft pulley
- Timing belt cover and the timing belt
- Camshaft sprocket
- Cylinder head cover
- Distributor and distributor case, if equipped
- Ignition coils and Camshaft Position (CMP) sensor case, with the CMP sensor housing coupler disconnected, if equipped

4. Loosen the valve adjusting screw locknuts and screws to allow all the valves to close.

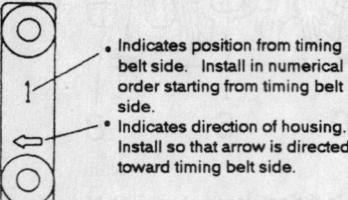

- Indicates position from timing belt side. Install in numerical order starting from timing belt side.
- Indicates direction of housing. Install so that arrow is directed toward timing belt side.

7923UG21

Camshaft bearing cap identification—1.6L engine

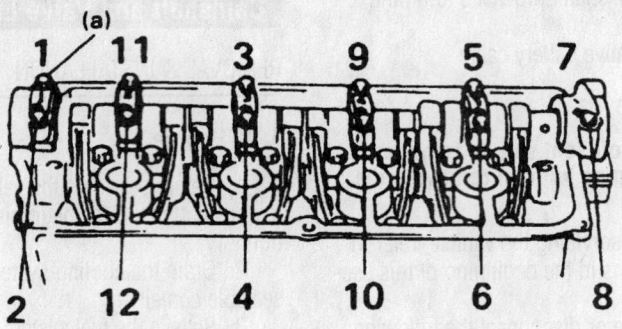

Camshaft housing bolt tightening sequence—1.6L engine

5. Remove or disconnect the following:
- Camshaft housing bolts
- Housings
- Camshaft

※※ CAUTION

The camshaft housing bolts must be removed in the reverse order of installation or damage to the camshaft may occur.

To install:

6. Lubricate the lobes and journals of the camshaft with clean engine oil.

7. Install or connect the following:
- Camshaft
- Camshaft housing on the camshaft and cylinder head, starting with the number 1 housing

➡**Embossed marks are provided on each camshaft housing, indicating position and direction for installation.**

8. Apply engine oil to the camshaft journal sliding surface of each housing. Apply sealant to the mating surface of the number 6 housing which will mate with the cylinder head.

9. Apply engine oil to the housing bolts, and hand-tighten the bolts into the housing. Follow the tightening sequence in 3 to 4 even stages, finishing with a final torque of 96 inch lbs. (11 Nm).

※※ CAUTION

The camshaft housing bolts must be tightened in the correct order or damage to the camshaft may occur.

10. Apply engine oil to the camshaft oil seal lip. Install the camshaft oil seal until the surface becomes flush.

11. Install or connect the following:
- Camshaft sprocket
- Timing belt
- Timing belt cover

- Crankshaft pulley
- Water pump pulley
- Water pump belt. Be sure the pin on the camshaft fits into the slot at the **E** mark on the camshaft sprocket. Tighten the sprocket bolt to 41–46 ft. lbs. (56–64 Nm).

12. Prior to installation on models equipped with a distributor, apply sealant to the area of the distributor housing that covers the rear of the rocker arm shaft on the cylinder head. Install the distributor and distributor housing to the cylinder head. Be sure the distributor is facing the correct firing position.

13. Prior to installation on models equipped with a Distributorless Ignition System (DIS), apply sealant to the area of the housing that covers the rear of the rocker arm shaft on the cylinder head. Install the CMP sensor and housing to the cylinder head and tighten the bolts to 96 inch lbs. (11 Nm).
- Ignition coil assembly, if equipped

14. Adjust the valve lash.

15. Install or connect the following:
- New cylinder head cover gasket and the cover
- Negative battery cable

16. Start the engine and check for any water or oil leaks when finished.

17. Check and/or adjust the ignition timing as necessary.

Valve Lash

ADJUSTMENT

1.3L Engines

Hydraulic valve lash adjusters are used to adjust the valve clearance to **0** lash automatically at all times. Adjustment is not required.

1.6L engine

1. Before servicing the vehicle, refer to the precautions in the beginning of this section.

2. Disconnect the negative battery cable.

3. Remove the cylinder head cover.

4. Remove the right front wheel and fender apron.

5. Turn the crankshaft pulley clockwise until the **V** mark on the pulley is aligned with the **0** calibration mark on the timing belt cover.

6. Remove the Camshaft Position (CMP) sensor and look in the hole for the notch on the camshaft rotor gear. If the notch is not visible in the hole, rotate the crankshaft 1 complete revolution until the notch is visible.

7. The valve lash is measured between the rocker arm adjusting screw and the valve stem. Use a thickness gauge to measure the gap.

8. Check the valve lash for the following valves:
- Intake valve of cylinder number 1 (ID 1)
- Intake valve of cylinder number 2 (ID 2)
- Exhaust valve of cylinder number 1 (ID 5)
- Exhaust valve of cylinder number 3 (ID 7)

9. If the valve lash is out of specification, adjust the specification after loosening the locknut and turning the adjusting screw. Hold the screw stationary while tightening the locknut. Recheck the specification after tightening the locknut.

10. Rotate the crankshaft 1 rotation clockwise and realign the timing marks.

11. Check the valve lash for the following valves:
- Intake valve of cylinder number 3 (ID 3)
- Intake valve of cylinder number 4 (ID 4)
- Exhaust valve of cylinder number 2 (ID 6)
- Exhaust valve of cylinder number 4 (ID 8)

12. Adjust the valves that are out of specification and recheck after tightening the locknut.

13. Install the remaining components.

14. Connect the negative battery cable.

Starter Motor

REMOVAL & INSTALLATION

1. Remove or disconnect the following:
- Negative battery cable
- Starter electrical connections
- 2 starter mounting bolts
- Starter motor

To install:

2. Install or connect the following:
- Starter
- 2 starter mounting bolts and tighten to 96 inch lbs. (10 Nm)
- Starter electrical connections
- Negative battery cable

Oil Pan

REMOVAL & INSTALLATION

1. Before servicing the vehicle, refer to the precautions in the beginning of this section.

2. Remove or disconnect the following:
- Negative battery cable
- Engine undercovers on 1.6L engines

3. Drain the engine oil into a suitable container.
- Lower plate on 1.3L models, from the clutch housing if equipped with a manual transaxle, or from the torque converter housing if equipped with a automatic transaxle.

4. On 1.6L models, remove or disconnect the following:
- Front exhaust pipe from the vehicle
- Transaxle stiffener plate from the engine and transaxle

5. Support the transmission and engine.

6. Remove or disconnect the following:
- Vehicle center member by removing the 7 nuts and 4 bolts from the center member, on 1.6L models
- Crankshaft Position (CKP) sensor from the oil pan, if equipped
- Oil pan retainer bolts
- Oil pan from the cylinder block
- Oil pump pick-up

To install:

7. Clean the mating surfaces of the oil pan and the engine block.

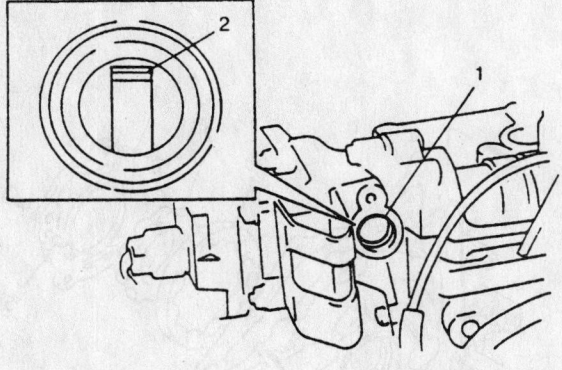

1. Hole for CMP sensor
2. Camshaft rotor gear

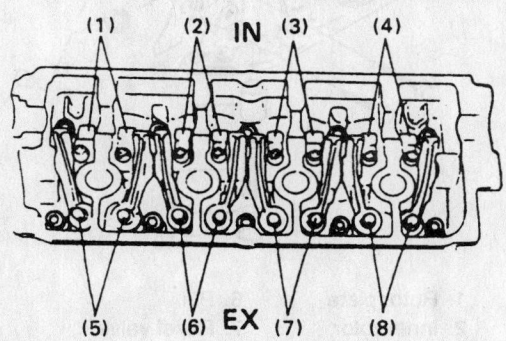

Valve numbered locations and camshaft rotor gear mark—1.6L engines

7923UG23

For Accessory Drive Belt illustrations, see Section 1 of this manual

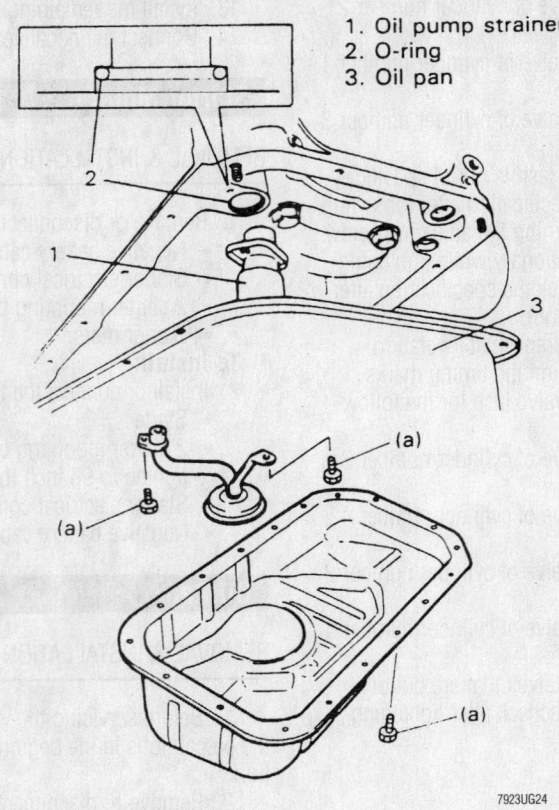

1. Oil pump strainer
2. O-ring
3. Oil pan

Oil pan mounting—1.3L and 1.6L engines

8. Install or connect the following:
   - Oil pump strainer. Tighten the strainer bolt first at the bracket. Tighten the bolts to 96 inch lbs. (11 Nm).
9. Apply silicon sealant to the oil pan mating surface in one continuous bead.
10. Install or connect the following:
   - Oil pan to the engine block and install the bolts. Start tightening the bolts at the center and move outward. Tighten the bolts to 96 inch lbs. (11 Nm).
   - CKP sensor on the oil pan
   - Drain plug and drain plug gasket on the oil pan
   - Lower plate on the clutch housing, on 1.3L models; if equipped with a manual transaxle, or on the torque converter housing if equipped with an automatic transaxle
11. On 1.6L models, install or connect the following:
   - Center member on the vehicle. Tighten the bolts and nuts at the engine mounts to 40 ft. lbs. (55 Nm) and all other nuts to 33 ft. lbs. (45 Nm).
   - Transaxle stiffener plate and tighten the bolts to 37 ft. lbs. (50 Nm)

   - Exhaust pipe and tighten the nuts and bolts to 37 ft. lbs. (50 Nm)
   - Engine undercovers
12. Refill the engine with oil.
13. Connect the negative battery cable.
14. Start the engine and check for leaks.

Oil Pump

REMOVAL & INSTALLATION

1. Before servicing the vehicle, refer to the precautions in the beginning of this section.
2. Disconnect the negative battery cable.
3. Drain the engine oil.
4. Remove or disconnect the following:
   - Right side fender apron clips by pushing the center pin

✶✶ WARNING

Do not push the center pin too far in, or it will fall off into the fender.

   - Power steering and air conditioning belt, if equipped
   - Water pump pulley bolts, loosen
   - Alternator drive belt

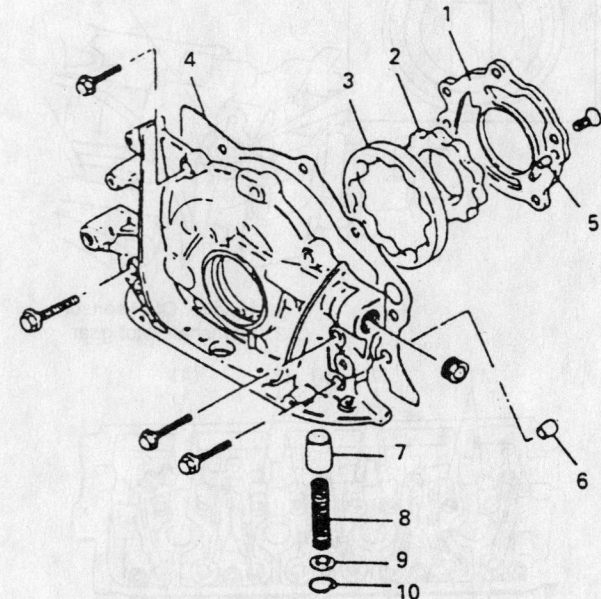

1. Rotor plate	6. Pin
2. Inner rotor	7. Relief valve
3. Outer rotor	8. Spring
4. Gasket	9. Retainer
5. Pin	10. Retainer ring

Exploded view of the oil pump—1.3L and 1.6L engine

- Water pump pulley and alternator bracket
- Crankshaft pulley
- Timing belt outside covers
- Timing belt guide and the timing belt
- Engine oil level gauge
- Air conditioning compressor bracket bolts, if equipped
- Timing belt and crankshaft pulley
- Oil pan and oil pump pick-up
- 7 bolts securing the oil pump to the engine block
- Oil pump

To install:

5. Clean the engine block where the oil pump mounts, then install the oil pump gasket and the 2 oil pump alignment pins.

6. To prevent damage to the oil seal when installing the oil pump, fit a guide sleeve on the crankshaft and apply a thin coating of engine oil to the special tool.

7. Install or connect the following:
- Oil pump. Install a long bolt in the lowest bolt hole on the intake manifold side of the engine. Install 2 long bolts in the 2 lowest bolt holes on the exhaust manifold side of the engine. Install the 4 short bolts in the other 4 bolt holes in the oil pump. Tighten all of the bolts to 96 inch lbs. (11 Nm). Check that the oil seal lip is not turned upward, then remove the special tool.

8. If the of the oil pump gasket bulges where the oil pan attaches cut the excess off with a sharp knife.

9. Install or connect the following:
- Oil pan and the oil pump pick-up
- Rubber seal between the oil pump and the water pump
- Crankshaft key
- Timing belt pulley
- Crankshaft pulley pin
- Pulley bolt and tighten to 80 ft. lbs. (110 Nm), with the crankshaft locked
- Timing belt guide so that the concave side faces the oil pump
- Crankshaft key and the timing belt pulley
- Remaining components in the reverse order of removal
- Negative battery cable

10. Start the engine and check the engine oil pressure.

11. Check that no leaks are present.

Rear Main Seal

REMOVAL & INSTALLATION

1. Before servicing the vehicle, refer to the precautions in the beginning of this section.

2. Remove or disconnect the following:
- Transaxle assembly
- Flexplate/flywheel from the crankshaft

3. Carefully pry the oil seal out of the retainer without scratching the sealing surface of the crankshaft.

To install:

4. Apply engine oil the lip of the new seal.

5. Install or connect the following:
- Seal in the retainer using a suitable seal driver
- Flexplate/flywheel
- Transaxle assembly

Piston and Ring

POSITIONING

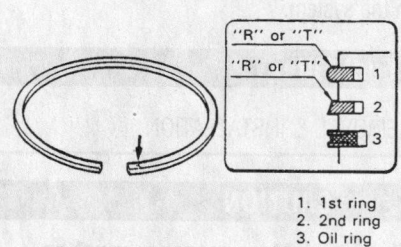

1. 1st ring
2. 2nd ring
3. Oil ring

7923AG84

Suzuki engines—piston ring positioning

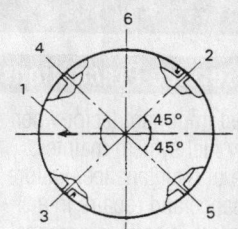

1. Arrow mark
2. 1st ring end gap
3. 2nd ring end gap and oil ring specer
4. Oil ring upper rail gap
5. Oil ring lower rail gap
6. Intake side
7. Exhaust side

7923AG85

Suzuki engines—piston ring end-gap spacing

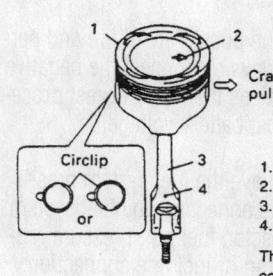

1. Piston
2. Arrow mark
3. Connecting rod
4. Oil hole

The oil hole should come on intake side

7923AG88

Suzuki engines—piston/connecting rod assembly-to-engine positioning

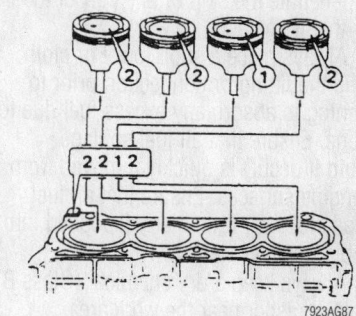

7923AG87

Suzuki engines—the piston ID number must match the number stamped in the engine block

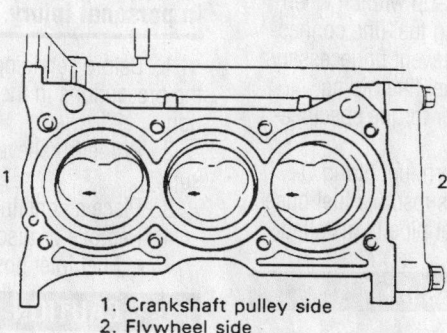

1. Crankshaft pulley side
2. Flywheel side

7923AG86

Suzuki engines—the directional arrow on the piston face must face the crankshaft pulley end of the engine

FUEL SYSTEM

Fuel System Service Precautions

Safety is the most important factor when performing not only fuel system maintenance, but any type of maintenance. Failure to conduct maintenance and repairs in a safe manner may result in serious personal injury or death. Maintenance and testing of the vehicle's fuel system components can be accomplished safely and effectively by adhering to the following rules and guidelines.

• To avoid the possibility of fire and personal injury, always disconnect the negative battery cable unless the repair or test procedure requires that battery voltage be applied.

• Always relieve the fuel system pressure prior to disconnecting any fuel system component (injector, fuel rail, pressure regulator, etc.), fitting or fuel line connection. Exercise extreme caution whenever relieving the fuel system pressure, to avoid exposing your skin, face and eyes to fuel spray. Please be advised that fuel under pressure may penetrate the skin or any part of the body that it contacts.

• Always place a shop towel or cloth around the fitting or connection prior to loosening to absorb any excess fuel due to spillage. Ensure that all fuel spillage (should it occur) is quickly removed from the engine surfaces. Ensure that all fuel soaked cloths or towels are deposited into a suitable waste container.

• Always keep a dry chemical (Class B) fire extinguisher near the work area.

• Do not allow fuel spray or fuel vapors to come into contact with a spark or open flame.

• Always use a back-up wrench when loosening and tightening fuel line connection fittings. This will prevent unnecessary stress and torsion on fuel line piping. Always follow the proper torque specifications.

• Always replace worn fuel fitting O-rings with new. Do not substitute fuel hose or equivalent, where fuel pipe is installed.

Fuel System Pressure

RELIEVING

❊❊ CAUTION

Care should be used when working around the fuel system. DO NOT smoke or expose the fuel system to

any open flames. Keep a fire extinguisher handy.

1. Before servicing the vehicle, refer to the precautions in the beginning of this section.
2. Disconnect the negative battery cable from the battery.
3. Place the vehicle in **PARK** for automatic transmission or **NEUTRAL** for manual transmission.
4. Remove the relay box cover.
5. Disconnect the fuel pump relay from the relay box.
6. Remove the fuel filler cap from the filler neck to release the fuel vapor pressure in the fuel tank.
7. Start the vehicle and allow the engine to run until it stalls.
8. Crank the engine 3 more revolutions to eliminate any remaining pressure in the fuel lines.
9. Disconnect the negative battery cable.
10. Connect the fuel pump relay to the relay box.
11. After servicing the fuel system, connect the negative battery cable.
12. Start the engine and check for leaks in the system.

Fuel Filter

REMOVAL & INSTALLATION

❊❊ CAUTION

The fuel system pressure must be relieved before disconnecting any fuel lines. Failure to do so may result in personal injury.

1. Before servicing the vehicle, refer to the precautions in the beginning of this section.
2. Properly relieve the fuel system pressure.
3. Place a container under the fuel filter.
4. Remove or disconnect the following:
   • Fuel inlet hose from the fuel filter

❊❊ CAUTION

A small amount of fuel may be released after the fuel hose is disconnected. Cover the hose and pipe with a shop towel.

   • Outlet hose from the fuel feed pipe
   • 2 fuel filter mounting bracket bolts

and remove the fuel filter from the frame with the outlet hose attached
   • Fuel filter from the bracket by unfastening the mounting bolt
   • Outlet hose from the fuel filter

To install:

5. Install or connect the following:
   • Fuel filter on the bracket
   • Mounting bolt
   • Remaining components in the reverse order of removal
   • Inlet and outlet hoses to the filter
   • Negative battery cable
6. With the ignition **ON** and the engine **OFF** check for leaks.

Fuel Pump

REMOVAL & INSTALLATION

1. Before servicing the vehicle, refer to the precautions in the beginning of this section.
2. Relieve the pressure from the fuel system.
3. Remove or disconnect the following:
4. Remove the rear seat cushion by removing or disconnecting the following:
   • Spare tire
   • Seat back by unfastening the 2 center mounting nuts and the 4 mounting screws
   • Fitting screws from the rear of the seat cushion
   • Lift the front of the seat cushion and remove the cushion
5. Remove or disconnect the following:
   • Fuel level gauge and the fuel pump lead wire couplers and detach the wire tape
   • Fuel filler hose from the fuel tank
   • Breather hose from the filler neck
6. Drain the fuel from the tank by pumping the fuel out through the fuel tank filler.

❊❊ CAUTION

Use a gasoline safe hand operated pump device to drain the fuel tank.

7. Remove or disconnect the following:
   • Fuel hoses from the fuel pipes, located near the fuel filter

❊❊ CAUTION

A small amount of fuel may be released after the fuel hose is disconnected. Cover the hose and pipe to be disconnected with a shop cloth.

8. Install a support (for example, a transmission jack) under the fuel tank.

9. Remove or disconnect the following:
- Fuel tank mounting hardware
- Tank from the vehicle
- Fuel lines from the fuel pump and sender assembly
- 12 screws that secure the fuel pump and fuel gauge assembly to the tank
- Pump and sender assembly
- Fuel pump electrical connectors
- Fuel strainer
- Fuel pump

To install:

10. Install or connect the following:
- Fuel pump on the fuel gauge assembly
- Fuel strainer on the fuel pump

➡**Always install a new fuel pump strainer when replacing the fuel pump.**

- Electrical connectors to the fuel pump
- Remaining components in the reverse order of removal
- Negative battery cable

11. Turn the ignition switch to the **ON** position, but leave the engine **OFF** and check for fuel leaks.

Fuel Injector

REMOVAL & INSTALLATION

Swift

1. Before servicing the vehicle, refer to the precautions in the beginning of this section.

2. Relieve the pressure from the fuel system.

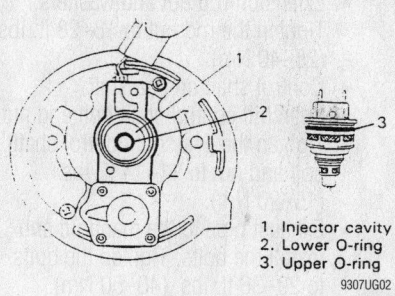

1. Injector cavity
2. Lower O-ring
3. Upper O-ring

9307UG02

Install the lower injector O-ring (2) to the injector cavity (1) and the upper O-ring (3) to the injector

3. Remove or disconnect the following:
- Air cleaner assembly
- Air cleaner mounting stay from the throttle body
- Injector cover
- Injector from the throttle body
- Injector O-rings and discard

To install:

4. Inspect the injector filter for dirt or contamination. If present, clean and check for dirt in the fuel lines and tank.

5. Apply a tin coat of spindle oil or gasoline to the new O-rings.

6. Install or connect the following:
- Lower injector O-ring to the injector cavity and the upper O-ring to the injector
- Injector by pushing straight into the injector cavity. Do not turn the injector while pushing.

7. Make sure the injector cover O-ring is free of contamination and has not deteriorated. Apply a coat of spindle oil or gasoline to the O-ring.

8. Install or connect the following:
- Injector cover and tighten the

retainers to 24–36 inch lbs. (3–4 Nm)
- Negative battery cable

9. Turn the ignition switch on and check for fuel leaks
- Air cleaner mounting stay to the throttle body
- Air cleaner assembly

Esteem

1. Before servicing the vehicle, refer to the precautions in the beginning of this section.

2. Relieve the pressure from the fuel system.

3. Remove or disconnect the following:
- Vacuum hose from the fuel pressure regulator
- Fuel injector electrical connections
- Fuel rail bolts
- Injector from the fuel rail
- Injector O-ring and discard

To install:

4. Coat the injector O-rings with gasoline.

5. Install or connect the following:
- Injector O-ring
- Injector to the fuel rail and intake manifold.

6. Make sure the injectors rotate smoothly. If not, the O-ring has been improperly installed and must be replaced.
- Fuel rail bolts. Once the bolts are tightened make sure the injectors rotate smoothly.
- Fuel injector electrical connections
- Vacuum hose to the fuel pressure regulator
- Negative battery cable

7. Start the vehicle and check for leaks.

DRIVE TRAIN

Transaxle Assembly

REMOVAL & INSTALLATION

Manual

1. Before servicing the vehicle, refer to the precautions in the beginning of this section.

2. Drain the transaxle oil.

3. Remove or disconnect the following:
- Battery cables
- Battery and tray

- Clutch cable adjusting nut, joint pin from the cable and cable from the bracket
- All the wiring harness clamps and connectors involved with the transaxle removal, tag if necessary for location to aid during installation
- Speedometer cable boot, case clip and cable from the case
- Radiator outlet pipe from the transmission side cover, on Swift models

- Transaxle retaining bolts
- Starter
- Fender apron extension on the left side
- Bolts connecting the exhaust pipe to the exhaust manifold and disconnect the joint

4. On Esteem models, with the engine supported, remove the vehicle center mounting member by removing the 7 nuts and 4 bolts.
- Gearshift control shaft nut and bolt

For Wheel Alignment specifications, see Section 1 of this manual

- Control shaft from the gearshift control shaft
- Extension rod nut and washers
- Clutch housing lower plate
- Sway bar
- Left and right front wheels
- Left and right ball joints
- Left and right halfshafts
- Transaxle stiffener
- Transmission to engine bolt and nut
- Engine rear mounting bracket bolts

5. Install an engine support.
6. Support the transaxle with a suitable jack.
7. Remove the left engine mounting bracket and stiffener.
8. Lower the transaxle with the engine attached. Pull the transaxle straight out toward the left side.
9. Lower and remove the transaxle.

To install:

10. Install or connect the following:
- Transaxle from the left side of the vehicle. Use care when inserting the pilot shaft into the clutch assembly. If the spline on the input shaft does not align with the clutch assembly spline, turn the crankshaft slightly to aid in spline alignment.

11. Raise the transaxle and engine.
- Left engine mounting bracket and stiffener. Tighten the bolts to 29–43 ft. lbs. (40–60 Nm).
- Rear engine mounting bracket bolts and tighten them to 29–43 ft. lbs. (40–60 Nm)

➡**Before installing the bolts into the rear mounting bracket, apply sealant to the bolt threads.**

- Transmission-to-engine bolt and nut. Tighten the nut and bolt to 29–43 ft. lbs. (40–60 Nm)

12. On Esteem models, install the center member on the vehicle and install the 7 nuts and 4 bolts. Tighten the bolts and nuts as follows:
 a. Center member-to-radiator support: 33 ft. lbs. (45 Nm).
 b. Center member-to-crossmember: 33 ft. lbs. (45 Nm).
 c. Engine mounts-to-center member nuts: 40 ft. lbs. (55 Nm).
13. Install or connect the following:
- Transaxle stiffener
14. Lower the transaxle supporting jack.
- Left and right halfshafts
- Ball joints
- Sway bar
- Clutch housing lower plate

- Extension rod nut and washers. Tighten the rod nut to 18–28 ft. lbs. (25–40 Nm).
- Control shaft on gearshift
- Gearshift control shaft bolt and nut. Tighten the gearshift control shaft bolt and nut to 11–14 ft. lbs. (15–20 Nm).
- Exhaust pipe to the manifold and install the bolts. Tighten the bolts to 29–36 ft. lbs. (40–50 Nm).
- Left fender apron extension

15. Refill the transaxle with the recommended lubricant.
16. Remove the engine support fixture.
17. Install or connect the following:
- Starter
- Transaxle retaining bolts. Tighten the retaining bolts to 29–43 ft. lbs. (40–60 Nm).
- Remaining components in the reverse order of removal
- Negative battery cable and the ground strap on the transaxle

Automatic

1. Before servicing the vehicle, refer to the precautions in the beginning of this section.
2. Drain the cooling system.
3. Drain the transaxle fluid from the transaxle.
4. Remove or disconnect the following:
- Negative battery cable from the battery and transaxle
- Speedometer cable

5. Disconnect the following electrical connectors:
- Solenoids
- Vehicle Speed Sensor (VSS), if equipped
- Shift lever switch
- Forward clutch cylinder revolution sensor

6. Remove or disconnect the following:
- Wiring harness from the clamps on the transaxle
- Select cable from the transaxle
- Cooling system pipe from the transaxle
- Top transaxle-to-engine bolts
- Starter
- Exhaust manifold cover
- Front exhaust pipe from the exhaust manifold

7. Support the engine.
- Engine undercovers, if equipped
- Transaxle cooler hoses

8. On Esteem models, with the engine supported, remove the vehicle center

mounting member by removing the 7 nuts and 4 bolts.
- Front exhaust pipe from the vehicle
- Transaxle stiffener plate
- Transaxle housing lower plate
- Torque converter bolts. To lock the drive plate, engage a flat bladed tool in the flywheel.
- Sway bar from the control arms
- Left and right front wheels
- Left and right ball joints from the steering knuckles
- Left and right halfshafts
- Engine rear mount and bracket

9. After removing the rear mount, remove the engine-to-transaxle bolt and nut located behind the rear bracket. Remove all bolts holding the engine to the transaxle.
10. Support the transaxle with a transaxle jack.
11. Remove or disconnect the following:
- Bolts from the engine left-hand mount
- Transaxle with the torque converter from the engine compartment

➡**When removing the transaxle from the engine, move it parallel with the crankshaft and use care so not to apply excessive force to the drive plate and torque converter.**

✴✴ CAUTION

Be sure to keep the transaxle with the torque converter horizontal or facing up throughout the work. Should it be tilted with converter down, the converter may fall off and cause personal injury.

To install:
12. Install or connect the following:
- Transaxle to the engine assembly
- Transaxle attaching nuts and bolts
- Left-hand mounting bolts and tighten to 40 ft. lbs. (55 Nm)
- Engine-to-transaxle bolt and nut before installing the rear transaxle mount. Tighten the nut to 65 ft. lbs. (90 Nm).
- All bolts for the transaxle. Tighten the bolts to 65 ft. lbs. (90 Nm).
- Left halfshafts
- Ball joints
- Sway bar to the control arms
- Torque converter bolts and tighten the bolts to 14 ft. lbs. (19 Nm)
- Transaxle housing lower plate
- Stiffener plate with the 4 bolts. Tighten the bolts to 40 ft. lbs. (55 Nm).
- Exhaust pipe on the center pipe and

tighten the bolts to 37 ft. lbs. (50 Nm)

13. On Esteem models, install the center member on the vehicle and install the 7 nuts and 4 bolts. Tighten the bolts and nuts as follows:

    a. Center member-to-radiator support: 33 ft. lbs. (45 Nm).

    b. Center member-to-crossmember: 33 ft. lbs. (45 Nm).

    c. Engine mounts-to-center member nuts: 40 ft. lbs. (55 Nm).

14. Install or connect the following:

- Oil hoses for the transaxle
- Engine undercovers

15. Remove the engine support.

- Exhaust pipe to the exhaust manifold and install the nuts
- Exhaust manifold cover
- Starter motor
- Remaining components in the reverse order of removal

16. Fill the cooling system and the transaxle. Check all fluids.

17. Connect the negative battery cable to the transaxle and battery.

Clutch

ADJUSTMENT

1. Before servicing the vehicle, refer to the precautions in the beginning of this section.

2. Depress the clutch pedal lightly until tension on the clutch cable can be felt.

3. Measure the clutch pedal free-play. It should be 0.6–0.8 in. (15–20mm).

4. Adjust the clutch pedal free-play by tightening or loosening the clutch cable adjustment nut.

5. Measure the clutch lever free-play. It should be 0–0.08 in. (0–2mm). If the clutch release lever free-play exceeds specification,

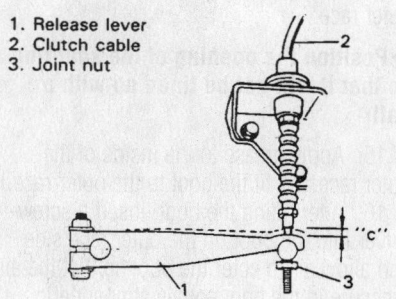

1. Release lever
2. Clutch cable
3. Joint nut

Clutch cable free-play adjustment

inspect the release shaft return spring for cracks or weakness.

➡**Be sure the marks on the clutch release lever and release shaft are aligned. If they are not, remove the lever from the shaft, align the marks and repeat the free-play adjustment procedure.**

REMOVAL & INSTALLATION

1. Before servicing the vehicle, refer to the precautions in the beginning of this section.

2. Remove the transaxle.

3. Hold the flywheel stationary.

4. Matchmark the pressure plate and flywheel for installation reference.

5. Loosen the pressure plate attaching bolts 1 turn at a time (evenly) until the spring pressure is released.

6. Remove the clutch disc and pressure plate.

To install:

7. Clean the flywheel mating surfaces of all oil, grease and metal deposits. Inspect flywheel for cracks, heat checking or other defects and replace or resurface as necessary.

8. Check the wear on the facings of the clutch disc by measuring the depth of each rivet head depression. Replace clutch disc

when rivet heads are 0.02 in. (0.5mm) below the surface of clutch surface.

9. Check the diaphragm spring and pressure plate for wear or damage. If the spring or plate is excessively worn, replace the pressure plate assembly.

10. Check the pilot bearing for smooth operation. If the bearing does not spin freely, replace it.

11. Position the clutch disc and pressure plate with the matchmarks aligned and install a clutch alignment tool.

12. Install the pressure plate bolts. Tighten the mounting bolts evenly and in a crisscross pattern to 13–20 ft. lbs. (18–28 Nm). Remove the alignment tool and the flywheel holding tool.

13. Lightly lubricate the transaxle input shaft splines, pilot bearing surface of the input shaft, and the release bearing with grease.

14. Install the transaxle.

15. Adjust the clutch cable.

Halfshaft

REMOVAL & INSTALLATION

1. Before servicing the vehicle, refer to the precautions in the beginning of this section.

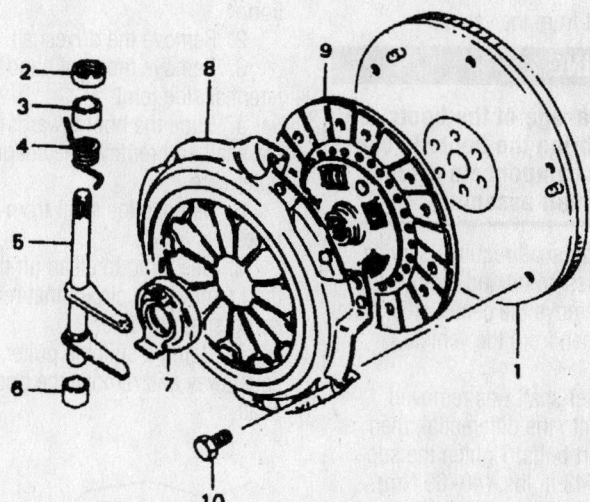

1. Flywheel
2. Release shaft seal
3. No. 2 bush
4. Return spring
5. Release shaft
6. No. 1 bush
7. Release bearing
8. Clutch cover
9. Clutch disc
10. Clutch cover bolt

Clutch component identification

1 Drive shaft joint (LH)
2 Pry tool

7923UG28

Disconnecting the left inboard joint

2. Remove or disconnect the following:

- Negative battery cable
- Undo the sealer on the halfshaft nut
- Halfshaft nut and washer

3. Drain the oil from the transmission.

4. On the left side shaft, use 2 large prybars to release the snapring fitting on the halfshaft inner joint from the differential.

5. On the right side shaft, use a plastic hammer to drive the shaft joint to release the snapring fitting on the halfshaft inner joint from the differential.

6. Remove or disconnect the following:

- Sway bar attaching nut, washer and bushing from the suspension arm
- Ball joint stud bolt and nut
- Inboard joint from the differential by pulling it
- Outer joint from the steering knuckle
- Halfshaft from the vehicle

✵✵ WARNING

To prevent breakage of the boots, be careful not to bring the boots in contact with other components when removing the shaft assembly.

7. If the center shaft requires service, drain the transmission oil and remove the support bolts. Remove the center shaft from the differential, then from the vehicle.

To install:

8. If the center shaft was removed, install the shaft into the differential, then install the support bolts. Tighten the support bolts to 29–43 ft. lbs. (40–60 Nm).

9. Clean the grease seal on the steering knuckle and apply a small amount of fresh grease.

10. Install or connect the following:

- Wheel side joint on the steering knuckle, then the differential side joint on the differential. Seat the differential joint by hand, making sure that the snapring is seated.
- Halfshaft nut on the outer joint loosely, to hold it in position.

✵✵ WARNING

Do not hit the joints with a hammer to seat them. Use hand pressure only or component damage may occur.

- Suspension arm to the steering knuckle
- Ball joint stud
- Sway bar on the suspension arm
- Sway bar bushing, washer and nut. Tighten the nut to 17–20 ft. lbs. (23–28 Nm).

11. If equipped with a manual transaxle fill the transaxle with the specified gear oil.

12. Tighten the halfshaft nut to 109–145 ft. lbs. (150–200 Nm).

13. Connect the negative battery cable.

14. If equipped with an automatic transaxle, fill the transaxle with the specified transmission oil.

CV-Joints

OVERHAUL

Double Offset Joint (DOJ) Type

The Double Offset Joint (DOJ) type is identified by the outside shape of the differential side joint which has no dent.

1. Before servicing the vehicle, refer to the precautions in the beginning of this section.

2. Remove the driveshaft.

3. Remove the boot band from the differential side joint.

4. Slide the boot towards the center of the shaft and remove the snapring from the outer race.

5. Remove the shaft from the outer race.

6. Use a rag to clean off the grease, then remove the circlip that retains the cage using snapring pliers.

7. Using a suitable puller, draw the cage away and remove the boot from the shaft.

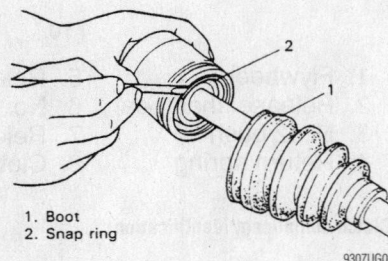

1. Boot
2. Snap ring

9307UG03

Remove the snapring from the outer race—DOJ type

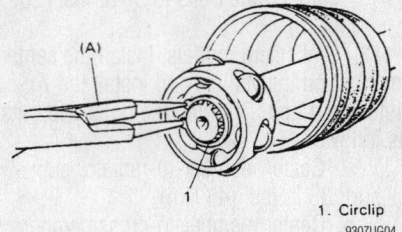

1. Circlip

9307UG04

Remove the circlip that retains the cage—DOJ type

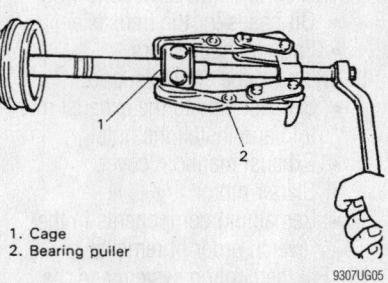

1. Cage
2. Bearing puller

9307UG05

Draw the cage away using a suitable puller—DOJ type

8. Inspect all components for wear and/or damage and replace as necessary.

To install:

9. Clean all components and allow them to completely dry.

10. Install the boot onto the shaft until its small diameter side fits to the shaft groove and attach it there with a boot band.

11. Using a pipe whose inner diameter is 0.906 in. (23mm) or more and outer diameter is 1.260 in. (32mm) or less, drive the cage into position. Install the cage using the smaller outside diameter side to the shaft end.

12. Install the circlip.

13. Apply grease to the entire surface of the cage.

14. Insert the cage into the outer race and fit the snapring into the groove of the outer race.

➡**Position the opening of the snapring so that it will not be lined up with a ball.**

15. Apply grease to the inside of the outer race and fit the boot to the outer race.

16. After fitting the boot, insert a screwdriver into the boot on the outer race side and allow air to enter the boot so that the air pressure in the boot equals atmospheric pressure

17. Fix the boot to the outer race with a boot band. Before tightening the band adjust the boot so that measurement D which is 8.10 in. (205.8mm) and measure-

"d" ⎫ Dimensions to use when fixing
"e" ⎭ boot with boot band.

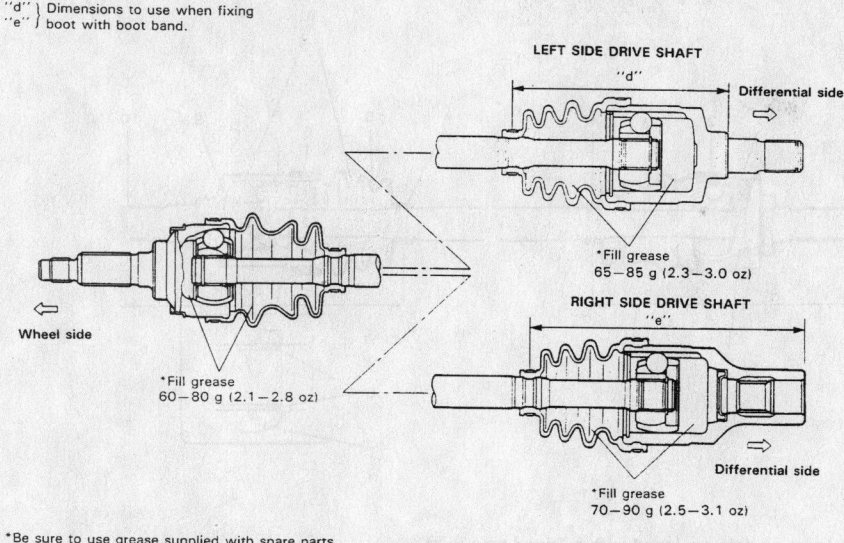

LEFT SIDE DRIVE SHAFT

"d"

Differential side ⟹

*Fill grease
65—85 g (2.3—3.0 oz)

RIGHT SIDE DRIVE SHAFT

"e"

Differential side ⟹

*Fill grease
70—90 g (2.5—3.1 oz)

Wheel side ⟸

*Fill grease
60—80 g (2.1—2.8 oz)

*Be sure to use grease supplied with spare parts.

9307UG06

Adjust the boot so that measurements D and E are as illustrated—DOJ type

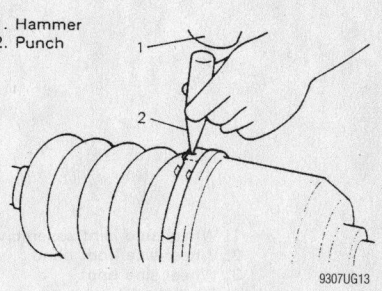

1. Hammer
2. Punch

9307UG13

Using a punch, caulk the center of the band folded over the fixture—Esteem models

ment E which is 7.34 in. (186.4mm) are as shown in the accompanying illustration.

18. Install the driveshaft in the vehicle.

Tripod Joint Type

The Tripod joint type can be identified by the 3 dent lines on the outside of the differential side joint.

1. Before servicing the vehicle, refer to the precautions in the beginning of this section.

2. Remove the driveshaft.

3. Remove the boot band and the Tripod joint housing.

4. Use a rag to clean off the grease, then remove the circlip using snapring pliers.

5. Remove the spider by using a suitable puller.

6. Remove the boot band and pull the differential side boot from the shaft.

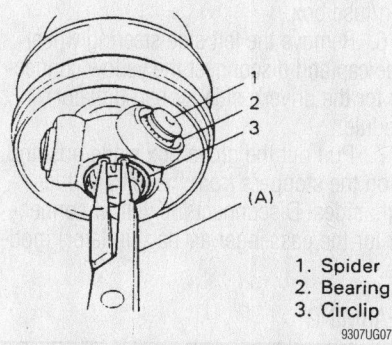

1
2
3

(A)

1. Spider
2. Bearing
3. Circlip

9307UG07

Remove the circlip that retains the spider assembly—Tripod type

7. Remove the dynamic damper band, then pull the damper through the shaft, if equipped.

8. Remove the boot bands from the wheel side joint boot and pull the boot through the shaft.

9. Inspect all components for wear and/or damage and replace as necessary.

To install:

10. Clean all components and allow them to completely dry.

11. On Swift models, perform the following:

a. Apply grease to the wheel side joint. Use black grease in the tube included with the wheel side boot kit.

b. Install the wheel side boot on the shaft.

c. Fill the inside of the boot with grease and fasten the boot with a band.

d. Install the dynamic damper on the shaft, if equipped.

e. Install the differential side boot on the shaft.

f. Apply grease to the Tripod joint. Use the yellow grease in the tube included in the differential side joint kit.

g. Using a pipe whose inner diameter is 0.906 in. (23mm) or more and outer diameter is 1.260 in. (32mm) or less, drive the spider into position. Face the chamfered side inward and fasten it in place with the circlip.

h. Fill the boot with grease, then install the housing and joint it with the boot.

i. Fasten the boot bands.

12. On Esteem models, perform the following:

a. Apply grease to the wheel side joint. Use black grease in the tube included with the wheel side boot kit.

b. Install the wheel side boot on the shaft.

c. Fill the inside of the boot with grease.

d. Fit the boot band into the groove in the boot.

e. Tighten the boot band until its outer diameter is 3.05 in. (77.5mm).

f. Fold the boot band over the metal fixture.

g. Using a punch, caulk the center of the band folded over the fixture.

h. Cut the band about 0.28 in. (7mm) from the clip.

i. Clamp the band with the clip.

j. Fix the small diameter side of the boot band into the groove in the boot.

k. Tighten the boot band until its outer diameter is 1.14 in. (29mm).

l. Fold the boot band over the metal fixture.

m. Using a punch, caulk the center of the band folded over the fixture.

n. Install the dynamic damper on the right side driveshaft.

o. Install the differential side boot on the shaft.

p. Apply grease to the Tripod joint. Use the yellow grease in the tube included in the differential side joint kit.

q. Install the spider into position. Face the chamfered side inward and fasten it in place with the circlip.

r. Fill the boot with grease, then install the housing and joint it with the boot.

s. Adjust the boot so that the when measured from the tip of the driveshaft (wheels side), to the inner (small) band; the measurement is 8 in. (204mm).

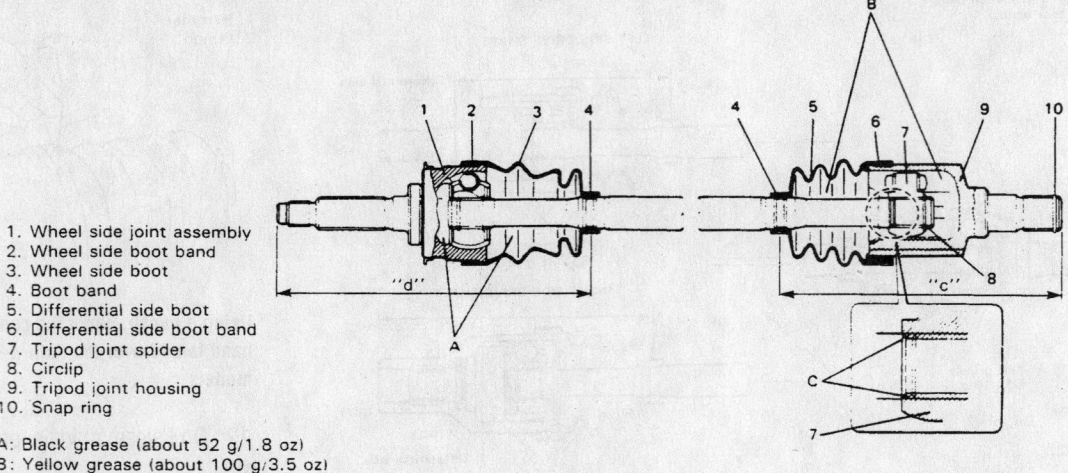

1. Wheel side joint assembly
2. Wheel side boot band
3. Wheel side boot
4. Boot band
5. Differential side boot
6. Differential side boot band
7. Tripod joint spider
8. Circlip
9. Tripod joint housing
10. Snap ring

A: Black grease (about 52 g/1.8 oz)
B: Yellow grease (about 100 g/3.5 oz)
C: Chamfered spline

9307UG14

Adjust the boot so that measurement D is correct—Esteem models equipped with a Tripod type joint

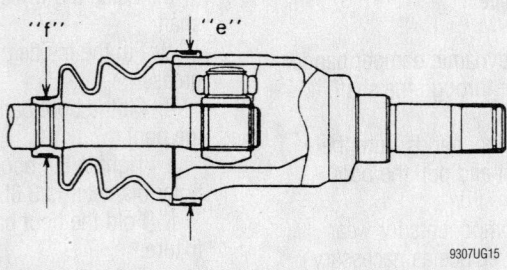

9307UG15

When tightening the bands, make sure that diameters E and F are correct—Esteem models equipped with a Tripod type joint

13. Insert a screwdriver into the boot on the outer race side and allow air to enter the boot so that the air pressure in the boot equals atmospheric pressure

a. Fasten the boot band so that diameter E which measures 3.11 in. (79mm) and diameter F which measures 1.14 in. (29mm) are correct. Refer to the accompanying illustration for the diameter locations.

14. Install the driveshaft.

STEERING AND SUSPENSION

Air Bag

✳✳ CAUTION

Most vehicles are equipped with an air bag system. The system must be disabled before performing service on or around system components, steering column, instrument panel components, wiring and sensors. Failure to follow safety and disabling procedures could result in accidental air bag deployment, possible personal injury and unnecessary system repairs.

PRECAUTIONS

Several precautions must be observed when handling the inflator module to avoid accidental deployment and possible personal injury.

• Never carry the inflator module by the wires or connector on the underside of the module.

• When carrying a live inflator module, hold securely with both hands, and ensure that the bag and trim cover are pointed away.

• Place the inflator module on a bench or other surface with the bag and trim cover facing up.

• With the inflator module on the bench, never place anything on or close to the module that may be thrown in the event of an accidental deployment.

DISARMING

✳✳ WARNING

When performing service on or around the air bag system components or wiring, disable the air bag system. Failure to follow the procedures could result in possible deployment, personal injury or unneeded system repairs.

1. Before servicing the vehicle, refer to the precautions in the beginning of this section.

2. Disconnect the negative battery cable.

3. Turn the steering wheel so the wheels are pointing straight ahead.

4. Turn the ignition switch to the **LOCK** position and remove the key.

5. Remove the **AIR BAG-IG** fuse from the air bag fuse box located near the junction/fuse box.

6. Remove the left side steering wheel side cap and disconnect the yellow connector for the driver's side air bag (inflator) module.

7. Pull out the glove box while pushing in on the stoppers from the left and the right sides. Disconnect the yellow connector for the passenger air bag (inflator) module.

REARMING

✳✳ WARNING

When performing service on or around the air bag system components or wiring, disable the air bag

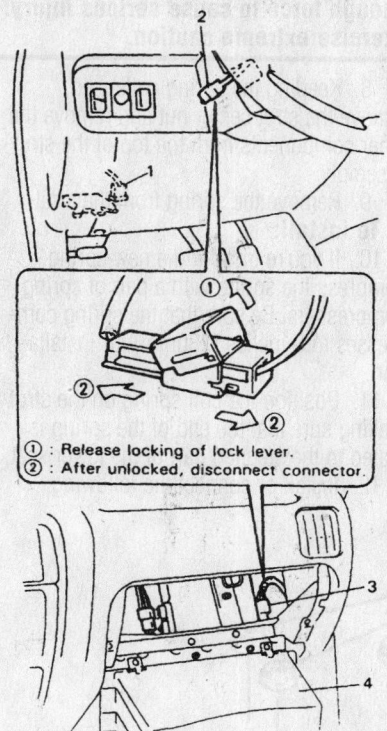

1 Air bag fuse box
2 Yellow connector of driver air bag (inflator)
   module
3 Yellow connectors of passenger air bag
   (inflator) module
4 Glove box

7923UG29

Air bag connector locations

system. Failure to follow the procedures could result in possible deployment, personal injury or unneeded
system repairs.

1. Before servicing the vehicle, refer to
the precautions in the beginning of this section.
2. Connect the negative battery cable.
3. Turn the ignition switch to the **LOCK**
position and remove the key.
4. Connect the yellow connector for the
passenger side air bag (inflator) module and
the yellow connector for the driver's side air
bag (inflator) module. Be sure to lock each
connector with the lock lever.
5. Install the glove box assembly.
6. Install the left side steering wheel
side cover.
7. Install the **AIR BAG-IG** fuse in the
air bag fuse box.
8. Turn the ignition **ON** and verify that
the **AIR BAG** warning lamp flashes 7
times, then turns off. If the system does not
operate as described, diagnosis and repairs
to the air bag system are necessary.

Rack and Pinion Steering Gear

REMOVAL & INSTALLATION

Manual

❊❊ WARNING

Be sure to set the front wheels
straight ahead and remove the ignition key from the cylinder before
starting repairs. If equipped with an
air bag, the contact coil of the air bag
system may be damaged if the key is
not removed and the wheels are not
straight ahead.

1. Before servicing the vehicle, refer to
the precautions in the beginning of this sec-
tion.
2. Disconnect the negative battery cable.
3. Slide the driver's seat back as far as
possible.
4. Remove or disconnect the following:
  • Pull back the front part of the floor
    mat on the driver's side
  • Steering shaft joint cover
  • Steering shaft upper joint bolt,
    loosen but do not remove it
  • Steering shaft lower joint bolt
  • Lower joint from the pinion
  • Front wheels
  • Tie rod ends from the steering
    knuckles
  • Steering gear mounting bolts and
    the brackets
  • Steering gear case from the vehicle

To install:
5. Install or connect the following:
  • Steering gear, brackets and mount-
    ing bolts. Tighten bolts to 14–21 ft.
    lbs. (20–30 Nm).
  • Tie rod ends to the steering knuck-
    les
6. Be sure the steering wheel is in the
straight-ahead position and the front wheels
are pointing straight ahead.
7. Install or connect the following:
  • Steering shaft to the steering gear
  • Lower steering shaft-to-steering
    gear clinch bolt and tighten both
    steering joint bolts (upper and
    lower) to 14–21 ft. lbs. (20–30
    Nm)
  • Remaining components in the
    reverse order of removal
  • Negative battery cable
8. Check and adjust the front wheel
alignment.

Power

❊❊ WARNING

Be sure to set the front wheels
straight ahead and remove the ignition key from the cylinder before
starting repairs. If equipped with an
air bag the contact coil of the air bag
system may be damaged if the key is
not removed and the wheels are not
straight ahead.

1. Before servicing the vehicle, refer to
the precautions in the beginning of this sec-
tion.
2. Remove or disconnect the following:
  • Negative battery cable
  • Steering column joint covers
  • Steering shaft upper joint bolt,
    loosen but do not remove
  • Steering shaft lower joint bolt
  • Lower joint from the pinion
  • Front wheels
  • Tie rod ends from the steering
    knuckles
  • Front exhaust pipe, on Swift mod-
    els
  • Engine rear torque rod with the
    torque rod and bracket, Swift mod-
    els equipped with an automatic
    transaxle
  • Gearshift control shaft and exten-
    sion rod from the transaxle, if
    equipped with a manual transaxle
  • Rear engine mount together with the
    bracket from the engine and sus-
    pension member, Esteem models
  • Mounting member from the sus-
    pension frame by removing the 2
    bolts, Esteem models
  • High and low pressure lines from
    the rack and pinion

➡**When the lines are disconnected
plug the lines or place an oil pan under
the vehicle.**

  • Cylinder lines from the rack and
    pinion
  • Rack and pinion mounting bolts
    and brackets
  • Rack and pinion case from the
    vehicle

To install:
3. Install or connect the following:
  • Rack and pinion, brackets and
    mounting bolts. Tighten the bolts to
    40 ft. lbs. (55 Nm) on Esteem mod-
    els or 22–28 ft. lbs. (30–40 Nm)
    on Swift models.

- Cylinder lines on the rack and pinion and tighten their fittings to 14–21 ft. lbs. (20–30 Nm)
- High and low pressure lines to the rack and pinion. Tighten the fittings to 22–28 ft. lbs. (30–40 Nm).
- Gearshift control shaft and extension rod to the transaxle, if equipped with a manual transaxle. Tighten the extension rod nut to 18–28 ft. lbs. (25–40 Nm) and tighten the control shaft nut and bolt to 11–14 ft. lbs. (15–20 Nm).
- Engine rear torque rod with the torque rod and bracket, Swift models equipped with an automatic transaxle
- Tie rod ends to the steering knuckles

4. Be sure the steering wheel is straight and the front wheels are pointing straight ahead.

- Steering shaft to the rack and pinion
- Lower steering shaft-to-rack and pinion clinch bolt and tighten both steering joint bolts (upper and lower) to 14–21 ft. lbs. (20–30 Nm)
- Front wheels
- Negative battery cable

5. Bleed the power steering system.
6. Lower the vehicle.
7. Check and adjust the front wheel alignment.

Strut

REMOVAL & INSTALLATION

Front

1. Before servicing the vehicle, refer to the precautions in the beginning of this section.
2. Remove or disconnect the following:
- Wheels
- Anti-Lock Brake (ABS) system wheel speed sensor, if equipped
- Brake hose clip, then the hose from the strut

3. Support the lower control arm with a floor jack.
4. Remove the strut bracket bolts.
5. Hold the strut by hand so it will not fall, and remove the upper strut support nuts from the engine compartment.

✳✳ WARNING

Do not loosen the center nut at this time or serious injury or vehicle damage may result.

6. Remove the strut assembly from the vehicle.
7. Install a pair of coil spring compressors on the coil spring on the strut assembly. Turn the spring compressors alternately until the spring tension is released from the strut assembly. If the spring can be turned slightly, then it has been collapsed enough.

✳✳ CAUTION

This procedure requires the use of a spring compressor. It cannot be performed without one. If you do not have access to this special tool, do not attempt to disassemble the strut. The coil spring is retained under considerable pressure. It can exert

enough force to cause serious injury. Exercise extreme caution.

8. Keeping the spring collapsed, remove the strut center nut and remove the other components from the top of the strut assembly.
9. Remove the spring from the strut.
To install:
10. If you're installing a new spring, compress the spring with a pair of spring compressors. Be sure that the spring compresses to 9 inches (230mm) for installation.
11. Position the coil spring on the strut making sure that the end of the spring is mated to the stepped part of the lower seat.
12. Install or connect the following:

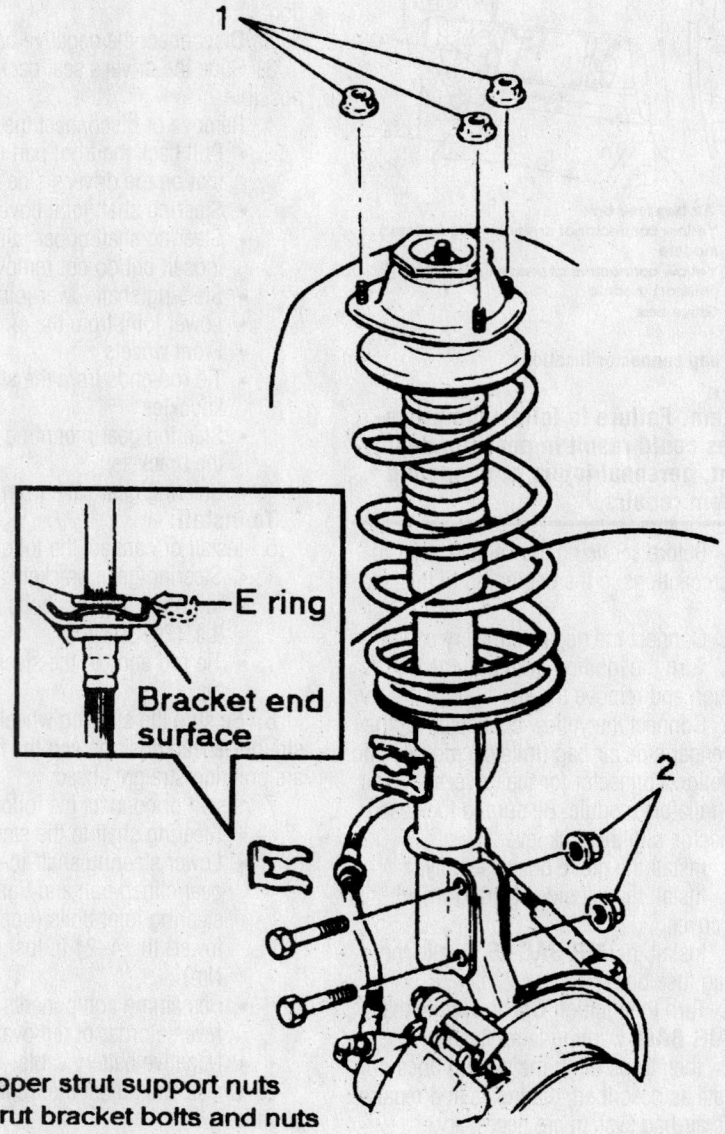

1. Upper strut support nuts
2. Strut bracket bolts and nuts

Strut assembly mounting

7923UG30

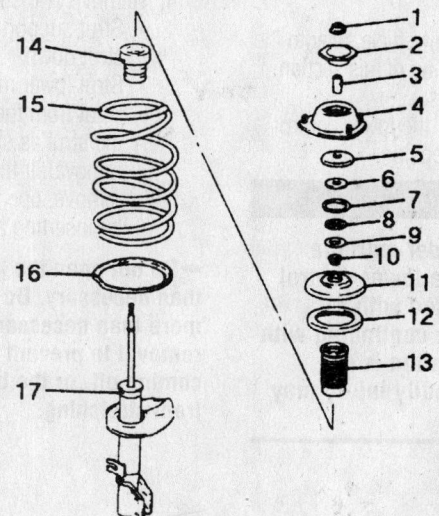

1. Nut
2. Stopper
3. Inner spacer
4. Support comp.
5. Bearing seat
6. Bearing upper washer
7. Bearing seal
8. Bearing
9. Bearing lower washer
10. Bearing spacer
11. Coil spring upper seat
12. Coil spring seat
13. Strut cover
14. Bump stopper
15. Coil spring
16. Coil spring lower seat
17. Strut

7923UG31

Exploded view of the strut assembly

- Bump stop on the strut rod
- Strut cover
- Spring seat
- Upper spring seat
- Bearing spacer. Align the strut bracket with the mark on the upper spring seat.

13. Clean the bearing lower washer and install it on the strut rod.

14. Clean the strut bearing and apply fresh grease to the bearing. Install the bearing on the lower washer.

15. Clean the bearing upper washer and install it.

16. Install these components in the following order:

- Bearing upper seal
- Bearing seat
- Strut support
- Inner spacer
- Washer
- Strut nut. Tighten the nut to 29–43 ft. lbs. (40–60 Nm). Apply a waterproof coating (paint or lacquer) to the nut and strut rod threads.

17. Loosen the spring compressors alternately, checking that the stepped part of the spring seat and spring are properly positioned.

18. Install or connect the following:

- Strut assembly onto the vehicle
- Upper support nuts loosely. Tighten the upper strut support nuts to 16–23 ft. lbs. (22–33 Nm) on Swift models or 20 ft. lbs. (28 Nm) on Esteem models.
- Strut bracket nuts and bolts and tighten to 51–65 ft. lbs. (70–90 Nm)
- ABS wheel speed sensor, if equipped
- Brake hose clip
- Wheels

19. Check the front end alignment.

Rear

ESTEEM

1. Before servicing the vehicle, refer to the precautions in the beginning of this section.

2. Remove wheel and tire assembly.

3. Place a jack under the lower control arm to support the suspension.

4. Remove or disconnect the following:

- Brake line from the brake hose at the strut
- E-ring securing the brake hose to the strut

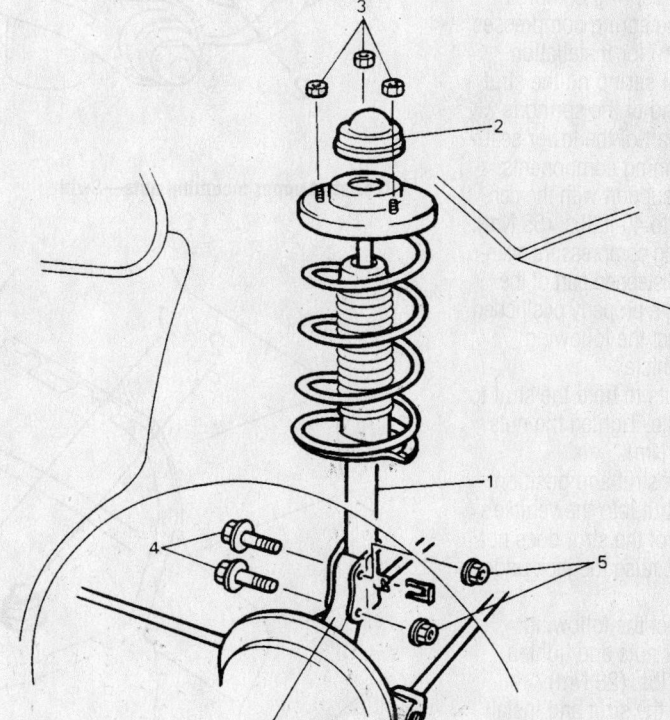

1. Strut assembly
2. Strut upper cap
3. Strut support nut
4. Strut bracket bolt
5. Strut bracket nut
6. Rear knuckle

7923UG32

Rear strut assembly mounting—Esteem

- Strut upper support nuts and push the strut down
- 2 bolts and nuts holding the strut to the rear knuckle
- Strut from the knuckle
- Strut from the vehicle

5. Install a pair of coil spring compressors on the coil spring on the strut assembly. Turn the spring compressors alternately until the spring tension is released from the strut assembly. If the spring can be turned slightly, then it has been collapsed enough.

※※ CAUTION

This procedure requires the use of a spring compressor. It cannot be performed without one. If you do not access to this special tool, do not attempt to disassemble the strut. The coil spring is retained under considerable pressure. Id can exert enough force to cause serious injury. Exercise extreme caution.

6. Keeping the spring collapsed, remove the strut center nut and remove the other components from the top of the strut assembly.

7. Remove the spring from the strut.

To install:

8. If installing a new spring, compress the spring with a pair of spring compressors. Make sure that the spring compresses to 11½ inches (290mm) for installation.

9. Position the coil spring on the strut making sure that the end of the spring is mated to the stepped part of the lower seat.

10. Install the remaining components.

11. Install the strut support with the center nut. Tighten the nut to 40 ft. lbs. (55 Nm).

12. Loosen the spring compressors alternately, checking that the stepped part of the spring seat and spring are properly positioned.

13. Install or connect the following:
- Strut in the vehicle
- 2 bolts and nuts to hold the strut to the rear knuckle. Tighten the nuts 65 ft. lbs. (90 Nm).

14. Fully extend the strut and position the upper part of the strut into the vehicle's body. If the upper part of the strut does not reach the vehicle body, raise the jack under the control arm a little.

15. Install or connect the following:
- Upper support nuts and tighten them to 20 ft. lbs. (28 Nm)
- Brake hose to the strut and install the E-clip
- Brake hose to the brake line
- Wheels

16. Fill the master cylinder with brake fluid and bleed the brake system.

SWIFT

1. Before servicing the vehicle, refer to the precautions in the beginning of this section.

2. Remove the wheels.

3. Place a jack under the lower control arm to support the suspension.

※※ CAUTION

The coil spring is under extreme pressure. Be sure the lower control arm is firmly supported with a hydraulic jack before continuing with procedure. If this caution is not observed, serious bodily injury may result.

4. Remove or disconnect the following:
- Strut support nuts and push the strut down
- Strut lower mounting bolt
- Strut from the knuckle. Compress the strut as short as possible for removal. If the strut is hard to remove, open the slit of the knuckle by inserting a wedge.

→Do not open the knuckle slit wider than necessary. Do not lower the jack more than necessary during the strut removal to prevent the coil spring from coming off, or the brake flexible hose from stretching.

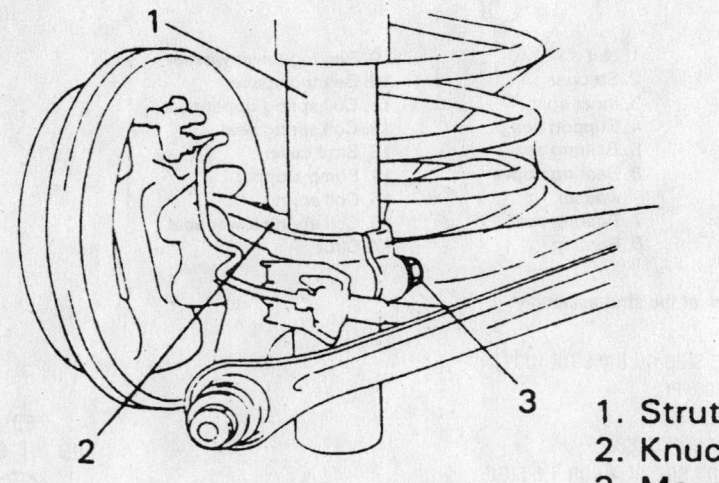

1. Strut
2. Knuckle
3. Mount bolt

7923UG33

Rear strut upper mounting nuts—Swift

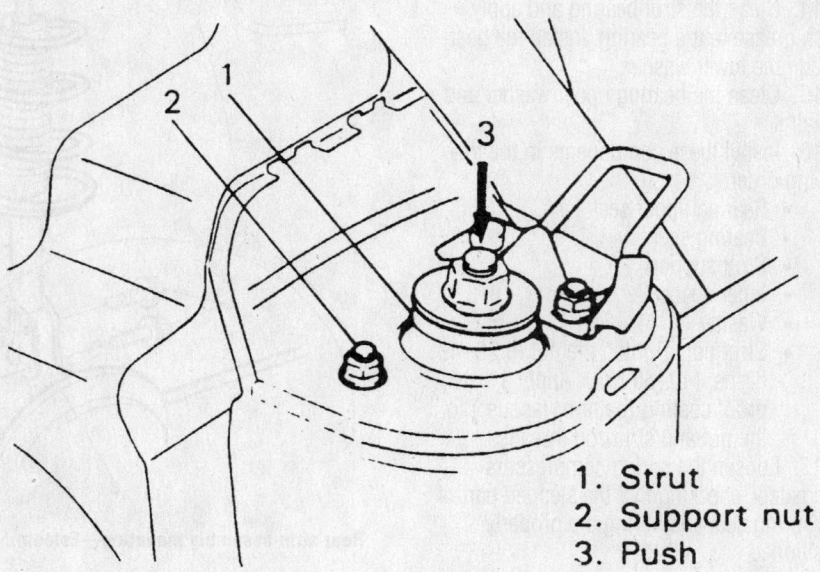

1. Strut
2. Support nut
3. Push

7923UG34

Rear strut lower mounting—Swift

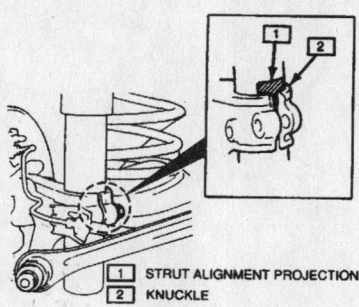

STRUT ALIGNMENT PROJECTION

KNUCKLE

7923UG35

Shock-to-steering knuckle alignment—Swift

To install:

5. Install or connect the following:
   - Strut in the vehicle. Position the bottom of the alignment projection inside the knuckle opening.
   - Strut lower mounting bolt and tighten it to 36–50 ft. lbs. (50–70 Nm)
6. Fully extend the strut and position the upper part of the strut into the vehicle's body. If the upper part of the strut does not reach the vehicle's body, raise the jack under the control arm a little.
7. Install or connect the following:
   - Upper support nuts and tighten them to 20–27 ft. lbs. (28–38 Nm)
   - Wheels

Coil Spring

REMOVAL & INSTALLATION

➡For coil spring service on the esteem, refer to the strut removal and installation procedure.

Swift

1. Before servicing the vehicle, refer to the precautions in the beginning of this section.
2. Remove or disconnect the following:
   - Rear wheels

➡To facilitate the toe-in adjustment after reinstallation, confirm which one of the lines stamped on the washer is in the closest alignment with the stamped line on the control rod. If not marked, add matchmarks.

   - Control rod inside bolt, body center side
   - Outside (wheel side) of the control rod from the rear knuckle stud
   - Control rod from the knuckle

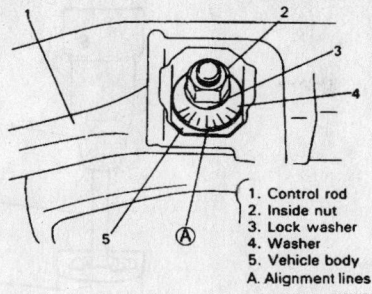

1. Control rod
2. Inside nut
3. Lock washer
4. Washer
5. Vehicle body
A. Alignment lines

7923UG36

Control rod inside mount—Swift

- Nuts, washers, and bushings connecting the rear sway bar to the rear lower control arms
- Rear mount nut on the control arm, loosen but do not remove the bolt
- Front nut of the control arm, loosen only at this time
- Wheel speed sensor, if equipped with Anti-Lock Brakes (ABS)

❋❋ CAUTION

The coil spring is under extreme pressure. Be sure that the control arm is firmly supported with a hydraulic jack before continuing with procedure. If this precaution is not observed, serious bodily injury may result.

3. Loosen the lower mount nut on the knuckle. Place a jack under the control arm

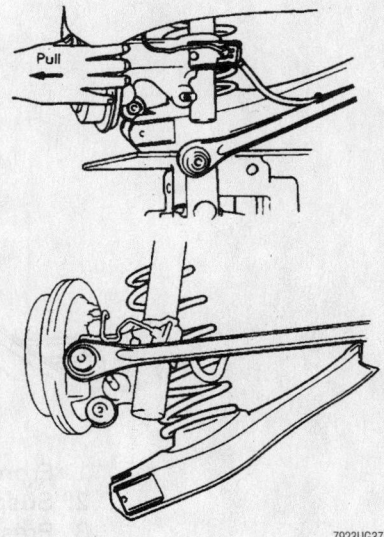

Pull

7923UG37

Disconnecting the knuckle and removing the coil spring

to prevent it from lowering and remove the lower mount nut on the knuckle.

4. Raise the jack placed under the control arm enough to allow the removal of the lower mount bolt of the knuckle.
5. Move the brake drum/backing plate toward the outside of the vehicle body so as to separate the lower mount of the knuckle from the control arm. Then, lower the jack gradually and remove the coil spring.

To install:

6. Place the jack under the control arm.
7. Install or connect the following:
   - Coil spring on the spring seat of the control arm. Raise the control arm. When seating the coil spring, mate the spring end with the stepped part of the control arm.
   - Lower knuckle mount bolt. Tighten the bolt to 29–33 ft. lbs. (40–45 Nm)
8. Remove the jack from under the suspension arm.
9. Install or connect the following:
   - Rear sway bar joints to the rear control arms and install the bushings, washers, and nuts. Tighten the nuts to 16–20 ft. lbs. (22–28 Nm).
   - Control rod
   - Inside control rod bolt and the outside control rod nut, but do not tighten them at this time
   - Wheel speed sensor, if equipped
   - Wheels and lower the vehicle
10. Tighten the control rod inside and outside nuts to 51–65 ft. lbs. (70–90 Nm).

➡**When tightening the nuts, the vehicle should be off the hoist and in a non-loaded state. Also when tightening the inside nut, align the line stamped on the body with the line on the washer as confirmed before removal or align the matchmarks if marked.**

11. Tighten the suspension arm front nuts to 36–50 ft. lbs. (50–70 Nm) and the rear nuts to 29–33 ft. lbs. (40–45 Nm). After tightening the suspension arm front nut, be sure that the washer is not tilted.
12. Check the rear wheel alignment.

Lower Ball Joint

REMOVAL & INSTALLATION

The lower ball joint is an integral part of the lower control arm assembly. If the ball joint is found to be defective the whole lower control arm assembly must be replaced.

Timing belt service is covered in Section 3 of this manual

Lower Control Arm

REMOVAL & INSTALLATION

The lower control arm and ball joint are a complete unit that will not separate.

1. Before servicing the vehicle, refer to the precautions in the beginning of this section.

2. Remove or disconnect the following:
   - Front wheels
   - Sway bar link nut, washer, and cushion
   - Ball joint stud bolt and nut
   - Lower control arm front bushing bolt
   - 2 bolts holding the lower control arm bracket to the vehicle
   - Lower control arm from the steering knuckle
   - Lower control arm from the vehicle

To install:

3. Install or connect the following:
   - Ball joint on the knuckle and secure it with the nut and bolt
   - Lower control arm rear bracket and bolts
   - Lower control arm front mounting bolt
   - Sway bar link to the control arm and install the cushion, washer and nut

4. Tighten the retainers to their proper torque specifications as follows:

   a. Control arm rear mounting bracket bolts: 27 ft. lbs. (37 Nm).

   b. Front mounting bolt: 65 ft. lbs. (90 Nm).

   c. Ball joint nut and bolt: 44 ft. lbs. (60 Nm).

   d. Sway bar link nut: 18 ft. lbs. (26 Nm).

5. Install the wheels.

6. Check the front wheel alignment.

CONTROL ARM BUSHING REPLACEMENT

1. Before servicing the vehicle, refer to the precautions in the beginning of this section.

2. Remove the lower control arm.

3. Use a hydraulic press to remove the rear bushing.

4. Cut the flange from the front bushing.

5. Use a hydraulic press to remove the front bushing.

To install:

6. Apply a solution of soapy water to the outer diameter of the front bushing, this will aid in installation.

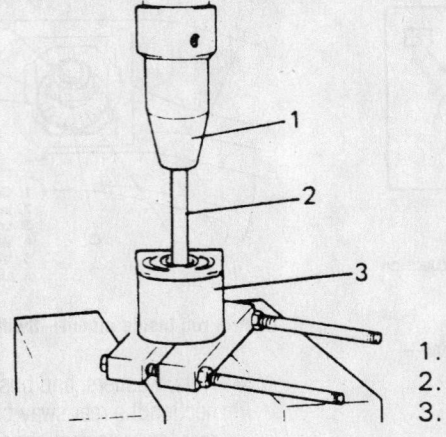

1. Press
2. Rod
3. Rear bushing

9307UG08

Use a suitable hydraulic press to remove the rear lower control arm bushing

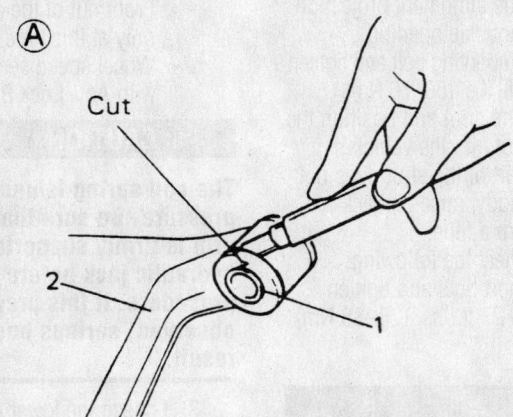

Ⓐ

Cut

2

1

Ⓑ

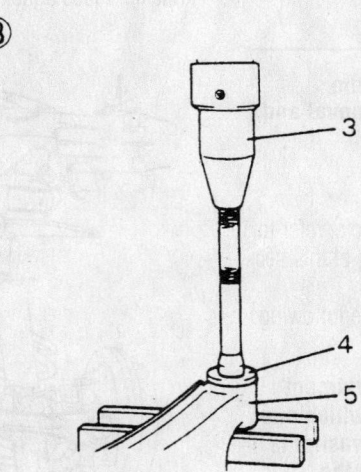

3

4

5

1. Front bushing
2. Suspension arm
3. Press
4. Front bushing
5. Suspension arm

9307UG09

Cut the flange from the front bushing, then using a suitable hydraulic press; remove the rear lower control arm bushing

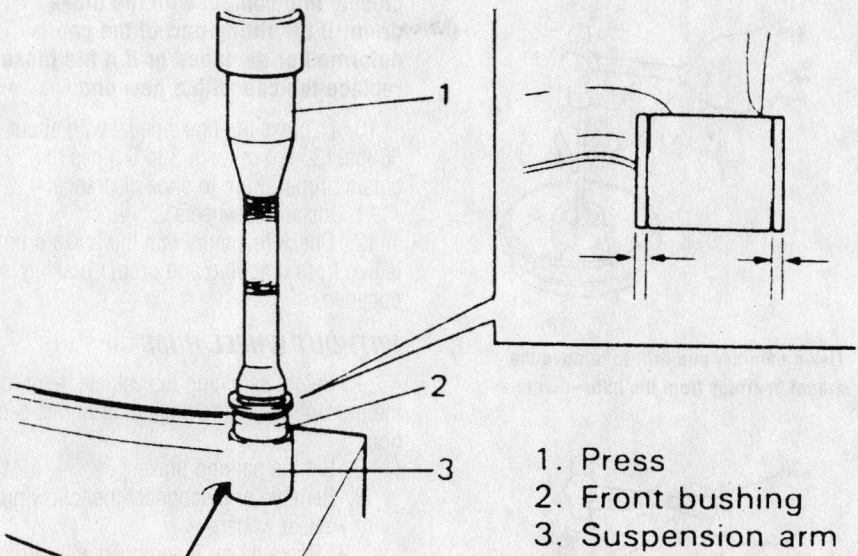

1. Press
2. Front bushing
3. Suspension arm

9307UG10

The front bushing should be positioned equally as shown after being pressed into position

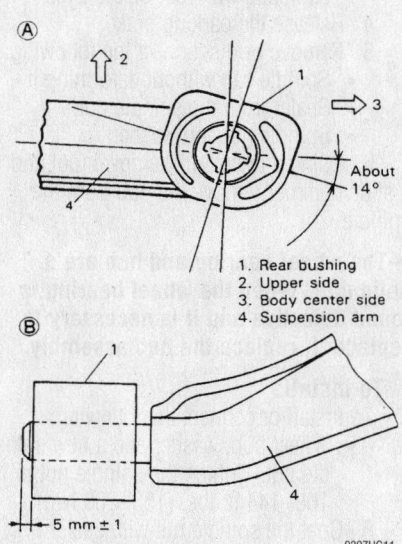

1. Rear bushing
2. Upper side
3. Body center side
4. Suspension arm

9307UG11

Install the rear bushing should be positioned as shown (A) before driving it into position (B)

7. Press the front bushing into its bore using a hydraulic press until the bushing is equal on the right and left of the arm as shown in the accompanying illustration.

8. Apply a solution of soapy water to the outer diameter of the rear bushing, this will aid in installation.

9. Install the rear bushing into the lower control arm in the direction and angle shown in figure A of the accompanying illustration.

10. Drive the bushing into position as shown in figure B of the accompanying illustration.

Wheel Bearings

ADJUSTMENT

The front and rear wheel bearings are a cartridge type design and cannot be adjusted.

REMOVAL & INSTALLATION

Front

➡**Always replace bearing races as a complete set.**

1. Before servicing the vehicle, refer to the precautions in the beginning of this section.

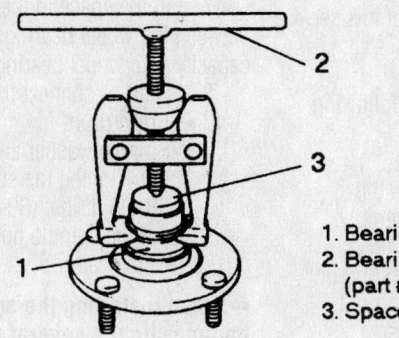

1. Bearing inner race
2. Bearing puller
   (part # 09913 - 61110)
3. Spacer to protect hub

7923UG38

Use a 2- or 3-jaw puller to remove the bearing race from the hub

2. Remove or disconnect the following:
   • Front wheel
   • Brake caliper, carrier and disc from the steering knuckle
   • Wheel speed sensor, if equipped with Anti-Lock Brakes (ABS)
   • Tie rod from the steering knuckle
   • Hub from the steering knuckle
   • Steering knuckle
   • Wheel bearing outside race from the hub using a suitable bearing puller
   • Outside oil seal, snapring, outside bearing, inside oil seal and the inside bearing in that order

➡**Once the bearing outer race is removed, the bearing set (outer race, bearings and inner races) should be replaced.**

   • Bearing outer race from the knuckle by pressing the race out of the knuckle

To install

➡**When installing the oil seals, be careful not to deform or tilt. Damage to the rubber part of the seal may occur.**

3. Install or connect the following:
   • New bearing outer race into the knuckle using a press and the following tools: Bearing Installer Handle 09924–74510, Bearing And Oil Seal Installer 09944–68210 and a Bearing Installer Support 0994–78210

4. Apply lithium grease to the bearing races, bearings and oil seal lips.

5. Install or connect the following:
   • Outside bearing on the steering knuckle
   • Snapring to hold the outside bearing in place, then the outside oil seal

Heater Core replacement is covered in Section 2 of this manual

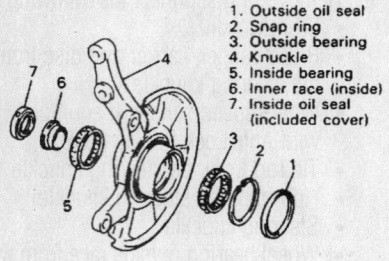

1. Outside oil seal
2. Snap ring
3. Outside bearing
4. Knuckle
5. Inside bearing
6. Inner race (inside)
7. Inside oil seal (included cover)

7923UG39

Steering knuckle bearing components—Swift

- Inside bearing to the steering knuckle
- Inside race and oil seal to the steering knuckle. When installing the inside oil seal, drive the oil seal in until it contacts the steering knuckle.

✴✴ CAUTION

If equipped with ABS use caution when installing the oil seal, because the seal has a hole that must align with the speed sensor position.

- Outside race to the wheel hub using a bearing installer

6. Using a press and the proper tools, press the hub into the steering knuckle. After installation, be sure the hub is installed straight and turns freely.

7. Install or connect the following:
- Steering knuckle in the vehicle
- Tie rod end
- Brake caliper, carrier and disc on the steering knuckle
- Front wheel

Rear

WITH WHEEL HUBS

1. Before servicing the vehicle, refer to the precautions in the beginning of this section.
2. Set the parking brake.
3. Remove or disconnect the following:
- Rear wheels
- Rear brake drums
4. Use a brass drift and knock the wheel bearings from the drum assembly.

To install:

5. Position the inner wheel bearing on the drum with the sealed side facing out. Using a rear wheel Bearing Installer 09913–76010, install the rear wheel bear-

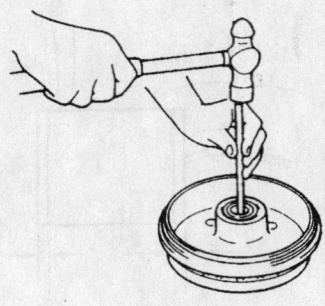

7923UG40

Use a hammer and drift to remove the wheel bearings from the hub—Swift

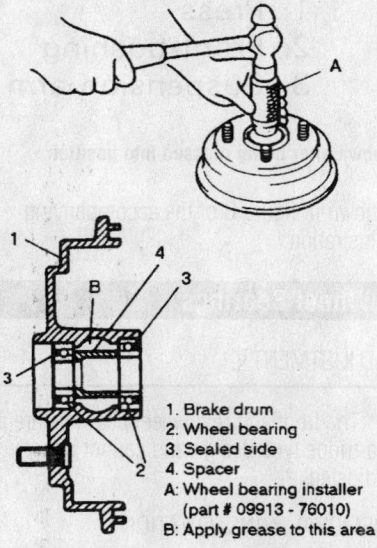

1. Brake drum
2. Wheel bearing
3. Sealed side
4. Spacer
A: Wheel bearing installer (part # 09913 - 76010)
B: Apply grease to this area

7923UG41

Wheel bearing installation—Swift

ing. Install the wheel bearing spacer into the drum.

6. Install or connect the following:
- Outer wheel bearing with the sealed side facing out, using a wheel bearing installer

7. Fill the space in the brake drum in between the wheel bearings to about 40% capacity with wheel bearing grease.

8. Install or connect the following:
- Brake drum
- Spindle washer and a new spindle nut. Tighten the spindle nut to 58–86 ft. lbs. (80–120 Nm).

9. Coat the spindle nut and the spindle dust cap with sealer.

➡**When installing the spindle cap, hammer lightly several times on the collar of the cap until the collar comes**

closely into contact with the brake drum. If the fitting part of the cap is deformed or damaged or if it fits loose, replace the cap with a new one.

10. Depress the brake pedal with about 66 lbs. (30 kg) of force 3 to 5 times to obtain proper drum to shoe clearance.
11. Install the wheels.
12. Check to ensure that the brake drum is free from dragging and proper braking is obtained.

WITHOUT WHEEL HUBS

1. Before servicing the vehicle, refer to the precautions in the beginning of this section.
2. Set the parking brake.
3. Remove or disconnect the following:
- Rear wheels
- Brake drum, if equipped with drum brakes
- Caliper, carrier and disc, if equipped with rear disc brakes
4. Release the parking brake.
5. Remove or disconnect the following:
- Spindle cap without deforming it
- Sealer from the spindle nut
- Spindle nut and washer
6. Using a brake hub removal tool and a slide hammer remove the hub from the spindle.

➡**The wheel bearing and hub are a solid unit. When the wheel bearing is found defective and it is necessary to replace it, replace the hub assembly.**

To install:

7. Install or connect the following:
- Wheel hub, washer and a new spindle nut. Tighten the spindle nut to 108–144 ft. lbs. (150–200 Nm).
8. Coat the spindle nut with sealer
9. Install or connect the following:
- Spindle cap
- Brake drums, if equipped with rear drum brakes
- Brake caliper carrier and disc, if equipped with rear disc brakes
10. Depress the brake pedal with about 66 lbs. (30 kg) of force 3 to 5 times to obtain proper drum/rotor to shoe/pad clearance.
11. Install the wheel and tighten the lug nuts to 36–58 ft. lbs. (50–80 Nm).
12. Check to ensure that the brakes are free from dragging and that proper braking is obtained.

PRECAUTIONS

Before servicing any vehicle, please be sure to read all of the following precautions, which deal with personal safety, prevention of component damage, and important points to take into consideration when servicing a motor vehicle:

• Never open, service or drain the radiator or cooling system when the engine is hot; serious burns can occur from the steam and hot coolant.

• Observe all applicable safety precautions when working around fuel. Whenever servicing the fuel system, always work in a well-ventilated area. Do not allow fuel spray or vapors to come in contact with a spark, open flame, or excessive heat (a hot drop light, for example). Keep a dry chemical fire extinguisher near the work area. Always keep fuel in a container specifically designed for fuel storage; also, always properly seal fuel containers to avoid the possibility of fire or explosion. Refer to the additional fuel system precautions later in this section.

• Fuel injection systems often remain pressurized, even after the engine has been turned **OFF**. The fuel system pressure must be relieved before disconnecting any fuel lines. Failure to do so may result in fire and/or personal injury.

• Brake fluid often contains polyglycol ethers and polyglycols. Avoid contact with the eyes and wash your hands thoroughly after handling brake fluid. If you do get brake fluid in your eyes, flush your eyes with clean, running water for 15 minutes. If eye irritation persists, or if you have taken brake fluid internally, IMMEDIATELY seek medical assistance.

• The EPA warns that prolonged contact with used engine oil may cause a number of skin disorders, including cancer! You should make every effort to minimize your exposure to used engine oil. Protective gloves should be worn when changing oil. Wash your hands and any other exposed skin areas as soon as possible after exposure to used engine oil. Soap and water, or waterless hand cleaner should be used.

• All new vehicles are now equipped with an air bag system. The system must be disabled before performing service on or around system components, steering column, instrument panel components, wiring and sensors. Failure to follow safety and disabling procedures could result in accidental air bag deployment, possible personal injury and unnecessary system repairs.

• Always wear safety goggles when working with, or around, the air bag system. When carrying a non-deployed air bag, be sure the bag and trim cover are pointed away from your body. When placing a non-deployed air bag on a work surface, always face the bag and trim cover upward, away from the surface. This will reduce the motion of the module if it is accidentally deployed. Refer to the additional air bag system precautions later in this section.

• Clean, high quality brake fluid from a sealed container is essential to the safe and proper operation of the brake system. You should always buy the correct type of brake fluid for your vehicle. If the brake fluid becomes contaminated, completely flush the system with new fluid. Never reuse any brake fluid. Any brake fluid that is removed from the system should be discarded. Also, do not allow any brake fluid to come in contact with a painted surface; it will damage the paint.

• Never operate the engine without the proper amount and type of engine oil; doing so WILL result in severe engine damage.

• Timing belt maintenance is extremely important! Many models utilize an interference-type, non-freewheeling engine. If the timing belt breaks, the valves in the cylinder head may strike the pistons, causing potentially serious (also time-consuming and expensive) engine damage. Refer to the maintenance interval charts in the front of this manual for the recommended replacement interval for the timing belt, and to the timing belt section for belt replacement and inspection.

• Disconnecting the negative battery cable on some vehicles may interfere with the functions of the on-board computer system(s) and may require the computer to undergo a relearning process once the negative battery cable is reconnected.

• When servicing drum brakes, only disassemble and assemble one side at a time, leaving the remaining side intact for reference.

ENGINE REPAIR

Distributor

REMOVAL

1. Before servicing the vehicle, refer to the precautions in the beginning of this section.
2. Remove or disconnect the following:
   • Negative battery cable. On vehicles equipped with an air bag, wait at least 90 seconds before proceeding.
   • Electrical connector from the air flow meter. Disconnect the air cleaner hose from the throttle body and remove the air cleaner cover, air flow meter and air duct as 1 unit.
   • Air intake and intercooler to provide clearance. Disconnect the air temperature sensor, cruise control actuator cable, and the air cleaner hose, if more clearance is necessary.
   • Spark plug wires from the distributor cap.
   • Distributor connector.
   • Distributor mounting bolt
   • O-ring from the distributor housing

➡ **The marks on the distributor drive gear and distributor housing should be aligned. If the marks are not aligned, mark the distributor housing and rotor position.**

INSTALLATION

Engine Disturbed

1. On the 2JZ-GE engine, remove the No. 3 timing belt cover.
2. Bring No. 1 cylinder to TDC
3. Install or connect the following:

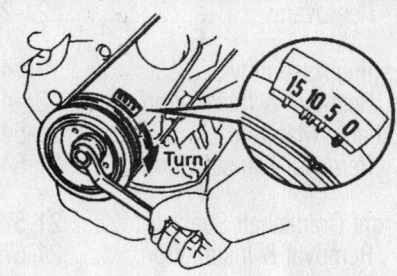

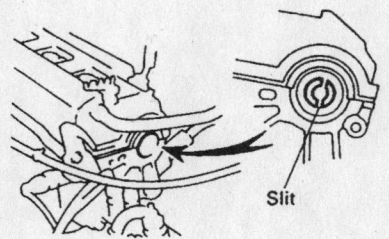

Common method of properly positioning the distributor drive mechanism to TDC

7923VG01

- O-ring
- Distributor. Tighten the mounting bolts
- Intercooler or the air intake
- Spark plug wires
- On the 2JZ-GE engine, reinstall the No. 3 timing belt cover
- Air flow meter, air hose and the air cleaner cover if removed
- Distributor and air flow meter connectors.
- Negative battery cable
- Check the ignition timing

Engine Not Disturbed

1. Install or connect the following:
   - O-ring
   - Distributor. Tighten the mounting bolts

➡ **Be sure the mark on the engine aligns with the mark on the distributor made during removal. Also be sure the rotor is in the same position as removal**

- Intercooler or the air intake
- Distributor cap
- Spark plug wires
- Air flow meter, air hose and the air cleaner cover
- Distributor and air flow meter connectors
- Negative battery cable
- Check the ignition timing.

Alternator

REMOVAL & INSTALLATION

Avalon, Camry

1. Before servicing the vehicle, refer to the precautions in the beginning of this section.
2. Remove or disconnect the following:
   - Negative battery cable. On models with an airbag, wait at least 90 seconds from the time that the ignition switch is turned to the LOCK position and the battery is disconnected before performing any further work.
   - On the 5S-FE, the two bolts and the No. 3 right hand engine mounting stay
   - On 1MZ-FE, the bolt and nut and then remove the No. 2 right hand engine mounting stay
   - Harness and wire (and nut) from the alternator

- Air cleaner, if necessary
- Drive belt from the pulley
- Alternator

To install:
3. Install or connect the following:
   - Alternator
   - Drive belt
   - Wiring
   - Negative and starter battery cables

Celica, GT-S, Supra

1. Before servicing the vehicle, refer to the precautions in the beginning of this section.
2. Remove or disconnect the following:
   - Negative battery cable

➡ **On some models it is necessary to remove the fuse/relay block located near the alternator.**

- Drive belt
- Wiring
- Alternator

To install:
3. Install or connect the following:
   - Alternator
   - Drive belt. Torque the pivot bolt and adjusting bolts: 5S-FE: Pivot bolt—40 ft. lbs. (54 Nm); Adjusting bolt—14 ft. lbs. (19 Nm).
   - Wiring
   - Negative battery cable

Corolla

1. Before servicing the vehicle, refer to the precautions in the beginning of this section.

➡ **It may be necessary to remove the gravel shield and work from underneath the car in order to gain access to the alternator retaining bolts.**

2. Remove or disconnect the following:
   - Negative battery cable

- Wiring from the alternator
- Drive belt
- Alternator

To install:
3. Install or connect the following:
   - Alternator. Torque the 14mm bolt to 18 ft. lbs. (24 Nm) and the 17 mm bolt to 40 ft. lbs. (54 Nm).
   - Alternator connector and wiring
   - Drive belt
   - Negative battery cable

Echo

1. Before servicing the vehicle, refer to the precautions in the beginning of this section.
2. Remove or disconnect the following:
   - Negative battery cable
   - Wire clamp from the rectifier
   - Alternator harness
   - 2 bolts
   - Alternator

To install:
3. Install or connect the following:
   - Alternator and hand tighten the bolts
   - Drive belt and adjust if necessary. Torque the 14 mm bolt to 14 ft. lbs. (18 Nm) and the 17 mm bolt to 40 ft. lbs. (54 Nm).
   - Alternator connector and clamp
   - Alternator electrical wiring
   - Negative battery cable

Ignition Timing

ADJUSTMENT

➡ **The timing on engines equipped with DIS is not adjustable.**

1. Before servicing the vehicle, refer to the precautions in the beginning of this section.

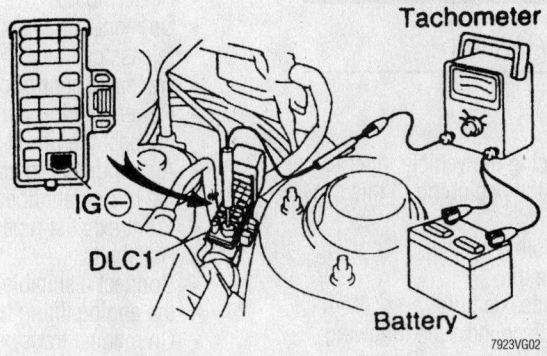

Attach the tachometer test probe to the IG terminal of the data link connector

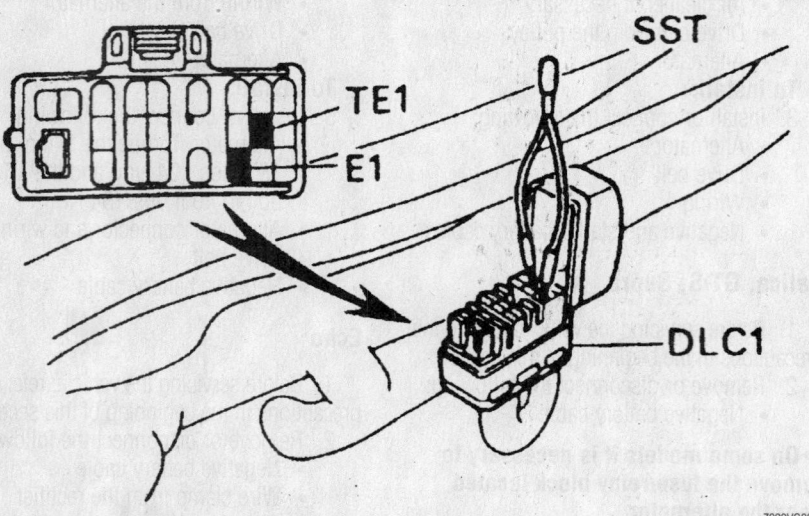

7923VG03

A jumper wire is used to connect terminals TE1 and E1 of the DLC

2. Perform the following:
   • Start the engine and let it reach operating temperature.
   • Connect the tachometer tester probe to terminal IG of the DLC1.

➡ **Never allow the tachometer test probe to touch ground as it could result in damage to the igniter and or ignition coil.**

➡ **Not all tachometers are compatible with this system. Be sure to confirm compatibility of your unit before use.**

   • Using a jumper connector, connect terminals TE1 and E1 of the DLC1.
   • Check the timing. Timing should be 8–12 degrees BTDC. Set the timing as needed.
   • Tighten the hold-down bolt.
   • Check that the ignition timing advances.

Engine Assembly

REMOVAL & INSTALLATION

5E-FE Engine

1. Before servicing the vehicle, refer to the precautions in the beginning of this section.
2. Drain the cooling system. Drain the engine oil.
3. Drain the transaxle fluid.
4. Remove or disconnect the following:
   • Negative battery cable. On vehicles equipped with an air bag, wait at least 90 seconds before proceeding.

   • Hood
   • Undercovers
   • Radiator
   • Accelerator cable
   • Throttle cable
   • Air cleaner assembly and bracket.
   • Charcoal canister
   • Fuel return and inlet hoses
   • Speedometer cable
   • Idle up air hoses from the power steering air control valve
   • Oxygen sensor wire, the oil pressure switch wire, the coolant fan switch wire, the water temperature gauge wire, the back-up light switch and neutral safety switch wires
   • All remaining wiring harnesses connected to the engine
   • All necessary vacuum hoses
   • Starter wires
   • If equipped with cruise control, actuator assembly
   • Heater hoses
   • On manual transaxles, clutch release cylinder
   • Transaxle control cables
   • Power steering pump. Do not disconnect the power steering hoses.
   • Air conditioning compressor. Leave the refrigerant lines connected.
   • Front exhaust pipe
   • Halfshafts
   • Connect a suitable lifting device to the engine lifting hooks.
   • On manual transaxles, rear mounting through-bolt and the rear mounting assembly. On automatic transaxle, front mounting through-bolt and front mounting assembly.

   • Right and left side mounting bolts and brackets.
   • Lift the engine/transaxle assembly out of the vehicle.
   • Starter
   • For automatic transaxles, torque converter clutch mounting bolts.
   • Engine from the transaxle by removing the bolts.

To install:

5. Install or connect the following:
   • Transaxle to the engine. On automatic transaxles, install the torque converter clutch and mounting bolts. Install the gray bolt first. Bolts: 20 ft. lbs. (27 Nm).
   • Starter. Bolts: 29 ft. lbs. (39 Nm).
   • Lower the engine and transaxle assembly in the vehicle.
   • Right side mounting insulator and temporarily install the through-bolt
   • Left side mounting bracket. Bolts marked (NT): 47 ft. lbs. (64 Nm), Bolts marked (7T): 35 ft. lbs. (48 Nm).
   • Ground strap. Tighten to 35 ft. lbs. (49 Nm).
   • Rear mounting bracket. Bolts: 35 ft. lbs. (48 Nm).
   • Rear insulator through-bolt. Bolt: 47 ft. lbs. (64 Nm).
   • Halfshafts
   • 2 bolts, nut and through-bolt of the right side mounting insulator. Bolts and nut: 47 ft. lbs. (64 Nm). Through-bolt: 54 ft. lbs. (73 Nm).
   • Front exhaust pipe
   • Air conditioning compressor
   • Power steering pump
   • Transaxle control cables
   • On manual transaxles, clutch release cylinder. Bolts: 108 inch lbs. (12 Nm).
   • Idle up air hoses
   • Speedometer cable
   • Heater hoses.
   • If equipped with cruise control, actuator assembly.
   • Starter wires and nut.
   • Vacuum hoses
   • Oxygen sensor wire, oil pressure switch, coolant fan switch wire, water temperature gauge wire, back-up light switch, and the neutral safety switch wires.
   • All wiring
   • Fuel line
   • Charcoal canister
   • Air cleaner assembly and bracket
   • Radiator, coolant hoses and on vehicles with automatic transmission, cooling lines

- All fluids
- Undercovers
- Hood
- Negative battery cable.

1NZ-FE

1. Before servicing the vehicle, refer to the precautions in the beginning of this section.
2. Drain the cooling system.
3. Drain the engine oil.
4. Drain the transaxle fluid.
5. Remove or disconnect the following:
- Hood
- Battery and tray
- Outer front cowl top panel
- Engine under covers
- Accelerator cable
- Air cleaner cap and related parts
- Air cleaner case and related parts
- All tubes, hoses and connectors attached to the engine
- Accessory drive belts
- Alternator
- Radiator
- With A/C, position the compressor out of the way
- With MT, the clutch release cylinder
- Transaxle control cables
- Fusible link
- Power steering pump
- Center exhaust pipe
- Halfshafts
- Suspension crossmember
- Rear engine mount
6. Attach a shop crane to the engine hangers.
7. Remove all remaining engine mount bolts/nuts.
8. Remove the engine/transaxle assembly.

To install:

9. Install or connect the following:
- Engine/transaxle assembly into position
- Right engine mount insulator. Torque the bolts to 35 ft. lbs. (47 Nm).
- Left engine mount. Torque the bolts to 35 ft. lbs. (47 Nm).
- Rear engine mount bracket. Torque the bolts to 35 ft. lbs. (47 Nm).
- Rear engine mount insulator. Torque the bolt to 47 ft. lbs. (64 Nm).
- Suspension crossmember. Torque the rear mount bolts to 86 ft. lbs. (116 Nm); the front mount bolts to 52 ft. lbs. (70 Nm).
- Halfshafts

- Center exhaust pipe
- Power steering pump
- Fusible link
- Transaxle control cables
- Clutch release cylinder. Torque the 2 bolts to 10 ft. lbs. (13 Nm).
- Compressor. Torque the 4 bolts to 18 ft. lbs. (25 Nm).
- Radiator
- Alternator
- All remaining tubes, hoses and connectors
- Air cleaner case
- Air cleaner cap
- Accelerator cable
- Outer front cowl top panel
- Engine under covers
- Battery and tray
- Coolant
- Engine oil
- Transaxle oil
- Hood

10. Start the vehicle, check for leaks and repair if necessary.

1ZZ-FE Engines

COROLLA

1. Before servicing the vehicle, refer to the precautions in the beginning of this section.
2. Relieve the fuel system pressure.
3. Drain the cooling system.
4. Drain the engine oil.
5. Drain the transaxle fluid.
6. Remove or disconnect the following:
- Negative battery cable. On vehicles equipped with an air bag, wait at least 90 seconds before proceeding.
- Battery
- Hood
- Undercover
- Accelerator cable
- With automatic transmission, throttle cable from the accelerator cable.
- Radiator and cooling fan
- Air cleaner assembly
- Coolant reservoir tank stay
- Electrical connector, the hose, the mounting bolt, and remove the washer tank
- Cruise control actuator
- The Manifold Absolute Pressure (MAP) sensor vacuum hose from the gas filter on the intake manifold
- The brake booster vacuum hose from the intake manifold

- With air conditioning: the air conditioning vacuum hose from the actuator
- With power steering: the air hose from the air pipe
- With air conditioning: the air conditioning actuator connector
7. Install the following wires and connectors from the right-hand fender apron:
a. The ground strap connector
b. The MAP sensor connector
c. With air conditioning: the air conditioning pressure switch
d. The engine wiring harness from the fender apron
8. Remove or disconnect the following:
- Data Link Connector 1(DLC1) connector and ground strap from the left-hand fender apron.
- Engine relay box and 4 connectors.
- Charcoal canister
- Heater hoses from water inlet housing
- Fuel inlet and return hoses
- With manual transmission, clutch release cylinder without disconnecting the pipe
- Transaxle control cable(s)
9. To disconnect the engine wiring harness, disconnect or remove the following components:
- Left-hand and right-hand front door scuff plate
- Lower finish panel
- Lower panel with the glove compartment
- Radio and center cluster finish panel
- Rear console box
- On manual transmission, shift lever knob
- On automatic transmission, shifting hole bezel
- Lower center finish panel
- Floor carpet bracket
- The 3 ECM connectors and cowl wire connector
10. Remove or disconnect the remaining components:
- Air conditioning compressor
- Front exhaust pipe
- Halfshafts
- Power steering pump
- Engine mounting center member
- Through-bolt and nut holding the mounting insulator to the mounting bracket
- Engine and transaxle assembly
- Front and rear engine mounting bracket

- Starter
- Separate the transaxle from the engine

To install:

11. Install or connect the following:
    - Engine to the transaxle
    - Starter
    - Rear engine mounting bracket bolts: 57 ft. lbs. (77 Nm).
    - Front engine mounting bracket. bolts: 57 ft. lbs. (77 Nm).
    - Engine and transaxle assembly into the vehicle
    - Engine mounting center member.
    - Front engine mounting insulator through-bolt and nut. Torque the bolt to 64 ft. lbs. (87 Nm).
    - Halfshafts
    - Front exhaust pipe
    - Power steering pump. Torque the bolts to 29 ft. lbs. (39 Nm).
    - Drive belt
    - Air conditioner compressor. Torque the bolts to 18 ft. lbs. (25 Nm).
    - Drive belt and reconnect the connector.

12. To install and connect the engine wiring harness, perform the following:
    - Push the wire through the cowl
    - Connect the 3 ECM connectors
    - Attach the cowl wire connector
    - Floor carpet bracket
    - Center lower finish panel
    - With automatic transmission, install the shifting hole bezel, with manual transmission, install the shift lever knob
    - Rear console box
    - Center cluster finish panel and the radio
    - Lower panel with the glove compartment door
    - Right and left-hand door scuff plates
    - Lower finish panel

13. Install or connect the following:
    - With manual transmission, clutch release cylinder
    - Transaxle control cable(s)
    - Fuel return and inlet hose. Torque the bolt to 22 ft. lbs. (29 Nm).
    - Heater hoses to the water inlet housing
    - Charcoal canister

14. Connect the following wires and connectors on the left-hand fender apron:
    - The 4 connectors to the engine relay box
    - Engine relay box
    - The DLC1 connector
    - The connector on the fender apron
    - The ground strap on the fender apron

15. Install or connect on the right-hand fender apron:
    - The ground strap connector
    - The MAP sensor connector
    - With air conditioning, the air conditioning pressure switch
    - The engine wire from the fender apron

16. Install or connect the following:
    - With A/C, the actuator connector
    - With power steering, the air hoses to the air pipe
    - The vacuum hose from the MAP sensor to the gas filter to the intake chamber
    - The brake booster vacuum hose to the air intake chamber
    - With A/C, the vacuum hose from the actuator
    - With cruise control, actuator, actuator cable and cover
    - Electrical connector and vinyl hose
    - Washer tank with the bolt
    - Coolant reservoir tank stay
    - Air cleaner
    - Radiator and cooling fan
    - With automatic transmission, connect the throttle cable.
    - Accelerator cable
    - All fluids
    - Negative battery cable
    - Undercovers and hood

17. Start the vehicle, check for leaks and repair if necessary.

CELICA

1. Before servicing the vehicle, refer to the precautions in the beginning of this section.
2. Release the fuel system pressure.
3. Drain the engine oil.
4. Drain the cooling system.
5. Drain the transaxle fluid.
6. Remove or disconnect the following:
   - Negative battery cable. On vehicles equipped with an air bag, wait at least 90 seconds before proceeding.
   - Battery
   - Hood
   - Undercover
   - Accelerator cable, cable bracket, and clamps.
   - Air cleaner
   - Cruise control actuator cable
   - Radiator
   - MAP sensor vacuum hose from the gas filter on the intake manifold
   - Power steering air hose from the intake manifold
   - Power steering hose from the air the air pipe

- Brake booster vacuum hose from the intake manifold
- Air conditioning idle-up valve
- Air conditioning idle-up valve hose from the intake manifold
- Air conditioning idle-up valve hose from the air pipe
- DLC1 from the bracket
- Engine wiring harness from the bracket
- Ground cable from the body and the ground strap from the body
- Heater hoses from the water outlet
- Heater hose from the water bypass pipe
- Fuel inlet hose from the fuel filter and the fuel return hose from the return pipe
- EVAP hose from the charcoal canister
- Engine wiring harness from the engine compartment relay box

7. To remove the engine wiring harness from the passenger's compartment, remove or disconnect the following:
   - The scuff plate
   - The cowl side trim
   - The finish panel from the lower instrument panel
   - Front side of the floor carpet
   - The wiring harness from the clamp of the ECM bracket
   - The 3 ECM connectors
   - The circuit opening relay connector
   - The 3 connectors from the connectors on the bracket
   - The A/C amplifier connector
   - The MAP sensor connector
   - The MAP sensor wire from the clamp on the bracket
   - The wire clamp from the bracket
   - The 2 nuts holding the engine wiring harness to the cowl

8. Remove or disconnect the following:
   - Front exhaust pipe
   - Halfshafts.
   - Alternator drive belt
   - Air conditioning drive belt, compressor connector, and compressor. Do not disconnect the air conditioning lines
   - Remove the drive belt and remove the 4 bolts that secure the power steering pump. Without disconnecting the lines, securely hang the pump out of the way
   - A/C relay box
   - On manual transmission, clutch release cylinder from the transaxle
   - Transaxle control cable(s)
   - On automatic transmission,

transaxle control cable from the engine mounting center member.
- Exhaust pipe support bracket.

9. To remove the engine mounting center member, remove the following components:
- The 2 dust covers from the rear side of the member
- The A/C pipe from the bracket
- The bolt and nut holding the front engine mounting bracket to the mounting insulator
- The bolt holding the rear engine mounting bracket to the insulator
- The bolt and 2 nuts holding the rear engine mounting insulator to the front suspension member
- The 2 bolts and the rear engine mounting bracket, and the center member with the rear mounting insulator

10. Remove or disconnect the remaining components:
- Attach an engine chain hoist to the engine hangers
- Left-hand engine mounting bracket from the mounting insulator
- Through-bolt and the left-hand mounting insulator
- Ground strap connector
- Right-hand engine mounting bracket from the mounting insulator
- Lift the engine and transaxle assembly from the vehicle
- Transaxle from the engine assembly

To install:

11. Install or connect the following:
- Transaxle to the engine assembly.
- Engine into the engine compartment
- Right-hand engine mounting bracket to the mounting insulator. Temporarily install the 3 nuts.
- Left-hand engine mounting insulator to the body with the through-bolt
- Left-hand engine mounting bracket to the mounting insulator and install the 2 bolts and nut. Bolts and nut: 47 ft. lbs. (64 Nm).
- Left-hand engine mounting through-bolt to the body. Bolt: 54 ft. lbs. (73 Nm).
- Right-hand mounting bracket to the insulator. 12mm nut: 21 ft. lbs. (28 Nm), 14mm nut: 38 ft. lbs. (52 Nm).
- Engine ground strap connector and remove the engine hoist.

12. To install the engine mounting center member, perform the following:
- Attach the center member together with the rear engine mounting insulator to the front suspension member
- Temporarily install the 2 bolts and nut holding the center member to the body
- Install the rear engine mounting bracket. Bolts: 58 ft. lbs. (78 Nm)
- Temporarily install the bolt and 2 nuts holding the rear engine mounting insulator to the front suspension member
- Temporarily install the bolt holding the rear engine mounting bracket to the insulator
- Temporarily install the bolt and nut holding the front engine mounting bracket to the insulator
- Tighten the 2 bolts holding the center member to the body to 26 ft. lbs. (35 Nm)
- Tighten the bolt and 2 nuts holding the rear mounting insulator to the front suspension member to 59 ft. lbs. (80 Nm)
- Tighten the bolt holding the rear engine mounting bracket to the insulator to 65 ft. lbs. (88 Nm)
- Tighten the bolt and nut holding the front engine mounting bracket to the insulator to 65 ft. lbs. (88 Nm
- Install the air conditioning pipe to the bracket and install the 2 dust covers to the center member

13. Install or connect the following:
- Exhaust pipe support bracket. Bolts to 14 ft. lbs. (19 Nm).
- Transaxle control cable(s)
- On automatic transmission, transaxle control cable to the engine mounting center member.
- On manual transmission, clutch release cylinder. Bolts: 108 inch lbs. (12 Nm), then attach the bracket with the bolt.
- A/C relay box to the body.
- Power steering pump. 12mm bolts: 14 ft. lbs. (19 Nm). 14mm bolts: 29 ft. lbs. (39 Nm). Install the drive belt. Adjusting bolt: 29 ft. lbs. (39 Nm).
- A/C compressor. Bolts: 18 ft. lbs. (25 Nm).
- A/C drive belt with the adjusting bolt. Torque the locknut to 29 ft.

lbs. (39 Nm). Connect the connector.
- Alternator drive belt
- Halfshafts
- Front exhaust pipe

14. To install the engine wiring harness in the passenger compartment, perform the following:
- Push the harness through the cowl panel, install the retainer to the cowl with the 2 nuts and install the wire clamp to the bracket
- Connect the harness to the clamp on the ECM
- Connect the 3 ECM connectors and the circuit opening relay connector
- Connect the 3 connectors to the connectors on the bracket
- Connect the A/C amplifier connector
- Install the floor carpet, the lower instrument panel finish panel, the cowl side trim panel, and the scuff plate

15. Install or connect the following:
- Engine wiring harness with the 2 connectors to the engine compartment relay box and install the relay box covers
- MAP sensor connector
- MAP sensor wire to the clamp on the bracket
- MAP sensor vacuum hose to the gas filter on the intake manifold
- Brake booster vacuum hose to the intake manifold
- A/C idle-up valve connector
- A/C idle-up valve hose to the intake manifold
- A/C idle-up valve hose to the air pipe
- DLC1 to the bracket
- Engine harness protector to the bracket
- Ground cable and the ground strap
- Heater hose to the water outlet and the heater hose to the water bypass pipe
- Fuel inlet hose to the fuel filter
- Fuel inlet hose with 2 new gaskets and the union bolt. Bolt: 22 ft. lbs. (30 Nm).
- Fuel return hose to the return pipe and connect the EVAP hose to the charcoal canister
- Power steering air hoses to the intake manifold and the air pipe.
- Radiator
- On models equipped with cruise

Timing belt service is covered in Section 3 of this manual

control, install the actuator cable to the clamps.
- Accelerator cable to the throttle body, cable bracket, and the clamps.
- Air cleaner
- Battery tray and battery
- Hood
- Refill the transaxle assembly, engine oil and coolant.
- Negative battery cable
- Engine undercover

16. Start the vehicle, check for leaks and repair if necessary.

2ZZ-GE Engine

1. Before servicing the vehicle, refer to the precautions in the beginning of this section.
2. Release the fuel system pressure.
3. Drain the cooling system.
4. Drain the engine oil.
5. Drain the transaxle fluid.
6. Remove or disconnect the following:
- Negative battery cable. On vehicles equipped with an air bag, wait at least 90 seconds before proceeding.
- Battery
- Hood
- Undercover
- Accelerator cable, cable bracket, and clamps.
- Air cleaner
- ECM box
- Cruise control actuator cable
- Radiator
- MAP sensor vacuum hose from the gas filter on the intake manifold
- Power steering air hose from the intake manifold
- Power steering hose from the air the air pipe
- Brake booster vacuum hose from the intake manifold
- A/C idle-up valve
- Ai/C idle-up valve hose from the intake manifold
- A/C idle-up valve hose from the air pipe
- DLC1 from the bracket
- Engine wiring harness from the bracket
- Ground cable from the body and the ground strap from the body
- Heater hoses from the water outlet
- Heater hose from the water bypass pipe
- Fuel inlet hose from the fuel filter and the fuel return hose from the return pipe

- EVAP hose from the charcoal canister
- Engine wiring harness from the engine compartment relay box

7. To remove the engine wiring harness from the passenger's compartment, remove or disconnect the following:
- The scuff plate
- The cowl side trim
- The finish panel from the lower instrument panel
- Remove the front side of the floor carpet
- The wiring harness from the clamp of the ECM bracket
- The 3 ECM connectors
- The circuit opening relay connector
- The 3 connectors from the connectors on the bracket
- The A/C amplifier connector
- The MAP sensor connector
- The MAP sensor wire from the clamp on the bracket
- The wire clamp from the bracket
- The 2 nuts holding the engine wiring harness to the cowl

8. Remove or disconnect the following:
- Front exhaust pipe
- Exhaust manifold
- Halfshafts
- Alternator drive belt
- A/C drive belt, compressor connector, and compressor. Do not disconnect the A/C lines
- Remove the drive belt and remove the 4 bolts that secure the power steering pump. Without disconnecting the lines, securely hang the pump out of the way.
- A/C relay box
- On manual transmission, clutch release cylinder from the transaxle
- Transaxle control cable(s)
- On automatic transmission, transaxle control cable from the engine mounting center member.
- Exhaust pipe support bracket

9. To remove the engine mounting center member, remove the following components:
- The 2 dust covers from the rear side of the member
- The air conditioning pipe from the bracket
- The bolt and nut holding the front engine mounting bracket to the mounting insulator
- The bolt holding the rear engine mounting bracket to the insulator
- The bolt and 2 nuts holding the rear engine mounting insulator to the front suspension member

- The 2 bolts and the rear engine mounting bracket, and the center member with the rear mounting insulator

10. Remove or disconnect the remaining components:
- Attach an engine chain hoist to the engine hangers
- Left-hand engine mounting bracket from the mounting insulator
- Through-bolt and the left-hand mounting insulator
- Ground strap connector
- Right-hand engine mounting bracket from the mounting insulator
- Lift the engine and transaxle assembly from the vehicle
- Transaxle from the engine assembly

To install:

11. Install or connect the following:
- Transaxle to the engine assembly
- Engine into the engine compartment
- Right-hand engine mounting bracket to the mounting insulator. Temporarily install the 3 nuts.
- Left-hand engine mounting insulator to the body with the through-bolt
- Left-hand engine mounting bracket to the mounting insulator and install the 2 bolts and nut. Bolts and nut: 47 ft. lbs. (64 Nm).
- Left-hand engine mounting through-bolt to the body. Bolt: 54 ft. lbs. (73 Nm).
- Right-hand mounting bracket to the insulator. 12mm nut: 21 ft. lbs. (28 Nm), 14mm nut: 38 ft. lbs. (52 Nm).
- Engine ground strap connector and remove the engine hoist

12. To install the engine mounting center member, perform the following:
- Attach the center member together with the rear engine mounting insulator to the front suspension member
- Temporarily install the 2 bolts and nut holding the center member to the body
- Install the rear engine mounting bracket. Bolts: 58 ft. lbs. (78 Nm)
- Temporarily install the bolt and 2 nuts holding the rear engine mounting insulator to the front suspension member
- Temporarily install the bolt holding the rear engine mounting bracket to the insulator
- Temporarily install the bolt and nut

holding the front engine mounting bracket to the insulator
- Tighten the 2 bolts holding the center member to the body to 26 ft. lbs. (35 Nm)
- Tighten the bolt and 2 nuts holding the rear mounting insulator to the front suspension member to 59 ft. lbs. (80 Nm)
- Tighten the bolt holding the rear engine mounting bracket to the insulator to 65 ft. lbs. (88 Nm)
- Tighten the bolt and nut holding the front engine mounting bracket to the insulator to 65 ft. lbs. (88 Nm
- Install the A/C pipe to the bracket and install the 2 dust covers to the center member
- Exhaust manifold

13. Install or connect the following:
- Exhaust pipe support bracket. Bolts to 14 ft. lbs. (19 Nm).
- Transaxle control cable(s)
- On automatic transmission, transaxle control cable to the engine mounting center member.
- On manual transmission, clutch release cylinder. Bolts: 108 inch lbs. (12 Nm), then attach the bracket with the bolt
- A/C relay box to the body
- Power steering pump. 12mm bolts: 14 ft. lbs. (19 Nm). 14mm bolts: 29 ft. lbs. (39 Nm). Install the drive belt. Adjusting bolt: 29 ft. lbs. (39 Nm).
- A/C compressor. Bolts: 18 ft. lbs. (25 Nm).
- A/C drive belt with the adjusting bolt and tighten the idler pulley locknut to 29 ft. lbs. (39 Nm). Connect the connector.
- Alternator drive belt
- Halfshafts
- Front exhaust pipe

14. To install the engine wiring harness in the passenger compartment, perform the following:
- Push the harness through the cowl panel, install the retainer to the cowl with the 2 nuts and install the wire clamp to the bracket
- Connect the harness to the clamp on the ECM
- Connect the 3 ECM connectors and the circuit opening relay connector
- Connect the 3 connectors to the connectors on the bracket
- A/C amplifier connector

- Install the floor carpet, the lower instrument panel finish panel, the cowl side trim panel, and the scuff plate

15. Install or connect the following:
- Engine wiring harness with the 2 connectors to the engine compartment relay box and install the relay box covers.
- MAP sensor connector
- MAP sensor wire to the clamp on the bracket
- MAP sensor vacuum hose to the gas filter on the intake manifold
- Brake booster vacuum hose to the intake manifold
- A/C idle-up valve connector
- A/C idle-up valve hose to the intake manifold
- A/C idle-up valve hose to the air pipe
- DLC1 to the bracket
- Engine harness protector to the bracket
- Ground cable and the ground strap
- Heater hose to the water outlet and the heater hose to the water bypass pipe
- Fuel inlet hose to the fuel filter
- Fuel inlet hose with 2 new gaskets and the union bolt. Bolt: 22 ft. lbs. (30 Nm).
- Fuel return hose to the return pipe and connect the EVAP hose to the charcoal canister
- Power steering air hoses to the intake manifold and the air pipe.
- Radiator
- On models equipped with cruise control, install the actuator cable to the clamps.
- Accelerator cable to the throttle body, cable bracket, and the clamps.
- Air cleaner
- Battery tray and battery
- Hood
- Refill the transaxle assembly, engine oil and coolant.
- Negative battery cable
- Engine undercover

16. Start the vehicle, check for leaks and repair if necessary.

5S-FE Engine

CAMRY AND CAMRY SOLARA

1. Before servicing the vehicle, refer to the precautions in the beginning of this section.

2. Drain the cooling system.
3. Drain the engine oil.
4. Drain the transaxle fluid.
5. Remove or disconnect the following:
- Negative battery cable. On vehicles equipped with an air bag, wait at least 90 seconds before proceeding.
- On the Camry Solara, strut tower brace
- Battery and battery tray
- Hood
- Engine undercover
- Accelerator cable from the throttle body. With automatic transmission, throttle cable
- Air cleaner, resonator, and air intake hose
- On models with cruise control, actuator cover, actuator with the bracket.
- Ground strap at the battery carrier
- Radiator and coolant reservoir hose
- Washer tank, electrical lead and hose

6. To disconnect the wiring harness, tag and disconnect the following:
- The 5 connectors to the engine relay box
- The igniter connector
- The noise filter connector
- The connector at the left-hand fender apron
- The 2 ground straps from the left-hand and right-hand fender aprons
- The DLC1
- Disconnect the MAP sensor connector

7. Remove or disconnect the following:
- Dash panel undercover, glove compartment door, glove compartment, cowl harness connectors and the 2 ECM connectors
- Heater hoses, fuel return hose, and fuel inlet hose
- With manual transmission, starter and the clutch release cylinder. Don't disconnect the hydraulic line, simply hang the cylinder out of the way
- Transaxle control cables at the transaxle
- Tag and disconnect all remaining vacuum hoses and connectors
- Engine wire from the cowl panel
- Without disconnecting the refrigerant lines, A/C compressor
- Front exhaust pipe bracket and the front pipe from the exhaust manifold

Heater Core replacement is covered in Section 2 of this manual

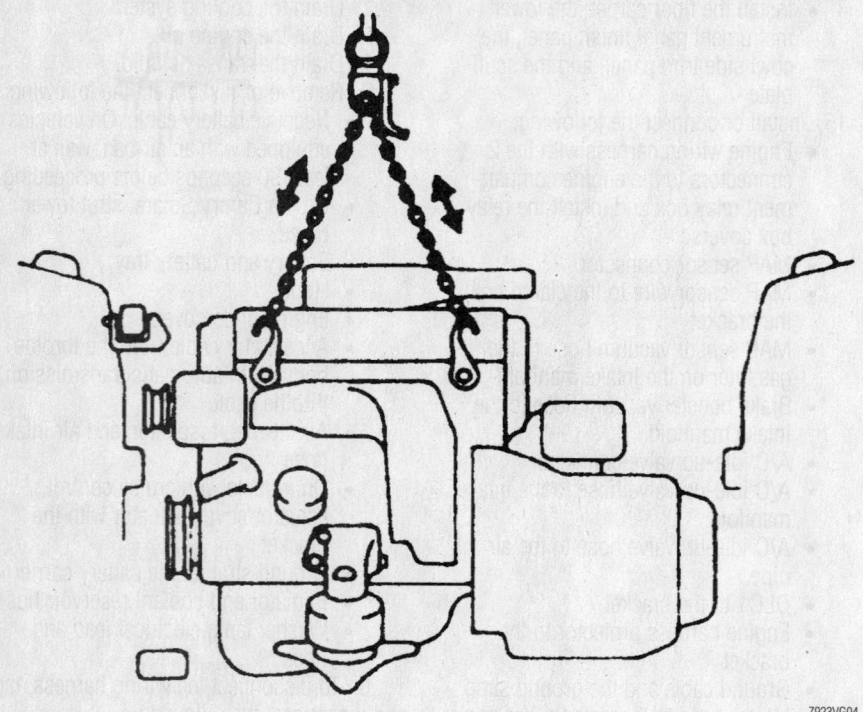

Use a hoist to remove the engine assembly—Camry 5S-FE engine

- Halfshafts
- Without disconnecting the hydraulic lines, power steering pump
- Left engine mounting insulator
- Right rear engine mounting insulator
- Front right engine mounting insulator

8. Attach an engine lifting device to the lift hooks. Remove the 3 bolts and disconnect the control rod. Slowly and carefully, lift the engine/transaxle assembly out of the engine compartment.

9. If equipped with automatic transmission, remove the starter. Separate the engine assembly from the transaxle.

To install:

10. Install or connect the following:
- Engine assembly to the transaxle
- With automatic transmission, starter
- Engine control rod. Bolts: 47 ft. lbs. (64 Nm).
- Front right engine mount. Bolts: 59 ft. lbs. (80 Nm).
- Rear engine mount. Nuts: 48 ft. lbs. (66 Nm).
- Left mount. Bolts (3 or 4): 47 ft. lbs. (64 Nm).
- Power steering pump. Bolts: 31 ft. lbs. (43 Nm). Install the drive belt and connect the 2 air hoses to the air pipe.
- Halfshafts

- Front pipe to the manifold. Nuts: 46 ft. lbs. (62 Nm).
- A/C compressor. Bolts: 20 ft. lbs. (27 Nm).

11. Feed the engine harness through the cowl and reattach the clamp to the cowl. Make the following connections:
- The 2 ECM connectors
- The 2 cowl wire connectors
- Install the glove compartment and door
- Install the lower instrument panel and the undercover

12. Install or connect the following:
- Vacuum hoses and the transaxle control cables
- On manual transmission vehicles, release cylinder and the starter
- Fuel inlet hose and tighten it to 22 ft. lbs. (29 Nm). Connect the return hose and the 2 heater hoses.
- Attach the 5 connectors to the relay box
- The connectors from the left-hand fender apron
- Install the engine relay box
- The igniter connector
- On California models, the ignition coil connector
- The noise filter connector
- The 2 ground straps from the left-hand and right-hand fender apron
- The DLC1
- The MAP sensor connector.

- Washer tank and connect the electrical lead and hose
- Coolant reservoir hose and the radiator
- With cruise control, actuator and bracket. Connect the actuator connector and install the cover.
- Ground strap to the battery carrier.
- Air cleaner assembly
- On California models, air hose to the air cleaner assembly and connect the air intake temperature sensor connector
- With automatic transmission, throttle cable
- Accelerator cable
- Battery tray and battery
- On Camry Solara models, strut tower brace
- Hood
- Oil and coolant.
- Negative battery cable
- Undercover

13. Start the vehicle, check for leaks and repair if necessary.

CELICA

1. Before servicing the vehicle, refer to the precautions in the beginning of this section.

2. Properly relieve the fuel system pressure.

3. Drain the cooling system.

4. Drain the engine oil.

5. Drain the transaxle fluid.

6. Remove or disconnect the following:
- Negative battery cable. On vehicles equipped with an air bag, wait at least 90 seconds before proceeding.
- Battery
- Hood
- Undercover
- Accelerator cable from the throttle body, the cable bracket and clamps
- Air cleaner assembly
- With cruise control, actuator from the bracket
- Radiator
- The MAP sensor
- The MAP sensor wire from the clamp
- The MAP sensor vacuum hose from the gas filter on the intake manifold
- The brake booster vacuum hose from the intake manifold
- The DLC1 from the bracket
- The igniter connector
- On California models, the ignition coil connector
- On California models, the ignition coil high tension wire, the noise fil-

ter, the wire clamp from bracket, the ignition coil and igniter assembly, and disconnect the wire from the bracket

- The ground cable from the body
- The ground strap from the body
- The heater hose from the water outlet and bypass pipe
- The fuel inlet hose from the fuel filter and the fuel return hose from the return pipe
- The EVAP hose from the charcoal canister
- Engine wiring harness from the engine compartment relay box.

7. To remove the engine wiring harness from the passenger's compartment, remove or disconnect the following:

- The scuff plate
- The cowl side trim
- The finish panel from the lower instrument panel
- Remove the front side of the floor carpet
- The wiring harness from the clamp of the ECM bracket
- The 3 ECM connectors
- The circuit opening relay connector
- The 3 connectors from the connectors on the bracket
- The air conditioning amplifier connector
- The wire clamp from the bracket
- The 2 nuts holding the engine wiring harness to the cowl

8. Remove or disconnect the following:

- Front exhaust pipe
- Halfshafts
- Alternator drive belt
- A/C compressor and suspend it securely out of the way
- Power steering pump. Securely hang the pump out of the way
- On manual transmission, the starter
- On manual transmission, the clutch release cylinder and associated components
- Transaxle control cable(s) from the transaxle
- On automatic transmission model, transaxle control cable from the engine mounting center member.
- Exhaust pipe support bracket

9. To remove the engine mounting center member, remove the following components:

- The 2 dust covers from the rear side of the member

- The air conditioning pipe from the bracket
- The bolt and nut holding the front engine mounting bracket to the mounting insulator
- The bolt holding the rear engine mounting bracket to the insulator
- The bolt and 2 nuts holding the rear engine mounting insulator to the front suspension member
- The 2 bolts (automatic transmission) or 3 bolts (manual transmission) and rear engine mounting bracket.
- Center member with the rear mounting insulator

10. Attach an engine chain hoist to the engine hangers

11. Remove or disconnect the following:

- Left-hand engine mount bracket from the mounting insulator
- Left-hand mounting insulator
- Ground strap connector
- Right-hand engine mount bracket from the mounting insulator
- Halfshaft bearing bracket
- Transaxle

To install:

12. Install or connect the following:

- Transaxle to the engine assembly and reattach the halfshaft bearing bracket. Bolts: 47 ft. lbs. (64 Nm).
- Transaxle/engine assembly into engine compartment.
- Right-hand engine mount bracket to the mounting insulator, and temporarily install the bolt and 2 nuts.
- Left-hand engine mounting insulator to the body with the through-bolt
- Left-hand engine mount bracket to the mounting insulator and install the 2 nuts and bolt (manual transmission) or the 2 bolts (automatic transmission). Bolts and nuts: 47 ft. lbs. (64 Nm).

13. Tighten the left-hand engine mounting through-bolt to the body: 54 ft. lbs. (73 Nm). Tighten the bolt and 2 nuts holding the right-hand mounting bracket to the insulator: 27 ft. lbs. (37 Nm) and nut: 38 ft. lbs. (52 Nm).

14. Install the engine ground strap connector

15. To install the engine mounting center member, perform the following:

- Attach the center member together with the rear engine mounting insulator to the front suspension member

- Temporarily install the 2 bolts holding the center member to the body
- Install the rear engine mounting bracket with the 2 bolts (automatic transmission) or the 3 bolts (manual transmission). Bolt: 58 ft. lbs. (79 Nm).
- Temporarily install the bolt and 2 nuts holding the rear engine mounting insulator to the front suspension member
- Temporarily install the bolt holding the rear engine mounting bracket to the insulator
- Temporarily install the bolt and nut holding the front engine mounting bracket to the insulator
- Tighten the 2 bolts holding the center member to the body to 26 ft. lbs. (35 Nm).
- Tighten the bolt and 2 nuts holding the rear mounting insulator to the front suspension member to 59 ft. lbs. (80 Nm).
- Tighten the bolt holding the rear engine mounting bracket to the insulator to 65 ft. lbs. (88 Nm).
- Tighten the bolt and nut holding the front engine mounting bracket to the insulator to 65 ft. lbs. (88 Nm).
- Install the A/C pipe to the bracket and install the 2 dust covers to the center member

16. Install or connect the following:

- Exhaust pipe support bracket. Bolts: 14 ft. lbs. (19 Nm).
- Transaxle control cable(s) to the transaxle.

17. On the automatic transmission vehicles, transaxle control cable to the engine mounting center member.

18. On manual transmission model only, install and/or connect the following:

- The clutch release cylinder. Bolts: 108 inch lbs. (12 Nm).
- The bracket with the bolt and the tube to the bracket with the clamp
- The tube with the clamp and bolt
- The back-up switch connector

19. Install or connect the following:

- On manual transmission model only, starter, starter wire, and starter connector.
- Power steering pump. 3 mounting bolts: 32 ft. lbs. (44 Nm); adjusting bolt: 29 ft. lbs. (39 Nm); pivot bolt: 32 ft. lbs. (44 Nm)
- Hoses to the air tube

Brake service is covered in Section 4 of this manual

- A/C compressor. Bolts: 18 ft. lbs. (24 Nm).
- Alternator drive belt
- Halfshafts
- Front exhaust pipe

20. To install the engine wiring harness to the passenger's compartment, perform the following:

- Push the harness through the cowl panel, install the retainer to the cowl with the 2 nuts and install the wire clamp to the bracket
- Connect the harness to the clamp on the ECM
- Connect the 3 ECM connectors and the circuit opening relay connector
- Connect the 3 connectors to the connectors on the bracket
- Connect the A/C amplifier connector
- Install the floor carpet, the lower instrument panel finish panel, the cowl side trim panel, and the scuff plate

21. Install or connect the following:

- Engine wiring harness with the 2 connectors to the engine compartment relay box and install the relay box covers
- The MAP sensor connector
- The MAP sensor wire to the clamp on the bracket
- The MAP sensor vacuum hose to the gas filter on the intake manifold
- The brake booster vacuum hose to the intake manifold
- The DLC1 to the bracket
- The engine harness protector to the bracket
- On California models, the engine harness clamp to the bracket and the ignition coil and igniter assembly with the 3 bolts, and install the harness to the bracket
- The igniter connector
- On California model, the ignition coil connector, high tension wire to the coil, and the noise filter
- The ground cable and the ground strap to the body
- The heater hose to the water outlet and the heater hose to the water bypass pipe
- Fuel inlet hose to the fuel filter
- Fuel inlet hose with 2 new gaskets and the union bolt. Bolt: 22 ft. lbs. (30 Nm).
- Fuel return hose to the return pipe and connect the EVAP hose to the charcoal canister
- Radiator

- With cruise control, actuator and connect the connector
- Accelerator cable to the throttle body, the cable bracket, and the clamps.
- Air cleaner
- Battery tray and battery
- Hood
- Oil and coolant
- Transaxle fluid to the proper level
- Negative battery cable
- Undercover

22. Start the vehicle, check for leaks and repair if necessary.

1MZ-FE Engine

AVALON, CAMRY AND CAMRY SOLARA

1. Before servicing the vehicle, refer to the precautions in the beginning of this section.
2. Properly relieve the fuel system pressure.
3. Drain the cooling system.
4. Drain the engine oil.
5. Drain the transaxle fluid.
6. Remove or disconnect the following:

- Negative battery cable. On vehicles equipped with an air bag, wait at least 90 seconds before proceeding.
- Hood
- Battery and battery tray
- Accelerator and throttle cables
- Cruise control actuator, if equipped
- Air cleaner assembly, mass air flow meter and air cleaner hose
- Radiator
- Engine relay box
- 2 igniter connectors
- Noise filter connector
- Connector from the left-hand fender apron
- 2 ground straps and any other electrical connections keeping them from being removed.
- Vacuum hoses from the engine.
- Fuel inlet and return hoses
- Heater hoses
- Transaxle control cable from the transaxle
- Instrument panel undercover, the lower instrument panel and glove box assembly
- 3 ECM connectors, the 5 cowl wire connectors, and the cooling fan ECM connector. Push the engine wire through the cowl panel
- Front exhaust pipe
- Halfshafts
- Power steering pressure tube
- Power steering pump

- A/C compressor without disconnecting the hoses
- Left-hand engine mounting insulator
- Right-hand engine mounting insulator
- Engine mounting shock absorber
- Front right engine mounting insulator

7. Attach a hoist chain to the engine hangers.
8. Remove or disconnect the following:

- Coolant reservoir hose and reservoir tank
- Right-side engine mounting stay bracket
- Engine control rod and bracket assembly

➡ **Make certain all wires, connectors and hoses are cleared from the engine.**

- Engine/transaxle assembly from the vehicle

To install:

9. Carefully lower the engine position. Keep the engine level while aligning the engine mounts.

10. Install or connect the following:

- Engine control rod and bracket. Tighten to 47 ft. lbs. (64 Nm).
- Right engine mount stay bracket. Tighten to 23 ft. lbs. (31 Nm).
- Engine ground straps.
- Coolant reservoir tank
- Front engine insulator. Tighten to 48 ft. lbs. (66 Nm).
- Engine mounting shock absorber. Tighten to 35 ft. lbs. (48 Nm).
- Left and right engine mounts. Tighten to 48 ft. lbs. (66 Nm).
- Power steering pump and A/C compressor
- Power steering pressure tube
- Halfshafts and front exhaust pipe
- Engine wires and connectors
- Transaxle control cable to the transaxle
- Fuel hoses and heater hoses
- All vacuum hoses, wiring and connectors
- Radiator
- Cruise control actuator
- Throttle cable and accelerator cable
- MAF meter, the air cleaner assembly, and air cleaner hose
- Coolant and engine oil
- Battery tray and battery.
- Transaxle fluid to the proper level
- Hood
- Negative battery cable

11. Start the vehicle, check for leaks and repair if necessary.

2JZ-GE and 2JZ-GTE Engines

1. Before servicing the vehicle, refer to the precautions in the beginning of this section.

2. Properly relieve the fuel system pressure.

3. Drain the cooling system.

4. Drain the engine oil.

5. Remove or disconnect the following:

- Negative battery cable. Do not start any work for at least 90 seconds to prevent accidental deployment of the air bag.
- Hood
- Radiator

❊❊ CAUTION

The fuel injection system remains under pressure even after the engine has been turned OFF. The fuel system pressure must be relieved before disconnecting any fuel lines. Failure to do so may result in fire and/or personal injury.

- Accelerator cable and cruise control actuator
- Air cleaner assembly, volume air flow meter, and the air intake hose
- Drive belt by turning the tensioner clockwise
- Fan, fan clutch, and the water pump pulley
- Charcoal canister
- Heater water hoses
- Brake booster vacuum hose
- EVAP hose
- Noise filter connector
- Ignition coil connector
- Engine wire from the wire clamp
- Rubber cap, nut and wire from the alternator
- Engine room main wire
- Igniter connector
- Theft deterrent horn connector
- Engine wire from the 2 wire clamps
- Wire clamp and power steering solenoid valve connector
- Ground strap from the cylinder block by removing the bolt
- Rubber cap, nut, and the wire from the starter
- Fuel inlet hose from the engine. Suspend the hose union upward
- Fuel return hose from the fuel return guide
- Fuel return hose from the fuel return hose. Plug the hose end

- Engine wire from the intake manifold stay
- Power steering pump
- Power steering pressure tube
- A/C compressor without disconnecting the hoses
- Engine wire from the cowl panel
- Remove the scuff plate from the right door
- Take out the front side of the floor carpet
- Remove the 2 nuts and ECM protector
- Remove the nut and disconnect the ECM from the floor panel
- Disconnect the 2 connectors from the ECM
- Disconnect the connector from the instrument panel wire
- Disconnect the connector from the connector cassette
- Pull out the engine wire from the cabin
- With manual transmission, upper console panel, shift lever boots, and holding bolts
- With manual transmission, clutch release cylinder and the ground strap from the transmission.
- No. 2 front exhaust pipe
- Exhaust pipe heat insulator
- Driveshaft
- With automatic transmissions, control rod from the shift lever by removing the nut.

6. Support the transmission with a jack.

7. Remove the rear support member by removing the 8 bolts.

8. Attach the engine hoist chain to the engine and raise the engine slightly.

9. Remove or disconnect the following:

- Nuts holding the engine front mounting insulators to the front suspension crossmember.
- Engine from the vehicle
- Oil dipstick guide from the transmission
- Engine wire from the transmission
- Starter connector, 2 bolts, engine wire bracket, and the starter
- With automatic transmissions, oil cooler tubes from the transmission
- With automatic transmissions, torque converter clutch mounting bolts
- Transmission

To install:

10. Assemble the engine and transmission. Tighten the bolts as follows:

- 14mm: 29 ft. lbs. (39 Nm)
- 17mm: 43 ft. lbs. (72 Nm)

11. Install or connect the following:

- With automatic transmission, torque converter clutch mounting bolts by first installing the gray bolt, then install the other 5 bolts. Bolts: 25 ft. lbs. (33 Nm).
- With automatic transmission, oil cooler tubes to the transmission. Union nuts: 25 ft. lbs. (33 Nm).
- Starter
- Engine wire to the transmission
- With automatic transmission, oil dipstick guide and dipstick
- Engine and transmission as an assembly to the vehicle. Keep slight tension on the engine until the mounting bolts and nuts are installed
- Engine front mounting insulators to the front suspension crossmember. Nuts: 43 ft. lbs. (59 Nm).
- Bolts holding the support member to the body. Bolts: 19 ft. lbs. (25 Nm).
- Nuts holding the support member to the engine rear mounting insulator. Nuts: 10 ft. lbs. (13 Nm).
- Driveshaft

12. With automatic transmissions, transmission control rod as follows:

- Shift the shift lever to the **N** position.
- Fully turn the control shaft lever back and return 2 notches. The control shaft is now in the neutral position
- Connect the control rod to the shift lever with the nut. Tighten the nut to 108 inch lbs. (13 Nm).

13. Install or connect the following:

- Exhaust pipe heat insulator
- No. 2 front exhaust pipe
- With manual transmission, clutch release cylinder and ground strap. Tighten the clutch release cylinder bolts to 108 inch lbs. (13 Nm) and the ground strap bolt to 27 ft. lbs. (37 Nm).
- With manual transmission, upper console panel, shift lever boots, and holding bolts.
- Engine wire through the cowl panel.
- Connect the connector to the connector cassette
- Connect the connector to the instrument panel wire connector

For complete Engine Mechanical specifications, see Section 1 of this manual

- Connect the 2 connectors to the ECM
- Insert the ECM bracket into the stay on the floor panel
- Install the ECM with the nut
- Install the ECM protector with the 2 nuts
- Install the floor carpet
- Install the scuff plate
- Engine wire to the cowl panel
- Air conditioning compressor to the engine
- Power steering tube with the 2 clamp bolts
- Power steering pump. Lower bolt: 43 ft. lbs. (58 Nm); upper bolt: 29 ft. lbs. (39 Nm); rear stay bolts: 29 ft. lbs. (39 Nm); front bracket bolts: 43 ft. lbs. (58 Nm).
- Engine wire bracket
- Fuel return hose to the fuel return pipe
- Fuel return hose to the clamp of the oil dipstick guide
- Fuel inlet hose with the 2 new gaskets and the union bolt. Bolt: 22 ft. lbs. (29 Nm).
- Wires and connectors
- EVAP hose
- Brake booster vacuum hose
- Heater hoses
- Charcoal canister
- Water pump pulley, fan, and the fan clutch
- Drive belt to the engine
- Air cleaner, VAF meter, and the intake air connector pipe.
- Control cables to the throttle body
- Oil
- Radiator
- Cooling system
- Negative battery cable
- Hood

14. Start the vehicle, check for leaks and repair if necessary.

Water Pump

REMOVAL & INSTALLATION

5E-FE Engine

1. Before servicing the vehicle, refer to the precautions in the beginning of this section.
2. Drain the cooling system.
3. Remove or disconnect the following:
   - Negative battery cable. On vehicles equipped with an air bag, wait at least 90 seconds before proceeding.
   - Alternator

- With distributorless ignition, intake manifold stay bracket by disconnecting the wire clamps and removing the 2 nuts
- With distributor ignition, remove the intake manifold stay bracket by removing the 2 nuts and 2 bolts.
- Water inlet pipe
- Oil dipstick guide
- Alternator adjusting bar
- Water pump attaching bolt and nuts
- Water pump

To install:

4. Install or connect the following:
   - 0.08–0.12 in. (2–3mm) bead of sealant to the groove in the pump.
   - O-ring, lubricated with a little soap and water, on the water inlet pipe.
   - Water pump. Bolts: 13 ft. lbs. (17 Nm).
   - Oil dipstick guide
   - Alternator adjusting bar
   - Dipstick guide clamp bolt
   - Water inlet pipe. Bolt: 65 inch lbs. (7.5 Nm).
   - Water bypass, heater inlet and water inlet hoses
   - For distributorless ignition engines, install the intake manifold bracket by installing the 2 bolts and the wire clamp. Bolts: 15 ft. lbs. (20 Nm).
   - For distributor ignition, install the intake manifold bracket by installing the 2 bolts. Bolts: 15 ft. lbs. (20 Nm).
   - Alternator and belt
   - Negative battery cable

5. Fill the coolant system to the proper level.
6. Start the vehicle, check for leaks and repair if necessary.

1NZ-FE Engine

1. Before servicing the vehicle, refer to the precautions in the beginning of this section.
2. Drain the cooling system.
3. Remove or disconnect the following:
   - Negative battery cable
   - Accessory drive belt
   - Water pump pulley, using a holding tool
   - Water pump and gasket (3 bolts; 2 nuts)

To install:

4. Install or connect:
   - Water pump, with a new gasket. Torque the nuts and bolts to 96 inch lbs. (11 Nm).
   - Pulley. Torque the bolts to 11 ft. lbs. (15 Nm).
   - Drive belt
   - Negative battery cable

5. Fill the cooling system to the proper level.
6. Start the vehicle, check for leaks and repair if necessary.

1ZZ-FE Engine

1. Before servicing the vehicle, refer to the precautions in the beginning of this section.
2. Drain the cooling system.

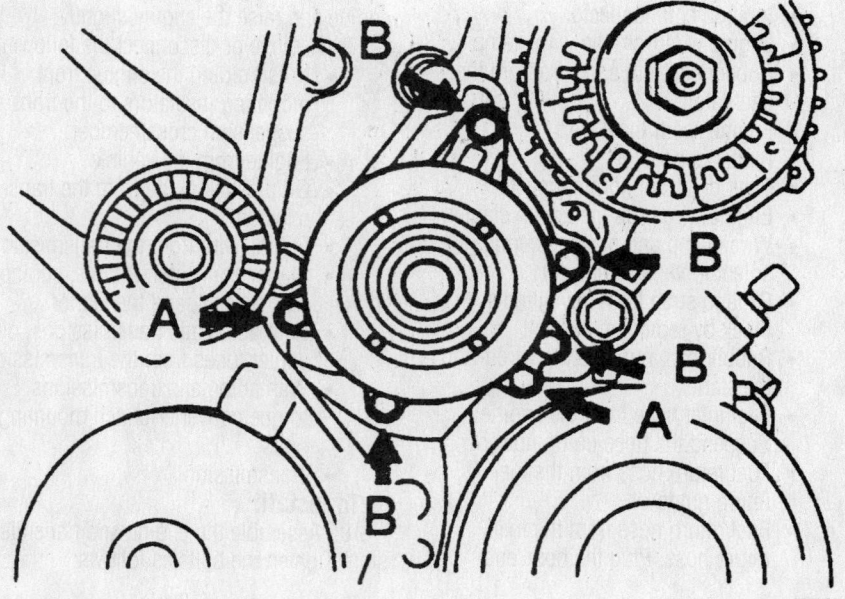

Water pump bolt identification—1.8L (1ZZ-FE) engine

7923VG06

3. Remove or disconnect the following:
- Negative battery cable
- Right-hand engine under cover
- Drive belt
- Water pump

To install:

4. Install or connect the following:
- Water pump. Bolts marked **A** (short): 80 inch lbs. (9 Nm). Bolts marked **B** (long): 96 inch lbs. (11 Nm).
- Drive belt
- Right engine under cover
- Negative battery cable

5. Fill the cooling system to the proper level.

6. Start the vehicle, check for leaks and repair if necessary.

2ZZ-GE Engine

1. Before servicing the vehicle, refer to the precautions in the beginning of this section.

2. Drain the cooling system.

3. Remove or disconnect the following:
- Negative battery cable

- Right-hand engine under cover
- Drive belt
- Water pump pulley
- Water pump and O-ring

To install:

4. Install or connect the following:
- Water pump with new O-ring. Bolts: 80 inch lbs. (9 Nm).
- Water pump pulley. Bolts: 11 ft. lbs. (15 Nm).
- Drive belt
- Right engine under cover
- Negative battery cable

5. Fill the cooling system to the proper level.

6. Start the vehicle, check for leaks and repair if necessary.

5S-FE Engine

1. Before servicing the vehicle, refer to the precautions in the beginning of this section.

2. Drain the cooling system.

3. Remove or disconnect the following:

- Negative battery cable. On vehicles

equipped with an air bag, wait at least 90 seconds before proceeding.
- Right engine undercover
- Lower radiator hose from the water outlet
- Timing belt, timing belt tension spring, and the No. 2 idler pulley
- Alternator, drive belt and the adjusting bar if necessary
- 2 nuts holding the water pump to the water bypass pipe and remove the 3 bolts in sequence.
- Water pump cover assembly
- Gasket and 2 O-rings from the water pump and the bypass pipe.
- Water pump from the water pump cover by removing the 3 bolts in sequence.

To install:

4. Install or connect the following:
- Water pump to the water pump cover. Bolts: 78 inch lbs. (9 Nm) in proper sequence.
- Water pump cover to the water bypass pipe, but do not install the nuts yet.

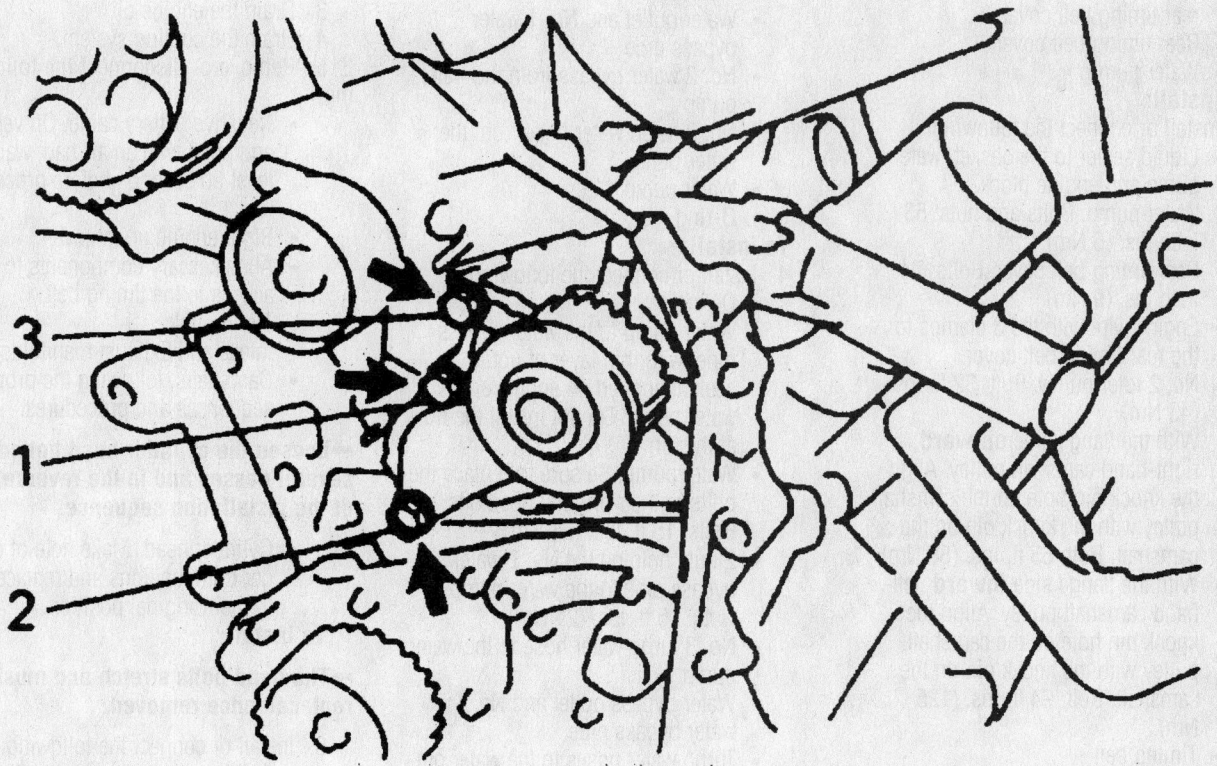

Install the 3 water pump bolts in this sequence—2.2L (5S-FE) engine

7923VG05

For Accessory Drive Belt illustrations, see Section 1 of this manual

- Water pump and tighten the 3 bolts in sequence. Bolts: 78 inch lbs. (9 Nm). Nuts: 82 inch lbs. (9 Nm).
- Alternator drive belt adjusting bar. Bolt: 13 ft. lbs. (18 Nm).
- No. 2 idler pulley and the timing belt tension spring
- Lower radiator hose
- Timing belt
- Right engine undercover
- Negative battery cable

5. Fill the cooling system to the proper level.

6. Start the vehicle, check for leaks and repair if necessary.

1MZ-FE Engine

1. Before servicing the vehicle, refer to the precautions in the beginning of this section.

2. Drain the cooling system.

3. Remove or disconnect the following:
- Negative battery cable. On vehicles equipped with an air bag, wait at least 90 seconds before proceeding
- Timing belt
- No. 2 idler pulley
- 3 clamps and engine wire from the rear timing belt cover
- Rear timing belt cover
- Water pump

To install:

4. Install or connect the following:
- Liquid sealer to the gasket, water pump and engine block.
- Water pump. Bolts and nuts: 53 inch lbs. (6 Nm).
- Rear timing belt cover. Bolts: 74 inch lbs. (9 Nm).
- Engine wire with the 3 clamps to the rear timing belt cover.
- No. 2 idler pulley. Bolt: 32 ft. lbs. (43 Nm).
- With the flange side **outward**, right-hand camshaft pulley. Align the knock pin hole on the camshaft pulley with the knock pin on the camshaft. Bolt: 65 ft. lbs. (88 Nm).
- With the flange side **inward**, left-hand camshaft pulley. Align the knock pin hole on the camshaft pulley with the knock pin on the camshaft. Bolt: 94 ft. lbs. (125 Nm).
- Timing belt
- Negative battery cable

5. Fill the cooling system to the proper level.

6. Start the vehicle, check for leaks and repair if necessary.

2JZ-GE and 2JZ-GTE Engines

1. Before servicing the vehicle, refer to the precautions in the beginning of this section.

2. Drain the cooling system.

3. Remove or disconnect the following:
- Negative battery cable. On vehicles equipped with an air bag, wait at least 90 seconds before proceeding.
- Air cleaner and MAF meter assembly
- Radiator
- With manual transmission, drive belt tensioner damper.
- Loosen the 4 nuts holding the fan clutch to the water pump.
- Drive belt from the engine
- Fan, fan clutch, and the water pump pulley
- Water inlet, lower radiator hose assembly, and the thermostat
- Timing belt
- Alternator
- On turbo models, turbo water hoses from the water outlet
- Except for the California vehicles, exhaust manifold heat insulator
- Water outlet and No. 1 water bypass pipe
- No. 2 water bypass from the water pump
- No. 3 turbo water hose from the water pump
- Water pump
- O-ring

To install:

4. Install or connect the following:
- O-ring
- Water pump to the water bypass pipe, with thin layer of liquid sealant applied on engine and water pump. Do not install the nut at this time.
- Water pump. Be sure to replace the bolts to their original positions. Bolts: 15 ft. lbs. (21 Nm).
- 2 nuts holding the No. 2 water bypass pipe to the water pump. Nuts: 15 ft. lbs. (21 Nm).
- No. 3 turbo water hose to the water pump.
- Water bypass outlet and No. 1 water bypass pipe.
- Turbo water hoses to the water outlet.
- Alternator
- Except for California vehicle, exhaust manifold heat insulator
- Engine wire bracket
- Timing belt

- Thermostat, water inlet, and the lower radiator hose assembly.
- Water pump pulley, fan, fluid clutch assembly, and the drive belt. Fan nuts: 12 ft. lbs. (16 Nm).
- With manual transmission, drive belt tensioner damper.
- Radiator
- Air cleaner and MAF meter assembly.
- No. 1 air hose
- Negative battery cable

5. Fill the cooling system to the proper level.

6. Start the vehicle, check for leaks and repair if necessary.

Cylinder Head

REMOVAL & INSTALLATION

5E-FE Engine

1. Before servicing the vehicle, refer to the precautions in the beginning of this section.

2. Properly relieve the fuel system pressure.

3. Drain the engine oil.

4. Drain the cooling system.

5. Remove or disconnect the following:
- Negative battery cable. On vehicles equipped with an air bag, wait at least 90 seconds before proceeding.
- Right engine undercover
- All necessary components to gain access to the timing belt
- Timing belt
- Intake and Exhaust manifolds
- Camshafts, following the proper sequences and procedures.

➡**Loosen the cylinder head bolts in several passes and in the reverse order of the installation sequence.**

- Cylinder head. Make note of cylinder bolt positions and replace them in their original position.

To install:

➡**The head bolts stretch and must be replaced once removed.**

6. Install or connect the following:
- 2 different size head bolts, lightly oiled, in their correct positions and tighten them in several passes in the proper sequence, evenly, to 33 ft. lbs. (44 Nm). Mark each bolt with a reference mark and tighten

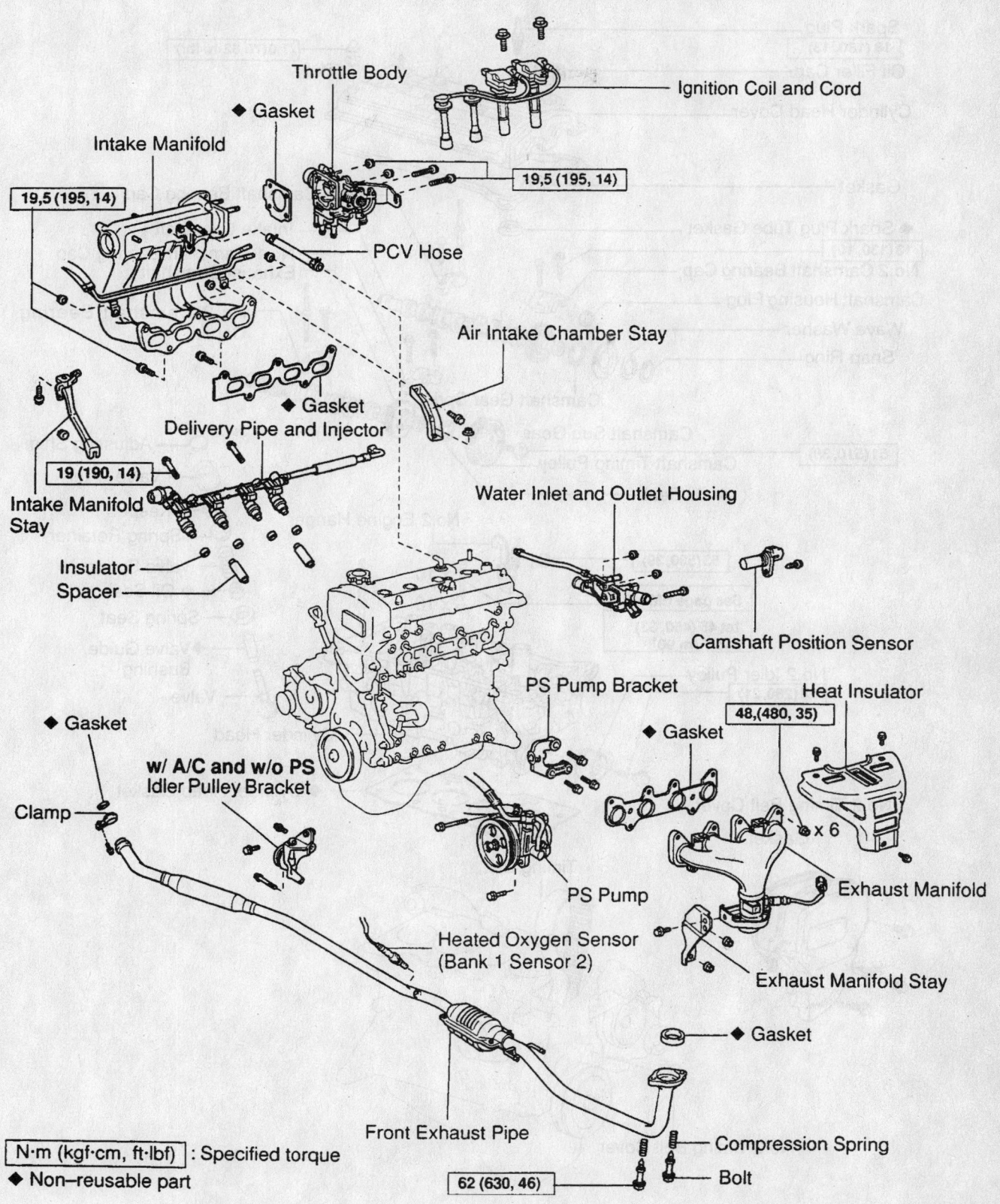

N·m (kgf·cm, ft·lbf) : Specified torque

◆ Non–reusable part

Removal of intake manifold, throttle body, exhaust front pipe and manifold—1.5L (5E-FE) engine

9301WG01

For Tire, Wheel and Ball Joint specifications, see Section 1 of this manual

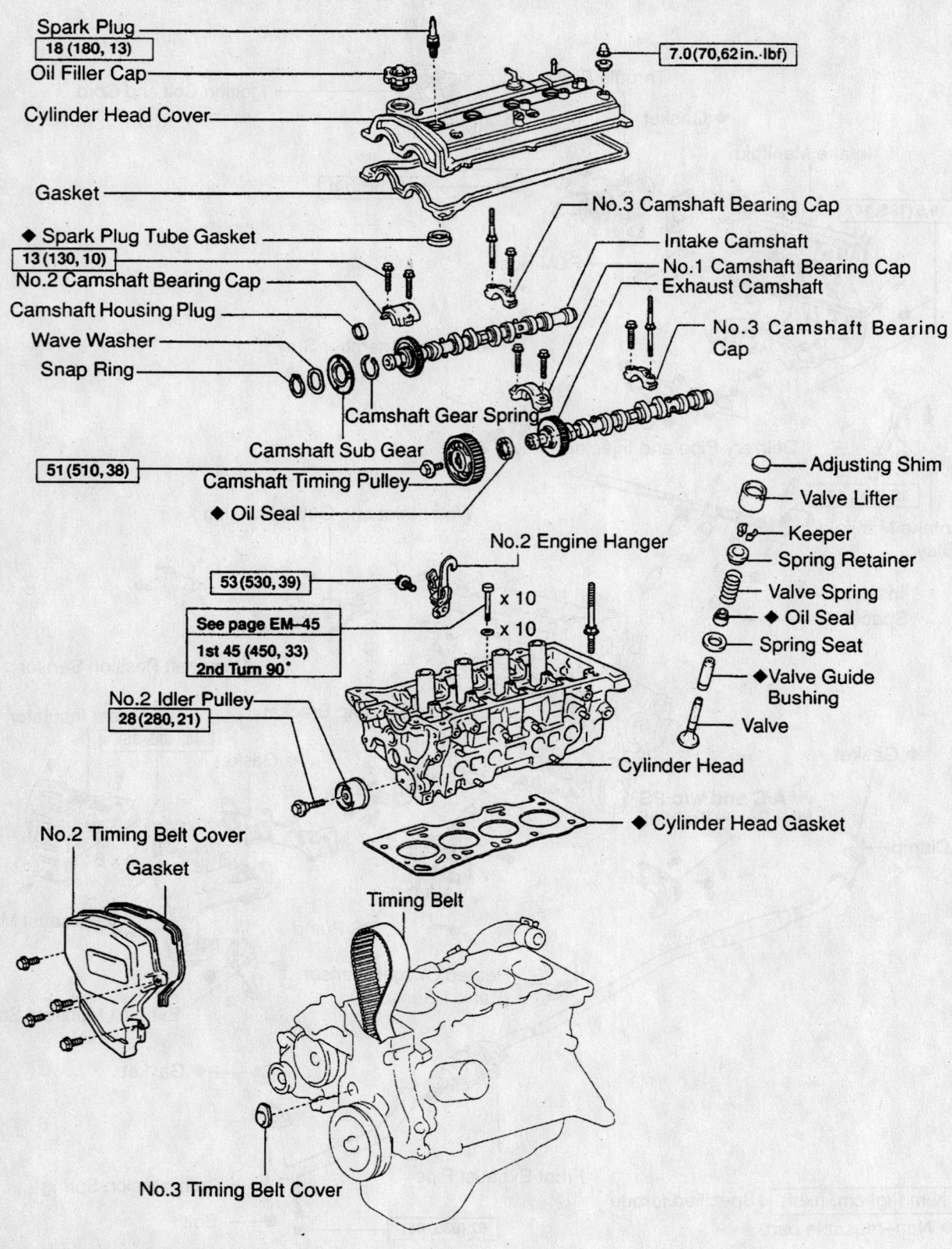

Spark Plug
18 (180, 13)
Oil Filler Cap
Cylinder Head Cover
Gasket
◆ Spark Plug Tube Gasket
13 (130, 10)
No.2 Camshaft Bearing Cap
Camshaft Housing Plug
Wave Washer
Snap Ring
Camshaft Gear Spring
Camshaft Sub Gear
51 (510, 38)
Camshaft Timing Pulley
◆ Oil Seal

7.0 (70, 62 in.·lbf)

No.3 Camshaft Bearing Cap
Intake Camshaft
No.1 Camshaft Bearing Cap
Exhaust Camshaft
No.3 Camshaft Bearing Cap

Adjusting Shim
Valve Lifter
Keeper
Spring Retainer
Valve Spring
◆ Oil Seal
Spring Seat
◆ Valve Guide Bushing
Valve

No.2 Engine Hanger
53 (530, 39)
See page EM–45
1st 45 (450, 33)
2nd Turn 90°
x 10
x 10

No.2 Idler Pulley
28 (280, 21)

Cylinder Head
◆ Cylinder Head Gasket

No.2 Timing Belt Cover
Gasket

Timing Belt

No.3 Timing Belt Cover

N·m (kgf·cm, ft·lbf) : Specified torque
◆ Non–reusable part

Exploded view of cylinder head disassembly—1.5L (5E-FE) engine

9301WG02

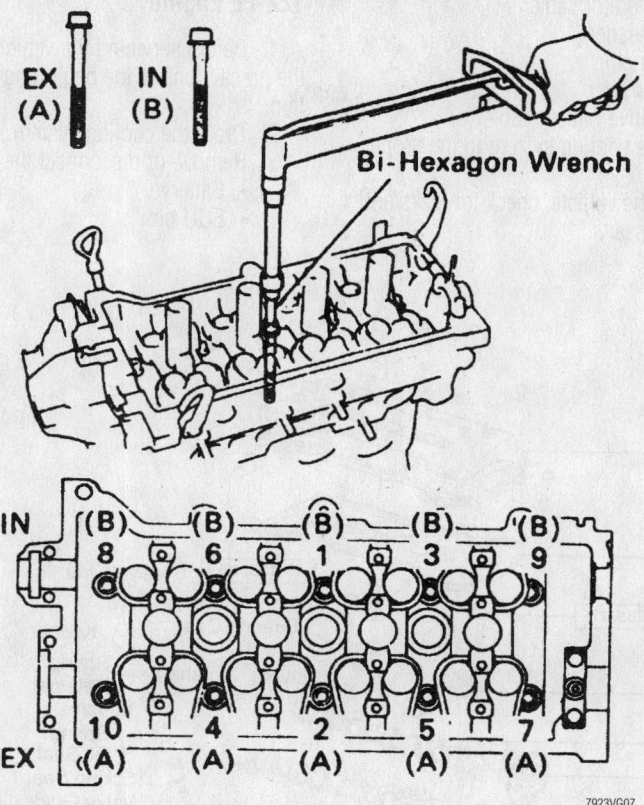

Cylinder head bolt tightening sequence—5E-FE engine

each bolt in sequence an additional 90 degrees.
- Camshafts and all other components following the proper sequences and procedures.

1NZ-FE Engine

1. Before servicing the vehicle, refer to the precautions in the beginning of this section.
2. Drain the cooling system.
3. Remove or disconnect the following:
- Negative battery cable
- Water filler
- Outer front cowl top panel
- Alternator
- Air cleaner
- Accelerator cable
- Center exhaust pipe
- Exhaust manifold support
- Exhaust manifold
- Ignition coil
- Spark plugs
- PCV hoses
- Throttle body
- Engine wiring harness at the head
- Intake manifold
- Camshaft position sensor

- ECT sensor
- Oil control valve
- PCV valve
- Oil filer cap
- Cylinder head cover
- Fuel injectors
- Timing chain cover

- Camshaft sprockets and valve timing control assembly
- Camshafts
- Cylinder head. Remove the bolts in a circular pattern, in several stages, starting from the ends and working towards the center

To install:
4. Install or connect the following:
- Cylinder head, using a new gasket. The Lod. No. on the gasket faces UP
5. Torque the cylinder head bolts, in sequence, in 3 steps:
   a. Step 1: 22 ft. lbs. (29 Nm).
   b. Step 2: Plus a 90 degree turn.
   c. Step 3: Plus a 90 degree turn.
6. Install or connect the following:
- Water bypass pipe. Torque the bolt to 80 inch lbs. (9 Nm).
- Camshafts
7. Camshaft bearing caps in 2 stages:
   a. Step: 10 ft. lbs. (13 Nm).
   b. Step 2: 17 ft. lbs. (23 Nm).
8. Install or connect the following:
- Sprockets and valve timing controller assembly, aligning the knock pin and hole. Torque the bolts to 47 ft. lbs. (64 Nm).
- Check and adjust the valves
- Cylinder head cover
- Oil filler cap
- PCV valve
- ECT sensor
- Camshaft position sensor
- Timing chain cover
- Intake manifold
- Engine wiring harness

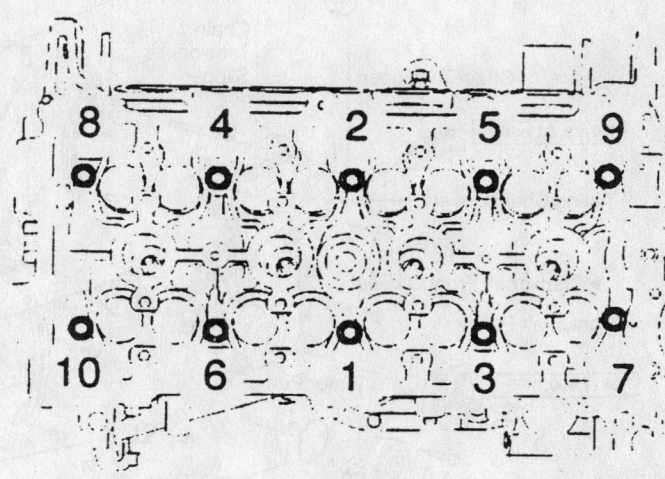

Cylinder head bolt tightening sequence—1NZ-FE engine

- Throttle body
- PCV hoses
- Spark plugs
- Ignition coils
- Exhaust manifold
- Exhaust manifold support. Torque the bolts to 27 ft. lbs. (37 Nm).
- Front exhaust pipe. Torque the nuts to 46 ft. lbs. (62 Nm).

- Accelerator cable
- Air cleaner
- Alternator
- Water filler
- Negative battery cable

9. Fill the cooling system to the proper level.

10. Start the vehicle, check for leaks and repair if necessary.

1ZZ-FE Engine

1. Before servicing the vehicle, refer to the precautions in the beginning of this section.

2. Drain the cooling system.

3. Remove or disconnect the following:
- Battery
- ECU box

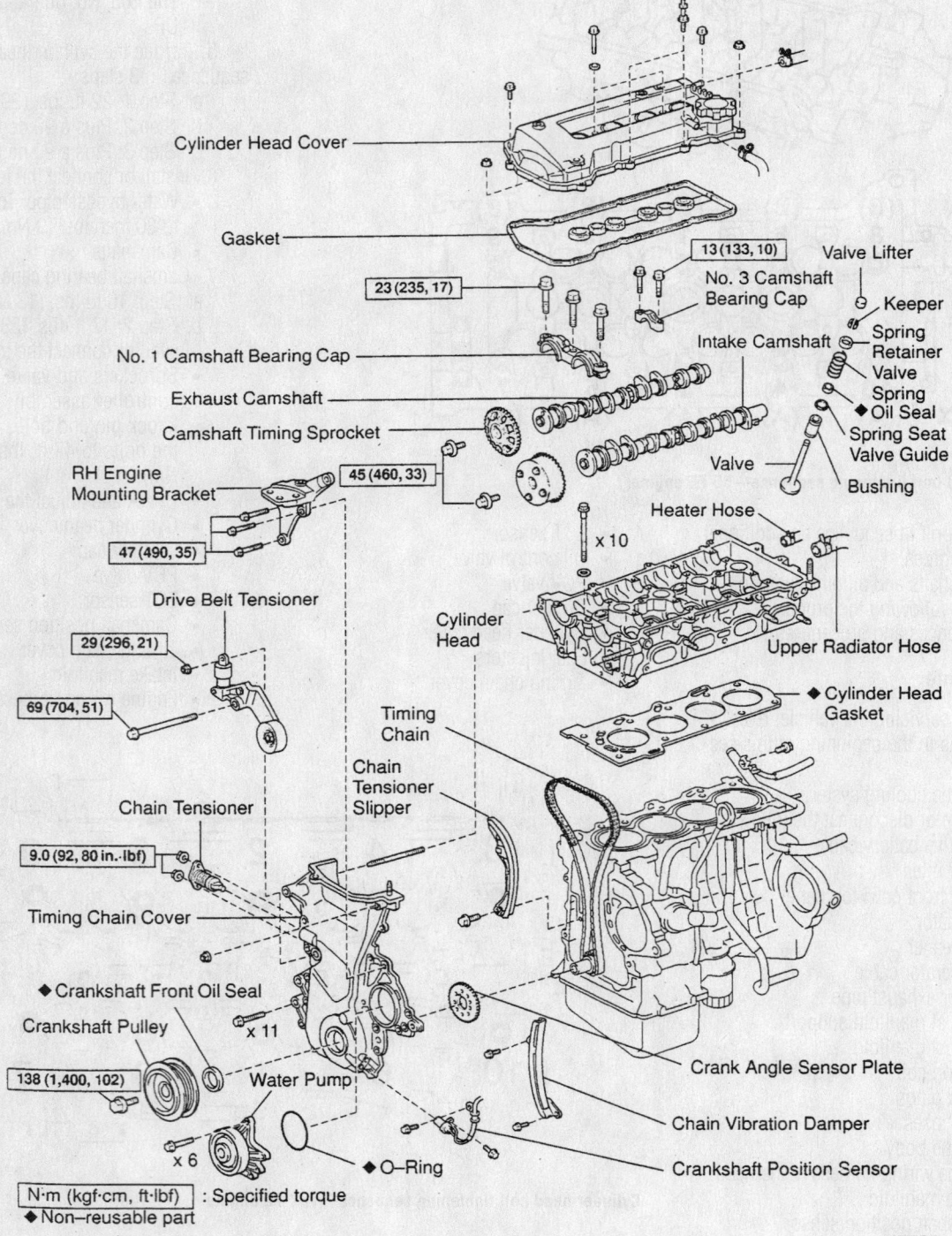

N·m (kgf·cm, ft·lbf) : Specified torque
◆ Non-reusable part

9307WG92

Cylinder head component removal—1ZZ-FE engine

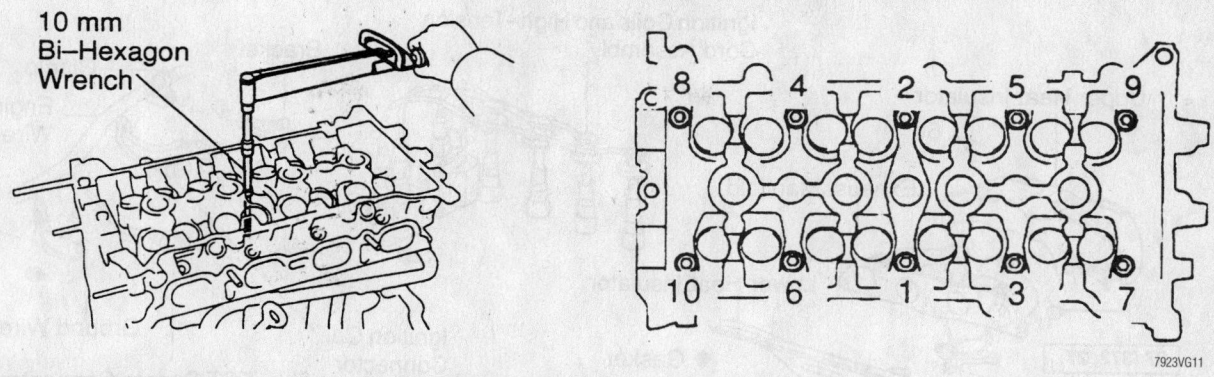

10 mm
Bi–Hexagon
Wrench

8 4 2 5 9

10 6 1 3 7

7923VG11

Cylinder head bolt tightening sequence—1ZZ-FE and 2ZZ-GE engines

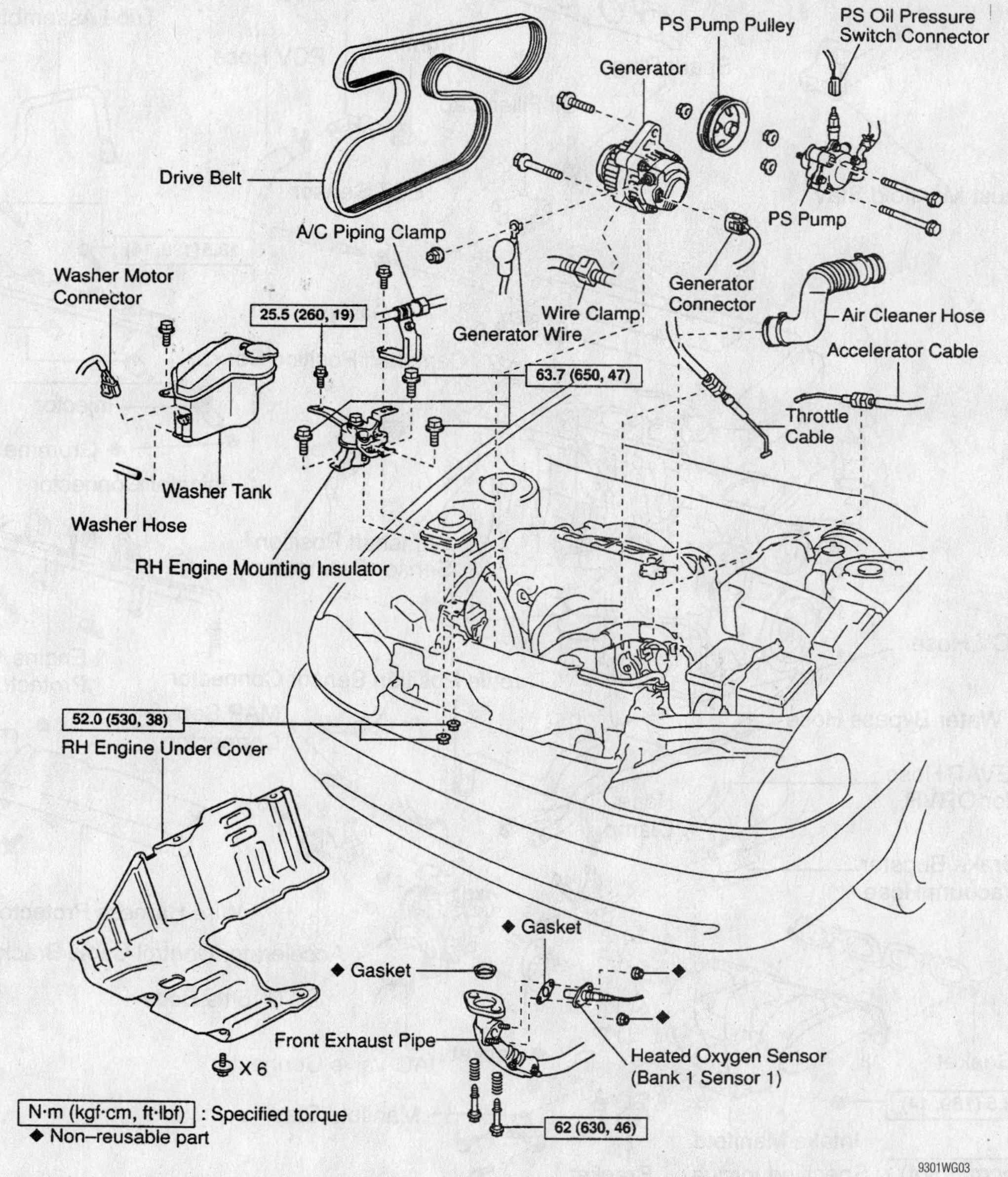

PS Pump Pulley

PS Oil Pressure
Switch Connector

Generator

Drive Belt

A/C Piping Clamp

PS Pump

Washer Motor
Connector

Generator Wire

Wire Clamp

Generator
Connector

Air Cleaner Hose

Accelerator Cable

25.5 (260, 19)

63.7 (650, 47)

Throttle
Cable

Washer Tank

Washer Hose

RH Engine Mounting Insulator

52.0 (530, 38)
RH Engine Under Cover

◆ Gasket

◆ Gasket

Front Exhaust Pipe

Heated Oxygen Sensor
(Bank 1 Sensor 1)

🔩 X 6

N·m (kgf·cm, ft·lbf) : Specified torque
◆ Non–reusable part

62 (630, 46)

9301WG03

Exploded view of engine accessories and right-hand engine under cover—1.8L (1ZZ-FE) Engines.

For Maintenance Interval recommendations, see Section 1 of this manual

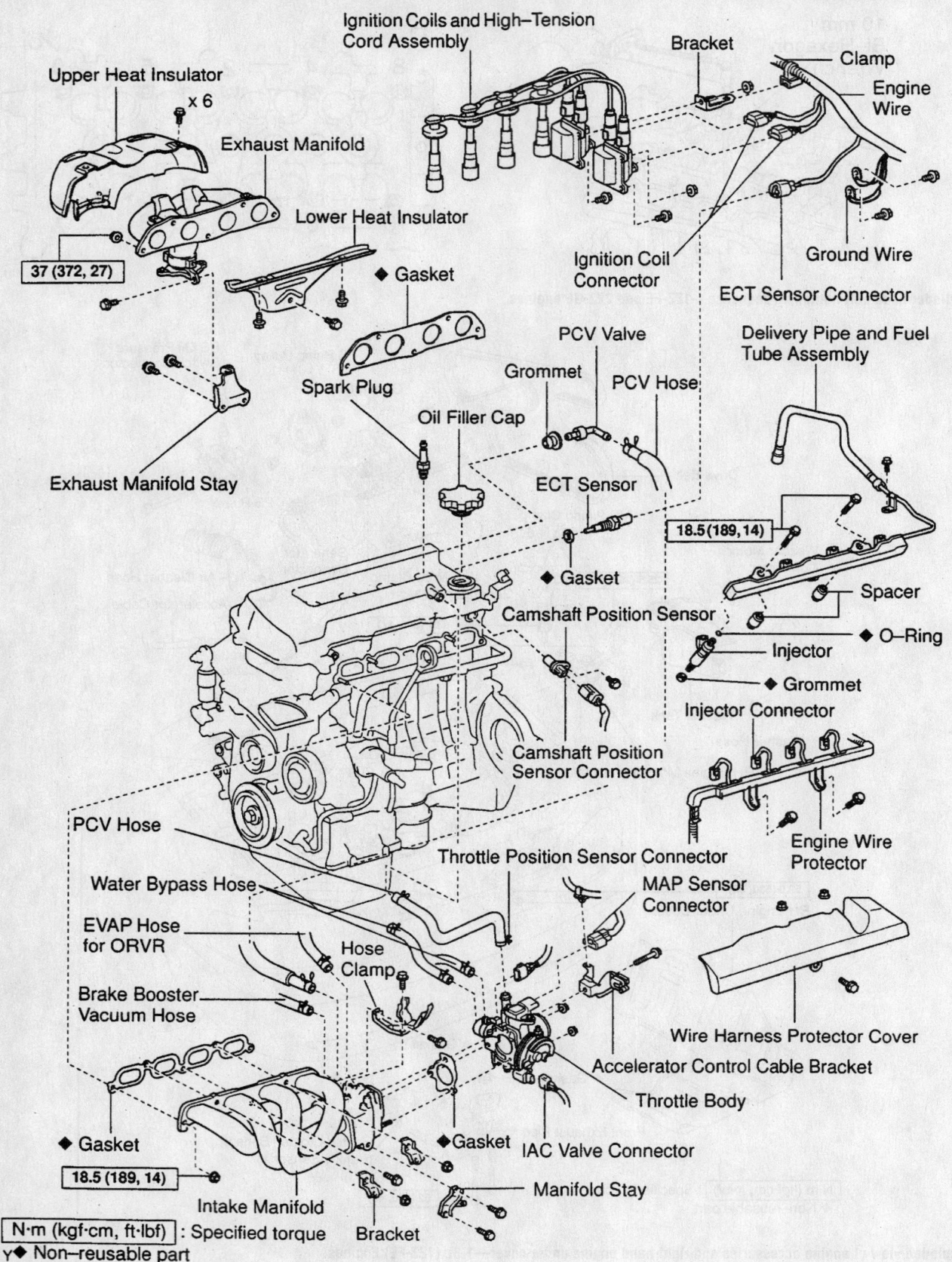

Ignition Coils and High–Tension Cord Assembly

Bracket

Clamp

Engine Wire

Upper Heat Insulator

x 6

Exhaust Manifold

Ground Wire

Ignition Coil Connector

ECT Sensor Connector

Lower Heat Insulator

37 (372, 27)

◆ Gasket

Spark Plug

PCV Valve

Grommet

PCV Hose

Delivery Pipe and Fuel Tube Assembly

Exhaust Manifold Stay

Oil Filler Cap

ECT Sensor

◆ Gasket

18.5 (189, 14)

Spacer

Camshaft Position Sensor

◆ O–Ring

Injector

◆ Grommet

Injector Connector

Camshaft Position Sensor Connector

Engine Wire Protector

PCV Hose

Throttle Position Sensor Connector

MAP Sensor Connector

Water Bypass Hose

EVAP Hose for ORVR

Hose Clamp

Wire Harness Protector Cover

Brake Booster Vacuum Hose

Accelerator Control Cable Bracket

Throttle Body

◆ Gasket

◆ Gasket

IAC Valve Connector

18.5 (189, 14)

Manifold Stay

Intake Manifold

N·m (kgf·cm, ft·lbf) : Specified torque

Bracket

Y◆ Non–reusable part

Spark Plug

View of engine intake, exhaust, ignition, and fuel system location—1.8L (1ZZ-FE) engines.

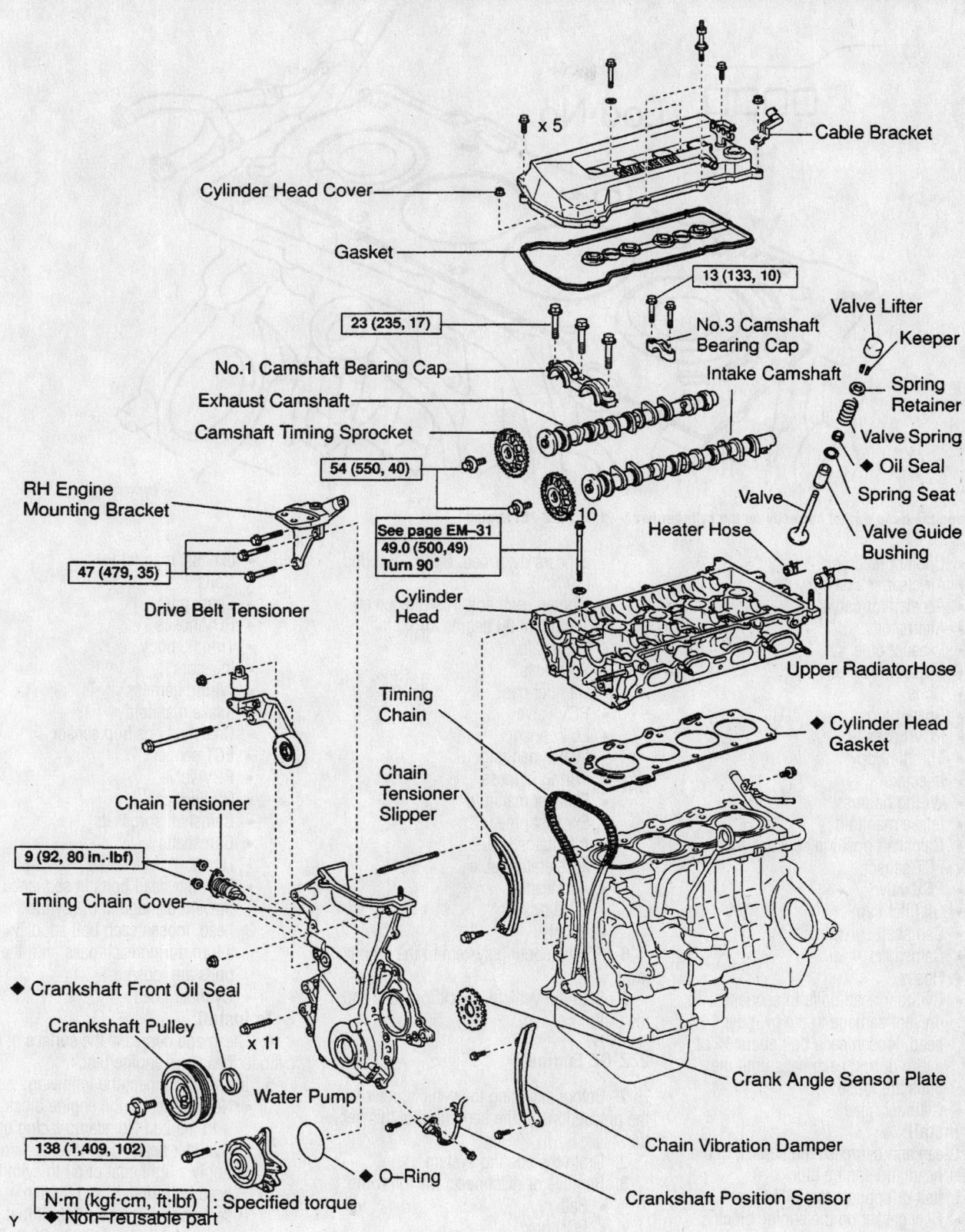

Cable Bracket

x 5

Cylinder Head Cover

Gasket

13 (133, 10)

23 (235, 17)

Valve Lifter

Keeper

No.3 Camshaft
Bearing Cap

No.1 Camshaft Bearing Cap

Intake Camshaft

Spring
Retainer

Exhaust Camshaft

Camshaft Timing Sprocket

Valve Spring

54 (550, 40)

◆ Oil Seal

RH Engine
Mounting Bracket

Spring Seat

Valve

See page EM–31
49.0 (500,49)
Turn 90°

Heater Hose

Valve Guide
Bushing

47 (479, 35)

Cylinder
Head

Drive Belt Tensioner

Upper RadiatorHose

Timing
Chain

◆ Cylinder Head
Gasket

Chain Tensioner

Chain
Tensioner
Slipper

9 (92, 80 in.·lbf)

Timing Chain Cover

◆ Crankshaft Front Oil Seal

Crankshaft Pulley

x 11

Crank Angle Sensor Plate

Water Pump

Chain Vibration Damper

138 (1,409, 102)

◆ O–Ring

Crankshaft Position Sensor

N·m (kgf·cm, ft·lbf) : Specified torque
Y ◆ Non–reusable part

9301WG05

Illustration of disassembled cylinder head assembly—1.8L (1ZZ-FE) Engines.

For Tune-up, Capacities and Firing orders, see Section 1 of this manual

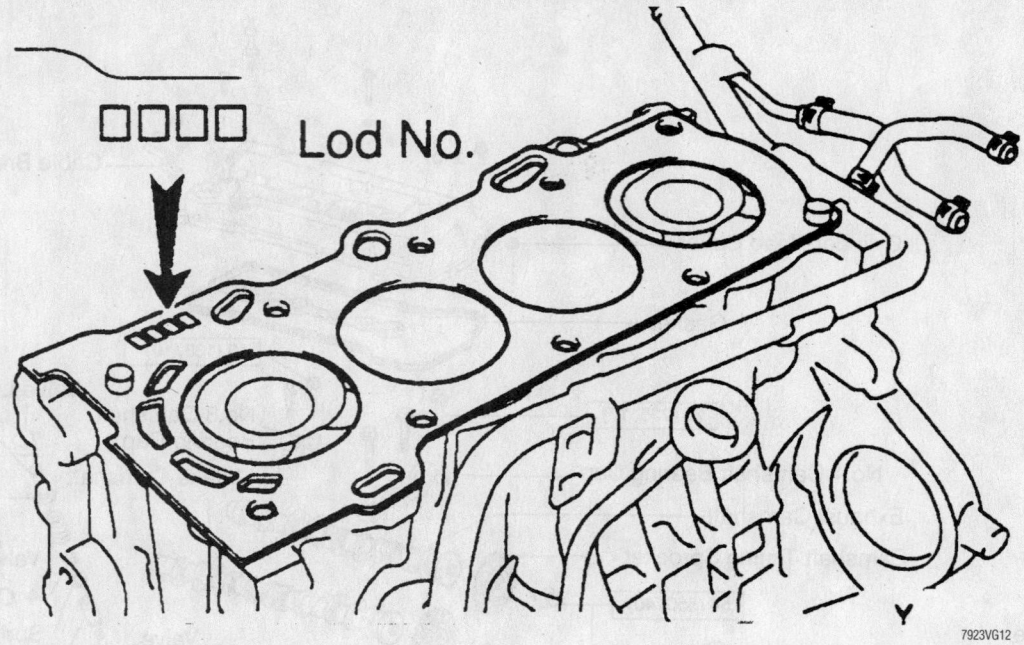

Position the head gasket correctly on the cylinder head—1.8L (1ZZ-FE) engine

- Coolant reservoir
- Air cleaner assembly
- Accelerator cable
- Alternator
- Exhaust pipe
- Exhaust manifold
- Coils
- Spark plugs
- PCV hoses
- Throttle body
- Injectors
- Wiring harness
- Intake manifold
- Camshaft position sensor
- ECT sensor
- PCV valve
- Oil filler cap
- Camshaft sprockets
- Camshafts
- Hoses
- Cylinder head bolts in sequence. To prevent damage to the cylinder head, loosen each bolt about ¼ of a turn during each pass until the bolts are loose.
- Cylinder head

To install:

4. Clean and degrease the surface of the cylinder head and engine block.

5. Install or connect the following:
- New gasket on the engine block with the Lod No. stamp facing up.
- Cylinder head
- Apply a light coat of oil to cylinder head bolt threads and tighten in sequence. Replace any bolt that appears deformed. Bolts: 36 ft. lbs. (49 Nm).
- Tighten each bolt in sequence an additional 90 degree turn.
- Camshafts
- Sprockets
- Oil filler cap
- PCV valve
- ECT sensor
- Intake manifold
- Wiring harness
- Exhaust manifold
- Exhaust pipe
- Alternator
- accelerator cable
- Air cleaner
- ECM box
- Battery

6. Fill the cooling system to the proper level.

7. Start the vehicle, check for leaks and repair if necessary.

2ZZ-GE Engine

1. Before servicing the vehicle, refer to the precautions in the beginning of this section.

2. Drain the cooling system.

3. Remove or disconnect the following:
- Battery
- ECU box
- Coolant reservoir
- Air cleaner assembly
- Accelerator cable
- Alternator
- Exhaust pipe
- Exhaust manifold
- Coils
- Spark plugs
- PCV hoses
- Throttle body
- Injectors
- Wiring harness
- Intake manifold
- Camshaft position sensor
- ECT sensor
- PCV valve
- Oil filler cap
- Camshaft sprockets
- Camshafts
- Hoses
- Cylinder head bolts in sequence. To prevent damage to the cylinder head, loosen each bolt about ¼ of a turn during each pass until the bolts are loose.
- Cylinder head

To install:

4. Clean and degrease the surface of the cylinder head and engine block.

5. Install or connect the following:
- New gasket on the engine block with the Lod No. stamp facing up.
- Cylinder head
- Apply a light coat of oil to cylinder head bolt threads and tighten in sequence. Replace any bolt that appears deformed. Bolts: 26 ft. lbs. (49 Nm).
- Torque each bolt in sequence an additional 180 degree turn.
- Camshafts

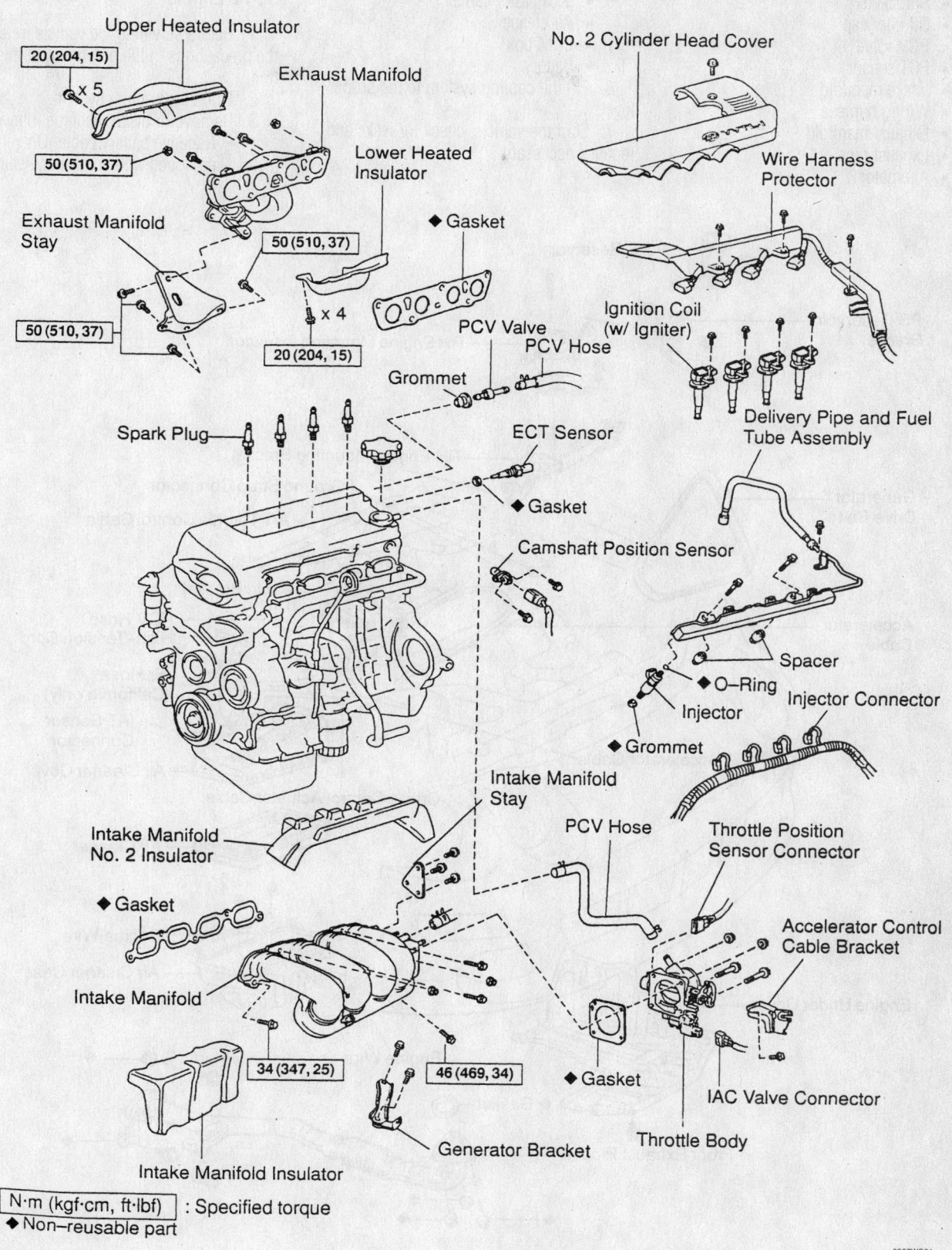

Cylinder head component exploded view—2ZZ-GE engine

- Sprockets
- Oil filler cap
- PCV valve
- ECT sensor
- Intake manifold
- Wiring harness
- Exhaust manifold
- Exhaust pipe
- Alternator

- accelerator cable
- Air cleaner
- ECM box
- Battery

6. Fill the cooling system to the proper level.

7. Start the vehicle, check for leaks and repair if necessary.

5S-FE Engine

1. Before servicing the vehicle, refer to the precautions in the beginning of this section.

2. Drain the cooling system.

3. Remove or disconnect the following:
- Negative battery cable. On vehicles equipped with an air bag, wait at

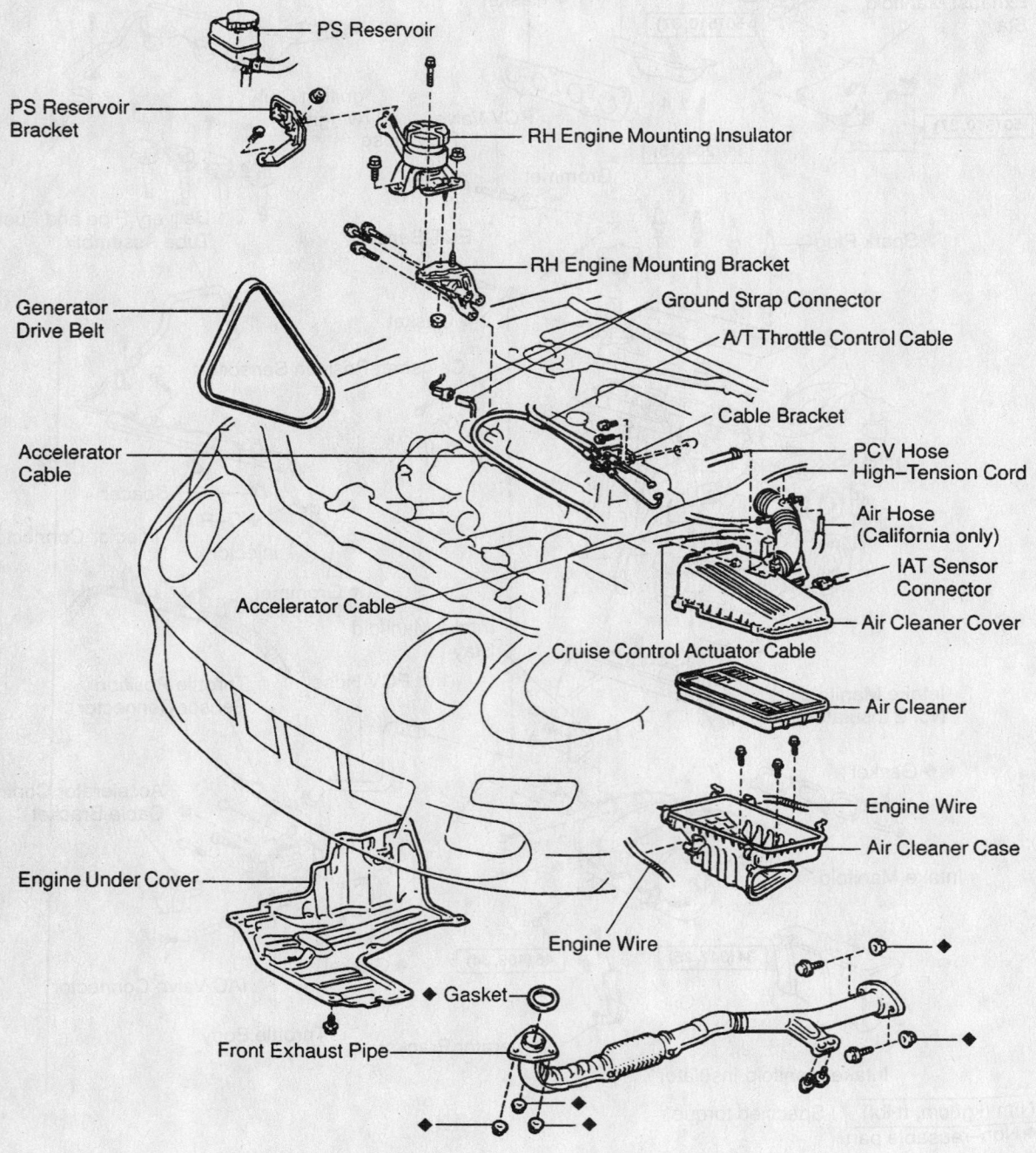

PS Reservoir

PS Reservoir Bracket

RH Engine Mounting Insulator

RH Engine Mounting Bracket

Ground Strap Connector

A/T Throttle Control Cable

Generator Drive Belt

Cable Bracket

Accelerator Cable

PCV Hose

High-Tension Cord

Air Hose (California only)

IAT Sensor Connector

Air Cleaner Cover

Accelerator Cable

Cruise Control Actuator Cable

Air Cleaner

Engine Wire

Air Cleaner Case

Engine Under Cover

Engine Wire

Gasket

Front Exhaust Pipe

◆ Non–reusable part

9301WG09

Exploded view of engine accessories and right-hand engine under cover—1.8L (1ZZ-FE) Engines.

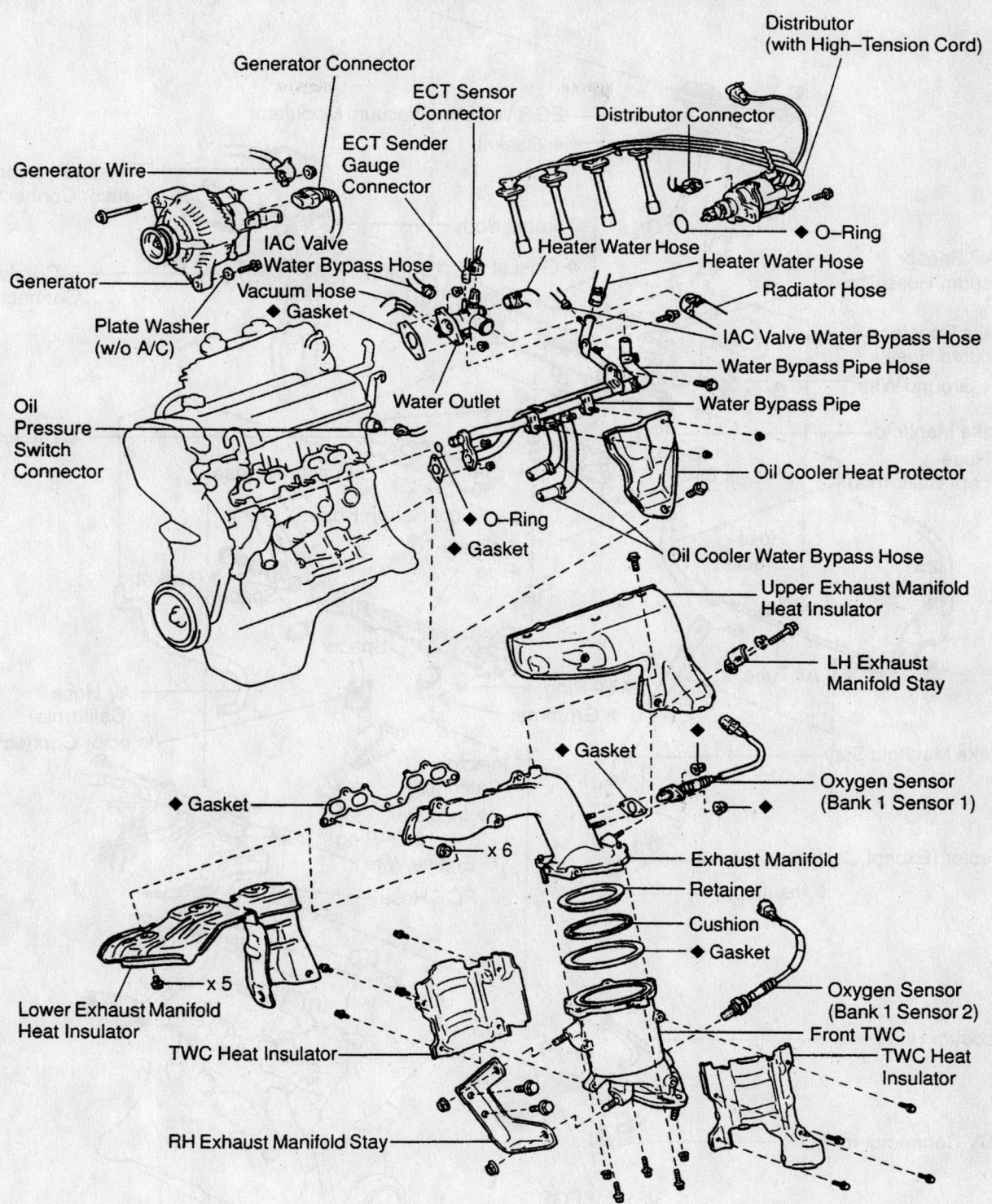

◆ Non–reusable part

9301WG10

View of engine, exhaust, ignition and fuel system location—1.8L (1ZZ-FE) Engines.

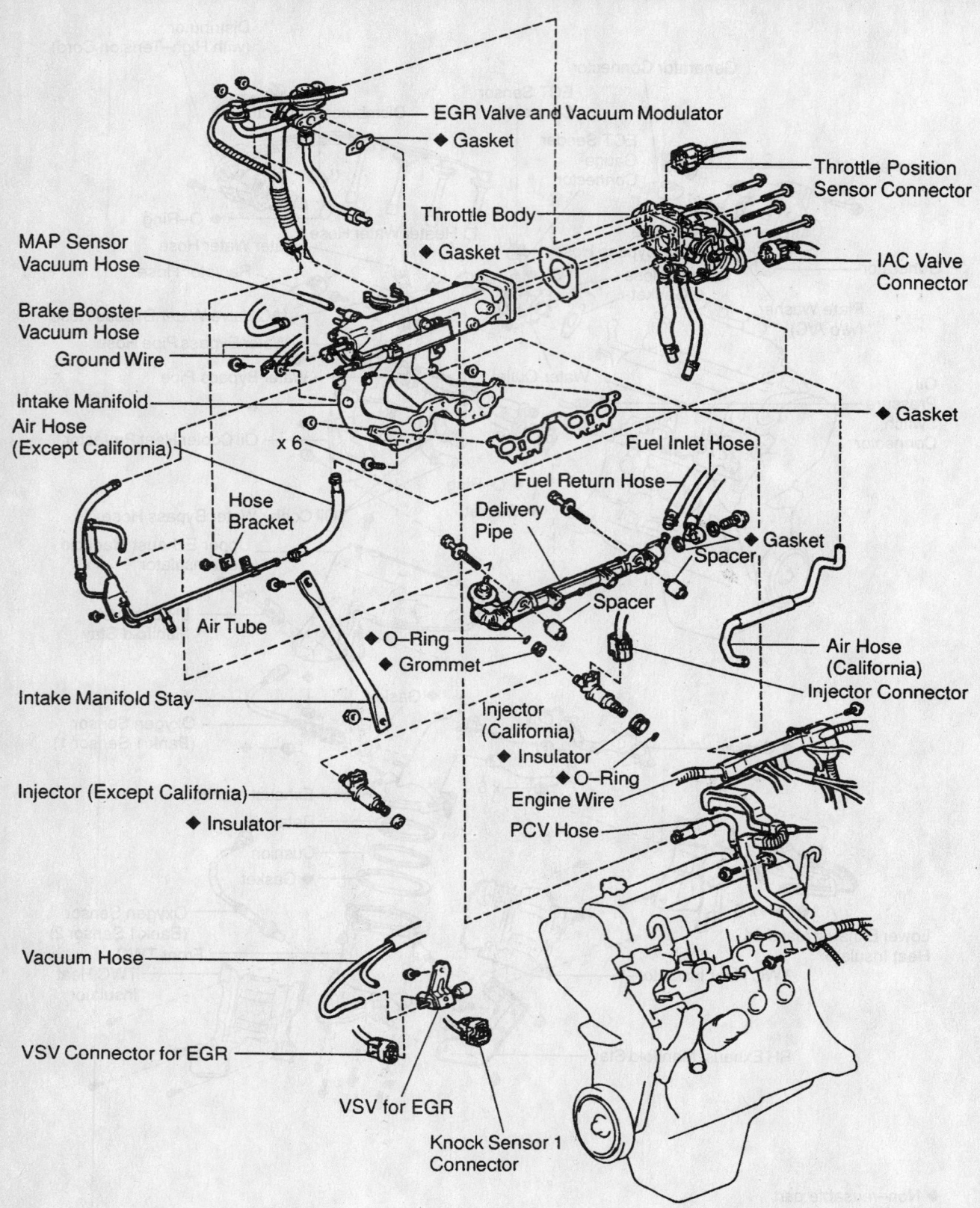

EGR Valve and Vacuum Modulator

◆ Gasket

Throttle Position Sensor Connector

Throttle Body

◆ Gasket

IAC Valve Connector

MAP Sensor Vacuum Hose

Brake Booster Vacuum Hose

Ground Wire

◆ Gasket

Intake Manifold Air Hose (Except California)

Fuel Inlet Hose

Fuel Return Hose

x 6

Hose Bracket

Delivery Pipe

Spacer

◆ Gasket

Spacer

Air Tube

◆ O–Ring

◆ Grommet

Air Hose (California)

Injector Connector

Intake Manifold Stay

Injector (California)

◆ Insulator

◆ O–Ring

Engine Wire

PCV Hose

Injector (Except California)

◆ Insulator

Vacuum Hose

VSV Connector for EGR

VSV for EGR

Knock Sensor 1 Connector

◆ Non–reusable part

View of engine, intake, ignition and fuel system location—1.8L (1ZZ-FE) Engines.

9301WG11

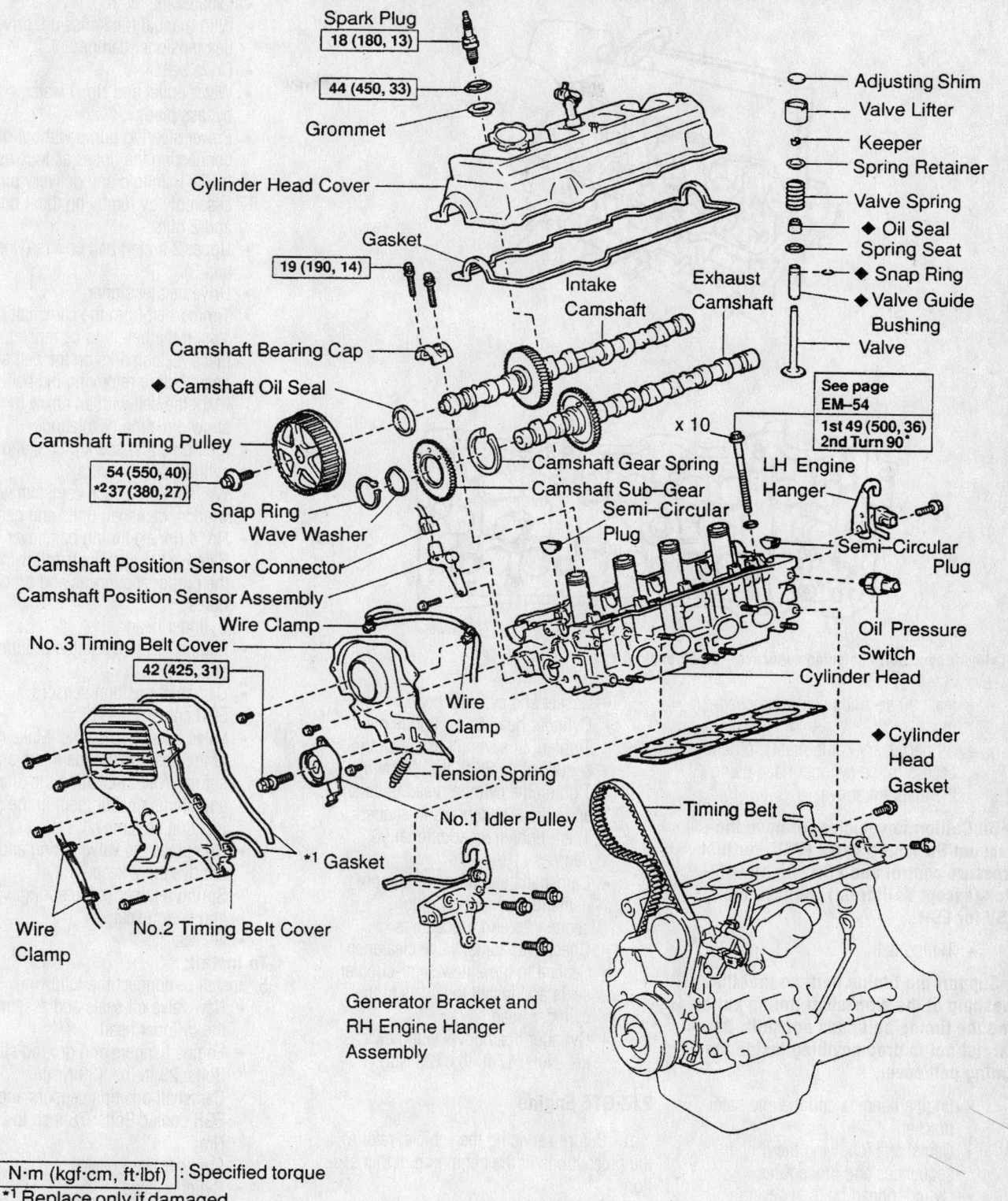

Spark Plug
18 (180, 13)

44 (450, 33)

Grommet

Cylinder Head Cover

Gasket

19 (190, 14)

Camshaft Bearing Cap

◆ Camshaft Oil Seal

Camshaft Timing Pulley

54 (550, 40)
*2 37 (380, 27)

Snap Ring

Wave Washer

Camshaft Position Sensor Connector

Camshaft Position Sensor Assembly

Wire Clamp

No. 3 Timing Belt Cover

42 (425, 31)

Wire Clamp

Wire Clamp

No.2 Timing Belt Cover

*1 Gasket

Tension Spring

No.1 Idler Pulley

Generator Bracket and RH Engine Hanger Assembly

Intake Camshaft

Exhaust Camshaft

Adjusting Shim
Valve Lifter
Keeper
Spring Retainer
Valve Spring
◆ Oil Seal
Spring Seat
◆ Snap Ring
◆ Valve Guide Bushing
Valve

See page EM-54
1st 49 (500, 36)
2nd Turn 90°

Camshaft Gear Spring

Camshaft Sub-Gear

Semi-Circular Plug

x 10

LH Engine Hanger

Semi-Circular Plug

Oil Pressure Switch

Cylinder Head

◆ Cylinder Head Gasket

Timing Belt

N·m (kgf·cm, ft·lbf) : Specified torque
*1 Replace only if damaged
*2 For use with SST
◆ Non-reusable part

9301WG12

Illustration of disassembled cylinder head assembly—1.8L (1ZZ-FE) Engines.

Timing belt service is covered in Section 3 of this manual

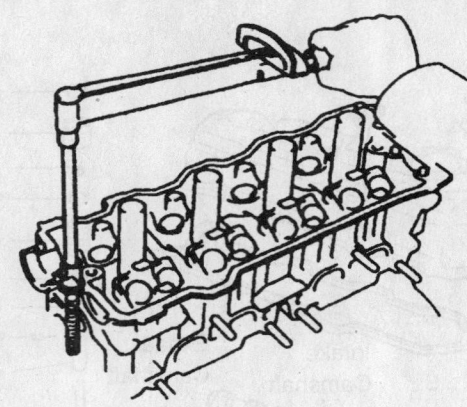

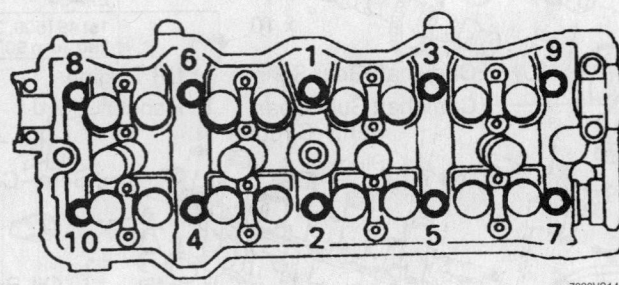

7923VG14

Cylinder head bolt tightening sequence—5S-FE engine

least 90 seconds before proceeding.
- All necessary components to gain access to the cylinder head using the diagram above.

➡**On California vehicles, remove the Vacuum Switching Valve (VSV) for fuel pressure control and EGR. On all vehicles (except California), remove the VSV for EGR.**

- Timing belt

➡**Support the timing belt, so that the meshing of the crankshaft timing pulley and the timing belt does not shift. Be careful not to drop anything inside the timing belt cover.**

- Engine hangers and the alternator bracket.
- Camshafts following the proper sequences and procedures.
- Cylinder head bolts in several passes and in the reverse order of the installation sequence.
- Cylinder head from the cylinder block, disengaging the cylinder head from the block dowel pins.

To install:
4. Install or connect the following:

- Gasket and cylinder head
- Cylinder head bolts, lightly oiled. Tighten, in several passes and in sequence, to 36 ft. lbs. (49 Nm). Tighten the cylinder head bolts an additional 90 degrees in sequence. Then, tighten an additional 90 degrees.
- Camshafts and all other components following the proper sequences and procedures.
- Check and adjust valve clearance.
- Sealant to the 2 new semi-circular seals and install the seals to the cylinder head.
- Cylinder head cover with new gasket. Nuts: 17 ft. lbs. (23 Nm).

2JZ-GTE Engine

1. Before servicing the vehicle, refer to the precautions in the beginning of this section.
2. Drain the cooling system.
3. Drain the engine oil.
4. Remove or disconnect the following:
- Negative battery cable from the battery. On vehicles equipped with an air bag, wait at least 90 seconds before proceeding.
- Turbocharger

- Exhaust manifold
- With manual transmission, drive belt tensioner damper.
- Drive belt
- Water outlet and No. 1 water bypass pipe.
- Power steering pump without disconnecting the hoses as follows:
- Intake manifold and delivery pipe assembly by removing the 4 bolts and 2 nuts.
- Upper 2 timing belt covers (Nos. 2 and 3).
- Drive belt tensioner
- Timing belt from the camshaft pulleys. If the belt is to be reused, place matchmarks on the belt and gears before removing the belt. Mark the belt with an arrow to show direction of rotation
- Ignition coils, spark plugs, and cylinder head covers
- While holding each camshaft with a wrench, camshaft bolts and gears
- No. 4 (inner) timing belt cover
- Remove the camshafts following the proper sequences and procedures
- Cylinder head
- Engine hangers and the ground strap
- Camshaft position sensors
- EGR cooler
- Valve lifters and shims. Make note of the positions of the lifters and shims. When installing the lifters and shims, install them in the same position as removal.
- Compress the valve spring and remove the 2 keepers
- Spring retainer, valve spring, valve and spring seat
- Oil seal

To install:
5. Install or connect the following:
- New valve oil seals and assemble the cylinder head
- Engine hangers and ground strap. Bolts: 29 ft. lbs. (39 Nm).
- Camshaft position sensors and EGR cooler. Bolts: 78 inch lbs. (9 Nm).
- Cylinder head gasket
- Cylinder head
- Head bolts, lightly oiled and plate washers. Uniformly tighten the head bolts in several passes in the correct order, to 25 ft. lbs. (34 Nm). Tighten each bolt an additional 90 degrees. Again, tighten the bolts 90 degrees.

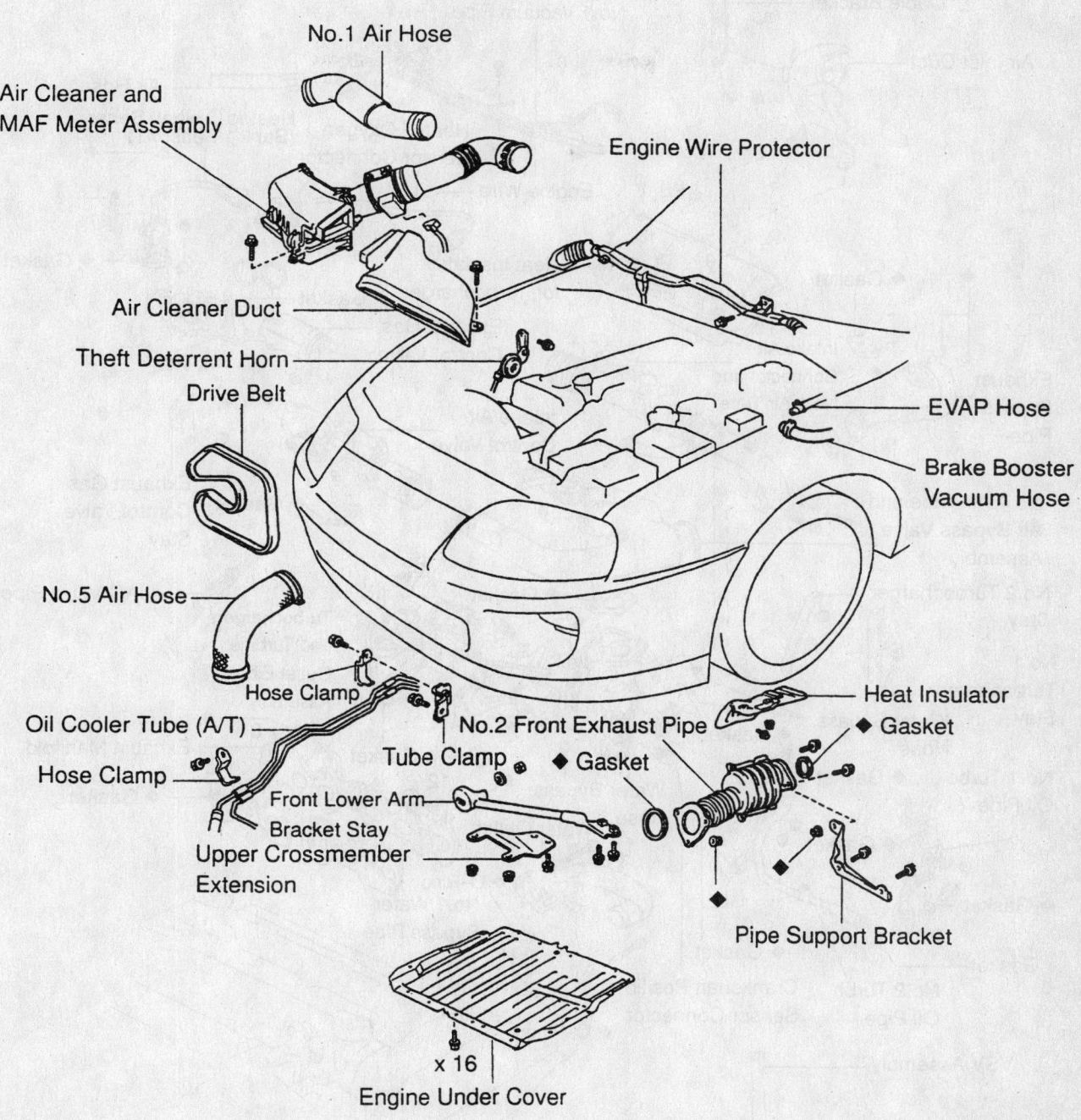

No.1 Air Hose

Air Cleaner and
MAF Meter Assembly

Engine Wire Protector

Air Cleaner Duct

Theft Deterrent Horn

Drive Belt

EVAP Hose

Brake Booster
Vacuum Hose

No.5 Air Hose

Hose Clamp

Oil Cooler Tube (A/T)

Hose Clamp

Tube Clamp

No.2 Front Exhaust Pipe

♦ Gasket

Front Lower Arm
Bracket Stay

Upper Crossmember
Extension

Heat Insulator

♦ Gasket

Pipe Support Bracket

x 16

Engine Under Cover

♦ Non–reusable part

9301WG17

Removal of air box, engine under cover and front exhaust pipe—3.0L (2JZ-GTE) engine

Heater Core replacement is covered in Section 2 of this manual

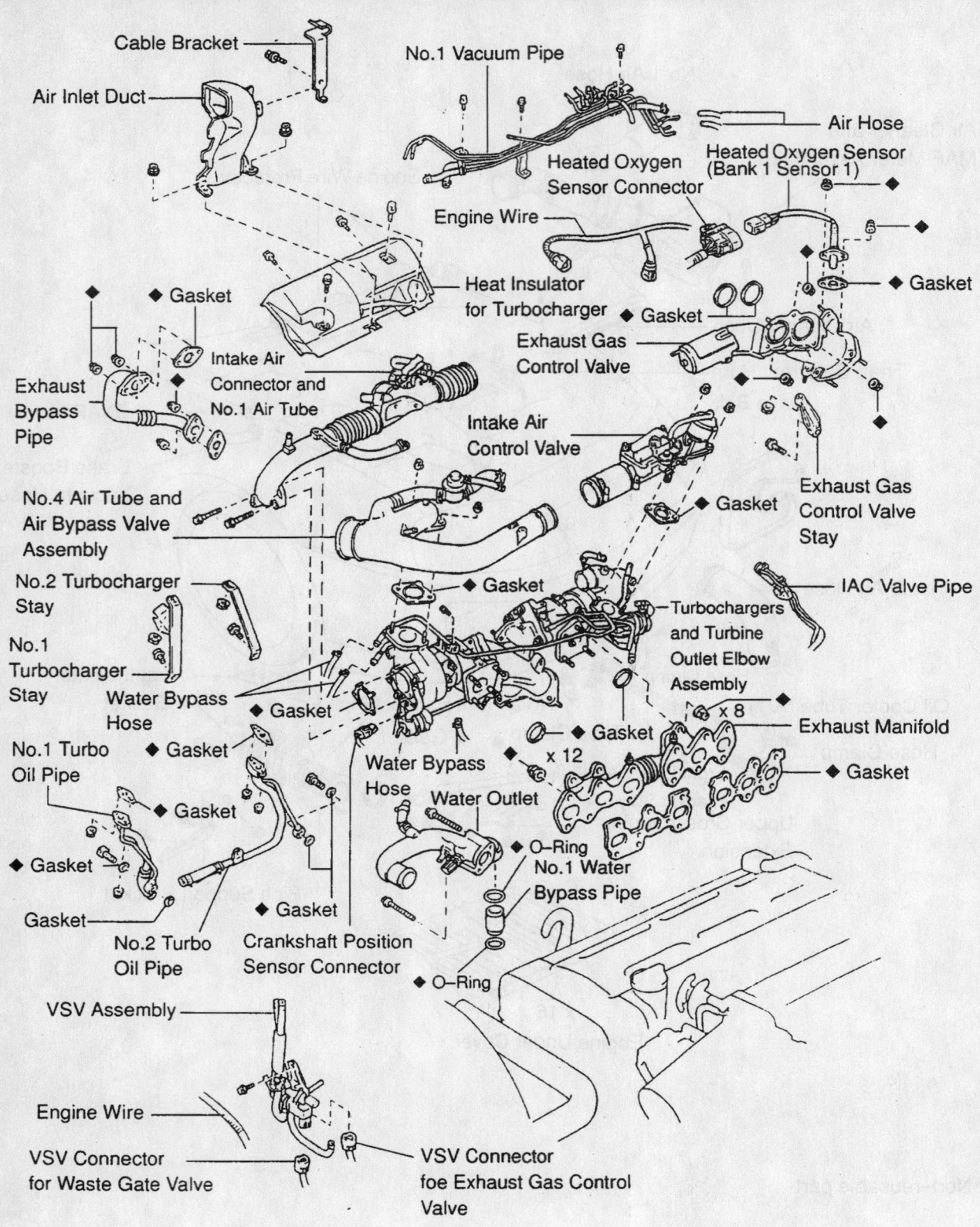

Cable Bracket

Air Inlet Duct

No.1 Vacuum Pipe

Air Hose

Heated Oxygen Sensor Connector

Heated Oxygen Sensor (Bank 1 Sensor 1)

Engine Wire

◆ Gasket

◆ Gasket

◆ Gasket

◆ Gasket

Exhaust Bypass Pipe

Intake Air Connector and No.1 Air Tube

Heat Insulator for Turbocharger

Exhaust Gas Control Valve

Intake Air Control Valve

Exhaust Gas Control Valve Stay

No.4 Air Tube and Air Bypass Valve Assembly

◆ Gasket

No.2 Turbocharger Stay

No.1 Turbocharger Stay

Water Bypass Hose

◆ Gasket

◆ Gasket

IAC Valve Pipe

Turbochargers and Turbine Outlet Elbow Assembly

x 8

Exhaust Manifold

No.1 Turbo Oil Pipe

◆ Gasket

◆ Gasket

Water Bypass Hose

x 12

◆ Gasket

◆ Gasket

◆ Gasket

◆ Gasket

Gasket

No.2 Turbo Oil Pipe

◆ Gasket

Crankshaft Position Sensor Connector

Water Outlet

O–Ring No.1 Water Bypass Pipe

◆ O–Ring

VSV Assembly

Engine Wire

VSV Connector for Waste Gate Valve

VSV Connector foe Exhaust Gas Control Valve

◆ Non–reusable part

Diagram of intake and exhaust manifolds—3.0L (2JZ-GTE) engine

9301WG18

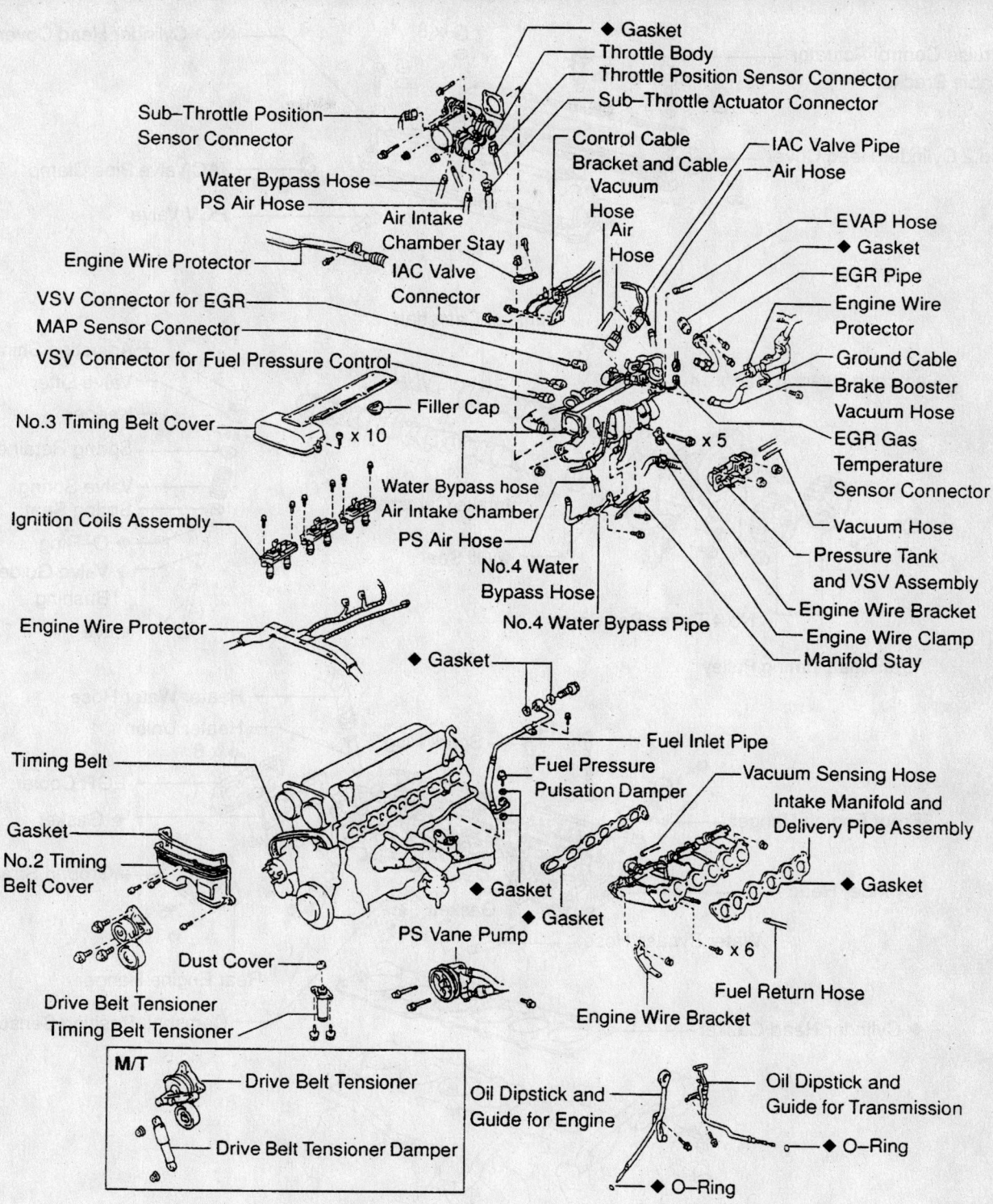

◆ Non–reusable part

Exploded view of intake manifold and throttle body—3.0L (2JZ-GTE) engine

9301WG19

Brake service is covered in Section 4 of this manual

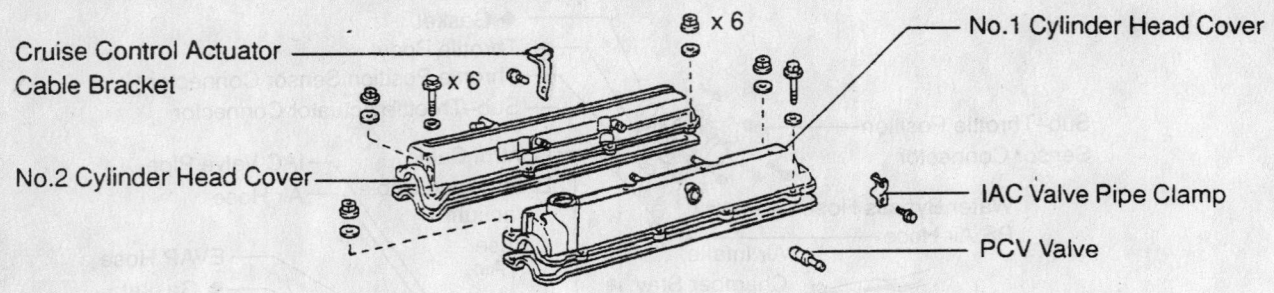

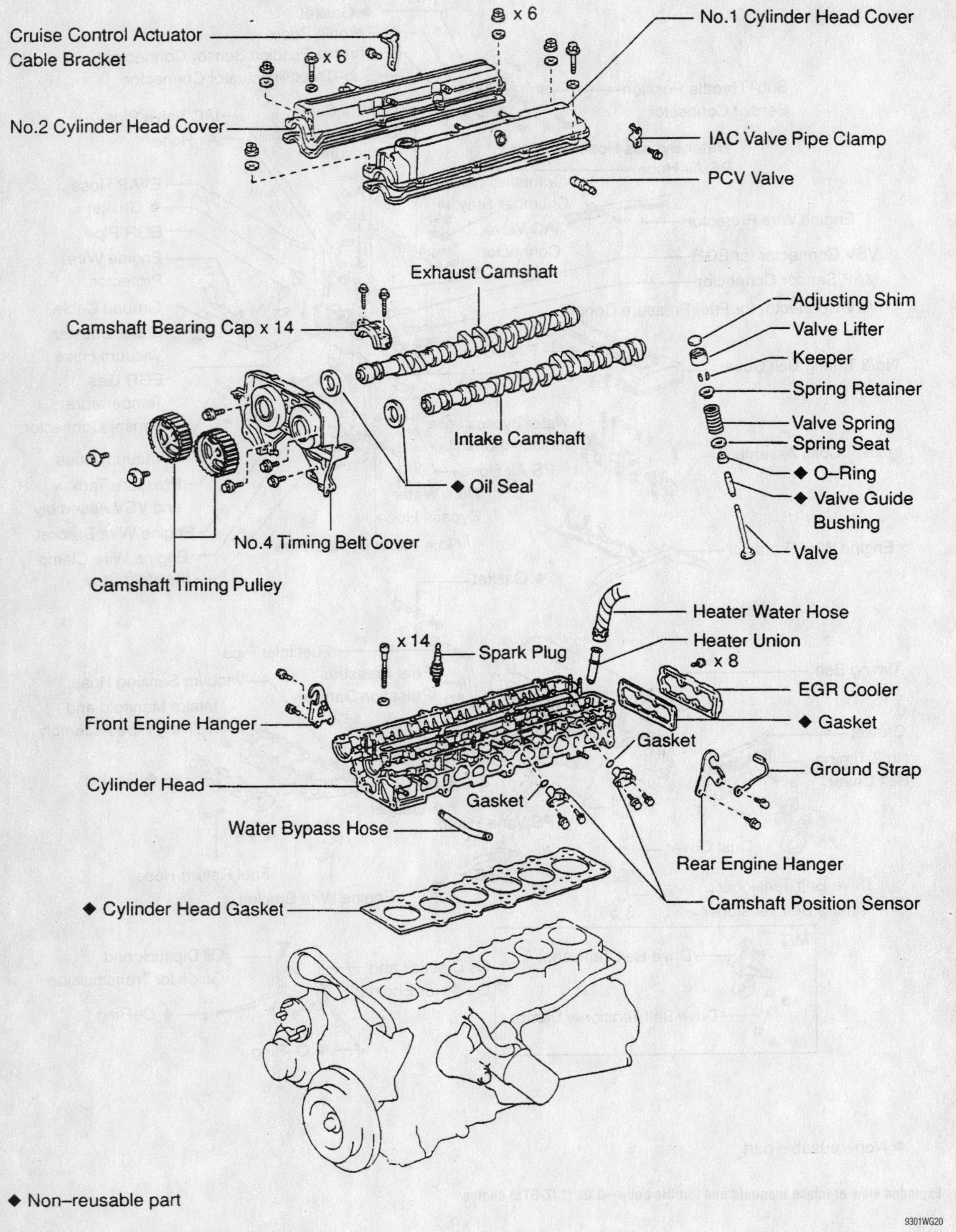

Cruise Control Actuator
Cable Bracket

No.2 Cylinder Head Cover

× 6

× 6

No.1 Cylinder Head Cover

IAC Valve Pipe Clamp

PCV Valve

Exhaust Camshaft

Camshaft Bearing Cap x 14

Intake Camshaft

Adjusting Shim
Valve Lifter
Keeper
Spring Retainer
Valve Spring
Spring Seat
◆ O–Ring
◆ Valve Guide
Bushing
Valve

Oil Seal

No.4 Timing Belt Cover

Camshaft Timing Pulley

× 14

Spark Plug

Heater Water Hose

Heater Union

× 8

EGR Cooler

◆ Gasket

Front Engine Hanger

Cylinder Head

Gasket

Gasket

Ground Strap

Water Bypass Hose

Rear Engine Hanger

Camshaft Position Sensor

◆ Cylinder Head Gasket

◆ Non–reusable part

Exploded view of cylinder head—3.0L (2JZ-GTE) engine

9301WG20

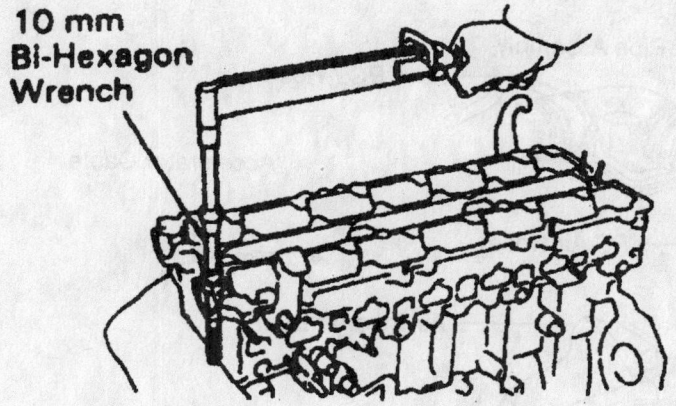

10 mm
Bi-Hexagon
Wrench

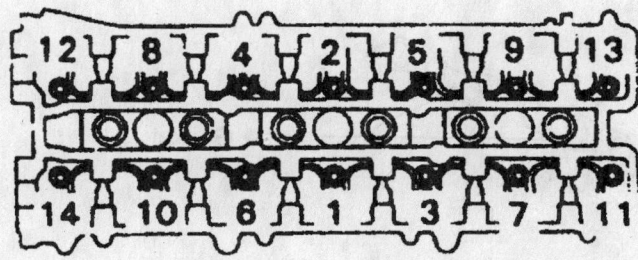

Cylinder head bolt tightening sequence—2JZ-GE and 2JZ-GTE engines

7923VG15

➡**Correct bolt torque must be achieved in 3 steps; do not attempt to shorten the procedure by combining the 2, 90 degree steps.**

- Cylinder head covers, spark plugs, ignition coils, and timing belt
- Intake manifold and delivery pipe assembly by installing a new gasket and engine wire. Bolts and nuts: 20 ft. lbs. (27 Nm)
- Fuel inlet pipe by installing a new gasket and the union bolt. Bolt: 30 ft. lbs. (42 Nm).
- Fuel inlet pipe clamp bolt to the intake manifold
- All other components in the reverse of installation
- Power steering pump. Bolts: 43 ft. lbs. (58 Nm).
- Water outlet and No. 1 water bypass pipe. Bolts: 15 ft. lbs. (21 Nm).
- ECT sensor and sender gauge connectors
- Upper radiator hose to the water outlet
- Drive belt
- With manual transmission torque

the belt tensioner damper nuts to 14 ft. lbs. (20 Nm).
- 2 new gaskets to the cylinder head
- Exhaust manifold. Bolts: 29 ft. lbs. (39 Nm).
- Turbocharger
- Negative battery cable

6. Fill the cooling system to the proper level

7. Fill the engine with clean oil.

8. Start the vehicle, check for leaks and repair if necessary.

2JZ-GE Engine

1. Before servicing the vehicle, refer to the precautions in the beginning of this section.

2. Relieve the fuel pressure in the fuel lines.

3. Drain the engine oil.

4. Drain the cooling system.

5. Remove or disconnect the following:

- Negative battery cable. Wait at least 90 seconds before performing any other work
- Undercovers
- Accelerator, throttle control (auto-

matic transmission only) and cruise control cables from the throttle body
- Air cleaner duct
- Air cleaner, airflow meter, and the intake air pipe
- Drive belt, the fan and fluid coupling, and the water pump pulley
- Front exhaust pipe
- Except for California vehicles, the manifold heat insulator
- Oxygen sensor connector(s)
- Exhaust manifolds and gaskets
- Power steering pump without disconnecting the hoses
- Brake booster vacuum hose
- EVAP hose
- Throttle body and intake air connector assembly
- Air intake chamber stays
- No. 2 vacuum pipe and VSV assembly
- No. 3 timing belt cover by removing the oil filler cap
- Cylinder head rear cover
- Spark plug wires from the cylinder head covers
- Distributor and wires
- Spark plugs
- Timing belt
- Water bypass outlet and No. 1 bypass pipe
- Fuel return hose
- Engine wire bracket from the intake manifold
- Oil dipstick and guide
- Starter
- Air intake chamber
- Vacuum control valve set
- Intake manifold and gaskets
- Cylinder head covers (valve covers)
- Camshaft timing pulleys. Hold the hexagon portion of the camshaft with a wrench and remove the pulley mounting bolt and camshaft pulley
- No. 4 timing belt cover
- Camshafts following the proper sequences and procedures
- Cylinder head bolts in the reverse order of the installation sequence
- Cylinder head

To install:

6. Install or connect the following:
- New gasket on the cylinder block
- Plate washers and cylinder head bolts, lightly coated with engine oil. Tighten the bolts to 25 ft. lbs. (34 Nm). Tighten each bolt an addi-

For complete Engine Mechanical specifications, see Section 1 of this manual

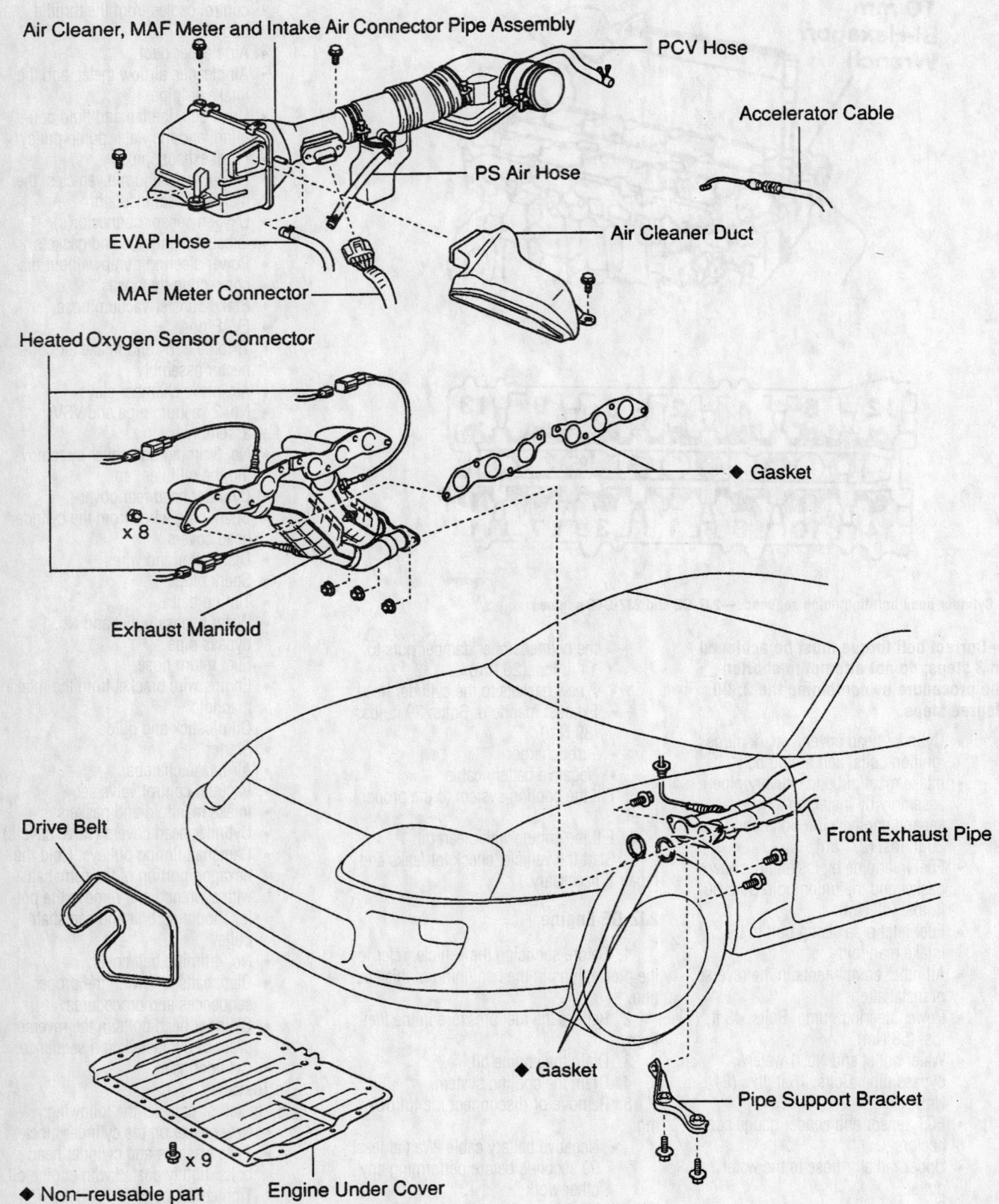

Air Cleaner, MAF Meter and Intake Air Connector Pipe Assembly

PCV Hose

Accelerator Cable

PS Air Hose

EVAP Hose

Air Cleaner Duct

MAF Meter Connector

Heated Oxygen Sensor Connector

◆ Gasket

x 8

Exhaust Manifold

Drive Belt

Front Exhaust Pipe

◆ Gasket

Pipe Support Bracket

x 9

◆ Non-reusable part

Engine Under Cover

9301WG13

Removal of the air duct assembly, exhaust manifold, front exhaust pipe and engine under cover—3.0L (2JZ-GE) engine

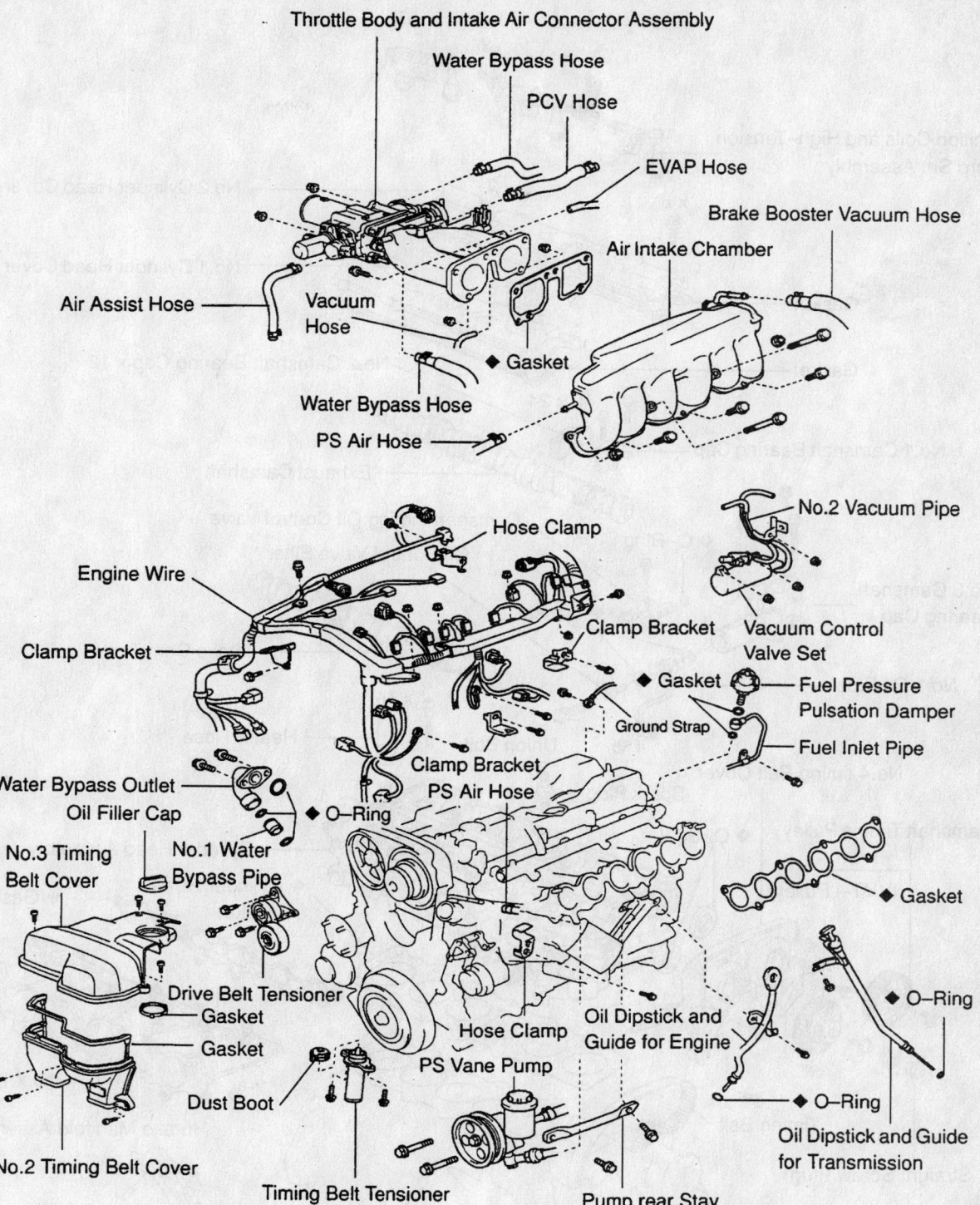

Throttle Body and Intake Air Connector Assembly

Water Bypass Hose

PCV Hose

EVAP Hose

Brake Booster Vacuum Hose

Air Intake Chamber

Air Assist Hose

Vacuum Hose

◆ Gasket

Water Bypass Hose

PS Air Hose

No.2 Vacuum Pipe

Hose Clamp

Engine Wire

Clamp Bracket

Vacuum Control Valve Set

◆ Gasket

Fuel Pressure Pulsation Damper

Ground Strap

Fuel Inlet Pipe

Clamp Bracket
PS Air Hose

Water Bypass Outlet
Oil Filler Cap

◆ O–Ring

No.1 Water Bypass Pipe

No.3 Timing Belt Cover

◆ Gasket

Drive Belt Tensioner
Gasket

Gasket

Hose Clamp

PS Vane Pump

Oil Dipstick and Guide for Engine

◆ O–Ring

Dust Boot

◆ O–Ring

Oil Dipstick and Guide for Transmission

No.2 Timing Belt Cover

Timing Belt Tensioner

Pump rear Stay

◆ Non–reusable part

9301WG14

Removal of the intake manifold and throttle body—3.0L (2JZ-GE) engine

For Accessory Drive Belt illustrations, see Section 1 of this manual

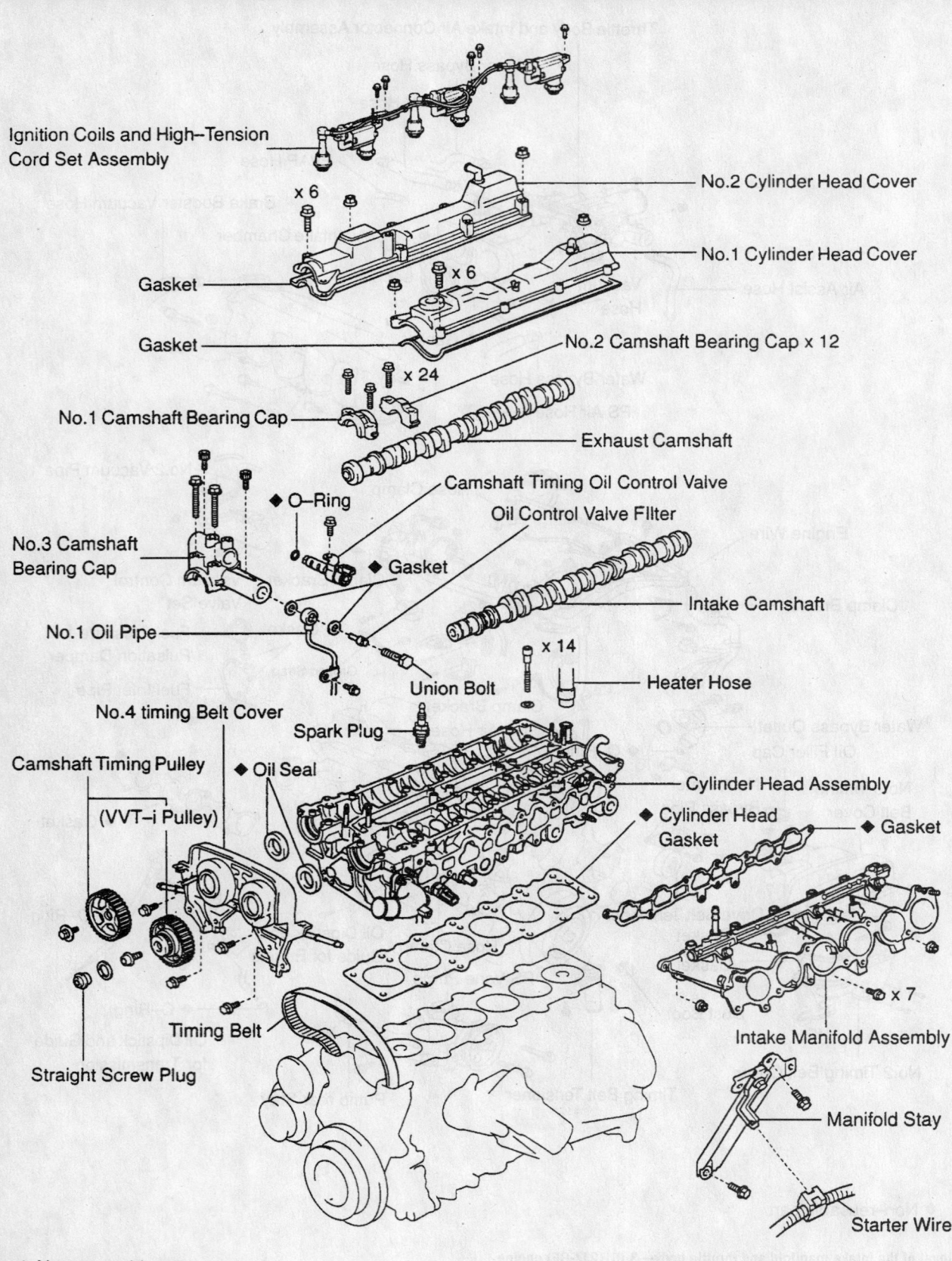

Ignition Coils and High–Tension Cord Set Assembly

No.2 Cylinder Head Cover

x 6

Gasket

No.1 Cylinder Head Cover

x 6

Gasket

No.2 Camshaft Bearing Cap x 12

x 24

No.1 Camshaft Bearing Cap

Exhaust Camshaft

◆ O–Ring

Camshaft Timing Oil Control Valve

Oil Control Valve Filter

No.3 Camshaft Bearing Cap

◆ Gasket

No.1 Oil Pipe

Intake Camshaft

Union Bolt

x 14

Heater Hose

No.4 timing Belt Cover

Spark Plug

Camshaft Timing Pulley

◆ Oil Seal

Cylinder Head Assembly

(VVT–i Pulley)

◆ Cylinder Head Gasket

◆ Gasket

Timing Belt

x 7

Straight Screw Plug

Intake Manifold Assembly

Manifold Stay

Starter Wire

◆ **Non–reusable part**

9301WG15

View of cylinder head covers, camshafts, and the cylinder head—3.0L (2JZ-GE) engine

tional 90 degrees. Again, tighten the bolts another 90 degrees of rotation.

- Position camshafts in the cylinder head with the cam lobes and the knock pins in the correct position.
- Position the No. 3 and No. 7 bearing caps in place, then uniformly and alternately tighten them temporarily
- New oil seals over the camshafts

7. Clean the surfaces of the No. 1 bearing cap and cylinder head with cleaner. Apply seal packing to the No. 1 bearing cap.

8. Install camshafts following the proper sequences and procedures.

9. Press the 2 oil seals in as far as they will go.

10. Rotate each camshaft until the forward straight (knock) pin is straight up. Loosen exhaust No. 1, 2, and 6 bearing cap bolts until they can be turned by hand; tighten, in several passes, to 14 ft. lbs. (20 Nm). Loosen intake No. 1, 2, and 5 and tighten, in several passes, to 14 ft. lbs. (20 Nm).

11. Turn each camshaft ⅓ of a revolution (120 degrees). Loosen exhaust Nos. 4 and 7 bearing cap bolts; tighten, in several passes, to 14 ft. lbs. (20 Nm). Loosen intake No. 4 and 6 bearing cap bolts; tighten, in several passes, to 14 ft. lbs. (20 Nm).

12. Turn each camshaft an additional ⅓ of a revolution, loosen exhaust bearing cap bolts Nos. 3 and 5, then tighten them, in several passes, to 14 ft. lbs. (20 Nm). Loosen intake bearing cap bolts No. 3 and 7, then tighten them, in several passes to 14 ft. lbs. (20 Nm).

13. Check and adjust the valve clearance.

14. Install or connect the following:
- No. 4 timing belt cover. Bolts: 78 inch lbs. (9 Nm).
- Camshaft timing pulleys. Align the shaft pin with the pulley groove and slide the pulley on. Install the bolt temporarily. Hold the hex portion of the camshaft with a wrench and tighten the pulley bolt to 59 ft. lbs. (79 Nm).
- Cylinder head covers
- Intake manifold with a new gasket. Bolts: 20 ft. lbs. (27 Nm).
- Injectors
- Delivery pipe. Bolts: 20 ft. lbs. (27 Nm).
- Fuel inlet pipe. Bolt: 30 ft. lbs. (42 Nm).

- Clamp bolt to the intake manifold
- Fuel pressure pulsation damper
- Intake manifold stay. Bolts: 29 ft. lbs. (39 Nm).
- Water outlet and No. 1 bypass hose assembly
- Engine wire protector to the intake manifold with the 3 nuts
- 6 injector connectors
- ECT sensor connector and sender gauge
- 2 wire clamps and the 2 ground straps to the intake manifold with the bolts
- Engine wire bracket to the water pump with the bolt
- Vacuum control valve set. Bolts: 15 ft. lbs. (21 Nm)
- VSV connector
- Air intake chamber as follows. Bolts and nut: 20 ft. lbs. (27 Nm).
- Bolt holding the engine wire protector to the air intake chamber
- Vacuum sensing hose to the fuel pressure regulator (except for California vehicles)
- Starter
- Oil and transmission dipstick tubes
- Engine wire bracket
- Fuel return hose
- Water bypass outlet and No. 1 water bypass pipe

15. Install the timing belt as follows:

a. Turn the crankshaft pulley and align its groove with the timing mark, **0**, on the No. 1 timing belt cover.

b. Align the timing marks on the camshaft timing gears and the No. 4 timing belt cover.

c. Install the timing belt.

d. Double check that all the timing marks for the crankshaft pulley and the camshaft gears are aligned as they were during disassembly.

e. Set the timing belt tensioner:

f. Use a press to slowly push in the pushrod on the tensioner. This will require between 220–2200 pounds of pressure.

g. Align the holes of the pushrod and housing. Place a 1.5mm hex wrench through the holes to keep the pushrod retracted.

h. Release the press and install the dust boot onto the tensioner.

i. Install the tensioner; alternately tighten the bolts: 20 ft. lbs. (26 Nm).

j. Remove the hex wrench from the tensioner with a pair of pliers.

k. Turn the crankshaft pulley 2 full turns clockwise. Check that each pulley's timing marks align correctly after the 2 turns. If any mark does not align, remove the timing belt and reinstall it.

16. Install or connect the following:
- Drive belt tensioner. Bolts: 15 ft. lbs. (21 Nm).
- No. 2 timing belt cover
- Spark plugs
- Distributor and spark plug wires
- No. 3 timing belt cover
- Cylinder head rear cover
- No. 2 vacuum pipe and VSV assembly
- Air intake chamber stays. Bolt and nut: 13 ft. lbs. (18 Nm).
- Throttle body and intake air connector assembly
- EVAP hose
- Brake booster vacuum hose
- Power steering pump
- Exhaust manifolds
- No. 2 front exhaust pipe
- Water pump pulley, fan, fluid coupling assembly, and the drive belt. Pulley bolts: 12 ft. lbs. (16 Nm).
- Air cleaner, VAF meter, and the intake air connector pipe assembly
- Air cleaner duct
- Cruise control cable, throttle control, and accelerator cables
- Negative battery cable
- Engine undercover

17. Fill the cooling system to the proper level.

18. Fill the engine with clean oil.

19. Start the vehicle, check for leaks and repair if necessary.

1MZ-FE Engine

1. Before servicing the vehicle, refer to the precautions in the beginning of this section.

2. Drain the engine oil.

3. Drain the cooling system.

4. Remove or disconnect the following:

- Negative battery cable. On vehicles equipped with an air bag, wait at least 90 seconds before proceeding.
- Accelerator cable and the throttle cable on vehicles equipped with an automatic transaxle.
- Air cleaner cover, air flow meter, air duct, cruise control actuator and bracket (if equipped), the 2 engine ground straps, right engine mount-

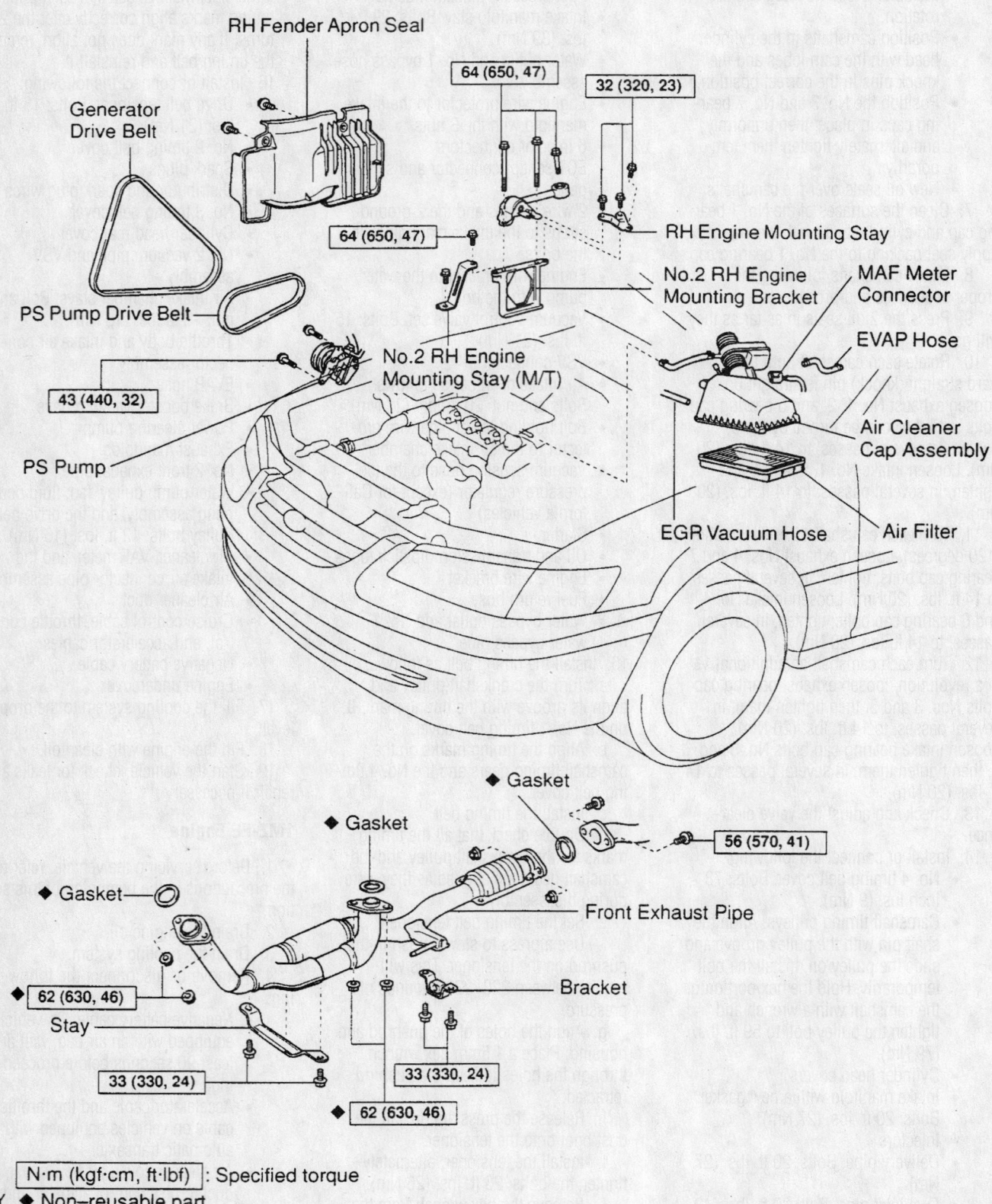

RH Fender Apron Seal

64 (650, 47)

32 (320, 23)

Generator Drive Belt

RH Engine Mounting Stay

64 (650, 47)

No.2 RH Engine Mounting Bracket

MAF Meter Connector

PS Pump Drive Belt

EVAP Hose

No.2 RH Engine Mounting Stay (M/T)

43 (440, 32)

Air Cleaner Cap Assembly

PS Pump

EGR Vacuum Hose

Air Filter

◆ Gasket

56 (570, 41)

◆ Gasket

◆ Gasket

Front Exhaust Pipe

Bracket

◆ 62 (630, 46)

Stay

33 (330, 24)

33 (330, 24)

◆ 62 (630, 46)

N·m (kgf·cm, ft·lbf) : Specified torque

Y ◆ Non–reusable part

9301WG21

Removal of the right-hand fender apron, front exhaust pipe, and air cleaner assembly—3.0L (1MZ-FE) engine

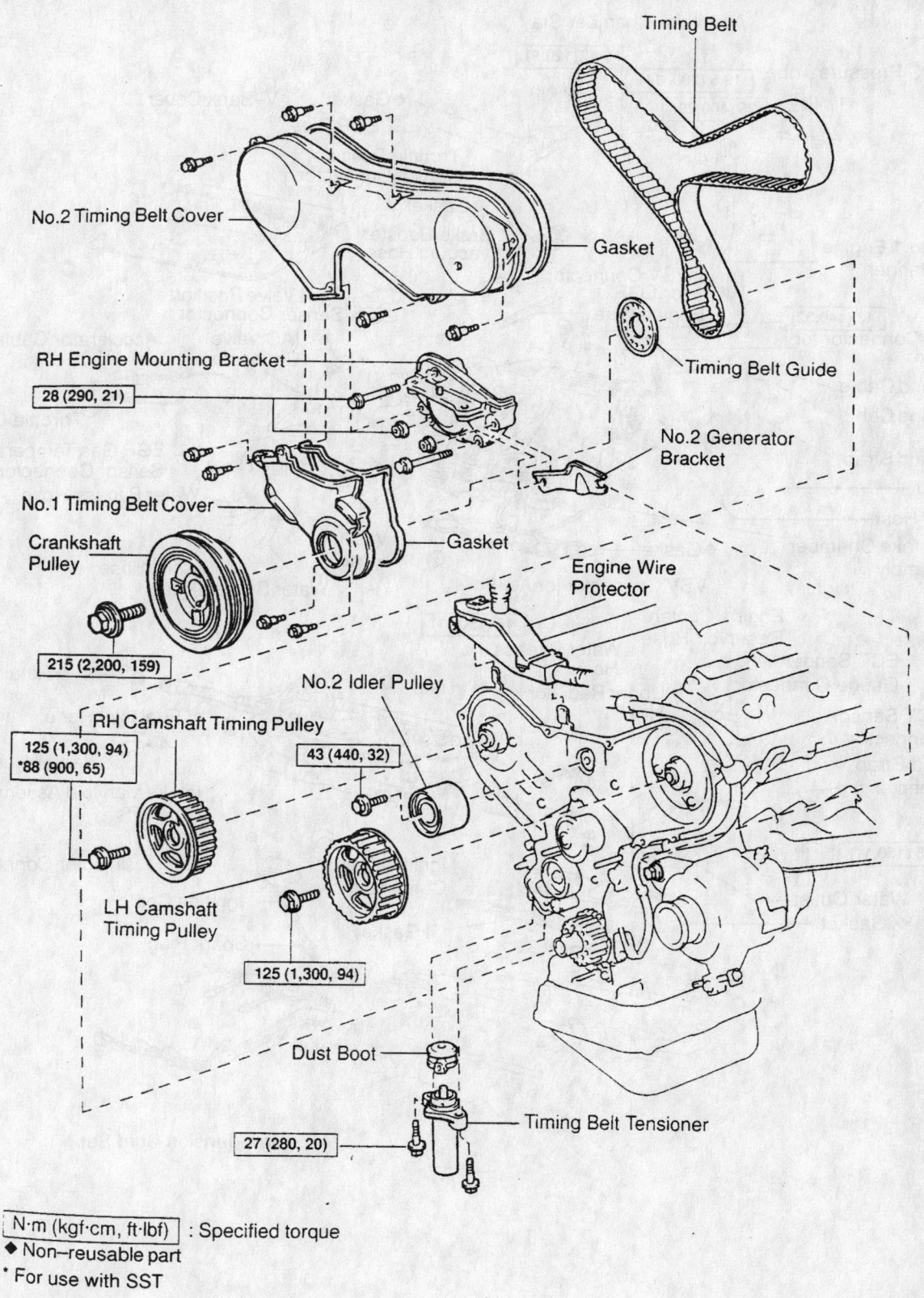

Timing Belt

No.2 Timing Belt Cover

Gasket

Timing Belt Guide

RH Engine Mounting Bracket

28 (290, 21)

No.2 Generator Bracket

No.1 Timing Belt Cover

Gasket

Crankshaft Pulley

215 (2,200, 159)

Engine Wire Protector

No.2 Idler Pulley

RH Camshaft Timing Pulley

125 (1,300, 94)
*88 (900, 65)

43 (440, 32)

LH Camshaft Timing Pulley

125 (1,300, 94)

Dust Boot

Timing Belt Tensioner

27 (280, 20)

N·m (kgf·cm, ft·lbf) : Specified torque
◆ Non−reusable part
* For use with SST

9301WG23

Timing belt and pulleys—3.0L (1MZ-FE) engine

For Wheel Alignment specifications, see Section 1 of this manual

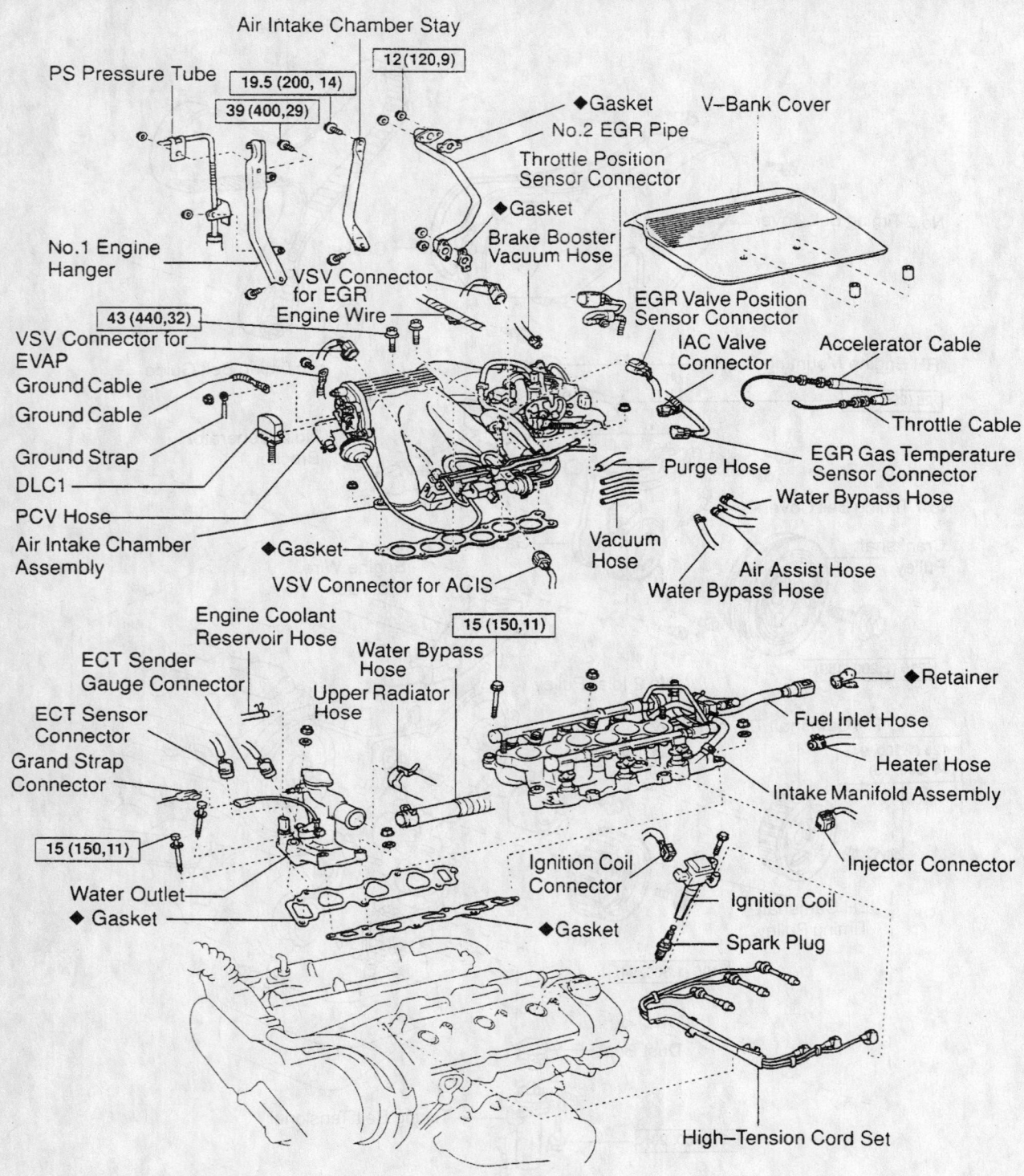

Air Intake Chamber Stay

12 (120,9)

PS Pressure Tube

19.5 (200, 14)

39 (400,29)

◆Gasket

No.2 EGR Pipe

V–Bank Cover

Throttle Position Sensor Connector

No.1 Engine Hanger

◆Gasket

Brake Booster Vacuum Hose

VSV Connector for EGR

Engine Wire

EGR Valve Position Sensor Connector

43 (440,32)

VSV Connector for EVAP

IAC Valve Connector

Accelerator Cable

Ground Cable

Ground Cable

Throttle Cable

Ground Strap

DLC1

Purge Hose

EGR Gas Temperature Sensor Connector

PCV Hose

Vacuum Hose

Water Bypass Hose

Air Intake Chamber Assembly

◆Gasket

Air Assist Hose

VSV Connector for ACIS

Water Bypass Hose

Engine Coolant Reservoir Hose

15 (150,11)

ECT Sender Gauge Connector

Water Bypass Hose

◆Retainer

ECT Sensor Connector

Upper Radiator Hose

Fuel Inlet Hose

Grand Strap Connector

Heater Hose

Intake Manifold Assembly

15 (150,11)

Injector Connector

Water Outlet

Ignition Coil Connector

◆ Gasket

◆Gasket

Ignition Coil

Spark Plug

High–Tension Cord Set

N·m (kgf·cm, ft·lbf) : Specified torque

◆ Non–reusable part

9301WG22

Intake manifold removal—3.0L (1MZ-FE) engine

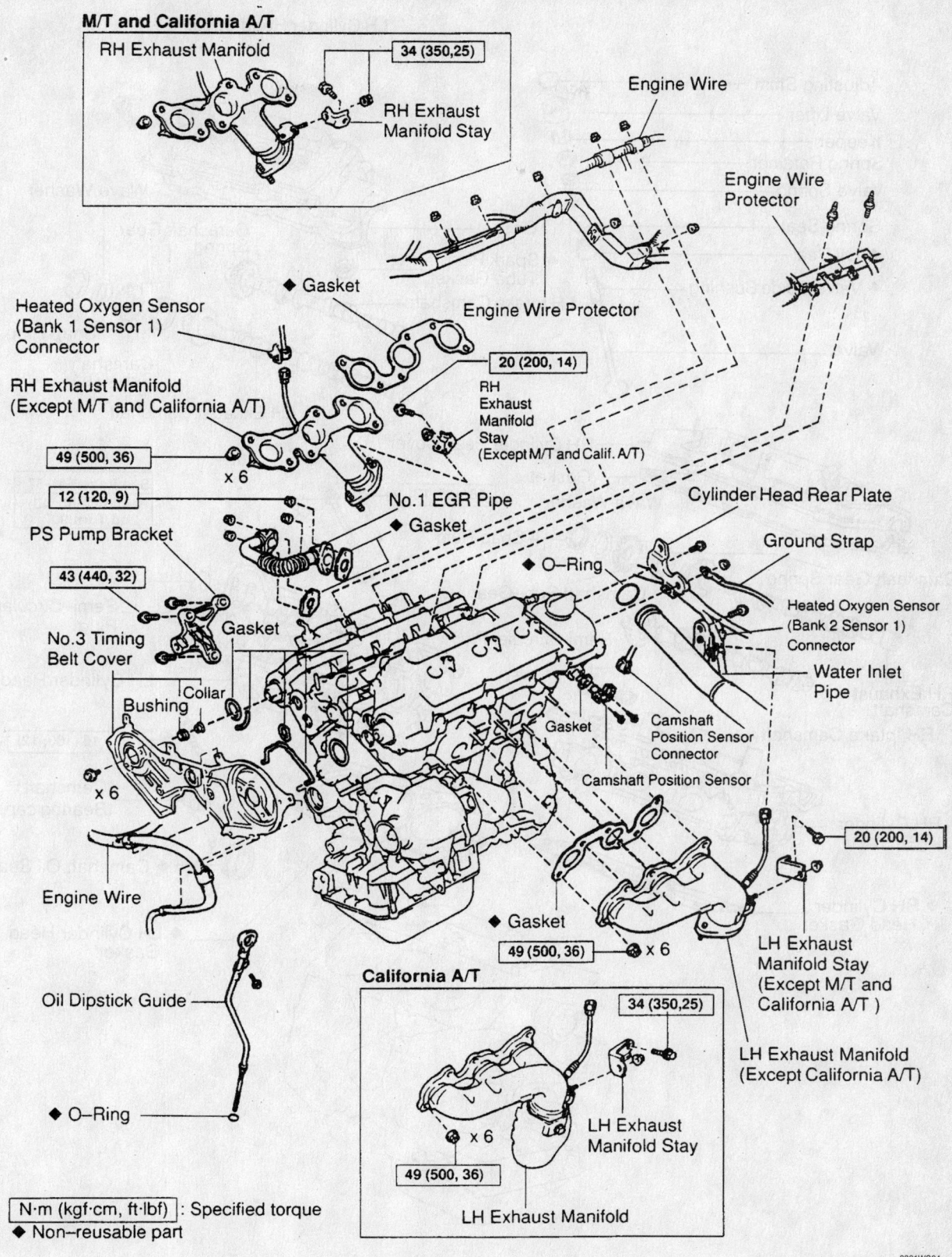

M/T and California A/T

RH Exhaust Manifold

34 (350,25)

RH Exhaust Manifold Stay

Engine Wire

Engine Wire Protector

◆ Gasket

Heated Oxygen Sensor (Bank 1 Sensor 1) Connector

Engine Wire Protector

RH Exhaust Manifold (Except M/T and California A/T)

20 (200, 14)

RH Exhaust Manifold Stay (Except M/T and Calif. A/T)

49 (500, 36)

x 6

12 (120, 9)

No.1 EGR Pipe

◆ Gasket

Cylinder Head Rear Plate

PS Pump Bracket

◆ O-Ring

Ground Strap

43 (440, 32)

Heated Oxygen Sensor (Bank 2 Sensor 1) Connector

No.3 Timing Belt Cover

Gasket

Water Inlet Pipe

Bushing

Collar

Gasket

Camshaft Position Sensor Connector

x 6

Camshaft Position Sensor

Engine Wire

◆ Gasket

20 (200, 14)

49 (500, 36)

x 6

LH Exhaust Manifold Stay (Except M/T and California A/T)

Oil Dipstick Guide

California A/T

34 (350,25)

LH Exhaust Manifold (Except California A/T)

◆ O-Ring

LH Exhaust Manifold Stay

x 6

49 (500, 36)

LH Exhaust Manifold

N·m (kgf·cm, ft·lbf) : Specified torque
◆ Non-reusable part

9301WG24

Exhaust manifold removal—3.0L (1MZ-FE) engine

For Maintenance Interval recommendations, see Section 1 of this manual

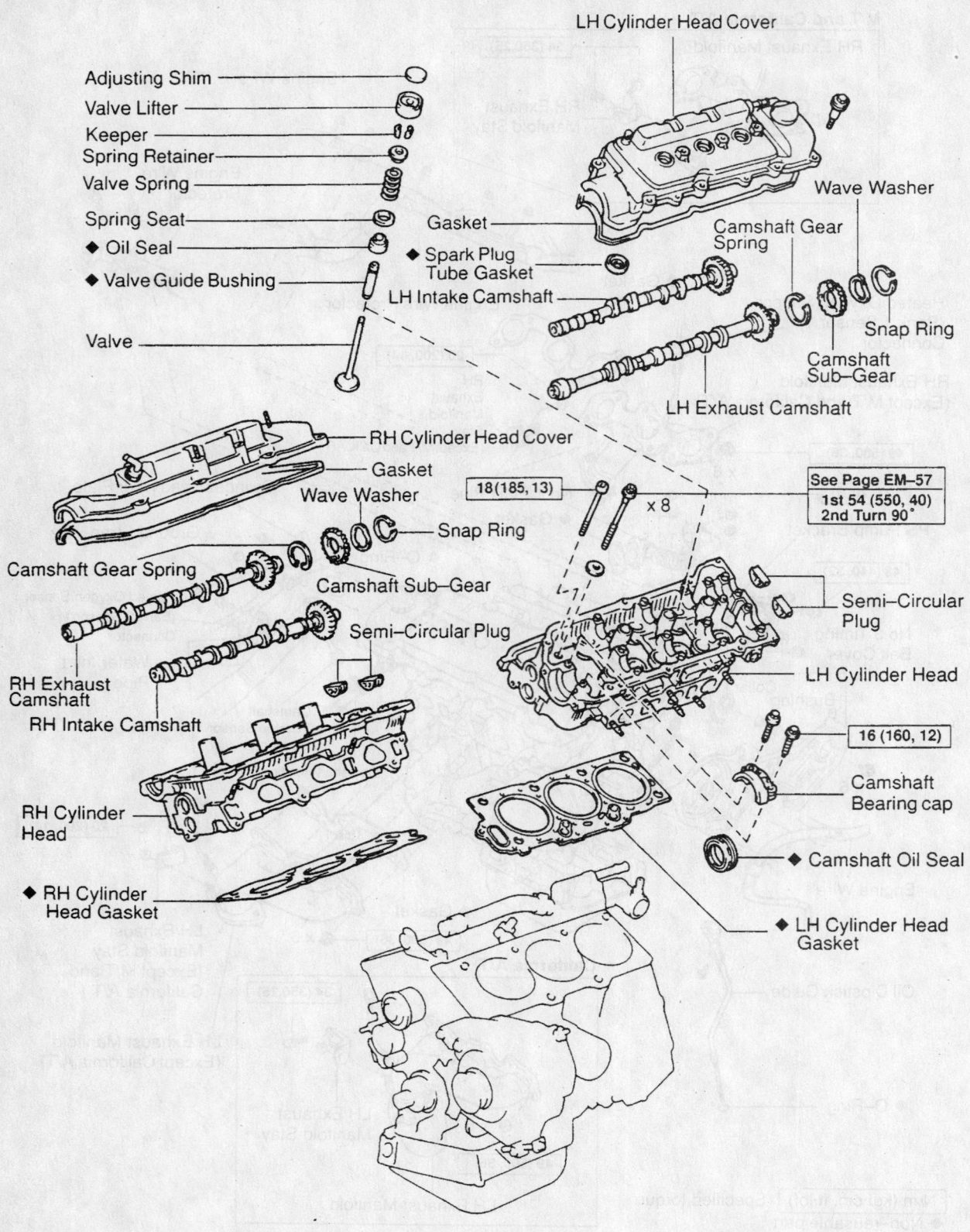

Adjusting Shim
Valve Lifter
Keeper
Spring Retainer
Valve Spring
Spring Seat
◆ Oil Seal
◆ Valve Guide Bushing
Valve

LH Cylinder Head Cover

Gasket
◆ Spark Plug Tube Gasket
LH Intake Camshaft

Wave Washer
Camshaft Gear Spring
Snap Ring
Camshaft Sub-Gear
LH Exhaust Camshaft

RH Cylinder Head Cover
Gasket
Wave Washer
Snap Ring
Camshaft Gear Spring
Camshaft Sub-Gear
Semi-Circular Plug
RH Exhaust Camshaft
RH Intake Camshaft
RH Cylinder Head
◆ RH Cylinder Head Gasket

18(185, 13)   x 8

See Page EM-57
1st 54 (550, 40)
2nd Turn 90°

Semi-Circular Plug
LH Cylinder Head

16 (160, 12)

Camshaft Bearing cap

◆ Camshaft Oil Seal

◆ LH Cylinder Head Gasket

N·m (kgf·cm, ft·lbf) : Specified torque
◆ Non-reusable part

9301WG25

Cylinder head gasket position and torque specification—3.0L (1MZ-FE) engine

ing support, radiator hoses, and 2 heater hoses
- Fuel feed and return lines and plug the lines
- Pressure hose, plug, and remove from V-bank cover
- Vacuum hoses from: Fuel pressure VSV; Fuel pressure regulator; Cylinder head rear plate; Intake air control valve VSV; EGR vacuum modulator; EGR valve
- All necessary connectors and hoses
- Ground straps and the hydraulic motor pressure hose
- 2 nuts and the power steering pressure tube
- Engine hanger and the intake chamber support
- Ignition coils and the spark plugs, timing belt, camshaft pulleys and the timing belt rear cover, cylinder head rear plate, water inlet pipe, and water outlet
- Intake manifold and fuel rail assembly
- EGR pipe

- Exhaust manifolds
- Dipstick assembly and the power steering pump bracket
- Valve covers and the camshaft position sensor
- Camshafts following the proper sequences and procedures
- 2 (1 on each head) 8mm recessed hex bolts. Loosen and remove the 8 head bolts evenly, in 3 passes, in the reverse order of the installation sequence
- Cylinder head gasket

To install:
5. Install or connect the following:
- New cylinder head gasket on cylinder block. Place the cylinder head onto the gasket.
- Cylinder head bolts into the cylinder head. Bolts: 40 ft. lbs. (54 Nm), then an additional 90 degrees
- 2 remaining 8mm bolts. Bolts: 13 ft. lbs. (18 Nm).
- Camshafts following the proper sequences and procedures
- Check and adjust the valves
- Sealant to the cylinder heads where

the camshaft supports meet the cylinder heads
- All remaining components in reverse of removal using the diagrams for the proper torque figure.

Turbocharger

REMOVAL & INSTALLATION

3.0L (2JZ-GTE) Engine

1. Before servicing the vehicle, refer to the precautions in the beginning of this section.
2. Drain the cooling system.
3. Remove or disconnect the following:
- Negative battery cable
- Cruise control actuator cable
- No. 1 air hose
- Air cleaner and MAF meter assembly as follows:
- Theft deterrent horn
- Front lower arm bracket stay
- Front upper crossmember extension
- No. 2 front exhaust pipe as follows:
- Heat insulator for the No. 2 front exhaust pipe
- With automatic transmission, automatic transmission oil cooler tubes
- Engine wire protector
- Heater hose from the No. 3 water bypass pipe
- EVAP hose from the No. 1 vacuum pipe
- IAC valve pipe from the No. 2 air tube
- No. 1 vacuum pipe from the air tubes
- VSV assembly
- Crankshaft position sensor connector from the clamp
- Water bypass hose (from the water pump) from the No.1 turbo water pipe
- Water bypass hose (from the water outlet) from the No. 1 turbo water pipe
- Water bypass hose (from the water outlet) from the No. 2 turbo water pipe
- No. 2 turbo water pipe from the No. 4 air tube
- No. 1 air tube from the No. 1 turbocharger by removing the 2 bolts
- 2 bolts holding the No. 4 air tube to the No. 1 turbocharger
- Air hose from the No. 4 air tube

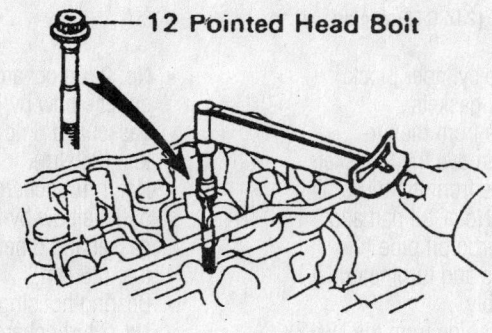

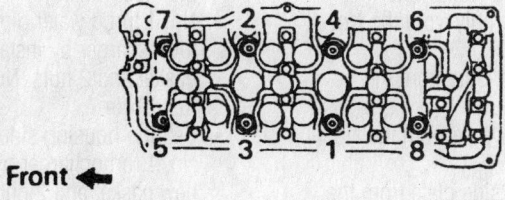

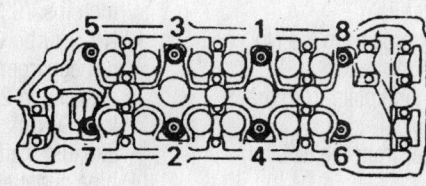

Cylinder head bolt tightening sequence—1MZ-FE engine

7923VG16

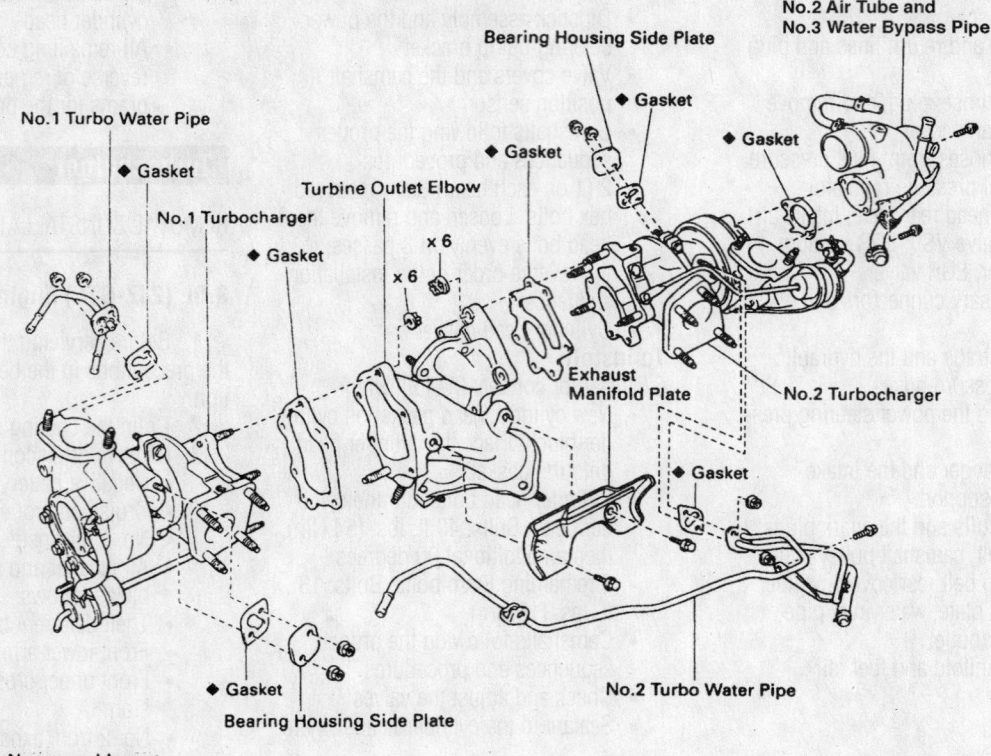

No.2 Air Tube and
No.3 Water Bypass Pipe

Bearing Housing Side Plate

◆ Gasket

No.1 Turbo Water Pipe

◆ Gasket

◆ Gasket

◆ Gasket

Turbine Outlet Elbow

No.1 Turbocharger

◆ Gasket

× 6

× 6

Exhaust
Manifold Plate

No.2 Turbocharger

◆ Gasket

No.2 Turbo Water Pipe

◆ Gasket

Bearing Housing Side Plate

◆ **Non-reusable part**

7923VG17

Exploded view of the turbocharger component assembly—Supra with 3.0L (2JZ-GTE) engine

- Air hose from the intake air connector
- No. 4 air tube and air bypass valve assembly
- Intake air control valve and gasket by removing the 2 nuts
- Air hose from the No. 2 air tube
- PCV hose from the No. 2 cylinder head cover
- Intake air connector and No. 1 air tube assembly
- Air inlet duct and cable bracket
- Heat insulator for the turbocharger
- Exhaust bypass pipe and gasket
- Exhaust gas control valve stay
- Main heated oxygen sensor
- Exhaust gas control valve
- No. 1 turbocharger stay
- No. 2 turbocharger stay
- Union bolt holding the No. 1 turbo oil pipe to the cylinder block. Remove the 2 gaskets
- 2 nuts and disconnect the turbo oil pipe from the turbocharger. Remove the gaskets
- Turbo oil hose from the turbo oil outlet on the No. 1 oil pan. Remove the No. 1 turbo oil pipe
- Union bolt holding the No. 2 turbo

oil pipe to the cylinder block. Remove the 2 gaskets
- Turbo oil pipe from the turbocharger. Remove the 2 gaskets
- Turbo oil hose from the turbo oil outlet on the No. 1 oil pan and remove the turbo oil pipe
- Turbochargers and turbine outlet elbow assembly
- No. 1 vacuum pipe from the No. 2 turbocharger
- No. 2 air tube and No. 3 water bypass pipe assembly from the No. 2 turbocharger
- Exhaust manifold plate from the turbine outlet elbow
- No. 2 turbo water pipe from the No. 2 turbocharger
- Bearing housing side plate from the No. 1 turbocharger
- No. 1 turbo water pipe from the No. 1 turbocharger
- Bearing housing side plate from the No. 2 turbocharger
- No. 1 turbocharger from the turbine outlet elbow
- No. 2 turbocharger from the turbine outlet elbow

To install:
4. Install or connect the following:

- No. 2 turbocharger to the turbine outlet elbow by installing a new gasket and 6 new nuts. Nuts: 18 ft. lbs. (25 Nm).
- No. 1 turbocharger to the turbine outlet elbow by installing a new gasket and 6 new nuts. Nuts: 18 ft. lbs. (25 Nm).
- Bearing housing side plate to the No. 2 turbocharger by installing a new gasket and 2 nuts. Nuts: 78 inch lbs. (9 Nm).
- No. 1 turbo water pipe to the No. 1 turbocharger by installing a new gasket and 2 nuts. Nuts: 78 inch lbs. (9 Nm).
- Bearing housing side plate to the No. 1 turbocharger by installing a new gasket and 2 nuts. Nuts: 78 inch lbs. (9 Nm).
- No. 2 turbo water pipe to the No. 2 turbocharger by installing a new gasket and 2 nuts. Nuts: 78 inch lbs. (9 Nm).
- Exhaust manifold plate to the turbine outlet elbow
- No. 2 air tube and No. 3 water bypass pipe assembly to the No. 2 turbocharger. Bolts: 15 ft. lbs. (21 Nm).

- No. 1 vacuum pipe to the No. 2 turbocharger.
- Turbochargers and turbine outlet elbow assembly. Install 8 **new** nuts and uniformly tighten the nuts in several passes. Nuts: 40 ft. lbs. (54 Nm).
- Water bypass hose (from the No. 2 turbo water pipe) to the No. 2 water bypass pipe
- Heater hose (from the No. 3 water bypass pipe) to the No. 2 water bypass pipe
- Install the No. 2 turbo oil pipe. Nuts: 15 ft. lbs. (21 Nm).
- Union bolt to hold the turbo oil pipe to the cylinder block. Bolt: 29 ft. lbs. (39 Nm).
- No. 1 turbo oil pipe. Nuts: 15 ft. lbs. (21 Nm).
- Union bolt to hold the turbo oil pipe to the cylinder block. Bolts: 29 ft. lbs. (39 Nm).
- No. 2 turbocharger stay. Nut and bolt: 32 ft. lbs. (43 Nm).
- No. 1 turbocharger stay. Nut and bolt: 32 ft. lbs. (43 Nm).
- Exhaust gas control valve by installing 2 new gaskets and the 3 nuts. Nuts: 51 ft. lbs. (69 Nm).
- Main heated oxygen sensor. Nuts: 14 ft. lbs. (20 Nm).
- Exhaust gas control valve stay. Bolt and nut: 32 ft. lbs. (43 Nm).
- Exhaust bypass pipe by installing 2 new gaskets and 4 new nuts. Nuts: 18 ft. lbs. (25 Nm).
- Heat insulator for the turbocharger
- Air inlet duct by installing the cable bracket, bolt, and 2 nuts
- Intake air connector and No. 1 air tube assembly
- PCV hose to the No. 2 cylinder head cover
- Air hose to the No. 2 air tube
- Intake air control valve and gasket. Nuts: 15 ft. lbs. (21 Nm)
- No. 4 air tube and air bypass valve assembly
- Air hose to the intake air connector
- Air hose to the No. 4 air tube
- 2 bolts holding the No. 4 air tube to the No. 1 turbocharger. Bolts: 15 ft. lbs. (21 Nm).
- No. 1 air tube to the No. 1 turbocharger. Bolts: 15 ft. lbs. (21 Nm).
- No. 2 turbo water pipe to the No. 4 air tube by installing the bolt

- Water bypass hose (from the water outlet) to the No. 2 turbo water pipe
- Water hose (from the water outlet) to the No. 1 turbo water pipe
- Water bypass hose (from the water pump) to the No. 1 turbo water pipe
- Crankshaft position sensor connector to the clamp
- VSV assembly and connect the 2 VSV connectors
- Engine wire to the wire clamp
- Air hose to the hose clamp
- Air hose to the actuator for the exhaust gas control valve
- Air hose to the actuator for the waste gate valve
- Install the No. 1 vacuum pipe to the air tubes and install the 3 bolts
- 2 air hoses (from the vacuum pressure tank) to the vacuum pipe
- Air hose to the VSV for the intake air control valve
- Air hose (from the No. 2 air tube) from the vacuum pipe
- 2 air hoses (from the VSV for exhaust bypass valve) to the vacuum pipe
- Vacuum hose (from the air bypass valve) to the No. 1 air tube
- Air hose (from the VSV for exhaust gas control valve) to the vacuum pipe
- Air hose (from the VSV for waste gate valve) to the vacuum pipe
- Air hose to the No. 1 air tube
- Air hose to the No. 4 air tube
- Engine wire to the 3 clamps
- VSV connector for the exhaust bypass valve
- VSV connector for the intake air control valve
- IAC valve pipe to the clamp
- Air hose to the No. 2 air tube
- Air hose (from the No. 1 vacuum pipe) to the IAC valve pipe
- Engine wire to the clamp
- EVAP hose to the No. 1 vacuum pipe
- Heater hose to the No. 3 water bypass pipe
- Engine wire protector
- Automatic transmission oil cooler tubes
- Heat insulator for the No. 2 front exhaust pipe
- No. 2 front exhaust pipe. Nuts: 46 ft. lbs. (62 Nm).

- Pipe support bracket. Bolts: 32 ft. lbs. (43 Nm).
- 2 bolts and nuts to hold the front exhaust pipe to the No. 2 front exhaust pipe. Bolts and nuts: 43 ft. lbs. (58 Nm).
- Upper front crossmember extension. Bolts: 22 ft. lbs. (29 Nm). Nuts: 25 ft. lbs. (33 Nm).
- Front lower arm bracket stay. Bolts: 33 ft. lbs. (44 Nm). Nut: 43 ft. lbs. (59 Nm).
- Theft deterrent horn
- Air cleaner and MAF meter assembly as follows:
- Air cleaner duct
- No. 1 air hose
- Cruise control actuator cable
- Undercover
- Negative battery cable

5. Fill the cooling system to the proper level.

6. Start the vehicle, check for leaks and repair if necessary.

Intake Manifold

REMOVAL & INSTALLATION

5E-FE Engine

1. Before servicing the vehicle, refer to the precautions in the beginning of this section.

2. Drain the cooling system.

3. Remove or disconnect the following:
- Negative battery cable. On vehicles equipped with an air bag, wait at least 90 seconds before proceeding.
- Accelerator cable
- With automatic transmission, throttle cable
- PCV hose
- Air cleaner
- All electrical wires and vacuum hoses that interfere with removal of the intake manifold
- EGR pipe and EGR valve
- Throttle body assembly
- Air intake chamber stay
- With distributor ignition, air pipe
- Engine wire clamps
- Intake manifold stay
- Intake manifold
- Intake manifold gasket

To install:

4. Install or connect the following:
- Using a new gasket, intake mani-

fold. Nuts and bolts: 14 ft. lbs. (19 Nm).
- Intake manifold stay. Bolt(s) and nut(s): 15 ft. lbs. (20 Nm).
- With distributor ignition, vacuum hoses, then the air pipe. Bolts: 48 inch lbs. (6 Nm).
- Air intake chamber stay
- Throttle body, using a new gasket. Nuts and bolts: 108 inch lbs. (13 Nm).
- All components to the throttle body
- EGR pipe and EGR valve
- All electrical wires and vacuum hoses removed from the intake manifold
- Air cleaner hose
- Negative battery cable

5. Fill the cooling system.
6. Start the vehicle, check for leaks and repair if necessary.

1NZ-FE Engine

1. Before servicing the vehicle, refer to the precautions in the beginning of this section.
2. Drain the cooling system.
3. Remove or disconnect the following:
   - Negative battery cable
   - Water filler
   - Outer front cowl top panel
   - Alternator
   - Air cleaner
   - Accelerator cable
   - Ignition coil
   - PCV hoses
   - Throttle body
   - Intake manifold and discard the gasket

To install:

4. Install or connect the following:
   - Intake manifold with a new gasket. Uniformly tighten the bolts and nuts, in several passes, from the ends, working towards the center, to 22 ft. lbs. (30 Nm).
   - Engine wiring harness
   - Throttle body
   - PCV hoses
   - Ignition coils
   - Accelerator cable
   - Air cleaner
   - Alternator
   - Water filler
   - Negative battery cable

5. Fill the cooling system.
6. Start the vehicle, check for leaks and repair if necessary.

1ZZ-FE Engine

1. Before servicing the vehicle, refer to the precautions in the beginning of this section.

2. Drain the cooling system.
3. Remove or disconnect the following:
   - Negative battery cable
   - Drive belt and alternator
   - Air intake duct
   - Accelerator cable
   - Exhaust pipe from the manifold.
   - Exhaust manifold support bracket
   - Spark plug wires, then ignition coils
   - Spark plugs
   - PCV hoses
   - Throttle body assembly
   - 2 bolts securing the wiring harness protector
   - Wiring connectors and ground wires
   - Intake manifold support bracket
   - Intake manifold and gasket

To install:

4. Install or connect the following:
   - Intake manifold with a new gasket. Torque the bolts to 14 ft. lbs. (18.5 Nm).
   - Harness wiring to the cylinder head and harness protector
   - Fuel injectors, throttle body and the PCV hoses
   - Spark plugs and ignition coils. Bolts and nuts: 80 inch lbs. (9 Nm).
   - Exhaust manifold and support bracket. Bolts: 37 ft. lbs. (49 Nm).
   - Front exhaust pipe to the manifold. Bolts: 46 ft. lbs. (62 Nm).
   - Oxygen sensor. Nuts: 14 ft. lbs. (20 Nm).
   - Accelerator cable and air intake duct

- Alternator and drive belt
- Negative battery cable

5. Fill the cooling system.
6. Start the vehicle, check for leaks and repair if necessary.

2ZZ-GE Engine

1. Before servicing the vehicle, refer to the precautions in the beginning of this section.
2. Drain the cooling system.
3. Remove or disconnect the following:
   - Negative battery cable
   - Drive belt and alternator
   - Air intake duct
   - Accelerator cable
   - Spark plug wires, then ignition coils
   - Spark plugs
   - PCV hoses
   - Throttle body assembly
   - Wiring harness
   - Hoses and tubes connected to the head
   - Intake manifold support bracket
   - Intake manifold and gasket

To install:

4. Install or connect the following:
   - Intake manifold with a new gasket. Bolts A: 20 ft. lbs. (27 Nm); bolt B: 34 ft. lbs. (46 Nm)
   - Harness wiring to the cylinder head and harness protector
   - Fuel injectors, throttle body and the PCV hoses
   - Spark plugs and ignition coils. Bolts and nuts: 80 inch lbs. (9 Nm).

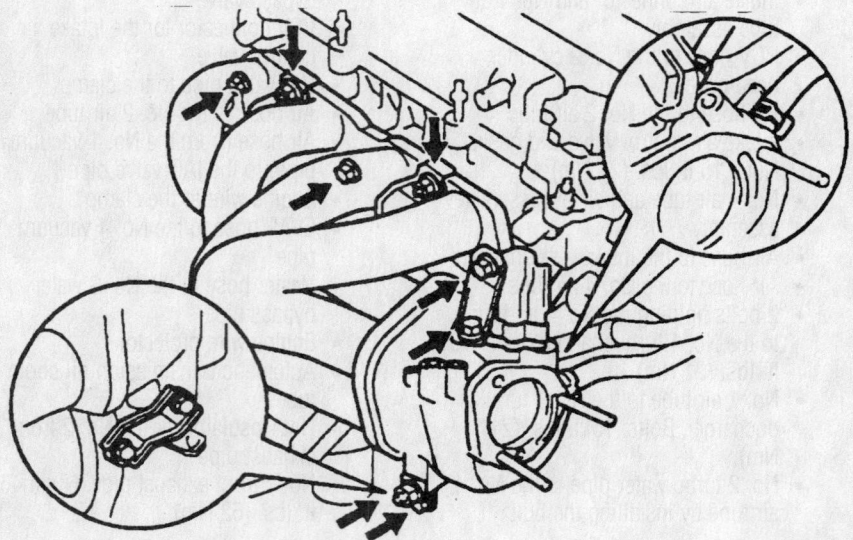

Intake manifold mounting fastener locations—1.8L (1ZZ-FE) engine

7923VG19

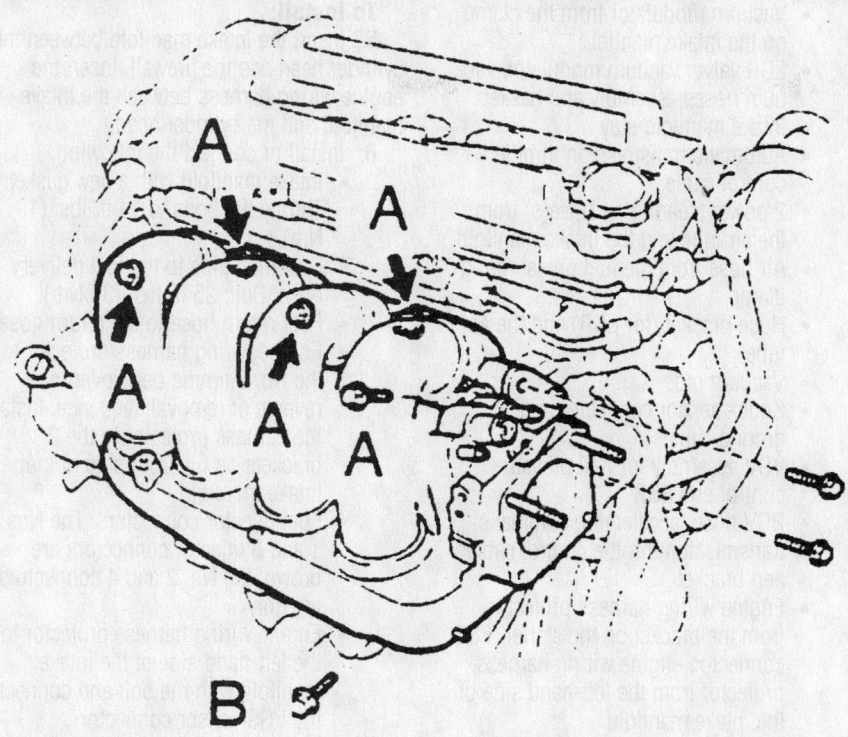

Intake manifold bolt installation—2ZZ-GE engine

- Oxygen sensor. Nuts: 14 ft. lbs. (20 Nm).
- Accelerator cable and air intake duct
- Alternator and drive belt
- Negative battery cable
5. Fill the cooling system.
6. Start the vehicle, check for leaks and repair if necessary.

5S-FE Engine

CAMRY AND CAMRY SOLARA

1. Before servicing the vehicle, refer to the precautions in the beginning of this section.
2. Drain the cooling system.
3. Remove or disconnect the following:
- Negative battery cable. On vehicles equipped with an air bag, wait at least 90 seconds before proceeding
- Accelerator cable. With automatic transmission, throttle cable from the throttle body
- Intake air temperature sensor connector
- On California models, air cleaner hose
- Air cleaner hose clamp bolt, air cleaner cap clips, air hose from the throttle body, and the air cleaner cap together with the resonator and air cleaner hose
- Electrical connections and hoses from the throttle body

- Throttle body. Type A throttle bodies are secured with 4 bolts and Type B throttle bodies are secured with 2 bolts and 2 nuts
- Vacuum hose bracket and the engine wiring harness
- EGR valve
- No. 1 air intake chamber, and manifold stays
- Intake manifold and gasket

To install:
4. Install or connect the following:
- Intake manifold with a new gasket. Torque the bolts to 14 ft. lbs. (19 Nm).
- Wire clamps to the wire brackets on the intake manifold
- Vacuum hose bracket and engine wiring harness
- No. 1 air intake chamber and manifold stays. 14mm bolts: 31 ft. lbs. (42 Nm). 12mm bolts: 16 ft. lbs. (22 Nm).
- EGR valve
- Throttle body with a new gasket
- Hoses and electrical connections to the throttle body

➡ **The protrusion on the gasket should be facing down and the water hose connections on the throttle body should also face down.**

- On type A throttle body, Bolts: 14 ft. lbs. (19 Nm). Bolt A is 45mm in length and bolt B is 55mm.

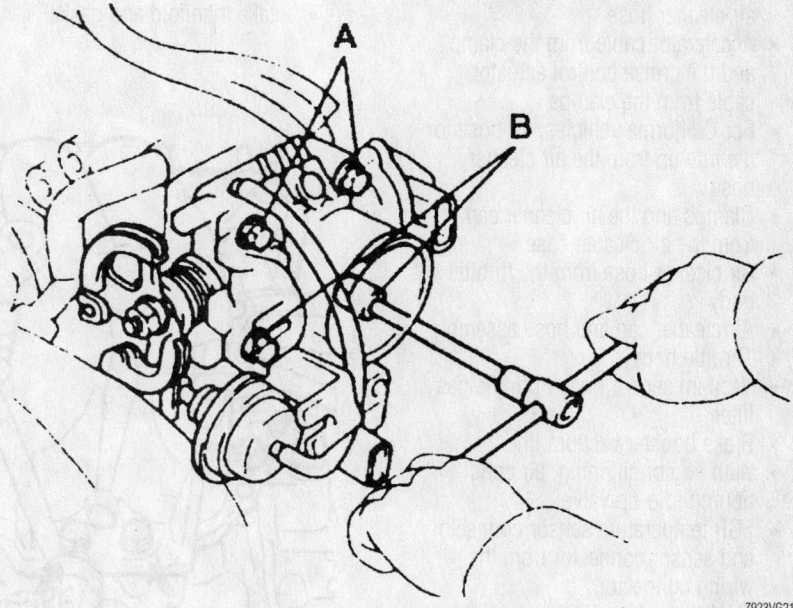

Mounting bolt length identification for the throttle body—2.2L (5S-FE) engine

- On type B throttle body, bolts and nuts: 14 ft. lbs. (19 Nm).
- PCV hose
- 2 vacuum hoses to the EGR modulator
- Vacuum hose to the TVV for EVAP
- IAC valve connector
- Throttle position sensor connector
- Air cleaner hose to the throttle body
- Air cleaner cap with the resonator and air cleaner hose
- Intake air temperature sensor connector
- On California models, air hose to the air cleaner hose
- With automatic transmission, throttle cable
- Accelerator cable
- Negative battery cable

5. Fill the cooling system.

6. Start the vehicle, check for leaks and repair if necessary

CELICA

1. Before servicing the vehicle, refer to the precautions in the beginning of this section.

2. Relieve the fuel system pressure.

3. Drain the cooling system.

4. Remove or disconnect the following:
- Negative battery cable. On vehicles equipped with an air bag, wait at least 90 seconds before proceeding.
- IAT sensor connector
- High tension spark plug wire from air cleaner hose
- Accelerator cable from the clamp and the cruise control actuator cable from the clamps
- For California vehicles, air hose for the idle up from the air cleaner hose
- Clamps and the air cleaner cap from the air cleaner case
- Air cleaner hose from the throttle body
- Air cleaner cap and hose assembly
- Throttle body
- Vacuum sensor hose from the gas filter
- Brake booster vacuum line
- With air conditioning, air conditioning idle-up valve
- EGR temperature sensor connector and sensor connector from the wiring connector
- EVAP hose from the charcoal canister and hose clamp from the bracket on the air tube
- Vacuum hoses from the VSV for the EGR

- Vacuum modulator from the clamp on the intake manifold
- EGR valve, vacuum modulator, vacuum hoses assembly and gasket
- Intake manifold stay
- Automatic transmission throttle control cable
- 2 power steering air hose(s) from the air tube and the intake manifold
- Air hose from the fuel pressure regulator
- Hose bracket (for EGR) and the air tube
- Vacuum pipe
- Knock sensor connector and ground cables
- VSV assembly for fuel pressure control and EGR
- PCV hose, accelerator, automatic transmission throttle control cables and bracket
- Engine wiring harness protector from the bracket on the starter, VSS connector, engine wiring harness protector from the left-hand side of the intake manifold
- Fuel injector connectors and engine wiring harness protector from the front side of the intake manifold.
- Engine wiring harness protector from the No. 2 timing belt cover
- Fuel inlet hose from the delivery pipe
- Fuel return hose from the return pipe
- Engine wiring harness between the intake manifold and the cylinder head
- Intake manifold and gasket

To install:

5. Insert the intake manifold between the cylinder head and the firewall. Insert the engine wiring harness between the intake manifold and the cylinder head.

6. Install or connect the following:
- Intake manifold with a new gasket. Torque the bolts to 14 ft. lbs. (19 Nm).
- Fuel inlet pipe to the fuel delivery hose. Bolt: 25 ft. lbs. (34 Nm).
- Fuel return hose to the return hose
- Engine wiring harness protector to the No. 2 timing belt cover in reverse of removal sequence. Install the harness protector to the 2 brackets on the front side of the intake manifold
- Fuel injector connectors. The Nos. 1 and 3 injector connectors are brown; the No. 2 and 4 connectors are gray
- Engine wiring harness protector to the left-hand side of the intake manifold with the bolt and connect the VSS sensor connector
- Engine wiring harness protector to the starter bracket
- Cable bracket to the intake manifold, and the accelerator and automatic transmission throttle control cables
- PCV hose to the intake manifold
- VSV for fuel pressure control and EGR
- Knock sensor connector and ground cable to the intake manifold

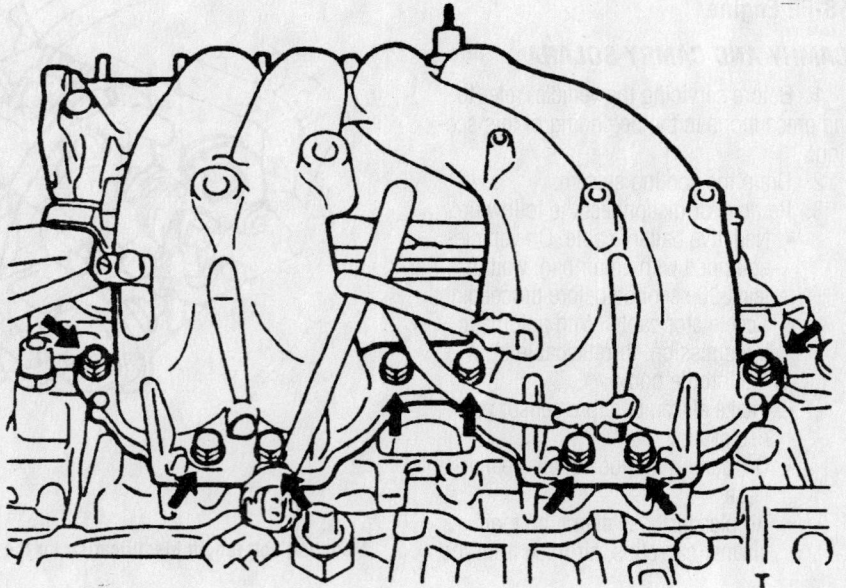

Intake manifold mounting fastener locations—2.2L (5S-FE) engine

7923VG20

- Air tube and hose bracket (for EGR)
- Power steering air hose(s) to the air tube and the intake manifold
- Vacuum sensing hose to the fuel pressure regulator
- Automatic transmission throttle control cable to the clamp on the rear side of the intake manifold.
- Intake manifold stay bolt. Bolt: 15 ft. lbs. (21 Nm). Nut: 32 ft. lbs. (44 Nm).
- EGR valve. Bolt: 43 ft. lbs. (59 Nm). Nuts: 108 inch lbs. (13 Nm).
- Vacuum modulator
- Vacuum hoses to the VSV for the EGR and hose clamp to the bracket on the air tube
- With A/C, the idle-up valve
- EGR gas temperature sensor connector and the sensor connector to the bracket on the intake manifold.
- EVAP hose to the charcoal canister
- Vacuum sensor hose to the gas filter and the brake booster vacuum hose
- Throttle body assembly
- Air filter, air cleaner cap, hose assembly, air cleaner hose to the throttle body, and secure the air cleaner cap with the 4 clamps
- Accelerator and cruise control cables to the clamps
- High tension spark plug wire to the air cleaner hose
- IAT sensor connector
- Negative battery cable

7. Fill the cooling system.

8. Start the vehicle, check for leaks and repair if necessary.

1MZ-FE Engine

1. Before servicing the vehicle, refer to the precautions in the beginning of this section.

2. Relieve the fuel pressure from the fuel lines.

3. Remove or disconnect the following:
- Negative battery terminal. If equipped with an air bag system, wait at least 90 seconds or longer before performing any other work
- Air cleaner hose
- V-bank cover
- Air cleaner chamber assembly
- Accelerator cable
- automatic transmission throttle cable
- TPS connector
- IAC valve connector

- EGR gas temperature sensor connector
- A/C idle up valve connector
- VSV connector for the Acoustic Control Induction System (ACIS)
- VSV connector for the fuel pressure control
- Disconnect the VSV for the EVAP
- VSV connector for the EGR
- DLC1 from the bracket on the intake air control valve
- Power steering pressure tube from the No. 1 engine hanger by removing the 2 bolts
- Brake booster vacuum hose from the intake air control valve for the ACIS
- PCV hose from the PCV valve on the right-hand cylinder head
- Ground strap and cable from the air intake air control valve from the ACIS
- Ground cable from the air intake chamber
- Vacuum hose clamp from fuel pipe
- 2 bypass hoses from the throttle body
- 2 power steering air hoses to the air intake chamber
- Air assist hose from the throttle body
- Remove the EVAP hose from the pipe on emission control valve set
- 2 vacuum hoses from the pipes on the cylinder head rear plate
- Vacuum sensing hose from the fuel pressure regulator
- Engine wire clamp from emission control valve set
- 2 bolts and the No. 1 engine hanger
- 2 bolts and the air intake chamber stay
- No. 2 EGR pipe and 2 gaskets by removing the 4 nuts.
- Hose from the VSV from the EVAP
- Air intake chamber assembly and gasket
- Fuel injector connectors
- Air assist hoses and pipe
- Fuel return hose from the No. 1 fuel pipe
- Fuel inlet hose from the fuel filter
- Delivery pipes and injectors from the engine
- Heater hoses
- Intake manifold and gasket

To install:

4. Thoroughly clean the intake manifold and cylinder head surfaces.

5. Install or connect the following:
- Intake manifold with a new gasket. Torque the bolts to 11 ft. lbs. (15 Nm).
- Heater hoses to the intake manifold
- 2 new grommets to each injector
- 2 new O-rings, lightly oiled, on each injector
- Fuel injectors

➡ **Be sure to position the injector electrical connector outward**

- 4 spacers in position on the intake manifold
- Right-hand delivery pipe and the No. 1 fuel pipe together with the 3 injectors in position on the intake manifold
- 2 bolts, temporarily, holding the right-hand delivery pipe to the intake manifold
- Bolt, temporarily, holding the No. 1 fuel pipe to the intake manifold
- Left-hand delivery pipe and the No. 2 fuel pipe together with the 3 injectors in position
- Fuel return hose to the fuel pressure regulator
- 2 bolts, temporarily, holding the left-hand delivery pipe to the intake manifold
- No. 2 fuel pipe, temporarily, to the left and right-hand delivery pipes with the union bolts and 2 new gaskets

➡ **Check that the injectors rotate smoothly. If the injectors do not rotate smoothly, the probable cause is incorrect installation of the O-rings. Replace the O-rings.**

6. Tighten the 4 bolts holding the delivery pipes to the intake manifold. Bolts: 84 inch lbs. (10 Nm).

7. Tighten the bolt holding the No. 1 fuel pipe to the intake manifold. Bolts: 14 ft. lbs. (20 Nm).

8. Tighten the 2 union bolts holding the No. 2 fuel pipe to the delivery pipes. Bolts: 24 ft. lbs. (33 Nm).

9. Install or connect the following:
- Fuel inlet hose to the fuel filter. Use 2 new gaskets when installing the union bolt
- Fuel return hose to the No. 1 fuel pipe. When routing the fuel return hose, pass the hose under the heater hoses
- Air assist hoses to the intake mani-

fold, then the air assist pipe to the bracket on the No. 1 fuel pipe
- Injector connectors
- Air intake chamber assembly. Bolts and nuts: 32 ft. lbs. (43 Nm).
- Hose to the VSV for the EVAP system
- 2 new gaskets and No. 2 EGR pipe with the 4 nuts. Nuts: 108 inch lbs. (12 Nm).
- No. 1 engine hanger with the 2 bolts. Bolts: 19 ft. lbs. (39 Nm).
- Air intake chamber stay with the 2 bolts. Bolts: 14 ft. lbs. (20 Nm).
- Brake booster vacuum hose to the intake air control valve for the ACIS
- PCV hose to the PCV valve on the right-hand cylinder head
- Ground strap and cable to the intake air control valve for the ACIS
- Connect the ground cable and strap with the nut. Nut: 10 ft. lbs. (15 Nm).
- Ground cable to air intake chamber
- Vacuum hose clamp to fuel pipe
- 2 water bypass hoses to the throttle body
- Air assist hose to the throttle body
- 2 power steering air hoses to the air intake chamber
- Connect the EVAP hose to the pipe on the emission control valve set
- 2 vacuum hoses to the pipes on the cylinder head rear plate
- Vacuum sensing hose to the fuel pressure regulator
- Engine wire clamp to the emission control valve set
- Power steering pressure tube with the 2 nuts
- TPS sensor connector
- IAC valve connector
- EGR gas temperature sensor connector
- A/C idle up valve connector
- VSV connector for the ACIS
- VSV connector for the fuel pressure control
- For California vehicles, install the VSV connector for the EVAP
- VSV connector for the EGR
- DLC1 to the bracket on the intake air control valve
- Accelerator cable
- automatic transmission throttle cable
- V-bank cover
- Air cleaner hose
- Negative battery cable

10. Fill the cooling system.
11. Start the vehicle, check for leaks and repair if necessary.

2JZ-GE Engine

1. Before servicing the vehicle, refer to the precautions in the beginning of this section.
2. Drain the cooling system.
3. Remove or disconnect the following:
- Negative battery cable. Wait at least 90 seconds before proceeding with any other work.
- VSV connector
- Vacuum sensing hose from the fuel pressure control and the air intake chamber
- With automatic transmissions, transmission dipstick and guide
- EGR pipe
- EGR gas temperature sensor connector from the No. 2 vacuum pipe and wiring connector
- No. 2 vacuum pipe from the air intake chamber and from the intake manifold
- 2 nuts holding the throttle body bracket to the cylinder head
- Intake air connector from the air intake chamber by removing the 4 bolts and 2 nuts
- Engine wire protector from the air intake chamber
- 3 vacuum hoses from the No. 1 vacuum pipe
- Vacuum hose from the air intake chamber
- Power steering air hose from the air intake chamber
- Brake booster vacuum hose from the air intake chamber by removing the union bolt and 2 gaskets
- Vacuum hose (from the actuator for ACIS) from the No. 1 vacuum pipe
- Fuel pressure regulator vacuum sensing hose from the air intake chamber, (except California)
- 2 nuts holding the air intake chamber stays to the cylinder head
- 2 bolts and disconnect the 2 air intake chamber stays from the air intake chamber
- Air intake chamber by removing the nut and 5 bolts

⁕⁕ CAUTION

Relieve the fuel pressure in the fuel line before disconnecting any fuel lines. Failure to do so may result in fire and/or explosion.

- Fuel return hose
- Oil dipstick and guide
- Engine wire bracket from the water pump

- Ground straps from the intake manifold
- Wire clamps from the intake manifold
- 6 injector connectors
- ECT sensor and the ECT sender gauge connectors
- Engine wire protector from the intake manifold
- Intake manifold stay
- Fuel inlet pipe clamp bolt from the intake manifold
- Union bolt and disconnect the fuel inlet pipe
- Intake manifold and gaskets

To install:

4. Thoroughly clean the contact surfaces of the head and manifold.
5. Install or connect the following:
- New gasket and intake manifold. Torque the bolts to 20 ft. lbs. (27 Nm).
- Inlet pipe with new gaskets. Bolt: 30 ft. lbs. (42 Nm).
- clamp bolt to the intake manifold
- Intake manifold stay. Bolts: 29 ft. lbs. (39 Nm).
- Engine wire protector to the intake manifold with the 3 nuts
- 6 injector connectors
- ECT sender gauge and connectors
- 2 wire clamps to the intake manifold with the bolts
- 2 ground straps to the intake manifold with the bolts
- Engine wire bracket to the water pump with the bolt
- New gasket on the air intake chamber

➡ **The protrusion on the gasket faces rearward**

- Intake air chamber, nut and 5 bolts
- 2 air intake chamber stays to the air intake chamber. Bolts: 13 ft. lbs. (18 Nm).
- 2 nuts holding the air intake chamber to the cylinder head. Nuts: 13 ft. lbs. (18 Nm).
- Except California vehicles, connect the vacuum sensing hose (from the fuel pressure regulator) to the air intake chamber
- Vacuum hose (from the actuator for ACIS) to the No. 1 vacuum pipe
- Brake booster vacuum hose to the air intake chamber. Tighten the union bolt to 22 ft. lbs. (29 Nm).
- Power steering air hose to the air intake chamber
- Vacuum hose (from the No. 2 vacuum pipe) to the air intake chamber

- 3 vacuum hoses (from the No. 2 vacuum pipe) to the No. 1 vacuum pipe
- Engine wire protector to the air intake chamber and install the bolt
- 4 bolts and 2 nuts holding the intake air connector to the air intake chamber
- Throttle body bracket to the cylinder head. Nuts: 15 ft. lbs. (21 Nm).
- No. 2 vacuum pipe to the air intake chamber and intake manifold. Nuts: 20 ft. lbs. (27 Nm).
- EGR gas temperature sensor connector
- EGR pipe. Union nut: 47 ft. lbs. (64 Nm). Bolts: 20 ft. lbs. (27 Nm).
- Oil dipstick guide and dipstick
- Fuel return hose
- With automatic transmission, transmission oil dipstick guide and dipstick
- VSV for the fuel pressure control
- Negative battery cable

6. Fill the cooling system.
7. Start the vehicle, check for leaks and repair if necessary.

2JZ-GTE Engine

1. Before servicing the vehicle, refer to the precautions in the beginning of this section.
2. Relieve fuel system pressure.
3. Drain the engine coolant.
4. Remove or disconnect the following:
   - Negative battery cable. On vehicles equipped with an air bag, wait at least 90 seconds before proceeding
   - Undercover
   - Accelerator and cruise control actuator cables
   - Throttle position sensor connector
   - Sub throttle position sensor connector
   - Sub throttle actuator connector
   - Throttle body from the air intake chamber by removing the 2 bolts and 2 nuts
   - Gasket from the throttle body
   - EVAP hose
   - Water bypass hose (from the No. 4 water bypass pipe)
   - Power steering air hose
   - Throttle body
   - With automatic transmissions, transmission dipstick and guide
   - Oil dipstick and guide
   - Air intake chamber stay
   - Control cable bracket from the air intake chamber

- IAC valve connector
- Turbo pressure sensor connector
- VSV connector for the fuel pressure control
- VSV connector for the EGR valve
- Engine wire protector
- IAC valve pipe from the clamp on the cylinder head cover and disconnect the air hose form the IAC valve
- Air hose (from the air intake chamber) from the vacuum pipe on the IAC valve pipe
- Air hose for the EGR from the valve pipe
- PCV hose from the PCV valve
- Vacuum sensing hose from the fuel pressure regulator
- Water bypass hose (from the IAC valve) from the No. 4 water bypass pipe
- EVAP hose (from the air intake chamber) from the vacuum pipe on the manifold stay
- EVAP hose (from the vacuum pipe on the No. 4 water bypass pipe) from the No. 2 vacuum pipe
- EVAP hose (from the charcoal canister) from the No. 2 vacuum pipe
- Power steering air hose from the air intake chamber
- Brake booster vacuum hose from the union on the air intake chamber
- EGR gas temperature sensor connector from the No. 2 vacuum pipe and wiring connector.
- EGR pipe
- No. 4 water bypass pipe
- Intake manifold stay
- Ground cable from the intake manifold
- Engine wire protector from the intake manifold
- Engine wire protector from the brackets
- Engine wire bracket.
- Air intake air chamber assembly from the intake manifold
- Water bypass hose from the IAC valve
- 6 injector connectors
- Camshaft position sensor connectors
- Engine wire clamps from the injector holders
- Fuel inlet pipe from the delivery pipe
- Fuel return pipe from the fuel pressure regulator

- Intake manifold and delivery pipe assembly

To install:

5. Install or connect the following:
   - New gasket, the intake manifold and delivery pipe assembly. Bolts and nuts: 20 ft. lbs. (27 Nm).
   - Fuel return pipe to the fuel pressure regulator. Union bolts: 20 ft. lbs. (27 Nm).
   - Fuel inlet pipe to the delivery pipe. Union bolt: 30 ft. lbs. (41 Nm).
   - Engine wire clamps to the injector holders
   - Camshaft position sensor connectors
   - Injector connectors. The No. 1, 3, and 5 injector connectors are dark gray; the No. 2, 4, and 6 injector connectors are gray
   - Air intake chamber gasket
   - Water bypass hose to the IAC valve
   - Air intake chamber assembly to the intake manifold and install the 5 bolts and 2 nuts. Tighten the nuts and bolts in several passes to 20 ft. lbs. (27 Nm).
   - Engine wire protector
   - Ground cable to the intake manifold
   - Intake manifold stay. Bolts: 29 ft. lbs. (39 Nm).
   - No. 4 water bypass pipe
   - EGR pipe and gasket with the 2 bolts. Bolts: 20 ft. lbs. (27 Nm).
   - Union bolt holding the EGR pipe to the EGR valve. Bolt: 47 ft. lbs. (64 Nm).
   - EGR gas temperature sensor connector
   - Brake booster vacuum hose to the union on the air intake chamber
   - Power steering air hose to the air intake chamber
   - EVAP hose (from the charcoal canister) to the No. 2 vacuum pipe
   - EVAP hose (from the vacuum pipe on the No. 4 water bypass pipe) to the No. 2 vacuum pipe
   - EVAP hose (from the air intake chamber) to the vacuum pipe on the manifold stay
   - Water bypass hose (from the IAC valve) to the No. 4 water bypass pipe
   - Vacuum sensing hose to the fuel pressure regulator
   - PCV hose to the PCV valve
   - Air hose for the EGR to the valve pipe

Heater Core replacement is covered in Section 2 of this manual

- Air hose (from the air intake chamber) to the vacuum pipe on the IAC valve pipe
- Connect the IAC valve pipe to the clamp on the cylinder head cover
- Connect the air hose to the IAC valve
- Engine wire protector to the vehicle body with the bolt
- VSV connector for the EGR valve and the fuel pressure control
- Turbo pressure sensor and the IAC valve connectors
- Control cable bracket to the intake chamber. Bolts: 14 ft. lbs. (19 Nm).
- Intake chamber stay. Nut and bolt: 14 ft. lbs. (19 Nm).
- Oil dipstick guide and dipstick. Use a new O-ring for the dipstick guide
- Transmission oil dipstick guide and dipstick. Use a new O-ring for the dipstick guide
- Throttle body
- Power steering air hose
- Water bypass hose (from the cylinder head)
- Water bypass hose (from the No. 4 water bypass pipe)
- EVAP hose
- Gasket and throttle body to the air intake chamber. Nuts and bolts: 15 ft. lbs. (21 Nm).
- Sub throttle actuator connector
- Sub throttle position sensor connector
- Throttle position sensor connector
- Cruise control actuator and the accelerator cables
- Air hose
- Undercover
- Negative battery cable
6. Fill the cooling system.
7. Start the vehicle, check for leaks and repair if necessary.

Exhaust Manifold

REMOVAL & INSTALLATION

5E-FE Engine

1. Before servicing the vehicle, refer to the precautions in the beginning of this section.
2. Remove or disconnect the following:
- Negative battery cable from the battery. On vehicles equipped with an air bag, wait at least 90 seconds before proceeding.
- All electrical wires and vacuum hoses that interfere with removal of the exhaust manifold.

- Exhaust heat insulator
- Exhaust pipe stay
- Exhaust pipe from the manifold by removing the 2 bolts and 2 compression springs.
- Exhaust manifold and gasket

To install:
3. Clean the gasket surfaces
4. Install or connect the following:
- Exhaust manifold with a new gasket. Bolts: 35 ft. lbs. (47 Nm).
- Exhaust stay to the engine and exhaust manifold. Bolt and nuts: 29 ft. lbs. (40 Nm).
- Exhaust manifold heat insulator. Bolts: 69 inch lbs. (8 Nm).
- Exhaust pipe to the exhaust manifold with the 2 compression springs and 2 bolts. Bolts: 46 ft. lbs. (62 Nm).
- All electrical wires and vacuum hoses that were disconnected for removal of the exhaust manifold.
- Negative battery cable

1NZ-FE

1. Before servicing the vehicle, refer to the precautions in the beginning of this section.
2. Remove or disconnect the following:
- Negative battery cable from the battery. On vehicles equipped with an air bag, wait at least 90 seconds before proceeding.

- All electrical wires and vacuum hoses that interfere with removal of the exhaust manifold.
- Exhaust heat insulator
- Exhaust pipe stay
- Exhaust pipe from the manifold by removing the 2 bolts and 2 compression springs.
- Exhaust manifold and gasket

To install:
3. Clean the gasket surfaces
4. Install or connect the following:
- Exhaust manifold with a new gasket. Bolts: 20 ft. lbs. (27 Nm).
- Exhaust stay to the engine and exhaust manifold. Bolt and nuts: 29 ft. lbs. (40 Nm).
- Exhaust manifold heat insulator. Bolts: 71 inch lbs. (8 Nm).
- Exhaust pipe to the exhaust manifold with the 2 compression springs and 2 bolts. Bolts: 46 ft. lbs. (62 Nm).
- All electrical wires and vacuum hoses that were disconnected for removal of the exhaust manifold.
- Negative battery cable

1ZZ-FE Engine

1. Before servicing the vehicle, refer to the precautions in the beginning of this section.

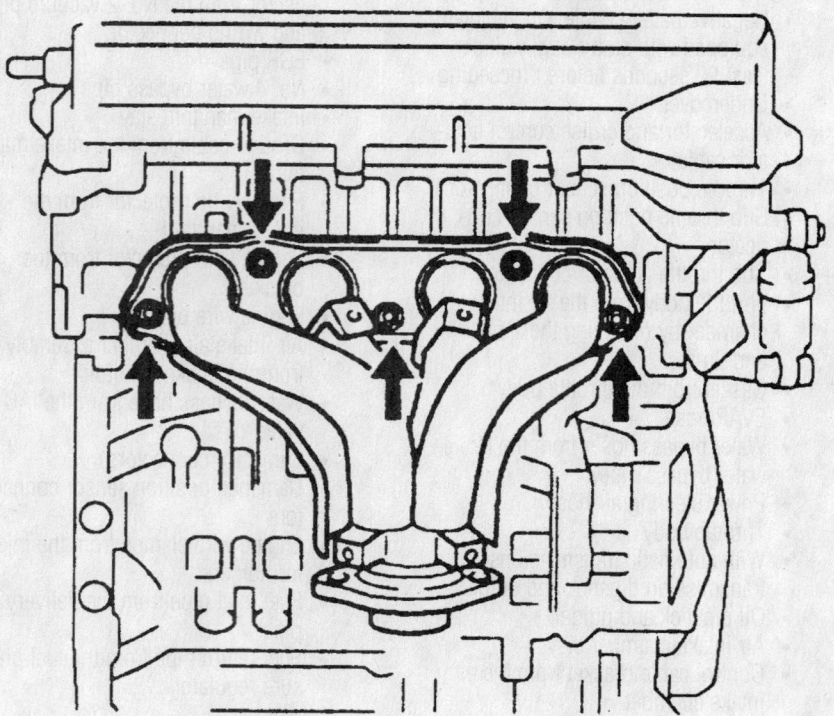

Exhaust manifold mounting nut locations—1.8L (1ZZ-FE) engine

7923VG22

2. Drain the cooling system.
3. Remove or disconnect the following:
   - Negative battery cable
   - Drive belt and alternator
   - Air intake duct
   - Accelerator cable
   - Exhaust pipe from the manifold
   - Exhaust manifold support bracket
   - Heat insulator from the dash panel
   - Upper heat insulator
   - Exhaust manifold and gasket
   - If necessary, the lower heat insulator from the exhaust manifold.

To install:

4. Install or connect the following:
   - Lower heat insulator on the exhaust manifold. Bolts: 108 inch lbs. (12 Nm).
   - Exhaust manifold using a new gasket. Nuts, tightened several passes: 27 ft. lbs. (37 Nm).
   - Upper heat insulator. Bolts: 108 inch lbs. (12 Nm).
   - Heat insulator on the dash panel
   - Exhaust manifold support bracket. Bolts in an alternating pattern: 37 ft. lbs. (49 Nm).
   - Front exhaust pipe to the manifold. Bolts: 46 ft. lbs. (62 Nm).
   - Oxygen sensor, using new gasket and nuts. Nuts: 14 ft. lbs. (20 Nm).
   - Accelerator cable and air intake duct
   - Alternator and drive belt
   - Negative battery cable
5. Fill the cooling system.
6. Start the vehicle, check for leaks and repair if necessary.

2ZZ-GE Engine

1. Before servicing the vehicle, refer to the precautions in the beginning of this section.
2. Drain the cooling system.
3. Remove or disconnect the following:
   - Negative battery cable
   - Drive belt and alternator
   - Air intake duct
   - Accelerator cable
   - Exhaust pipe from the manifold
   - Exhaust manifold support bracket
   - Heat insulator from the dash panel.
   - Upper heat insulator
   - Exhaust manifold and gasket
   - If necessary, the lower heat insulator from the exhaust manifold.

To install:

4. Install or connect the following:
   - Lower heat insulator on the

exhaust manifold. Bolts: 15 ft. lbs. (20 Nm).
   - Exhaust manifold using a new gasket. Nuts, tightened several passes: 37 ft. lbs. (50 Nm).
   - Upper heat insulator. Bolts: 15 ft. lbs. (20 Nm).
   - Heat insulator on the dash panel.
   - Exhaust manifold support bracket. Bolts: 37 ft. lbs. (49 Nm).
   - Front exhaust pipe to the manifold. Bolts: 46 ft. lbs. (62 Nm).
   - Oxygen sensor, using new gasket and nuts. Nuts: 14 ft. lbs. (20 Nm).
   - Accelerator cable and air intake duct.
   - Alternator and drive belt.
   - Negative battery cable
5. Fill the cooling system.
6. Start the vehicle, check for leaks and repair if necessary.

5S-FE Engine

1. Before servicing the vehicle, refer to the precautions in the beginning of this section.
2. Remove or disconnect the following:
   - Negative battery cable. On vehicles equipped with an air bag, wait at least 90 seconds before proceeding
   - Bolts holding the front exhaust pipe to the mounting bracket
   - Front exhaust pipe from the manifold
   - Main oxygen sensor connector and the sub oxygen sensor connector
   - Left-hand exhaust manifold stay
   - Upper manifold heat insulator
   - Right-hand exhaust manifold stay
   - Exhaust manifold, and the 3-way catalytic converter assembly
   - Oxygen sensor and gasket from the exhaust manifold
   - Sub oxygen sensor from the 3-way catalytic converter
   - Lower heat insulator from the exhaust manifold
   - 3-way catalytic converter heat insulators
   - 3-way Catalytic Converter (TWC), gasket, retainer and cushion from the exhaust manifold

To install:

3. Place the cushion, retainer, and a new gasket on the TWC and reinstall it to the exhaust manifold with the bolts and the nuts. Bolts and nuts: 22 ft. lbs. (30 Nm).
4. Install or connect the following:
   - Lower manifold heat insulator and

TWC heat insulators with the bolts.
   - Main oxygen sensor to the exhaust manifold with a new gasket and new nuts. Nuts: 14 ft. lbs. (19 Nm).
   - Sub oxygen sensor to the front TWC and tighten to 33 ft. lbs. (45 Nm).
   - Exhaust manifold, using new gasket and front TWC assembly to the engine with the nuts. Nuts, tighten uniformly in several passes: 36 ft. lbs. (49 Nm).
   - Right-hand exhaust manifold stay. Bolts and nuts: 31 ft. lbs. (42 Nm).
   - Upper heat insulator
   - Left-hand exhaust manifold stay. Bolt: 29 ft. lbs. (39 Nm). Nut: 31 ft. lbs. (42 Nm).
   - Main and the sub oxygen sensors connectors.
   - Front exhaust pipe with a new gasket to the TWC. Nuts: 46 ft. lbs. (62 Nm).
   - Front exhaust pipe to the exhaust pipe bracket. Bolts: 14 ft. lbs. (19 Nm).
   - Negative battery cable

1MZ-FE Engine

1. Before servicing the vehicle, refer to the precautions in the beginning of this section.
2. Remove or disconnect the following:
   - Negative battery cable
   - Undercovers
   - Front exhaust pipe from the exhaust manifold
   - EGR pipe from the exhaust manifold
   - Heated oxygen sensor connector from the right exhaust manifold
   - Exhaust manifold stay
   - Exhaust manifold and gasket

To install:

3. Install or connect the following:
   - Exhaust manifold using new gasket. Bolts, uniformly tightened: 36 ft. lbs. (49 Nm).
   - Exhaust manifold stay. Bolt and nut: 15 ft. lbs. (20 Nm).
   - Heated oxygen sensor connector to the right exhaust manifold
   - EGR pipe, using new gaskets, to the exhaust manifold and the engine. Nuts: 108 inch lbs. (12 Nm).
   - Front exhaust pipe, using new gas-

ket, to the exhaust manifold. Nuts: 46 ft. lbs. (62 Nm).

- Undercovers
- Negative battery cable

2JZ-GTE Engine

1. Before servicing the vehicle, refer to the precautions in the beginning of this section.

2. Drain the cooling system.

3. Remove or disconnect the following:

- Negative battery cable. On vehicles equipped with an air bag, wait at least 90 seconds before proceeding.
- Cruise control actuator cable from the throttle body
- No. 1 air hose
- Air cleaner and MAF meter assembly
- Theft deterrent horn from the body.
- Undercover
- Front lower arm bracket stay
- Front upper crossmember extension
- No. 2 front exhaust pipe
- Heat insulator for the No. 2 front exhaust pipe
- With automatic transmission, automatic transmission oil cooler tubes from the engine
- Engine wire protector from the body
- Heater hose from the No. 3 water bypass pipe
- EVAP hose from the No. 1 vacuum pipe
- IAC valve pipe from the No. 2 air tube
- No. 1 vacuum pipe from the air tubes
- VSV assembly
- Crankshaft position sensor connector from the clamp
- Water bypass hose (from the water pump) from the No.1 turbo water pipe
- Water bypass hose (from the water outlet) from the No. 1 turbo water pipe
- Water bypass hose (from the water outlet) from the No. 2 turbo water pipe
- No. 2 turbo water pipe from the No. 4 air tube
- No. 1 air tube from the No. 1 turbocharger by removing the 2 bolts
- 2 bolts holding the No. 4 air tube to the No. 1 turbocharger
- Air hose from the No. 4 air tube
- Air hose from the intake air connector

- No. 4 air tube and air bypass valve assembly
- Intake air control valve and gasket by removing the 2 nuts
- Air hose from the No. 2 air tube
- PCV hose from the No. 2 cylinder head cover
- Intake air connector and No. 1 air tube assembly
- Air inlet duct and cable bracket
- Heat insulator for the turbocharger
- Exhaust bypass pipe and gasket
- Exhaust gas control valve stay
- Main heated oxygen sensor by disconnecting the electrical connector and 2 nuts
- Exhaust gas control valve
- No. 1 turbocharger stay
- No. 2 turbocharger stay
- Union bolt holding the No. 1 turbo oil pipe to the cylinder block. Remove the 2 gaskets
- 2 nuts and disconnect the turbo oil pipe from the turbocharger. Remove the gaskets
- Turbo oil hose from the turbo oil outlet on the No. 1 oil pan. Remove the turbo oil pipe
- Union bolt holding the No. 2 turbo oil pipe to the cylinder block. Remove the 2 gaskets
- Turbo oil pipe from the turbocharger. Remove the 2 gaskets
- Turbo oil hose from the turbo oil outlet on the No. 1 oil pan and remove the turbo oil pipe
- Heater hose (from the No. 3 water bypass pipe) from the No. 2 water bypass pipe.

- Water bypass hose (from the No. 2 turbo water pipe) from the No. 2 water bypass pipe
- 8 nuts holding the turbochargers to the exhaust manifold
- 2 turbochargers and turbine outlet elbow assembly
- 2 gaskets
- Exhaust manifold

To install:

4. Install or connect the following:

- Exhaust manifold with 2 new gaskets. Torque the nuts, in several passes and in sequence: 29 ft. lbs. (39 Nm).
- 2 new gaskets and the turbochargers and turbine outlet elbow assembly to the exhaust manifold
- 8 **new** nuts and uniformly tighten the nuts in several passes. Nuts: 40 ft. lbs. (54 Nm).
- Water bypass hose (from the No. 2 turbo water pipe) to the No. 2 water bypass pipe
- Heater hose (from the No. 3 water bypass pipe) to the No. 2 water bypass pipe
- No. 2 turbo oil pipe. Nuts: 15 ft. lbs. (21 Nm). Union bolt: 29 ft. lbs. (39 Nm).
- No. 1 turbo oil pipe. Nuts: 15 ft. lbs. (21 Nm). Union bolt: 29 ft. lbs. (39 Nm).
- No. 2 turbocharger stay. Nut and bolt: 32 ft. lbs. (43 Nm).
- No. 1 turbocharger stay. Nut and bolt: 32 ft. lbs. (43 Nm).
- Exhaust gas control valve by

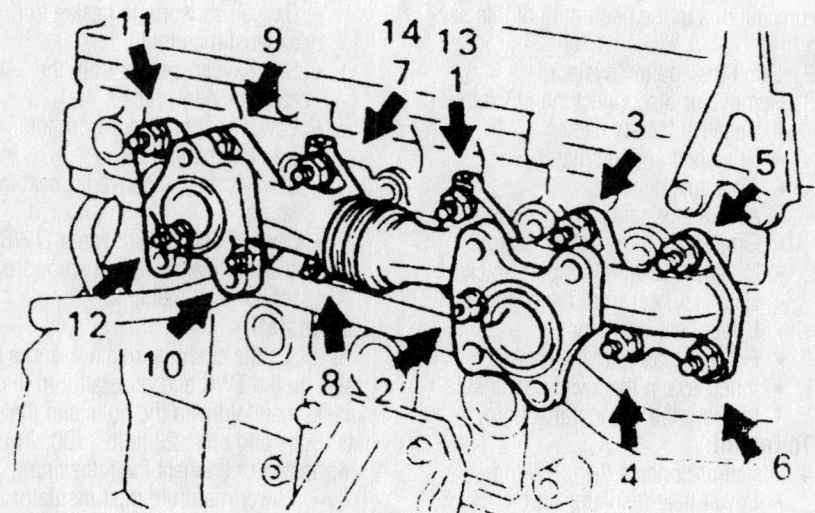

Exhaust manifold bolt installation sequence—3.0L (2JZ-GTE) engine

7923VG23

installing 2 new gaskets and the 3 nuts. Nuts: 51 ft. lbs. (69 Nm).
- Main heated oxygen sensor. Nuts: 14 ft. lbs. (20 Nm).
- Exhaust gas control valve stay. Bolt and nut: 32 ft. lbs. (43 Nm).
- Exhaust bypass pipe, with 2 new gaskets. Nuts: 18 ft. lbs. (25 Nm).
- Heat insulator for the turbocharger
- Air inlet duct
- Intake air connector and No. 1 air tube assembly
- PCV hose to the No. 2 cylinder head cover
- Air hose to the No. 2 air tube
- Intake air control valve and gasket. Nut: 15 ft. lbs. (21 Nm).
- No. 4 air tube and air bypass valve assembly
- Air hose to the intake air connector
- Air hose to the No. 4 air tube
- 2 bolts holding the No. 4 air tube to the No. 1 turbocharger. Bolt: 15 ft. lbs. (21 Nm).
- No. 1 air tube to the No. 1 turbocharger and install the 2 bolts. Bolts: 15 ft. lbs. (21 Nm).
- No. 2 turbo water pipe to the No. 4 air tube
- Water bypass hose (from the water outlet) to the No. 2 turbo water pipe
- Water hose (from the water outlet) to the No. 1 turbo water pipe
- Water bypass hose (from the water pump) to the No. 1 turbo water pipe
- Crankshaft position sensor connector to the clamp
- VSV assembly
- No. 1 vacuum pipe to the air tubes and install the 3 bolts
- 2 air hoses (from the vacuum pressure tank) to the vacuum pipe
- Air hose to the VSV for the intake air control valve
- Air hose (from the No. 2 air tube) from the vacuum pipe
- 2 air hoses (from the VSV for exhaust bypass valve) to the vacuum pipe
- Vacuum hose (from the air bypass valve) to the No. 1 air tube
- Air hose (from the VSV for exhaust gas control valve) to the vacuum pipe
- Air hose (from the VSV for waste gate valve) to the vacuum pipe
- Air hose to the No. 1 air tube
- Air hose to the No. 4 air tube

- Engine wire to the 3 clamps
- VSV connector for the exhaust bypass valve
- VSV connector for the intake air control valve
- IAC valve pipe to the No. 2 air tube
- EVAP hose to the No. 1 vacuum pipe
- Heater hose to the No. 3 water bypass pipe
- Engine wire protector to the body
- Automatic transmission oil cooler tubes to the engine
- Heat insulator for the No. 2 front exhaust pipe
- No. 2 front exhaust pipe, using new gasket. Nuts: 46 ft. lbs. (62 Nm).
- Front exhaust pipe to the No. 2 exhaust pipe and install the pipe support bracket. Bolts: 32 ft. lbs. (43 Nm).
- Bolts and nuts to hold the front exhaust pipe to the No. 2 front exhaust pipe. Tighten to 43 ft. lbs. (58 Nm).
- Upper front crossmember extension. Bolts to 22 ft. lbs. (29 Nm). Nuts: 25 ft. lbs. (33 Nm).
- Front lower arm bracket stay. Bolts: 33 ft. lbs. (44 Nm). Nut: 43 ft. lbs. (59 Nm).
- Theft deterrent horn
- Air cleaner and MAF meter
- Air cleaner duct
- No. 1 air hose
- Cruise control actuator cable to the throttle body
- Undercover
- Negative battery cable

5. Fill the cooling system to the proper level.

6. Start the vehicle, check for leaks and repair if necessary.

2JZ-GE Engine

1. Before servicing the vehicle, refer to the precautions in the beginning of this section.

2. Remove or disconnect the following:
- Negative battery cable. Wait at least 90 seconds before performing any other work.
- O2 sensors at the exhaust manifolds.
- If equipped, outside heat insulator

3. Remove the 4 nuts to disconnect the manifold from the front exhaust pipe. Loosen the mounting nuts to the exhaust manifolds and remove the 2 manifolds and the gasket.

To install:

4. Scrape the mating surfaces of all old gasket material.

5. Install or connect the following:
- Manifolds with a new gasket. Nuts: 29 ft. lbs. (39 Nm).
- Front exhaust pipe, using new gasket, to the exhaust manifolds. Nuts: 46 ft. lbs. (62 Nm).
- Outer heat insulator. Nuts: 13 ft. lbs. (18 Nm).
- O2 sensors
- Undercovers
- Negative battery cable

Front Crankshaft Seal

REMOVAL & INSTALLATION

➡**The following procedures apply to engines using a timing belt. The procedures for front cover seals can be found later in this section.**

5E-FE Engine

➡**The front oil seal can be removed from the engine without removing the oil pump.**

1. Before servicing the vehicle, refer to the precautions in the beginning of this section.

2. Remove or disconnect the following:
- Negative battery cable from the battery. On vehicles equipped with an air bag, wait at least 90 seconds before proceeding.
- Front covers and the timing belt
- Crankshaft timing belt sprocket

❄❄ WARNING

When removing the front seal, be extremely careful not to damage the crankshaft.

3. Using a knife, cut off the oil seal lip

4. Tape the end of a flat bladed tool to avoid damaging crankshaft. Pry out the oil seal using the taped end of the tool

5. Inspect the oil seal riding surface on the crankshaft for signs of wear or damage

To install:

6. Wipe the seal bore with a clean rag

7. Apply multipurpose grease the lip of a new oil seal

8. Drive the oil seal into place. Be sure the seal surface is flush with the oil pump case edge. Work from the front of the cover. Be extremely careful not to damage the seal

---

For complete Engine Mechanical specifications, see Section 1 of this manual

9. Install or connect the following:
- Sprocket without disturbing the Woodruff key
- Timing belt and front covers
- Negative battery cable

5S-FE Engine

1. Before servicing the vehicle, refer to the precautions in the beginning of this section.
2. Remove or disconnect the following:
- Negative battery cable. On vehicles equipped with an air bag, wait at least 90 seconds before proceeding

➡ **The front oil seal can be removed from the engine without removing the oil pump.**

- Timing belt covers and the timing belt
- Front crankshaft gear from the crankshaft. Be sure not to damage any part of the crankshaft
3. Using a knife, cut off the oil seal lip.
4. Pry out the oil seal. Wrap the edge of the tool with a rag or tape to prevent damaging the crankshaft. Be careful not to damage the crankshaft.

To install:
5. Using a new seal, apply a thin layer of liquid sealer to the outside of the seal.
6. Apply multi purpose grease to the new oil seal lip.
7. Oil seal until its surface is flush with the oil pump body edge.
8. Install or connect the following:
- Timing belt and the timing belt covers
- All other components
- Negative battery cable

1MZ-FE Engine

AVALON, CAMRY AND CAMRY SOLARA

1. Before servicing the vehicle, refer to the precautions in the beginning of this section.
2. Remove or disconnect the following:
- Negative battery cable. On vehicles equipped with an air bag, wait at least 90 seconds before proceeding.
- Timing belt
- Crankshaft timing gear
3. Cut out the lip portion of the oil seal.
4. Tape the end of a suitable prybar to protect the crankshaft and carefully remove the oil seal.

❋❋ **WARNING**

Be careful not to damage the crankshaft sealing surface.

To install:
5. Apply multi-purpose grease to the lip of a new oil seal. Also apply a light coating of liquid sealant to the outside of the oil seal.
6. Oil seal, until its surface is flush with the oil pump case edge.
7. Install or connect the following:
- Crankshaft timing gear
- Timing belt
- Negative battery cable

2JZ-GE and 2JZ-GTE Engines

1. Before servicing the vehicle, refer to the precautions in the beginning of this section.
2. Remove or disconnect the following:
- Timing belt
- Crankshaft timing pulley
3. Cut the oil seal lip.
4. Tape the end of a small prying tool and remove the oil seal.

❋❋ **WARNING**

Be careful not to damage the crankshaft.

To install:
5. Apply multi-purpose grease to a new oil seal lip.
6. Oil seal, until its surface is flush with the oil pump case edge.
7. Install or connect the following:
- Crankshaft timing pulley
- Timing belt

Camshaft(s)

REMOVAL & INSTALLATION

5E-FE Engine

1. Before servicing the vehicle, refer to the precautions in the beginning of this section.
2. Remove or disconnect the following:
- Negative battery cable. On vehicles equipped with an air bag, wait at least 90 seconds before proceeding.
- Valve cover
- Timing belt assembly
- Camshaft timing sprocket

❋❋ **WARNING**

Due to the relatively small amount of camshaft thrust clearance, the camshaft must be kept level during removal. If the camshaft is not level on removal, the portion of the head receiving the thrust may crack or be damaged.

3. Set the intake camshaft so the service bolt holes of the intake camshaft gears are directly up.
4. Remove each front bearing cap of the intake and exhaust camshafts.
5. Secure the intake camshaft sub-gear to the main gear with a 6mm diameter bolt, 16–20mm long and with a pitch of 1.0mm. Be sure that the torsional spring force of the subgear has been eliminated by the above operation.

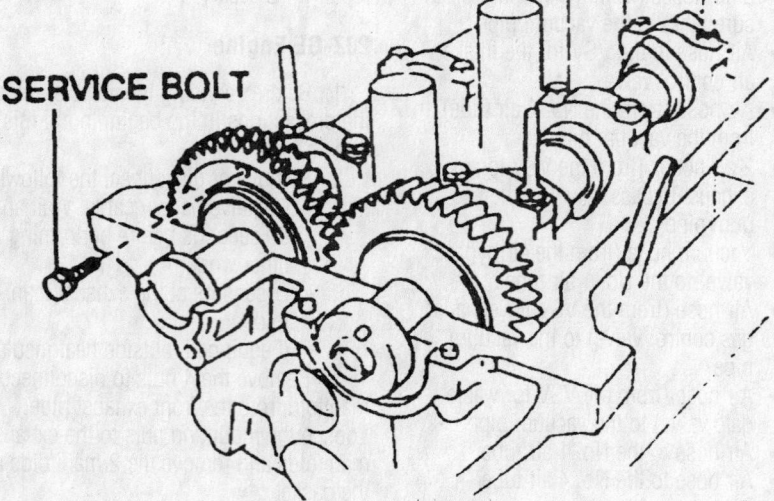

SERVICE BOLT

Securing the intake camshaft subgear to the main gear—1.5L (5E-FE) engine

7923VG24

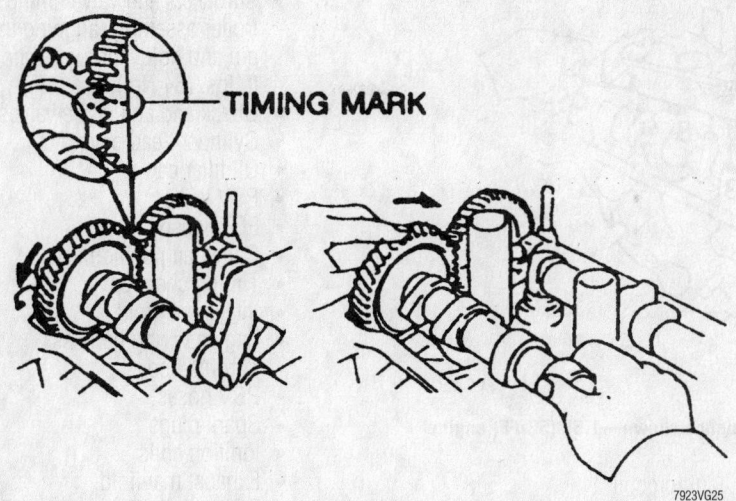

TIMING MARK

7923VG25

Exhaust and intake camshaft gear timing marks—1.5L (5E-FE) engine

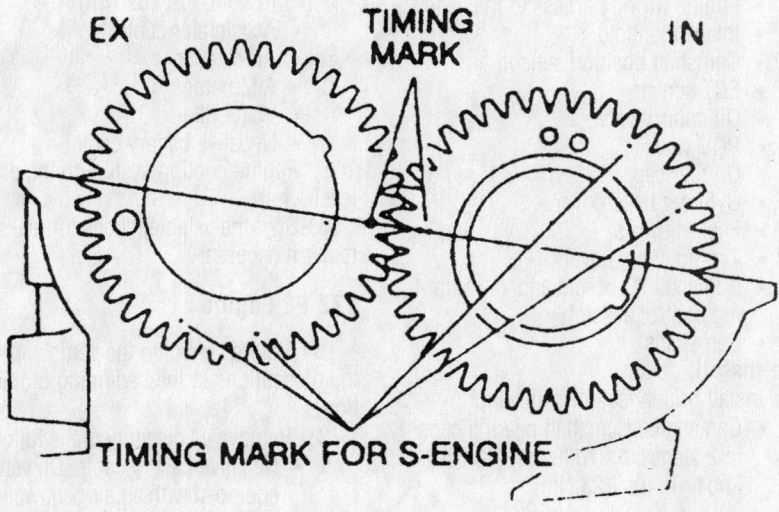

EX   TIMING
MARK   IN

TIMING MARK FOR S-ENGINE

7923VG26

Correct exhaust and intake camshaft gear timing mark alignment—1.5L (5E-FE) engine

7923VG27

Tighten the intake camshaft bearing cap bolts in sequence—1.5L (5E-FE) engine

6. In correct sequence, loosen and remove the 8 bolts of the 4 bearing caps of the exhaust camshaft and remove the exhaust camshaft. If the camshaft is not being lifted out straight, install the middle bearing cap and loosen it evenly to keep the camshaft straight.

7. In the reverse order of the installation sequence, loosen and remove the 8 bolts of the 4 bearing caps of the intake camshaft and remove the intake camshaft. If the camshaft is not being lifted out straight, install the middle bearing cap and loosen it evenly to keep the camshaft straight.

8. Remove the adjusting shim.

To install:

9. Install the valve shim to the engine.

10. Apply engine oil to the surface of the intake camshaft.

11. Place the intake camshaft on the cylinder head so the service bolt points directly up.

12. Install the 4 rearward bearing caps in their original order and temporarily tighten them evenly. Do not tighten the bolts at this time.

13. Apply engine oil to the portion of the exhaust camshaft.

14. Engage the exhaust camshaft gear to the intake camshaft gear by matching the proper timing marks on each gear.

15. Roll down the exhaust camshaft onto the bearing journals while engaging the gears with each other.

➡There are other marks present for the "S" engine. Do not use these marks.

16. Install the 4 rearward intake bearing caps in their original order and tighten evenly. Do not tighten the bolts at this time.

17. Remove the service bolt.

18. Clean the mating surfaces of the No. 2 bearing cap and apply sealer. Install and temporarily tighten the bolts. Install the camshaft housing plug.

19. Tighten the intake camshaft cap bolts to 108 inch lbs. (13 Nm), in the proper sequence.

20. Apply grease to a new camshaft oil seal lip and install it as far as the deepest part of the cylinder head.

21. Install the No. 1 bearing cap and temporarily tighten the bolts.

22. Tighten the exhaust camshaft bearing cap bolts to 108 inch lbs. (13 Nm) evenly and in the proper sequence.

23. Turn the camshaft 1 revolution and check that the timing marks of the camshaft gears are aligned.

For Accessory Drive Belt illustrations, see Section 1 of this manual

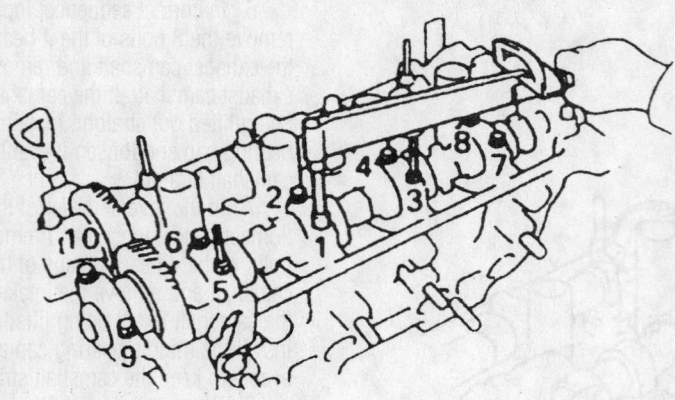

7923VG28

Tighten the exhaust bearing cap bolts according to the sequence shown—1.5L (5E-FE) engine

24. Check and adjust the valve clearance.
25. Install or connect the following:
   - Valve cover
   - Camshaft timing sprocket. Bolt: 37 ft. lbs. (50 Nm).
   - Timing belt
   - Negative battery cable

1NZ-FE Engine

1. Before servicing the vehicle, refer to the precautions in the beginning of this section.
2. Drain the cooling system.
3. Remove or disconnect the following:
   - Negative battery cable
   - Water filler
   - Outer front cowl top panel
   - Alternator
   - Air cleaner
   - Accelerator cable
   - Center exhaust pipe
   - Exhaust manifold support
   - Exhaust manifold

   - Ignition coil
   - Spark plugs
   - PCV hoses
   - Throttle body
   - Engine wiring harness at the head
   - Intake manifold
   - Camshaft position sensor
   - ECT sensor
   - Oil control valve
   - PCV valve
   - Oil filer cap
   - Cylinder head cover
   - Fuel injectors
   - Timing chain cover
   - Camshaft sprockets and valve timing control assembly
   - Camshafts

To install:

4. Install or connect the following:
   - Camshafts. Camshaft bearing caps in 2 stages: 1st 10 ft. lbs. (13 Nm); 2nd 17 ft. lbs. (23 Nm).

- Sprockets and valve timing controller assembly, aligning the knock pin and hole. Torque the bolts to 47 ft. lbs. (64 Nm).
- Check and adjust the valves.
- Cylinder head cover
- Oil filler cap
- PCV valve
- ECT sensor
- Camshaft position sensor
- Timing chain cover
- Intake manifold
- Engine wiring harness
- Throttle body
- PCV hoses
- Spark plugs
- Ignition coils
- Exhaust manifold
- Exhaust manifold support. Torque the bolts to 27 ft. lbs. (37 Nm).
- Front exhaust pipe. Torque the nuts to 46 ft. lbs. (62 Nm).
- Accelerator cable
- Air cleaner
- Alternator
- Water filler
- Negative battery cable

5. Fill the cooling system to the proper level.
6. Start the vehicle, check for leaks and repair if necessary.

1ZZ-FE Engine

1. Before servicing the vehicle, refer to the precautions in the beginning of this section.
2. Remove or disconnect the following:
   - Negative battery cable. On vehicles equipped with an air bag, wait at least 90 seconds before proceeding.
   - Cylinder head cover

3. Turn the crankshaft so that the No. 1 piston is at TDC on the compression stroke. Check to see that the point marks on the camshaft sprockets are facing each other, if not, rotate the crankshaft 1 full revolution.
4. Tie the timing chain to each sprocket with string or wire to maintain correct valve timing.
5. Hold the camshafts with a wrench and remove the bolts securing the sprockets to the camshafts.
6. Using several passes, gradually remove the bearing cap bolts in the proper sequence. Then, remove the camshafts
To install:
7. Lubricate the camshafts with clean engine oil and place them on the cylinder head. Be sure to position the lobes for the No. 1 cylinder as shown in the illustration.
8. Install the bearing caps in their origi-

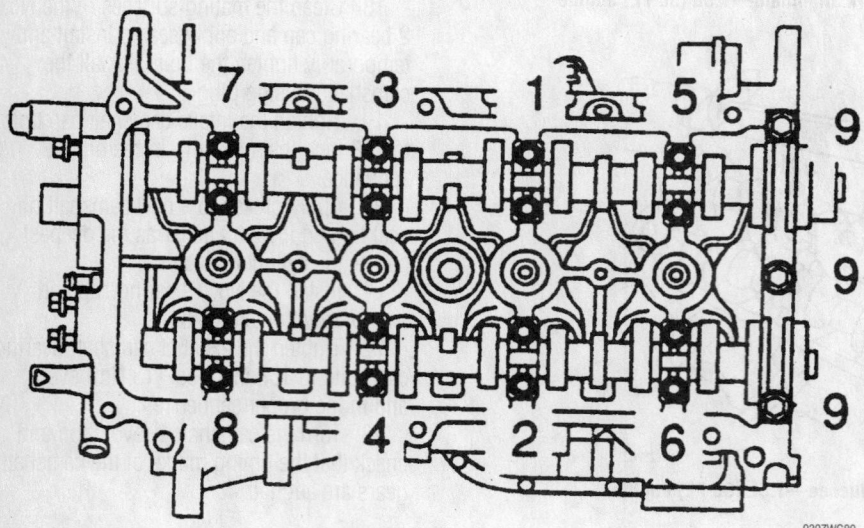

9307WG80

Camshaft bolt torque sequence—1NZ-FE engine

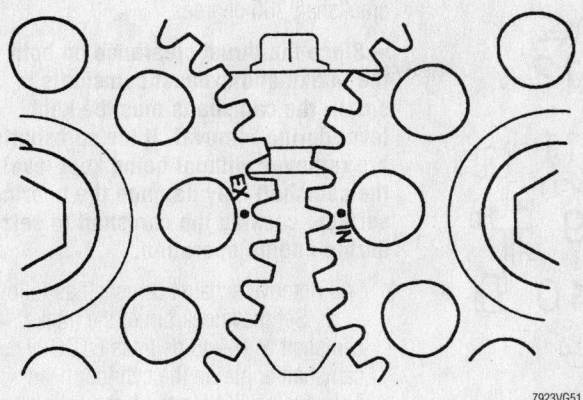

7923VG51

The sprocket marks will align when the No. 1 piston is at TDC on the compression stroke—1.8L (1ZZ-FE) engine

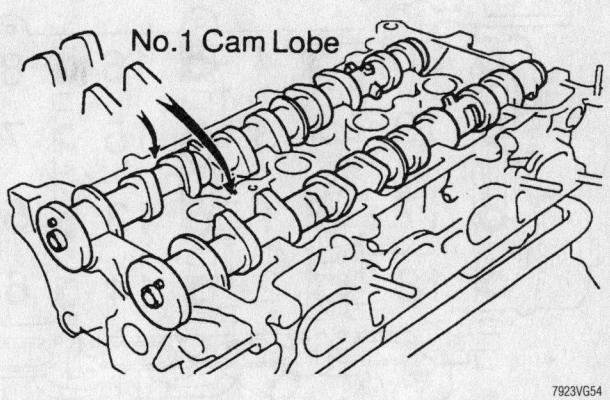

7923VG54

When installing the camshafts, position the lobes for the No. 1 cylinder as shown—1.8L (1ZZ-FE) engine

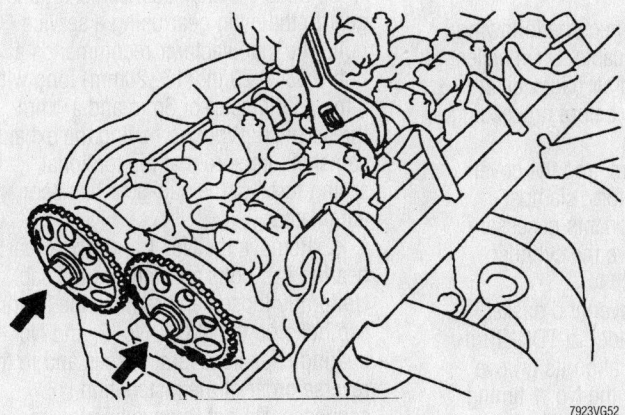

7923VG52

Hold the camshaft with a wrench while removing the sprocket bolt—1.8L (1ZZ-FE) engine

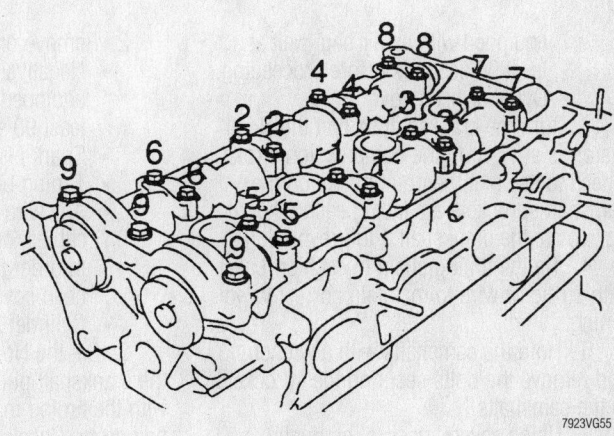

7923VG55

Camshaft bearing cap bolt tightening sequence—1.8L (1ZZ-FE) engine

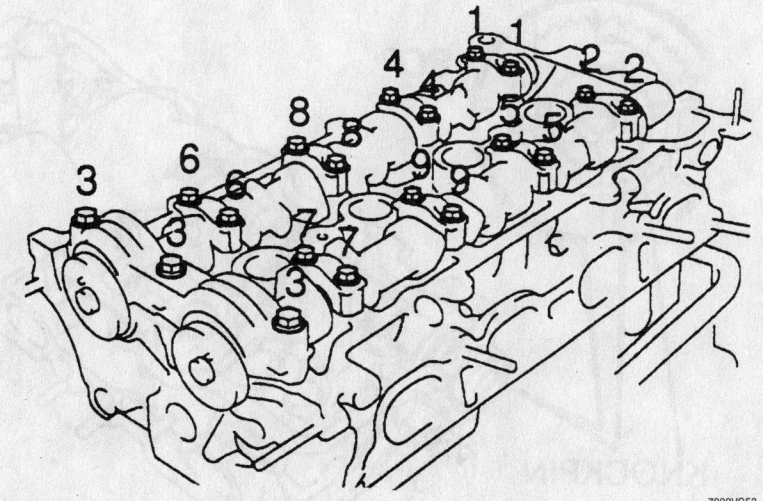

7923VG53

Camshaft bearing cap bolt removal sequence—1.8L (1ZZ-FE) engine

nal positions. Apply clean engine oil to the threads and under the heads of the bearing cap bolts. After tightening the bolts on the No. 1 bearing cap to 17 ft. lbs. (23 Nm), tighten the remaining bolts in sequence using several passes to 10 ft. lbs. (13 Nm).

9. Check the valve clearance and make adjustments as needed.

10. Install or connect the following:
- Camshaft sprockets and the chain
- Cylinder head cover
- Negative battery cable

2ZZ-GE Engine

1. Before servicing the vehicle, refer to the precautions in the beginning of this section.

2. Remove or disconnect the following:
- Negative battery cable. On vehicles

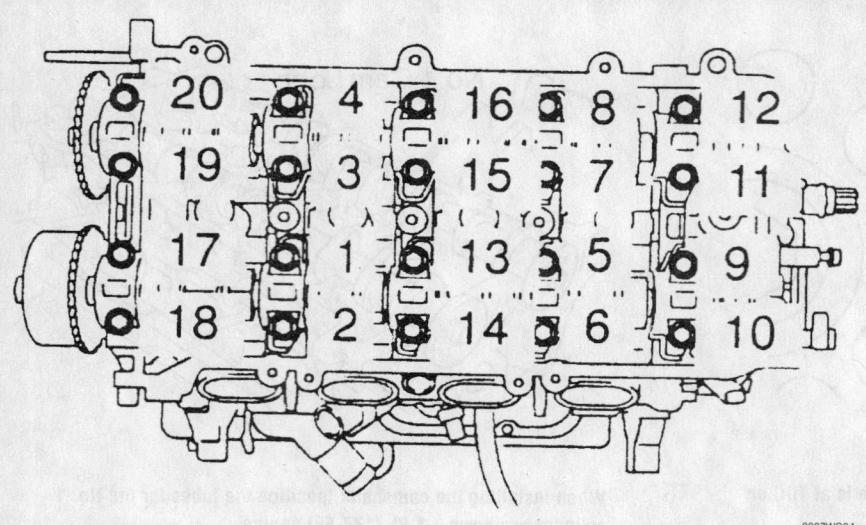

Camshaft bearing cap torque sequence—2ZZ-GE engines

equipped with an air bag, wait at least 90 seconds before proceeding.

• Cylinder head cover

3. Turn the crankshaft so that the No. 1 piston is at TDC on the compression stroke. Check to see that the point marks on the camshaft sprockets are facing each other, if not, rotate the crankshaft 1 full revolution.

4. Tie the timing chain to each sprocket with string or wire to maintain correct valve timing.

5. Hold the camshafts with a wrench and remove the bolts securing the sprockets to the camshafts.

6. Using several passes, gradually remove the bearing cap bolts in the proper sequence. Then, remove the camshafts

To install:

7. Lubricate the camshafts with clean engine oil and place them on the cylinder head. Be sure to position the lobes for the No. 1 cylinder as shown in the illustration.

8. Install the bearing caps in their original positions. Apply clean engine oil to the threads and under the heads of the bearing cap bolts. After tightening the bolts on the No. 1 bearing cap to 14 ft. lbs. (18 Nm), tighten the remaining bolts in sequence using several passes to 14 ft. lbs. (18 Nm).

9. Check the valve clearance and make adjustments as needed.

10. Install or connect the following:

• Camshaft sprockets and the chain
• Cylinder head cover
• Negative battery cable

5S-FE Engine

1. Before servicing the vehicle, refer to the precautions in the beginning of this section.

2. Remove or disconnect the following:

• Negative battery cable. On vehicles equipped with an air bag, wait at least 90 seconds before proceeding.
• Spark plug wires
• Timing belt, gears, and the covers
• Any wire connectors, clamps, cables, or components necessary in order to remove the cylinder head cover.
• Cylinder head cover and gasket

3. Set the No. 1 cylinder to TDC. Turn the crankshaft pulley and align its groove with the timing mark **0** of the No. 1 timing belt cover. Check that the valve lifters on the No. 1 cylinder are loose and valve lifters on

the No. 4 cylinder are tight. If not, rotate the crankshaft 360 degrees.

➡ Since the thrust clearance on both the intake and exhaust camshafts is small, the camshafts must be kept level during removal. If the camshafts are removed without being kept level, the camshaft may damage the bearing surface, causing the camshaft to seize during engine operation.

4. Remove exhaust camshaft as follows:

a. Set the knock pin of the intake camshaft at 10–45 degrees BTDC of camshaft angle on the cylinder head. This angle will help to lift the exhaust camshaft level and evenly by pushing the No. 2 and No. 4 cylinder camshaft lobes of the exhaust camshaft toward their valve lifters.

b. Secure the exhaust camshaft sub-gear to the main gear using a service bolt. The manufacturer recommends a bolt 0.63–0.79 in. (16–20mm) long with a thread diameter of 6mm and a 1mm thread pitch. When removing the exhaust camshaft, be sure that the torsional spring force of the sub-gear has been eliminated.

c. Remove the No. 1 and No. 2 rear bearing cap bolts and remove the cap. Uniformly loosen and remove the bearing cap bolts on the No. 1, No. 2, and No. 4 bearing caps in several passes and in the reverse order of the installation sequence. Do not remove bearing cap bolts to No. 3 bearing cap at this time. Remove the No. 1, 2, and 4 bearing caps.

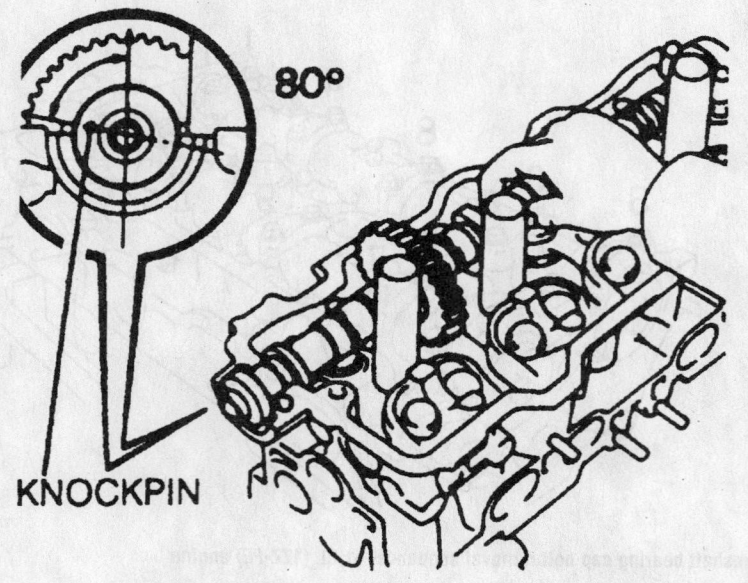

Intake camshaft removal and installation positioning—2.2L (5S-FE) engine

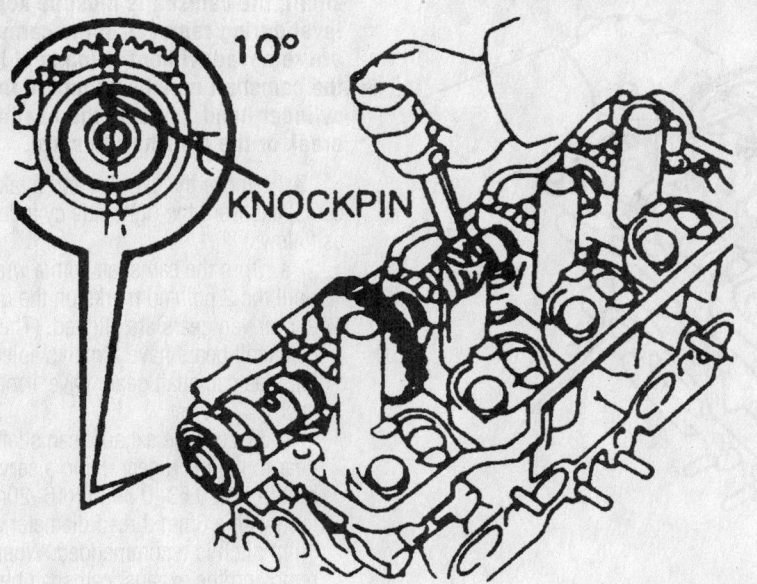

Exhaust camshaft removal and installation positioning—2.2L (5S-FE) engine

d. Alternately loosen and remove the bearing cap bolts on the No. 3 bearing cap. As these bolts are loosened check to see that the camshaft is being lifted out straight and level.

➡ **If the camshaft is not lifted out straight and level, tighten the No. 3 bearing cap bolts. Reverse the order of Steps 7c through 7a and reset the intake camshaft knock pin to 10–45 degrees BTDC, then repeat Steps 7a through 7c. Do not attempt to pry the camshaft from its mounting.**

e. Remove the No. 3 bearing cap and exhaust camshaft from the engine.

5. Remove intake camshaft as follows:

a. Set the knock pin of the intake camshaft at 80–115 degrees BTDC of the camshaft angle on the cylinder head. This angle will help to lift the intake camshaft level and evenly by pushing No. 1 and No. 3 cylinder camshaft lobes of the intake camshaft toward their valve lifters.

b. Remove the 2 front bearing cap bolts and remove the front bearing cap and oil seal. If the cap will not come apart easily, leave it in place without the bolts.

c. Uniformly loosen and remove the bearing cap bolts to No. 1, No. 3, and the No. 4 bearing caps in several phases and in the reverse order of the installation sequence. Do not remove bearing cap

bolts to the No. 2 bearing cap at this time. Remove No. 1, 3, and 4 bearing caps.

d. Alternately loosen and remove bearing cap bolts to the No. 2 bearing cap. As these bolts are loosened and after breaking the adhesion on the front

bearing cap, check to see that the camshaft is being lifted out straight and level.

➡ **If the camshaft is not lifting out straight and level tighten the No. 2 bearing cap bolts. Reverse Steps 8b through 8d, then start over from Step 8b. Do not attempt to pry the camshaft from its mounting.**

e. Remove the No. 2 bearing cap with the intake camshaft from the engine.

6. Remove the valve lifter shims and hydraulic lifters. Identify each lifter and shim as it is removed so it can be reinstalled in the same position. If the lifters are to be reused, store them upside down in a sealed container.

To install:

7. Install the valve lifters into their original positions and shims.

➡ **Before installing the intake camshaft, apply multi-purpose grease to the camshaft.**

8. Install the intake camshaft, as follows:

a. Position the camshaft at 80–115 degrees BTDC of camshaft angle on the cylinder head.

b. Apply sealant to the front bearing cap.

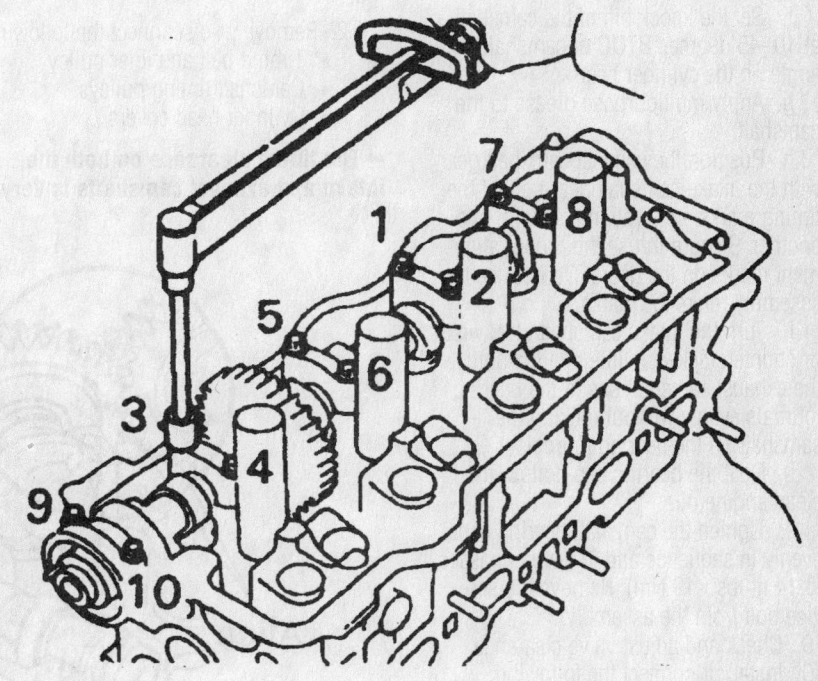

Intake camshaft bearing cap bolt tightening sequence—2.2L (5S-FE) engine

For Wheel Alignment specifications, see Section 1 of this manual

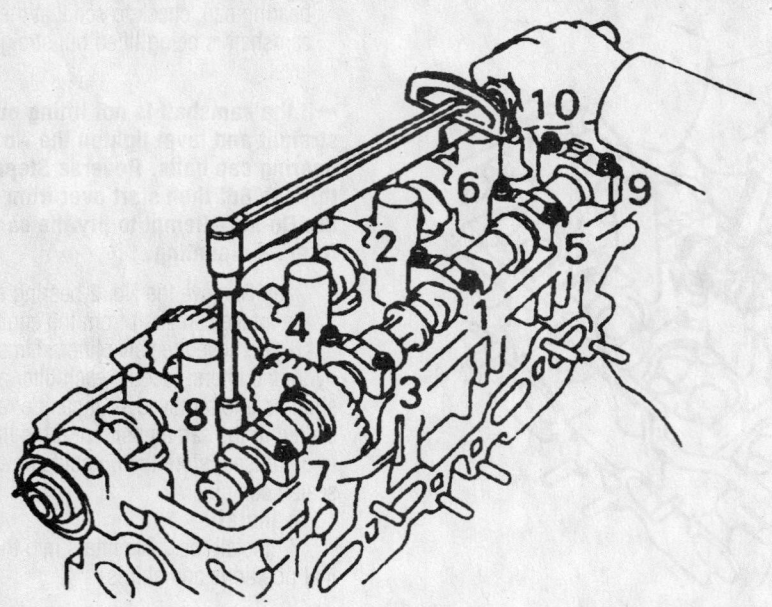

7923VG38

Exhaust camshaft bearing cap bolt tightening sequence—2.2L (5S-FE) engine

c. Coat the bearing cap bolts with clean engine oil.

d. Tighten the camshaft bearing caps evenly in sequence and in several passes to 14 ft. lbs. (19 Nm).

e. Apply MP grease to a new oil seal lip, and by using a suitable tool, tap a new oil seal into place.

• Exhaust camshaft as follows:

f. Set the knock pin of the camshaft at 10–45 degrees BTDC of camshaft angle on the cylinder head.

g. Apply multipurpose grease to the camshaft.

h. Position the exhaust camshaft gear with the intake camshaft gear so that the timing marks are in alignment with one another. Be sure to use the proper alignment marks on the gears. Do not use the assembly reference marks.

i. Turn the intake camshaft clockwise or counterclockwise little by little until the exhaust camshaft sits in the bearing journals evenly without rocking the camshaft on the bearing journals.

j. Coat the bearing cap bolts with clean engine oil.

k. Tighten the camshaft bearing caps evenly in sequence and in several passes to 14 ft. lbs. (19 Nm). Remove the service bolt from the assembly.

9. Check and adjust valve clearance.

10. Install or connect the following:
• Cylinder head cover
• Timing belt and related components.
• Electrical connectors, cables,

brackets, and components attached to the cylinder head cover.
• Spark plug wires
• Negative battery cable

1MZ-FE Engine

1. Before servicing the vehicle, refer to the precautions in the beginning of this section.

2. Remove or disconnect the following:
• Timing belt and idler pulley
• Camshaft timing pulleys
• Cylinder head covers

➡**The thrust clearance on both the intake and exhaust camshafts is very** small; the camshafts must be kept level during removal. If the camshafts are removed without being kept level, the camshaft may be caught in the cylinder head, causing the head to break or the camshaft to seize.

3. Remove the exhaust and intake camshafts from the right side cylinder head, as follows:

a. Turn the camshaft with a wrench until the 2 pointed marks on the drive and driven gears are aligned. (The right camshaft gears have 2 marks apiece; the left side camshaft gears have 1 mark each.)

b. Secure the exhaust camshaft sub-gear to the main gear using a service bolt. A bolt 0.63–0.79 in. (16–20mm) long with a 6mm thread diameter and a 1mm pitch is recommended. When removing the exhaust camshaft be sure the sub-gear is not loaded; all the force must be eliminated.

c. Uniformly loosen and remove the exhaust camshaft bearing cap bolts in several passes and in the proper sequence. Remove the 8 bearing cap bolts and remove the caps, keeping them in the correct order.

d. Remove the exhaust camshaft from the engine.

e. Uniformly loosen and remove the 10 bearing cap bolts in several passes, in the proper sequence. Remove the bearing caps, keeping them in order, remove the oil seal, then, lift out the intake camshaft.

4. Remove the exhaust and intake camshafts from the left side cylinder head, as follows:

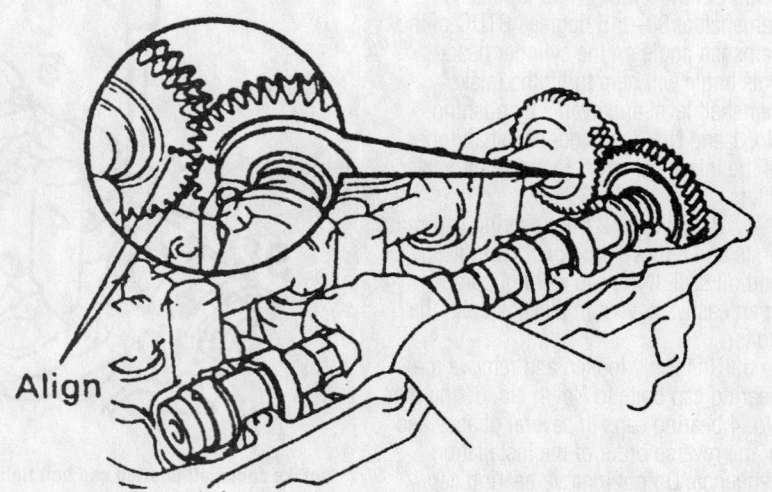

Align

7923VG40

Aligning the camshaft gear timing marks for the right camshafts—3.0L (1MZ-FE) engine

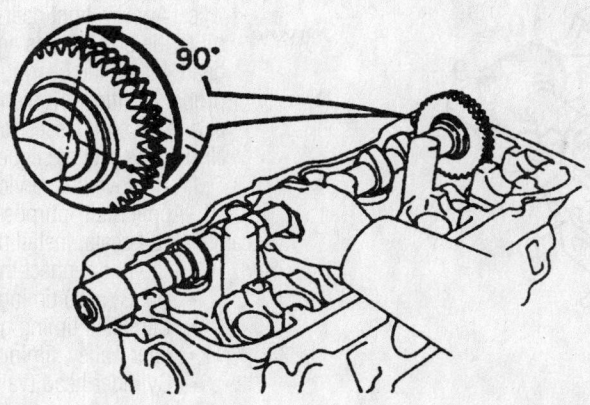

7923VG41

Camshaft installation for the right exhaust camshaft—3.0L (1MZ-FE) engine

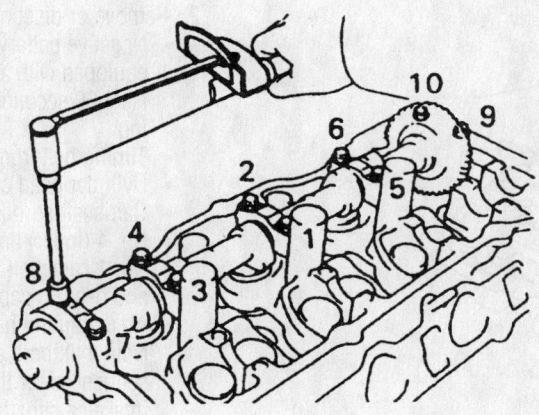

7923VG42

Bearing cap bolt tightening sequence for the right exhaust camshaft—3.0L (1MZ-FE) engine

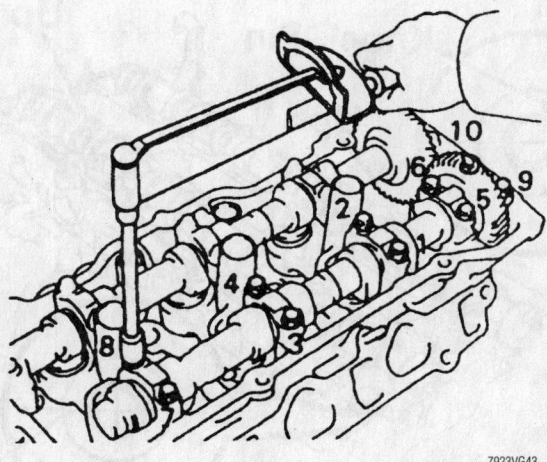

7923VG43

Bearing cap bolt tightening sequence for the right intake camshaft—3.0L (1MZ-FE) engine

a. Turn the camshaft with a wrench until the pointed marks on the drive and driven gears are aligned. (The right camshaft gears have 2 marks apiece; the left side camshaft gears have 1 mark each.)

b. Secure the exhaust camshaft sub-gear to the main gear using a service bolt. A bolt 0.63–0.79 in. (16–20mm) long with a 6mm thread diameter and a 1mm pitch is recommended. When removing the exhaust camshaft be sure the sub-gear is not loaded; all the force must be eliminated.

c. Uniformly loosen and remove the exhaust camshaft bearing cap bolts in several passes and in the proper sequence. Remove the 8 bearing cap bolts and remove the caps. Keep the caps in the correct order.

d. Remove the exhaust camshaft from the engine.

e. Uniformly loosen and remove the 10 bearing cap bolts in several passes, in the reverse order of the installation sequence. Remove the bearing caps, keeping them in order, remove the oil seal, then lift out the intake camshaft.

5. Remove the valve lifter shims and hydraulic lifters. If the lifters are to be reused, store them upside down in a sealed container.

To install:

6. Install the valve lifters into their original positions and shims. Check the valve clearance and replace the shims as necessary.

➡**Before installing the camshafts in either cylinder head, apply multi-purpose grease to each camshaft.**

7. Install the right camshafts, as follows:

a. Position the intake camshaft on the head so that the alignment marks are at a 90 degrees angle from vertical. The mark should be at the "3 o'clock" position.

b. Apply sealant to the No. 1 bearing cap.

c. Apply a light coat of clean engine oil to the bolt threads and under the bolt head. Install the bearing caps to their proper position. Tighten the bolts evenly and in several passes to 12 ft. lbs. (16 Nm) in the proper sequence.

d. Position the exhaust camshaft on the head so that the alignment marks are at a 90 degrees angle from vertical. The mark must align with the marks on the other gear.

e. Apply a light coat of clean engine oil to the bolt threads and under the bolt

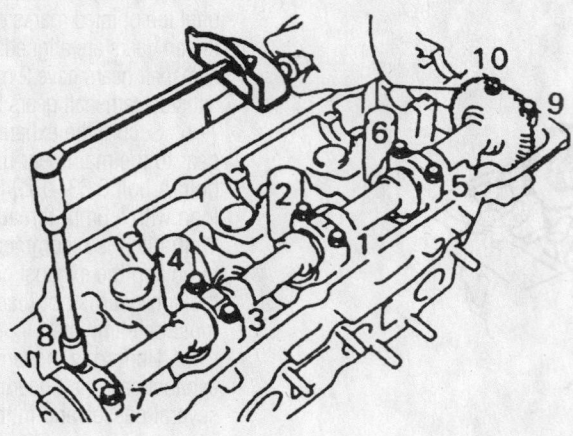

7923VG44

Bearing cap bolt tightening sequence for the left exhaust camshaft—3.0L (1MZ-FE) engine

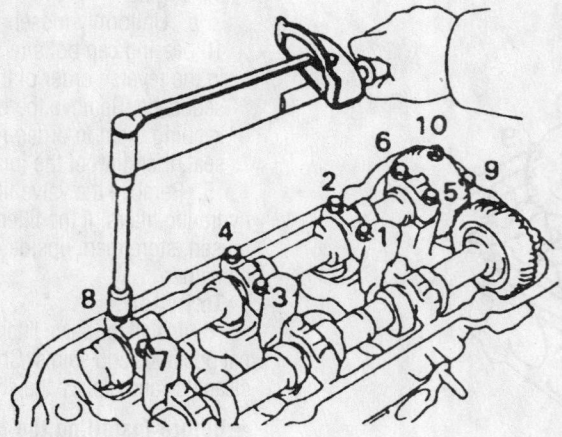

7923VG45

Bearing cap bolt tightening sequence for the left intake camshaft—3.0L (1MZ-FE) engine

tion and must align with the marks on the other gear.

e. Apply a light coat of clean engine oil to the bolt threads and under the bolt head. Install the bearing caps to their proper position. Tighten the bolts evenly and in several passes to 12 ft. lbs. (16 Nm) in the proper sequence.

f. Remove the service bolt.

9. Apply multi-purpose grease to new camshaft oil seals. Install the seals.

10. Install or connect the following:
- No. 3 (rear) timing belt cover
- Camshaft timing gears
- Idler pulley, timing belt and covers
- Cylinder head (valve) covers

2JZ-GTE and 2JZ-GE Engines

1. Before servicing the vehicle, refer to the precautions in the beginning of this section.

2. Remove or disconnect the following:
- Negative battery cable. On vehicles equipped with an air bag, wait at least 90 seconds before proceeding.
- Timing belt from the engine
- Cylinder head covers
- Camshaft sprocket
- No. 4 (inner) timing belt cover
- No. 1 camshaft bearing cap bolts and bearing caps.
- All remaining bearing cap bolts. Note that there are separate sequences for the exhaust and intake camshafts. Lift off all 12 bearing caps.
- Exhaust and intake camshafts

head. Install the bearing caps to their proper position. Tighten the bolts evenly and in several passes to 12 ft. lbs. (16 Nm) in the proper sequence.

f. Remove the service bolt.

8. Install the left camshaft, as follows:

a. Position the intake camshaft on the head so that the alignment mark is at a 90degrees angle from vertical. The mark should be at the "9 o'clock" position.

b. Apply sealant to the No. 1 bearing cap.

c. Apply a light coat of clean engine oil to the bolt threads and under the bolt head. Install the bearing caps to their proper position. Tighten the bolts evenly and in several passes to 12 ft. lbs. (16 Nm) in the proper sequence.

d. Position the exhaust camshaft on the head so that the alignment marks are at a 90 degree angle from vertical. The mark should be at the "3 o'clock" posi-

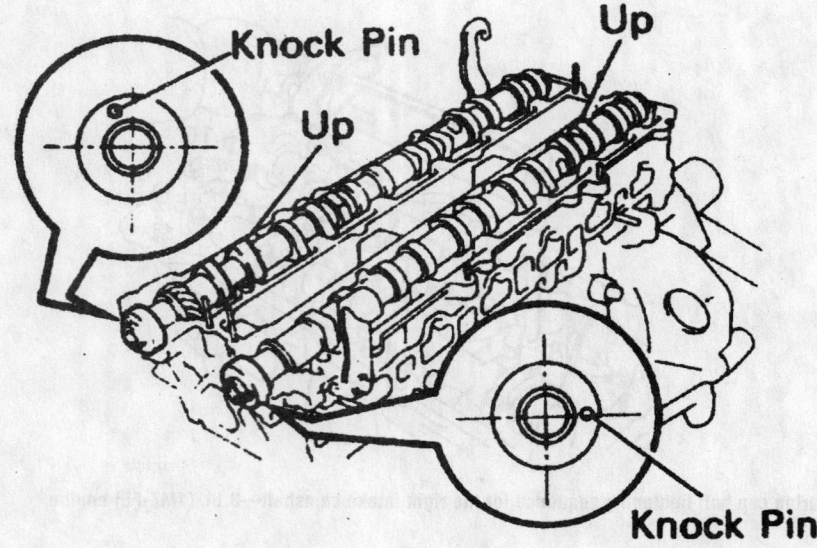

7923VG46

During installation, position the knock pins as shown—3.0L (2JZ-GE and 2JZ-GTE) engines

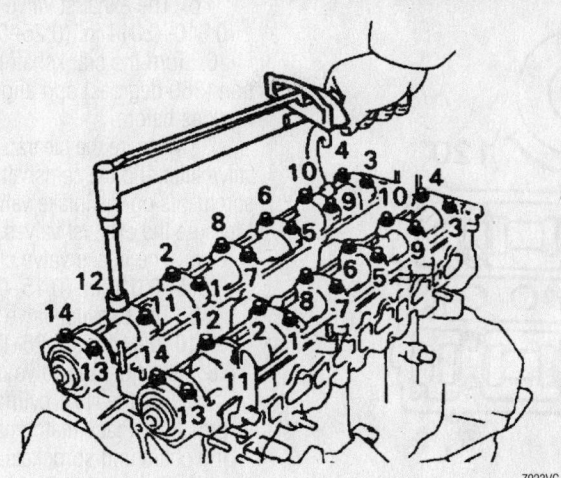

Camshaft bearing cap bolt tightening sequence—3.0L (2JZ-GE and 2JZ-GTE) engines

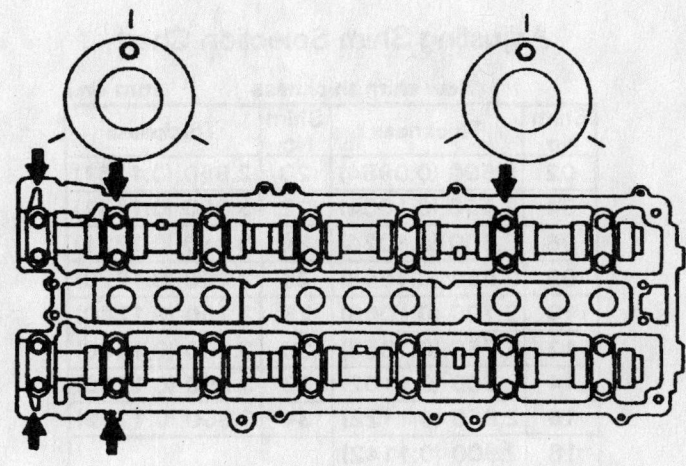

Tightening the camshafts (Step 1)—3.0L (2JZ-GE and 2JZ-GTE) engines

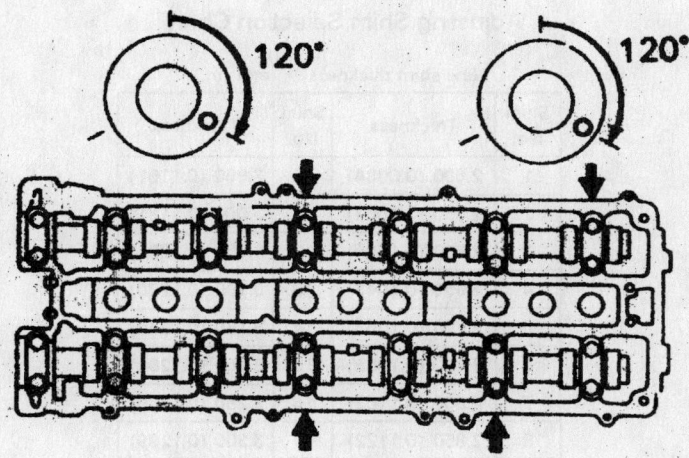

Tightening the camshafts (Step 2)—3.0L (2JZ-GE and 2JZ-GTE) engines

- Valve lifter shims and hydraulic lifters. Identify each lifter and shim as it is removed so it can be reinstalled in the same position. If the lifters are to be reused, store them upside down in a sealed container.

To install:

3. Install or connect the following:
   - Valve lifters into their original positions
   - Shims. Check the valve clearance and replace the shims as necessary

➡**When reinstalling, remember that the camshafts must be handled carefully and kept straight and level to avoid damage.**

- Camshafts, coated with engine oil
- No. 3 and No. 7 bearing caps in place
- Cap bolts, coated with oil. Then tighten them temporarily
- New oil seals, coated with multipurpose grease, over the camshafts
- 2 No. 1 bearing caps. The mating surfaces should be coated with sealant. Install the bolts
- All remaining bearing caps
- All remaining cap bolts, coated with clean oil. Tighten them, in several passes, in the correct sequence, to 14 ft. lbs. (20 Nm). Note that there are separate sequences for the intake and exhaust sides

4. Install the oil seal, as far as it will go.

5. Rotate each camshaft until the forward straight (knock) pin is straight up. Loosen exhaust No. 1, 2 and 6 bearing cap bolts until they can be turned by hand; retighten them to 14 ft. lbs. (20 Nm). Loosen intake No. 1 and 2 bolts and retighten to 14 ft. lbs. (20 Nm).

6. Turn each camshaft ⅓ of a revolution (120 degrees). Loosen exhaust Nos. 4 and 7 bearing cap bolts; tighten them to 14 ft. lbs. (20 Nm). Loosen intake Nos. 4 and 6 bearing cap bolts; tighten them to 14 ft. lbs. (20 Nm).

7. Turn each camshaft an additional ⅓ of a revolution, loosen exhaust bearing cap bolts No. 3 and 5, then tighten them to 14 ft. lbs. (20 Nm). Loosen intake bearing cap bolts No. 3 and 7, then tighten them to 14 ft. lbs. (20 Nm).

8. Check and adjust the valve clearance.

9. Install or connect the following:
   - No 4. inside timing belt cover and the camshaft pulleys. Align the

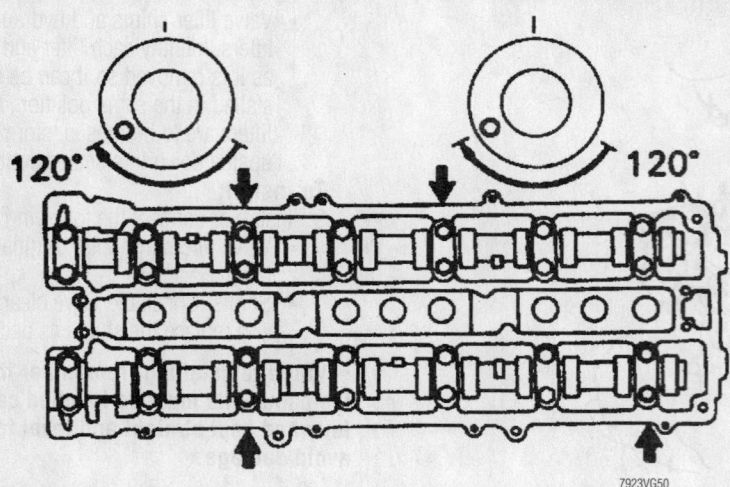

7923VG50

Tightening the camshafts (Step 3)—3.0L (2JZ-GE and 2JZ-GTE) engines

shaft pin with the pulley groove and slide the pulley on. Install the bolt temporarily. Hold the hex portion of the camshaft with a wrench; tighten the pulley bolt to 59 ft. lbs. (79 Nm).

- Cylinder head covers
- Timing belt
- Negative battery cable

Valve Lash

ADJUSTMENT

1NZ-FE Engine

➡ **Adjust the valve clearance when the engine is cold.**

1. Before servicing the vehicle, refer to the precautions in the beginning of this section.

2. Remove or disconnect the following:
   - Negative battery cable. On vehicles equipped with an air bag, wait at least 90 seconds before proceeding.
   - Cylinder head cover

3. Turn the crankshaft pulley and align its groove with the timing mark **0** of the No. 1 timing cover.

4. Check that the timing marks on the camshaft sprockets and valve timing controller are facing up (12 o'clock). If not, turn the crankshaft 1 complete revolution (360 degrees).

5. Measure the clearance between the valve lifter and the camshaft. Record the measurements on the intake valves No. 1 and 2. Measure the exhaust valves at No. 1 and 3.

   a. The intake valve clearance cold is 0.006–0.010 in. (0.15–0.25mm).

b. The exhaust valve clearance cold is 0.010–0.014 in. (0.25–0.35mm).

6. Turn the crankshaft pulley 1 revolution (360 degrees) and align the timing mark as before.

7. Measure the clearance between the valve lifter and the camshaft. Record the measurements on the intake valves No. 3 and 4. Measure the exhaust valves at No. 2 and 4.

   a. The intake valve clearance cold is 0.006–0.010 in. (0.15–0.25mm).

   b. The exhaust valve clearance cold is 0.010–0.014 in. (0.25–0.35mm).

8. To adjust the valve clearance:

   a. Set the No.1 cylinder at TDC compression. Place matchmarks on the timing chain and sprockets.

   b. Remove the 2 plugs from the timing chain cover.

Adjusting Shim Selection Chart

Shim No.	Thickness	Shim No.	Thickness
02	2.500 (0.0984)	20	2.950 (0.1161)
04	2.550 (0.1004)	22	3.000 (0.1181)
06	2.600 (0.1024)	24	3.050 (0.1201)
08	2.650 (0.1043)	26	3.100 (0.1220)
10	2.700 (0.1063)	28	3.150 (0.1240)
12	2.750 (0.1083)	30	3.200 (0.1260)
14	2.800 (0.1102)	32	3.250 (0.1280)
16	2.850 (0.1122)	34	3.300 (0.1299)
18	2.900 (0.1142)		

(New shim thickness — mm (in.))

7923VG56

Adjusting shim chart (intake and exhaust)—5E-FE engine

Adjusting Shim Selection Chart

Shim No.	Thickness	Shim No.	Thickness
1	2.500 (0.0984)	10	2.950 (0.1161)
2	2.550 (0.1004)	11	3.000 (0.1181)
3	2.600 (0.1024)	12	3.050 (0.1201)
4	2.650 (0.1043)	13	3.100 (0.1220)
5	2.700 (0.1063)	14	3.150 (0.1240)
6	2.750 (0.1083)	15	3.200 (0.1260)
7	2.800 (0.1102)	16	3.250 (0.1280)
8	2.850 (0.1122)	17	3.300 (0.1299)
9	2.900 (0.1142)		

(New shim thickness — mm (in.))

HINT: New shims have the thickness in millimeters imprinted on the face.

7923VG57

Adjusting shim chart (intake and exhaust)— 5S-FE, 1MZ-FE, 2JZ-GE and 2JZ-GTE engines

1ZZ–FE: Valve Lifter Selection Chart (Intake)

New lifter thickness mm (in.)

Lifter No.	Thickness	Lifter No.	Thickness	Lifter No.	Thickness
06	5.060 (0.1992)	30	5.300 (0.2087)	54	5.540 (0.2181)
08	5.080 (0.2000)	32	5.320 (0.2094)	56	5.560 (0.2189)
10	5.100 (0.2008)	34	5.340 (0.2102)	58	5.580 (0.2197)
12	5.120 (0.2016)	36	5.360 (0.2110)	60	5.600 (0.2205)
14	5.140 (0.2024)	38	5.380 (0.2118)	62	5.620 (0.2213)
16	5.160 (0.2031)	40	5.400 (0.2126)	64	5.640 (0.2220)
18	5.180 (0.2039)	42	5.420 (0.2134)	66	5.660 (0.2228)
20	5.200 (0.2047)	44	5.440 (0.2142)	68	5.680 (0.2236)
22	5.220 (0.2055)	46	5.460 (0.2150)	70	5.700 (0.2244)
24	5.240 (0.2063)	48	5.480 (0.2157)	72	5.720 (0.2252)
26	5.260 (0.2071)	50	5.500 (0.2165)	74	5.740 (0.2260)
28	5.280 (0.2079)	52	5.520 (0.2173)		

Intake valve clearance (Cold):
0.15 – 0.25 mm (0.006 – 0.010 in.)

EXAMPLE: The 5.250 mm (0.2067 in.) lifter is installed, and the measured clearance is 0.400 mm (0.0157 in.). Replace the 5.250 mm (0.2067 in.) lifter with a new No. 48 lifter.

Adjusting shim chart (intake)—1ZZ-FE engine

9307WG70

The Valve Lifter Selection Chart cross-references **Measured clearance mm (in.)** (rows) against **Installed lifter thickness mm (in.)** (columns) to give the new lifter number.

Installed lifter thickness (column headers): 5.060 (0.1992), 5.080 (0.2000), 5.100 (0.2008), 5.120 (0.2016), 5.140 (0.2024), 5.160 (0.2031), 5.180 (0.2039), 5.200 (0.2047), 5.210 (0.2051), 5.220 (0.2055), 5.230 (0.2059), 5.240 (0.2063), 5.250 (0.2067), 5.260 (0.2071), 5.270 (0.2075), 5.280 (0.2079), 5.290 (0.2083), 5.310 (0.2091), 5.320 (0.2094), 5.330 (0.2098), 5.340 (0.2102), 5.360 (0.2110), 5.370 (0.2114), 5.380 (0.2118), 5.390 (0.2122), 5.400 (0.2126), 5.410 (0.2130), 5.420 (0.2134), 5.430 (0.2138), 5.440 (0.2142), 5.450 (0.2146), 5.460 (0.2150), 5.470 (0.2154), 5.480 (0.2157), 5.490 (0.2161), 5.500 (0.2165), 5.510 (0.2169), 5.520 (0.2173), 5.530 (0.2177), 5.540 (0.2181), 5.550 (0.2185), 5.560 (0.2189), 5.570 (0.2193), 5.580 (0.2197), 5.590 (0.2201), 5.600 (0.2213), 5.620 (0.2213), 5.640 (0.2220), 5.660 (0.2228), 5.680 (0.2236), 5.700 (0.2244), 5.720 (0.2252), 5.740 (0.2260)

Measured clearance (row headers):
- 0.000 – 0.030 (0.0000 – 0.0012)
- 0.031 – 0.050 (0.0012 – 0.0020)
- 0.051 – 0.070 (0.0020 – 0.0028)
- 0.071 – 0.090 (0.0028 – 0.0035)
- 0.091 – 0.110 (0.0036 – 0.0043)
- 0.111 – 0.130 (0.0044 – 0.0051)
- 0.131 – 0.140 (0.0052 – 0.0059)
- 0.150 – 0.250 (0.0059 – 0.0098)
- 0.251 – 0.270 (0.0099 – 0.0105)
- 0.271 – 0.290 (0.0107 – 0.0114)
- 0.291 – 0.310 (0.0115 – 0.0122)
- 0.311 – 0.330 (0.0122 – 0.0130)
- 0.331 – 0.350 (0.0130 – 0.0138)
- 0.351 – 0.370 (0.0138 – 0.0146)
- 0.371 – 0.390 (0.0146 – 0.0154)
- 0.391 – 0.410 (0.0154 – 0.0161)
- 0.411 – 0.430 (0.0162 – 0.0169)
- 0.431 – 0.450 (0.0170 – 0.0177)
- 0.451 – 0.470 (0.0178 – 0.0185)
- 0.471 – 0.490 (0.0185 – 0.0193)
- 0.491 – 0.510 (0.0193 – 0.0201)
- 0.511 – 0.530 (0.0201 – 0.0209)
- 0.531 – 0.550 (0.0209 – 0.0217)
- 0.551 – 0.570 (0.0217 – 0.0224)
- 0.571 – 0.590 (0.0225 – 0.0232)
- 0.591 – 0.610 (0.0233 – 0.0240)
- 0.611 – 0.630 (0.0241 – 0.0248)
- 0.631 – 0.650 (0.0248 – 0.0256)
- 0.651 – 0.670 (0.0256 – 0.0264)
- 0.671 – 0.690 (0.0264 – 0.0272)
- 0.691 – 0.710 (0.0272 – 0.0280)
- 0.711 – 0.730 (0.0280 – 0.0287)
- 0.731 – 0.750 (0.0288 – 0.0295)
- 0.751 – 0.770 (0.0296 – 0.0303)
- 0.771 – 0.790 (0.0304 – 0.0311)
- 0.791 – 0.810 (0.0311 – 0.0319)
- 0.811 – 0.830 (0.0319 – 0.0327)
- 0.831 – 0.850 (0.0327 – 0.0335)
- 0.851 – 0.870 (0.0335 – 0.0343)
- 0.871 – 0.890 (0.0343 – 0.0350)
- 0.891 – 0.910 (0.0351 – 0.0358)
- 0.911 – 0.930 (0.0359 – 0.0366)

1ZZ–FE: Valve Lifter Selection Chart (Exhaust)

New lifter thickness   mm (in.)

Lifter No.	Thickness	Lifter No.	Thickness	Lifter No.	Thickness
06	5.060 (0.1992)	30	5.300 (0.2087)	54	5.540 (0.2181)
08	5.080 (0.2000)	32	5.320 (0.2094)	56	5.560 (0.2189)
10	5.100 (0.2008)	34	5.340 (0.2102)	58	5.580 (0.2197)
12	5.120 (0.2016)	36	5.360 (0.2110)	60	5.600 (0.2205)
14	5.140 (0.2024)	38	5.380 (0.2118)	62	5.620 (0.2213)
16	5.160 (0.2031)	40	5.400 (0.2126)	64	5.640 (0.2220)
18	5.180 (0.2039)	42	5.420 (0.2134)	66	5.660 (0.2228)
20	5.200 (0.2047)	44	5.440 (0.2142)	68	5.680 (0.2236)
22	5.220 (0.2055)	46	5.460 (0.2150)	70	5.700 (0.2244)
24	5.240 (0.2063)	48	5.480 (0.2157)	72	5.720 (0.2252)
26	5.260 (0.2071)	50	5.500 (0.2165)	74	5.740 (0.2260)
28	5.280 (0.2079)	52	5.520 (0.2173)		

Exhaust valve clearance (Cold):
0.25 – 0.35 mm (0.010 – 0.014 in.)

EXAMPLE: The 5.340 mm (0.2102 in.) lifter is installed, and the measured clearance is 0.440 mm (0.0173 in.). Replace the 5.340 mm (0.2102 in.) lifter with a new No. 48 lifter.

Adjusting shim chart (exhaust)—1ZZ-FE engine

9307WG71

2ZZ-GE: Valve Shim Selection Chart (Intake)

Measure clearance mm (in.) / Installed lifter thickness mm (in.)	2.000 (0.0787)	2.020 (0.0795)	2.040 (0.0803)	2.060 (0.0811)	2.080 (0.0819)	2.100 (0.0827)	2.120 (0.0835)	2.140 (0.0843)	2.160 (0.0850)	2.180 (0.0858)	2.190 (0.0858)	2.200 (0.0862)	2.210 (0.0866)	2.220 (0.0870)	2.230 (0.0878)	2.240 (0.0882)	2.250 (0.0886)	2.260 (0.0890)	2.270 (0.0894)	2.280 (0.0898)	2.290 (0.0902)	2.300 (0.0906)	2.310 (0.0909)	2.320 (0.0913)	2.330 (0.0917)	2.340 (0.0921)
0.000-0.030 (0.0000-0.0012)									00	00	00	00	02	02	04	04	06	06	08	08	10	10	12	12	14	14
0.031-0.050 (0.0012-0.0020)							00	00	00	00	02	02	04	04	06	06	08	08	10	10	12	12	14	14	16	16
0.051-0.070 (0.0020-0.0028)						00	00	00	00	02	04	04	06	06	08	08	10	10	12	12	14	14	16	16	18	18
0.071-0.090 (0.0028-0.0035)					00	00	00	00	02	04	06	06	08	08	10	10	12	12	14	14	16	16	18	18	20	20
0.091-0.110 (0.0035-0.0043)				00	00	00	00	02	04	06	08	08	10	10	12	12	14	14	16	16	18	18	20	20	22	22
0.111-0.130 (0.0043-0.0051)			00	00	00	00	02	04	06	08	10	10	12	12	14	14	16	16	18	18	20	20	22	22	24	24
0.131-0.149 (0.0051-0.0059)		00	00	00	00	02	04	06	08	10	12	12	14	14	16	16	18	18	20	20	22	22	24	24	26	26
0.150-0.250 (0.0059-0.0098)																										
0.251-0.270 (0.0099-0.0106)	06	08	10	12	14	16	18	20	22	24	26	26	28	28	30	30	32	32	34	34	36	36	38	38	40	40
0.271-0.290 (0.0107-0.0114)	08	10	12	14	16	18	20	22	24	26	28	28	30	30	32	32	34	34	36	36	38	38	40	40	42	42
0.291-0.310 (0.0115-0.0122)	10	12	14	16	18	20	22	24	26	28	30	30	32	32	34	34	36	36	38	38	40	40	42	42	44	44
0.311-0.330 (0.0122-0.0130)	12	14	16	18	20	22	24	26	28	30	32	32	34	34	36	36	38	38	40	40	42	42	44	44	46	46
0.331-0.350 (0.0130-0.0138)	14	16	18	20	22	24	26	28	30	32	34	34	36	36	38	38	40	40	42	42	44	44	46	46	48	48
0.351-0.370 (0.0138-0.0146)	16	18	20	22	24	26	28	30	32	34	36	36	38	38	40	40	42	42	44	44	46	46	48	48	50	50
0.371-0.390 (0.0146-0.0154)	18	20	22	24	26	28	30	32	34	36	38	38	40	40	42	42	44	44	46	46	48	48	50	50	52	52
0.391-0.410 (0.0154-0.0161)	20	22	24	26	28	30	32	34	36	38	40	40	42	42	44	44	46	46	48	58	50	50	52	52	54	54
0.411-0.430 (0.0162-0.0169)	22	24	26	28	30	32	34	36	38	40	42	42	44	44	46	46	48	48	50	50	52	52	54	54	56	56
0.431-0.450 (0.0170-0.0177)	24	26	28	30	32	34	36	38	40	42	44	44	46	46	48	48	50	50	52	52	54	54	56	56	58	58
0.451-0.470 (0.0178-0.0185)	26	28	30	32	34	36	38	40	42	44	46	46	48	48	50	50	52	52	54	54	56	56	58	58	60	60
0.471-0.490 (0.0185-0.0193)	28	30	32	34	36	38	40	42	44	46	48	48	50	50	52	52	54	54	56	56	58	58	60	60	62	62
0.491-0.510 (0.0193-0.0201)	30	32	34	36	38	40	42	44	46	48	50	50	52	52	54	54	56	56	58	58	60	60	62	62	64	64
0.511-0.530 (0.0201-0.0209)	32	34	36	38	40	42	44	46	48	50	52	52	54	54	56	56	58	58	60	60	62	62	64	64	66	66
0.531-0.550 (0.0209-0.0217)	34	36	38	40	42	44	46	48	50	52	54	54	56	56	58	58	60	60	62	62	64	64	66	66	68	68
0.551-0.570 (0.0217-0.0224)	36	38	40	42	44	46	48	50	52	54	56	56	58	58	60	60	62	62	64	64	66	66	68	68	70	70
0.571-0.590 (0.0225-0.0232)	38	40	42	44	46	48	50	52	54	56	58	58	60	60	62	62	64	64	66	66	68	68	70	70	72	72
0.591-0.610 (0.0233-0.0240)	40	42	44	46	48	50	52	54	56	58	60	60	62	62	64	64	66	66	68	68	70	70	72	72	74	74
0.611-0.630 (0.0241-0.0248)	42	44	46	48	50	52	54	56	58	60	62	62	64	64	66	66	68	68	70	70	72	72	74	74	76	76
0.631-0.650 (0.0248-0.0256)	44	46	48	50	52	54	56	58	60	62	64	64	66	66	68	68	70	70	72	72	74	74	76	76	78	78
0.651-0.670 (0.0256-0.0264)	46	48	50	52	54	56	58	60	62	64	66	66	68	68	70	70	72	72	74	74	76	76	78	78	80	80
0.671-0.690 (0.0264-0.0272)	48	50	52	54	56	58	60	62	64	66	68	68	70	70	72	72	74	74	76	76	78	78	80	80	80	80
0.691-0.710 (0.0272-0.0280)	50	52	54	56	58	60	62	64	66	68	70	70	72	72	74	74	76	76	78	78	80	80	80	80	80	80
0.711-0.730 (0.0280-0.0287)	52	54	56	58	60	62	64	66	68	70	72	72	74	74	76	76	78	78	80	80	80	80	80	80		
0.731-0.750 (0.0288-0.0295)	54	56	58	60	62	64	66	68	70	72	74	74	76	76	78	78	80	80	80	80	80	80				
0.751-0.770 (0.0296-0.0303)	56	58	60	62	64	66	68	70	72	74	76	76	78	78	80	80	80	80	80	80						
0.771-0.790 (0.0304-0.0311)	58	60	62	64	66	68	70	72	74	76	78	78	80	80	80	80	80	80								
0.791-0.810 (0.0311-0.0319)	60	62	64	66	68	70	72	74	76	78	80	80	80	80	80	80										
0.811-0.830 (0.0319-0.0327)	62	64	66	68	70	72	74	76	78	80	80	80	80	80												
0.831-0.850 (0.0327-0.0335)	64	66	68	70	72	74	76	78	80	80	80	80														
0.851-0.870 (0.0335-0.0343)	66	68	70	72	74	76	78	80	80	80																
0.871-0.890 (0.0343-0.0350)	68	70	72	74	76	78	80	80	80																	
0.891-0.910 (0.0351-0.0358)	70	72	74	76	78	80	80	80																		
0.911-0.930 (0.0359-0.0366)	72	74	76	78	80	80	80																			
0.931-0.950 (0.0367-0.0374)	74	76	78	80	80	80																				
0.951-0.970 (0.0374-0.0382)	76	78	80	80	80																					
0.971-0.990 (0.0382-0.0390)	78	80	80	80																						
0.991-1.010 (0.0390-0.0398)	80	80	80																							
1.011-1.030 (0.0398-0.0406)	80	80																								
1.031-1.050 (0.0406-0.0413)	80																									

Measure clearance mm (in.) / Installed lifter thickness mm (in.)	2.350 (0.0925)	2.360 (0.0929)	2.370 (0.0933)	2.380 (0.0937)	2.390 (0.0941)	2.400 (0.0945)	2.410 (0.0949)	2.420 (0.0953)	2.430 (0.0957)	2.440 (0.0961)	2.450 (0.0965)	2.460 (0.0969)	2.470 (0.0972)	2.480 (0.0976)	2.490 (0.0980)	2.500 (0.0984)	2.510 (0.0988)	2.520 (0.0992)	2.530 (0.0996)	2.540 (0.1000)	2.550 (0.1004)	2.560 (0.1008)	2.580 (0.1016)	2.600 (0.1024)	2.620 (0.1031)	2.640 (0.1039)	2.660 (0.1047)	2.680 (0.1055)	2.700 (0.1063)	2.720 (0.1071)	2.740 (0.1079)	2.760 (0.1087)	2.780 (0.1094)	2.800 (0.1102)
0.000-0.030 (0.0000-0.0012)	16	16	18	18	20	20	22	22	24	24	26	26	28	28	30	30	32	32	34	34	36	36	38	40	42	44	46	48	50	52	54	56	58	60
0.031-0.050 (0.0012-0.0020)	18	18	20	20	22	22	24	24	26	26	28	28	30	30	32	32	34	34	36	36	38	38	40	42	44	46	48	50	52	54	56	58	60	62
0.051-0.070 (0.0020-0.0028)	20	20	22	22	24	24	26	26	28	28	30	30	32	32	34	34	36	36	38	38	40	40	42	44	46	48	50	52	54	56	58	60	62	64
0.071-0.090 (0.0028-0.0035)	22	22	24	24	26	26	28	28	30	30	32	32	34	34	36	36	38	38	40	40	42	42	44	46	48	50	52	54	56	58	60	62	64	66
0.091-0.110 (0.0035-0.0043)	24	24	26	26	28	28	30	30	32	32	34	34	36	36	38	38	40	40	42	42	44	44	46	48	50	52	54	56	58	60	62	64	66	68
0.111-0.130 (0.0043-0.0051)	26	26	28	28	30	30	32	32	34	34	36	36	38	38	40	40	42	42	44	44	46	46	48	50	52	54	56	58	60	62	64	66	68	70
0.131-0.149 (0.0051-0.0059)	28	28	30	30	32	32	34	34	36	36	38	38	40	40	42	42	44	44	46	46	48	48	50	52	54	56	58	60	62	64	66	68	70	72
0.150-0.250 (0.0059-0.0098)																																		
0.251-0.270 (0.0099-0.0106)	42	42	44	44	46	46	48	48	50	50	52	52	54	54	56	56	58	58	60	60	62	64	66	68	70	72	74	76	78	80	80	80	80	
0.271-0.290 (0.0107-0.0114)	44	44	46	46	48	48	50	50	52	52	54	54	56	56	58	58	60	60	62	62	64	64	66	68	70	72	74	76	78	80	80	80		
0.291-0.310 (0.0115-0.0122)	46	46	48	48	50	50	52	52	54	54	56	56	58	58	60	60	62	62	64	64	66	66	68	70	72	74	76	78	80	80	80			
0.311-0.330 (0.0122-0.0130)	48	48	50	50	52	52	54	54	56	56	58	58	60	60	62	62	64	64	66	66	68	68	70	72	74	76	78	80	80	80				
0.331-0.350 (0.0130-0.0138)	50	50	52	52	54	54	56	56	58	58	60	60	62	62	64	64	66	66	68	68	70	70	72	74	76	78	80	80	80					
0.351-0.370 (0.0138-0.0146)	52	52	54	54	56	56	58	58	60	60	62	62	64	64	66	66	68	68	70	70	72	72	74	76	78	80	80	80						
0.371-0.390 (0.0146-0.0154)	54	54	56	56	58	58	60	60	62	62	64	64	66	66	68	68	70	70	72	72	74	74	76	78	80	80	80							
0.391-0.410 (0.0154-0.0161)	56	56	58	58	60	60	62	62	64	64	66	66	68	68	70	70	72	72	74	74	76	76	78	80	80	80								
0.411-0.430 (0.0162-0.0169)	58	58	60	60	62	62	64	64	66	66	68	68	70	70	72	72	74	74	76	76	78	78	80	80	80									
0.431-0.450 (0.0170-0.0177)	60	60	62	62	64	64	66	66	68	68	70	70	72	72	74	74	76	76	78	78	80	80	80	80										
0.451-0.470 (0.0178-0.0185)	62	62	64	64	66	66	68	68	70	70	72	72	74	74	76	76	78	78	80	80	80	80	80											
0.471-0.490 (0.0185-0.0193)	64	64	66	66	68	68	70	70	72	72	74	74	76	76	78	78	80	80	80	80	80	80												
0.491-0.510 (0.0193-0.0201)	66	66	68	68	70	70	72	72	74	74	76	76	78	78	80	80	80	80	80	80														
0.511-0.530 (0.0201-0.0209)	68	68	70	70	72	72	74	74	76	76	78	78	80	80	80	80	80	80																
0.531-0.550 (0.0209-0.0217)	70	70	72	72	74	74	76	76	78	78	80	80	80	80	80	80																		
0.551-0.570 (0.0217-0.0224)	72	72	74	74	76	76	78	78	80	80	80	80	80	80																				
0.571-0.590 (0.0225-0.0232)	74	74	76	76	78	78	80	80	80	80	80	80																						
0.591-0.610 (0.0233-0.0240)	76	76	78	78	80	80	80	80	80	80																								
0.611-0.630 (0.0241-0.0248)	78	78	80	80	80	80	80	80																										
0.631-0.650 (0.0248-0.0256)	80	80	80	80	80	80																												
0.651-0.670 (0.0256-0.0264)	80	80	80	80																														
0.671-0.690 (0.0264-0.0272)	80	80																																
0.691-0.710 (0.0272-0.0280)																																		
0.711-0.730 (0.0280-0.0287)																																		
0.731-0.750 (0.0288-0.0295)																																		
0.751-0.770 (0.0296-0.0303)																																		
0.771-0.790 (0.0304-0.0311)																																		
0.791-0.810 (0.0311-0.0319)																																		
0.811-0.830 (0.0319-0.0327)																																		
0.831-0.850 (0.0327-0.0335)																																		
0.851-0.870 (0.0335-0.0343)																																		
0.871-0.890 (0.0343-0.0350)																																		
0.891-0.910 (0.0351-0.0358)																																		
0.911-0.930 (0.0359-0.0366)																																		
0.931-0.950 (0.0367-0.0374)																																		
0.951-0.970 (0.0374-0.0382)																																		
0.971-0.990 (0.0382-0.0390)																																		
0.991-1.010 (0.0390-0.0398)																																		
1.011-1.030 (0.0398-0.0406)																																		
1.031-1.050 (0.0406-0.0413)																																		

New Shim thickness mm (in.)

Shim No.	Thickness	Shim No.	Thickness	Shim No.	Thickness
00	2.000 (0.0787)	28	2.280 (0.0898)	56	2.560 (0.1008)
02	2.020 (0.0795)	30	2.300 (0.0906)	58	2.580 (0.1016)
04	2.040 (0.0803)	32	2.320 (0.0913)	60	2.600 (0.1024)
06	2.060 (0.0811)	34	2.340 (0.0921)	62	2.620 (0.1031)
08	2.080 (0.0819)	36	2.360 (0.0929)	64	2.640 (0.1039)
10	2.100 (0.0827)	38	2.380 (.0.0937)	66	2.660 (0.1047)
12	2.120 (0.0835)	40	2.400 (0.0945)	68	2.680 (0.1055)
14	2.140 (0.0843)	42	2.420 (0.0953)	70	2.700 (0.1063)
16	2.160 (0.0850)	44	2.440 (0.0961)	72	2.720 (0.1071)
18	2.180 (0.0858)	46	2.460 (0.0969)	74	2.740 (0.1079)
20	2.200 (0.0866)	48	2.480 (0.0976)	76	2.760 (0.1087)
22	2.220 (0.0874)	50	2.500 (0.0984)	78	2.780 (0.1094)
24	2.240 (0.0882)	52	2.520 (0.0992)	80	2.800 (0.1102)
26	2.260 (0.0890)	54	2.540 (0.1000)		

Intake valve clearance (Cold):
0.15 – 0.25 mm (0.006 – 0.010 in.)

EXAMPLE: The 2.200 mm (0.0826 in.) shim is installed, and the measured clearance is 0.400 mm (0.0157 in.). Replace the 2.400 mm (0.0945 in.) shim with a new No. 40 shim.

9307WG72

Adjusting shim chart (intake)—2ZZ-GE engine

2ZZ-GE: Valve Shim Selection Chart (Exhaust)

Measure clearance mm (in.) / Installed lifter thickness mm (in.)	2.000 (0.0787)	2.020 (0.0795)	2.040 (0.0803)	2.060 (0.0811)	2.080 (0.0819)	2.100 (0.0827)	2.120 (0.0835)	2.140 (0.0843)	2.160 (0.0850)	2.180 (0.0858)	2.190 (0.0862)	2.200 (0.0866)	2.210 (0.0870)	2.220 (0.0874)	2.230 (0.0878)	2.240 (0.0882)	2.250 (0.0886)	2.260 (0.0890)	2.270 (0.0894)	2.280 (0.0898)	2.290 (0.0902)	2.300 (0.0906)	2.310 (0.0909)	2.320 (0.0913)	2.330 (0.0917)	2.340 (0.0921)	2.350 (0.0925)	2.360 (0.0929)	2.370 (0.0933)	2.380 (0.0937)
0.000-0.030 (0.0000-0.0012)																											00	00	00	00
0.031-0.050 (0.0012-0.0020)																								00	00	00	00	00	00	00
0.051-0.070 (0.0020-0.0028)																						00	00	00	00	00	00	00	02	02
0.071-0.090 (0.0028-0.0035)																				00	00	00	00	00	00	00	02	02	04	04
0.091-0.110 (0.0036-0.0043)																		00	00	00	00	00	00	00	02	02	04	04	06	06
0.111-0.130 (0.0044-0.0051)																00	00	00	00	00	00	00	02	02	04	04	06	06	08	08
0.131-0.150 (0.0052-0.0059)														00	00	00	00	00	00	00	02	02	04	04	06	06	08	08	10	10
0.151-0.170 (0.0059-0.0067)												00	00	00	00	00	00	00	02	02	04	04	06	06	08	08	10	10	12	12
0.171-0.190 (0.0067-0.0075)										00	00	00	00	00	00	00	02	02	04	04	06	06	08	08	10	10	12	12	14	14
0.191-0.210 (0.0075-0.0083)									00	00	00	00	00	00	02	02	04	04	06	06	08	08	10	10	12	12	14	14	16	16
0.211-0.230 (0.0083-0.0091)								00	00	00	00	00	02	02	04	04	06	06	08	08	10	10	12	12	14	14	16	16	18	18
0.231-0.250 (0.0091-0.0098)							00	00	00	00	02	02	04	04	06	06	08	08	10	10	12	12	14	14	16	16	18	18	20	20
0.251-0.270 (0.0099-0.0106)						00	00	00	00	02	04	04	06	06	08	08	10	10	12	12	14	14	16	16	18	18	20	20	22	22
0.271-0.290 (0.0107-0.0114)					00	00	00	00	02	04	06	06	08	08	10	10	12	12	14	14	16	16	18	18	20	20	22	22	24	24
0.291-0.310 (0.0115-0.0122)				00	00	00	00	02	04	06	08	08	10	10	12	12	14	14	16	16	18	18	20	20	22	22	24	24	26	26
0.311-0.330 (0.0122-0.0130)			00	00	00	00	02	04	06	08	10	10	12	12	14	14	16	16	18	18	20	20	22	22	24	24	26	26	28	28
0.331-0.349 (0.0130-0.0137)		00	00	00	00	02	04	06	08	10	12	12	14	14	16	16	18	18	20	20	22	22	24	24	26	26	28	28	30	30
0.350-0.450 (0.0138-0.0177)																														
0.451-0.470 (0.0178-0.0185)	06	08	10	12	14	16	18	20	22	24	26	26	28	28	30	30	32	32	34	34	36	36	38	38	40	42	42	44	44	46
0.471-0.490 (0.0185-0.0193)	08	10	12	14	16	18	20	22	24	26	28	28	30	30	32	32	34	34	36	36	38	38	40	42	42	44	44	46	46	48
0.491-0.510 (0.0193-0.0201)	10	12	14	16	18	20	22	24	26	28	30	30	32	32	34	34	36	36	38	38	40	42	42	44	44	46	46	48	48	50
0.511-0.530 (0.0201-0.0209)	12	14	16	18	20	22	24	26	28	30	32	32	34	34	36	36	38	38	40	42	42	44	44	46	46	48	48	50	50	52
0.531-0.550 (0.0209-0.0217)	14	16	18	20	22	24	26	28	30	32	34	34	36	36	38	38	40	42	42	44	44	46	46	48	48	50	50	52	52	54
0.551-0.570 (0.0217-0.0224)	16	18	20	22	24	26	28	30	32	34	36	36	38	38	40	42	42	44	44	46	46	48	48	50	50	52	52	54	54	56
0.571-0.590 (0.0225-0.0232)	18	20	22	24	26	28	30	32	34	36	38	38	40	42	42	44	44	46	46	48	48	50	50	52	52	54	54	56	56	58
0.591-0.610 (0.0223-0.0240)	20	22	24	26	28	30	32	34	36	38	40	42	42	44	44	46	46	48	48	50	50	52	52	54	54	56	56	58	58	60
0.611-0.630 (0.0241-0.0248)	22	24	26	28	30	32	34	36	38	40	42	44	44	46	46	48	48	50	50	52	52	54	54	56	56	58	58	60	60	62
0.631-0.650 (0.0248-0.0256)	24	26	28	30	32	34	36	38	40	42	44	46	46	48	48	50	50	52	52	54	54	56	56	58	58	60	60	62	62	64
0.651-0.670 (0.0256-0.0264)	23	28	30	32	34	36	38	40	42	44	46	48	48	50	50	52	52	54	54	56	56	58	58	60	60	62	62	64	64	66
0.671-0.690 (0.0264-0.0272)	28	30	32	34	36	38	40	42	44	46	48	50	50	52	52	54	54	56	56	58	58	60	60	62	62	64	64	66	66	68
0.691-0.710 (0.0272-0.0280)	30	32	34	36	38	40	42	44	46	48	50	52	52	54	54	56	56	58	58	60	60	62	62	64	64	66	66	68	68	70
0.711-0.730 (0.0280-0.0287)	32	34	36	38	40	42	44	46	48	50	52	54	54	56	56	58	58	60	60	62	62	64	64	66	66	68	68	70	70	72
0.731-0.750 (0.0288-0.0295)	34	36	38	40	42	44	46	48	50	52	54	56	56	58	58	60	60	62	62	64	64	66	66	68	68	70	70	72	72	74
0.751-0.770 (0.0296-0.0303)	36	38	40	42	44	46	48	50	52	54	56	58	58	60	60	62	62	64	64	66	66	68	68	70	70	72	72	74	74	76
0.771-0.790 (0.0304-0.0311)	38	40	42	44	46	48	50	52	54	56	58	60	60	62	62	64	64	66	66	68	68	70	70	72	72	74	74	76	76	78
0.791-0.810 (0.0311-0.0319)	40	42	44	46	48	50	52	54	56	58	60	62	62	64	64	66	66	68	68	70	70	72	72	74	74	76	76	78	78	80
0.811-0.830 (0.0319-0.0327)	42	44	46	48	50	52	54	56	58	60	62	64	64	66	66	68	68	70	70	72	72	74	74	76	76	78	78	80	80	80
0.831-0.850 (0.0327-0.0335)	44	46	48	50	52	54	56	58	60	62	64	66	66	68	68	70	70	72	72	74	74	76	76	78	78	80	80	80	80	80
0.851-0.870 (0.0335-0.0343)	46	48	50	52	54	56	58	60	62	64	66	68	68	70	70	72	72	74	74	76	76	78	78	80	80	80	80	80	80	80
0.871-0.890 (0.0343-0.0350)	48	50	52	54	56	58	60	62	64	66	68	70	70	72	72	74	74	76	76	78	78	80	80	80	80	80	80	80		
0.891-0.910 (0.0351-0.0358)	50	52	54	56	58	60	62	64	66	68	70	72	72	74	74	76	76	78	78	80	80	80	80	80	80	80				
0.911-0.930 (0.0359-0.0366)	52	54	56	58	60	62	64	66	68	70	72	74	74	76	76	78	78	70	80	80	80	80	80	80						
0.931-0.950 (0.0367-0.0374)	54	56	58	60	62	64	66	68	70	72	74	76	76	78	78	80	80	80	80	80	80	80								
0.951-0.970 (0.0374-0.0382)	56	58	60	62	64	66	68	70	72	74	76	78	78	80	80	80	80	80	80	80										
0.971-0.990 (0.0382-0.0390)	58	60	62	64	66	68	70	72	74	76	78	80	80	80	80	80	80	80												
0.991-1.010 (0.0390-0.0398)	60	62	64	66	68	70	72	74	76	78	80	80	80	80	80	80														
1.011-1.030 (0.0398-0.0406)	62	64	66	68	[illegible]	72	74	76	78	80	80	80	80	80																
1.031-1.050 (0.0406-0.0413)	64	66	68	70	72	74	76	78	80	80	80	80																		
1.051-1.070 (0.0414-0.0421)	66	68	70	72	74	76	78	80	80	80																				
1.071-1.090 (0.0422-0.0429)	68	70	72	74	76	78	80	80	80																					
1.091-1.110 (0.0430-0.0437)	70	72	74	76	78	80	80	80																						
1.111-1.130 (0.0437-0.0445)	72	74	76	78	80	80	80																							
1.131-1.150 (0.0445-0.0453)	74	76	78	80	80	80																								
1.151-1.170 (0.0453-0.0461)	76	78	80	80	80																									
1.171-1.190 (0.0461-0.0469)	78	80	80	80																										
1.191-1.210 (0.0469-0.0476)	80	80	80																											
1.211-1.230 (0.0477-0.0484)	80	80																												
1.231-1.250 (0.0485-0.0492)	80																													

Measure clearance mm (in.) / Installed lifter thickness mm (in.)	2.390 (0.0941)	2.400 (0.0945)	2.410 (0.0949)	2.420 (0.0953)	2.430 (0.0957)	2.440 (0.0961)	2.450 (0.0965)	2.460 (0.0969)	2.470 (0.0972)	2.480 (0.0976)	2.490 (0.0980)	2.500 (0.0984)	2.510 (0.0988)	2.520 (0.0992)	2.530 (0.0996)	2.540 (0.1000)	2.550 (0.1004)	2.560 (0.1008)	2.580 (0.1016)	2.600 (0.1024)	2.620 (0.1031)	2.640 (0.1039)	2.660 (0.1047)	2.680 (0.1055)	2.700 (0.1063)	2.720 (0.1071)	2.740 (0.1079)	2.760 (0.1087)	2.780 (0.1094)	2.800 (0.1102)
0.000-0.030 (0.0000-0.0012)	00	00	02	02	04	04	06	06	08	08	10	10	12	12	14	14	16	16	18	20	22	24	26	28	30	32	34	36	38	40
0.031-0.050 (0.0012-0.0020)	02	02	04	04	06	06	08	08	10	10	12	12	14	14	16	16	18	18	20	22	24	26	28	30	32	34	36	38	40	42
0.051-0.070 (0.0020-0.0028)	04	04	06	06	08	08	10	10	12	12	14	14	16	16	18	18	20	20	22	24	26	28	30	32	34	36	38	40	42	44
0.071-0.090 (0.0028-0.0035)	06	06	08	08	10	10	12	12	14	14	16	16	18	18	20	20	22	22	24	26	28	30	32	34	36	38	40	42	44	46
0.091-0.110 (0.0036-0.0043)	08	08	10	10	12	12	14	14	16	16	18	18	20	20	22	22	24	24	26	28	30	32	34	36	38	40	42	44	46	48
0.111-0.130 (0.0044-0.0051)	10	10	12	12	14	14	16	16	18	18	20	20	22	22	24	24	26	36	28	30	32	34	36	38	40	42	44	46	48	50
0.131-0.150 (0.0052-0.0059)	12	12	14	14	16	16	18	18	20	20	22	22	24	24	26	26	28	28	30	32	34	36	38	40	42	44	46	48	50	52
0.151-0.170 (0.0059-0.0067)	14	14	16	16	18	18	20	20	22	22	24	24	26	26	28	28	30	30	32	34	36	38	40	42	44	46	48	50	52	54
0.171-0.190 (0.0067-0.0075)	16	16	18	18	20	20	22	22	24	24	26	26	28	28	30	30	32	32	34	36	38	40	42	44	46	48	50	52	54	56
0.191-0.210 (0.0075-0.0083)	18	18	20	20	22	22	24	24	26	26	28	28	30	30	32	32	34	34	36	38	40	42	44	46	48	50	52	54	56	58
0.211-0.230 (0.0083-0.0091)	20	20	22	22	24	24	26	26	28	28	30	30	32	32	34	34	36	36	38	40	42	44	46	48	50	52	54	56	58	60
0.231-0.250 (0.0091-0.0098)	22	22	24	24	26	26	28	28	30	30	32	32	34	34	36	36	38	38	40	42	44	46	48	50	52	54	56	58	60	62
0.251-0.270 (0.0099-0.0106)	24	24	26	26	28	28	30	30	32	32	34	34	36	36	38	38	40	40	42	44	46	48	50	52	54	56	58	60	62	64
0.271-0.290 (0.0107-0.0114)	26	26	28	28	30	30	32	32	34	34	36	36	38	38	40	40	42	42	44	46	48	50	52	54	56	58	60	62	64	66
0.291-0.310 (0.0115-0.0122)	28	28	30	30	32	32	34	34	36	36	38	38	40	40	42	42	44	44	46	48	50	52	54	56	58	60	62	64	66	68
0.311-0.330 (0.0122-0.0130)	30	30	32	32	34	34	36	36	38	38	40	40	42	42	44	44	46	46	48	50	52	54	56	58	60	62	64	66	68	70
0.331-0.349 (0.0130-0.0137)	32	32	34	34	36	36	38	38	40	40	42	42	44	44	46	46	48	48	50	52	54	56	58	60	62	64	66	68	70	72
0.350-0.450 (0.0138-0.0177)																														
0.451-0.470 (0.0178-0.0185)	46	48	48	50	50	52	52	54	54	56	56	58	58	60	60	62	62	64	66	68	70	72	74	76	78	80	80	80	80	
0.471-0.490 (0.0185-0.0193)	48	50	50	52	52	54	54	56	56	56	60	60	60	62	62	64	64	66	68	70	62	74	76	78	80	80	80	80		
0.491-0.510 (0.0193-0.0201)	50	52	52	54	54	56	56	58	58	60	60	62	62	64	64	66	66	68	70	72	74	76	78	80	80	80	80			
0.511-0.530 (0.0201-0.0209)	52	54	54	56	56	58	68	60	60	60	64	64	64	66	66	68	68	70	72	74	76	78	80	80	80	80				
0.531-0.550 (0.0209-0.0217)	54	56	56	58	58	60	60	62	62	64	64	66	66	68	68	70	70	72	74	76	78	80	80	80	80					
0.551-0.570 (0.0217-0.0224)	56	58	58	60	60	62	62	64	64	64	68	68	68	70	70	72	72	74	76	78	80	80	80	80						
0.571-0.590 (0.0225-0.0232)	68	60	60	62	62	64	64	66	66	68	68	70	70	72	72	74	74	76	78	80	80	80	80							
0.591-0.610 (0.0223-0.0240)	60	62	62	64	64	66	66	68	68	68	72	72	72	74	74	76	76	78	80	80	80	80								
0.611-0.630 (0.0241-0.0248)	62	64	64	66	66	68	68	70	70	72	72	74	74	76	76	78	78	80	80	80	80									
0.631-0.650 (0.0248-0.0256)	64	66	66	68	68	70	70	72	72	72	76	76	76	78	78	80	80	80	80	80										
0.651-0.670 (0.0256-0.0264)	66	68	68	70	70	72	72	74	74	76	76	78	78	80	80	80	80	80	80											
0.671-0.690 (0.0264-0.0272)	68	70	70	72	72	74	74	76	76	76	80	80	80	80	80	80	80	80												
0.691-0.710 (0.0272-0.0280)	70	72	72	74	74	76	76	78	78	80	80	80	80	80	80	80														
0.711-0.730 (0.0280-0.0287)	72	74	74	76	76	78	78	80	80	80	80	80	80	80																
0.731-0.750 (0.0288-0.0295)	74	76	76	78	78	80	80	80	80	80	80	80																		
0.751-0.770 (0.0296-0.0303)	76	78	78	80	80	80	80	80	80	80																				
0.771-0.790 (0.0304-0.0311)	78	80	80	80	80	80	80	80																						
0.791-0.810 (0.0311-0.0319)	80	80	80	80	80	80																								
0.811-0.830 (0.0319-0.0327)	80	80	80	80																										
0.831-0.850 (0.0327-0.0335)	80	80																												

New Shim thickness mm (in.)

Shim No.	Thickness	Shim No.	Thickness	Shim No.	Thickness
00	2.000(0.0787)	28	2.280 (0.0898)	56	2.560 (0.1008)
02	2.020 (0.0795)	30	2.300 (0.0906)	58	2.580 (0.1016)
04	2.040 (0.0803)	32	2.320 (0.0913)	60	2.600 (0.1024)
06	2.060 (0.0811)	34	2.340 (0.0921)	62	2.620 (0.1031)
08	2.080 (0.0819)	36	2.360 (0.0929)	64	2.640 (0.1039)
10	2.100 (0.0827)	38	2.380 (.0.0937)	66	2.660 (0.1047)
12	2.120 (0.0835)	40	2.400 (0.0945)	68	2.680 (0.1055)
14	2.140 (0.0843)	42	2.420 (0.0953)	70	2.700 (0.1063)
16	2.160 (0.0850)	44	2.440 (0.0961)	72	2.720 (0.1071)
18	2.180(0.0858)	46	2.460 (0.0969)	74	2.740 (0.1079)
20	2.200 (0.0866)	48	2.480 (0.0976)	76	2.760 (0.1087)
22	2.220 (0.0874)	50	2.500 (0.0984)	78	2.780 (0.1094)
24	2.240 (0.0882)	52	2.520 (0.0992)	80	2.800 (0.1102)
26	2.260 (0.0890)	54	2.540 (0.1000)		

Exhaust valve clearance (Cold):
0.35 – 0.45 mm (0.014 – 0.018 in.)

EXAMPLE: The 2.200 mm (0.0862 in.) shim is installed, and the measured clearance is 0.500 mm (0.0197 in.).
Replace the 2.300 mm (0.0906 in.) shim with a new No. 30 shim.

9307WG73

Adjusting shim chart (exhaust)—2ZZ-GE engine

Adjusting Shim Selection Chart

New shim thickness mm (in.)

Shim No.	Thickness	Shim No.	Thickness
02	2.500 (0.0984)	20	2.950 (0.1161)
04	2.550 (0.1004)	22	3.000 (0.1181)
06	2.600 (0.1024)	24	3.050 (0.1201)
08	2.650 (0.1043)	26	3.100 (0.1220)
10	2.700 (0.1063)	28	3.150 (0.1240)
12	2.750 (0.1083)	30	3.200 (0.1260)
14	2.800 (0.1102)	32	3.250 (0.1280)
16	2.850 (0.1122)	34	3.300 (0.1299)
18	2.900 (0.1142)		

7923VG56

Adjusting shim chart (exhaust)—1NZ-FE engine

Adjusting Shim Selection Chart

New shim thickness mm (in.)

Shim No.	Thickness	Shim No.	Thickness
02	2.500 (0.0984)	20	2.950 (0.1161)
04	2.550 (0.1004)	22	3.000 (0.1181)
06	2.600 (0.1024)	24	3.050 (0.1201)
08	2.650 (0.1043)	26	3.100 (0.1220)
10	2.700 (0.1063)	28	3.150 (0.1240)
12	2.750 (0.1083)	30	3.200 (0.1260)
14	2.800 (0.1102)	32	3.250 (0.1280)
16	2.850 (0.1122)	34	3.300 (0.1299)
18	2.900 (0.1142)		

7923VG56

Adjusting shim chart (intake)—1NZ-FE engine

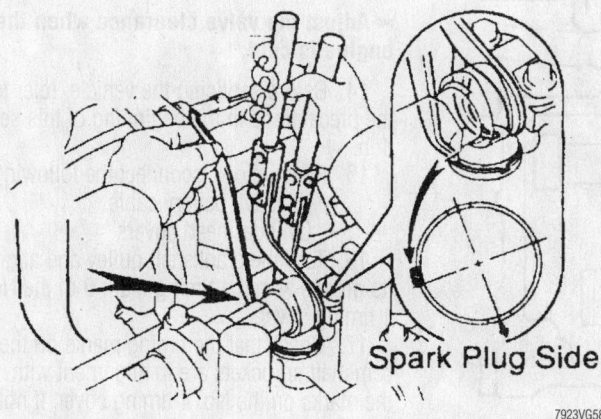

Spark Plug Side

7923VG58

Common method of removing valve shims

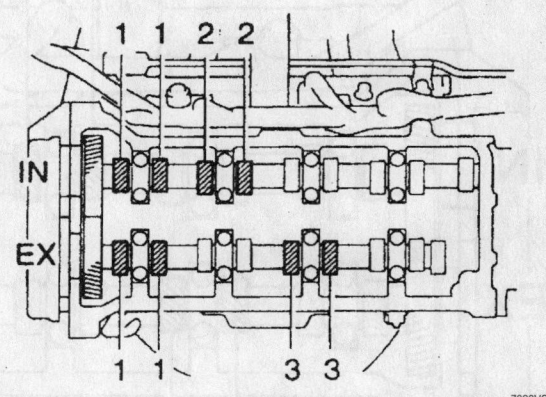

7923VG59

Intake valves (1 and 2) and exhaust valves (1 and 3)—1NZ-FE engine

c. Turn the exhaust camshaft clockwise slightly while rotating the stopper plate on the tensioner downward. Push in on the tension plunger. When the stopper plate cannot be easily lowered, rotate the exhaust camshaft clockwise and counterclockwise slightly. Insert a 3mm bar into the holes in the stopper plate and tensioner to lock the tensioner. Remove the timing chain.

d. Remove the valve timing controller assembly.

e. Remove the lifters.

9. Determine the replacement adjusting shim size by either using the chart or the following formula:

- Intake: N = T + A—0.008 in. (0.20mm)
- Exhaust: N = T + A—0.012 in. (0.30mm)

- T = Thickness of removed shim
- A = Measured valve clearance
- N = Thickness of new shim

10. Install a new shim.
11. Recheck the valve clearance.
12. Install the cylinder head covers.
13. Connect the negative battery cable.

5E-FE

➡**Adjust the valve clearance when the engine is cold.**

1. Before servicing the vehicle, refer to the precautions in the beginning of this section.

2. Remove or disconnect the following:

- Negative battery cable. On vehicles equipped with an air bag, wait at least 90 seconds before proceeding.
- Cylinder head covers

3. Turn the crankshaft pulley and align its groove with the timing mark **0** of the No. 1 timing cover.

4. Check that the timing marks on the camshaft sprockets are in alignment with the marks on the No. 4 timing cover. If not, turn the crankshaft 1 complete revolution (360 degrees).

5. Measure the clearance between the valve lifter and the camshaft. Record the measurements on the intake valves No. 1 and 2. Measure the exhaust valves at No. 1 and 3.

a. The intake valve clearance cold is 0.006–0.010 in. (0.15–0.25mm).

b. The exhaust valve clearance cold is 0.012–0.016 in. (0.31–0.41mm).

6. Turn the crankshaft pulley 1 revolution (360 degrees) and align the groove with the timing mark **0** of the No.1 timing belt cover.

Timing belt service is covered in Section 3 of this manual

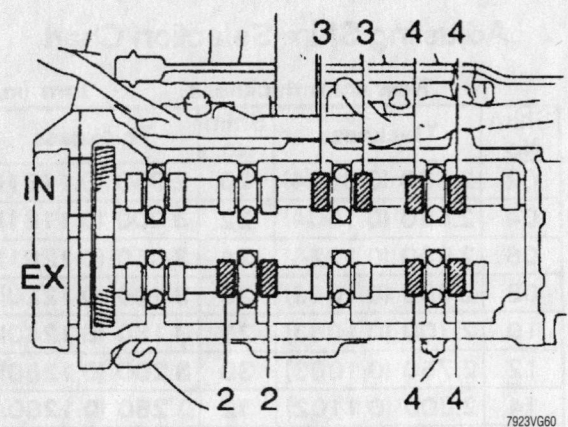

Intake valves (3 and 4) and exhaust valves (2 and 4)—5E-FE engine

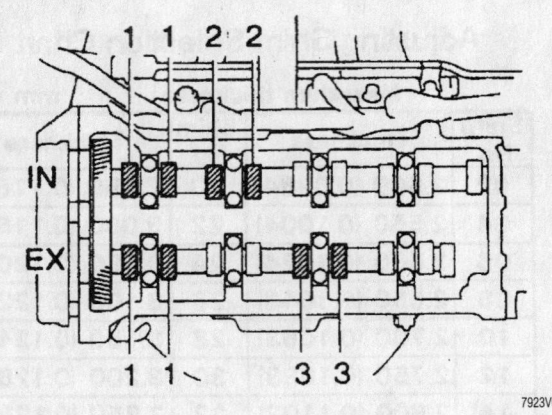

Intake valves (1 and 2) and exhaust valves (1 and 3)—5E-FE engine

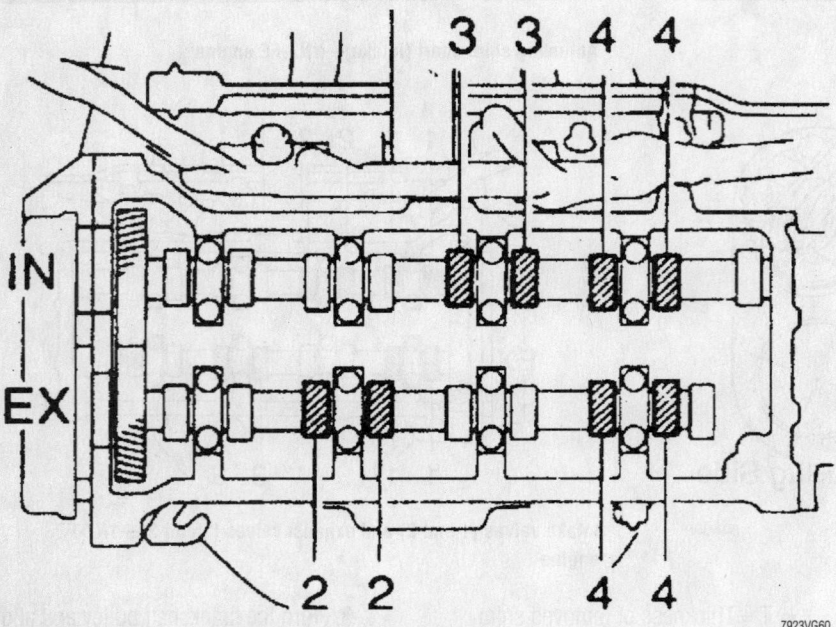

Intake valves (3 and 4) and exhaust valves (2 and 4)—5E-FE engine

7. Measure the clearance between the valve lifter and the camshaft. Record the measurements on the intake valves No. 3 and 4. Measure the exhaust valves at No. 2 and 4.

a. The intake valve clearance cold is 0.006–0.010 in. (0.15–0.25mm).

b. The exhaust valve clearance cold is 0.012–0.016 in. (0.31–0.41mm).

8. To adjust the valve clearance:

a. Remove the adjusting shim and turn the crankshaft to position the cam lobe of the camshaft on the valve to be adjusted upward.

b. Turn the valve lifter so that the notch is perpendicular to the camshaft and facing the spark plug side.

c. Using SST 09248-55040 (valve lifter press), or equivalent, hold the camshaft in place.

d. Using SST 09248-55040 (valve lifter press), or equivalent, press down the valve lifter and place SST 09248-05420 (valve lifter stopper), or equivalent between the camshaft and valve lifter.

e. Remove the SST 09248-44040 tool.

f. Using a small screwdriver and a magnetic finger, remove the adjusting shim.

9. Determine the replacement adjusting shim size by either using the chart or the following formula:

- Intake: N = T + A—0.008 in. (0.20mm)
- Exhaust: N = T + A—0.014 in. (0.36mm)
- T = Thickness of removed shim
- A = Measured valve clearance
- N = Thickness of new shim

10. Install a new shim.

11. Recheck the valve clearance.

12. Install the cylinder head covers.

13. Connect the negative battery cable.

➡ **Adjust the valve clearance when the engine is cold.**

14. Before servicing the vehicle, refer to the precautions in the beginning of this section.

15. Remove or disconnect the following:
- Negative battery cable.
- Cylinder head covers

16. Turn the crankshaft pulley and align its groove with the timing mark **0** of the No. 1 timing cover.

17. Check that the timing marks on the camshaft sprockets are in alignment with the marks on the No. 4 timing cover. If not, turn the crankshaft 1 complete revolution (360 degrees).

18. Measure the clearance between the valve lifter and the camshaft. Record the measurements on the intake valves No. 1 and 2. Measure the exhaust valves at No. 1 and 3.

a. The intake valve clearance cold is 0.006–0.010 in. (0.15–0.25mm).

b. The exhaust valve clearance cold is 0.010–0.014 in. (0.25–0.35mm).

19. Turn the crankshaft pulley 1 revolution (360 degrees) and align the groove with the timing mark **0** of the No.1 timing belt cover.

20. Measure the clearance between the valve lifter and the camshaft. Record the measurements on the intake valves No. 3 and 4. Measure the exhaust valves at No. 2 and 4.

a. The intake valve clearance cold is 0.006–0.010 in. (0.15–0.25mm).

b. The exhaust valve clearance cold is 0.010–0.014 in. (0.25–0.35mm).

21. To adjust the intake valve clearance:

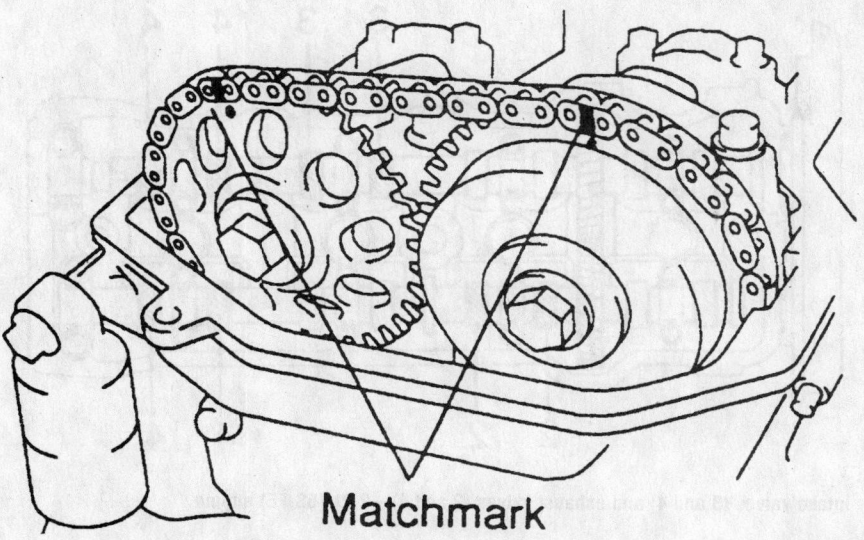

Matchmark

9307WG88

Timing marks at TDC compression—1ZZ-FE engine

a. Remove the intake camshaft.

b. Using a small screwdriver and a magnetic finger, remove the adjusting shim.

c. Determine the replacement adjusting shim size by either using the chart or the following formula:

- Intake: $N = T + A$—0.008 in. (0.20mm)
- T = Thickness of removed shim
- A = Measured valve clearance
- N = Thickness of new shim

d. Install a new shim.

e. Install intake camshaft.

f. Recheck the valve clearance.

22. To adjust the exhaust valve clearance:

a. Turn the crankshaft to position the cam lobe of the camshaft on the valve to be adjusted, upward.

b. Turn the valve lifter so that the notch is perpendicular to the camshaft and facing the spark plug side.

c. Using SST 09248–55040 (valve lifter press), or equivalent, hold the camshaft in place.

d. Using SST 09248–55040 (valve lifter press), or equivalent, press down the valve lifter and place SST 09248–05420 (valve lifter stopper), or equivalent between the camshaft and valve lifter.

e. Remove the SST 09248–44040 tool.

f. Using a small screwdriver and a magnetic finger, remove the adjusting shim.

23. Determine the replacement adjusting shim size by either using the chart or the following formula:

- Exhaust: $N = T + A$—0.014 in. (0.36mm)
- T = Thickness of removed shim
- A = Measured valve clearance
- N = Thickness of new shim

24. Install a new shim.

25. Recheck the valve clearance.

26. Install or connect the following:

- Cylinder head covers
- Negative battery cable

2ZZ-GE Engine

➡ **Adjust the valve clearance when the engine is cold.**

1. Before servicing the vehicle, refer to the precautions in the beginning of this section.

2. Remove or disconnect the following:

- Negative battery cable.
- Cylinder head covers

3. Turn the crankshaft pulley and align its groove with the timing mark **0** of the No. 1 timing cover.

4. Check that the timing marks on the camshaft sprockets are in alignment with the upper edge of the timing cover. If not, turn the crankshaft 1 complete revolution (360 degrees).

5. Measure the clearance between the valve lifter and the camshaft. Record the measurements on the intake valves No. 1 and 2. Measure the exhaust valves at No. 1 and 3.

a. The intake valve clearance cold is 0.006–0.010 in. (0.15–0.25mm).

b. The exhaust valve clearance cold is 0.014–0.018 in. (0.36–0.45mm).

6. Turn the crankshaft pulley 1 revolution (360 degrees) and align the groove with the timing mark **0** of the No.1 timing belt cover.

7. Measure the clearance between the valve lifter and the camshaft. Record the measurements on the intake valves No. 3 and 4. Measure the exhaust valves at No. 2 and 4.

a. The intake valve clearance cold is 0.006–0.010 in. (0.15–0.25mm).

b. The exhaust valve clearance cold is 0.014–0.018 in. (0.36–0.45mm).

8. To adjust the intake valve clearance:

a. Remove the intake camshaft.

b. Using a small screwdriver and a magnetic finger, remove the adjusting shim.

c. Determine the replacement adjusting shim size by either using the chart or the following formula:

- Intake: $N = T + A$—0.008 in. (0.20mm)
- T = Thickness of removed shim
- A = Measured valve clearance
- N = Thickness of new shim

d. Install a new shim.

e. Install intake camshaft.

f. Recheck the valve clearance.

9. To adjust the exhaust valve clearance:

a. Turn the crankshaft to position the cam lobe of the camshaft on the valve to be adjusted, upward.

b. Turn the valve lifter so that the notch is perpendicular to the camshaft and facing the spark plug side.

c. Using SST 09248–55040 (valve lifter press), or equivalent, hold the camshaft in place.

d. Using SST 09248–55040 (valve lifter press), or equivalent, press down the valve lifter and place SST 09248–05420 (valve lifter stopper), or equivalent between the camshaft and valve lifter.

e. Remove the SST 09248–44040 tool.

f. Using a small screwdriver and a magnetic finger, remove the adjusting shim.

10. Determine the replacement adjusting shim size by either using the chart or the following formula:

- Exhaust: $N = T + A$—0.014 in. (0.36mm)

Heater Core replacement is covered in Section 2 of this manual

- T = Thickness of removed shim
- A = Measured valve clearance
- N = Thickness of new shim
11. Install a new shim.
12. Recheck the valve clearance.
13. Install or connect the following:
   - Cylinder head covers
   - Negative battery cable

5S-FE Engine

➡ **Adjust the valve clearance when the engine is cold.**

1. Before servicing the vehicle, refer to the precautions in the beginning of this section.

2. Remove or disconnect the following:
   - Negative battery cable. On vehicles equipped with an air bag, wait at least 90 seconds before proceeding.
   - Cylinder head covers

3. Turn the crankshaft pulley and align its groove with the timing mark **0** of the No. 1 timing cover.

4. Check that the timing marks on the camshaft sprockets are in alignment with the marks on the No. 4 timing cover. If not, turn the crankshaft 1 complete revolution (360 degrees).

5. Measure the clearance between the valve lifter and the camshaft. Record the measurements on the intake valves No. 1 and 2. Measure the exhaust valves at No. 1 and 3.
   a. The intake valve clearance cold is 0.007–0.011 in. (0.19–0.29mm).
   b. The exhaust valve clearance cold is 0.011–0.015 in. (0.28–0.38mm).

6. Turn the crankshaft pulley 1 revolution (360 degrees) and align the groove with the timing mark **0** of the No.1 timing belt cover.

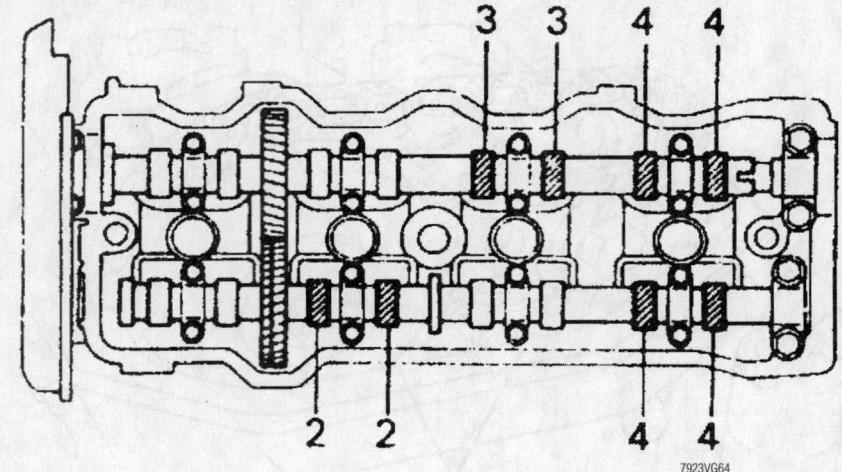

Intake valves (3 and 4) and exhaust valves (2 and 4)—2.2L (5S-FE) engine

7923VG64

7. Measure the clearance between the valve lifter and the camshaft. Record the measurements on the intake valves No. 3 and 4. Measure the exhaust valves at 2 and 4.
   a. The intake valve clearance cold is 0.007–0.011 in. (0.19–0.29mm).
   b. The exhaust valve clearance cold is 0.011–0.015 in. (0.29–0.38mm).

8. To adjust the valve clearance:
   a. Turn the crankshaft to position the cam lobe of the camshaft on the valve to be adjusted, upward.
   b. Turn the valve lifter so that the notch is perpendicular to the camshaft and facing the spark plug side.
   c. Using SST 09248–55040 (valve lifter press), or equivalent, hold the camshaft in place.
   d. Using SST 09248–55040 (valve lifter press), or equivalent, press down the valve lifter and place SST

09248–05420 (valve lifter stopper), or equivalent between the camshaft and valve lifter.
   e. Remove the SST 09248–44040 tool.
   f. Using a small screwdriver and a magnetic finger, remove the adjusting shim.

9. Determine the replacement adjusting shim size by either using the chart or the following formula:
   - Intake: N = T + A—0.009 in. (24mm)
   - Exhaust: N = T + A—0.013 in. (0.33mm)
   - T = Thickness of removed shim
   - A = Measured valve clearance
   - N = Thickness of new shim
10. Install a new shim.
11. Recheck the valve clearance.
12. Install or connect the following:
   - Cylinder head covers
   - Negative battery cable

1MZ-FE engine

➡ **Adjust the valve clearance when the engine is cold.**

1. Before servicing the vehicle, refer to the precautions in the beginning of this section.

2. Remove or disconnect the following:
   - Negative battery cable. On vehicles equipped with an air bag, wait at least 90 seconds before proceeding.
   - Accelerator/throttle cable from the throttle linkage
   - Air cleaner cover, air flow meter, and air duct assembly
   - V-bank cover
   - Emission control valve set

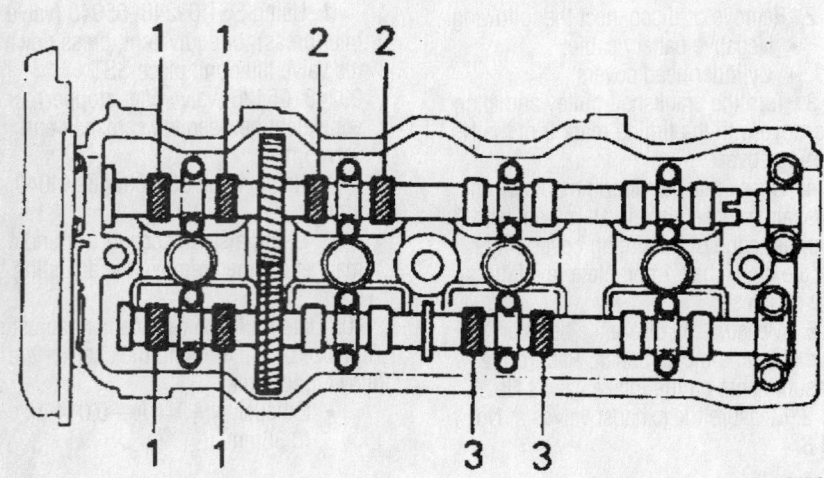

7923VG63

Intake valves (1 and 2) and exhaust valves (1 and 3)—2.2L (5S-FE) engine

- Air intake chamber
- Engine harness from the injectors and the ignition coils
- Ignition coils and keep them in order for reassembly
- Spark plugs
- Cylinder head covers

3. Turn the crankshaft pulley and align its groove with the timing mark **0** of the No. 1 timing cover.

4. Check that the valve lifters on the No. 1 intake are loose and the No. 1 exhaust are tight. If not, turn the crankshaft 1 complete revolution (360 degrees).

➥**All measurements should be written down. These recorded measurements will need to be used in conjunction with a mathematical formula to determine the thickness of the replacement shims.**

5. Measure the clearance between the valve lifters and the camshaft. Record the measurements on valves No. 1 and 6 intake; No. 2 and 3 exhaust.

   a. The intake valve clearance cold is 0.006–0.010 in. (0.15–0.25mm).

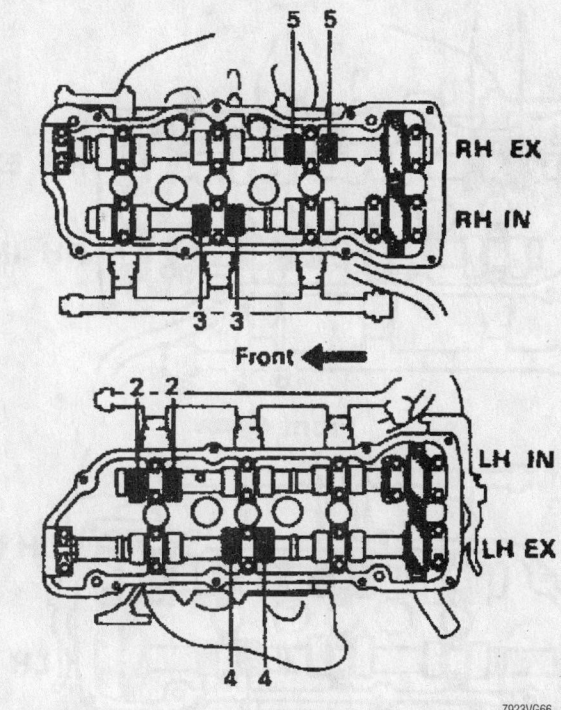

Adjust these valves during the 2nd step—3.0L (1MZ-FE) engine

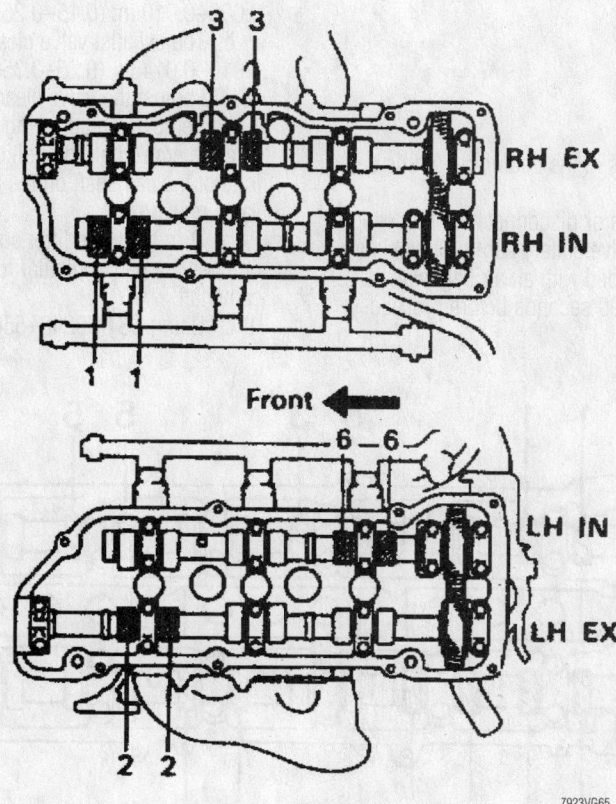

Adjust these valves during the 1st step—3.0L (1MZ-FE) engine

   b. The exhaust valve clearance cold is 0.010–0.014 in. (0.25–0.35mm).

6. Turn the crankshaft ⅔ of a revolution (240 degrees). Record the measurements on valves No. 2 and 3 intake; No. 4 and 5 exhaust.

7. Turn the crankshaft another ⅔ of a revolution. Record the measurements on valves No. 4 and 5 intake; No. 1 and 6 exhaust.

8. Remove the adjusting shim by turning the crankshaft to position the cam lobe of the camshaft in the up position on the valve to be adjusted. Using a small thin flat bladed tool, turn the valve lifter so that the notches are perpendicular to the camshaft. Press down the valve lifter with SST 09248–55010 part A, or equivalent. Place SST 09248–55010 part B between the camshaft and the valve lifter; remove part A.

9. Remove the adjusting shim with a magnet and a small screwdriver.

10. Determine the replacement adjusting shim size by either using the charts or the following formulas:

- Intake: N = T + (A—0.008 in./0.020mm)
- Exhaust: N = T + (A—0.012 in./0.30mm)
- T = Thickness of removed shim
- A = Measured valve clearance

Brake service is covered in Section 4 of this manual

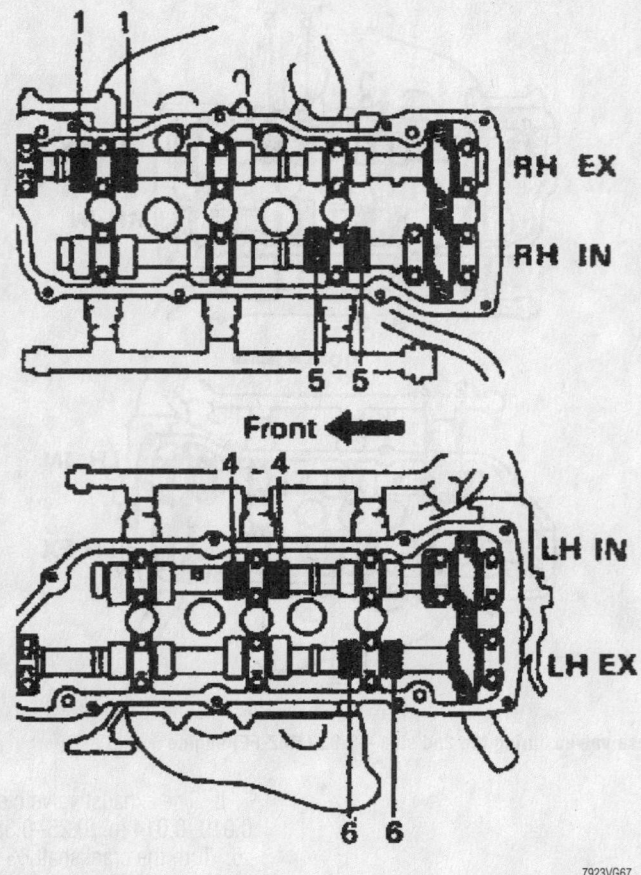

7923VG67

Adjust these valves during the 3rd step—3.0L (1MZ-FE) engine

• N = Thickness of new shim

11. Select a new shim with a thickness as close as possible to the calculated value. Install the new replacement shim.

➡**Shims are available in 17 sizes in increments of 0.0020 in. (0.050mm), from 0.0984 in. (2.500mm) to 0.1299 in. (3.300mm).**

12. Recheck the valve clearance.
13. Install or connect the following:
  • Cylinder head covers
  • Spark plugs and the ignition coils
  • Engine wiring harness to the injectors and the coils
  • Intake chamber
  • Emission control valve set
  • V-bank cover
  • Air flow meter, air duct, and air cleaner cover
  • Negative battery cable

2JZ-GE and 2JZ-GTE Engines

➡**Adjust the valve clearance when the engine is cold.**

1. Before servicing the vehicle, refer to the precautions in the beginning of this section.

2. Remove or disconnect the following:
  • Negative battery cable. On vehicles equipped with an air bag, wait at least 90 seconds before proceeding.

  • Cylinder head covers

3. Turn the crankshaft pulley and align its groove with the timing mark **0** of the No. 1 timing cover.

4. Check that the timing marks on the camshaft sprockets are in alignment with the marks on the No. 4 timing cover. If not, turn the crankshaft 1 complete revolution (360 degrees).

5. Measure the clearance between the valve lifter and the camshaft. Record the measurements on the intake valves No. 1, 2 and 4. Measure the exhaust valves at No. 1, 3 and 5.

  a. The intake valve clearance cold is 0.006–0.010 in. (0.15–0.25mm).

  b. The exhaust valve clearance cold is 0.010–0.014 in. (0.25–0.35mm).

6. Turn the crankshaft pulley 1 revolution (360 degrees) and align the groove with the timing mark **0** of the No.1 timing belt cover.

7. Measure the clearance between the valve lifter and the camshaft. Record the measurements on the intake valves No. 3, 5 and 6. Measure the exhaust valves at No. 2, 4, and 6.

  a. The intake valve clearance cold is 0.006–0.010 in. (0.15–0.25mm).

  b. The exhaust valve clearance cold is 0.010–0.014 in. (0.25–0.35mm).

8. To adjust the valve clearance:

  a. Remove the adjusting shim and turn the crankshaft to position the cam lobe of the camshaft on the adjusting valve upward.

  b. Turn the valve lifter so that the notches are perpendicular to the camshaft.

  c. Using SST 09248–55040 (valve

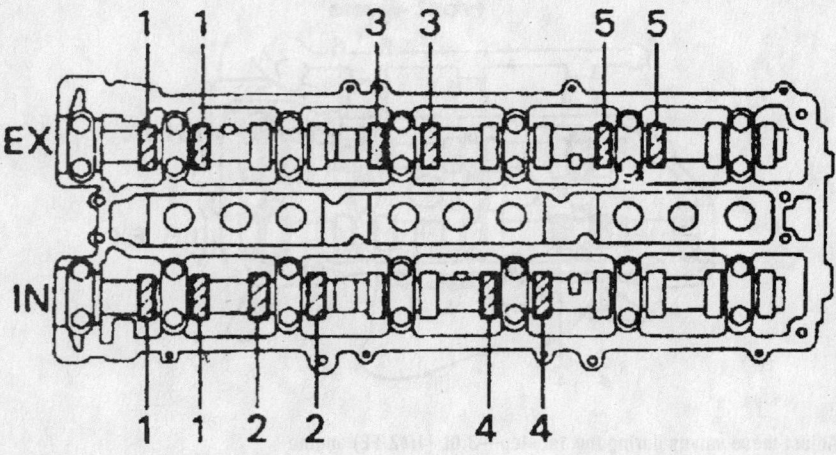

7923VG68

Valve clearance inspection (before turning crankshaft 360 degrees)—2JZ-GE and 2JZ-GTE engines

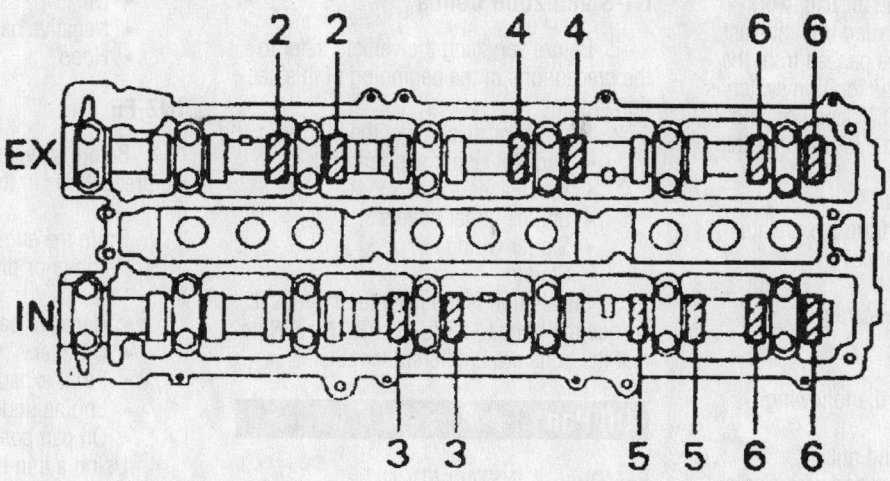

Valve clearance inspection (after turning crankshaft 360 degrees)—2JZ-GE and 2JZ-GTE engines

lifter press), or equivalent, hold the camshaft in place.

   d. Using SST 09248–55040 (valve lifter press), or equivalent, press down the valve lifter and place SST 09248–05420 (valve lifter stopper), or equivalent between the camshaft and valve lifter.

   e. Remove the SST 09248–44040 tool.

   f. Using a small screwdriver and a magnetic finger, remove the adjusting shim.

9. Determine the replacement adjusting shim size by either using the chart or the following formula:

- Intake: N = T + (A—0.008 in./0.20mm)
- Exhaust: N = T + (A—0.012 in./0.30mm)
- T = Thickness of removed shim
- A = Measured valve clearance
- N = Thickness of new shim

10. Recheck the valve clearance.
11. Install or connect the following:
- Cylinder head covers
- Negative battery cable

Starter

REMOVAL & INSTALLATION

Tercel

1. Before servicing the vehicle, refer to the precautions in the beginning of this section.
2. Remove or disconnect the following:
- Negative battery cable. On models equipped with an air bag, work

must NOT be started until at least 90 seconds have passed from the time that both the ignition switch is turned to the LOCK position and the negative cable is disconnected from the battery.
- Wiring from the starter terminals. On some models, it may be necessary to remove the transaxle cable and bracket from the transaxle
- Heat insulator, if equipped, and the starter mounting bolts
- Starter

3. Installation is the reverse of the removal procedure

Echo

1. Before servicing the vehicle, refer to the precautions in the beginning of this section.
2. Remove or disconnect the following:
- Negative battery cable. On models equipped with an air bag, work must NOT be started until at least 90 seconds have passed from the time that both the ignition switch is turned to the LOCK position and the negative cable is disconnected from the battery
- Engine under covers
- Wiring from the starter terminals
- Starter mounting bolts
- Starter

To install:
3. Install or connect the following:
- Starter

- Starter wiring
- Engine under covers

Avalon and Camry

1. Before servicing the vehicle, refer to the precautions in the beginning of this section.
2. Remove or disconnect the following:
- Negative battery cable. On models equipped with an air bag, work must NOT be started until at least 90 seconds have passed from the time that both the ignition switch is turned to the LOCK position and the negative cable is disconnected from the battery.
- With cruise control, the battery
- With cruise control, the actuator and bracket
- Wiring from the starter
- Starter

To install:
3. Install or connect the following:
- Starter. Torque the bolts to 27 ft. lbs. (37 Nm).
- Starter cable
- Electrical connectors
- Actuator connector and clamp, if equipped with cruise control
- Battery and tray, if equipped with cruise control

Corolla

1. Before servicing the vehicle, refer to the precautions in the beginning of this section.
2. Remove or disconnect the following:
- Battery and tray. On models

For complete Engine Mechanical specifications, see Section 1 of this manual

equipped with an air bag, work must NOT be started until at least 90 seconds have passed from the time that both the ignition switch is turned to the LOCK position and the negative cable is disconnected from the battery.

- Coolant reservoir
- Reservoir hose from the radiator
- Right side under cover
- Wiring clamp
- Wiring from the starter
- Starter

To install:

3. Install or connect the following:
- Starter
- Starter wiring and nut
- Starter connector and wire clamp
- Right side engine cover
- Radiator reservoir and hose
- Battery and tray

4. Check the cooling system and top off if necessary.

Supra and 1999 Celica

1. Before servicing the vehicle, refer to the precautions in the beginning of this section.

2. Remove or disconnect the following:

- Negative battery cable. On models equipped with an air bag, work must NOT be started until at least 90 seconds have passed from the time that both the ignition switch is turned to the LOCK position and the negative cable is disconnected from the battery.
- Air cleaner assembly
- On the 5S-FE, the battery if necessary
- With cruise control, the cruise control actuator from the body bracket
- The starter wiring
- The starter bolts
- The oxygen sensor wiring harness from the engine wire brackets and starter
- The starter

To install:

3. Install or connect the following:
- Starter and oxygen sensor wiring and brackets. Bolts: 29 ft. lbs. (39 Nm).
- Starter cable and tighten the nut
- All other starter wiring
- With cruise control, the actuator
- If removed, the battery
- Air cleaner
- Negative battery cable

GT-S and 2000 Celica

1. Before servicing the vehicle, refer to the precautions in the beginning of this section.

2. Remove or disconnect the following:
- Radiator upper support seal
- Air cleaner
- Engine under covers
- Starter wiring
- Starter

3. Installation is the reverse of removal. Torque the bolts to 28 ft. lbs. (37 Nm). Note that the bolts are 2 different lengths.

Oil Pan

REMOVAL & INSTALLATION

5E-FE

1. Before servicing the vehicle, refer to the precautions in the beginning of this section.

2. Drain the engine oil.

3. Remove or disconnect the following:
- Negative battery cable. On vehicles equipped with an air bag, wait at least 90 seconds before proceeding.
- Hood
- Oil dipstick
- Timing belt and suspend the engine with a hoist

➡ **Do not raise the engine more than necessary, since wiring and other components can be damaged.**

- Crankshaft timing sprocket and oil pump sprocket
- Air conditioning compressor and mounting bracket
- With distributor ignition, exhaust pipe stay
- Oxygen sensor
- Front exhaust pipe
- Oil pan

To install:

4. Clean all gasket mating surfaces.

5. Install or connect the following:
- Oil pan, with new gasket and sealer. Torque the bolts to 10 ft. lbs. (13 Nm).
- Front exhaust pipe using a new gasket. Bolts: 46 ft. lbs. (62 Nm).
- Oxygen sensor
- Exhaust pipe stay. Bolts: 14 ft. lbs. (19 Nm).
- Crankshaft timing and oil pump sprockets
- Timing belt
- Oil dipstick

- Oil
- Negative battery cable
- Hood

1NZ-FE

1. Before servicing the vehicle, refer to the precautions in the beginning of this section.

2. Drain the engine oil.

3. Remove or disconnect the following:

- Negative battery cable
- Oil filter
- Front exhaust pipe
- Engine under covers
- Oil pan bolts

4. Using a thin blade, cut the sealer holding the oil pan and lower the pan.

5. Installation is the reverse of removal. Use RTV sealer. Torque the bolts, in a criss-cross pattern, to 80 inch lbs. (9 Nm).

1ZZ-FE Engine

1. Before servicing the vehicle, refer to the precautions in the beginning of this section.

2. Drain the engine oil.

3. Remove or disconnect the following:
- Negative battery cable. On vehicles equipped with an air bag, wait at least 90 seconds before proceeding.
- Undercovers
- Front exhaust pipe
- Oil pan mounting bolts and nuts
- Oil pan, cutting off the applied sealer.

To install:

4. Remove any old sealant from the oil pan flange and thoroughly clean the sealing surface.

5. Install or connect the following:
- Oil pan. Tighten the bolts and nuts in several passes. Bolts and nuts: 80 inch lbs. (9 Nm).
- Front exhaust pipe
- Negative battery cable
- Undercovers

6. Fill the engine with clean oil.

7. Start the vehicle, check for leaks and repair if necessary.

2ZZ-GE Engine

1. Before servicing the vehicle, refer to the precautions in the beginning of this section.

2. Drain the engine oil.

3. Remove or disconnect the following:
- Negative battery cable. On vehicles equipped with an air bag, wait at

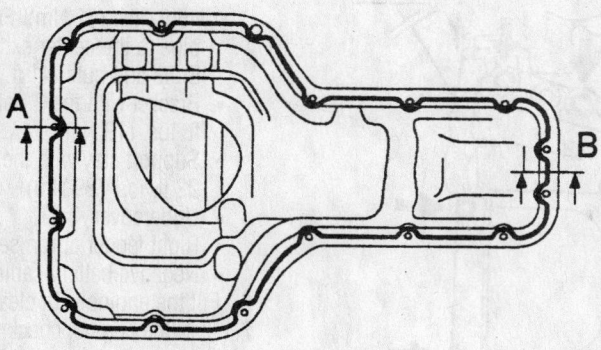

**Seal Width
4 – 5 mm**

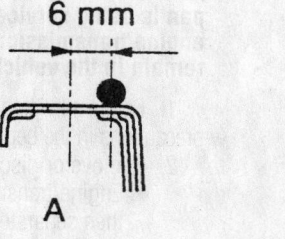
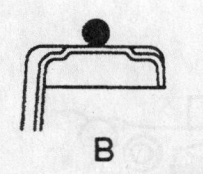

7923VG72

Apply sealant to the oil pan as shown—1.8L (1ZZ-FE) engine

least 90 seconds before proceeding.
- Undercovers
- Front exhaust pipe
- Oil pan mounting bolts and nuts
- Oil pan, cutting off the applied sealer.

To install:

4. Remove any old sealant from the oil pan flange and thoroughly clean the sealing surface.

5. Install or connect the following:
- Oil pan. Tighten the bolts and nuts in several passes. Bolts and nuts: 80 inch lbs. (9 Nm).
- Front exhaust pipe
- Negative battery cable
- Undercovers

6. Fill the engine with clean oil.

7. Start the vehicle, check for leaks and repair if necessary.

5S-FE Engine

1. Before servicing the vehicle, refer to the precautions in the beginning of this section.

2. Drain the engine oil.

3. Remove or disconnect the following:

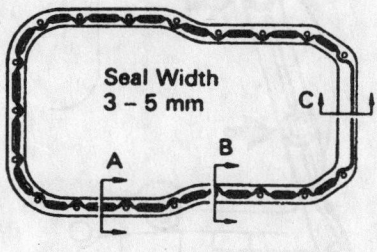

**Seal Width
3 – 5 mm**

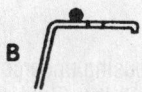

5 mm (0.20 in.)

7923VG73

Oil pan sealing diagram—2.2L (5S-FE) engine

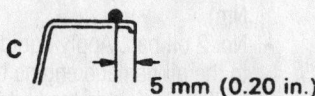

- Negative battery cable. On vehicles equipped with an air bag, wait at least 90 seconds before proceeding.
- Undercovers
- Front exhaust pipe
- Engine mounting center member
- On Celica, the 3-way Catalytic Converter (TWC)
- Rear end stiffener plate
- Oil dipstick
- Oil pan bolts and nuts. Cut off the applied sealer.

➡**Do not use the cutting tool for the oil pump body side and rear oil seal retainer.**

To install:

4. Remove any old sealant from the oil pan flange and thoroughly clean both sealing surfaces.

5. Install or connect the following:
- Oil pan. Uniformly tighten the bolts and nuts in several passes. Bolts and nuts: 48 inch lbs. (5.4 Nm)
- Oil dipstick
- Rear end stiffener plate
- On Celica, the TWC.
- Engine mounting center member
- Front exhaust pipe
- Negative battery cable
- Undercovers

6. Fill the engine with clean oil.

7. Start the vehicle, check for leaks and repair if necessary.

1MZ-FE Engine

1. Before servicing the vehicle, refer to the precautions in the beginning of this section.

2. Drain the engine oil.

3. Remove or disconnect the following:

- Negative battery cable. On vehicles equipped with an air bag, wait at least 90 seconds before proceeding.
- Fender apron seal
- Undercover
- Front exhaust pipe
- Front exhaust pipe bracket from the No. 1 oil pan.
- Flywheel housing undercover
- Bolts and nuts from the No. 2 oil pan.
- Oil strainer and gasket
- Remove the No.1 oil pan
- Baffle plate from the No. 1 oil pan

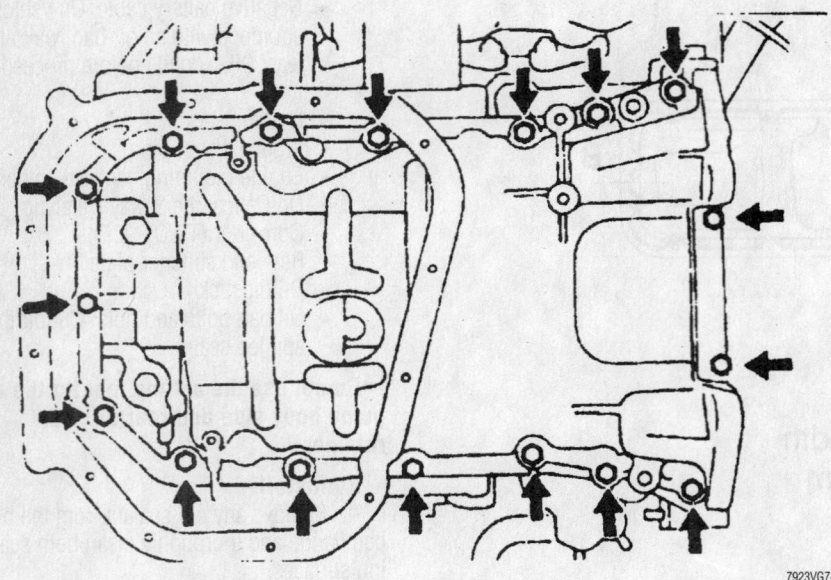

No. 1 oil pan mounting bolt locations—3.0L (1MZ-FE) engine

7923VG74

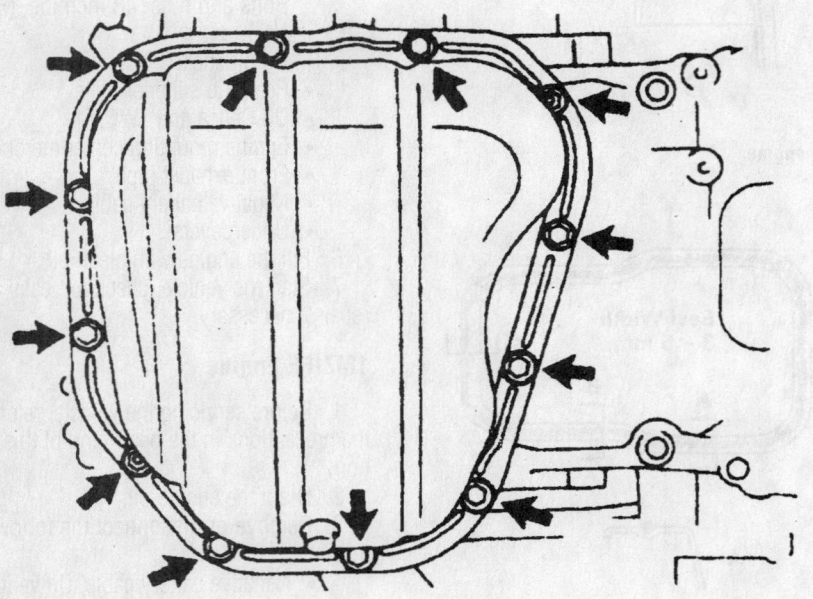

No. 2 oil pan mounting bolt locations—3.0L (1MZ-FE) engine

7923VG75

To install:

4. Clean all mating surfaces of the oil pans.

5. Install or connect the following:
- Baffle plate to the No. 1 oil pan and tighten: 69 inch lbs. (8 Nm).
- Install the No. 1 oil pan. with liquid sealant. Uniformly tighten the bolts and nuts in several passes. 10mm head bolt: 69 inch lbs. (8 Nm). 12mm head bolt: 14 ft. lbs. (19.5 Nm). 14mm head bolt-27 ft. lbs. (37.2 Nm)
- Flywheel housing undercover. Bolts: 69 inch lbs. (7.8 Nm).
- Oil strainer. Nuts: 69 inch lbs. (7.8 Nm).
- No. 2 oil pan. Apply liquid sealant to the oil pan and engine block. Uniformly tighten the bolts and nuts in several passes. Bolts: 69 inch lbs. (7.8 Nm).
- Flywheel housing undercover
- Front exhaust pipe bracket to the No. 1 oil pan. Bolts: 15 ft. lbs. (21 Nm).
- Front exhaust pipe. Exhaust mani-

folds to the front exhaust pipe nuts: 46 ft. lbs. (62 Nm). Front exhaust pipe to the center exhaust pipe. Bolts and nuts: 41 ft. lbs. (56 Nm).
- Bracket with the 2 bolts. Bolts: 14 ft. lbs. (19 Nm).
- Support stay with the 2 bolts. Bolts: 22 ft. lbs. (29 Nm).
- Undercover
- Right fender apron seal
- Negative battery cable

6. Fill the engine with clean oil.

7. Start the engine, check for leaks and repair if necessary.

2JZ-GE and 2JZ-GTE Engines

➡The No. 1 oil pan can not be removed with the engine in the vehicle. The engine/transmission assembly must be removed. If only the No. 2 oil pan is being serviced, the engine/transmission assembly can remain in the vehicle.

1. Before servicing the vehicle, refer to the precautions in the beginning of this section.

2. Remove or disconnect the following:
- Engine/transmission assembly, then separate the transmission from the engine.
- Timing belt
- Idler pulley
- Crankshaft timing pulley
- Oil dipstick and guide
- Oil sensor lead, 4 attaching bolts and oil level sensor.
- Lower (No. 2) oil pan.
- Oil strainer and gasket
- Baffle plate
- On turbocharged engines, turbo oil outlet pipe
- Upper (No. 1) oil pan.
- O-ring

To install:

3. Install or connect the following:
- New O-ring in the block. Apply a ⅛ inch (3–4mm) sealant bead to the pan mating surface.
- Upper pan. 12mm bolts to 15 ft. lbs. (21 Nm). 14mm bolts to 29 ft. lbs. (39 Nm).
- On turbocharged engines, the turbo oil outlet pipe and gasket. Nuts: 20 ft. lbs. (27 Nm).
- 2 turbo oil outlet hoses.
- Baffle plate and oil strainer. Tighten both to 78 inch lbs. (9 Nm).
- Lower oil pan in the same manner as the upper pan. Bolts to 78 inch lbs. (9 Nm).
- Oil level sensor, using a new gasket. Tighten to 48 inch lbs. (5.4 Nm).

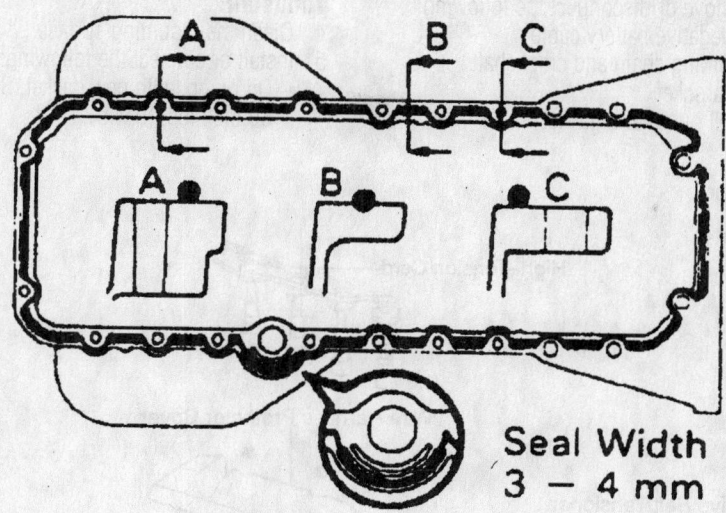

Seal Width
3 – 4 mm

Upper oil pan sealant application—3.0L (2JZ-GE and 2JZ-GTE) engines

7923VG76

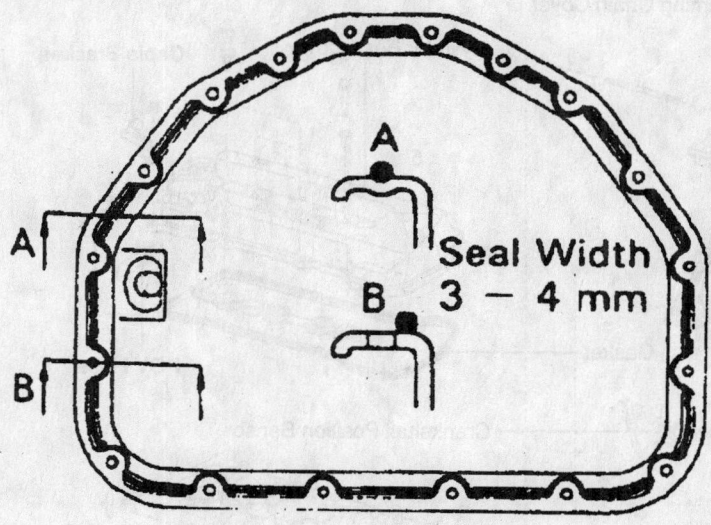

Seal Width
3 – 4 mm

Lower oil pan sealant application—3.0L (2JZ-GE and 2JZ-GTE) engines

7923VG77

- Oil dipstick and guide, the timing pulleys and belt, and transmission to the engine.
- Engine and transmission
- All fluids

Oil Pump

REMOVAL & INSTALLATION

5E-FE Engine

1. Before servicing the vehicle, refer to the precautions in the beginning of this section.

2. Drain the engine oil.
3. Remove or disconnect the following:
    - Negative battery cable. On vehicles equipped with an air bag, wait at least 90 seconds before proceeding.
    - Hood
    - Oil dipstick
    - Timing belt
    - Oil pan
    - Oil strainer with the O-ring
    - Pressure regulator valve
    - Oil pump and tension spring bracket
    - Nut and oil pump pulley

To install:
4. Clean all gasket mating surfaces.
5. Install or connect the following:
    - Oil pump pulley by first placing the driven rotors into the pump body with the marks facing the front
    - Oil pump pulley. Nut: 27 ft. lbs. (37 Nm).
    - Pressure regulator valve. Tighten to 22 ft. lbs. (30 Nm).
    - Oil strainer, using a new O-ring. Bolts: 84 inch lbs. (10 Nm).
    - Oil pan, using a new gasket and sealer
    - All remaining components in the reverse order they were removed.
    - Negative battery cable
    - Hood
6. Fill the engine with clean oil.
7. Start the vehicle, check for leaks and repair if necessary.

1NZ-FE

1. Before servicing the vehicle, refer to the precautions in the beginning of this section.
2. Drain the engine oil.
3. Remove or disconnect the following:
    - Negative battery cable
    - Timing chain cover
    - 2 bolts, 3 screws and the oil pump cover from the timing chain cover
    - Drive and driven rotors
    - Plug, spring and relief valve
4. Inspect the relief valve motion
5. Check rotor side clearance: 0.0012–0.0035 in. (0.03–0.09mm).
6. Check rotor tip clearance: 0.0024–0.0071 in. (0.06–0.18mm).
7. Check rotor-to-body clearance: 0.0098–0.0128 in. (0.250–0.325mm).

To install:
8. Install or connect the following:
    - Relief valve and spring. Torque the plug to 18 ft. lbs. (24 Nm).
    - Drive and driven rotors with the marks on the cover side
    - Cover. Torque the bolts to 80 inch lbs. (9 Nm); the screws to 96 inch lbs. (11 Nm).
    - Timing chain cover
    - Negative battery cable
9. Fill the engine with clean oil.
10. Start the vehicle, check for leaks and repair if necessary.

For Tire, Wheel and Ball Joint specifications, see Section 1 of this manual

1ZZ-FE Engine

1. Before servicing the vehicle, refer to the precautions in the beginning of this section.

2. Drain the engine oil.

3. Remove or disconnect the following:
   • Negative battery cable
   • Timing chain and crankshaft sprocket
   • Oil pump and gasket

To install:

4. Clean the mounting surface.

5. Install or connect the following:
   • Oil pump, with new gasket. Bolts: 80 inch lbs. (9 Nm).

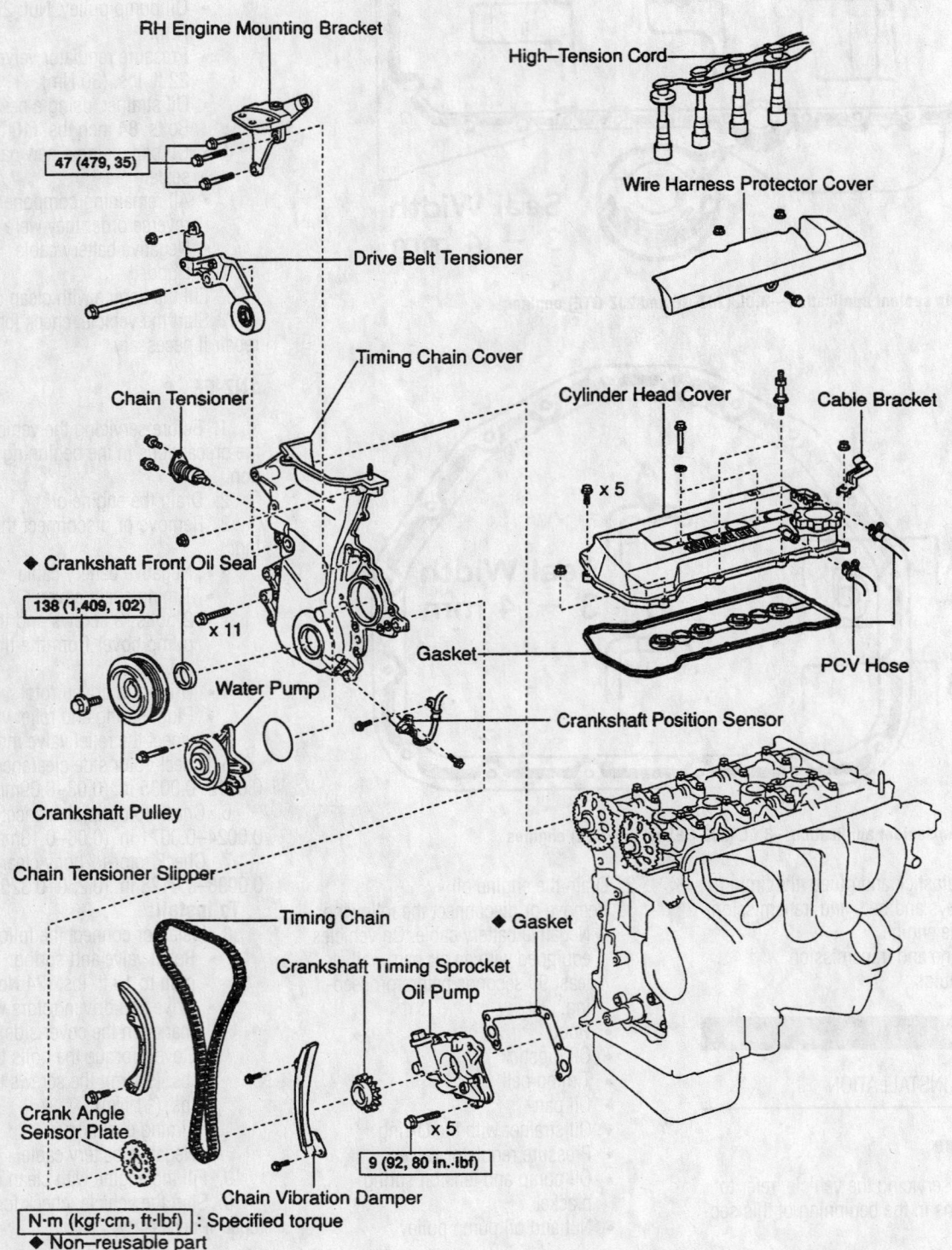

RH Engine Mounting Bracket

High–Tension Cord

47 (479, 35)

Wire Harness Protector Cover

Drive Belt Tensioner

Timing Chain Cover

Chain Tensioner

Cylinder Head Cover

Cable Bracket

x 5

◆ Crankshaft Front Oil Seal

138 (1,409, 102)

x 11

Gasket

PCV Hose

Water Pump

Crankshaft Position Sensor

Crankshaft Pulley

Chain Tensioner Slipper

Timing Chain

◆ Gasket

Crankshaft Timing Sprocket

Oil Pump

Crank Angle Sensor Plate

x 5

9 (92, 80 in.·lbf)

Chain Vibration Damper

N·m (kgf·cm, ft·lbf) : Specified torque
◆ Non–reusable part

7923VGB0

Exploded view of the oil pump mounting—1.8L (1ZZ-FE) engine

- Crankshaft sprocket and timing chain
- Negative battery cable
6. Fill the engine with clean oil.
7. Start the vehicle, check for leaks and repair if necessary.

2ZZ-GE Engine

1. Before servicing the vehicle, refer to the precautions in the beginning of this section.
2. Drain the engine oil.
3. Remove or disconnect the following:
- Negative battery cable
- Timing chain and crankshaft sprocket
- Oil pump and gasket

To install:
4. Clean the mounting surface.
5. Install or connect the following:
- Oil pump, with new gasket. Bolts: 80 inch lbs. (9 Nm).
- Crankshaft sprocket and timing chain
- Negative battery cable
6. Fill the engine with clean oil.
7. Start the vehicle, check for leaks and repair if necessary.

5S-FE Engine

CAMRY AND CAMRY SOLARA

1. Before servicing the vehicle, refer to the precautions in the beginning of this section.
2. Drain the engine oil.
3. Remove or disconnect the following:
- Negative battery cable. On vehicles equipped with an air bag, wait at least 90 seconds before proceeding.
- Hood
- Front exhaust pipe
- Rear end stiffener plate
- Oil dipstick and oil pan
- Oil strainer and gasket
- Timing belt and pulleys
- Crankshaft position sensor
- Oil pump and gasket

To install:
4. Install or connect the following:
- Oil pump with new gasket. Bolts: 82 inch lbs. (9 Nm).

➡**The long bolts are 1.38 in. (35mm) and all the others are 0.98 in. (25mm).**

- Crankshaft position sensor
- Timing belt and pulleys
- Oil strainer with new gasket.

Tighten to 48 inch lbs. (5 Nm).
- Oil pan and dipstick

➡**The pan must be installed within 5 minutes of sealant application or the procedure will have to be repeated.**

- Rear end stiffener plate. Bolts: 27 ft. lbs. (37 Nm).
- Front exhaust pipe
- Negative battery cable
- Hood
5. Fill the engine with clean oil.
6. Start the vehicle, check for leaks and repair if necessary.

CELICA

1. Before servicing the vehicle, refer to the precautions in the beginning of this section.
2. Drain the engine oil.
3. Remove or disconnect the following:
- Negative battery cable. On vehicles equipped with an air bag, wait at least 90 seconds before proceeding.
- Undercovers
- Front exhaust pipe
- Engine mounting center member
- Sub oxygen sensor wiring.
- Right-hand side exhaust manifold stay, TWC with gasket, retainer, and cushion.
- Rear end stiffener plate
- Oil dipstick
- Oil pan, cutting off the applied sealer.

➡**Do not use the tool for the oil pump body side and rear oil seal retainer.**

- Oil strainer, baffle plate, and the gasket
- Timing belt
- No. 2 idler pulley and crankshaft timing pulley.
- Oil pump pulley
- Oil pump and gasket

To install:
4. Install or connect the following:
- Oil pump, with new gasket. Bolts, uniformly tightened in several passes: 78 inch lbs. (9 Nm). Bolt A is 0.98 in. (25mm) long and bolt B is 1.38 in. (35mm) long.
- Oil pump pulley. Nut: 18 ft. lbs. (24 Nm).
- Crankshaft timing pulley and the No. 2 idler pulley.
- Timing belt
- New gasket, oil strainer and baffle plate. Bolts and nuts: 48 inch lbs. (5 Nm).
- Oil pan and dipstick.
- Rear end stiffener plate
- Cushion, retainer, and a new gasket to the front TWC. Bolts and nuts: 21 ft. lbs. (29 Nm).
- Right-hand side exhaust manifold stay. Bolts and nuts: 31 ft. lbs. (42 Nm)
- Sub oxygen sensor connector
- Engine mounting center member
- Front exhaust pipe

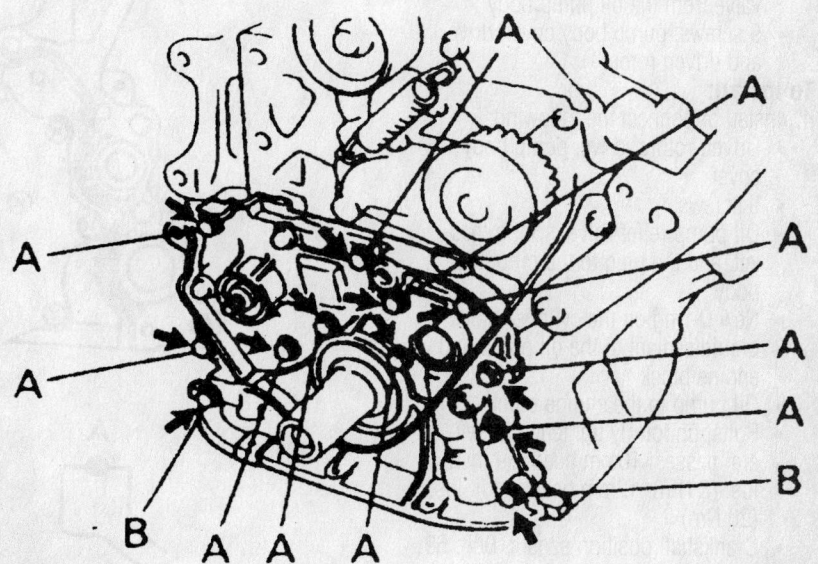

Oil pump mounting bolt identification—2.2L (5S-FE) engine

7923VG78

- Negative battery cable
- Undercovers

5. Fill the engine with clean oil.

6. Start the vehicle, check for leaks and repair if necessary.

1MZ-FE Engine

1. Before servicing the vehicle, refer to the precautions in the beginning of this section.

2. Drain the engine oil.

3. Remove or disconnect the following:
- Negative battery cable. On vehicles equipped with an air bag, wait at least 90 seconds before proceeding.
- Fender apron seal
- Undercover
- Front exhaust pipe
- Front exhaust pipe bracket from the No. 1 oil pan
- Alternator drive belt
- A/C compressor
- Power steering pump drive belt and adjusting strut
- Timing belt and belt pulleys
- Rear timing belt cover
- A/C compressor housing bracket
- No. 2 oil pan, oil strainer, No.1 oil pan and baffle plate
- Crankshaft position sensor
- 9 oil pump bolts. Make a note of the position of the each bolt.
- Oil pump body by prying between the oil pump and main bearing cap
- O-ring from the cylinder block
- Plug, gasket, spring, and relief valve from the oil pump body
- 9 screws, pump body cover, drive, and driven rotors

To install:

4. Install or connect the following:
- Driven rotors, drive, pump body cover
- 9 screws
- Oil pump relief valve, spring, gasket, and the plug to the oil pump body
- New O-ring on the cylinder block
- Liquid sealant to the oil pump and engine block
- Oil pump to the engine block
- Bolts, uniformly tightened in several passes: 10mm head: 69 inch lbs. (8 Nm). 12mm head: 14 ft. lbs. (20 Nm)
- Crankshaft position sensor. Bolt: 69 inch lbs. (8 Nm).
- Baffle plate to the No. oil pan. Tighten to 69 inch lbs. (8 Nm).
- No. 1 oil pan, oil strainer and No. 2 oil pan.

- A/C compressor housing bracket. Bolts: 18 ft. lbs. (25 Nm).
- Rear timing belt cover. Bolts: 74 inch lbs. (9 Nm).
- Timing belt pulleys
- Timing belt
- Adjusting strut and power steering drive belt. Bolt and nut: 32 ft. lbs. (43 Nm).
- A/C compressor
- Alternator drive belt
- Front exhaust pipe bracket to the No. 1 oil pan. Bolts: 15 ft. lbs. (21 Nm).
- Front exhaust pipe
- Undercover
- Right fender apron seal
- Negative battery cable

5. Fill the engine with clean oil.

6. Start the vehicle, check for leaks and repair if necessary.

2JZ-GE and 2JZ-GTE Engines

1. Before servicing the vehicle, refer to the precautions in the beginning of this section.

2. Remove or disconnect the following:
- Negative battery cable. On vehicles equipped with an air bag, wait at least 90 seconds before proceeding.

- Engine and transmission. Separate the transmission from the engine.
- Timing belt
- Idler pulley
- Crankshaft timing pulley
- Oil dipstick and tube, oil level sensor, No. 2 (lower) oil pan, oil strainer and oil baffle plate.
- With turbocharger, turbo oil outlet pipe.
- No.1 (upper) oil pan
- Oil pump, by driving the pump off the cylinder block using a brass drift.
- O-rings

To install:

3. Install or connect the following:
- 2 new O-rings in the cylinder block
- A ⅛ inch (3–4mm) bead of sealant around the pump mating surface
- Oil pump. Bolts: 15 ft. lbs. (21 Nm)
- O-ring on the block
- A bead of sealant around the No. 1 oil pan
- No.1 oil pan. Bolts: 12mm heads to 15 ft. lbs. (21 Nm). 14mm heads to 29 ft. lbs. (39 Nm).
- With turbocharger, turbo oil outlet pipe. Bolts: 20 ft. lbs. (27 Nm).
- Oil baffle plate, oil strainer, No. 2

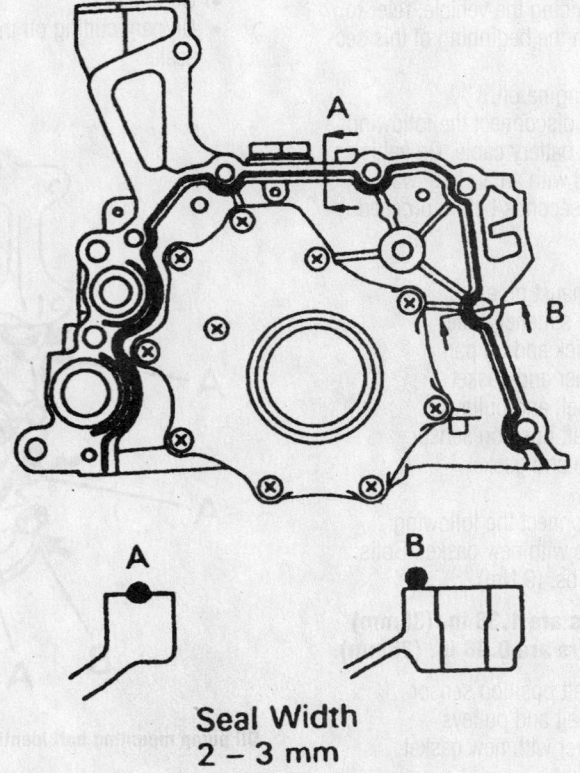

Seal Width 2 – 3 mm

7923VG79

Oil pump sealant application—3.0L (2JZ-GE and 2JZ-GTE) engines

oil pan, oil lever sensor and oil dipstick with a new O-ring.
- Crankshaft and idler pulley
- Timing belt
- Engine and transmission
- Negative battery cable

Rear Main Seal

REMOVAL & INSTALLATION

5S-FE, 1ZZ-FE, 2ZZ-GE and 1NZ-FE Engines

1. Remove or disconnect the following:
   - Transaxle
   - Clutch assembly
   - Flywheel or flexplate
2. Use a small sharp knife to cut off the lip of the oil seal. Take great care not to score any metal with the knife.
3. Use a small prytool to pry the old seal from the retaining plate. Be careful not to damage the plate. Protect the tip of the tool with tape and pad the fulcrum point with cloth.
4. Inspect the crankshaft and seal lip contact surfaces for any sign of damage.

To install:

5. Apply a light coat of multi-purpose grease to the lip of a new oil seal. Loosely fit the seal into place by hand, making sure it is not crooked.
6. Use a seal driver of the correct size to install the seal. Tap it into place until the surface of the seal is flush with the edge of the housing.

1MZ-FE Engine

1. Remove or disconnect the following:
   - Transaxle
   - Clutch cover assembly and flywheel or driveplate
   - Remove the rear end plate.
   - Oil seal retainer and gasket. Discard the gasket or sealant.
2. Use a small prybar to pry the oil seal from the retaining plate. Be careful not to damage the plate.

To install:

3. Clean the retainer contact surfaces thoroughly and lubricate the new oil seal with multi-purpose grease.
4. Drive the oil seal into the retainer until its surface is flush with the edge of the retainer. Make sure that the seal is installed evenly in the retainer to ensure proper sealing.
5. Apply a ⅛ inch bead of sealant to the oil seal retainer. Install the retainer and install the dust seal. Tighten the bolts to 69 inch lbs. (8 Nm).
6. Install the rear end plate.
7. On automatic transaxle equipped vehicles, install the driveplate.
8. On manual transaxle equipped vehicles, install the clutch disc and clutch cover.
9. Install the transaxle.

5E-FE Engine

❋❋ CAUTION

On models equipped with a Supplemental Restraint System (SRS), work must NOT be started until at least 90 seconds have passed from the time the ignition switch is turned to the LOCK position and the negative cable is disconnected from the battery.

1. Remove or disconnect the following:
   - Transaxle
   - With a manual transaxle, the clutch assembly and flywheel
   - With an automatic transaxle, the flexplate
   - The rear endplate
   - Rear oil seal retainer
2. Using a small prpbar, pry the rear oil seal retainer from the mating surfaces.
3. Using a drive punch or a hammer and a small prytool, drive the oil seal from the rear bearing retainer.

➡ **When removing the rear oil seal, be careful not to damage the seal mounting surface.**

4. Using a putty knife, clean the gasket mounting surfaces. Make certain that the contact surfaces are completely free of oil and foreign matter.

To install:

5. Clean the oil seal mounting surface.
6. Using multi-purpose grease, lubricate the new seal lips.
7. Using a seal installation tool or a smooth, round driver, tap the seal straight into the bore of the retainer.
8. Position a new gasket on the retainer and coat it lightly with gasket sealer. Fit the seal retainer into place on the motor; be careful when installing the oil seal over the crankshaft.

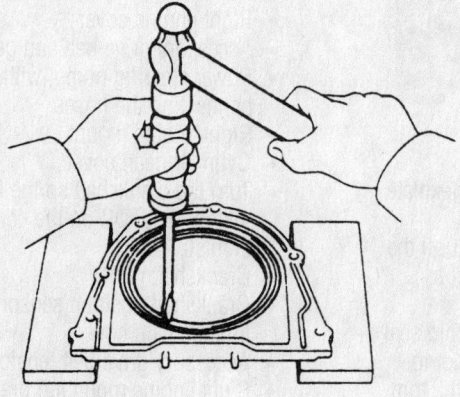

89553GG1

Always place the seal on blocks of wood, then tap the seal from the retainer—1MZ-FE engine

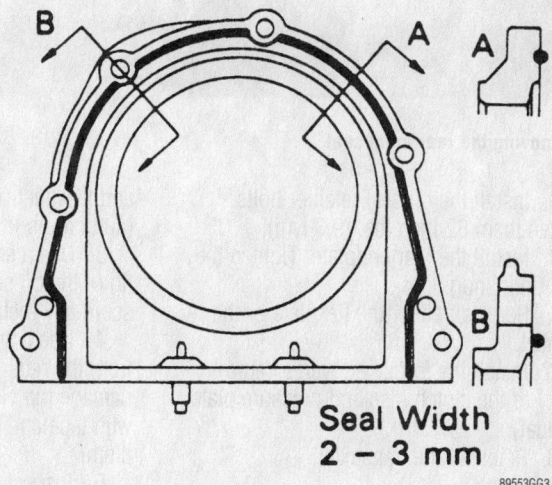

Seal Width 2 – 3 mm

89553GG3

Apply sealant to the rear oil seal retainer—1MZ-FE engine

Removing the rear endplate

85993234

Loosening the rear main seal attaching bolts

85993235

Removing the rear main seal

85993236

9. Install the oil seal retainer bolts. Tighten them 82 inch lbs. (9.3 Nm).

10. Install the rear endplate. Tighten the bolts until snug.

11. Reinstall either the flexplate or the flywheel.

12. Install the torque converter (automatic) or the clutch disc and pressure plate (manual).

13. Reinstall the transaxle.

2JZ-GE and 2JZ-GTE Engines

1. Remove the transmission.
2. Remove the clutch cover assembly

and flywheel (manual trans.) or the flexplate (auto. trans.).

3. Use a small, sharp knife to cut off the lip of the oil seal. Take great care not to score any metal with the knife.

4. Use a small prybar to pry the old seal from the retaining plate. Be careful not to damage the plate. Protect the tip of the tool with tape and pad the fulcrum point with cloth.

5. Inspect the crankshaft and seal lip contact surfaces for any sign of damage.

To install:

6. Apply a light coat of multi-purpose grease to the lip of a new oil seal. Loosely

fit the seal into place by hand, making sure it is not crooked.

7. Use a seal driver. Tap it into place until the surface of the seal is flush with the edge of the housing.

Timing Chain, Sprockets, Front Cover and seal

REMOVAL & INSTALLATION

1ZZ-FE Engine

1998–99

1. Before servicing the vehicle, refer to the precautions in the beginning of this section.

2. Drain the cooling system.

3. Remove or disconnect the following:
- Negative battery cable
- Right engine cover
- Accessory drive belt and generator
- Power steering pump, without disconnecting the hoses
- Right engine mount
- Cylinder head cover
- Turn the crankshaft so the No. 1 piston is at TDC on the compression stroke
- Crankshaft pulley
- Crankshaft position sensor from the timing chain cover
- Accessory drive belt tensioner
- Right engine mounting bracket
- Chain tensioner
- Water pump
- Timing chain cover
- Crankshaft angle sensor plate
- Timing chain tensioner slipper
- Timing chain and crankshaft timing sprocket

Align the No. 1 mark link with the mark on the crankshaft sprocket—1.8L (1ZZ-FE) engine

7923VG80

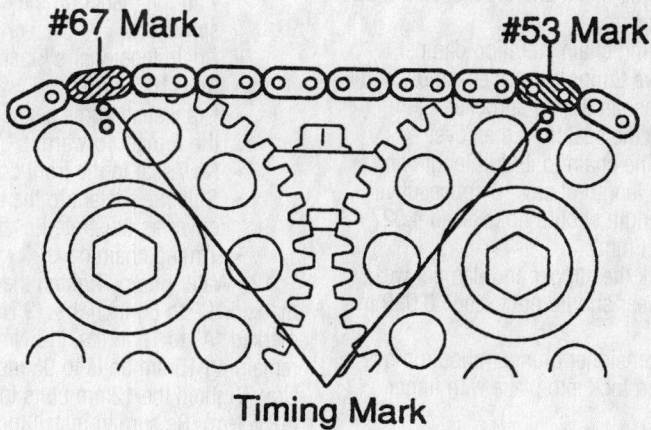

Align mark links Nos. 53 and 67 with the marks on the camshaft sprockets as shown—1.8L (1ZZ-FE) engine

7923VG81

To install:

4. Install or connect the following:
   - Crankshaft sprocket with timing chain. Be sure to align the No. 1 mark link with the mark on the sprocket
   - Timing chain on the camshaft sprockets. Align the Nos. 53 and 67 mark links with the marks on the camshaft sprockets
   - Chain tensioner slipper. Bolt: 14 ft. lbs. (18.5 Nm).
   - Crankshaft angle sensor plate
   - New seal in the front cover
   - Silicone sealant to the timing chain cover as illustrated
   - Timing chain cover

5. Water pump. Tighten the bolts marked "C" to 80 inch lbs. (9 Nm) and tighten the remaining bolts to 14 ft. lbs.

(18.5 Nm). Be sure to install the bolts in their original locations. Bolt lengths:
   a. A: 1.77 in. (45mm)
   b. B: 1.38 in. (35mm)
   c. C: 1.18 in. (30mm)
   d. D: 0.98 in. (25mm)

6. Install or connect the following:
   - Right engine mounting bracket. Bolts, with sealant applied: 35 ft. lbs. (47 Nm).
   - Drive belt tensioner. Bolt: 51 ft. lbs. (69 Nm). Nut: 21 ft. lbs. (29 Nm).
   - Crankshaft position sensor. Tighten to 80 inch lbs. (9 Nm).
   - Crankshaft pulley. Bolt: 102 ft. lbs. (138 Nm).

7. Release the ratchet pawl and compress the chain tensioner. Place the hook on the pin to keep the tensioner compressed.

8. Install the tensioner, using a new O-ring. Torque the bolts to 80 inch lbs. (9 Nm).

9. Turn the crankshaft counterclockwise and remove the hook from the pin. Turn the crankshaft clockwise and be sure the slipper is pushed by the plunger.

10. Check the valve timing by turning the crankshaft clockwise until the mark of the pulley is aligned with the mark on the timing chain cover. The marks on the camshaft sprockets should be facing each other as shown.

11. Install or connect the following:
    - Silicone sealant to the 2 areas where the timing chain cover meets the cylinder head.
    - Cylinder head cover. Bolts with washers in the sequence shown: 80

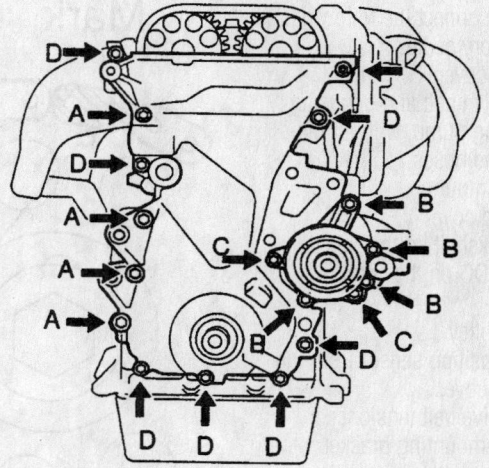

7923VG82

Timing chain cover bolt identification—1.8L (1ZZ-FE) engine

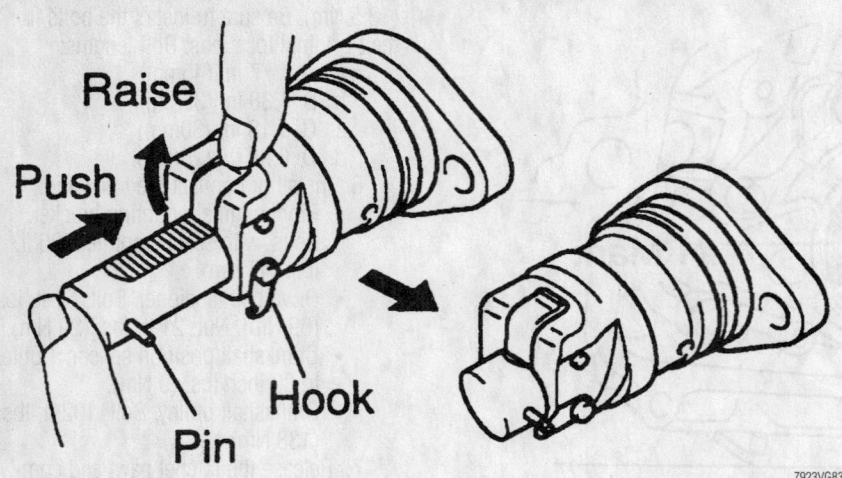

Compress the timing chain tensioner and place the hook on the pin—1.8L (1ZZ-FE) engine

inch lbs. (9 Nm). Bolts without washers: 96 inch lbs. (11 Nm).
- Right engine mount. Bolts: Marked **A**: 47 ft. lbs. (64 Nm). Marked **B**: 19 ft. lbs. (26 Nm). Nut: 38 ft. lbs. (52 Nm).
- Power steering pump
- Alternator and drive belt
- Right engine undercover
- Negative battery cable
- Washer tank

12. Fill the cooling system to the proper level.

13. Start the vehicle, check for leaks and repair if necessary.

2000

1. Before servicing the vehicle, refer to the precautions in the beginning of this section.

2. Drain the cooling system.

3. Remove or disconnect the following:
- Negative battery cable
- Right engine cover
- Accessory drive belt and generator
- Power steering pump, without disconnecting the hoses.
- Right engine mount
- Cylinder head cover
- Turn the crankshaft so the No. 1 piston is at TDC on the compression stroke.
- Crankshaft pulley
- Crankshaft position sensor from the timing chain cover.
- Accessory drive belt tensioner.
- Right engine mounting bracket
- Chain tensioner
- Water pump
- Timing chain cover
- Crankshaft angle sensor plate
- Timing chain tensioner slipper

- Timing chain and crankshaft timing sprocket.
- Timing chain vibration damper
- Valve timing control assembly and camshaft timing sprocket

4. Drive the seal from the cover.

5. Pull the chain to its full length and measure the length of any 16 consecutive links. The length should not exceed 4.827 inches (122.6mm).

6. Check the slipper and damper wear. Maximum wear should not exceed 0.039 in. (1mm).

7. The tensioner plunger should move smoothly and lock into place with finger pressure.

To install:

8. Apply engine oil from the tip of the intake camshaft, back to 16mm.

9. Align the timing mark on the valve timing controller with the knock pin and gently push the valve timing controller onto the camshaft.

10. Set the No.1 piston to TDC compression. The key on the crankshaft should be at 12 o'clock.

11. Install or connect the following:
- Sprockets. Torque the bolt to 33 ft. lbs. (45 Nm). Turn the camshafts to align the point marks on the sprockets.
- Chain damper. Bolts: 96 inch lbs. (11 Nm).
- Timing chain and crankshaft sprocket. Be sure to align the yellow chain link with the mark on the crankshaft sprocket.
- Timing chain on the camshaft sprockets. Align the yellow links with the marks on the camshaft sprockets.
- Chain tensioner slipper. Bolt: 14 ft. lbs. (18.5 Nm).
- Crankshaft angle sensor plate with the **F** mark forward
- New seal in the front cover
- Silicone sealant to the timing chain cover as illustrated
- Timing chain cover

12. Water pump. Tighten the 10mm bolts marked "C" to 80 inch lbs. (9 Nm), those marked "A" to 10 ft. lbs. (13 Nm), and the remaining 10mm bolts to 96 inch lbs. (11 Nm). Tighten the 12mm bolts to 14 ft. lbs. (18.5 Nm). Be sure to install the bolts in their original locations. Bolt lengths:

a. A: 1.77 in. (45mm)

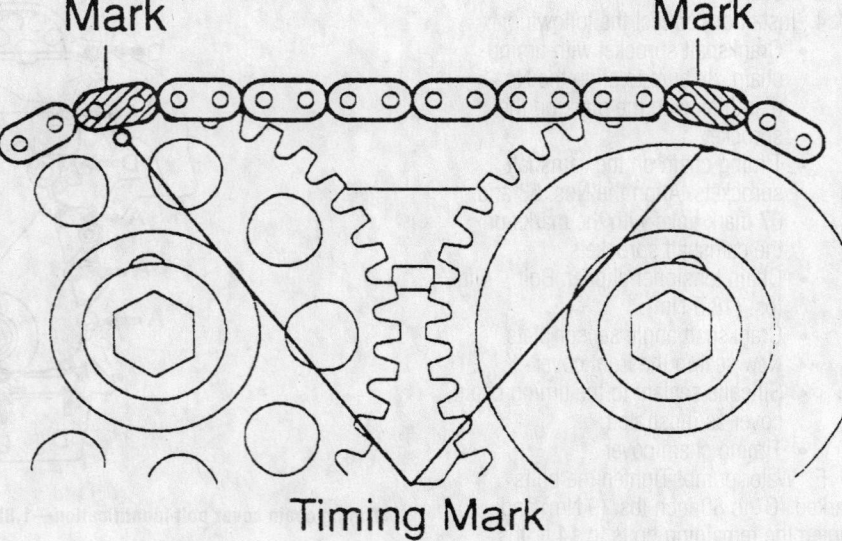

Timing chain link marks—1ZZ-FE and 2ZZ-GE engines

b. B: 1.38 in. (35mm)

c. C: 1.18 in. (30mm)

d. D: 0.98 in. (25mm)

13. With a Torx wrench, tighten the stud bolt to 82 inch lbs. (9.3 Nm).

14. Install or connect the following:

- Right engine mounting bracket. Bolts, with sealant applied: 35 ft. lbs. (47 Nm).
- Drive belt tensioner. Bolt: 51 ft. lbs. (69 Nm). Nut: 21 ft. lbs. (29 Nm).
- Crankshaft position sensor. Tighten to 80 inch lbs. (9 Nm).
- Crankshaft pulley. Bolt: 102 ft. lbs. (138 Nm).

15. Release the ratchet pawl and compress the chain tensioner. Place the hook on the pin to keep the tensioner compressed.

16. Install the tensioner, using a new O-ring. Torque the bolts to 80 inch lbs. (9 Nm).

17. Turn the crankshaft counterclockwise and remove the hook from the pin. Turn the crankshaft clockwise and be sure the slipper is pushed by the plunger.

18. Check the valve timing by turning the crankshaft clockwise until the mark of the pulley is aligned with the mark on the timing chain cover. The marks on the camshaft sprockets should be facing each other as shown.

19. Install or connect the following:

- Silicone sealant to the 2 areas where the timing chain cover meets the cylinder head.
- Cylinder head cover. Bolts with washers in the sequence shown: 80 inch lbs. (9 Nm). Bolts without washers: 96 inch lbs. (11 Nm).
- Right engine mount. Bolts and nuts: 38 ft. lbs. (52 Nm).
- Power steering pump
- Alternator and drive belt
- Right engine undercover
- Negative battery cable
- Washer tank

20. Fill the cooling system to the proper level.

21. Start the vehicle, check for leaks and repair if necessary.

1NZ-FE

1. Before servicing the vehicle, refer to the precautions in the beginning of this section.

2. Drain the cooling system.

3. Remove or disconnect the following:

- Negative battery cable
- Right front wheel
- Alternator

- Power steering pump
- Right engine mount insulator. Use a jack and wood block for support.
- With A/C, the bolt holding the liquid tube to the insulator
- Ignition coils
- Cylinder head cover

4. Place No.1 cylinder on TDC compression. Make sure that the timing marks on the camshaft sprockets and valve timing controller assembly are facing UP (12 o'clock). If not, turn the crankshaft 360 degrees to align the marks.

5. Remove or disconnect the following:

- Crankshaft pulley bolt
- Pulley and pin
- Crankshaft position sensor
- Right engine mount bracket
- Water pump
- Oil control valve
- 13 bolts, 1 nut and 1 stud bolt. Pry the cover off.
- 2 O-rings from the block and pan
- Chain tensioner
- Tensioner slipper
- Chain vibration damper
- Chain

6. Drive the seal from the cover.

7. Pull the chain to its full length and measure the length of any 16 consecutive links. The length should not exceed 4.85 inches (123.2mm).

8. Check the slipper and damper wear. Maximum wear should not exceed 0.039 in. (1mm).

9. The tensioner plunger should move smoothly and lock into place with finger pressure.

To install:

10. Set the crankshaft at 140 degrees ATDC. Set the camshaft sprockets at 20 degrees ATDC; then, reset the crankshaft to 20 degrees ATDC.

11. Install or connect the following:

- New seal, driven into place until flush with the cover edge
- Vibration damper. Torque the bolts to 80 inch lbs. (9 Nm).
- Timing chain

➡**A new chain will have 3 marked links to align with the 3 sprockets, as shown in the accompanying illustration.**

- Slipper

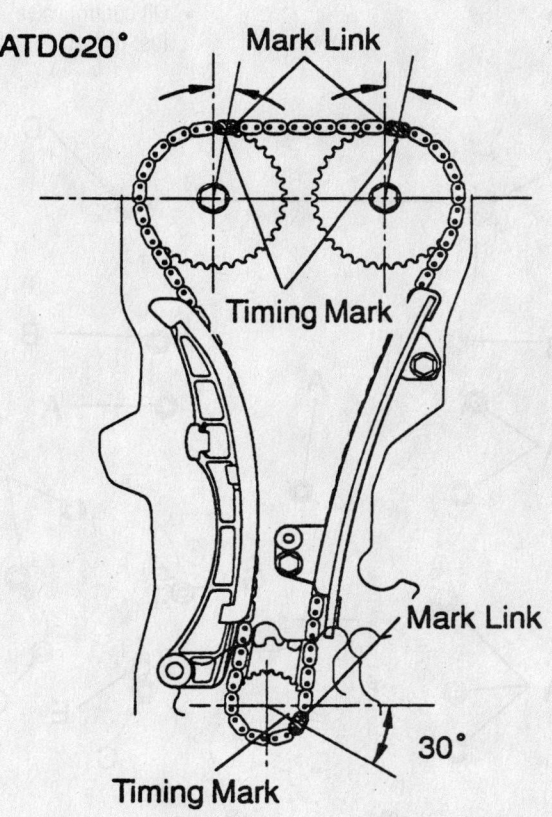

Timing chain installation—1NZ-FE engine

9307WG78

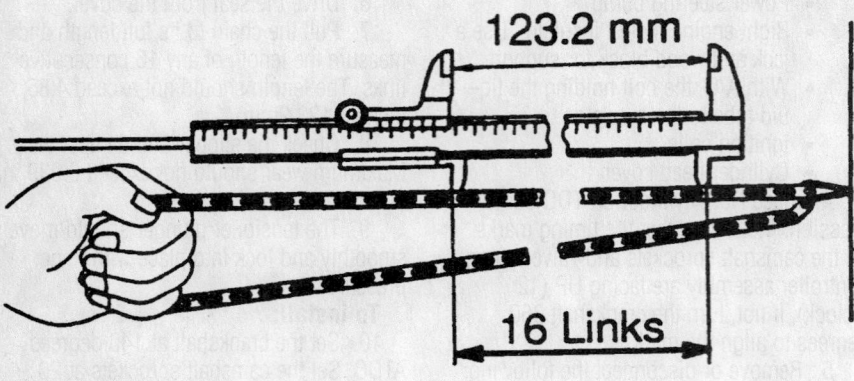

123.2 mm

16 Links

9307WG77

Measuring the timing chain—1NZ-FE engine

- Tensioner. Torque the bolts to 80 inch lbs. (9 Nm).
- Cover, using a 4–5mm bead of RTV sealer and new O-rings to the block and pan

12. Uniformly tighten the bolts in several passes, using the accompanying illustration, to:
- A, C, E and G: 96 inch lbs. (11 Nm)
- B, D and F: 18 ft. lbs. (24 Nm)

13. Bolt lengths are as follows:
- A: 20mm
- B: 30mm
- C: 35mm
- D: 20mm
- E: 35mm

14. Install or connect the following:
- Right engine mount bracket. Coat all but the end 2 threads of the bolt with RTV sealer. Torque the bolt to 41 ft. lbs. (55 Nm).
- Crankshaft position sensor. Torque bolt A to 66 inch lbs. (7.5 Nm); bolts B to 96 inch lbs. (11 Nm).
- Oil control valve. Torque: 71 inch lbs. (8 Nm).

- Crankshaft pulley and pin. Torque the bolt to 94 ft. lbs. (128 Nm).
- Cylinder head cover with RTV gasket material at the 2 locations shown. Uniformly tighten the bolts and nuts, in several passes, to 84 inch lbs. (10 Nm).
- PCV hoses
- Ignition coils
- Right engine mount insulator. Torque the bolts and nuts to 35 ft. lbs. (47 Nm).
- Power steering pump
- Alternator
- Right under cover
- Wheel
- Negative battery cable

15. Fill the cooling system to the proper level.

16. Start the vehicle, check for leaks and repair if necessary.

2ZZ-GE Engine

1. Before servicing the vehicle, refer to the precautions in the beginning of this section.

2. Drain the cooling system.

3. Remove or disconnect the following:
- Negative battery cable
- Right engine cover
- Accessory drive belt and generator
- Power steering pump, without disconnecting the hoses.
- Right engine mount
- Cylinder head cover
- Turn the crankshaft so the No. 1 piston is at TDC on the compression stroke.
- Crankshaft pulley
- Crankshaft position sensor from the timing chain cover.
- Accessory drive belt tensioner.
- Right engine mounting bracket
- Chain tensioner
- Water pump
- Timing chain cover
- Crankshaft angle sensor plate
- Timing chain tensioner slipper
- Timing chain and crankshaft timing sprocket.
- Timing chain vibration damper
- Valve timing control assembly and camshaft timing sprocket

4. Drive the seal from the cover.

5. Pull the chain to its full length and measure the length of any 16 consecutive links. The length should not exceed 4.827 inches (122.6mm).

6. Check the slipper and damper wear. Maximum wear should not exceed 0.039 in. (1mm).

F   D

B   B

A   A

A

E

G

A

E   G

E

C

A

9307WG79

Timing cover bolt installation—1NZ-FE engine

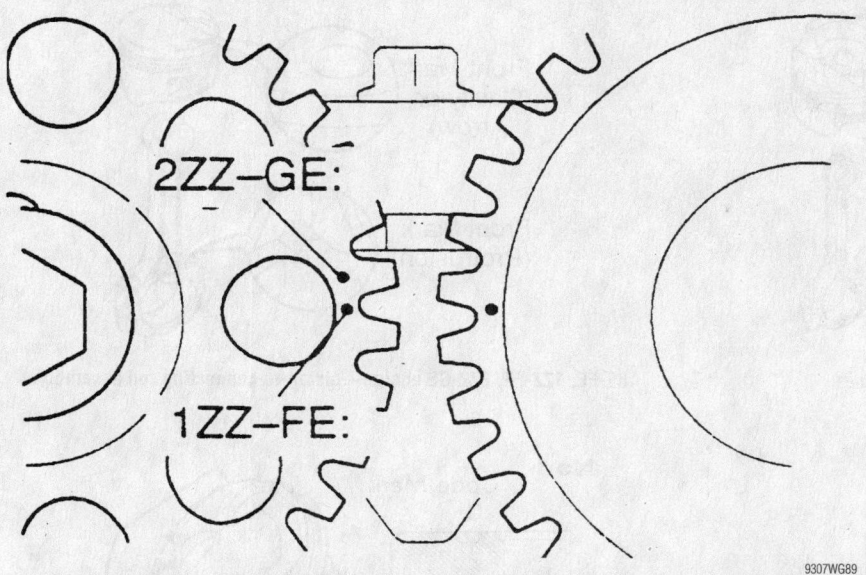

Timing mark identification at TDC compression—1ZZ-FE and 2ZZ-GE engines

7. The tensioner plunger should move smoothly and lock into place with finger pressure.

To install:

8. Apply engine oil from the tip of the intake camshaft, back to 16mm.

9. Align the timing mark on the valve timing controller with the knock pin and gently push the valve timing controller onto the camshaft.

10. Set the No.1 piston to TDC compression. The key on the crankshaft should be at 12 o'clock.

11. Install or connect the following:
- Sprockets. Torque the bolt to 33 ft. lbs. (45 Nm). Turn the camshafts to align the point marks on the sprockets.
- Chain damper. Bolts: 96 inch lbs. (11 Nm).
- Timing chain and crankshaft sprocket. Be sure to align the yellow chain link with the mark on the crankshaft sprocket.
- Timing chain on the camshaft sprockets. Align the yellow links with the marks on the camshaft sprockets.
- Chain tensioner slipper. Bolt: 14 ft. lbs. (18.5 Nm).
- Crankshaft angle sensor plate with the **F** mark forward
- New seal in the front cover
- Silicone sealant to the timing chain cover as illustrated
- Timing chain cover

12. Install the water pump. Tighten the 10mm bolts marked "C" to 80 inch lbs. (9 Nm), those marked "A" to 10 ft. lbs. (13 Nm), and the remaining 10mm bolts to 96 inch lbs. (11 Nm). Tighten the 12mm bolts to 14 ft. lbs. (18.5 Nm). Be sure to install the bolts in their original locations. Bolt lengths:
- a. A: 1.77 in. (45mm)
- b. B: 1.38 in. (35mm)
- c. C: 1.18 in. (30mm)
- d. D: 0.98 in. (25mm)

13. With a Torx wrench, tighten the stud bolt to 82 inch lbs. (9.3 Nm).

14. Install or connect the following:
- Right engine mounting bracket. Bolts, with sealant applied: 35 ft. lbs. (47 Nm).
- Drive belt tensioner. Bolt: 51 ft. lbs. (69 Nm). Nut: 21 ft. lbs. (29 Nm).

- Crankshaft position sensor. Tighten to 80 inch lbs. (9 Nm).
- Crankshaft pulley. Bolt: 102 ft. lbs. (138 Nm).

15. Release the ratchet pawl and compress the chain tensioner. Place the hook on the pin to keep the tensioner compressed.

16. Install the tensioner, using a new O-ring. Torque the bolts to 80 inch lbs. (9 Nm).

17. Turn the crankshaft counterclockwise and remove the hook from the pin. Turn the crankshaft clockwise and be sure the slipper is pushed by the plunger.

18. Check the valve timing by turning the crankshaft clockwise until the mark of the pulley is aligned with the mark on the timing chain cover. The marks on the camshaft sprockets should be facing each other as shown.

19. Install or connect the following:
- Silicone sealant to the 2 areas where the timing chain cover meets the cylinder head.
- Cylinder head cover. Bolts with washers in the sequence shown: 80 inch lbs. (9 Nm). Bolts without washers: 96 inch lbs. (11 Nm).
- Right engine mount. Bolts and nuts: 38 ft. lbs. (52 Nm).
- Power steering pump
- Alternator and drive belt
- Right engine undercover
- Negative battery cable
- Washer tank

20. Fill the cooling system to the proper level.

21. Start the vehicle, check for leaks and repair if necessary.

Piston and Ring Positioning

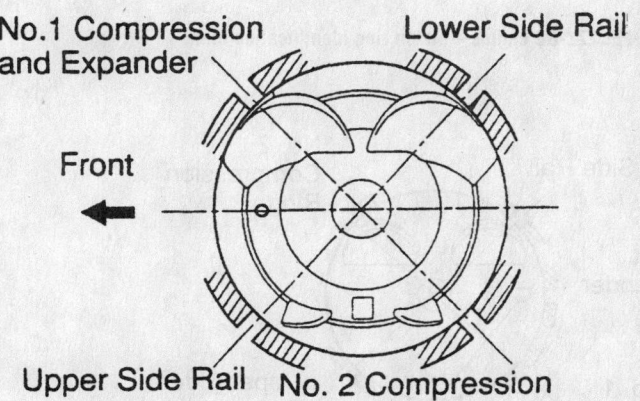

Piston ring positioning—1NZ-FE engine

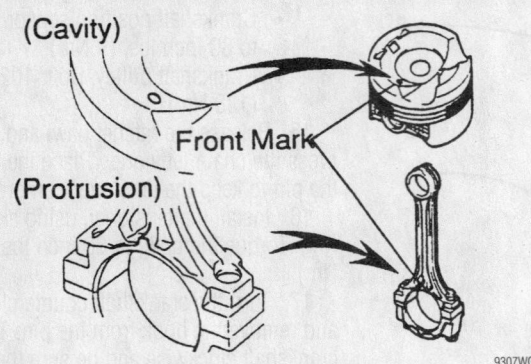

Piston and connecting rod positioning—1NZ-FE engine

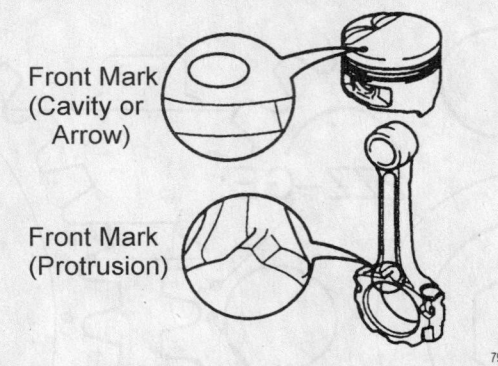

5E-FE, 1ZZ-FE, 2ZZ-GE engine—piston-to-connecting rod assembly

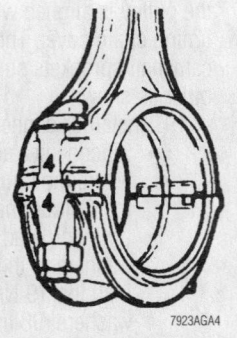

Before removing the caps from the connecting rods, be sure to matchmark them as shown

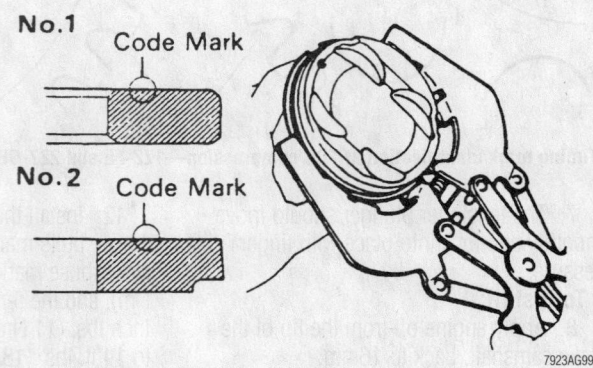

3S-GTE engine—compression ring positioning

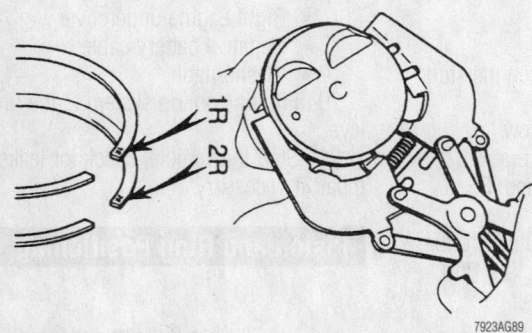

5E-FE, 1ZZ-FE, 2ZZ-GE engine—piston ring identification mark locations

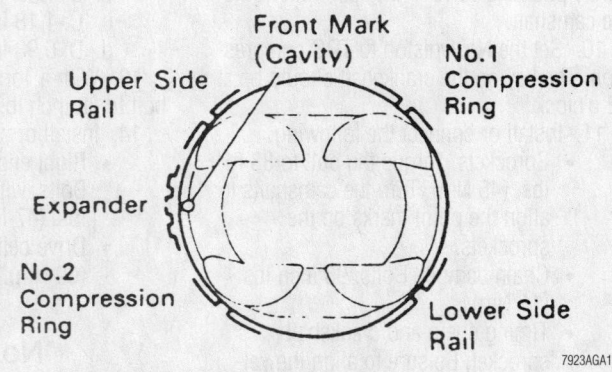

3S-GTE engine—piston ring end-gap spacing

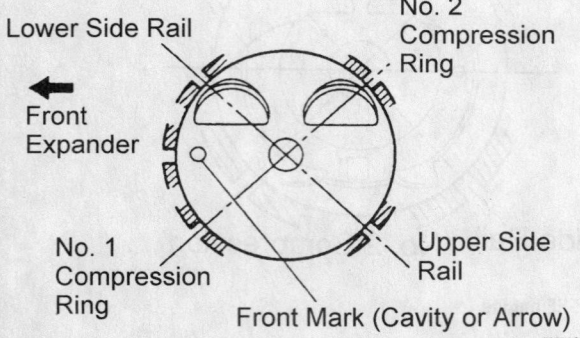

5E-FE, 1ZZ-FE, 2ZZ-GE engine—piston ring end-gap spacing

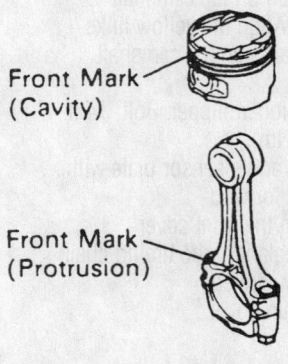

3S-GTE and 5S-FE engines—piston-to-connecting rod assembly

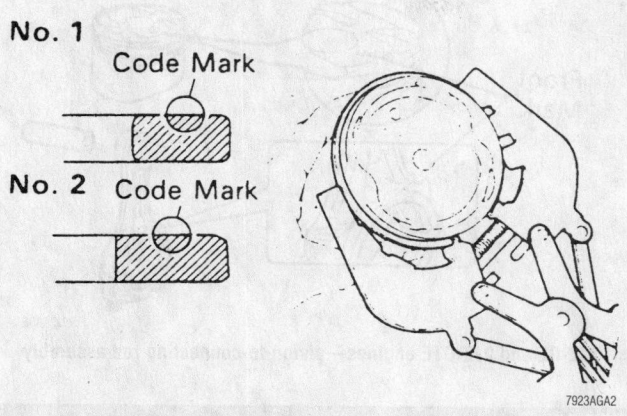

5S-FE engine—compression ring positioning

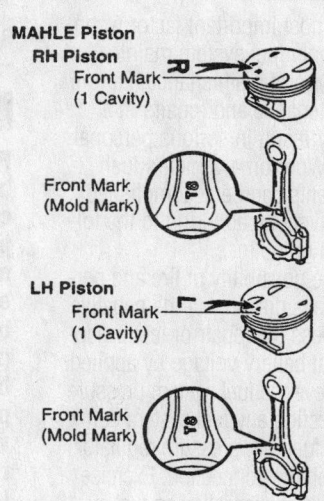

1MZ-FE engine—piston ring end-gap spacing

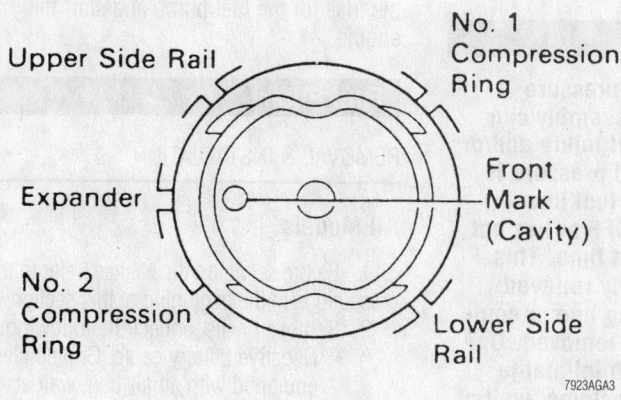

5S-FE engine—piston ring end-gap spacing

Avalon 1MZ-FE engine—piston-to-connecting rod assembly

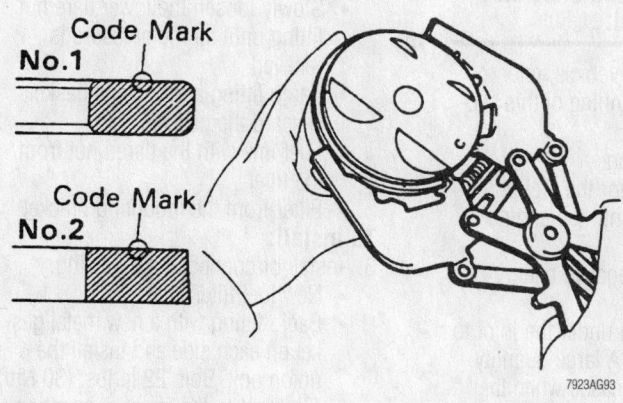

1MZ-FE, 2JZ-GE and 2JZ-GTE engines—compression ring positioning

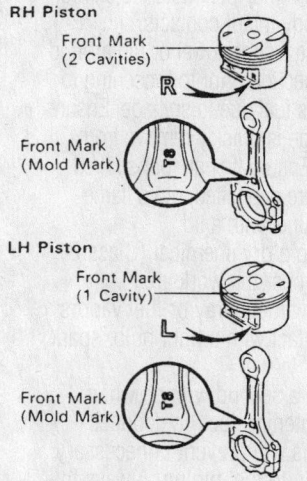

Camry 1MZ-FE engine—piston-to-connecting rod assembly

Timing belt service is covered in Section 3 of this manual

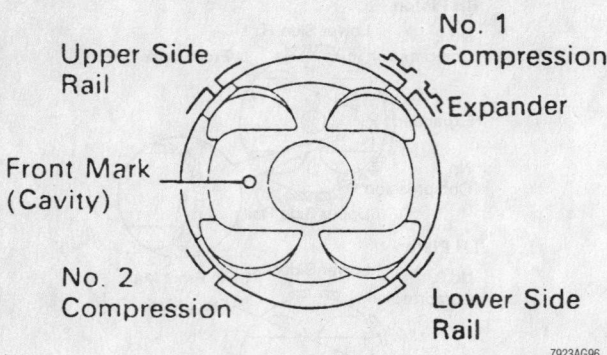

2JZ-GE and 2JZ-GTE engines—piston ring end-gap spacing

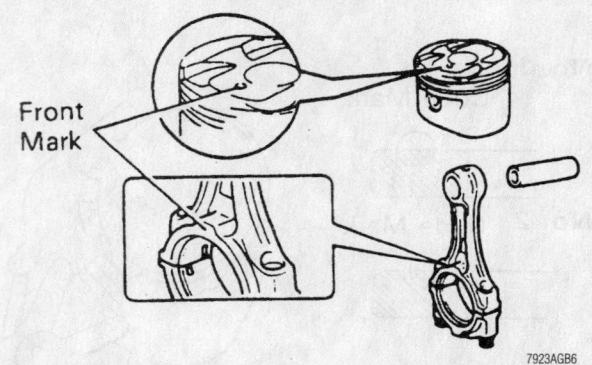

2JZ-GE and 2JZ-GTE engines—piston-to-connecting rod assembly

FUEL SYSTEM

Fuel System Service Precautions

Safety is the most important factor when performing not only fuel system maintenance, but any type of maintenance. Failure to conduct maintenance and repairs in a safe manner may result in serious personal injury or death. Work on a vehicle's fuel system components can be accomplished safely and effectively by adhering to the following rules and guidelines.

- To avoid the possibility of fire and personal injury, always disconnect the negative battery cable unless the repair or test procedure requires that battery voltage by applied.
- Always relieve the fuel system pressure prior to disconnecting any fuel system component (injector, fuel rail, pressure regulator, etc.) fitting or fuel line connection. Exercise extreme caution whenever relieving fuel system pressure, to avoid exposing skin, face and eyes to fuel spray. Please be advised that fuel under pressure may penetrate the skin or any part of the body that it contacts.
- Always place a shop towel or rag around the fitting or connection prior to loosening to absorb any excess fuel due to spillage. Ensure that all fuel spillage is quickly remove from engine surfaces. Ensure that all fuel-soaked cloths or towels are deposited into a flame-proof waste container with a lid.
- Always keep a dry chemical (Class B) fire extinguisher near the work area.
- Do not allow fuel spray or fuel vapors to come into contact with a light bulb, spark or open flame.
- Always use a second wrench when loosening or tightening fuel line connections fittings. This will prevent unnecessary stress and torsion to fuel piping. Always follow the proper torque specifications.
- Always replace worn fuel fitting O-rings with new ones. Do not substitute fuel hose where rigid pipe is installed.

Fuel System Pressure

RELIEVING

✳✳ CAUTION

Failure to relieve fuel pressure before repairs or disassembly can cause serious personal injury and/or property damage. Fuel pressure is maintained within the fuel lines, even if the engine is OFF or has not been run in a period of time. This pressure must be safely relieved before any fuel-bearing line or component is loosened or removed. On vehicles equipped with inflatable restraints or air bag systems, wait at least 90 seconds after disconnecting the battery cable before performing any other work. The back-up power will keep the restraint system energized for a period of time after the battery is disconnected.

1. Before servicing the vehicle, refer to the precautions in the beginning of this section.
2. Perform the following:
- Remove the fuse for the fuel pump
- Start the engine until the engine stalls
- Disconnect the negative battery cable
- Place a catch-pan under the joint to be disconnected. A large quantity of fuel may be released when the joint is opened
- Wear eye or full face protection
- Place a shop towel over the area and slowly release the joint using a wrench of the correct size.
- Allow the any fuel left in the line to bleed off slowly before fully disconnecting the joint.
- Plug the opened lines
3. After connecting fuel lines, install the fuse for the fuel pump and start the engine.

Fuel Filter

REMOVAL & INSTALLATION

All Models

1. Before servicing the vehicle, refer to the precautions in the beginning of this section.
2. Remove or disconnect the following:
- Negative battery cable. On vehicles equipped with an air bag, wait at least 90 seconds before proceeding
- Protective shield for the fuel filter
- If necessary, air cleaner hose and cap
- If necessary, charcoal canister.
- Slowly loosen the lower flare nut fitting until all the pressure is relieved
- Banjo fitting and 2 metal gaskets. Discard the gaskets
- Fuel line with the flared nut from the filter
- Filter from the mounting bracket

To install:
3. Install or connect the following:
- New fuel filter
- Banjo fitting with a new metal gasket on each side and install the union bolt. Bolt: 22 ft. lbs. (30 Nm).
- Flare nut to the lower connection. Nut: 22 ft. lbs. (30 Nm).
- Charcoal canister
- Air cleaner hose and cap
- Protective shield
- Negative battery cable

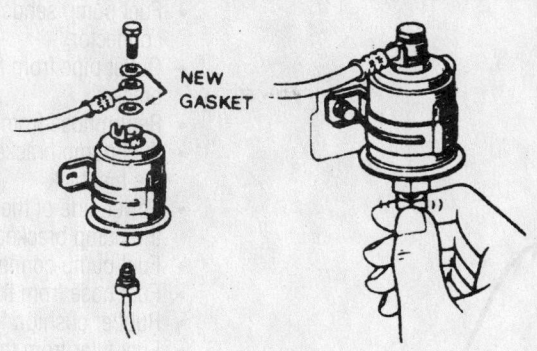

7923VG84

Always use new gaskets when replacing a fuel filter—Tercel shown

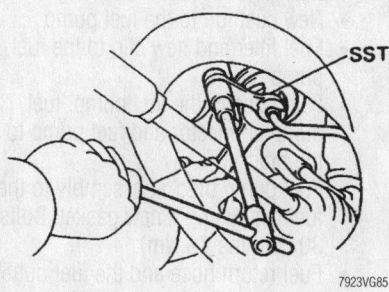

7923VG85

A line wrench with an extension may be needed to loosen the inlet line at the filter—Corolla

Fuel Pump

REMOVAL & INSTALLATION

Avalon, Camry and Camry Solara

1. Before servicing the vehicle, refer to the precautions in the beginning of this section.
2. Relieve the fuel system pressure.
3. Remove or disconnect the following:
   • Negative battery cable. On vehicles equipped with an air bag, wait at least 90 seconds before proceeding.
   • Fuel tank

➡**On some models it will be necessary to remove the rear seat cushion.**

   • Floor service hole cover
   • Electrical connector at the fuel pump assembly
   • Fuel outlet pipe from the fuel pump bracket
   • Return hose from the fuel pump bracket
   • Fuel pump bracket assembly from the fuel tank by removing the bolts
   • Pump bracket assembly
   • Fuel pump from the fuel bracket

To install:
4. Install or connect the following:
   • Fuel pump to the fuel bracket
   • Pump bracket assembly

   • Fuel pump to the fuel tank. Bolts: 35 inch lbs. (4 Nm).
   • Return hose to the fuel pump bracket
   • Outlet pipe to the fuel pump bracket. Tighten to 21 ft. lbs. (28 Nm).
   • Service hole cover to the fuel tank
   • Fuel pump connector
   • Rear seat cushion
   • Fuel tank
   • Negative battery cable

Echo

1. Before servicing the vehicle, refer to the precautions in the beginning of this section.
2. Relieve the fuel system pressure.
3. Remove or disconnect the following:

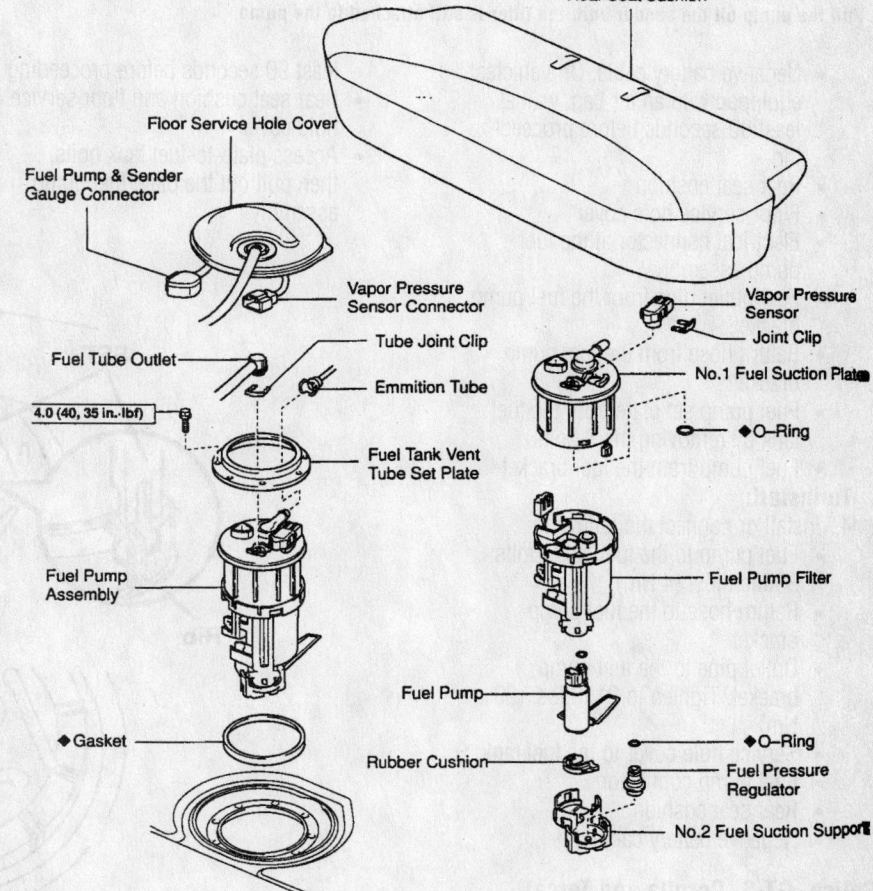

Rear Seat Cushion

Floor Service Hole Cover

Fuel Pump & Sender Gauge Connector

Vapor Pressure Sensor Connector

Fuel Tube Outlet

Tube Joint Clip

4.0 (40, 35 in. lbf)

Emmition Tube

Fuel Tank Vent Tube Set Plate

Fuel Pump Assembly

◆Gasket

Vapor Pressure Sensor

Joint Clip

No.1 Fuel Suction Plate

◆O–Ring

Fuel Pump Filter

Fuel Pump

Rubber Cushion

◆O–Ring

Fuel Pressure Regulator

No.2 Fuel Suction Support

N·m (kgf·cm, ft·lbf) : Specified torque
◆ Non–reusable part

9307WG83

Fuel pump removal—Echo

Heater Core replacement is covered in Section 2 of this manual

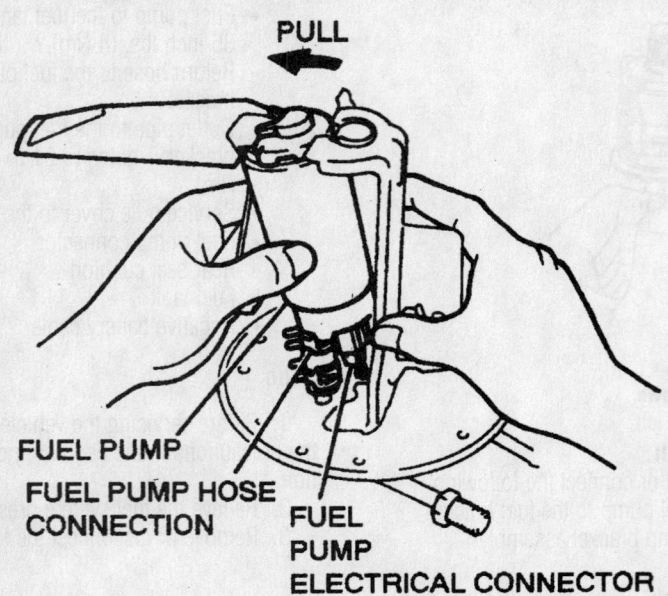

PULL

FUEL PUMP

FUEL PUMP HOSE
CONNECTION

FUEL
PUMP
ELECTRICAL CONNECTOR

7923VG86

Pull the pump off the sender unit; the filter is still attached to the pump

- Negative battery cable. On vehicles equipped with an air bag, wait at least 90 seconds before proceeding.
- Rear seat cushion
- Floor service hole cover
- Electrical connector at the fuel pump assembly
- Fuel outlet pipe from the fuel pump bracket
- Return hose from the fuel pump bracket.
- Fuel pump set plate from the fuel tank by removing the 8 bolts
- Fuel pump from the fuel bracket

To install:

4. Install or connect the following:
- Fuel pump to the fuel tank. Bolts: 35 inch lbs. (4 Nm).
- Return hose to the fuel pump bracket
- Outlet pipe to the fuel pump bracket. Tighten to 21 ft. lbs. (28 Nm).
- Service hole cover to the fuel tank
- Fuel pump connector
- Rear seat cushion
- Negative battery cable

Celica, GT-S, Corolla and Tercel

1. Before servicing the vehicle, refer to the precautions in the beginning of this section.

2. Relieve the fuel system pressure.

3. Remove or disconnect the following:
- Negative battery cable. On vehicles equipped with an air bag, wait at least 90 seconds before proceeding
- Rear seat cushion and floor service hole cover
- Access plate-to-fuel tank bolts, then pull out the plate/fuel pump assembly

- Fuel pump sender and fuel pump connector
- Outlet pipe from the fuel pump bracket
- Return hose from the pump bracket
- Fuel pump bracket assembly from the fuel tank
- Lower side of the fuel pump from the pump bracket
- Fuel pump connector
- Fuel hose from the fuel pump
- Rubber cushion from the pump
- Fuel filter from the pump by removing the small clip

To install:

4. Install or connect the following:
- New cushion to the fuel pump
- Fuel filter and new clip to the fuel pump
- Fuel hose to the fuel pump, fuel pump connector and fuel pump to the bracket
- Fuel pump bracket assembly to the fuel tank using a new gasket. Bolts: 30 inch lbs. (3 Nm).
- Fuel return hose and the fuel outlet pipe to the fuel pump bracket
- Fuel pump and fuel pump sender connector
- Fuel tank

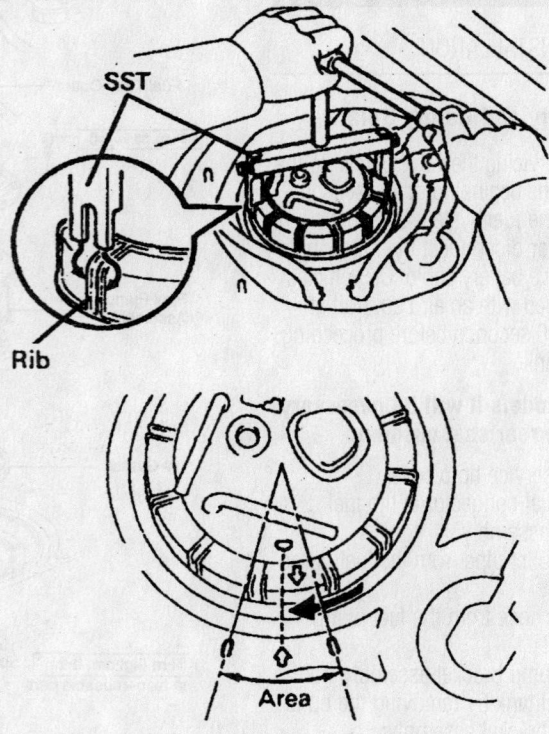

SST

Rib

Area

7923VG87

Tighten the fuel pump retainer until the arrow mark on the retainer is within the lines on the fuel tank—Supra

- Negative battery cable
- Floor service hole cover and rear seat cushion

Supra

1. Before servicing the vehicle, refer to the precautions in the beginning of this section.
2. Relieve the fuel system pressure.
3. Remove or disconnect the following:
   - Negative battery cable. On vehicles equipped with an air bag, wait at least 90 seconds before proceeding
   - Floor carpet, spare wheel cover, spare wheel, service hole cover
   - Fuel pump electrical connector from the fuel pump
   - Outlet hose from the fuel pump
   - Fuel return hose from the fuel pump
   - Fuel breather clamp
   - Loosen retainer to the fuel pump
   - Fuel return hose from the return port of the fuel pump bracket
   - Retainer, fuel pump, sender gauge assembly, and the gasket as a unit

To install:

4. Install or connect the following:
   - New gasket to the fuel pump
   - Fuel pump and sender gauge assembly into the fuel tank
   - Align the arrow marks of the fuel pump bracket and the fuel tank
   - Fuel pump retainer until the arrow mark on the retainer is within the lines on the fuel tank
   - Retainer clamp to the fuel pump
   - Fuel pump electrical connector to the fuel pump
   - Outlet hose with 2 new gaskets. Bolt: 22 ft. lbs. (29 Nm).
   - Fuel return hose to the fuel pump
   - Fuel breather hose to the fuel pump
   - Negative battery cable
   - Service hole cover
   - Spare wheel, spare wheel cover, floor carpet

Fuel Injectors

REMOVAL & INSTALLATION

Avalon and Camry

5S-FE ENGINE

1. Before servicing the vehicle, refer to the precautions in the beginning of this section.
2. Drain the cooling system.

Loosening and removing the pulsation damper—5S-FE engine

3. Remove or disconnect the following:
   - Negative battery cable
   - Accelerator cable from the throttle body
   - If equipped with automatic transmission, the throttle cable
   - Air intake temperature sensor connector
   - Cruise control actuator cable from the clamp on the resonator
   - Air cleaner
   - Wiring from the throttle position sensor and the ISC valve
   - Hoses from the PCV, EGR vacuum modulator and EVAP VSV
   - Throttle body
   - PS vacuum hoses
   - Hoses from the EVAP Bi-metal Vacuum Switching Valve (BVSV)
   - EGR valve and the vacuum modulator
   - Vacuum sensor hose at the air intake chamber
   - Brake booster vacuum hose and the vacuum sensing hose
   - With air conditioning, the magnet switch VSV wiring
   - Ground straps from the intake manifold
   - Knock sensor and EGR Vacuum Switching Valve (VSV) wiring
   - Engine wire harness
   - Stays or supports holding the air intake chamber and the intake manifold
   - Intake manifold
   - Wiring from each injector
   - Fuel inlet pipe

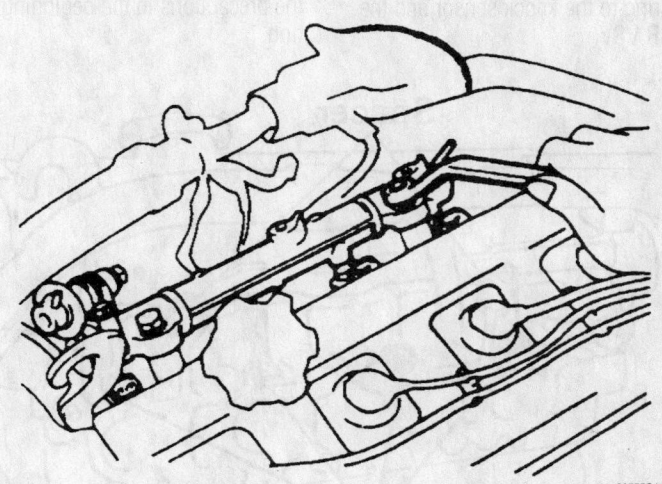

Unbolt and remove the fuel delivery pipe and injectors as an assembly

Brake service is covered in Section 4 of this manual

- Fuel return hose
- Delivery pipe or fuel rail along with the injectors
- Insulators and rail spacers from the head
- Injectors from the fuel rail
- O-ring and grommet from each injector

To install:

4. Install or connect the following:
- New grommet on each injector
- New O-rings, coated with gasoline, on each injector
- Injectors into the fuel rail while turning each left and right. After installation, check that the injectors turn freely in place; if not, remove the injector and inspect the O-ring for damage or deformation.
- New insulators and spacers on the head
- Fuel rail and injectors; check that the injectors still turn freely in position. Position the injector connectors upward
- Retaining bolts, tightening them to 9 ft. lbs. (13 Nm)
- Fuel return hose
- Fuel inlet pipe and pulsation damper to the delivery pipe. Use new gaskets; tighten the union bolt to 25 ft. lbs. (34 Nm).
- Wiring to each injector
- Intake manifold. Nuts and bolts evenly in several passes to 14 ft. lbs. (19 Nm).
- Air chamber and manifold stays. 14mm bolt: 31 ft. lbs. (42 Nm) and the 12mm bolt to 16 ft. lbs. (22 Nm).
- Engine wire harness
- Wiring to the knock sensor and the EGR VSV

- Both engine ground straps to the intake manifold
- A/C magnet switch wiring
- Hoses for the vacuum sensor, brake booster and vacuum sensing hose
- EGR valve and vacuum modulator. Use new gaskets. Tighten the union nut to 43 ft. lbs. (59 Nm) and the bolt to 9 ft. lbs. (13 Nm).
- Hoses to the charcoal canister and EGR VSV
- Wiring to the EGR temperature sensor if it was removed
- Vacuum hoses to the EVAP BVSV
- PS vacuum hoses
- Throttle body. Bolts, evenly and alternately, to 14 ft. lbs. (19 Nm).

➡ **The upper mounting bolts are shorter than the lower mounting bolts. Make certain the bolts are correctly placed before tightening.**

- PCV, EGR vacuum modulator and EGR VSV hoses to the throttle body
- Wiring for the throttle position sensor
- Air cleaner cap, resonator and intake hose
- Wiring to the air intake temperature sensor
- Cruise control actuator cable
- Throttle control cable
- Accelerator cable
- Negative battery cable

5. Fill the cooling system to the proper level.

6. Start the vehicle, check for leaks and repair if necessary.

1MZ-FE ENGINE

1. Before servicing the vehicle, refer to the precautions in the beginning of this section.

2. Properly relieve the fuel system pressure.

3. Drain the cooling system.

4. Remove or disconnect the following:
- Negative battery cable. Work must be started approximately 90 seconds or longer after the negative battery cable has been disconnected, if equipped with an air bag
- Accelerator and throttle cables
- Air cleaner assembly
- V-bank cover
- Emission valve control set
- No. 2 EGR pipe
- Hydraulic motor pressure pipe from the water inlet and air inlet chamber
- Air intake chamber assembly
- Injector wiring
- Air assist pipe from the bracket on the No. 1 fuel pipe
- Air assist hoses from the intake manifold
- Fuel return hose from the No. 1 fuel pipe
- Fuel inlet hose for the fuel filter
- 2 union bolts holding the No. 2 fuel pipe to the delivery pipes
- Fuel return hose from the fuel pressure regulator
- Union bolt for the right hand delivery pipe, 2 gaskets, 2 bolts, left hand delivery pipe together with the 3 injectors and the No. 2 fuel pipe
- Union bolt for the delivery pipe and 2 gaskets from the No. 2 fuel pipe
- The 3 bolts, right hand delivery pipe together with the 3 injectors and the No. 1 fuel pipe
- The 4 spacers from the intake manifold
- The 6 injectors from the delivery pipes
- The two O-rings and two grommets from each injector

To install:

5. Install or connect the following:
- 2 new grommets to each injector
- New O-rings, with a light coat of fuel, to each injector
- Injectors
- The 4 spacers on the intake manifold
- Right hand delivery pipe and the No. 1 fuel pipe together with the 3 injectors in position on the intake manifold
- Bolt holding the right side delivery pipe, temporarily, to the intake manifold
- Left hand delivery pipe and the No. 2 fuel pipe together with the 3

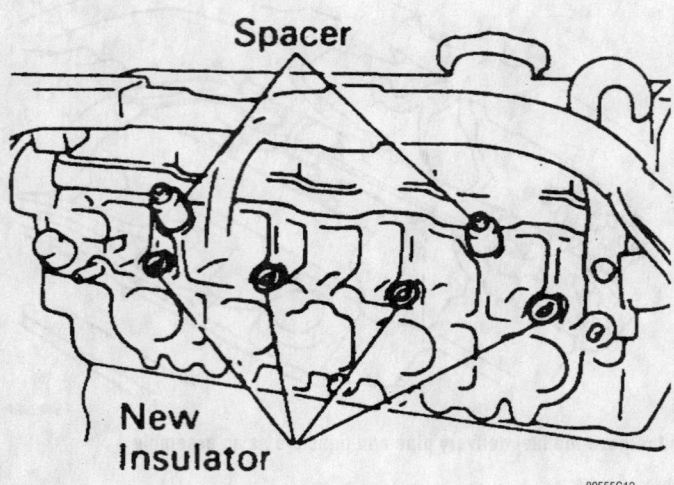

Spacer

New Insulator

89555G12

Place new injector seals and spacers in position on the cylinder head—5S-FE engine shown

injectors in position on the intake manifold

- Fuel return hose to the fuel pressure regulator

6. Temporarily install the 2 bolts holding the left hand delivery pipe to the intake manifold.

7. Temporarily install the No. 2 fuel pipe to the left side delivery pipe with the union bolt and 2 new gaskets.

8. Check that the injectors rotate smoothly. If they do not, Replace the O-rings.

9. Position the injector connector outward. Tighten the 4 bolts holding the delivery pipes to the intake manifold and tighten to 7 ft. lbs. (10 Nm). Tighten the bolt holding the No. 1 fuel pipe to the intake manifold to 14 ft. lbs. (20 Nm).

Tighten the 2 union bolts holding the no. 2 fuel pipe to the delivery pipes to 24 ft. lbs. (32 Nm).

10. Install or connect the following:
- Fuel inlet and return hoses. Union bolt: 22 ft. lbs. (30 Nm).
- Fuel return hose to the No. 1 fuel pipe. Pass the fuel return hose under the heater hoses
- Air assist hoses to the intake manifold
- Air assist pipe to the bracket on the No. 1 fuel pipe
- Fuel injector wiring connectors
- Air intake chamber assembly
- Hydraulic motor pressure pipe to the intake chamber. Bolts: 69 inch lbs. (8 Nm)

- No. 2 EGR pipe with new gaskets, tighten to 9 ft. lbs. (12 Nm)
- Emission control valve set
- V-bank cover
- Air cleaner hose
- Throttle and accelerator cables
- Negative battery cable

11. Fill the cooling system to the proper level.

12. Start the vehicle, check for leaks and repair if necessary.

Celica, Supra, GT-S and Tercel

5S-FE ENGINE

1. Before servicing the vehicle, refer to the precautions in the beginning of this section.

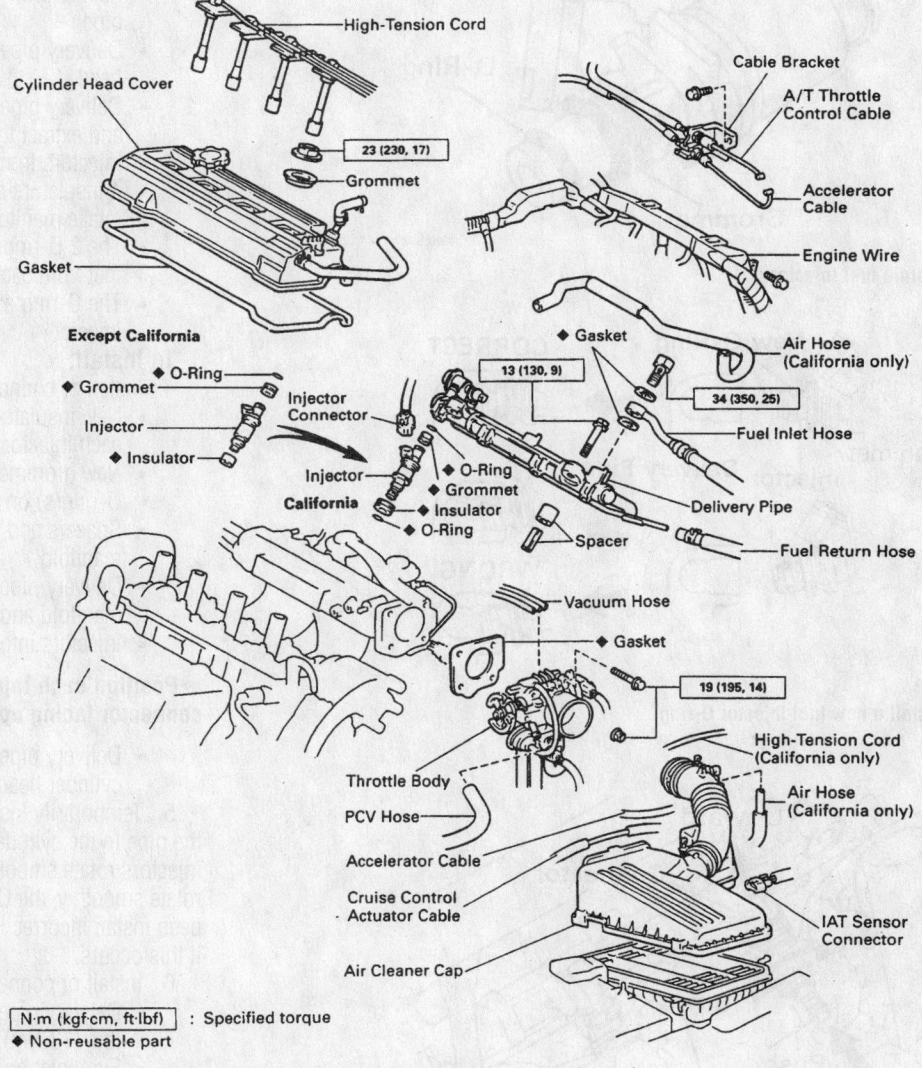

High-Tension Cord
Cylinder Head Cover
23 (230, 17)
Grommet
Gasket
Except California
O-Ring
Grommet
Injector
Injector Connector
Insulator
Injector
O-Ring
California
Grommet
Insulator
O-Ring
Spacer
13 (130, 9)
Gasket
Air Hose (California only)
34 (350, 25)
Fuel Inlet Hose
Delivery Pipe
Fuel Return Hose
Cable Bracket
A/T Throttle Control Cable
Accelerator Cable
Engine Wire
Vacuum Hose
Gasket
19 (195, 14)
High-Tension Cord (California only)
Air Hose (California only)
Throttle Body
PCV Hose
Accelerator Cable
Cruise Control Actuator Cable
Air Cleaner Cap
IAT Sensor Connector

N·m (kgf·cm, ft·lbf) : Specified torque
♦ Non-reusable part

89595G28

View of the fuel injector, delivery pipe and related components—5S-FE engine

For complete Engine Mechanical specifications, see Section 1 of this manual

California

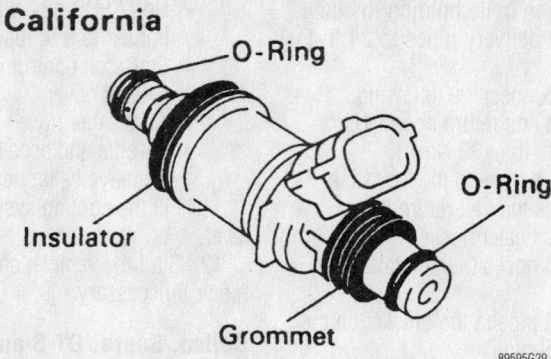

The California fuel injector has 2 O-rings a insulator and grommet

Except California

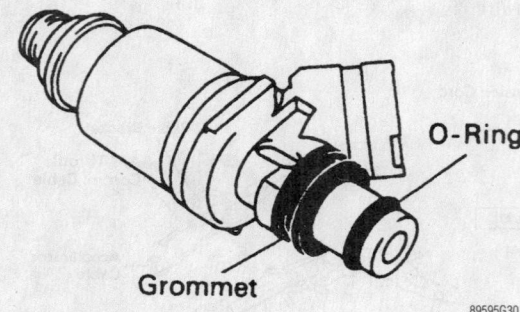

View of the non-California fuel injector

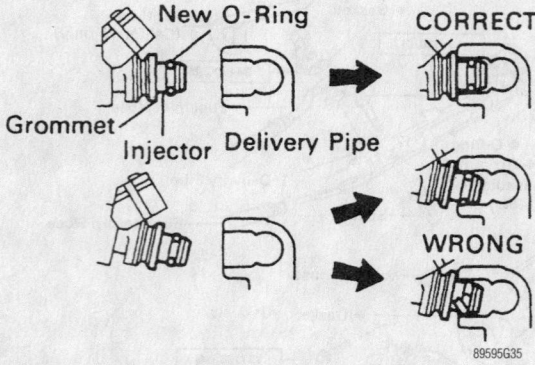

The correct way to install a new fuel injector O-ring

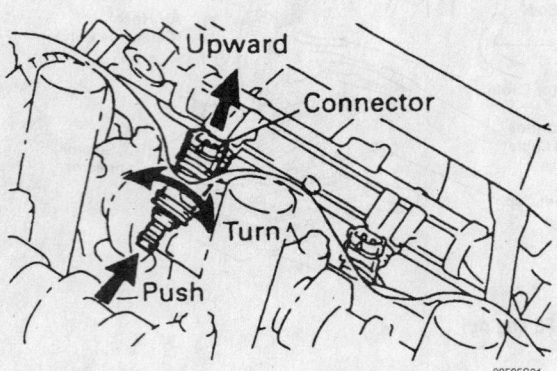

When inserting the injector, turn left and right making sure the connector is facing upward

2. Properly relieve the fuel system pressure.
3. Remove or disconnect the following:
   - Negative battery cable
   - Air cleaner cap and hose
   - Throttle body assembly
   - Valve cover assembly
   - Vacuum sensing hose from the fuel pressure regulator
   - Air hose from the air assist system at the intake manifold port
   - Engine wire protector from the left side of the intake manifold
   - Injector wiring
   - Engine wiring protector from the 2 brackets on the front of the intake manifold
   - Union bolt, 2 gaskets, and the fuel inlet hoses from the delivery pipe
   - Fuel return hose from the return pipe
   - Delivery pipe from the cylinder head
   - Delivery pipe from the 4 injectors and extract the pipe
   - Injectors from the pipe
   - 4 insulators and 2 spacers from the intake manifold
   - The 2 O-rings, insulator and grommet from each injector
   - The O-ring and grommet from each injector

To install:

4. Install or connect the following:
   - New insulator and grommet on each injector
   - New grommet on each insulator
   - O-ring(s) on each injector
   - Spacers and insulators to the intake manifold
   - Delivery pipe between the intake manifold and cylinder head
   - Injectors into the delivery pipe

➡**Position each injector in with the connector facing upward.**

   - Delivery pipe assembly to the cylinder head

5. Temporarily install the bolts holding the pipe to the cylinder head. Check that the injectors rotate smoothly. If they do not rotate smoothly, the O-ring(s) may have been install incorrectly. Replace the O-ring if this occurs.
6. Install or connect the following:
   - Delivery pipe retaining bolts: 9 ft. lbs. (13 Nm).
   - Fuel inlet hose to the delivery pipe with new gaskets and union bolt. Tighten the assembly to 25 ft. lbs. (34 Nm).

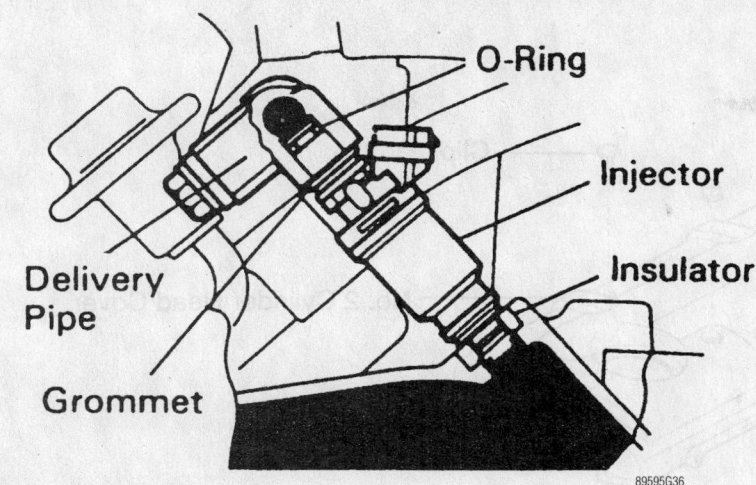

The injector will seat in the fuel rail and intake manifold as shown

89595G36

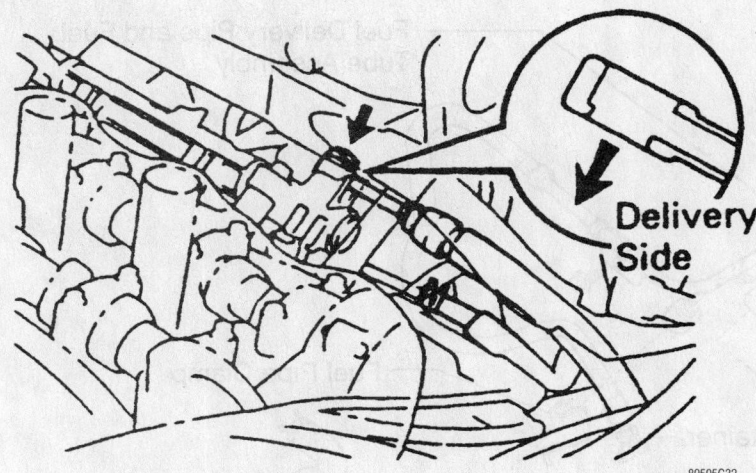

Once the injector is in check for it to rotate smoothly

89595G32

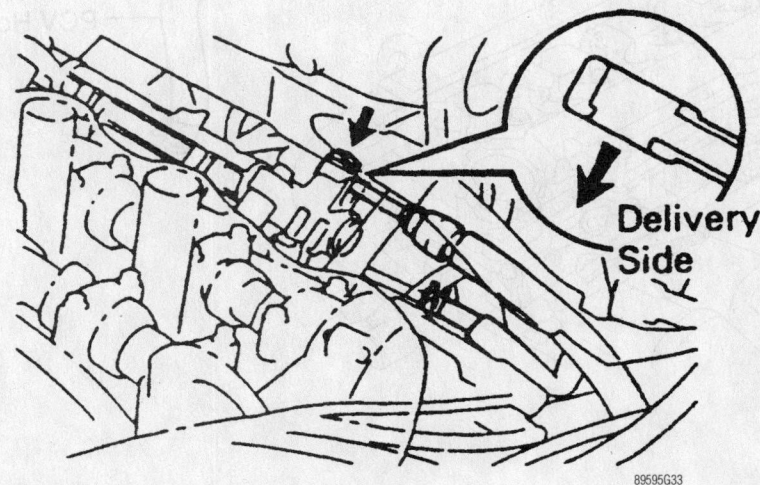

Place the fuel inlet hose on in the correct direction

89595G33

➡**Be careful of the fuel inlet hose installation direction.**

- Fuel return hose to the return pipe
- Engine wire protector to the brackets on the front side of the intake manifold
- Injector wiring harnesses

➡**The No. 1 and No. 3 injector wiring are brown and the No. 2 and No. 4 are gray.**

- Engine wire protector to the left side of the intake manifold
- Vacuum sensing hose to the fuel pressure regulator
- Air hose for the air assist system to the intake manifold port
- Valve cover assembly
- Throttle body
- Control cables to the throttle body
- Air cleaner cap and hose
- Negative battery cable

1ZZ-FE ENGINE

1. Before servicing the vehicle, refer to the precautions in the beginning of this section.

2. Properly relieve the fuel system pressure.

3. Remove or disconnect the following:
- No. 2 cylinder head cover
- PCV hose
- Fuel tube from the fuel pipe
- Injector connectors
- Delivery pipe and injectors
- Spacers from the head
- Injectors from the delivery pipe
- O-ring and grommet from each injector

To install:

4. Install or connect the following:
- New grommets
- New O-rings coated with light machine oil
- Injectors on the delivery pipe

➡**Coat the contact point on the pipe with light machine oil and twist the injectors into place. The connector should face outward.**

- Spacers

➡**Coat the seats in the head where the injectors contact, with light machine oil.**

- Delivery pipe and injectors

5. Loosely install the hold-down bolts and check that the injectors rotate smoothly. If they don't, the probable cause is incorrect O-ring installation.

For Accessory Drive Belt illustrations, see Section 1 of this manual

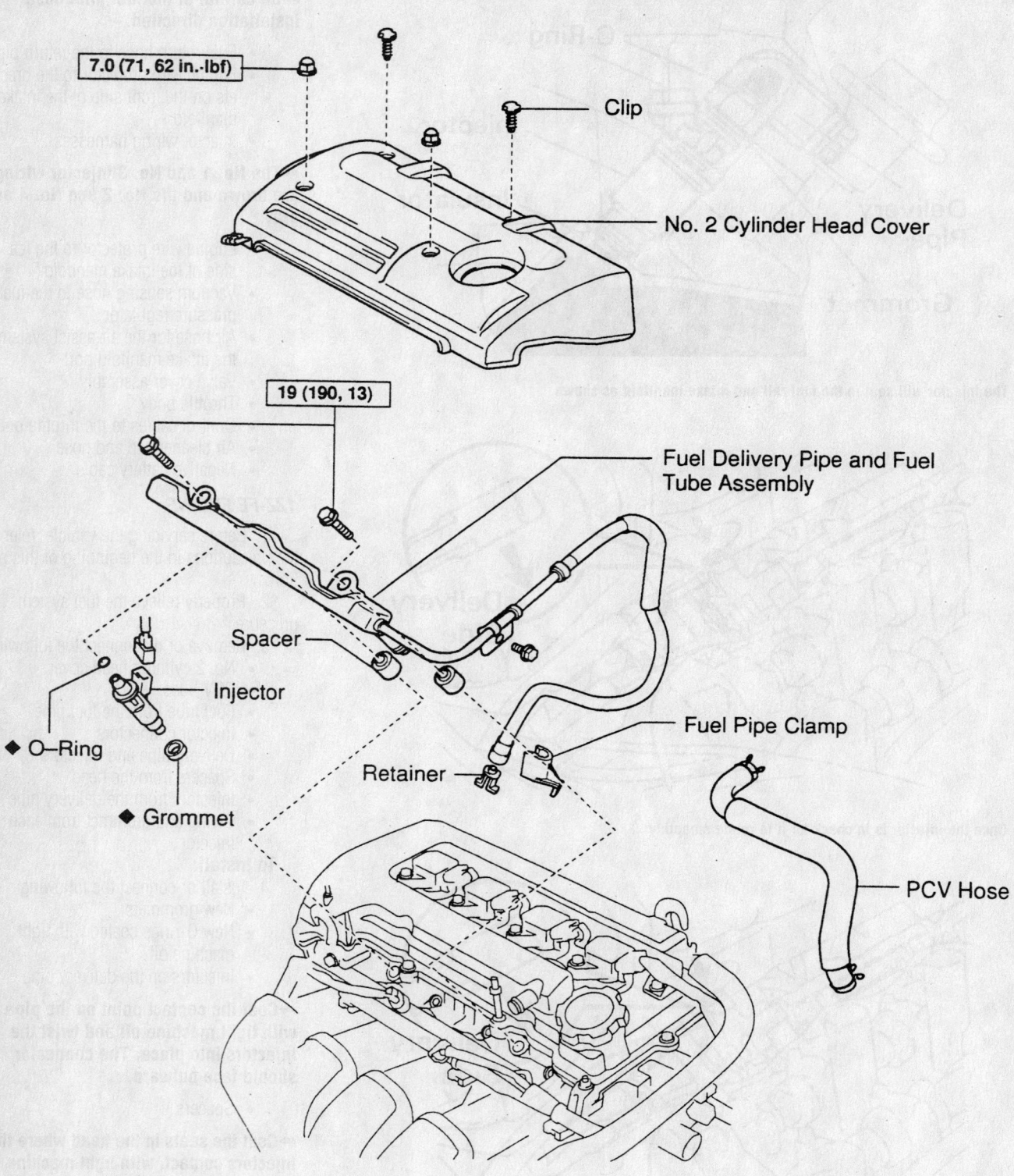

7.0 (71, 62 in.-lbf)

Clip

No. 2 Cylinder Head Cover

19 (190, 13)

Fuel Delivery Pipe and Fuel Tube Assembly

Spacer

Injector

◆ O–Ring

◆ Grommet

Retainer

Fuel Pipe Clamp

PCV Hose

N·m (kgf·cm, ft·lbf) : Specified torque

◆ Non–reusable part

Fuel injector removal and installation—1ZZ-FE engine

9307WG95

6. Torque the hold-down bolts to 14 ft. lbs. (19 Nm).

7. Torque the fuel pipe bolt to 84 inch lbs. (9 Nm).

8. Connect the fuel line.

9. Install the PCV hose.

10. Install the No. 2 cover.

2ZZ-GE ENGINE

1. Before servicing the vehicle, refer to the precautions in the beginning of this section.

2. Properly relieve the fuel system pressure.

3. Remove or disconnect the following:

- No. 2 cylinder head cover
- Fuel tube from the fuel pipe
- Injector connectors
- Delivery pipe and injectors
- Spacers from the head
- Injectors from the delivery pipe
- O-ring and grommet from each injector

To install:

4. Install or connect the following:

- New grommets
- New O-rings coated with light machine oil
- Injectors on the delivery pipe

➡ **Coat the contact point on the pipe with light machine oil and twist the injectors into place. The connector should face outward.**

- Spacers

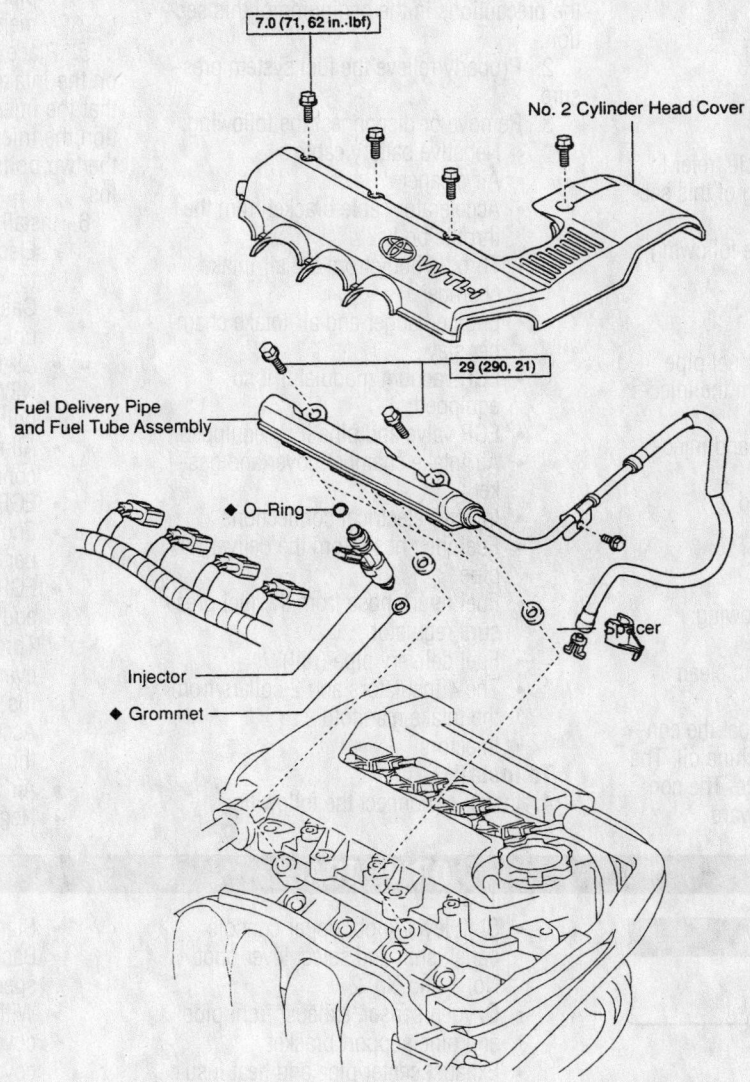

7.0 (71, 62 in.·lbf)

No. 2 Cylinder Head Cover

29 (290, 21)

Fuel Delivery Pipe and Fuel Tube Assembly

◆ O–Ring

Spacer

Injector

◆ Grommet

N·m (kgf·cm, ft·lbf) : Specified torque
◆ Non–reusable part

9307WG96

Fuel injector removal and installation—2ZZ-GE engine

➡**Coat the seats in the head where the injectors contact, with light machine oil.**

- Delivery pipe and injectors

5. Loosely install the hold-down bolts and check that the injectors rotate smoothly. If they don't, the probable cause is incorrect O-ring installation.

6. Torque the hold-down bolts to 21 ft. lbs. (29 Nm).

7. Torque the fuel pipe bolt to 84 inch lbs. (9 Nm).

8. Connect the fuel line.

9. Install the PCV hose.

10. Install the No. 2 cover.

Echo

1NZ-FE ENGINE

1. Before servicing the vehicle, refer to the precautions in the beginning of this section.

2. Remove or disconnect the following:
- Negative battery cable
- Cylinder head cover
- Fuel pipe clamp
- Fuel inlet line from the fuel pipe
- Injector connectors from the injectors
- Delivery pipe (3 bolts) and injectors
- 2 spacers from the head
- Injectors from the pipe
- O-rings and grommets

To install:

3. Install or connect the following:
- New grommets
- New O-rings coated with clean engine oil
- Injectors to the pipe. Coat the contact area with light machine oil. The injectors twist into place. The connector should face outward

- Delivery pipe bolts. Torque: 14 ft. lbs. (19 Nm).
- Fuel pipe bolt. Torque: 80 inch lbs. (9 Nm).
- Fuel hose to fuel pipe
- Pipe clamp
- Wire harness cover
- PCV hose
- Negative battery cable

Corolla

1ZZ-FE ENGINES

1. Before servicing the vehicle, refer to the precautions in the beginning of this section.

2. Properly relieve the fuel system pressure.

3. Remove or disconnect the following:
- Negative battery cable
- Air cleaner
- Accelerator cable bracket from the throttle body
- Throttle body from the air intake chamber
- Engine hanger and air intake chamber stay
- EGR vacuum modulator if so equipped
- EGR valve and pipe if so equipped
- Air intake chamber cover and gasket
- Injector electrical connections
- Fuel inlet hose from the delivery pipe
- Fuel return hose from the fuel pressure regulator
- Fuel delivery pipe (rail)
- The 4 insulators and 2 collars from the intake manifold
- Injectors

To install:

4. Install or connect the following:

➡**Before installing the injectors back into the fuel rail, install a NEW O-ring on each injector, coated with a light coat of gasoline (NEVER use oil of any sort).**

- Injectors

➡**Make certain each injector can be smoothly rotated. If they do not rotate smoothly, the O-ring is not in its correct position.**

- Insulators into each injector hole
- The two spacers on the delivery pipe mounting holes in the intake manifold

5. Place the delivery pipe and injectors on the intake manifold and again check that the injectors rotate smoothly. Position the injector connector upward. Install the two bolts and tighten them to 11 ft. lbs.

6. Install or connect the following:
- Electrical connectors to each injector
- Gaskets, the inlet pipe and fuel union bolt. Bolt to 22 ft. lbs.
- Air intake chamber cover with a NEW gasket. Torque the retaining bolts in steps to 14 ft. lbs.
- All necessary hoses and electrical connections
- EGR valve and pipe if so equipped
- Engine hanger and air intake chamber stay
- EGR vacuum modulator if so equipped
- Throttle body. Torque the bolts evenly (in a X-pattern) to 16 ft. lbs.
- Accelerator cable bracket to the throttle body
- Air cleaner hose and cap
- Negative battery cable

DRIVE TRAIN

Transmission Assembly

REMOVAL & INSTALLATION

Supra

MANUAL

1. Before servicing the vehicle, refer to the precautions in the beginning of this section.

2. Drain the transmission fluid.

3. Remove or disconnect the following:
- Negative battery cable. On vehicles equipped with an air bag, wait at least 90 seconds before proceeding
- Fan shroud set bolts

- Shift lever knob, upper console panel, shift and select lever boot No. 1 and No. 2
- Oxygen sensor, exhaust front pipe and pipe support bracket
- Exhaust center pipe and heat insulator
- Center floor crossmember brace
- Driveshafts. Cap the end of the transmission to prevent leakage
- Transmission lever bolt and nut, remove the shift lever
- Clutch release cylinder. Remove the bolt, ground cable and flexible hose bracket

- Starter connector
- Back-up light switch and vehicle speed sensor connectors
- With a V160 transmission, clutch cover set bolts and service hole cover. Place matchmarks on the flywheel and clutch cover. Remove the 6 bolts
- Rear engine mounting member
- Starter
- Remaining transmission mounting bolts and remove the transmission from the engine
- With a V160 transmission, shift lever retainer from the transmission

- Rear engine mounting from the transmission

To install:

4. Install or connect the following:
- Rear engine mount to the transmission. Bolts: 18 ft. lbs. (25 Nm).
- With a V160 transmission, shift lever retainer. Bolts: 14 ft. lbs. (19 Nm). Through-bolt and nut: 18 ft. lbs. (25 Nm).
- Transmission to the engine. Bolts: 53 ft. lbs. (72 Nm).
- Starter. Bolts: 29 ft. lbs. (39 Nm).
- Rear engine mounting member. Nuts: 10 ft. lbs. (13 Nm). Bolts: 19 ft. lbs. (25 Nm).
- With a V160 transmission, clutch cover set bolts. Bolts: 14 ft. lbs. (19 Nm). Install the service hole cover. Bolts: 108 inch lbs. (12 Nm).
- Vehicle speed sensor and back-up light switch connectors
- Starter wire connector
- Clutch release cylinder. Bolts: 108 inch lbs. (12 Nm). Install bolt with the clamp and ground cable. Bolt: 53 ft. lbs. (72 Nm).
- Transmission shift lever. Bolts: 14 ft. lbs. (19 Nm).
- Driveshafts, with grease applied to the flexible coupling centering bushings
- Center floor crossmember brace. Bolts: 108 inch lbs. (13 Nm).
- Heat insulator. Bolts: 48 inch lbs. (5 Nm).
- Center pipe, using new gaskets. Bolts: 14 ft. lbs. (19 Nm).
- Front pipe, using new gaskets. Bolts: 43 ft. lbs. (58 Nm).
- Pipe support bracket. Bolts: 27 ft. lbs. (37 Nm).
- Front pipe set bolts and nuts. Bolts and nuts: 32 ft. lbs. (43 Nm).
- Oxygen sensor and cover using a new gasket. Tighten to 13 ft. lbs. (18 Nm).
- Shift lever mount bolts. Bolts: 69 inch lbs. (8 Nm).
- Shift and select lever boot No. 1 and No. 2.
- Upper console panel to the console box with the 4 clips.
- Shift lever knob
- Fan shroud set bolts
- Negative battery cable

5. Fill the transmission fluid to the proper level.

6. Start the vehicle, check for leaks and repair if necessary.

AUTOMATIC

1. Before servicing the vehicle, refer to the precautions in the beginning of this section.

2. Remove or disconnect the following:
- Negative battery cable. On vehicles equipped with an air bag, wait at least 90 seconds before proceeding.
- Transmission oil level gauge and filler pipe.
- Undercover
- Oxygen sensor
- Exhaust pipe
- Heat insulator by removing the 4 bolts (normal roof) or 6 bolts (sport roof).
- Rear center floor crossmember brace
- Driveshaft
- Control rod from the shift lever
- Shift control rod from the park/neutral position switch by removing the nut.
- No. 1 vehicle speed sensor
- No. 2 vehicle speed sensor
- Solenoid wire
- Park/neutral position switch
- Automatic transmission fluid temperature sensor
- With overdrive, overdrive direct clutch speed sensor

- Starter electrical connector to the starter, then the nut and cable from the starter
- Transmission oil cooler pipes
- With turbocharged models, intercooler pipe
- Torque converter clutch mounting bolts
- Rear mounting by removing the 4 outer bolts
- 4 wiring harness clamps from the retainer
- Starter and transmission set bolts
- Transmission

To install:

3. Install or connect the following:
- Transmission and starter. Bolts: 14mm bolts: 27 ft. lbs. (37 Nm.). 17mm bolts: 53 ft. lbs. (72 Nm)
- Starter cable and nut, then electrical connector to the starter
- 4 wiring harness clamps to the retainer
- Rear mounting bolts. Bolts: 19 ft. lbs. (25 Nm).
- Torque converter clutch mounting bolts
- With turbocharged models, intercooler pipe
- Transmission oil cooler pipes. Cooler union nuts: 25 ft. lbs. (34 Nm). Cooler pipe bracket bolt: 84 inch lbs. (10 Nm).
- Center and rear oil cooler pipe brackets. Bolts: 84 inch lbs. (34 Nm).

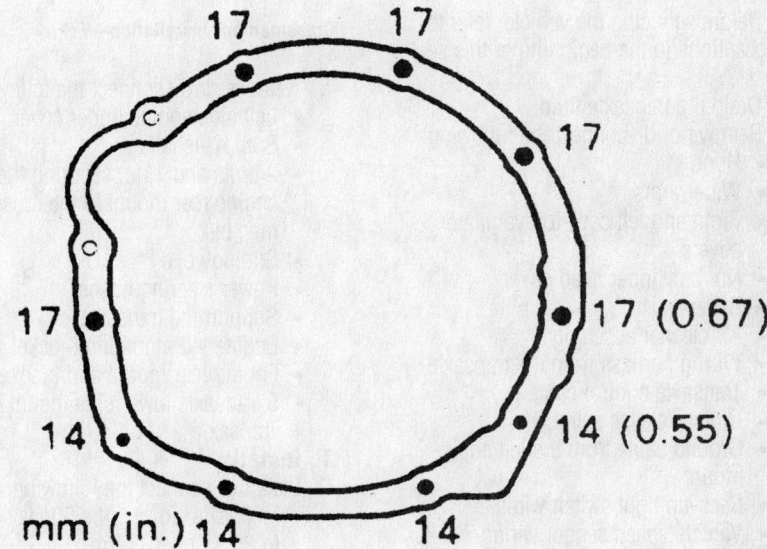

Transmission bolt identification—Supra

7923VG99

- Automatic transmission fluid temperature sensor
- Park/neutral position switch
- Solenoid wire
- No. 2 vehicle speed sensor
- No. 1 vehicle speed sensor
- Overdrive direct clutch speed sensor
- Shift control rod to the park/neutral position switch. Nut: 12 ft. lbs. (16 Nm).
- Shift control rod to the shift lever. Nut: 108 inch lbs. (13 Nm). Inspect and adjust the park/neutral position switch as needed
- Driveshaft
- Rear center floor crossmember brace. Bolts: 108 inch lbs. 913 Nm).
- Heat insulator with the bolts and tighten to 48 inch lbs. (5.4 Nm).
- Exhaust pipe. Bracket to transmission housing: 27 ft. lbs. (37 Nm); No. 2 exhaust pipe to center exhaust pipe: 43 ft. lbs. (58 Nm).
- Oxygen sensor
- Undercover
- Transmission filler pipe and level gauge
- Negative battery cable

Manual Transaxle Assembly

REMOVAL & INSTALLATION

Echo

1. Before servicing the vehicle, refer to the precautions in the beginning of this section.
2. Drain the transaxle fluid.
3. Remove or disconnect the following:
   - Hood
   - Wiper arms
   - Right and left cowl top ventilator covers
   - No. 2 cylinder head cover
   - Battery
   - Air cleaner assembly
   - Wiring harness from the transaxle
   - Transaxle control cable
   - Clutch release cylinder
   - Ground cable from the left engine mount
   - Back-up light switch wires
   - Vehicle speed sensor wiring
   - 2 transaxle upper side mounting bolts
   - Starter
4. At this point, attach an engine crane to support the engine.

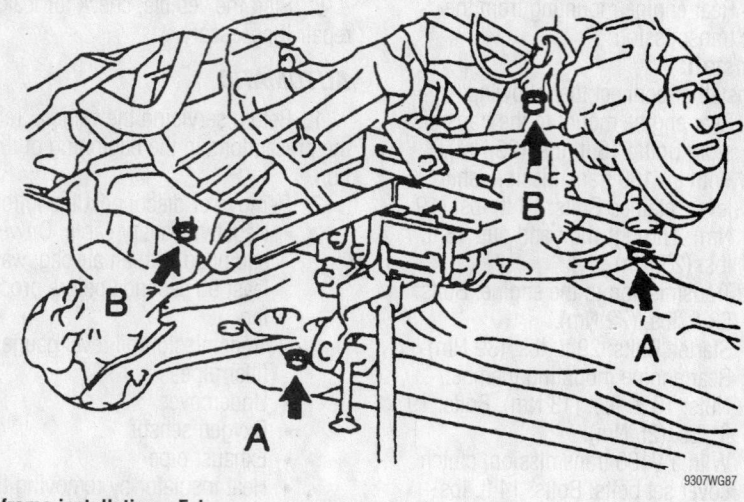

Sub-frame installation—Echo

9307WG87

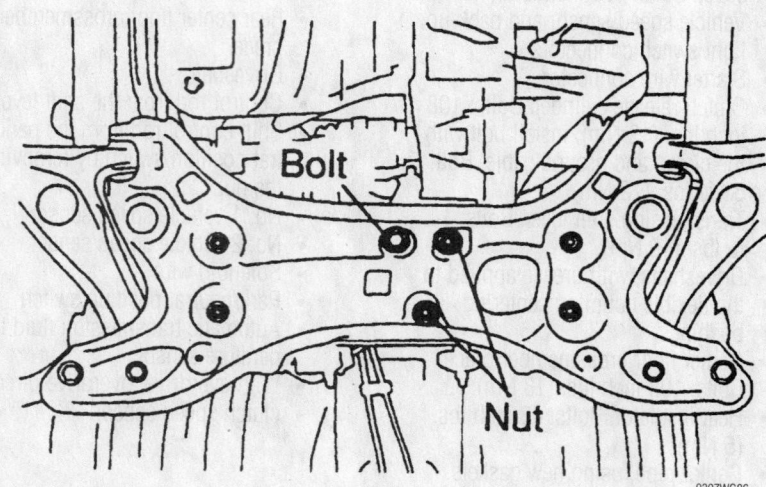

Crossmember installation—Echo

9307WG86

5. Remove or disconnect the following:
   - Left side engine under cover
   - Both halfshafts
   - 2 bolts and 1 nut securing the engine rear mount to the crossmember
   - Sliding yoke
   - Power steering hoses
   - Support the transaxle
   - Engine left mounting bracket
   - Engine rear mount and bracket
   - 5 transaxle lower side mount bolts
   - Transaxle

To install:
6. Install or connect the following:
   - Transaxle. Torque the 5 lower bolts to 25 ft. lbs. (33 Nm).
   - Engine rear mount and bracket. Torque the mount bolt and nut to 47 ft. lbs. (64 Nm); the bracket bolts to 36 ft. lbs. (49 Nm).
   - Engine left mounting bracket.

Torque the bolts to 36 ft. lbs. (49 Nm).
   - Power steering hoses
   - Sliding yoke
   - 2 bolts and 1 nut securing the engine rear mount to the crossmember. Torque the bolts to 36 ft. lbs. (49 Nm).
   - Both halfshafts
   - Left side engine under cover
   - Starter. Torque the bolts to 29 ft. lbs. (39 Nm).
   - 2 transaxle upper side mounting bolts. Torque the bolts to 25 ft. lbs. (33 Nm).
   - Vehicle speed sensor wiring
   - Back-up light switch wires
   - Ground cable from the left engine mount
   - Clutch release cylinder
   - Transaxle control cable
   - Wiring harness from the transaxle

- Air cleaner assembly
- Battery
- No. 2 cylinder head cover
- Right and left cowl top ventilator covers
- Wiper arms
- Hood

7. Fill the transaxle to the proper level.

8. Start the vehicle, check for leaks and repair if necessary.

Camry and Camry Solara

1. Before servicing the vehicle, refer to the precautions in the beginning of this section.

2. Drain the transaxle fluid.

3. Remove or disconnect the following:
- Negative battery cable. On vehicles equipped with an air bag, wait at least 90 seconds before proceeding.
- Air cleaner
- With cruise control, cruise control actuator
- Clutch release cylinder and tube clamp
- Starter
- Back-up light switch connector and ground strap
- Wires clamp
- Clips and washers that attach the transaxle control cables to the control levers
- Transaxle control cables.
- Speed sensor connector
- Undercovers
- Left and right halfshafts
- 4 steering gear housing bolts.
- Stabilizer bar bushing bracket
- 2 set bolts and nuts
- Steering gear box from the suspension member and suspend it securely

- Exhaust pipe
- Stiffener plate
- Engine front mounting from the suspension member
- Engine rear mounting from the front suspension member
- Left engine mounting
- Steering cooler pipe from the suspension member
- 2 fender liner set screws
- The 2 bolts and 4 nuts located on the outside of the suspension member brackets
- The 4 larger bolts holding the suspension member to the vehicle body
- The 2 front lower braces, rear braces, and the front suspension member.
- Transaxle

To install:

4. Move the transaxle into position so that the input shaft spline is aligned with the clutch disc.

5. Install or connect the following:
- Transaxle into the engine and secure with the lower mounting bolts. Bolts: 10mm mounting bolts: 47 ft. lbs. (63 Nm). 12mm bolts to 34 ft. lbs. (46 Nm).
- Front suspension member and the 2 front lower braces and rear lower braces. 4 large bolts that hold the suspension member to the vehicle: 134 ft. lbs. (181 Nm); 2 outside bolts and 4 outside nuts: 24 ft. lbs. (32 Nm).
- 2 fender liner set screws
- Steering cooler pipe to the suspension member
- Engine left mount. Bolts: 38 ft. lbs. (52 Nm); 2 nuts and 2 grommets. Nuts: 59 ft. lbs. (80 Nm).

- Engine rear mounting to the front suspension member. Nuts: 59 ft. lbs. (80 Nm).
- Engine front mounting to the suspension member. Bolt: 59 ft. lbs. (80 Nm).
- Stiffener plate. Bolts: 27 ft. lbs. (37 Nm).
- Exhaust pipe
- Steering gear housing to the front suspension member. Bolts and nuts: 134 ft. lbs. (181 Nm).
- Stabilizer bar bushing bracket. 4 bolts: 14 ft. lbs. (19 Nm).
- Right and left halfshafts
- Undercovers
- Vehicle speed sensor
- Control cables by installing the washers and clips
- Clamp that retains the wires to the transaxle
- Back-up light switch connector and ground cables
- Starter. Bolts: 29 ft. lbs. (39 Nm).
- Pipe clamp and clutch release cylinder to the transaxle. Bolts: 108 inch lbs. (13 Nm).
- Cruise control actuator
- Air cleaner
- Negative battery cable

6. Fill the transaxle fluid to the proper level.

7. Start the vehicle, check for leaks and repair if necessary.

Celica

1. Before servicing the vehicle, refer to the precautions in the beginning of this section.

2. Drain the transaxle fluid.

3. Remove or disconnect the following:
- Negative battery cable. On vehicles equipped with an air bag, wait at least 90 seconds before proceeding

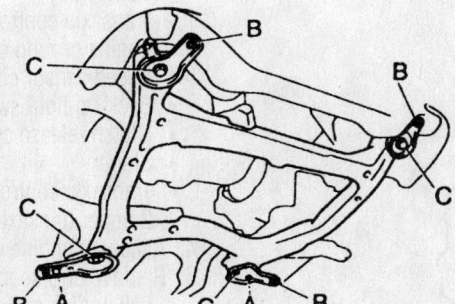

Bolt A: 32 N·m (330 kgf·cm, 24 ft·lbf)
Nut B: 36 N·m (370 kgf·cm, 27 ft·lbf)
Bolt C: 181 N·m (1,850 kgf·cm, 134 ft·lbf)

7923VG93

Front suspension member and fastener locations—Camry

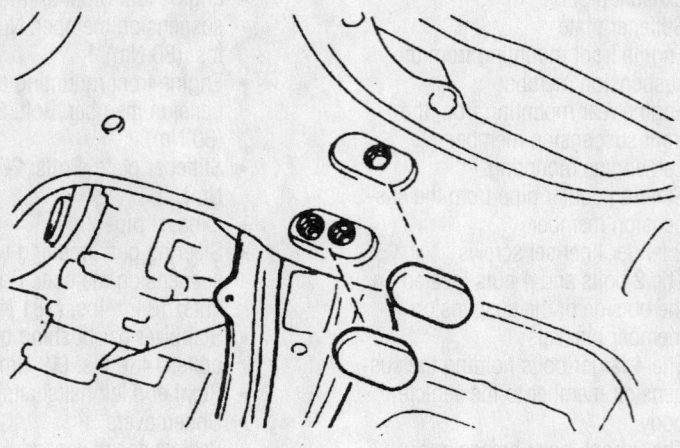

7923VG91

Rear mounting insulator set bolt locations—Celica

- Air cleaner case assembly with hose
- Release cylinder tube bracket
- Clutch release cylinder
- Back-up light switch connector
- Ground cable on the transaxle
- Shift cables from the transaxle
- Vehicle speed sensor connector or the speedometer cable
- Engine wire clamps
- Starter set bolt from the transaxle upper side
- Undercovers
- Halfshafts
- Front exhaust pipe and support bracket
- Starter
- Engine center support member
- Engine rear mounting
- Engine front mounting bracket and insulator
- Engine left mounting bracket

- Transaxle mounting bolts from the engine rear end plate side.
- Transaxle case protector
- Engine left side and remove the 3 upper transaxle bolts.
- Transaxle

To install:

4. Position the transaxle to the engine and raise the engine right side. Align the input shaft with the clutch disc

5. Install or connect the following:
- Transaxle to the engine. 3 upper transaxle bolts: 47 ft. lbs. (64 Nm).
- Transaxle case protector. Bolts: 108 inch lbs. (13 Nm).
- 4 transaxle lower bolts. Bolt A: 17 ft. lbs. (23 Nm); Bolt B: 34 ft. lbs. (46 Nm).
- Left engine mounting bracket to the engine left mounting insulator. Bolts: 47 ft. lbs. (64 Nm).
- Engine front mounting bracket and

insulator. 2 bracket bolts: 57 ft. lbs. (77 Nm). Through-bolt: 64 ft. lbs. (87 Nm).
- Engine rear mounting bracket and insulator. Bracket bolts: 57 ft. lbs. (77 Nm). Through-bolt: 64 ft. lbs. (87 Nm).
- Engine center support member
- Starter
- Front exhaust pipe and support bracket
- Halfshafts
- Transaxle oil
- Undercovers
- Engine support fixture
- Starter set bolt to the transaxle upper side. Bolt: 29 ft. lbs. (39 Nm).
- Engine wire clamps
- Vehicle speed sensor connector or the speedometer cable
- Transaxle shift cables and ground cable
- Back-up light switch connector
- Release cylinder
- Air cleaner case assembly
- Negative battery cable

6. Fill the transaxle fluid to the proper level.

7. Start the vehicle, check for leaks and repair if necessary.

Celica GT-S

1. Before servicing the vehicle, refer to the precautions in the beginning of this section.

2. Drain the transaxle fluid.

3. Remove or disconnect the following:
- Hood
- No. 2 cylinder head cover
- Radiator overflow bottle
- Battery
- Air cleaner assembly
- ECM box
- Wiring harness
- Battery tray
- Transaxle control cable
- Engine ground cable
- Speed sensor connector
- Back-up light switch connector
- Clutch release cylinder
- Starter
- Transaxle control cable bracket
- 2 upper transaxle mounting bolts

4. Attach an engine crane to the engine.

5. Remove or disconnect the following:
- Left engine mount bracket
- Engine under covers
- Both halfshafts
- Front exhaust pipe

6. Unbolt the steering rack from the sub-frame and suspend it.

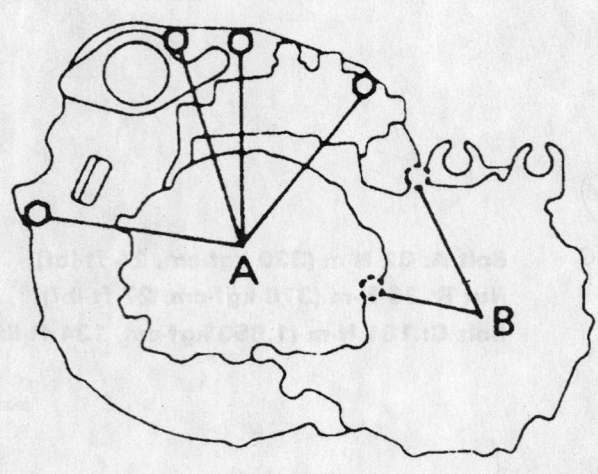

7923VG92

Upper transaxle mounting bolt locations—Celica

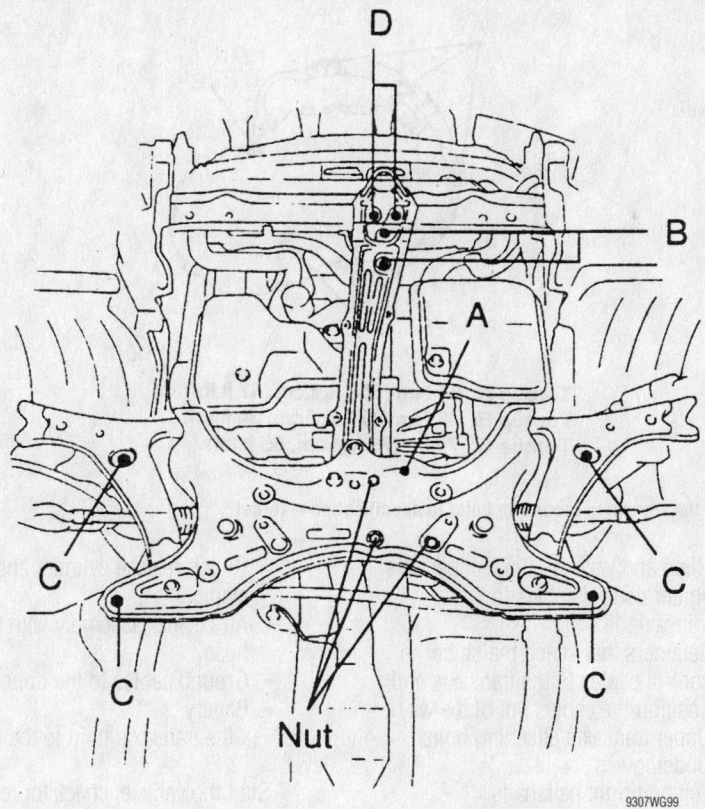

Celica sub-frame torques

9307WG99

7. Remove or disconnect the following:
- Stabilizer bar
- Wheels
- Ball joints from the lower arms
- Lower arms and sub-frame
- Engine rear mount and bracket

8. Raise the engine slightly.

9. Remove or disconnect the following:
- 4 lower transaxle side bolts
- Transaxle from the engine

To install:

10. Install or connect the following:
- Transaxle to the engine. Install the 4 lower side bolts. Torque the 2 bottom bolts to 17 ft. lbs. (23 Nm); the 2 side bolts to 35 ft. lbs. (48 Nm).
- Rear mount and bracket. Torque the bracket bolts to 47 ft. lbs. (64 Nm); the mount bolt to 64 ft. lbs. (87 Nm).
- Lower control arms-to-frame. Bolts: 101 ft. lbs. (137 Nm); tie rod ball stud nuts: 36 ft. lbs. (49 Nm).

11. Install the sub-frame. Torque the bolts as illustrated:
- a. A and B: 38 ft. lbs. (52 Nm)
- b. C: 116 ft. lbs. (157 Nm)
- c. D: 29 ft. lbs. (39 Nm)

12. Install or connect the following:
- Steering rack. Bolts: 43 ft. lbs. (58 Nm).
- Front exhaust pipe. Bolts: 32 ft. lbs. (43 Nm).
- Halfshafts
- Under covers
- Left mount bracket. Bolt: 44 ft. lbs. (60 Nm).
- Left mount. Bolts: 44 ft. lbs. (60 Nm); nut: 59 ft. lbs. (80 Nm).
- Transaxle upper side mount bolts. Torque: 47 ft. lbs. (64 Nm).
- Control cable bracket. Bolts: 18 ft. lbs. (25 Nm).
- Starter. Bolts: 28 ft. lbs. (37 Nm).
- Clutch release cylinder
- Wiring
- Control cable
- Battery tray
- ECM
- Air cleaner
- Battery
- Reservoir
- No. 2 cylinder head cover
- Hood

13. Fill the transaxle fluid to the proper level.

14. Start the vehicle, check for leaks and repair if necessary.

Corolla

1. Before servicing the vehicle, refer to the precautions in the beginning of this section.

2. Drain the transaxle fluid.

3. Remove or disconnect the following:
- Negative battery cable. On vehicles equipped with an air bag, wait at least 90 seconds before proceeding.
- Air cleaner case assembly with hose
- Coolant reservoir tank
- Release cylinder tube bracket
- Clutch release cylinder
- Back-up light switch connector
- Ground cable
- Shift cables from the transaxle
- Vehicle speed sensor connector or the speedometer cable
- Engine wire clamps
- Starter set bolt from the transaxle upper side
- 2 transaxle upper mounting bolts
- Engine left mounting stay
- Engine left mounting set bolt from the rear side
- Undercovers
- Lower ball joint from the lower arm.
- Halfshafts
- Front exhaust pipe
- Hole cover
- Engine front mounting set bolts
- Engine rear mounting
- Engine center support member
- Starter
- Transaxle mounting bolts from the engine rear end plate side
- Engine left mounting set bolts from the front side
- Transaxle mounting bolts from the engine front side, then engine rear side
- Transaxle

To install:

4. Align the input shaft with the clutch disc and install the transaxle to the engine. 12mm bolts: 47 ft. lbs. (64 Nm). 10mm bolts: 34 ft. lbs. (46 Nm).

5. Install or connect the following:
- Left engine mount. Bolts: 41 ft. lbs. (56 Nm).
- Transaxle mounting bolts to the engine rear end plate side. Bolts: 17 ft. lbs. (23 Nm).

- Starter, lower bolt and electrical connector to the starter. Bolt: 29 ft. lbs. (39 Nm).
- Engine center support member. Radiator support bolts: 45 ft. lbs. (61 Nm). Frame bolts: 152 ft. lbs. (206 Nm).
- Engine rear mounting. Bolts: 35 ft. lbs. (48 Nm).
- Engine front mounting. Bolts: 47 ft. lbs. (64 Nm).
- Hole covers
- Front exhaust pipe
- Halfshafts
- Lower ball joint to lower arm. Bolt and nuts: 105 ft. lbs. (142 Nm).
- Undercovers
- Engine left mounting set bolt to the rear side. Bolt: 41 ft. lbs. (56 Nm).
- Engine left mounting stay. Bolt: 15 ft. lbs. (21 Nm).
- 2 transaxle upper side mounting bolts. Bolts: 29 ft. lbs. (39 Nm).
- Starter set bolt to the transaxle upper side. Bolt: 29 ft. lbs. (39 Nm).
- Engine wire clamps
- Vehicle speed sensor connector or the speedometer cable.
- Transaxle shift cables and ground cable.
- Back-up light switch connector.
- Release cylinder and release cylinder tube bracket. Bolts: 108 inch lbs. (12 Nm).
- Coolant reservoir tank
- Air cleaner case assembly
- Negative battery cable

6. Fill the transaxle fluid to the proper level.

7. Start the vehicle, check for leaks and repair if necessary.

Tercel

1. Before servicing the vehicle, refer to the precautions in the beginning of this section.
2. Drain the transaxle fluid.
3. Remove or disconnect the following:
   - Negative battery cable. On vehicles equipped with an air bag, wait at least 90 seconds before proceeding
   - Battery
   - Air cleaner case assembly with the air hose
   - Clutch release cylinder and tube clamp
   - Ground cable from the transaxle side and engine left mounting bracket side
   - Back-up light switch electrical connector

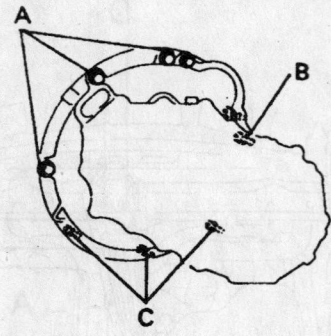

Torque A: 64 Nm (650 kgf.cm, 47 ft.lbf)
Torque B: 46 Nm (470 kgf.cm, 34 ft.lbf)
Torque C: 7 Nm (75 kgf.cm, 65 ft.lbf)

7923VG88

Tighten the transaxle mounting bolts to specification—Tercel

- Clips and washers that connect the shifter control cables to the transaxle
- Retainers that attach the shifter control cables to the transaxle and position the cables out of the way
- Upper transaxle attaching bolts
- Undercovers
- Left and right halfshafts
- Front exhaust pipe, if necessary
- Speedometer cable
- Starter
- Rear engine mounting insulator and bracket
- Left engine mount
- Transaxle attachment bolts
- Transaxle

To install:

4. Align the input shaft spline with the clutch disc and install the transaxle to the engine. Install the transaxle attaching bolts and tighten the bolts to specification.

5. Install or connect the following:
   - Left engine mounting bracket. Bolts: 36 ft. lbs. (48 Nm).
   - Rear mounting insulator. A: Body side bolt through brackets to 58 ft. lbs. (78 Nm); B: Through-bolt to 47 ft. lbs. (64 Nm); C: Body side bolt though insulator to 67 ft. lbs. (90 Nm).
   - Starter and electrical connectors. Bolts: 29 ft. lbs. (39 Nm).
   - Speedometer to the transaxle
   - Front exhaust pipe
   - Right and left halfshafts
   - Transaxle oil
   - Undercovers
   - Retainers, the washers and the clips for the shift control cables
   - Back-up light switch electrical connector

- Clutch release cylinder and tube clamp
- Air cleaner assembly with the air hose
- Ground cables to the transaxle
- Battery

6. Fill the transaxle fluid to the proper level.

7. Start the vehicle, check for leaks and repair if necessary.

Automatic Transaxle

REMOVAL & INSTALLATION

Echo

1. Before servicing the vehicle, refer to the precautions in the beginning of this section.
2. Drain the transaxle fluid.
3. Remove or disconnect the following:
   - Hood
   - Right and left cowl top ventilator covers
   - No. 2 cylinder head cover
   - Battery
   - Air cleaner bracket
   - Wiring harness from the transaxle
   - Transaxle shift cable
   - Clutch release cylinder
   - Ground cable from the left engine mount
   - Park/Neutral switch wiring
   - Solenoid wiring
   - Direct Clutch Speed Sensor wiring
   - Filler pipe hose
   - 2 transaxle upper side mounting bolts
   - Engine under covers
   - Starter

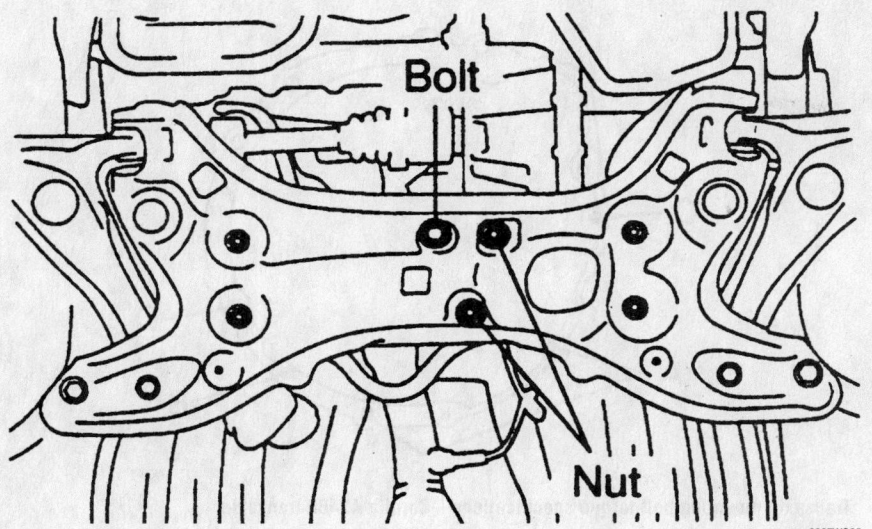

Rear mount installation—Echo

9307WG85

- Oil cooler hose
- Both halfshafts
- Torque converter access plug
- Torque converter bolts and attach an engine crane to support the engine
- Front suspension subframe (lower arms, steering gear and stabilizer)
- 5 transaxle bolts
- 2 transaxle mount bolts
- Transaxle

To install:

4. Install or connect the following:
   - Transaxle. Torque the 5 bolts to 22 ft. lbs. (30 Nm).
   - 2 transaxle mount bolts. Torque the mount bolts to 36 ft. lbs. (49 Nm).
   - Torque converter bolts: 20 ft. lbs. (27 Nm).
   - Access plug
   - Subframe assembly
   - Oil cooler hoses
   - Starter. Torque the bolts to 29 ft. lbs. (39 Nm).
   - 2 transaxle upper side mounting bolts. Torque the bolts to 22 ft. lbs. (30 Nm).
   - Both halfshafts
   - Engine under covers
   - Filler hose
   - Transaxle control cable
   - Ground cable from the left engine mount
   - Wiring
   - Air cleaner bracket
   - No. 2 cylinder head cover

- Right and left cowl top ventilator covers
- Battery
- Hood

5. Fill the transaxle fluid to the proper level.

6. Start the vehicle, check for leaks and repair if necessary.

Avalon, Camry and Camry Solara

1. Before servicing the vehicle, refer to the precautions in the beginning of this section.
2. Drain the transaxle fluid.
3. Remove or disconnect the following:
   - Negative battery cable. On vehicles equipped with an air bag, wait at least 90 seconds before proceeding
   - Battery
   - Air cleaner assembly
   - Throttle cable from the throttle body

- Cruise control actuator cover and connector, if equipped
- Ground wire
- Starter
- Speed sensor connectors, direct clutch speed sensor, and the park/neutral position switch connector on the transaxle
- Solenoid connector on the transaxle
- Shift control cable
- Oil cooler hoses
- Front side transaxle mounting bolts
- Front engine mounting bolts
- Oil cooler line mounting bolts from the front frame
- Upper transaxle to engine mounting bolts
- Front exhaust pipe
- Engine side covers and undercovers
- Both halfshafts
- Front side engine mounting nut
- Rear side engine mounting bolts (remove hole plugs)
- Left side transaxle mounting bolts
- Steering gear housing
- Front frame assembly
- Rear end plate mounting bolts
- Torque converter cover
- Torque converter retaining bolts
- Remaining transaxle mounting bolts
- Transaxle

To install:

4. Install or connect the following:
   - Transaxle aligning the 2 dowel pins on the block with the converter housing. 10mm bolts: 34 ft. lbs. (46 Nm); 12mm bolts: 47 ft. lbs. (64 Nm)
   - Torque converter bolts coated with sealer. Install the bolts starting with

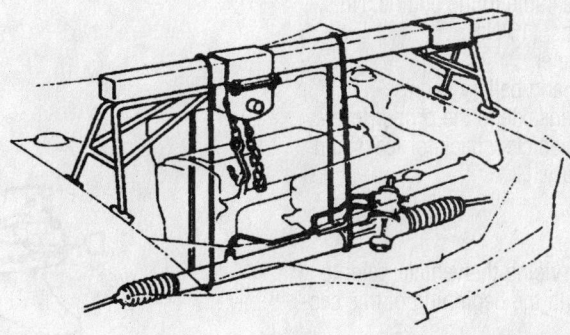

7923VG98

Tie the steering rack to the engine support fixture components, as shown—Avalon and Camry

the green bolt followed by the rest. Bolts: 20 ft. lbs. (27 Nm).

- Rear end plate. Bolts: 27 ft. lbs. (37 Nm).
- Front frame assembly. 12mm bolts: 24 ft. lbs. (32 Nm); 19mm bolts: 134 ft. lbs. (181 Nm); Nut: 27 ft. lbs. (36 Nm).
- Fender liner set screws
- Steering gear to the frame. Bolts and nuts: 134 ft. lbs. (181 Nm).
- Sway bar brackets. Bolts: 14 ft. lbs. (19 Nm).
- Left transaxle mounting bolts. Bolts: 38 ft. lbs. (52 Nm).
- Rear side mounting bolts and nuts. Bolts and nuts: 48 ft. lbs. (66 Nm). Install the plugs.
- Front engine mounting nut. Nut: 59 ft. lbs. (80 Nm).
- Halfshafts
- Right and left engine side covers
- Lower engine cover
- Exhaust pipe. Nuts: 46 ft. lbs. (62 Nm).
- Exhaust pipe to the converter. Nuts and bolts: 32 ft. lbs. (43 Nm).
- Upper transaxle mounting bolts. Bolts: 47 ft. lbs. (64 Nm).
- Oil cooler clamping bolts to the front frame.
- Front side engine mounting bolts. Bolts: 59 ft. lbs. (80 Nm).
- Front side transaxle mounting bolts. Bolts: 59 ft. lbs. (80 Nm).
- Oil cooler hoses
- Shift control cable
- Solenoid electrical connector
- Park/neutral switch electrical connector
- Speed sensor and the direct clutch speed sensor connectors.
- Starter
- Ground strap
- Cruise control actuator and cover
- Throttle cable to the engine. Nuts: 11 ft. lbs. (15 Nm).
- Air cleaner
- Battery and battery cables.

5. Fill the transaxle to the proper level.
6. Start the vehicle, check for leaks and repair if necessary.

Corolla

1. Before servicing the vehicle, refer to the precautions in the beginning of this section.
2. Drain the transaxle fluid.
3. Remove or disconnect the following:
- Negative battery cable. On vehicles equipped with an air bag, wait at

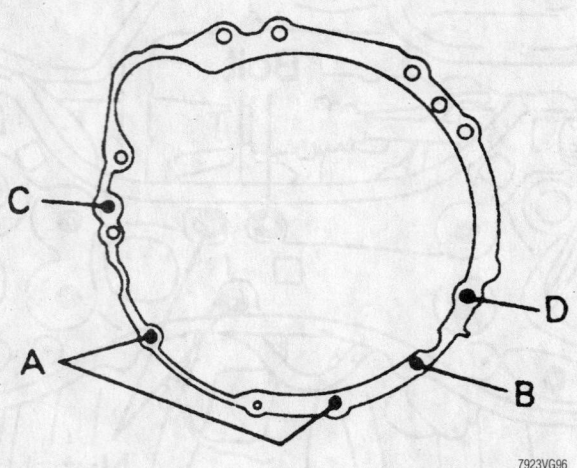

Transaxle mounting bolt torque specifications—Corolla A245E transaxle

least 90 seconds before proceeding
- Negative battery cable from the transaxle
- Transaxle level gauge
- With A245E transaxle, reservoir tank and air cleaner assembly
- Throttle cable from the bracket
- Engine left mounting upper side bolts
- Engine left mounting stay
- Ground cable from the transaxle
- Wiring harness clamp and throttle cable clamp
- With A245E transaxle, starter upper side mounting bolt and the 2 transaxle mounting bolts from transaxle side
- Undercovers
- Left and right halfshafts
- Front exhaust pipe
- Engine support fixture
- Suspension member

- Starter
- Vehicle speed sensor connector
- Solenoid connector and park/neutral position switch connector. Remove the wiring harness clamps
- Nut from the manual shift lever, then the control cable from the bracket by removing the clip
- Oil cooler hoses
- Transaxle filler tube
- With A131L transaxle, stiffener plate
- With A245E transaxle, converter cover
- Torque converter bolts
- 5 (A245E) or 4 (A131L) transaxle mounting bolts
- Transaxle

To install:

4. Install or connect the following:
- A245E Transaxle. Install the 5 lower transaxle bolts. Tighten the bolts as

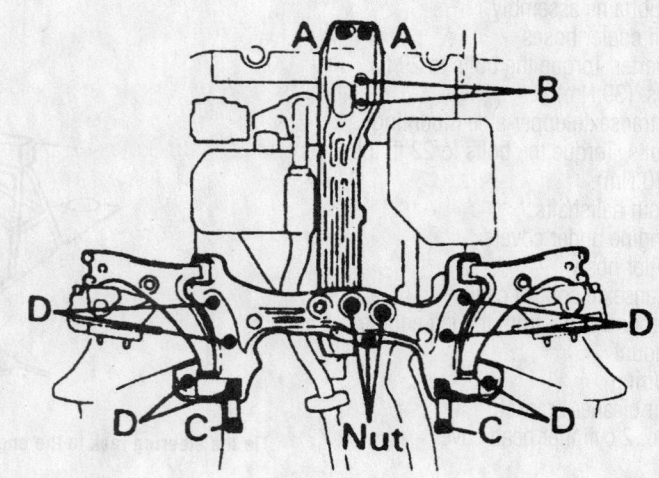

Suspension member fastener identification—Corolla

follows: Bolt A: 17 ft. lbs. (23 Nm); Bolt B: 18 ft. lbs. (25 Nm); Bolt C: 34 ft. lbs. (46 Nm); Bolt D: 47 ft. lbs. (64 Nm)
- A131L transaxle. Install the 4 lower transaxle bolts. Bolts: 47 ft. lbs. (64 Nm).
- Torque converter bolts to the transaxle. Bolts: 18 ft. lbs. (25 Nm).
- With the A131L transaxle, stiffener plate. Bolts: 13 ft. lbs. (18 Nm).
- With the A245E transaxle, torque converter cover.
- Transaxle filler pipe
- Oil cooler hoses and replace the clips to their original positions
- Control cable for the transaxle to the bracket and install the clip
- Control cable to the manual shaft lever by installing the nut
- Solenoid connector and park/neutral position switch connector. Connect the wiring to the clamps
- Vehicle speed sensor wiring
- Starter. Bolt: 29 ft. lbs. (39 Nm).
- Suspension member. Bolt A: 45 ft. lbs. (61 Nm); Bolt B: 47 ft. lbs. (64 Nm); Bolt C: 152 ft. lbs. (206 Nm); Bolt D: 152 ft. lbs. (206 Nm); Nuts: 42 ft. lbs. (57 Nm).
- Front exhaust pipe
- Left and right halfshafts
- Undercovers
- Transaxle mounting bolts to the transaxle side
- Starter upper side mounting bolt. Bolt: 29 ft. lbs. (39 Nm).
- Wiring harness clamp and throttle cable clamp
- Ground cable. Bolt: 13 ft. lbs. (18 Nm).
- Engine left mounting stay. Bolts: 15 ft. lbs. (21 Nm).
- Engine left mounting upper side bolts. Bolts: 38 ft. lbs. (52 Nm).
- Throttle cable
- Air cleaner and the reservoir tank
- Transaxle level gauge
- Negative battery cable

5. Fill the transaxle fluid to the proper level.
6. Start the vehicle, check for leaks and repair if necessary.

Celica

1. Before servicing the vehicle, refer to the precautions in the beginning of this section.
2. Drain the transaxle fluid.

3. Remove or disconnect the following:
- Negative battery cable. On vehicles equipped with an air bag, wait at least 90 seconds before proceeding.
- Throttle cable from the engine
- Cruise control actuator. The cruise control actuator and bracket should be removed as an assembly.
- Air cleaner assembly and battery.
- Vehicle speed sensor and the transaxle ground strap.
- Engine left mounting upper side bolt
- Starter
- Park/neutral position switch connector
- Solenoid connectors
- Upper transaxle retaining bolts
- Transaxle oil cooler hoses
- Undercover
- Both halfshafts
- Shift control cable from the control shaft lever and body bracket
- Engine rear mounting through-bolt
- Front exhaust pipe
- Air conditioner pipe bracket by removing the bolt
- Shift cable from the suspension member
- The 2 power steering gear assembly set bolts and nuts
- The 3 grommets from the center crossmember
- The 13 bolts and 2 nuts holding the suspension and center crossmembers
- Crossmembers from the vehicle
- No. 1 manifold stay
- Stiffener plate
- Torque converter bolts
- Transaxle

To install:
4. Install or connect the following:
- Transaxle. 10mm bolt: 34 ft. lbs. (46 Nm); 12mm bolt: 47 ft. lbs. (64 Nm).
- Torque converter bolts, with silicone applied to threads. Bolts: 18 ft. lbs. (25 Nm).
- Stiffener plate. Bolts, alternately tightening: 12mm bolts: 15 ft. lbs. (21 Nm); 14mm bolts: 32 ft. lbs. (43 Nm).
- No. 1 manifold stay. Bolt: 15 ft. lbs. (21 Nm). Nut: 32 ft. lbs. (43 Nm).
- Raise the suspension member into position and install the 2 bolts to hold the suspension to the body. Bolts to 94 ft. lbs. (127 Nm).

- The 3 bolts to hold the rear of the lower control arms to the subframe and body. Torque the bolt that goes through the lower control arm to 123 ft. lbs. (167 Nm) and the other 2 bolts to 130 ft. lbs. (175 Nm).
- Side bolts
- Center member
- Engine rear mount and bracket. Nuts: 59 ft. lbs. (80 Nm); Bolt: 65 ft. lbs. (88 Nm).
- Engine front mount. Bolts: 59 ft. lbs. (80 Nm).
- The 2 front bolts connecting the center mount to the radiator support. Bolts: 26 ft. lbs. (35 Nm).
- Grommets to the center member
- The 2 power steering gear assembly set bolts and nuts. Nuts and bolts: 94 ft. lbs. (127 Nm).
- The 2 shift cable mounting bolts
- Air conditioner pipe bracket
- Front exhaust pipe
- Engine rear mounting bolt. Bolt: 64 ft. lbs. (88 Nm).
- Shift control cable to the control shaft lever and body bracket. Install the clips
- Left and right halfshafts
- Undercovers
- Oil cooler hoses with the 2 clips
- Upper transaxle mounting bolts. Bolts: 47 ft. lbs. (64 Nm).
- Solenoid connectors
- Park/neutral position switch connector
- Vehicle speed sensor connector and the ground strap to the transaxle
- Starter. Bolts: 29 ft. lbs. (39 Nm).
- Left mounting upper side bolt. Bolt: 47 ft. lbs. (64 Nm).
- Air cleaner assembly
- Cruise control actuator
- Battery and cables
- Throttle cable

5. Fill the transaxle fluid to the proper level.
6. Start the vehicle, check for leaks and repair if necessary.

Celica GT-S

1. Before servicing the vehicle, refer to the precautions in the beginning of this section.
2. Drain the cooling system.
3. Drain the transaxle fluid.
4. Remove or disconnect the following:

- Hood
- Battery
- ECM
- Air cleaner assembly
- No. 2 cylinder head cover
- Ground cables
- Control cable bracket
- Input speed turbine sensor
- Vehicle speed sensor
- Solenoid connector
- Park/neutral switch connector
- Control cable
- Coolant reservoir
- Starter
- Engine under covers
- Halfshafts
- Stabilizer bar end links

5. Unbolt the steering rack and suspend it

- 9 bolts and 3 nuts and lower the suspension member
- Engine rear mount and attach and engine crane to support the engine.
- Upper left side engine
- Fluid cooler hoses and support the transaxle with a jack.
- Torque converter access plug
- Torque converter bolts
- Transaxle bolts
- Transaxle

To install:

6. Install or connect the following:
- Transaxle. 2 lower bolts: 17 ft. lbs. (23 Nm); 2 lower side bolts: 34 ft. lbs. (46 Nm).
- Torque converter bolts. Torque: 25 ft. lbs. (41 Nm).
- Oil cooler hoses
- Upper left side mount bolt and nut. Torque: 59 ft. lbs. (80 Nm).
- Engine rear mount insulator. Bolt: 64 ft. lbs. (87 Nm).

7. Install the suspension member. Torque the bolts as illustrated:
   a. A: 116 ft. lbs. (157 Nm).
   b. B: 38 ft. lbs. (52 Nm).
   c. C: 29 ft. lbs. (39 Nm).
   d. Nut: 38 ft. lbs. (52 Nm).

8. Install or connect the following:
- Steering rack. Bolts: 33 ft. lbs. (45 Nm).
- Stabilizer bar end links. Torque: 32 ft. lbs. (44 Nm).
- Halfshafts
- Under covers
- Starter. Torque: 28 ft. lbs. (37 Nm).
- Coolant reservoir
- Control cable
- Wiring
- Upper transaxle bolts. Torque: 47 ft. lbs. (64 Nm).
- Control cable clamp

- No. 2 cylinder head cover. Torque: 62 inch lbs. (7 Nm).
- Air cleaner
- ECM
- Battery
- Hood

9. Fill the cooling system.
10. Fill the transaxle to the proper level.
11. Start the vehicle, check for leaks and repair if necessary.

Tercel

1. Before servicing the vehicle, refer to the precautions in the beginning of this section.
2. Drain the transaxle/differential fluid.
3. Remove or disconnect the following:
- Negative battery cable. On vehicles equipped with an air bag, wait at least 90 seconds before proceeding.
- Positive battery cable, battery hold down, and the battery
- Air cleaner
- Throttle cable from the throttle link
- Ground cable and bracket from the transaxle
- Upper side mounting bolts from the transaxle
- Starter wiring
- Undercovers
- Right and left halfshafts from the transaxle
- Speedometer cable from the transaxle
- Shift control cable from the control lever
- Park/neutral position switch electrical connector
- Overdrive solenoid connector, if equipped

- Oil cooler hose from the transaxle
- Starter
- Front exhaust pipe
- Transaxle converter cover
- Torque converter bolts
- Bolts holding the left-hand engine mounting bracket to the body
- Rear mounting insulator through-bolt
- Rear mounting insulator
- Transaxle
- Torque converter from the transaxle

To install:

4. Remove or disconnect the following:
- Torque converter to the transaxle
- Transaxle assembly to the engine
- Left-hand engine mounting bracket. Bolts: 35 ft. lbs. (48 Nm).
- Rear engine mounting bracket. Outside bolts through brackets: 58 ft. lbs. (78 Nm); Inside bolts though rear mounting: 69 ft. lbs. (92 Nm)
- Rear mounting insulator through-bolt. Bolt: 47 ft. lbs. (64 Nm).
- Torque converter mounting bolts. Bolts: 20 ft. lbs. (27 Nm).
- Torque converter cover
- Front exhaust pipe
- Starter
- Oil cooler hose to the transaxle.
- Overdrive solenoid connector
- Park/neutral switch electrical connector
- Shift control cable to the lever.
- Shift control cable to the body.
- Speedometer cable to the transaxle
- Left and right halfshafts
- Undercovers
- Ground cable and bracket to the transaxle.
- Starter upper bolt. Bolt to 29 ft. lbs. (39 Nm).

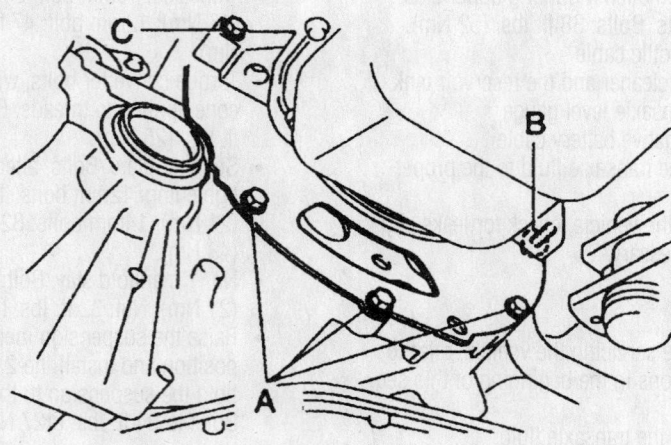

Tighten the lower transaxle mounting bolts to the proper specification—Tercel

7923VG94

- Transaxle upper side mounting bolts. Bolts: 47 ft. lbs. (64 Nm).
- Throttle cable
- Air cleaner
- Battery and battery cables

5. Fill the transmission and differential with the proper fluid, Dexron®II ATF.

6. Start the vehicle, check for leaks and repair if necessary.

Clutch

ADJUSTMENTS

Hydraulic clutch actuating systems used in Toyota vehicles do not require adjustment.

REMOVAL & INSTALLATION

Camry, Camry, Echo, Solara, Tercel and Supra

1. Before servicing the vehicle, refer to the precautions in the beginning of this section.

2. Remove or disconnect the following:
- Negative battery cable. On vehicles equipped with an air bag, wait at

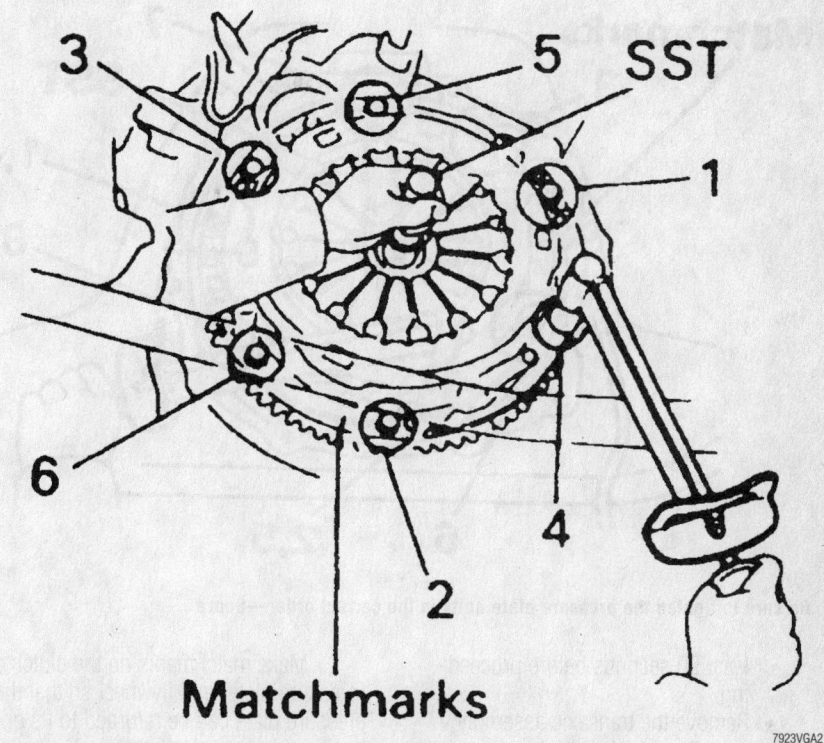

Matchmarks

Tighten the pressure plate bolts according to the sequence shown—Camry

7923VGA2

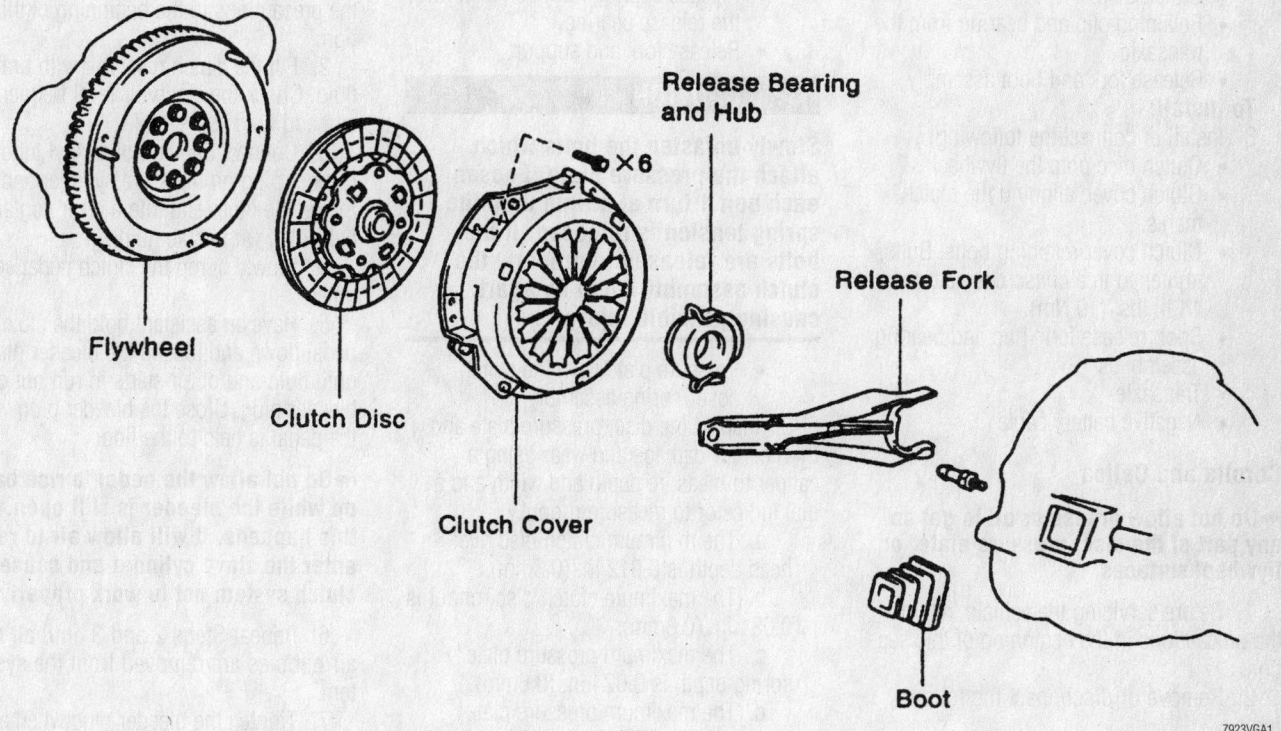

Clutch component assembly—Camry shown, others similar

7923VGA1

Timing belt service is covered in Section 3 of this manual

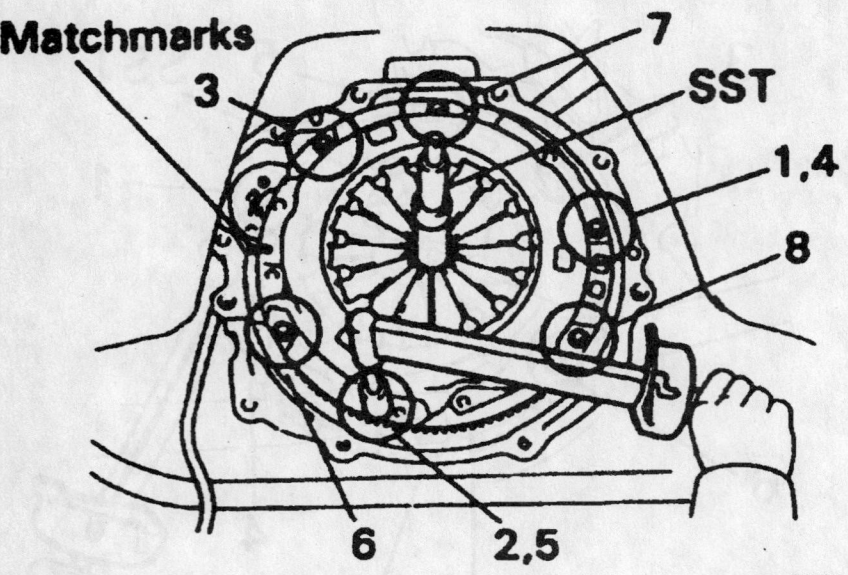

Matchmarks

Be sure to tighten the pressure plate bolts in the correct order—Supra

7923VGA3

least 90 seconds before proceeding.
- Remove the transaxle assembly from the vehicle
- Clutch pressure plate retaining bolts
- Clutch cover
- Clutch disc
- Retaining clip and bearing from the transaxle
- Release fork and boot assembly

To install:

3. Install or connect the following:
- Clutch disc onto the flywheel
- Clutch cover, aligning the matchmarks
- Clutch cover retaining bolts. Bolts, tightened in a crisscross pattern: 14 ft. lbs. (19 Nm).
- Boot, release fork, hub and bearing assemblies
- Transaxle
- Negative battery cable

Corolla and Celica

➡Do not allow grease or oil to get on any part of the disc, pressure plate, or flywheel surfaces.

1. Before servicing the vehicle, refer to the precautions in the beginning of this section.

2. Remove or disconnect the following:
- Negative battery cable. On vehicles equipped with an air bag, wait at least 90 seconds before proceeding
- Transaxle assembly

3. Make matchmarks on the clutch cover (pressure plate) and flywheel so that the pressure plate can be returned to its original position during installation.

4. Remove or disconnect the following:
- Release fork bearing clips
- Release bearing hub, complete with the release bearing
- Release fork and support

✳✳ CAUTION

Slowly unfasten the bolts which attach the pressure plate. Loosen each bolt 1 turn at a time until the spring tension is released. If the bolts are released improperly the clutch assembly could fly apart, causing possible injury.

- Pressure plate from the clutch cover/spring assembly

5. Inspect the disc, pressure plate and flywheel for damage and wear using a caliper to measure depth and width and a dial indicator to measure runout.
   a. The minimum clutch disc rivet head depth is 0.012 in. (0.3mm).
   b. The maximum clutch disc runout is 0.031 in. (0.8mm).
   c. The maximum pressure plate spring depth is 0.024 in. (0.6mm).
   d. The maximum pressure plate spring width is 0.197 in. (5.0mm).
   e. The maximum flywheel runout is 0.004 in. (0.1mm).

6. Replace or machine parts as necessary.

To install:

7. When reassembling, apply a thin coating of multipurpose grease to the release bearing hub and release fork contact points. Also, pack the groove inside the clutch hub with multipurpose grease and lubricate the pivot points of the release fork.

8. Install or connect the following:
- Clutch disc and pressure plate. The bolts should be tightened in 2 or 3 steps, gradually and evenly. Final bolt torque is 14 ft. lbs. (19 Nm).
- Release bearing, fork and boot
- Transaxle assembly
- Negative battery cable

Hydraulic Clutch System

BLEEDING

➡If any maintenance on the clutch system was performed or the system is suspected of containing air, bleed the system. Use care; brake fluid will remove the paint from any surface. If the brake fluid spills onto any painted surface, wash it off immediately with soap and water.

1. Before servicing the vehicle, refer to the precautions in the beginning of this section.

2. Fill the clutch reservoir with brake fluid. Check the reservoir level frequently and add fluid as needed.

3. Connect one end of a vinyl tube to the bleeder plug on the slave cylinder and submerge the other end into a clear container half-filled with brake fluid.

4. Slowly pump the clutch pedal several times.

5. Have an assistant hold the clutch pedal down and loosen the bleeder plug until fluid and/or air starts to run out of the bleeder plug. Close the bleeder plug while the pedal is held to the floor.

➡Do not allow the pedal to rise back-up while the bleeder is still open. If this happens, it will allow air to re-enter the slave cylinder and cause the clutch system not to work properly.

6. Repeat Steps 2 and 3 until all the air bubbles are removed from the system.

7. Tighten the bleeder plug when all the air is gone.

8. Refill the master cylinder to the proper level as required.

9. Check the system for leaks.

Halfshaft

REMOVAL & INSTALLATION

Avalon, Camry and Camry Solara

1. Before servicing the vehicle, refer to the precautions in the beginning of this section.
2. Drain the transaxle fluid.
3. Remove or disconnect the following:
   - Negative battery cable. On vehicles equipped with an air bag, wait at least 90 seconds before proceeding.
   - Front fender apron seal
   - Tie rod end from the steering knuckle by removing the cotter pin and nut. Separate the tie rod from the steering knuckle
   - Stabilizer bar link from the lower control arm. Make note of the washers and cushions positions.
   - Lower ball joint from the steering knuckle. Push down on the lower control arm and separate the steering knuckle from the ball joint
   - Cotter pin, lock cap and locknut holding the halfshaft to the steering knuckle
   - Halfshaft from the steering knuckle
   - Left halfshaft from the transaxle
   - Snapring from the halfshaft.
   - Right halfshaft from the transaxle

➡ **The lockbolt is located in the center of the halfshaft, near the dampener.**

To install:

4. Install the right halfshaft to the transaxle, as follows:
   a. Coat the side gear shaft and differential case sliding surface with gear oil.
   b. Using snapring pliers, install the snapring to the halfshaft.
   c. Install the halfshaft and the bearing lockbolt. Lockbolt: 24 ft. lbs. (32 Nm).
5. Install the left halfshaft to the transaxle, as follows:
   a. Install a new snapring to the inner spline of the halfshaft.
   b. Coat the side gear shaft and differential case sliding surface with gear oil.
   c. Install the halfshaft to the transaxle with the snapring opening facing down. The halfshaft should click into place when installing.
   d. After installation of the halfshaft, check that the halfshaft cannot be removed by hand.

6. Install or connect the following:
   - Halfshaft to the steering knuckle, then install the locknut. Locknut: 217 ft. lbs. (294 Nm).
   - Lock cap and new cotter pin to the halfshaft
   - Steering knuckle to the lower ball joint. Nuts and bolt: 94 ft. lbs. (127 Nm).
   - Stabilizer bar link to the lower control arm. Nut: 29 ft. lbs. (39 Nm).
   - Tie rod to the steering knuckle. Nut: 36 ft. lbs. (49 Nm).
   - New cotter pin to the tie rod end
   - Front fender apron seal
   - Front wheels
   - Negative battery cable
7. Fill the transaxle fluid to the proper level.
8. Start the vehicle, check for leaks and repair if necessary.

Echo

1. Before servicing the vehicle, refer to the precautions in the beginning of this section.
2. Drain the transaxle fluid.
3. Remove or disconnect the following:
   - Negative battery cable. On vehicles equipped with an air bag, wait at least 90 seconds before proceeding.
   - Both front wheels
   - Locknut holding the halfshaft to the steering knuckle
   - Tie rod end from the steering knuckle by removing the cotter pin and nut. Separate the tie rod from the steering knuckle

   - Stabilizer bar link from the lower control arm. Make note of the washers and cushions positions
   - Lower ball joint from the steering knuckle. Push down on the lower control arm and separate the steering knuckle from the ball joint
   - Halfshaft from the steering knuckle
   - Left halfshaft from the transaxle
   - Snapring from the halfshaft
   - Right halfshaft from the transaxle

➡ **The lockbolt is located in the center of the halfshaft, near the dampener.**

To install:

4. Install the right halfshaft to the transaxle, as follows:
   a. Coat the side gear shaft and differential case sliding surface with gear oil.
   b. Using snapring pliers, install the snapring to the halfshaft.
   c. Install the halfshaft
5. Install the left halfshaft to the transaxle, as follows:
   a. Install a new snapring to the inner spline of the halfshaft.
   b. Coat the side gear shaft and differential case sliding surface with gear oil.
   c. Install the halfshaft to the transaxle with the snapring opening facing down. The halfshaft should click into place when installing.
   d. After installation of the halfshaft, check that the halfshaft cannot be removed by hand.
6. Install or connect the following:
   - Halfshaft to the steering knuckle,

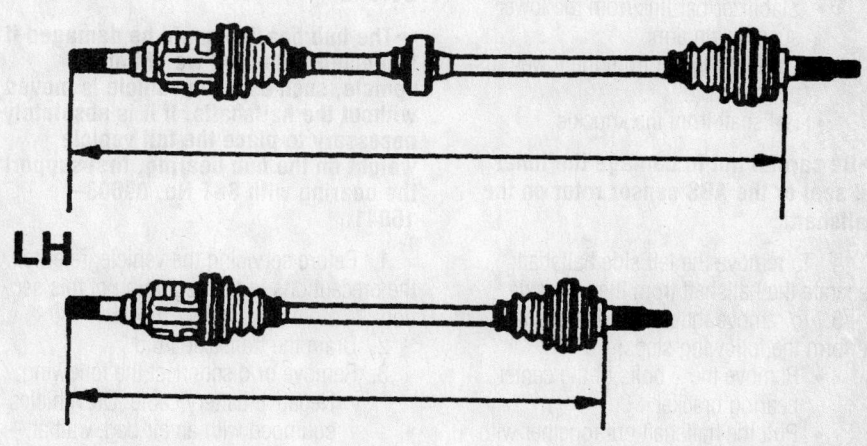

RH

LH

Measuring halfshaft length—Echo

9307WG84

Heater Core replacement is covered in Section 2 of this manual

then install the locknut. Locknut: 159 ft. lbs. (216 Nm).

- Lock cap and new cotter pin to the halfshaft
- Steering knuckle to the lower ball joint. Nut: 72 ft. lbs. (98 Nm).
- Stabilizer bar link to the lower control arm. Nut: 29 ft. lbs. (39 Nm).
- Tie rod to the steering knuckle. Nut: 36 ft. lbs. (49 Nm).
- New cotter pin to the tie rod end
- Both front wheels
- Negative battery cable

7. Fill the transaxle fluid to the proper level.

8. Start the vehicle, check for leaks and repair if necessary.

Celica and Celica GT-S

➡The hub bearing could be damaged if subjected to the full weight of the vehicle, such as if the vehicle is moved without the halfshafts. If it is absolutely necessary to place the full vehicle weight on the hub bearing, first support the bearing with SST No. 09608–16041.

1. Before servicing the vehicle, refer to the precautions in the beginning of this section.

2. Drain the transaxle fluid.

3. Remove or disconnect the following:
- Negative battery cable. On vehicles equipped with an air bag, wait at least 90 seconds before proceeding
- Both front wheels
- Cotter pin, locknut cap, and the bearing locknut
- Undercovers
- Tie rod ball joint from the steering knuckle
- Stabilizer bar link from the lower suspension arm
- Lower ball joint from the lower suspension arm
- Halfshaft from the knuckle

➡Be careful not to damage the inner oil seal or the ABS sensor rotor on the halfshaft.

4. To remove the left side halfshaft, separate the halfshaft from the transaxle.

5. To remove the right side halfshaft perform the following steps:
- Remove the 2 bolts of the center bearing bracket
- Pull the halfshaft out together with the center bearing case and the center halfshaft.
- Remove the center shaft with the right-hand halfshaft from the

transaxle through the bearing bracket.

➡Do not damage the oil seal lip.

To install:

6. Install or connect the following:
- Snapring opening side facing downward, on the oiled inboard joint tulip
- Left side halfshaft into the transaxle
- Right side halfshaft, with the bearing case and center shaft, into the transaxle
- Center bearing case (right side). Bolts: 47 ft. lbs. (64 Nm)

7. After installing either halfshaft, check that there is 0.08–0.12 in. (2–3mm) of axial play. Check that the halfshaft is making contact with the pinion shaft and that the halfshaft cannot be pulled out.

8. Install or connect the following:
- Halfshaft into the knuckle
- Lower suspension arm to the lower ball joint. Bolt and nuts: 94 ft. lbs. (127 Nm).
- Tie rod end to the steering knuckle. Nut: 36 ft. lbs. (49 Nm).
- Stabilizer bar link to the lower suspension arm. Nuts: 33 ft. lbs. (44 Nm).
- Front wheels
- Locknut and washer. Locknut: 159 ft. lbs. (216 Nm).
- Negative battery cable
- Locknut cap and a new cotter pin.
- Undercover

9. Fill the transaxle fluid to the proper level

10. Start the vehicle, check for leaks and repair if necessary.

Corolla

➡The hub bearing could be damaged if subjected to the full weight of the vehicle, such as if the vehicle is moved without the halfshafts. If it is absolutely necessary to place the full vehicle weight on the hub bearing, first support the bearing with SST No. 09608–16041.

1. Before servicing the vehicle, refer to the precautions in the beginning of this section.

2. Drain the transaxle fluid.

3. Remove or disconnect the following:
- Negative battery cable. On vehicles equipped with an air bag, wait at least 90 seconds before proceeding
- Cotter pin, locknut cap, and bearing locknut
- Front wheels

- Undercovers
- With ABS, speed sensor
- Tie rod ball joint from the steering knuckle
- Lower ball joint from the lower suspension arm

4. Drive the halfshaft from the knuckle.

➡Most halfshafts can be separated from the knuckle using a brass or plastic hammer; some others may require the use of a puller.

5. Remove the halfshaft from the transaxle

To install:

6. Install or connect the following:
- Snapring, opening side facing downward, to the inboard, oiled, joint tulip
- Halfshaft into the transaxle. After installing the halfshaft to the transaxle, check that there is 0.08–0.12 in. (2–3mm) of axial play. Check that the halfshaft is making contact with the pinion shaft and that the halfshaft cannot be pulled out
- Halfshaft into the knuckle
- Lower suspension arm to the steering knuckle. Nuts and bolts: 105 ft. lbs. (142 Nm).
- Tie rod end to the steering knuckle. Nut: 36 ft. lbs. (49 Nm).
- ABS speed sensor
- Hub locknut and washer
- Negative battery cable
- Wheels
- Locknut cap and NEW cotter pin.
- Undercovers

7. Fill the transaxle fluid to the proper level.

8. Start the vehicle, check for leaks and repair if necessary.

Tercel

1. Before servicing the vehicle, refer to the precautions in the beginning of this section.

2. Drain the transaxle fluid.

3. Remove or disconnect the following:
- Negative battery cable. On vehicles equipped with an air bag, wait at least 90 seconds before proceeding.
- Left engine undercover
- Front wheels
- With ABS, ABS speed sensor
- Cotter pin and lock cap from the halfshaft
- Hub nut
- Tie rod from the steering knuckle

- Lower ball joint from the lower control arm
- Halfshaft from the axle hub
- Halfshaft from the transaxle

To install:

4. Install or connect the following:
- Halfshaft into the transaxle until it clicks into position Use a new snapring
- Outer joint into the axle hub
- Lower ball joint to the lower arm. Bolt and nuts: 59 ft. lbs. (80 Nm).
- Tie rod to the knuckle. Nut: 36 ft. lbs. (49 Nm).
- Hub nut: 152 ft. lbs. (206 Nm).
- New cotter pin
- Undercover and wheels
- Negative battery cable

5. Fill the transaxle fluid to the proper level.

6. Start the vehicle, check for leaks and repair if necessary.

Supra

1. Before servicing the vehicle, refer to the precautions in the beginning of this section.

2. Remove or disconnect the following:
- Negative battery cable. On vehicles equipped with an air bag, wait at least 90 seconds before proceeding.
- Rear exhaust assembly
- Cotter pin, locknut cap, and the locknut holding the halfshaft to the rear axle carrier.

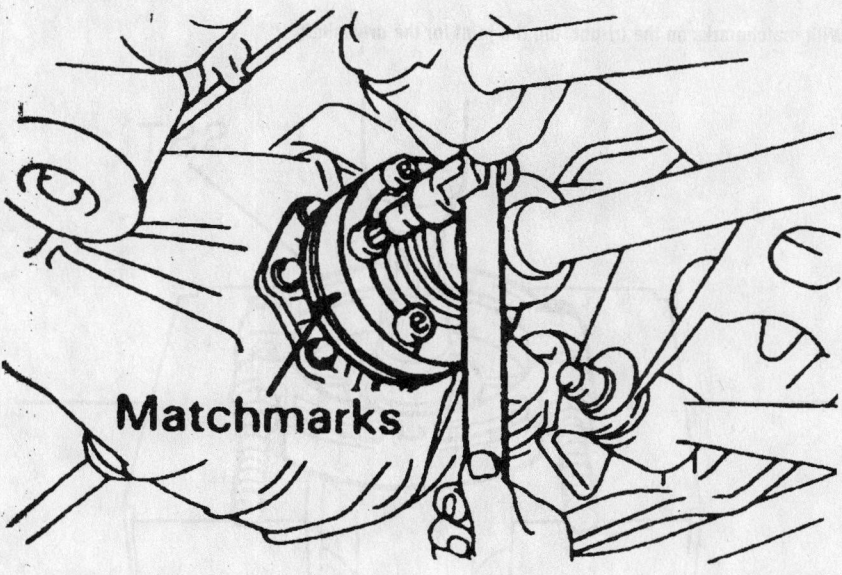

Matchmarks

Whenever removing the halfshaft, be sure to matchmark it to the differential side gear—Supra

---

- Lower suspension arm brace

3. Place matchmarks on the halfshaft and the differential side gear shaft.

4. Remove or disconnect the following:
- Hexagon bolts and washers
- Halfshaft

To install:

5. Install or connect the following:
- Halfshaft
- Hexagon bolts, oiled. Bolts: 61 ft. lbs. (83 Nm).
- Lower suspension arm brace. Bolts: 13 ft. lbs. (18 Nm).
- Bearing locknut: 213 ft. lbs. (289 Nm).
- Locknut cap and new cotter pin
- Rear exhaust assembly
- Negative battery cable

CV-Joints

OVERHAUL

Avalon and Camry

1. Before servicing the vehicle, refer to the precautions in the beginning of this section.

2. Remove or disconnect the following:
- Boot clamps
- Inboard joint tulip
- Snapring from the tri-pot

3. Place matchmarks on the driveshaft and tri-pot

---

4. Remove or disconnect the following:
- Tri-pot joint off the driveshaft without hitting the joint roller
- Inboard joint boot
- On the right side, the clamp and dynamic damper
- Clamps and the outboard drive boot. DO NOT disassemble the outboard joint
- Dust cover from the inboard joint using a press

➡ **If equipped, be careful not to damage the ABS speed sensor rotor.**

To assemble:

5. Install or connect the following:
- Using a press, the inboard joint tulip into a new dust cover
- On the right side, the clamp position in line with the groove of the halfshaft
- Seal packing to the inboard joint cover
- Inboard joint cover
- Bolts and nuts and washers to keep the joint together. Tighten the bolts by hand
- Outboard tulip joint and the outboard boot with about 4.8–5.5 ounces of grease
- Boot onto the outboard joint

6. Pack the inboard tulip joint and boot with grease that was supplied with the boot kit.

7. Install or connect the following:
- Inboard tulip joint onto the halfshaft
- Boot onto the halfshaft

8. Before checking the standard length, bend the band and lock it.

9. Make sure that the boot is not stretched or squashed when the driveshaft is at standard length. Standard driveshaft length: 34.14 +/- 0.099 inch (867.3 +/- 2.5mm)

10. After making sure that the boots are in the shaft groove, secure the band

11. Pack in grease to the center driveshaft or side gear shaft. Use 5160 g (1.82.1 oz.)

12. Connect the driveshaft and the center driveshaft or the side gear shaft, placing a new gasket on the inboard joint without compressing the inboard boot.

13. Check to see that there is no play in the inboard and outboard joints and that the inboard joint slides smoothly in the thrust direction.

7923VGA4

Brake service is covered in Section 4 of this manual

Echo

1. Before servicing the vehicle, refer to the precautions in the beginning of this section.

➡ **The outboard joint cannot be disassembled.**

2. Remove or disconnect the following:
   • The 2 inboard boot clamps and slide the clamp down the shaft

➡ **Paint-mark the inboard joint shaft and tri-pot joint.**

   • Inboard joint shaft from the outboard joint shaft
   • Snapring from the tri-pot joint, and paint-mark the tri-pot and outboard joint shaft
   • Tri-pot joint from the shaft with a brass hammer
   • Inboard boot and clamps
   • Damper (right shaft)
   • Outboard joint boot
   • Dust cover from the inboard joint shaft (press)

To assemble:

3. Install or connect the following:
   • Dust cover
   • Boots and clamps, loosely
   • Damper (right shaft)
   • Tri-pot joint, beveled edge toward the outboard shaft. Align the matchmarks and drive it into place
   • New snapring
   • Outboard boot. The grease capacity is 5.5–6 ounces (155–170 g).
   • Inboard shaft to outboard shaft
   • Inboard boot. The grease capacity is 4.4–4.8 ounces (125–135 g).

➡ **Toyota recommends different types of grease for the joints. OEM replacement boot kits have the grease color coded. The outboard grease is black; the inboard, yellow.**

4. Check the boots at standard halfshaft length. The right shaft should be 32.02 inches +/- 0.197 in. (813.3mm +/- 5mm); the left shaft should be 22.61 inches +/- 0.197 in. (574.3mm +/- 5mm).

5. Check the position of the damper before installing the clamp. Distance from the outer face of the damper to the outer face of the outer joint should be 16.835 inches +/- 0.079 in. (427.6mm +/- 2mm).

Celica, Supra, GT-S

5S-FE, 1ZZ-FE AND 2ZZ-GE ENGINES

1. Before servicing the vehicle, refer to the precautions in the beginning of this section.

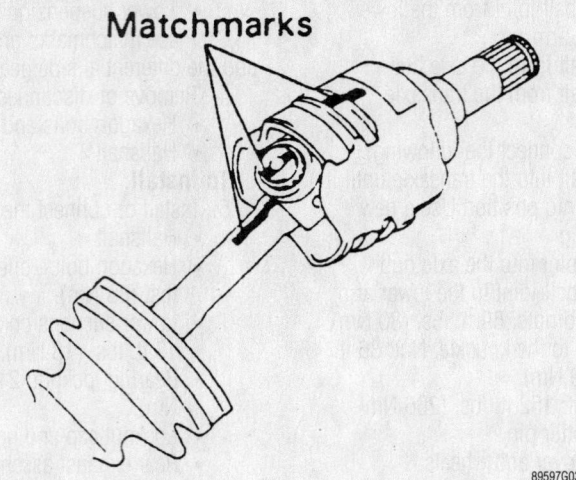

Place matchmarks on the tri-pot and outboard joints

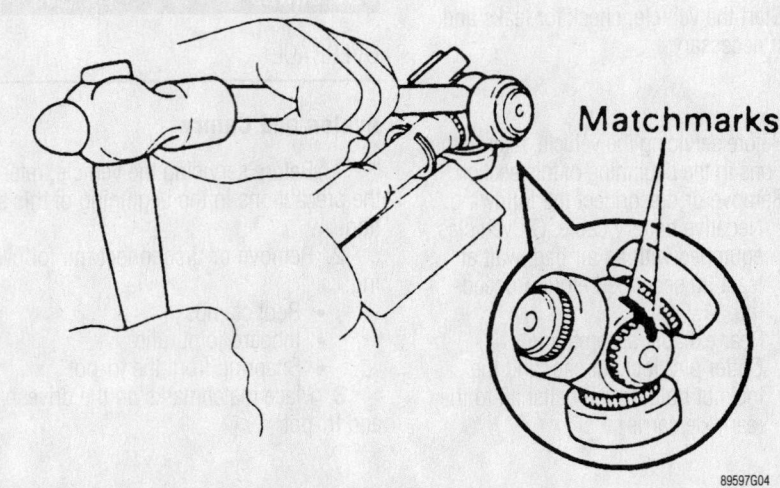

With matchmarks on the tri-pot, tap the joint for the driveshaft

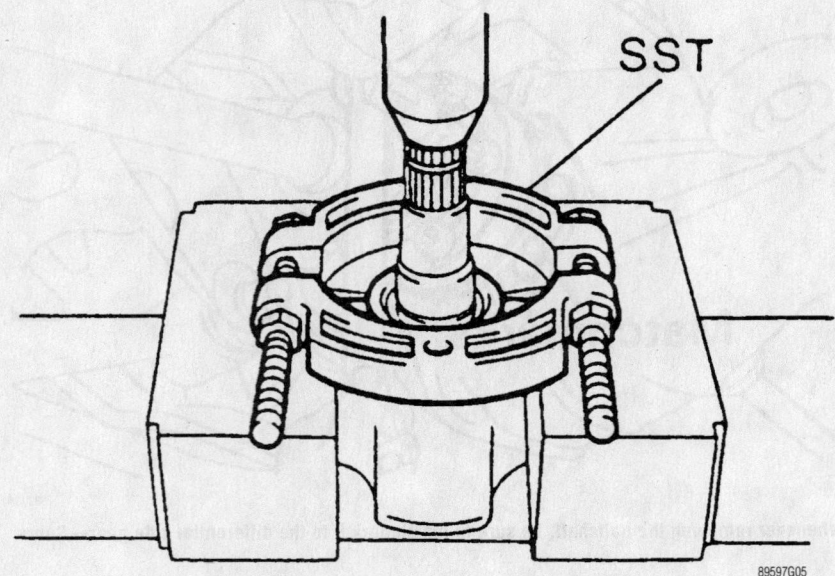

Using a press, remove the dust cover from the center driveshaft on the left side

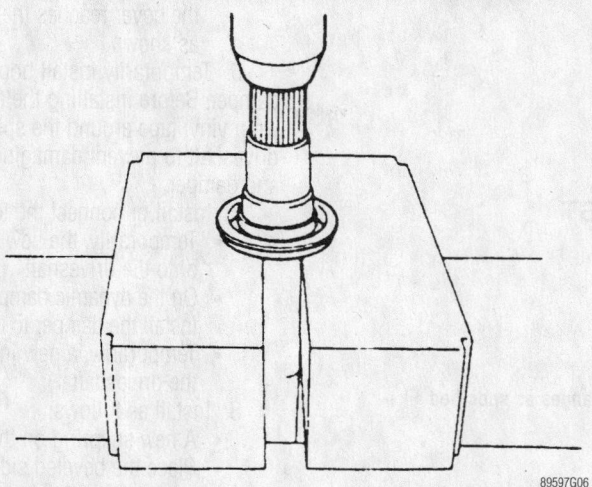

Use a press to drive the old dust cover out of the right side center driveshaft

89597G06

w/o ABS   w/ ABS

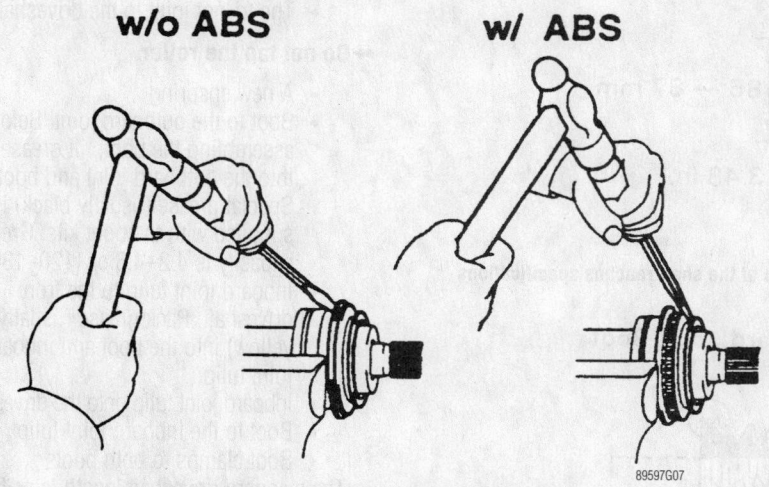

89597G07

Removing the no. 2 dust deflector on models with and without ABS

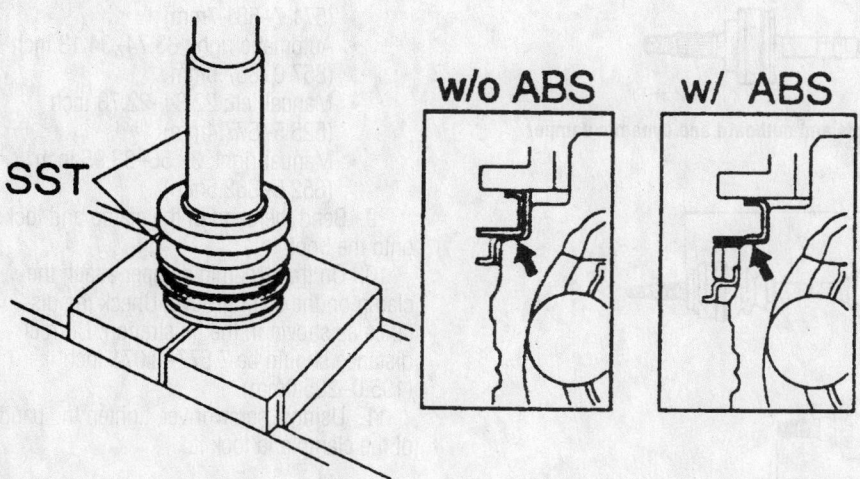

Press the new No. 2 deflector and seat it properly

89597G08

2. Remove or disconnect the following:
- Halfshaft
- On the inboard joint tulip, the boot clamps. Slide the inboard joint boot toward the outboard joint.

➡ **Place matchmarks on the tri-pot and inboard joint tulip or center driveshaft. Do not punch the marks.**

- Inboard joint tulip or center driveshaft from the driveshaft

3. On the tri-pot, use snapring expander, temporarily, slide the snapring toward the outboard joint side.

➡ **Place matchmarks on the driveshaft and tri-pot. Do not punch the marks.**

4. Remove or disconnect the following:
- Tri-pot from the driveshaft
- Snapring
- Inboard joint boot
- On the right side with dynamic damper, the clamps from the damper and extract the unit
- Clamps from the outboard joint
- Boot from the outboard joint

➡ **Do not disassemble the outboard joint.**

- On the left side, the dust cover from the inboard joint tulip
- On the right side, the dust cover from the center driveshaft
- On the right side, the snapring and bearing from the case
- The dust cover and snapring
- The bearing from the shaft and the other snapring
- The straight pin
- The No. 2 dust deflector

To assemble:

5. Install or connect the following:
- On the right side, assemble the center driveshaft by inserting the straight pin into the bearing case
- A new bearing into the case
- A new snapring
- A new bearing into the case assembly on the center driveshaft
- A new snapring
- New dust cover. The clearance between the dust cover and bearing should be kept in the ranges in the illustration.
- On the left side, use a press and insert the new dust cover
- On the right side, using a steel plate and press, press in a new dust cover until the distance from the tip of the center driveshaft to

For complete Engine Mechanical specifications, see Section 1 of this manual

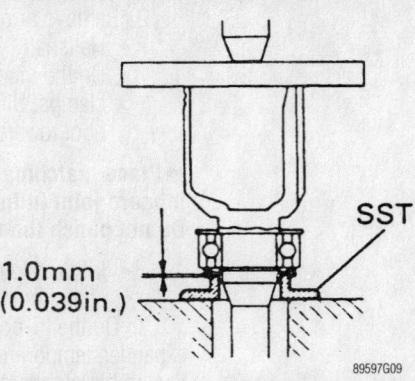

1.0mm (0.039in.)

SST

89597G09

When installing the dust cover, the clearance should be within the ranges as specified

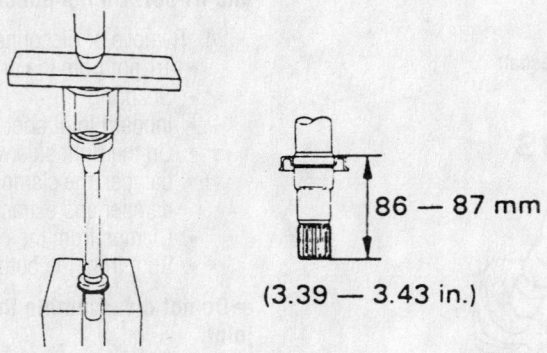

86 — 87 mm

(3.39 — 3.43 in.)

89597G10

On the right side driveshaft, press the dust cover till the tip of the shaft reaches specifications

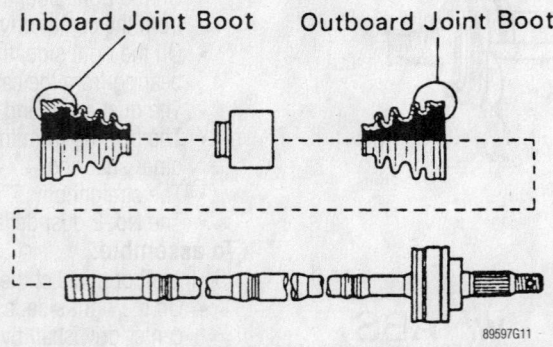

Inboard Joint Boot   Outboard Joint Boot

89597G11

Temporarily install the driveshaft boots on the inboards and outboard and dynamic damper

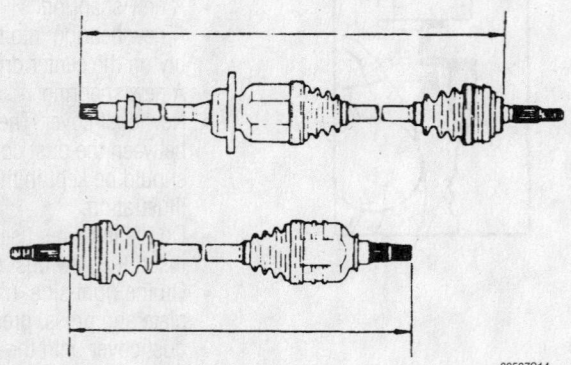

89597G14

Measure the driveshaft length at these points and make sure it reaches specifications prior to installation

the cover reaches the specification as shown

6. Temporarily install boots and the damper. Before installing the outboard joint, wrap vinyl tape around the spline of the driveshaft to prevent damaging the boots and damper.

7. Install or connect the following:
- Temporarily, the new outboard boot onto the driveshaft
- On the dynamic damper, temporarily install the damper to the driveshaft.
- Temporarily, a new inboard joint to the driveshaft

8. Install as follows:
- A new snapring on the tri-pot joint
- Place the beveled side of the tri-pot joint axial spline toward the outboard joint. Align the matchmarks placed before removal
- The tri-pot joint to the driveshaft

➡ **Do not tap the roller.**

- A new snapring
- Boot to the outboard joint. Before assembling the boot, fill grease into the outboard joint and boot. Special grease (usually black) is supplied with the boot kits. Grease capacity is 4.2–4.6 oz (120–130 g)
- Inboard joint tulip to the front driveshaft. Pack grease (usually yellow) into the boot and inboard joint tulip.
- Inboard joint tulip onto the driveshaft
- Boot to the inboard joint tulip
- Boot clamps to both boots

The standard driveshaft length is as follows:

- Automatic left: 22.516–22.910 inch (571.7–581.7mm)
- Automatic right: 33.74–34.13 inch (857.0–867.0mm
- Manual left: 22.34–22.73 inch (525.5–577.4mm)
- Manual right: 33.56–33.95 inch (852.5–862.5mm)

9. Bend the band of the clamp and lock onto the boot.

10. On the dynamic damper attach the clamp on the shaft groove. Check the distance as shown in the illustration. Correct distance should be 7.677–8.071 inch (195.0–205.0mm).

11. Using a screwdriver, tighten the band of the clamp and lock it.

2JZ-GE ENGINES

1. Before servicing the vehicle, refer to the precautions in the beginning of this section.

2. Remove or disconnect the following:

- Halfshaft
- Inboard joint boot clamps
- Inboard joint tulip
- Tri-pot joint
- Inboard joint boot
- On the dynamic damper (right side), the clamp from the damper
- Dynamic damper
- Outboard joint boot, but do not disassemble the joint.
- Dust cover from the inboard joint tulip using a press

To assemble:

3. Install or connect the following:
- New dust cover onto the inboard joint tulip
- Temporarily, boots and the right damper. Before installing the outboard joint boot, wrap vinyl tape around the spline of the driveshaft to prevent damaging the boots and damper
- Temporarily, a new outboard boot to the driveshaft
- On the right side, the dynamic damper to the driveshaft
- Temporarily, the new inboard joint boot to the driveshaft

4. Place the beveled side of the tri-pot joint axial spline toward the outboard joint. Align the matchmarks placed before removal.

5. Install or connect the following:
- Using a brass bar and hammer, the tri-pot joint to the driveshaft. Do not tap the roller.
- A new snapring
- Boot to the outboard joint

➡**Before assembling the boot, fill it with the grease (black), supplied with** the kit, into the outboard joint and boot.

- The inboard joint tulip to the front driveshaft. Pack grease into the inboard joint tulip and boot. The grease should be yellow in color and supplied with the boot kit
- The inboard joint tulip to the driveshaft
- Temporarily, the boot to the inboard joint tulip
- The boot clamps on both boots

6. Be sure the boots are on the shaft groove. Set the driveshaft length to: left—21.81–22.21 inch (554.2–564.2mm); right—34.33–34.73 inch (872.25–882.2mm)

7. Using a screwdriver, bend the band and lock it.

8. On the right side driveshaft, place the clamp on in the shaft groove. Check that the distance is within specifications. Distance should be 15.25–15.65 inch (387.5–397.5mm).

9. Using a screwdriver, bend the band and lock it.

Corolla and Tercel

1. Before servicing the vehicle, refer to the precautions in the beginning of this section.

2. Remove or disconnect the following:
- Inboard joint boot clips
- Inboard joint tulip from the driveshaft
- Snapring
- Using a brass rod and hammer, the tri-pot joint off the driveshaft without hitting the joint roller

- Inboard joint boot
- Clamp and driveshaft damper
- Clamps and the outboard drive boot. DO NOT disassemble the outboard joint.

To assemble:

3. Install or connect the following:

➡**Before installing the boot, wrap the spline end of the shaft with masking tape to prevent damage to the boot.**

- Driveshaft damper with a new clamp
- Temporarily, the inboard boot with new clamp to the drive joint

➡**The inboard boot and clamp are larger than those of the outboard boot.**

- The tri-pot onto the driveshaft with a brass rod and hammer without hitting the joint roller
- The snapring

4. Pack the outboard tulip joint and the outboard boot with about 0.26–0.33 lbs. ounces of grease that was supplied with the boot kit

5. Install or connect the following:
- Boot onto the outboard joint
- Inboard tulip joint and boot with ½ lb. of grease that was supplied with the boot kit
- Inboard tulip joint onto the driveshaft
- Boot onto the driveshaft

6. Before checking the standard length, bend the band and lock it. Make sure that the boot is not stretched or squashed when the driveshaft is at standard length. Standard driveshaft length: LH: 540.2 mm (21.268 in.); RH: 857.4 mm (33.756 in.).

STEERING AND SUSPENSION

Air Bag

PRECAUTIONS

Several precautions must be observed when handling the inflator module to avoid accidental deployment and possible personal injury.

• Never carry the inflator module by the wires or connector on the underside of the module.

• When carrying a live inflator module, hold securely with both hands, and ensure that the bag and trim cover are pointed away.

• Place the inflator module on a bench or other surface with the bag and trim cover facing up.

• With the inflator module on the bench, never place anything on or close to the module that may be thrown in the event of an accidental deployment.

DISARMING

To avoid personal injury when working on vehicles equipped with an air bag, the negative battery cable must be disconnected and at least 90 seconds must elapse before working on the system. Failure to do so may result in deployment of the air bag.

REARMING

After vehicle service is completed, reattach the battery cables (positive cable first!) to rearm the air bag system.

Rack and Pinion Steering Gear

REMOVAL & INSTALLATION

Manual

TERCEL

1. Before servicing the vehicle, refer to the precautions in the beginning of this section.
2. Position the front wheels straight ahead.
3. Remove or disconnect the following:
   • Negative battery cable. On vehicles equipped with an air bag, wait at least 90 seconds before proceeding

➡ **If equipped with an air bag, disable the system and secure the steering wheel.**

   • Sliding yoke on rack and pinion assembly

   • Front wheels
   • Tie rod ends
   • Rack and pinion assembly bracket bolts and rack and pinion assembly

To install:
4. Install or connect the following:
   • Rack and pinion assembly. Bolts: 43 ft. lbs. (58 Nm).
   • Tie rod ends to the knuckle arms
   • Sliding yoke onto rack and pinion assembly. Torque the bolts to 19 ft. lbs. (26 Nm).
   • Front wheels
   • Negative battery cable.

ECHO

1. Before servicing the vehicle, refer to the precautions in the beginning of this section.
2. Place the wheels in the straight-ahead position.
3. Remove or disconnect the following:
   • Steering wheel
   • Engine under covers
   • Tie rod ends
   • No. 2 column hole cover
   • Sliding yoke
   • Hood
4. Attach and engine crane to the engine for support.
5. Remove or disconnect the following:
   • Lower arms from the knuckles
   • Rear engine mount insulator
   • Front crossmember with the steering rack
   • Column hole cover sub-assembly
   • 4 bolts and nuts attaching the gear to the crossmember

➡ **Because the nut has its own stopper, do not turn the nut with the bolt tight.**

To install:
6. Install or connect the following:
   • Steering rack to crossmember. Torque the nuts to 54 ft. lbs. (74 Nm).
   • Column hole cover sub-assembly
   • Front suspension assembly. Front bolts: 52 ft. lbs. (70 Nm); rear bolts: 86 ft. lbs. (116 Nm).
   • Rear mount insulator. Torque the bolt and 2 nuts to 59 ft. lbs. (80 Nm).
   • Lower arm to knuckle. Horizontal bolt: 65 ft. lbs. (88 Nm); vertical bolt: 97 ft. lbs. (132 Nm).
   • Hod
   • Sliding yoke
   • No. 2 column hole cover
   • Tie rods

   • Under covers
   • Steering wheel. Nut: 25 ft. lbs. (34 Nm).

Power

ECHO

1. Before servicing the vehicle, refer to the precautions in the beginning of this section.
2. Place the wheels in the straight-ahead position.
3. Remove or disconnect the following:
   • Steering wheel
   • Engine under covers
   • Tie rod ends
   • No. 2 column hole cover
   • Sliding yoke
   • Pressure and return lines
   • Hood
4. Attach and engine crane to the engine for support.
5. Remove or disconnect the following:
   • Lower arms from the knuckles
   • Rear engine mount insulator
   • Front crossmember with the steering rack
   • Stabilizer bar
   • Heat insulator
   • Damper
   • 4 bolts and nuts attaching the gear to the crossmember

➡ **Because the nut has its own stopper, do not turn the nut with the bolt tight.**

To install:
6. Install or connect the following:
   • Steering rack to crossmember. Torque the nuts to 54 ft. lbs. (74 Nm).
   • Damper. Bolts: 13 ft. lbs. (18 Nm).
   • Heat insulator. Bolt: 26 ft. lbs. (35 Nm).
   • Stabilizer bar
   • Front suspension assembly. Front bolts: 52 ft. lbs. (70 Nm); rear bolts: 86 ft. lbs. (116 Nm).
   • Rear mount insulator. Torque the bolt and 2 nuts to 59 ft. lbs. (80 Nm).
   • Lower arm to knuckle. Horizontal bolt: 65 ft. lbs. (88 Nm); vertical bolt: 97 ft. lbs. (132 Nm).
   • Hood
   • Sliding yoke
   • No. 2 column hole cover
   • Tie rods
   • Pressure and return lines
   • Under covers
   • Steering wheel. Nut: 25 ft. lbs. (34 Nm).

AVALON, CAMRY AND CAMRY SOLARA

1. Before servicing the vehicle, refer to the precautions in the beginning of this section.

2. Position the front wheels straight ahead.

3. Remove or disconnect the following:
- Negative battery cable. On vehicles equipped with an air bag, wait at least 90 seconds before proceeding.

➡**If equipped with an air bag, disable the system and secure the steering wheel.**

- Front wheels
- Left and right front fender apron seals.
- Cotter pin and nut holding the steering knuckle to the tie rod end.
- Tie rod end from the steering knuckle.

➡**Place matchmarks on the intermediate shaft and the control valve shaft.**

- Lower bolt holding the control valve shaft to the intermediate shaft.
- Intermediate shaft from steering rack housing.
- Tube clamp
- Return line and the pressure line from the control valve housing.
- Stabilizer bar bolts and nuts. Do not remove the bar from the vehicle.
- If necessary, rear engine mounting and bracket for additional clearance.
- On the V6 engine, oxygen sensor.
- Steering gear mounting bolts and nuts
- Steering gear

To install:

4. Install or connect the following:
- Steering gear. Nuts and bolts: 134 ft. lbs. (181 Nm).
- On the V6, oxygen sensor.
- Rear engine mounting bracket. Bolts: 38 ft. lbs. (52 Nm).
- Stabilizer bar bolts and nuts.
- Tube clamp. Nut: 84 inch lbs. (10 Nm).
- Intermediate shaft to the steering rack. Bolts: 26 ft. lbs. (35 Nm).
- Tie rods to the steering knuckles with the castellated nuts.
- Front fender apron seals
- Front wheels

- Negative battery cable
- Power steering fluid

CELICA AND GT-S

1. Before servicing the vehicle, refer to the precautions in the beginning of this section.

2. Remove or disconnect the following:

- Negative battery cable. On vehicles equipped with an air bag, wait at least 90 seconds before proceeding.
- Right and left-hand engine undercovers.
- Left and right-hand tie rod ends. Separate the tie rod using a puller.
- Oxygen sensor
- Front exhaust pipe and brackets.

➡**Place matchmarks on the steering column intermediate shaft and the steering gear control valve shaft.**

- Lower bolt to the intermediate shaft
- Intermediate shaft from the control valve shaft.
- Pressure feed and return tubes from the steering rack.
- Tube clamp bracket

3. Support the engine and transaxle with a support fixture.

4. Remove or disconnect the following:
- Lower control arms from the lower ball joints.
- Through-bolts to the front and rear mounting insulators.
- Bolts holding the rear of the lower control arm to the sub-frame and body. Remove the bolts on both sides of the sub-frame.

5. Support the front sub-frame with a jack.

6. Remove or disconnect the following:
- Bolts holding the sub-frame to the body. Lower the front sub-frame with the lower suspension arms and steering gear.
- Power steering gear assembly by removing the set bolts and nuts from the sub-frame.

To install:

7. Install or connect the following:
- Power steering gear assembly to the sub-frame. Bolts and nuts: 94 ft. lbs. (127 Nm).
- Sub-frame to the body. Bolts: 94 ft. lbs. (127 Nm).
- Bolts to hold the rear of the lower control arms to the sub-frame and body. Bolt that goes through the

lower control arm to 123 ft. lbs. (167 Nm) and the other 2 bolts to 130 ft. lbs. (175 Nm). Install the bolts to both sides
- Through-bolts to the front and rear mounting insulators. Bolts: 64 ft. lbs. (88 Nm).
- Lower control arms to the lower ball joints. Nuts and bolts: 94 ft. lbs. (127 Nm).
- Pressure feed and return tubes to the steering rack. Tubes: 26 ft. lbs. (36 Nm).
- Tube clamp bracket. Bolts: 108 inch lbs. (13 Nm).

8. Align the matchmarks on the intermediate shaft and control valve shaft.

9. Install or connect the following:
- Lower bolt. Upper and lower bolts: 26 ft. lbs. (35 Nm).
- Front exhaust pipe
- Oxygen sensor
- Tie rod ends to the steering knuckles
- Right and left-hand engine undercovers
- Negative battery cable
- Power steering fluid

COROLLA

1. Before servicing the vehicle, refer to the precautions in the beginning of this section.

2. Position the front wheels straight ahead.

3. Remove or disconnect the following:
- Negative battery cable. On vehicles equipped with an air bag, wait at least 90 seconds before proceeding

➡**If equipped with an air bag, disable the system and secure the steering wheel.**

- Steering column hole cover and loosen the upper pinch bolt on the sliding yoke
- Lower pinch bolt at the pinion shaft
- Front wheels
- Left and right engine undercovers.
- Left and right tie rod ends.

4. Install an engine support and tension it to support the engine without raising it.

✳✳ CAUTION

The engine hoist is now in place and under tension. Use care when repositioning the vehicle and make necessary adjustments to the engine support.

Timing belt service is covered in Section 3 of this manual

5. Remove or disconnect the following:
- Lower control arms from the ball joints
- If equipped with a stabilizer bar, stabilizer bar links from both lower control arms
- Right rear control arm bushing retaining bracket. Do this for both lower control arms
- Stabilizer bar
- Grommet in the crossmember
- Bolt and nuts holding in the middle of the crossmember and support the crossmember with a jack
- Bolts from the outer side of the suspension crossmember
- Suspension crossmember with the lower suspension arms
- Exhaust front pipe support
- Engine rear mount insulator
- Engine rear mount bracket
- Pressure feed and return tubes
- Brackets and grommets to the power steering rack

6. Slide the power steering gear assembly to the right side of the vehicle.

To install:

7. Install or connect the following:
- Power steering assembly
- Grommets and brackets. Nuts and bolts: 43 ft. lbs. (59 Nm).
- Pressure feed and return tubes. Nuts: 26 ft. lbs. (36 Nm).
- Engine rear mount bracket. Bolts: 57 ft. lbs. (77 Nm).
- Engine rear mount insulator. Bolt: 64 ft. lbs. (87 Nm).
- Exhaust front pipe support. Bolts: 14 ft. lbs. (19 Nm).
- Outer 6 bolts to hold the crossmember to the vehicle. Bolts: 152 ft. lbs. (206 Nm).
- Center crossmember-to-radiator support bolts: 45 ft. lbs. (61 Nm).
- Lower A frame-to-center bolts: 161 ft. lbs. (218 Nm).
- Lower A frame-to-outer bolts: 109 ft. lbs. (147 Nm).
- Front, center and rear mount bolts: 45 ft. lbs. (61 Nm).
- Grommet to the crossmember
- Stabilizer bar
- Lower control arm bushing retaining bracket. Do not tighten the bolts or nut at this time
- Lower control arm to the lower ball joint. Bolt and nuts: 105 ft. lbs. (142 Nm). Connect both lower control arms to the ball joints
- Stabilizer bar links to the lower control arms. Nuts: 33 ft. lbs. (44 Nm).

- Sliding yoke to the pinion shaft. Lower bolt: 26 ft. lbs. (35 Nm). Tighten upper bolt: 20 ft. lbs. (27 Nm).
- Steering column hole cover. Bolts: 43 inch lbs. (5 Nm).
- Left and right-hand tie rod ends. Nuts: 36 ft. lbs. (49 Nm).
- Front wheels
- Control arm bracket bolts: 108 ft. lbs. (147 Nm).
- Stabilizer bar bracket bolt: 37 ft. lbs. (50 Nm).
- Bracket nut: 14 ft. lbs. (19 Nm).
- Negative battery cable

8. Check and top off the power steering fluid.

9. Check and adjust the alignment, if needed.

SUPRA

1. Before servicing the vehicle, refer to the precautions in the beginning of this section.

2. Position the front wheels straight ahead.

3. Remove or disconnect the following:
- Negative battery cable. On vehicles equipped with an air bag, wait at least 90 seconds before proceeding

➡**If equipped with an air bag, disable the system and secure the steering wheel.**

- Front wheels
- Undercovers and front suspension member protection

4. Place matchmarks on the universal joint and the control valve shaft.

5. Loosen the bolt on the upper side of the intermediate shaft.

6. Remove or disconnect the following:
- Bolt on the lower side of the intermediate shaft and disconnect the universal joint from the steering rack
- Hydraulic lines to the rack assembly by removing the union bolts and gaskets
- Tie rod ends from the steering knuckles
- Solenoid wiring from the rack and pinion unit
- Bolts and brackets holding the steering rack to the frame
- Rack assembly

To install:

7. Install or connect the following:
- Rack. Mounting bracket bolts: 55 ft. lbs. (75 Nm).
- Solenoid wiring

- Tie rod ends to the steering knuckles
- Fluid lines to the rack and pinion with new washers. Union bolts: 36 ft. lbs. (49 Nm).

8. Align the matchmarks on the universal joint and the control valve shaft. Tighten the upper and lower bolts to 26 ft. lbs. (35 Nm).

9. Install or connect the following:
- Front suspension member protection and the engine undercovers
- Front wheels
- Negative battery cable

10. Chedck and top off the power steering fluid.

11. Check and adjust the alignment, if necessary.

TERCEL

1. Before servicing the vehicle, refer to the precautions in the beginning of this section.

2. Position the front wheels straight ahead.

3. Remove or disconnect the following:

4. Negative battery cable. On vehicles equipped with an air bag, wait at least 90 seconds before proceeding

➡**If equipped with an air bag, disable the system and secure the steering wheel.**

- Tie rod ends from the knuckle arm
- Column hole cover

➡**Matchmark the sliding yoke and control valve shaft for installation.**

5. Loosen the upper bolt and disconnect the lower bolt to the control valve shaft. Slide the shaft upward and disconnect the control valve shaft from the steering rack.

6. Remove or disconnect the following:
- Oxygen sensor and exhaust pipe
- Stabilizer bar, if necessary for additional access
- Engine rear mount insulator bolts
- Vacuum hoses
- With a manual transaxle, transmission control cables
- Power steering hoses
- Tube clamp
- Brackets and grommets
- Housing-to-frame retaining bolts and remove the assembly. Slide the housing out the left-hand side of the vehicle

To install:

7. Line up the steering splines, then install the unit.

8. Install or connect the following:
- Grommets and brackets. Bolts and nuts: 43 ft. lbs. (58 Nm).

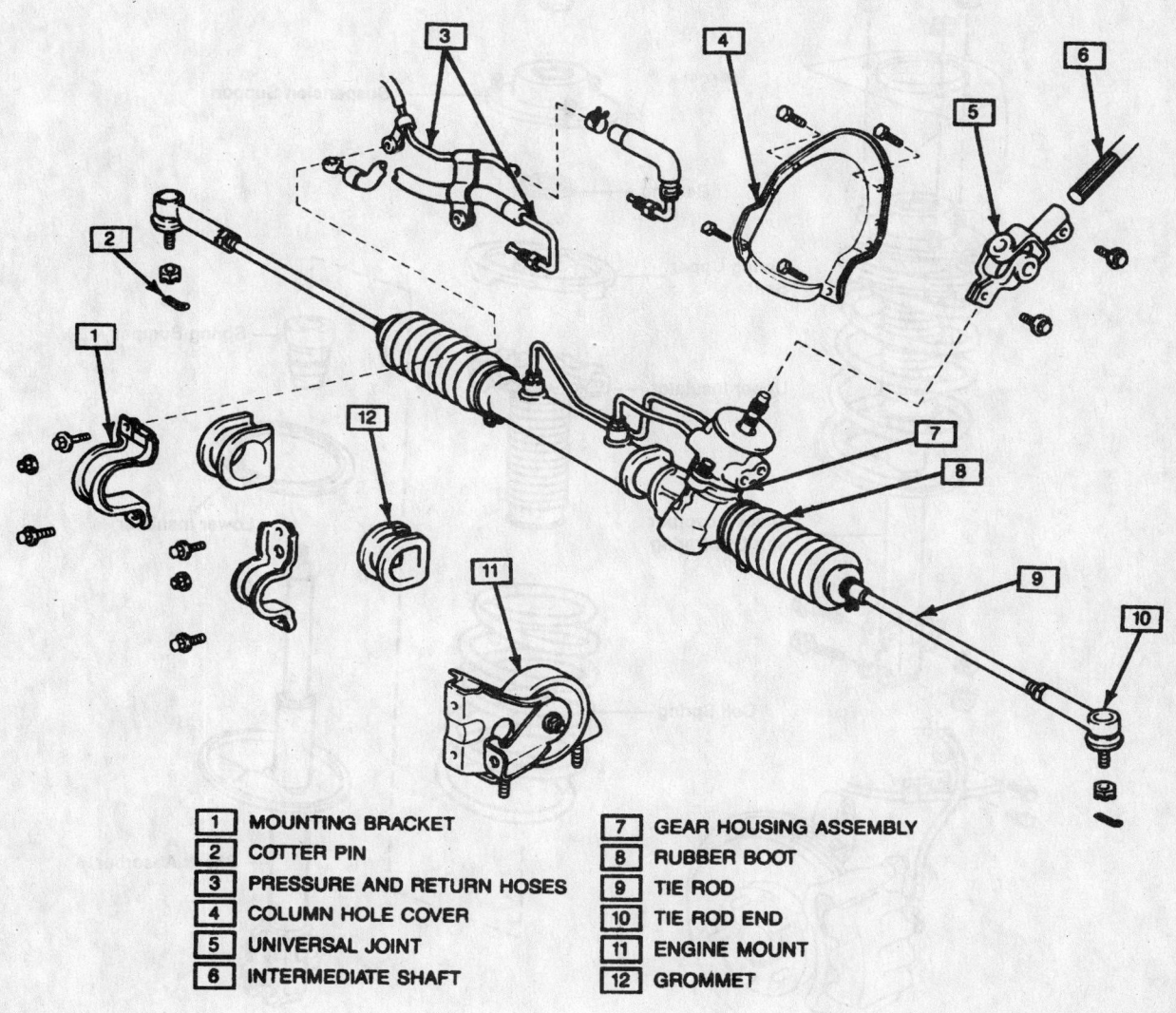

1	MOUNTING BRACKET	**7**	GEAR HOUSING ASSEMBLY
2	COTTER PIN	**8**	RUBBER BOOT
3	PRESSURE AND RETURN HOSES	**9**	TIE ROD
4	COLUMN HOLE COVER	**10**	TIE ROD END
5	UNIVERSAL JOINT	**11**	ENGINE MOUNT
6	INTERMEDIATE SHAFT	**12**	GROMMET

7923VGA5

Exploded view of a typical power rack and pinion steering gear unit

- Pressure feed and return tubes. Tighten to 26 ft. lbs. (36 Nm).
- Tube clamp. Bolt: 108 inch lbs. (13 Nm).
- Vacuum hoses
- Rear brackets. Bolts: 35 ft. lbs. (48 Nm).
- Engine rear mount insulator. Through-bolt: 47 ft. lbs. (64 Nm). Support bracket bolts: 58 ft. lbs. (78 Nm).
- Stabilizer bar
- Sliding yoke to the control valve shaft. Bolts: 21 ft. lbs. (28 Nm).
- Column hole cover. Nuts: 43 inch lbs. (5 Nm).
- Right and left tie rod ends

- Wheels
- Negative battery cable

9. Check and top off the power steering fluid.

10. Check and adjust the alignment, if necessary.

Strut and Coil Spring

REMOVAL & INSTALLATION

Front

EXCEPT SUPRA

1. Before servicing the vehicle, refer to the precautions in the beginning of this section.

2. Remove or disconnect the following:
- Negative battery cable. On vehicles equipped with an air bag, wait at least 90 seconds before proceeding

✳✳ WARNING

Do not support the weight of the vehicle on the suspension arm; the arm will deform under its weight.

- Wheel
- Bolt, and disconnect the brake hose from the strut
- With ABS brakes, wiring harness from the strut
- Bolts and strut from the steering knuckle

Heater Core replacement is covered in Section 2 of this manual

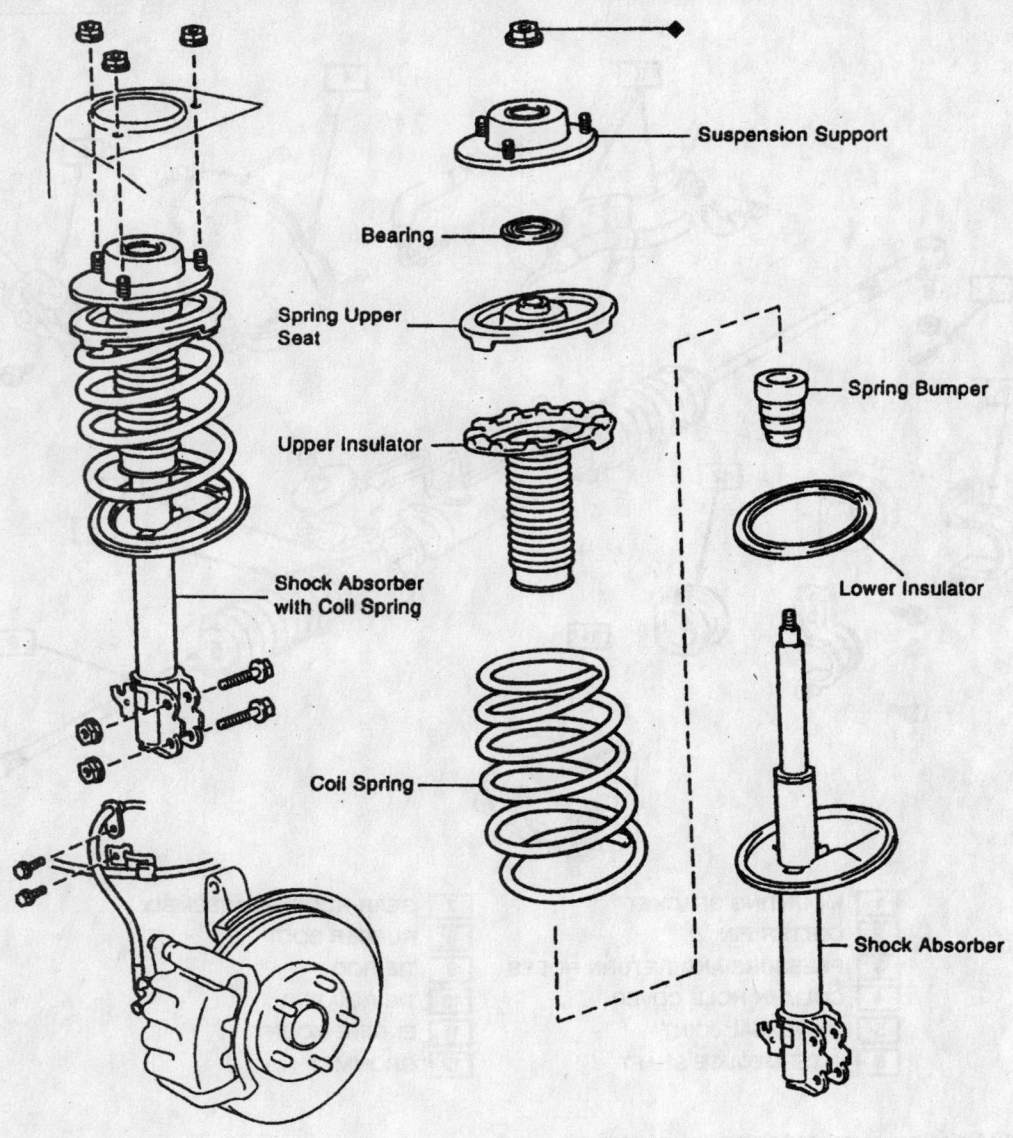

- Suspension Support
- Bearing
- Spring Upper Seat
- Upper Insulator
- Shock Absorber with Coil Spring
- Coil Spring
- Spring Bumper
- Lower Insulator
- Shock Absorber

◆ Non-reusable part

7923VGA6

Common coil spring and strut component assembly—except Supra

- Strut
3. To disassemble the strut:
  - Install a bolt and 2 nuts to the bracket at the lower portion of the strut shell and secure it in a vise
  - Compress the coil spring
  - Remove the dust cover and hold the spring seat so that it will not turn. Remove the nut on the top of the strut
  - Remove the suspension support, bearing, dust seal, spring seat, spring, insulators and bumper

To install:
4. To assemble the strut:

- Install the spring bumper to piston
- Using a spring compressor, compress the spring
- Install the coil spring to the strut. Fit the lower end of the coil spring into the gap of the lower seat
- Install the spring seat with the insulator
- Install the dust seal on the spring seat
- Install the suspension support and tighten 35 ft. lbs. (47 Nm). After the nut has been tighten, release the compressor tool tension
- Pack multipurpose grease into the

suspension support. Install the dust cover

➡Do not use an impact wrench to tighten the nut. Also, check that the bearing fits into the recess in the suspension support.

5. Install or connect the following:
  - Nuts holding the strut to the strut tower. Nuts to 29 ft. lbs. (39 Nm), except on Avalon, Camry and Celica: 59 ft. lbs. (80 Nm)
  - Steering knuckle to the strut lower bracket
6. Insert the 2 bolts from the rear side

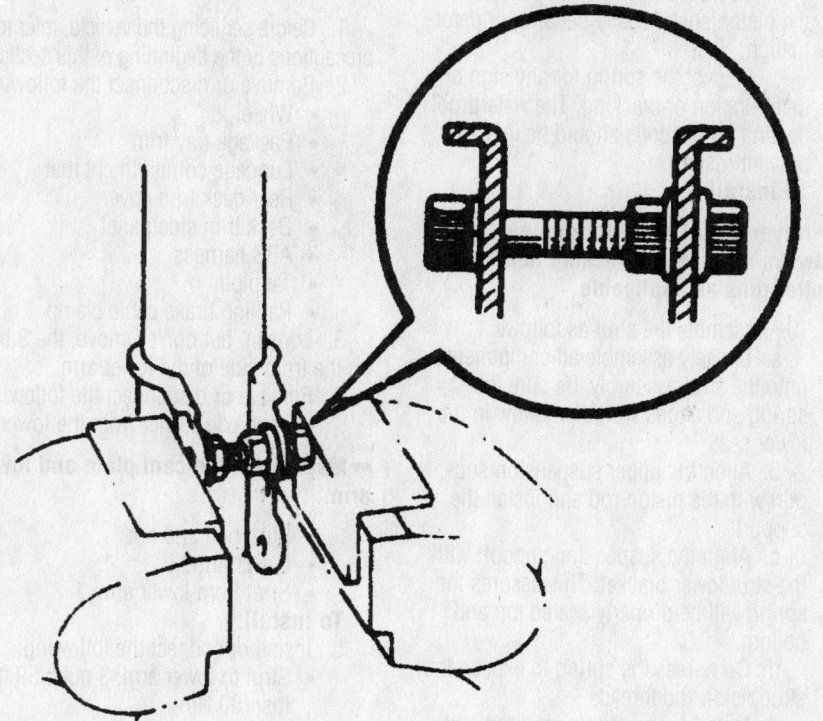

Proper method of supporting the strut in a vise

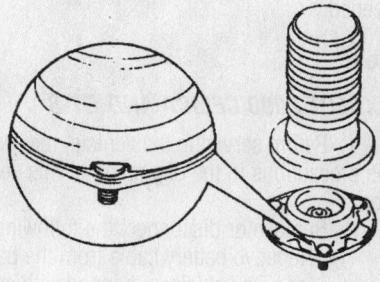

Aligning the insulator to the support—
Supra

and tighten the strut-to-steering knuckle arm bolts. Tighten as follows:

- Tercel: 166 ft. lbs. (226 Nm).
- Echo: 97 ft. lbs. (132 Nm).
- Corolla: 203 ft. lbs. (275 Nm).
- Celica: 113 ft. lbs. (153 Nm).
- Avalon and Camry: 156 ft. lbs. (211 Nm).

7. Install or connect the following:
- Brake line to the steering knuckle
- If equipped with ABS, secure the wiring harness
- Wheel
- Negative battery cable

8. Check and adjust the alignment, if needed.

SUPRA

1. Before servicing the vehicle, refer to the precautions in the beginning of this section.

2. Remove or disconnect the following:
- Negative battery cable. On vehicles equipped with an air bag, wait at least 90 seconds before proceeding
- Wheel
- Brake caliper support bracket. Suspend it with a piece of wire
- Fender apron, engine undercover,

and the front fender wheel opening molding
- Windshield washer tank, if removing the left side strut
- ABS speed sensor at the steering knuckle and wiring harness clamp
- Plug from the upper strut mount. Do not remove the center bolt

✳✳ CAUTION

Do not remove the center bolt to the strut at this time. The spring on the strut is under high pressure and can cause serious injury or vehicle damage.

- Upper control arm through-bolt from the sub-frame
- Upper control arm and turn the control arm completely around. It is not necessary to remove the upper ball joint
- Strut at the lower control arm
- Upper mounting nuts and strut assembly with the coil spring

3. Compress the coil spring.
4. Remove or disconnect the following:
- Piston rod locknut
- Suspension support, coil spring and bumper

To install:

5. Match the bolt of the suspension support with the cut out portion of the insulator.

6. Install or connect the following:
- Spring bumper
- Compressed coil spring, matching the end of the coil into the recess of the strut spring seat
- Suspension support to the rod and temporarily install a new nut

7. Turn the suspension support so one of the bolts on the support faces the same direction as shown in the illustration.

➡️**Align the bolt so a line drawn between the rod and bolt would be at 90 degrees to the direction of the lower bushing.**

8. Remove the spring compressor.
9. Install or connect the following:
- Strut and tighten the 3 upper strut mount nuts to 26 ft. lbs. (35 Nm). Tighten the middle nut to 22 ft. lbs. (29 Nm) and install the plug
- Lower end of the strut to the lower control arm. Do not tighten the bolt at this time
- Upper control arm and through-bolt and nut. Do not tighten the bolt at this time
- Speed sensor, wiring harness, and the washer tank
- Fender apron and engine under-cover
- Caliper support bracket. Bolts: 87 ft. lbs. (118 Nm).
- Wheel

10. Bounce the vehicle a few times to stabilize the suspension, then tighten the strut-to-lower arm bolt to 106 ft. lbs. (143 Nm). Tighten the upper arm to 121 ft. lbs. (164 Nm).

11. Reconnect negative battery cable.

Brake service is covered in Section 4 of this manual

12. Check and adjust the alignment, if needed.

Rear

EXCEPT 2000 CELICA AND GT-S

1. Before servicing the vehicle, refer to the precautions in the beginning of this section.
2. Remove or disconnect the following:
   • Negative battery cable from the battery. On vehicles equipped with an air bag, wait at least 90 seconds before proceeding
   • Rear seat cushion and any trim necessary to access the strut towers
3. Support the axle beam with a jack.
4. Remove or disconnect the following:
   • Wheel
   • On Supra, brake caliper support bracket. Leave the brake line connected
   • With ABS, sensor wire from the strut
   • Stabilizer bar
   • On Camry, LSPV from the lower control arm
5. Loosen the fasteners securing the strut to the axle carrier. Do not remove the bolts at this time.
6. Support the axle carrier with a jack.
7. Remove or disconnect the following:
   • Strut-to-strut tower nuts

✻✻ CAUTION

Do not loosen the center nut on the top of the strut piston.

   • Strut
8. To disassemble the strut:
   a. Place the strut assembly in a pipe vise or strut vise.

➡**Do not attempt to clamp the strut assembly in a flat jaw vise as this will result in damage to the strut tube.**

   b. Compress the spring until the upper suspension support is free of any spring tension. Do not over-compress the spring.
   c. Hold the upper support, then remove the nut on the end of the shock piston rod.
   d. Remove the support, coil spring, insulator, and bumper.
9. Inspect the strut as follows:
   a. Check the shock absorber by moving the piston shaft through its full range of travel. It should move smoothly and evenly throughout its entire travel without any trace of binding or notching.

   b. Use a small straightedge to check the piston shaft for any bending or deformation.
   c. Inspect the spring for any sign of deterioration or cracking. The waterproof coating on the coils should be intact to prevent rusting.

To install:

➡**Never reuse a self-locking nut. Always replace self-locking nuts and cotter pins as applicable.**

10. Assemble the strut as follows:
    a. Loosely assemble all components onto the strut assembly. Be sure the spring end aligns with the hollow in the lower seat.
    b. Align the upper suspension support with the piston rod and install the support.
    c. Align the suspension support with the strut lower bracket. This assures the spring will be properly seated top and bottom.
    d. Compress the spring to expose the strut piston rod threads.
    e. Install a new strut piston nut and tighten to the following:
       • Tercel: 25 ft. lbs. (34 Nm).
       • Corolla, Celica, Camry and Avalon: 36 ft. lbs. (49 Nm).
       • Supra: 20 ft. lbs. (27 Nm).
    f. Remove the spring compressor. Be sure the paint mark on the upper support faces the outside of the strut.
11. Place the strut on the vehicle and install the nuts to hold the strut to the strut tower. Nuts: 29 ft. lbs. (39 Nm), except on Supra. On Supra models: 19 ft. lbs. (26 Nm).
12. Install or connect the following:
    • Strut to the axle carrier and install the bolt and nut. Do not tighten at this time
    • Stabilizer link to the strut
    • On Camry, load sensing proportioning valve to the control arm. Tighten to 94 ft. lbs. (130 Nm)
    • On Supra, brake caliper
    • Wheel. Bounce the vehicle up and down to stabilize the suspension
13. With the vehicle weight on the suspension, tighten the bolt holding the strut to the axle carrier as follows:
    • Tercel: 50 ft. lbs. (68 Nm).
    • Corolla: 105 ft. lbs. (142 Nm).
    • Celica, Camry and Avalon: 188 ft. lbs. (255 Nm)
    • Supra: 106 ft. lbs. (143 Nm).
14. Install or connect the following:
    • Rear seat cushion and any trim
    • Negative battery cable

2000 CELICA AND GT-S

1. Before servicing the vehicle, refer to the precautions in the beginning of this section.
2. Remove or disconnect the following:
   • Wheel
   • Package tray trim
   • Luggage compartment mat
   • Rear deck trim cover
   • Deck trim side panel
   • ABS harness
   • Tailpipe
   • Parking brake cable clamp
3. Loosen, but don't remove, the 3 bolts on the front side of the lower arm.
4. Remove or disconnect the following:
   • Rear axle carrier from the lower arm

➡**Matchmark the cam plate and lower arm.**

   • Strut bolt and nut
   • Lower arm
   • Strut from lower arm

To install:
5. Install or connect the following:
   • Strut to lower arm. 3 nuts: 59 ft. lbs. (80 Nm).
   • Lower arm. Bolts: 55 ft. lbs. (74 Nm). Strut upper bolt: 103 ft. lbs. (140 Nm), while holding the nut.
   • Rear axle carrier. Side bolt: 55 ft. lbs. (74 Nm); cam bolt: 55 ft. lbs. (Nm).
   • Lower suspension arm bolts: 48 ft. lbs. (65 Nm).
   • Parking brake cable clamp: 48 inch lbs. (5.4 Nm).
   • Tailpipe
   • ABS sensor harness
   • Wheel

Rear Shock Absorber and Spring

REMOVAL & INSTALLATION

Echo

1. Before servicing the vehicle, refer to the precautions in the beginning of this section.
2. Support the rear axle beam.
3. Remove or disconnect the following:
   • Wheels
   • Package tray trim
   • Rear seat
   • Door scuff plate and door opening trim
   • Quarter panel trim
   • Roof side inner garnish
   • Partition board
   • Shock absorber upper nuts and lower bolt

4. Lower the axle slowly and remove the spring

To install:

5. Position the upper insulator so that its gap fits into the end of the spring.

6. Place the lower insulator on the axle.

7. Raise the axle while positioning the shock. Torque the lower bolt to 36 ft. lbs. (49 Nm).

8. Position the lower nut on the rod so that the rod protrudes 15–18mm above the nut.

9. Install or connect the following:
- Upper nut. Torque the nut to 18 ft. lbs. (25 Nm).
- Partition board
- Roof side inner garnish
- Quarter panel
- Door scuff plate and trim
- Rear seat
- Package tray trim
- Wheels

Upper Ball Joint

REMOVAL & INSTALLATION

The upper ball joint (used only on the Supra) is an integral part of the upper arm and is not replaced separately. The upper ball joint replacement is accomplished by replacing the upper arm, as follows:

Upper Control Arm

REMOVAL & INSTALLATION

Supra

1. Before servicing the vehicle, refer to the precautions in the beginning of this section.

2. Remove or disconnect the following:
- Negative battery cable. On vehicles equipped with an air bag, wait at least 90 seconds before proceeding
- Wheel
- Caliper support bracket. Leave the brake line connected and suspend it aside
- Rotor
- Front fender splash shield, fender liner, and wheel opening molding
- If removing the left side arm, the washer tank
- ABS speed sensor from the steering knuckle. Remove the 3 bolts and disconnect the wire harness clamp

- Cotter pin and the nut from the upper ball joint; press the upper ball joint from the knuckle
- Upper control arm

To install:

3. Install or connect the following:
- Upper control arm. Connect the upper control arm to the sub-frame and install the through-bolt. Do not tighten the bolt at this time

➡**The upper control arm mounting bolts are not tightened until the suspension has been assembled and vehicle is on the ground.**

- Ball joint to the knuckle. Nut: 76 ft. lbs. (103 Nm). Install a new cotter pin
- Wire harness and ABS speed sensor
- Washer tank, the fender liner, splash shield, and molding
- Rotor
- Caliper support bracket. Bolts: 87 ft. lbs. (118 Nm).
- Wheel

4. Support the lower arm and tighten the upper control arm through-bolt and nut to 121 ft. lbs. (164 Nm).

5. Connect the negative battery cable

6. Check and adjust the alignment, if needed.

Lower Ball Joint

REMOVAL & INSTALLATION

Except Echo

➡**On the Supra, the lower ball joint is not replaceable. If the lower ball joint is defective, replace the lower arm and ball joint as an assembly.**

1. Before servicing the vehicle, refer to the precautions in the beginning of this section.

2. Remove or disconnect the following:
- Negative battery cable. On vehicles equipped with an air bag, wait at least 90 seconds before proceeding
- Front wheels
- Cotter pin from the bearing locknut cap, then remove the cap

3. Depress the brake pedal and loosen the axle nut

4. Remove or disconnect the following:
- Brake caliper attaching hardware, position the caliper aside with the hydraulic line still attached and suspend it with a wire
- ABS speed sensor, if equipped
- Rotor

5. Loosen the 2 nuts holding the strut to the steering knuckle assembly. Do not remove at this time.

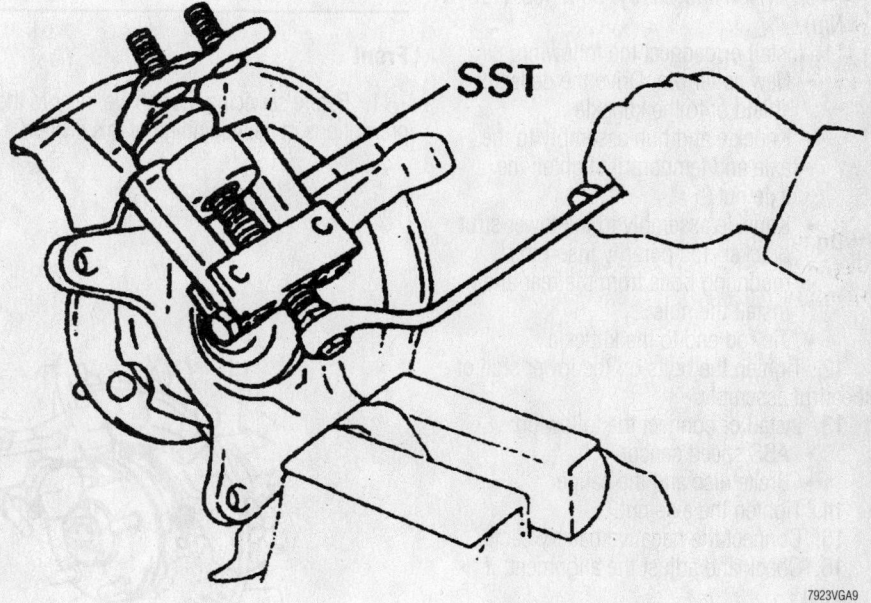

Removing the ball joint from the knuckle

7923VGA9

For complete Engine Mechanical specifications, see Section 1 of this manual

6. Remove or disconnect the following:
- Cotter pin and nut from the tie rod end. Using a tie rod end removal tool, separate the tie rod end from the steering knuckle
- Steering knuckle from the strut assembly
- Axle nut and grasp the hub and knuckle assembly. With a plastic hammer tap the axle shaft to remove knuckle and hub

➡**Cover the halfshaft boot with a shop rag to protect it from any damage.**

7. Clamp the steering knuckle in a vise and remove the dust deflector. Remove the nut holding the steering knuckle to the ball joint. Press the ball joint out of the steering knuckle.

8. Remove the ball joint from the arm.

To install:

9. Install the Lower ball joint to the lower arm. Tighten the fasteners to:
   a. Tercel: 59 ft. lbs. (79 Nm).
   b. Corolla, Avalon, Camry and 1998–99 Celica,: 94 ft. lbs. (127 Nm).
   c. GT-S and 2000 Celica: 105 ft. lbs. (142 Nm).

10. Install the ball joint to the steering knuckle. Tighten the ball joint-to-steering knuckle nut to:
   a. Tercel: 72 ft. lbs. (97 Nm).
   b. Corolla and 1998–99 Celica: 87 ft. lbs. (118 Nm).
   c. GT-S and 2000 Celica: 76 ft. lbs. (103 Nm).
   d. Avalon and Camry: 90 ft. lbs. (123 Nm).

11. Install or connect the following:
- New cotter pin. Drive the deflector shield onto the knuckle
- Knuckle and hub assembly to the axle and temporarily tighten the axle nut
- Knuckle assembly to the lower strut bracket. Temporarily insert the mounting bolts from the rear and install the nuts
- Tie rod end to the knuckle

12. Tighten the bolts on the lower side of the strut assembly.

13. Install or connect the following:
- ABS speed sensor
- Brake disc and the caliper

14. Tighten the axle nut.

15. Connect the negative battery cable.

16. Check and adjust the alignment, if needed.

Echo

1. Before servicing the vehicle, refer to the precautions in the beginning of this section.

➡**The ball joint is not replaceable.**

2. Remove or disconnect the following:
- Wheels
- Stabilizer bar end link
- Lower arm from the knuckle

3. For the right arm, jack up the engine slightly and disconnect the rear engine mount.

4. On either side, remove the 2 bolts and 1 nut and remove the arm.

5. Flip the ball stud back and forth a few times. Install the nut. Using a torque wrench, rotate the ball stud at the rate of 1 turn in 2–4 seconds. Take a torque reading on the 5th revolution. Turning torque should be at least 5 inch lbs. (0.59 Nm), and not more than 30 inch lbs. (3.4 Nm).

To install:

6. Install or connect the following:
- Lower control arm. Torque the bolts to A: 65 ft. lbs. (88 Nm); B: 97 ft. lbs. (132 Nm).

➡**DO NOT turn the nut.**

- On the right side, the engine rear mount. Bolt and nuts: 59 ft. lbs. (80 Nm).
- Lower arm to knuckle. Torque the nut to 72 ft. lbs. (98 Nm).
- Stabilizer bar
- Wheel

Wheel Bearings

ADJUSTMENT

Front

1. Before servicing the vehicle, refer to the precautions in the beginning of this section.

2. All models use a non-adjustable wheel bearing. To determine the condition of the wheel bearing, check the backlash in bearing shaft direction and the axle hub deviation. Maximum for backlash should be as follows:
- Corolla, Echo, Supra, Avalon, Camry and Camry Solara: 0.0020 in. (0.05mm)
- Celica: 0.0031 in. (0.08mm)

3. Maximum axle hub deviation is:
- Corolla: 0.0028 in. (0.07mm)
- Celica: 0.0028 in. (0.07mm)
- Supra, Echo, Avalon and Camry: 0.0020 in. (0.05mm)

4. If the wheel bearing is out of specifications, replace the wheel bearing.

Rear

TERCEL

1. Before servicing the vehicle, refer to the precautions in the beginning of this section.

2. Remove or disconnect the following:
- Negative battery cable. On vehicles equipped with an air bag, wait at least 90 seconds before proceeding
- Rear wheels
- Locknut cap, cotter pin and locknut

3. Install the bearing locknut and tighten it to 22 ft. lbs. (29 Nm) while spinning the drum.

4. Spin the brake drum several times to snug down the bearing, then loosen the bearing locknut until it can be turned by hand.

➡**There must be absolutely no brake drag at this time.**

5. Retighten the bearing locknut until there is a bearing preload of 0.9–2.2 lbs.

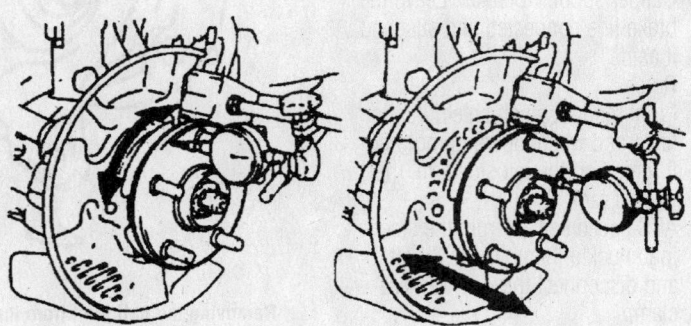

Checking the wheel bearings for deviation and free-play

7923VGB1

(3.2–9.8 N) while turning the wheel. Measure with a spring scale hooked to one of the studs.

6. Install or connect the following:
- Locknut lock, a new cotter pin, and the cap. If the cotter pin hole does not align properly, align the holes by tightening the nut to the next hole. Do not loosen the nut
- Rear wheel
- Negative battery cable

EXCEPT TERCEL

Check the backlash in bearing shaft direction and the axle hub deviation. Maximum for backlash should be 0.0020 in. (0.05mm). Maximum axle hub deviation is 0.0028 in. (0.07mm), except on Supra, which is 0.020 in. (0.05mm).

➡**The wheel bearing is non-adjustable. If the wheel bearing is out of specifications, replace the wheel bearing.**

REMOVAL & INSTALLATION

Front

EXCEPT SUPRA

1. Before servicing the vehicle, refer to the precautions in the beginning of this section.
2. Remove or disconnect the following:
- Negative battery cable. On vehicles

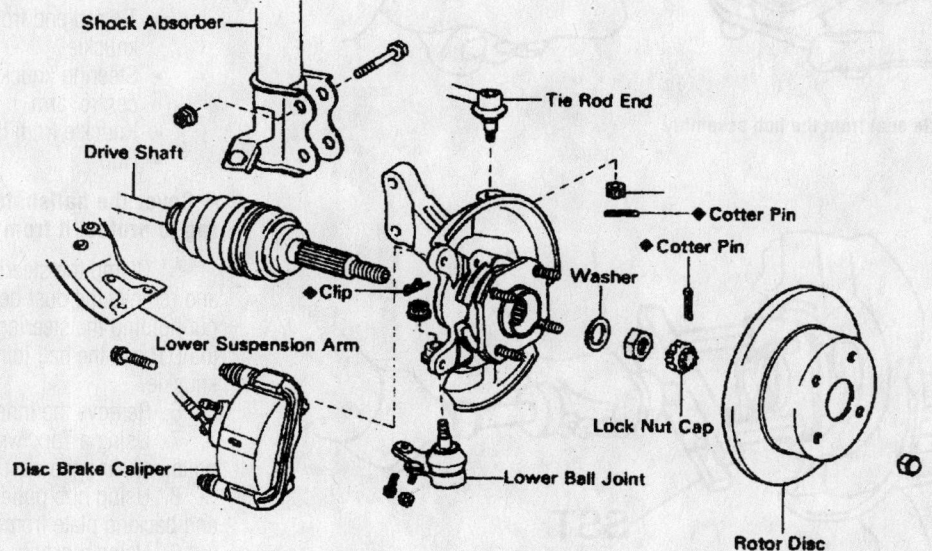

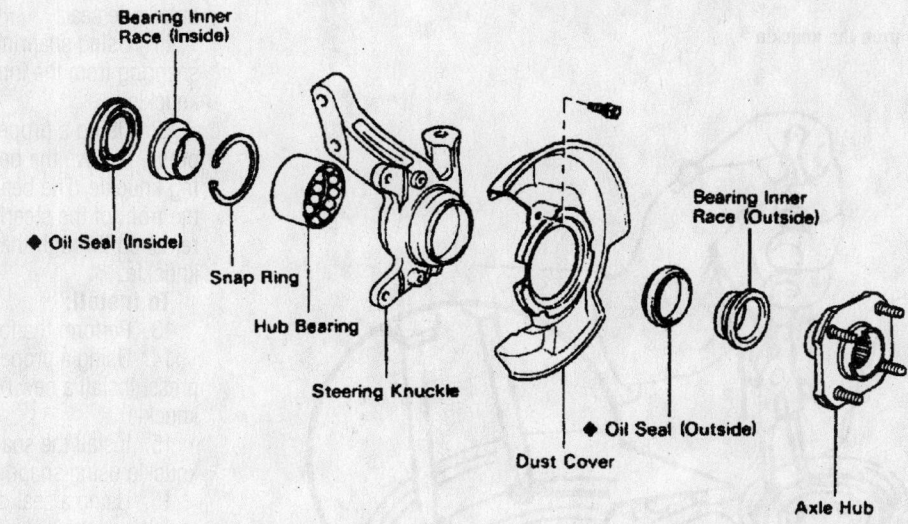

Steering knuckle and hub assembly—except Supra

◆ Non-reusable part

7923VGB2

For Accessory Drive Belt illustrations, see Section 1 of this manual

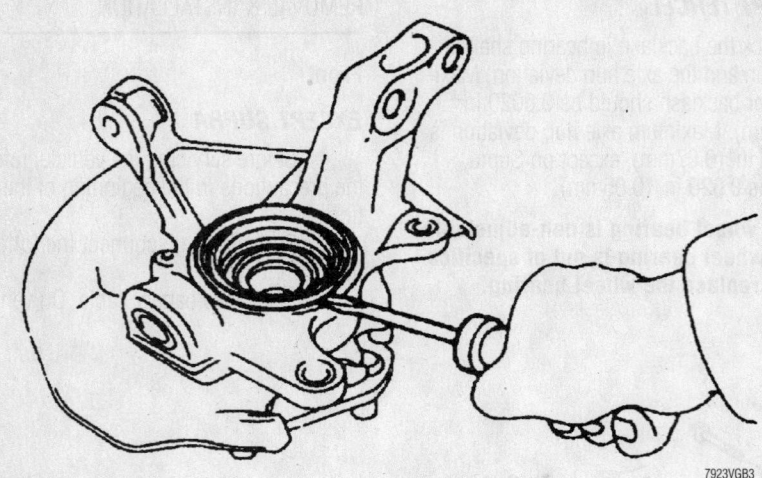

Removing the inner axle seal from the hub assembly

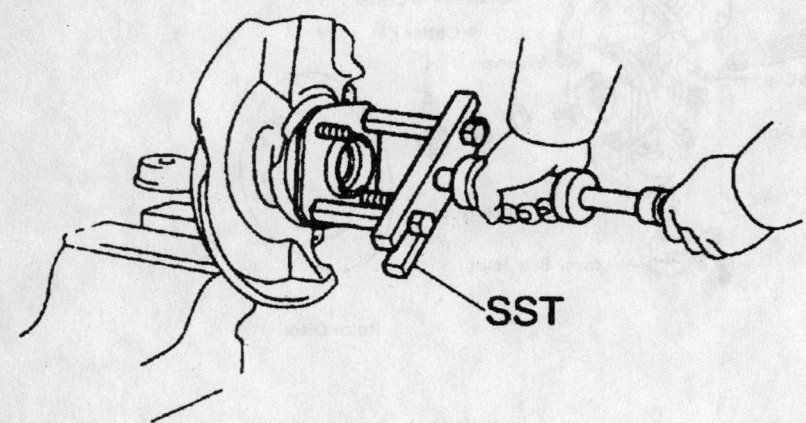

SST

Removing the axle hub from the knuckle

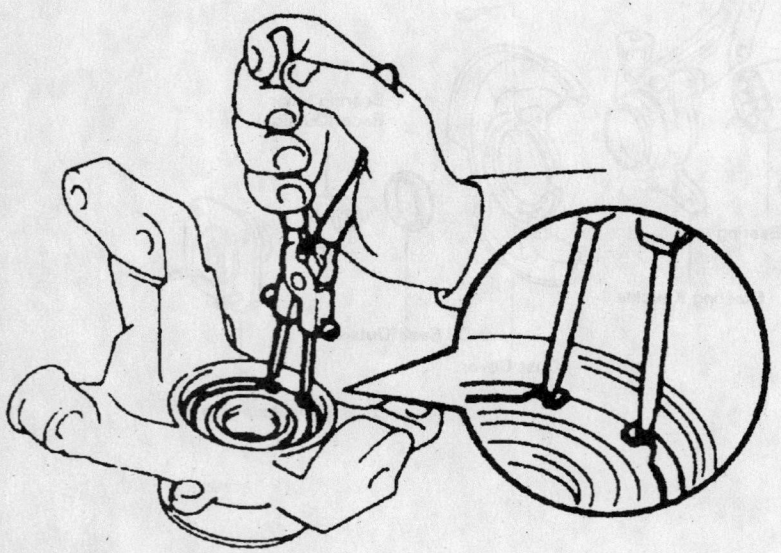

Removing the snapring from the knuckle before pressing out the bearing

equipped with an air bag, wait at least 90 seconds before proceeding
- Wheels
- Axle nut cap
- Axle nut
- Caliper. Position the caliper aside with the hydraulic line still attached and suspend it with a wire.
- ABS speed sensor
- Rotor

3. Loosen the nuts on the lower side of the strut assembly. Do not remove at this time.

4. Remove or disconnect the following:
- Tie rod end from the steering knuckle
- Steering knuckle from the lower control arm
- Knuckle from the strut assembly
- Hub

➡**Cover the halfshaft boot with a shop rag to protect it from any damage.**

5. Clamp the steering knuckle in a vise and remove the dust deflector. Remove the nut holding the steering knuckle to the ball joint. Press the ball joint out of the steering knuckle.

6. Remove the inner axle seal.

7. Using a Torx®wrench, remove the bolts securing the dust cover.

8. Using hub puller, remove the hub and backing plate from the steering knuckle.

9. Using a proper sized driver and a press, remove the inner hub race from the axle hub.

10. Using seal removal tool, remove the outer axle seal.

11. Using snapring pliers, remove the snapring from the inner side of the steering knuckle.

12. Using a proper sized driver and a press, remove the bearing from the steering knuckle. The bearing is pressed from the front of the steering knuckle and is removed through the back of the steering knuckle.

To install:

13. Perform the following:

14. Using a proper sized driver and a press, install a new bearing to the steering knuckle.

15. Install the snapring to the steering knuckle using snapring pliers.

16. Using a seal driver and a hammer, install a new outer oil seal. Apply multipurpose grease to the oil seal lip.

17. Place the dust cover on the steering knuckle. Bolts: 78 inch lbs. (9 Nm).

18. Using a press and a proper sized driver, install the axle hub to the steering knuckle.

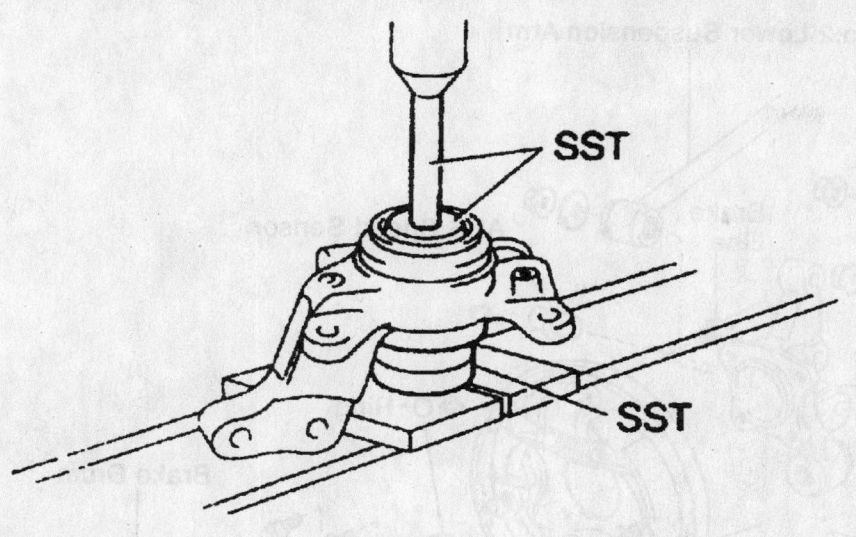

Removing the bearing from the steering knuckle using a press

19. Attach the ball joint to the steering knuckle. Install a new cotter pin.

20. Using a seal driver and a hammer, install a new inner oil seal. Apply multipurpose grease to the oil seal lip.

21. Install the knuckle and hub assembly to the axle and temporarily tighten the axle nut.

22. Connect the knuckle assembly to the lower strut bracket. Temporarily insert the mounting bolts from the rear and install the nuts making sure the matchmarks made earlier are in alignment.

23. Connect the lower ball joint to lower arm.

24. Connect the tie rod end to the knuckle.

25. Tighten on the lower side of the strut assembly.

26. If equipped, install the ABS speed sensor.

27. Install the brake disc and the caliper.

28. Tighten the axle nut while someone depresses the brake pedal. Install the adjusting nut cap and insert a new cotter pin.

29. Install the wheels to the vehicle. Verify that the wheel turns freely.

30. Connect the negative battery cable to the battery.

31. Check alignment.

SUPRA

1. Before servicing the vehicle, refer to the precautions in the beginning of this section.

2. Remove or disconnect the following:

- Negative battery cable. On vehicles equipped with an air bag, wait at least 90 seconds before proceeding
- Front tire and wheel assembly
- Brake caliper support bracket, leaving the brake line connected and support it using a piece of wire

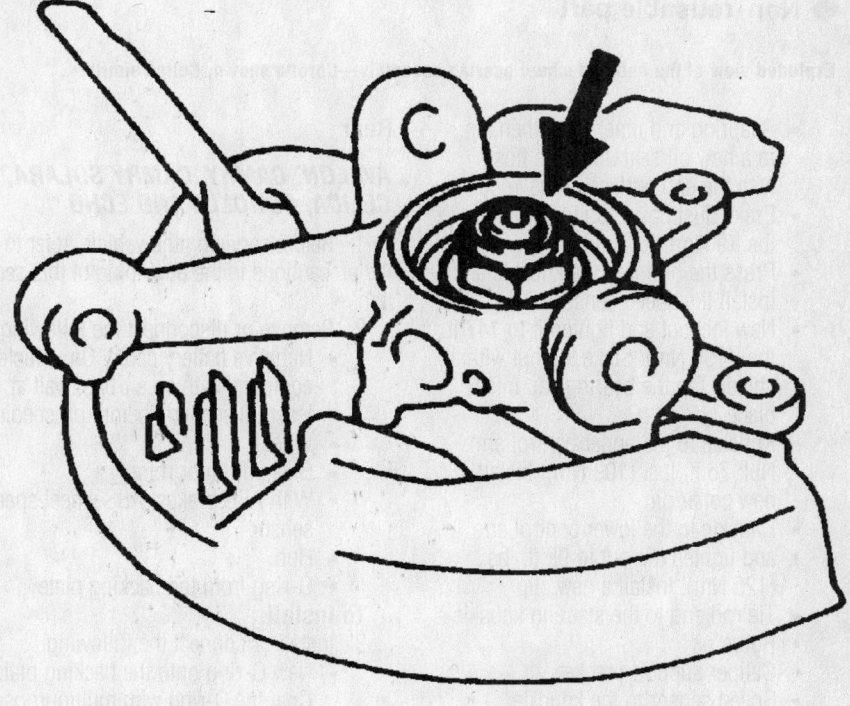

Axle hub locknut location—Supra

- Rotor
- ABS speed sensor
- Cotter pin, nut and tie rod from the steering knuckle
- Cotter pin, nut and steering knuckle from the upper control arm
- Clip and nut and press the knuckle off the lower control arm
- Steering knuckle
- Hub bearing cap from the steering knuckle. Using a hammer and chisel, loosen the staked part of the hub nut and remove it
- ABS sensor rotor

3. Remove the 4 bolts and shift the brake dust shield toward the hub (outside).

4. Using a 2 armed puller, remove the axle hub from the knuckle.

5. With a puller, remove the inner bearing race from the axle hub. Pry out the oil seal.

6. Remove the bearing snapring, then position the inner race above the bearing on the inner side. Press the bearing out.

To install:

7. Press the bearing into the knuckle. If the inner race and balls come loose from the outer race, be sure to install them on the same side as before.

8. Install or connect the following:

For Tire, Wheel and Ball Joint specifications, see Section 1 of this manual

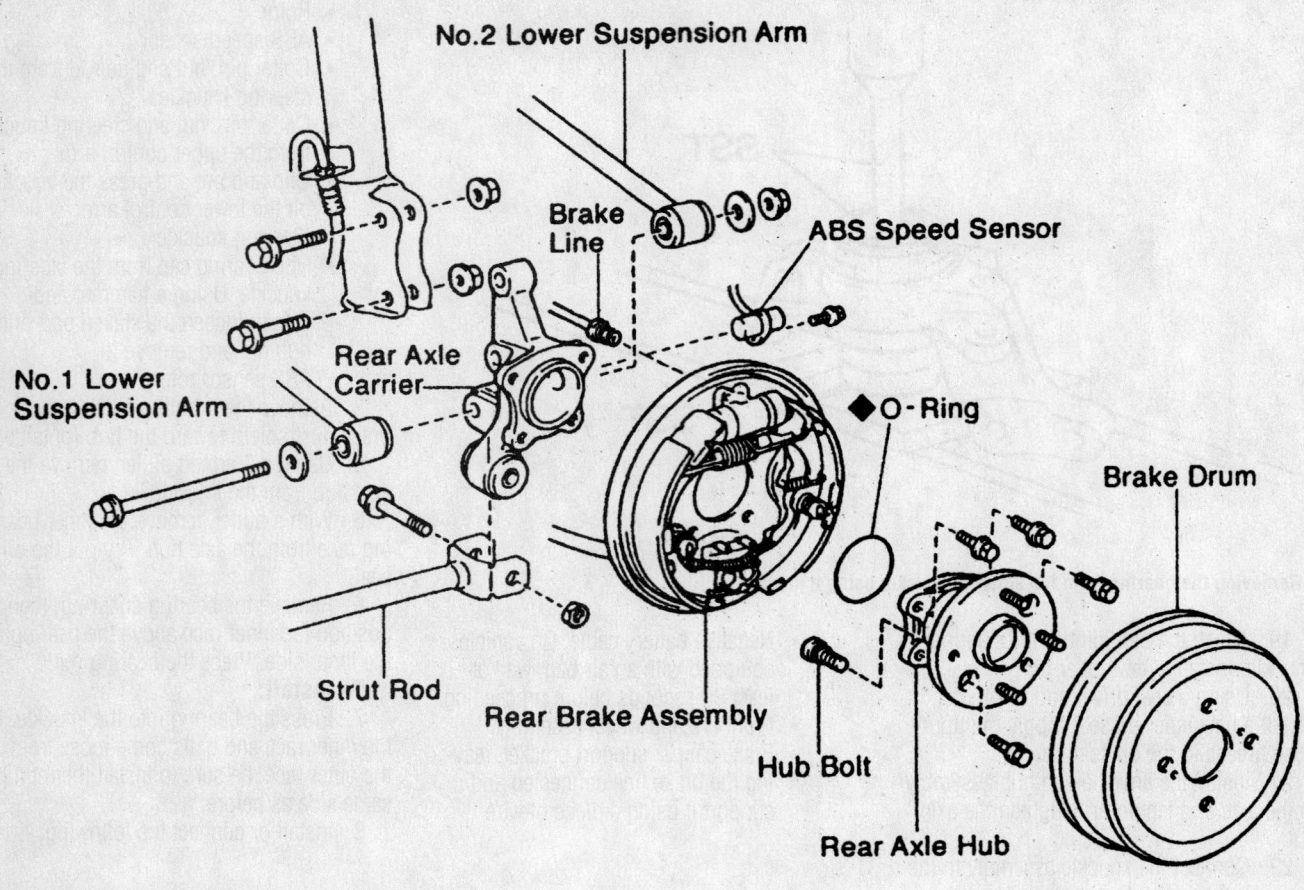

◆ **Non-reusable part**

Exploded view of the hub and wheel bearing assembly—Corolla shown, Celica similar

- Snapring and inner race, then tap in a new oil seal until it is flush with the end surface of the knuckle
- Brake dust cover. Bolts: 74 inch lbs. (9 Nm).
- Press the hub into the knuckle and install the speed sensor
- New locknut and tighten it to 147 ft. lbs. (199 Nm). Stake the nut with a chisel. Tap the bearing cap into place
- Knuckle to the upper control arm. Nut: 76 ft. lbs. (103 Nm). Install a new cotter pin
- Knuckle to the lower control arm and tighten the nut to 92 ft. lbs. (125 Nm). Install a new clip
- Tie rod end to the steering knuckle
- Rotor
- Caliper support bracket
- Speed sensor to the knuckle
- Front wheel
- Negative battery cable

9. Check and adjust the alignment, if needed.

Rear

AVALON, CAMRY, CAMRY SOLARA, CELICA, COROLLA AND ECHO

1. Before servicing the vehicle, refer to the precautions in the beginning of this section.
2. Remove or disconnect the following:
   - Negative battery cable. On vehicles equipped with an air bag, wait at least 90 seconds before proceeding
   - Wheel
   - Brake drum or rotor
   - With ABS brakes, ABS wheel speed sensor
   - Hub
   - O-ring from the backing plate

To install:

3. Install or connect the following:
   - New O-ring onto the backing plate. Coat the O-ring with multipurpose grease
   - Hub to the knuckle. Bolts: Except Echo, 59 ft. lbs. (80 Nm); Echo, 38 ft. lbs. (52 Nm).

- With ABS brakes, ABS wheel speed sensor
- Brake drum or rotor
- Wheel
- Negative battery cable

4. Check and adjust the alignment, if needed.

TERCEL

1. Before servicing the vehicle, refer to the precautions in the beginning of this section.
2. Remove or disconnect the following:
   - Negative battery cable. On vehicles equipped with an air bag, wait at least 90 seconds before proceeding
   - Rear wheels
   - Locknut cap, cotter pin and locknut
   - Brake drum along with the outer wheel bearing and thrust washer
   - Inner bearing oil seal out of the brake drum assembly
   - Inner bearing
   - Bearing races, as required

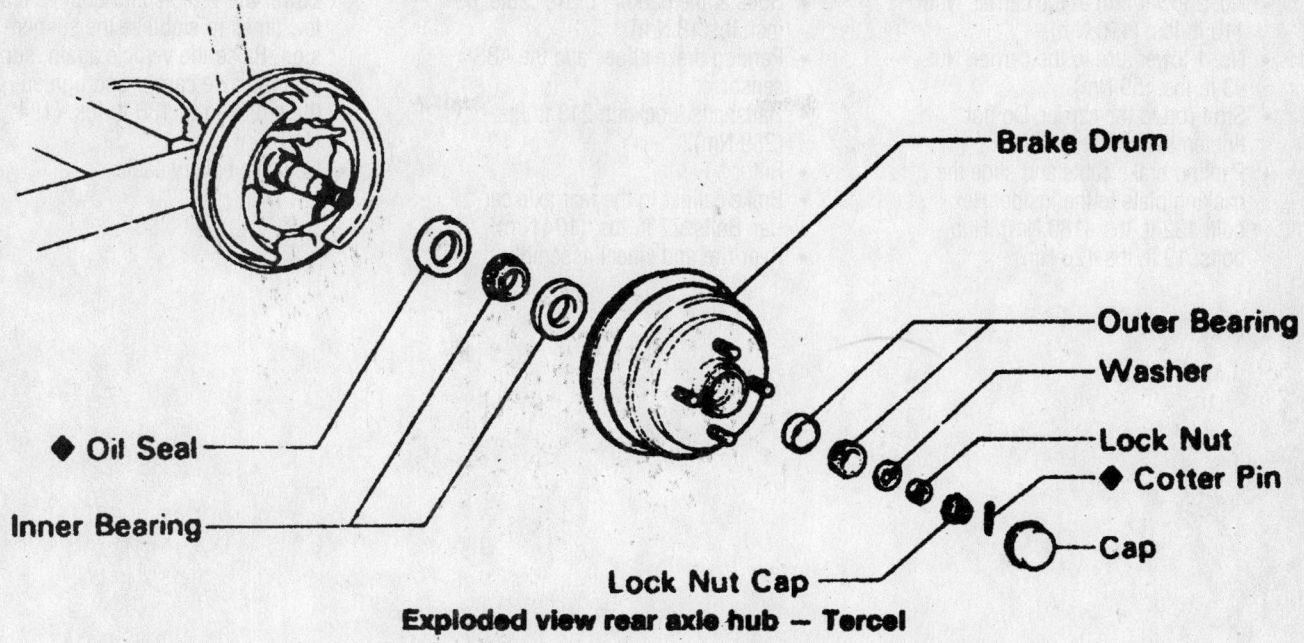

Brake Drum

Outer Bearing

Washer

Lock Nut

◆ **Cotter Pin**

Cap

◆ Oil Seal

Inner Bearing

Lock Nut Cap

Exploded view rear axle hub — Tercel

7923VGB8

Exploded view of the rear wheel bearing assembly—Tercel shown

To install:

3. Clean and pack the bearings with grease.

4. Install or connect the following:
- New outer bearing races into the brake drum and add a liberal amount of bearing grease to the inside of the hub and the bearing cap
- Inner bearing into the brake drum
- New oil seal
- Brake drum onto the axle shaft
- Outer bearing and position the thrust washer
- Bearing locknut and tighten it to 22 ft. lbs. (29 Nm) while spinning the drum

5. Spin the brake drum several times to snug down the bearing, then loosen the bearing locknut until it can be turned by hand.

➡**There must be absolutely no brake drag at this time.**

6. Retighten the bearing locknut until there is a bearing preload of 0.9–2.2 lbs. (3.2–9.8 N) while turning the wheel. Measure with a spring scale hooked to one of the studs.

7. Install or connect the following:
- Locknut lock, a new cotter pin, and the cap. If the cotter pin hole does not align properly, align the holes

by tightening the nut to the next hole. Do not loosen the nut.
- Rear wheel
- Negative battery cable

SUPRA

1. Before servicing the vehicle, refer to the precautions in the beginning of this section.

2. Remove or disconnect the following:

- Negative battery cable. On vehicles equipped with an air bag, wait at least 90 seconds before proceeding
- Rear tire and wheel assembly
- Brake caliper support bracket from the rear axle carrier and support it with a piece of wire

➡**Matchmark the disc brake rotor and the axle hub.**

- Rotor
- Speed sensor
- Rear halfshaft
- Parking brake shoes
- Parking brake cable
- Strut rod at the axle carrier

3. Perform the following:
   a. Remove the nut, then press out the No. 1 lower suspension arm.
   b. Remove the nut, then press out the No. 2 lower suspension arm.

c. Remove the nut, then press out the upper suspension arm. Remove the axle carrier.

d. Remove the dust deflector and pull out the oil seal.

e. Using a 2 arm puller, remove the axle hub from the carrier.

f. Remove the backing plate.

g. Press the inner race (outside) from the hub. Then, remove the oil seal and the snapring.

h. Place the inner race (outside) over the bearing and tap out the bearing and inner race (inside).

To install:

4. Install or connect the following:
- Bearing to the axle carrier.

➡**If the inner races come loose from the bearing outer race, be sure to install them on the same side as before.**

- Snapring, the inner race (outside), and a new oil seal
- Backing plate. Install the inner race (inside) and press in the axle hub with the proper tools
- Inner oil seal. Align the holes for the speed sensor in the dust deflector and axle carrier. Install the dust deflector
- Upper arm to the axle carrier. Nut and bolt: 80 ft. lbs. (109 Nm).

For Wheel Alignment specifications, see Section 1 of this manual

- No. 2 lower arm to the carrier. Nut: 110 ft. lbs. (150 Nm).
- No. 1 lower arm to the carrier. Nut: 43 ft. lbs. (59 Nm).
- Strut rod to the carrier. Do not tighten the bolt at this time.
- Parking brake cable and slide the backing plate to the inside. Hex bolt: 132 ft. lbs. (180 Nm). Hub bolts: 19 ft. lbs. (26 Nm).

- Bolts at the parking brake cable: 69 inch lbs. (8 Nm).
- Parking brake shoes and the ABS sensor
- Halfshafts. Locknut: 213 ft. lbs. (289 Nm).
- Rotor
- Brake caliper to the rear axle carrier. Bolts: 77 ft. lbs. (104 Nm).
- Rear tire and wheel assembly.

Lower the vehicle and bounce it a few times to stabilize the suspension. Raise the vehicle again, support the axle carrier and tighten the strut rod to 136 ft. lbs. (184 Nm).

- Negative battery cable

VOLKSWAGEN

22

1998–00
Cabrio • Golf • GTI • Jetta • New Beetle • Passat

PRECAUTIONS

Before servicing any vehicle, please be sure to read all of the following precautions, which deal with personal safety, prevention of component damage, and important points to take into consideration when servicing a motor vehicle:

• Never open, service or drain the radiator or cooling system when the engine is hot; serious burns can occur from the steam and hot coolant.

• Observe all applicable safety precautions when working around fuel. Whenever servicing the fuel system, always work in a well-ventilated area. Do not allow fuel spray or vapors to come in contact with a spark, open flame or excessive heat (a hot drop light, for example). Keep a dry chemical fire extinguisher near the work area. Always keep fuel in a container specifically designed for fuel storage; also, always properly seal fuel containers to avoid the possibility of fire or explosion. Refer to the additional fuel system precautions later in this section.

• Fuel injection systems often remain pressurized, even after the engine has been turned **OFF**. The fuel system pressure must be relieved before disconnecting any fuel lines. Failure to do so may result in fire and/or personal injury.

• Brake fluid often contains polyglycol ethers and polyglycols. Avoid contact with the eyes and wash your hands thoroughly after handling brake fluid. If you do get brake fluid in your eyes, flush your eyes with clean, running water for 15 minutes. If eye irritation persists, or if you have taken brake fluid internally, IMMEDIATELY seek medical assistance.

• The EPA warns that prolonged contact with used engine oil may cause a number of skin disorders, including cancer! You should make every effort to minimize your exposure to used engine oil. Protective gloves should be worn when changing oil. Wash your hands and any other exposed skin areas as soon as possible after exposure to used engine oil. Soap and water, or waterless hand cleaner should be used.

• All new vehicles are now equipped with an air bag system. The system must be disabled before performing service on or around system components, steering column, instrument panel components, wiring and sensors. Failure to follow safety and disabling procedures could result in accidental air bag deployment, possible personal injury and unnecessary system repairs.

• Always wear safety goggles when working with, or around, the air bag system. When carrying a non-deployed air bag, be sure the bag and trim cover are pointed away from your body. When placing a non-deployed air bag on a work surface, always face the bag and trim cover upward, away from the surface. This will reduce the motion of the module if it is accidentally deployed. Refer to the additional air bag system precautions later in this section.

• Clean, high quality brake fluid from a sealed container is essential to the safe and proper operation of the brake system. You should always buy the correct type of brake fluid for your vehicle. If the brake fluid becomes contaminated, completely flush the system with new fluid. Never reuse any brake fluid. Any brake fluid that is removed from the system should be discarded. Also, do not allow any brake fluid to come in contact with a painted surface; it will damage the paint.

• Never operate the engine without the proper amount and type of engine oil; doing so WILL result in severe engine damage.

• Timing belt maintenance is extremely important! Many models utilize an interference-type, non-freewheeling engine. If the timing belt breaks, the valves in the cylinder head may strike the pistons, causing potentially serious (also time-consuming and expensive) engine damage. Refer to the maintenance interval charts in the front of this manual for the recommended replacement interval for the timing belt, and to the timing belt section for belt replacement and inspection.

• Disconnecting the negative battery cable on some vehicles may interfere with the functions of the on-board computer system(s) and may require the computer to undergo a relearning process once the negative battery cable is reconnected.

• When servicing drum brakes, only disassemble and assemble one side at a time, leaving the remaining side intact for reference.

• Only an MVAC-trained, EPA-certified automotive technician should service the air conditioning system or its components.

GASOLINE ENGINE REPAIR

Distributor

The 1998–00 4-cylinder and all 6-cylinder engines are equipped with a Distributorless Ignition System (DIS).

REMOVAL

1. Before servicing the vehicle, refer to the precautions in the beginning of this section.
2. Remove or disconnect:
    - Coil high tension wire and the connector plug at the distributor.
    - All vacuum lines, if equipped.
    - Cap and static shield as a unit.
3. At the front crankshaft pulley bolt, turn the engine to TDC on the No. 1 piston. Matchmark the rotor points to the rim of the distributor; some vehicles already have a notch there. Also matchmark the distributor to the engine block or head.
4. Remove the bolt and distributor clamp, and lift the distributor straight out.

INSTALLATION

Timing Not Disturbed

1. Before servicing the vehicle, refer to the precautions in the beginning of this section.
2. On some vehicles, the distributor engages its drive with an offset slot and is easy to reinstall in the reverse order of removal, even if the crankshaft or camshaft has been turned. Gently rotate the rotor while pushing the distributor into place. Install the hold-down bolt and adjust the ignition timing.
3. On engines with the drive gear on the distributor, be sure the engine is still at TDC and insert the distributor with the matchmarks aligned.
4. Install the hold-down clamp and bolt, connector plug, cap and static shield, and high tension wires.
5. Check and adjust the ignition timing.

Timing Disturbed

1. Before servicing the vehicle, refer to the precautions in the beginning of this section.
2. Rotate the crankshaft to TDC of No. 1 piston.
3. With a suitable tool, turn the oil pump drive so it is parallel with the crankshaft.

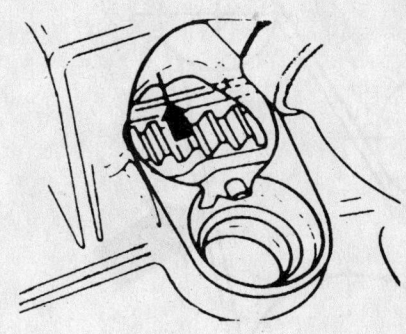

7923WG01

Timing mark locations on the bell housing

4. Install or connect:
    - Rotor onto the distributor and align it with the No. 1 mark on the rim of the body.
    - Distributor, making sure the rotor still aligns with the mark when the distributor is all the way in.
    - Hold-down clamp and bolt, connector plug, cap and static shield, and high tension wires.
5. Check and adjust the ignition timing.

Alternator

REMOVAL & INSTALLATION

Before purchasing a replacement alternator, read the specification plate on the housing. The number 14V will appear to indicate maximum voltage rating. On the same line will be two more digits followed by the letter **A**. This is the maximum amperage output. Be sure to purchase an alternator with the same rating. The regulator can be replaced without removing the alternator.

Fox

1. Disconnect the negative battery cable.
2. Disconnect the wiring from the alternator. If necessary, mark the wires with tape or other means to ensure they are connected properly upon installation.
3. Loosen the alternator adjustment bolt, and remove the drive belt from the alternator.
4. Remove the alternator adjustment bolt, followed by the pivot bolt.
5. Carefully lift the alternator from the bracket.

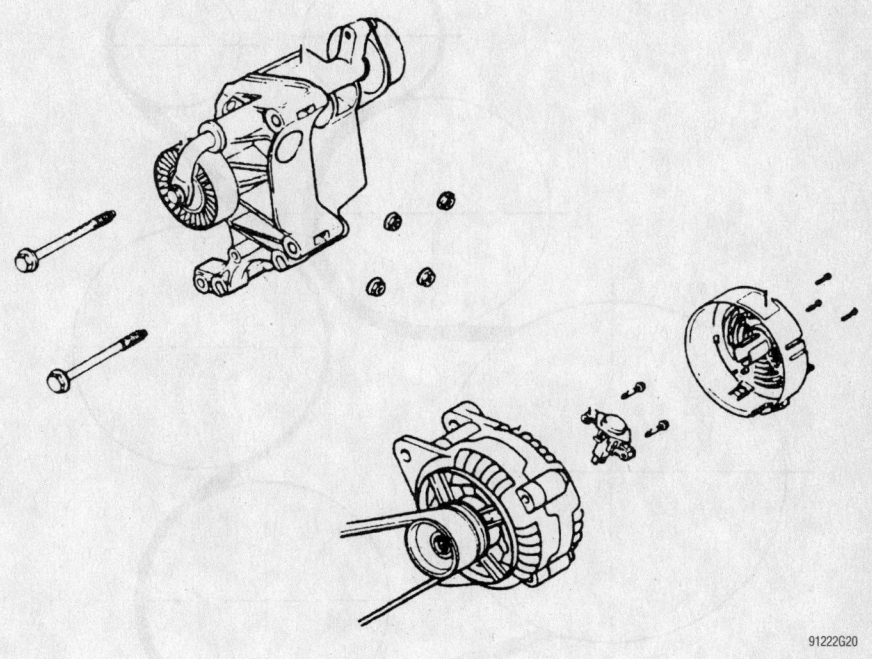

91222G20

Alternator mounting detail—2.0L engines

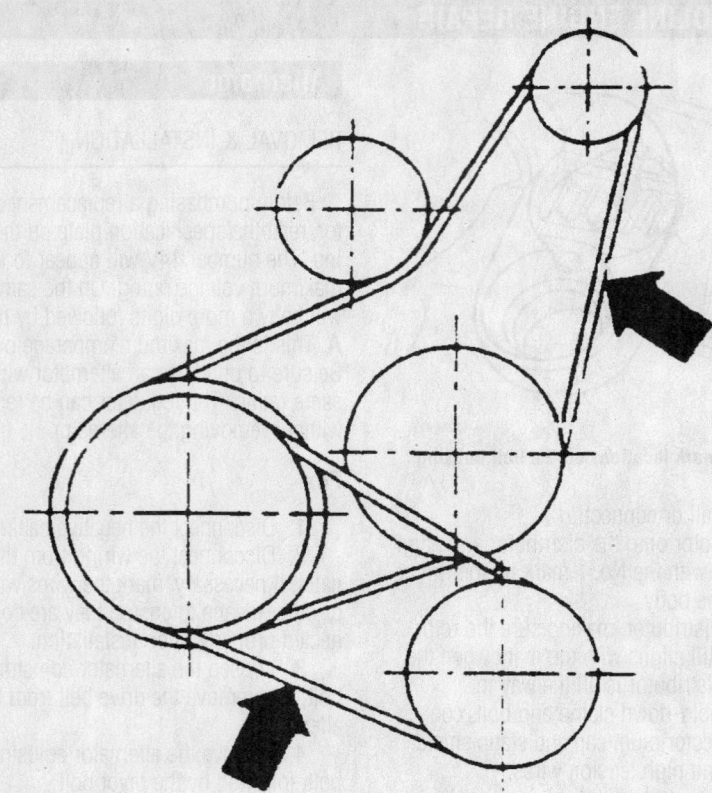

Alternator belt routing—all except VR6, with power steering, no A/C

91222G22

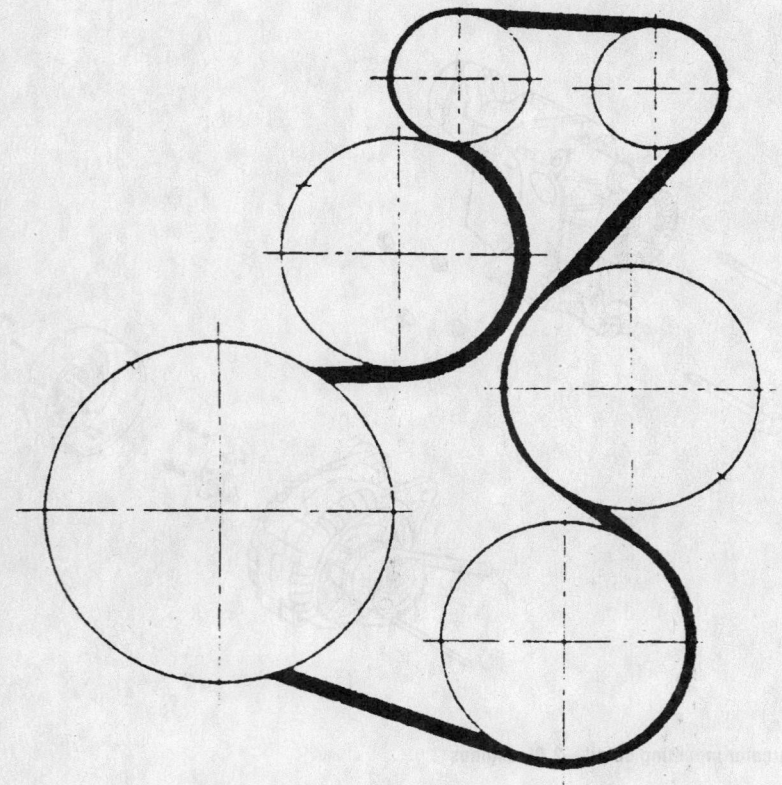

Alternator belt mounting detail—VR6 engines

91222G23

To install:

6. Hold the alternator in position on the mounting bracket, and install the pivot bolt.

7. Install (but do not tighten) the alternator adjustment bolt.

8. Place the alternator drive belt on the pulley.

9. Adjust belt tension and tighten the mounting and adjustment bolts as necessary.

10. Connect the wiring to the alternator. If tape was used to identify the wires, make sure it is removed once the wires are connected.

11. Connect the negative battery cable.

All other vehicles

1. Disconnect the negative battery cable.

2. Disconnect the wiring from the alternator. If necessary, mark the wires with tape or other means to ensure they are connected properly upon installation.

3. Loosen the belt tension, and remove the drive belt from the alternator. On VR6 (AAA) engines only, remove the belt tensioner from the cylinder head.

4. Remove the alternator adjustment bolt, followed by the pivot bolts.

5. Carefully lift the alternator from the bracket.

To install:

6. Hold the alternator in position on the mounting bracket, and install the pivot bolt. On later engines with automatic belt tensioners, install the upper mounting bolt.

7. Install (but do not tighten) the alternator adjustment bolt (earlier models only).

8. Place the alternator drive belt on the pulley.

9. Adjust belt tension and tighten the mounting and adjustment bolts as necessary.

10. Connect the wiring to the alternator. If tape was used to identify the wires, make sure it is removed once the wires are connected.

11. Connect the negative battery cable.

Ignition Timing

ADJUSTMENT

➡The ignition timing is controlled by the engine control module and is not adjustable. However the timing can be monitored on a scan tool connected to the DLC in the vehicle. No specification has been given by the manufacturer.

Engine Assembly

REMOVAL & INSTALLATION

Cabrio, Golf, Jetta

1. Before servicing the vehicle, refer to the precautions in the beginning of this section.

➡**The engine and transaxle are lifted from the vehicle as an assembly.**

2. Remove or disconnect:
   - Battery cables and battery.
   - Fuel pressure, by opening the fuel filler cap, loosening the fuel filter fitting. Be sure to take the appropriate fire safety precautions.
   - Air filter
   - Accelerator cable from the injection pump.
   - Radiator cap. Turn the heater temperature control all the way towards warm and remove the thermostat housing to drain the coolant.
   - Upper radiator hose and wiring from the radiator fan motor and switches.
   - Radiator and fan shroud as an assembly.
   - Electrical connections and vacuum lines, carefully labeling each one.
   - With power steering, power steering pump and secure it to the body. Do not disconnect the hydraulic lines.
   - With air conditioning, air conditioning compressor and secure it aside without disconnecting the lines.
   - Fuel inlet and outlet lines from the injection pump and plug the holes to keep the pump clean. Note the outlet fitting has a special orifice.
   - On turbocharged engines, exhaust pipe and the oil lines from the turbocharger and cap the oil line fittings on the turbocharger. Unbolt the turbocharger and lift it out of the engine.
   - With an automatic transaxle, selector cable at the transaxle, with selector lever in P (park).
   - On manual transaxle shift linkage, 2 rods with the plastic socket ends and remaining linkage from the case as required. Disconnect the clutch cable, lift it out of the case and set it aside.
   - Wiring from the starter and the back-up light switch and the ground cable from the transaxle. Remove the speedometer cable from the transaxle and plug the hole in the case.

3. Attach an engine sling tool VW-2024A , to the engine and attach the sling to a suitable lifting device.

4. Remove or disconnect:
   - Exhaust pipe from the manifold or turbocharger.
   - Halfshafts from the flanges and hang them from the body with wire.
   - Starter along with the front mount.

5. With all mounts unbolted, slightly lower the engine/transaxle assembly and tilt it towards the transaxle side. Then, carefully lift the assembly out of the vehicle.

To install:

6. Carefully install the engine/transaxle assembly and be sure all mounts are securely bolted to the engine/transaxle. Start all nuts and bolts that secure the mounts to the body but don't tighten them yet.

7. With all mounts installed and the engine safely in the vehicle, allow some slack in the lifting equipment. With the vehicle safely supported, shake the engine/transaxle as a unit to settle it in the mounts. Tighten all mounting bolts, starting at the rear and working forward. Tighten to 33 ft. lbs. (41 Nm) for 10mm bolts or 54 ft. lbs. (73 Nm) for 12mm bolts.

8. Install or connect:
   - Starter. Bolts: 33 ft. lbs. (45 Nm).
   - Halfshafts to the flanges. Bolts: 33 ft. lbs. (45 Nm).
   - Exhaust pipe and use new self-locking nuts to secure the flange. Nuts: 30 ft. lbs. (40 Nm). If equipped with spring clamps, the clamps can be used again.
   - Shift linkage and the clutch cable, if equipped.
   - Fuel injector lines and tighten to 18 ft. lbs. (25 Nm). Be careful not to over-tighten the line nuts. If a line is damaged or clogged, replace all lines as a set.
   - Inlet and outlet lines to the injector pump, using new gaskets. Note the special outlet fitting has the word "OUT" printed on the top.
   - Air conditioning compressor and/or power steering pump, if equipped. Install and adjust the drive belts.
   - Wiring and vacuum hoses.
   - Radiator, fan and heater hoses. Use a new O-ring on the thermostat and tighten the thermostat housing bolts to 84 inch lbs. (10 Nm).
   - Coolant
   - Air filter.
   - Battery and battery cables.

1998–00 Passat

➡**Tag all hoses and wiring during removal, to use as reference during reassembly.**

1. Lock the carrier into service position as follows:
   a. Remove the front bumper.
   b. Tag and remove any wiring or connector that would inhibit locking the carrier.
   c. Remove the 3 quick-release screws on the front noise insulation panel.
   d. Unbolt the air guide between the lock carrier and the air filter.
   e. If installed, remove the retaining clamps for the wiring harness at the left side of the radiator frame.
   f. Remove the No. 2 bolts and install Support tool 3369 .
   g. Remove the remaining bolts and pull the lock carrier out to the stop.
   h. To secure the lock carrier, install the appropriate M6 bolts into the rear of the lock carrier and fender.

2. Position the wipers to the vertical position.

3. Properly relieve the fuel system pressure.

4. Remove or disconnect:
   - Negative battery cable.
   - Engine under cover.
   - Coolant.
   - Front bumper.
   - Power steering cooling coil from the radiator, leaving it connected and hanging.
   - If equipped, transaxle oil cooling lines.
   - Electric cooling fan thermal switch at the lower left of the radiator.
   - Air intake duct and assembly.
   - Headlight height adjuster wiring harness.
   - Turn signal bulbs from the light housing.
   - Coolant hose from the radiator at the upper coolant pipe.
   - Hood release cable at the carrier lock.
   - Power steering fluid reservoir cap/dipstick.
   - Wiring harness for the ABS hydraulic unit.

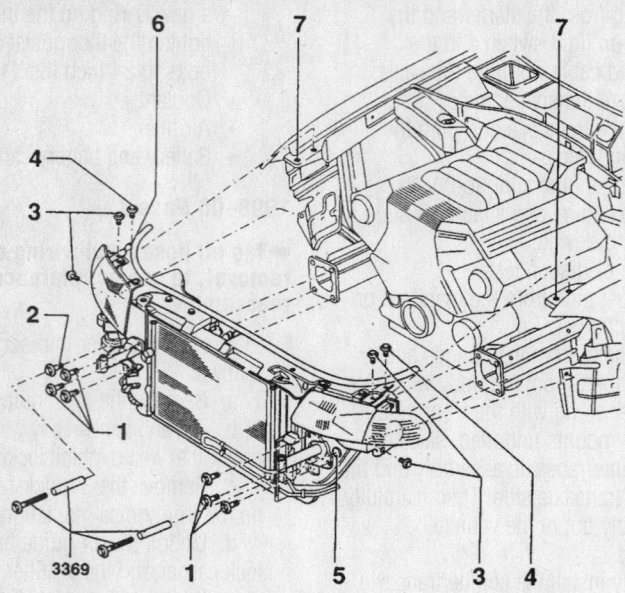

1. Bolts 33 ft. lbs. (45 Nm)
2. Bolts 33 ft. lbs. (45 Nm)
3. Bolts 7 ft. lbs. (10 Nm)
4. Bolts 7 ft. lbs. (10 Nm)
5. Bore for support tool
6. Lock carrier bore
7. Fender bore

7923CG12

Moving the lock carrier into the service position—1998–00 Passat

- Horn electrical connectors.
- Air guides at the left and right sides of the radiator.
- Air conditioning condenser retaining fasteners.
- Air conditioning low pressure switch.
- Condenser, and position it over the fender.
- Green harness connector from the air conditioning compressor magnetic clutch.
- Engine covers.
- Wiring harness connectors for the wastegate bypass regulator valve, the EVAP canister purge regulator valve, the power output stage, and the MAF sensor.
- ECL warning switch.
- Coolant hoses at the expansion tank, then remove the tank and position it aside.
- With cruise control, actuating rod from the throttle valve control module, then remove the vacuum hose from the vacuum unit.
- Accelerator pedal cable from the throttle valve control module.
- Hose for the Leak Detection Pump (LDP).

- Fuel supply and return lines.
- Brake booster vacuum hose.
- Vacuum hose for the EVAP canister purge regulator valve.
- ECM retaining bracket.
- Wiring harness to the ECM.
- With an automatic transaxle, the kickdown switch connector.
- Heated oxygen sensor wiring harness.
- Ground connection at the plenum chamber.
- Heater hoses from the heater core.
- VSS from the transaxle and position it aside.
- With a manual transaxle, back-up light switch from the transaxle.
- Engine driven cooling fan.
- Accessory drive belts.
- Air conditioning compressor from the mounting bracket and position it aside using wire.
- Power steering pump and position it aside leaving the hoses attached.

❋❋ WARNING

The flexpipe at the front exhaust pipe must not be bent more than 10°, otherwise it may be damaged.

- Catalytic converter from the turbocharger.
- Starter, and the ground strap at the right engine mount.
- With an automatic transaxle, 3 torque converter-to-driveplate mounting bolts through the opening left by the starter.

5. Loosen the upper nuts for the left and right engine mounts, matchmark the threaded bolt and centering sleeves at the bottom of the left and right engine mounts, then remove the mounting nuts.

6. Remove or disconnect:
- Lower engine-to-transaxle mounting bolts.
- With an automatic transaxle, ATF cooler line bracket form the left side of the engine.
- Upper nuts from the engine mounts.

7. Position an Engine Support Bridge 10-222A to the bolted flanges of the fenders with the spindle facing forward.

8. Attach the Engine Support Adapter 3147 or equivalent to a bolt hole in the transaxle bell housing.

9. Connect the Engine Support Adapter 3147 and the Engine Support Bridge 10-222A using Adapter 2024A/1 and Extension 2024A/2.

10. Attach an engine sling between the engine and the hoist.

11. Remove the upper engine-to-transaxle mounting bolts.

12. Separate the engine from the transaxle, then slowly lift the engine up and out the front of the engine compartment.

13. If equipped with an automatic transaxle, secure the torque converter to prevent it from falling out.

To install:

➡**Be sure that the centering sleeves for the engine-to-transaxle are correctly installed in the cylinder block.**

14. Verify that the intermediate plate is over the centering sleeves.

15. Install or connect the following:
- Engine into the vehicle
- Upper engine-to-transaxle mounting bolts

16. Lower the engine into position, then remove the engine sling and hoist and the transaxle support apparatus from the vehicle.

17. Install or connect:
- Engine mounting fasteners without any tension or pre-load.
- With an automatic transaxle, ATF cooler line bracket to the left side of the engine.

- Lower engine-to-transaxle mounting bolts. M12 bolts: 48 ft. lbs. (65 Nm), M10 bolts: 33 ft. lbs. (45 Nm)
- Engine mounting nuts/bolts: 18 ft. lbs. (25 Nm).
- With an automatic transaxle, driveplate-to-torque converter mounting bolts through the starter opening. Bolts: 63 ft. lbs. (85 Nm).
- Starter and attach the ground strap to the right engine mount.
- Catalytic converter to the turbocharger. Mounting bolts: 22 ft. lbs. (30 Nm).
- Power steering pump, the air conditioning compressor, and the engine cooling fan, then the accessory drive belts.
- With a manual transaxle, back-up light switch to the transaxle.
- VSS to the transaxle.
- Heater hoses to the heater core.
- Ground connection at the plenum chamber.
- Heated oxygen sensor wiring harness.
- With an automatic transaxle, kickdown switch connector.
- Wiring harness to the ECM.
- ECM retaining bracket and cover the E-box.
- Vacuum hose for the EVAP canister purge regulator valve.
- Fuel supply and return lines.
- Brake booster vacuum hose.

18. The completion of the installation procedure is the reverse of the removal.

19. If equipped with an automatic transaxle, check the ATF level

20. Fill the engine with coolant

21. Fully close all power windows, operate all window switches for at least 1 second in the close direction to activate the one-touch opening/closing function

22. Connect the negative battery cable.

➡DTCs are stored when harness connectors are detached.

23. Read the DTCs and clear the fault codes.

New Beetle

The engine and transaxle are removed from under the vehicle as an assembly.

1. Before servicing the vehicle, refer to the precautions in the beginning of this section.

2. Remove or disconnect:
- Negative battery cable.

- Engine cover.
- Power steering reservoir from the battery support leaving the hoses attached, and positioning it aside.
- Battery and bracket.
- Fuel supply and return lines.
- Air cleaner and intake air duct.
- Throttle cable.
- With a manual transaxle, transaxle range selector mechanism and the clutch slave cylinder.
- With an automatic transaxle, range selector lever cable at the transaxle.
- Engine undercover.
- Coolant.
- Accessory drive belt.
- If necessary, right side cooling fan.
- Power steering pump and bracket and position it aside leaving the hoses connected.
- Air conditioner compressor and position it aside leaving the hoses connected.
- Coolant, vacuum and intake hoses.
- Secondary air injection pump and bracket.
- All wires from the transaxle, starter and generator, position the wires out of the way.
- Any remaining wires or connectors that would interfere with the engine removal.
- Engine pendulum support.
- Right-hand halfshaft and disconnect the left-hand halfshaft at the transaxle.
- Front exhaust pipe from the manifold.

3. Install Engine Bracket T10012, or equivalent, in Engine/transaxle Jack VAG 1383 A or equivalent.

4. Remove the bracket for the coolant hose under the engine block.

5. Attach the Engine Bracket T10012 to the engine block using the threaded holes at the corners of the engine block.

6. Raise the engine/transaxle jack to relieve the tension on the mounts.

7. Disconnect the engine and transaxle mounts from inside the engine compartment.

8. Lower the engine/transaxle assembly from the vehicle, be sure to guide the power steering pressure hose past the transaxle.

To install:

9. Raise the engine/transaxle assembly into the vehicle, be sure to guide the power steering hose around the transaxle.

10. Using new bolts, connect the engine/transaxle mounts following the accompanying illustration.

11. Remove the engine/transaxle jack and Engine Bracket T10012.

12. Install or connect:
- Bracket for the coolant hose under the engine block.
- Front exhaust pipe to the manifold.
- Left halfshaft and the right halfshaft.

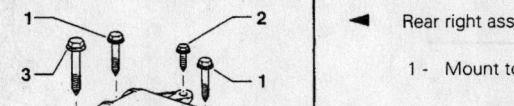

Engine mount fastener locations—New Beetle

Engine / transmission mounts

Tightening torques

Front assembly mounting

1 -	Mount to body [1]	40 Nm (30 ft lb) + 90° ($^1/_4$ turn)
2 -	Mount/bracket to body	25 Nm (18 ft lb)
3 -	Mount to engine bracket [1]	60 Nm (44 ft lb) + 90° ($^1/_4$ turn)

[1] Replace bolts

Rear right assembly mounting

1 -	Mount to body [1]	40 Nm (30 ft lb) + 90° ($^1/_4$ turn)
2 -	Mount to body	25 Nm (18 ft lb)
3 -	Bearing on transmission console [1]	60 Nm (44 ft lb) + 90° ($^1/_4$ turn)

[1] Replace bolts

9301WG02

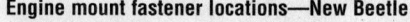

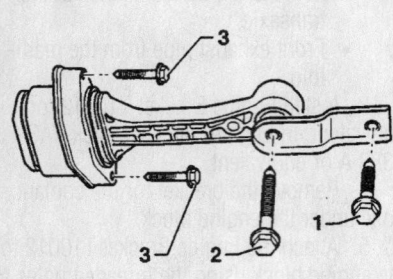

Pendulum support mounting bolt locations—New Beetle

9301WG01

- Engine pendulum support. Bolts (1) and (2): 30 ft. lbs. (40 Nm) plus 90 degrees, and bolts (3): 15 ft. lbs. (20 Nm) plus 90 degrees.
- Any wiring that was removed.
- Secondary air injection pump and bracket.
- Any hoses that were removed.
- Air conditioner compressor.
- Power steering bracket and pump.
- If removed, right side cooling fan.
- Accessory drive belt and fill the cooling system.
- With an automatic transaxle, the transaxle range selector cable.
- With a manual transaxle, clutch slave cylinder and connect the range selector lever.
- Throttle cable.
- Intake air duct and air cleaner assembly.
- Fuel supply and return lines.
- Battery bracket.
- Power steering reservoir.
- Engine oil.
- Battery cables.

13. Check and clear any DTCs, then match the ECM to the TCM.

14. Install the engine covers.

Water Pump

REMOVAL & INSTALLATION

1.8L Engine

➡**The coolant pump is bolted to the brackets for the generator, power steering pump, and cooling fan.**

1. Before servicing the vehicle, refer to the precautions in the beginning of this section.

2. Position the lock carrier into the service position as follows:

   a. Remove or disconnect:

- Front bumper.
- Any wiring or connector that would inhibit locking the carrier.
- 3 quick-release screws on the front noise insulation panel.
- Air guide between the lock carrier and the air filter.
- Retaining clamps for the wiring harness at the left side of the radiator frame.
- No. 2 bolts and install Support tool 3369 or equivalent.
- Remaining bolts and pull the lock carrier out to the stop.

   b. To secure the lock carrier, install the appropriate M6 bolts into the rear of the lock carrier and fender.

3. Remove or disconnect

- Negative battery cable.
- Accessory drive belt, then the engine driven cooling fan.
- Coolant.
- Clamps for the coolant hoses at the water pump.
- Intake air duct between the intake manifold and the charge air cooler.
- Generator mounting bolts and slide it forward.
- Wiring from the generator once it is removed.

4. Unbolt the following supports and brackets for the generator, power steering pump, and engine cooling fan:

   a. Intake manifold support

   b. Support for the engine torque bracket

   c. Brace to the cylinder block (remove completely)

5. Remove or disconnect:

- Brackets for the generator, power steering pump, and engine cooling fan, positioning the brackets to the left side using a piece of wire
- Coolant hoses from the pump and thermostat housing
- Coolant pump housing from the timing belt cover
- Coolant pump mounting bolts, then the pump
- Impeller housing from the pump housing

To install:

6. Using a new gasket, mount the new coolant pump to the pump housing. Mounting bolts: 84 inch lbs. (10 Nm).

7. Using a new gasket and O-ring, install the coolant pump. Mounting bolts in alphabetical sequence: 18 ft. lbs. (25 Nm).

8. Tighten the coolant pump housing to the timing belt cover to 84 inch lbs. (10 Nm).

9. Install or connect:

- Coolant hoses to the pump and thermostat housing.
- Brackets that were removed. Mounting bolts: 18 ft. lbs. (25 Nm).
- Wires to the generator, then install the generator.
- Air intake duct between the intake manifold and the charge air cooler.

10. The remaining steps are the reverse of the removal.

11. Fill the engine with coolant

12. Fully close all power windows to stop, operate all window switches for at

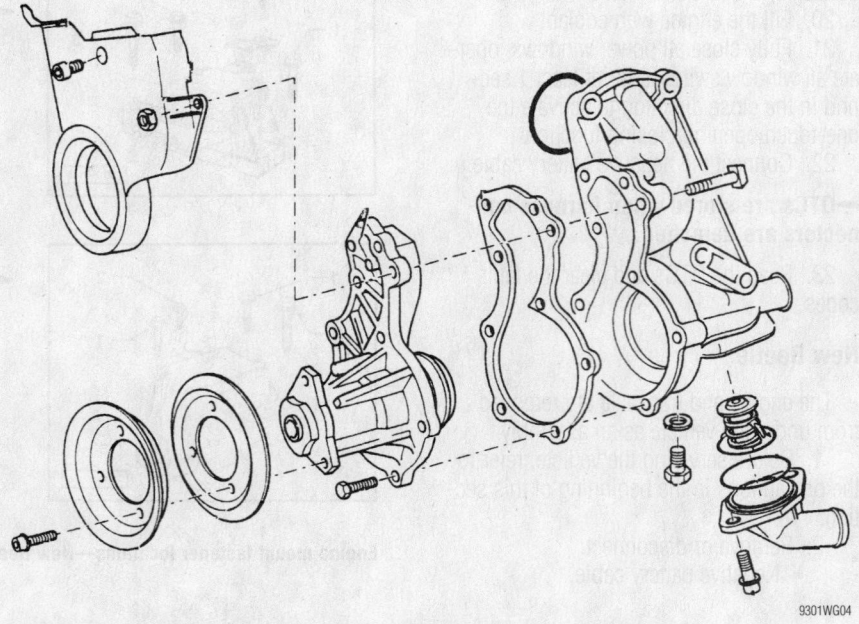

Exploded view of the water pump, housing and related components—1.8L engine

9301WG04

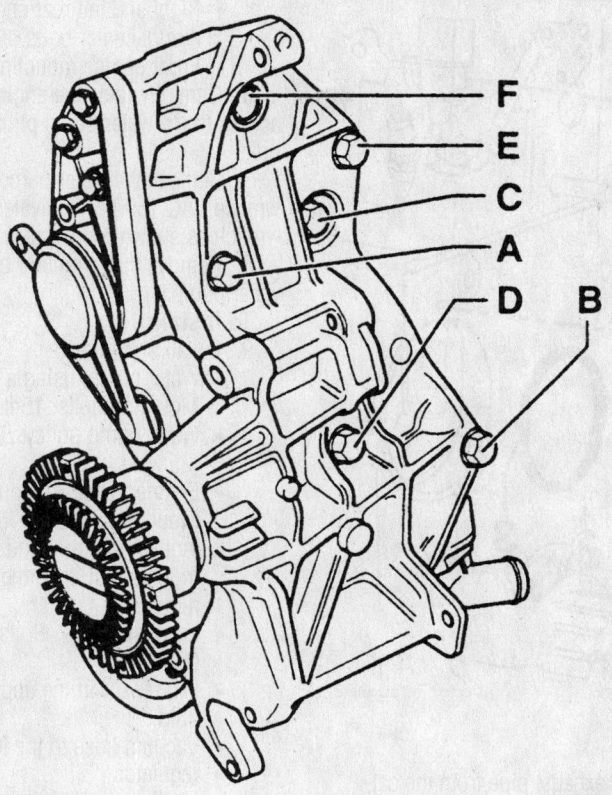

Assembly bracket with water pump mounting bolt tightening sequence—1.8L engine

least 1 second in the close direction to activate the one-touch opening/closing function

13. After installing the lock carrier, check the wiring for proper routing near the cooling fan

2.0L (ABA) Engine

1. Before servicing the vehicle, refer to the precautions in the beginning of this section.

2. Remove or disconnect:
   • Coolant
   • Thermostat housing from under the water pump housing.

3. Loosen but don't remove the bolts holding the pulley to the water pump.

4. Remove or disconnect:
   • Timing belt cover.
   • Alternator and/or steering pump as required to remove the water pump drive belt.
   • Water pump pulley. On some vehicles, the crankshaft pulley must also be removed by removing the bolts holding the pulley to the timing belt sprocket.
   • Water pump from its housing.

To install:

5. Be sure to clean the housing before installing the new gasket. Install the pump into the housing. Pump-to-housing bolts: 84 inch lbs. (10 Nm).

6. Install the water pump drive pulley. Bolts: 15 ft. lbs. (20 Nm). Install crankshaft drive pulley. Bolts: 15 ft. lbs. (20 Nm).

7. Adjust drive belt tension and install the thermostat and housing. Bolts: 84 inch lbs. (10 Nm).

2.0L (AEG) Engine

1. Before servicing the vehicle, refer to the precautions in the beginning of this section.

2. Remove or disconnect:
   • Negative battery cable.
   • Coolant.
   • Accessory drive belt and tensioner.
   • Upper and center timing belt covers.

3. Position the engine so that the No. 1 cylinder is at TDC.

> ※※ **WARNING**
>
> **Cover the timing belt to protect it from being contaminated with coolant.**

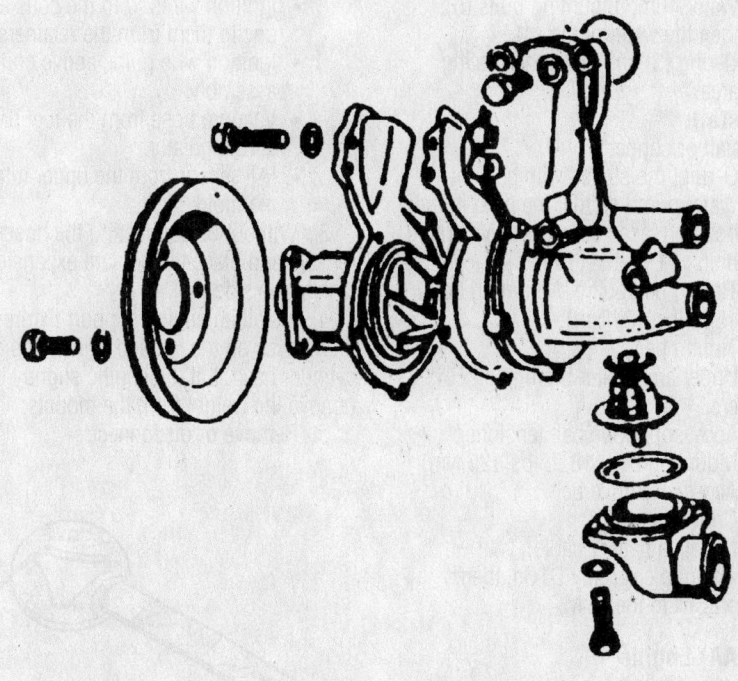

Water pump and thermostat housing—2.0L (ABA) engine

Heater Core replacement is covered in Section 2 of this manual

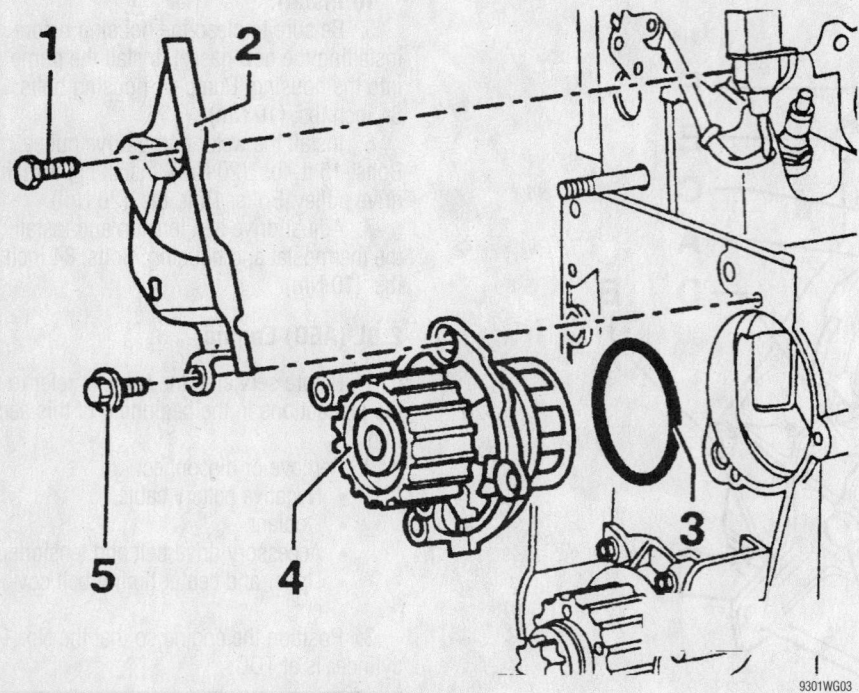

Exploded view of the water pump mounting—2.0L (AEG) engine

4. Loosen the timing belt tension and slide the belt off the water pump sprocket.

5. Remove or disconnect:
- Timing belt guard (2) mounting bolt (1).
- Water pump mounting bolts (5), then the water pump (4).
- O-ring (3) and clean the seating area.

To install:

6. Install or connect:
- O-ring, moistened with coolant.
- Water pump so that the plug in the housing faces downward. Mounting bolts: 11 ft. lbs. (15 Nm).
- Timing belt guard. Mounting bolt: 15 ft. lbs. (20 Nm).
- Timing belt.
- Upper and center timing belt covers.
- Accessory drive belt tensioner. Mounting bolt: 18 ft. lbs. (25 Nm).
- Accessory drive belt.
- Coolant.
- Negative battery cable.

7. Check and clear any DTCs, then match the ECM to the TCM.

2.8L (AAA) Engine

1. Before servicing the vehicle, refer to the precautions in the beginning of this section.

2. Remove or disconnect:
- Negative battery cable.

- Front exhaust pipe from the catalytic converter.
- Coolant.
- Accessory drive belt.
- Air intake duct.
- Ignition wires from the coils and unclip them from the retainers.
- Ignition wire guide above coil assembly.
- Vacuum hose from the fuel pressure regulator.
- IAT sensor from the upper intake manifold.

3. Without disconnecting the hoses, remove and place the coolant expansion tank to the side.

4. Install an engine support fixture to the lifting eyes on the left and right sides of the cylinder head. Lift the engine slightly to remove the weight from the mounts.

5. Remove or disconnect:

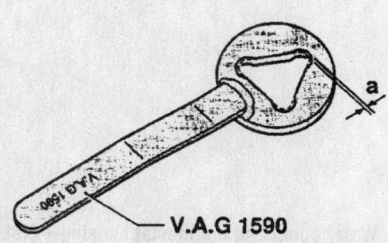

V.A.G 1590

Modify wrench VAG 1590 or equivalent to fit the water pump pulley bolt as needed

- Right and left rear engine/transaxle mount center bolts.
- Front engine mounting center bolts.

6. Carefully raise the engine to gain access to the water pump pulley mounting bolts.

7. Remove the water pump pulley using wrench VAG 1590 or equivalent. Modify the wrench as shown if necessary to fit the bolt.

8. Remove the mounting bolts and the water pump.

To install:

9. Install or connect:
- Water pump, using a new O-ring. Mounting bolts: 15 ft. lbs. (20 Nm).
- Water pump pulley. Bolt: 18 ft. lbs. (25 Nm).
- Engine/transaxle mount bolts. Mounting bolts: 44 ft. lbs. (60 Nm). Tighten the front and right rear mounts first, then the left rear mount.
- Expansion tank. Bolts: 84 inch lbs. (10 Nm).
- IAT sensor in the upper intake manifold.
- Vacuum hose to the fuel pressure regulator.
- Ignition wires and the wire guide.
- Air duct and accessory drive belt.
- Coolant.

2.8L (AHA) Engine

1. Before servicing the vehicle, refer to the precautions in the beginning of this section.

2. Disconnect the negative battery cable.

3. Lock the carrier into service position as follows:

   a. Remove or disconnect:
   - Front bumper.
   - Any wiring or connector that would inhibit locking the carrier.
   - 3 quick-release screws on the front noise insulation panel.
   - Air guide between the lock carrier and the air filter.
   - Retaining clamps for the wiring harness at the left side of the radiator frame.
   - No. 2 bolts and install Support 3369 or equivalent.
   - Remaining bolts and pull the lock carrier out to the stop.

   b. To secure the lock carrier, install the appropriate M6 bolts into the rear of the lock carrier and fender.

4. Remove or disconnect:
- Accessory drive belt.
- Timing belt.
- Coolant.

- Timing belt tensioner and idler rollers.
- Water pump mounting bolts, then the pump.

To install:

5. Install or connect:
   - New gasket on the water pump flange.
   - Water pump. Mounting bolts: 10 ft. lbs. (15 Nm).
   - Timing belt idler roller. Mounting bolt: 33 ft. lbs. (45 Nm).
   - Timing belt tensioner roller. Mounting bolt: 15 ft. lbs. (20 Nm).
   - Timing belt and accessory drive belt.
   - Coolant.
6. Return the lock carrier to the normal position.
7. Connect the negative battery cable.

Cylinder Head

REMOVAL & INSTALLATION

➡**Cylinder head bolt torque sequence illustrations are found in Section 1, following the Torque Charts.**

1.8L Engine

1. Before servicing the vehicle, refer to the precautions in the beginning of this section.
2. Place the lock carrier into the service position as follows:
   a. Remove or disconnect:
   - Front bumper.
   - Any wiring or connector that would inhibit locking the carrier.
   - 3 quick-release screws on the front noise insulation panel.
   - Air guide between the lock carrier and the air filter.
   - If installed, retaining clamps for the wiring harness at the left side of the radiator frame.
   - No. 2 bolts and install Support 3369 or equivalent.
   - Remaining bolts and pull the lock carrier out to the stop.
   b. To secure the lock carrier, install the appropriate M6 bolts into the rear of the lock carrier and fender.
3. Remove or disconnect:
   - Negative battery cable.
   - Accessory drive belt, then the engine driven cooling fan.
   - Coolant.
   - Intake manifold.
   - Accessory drive belts.

4. Label and detach the following lines and electrical connectors:
   - Wastegate bypass regulator valve
   - EVAP canister purge regulator valve
   - Power outage stage
   - MAF sensor
5. Remove or disconnect:
   - Air cleaner housing.
   - ECT and the Temperature II sensor harness connector.
   - All connections from the cylinder head and position them aside.
   - Crankcase breather line.
   - Fuel supply and return lines.
   - Oil supply line at the cylinder head.
   - Exhaust manifold heat shield.
   - Turbocharger from the exhaust manifold.
   - Coolant hose to the heat exchanger at the rear of the cylinder head.
   - Upper timing belt cover.
6. Turn the crankshaft, in the direction of rotation (clockwise), until the No. 1 cylinder is at TDC.
7. Using a T45 Torx® wrench, loosen the timing belt tensioner, and remove the belt from the camshaft gear.
8. Remove:
   - The Torx® bolt and swing the tensioner assembly bracket forward.
   - Valve cover.
   - Cylinder head bolts in sequence, as shown.

- Cylinder head, then clean the gasket mating surfaces.
9. Clean and dry out the cylinder head bolt holes.

To install:

✳✳ WARNING

Always replace the cylinder head bolts. Always replace self-locking nuts, bolts, gaskets and O-rings. Do not remove the new head gasket from the package until immediately before installing.

10. Before installing the cylinder head, be sure NO pistons are at TDC.
11. Loosen the turbocharger support bracket to reduce the likelihood of any tension while installing the cylinder head.
12. Install or connect:
    - Head gasket with the part number visible from the intake side.
    - Cylinder head.
    - New cylinder head bolts and tighten by hand.
13. Tighten the new cylinder head bolts in sequence in 2 steps:
    a. Tighten all of the bolts to 44 ft. lbs. (60 Nm).
    b. Tighten all of the bolts an additional ½ turn (180°).

➡**It is not necessary to retighten the cylinder head bolts.**

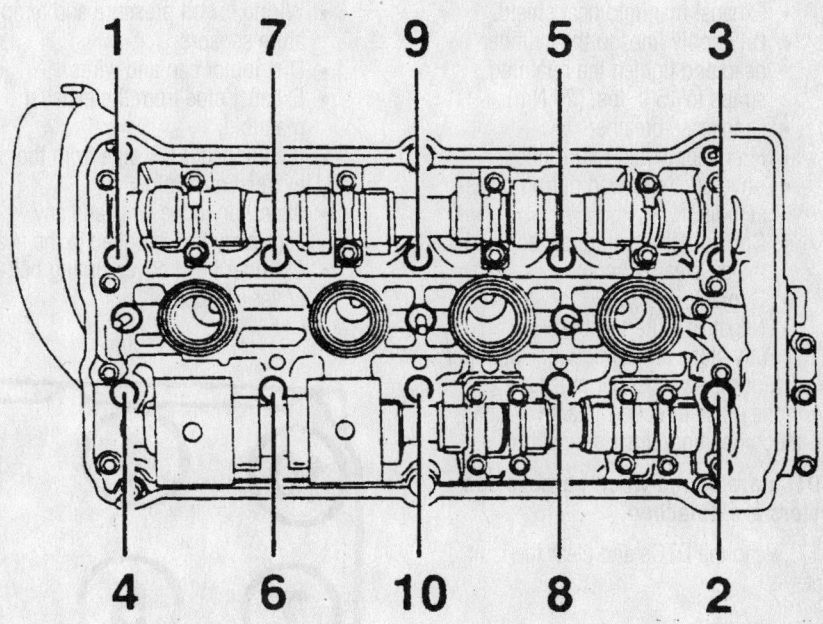

Cylinder head bolt removal sequence—1.8L engine

7923CG14

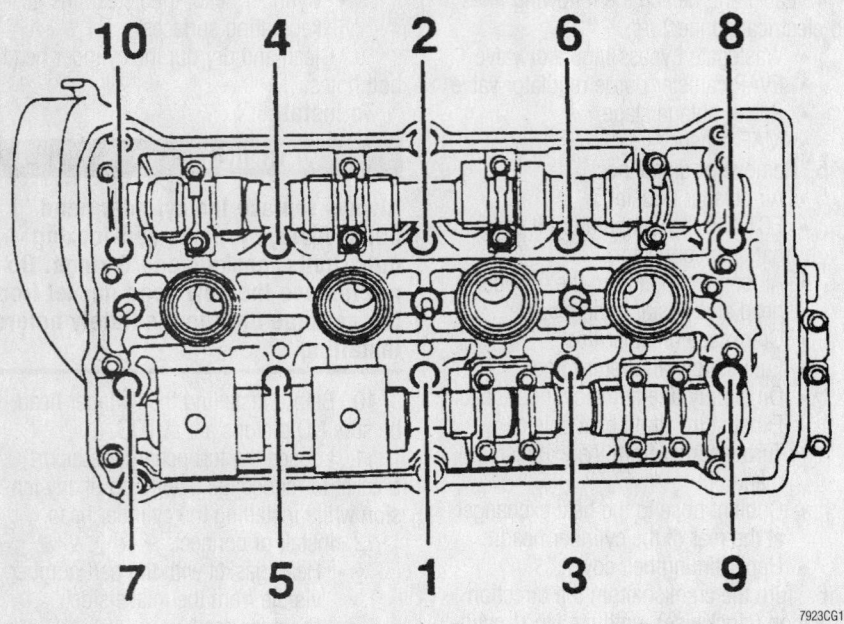

Cylinder head bolt tightening sequence—1.8L engine

14. Using new gaskets, install the turbocharger to the exhaust manifold , coat the bolts with Hot Bolt Paste G 052 112 A3 (or equivalent). Mounting bolts: 26 ft. lbs. (35 Nm). Turbo support bracket mounting bolts: 33 ft. lbs. (40 Nm).

15. Install or connect:
- Valve cover.
- Timing belt.
- Accessory drive belts.
- Exhaust manifold heat shield.
- Oil supply lines to the cylinder head and tighten the retaining straps to 15 ft. lbs. (20 Nm).
- Crankcase breather.
- Fuel supply and return lines.
- Any items removed during disassembly.
- Coolant temperature sensors, and the air cleaner housing.
- Coolant.
- Negative battery cable.

16. Fully close all power windows, operate all window switches for at least 1 second in the close direction to activate the one-touch opening/closing function

➡ DTCs are stored when harness connectors are detached.

17. Read the DTCs and clear the fault codes.

2.0L (ABA) Engine

1. Before servicing the vehicle, refer to the precautions in the beginning of this section.

2. Remove or disconnect:
- Negative battery cable.
- Coolant.
- Throttle cable.
- All wiring and vacuum lines from the intake manifold.
- Upper intake manifold.
- Fuel supply and return lines.
- Radiator and heater hoses from the cylinder head.
- Wiring for oil pressure and temperature sensors.
- Distributor cap and wires.
- Exhaust pipe from the exhaust manifold.
- if equipped, EGR pipe from the exhaust manifold.
- Accessory drive belts and any accessory that is bolted to the head.
- Cylinder head cover, timing belt cover and belt.

3. Loosen the cylinder head bolts in the reverse of the tightening sequence.

4. Remove the bolts and lift the head straight off.

To install:

5. Before reinstalling the head, check the flatness of the head and block in both width and length, then diagonally from each corner.

6. Install the new cylinder head gasket with the word TOP or OBEN facing upward; do not use any sealing compound.

7. To align the cylinder head, install Guide Pins from tool 3070, or equivalent, into the holes for cylinder head bolts 8 and 10.

8. Install:
- Cylinder head.
- Cylinder head bolts, except 8 and 10, by hand.

9. Remove the Guide Pins using tool 3070 or equivalent, then install head bolts 8 and 10.

10. Tighten the head bolts in sequence in the following steps:
- 30 ft. lbs. (40 Nm)
- 44 ft. lbs. (60 Nm)
- + 90°
- + an additional 90°

11. Install or connect:
- Camshaft drive belt and adjust the tension.
- Exhaust pipe to the manifold. Use new gaskets and self-locking nuts. Nuts: 18 ft. lbs. (25 Nm).
- EGR pipe, if equipped.
- Wiring to the oil pressure and temperature sensors.
- Ignition system components.
- Radiator and heater hoses.
- Throttle cable and all wiring and vacuum lines.
- Thermostat with a new O-ring. Housing bolts: 84 inch lbs. (10 Nm).
- Coolant.

Tighten the cylinder head bolts in the sequence shown—2.0L (ABA) engine

- Accessory drive belts and adjust the tension.
- Upper intake manifold.
- All wiring and vacuum lines disconnected from the intake manifold. Connect the throttle cable.
- Fuel supply and return lines.
- Negative battery cables.

2.0L (AEG) Engine

1. Before servicing the vehicle, refer to the precautions in the beginning of this section.
2. Remove or disconnect:
   - Negative battery cable.
   - Engine cover.
   - Coolant.
   - Fuel supply and return lines.
   - All vacuum lines related to the cylinder head removal.
   - Air cleaner.
   - Accelerator cable from the TCM.
   - All electrical connectors that would interfere with the cylinder head removal.
   - Front bolts for the upper and lower intake manifold.
   - Warm air deflector.
   - Any coolant hoses attached to the cylinder head and intake manifold.
   - Accessory drive belt and tensioner.
   - Timing belt upper cover.
   - Timing belt off the camshaft sprocket.

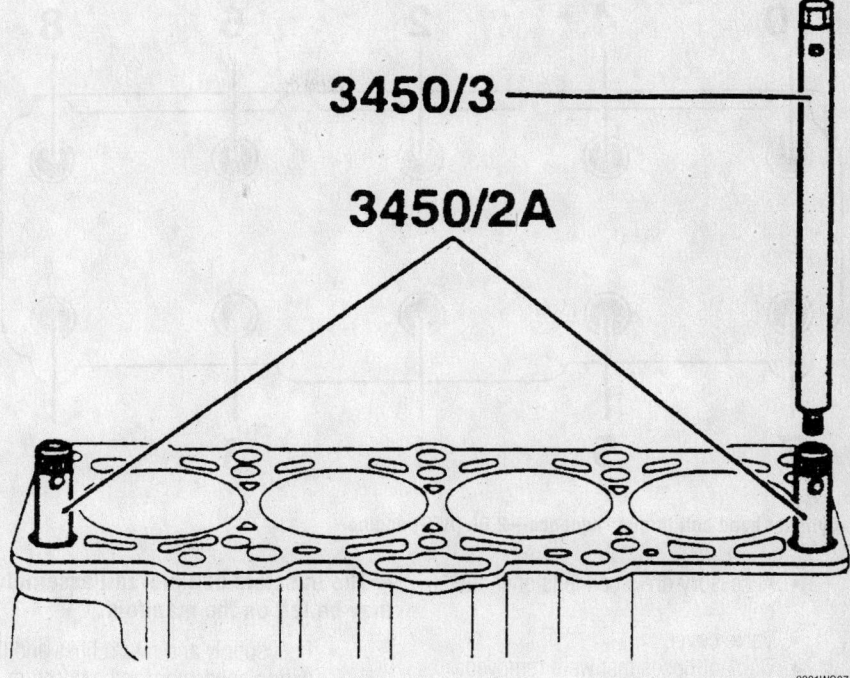

To properly install the head gasket and cylinder head, install the tools as shown—2.0L (AEG) engine

- Upper bolt for the rear timing belt guide.
- Front exhaust pipe from the manifold.
- Valve cover.

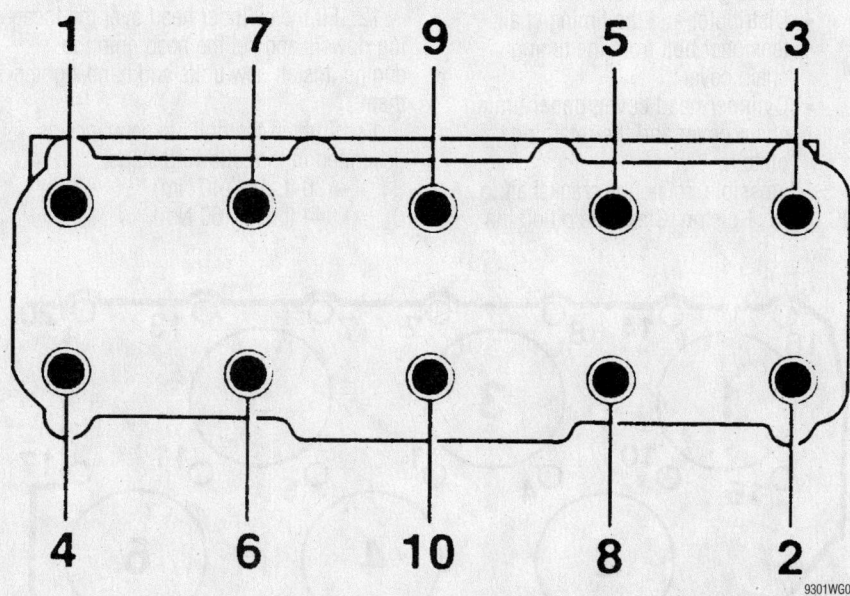

Cylinder head bolt removal sequence—2.0L (AEG) engine

- Cylinder head bolts in sequence.
3. Carefully lift the cylinder head off the engine block.
4. Clean the gasket sealing surfaces.
To install:
5. Install:
   - Guide Pins 3450/2A, or equivalent, into the head bolt holes as shown.

➡**The head gasket part number must be readable when installed on the engine block.**

   - Head gasket.
   - Cylinder head and tighten the 8 remaining head bolts by hand.
6. Remove the Guide Pins 3450/2A using the Pin Removal tool 3450/3, or equivalent and install the last 2 head bolts and tighten by hand.
7. Tighten all the cylinder head bolts in sequence as follows:
   - 30 ft. lbs. (40 Nm)
   - + 90°
   - + an additional 90°.
8. Install or connect:
   - Upper bolt for the rear timing belt cover and tighten to 15 ft. lbs. (20 Nm).
   - Timing belt and covers.
   - Front exhaust pipe to the manifold.

For complete Engine Mechanical specifications, see Section 1 of this manual

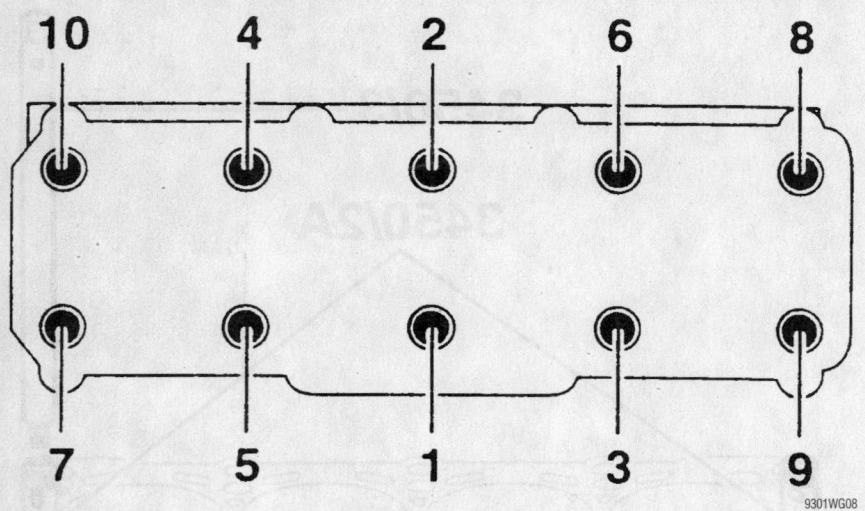

Cylinder head bolt torque sequence—2.0L (AEG) engine

- Accessory drive belt tensioner and belt.
- Valve cover.
- Coolant hoses that were removed.
- Warm air deflector. Mounting bolts: 18 ft. lbs. (25 Nm).
- Any electrical connectors that were removed.
- Throttle cable to the TCM.
- Air cleaner and any vacuum hoses that were removed.
- Coolant and engine oil.
- Negative battery cable.

9. Check and clear and DTCs, then match the ECM to the TCM.

2.8L (AAA) Engine

This procedure requires special tool 3268 or equivalent. This is a setting tool that holds the camshafts in the correct position for installing the timing chains. Before removing the cylinder head, be sure new bolts are available. The cylinder head bolts are made to stretch and cannot be used again.

1. Before servicing the vehicle, refer to the precautions in the beginning of this section.

2. Remove or disconnect:
- Battery cables and battery.
- Wiring and vacuum lines as required to remove the air cleaner, MAF sensor and duct.
- Coolant.
- Engine trim cover.
- Distributor cap, ignition wires and wire guide as an assembly.
- Throttle cable.
- Wiring and vacuum lines from the intake manifold and remove the upper manifold.

➡ The injectors and fuel rail assembly may be left on the manifold.

- Fuel supply and return lines and the wiring connector for the injectors.
- Radiator and heater hoses.

3. Thread a long 8 x 10mm bolt into the accessory drive belt tensioner to release the tension. Move the tensioner only as required to remove the belt.

4. Remove or disconnect:
- Alternator and belt tensioner.
- Heat shield and the bolts to disconnect the 2 piece exhaust manifold from the engine. Note the position of the gaskets.
- Distributor and the timing chain tensioner bolt from the timing chain cover.
- Cylinder head cover, upper timing chain cover and the retaining plate.

5. If possible, rotate the crankshaft to TDC of No. 1 piston. Clean the oil off the

chain and sprockets and mark the direction of rotation for assembly.

6. Hold the camshafts at the flats with a 24mm wrench and remove the bolts to remove the sprockets and chain. Note the position of the distributor drive on the short camshaft.

❋❋ WARNING

Do not use the setting tool to hold the camshafts when removing or installing the sprocket bolts. The camshafts and the tool will be damaged.

7. Carefully check to be sure all necessary wires, hoses and brackets and components have been removed.

8. Loosen the cylinder head bolts in the reverse of the torque sequence. Remove and discard the bolts.

9. Remove the cylinder head.

To install:

10. Carefully clean the old gasket material from the head and the block. Before reinstalling the head, check the flatness of the head and block in both width and length, then diagonally from each corner. Maximum allowable distortion is 0.004 in. (0.1mm).

11. If the new head gasket already has sealant in the small holes at the timing chain end, remove the sealant. Apply a silicone sealer to the timing chain end and install the gasket onto the block with the word TOP or OBEN facing up.

12. Fit the cylinder head over the locating dowels and set the head onto the engine. Install new bolts and hand-tighten them.

13. Tighten the bolts in sequence as described in the following steps:
- 30 ft. lbs. (40 Nm)
- 44 ft. lbs. (60 Nm)

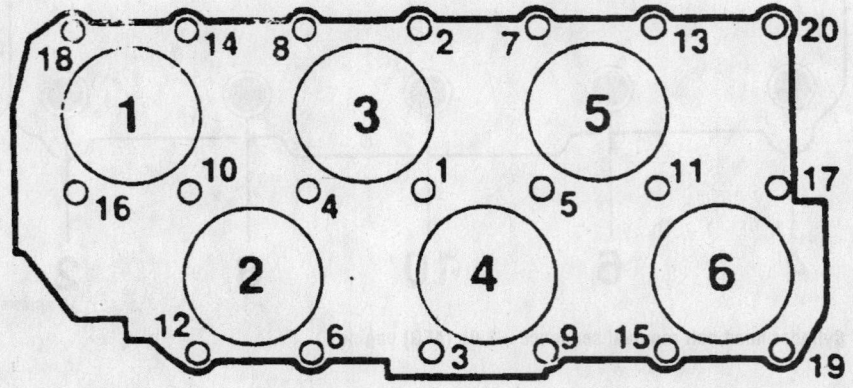

Be sure to tighten the cylinder head bolts in the sequence shown—2.8L (AAA) engine

- +90°
- + an additional 90°

14. Be sure the crankshaft is at TDC on No. 1 piston. Install the setting tool to lock the camshafts in place, then install the timing chain and sprockets. Be sure they are positioned to rotate in the original direction.

15. Hold the camshaft with a 24mm wrench and install the sprocket bolt. Be sure the distributor drive is correctly positioned. Bolts: 74 ft. lbs. (100 Nm).

16. Install the tensioner shoe and temporarily install the upper timing chain cover. Install the tensioner bolt and remove the setting tool. Rotate the crankshaft 4 full turns and stop at TDC of No. 1 piston. The setting tool should fit into the camshafts.

17. Remove the tensioner bolt and upper timing chain cover again. Apply new sealant as required, install the cover. Bolts: 82 inch lbs. (10 Nm). Tensioner bolt: 15 ft. lbs. (20 Nm).

18. Install or connect:
- Cylinder head cover.
- Intake and exhaust manifolds, using new gaskets. Nuts and bolts: 18 ft. lbs. (25 Nm).
- Alternator belt and adjust tension.
- Accessory drive belt and adjust tension.
- Radiator and heater hoses.
- Injectors, fuel rail assembly and manifold. Connect the fuel supply and return lines. Connect the wiring connector for the injectors.
- Upper manifold. Bolts: 18 ft. lbs. (25 Nm).
- Wiring and vacuum lines disconnected from the intake manifold.
- Throttle cable.
- Distributor cap, ignition wires and wire guide as an assembly.
- Trim cover.
- Battery cables and battery.
- Coolant.

2.8L (AHA) Engine

➡**This procedure is for removing the left cylinder head, the right cylinder head service is similar.**

1. Before servicing the vehicle, refer to the precautions in the beginning of this section.

2. Place the lock carrier into the service position as follows:
   a. Remove or disconnect:
   - Front bumper.
   - Any wiring or connector that would inhibit locking the carrier.

- 3 quick-release screws on the front noise insulation panel.
- Air guide between the lock carrier and the air filter.
- If installed, retaining clamps for the wiring harness at the left side of the radiator frame.
- No. 2 bolts and install Support 3369 or equivalent.
- Remaining bolts and pull the lock carrier out to the stop.
   b. To secure the lock carrier, install the appropriate M6 bolts into the rear of the lock carrier and fender.

3. Remove or disconnect:
- Negative battery cable.
- Engine cover and under cover.
- Accessory drive belt.
- Timing belt.
- Left and right front exhaust pipes from the manifold.
- Exhaust system from the transaxle bracket.
- Coolant.
- Coolant expansion tank.
- Crankshaft housing ventilation hose from the valve cover.
- Intake air duct between the MAF sensor and elbow.
- Intake Manifold Tuning (IMT) valve and IAT sensor electrical connectors.
- Vacuum hoses from the cruise control diaphragm and fuel pressure regulator.
- Cruise control vacuum diaphragm and throttle control linkage from the throttle body.
- ECT and CMP sensor electrical connector.
- Tie wraps for the wiring harnesses and position the wires to the rear.
- Any remaining connectors and hoses that would interfere with the cylinder head removal.
- Fuel supply and return lines.
- 4 bolts mounting the fuel rail to the intake manifold.
- Fuel rail from the intake manifold.

4. Install Camshaft Locator Bar 3391, or equivalent, onto the camshaft sprocket.

5. Loosen the 2 bolts for the camshaft and back out approximately 5 turns.

6. Remove the Camshaft Locating Bar 3391.

7. Using Camshaft Gear Puller T40001, or equivalent, remove the camshaft sprocket.

8. Remove or disconnect:

- Spark plug connectors.
- Ignition coil bracket and position it aside with the coils.
- Intake manifold.
- Timing belt rear cover.
- Coolant hoses related to the cylinder head service.
- Valve cover.

9. Using Polydrive Special tool 3452, or equivalent, loosen, then remove the cylinder head bolts in the reverse order of the tightening sequence.

To install:

➡**The bolt holes in the engine block MUST be clean and free of debris and fluid. Always replace the cylinder head bolts.**

10. Set the crankshaft and camshaft to TDC for cylinder No. 3.

11. Install or connect:
- Head gasket, making sure that the part No. or the word "OBEN" is facing the cylinder head.
- Cylinder head and tighten the head bolt hand-tight.
- New cylinder head bolts in sequence in 2 steps:
   a. Tighten all bolts to 44 ft. lbs. (60 Nm).
   b. Tighten all bolts an additional ½ turn (180°).

➡**It is not necessary to retighten the cylinder head bolts.**

- Timing belt rear cover.
- Any coolant hoses that were removed.
- Intake manifold.
- Ignition coil bracket with the coil attached.
- Camshaft sprocket. Retaining bolt:

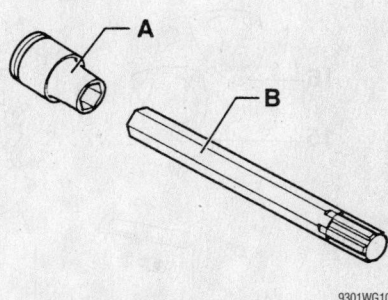

9301WG10

To remove and install the cylinder head bolts, use the Polydrive Special tool 3452 with a 10mm socket—2.8L (AHA) engine

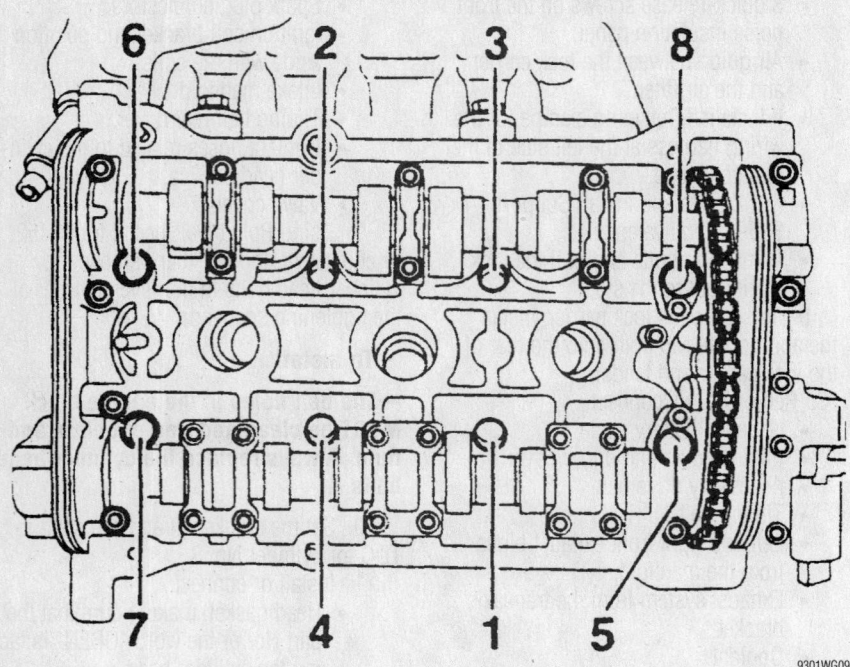

Cylinder head bolt torque sequence—2.8L (AHA) engine

9301WG09

41 ft. lbs. (55 Nm).
- Camshaft Locating Bar 3391, tighten the 2 bolts for the camshafts, then remove the tool.

- Fuel rail to the intake manifold.
- Fuel supply and return lines.
- All hoses, vacuum lines and electrical connectors.

- Any tie wraps that were cut during the removal steps.
- Exhaust system.
- Coolant expansion tank and fill the cooling system.
- Timing belt.
- Accessory drive belt.

12. Return the lock carrier to the normal position.

13. Connect the negative battery cable.

14. Install the engine cover and under cover.

Turbocharger

REMOVAL & INSTALLATION

1.8L Engine

1. Before servicing the vehicle, refer to the precautions in the beginning of this section.

2. Remove or disconnect:
- Negative battery cable.
- Engine undercover, and unbolt the air conditioning compressor.
- Turbocharger support bracket.
- Oil return line at the turbocharger.
- Air hoses from the turbocharger.
- Oil feed line at the turbocharger.

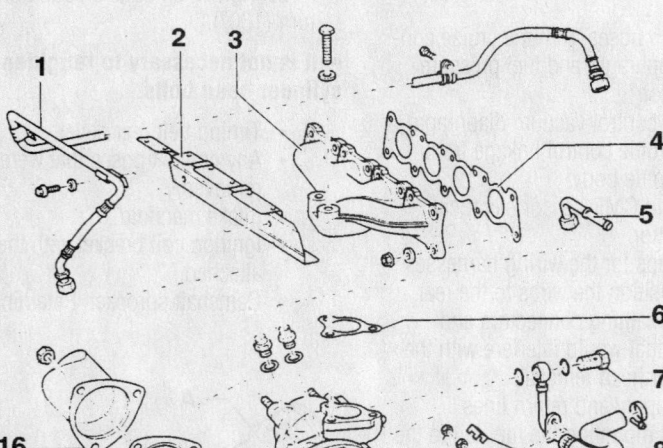

1. Oil supply line
2. Heat shield
3. Exhaust manifold
4. Exhaust manifold gasket
5. Coolant return line
6. Exhaust manifold-to-turbo gasket
7. Banjo bolt
8. Coolant supply hose
9. Fuse
10. Vacuum diaphragm for the wastegate
11. Gasket
12. Oil return line
13. Turbocharger
14. Support bracket
15. Gasket
16. Three Way Catalytic Converter (TWC)

Exploded view of the turbocharger and related components—1.8L engine

7923WG07

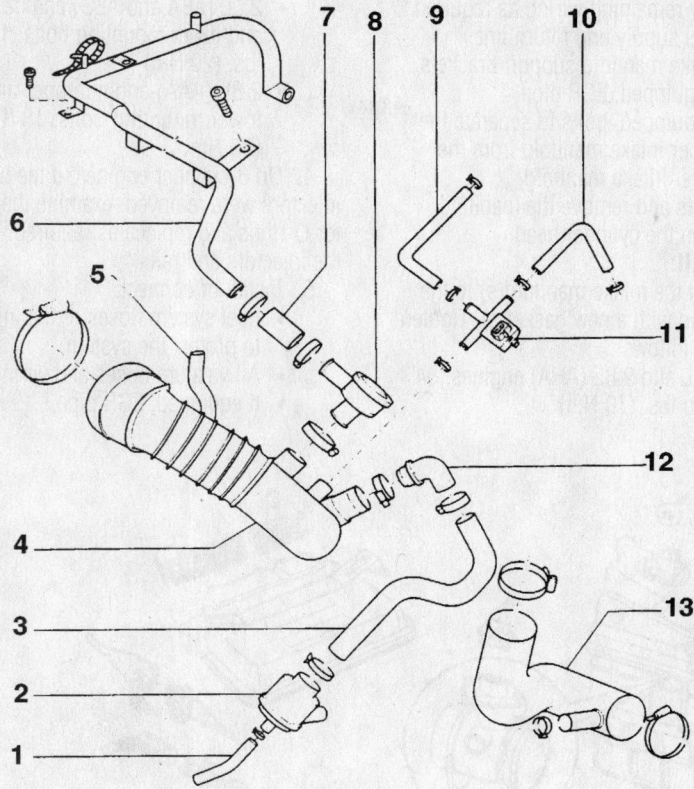

1. Vacuum hose
2. Boost pressure recirculation valve
3. Hose
4. Intake air duct
5. EVAP hose
6. Crankcase ventilation hose
7. Crankcase ventilation hose
8. PCV valve
9. Hose
10. Wastegate vacuum hose
11. Wastegate bypass regulator valve
12. Elbow
13. Hose to the turbocharger

7923WG08

Exploded view of the hoses related to the turbocharger—1.8L engine

- Hose for the boost pressure regulation valve vacuum diaphragm.
- Bracket for the coolant supply line at the boost pressure regulation valve vacuum diaphragm, and using Clamp 3094 or equivalent, pinch off the coolant supply hose.
- Intake air duct between the cowl and the air cleaner housing.
- Air cleaner housing cover.

3. Label and detach the following lines and electrical connectors:
- Wastegate bypass regulator valve
- EVAP canister purge regulator valve
- Power outage stage
- MAF sensor

4. Remove or disconnect:
- Air cleaner housing and the engine cover.
- Crankcase breather hose at the valve cover.
- Oil supply line at the turbocharger.
- Heat shield, and sleeve from the coolant return hose, and using Clamp 3094 or equivalent, pinch off the coolant return hose, then remove the hose.

❄❄ WARNING

The exhaust flexpipe may be damaged if bent more than 10 degrees.

- TWC from the turbocharger.
- Turbocharger from the exhaust manifold.
- Coolant supply banjo fitting.
- Turbocharger.

To install:
5. Install or connect:
- Turbocharger.
- Coolant supply banjo fitting and tighten to 18 ft. lbs. (25 Nm).
- Turbocharger to the exhaust manifold, using new gaskets. Mounting bolts, coating the bolts with Hot Bolt Paste G 052 112 A3 (or equivalent): 26 ft. lbs. (35 Nm). Turbocharger support bracket mounting bolts: 33 ft. lbs. (45 Nm).
- TWC to the turbo.
- Coolant supply hose.
- Sleeve and heat shield to the return hose.
- Oil return hose.

6. Add oil to the turbocharger through the oil feed line.
7. Connect:
- Oil supply line to the turbocharger and tighten to 18 ft. lbs. (25 Nm).
- Crankcase breather, and install the engine cover and air cleaner housing.

8. Attach the following lines and electrical connectors:
- MAF sensor
- Power outage stage
- Wastegate bypass regulator valve

9. Install or connect:
- Hoses and brackets for the boost pressure regulation valve vacuum diaphragm.
- Air hoses to the air cleaner assembly and the turbo.
- Air conditioning compressor and engine undercovers.
- Coolant.
- Negative battery cable.

10. Start the vehicle and check for leaks, then let the engine idle for approximately 1 minute without increasing the engine speed. This ensures adequate oil supply to the turbocharger.

For Tire, Wheel and Ball Joint specifications, see Section 1 of this manual

Intake Manifold

REMOVAL & INSTALLATION

1. Before servicing the vehicle, refer to the precautions in the beginning of this section.

2. Remove or disconnect:
   - Negative battery cable.
   - Engine cover as required.
   - Air duct from the throttle body and disconnect the accelerator cable.
   - Vacuum and coolant hoses as required.
   - Any remaining wiring as required.
   - Fuel supply and return lines.
   - Intake manifold support brackets.
   - If equipped, EGR pipe.
   - If equipped, bolts to separate the upper intake manifold from the lower intake manifold.
   - Bolts and remove the manifold from the cylinder head.

To install:

3. Install the intake manifold(s) to the cylinder head with a new gasket(s). Tighten the bolts as follows:
   - 1.8L and 2.8L (AHA) engines: 84 inch lbs. (10 Nm)
   - 2.0L (ABA and AEG) engine upper and lower mounting bolts: 15 ft. lbs. (20 Nm)
   - 2.8L (AAA) engine upper and lower mounting bolts: 18 ft. lbs. (25 Nm)

4. On 6-cylinder engines, if the fuel injectors were removed, examine the injector O-rings and replace as required. Install the injectors and rail.

5. Install or connect:
   - Fuel system hoses or the injectors to protect the system.
   - All vacuum hoses and wiring.
   - If equipped, EGR pipe.

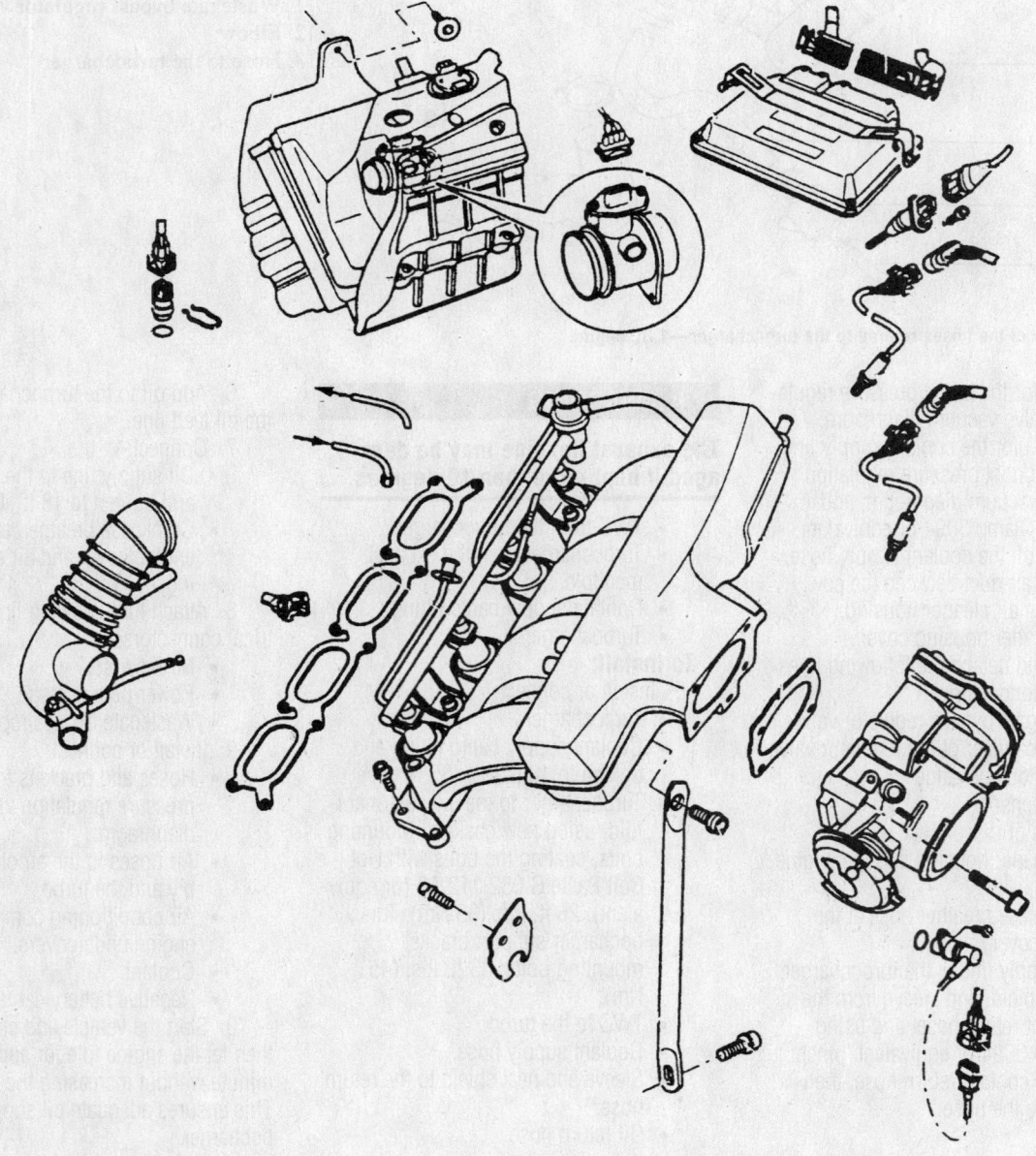

9301WG11

Exploded view of the intake manifold mounting and related components—1.8L engine

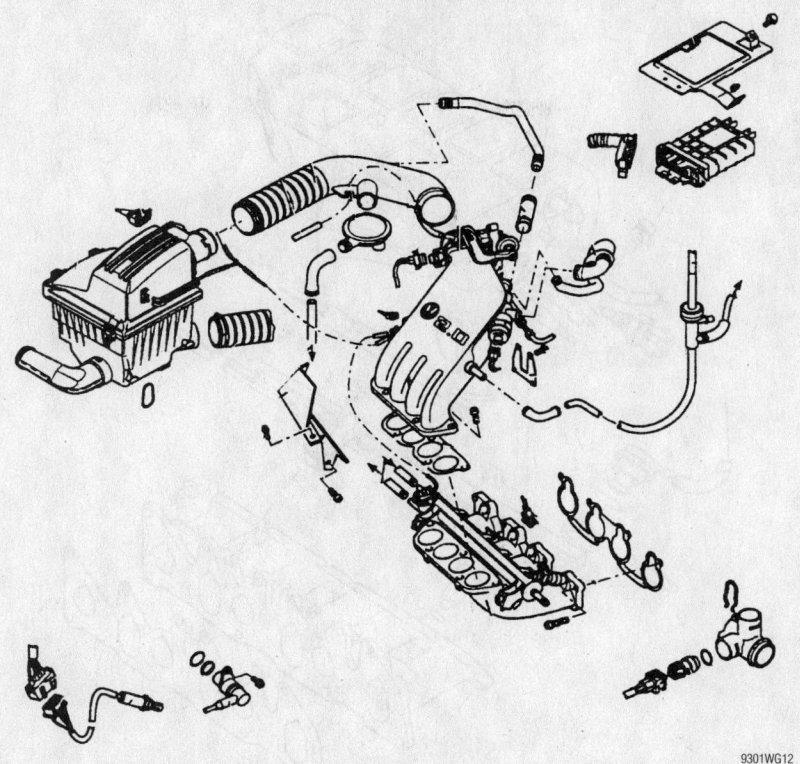

9301WG12

Exploded view of the intake manifold mounting and related components—2.0L (ABA) engine

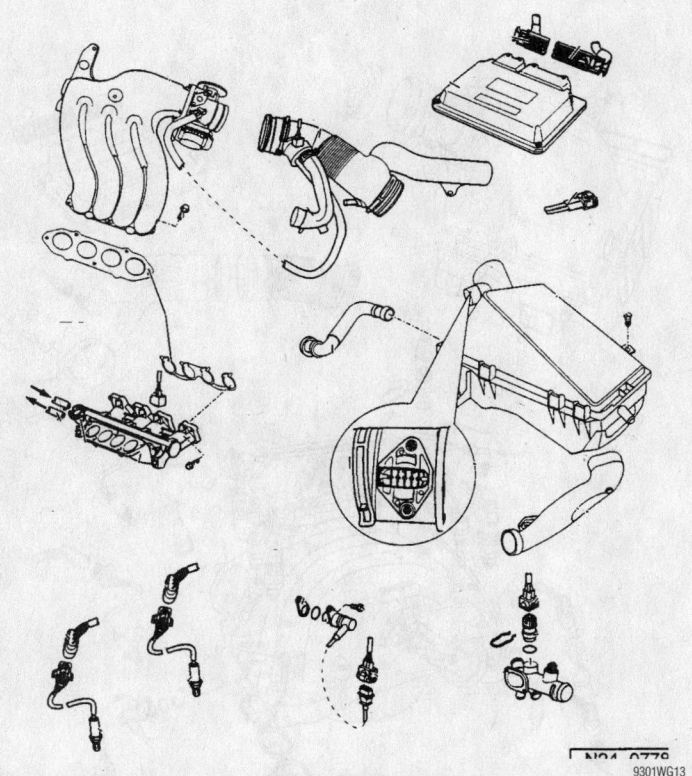

N24 0778
9301WG13

Exploded view of the intake manifold mounting and related components—2.0L (AEG) engine

For Wheel Alignment specifications, see Section 1 of this manual

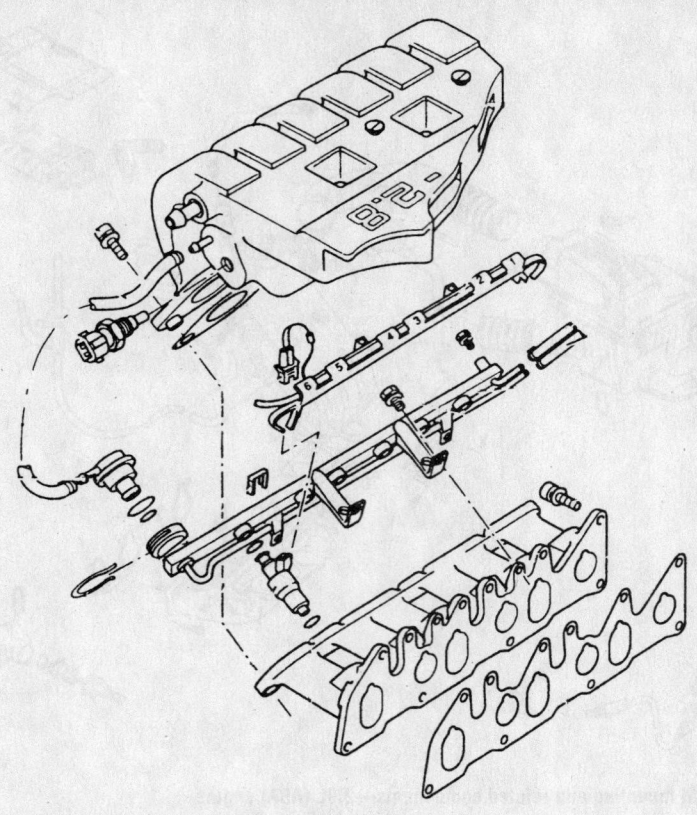

9301WG14

Exploded view of the intake manifold mounting and related components—2.8L (AAA) engine

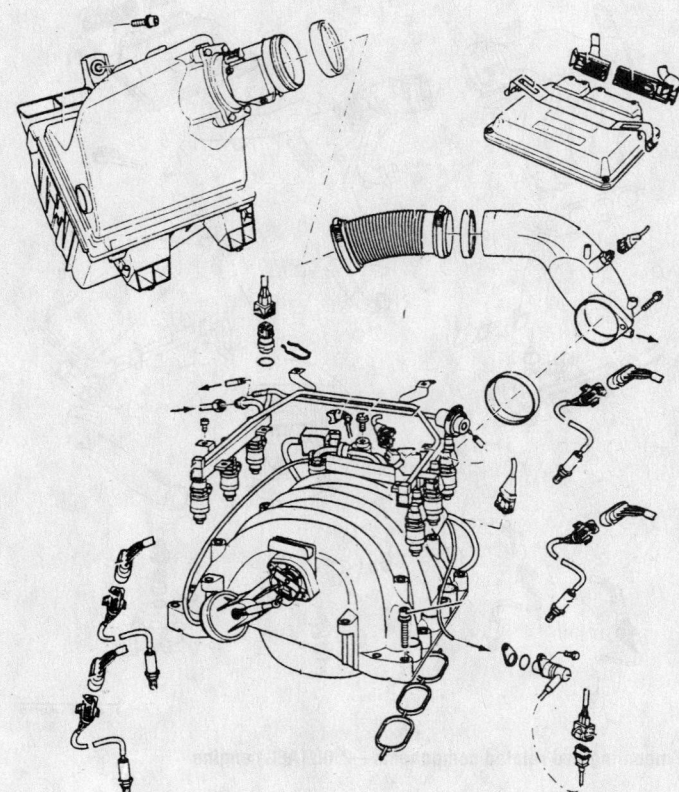

9301WG15

Exploded view of the intake manifold mounting and related components—2.8L (AHA) engine

- Throttle cable.
- Remaining components.

Exhaust Manifold

REMOVAL & INSTALLATION

1. Before servicing the vehicle, refer to the precautions in the beginning of this section.
2. Remove or disconnect:
   - O_2 sensor wiring and remove any heat shields that may be in the way.
   - Exhaust support brackets as necessary.
   - Front exhaust pipe from the manifold or turbocharger.
   - If equipped, the turbocharger.
   - Self-locking nuts or bolts and the manifold.

To install:

3. Installation is the reverse of removal. Use new gaskets and self-locking nuts. Nuts: 18 ft. lbs. (25 Nm).
4. If the exhaust pipe is bolted to the manifold, install a new gasket and use new self-locking nuts. Nuts: 30 ft. lbs. (40 Nm).
5. Check and clear any DTCs.

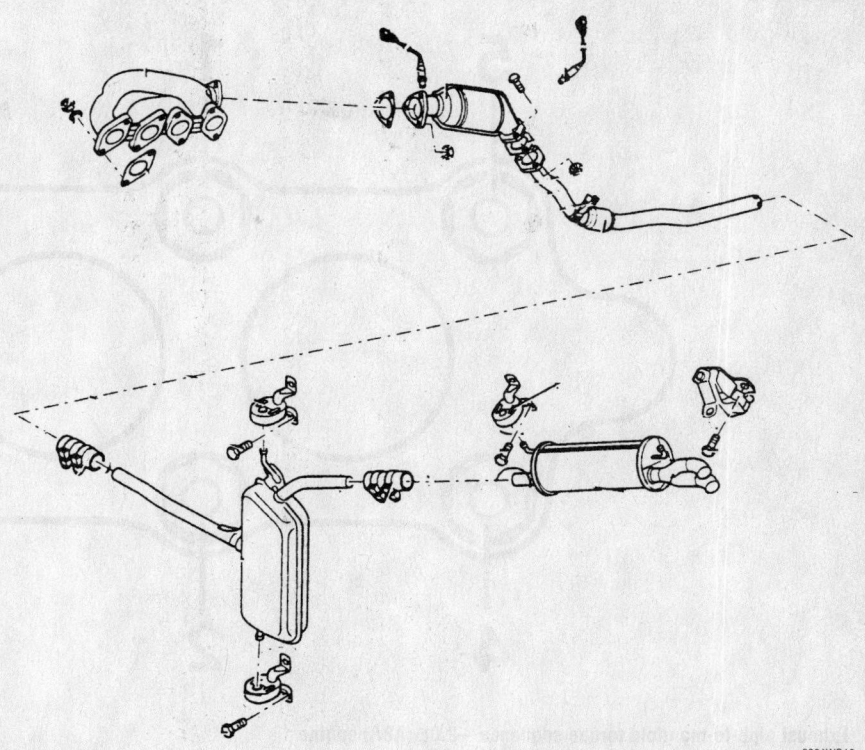

9301WG16

Exploded view of the exhaust manifold mounting and related components—1.8L engine

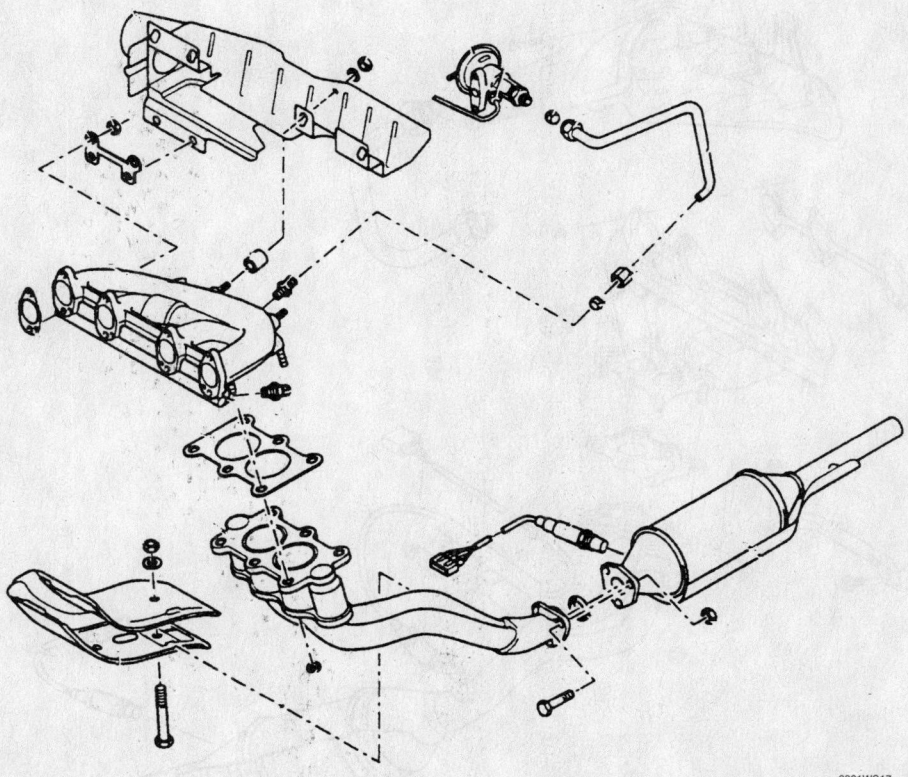

9301WG17

Exploded view of the exhaust manifold mounting and related components—2.0L (ABA) engine

For Maintenance Interval recommendations, see Section 1 of this manual

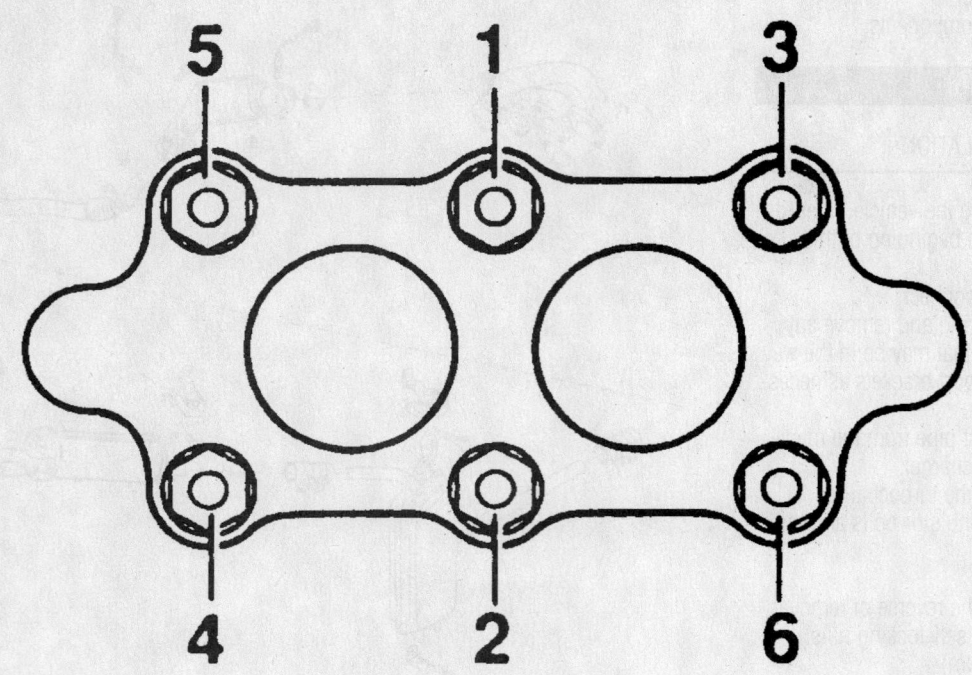

9301WG18

Exhaust pipe-to-manifold torque sequence—2.0L (ABA) engine

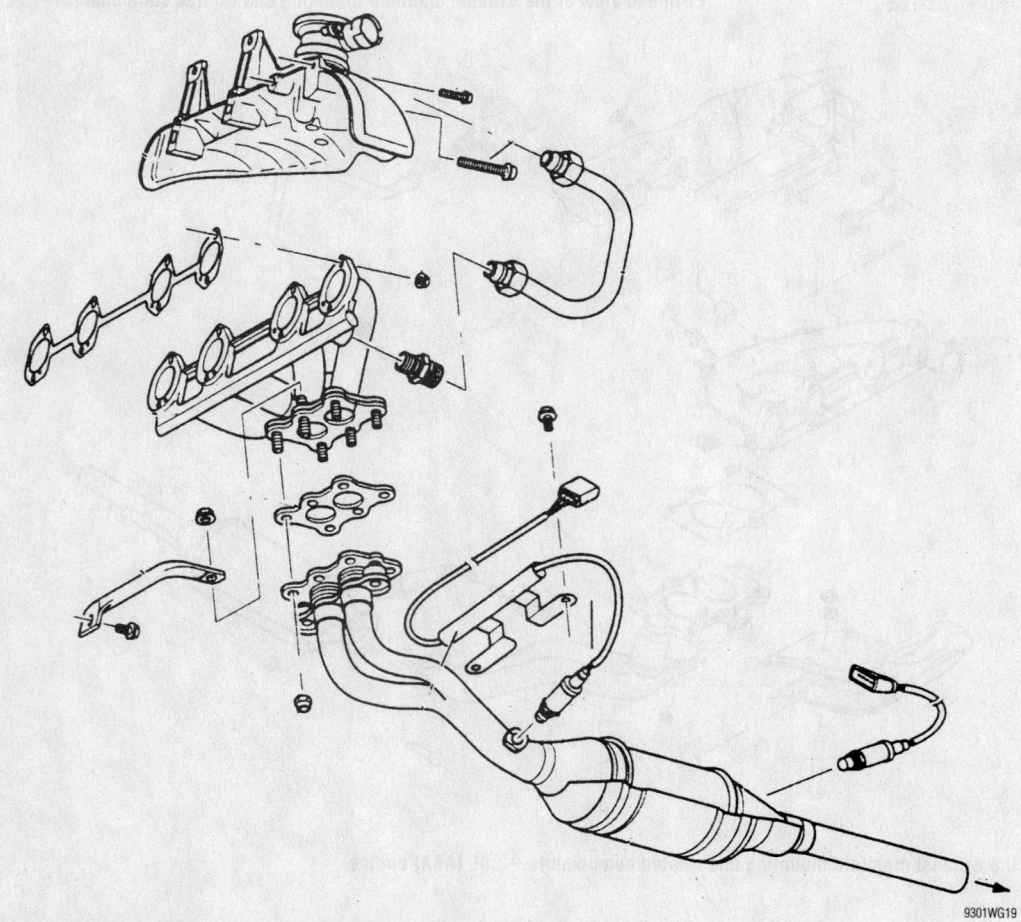

9301WG19

Exploded view of the exhaust manifold mounting and related components—2.0L (AEG) engine

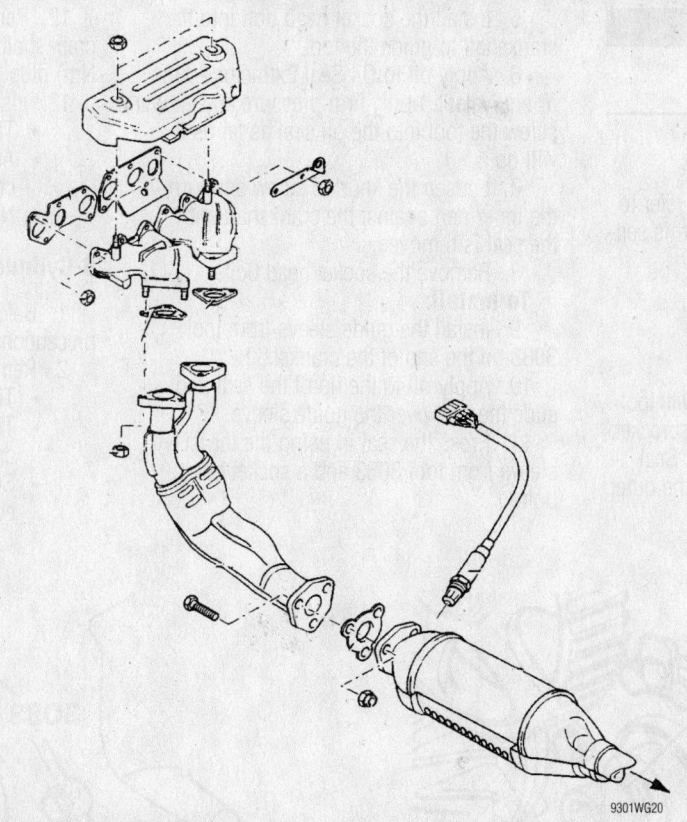

9301WG20

Exploded view of the exhaust manifold mounting and related components—2.8L (AAA) engine

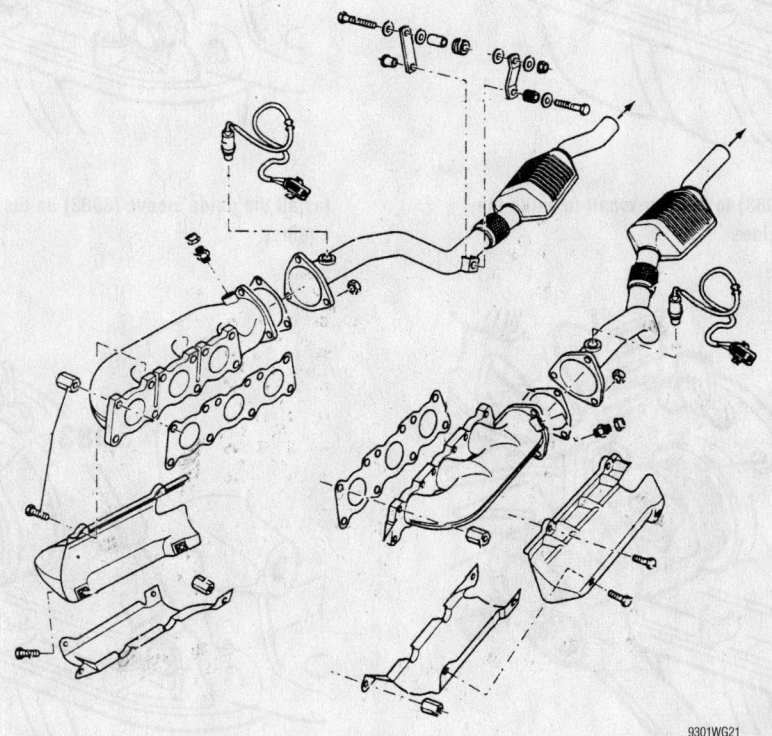

9301WG21

Exploded view of the exhaust manifold mounting and related components—2.8L (AHA) engine

For Tune-up, Capacities and Firing orders, see Section 1 of this manual

Front Crankshaft Seal

REMOVAL & INSTALLATION

4-Cylinder Engines

1. Before servicing the vehicle, refer to the precautions in the beginning of this section.
2. Remove or disconnect:
   - Negative battery cable.
   - Accessory drive belts.
   - Timing belt.
3. Hold the crankshaft sprocket with tool 3099 and remove the center bolt and sprocket.
4. Unscrew the inner part of Oil Seal Extractor 2085 or equivalent out of the outer part about 2 turns.

5. Install the socket head bolt into the crankshaft to guide the tool.
6. Apply oil to Oil Seal Extractor 2085 or equivalent. Apply firm pressure and screw the tool into the oil seal as far as it will go.
7. Loosen the knurled screw and turn the inner part against the crankshaft until the seal is removed.
8. Remove the socket head bolt.

To install:

9. Install the guide sleeve from tool 3083 on the end of the crankshaft.
10. Apply oil to the lip of the seal and slide the seal over the guide sleeve.
11. Press the seal in using the thrust sleeve from tool 3083 and a socket head bolt.

12. Remove the tools and install the crankshaft sprocket. Bolt: 66 ft. lbs. (90 Nm) plus ¼ turn.
13. Install or connect:
    - Timing belt.
    - Accessory drive belt and remaining components.
    - Negative battery cable.

6-Cylinder Engines

1. Before servicing the vehicle, refer to the precautions in the beginning of this section.
2. Remove or disconnect:
   - Timing belt.
   - Timing belt sprocket from the crankshaft.
   - Seal with Seal Remover 3203 or equivalent.

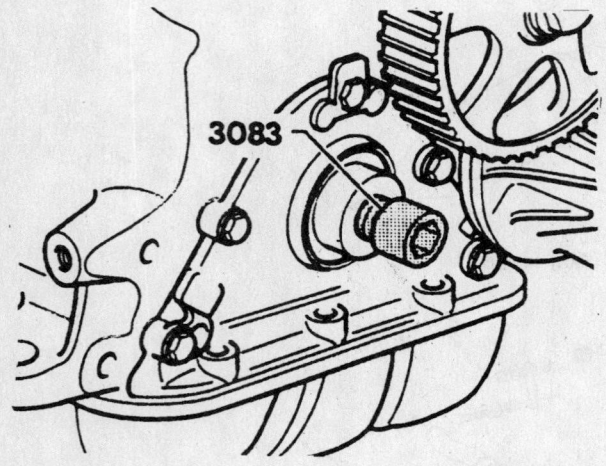

Install the socket head bolt (3083) in the crankshaft to guide the Seal Extractor—4-cylinder engines

7923WG09

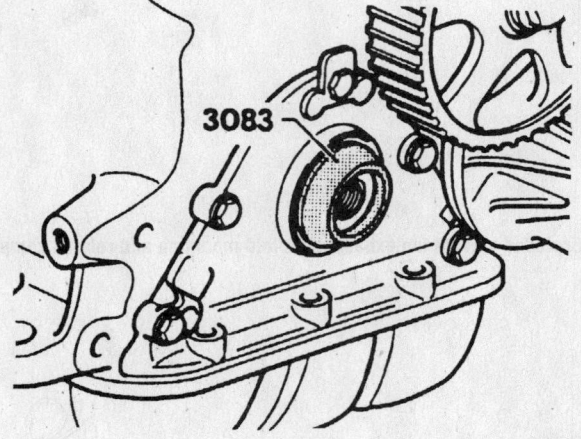

Install the guide sleeve (3083) on the crankshaft—4-cylinder engines

7923WG11

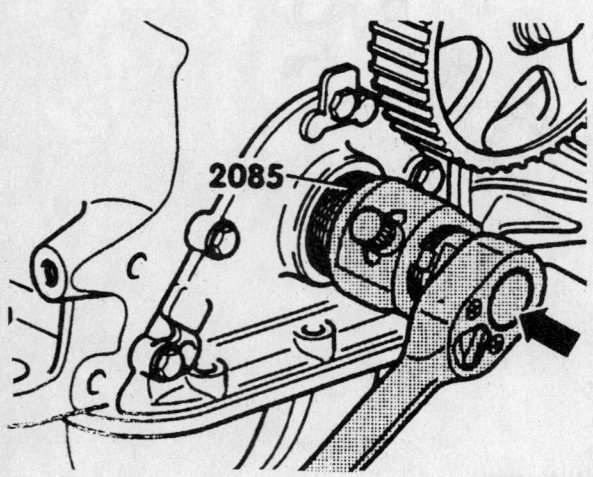

Screw the seal Extractor (2085) into the oil seal while applying pressure—4-cylinder engines

7923WG10

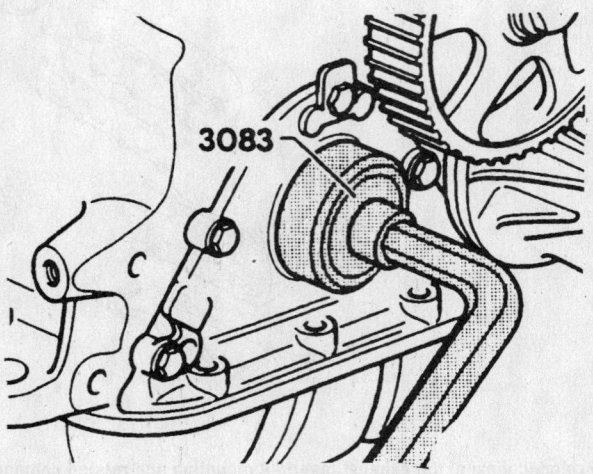

Tighten the socket head bolt to press the seal into place—4-cylinder engines

7923WG12

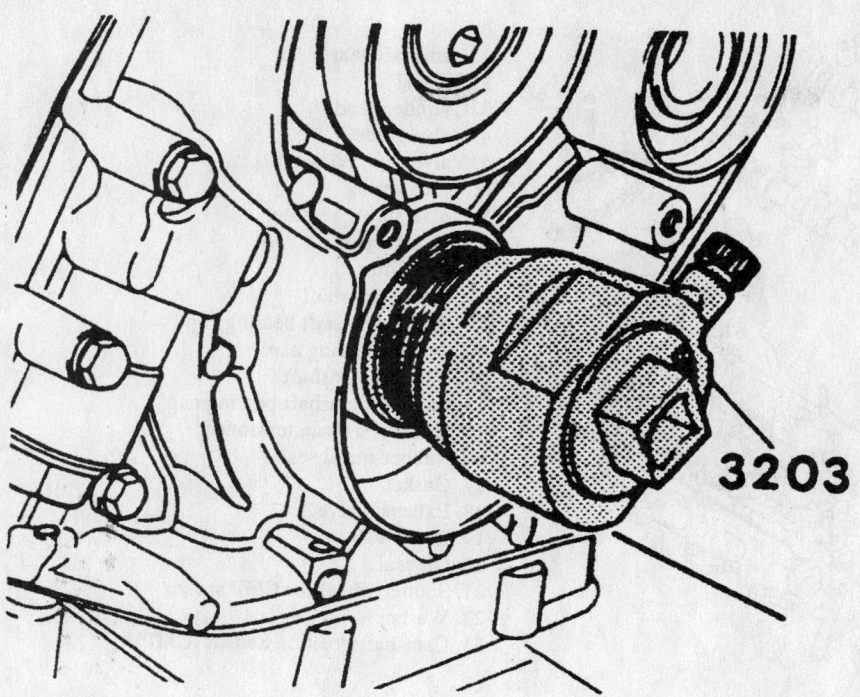

Removing the seal using Seal Remover 3203—6-cylinder engines

7923CG09

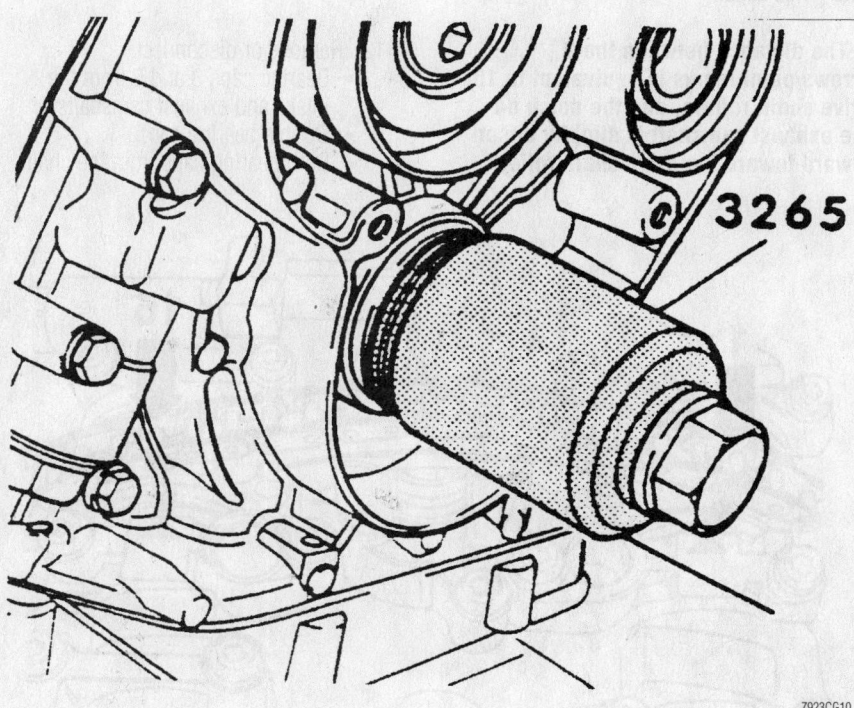

Installing the seal using the seal installer and the crankshaft center bolt—6-cylinder engines

7923CG10

3. Clean the running and sealing surfaces.

To install:

4. Slide the seal over the Installing Sleeve 3202 or equivalent.

5. Press the oil seal flush with Seal Installer 3265, or equivalent, and the center crankshaft bolt.

6. Install the timing belt sprocket. Center crankshaft bolt: 66 ft. lbs. (90 Nm) plus ¼ turn.

7. Install the timing belt and all remaining components.

Camshaft and Valve Lifters

REMOVAL & INSTALLATION

1.8L Engine

1. Place the lock carrier into the service position as follows:

a. Remove or disconnect:
- Front bumper.
- Any wiring or connector that would inhibit locking the carrier.
- 3 quick-release screws on the front noise insulation panel.
- Air guide between the lock carrier and the air filter.
- If installed, retaining clamps for the wiring harness at the left side of the radiator frame.
- No. 2 bolts and install Support 3369 or equivalent.
- Remaining bolts and pull the lock carrier out to the stop.

b. To secure the lock carrier, install the appropriate M6 bolts into the rear of the lock carrier and fender.

2. Remove or disconnect:
- Negative battery cable.
- Accessory drive belts.
- Engine covers.
- Timing belt upper cover.

3. Turn the crankshaft, in the direction of rotation (clockwise), until the No. 1 cylinder is at TDC.

4. Using a T45 Torx® wrench, loosen the timing belt tensioner.

5. Push down on the tensioner, and remove the belt from the camshaft gear.

6. Remove or disconnect:
- Torx® bolt and swing the tensioner assembly bracket forward.
- Valve cover.

7. Using Retainer 3036 or equivalent, loosen the cam gear retaining bolt.

8. Remove:

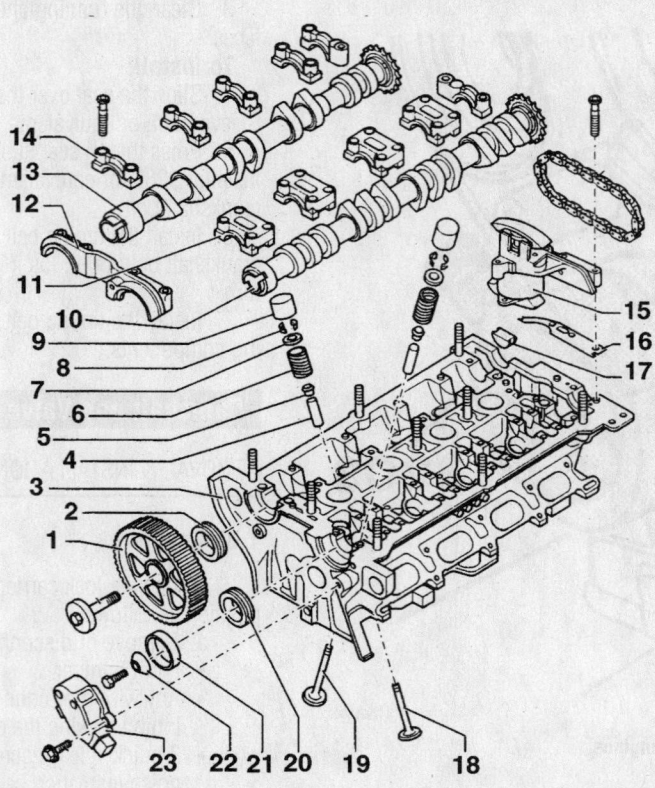

1. Camshaft gear
2. Oil seal
3. Cylinder head
4. Valve guide
5. Valve stem oil seal
6. Valve spring
7. Valve spring retainer
8. Valve keeper
9. Valve lifter
10. Intake camshaft
11. Intake camshaft bearing cap
12. Double bearing cap
13. Exhaust camshaft
14. Exhaust camshaft bearing cap
15. Hydraulic chain tensioner
16. Rubber/metal seal
17. Gasket
18. Exhaust valve
19. Intake valve
20. Oil seal
21. Shutter wheel for CMP sensor
22. Washer
23. Camshaft Position Sensor (CMP)

7923WG14

Exploded view of the camshaft and related components—1.8L engine

- Camshaft gear.
- Housing for CMP sensor and shutter wheel.

9. Secure the hydraulic chain tensioner with Bracket-Tensioner 3366 or equivalent.

10. Verify that the camshafts are at TDC for the No. 1 cylinder. Both camshaft markings must align with arrows on the bearing caps.

11. Clean the drive chain and the cam chain gears opposite both arrows on the bearing caps. Matchmark the installed position using paint.

➡ **The distance between the 2 arrows/paint marks is equivalent to 16 drive chain rollers, and the notch on the exhaust camshaft is slightly offset inward toward the drive chain roller.**

12. Remove or disconnect:
- Bearing caps 3 and 5 from the intake and exhaust camshafts.
- Double bearing cap.
- Both bearing caps from the chain

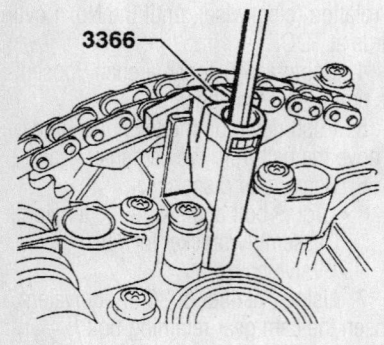

Over-tightening will damage the chain tensioner (3366)—1.8L engine

7923WG15

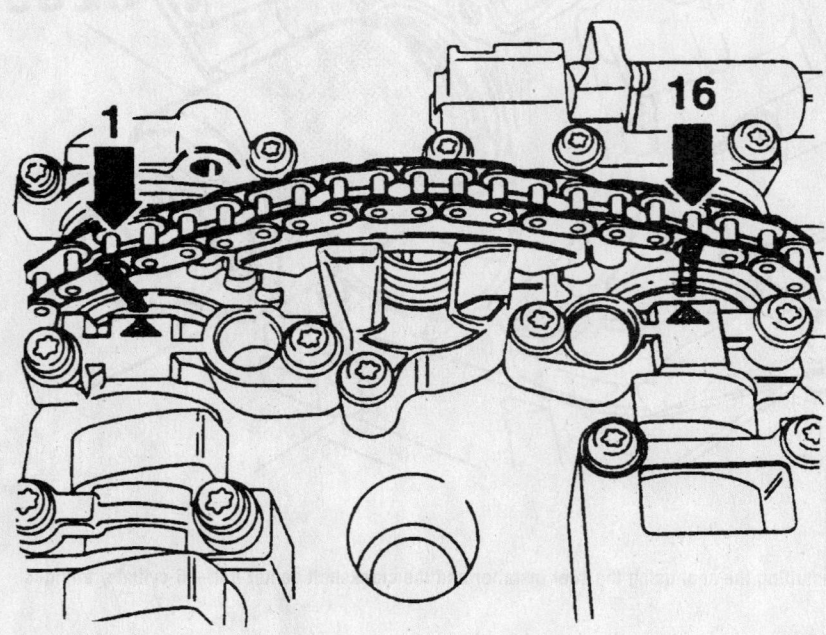

7923WG16

To ensure proper installation, matchmark the chain-to-camshaft position—1.8L engine

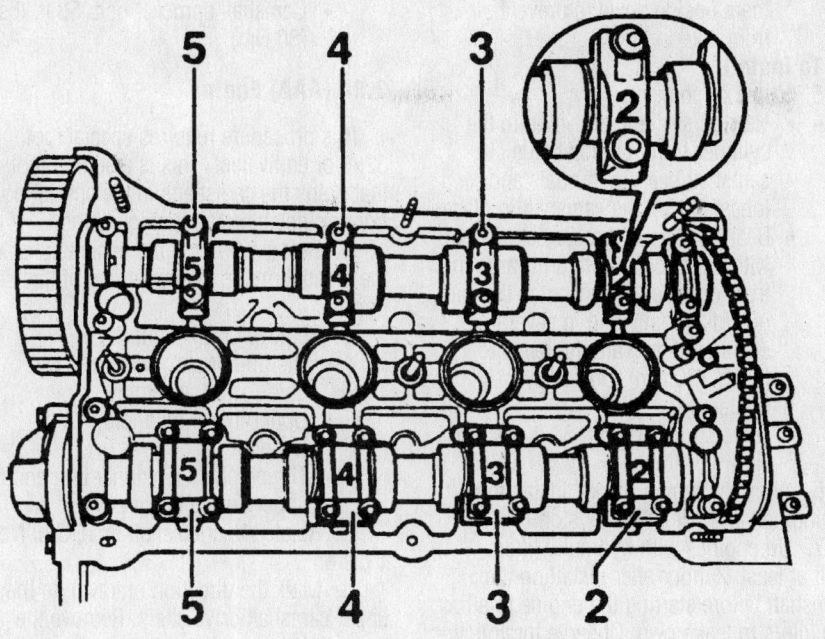

Camshaft bearing cap identification—1.8L engine

gears on the intake and exhaust camshafts.
- Hydraulic chain tensioner retaining bolts.

13. In an alternating and diagonal sequence, loosen the bearing caps 2 and 4 of the intake and exhaust manifold, then remove.

14. Remove the camshafts with the hydraulic chain tensioner.

To install:

※ WARNING

After installing the lifters or the camshaft(s), the engine must NOT be started for at least 30 minutes. Otherwise the valves could strike the pistons. Rotate the engine by hand, at least 2 revolutions, to ensure that the valves do not strike the pistons.

15. Replace the rubber/metal chain tensioner gasket and apply sealant to the hatched area, as shown.
16. Install the drive chain on the camshaft as follows:
   a. If installing the old chain, align the paint marks with the camshaft marks.
   b. If installing a new chain, the distance between the notches **A** and **B** on the camshafts must equal the distance between 16 drive chain rollers.
17. Slide the hydraulic chain tensioner between the drive chain.
18. Install the camshafts with the chain tensioner into the cylinder head.
19. Oil the camshaft contact surfaces.

➡ When installing the bearing caps, verify the markings on the caps are readable from the intake side of the cylinder head.

20. Tighten the bearing caps 2 and 4 of the intake and exhaust camshafts in an alternating diagonal sequence to 84 inch lbs. (10 Nm).
21. Install both bearing caps on the chain sprockets of the intake and exhaust camshafts and tighten to 84 inch lbs. (10 Nm).
22. Verify the correct positions of the camshafts.
23. Remove the Bracket-Tensioner 3366.
24. Lightly coat the cylinder head mating surface of the double bearing cap with sealant, then install.
25. Install:
   - Remaining bearing caps: 84 inch lbs. (10 Nm).
   - Camshaft gear. Retaining bolt: 48 ft. lbs. (65 Nm).
   - CMP shutter wheel and housing cover.
   - Valve cover.
26. Align the camshaft gear and the vibration damper with the TDC markings.
27. Install or connect:
   - Timing belt.
   - Accessory drive belts, then the engine cover.
   - Lock carrier.
   - Negative battery cable.

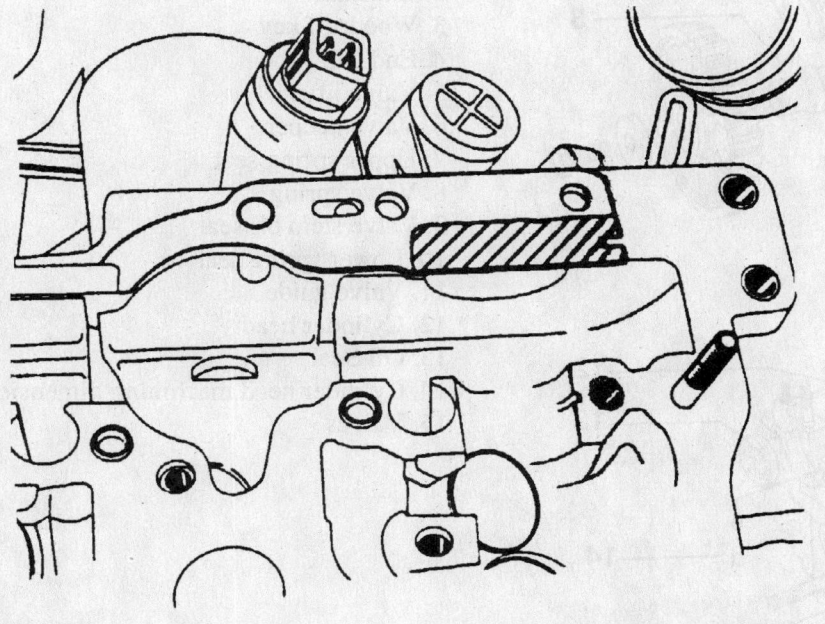

To ensure a proper seal, be sure to apply sealant to the hatched area—1.8L engine

2.0L Engines

1. Before servicing the vehicle, refer to the precautions in the beginning of this section.

2. Remove or disconnect:
- Negative battery cable.
- Timing belt cover(s), the timing belt
- Camshaft sprocket and cylinder head cover.

3. Number the bearing caps from front-to-back. Scribe an arrow pointing towards the front of the engine. The caps are offset and must be installed correctly. Factory numbers on the caps are not always on the same side.

4. Remove or disconnect:
- Front and rear bearing caps. Loosen the remaining bearing cap nuts diagonally, in several steps, starting from the outside caps near the ends of the head and working toward the center.
- Bearing caps and the camshaft.
- If required, lifters from the valves. Keep them in order so they can be installed in their original positions. Place them in a bath of oil or place them upside down to prevent air from entering them.

To install:

5. Install or connect:
- New oil seal and end-plug in the cylinder head. Lubricate the camshaft bearing journals and lobes and set the camshaft in place.
- Bearing caps in the correct position with the arrow pointing towards the front of the engine. Tighten the cap nuts diagonally and in several steps until they are tightened to 15 ft. lbs. (20 Nm). Do not over-tighten.
- Drive sprocket. Bolt: 58 ft. lbs. (80 Nm).

6. Align the timing marks, install the timing belt and adjust the tension.

7. On engines with hydraulic lifters, wait at least ½ hour after installing the camshaft before starting the engine to allow the lifters to leak down. Observe the following values:
- Camshaft shaft end-play: 0.006 in. (0.15mm)
- Bearing cap bolts: 15 ft. lbs. (20 Nm)
- Camshaft sprocket bolt: 58 ft. lbs. (80 Nm)

2.8L (AAA) Engine

This procedure requires special tool 3268 or equivalent. This is a setting tool that holds the camshafts in the correct position for installing the timing chains.

1. Before servicing the vehicle, refer to the precautions in the beginning of this section.

2. Remove or disconnect:
- Distributor cap, wires and wire guide as an assembly.
- Upper intake manifold.
- Cylinder head cover.
- Timing chain tensioner bolt and the upper timing chain cover.

3. Rotate the crankshaft to TDC of No. 1 piston.

4. Mark the direction of travel on the upper camshaft drive chain. Remove the tensioner shoe and the chain.

5. Hold the camshafts at the flats with a 24mm wrench and remove the bolts to remove the sprockets. Note the position of the distributor drive on the short camshaft.

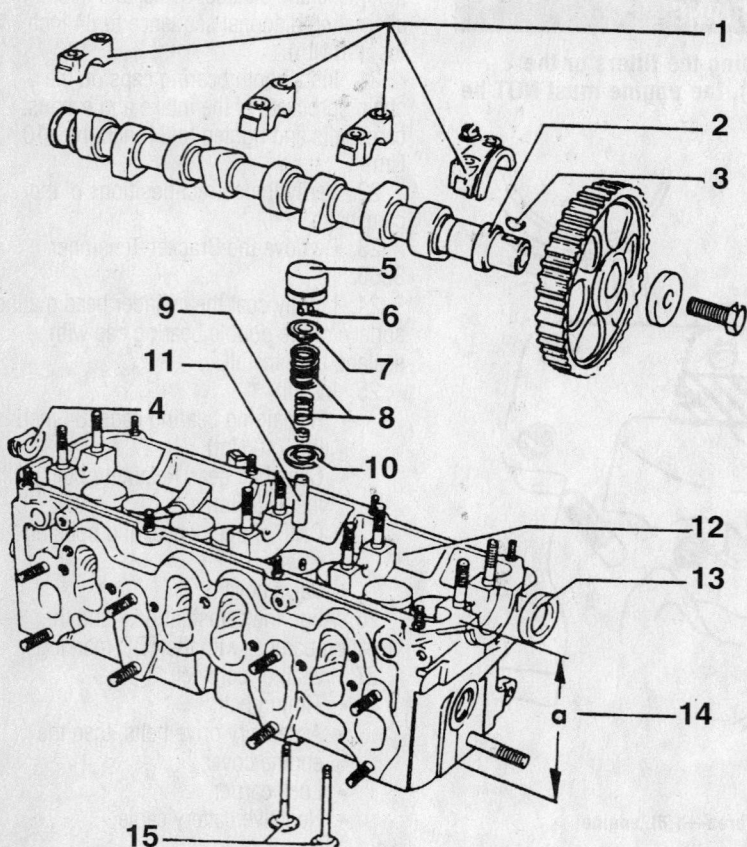

1. Bearing cap
2. Camshaft
3. Woodruff key
4. End cap
5. Valve lifter
6. Valve keeper
7. Upper spring seat
8. Valve spring
9. Valve stem oil seal
10. Lower spring seat
11. Valve guide
12. Cylinder head
13. Oil Seal
14. Cylinder head machining dimension
15. Valves

Exploded view of the camshaft and related components—2.0L engines

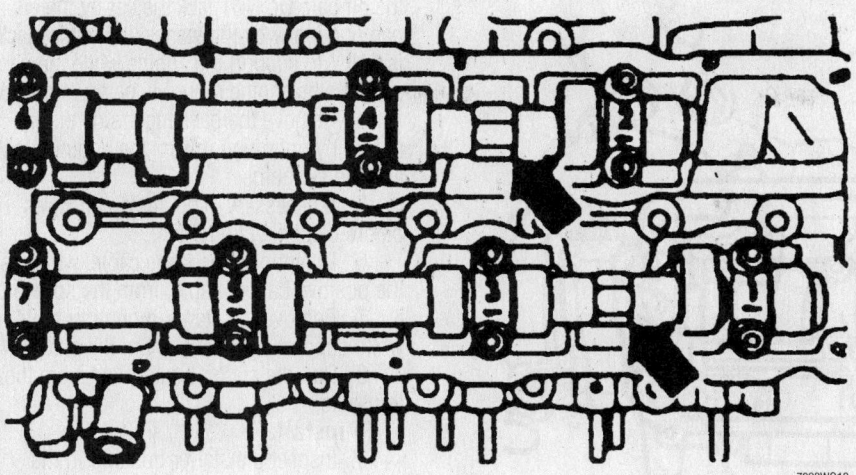

7923WG19

Hold the camshafts with an open end wrench on the flats (arrows) when removing or installing the sprocket bolts

✳✳ WARNING

Do not use the setting tool to hold the camshafts when removing or installing the sprocket bolts. The camshafts or the tool will be damaged.

6. On the long camshaft, remove the end bearing caps. Loosen the center cap nuts in a diagonal pattern 2 turns at a time until the valve springs are relieved. Remove the camshaft.

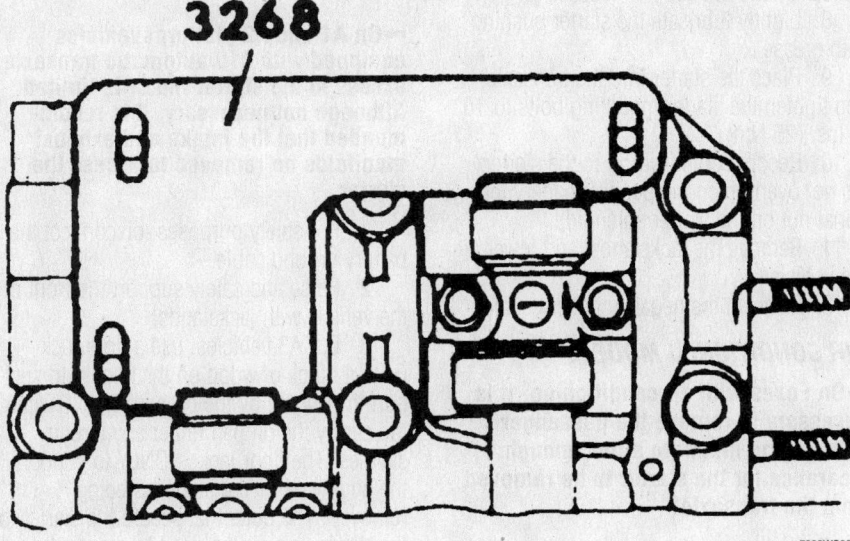

3268

7923WG20

The camshaft setting tool holds the camshafts in place when installing the chain—do not use this tool to loosen or tighten the sprocket bolts

7. On the short camshaft, remove the center bearing cap and loosen the nuts on the end-caps in a diagonal pattern 2 turns at a time. When the valve springs are relieved, remove the camshaft.

To install:

8. Lubricate the long camshaft and the cylinder head bearing surfaces and set the camshaft in place. Install bearing caps 3 and 5 and tighten the bolts 2 turns at a time in a diagonal pattern to draw the camshaft down against the valve springs.

9. Install the other bearing caps. Nuts: 15 ft. lbs. (20 Nm).

10. Repeat the process with the short camshaft, using bearing caps 2 and 6 to draw the camshaft down against the springs.

11. Hold the camshaft with a 24mm wrench and install the sprockets. Be sure the distributor drive is correctly positioned. Bolts: 74 ft. lbs. (100 Nm).

12. Be sure the crankshaft is at TDC on No. 1 piston. Install the setting tool and install the timing chain.

13. Install the tensioner shoe and temporarily install the upper timing chain cover. Install the tensioner bolt and remove the setting tool. Rotate the crankshaft 4 full turns and stop at TDC of No. 1 piston. The setting tool should fit into the camshafts.

14. Remove the tensioner bolt and upper timing chain cover again. Clean the old sealant off the cylinder head gasket and apply new sealant.

15. Install:
- Upper timing chain cover. Bolts: 82 inch lbs. (10 Nm).
- Tensioner bolt and tighten to 15 ft. lbs. (20 Nm).
- Cylinder head cover, upper intake manifold and ignition system components.

2.8L (AHA) Engine

1. Remove or disconnect:
- Timing belt.
- Valve cover(s).
- On the left cylinder head, Camshaft Position (CMP) sensor.
- On the right cylinder head, plug/cover on the head.
- Camshaft timing belt sprocket.

➡**DO NOT allow the bearing caps to become mixed up.**

- Camshaft bearing caps 2 and 3.

2. Gradually and evenly, loosen the nuts for the camshaft bearing caps 1 and 4, in a diagonal sequence.

3. Remove the camshaft and lift out the valve lifters. If it is to be reused, it must go in the bore from which it was removed.

To install:

4. Install or connect:
- Lifters into their respective bore.
- Bearing caps 1 and 4 in a alternating and diagonal sequence.
- Bearing caps 2 and 3. All bearing caps: 15 ft. lbs. (20 Nm).

Timing belt service is covered in Section 3 of this manual

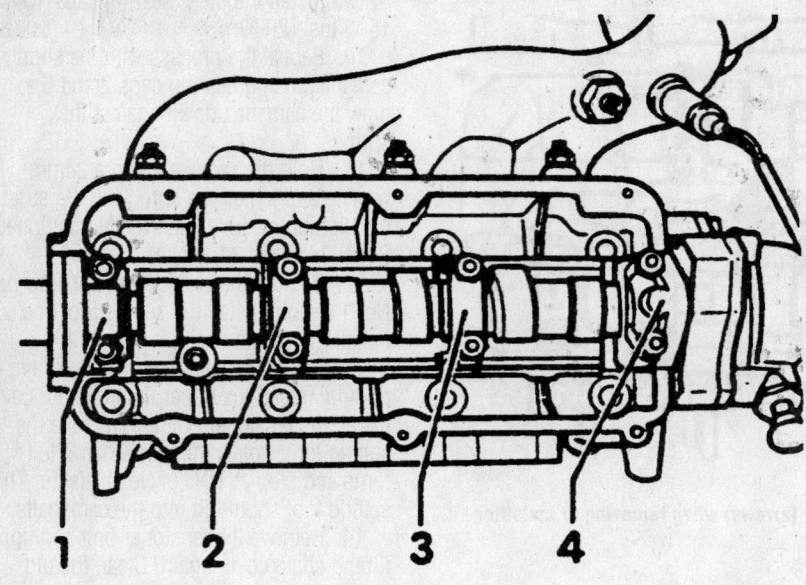

Camshaft bearing cap identification—2.8L (AHA) engine

7923CG11

✳✳ WARNING

After installing the lifters or the camshaft(s), the engine must NOT be started for at least 30 minutes. Otherwise the valves could strike the pistons. Rotate the engine by hand, at least 2 revolutions, to ensure that the valves do not strike the pistons.

- Camshaft timing belt sprocket. Mounting bolt: 52 ft. lbs. (71 Nm).
- On the right cylinder head, plug/cover on the head.
- On the left cylinder head, CMP sensor. Mounting bolts: 89 inch lbs. (10 Nm).
- Valve cover.
- Timing belt.

Valve Lash

ADJUSTMENT

All engines are equipped with hydraulic valve lash adjusters. No periodic valve lash adjustment is necessary.

Starter

REMOVAL & INSTALLATION

Fox

NON-AIR CONDITIONED MODELS

1. Disconnect the negative battery cable for safety purposes.

2. Raise and safely support the front of the vehicle with jackstands.

3. If necessary, label the small wires before disconnecting them.

4. Disconnect the large cable, which is the positive battery cable, from the solenoid.

5. Remove the starter mounting bolts, while supporting the weight of the starter.

6. Pull the starter straight out from the transaxle.

To install:

7. Inspect the starter bushing in the transaxle, and replace it if necessary.

8. Lightly lubricate the starter bushing with grease.

9. Place the starter into the transaxle, and tighten the starter mounting bolts to 18 ft. lbs. (25 Nm).

10. Reconnect the wiring to the starter. Do not overtighten the positive battery terminal nut on the starter solenoid.

11. Remove the jackstands, and lower the vehicle.

12. Connect the negative battery cable.

AIR CONDITIONED MODELS

➡**On Foxes with air conditioning, it is necessary to remove the passenger side engine mount to allow enough clearance for the starter to be removed from the transaxle.**

1. Disconnect the negative battery cable for safety purposes.

2. Raise and safely support the front of the vehicle with jackstands.

3. Using a floor jack, (with a block of wood on the chock) support the engine by

the oil pan. Do NOT jack the car by the oil pan under any circumstances! The floor jack is ONLY to support the engine while the passenger side engine mount is being removed.

4. Remove the passenger side engine mount by unbolting it from the engine block and the subframe.

5. If necessary, label the small wires before disconnecting them.

6. Disconnect the large cable, which is the positive battery cable, from the solenoid.

7. Remove the starter mounting bolts, while supporting the weight of the starter.

8. Pull the starter straight out from the transaxle.

To install:

9. Inspect the starter bushing in the transaxle, and replace it if necessary.

10. Lightly lubricate the starter bushing with grease.

11. Place the starter into the transaxle, and tighten the starter mounting bolts to 18 ft. lbs. (25 Nm).

12. Reconnect the wiring to the starter. Do not overtighten the positive battery terminal nut on the starter solenoid.

13. Install the engine mount. Tighten the bracket-to-engine bolts to 26 ft. lbs. (35 Nm) and the engine mount-to-subframe nut to 30 ft. lbs. (40 Nm).

14. Remove the floor jack from the oil pan, and place it on the subframe. Raise the vehicle slightly to remove the jackstands, and lower the vehicle.

15. Connect the negative battery cable.

All others

➡**On A1 and A2 platform vehicles equipped with 010 automatic transaxle, access to the starter motor is limited. Although not necessary, it is recommended that the intake and exhaust manifolds be removed to access the starter.**

1. For safety purposes, disconnect the battery ground cable.

2. Raise and safely support the front of the vehicle with jackstands.

3. For A3 vehicles, use a floor jack (with a block of wood on the chock) to support the engine by the oil pan. Do NOT jack the car by the oil pan under any circumstances! The floor jack is ONLY to support the engine while the starter is being removed. The bolts that secure the starter to the transaxle are also used to secure the front engine mount to the transaxle.

4. If necessary, label the small wires before disconnecting them.

5. Disconnect the large cable, which is the positive battery cable, from the solenoid.

6. On 010 transaxle-equipped vehicles, remove the bracket that secures the starter to the engine.

7. Remove the starter mounting bolts, while supporting the weight of the starter.

8. Pull the starter straight out from the transaxle.

To install:

➡**On vehicles with a manual transaxle, there is a bushing where the starter shaft fits into the bell housing. If the shaft or bushing are worn or if the starter has been jamming, the bushing should be replaced. There is a special bushing removal tool available but a small inside bearing removal tool is usually sufficient.**

9. Install the starter into the transaxle.

10. Tighten the starter mounting bolts as follows:

  a. All vehicles except 010 transaxle:
- M8 nut: 89 inch lbs. (10 Nm)
- M10 nut and bolt: 44 ft. lbs. (60 Nm)
- M12 bolt: 33 ft. lbs. (45 Nm)

  b. 010 transaxle only:
- Mounting flange bolt: 15 ft. lbs. (20 Nm)
- Mounting bracket bolt: 18 ft. lbs. (25 Nm)

11. Attach the electrical connections to the starter.

➡**Be careful not to over tighten the battery cable connection. The metal is soft and the threads will strip easily.**

12. Lower the vehicle from the jackstands.

13. Connect the negative battery cable.

Oil Pan

REMOVAL & INSTALLATION

➡**The oil pan can be removed with the engine in the vehicle. On some vehicles, it may be necessary to lower the subframe to service the oil pan.**

1. Before servicing the vehicle, refer to the precautions in the beginning of this section.

2. Remove or disconnect:
- Engine oil.
- Bolts retaining the oil pan.
- Oil pan.

To install:

3. Be sure the gasket surface is flat and install the pan with a new gasket.

4. Tighten the retaining bolts in a criss-cross pattern as follows:
- 2.0L (AEB and AEG) engines: 11 ft. lbs. (15 Nm)
- 2.0L (ABA) engine: 15 ft. lbs. (20 Nm)
- 1.8L (AEB) and 2.8L (AAA) engine: 11 ft. lbs. (15 Nm)
- 2.8L (AHA) engine: 84 inch lbs. (10 Nm)

5. Refill the engine with oil. Start the engine and check for leaks.

Oil Pump

REMOVAL & INSTALLATION

1.8L, 2.0L (ABA), and 2.8L (AAA) Engines

1. Before servicing the vehicle, refer to the precautions in the beginning of this section.

2. Remove or disconnect:

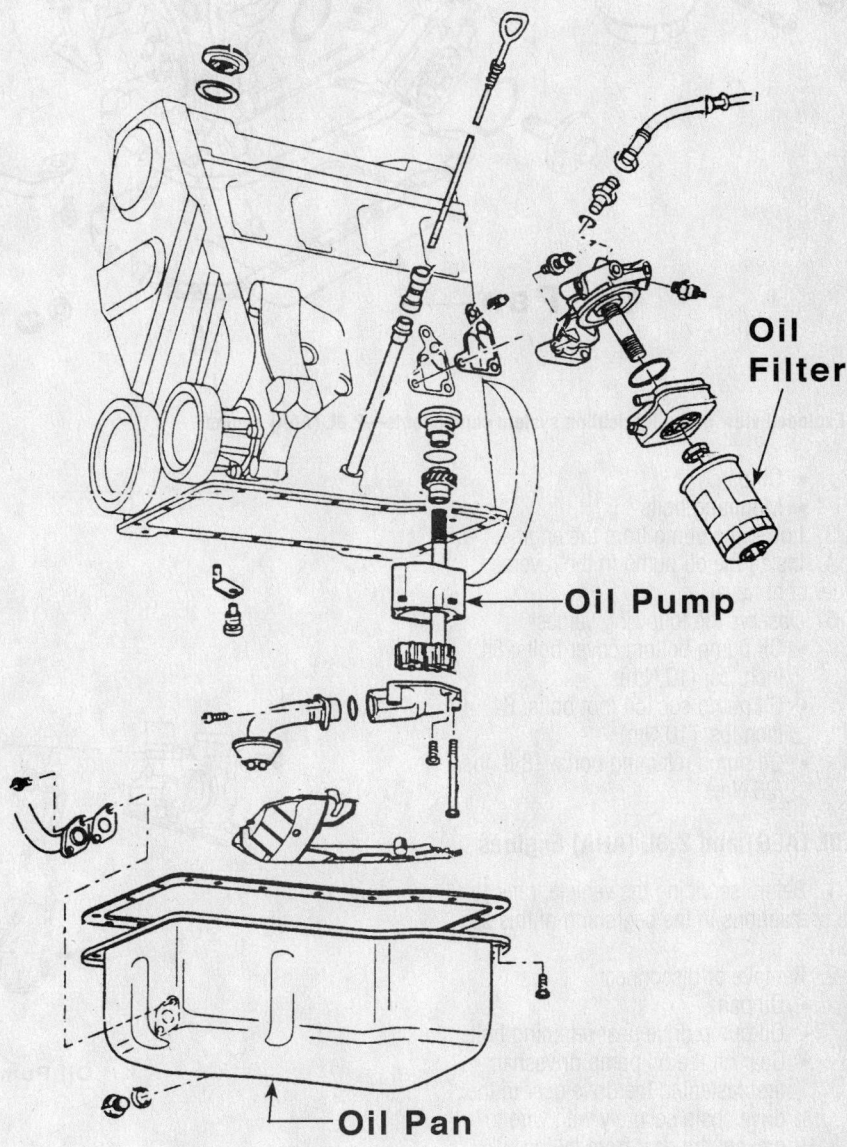

Exploded view of the lubrication system components—1.8L engine, 2.0L (ABA) engine similar

9301WG22

Heater Core replacement is covered in Section 2 of this manual

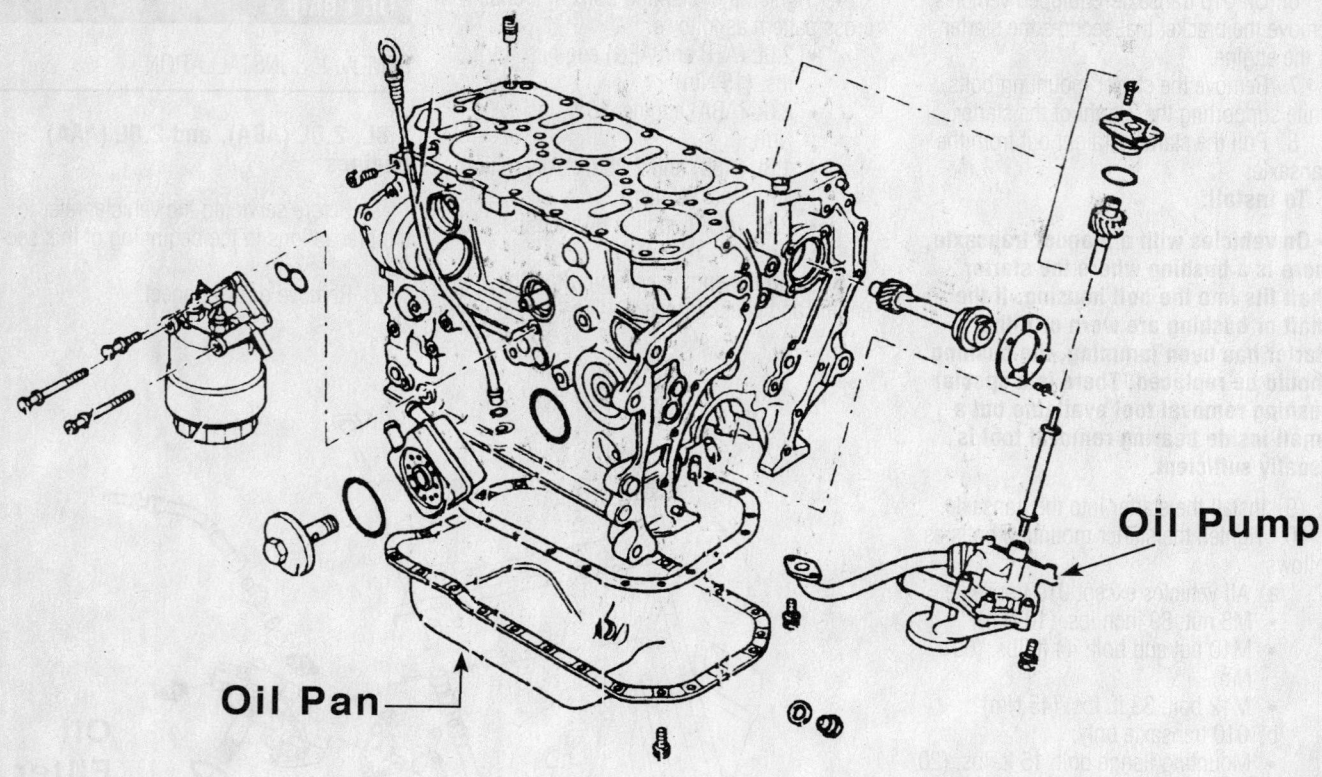

Exploded view of the lubrication system components—2.8L (AAA) engine

- • Oil pan.
- • Mounting bolts
3. Lower the pump from the engine.
4. Install the oil pump in the reverse order of removal.
5. Observe the following values:
   - • Oil pump bottom cover bolts: 84 inch lbs. (10 Nm)
   - • Oil pump suction foot bolts: 84 inch lbs. (10 Nm)
   - • Oil pump retaining bolts: 18 ft. lbs. (25 Nm)

2.0L (AEG) and 2.8L (AHA) Engines

1. Before servicing the vehicle, refer to the precautions in the beginning of this section.
2. Remove or disconnect:
   - • Oil pan.
   - • Oil pump drive gear retaining bolt.
   - • Gear off the oil pump driveshaft, first fastening the drive gear to the drive chain securely with wire to prevent the gear from falling.
   - • Oil pump mounting bolts.
   - • Oil pump.

To install:

3. Be sure the oil pump dowel sleeves are located on the engine block.
4. Install:

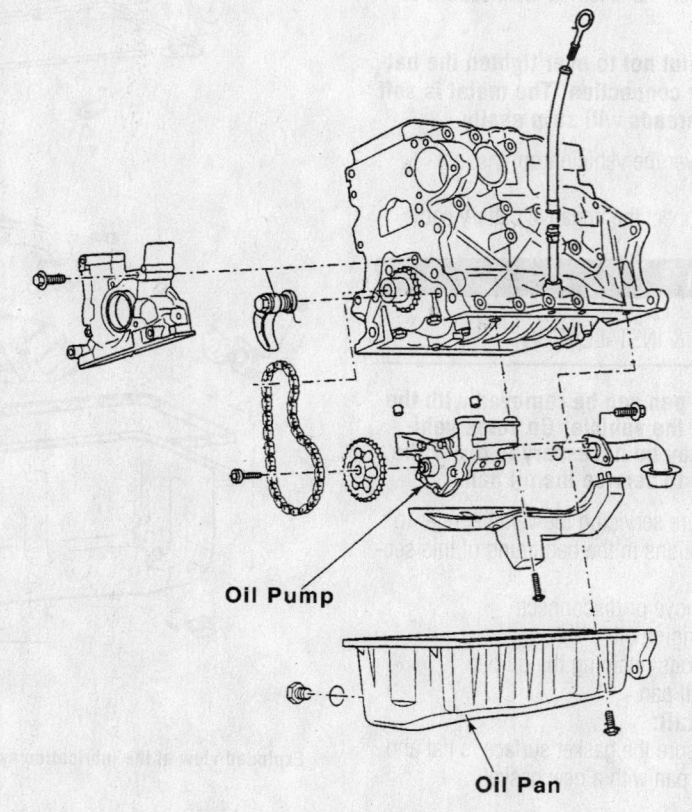

Exploded view of the lubrication system components—2.0L (AEG) engine

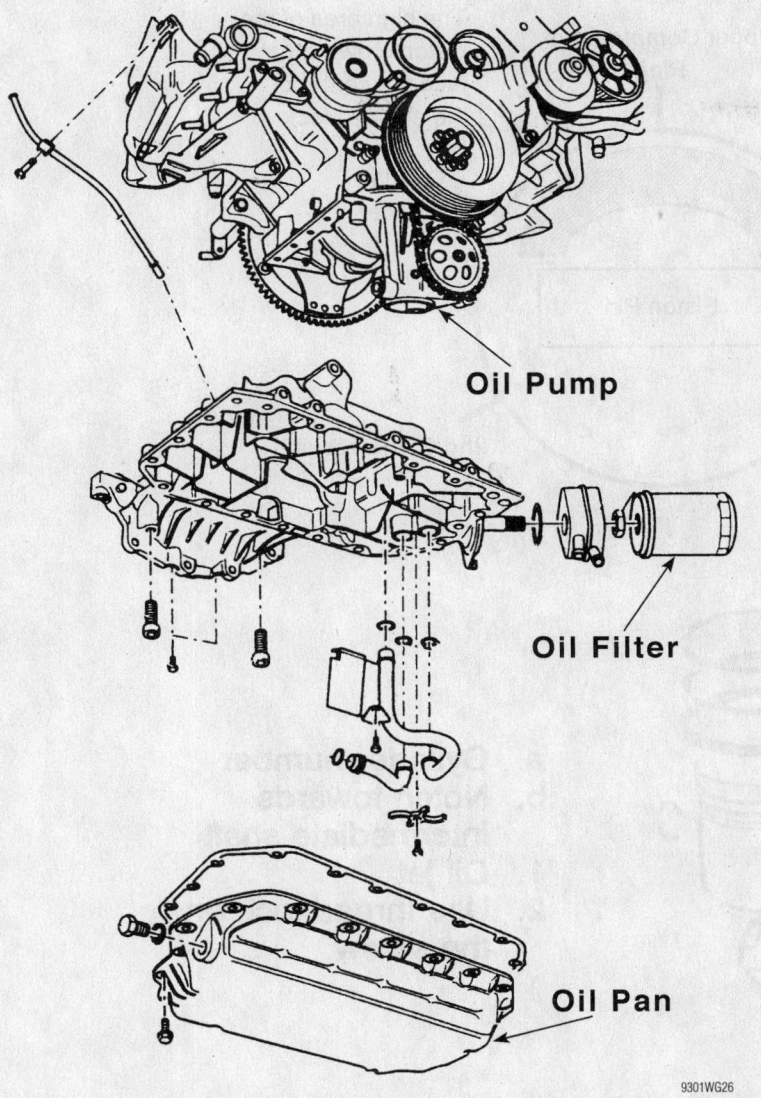

Oil Pump

Oil Filter

Oil Pan

9301WG26

Exploded view of the lubrication system components—2.8L (AHA) engine

- Oil pump. Mounting bolts: 18 ft. lbs. (25 Nm).
- Drive gear and remove the string, cord, wire or tie.
- Drive gear retaining bolt. Bolt: 18 ft. lbs. (25 Nm).
- Oil pan.

Rear Main Seal

REMOVAL & INSTALLATION

The rear main oil seal is located in a housing on the rear of the cylinder block. To replace the seal on all vehicles it is necessary to remove the transaxle and flywheel.

1. Before servicing the vehicle, refer to the precautions in the beginning of this section.
2. Remove or disconnect:
   - Transaxle and flywheel.
   - On 6-cylinder engines, old seal out of the support ring.
   - On 4-cylinder engines, oil seal with the mounting flange as a complete unit.

To install:

3. On 6-cylinder engines, oil the new seal and press it into place using tool VW-2003/2A, or equivalent, to start the seal and tool VW-2003/1, or equivalent, to seat the seal. Be careful not to damage the seal or score the crankshaft.
4. On 4-cylinder engines, install a new mounting flange with seal using a new gasket. Mounting flange bolts to 84 inch lbs. (10 Nm).
5. Install the flywheel and transaxle.

Piston and Ring Positioning

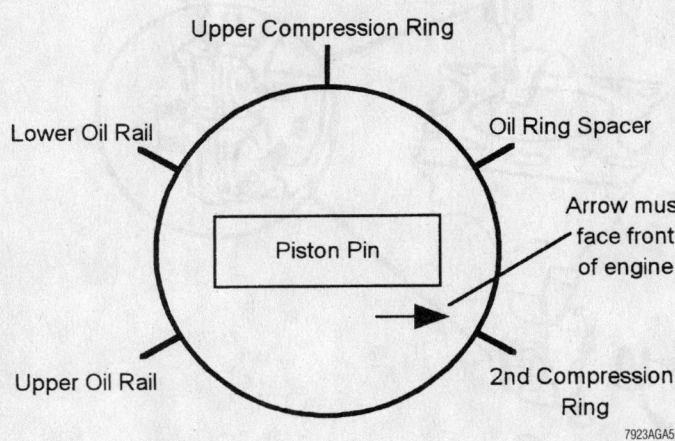

7923AGA5

1.8L (ABE), and 2.0L (ABA) engines—piston ring end-gap spacing

Brake service is covered in Section 4 of this manual

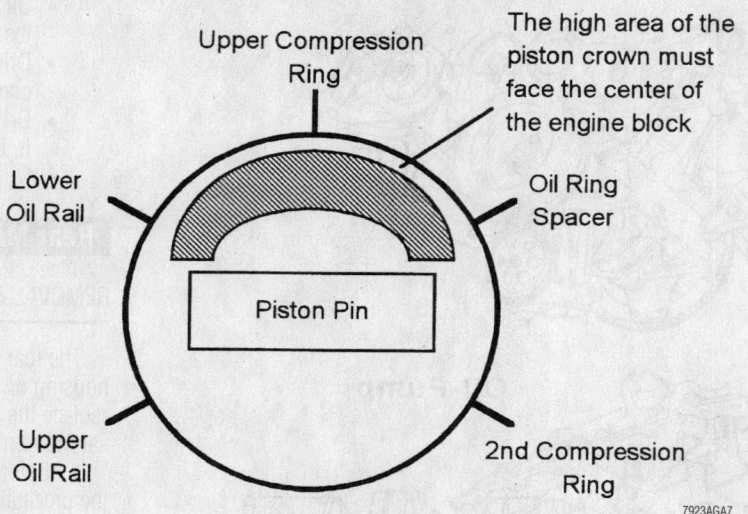

Upper Compression Ring

The high area of the piston crown must face the center of the engine block

Lower Oil Rail

Oil Ring Spacer

Piston Pin

Upper Oil Rail

2nd Compression Ring

7923AGA7

2.8L (AAA) engine—piston ring end-gap spacing

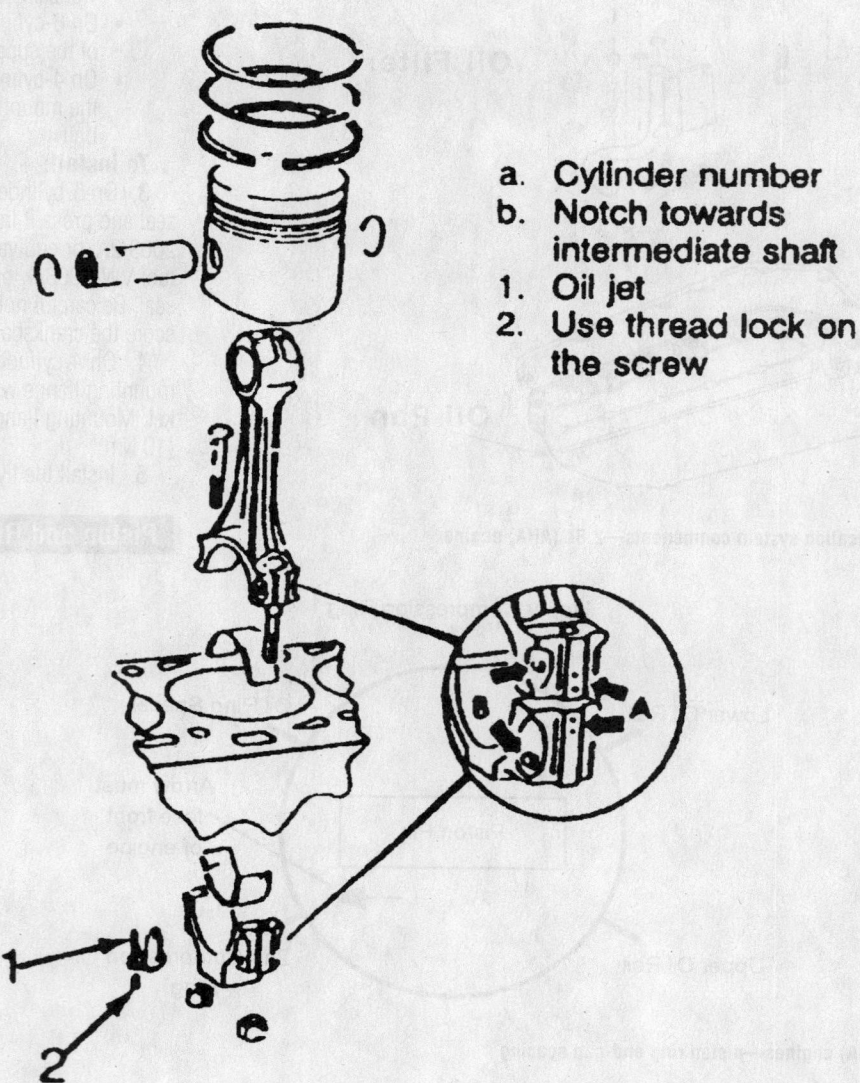

a. **Cylinder number**
b. **Notch towards intermediate shaft**
1. **Oil jet**
2. **Use thread lock on the screw**

7923AGA6

Piston and connecting rod installation

DIESEL ENGINE REPAIR

Alternator

REMOVAL & INSTALLATION

1. Disconnect the negative battery cable.
2. Disconnect the wiring from the alternator. If necessary, mark the wires with tape or other means to ensure they are connected properly upon installation.
3. Loosen the belt tension, and remove the drive belt from the alternator.
4. Remove the alternator adjustment bolt, followed by the pivot bolts.
5. Carefully lift the alternator from the bracket.

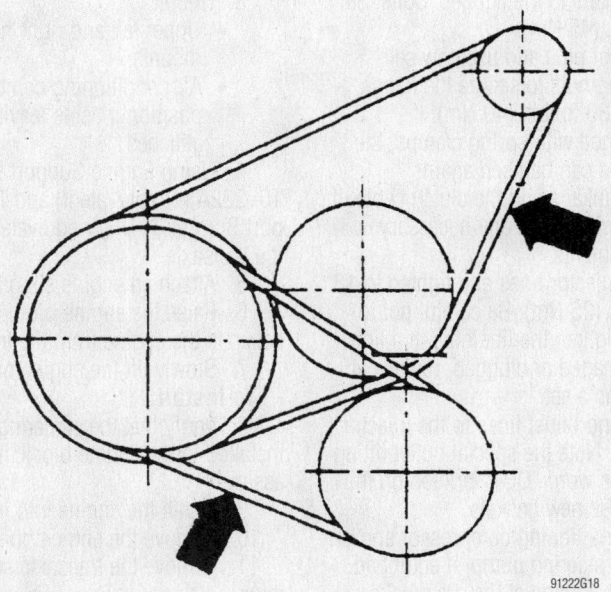

Alternator belt mounting detail—diesel engines

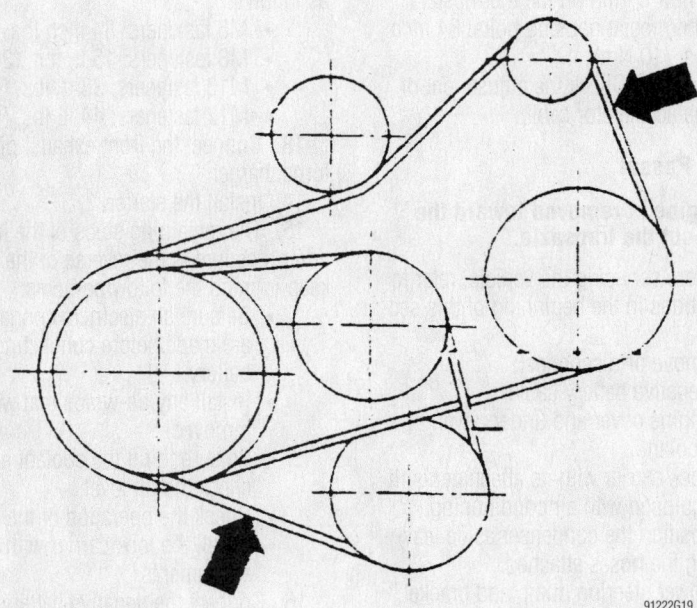

Diesel engine belt alternator belt routing

To install:

6. Hold the alternator in position on the mounting bracket, and install the pivot bolt. On later engines with automatic belt tensioners, install the upper mounting bolt.
7. Install (but do not tighten) the alternator adjustment bolt (earlier models only).
8. Place the alternator drive belt on the pulley.
9. Adjust belt tension and tighten the mounting and adjustment bolts as necessary.
10. Connect the wiring to the alternator. If tape was used to identify the wires, make sure it is removed once the wires are connected.
11. Connect the negative battery cable.

Engine Assembly

REMOVAL & INSTALLATION

Golf, Jetta

1. Before servicing the vehicle, refer to the precautions in the beginning of this section.
2. The engine and transaxle are lifted from the vehicle as an assembly.
3. Remove or disconnect:
   - Battery cables and battery.
   - Fuel pressure. Open the fuel filler cap to relieve tank pressure, then loosen the fuel filter fitting. Be sure to take the appropriate fire safety precautions.
   - Air filter and accelerator cable from the injection pump.
   - Coolant.
   - Upper radiator hose and wiring from the radiator fan motor and switches. Remove the mounting nuts or bolts and lift out the radiator and fan shroud as an assembly.
   - Electrical connections and vacuum lines, carefully labeling each one. Don't forget ground connections that are attached to the body.
   - With power steering, power steering pump and secure it to the body. Do not disconnect the hydraulic lines.
   - With air conditioning, air conditioning compressor and secure it aside without disconnecting the lines.
   - Fuel inlet and outlet lines from the

For complete Engine Mechanical specifications, see Section 1 of this manual

injection pump and plug the holes to keep the pump clean. Note the outlet fitting has a special orifice.

- On turbocharged engines, exhaust pipe and the oil lines from the turbocharger and cap the oil line fittings on the turbocharger. Unbolt the turbocharger and lift it out of the engine.
- With an automatic transaxle, selector cable at the transaxle.
- On manual transaxle shift linkage, the 2 rods with the plastic socket ends and unbolt the remaining linkage from the case as required.
- Clutch cable, lift it out of the case and set it aside.
- Wiring from the starter, the back-up light switch and the ground cable from the transaxle.
- Speedometer cable from the transaxle and plug the hole in the case.

4. Attach an engine sling tool VW-2024A or equivalent, to the engine and attach the sling to a suitable lifting device.

5. Remove or disconnect:
- Nuts or spring clamps holding the exhaust pipe to the manifold or turbocharger.

✳✳ CAUTION

On some models, special tools are required for removing and installing the exhaust pipe-to-manifold spring clamps: VW3140/1 and /2 or equivalent. This is a set of different sized wedges for spreading the spring clamps in steps. The installed spring clamp has considerable tension and could cause damage or injury if not properly removed. Clamps with wedges installed are also under high tension and should be handled carefully.

- Halfshafts from the flanges and hang them from the body with wire.

6. Unbolt the mounts. Remove the starter first and the front mount with it.

7. With all mounts unbolted, slightly lower the engine/transaxle assembly and tilt it towards the transaxle side. Then, carefully lift the assembly out of the vehicle.

To install:

8. Carefully install the engine/transaxle assembly and be sure all mounts are securely bolted to the engine/transaxle. Start all nuts and bolts that secure the mounts to the body but don't tighten them yet.

9. With all mounts installed and the engine safely in the vehicle, allow some slack in the lifting equipment. With the vehicle safely supported, shake the engine/transaxle as a unit to settle it in the mounts. Mounting bolts, starting at the rear and working forward: 33 ft. lbs. (41 Nm) for 10mm bolts or 54 ft. lbs. (73 Nm) for 12mm bolts.

10. Install or connect:
- Starter. Bolts: 33 ft. lbs. (45 Nm).
- Halfshafts to the flanges. Bolts: 33 ft. lbs. (45 Nm).
- Exhaust pipe and use new self-locking nuts to secure the flange. Nuts: 30 ft. lbs. (40 Nm). If equipped with spring clamps, the clamps can be used again.
- Shift linkage and the clutch cable, if equipped. Make any necessary adjustments.
- Fuel injector lines and tighten to 18 ft. lbs. (25 Nm). Be careful not to over-tighten the line nuts. If a line is damaged or clogged, replace all lines as a set.
- Inlet and outlet lines to the injector pump. Note the special outlet fitting has the word "OUT" printed on the top. Use new gaskets.
- Air conditioning compressor and/or power steering pump, if equipped. Install and adjust the drive belts.
- Wiring and vacuum hoses.
- Radiator, fan and heater hoses. Use a new O-ring on the thermostat. Thermostat housing bolts: 84 inch lbs. (10 Nm).
- Coolant. Check the adjustment of the accelerator cable.

1998–00 Passat

➡**The engine is removed toward the front without the transaxle.**

1. Before servicing the vehicle, refer to the precautions in the beginning of this section.

2. Remove or disconnect:
- Negative battery cable.
- Engine cover and under cover.
- Coolant.
- Lock carrier with its attachments. If equipped with air conditioning, position the condenser aside leaving the hoses attached.
- Power steering pump and bracket, and position it aside leaving the hoses attached.
- Fuel supply and return lines at the fuel filter.

- All vacuum, coolant, and intake air hoses that would interfere with engine removal.
- Any electrical connections that would interfere with engine removal.
- Starter.
- Front exhaust pipe from the turbocharger.
- Engine-to-transaxle mounting bolts.

3. Unbolt:
- Upper left and right-hand motor mounts.
- Air conditioning compressor and position it aside leaving the hoses attached.

4. Using Engine Support Bracket 10–222A (or equivalent) and Transaxle Support Bracket 3147 (or equivalent) support the transaxle.

5. Attach an engine sling and hoist.

6. Raise the engine off the mounts and separate the engine from the transaxle.

7. Slowly lift the engine out forward.

To install:

8. Verify that the centering dowels are installed in the engine block, install if necessary.

9. Install the engine into the vehicle.

10. Remove the engine hoist and sling.

11. Remove the transaxle support apparatus.

12. Install the engine-to-transaxle and motor mount attaching hardware and tighten as follows:
- M6 fasteners: 84 inch lbs. (10 Nm)
- M8 fasteners: 15 ft. lbs. (20 Nm)
- M10 fasteners: 33 ft. lbs. (45 Nm)
- M12 fasteners: 44 ft. lbs. (60 Nm)

13. Connect the front exhaust pipe to the turbocharger.

14. Install the starter.

15. The remaining steps of the installation procedure is the reverse of the removal, keep in mind the following items:
- Be sure all electrical connections are made before connecting the battery.
- Install any tie-wraps that were removed.
- Be sure to fill the coolant and check the oil level.
- Check the operation of the clutch.
- Install the lock carrier with its attachments.

16. Connect the negative battery cable.

17. Install the engine cover and undercover.

18. Read and clear any DTCs.

New Beetle

The engine and transaxle are removed from beneath the vehicle as an assembly.

1. Before servicing the vehicle, refer to the precautions in the beginning of this section.

2. Remove or disconnect:
- Negative battery cable.
- Engine cover.
- Power steering reservoir from the battery support leaving the hoses attached, and position it aside.
- Battery and bracket.
- Fuel supply and return lines at the fuel filter.
- Air cleaner and intake air duct.
- Throttle cable.
- With a manual transaxle, transaxle range selector mechanism and the clutch slave cylinder.
- With an automatic transaxle, range selector lever cable at the transaxle.
- Engine undercover.
- Coolant.
- Accessory drive belt.
- If necessary, right side cooling fan.
- Power steering pump and bracket and position it aside leaving the hoses connected.
- Air conditioner compressor and position it aside leaving the hoses connected.
- Any coolant, vacuum and intake hoses.
- All wires from the transaxle, starter and generator; position the wires out of the way.
- Any remaining wires or connectors that would interfere with the engine removal.
- Engine pendulum support.
- Right-hand halfshaft
- Left-hand halfshaft at the transaxle.
- Front exhaust pipe from the manifold.

3. Install Engine Bracket T10012, or equivalent, in the Engine/Transaxle Jack VAG 1383 A or equivalent.

4. Remove the bracket for the coolant hose under the engine block.

5. Attach the Engine Bracket T10012 to the engine block using the threaded holes at the corners of the engine block.

6. Raise the engine/transaxle jack to relieve the tension on the mounts.

7. Disconnect the engine and transaxle mounts from inside the engine compartment.

8. Lower the engine/transaxle assembly

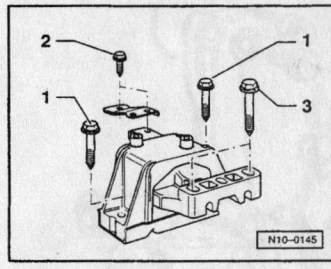

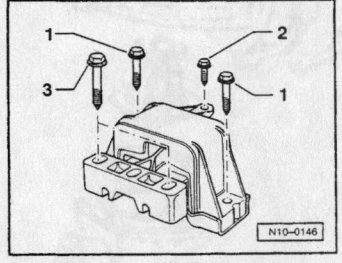

Exploded view of the engine mounts—New Beetle

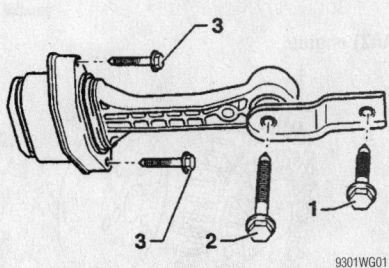

Pendulum support mounting bolt locations—New Beetle

from the vehicle, be sure to guide the power steering pressure hose past the transaxle.

To install:

9. Raise the engine/transaxle assembly into the vehicle, be sure to guide the power steering hose around the transaxle.

10. Using new bolts, connect the engine/transaxle mounts following the accompanying illustration.

11. Remove the engine/transaxle jack and Engine Bracket T10012.

12. Install or connect:
- Bracket for the coolant hose under the engine block.
- Front exhaust pipe to the manifold.
- Left halfshaft and right halfshaft.
- Engine pendulum support. Bolts (1) and (2) to 30 ft. lbs. (40 Nm) plus 90 degrees, and bolts (3) to 15 ft. lbs. (20 Nm) plus 90 degrees.

Engine / transmission mounts

Tightening torques

Front assembly mounting

1 -	Mount to body [1]	40 Nm (30 ft lb) + 90° (¼ turn)
2 -	Mount/bracket to body	25 Nm (18 ft lb)
3 -	Mount to engine bracket [1]	60 Nm (44 ft lb) + 90° (¼ turn)

[1] Replace bolts

Rear right assembly mounting

1 -	Mount to body [1]	40 Nm (30 ft lb) + 90° (¼ turn)
2 -	Mount to body	25 Nm (18 ft lb)
3 -	Bearing on transmission console [1]	60 Nm (44 ft lb) + 90° (¼ turn)

[1] Replace bolts

- Any wiring that was removed.
- Any hoses that were removed.
- Air conditioner compressor.
- Power steering bracket and pump.
- If removed, right side cooling fan.
- Accessory drive belt and fill the cooling system.
- With an automatic transaxle, transaxle range selector cable.
- With a manual transaxle, clutch slave cylinder and connect the range selector lever.
- Throttle cable.
- Intake air duct and air cleaner assembly.
- Fuel supply and return lines to the filter housing.
- Battery bracket.
- Power steering reservoir.
- Battery cables.
- Engine covers.

Water Pump

REMOVAL & INSTALLATION

1.9L (AAZ) Engine

On some Diesel engines, the belt tension is adjusted with shims between the outer and inner halves of the pulley. On others, the alternator swivels to adjust belt tension.

1. Before servicing the vehicle, refer to the precautions in the beginning of this section.

For Accessory Drive Belt illustrations, see Section 1 of this manual

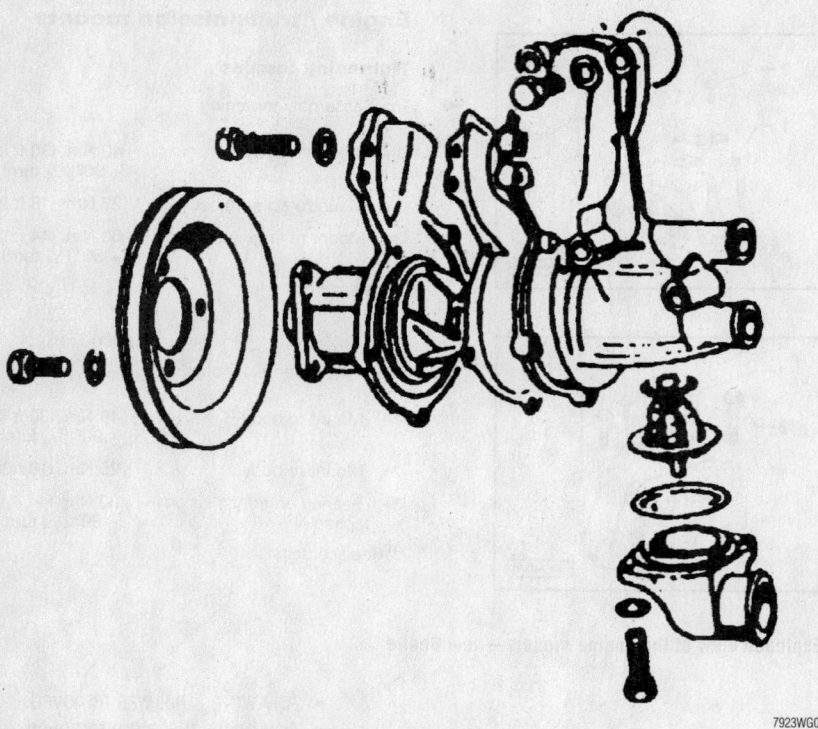

Exploded view of the water pump assembly—1.9L (AAZ) engine

2. Drain the coolant.

3. Loosen but don't remove the bolts holding the pulley to the water pump.

4. With a movable alternator, loosen the alternator and remove the drive belt.

5. Remove the water pump pulley and water pump.

To install:

6. Installation is the reverse of removal. Be sure to clean the pump housing before installing the new gasket. Tighten the following:

- Water pump-to-housing—84 inch lbs. (10 Nm)
- Water pump drive pulley—15 ft. lbs. (20 Nm)
- Thermostat housing—84 inch lbs. (10 Nm)
- Alternator mounting bolts—18 ft. lbs. (25 Nm)

1.9L (AHH) Engine

1. Before servicing the vehicle, refer to the precautions in the beginning of this section.

2. Remove or disconnect:
- Negative battery cable.
- Engine cover and under cover.
- Coolant.

3. Position the lock carrier into the service position, as follows:

   a. Remove or disconnect:
   - Front bumper.

- Any wiring or connector that would inhibit locking the carrier.
- 3 quick-release screws on the front noise insulation panel.
- Air guide between the lock carrier and the air filter.
- If installed, retaining clamps for the wiring harness at the left side of the radiator frame.
- No. 2 bolts and install Support tool 3369 or equivalent.
- Remaining bolts and pull the lock carrier out to the stop.

   b. To secure the lock carrier, install the appropriate M6 bolts into the rear of the lock carrier and fender.

4. Remove or disconnect:
- Accessory drive belt, fan clutch and belt pulleys.
- If necessary, drive belt tensioner.
- Coolant pump assembly bracket.
- Coolant pump from the bracket.

To install:

5. Using a new O-ring, install the coolant pump onto the bracket and tighten the mounting bolts to 84 inch lbs. (10 Nm).

6. Using a new gasket and O-ring,

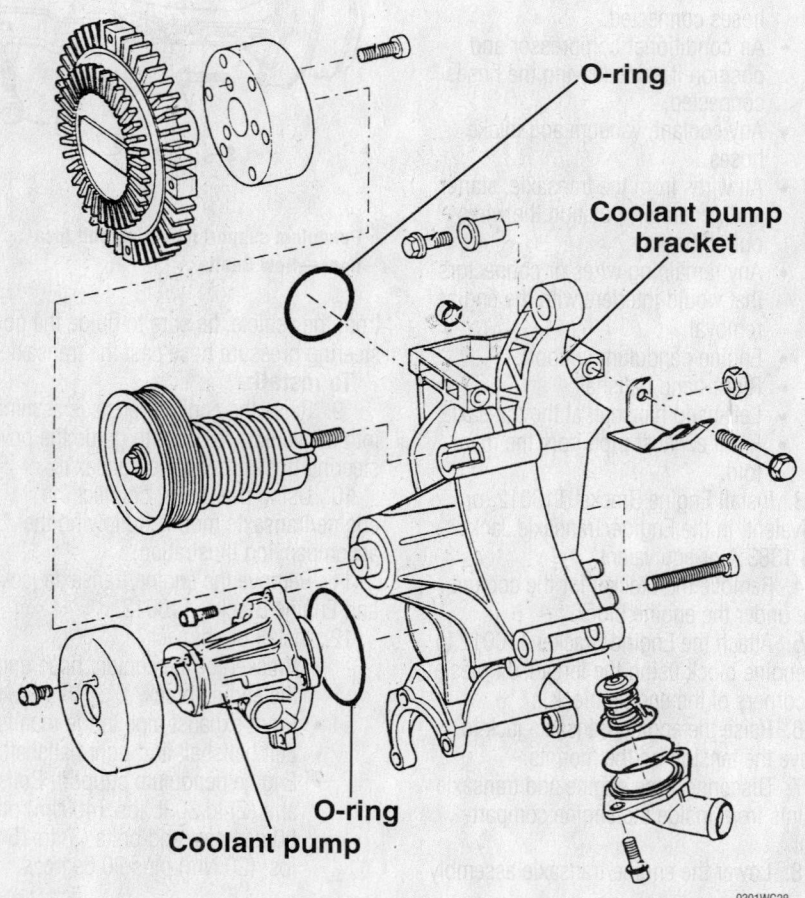

Exploded view of the coolant pump, bracket and related components—1.9L (AHH) engine

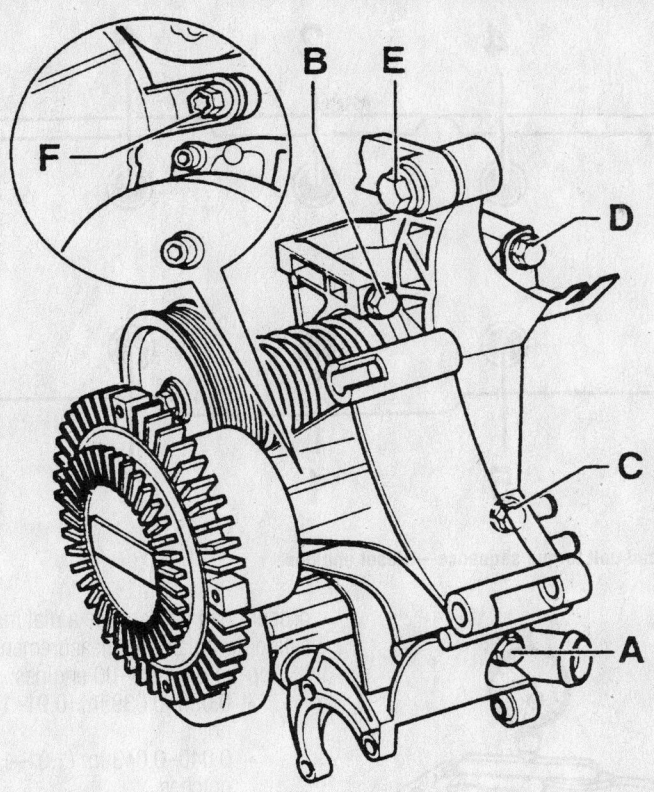

Coolant pump bracket torque sequence—1.9L (AHH) engine

- Hose between the charge air cooler and the intake manifold.
- Fuel filter and bracket.
- Timing belt upper and center covers.
- Timing belt tension, then remove the belt from the camshaft sprocket, injection pump sprocket and coolant pump sprocket.
- Coolant pump mounting bolts, then the pump.
- O-ring and clean the seating area.

To install:

3. Moisten a new O-ring with coolant, then install.

4. Install or connect:
- Water pump, so that the plug in the housing faces downward. Mounting bolts: 11 ft. lbs. (15 Nm).
- Timing belt.
- Upper and center timing belt covers.
- Fuel filter and bracket.
- Fuel lines.
- Hose between the charge air cooler and the intake manifold.
- Accessory drive belt tensioner. Mounting bolt: 18 ft. lbs. (25 Nm).
- Accessory drive belt.
- Coolant.
- Negative battery cable.

install the coolant pump bracket. Mounting bolts, in alphabetical sequence: 18 ft. lbs. (25 Nm).

7. Tighten the coolant pump housing to the timing belt cover mounting bolt (F) to 84 inch lbs. (10 Nm).

8. If removed, install the drive belt tensioner.

9. Install the belt pulleys, fan clutch and accessory drive belt.

10. Return the lock carrier into the normal position.

11. Install or connect:
- Coolant.
- Negative battery cable.
- Engine cover and under cover.

1.9L (ALH) Engine

1. Before servicing the vehicle, refer to the precautions in the beginning of this section.

2. Remove or disconnect:
- Negative battery cable.
- Engine cover and under cover.
- Coolant.
- Accessory drive belt.
- Fuel supply and return lines at the filter.

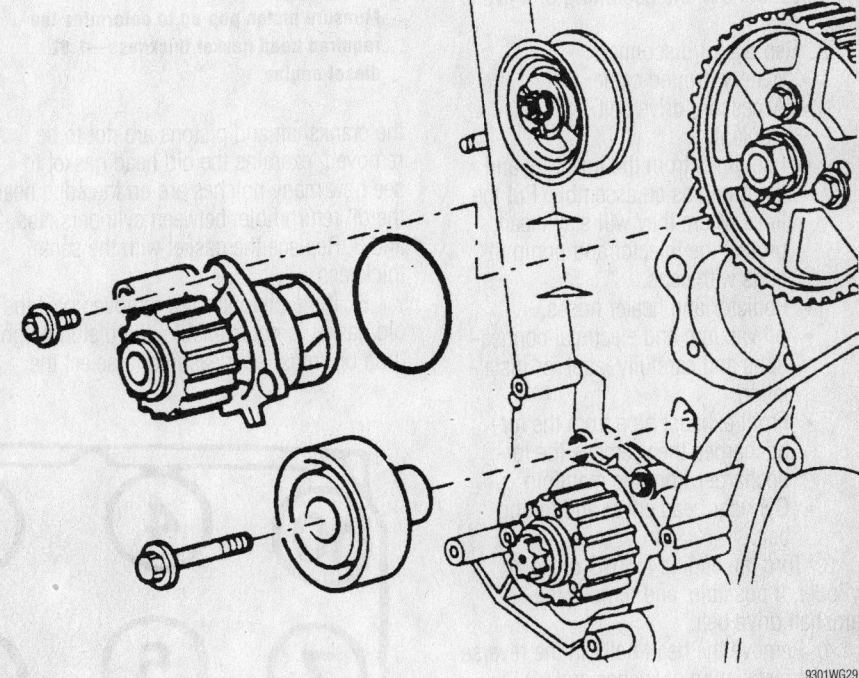

Exploded view of the water pump mounting—1.9L (ALH) engine

For Tire, Wheel and Ball Joint specifications, see Section 1 of this manual

Diesel Glow Plugs

REMOVAL & INSTALLATION

1. Before servicing the vehicle, refer to the precautions in the beginning of this section.

2. Remove the bus bar connecting the glow plugs and determine which plugs need replacement.

3. Remove the defective plugs.

4. When installing new plugs, tighten them to 11 ft. lbs. (15 Nm).

➡Diesel glow plugs have an air gap much like a spark plug to prevent overheating of the plug. Over-tightening the glow plug will close the gap and cause the plug to burn out.

Cylinder Head

REMOVAL & INSTALLATION

➡Cylinder head bolt torque sequence illustrations are found in Section 1, following the Torque Charts.

➡The cylinder head bolts on all diesel vehicles are stretch bolts and must be replaced when removed.

1. Before servicing the vehicle, refer to the precautions in the beginning of this section.

2. Remove or disconnect:
   • Battery ground cable.
   • Accessory drive belt.
   • Coolant.
   • Fuel lines from the injectors and the pump as an assembly. Put the lines where they will stay clean; protect the injector and pump fittings with caps.
   • Radiator and heater hoses.
   • All vacuum and electrical connections and carefully label for installation.
   • Front exhaust pipe from the turbocharger, then remove the turbocharger from the manifold.
   • Cylinder head cover and timing belt.

3. Turn the engine to TDC of No. 1 cylinder, if possible, and remove the camshaft drive belt.

4. Remove the head bolts in the reverse order of installation sequence and lift the head out of the vehicle.

To install:

5. On these engines, the pistons actually project above the deck of the block. If

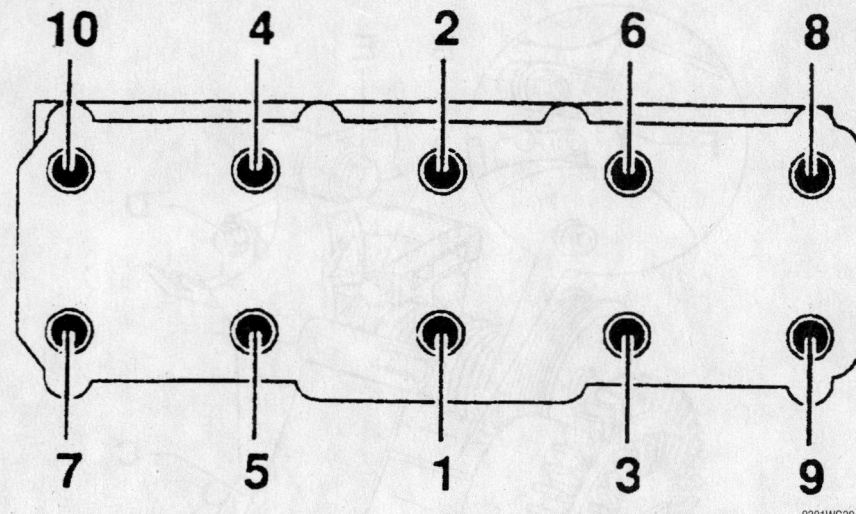

Cylinder head bolt torque sequence—Diesel engines

9301WG39

Measure piston pop-up to determine the required head gasket thickness—1.9L diesel engine

7923WG23

the crankshaft and pistons are not to be removed, examine the old head gasket to see how many notches are on the edge near the oil return hole, between cylinders Nos. 2 and 3. Replace the gasket with the same thickness.

6. If the pistons were removed or if the old gasket is not available, the piston height (pop up) must be measured to select the

proper head gasket. Use a dial indicator or caliper to obtain the measurement.

Pop-up on 1998–00 engines:
   • 0.036–0.039 in. (0.91–1.00mm): 1 notch
   • 0.040–0.043 in. (1.01–1.10mm): 2 notches
   • 0.044–0.047 in. (1.11–1.20mm): 3 notches

7. Install the new cylinder head gasket with the word TOP or OBEN facing upward. Do not use any sealing compound.

8. Turn the crankshaft to TDC of No. 1 cylinder, then back about ¼ turn to bring all pistons about even.

9. Carefully lower the head on and install new head bolts into No. 8 and 10 first. These holes are smaller and will properly locate the gasket and cylinder head.

10. Install the remaining bolts and tighten in the proper sequence in 3 steps: 30 ft. lbs. (40 Nm), 44 ft. lbs. (60 Nm), then a full ½ turn more. 2 quarter turns are allowed.

Tighten the cylinder head bolts in the correct order as shown—1.9L diesel engine

7923WG04

11. Installation of the remaining parts is the reverse of removal, be sure to change the oil and filter. Install the camshaft drive belt and set injection pump timing.

12. Install the fuel injector lines and tighten to 18 ft. lbs. (25 Nm). Be careful not to over tighten the line nuts. If a line is damaged or clogged, replace all lines as a set.

13. Connect the negative battery cable.

Turbocharger

REMOVAL & INSTALLATION

1. Before servicing the vehicle, refer to the precautions in the beginning of this section.

2. Remove or disconnect:
- Negative battery cable.
- Exhaust pipe from the turbocharger outlet.
- Oil supply line and bracket, after cleaning the oil supply fitting on the top of the turbocharger.
- Inlet air hose.
- Oil return line and the turbocharger mounting bracket.
- Turbo-to-manifold bolts. Lift the turbocharger out from the top.

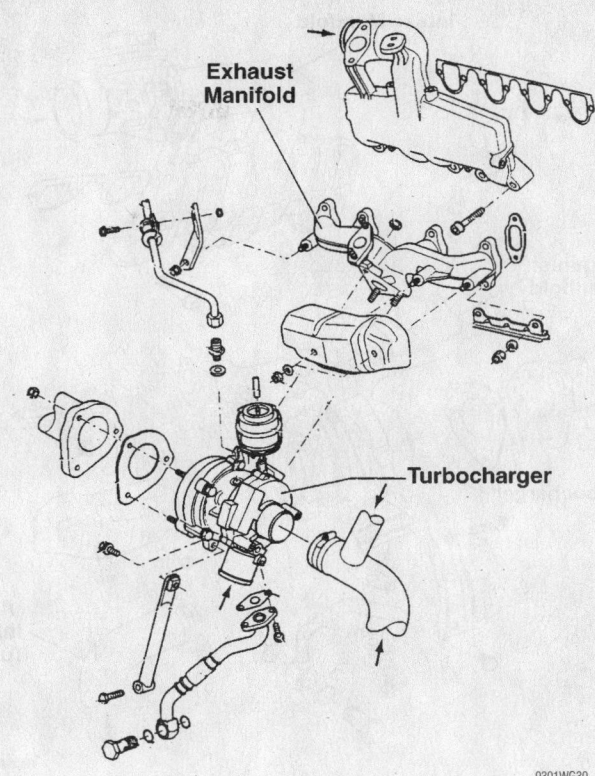

9301WG30

Exploded view of the intake and exhaust manifold, turbocharger and related components—1.9L (AHH) engine

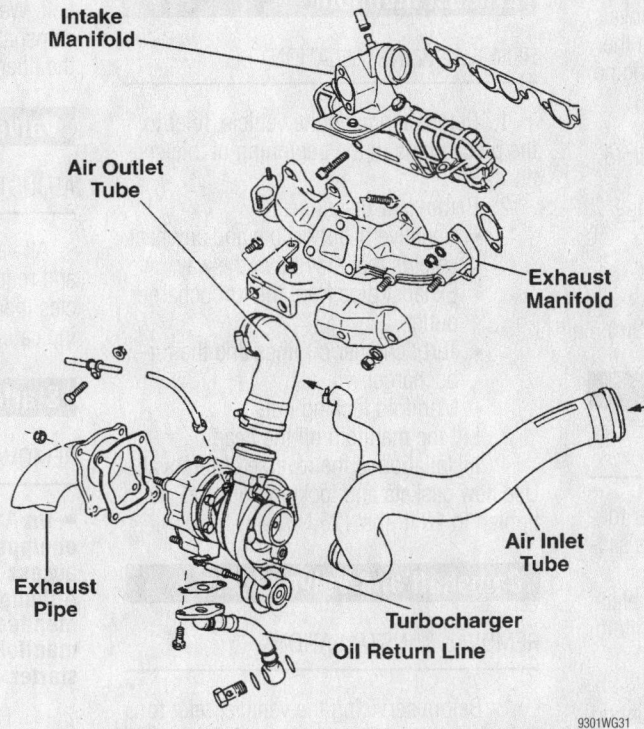

9301WG31

Exploded view of the intake and exhaust manifold, turbocharger and related components—1.9L (AAA) engine

For Wheel Alignment specifications, see Section 1 of this manual

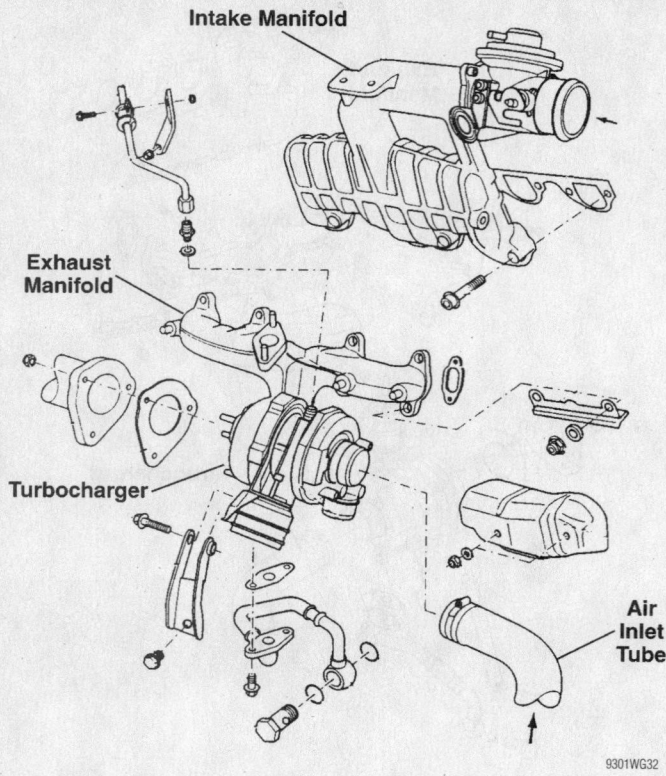

Intake Manifold

Exhaust Manifold

Turbocharger

Air Inlet Tube

9301WG32

Exploded view of the intake and exhaust manifold, turbocharger and related components—1.9L (ALH) engine

To install:

3. Installation is the reverse of removal. Before installing the oil supply line, fill the connection on the turbocharger with engine oil.

4. Tighten the following:
- Turbocharger-to-exhaust manifold: 33 ft. lbs. (45 Nm)
- Mounting bracket nuts: 18 ft. lbs. (25 Nm)
- Turbocharger outlet nuts: 18 ft. lbs. (25 Nm)
- Oil return line: 22 ft. lbs. (30 Nm)

Intake Manifold

REMOVAL & INSTALLATION

1. Before servicing the vehicle, refer to the precautions in the beginning of this section.

2. Label and detach any hoses and electrical connections related to intake manifold service.

3. Disconnect the air inlet hose.

4. Remove the intake manifold mounting bolts.

5. Installation is the reverse of removal. Use a new gasket and tighten the bolts to 18 ft. lbs. (25 Nm).

Exhaust Manifold

REMOVAL & INSTALLATION

1. Before servicing the vehicle, refer to the precautions in the beginning of this section.

2. Remove or disconnect:
- Negative battery cable and any heat shields that may be in the way.
- Exhaust pipe from the turbocharger outlet.
- Turbocharger oil lines and the turbocharger.
- Manifold locking nuts

3. Lift the manifold off the head.

4. Installation is the reverse of removal. Use new gaskets and locking nuts and tighten to 18 ft. lbs. (25 Nm).

Camshaft and Valve Lifters

REMOVAL & INSTALLATION

1. Before servicing the vehicle, refer to the precautions in the beginning of this section.

2. Remove or disconnect:
- Negative battery cable.

- Timing belt cover(s), the timing belt, cylinder head cover and the camshaft sprocket.

3. Number the bearing caps from front-to-back. If the cap does not already have one, scribe an arrow pointing towards the front of the engine. The caps are offset and must be installed correctly. Factory numbers on the caps are not always on the same side.

4. Remove:
- Front and rear bearing caps. Loosen the remaining bearing cap nuts a little at a time to avoid bending the camshaft. Start from the outside caps near the ends of the head and work toward the center.
- Bearing caps and the camshaft.

To install:

5. Install a new oil seal and end-plug in the cylinder head. Lubricate the camshaft bearing journals and lobes and set the camshaft in place.

6. Install the bearing caps in the correct position with the arrow pointing towards the front of the engine. Cap nuts, diagonally and in several steps: 15 ft. lbs. (20 Nm). Do not over tighten. Camshaft shaft end-play should be about 0.006 in. (0.15mm).

7. Install the drive sprocket and timing belt. Wait at least ½ hour after installing the camshaft before starting the engine to allow the lifters to leak down.

Valve Lash

ADJUSTMENT

All vehicles have hydraulic valve lifters and require no adjustment. On these vehicles there will be a sticker under the hood indicating hydraulic lifters.

Starter

REMOVAL & INSTALLATION

➡**On A1 and A2 platform vehicles equipped with 010 automatic transaxle, access to the starter motor is limited. Although not necessary, it is recommended that the intake and exhaust manifolds be removed to access the starter.**

1. For safety purposes, disconnect the battery ground cable.

2. Raise and safely support the front of the vehicle with jackstands.

3. For A3 vehicles, use a floor jack

(with a block of wood on the chock) to support the engine by the oil pan. Do NOT jack the car by the oil pan under any circumstances! The floor jack is ONLY to support the engine while the starter is being removed. The bolts that secure the starter to the transaxle are also used to secure the front engine mount to the transaxle.

4. If necessary, label the small wires before disconnecting them.

5. Disconnect the large cable, which is the positive battery cable, from the solenoid.

6. On 010 transaxle-equipped vehicles, remove the bracket that secures the starter to the engine.

7. Remove the starter mounting bolts, while supporting the weight of the starter.

8. Pull the starter straight out from the transaxle.

To install:

➡ **On vehicles with a manual transaxle, there is a bushing where the starter shaft fits into the bell housing. If the shaft or bushing are worn or if the starter has been jamming, the bushing should be replaced. There is a special bushing removal tool available but a small inside bearing removal tool is usually sufficient.**

9. Install the starter into the transaxle.

10. Tighten the starter mounting bolts as follows:

  a. All vehicles except 010 transaxle:
- M8 nut: 89 inch lbs. (10 Nm)
- M10 nut and bolt: 44 ft. lbs. (60 Nm)
- M12 bolt: 33 ft. lbs. (45 Nm)

  b. 010 transaxle only:
- Mounting flange bolt: 15 ft. lbs. (20 Nm)
- Mounting bracket bolt: 18 ft. lbs. (25 Nm)

11. Attach the electrical connections to the starter.

➡ **Be careful not to over tighten the battery cable connection. The metal is soft and the threads will strip easily.**

12. Lower the vehicle from the jackstands.

13. Connect the negative battery cable.

Oil Pan

REMOVAL & INSTALLATION

The oil pan can be removed with the engine in the vehicle.

1. Before servicing the vehicle, refer to the precautions in the beginning of this section.

2. Remove or disconnect:
- Engine oil.
- Bolts retaining the oil pan.
- Oil pan.

To install:

3. Be sure the gasket surface is flat and install the pan with a new gasket.

4. Tighten the retaining bolts in a criss-cross pattern to 15 ft. lbs. (20 Nm). Do not over-tighten.

5. Refill the engine with oil.

Oil Pump

REMOVAL & INSTALLATION

1. Before servicing the vehicle, refer to the precautions in the beginning of this section.

2. Remove or disconnect:
- Oil pan.
- Mounting bolts and lower the pump from the engine.
- Bottom cover and disassemble the pump. The pressure relief valve is in the bottom cover.

3. After reassembling the pump, prime it with oil and install it in the reverse order of removal.

4. Observe the following values:
- Oil pump bottom cover bolts: 84 inch lbs. (10 Nm)
- Oil pump suction foot bolts: 84 inch lbs. (10 Nm)
- Oil pump retaining bolts: 18 ft. lbs. (25 Nm)

Rear Main Seal

REMOVAL & INSTALLATION

The rear main oil seal is located in a housing on the rear of the cylinder block.

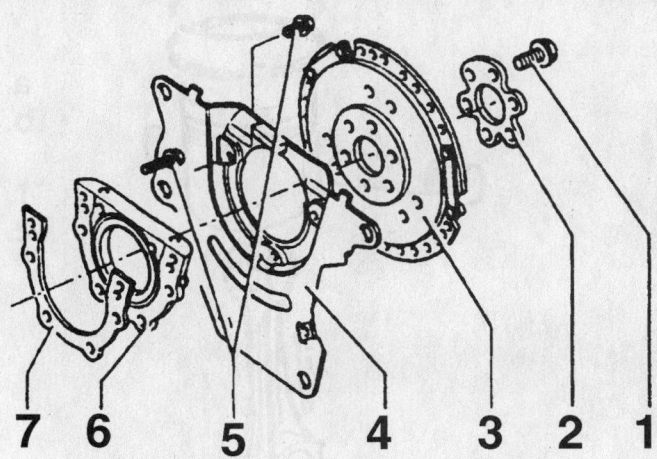

7 6 5 4 3 2 1

1. Bolt
2. Washer
3. Drive plate
4. Intermediate plate
5. Intermediate plate mounting bolt
6. Mounting flange with seal
7. Gasket

7923WG22

Rear main oil seal and related components—1.9L diesel engine

To replace the seal on all vehicles it is necessary to remove the transaxle and flywheel.

1. Before servicing the vehicle, refer to the precautions in the beginning of this section.

2. Remove the transaxle and flywheel.

3. Remove the oil seal with the mounting flange as a complete unit.

To install:

4. Install a new mounting flange with seal using a new gasket. Mounting flange bolts: 84 inch lbs. (10 Nm).

5. Install the flywheel and transaxle.

Piston and Ring Positioning

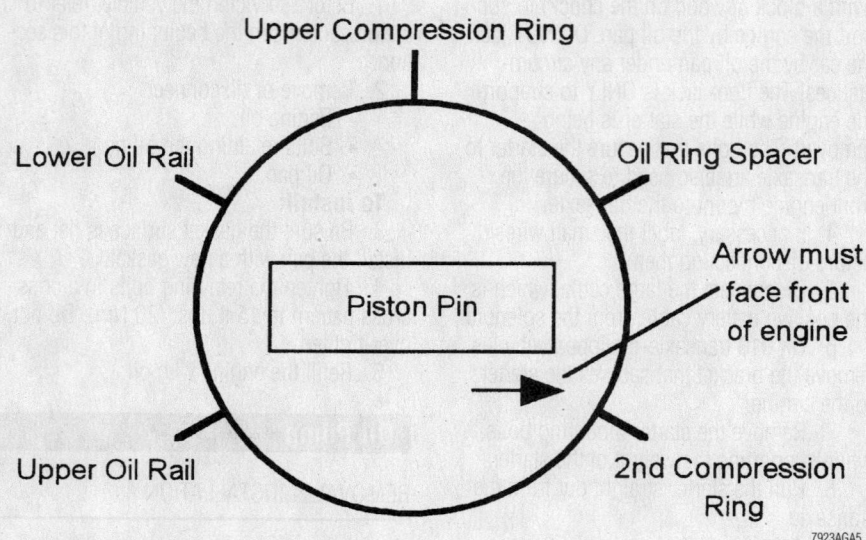

Upper Compression Ring

Lower Oil Rail

Oil Ring Spacer

Piston Pin

Arrow must face front of engine

Upper Oil Rail

2nd Compression Ring

7923AGA5

Diesel engines—piston ring end-gap spacing

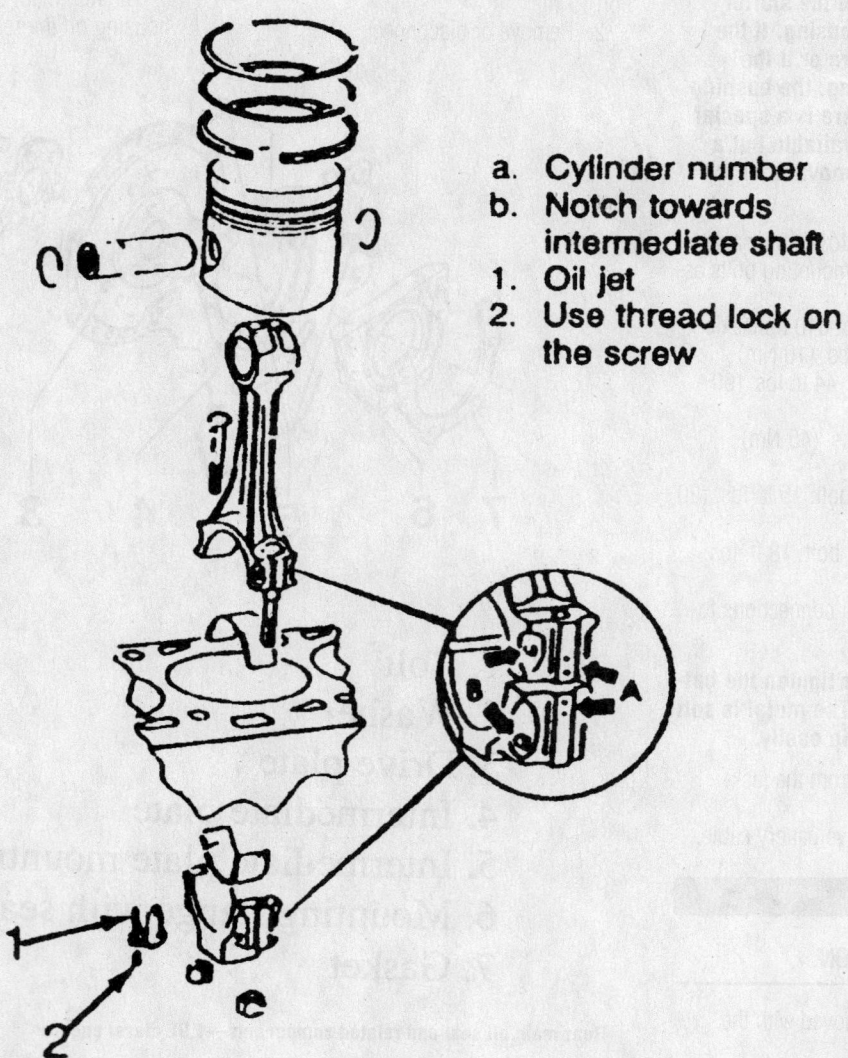

a. **Cylinder number**
b. **Notch towards intermediate shaft**
1. **Oil jet**
2. **Use thread lock on the screw**

7923AGA6

Piston and connecting rod installation

GASOLINE FUEL SYSTEM

Fuel System Service Precautions

Whenever working on or around gasoline or the fuel delivery system, heed the following precautions:

• Do not allow fuel spray or fuel vapors to come into contact with a heating element or open flame. Do not smoke while working on the fuel system.

• Always disconnect the negative battery cable unless the repair or test procedure requires that battery voltage be applied.

• Always relieve the fuel system pressure prior to disconnecting any fitting or fuel line connection.

• To control fuel spray when relieving system pressure, place a shop towel around the fitting prior to loosening to catch the spray. Ensure that all fuel spillage is quickly wiped up and that all fuel soaked rags are deposited into a proper fire safety container.

• Always keep a dry chemical (Class B) fire extinguisher near the work area.

• Always use a back-up wrench when loosening and tightening fuel line fittings.

• Do not re-use fuel system gaskets and O-rings, replace with new ones. Do not substitute fuel hose where fuel pipe is installed.

➡**Before servicing the vehicle, also refer to the precautions in the beginning of this section.**

Fuel System Pressure

RELIEVING

The fuel injection system operates under high pressure. This makes it necessary to first relieve the system of pressure before servicing. The pressurized fuel, when released, may ignite or cause personal injury.

1. Before servicing the vehicle, refer to the precautions in the beginning of this section.
2. Remove or disconnect:
   • Power to the fuel pump by removing the relay or the fuel pump fuse. Check the list on the fuse box lid to be sure. The fuse can be removed to stop the fuel pump from running. With the engine operating at idle,

wait until the engine stalls from fuel starvation.
   • Negative battery cable.
3. Carefully loosen the fuel line on the control pressure regulator or component to be serviced.
4. Wrap a clean rag around the connection, while loosening, to catch any fuel.
5. After service is complete, discard the fuel soaked rag in the proper manner and reconnect the negative battery cable, relay or fuses.

Fuel Filter

REMOVAL & INSTALLATION

Most vehicles use a fuel filter mounted under the vehicle, below the fuel tank. An arrow should be on the filter indicating fuel flow direction. Install with arrow pointing to engine. Use care not to mix up fuel supply or return lines. Fuel pressure applied to the return side of the system will cause damage.

In addition, some vehicles use a filter in the engine compartment near the fuel distributor. If equipped, use the following procedure:

1. Before servicing the vehicle, refer to the precautions in the beginning of this section.
2. Make certain to follow the precautions and relieve fuel the pressure.
3. Disconnect the fuel lines leading into and out of the fuel filter.

4. Unscrew the filter retaining bracket and remove the filter.
5. Install a new filter in the bracket and reattach the bracket. Be sure the arrows are pointing in the direction of the fuel flow.
6. Reconnect the fuel lines, start the engine and check for leaks.

Fuel Pump

REMOVAL & INSTALLATION

1. Before servicing the vehicle, refer to the precautions in the beginning of this section.
2. The main fuel pump is located under the vehicle in front of the rear axle or in front of the tank on the right side. Disconnect the negative battery cable.
3. Remove or disconnect:
   • Electrical connector.
   • Fuel system pressure.
   • Mounting bolts
   • Fuel pump.
4. Installation is the reverse of removal. Be sure to use new sealing rings and/or gaskets.

Fuel Injector(s)

REMOVAL & INSTALLATION

1. Relieve the fuel system pressure.
2. On CIS-E Motronic systems pull the

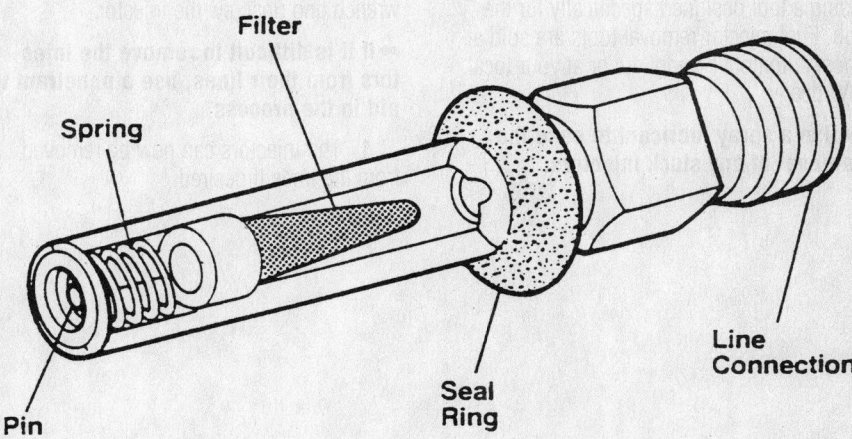

Filter

Spring

Pin

Seal Ring

Line Connection

91225G06

Inner view of a mechanical fuel injector

Pull the fuel injector wiring harness from the injector

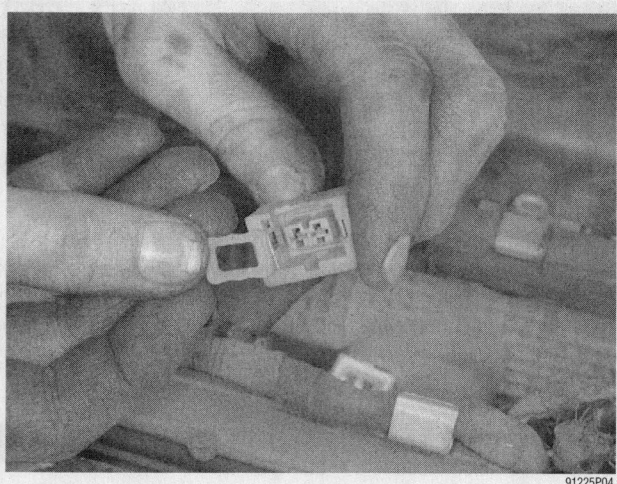

Close up of injector wiring harness

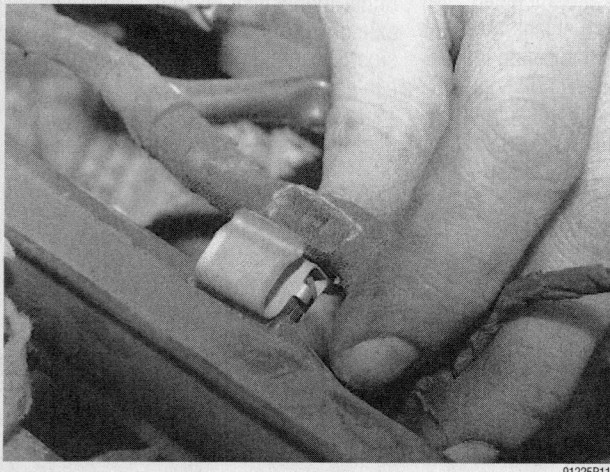

Observe the location of the injector wiring harness spring clips

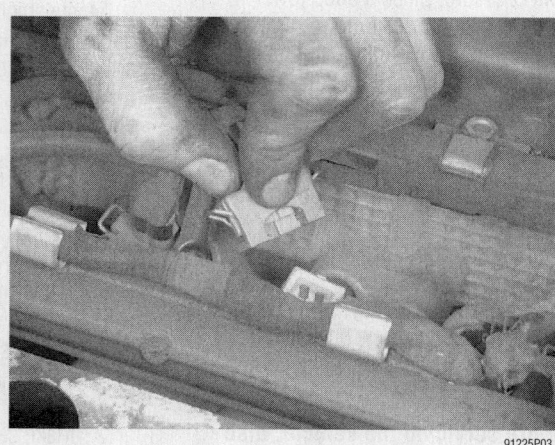

Notice the arch in the injector wiring harness connector. It will only slide over the injector in one direction

injectors straight out of the intake manifold using a tool designed specifically for the job. Fuel injector removal tools are sold at most automotive suppliers or at your local VW dealer.

➡**Use a spray lubricant to ease the removal of any stuck injectors.**

3. Now hold the fitting with a line wrench and unscrew the injector.

➡**If it is difficult to remove the injectors from their lines, use a penetrant to aid in the process.**

4. The injectors can now be removed from the lines if desired.

To install:

5. Install the fuel injectors on the lines.

6. Lubricate the injector O-rings with a spray lubricant or gasoline.

7. Install the injectors.

DIESEL FUEL SYSTEM

Fuel System Service Precautions

Although Diesel fuel is not as flammable as gasoline, whenever working on or around diesel fuel or the fuel delivery system, heed the following precautions:

• Do not allow fuel spray or fuel vapors to come into contact with a heating element or open flame. Do not smoke while working on the fuel system.

• Always disconnect the negative battery cable unless the repair or test procedure requires that battery voltage be applied.

• Always relieve the fuel system pressure prior to disconnecting any fitting or fuel line connection.

• To control fuel spray when relieving system pressure, place a shop towel around the fitting prior to loosening to catch the spray. Ensure that all fuel spillage is quickly wiped up and that all fuel soaked rags are deposited into a proper fire safety container.

• Always keep a dry chemical (Class B) fire extinguisher near the work area.

• Always use a back-up wrench when loosening and tightening fuel line fittings.

• Do not re-use fuel system gaskets and O-rings. Replace with new ones. Do not substitute fuel hose where fuel pipe is installed.

Idle Speed

ADJUSTMENT

Diesel engines have both an idle speed and a maximum speed adjustment. The maximum speed adjustment is a high idle speed that prevents the engine from over-revving when the control lever is in the full speed position but there is no load on the engine. No increase in power is available through this adjustment. The control lever idle stop screw is no longer used for idle speed adjustment. The idle speed boost linkage includes an adjustment for basic idle speed.

1. Before servicing the vehicle, refer to the precautions in the beginning of this section.

2. If the vehicle has no tachometer, connect a suitable diesel engine tachometer as per the manufacturer's instructions.

3. Run the engine to normal operating temperature.

4. Be sure the manual cold start/idle speed boost knob is pushed in all the way.

5. Turn the linkage cap nut to adjust idle speed to 820–880 rpm, at a point where there is the least vibration.

6. Advance the control lever to full speed. The high idle speed is 5300–5400 rpm. Adjust as needed and secure the locknut with sealer.

Fuel Filter/Water Separator

DRAINING WATER

Although diesel fuel and water do not readily mix, fuel does tend to entrap moisture from the air each time it is moved from one container to another. Eventually every diesel fuel system collects enough water to become a potential hazard. Fortunately, when it's allowed to settle, the water will always drop to the bottom of the tank or filter housing. Some diesel fuel filters are equipped with a water drain; a bolt or petcock at the bottom of the housing.

At The Water Separator

1. Before servicing the vehicle, refer to the precautions in the beginning of this section.

2. Remove the fuel filler cap.

3. At the separator, connect a hose from the separator drain to a catch pan.

4. Open the drain valve (3 turns) and drain the separator until a steady stream of fuel flows from the separator, then close the valve.

➡ **Don't forget to install the filler cap.**

At The Filter

1. Before servicing the vehicle, refer to the precautions in the beginning of this section.

2. If the filter is equipped with a water drain at the bottom, place a pan under the drain to catch the water and fuel.

3. If equipped, loosen the vent bolt on the filter base. If there is no vent, loosen the

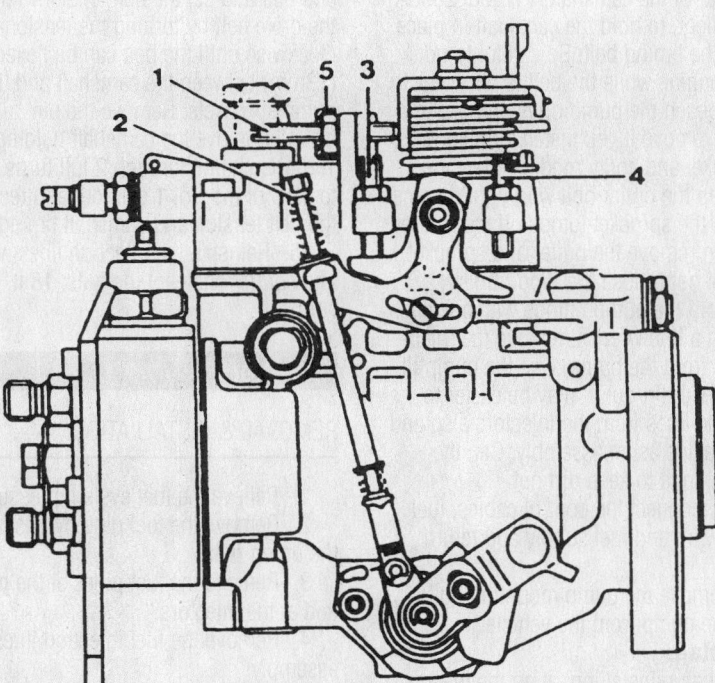

1. Previous idle adjustment screw
2. Linkage with cap nut for idle adjustment
3. Stop screw for minimum idle speed
4. Stop screw for idle speed boost
5. Tamper-proof cap

7923WG27

Low idle speed adjustment is made at the linkage cap—diesel engines

return line at the pump (the line not connected to the filter).

4. Loosen the bolt or valve. When fuel flows in a clean stream, close the drain and tighten the vent or return line.

REMOVAL & INSTALLATION

❊❊ WARNING

Do not allow diesel fuel to contact the coolant hoses. If this happens, wipe it off and wash the hoses with soap and water immediately.

1. Before servicing the vehicle, refer to the precautions in the beginning of this section.
2. Remove or disconnect:
   - Retaining clip (5).
   - Control valve from the filter with the fuel lines attached.
   - Hoses from connections (1) and (2).
   - Filter assembly.

To install:

3. Use a new O-ring and install the control valve on the filter.
4. Install the retaining clip (5).
5. Connect the hoses to connections (1) and (2) and secure them with clamps.

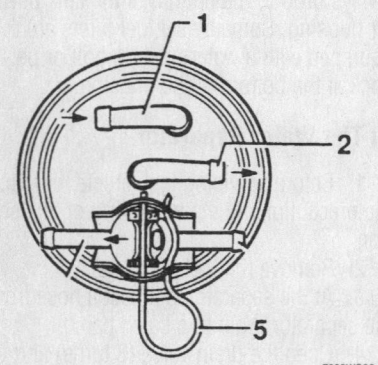

Fuel filter assembly—diesel engines

Diesel Injection Pump

REMOVAL & INSTALLATION

➡**Special tools are required for injection pump installation. Do not remove the pump without these tools on hand.**

1. Before servicing the vehicle, refer to the precautions in the beginning of this section.
2. Remove or disconnect:
   - Negative battery cable
   - Air cleaner

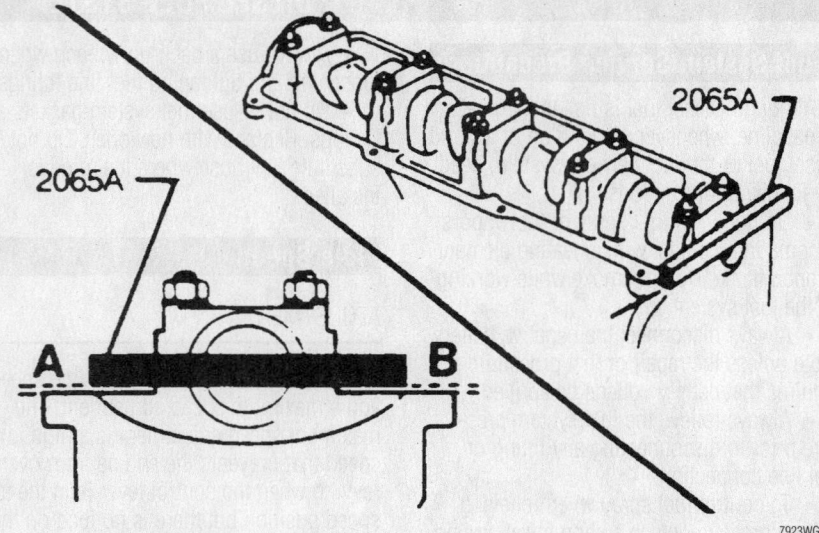

Install the bar to hold the camshaft in position during diesel injection pump service

- Cylinder head cover
- Timing belt cover.

3. Turn the engine to TDC of No. 1 cylinder and insert a setting bar into the slot on the rear of the camshaft, VW tool 2065A or equivalent, to hold the camshaft in place. Remove the timing belt. Be careful to not turn the engine while the belt is removed.
4. Loosen the pump drive sprocket nut but don't remove it yet. Install a puller on the sprocket and apply moderate tension.
5. Rap the puller bolt with light hammer taps until the sprocket jumps off the tapered shaft, then remove the puller and sprocket. Be careful not to lose the Woodruff key.
6. Hold the pump fittings with a wrench and using a line wrench, remove the injection lines from the pump. Cap the pump fittings to keep dirt out. It may be easier to remove the lines from the injectors also and set them aside as an assembly. Cap the injector fittings to keep dirt out.
7. Disconnect the control cables, fuel solenoid wire and fuel supply and return lines.
8. Remove the pump mounting bolts and lift the pump from the vehicle.

To install:

9. When reinstalling, align the marks on the top of the mounting flange and the pump. Mounting bolts: 18 ft. lbs. (25 Nm).
10. Install the Woodruff key and sprocket. Nut: 33 ft. lbs. (45 Nm).
11. When reinstalling the supply and return lines, be sure the fitting marked OUT is used for the return line. This fitting has an orifice and must be in the correct place. Use new gaskets.
12. Turn the pump sprocket so the mark aligns with the mark on the side of the mounting flange and insert a pin through the hole in the sprocket to hold it in place.
13. Install the camshaft drive sprocket and belt and set the belt tension. Tension the drive belt by turning the tensioner pulley clockwise until the belt can be flexed ½ in. (13mm) between the camshaft and the pump sprockets. Remove the pin.
14. Remove the camshaft holding bar. Turn the engine through 2 full turns, return to TDC of the No. 1 cylinder and recheck the belt tension and camshaft timing.
15. Reinstall the injection lines, wiring and control cables. Line nuts: 18 ft. lbs. (25 Nm).

Injectors

REMOVAL & INSTALLATION

1. Relieve the fuel system pressure.
2. Remove the fuel pipe by unscrewing the union nuts.
3. Remove the fuel pipes at the pump and at the injectors.
4. Remove the fuel injection lines as an assembly.
5. Unscrew the injector from the cylinder head.

To install:

➡**Replace the heat shields at each injector hole every time a new injector is installed.**

6. Screw the new injector into the cylinder head and tighten it.
7. Install the fuel line assembly.

Transaxle Assembly

REMOVAL & INSTALLATION

Manual

CABRIO, GOLF, AND JETTA

1. Before servicing the vehicle, refer to the precautions in the beginning of this section.
2. Remove or disconnect:
   - Negative battery cable.
   - Back-up light switch connector and the speedometer cable from the transaxle; plug the speedometer cable hole.
   - Upper engine-to-transaxle bolts.
   - 3 right side engine mount bolts, between engine and firewall.
   - Shift linkage as follows: Pry open the ball joint ends and remove the shift and relay shaft rods.
   - Center bolt from the left transaxle mount.
   - Front wheels. Connect the engine sling tool VW-10–222A or equivalent, to the loop in the cylinder head and just take the weight of the engine off the mounts. On 16V engine, the idle stabilizer valve must be removed to attach the tool. Do not try to support the engine from below.
   - Oil from the transaxle.
   - Left inner fender liner.
   - Halfshafts from the inner drive flanges and hang them from the body.
   - Clutch cover plate and the small plate behind the right halfshaft flange.
   - Starter and front engine mount.
   - Clutch cable and remove it from the transaxle housing.
   - Remaining transaxle mount bolts and mounts.
3. Place a jack under the transaxle and remove the last bolts holding it to the engine. Carefully pry the transaxle away from the engine and lower it from the vehicle.

To install:

4. Coat the input shaft lightly with molybdenum grease and carefully fit the transaxle in place. If necessary, put the transaxle in any gear and turn an output

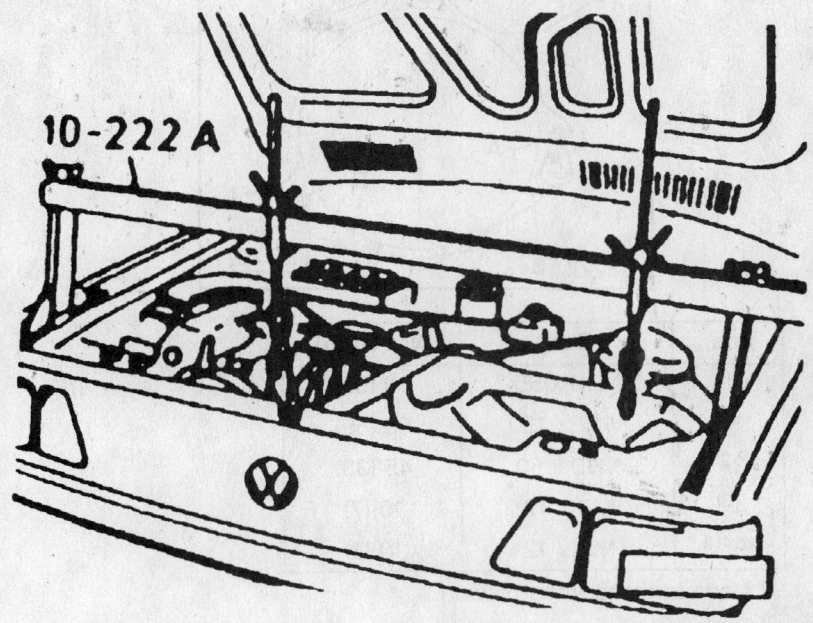

10-222A

Supporting the engine to remove the transaxle

7923WG30

flange to align the input shaft spline with the clutch spline.

5. Install the engine-to-transaxle bolts. Bolts: 55 ft. lbs. (75 Nm).

6. When installing the mounts to the transaxle, tighten the rear bracket-to-engine bolts and the transaxle support bolts to 18 ft. lbs. (25 Nm). Tighten the left bracket-to-transaxle bolts to 25 ft. lbs. (35 Nm) and the remaining mounting bolts to 44 ft. lbs. (60 Nm). Install but do not tighten the bolts that go into the rubber mounts.

7. Install the starter and front mount.

8. With all mounts installed and the transaxle safely in the vehicle, allow some slack in the lifting equipment. With the vehicle safely supported, shake the engine/transaxle as a unit to settle it in the mounts. Tighten all mounting bolts, starting at the rear and working forward. Tighten the bolts that go into the rubber mounts to 44 ft. lbs. (60 Nm).

9. Remove or disconnect:
   - Halfshafts. Bolts: 33 ft. lbs. (45 Nm).
   - Clutch cover plates.
   - Shift linkage and clutch cable and adjust as required.
   - Inner fender and complete the remaining installation.
   - Transaxle with oil.

NEW BEETLE

1. Before servicing the vehicle, refer to the precautions in the beginning of this section.
2. Remove or disconnect:
   - Engine cover and undercover.
   - Negative, then the positive battery cable.
   - Power steering reservoir from the battery tray and position it aside leaving the hoses attached.
   - Battery and tray.
   - Air cleaner and intake air hose.
   - Connector for the reverse light switch and VSS.
   - Gear selector cable from the transaxle.
   - Slave cylinder and position it side leaving the hydraulic hose attached.
   - Cable retainer near the starter.
   - Upper starter mounting bolt.
   - Ground strap at the engine-to-transaxle top mounting bolt.
   - Upper engine-to-transaxle mounting bolts.
3. Reposition any wiring or hoses that would interfere with installing an engine support tool.
4. Install a suitable engine support tool.
5. Remove or disconnect:
   - Starter.

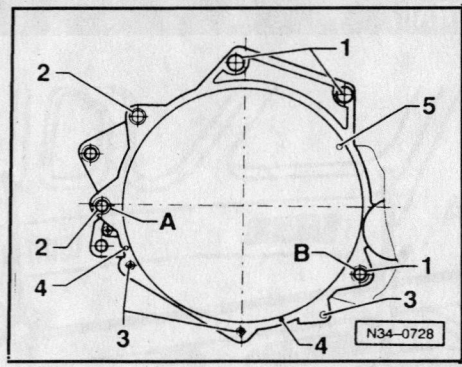

Pos.	Bolt	Nm (ft lb)
1	M12 x 55	80 (59)
2 [1]	M12 x 150	80 (59)
3 [2]	M10 x 50	45 (33)
4 [3]	M7 x 12	10 (7)
5 [4]	M7 x 12	10 (7)

[1] Also starter to transmission

[2] Only on engines with an aluminium oil pan.

[3] Large cover plate for flywheel only on engines with sheet steel oil pan (painted black)

[4] Small cover plate for flywheel

Pos. A + B = Dowel sleeves

Engine-to-transaxle (automatic and manual) mounting bolt torque specification and locations—New Beetle

- Halfshafts at the transaxle, after turning the wheels to the left lock, and tie them up out of the way.
- Flywheel cover plates.
- Front exhaust pipe from the manifold.
- Pendulum and transaxle support.
- Bolts for the transaxle mount.
- If necessary, right side cooling fan.

6. Install a suitable transaxle jack and support the weight of the transaxle.

7. Remove the lower engine-to-transaxle mounting bolts and lower the transaxle out of the vehicle.

To install:

8. Clean the input shaft and lubricate lightly with grease.

9. Raise the transaxle into the vehicle.

10. Install the lower engine-to-transaxle mounting bolts.

11. Remove the transaxle jack.

12. Install or connect:
- Upper engine-to-transaxle mounting bolts.
- Transaxle mount, using new bolts. Mounting bolt: 30 ft. lbs. (40 Nm) plus 90°.
- Pendulum and transaxle supports.
- Exhaust system.
- Install the flywheel cover plates.
- Halfshafts to the transaxle.
- Starter and clutch slave cylinder.

13. The remaining steps of the installation are the reverse of the removal, keeping in mind the following items:
- Attach any ground cables that were removed
- Remove the engine support bracket
- Be sure all connectors are attached
- Check the transaxle
- Install engine covers
- Read and clear any DTCs

1998–00 PASSAT

1. Before servicing the vehicle, refer to the precautions in the beginning of this section.

2. Remove or disconnect:
- Negative battery cable.
- Engine undercover.
- Front exhaust pipes from the engine. Loosen the U-bolt and push to the rear.
- Starter.
- Shift rod from the transaxle.
- Speed sensor and left back-up light connectors from the transaxle.

3. Support the transaxle with a jack.

4. Remove:
- Right and left transaxle mounts.
- Right and left halfshaft from the transaxle.

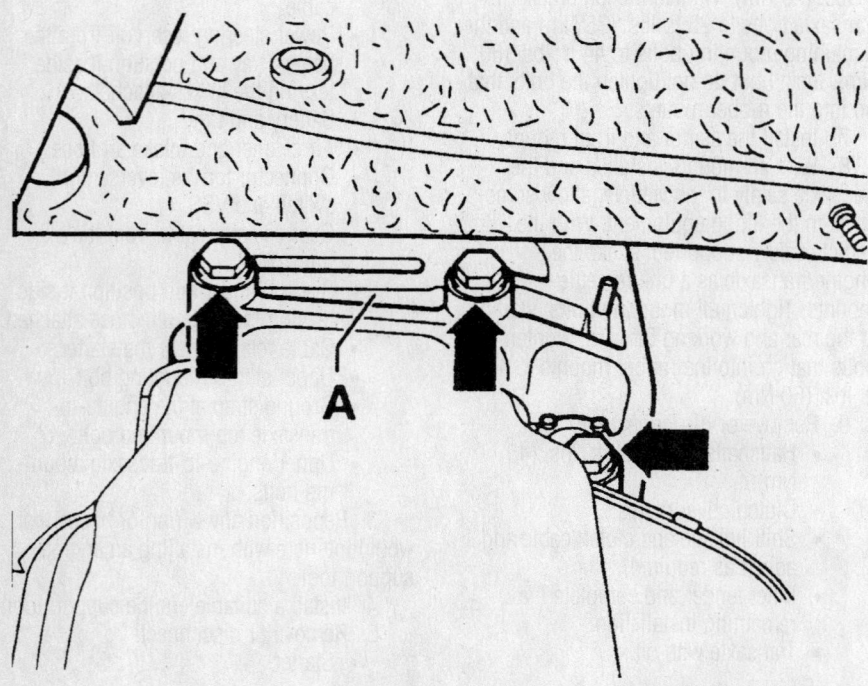

When installing the transaxle mount (A), be sure to install new bolts (arrows)—New Beetle

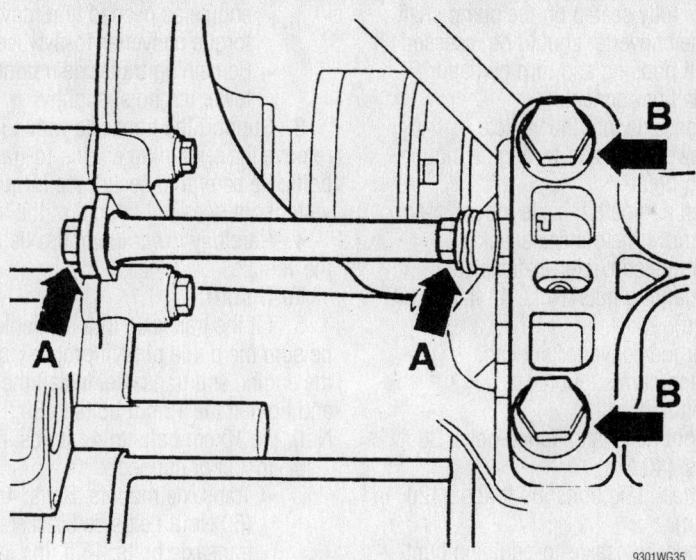

9301WG35

Be sure to tighten bolts (A) to 18 ft. lbs. (25 Nm) and bolts (B) to 44 ft. lbs. (60 Nm) plus 90°— New Beetle

- Halfshaft shield.
- Transaxle-to-engine mounting bolts.

5. Pry the transaxle from the engine and lower it about 6 in. (13cm) to access the slave cylinder.

6. Remove the slave cylinder with bracket without disconnecting the fluid line.

7. Lower and remove the transaxle assembly.

To install:

8. Clean the input shaft and apply a thin film of No. 000 100 high-performance grease or equivalent to the splines.

9. Lubricate the plunger contact surface on the release lever with Dow Corning® CU-7439 Plus Copper Paste or equivalent.

10. Raise the transaxle into position and install the slave cylinder. Mounting bolts: 18 ft. lbs. (25 Nm).

11. Install or connect:
- Transaxle-to-engine bolts. M8 bolts: 18 ft. lbs. (25 Nm), M10 bolts: 33 ft. lbs. (45 Nm) and the M12 bolts: 48 ft. lbs. (65 Nm).
- Transaxle mounts. Mounting bolts: 30 ft. lbs. (40 Nm).
- Halfshafts. M8 bolts: 33 ft. lbs. (45 Nm) and M10 bolts: 59 ft. lbs. (80 Nm).
- Halfshaft shield.
- Shift rod. Bolts: 15 ft. lbs. (20 Nm).
- Starter.
- Exhaust system.
- Engine undercover.
- Negative battery cable.

Automatic

CABRIO, GOLF, AND JETTA

1. Before servicing the vehicle, refer to the precautions in the beginning of this section.

2. Remove or disconnect:
- Battery and the speedometer drive and plug the hole in the transaxle.
- On the Golf and Jetta models, front axle nuts.

✻✻ CAUTION

When loosening or tightening an axle nut, be sure the vehicle is on the ground. Axle nut torque is high enough that attempting to loosen it may cause the vehicle to fall.

- Femove the front wheels. Connect the engine sling tool VW-10–222A or equivalent, to the cylinder head and just take the weight of the engine off the mounts. On 16V engine, the idle stabilizer valve must be removed to attach the tool. Do not try to support the engine from below.
- Driver's side rear transaxle mount and support bracket.
- On the Golf and Jetta models, front mount bolts from the transaxle and from the body and mount as a complete assembly.
- Selector and accelerator cables from the transaxle lever but leave them attached to the bracket.

Remove the bracket assembly to save the adjustment.
- Halfshafts from the drive flanges. On the Golf and Jetta models, the shafts must be removed, which may require separating the ball joints from the wheel bearing housing to gain the necessary clearance. Remove the ball joint clamping bolt.
- Heat shield and brackets and remove the starter. On the Cabrio models, the front mount comes off with the starter.
- Torque converter-to-flywheel bolts, turning the engine as needed.
- Remaining transaxle mounts, on the Golf and Jetta models, the subframe bolts and allow the subframe to hang free.

3. Support the transaxle with a jack and remove the remaining engine-to-transaxle bolts. Be careful to secure the torque converter so it does not fall out of the transaxle.

4. Carefully lower the transaxle from the vehicle.

To install:

5. When reinstalling, be sure the torque converter is fully seated on the pump shaft splines. The converter should be recessed into the bell housing and turn by hand. Keep checking that it still turns while drawing the engine and transaxle together with the bolts.

6. Install or connect:
- Engine-to-transaxle bolts. Bolts: 55 ft. lbs. (75 Nm).
- All mount and subframe bolts before tightening any on them. Tighten the bolts starting at the rear and work forward. Smaller bolts: 25 ft. lbs. (34 Nm) and larger bolts: 58 ft. lbs. (80 Nm). Remove the lifting equipment when all mounts are installed.
- Torque converter-to-flywheel bolts. Bolts: 26 ft. lbs. (35 Nm).
- Starter. Bolts: 14 ft. lbs. (20 Nm).
- Heat shields.

➥**If the halfshafts were removed, be sure the splines are clean and apply a thread-locking compound to the splines before sliding it into the hub.**

- Halfshafts to the drive flanges. Bolts: 37 ft. lbs. (50 Nm). Install new axle nuts but do not fully tighten them until the vehicle is on the ground.
- If removed, fit the ball joints to the

Timing belt service is covered in Section 3 of this manual

control arm. Clamping bolt: 37 ft. lbs. (50 Nm).
- Shift linkage. Adjust as required.

7. When assembly is complete and the vehicle is on its wheels, tighten the axle nuts to 195 ft. lbs. (265 Nm).

NEW BEETLE

1. Before servicing the vehicle, refer to the precautions in the beginning of this section.

2. Remove or disconnect:
- Engine cover and under cover.
- Negative, then the positive battery cable.
- Power steering reservoir from the battery tray and position it aside leaving the hoses attached.
- Battery and tray.
- Air cleaner and intake air hose.
- Connector for the reverse light switch and VSS.
- Electrical connections at the transaxle.
- Selector lever at the transaxle.
- Ground cable from the upper transaxle-to-engine mounting bolt.
- Starter.
- After clamping off the ATF cooler lines at the transaxle, the cooler.
- Upper transaxle-to-engine mounting bolts.

3. Install a suitable engine support fixture and slightly lift the engine.

4. Remove or disconnect:
- Left front wheel.
- All engine under covers
- Halfshafts at the transaxle.
- Right halfshaft and position it out of the way.
- Left halfshaft.
- Pendulum support.
- If necessary, right side cooling fan.
- Cap for the torque converter nut cover.
- 3 torque converter nuts.
- Left engine/transaxle mount.
- Transaxle jack, then support the weight of the transaxle.
- Lower transaxle-to-engine mounting bolts.

5. Separate the transaxle from the engine while pushing the torque converter out of the driveplate.

6. Push the torque converter against the ATF pump.

7. While lowering the transaxle, guide the power steering hoses past the transaxle.

8. Secure the torque converter to prevent it from falling out.

To install:

9. When reinstalling, be sure the torque converter is fully seated on the pump shaft splines. The converter should be recessed into the bell housing and turn by hand.

10. Install or connect:
- Transaxle into the vehicle.
- Lower transaxle-to-engine mounting bolts.
- Left engine/transaxle mount, then remove the transaxle jack.
- Torque converter-to-flexplate mounting nuts. Nuts: 44 ft. lbs. (60 Nm).
- Torque converter nut cap.
- Pendulum support.
- Left halfshaft.
- Right halfshaft. Flange bolts: 30 ft. lbs. (40 Nm).
- Wheel. Lug bolts: 89 ft. lbs. (120 Nm).
- Upper transaxle-to-engine mounting bolts.
- ATF cooler using new O-rings and remove the clamps.
- Starter.
- Ground cable at the upper transaxle mounting bolt.
- Transaxle selector lever.
- Electrical connections at the transaxle.
- Engine support fixture.
- Intake air hose and air cleaner assembly.
- Battery tray and battery.
- Power steering reservoir to the battery.
- Battery cables.
- All covers that were removed.

PASSAT

1. Before servicing the vehicle, refer to the precautions in the beginning of this section.

2. Remove or disconnect:
- Battery.
- Wiring from the transaxle.
- Upper engine-to-transaxle bolts.
- Front wheels. Connect the engine sling tool VW-10-222A or equivalent, to the cylinder head and just take the weight of the engine off the mounts. The idle stabilizer valve must be removed to attach the tool. Do not try to support the engine from below.
- Shift cable.
- Hoses at the transaxle cooler.
- Starter and the engine's left and right mounts.
- Skid plate
- Halfshafts from the drive flanges. Hang them from the body with wire.
- Torque converter plate and turn the

engine as needed to remove the torque converter-to-flywheel bolts.
- Remaining transaxle mounts and lower the hoist slightly.

3. Support the transaxle with a jack and remove the remaining engine-to-transaxle bolts. Be careful to secure the torque converter so it does not fall out of the transaxle.

4. Carefully lower the transaxle out of the vehicle.

To install:

5. Fit the transaxle into the vehicle and be sure the guide pins fit properly between the engine and transaxle. Install the bolts and tighten the 12mm bolts to 59 ft. lbs. (80 Nm), the 10mm bolts to 44 ft. lbs. (60 Nm).

6. Install or connect:
- Transaxle mounts. Bolts: 44 ft. lbs. (60 Nm). Left side bracket-to-transaxle bolts: 18 ft. lbs. (25 Nm).
- Torque converter bolts: 44 ft. lbs. (60 Nm).
- Halfshafts. Bolts: 33 ft. lbs. (45 Nm).
- Shift linkage. Adjust as required.

7. Install the remaining parts and check the fluid level in the transaxle.

Clutch

ADJUSTMENT

All vehicles use a hydraulic clutch release mechanism. No free-play adjustment is required or possible. If the clutch does not release or engage properly, bleed the system before moving on to more extensive repairs.

REMOVAL & INSTALLATION

1. Before servicing the vehicle, refer to the precautions in the beginning of this section.

2. Remove the transaxle.

3. Matchmark the flywheel and pressure plate if the pressure plate is going to be reused.

4. Gradually loosen the pressure plate bolts 1-2 turns at a time in a crisscross pattern to prevent distortion.

5. Remove the pressure plate and disc.

6. Check the clutch disc for uneven or excessive lining wear. Examine the pressure plate for cracking, scorching or scoring. Replace any questionable components.

To install:

7. Install the clutch disc and pressure plate with the springs on the disc towards the plate. Use an alignment tool to keep the clutch disc centered.

8. Gradually tighten the pressure plate-

to-flywheel bolts in a crisscross pattern. Bolts: 18 ft. lbs. (24 Nm).

9. Install the clutch release bearing.

10. Install the transaxle.

Hydraulic Clutch System

BLEEDING

1. Before servicing the vehicle, refer to the precautions in the beginning of this section.

2. The clutch and brakes share the same reservoir. Clean all dirt and grease from the cap to be sure no foreign substances enter the system.

3. Remove the cap and diaphragm and fill the reservoir to the top with the approved DOT 3 or 4 brake fluid. Fully loosen the bleed screw which is in the slave cylinder body next to the inlet connection.

4. At this point bubbles of air will appear at the bleed screw outlet. When the slave cylinder is full and a steady stream of fluid comes out of the slave cylinder bleeder, tighten the bleed screw.

5. Refill the reservoir and cap it. Exert a light load of about 20 lbs. (9 kg) to the slave cylinder piston by pushing the release lever towards the cylinder and loosen the bleed screw. Maintain a constant light load; fluid and any air that is left will be expelled through the bleed port. Tighten the bleed screw when a steady flow of fluid with no air is being expelled.

6. Fill the reservoir fluid level back to normal capacity, if necessary repeat Step 4.

7. Exert a light load to the release lever but do not open the bleeder screw as the piston in the slave cylinder will move slowly down the bore. Repeat this operation 2–3 times; the fluid movement will force any air left in the system into the reservoir. The hydraulic system should now be fully bled.

8. Check the operation of the clutch hydraulic system and repeat this procedure, if necessary.

Halfshaft

REMOVAL & INSTALLATION

Except Passat

✳✳ CAUTION

When loosening or tightening axle nuts, be sure the vehicle is on the ground. Axle nut torque is high enough that attempting to loosen it may cause the vehicle to fall.

1. Before servicing the vehicle, refer to the precautions in the beginning of this section.

2. Remove or disconnect:
- Front axle nut.
- Front wheels.
- Socket head bolts retaining the halfshaft to the transaxle flange.

3. Matchmark the installed position of the ball joint, then detach the ball joint from the control arm.

4. Remove the transaxle side of the halfshaft from the drive flange and secure it out of the way. Do not let it hang unsupported.

5. Push the halfshaft out of the hub. A wheel puller may be required.

To install:

6. Fit the halfshaft to the drive flange and install the bolts. It is not necessary to tighten them yet.

7. Apply a thread-locking compound to the outer ¼ in. (6mm) of the spline. Slip the spline through the hub and loosely install a new axle nut.

8. Assemble the front suspension, being careful to align the matchmarks.

a. On New Beetle models, tighten the ball joint bolts to 15 ft. lbs. (20 Nm) plus 90°.

b. On Cabrio, Golf, GTI, Jetta and Passat models, tighten the ball joint clamping bolt to 33 ft. lbs. (45 Nm).

9. Install the wheel and hold it to keep the axle from turning. Inner axle bolts: 33 ft. lbs. (45 Nm).

10. With the vehicle on the ground, tighten the axle nut as follows:
- New Beetle—221 ft. lbs. (300 Nm), loosen 1 turn, tighten to 37 ft. lbs. (50 Nm) then an additional 30°
- Cabrio, Golf, GTI, Jetta and Passat—66 ft. lbs. (90 Nm) plus 45°

11. Check and adjust the front wheel alignment.

PASSAT

✳✳ CAUTION

When loosening or tightening axle nut or bolt, be sure the vehicle is on the ground. Axle nut torque is high enough that attempting to loosen it may cause the vehicle to fall.

1. Remove the hub cap or center cap.

2. Loosen the hex collar bolt.

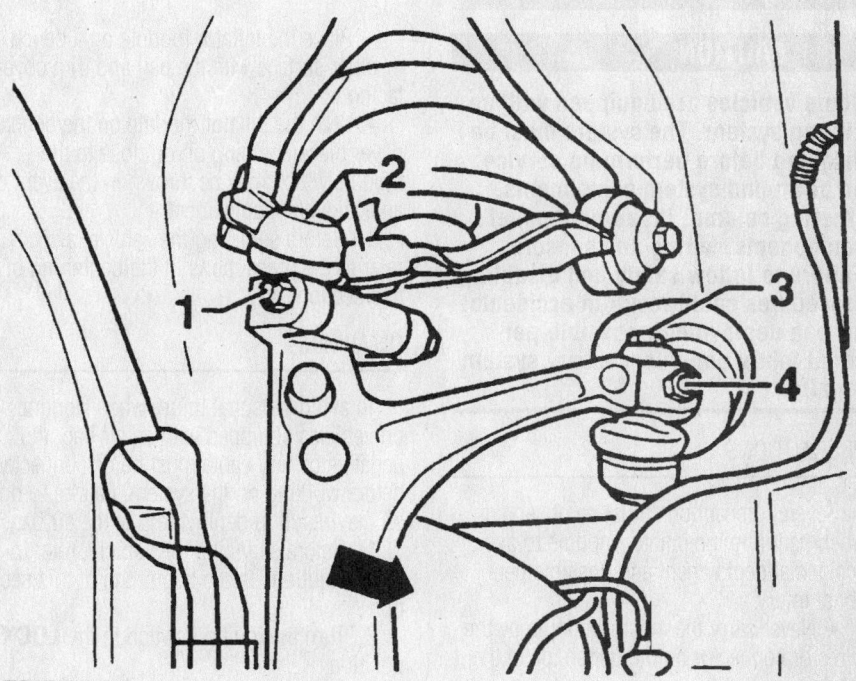

7923CG28

Loosen nut (1), remove the hex bolt, and pull both arms (2) upward and out—1998–00 Passat

Heater Core replacement is covered in Section 2 of this manual

3. Remove or disconnect:
- Front wheels.
- Halfshaft-to-transaxle flange bolts, then the hex collar bolt.
- ABS wheel speed sensor cable from the brake caliper bracket.

4. Slide the ABS speed sensor partly out of its mount.

5. Remove nut/bolt No. 1, as shown, then pull both arms up and out of the swing arm.

➡ **The slots in the swing arm must not be widened. Do not loosen the bolts No. 3 and 4, otherwise the axle geometry must be checked.**

6. Tilt the swing arm out and to the rear of the vehicle, then remove the halfshaft.

To install:

7. Install or connect:
- Halfshaft into the wheel hub.
- Swing arm bolt. Bolt: 30 ft. lbs. (40 Nm).
- Halfshaft-to-transaxle flange. M8 bolts: 30 ft. lbs. (40 Nm) and M10 bolts: 57 ft. lbs. (77 Nm).
- ABS wheel speed sensor, and the sensor cable into the caliper bracket.

8. With the wheels installed and the vehicle on the ground, tighten the axle bolt as follows: M14 bolt 85 ft. lbs. (115 Nm) plus an additional ¼ (90°) turn, M16 bolt 140 ft. lbs. (190 Nm) plus an additional ¼ (90°) turn.

CV-JOINT AND BOOT OVERHAUL

The constant velocity joints (CV-joints) can be disassembled for cleaning and inspection but they cannot be repaired. All parts are machined to a matched tolerance and the entire CV-joint must be replaced as a unit. On Golf and Jetta, the CV-joints are different on the left and right sides and cannot be interchanged.

1. Raise and safely support the vehicle and remove the halfshaft.

2. Pry open and remove the boot clamps with a pair of wire cutters.

3. With the halfshaft securely clamped in a vise, the outer CV-joint can be removed by sharply rapping out on the joint with a plastic hammer. The joint will snap off of the circlip and slide off the axle.

4. To remove the inner joint, remove the circlip from the center and slide the joint off the axle.

5. Both boots can be removed after removing the CV-joint

To install:

6. Always replace both circlips and make sure the CV-joint is clean before installation. Wrap a piece of black electrical tape around the shaft splines and slip the inner clamp and the boot onto the shaft.

7. Remove the tape and install the dished washer with the concave side out so it acts as a spring pushing the CV-joint out. On the outer joint, install the thrust washer and a new circlip.

8. To install the outer joint, place it onto the spline and carefully tap straight in on the end with a plastic hammer. The joint will click into place over the circlip.

9. To install the inner joint, slide it onto the spline and push in enough to allow the circlip to fit into the groove in the axle shaft.

10. Pack the CV-joint with special CV-joint grease. DO NOT use any other type of grease.

11. Pack any remaining grease into the boot and install the clamps on the outer boot.

STEERING AND SUSPENSION

Air Bag

※ CAUTION

Some vehicles are equipped with an air bag system. The system must be disabled before performing service on or around system components, steering column, instrument panel components, wiring and sensors. Failure to follow safety and disabling procedures could result in accidental air bag deployment, possible personal injury and unnecessary system repairs.

PRECAUTIONS

Several precautions must be observed when handling the inflator module to avoid accidental deployment and possible personal injury.

- Never carry the inflator module by the wires or connector on the underside of the module.
- When carrying a live inflator module, hold securely with both hands, and ensure that the bag and trim cover are pointed away.
- Place the inflator module on a bench or other surface with the bag and trim cover facing up.
- With the inflator module on the bench, never place anything on or close to the module which may be thrown in the event of an accidental deployment.

1. Before servicing the vehicle, also refer to the precautions in the beginning of this section.

DISARMING

To avoid personal injury when working on vehicles equipped with an air bag, the negative battery cable must be disconnected before working on the system. Failure to do so may result in deployment of the air bag.

1. Before servicing the vehicle, refer to the precautions in the beginning of this section.

2. Turn the ignition switch to the **LOCK** position.

3. Disconnect the negative battery cable.

4. Wait 10 minutes for the battery back-up power to discharge.

Rack and Pinion Steering Gear

REMOVAL & INSTALLATION

Except Passat

1. Before servicing the vehicle, refer to the precautions in the beginning of this section.

2. Remove or disconnect:
- Front wheels and disengage both tie rod ends.
- Low pressure (suction) hose from the pump and drain the system into a catch pan.
- At the steering column, the boot clamp, push the boot towards the body and remove the clamp bolt from the universal joint.
- On Cabrio models, exhaust manifold and shift linkage bracket.
- Rack mounting clamp nuts and clamps.

3. At this point on some vehicles, the rack cannot be removed from the body. Support the engine/transaxle and remove the subframe bolts to allow the rack to move

towards the rear. On Cabrio models, remove the transaxle mount and bracket.

4. Disconnect the power steering hydraulic lines and remove the rack toward the right.

To install:

5. Be sure the mounting bushings are in good condition. Fit the rack assembly into place and tighten the clamp nuts as follows:
- Except New Beetle: 22 ft. lbs. (32 Nm)
- New Beetle: 15 ft. lbs. (20 Nm) plus 90°

6. Install any subframe bolts that were removed.

7. Connect the hydraulic lines and install the steering column universal joint bolt.

8. Fill the system with new fluid and run the engine to check for leaks and bleed the system.

1998–00 Passat

1. Before servicing the vehicle, refer to the precautions in the beginning of this section.

2. Remove or disconnect:

- Battery, then the battery box.
- Bolt at the steering column U-joint.

3. Release the eccentric by turning the T50 Torx® bolt clockwise, then remove the bolt.

4. Before removing the steering column form the steering gear, secure the steering column with safety wire.

✳✳ WARNING

Be sure to lock the steering wheel, otherwise the air bag unit coil spring may be damaged.

5. Lock the steering wheel in the center position and do not move during the repairs.

➡**The splines between the top and bottom part of the steering column must not be separated.**

6. Move the U-joint down and out of the way.

7. Using Hose Clamps 3094 or equivalent, pinch off the suction and return lines to the steering gear.

8. Remove or disconnect:

- Front wheels.
- Left and right tie rods.
- Tie rod opening cover.

➡**Place a drip tray under the vehicle to catch any residual power steering fluid.**

- Banjo bolts for the steering gear suction and return hydraulic hoses.
- Steering gear mounting bolts.
- Steering gear through the left side wheel opening.

To install:

9. Remove the screw plug to lock the steering gear in the center position with Locking tool VAG 1907 or equivalent, and tighten to 13 ft. lbs. (18 Nm).

10. Insert the steering gear into the vehicle through the left side wheel opening.

11. Hand-tighten mounting bolts 1 and 2.

12. Install bolt 3 and tighten to 48 ft. lbs. (65 Nm), then tighten bolts 1 and 2 to 48 ft. lbs. (65 Nm).

13. Using new sealing gaskets, install the return hose. Banjo bolt: 37 ft. lbs. (50 Nm). Suction hose banjo bolt: 30 ft. lbs. (40 Nm).

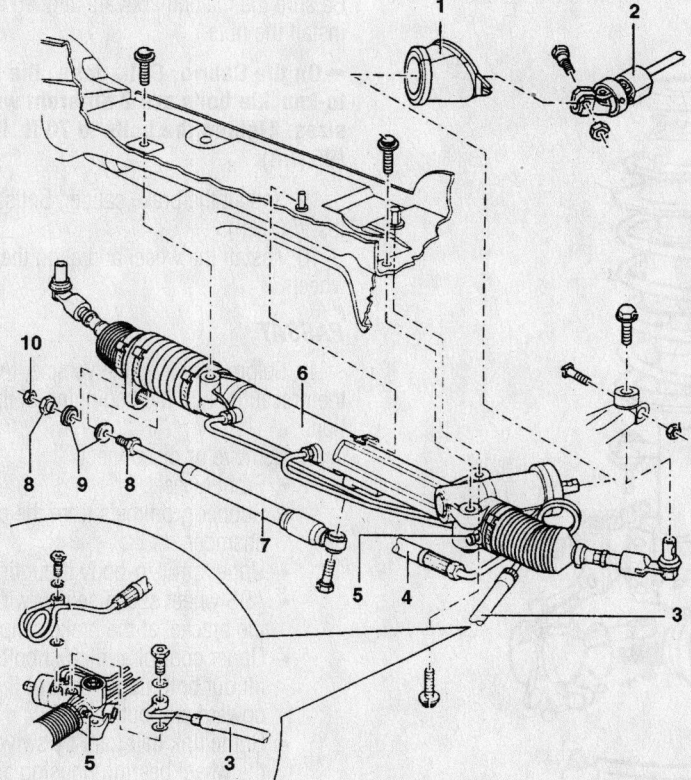

1. Boot seal
2. Steering column
3. Return hose
4. Flexible hose
5. Screw plug for centering the steering wheel
6. Rack and pinion steering gear
7. Steering damper
8. Bushing
9. Two-piece rubber bushing
10. Nut

7923CG33

Exploded view of the steering gear mounting—Passat

14. Connect the left and right tie rods. Mounting through-bolt: 33 ft. lbs. (45 Nm).

15. Install the tie rod opening cover.

16. Fit the J-joint to the steering gear, then insert the Torx® adjusting bolt by turning it clockwise.

17. Remove the Locking tool VAG 1907, then reinstall the screw plug and tighten to 13 ft. lbs. (18 Nm).

18. Tighten the adjusting bolt nut to 30 ft. lbs. (40 Nm).

19. Remove the steering wheel lock.

20. Remove the hose clamps 3094, and check the hydraulic fluid.

21. Install the battery tray, then connect the battery.

Strut

REMOVAL & INSTALLATION

Front

EXCEPT PASSAT

The upper strut-to-steering knuckle bolt may have an eccentric washer for adjusting wheel camber. Use a wire brush to clean the area and use a cold chisel to mark a fine line on the washer and the strut together.

Cut away socket for removing the upper strut nut

This matchmark may be enough to preserve the front wheel camber adjustment. It will at least be accurate enough to allow driving the vehicle until a proper front wheel alignment can be performed. If there is no eccentric washer, a new bolt and eccentric washer can be substituted. The parts are available through the dealer.

A special tool is required to remove the upper strut nut. If necessary, it can be made by cutting away part of a 22mm socket.

1. Before servicing the vehicle, refer to the precautions in the beginning of this section.

2. Remove or disconnect:
   • Front wheels.
   • If equipped, ABS wheel speed sensor.
   • Brake line from the strut and

remove the caliper. DO NOT let the caliper hang by the hydraulic line, hang it from the body with wire.
   • On the New Beetle, strut-to-wheel bearing housing pinch bolt and, slightly, spread the joint.

3. Clean and matchmark the position of the strut-to-steering knuckle bolt.

4. Remove the bolts and push the steering knuckle down away from the strut. Support the knuckle so it is not hanging on the outer CV-joint.

➡**On the New Beetle, it may be necessary to remove the wiper arms and external plenum chamber to access the upper strut mounting.**

5. Use a hex wrench to hold the shock absorber rod and use the cut away socket to remove the upper nut. Lower the strut from the vehicle.

   To install:

6. Place the strut into the fender and install the nuts. Install a new center nut and tighten it to 44 ft. lbs. (60 Nm).

7. On the New Beetle, tighten the pinch bolt to 37 ft. lbs. (50 Nm) plus 90°.

8. On the Cabrio, Golf, Jetta, fit the knuckle into the strut and install the bolts. Be sure the matchmarks are aligned and install the nuts.

➡**On the Cabrio, Golf, Jetta, the strut-to-knuckle bolts are 2 different wrench sizes. Tighten the bolts to 70 ft. lbs. (95 Nm).**

9. Install the brake caliper. Bolts: 44 ft. lbs. (60 Nm).

10. Install the wheel and align the front wheels.

PASSAT

1. Before servicing the vehicle, refer to the precautions in the beginning of this section.

2. Remove or disconnect:
   • Front wheels.
   • Rubber grommets from the plenum chamber.
   • Upper strut-to-body mounting nuts.
   • ABS wheel speed sensor wire from the bracket at the brake caliper.
   • Upper control arm pinchbolt, then lift out both upper control links upward and out.
   • Guide link ball joint by swiveling the wheel bearing housing aside.
   • Lower strut mounting bolt.

➡**When removing the strut be sure not to damage the CV joint boot.**

3. Remove the strut downward.

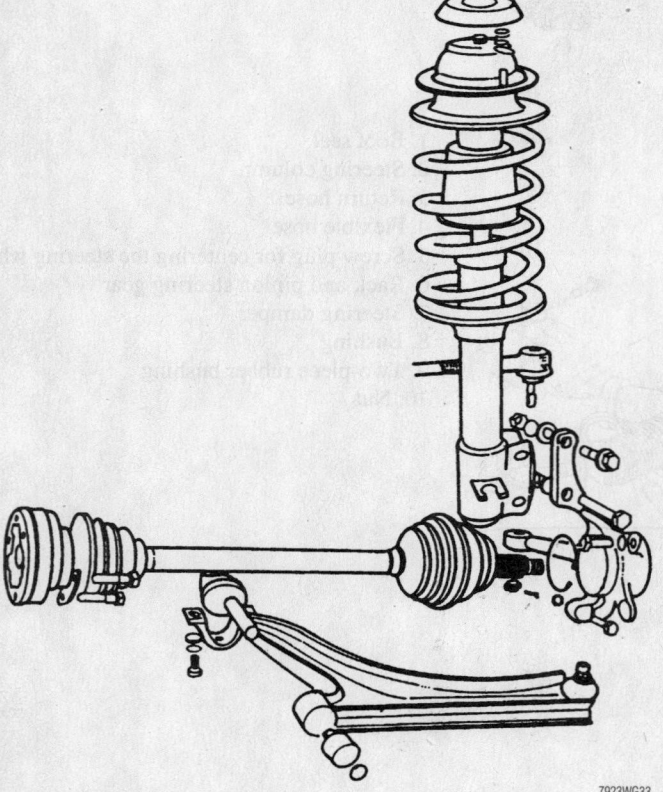

The MacPherson strut is mounted between the steering knuckle and body

To install:

✳✳ WARNING

The bonded rubber bushing can only turned to a limited extent. The bolted connections between the suspension strut and the lower track control links should therefore only be tightened when the vehicle is standing on the ground.

4. Install or connect:
- Strut into the vehicle and position it so that the hole in the spring plate faces the middle of the vehicle.
- Lower mounting bolt. Bolt: 66 ft. lbs. (90 Nm).
- Upper control links to the wheel bearing housing. Pinchbolt: 30 ft. lbs. (40 Nm).

➡ **It may be necessary to hold the ball joint stud with a 4mm hex wrench.**

- Ball joint nut. Nut: 74 ft. lbs. (100 Nm).
- ABS wheel speed sensor wire into the holder at the brake caliper.
- Upper strut-to-body mounting nuts. Nuts: 15 ft. lbs. (20 Nm).
- Rubber grommets into the plenum chamber.
- Wheels. Lug bolts to 89 ft. lbs. (120 Nm).

Rear

✳✳ WARNING

Do not remove both suspension struts at the same time or the axle beam will be hanging on the brake lines.

1. Before servicing the vehicle, refer to the precautions in the beginning of this section.
2. On Cabrio, Golf, Jetta, perform the following:
   a. Working inside the vehicle, remove the cap from the top shock mount and remove the top nut, washer and rubber bushings.
   b. Remove the second nut.
   c. Slowly lift the vehicle and safely support it. Do not place supports under the axle beam.
3. On the New Beetle and 1998–00 Passat, raise the vehicle and remove the upper mounting bolts through the wheel opening.
4. Unbolt the strut from the axle and carefully remove the strut and spring from

the vehicle. It may be necessary to press the axle down slightly.

To install:
5. Install the shock on the axle assembly. Do not tighten the nut until the vehicle is on the floor at normal riding height.
6. On the Cabrio, Golf, Jetta, perform the following:
   a. Install the upper end of the strut to the body. Lower nut: 11 ft. lbs. (15 Nm) and Upper nut: 18 ft. lbs. (25 Nm).
   b. Install the wheel and lower the vehicle to the floor.
   c. Tighten the lower strut mounting nut to 52 ft. lbs. (70 Nm).
7. On the New Beetle and 1998–00 Passat, perform the following:
   a. Install the upper end of the strut to the body and tighten the attaching bolt to 55 ft. lbs. (75 Nm).
   b. Install the wheel.
   c. Tighten the lower strut mounting nut to 44 ft. lbs. (60 Nm).

Coil Spring

REMOVAL & INSTALLATION

1. Before servicing the vehicle, refer to the precautions in the beginning of this section.
2. Remove the strut from the vehicle.
3. Clamp the Spring Compressor VAG 1752/2 or equivalent in a vise.
4. Install the strut into the spring compressor.
5. Pry off the mounting bolt cap.
6. Compress the coil spring and remove the self-locking nut from the piston rod.
7. Matchmark the position of the spring retainer and spring mount.
8. Remove:
- Spring seat and related components noting the order of removal.

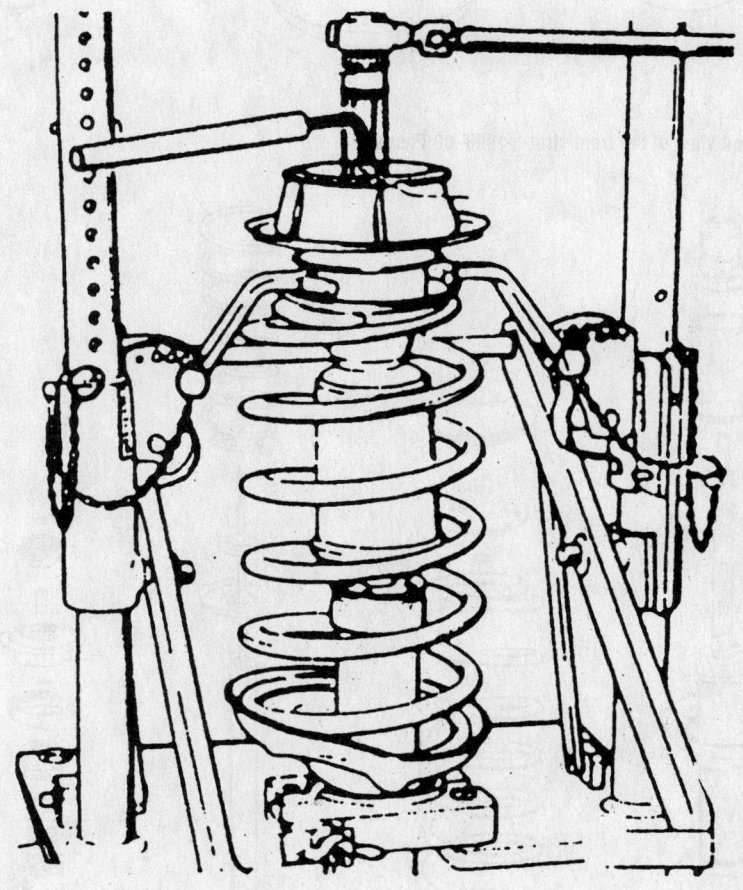

7923WG35

Compress the coil spring before removing the upper strut rod nut

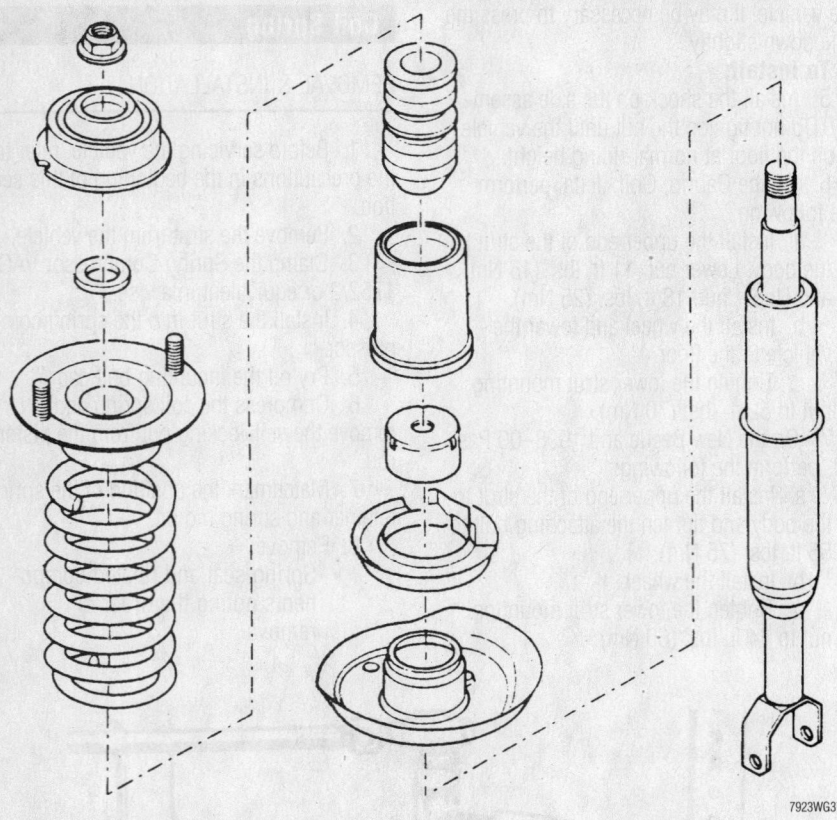

Exploded view of the front strut—1998–00 Passat

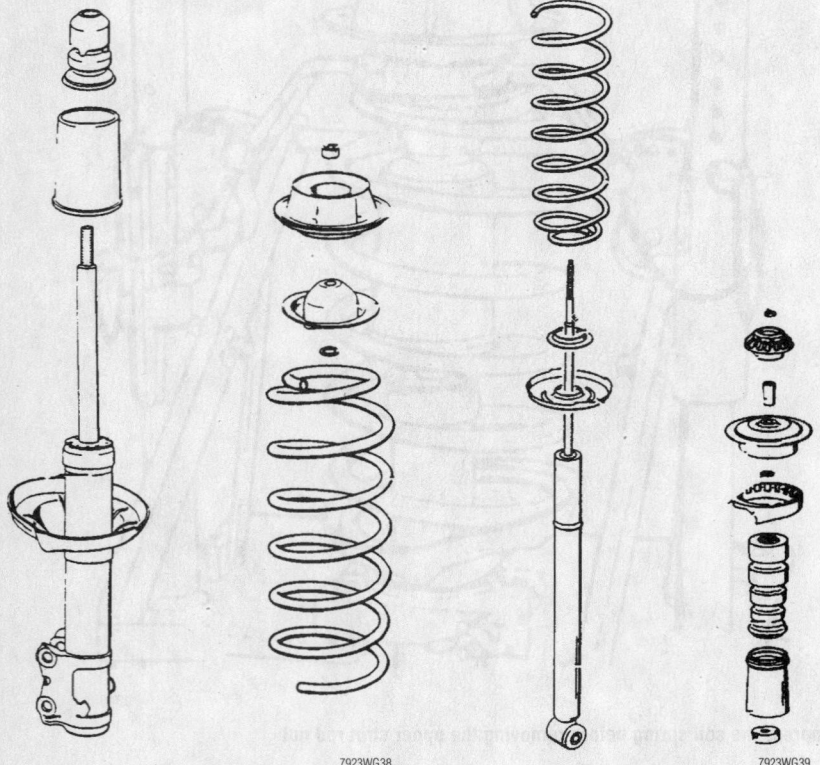

Exploded view of the front strut—except passat

Exploded view of the rear strut—except Passat

- Strut from the spring compressor.
- Spring out of the compressor.

To install:

9. Install the new spring into the compressor.

10. Compress the spring and insert the strut through the spring.

11. Install the spring seat and related components in the reverse order as they were removed and aligning the matchmarks.

12. Install a new self-locking nut.

13. Reinstall the mounting bolt cap.

14. Release the spring compressor and install the strut into the vehicle.

Upper Ball Joint

REMOVAL & INSTALLATION

Passat

The 1998–00 Passat front suspension is equipped with 2 separate upper ball joints that are not replaceable, the upper link (front or rear) must be replaced as follows:

1. Before servicing the vehicle, refer to the precautions in the beginning of this section.

2. Remove or disconnect:
- Front wheels.
- Clip No. 1 as shown. The clip does not have to be replaced.
- Pinchbolt and pull both control arms upward and out.

3. Cover the steering gear boot.

4. Remove or disconnect:
- Guide link ball joint and press off the joint.
- ABS wheel speed sensor wire front the bracket on the brake caliper.

5. Support the suspension from excessive rebound travel.

6. Remove or disconnect:
- Lower strut mounting bolt and swing the wheel bearing housing aside.
- Rubber grommets from the plenum chamber.
- Upper strut-to-body mounting nuts.
- Strut together with the mounting bracket.

7. Clamp the strut in a vise with the protective jaw covers.

8. Remove:
- Upper link bolts and detach both of the links.
- Bracket-to-strut mounting nuts, then separate.

To install:

9. Position the brackets and links as shown, and tighten the bracket-to-strut mounting nuts to 15 ft. lbs. (20 Nm).

10. Align the links as shown, then tighten to 37 ft. lbs. (50 Nm) plus ¼ turn (90°).

11. Install or connect:
- Strut with mounting bracket into the vehicle. Upper strut-to-body mounting nuts: 48 ft. lbs. (75 Nm).
- Lower strut mounting bolt: 66 ft. lbs. (90 Nm).
- Nut on the ball joint: 74 ft. lbs. (100 Nm).
- Upper links to the wheel bearing housing. Pinchbolt: 30 ft. lbs. (40 Nm).
- ABS wiring to the brake caliper bracket.
- Wheel.

Lower Ball Joint

REMOVAL & INSTALLATION

1. Before servicing the vehicle, refer to the precautions in the beginning of this section.

2. Remove:
- Front wheels.
- Ball joint clamping bolt.
- Ball joint from the steering knuckle, by prying the lower control arm down.
- Ball joint-to-lower control arm retaining nuts and bolts or drill out the rivets with a ¼ in. (6mm) drill.
- Ball joint.

To install:

3. Install the new ball joint in the reverse order of removal. If no parts were installed other than the ball joint, no camber adjustment is necessary. Tighten the 2 control arm-to-ball joint bolts to 18 ft. lbs. (25 Nm) and the ball joint clamping bolt to 37 ft. lbs. (50 Nm).

Lower Control Arm

REMOVAL & INSTALLATION

Fox

1. Disconnect the negative battery cable.

2. Raise the vehicle and support it on jackstands. Remove the wheels.

3. Remove the nut and bolt attaching the ball joint to the hub (wheel bearing housing), then pry the joint down and out of the hub.

4. Remove the stabilizer bar.

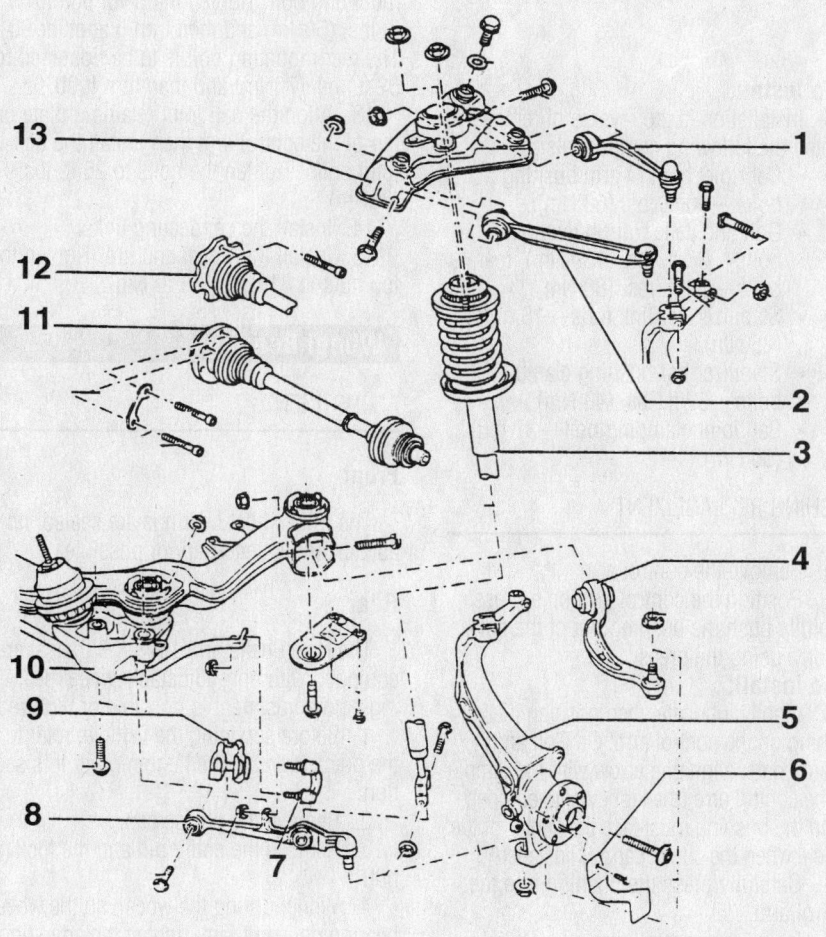

1. Upper link, rear
2. Upper link, front
3. Suspension strut
4. Guide link
5. Wheel bearing housing
6. Splash shield
7. Connecting link
8. Lower track control link
9. Clamp
10. Subframe
11. Halfshaft w/CV joint
12. Halfshaft w/triple-rotor joint
13. Mounting bracket

Exploded view of the front suspension—1998–00 Passat

7923WG41

91227P28

Removal of the lower control arm

5. Detach the control arm mounting bolts from the frame.

6. Remove the bolts securing the control arm, then remove the arm.

To install:

7. Installation is the reverse of removal. Observe the following torque's:

- Fox control arm-to-subframe bolts—50 ft. lbs. (37 Nm).
- Fox ball joint-to-hub bolt—25 ft. lbs. (34 Nm) for M8 nuts or 36 ft. lbs. (49 Nm) for M10 nuts.
- Stabilizer bar link rods—18 ft. lbs. (25 Nm).
- Stabilizer bar bushing clamp bolts—32 ft. lbs. (43 Nm).

Cabrio, Golf and Jetta

When removing the driver's side control arm on Golf and Jetta equipped with an automatic transaxle, it may be necessary to lift the engine/transaxle. First support the engine from above or below. Remove the front left engine mounting nut and bolt, remove the rear mount and raise the engine to expose the front control arm bolt.

1. Raise and safely support the vehicle and remove the wheels.

2. Remove the ball joint clamping bolt and pry the control arm down.

3. Remove the rubber bushings to unfasten the stabilizer bar.

4. Remove the control arm mounting bolts and remove the control arm.

To install:

5. Installation is the reverse of removal. Tighten the following components:

- Cabriolet control arm bushing bolts—50 ft. lbs. (68 Nm).
- Golf and Jetta front bushing bolts—96 ft. lbs. (130 Nm), rear bolts—59 ft. lbs. (80 Nm).
- Stabilizer bar link rods—18 ft. lbs. (25 Nm).
- Stabilizer bar bushing clamp bolts—32 ft. lbs. (43 Nm).
- Ball joint clamping bolt—37 ft. lbs. (50 Nm).

BUSHING REPLACEMENT

1. Remove the control arm.

2. Position the control arm on a press. Carefully push the bushing out of the control arm using the press.

To install:

3. Lightly lubricate, then position the bushing on the control arm. On Golf and Jetta models, align one arrow with the dimple on the control arm (the kidney shaped opening in the bushing must face the center of the vehicle when the control arm is installed).

4. Carefully press the bushing into control arm.

5. Install the control arm.

Cabrio, Golf and Jetta

1. Chock the rear wheels and then lift the vehicle supporting it on jack stands.

2. Remove the front wheel.

3. Base suspension models require the removal of the connecting link to the control arm.

4. Plus suspension models require the removal of the nut that attaches the stabilizer bar to the control arm.

5. Separate the link from the arm.

6. Matchmark the correct installed position of the ball joint into the control arm.

7. Remove the ball joint.

8. Remove the pivot bolt from the control arm.

9. Remove the rear control arm mounting bolt.

10. Pull the rear control arm from the vehicle.

To install:

11. Install the control arm and slide the ball joint into it.

12. Push the control arm pivot bolt through the control arm. Also install the rear mounting bolt. Tighten the pivot bolt to 37 ft. lbs. (50Nm) and then turn it another 90°. The rear mounting bolt is to be tightened to 52 ft. lbs. (70Nm) and then turn it 90°.

13. Align the ball joint retaining plate on top of the control arm then install the ball joint bolts. Tighten the bolts to 26 ft. lbs. (35Nm).

14. Install the connecting links.

15. Install the wheel and tire. Tighten the lug nuts to 81 ft. lbs. (110 Nm).

Wheel Bearings

ADJUSTMENT

Front

The front wheel bearings are sealed, no adjustment is necessary or possible.

Rear

The New Beetle and 1998–00 Passat are equipped with non-adjustable wheel bearings, no adjustment is possible or required.

1. Before servicing the vehicle, refer to the precautions in the beginning of this section.

2. Remove the grease cap.

3. Remove the cotter pin and the locking nut.

4. While turning the wheel, so the wheel bearing does not jam, tighten the adjusting nut firmly.

5. Back the nut off slightly. The nut is properly adjusted when it is possible to pry the thrust washer side to side with some drag by using finger pressure on the tool.

6. Install the locking nut and a new cotter pin. When installing the cap, be sure it is securely in place.

REMOVAL & INSTALLATION

Front

EXCEPT PASSAT

➡ **The hub and bearing are pressed into the knuckle and the bearing cannot be reused once the hub has been removed.**

1. Before servicing the vehicle, refer to the precautions in the beginning of this section.

2. With the vehicle on the ground, remove the front axle nut.

3. Remove the steering knuckle.

4. Clamp the upper knuckle-to-strut bolt boss in a vice.

5. Install the special press tool onto the hub as shown and press the hub out of the bearing.

6. If the inner bearing race stayed on the hub, clamp the hub in a vise and use a bearing puller to remove it.

7. On the knuckle, remove the splash shield and internal snaprings from the bearing housing.

8. After removing the snapring, the same press tool can be used to push the bearing out of the knuckle.

9. Clean the bearing housing and hub with a wire brush and inspect all parts. Replace parts that have been distorted or discolored from heat. If the hub is not absolutely prefect where it contacts the inner bearing race, the new bearing will fail quickly.

To install:

10. The new bearing is pressed in from the hub side using a regular arbor press. Install the snapring and support the steering knuckle on the press.

11. Using the old bearing as a press tool, press the new bearing into the housing up against the snapring. Be sure the press tool contacts only the outer race of the bearing.

12. Install the outer snapring and splash shield. If removed, install the speed sensor rotor onto the hub.

13. Support the inner race on the press and press the hub into the bearing. Be sure the inner race is supported or the bearing fail quickly.

14. Install the steering knuckle and be sure to tighten the axle nut correctly before allowing the vehicle to roll.

PASSAT

1. Before servicing the vehicle, refer to the precautions in the beginning of this section.

2. Loosen the halfshaft retaining bolt.

3. Remove or disconnect:
   • Front wheel.
   • ABS wheel speed sensor.

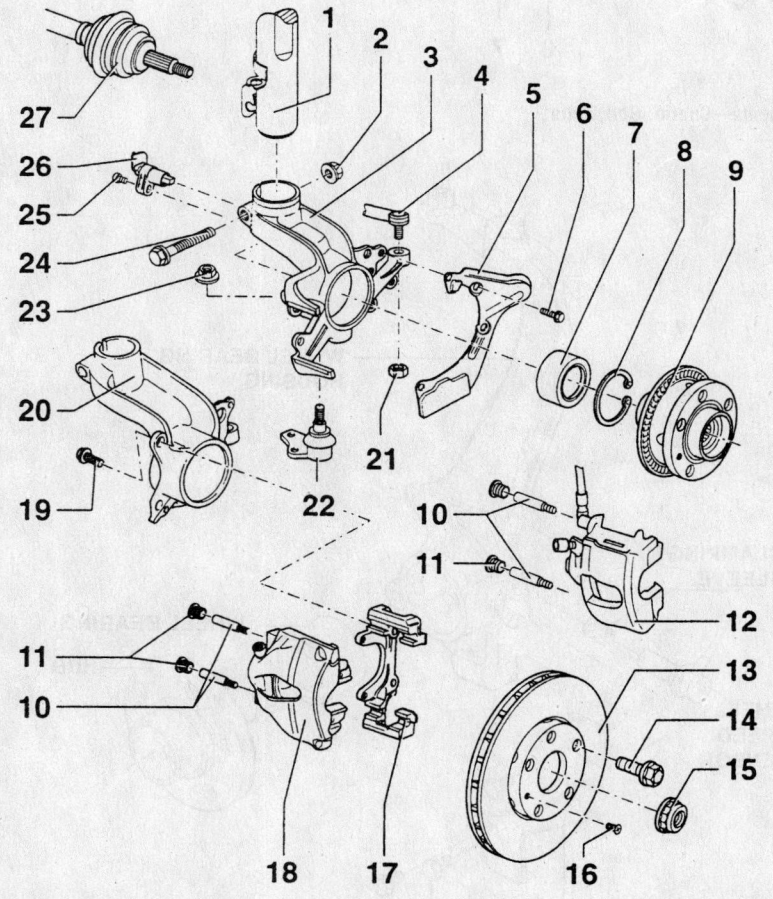

1. Strut
2. Self-locking nut (37 ft. lbs. (50 Nm) plus 90 degrees)
3. Wheel bearing housing (1.9L and 2.0L engines)
4. Tie rod end
5. Splash plate
6. Bolt (7 ft. lbs. (10 Nm))
7. Wheel bearing
8. Snapring
9. Wheel hub with ABS wheel speed sensor rotor
10. Guide pins (20 ft. lbs. (28 Nm))
11. Cap
12. Brake caliper (1.9L and 2.0L engines)
13. Brake disk
14. Wheel bolts (89 ft. lbs. (120 Nm))
15. Self-locking 12-point nut
16. Phillips-head screw
17. Brake carrier
18. Brake caliper (1.8L engine)
19. Self-locking bolt (92 ft. lbs. (125 Nm))
20. Wheel bearing (1.8L engine)
21. Self-locking nut (33 ft. lbs. (45 Nm))
22. Ball joint
23. Self-locking nut (33 ft. lbs. (45 Nm))
24. Bolt
25. Bolt (71 inch lbs. (8 Nm))
26. ABS wheel speed sensor
27. Halfshaft

9301WG36

Exploded view of the front wheel bearing and related components—New Beetle

For Tire, Wheel and Ball Joint specifications, see Section 1 of this manual

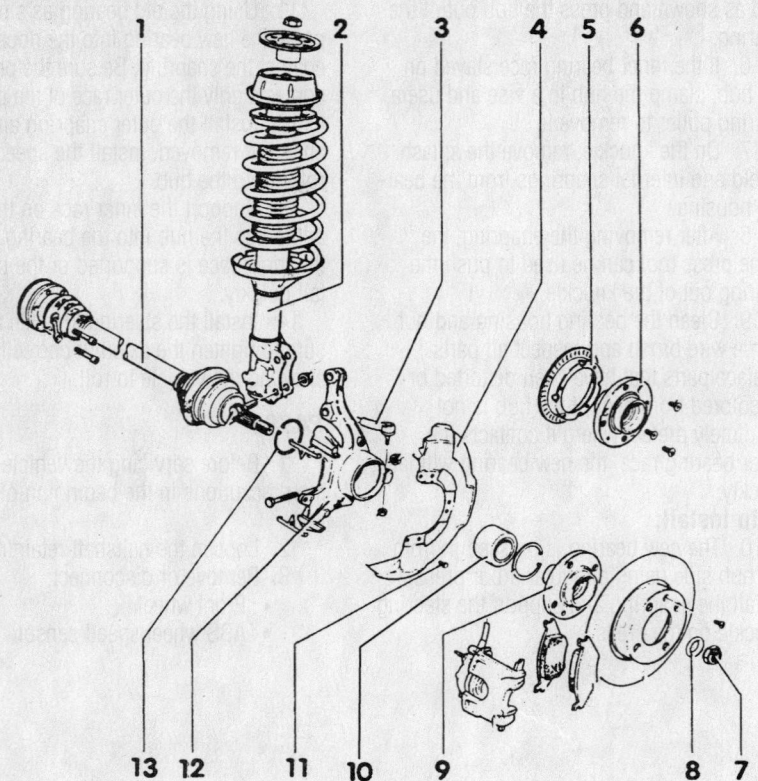

1. Halfshaft
2. Strut
3. Tie rod end
4. Splash plate
5. ABS wheel speed sensor
6. Wheel hub w/ABS
7. Axle nut (195 ft. lbs. (265 Nm))
8. Washer
9. Wheel hub wo/ABS
10. Snapring
11. Wheel bearing
12. Wheel bearing housing
13. Snapring

9301WG37

Exploded view of the front wheel bearing and related components—Cabrio, Golf, Jetta

- Caliper bracket mounting bolts, then the rotor.
- Brake splash guard.

4. Loosen the mounting nuts for the lower guide and track links.

5. Remove or disconnect:
- Tie rod end from the wheel bearing housing.
- Mounting nuts for the lower guide and track links and press out the joints.
- Upper control arm pinchbolt and disconnect the arms.
- Wheel bearing housing.

6. Place the wheel bearing housing on a press.

7. Drive out the hub with the wheel bearing.

8. Using a bearing separator and press, drive hub out of the bearing.

To install:

9. Press the new wheel bearing into the bearing housing using the appropriate bearing driver.

10. Press the hub into the wheel bearing using the appropriate bearing driver.

11. Install the wheel bearing housing.

12. Slide the CV-joint through the wheel hub and hand-tighten the new nut.

13. Install or connect:

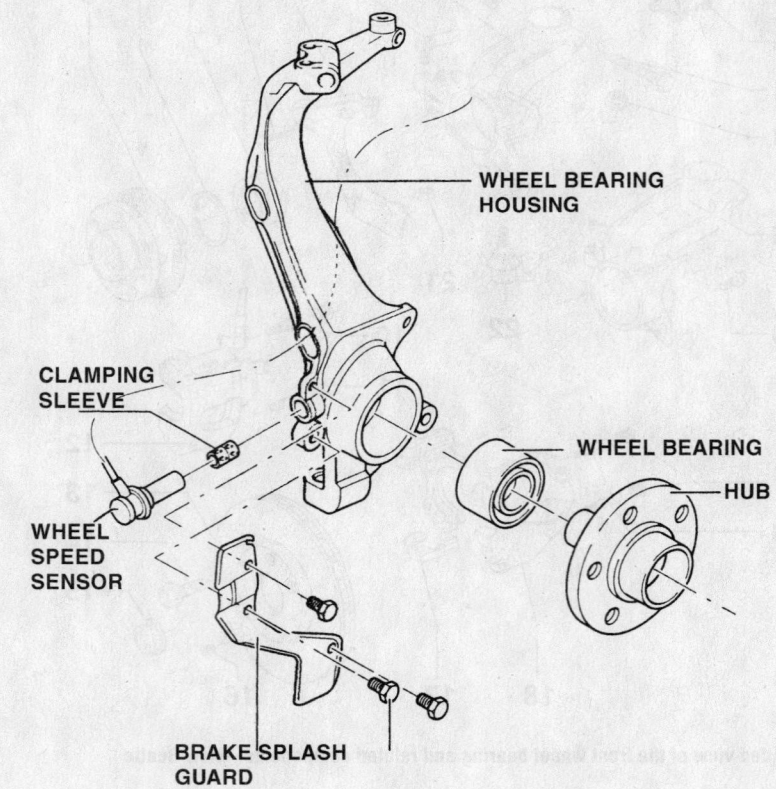

WHEEL BEARING HOUSING

CLAMPING SLEEVE

WHEEL SPEED SENSOR

WHEEL BEARING

HUB

BRAKE SPLASH GUARD

7923CG44

Exploded view of the front wheel bearing housing—1998–00 Passat

- Lower track control and guide link. New self-locking nut: 74 ft. lbs. (100 Nm).
- Both of the upper link ball joints into the wheel bearing. Pinchbolt: 30 ft. lbs. (40 Nm).
- Tie rod end. New self-locking nut: 37 ft. lbs. (50 Nm). Bolt: 44 inch lbs. (5 Nm).
- ABS wheel speed sensor.
- Brake splash guard. Retaining bolts: 84 inch lbs. (10 Nm).
- Brake rotor.
- Brake caliper. Retaining bolt: 89 ft. lbs. (120 Nm).
- Wheels. Lug bolts: 89 ft. lbs. (120 Nm).

14. Tighten the halfshaft retaining bolt as follows:
- If equipped with a M14 bolt, tighten to 85 ft. lbs. (115 Nm) plus ½ turn (180°)
- If equipped with a M16 bolt, tighten to 140 ft. lbs. (190 Nm) plus ½ turn (180°)

15. Check the front suspension alignment, if necessary, adjust.

Rear

NEW BEETLE AND PASSAT

The Beetle and 1998–00 Passat are equipped with a sealed bearing. The wheel bearing, wheel hub and wheel speed sensor are replaced as an assembly.

1. Before servicing the vehicle, refer to the precautions in the beginning of this section.
2. Remove the rear wheels.
3. On the Passat model, perform the following:
   a. Remove the caliper and rotor.
   b. Slightly, withdraw the ABS wheel speed sensor.
   c. Remove the wheel hub-to-axle beam mounting bolts through the openings in the wheel hub flange.
   d. Remove the wheel hub.
4. On the New Beetle model, perform the following:
   a. Slightly, withdraw the ABS wheel speed sensor.
   b. If equipped with drum brakes, remove the drum.
   c. If equipped with disk brakes, remove the caliper and rotor.
   d. Remove the wheel hub center dust cover.
   e. Remove the self-locking 12-point nut.
   f. Using a multiple jaw puller, withdraw the wheel hub off of the stub axle.

To install:

5. Install the new wheel hub and tighten the mounting fasteners as follows:
- Passat—44 ft. lbs. (60 Nm)
- New Beetle—129 ft. lbs. (175 Nm)
6. Install:
- Drum or caliper and rotor.
- If equipped, center dust cap.
- If removed, ABS wheel speed sensor.
- Wheels.
7. If necessary, check and adjust the wheel alignment.

CABRIO, GOLF, JETTA

1. Before servicing the vehicle, refer to the precautions in the beginning of this section.
2. Remove the rear wheels.
3. On drum brakes, insert a small pry-tool through a wheel bolt hole and push up

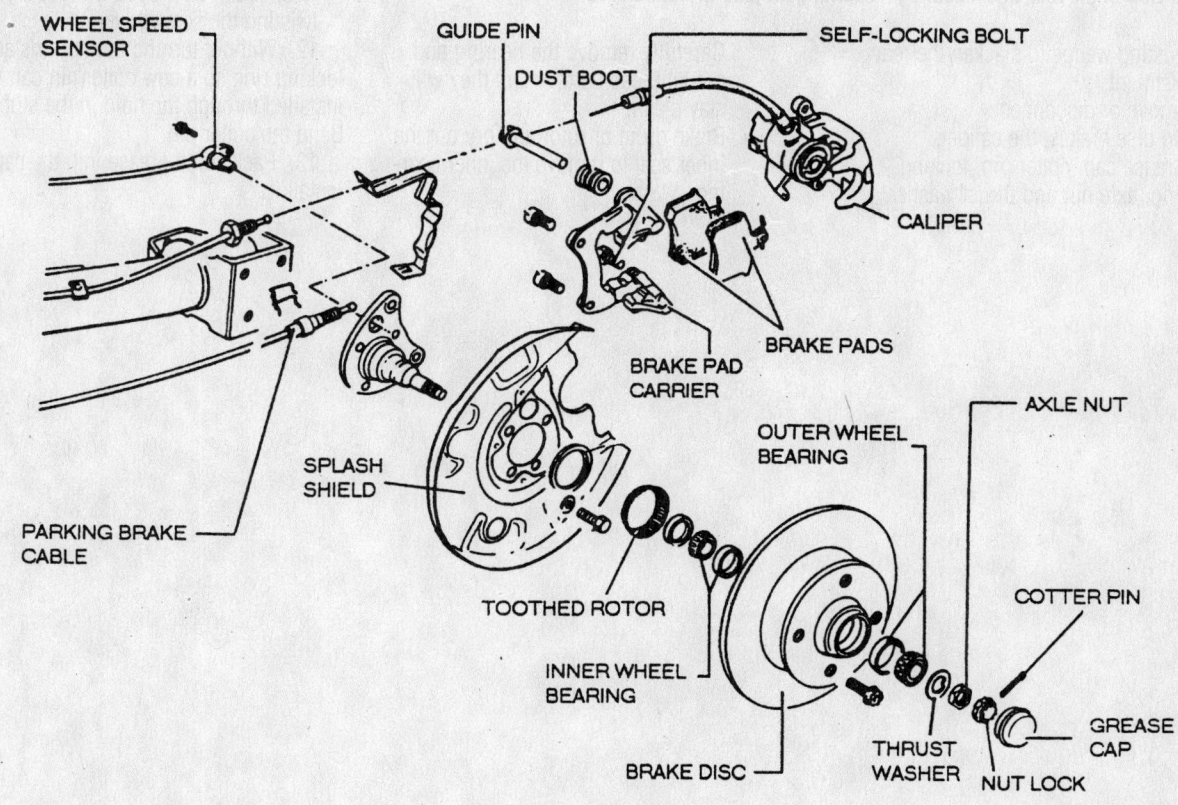

WHEEL SPEED SENSOR

GUIDE PIN

DUST BOOT

SELF-LOCKING BOLT

CALIPER

BRAKE PADS

BRAKE PAD CARRIER

SPLASH SHIELD

PARKING BRAKE CABLE

TOOTHED ROTOR

INNER WHEEL BEARING

BRAKE DISC

OUTER WHEEL BEARING

THRUST WASHER

NUT LOCK

AXLE NUT

COTTER PIN

GREASE CAP

Exploded view of the rear wheel bearing—Cabrio, golf, jetta w/disk brakes

7923CG46

For Wheel Alignment specifications, see Section 1 of this manual

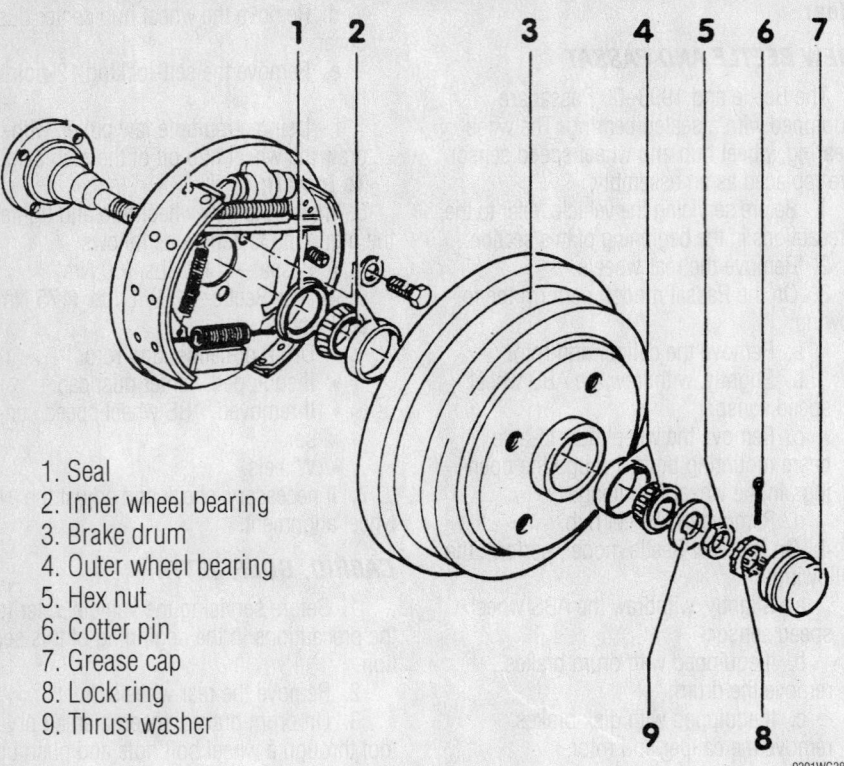

1. Seal
2. Inner wheel bearing
3. Brake drum
4. Outer wheel bearing
5. Hex nut
6. Cotter pin
7. Grease cap
8. Lock ring
9. Thrust washer

9301WG38

Exploded view of the rear wheel bearing—Cabrio, golf, jetta w/drum brakes

on the adjusting wedge to slacken the rear brake adjustment.

4. Remove or disconnect:
   • On disc brakes, the caliper.
   • Grease cap, cotter pin, locking ring, axle nut and thrust washer.

Carefully remove the bearing and put all these parts where they will stay clean.
   • Brake drum or rotor and pry out the inner seal to remove the inner bearing.

5. Clean all the grease off the bearings using solvent. If the bearings appear worn or damaged, they must be replaced.

6. To remove the bearing races, support the drum or rotor and carefully drive the race out with a long drift pin. They can also be removed on a press.

To install:

7. Carefully press the new race into the drum or rotor. The old race can be used as a press tool but be sure it does not become stuck in the hub.

8. Pack the inner bearing with clean wheel bearing grease and fit it into the inner race. Press a new axle seal into place by hand.

9. Lightly coat the stub axle with grease and install the drum or rotor. Be careful not to damage the axle seal.

10. Pack the outer bearing and install the bearing, thrust washer and nut.

11. To adjust the bearing pre-load:
    a. Begin tightening the nut while turning the drum or rotor.
    b. When the nut is snug, try to move the thrust washer with a screwdriver.
    c. Back the nut off until the thrust washer can be moved without prying or twisting the screwdriver.

12. Without turning the nut, install the locking ring so a new cotter pin can be installed through the hold in the stub axle. Bend the cotter pin.

13. Pack some grease into the cap and install it.

PRECAUTIONS

Before servicing any vehicle, please be sure to read all of the following precautions, which deal with personal safety, prevention of component damage, and important points to take into consideration when servicing a motor vehicle:

• Never open, service or drain the radiator or cooling system when the engine is hot; serious burns can occur from the steam and hot coolant.

• Observe all applicable safety precautions when working around fuel. Whenever servicing the fuel system, always work in a well-ventilated area. Do not allow fuel spray or vapors to come in contact with a spark, open flame, or excessive heat (a hot drop light, for example). Keep a dry chemical fire extinguisher near the work area. Always keep fuel in a container specifically designed for fuel storage; also, always properly seal fuel containers to avoid the possibility of fire or explosion. Refer to the additional fuel system precautions later in this section.

• Fuel injection systems often remain pressurized, even after the engine has been turned **OFF**. The fuel system pressure must be relieved before disconnecting any fuel lines. Failure to do so may result in fire and/or personal injury.

• Brake fluid often contains polyglycol ethers and polyglycols. Avoid contact with the eyes and wash your hands thoroughly after handling brake fluid. If you do get brake fluid in your eyes, flush your eyes with clean, running water for 15 minutes. If eye irritation persists, or if you have taken brake fluid internally, IMMEDIATELY seek medical assistance.

• The EPA warns that prolonged contact with used engine oil may cause a number of skin disorders, including cancer. You should make every effort to minimize your exposure to used engine oil. Protective gloves should be worn when changing oil. Wash your hands and any other exposed skin areas as soon as possible after exposure to used engine oil. Soap and water, or waterless hand cleaner should be used.

• All new vehicles are now equipped with an air bag system, often referred to as a Supplemental Restraint System (SRS) or Supplemental Inflatable Restraint (SIR) system. The system must be disabled before performing service on or around system components, steering column, instrument panel components, wiring and sensors. Failure to follow safety and disabling procedures could result in accidental air bag deployment, possible personal injury, and unnecessary system repairs.

• Always wear safety goggles when working with, or around, the air bag system. When carrying a non-deployed air bag, be sure the bag and trim cover are pointed away from your body. When placing a non-deployed air bag on a work surface, always face the bag and trim cover upward, away from the surface. This will reduce the motion of the module if it is accidentally deployed. Refer to the additional air bag system precautions later in this section.

• Clean, high quality brake fluid from a sealed container is essential to the safe and proper operation of the brake system. You should always buy the correct type of brake fluid for your vehicle. If the brake fluid becomes contaminated, completely flush the system with new fluid. Never reuse any brake fluid. Any brake fluid that is removed from the system should be discarded. Also, do not allow any brake fluid to come in contact with a painted surface; it will damage the paint.

• Never operate the engine without the proper amount and type of engine oil; doing so will result in severe engine damage.

• Timing belt maintenance is extremely important. Many models utilize an interference-type, non-freewheeling engine. If the timing belt breaks, the valves in the cylinder head may strike the pistons, causing potentially serious (also time-consuming and expensive) engine damage. Refer to the maintenance interval charts in the front of this manual for the recommended replacement interval for the timing belt, and to the timing belt section for belt replacement and inspection.

• Disconnecting the negative battery cable on some vehicles may interfere with the functions of the on-board computer system(s) and may require the computer to undergo a relearning process once the negative battery cable is reconnected.

• When servicing drum brakes, only disassemble and assemble one side at a time, leaving the remaining side intact for reference.

ENGINE REPAIR

➡ **Disconnecting the negative battery cable on some vehicles may interfere with the functions of the on board computer system. The computer may undergo a relearning process once the negative battery cable is reconnected.**

Distributor

REMOVAL AND INSTALLATION

6-Cylinder Engine and 1999–00 5-Cylinder Engines

These models use a Distributorless Ignition System (DIS) controlled by the Powertrain Control Module (PCM).

1998 5 Cylinder Engines

1. Before servicing the vehicle, refer to the precautions in the beginning of this section.
2. Remove or disconnect the following:
   • Negative battery cable
   • Electrical leads from the distributor
   • Distributor cap. Mark the position of the rotor in relation to the cylinder head.
   • Distributor retaining bolts and the distributor

To install:
3. Install or connect the following:
   • Distributor with the matchmarks aligned
   • Distributor cap

   • Electrical leads
   • Negative battery cable

Alternator

REMOVAL

5 Cylinder Engines

1. Before servicing the vehicle, refer to the precautions in the beginning of this section.
2. Remove or disconnect the following:
   • Negative battery cable
   • Accessory drive belt
   • Power steering pump
   • Alternator harness connectors
   • Alternator

6 Cylinder Engines

1. Before servicing the vehicle, refer to the precautions in the beginning of this section.
2. Remove or disconnect the following:
   - Negative battery cable
   - Accessory drive belt
   - Alternator harness connectors
   - Alternator

INSTALLATION

5 Cylinder Engines

Install or connect the following:
- Alternator
- Alternator harness connectors
- Power steering pump
- Accessory drive belt
- Negative battery cable

6 Cylinder Engines

Install or connect the following:
- Alternator
- Alternator harness connectors
- Accessory drive belt
- Negative battery cable

Ignition Timing

ADJUSTMENT

All engines are equipped with a Motronic 4.3 or 4.4 control system. Manual adjustment of the ignition timing is not possible.

Engine Assembly

REMOVAL & INSTALLATION

Front Wheel Drive and All Wheel Drive

1. Before servicing the vehicle, refer to the precautions in the beginning of this section.
2. Drain the cooling system.
3. Remove or disconnect the following:
   - Battery and tray
   - Lower air baffle, if equipped
   - Front wheels
   - Brake hose bracket
   - Rear driveshaft and transfer case, if equipped
   - Lower ball joints
   - Track rods
   - Axle halfshafts
   - Exhaust front pipe

- Right engine mount
- Torque rod
- Speedometer cable
- Front and rear lower engine mounts
- Air intake assembly
- Ignition coil connections
- Accelerator cable
- Idle Air Control (IAC) valve hose
- Positive Crankcase Ventilation (PCV) valve and hose
- Preheat hose
- Mass Air Flow (MAF) sensor connector
- Firewall ground cable
- Heated Oxygen (HO2S) sensor connector
- Distributor connector, if equipped
- Crankshaft Position (CKP) sensor connector
- Camshaft Position (CMP) sensor connector
- Brake booster vacuum hose
- Cooling fan
- Radiator hoses
- Radiator
- Coolant overflow tank
- Turbocharger vacuum hose, if equipped
- Exhaust Gas Recirculation (EGR) valve regulator vacuum hose
- Clutch slave cylinder, if equipped
- Gear select cable
- Accessory drive belts
- A/C compressor
- Starter motor
- Turbocharger oil cooler lines and valve, if equipped
- Fuel supply manifold covers
- Fuel line retainers

4. Install fuel injector holders on the injectors.
5. Remove or disconnect the following:
   - Fuel pressure regulator vacuum hose
   - Fuel supply manifold with injectors attached
   - Engine control wiring harness
   - Air pump
   - Front engine mount

6. Lift the engine and transmission out of the vehicle.

To install:

7. Lower the drivetrain into the vehicle.
8. Install or connect the following:
   - Front engine mount. Tighten the bolt to 37 ft. lbs. (50 Nm).
   - Air pump
   - Engine control wiring harness
   - Fuel supply manifold with injectors

attached and remove the injector holders
- Fuel pressure regulator vacuum hose
- Fuel line retainers
- Fuel supply manifold covers
- Turbocharger oil cooler lines and valve, if equipped
- Starter motor
- A/C compressor
- Accessory drive belts
- Gear select cable
- Clutch slave cylinder, if equipped
- EGR valve regulator vacuum hose
- Turbocharger vacuum hose, if equipped
- Coolant overflow tank
- Radiator
- Radiator hoses
- Cooling fan
- Brake booster vacuum hose
- CMP sensor connector
- CKP sensor connector
- Distributor connector, if equipped
- HO2S sensor connector
- Firewall ground cable
- MAF sensor connector
- Preheat hose
- PCV valve and hose
- IAC valve hose
- Accelerator cable
- Ignition coil connections
- Air intake assembly
- Front and rear lower engine mounts
- Speedometer cable
- Torque rod with new fasteners. Tighten to 26 ft. lbs. (35 Nm) plus 90 degrees.
- Right engine mount. Tighten the bolt to 37 ft. lbs. (50 Nm).
- Exhaust front pipe
- Axle halfshafts
- Track rods
- Lower ball joints
- Rear driveshaft and transfer case, if equipped
- Brake hose bracket
- Front wheels
- Lower air baffle, if equipped
- Battery and tray

9. Fill the cooling system.
10. Start the engine and check for leaks.

Rear Wheel Drive

1. Before servicing the vehicle, refer to the precautions in the beginning of this section.
2. Drain the cooling system.
3. Drain the engine oil.

4. Relieve the fuel system pressure.
5. Remove or disconnect the following:
- Battery
- Body ground cable
- Accessory drive belts
- Cooling fan
- Battery relay connector
- Right ground cable
- Radiator hoses
- Transmission cooler lines
- Left and right motor mounts
- Positive Crankcase Ventilation (PCV) valve and hose
- Idle Air Control (IAC) valve connector and hose
- Evaporative Emissions (EVAP) valve hoses
- Mass Air Flow (MAF) sensor connector
- Preheat hose
- Servo pump
- Fuel lines
- Accelerator cable
- Cruise control cable and vacuum line
- Engine control wiring harness
- Heater hoses
- Brake booster vacuum line
- Camshaft Position (CMP) sensor connector
- Crankshaft Position (CKP) sensor connector
- Engine splash shield and air baffle
- Radiator
- Oil thermostat hose
- A/C compressor
- Heated Oxygen (HO2S) sensor connectors
- Exhaust heat shield
- Exhaust front pipe
- Gear selector rods
- Driveshaft
- Transmission mount and cross-member

6. Lift the engine and transmission out of the vehicle.

To install:
7. Lower the drivetrain into the vehicle.
8. Install or connect the following:
- Transmission mount and cross-member. Tighten to 37 ft. lbs. (50 Nm).
- Driveshaft
- Gear selector rods
- Exhaust front pipe
- Exhaust heat shield
- HO2S sensor connectors
- A/C compressor
- Oil thermostat hose
- Radiator
- Engine splash shield and air baffle
- CKP sensor connector

- CMP sensor connector
- Brake booster vacuum line
- Heater hoses
- Engine control wiring harness
- Cruise control cable and vacuum line
- Accelerator cable
- Fuel lines
- Servo pump
- Preheat hose
- MAF sensor connector
- EVAP valve hoses
- IAC valve connector and hose
- PCV valve and hose
- Left and right motor mounts. Tighten the nuts to 37 ft. lbs. (50 Nm).
- Transmission cooler lines
- Radiator hoses
- Right ground cable
- Battery relay connector
- Cooling fan
- Accessory drive belts
- Body ground cable
- Battery

9. Fill the crankcase to the correct level.
10. Fill the cooling system.
11. Start the engine and check for leaks.

Water Pump

REMOVAL & INSTALLATION

Except Rear Wheel Drive

1. Before servicing the vehicle, refer to the precautions in the beginning of this section.
2. Drain the cooling system.
3. Remove or disconnect the following:
- Negative battery cable
- Spark plug cover
- Fuel line clips
- Expansion tank
- Accessory drive belts
- Vibration damper guard
- Front wheel
- Inner fender panel
- Front timing cover
- Timing belt
- Water pump

To install:
4. Install or connect the following:
- Water pump with a new gasket. Tighten the bolts to 15 ft. lbs. (20 Nm).
- Timing belt
- Front timing cover
- Inner fender panel
- Front wheel
- Vibration damper guard

- Accessory drive belts
- Expansion tank
- Fuel line clips
- Spark plug cover
- Negative battery cable

5. Fill the cooling system.
6. Start the engine and check for leaks.

Rear Wheel Drive

1. Before servicing the vehicle, refer to the precautions in the beginning of this section.
2. Drain the cooling system.
3. Remove or disconnect the following:
- Negative battery cable
- Accessory drive belts
- Front cover
- Timing belt
- Water pump

To install:
- Water pump with a new gasket. Tighten the bolts to 15 ft. lbs. (20 Nm).
- Timing belt
- Front cover
- Accessory drive belts
- Negative battery cable

4. Fill the cooling system.
5. Start the engine and check for leaks.

Cylinder Head

REMOVAL & INSTALLATION

5-Cylinder Engines

1. Before servicing the vehicle, refer to the precautions in the beginning of this section.
2. Drain the cooling system.
3. Remove or disconnect the following:
- Negative battery cable
- Air intake assembly
- Accessory drive belts
- Front cover
- Timing belt
- Exhaust manifold
- Fuel supply manifold covers
- Fuel line retainers

4. Install fuel injector holders on the injectors.
- Fuel pressure regulator vacuum hose
- Fuel supply manifold with injectors attached
- Cooling fan
- Intake manifold
- Upper radiator hose
- Camshaft sprockets
- Rear timing cover
- Distributor, if equipped

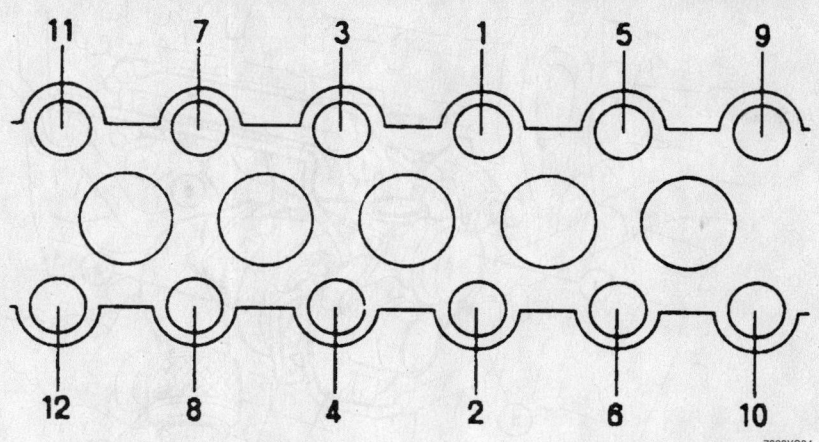

Cylinder head torque sequence—5–cylinder engines

- Camshaft Position (CMP) sensor, if equipped
- Extension arm and brackets
- Upper cylinder head
- Camshafts
- Coolant pipe
- Lower cylinder head

To install:

5. Install the lower cylinder head with a new gasket. Tighten the bolts in sequence as follows:
   a. Step 1: 15 ft. lbs. (20 Nm)
   b. Step 2: 44 ft. lbs. (60 Nm)
   c. Step 3: Plus 130 degrees

6. Install or connect the following:
   - Coolant pipe
   - Camshafts
   - Upper cylinder head. Tighten the bolts to 13 ft. lbs. (17 Nm).
   - Extension arm and brackets
   - CMP sensor, if equipped

- Distributor, if equipped
- Rear timing cover
- Camshaft sprockets
- Upper radiator hose
- Intake manifold
- Cooling fan
- Fuel supply manifold with injectors attached and remove the injector holders
- Fuel pressure regulator vacuum hose
- Fuel line retainers
- Fuel supply manifold covers
- Exhaust manifold
- Timing belt
- Front cover
- Accessory drive belts
- Air intake assembly
- Negative battery cable

7. Fill the cooling system.
8. Start the engine and check for leaks.

6-Cylinder Engine

1. Before servicing the vehicle, refer to the precautions in the beginning of this section.
2. Drain the cooling system.
3. Remove or disconnect the following:
   - Negative battery cable
   - Exhaust front pipe
   - Exhaust manifold heat shield
   - Exhaust manifold
   - Accessory drive belt
   - Coolant pipe
   - Front cover
   - Timing belt
   - Transmission mounting plate bolt
   - Mass Air Flow (MAF) sensor
   - Air intake assembly
   - Accelerator cable and bracket
   - Throttle Position (TP) sensor connector
   - Idle Air Control (IAC) valve hoses
   - Throttle body vacuum hoses
   - Cruise control servo vacuum line
   - Fuel supply manifold covers
   - Fuel line retainers

4. Install fuel injector holders on the injectors.
   - Fuel pressure regulator vacuum hose
   - Fuel supply manifold with injectors attached
   - Intake manifold
   - Ignition coils
   - Camshaft sprockets
   - Camshaft Position (CMP) sensor
   - Engine Coolant Temperature (ECT) sensor connector
   - Upper radiator hose
   - Upper cylinder head
   - Camshafts
   - Lower cylinder head

To install:

5. Install the lower cylinder head with a new gasket. Tighten the bolts in sequence as follows:
   a. Step 1: 15 ft. lbs. (20 Nm)
   b. Step 2: 44 ft. lbs. (60 Nm)
   c. Step 3: Plus 130 degrees

6. Install or connect the following:
   - Camshafts
   - Upper cylinder head. Tighten the bolts to 13 ft. lbs. (17 Nm).
   - Upper radiator hose
   - ECT sensor connector
   - CMP sensor
   - Camshaft sprockets
   - Ignition coils
   - Intake manifold
   - Fuel supply manifold with injectors attached

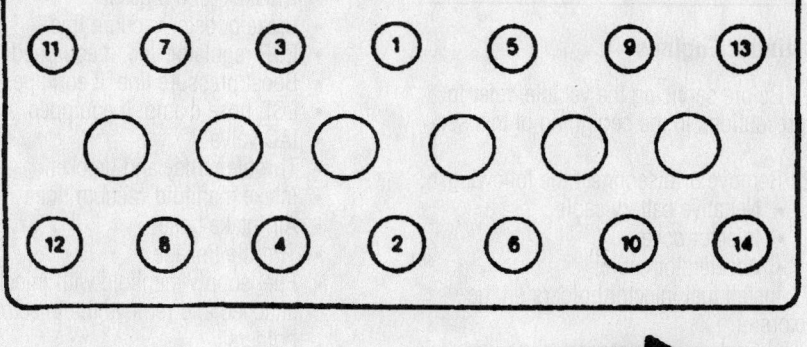

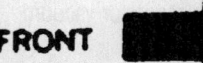

FRONT

7923XG03

Cylinder head torque sequence—6–cylinder engines

- Fuel pressure regulator vacuum hose
- Fuel line retainers
- Fuel supply manifold covers
- Cruise control servo vacuum line
- Throttle body vacuum hoses
- IAC valve hoses
- TP sensor connector
- Accelerator cable and bracket
- Air intake assembly
- MAF sensor
- Transmission mounting plate bolt
- Timing belt
- Front cover
- Coolant pipe
- Accessory drive belt
- Exhaust manifold
- Exhaust manifold heat shield
- Exhaust front pipe
- Negative battery cable

7. Fill the cooling system.
8. Start the engine and check for leaks.

Rocker Arms/Shafts

REMOVAL & INSTALLATION

The covered vehicles are not equipped with rocker arms or shafts. The valves are directly actuated by the camshaft.

Turbocharger

REMOVAL & INSTALLATION

1. Before servicing the vehicle, refer to the precautions in the beginning of this section.
2. Drain the cooling system.
3. Remove or disconnect the following:
   - Negative battery cable
   - Exhaust manifold heat shield
   - Upper air charge pipe
   - Inner heat shield
   - Air intake hose
   - Turbo coolant hoses
   - Turbo oil lines
   - Exhaust front pipe and bracket
   - Red boost pressure hose
   - White bypass valve hose
   - Yellow pressure regulator hose
   - Turbocharger

To install:

4. Install new pin bolts with thread locking compound. Tighten them to 15 ft. lbs. (20 Nm).
5. Install or connect the following:
   - Turbocharger and tighten the fasteners to 22 ft. lbs. (30 Nm)
   - Yellow pressure regulator hose
   - White bypass valve hose

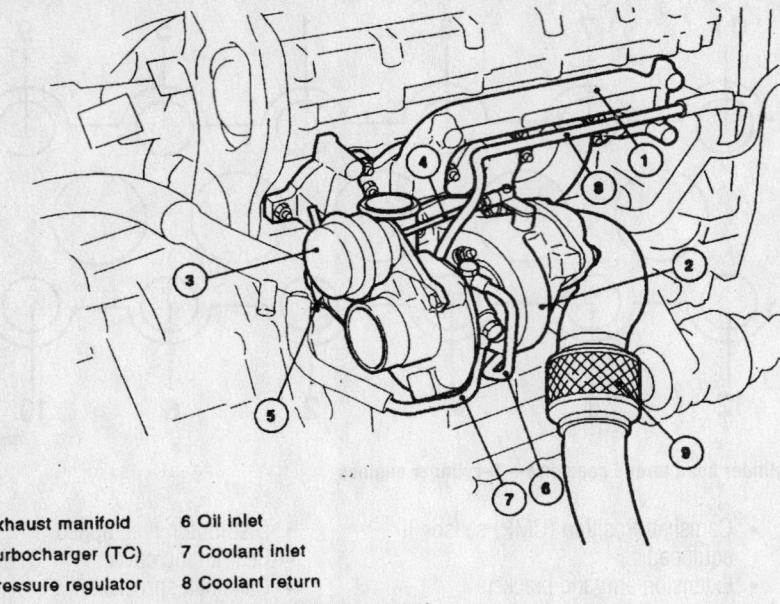

1 Exhaust manifold	6 Oil inlet
2 Turbocharger (TC)	7 Coolant inlet
3 Pressure regulator	8 Coolant return
4 Link	9 Flexible joint (bellows type)
5 Bypass valve	

7923XG06

Turbocharger and exhaust manifold component identification—5-cylinder engines

- Red boost pressure hose
- Exhaust front pipe and bracket
- Turbo oil lines
- Turbo coolant hoses
- Air intake hose
- Inner heat shield
- Upper air charge pipe
- Exhaust manifold heat shield
- Negative battery cable

6. Fill the cooling system.
7. Start the engine and check for leaks.

Intake Manifold

REMOVAL & INSTALLATION

5-Cylinder Engines

1. Before servicing the vehicle, refer to the precautions in the beginning of this section.
2. Remove or disconnect the following:
   - Negative battery cable
   - Injector cover
   - Accelerator cable
3. Install fuel injector holders on the injectors.
   - Fuel pressure regulator vacuum hose
   - Fuel supply manifold with injectors attached
   - Throttle linkage
   - Air intake hose
   - Intake manifold vacuum lines

- Throttle pulley and bracket
- Idle Air Control (IAC) valve
- Exhaust Gas Recirculation (EGR) hose clamp, if equipped
- Boost pressure line, if equipped
- EGR regulator line, if equipped
- Brake booster vacuum line
- Dipstick tube bracket
- Intake manifold lower bracket
- Intake manifold

To install:

4. Install or connect the following:
   - Intake manifold with a new gasket. Tighten the bolts to 15 ft. lbs. (20 Nm).
   - Intake manifold lower bracket
   - Dipstick tube bracket
   - Brake booster vacuum line
   - EGR regulator line, if equipped
   - Boost pressure line, if equipped
   - EGR hose clamp, if equipped
   - IAC valve
   - Throttle pulley and bracket
   - Intake manifold vacuum lines
   - Air intake hose
   - Throttle linkage
   - Fuel supply manifold with injectors attached and remove the injector holders
   - Fuel pressure regulator vacuum hose
   - Accelerator cable
   - Injector cover
   - Negative battery cable

5. Start the engine and check for leaks.

6-Cylinder Engine

1. Before servicing the vehicle, refer to the precautions in the beginning of this section.
2. Remove or disconnect the following:
   - Negative battery cable
   - Mass Air Flow (MAF) sensor connector
   - Idle Air Control (IAC) valve connector and hose
   - Air intake hose
   - Throttle pulley cover
   - Throttle Position (TP) sensor connector
   - Accelerator cable and bracket
   - Cruise control servo
   - Throttle body vacuum hoses
   - Injector cover
3. Install fuel injector holders on the injectors.
   - Fuel pressure regulator vacuum hose
   - Fuel supply manifold with injectors attached
   - Air preheater hose
   - Left and right power stage connectors
   - Intake manifold bracket
   - Brake booster vacuum line
   - Positive Crankcase Ventilation (PCV) valve and hose
   - Intake manifold outer section
   - Intake manifold

To install:
4. Install or connect the following:
   - Intake manifold with a new gasket and tighten the bolts to 15 ft. lbs. (20 Nm)
   - Intake manifold outer section with new clamps
   - PCV valve and hose
   - Brake booster vacuum line
   - Intake manifold bracket
   - Left and right power stage connectors
   - Air preheater hose
   - Fuel supply manifold with injectors attached
   - Fuel pressure regulator vacuum hose
   - Injector cover
   - Throttle body vacuum hoses
   - Cruise control servo
   - Accelerator cable and bracket
   - TP sensor connector
   - Throttle pulley cover
   - Air intake hose
   - IAC valve connector and hose
   - MAF sensor connector

   - Negative battery cable
5. Start the engine and check for leaks.

Exhaust Manifold

REMOVAL & INSTALLATION

1. Before servicing the vehicle, refer to the precautions in the beginning of this section.
2. Remove or disconnect the following:
   - Negative battery cable
   - Exhaust manifold heat shield
   - Exhaust front pipe
   - Turbocharger, if equipped
   - Exhaust manifold

To install:
3. Install or connect the following:
   - Exhaust manifold and tighten the fasteners to 18 ft. lbs. (25 Nm)
   - Turbocharger, if equipped
   - Exhaust front pipe and tighten the fasteners to 44 ft. lbs. (60 Nm)
   - Exhaust manifold heat shield and tighten the fasteners to 86 inch lbs. (10 Nm)
4. Loosen the joint at the catalytic converter and re-tighten to 18 ft. lbs. (25 Nm). This is necessary to prevent stresses in the system.
5. Connect the negative battery cable.
6. Start the engine and check for leaks.

Front Crankshaft Seal

REMOVAL & INSTALLATION

5-Cylinder Engines

1. Before servicing the vehicle, refer to the precautions in the beginning of this section.
2. Remove or disconnect the following:
   - Negative battery cable
   - Fuel line clips
   - Coolant recovery tank
   - Accessory drive belts
   - Right front wheel
   - Inner fender liner
   - Front cover
   - Timing belt
   - Crankshaft timing sprocket
   - Front crankshaft seal

To install:
3. Install or connect the following:
   - Front crankshaft seal so that it is flush with the oil pump housing
   - Crankshaft timing sprocket and tighten the nut to 133 ft. lbs. (180 Nm)

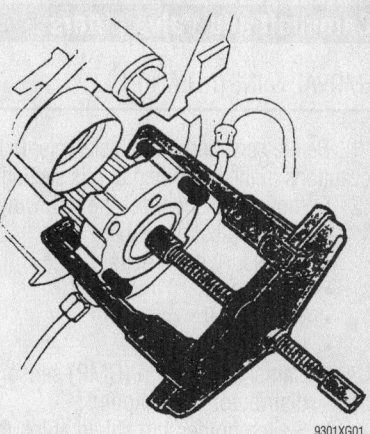

9301XG01

Removing the crankshaft timing belt sprocket

   - Timing belt
   - Front cover
   - Inner fender liner
   - Right front wheel
   - Accessory drive belts
   - Coolant recovery tank
   - Fuel line clips
   - Negative battery cable
4. Start the engine and check for leaks.

6-Cylinder Engine

1. Before servicing the vehicle, refer to the precautions in the beginning of this section.
2. Drain the cooling system.
3. Remove or disconnect the following:
   - Negative battery cable
   - Accessory drive belts
   - Cooling fan and shroud
   - Radiator
   - Front cover
   - Timing belt
   - Crankshaft timing sprocket
   - Front crankshaft seal

To install:
4. Install or connect the following:
   - Front crankshaft seal so that it is flush with the oil pump housing
   - Crankshaft timing sprocket and tighten the nut to 221 ft. lbs. (300 Nm)
   - Timing belt
   - Front cover
   - Radiator
   - Cooling fan and shroud
   - Accessory drive belts
   - Negative battery cable
5. Fill the cooling system.
6. Start the engine and check for leaks.

Timing belt service is covered in Section 3 of this manual

Camshaft and Valve Lifters

REMOVAL & INSTALLATION

1. Before servicing the vehicle, refer to the precautions in the beginning of this section.
2. Remove or disconnect the following:
   - Negative battery cable
   - Accessory drive belts
   - Front cover
   - Timing belt
   - Ignition coil cover
   - Camshaft Position (CMP) sensor or distributor, as equipped
   - Switch holder and shield at left rear of the engine
   - Ignition coils
   - Camshaft sprockets
   - Upper cylinder head
   - Camshafts
   - Hydraulic lash adjusters

➡**Keep all valvetrain components in order for assembly.**

To install:
3. Install or connect the following:
   - Hydraulic lash adjusters
   - Camshafts
   - Upper cylinder head and tighten the bolts to 13 ft. lbs. (17 Nm)
   - Camshaft sprockets and tighten the bolts to 15 ft. lbs. (20 Nm)
   - Ignition coils
   - Switch holder and shield at left rear of the engine
   - CMP sensor or distributor, as equipped
   - Ignition coil cover
   - Timing belt
   - Front cover
   - Accessory drive belts
   - Negative battery cable

Valve Lash

ADJUSTMENT

All engines covered use hydraulic lash adjusters. No adjustment is necessary.

Starter Motor

REMOVAL & INSTALLATION

1. Before servicing the vehicle, refer to the precautions in the beginning of this section.
2. Remove or disconnect the following:
   - Negative battery cable
   - Starter motor harness connectors
   - Starter motor

To install:
3. Install or connect the following:
   - Starter motor and tighten the bolts to 25 ft. lbs. (34 Nm)
   - Starter motor harness connectors
   - Negative battery cable

Oil Pan

REMOVAL & INSTALLATION

5-Cylinder Engines

1. Before servicing the vehicle, refer to the precautions in the beginning of this section.
2. Remove the engine from the vehicle and mount on a suitable work stand.
   - Oil filter
   - Oil pan
   - Oil passage O-rings

To install:
3. Install or connect the following:
   - New oil passage O-rings
   - Oil pan and tighten the bolts to 13 ft. lbs. (17 Nm)
   - New oil filter
4. Install the engine into the vehicle.

6-Cylinder Engine

1. Before servicing the vehicle, refer to the precautions in the beginning of this section.
2. Drain the engine oil.
3. Remove or disconnect the following:
   - Negative battery cable
   - Splash guard, if equipped
   - Left and right motor mounts
4. Raise and support the engine.
   - Front axle crossmember
   - Oil pan support bracket
   - Oil pan

To install:
5. Install or connect the following:
   - Oil pan with a new gasket and tighten the bolts to 96 ft. lbs. (11 Nm)
   - Oil pan support bracket
   - Front axle crossmember
   - Left and right motor mounts
   - Splash guard, if equipped
   - Negative battery cable
6. Start the engine and check for leaks.

Oil Pump

REMOVAL & INSTALLATION

5-Cylinder Engines

1. Before servicing the vehicle, refer to the precautions in the beginning of this section.

2. Remove or disconnect the following:
   - Negative battery cable
   - Fuel line clips
   - Coolant recovery tank
   - Accessory drive belts
   - Right front wheel
   - Inner fender liner
   - Front cover
   - Timing belt
   - Crankshaft timing sprocket
   - Front crankshaft seal
   - Oil pump

To install:
3. Install the oil pump using special tool 999-5455. Use the oil pump bolts to guide the pump. Use the crankshaft nut to press the pump in until it is seated fully. Tighten the oil pump bolts to 84 inch lbs. (10 Nm).
4. Remove the crankshaft nut and the press tool.
5. Install or connect the following:
   - Front crankshaft seal
   - Crankshaft timing sprocket and tighten the nut to 133 ft. lbs. (180 Nm)
   - Timing belt
   - Front cover
   - Inner fender liner
   - Right front wheel
   - Accessory drive belts
   - Coolant recovery tank
   - Fuel line clips
   - Negative battery cable
6. Start the engine and check for leaks.

6-Cylinder Engine

1. Before servicing the vehicle, refer to the precautions in the beginning of this section.
2. Drain the cooling system.
3. Remove or disconnect the following:
   - Negative battery cable
   - Accessory drive belts
   - Cooling fan and shroud
   - Radiator
   - Front cover
   - Timing belt
   - Crankshaft timing sprocket
   - Front crankshaft seal
   - Oil pump

To install:
4. Install the oil pump using special tool 999-5455. Use the oil pump bolts to guide the pump. Use the crankshaft nut to press the pump in until it is seated fully. Tighten the oil pump bolts to 84 inch lbs. (10 Nm).
5. Remove the crankshaft nut and the press tool.
6. Install or connect the following:

- Front crankshaft seal
- Crankshaft timing sprocket and tighten the nut to 221 ft. lbs. (300 Nm)
- Timing belt
- Front cover
- Radiator
- Cooling fan and shroud
- Accessory drive belts
- Negative battery cable
7. Fill the cooling system.
8. Start the engine and check for leaks.

Rear Main Seal

REMOVAL & INSTALLATION

1. Before servicing the vehicle, refer to the precautions in the beginning of this section.
2. Remove or disconnect the following:
   - Negative battery cable
   - Transmission
   - Clutch, if equipped
   - Flexplate or flywheel
   - Rear main seal

To install:

3. Install the rear main seal so that it is flush with the cylinder block.
4. Install or connect the following:
   - Flexplate or flywheel. Tighten the bolts to 33 ft. lbs. (45 Nm) plus 50 degrees.

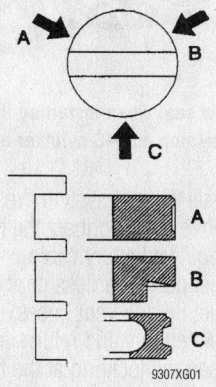

Ring identification and end-gap spacing—all engines

- Clutch, if equipped
- Transmission
- Negative battery cable
5. Start the engine and check for leaks.

Piston and Ring

POSITIONING

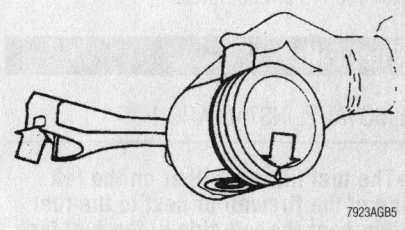

**Piston and rod positioning—all engines
The notch on the piston crown faces the front of the engine**

FUEL SYSTEM

Fuel System Service Precautions

Safety is the most important factor when performing not only fuel system maintenance but any type of maintenance. Failure to conduct maintenance and repairs in a safe manner may result in serious personal injury or death. Maintenance and testing of the vehicle fuel system components can be accomplished safely and effectively by adhering to the following rules and guidelines.

• To avoid the possibility of fire and personal injury, always disconnect the negative battery cable unless the repair or test procedure requires that battery voltage be applied.

• Always relieve the fuel system pressure prior to disconnecting any fuel system component (injector, fuel rail, pressure regulator, etc.), fitting or fuel line connection. Exercise extreme caution whenever relieving fuel system pressure to avoid exposing skin, face and eyes to fuel spray. Please be advised that fuel under pressure may penetrate the skin or any part of the body that it contacts.

• Always place a shop towel or cloth around the fitting or connection prior to loosening to absorb any excess fuel due to spillage. Ensure that all fuel spillage

(should it occur) is quickly removed from engine surfaces. Ensure that all fuel soaked cloths or towels are deposited into a suitable waste container.

• Always keep a dry chemical (Class B) fire extinguisher near the work area.

• Do not allow fuel spray or fuel vapors to come into contact with a spark or open flame.

• Always use a back-up wrench when loosening and tightening fuel line connection fittings. This will prevent unnecessary stress and torsion to fuel line piping.

Always follow the proper tighten specifications.

• Always replace worn fuel fitting O-rings with new. Do not substitute fuel hose or equivalent, where fuel pipe is installed.

Fuel System Pressure

RELIEVING

1. Before servicing the vehicle, refer to the precautions in the beginning of this section.

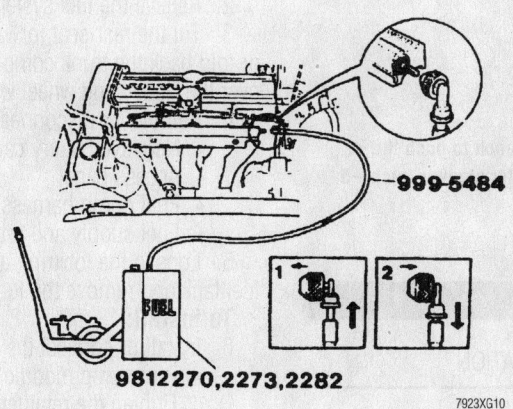

Connecting the adapter and fuel drainage unit—5-cylinder engine shown

Heater Core replacement is covered in Section 2 of this manual

2. Disconnect the negative battery cable.

3. Remove protective cap from the valve on the fuel rail.

4. Connect a hose to adapter 999-5484 and place the other end in a clean container.

5. Connect the adapter in the locked or closed position to the valve on the fuel rail.

6. Unlock or open the adapter valve.

7. After the fuel system pressure is relieved, remove the adapter and hose and replace the protective cap.

8. Connect the negative battery cable when repairs are complete.

Fuel Filter

REMOVAL & INSTALLATION

➡The fuel filter is either on the left side of the firewall or next to the fuel pump near the left side of the fuel tank.

1. Before servicing the vehicle, refer to the precautions in the beginning of this section.

2. Relieve the fuel system pressure.

3. Remove or disconnect the following:
- Negative battery cable
- Fuel filler cap
- Fuel lines from the fuel filter
- Fuel filter from the bracket

To install:

4. Install or connect the following:
- Fuel filter to the bracket
- Fuel lines to the fuel filter
- Fuel filler cap
- Negative battery cable

5. Start the engine and check for leaks.

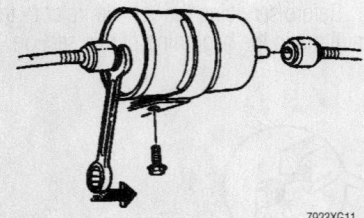

7923XG11

Using an open end wrench to push the quick disconnect coupler sleeves back—5 cylinder models

Fuel Pump

REMOVAL & INSTALLATION

6-Cylinder Engine

1. Before servicing the vehicle, refer to the precautions in the beginning of this section.

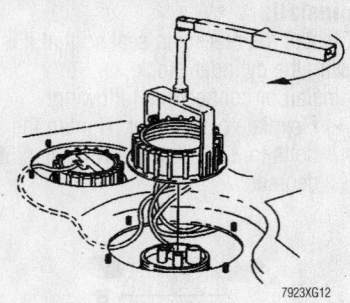

7923XG12

Use a new seal when installing the fuel pump retaining nut—5-cylinder engines

2. Relieve the fuel system pressure.

3. Remove or disconnect the following:
- Negative battery cable
- Fuel pump harness connections
- Fuel fill and vent hoses
- Fuel supply and return lines

4. Loosen the lockring at the top of the fuel tank and remove the sending unit with the transfer pump attached. Note the direction of the float in the tank.

5. Remove the transfer pump from the sending unit.

To install:

6. Install the transfer pump on the sending unit. Install the sending unit in the fuel tank and tighten the lockring.

7. Install or connect the following:
- Fuel supply and return lines
- Fuel fill and vent hoses
- Fuel pump harness connections
- Negative battery cable

8. Start the engine and check for leaks.

5-Cylinder Engines

1. Before servicing the vehicle, refer to the precautions in the beginning of this section.

2. Relieve the fuel system pressure.

3. Tilt the rear seat forward and remove or fold back the trunk compartment carpet over the right-hand wheel well panel.

4. Remove or disconnect the following:
- Negative battery cable
- Access panel
- Fuel pump harness connectors
- Fuel supply and return lines

5. Loosen the lockring at the top of the fuel tank and remove the fuel pump module.

To install:

6. Install or connect the following:
- Fuel pump module with a new seal. Tighten the retainer to 30 ft. lbs. (40 Nm).
- Fuel supply and return lines
- Access panel
- Trunk carpeting

- Rear seat
- Negative battery cable

7. Start the engine and check for leaks.

Fuel Injector

REMOVAL & INSTALLATION

5-Cylinder Engines

1. Before servicing the vehicle, refer to the precautions in the beginning of this section.

2. Remove or disconnect the following:
- Negative battery cable
- Injector cover
- Accelerator cable

3. Install fuel injector holders on the injectors.
- Fuel pressure regulator vacuum hose
- Fuel supply manifold with injectors attached

4. Remove the injector holders and separate the injectors from the fuel supply manifold.

To install:

5. Install or connect the following:
- Injectors to the fuel supply manifold with new O-ring seals
- Injector holders
- Fuel supply manifold with injectors attached. Remove the injector holders.
- Fuel pressure regulator vacuum hose
- Accelerator cable
- Injector cover
- Negative battery cable

6. Start the engine and check for leaks.

6 Cylinder Engines

1. Before servicing the vehicle, refer to the precautions in the beginning of this section.

2. Remove or disconnect the following:
- Negative battery cable
- Mass Air Flow (MAF) sensor connector
- Idle Air Control (IAC) valve connector and hose
- Air intake hose
- Throttle pulley cover
- Throttle Position (TP) sensor connector
- Accelerator cable and bracket
- Cruise control servo
- Throttle body vacuum hoses
- Injector cover

3. Install fuel injector holders on the injectors.
- Fuel pressure regulator vacuum hose

- Fuel supply manifold with injectors attached

4. Remove the injector holders and separate the injectors from the fuel supply manifold.

To install:

5. Install or connect the following:
- Injectors to the fuel supply manifold with new O-ring seals

- Injector holders
- Fuel supply manifold with injectors attached. Remove the injector holders.
- Fuel pressure regulator vacuum hose
- Injector cover
- Throttle body vacuum hoses
- Cruise control servo

- Accelerator cable and bracket
- TP sensor connector
- Throttle pulley cover
- Air intake hose
- IAC valve connector and hose
- MAF sensor connector
- Negative battery cable

6. Start the engine and check for leaks.

DRIVE TRAIN

Transmission Assembly

REMOVAL & INSTALLATION

Rear Wheel Drive

MANUAL

1. Before servicing the vehicle, refer to the precautions in the beginning of this section.
2. Drain the transmission fluid.
3. Remove or disconnect the following:
- Negative battery cable
- Reverse light connector at the firewall
- Shift linkage
- Gearshift boot
- Reverse detent fork
- Overdrive switch connector, if equipped
- Clutch slave cylinder
- Exhaust front pipe bracket
- Driveshaft
- Speedometer cable
- Starter motor
- Transmission crossmember and support the transmission with a jack
- Transmission flange bolts
- Transmission

To install:

4. Install or connect the following:
- Transmission and tighten the flange bolts to 30 ft. lbs. (40 Nm)
- Transmission crossmember
- Starter motor
- Speedometer cable
- Driveshaft
- Exhaust front pipe bracket
- Clutch slave cylinder
- Overdrive switch connector, if equipped
- Reverse detent fork
- Gearshift boot
- Shift linkage
- Reverse light connector at the firewall

- Negative battery cable

5. Fill the transmission to the correct level.

AUTOMATIC

1. Before servicing the vehicle, refer to the precautions in the beginning of this section.
2. Drain the transmission fluid.
3. Remove or disconnect the following:
- Negative battery cable
- Throttle Valve (TV) cable
- Oil dipstick tube
- Shift control and reaction rods
- Driveshaft
- Exhaust front pipe
- Starter motor
- Transmission crossmember and support the transmission with a jack
- Flywheel access cover
- Torque converter
- Transmission oil cooler lines
- Transmission flange bolts
- Transmission

To install:

4. Install or connect the following:
- Transmission and tighten the flange bolts to 65 ft. lbs. (85 Nm)
- Transmission oil cooler lines
- Torque converter and tighten the bolts to 32 ft. lbs. (42 Nm)
- Flywheel access cover
- Transmission crossmember and support the transmission with a jack
- Starter motor
- Exhaust front pipe
- Driveshaft
- Shift control and reaction rods
- Oil dipstick tube
- TV cable
- Negative battery cable

5. Fill the transmission to the correct level.

6. Start the engine and check for leaks.

Transaxle Assembly

REMOVAL & INSTALLATION

Front Wheel Drive and AWD

MANUAL

1. Before servicing the vehicle, refer to the precautions in the beginning of this section.
2. Install a support fixture to the engine lifting eyes.
3. Drain the transaxle fluid.
4. Remove or disconnect the following:
- Battery and tray
- Air intake ducts
- Gear select cables and link plate
- Reverse light switch connector
- Turbocharger inlet pipe, if equipped
- Upper engine oil cooler hose
- Clutch slave cylinder
- Transaxle ground cable
- Rear engine mount
- Starter motor
- Firewall ground cable
- Torque arm
- Front wheels
- Wheel speed sensors
- Brake line brackets
- Wheel speed sensor wiring brackets
- Inner fender liners
- Lower ball joints
- Stabilizer bar links
- Evaporative Emissions (EVAP) canister hoses
- Exhaust front pipe
- Axle halfshafts
- Heated Oxygen (HO2S) sensor harness clamps
- Vehicle Speed (VSS) sensor connector
- Transaxle mount
- Transaxle flange bolts

5. Loosen the 2 right side subframe-to-body bolts approximately ½ inch. Support the subframe with a jack and remove the subframe-to-body bolts on the left side.

Brake service is covered in Section 4 of this manual

6. Lower the jack and let the frame hang down from the right side bolts.

7. Tie the left side of the steering gear to the left side frame rail for support. Remove the steering gear engine mount.

8. Lower the engine and transaxle with the lifting hook.

9. Pull the transaxle away from the engine. Lower the transaxle from the vehicle.

To install:

10. Install or connect the following:
- Transaxle and tighten the flange bolts to 37 ft. lbs. (50 Nm)
- Transaxle mount and tighten the bolts to 37 ft. lbs. (50 Nm)
- VSS sensor connector
- HO2S sensor harness clamps

11. Install the subframe using new bolts. Starting on the left side, lift the frame with a jack. Mount the support brackets on both sides. Tighten the frame bolts to 78 ft. lbs. (105 Nm) plus 120 degrees. Tighten the bracket bolts to 37 ft. lbs. (50 Nm). Repeat the procedure for the right side. Remove the subframe jack.

12. Install or connect the following:
- Steering gear engine mount and tighten the bolts to 37 ft. lbs. (50 Nm)
- Axle halfshafts
- Exhaust front pipe
- EVAP canister hoses
- Stabilizer bar links
- Lower ball joints
- Inner fender liners
- Wheel speed sensor wiring brackets
- Brake line brackets
- Wheel speed sensors
- Front wheels
- Torque arm and tighten the bolts to 26 ft. lbs. (35 Nm) plus 40 degrees
- Firewall ground cable
- Starter motor
- Rear engine mount and tighten the bolts to 37 ft. lbs. (50 Nm)
- Transaxle ground cable
- Clutch slave cylinder
- Upper engine oil cooler hose
- Turbocharger inlet pipe, if equipped
- Reverse light switch connector
- Gear select cables and link plate
- Air intake ducts
- Battery and tray

13. Fill the transaxle to the correct level.

14. Start the engine and check for leaks.

AUTOMATIC

1. Before servicing the vehicle, refer to the precautions in the beginning of this section.

2. Drain the transaxle fluid.

3. Attach a support fixture to the engine lifting eyes.

4. Remove or disconnect the following:
- Battery and tray
- Air intake assembly
- Turbo control valve
- Turbocharger inlet tube
- Gear select cable
- Transaxle harness connector
- Transaxle ground cable
- Heated Oxygen (HO2S) sensor connector
- Transaxle oil cooler lines
- Transaxle dipstick tube
- Exhaust Gas Recirculation (EGR) valve hoses, if equipped
- Starter motor
- Coolant recovery tank
- Torque rod extension arm
- Front wheels
- Wheel speed sensors
- Brake line brackets
- Wheel speed sensor wiring brackets
- Inner fender liners
- Axle halfshafts
- Transfer case, if equipped
- Splash guards
- Lower ball joints
- Stabilizer bar links
- Evaporative Emissions (EVAP) canister and hoses
- Exhaust front pipe
- Oil line bracket
- Steering gear engine mount
- Steering gear mounting nuts
- Vehicle Speed (VSS) sensor connector

- Transaxle mount
- Torque converter

5. Loosen the 2 right side subframe-to-body bolts approximately ½ inch. Support the subframe with a jack and remove the subframe-to-body bolts on the left side.

6. Lower the jack and let the frame hang down from the right side bolts.

7. Tie the left side of the steering gear to the left side frame rail for support. Remove the steering gear engine mount.

8. Lower the engine and transaxle with the lifting hook.

9. Install transaxle fixture 5463 on the transaxle jack, using the torque rod mounting bolts to hold it in place. At the same time, fit tool 5463-1 support plate and raise the jack so that it is making light contact.

10. Remove the transaxle flange bolts.

11. Remove the transaxle.

To install:

➡ **Use new fasteners where indicated.**

12. Install or connect the following:
- Transaxle and tighten the bolts to 37 ft. lbs. (50 Nm)
- Torque converter. Use new bolts and tighten them to 22 ft. lbs. (30 Nm).
- Rear transaxle mount and torque the bolts to 37 ft. lbs. (50 Nm)
- HO2S sensor connector
- VSS sensor connector

13. Install the subframe using new bolts. Starting on the left side, lift the frame with a jack. Mount the support brackets on both sides. Tighten the frame bolts to 78 ft. lbs. (105 Nm) plus 120 degrees. Tighten the bracket bolts to 37 ft. lbs. (50 Nm). Repeat

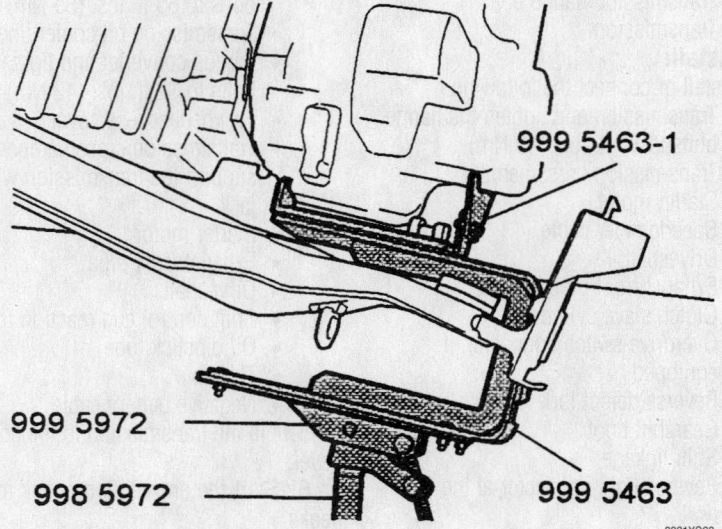

999 5463-1

999 5972

998 5972

999 5463

Using the transaxle fixture

9301XG03

the procedure for the right side. Remove the subframe jack.

14. Install or connect the following:
- Steering gear. Use new mounting nuts and tighten them to 37 ft. lbs. (50 Nm).
- Steering gear engine mount and tighten the bolts to 37 ft. lbs. (50 Nm)
- Oil line bracket
- Torque rod extension arm and tighten the bolts to 26 ft. lbs. (35 Nm) plus 40 degrees
- Exhaust front pipe
- EVAP canister and hoses
- Stabilizer bar links
- Lower ball joints
- Splash guards
- Transfer case, if equipped
- Axle halfshafts
- Inner fender liners
- Wheel speed sensor wiring brackets
- Brake line brackets
- Wheel speed sensors
- Front wheels
- Coolant recovery tank
- Starter motor
- EGR valve hoses, if equipped
- Transaxle dipstick tube
- Transaxle oil cooler lines
- Transaxle ground cable
- Transaxle harness connector
- Gear select cable
- Turbocharger inlet tube
- Turbo control valve
- Air intake assembly
- Battery and tray

15. Fill the transaxle to the correct level.

Clutch

ADJUSTMENT

The vehicles covered are equipped with a hydraulic clutch system that is self-adjusting.

REMOVAL & INSTALLATION

1. Before servicing the vehicle, refer to the precautions in the beginning of this section.
2. Remove the transmission or transaxle.
3. Remove the throw-out bearing from the sleeve and fork.
4. Unbolt the pressure plate in a crossing pattern and in several passes.
5. Remove the pressure plate and clutch disk.

To install:
6. Install the clutch disk and pressure plate. Tighten the bolts in a crossing pattern and in several passes to 18 ft. lbs. (25 Nm).
7. Install the throw-out bearing.
8. Install the transmission or transaxle.

Hydraulic Clutch System

BLEEDING

✳✳ CAUTION

Use only DOT 4 brake fluid.

1. Before servicing the vehicle, refer to the precautions in the beginning of this section.
2. Depress the clutch pedal a few times to purge the air bubbles in the master cylinder.
3. Connect a hose from a drain bottle to the nipple on the slave cylinder.
4. While the clutch pedal is depressed to the floor, open the bleed nipple.
5. Hold the pedal to the floor to allow brake fluid and air bubble to exit through the hose. Close the nipple.
6. Repeat this procedure until no air bubbles are visible in the escaping fluid.
7. Pump the clutch pedal a few times to build pressure in the system.
8. Check the fluid reservoir. The fluid level should not be above the MAX level.

Transfer Case Assembly

REMOVAL & INSTALLATION

1. Before servicing the vehicle, refer to the precautions in the beginning of this section.
2. Remove or disconnect the following:
- Right front wheel
- Right axle halfshaft
- Transfer case vibration damper and bracket
- Rear driveshaft
- Transfer case

To install:
3. Install or connect the following:
- Transfer case and tighten the bolts to 37 ft. lbs. (50 Nm)
- Rear driveshaft
- Transfer case vibration damper and bracket
- Right axle halfshaft
- Right front wheel

Halfshaft

REMOVAL & INSTALLATION

Front

1. Before servicing the vehicle, refer to the precautions in the beginning of this section.
2. Remove or disconnect the following:
- Front wheel
- Wheel speed sensor and wiring bracket
- Brake hose bracket
- Stabilizer bar link
- Splash guards
- Lower ball joint
- Hub retainer nut
3. Pull the hub off of the stub shaft.
4. Pry the inner joint out of the transaxle and remove the axle halfshaft.

To install:

➡**Use new fasteners for assembly.**

5. Install the axle halfshaft so that the circlip is felt to seat in the retainer groove.
6. Guide the stub shaft into the hub.
7. Install or connect the following:
- Hub retainer nut and tighten the nut to 89 ft. lbs. (120 Nm) plus 60 degrees
- Lower ball joint
- Splash guards
- Stabilizer bar link
- Brake hose bracket
- Wheel speed sensor and wiring bracket
- Front wheel

Rear

1. Before servicing the vehicle, refer to the precautions in the beginning of this section.
2. Remove or disconnect the following:
- Rear wheel
- Hub retainer nut
- Final drive subframe section
- Inner axle halfshaft bolts
- Axle halfshaft

To install:

➡**Use new fasteners for assembly.**

3. Fit the outer joint stub shaft into the wheel hub. Tighten the inner halfshaft bolts to 70 ft. lbs. (91 Nm).
4. Install or connect the following:
- Final drive subframe section and tighten the bolts to 52 ft. lbs. (68 Nm) plus 30 degrees
- Hub retainer nut and tighten to 103 ft. lbs. (134 Nm) plus 60 degrees
- Rear wheel

For complete Engine Mechanical specifications, see Section 1 of this manual

CV-Joints

OVERHAUL

Inner CV-Joint

The inner CV-joint is serviced with the axle shaft as an assembly. The inner CV-joint boot can be serviced by removing the outer CV-joint.

Air Bag

❊❊ CAUTION

Some vehicles are equipped with an air bag system. The system must be disarmed before performing service on, or around, system components, the steering column, instrument panel components, wiring and sensors. Failure to follow the safety precautions and the disarming procedure could result in accidental air bag deployment, possible injury and unnecessary system repairs.

PRECAUTIONS

Several precautions must be observed when handling the inflator module to avoid accidental deployment and possible personal injury.

• Never carry the inflator module by the wires or connector on the underside of the module.

• When carrying a live inflator module, hold securely with both hands, and ensure that the bag and trim cover are pointed away.

• Place the inflator module on a bench or other surface with the bag and trim cover facing up.

• With the inflator module on the bench, never place anything on or close to the module which may be thrown in the event of an accidental deployment.

DISARMING

1. Before servicing the vehicle, refer to the precautions in the beginning of this section.

2. Disconnect the negative battery cable.

3. Wait at least 1 minute before working on the vehicle. The air bag system is designed to retain enough power to deploy the air bag for a short time after the battery has been disconnected.

Outer CV-Joint

1. Before servicing the vehicle, refer to the precautions in the beginning of this section.

2. Remove the halfshaft from the vehicle.

3. Remove the grease boot clamps and push the boot away from the joint.

4. Expand the inner race circlip and pull the CV-joint off of the axle shaft.

5. Disassemble the inner race, cage and balls for cleaning and inspection.

STEERING AND SUSPENSION

4. After repairs are complete, connect the negative battery cable. Turn the ignition switch to the **ON** position and check the SRS light for proper operation.

Rack and Pinion Steering Gear

REMOVAL & INSTALLATION

850, C70, S70, and V70 Models

1. Before servicing the vehicle, refer to the precautions in the beginning of this section.

2. Install a support fixture to the engine lifting eyes.

3. Remove or disconnect the following:
• Negative battery cable
• Front wheels
• Outer tie rod ends
• Splash guards
• Power steering hose brackets
• Steering gear mounting nuts
• Subframe brackets

4. Support the subframe with a jack.

5. Loosen the front subframe bolts and remove the rear subframe bolts.

6. Remove or disconnect the following:
• Power steering hoses
• Steering shaft coupler
• Steering gear

To install:

7. Install or connect the following:
• Steering gear
• Steering shaft coupler and tighten the bolt to 15 ft. lbs. (20 Nm)
• Power steering hoses
• Subframe. Use new bolts and tighten them to 77 ft. lbs. (105 Nm) plus 120 degrees.
• Subframe brackets and tighten the bolts to 37 ft. lbs. (50 Nm)
• Steering gear mounting nuts. Use new nuts and tighten them to 37 ft. lbs. (50 Nm).
• Power steering hose brackets
• Splash guards
• Outer tie rod ends

To install:

6. Assemble the inner race, cage and balls into the outer joint housing.

7. Expand the circlip and push the joint on to the axle shaft.

8. Fill the outer race and the grease boot with CV-joint grease and tighten the boot clamps.

9. Install the axle halfshaft.

• Front wheels
• Negative battery cable

8. Fill the power steering fluid reservoir with fluid.

9. Check the wheel alignment and adjust as necessary.

960, S90 and V90 Models

1. Before servicing the vehicle, refer to the precautions in the beginning of this section.

2. Remove or disconnect the following:
• Negative battery cable
• Splash guards
• Front wheels
• Steering shaft coupler
• Outer tie rod ends
• Power steering hoses
• Stabilizer bar brackets
• Steering gear

To install:

3. Install or connect the following:
• Steering gear and tighten the bolts to 32 ft. lbs. (44 Nm)
• Stabilizer bar brackets
• Power steering hoses
• Outer tie rod ends and tighten the nuts to 44 ft. lbs. (60 Nm)
• Steering shaft coupler and tighten the bolts to 15 ft. lbs. (20 Nm)
• Front wheels
• Splash guards
• Negative battery cable

4. Fill the power steering pump reservoir.

5. Check the wheel alignment and adjust as necessary.

Strut

REMOVAL & INSTALLATION

Front

960, S90 AND V90 MODELS

1. Before servicing the vehicle, refer to the precautions in the beginning of this section.

2. Remove or disconnect the following:
- Front wheel
- Outer tie rod end
- Wheel speed sensor and wiring bracket
- Stabilizer bar link
- Brake caliper and rotor
- Brake hose bracket
- Lower ball joint
- Upper strut mount
- Strut assembly

To install:

3. Install or connect the following:
- Strut assembly and tighten the upper mount nuts to 30 ft. lbs. (40 Nm)
- Lower ball joint and tighten the nut to 44 ft. lbs. (60 Nm)
- Brake hose bracket
- Brake caliper and rotor
- Stabilizer bar link
- Wheel speed sensor and wiring bracket
- Outer tie rod end and tighten the nut to 44 ft. lbs. (60 Nm)
- Front wheel

4. Check the wheel alignment and adjust as necessary.

850, C70, S70 AND V70 MODELS

1. Before servicing the vehicle, refer to the precautions in the beginning of this section.

2. Support the lower control arm with a jack stand.

3. Remove or disconnect the following:
- Front wheel
- Stabilizer bar link
- Wheel speed sensor and wiring
- Strut bracket bolts
- Upper strut mount nuts
- Strut assembly

To install:

➡**Use new fasteners for assembly.**

4. Install or connect the following:
- Strut assembly and tighten the upper mount nuts to 18 ft. lbs. (25 Nm)
- Strut bracket bolts and tighten the bolts to 48 ft. lbs. (65 Nm) plus 90 degrees
- Wheel speed sensor and wiring
- Stabilizer bar link
- Front wheel

5. Check the wheel alignment and adjust as necessary.

Shock Absorber

REMOVAL & INSTALLATION

Rear

960 WAGON AND V90 MODELS

1. Before servicing the vehicle, refer to the precautions in the beginning of this section.

2. Remove or disconnect the following:
- Upper shock absorber retainer bolt
- Lower shock absorber bolt
- Shock absorber

To install:

3. Install the shock absorber and tighten the bolts to 63 ft. lbs. (85 Nm).

960 SEDAN, S90 AND AWD V70 MODELS

1. Before servicing the vehicle, refer to the precautions in the beginning of this section.

2. Remove or disconnect the following:
- Rear wheel
- Upper shock absorber retainer bolt
- Lower shock absorber bolt
- Shock absorber

To install:

3. Install the shock absorber and tighten the bolts to 59 ft. lbs. (80 Nm).

4. Install the rear wheel.

850, C70, S70 AND FWD V70 MODELS

1. Before servicing the vehicle, refer to the precautions in the beginning of this section.

2. Remove or disconnect the following:
- Access panel
- Upper shock mount bolts
- Lower shock mount bolt
- Shock absorber

To install:

3. Install the shock absorber. Tighten the lower bolt to 59 ft. lbs. (80 Nm) and the upper bolts to 18 ft. lbs. (25 Nm).

4. Install the access panel.

Coil Spring

REMOVAL & INSTALLATION

Front

1. Before servicing the vehicle, refer to the precautions in the beginning of this section.

2. Remove the strut from the vehicle.

3. Install a spring compressor and tighten until the strut is loose inside the spring.

4. Remove or disconnect the following:
- Upper strut mount
- Spring retainer
- Coil spring

To install:

5. Install or connect the following:
- Coil spring
- Spring retainer
- Upper strut mount. Tighten the nut to 52 ft. lbs. (70 Nm).

6. Remove the spring compressor.

7. Install the strut to the vehicle.

8. Check the wheel alignment and adjust as necessary.

Rear

850, C70, S70, AND FWD V70 MODELS

1. Before servicing the vehicle, refer to the precautions in the beginning of this section.

2. Support the vehicle at the frame with jack stands.

3. Raise the trailing arm to curb height with a floor jack.

4. Remove or disconnect the following:
- Rear wheel
- Lower shock absorber mount bolt
- Spring mounting nut

5. Lower the jack and remove the coil spring.

To install:

6. Install or connect the following:
- Coil spring and tighten the mounting nut to 30 ft. lbs. (40 Nm)
- Lower shock absorber mount bolt
- Rear wheel

960, S90, V90 AND AWD V70 MODELS

1. Before servicing the vehicle, refer to the precautions in the beginning of this section.

2. Support the vehicle at the frame with jackstands.

3. Support the lower control arm with a jack.

4. Remove or disconnect the following:
- Rear wheel
- Stabilizer bar
- Shock absorber
- Control arm mounting bolts and nuts

For Accessory Drive Belt illustrations, see Section 1 of this manual

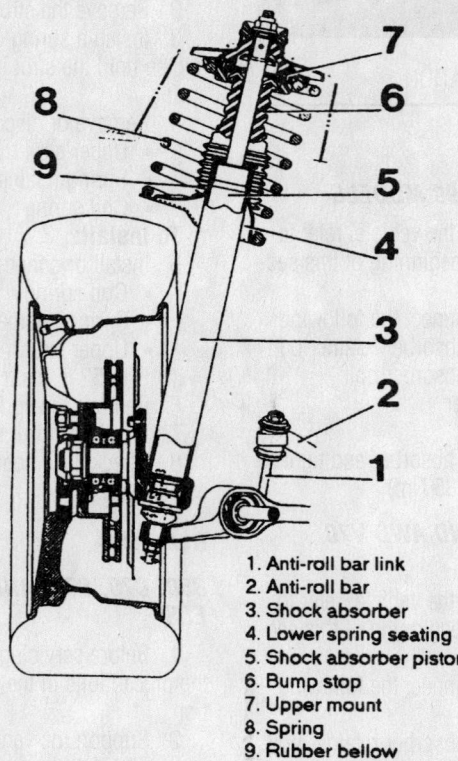

1. Anti-roll bar link
2. Anti-roll bar
3. Shock absorber
4. Lower spring seating
5. Shock absorber piston
6. Bump stop
7. Upper mount
8. Spring
9. Rubber bellow

7923XG14

Spring strut assembly—960, S90 and V90 models

- Jacking point
- Front support arm bracket

5. Lower the control arm and remove the coil spring.

To install:

6. Install the coil spring and raise the control arm.

7. Install or connect the following:
- Front support arm bracket and tighten the bolts to 48 ft. lbs. (65 Nm)
- Jacking point and tighten the bolts to 78 ft. lbs. (105 Nm)
- Control arm mounting bolts and nuts. Tighten the nuts to 59 ft. lbs. (80 Nm).
- Shock absorber
- Stabilizer bar
- Rear wheel

Lower Ball Joint

REMOVAL & INSTALLATION

960, S90 and V90 Models

1. Before servicing the vehicle, refer to the precautions in the beginning of this section.

2. Remove or disconnect the following:

- Front wheel
- Stabilizer bar link
- Lower ball joint from the lower control arm
- Lower ball joint from the strut

To install:

3. Install or connect the following:
- Lower ball joint. Tighten the strut bolts to 22 ft. lbs. (30 Nm) plus 90 degrees.
- Lower control arm. Tighten the nut to 44 ft. lbs. (60 Nm).
- Stabilizer bar link
- Front wheel

850, C70, S70 and V70 Models

1. Before servicing the vehicle, refer to the precautions in the beginning of this section.

2. Remove or disconnect the fol_ lowing:

- Front wheel
- Steering knuckle pinch bolt
- Lower control arm bolts
- Lower ball joint

To install:

3. Remove or disconnect the fol_ lowing:

- Lower ball joint. Tighten the pinch bolt to 37 ft. lbs. (50 Nm).

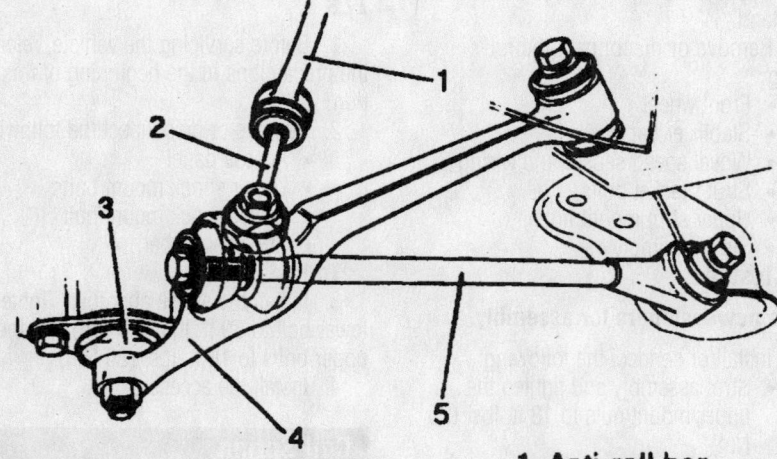

1 Anti-roll bar
2 Anti-roll bar link
3 Ball joint
4 Control arm
5 Control arm strut

7923XG15

Front suspension assembly—960, S90 and V90 models

- Lower control arm bolts. Tighten the bolts to 13 ft. lbs. (18 Nm) plus 120 degrees.
- Front wheel

Lower Control Arm

REMOVAL & INSTALLATION

960, S90 and V90 Models

1. Before servicing the vehicle, refer to the precautions in the beginning of this section.
2. Remove or disconnect the following:
   - Front wheel
   - Lower ball joint
   - Stabilizer bar link
   - Locator rod
   - Control arm

To install:

3. Install or connect the following:
   - Control arm. Tighten the inner bolt to 63 ft. lbs. (85 Nm).
   - Locator rod. Tighten the inner bolt to 63 ft. lbs. (85 Nm) and the control arm bolt to 70 ft. lbs. (95 Nm).
   - Stabilizer bar link.
   - Lower ball joint. Tighten the nut to 44 ft. lbs. (60 Nm).
   - Front wheel
4. Check the wheel alignment and adjust as necessary.

850, C70, S70 and V70 Models

1. Before servicing the vehicle, refer to the precautions in the beginning of this section.
2. Remove or disconnect the following:
   - Front wheel
   - Lower ball joint
   - Lower control arm

To install:

3. Install or connect the following:
   - Lower control arm. Tighten the subframe bolts to 48 ft. lbs. (65 Nm) plus 120 degrees
   - Lower ball joint. Tighten the bolts to 13 ft. lbs. (18 Nm) plus 90 degrees.
   - Front wheel
4. Check the wheel alignment and adjust as necessary.

CONTROL ARM BUSHING REPLACEMENT

The control arm bushings are serviced with the control arm as an assembly.

Wheel Bearing

ADJUSTMENT

Front

960, S90 AND V90 MODELS

The front wheel bearings are not adjustable. If the lateral run-out on the hub with the disc removed exceeds 0.0012 in. (0.030mm), the hub must be replaced.

850, C70, S70 AND V70 MODELS

The front wheel bearings are not adjustable. If the lateral run-out on the hub with the disc removed exceeds 0.0007 in. (0.020mm), the hub must be replaced.

Rear

The rear wheel bearings are sealed, pressed-in units, and no adjustment is possible.

REMOVAL & INSTALLATION

Front

960, S90 AND V90 MODELS

1. Before servicing the vehicle, refer to the precautions in the beginning of this section.
2. Remove or disconnect the following:
   - Front wheel
   - Brake caliper and rotor
   - Split pin
   - Spindle nut
   - Hub and bearing assembly

To install:

➡**Use a new spindle nut and split pin for assembly.**

3. Install or connect the following:
   - Hub and bearing assembly
   - Spindle nut. Tighten the nut to 74 ft. lbs. (100 Nm) plus 45 degrees.
   - Split pin
   - Brake caliper and rotor
   - Front wheel

850, C70, S70 AND V70 MODELS

1. Before servicing the vehicle, refer to the precautions in the beginning of this section.
2. Remove or disconnect the following:
   - Front wheel
   - Wheel speed sensor
   - Brake caliper and rotor

- Spindle nut
- Hub retainer bolts
- Hub and bearing assembly

To install:

3. Install or connect the following:
   - Hub and bearing assembly. Tighten the bolts to 33 ft. lbs. (45 Nm) plus 60 degrees.
   - Spindle nut. Tighten the nut to 89 ft. lbs. (120 Nm) plus 60 degrees.
   - Brake caliper and rotor
   - Wheel speed sensor
   - Front wheel

Rear

960, S90, V90 AND AWD V70 MODELS

1. Before servicing the vehicle, refer to the precautions in the beginning of this section.
2. Remove or disconnect the following:
   - Rear wheel
   - Brake caliper and rotor
   - Parking brake shoes
   - Parking brake cable
   - Support arm
   - Lower link arm
   - Track rod
   - Axle nut
   - Wheel bearing housing
   - Housing circlip
3. Press the hub out of the bearing and press the bearing out of the housing.

To install:

4. Press the new bearing into the housing. Install the circlip.
5. Support the inner race and press the hub into the bearing.
6. Install or connect the following:
   - Wheel bearing housing
   - Axle nut. Tighten the nut to 103 ft. lbs. (140 Nm) plus 60 degrees.
   - Track rod. Tighten the bolt to 63 ft. lbs. (85 Nm).
   - Lower link arm. Tighten the bolt to 45 ft. lbs. (60 Nm) plus 90 degrees
   - Support arm. Tighten the nut to 85 ft. lbs. (115 Nm).
   - Parking brake cable
   - Parking brake shoes
   - Brake caliper and rotor
   - Rear wheel
7. Check the wheel alignment and adjust as necessary.

850, C70, S70 AND FWD V70 MODELS

1. Before servicing the vehicle, refer to the precautions in the beginning of this section.

For Tire, Wheel and Ball Joint specifications, see Section 1 of this manual

2. Remove or disconnect the following:
    - Rear wheel
    - Brake caliper and rotor
    - Dust cap

- Hub nut
- Hub and bearing assembly

To install:
3. Install or connect the following:
    - Hub and bearing assembly. Use a

new hub nut and tighten it to 89 ft. lbs. (120 Nm) plus 30 degrees.
- Dust cap
- Brake caliper and rotor
- Rear wheel

GLOSSARY

ABS: Anti-lock braking system. An electro-mechanical braking system which is designed to minimize or prevent wheel lock-up during braking.

ABSOLUTE PRESSURE: Atmospheric (barometric) pressure plus the pressure gauge reading.

ACCELERATOR PUMP: A small pump located in the carburetor that feeds fuel into the air/fuel mixture during acceleration.

ACCUMULATOR: A device that controls shift quality by cushioning the shock of hydraulic oil pressure being applied to a clutch or band.

ACTUATING MECHANISM: The mechanical output devices of a hydraulic system, for example, clutch pistons and band servos.

ACTUATOR: The output component of a hydraulic or electronic system.

ADVANCE: Setting the ignition timing so that spark occurs earlier before the piston reaches top dead center (TDC).

ADAPTIVE MEMORY (ADAPTIVE STRATEGY): The learning ability of the TCM or PCM to redefine its decision-making process to provide optimum shift quality.

AFTER TOP DEAD CENTER (ATDC): The point after the piston reaches the top of its travel on the compression stroke.

AIR BAG: Device on the inside of the car designed to inflate on impact of crash, protecting the occupants of the car.

AIR CHARGE TEMPERATURE (ACT) SENSOR: The temperature of the airflow into the engine is measured by an ACT sensor, usually located in the lower intake manifold or air cleaner. ALDL (assembly line diagnostic link): **Electrical connector for scanning ECM/PCM/TCM input and output devices.**

AIR CLEANER: An assembly consisting of a housing, filter and any connecting ductwork. The filter element is made up of a porous paper, sometimes with a wire mesh screening, and is designed to prevent airborne particles from entering the engine through the carburetor or throttle body.

AIR INJECTION: One method of reducing harmful exhaust emissions by injecting air into each of the exhaust ports of an engine. The fresh air entering the hot exhaust manifold causes any remaining fuel to be burned before it can exit the tailpipe.

AIR PUMP: An emission control device that supplies fresh air to the exhaust manifold to aid in more completely burning exhaust gases.

AIR/FUEL RATIO: The ratio of air-to-gasoline by weight in the fuel mixture drawn into the engine.

ALIGNMENT RACK: A special drive-on vehicle lift apparatus/measuring device used to adjust a vehicle's toe, caster and camber angles.

ALL WHEEL DRIVE: Term used to describe a full time four wheel drive system or any other vehicle drive system that continuously delivers power to all four wheels. This system is found primarily on station wagon vehicles and SUVs not utilized for significant off road use.

ALTERNATING CURRENT (AC): Electric current that flows first in one direction, then in the opposite direction, continually reversing flow.

ALTERNATOR: A device which produces AC (alternating current) which is converted to DC (direct current) to charge the car battery.

AMMETER: An instrument, calibrated in amperes, used to measure the flow of an electrical current in a circuit. Ammeters are always connected in series with the circuit being tested.

AMPERAGE: The total amount of current (amperes) flowing in a circuit.

AMPLIFIER: A device used in an electrical circuit to increase the voltage of an output signal.

AMP/HR. RATING (BATTERY): Measurement of the ability of a battery to deliver a stated amount of current for a stated period of time. The higher the amp/hr. rating, the better the battery.

AMPERE: The rate of flow of electrical current present when one volt of electrical pressure is applied against one ohm of electrical resistance.

ANALOG COMPUTER: Any microprocessor that uses similar (analogous) electrical signals to make its calculations.

ANODIZED: A special coating applied to the surface of aluminum valves for extended service life.

ANTIFREEZE: A substance (ethylene or propylene glycol) added to the coolant to prevent freezing in cold weather.

ANTI-FOAM AGENTS: Minimize fluid foaming from the whipping action encountered in the converter and planetary action.

ANTI-WEAR AGENTS: Zinc agents that control wear on the gears, bushings, and thrust washers.

ANTI-LOCK BRAKING SYSTEM: A supplementary system to the base hydraulic system that prevents sustained lock-up of the wheels during braking as well as automatically controlling wheel slip.

ANTI-ROLL BAR: See stabilizer bar.

ARC: A flow of electricity through the air between two electrodes or contact points that produces a spark.

ARMATURE: A laminated, soft iron core wrapped by a wire that converts electrical energy to mechanical energy as in a motor or relay. When rotated in a magnetic field, it changes mechanical energy into electrical energy as in a generator.

ATDC: After Top Dead Center.

ATF: Automatic transmission fluid.

ATMOSPHERIC PRESSURE: The pressure on the Earth's surface caused by the weight of the air in the atmosphere. At sea level, this pressure is 14.7 psi at 32°F (101 kPa at 0°C).

ATOMIZATION: The breaking down of a liquid into a fine mist that can be suspended in air.

AUXILIARY ADD-ON COOLER: A supplemental transmission fluid cooling device that is installed in series with the heat exchanger (cooler), located inside the radiator, to provide additional support to cool the hot fluid leaving the torque converter.

AUXILIARY PRESSURE: An added fluid pressure that is introduced into a regulator or balanced valve system to control valve movement. The auxiliary pressure itself can be either a fixed or a variable value. (See balanced valve; regulator valve.)

AWD: All wheel drive.

AXIAL FORCE: A side or end thrust force acting in or along the same plane as the power flow.

AXIAL PLAY: Movement parallel to a shaft or bearing bore.

AXLE CAPACITY: The maximum load-carrying capacity of the axle itself, as specified by the manufacturer. This is usually a higher number than the GAWR.

AXLE RATIO: This is a number (3.07:1, 4.56:1, for example) expressing the ratio between driveshaft revolutions and wheel revolutions. A low numerical ratio allows the engine to work easier because it doesn't have to turn as fast. A high numerical ratio means that the engine has to turn more rpm's to move the wheels through the same number of turns.

BACKFIRE: The sudden combustion of gases in the intake or exhaust system that results in a loud explosion.

BACKLASH: The clearance or play between two parts, such as meshed gears.

BACKPRESSURE: Restrictions in the exhaust system that slow the exit of exhaust gases from the combustion chamber.

BAKELITE®: A heat resistant, plastic insulator material commonly used in printed circuit boards and transistorized components.

BALANCED VALVE: A valve that is positioned by opposing auxiliary hydraulic pressures and/or spring force. Examples include mainline regulator, throttle, and governor valves. (See regulator valve.)

BAND: A flexible ring of steel with an inner lining of friction material. When tightened around the outside of a drum, a planetary member is held stationary to the transmission/transaxle case.

BALL BEARING: A bearing made up of hardened inner and outer races between which hardened steel balls roll.

BALL JOINT: A ball and matching socket connecting suspension components (steering knuckle to lower control arms). It permits rotating movement in any direction between the components that are joined.

BARO (BAROMETRIC PRESSURE SENSOR): Measures the change in the intake manifold pressure caused by changes in altitude.

BAROMETRIC MANIFOLD ABSOLUTE PRESSURE (BMAP) SEN-

SOR: Operates similarly to a conventional MAP sensor; reads intake manifold pressure and is also responsible for determining altitude and barometric pressure prior to engine operation.

BAROMETRIC PRESSURE: (See atmospheric pressure.)

BALLAST RESISTOR: A resistor in the primary ignition circuit that lowers voltage after the engine is started to reduce wear on ignition components.

BATTERY: A direct current electrical storage unit, consisting of the basic active materials of lead and sulfuric acid, which converts chemical energy into electrical energy. Used to provide current for the operation of the starter as well as other equipment, such as the radio, lighting, etc.

BEAD: The portion of a tire that holds it on the rim.

BEARING: A friction reducing, supportive device usually located between a stationary part and a moving part.

BEFORE TOP DEAD CENTER (BTDC): The point just before the piston reaches the top of its travel on the compression stroke.

BELTED TIRE: Tire construction similar to bias-ply tires, but using two or more layers of reinforced belts between body plies and the tread.

BEZEL: Piece of metal surrounding radio, headlights, gauges or similar components; sometimes used to hold the glass face of a gauge in the dash.

BIAS-PLY TIRE: Tire construction, using body ply reinforcing cords which run at alternating angles to the center line of the tread.

BI-METAL TEMPERATURE SENSOR: Any sensor or switch made of two dissimilar types of metal that bend when heated or cooled due to the different expansion rates of the alloys. These types of sensors usually function as an on/off switch.

BLOCK: See Engine Block.

BLOW-BY: Combustion gases, composed of water vapor and unburned fuel, that leak past the piston rings into the crankcase during normal engine operation. These gases are removed by the PCV system to prevent the buildup of harmful acids in the crankcase.

BOOK TIME: See Labor Time.

BOOK VALUE: The average value of a car, widely used to determine trade-in and resale value.

BOOST VALVE: Used at the base of the regulator valve to increase mainline pressure.

BORE: Diameter of a cylinder.

BRAKE CALIPER: The housing that fits over the brake disc. The caliper holds the brake pads, which are pressed against the discs by the caliper pistons when the brake pedal is depressed.

BRAKE HORSEPOWER(BHP): The actual horsepower available at the engine flywheel as measured by a dynamometer.

BRAKE FADE: Loss of braking power, usually caused by excessive heat after repeated brake applications.

BRAKE HORSEPOWER: Usable horsepower of an engine measured at the crankshaft.

BRAKE PAD: A brake shoe and lining assembly used with disc brakes.

BRAKE PROPORTIONING VALVE: A valve on the master cylinder which restricts hydraulic brake pressure to the wheels to a specified amount, preventing wheel lock-up.

BREAKAWAY: Often used by Chrysler to identify first-gear operation in D and 2 ranges. In these ranges, first-gear operation depends on a one-way roller clutch that holds on acceleration and releases (breaks away) on deceleration, resulting in a freewheeling coast-down condition.

BRAKE SHOE: The backing for the brake lining. The term is, however, usually applied to the assembly of the brake backing and lining.

BREAKER POINTS: A set of points inside the distributor, operated by a cam, which make and break the ignition circuit.

BRINNELLING: A wear pattern identified by a series of indentations at regular intervals. This condition is caused by a lack of lube, overload situations, and/or vibrations.

BTDC: Before Top Dead Center.

BUMP: Sudden and forceful apply of a clutch or band.

BUSHING: A liner, usually removable, for a bearing; an anti-friction liner used in place of a bearing.

CALIFORNIA ENGINE: An engine certified by the EPA for use in California only; conforms to more stringent emission regulations than Federal engine.

CALIPER: A hydraulically activated device in a disc brake system, which is mounted straddling the brake rotor (disc). The caliper contains at least one piston and two brake pads. Hydraulic pressure on the piston(s) forces the pads against the rotor.

CAPACITY: The quantity of electricity that can be delivered from a unit, as from a battery in ampere-hours, or output, as from a generator.

CAMBER: One of the factors of wheel alignment. Viewed from the front of the car, it is the inward or outward tilt of the wheel. The top of the tire will lean outward (positive camber) or inward (negative camber).

CAMSHAFT: A shaft in the engine on which are the lobes (cams) which operate the valves. The camshaft is driven by the crankshaft, via a belt, chain or gears, at one half the crankshaft speed.

CAPACITOR: A device which stores an electrical charge.

CARBON MONOXIDE (CO): A colorless, odorless gas given off as a normal byproduct of combustion. It is poisonous and extremely dangerous in confined areas, building up slowly to toxic levels without warning if adequate ventilation is not available.

CARBURETOR: A device, usually mounted on the intake manifold of an engine, which mixes the air and fuel in the proper proportion to allow even combustion.

CASTER: The forward or rearward tilt of an imaginary line drawn through the upper ball joint and the center of the wheel. Viewed from the sides, positive caster (forward tilt) lends directional stability, while negative caster (rearward tilt) produces instability.

CATALYTIC CONVERTER: A device installed in the exhaust system, like a muffler, that converts harmful byproducts of combustion into carbon dioxide and water vapor by means of a heat-producing chemical reaction.

CENTRIFUGAL ADVANCE: A mechanical method of advancing the spark timing by using flyweights in the distributor that react to centrifugal force generated by the distributor shaft rotation.

CENTRIFUGAL FORCE: The outward pull of a revolving object, away from the center of revolution. Centrifugal force increases with the speed of rotation.

CETANE RATING: A measure of the ignition value of diesel fuel. The higher the cetane rating, the better the fuel. Diesel fuel cetane rating is roughly comparable to gasoline octane rating.

CHECK VALVE: Any one-way valve installed to permit the flow of air, fuel or vacuum in one direction only.

CHOKE: The valve/plate that restricts the amount of air entering an engine on the induction stroke, thereby enriching the air/fuel ratio.

CHUGGLE: Bucking or jerking condition that may be engine related and may be most noticeable when converter clutch is engaged; similar to the feel of towing a trailer.

CIRCLIP: A split steel snapring that fits into a groove to hold various parts in place.

CIRCUIT BREAKER: A switch which protects an electrical circuit from overload by opening the circuit when the current flow exceeds a pre-determined level. Some circuit breakers must be reset manually, while most reset automatically.

CIRCUIT: Any unbroken path through which an electrical current can flow. Also used to describe fuel flow in some instances.

CIRCUIT, BYPASS: Another circuit in parallel with the major circuit through which power is diverted.

CIRCUIT, CLOSED: An electrical circuit in which there is no interruption of current flow.

CIRCUIT, GROUND: The non-insulated portion of a complete circuit used as a common potential point. In automotive circuits, the ground is composed of metal parts, such as the engine, body sheet metal, and frame and is usually a negative potential.

CIRCUIT, HOT: That portion of a circuit not at ground potential. The hot circuit is usually insulated and is connected to the positive side of the battery.

CIRCUIT, OPEN: A break or lack of contact in an electrical circuit, either intentional (switch) or unintentional (bad connection or broken wire).

CIRCUIT, PARALLEL: A circuit having two or more paths for current flow with common positive and negative tie points. The same voltage is applied to each load device or parallel branch.

CIRCUIT, SERIES: An electrical system in which separate parts are connected end to end, using one wire, to form a single path for current to flow.

CIRCUIT, SHORT: A circuit that is accidentally completed in an electrical path for which it was not intended.

CLAMPING (ISOLATION) DIODES: Diodes positioned in a circuit to prevent self-induction from damaging electronic components.

CLEARCOAT: A transparent layer which, when sprayed over a vehicle's paint job, adds gloss and depth as well as an additional protective coating to the finish.

CLUTCH: Part of the power train used to connect/disconnect power to the rear wheels.

CLUTCH, FLUID: The same as a fluid coupling. A fluid clutch or coupling performs the same function as a friction clutch by utilizing fluid friction and inertia as opposed to solid friction used by a friction clutch. (See fluid coupling.)

CLUTCH, FRICTION: A coupling device that provides a means of smooth and positive engagement and disengagement of engine torque to the vehicle powertrain. Transmission of power through the clutch is accomplished by bringing one or more rotating drive members into contact with complementing driven members.

COAST: Vehicle deceleration caused by engine braking conditions.

COEFFICIENT OF FRICTION: The amount of surface tension between two contacting surfaces; identified by a scientifically calculated number.

COIL: Part of the ignition system that boosts the relatively low voltage supplied by the car's electrical system to the high voltage required to fire the spark plugs.

COMBINATION MANIFOLD: An assembly which includes both the intake and exhaust manifolds in one casting.

COMBINATION VALVE: A device used in some fuel systems that routes fuel vapors to a charcoal storage canister instead of venting them into the atmosphere. The valve relieves fuel tank pressure and allows fresh air into the tank as the fuel level drops to prevent a vapor lock situation.

COMBUSTION CHAMBER: The part of the engine in the cylinder head where combustion takes place.

COMPOUND GEAR: A gear consisting of two or more simple gears with a common shaft.

COMPOUND PLANETARY: A gearset that has more than the three elements found in a simple gearset and is constructed by combining members of two planetary gearsets to create additional gear ratio possibilities.

COMPRESSION CHECK: A test involving removing each spark plug and inserting a gauge. When the engine is cranked, the gauge will record a pressure reading in the individual cylinder. General operating condition can be determined from a compression check.

COMPRESSION RATIO: The ratio of the volume between the piston and cylinder head when the piston is at the bottom of its stroke (bottom dead center) and when the piston is at the top of its stroke (top dead center).

COMPUTER: An electronic control module that correlates input data according to prearranged engineered instructions; used for the management of an actuator system or systems.

CONDENSER:

An electrical device which acts to store an electrical charge, preventing voltage surges.

2. A radiator-like device in the air conditioning system in which refrigerant gas condenses into a liquid, giving off heat.

CONDUCTOR: Any material through which an electrical current can be transmitted easily.

CONNECTING ROD: The connecting link between the crankshaft and piston.

CONSTANT VELOCITY JOINT: Type of universal joint in a halfshaft assembly in which the output shaft turns at a constant angular velocity without variation, provided that the speed of the input shaft is constant.

CONTINUITY: Continuous or complete circuit. Can be checked with an ohmmeter.

CONTROL ARM: The upper or lower suspension components which are mounted on the frame and support the ball joints and steering knuckles.

CONVENTIONAL IGNITION: Ignition system which uses breaker points.

CONVERTER: (See torque converter.)

CONVERTER LOCKUP: The switching from hydrodynamic to direct mechanical drive, usually through the application of a friction element called the converter clutch.

COOLANT: Mixture of water and anti-freeze circulated through the engine to carry off heat produced by the engine.

CORROSION INHIBITOR: An inhibitor in ATF that prevents corrosion of bushings, thrust washers, and oil cooler brazed joints.

COUNTERSHAFT: An intermediate shaft which is rotated by a mainshaft and transmits, in turn, that rotation to a working part.

COUPLING PHASE: Occurs when the torque converter is operating at its greatest hydraulic efficiency. The speed differential between the impeller and the turbine is at its minimum. At this point, the stator freewheels, and there is no torque multiplication.

CRANKCASE: The lower part of an engine in which the crankshaft and related parts operate.

CRANKSHAFT: Engine component (connected to pistons by connecting rods) which converts the reciprocating (up and down) motion of pistons to rotary motion used to turn the driveshaft.

CURB WEIGHT: The weight of a vehicle without passengers or payload, but including all fluids (oil, gas, coolant, etc.) and other equipment specified as standard.

CURRENT: The flow (or rate) of electrons moving through a circuit. Current is measured in amperes (amp).

CURRENT FLOW CONVENTIONAL: Current flows through a circuit from the positive terminal of the source to the negative terminal (plus to minus).

CURRENT FLOW, ELECTRON: Current or electrons flow from the negative terminal of the source, through the circuit, to the positive terminal (minus to plus).

CV-JOINT: Constant velocity joint.

CYCLIC VIBRATIONS: The off-center movement of a rotating object that is affected by its initial balance, speed of rotation, and working angles.

CYLINDER BLOCK: See engine block.

CYLINDER HEAD: The detachable portion of the engine, usually fastened to the top of the cylinder block and containing all or most of the combustion chambers. On overhead valve engines, it contains the valves and their operating parts. On overhead cam engines, it contains the camshaft as well.

CYLINDER: In an engine, the round hole in the engine block in which the piston(s) ride.

DATA LINK CONNECTOR (DLC): Current acronym/term applied to the federally mandated, diagnostic junction connector that is used to monitor ECM/PC/TCM inputs, processing strategies, and outputs including diagnostic trouble codes (DTCs).

DEAD CENTER: The extreme top or bottom of the piston stroke.

DECELERATION BUMP: When referring to a torque converter clutch in the applied position, a sudden release of the accelerator pedal causes a forceful reversal of power through the drivetrain (engine braking), just prior to the apply plate actually being released.

DELAYED (LATE OR EXTENDED): Condition where shift is expected but does not occur for a period of time, for example, where clutch or band engagement does not occur as quickly as expected during part throttle or wide open throttle apply of accelerator or when manually downshifting to a lower range.

DETENT: A spring-loaded plunger, pin, ball, or pawl used as a holding device on a ratchet wheel or shaft. In automatic transmissions, a detent mechanism is used for locking the manual valve in place.

DETENT DOWNSHIFT: (See kickdown.)

DETERGENT: An additive in engine oil to improve its operating characteristics.

DETONATION: An unwanted explosion of the air/fuel mixture in the combustion chamber caused by excess heat and compression, advanced timing, or an overly lean mixture. Also referred to as "ping".

DEXRON®: A brand of automatic transmission fluid.

DIAGNOSTIC TROUBLE CODES (DTCs): A digital display from the control module memory that identifies the input, processor, or output device circuit that is related to the powertrain emission/driveability malfunction detected. Diagnostic trouble codes can be read by the MIL to flash any codes or by using a handheld scanner.

DIAPHRAGM: A thin, flexible wall separating two cavities, such as in a vacuum advance unit.

DIESELING: The engine continues to run after the car is shut off; caused by fuel continuing to be burned in the combustion chamber.

DIFFERENTIAL: A geared assembly which allows the transmission of motion between drive axles, giving one axle the ability to rotate faster than the other, as in cornering.

DIFFERENTIAL AREAS: When opposing faces of a spool valve are acted upon by the same pressure but their areas differ in size, the face with the larger area produces the differential force and valve movement. (See spool valve.)

DIFFERENTIAL FORCE: (See differential areas)

DIGITAL READOUT: A display of numbers or a combination of numbers and letters.

DIGITAL VOLT OHMMETER: An electronic diagnostic tool used to measure voltage, ohms and amps as well as several other functions, with the readings displayed on a digital screen in tenths, hundredths and thousandths.

DIODE: An electrical device that will allow current to flow in one direction only.

DIRECT CURRENT (DC): Electrical current that flows in one direction only.

DIRECT DRIVE: The gear ratio is 1:1, with no change occurring in the torque and speed input/output relationship.

DISC BRAKE: A hydraulic braking assembly consisting of a brake disc, or rotor, mounted on an axle shaft, and a caliper assembly containing, usually two brake pads which are activated by hydraulic pressure. The pads are forced against the sides of the disc, creating friction which slows the vehicle.

DISPERSANTS: Suspend dirt and prevent sludge buildup in a liquid, such as engine oil.

DOUBLE BUMP (DOUBLE FEEL): Two sudden and forceful applies of a clutch or band.

DISPLACEMENT: The total volume of air that is displaced by all pistons as the engine turns through one complete revolution.

DISTRIBUTOR: A mechanically driven device on an engine which is responsible for electrically firing the spark plug at a pre-determined point of the piston stroke.

DOHC: Double overhead camshaft.

DOUBLE OVERHEAD CAMSHAFT: The engine utilizes two camshafts mounted in one cylinder head. One camshaft operates the exhaust valves, while the other operates the intake valves.

DOWEL PIN: A pin, inserted in mating holes in two different parts allowing those parts to maintain a fixed relationship.

DRIVELINE: The drive connection between the transmission and the drive wheels.

DRIVE TRAIN: The components that transmit the flow of power from the engine to the wheels. The components include the clutch, transmission, driveshafts (or axle shafts in front wheel drive), U-joints and differential.

DRUM BRAKE: A braking system which consists of two brake shoes and one or two wheel cylinders, mounted on a fixed backing plate, and a brake drum, mounted on an axle, which revolves around the assembly.

DRY CHARGED BATTERY: Battery to which electrolyte is added when the battery is placed in service.

DVOM: Digital volt ohmmeter

DWELL: The rate, measured in degrees of shaft rotation, at which an electrical circuit cycles on and off.

DYNAMIC: An application in which there is rotating or reciprocating motion between the parts.

EARLY: Condition where shift occurs before vehicle has reached proper speed, which tends to labor engine after upshift.

EBCM: See Electronic Control Unit (ECU).

ECM: See Electronic Control Unit (ECU).

ECU: Electronic control unit.

ELECTRODE: Conductor (positive or negative) of electric current.

ELECTROLYSIS: A surface etching or bonding of current conducting

transmission/transaxle components that may occur when grounding straps are missing or in poor condition.

ELECTROLYTE: A solution of water and sulfuric acid used to activate the battery. Electrolyte is extremely corrosive.

ELECTROMAGNET: A coil that produces a magnetic field when current flows through its windings.

ELECTROMAGNETIC INDUCTION: A method to create (generate) current flow through the use of magnetism.

ELECTROMAGNETISM: The effects surrounding the relationship between electricity and magnetism.

ELECTROMOTIVE FORCE (EMF): The force or pressure (voltage) that causes current movement in an electrical circuit.

ELECTRONIC CONTROL UNIT: A digital computer that controls engine (and sometimes transmission, brake or other vehicle system) functions based on data received from various sensors. Examples used by some manufacturers include Electronic Brake Control Module (EBCM), Engine Control Module (ECM), Powertrain Control Module (PCM) or Vehicle Control Module (VCM).

ELECTRONIC IGNITION: A system in which the timing and firing of the spark plugs is controlled by an electronic control unit, usually called a module. These systems have no points or condenser.

ELECTRONIC PRESSURE CONTROL (EPC) SOLENOID: A specially designed solenoid containing a spool valve and spring assembly to control fluid mainline pressure. A variable current flow, controlled by the ECM/PCM, varies the internal force of the solenoid on the spool valve and resulting mainline pressure. (See variable force solenoid.)

ELECTRONICS: Miniaturized electrical circuits utilizing semiconductors, solid-state devices, and printed circuits. Electronic circuits utilize small amounts of power.

ELECTRONIFICATION: The application of electronic circuitry to a mechanical device. Regarding automatic transmissions, electrification is incorporated into converter clutch lockup, shift scheduling, and line pressure control systems.

ELECTROSTATIC DISCHARGE (ESD): An unwanted, high-voltage electrical current released by an individual who has taken on a static charge of electricity. Electronic components can be easily damaged by ESD.

ELEMENT: A device within a hydrodynamic drive unit designed with a set of blades to direct fluid flow.

ENAMEL: Type of paint that dries to a smooth, glossy finish.

END BUMP (END FEEL OR SLIP BUMP): Firmer feel at end of shift when compared with feel at start of shift.

END-PLAY: The clearance/gap between two components that allows for expansion of the parts as they warm up, to prevent binding and to allow space for lubrication.

ENERGY: The ability or capacity to do work.

ENGINE: The primary motor or power apparatus of a vehicle, which converts liquid or gas fuel into mechanical energy.

ENGINE BLOCK: The basic engine casting containing the cylinders, the crankshaft main bearings, as well as machined surfaces for the mounting of other components such as the cylinder head, oil pan, transmission, etc..

ENGINE BRAKING: Use of engine to slow vehicle by manually downshifting during zero-throttle coast down.

ENGINE CONTROL MODULE (ECM): Manages the engine and incorporates output control over the torque converter clutch solenoid. (Note: Current designation for the ECM in late model vehicles is PCM.)

ENGINE COOLANT TEMPERATURE (ECT) SENSOR: Prevents converter clutch engagement with a cold engine; also used for shift timing and shift quality.

EP LUBRICANT: EP (extreme pressure) lubricants are specially formulated for use with gears involving heavy loads (transmissions, differentials, etc.).

ETHYL: A substance added to gasoline to improve its resistance to knock, by slowing down the rate of combustion.

ETHYLENE GLYCOL: The base substance of antifreeze.

EXHAUST MANIFOLD: A set of cast passages or pipes which conduct exhaust gases from the engine.

FAIL-SAFE (BACKUP) CONTROL: A substitute value used by the PCM/TCM to replace a faulty signal from an input sensor. The temporary value allows the vehicle to continue to be operated.

FAST IDLE: The speed of the engine when the choke is on. Fast idle speeds engine warm-up.

FEDERAL ENGINE: An engine certified by the EPA for use in any of the 49 states (except California).

FEEDBACK: A circuit malfunction whereby current can find another path to feed load devices.

FEELER GAUGE: A blade, usually metal, of precisely predetermined thickness, used to measure the clearance between two parts.

FILAMENT: The part of a bulb that glows; the filament creates high resistance to current flow and actually glows from the resulting heat.

FINAL DRIVE: An essential part of the axle drive assembly where final gear reduction takes place in the powertrain. In RWD applications and north-south FWD applications, it must also change the power flow direction to the axle shaft by ninety degrees. (Also see axle ratio).

FIRING ORDER: The order in which combustion occurs in the cylinders of an engine. Also the order in which spark is distributed to the plugs by the distributor.

FIRM: A noticeable quick apply of a clutch or band that is considered normal with medium to heavy throttle shift; should not be confused with harsh or rough.

FLAME FRONT: The term used to describe certain aspects of the fuel explosion in the cylinders. The flame front should move in a controlled pattern across the cylinder, rather than simply exploding immediately.

FLARE (SLIPPING): A quick increase in engine rpm accompanied by momentary loss of torque; generally occurs during shift.

FLAT ENGINE: Engine design in which the pistons are horizontally opposed. Porsche, Subaru and some old VW are common examples of flat engines.

FLAT RATE: A dealership term referring to the amount of money paid to a technician for a repair or diagnostic service based on that particular service versus dealership's labor time (NOT based on the actual time the technician spent on the job).

FLAT SPOT: A point during acceleration when the engine seems to lose power for an instant.

FLOODING: The presence of too much fuel in the intake manifold and combustion chamber which prevents the air/fuel mixture from firing, thereby causing a no-start situation.

FLUID: A fluid can be either liquid or gas. In hydraulics, a liquid is used for transmitting force or motion.

FLUID COUPLING: The simplest form of hydrodynamic drive, the fluid coupling consists of two look-alike members with straight radial varies referred to as the impeller (pump) and the turbine. Input torque is always equal to the output torque.

FLUID DRIVE: Either a fluid coupling or a fluid torque converter. (See hydrodynamic drive units.)

FLUID TORQUE CONVERTER: A hydrodynamic drive that has the ability to act both as a torque multiplier and fluid coupling. (See hydrodynamic drive units; torque converter.)

FLUID VISCOSITY: The resistance of a liquid to flow. A cold fluid (oil) has greater viscosity and flows more slowly than a hot fluid (oil).

FLYWHEEL: A heavy disc of metal attached to the rear of the crankshaft. It smoothes the firing impulses of the engine and keeps the crankshaft turning during periods when no firing takes place. The starter also engages the flywheel to start the engine.

FOOT POUND (ft. lbs., lbs. ft. or sometimes, ft. lb.): The amount of energy or work needed to raise an item weighing one pound, a distance of one foot.

FREEZE PLUG: A plug in the engine block which will be pushed out if the coolant freezes. Sometimes called expansion plugs, they protect the block from cracking should the coolant freeze.

FRICTION: The resistance that occurs between contacting surfaces. This relationship is expressed by a ratio called the coefficient of friction (CL).

FRICTION, COEFFICIENT OF: The amount of surface tension between two contacting surfaces; expressed by a scientifically calculated number.

FRONT END ALIGNMENT: A service to set caster, camber and toe-in to the correct specifications. This will ensure that the car steers and handles properly and that the tires wear properly.

FRICTION MODIFIER: Changes the coefficient of friction of the fluid between the mating steel and composition clutch/band surfaces during the engagement process and allows for a certain amount of intentional slipping for a good "shift-feel".

FRONTAL AREA: The total frontal area of a vehicle exposed to air flow.

FUEL FILTER: A component of the fuel system containing a porous paper element used to prevent any impurities from entering the engine through the fuel system. It usually takes the form of a canister-like housing, mounted in-line with the fuel hose, located anywhere on a vehicle between the fuel tank and engine.

FUEL INJECTION: A system replacing the carburetor that sprays fuel into the cylinder through nozzles. The amount of fuel can be more precisely controlled with fuel injection.

FULL FLOATING AXLE: An axle in which the axle housing extends through the wheel giving bearing support on the outside of the housing. The front axle of a four-wheel drive vehicle is usually a full floating axle, as are the rear axles of many larger (1 ton and over) pick-ups and vans.

FULL-TIME FOUR-WHEEL DRIVE: A four-wheel drive system that continuously delivers power to all four wheels. A differential between the front and rear driveshafts permits variations in axle speeds to control gear wind-up without damage.

FULL THROTTLE DETENT DOWNSHIFT: A quick apply of accelerator pedal to its full travel, forcing a downshift.

FUSE: A protective device in a circuit which prevents circuit overload by breaking the circuit when a specific amperage is present. The device is constructed around a strip or wire of a lower amperage rating than the circuit it is designed to protect. When an amperage higher than that stamped on the fuse is present in the circuit, the strip or wire melts, opening the circuit.

FUSIBLE LINK: A piece of wire in a wiring harness that performs the same job as a fuse. If overloaded, the fusible link will melt and interrupt the circuit.

FWD: Front wheel drive.

GAWR: (Gross axle weight rating) the total maximum weight an axle is designed to carry.

GCW: (Gross combined weight) total combined weight of a tow vehicle and trailer.

GARAGE SHIFT: initial engagement feel of transmission, neutral to reverse or neutral to a forward drive.

GARAGE SHIFT FEEL: A quick check of the engagement quality and responsiveness of reverse and forward gears. This test is done with the vehicle stationary.

GEAR: A toothed mechanical device that acts as a rotating lever to transmit power or turning effort from one shaft to another. (See gear ratio.)

GEAR RATIO: A ratio expressing the number of turns a smaller gear will make to turn a larger gear through one revolution. The ratio is found by dividing the number of teeth on the smaller gear into the number of teeth on the larger gear.

GEARBOX: Transmission

GEAR REDUCTION: Torque is multiplied and speed decreased by the factor of the gear ratio. For example, a 3:1 gear ratio changes an input torque of 180 ft. lbs. and an input speed of 2700 rpm to 540 Ft. lbs. and 900 rpm, respectively. (No account is taken of frictional losses, which are always present.)

GEARTRAIN: A succession of intermeshing gears that form an assembly and provide for one or more torque changes as the power input is transmitted to the power output.

GEL COAT: A thin coat of plastic resin covering fiberglass body panels.

GENERATOR: A device which produces direct current (DC) necessary to charge the battery.

GOVERNOR: A device that senses vehicle speed and generates a hydraulic oil pressure. As vehicle speed increases, governor oil pressure rises.

GROUND CIRCUIT: (See circuit, ground.)

GROUND SIDE SWITCHING: The electrical/electronic circuit control switch is located after the circuit load.

GVWR: (Gross vehicle weight rating) total maximum weight a vehicle is designed to carry including the weight of the vehicle, passengers, equipment, gas, oil, etc.

HALOGEN: A special type of lamp known for its quality of brilliant white light. Originally used for fog lights and driving lights.

HARD CODES: DTCs that are present at the time of testing; also called continuous or current codes.

HARSH(ROUGH): An apply of a clutch or band that is more noticeable than a firm one; considered undesirable at any throttle position.

HEADER TANK: An expansion tank for the radiator coolant. It can be located remotely or built into the radiator.

HEAT RANGE: A term used to describe the ability of a spark plug to carry away heat. Plugs with longer nosed insulators take longer to carry heat off effectively.

HEAT RISER: A flapper in the exhaust manifold that is closed when the engine is cold, causing hot exhaust gases to heat the intake manifold providing better cold engine operation. A thermostatic spring opens the flapper when the engine warms up.

HEAVY THROTTLE: Approximately three-fourths of accelerator pedal travel.

HEMI: A name given an engine using hemispherical combustion chambers.

HERTZ (HZ): The international unit of frequency equal to one cycle per second (10,000 Hertz equals 10,000 cycles per second).

HIGH-IMPEDANCE DVOM (DIGITAL VOLT-OHMMETER): This styled device provides a built-in resistance value and is capable of limiting circuit current flow to safe milliamp levels.

HIGH RESISTANCE: Often refers to a circuit where there is an excessive amount of opposition to normal current flow.

HORSEPOWER: A measurement of the amount of work; one horsepower is the amount of work necessary to lift 33,000 lbs. one foot in one minute. Brake horsepower (bhp) is the horsepower delivered by an engine on a dynamometer. Net horsepower is the power remaining (measured at the flywheel of the engine) that can be used to turn the wheels after power is consumed through friction and running the engine accessories (water pump, alternator, air pump, fan etc.)

HOT CIRCUIT: (See circuit, hot; hot lead.) hot lead: **A wire or conductor in the power side of the circuit. (See circuit, hot.)**

HOT SIDE SWITCHING: The electrical/electronic circuit control switch is located before the circuit load.

HUB: The center part of a wheel or gear.

HUNTING (BUSYNESS): Repeating quick series of up-shifts and downshifts that causes noticeable change in engine rpm, for example, as in a 4-3-4 shift pattern.

HYDRAULICS: The use of liquid under pressure to transfer force of motion.

HYDROCARBON (HC): Any chemical compound made up of hydrogen and carbon. A major pollutant formed by the engine as a by-product of combustion.

HYDRODYNAMIC DRIVE UNITS: Devices that transmit power solely by the action of a kinetic fluid flow in a closed recirculating path. An impeller energizes the fluid and discharges the high-speed jet stream into the turbine for power output.

HYDROMETER: An instrument used to measure the specific gravity of a solution.

HYDROPLANING: A phenomenon of driving when water builds up under the tire tread, causing it to lose contact with the road. Slowing down will usually restore normal tire contact with the road.

HYPOID GEARSET: The drive pinion gear may be placed below or above the centerline of the driven gear; often used as a final drive gearset.

IDLE MIXTURE: The mixture of air and fuel (usually about 14:1) being fed to the cylinders. The idle mixture screw(s) are sometimes adjusted as part of a tune-up.

IDLER ARM: Component of the steering linkage which is a geometric duplicate of the steering gear arm. It supports the right side of the center steering link.

IMPELLER: Often called a pump, the impeller is the power input (drive) member of a hydrodynamic drive. As part of the torque converter cover, it acts as a centrifugal pump and puts the fluid in motion.

INCH POUND (inch lbs.; sometimes in. lb. or in. lbs.): One twelfth of a foot pound.

INDUCTANCE: The force that produces voltage when a conductor is passed through a magnetic field.

INDUCTION: A means of transferring electrical energy in the form of a magnetic field. Principle used in the ignition coil to increase voltage.

INITIAL FEEL: A distinct firmer feel at start of shift when compared with feel at finish of shift.

INJECTOR: A device which receives metered fuel under relatively low pressure and is activated to inject the fuel into the engine under relatively high pressure at a predetermined time.

INPUT: In an automatic transmission, the source of power from the engine is absorbed by the torque converter, which provides the power input into the transmission. The turbine drives the input(turbine)shaft.

INPUT SHAFT: The shaft to which torque is applied, usually carrying the driving gear or gears.

INTAKE MANIFOLD: A casting of passages or pipes used to conduct air or a fuel/air mixture to the cylinders.

INTERNAL GEAR: The ring-like outer gear of a planetary gearset with the gear teeth cut on the inside of the ring to provide a mesh with the planet pinions.

ISOLATION (CLAMPING) DIODES: Diodes positioned in a circuit to prevent self-induction from damaging electronic components.

IX ROTARY GEAR PUMP: Contains two rotating members, one shaped with internal gear teeth and the other with external gear teeth. As the gears separate, the fluid fills the gaps between gear teeth, is pulled across a crescent-shaped divider, and then is forced to flow through the outlet as the gears mesh.

IX ROTARY LOBE PUMP: Sometimes referred to as a gerotor type pump. Two rotating members, one shaped with internal lobes and the other with external lobes, separate and then mesh to cause fluid to flow.

JOURNAL: The bearing surface within which a shaft operates.

JUMPER CABLES: Two heavy duty wires with large alligator clips used to provide power from a charged battery to a discharged battery mounted in a vehicle.

JUMPSTART: Utilizing the sufficiently charged battery of one vehicle to start the engine of another vehicle with a discharged battery by the use of jumper cables.

KEY: A small block usually fitted in a notch between a shaft and a hub to prevent slippage of the two parts.

KICKDOWN: Detent downshift system; either linkage, cable, or electrically controlled.

KILO: A prefix used in the metric system to indicate one thousand.

KNOCK: Noise which results from the spontaneous ignition of a portion of the air-fuel mixture in the engine cylinder caused by overly advanced ignition timing or use of incorrectly low octane fuel for that engine.

KNOCK SENSOR: An input device that responds to spark knock, caused by over advanced ignition timing.

LABOR TIME: A specific amount of time required to perform a certain repair or diagnostic service as defined by a vehicle or after-market manufacturer.

LACQUER: A quick-drying automotive paint.

LATE: Shift that occurs when engine is at higher than normal rpm for given amount of throttle.

LIGHT-EMITTING DIODE (LED): A semiconductor diode that emits light as electrical current flows through it; used in some electronic display devices to emit a red or other color light.

LIGHT THROTTLE: Approximately one-fourth of accelerator pedal travel.

LIMITED SLIP: A type of differential which transfers driving force to the wheel with the best traction.

LIMP-IN MODE: Electrical shutdown of the transmission/ transaxle output solenoids, allowing only forward and reverse gears that are hydraulically energized by the manual valve. This permits the vehicle to be driven to a service facility for repair.

LIP SEAL: Molded synthetic rubber seal designed with an outer sealing edge (lip) that points into the fluid containing area to be sealed. This type of seal is used where rotational and axial forces are present.

LITHIUM-BASE GREASE: Chassis and wheel bearing grease using lithium as a base. Not compatible with sodium-base grease.

LOAD DEVICE: A circuit's resistance that converts the electrical energy into light, sound, heat, or mechanical movement.

LOAD RANGE: Indicates the number of plies at which a tire is rated. Load range B equals four-ply rating; C equals six-ply rating; and, D equals an eight-ply rating.

LOAD TORQUE: The amount of output torque needed from the transmission/transaxle to overcome the vehicle load.

LOCKING HUBS: Accessories used on part-time four-wheel drive systems that allow the front wheels to be disengaged from the drive train when four-wheel drive is not being used. When four-wheel drive is desired, the hubs are engaged, locking the wheels to the drive train.

LOCKUP CONVERTER: A torque converter that operates hydraulically and mechanically. When an internal apply plate (lockup plate) clamps to the torque converter cover, hydraulic slippage is eliminated.

LOCK RING: See Circlip or Snapring

MAGNET: Any body with the property of attracting iron or steel.

MAGNETIC FIELD: The area surrounding the poles of a magnet that is affected by its attraction or repulsion forces.

MAIN LINE PRESSURE: Often called control pressure or line pressure, it refers to the pressure of the oil leaving the pump and is controlled by the pressure regulator valve.

MALFUNCTION INDICATOR LAMP (MIL): Previously known as a check engine light, the dash-mounted MIL illuminates and signals the driver that an emission or driveability problem with the powertrain has been detected by the ECM/PCM. When this occurs, at least one diagnostic trouble code (DTC) has been stored into the control module memory.

MANIFOLD ABSOLUTE PRESSURE (MAP) SENSOR: Reads the amount of air pressure (vacuum) in the engine's intake manifold system; its signal is used to analyze engine load conditions.

MANIFOLD VACUUM: Low pressure in an engine intake manifold formed just below the throttle plates. Manifold vacuum is highest at idle and drops under acceleration.

MANIFOLD: A casting of passages or set of pipes which connect the cylinders to an inlet or outlet source.

MANUAL LEVER POSITION SWITCH (MLPS): A mechanical switching unit that is typically mounted externally to the transmission/transaxle to inform the PCM/ECM which gear range the driver has selected.

MANUAL VALVE: Located inside the transmission/transaxle, it is directly connected to the driver's shift lever. The position of the manual valve determines which hydraulic circuits will be charged with oil pressure and the operating mode of the transmission.

MANUAL VALVE LEVER POSITION SENSOR (MVLPS): The input from this device tells the TCM what gear range was selected.

MASS AIR FLOW (MAF) SENSOR: Measures the airflow into the engine.

MASTER CYLINDER: The primary fluid pressurizing device in a hydraulic system. In automotive use, it is found in brake and hydraulic clutch systems and is pedal activated, either directly or, in a power brake system, through the power booster.

MacPherson STRUT: A suspension component combining a shock absorber and spring in one unit.

MEDIUM THROTTLE: Approximately one-half of accelerator pedal travel.

MEGA: A metric prefix indicating one million.

MEMBER: An independent component of a hydrodynamic unit such as an impeller, a stator, or a turbine. It may have one or more elements.

MERCON: A fluid developed by Ford Motor Company in 1988. It contains a friction modifier and closely resembles operating characteristics of Dexron.

METAL SEALING RINGS: Made from cast iron or aluminum, their primary application is with dynamic components involving pressure sealing circuits of rotating members. These rings are designed with either butt or hook lock end joints.

METER (ANALOG): A linear-style meter representing data as lengths; a needle-style instrument interfacing with logical numerical increments. This style of electrical meter uses relatively low impedance internal resistance and cannot be used for testing electronic circuitry.

METER(DIGITAL): Uses numbers as a direct readout to show values. Most meters of this style use high impedance internal resistance and must be used for testing low current electronic circuitry.

MICRO: A metric prefix indicating one-millionth (0.000001).

MILLI: A metric prefix indicating one-thousandth (0.001).

MINIMUM THROTTLE: The least amount of throttle opening required for upshift; normally close to zero throttle.

MISFIRE: Condition occurring when the fuel mixture in a cylinder fails to ignite, causing the engine to run roughly.

MODULE: Electronic control unit, amplifier or igniter of solid state or integrated design which controls the current flow in the ignition primary circuit based on input from the pick-up coil. When the module opens the primary circuit, high secondary voltage is induced in the coil.

MODULATED: In an electronic-hydraulic converter clutch system (or shift valve system), the term modulated refers to the pulsing of a solenoid, at a variable rate. This action controls the buildup of oil pressure in the hydraulic circuit to allow a controlled amount of clutch slippage.

MODULATED CONVERTER CLUTCH CONTROL (MCCC): A pulse width duty cycle valve that controls the converter lockup apply pressure and maximizes smoother transitions between lock and unlock conditions.

MODULATOR PRESSURE (THROTTLE PRESSURE): A hydraulic signal oil pressure relating to the amount of engine load, based on either the amount of throttle plate opening or engine vacuum.

MODULATOR VALVE: A regulator valve that is controlled by engine vacuum, providing a hydraulic pressure that varies in relation to engine torque. The hydraulic torque signal functions to delay the shift pattern and provide a line pressure boost. (See throttle valve.)

MOTOR: An electromagnetic device used to convert electrical energy into mechanical energy.

MULTIPLE-DISC CLUTCH: A grouping of steel and friction lined plates that, when compressed together by hydraulic pressure acting upon a piston, lock or unlock a planetary member.

MULTI-WEIGHT: Type of oil that provides adequate lubrication at both high and low temperatures.

needed to move one amp through a resistance of one ohm.

MUSHY: Same as soft; slow and drawn out clutch apply with very little shift feel.

MUTUAL INDUCTION: The generation of **Current from one wire circuit to another by movement of the magnetic field surrounding a current-carrying circuit as its ampere flow increases or decreases.**

NEEDLE BEARING: A bearing which consists of a number (usually a large number) of long, thin rollers.

NITROGEN OXIDE (NOx): One of the three basic pollutants found in the exhaust emission of an internal combustion engine. The amount of NOx usually varies in an inverse proportion to the amount of HC and CO.

NONPOSITIVE SEALING: A sealing method that allows some minor leakage, which normally assists in lubrication.

O2 SENSOR: Located in the engine's exhaust system, it is an input device to the ECM/PCM for managing the fuel delivery and ignition system. A scanner can be used to observe the fluctuating voltage readings produced by an O2 sensor as the oxygen content of the exhaust is analyzed.

O-RING SEAL: Molded synthetic rubber seal designed with a circular cross-section. This type of seal is used primarily in static applications.

OBD II (ON-BOARD DIAGNOSTICS, SECOND GENERATION): Refers to the federal law mandating tighter control of 1996 and newer vehicle emissions, active monitoring of related devices, and standardization of terminology, data link connectors, and other technician concerns.

OCTANE RATING: A number, indicating the quality of gasoline based on its ability to resist knock. The higher the number, the better the quality. Higher compression engines require higher octane gas.

OEM: Original Equipment Manufactured. OEM equipment is that furnished standard by the manufacturer.

OFFSET: The distance between the vertical center of the wheel and the mounting surface at the lugs. Offset is positive if the center is outside the lug circle; negative offset puts the center line inside the lug circle.

OHM'S LAW: A law of electricity that states the relationship between voltage, current, and resistance. Volts = amperes x ohms

OHM: The unit used to measure the resistance of conductor-to-electrical flow. One ohm is the amount of resistance that limits current flow to one ampere in a circuit with one volt of pressure.

OHMMETER: An instrument used for measuring the resistance, in ohms, in an electrical circuit.

ONE-WAY CLUTCH: A mechanical clutch of roller or sprag design that resists torque or transmits power in one direction only. It is used to either hold or drive a planetary member.

ONE-WAY ROLLER CLUTCH: A mechanical device that transmits or holds torque in one direction only.

OPENCIRCUIT: A break or lack of contact in an electrical circuit, either intentional (switch) or unintentional (bad connection or broken wire).

ORIFICE: Located in hydraulic oil circuits, it acts as a restriction. It slows down fluid flow to either create back pressure or delay pressure buildup downstream.

OSCILLOSCOPE: A piece of test equipment that shows electric impulses as a pattern on a screen. Engine performance can be analyzed by interpreting these patterns.

OUTPUT SHAFT: The shaft which transmits torque from a device, such as a transmission.

OUTPUT SPEED SENSOR (OSS): Identifies transmission/transaxle output shaft speed for shift timing and may be used to calculate TCC slip; often functions as the VSS (vehicle speed sensor).

OVERDRIVE: (1.) A device attached to or incorporated in a transmission/transaxle that allows the engine to turn less than one full revolution for every complete revolution of the wheels. The net effect is to reduce engine rpm, thereby using less fuel. A typical overdrive gear ratio would be .87:1, instead of the normal 1:1 in high gear. (2.) A gear assembly which produces more shaft revolutions than that transmitted to it.

OVERDRIVE PLANETARY GEARSET: A single planetary gearset designed to provide a direct drive and overdrive ratio. When coupled to a three-speed transmission/transaxle configuration, a four-speed/overdrive unit is present.

OVERHEAD CAMSHAFT (OHC): An engine configuration in which the camshaft is mounted on top of the cylinder head and operates the valve either directly or by means of rocker arms.

OVERHEAD VALVE (OHV): An engine configuration in which all of the valves are located in the cylinder head and the camshaft is located in the cylinder block. The camshaft operates the valves via lifters and pushrods.

OVERRUNCLUTCH: Another name for a one-way mechanical clutch. Applies to both roller and sprag designs.

OVERSTEER: The tendency of some vehicles, when steering into a turn, to over-respond or steer more than required, which could result in excessive slip of the rear wheels. Opposite of under-steer.

OXIDATION STABILIZERS: Absorb and dissipate heat. Automatic transmission fluid has high resistance to varnish and sludge buildup that occurs from excessive heat that is generated primarily in the torque converter. Local temperatures as high as 6000F (3150C) can occur at the clutch plates during engagement, and this heat must be absorbed and dissipated. If the fluid cannot withstand the heat, it burns or oxidizes, resulting in an almost immediate destruction of friction materials, clogged filter screen and hydraulic passages, and sticky valves.

OXIDES OF NITROGEN: See nitrogen oxide (NOx).

OXYGEN SENSOR: Used with a feedback system to sense the presence of oxygen in the exhaust gas and signal the computer which can use the voltage signal to determine engine operating efficiency and adjust the air/fuel ratio.

PARALLEL CIRCUIT: (See circuit, parallel.)

PARTS WASHER: A basin or tub, usually with a built-in pump mechanism and hose used for circulating chemical solvent for the purpose of cleaning greasy, oily and dirty components.

PART-TIME FOUR WHEEL DRIVE: A system that is normally in the two wheel drive mode and only runs in four-wheel drive when the system is manually engaged because more traction is desired. Two or four wheel drive is normally selected by a lever to engage the front axle, but if locking hubs are used, these must also be manually engaged in the Lock position. Otherwise, the front axle will not drive the front wheels.

PASSIVE RESTRAINT: Safety systems such as air bags or automatic seat belts which operate with no action required on the part of the driver or passenger. Mandated by Federal regulations on all vehicles sold in the U.S. after 1990.

PAYLOAD: The weight the vehicle is capable of carrying in addition to its own weight. Payload includes weight of the driver, passengers and cargo, but not coolant, fuel, lubricant, spare tire, etc.

PCM: Powertrain control module.

PCV VALVE: A valve usually located in the rocker cover that vents crankcase vapors back into the engine to be reburned.

PERCOLATION: A condition in which the fuel actually "boils," due to excessive heat. Percolation prevents proper atomization of the fuel causing rough running.

PICK-UP COIL: The coil in which voltage is induced in an electronic ignition.

PINION GEAR: The smallest gear in a drive gear assembly. piston: **A disc or cup that fits in a cylinder bore and is free to move. In hydraulics, it provides the means of converting hydraulic pressure into a usable force. Examples of piston applications are found in servo, clutch, and accumulator units.**

PING: A metallic rattling sound produced by the engine during acceleration. It is usually due to incorrect ignition timing or a poor grade of gasoline.

PINION: The smaller of two gears. The rear axle pinion drives the ring gear which transmits motion to the axle shafts.

PISTON RING: An open-ended ring which fits into a groove on the outer diameter of the piston. Its chief function is to form a seal between the piston and cylinder wall. Most automotive pistons have three rings: two for compression sealing; one for oil sealing.

PITMAN ARM: A lever which transmits steering force from the steering gear to the steering linkage.

PLANET CARRIER: A basic member of a planetary gear assembly that carries the pinion gears.

PLANET PINIONS: Gears housed in a planet carrier that are in constant mesh with the sun gear and internal gear. Because they have their own independent rotating centers, the pinions are capable of rotating around the sun gear or the inside of the internal gear.

PLANETARY GEAR RATIO: The reduction or overdrive ratio developed by a planetary gearset.

PLANETARY GEARSET: In its simplest form, it is made up of a basic assembly group containing a sun gear, internal gear, and planet carrier. The gears are always in constant mesh and offer a wide range of gear ratio possibilities.

PLANETARY GEARSET(COMPOUND): Two planetary gearsets combined together.

PLANETARY GEARSET(SIMPLE): An assembly of gears in constant mesh consisting of a sun gear, several pinion gears mounted in a carrier, and a ring gear. It provides gear ratio and direction changes, in addition to a direct drive and a neutral.

PLY RATING: A. rating given a tire which indicates strength (but not necessarily actual plies). A two-ply/four-ply rating has only two plies, but the strength of a four-ply tire.

POLARITY: Indication (positive or negative) of the two poles of a battery.

PORT: An opening for fluid intake or exhaust.

POSITIVE SEALING: A sealing method that completely prevents leakage.

POTENTIAL: Electrical force measured in volts; sometimes used interchangeably with voltage.

POWER: The ability to do work per unit of time, as expressed in horsepower; one horsepower equals 33,000 ft. lbs. of work per minute, or 550 ft. lbs. of work per second.

POWER FLOW: The systematic flow or transmission of power through the gears, from the input shaft to the output shaft.

POWER-TO-WEIGHT RATIO: Ratio of horsepower to weight of car.

POWERTRAIN: See Drivetrain.

POWERTRAIN CONTROL MODULE(PCM): Current designation for the engine control module (ECM). In many cases, late model vehicle control units manage the engine as well as the transmission. In other settings, the PCM controls the engine and is interfaced with a TCM to control transmission functions.

Ppm: Parts per million; unit used to measure exhaust emissions.

PREIGNITION: Early ignition of fuel in the cylinder, sometimes due to glowing carbon deposits in the combustion chamber. Preignition can be damaging since combustion takes place prematurely.

PRELOAD: A predetermined load placed on a bearing during assembly or by adjustment.

PRESS FIT: The mating of two parts under pressure, due to the inner diameter of one being smaller than the outer diameter of the other, or vice versa; an interference fit.

PRESSURE: The amount of force exerted upon a surface area.

PRESSURE CONTROL SOLENOID (PCS): An output device that provides a boost oil pressure to the mainline regulator valve to control line pressure. Its operation is determined by the amount of current sent from the PCM.

PRESSURE GAUGE: An instrument used for measuring the fluid pressure in a hydraulic circuit.

PRESSURE REGULATOR VALVE: In automatic transmissions, its purpose is to regulate the pressure of the pump output and supply the basic fluid pressure necessary to operate the transmission. The regulated fluid pressure may be referred to as mainline pressure, line pressure, or control pressure.

PRESSURE SWITCH ASSEMBLY (PSA): Mounted inside the transmission, it is a grouping of oil pressure switches that inputs to the PCM when certain hydraulic passages are charged with oil pressure.

PRESSURE PLATE: A spring-loaded plate (part of the clutch) that transmits power to the driven (friction) plate when the clutch is engaged.

PRIMARY CIRCUIT: The low voltage side of the ignition system which consists of the ignition switch, ballast resistor or resistance wire, bypass, coil, electronic control unit and pick-up coil as well as the connecting wires and harnesses.

PROFILE: Term used for tire measurement (tire series), which is the ratio of tire height to tread width.

PROM (PROGRAMMABLE READ-ONLY MEMORY): The heart of the computer that compares input data and makes the engineered program or strategy decisions about when to trigger the appropriate output based on stored computer instructions.

Pulse generator: A two-wire pickup sensor used to produce a fluctuating electrical signal. This changing signal is read by the controller to determine the speed of the object and can be used to measure transmission/transaxle input speed, output speed, and vehicle speed.

PSI: Pounds per square inch; a measurement of pressure.

PULSE WIDTH DUTY CYCLE SOLENOID (PULSE WIDTH MODULATED SOLENOID): A computer-controlled solenoid that turns on and off at a variable rate producing a modulated oil pressure; often referred to as a pulse width modulated (PWM) solenoid. Employed in many electronic automatic transmissions and transaxles, these solenoids are used to manage shift control and converter clutch hydraulic circuits.

PUSHROD: A steel rod between the hydraulic valve lifter and the valve rocker arm in overhead valve (OHV) engines.

PUMP: A mechanical device designed to create fluid flow and pressure buildup in a hydraulic system.

QUARTER PANEL: General term used to refer to a rear fender. Quarter panel is the area from the rear door opening to the tail light area and from rear wheel well to the base of the trunk and roof-line.

RACE: The surface on the inner or outer ring of a bearing on which the balls, needles or rollers move.

RACK AND PINION: A type of automotive steering system using a pinion gear attached to the end of the steering shaft. The pinion meshes with a long rack attached to the steering linkage.

RADIAL TIRE: Tire design which uses body cords running at right angles to the center line of the tire. Two or more belts are used to give tread strength. Radials can be identified by their characteristic sidewall bulge.

RADIATOR: Part of the cooling system for a water-cooled engine, mounted in the front of the vehicle and connected to the engine with rubber hoses. Through the radiator, excess combustion heat is dissipated into the atmosphere through forced convection using a water and glycol based mixture that circulates through, and cools, the engine.

RANGE REFERENCE AND CLUTCH/BAND APPLY CHART: A guide that shows the application of clutches and bands for each gear, within the selector range positions. These charts are extremely useful for understanding how the unit operates and for diagnosing malfunctions.

RAVIGNEAUX GEARSET: A compound planetary gearset that features matched dual planetary pinions (sets of two) mounted in a single planet carrier. Two sun gears and one ring mesh with the carrier pinions.

REACTION MEMBER: The stationary planetary member, in a planetary gearset, that is grounded to the transmission/transaxle case through the use of friction and wedging devices known as bands, disc clutches, and one-way clutches.

REACTION PRESSURE: The fluid pressure that moves a spool valve against an opposing force or forces; the area on which the opposing force acts. The opposing force can be a spring or a combination of spring force and auxiliary hydraulic force.

REACTOR, TORQUE CONVERTER: The reaction member of a fluid torque converter, more commonly called a stator. (See stator.)

REAR MAIN OIL SEAL: A synthetic or rope-type seal that prevents oil from leaking out of the engine past the rear main crankshaft bearing.

RECIRCULATING BALL: Type of steering system in which recirculating steel balls occupy the area between the nut and worm wheel, causing a reduction in friction.

RECTIFIER: A device (used primarily in alternators) that permits electrical current to flow in one direction only.

REDUCTION: (See gear reduction.)

REFRIGERANT 12 (R-12) or 134 (R-134): The generic name of the refrigerant used in automotive air conditioning systems.

REGULATOR: A device which maintains the amperage and/or voltage levels of a circuit at predetermined values.

REGULATOR VALVE: A valve that changes the pressure of the oil in a hydraulic circuit as the oil passes through the valve by bleeding off (or exhausting) some of the volume of oil supplied to the valve.

RELAY: A switch which automatically opens and/or closes a circuit.

RELAY VALVE: A valve that directs flow and pressure. Relay valves simply connect or disconnect interrelated passages without restricting the fluid flow or changing the pressure.

RELIEF VALVE: A spring-loaded, pressure-operated valve that limits oil pressure buildup in a hydraulic circuit to a predetermined maximum value.

RELUCTOR: A wheel that rotates inside the distributor and triggers the release of voltage in an electronic ignition.

RESERVOIR: The storage area for fluid in a hydraulic system; often called a sump.

RESIN: A liquid plastic used in body work.

RESIDUAL MAGNETISM: The magnetic strength stored in a material after a magnetizing field has been removed.

RESISTANCE: The opposition to the flow of current through a circuit or electrical device, and is measured in ohms. Resistance is equal to the voltage divided by the amperage.

RESISTOR SPARK PLUG: A spark plug using a resistor to shorten the spark duration. This suppresses radio interference and lengthens plug life.

RESISTOR: A device, usually made of wire, which offers a preset amount of resistance in an electrical circuit.

RESULTANT FORCE: The single effective directional thrust of the fluid force on the turbine produced by the vortex and rotary forces acting in different planes.

RETARD: Set the ignition timing so that spark occurs later (fewer degrees before TDC).

RHEOSTAT: A device for regulating a current by means of a variable resistance.

RING GEAR: The name given to a ring-shaped gear attached to a differential case, or affixed to a flywheel or as part of a planetary gear set.

ROADLOAD: grade.

ROCKER ARM: A lever which rotates around a shaft pushing down (opening) the valve with an end when the other end is pushed up by the pushrod. Spring pressure will later close the valve.

ROCKER PANEL: The body panel below the doors between the wheel opening.

ROLLER BEARING: A bearing made up of hardened inner and outer races between which hardened steel rollers move.

ROLLER CLUTCH: A type of one-way clutch design using rollers and springs mounted within an inner and outer cam race assembly.

ROTARY FLOW: The path of the fluid trapped between the blades of the members as they revolve with the rotation of the torque converter cover (rotational inertia).

ROTOR: (1.) The disc-shaped part of a disc brake assembly, upon which the brake pads bear; also called, brake disc. (2.) The device mounted atop the distributor shaft, which passes current to the distributor cap tower contacts.

ROTARY ENGINE: See Wankel engine.

RPM: Revolutions per minute (usually indicates engine speed).

RTV: A gasket making compound that cures as it is exposed to the atmosphere. It is used between surfaces that are not perfectly machined to one another, leaving a slight gap that the RTV fills and in which it hardens. The letters RTV represent room temperature vulcanizing.

RUN-ON: Condition when the engine continues to run, even when the key is turned off. See dieseling.

SEALED BEAM: A automotive headlight. The lens, reflector and filament from a single unit.

SEATBELT INTERLOCK: A system whereby the car cannot be started unless the seatbelt is buckled.

SECONDARY CIRCUIT: The high voltage side of the ignition system, usually above 20,000 volts. The secondary includes the ignition coil, coil wire, distributor cap and rotor, spark plug wires and spark plugs.

SELF-INDUCTION: The generation of voltage in a current-carrying wire by changing the amount of current flowing within that wire.

SEMI-CONDUCTOR: A material (silicon or germanium) that is neither a good conductor nor an insulator; used in diodes and transistors.

SEMI-FLOATING AXLE: In this design, a wheel is attached to the axle shaft, which takes both drive and cornering loads. Almost all solid axle passenger cars and light trucks use this design.

SENDING UNIT: A mechanical, electrical, hydraulic or electromagnetic device which transmits information to a gauge.

SENSOR: Any device designed to measure engine operating conditions or ambient pressures and temperatures. Usually electronic in nature and designed to send a voltage signal to an on-board computer, some sensors may operate as a simple on/off switch or they may provide a variable voltage signal (like a potentiometer) as conditions or measured parameters change.

SERIES CIRCUIT: (See circuit, series.)

SERPENTINE BELT: An accessory drive belt, with small multiple v-ribs, routed around most or all of the engine-powered accessories such as the alternator and power steering pump. Usually both the front and the back side of the belt comes into contact with various pulleys.

SERVO: In an automatic transmission, it is a piston in a cylinder assembly that converts hydraulic pressure into mechanical force and movement; used for the application of the bands and clutches.

SHIFT BUSYNESS: When referring to a torque converter clutch, it is the frequent apply and release of the clutch plate due to uncommon driving conditions.

SHIFT VALVE: Classified as a relay valve, it triggers the automatic shift in response to a governor and a throttle signal by directing fluid to the appropriate band and clutch apply combination to cause the shift to occur.

SHIM: Spacers of precise, predetermined thickness used between parts to establish a proper working relationship.

SHIMMY: Vibration (sometimes violent) in the front end caused by misaligned front end, out of balance tires or worn suspension components.

SHORT CIRCUIT: An electrical malfunction where current takes the path of least resistance to ground (usually through damaged insulation). Current flow is excessive from low resistance resulting in a blown fuse.

SHUDDER: Repeated jerking or stick-slip sensation, similar to chuggle but more severe and rapid in nature, that may be most noticeable during certain ranges of vehicle speed; also used to define condition after converter clutch engagement.

SIMPSON GEARSET: A compound planetary gear train that integrates two simple planetary gearsets referred to as the front planetary and the rear planetary.

SINGLE OVERHEAD CAMSHAFT: See overhead camshaft.

SKIDPLATE: A metal plate attached to the underside of the body to protect the fuel tank, transfer case or other vulnerable parts from damage.

SLAVE CYLINDER: In automotive use, a device in the hydraulic clutch system which is activated by hydraulic force, disengaging the clutch.

SLIPPING: Noticeable increase in engine rpm without vehicle speed increase; usually occurs during or after initial clutch or band engagement.

SLUDGE: Thick, black deposits in engine formed from dirt, oil, water, etc. It is usually formed in engines when oil changes are neglected.

SNAP RING: A circular retaining clip used inside or outside a shaft or part to secure a shaft, such as a floating wrist pin.

SOFT: Slow, almost unnoticeable clutch apply with very little shift feel.

SOFTCODES: DTCs that have been set into the PCM memory but are not present at the time of testing; often referred to as history or intermittent codes.

SOHC: Single overhead camshaft.

SOLENOID: An electrically operated, magnetic switching device.

SPALLING: A wear pattern identified by metal chips flaking off the hardened surface. This condition is caused by foreign particles, overloading situations, and/or normal wear.

SPARK PLUG: A device screwed into the combustion chamber of a spark ignition engine. The basic construction is a conductive core inside of a ceramic insulator, mounted in an outer conductive base. An electrical charge from the spark plug wire travels along the conductive core and jumps a preset air gap to a grounding point or points at the end of the conductive base. The resultant spark ignites the fuel/air mixture in the combustion chamber.

SPECIFIC GRAVITY (BATTERY): The relative weight of liquid (battery electrolyte) as compared to the weight of an equal volume of water.

SPLINES: Ridges machined or cast onto the outer diameter of a shaft or inner diameter of a bore to enable parts to mate without rotation.

SPLIT TORQUE DRIVE: In a torque converter, it refers to parallel paths of torque transmission, one of which is mechanical and the other hydraulic.

SPONGY PEDAL: A soft or spongy feeling when the brake pedal is depressed. It is usually due to air in the brake lines.

SPOOLVALVE: A precision-machined, cylindrically shaped valve made up of lands and grooves. Depending on its position in the valve bore, various interconnecting hydraulic circuit passages are either opened or closed.

SPRAG CLUTCH: A type of one-way clutch design using cams or contoured-shaped sprags between inner and outer races. (See one-way clutch.)

SPRUNG WEIGHT: The weight of a car supported by the springs.

SQUARE-CUT SEAL: Molded synthetic rubber seal designed with a square- or rectangular-shaped cross-section. This type of seal is used for both dynamic and static applications.

SRS: Supplemental restraint system

STABILIZER (SWAY) BAR: A bar linking both sides of the suspension. It resists sway on turns by taking some of added load from one wheel and putting it on the other.

STAGE: The number of turbine sets separated by a stator. A turbine set may be made up of one or more turbine members. A three-element converter is classified as a single stage.

STALL: In fluid drive transmission/transaxle applications, stall refers to engine rpm with the transmission/transaxle engaged and the vehicle stationary; throttle valve can be in any position between closed and wide open.

STALL SPEED: In fluid drive transmission/transaxle applications, stall speed refers to the maximum engine rpm with the transmission/transaxle engaged and vehicle stationary, when the throttle valve is wide open. (See stall; stall test.)

STALL TEST: A procedure recommended by many manufacturers to help determine the integrity of an engine, the torque converter stator, and certain clutch and band combinations. With the shift lever in each of the forward and reverse positions and with the brakes firmly applied, the accelerator pedal is momentarily pressed to the wide open throttle (WOT) position. The engine rpm reading at full throttle can provide clues for diagnosing the condition of the items listed above.

STALL TORQUE: The maximum design or engineered torque ratio of a fluid torque converter, produced under stall speed conditions. (See stall speed.)

STARTER: A high-torque electric motor used for the purpose of starting the engine, typically through a high ratio geared drive connected to the flywheel ring gear.

STATIC: A sealing application in which the parts being sealed do not move in relation to each other.

STATOR (REACTOR): The reaction member of a fluid torque converter that changes the direction of the fluid as it leaves the turbine to enter the impeller vanes. During the torque multiplication phase, this action assists the impeller's rotary force and results in an increase in torque.

STEERING GEOMETRY: Combination of various angles of suspension components (caster, camber, toe-in); roughly equivalent to front end alignment.

STRAIGHT WEIGHT: Term designating motor oil as suitable for use within a narrow range of temperatures. Outside the narrow temperature range its flow characteristics will not adequately lubricate.

STROKE: The distance the piston travels from bottom dead center to top dead center.

SUBSTITUTION: Replacing one part suspected of a defect with a like part of known quality.

SUMP: The storage vessel or reservoir that provides a ready source of fluid to the pump. In an automatic transmission, the sump is the oil pan. All fluid eventually returns to the sump for recycling into the hydraulic system.

SUN GEAR: In a planetary gearset, it is the center gear that meshes with a cluster of planet pinions.

SUPERCHARGER: An air pump driven mechanically by the engine through belts, chains, shafts or gears from the crankshaft. Two general types of supercharger are the positive displacement and centrifugal type, which pump air in direct relationship to the speed of the engine.

SUPPLEMENTAL RESTRAINT SYSTEM: See air bag.

SURGE: Repeating engine-related feeling of acceleration and deceleration that is less intense than chuggle.

SWITCH: A device used to open, close, or redirect the current in an electrical circuit.

SYNCHROMESH: A manual transmission/transaxle that is equipped with devices (synchronizers) that match the gear speeds so that the transmission/transaxle can be downshifted without clashing gears.

SYNTHETIC OIL: Non-petroleum based oil.

TACHOMETER: A device used to measure the rotary speed of an engine, shaft, gear, etc., usually in rotations per minute.

TDC: Top dead center. The exact top of the piston's stroke.

TEFLON SEALING RINGS: Teflon is a soft, durable, plastic-like material that is resistant to heat and provides excellent sealing. These rings are designed with either scarf-cut joints or as one-piece rings. Teflon sealing rings have replaced many metal ring applications.

TERMINAL: A device attached to the end of a wire or cable to make an electrical connection.

TEST LIGHT, CIRCUIT-POWERED: Uses available circuit voltage to test circuit continuity.

TEST LIGHT, SELF-POWERED: Uses its own battery source to test circuit continuity.

THERMISTOR: A special resistor used to measure fluid temperature; it decreases its resistance with increases in temperature.

THERMOSTAT: A valve, located in the cooling system of an engine, which is closed when cold and opens gradually in response to engine heating, controlling the temperature of the coolant and rate of coolant flow.

THERMOSTATIC ELEMENT: A heat-sensitive, spring-type device that controls a drain port from the upper sump area to the lower sump. When the transaxle fluid reaches operating temperature, the port is closed and the upper sump fills, thus reducing the fluid level in the lower sump.

THROTTLE POSITION (TP) SENSOR: Reads the degree of throttle opening; its signal is used to analyze engine load conditions. The ECM/PCM decides to apply the TCC, or to disengage it for coast or load conditions that need a converter torque boost.

THROTTLE PRESSURE/MODULATOR PRESSURE: A hydraulic signal oil pressure relating to the amount of engine load, based on either the amount of throttle plate opening or engine vacuum.

THROTTLE VALVE: A regulating or balanced valve that is controlled mechanically by throttle linkage or engine vacuum. It sends a hydraulic signal to the shift valve body to control shift timing and shift quality. (See balanced valve; modulator valve.)

THROW-OUT BEARING: As the clutch pedal is depressed, the throwout bearing moves against the spring fingers of the pressure plate, forcing the pressure plate to disengage from the driven disc.

TIE ROD: A rod connecting the steering arms. Tie rods have threaded ends that are used to adjust toe-in.

TIE-UP: Condition where two opposing clutches are attempting to apply at same time, causing engine to labor with noticeable loss of engine rpm.

TIMING BELT: A square-toothed, reinforced rubber belt that is driven by the crankshaft and operates the camshaft.

TIMING CHAIN: A roller chain that is driven by the crankshaft and operates the camshaft.

TIRE ROTATION: Moving the tires from one position to another to make the tires wear evenly.

TOE-IN (OUT): A term comparing the extreme front and rear of the front tires. Closer together at the front is toe-in; farther apart at the front is toe-out.

TOP DEAD CENTER (TDC): The point at which the piston reaches the top of its travel on the compression stroke.

TORQUE: Measurement of turning or twisting force, expressed as foot-pounds or inch-pounds.

TORQUE CONVERTER: A turbine used to transmit power from a driving member to a driven member via hydraulic action, providing changes in drive ratio and torque. In automotive use, it links the driveplate at the rear of the engine to the automatic transmission.

TORQUE CONVERTER CLUTCH: The apply plate (lockup plate) assembly used for mechanical power flow through the converter.

TORQUE PHASE: Sometimes referred to as slip phase or stall phase, torque multiplication occurs when the turbine is turning at a slower speed than the impeller, and the stator is reactionary (stationary). This sequence generates a boost in output torque.

TORQUE RATING (STALL TORQUE): The maximum torque multiplication that occurs during stall conditions, with the engine at wide open throttle (WOT) and zero turbine speed.

TORQUE RATIO: An expression of the gear ratio factor on torque effect. A 3:1 gear ratio or 3:1 torque ratio increases the torque input by the ratio factor of 3. Input torque (100 ft. lbs.)x 3 = output torque (300 ft. lbs.)

TRACTION: The amount of usable tractive effort before the drive wheels slip on the road contact surface.

TORSION BAR SUSPENSION: Long rods of spring steel which take the place of springs. One end of the bar is anchored and the other arm (attached to the suspension) is free to twist. The bars' resistance to twisting causes springing action.

TRACK: Distance between the centers of the tires where they contact the ground.

TRACTION CONTROL: A control system that prevents the spinning of a vehicle's drive wheels when excess power is applied.

TRACTIVE EFFORT: The amount of force available to the drive wheels, to move the vehicle.

TRANSAXLE: A single housing containing the transmission and differential. Transaxles are usually found on front engine/front wheel drive or rear engine/rear wheel drive cars.

TRANSDUCER: A device that changes energy from one form to another. For example, a transducer in a microphone changes sound energy to electrical energy. In automotive air-conditioning controls used in automatic temperature systems, a transducer changes an electrical signal to a vacuum signal, which operates mechanical doors.

TRANSMISSION: A powertrain component designed to modify torque and speed developed by the engine; also provides direct drive, reverse, and neutral.

TRANSMISSION CONTROL MODULE (TCM): Manages transmission functions. These vary according to the manufacturer's product design but may include converter clutch operation, electronic shift scheduling, and mainline pressure.

TRANSMISSION FLUID TEMPERATURE (TFT)SENSOR: Originally called a transmission oil temperature (TOT) sensor, this input device to the ECM/PCM senses the fluid temperature and provides a resistance value. It operates on the thermistor principle.

TRANSMISSION INPUT SPEED (TIS) SENSOR: Measures turbine shaft (input shaft) rpm's and compares to engine rpm's to determine torque converter slip. When compared to the transmission output speed sensor or VSS, gear ratio and clutch engagement timing can be determined.

TRANSMISSION OIL TEMPERATURE (TOT) SENSOR: (See transmission fluid temperature (TFT) sensor.)

TRANSMISSION RANGE SELECTOR (TRS) SWITCH: Tells the module which gear shift position the driver has chosen. turbine: The output (driven) member of a fluid coupling or fluid torque converter. It is splined to the input (turbine) shaft of the transmission.

TRANSFER CASE: A gearbox driven from the transmission that delivers power to both front and rear driveshafts in a four-wheel drive system. Transfer cases usually have a high and low range set of gears, used depending on how much pulling power is needed.

TRANSISTOR: A semi-conductor component which can be actuated by a small voltage to perform an electrical switching function.

TREAD WEAR INDICATOR: Bars molded into the tire at right angles to the tread that appear as horizontal bars when 1/16 in. of tread remains.

TREAD WEAR PATTERN: The pattern of wear on tires which can be "read" to diagnose problems in the front suspension.

TUNE-UP: A regular maintenance function, usually associated with the replacement and adjustment of parts and components in the electrical and fuel systems of a vehicle for the purpose of attaining optimum performance.

TURBOCHARGER: An exhaust driven pump which compresses intake air and forces it into the combustion chambers at higher than atmospheric pressures. The increased air pressure allows more fuel to be burned and results in increased horsepower being produced.

TURBULENCE: The interference of molecules of a fluid (or vapor) with each other in a fluid flow.

TYPE F: Transmission fluid developed and used by Ford Motor Company up to 1982. This fluid type provides a high coefficient of friction.

TYPE 7176: The preferred choice of transmission fluid for Chrysler automatic transmissions and transaxles. Developed in 1986, it closely resembles Dexron and Mercon. Type 7176 is the recommended service fill fluid for all Chrysler products utilizing a lockup torque converter dating back to 1978.

U-JOINT (UNIVERSAL JOINT): A flexible coupling in the drive train that allows the driveshafts or axle shafts to operate at different angles and still transmit rotary power.

UNDERSTEER: The tendency of a car to continue straight ahead while negotiating a turn.

UNIT BODY: Design in which the car body acts as the frame.

UNLEADED FUEL: Fuel which contains no lead (a common gasoline additive). The presence of lead in fuel will destroy the functioning elements of a catalytic converter, making it useless.

UNSPRUNG WEIGHT: The weight of car components not supported by the springs (wheels, tires, brakes, rear axle, control arms, etc.).

UPSHIFT: A shift that results in a decrease in torque ratio and an increase in speed.

VACUUM: A negative pressure; any pressure less than atmospheric pressure.

VACUUM ADVANCE: A device which advances the ignition timing in response to increased engine vacuum.

VACUUM GAUGE: An instrument used for measuring the existing vacuum in a vacuum circuit or chamber. The unit of measure is inches (of mercury in a barometer).

VACUUM MODULATOR: Generates a hydraulic oil pressure in response to the amount of engine vacuum.

VALVES: Devices that can open or close fluid passages in a hydraulic system and are used for directing fluid flow and controlling pressure.

VALVE BODY ASSEMBLY: The main hydraulic control assembly of the transmission/transaxle that contains numerous valves, check balls, and other components to control the distribution of pressurized oil throughout the transmission.

VALVE CLEARANCE: The measured gap between the end of the valve stem and the rocker arm, cam lobe or follower that activates the valve.

VALVE GUIDES: The guide through which the stem of the valve passes. The guide is designed to keep the valve in proper alignment.

VALVE LASH (clearance): The operating clearance in the valve train.

VALVE TRAIN: The system that operates intake and exhaust valves, consisting of camshaft, valves and springs, lifters, pushrods and rocker arms.

VAPOR LOCK: Boiling of the fuel in the fuel lines due to excess heat. This will interfere with the flow of fuel in the lines and can completely stop the flow. Vapor lock normally only occurs in hot weather.

VARIABLE DISPLACEMENT (VARIABLE CAPACITY) VANE PUMP: Slipper-type vanes, mounted in a revolving rotor and contained within the bore of a movable slide, capture and then force fluid to flow. Movement of the slide to various positions changes the size of the vane chambers and the amount of fluid flow. Note: **GM refers to this pump design as variable displacement, and Ford terms it variable capacity.**

VARIABLE FORCE SOLENOID (VFS): Commonly referred to as the electronic pressure control (EPC) solenoid, it replaces the cable/linkage style of TV system control and is integrated with a spool valve and spring assembly to control pressure. A variable computer-controlled current flow varies the internal force of the solenoid on the spool valve and resulting control pressure.

VARIABLE ORIFICE THERMAL VALVE: Temperature-sensitive hydraulic oil control device that adjusts the size of a circuit path opening. By altering the size of the opening, the oil flow rate is adapted for cold to hot oil viscosity changes.

VARNISH: Term applied to the residue formed when gasoline gets old and stale.

VCM: See Electronic Control Unit (ECU).

VEHICLE SPEED SENSOR (VSS): Provides an electrical signal to the computer module, measuring vehicle speed, and affects the torque converter clutch engagement and release.

VESPEL SEALING RINGS: Hard plastic material that produces excellent sealing in dynamic settings. These rings are found in late versions of the 4T60 and in all 4T60-E and 4T80-E transaxles.

VISCOSITY: The ability of a fluid to flow. The lower the viscosity rating, the easier the fluid will flow. 10 weight motor oil will flow much easier than 40 weight motor oil.

VISCOSITY INDEX IMPROVERS: Keeps the viscosity nearly constant with changes in temperature. This is especially important at low temperatures, when the oil needs to be thin to aid in shifting and for cold-weather starting. Yet it must not be so thin that at high temperatures it will cause excessive hydraulic leakage so that pumps are unable to maintain the proper pressures.

VISCOUS CLUTCH: A specially designed torque converter clutch apply plate that, through the use of a silicon fluid, clamps smoothly and absorbs torsional vibrations.

VOLT: Unit used to measure the force or pressure of electricity. It is defined as the pressure

VOLTAGE: The electrical pressure that causes current to flow. Voltage is measured in volts (V).

VOLTAGE, APPLIED: The actual voltage read at a given point in a circuit. It equals the available voltage of the power supply minus the losses in the circuit up to that point.

VOLTAGE DROP: The voltage lost or used in a circuit by normal loads such as a motor or lamp or by abnormal loads such as a poor (high-resistance) lead or terminal connection.

VOLTAGE REGULATOR: A device that controls the current output of the alternator or generator.

VOLTMETER: An instrument used for measuring electrical force in units called volts. Voltmeters are always connected parallel with the circuit being tested.

VORTEX FLOW: The crosswise or circulatory flow of oil between the blades of the members caused by the centrifugal pumping action of the impeller.

WANKEL ENGINE: An engine which uses no pistons. In place of pistons, triangular-shaped rotors revolve in specially shaped housings.

WATER PUMP: A belt driven component of the cooling system that mounts on the engine, circulating the coolant under pressure.

WATT: The unit for measuring electrical power. One watt is the product of one ampere and one volt (watts equals amps times volts). Wattage is the horsepower of electricity (746 watts equal one horsepower).

WHEEL ALIGNMENT: Inclusive term to describe the front end geometry (caster, camber, toe-in/out).

WHEEL CYLINDER: Found in the automotive drum brake assembly, it is a device, actuated by hydraulic pressure, which, through internal pistons, pushes the brake shoes outward against the drums.

WHEEL WEIGHT: Small weights attached to the wheel to balance the wheel and tire assembly. Out-of-balance tires quickly wear out and also give erratic handling when installed on the front.

WHEELBASE: Distance between the center of front wheels and the center of rear wheels.

WIDE OPEN THROTTLE (WOT): Full travel of accelerator pedal.

WORK: The force exerted to move a mass or object. Work involves motion; if a force is exerted and no motion takes place, no work is done. Work per unit of time is called power. Work = force x distance = ft. lbs. 33,000 ft. lbs. in one minute = 1 horsepower

ZERO-THROTTLE COAST DOWN: A full release of accelerator pedal while vehicle is in motion and in drive range.

Commonly Used Abbreviations

2

2WD	Two Wheel Drive

4

4WD	Four Wheel Drive

A

A/C	Air Conditioning
ABDC	After Bottom Dead Center
ABS	Anti-lock Brakes
AC	Alternating Current
ACL	Air cleaner
ACT	Air Charge Temperature
AIR	Secondary Air Injection
ALCL	Assembly Line Communications Link
ALDL	Assembly Line Diagnostic Link
AT	Automatic Transaxle/Transmission
ATDC	After Top Dead Center
ATF	Automatic Transmission Fluid
ATS	Air Temperature Sensor
AWD	All Wheel Drive

B

BAP	Barometric Absolute Pressure
BARO	Barometric Pressure
BBDC	Before Bottom Dead Center
BCM	Body Control Module
BDC	Bottom Dead Center
BPT	Backpressure Transducer
BTDC	Before Top Dead Center
BVSV	Bimetallic Vacuum Switching Valve

C

CAC	Charge Air Cooler
CARB	California Air Resources Board
CAT	Catalytic Converter
CCC	Computer Command Control
CCCC	Computer Controlled Catalytic Converter
CCCI	Computer Controlled Coil Ignition
CCD	Computer Controlled Dwell
CDI	Capacitor Discharge Ignition
CEC	Computerized Engine Control
CFI	Continuous Fuel Injection
CIS	Continuous Injection System
CIS-E	Continuous Injection System - Electronic
CKP	Crankshaft Position
CL	Closed Loop
CMP	Camshaft Position
CPP	Clutch Pedal Position
CTOX	Continuous Trap Oxidizer System
CTP	Closed Throttle Position
CVC	Constant Vacuum Control
CYL	Cylinder

D

DBC	Dual Bed Catalyst
DC	Direct Current
DFI	Direct Fuel Injection
DIS	Distributorless Ignition System
DLC	Data Link Connector
DMM	Digital Multimeter
DOHC	Double Overhead Camshaft
DRB	Diagnostic Readout Box
DTC	Diagnostic Trouble Code
DTM	Diagnostic Test Mode
DVOM	Digital Volt/Ohmmeter

E

EBCM	Electronic Brake Control Module
ECM	Engine Control Module
ECT	Engine Coolant Temperature
ECU	Engine Control Unit or Electronic Control Unit
EDIS	Electronic Distributorless Ignition System
EEC	Electronic Engine Control
EEPROM	Electrically Erasable Programmable Read Only Memory
EFE	Early Fuel Evaporation
EGR	Exhaust Gas Recirculation
EGRT	Exhaust Gas Recirculation Temperature
EGRVC	EGR Valve Control
EPROM	Erasable Programmable Read Only Memory
EVAP	Evaporative Emissions
EVP	EGR Valve Position

F

FBC	Feedback Carburetor
FEEPROM	Flash Electrically Erasable Programmable Read Only Memory
FF	Flexible Fuel
FI	Fuel Injection
FT	Fuel Trim
FWD	Front Wheel Drive

G

GND	Ground

H

HAC	High Altitude Compensation
HEGO	Heated Exhaust Gas Oxygen sensor
HEI	High Energy Ignition
HO2 Sensor	Heated Oxygen Sensor

I

IAC	Idle Air Control
IAT	Intake Air Temperature
ICM	Ignition Control Module
IFI	Indirect Fuel Injection
IFS	Inertia Fuel Shutoff
ISC	Idle Speed Control
IVSV	Idle Vacuum Switching Valve

Commonly Used Abbreviations

K

KOEO	Key On, Engine Off
KOER	Key ON, Engine Running
KS	Knock Sensor

M

MAF	Mass Air Flow
MAP	Manifold Absolute Pressure
MAT	Manifold Air Temperature
MC	Mixture Control
MDP	Manifold Differential Pressure
MFI	Multiport Fuel Injection
MIL	Malfunction Indicator Lamp or Maintenance
MST	Manifold Surface Temperature
MVZ	Manifold Vacuum Zone

N

NVRAM	Nonvolatile Random Access Memory

O

O2 Sensor	Oxygen Sensor
OBD	On-Board Diagnostic
OC	Oxidation Catalyst
OHC	Overhead Camshaft
OL	Open Loop

P

P/S	Power Steering
PAIR	Pulsed Secondary Air Injection
PCM	Powertrain Control Module
PCS	Purge Control Solenoid
PCV	Positive Crankcase Ventilation
PIP	Profile Ignition Pick-up
PNP	Park/Neutral Position
PROM	Programmable Read Only Memory
PSP	Power Steering Pressure
PTO	Power Take-Off
PTOX	Periodic Trap Oxidizer System

R

RABS	Rear Anti-lock Brake System
RAM	Random Access Memory
ROM	Read Only Memory
RPM	Revolutions Per Minute
RWAL	Rear Wheel Anti-lock Brakes
RWD	Rear Wheel Drive

S

SBC	Single Bed Converter
SBEC	Single Board Engine Controller
SC	Supercharger
SCB	Supercharger Bypass
SFI	Sequential Multiport Fuel Injection
SIR	Supplemental Inflatible Restraint
SOHC	Single Overhead Camshaft
SPL	Smoke Puff Limiter
SPOUT	Spark Output
SRI	Service Reminder Indicator
SRS	Supplemental Restraint System
SRT	System Readiness Test
SSI	Solid State Ignition
ST	Scan Tool
STO	Self-Test Output

T

TAC	Thermostatic Air Cleaner
TBI	Throttle Body Fuel Injection
TC	Turbocharger
TCC	Torque Converter Clutch
TCM	Transmission Control Module
TDC	Top Dead Center
TFI	Thick Film Ignition
TP	Throttle Position
TR Sensor	Transaxle/Transmission Range Sensor
TVV	Thermal Vacuum Valve
TWC	Three-way Catalytic Converter

V

VAF	Volume Air Flow, or Vane Air Flow
VAPS	Variable Assist Power Steering
VRV	Vacuum Regulator Valve
VSS	Vehicle Speed Sensor
VSV	Vacuum Switching Valve

W

WOT	Wide Open Throttle
WU-TWC	Warm Up Three-way Catalytic Converter

ENGLISH TO METRIC CONVERSION: TORQUE

To convert foot-pounds (ft. lbs.) to Newton-meters (Nm), multiply the number of ft. lbs. by 1.36
To convert Newton-meters (Nm) to foot-pounds (ft. lbs.), multiply the number of Nm by 0.7376

ft. lbs.	Nm	ft. lbs.	Nm	ft. lbs.	Nm	ft. lbs.	Nm
0.1	0.1	34	46.2	76	103.4	118	160.5
0.2	0.3	35	47.6	77	104.7	119	161.8
0.3	0.4	36	49.0	78	106.1	120	163.2
0.4	0.5	37	50.3	79	107.4	121	164.6
0.5	0.7	38	51.7	80	108.8	122	165.9
0.6	0.8	39	53.0	81	110.2	123	167.3
0.7	1.0	40	54.4	82	111.5	124	168.6
0.8	1.1	41	55.8	83	112.9	125	170.0
0.9	1.2	42	57.1	84	114.2	126	171.4
1	1.4	43	58.5	85	115.6	127	172.7
2	2.7	44	59.8	86	117.0	128	174.1
3	4.1	45	61.2	87	118.3	129	175.4
4	5.4	46	62.6	88	119.7	130	176.8
5	6.8	47	63.9	89	121.0	131	178.2
6	8.2	48	65.3	90	122.4	132	179.5
7	9.5	49	66.6	91	123.8	133	180.9
8	10.9	50	68.0	92	125.1	134	182.2
9	12.2	51	69.4	93	126.5	135	183.6
10	13.6	52	70.7	94	127.8	136	185.0
11	15.0	53	72.1	95	129.2	137	186.3
12	16.3	54	73.4	96	130.6	138	187.7
13	17.7	55	74.8	97	131.9	139	189.0
14	19.0	56	76.2	98	133.3	140	190.4
15	20.4	57	77.5	99	134.6	141	191.8
16	21.8	58	78.9	100	136.0	142	193.1
17	23.1	59	80.2	101	137.4	143	194.5
18	24.5	60	81.6	102	138.7	144	195.8
19	25.8	61	83.0	103	140.1	145	197.2
20	27.2	62	84.3	104	141.4	146	198.6
21	28.6	63	85.7	105	142.8	147	199.9
22	29.9	64	87.0	106	144.2	148	201.3
23	31.3	65	88.4	107	145.5	149	202.6
24	32.6	66	89.8	108	146.9	150	204.0
25	34.0	67	91.1	109	148.2	151	205.4
26	35.4	68	92.5	110	149.6	152	206.7
27	36.7	69	93.8	111	151.0	153	208.1
28	38.1	70	95.2	112	152.3	154	209.4
29	39.4	71	96.6	113	153.7	155	210.8
30	40.8	72	97.9	114	155.0	156	212.2
31	42.2	73	99.3	115	156.4	157	213.5
32	43.5	74	100.6	116	157.8	158	214.9
33	44.9	75	102.0	117	159.1	159	216.2

METRIC TO ENGLISH CONVERSION: TORQUE

To convert foot-pounds (ft. lbs.) to Newton-meters (Nm), multiply the number of ft. lbs. by 1.36

To convert Newton-meters (Nm) to foot-pounds (ft. lbs.), multiply the number of Nm by 0.7376

Nm	ft. lbs.	Nm	ft. lbs.	Nm	ft. lbs.	Nm	ft. lbs.	Nm	ft. lbs.
0.1	0.1	34	25.0	76	55.9	118	86.8	160	117.6
0.2	0.1	35	25.7	77	56.6	119	87.5	161	118.4
0.3	0.2	36	26.5	78	57.4	120	88.2	162	119.1
0.4	0.3	37	27.2	79	58.1	121	89.0	163	119.9
0.5	0.4	38	27.9	80	58.8	122	89.7	164	120.6
0.6	0.4	39	28.7	81	59.6	123	90.4	165	121.3
0.7	0.5	40	29.4	82	60.3	124	91.2	166	122.1
0.8	0.6	41	30.1	83	61.0	125	91.9	167	122.8
0.9	0.7	42	30.9	84	61.8	126	92.6	168	123.5
1	0.7	43	31.6	85	62.5	127	93.4	169	124.3
2	1.5	44	32.4	86	63.2	128	94.1	170	125.0
3	2.2	45	33.1	87	64.0	129	94.9	171	125.7
4	2.9	46	33.8	88	64.7	130	95.6	172	126.5
5	3.7	47	34.6	89	65.4	131	96.3	173	127.2
6	4.4	48	35.3	90	66.2	132	97.1	174	127.9
7	5.1	49	36.0	91	66.9	133	97.8	175	128.7
8	5.9	50	36.8	92	67.6	134	98.5	176	129.4
9	6.6	51	37.5	93	68.4	135	99.3	177	130.1
10	7.4	52	38.2	94	69.1	136	100.0	178	130.9
11	8.1	53	39.0	95	69.9	137	100.7	179	131.6
12	8.8	54	39.7	96	70.6	138	101.5	180	132.4
13	9.6	55	40.4	97	71.3	139	102.2	181	133.1
14	10.3	56	41.2	98	72.1	140	102.9	182	133.8
15	11.0	57	41.9	99	72.8	141	103.7	183	134.6
16	11.8	58	42.6	100	73.5	142	104.4	184	135.3
17	12.5	59	43.4	101	74.3	143	105.1	185	136.0
18	13.2	60	44.1	102	75.0	144	105.9	186	136.8
19	14.0	61	44.9	103	75.7	145	106.6	187	137.5
20	14.7	62	45.6	104	76.5	146	107.4	188	138.2
21	15.4	63	46.3	105	77.2	147	108.1	189	139.0
22	16.2	64	47.1	106	77.9	148	108.8	190	139.7
23	16.9	65	47.8	107	78.7	149	109.6	191	140.4
24	17.6	66	48.5	108	79.4	150	110.3	192	141.2
25	18.4	67	49.3	109	80.1	151	111.0	193	141.9
26	19.1	68	50.0	110	80.9	152	111.8	194	142.6
27	19.9	69	50.7	111	81.6	153	112.5	195	143.4
28	20.6	70	51.5	112	82.4	154	113.2	196	144.1
29	21.3	71	52.2	113	83.1	155	114.0	197	144.9
30	22.1	72	52.9	114	83.8	156	114.7	198	145.6
31	22.8	73	53.7	115	84.6	157	115.4	199	146.3
32	23.5	74	54.4	116	85.3	158	116.2	200	147.1
33	24.3	75	55.1	117	86.0	159	116.9	201	147.8

ENGLISH/METRIC CONVERSION: TEMPERATURE

To convert Fahrenheit (F°) to Celsius (C°), take F° temperature and subtract 32, multiply the result by 5 and divide the result by 9
To convert Celsius (C°) to Fahrenheit (F°), take C° temperature and multiply it by 9, divide the result by 5 and add 32

F°	C°	F°	C°	C°	F°	C°	F°
-40	-40.0	150	65.6	-38	-36.4	46	114.8
-35	-37.2	155	68.3	-36	-32.8	48	118.4
-30	-34.4	160	71.1	-34	-29.2	50	122
-25	-31.7	165	73.9	-32	-25.6	52	125.6
-20	-28.9	170	76.7	-30	-22	54	129.2
-15	-26.1	175	79.4	-28	-18.4	56	132.8
-10	-23.3	180	82.2	-26	-14.8	58	136.4
-5	-20.6	185	85.0	-24	-11.2	60	140
0	-17.8	190	87.8	-22	-7.6	62	143.6
1	-17.2	195	90.6	-20	-4	64	147.2
2	-16.7	200	93.3	-18	-0.4	66	150.8
3	-16.1	205	96.1	-16	3.2	68	154.4
4	-15.6	210	98.9	-14	6.8	70	158
5	-15.0	212	100.0	-12	10.4	72	161.6
10	-12.2	215	101.7	-10	14	74	165.2
15	-9.4	220	104.4	-8	17.6	76	168.8
20	-6.7	225	107.2	-6	21.2	78	172.4
25	-3.9	230	110.0	-4	24.8	80	176
30	-1.1	235	112.8	-2	28.4	82	179.6
35	1.7	240	115.6	0	32	84	183.2
40	4.4	245	118.3	2	35.6	86	186.8
45	7.2	250	121.1	4	39.2	88	190.4
50	10.0	255	123.9	6	42.8	90	194
55	12.8	260	126.7	8	46.4	92	197.6
60	15.6	265	129.4	10	50	94	201.2
65	18.3	270	132.2	12	53.6	96	204.8
70	21.1	275	135.0	14	57.2	98	208.4
75	23.9	280	137.8	16	60.8	100	212
80	26.7	285	140.6	18	64.4	102	215.6
85	29.4	290	143.3	20	68	104	219.2
90	32.2	295	146.1	22	71.6	106	222.8
95	35.0	300	148.9	24	75.2	108	226.4
100	37.8	305	151.7	26	78.8	110	230
105	40.6	310	154.4	28	82.4	112	233.6
110	43.3	315	157.2	30	86	114	237.2
115	46.1	320	160.0	32	89.6	116	240.8
120	48.9	325	162.8	34	93.2	118	244.4
125	51.7	330	165.6	36	96.8	120	248
130	54.4	335	168.3	38	100.4	122	251.6
135	57.2	340	171.1	40	104	124	255.2
140	60.0	345	173.9	42	107.6	126	258.8
145	62.8	350	176.7	44	111.2	128	262.4

LENGTH CONVERSION

To convert inches (in.) to millimeters (mm), multiply the number of inches by 25.4

To convert millimeters (mm) to inches (in.), multiply the number of millimeters by 0.04

Inches	Millimeters	Inches	Millimeters	Inches	Millimeters	Inches	Millimeters
0.0001	0.00254	0.005	0.1270	0.09	2.286	4	101.6
0.0002	0.00508	0.006	0.1524	0.1	2.54	5	127.0
0.0003	0.00762	0.007	0.1778	0.2	5.08	6	152.4
0.0004	0.01016	0.008	0.2032	0.3	7.62	7	177.8
0.0005	0.01270	0.009	0.2286	0.4	10.16	8	203.2
0.0006	0.01524	0.01	0.254	0.5	12.70	9	228.6
0.0007	0.01778	0.02	0.508	0.6	15.24	10	254.0
0.0008	0.02032	0.03	0.762	0.7	17.78	11	279.4
0.0009	0.02286	0.04	1.016	0.8	20.32	12	304.8
0.001	0.0254	0.05	1.270	0.9	22.86	13	330.2
0.002	0.0508	0.06	1.524	1	25.4	14	355.6
0.003	0.0762	0.07	1.778	2	50.8	15	381.0
0.004	0.1016	0.08	2.032	3	76.2	16	406.4

ENGLISH/METRIC CONVERSION: LENGTH

To convert inches (in.) to millimeters (mm), multiply the number of inches by 25.4
To convert millimeters (mm) to inches (in.), multiply the number of millimeters by 0.04

Inches		Millimeters	Inches		Millimeters	Inches		Millimeters
Fraction	Decimal	Decimal	Fraction	Decimal	Decimal	Fraction	Decimal	Decimal
1/64	0.016	0.397	11/32	0.344	8.731	11/16	0.688	17.463
1/32	0.031	0.794	23/64	0.359	9.128	45/64	0.703	17.859
3/64	0.047	1.191	3/8	0.375	9.525	23/32	0.719	18.256
1/16	0.063	1.588	25/64	0.391	9.922	47/64	0.734	18.653
5/64	0.078	1.984	13/32	0.406	10.319	3/4	0.750	19.050
3/32	0.094	2.381	27/64	0.422	10.716	49/64	0.766	19.447
7/64	0.109	2.778	7/16	0.438	11.113	25/32	0.781	19.844
1/8	0.125	3.175	29/64	0.453	11.509	51/64	0.797	20.241
9/64	0.141	3.572	15/32	0.469	11.906	13/16	0.813	20.638
5/32	0.156	3.969	31/64	0.484	12.303	53/64	0.828	21.034
11/64	0.172	4.366	1/2	0.500	12.700	27/32	0.844	21.431
3/16	0.188	4.763	33/64	0.516	13.097	55/64	0.859	21.828
13/64	0.203	5.159	17/32	0.531	13.494	7/8	0.875	22.225
7/32	0.219	5.556	35/64	0.547	13.891	57/64	0.891	22.622
15/64	0.234	5.953	9/16	0.563	14.288	29/32	0.906	23.019
1/4	0.250	6.350	37/64	0.578	14.684	59/64	0.922	23.416
17/64	0.266	6.747	19/32	0.594	15.081	15/16	0.938	23.813
9/32	0.281	7.144	39/64	0.609	15.478	61/64	0.953	24.209
19/64	0.297	7.541	5/8	0.625	15.875	31/32	0.969	24.606
5/16	0.313	7.938	41/64	0.641	16.272	63/64	0.984	25.003
21/64	0.328	8.334	21/32	0.656	16.669	1/1	1.000	25.400
			43/64	0.672	17.066			

Chilton 2003 Labor Guide Manual

Manual: 0-8019-9360-1/Part No. 9360
CD-ROM: 0-8019-9361-X/Part No. 9361

Designed for today's Professional Technician, *Chilton Labor Guide Manual* provides estimated repair times for nearly every automotive repair procedure imaginable. Available in both hardcover and CD-ROM, this manual covers 22 years of labor operations for virtually every domestic and import vehicle on the road offering the following special features:

KEY FEATURES
(OF HARDCOVER MANUAL)

- includes vehicles from 1998 through 2003
- labor times which take into account vehicle age as well as parts wear and tear, plus labor times reflecting repairs to vehicles subjected to Severe Service
- separate indexed tabs for each major section
- easy-to-use model-to-page indexing
- testing/diagnostic times preceding each labor section, where applicable
- transaxle/transmission service and repair times – no need to buy additional manuals
- operation descriptions written for easy reference
- wide acceptance by extended warranty companies
- add-on, combination, and Severe Service times in italics
- table of contents with manufacturer/ model listing to speed lookup
- a 'How to Use' page to guide new users

2,240 pp *8⅛" x 11"*, HC, 1-Color, ©2003

PREVIOUS YEAR
Chilton Labor Guide Manual,
1981-2002,
0-8019-9349-0
Part No. 9349
2336 pp

KEY FEATURES
(OF CD-ROM)

- labor times which take into account vehicle age as well as parts wear and tear, plus labor times reflecting repairs to vehicles subjected to Severe Service
- improved navigation
- enhanced functionality
- easy-to-use quick lookup function
- testing/diagnostic times, where applicable
- transaxle/transmission service and repair times
- 'Customer' and 'Estimates' drop-down screens
- 'Specials' box allows users to create their own service offerings
- labor selection 'highlighting' feature
- wide acceptance by extended warranty companies
- 'How to Use' drop-down screens to guide new users

PREVIOUS YEAR
Chilton Labor Guide CD-ROM,
1981-2002
0-8019-9350-4
Part No. 9350

Chilton Hardcover Reference Manuals

Chilton Hardcover Reference Manuals are perfect for enthusiasts of vintage autos. These manuals contain repair and maintenance information for all major systems that may not be available elsewhere. Included are repair and overhaul procedures, using thousands of illustrations, as well as troubleshooting and diagnosis. They offer a wide range of repair information on domestic and import cars, trucks and vans from 1979 to 2002.

CHILTON AUTO REPAIR MANUALS

1998-2002
0-8019-9362-8/Part No. 9362
1,426 pp
Covers all popular American and Canadian cars. Added features include Maintenance Labor Times, Technical Service Bulletin Indexes, Scheduled Maintenance Intervals Charts, and much more.

1993-1997
0-8019-7919-6/Part No. 7919
2,064 pp
Covers all popular American and Canadian cars.

1988-1992
0-8019-7906-4/Part No. 7906
1,284 pp
Covers all popular American and Canadian cars.

1980-1987
0-8019-7670-7/Part No. 7670
1,344 pp
Covers all popular American and Canadian cars.

CHILTON IMPORT AUTO REPAIR MANUALS

1998-2002
0-8019-9363-6/Part No. 9363
1,792 pp
Covers all popular Import cars. Added features include Maintenance Labor Times, Technical Service Bulletin Indexes, Scheduled Maintenance Intervals Charts, and much more.

1993-1997
0-8019-7920-X/Part No. 7920
2,080 pp
Covers all popular Import cars. Also features Frequent Maintenance Labor Times, Technical Service Bulletin Indexes, Scheduled Maintenance Interval Charts and more.

1988-1992
0-8019-7907-2/Part No. 7907
1,632 pp
Covers all popular Import cars.

1980-1987
0-8019-7672-3/Part No. 7672
1,488 pp
Covers all popular Import cars.

CHILTON TRUCK AND VAN REPAIR MANUALS

1998-2002
0-8019-9364-4/Part No. 9364
1,408 pp
Covers popular U.S., Canadian, and Import Pick-Ups, Vans, and 4WDs. Added features include Maintenance Labor Times, Technical Service Bulletin Indexes, Scheduled Maintenance Intervals Charts, and much more.

1993-1997
0-8019-7921-8/Part No. 7921
2,096 pp
Covers popular U.S., Canadian and Import Pick-Ups, Sport-Utilities, Vans, RVs and 4 wheel drives.

1991-1995
0-8019-7911-0/Part No. 7911
1,664 pp

1986-1990
0-8019-7902-1/Part No. 7902
1,536 pp

1979-1986
0-8019-7655-3/Part No. 7655
1,440 pp

CHILTON SUV REPAIR MANUAL

1998-2002
0-8019-9365-2/Part No. 9365
1,600 pp
Covers popular U.S., Canadian, and Import SUVs. Added features include Maintenance Labor Times, Technical Service Bulletin Indexes, Scheduled Maintenance Intervals Charts, and much more.

COLLECTOR'S SERIES

CHILTON'S AUTO REPAIR MANUALS
1940-1953
0-8019-5631-5/Part No. 5631
1964-1971
0-8019-5974-8/Part No. 5974

CHILTON'S TRUCK AND VAN REPAIR MANUALS
1961-1971
0-8019-6198-X/Part No. 6198
1971-1978
0-8019-7012-1/Part No. 7012

Chilton Hardcover Professional Service Manuals

Designed for today's Professional Technician, *Chilton Professional Service Manuals* on automotive repair are comprehensive and technically detailed, offering TOTAL maintenance, service and repair information for the automotive professional. Information is provided in an easy-to-read format, supported by quick-reference sections as well as exploded-view illustrations, diagrams and charts. Complete coverage of repair procedures is offered, from drive train to chassis and all associated components. All this at the user's fingertips insures fast, accurate repairs.

KEY FEATURES

- exploded-view illustrations, diagrams, and charts
- simple-to-follow removal and installation instructions for heater core and related components
- complete coverage of suspension components
- specifications for ball joint inspection and OEM tire size
- complete coverage of repair procedures from drive train to chassis and all associated components

Chilton Auto Service Manual, 1999-2003
0-8019-9356-3
Part No. 9356
1,668 pp

Chilton Import Auto Service Manual 1999-2003
0-8019-9357-1
Part No. 9357
2,032 pp

Chilton Truck and Van Service Manual 1999-2003
0-8019-9358-X
Part No. 9358
2,240 pp

Chilton SUV Service Manual, 1999-2003
0-8019-9359-8
Part No. 9359
2,240 pp

8½" x 11", HC, 1-Color, ©2003

Quick-Reference Manuals

The Chilton Professional Series offers *Quick-Reference Manuals* for the automotive professional, providing complete coverage on repair and maintenance, adjustments, and diagnostic procedures for specific systems and components. Each soft-cover manual contains model-specific procedures supported by countless photos and illustrations, covering virtually all domestic and import cars, trucks, and vans.

KEY FEATURES

- step-by-step procedures
- detailed illustrations and exploded views
- easy-to-use manufacturer and model indexing
- handy specifications or data charts

Timing Belts 1980-2000
0-8019-9305-9
Part No. 9305
320 pp

Specifications and Maintenance Intervals, 1990-2000
0-8019-9310-5
Part No. 9310

Heater Core Service 1990-2000
0-8019-9311-3
Part No. 9311
560 pp

Brake Specifications and Service, 1990-2000
0-8019-9312-1
Part No. 9312
520 pp

Electric Cooling Fans, Accessory Drive Belts & Water Pumps, 1995-1999
0-8019-9126-9
Part No. 9126
312 pp

Powertrain Codes & Oxygen Sensors, 1990-1999
0-8019-9127-7
Part No. 9127
400 pp, SC

8½" x 11", SC, 1-Color

Delmar's Automotive Dictionary
David W. South & Boyce Dwiggins
0-8273-7405-4

This handy, ready-reference dictionary provides the automotive engineer, technician, mechanic, student, enthusiast or layperson with a single source for the most up-to-date definitions available of technical, professional and informal terminology used in today s automotive world. It is descriptive and covers the wide scope of terms pertinent to the automotive field. With multiple definitions and aids, and proper pronunciation of terms, this dictionary is a must for all!

KEY FEATURES

- over 3000 terms comprehensively covering more than 100 subject areas
- enhanced by a list of acronyms and abbreviations
- up-to-date definitions of today's automotive terminology
- aids for proper pronunciation
- each term has multiple definitions

281 pp, 6" x 9", SC, 1-Color, ©1997

Practical Problems in Mathematics for Automotive Technicians, 5E
George Morre, Todd Sformo & Larry Sformo
0-8273-7944-7

By showing how to apply math solutions to everyday problems, this all-in-one math reference transforms the "remove it and replace it" mechanic into a complete automotive technician. The book builds from math basics to cover more complex topics--not to mention such workplace issues as invoices and scale reading of test meters. Each easy-to-read chapter features step-by-step instructions, diagrams, charts and examples to make the problem-solving process a snap.

256 pp, 7⅞" x 9¼", SC, 1-Color, ©1998
Instructor's Manual   0-8273-7945-5

Math for the Automotive Trade, 3E
John C. Peterson & William deKryger
0-8273-6712-0

Math for Automotive Trades, 3E provides excellent examples and problems that reflect technological requirements of workers in automotive technology. The text has three parts: review of basic mathematics skills, math applications to specific automotive situations, and an examination of measurement aspects beginning with angle and linear measurements and ending with an extensive look at measurement tools used in the automotive trade.

345 pp, 8½" x 11", SC, 1-Color, ©1995
Instructor's Manual   0-8273-6713-9

ASE Test Preparation Series
Delmar Learning
0-7668-3877-3
Part No. 13877
(Complete Set)

Delmar Learning has developed comprehensive ASE Test Preparation Manuals to help automotive technicians increase their success on these certification programs. The material covers the topics one might find during the test process. The booklets include many review questions and answers, as well as detailed descriptions of the repairs involved. Designed to look like the actual test, participants will feel more comfortable with practice, which will translate into greater success in taking the actual tests. The design of the Delmar Learning product also includes helpful test taking hints and student preparation ideas designed to enhance success.

KEY FEATURES
- The history of the ASE
- Test-taking strategies
- Tasks lists and overview
- Sample test questions
- ASE-style exams
- Explanations to the answers (right and wrong)
- Glossary of terms

(A1) Automotive Engine Repair, 2E
0-7668-3424-7
Part No. 13424
General Engine Diagnosis, Cylinder Head and Valve Train Diagnosis and Repair, Engine Block Diagnosis and Repair, Lubrication and Cooling Systems Diagnosis and Repair, and Fuel, Electrical, Ignition and Exhaust Systems Inspection and Service.

(A2) Automotive Transmissions and Transaxles, 2E
0-7668-3425-5
Part No. 13425
General Transmission/ Transaxle Diagnosis (Mechanical/Hydraulic Systems and Electronic Systems), Transmission/Transaxle Maintenance and Adjustment, In-Vehicle Transmission/Transaxle Repair, Off-Vehicle Transmission/Transaxle Repair.

(A3) Automotive Manual Drive Trains and Axles, 2E
0-7668-3426-3
Part No. 13426
Clutch Diagnosis and Repair, Transmission Diagnosis and Repair, Transaxle Diagnosis and Repair, Drive Shaft/Half Shaft and Universal Joint/Constant Velocity (CV) Joint Diagnosis and Repair (Front and Rear Wheel Drive), Rear Axle Diagnosis and Repair, Four Wheel Drive/All Wheel Drive Component Diagnosis and Repair.

(A4) Automotive Suspension and Steering, 2E
0-7668-3427-1
Part No. 13427
Steering Systems Diagnosis and Repair (Steering Columns and Manual Steering Gears, Power Assisted Steering Units, Steering Linkage), Suspension Systems Diagnosis and Repair (Front Suspensions, Rear Suspensions, Miscellaneous Services), Wheel Alignment Diagnosis, Adjustment and Repair, and Wheel and Tire Diagnosis and Repair.

(A5) Automotive Brakes, 2E
0-7668-3428-X
Part No. 13428
Hydraulic System Diagnosis and Repair, Drum Brake Diagnosis and Repair, Disc Brake Diagnosis and Repair, Power Assist Units Diagnosis and Repair, Miscellaneous Systems Diagnosis and Repair, Antilock Brake Systems (ABS) Diagnosis and Repair.

(A6) Automotive Electrical-Electronic Systems, 2E
0-7668-3429-8
Part No. 13429
General Electrical/Electronic Systems Diagnosis, Battery Diagnosis and Service, Starting Systems Diagnosis and Repair, Charging Systems Diagnosis and Repair, Lighting Systems Diagnosis and Repair, Gauges, Warning Devices and Driver Information Systems Diagnosis and Repair, Horn and Wiper/Washer Diagnosis and Repair, Accessories

Diagnosis and Repair.

(A7) Automotive Heating and Air Conditioning, 2E
0-7668-3430-1
Part No. 13430
The manual for A7 includes the following topics: A/C System Diagnosis and Repair, Refrigeration System Component Diagnosis and Repair, Heating and Engine Cooling Systems Diagnosis and Repair, Operating Systems and Related Controls Diagnosis and Repair, Refrigerant Recovery, Recycling, Handling and Retrofit.

(A8) Automotive Engine Performance, 2E
0-7668-3431-X
Part No. 13431
The manual for A8 includes the following topics: General Engine Diagnosis, Ignition System Diagnosis and Repair, Fuel, Air Induction, and Exhaust Systems Diagnosis and Repair, Emissions Control Systems Diagnosis and Repair (Including OBDII), Computerized Engine controls Diagnosis and Repair (Including OBDII), Engine Electrical Systems diagnosis and Repair.

(L1) Automotive Advance Engine Performance, 2E
0-7668-3432-8
Part No. 13432
The manual for L1 includes the following topics: General Powertrain Diagnosis, Computerized Powertrain Controls Diagnosis (Including OBDII), Ignition System Diagnosis, Fuel Systems and Air Induction Systems Diagnosis, Emission Control Systems Diagnosis, I/M Failure Diagnosis.

(P2) Automobile Parts Specialist, 2E
0-7668-3433-6
Part No. 13433
The manual for P2 includes the following topics: General Operations, Customer Relations and Sales Skills, Vehicle Systems Knowledge, Vehicle Identification, Cataloging Skills, Inventory Management, Merchandising.

(X1) Exhaust Systems
0-7668-3434-4
Part No. 13434
Exhaust Systems includes the following topics: Exhaust Systems Inspection and Repair, Emissions Systems Diagnosis, Exhaust System Fabrication, Exhaust System Installation, Exhaust System Repair Regulations.

ASE Test Preparation Series in Español!
1-4018-1530-8 *(Complete Set)*
Now available in Español – the first of its kind for Spanish-speaking technicians! This comprehensive package of ASE test preparation booklets are intended for any Spanish-speaking automotive technician who is preparing to take an ASE examination. The series includes questions that relate to each competency required for certification by ASE. In addition to a multitude of questions, the reason why each answer is right or wrong is explained, along with task lists and overview, test-taking strategies, and more.

(A1) Reparación de Motores, 2A Edición
1-4018-1014-4/Part No. 21014

(A2) Transmision Automática/ Eje de Transmisión Automática, 2A Edición
1-4018-1015-2/Part No. 21015

(A3) Tren de y Mando Ejes Manuales, 2A Edición
1-4018-1016-0/Part No. 21016

(A4) Suspensión y Dirección, 2A Edición
1-4018-1017-9/Part No. 21017

(A5) Frenos, 2A Edición
1-4018-1018-7/Part No. 21018

(A6) Sistemas Eléctricos/ Electrónicos, 2A Edición
1-4018-1019-5/Part No. 21019

(A7) Calefacción y Aire Acondicionado, 2A Edición
1-4018-1020-9/Part No. 21020

(A8) Funcionamiento de Motores, 2A Edición
1-4018-1021-7/Part No. 21021

(L1) Especialista en el Funciommiato Avansado de Motores, 2A Edición
1-4018-1022-5/Part No. 21022

(P2) Especialista en Partes de Automovil, 2A Edición
1-4018-1023-3/Part No. 21023

(X1) Sistemas de Escape, 2A Edición
1-4018-1024-1/Part No. 21024

Engine Machinist **NEW!**
0-7668-6283-6/Part No. 16283
Complete Set (M1-M3)

(M1)
Cylinder Head Specialist
0-7668-6280-1/Part No. 16280

(M2)
Cylinder Block Specialist
0-7668-6281-X/Part No. 16281

(M3)
Assembly Specialist
0-7668-6282-8/Part No. 16282

Automotive ASE Test Preparation Video Series

Delmar Learning

0-7668-3168-X *(Complete Set of 12 Tapes)*
0-7668-8042-7 *(Complete Set of 3 CD-ROMs)*

Delmar's Automotive ASE Test Prep Videos present test takers with a review of the A1-A8, L1, and P2 tests prior to taking the exam. Each tape summarizes key topics and key task areas through live action and animation. Actual technicians, authentic automotive shops, and late-model vehicles are featured for an up-to-date look and feel. Safety is emphasized throughout each tape. An overview tape introduces test takers to the ASE testing style.

FEATURES OF THE VIDEO SERIES

- lively, easy to follow videos emphasize safety throughout
- covers major task areas and topics for each of the ASE exams
- accompanying Instructor's Guide helps users comprehend and retain information presented

Complete Set of 12 Tapes (with Instructor's Guide), ©2001

Tape 1: Overview of ASE, 0-7668-2484-5
Tape 2: A1 Engine Repair, 0-7668-2485-3
Tape 3: A2 Automatic Transmission, 0-7668-2498-5
Tape 4: A3 Manual Transmission, 0-7668-2499-3
Tape 5: A4 Steering and Suspension, 0-7668-2500-0
Tape 6: A5 Automotive Brakes, 0-7668-2501-9
Tape 7: A6 Electricity/Electronics, 0-7668-2493-4
Tape 8: A7 Air Conditioning, 0-7668-2486-1
Tape 9: A8 Engine Performance, 0-7668-2494-2
Tape 10: P2 Parts Specialist, 0-7668-2487-X
Tape 11: L1 Advanced Engine Performance (Part 1), 0-7668-2491-8
Tape 12: L1 Advanced Engine Performance (Part 2), 0-7668-2492-6

BUNDLES

Bundle 1: Specialty Topics (Set of 4 Tapes) includes Overview of ASE, A1 Engine Repair, A7 Air Conditioning, and P2 Parts Specialist, 0-7668-2483-7
Bundle 2: Engine Performance/Electronics (Set of 4 Tapes) includes L1 Part 1, L1 Part 2, A6 Electricity/ Electronics, and A8 Engine Performance, 0-7668-2490-X
Bundle 3: Undercar (Set of 4 Tapes) includes A2 Automatic Transmissions, A3 Manual Transmissions, A4 Steering and Suspension, and A5 Automotive Brakes, 0-7668-2497-7

CD-ROM COURSEWARE

Based on the ASE Test Prep Series, the CD-ROMs offer the following in addition to the video content:

- Gradebook
- Pre-test/Post-test
- Ability to modify
- Video Glossary
- Variety of question types
- Automatic remediation
- Video File Server compatible

See Inside Front Cover for System Requirements

CD-ROM 1: Specialty Topics CD-ROM includes Overview of ASE, A1 Engine Repair, A7 Air Conditioning, and P2 Parts Specialist, 0-7668-2489-6
CD-ROM 2: Engine Performance/Electronics CD-ROM includes L1 Part 1, L1 Part 2, A6 Electricity/ Electronics, and A8 Engine Performance, 0-7668-2496-9
CD-ROM 3: Undercar CD-ROM includes A2 Automatic Transmissions, A3 Manual Transmissions, A4 Steering and Suspension, and A5 Automotive Brakes, 0-7668-2503-5

The ASE "Passing Lane" Package

Delmar Learning

0-7668-4338-6

(Complete Set)

The most comprehensive test preparation for Automotive Tests A1-A8, L1, and P2. Combining the most thorough ASE Test Preparation books with the latest in ASE videos, this package provides a program of self-study for the automotive ASE Tests.

EACH BOOK IN THE SERIES FEATURES:

- test-taking strategies
- tasks lists and overview
- sample test questions
- ASE-style exams
- explanations to the answers
- glossary of terms

EACH VIDEO IN THE SERIES FEATURES:

- lively, easy to follow videos emphasize safety throughout
- covers major task areas and topics for each of the ASE exams
- accompanying Activity Sheets help comprehend and retain information

(A1) Automotive Engine Repair Book/Video,
0-7668-4181-2
(A2) Automotive Transmissions and Transaxles Book/Video,
0-7668-4182-0
(A3) Automotive Manual Drive Trains and Axles Book/Video,
0-7668-4183-9
(A4) Automotive Suspension and Steering Book/Video,
0-7668-4184-7
(A5) Automotive Brakes Book/Video,
0-7668-4185-5
(A6) Automotive Electrical-Electronics Systems Book/Video,
0-7668-4186-3
(A7) Automotive Heating and Air Conditioning Book/Video,
0-7668-4187-1
(A8) Automotive Engine Performance Book/Video,
0-7668-4188-X
(L1) Automotive Advanced Engine Performance Book/Video,
0-7668-4189-8
(P2) Automobile Parts Specialist Book/Video,
0-7668-4190-1

Automotive Technician Certification Test Preparation Manual, 2E

Don Knowles

0-7668-1948-5

The second edition of Certified ASE Master Technician Don Knowles' popular ASE test preparation book adds coverage of the L1 Advanced Engine Performance test to its coverage of automotive tests A1 through A8. All nine tests covered in this book reflect year 2000 task lists, including the updated composite vehicle in the L1 test. This revised edition contains at least one practice question for every ASE task in the tests. Also included is the updated and expanded coverage of electronic automatic transmissions, electronically controlled automatic transmissions, electronically controlled 4 wheel drive and steering, ABS systems, wiring diagrams, and repairing electronic components.

KEY FEATURES

- a new section has been added on computer-controlled automatic transmissions and transaxles including those used in OBD II vehicles
- new information has been included on electronically-controlled 4WD systems and ABS systems
- the chapter on Electrical/Electronic Systems has been expanded to include information on reading wiring diagrams and inspecting, testing, and repairing electronic components
- a complete chapter has been added to prepare technicians for the Advanced Engine Performance (L1) test

CONTENTS

Engine Repair Automatic Transmission/Transaxle. Manual Drive Train and Axles. Suspension and Steering. Brakes. Electrical/Electronic Systems. Heating, Ventilation, and Air Conditioning Systems. Engine Performance. Advanced Engine Performance.

788 pp, 8½" x 11", SC, 1-Color, ©2001

Prepare to Pass the ASE Exam Online

ATCChallenge.com
Delmar Learning

Delmar Learning's online ASE test preparation web site has been carefully reviewed and researched by ASE master technicians to include fully updated content on tests A1-A8, L1, P2, and X1. The site offers two different study options so users can choose their study method each time they sign on.

- practice questions provide helpful hints, insight into right and wrong answers, and links back to further reading for each task area
- sample tests prepare users for test day by using ASE-style questions - reflecting the type of questions and task areas on the actual exam - making this the most up-to-date and realistic ASE test preparation study aid available
- both study options offer detailed reports that provide accurate test results, feedback, and links on areas of further study down to task areas
- one year secure access through any web-enabled computer
- a complete task list including an overview of each task to further enhance study
- includes coverage of the new ASE Exhaust Systems exam (Test X1)
- automotive dictionary with more than 5,000 terms and Spanish translation
- technical support provided by Delmar Learning

ATCChallenge.com Plus
Delmar Learning

New from Delmar Learning, ATCChallenge.com Plus is the ideal way to gain the expertise required to pass the A5 (brakes) and A7 (heating and air conditioning) ASE exams. Thoroughly reviewed and researched by master automotive technicians, this total online courseware solution may be used effectively in professional training and education courses as well as by individuals preparing for selected ASE exams.

While this version includes all the features that ATCChallenge.com contains, it also brings with it a variety of new tools for use by students and educators. The biggest addition from the original version is the immediate remediation to the ASE-style questions that brings the user to a file or a video clip that explains the answer to each question.

- combines the Today's Technician Series, Erjavec's Automotive Technology, and Delmar Learning's Automotive Video Series to bring technicians the most comprehensive ASE coverage available in one place
- individuals can track their scores by ASE task, by test, or by question; create personalized testbanks; and generate their own progress reports; making ATCChallenge.com Plus an ideal self-study guide and test preparation tool
- a single training director or administrator can easily track average and/or raw scores at the shop-level, by region, or system-wide to gain a measure of learning achievement and program effectiveness
- a complete task list, with an overview of each task to enhances the learning experience, ensures that users are 100% prepared to pass each ASE test

Call Your Delmar Sales Rep for Part Numbers & Pricing

> ## Visit ATCChallenge.com to see the latest modules and a free demo!

ATC Challenge 3.0 CD-ROM
Delmar Learning
0-7668-2982-0

These exciting interactive CD-ROMs have been designed to prepare technicians for successful completion of the Automotive ASE task areas (A1-A8). This multimedia software assesses strengths and weaknesses by identifying topics needing further study while allowing users to review ASE task areas at their own pace. Explanations, hints, notes, and a glossary aid the user in comprehension, critical thinking and retention. These CD-ROMs offer hundreds of ASE-style questions, a test taking strategy section and LAN compatibility. Not only is ATC Challenge 3.0 the ultimate in test preparation, but it is also an excellent learning tool!

CD-ROM, ©2001

Site License Available for Multiple Unit Purchases or Multiple Workstations for ATC Challenge 3.0:
User 1: Full Price (List or Net)
Users 2-5:
$80/workstation + Full Price
Users 6-10:
$70/workstation + Full Price
Users 11-20:
$60/workstation + Full Price
Users 21+:
$50/workstation + Full Price

ATC Challenge for P2
Delmar Learning
0-7668-1827-6

This interactive CD-ROM contains material that will help prepare technicians for the Automotive Parts Specialist (P2) certification exam.

CD-ROM, ©2000

NATEF Standards Job Sheets
Delmar Learning
0-7668-6375-1
(Complete Set, Tests A1-A8)

Each of our eight NATEF (National Automotive Technicians Education Foundation) Standards Job Sheets workbooks has been thoughtfully designed to assist users in gaining valuable job preparedness skills and mastering specific technical competencies required for success as a professional automotive technician. The entire series is based on current NATEF standards.

Central to each manual are well-designed and easy-to-read job sheets, each of which contains specific, performance-based objectives, lists of required tools and materials, safety precautions, plus step-by-step procedures to lead users to completion of shop activities.

KEY FEATURES

- easy to use in any automotive education or training program in which NATEF coverage is desired
- completed Job Sheets may be kept as records, providing tangible evidence that instructors are addressing all NATEF tasks while paving the way for program certification

JOB SHEETS AVAILABLE FOR:
(A1) Automotive Engine Repair, 0-7668-6367-0
(A2) Automatic Transmissions and Transaxles, 0-7668-6368-9
(A3) Manual Drive Trains and Axles, 0-7668-6369-7
(A4) Automotive Suspension and Steering, 0-7668-6370-0
(A5) Automotive Brakes, 0-7668-6371-9
(A6) Automotive Electrical and Electronic Systems, 0-7668-6372-7
(A7) Automotive Heating and Air Conditioning, 0-7668-6373-5
(A8) Automotive Engine Performance, 0-7668-6374-3

All share the following information:
8½" x 11", SC, 1-Color, ©2002

This pioneering eight-book series offers automotive repair shop owners and those wanting to be shop owners the necessary business and customer service skills to run a successful automotive service facility.

The series covers three main topical areas: personnel management, business management, and sales and marketing. Each book provides a framework to help technicians make consistent, high-quality, and productive service a part of every day shop operations. According to the author, "Great performance coupled with increased customer loyalty, trust, and operational excellence will almost always reult in increased profits."

Automotive Service Management Series Benefits:

- real-world approach reflects author's experience as a fourth generation technician, a repair & service company owner, and an automotive industry trainer
- all-inclusive coverage spans from designing an automotive repair facility floor plan through financial management techniques, customer/staff relations, and more
- length of each book makes it easy to incorporate this series into workshops, seminars, and training/education courses
- information is available "as is" or for customization

Total Customer Relationship Management
 Part # 22657, ISBN 1-4018-2657-1
From Intent to Implementation
 Part # 22658, ISBN 1-4018-2658-X
Operational Excellence
 Part # 22659, ISBN 1-4018-2659-8
Building a Team
 Part # 22660, ISBN 1-4018-2660-1
The High Performance Shop
 Part # 22661, ISBN 1-4018-2661-X
Safety Communications
 Part # 22662, ISBN 1-4018-2662-8
Managing Dollars with Sense
 Part # 22663, ISBN 1-4018-2663-6
Operations Management
 Part # 22665, ISBN 1-4018-2665-2
Entire Set of 8 Books
 Part # 22499, ISBN 1-4018-2499-4

Softcover manuals are 8 1/2" x 11", printed in 1-Color, ©2003

ABOUT THE AUTHOR

Mitch Schneider is a fourth generation mechanic/technician and is a frequent speaker at major conventions and meetings of automotive industry trade organizations. Schneider is also an award-winning journalist and is a regular contributor and senior contributing editor for *Motor Age* magazine. He provides commentary on the evolving relationship between service dealers, jobbers, warehouse directors and manufacturers.

Schneider has also appeared on the TNN cable show "Truckin' USA" where he hosted the "Tech Tips" segment. In addition to operating the award-winning Schneider's Automotive for 22 years in Simi Valley, CA, he is also the president and founder of Schneider's Future-Tech, a service company specializing in conducting management seminars for automotive service dealers, jobbers, warehouse distribution companies, and manufacturers.

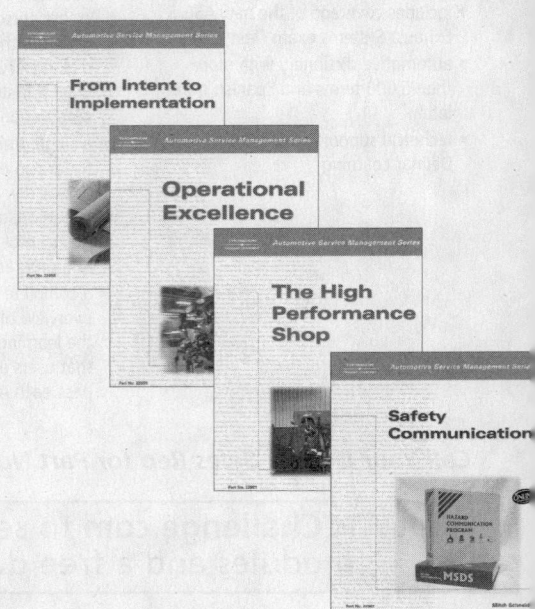